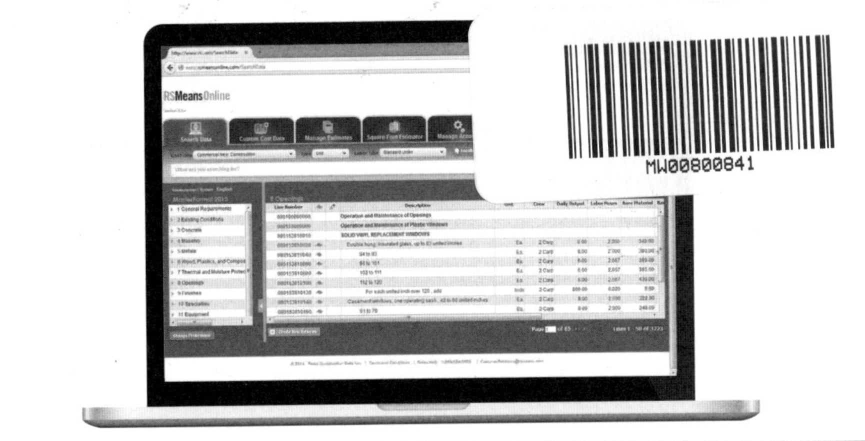

WIN the complete library of RSMeans Online data!

Register your book below to receive your FREE quarterly updates, PLUS you will be entered into a quarterly drawing to win the COMPLETE RSMeans Online library of 2015 data!

Be sure to keep up-to-date in 2015!

Fill out the card below for RSMeans' free quarterly updates, as well as a chance to win the complete RSMeans Online library. Please provide your name, address, and email below and return this card by mail, or to register online:

http://info.thegordiangroup.com/RSMeans.html

Name ____

email ____________________________ Title ____________________

Company ____

Street ____

City/Town ____________________ State/Prov. _____ Zip/Postal Code __________

RSMeans

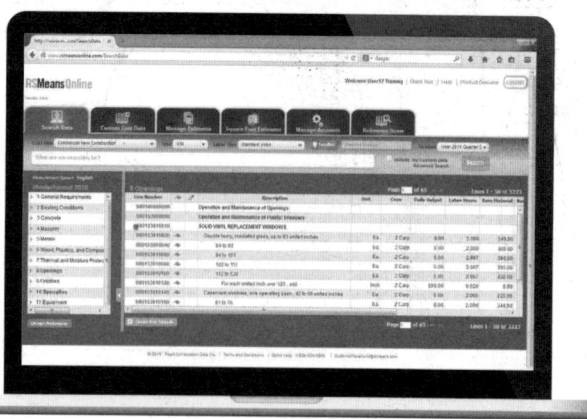

RSMeans Facilities Construction Cost Data

2015 30th annual edition

Melville J. Mossman, PE, Senior Editor

**Numbers in italics are the divisional responsibilities for each editor. Please contact the designated editor directly with any questions.*

RSMeans
Construction Publishers & Consultants
700 Longwater Drive
Norwell, MA 02061
United States of America
1-877-792-2083
www.rsmeans.com

Printed in the United States of America
ISSN 1540-6202
ISBN 978-1-940238-55-5

0205A $524.95 per copy (in United States)
Price is subject to change without prior notice.

This book is printed on 50% recycled paper using soy-based ink and is entirely recyclable.

Related RSMeans Products and Services

The engineers at RSMeans suggest the following products and services as companion information resources to *RSMeans Facilities Constructi Cost Data:*

Construction Cost Data Books

Assemblies Cost Data 2015
Square Foot Costs 2015

Reference Books

Estimating Building Costs
RSMeans Estimating Handbook
Green Building: Project Planning & Estimating
How to Estimate with RSMeans Data
Plan Reading & Material Takeoff
Project Scheduling & Management for Construction
Universal Design Ideas for Style, Comfort & Safety

Seminars and In-House Training

Unit Price Estimating
RSMeans Online® Training
Practical Project Management for Construction Professionals
Scheduling with MSProject for Construction Professionals
Mechanical & Electrical Estimating

RSMeans Online Bookstore

Visit RSMeans at **www.rsmeans.com** for a one-stop portal for the most reliable and current resources available on the market. Here you can learn about our more than 20 annual cost data books and eBooks, also available on CD and online versions, as well as our libary of reference books and RSMeans seminars.

RSMeans Electronic Data

Get the information found in RSMeans cost books electronically at **RSMeansOnline.com**.

RSMeans Business Solutions

Cost engineers and analysts conduct facility life cycle and benchmark research studies and predictive cost modeling, as well as offer consultation services for conceptual estimating and real property management. Research studies are designed with the application of proprietary cost and project data from RSMeans' extensive North American databases. Analysts offer building produ manufacturers qualitative and quantitative market research, as wel as time/motion studies for new products, and Web-based dashboar for market opportunity analysis. Clients are from both the public ar private sectors, including federal, state, and municipal agencies; corporations; institutions; construction management firms; hospita and associations.

Construction Costs for Software Applications

More than 25 unit price and assemblies cost databases are availab through a number of leading estimating and facilities management software partners. For more information see the "Other RSMeans Products and Services" pages at the back of this publication.

RSMeans data is also available to federal, state, and local government agencies as multi-year, multi-seat licenses.

For information on our current partners, visit: **www.rsmeans.com/partners** or call 1-877-792-2083.

Table of Contents

Foreword

Our Mission

Since 1942, RSMeans has been actively engaged in construction cost publishing and consulting throughout North America.

Today, more than 70 years after RSMeans began, our primary objective remains the same: to provide you, the construction and facilities professional, with the most current and comprehensive construction cost data possible.

Whether you are a contractor, owner, architect, engineer, facilities manager, or anyone else who needs a reliable construction cost estimate, you'll find this publication to be a highly useful and necessary tool.

With the constant flow of new construction methods and materials today, it's difficult to find the time to look at and evaluate all the different construction cost possibilities. In addition, because labor and material costs keep changing, last year's cost information is not a reliable basis for today's estimate or budget.

That's why so many construction professionals turn to RSMeans. We keep track of the costs for you, along with a wide range of other key information, from city cost indexes . . . to productivity rates . . . to crew composition . . . to contractor's overhead and profit rates.

RSMeans performs these functions by collecting data from all facets of the industry and organizing it in a format that is instantly accessible to you. From the preliminary budget to the detailed unit price estimate, you'll find the data in this book useful for all phases of construction cost determination.

The Staff, the Organization, and Our Services

When you purchase one of RSMeans' publications, you are, in effect, hiring the services of a full-time staff of construction and engineering professionals.

Our thoroughly experienced and highly qualified staff works daily at collecting, analyzing, and disseminating comprehensive cost information for your needs. These staff members have years of practical construction experience and engineering training prior to joining the firm. As a result, you can count on them not only for accurate cost figures, but also for additional background reference information that will help you create a realistic estimate.

The RSMeans organization is always prepared to help you solve construction problems through its variety of data solutions, including online, CD, and print book formats, as well as cost estimating expertise available via our business solutions, training, and seminars.

Besides a full array of construction cost estimating books, RSMeans also publishes a number of other reference works for the construction industry. Subjects include construction estimating and project and business management, green building, and a library of facility management references.

In addition, you can access all of our construction cost data electronically in convenient CD format or on the web. Visit **RSMeansOnline.com** for more information on our 24/7 online cost data.

What's more, you can increase your knowledge and improve your construction estimating and management performance with an RSMeans construction seminar or in-house training program. These two-day seminar programs offer unparalleled opportunities for everyone in your organization to become updated on a wide variety of construction-related issues.

RSMeans is also a worldwide provider of construction cost management and analysis services for commercial and government owners.

In short, RSMeans can provide you with the tools and expertise for constructing accurate and dependable construction estimates and budgets in a variety of ways.

Robert Snow Means Established a Tradition of Quality That Continues Today

Robert Snow Means spent years building RSMeans, making certain he always delivered a quality product.

Today, at RSMeans, we do more than talk about the quality of our data and the usefulness of our books. We stand behind all of our data, from historical cost indexes to construction materials and techniques to current costs.

If you have any questions about our products or services, please call us toll-free at 1-877-792-2083. Our customer service representatives will be happy to assist you. You can also visit our website at **www.rsmeans.com**

How the Book Is Built: An Overview

The Construction Specifications Institute (CSI) and Construction Specifications Canada (CSC) have produced the 2014 edition of MasterFormat®, a system of titles and numbers used extensively to organize construction information.

All unit price data in the RSMeans cost data books is now arranged in the 50-division MasterFormat® 2014 system.

A Powerful Construction Tool

You have in your hands one of the most powerful construction tools available today. A successful project is built on the foundation of an accurate and dependable estimate. This book will enable you to construct just such an estimate.

For the casual user the book is designed to be:

- quickly and easily understood so you can get right to your estimate.
- filled with valuable information so you can understand the necessary factors that go into the cost estimate.

For the regular user, the book is designed to be:

- a handy desk reference that can be quickly referred to for key costs.
- a comprehensive, fully reliable source of current construction costs and productivity rates so you'll be prepared to estimate any project.
- a source book for preliminary project cost, product selections, and alternate materials and methods.

To meet all of these requirements, we have organized the book into the following clearly defined sections.

Quick Start

See our "Quick Start" instructions on the following page to get started right away.

Estimating with RSMeans Unit Price Cost Data

Please refer to these steps for guidance on completing an estimate using RSMeans unit price cost data.

How to Use the Book: The Details

This section contains an in-depth explanation of how the book is arranged . . . and how you can use it to determine a reliable construction cost estimate. It includes information about how we develop our cost figures and how to completely prepare your estimate.

Unit Price Section

All cost data has been divided into the 50 divisions according to the MasterFormat system of classification and numbering. For a listing of these divisions and an outline of their subdivisions, see the Unit Price Section Table of Contents.
Estimating tips are included at the beginning of each division.

Assemblies Section

The cost data in this section has been organized in an "Assemblies" format. These assemblies are the functional elements of a building and are arranged according to the 7 elements of the UNIFORMAT II classification system. For a complete explanation of a typical "Assemblies" page, see "How RSMeans Assemblies Data Works."

Reference Section

This section includes information on Equipment Rental Costs, Crew Listings, Historical Cost Indexes, City Cost Indexes, Location Factors, Reference Tables, Change Orders, Square Foot Costs, and a listing of Abbreviations.

Equipment Rental Costs: This section contains the average costs to rent and operate hundreds of pieces of construction equipment.

Crew Listings: This section lists all of the crews referenced in the book. For the purposes of this book, a crew is composed of more than one trade classification and/or the addition of power equipment to any trade classification. Power equipment is included in the cost of the crew. Costs are shown both with bare labor rates and with the installing contractor's overhead and profit added. For each, the total crew cost per eight-hour day and the composite cost per labor-hour are listed.

Historical Cost Indexes: These indexes provide you with data to adjust construction costs over time.

City Cost Indexes: All costs in this book are U.S. national averages. Costs vary because of the regional economy. You can adjust costs by CSI Division to over 700 locations throughout the U.S. and Canada by using the data in this section.

Location Factors: You can adjust total project costs to over 900 locations throughout the U.S. and Canada by using the data in this section.

Reference Tables: At the beginning of selected major classifications in the Unit Price Section are reference numbers shown in a shaded box. These numbers refer you to related information in the Reference Section. In this section, you'll find reference tables, explanations, estimating information that support how we develop the unit price data, technical data, and estimating procedures.

Change Orders: This section includes information on the factors that influence the pricing of change orders.

Square Foot Costs: This section contains costs for 59 different building types that allow you to make a rough estimate for the overall cost of a project or its major components.

Abbreviations: A listing of abbreviations used throughout this book, along with the terms they represent, is included in this section.

Index

A comprehensive listing of all terms and subjects in this book will help you quickly find what you need when you are not sure where it occurs in MasterFormat.

The Scope of This Book

This book is designed to be as comprehensive and as easy to use as possible. To that end we have made certain assumptions and limited its scope in two key ways:

1. We have established material prices based on a national average.
2. We have computed labor costs based on a 30-city national average of union wage rates.

For a more detailed explanation of how the cost data is developed, see "How To Use the Book: The Details."

Project Size/Type

The material prices in RSMeans cost data books are "contractor's prices." They are the prices that contractors can expect to pay at the lumberyards, suppliers'/distributors' warehouses, etc. Small orders of speciality items would be higher than the costs shown, while very large orders, such as truckload lots, would be less. The variation would depend on the size, timing, and negotiating power of the contractor. The labor costs are primarily for new construction or major renovation rather than repairs or minor alterations.
With reasonable exercise of judgment, the figures can be used for any building work.

Absolute Essentials for a Quick Start

If you feel you are ready to use this book and don't think you will need the detailed instructions that begin on the following page, this Absolute Essentials for a Quick Start page is for you. These steps will allow you to get started estimating in a matter of minutes.

1 Scope

Think through the project that you will be estimating, and identify the many individual work tasks that will need to be covered in your estimate.

2 Quantify

Determine the number of units that will be required for each work task that you identified.

3 Pricing

Locate individual Unit Price line items that match the work tasks you identified. The Unit Price Section Table of Contents that begins on page 1 and the Index in the back of the book will help you find these line items.

4 Multiply

Multiply the Total Incl O&P cost for a Unit Price line item in the book by your quantity for that item. The price you calculate will be an estimate for a completed item of work performed by a subcontractor. Keep adding line items in this manner to build your estimate.

5 Project Overhead

Include project overhead items in your estimate. These items are needed to make the job run and are typically, but not always, provided by the General Contractor. They can be found in Division 1. An alternate method of estimating project overhead costs is to apply a percentage of the total project cost.

Include rented tools not included in crews, waste, rubbish handling, and cleanup.

6 Estimate Summary

Include General Contractor's markup on subcontractors, General Contractor's office overhead and profit, and sales tax on materials and equipment.

Adjust your estimate to the project's location by using the City Cost Indexes or Location Factors found in the Reference Section.

Editors' Note: We urge you to spend time reading and understanding the supporting material in the front of this book. An accurate estimate requires experience, knowledge, and careful calculation. The more you know about how we at RSMeans developed the data, the more accurate your estimate will be. In addition, it is important to take into consideration the reference material in the back of the book such as Equipment Listings, Crew Listings, City Cost Indexes, Location Factors, and Reference Tables.

Estimating with RSMeans Unit Price Cost Data

ollowing these steps will allow you to omplete an accurate estimate using SMeans Unit Price cost data.

Scope Out the Project

- Think through the project and identify those CSI Divisions needed in your estimate.
- Identify the individual work tasks that will need to be covered in your estimate.
- The Unit Price data in this book has been divided into 50 Divisions according to the CSI MasterFormat® 2014—their titles are listed on the back cover of your book.
- The Unit Price Section Table of Contents on page 1 may also be helpful when scoping out your project.
- Experienced estimators find it helpful to begin with Division 2 and continue through completion. Division 1 can be estimated after the full project scope is known.

Quantify

- Determine the number of units required for each work task that you identified.
- Experienced estimators include an allowance for waste in their quantities. (Waste is not included in RSMeans Unit Price line items unless so stated.)

3 Price the Quantities

- Use the Unit Price Table of Contents, and the Index, to locate individual Unit Price line items for your estimate.
- Reference Numbers indicated within a Unit Price section refer to additional information that you may find useful.
- The crew indicates who is performing the work for that task. Crew codes are expanded in the Crew Listings in the Reference Section to include all trades and equipment that comprise the crew.
- The Daily Output is the amount of work the crew is expected to do in one day.
- The Labor-Hours value is the amount of time it will take for the crew to install one unit of work.
- The abbreviated Unit designation indicates the unit of measure upon which the crew, productivity, and prices are based.
- Bare Costs are shown for materials, labor, and equipment needed to complete the Unit Price line item. Bare costs do not include waste, project overhead, payroll insurance, payroll taxes, main office overhead, or profit.
- The Total Incl O&P cost is the billing rate or invoice amount of the installing contractor or subcontractor who performs the work for the Unit Price line item.

4 Multiply

- Multiply the total number of units needed for your project by the Total Incl O&P cost for each Unit Price line item.
- Be careful that your take off unit of measure matches the unit of measure in the Unit column.
- The price you calculate is an estimate for a completed item of work.
- Keep scoping individual tasks, determining the number of units required for those tasks, matching each task with individual Unit Price line items in the book, and multiply quantities by Total Incl O&P costs.
- An estimate completed in this manner is priced as if a subcontractor, or set of subcontractors, is performing the work. The estimate does not yet include Project Overhead or Estimate Summary components such as General Contractor markups on subcontracted work, General Contractor office overhead and profit, contingency, and location factor.

5 Project Overhead

- Include project overhead items from Division 1 – General Requirements.
- These items are needed to make the job run. They are typically, but not always, provided by the General Contractor. Items include, but are not limited to, field personnel, insurance, performance bond, permits, testing, temporary utilities, field office and storage facilities, temporary scaffolding and platforms, equipment mobilization and demobilization, temporary roads and sidewalks, winter protection, temporary barricades and fencing, temporary security, temporary signs, field engineering and layout, final cleaning and commissioning.
- Each item should be quantified, and matched to individual Unit Price line items in Division 1, and then priced and added to your estimate.
- An alternate method of estimating project overhead costs is to apply a percentage of the total project cost, usually 5% to 15% with an average of 10% (see General Conditions, page ix).
- Include other project related expenses in your estimate such as:
  - Rented equipment not itemized in the Crew Listings
  - Rubbish handling throughout the project (see 02 41 19.19)

6 Estimate Summary

- Include sales tax as required by laws of your state or county.
- Include the General Contractor's markup on self-performed work, usually 5% to 15% with an average of 10%.
- Include the General Contractor's markup on subcontracted work, usually 5% to 15% with an average of 10%.
- Include General Contractor's main office overhead and profit:
  - RSMeans gives general guidelines on the General Contractor's main office overhead (see section 01 31 13.60 and Reference Number R013113-50).
  - RSMeans gives no guidance on the General Contractor's profit.
  - Markups will depend on the size of the General Contractor's operations, his projected annual revenue, the level of risk he is taking on, and on the level of competition in the local area and for this project in particular.
- Include a contingency, usually 3% to 5%, if appropriate.
- Adjust your estimate to the project's location by using the City Cost Indexes or the Location Factors in the Reference Section:
  - Look at the rules on the pages for "How to Use the City Cost Indexes" to see how to apply the Indexes for your location.
  - When the proper Index or Factor has been identified for the project's location, convert it to a multiplier by dividing it by 100, and then multiply that multiplier by your estimated total cost. The original estimated total cost will now be adjusted up or down from the national average to a total that is appropriate for your location.

Editors' Notes:
We urge you to spend time reading and understanding the supporting material in the front of this book. An accurate estimate requires experience, knowledge, and careful calculation. The more you know about how we at RSMeans developed the data, the more accurate your estimate will be. In addition, it is important to take into consideration the reference material in the back of the book such as Equipment Listings, Crew Listings, City Cost Indexes, Location Factors, and Reference Tables.

How to Use the Book: The Details

hat's Behind the Numbers? he Development of Cost Data

ne staff at RSMeans continually monitors evelopments in the construction industry order to ensure reliable, thorough, and p-to-date cost information. While overall onstruction costs may vary relative to general conomic conditions, price fluctuations ithin the industry are dependent upon many ctors. Individual price variations may, in ct, be opposite to overall economic trends. herefore, costs are constantly tracked and omplete updates are published yearly. Also, ew items are frequently added in response changes in materials and methods.

osts—$ (U.S.)

ll costs represent U.S. national averages and e given in U.S. dollars. The RSMeans City ost Indexes can be used to adjust costs to a articular location. The City Cost Indexes for anada can be used to adjust U.S. national verages to local costs in Canadian dollars. o exchange rate conversion is necessary.

G The processes or products identified by the green symbol in our publications ave been determined to be environmentally esponsible and/or resource-efficient solely by he RSMeans engineering staff. The inclusion f the green symbol does not represent ompliance with any specific industry ssociation or standard.

aterial Costs

he RSMeans staff contacts manufacturers, ealers, distributors, and contractors all cross the U.S. and Canada to determine ational average material costs. If you have ccess to current material costs for your pecific location, you may wish to make djustments to reflect differences from the ational average. Included within material osts are fasteners for a normal installation. SMeans engineers use manufacturers' ecommendations, written specifications, nd/or standard construction practice for ize and spacing of fasteners. Adjustments o material costs may be required for our specific application or location. The manufacturer's warranty is assumed. Extended warranties are not included in the material costs. Material costs do not include sales tax.

Labor Costs

Labor costs are based on the average of wage rates from 30 major U.S. cities. Rates are determined from labor union agreements or prevailing wages for construction trades for the current year. Rates, along with overhead and profit markups, are listed on the inside back cover of this book.

- If wage rates in your area vary from those used in this book, or if rate increases are expected within a given year, labor costs should be adjusted accordingly.

Labor costs reflect productivity based on actual working conditions. In addition to actual installation, these figures include time spent during a normal weekday on tasks such as, material receiving and handling, mobilization at site, site movement, breaks, and cleanup.

Productivity data is developed over an extended period so as not to be influenced by abnormal variations and reflects a typical average.

Equipment Costs

Equipment costs include not only rental, but also operating costs for equipment under normal use. The operating costs include parts and labor for routine servicing such as repair and replacement of pumps, filters, and worn lines. Normal operating expendables, such as fuel, lubricants, tires, and electricity (where applicable), are also included. Extraordinary operating expendables with highly variable wear patterns, such as diamond bits and blades, are excluded. These costs are included under materials. Equipment rental rates are obtained from industry sources throughout North America—contractors, suppliers, dealers, manufacturers, and distributors.

Rental rates can also be treated as reimbursement costs for contractor-owned equipment. Owned equipment costs include depreciation, loan payments, interest, taxes, insurance, storage and major repairs.

Equipment costs do not include operators' wages; nor do they include the cost to move equipment to a job site (mobilization) or from a job site (demobilization).

Equipment Cost/Day—The cost of power equipment required for each crew is included in the Crew Listings in the Reference Section (small tools that are considered as essential everyday tools are not listed out separately). The Crew Listings itemize specialized tools and heavy equipment along with labor trades. The daily cost of itemized equipment included in a crew is based on dividing the weekly bare rental rate by 5 (number of working days per week) and then adding the hourly operating cost times 8 (the number of hours per day). This Equipment Cost/Day is shown in the last column of the Equipment Rental Cost pages in the Reference Section.

Mobilization/Demobilization—The cost to move construction equipment from an equipment yard or rental company to the job site and back again is not included in equipment costs. Mobilization (to the site) and demobilization (from the site) costs can be found in the Unit Price Section. If a piece of equipment is already at the job site, it is not appropriate to utilize mob./demob. costs again in an estimate.

Overhead and Profit

Total Cost including O&P for the *Installing Contractor* is shown in the last column on the Unit Price and/or the Assemblies pages of this book. This figure is the sum of the bare material cost plus 10% for profit, the bare labor cost plus total overhead and profit, and the bare equipment cost plus 10% for profit. Details for the calculation of Overhead and Profit on labor are shown on the inside back cover and in the Reference Section of this book. (See "How RSMeans Unit Price Data Works" for an example of this calculation.)

General Conditions

Cost data in this book is presented in two ways: Bare Costs and Total Cost including O&P (Overhead and Profit). General Conditions, or General Requirements, of the contract should also be added to the Total Cost including O&P when applicable. Costs for General Conditions are listed in Division 1 of the Unit Price Section and the Reference

Section of this book. General Conditions for the *Installing Contractor* may range from 0% to 10% of the Total Cost including O&P. For the *General* or *Prime Contractor*, costs for General Conditions may range from 5% to 15% of the Total Cost including O&P, with a figure of 10% as the most typical allowance. If applicable, the Assemblies and Models sections of this book use costs that include the installing contractor's overhead and profit (O&P).

Factors Affecting Costs

Costs can vary depending upon a number of variables. Here's how we have handled the main factors affecting costs.

Quality—The prices for materials and the workmanship upon which productivity is based represent sound construction work. They are also in line with U.S. government specifications.

Overtime—We have made no allowance for overtime. If you anticipate premium time or work beyond normal working hours, be sure to make an appropriate adjustment to your labor costs.

Productivity—The productivity, daily output, and labor-hour figures for each line item are based on working an eight-hour day in daylight hours in moderate temperatures. For work that extends beyond normal work hours or is performed under adverse conditions, productivity may decrease. (See "How RSMeans Unit Price Data Works" for more on productivity.)

Size of Project—The size, scope of work, and type of construction project will have a significant impact on cost. Economies of scale can reduce costs for large projects. Unit costs can often run higher for small projects.

Location—Material prices in this book are for metropolitan areas. However, in dense urban areas, traffic and site storage limitations may increase costs. Beyond a 20-mile radius of large cities, extra trucking or transportation charges may also increase the material costs slightly. On the other hand, lower wage rates may be in effect. Be sure to consider both of these factors when preparing an estimate, particularly if the job site is located in a central city or remote rural location.
In addition, highly specialized subcontract items may require travel and per-diem expenses for mechanics.

Other Factors—

- season of year
- contractor management
- weather conditions
- local union restrictions
- building code requirements
- availability of:
  - adequate energy
  - skilled labor
  - building materials
- owner's special requirements/restrictions
- safety requirements
- environmental considerations

Unpredictable Factors—General business conditions influence "in-place" costs of all items. Substitute materials and construction methods may have to be employed. These may affect the installed cost and/or life cycle costs. Such factors may be difficult to evaluate and cannot necessarily be predicted on the basis of the job's location in a particular section of the country. Thus, where these factors apply, you may find significant but unavoidable cost variations for which you will have to apply a measure of judgment to your estimate.

Rounding of Costs

In general, all unit prices in excess of $5.00 have been rounded to make them easier to use and still maintain adequate precision of the results. The rounding rules we have chosen are in the following table.

Prices from . . .	Rounded to the nearest . . .
$.01 to $5.00	$.01
$5.01 to $20.00	$.05
$20.01 to $100.00	$.50
$100.01 to $300.00	$1.00
$300.01 to $1,000.00	$5.00
$1,000.01 to $10,000.00	$25.00
$10,000.01 to $50,000.00	$100.00
$50,000.01 and above	$500.00

How Subcontracted Items Affect Costs

A considerable portion of all large construction jobs is usually subcontracted. In fact, the percentage done by subcontractors is constantly increasing and may run over 90%. Since the workers employed by these companies do nothing else but install their particular product, they soon become expert in that line. The result is, installation by these firms is accomplished so efficiently that the total in-place cost, even adding the general contractor's overhead and profit, is no more, and often less, than if the principal contractor had handled the installation himself/herself. Companies that deal with construction specialties are anxious to have their product perform well and, consequently, the installation will be the best possible.

Contingencies

The allowance for contingencies generally provides for unforeseen construction difficulties. On alterations or repair jobs, 10% is not too much. If drawings are final and only field contingencies are being considered, 2% or 3% is probably sufficient, and often nothing need be added. Contractually, changes in plans will be covered by extras. The contractor should consider inflationary price trends and possible material shortages during the course of the job. These escalation factors are dependent upon both economic conditions and the anticipated time between the estimate and actual construction. If drawings are not complete or approved, or a budget cost is wanted, it is wise to add 5% to 10%. Contingencies, then, are a matter of judgment. Additional allowances for contingencies are shown in Division 1.

Important Estimating Considerations

The productivity or daily output of each craftsman or crew assumes a well-managed job where tradesmen with the proper tools and equipment, along with the appropriate construction materials, are present. Included are daily mobilization and cleanup time, break time and plan layout time. Unless otherwise indicated, time for material movement on site (for items that can be transported by hand) of up to 200' into the building and to the first or second floor is also included. If material has to be transported by other means, over greater distances, or to higher floors, an additional allowance should be considered by the estimator.

While horizontal movement is typically a sole function of distances, vertical transport introduces other variables that can significantly impact productivity. In an occupied building, the use of elevators (assuming access, size, and required protective measures are acceptable) must be understood at the time of the estimate. For new construction, hoist wait and cycle times can easily be 15 minutes and may result in scheduled access extending beyond the normal work day. Finally, all vertical transport will impose strict weight limits likely to preclude the use of any motorized material handling.

The productivity, or daily output, also assumes installation that meets manufacturer/designer/standard specifications. A time allowance for quality control checks, minor adjustments and any task required to ensure the proper function or operation is also included. For items that require connections to services, time is included for positioning, leveling, securing the unit and for making all the necessary connections (and start up where applicable) ensuring a complete installation. Estimating of the services themselves (electrical, plumbing, water, steam, hydraulics, dust collection, etc.) is separate.

In some cases, the estimator must consider the use of a crane and an appropriate crew for the installation of large or heavy items. For those situations where a crane is not included in the assigned crew and as part of the line item cost, equipment rental costs, mobilization and demobilization costs, and operator and support personnel costs must be considered.

Labor-Hours

The labor-hours expressed in this publication are derived by dividing the total daily labor hours for the crew by the daily output. Based on average installation time and the assumptions listed above, the labor-hours include: direct labor, indirect labor, and nonproductive time. A typical day for a craftsman might include, but not be limited to:

- Direct Work
  - Measuring and layout
  - Preparing materials
  - Actual installation
  - Quality assurance/quality control
- Indirect Work
  - Reading plans or specifications
  - Preparing space
  - Receiving materials
  - Material movement
  - Giving or receiving instruction
  - Miscellaneous
- Non-Work
  - Chatting
  - Personal issues
  - Breaks
  - Interruptions (i.e., sickness, weather, material or equipment shortages, etc.)

If any of the items for a typical day do not apply to the particular work or project situation, the estimator should make any necessary adjustments.

Final Checklist

Estimating can be a straightforward process provided you remember the basics. Here's a checklist of some of the steps you should remember to complete before finalizing your estimate.

Did you remember to . . .

- factor in the City Cost Index for your locale?
- take into consideration which items have been marked up and by how much?
- mark up the entire estimate sufficiently for your purposes?
- read the background information on techniques and technical matters that could impact your project time span and cost?
- include all components of your project in the final estimate?
- double check your figures for accuracy?
- call RSMeans if you have any questions about your estimate or the data you've found in our publications? Remember, RSMeans stands behind its publications. If you have any questions about your estimate . . . about the costs you've used from our books . . . or even about the technical aspects of the job that may affect your estimate, feel free to call the RSMeans editors at 1-877-792-2083.

Free Quarterly Updates

Stay up-to-date throughout 2015 with RSMeans' free cost data updates four times a year. Sign up online to make sure you have access to the newest data. Every quarter we provide the city cost adjustment factors for hundreds of cities and key materials. Register at: **http://info.thegordiangroup.com/RSMeans.html**.

Using Minimum Labor/Equipment Charges for Small Quantities

Estimating small construction or repair tasks often creates situations in which the quantity of work to be performed is very small. When this occurs, the labor and/or equipment costs to perform the work may be too low to allow for the crew to get to the job, receive instructions, find materials, get set up, perform the work, clean up, and get to the next job. In these situations, the estimator should compare the developed labor and/or equipment costs for performing the work (e.g., quantity × labor and/or equipment costs) with the *minimum labor/equipment charge* within that Unit Price section of the book.(These minimum labor/equipment charge line items appear only in *Facilities Construction Cost Data* and *Commercial Renovation Cost Data.*)

If the labor and/or equipment costs developed by the estimator are LOWER THAN the *minimum labor/equipment charge* listed at the bottom of specific sections of Unit Price costs, the estimator should adjust the developed costs upward to the *minimum labor/equipment charge*. The proper use of a *minimum labor/equipment charge* results in having enough money in the estimate to cover the contractor's higher cost of performing a very small amount of work during a partial workday.

A *minimum labor/equipment charge* should be used only when the task being estimated is the only task the crew will perform at the job site that day. **If, however, the crew will be able to perform other tasks at the job site that day, the use of a minimum labor/equipment charge is not appropriate.**

08 52 Wood Windows

08 52 10 – Plain Wood Windows

08 52 10.40 Casement Window		Crew	Daily Output	Labor-Hours	Unit	Material	2015 Bare Costs Labor	Equipment	Total	Total Incl O&P
0010	**CASEMENT WINDOW**, including frame, screen and grilles									
0100	2'-0" x 3'-0" H, dbl. insulated glass G	1 Carp	10	.800	Ea.	268	37.50		305.50	355
0150	Low E glass G		10	.800		259	37.50		296.50	345
0200	2'-0" x 4'-6" high, double insulated glass G		9	.889		360	41.50		401.50	465
0250	Low E glass G		9	.889		370	41.50		411.50	475
9000	Minimum labor/equipment charge	1 Carp	3	2.667	Job		125		125	205

Example:

Establish the bid price to install two casement windows. Assume installation of 2'×4'-6" metal clad windows with insulating glass [Unit Price line number 08 52 10.40 0250], and that this is the only task this crew will perform at the job site that day.

Solution:

Step One — Develop the Bare Labor Cost for this task:
Bare Labor Cost = 2 windows @ $41.50/each = $83.00

Step Two — Evaluate the *minimum labor/equipment charge* for this Unit Price section against the developed Bare Labor Cost for this task:
Minimum labor/equipment charge = $125.00 (compare with $83.00)

Step Three — Choose to adjust the developed labor cost upward to the *minimum labor/equipment charge.*

Step Four — Develop the bid price for this task (including O&P):
Add together the marked-up Bare Material Cost for this task and the marked-up *minimum labor/equipment charge* for this Unit Price section.

2 × ($370.00 + 10%) + ($125.00 + 63.9%)

= 2 × ($370.00 + $37.00) + ($125.00 + $79.88)

= 2 × ($407.00) + $204.88

= $814.00 + $204.88

= $1,018.88

ANSWER: $1,018.88 is the correct bid price to use. This sum takes into consideration the Material Cost (with 10% for profit) for these two windows, plus the *minimum labor/ equipment charge* (with O&P included) for this section of the Unit Price.

Unit Price Section

Table of Contents

Table of Contents (cont.)

Table of Contents (cont.)

How RSMeans Unit Price Data Works

All RSMeans unit price data is organized in the same way.

It is important to understand the structure, so that you can find information easily and use it correctly.

RSMeans **Line Numbers** consist of 12 characters, which identify a unique location in the database for each task. The first 6 or 8 digits conform to the Construction Specifications Institute MasterFormat® 2014. The remainder of the digits are a further breakdown by RSMeans in order to arrange items in understandable groups of similar tasks. Line numbers are consistent across all RSMeans publications, so a line number in any RSMeans product will always refer to the same unit of work.

RSMeans engineers have created **reference** information to assist you in your estimate. If there is information that applies to a section, it will be indicated at the start of the section. In this case, R033105-10 provides information on the proportionate quantities of formwork, reinforcing, and concrete used in cast-in-place concrete items such as footings, slabs, beams, and columns. The Reference Section is located in the back of the book on the pages with a gray edge.

RSMeans **Descriptions** are shown in a hierarchical structure to make them readable. In order to read a complete description, read up through the indents to the top of the section. Include everything that is above and to the left that is not contradicted by information below. For instance, the complete description for line 03 30 53.40 3550 is "Concrete in place, including forms (4 uses), Grade 60 rebar, concrete (Portland cement Type 1), placement and finishing unless otherwise indicated; Equipment pad (3000 psi), 4' x 4' x 6" thick."

When using **RSMeans data**, it is important to read through an entire section to ensure that you use the data that most closely matches your work. Note that sometimes there is additional information shown in the section that may improve your price. There are frequently lines that further describe, add to, or adjust data for specific situations.

03 30 Cast-In-Place Concrete

03 30 53 – Miscellaneous Cast-In-Place Concrete

03 30 53.40 Concrete In Place		
0010	**CONCRETE IN PLACE**	R033105-10
0020	Including forms (4 uses), Grade 60 rebar, concrete (Portland cement	R033105-20
0050	Type I), placement and finishing unless otherwise indicated	R033105-70
0300	Beams (3500 psi), 5 kip per L.F., 10′ span	R033105-85
0350	25′ span	
0500	Chimney foundations (5000 psi), over 5 C.Y.	
0510	(3500 psi), under 5 C.Y.	
0700	Columns, square (4000 psi), 12″ x 12″, less than 2% reinforcing	
3100	Elevated slabs, flat plate, including finish, not	
3400	Cellular concrete, 1-5/8″ fill, under 5000 S.F.	
3450	Over 10,000 S.F.	
3500	Add per floor for 3 to 6 stories high	
3520	For 7 to 20 stories high	
3540	Equipment pad (3000 psi), 3′ x 3′ x 6″ thick	
3550	4′ x 4′ x 6″ thick	
3560	5′ x 5′ x 8″ thick	
3570	6′ x 6′ x 8″ thick	

The data published in RSMeans print books represents a "national average" cost. This data should be modified to the project location using the **City Cost Indexes** or **Location Factors** tables found in the Reference Section (see pages 1390–1438). Use the location factors to adjust estimate totals if the project covers multiple trades. Use the city cost indexes (CCI) for single trade projects or projects where a more detailed analysis is required. All figures in the two tables are derived from the same research. The last row of data in the CCI, the weighted average, is the same as the numbers reported for each location in the location factor table.

Crews include labor or labor and equipment necessary to accomplish each task. In this case, Crew C-14H is used. RSMeans selects a crew to represent the workers and equipment that are typically used for that task. In this case, Crew C-14H consists of one carpenter foreman (outside), two carpenters, one rodman, one laborer, one cement finisher, and one gas engine vibrator. Details of all crews can be found in the reference section.

Crews

Crew No.	Bare Costs		Incl. Subs O&P		Cost Per Labor-Hour	
Crew C-14C	**Hr.**	**Daily**	**Hr.**	**Daily**	**Bare Costs**	**Incl. O&P**
1 Carpenter Foreman (outside)	$48.95	$391.60	$80.25	$642.00	$45.08	$73.66
6 Carpenters	46.95	2253.60	76.95	3693.60		
2 Rodmen (reinf.)	52.55	840.80	85.75	1372.00		
4 Laborers	37.60	1203.20	61.65	1972.80		
1 Cement Finisher	45.00	360.00	71.15	569.20		
1 Gas Engine Vibrator		31.20		34.32	.28	.31
112 L.H., Daily Totals		$5080.40		$8283.92	$45.36	$73.96

The **Daily Output** is the amount of work that the crew can do in a normal 8-hour workday, including mobilization, layout, movement of materials, and cleanup. In this case, crew C-14H can install thirty 4' x 4' x 6" thick concrete pads in a day. Daily output is variable, based on many factors, including the size of the job, location, and environmental conditions. RSMeans data represents work done in daylight (or adequate lighting) and temperate conditions.

Bare Costs are the costs of materials, labor, and equipment that the installing contractor pays. They represent the cost, in U.S. dollars, for one unit of work. They do not include any markups for profit or labor burden.

Crew	Daily Output	Labor-Hours	Unit	Material	2015 Bare Costs Labor	Equipment	Total	Total Incl O&P
C-14A	15.62	12.804	C.Y.	320	605	48	973	1,400
"	18.55	10.782		335	510	40.50	885.50	1,250
C-14C	32.22	3.476		150	157	.97	307.97	420
"	23.71	4.724		177	213	1.32	391.32	545
C-14A	11.96	16.722		365	790	63	1,218	1,775
	2000	.028		.99	1.17	.36	2.52	3.38
	2200	.025		.94	1.07	.33	2.34	3.12
	31800	.002			.07	.02	.09	.14
	21200	.003			.11	.03	.14	.22
C-14H	45	1.067	Ea.	47	49.50	.69	97.19	133
	30	1.600		69.50	74	1.04	144.54	198
	18	2.667		122	124	1.73	247.73	335
	14	3.429		164	159	2.23	325.23	440

The **Total Incl O&P column** is the total cost, including overhead and profit, that the installing contractor will charge the customer. This represents the cost of materials plus 10% profit, the cost of labor plus labor burden and 10% profit, and the cost of equipment plus 10% profit. It does not include the general contractor's overhead and profit. Note: See the inside back cover for details of how RSMeans calculates labor burden.

The **Total column** represents the total bare cost for the installing contractor, in U.S. dollars. In this case, the sum of $69.50 for material + $74.00 for labor + $1.04 for equipment is $144.54.

The figure in the **Labor Hours** column is the amount of labor required to perform one unit of work—in this case the amount of labor required to construct one 4' x 4' equipment pad. This figure is calculated by dividing the number of hours of labor in the crew by the daily output (48 labor hours divided by 30 pads = 1.6 hours of labor per pad). Multiply 1.600 times 60 to see the value in minutes: 60 x 1.6 = 96 minutes. Note: the labor hour figure is not dependent on the crew size. A change in crew size will result in a corresponding change in daily output, but the labor hours per unit of work will not change.

All RSMeans unit cost data includes the typical **Unit of Measure** used for estimating that item. For concrete-in-place the typical unit is cubic yards (C.Y.) or each (Ea.). For installing broadloom carpet it is square yard, and for gypsum board it is square foot. The estimator needs to take special care that the unit in the data matches the unit in the take-off. Unit conversions may be found in the Reference Section.

How RSMeans Unit Price Works (Continued)

Sample Estimate

This sample demonstrates the elements of an estimate, including a tally of the RSMeans data lines, and a summary of the markups on a contractor's work to arrive at a total cost to the owner. The RSMeans Location Factor is added at the bottom of the estimate to adjust the cost of the work to a specific location.

Work Performed: The body of the estimate shows the RSMeans data selected, including line number, a brief description of each item, its take-off unit and quantity, and the bare costs of materials, labor, and equipment. This estimate also includes a column titled "SubContract." This data is taken from the RSMeans column "Total Incl O&P," and represents the total that a subcontractor would charge a general contractor for the work, including the sub's markup for overhead and profit.

Division 1, General Requirements: This is the first division numerically, but the last division estimated. Division 1 includes project-wide needs provided by the general contractor. These requirements vary by project, but may include temporary facilities and utilities, security, testing, project cleanup, etc. For small projects a percentage can be used, typically between 5% and 15% of project cost. For large projects the costs may be itemized and priced individually.

Bonds: Bond costs should be added to the estimate. The figures here represent a typical performance bond, ensuring the owner that if the general contractor does not complete the obligations in the construction contract the bonding company will pay the cost for completion of the work.

Location Adjustment: RSMeans published data is based on national average costs. If necessary, adjust the total cost of the project using a location factor from the "Location Factor" table or the "City Cost Index" table. Use location factors if the work is general, covering multiple trades. If the work is by a single trade (e.g., masonry) use the more specific data found in the "City Cost Indexes."

Project Name: Pre-Engineered Steel Building		Architect: As Shown		
Location:	Anywhere, USA			
Line Number	Description	Qty	Unit	Materia
03 30 53.40 3940	Strip footing, 12" x 24", reinforced	34	C.Y.	$4,726.0
03 30 53.40 3950	Strip footing, 12" x 36", reinforced	15	C.Y.	$1,995.0
03 11 13.65 3000	Concrete slab edge forms	500	L.F.	$175.0
03 22 11.10 0200	Welded wire fabric reinforcing	150	C.S.F.	$2,580.0
03 31 13.35 0300	Ready mix concrete, 4000 psi for slab on grade	278	C.Y.	$29,746.0
03 31 13.70 4300	Place, strike off & consolidate concrete slab	278	C.Y.	$0.0
03 35 13.30 0250	Machine float & trowel concrete slab	15,000	S.F.	$0.0
03 15 16.20 0140	Cut control joints in concrete slab	950	L.F.	$57.0
03 39 23.13 0300	Sprayed concrete curing membrane	150	C.S.F.	$1,657.5
Division 03	Subtotal			$40,936.5
08 36 13.10 2650	Manual 10' x 10' steel sectional overhead door	8	Ea.	$8,800.0
08 36 13.10 2860	Insulation and steel back panel for OH door	800	S.F.	$3,800.0
Division 08	Subtotal			$12,600.0
13 34 19.50 1100	Pre-Engineered Steel Building, 100' x 150' x 24'	15,000	SF Flr.	$0.0
13 34 19.50 6050	Framing for PESB door opening, 3' x 7'	4	Opng.	$0.0
13 34 19.50 6100	Framing for PESB door opening, 10' x 10'	8	Opng.	$0.0
13 34 19.50 6200	Framing for PESB window opening, 4' x 3'	6	Opng.	$0.0
13 34 19.50 5750	PESB door, 3' x 7', single leaf	4	Opng.	$2,340.0
13 34 19.50 7750	PESB sliding window, 4' x 3' with screen	6	Opng.	$2,280.0
13 34 19.50 6550	PESB gutter, eave type, 26 ga., painted	300	L.F.	$2,085.0
13 34 19.50 8650	PESB roof vent, 12" wide x 10' long	15	Ea.	$540.0
13 34 19.50 6900	PESB insulation, vinyl faced, 4" thick	27,400	S.F.	$11,508.0
Division 13	Subtotal			$18,753.0
		Subtotal		$72,289.50
Division 01	General Requirements @ 7%			5,060.27
		Estimate Subtotal		$77,349.77
		Sales Tax @ 5%		3,867.49
		Subtotal A		81,217.25
		GC O & P		8,121.73
		Subtotal B		89,338.98
		Contingency @ 5%		
		Subtotal C		
		Bond @ $12/1000 +10% O&P		
		Subtotal D		
		Location Adjustment Factor		
		Grand Total		

This estimate is based on an interactive spreadsheet. A copy of this spreadsheet is located on the RSMeans website at **http:/www.reedconstructiondata.com/rsmeans/extras/546011.** You are free to download it and adjust it to your methodology.

his example shows the cost to construct a pre-engineered steel building. The foundation, doors, windows, and insulation will be stalled by the general contractor. A subcontractor will install the structural steel, roofing, and siding.

		01/01/15	R+R	
Labor	Equipment	SubContract	Estimate Total	
3,570.00	$22.10	$0.00		
1,260.00	$7.80	$0.00		
$1,190.00	$0.00	$0.00		
4,050.00	$0.00	$0.00		
$0.00	$0.00	$0.00		
4,753.80	$158.46	$0.00		
9,000.00	$450.00	$0.00		
$380.00	$85.50	$0.00		
$952.50	$0.00	$0.00		
5,156.30	$723.86	$0.00	$66,816.66	Division 03
$3,320.00	$0.00	$0.00		
$0.00	$0.00	$0.00		
3,320.00	$0.00	$0.00	$15,920.00	Division 08
$0.00	$0.00	$360,000.00		
$0.00	$0.00	$2,340.00		
$0.00	$0.00	$9,400.00		
$0.00	$0.00	$3,420.00		
$672.00	$0.00	$0.00		
$576.00	$67.20	$0.00		
$789.00	$0.00	$0.00		
$3,165.00	$0.00	$0.00		
$9,042.00	$0.00	$0.00		
4,244.00	$67.20	$375,160.00	$408,224.20	Division 13
42,720.30	$791.06	$375,160.00	$490,960.86	Subtotal
2,990.42	55.37	26,261.20		Gen. Requirements
$45,710.72	$846.43	$401,421.20	$490,960.86	Estimate Subtotal
	42.32	10,035.53		Sales tax
45,710.72	888.76	411,456.73		Subtotal
28,569.20	88.88	41,145.67		GC O&P
74,279.92	977.63	452,602.40	$617,198.93	Subtotal
			30,859.95	Contingency
			$648,058.88	Subtotal
			8,554.38	Bond
			$656,613.26	Subtotal
102.30			15,102.10	Location Adjustment
			$671,715.36	Grand Total

Sales Tax: If the work is subject to state or local sales taxes, the amount must be added to the estimate. Sales tax may be added to material costs, equipment costs, and subcontracted work. In this case, sales tax was added in all three categories. It was assumed that approximately half the subcontracted work would be material cost, so the tax was applied to 50% of the subcontract total.

GC O&P: This entry represents the general contractor's markup on material, labor, equipment, and subcontractor costs. RSMeans' standard markup on materials, equipment, and subcontracted work is 10%. In this estimate, the markup on the labor performed by the GC's workers uses "Skilled Workers Average" shown in Column F on the table "Installing Contractor's Overhead & Profit," which can be found on the inside-back cover of the book.

Contingency: A factor for contingency may be added to any estimate to represent the cost of unknowns that may occur between the time that the estimate is performed and the time the project is constructed. The amount of the allowance will depend on the stage of design at which the estimate is done, and the contractor's assessment of the risk involved. Refer to section 01 21 16.50 for contigency allowances.

Division 1 – General Requirements

Estimating Tips

01 20 00 Price and Payment Procedures

- Allowances that should be added to estimates to cover contingencies and job conditions that are not included in the national average material and labor costs are shown in section 01 21.
- When estimating historic preservation projects (depending on the condition of the existing structure and the owner's requirements), a 15%–20% contingency or allowance is recommended, regardless of the stage of the drawings.

01 30 00 Administrative Requirements

- Before determining a final cost estimate, it is a good practice to review all the items listed in Subdivisions 01 31 and 01 32 to make final adjustments for items that may need customizing to specific job conditions.
- Requirements for initial and periodic submittals can represent a significant cost to the General Requirements of a job. Thoroughly check the submittal specifications when estimating a project to determine any costs that should be included.

01 40 00 Quality Requirements

- All projects will require some degree of Quality Control. This cost is not included in the unit cost of construction listed in each division. Depending upon the terms of the contract, the various costs of inspection and testing can be the responsibility of either the owner or the contractor. Be sure to include the required costs in your estimate.

01 50 00 Temporary Facilities and Controls

- Barricades, access roads, safety nets, scaffolding, security, and many more requirements for the execution of a safe project are elements of direct cost. These costs can easily be overlooked when preparing an estimate. When looking through the major classifications of this subdivision, determine which items apply to each division in your estimate.
- Construction Equipment Rental Costs can be found in the Reference Section in section 01 54 33. Operators' wages are not included in equipment rental costs.
- Equipment mobilization and demobilization costs are not included in equipment rental costs and must be considered separately in section 01 54 36.50.
- The cost of small tools provided by the installing contractor for his workers is covered in the "Overhead" column on the "Installing Contractor's Overhead and Profit" table that lists labor trades, base rates and markups and, therefore, is included in the "Total Incl. O&P" cost of any Unit Price line item.

01 70 00 Execution and Closeout Requirements

- When preparing an estimate, thoroughly read the specifications to determine the requirements for Contract Closeout. Final cleaning, record documentation, operation and maintenance data, warranties and bonds, and spare parts and maintenance materials can all be elements of cost for the completion of a contract. Do not overlook these in your estimate.

Reference Numbers

Reference numbers are shown in shaded boxes at the beginning of some major classifications. These numbers refer to related items in the Reference Section. The reference information may be an estimating procedure, an alternate pricing method, or technical information.

Note: Not all subdivisions listed here necessarily appear in this publication. ■

01 11 Summary of Work

01 11 31 – Professional Consultants

01 11 31.10 Architectural Fees

	01 11 31.10 Architectural Fees	Crew	Daily Output	Labor-Hours	Unit	Material	Labor (2015 Bare Costs)	Equipment	Total	Total Incl O&P
0010	**ARCHITECTURAL FEES** R011110-10									
0020	For new construction									
0060	Minimum				Project				4.90%	4.90%
0090	Maximum								16%	16%
0100	For alteration work, to $500,000, add to new construction fee								50%	50%
0150	Over $500,000, add to new construction fee								25%	25%
2000	For "Greening" of building G								3%	3%

01 11 31.20 Construction Management Fees

		Crew	Daily Output	Labor-Hours	Unit	Material	Labor	Equipment	Total	Total Incl O&P
0010	**CONSTRUCTION MANAGEMENT FEES**									
0060	For work to $100,000				Project				10%	10%
0070	To $250,000								9%	9%
0090	To $1,000,000								6%	6%
0100	To $5,000,000								5%	5%
0110	To $10,000,000								4%	4%
0300	$50,000,000 job, minimum								2.50%	2.50%
0350	Maximum								4%	4%

01 11 31.30 Engineering Fees

		Crew	Daily Output	Labor-Hours	Unit	Material	Labor	Equipment	Total	Total Incl O&P
0010	**ENGINEERING FEES** R011110-30									
0020	Educational planning consultant, minimum				Project				.50%	.50%
0100	Maximum				"				2.50%	2.50%
0200	Electrical, minimum				Contrct				4.10%	4.10%
0300	Maximum								10.10%	10.10%
0400	Elevator & conveying systems, minimum								2.50%	2.50%
0500	Maximum								5%	5%
0600	Food service & kitchen equipment, minimum								8%	8%
0700	Maximum								12%	12%
0800	Landscaping & site development, minimum								2.50%	2.50%
0900	Maximum								6%	6%
1000	Mechanical (plumbing & HVAC), minimum								4.10%	4.10%
1100	Maximum								10.10%	10.10%
1200	Structural, minimum				Project				1%	1%
1300	Maximum				"				2.50%	2.50%

01 11 31.50 Models

		Crew	Daily Output	Labor-Hours	Unit	Material	Labor	Equipment	Total	Total Incl O&P
0010	**MODELS**									
0500	2 story building, scaled 100′ x 200′, simple materials and details				Ea.	5,000			5,000	5,500
0510	Elaborate materials and details				"	30,000			30,000	33,000

01 11 31.75 Renderings

		Crew	Daily Output	Labor-Hours	Unit	Material	Labor	Equipment	Total	Total Incl O&P
0010	**RENDERINGS** Color, matted, 20" x 30", eye level,									
0020	1 building, minimum				Ea.	2,050			2,050	2,275
0050	Average					2,875			2,875	3,175
0100	Maximum					4,625			4,625	5,100
1000	5 buildings, minimum					4,125			4,125	4,525
1100	Maximum					8,250			8,250	9,075
2000	Aerial perspective, color, 1 building, minimum					3,000			3,000	3,275
2100	Maximum					8,150			8,150	8,950
3000	5 buildings, minimum					5,875			5,875	6,450
3100	Maximum					11,700			11,700	12,900

01 21 Allowances

01 21 16 – Contingency Allowances

01 21 16.50 Contingencies		Crew	Daily Output	Labor-Hours	Unit	2015 Bare Costs Material	Labor	Equipment	Total	Total Incl O&P
0010	**CONTINGENCIES**, Add to estimate									
0020	Conceptual stage				Project				20%	20%
0050	Schematic stage								15%	15%
0100	Preliminary working drawing stage (Design Dev.)								10%	10%
0150	Final working drawing stage								3%	3%

01 21 53 – Factors Allowance

01 21 53.50 Factors

		Crew	Daily Output	Labor-Hours	Unit	Material	Labor	Equipment	Total	Total Incl O&P
0010	**FACTORS** Cost adjustments R012153-10									
0100	Add to construction costs for particular job requirements									
0500	Cut & patch to match existing construction, add, minimum				Costs	2%	3%			
0550	Maximum					5%	9%			
0800	Dust protection, add, minimum					1%	2%			
0850	Maximum					4%	11%			
1100	Equipment usage curtailment, add, minimum					1%	1%			
1150	Maximum					3%	10%			
1400	Material handling & storage limitation, add, minimum					1%	1%			
1450	Maximum					6%	7%			
1700	Protection of existing work, add, minimum					2%	2%			
1750	Maximum					5%	7%			
2000	Shift work requirements, add, minimum						5%			
2050	Maximum						30%			
2300	Temporary shoring and bracing, add, minimum					2%	5%			
2350	Maximum					5%	12%			
2400	Work inside prisons and high security areas, add, minimum						30%			
2450	Maximum						50%			

01 21 55 – Job Conditions Allowance

01 21 55.50 Job Conditions

		Crew	Daily Output	Labor-Hours	Unit	Material	Labor	Equipment	Total	Total Incl O&P
0010	**JOB CONDITIONS** Modifications to applicable									
0020	cost summaries									
0100	Economic conditions, favorable, deduct				Project				2%	2%
0200	Unfavorable, add								5%	5%
0300	Hoisting conditions, favorable, deduct								2%	2%
0400	Unfavorable, add								5%	5%
0500	General Contractor management, experienced, deduct								2%	2%
0600	Inexperienced, add								10%	10%
0700	Labor availability, surplus, deduct								1%	1%
0800	Shortage, add								10%	10%
0900	Material storage area, available, deduct								1%	1%
1000	Not available, add								2%	2%
1100	Subcontractor availability, surplus, deduct								5%	5%
1200	Shortage, add								12%	12%
1300	Work space, available, deduct								2%	2%
1400	Not available, add								5%	5%

01 21 57 – Overtime Allowance

01 21 57.50 Overtime

		Crew	Daily Output	Labor-Hours	Unit	Material	Labor	Equipment	Total	Total Incl O&P
0010	**OVERTIME** for early completion of projects or where R012909-90									
0020	labor shortages exist, add to usual labor, up to				Costs		100%			

01 21 Allowances

01 21 61 – Cost Indexes

01 21 61.20 Historical Cost Indexes		Crew	Daily Output	Labor-Hours	Unit	Material	2015 Bare Costs Labor	Equipment	Total	Total Incl O&P
0010	**HISTORICAL COST INDEXES** (See Reference Section)									

01 21 63 – Taxes

01 21 63.10 Taxes

		Crew	Daily Output	Labor-Hours	Unit	Material	Labor	Equipment	Total	Total Incl O&P
0010	**TAXES** R012909-80									
0020	Sales tax, State, average				%	5.04%				
0050	Maximum R012909-85					7.50%				
0200	Social Security, on first $117,000 of wages						7.65%			
0300	Unemployment, combined Federal and State, minimum						.60%			
0350	Average						7.80%			
0400	Maximum						12.87%			

01 31 Project Management and Coordination

01 31 13 – Project Coordination

01 31 13.20 Field Personnel

		Crew	Daily Output	Labor-Hours	Unit	Material	Labor	Equipment	Total	Total Incl O&P
0010	**FIELD PERSONNEL**									
0020	Clerk, average				Week		450		450	730
0100	Field engineer, minimum						1,075		1,075	1,725
0120	Average						1,400		1,400	2,275
0140	Maximum						1,575		1,575	2,575
0160	General purpose laborer, average						1,500		1,500	2,475
0180	Project manager, minimum						1,975		1,975	3,225
0200	Average						2,275		2,275	3,700
0220	Maximum						2,600		2,600	4,225
0240	Superintendent, minimum						1,925		1,925	3,125
0260	Average						2,125		2,125	3,450
0280	Maximum						2,400		2,400	3,925
0290	Timekeeper, average						1,225		1,225	2,000

01 31 13.30 Insurance

		Crew	Daily Output	Labor-Hours	Unit	Material	Labor	Equipment	Total	Total Incl O&P
0010	**INSURANCE** R013113-40									
0020	Builders risk, standard, minimum				Job				.24%	.24%
0050	Maximum R013113-60								.64%	.64%
0200	All-risk type, minimum								.25%	.25%
0250	Maximum								.62%	.62%
0400	Contractor's equipment floater, minimum				Value				.50%	.50%
0450	Maximum				"				1.50%	1.50%
0800	Workers' compensation & employer's liability, average									
0850	by trade, carpentry, general				Payroll		14.93%			
0900	Clerical						.50%			
0950	Concrete						12.93%			
1000	Electrical						5.76%			
1050	Excavation						9.70%			
1100	Glazing						13.29%			
1150	Insulation						11.66%			
1200	Lathing						8.38%			
1250	Masonry						13.68%			
1300	Painting & decorating						11.72%			
1350	Pile driving						14.31%			
1400	Plastering						10.90%			
1450	Plumbing						6.98%			
1500	Roofing						31.69%			

01 31 13 – Project Coordination

01 31 13.30 Insurance		Crew	Daily Output	Labor-Hours	Unit	Material	2015 Bare Costs Labor	Equipment	Total	Total Incl O&P
1550	Sheet metal work (HVAC)				Payroll		8.82%			
1600	Steel erection, structural						31.68%			
1650	Tile work, interior ceramic						8.86%			
1700	Waterproofing, brush or hand caulking						6.56%			
1800	Wrecking						26.14%			
2000	Range of 35 trades in 50 states, excl. wrecking, min.						1.31%			
2100	Average						13.50%			
2200	Maximum				↓		132%			

01 31 13.40 Main Office Expense		Crew	Daily Output	Labor-Hours	Unit	Material	Labor	Equipment	Total	Total Incl O&P
0010	**MAIN OFFICE EXPENSE** Average for General Contractors R013113-50									
0020	As a percentage of their annual volume									
0030	Annual volume to $300,000, minimum				% Vol.				20%	
0040	Maximum								30%	
0060	To $500,000, minimum								17%	
0070	Maximum								22%	
0080	To $1,000,000, minimum								16%	
0090	Maximum								19%	
0110	To $3,000,000, minimum								14%	
0120	Maximum								16%	
0130	To $5,000,000, minimum								8%	
0140	Maximum				↓				10%	

01 31 13.50 General Contractor's Mark-Up		Crew	Daily Output	Labor-Hours	Unit	Material	Labor	Equipment	Total	Total Incl O&P
0010	**GENERAL CONTRACTOR'S MARK-UP** on Change Orders									
0200	Extra work, by subcontractors, add				%				10%	10%
0250	By General Contractor, add								15%	15%
0400	Omitted work, by subcontractors, deduct all but								5%	5%
0450	By General Contractor, deduct all but								7.50%	7.50%
0600	Overtime work, by subcontractors, add								15%	15%
0650	By General Contractor, add				↓				10%	10%

01 31 13.70 Overhead		Crew	Daily Output	Labor-Hours	Unit	Material	Labor	Equipment	Total	Total Incl O&P
0010	**OVERHEAD** As a percent of installing contractors direct costs R013113-50									
0040	Includes an allowance for home office expenses, FICA,									
0060	Risk & public liability insur. and unemploy.; minimum				%				5%	
0080	Average								15%	
0100	Maximum								30%	
0120	With profit allowance, by size of project; under $50,000								40%	
0140	$50,000 to $100,000								35%	
0160	$100,000 to $500,000								25%	
0180	$500,000 to $1,000,000				↓				20%	
0200	Mark-up on general contractors total incl. O&P for									
0220	Handling subcontract; minimum				%				5%	5%
0240	Average								10%	10%
0260	Maximum				↓				15%	15%

01 31 13.90 Performance Bond		Crew	Daily Output	Labor-Hours	Unit	Material	Labor	Equipment	Total	Total Incl O&P
0010	**PERFORMANCE BOND** R013113-80									
0020	For buildings, minimum				Job				.60%	.60%
0100	Maximum				"				2.50%	2.50%

01 32 Construction Progress Documentation

01 32 13 – Scheduling of work

01 32 13.50 Scheduling		Crew	Daily Output	Labor-Hours	Unit	Material	Labor	Equipment	Total	Total Incl O&P
						2015 Bare Costs				
0010	**SCHEDULING**									
0020	Critical path, as % of architectural fee, minimum				%				.50%	.50%
0100	Maximum				"				1%	1%
0300	Computer-update, micro, no plots, minimum				Ea.				455	500
0400	Including plots, maximum				"				1,450	1,600
0600	Rule of thumb, CPM scheduling, small job ($10 Million)				Job				.05%	.05%
0650	Large job ($50 Million +)								.03%	.03%
0700	Including cost control, small job								.08%	.08%
0750	Large job				↓				.04%	.04%

01 32 33 – Photographic Documentation

01 32 33.50 Photographs		Crew	Daily Output	Labor-Hours	Unit	Material	Labor	Equipment	Total	Total Incl O&P
0010	**PHOTOGRAPHS**									
0020	8" x 10", 4 shots, 2 prints ea., std. mounting				Set	475			475	525
0100	Hinged linen mounts					530			530	580
0200	8" x 10", 4 shots, 2 prints each, in color					425			425	470
0300	For I.D. slugs, add to all above					5.30			5.30	5.85
0500	Aerial photos, initial fly-over, 5 shots, digital images					400			400	440
0550	10 shots, digital images, 1 print					450			450	495
0600	For each addtional print from fly-over					200			200	220
0700	For full color prints, add				↓	40%				
0750	Add for traffic control area				Ea.	305			305	335
0900	For over 30 miles from airport, add per				Mile	6			6	6.60
1500	Time lapse equipment, camera and projector, buy				Ea.	2,650			2,650	2,925
1550	Rent per month				"	1,600			1,600	1,750
1700	Cameraman and film, including processing, B.&W.				Day	1,225			1,225	1,350
1720	Color				"	1,425			1,425	1,550

01 41 Regulatory Requirements

01 41 26 – Permit Requirements

01 41 26.50 Permits		Crew	Daily Output	Labor-Hours	Unit	Material	Labor	Equipment	Total	Total Incl O&P
0010	**PERMITS**									
0020	Rule of thumb, most cities, minimum				Job				.50%	.50%
0100	Maximum				"				2%	2%

01 45 Quality Control

01 45 23 – Testing and Inspecting Services

01 45 23.50 Testing		Crew	Daily Output	Labor-Hours	Unit	Material	Labor	Equipment	Total	Total Incl O&P
0010	**TESTING** and Inspecting Services									
0015	For concrete building costing $1,000,000, minimum				Project				4,725	5,200
0020	Maximum								38,000	41,800
0050	Steel building, minimum								4,725	5,200
0070	Maximum								14,800	16,300
0100	For building costing, $10,000,000, minimum								30,100	33,100
0150	Maximum				↓				48,200	53,000
0200	Asphalt testing, compressive strength Marshall stability, set of 3				Ea.				145	165
0220	Density, set of 3								86	95
0250	Extraction, individual tests on sample								136	150
0300	Penetration								41	45
0350	Mix design, 5 specimens				↓				182	200

01 45 23 – Testing and Inspecting Services

01 45 23.50 Testing		Crew	Daily Output	Labor-Hours	Unit	Material	2015 Bare Costs Labor	Equipment	Total	Total Incl O&P
0360	Additional specimen				Ea.				36	40
0400	Specific gravity								41	45
0420	Swell test								64	70
0450	Water effect and cohesion, set of 6								182	200
0470	Water effect and plastic flow								64	70
0600	Concrete testing, aggregates, abrasion, ASTM C 131								136	150
0650	Absorption, ASTM C 127								42	46
0800	Petrographic analysis, ASTM C 295								775	850
0900	Specific gravity, ASTM C 127								50	55
1000	Sieve analysis, washed, ASTM C 136								59	65
1050	Unwashed								59	65
1200	Sulfate soundness								114	125
1300	Weight per cubic foot								36	40
1500	Cement, physical tests, ASTM C 150								320	350
1600	Chemical tests, ASTM C 150								245	270
1800	Compressive test, cylinder, delivered to lab, ASTM C 39								12	13
1900	Picked up by lab, minimum								14	15
1950	Average								18	20
2000	Maximum								27	30
2200	Compressive strength, cores (not incl. drilling), ASTM C 42								36	40
2250	Core drilling, 4" diameter (plus technician)				Inch				23	25
2260	Technician for core drilling				Hr.				45	50
2300	Patching core holes				Ea.				22	24
2400	Drying shrinkage at 28 days								236	260
2500	Flexural test beams, ASTM C 78								59	65
2600	Mix design, one batch mix								259	285
2650	Added trial batches								120	132
2800	Modulus of elasticity, ASTM C 469								164	180
2900	Tensile test, cylinders, ASTM C 496								45	50
3000	Water-Cement ratio curve, 3 batches								141	155
3100	4 batches								186	205
3300	Masonry testing, absorption, per 5 brick, ASTM C 67								45	50
3350	Chemical resistance, per 2 brick								50	55
3400	Compressive strength, per 5 brick, ASTM C 67								68	75
3420	Efflorescence, per 5 brick, ASTM C 67								68	75
3440	Imperviousness, per 5 brick								87	96
3470	Modulus of rupture, per 5 brick								86	95
3500	Moisture, block only								32	35
3550	Mortar, compressive strength, set of 3								23	25
4100	Reinforcing steel, bend test								55	61
4200	Tensile test, up to #8 bar								36	40
4220	#9 to #11 bar								41	45
4240	#14 bar and larger								64	70
4400	Soil testing, Atterberg limits, liquid and plastic limits								59	65
4510	Hydrometer analysis								109	120
4530	Specific gravity, ASTM D 354								44	48
4600	Sieve analysis, washed, ASTM D 422								55	60
4700	Unwashed, ASTM D 422								59	65
4710	Consolidation test (ASTM D2435), minimum								250	275
4715	Maximum								430	475
4720	Density and classification of undisturbed sample								73	80
4735	Soil density, nuclear method, ASTM D2922								35	38.50
4740	Sand cone method ASTM D1556								27	30

01 45 Quality Control

01 45 23 – Testing and Inspecting Services

01 45 23.50 Testing		Crew	Daily Output	Labor-Hours	Unit	Material	Labor	Equipment	Total	Total Incl O&P
						2015 Bare Costs				
4750	Moisture content, ASTM D 2216				Ea.				9	10
4780	Permeability test, double ring infiltrometer								500	550
4800	Permeability, var. or constant head, undist., ASTM D 2434								227	250
4850	Recompacted								250	275
4900	Proctor compaction, 4" standard mold, ASTM D 698								123	135
4950	6" modified mold								68	75
5100	Shear tests, triaxial, minimum								410	450
5150	Maximum								545	600
5300	Direct shear, minimum, ASTM D 3080								320	350
5350	Maximum								410	450
5550	Technician for inspection, per day, earthwork								320	350
5570	Concrete								280	310
5650	Bolting								400	440
5750	Roofing								480	530
5790	Welding								480	530
5820	Non-destructive metal testing, dye penetrant				Day				310	340
5840	Magnetic particle								310	340
5860	Radiography								450	495
5880	Ultrasonic								310	340
6000	Welding certification, minimum				Ea.				91	100
6100	Maximum				"				250	275
7000	Underground storage tank									
7500	Volumetric tightness test ,<=12,000 gal.				Ea.				435	480
7510	<=30,000 gal.				"				615	675
7600	Vadose zone (soil gas) sampling, 10-40 samples, min.				Day				1,375	1,500
7610	Maximum				"				2,275	2,500
7700	Ground water monitoring incl. drilling 3 wells, min.				Total				4,550	5,000
7710	Maximum				"				6,375	7,000
8000	X-ray concrete slabs				Ea.				182	200
9000	Thermographic testing, for bldg envelope heat loss, average 2,000 S.F. G				"				500	500

01 51 Temporary Utilities

01 51 13 – Temporary Electricity

01 51 13.80 Temporary Utilities

		Crew	Daily Output	Labor-Hours	Unit	Material	Labor	Equipment	Total	Total Incl O&P
0010	**TEMPORARY UTILITIES**									
0350	Lighting, lamps, wiring, outlets, 40,000 SF building, 8 strings	1 Elec	34	.235	CSF Flr	4.91	12.85		17.76	25.50
0360	16 strings	"	17	.471		9.80	25.50		35.30	51
0400	Power for temp lighting only, 6.6 KWH, per month								.92	1.01
0430	11.8 KWH, per month								1.65	1.82
0450	23.6 KWH, per month								3.30	3.63
0600	Power for job duration incl. elevator, etc., minimum								47	51.50
0650	Maximum								110	121
0700	Temporary construction water bill per mo. average				Month	68			68	75
1000	Toilet, portable, see Equip. Rental 01 54 33 in Reference Section									

01 52 Construction Facilities

01 52 13 – Field Offices and Sheds

01 52 13.20 Office and Storage Space

	01 52 13.20 Office and Storage Space	Crew	Daily Output	Labor-Hours	Unit	2015 Bare Costs Material	Labor	Equipment	Total	Total Incl O&P
0010	OFFICE AND STORAGE SPACE									
0020	Office trailer, furnished, no hookups, 20′ x 8′, buy	2 Skwk	1	16	Ea.	10,400	780		11,180	12,700
0250	Rent per month					188			188	206
0300	32′ x 8′, buy	2 Skwk	.70	22.857		15,500	1,100		16,600	18,900
0350	Rent per month					239			239	262
0400	50′ x 10′, buy	2 Skwk	.60	26.667		24,100	1,300		25,400	28,600
0450	Rent per month					340			340	375
0500	50′ x 12′, buy	2 Skwk	.50	32		29,400	1,550		30,950	34,800
0550	Rent per month					410			410	455
0700	For air conditioning, rent per month, add					48.50			48.50	53.50
0800	For delivery, add per mile				Mile	11			11	12.10
0890	Delivery each way				Ea.	200			200	220
0900	Bunk house trailer, 8′ x 40′ duplex dorm with kitchen, no hookups, buy	2 Carp	1	16		37,300	750		38,050	42,200
0910	9 man with kitchen and bath, no hookups, buy		1	16		38,000	750		38,750	43,000
0920	18 man sleeper with bath, no hookups, buy		1	16		49,000	750		49,750	55,000
1000	Portable buildings, prefab, on skids, economy, 8′ x 8′		265	.060	S.F.	25	2.83		27.83	32
1100	Deluxe, 8′ x 12′		150	.107	"	20	5		25	30
1200	Storage boxes, 20′ x 8′, buy	2 Skwk	1.80	8.889	Ea.	2,775	430		3,205	3,750
1250	Rent per month					82.50			82.50	91
1300	40′ x 8′, buy	2 Skwk	1.40	11.429		3,850	555		4,405	5,125
1350	Rent per month					101			101	111
5000	Air supported structures, see Section 13 31 13.13									

01 52 13.40 Field Office Expense

	01 52 13.40 Field Office Expense	Crew	Daily Output	Labor-Hours	Unit	Material	Labor	Equipment	Total	Total Incl O&P
0010	FIELD OFFICE EXPENSE									
0100	Office equipment rental average				Month	200			200	220
0120	Office supplies, average				"	80			80	88
0125	Office trailer rental, see Section 01 52 13.20									
0140	Telephone bill; avg. bill/month incl. long dist.				Month	85			85	93.50
0160	Lights & HVAC				"	160			160	176

01 54 Construction Aids

01 54 09 – Protection Equipment

01 54 09.50 Personnel Protective Equipment

	01 54 09.50 Personnel Protective Equipment	Crew	Daily Output	Labor-Hours	Unit	Material	Labor	Equipment	Total	Total Incl O&P
0010	PERSONNEL PROTECTIVE EQUIPMENT									
0015	Hazardous waste protection									
0020	Respirator mask only, full face, silicone				Ea.	335			335	370
0030	Half face, silicone					62.50			62.50	68.50
0040	Respirator cartridges, 2 req'd/mask, dust or asbestos					4.03			4.03	4.43
0050	Chemical vapor					2.88			2.88	3.16
0060	Combination vapor and dust					8.60			8.60	9.45
0100	Emergency escape breathing apparatus, 5 minutes					720			720	790
0110	10 minutes					830			830	915
0150	Self contained breathing apparatus with full face piece, 30 minutes					2,375			2,375	2,600
0160	60 minutes					4,050			4,050	4,450
0200	Encapsulating suits, limited use, level A					1,900			1,900	2,100
0210	Level B					435			435	480
0300	Over boots, latex				Pr.	7.80			7.80	8.60
0310	PVC					29.50			29.50	32.50
0320	Neoprene					40.50			40.50	44.50
0400	Gloves, nitrile/PVC					38			38	41.50
0410	Neoprene coated					34			34	37.50

01 54 Construction Aids

01 54 09 – Protection Equipment

01 54 09.50 Personnel Protective Equipment

		Crew	Daily Output	Labor-Hours	Unit	2015 Bare Costs Material	Labor	Equipment	Total	Total Incl O&P
0500	Fire protection equipment, shoes/boots				Pr.	247			247	272
0510	Hard-hats				Ea.	254			254	280
0520	Gloves				Pr.	65.50			65.50	72

01 54 09.60 Safety Nets

		Crew	Daily Output	Labor-Hours	Unit	Material	Labor	Equipment	Total	Total Incl O&P
0010	**SAFETY NETS**									
0020	No supports, stock sizes, nylon, 3 1/2" mesh				S.F.	2.91			2.91	3.20
0100	Polypropylene, 6" mesh					1.59			1.59	1.75
0200	Small mesh debris nets, 1/4" mesh, stock sizes					.67			.67	.74
0220	Combined 3 1/2" mesh and 1/4" mesh, stock sizes					4.65			4.65	5.10
0300	Monthly rental, 4" mesh, stock sizes, 1st month					.50			.50	.55
0320	2nd month rental					.25			.25	.28
0340	Maximum rental/year				↓	1.15			1.15	1.27

01 54 16 – Temporary Hoists

01 54 16.50 Weekly Forklift Crew

		Crew	Daily Output	Labor-Hours	Unit	Material	Labor	Equipment	Total	Total Incl O&P
0010	**WEEKLY FORKLIFT CREW**									
0100	All-terrain forklift, 45' lift, 35' reach, 9000 lb. capacity	A-3P	.20	40	Week		1,950	2,625	4,575	5,975

01 54 19 – Temporary Cranes

01 54 19.50 Daily Crane Crews

		Crew	Daily Output	Labor-Hours	Unit	Material	Labor	Equipment	Total	Total Incl O&P
0010	**DAILY CRANE CREWS** for small jobs, portal to portal									
0100	12-ton truck-mounted hydraulic crane	A-3H	1	8	Day		415	860	1,275	1,600
0200	25-ton	A-3I	1	8			415	990	1,405	1,750
0300	40-ton	A-3J	1	8			415	1,225	1,640	2,000
0400	55-ton	A-3K	1	16			775	1,625	2,400	3,025
0500	80-ton	A-3L	1	16			775	2,350	3,125	3,800
0600	100-ton	A-3M	1	16	↓		775	2,325	3,100	3,800
0900	If crane is needed on a Saturday, Sunday or Holiday									
0910	At time-and-a-half, add				Day		50%			
0920	At double time, add				"		100%			

01 54 19.60 Monthly Tower Crane Crew

		Crew	Daily Output	Labor-Hours	Unit	Material	Labor	Equipment	Total	Total Incl O&P
0010	**MONTHLY TOWER CRANE CREW**, excludes concrete footing									
0100	Static tower crane, 130' high, 106' jib, 6200 lb. capacity	A-3N	.05	176	Month		9,100	24,800	33,900	41,700

01 54 23 – Temporary Scaffolding and Platforms

01 54 23.60 Pump Staging

		Crew	Daily Output	Labor-Hours	Unit	Material	Labor	Equipment	Total	Total Incl O&P
0010	**PUMP STAGING**, Aluminum R015423-20									
0200	24' long pole section, buy				Ea.	410			410	450
0300	18' long pole section, buy					320			320	350
0400	12' long pole section, buy					215			215	236
0500	6' long pole section, buy					113			113	124
0600	6' long splice joint section, buy					84			84	92.50
0700	Pump jack, buy					172			172	189
0900	Foldable brace, buy					67.50			67.50	74
1000	Workbench/back safety rail support, buy					91			91	100
1100	Scaffolding planks/workbench, 14" wide x 24' long, buy					775			775	850
1200	Plank end safety rail, buy					345			345	380
1250	Safety net, 22' long, buy				↓	410			410	450
1300	System in place, 50' working height, per use based on 50 uses	2 Carp	84.80	.189	C.S.F.	6.95	8.85		15.80	22
1400	100 uses		84.80	.189		3.46	8.85		12.31	18.30
1500	150 uses	↓	84.80	.189	↓	2.32	8.85		11.17	17.05

01 54 Construction Aids

01 54 23 – Temporary Scaffolding and Platforms

01 54 23.70 Scaffolding		Crew	Daily Output	Labor-Hours	Unit	2015 Bare Costs Material	Labor	Equipment	Total	Total Incl O&P
0010	**SCAFFOLDING** R015423-10									
0015	Steel tube, regular, no plank, labor only to erect & dismantle									
0090	Building exterior, wall face, 1 to 5 stories, 6′-4″ x 5′ frames	3 Carp	8	3	C.S.F.		141		141	231
0200	6 to 12 stories	4 Carp	8	4			188		188	310
0301	13 to 20 stories	5 Clab	8	5			188		188	310
0460	Building interior, wall face area, up to 16′ high	3 Carp	12	2			94		94	154
0560	16′ to 40′ high		10	2.400			113		113	185
0800	Building interior floor area, up to 30′ high		150	.160	C.C.F.		7.50		7.50	12.30
0900	Over 30′ high	4 Carp	160	.200	″		9.40		9.40	15.40
0906	Complete system for face of walls, no plank, material only rent/mo				C.S.F.	35.50			35.50	39.50
0908	Interior spaces, no plank, material only rent/mo				C.C.F.	4.33			4.33	4.76
0910	Steel tubular, heavy duty shoring, buy									
0920	Frames 5′ high 2′ wide				Ea.	81.50			81.50	90
0925	5′ high 4′ wide					93			93	102
0930	6′ high 2′ wide					93.50			93.50	103
0935	6′ high 4′ wide					109			109	120
0940	Accessories									
0945	Cross braces				Ea.	15.50			15.50	17.05
0950	U-head, 8″ x 8″					19.10			19.10	21
0955	J-head, 4″ x 8″					13.90			13.90	15.30
0960	Base plate, 8″ x 8″					15.50			15.50	17.05
0965	Leveling jack					33.50			33.50	36.50
1000	Steel tubular, regular, buy									
1100	Frames 3′ high 5′ wide				Ea.	75			75	82.50
1150	5′ high 5′ wide					86.50			86.50	95
1200	6′-4″ high 5′ wide					118			118	129
1350	7′-6″ high 6′ wide					158			158	173
1500	Accessories cross braces					15.50			15.50	17.05
1550	Guardrail post					16.40			16.40	18.05
1600	Guardrail 7′ section					8.70			8.70	9.60
1650	Screw jacks & plates					21.50			21.50	24
1700	Sidearm brackets					30.50			30.50	33.50
1750	8″ casters					31			31	34
1800	Plank 2″ x 10″ x 16′-0″					57.50			57.50	63.50
1900	Stairway section					286			286	315
1910	Stairway starter bar					32			32	35
1920	Stairway inside handrail					58.50			58.50	64
1930	Stairway outside handrail					86.50			86.50	95
1940	Walk-thru frame guardrail					41.50			41.50	46
2000	Steel tubular, regular, rent/mo.									
2100	Frames 3′ high 5′ wide				Ea.	5			5	5.50
2150	5′ high 5′ wide					5			5	5.50
2200	6′-4″ high 5′ wide					5.50			5.50	6.05
2250	7′-6″ high 6′ wide					10			10	11
2500	Accessories, cross braces					1			1	1.10
2550	Guardrail post					1			1	1.10
2600	Guardrail 7′ section					1			1	1.10
2650	Screw jacks & plates					2			2	2.20
2700	Sidearm brackets					2			2	2.20
2750	8″ casters					8			8	8.80
2800	Outrigger for rolling tower					3			3	3.30
2850	Plank 2″ x 10″ x 16′-0″					10			10	11

01 54 23 – Temporary Scaffolding and Platforms

01 54 23.70 Scaffolding		Crew	Daily Output	Labor-Hours	Unit	Material	2015 Bare Costs Labor	Equipment	Total	Total Incl O&P
2900	Stairway section				Ea.	35			35	38.50
2940	Walk-thru frame guardrail					2.50			2.50	2.75
3000	Steel tubular, heavy duty shoring, rent/mo.									
3250	5′ high 2′ & 4′ wide				Ea.	8.50			8.50	9.35
3300	6′ high 2′ & 4′ wide					8.50			8.50	9.35
3500	Accessories, cross braces					1			1	1.10
3600	U - head, 8″ x 8″					2.50			2.50	2.75
3650	J - head, 4″ x 8″					2.50			2.50	2.75
3700	Base plate, 8″ x 8″					1			1	1.10
3750	Leveling jack					2.50			2.50	2.75
4000	Scaffolding, stl. tubular, reg., no plank, labor only to erect & dismantle									
4100	Building exterior 2 stories	3 Carp	8	3	C.S.F.		141		141	231
4150	4 stories	"	8	3			141		141	231
4200	6 stories	4 Carp	8	4			188		188	310
4250	8 stories		8	4			188		188	310
4300	10 stories		7.50	4.267			200		200	330
4350	12 stories		7.50	4.267			200		200	330
5700	Planks, 2″ x 10″ x 16′-0″, labor only to erect & remove to 50′ H	3 Carp	72	.333	Ea.		15.65		15.65	25.50
5800	Over 50′ high	4 Carp	80	.400	"		18.80		18.80	31
6000	Heavy duty shoring for elevated slab forms to 8′-2″ high, floor area									
6100	Labor only to erect & dismantle	4 Carp	16	2	C.S.F.		94		94	154
6110	Materials only, rent.mo				"	44.50			44.50	49
6500	To 14′-8″ high									
6600	Labor only to erect & dismantle	4 Carp	10	3.200	C.S.F.		150		150	246
6610	Materials only, rent/mo				"	65			65	71.50

01 54 23.75 Scaffolding Specialties		Crew	Daily Output	Labor-Hours	Unit	Material	Labor	Equipment	Total	Total Incl O&P
0010	**SCAFFOLDING SPECIALTIES**									
1200	Sidewalk bridge, heavy duty steel posts & beams, including									
1210	parapet protection & waterproofing (material cost is rent/month)									
1220	8′ to 10′ wide, 2 posts	3 Carp	15	1.600	L.F.	45	75		120	173
1230	3 posts	"	10	2.400	"	69.50	113		182.50	262
1500	Sidewalk bridge using tubular steel scaffold frames including									
1510	planking (material cost is rent/month)	3 Carp	45	.533	L.F.	8.55	25		33.55	50.50
1600	For 2 uses per month, deduct from all above					50%				
1700	For 1 use every 2 months, add to all above					100%				
1900	Catwalks, 20″ wide, no guardrails, 7′ span, buy				Ea.	166			166	183
2000	10′ span, buy					233			233	256
3720	Putlog, standard, 8′ span, with hangers, buy					162			162	178
3730	Rent per month					20			20	22
3750	12′ span, buy					203			203	223
3755	Rent per month					25			25	27.50
3760	Trussed type, 16′ span, buy					365			365	400
3770	Rent per month					30			30	33
3790	22′ span, buy					420			420	465
3795	Rent per month					40			40	44
3800	Rolling ladders with handrails, 30″ wide, buy, 2 step					267			267	294
4000	7 step					760			760	835
4050	10 step					1,075			1,075	1,175
4100	Rolling towers, buy, 5′ wide, 7′ long, 10′ high					1,200			1,200	1,325
4200	For 5′ high added sections, to buy, add					204			204	224
4300	Complete incl. wheels, railings, outriggers,									
4350	21′ high, to buy				Ea.	2,025			2,025	2,225

01 54 23 – Temporary Scaffolding and Platforms

01 54 23.75 Scaffolding Specialties

	01 54 23.75 Scaffolding Specialties	Crew	Daily Output	Labor-Hours	Unit	2015 Bare Costs Material	Labor	Equipment	Total	Total Incl O&P
4400	Rent/month = 5% of purchase cost				Ea.	200			200	220
5000	Motorized work platform, mast climber									
5050	Base unit, 50′ W, less than 100′ tall, rent/mo				Ea.	2,900			2,900	3,175
5100	Less than 200′ tall, rent/mo					3,700			3,700	4,075
5150	Less than 300′ tall, rent/mo					4,375			4,375	4,825
5200	Less than 400′ tall, rent/mo					5,000			5,000	5,500
5250	Set up and demob, per unit, less than 100′ tall	B-68F	16.60	1.446	C.S.F.		71.50	20	91.50	138
5300	Less than 200′ tall		25	.960			47.50	13.30	60.80	91.50
5350	Less than 300′ tall		30	.800			39.50	11.05	50.55	76
5400	Less than 400′ tall		33	.727			36	10.05	46.05	69.50
5500	Mobilization (price includes freight in and out) per unit				Ea.				1,100	1,225

01 54 23.80 Staging Aids

	01 54 23.80 Staging Aids	Crew	Daily Output	Labor-Hours	Unit	Material	Labor	Equipment	Total	Total Incl O&P
0010	**STAGING AIDS** and fall protection equipment									
0100	Sidewall staging bracket, tubular, buy				Ea.	55			55	60.50
0110	Cost each per day, based on 250 days use				Day	.22			.22	.24
0200	Guard post, buy				Ea.	50.50			50.50	55.50
0210	Cost each per day, based on 250 days use				Day	.20			.20	.22
0300	End guard chains, buy per pair				Pair	36			36	39.50
0310	Cost per set per day, based on 250 days use				Day	.22			.22	.24
1000	Roof shingling bracket, steel, buy				Ea.	9.80			9.80	10.80
1010	Cost each per day, based on 250 days use				Day	.04			.04	.04
1100	Wood bracket, buy				Ea.	16.80			16.80	18.50
1110	Cost each per day, based on 250 days use				Day	.07			.07	.07
2000	Ladder jack, aluminum, buy per pair				Pair	118			118	130
2010	Cost per pair per day, based on 250 days use				Day	.47			.47	.52
2100	Steel siderail jack, buy per pair				Pair	97.50			97.50	107
2110	Cost per pair per day, based on 250 days use				Day	.39			.39	.43
3000	Laminated wood plank, 2″ x 10″ x 16′, buy				Ea.	57.50			57.50	63.50
3010	Cost each per day, based on 250 days use				Day	.23			.23	.25
3100	Aluminum scaffolding plank, 20″ wide x 24′ long, buy				Ea.	755			755	830
3110	Cost each per day, based on 250 days use				Day	3.01			3.01	3.31
4000	Nylon full body harness, lanyard and rope grab				Ea.	172			172	189
4010	Cost each per day, based on 250 days use				Day	.69			.69	.76
4100	Rope for safety line, 5/8″ x 100′ nylon, buy				Ea.	56			56	61.50
4110	Cost each per day, based on 250 days use				Day	.22			.22	.25
4200	Permanent U-Bolt roof anchor, buy				Ea.	35.50			35.50	39.50
4300	Temporary (one use) roof ridge anchor, buy				"	32			32	35.50
5000	Installation (setup and removal) of staging aids									
5010	Sidewall staging bracket	2 Carp	64	.250	Ea.		11.75		11.75	19.25
5020	Guard post with 2 wood rails	"	64	.250			11.75		11.75	19.25
5030	End guard chains, set	1 Carp	64	.125			5.85		5.85	9.60
5100	Roof shingling bracket		96	.083			3.91		3.91	6.40
5200	Ladder jack		64	.125			5.85		5.85	9.60
5300	Wood plank, 2″ x 10″ x 16′	2 Carp	80	.200			9.40		9.40	15.40
5310	Aluminum scaffold plank, 20″ x 24′	"	40	.400			18.80		18.80	31
5410	Safety rope	1 Carp	40	.200			9.40		9.40	15.40
5420	Permanent U-Bolt roof anchor (install only)	2 Carp	40	.400			18.80		18.80	31
5430	Temporary roof ridge anchor (install only)	1 Carp	64	.125			5.85		5.85	9.60

01 54 Construction Aids

01 54 26 – Temporary Swing Staging

01 54 26.50 Swing Staging		Crew	Daily Output	Labor-Hours	Unit	2015 Bare Costs Material	Labor	Equipment	Total	Total Incl O&P
0010	**SWING STAGING**, 500 lb. cap., 2′ wide to 24′ long, hand operated									
0020	steel cable type, with 60′ cables, buy				Ea.	4,900			4,900	5,375
0030	Rent per month				"	490			490	540
0600	Lightweight (not for masons) 24′ long for 150′ height,									
0610	manual type, buy				Ea.	10,200			10,200	11,200
0620	Rent per month					1,025			1,025	1,125
0700	Powered, electric or air, to 150′ high, buy					25,600			25,600	28,200
0710	Rent per month					1,800			1,800	1,975
0780	To 300′ high, buy					26,000			26,000	28,600
0800	Rent per month					1,825			1,825	2,000
1000	Bosun's chair or work basket 3′ x 3.5′, to 300′ high, electric, buy					10,400			10,400	11,400
1010	Rent per month					730			730	800
2200	Move swing staging (setup and remove)	E-4	2	16	Move		850	73	923	1,600

01 54 36 – Equipment Mobilization

01 54 36.50 Mobilization

Line	Description	Crew	Daily Output	Labor-Hours	Unit	Material	Labor	Equipment	Total	Total Incl O&P
0010	**MOBILIZATION** (Use line item again for demobilization) R015436-50									
0015	Up to 25 mi. haul dist. (50 mi. RT for mob/demob crew)									
1200	Small equipment, placed in rear of, or towed by pickup truck	A-3A	4	2	Ea.		97	39	136	197
1300	Equipment hauled on 3-ton capacity towed trailer	A-3Q	2.67	3			146	67	213	305
1400	20-ton capacity	B-34U	2	8			355	237	592	825
1500	40-ton capacity	B-34N	2	8			365	375	740	995
1600	50-ton capacity	B-34V	1	24			1,125	1,050	2,175	2,950
1700	Crane, truck-mounted, up to 75 ton (driver only)	1 Eqhv	4	2			103		103	164
1800	Over 75 ton (with chase vehicle)	A-3E	2.50	6.400			294	62.50	356.50	540
2400	Crane, large lattice boom, requiring assembly	B-34W	.50	144			6,275	7,500	13,775	18,400
2500	For each additional 5 miles haul distance, add						10%	10%		
3000	For large pieces of equipment, allow for assembly/knockdown									
3001	For mob/demob of vibrofloatation equip, see Section 31 45 13.10									
3100	For mob/demob of micro-tunneling equip, see Section 33 05 23.19									
3200	For mob/demob of pile driving equip, see Section 31 62 19.10									
3300	For mob/demob of caisson drilling equip, see Section 31 63 26.13									

01 55 Vehicular Access and Parking

01 55 23 – Temporary Roads

01 55 23.50 Roads and Sidewalks

Line	Description	Crew	Daily Output	Labor-Hours	Unit	Material	Labor	Equipment	Total	Total Incl O&P
0010	**ROADS AND SIDEWALKS** Temporary									
0050	Roads, gravel fill, no surfacing, 4″ gravel depth	B-14	715	.067	S.Y.	4.04	2.67	.51	7.22	9.35
0100	8″ gravel depth	"	615	.078	"	8.10	3.10	.59	11.79	14.60
1000	Ramp, 3/4″ plywood on 2″ x 6″ joists, 16″ O.C.	2 Carp	300	.053	S.F.	1.58	2.50		4.08	5.85
1100	On 2″ x 10″ joists, 16″ O.C.	"	275	.058	"	2.24	2.73		4.97	6.95

01 56 Temporary Barriers and Enclosures

01 56 13 – Temporary Air Barriers

01 56 13.60 Tarpaulins		Crew	Daily Output	Labor-Hours	Unit	Material	2015 Bare Costs Labor	Equipment	Total	Total Incl O&P
0010	**TARPAULINS**									
0020	Cotton duck, 10 oz. to 13.13 oz. per S.Y., 6'x8'				S.F.	.83			.83	.91
0050	30'x30'					.59			.59	.65
0100	Polyvinyl coated nylon, 14 oz. to 18 oz., minimum					1.36			1.36	1.50
0150	Maximum					1.36			1.36	1.50
0200	Reinforced polyethylene 3 mils thick, white					.04			.04	.04
0300	4 mils thick, white, clear or black					.09			.09	.10
0400	5.5 mils thick, clear					.18			.18	.20
0500	White, fire retardant					.44			.44	.48
0600	12 mils, oil resistant, fire retardant					.39			.39	.43
0700	8.5 mils, black					.57			.57	.63
0710	Woven polyethylene, 6 mils thick					.18			.18	.20
0720	Steel reinforced polyethylene, 20 mils thick					.52			.52	.57
0730	Polyester reinforced w/integral fastening system 11 mils thick					.21			.21	.23
0740	Mylar polyester, non-reinforced, 7 mils thick					1.17			1.17	1.29

01 56 13.90 Winter Protection		Crew	Daily Output	Labor-Hours	Unit	Material	Labor	Equipment	Total	Total Incl O&P
0010	**WINTER PROTECTION**									
0100	Framing to close openings	2 Clab	500	.032	S.F.	.46	1.20		1.66	2.47
0200	Tarpaulins hung over scaffolding, 8 uses, not incl. scaffolding		1500	.011		.24	.40		.64	.92
0250	Tarpaulin polyester reinf. w/integral fastening system 11 mils thick		1600	.010		.21	.38		.59	.85
0300	Prefab fiberglass panels, steel frame, 8 uses		1200	.013		2.40	.50		2.90	3.46

01 56 16 – Temporary Dust Barriers

01 56 16.10 Dust Barriers, Temporary		Crew	Daily Output	Labor-Hours	Unit	Material	Labor	Equipment	Total	Total Incl O&P
0010	**DUST BARRIERS, TEMPORARY**									
0020	Spring loaded telescoping pole & head, to 12', erect and dismantle	1 Clab	240	.033	Ea.		1.25		1.25	2.05
0025	Cost per day (based upon 250 days)				Day	.26			.26	.29
0030	To 21', erect and dismantle	1 Clab	240	.033	Ea.		1.25		1.25	2.05
0035	Cost per day (based upon 250 days)				Day	.44			.44	.48
0040	Accessories, caution tape reel, erect and dismantle	1 Clab	480	.017	Ea.		.63		.63	1.03
0045	Cost per day (based upon 250 days)				Day	.36			.36	.40
0060	Foam rail and connector, erect and dismantle	1 Clab	240	.033	Ea.		1.25		1.25	2.05
0065	Cost per day (based upon 250 days)				Day	.10			.10	.11
0070	Caution tape	1 Clab	384	.021	C.L.F.	2.70	.78		3.48	4.25
0080	Zipper, standard duty		60	.133	Ea.	8	5		13	17
0090	Heavy duty		48	.167	"	9.50	6.25		15.75	21
0100	Polyethylene sheet, 4 mil		37	.216	Sq.	2.90	8.15		11.05	16.55
0110	6 mil		37	.216	"	4.02	8.15		12.17	17.75
1000	Dust partition, 6 mil polyethylene, 1" x 3" frame	2 Carp	2000	.008	S.F.	.30	.38		.68	.94
1080	2" x 4" frame	"	2000	.008	"	.35	.38		.73	1.01

01 56 23 – Temporary Barricades

01 56 23.10 Barricades		Crew	Daily Output	Labor-Hours	Unit	Material	Labor	Equipment	Total	Total Incl O&P
0010	**BARRICADES**									
0020	5' high, 3 rail @ 2" x 8", fixed	2 Carp	20	.800	L.F.	6	37.50		43.50	68
0150	Movable	"	30	.533	"	5	25		30	46.50
0300	Stock units, 6' high, 8' wide, plain, buy				Ea.	390			390	430
0350	With reflective tape, buy				"	405			405	445
0400	Break-a-way 3" PVC pipe barricade									
0410	with 3 ea. 1' x 4' reflectorized panels, buy				Ea.	106			106	117
0500	Plywood with steel legs, 24" wide					59			59	65
0600	Warning signal flag tree, 11' high, 2 flags, buy					257			257	283
0800	Traffic cones, PVC, 18" high					12.50			12.50	13.75

01 56 Temporary Barriers and Enclosures

01 56 23 – Temporary Barricades

01 56 23.10 Barricades		Crew	Daily Output	Labor-Hours	Unit	2015 Bare Costs Material	Labor	Equipment	Total	Total Incl O&P
0850	28" high				Ea.	19.50			19.50	21.50
1000	Guardrail, wooden, 3' high, 1" x 6", on 2" x 4" posts	2 Carp	200	.080	L.F.	1.27	3.76		5.03	7.55
1100	2" x 6", on 4" x 4" posts	"	165	.097		2.46	4.55		7.01	10.15
1200	Portable metal with base pads, buy					15.55			15.55	17.10
1250	Typical installation, assume 10 reuses	2 Carp	600	.027		2.55	1.25		3.80	4.86
1300	Barricade tape, polyethylene, 7 mil, 3" wide x 500' long roll				Ea.	25			25	27.50
3000	Detour signs, set up and remove									
3010	Reflective aluminum, MUTCD, 24" x 24", post mounted	1 Clab	20	.400	Ea.	2.46	15.05		17.51	27
4000	Roof edge portable barrier stands and warning flags, 50 uses	1 Rohe	9100	.001	L.F.	.07	.03		.10	.12
4010	100 uses	"	9100	.001	"	.03	.03		.06	.09
5000	Barricades, see Section 01 54 33.40									

01 56 26 – Temporary Fencing

01 56 26.50 Temporary Fencing		Crew	Daily Output	Labor-Hours	Unit	Material	Labor	Equipment	Total	Total Incl O&P
0010	**TEMPORARY FENCING**									
0020	Chain link, 11 ga., 4' high	2 Clab	400	.040	L.F.	2.95	1.50		4.45	5.70
0100	6' high		300	.053		2.95	2.01		4.96	6.55
0200	Rented chain link, 6' high, to 1000' (up to 12 mo.)		400	.040		4.29	1.50		5.79	7.20
0250	Over 1000' (up to 12 mo.)		300	.053		4.29	2.01		6.30	8
0350	Plywood, painted, 2" x 4" frame, 4' high	A-4	135	.178		6.20	7.95		14.15	19.80
0400	4" x 4" frame, 8' high	"	110	.218		12	9.75		21.75	29
0500	Wire mesh on 4" x 4" posts, 4' high	2 Carp	100	.160		9.85	7.50		17.35	23
0550	8' high	"	80	.200		14.90	9.40		24.30	32
0600	Plastic safety fence, light duty, 4' high, posts at 10'	B-1	500	.048		1.17	1.84		3.01	4.29
0610	Medium duty, 4' high, posts at 10'		500	.048		1.58	1.84		3.42	4.75
0620	Heavy duty, 4' high, posts at 10'		500	.048		2.19	1.84		4.03	5.40
0630	Reflective heavy duty, 4' high, posts at 10'		500	.048		4.24	1.84		6.08	7.65

01 56 29 – Temporary Protective Walkways

01 56 29.50 Protection		Crew	Daily Output	Labor-Hours	Unit	Material	Labor	Equipment	Total	Total Incl O&P
0010	**PROTECTION**									
0020	Stair tread, 2" x 12" planks, 1 use	1 Carp	75	.107	Tread	5.10	5		10.10	13.80
0100	Exterior plywood, 1/2" thick, 1 use		65	.123		1.97	5.80		7.77	11.60
0200	3/4" thick, 1 use		60	.133		2.78	6.25		9.03	13.30
2200	Sidewalks, 2" x 12" planks, 2 uses		350	.023	S.F.	.85	1.07		1.92	2.70
2300	Exterior plywood, 2 uses, 1/2" thick		750	.011		.33	.50		.83	1.18
2500	3/4" thick		600	.013		.46	.63		1.09	1.54

01 56 32 – Temporary Security

01 56 32.50 Watchman		Crew	Daily Output	Labor-Hours	Unit	Material	Labor	Equipment	Total	Total Incl O&P
0010	**WATCHMAN**									
0020	Service, monthly basis, uniformed person, minimum				Hr.				25	27.50
0100	Maximum								45.50	50
0200	Person and command dog, minimum								31	34
0300	Maximum								54.50	60
0500	Sentry dog, leased, with job patrol (yard dog), 1 dog				Week				290	320
0600	2 dogs				"				390	430
0800	Purchase, trained sentry dog, minimum				Ea.				1,375	1,500
0900	Maximum				"				2,725	3,000

01 58 Project Identification

01 58 13 – Temporary Project Signage

01 58 13.50 Signs		Crew	Daily Output	Labor-Hours	Unit	Material	2015 Bare Costs Labor	Equipment	Total	Total Incl O&P
0010	**SIGNS**									
0020	High intensity reflectorized, no posts, buy				Ea.	25			25	27.50

01 71 Examination and Preparation

01 71 23 – Field Engineering

01 71 23.13 Construction Layout

		Crew	Daily Output	Labor-Hours	Unit	Material	Labor	Equipment	Total	Total Incl O&P
0010	**CONSTRUCTION LAYOUT**									
1100	Crew for layout of building, trenching or pipe laying, 2 person crew	A-6	1	16	Day		750	55	805	1,275
1200	3 person crew	A-7	1	24			1,225	54.50	1,279.50	2,050
1400	Crew for roadway layout, 4 person crew	A-8	1	32	↓		1,600	54.50	1,654.50	2,625

01 71 23.19 Surveyor Stakes

		Crew	Daily Output	Labor-Hours	Unit	Material	Labor	Equipment	Total	Total Incl O&P
0010	**SURVEYOR STAKES**									
0020	Hardwood, 1" x 1" x 48" long				C	70			70	77
0100	2" x 2" x 18" long					78			78	86
0150	2" x 2" x 24" long				↓	140			140	154

01 74 Cleaning and Waste Management

01 74 13 – Progress Cleaning

01 74 13.20 Cleaning Up

		Crew	Daily Output	Labor-Hours	Unit	Material	Labor	Equipment	Total	Total Incl O&P
0010	**CLEANING UP**									
0020	After job completion, allow, minimum				Job				.30%	.30%
0040	Maximum				"				1%	1%
0042	Rubbish removal, see Section 02 41 19.19									
0052	Cleanup of floor area, continuous, per day, during const.	A-5	16	1.125	M.S.F.	2.18	42.50	3.84	48.52	76
0100	Final by GC at end of job	"	11.50	1.565	"	2.31	59	5.35	66.66	105

01 76 Protecting Installed Construction

01 76 13 – Temporary Protection of Installed Construction

01 76 13.20 Temporary Protection

		Crew	Daily Output	Labor-Hours	Unit	Material	Labor	Equipment	Total	Total Incl O&P
0010	**TEMPORARY PROTECTION**									
0020	Flooring, 1/8" tempered hardboard, taped seams	2 Carp	1500	.011	S.F.	.41	.50		.91	1.27
0030	Peel away carpet protection	1 Clab	3200	.003	"	.11	.09		.20	.27

01 91 Commissioning

01 91 13 – General Commissioning Requirements

01 91 13.50 Building Commissioning

		Crew	Daily Output	Labor-Hours	Unit	Material	Labor	Equipment	Total	Total Incl O&P
0010	**BUILDING COMMISSIONING**									
0100	Basic building commissioning, minimum				%				.25%	.25%
0150	Maximum								.50%	.50%
0200	Enhanced building commissioning, minimum								.50%	.50%
0250	Maximum				↓				1%	1%

01 93 08 – Facility Maintenance Equipment

01 93 08.50 Equipment		Crew	Daily Output	Labor-Hours	Unit	Material	Labor	Equipment	Total	Total Incl O&P
0010	**EQUIPMENT**, Purchase									
0090	Carpet care equipment									
0110	Dual motor vac, 1 H.P., 16" brush				Ea.	1,175			1,175	1,300
0120	Upright vacuum, 12" brush					221			221	243
0130	14" brush					390			390	430
0140	Soil extractor, hot water 5' wand, 12" head					3,075			3,075	3,375
0150	Dry foam, 13" brush					1,700			1,700	1,875
0160	24" brush				↓	3,075			3,075	3,375
0240	Floor care equipment									
0260	Polishing, buffing, waxing machine, 175 RPM									
0270	.33 H.P., 20" diam. brush				Ea.	720			720	790
0280	1 H.P., 16" diam. brush					1,150			1,150	1,250
0290	17" diam. brush					1,275			1,275	1,425
0300	1.5 H.P., 20" diam. brush					1,350			1,350	1,475
0310	20" diam. brush, 1500 RPM				↓	1,100			1,100	1,200
0330	Scrubber, automatic, 2 stage 1 H.P. vacuum motor									
0340	20" diam. brush				Ea.	12,100			12,100	13,300
0350	28" diam. brush				"	17,500			17,500	19,200
1200	Plumbing maintenance equipment									
1220	Kinetic water ram				Ea.	385			385	420
1240	Cable pipe snake, 104 ft., self feed, electric, .5 H.P.				"	2,625			2,625	2,900
1500	Specialty equipment									
1550	Litterbuggy, 9 H.P. gasoline				Ea.	5,075			5,075	5,575
1560	Propane					3,550			3,550	3,900
1570	Pressure cleaner, hot water, 1000 psi					3,550			3,550	3,900
1580	3500 psi				↓	7,025			7,025	7,725
1590	Vacuum cleaners, steel canister									
1600	Dry only, two stage, 1 H.P.				Ea.	330			330	365
1610	Wet/dry, two stage, 2 H.P.					655			655	720
1620	4 H.P.					1,025			1,025	1,125
1630	Wet/dry, three stage, 1.5 H.P.					655			655	720
1670	Squeegee wet only, 2 stage, 1 H.P.					1,175			1,175	1,300
1680	2 H.P.					1,700			1,700	1,850
1690	Asbestos and hazardous waste dry vacs, 1 H.P.					925			925	1,025
1700	2 H.P.					1,550			1,550	1,700
1710	Three stage, 1 H.P.					545			545	600
1800	Upholstery soil extractor, dry-foam				↓	3,400			3,400	3,750
1900	Snow removal equipment									
1920	Thrower, 3 H.P., 20", single stage, gas				Ea.	625			625	690
1930	Electric start					725			725	800
1940	5 H.P., 24", 2 stage, gas					1,200			1,200	1,325
1950	11 H.P., 36"				↓	2,700			2,700	2,975
1970	Trash receptacles and closures; see Section 12 93 23.20									
2000	Tube cleaning equip., for heat exchangers, condensers, evap.									
2010	Tubes/pipes 1/4"-1", 115 volt, .5 H.P. drive				Ea.	1,650			1,650	1,825
2020	Air powered					1,700			1,700	1,875
2040	1" and up, 115 volt, 1 H.P. drive				↓	1,800			1,800	1,975

Labor, Equipment and Total columns above are under "2015 Bare Costs".

01 93 09.50 Moving Equipment		Crew	Daily Output	Labor-Hours	Unit	Material	2015 Bare Costs Labor	Equipment	Total	Total Incl O&P
0010	**MOVING EQUIPMENT**, Remove and reset, 100′ distance,									
0020	No obstructions, no assembly or leveling unless noted									
0100	Annealing furnace, 24′ overall	B-67	4	4	Ea.		196	78	274	395
0200	Annealing oven, small		14	1.143			56	22	78	113
0240	Very large		1	16			780	310	1,090	1,575
0400	Band saw, small		12	1.333			65	26	91	132
0440	Large		8	2			98	39	137	197
0500	Blue print copy machine		7	2.286			112	44.50	156.50	225
0600	Bonding mill, 6″		7	2.286			112	44.50	156.50	225
0620	12″		6	2.667			130	52	182	263
0640	18″		4	4			196	78	274	395
0660	24″		2	8			390	156	546	790
0700	Boring machine (jig)	B-68	7	3.429			168	44.50	212.50	315
0800	Bridgeport mill, standard	B-67	14	1.143			56	22	78	113
1000	Calibrator, 6 unit	″	14	1.143			56	22	78	113
1100	Comparitor, bench top	2 Clab	14	1.143			43		43	70.50
1140	Floor mounted	B-67	7	2.286			112	44.50	156.50	225
1200	Computer, desk top	2 Clab	25	.640			24		24	39.50
1300	Copy machine	″	25	.640			24		24	39.50
1500	Deflasher	B-67	14	1.143			56	22	78	113
1600	Degreaser, small		14	1.143			56	22	78	113
1640	Large 24′ overall		1	16			780	310	1,090	1,575
1700	Desk with chair	2 Clab	25	.640			24		24	39.50
1725	Desk only	″	22	.727			27.50		27.50	45
1750	Chair only	1 Clab	300	.027			1		1	1.64
1800	Dial press	B-67	7	2.286			112	44.50	156.50	225
1900	Drafting table	2 Clab	14	1.143			43		43	70.50
2000	Drill press, bench top	″	14	1.143			43		43	70.50
2040	Floor mounted	B-67	14	1.143			56	22	78	113
2080	Industrial radial	″	7	2.286			112	44.50	156.50	225
2100	Dust collector, portable	2 Clab	25	.640			24		24	39.50
2140	Stationary, small	B-67	7	2.286			112	44.50	156.50	225
2180	Stationary, large	″	2	8			390	156	546	790
2300	Electric discharge machine	B-68	7	3.429			168	44.50	212.50	315
2400	Environmental chamber walls, including assembly	4 Clab	18	1.778	L.F.		67		67	110
2600	File cabinet	2 Clab	25	.640	Ea.		24		24	39.50
2800	Grinder/sander, pedestal mount	B-67	14	1.143			56	22	78	113
3000	Hack saw, power	2 Clab	24	.667			25		25	41
3100	Hydraulic press	B-67	14	1.143			56	22	78	113
3500	Laminar flow tables	″	14	1.143			56	22	78	113
3600	Lathe, bench	2 Clab	14	1.143			43		43	70.50
3640	6″	B-67	14	1.143			56	22	78	113
3680	10″		13	1.231			60	24	84	122
3720	12″		12	1.333			65	26	91	132
4000	Milling machine		8	2			98	39	137	197
4100	Molding press, 25 ton		5	3.200			156	62	218	315
4140	60 ton		4	4			196	78	274	395
4180	100 ton		2	8			390	156	546	790
4220	150 ton		1.50	10.667			520	207	727	1,050
4260	200 ton		1	16			780	310	1,090	1,575
4300	300 ton		.75	21.333			1,050	415	1,465	2,100
4700	Oil pot stand		14	1.143			56	22	78	113

01 93 09.50 Moving Equipment		Crew	Daily Output	Labor-Hours	Unit	Material	2015 Bare Costs Labor	Equipment	Total	Total Incl O&P
5000	Press, 10 ton	B-67	14	1.143	Ea.		56	22	78	113
5040	15 ton		12	1.333			65	26	91	132
5080	20 ton		10	1.600			78	31	109	158
5120	30 ton		8	2			98	39	137	197
5160	45 ton		6	2.667			130	52	182	263
5200	60 ton		4	4			196	78	274	395
5240	75 ton		2.50	6.400			315	124	439	630
5280	100 ton		2	8			390	156	546	790
5500	Raised floor, including assembly	2 Carp	250	.064	S.F.		3		3	4.92
5600	Rolling mill, 6"	B-67	7	2.286	Ea.		112	44.50	156.50	225
5640	9"		6	2.667			130	52	182	263
5680	12"		4	4			196	78	274	395
5720	13"		3.50	4.571			223	89	312	455
5760	18"		2	8			390	156	546	790
5800	25"		1	16			780	310	1,090	1,575
6000	Sander, floor stand		14	1.143			56	22	78	113
6100	Screw machine		7	2.286			112	44.50	156.50	225
6200	Shaper, 16"		14	1.143			56	22	78	113
6300	Shear, power assist		4	4			196	78	274	395
6400	Slitter, 6"		14	1.143			56	22	78	113
6440	8"		13	1.231			60	24	84	122
6480	10"		12	1.333			65	26	91	132
6520	12"		11	1.455			71	28.50	99.50	143
6560	16"		10	1.600			78	31	109	158
6600	20"		8	2			98	39	137	197
6640	24"		6	2.667			130	52	182	263
6800	Snag and tap machine		7	2.286			112	44.50	156.50	225
6900	Solder machine (auto)		7	2.286			112	44.50	156.50	225
7000	Storage cabinet metal, small	2 Clab	36	.444			16.70		16.70	27.50
7040	Large		25	.640			24		24	39.50
7100	Storage rack open, small		14	1.143			43		43	70.50
7140	Large		7	2.286			86		86	141
7200	Surface bench, small	B-67	14	1.143			56	22	78	113
7240	Large	"	5	3.200			156	62	218	315
7300	Surface grinder, large wet	B-68	5	4.800			235	62.50	297.50	440
7500	Time check machine	2 Clab	14	1.143			43		43	70.50
8000	Welder, 30 kVA (bench)		14	1.143			43		43	70.50
8100	Work bench with chair		25	.640			24		24	39.50
8110	For haul up to 200 feet, add				%		25%			
8200	Storage rack, relocate, disassemble, xport & reassemble, 8' x 36"	B-68A	600	.040	S.F.		1.97	.31	2.28	3.44
8205	8' x 42"		600	.040			1.97	.31	2.28	3.44
8210	8' x 48"		600	.040			1.97	.31	2.28	3.44
8215	Excess or store, disassemble and deliver, 8' x 36"		1200	.020			.99	.15	1.14	1.72
8220	8' x 42"		1200	.020			.99	.15	1.14	1.72
8225	8' x 48"		1200	.020			.99	.15	1.14	1.72
8230	Work bench, remove, transport, reinstall	B-67B	4	4	Ea.		198		198	310
8235	Excess or store, remove and deliver	"	8	2			99		99	155
8240	Clean bench, relocate, remove, transport, and set in place	B-68A	4	6			296	46.50	342.50	515
8245	Excess or store, remove and deliver	"	8	3			148	23	171	259
8250	Storage cabinet, relocate, remove, transport, and set in place	B-67B	8	2			99		99	155
8255	Excess or store, remove and deliver		10	1.600			79		79	124
8260	Bookcase cabinet, relocate, remove, transport, and set in place		16	1			49.50		49.50	77.50
8265	Excess or store, remove and deliver		20	.800			39.50		39.50	62

01 93 Facility Maintenance

01 93 09 – Facility Equipment

01 93 09.50 Moving Equipment		Crew	Daily Output	Labor-Hours	Unit	Material	2015 Bare Costs Labor	Equipment	Total	Total Incl O&P
8270	Bench top equip, relocate, remove, transport, and set in place	B-67B	8	2	Ea.		99		99	155
8275	Excess or store, remove and deliver		12.80	1.250			62		62	97
8280	Floor mounted equipment									
8285	Small, relocate, generally less than 2000 lb.	B-68A	3	8	Ea.		395	62	457	690
8290	Excess or store, less than 2000 lb.		6	4			197	31	228	345
8295	Medium, relocate, 2000 to 6000 lb.		1.04	23.077			1,150	178	1,328	2,000
8300	Excess or store, 2000 to 6000 lb.		1.68	14.286			705	110	815	1,225
8305	Large, relocate, 6000 lb. or more		.40	60			2,950	465	3,415	5,150
8310	Excess or store, 6000 lb. or more		.64	37.500			1,850	290	2,140	3,225
8315	Bridge crane to 40′ length									
8320	1-5 Ton	B-68E	1.36	29.412	Ea.		1,550	136	1,686	2,950
8325	5-10 Ton		1.04	38.462	″		2,025	178	2,203	3,875
8330	Rails, both sides per L.F., removal, excludes support steel		8	5	L.F.		264	23	287	500
8335	Jib crane, remove and store, including support steel, rail track		2	20	Ea.		1,050	92.50	1,142.50	2,000
8340	Office desk, relocate, remove, transport, and set in place	B-67B	8	2			99		99	155
8345	Excess or store, remove and deliver	B-68A	10.64	2.256			111	17.40	128.40	194
8350	Office file cabinet, relocate, remove, transport, and set in place									
8355	2 drawer	B-67B	8	2	Ea.		99		99	155
8360	4 drawer	″	8	2	″		99		99	155
8365	Excess or store, remove and deliver									
8370	2 drawer	B-67B	16	1	Ea.		49.50		49.50	77.50
8375	4 drawer		10.64	1.504			74.50		74.50	117
8380	Office partition, excess or store, remove and deliver		16	1			49.50		49.50	77.50
8385	Gas bottle rack, relocate, remove, transport, and set in place	B-68A	8	3			148	23	171	259
8390	Excess or store, remove and deliver	″	10.64	2.256			111	17.40	128.40	194
8395	Fume hood & filter, bench mounted, relocate, remove, xport & set in place	B-67B	1.36	11.765			580		580	915
8400	Excess or store, remove and deliver	″	2	8			395		395	620
8405	Bench top microscope, relocate, remove, transport, and set in place	1 Mill	8	1			49		49	77.50
8410	Excess or store, remove and deliver	″	8	1			49		49	77.50
8415	Modular work station									
8420	Includes 2 desktops, 4 pedestal files	B-68A	1	24	Ea.		1,175	185	1,360	2,050
8425	Install, purchase new		1.60	15			740	116	856	1,300
8430	Relocate		1.60	15			740	116	856	1,300
8435	Excess or store		2.64	9.091			450	70	520	780
8440	Existing, install, includes 2 workstations		1.60	15			740	116	856	1,300
8445	Wall mounted shelf, relocate, remove, transport and install	B-67B	8	2			99		99	155
8450	Excess or store, remove and deliver		12	1.333			66		66	104
8455	Wall mounted coat rack, excess or store		16	1			49.50		49.50	77.50
8460	Office or work table, excess or store		10.64	1.504			74.50		74.50	117
8465	Floor stand testers									
8470	Small, relocate, footprint generally less than 12 S.F.	B-67B	8	2	Ea.		99		99	155
8475	Excess or store, footprint generally less than 12 S.F.		8	2			99		99	155
8480	Medium, relocate, footprint between 12 S.F. & 20 S.F.		2.64	6.061			299		299	470
8485	Excess or store, footprint between 12 S.F. & 20 S.F.		8	2			99		99	155
8490	Large, relocate, footprint generally greater than 20 S.F.	B-68A	1.04	23.077			1,150	178	1,328	2,000
8495	Excess or store, footprint greater than 20 S.F.		2	12			590	92.50	682.50	1,025
8500	Office safe, relocate, remove, transport, and set in place		8	3			148	23	171	259
8505	Excess or store, remove and deliver		10.64	2.256			111	17.40	128.40	194
8510	Screen, relocate	1 Mill	16	.500			24.50		24.50	38.50
8515	Excess or store, remove and deliver	″	16	.500			24.50		24.50	38.50
8520	Equipment pieces									
8525	Interconnect	L-1	8	2	Ea.		113		113	176
8530	Disconnect	L-1	8	4			227		227	355

01 93 09 – Facility Equipment

01 93 09.50 Moving Equipment		Crew	Daily Output	Labor-Hours	Unit	Material	2015 Bare Costs Labor	Equipment	Total	Total Incl O&P
8535	Vidmar cabinet, floor mounted, relocate, remove, xport and set in place	B-68A	4	6	Ea.		296	46.50	342.50	515
8540	Excess or store, remove and deliver	″	8	3			148	23	171	259
8545	Work bench, relocate	B-67B	8	2			99		99	155
8550	Excess or store, remove and deliver		10.64	1.504			74.50		74.50	117
8555	Steel wire shelves, relocate, 18″ x 48″ x 5′ high		4	4			198		198	310
8560	Excess or store, remove and deliver, 18″ x 48″ x 5′ high		8	2			99		99	155
8565	Coat rack, relocate	1 Mill	16	.500			24.50		24.50	38.50
8570	Excess or store, remove and deliver	″	8	1			49		49	77.50
8575	Pallet racks, 8′ x 36″, disassemble & reassemble in another location	B-68A	504	.048	S.F.		2.35	.37	2.72	4.09
8580	Install clean room equipment	B-68B	1	56	Ea.		3,000	330	3,330	5,050
8585	Small	B-68C	2	16			850	166	1,016	1,500
8590	Large	B-68B	.24	233			12,500	1,375	13,875	21,000
8595	Install large 10′ x 25′ x 10′ equipment		.08	700			37,500	4,150	41,650	63,000
8600	Install large fume hood		.16	350			18,700	2,075	20,775	31,600
8605	Install equipment									
8610	Large	B-68B	.72	77.778	Ea.		4,175	460	4,635	7,000
8615	Medium		1.20	46.667			2,500	277	2,777	4,200
8620	Small		2.40	23.333			1,250	138	1,388	2,100
8900	For moving distance to 200 feet, add						25%			

01 93 13 – Facility Maintenance Procedures

01 93 13.03 Concrete Facilities Maintenance		Crew	Daily Output	Labor-Hours	Unit	Material	Labor	Equipment	Total	Total Incl O&P
0010	**CONCRETE FACILITIES MAINTENANCE**									
0200	Resurface concrete floor with epoxy/silica finish									
0205	For new cured concrete floors with open surface texture:									
0210	Clean, roll base coat(s), broadcast silica, roll top coat									
0220	1/16″ thick (1 base coat) plus top coat	2 Cefi	600	.027	S.F.	1.77	1.20		2.97	3.85
0230	1/8″ thick (2 base coats) plus top coat, rolled	″	350	.046	″	2.37	2.06		4.43	5.85
0305	For old or coated concrete floors with open surface texture:									
0310	Medium shotblast, roll base coat(s), broadcast silica, roll top coat									
0320	1/16″ thick (1 base coat) plus top coat	2 Cefi	400	.040	S.F.	1.19	1.80		2.99	4.16
0330	1/8″ thick (2 base coats) plus top coat, rolled	″	300	.053	″	1.79	2.40		4.19	5.75
1000	Patching concrete, see Section 03 01 30.62									
1037	Stair cleaning, sweep	1 Clab	50	.160	Flight		6		6	9.85
1039	Mop or scrub	″	5	1.600	″	.16	60		60.16	98.50
1040	Sealing concrete floor, spray, oil or urethane base, 1 coat	1 Cefi	8000	.001	S.F.	.10	.05		.15	.18
1050	2 coats	″	4000	.002	″	.20	.09		.29	.37
1080	Sweep and damp mop concrete floor	1 Clab	20	.400	M.S.F.	3.63	15.05		18.68	28.50

01 93 13.04 Masonry Facilities Maintenance		Crew	Daily Output	Labor-Hours	Unit	Material	Labor	Equipment	Total	Total Incl O&P
0010	**MASONRY FACILITIES MAINTENANCE**									
0040	Caulking masonry, no staging included									
0050	Re-caulk only, oil base, 1/2″ x 1/2″	1 Bric	225	.036	L.F.	.94	1.64		2.58	3.70
0060	Butyl		205	.039		.45	1.80		2.25	3.43
0070	Polysulfide		200	.040		.94	1.85		2.79	4.03
0080	Silicone		195	.041		.64	1.89		2.53	3.78
0100	Cut out and re-caulk, oil base		145	.055		.94	2.55		3.49	5.15
0110	Butyl		130	.062		.45	2.84		3.29	5.10
0120	Polysulfide		125	.064		.47	2.95		3.42	5.30
0130	Silicone		125	.064		.64	2.95		3.59	5.50
0800	Cleaning masonry, see Section 04 01 30.20									
3000	Pointing masonry, see Section 04 01 20.20									

01 93 Facility Maintenance

01 93 13 – Facility Maintenance Procedures

01 93 13.05 Metals Facilities Maintenance

	01 93 13.05 Metals Facilities Maintenance	Crew	Daily Output	Labor-Hours	Unit	2015 Bare Costs Material	Labor	Equipment	Total	Total Incl O&P
0010	**METALS FACILITIES MAINTENANCE**									
2000	Steel surface treat., sand blast & paint, see Section 05 01 10.51									
2010	Steel siding painting, see Section 09 91 13									

01 93 13.07 Moisture-Thermal Control Facilities Maintenance

	Description	Crew	Daily Output	Labor-Hours	Unit	Material	Labor	Equipment	Total	Total Incl O&P
0010	**MOISTURE-THERMAL CONTROL FACILITIES MAINTENANCE**									
0100	Caulking and sealants see Section 07 92									
0110	Caulking around exterior doors and windows, silicone	1 Carp	400	.020	L.F.	.16	.94		1.10	1.72
2000	Roofing & siding demolition see Section 07 05 05.10									
2500	Single-ply or built-up roofing maintenance									
2510	Walk roof, clean drain trap, pick-up trash									
2520	Two inspections per year				S.F.				.02	.02
2530	Six inspections per year				″				.06	.06
3000	Roof non-destructive testing, infrared analysis system									
3040	One test per year				S.F.				.04	.04
3050	One test every 8 months				″				.06	.06

01 93 13.08 Door and Window Facilities Maintenance

	Description	Crew	Daily Output	Labor-Hours	Unit	Material	Labor	Equipment	Total	Total Incl O&P
0010	**DOOR & WINDOW FACILITIES MAINTENANCE**									
0250	Cylinder dead bolt lock set, repair	1 Skwk	11	.727	Ea.		35.50		35.50	57.50
0300	Demolition, windows see Section 08 05 05.20									
0500	Door closer, overhaul	1 Skwk	5	1.600	Ea.		78		78	126
0510	Repair concealed door check or closer		4	2			97.50		97.50	158
0520	Adjust		26	.308			14.95		14.95	24.50
0530	Replace, see Section 08 71 20.30									
1500	Remove and reset window, no interior/exterior trim or painting	1 Carp	4	2	Ea.		94		94	154
1510	Interior trim and painting only		2	4			188		188	310
1520	Interior/exterior trim and painting		1.33	6.015			282		282	465
1900	Washing door, 3′ x 7′, damp wipe	1 Clab	145	.055		.19	2.07		2.26	3.61
2000	Washing windows, at ground level w/ladder, sponge & squeegee									
2010	Both sides, 4′ x 7′ 2 pane	1 Clab	60	.133	Ea.	.24	5		5.24	8.45
2020	4′ x 6′, 4 pane		55	.145		.21	5.45		5.66	9.20
2030	3.5′ x 5.5′, 8 pane		45	.178		.17	6.70		6.87	11.15
2040	2.5′ x 5.7′, 12 pane		40	.200		.12	7.50		7.62	12.50
2050	4′ x 6′, 16 pane		35	.229		.21	8.60		8.81	14.35
2060	4′ x 7′, industrial, 20 pane		30	.267		.24	10.05		10.29	16.70
2070	6′ x 7′, austral casement, 6 pane		35	.229		.36	8.60		8.96	14.50
2080	Clear glass partition, 8 sq. ft. per unit		3	2.667	M.S.F.	4.29	100		104.29	169
2090	Opaque, 20 sq. ft. per unit		3	2.667	″	4.29	100		104.29	169
2100	Alternate pricing method by window area to be washed									
2110	Minimum productivity	1 Clab	1	8	M.S.F.	4.29	300		304.29	500
2120	Average productivity		2.50	3.200		4.29	120		124.29	202
2130	Maximum productivity		4	2		4.29	75		79.29	128
2200	Weatherstripping see Section 08 75 30.10									

01 93 13.09 Finishes, Facilities Maintenance

	Description	Crew	Daily Output	Labor-Hours	Unit	Material	Labor	Equipment	Total	Total Incl O&P
0010	**FINISHES, FACILITIES MAINTENANCE** R019313-10									
0100	Ceiling maintenance									
0115	Acoustic tile cleaning, chemical spray, including masking									
0120	Unobstructed floor R019313-20	4 Clab	5400	.006	S.F.	.06	.22		.28	.44
0125	Obstructed floor	″	4000	.008	″	.06	.30		.36	.56
0128	Acoustic tile cleaning & coating, chemical, including masking									
0130	Unobstructed floor	4 Clab	4800	.007	S.F.	.13	.25		.38	.56
0135	Obstructed floor	″	3400	.009	″	.13	.35		.48	.73

01 93 13.09 Finishes, Facilities Maintenance

	01 93 13.09 Finishes, Facilities Maintenance	Crew	Daily Output	Labor-Hours	Unit	Material	2015 Bare Costs Labor	Equipment	Total	Total Incl O&P
0400	Floor maintenance R019313-30									
0500	Carpet fiber sealant, anti-soilant, commercial				S.F.				.28	.28
0510	Residential				″				.37	.37
0530	Carpet cleaning, vacuum, dry pick-up									
0540	Unobstructed	1 Clab	30	.267	M.S.F.		10.05		10.05	16.45
0550	Obstructed		22	.364			13.65		13.65	22.50
0560	Carpet cleaning, portable extractor with floor wand		20	.400		8	15.05		23.05	33.50
0570	Deep carpet cleaning, self-contained extractor									
0572	24″ path at 100 fpm	1 Clab	64	.125	M.S.F.	10.05	4.70		14.75	18.75
0574	20″ path		54	.148		10.05	5.55		15.60	20
0576	16″ path		43	.186		10.05	7		17.05	22.50
0578	12″ path		32	.250		10.05	9.40		19.45	26.50
0580	Carpet pile lifting, self-contained extractor									
0582	24″ path at 90 fpm	1 Clab	60	.133	M.S.F.	11.05	5		16.05	20.50
0584	20″ path		50	.160		11.05	6		17.05	22
0586	16″ path		40	.200		11.05	7.50		18.55	24.50
0588	12″ path		30	.267		11.05	10.05		21.10	28.50
0590	Carpet restoration cleaning, self-contained extractor									
0592	24″ path at 50 fpm	1 Clab	32	.250	M.S.F.	12.05	9.40		21.45	28.50
0594	20″ path		27	.296		12.05	11.15		23.20	31.50
0596	16″ path		21	.381		12.05	14.30		26.35	37
0598	12″ path		16	.500		12.05	18.80		30.85	44.50
0700	Composition, resilient or wood flooring									
0720	Dust mop, unobstructed	1 Clab	60	.133	M.S.F.		5		5	8.20
0730	Obstructed		35	.229			8.60		8.60	14.10
0740	Damp mop, unobstructed		26	.308		5	11.55		16.55	24.50
0750	Obstructed		16	.500		5	18.80		23.80	36.50
0780	Hand scrub, unobstructed		1.80	4.444		5	167		172	280
0790	Obstructed		1.30	6.154		5	231		236	385
0800	Machine scrub, unobstructed		17	.471		5	17.70		22.70	34.50
0810	Obstructed		12	.667		5	25		30	46.50
0820	Machine polish, unobstructed		28	.286			10.75		10.75	17.60
0830	Obstructed		14	.571			21.50		21.50	35
0850	Refinish old wood floors, minimum	1 Carp	400	.020	S.F.	.04	.94		.98	1.58
0860	Maximum cost	″	130	.062		.06	2.89		2.95	4.80
0862	Light sanding, clean and 1 coat of polyurethane	J-7	700	.023		.13	.92	.04	1.09	1.66
0863	Heavy sanding, 1 coat sealer, 2 coats polyurethane	″	400	.040		.38	1.61	.07	2.06	3.07
0870	Strip & rewax/polish, unobstructed	1 Clab	4	2	M.S.F.	39.50	75		114.50	166
0880	Obstructed		3	2.667		39.50	100		139.50	207
0890	Sweeping, unobstructed		47	.170			6.40		6.40	10.50
0900	Obstructed		35	.229			8.60		8.60	14.10
0950	Painting concrete or wood floors, see Section 09 91 23.52									
1200	Paint removal, see Section 02 83 19.26									
1300	Scrape after fire damage, see Section 09 91 03.41									
2000	Wall maintenance									
2400	Washing enamel finish walls with mild cleanser	1 Clab	16	.500	M.S.F.	4.58	18.80		23.38	36

01 93 13.10 Specialties, Facilities Maintenance

	01 93 13.10 Specialties, Facilities Maintenance	Crew	Daily Output	Labor-Hours	Unit	Material	Labor	Equipment	Total	Total Incl O&P
0010	**SPECIALTIES, FACILITIES MAINTENANCE**									
0100	Bathroom accessories									
0120	Accessory, replacement see Section 10 28 13.13									
0130	All purpose cleaner, concentrate				Gal.	18.30			18.30	20
0150	Clean mirror, 36″ x 24″	1 Clab	840	.010	Ea.	.03	.36		.39	.62

01 93 13.10 Specialties, Facilities Maintenance		Crew	Daily Output	Labor-Hours	Unit	2015 Bare Costs Material	Labor	Equipment	Total	Total Incl O&P
0160	48" x 24"	1 Clab	630	.013	Ea.	.04	.48		.52	.82
0170	72" x 24"		420	.019		.05	.72		.77	1.23
0200	General cleaning of fixtures (basins, water closets, urinals)									
0210	Including shelves, partitions and dispenser servicing	1 Clab	80	.100	Fixture	.03	3.76		3.79	6.20
1500	Partition replacement, hospital, office, toilet see Section 10 21 13.13									

01 93 13.11 Architectural Equipment, Facilities Maintenance

Line	Description	Crew	Daily Output	Labor-Hours	Unit	Material	Labor	Equipment	Total	Total Incl O&P
0010	**ARCHITECTURAL EQUIPMENT, FACILITIES MAINTENANCE**									
0100	Bird control needle strips									
0110	Anchor clip mounted	2 Clab	500	.032	L.F.	6.20	1.20		7.40	8.75
0120	Adhesive mounted	"	550	.029	"	8	1.09		9.09	10.60
0500	Laundry equipment repair and maintenance									
0520	Dryer, domestic, remove or reinstall basket	1 Skwk	6	1.333	Ea.		65		65	105
0530	Commercial, disassemble, clean & reassemble		2	4			195		195	315
0550	Extractor, replace belts (2)		6	1.333			65		65	105
0560	Replace pressure pad		2	4			195		195	315
0570	Flatwork ironer, replace feed ribbons		2	4			195		195	315
0580	Press, overhaul air system		4	2			97.50		97.50	158
0590	Tumbler, extractor, replace bearings and seal		2	4			195		195	315
0600	Washer, domestic, replace bearings and seal		3	2.667			130		130	211
0610	Remove or reinstall tub		4	2			97.50		97.50	158
0700	Medical and ward equipment repair and maintenance									
0710	Aspirator-overhaul	1 Skwk	6	1.333	Ea.		65		65	105
0720	Autoclave-overhaul		4	2	"		97.50		97.50	158
0730	Bed rails, assemble or repair		10	.800	Pr.		39		39	63
0770	Cart, medication, ice, etc. - replace wheels		6	1.333	Carton		65		65	105
0780	Replace rubber bumper		6	1.333	"		65		65	105
0800	Centrifuge, repair or replace timer or switch		11	.727	Ea.		35.50		35.50	57.50
0850	Litter, attach safety belt		16	.500			24.50		24.50	39.50
0880	Operating table, overhaul and lubricate		6	1.333			65		65	105
0900	Oxygen tent apparatus, overhaul		2	4			195		195	315
0950	Still, clean		3	2.667			130		130	211
0970	Striker saw, overhaul		5	1.600			78		78	126
1100	Wheelchair, complete overhaul		3	2.667			130		130	211
1110	Replace wheel		11	.727			35.50		35.50	57.50
2000	Other equipment repair and maintenance									
2040	Cart or dolly, assemble or disassemble axles and wheels	1 Skwk	6	1.333	Carton		65		65	105
2090	Forge and harden tool		13	.615	Ea.		30		30	48.50
3100	Office furniture, repair		7	1.143			55.50		55.50	90.50
3120	Portable power tool, assemble or disassemble		6	1.333			65		65	105
3130	Repair or replace part		6	1.333			65		65	105
3180	Replace tool handle		16	.500			24.50		24.50	39.50
3200	Sharpen hand tool		16	.500			24.50		24.50	39.50

01 93 13.12 Furnishings Facilities Maintenance

Line	Description	Crew	Daily Output	Labor-Hours	Unit	Material	Labor	Equipment	Total	Total Incl O&P
0010	**FURNISHINGS FACILITIES MAINTENANCE**									
1000	Reupholster, side chair	1 Skwk	2.50	3.200	Ea.		156		156	253
1040	Tubular frame chair seat and back		3	2.667			130		130	211
1080	Sofa		2	4			195		195	315
1500	Upholstery fabric protection/cleaner, minimum cost				S.Y.				10	10
1510	Maximum cost				"				26	26
1540	Vacuum, large divan	1 Clab	80	.100	Ea.		3.76		3.76	6.15

01 93 13.14 Conveying Systems Facilities Maintenance

	01 93 13.14 Conveying Systems Facilities Maintenance	Crew	Daily Output	Labor-Hours	Unit	Material	2015 Bare Costs Labor	Equipment	Total	Total Incl O&P
0010	**CONVEYING SYSTEMS FACILITIES MAINTENANCE**									
0900	Elevator, general cleaning, vacuum, dust, mop, polish	1 Clab	16	.500	Ea.	.88	18.80		19.68	32
0920	Motor generator set, service	1 Elec	8	1			54.50		54.50	84.50
0930	Repair, minor		2	4			219		219	340
0940	Limit switch, repair		11	.727			40		40	61.50
0950	Troubleshoot, main control panel	2 Elec	2.50	6.400			350		350	540
1000	Escalator cleaning, dusting, polishing	1 Clab	20	.400	Flight	.92	15.05		15.97	25.50
1500	Vehicle maintenance, grease, light vehicle	1 Skwk	20	.400	Ea.		19.45		19.45	31.50
1510	Heavy vehicle		7	1.143			55.50		55.50	90.50
1520	Wash, light vehicle		8	1		.76	48.50		49.26	80
1530	Heavy vehicle		6	1.333		1.36	65		66.36	107
1540	Change oil and filter, light vehicle		10	.800		30	39		69	96
1570	Rotate tires, light vehicle		18	.444			21.50		21.50	35

01 93 13.15 Mechanical Facilities Maintenance

	01 93 13.15 Mechanical Facilities Maintenance		Crew	Daily Output	Labor-Hours	Unit	Material	Labor	Equipment	Total	Total Incl O&P
0010	**MECHANICAL FACILITIES MAINTENANCE**										
0100	Air conditioning system maintenance										
0130	Belt, replace		1 Stpi	15	.533	Ea.		32		32	49.50
0170	Fan, clean			16	.500			30		30	46.50
0180	Filter, remove, clean, replace			12	.667			40		40	62
0190	Flexible coupling alignment, inspect			40	.200			11.95		11.95	18.65
0200	Gas leak locate and repair			4	2			120		120	186
0250	Pump packing gland, remove and replace			11	.727			43.50		43.50	68
0270	Tighten			32	.250			14.95		14.95	23.50
0290	Pump disassemble and assemble			4	2			120		120	186
0300	Air pressure regulator, disassemble, clean, assemble		1 Skwk	4	2			97.50		97.50	158
0310	Repair or replace part		"	6	1.333			65		65	105
0320	Purging system		1 Stpi	16	.500			30		30	46.50
0400	Compressor, air, remove or install fan wheel		1 Skwk	20	.400			19.45		19.45	31.50
0410	Disassemble or assemble 2 cylinder, 2 stage			4	2			97.50		97.50	158
0420	4 cylinder, 4 stage			1	8			390		390	630
0430	Repair or replace part			2	4			195		195	315
0700	Demolition, for mech. demolition see Section 23 05 05.10 or 22 05 05.10										
0800	Ductwork, clean										
0810	Rectangular										
0820	6″	G	1 Shee	187.50	.043	L.F.		2.39		2.39	3.77
0830	8″	G		140.63	.057			3.18		3.18	5
0840	10″	G		112.50	.071			3.98		3.98	6.30
0850	12″	G		93.75	.085			4.77		4.77	7.55
0860	14″	G		80.36	.100			5.55		5.55	8.80
0870	16″	G		70.31	.114			6.35		6.35	10.05
0900	Round										
0910	4″	G	1 Shee	358.10	.022	L.F.		1.25		1.25	1.97
0920	6″	G		238.73	.034			1.87		1.87	2.96
0930	8″	G		179.05	.045			2.50		2.50	3.95
0940	10″	G		143.24	.056			3.12		3.12	4.93
0950	12″	G		119.37	.067			3.75		3.75	5.90
0960	16″	G		89.52	.089			5		5	7.90
1200	Fire protection equipment										
1220	Fire hydrant, replace		Q-1	3	5.333	Ea.		282		282	440
1230	Service, lubricate, inspect, flush, clean		1 Plum	7	1.143			67		67	105
1240	Test		"	11	.727			42.50		42.50	66.50
1310	Inspect valves, pressure, nozzle,		1 Spri	4	2	System		112		112	176

01 93 Facility Maintenance

01 93 13 – Facility Maintenance Procedures

01 93 13.15 Mechanical Facilities Maintenance		Crew	Daily Output	Labor-Hours	Unit	2015 Bare Costs Material	Labor	Equipment	Total	Total Incl O&P
1500	Instrumentation preventive maintenance, rule of thumb									
1510	Annual cost per instrument for calibration & trouble shooting									
1520	Includes flow meters, analytical transmitters, flow control									
1530	chemical feed systems, recorders and distrib. control sys.				Ea.				325	325
1800	Plumbing fixtures, for installation see Section 22 41 00									
1840	Open drain with snake	1 Plum	13	.615	Ea.		36		36	56.50
1850	Open drain with toilet auger		16	.500			29.50		29.50	46
1860	Grease trap, clean		2	4			235		235	365
1870	Plaster trap, clean		6	1.333			78.50		78.50	122
1900	Relief valve, test and adjust	1 Stpi	20	.400			24		24	37.50
2000	Repair or replace-steam trap		8	1			60		60	93
2020	Y-type or bell strainer		6	1.333			79.50		79.50	124
2040	Water trap or vacuum breaker, screwed joints		13	.615			37		37	57.50
2100	Steam specialties, clean									
2120	Air separator with automatic trap, 1" fittings	1 Stpi	12	.667	Ea.		40		40	62
2130	Bucket trap, 2" pipe		7	1.143			68.50		68.50	107
2150	Drip leg, 2" fitting		45	.178			10.60		10.60	16.55
2200	Thermodynamic trap, 1" fittings		50	.160			9.55		9.55	14.90
2210	Thermostatic		65	.123			7.35		7.35	11.45
2240	Screen and seat in y-type strainer, plug type.		25	.320			19.10		19.10	30
2242	Screen and seat in y-type strainer, flange type.		12	.667			40		40	62
2500	Valve, replace broken handwheel		24	.333			19.90		19.90	31
3000	Valve, overhaul, regulator, relief, flushometer, mixing									
3040	Cold water, gas	1 Stpi	5	1.600	Ea.		95.50		95.50	149
3050	Hot water, steam		3	2.667			159		159	249
3080	Globe, gate, check up to 4" cold water, gas		10	.800			48		48	74.50
3090	Hot water, steam		5	1.600			95.50		95.50	149
3100	Over 4" ID hot or cold line		1.40	5.714			340		340	535
3120	Remove and replace, gate, globe or check up to 4"	Q-5	6	2.667			143		143	224
3130	Over 4"	"	2	8			430		430	670
3150	Repack up to 4"	1 Stpi	13	.615			37		37	57.50
3160	Over 4"	"	4	2			120		120	186

01 93 13.16 Electrical Facilities Maintenance

01 93 13.16 Electrical Facilities Maintenance		Crew	Daily Output	Labor-Hours	Unit	Material	Labor	Equipment	Total	Total Incl O&P
0010	**ELECTRICAL FACILITIES MAINTENANCE**									
0700	Cathodic protection systems									
0720	Check and adjust reading on rectifier	1 Elec	20	.400	Ea.		22		22	34
0730	Check pipe to soil potential		20	.400			22		22	34
0740	Replace lead connection		4	2			109		109	169
0800	Control device, install		5.70	1.404			77		77	119
0810	Disassemble, clean and reinstall		7	1.143			62.50		62.50	97
0820	Replace		10.70	.748			41		41	63.50
0830	Trouble shoot		10	.800			44		44	68
0900	Demolition, for electrical demolition see Section 26 05 05.10									
1000	Distribution systems and equipment install or repair a breaker									
1010	In power panels up to 200 amps	1 Elec	7	1.143	Ea.		62.50		62.50	97
1020	Over 200 amps		2	4			219		219	340
1030	Reset breaker or replace fuse		20	.400			22		22	34
1100	Megger test MCC (each stack)		4	2			109		109	169
1110	MCC vacuum and clean (each stack)		5.30	1.509			82.50		82.50	128
2500	Remove and replace or maint., road fixture & lamp		3	2.667		925	146		1,071	1,250
2510	Fluorescent fixture		7	1.143		67	62.50		129.50	171
2515	Relamp (fluor.) facility area each tube, spot	1 Clab	24	.333		3.38	12.55		15.93	24

01 93 13 – Facility Maintenance Procedures

01 93 13.16 Electrical Facilities Maintenance		Crew	Daily Output	Labor-Hours	Unit	Material	2015 Bare Costs Labor	Equipment	Total	Total Incl O&P
2516	Group	1 Clab	100	.080	Ea.	3.38	3.01		6.39	8.65
2518	Fluorescent fixture, clean (area)		44	.182		.11	6.85		6.96	11.30
2520	Incandescent fixture	1 Elec	11	.727		61.50	40		101.50	130
2530	Lamp (incandescent or fluorescent)	1 Clab	60	.133		3.38	5		8.38	11.90
2535	Replace cord in socket lamp	1 Elec	13	.615		1.66	33.50		35.16	54
2540	Ballast electronic type for two tubes		8	1		37	54.50		91.50	126
2541	Starter		30	.267		.37	14.60		14.97	23
2545	Replace other lighting parts		11	.727		14.20	40		54.20	77
2550	Switch		11	.727		9.85	40		49.85	72.50
2555	Receptacle		11	.727		7.90	40		47.90	70
2560	Floodlight		4	2		273	109		382	470
2570	Christmas lighting, indoor, per string	1 Clab	16	.500			18.80		18.80	31
2580	Outdoor		13	.615			23		23	38
2590	Test battery operated emergency lights		40	.200			7.50		7.50	12.35
2600	Repair and replace component in communication system	1 Elec	6	1.333		53	73		126	171
2700	Repair misc. appliances (incl. clocks, vent fan, blower, etc.)	"	6	1.333			73		73	113
2710	Reset clocks & timers	1 Clab	50	.160			6		6	9.85
2720	Adjust time delay relays	1 Elec	16	.500			27.50		27.50	42.50
2730	Test specific gravity of lead-acid batteries	1 Clab	80	.100			3.76		3.76	6.15
3000	Motors and generators									
3020	Disassemble, clean and reinstall motor, up to 1/4 HP	1 Elec	4	2	Ea.		109		109	169
3030	Up to 3/4 HP		3	2.667			146		146	226
3040	Up to 10 HP		2	4			219		219	340
3050	Replace part, up to 1/4 HP		6	1.333			73		73	113
3060	Up to 3/4 HP		4	2			109		109	169
3070	Up to 10 HP		3	2.667			146		146	226
3080	Megger test motor windings		5.33	1.501			82		82	127
3082	Motor vibration check		16	.500			27.50		27.50	42.50
3084	Oil motor bearings		25	.320			17.50		17.50	27
3086	Run test emergency generator for 30 minutes		11	.727			40		40	61.50
3090	Rewind motor, up to 1/4 HP		3	2.667			146		146	226
3100	Up to 3/4 HP		2	4			219		219	340
3110	Up to 10 HP		1.50	5.333			292		292	450
3150	Generator, repair or replace part		4	2			109		109	169
3160	Repair DC generator		2	4			219		219	340
4000	Stub pole, install or remove		3	2.667			146		146	226
4500	Transformer maintenance up to 15 kVA		2.70	2.963			162		162	251

Estimating Tips

02 30 00 Subsurface Investigation

In preparing estimates on structures involving earthwork or foundations, all information concerning soil characteristics should be obtained. Look particularly for hazardous waste, evidence of prior dumping of debris, and previous stream beds.

02 40 00 Demolition and Structure Moving

The costs shown for selective demolition do not include rubbish handling or disposal. These items should be estimated separately using RSMeans data or other sources.

- Historic preservation often requires that the contractor remove materials from the existing structure, rehab them, and replace them. The estimator must be aware of any related measures and precautions that must be taken when doing selective demolition and cutting and patching. Requirements may include special handling and storage, as well as security.
- In addition to Subdivision 02 41 00, you can find selective demolition items in each division. Example: Roofing demolition is in Division 7.

02 40 00 Building Deconstruction

This section provides costs for the careful dismantling and recycling of most of low-rise building materials.

02 50 00 Containment of Hazardous Waste

This section addresses on-site hazardous waste disposal costs.

02 80 00 Hazardous Material Disposal/ Remediation

This subdivision includes information on hazardous waste handling, asbestos remediation, lead remediation, and mold remediation. See reference R028213-20 and R028319-60 for further guidance in using these unit price lines.

02 90 00 Monitoring Chemical Sampling, Testing Analysis

This section provides costs for on-site sampling and testing hazardous waste.

Reference Numbers

Reference numbers are shown in shaded boxes at the beginning of some major classifications. These numbers refer to related items in the Reference Section. The reference information may be an estimating procedure, an alternate pricing method, or technical information.

Note: Not all subdivisions listed here necessarily appear in this publication. ■

02 21 Surveys

02 21 13 – Site Surveys

02 21 13.09 Topographical Surveys

	02 21 13.09 Topographical Surveys	Crew	Daily Output	Labor-Hours	Unit	2015 Bare Costs Material	Labor	Equipment	Total	Total Incl O&P
0010	**TOPOGRAPHICAL SURVEYS**									
0020	Topographical surveying, conventional, minimum	A-7	3.30	7.273	Acre	20	375	16.60	411.60	640
0100	Maximum	A-8	.60	53.333	"	60	2,675	91	2,826	4,475

02 21 13.13 Boundary and Survey Markers

	02 21 13.13 Boundary and Survey Markers	Crew	Daily Output	Labor-Hours	Unit	Material	Labor	Equipment	Total	Total Incl O&P
0010	**BOUNDARY AND SURVEY MARKERS**									
0300	Lot location and lines, large quantities, minimum	A-7	2	12	Acre	35	620	27.50	682.50	1,075
0320	Average	"	1.25	19.200		55	990	44	1,089	1,700
0400	Small quantities, maximum	A-8	1	32		75	1,600	54.50	1,729.50	2,725
0600	Monuments, 3′ long	A-7	10	2.400	Ea.	40	124	5.45	169.45	249
0800	Property lines, perimeter, cleared land	"	1000	.024	L.F.	.05	1.24	.05	1.34	2.11
0900	Wooded land	A-8	875	.037	"	.07	1.83	.06	1.96	3.09

02 21 13.16 Aerial Surveys

	02 21 13.16 Aerial Surveys	Crew	Daily Output	Labor-Hours	Unit	Material	Labor	Equipment	Total	Total Incl O&P
0010	**AERIAL SURVEYS**									
1500	Aerial surveying, including ground control, minimum fee, 10 acres				Total				4,700	4,700
1510	100 acres								9,400	9,400
1550	From existing photography, deduct								1,625	1,625
1600	2′ contours, 10 acres				Acre				470	470
1850	100 acres				"				94	94

02 32 Geotechnical Investigations

02 32 13 – Subsurface Drilling and Sampling

02 32 13.10 Boring and Exploratory Drilling

	02 32 13.10 Boring and Exploratory Drilling	Crew	Daily Output	Labor-Hours	Unit	Material	Labor	Equipment	Total	Total Incl O&P
0010	**BORING AND EXPLORATORY DRILLING**									
0020	Borings, initial field stake out & determination of elevations	A-6	1	16	Day		750	55	805	1,275
0100	Drawings showing boring details				Total		335		335	425
0200	Report and recommendations from P.E.						775		775	970
0300	Mobilization and demobilization	B-55	4	6			229	271	500	675
0350	For over 100 miles, per added mile		450	.053	Mile		2.03	2.41	4.44	5.95
0600	Auger holes in earth, no samples, 2-1/2″ diameter		78.60	.305	L.F.		11.65	13.75	25.40	34
0650	4″ diameter		67.50	.356			13.55	16.05	29.60	39.50
0800	Cased borings in earth, with samples, 2-1/2″ diameter		55.50	.432		14	16.50	19.50	50	64
0850	4″ diameter		32.60	.736		18	28	33	79	102
1000	Drilling in rock, "BX" core, no sampling	B-56	34.90	.458			19.75	45.50	65.25	82
1050	With casing & sampling		31.70	.505		14	22	50.50	86.50	106
1200	"NX" core, no sampling		25.92	.617			26.50	61.50	88	111
1250	With casing and sampling		25	.640		15	27.50	63.50	106	131
1400	Borings, earth, drill rig and crew with truck mounted auger	B-55	1	24	Day		915	1,075	1,990	2,700
1450	Rock using crawler type drill	B-56	1	16	"		690	1,600	2,290	2,850
1500	For inner city borings add, minimum								10%	10%
1510	Maximum								20%	20%

02 32 19 – Exploratory Excavations

02 32 19.10 Test Pits

	02 32 19.10 Test Pits	Crew	Daily Output	Labor-Hours	Unit	Material	Labor	Equipment	Total	Total Incl O&P
0010	**TEST PITS**									
0020	Hand digging, light soil	1 Clab	4.50	1.778	C.Y.		67		67	110
0100	Heavy soil	"	2.50	3.200			120		120	197
0120	Loader-backhoe, light soil	B-11M	28	.571			25	14	39	56
0130	Heavy soil	"	20	.800			35.50	19.60	55.10	78.50
1000	Subsurface exploration, mobilization				Mile				6.75	8.40
1010	Difficult access for rig, add				Hr.				260	320
1020	Auger borings, drill rig, incl. samples				L.F.				26.50	33

02 32 Geotechnical Investigations

02 32 19 – Exploratory Excavations

02 32 19.10 Test Pits		Crew	Daily Output	Labor-Hours	Unit	Material	2015 Bare Costs Labor	Equipment	Total	Total Incl O&P
1030	Hand auger				L.F.				31.50	40
1050	Drill and sample every 5′, split spoon				↓				31.50	40
1060	Extra samples				Ea.				36	45.50

02 41 Demolition

02 41 13 – Selective Site Demolition

02 41 13.15 Hydrodemolition

		Crew	Daily Output	Labor-Hours	Unit	Material	Labor	Equipment	Total	Total Incl O&P
0010	**HYDRODEMOLITION** R024119-10									
0015	Hydrodemolition, concrete pavement									
0120	20,000 PSI, Crew to include loader / vacuum truck as required									
0130	2″ depth	B-5	1000	.056	S.F.		2.33	1.42	3.75	5.35
0410	4″ depth		800	.070			2.91	1.78	4.69	6.65
0420	6″ depth	↓	600	.093	↓		3.88	2.37	6.25	8.90

02 41 13.17 Demolish, Remove Pavement and Curb

		Crew	Daily Output	Labor-Hours	Unit	Material	Labor	Equipment	Total	Total Incl O&P
0010	**DEMOLISH, REMOVE PAVEMENT AND CURB** R024119-10									
5010	Pavement removal, bituminous roads, up to 3″ thick	B-38	690	.058	S.Y.		2.48	1.85	4.33	6.05
5050	4″ to 6″ thick		420	.095			4.08	3.04	7.12	9.95
5100	Bituminous driveways		640	.063			2.68	1.99	4.67	6.50
5200	Concrete to 6″ thick, hydraulic hammer, mesh reinforced		255	.157			6.70	5	11.70	16.35
5300	Rod reinforced		200	.200	↓		8.55	6.40	14.95	21
5400	Concrete, 7″ to 24″ thick, plain		33	1.212	C.Y.		52	38.50	90.50	127
5500	Reinforced		24	1.667	″		71.50	53	124.50	174
5590	Minimum labor/equipment charge	↓	6	6.667	Job		285	213	498	695
5600	With hand held air equipment, bituminous, to 6″ thick	B-39	1900	.025	S.F.		1	.12	1.12	1.77
5700	Concrete to 6″ thick, no reinforcing		1600	.030			1.19	.15	1.34	2.10
5800	Mesh reinforced		1400	.034			1.36	.17	1.53	2.40
5900	Rod reinforced	↓	765	.063	↓		2.50	.30	2.80	4.39
5990	Minimum labor/equipment charge	B-38	6	6.667	Job		285	213	498	695
6000	Curbs, concrete, plain	B-6	360	.067	L.F.		2.75	1.01	3.76	5.55
6100	Reinforced		275	.087			3.60	1.32	4.92	7.30
6200	Granite		360	.067			2.75	1.01	3.76	5.55
6300	Bituminous		528	.045	↓		1.88	.69	2.57	3.80
6390	Minimum labor/equipment charge		6	4	Job		165	60.50	225.50	335
6500	Site demo, berms under 4″ in height, bituminous		528	.045	L.F.		1.88	.69	2.57	3.80
6600	4″ or over in height	↓	300	.080	″		3.30	1.21	4.51	6.70

02 41 13.20 Selective Demo, Highway Guard Rails & Barriers

		Crew	Daily Output	Labor-Hours	Unit	Material	Labor	Equipment	Total	Total Incl O&P
0010	**SELECTIVE DEMOLITION, HIGHWAY GUARD RAILS & BARRIERS**									
0100	Guard rail, corrugated steel	B-6	600	.040	L.F.		1.65	.61	2.26	3.34
0200	End sections		40	.600	Ea.		25	9.10	34.10	50
0300	Wrap around		40	.600	″		25	9.10	34.10	50
0400	Timber 4″ x 8″		600	.040	L.F.		1.65	.61	2.26	3.34
0500	Three 3/4″ cables		600	.040	″		1.65	.61	2.26	3.34
0600	Wood posts	↓	240	.100	Ea.		4.13	1.52	5.65	8.35
0700	Guide rail, 6″ x 6″ box beam	B-80B	120	.267	L.F.		10.75	2.03	12.78	19.70
0800	Median barrier, 6″ x 8″ box beam		240	.133			5.40	1.02	6.42	9.85
0850	Precast concrete 3′-6″ high x 2′ wide	↓	300	.107	↓		4.30	.81	5.11	7.90
0900	Impact barrier, UTMCD, barrel type	B-16	60	.533	Ea.		20.50	11.50	32	46
1000	Resilient guide fence and light shield 6′ high	″	120	.267	L.F.		10.30	5.75	16.05	23
1100	Concrete posts, 6′-5″ triangular	B-6	200	.120	Ea.		4.95	1.82	6.77	10
1200	Speed bumps 10-1/2″ x 2-1/4″ x 48″	↓	300	.080	L.F.		3.30	1.21	4.51	6.70

02 41 Demolition

02 41 13 – Selective Site Demolition

02 41 13.20	Selective Demo, Highway Guard Rails & Barriers	Crew	Daily Output	Labor-Hours	Unit	Material	2015 Bare Costs Labor	Equipment	Total	Total Incl O&P
1300	Pavement marking channelizing	B-6	200	.120	Ea.		4.95	1.82	6.77	10
1400	Barrier and curb delineators		300	.080			3.30	1.21	4.51	6.70
1500	Rumble strips 24" x 3-1/2" x 1/2"		150	.160			6.60	2.43	9.03	13.35

02 41 13.23 Utility Line Removal

Line	Description	Crew	Daily Output	Labor-Hours	Unit	Material	Labor	Equipment	Total	Total Incl O&P
0010	**UTILITY LINE REMOVAL**									
0015	No hauling, abandon catch basin or manhole	B-6	7	3.429	Ea.		142	52	194	286
0020	Remove existing catch basin or manhole, masonry		4	6			248	91	339	500
0030	Catch basin or manhole frames and covers, stored		13	1.846			76	28	104	154
0040	Remove and reset		7	3.429			142	52	194	286
0900	Hydrants, fire, remove only	B-21A	5	8			375	94.50	469.50	700
0950	Remove and reset	"	2	20			940	237	1,177	1,750
0990	Minimum labor/equipment charge	2 Plum	2	8	Job		470		470	730
2900	Pipe removal, sewer/water, no excavation, 12" diameter	B-6	175	.137	L.F.		5.65	2.08	7.73	11.45
2930	15"-18" diameter	B-12Z	150	.160			6.75	10.70	17.45	22.50
2960	21"-24" diameter		120	.200			8.45	13.40	21.85	28.50
3000	27"-36" diameter		90	.267			11.30	17.85	29.15	38
3200	Steel, welded connections, 4" diameter	B-6	160	.150			6.20	2.28	8.48	12.50
3300	10" diameter		80	.300			12.40	4.55	16.95	25
3390	Minimum labor/equipment charge		3	8	Job		330	121	451	670

02 41 13.30 Minor Site Demolition

Line	Description	Crew	Daily Output	Labor-Hours	Unit	Material	Labor	Equipment	Total	Total Incl O&P
0010	**MINOR SITE DEMOLITION** R024119-10									
0100	Roadside delineators, remove only	B-80	175	.183	Ea.		7.55	4.12	11.67	16.75
0110	Remove and reset	"	100	.320	"		13.20	7.20	20.40	29.50
0400	Minimum labor/equipment charge	B-6	4	6	Job		248	91	339	500
0800	Guiderail, corrugated steel, remove only	B-80A	100	.240	L.F.		9	3.03	12.03	18.15
0850	Remove and reset	"	40	.600	"		22.50	7.60	30.10	45.50
0860	Guide posts, remove only	B-80B	120	.267	Ea.		10.75	2.03	12.78	19.70
0870	Remove and reset	B-55	50	.480	"		18.30	21.50	39.80	54
0890	Minimum labor/equipment charge	2 Clab	4	4	Job		150		150	247
1000	Masonry walls, block, solid	B-5	1800	.031	C.F.		1.29	.79	2.08	2.97
1200	Brick, solid		900	.062			2.59	1.58	4.17	5.95
1400	Stone, with mortar		900	.062			2.59	1.58	4.17	5.95
1500	Dry set		1500	.037			1.55	.95	2.50	3.56
1600	Median barrier, precast concrete, remove and store	B-3	430	.112	L.F.		4.57	6	10.57	14
1610	Remove and reset	"	390	.123	"		5.05	6.60	11.65	15.40
1650	Minimum labor/equipment charge	A-1	4	2	Job		75	20.50	95.50	146
4000	Sidewalk removal, bituminous, 2" thick	B-6	350	.069	S.Y.		2.83	1.04	3.87	5.70
4010	2-1/2" thick		325	.074			3.05	1.12	4.17	6.15
4050	Brick, set in mortar		185	.130			5.35	1.97	7.32	10.80
4100	Concrete, plain, 4"		160	.150			6.20	2.28	8.48	12.50
4110	Plain, 5"		140	.171			7.05	2.60	9.65	14.30
4120	Plain, 6"		120	.200			8.25	3.04	11.29	16.70
4200	Mesh reinforced, concrete, 4 "		150	.160			6.60	2.43	9.03	13.35
4210	5 " thick		131	.183			7.55	2.78	10.33	15.30
4220	6 " thick		112	.214			8.85	3.25	12.10	17.90
4290	Minimum labor/equipment charge	B-39	12	4	Job		159	19.40	178.40	281
6850	Runways, remove rubber skid marks, 4-6 passes	B-59A	35	.686	M.S.F.	37.50	26.50	14.30	78.30	100
6860	6-10 passes	"	35	.686	"	56.50	26.50	14.30	97.30	121

02 41 13.33 Railtrack Removal

Line	Description	Crew	Daily Output	Labor-Hours	Unit	Material	Labor	Equipment	Total	Total Incl O&P
0010	**RAILTRACK REMOVAL**									
3500	Railroad track removal, ties and track	B-13	330	.170	L.F.		6.95	2.23	9.18	13.75
3600	Ballast	B-14	500	.096	C.Y.		3.82	.73	4.55	7

02 41 13 – Selective Site Demolition

02 41 13.33 Railtrack Removal		Crew	Daily Output	Labor-Hours	Unit	Material	2015 Bare Costs Labor	Equipment	Total	Total Incl O&P
3700	Remove and re-install, ties & track using new bolts & spikes	B-14	50	.960	L.F.		38	7.30	45.30	70
3800	Turnouts using new bolts and spikes		1	48	Ea.		1,900	365	2,265	3,500
3890	Minimum labor/equipment charge		5	9.600	Job		380	73	453	700

02 41 13.34 Selective Demolition, Utility Materials

Line	Description	Crew	Daily Output	Labor-Hours	Unit	Material	Labor	Equipment	Total	Total Incl O&P
0010	SELECTIVE DEMOLITION, UTILITY MATERIALS R024119-10									
0015	Excludes excavation									
0020	See other utility items in Section 02 41 13.33									
0100	Fire Hydrant extensions	B-20	14	1.714	Ea.		72		72	117
0200	Precast Utility boxes up to 8′ x 14′ x 7′	B-13	2	28			1,150	370	1,520	2,250
0300	Handholes and meter pits	B-6	2	12			495	182	677	1,000
0400	Utility valves 4″-12″	B-20	4	6			252		252	410
0500	14″-24″	B-21	2	14			605	69	674	1,050

02 41 13.36 Selective Demolition, Utility Valves and Accessories

Line	Description	Crew	Daily Output	Labor-Hours	Unit	Material	Labor	Equipment	Total	Total Incl O&P
0010	SELECTIVE DEMOLITION, UTILITY VALVES & ACCESSORIES									
0015	Excludes excavation									
0100	Utility valves 4″-12″ diam.	B-20	4	6	Ea.		252		252	410
0200	14″-24″ diam.	B-21	2	14			605	69	674	1,050
0300	Crosses 4″-12″	B-20	8	3			126		126	206
0400	14″-24″	B-21	4	7			305	34.50	339.50	535
0500	Utility cut-in valves 4″-12″ diam.	B-20	20	1.200			50.50		50.50	82
0600	Curb boxes	″	20	1.200			50.50		50.50	82

02 41 13.38 Selective Demo., Water & Sewer Piping & Fittings

Line	Description	Crew	Daily Output	Labor-Hours	Unit	Material	Labor	Equipment	Total	Total Incl O&P
0010	SELECTIVE DEMOLITION, WATER & SEWER PIPING AND FITTINGS									
0015	Excludes excavation									
0090	Concrete pipe 4″-10″ dia	B-6	250	.096	L.F.		3.96	1.46	5.42	8
0100	42″-48″ diameter	B-13B	96	.583			24	11.75	35.75	52
0200	60″-84″ diameter	″	80	.700			28.50	14.10	42.60	62
0300	96″ diameter	B-13C	80	.700			28.50	21	49.50	69.50
0400	108″-144″ diameter	″	64	.875			36	26	62	86.50
0450	Concrete fittings 12″ diameter	B-6	24	1	Ea.		41.50	15.20	56.70	83.50
0480	Concrete end pieces 12″ diameter		200	.120	L.F.		4.95	1.82	6.77	10
0485	15″ diameter		150	.160			6.60	2.43	9.03	13.35
0490	18″ diameter		150	.160			6.60	2.43	9.03	13.35
0500	24″ -36″ diameter		100	.240			9.90	3.64	13.54	20
0600	Concrete fittings 24″-36″ diameter		12	2	Ea.		82.50	30.50	113	168
0700	48″-84″ diameter	B-13B	12	4.667			191	94	285	415
0800	96″ diameter	″	8	7			287	141	428	620
0900	108″-144″ diameter	B-13C	4	14			575	415	990	1,400
1000	Ductile iron pipe 4″ diameter	B-21B	200	.200	L.F.		8.15	3.27	11.42	16.90
1100	6″-12″ diameter		175	.229			9.35	3.74	13.09	19.25
1200	14″-24″ diameter		120	.333			13.60	5.45	19.05	28
1300	Ductile iron fittings 4″-12″ diameter		24	1.667	Ea.		68	27.50	95.50	141
1400	14″-16″ diameter		18	2.222			90.50	36.50	127	188
1500	18″-24″ diameter		12	3.333			136	54.50	190.50	281
1600	Plastic pipe 3/4″-4″ diameter	B-6	700	.034	L.F.		1.42	.52	1.94	2.86
1700	6″-8″ diameter		500	.048			1.98	.73	2.71	4.01
1800	10″-18″ diameter		300	.080			3.30	1.21	4.51	6.70
1900	20″-36″ diameter		200	.120			4.95	1.82	6.77	10
1910	42″-48″ diameter		180	.133			5.50	2.02	7.52	11.15
1920	54″-60″ diameter		160	.150			6.20	2.28	8.48	12.50
2000	Plastic fittings 4″- 8″ diameter		75	.320	Ea.		13.20	4.86	18.06	27
2100	10″-14″ diameter		50	.480			19.80	7.30	27.10	40

02 41 13.38 Selective Demo., Water & Sewer Piping & Fittings		Crew	Daily Output	Labor-Hours	Unit	Material	2015 Bare Costs Labor	Equipment	Total	Total Incl O&P
2200	16"-24" diameter	B-6	20	1.200	Ea.		49.50	18.20	67.70	100
2210	30"-36" diameter		15	1.600			66	24.50	90.50	134
2220	42"- 48" diameter		12	2			82.50	30.50	113	168
2300	Copper pipe 3/4"-2" diameter	Q-1	500	.032	L.F.		1.69		1.69	2.64
2400	2 1/2" - 3" diameter		300	.053			2.82		2.82	4.39
2500	4"- 6" diameter		200	.080			4.23		4.23	6.60
2600	Copper fittings 3/4"- 2" diameter		15	1.067	Ea.		56.50		56.50	88
2700	Cast iron pipe 4" diameter		200	.080	L.F.		4.23		4.23	6.60
2800	5"- 6" diameter	Q-2	200	.120			6.55		6.55	10.25
2900	8"- 12" diameter	Q-3	200	.160			8.95		8.95	13.95
3000	Cast iron fittings 4" diameter	Q-1	30	.533	Ea.		28		28	44
3100	5"- 6" diameter	Q-2	30	.800			44		44	68.50
3200	8"- 15" diameter	Q-3	30	1.067			59.50		59.50	93
3300	Vent cast iron pipe 4"- 8" diameter	Q-1	200	.080	L.F.		4.23		4.23	6.60
3400	10"- 15" diameter	Q-3	200	.160	"		8.95		8.95	13.95
3500	Vent cast iron fittings 4"- 8" diameter	Q-2	30	.800	Ea.		44		44	68.50
3600	10"- 15" diameter	"	20	1.200	"		65.50		65.50	103

02 41 13.40 Selective Demolition, Metal Drainage Piping

		Crew	Daily Output	Labor-Hours	Unit	Material	Labor	Equipment	Total	Total Incl O&P
0010	**SELECTIVE DEMOLITION, METAL DRAINAGE PIPING**									
0015	Excludes excavation									
0100	CMP pipe, aluminum, 6"-10" dia	B-21	800	.035	L.F.		1.52	.17	1.69	2.66
0110	12" dia		600	.047			2.02	.23	2.25	3.54
0120	18" dia		600	.047			2.02	.23	2.25	3.54
0140	Steel, 6"-10" dia		800	.035			1.52	.17	1.69	2.66
0150	12" dia		600	.047			2.02	.23	2.25	3.54
0160	18" dia		400	.070			3.03	.34	3.37	5.30
0170	24" dia	B-13	300	.187			7.65	2.45	10.10	15.10
0180	30" - 36" dia		250	.224			9.20	2.95	12.15	18.15
0190	48" - 60" dia		200	.280			11.50	3.68	15.18	22.50
0200	72" dia	B-13B	100	.560			23	11.30	34.30	49.50
0210	CMP end sections, steel, 10"-18" dia	B-21	40	.700	Ea.		30.50	3.44	33.94	53.50
0220	24"-36" dia	B-13	30	1.867			76.50	24.50	101	151
0230	48" dia		20	2.800			115	37	152	227
0240	60" dia		10	5.600			230	73.50	303.50	450
0250	72" dia	B-13B	10	5.600			230	113	343	495
0260	CMP fittings, 8"-12" dia	B-21	60	.467			20	2.30	22.30	35.50
0270	18" dia	"	40	.700			30.50	3.44	33.94	53.50
0280	24"-48" dia	B-13	30	1.867			76.50	24.50	101	151
0290	60" dia	"	20	2.800			115	37	152	227
0300	72" dia	B-13B	10	5.600			230	113	343	495
0310	Oval arch 17" x 13", 21" x 15", 15-18" equivalent	B-21	400	.070	L.F.		3.03	.34	3.37	5.30
0320	28" x 20", 24" equivalent	B-13	300	.187			7.65	2.45	10.10	15.10
0330	35" x 24", 42" x 29", 30-36" equivalent		250	.224			9.20	2.95	12.15	18.15
0340	49" x 33", 57" x 38", 42-48" equivalent		200	.280			11.50	3.68	15.18	22.50
0350	Oval arch 17" x 13" end piece, 15" equivalent	B-21	40	.700	Ea.		30.50	3.44	33.94	53.50
0360	42" x 29" end piece, 36" equivalent	B-13	30	1.867	"		76.50	24.50	101	151

02 41 13.42 Selective Demolition, Manholes and Catch Basins

		Crew	Daily Output	Labor-Hours	Unit	Material	Labor	Equipment	Total	Total Incl O&P
0010	**SELECTIVE DEMOLITION, MANHOLES & CATCH BASINS**									
0015	Excludes excavation									
0100	Manholes, precast or brick over 8' deep	B-6	8	3	V.L.F.		124	45.50	169.50	250
0200	Cast in place 4'-8' deep	B-9	127	.315	SF Face		11.95	1.83	13.78	21.50
0300	Over 8' deep	"	100	.400	"		15.20	2.33	17.53	27.50

02 41 Demolition

02 41 13 – Selective Site Demolition

02 41 13.42 Selective Demolition, Manholes and Catch Basins		Crew	Daily Output	Labor-Hours	Unit	Material	2015 Bare Costs Labor	Equipment	Total	Total Incl O&P
0400	Top, precast, 8" thick, 4'-6' dia	B-6	8	3	Ea.		124	45.50	169.50	250
0500	Steps	1 Clab	60	.133	"		5		5	8.20

02 41 13.43 Selective Demolition, Box Culvert		Crew	Daily Output	Labor-Hours	Unit	Material	Labor	Equipment	Total	Total Incl O&P
0010	**SELECTIVE DEMOLITION, BOX CULVERT**									
0015	Excludes excavation									
0100	Box culvert 8' x 6' x 3' to 8' x 8' x 8'	B-69	300	.160	L.F.		6.65	5.40	12.05	16.70
0200	8' x 10' x 3' to 8' x 12' x 8'	"	200	.240	"		9.95	8.10	18.05	25

02 41 13.44 Selective Demolition, Septic Tanks and Related Components		Crew	Daily Output	Labor-Hours	Unit	Material	Labor	Equipment	Total	Total Incl O&P
0010	**SELECTIVE DEMOLITION, SEPTIC TANKS & RELATED COMPONENTS**									
0020	Excludes excavation									
0100	Septic tanks, precast, 1000-1250 gal.	B-21	8	3.500	Ea.		152	17.20	169.20	266
0200	1500 gal.		7	4			173	19.70	192.70	305
0300	2000-2500 gal.		5	5.600			243	27.50	270.50	425
0400	4000 gal.		4	7			305	34.50	339.50	535
0500	Precast, 5000 gal., 4 piece	B-13	3	18.667			765	245	1,010	1,525
0600	15,000 gal.	B-13B	1.70	32.941			1,350	665	2,015	2,925
0700	25,000 gal.		1.10	50.909			2,075	1,025	3,100	4,500
0800	40,000 gal.		.80	70			2,875	1,400	4,275	6,200
0900	Precast, 50,000 gal., 5 piece	B-13C	.60	93.333			3,825	2,775	6,600	9,250
1000	Cast-in-place, 75,000 gal.	B-9	.06	666			25,300	3,875	29,175	45,800
1100	100,000 gal.	"	.05	800			30,400	4,650	35,050	55,000
1200	HDPE, 1000 gal.	B-21	9	3.111			135	15.30	150.30	236
1300	1500 gal.		8	3.500			152	17.20	169.20	266
1400	Galley, 4' x 4' x 4'		16	1.750			76	8.60	84.60	132
1500	Distribution boxes, concrete, 7 outlets	2 Clab	16	1			37.50		37.50	61.50
1600	9 outlets	"	8	2			75		75	123
1700	Leaching chambers 13' x 3'-7" x 1'-4", standard	B-13	16	3.500			143	46	189	284
1800	8' x 4' x 1'-6", heavy duty		14	4			164	52.50	216.50	325
1900	13' x 3'-9" x 1'-6"		12	4.667			191	61.50	252.50	380
2100	20' x 4' x 1'-6"		5	11.200			460	147	607	905
2200	Leaching pit 6'-6" x 6' deep	B-21	5	5.600			243	27.50	270.50	425
2300	6'-6" x 8' deep		4	7			305	34.50	339.50	535
2400	8' x 6' deep H20		4	7			305	34.50	339.50	535
2500	8' x 8' deep H20		3	9.333			405	46	451	710
2600	Velocity reducing pit, precast 6' x 3' deep		4.70	5.957			258	29.50	287.50	450

02 41 13.46 Selective Demolition, Steel Pipe With Insulation		Crew	Daily Output	Labor-Hours	Unit	Material	Labor	Equipment	Total	Total Incl O&P
0010	**SELECTIVE DEMOLITION, STEEL PIPE WITH INSULATION**									
0020	Excludes excavation									
0100	Steel pipe, with insulation, 3/4"-4"	B-1A	400	.060	L.F.		2.30	.82	3.12	4.66
0200	5"-10"	B-1B	360	.089			3.70	2.72	6.42	9
0300	12"-16"		240	.133			5.55	4.09	9.64	13.50
0400	18"-24"		160	.200			8.35	6.15	14.50	20.50
0450	26"-36"		100	.320			13.30	9.80	23.10	32.50
0500	Steel gland seal, with insulation, 3/4"-4"	B-1A	100	.240	Ea.		9.20	3.27	12.47	18.65
0600	5"-10"	B-1B	75	.427			17.75	13.05	30.80	43.50
0700	12"-16"		60	.533			22	16.35	38.35	54
0800	18"-24"		50	.640			26.50	19.60	46.10	64.50
0850	26"-36"		40	.800			33.50	24.50	58	81
0900	Demo steel fittings with insulation 3/4"-4"	B-1A	60	.400			15.30	5.45	20.75	31
1000	5"-10"	B-1B	40	.800			33.50	24.50	58	81
1100	12"-16"		30	1.067			44.50	32.50	77	108
1200	18"-24"		20	1.600			66.50	49	115.50	162

02 41 13 – Selective Site Demolition

02 41 13.46 Selective Demolition, Steel Pipe With Insulation		Crew	Daily Output	Labor-Hours	Unit	Material	2015 Bare Costs Labor	Equipment	Total	Total Incl O&P
1300	26"-36"	B-1B	15	2.133	Ea.		89	65.50	154.50	216
1400	Steel pipe anchors, 5"-10"		40	.800			33.50	24.50	58	81
1500	12"-16"		30	1.067			44.50	32.50	77	108
1600	18"-24"		20	1.600			66.50	49	115.50	162
1700	26"-36"		15	2.133			89	65.50	154.50	216

02 41 13.48 Selective Demolition, Gasoline Containment Piping

		Crew	Daily Output	Labor-Hours	Unit	Material	Labor	Equipment	Total	Total Incl O&P
0010	SELECTIVE DEMOLITION, GASOLINE CONTAINMENT PIPING									
0020	Excludes excavation									
0030	Excludes environmental site remediation									
0100	Gasoline plastic primary containment piping 2" to 4"	Q-6	800	.030	L.F.		1.67		1.67	2.61
0200	Fittings 2" to 4"		40	.600	Ea.		33.50		33.50	52
0300	Gasoline plastic secondary containment piping 3" to 6"		800	.030	L.F.		1.67		1.67	2.61
0400	Fittings 3" to 6"		40	.600	Ea.		33.50		33.50	52

02 41 13.50 Selective Demolition, Natural Gas, PE Pipe

		Crew	Daily Output	Labor-Hours	Unit	Material	Labor	Equipment	Total	Total Incl O&P
0010	SELECTIVE DEMOLITION, NATURAL GAS, PE PIPE									
0020	Excludes excavation									
0100	Natural gas coils, PE, 1 1/4" to 3"	Q-6	800	.030	L.F.		1.67		1.67	2.61
0200	Joints, 40', PE, 3" - 4"		800	.030			1.67		1.67	2.61
0300	6" - 8"		600	.040			2.23		2.23	3.48

02 41 13.51 Selective Demolition, Natural Gas, Steel Pipe

		Crew	Daily Output	Labor-Hours	Unit	Material	Labor	Equipment	Total	Total Incl O&P
0010	SELECTIVE DEMOLITION, NATURAL GAS, STEEL PIPE									
0020	Excludes excavation									
0100	Natural gas steel pipe 1" - 4"	B-1A	800	.030	L.F.		1.15	.41	1.56	2.33
0200	5" - 10"	B-1B	360	.089			3.70	2.72	6.42	9
0300	12" - 16"		240	.133			5.55	4.09	9.64	13.50
0400	18" - 24"		160	.200			8.35	6.15	14.50	20.50
0500	Natural gas steel fittings 1" - 4"	B-1A	160	.150	Ea.		5.75	2.04	7.79	11.65
0600	5" - 10"	B-1B	160	.200			8.35	6.15	14.50	20.50
0700	12" - 16"		108	.296			12.35	9.10	21.45	30
0800	18" - 24"		70	.457			19.05	14	33.05	46.50

02 41 13.52 Selective Demo, Natural Gas, Valves, Fittings, Regulators

		Crew	Daily Output	Labor-Hours	Unit	Material	Labor	Equipment	Total	Total Incl O&P
0010	SELECTIVE DEMO, NATURAL GAS, VALVES, FITTINGS, REGULATORS									
0100	Gas stops 1 1/4" - 2"	1 Plum	22	.364	Ea.		21.50		21.50	33.50
0200	Gas regulator 1 1/2" - 2"	"	22	.364			21.50		21.50	33.50
0300	3" - 4"	Q-1	22	.727			38.50		38.50	60
0400	Gas plug valve, 3/4" - 2"	1 Plum	22	.364			21.50		21.50	33.50
0500	2 1/2" - 3"	Q-1	10	1.600			84.50		84.50	132

02 41 13.54 Selective Demolition, Electric Ducts and Fittings

		Crew	Daily Output	Labor-Hours	Unit	Material	Labor	Equipment	Total	Total Incl O&P
0010	SELECTIVE DEMOLITION, ELECTRIC DUCTS & FITTINGS									
0020	Excludes excavation									
0100	Plastic conduit, 1/2" - 2"	1 Elec	600	.013	L.F.		.73		.73	1.13
0200	3" - 6"	2 Elec	400	.040	"		2.19		2.19	3.39
0300	Fittings, 1/2" - 2"	1 Elec	50	.160	Ea.		8.75		8.75	13.55
0400	3" - 6"	"	40	.200	"		10.95		10.95	16.95

02 41 13.56 Selective Demolition, Electric Duct Banks

		Crew	Daily Output	Labor-Hours	Unit	Material	Labor	Equipment	Total	Total Incl O&P
0010	SELECTIVE DEMOLITION, ELECTRIC DUCT BANKS									
0020	Excludes excavation									
0100	Hand holes sized to 4' x 4' x 4'	R-3	7	2.857	Ea.		155	19.70	174.70	263
0200	Manholes sized to 6' x 10' x 7'	B-13	6	9.333	"		385	123	508	755
0300	Conduit 1 @ 2" diameter, EB plastic, no concrete	2 Elec	1000	.016	L.F.		.88		.88	1.36
0400	2 @ 2" diameter		500	.032			1.75		1.75	2.71

02 41 13.56 Selective Demolition, Electric Duct Banks		Crew	Daily Output	Labor-Hours	Unit	Material	2015 Bare Costs Labor	Equipment	Total	Total Incl O&P
0500	4 @ 2" diameter	2 Elec	250	.064	L.F.		3.50		3.50	5.40
0600	1 @ 3" diameter		800	.020			1.09		1.09	1.69
0700	2 @ 3" diameter		400	.040			2.19		2.19	3.39
0800	4 @ 3" diameter		200	.080			4.38		4.38	6.80
0900	1 @ 4" diameter		800	.020			1.09		1.09	1.69
1000	2 @ 4" diameter		400	.040			2.19		2.19	3.39
1100	4 @ 4" diameter		200	.080			4.38		4.38	6.80
1200	6 @ 4" diameter		100	.160			8.75		8.75	13.55
1300	1 @ 5" diameter		500	.032			1.75		1.75	2.71
1400	2 @ 5" diameter		250	.064			3.50		3.50	5.40
1500	4 @ 5" diameter		160	.100			5.45		5.45	8.45
1600	6 @ 5" diameter		100	.160			8.75		8.75	13.55
1700	1 @ 6" diameter		500	.032			1.75		1.75	2.71
1800	2 @ 6" diameter		250	.064			3.50		3.50	5.40
1900	4 @ 6" diameter		160	.100			5.45		5.45	8.45
2000	6 @ 6" diameter		100	.160			8.75		8.75	13.55
2100	Conduit 1 EB plastic, with concrete, 0.92 C.F./L.F.	B-9	150	.267			10.15	1.55	11.70	18.30
2200	2 EB plastic 1.52 C.F./L.F.		100	.400			15.20	2.33	17.53	27.50
2300	2 x 2 EB plastic 2.51 C.F./L.F.		80	.500			19	2.91	21.91	34
2400	2 x 3 EB plastic 3.56 C.F./L.F.		60	.667			25.50	3.88	29.38	46
2500	Conduit 2 @ 2" diameter, steel, no concrete	2 Elec	400	.040			2.19		2.19	3.39
2600	4 @ 2" diameter		200	.080			4.38		4.38	6.80
2700	2 @ 3" diameter		200	.080			4.38		4.38	6.80
2800	4 @ 3" diameter		100	.160			8.75		8.75	13.55
2900	2 @ 4" diameter		200	.080			4.38		4.38	6.80
3000	4 @ 4" diameter		100	.160			8.75		8.75	13.55
3100	6 @ 4" diameter		50	.320			17.50		17.50	27
3200	2 @ 5" diameter		120	.133			7.30		7.30	11.30
3300	4 @ 5" diameter		60	.267			14.60		14.60	22.50
3400	6 @ 5" diameter		40	.400			22		22	34
3500	2 @ 6" diameter		120	.133			7.30		7.30	11.30
3600	4 @ 6" diameter		60	.267			14.60		14.60	22.50
3700	6 @ 6" diameter		40	.400			22		22	34
3800	Conduit 2 steel, with concrete, 1.52 C.F./L.F.	B-9	80	.500			19	2.91	21.91	34
3900	2 x 2 EB steel 2.51 C.F./L.F.		60	.667			25.50	3.88	29.38	46
4000	2 x 3 steel 3.56 C.F./L.F.		50	.800			30.50	4.66	35.16	55
4100	Conduit fittings, PVC type EB, 2" - 3"	1 Elec	30	.267	Ea.		14.60		14.60	22.50
4200	4" - 6"	"	20	.400	"		22		22	34

02 41 13.60 Selective Demolition Fencing		Crew	Daily Output	Labor-Hours	Unit	Material	Labor	Equipment	Total	Total Incl O&P
0010	**SELECTIVE DEMOLITION FENCING** R024119-10									
0700	Snow fence, 4' high	B-6	1000	.024	L.F.		.99	.36	1.35	2
1600	Fencing, barbed wire, 3 strand	2 Clab	430	.037			1.40		1.40	2.29
1650	5 strand	"	280	.057			2.15		2.15	3.52
1700	Chain link, posts & fabric, 8' to 10' high, remove only	B-6	445	.054			2.23	.82	3.05	4.50
1750	Remove and reset	"	70	.343			14.15	5.20	19.35	28.50

02 41 13.62 Selective Demo., Chain Link Fences & Gates		Crew	Daily Output	Labor-Hours	Unit	Material	Labor	Equipment	Total	Total Incl O&P
0010	**SELECTIVE DEMOLITION, CHAIN LINK FENCES & GATES**									
0100	Chain link, gates, 3' - 4' width	B-6	30	.800	Ea.		33	12.15	45.15	67
0200	10-12' width		16	1.500			62	23	85	125
0300	14' width		15	1.600			66	24.50	90.50	134
0400	20' width		10	2.400			99	36.50	135.50	200
0500	18' width with overhead & cantilever		80	.300	L.F.		12.40	4.55	16.95	25

02 41 13.62 Selective Demo., Chain Link Fences & Gates		Crew	Daily Output	Labor-Hours	Unit	Material	2015 Bare Costs Labor	Equipment	Total	Total Incl O&P
0510	Sliding	B-6	80	.300	L.F.		12.40	4.55	16.95	25
0520	Cantilever to 40′ wide		80	.300			12.40	4.55	16.95	25
0530	Motor operators	2 Skwk	1	16	Ea.		780		780	1,275
0540	Transmitter systems	″	15	1.067	″		52		52	84.50
0600	Chain link, fence, 5′ high	B-6	890	.027	L.F.		1.11	.41	1.52	2.25
0650	3′ - 4′ high		1000	.024			.99	.36	1.35	2
0675	12′ high		400	.060			2.48	.91	3.39	5
0800	Chain link, fence, braces	B-1	2000	.012	Ea.		.46		.46	.75
0900	Privacy slats	″	2000	.012			.46		.46	.75
1000	Fence posts, steel, in concrete	B-6	80	.300			12.40	4.55	16.95	25
1100	Fence fabric & accessories, fabric to 8′ high		800	.030	L.F.		1.24	.46	1.70	2.50
1200	Barbed wire		5000	.005	″		.20	.07	.27	.40
1300	Extension arms & eye tops		300	.080	Ea.		3.30	1.21	4.51	6.70
1400	Fence rails		2000	.012	L.F.		.50	.18	.68	1
1500	Reinforcing wire		5000	.005	″		.20	.07	.27	.40

02 41 13.64 Selective Demolition, Vinyl Fences and Gates

		Crew	Daily Output	Labor-Hours	Unit	Material	Labor	Equipment	Total	Total Incl O&P
0010	SELECTIVE DEMOLITION, VINYL FENCES & GATES									
0100	Vinyl fence up to 6′ high	B-6	1000	.024	L.F.		.99	.36	1.35	2
0200	Gates, up to 6′ high	″	40	.600	Ea.		25	9.10	34.10	50

02 41 13.66 Selective Demolition, Misc Metal Fences and Gates

		Crew	Daily Output	Labor-Hours	Unit	Material	Labor	Equipment	Total	Total Incl O&P
0010	SELECTIVE DEMOLITION, MISC METAL FENCES & GATES									
0100	Misc steel mesh fences, 4′ - 6′ high	B-6	600	.040	L.F.		1.65	.61	2.26	3.34
0200	Kennels, 6′ - 12′ long	2 Clab	8	2	Ea.		75		75	123
0300	Tops, 6′ - 12′ long	″	30	.533	″		20		20	33
0400	Security fences, 12′ - 16′ high	B-6	100	.240	L.F.		9.90	3.64	13.54	20
0500	Metal tubular picket fences 4′ - 6′ high		500	.048	″		1.98	.73	2.71	4.01
0600	Gates 3′ - 4′ wide		20	1.200	Ea.		49.50	18.20	67.70	100

02 41 13.68 Selective Demolition, Wood Fences and Gates

		Crew	Daily Output	Labor-Hours	Unit	Material	Labor	Equipment	Total	Total Incl O&P
0010	SELECTIVE DEMOLITION, WOOD FENCES & GATES									
0100	Wood fence gates 3′ - 4′ wide	2 Clab	20	.800	Ea.		30		30	49.50
0200	Wood fence, open rail, to 4′ high		2560	.006	L.F.		.24		.24	.39
0300	to 8′ high		368	.043	″		1.63		1.63	2.68
0400	Post, in concrete		50	.320	Ea.		12.05		12.05	19.75

02 41 13.70 Selective Demolition, Rip-Rap and Rock Lining

		Crew	Daily Output	Labor-Hours	Unit	Material	Labor	Equipment	Total	Total Incl O&P
0010	SELECTIVE DEMOLITION, RIP-RAP & ROCK LINING R024119-10									
0100	Slope protection, broken stone	B-13	62	.903	C.Y.		37	11.90	48.90	73
0200	3/8 to 1/4 C.Y. pieces		60	.933	S.Y.		38.50	12.25	50.75	75.50
0300	18 inch depth		60	.933	″		38.50	12.25	50.75	75.50
0400	Dumped stone		93	.602	Ton		24.50	7.90	32.40	48.50
0500	Gabions, 6 – 12 inches deep		60	.933	S.Y.		38.50	12.25	50.75	75.50
0600	18 – 36 inches deep		30	1.867	″		76.50	24.50	101	151

02 41 13.72 Selective Demo., Shore Protect/Mooring Struct.

		Crew	Daily Output	Labor-Hours	Unit	Material	Labor	Equipment	Total	Total Incl O&P
0010	SELECTIVE DEMOLITION, SHORE PROTECT/MOORING STRUCTURES									
0100	Breakwaters, bulkheads, concrete, maximum	B-9	12	3.333	L.F.		127	19.40	146.40	230
0200	Breakwaters, bulkheads, concrete, 12′, minimum		10	4			152	23.50	175.50	275
0300	Maximum		9	4.444			169	26	195	305
0400	Steel, from shore	B-40B	54	.889			37	21.50	58.50	83.50
0500	from barge	B-76A	30	2.133			86.50	69	155.50	217
0600	Jetties, docks, floating	B-21B	600	.067	S.F.		2.72	1.09	3.81	5.65
0700	Pier supported, 3″ - 4″ decking		300	.133			5.45	2.18	7.63	11.25
0800	Floating, prefab, small boat, minimum		600	.067			2.72	1.09	3.81	5.65

02 41 Demolition

02 41 13 – Selective Site Demolition

02 41 13.72 Selective Demo., Shore Protect/Mooring Struct.		Crew	Daily Output	Labor-Hours	Unit	Material	2015 Bare Costs Labor	Equipment	Total	Total Incl O&P
0900	Maximum	B-21B	300	.133	S.F.		5.45	2.18	7.63	11.25
1000	Floating, prefab, per slip, minimum		3.20	12.500	Ea.		510	204	714	1,050
1010	Maximum		2.80	14.286	"		585	234	819	1,200

02 41 13.74 Selective Demolition, Piles

Line	Description	Crew	Daily Output	Labor-Hours	Unit	Material	Labor	Equipment	Total	Total Incl O&P
0010	SELECTIVE DEMOLITION, PILES									
0100	Cast in place piles, corrugated, 8" - 10"	B-19	600	.107	V.L.F.		5.10	2.90	8	11.40
0200	12"-14"		500	.128			6.10	3.47	9.57	13.65
0300	16"		400	.160			7.60	4.34	11.94	17.10
0400	fluted, 12"		600	.107			5.10	2.90	8	11.40
0500	14"-18"		500	.128			6.10	3.47	9.57	13.65
0600	end bearing, 12"		600	.107			5.10	2.90	8	11.40
0700	14"-18"		500	.128			6.10	3.47	9.57	13.65
0800	Precast prestressed piles, 12"-14" dia		700	.091			4.36	2.48	6.84	9.80
0900	16"-24" dia		600	.107			5.10	2.90	8	11.40
1000	36"-66" dia		300	.213			10.15	5.80	15.95	23
1100	10"-14" thick		600	.107			5.10	2.90	8	11.40
1200	16"-24" thick		500	.128			6.10	3.47	9.57	13.65
1300	Pressure grouted pile, 5"		150	.427			20.50	11.60	32.10	46
1400	Steel piles, 8"-12" tip		600	.107			5.10	2.90	8	11.40
1500	H sections HP8 to HP12		600	.107			5.10	2.90	8	11.40
1600	HP14	B-19A	600	.107			5.10	3.61	8.71	12.15
1700	Steel pipe piles, 8"-12"	B-19	600	.107			5.10	2.90	8	11.40
1800	14"-18" plain		500	.128			6.10	3.47	9.57	13.65
1900	concrete filled, 14"-18"		400	.160			7.60	4.34	11.94	17.10
2000	Timber piles to 14" dia		600	.107			5.10	2.90	8	11.40

02 41 13.76 Selective Demolition, Water Wells

Line	Description	Crew	Daily Output	Labor-Hours	Unit	Material	Labor	Equipment	Total	Total Incl O&P
0010	SELECTIVE DEMOLITION, WATER WELLS									
0100	Well, 40' deep with casing & gravel pack, 24"-36" dia	B-23	.25	160	Ea.		6,075	11,400	17,475	22,500
0200	Riser pipe, 1-1/4", for observation well	"	300	.133	L.F.		5.05	9.50	14.55	18.75
0300	Pump, 1/2 to 5 HP up to 100' depth	Q-1	3	5.333	Ea.		282		282	440
0400	Up to 150' well 25 HP pump	Q-22	1.50	10.667			565	435	1,000	1,350
0500	Up to 500' well 30 HP pump	"	1	16			845	655	1,500	2,050
0600	Well screen 2" to 8"	B-23	300	.133	V.L.F.		5.05	9.50	14.55	18.75
0700	10" to 16"		200	.200			7.60	14.25	21.85	28
0800	18" to 26"		150	.267			10.15	18.95	29.10	37.50
0900	Slotted PVC for wells 1-1/4"-8"		600	.067			2.53	4.74	7.27	9.35
1000	Well screen and casing 6" to 16"		300	.133			5.05	9.50	14.55	18.75
1100	18" to 26"		150	.267			10.15	18.95	29.10	37.50
1200	30" to 36"		100	.400			15.20	28.50	43.70	56.50

02 41 13.78 Selective Demolition, Radio Towers

Line	Description	Crew	Daily Output	Labor-Hours	Unit	Material	Labor	Equipment	Total	Total Incl O&P
0010	SELECTIVE DEMOLITION, RADIO TOWERS									
0100	Radio tower,guyed, 50'	2 Skwk	16	1	Ea.		48.50		48.50	79
0200	190', 40 lb. section	K-2	.70	34.286			1,675	435	2,110	3,400
0300	200', 70 lb. section		.70	34.286			1,675	435	2,110	3,400
0400	300', 70 lb. section		.40	60			2,925	760	3,685	5,975
0500	270', 90 lb. section		.40	60			2,925	760	3,685	5,975
0600	400'		.30	80			3,900	1,000	4,900	7,950
0700	Self supported, 60'		.90	26.667			1,300	335	1,635	2,650
0800	120'		.80	30			1,475	380	1,855	3,000
0900	190'		.40	60			2,925	760	3,685	5,975

02 41 13 – Selective Site Demolition

02 41 13.80 Selective Demo., Utility Poles and Cross Arms

Line	02 41 13.80 Selective Demo., Utility Poles and Cross Arms	Crew	Daily Output	Labor-Hours	Unit	Material	2015 Bare Costs Labor	Equipment	Total	Total Incl O&P
0010	SELECTIVE DEMOLITION, UTILITY POLES & CROSS ARMS									
0100	Utility poles, wood, 20′ - 30′ high	R-3	6	3.333	Ea.		181	23	204	305
0200	35′ - 45′ high	"	5	4			217	27.50	244.50	370
0300	Cross arms, wood, 4′ - 6′ long	1 Elec	5	1.600			87.50		87.50	136

02 41 13.82 Selective Removal, Pavement Lines and Markings

Line	Description	Crew	Daily Output	Labor-Hours	Unit	Material	Labor	Equipment	Total	Total Incl O&P
0010	SELECTIVE REMOVAL, PAVEMENT LINES & MARKINGS									
0015	Does not include traffic control costs									
0020	See other items in Section 32 17 23.13									
0100	Remove permanent painted traffic lines and markings	B-78A	500	.016	C.L.F.		.78	1.64	2.42	3.04
0200	Temporary traffic line tape	2 Clab	1500	.011	L.F.		.40		.40	.66
0300	Thermoplastic traffic lines and markings	B-79A	500	.024	C.L.F.		1.17	2.56	3.73	4.66
0400	Painted pavement markings	B-78B	500	.036	S.F.		1.40	.75	2.15	3.11

02 41 13.84 Selective Demolition, Walks, Steps and Pavers

Line	Description	Crew	Daily Output	Labor-Hours	Unit	Material	Labor	Equipment	Total	Total Incl O&P
0010	SELECTIVE DEMOLITION, WALKS, STEPS AND PAVERS									
0100	Splash blocks	1 Clab	300	.027	S.F.		1		1	1.64
0200	Tree grates	"	50	.160	Ea.		6		6	9.85
0300	Walks, limestone pavers	2 Clab	150	.107	S.F.		4.01		4.01	6.60
0400	Redwood sections		600	.027			1		1	1.64
0500	Redwood planks		480	.033			1.25		1.25	2.05
0600	Shale paver		300	.053			2.01		2.01	3.29
0700	Tile thinset paver		675	.024			.89		.89	1.46
0800	Wood round	B-1	350	.069	Ea.		2.62		2.62	4.30
0900	Asphalt block	2 Clab	450	.036	S.F.		1.34		1.34	2.19
1000	Bluestone		450	.036			1.34		1.34	2.19
1100	Slate, 1″ or thinner		675	.024			.89		.89	1.46
1200	Granite blocks		300	.053			2.01		2.01	3.29
1300	Precast patio blocks		450	.036			1.34		1.34	2.19
1400	Planter blocks		600	.027			1		1	1.64
1500	Brick paving, dry set		300	.053			2.01		2.01	3.29
1600	Mortar set		180	.089			3.34		3.34	5.50
1700	Dry set on edge		240	.067			2.51		2.51	4.11
1800	Steps, brick		200	.080	L.F.		3.01		3.01	4.93
1900	Railroad tie		150	.107			4.01		4.01	6.60
2000	Bluestone		180	.089			3.34		3.34	5.50
2100	Wood/steel edging for steps		1000	.016			.60		.60	.99
2200	Timber or railroad tie edging for steps		400	.040			1.50		1.50	2.47

02 41 13.86 Selective Demolition, Athletic Surfaces

Line	Description	Crew	Daily Output	Labor-Hours	Unit	Material	Labor	Equipment	Total	Total Incl O&P
0010	SELECTIVE DEMOLITION, ATHLETIC SURFACES									
0100	Synthetic grass	2 Clab	2000	.008	S.F.		.30		.30	.49
0200	Surface coat latex rubber	"	2000	.008	"		.30		.30	.49
0300	Tennis court posts	B-11C	16	1	Ea.		44	23	67	96

02 41 13.88 Selective Demolition, Lawn Sprinkler Systems

Line	Description	Crew	Daily Output	Labor-Hours	Unit	Material	Labor	Equipment	Total	Total Incl O&P
0010	SELECTIVE DEMOLITION, LAWN SPRINKLER SYSTEMS									
0100	Golf course sprinkler system, 9 hole	4 Skwk	.10	320	Ea.		15,600		15,600	25,300
0200	Sprinkler system, 24′ diam. @ 15′ O.C., per head	B-20	110	.218	Head		9.15		9.15	14.95
0300	60′ diam. @ 24′ O.C., per head	"	52	.462	"		19.35		19.35	31.50
0400	Sprinkler heads, plastic	2 Skwk	150	.107	Ea.		5.20		5.20	8.45
0500	Impact circle pattern, 28′ - 76′ diam.		75	.213			10.40		10.40	16.85
0600	Pop-up, 42′ - 76′ diam.		50	.320			15.55		15.55	25.50
0700	39′ - 99′ diameter		50	.320			15.55		15.55	25.50
0800	Sprinkler valves		40	.400			19.45		19.45	31.50

02 41 13 – Selective Site Demolition

02 41 13.88 Selective Demolition, Lawn Sprinkler Systems		Crew	Daily Output	Labor-Hours	Unit	Material	2015 Bare Costs Labor	Equipment	Total	Total Incl O&P
0900	Valve boxes	2 Skwk	40	.400	Ea.		19.45		19.45	31.50
1000	Controls		2	8			390		390	630
1100	Backflow preventer		4	4			195		195	315
1200	Vacuum breaker		4	4			195		195	315

02 41 13.90 Selective Demolition, Retaining Walls

		Crew	Daily Output	Labor-Hours	Unit	Material	Labor	Equipment	Total	Total Incl O&P
0010	SELECTIVE DEMOLITION, RETAINING WALLS									
0100	Concrete retaining wall, 6′ high, no reinforcing	B-13K	200	.080	L.F.		4.14	8.10	12.24	15.45
0200	8′ high		150	.107			5.50	10.80	16.30	20.50
0300	10′ high		150	.107			5.50	10.80	16.30	20.50
0400	With reinforcing, 6′ high		200	.080			4.14	8.10	12.24	15.45
0500	8′ high		150	.107			5.50	10.80	16.30	20.50
0600	10′ high		120	.133			6.90	13.50	20.40	26
0700	20′ high		60	.267			13.80	27	40.80	51.50
0800	Concrete cribbing, 12′ high, open/closed face		150	.107	S.F.		5.50	10.80	16.30	20.50
0900	Interlocking segmental retaining wall	B-62	800	.030			1.24	.22	1.46	2.24
1000	Wall caps	"	600	.040			1.65	.29	1.94	2.99
1100	Metal bin retaining wall, 10′ wide, 4-12′ high	B-13	1200	.047			1.91	.61	2.52	3.78
1200	10′ wide, 16-28′ high		1000	.056			2.30	.74	3.04	4.53
1300	Stone filled gabions, 6′ x 3′ x 1′		170	.329	Ea.		13.50	4.33	17.83	27
1400	6′ x 3′ x 1′-6″		75	.747			30.50	9.80	40.30	60.50
1500	6′ x 3′ x 3′		25	2.240			92	29.50	121.50	182
1600	9′ x 3′ x 1′		75	.747			30.50	9.80	40.30	60.50
1700	9′ x 3′ x 1′-6″		33	1.697			69.50	22.50	92	138
1800	9′ x 3′ x 3′		12	4.667			191	61.50	252.50	380
1900	12′ x 3′ x 1′		42	1.333			54.50	17.55	72.05	108
2000	12′ x 3′ x 1′-6″		20	2.800			115	37	152	227
2100	12′ x 3′ x 3′		6	9.333			385	123	508	755

02 41 13.92 Selective Demolition, Parking Appurtenances

		Crew	Daily Output	Labor-Hours	Unit	Material	Labor	Equipment	Total	Total Incl O&P
0010	SELECTIVE DEMOLITION, PARKING APPURTENANCES									
0100	Bumper rails, garage, 6″ wide	B-6	300	.080	L.F.		3.30	1.21	4.51	6.70
0200	12″ channel rail		300	.080			3.30	1.21	4.51	6.70
0300	Parking bumper, timber		1000	.024			.99	.36	1.35	2
0400	Folding, with locks	B-1	100	.240	Ea.		9.20		9.20	15.05
0500	Flexible fixed garage stanchion	B-6	150	.160			6.60	2.43	9.03	13.35
0600	Wheel stops, precast concrete		120	.200			8.25	3.04	11.29	16.70
0700	Thermoplastic		120	.200			8.25	3.04	11.29	16.70
0800	Pipe bollards, 6″ - 12″ dia		80	.300			12.40	4.55	16.95	25

02 41 13.93 Selective Demolition, Site Furnishings

		Crew	Daily Output	Labor-Hours	Unit	Material	Labor	Equipment	Total	Total Incl O&P
0010	SELECTIVE DEMOLITION, SITE FURNISHINGS									
0100	Benches, all types	2 Clab	10	1.600	Ea.		60		60	98.50
0200	Trash receptacles, all types		80	.200			7.50		7.50	12.35
0300	Trash enclosures, steel or wood		10	1.600			60		60	98.50

02 41 13.94 Selective Demolition, Athletic Screening

		Crew	Daily Output	Labor-Hours	Unit	Material	Labor	Equipment	Total	Total Incl O&P
0010	SELECTIVE DEMOLITION, ATHLETIC SCREENING									
0020	See other items in Section 02 41 13.60 & 02 41 13.62									
0100	Baseball backstops	B-6	2	12	Ea.		495	182	677	1,000
0200	Basketball goal, single		6	4			165	60.50	225.50	335
0300	Double		4	6			248	91	339	500
0400	Tennis wire mesh, pair ends		5	4.800	Set		198	73	271	400
0500	Enclosed court		3	8	Ea.		330	121	451	670
0600	Wood or masonry handball court		1	24	"		990	365	1,355	2,000

02 41 13.95 Selective Demo., Athletic/Playground Equipment		Crew	Daily Output	Labor-Hours	Unit	Material	2015 Bare Costs Labor	Equipment	Total	Total Incl O&P
0010	**SELECTIVE DEMO., ATHLETIC/PLAYGROUND EQUIPMENT**									
0020	See other items in Section 02 41 13.96									
0100	Bike rack	B-6	32	.750	Ea.		31	11.40	42.40	62.50
0200	Climber arch		16	1.500			62	23	85	125
0300	Fitness trail, 9 to 10 stations wood or metal		.50	48			1,975	730	2,705	4,000
0400	16 to 20 stations wood or metal		.25	96			3,950	1,450	5,400	8,025
0500	Monkey bars 14′ long		8	3			124	45.50	169.50	250
0600	Parallel bars 10′ long		8	3			124	45.50	169.50	250
0700	Post tether ball		32	.750			31	11.40	42.40	62.50
0800	Poles 10′-6″ long		32	.750			31	11.40	42.40	62.50
0900	Pole ground sockets		80	.300			12.40	4.55	16.95	25
1000	See-saw, 2 units		12	2			82.50	30.50	113	168
1100	4 units		10	2.400			99	36.50	135.50	200
1200	6 units		8	3			124	45.50	169.50	250
1300	Shelter, fiberglass, 3 person		10	2.400			99	36.50	135.50	200
1400	Slides, 12′ long		8	3			124	45.50	169.50	250
1500	20′ long		6	4			165	60.50	225.50	335
1600	Swings, 4 seats		6	4			165	60.50	225.50	335
1700	8 seats		4	6			248	91	339	500
1800	Whirlers, 8 – 10′ diameter		6	4			165	60.50	225.50	335
1900	Football or football/soccer goal posts		3	8	Pair		330	121	451	670
2000	Soccer goal posts		4	6	″		248	91	339	500
2100	Playground surfacing, 4″ depth	2 Clab	900	.018	S.F.		.67		.67	1.10
2200	2″ topping	″	1800	.009	″		.33		.33	.55
2300	Platform/paddle tennis court, alum. deck & frame	B-1	.20	120	Court		4,600		4,600	7,525
2400	Aluminum deck and wood frame		.25	96	″		3,675		3,675	6,025
2500	Heater		2	12	Ea.		460		460	755
2600	Wood deck & frame		.25	96	Court		3,675		3,675	6,025
2700	Wood deck & steel frame		.25	96			3,675		3,675	6,025
2800	Steel deck & wood frame		.25	96			3,675		3,675	6,025

02 41 13.96 Selective Demo., Modular Playground Equipment		Crew	Daily Output	Labor-Hours	Unit	Material	2015 Bare Costs Labor	Equipment	Total	Total Incl O&P
0010	**SELECTIVE DEMOLITION, MODULAR PLAYGROUND EQUIPMENT**									
0100	Modular playground, 48″ deck, square or triangle shape	B-1	2	12	Ea.		460		460	755
0200	Various posts		18	1.333			51		51	83.50
0300	Roof 54″ x 54″		18	1.333			51		51	83.50
0400	Wheel chair transfer module		6	4			153		153	251
0500	Guardrail pipe		90	.267	L.F.		10.20		10.20	16.75
0600	Steps, 3 risers		16	1.500	Ea.		57.50		57.50	94
0700	Activity panel crawl through		8	3			115		115	188
0800	Alphabet/spelling panel		8	3			115		115	188
0900	Crawl tunnel, each 4′ section		8	3			115		115	188
1000	Slide tunnel		16	1.500			57.50		57.50	94
1100	Spiral slide tunnel		12	2			76.50		76.50	125
1200	Ladder, horizontal or vertical		10	2.400			92		92	151
1300	Corkscrew climber		6	4			153		153	251
1400	Fire pole		12	2			76.50		76.50	125
1500	Bridge, ring chamber		8	3			115		115	188
1600	Bridge, suspension		8	3			115		115	188

02 41 13.97 Selective Demolition, Traffic Light Systems		Crew	Daily Output	Labor-Hours	Unit	Material	2015 Bare Costs Labor	Equipment	Total	Total Incl O&P
0010	**SELECTIVE DEMOLITION, TRAFFIC LIGHT SYSTEMS**									
0100	Traffic signal, pedestrian cross walk	R-2	.50	112	Total		5,475	550	6,025	9,225
0200	8 signal intersection, 2 each direction	″	.25	224			10,900	1,100	12,000	18,400

02 41 Demolition

02 41 13 – Selective Site Demolition

02 41 13.97 Selective Demolition, Traffic Light Systems		Crew	Daily Output	Labor-Hours	Unit	Material	2015 Bare Costs Labor	Equipment	Total	Total Incl O&P
0300	Each traffic phase controller	L-9	3	12	Total		515		515	860
0400	Each semi-actuated detector		1.50	24			1,025		1,025	1,725
0500	Each fully-actuated detector		1.50	24			1,025		1,025	1,725
0600	Pedestrian push button system		2	18			775		775	1,300
0650	Each optically programmed head		3	12			515		515	860
0700	School signal flashing system		1	36	Signal		1,550		1,550	2,575
0800	Complete traffic light intersection system, without lane control	R-2	.25	224	Ea.		10,900	1,100	12,000	18,400
0900	With lane control		.20	280			13,700	1,375	15,075	23,000
1000	Traffic light, each left turn protection system		.66	84.848			4,150	415	4,565	6,975

02 41 13.98 Select. Demo., Sod, Edging, Planters & Tree Guying		Crew	Daily Output	Labor-Hours	Unit	Material	Labor	Equipment	Total	Total Incl O&P
0010	SELECTIVE DEMOLITION, SOD, EDGING, PLANTERS & TREE GUYING									
0100	Remove sod 1″ deep, 3 M.S.F. or greater, level ground	B-63	40	1	M.S.F.		40	4.34	44.34	70
0200	Less than 3 M.S.F., level ground		27	1.481			59	6.45	65.45	103
0300	3 M.S.F. or greater, sloped ground		12	3.333			133	14.45	147.45	232
0400	Less than 3 M.S.F., sloped ground		8	5			199	21.50	220.50	350
0500	Edging, aluminum	B-1	800	.030	L.F.		1.15		1.15	1.88
0600	Brick horizontal		400	.060			2.30		2.30	3.76
0700	Brick vertical		200	.120			4.59		4.59	7.55
0800	Corrugated aluminum		1200	.020			.77		.77	1.25
0900	Granite 5″ x 16″		600	.040			1.53		1.53	2.51
1000	Polyethylene grass barrier		1200	.020			.77		.77	1.25
1100	Precast 2″ x 8″ x 16″		800	.030			1.15		1.15	1.88
1200	Railroad ties		200	.120			4.59		4.59	7.55
1300	Wood		400	.060			2.30		2.30	3.76
1400	Steel strips including stakes		600	.040			1.53		1.53	2.51
1500	Planters, concrete, 48″ dia	2 Clab	30	.533	Ea.		20		20	33
1600	Concrete 7′ dia		20	.800			30		30	49.50
1700	Fiberglass 36″ dia		30	.533			20		20	33
1800	60″ dia		20	.800			30		30	49.50
1900	24″ square		30	.533			20		20	33
2000	48″ square		30	.533			20		20	33
2100	Wood 48″ square		30	.533			20		20	33
2200	48″ dia		20	.800			30		30	49.50
2300	72″ dia		20	.800			30		30	49.50
2400	Planter/bench, fiberglass 72″ - 96″ square		10	1.600			60		60	98.50
2500	Wood, 72″ square		10	1.600			60		60	98.50
2600	Tree guying, stakes, 3″ - 4″ caliper		40	.400			15.05		15.05	24.50
2700	anchors, 3″ caliper		40	.400			15.05		15.05	24.50
2800	3″ - 6″ caliper		30	.533			20		20	33
2900	6″ caliper		20	.800			30		30	49.50
3000	8″ caliper		15	1.067			40		40	66

02 41 16 – Structure Demolition

02 41 16.13 Building Demolition		Crew	Daily Output	Labor-Hours	Unit	Material	Labor	Equipment	Total	Total Incl O&P
0010	BUILDING DEMOLITION Large urban projects, incl. 20 mi. haul R024119-10									
0011	No foundation or dump fees, C.F. is vol. of building standing									
0020	Steel	B-8	21500	.003	C.F.		.13	.15	.28	.37
0050	Concrete		15300	.004			.18	.22	.40	.53
0080	Masonry		20100	.003			.14	.16	.30	.40
0100	Mixture of types		20100	.003			.14	.16	.30	.40
0500	Small bldgs, or single bldgs, no salvage included, steel	B-3	14800	.003			.13	.17	.30	.40
0600	Concrete		11300	.004			.17	.23	.40	.53
0650	Masonry		14800	.003			.13	.17	.30	.40

02 41 Demolition

02 41 16 – Structure Demolition

02 41 16.13 Building Demolition		Crew	Daily Output	Labor-Hours	Unit	Material	2015 Bare Costs Labor	Equipment	Total	Total Incl O&P
0700	Wood	B-3	14800	.003	C.F.		.13	.17	.30	.40
0750	For buildings with no interior walls, deduct								50%	50%
1000	Demoliton single family house, one story, wood 1600 S.F.	B-3	1	48	Ea.		1,975	2,575	4,550	6,000
1020	3200 S.F.		.50	96			3,925	5,150	9,075	12,000
1200	Demoliton two family house, two story, wood 2400 S.F.		.67	71.964			2,950	3,850	6,800	9,025
1220	4200 S.F.		.38	128			5,250	6,850	12,100	16,000
1300	Demoliton three family house, three story, wood 3200 S.F.		.50	96			3,925	5,150	9,075	12,000
1320	5400 S.F.		.30	160			6,550	8,575	15,125	20,000
5000	For buildings with no interior walls, deduct								50%	50%

02 41 16.17 Building Demolition Footings and Foundations		Crew	Daily Output	Labor-Hours	Unit	Material	Labor	Equipment	Total	Total Incl O&P
0010	BUILDING DEMOLITION FOOTINGS AND FOUNDATIONS R024119-10									
0200	Floors, concrete slab on grade,									
0240	4" thick, plain concrete	B-13L	5000	.003	S.F.		.17	.41	.58	.71
0280	Reinforced, wire mesh		4000	.004			.21	.51	.72	.90
0300	Rods		4500	.004			.18	.46	.64	.79
0400	6" thick, plain concrete		4000	.004			.21	.51	.72	.90
0420	Reinforced, wire mesh		3200	.005			.26	.64	.90	1.12
0440	Rods		3600	.004			.23	.57	.80	.99
1000	Footings, concrete, 1' thick, 2' wide	B-5	300	.187	L.F.		7.75	4.74	12.49	17.80
1080	1'-6" thick, 2' wide		250	.224			9.30	5.70	15	21.50
1120	3' wide		200	.280			11.65	7.10	18.75	26.50
1140	2' thick, 3' wide		175	.320			13.30	8.10	21.40	30.50
1200	Average reinforcing, add								10%	10%
1220	Heavy reinforcing, add								20%	20%
2000	Walls, block, 4" thick	B-13L	8000	.002	S.F.		.10	.26	.36	.44
2040	6" thick		6000	.003			.14	.34	.48	.60
2080	8" thick		4000	.004			.21	.51	.72	.90
2100	12" thick		3000	.005			.28	.68	.96	1.19
2200	For horizontal reinforcing, add								10%	10%
2220	For vertical reinforcing, add								20%	20%
2400	Concrete, plain concrete, 6" thick	B-13L	4000	.004			.21	.51	.72	.90
2420	8" thick		3500	.005			.24	.59	.83	1.03
2440	10" thick		3000	.005			.28	.68	.96	1.19
2500	12" thick		2500	.006			.33	.82	1.15	1.43
2600	For average reinforcing, add								10%	10%
2620	For heavy reinforcing, add								20%	20%
4300	Footings & foundation removal, with expansion grout, no hauling	B-9	5.68	7.042	C.Y.		268	41	309	485
9000	Minimum labor/equipment charge	A-1	2	4	Job		150	40.50	190.50	292

02 41 16.33 Bridge Demolition		Crew	Daily Output	Labor-Hours	Unit	Material	Labor	Equipment	Total	Total Incl O&P
0010	BRIDGE DEMOLITION									
0100	Bridges, pedestrian, precast, 60' to 150' long	B-21C	250	.224	S.F.		9.20	7.35	16.55	23
0200	Steel, 50' to 160' long, 8' to 10' wide	"	500	.112			4.59	3.68	8.27	11.50
0300	Laminated wood, 80' to 130' long	C-12	300	.160			7.45	2.18	9.63	14.55

02 41 19 – Selective Demolition

02 41 19.13 Selective Building Demolition	
0010	SELECTIVE BUILDING DEMOLITION
0020	Costs related to selective demolition of specific building components
0025	are included under Common Work Results (XX 05)
0030	in the component's appropriate division.

02 41 Demolition

02 41 19 – Selective Demolition

02 41 19.16 Selective Demolition, Cutout

	02 41 19.16 Selective Demolition, Cutout	Crew	Daily Output	Labor-Hours	Unit	Material	2015 Bare Costs Labor	Equipment	Total	Total Incl O&P
0010	SELECTIVE DEMOLITION, CUTOUT R024119-10									
0020	Concrete, elev. slab, light reinforcement, under 6 C.F.	B-9	65	.615	C.F.		23.50	3.58	27.08	42.50
0050	Light reinforcing, over 6 C.F.		75	.533	"		20.50	3.10	23.60	36.50
0200	Slab on grade to 6" thick, not reinforced, under 8 S.F.		85	.471	S.F.		17.90	2.74	20.64	32.50
0250	8 – 16 S.F.		175	.229	"		8.70	1.33	10.03	15.70
0255	For over 16 S.F. see Line 02 41 16.17 0400									
0600	Walls, not reinforced, under 6 C.F.	B-9	60	.667	C.F.		25.50	3.88	29.38	46
0650	6 – 12 C.F.	"	80	.500	"		19	2.91	21.91	34
0655	For over 12 C.F. see Line 02 41 16.17 2500									
1000	Concrete, elevated slab, bar reinforced, under 6 C.F.	B-9	45	.889	C.F.		34	5.15	39.15	61
1050	Bar reinforced, over 6 C.F.		50	.800	"		30.50	4.66	35.16	55
1200	Slab on grade to 6" thick, bar reinforced, under 8 S.F.		75	.533	S.F.		20.50	3.10	23.60	36.50
1250	8 – 16 S.F.		150	.267	"		10.15	1.55	11.70	18.30
1255	For over 16 S.F. see Line 02 41 16.17 0440									
1400	Walls, bar reinforced, under 6 C.F.	B-9	50	.800	C.F.		30.50	4.66	35.16	55
1450	6 – 12 C.F.	"	70	.571	"		21.50	3.33	24.83	39
1455	For over 12 C.F. see Lines 02 41 16.17 2500 and 2600									
2000	Brick, to 4 S.F. opening, not including toothing									
2040	4" thick	B-9	30	1.333	Ea.		50.50	7.75	58.25	91.50
2060	8" thick		18	2.222			84.50	12.95	97.45	152
2080	12" thick		10	4			152	23.50	175.50	275
2400	Concrete block, to 4 S.F. opening, 2" thick		35	1.143			43.50	6.65	50.15	78.50
2420	4" thick		30	1.333			50.50	7.75	58.25	91.50
2440	8" thick		27	1.481			56.50	8.60	65.10	102
2460	12" thick		24	1.667			63.50	9.70	73.20	115
2600	Gypsum block, to 4 S.F. opening, 2" thick		80	.500			19	2.91	21.91	34
2620	4" thick		70	.571			21.50	3.33	24.83	39
2640	8" thick		55	.727			27.50	4.23	31.73	50
2800	Terra cotta, to 4 S.F. opening, 4" thick		70	.571			21.50	3.33	24.83	39
2840	8" thick		65	.615			23.50	3.58	27.08	42.50
2880	12" thick		50	.800			30.50	4.66	35.16	55
4000	For toothing masonry, see Section 04 01 20.50									
6000	Walls, interior, not including re-framing,									
6010	openings to 5 S.F.									
6100	Drywall to 5/8" thick	1 Clab	24	.333	Ea.		12.55		12.55	20.50
6200	Paneling to 3/4" thick		20	.400			15.05		15.05	24.50
6300	Plaster, on gypsum lath		20	.400			15.05		15.05	24.50
6340	On wire lath		14	.571			21.50		21.50	35
7000	Wood frame, not including re-framing, openings to 5 S.F.									
7200	Floors, sheathing and flooring to 2" thick	1 Clab	5	1.600	Ea.		60		60	98.50
7310	Roofs, sheathing to 1" thick, not including roofing		6	1.333			50		50	82
7410	Walls, sheathing to 1" thick, not including siding		7	1.143			43		43	70.50
8500	Minimum labor/equipment charge		4	2	Job		75		75	123

02 41 19.18 Selective Demolition, Disposal Only

	02 41 19.18 Selective Demolition, Disposal Only	Crew	Daily Output	Labor-Hours	Unit	Material	2015 Bare Costs Labor	Equipment	Total	Total Incl O&P
0010	SELECTIVE DEMOLITION, DISPOSAL ONLY R024119-10									
0015	Urban bldg w/salvage value allowed									
0020	Including loading and 5 mile haul to dump									
0200	Steel frame	B-3	430	.112	C.Y.		4.57	6	10.57	14
0300	Concrete frame		365	.132			5.40	7.05	12.45	16.45
0400	Masonry construction		445	.108			4.41	5.80	10.21	13.50
0500	Wood frame		247	.194			7.95	10.40	18.35	24.50

02 41 19.19 Selective Demolition		Crew	Daily Output	Labor-Hours	Unit	Material	Labor (2015 Bare Costs)	Equipment (2015 Bare Costs)	Total (2015 Bare Costs)	Total Incl O&P
0010	**SELECTIVE DEMOLITION,** Rubbish Handling R024119-10									
0020	The following are to be added to the demolition prices									
0050	The following are components for a complete chute system									
0100	Top chute circular steel, 4′ long, 18″ diameter R024119-30	B-1C	15	1.600	Ea.	266	61	29	356	425
0102	23″ diameter		15	1.600		288	61	29	378	445
0104	27″ diameter		15	1.600		310	61	29	400	470
0106	30″ diameter		15	1.600		330	61	29	420	495
0108	33″ diameter		15	1.600		355	61	29	445	520
0110	36″ diameter		15	1.600		375	61	29	465	545
0112	Regular chute, 18″ diameter		15	1.600		199	61	29	289	350
0114	23″ diameter		15	1.600		222	61	29	312	375
0116	27″ diameter		15	1.600		243	61	29	333	400
0118	30″ diameter		15	1.600		266	61	29	356	425
0120	33″ diameter		15	1.600		288	61	29	378	445
0122	36″ diameter		15	1.600		310	61	29	400	470
0124	Control door chute, 18″ diameter		15	1.600		375	61	29	465	545
0126	23″ diameter		15	1.600		400	61	29	490	570
0128	27″ diameter		15	1.600		420	61	29	510	595
0130	30″ diameter		15	1.600		445	61	29	535	620
0132	33″ diameter		15	1.600		465	61	29	555	640
0134	36″ diameter		15	1.600		490	61	29	580	665
0136	Chute liners, 14 ga., 18-30″ diameter		15	1.600		223	61	29	313	380
0138	33-36″ diameter		15	1.600		280	61	29	370	440
0140	17% thinner chute, 30″ diameter		15	1.600		222	61	29	312	375
0142	33% thinner chute, 30″ diameter		15	1.600		167	61	29	257	315
0144	Top chute cover	1 Clab	24	.333		153	12.55		165.55	189
0146	Door chute cover	″	24	.333		153	12.55		165.55	189
0148	Top chute trough	2 Clab	12	1.333		510	50		560	640
0150	Bolt down frame & counter weights, 250 lb.	B-1	4	6		4,250	230		4,480	5,025
0152	500 lb.		4	6		6,200	230		6,430	7,200
0154	750 lb.		4	6		9,100	230		9,330	10,400
0156	1000 lb.		2.67	8.989		10,200	345		10,545	11,800
0158	1500 lb.		2.67	8.989		13,000	345		13,345	14,900
0160	Chute warning light system, 5 stories	B-1C	4	6		8,975	230	109	9,314	10,400
0162	10 stories	″	2	12		14,300	460	218	14,978	16,700
0164	Dust control device for dumpsters	1 Clab	8	1		102	37.50		139.50	174
0166	Install or replace breakaway cord		8	1		27	37.50		64.50	91
0168	Install or replace warning sign		16	.500		11.45	18.80		30.25	43.50
0600	Dumpster, weekly rental, 1 dump/week, 6 C.Y. capacity (2 Tons)				Week	415			415	455
0700	10 C.Y. capacity (3 Tons)					480			480	530
0725	20 C.Y. capacity (5 Tons) R024119-20					565			565	625
0800	30 C.Y. capacity (7 Tons)					730			730	800
0840	40 C.Y. capacity (10 Tons)					775			775	850
0900	Alternate pricing for dumpsters									
0910	Delivery, average for all sizes				Ea.	75			75	82.50
0920	Haul, average for all sizes					235			235	259
0930	Rent per day, average for all sizes					20			20	22
0940	Rent per month, average for all sizes					80			80	88
0950	Disposal fee per ton, average for all sizes				Ton	88			88	97
2000	Load, haul, dump and return, 0 – 50′ haul, hand carried	2 Clab	24	.667	C.Y.		25		25	41
2005	Wheeled		37	.432			16.25		16.25	26.50
2040	0 – 100′ haul, hand carried		16.50	.970			36.50		36.50	60

02 41 Demolition

02 41 19 – Selective Demolition

02 41 19.19 Selective Demolition		Crew	Daily Output	Labor-Hours	Unit	2015 Bare Costs Material	Labor	Equipment	Total	Total Incl O&P
2045	Wheeled	2 Clab	25	.640	C.Y.		24		24	39.50
2080	Haul and return, add per each extra 100′ haul, hand carried		35.50	.451			16.95		16.95	28
2085	Wheeled		54	.296			11.15		11.15	18.25
2120	For travel in elevators, up to 10 floors, add		140	.114			4.30		4.30	7.05
2130	0 – 50′ haul, incl. up to 5 riser stair, hand carried		23	.696			26		26	43
2135	Wheeled		35	.457			17.20		17.20	28
2140	6 – 10 riser stairs, hand carried		22	.727			27.50		27.50	45
2145	Wheeled		34	.471			17.70		17.70	29
2150	11 – 20 riser stairs, hand carried		20	.800			30		30	49.50
2155	Wheeled		31	.516			19.40		19.40	32
2160	21 – 40 riser stairs, hand carried		16	1			37.50		37.50	61.50
2165	Wheeled		24	.667			25		25	41
2170	0 – 100′ haul, incl. 5 riser stair, hand carried		15	1.067			40		40	66
2175	Wheeled		23	.696			26		26	43
2180	6 – 10 riser stair, hand carried		14	1.143			43		43	70.50
2185	Wheeled		21	.762			28.50		28.50	47
2190	11 – 20 riser stair, hand carried		12	1.333			50		50	82
2195	Wheeled		18	.889			33.50		33.50	55
2200	21 – 40 riser stair, hand carried		8	2			75		75	123
2205	Wheeled		12	1.333			50		50	82
2210	Haul and return, add per each extra 100′ haul, hand carried		35.50	.451			16.95		16.95	28
2215	Wheeled		54	.296			11.15		11.15	18.25
2220	For each additional flight of stairs, up to 5 risers, add		550	.029	Flight		1.09		1.09	1.79
2225	6 – 10 risers, add		275	.058			2.19		2.19	3.59
2230	11 – 20 risers, add		138	.116			4.36		4.36	7.15
2235	21 – 40 risers, add		69	.232			8.70		8.70	14.30
3000	Loading & trucking, including 2 mile haul, chute loaded	B-16	45	.711	C.Y.		27.50	15.35	42.85	62
3040	Hand loading truck, 50′ haul	"	48	.667			26	14.40	40.40	58
3080	Machine loading truck	B-17	120	.267			10.90	6.45	17.35	25
5000	Haul, per mile, up to 8 C.Y. truck	B-34B	1165	.007			.28	.59	.87	1.09
5100	Over 8 C.Y. truck	"	1550	.005			.21	.45	.66	.82

02 41 19.20 Selective Demolition, Dump Charges		Crew	Daily Output	Labor-Hours	Unit	Material	Labor	Equipment	Total	Total Incl O&P
0010	**SELECTIVE DEMOLITION, DUMP CHARGES** R024119-10									
0020	Dump charges, typical urban city, tipping fees only									
0100	Building construction materials				Ton	74			74	81
0200	Trees, brush, lumber					63			63	69.50
0300	Rubbish only					63			63	69.50
0500	Reclamation station, usual charge					74			74	81

02 41 19.21 Selective Demolition, Gutting		Crew	Daily Output	Labor-Hours	Unit	Material	Labor	Equipment	Total	Total Incl O&P
0010	**SELECTIVE DEMOLITION, GUTTING** R024119-10									
0020	Building interior, including disposal, dumpster fees not included									
0500	Residential building									
0560	Minimum	B-16	400	.080	SF Flr.		3.10	1.73	4.83	6.95
0580	Maximum	"	360	.089	"		3.44	1.92	5.36	7.70
0900	Commercial building									
1000	Minimum	B-16	350	.091	SF Flr.		3.54	1.97	5.51	7.95
1020	Maximum		250	.128	"		4.95	2.76	7.71	11.15
3000	Minimum labor/equipment charge		4	8	Job		310	173	483	695

02 41 19.25 Selective Demolition, Saw Cutting		Crew	Daily Output	Labor-Hours	Unit	Material	Labor	Equipment	Total	Total Incl O&P
0010	**SELECTIVE DEMOLITION, SAW CUTTING** R024119-10									
0015	Asphalt, up to 3″ deep	B-89	1050	.015	L.F.	.12	.67	.46	1.25	1.71
0020	Each additional inch of depth	"	1800	.009		.04	.39	.27	.70	.96

02 41 Demolition

02 41 19 – Selective Demolition

02 41 19.25 Selective Demolition, Saw Cutting		Crew	Daily Output	Labor-Hours	Unit	2015 Bare Costs Material	Labor	Equipment	Total	Total Incl O&P
1200	Masonry walls, hydraulic saw, brick, per inch of depth	B-89B	300	.053	L.F.	.04	2.34	2.86	5.24	6.95
1220	Block walls, solid, per inch of depth	"	250	.064		.04	2.81	3.43	6.28	8.30
2000	Brick or masonry w/hand held saw, per inch of depth	A-1	125	.064		.05	2.41	.65	3.11	4.71
5000	Wood sheathing to 1" thick, on walls	1 Carp	200	.040			1.88		1.88	3.08
5020	On roof	"	250	.032	↓		1.50		1.50	2.46
9000	Minimum labor/equipment charge	A-1	2	4	Job		150	40.50	190.50	292

02 41 19.27 Selective Demolition, Torch Cutting		Crew	Daily Output	Labor-Hours	Unit	Material	Labor	Equipment	Total	Total Incl O&P
0010	**SELECTIVE DEMOLITION, TORCH CUTTING** R024119-10									
0020	Steel, 1" thick plate	E-25	333	.024	L.F.	.84	1.31	.03	2.18	3.33
0040	1" diameter bar	"	600	.013	Ea.	.14	.73	.02	.89	1.49
1000	Oxygen lance cutting, reinforced concrete walls									
1040	12" to 16" thick walls	1 Clab	10	.800	L.F.		30		30	49.50
1080	24" thick walls	"	6	1.333	"		50		50	82
1090	Minimum labor/equipment charge	E-25	2	4	Job		219	5.70	224.70	400
1100	See Section 05 05 21.10									

02 42 Removal and Salvage of Construction Materials

02 42 10 – Building Deconstruction

02 42 10.10 Estimated Salvage Value or Savings		Crew	Daily Output	Labor-Hours	Unit	Material	Labor	Equipment	Total	Total Incl O&P
0010	**ESTIMATED SALVAGE VALUE OR SAVINGS**									
0015	Excludes material handling, packaging, container costs and									
0020	transportation for salvage or disposal									
0050	All Items in Section 02 42 10.10 are credit deducts and not costs									
0100	Copper Wire Salvage Value G				Lb.				1.60	1.60
0110	Disposal Savings G								.04	.04
0200	Copper Pipe Salvage Value G								2.50	2.50
0210	Disposal Savings G								.05	.05
0300	Steel Pipe Salvage Value G								.06	.06
0310	Disposal Savings G								.03	.03
0400	Cast Iron Pipe Salvage Value G								.03	.03
0410	Disposal Savings G								.01	.01
0500	Steel Doors or Windows Salvage Value G								.06	.06
0510	Aluminum G								.55	.55
0520	Disposal Savings G								.03	.03
0600	Aluminum Siding Salvage Value G								.49	.49
0630	Disposal Savings G				↓				.03	.03
0640	Wood Siding (no lead or asbestos) G				C.Y.				12	12
0800	Clean Concrete Disposal Savings G				Ton				62	62
0850	Asphalt Shingles Disposal Savings G				"				60	60
1000	Wood wall framing clean salvage value G				M.B.F.				55	55
1010	Painted G								44	44
1020	Floor framing G								55	55
1030	Painted G								44	44
1050	Roof framing G								55	55
1060	Painted G								44	44
1100	Wood beams salvage value G				↓				55	55
1200	Wood framing and beams disposal savings G				Ton				66	66
1220	Wood sheating and sub-base flooring G								72.50	72.50
1230	Wood wall paneling (1/4 inch thick) G				↓				66	66
1300	Wood panel 3/4-1 inch thick low salvage value G				S.F.				.55	.55
1350	high salvage value G				"				2.20	2.20

02 42 10.10 Estimated Salvage Value or Savings			Crew	Daily Output	Labor-Hours	Unit	Material	2015 Bare Costs Labor	Equipment	Total	Total Incl O&P
1400	Disposal savings	G				Ton				66	66
1500	Flooring tongue and groove 25/32 inch thick low salvage value	G				S.F.				.55	.55
1530	High salvage value	G				"				1.10	1.10
1560	Disposal savings	G				Ton				66	66
1600	Drywall or sheet rock salvage value	G								22	22
1650	Disposal savings	G				↓				66	66

02 42 10.20 Deconstruction of Building Components			Crew	Daily Output	Labor-Hours	Unit	Material	Labor	Equipment	Total	Total Incl O&P
0010	**DECONSTRUCTION OF BUILDING COMPONENTS**										
0012	Buildings one or two stories only										
0015	Excludes material handling, packaging, container costs and										
0020	transportation for salvage or disposal										
0050	Deconstruction of Plumbing Fixtures										
0100	Wall hung or countertop lavatory	G	2 Clab	16	1	Ea.		37.50		37.50	61.50
0110	Single or double compartment kitchen sink	G		14	1.143			43		43	70.50
0120	Wall hung urinal	G		14	1.143			43		43	70.50
0130	Floor mounted	G		8	2			75		75	123
0140	Floor mounted water closet	G		16	1			37.50		37.50	61.50
0150	Wall hung	G		14	1.143			43		43	70.50
0160	Water fountain, free standing	G		16	1			37.50		37.50	61.50
0170	Wall hung or deck mounted	G		12	1.333			50		50	82
0180	Bathtub, steel or fiberglass	G		10	1.600			60		60	98.50
0190	Cast iron	G		8	2			75		75	123
0200	Shower, single	G		6	2.667			100		100	164
0210	Group	G	↓	7	2.286	↓		86		86	141
0300	Deconstruction of Electrical Fixtures										
0310	Surface mount incandescent fixtures	G	2 Clab	48	.333	Ea.		12.55		12.55	20.50
0320	Fluorescent, 2 lamp	G		32	.500			18.80		18.80	31
0330	4 lamp	G		24	.667			25		25	41
0340	Strip Fluorescent, 1 lamp	G		40	.400			15.05		15.05	24.50
0350	2 lamp	G		32	.500			18.80		18.80	31
0400	Recessed drop-in fluorescent fixture, 2 lamp	G		27	.593			22.50		22.50	36.50
0410	4 lamp	G	↓	18	.889	↓		33.50		33.50	55
0500	Deconstruction of appliances										
0510	Cooking stoves	G	2 Clab	26	.615	Ea.		23		23	38
0520	Dishwashers	G	"	26	.615	"		23		23	38
0600	Deconstruction of millwork and trim										
0610	Cabinets, wood	G	2 Carp	40	.400	L.F.		18.80		18.80	31
0620	Countertops	G		100	.160	"		7.50		7.50	12.30
0630	Wall paneling, 1 inch thick	G		500	.032	S.F.		1.50		1.50	2.46
0640	Ceiling trim	G		500	.032	L.F.		1.50		1.50	2.46
0650	Wainscoting	G		500	.032	S.F.		1.50		1.50	2.46
0660	Base, 3/4" to 1" thick	G	↓	600	.027	L.F.		1.25		1.25	2.05
0700	Deconstruction of doors and windows										
0710	Doors, wrap, interior, wood, single, no closers	G	2 Carp	21	.762	Ea.	4.50	36		40.50	63.50
0720	Double	G		13	1.231		9	58		67	104
0730	Solid core, single, exterior or interior	G		10	1.600		4.50	75		79.50	128
0740	Double	G	↓	8	2	↓	9	94		103	164
0810	Windows, wrap, wood, single										
0812	with no casement or cladding	G	2 Carp	21	.762	Ea.	4.50	36		40.50	63.50
0820	with casement and/or cladding	G	"	18	.889	"	4.50	41.50		46	73.50
0900	Deconstruction of interior finishes										
0910	Drywall for recycling	G	2 Clab	1775	.009	S.F.		.34		.34	.56

02 42 Removal and Salvage of Construction Materials

02 42 10 – Building Deconstruction

02 42 10.20 Deconstruction of Building Components			Crew	Daily Output	Labor-Hours	Unit	Material	2015 Bare Costs Labor	Equipment	Total	Total Incl O&P
0920	Plaster wall, first floor	G	2 Clab	1775	.009	S.F.		.34		.34	.56
0930	Second floor	G		1330	.012			.45		.45	.74
1000	Deconstruction of roofing and accessories										
1010	Built-up roofs	G	2 Clab	570	.028	S.F.		1.06		1.06	1.73
1020	Gutters, facia and rakes	G	″	1140	.014	L.F.		.53		.53	.87
2000	Deconstruction of wood components										
2010	Roof sheeting	G	2 Clab	570	.028	S.F.		1.06		1.06	1.73
2020	Main roof framing	G		760	.021	L.F.		.79		.79	1.30
2030	Porch roof framing	G		445	.036			1.35		1.35	2.22
2040	Beams 4″ x 8″	G	B-1	375	.064			2.45		2.45	4.01
2050	4″ x 10″	G		300	.080			3.06		3.06	5
2055	4″ x 12″	G		250	.096			3.67		3.67	6
2060	6″ x 8″	G		250	.096			3.67		3.67	6
2065	6″ x 10″	G		200	.120			4.59		4.59	7.55
2070	6″ x 12″	G		170	.141			5.40		5.40	8.85
2075	8″ x 12″	G		126	.190			7.30		7.30	11.95
2080	10″ x 12″	G		100	.240			9.20		9.20	15.05
2100	Ceiling joists	G	2 Clab	800	.020			.75		.75	1.23
2150	Wall framing, interior	G		1230	.013			.49		.49	.80
2160	Sub-floor	G		2000	.008	S.F.		.30		.30	.49
2170	Floor joists	G		2000	.008	L.F.		.30		.30	.49
2200	Wood siding (no lead or asbestos)	G		1300	.012	S.F.		.46		.46	.76
2300	Wall framing, exterior	G		1600	.010	L.F.		.38		.38	.62
2400	Stair risers	G		53	.302	Ea.		11.35		11.35	18.60
2500	Posts	G		800	.020	L.F.		.75		.75	1.23
3000	Deconstruction of exterior brick walls										
3010	Exterior brick walls, first floor	G	2 Clab	200	.080	S.F.		3.01		3.01	4.93
3020	Second floor	G		64	.250	″		9.40		9.40	15.40
3030	Brick chimney	G		100	.160	C.F.		6		6	9.85
4000	Deconstruction of concrete										
4010	Slab on grade, 4″ thick, plain concrete	G	B-9	500	.080	S.F.		3.04	.47	3.51	5.50
4020	Wire mesh reinforced	G		470	.085			3.23	.50	3.73	5.85
4030	Rod reinforced	G		400	.100			3.80	.58	4.38	6.90
4110	Foundation wall, 6″ thick, plain concrete	G		160	.250			9.50	1.46	10.96	17.20
4120	8″ thick	G		140	.286			10.85	1.66	12.51	19.65
4130	10″ thick	G		120	.333			12.65	1.94	14.59	23
9000	Deconstruction process, support equipment as needed										
9010	Daily use, portal to portal, 12-ton truck-mounted hydraulic crane crew	G	A-3H	1	8	Day		415	860	1,275	1,600
9020	Daily use, skid steer and operator	G	A-3C	1	8			390	320	710	965
9030	Daily use, backhoe 48 H.P., operator and labor	G	″	1	8			390	320	710	965

02 42 10.30 Deconstruction Material Handling			Crew	Daily Output	Labor-Hours	Unit	Material	2015 Bare Costs Labor	Equipment	Total	Total Incl O&P
0010	**DECONSTRUCTION MATERIAL HANDLING**										
0012	Buildings one or two stories only										
0100	Clean and stack brick on pallet	G	2 Clab	1200	.013	Ea.		.50		.50	.82
0200	Haul 50′ and load rough lumber up to 2″ x 8″ size	G		2000	.008	″		.30		.30	.49
0210	Lumber larger than 2″ x 8″	G		3200	.005	B.F.		.19		.19	.31
0300	Finish wood for recycling stack and wrap per pallet	G		8	2	Ea.	36	75		111	163
0350	Light fixtures			6	2.667		65	100		165	236
0375	Windows			6	2.667		61	100		161	232
0400	Miscellaneous materials			8	2		18	75		93	143
1000	See Section 02 41 19.19 for bulk material handling										

02 43 Structure Moving

02 43 13 – Structure Relocation

02 43 13.13 Building Relocation		Crew	Daily Output	Labor-Hours	Unit	2015 Bare Costs Material	Labor	Equipment	Total	Total Incl O&P
0010	**BUILDING RELOCATION**									
0011	One day move, up to 24′ wide									
0020	Reset on existing foundation				Total				11,500	11,500
0040	Wood or steel frame bldg., based on ground floor area G	B-4	185	.259	S.F.		9.95	2.78	12.73	19.30
0060	Masonry bldg., based on ground floor area G	"	137	.350			13.45	3.75	17.20	26
0200	For 24′ to 42′ wide, add				↓				15%	15%

02 58 Snow Control

02 58 13 – Snow Fencing

02 58 13.10 Snow Fencing System		Crew	Daily Output	Labor-Hours	Unit	Material	Labor	Equipment	Total	Total Incl O&P
0010	**SNOW FENCING SYSTEM**									
7001	Snow fence on steel posts 10′ O.C., 4′ high	B-1	500	.048	L.F.	.93	1.84		2.77	4.03

02 65 Underground Storage Tank Removal

02 65 10 – Underground Tank and Contaminated Soil Removal

02 65 10.30 Removal of Underground Storage Tanks		Crew	Daily Output	Labor-Hours	Unit	Material	Labor	Equipment	Total	Total Incl O&P
0010	**REMOVAL OF UNDERGROUND STORAGE TANKS** R026510-20									
0011	Petroleum storage tanks, non-leaking									
0100	Excavate & load onto trailer									
0110	3000 gal. to 5000 gal. tank G	B-14	4	12	Ea.		475	91	566	875
0120	6000 gal. to 8000 gal. tank G	B-3A	3	13.333			535	345	880	1,250
0130	9000 gal. to 12000 gal. tank G	"	2	20	↓		805	515	1,320	1,875
0190	Known leaking tank, add				%				100%	100%
0200	Remove sludge, water and remaining product from tank bottom									
0201	of tank with vacuum truck									
0300	3000 gal. to 5000 gal. tank G	A-13	5	1.600	Ea.		78	153	231	291
0310	6000 gal. to 8000 gal. tank G		4	2			97	191	288	365
0320	9000 gal. to 12000 gal. tank G	↓	3	2.667	↓		130	254	384	485
0390	Dispose of sludge off-site, average				Gal.				6.25	6.80
0400	Insert inert solid CO_2 "dry ice" into tank									
0401	For cleaning/transporting tanks (1.5 lb./100 gal. cap) G	1 Clab	500	.016	Lb.	1.17	.60		1.77	2.28
0403	Insert solid carbon dioxide, 1.5 lb./100 gal. G	"	400	.020	"	1.17	.75		1.92	2.52
0503	Disconnect and remove piping G	1 Plum	160	.050	L.F.		2.94		2.94	4.58
0603	Transfer liquids, 10% of volume G	"	1600	.005	Gal.		.29		.29	.46
0703	Cut accessway into underground storage tank G	1 Clab	5.33	1.501	Ea.		56.50		56.50	92.50
0813	Remove sludge, wash and wipe tank, 500 gal. G	1 Plum	8	1			58.50		58.50	91.50
0823	3,000 gal. G		6.67	1.199			70.50		70.50	110
0833	5,000 gal. G		6.15	1.301			76.50		76.50	119
0843	8,000 gal. G		5.33	1.501			88		88	137
0853	10,000 gal. G		4.57	1.751			103		103	160
0863	12,000 gal. G	↓	4.21	1.900	↓		112		112	174
1020	Haul tank to certified salvage dump, 100 miles round trip									
1023	3000 gal. to 5000 gal. tank				Ea.				760	830
1026	6000 gal. to 8000 gal. tank								880	960
1029	9,000 gal. to 12,000 gal. tank				↓				1,050	1,150
1100	Disposal of contaminated soil to landfill									
1110	Minimum				C.Y.				145	160
1111	Maximum				"				400	440
1120	Disposal of contaminated soil to									

02 65 Underground Storage Tank Removal

02 65 10 – Underground Tank and Contaminated Soil Removal

02 65 10.30 Removal of Underground Storage Tanks		Crew	Daily Output	Labor-Hours	Unit	2015 Bare Costs Material	Labor	Equipment	Total	Total Incl O&P
1121	bituminous concrete batch plant									
1130	Minimum				C.Y.				80	88
1131	Maximum				"				115	125
1203	Excavate, pull, & load tank, backfill hole, 8,000 gal. + G	B-12C	.50	32	Ea.		1,425	2,350	3,775	4,875
1213	Haul tank to certified dump, 100 miles rt, 8,000 gal. + G	B-34K	1	8			320	950	1,270	1,575
1223	Excavate, pull, & load tank, backfill hole, 500 gal. G	B-11C	1	16			705	365	1,070	1,525
1233	Excavate, pull, & load tank, backfill hole, 3,000 – 5,000 gal. G	B-11M	.50	32			1,400	785	2,185	3,125
1243	Haul tank to certified dump, 100 miles rt, 500 gal. G	B-34L	1	8			390	245	635	885
1253	Haul tank to certified dump, 100 miles rt, 3,000 – 5,000 gal. G	B-34M	1	8	↓		390	305	695	950
2010	Decontamination of soil on site incl poly tarp on top/bottom									
2011	Soil containment berm, and chemical treatment									
2020	Minimum G	B-11C	100	.160	C.Y.	7.80	7.05	3.64	18.49	24
2021	Maximum G	"	100	.160		10.10	7.05	3.64	20.79	26.50
2050	Disposal of decontaminated soil, minimum								135	150
2055	Maximum				↓				400	440

02 81 Transportation and Disposal of Hazardous Materials

02 81 20 – Hazardous Waste Handling

02 81 20.10 Hazardous Waste Cleanup/Pickup/Disposal		Crew	Daily Output	Labor-Hours	Unit	Material	Labor	Equipment	Total	Total Incl O&P
0010	**HAZARDOUS WASTE CLEANUP/PICKUP/DISPOSAL**									
0100	For contractor rental equipment, i.e., Dozer,									
0110	Front end loader, Dump truck, etc., see 01 54 33 Reference Section									
1000	Solid pickup									
1100	55 gal. drums				Ea.				240	265
1120	Bulk material, minimum				Ton				190	210
1130	Maximum				"				595	655
1200	Transportation to disposal site									
1220	Truckload = 80 drums or 25 C.Y. or 18 tons									
1260	Minimum				Mile				3.95	4.45
1270	Maximum				"				7.25	7.35
3000	Liquid pickup, vacuum truck, stainless steel tank									
3100	Minimum charge, 4 hours									
3110	1 compartment, 2200 gallon				Hr.				140	155
3120	2 compartment, 5000 gallon				"				200	225
3400	Transportation in 6900 gallon bulk truck				Mile				7.95	8.75
3410	In teflon lined truck				"				10.20	11.25
5000	Heavy sludge or dry vacuumable material				Hr.				140	160
6000	Dumpsite disposal charge, minimum				Ton				140	155
6020	Maximum				"				415	455

02 82 13.39 Asbestos Remediation Plans and Methods		Crew	Daily Output	Labor-Hours	Unit	Material (2015 Bare Costs)	Labor (2015 Bare Costs)	Equipment (2015 Bare Costs)	Total (2015 Bare Costs)	Total Incl O&P
0010	**ASBESTOS REMEDIATION PLANS AND METHODS**									
0100	Building Survey-Commercial Building				Ea.				2,200	2,400
0200	Asbestos Abatement Remediation Plan				"				1,350	1,475

02 82 13.41 Asbestos Abatement Equipment

		Crew	Daily Output	Labor-Hours	Unit	Material	Labor	Equipment	Total	Total Incl O&P
0010	**ASBESTOS ABATEMENT EQUIPMENT** R028213-20									
0011	Equipment and supplies, buy									
0200	Air filtration device, 2000 CFM				Ea.	960			960	1,050
0250	Large volume air sampling pump, minimum					345			345	380
0260	Maximum					335			335	365
0300	Airless sprayer unit, 2 gun					4,500			4,500	4,950
0350	Light stand, 500 watt				↓	49			49	54
0400	Personal respirators									
0410	Negative pressure, 1/2 face, dual operation, min.				Ea.	26.50			26.50	29
0420	Maximum					29.50			29.50	32.50
0450	P.A.P.R., full face, minimum					124			124	137
0460	Maximum					165			165	182
0470	Supplied air, full face, incl. air line, minimum					168			168	185
0480	Maximum					405			405	445
0500	Personnel sampling pump					226			226	249
1500	Power panel, 20 unit, incl. GFI					600			600	660
1600	Shower unit, including pump and filters					1,275			1,275	1,425
1700	Supplied air system (type C)					3,425			3,425	3,750
1750	Vacuum cleaner, HEPA, 16 gal., stainless steel, wet/dry					435			435	475
1760	55 gallon					560			560	615
1800	Vacuum loader, 9 – 18 ton/hr.					94,500			94,500	104,000
1900	Water atomizer unit, including 55 gal. drum					281			281	310
2000	Worker protection, whole body, foot, head cover & gloves, plastic					8.65			8.65	9.55
2500	Respirator, single use					25			25	27.50
2550	Cartridge for respirator					5.75			5.75	6.30
2570	Glove bag, 7 mil, 50" x 64"					9.20			9.20	10.10
2580	10 mil, 44" x 60"					5.75			5.75	6.35
3000	HEPA vacuum for work area, minimum					305			305	335
3050	Maximum					815			815	900
6000	Disposable polyethylene bags, 6 mil, 3 C.F.					.82			.82	.90
6300	Disposable fiber drums, 3 C.F.					18.20			18.20	20
6400	Pressure sensitive caution labels, 3" x 5"					3.47			3.47	3.82
6450	11" x 17"					7.35			7.35	8.05
6500	Negative air machine, 1800 CFM				↓	845			845	930

02 82 13.42 Preparation of Asbestos Containment Area

		Crew	Daily Output	Labor-Hours	Unit	Material	Labor	Equipment	Total	Total Incl O&P
0010	**PREPARATION OF ASBESTOS CONTAINMENT AREA**									
0100	Pre-cleaning, HEPA vacuum and wet wipe, flat surfaces	A-9	12000	.005	S.F.	.02	.28		.30	.47
0200	Protect carpeted area, 2 layers 6 mil poly on 3/4" plywood	"	1000	.064		2.04	3.35		5.39	7.65
0300	Separation barrier, 2" x 4" @ 16", 1/2" plywood ea. side, 8' high	2 Carp	400	.040		3.32	1.88		5.20	6.75
0310	12' high		320	.050		3.32	2.35		5.67	7.50
0320	16' high		200	.080		2.30	3.76		6.06	8.70
0400	Personnel decontam. chamber, 2" x 4" @ 16", 3/4" ply ea. side		280	.057		4.34	2.68		7.02	9.15
0450	Waste decontam. chamber, 2" x 4" studs @ 16", 3/4" ply ea. side	↓	360	.044	↓	4.34	2.09		6.43	8.20
0500	Cover surfaces with polyethylene sheeting									
0501	Including glue and tape									
0550	Floors, each layer, 6 mil	A-9	8000	.008	S.F.	.04	.42		.46	.72
0551	4 mil		9000	.007		.03	.37		.40	.63
0560	Walls, each layer, 6 mil	↓	6000	.011	↓	.04	.56		.60	.95

02 82 13.42 Preparation of Asbestos Containment Area		Crew	Daily Output	Labor-Hours	Unit	2015 Bare Costs Material	Labor	Equipment	Total	Total Incl O&P
0561	4 mil	A-9	7000	.009	S.F.	.03	.48		.51	.80
0570	For heights above 12′, add						20%			
0575	For heights above 20′, add						30%			
0580	For fire retardant poly, add					100%				
0590	For large open areas, deduct					10%	20%			
0600	Seal floor penetrations with foam firestop to 36 sq. in.	2 Carp	200	.080	Ea.	7.75	3.76		11.51	14.65
0610	36 sq. in. to 72 sq. in.		125	.128		15.50	6		21.50	27
0615	72 sq. in. to 144 sq. in.		80	.200		31	9.40		40.40	49.50
0620	Wall penetrations, to 36 square inches		180	.089		7.75	4.17		11.92	15.35
0630	36 sq. in. to 72 sq. in.		100	.160		15.50	7.50		23	29.50
0640	72 sq. in. to 144 sq. in.		60	.267		31	12.50		43.50	54.50
0800	Caulk seams with latex	1 Carp	230	.035	L.F.	.17	1.63		1.80	2.87
0900	Set up neg. air machine, 1-2k CFM/25 M.C.F. volume	1 Asbe	4.30	1.860	Ea.		97.50		97.50	157
0950	Set up and remove portable shower unit	2 Asbe	4	4	"		209		209	335

02 82 13.43 Bulk Asbestos Removal		Crew	Daily Output	Labor-Hours	Unit	Material	Labor	Equipment	Total	Total Incl O&P
0010	**BULK ASBESTOS REMOVAL**									
0020	Includes disposable tools and 2 suits and 1 respirator filter/day/worker									
0100	Beams, W 10 x 19	A-9	235	.272	L.F.	.79	14.25		15.04	24
0110	W 12 x 22		210	.305		.88	15.95		16.83	26.50
0120	W 14 x 26		180	.356		1.03	18.65		19.68	31
0130	W 16 x 31		160	.400		1.15	21		22.15	35
0140	W 18 x 40		140	.457		1.32	24		25.32	40
0150	W 24 x 55		110	.582		1.68	30.50		32.18	51
0160	W 30 x 108		85	.753		2.17	39.50		41.67	66
0170	W 36 x 150		72	.889		2.56	46.50		49.06	78
0200	Boiler insulation		480	.133	S.F.	.45	7		7.45	11.75
0210	With metal lath, add				%				50%	50%
0300	Boiler breeching or flue insulation	A-9	520	.123	S.F.	.36	6.45		6.81	10.75
0310	For active boiler, add				%				100%	100%
0400	Duct or AHU insulation	A-10B	440	.073	S.F.	.21	3.82		4.03	6.40
0500	Duct vibration isolation joints, up to 24 sq. in. duct	A-9	56	1.143	Ea.	3.30	60		63.30	100
0520	25 sq. in. to 48 sq. in. duct		48	1.333		3.85	70		73.85	116
0530	49 sq. in. to 76 sq. in. duct		40	1.600		4.62	84		88.62	140
0600	Pipe insulation, air cell type, up to 4″ diameter pipe		900	.071	L.F.	.21	3.73		3.94	6.25
0610	4″ to 8″ diameter pipe		800	.080		.23	4.19		4.42	7
0620	10″ to 12″ diameter pipe		700	.091		.26	4.79		5.05	8
0630	14″ to 16″ diameter pipe		550	.116		.34	6.10		6.44	10.15
0650	Over 16″ diameter pipe		650	.098	S.F.	.28	5.15		5.43	8.60
0700	With glove bag up to 3″ diameter pipe		200	.320	L.F.	9.15	16.75		25.90	37
1000	Pipe fitting insulation up to 4″ diameter pipe		320	.200	Ea.	.58	10.50		11.08	17.50
1100	6″ to 8″ diameter pipe		304	.211		.61	11.05		11.66	18.40
1110	10″ to 12″ diameter pipe		192	.333		.96	17.45		18.41	29
1120	14″ to 16″ diameter pipe		128	.500		1.44	26		27.44	43.50
1130	Over 16″ diameter pipe		176	.364	S.F.	1.05	19.05		20.10	31.50
1200	With glove bag, up to 8″ diameter pipe		75	.853	L.F.	6.25	44.50		50.75	79
2000	Scrape foam fireproofing from flat surface		2400	.027	S.F.	.08	1.40		1.48	2.33
2100	Irregular surfaces		1200	.053		.15	2.80		2.95	4.66
3000	Remove cementitious material from flat surface		1800	.036		.10	1.86		1.96	3.11
3100	Irregular surface		1000	.064		.13	3.35		3.48	5.55
4000	Scrape acoustical coating/fireproofing, from ceiling		3200	.020		.06	1.05		1.11	1.75
5000	Remove VAT and mastic from floor by hand		2400	.027		.08	1.40		1.48	2.33
5100	By machine	A-11	4800	.013		.04	.70	.01	.75	1.17

02 82 Asbestos Remediation

02 82 13 – Asbestos Abatement

02 82 13.43 Bulk Asbestos Removal		Crew	Daily Output	Labor-Hours	Unit	Material	2015 Bare Costs Labor	Equipment	Total	Total Incl O&P
5150	For 2 layers, add				%				50%	50%
6000	Remove contaminated soil from crawl space by hand	A-9	400	.160	C.F.	.46	8.40		8.86	14
6100	With large production vacuum loader	A-12	700	.091	"	.26	4.79	1.09	6.14	9.20
7000	Radiator backing, not including radiator removal	A-9	1200	.053	S.F.	.15	2.80		2.95	4.66
8000	Cement-asbestos transite board and cement wall board	2 Asbe	1000	.016		.15	.84		.99	1.51
8100	Transite shingle siding	A-10B	750	.043		.21	2.24		2.45	3.83
8200	Shingle roofing	"	2000	.016		.08	.84		.92	1.44
8250	Built-up, no gravel, non-friable	B-2	1400	.029		.08	1.09		1.17	1.87
8260	Bituminous flashing	1 Rofc	300	.027		.08	1.07		1.15	2.02
8300	Asbestos millboard, flat board and VAT contaminated plywood	2 Asbe	1000	.016		.08	.84		.92	1.44
9000	For type B (supplied air) respirator equipment, add				%				10%	10%

02 82 13.44 Demolition In Asbestos Contaminated Area

Line	Description	Crew	Daily Output	Labor-Hours	Unit	Material	Labor	Equipment	Total	Total Incl O&P
0010	**DEMOLITION IN ASBESTOS CONTAMINATED AREA**									
0200	Ceiling, including suspension system, plaster and lath	A-9	2100	.030	S.F.	.09	1.60		1.69	2.67
0210	Finished plaster, leaving wire lath		585	.109		.32	5.75		6.07	9.55
0220	Suspended acoustical tile		3500	.018		.05	.96		1.01	1.60
0230	Concealed tile grid system		3000	.021		.06	1.12		1.18	1.87
0240	Metal pan grid system		1500	.043		.12	2.24		2.36	3.73
0250	Gypsum board		2500	.026		.07	1.34		1.41	2.24
0260	Lighting fixtures up to 2′ x 4′		72	.889	Ea.	2.56	46.50		49.06	78
0400	Partitions, non load bearing									
0410	Plaster, lath, and studs	A-9	690	.093	S.F.	.88	4.86		5.74	8.75
0450	Gypsum board and studs	"	1390	.046	"	.13	2.41		2.54	4.03
9000	For type B (supplied air) respirator equipment, add				%				10%	10%

02 82 13.45 OSHA Testing

Line	Description	Crew	Daily Output	Labor-Hours	Unit	Material	Labor	Equipment	Total	Total Incl O&P
0010	**OSHA TESTING**									
0100	Certified technician, minimum				Day				200	220
0110	Maximum								300	330
0120	Industrial hygienist, minimum								250	250
0130	Maximum								400	440
0200	Asbestos sampling and PCM analysis, NIOSH 7400, minimum	1 Asbe	8	1	Ea.	2.96	52.50		55.46	87.50
0210	Maximum		4	2		3.26	105		108.26	172
1000	Cleaned area samples		8	1		2.81	52.50		55.31	87
1100	PCM air sample analysis, NIOSH 7400, minimum		8	1		31.50	52.50		84	119
1110	Maximum		4	2		3.42	105		108.42	172
1200	TEM air sample analysis, NIOSH 7402, minimum								80	106
1210	Maximum								360	450

02 82 13.46 Decontamination of Asbestos Containment Area

Line	Description	Crew	Daily Output	Labor-Hours	Unit	Material	Labor	Equipment	Total	Total Incl O&P
0010	**DECONTAMINATION OF ASBESTOS CONTAINMENT AREA**									
0100	Spray exposed substrate with surfactant (bridging)									
0200	Flat surfaces	A-9	6000	.011	S.F.	.36	.56		.92	1.30
0250	Irregular surfaces		4000	.016	"	.31	.84		1.15	1.69
0300	Pipes, beams, and columns		2000	.032	L.F.	.56	1.68		2.24	3.32
1000	Spray encapsulate polyethylene sheeting		8000	.008	S.F.	.34	.42		.76	1.04
1100	Roll down polyethylene sheeting		8000	.008	"		.42		.42	.67
1500	Bag polyethylene sheeting		400	.160	Ea.	.81	8.40		9.21	14.40
2000	Fine clean exposed substrate, with nylon brush		2400	.027	S.F.		1.40		1.40	2.25
2500	Wet wipe substrate		4800	.013			.70		.70	1.12
2600	Vacuum surfaces, fine brush		6400	.010			.52		.52	.84
3000	Structural demolition									
3100	Wood stud walls	A-9	2800	.023	S.F.		1.20		1.20	1.93
3500	Window manifolds, not incl. window replacement		4200	.015			.80		.80	1.28

02 82 Asbestos Remediation

02 82 13 – Asbestos Abatement

02 82 13.46 Decontamination of Asbestos Containment Area		Crew	Daily Output	Labor-Hours	Unit	Material	2015 Bare Costs Labor	Equipment	Total	Total Incl O&P
3600	Plywood carpet protection	A-9	2000	.032	S.F.		1.68		1.68	2.70
4000	Remove custom decontamination facility	A-10A	8	3	Ea.	15.40	158		173.40	270
4100	Remove portable decontamination facility	3 Asbe	12	2	"	13.05	105		118.05	182
5000	HEPA vacuum, shampoo carpeting	A-9	4800	.013	S.F.	.07	.70		.77	1.20
9000	Final cleaning of protected surfaces	A-10A	8000	.003	"		.16		.16	.25

02 82 13.47 Asbestos Waste Pkg., Handling, and Disp.

		Crew	Daily Output	Labor-Hours	Unit	Material	Labor	Equipment	Total	Total Incl O&P
0010	**ASBESTOS WASTE PACKAGING, HANDLING, AND DISPOSAL**									
0100	Collect and bag bulk material, 3 C.F. bags, by hand	A-9	400	.160	Ea.	.82	8.40		9.22	14.40
0200	Large production vacuum loader	A-12	880	.073		.86	3.81	.87	5.54	8.05
1000	Double bag and decontaminate	A-9	960	.067		.82	3.49		4.31	6.50
2000	Containerize bagged material in drums, per 3 C.F. drum	"	800	.080		18.20	4.19		22.39	27
3000	Cart bags 50′ to dumpster	2 Asbe	400	.040	↓		2.09		2.09	3.37
5000	Disposal charges, not including haul, minimum				C.Y.				61	67
5020	Maximum				"				355	395
9000	For type B (supplied air) respirator equipment, add				%				10%	10%

02 82 13.48 Asbestos Encapsulation With Sealants

		Crew	Daily Output	Labor-Hours	Unit	Material	Labor	Equipment	Total	Total Incl O&P
0010	**ASBESTOS ENCAPSULATION WITH SEALANTS**									
0100	Ceilings and walls, minimum	A-9	21000	.003	S.F.	.31	.16		.47	.60
0110	Maximum		10600	.006		.45	.32		.77	1.01
0200	Columns and beams, minimum		13300	.005		.31	.25		.56	.75
0210	Maximum		5325	.012	↓	.51	.63		1.14	1.57
0300	Pipes to 12″ diameter including minor repairs, minimum		800	.080	L.F.	.43	4.19		4.62	7.20
0310	Maximum	↓	400	.160	"	1.12	8.40		9.52	14.75

02 83 Lead Remediation

02 83 19 – Lead-Based Paint Remediation

02 83 19.21 Lead Paint Remediation Plans and Methods

		Crew	Daily Output	Labor-Hours	Unit	Material	Labor	Equipment	Total	Total Incl O&P
0010	**LEAD PAINT REMEDIATION PLANS AND METHODS**									
0100	Building Survey-Commercial Building				Ea.				2,050	2,250
0200	Lead Abatement Remediation Plan								1,225	1,350
0300	Lead Paint Testing, AAS Analysis								51	56
0400	Lead Paint Testing, X-Ray Fluorescence				↓				51	56

02 83 19.22 Preparation of Lead Containment Area

		Crew	Daily Output	Labor-Hours	Unit	Material	Labor	Equipment	Total	Total Incl O&P
0010	**PREPARATION OF LEAD CONTAINMENT AREA**									
0020	Lead abatement work area, test kit, per swab	1 Skwk	16	.500	Ea.	4.21	24.50		28.71	44
0025	For dust barriers see Section 01 56 16.10									
0050	Caution sign	1 Skwk	48	.167	Ea.	8.35	8.10		16.45	22.50
0100	Pre-cleaning, HEPA vacuum and wet wipe, floor and wall surfaces	3 Skwk	5000	.005	S.F.	.01	.23		.24	.40
0105	Ceiling, 6′ - 11′ high		4100	.006		.08	.28		.36	.54
0108	12′ - 15′ high		3550	.007		.07	.33		.40	.61
0115	Over 15′ high	↓	3000	.008	↓	.09	.39		.48	.73
0500	Cover surfaces with polyethylene sheeting									
0550	Floors, each layer, 6 mil	3 Skwk	8000	.003	S.F.	.04	.15		.19	.29
0560	Walls, each layer, 6 mil	"	6000	.004	"	.04	.19		.23	.37
0570	For heights above 12′, add						20%			
2400	Post abatement cleaning of protective sheeting, HEPA vacuum & wet wipe	3 Skwk	5000	.005	S.F.	.01	.23		.24	.40
2450	Doff, bag and seal protective sheeting		12000	.002		.01	.10		.11	.17
2500	Post abatement cleaning, HEPA vacuum & wet wipe	↓	5000	.005	↓	.01	.23		.24	.40

02 83 19 – Lead-Based Paint Remediation

02 83 19.23 Encapsulation of Lead-Based Paint

02 83 19.23 Encapsulation of Lead-Based Paint		Crew	Daily Output	Labor-Hours	Unit	2015 Bare Costs Material	Labor	Equipment	Total	Total Incl O&P
0010	ENCAPSULATION OF LEAD-BASED PAINT									
0020	Interior, brushwork, trim, under 6"	1 Pord	240	.033	L.F.	2.30	1.34		3.64	4.69
0030	6" to 12" wide		180	.044		3.06	1.79		4.85	6.25
0040	Balustrades		300	.027		1.84	1.08		2.92	3.75
0050	Pipe to 4" diameter		500	.016		1.12	.65		1.77	2.27
0060	To 8" diameter		375	.021		1.48	.86		2.34	3.01
0070	To 12" diameter		250	.032		2.19	1.29		3.48	4.49
0080	To 16" diameter		170	.047		3.26	1.90		5.16	6.65
0090	Cabinets, ornate design		200	.040	S.F.	2.81	1.61		4.42	5.70
0100	Simple design		250	.032	"	2.24	1.29		3.53	4.54
0110	Doors, 3' x 7', both sides, incl. frame & trim									
0120	Flush	1 Pord	6	1.333	Ea.	28	54		82	118
0130	French, 10 – 15 lite		3	2.667		5.65	108		113.65	179
0140	Panel		4	2		34	80.50		114.50	168
0150	Louvered		2.75	2.909		31	117		148	224
0160	Windows, per interior side, per 15 S.F.									
0170	1 to 6 lite	1 Pord	14	.571	Ea.	19.40	23		42.40	58.50
0180	7 to 10 lite		7.50	1.067		21.50	43		64.50	93
0190	12 lite		5.75	1.391		29	56		85	122
0200	Radiators		8	1		69	40.50		109.50	141
0210	Grilles, vents		275	.029	S.F.	2.04	1.17		3.21	4.13
0220	Walls, roller, drywall or plaster		1000	.008		.56	.32		.88	1.14
0230	With spunbonded reinforcing fabric		720	.011		.63	.45		1.08	1.41
0240	Wood		800	.010		.69	.40		1.09	1.41
0250	Ceilings, roller, drywall or plaster		900	.009		.63	.36		.99	1.27
0260	Wood		700	.011		.78	.46		1.24	1.60
0270	Exterior, brushwork, gutters and downspouts		300	.027	L.F.	1.84	1.08		2.92	3.75
0280	Columns		400	.020	S.F.	1.38	.81		2.19	2.82
0290	Spray, siding		600	.013	"	.93	.54		1.47	1.88
0300	Miscellaneous									
0310	Electrical conduit, brushwork, to 2" diameter	1 Pord	500	.016	L.F.	1.12	.65		1.77	2.27
0320	Brick, block or concrete, spray		500	.016	S.F.	1.12	.65		1.77	2.27
0330	Steel, flat surfaces and tanks to 12"		500	.016		1.12	.65		1.77	2.27
0340	Beams, brushwork		400	.020		1.38	.81		2.19	2.82
0350	Trusses		400	.020		1.38	.81		2.19	2.82

02 83 19.26 Removal of Lead-Based Paint

02 83 19.26 Removal of Lead-Based Paint		Crew	Daily Output	Labor-Hours	Unit	Material	Labor	Equipment	Total	Total Incl O&P
0010	REMOVAL OF LEAD-BASED PAINT R028319-60									
0011	By chemicals, per application									
0050	Baseboard, to 6" wide	1 Pord	64	.125	L.F.	.72	5.05		5.77	8.90
0070	To 12" wide		32	.250	"	1.40	10.10		11.50	17.75
0200	Balustrades, one side		28	.286	S.F.	1.45	11.55		13	20
1400	Cabinets, simple design		32	.250		1.31	10.10		11.41	17.65
1420	Ornate design		25	.320		1.58	12.90		14.48	22.50
1600	Cornice, simple design		60	.133		1.50	5.40		6.90	10.30
1620	Ornate design		20	.400		5.35	16.15		21.50	32
2800	Doors, one side, flush		84	.095		1.88	3.84		5.72	8.25
2820	Two panel		80	.100		1.31	4.04		5.35	7.95
2840	Four panel		45	.178		1.40	7.15		8.55	13.10
2880	For trim, one side, add		64	.125	L.F.	.71	5.05		5.76	8.90
3000	Fence, picket, one side		30	.267	S.F.	1.31	10.75		12.06	18.75
3200	Grilles, one side, simple design		30	.267		1.33	10.75		12.08	18.75
3220	Ornate design		25	.320		1.43	12.90		14.33	22.50

02 83 Lead Remediation

02 83 19 – Lead-Based Paint Remediation

02 83 19.26 Removal of Lead-Based Paint		Crew	Daily Output	Labor-Hours	Unit	2015 Bare Costs Material	Labor	Equipment	Total	Total Incl O&P
3240	Handrails	1 Pord	90	.089	L.F.	1.33	3.59		4.92	7.20
4400	Pipes, to 4" diameter		90	.089		1.88	3.59		5.47	7.80
4420	To 8" diameter		50	.160		3.75	6.45		10.20	14.55
4440	To 12" diameter		36	.222		5.65	8.95		14.60	20.50
4460	To 16" diameter		20	.400		7.50	16.15		23.65	34.50
4500	For hangers, add		40	.200	Ea.	2.49	8.05		10.54	15.70
4800	Siding		90	.089	S.F.	1.24	3.59		4.83	7.10
5000	Trusses, open		55	.145	SF Face	1.96	5.85		7.81	11.60
6200	Windows, one side only, double hung, 1/1 light, 24" x 48" high		4	2	Ea.	23	80.50		103.50	156
6220	30" x 60" high		3	2.667		31	108		139	207
6240	36" x 72" high		2.50	3.200		37	129		166	249
6280	40" x 80" high		2	4		46.50	161		207.50	310
6400	Colonial window, 6/6 light, 24" x 48" high		2	4		46.50	161		207.50	310
6420	30" x 60" high		1.50	5.333		62	215		277	415
6440	36" x 72" high		1	8		93	325		418	620
6480	40" x 80" high		1	8		93	325		418	620
6600	8/8 light, 24" x 48" high		2	4		46.50	161		207.50	310
6620	40" x 80" high		1	8		93	325		418	620
6800	12/12 light, 24" x 48" high		1	8		93	325		418	620
6820	40" x 80" high		.75	10.667		124	430		554	825
6840	Window frame & trim items, included in pricing above									
7000	Hand scraping and HEPA vacuum, less than 4 S.F.	1 Pord	8	1	Ea.	.82	40.50		41.32	66
8000	Collect and bag bulk material, 3 C.F. bags, by hand		30	.267	"	.82	10.75		11.57	18.20
9000	Minimum labor/equipment charge		3	2.667	Job		108		108	173

02 85 Mold Remediation

02 85 16 – Mold Remediation Preparation and Containment

02 85 16.40 Mold Remediation Plans and Methods

		Crew	Daily Output	Labor-Hours	Unit	Material	Labor	Equipment	Total	Total Incl O&P
0010	**MOLD REMEDIATION PLANS AND METHODS**									
0020	Initial inspection, areas to 2500 S.F.				Total				268	295
0030	Areas to 5000 S.F.								440	485
0032	Areas to 10000 S.F.								440	485
0040	Testing, air sample each								126	139
0050	Swab sample								115	127
0060	Tape sample								125	140
0070	Post remediation air test								126	139
0080	Mold abatement plan, area to 2500 S.F.								1,225	1,325
0090	Areas to 5000 S.F.								1,625	1,775
0095	Areas to 10000 S.F.								2,550	2,800
0100	Packup & removal of contents, average 3 bedroom home, excl storage								8,150	8,975
0110	Average 5 bedroom home, excl storage								15,300	16,800
0600	For demolition in mold contaminated areas, see Section 02 85 33.50									
0610	For personal protection equipment, see Section 02 82 13.41									

02 85 16.50 Preparation of Mold Containment Area

		Crew	Daily Output	Labor-Hours	Unit	Material	Labor	Equipment	Total	Total Incl O&P
0010	**PREPARATION OF MOLD CONTAINMENT AREA**									
0100	Pre-cleaning, HEPA vacuum and wet wipe, flat surfaces	A-9	12000	.005	S.F.	.02	.28		.30	.47
0300	Separation barrier, 2" x 4" @ 16", 1/2" plywood ea. side, 8' high	2 Carp	400	.040		3.32	1.88		5.20	6.75
0310	12' high		320	.050		3.32	2.35		5.67	7.50
0320	16' high		200	.080		2.30	3.76		6.06	8.70
0400	Personnel decontam. chamber, 2" x 4" @ 16", 3/4" ply ea. side		280	.057		4.34	2.68		7.02	9.15
0450	Waste decontam. chamber, 2" x 4" studs @ 16", 3/4" ply each side		360	.044		4.34	2.09		6.43	8.20

02 85 Mold Remediation

02 85 16 – Mold Remediation Preparation and Containment

02 85 16.50 Preparation of Mold Containment Area		Crew	Daily Output	Labor-Hours	Unit	Material	2015 Bare Costs Labor	Equipment	Total	Total Incl O&P
0500	Cover surfaces with polyethylene sheeting									
0501	Including glue and tape									
0550	Floors, each layer, 6 mil	A-9	8000	.008	S.F.	.04	.42		.46	.72
0551	4 mil		9000	.007		.03	.37		.40	.63
0560	Walls, each layer, 6 mil		6000	.011		.04	.56		.60	.95
0561	4 mil		7000	.009		.03	.48		.51	.80
0570	For heights above 12′, add						20%			
0575	For heights above 20′, add						30%			
0580	For fire retardant poly, add					100%				
0590	For large open areas, deduct					10%	20%			
0600	Seal floor penetrations with foam firestop to 36 sq. in.	2 Carp	200	.080	Ea.	7.75	3.76		11.51	14.65
0610	36 sq. in. to 72 sq. in.		125	.128		15.50	6		21.50	27
0615	72 sq. in. to 144 sq. in.		80	.200		31	9.40		40.40	49.50
0620	Wall penetrations, to 36 square inches		180	.089		7.75	4.17		11.92	15.35
0630	36 sq. in. to 72 sq. in.		100	.160		15.50	7.50		23	29.50
0640	72 sq. in. to 144 sq. in.		60	.267		31	12.50		43.50	54.50
0800	Caulk seams with latex caulk	1 Carp	230	.035	L.F.	.17	1.63		1.80	2.87
0900	Set up neg. air machine, 1-2k CFM/25 M.C.F. volume	1 Asbe	4.30	1.860	Ea.		97.50		97.50	157

02 85 33 – Removal and Disposal of Materials with Mold

02 85 33.50 Demolition in Mold Contaminated Area		Crew	Daily Output	Labor-Hours	Unit	Material	Labor	Equipment	Total	Total Incl O&P
0010	**DEMOLITION IN MOLD CONTAMINATED AREA**									
0200	Ceiling, including suspension system, plaster and lath	A-9	2100	.030	S.F.	.09	1.60		1.69	2.67
0210	Finished plaster, leaving wire lath		585	.109		.32	5.75		6.07	9.55
0220	Suspended acoustical tile		3500	.018		.05	.96		1.01	1.60
0230	Concealed tile grid system		3000	.021		.06	1.12		1.18	1.87
0240	Metal pan grid system		1500	.043		.12	2.24		2.36	3.73
0250	Gypsum board		2500	.026		.07	1.34		1.41	2.24
0255	Plywood		2500	.026		.07	1.34		1.41	2.24
0260	Lighting fixtures up to 2′ x 4′		72	.889	Ea.	2.56	46.50		49.06	78
0400	Partitions, non load bearing									
0410	Plaster, lath, and studs	A-9	690	.093	S.F.	.88	4.86		5.74	8.75
0450	Gypsum board and studs		1390	.046		.13	2.41		2.54	4.03
0465	Carpet & pad		1390	.046		.13	2.41		2.54	4.03
0600	Pipe insulation, air cell type, up to 4″ diameter pipe		900	.071	L.F.	.21	3.73		3.94	6.25
0610	4″ to 8″ diameter pipe		800	.080		.23	4.19		4.42	7
0620	10″ to 12″ diameter pipe		700	.091		.26	4.79		5.05	8
0630	14″ to 16″ diameter pipe		550	.116		.34	6.10		6.44	10.15
0650	Over 16″ diameter pipe		650	.098	S.F.	.28	5.15		5.43	8.60
9000	For type B (supplied air) respirator equipment, add				%				10%	10%

Division Notes

		CREW	DAILY OUTPUT	LABOR-HOURS	UNIT	BARE COSTS MAT.	LABOR	EQUIP.	TOTAL	TOTAL INCL O&P

Estimating Tips

General

- Carefully check all the plans and specifications. Concrete often appears on drawings other than structural drawings, including mechanical and electrical drawings for equipment pads. The cost of cutting and patching is often difficult to estimate. See Subdivision 03 81 for Concrete Cutting, Subdivision 02 41 19.16 for Cutout Demolition, Subdivision 03 05 05.10 for Concrete Demolition, and Subdivision 02 41 19.19 for Rubbish Handling (handling, loading and hauling of debris).
- Always obtain concrete prices from suppliers near the job site. A volume discount can often be negotiated, depending upon competition in the area. Remember to add for waste, particularly for slabs and footings on grade.

03 10 00 Concrete Forming and Accessories

- A primary cost for concrete construction is forming. Most jobs today are constructed with prefabricated forms. The selection of the forms best suited for the job and the total square feet of forms required for efficient concrete forming and placing are key elements in estimating concrete construction. Enough forms must be available for erection to make efficient use of the concrete placing equipment and crew.
- Concrete accessories for forming and placing depend upon the systems used. Study the plans and specifications to ensure that all special accessory requirements have been included in the cost estimate, such as anchor bolts, inserts, and hangers.
- Included within costs for forms-in-place are all necessary bracing and shoring.

03 20 00 Concrete Reinforcing

- Ascertain that the reinforcing steel supplier has included all accessories, cutting, bending, and an allowance for lapping, splicing, and waste. A good rule of thumb is 10% for lapping, splicing, and waste. Also, 10% waste should be allowed for welded wire fabric.
- The unit price items in the subdivisions for Reinforcing In Place, Glass Fiber Reinforcing, and Welded Wire Fabric include the labor to install accessories such as beam and slab bolsters, high chairs, and bar ties and tie wire. The material cost for these accessories is not included; they may be obtained from the Accessories Division.

03 30 00 Cast-In-Place Concrete

- When estimating structural concrete, pay particular attention to requirements for concrete additives, curing methods, and surface treatments. Special consideration for climate, hot or cold, must be included in your estimate. Be sure to include requirements for concrete placing equipment, and concrete finishing.
- For accurate concrete estimating, the estimator must consider each of the following major components individually: forms, reinforcing steel, ready-mix concrete, placement of the concrete, and finishing of the top surface. For faster estimating, Subdivision 03 30 53.40 for Concrete-In-Place can be used; here, various items of concrete work are presented that include the costs of all five major components (unless specifically stated otherwise).

03 40 00 Precast Concrete
03 50 00 Cast Decks and Underlayment

- The cost of hauling precast concrete structural members is often an important factor. For this reason, it is important to get a quote from the nearest supplier. It may become economically feasible to set up precasting beds on the site if the hauling costs are prohibitive.

Reference Numbers

Reference numbers are shown in shaded boxes at the beginning of some major classifications. These numbers refer to related items in the Reference Section. The reference information may be an estimating procedure, an alternate pricing method, or technical information.

Note: Not all subdivisions listed here necessarily appear in this publication. ■

03 01 30.62 Concrete Patching

Line	Description	Crew	Daily Output	Labor-Hours	Unit	Material (2015 Bare Costs)	Labor (2015 Bare Costs)	Equipment (2015 Bare Costs)	Total (2015 Bare Costs)	Total Incl O&P
0010	**CONCRETE PATCHING**									
0100	Floors, 1/4" thick, small areas, regular grout	1 Cefi	170	.047	S.F.	1.46	2.12		3.58	4.95
0150	Epoxy grout	"	100	.080	"	8.05	3.60		11.65	14.60
2000	Walls, including chipping, cleaning and epoxy grout									
2100	1/4" deep	1 Cefi	65	.123	S.F.	7.65	5.55		13.20	17.20
2150	1/2" deep		50	.160		15.35	7.20		22.55	28.50
2200	3/4" deep		40	.200		23	9		32	40
9000	Minimum labor/equipment charge		4.50	1.778	Job		80		80	126

03 01 30.71 Concrete Crack Repair

Line	Description	Crew	Daily Output	Labor-Hours	Unit	Material (2015 Bare Costs)	Labor (2015 Bare Costs)	Equipment (2015 Bare Costs)	Total (2015 Bare Costs)	Total Incl O&P
0010	**CONCRETE CRACK REPAIR**									
1000	Structural repair of concrete cracks by epoxy injection (ACI RAP-1)									
1001	suitable for horizontal, vertical and overhead repairs									
1010	Clean/grind concrete surface(s) free of contaminants	1 Cefi	400	.020	L.F.		.90		.90	1.42
1015	Rout crack with v-notch crack chaser, if needed	C-32	600	.027		.03	1.10	.15	1.28	1.98
1020	Blow out crack with oil-free dry compressed air (1 pass)	C-28	3000	.003			.12	.01	.13	.20
1030	Install surface-mounted entry ports (spacing = concrete depth)	1 Cefi	400	.020	Ea.	1.39	.90		2.29	2.95
1040	Cap crack at surface with epoxy gel (per side/face)		400	.020	L.F.	.34	.90		1.24	1.80
1050	Snap off ports, grind off epoxy cap residue after injection		200	.040	"		1.80		1.80	2.85
1100	Manual injection with 2-part epoxy cartridge, excludes prep									
1110	Up to 1/32" (0.03125") wide x 4" deep	1 Cefi	160	.050	L.F.	.17	2.25		2.42	3.75
1120	6" deep		120	.067		.26	3		3.26	5
1130	8" deep		107	.075		.34	3.36		3.70	5.70
1140	10" deep		100	.080		.43	3.60		4.03	6.15
1150	12" deep		96	.083		.51	3.75		4.26	6.50
1210	Up to 1/16" (0.0625") wide x 4" deep		160	.050		.34	2.25		2.59	3.94
1220	6" deep		120	.067		.51	3		3.51	5.30
1230	8" deep		107	.075		.69	3.36		4.05	6.05
1240	10" deep		100	.080		.86	3.60		4.46	6.65
1250	12" deep		96	.083		1.03	3.75		4.78	7.10
1310	Up to 3/32" (0.09375") wide x 4" deep		160	.050		.51	2.25		2.76	4.13
1320	6" deep		120	.067		.77	3		3.77	5.60
1330	8" deep		107	.075		1.03	3.36		4.39	6.45
1340	10" deep		100	.080		1.29	3.60		4.89	7.10
1350	12" deep		96	.083		1.54	3.75		5.29	7.65
1410	Up to 1/8" (0.125") wide x 4" deep		160	.050		.69	2.25		2.94	4.31
1420	6" deep		120	.067		1.03	3		4.03	5.85
1430	8" deep		107	.075		1.37	3.36		4.73	6.80
1440	10" deep		100	.080		1.72	3.60		5.32	7.60
1450	12" deep		96	.083		2.06	3.75		5.81	8.20
1500	Pneumatic injection with 2-part bulk epoxy, excludes prep									
1510	Up to 5/32" (0.15625") wide x 4" deep	C-31	240	.033	L.F.	.24	1.50	1.55	3.29	4.34
1520	6" deep		180	.044		.36	2	2.07	4.43	5.85
1530	8" deep		160	.050		.48	2.25	2.32	5.05	6.65
1540	10" deep		150	.053		.61	2.40	2.48	5.49	7.20
1550	12" deep		144	.056		.73	2.50	2.58	5.81	7.60
1610	Up to 3/16" (0.1875") wide x 4" deep		240	.033		.29	1.50	1.55	3.34	4.39
1620	6" deep		180	.044		.44	2	2.07	4.51	5.90
1630	8" deep		160	.050		.58	2.25	2.32	5.15	6.75
1640	10" deep		150	.053		.73	2.40	2.48	5.61	7.30
1650	12" deep		144	.056		.87	2.50	2.58	5.95	7.75
1710	Up to 1/4" (0.1875") wide x 4" deep		240	.033		.39	1.50	1.55	3.44	4.50
1720	6" deep		180	.044		.58	2	2.07	4.65	6.05

03 01 Maintenance of Concrete

03 01 30 – Maintenance of Cast-In-Place Concrete

03 01 30.71 Concrete Crack Repair		Crew	Daily Output	Labor-Hours	Unit	Material	2015 Bare Costs Labor	Equipment	Total	Total Incl O&P
1730	8″ deep	C-31	160	.050	L.F.	.77	2.25	2.32	5.34	6.95
1740	10″ deep		150	.053		.97	2.40	2.48	5.85	7.60
1750	12″ deep		144	.056		1.16	2.50	2.58	6.24	8.05
2000	Non-structural filling of concrete cracks by gravity-fed resin (ACI RAP-2)									
2001	suitable for individual cracks in stable horizontal surfaces only									
2010	Clean/grind concrete surface(s) free of contaminants	1 Cefi	400	.020	L.F.		.90		.90	1.42
2020	Rout crack with v-notch crack chaser, if needed	C-32	600	.027		.03	1.10	.15	1.28	1.98
2030	Blow out crack with oil-free dry compressed air (1 pass)	C-28	3000	.003			.12	.01	.13	.20
2040	Cap crack with epoxy gel at underside of elevated slabs, if needed	1 Cefi	400	.020		.34	.90		1.24	1.80
2050	Insert backer rod into crack, if needed		400	.020		.02	.90		.92	1.44
2060	Partially fill crack with fine dry sand, if needed		800	.010		.02	.45		.47	.73
2070	Apply two beads of sealant alongside crack to form reservoir, if needed		400	.020		.25	.90		1.15	1.69
2100	Manual filling with squeeze bottle of 2-part epoxy resin									
2110	Full depth crack up to 1/16″ (0.0625″) wide x 4″ deep	1 Cefi	300	.027	L.F.	.09	1.20		1.29	2
2120	6″ deep		240	.033		.14	1.50		1.64	2.53
2130	8″ deep		200	.040		.19	1.80		1.99	3.06
2140	10″ deep		185	.043		.24	1.95		2.19	3.34
2150	12″ deep		175	.046		.28	2.06		2.34	3.56
2210	Full depth crack up to 1/8″ (0.125″) wide x 4″ deep		270	.030		.19	1.33		1.52	2.32
2220	6″ deep		215	.037		.28	1.67		1.95	2.96
2230	8″ deep		180	.044		.38	2		2.38	3.58
2240	10″ deep		165	.048		.47	2.18		2.65	3.97
2250	12″ deep		155	.052		.57	2.32		2.89	4.29
2310	Partial depth crack up to 3/16″ (0.1875″) wide x 1″ deep		300	.027		.07	1.20		1.27	1.98
2410	Up to 1/4″ (0.250″) wide x 1″ deep		270	.030		.09	1.33		1.42	2.21
2510	Up to 5/16″ (0.3125″) wide x 1″ deep		250	.032		.12	1.44		1.56	2.41
2610	Up to 3/8″ (0.375″) wide x 1″ deep		240	.033		.14	1.50		1.64	2.53

03 01 30.72 Concrete Surface Repairs		Crew	Daily Output	Labor-Hours	Unit	Material	Labor	Equipment	Total	Total Incl O&P
0010	CONCRETE SURFACE REPAIRS									
9000	Surface repair by Methacrylate flood coat (ACI RAP-13)									
9010	Suitable for healing and sealing horizontal surfaces only									
9020	Large cracks must previously have been repaired or filled									
9100	Shotblast entire surface to remove contaminants	A-1A	4000	.002	S.F.		.10	.05	.15	.22
9200	Blow off dust and debris with oil-free dry compressed air	C-28	16000	.001			.02		.02	.04
9300	Flood coat surface w/Methacrylate, distribute w/broom/squeegee, no prep.	3 Cefi	8000	.003		.98	.14		1.12	1.29
9400	Lightly broadcast even coat of dry silica sand while sealer coat is tacky	1 Cefi	8000	.001		.01	.05		.06	.08

03 05 Common Work Results for Concrete

03 05 05 – Selective Demolition for Concrete

03 05 05.10 Selective Demolition, Concrete		Crew	Daily Output	Labor-Hours	Unit	Material	Labor	Equipment	Total	Total Incl O&P
0010	SELECTIVE DEMOLITION, CONCRETE R024119-10									
0012	Excludes saw cutting, torch cutting, loading or hauling									
0050	Break into small pieces, reinf. less than 1% of cross-sectional area	B-9	24	1.667	C.Y.		63.50	9.70	73.20	115
0060	Reinforcing 1% to 2% of cross-sectional area		16	2.500			95	14.55	109.55	172
0070	Reinforcing more than 2% of cross-sectional area		8	5			190	29	219	340
0150	Remove whole pieces, up to 2 tons per piece	E-18	36	1.111	Ea.		58.50	26.50	85	132
0160	2 – 5 tons per piece		30	1.333			70	31.50	101.50	159
0170	5 – 10 tons per piece		24	1.667			87.50	39.50	127	199
0180	10 – 15 tons per piece		18	2.222			117	52.50	169.50	264
0250	Precast unit embedded in masonry, up to 1 C.F.	D-1	16	1			42		42	68.50
0260	1 – 2 C.F.		12	1.333			56		56	91.50

03 05 Common Work Results for Concrete

03 05 05 – Selective Demolition for Concrete

03 05 05.10 Selective Demolition, Concrete		Crew	Daily Output	Labor-Hours	Unit	Material	2015 Bare Costs Labor	Equipment	Total	Total Incl O&P
0270	2 – 5 C.F.	D-1	10	1.600	Ea.		67.50		67.50	110
0280	5 – 10 C.F.		8	2			84		84	137
0990	For hydrodemolition see Section 02 41 13.15									
1910	Minimum labor/equipment charge	B-9	2	20	Job		760	116	876	1,375

03 05 13 – Basic Concrete Materials

03 05 13.20 Concrete Admixtures and Surface Treatments		Crew	Daily Output	Labor-Hours	Unit	Material	Labor	Equipment	Total	Total Incl O&P
0010	**CONCRETE ADMIXTURES AND SURFACE TREATMENTS**									
0040	Abrasives, aluminum oxide, over 20 tons				Lb.	1.88			1.88	2.07
0050	1 to 20 tons					2.02			2.02	2.22
0070	Under 1 ton					2.10			2.10	2.31
0100	Silicon carbide, black, over 20 tons					2.87			2.87	3.16
0110	1 to 20 tons					3.04			3.04	3.35
0120	Under 1 ton					3.17			3.17	3.49
0200	Air entraining agent, .7 to 1.5 oz. per bag, 55 gallon drum				Gal.	14			14	15.40
0220	5 gallon pail					19.25			19.25	21
0300	Bonding agent, acrylic latex, 250 S.F. per gallon, 5 gallon pail					22.50			22.50	25
0320	Epoxy resin, 80 S.F. per gallon, 4 gallon case					61			61	67
0400	Calcium chloride, 50 lb. bags, TL lots				Ton	810			810	890
0420	Less than truckload lots				Bag	24			24	26.50
0500	Carbon black, liquid, 2 to 8 lb. per bag of cement				Lb.	9.10			9.10	10
0600	Colored pigments, integral, 2 to 10 lb. per bag of cement, subtle colors					2.40			2.40	2.64
0610	Standard colors					3.26			3.26	3.59
0620	Premium colors					5.40			5.40	5.95
0920	Dustproofing compound, 250 S.F./gal., 5 gallon pail				Gal.	6.65			6.65	7.30
1010	Epoxy based, 125 S.F./gal., 5 gallon pail				"	55.50			55.50	61
1100	Hardeners, metallic, 55 lb. bags, natural (grey)				Lb.	.70			.70	.76
1200	Colors					2.27			2.27	2.50
1300	Non-metallic, 55 lb. bags, natural grey					.43			.43	.47
1320	Colors					.94			.94	1.04
1550	Release agent, for tilt slabs, 5 gallon pail				Gal.	17.60			17.60	19.40
1570	For forms, 5 gallon pail					12.50			12.50	13.75
1590	Concrete release agent for forms, 100% biodegradeable, zero VOC, 5 gal pail G					19.95			19.95	22
1595	55 gallon drum G					16.80			16.80	18.50
1600	Sealer, hardener and dustproofer, epoxy-based, 125 S.F./gal., 5 gallon unit					55.50			55.50	61
1620	3 gallon unit					61			61	67.50
1630	Sealer, solvent-based, 250 S.F./gal., 55 gallon drum					24.50			24.50	27
1640	5 gallon pail					30.50			30.50	34
1650	Sealer, water based, 350 S.F., 55 gallon drum					22.50			22.50	25
1660	5 gallon pail					25.50			25.50	28.50
1900	Set retarder, 100 S.F./gal., 1 gallon pail					22.50			22.50	24.50
2000	Waterproofing, integral 1 lb. per bag of cement				Lb.	2.12			2.12	2.33
2100	Powdered metallic, 40 lb. per 100 S.F., standard colors					2.52			2.52	2.77
2120	Premium colors					3.53			3.53	3.88
3000	For integral colored pigments, 2500 psi (5 bag mix)									
3100	Standard colors, 1.8 lb. per bag, add				C.Y.	21.50			21.50	24
3200	9.4 lb. per bag, add					113			113	124
3400	Premium colors, 1.8 lb. per bag, add					29.50			29.50	32.50
3500	7.5 lb. per bag, add					122			122	134
3700	Ultra premium colors, 1.8 lb. per bag, add					49			49	53.50
3800	7.5 lb. per bag, add					203			203	224
6000	Concrete ready mix additives, recycled coal fly ash, mixed at plant G				Ton	58.50			58.50	64
6010	Recycled blast furnace slag, mixed at plant G				"	91			91	100

03 05 Common Work Results for Concrete

03 05 13 – Basic Concrete Materials

03 05 13.25 Aggregate

03 05 13.25 Aggregate		Crew	Daily Output	Labor-Hours	Unit	2015 Bare Costs Material	Labor	Equipment	Total	Total Incl O&P
0010	**AGGREGATE**									
0100	Lightweight vermiculite or perlite, 4 C.F. bag, C.L. lots G				Bag	23.50			23.50	26
0150	L.C.L. lots G				"	26.50			26.50	29
0250	Sand & stone, loaded at pit, crushed bank gravel				Ton	21.50			21.50	23.50
0350	Sand, washed, for concrete					19.35			19.35	21.50
0400	For plaster or brick					19.35			19.35	21.50
0450	Stone, 3/4" to 1-1/2"					17.80			17.80	19.60
0470	Round, river stone					26.50			26.50	29
0500	3/8" roofing stone & 1/2" pea stone					26			26	28.50
0550	For trucking 10-mile round trip, add to the above	B-34B	117	.068			2.74	5.90	8.64	10.90
0600	For trucking 30-mile round trip, add to the above	"	72	.111	↓		4.45	9.60	14.05	17.70
0850	Sand & stone, loaded at pit, crushed bank gravel				C.Y.	30			30	33
0950	Sand, washed, for concrete					27			27	29.50
1000	For plaster or brick					27			27	29.50
1050	Stone, 3/4" to 1-1/2"					34			34	37.50
1055	Round, river stone					36			36	39.50
1100	3/8" roofing stone & 1/2" pea stone					25.50			25.50	28
1150	For trucking 10-mile round trip, add to the above	B-34B	78	.103			4.11	8.85	12.96	16.35
1200	For trucking 30-mile round trip, add to the above	"	48	.167	↓		6.70	14.40	21.10	26.50
1310	Onyx chips, 50 lb. bags				Cwt.	34			34	37
1330	Quartz chips, 50 lb. bags					33.50			33.50	37
1410	White marble, 3/8" to 1/2", 50 lb. bags				↓	7.50			7.50	8.25
1430	3/4", bulk				Ton	117			117	129

03 05 13.30 Cement

Line	Description	Crew	Daily Output	Labor-Hours	Unit	Material	Labor	Equipment	Total	Total Incl O&P
0010	**CEMENT**									
0240	Portland, Type I/II, TL lots, 94 lb. bags				Bag	9.90			9.90	10.90
0250	LTL/LCL lots				"	11			11	12.10
0300	Trucked in bulk, per Cwt.				Cwt.	7.10			7.10	7.85
0400	Type III, high early strength, TL lots, 94 lb. bags				Bag	11.80			11.80	13
0420	L.T.L. or L.C.L. lots					12.10			12.10	13.30
0500	White, type III, high early strength, T.L. or C.L. lots, bags					23.50			23.50	26
0520	L.T.L. or L.C.L. lots					24			24	26.50
0600	White, type I, T.L. or C.L. lots, bags					25			25	27.50
0620	L.T.L. or L.C.L. lots				↓	26.50			26.50	29

03 05 13.80 Waterproofing and Dampproofing

Line	Description	Crew	Daily Output	Labor-Hours	Unit	Material	Labor	Equipment	Total	Total Incl O&P
0010	**WATERPROOFING AND DAMPPROOFING**									
0050	Integral waterproofing, add to cost of regular concrete				C.Y.	12.70			12.70	14

03 05 13.85 Winter Protection

Line	Description	Crew	Daily Output	Labor-Hours	Unit	Material	Labor	Equipment	Total	Total Incl O&P
0010	**WINTER PROTECTION**									
0012	For heated ready mix, add				C.Y.	4.05			4.05	4.46
0100	Temporary heat to protect concrete, 24 hours	2 Clab	50	.320	M.S.F.	535	12.05		547.05	605
0200	Temporary shelter for slab on grade, wood frame/polyethylene sheeting									
0201	Build or remove, light framing for short spans	2 Carp	10	1.600	M.S.F.	310	75		385	470
0210	Large framing for long spans	"	3	5.333	"	415	250		665	865
0500	Electrically heated pads, 110 volts, 15 watts per S.F., buy				S.F.	10.35			10.35	11.40
0600	20 watts per S.F., buy					13.80			13.80	15.15
0710	Electrically, heated pads, 15 watts/S.F., 20 uses				↓	.52			.52	.57

03 11 13.20 Forms In Place, Beams and Girders		Crew	Daily Output	Labor-Hours	Unit	Material	Labor	Equipment	Total	Total Incl O&P
						2015 Bare Costs				
0010	**FORMS IN PLACE, BEAMS AND GIRDERS**									
0500	Exterior spandrel, job-built plywood, 12" wide, 1 use R031113-60	C-2	225	.213	SFCA	3.12	9.75		12.87	19.45
0550	2 use		275	.175		1.64	8		9.64	14.90
0600	3 use		295	.163		1.25	7.45		8.70	13.55
0650	4 use		310	.155		1.01	7.10		8.11	12.70
1000	18" wide, 1 use		250	.192		2.78	8.80		11.58	17.45
1050	2 use		275	.175		1.53	8		9.53	14.80
1100	3 use		305	.157		1.11	7.20		8.31	13
1150	4 use		315	.152		.91	6.95		7.86	12.40
1500	24" wide, 1 use		265	.181		2.54	8.30		10.84	16.40
1550	2 use		290	.166		1.43	7.55		8.98	14
1600	3 use		315	.152		1.02	6.95		7.97	12.50
1650	4 use		325	.148		.83	6.75		7.58	11.95
2000	Interior beam, job-built plywood, 12" wide, 1 use		300	.160		3.67	7.30		10.97	16.05
2050	2 use		340	.141		1.77	6.45		8.22	12.55
2100	3 use		364	.132		1.47	6.05		7.52	11.50
2150	4 use		377	.127		1.19	5.80		6.99	10.85
2500	24" wide, 1 use		320	.150		2.60	6.85		9.45	14.10
2550	2 use		365	.132		1.47	6		7.47	11.45
2600	3 use		385	.125		1.03	5.70		6.73	10.50
2650	4 use		395	.122		.84	5.55		6.39	10
3000	Encasing steel beam, hung, job-built plywood, 1 use		325	.148		3.20	6.75		9.95	14.55
3050	2 use		390	.123		1.76	5.65		7.41	11.15
3100	3 use		415	.116		1.28	5.30		6.58	10.05
3150	4 use		430	.112		1.04	5.10		6.14	9.50
3500	Bottoms only, to 30" wide, job-built plywood, 1 use		230	.209		4.27	9.55		13.82	20.50
3550	2 use		265	.181		2.39	8.30		10.69	16.25
3600	3 use		280	.171		1.71	7.85		9.56	14.75
3650	4 use		290	.166		1.39	7.55		8.94	13.95
4000	Sides only, vertical, 36" high, job-built plywood, 1 use		335	.143		5.20	6.55		11.75	16.45
4050	2 use		405	.119		2.86	5.40		8.26	12.05
4100	3 use		430	.112		2.08	5.10		7.18	10.65
4150	4 use		445	.108		1.69	4.93		6.62	9.95
4500	Sloped sides, 36" high, 1 use		305	.157		5	7.20		12.20	17.30
4550	2 use		370	.130		2.80	5.95		8.75	12.80
4600	3 use		405	.119		2.01	5.40		7.41	11.10
4650	4 use		425	.113		1.63	5.15		6.78	10.25
5000	Upstanding beams, 36" high, 1 use		225	.213		6.35	9.75		16.10	23
5050	2 use		255	.188		3.53	8.60		12.13	18
5100	3 use		275	.175		2.56	8		10.56	15.90
5150	4 use	↓	280	.171	↓	2.08	7.85		9.93	15.15
9000	Minimum labor/equipment charge	2 Carp	2	8	Job		375		375	615

03 11 13.25 Forms In Place, Columns		Crew	Daily Output	Labor-Hours	Unit	Material	Labor	Equipment	Total	Total Incl O&P
0010	**FORMS IN PLACE, COLUMNS**									
0500	Round fiberglass, 4 use per mo., rent, 12" diameter	C-1	160	.200	L.F.	8.95	8.90		17.85	24.50
0550	16" diameter R031113-60		150	.213		10.65	9.50		20.15	27.50
0600	18" diameter		140	.229		11.95	10.20		22.15	30
0650	24" diameter		135	.237		14.85	10.55		25.40	33.50
0700	28" diameter		130	.246		16.60	11		27.60	36.50
0800	30" diameter		125	.256		17.35	11.40		28.75	38
0850	36" diameter		120	.267		23	11.90		34.90	45
1500	Round fiber tube, recycled paper, 1 use, 8" diameter G	↓	155	.206	↓	1.77	9.20		10.97	17.05

03 11 13.25 Forms In Place, Columns			Crew	Daily Output	Labor-Hours	Unit	2015 Bare Costs Material	Labor	Equipment	Total	Total Incl O&P
1550	10″ diameter	G	C-1	155	.206	L.F.	2.40	9.20		11.60	17.75
1600	12″ diameter	G		150	.213		2.83	9.50		12.33	18.70
1650	14″ diameter	G		145	.221		3.91	9.85		13.76	20.50
1700	16″ diameter	G		140	.229		4.88	10.20		15.08	22
1720	18″ diameter	G		140	.229		5.70	10.20		15.90	23
1750	20″ diameter	G		135	.237		7.60	10.55		18.15	26
1800	24″ diameter	G		130	.246		9.70	11		20.70	28.50
1850	30″ diameter	G		125	.256		14.50	11.40		25.90	34.50
1900	36″ diameter	G		115	.278		17.10	12.40		29.50	39.50
1950	42″ diameter	G		100	.320		44.50	14.30		58.80	72.50
2000	48″ diameter	G		85	.376		51.50	16.80		68.30	84.50
2200	For seamless type, add						15%				
3000	Round, steel, 4 use per mo., rent, regular duty, 14″ diameter	G	C-1	145	.221	L.F.	12.30	9.85		22.15	29.50
3050	16″ diameter	G		125	.256		12.50	11.40		23.90	32.50
3100	Heavy duty, 20″ diameter	G		105	.305		13.80	13.60		27.40	37.50
3150	24″ diameter	G		85	.376		15.05	16.80		31.85	44
3200	30″ diameter	G		70	.457		17.30	20.50		37.80	52.50
3250	36″ diameter	G		60	.533		18.60	24		42.60	59.50
3300	48″ diameter	G		50	.640		27.50	28.50		56	77.50
3350	60″ diameter	G		45	.711		34	31.50		65.50	89.50
4500	For second and succeeding months, deduct						50%				
5000	Job-built plywood, 8″ x 8″ columns, 1 use		C-1	165	.194	SFCA	2.73	8.65		11.38	17.20
5050	2 use			195	.164		1.56	7.30		8.86	13.70
5100	3 use			210	.152		1.09	6.80		7.89	12.35
5150	4 use			215	.149		.90	6.65		7.55	11.90
5500	12″ x 12″ columns, 1 use			180	.178		2.62	7.95		10.57	15.90
5550	2 use			210	.152		1.44	6.80		8.24	12.75
5600	3 use			220	.145		1.05	6.50		7.55	11.80
5650	4 use			225	.142		.85	6.35		7.20	11.35
6000	16″ x 16″ columns, 1 use			185	.173		2.62	7.70		10.32	15.55
6050	2 use			215	.149		1.41	6.65		8.06	12.45
6100	3 use			230	.139		1.05	6.20		7.25	11.30
6150	4 use			235	.136		.86	6.05		6.91	10.90
6500	24″ x 24″ columns, 1 use			190	.168		2.97	7.50		10.47	15.55
6550	2 use			216	.148		1.63	6.60		8.23	12.65
6600	3 use			230	.139		1.19	6.20		7.39	11.45
6650	4 use			238	.134		.96	6		6.96	10.90
7000	36″ x 36″ columns, 1 use			200	.160		2.31	7.15		9.46	14.25
7050	2 use			230	.139		1.31	6.20		7.51	11.60
7100	3 use			245	.131		.92	5.85		6.77	10.55
7150	4 use			250	.128		.75	5.70		6.45	10.20
7400	Steel framed plywood, based on 50 uses of purchased										
7420	forms, and 4 uses of bracing lumber										
7500	8″ x 8″ column		C-1	340	.094	SFCA	2.14	4.20		6.34	9.25
7550	10″ x 10″			350	.091		1.87	4.08		5.95	8.75
7600	12″ x 12″			370	.086		1.59	3.86		5.45	8.10
7650	16″ x 16″			400	.080		1.23	3.57		4.80	7.20
7700	20″ x 20″			420	.076		1.09	3.40		4.49	6.75
7750	24″ x 24″			440	.073		.79	3.24		4.03	6.15
7755	30″ x 30″			440	.073		1	3.24		4.24	6.40
7760	36″ x 36″			460	.070		.88	3.10		3.98	6.05
9000	Minimum labor/equipment charge		2 Carp	2	8	Job		375		375	615

03 11 Concrete Forming

03 11 13 – Structural Cast-In-Place Concrete Forming

03 11 13.30 Forms In Place, Culvert		Crew	Daily Output	Labor-Hours	Unit	Material	Labor	Equipment	Total	Total Incl O&P
						2015 Bare Costs				
0010	**FORMS IN PLACE, CULVERT**									
0015	5′ to 8′ square or rectangular, 1 use	C-1	170	.188	SFCA	4.03	8.40		12.43	18.20
0050	2 use R031113-60		180	.178		2.42	7.95		10.37	15.65
0100	3 use		190	.168		1.89	7.50		9.39	14.40
0150	4 use		200	.160		1.62	7.15		8.77	13.50

03 11 13.35 Forms In Place, Elevated Slabs		Crew	Daily Output	Labor-Hours	Unit	Material	Labor	Equipment	Total	Total Incl O&P
0010	**FORMS IN PLACE, ELEVATED SLABS**									
1000	Flat plate, job-built plywood, to 15′ high, 1 use	C-2	470	.102	S.F.	3.86	4.67		8.53	11.90
1050	2 use R031113-60		520	.092		2.13	4.22		6.35	9.25
1100	3 use		545	.088		1.55	4.03		5.58	8.30
1150	4 use		560	.086		1.26	3.92		5.18	7.80
1500	15′ to 20′ high ceilings, 4 use		495	.097		1.30	4.43		5.73	8.70
1600	21′ to 35′ high ceilings, 4 use		450	.107		1.60	4.88		6.48	9.75
2000	Flat slab, drop panels, job-built plywood, to 15′ high, 1 use		449	.107		4.42	4.89		9.31	12.85
2050	2 use		509	.094		2.43	4.31		6.74	9.70
2100	3 use		532	.090		1.77	4.13		5.90	8.70
2150	4 use		544	.088		1.43	4.04		5.47	8.20
2250	15′ to 20′ high ceilings, 4 use		480	.100		2.47	4.57		7.04	10.20
2350	20′ to 35′ high ceilings, 4 use		435	.110		2.77	5.05		7.82	11.30
3000	Floor slab hung from steel beams, 1 use		485	.099		2.73	4.53		7.26	10.40
3050	2 use		535	.090		2.12	4.10		6.22	9.05
3100	3 use		550	.087		1.92	3.99		5.91	8.65
3150	4 use		565	.085		1.81	3.89		5.70	8.35
3500	Floor slab, with 1-way joist pans, 1 use		415	.116		6.35	5.30		11.65	15.65
3550	2 use		445	.108		4.36	4.93		9.29	12.90
3600	3 use		475	.101		3.70	4.62		8.32	11.60
3650	4 use		500	.096		3.37	4.39		7.76	10.90
4500	With 2-way waffle domes, 1 use		405	.119		6.50	5.40		11.90	16.05
4520	2 use		450	.107		4.51	4.88		9.39	12.95
4530	3 use		460	.104		3.84	4.77		8.61	12
4550	4 use		470	.102		3.51	4.67		8.18	11.50
5000	Box out for slab openings, over 16″ deep, 1 use		190	.253	SFCA	4.57	11.55		16.12	24
5050	2 use		240	.200	″	2.51	9.15		11.66	17.75
5500	Shallow slab box outs, to 10 S.F.		42	1.143	Ea.	12.10	52.50		64.60	99
5550	Over 10 S.F. (use perimeter)		600	.080	L.F.	1.61	3.66		5.27	7.80
6000	Bulkhead forms for slab, with keyway, 1 use, 2 piece		500	.096		2.14	4.39		6.53	9.55
6100	3 piece (see also edge forms)		460	.104		2.31	4.77		7.08	10.35
6200	Slab bulkhead form, 4-1/2″ high, exp metal, w/keyway & stakes G	C-1	1200	.027		.92	1.19		2.11	2.96
6210	5-1/2″ high G		1100	.029		1.12	1.30		2.42	3.36
6215	7-1/2″ high G		960	.033		1.31	1.49		2.80	3.88
6220	9-1/2″ high G		840	.038		1.43	1.70		3.13	4.36
6500	Curb forms, wood, 6″ to 12″ high, on elevated slabs, 1 use		180	.178	SFCA	1.68	7.95		9.63	14.85
6550	2 use		205	.156		.93	6.95		7.88	12.40
6600	3 use		220	.145		.67	6.50		7.17	11.40
6650	4 use		225	.142		.55	6.35		6.90	11
7000	Edge forms to 6″ high, on elevated slab, 4 use		500	.064	L.F.	.21	2.86		3.07	4.91
7070	7″ to 12″ high, 1 use		162	.198	SFCA	1.27	8.80		10.07	15.85
7080	2 use		198	.162		.70	7.20		7.90	12.55
7090	3 use		222	.144		.51	6.45		6.96	11.10
7101	4 use		350	.091		.21	4.08		4.29	6.95
7500	Depressed area forms to 12″ high, 4 use		300	.107	L.F.	1.02	4.76		5.78	8.90
7550	12″ to 24″ high, 4 use		175	.183		1.38	8.15		9.53	14.85

03 11 Concrete Forming

03 11 13 – Structural Cast-In-Place Concrete Forming

03 11 13.35 Forms In Place, Elevated Slabs		Crew	Daily Output	Labor-Hours	Unit	Material	2015 Bare Costs Labor	Equipment	Total	Total Incl O&P
8000	Perimeter deck and rail for elevated slabs, straight	C-1	90	.356	L.F.	12.45	15.85		28.30	39.50
8050	Curved		65	.492		17.05	22		39.05	55
8500	Void forms, round plastic, 8" high x 3" diameter G		450	.071	Ea.	.71	3.17		3.88	6
8550	4" diameter G		425	.075		1.06	3.36		4.42	6.65
8600	6" diameter G		400	.080		1.75	3.57		5.32	7.80
8650	8" diameter G		375	.085		3.13	3.81		6.94	9.70
9000	Minimum labor/equipment charge	2 Carp	2	8	Job		375		375	615

03 11 13.40 Forms In Place, Equipment Foundations		Crew	Daily Output	Labor-Hours	Unit	Material	Labor	Equipment	Total	Total Incl O&P
0010	**FORMS IN PLACE, EQUIPMENT FOUNDATIONS**									
0020	1 use	C-2	160	.300	SFCA	3.45	13.70		17.15	26.50
0050	2 use R031113-60		190	.253		1.90	11.55		13.45	21
0100	3 use		200	.240		1.38	11		12.38	19.50
0150	4 use		205	.234		1.12	10.70		11.82	18.80
9000	Minimum labor/equipment charge	1 Carp	3	2.667	Job		125		125	205

03 11 13.45 Forms In Place, Footings		Crew	Daily Output	Labor-Hours	Unit	Material	Labor	Equipment	Total	Total Incl O&P
0010	**FORMS IN PLACE, FOOTINGS**									
0020	Continuous wall, plywood, 1 use	C-1	375	.085	SFCA	6.80	3.81		10.61	13.75
0050	2 use R031113-60		440	.073		3.74	3.24		6.98	9.40
0100	3 use		470	.068		2.72	3.04		5.76	7.95
0150	4 use		485	.066		2.22	2.94		5.16	7.25
0500	Dowel supports for footings or beams, 1 use		500	.064	L.F.	.94	2.86		3.80	5.70
1000	Integral starter wall, to 4" high, 1 use		400	.080		.98	3.57		4.55	6.95
1500	Keyway, 4 use, tapered wood, 2" x 4"	1 Carp	530	.015		.22	.71		.93	1.41
1550	2" x 6"		500	.016		.33	.75		1.08	1.59
2000	Tapered plastic		530	.015		1.31	.71		2.02	2.60
2250	For keyway hung from supports, add		150	.053		.94	2.50		3.44	5.15
3000	Pile cap, square or rectangular, job-built plywood, 1 use	C-1	290	.110	SFCA	2.91	4.92		7.83	11.25
3050	2 use		346	.092		1.60	4.13		5.73	8.50
3100	3 use		371	.086		1.16	3.85		5.01	7.60
3150	4 use		383	.084		.95	3.73		4.68	7.15
4000	Triangular or hexagonal, 1 use		225	.142		3.40	6.35		9.75	14.15
4050	2 use		280	.114		1.87	5.10		6.97	10.40
4100	3 use		305	.105		1.36	4.68		6.04	9.15
4150	4 use		315	.102		1.11	4.53		5.64	8.65
5000	Spread footings, job-built lumber, 1 use		305	.105		2.28	4.68		6.96	10.15
5050	2 use		371	.086		1.27	3.85		5.12	7.70
5100	3 use		401	.080		.91	3.56		4.47	6.85
5150	4 use		414	.077		.74	3.45		4.19	6.45
6000	Supports for dowels, plinths or templates, 2' x 2' footing		25	1.280	Ea.	6.05	57		63.05	100
6050	4' x 4' footing		22	1.455		12.10	65		77.10	119
6100	8' x 8' footing		20	1.600		24	71.50		95.50	144
6150	12' x 12' footing		17	1.882		32.50	84		116.50	174
7000	Plinths, job-built plywood, 1 use		250	.128	SFCA	3.38	5.70		9.08	13.05
7100	4 use		270	.119	"	1.11	5.30		6.41	9.85
9000	Minimum labor/equipment charge	1 Carp	3	2.667	Job		125		125	205

03 11 13.47 Forms In Place, Gas Station Forms		Crew	Daily Output	Labor-Hours	Unit	Material	Labor	Equipment	Total	Total Incl O&P
0010	**FORMS IN PLACE, GAS STATION FORMS**									
0050	Curb fascia, with template, 12 ga. steel, left in place, 9" high G	1 Carp	50	.160	L.F.	13.90	7.50		21.40	27.50
1000	Sign or light bases, 18" diameter, 9" high G		9	.889	Ea.	87.50	41.50		129	165
1050	30" diameter, 13" high G		8	1	"	139	47		186	230
1990	Minimum labor/equipment charge		2	4	Job		188		188	310
2000	Island forms, 10' long, 9" high, 3'- 6" wide G	C-1	10	3.200	Ea.	390	143		533	660

03 11 Concrete Forming

03 11 13 – Structural Cast-In-Place Concrete Forming

03 11 13.47 Forms In Place, Gas Station Forms			Crew	Daily Output	Labor-Hours	Unit	2015 Bare Costs Material	Labor	Equipment	Total	Total Incl O&P
2050	4′ wide	G	C-1	9	3.556	Ea.	400	159		559	700
2500	20′ long, 9″ high, 4′ wide	G		6	5.333		645	238		883	1,100
2550	5′ wide	G		5	6.400		670	286		956	1,200
9000	Minimum labor/equipment charge		1 Carp	3	2.667	Job		125		125	205

03 11 13.50 Forms In Place, Grade Beam

			Crew	Daily Output	Labor-Hours	Unit	Material	Labor	Equipment	Total	Total Incl O&P
0010	**FORMS IN PLACE, GRADE BEAM**										
0020	Job-built plywood, 1 use		C-2	530	.091	SFCA	3.14	4.14		7.28	10.25
0050	2 use	R031113-60		580	.083		1.73	3.78		5.51	8.10
0100	3 use			600	.080		1.25	3.66		4.91	7.40
0150	4 use			605	.079		1.02	3.63		4.65	7.05
9000	Minimum labor/equipment charge		2 Carp	2	8	Job		375		375	615

03 11 13.55 Forms In Place, Mat Foundation

			Crew	Daily Output	Labor-Hours	Unit	Material	Labor	Equipment	Total	Total Incl O&P
0010	**FORMS IN PLACE, MAT FOUNDATION**										
0020	Job-built plywood, 1 use		C-2	290	.166	SFCA	3.08	7.55		10.63	15.80
0050	2 use	R031113-60		310	.155		1.26	7.10		8.36	13
0100	3 use			330	.145		.81	6.65		7.46	11.80
0120	4 use			350	.137		.74	6.25		6.99	11.10

03 11 13.65 Forms In Place, Slab On Grade

			Crew	Daily Output	Labor-Hours	Unit	Material	Labor	Equipment	Total	Total Incl O&P
0010	**FORMS IN PLACE, SLAB ON GRADE**										
1000	Bulkhead forms w/keyway, wood, 6″ high, 1 use		C-1	510	.063	L.F.	1.04	2.80		3.84	5.75
1050	2 uses	R031113-60		400	.080		.57	3.57		4.14	6.50
1100	4 uses			350	.091		.34	4.08		4.42	7.05
1400	Bulkhead form for slab, 4-1/2″ high, exp metal, incl keyway & stakes	G		1200	.027		.92	1.19		2.11	2.96
1410	5-1/2″ high	G		1100	.029		1.12	1.30		2.42	3.36
1420	7-1/2″ high	G		960	.033		1.31	1.49		2.80	3.88
1430	9-1/2″ high	G		840	.038		1.43	1.70		3.13	4.36
2000	Curb forms, wood, 6″ to 12″ high, on grade, 1 use			215	.149	SFCA	2.44	6.65		9.09	13.60
2050	2 use			250	.128		1.35	5.70		7.05	10.85
2100	3 use			265	.121		.98	5.40		6.38	9.90
2150	4 use			275	.116		.79	5.20		5.99	9.35
3000	Edge forms, wood, 4 use, on grade, to 6″ high			600	.053	L.F.	.35	2.38		2.73	4.28
3050	7″ to 12″ high			435	.074	SFCA	.79	3.28		4.07	6.25
3060	Over 12″			350	.091	″	.91	4.08		4.99	7.70
3500	For depressed slabs, 4 use, to 12″ high			300	.107	L.F.	.75	4.76		5.51	8.65
3550	To 24″ high			175	.183		.99	8.15		9.14	14.45
4000	For slab blockouts, to 12″ high, 1 use			200	.160		.80	7.15		7.95	12.55
4050	To 24″ high, 1 use			120	.267		1.01	11.90		12.91	20.50
4100	Plastic (extruded), to 6″ high, multiple use, on grade			800	.040		8.20	1.78		9.98	11.95
5000	Screed, 24 ga. metal key joint, see Section 03 15 16.30										
5020	Wood, incl. wood stakes, 1″ x 3″		C-1	900	.036	L.F.	.77	1.59		2.36	3.44
5050	2″ x 4″			900	.036	″	.80	1.59		2.39	3.48
6000	Trench forms in floor, wood, 1 use			160	.200	SFCA	1.94	8.90		10.84	16.80
6050	2 use			175	.183		1.06	8.15		9.21	14.50
6100	3 use			180	.178		.77	7.95		8.72	13.85
6150	4 use			185	.173		.63	7.70		8.33	13.35
8760	Void form, corrugated fiberboard, 4″ x 12″, 4′ long	G		3000	.011	S.F.	2.87	.48		3.35	3.93
8770	6″ x 12″, 4′ long			3000	.011		3.42	.48		3.90	4.54
8780	1/4″ thick hardboard protective cover for void form		2 Carp	1500	.011		.58	.50		1.08	1.46
9000	Minimum labor/equipment charge		1 Carp	2	4	Job		188		188	310

03 11 13 – Structural Cast-In-Place Concrete Forming

03 11 13.85 Forms In Place, Walls		Crew	Daily Output	Labor-Hours	Unit	Material	2015 Bare Costs Labor	Equipment	Total	Total Incl O&P
0010	**FORMS IN PLACE, WALLS**									
0100	Box out for wall openings, to 16" thick, to 10 S.F.	C-2	24	2	Ea.	27	91.50		118.50	180
0150	Over 10 S.F. (use perimeter)	"	280	.171	L.F.	2.31	7.85		10.16	15.40
0250	Brick shelf, 4" w, add to wall forms, use wall area above shelf									
0260	1 use R031113-60	C-2	240	.200	SFCA	2.50	9.15		11.65	17.75
0300	2 use		275	.175		1.37	8		9.37	14.60
0350	4 use		300	.160		1	7.30		8.30	13.10
0500	Bulkhead, wood with keyway, 1 use, 2 piece		265	.181	L.F.	2.12	8.30		10.42	15.95
0600	Bulkhead forms with keyway, 1 piece expanded metal, 8" wall G	C-1	1000	.032		1.31	1.43		2.74	3.78
0610	10" wall G		800	.040		1.43	1.78		3.21	4.50
0620	12" wall G		525	.061		1.72	2.72		4.44	6.35
0700	Buttress, to 8' high, 1 use	C-2	350	.137	SFCA	4.28	6.25		10.53	15
0750	2 use		430	.112		2.35	5.10		7.45	10.95
0800	3 use		460	.104		1.72	4.77		6.49	9.70
0850	4 use		480	.100		1.41	4.57		5.98	9.05
1000	Corbel or haunch, to 12" wide, add to wall forms, 1 use		150	.320	L.F.	2.37	14.65		17.02	26.50
1050	2 use		170	.282		1.30	12.90		14.20	22.50
1100	3 use		175	.274		.95	12.55		13.50	21.50
1150	4 use		180	.267		.77	12.20		12.97	21
2000	Wall, job-built plywood, to 8' high, 1 use		370	.130	SFCA	2.74	5.95		8.69	12.70
2050	2 use		435	.110		1.75	5.05		6.80	10.15
2100	3 use		495	.097		1.27	4.43		5.70	8.65
2150	4 use		505	.095		1.03	4.35		5.38	8.25
2400	Over 8' to 16' high, 1 use		280	.171		3.03	7.85		10.88	16.20
2450	2 use		345	.139		1.34	6.35		7.69	11.90
2500	3 use		375	.128		.95	5.85		6.80	10.65
2550	4 use		395	.122		.78	5.55		6.33	9.95
2700	Over 16' high, 1 use		235	.204		2.71	9.35		12.06	18.30
2750	2 use		290	.166		1.49	7.55		9.04	14.05
2800	3 use		315	.152		1.09	6.95		8.04	12.60
2850	4 use		330	.145		.88	6.65		7.53	11.85
4000	Radial, smooth curved, job-built plywood, 1 use		245	.196		2.55	8.95		11.50	17.50
4050	2 use		300	.160		1.40	7.30		8.70	13.55
4100	3 use		325	.148		1.02	6.75		7.77	12.15
4150	4 use		335	.143		.83	6.55		7.38	11.65
4200	Below grade, job-built plywood, 1 use		225	.213		2.77	9.75		12.52	19.05
4210	2 use		225	.213		1.53	9.75		11.28	17.70
4220	3 use		225	.213		1.27	9.75		11.02	17.40
4230	4 use		225	.213		.90	9.75		10.65	17
4300	Curved, 2' chords, job-built plywood, to 8' high, 1 use		290	.166		2.16	7.55		9.71	14.75
4350	2 use		355	.135		1.19	6.20		7.39	11.45
4400	3 use		385	.125		.86	5.70		6.56	10.30
4450	4 use		400	.120		.70	5.50		6.20	9.75
4500	Over 8' to 16' high, 1 use		290	.166		.93	7.55		8.48	13.40
4525	2 use		355	.135		.51	6.20		6.71	10.70
4550	3 use		385	.125		.37	5.70		6.07	9.75
4575	4 use		400	.120		.31	5.50		5.81	9.35
4600	Retaining wall, battered, job-built plyw'd, to 8' high, 1 use		300	.160		2.03	7.30		9.33	14.25
4650	2 use		355	.135		1.12	6.20		7.32	11.40
4700	3 use		375	.128		.81	5.85		6.66	10.50
4750	4 use		390	.123		.66	5.65		6.31	9.95
4900	Over 8' to 16' high, 1 use		240	.200		2.22	9.15		11.37	17.45

03 11 Concrete Forming

03 11 13 – Structural Cast-In-Place Concrete Forming

03 11 13.85 Forms In Place, Walls		Crew	Daily Output	Labor-Hours	Unit	Material	2015 Bare Costs Labor	Equipment	Total	Total Incl O&P
4950	2 use	C-2	295	.163	SFCA	1.22	7.45		8.67	13.55
5000	3 use		305	.157		.89	7.20		8.09	12.80
5050	4 use	↓	320	.150		.72	6.85		7.57	12.05
5500	For gang wall forming, 192 S.F. sections, deduct					10%	10%			
5550	384 S.F. sections, deduct					20%	20%			
7500	Lintel or sill forms, 1 use	1 Carp	30	.267		3.26	12.50		15.76	24
7520	2 use		34	.235		1.79	11.05		12.84	20
7540	3 use		36	.222		1.30	10.45		11.75	18.55
7560	4 use	↓	37	.216	↓	1.06	10.15		11.21	17.80
7800	Modular prefabricated plywood, based on 20 uses of purchased									
7820	forms, and 4 uses of bracing lumber									
7860	To 8′ high	C-2	800	.060	SFCA	.94	2.74		3.68	5.55
8060	Over 8′ to 16′ high		600	.080		.99	3.66		4.65	7.10
8600	Pilasters, 1 use		270	.178		3.19	8.15		11.34	16.80
8620	2 use		330	.145		1.75	6.65		8.40	12.85
8640	3 use		370	.130		1.27	5.95		7.22	11.10
8660	4 use	↓	385	.125	↓	1.04	5.70		6.74	10.50
9010	Steel framed plywood, based on 50 uses of purchased									
9020	forms, and 4 uses of bracing lumber									
9060	To 8′ high	C-2	600	.080	SFCA	.74	3.66		4.40	6.80
9260	Over 8′ to 16′ high		450	.107		.74	4.88		5.62	8.80
9460	Over 16′ to 20′ high	↓	400	.120	↓	.74	5.50		6.24	9.80
9475	For elevated walls, add						10%			
9900	Minimum labor/equipment charge	2 Carp	2	8	Job		375		375	615

03 11 16 – Architectural Cast-in Place Concrete Forming

03 11 16.13 Concrete Form Liners		Crew	Daily Output	Labor-Hours	Unit	Material	Labor	Equipment	Total	Total Incl O&P
0010	**CONCRETE FORM LINERS**									
5750	Liners for forms (add to wall forms), ABS plastic									
5800	Aged wood, 4″ wide, 1 use	1 Carp	256	.031	SFCA	3.30	1.47		4.77	6.05
5820	2 use		256	.031		1.82	1.47		3.29	4.40
5830	3 use		256	.031		1.32	1.47		2.79	3.85
5840	4 use		256	.031		1.07	1.47		2.54	3.58
5900	Fractured rope rib, 1 use		192	.042		4.79	1.96		6.75	8.45
5925	2 use		192	.042		2.63	1.96		4.59	6.10
5950	3 use		192	.042		1.92	1.96		3.88	5.30
6000	4 use		192	.042		1.56	1.96		3.52	4.92
6100	Ribbed, 3/4″ deep x 1-1/2″ O.C., 1 use		224	.036		4.79	1.68		6.47	8
6125	2 use		224	.036		2.63	1.68		4.31	5.65
6150	3 use		224	.036		1.92	1.68		3.60	4.86
6200	4 use		224	.036		1.56	1.68		3.24	4.46
6300	Rustic brick pattern, 1 use		224	.036		3.30	1.68		4.98	6.40
6325	2 use		224	.036		1.82	1.68		3.50	4.75
6350	3 use		224	.036		1.32	1.68		3	4.20
6400	4 use		224	.036		1.07	1.68		2.75	3.93
6500	3/8″ striated, random, 1 use		224	.036		3.30	1.68		4.98	6.40
6525	2 use		224	.036		1.82	1.68		3.50	4.75
6550	3 use		224	.036		1.32	1.68		3	4.20
6600	4 use		224	.036		1.07	1.68		2.75	3.93
6850	Random vertical rustication, 1 use		384	.021		6.25	.98		7.23	8.50
6900	2 use		384	.021		3.45	.98		4.43	5.40
6925	3 use		384	.021		2.51	.98		3.49	4.36
6950	4 use	↓	384	.021	↓	2.04	.98		3.02	3.84

03 11 Concrete Forming

03 11 16 – Architectural Cast-in Place Concrete Forming

03 11 16.13 Concrete Form Liners		Crew	Daily Output	Labor-Hours	Unit	2015 Bare Costs Material	Labor	Equipment	Total	Total Incl O&P
7050	Wood, beveled edge, 3/4" deep, 1 use	1 Carp	384	.021	L.F.	.16	.98		1.14	1.78
7100	1" deep, 1 use		384	.021	"	.33	.98		1.31	1.96
7200	4" wide aged cedar, 1 use		256	.031	SFCA	3.30	1.47		4.77	6.05
7300	4" variable depth rough cedar		224	.036	"	4.79	1.68		6.47	8

03 11 19 – Insulating Concrete Forming

03 11 19.10 Insulating Forms, Left In Place		Crew	Daily Output	Labor-Hours	Unit	Material	Labor	Equipment	Total	Total Incl O&P
0010	**INSULATING FORMS, LEFT IN PLACE**									
0020	S.F. is for exterior face, but includes forms for both faces (total R22)									
2000	4" wall, straight block, 16" x 48" (5.33 S.F.) G	2 Carp	90	.178	Ea.	19.60	8.35		27.95	35
2010	90 corner block, exterior 16" x 38" x 22" (6.67 S.F.) G		75	.213		23	10		33	42
2020	45 corner block, exterior 16" x 34" x 18" (5.78 S.F.) G		75	.213		23	10		33	42
2100	6" wall, straight block, 16" x 48" (5.33 S.F.) G		90	.178		20	8.35		28.35	35.50
2110	90 corner block, exterior 16" x 32" x 24" (6.22 S.F.) G		75	.213		25	10		35	44
2120	45 corner block, exterior 16" x 26" x 18" (4.89 S.F.) G		75	.213		22.50	10		32.50	41
2130	Brick ledge block, 16" x 48" (5.33 S.F.) G		80	.200		25	9.40		34.40	43
2140	Taper top block, 16" x 48" (5.33 S.F.) G		80	.200		23.50	9.40		32.90	41.50
2200	8" wall, straight block, 16" x 48" (5.33 S.F.) G		90	.178		21.50	8.35		29.85	37
2210	90 corner block, exterior 16" x 34" x 26" (6.67 S.F.) G		75	.213		27.50	10		37.50	47
2220	45 corner block, exterior 16" x 28" x 20" (5.33 S.F.) G		75	.213		23.50	10		33.50	42.50
2230	Brick ledge block, 16" x 48" (5.33 S.F.) G		80	.200		26.50	9.40		35.90	44.50
2240	Taper top block, 16" x 48" (5.33 S.F.) G		80	.200		24.50	9.40		33.90	42.50

03 11 19.60 Roof Deck Form Boards		Crew	Daily Output	Labor-Hours	Unit	Material	Labor	Equipment	Total	Total Incl O&P
0010	**ROOF DECK FORM BOARDS** R051223-50									
0050	Includes bulb tee sub-purlins @ 32-5/8" O.C.									
0070	Non-asbestos fiber cement, 5/16" thick	C-13	2950	.008	S.F.	2.73	.41	.05	3.19	3.78
0100	Fiberglass, 1" thick		2700	.009		3.46	.45	.05	3.96	4.65
0500	Wood fiber, 1" thick G		2700	.009		2.24	.45	.05	2.74	3.31

03 11 23 – Permanent Stair Forming

03 11 23.75 Forms In Place, Stairs		Crew	Daily Output	Labor-Hours	Unit	Material	Labor	Equipment	Total	Total Incl O&P
0010	**FORMS IN PLACE, STAIRS**									
0015	(Slant length x width), 1 use	C-2	165	.291	S.F.	5.90	13.30		19.20	28.50
0050	2 use R031113-60		170	.282		3.35	12.90		16.25	24.50
0100	3 use		180	.267		2.50	12.20		14.70	23
0150	4 use		190	.253		2.08	11.55		13.63	21
1000	Alternate pricing method (1.0 L.F./S.F.), 1 use		100	.480	LF Rsr	5.90	22		27.90	42.50
1050	2 use		105	.457		3.35	21		24.35	38
1100	3 use		110	.436		2.50	19.95		22.45	35.50
1150	4 use		115	.417		2.08	19.10		21.18	34
2000	Stairs, cast on sloping ground (length x width), 1 use		220	.218	S.F.	2.36	10		12.36	18.95
2025	2 use		232	.207		1.30	9.45		10.75	16.90
2050	3 use		244	.197		.94	9		9.94	15.80
2100	4 use		256	.188		.77	8.55		9.32	14.90

03 15 05 – Concrete Forming Accessories

03 15 05.12 Chamfer Strips

03 15 05.12 Chamfer Strips		Crew	Daily Output	Labor-Hours	Unit	Material	Labor	Equipment	Total	Total Incl O&P
						2015 Bare Costs				
0010	**CHAMFER STRIPS**									
2000	Polyvinyl chloride, 1/2" wide with leg	1 Carp	535	.015	L.F.	.57	.70		1.27	1.78
2200	3/4" wide with leg		525	.015		.68	.72		1.40	1.92
2400	1" radius with leg		515	.016		.64	.73		1.37	1.90
2800	2" radius with leg		500	.016		1.22	.75		1.97	2.57
5000	Wood, 1/2" wide		535	.015		.15	.70		.85	1.32
5200	3/4" wide		525	.015		.16	.72		.88	1.35
5400	1" wide	↓	515	.016	↓	.33	.73		1.06	1.56

03 15 05.15 Column Form Accessories

Line	Description	Crew	Daily Output	Labor-Hours	Unit	Material	Labor	Equipment	Total	Total Incl O&P
0010	**COLUMN FORM ACCESSORIES**									
1000	Column clamps, adjustable to 24" x 24", buy G				Set	145			145	160
1100	Rent per month G					14.55			14.55	16
1300	For sizes to 30" x 30", buy G					170			170	187
1400	Rent per month G					17			17	18.70
1600	For sizes to 36" x 36", buy G					212			212	233
1700	Rent per month G					21			21	23.50
2000	Bar type with wedges, 36" x 36", buy G					112			112	123
2100	Rent per month G					7.85			7.85	8.60
2300	48" x 48", buy G					152			152	167
2400	Rent per month G					10.65			10.65	11.70
3000	Scissor type with wedges, 36" x 36", buy G					95			95	105
3100	Rent per month G					9.50			9.50	10.45
3300	60" x 60", buy G					135			135	149
3400	Rent per month G					13.50			13.50	14.85
4000	Friction collars 2'-6" diam., buy G					2,575			2,575	2,825
4100	Rent per month G					180			180	198
4300	4'-0" diam., buy G					3,150			3,150	3,450
4400	Rent per month G				↓	220			220	242

03 15 05.30 Hangers

Line	Description	Crew	Daily Output	Labor-Hours	Unit	Material	Labor	Equipment	Total	Total Incl O&P
0010	**HANGERS**									
0020	Slab and beam form									
0500	Banding iron									
0550	3/4" x 22 ga., 14 L.F. per lb. or 1/2" x 14 ga., 7 L.F. per lb. G				Lb.	1.34			1.34	1.47
1000	Fascia ties, coil type, to 24" long G				C	425			425	465
1500	Frame ties to 8-1/8" G					525			525	575
1550	8-1/8" to 10-1/8" G					550			550	605
5000	Snap tie hanger, to 30" overall length, 3000# G					425			425	465
5050	To 36" overall length G					480			480	525
5100	To 48" overall length G				↓	585			585	645
5500	Steel beam hanger									
5600	Flange to 8-1/8" G				C	525			525	575
5650	8-1/8" to 10-1/8" G				"	550			550	605
5900	Coil threaded rods, continuous, 1/2" diameter G				L.F.	1.36			1.36	1.50
6000	Tie hangers to 30" overall length, 4000# G				C	480			480	530
6100	To 36" overall length G					535			535	585
6500	Tie back hanger, up to 12-1/8" flange G				↓	1,400			1,400	1,550
8500	Wire, black annealed, 15 gage G				Cwt.	151			151	167
8600	16 ga				"	203			203	223

03 15 Concrete Accessories

03 15 05 – Concrete Forming Accessories

03 15 05.70 Shores			Crew	Daily Output	Labor-Hours	Unit	2015 Bare Costs Material	Labor	Equipment	Total	Total Incl O&P
0010	**SHORES**										
0020	Erect and strip, by hand, horizontal members										
0500	Aluminum joists and stringers	G	2 Carp	60	.267	Ea.		12.50		12.50	20.50
0600	Steel, adjustable beams	G		45	.356			16.70		16.70	27.50
0700	Wood joists			50	.320			15		15	24.50
0800	Wood stringers			30	.533			25		25	41
1000	Vertical members to 10′ high	G		55	.291			13.65		13.65	22.50
1050	To 13′ high	G		50	.320			15		15	24.50
1100	To 16′ high	G		45	.356			16.70		16.70	27.50
1500	Reshoring	G		1400	.011	S.F.	.58	.54		1.12	1.52
1600	Flying truss system	G	C-17D	9600	.009	SFCA		.43	.08	.51	.79
1760	Horizontal, aluminum joists, 6-1/4″ high x 5′ to 21′ span, buy	G				L.F.	15.45			15.45	17
1770	Beams, 7-1/4″ high x 4′ to 30′ span	G				″	18.10			18.10	19.90
1810	Horizontal, steel beam, W8x10, 7′ span, buy	G				Ea.	57			57	63
1830	10′ span	G					67.50			67.50	74.50
1920	15′ span	G					117			117	128
1940	20′ span	G					164			164	180
1970	Steel stringer, W8x10, 4′ to 16′ span, buy	G				L.F.	6.80			6.80	7.45
3000	Rent for job duration, aluminum joist @ 2′ O.C., per mo.	G				SF Flr.	.39			.39	.43
3050	Steel W8x10	G					.17			.17	.19
3060	Steel adjustable	G					.17			.17	.19
3500	#1 post shore, steel, 5′-7″ to 9′-6″ high, 10000# cap., buy	G				Ea.	148			148	163
3550	#2 post shore, 7′-3″ to 12′-10″ high, 7800# capacity	G					171			171	188
3600	#3 post shore, 8′-10″ to 16′-1″ high, 3800# capacity	G					187			187	206
5010	Frame shoring systems, steel, 12000#/leg, buy										
5040	Frame, 2′ wide x 6′ high	G				Ea.	93.50			93.50	103
5250	X-brace	G					15.50			15.50	17.05
5550	Base plate	G					15.50			15.50	17.05
5600	Screw jack	G					33.50			33.50	36.50
5650	U-head, 8″ x 8″	G					19.10			19.10	21

03 15 05.75 Sleeves and Chases			Crew	Daily Output	Labor-Hours	Unit	Material	Labor	Equipment	Total	Total Incl O&P
0010	**SLEEVES AND CHASES**										
0100	Plastic, 1 use, 12″ long, 2″ diameter		1 Carp	100	.080	Ea.	2.15	3.76		5.91	8.50
0150	4″ diameter			90	.089		6.05	4.17		10.22	13.50
0200	6″ diameter			75	.107		10.65	5		15.65	19.90
0250	12″ diameter			60	.133		31.50	6.25		37.75	45

03 15 05.80 Snap Ties			Crew	Daily Output	Labor-Hours	Unit	Material	Labor	Equipment	Total	Total Incl O&P
0010	**SNAP TIES**, 8-1/4″ L&W (Lumber and wedge)										
0100	2250 lb., w/flat washer, 8″ wall	G				C	90			90	99
0150	10″ wall	G					131			131	144
0200	12″ wall	G					136			136	150
0250	16″ wall	G					150			150	165
0300	18″ wall	G					156			156	172
0500	With plastic cone, 8″ wall	G					80			80	88
0550	10″ wall	G					83			83	91.50
0600	12″ wall	G					89			89	98
0650	16″ wall	G					98			98	108
0700	18″ wall	G					101			101	111
1000	3350 lb., w/flat washer, 8″ wall	G					163			163	179
1150	12″ wall	G					183			183	201
1200	16″ wall	G					210			210	231
1250	18″ wall	G					218			218	240

15 Concrete Accessories

15 05 – Concrete Forming Accessories

03 15 05.80 Snap Ties

03 15 05.80 Snap Ties			Crew	Daily Output	Labor-Hours	Unit	2015 Bare Costs Material	Labor	Equipment	Total	Total Incl O&P
1500	With plastic cone, 8″ wall	G				C	132			132	145
1600	12″ wall	G					149			149	164
1650	16″ wall	G					170			170	187
1700	18″ wall	G				↓	176			176	194

03 15 05.85 Stair Tread Inserts

03 15 05.85 Stair Tread Inserts			Crew	Daily Output	Labor-Hours	Unit	Material	Labor	Equipment	Total	Total Incl O&P
0010	**STAIR TREAD INSERTS**										
0105	Cast nosing insert, abrasive surface, pre-drilled, includes screws										
0110	Aluminum, 3″ wide x 3′ long		1 Cefi	32	.250	Ea.	55	11.25		66.25	78.50
0120	4′ long			31	.258		70	11.60		81.60	95.50
0130	5′ long		↓	30	.267	↓	87	12		99	114
0135	Extruded nosing insert, black abrasive strips, continuous anchor										
0140	Aluminum, 3″ wide x 3′ long		1 Cefi	64	.125	Ea.	35.50	5.65		41.15	48
0150	4′ long			60	.133		55	6		61	70
0160	5′ long		↓	56	.143	↓	62	6.45		68.45	78
0165	Extruded nosing insert, black abrasive strips, pre-drilled, incl. screws										
0170	Aluminum, 3″ wide x 3′ long		1 Cefi	32	.250	Ea.	49	11.25		60.25	72
0180	4′ long			31	.258		63.50	11.60		75.10	88.50
0190	5′ long		↓	30	.267	↓	74.50	12		86.50	101
9000	Minimum labor/equipment charge		1 Carp	4	2	Job		94		94	154

03 15 05.95 Wall and Foundation Form Accessories

03 15 05.95 Wall and Foundation Form Accessories			Crew	Daily Output	Labor-Hours	Unit	Material	Labor	Equipment	Total	Total Incl O&P
0010	**WALL AND FOUNDATION FORM ACCESSORIES**										
2000	Footings, turnbuckle form aligner	G				Ea.	16.20			16.20	17.85
2050	Spreaders for footer, adjustable	G				″	25			25	27.50
3000	Form oil, up to 1200 S.F. per gallon coverage					Gal.	13.55			13.55	14.90
3050	Up to 800 S.F. per gallon					″	20.50			20.50	22.50
3500	Form patches, 1-3/4″ diameter					C	27			27	29.50
3550	2-3/4″ diameter					″	47			47	51.50
4000	Nail stakes, 3/4″ diameter, 18″ long	G				Ea.	3.30			3.30	3.63
4050	24″ long	G					4.26			4.26	4.69
4200	30″ long	G					5.45			5.45	6
4250	36″ long	G				↓	6.60			6.60	7.25

03 15 13 – Waterstops

03 15 13.50 Waterstops

03 15 13.50 Waterstops		Crew	Daily Output	Labor-Hours	Unit	Material	Labor	Equipment	Total	Total Incl O&P
0010	**WATERSTOPS**, PVC and Rubber									
0020	PVC, ribbed 3/16″ thick, 4″ wide	1 Carp	155	.052	L.F.	1.35	2.42		3.77	5.45
0050	6″ wide		145	.055		2.27	2.59		4.86	6.75
0500	With center bulb, 6″ wide, 3/16″ thick		135	.059		1.99	2.78		4.77	6.75
0550	3/8″ thick		130	.062		3.67	2.89		6.56	8.80
0600	9″ wide x 3/8″ thick		125	.064		6.05	3		9.05	11.55
0800	Dumbbell type, 6″ wide, 3/16″ thick		150	.053		2.02	2.50		4.52	6.30
0850	3/8″ thick		145	.055		3.79	2.59		6.38	8.40
1000	9″ wide, 3/8″ thick, plain		130	.062		5.65	2.89		8.54	10.95
1050	Center bulb		130	.062		9.15	2.89		12.04	14.85
1250	Ribbed type, split, 3/16″ thick, 6″ wide		145	.055		1.92	2.59		4.51	6.35
1300	3/8″ thick		130	.062		4.38	2.89		7.27	9.55
2000	Rubber, flat dumbbell, 3/8″ thick, 6″ wide		145	.055		10.60	2.59		13.19	15.90
2050	9″ wide		135	.059		16.15	2.78		18.93	22.50
2500	Flat dumbbell split, 3/8″ thick, 6″ wide		145	.055		1.92	2.59		4.51	6.35
2550	9″ wide		135	.059		4.38	2.78		7.16	9.40
3000	Center bulb, 1/4″ thick, 6″ wide		145	.055		10.45	2.59		13.04	15.75
3050	9″ wide		135	.059		23	2.78		25.78	30
3500	Center bulb split, 3/8″ thick, 6″ wide	↓	145	.055	↓	14.80	2.59		17.39	20.50

03 15 Concrete Accessories

03 15 13 – Waterstops

03 15 13.50 Waterstops		Crew	Daily Output	Labor-Hours	Unit	2015 Bare Costs Material	Labor	Equipment	Total	Total Incl O&P
3550	9″ wide	1 Carp	135	.059	L.F.	25.50	2.78		28.28	32.50
5000	Waterstop fittings, rubber, flat									
5010	Dumbbell or center bulb, 3/8″ thick,									
5200	Field union, 6″ wide	1 Carp	50	.160	Ea.	35.50	7.50		43	52
5250	9″ wide		50	.160		47.50	7.50		55	65
5500	Flat cross, 6″ wide		30	.267		45	12.50		57.50	70
5550	9″ wide		30	.267		64.50	12.50		77	91.50
6000	Flat tee, 6″ wide		30	.267		43.50	12.50		56	68.50
6050	9″ wide		30	.267		59	12.50		71.50	85.50
6500	Flat ell, 6″ wide		40	.200		42.50	9.40		51.90	62
6550	9″ wide		40	.200		54.50	9.40		63.90	75.50
7000	Vertical tee, 6″ wide		25	.320		31.50	15		46.50	59.50
7050	9″ wide		25	.320		45	15		60	74
7500	Vertical ell, 6″ wide		35	.229		29.50	10.75		40.25	50
7550	9″ wide		35	.229		38	10.75		48.75	59

03 15 16 – Concrete Construction Joints

03 15 16.20 Control Joints, Saw Cut

		Crew	Daily Output	Labor-Hours	Unit	Material	Labor	Equipment	Total	Total Incl O&P
0010	**CONTROL JOINTS, SAW CUT**									
0100	Sawcut control joints in green concrete									
0120	1″ depth	C-27	2000	.008	L.F.	.04	.36	.08	.48	.70
0140	1-1/2″ depth		1800	.009		.06	.40	.09	.55	.79
0160	2″ depth		1600	.010		.08	.45	.10	.63	.92
0180	Sawcut joint reservoir in cured concrete									
0182	3/8″ wide x 3/4″ deep, with single saw blade	C-27	1000	.016	L.F.	.06	.72	.17	.95	1.38
0184	1/2″ wide x 1″ deep, with double saw blades		900	.018		.11	.80	.19	1.10	1.60
0186	3/4″ wide x 1-1/2″ deep, with double saw blades		800	.020		.23	.90	.21	1.34	1.91
0190	Water blast joint to wash away laitance, 2 passes	C-29	2500	.003			.12	.03	.15	.23
0200	Air blast joint to blow out debris and air dry, 2 passes	C-28	2000	.004			.18	.01	.19	.29
0300	For backer rod, see Section 07 91 23.10									
0340	For joint sealant, see Sections 03 15 16.30 or 07 92 13.20									
0900	For replacement of joint sealant, see Section 07 01 90.81									

03 15 16.30 Expansion Joints

			Crew	Daily Output	Labor-Hours	Unit	Material	Labor	Equipment	Total	Total Incl O&P
0010	**EXPANSION JOINTS**										
0020	Keyed, cold, 24 ga., incl. stakes, 3-1/2″ high	G	1 Carp	200	.040	L.F.	.81	1.88		2.69	3.97
0050	4-1/2″ high	G		200	.040		.92	1.88		2.80	4.09
0100	5-1/2″ high	G		195	.041		1.12	1.93		3.05	4.39
0150	7-1/2″ high	G		190	.042		1.31	1.98		3.29	4.68
0160	9-1/2″ high	G		185	.043		1.43	2.03		3.46	4.90
0300	Poured asphalt, plain, 1/2″ x 1″		1 Clab	450	.018		.69	.67		1.36	1.86
0350	1″ x 2″			400	.020		2.76	.75		3.51	4.27
0500	Neoprene, liquid, cold applied, 1/2″ x 1″			450	.018		2.88	.67		3.55	4.27
0550	1″ x 2″			400	.020		11.50	.75		12.25	13.90
0700	Polyurethane, poured, 2 part, 1/2″ x 1″			400	.020		1.28	.75		2.03	2.64
0750	1″ x 2″			350	.023		5.10	.86		5.96	7.05
0900	Rubberized asphalt, hot or cold applied, 1/2″ x 1″			450	.018		.40	.67		1.07	1.54
0950	1″ x 2″			400	.020		1.60	.75		2.35	2.99
1100	Hot applied, fuel resistant, 1/2″ x 1″			450	.018		.60	.67		1.27	1.76
1150	1″ x 2″			400	.020		2.40	.75		3.15	3.87
2000	Premolded, bituminous fiber, 1/2″ x 6″		1 Carp	375	.021		.44	1		1.44	2.12
2050	1″ x 12″			300	.027		1.99	1.25		3.24	4.24
2140	Concrete expansion joint, recycled paper and fiber, 1/2″ x 6″	G		390	.021		.42	.96		1.38	2.04
2150	1/2″ x 12″	G		360	.022		.83	1.04		1.87	2.63

03 15 Concrete Accessories

03 15 16 – Concrete Construction Joints

03 15 16.30 Expansion Joints		Crew	Daily Output	Labor-Hours	Unit	Material	2015 Bare Costs Labor	Equipment	Total	Total Incl O&P
2250	Cork with resin binder, 1/2" x 6"	1 Carp	375	.021	L.F.	1.11	1		2.11	2.86
2300	1" x 12"		300	.027		3.11	1.25		4.36	5.45
2500	Neoprene sponge, closed cell, 1/2" x 6"		375	.021		2.37	1		3.37	4.25
2550	1" x 12"		300	.027		8.25	1.25		9.50	11.15
2750	Polyethylene foam, 1/2" x 6"		375	.021		.68	1		1.68	2.39
2800	1" x 12"		300	.027		2.19	1.25		3.44	4.46
3000	Polyethylene backer rod, 3/8" diameter		460	.017		.03	.82		.85	1.37
3050	3/4" diameter		460	.017		.06	.82		.88	1.40
3100	1" diameter		460	.017		.09	.82		.91	1.44
3500	Polyurethane foam, with polybutylene, 1/2" x 1/2"		475	.017		1.14	.79		1.93	2.55
3550	1" x 1"		450	.018		2.88	.83		3.71	4.54
3750	Polyurethane foam, regular, closed cell, 1/2" x 6"		375	.021		.84	1		1.84	2.56
3800	1" x 12"		300	.027		3	1.25		4.25	5.35
4000	Polyvinyl chloride foam, closed cell, 1/2" x 6"		375	.021		2.28	1		3.28	4.15
4050	1" x 12"		300	.027		7.85	1.25		9.10	10.70
4250	Rubber, gray sponge, 1/2" x 6"		375	.021		1.92	1		2.92	3.75
4300	1" x 12"		300	.027		6.90	1.25		8.15	9.65
4400	Redwood heartwood, 1" x 4"		400	.020		1.16	.94		2.10	2.82
4450	1" x 6"		375	.021		1.75	1		2.75	3.57
5000	For installation in walls, add						75%			
5250	For installation in boxouts, add						25%			

03 15 19 – Cast-In Concrete Anchors

03 15 19.05 Anchor Bolt Accessories

		Crew	Daily Output	Labor-Hours	Unit	Material	Labor	Equipment	Total	Total Incl O&P
0010	**ANCHOR BOLT ACCESSORIES**									
0015	For anchor bolts set in fresh concrete, see Section 03 15 19.10									
8150	Anchor bolt sleeve, plastic, 1" diam. bolts	1 Carp	60	.133	Ea.	8.20	6.25		14.45	19.30
8500	1-1/2" diameter		28	.286		14.60	13.40		28	38
8600	2" diameter		24	.333		18.60	15.65		34.25	46
8650	3" diameter		20	.400		34	18.80		52.80	68.50

03 15 19.10 Anchor Bolts

		Crew	Daily Output	Labor-Hours	Unit	Material	Labor	Equipment	Total	Total Incl O&P
0010	**ANCHOR BOLTS**									
0015	Made from recycled materials									
0025	Single bolts installed in fresh concrete, no templates									
0030	Hooked w/nut and washer, 1/2" diameter, 8" long G	1 Carp	132	.061	Ea.	1.33	2.85		4.18	6.15
0040	12" long G		131	.061		1.48	2.87		4.35	6.35
0050	5/8" diameter, 8" long G		129	.062		2.98	2.91		5.89	8.05
0060	12" long G		127	.063		3.67	2.96		6.63	8.90
0070	3/4" diameter, 8" long G		127	.063		3.67	2.96		6.63	8.90
0080	12" long G		125	.064		4.59	3		7.59	9.95
0090	2-bolt pattern, including job-built 2-hole template, per set									
0100	J-type, incl. hex nut & washer, 1/2" diameter x 6" long G	1 Carp	21	.381	Set	5.20	17.90		23.10	35.50
0110	12" long G		21	.381		5.80	17.90		23.70	36
0120	18" long G		21	.381		6.70	17.90		24.60	37
0130	3/4" diameter x 8" long G		20	.400		10.20	18.80		29	42
0140	12" long G		20	.400		12	18.80		30.80	44
0150	18" long G		20	.400		14.75	18.80		33.55	47.50
0160	1" diameter x 12" long G		19	.421		20.50	19.75		40.25	55
0170	18" long G		19	.421		24	19.75		43.75	59
0180	24" long G		19	.421		28.50	19.75		48.25	64
0190	36" long G		18	.444		38	21		59	76
0200	1-1/2" diameter x 18" long G		17	.471		37.50	22		59.50	77.50
0210	24" long G		16	.500		44	23.50		67.50	87

03 15 Concrete Accessories

03 15 19 – Cast-In Concrete Anchors

03 15 19.10 Anchor Bolts			Crew	Daily Output	Labor-Hours	Unit	2015 Bare Costs Material	Labor	Equipment	Total	Total Incl O&P
0300	L-type, incl. hex nut & washer, 3/4" diameter x 12" long	G	1 Carp	20	.400	Set	12.90	18.80		31.70	45
0310	18" long	G		20	.400		15.75	18.80		34.55	48.50
0320	24" long	G		20	.400		18.60	18.80		37.40	51.50
0330	30" long	G		20	.400		23	18.80		41.80	56
0340	36" long	G		20	.400		25.50	18.80		44.30	59.50
0350	1" diameter x 12" long	G		19	.421		19.55	19.75		39.30	54
0360	18" long	G		19	.421		23.50	19.75		43.25	58.50
0370	24" long	G		19	.421		28.50	19.75		48.25	64
0380	30" long	G		19	.421		33	19.75		52.75	69
0390	36" long	G		18	.444		37.50	21		58.50	75.50
0400	42" long	G		18	.444		45	21		66	83.50
0410	48" long	G		18	.444		50	21		71	89
0420	1-1/4" diameter x 18" long	G		18	.444		30	21		51	67
0430	24" long	G		18	.444		35	21		56	72.50
0440	30" long	G		17	.471		40	22		62	80
0450	36" long	G		17	.471		45	22		67	85.50
0460	42" long	G	2 Carp	32	.500		50.50	23.50		74	94
0470	48" long	G		32	.500		57	23.50		80.50	102
0480	54" long	G		31	.516		67	24		91	113
0490	60" long	G		31	.516		73	24		97	120
0500	1-1/2" diameter x 18" long	G		33	.485		51	23		74	93.50
0510	24" long	G		32	.500		59	23.50		82.50	103
0520	30" long	G		31	.516		66	24		90	112
0530	36" long	G		30	.533		75.50	25		100.50	124
0540	42" long	G		30	.533		85.50	25		110.50	136
0550	48" long	G		29	.552		96	26		122	148
0560	54" long	G		28	.571		116	27		143	172
0570	60" long	G		28	.571		127	27		154	184
0580	1-3/4" diameter x 18" long	G		31	.516		74.50	24		98.50	122
0590	24" long	G		30	.533		87	25		112	137
0600	30" long	G		29	.552		101	26		127	154
0610	36" long	G		28	.571		114	27		141	170
0620	42" long	G		27	.593		128	28		156	187
0630	48" long	G		26	.615		141	29		170	203
0640	54" long	G		26	.615		174	29		203	239
0650	60" long	G		25	.640		187	30		217	256
0660	2" diameter x 24" long	G		27	.593		111	28		139	169
0670	30" long	G		27	.593		125	28		153	184
0680	36" long	G		26	.615		137	29		166	199
0690	42" long	G		25	.640		153	30		183	218
0700	48" long	G		24	.667		175	31.50		206.50	245
0710	54" long	G		23	.696		208	32.50		240.50	283
0720	60" long	G		23	.696		223	32.50		255.50	300
0730	66" long	G		22	.727		239	34		273	320
0740	72" long	G		21	.762		261	36		297	345
1000	4-bolt pattern, including job-built 4-hole template, per set										
1100	J-type, incl. hex nut & washer, 1/2" diameter x 6" long	G	1 Carp	19	.421	Set	7.55	19.75		27.30	41
1110	12" long	G		19	.421		8.75	19.75		28.50	42
1120	18" long	G		18	.444		10.55	21		31.55	45.50
1130	3/4" diameter x 8" long	G		17	.471		17.55	22		39.55	55.50
1140	12" long	G		17	.471		21	22		43	59.50
1150	18" long	G		17	.471		26.50	22		48.50	65.50
1160	1" diameter x 12" long	G		16	.500		38	23.50		61.50	80

03 15 Concrete Accessories

03 15 19 – Cast-In Concrete Anchors

03 15 19.10 Anchor Bolts			Crew	Daily Output	Labor-Hours	Unit	2015 Bare Costs Material	Labor	Equipment	Total	Total Incl O&P
1170	18″ long	G	1 Carp	15	.533	Set	45	25		70	90.50
1180	24″ long	G		15	.533		54	25		79	101
1190	36″ long	G		15	.533		73	25		98	122
1200	1-1/2″ diameter x 18″ long	G		13	.615		72	29		101	127
1210	24″ long	G		12	.667		85.50	31.50		117	146
1300	L-type, incl. hex nut & washer, 3/4″ diameter x 12″ long	G		17	.471		23	22		45	61.50
1310	18″ long	G		17	.471		28.50	22		50.50	67.50
1320	24″ long	G		17	.471		34.50	22		56.50	74
1330	30″ long	G		16	.500		43	23.50		66.50	85.50
1340	36″ long	G		16	.500		48.50	23.50		72	92
1350	1″ diameter x 12″ long	G		16	.500		36	23.50		59.50	78.50
1360	18″ long	G		15	.533		44.50	25		69.50	90
1370	24″ long	G		15	.533		54	25		79	101
1380	30″ long	G		15	.533		63.50	25		88.50	111
1390	36″ long	G		15	.533		72	25		97	121
1400	42″ long	G		14	.571		87	27		114	140
1410	48″ long	G		14	.571		97.50	27		124.50	151
1420	1-1/4″ diameter x 18″ long	G		14	.571		57	27		84	107
1430	24″ long	G		14	.571		67	27		94	118
1440	30″ long	G		13	.615		77	29		106	132
1450	36″ long	G	▼	13	.615		87	29		116	144
1460	42″ long	G	2 Carp	25	.640		98	30		128	158
1470	48″ long	G		24	.667		112	31.50		143.50	175
1480	54″ long	G		23	.696		131	32.50		163.50	198
1490	60″ long	G		23	.696		144	32.50		176.50	212
1500	1-1/2″ diameter x 18″ long	G		25	.640		99	30		129	159
1510	24″ long	G		24	.667		115	31.50		146.50	178
1520	30″ long	G		23	.696		129	32.50		161.50	196
1530	36″ long	G		22	.727		148	34		182	219
1540	42″ long	G		22	.727		169	34		203	241
1550	48″ long	G		21	.762		189	36		225	267
1560	54″ long	G		20	.800		230	37.50		267.50	315
1570	60″ long	G		20	.800		251	37.50		288.50	340
1580	1-3/4″ diameter x 18″ long	G		22	.727		146	34		180	217
1590	24″ long	G		21	.762		171	36		207	247
1600	30″ long	G		21	.762		199	36		235	277
1610	36″ long	G		20	.800		226	37.50		263.50	310
1620	42″ long	G		19	.842		254	39.50		293.50	345
1630	48″ long	G		18	.889		278	41.50		319.50	375
1640	54″ long	G		18	.889		345	41.50		386.50	450
1650	60″ long	G		17	.941		370	44		414	485
1660	2″ diameter x 24″ long	G		19	.842		220	39.50		259.50	305
1670	30″ long	G		18	.889		247	41.50		288.50	340
1680	36″ long	G		18	.889		272	41.50		313.50	370
1690	42″ long	G		17	.941		305	44		349	410
1700	48″ long	G		16	1		345	47		392	455
1710	54″ long	G		15	1.067		415	50		465	535
1720	60″ long	G		15	1.067		445	50		495	570
1730	66″ long	G		14	1.143		475	53.50		528.50	610
1740	72″ long	G	▼	14	1.143	▼	520	53.50		573.50	660
1990	For galvanized, add					Ea.	75%				

03 15 19 – Cast-In Concrete Anchors

03 15 19.20 Dovetail Anchor System

	03 15 19.20 Dovetail Anchor System		Crew	Daily Output	Labor-Hours	Unit	2015 Bare Costs Material	Labor	Equipment	Total	Total Incl O&P
0010	**DOVETAIL ANCHOR SYSTEM**										
0500	Dovetail anchor slot, galvanized, foam-filled, 26 ga.	G	1 Carp	425	.019	L.F.	.88	.88		1.76	2.42
0600	24 ga.	G		400	.020		1.64	.94		2.58	3.34
0625	22 ga.	G		400	.020		1.92	.94		2.86	3.65
0900	Stainless steel, foam-filled, 26 ga.	G		375	.021		1.41	1		2.41	3.19
1200	Dovetail brick anchor, corrugated, galvanized, 3-1/2" long, 16 ga.	G	1 Bric	10.50	.762	C	30.50	35		65.50	90.50
1300	12 ga.	G		10.50	.762		43.50	35		78.50	105
1500	Seismic, galvanized, 3-1/2" long, 16 ga.	G		10.50	.762		44.50	35		79.50	106
1600	12 ga.	G		10.50	.762		41.50	35		76.50	103
2000	Dovetail cavity wall, corrugated, galvanized, 5-1/2" long, 16 ga.	G		10.50	.762		38	35		73	99
2100	12 ga.	G		10.50	.762		55.50	35		90.50	119
3000	Dovetail furring anchors, corrugated, galvanized, 1-1/2" long, 16 ga.	G		10.50	.762		17.20	35		52.20	76
3100	12 ga.	G		10.50	.762		25.50	35		60.50	85
6000	Dovetail stone panel anchors, galvanized, 1/8" x 1" wide, 3-1/2" long	G		10.50	.762		99	35		134	166
6100	1/4" x 1" wide	G		10.50	.762		112	35		147	180

03 15 19.30 Inserts

	03 15 19.30 Inserts		Crew	Daily Output	Labor-Hours	Unit	Material	Labor	Equipment	Total	Total Incl O&P
0010	**INSERTS**										
1000	Inserts, slotted nut type for 3/4" bolts, 4" long	G	1 Carp	84	.095	Ea.	19.65	4.47		24.12	29
2100	6" long	G		84	.095		22	4.47		26.47	32
2150	8" long	G		84	.095		29.50	4.47		33.97	40
2200	Slotted, strap type, 4" long	G		84	.095		21	4.47		25.47	31
2250	6" long	G		84	.095		24	4.47		28.47	34
2300	8" long	G		84	.095		31.50	4.47		35.97	42.50
2350	Strap for slotted insert, 4" long	G		84	.095		11.55	4.47		16.02	20
4100	6" long	G		84	.095		12.55	4.47		17.02	21
4150	8" long	G		84	.095		15.65	4.47		20.12	24.50
4200	10" long	G		84	.095		19.05	4.47		23.52	28.50
7000	Loop ferrule type										
7100	1/4" diameter bolt	G	1 Carp	84	.095	Ea.	2.23	4.47		6.70	9.80
7350	7/8" diameter bolt	G	"	84	.095	"	10.35	4.47		14.82	18.75
9000	Wedge type										
9100	For 3/4" diameter bolt	G	1 Carp	60	.133	Ea.	9.20	6.25		15.45	20.50
9800	Cut washers, black										
9900	3/4" bolt	G				Ea.	1.17			1.17	1.29
9950	For galvanized inserts, add						30%				

03 15 19.45 Machinery Anchors

	03 15 19.45 Machinery Anchors		Crew	Daily Output	Labor-Hours	Unit	Material	Labor	Equipment	Total	Total Incl O&P
0010	**MACHINERY ANCHORS**, heavy duty, incl. sleeve, floating base nut,										
0020	lower stud & coupling nut, fiber plug, connecting stud, washer & nut.										
0030	For flush mounted embedment in poured concrete heavy equip. pads.										
0200	Stud & bolt, 1/2" diameter	G	E-16	40	.400	Ea.	72.50	21.50	3.65	97.65	123
0300	5/8" diameter	G		35	.457		80.50	24.50	4.17	109.17	138
0500	3/4" diameter	G		30	.533		93	28.50	4.86	126.36	159
0600	7/8" diameter	G		25	.640		101	34.50	5.85	141.35	179
0800	1" diameter	G		20	.800		117	43	7.30	167.30	215
0900	1-1/4" diameter	G		15	1.067		141	57	9.75	207.75	269

03 21 05.10 Rebar Accessories			Crew	Daily Output	Labor-Hours	Unit	2015 Bare Costs Material	Labor	Equipment	Total	Total Incl O&P
0010	REBAR ACCESSORIES										
0030	Steel & plastic made from recycled materials										
0100	Beam bolsters (BB), lower, 1-1/2" high, plain steel	G				C.L.F.	33			33	36.50
0102	Galvanized	G					39.50			39.50	43.50
0104	Stainless tipped legs	G					460			460	505
0106	Plastic tipped legs	G					49			49	54
0108	Epoxy dipped	G					84			84	92.50
0110	2" high, plain	G					42			42	46
0120	Galvanized	G					50.50			50.50	55.50
0140	Stainless tipped legs	G					465			465	515
0160	Plastic tipped legs	G					57			57	62.50
0162	Epoxy dipped	G					95			95	105
0200	Upper (BBU), 1-1/2" high, plain steel	G					81			81	89
0210	3" high	G					91			91	100
0500	Slab bolsters, continuous (SB), 1" high, plain steel	G					29			29	32
0502	Galvanized	G					35			35	38.50
0504	Stainless tipped legs	G					455			455	500
0506	Plastic tipped legs	G					42			42	46
0510	2" high, plain steel	G					36			36	39.50
0515	Galvanized	G					43			43	47.50
0520	Stainless tipped legs	G					495			495	545
0525	Plastic tipped legs	G					49			49	54
0530	For bolsters with wire runners (SBR), add	G					39			39	43
0540	For bolsters with plates (SBP), add	G				↓	92			92	101
0700	Bag ties, 16 ga., plain, 4" long	G				C	4			4	4.40
0710	5" long	G					5			5	5.50
0720	6" long	G					4			4	4.40
0730	7" long	G					5			5	5.50
1200	High chairs, individual (HC), 3" high, plain steel	G					59			59	65
1202	Galvanized	G					71			71	78
1204	Stainless tipped legs	G					485			485	530
1206	Plastic tipped legs	G					64			64	70.50
1210	5" high, plain	G					86			86	94.50
1212	Galvanized	G					103			103	114
1214	Stainless tipped legs	G					510			510	560
1216	Plastic tipped legs	G					94			94	103
1220	8" high, plain	G					126			126	139
1222	Galvanized	G					151			151	166
1224	Stainless tipped legs	G					550			550	605
1226	Plastic tipped legs	G					139			139	153
1230	12" high, plain	G					300			300	330
1232	Galvanized	G					360			360	395
1234	Stainless tipped legs	G					725			725	800
1236	Plastic tipped legs	G					330			330	360
1400	Individual high chairs, with plate (HCP), 5" high	G					179			179	197
1410	8" high	G					248			248	273
1500	Bar chair (BC), 1-1/2" high, plain steel	G					38			38	42
1520	Galvanized	G					43			43	47.50
1530	Stainless tipped legs	G					470			470	515
1540	Plastic tipped legs	G				↓	41			41	45
1700	Continuous high chairs (CHC), legs 8" O.C., 4" high, plain steel	G				C.L.F.	51			51	56
1705	Galvanized	G					61			61	67.50
1710	Stainless tipped legs	G				↓	475			475	525

03 21 05.10 Rebar Accessories			Crew	Daily Output	Labor-Hours	Unit	2015 Bare Costs Material	Labor	Equipment	Total	Total Incl O&P
1715	Plastic tipped legs	G				C.L.F.	69			69	76
1718	Epoxy dipped	G					89			89	98
1720	6" high, plain	G					70			70	77
1725	Galvanized	G					84			84	92.50
1730	Stainless tipped legs	G					495			495	545
1735	Plastic tipped legs	G					94			94	103
1738	Epoxy dipped	G					121			121	133
1740	8" high, plain	G					100			100	110
1745	Galvanized	G					120			120	132
1750	Stainless tipped legs	G					525			525	580
1755	Plastic tipped legs	G					119			119	131
1758	Epoxy dipped	G					153			153	169
1900	For continuous bottom wire runners, add	G					52			52	57
1940	For continuous bottom plate, add	G					214			214	235
2200	Screed chair base, 1/2" coil thread diam., 2-1/2" high, plain steel	G				C	325			325	360
2210	Galvanized	G					390			390	430
2220	5-1/2" high, plain	G					395			395	430
2250	Galvanized	G					470			470	520
2300	3/4" coil thread diam., 2-1/2" high, plain steel	G					410			410	450
2310	Galvanized	G					490			490	540
2320	5-1/2" high, plain steel	G					505			505	555
2350	Galvanized	G					605			605	665
2400	Screed holder, 1/2" coil thread diam. for pipe screed, plain steel, 6" long	G					395			395	435
2420	12" long	G					605			605	665
2500	3/4" coil thread diam. for pipe screed, plain steel, 6" long	G					565			565	625
2520	12" long	G					905			905	995
2700	Screw anchor for bolts, plain steel, 3/4" diameter x 4" long	G					540			540	595
2720	1" diameter x 6" long	G					890			890	975
2740	1-1/2" diameter x 8" long	G					1,125			1,125	1,250
2800	Screw anchor eye bolts, 3/4" x 3" long	G					2,825			2,825	3,125
2820	1" x 3-1/2" long	G					3,800			3,800	4,175
2840	1-1/2" x 6" long	G					11,600			11,600	12,800
2900	Screw anchor bolts, 3/4" x 9" long	G					1,625			1,625	1,775
2920	1" x 12" long	G					3,125			3,125	3,425
3001	Slab lifting inserts, single pickup, galv, 3/4" diam., 5" high	G					1,725			1,725	1,900
3010	6" high	G					1,750			1,750	1,925
3030	7" high	G					1,775			1,775	1,950
3100	1" diameter, 5-1/2" high	G					1,800			1,800	2,000
3120	7" high	G					1,875			1,875	2,050
3200	Double pickup lifting inserts, 1" diameter, 5-1/2" high	G					3,500			3,500	3,850
3220	7" high	G					3,875			3,875	4,250
3330	1-1/2" diameter, 8" high	G					4,875			4,875	5,375
3800	Subgrade chairs, #4 bar head, 3-1/2" high	G					38			38	42
3850	12" high	G					45			45	49.50
3900	#6 bar head, 3-1/2" high	G					38			38	42
3950	12" high	G					45			45	49.50
4200	Subgrade stakes, no nail holes, 3/4" diameter, 12" long	G					294			294	325
4250	24" long	G					360			360	395
4300	7/8" diameter, 12" long	G					375			375	410
4350	24" long	G					625			625	690
4500	Tie wire, 16 ga. annealed steel	G				Cwt.	203			203	223

03 21 Reinforcement Bars

03 21 05 – Reinforcing Steel Accessories

03 21 05.75 Splicing Reinforcing Bars			Crew	Daily Output	Labor-Hours	Unit	Material	2015 Bare Costs Labor	Equipment	Total	Total Incl O&P
0010	**SPLICING REINFORCING BARS**										
0020	Including holding bars in place while splicing										
0100	Standard, self-aligning type, taper threaded, #4 bars	G	C-25	190	.168	Ea.	5.95	6.95		12.90	18.35
0105	#5 bars	G		170	.188		7.30	7.75		15.05	21
0110	#6 bars	G		150	.213		8.40	8.80		17.20	24
0120	#7 bars	G		130	.246		9.80	10.15		19.95	28
0300	#8 bars	G		115	.278		16.60	11.50		28.10	38
0305	#9 bars	G	C-5	105	.533		18.15	27.50	7	52.65	72
0310	#10 bars	G		95	.589		20	30.50	7.75	58.25	80
0320	#11 bars	G		85	.659		21.50	34	8.65	64.15	88
0330	#14 bars	G		65	.862		32	44.50	11.35	87.85	119
0340	#18 bars	G		45	1.244		49	64.50	16.35	129.85	176
0500	Transition self-aligning, taper threaded, #18-14	G		45	1.244		51	64.50	16.35	131.85	178
0510	#18-11	G		45	1.244		52	64.50	16.35	132.85	179
0520	#14-11	G		65	.862		34	44.50	11.35	89.85	122
0540	#11-10	G		85	.659		23.50	34	8.65	66.15	90.50
0550	#10-9	G		95	.589		22	30.50	7.75	60.25	82.50
0560	#9-8	G	C-25	105	.305		20	12.60		32.60	43.50
0580	#8-7	G		115	.278		18.60	11.50		30.10	40
0590	#7-6	G		130	.246		11.75	10.15		21.90	30
0600	Position coupler for curved bars, taper threaded, #4 bars	G		160	.200		27.50	8.25		35.75	44
0610	#5 bars	G		145	.221		28.50	9.10		37.60	47
0620	#6 bars	G		130	.246		34.50	10.15		44.65	55.50
0630	#7 bars	G		110	.291		36.50	12		48.50	60.50
0640	#8 bars	G		100	.320		38	13.20		51.20	64
0650	#9 bars	G	C-5	90	.622		41.50	32	8.20	81.70	107
0660	#10 bars	G		80	.700		44.50	36	9.20	89.70	118
0670	#11 bars	G		70	.800		46.50	41.50	10.50	98.50	130
0680	#14 bars	G		55	1.018		58	52.50	13.40	123.90	164
0690	#18 bars	G		40	1.400		83.50	72.50	18.40	174.40	229
0700	Transition position coupler for curved bars, taper threaded, #18-14	G		40	1.400		85.50	72.50	18.40	176.40	232
0710	#18-11	G		40	1.400		86.50	72.50	18.40	177.40	233
0720	#14-11	G		55	1.018		60	52.50	13.40	125.90	166
0730	#11-10	G		70	.800		48.50	41.50	10.50	100.50	132
0740	#10-9	G		80	.700		46.50	36	9.20	91.70	120
0750	#9-8	G	C-25	90	.356		43.50	14.70		58.20	73
0760	#8-7	G		100	.320		40	13.20		53.20	66.50
0770	#7-6	G		110	.291		38.50	12		50.50	63
0800	Sleeve type w/grout filler, for precast concrete, #6 bars	G		72	.444		21.50	18.35		39.85	54.50
0802	#7 bars	G		64	.500		25.50	20.50		46	63.50
0805	#8 bars	G		56	.571		30.50	23.50		54	73.50
0807	#9 bars	G		48	.667		36.50	27.50		64	86.50
0810	#10 bars	G	C-5	40	1.400		43	72.50	18.40	133.90	185
0900	#11 bars	G		32	1.750		47.50	90.50	23	161	224
0920	#14 bars	G		24	2.333		74	121	30.50	225.50	310
1000	Sleeve type w/ferrous filler, for critical structures, #6 bars	G	C-25	72	.444		59	18.35		77.35	96
1210	#7 bars	G		64	.500		59.50	20.50		80	101
1220	#8 bars	G		56	.571		63	23.50		86.50	109
1230	#9 bars	G	C-5	48	1.167		64.50	60.50	15.35	140.35	185
1240	#10 bars	G		40	1.400		68.50	72.50	18.40	159.40	213
1250	#11 bars	G		32	1.750		83	90.50	23	196.50	263
1260	#14 bars	G		24	2.333		104	121	30.50	255.50	345

03 21 Reinforcement Bars

03 21 05 – Reinforcing Steel Accessories

03 21 05.75 Splicing Reinforcing Bars			Crew	Daily Output	Labor-Hours	Unit	2015 Bare Costs Material	Labor	Equipment	Total	Total Incl O&P
1270	#18 bars	G	C-5	16	3.500	Ea.	106	181	46	333	460
2000	Weldable half coupler, taper threaded, #4 bars	G	E-16	120	.133		8.90	7.15	1.22	17.27	24
2100	#5 bars	G		112	.143		10.50	7.65	1.30	19.45	27
2200	#6 bars	G		104	.154		16.60	8.25	1.40	26.25	34.50
2300	#7 bars	G		96	.167		19.30	8.95	1.52	29.77	39
2400	#8 bars	G		88	.182		20	9.75	1.66	31.41	41.50
2500	#9 bars	G		80	.200		22	10.75	1.82	34.57	46
2600	#10 bars	G		72	.222		22.50	11.90	2.03	36.43	48.50
2700	#11 bars	G		64	.250		24.50	13.40	2.28	40.18	53
2800	#14 bars	G		56	.286		28	15.35	2.61	45.96	61.50
2900	#18 bars	G		48	.333		45.50	17.90	3.04	66.44	86

03 21 11 – Plain Steel Reinforcement Bars

03 21 11.60 Reinforcing In Place			Crew	Daily Output	Labor-Hours	Unit	Material	Labor	Equipment	Total	Total Incl O&P
0010	**REINFORCING IN PLACE**, 50-60 ton lots, A615 Grade 60	R032110-10									
0020	Includes labor, but not material cost, to install accessories	R032110-80									
0030	Made from recycled materials										
0102	Beams & Girders, #3 to #7	G	4 Rodm	3200	.010	Lb.	.48	.53		1.01	1.39
0152	#8 to #18	G		5400	.006		.48	.31		.79	1.04
0202	Columns, #3 to #7	G		3000	.011		.48	.56		1.04	1.45
0252	#8 to #18	G		4600	.007		.48	.37		.85	1.13
0300	Spirals, hot rolled, 8" to 15" diameter	G		2.20	14.545	Ton	1,575	765		2,340	2,975
0320	15" to 24" diameter	G		2.20	14.545		1,500	765		2,265	2,900
0330	24" to 36" diameter	G		2.30	13.913		1,425	730		2,155	2,775
0340	36" to 48" diameter	G		2.40	13.333		1,350	700		2,050	2,650
0360	48" to 64" diameter	G		2.50	12.800		1,500	675		2,175	2,750
0380	64" to 84" diameter	G		2.60	12.308		1,575	645		2,220	2,775
0390	84" to 96" diameter	G		2.70	11.852		1,650	625		2,275	2,850
0402	Elevated slabs, #4 to #7	G		5800	.006	Lb.	.48	.29		.77	1
0502	Footings, #4 to #7	G		4200	.008		.48	.40		.88	1.18
0552	#8 to #18	G		7200	.004		.48	.23		.71	.91
0602	Slab on grade, #3 to #7	G		4200	.008		.48	.40		.88	1.18
0702	Walls, #3 to #7	G		6000	.005		.48	.28		.76	.99
0752	#8 to #18	G		8000	.004		.48	.21		.69	.87
0900	For other than 50 – 60 ton lots										
1000	Under 10 ton job, #3 to #7, add						25%	10%			
1010	#8 to #18, add						20%	10%			
1050	10 – 50 ton job, #3 to #7, add						10%				
1060	#8 to #18, add						5%				
1100	60 – 100 ton job, #3 to #7, deduct						5%				
1110	#8 to #18, deduct						10%				
1150	Over 100 ton job, #3 to #7, deduct						10%				
1160	#8 to #18, deduct						15%				
1200	Reinforcing in place, A615 Grade 75, add	G				Ton	92.50			92.50	102
1220	Grade 90, add						125			125	138
2000	Unloading & sorting, add to above		C-5	100	.560			29	7.35	36.35	55
2200	Crane cost for handling, 90 picks/day, up to 1.5 Tons/bundle, add to above			135	.415			21.50	5.45	26.95	40.50
2210	1.0 Ton/bundle			92	.609			31.50	8	39.50	60
2220	0.5 Ton/bundle			35	1.600			82.50	21	103.50	157
2400	Dowels, 2 feet long, deformed, #3	G	2 Rodm	520	.031	Ea.	.40	1.62		2.02	3.08
2410	#4	G		480	.033		.71	1.75		2.46	3.64
2420	#5	G		435	.037		1.11	1.93		3.04	4.37
2430	#6	G		360	.044		1.60	2.34		3.94	5.55

03 21 Reinforcement Bars

03 21 11 – Plain Steel Reinforcement Bars

03 21 11.60 Reinforcing In Place		Crew	Daily Output	Labor-Hours	Unit	Material	2015 Bare Costs Labor	Equipment	Total	Total Incl O&P
2450	Longer and heavier dowels, add G	2 Rodm	725	.022	Lb.	.53	1.16		1.69	2.48
2500	Smooth dowels, 12" long, 1/4" or 3/8" diameter G		140	.114	Ea.	.72	6		6.72	10.60
2520	5/8" diameter G		125	.128		1.26	6.75		8.01	12.40
2530	3/4" diameter G		110	.145		1.57	7.65		9.22	14.15
2600	Dowel sleeves for CIP concrete, 2-part system									
2610	Sleeve base, plastic, for 5/8" smooth dowel sleeve, fasten to edge form	1 Rodm	200	.040	Ea.	.53	2.10		2.63	4.01
2615	Sleeve, plastic, 12" long, for 5/8" smooth dowel, snap onto base		400	.020		1.18	1.05		2.23	3.02
2620	Sleeve base, for 3/4" smooth dowel sleeve		175	.046		.53	2.40		2.93	4.50
2625	Sleeve, 12" long, for 3/4" smooth dowel		350	.023		1.33	1.20		2.53	3.42
2630	Sleeve base, for 1" smooth dowel sleeve		150	.053		.67	2.80		3.47	5.30
2635	Sleeve, 12" long, for 1" smooth dowel		300	.027		1.40	1.40		2.80	3.83
2700	Dowel caps, visual warning only, plastic, #3 to #8	2 Rodm	800	.020		.26	1.05		1.31	2.01
2720	#8 to #18		750	.021		.63	1.12		1.75	2.52
2750	Impalement protective, plastic, #4 to #9		800	.020		1.22	1.05		2.27	3.06
9000	Minimum labor/equipment charge	1 Rodm	4	2	Job		105		105	172

03 21 13 – Galvanized Reinforcement Steel Bars

03 21 13.10 Galvanized Reinforcing		Crew	Daily Output	Labor-Hours	Unit	Material	Labor	Equipment	Total	Total Incl O&P
0010	GALVANIZED REINFORCING									
0150	Add to plain steel rebar pricing for galvanized rebar				Ton	460			460	505

03 21 16 – Epoxy-Coated Reinforcement Steel Bars

03 21 16.10 Epoxy-Coated Reinforcing		Crew	Daily Output	Labor-Hours	Unit	Material	Labor	Equipment	Total	Total Incl O&P
0010	EPOXY-COATED REINFORCING									
0100	Add to plain steel rebar pricing for epoxy-coated rebar				Ton	420			420	465

03 21 19 – Stainless Steel Reinforcement Bars

03 21 19.10 Stainless Steel Reinforcing		Crew	Daily Output	Labor-Hours	Unit	Material	Labor	Equipment	Total	Total Incl O&P
0010	STAINLESS STEEL REINFORCING									
0100	Add to plain steel rebar pricing for stainless steel rebar					300%				

03 21 21 – Composite Reinforcement Bars

03 21 21.11 Glass Fiber-Reinforced Polymer Reinforcement Bars		Crew	Daily Output	Labor-Hours	Unit	Material	Labor	Equipment	Total	Total Incl O&P
0010	GLASS FIBER-REINFORCED POLYMER REINFORCEMENT BARS									
0020	Includes labor, but not material cost, to install accessories									
0050	#2 bar, .043 lb./L.F.	4 Rodm	9500	.003	L.F.	.40	.18		.58	.73
0100	#3 bar, .092 lb./L.F.		9300	.003		.49	.18		.67	.84
0150	#4 bar, .160 lb./L.F.		9100	.004		.71	.19		.90	1.08
0200	#5 bar, .258 lb./L.F.		8700	.004		1.09	.19		1.28	1.52
0250	#6 bar, .372 lb./L.F.		8300	.004		1.41	.20		1.61	1.88
0300	#7 bar, .497 lb./L.F.		7900	.004		1.84	.21		2.05	2.37
0350	#8 bar, .620 lb./L.F.		7400	.004		2.38	.23		2.61	2.99
0400	#9 bar, .800 lb./L.F.		6800	.005		3.10	.25		3.35	3.81
0450	#10 bar, 1.08 lb./L.F.		5800	.006		3.75	.29		4.04	4.60
0500	For Bends, add per bend				Ea.	1.48			1.48	1.63

03 22 Fabric and Grid Reinforcing

03 22 11 – Plain Welded Wire Fabric Reinforcing

03 22 11.10 Plain Welded Wire Fabric			Crew	Daily Output	Labor-Hours	Unit	Material	2015 Bare Costs Labor	Equipment	Total	Total Incl O&P
0010	**PLAIN WELDED WIRE FABRIC** ASTM A185										
0020	Includes labor, but not material cost, to install accessories										
0030	Made from recycled materials										
0050	Sheets										
0100	6 x 6 - W1.4 x W1.4 (10 x 10) 21 lb. per C.S.F.	G	2 Rodm	35	.457	C.S.F.	14.50	24		38.50	55
0200	6 x 6 - W2.1 x W2.1 (8 x 8) 30 lb. per C.S.F.	G		31	.516		17.20	27		44.20	63.50
0300	6 x 6 - W2.9 x W2.9 (6 x 6) 42 lb. per C.S.F.	G		29	.552		22.50	29		51.50	72
0400	6 x 6 - W4 x W4 (4 x 4) 58 lb. per C.S.F.	G		27	.593		31.50	31		62.50	85.50
0500	4 x 4 - W1.4 x W1.4 (10 x 10) 31 lb. per C.S.F.	G		31	.516		20	27		47	66.50
0600	4 x 4 - W2.1 x W2.1 (8 x 8) 44 lb. per C.S.F.	G		29	.552		25	29		54	75
0650	4 x 4 - W2.9 x W2.9 (6 x 6) 61 lb. per C.S.F.	G		27	.593		40.50	31		71.50	95.50
0700	4 x 4 - W4 x W4 (4 x 4) 85 lb. per C.S.F.	G		25	.640		50.50	33.50		84	111
0750	Rolls										
0800	2 x 2 - #14 galv., 21 lb./C.S.F., beam & column wrap	G	2 Rodm	6.50	2.462	C.S.F.	41.50	129		170.50	257
0900	2 x 2 - #12 galv. for gunite reinforcing	G	"	6.50	2.462	"	62.50	129		191.50	280
9000	Minimum labor/equipment charge		1 Rodm	4	2	Job		105		105	172

03 22 13 – Galvanized Welded Wire Fabric Reinforcing

03 22 13.10 Galvanized Welded Wire Fabric			Crew	Daily Output	Labor-Hours	Unit	Material	Labor	Equipment	Total	Total Incl O&P
0010	**GALVANIZED WELDED WIRE FABRIC**										
0100	Add to plain welded wire pricing for galvanized welded wire					Lb.	.23			.23	.25

03 22 16 – Epoxy-Coated Welded Wire Fabric Reinforcing

03 22 16.10 Epoxy-Coated Welded Wire Fabric			Crew	Daily Output	Labor-Hours	Unit	Material	Labor	Equipment	Total	Total Incl O&P
0010	**EPOXY-COATED WELDED WIRE FABRIC**										
0100	Add to plain welded wire pricing for epoxy-coated welded wire					Lb.	.21			.21	.23

03 23 Stressed Tendon Reinforcing

03 23 05 – Prestressing Tendons

03 23 05.50 Prestressing Steel			Crew	Daily Output	Labor-Hours	Unit	Material	Labor	Equipment	Total	Total Incl O&P
0010	**PRESTRESSING STEEL**										
0100	Grouted strand, in beams, post-tensioned in field, 50′ span, 100 kip	G	C-3	1200	.053	Lb.	2.64	2.59	.09	5.32	7.20
0150	300 kip	G		2700	.024		1.14	1.15	.04	2.33	3.17
0300	100′ span, 100 kip	G		1700	.038		2.64	1.83	.07	4.54	5.95
0350	300 kip	G		3200	.020		2.27	.97	.04	3.28	4.12
0500	200′ span, 100 kip	G		2700	.024		2.64	1.15	.04	3.83	4.82
0550	300 kip	G		3500	.018		2.27	.89	.03	3.19	3.99
0800	Grouted bars, in beams, 50′ span, 42 kip	G		2600	.025		1.04	1.20	.04	2.28	3.15
0850	143 kip	G		3200	.020		1	.97	.04	2.01	2.72
1000	75′ span, 42 kip	G		3200	.020		1.06	.97	.04	2.07	2.79
1050	143 kip	G		4200	.015		.89	.74	.03	1.66	2.21
1200	Ungrouted strand, in beams, 50′ span, 100 kip	G	C-4	1275	.025		.63	1.33	.02	1.98	2.89
1250	300 kip	G		1475	.022		.63	1.15	.02	1.80	2.59
1400	100′ span, 100 kip	G		1500	.021		.63	1.13	.02	1.78	2.56
1450	300 kip	G		1650	.019		.63	1.03	.02	1.68	2.39
1600	200′ span, 100 kip	G		1500	.021		.63	1.13	.02	1.78	2.56
1650	300 kip	G		1700	.019		.63	1	.02	1.65	2.34
1800	Ungrouted bars, in beams, 50′ span, 42 kip	G		1400	.023		.48	1.21	.02	1.71	2.53
1850	143 kip	G		1700	.019		.48	1	.02	1.50	2.18
2000	75′ span, 42 kip	G		1800	.018		.48	.94	.02	1.44	2.09
2050	143 kip	G		2200	.015		.48	.77	.01	1.26	1.81
2220	Ungrouted single strand, 100′ elevated slab, 25 kip	G		1200	.027		.63	1.41	.03	2.07	3.03

03 23 Stressed Tendon Reinforcing

03 23 05 – Prestressing Tendons

03 23 05.50 Prestressing Steel			Crew	Daily Output	Labor-Hours	Unit	Material	2015 Bare Costs Labor	Equipment	Total	Total Incl O&P
2250	35 kip	G	C-4	1475	.022	Lb.	.63	1.15	.02	1.80	2.59
3000	Slabs on grade, 0.5-inch diam. non-bonded strands, HDPE sheathed,										
3050	attached dead-end anchors, loose stressing-end anchors										
3100	25′ x 30′ slab, strands @ 36″ O.C., placing		2 Rodm	2940	.005	S.F.	.60	.29		.89	1.13
3105	Stressing		C-4A	3750	.004			.22	.01	.23	.38
3110	42″ O.C., placing		2 Rodm	3200	.005		.53	.26		.79	1.02
3115	Stressing		C-4A	4040	.004			.21	.01	.22	.35
3120	48″ O.C., placing		2 Rodm	3510	.005		.47	.24		.71	.90
3125	Stressing		C-4A	4390	.004			.19	.01	.20	.32
3150	25′ x 40′ slab, strands @ 36″ O.C., placing		2 Rodm	3370	.005		.58	.25		.83	1.05
3155	Stressing		C-4A	4360	.004			.19	.01	.20	.32
3160	42″ O.C., placing		2 Rodm	3760	.004		.50	.22		.72	.92
3165	Stressing		C-4A	4820	.003			.17	.01	.18	.29
3170	48″ O.C., placing		2 Rodm	4090	.004		.45	.21		.66	.83
3175	Stressing		C-4A	5190	.003			.16	.01	.17	.27
3200	30′ x 30′ slab, strands @ 36″ O.C., placing		2 Rodm	3260	.005		.58	.26		.84	1.06
3205	Stressing		C-4A	4190	.004			.20	.01	.21	.34
3210	42″ O.C., placing		2 Rodm	3530	.005		.52	.24		.76	.97
3215	Stressing		C-4A	4500	.004			.19	.01	.20	.32
3220	48″ O.C., placing		2 Rodm	3840	.004		.47	.22		.69	.87
3225	Stressing		C-4A	4850	.003			.17	.01	.18	.29
3230	30′ x 40′ slab, strands @ 36″ O.C., placing		2 Rodm	3780	.004		.56	.22		.78	.98
3235	Stressing		C-4A	4920	.003			.17	.01	.18	.29
3240	42″ O.C., placing		2 Rodm	4190	.004		.49	.20		.69	.87
3245	Stressing		C-4A	5410	.003			.16	.01	.17	.26
3250	48″ O.C., placing		2 Rodm	4520	.004		.45	.19		.64	.79
3255	Stressing		C-4A	5790	.003			.15	.01	.16	.25
3260	30′ x 50′ slab, strands @ 36″ O.C., placing		2 Rodm	4300	.004		.53	.20		.73	.91
3265	Stressing		C-4A	5650	.003			.15	.01	.16	.25
3270	42″ O.C., placing		2 Rodm	4720	.003		.47	.18		.65	.81
3275	Stressing		C-4A	6150	.003			.14	.01	.15	.23
3280	48″ O.C., placing		2 Rodm	5240	.003		.42	.16		.58	.72
3285	Stressing		C-4A	6760	.002			.12	.01	.13	.21
3900	Minimum labor/equipment charge, placing		2 Rodm	2	8	Job		420		420	685
3905	Minimum labor/equipment charge, stressing		C-4A	2	8	"		420	20.50	440.50	710

03 24 Fibrous Reinforcing

03 24 05 – Reinforcing Fibers

03 24 05.30 Synthetic Fibers			Crew	Daily Output	Labor-Hours	Unit	Material	Labor	Equipment	Total	Total Incl O&P
0010	**SYNTHETIC FIBERS**										
0100	Synthetic fibers, add to concrete					Lb.	4.40			4.40	4.84
0110	1-1/2 lb. per C.Y.					C.Y.	6.80			6.80	7.50

03 24 05.70 Steel Fibers			Crew	Daily Output	Labor-Hours	Unit	Material	Labor	Equipment	Total	Total Incl O&P
0010	**STEEL FIBERS**										
0140	ASTM A850, Type V, continuously deformed, 1-1/2″ long x 0.045″ diam.										
0150	Add to price of ready mix concrete	G				Lb.	1.13			1.13	1.24
0205	Alternate pricing, dosing at 5 lb. per C.Y., add to price of RMC	G				C.Y.	5.65			5.65	6.20
0210	10 lb. per C.Y.	G					11.30			11.30	12.45
0215	15 lb. per C.Y.	G					16.95			16.95	18.65
0220	20 lb. per C.Y.	G					22.50			22.50	25
0225	25 lb. per C.Y.	G					28.50			28.50	31

03 24 Fibrous Reinforcing

03 24 05 – Reinforcing Fibers

03 24 05.70 Steel Fibers		Crew	Daily Output	Labor-Hours	Unit	Material	2015 Bare Costs Labor	Equipment	Total	Total Incl O&P
0230	30 lb. per C.Y. [G]				C.Y.	34			34	37.50
0235	35 lb. per C.Y. [G]					39.50			39.50	43.50
0240	40 lb. per C.Y. [G]					45			45	49.50
0250	50 lb. per C.Y. [G]					56.50			56.50	62
0275	75 lb. per C.Y. [G]					85			85	93
0300	100 lb. per C.Y. [G]					113			113	124

03 30 Cast-In-Place Concrete

03 30 53 – Miscellaneous Cast-In-Place Concrete

03 30 53.40 Concrete In Place

Code	Description	Crew	Daily Output	Labor-Hours	Unit	Material	Labor	Equipment	Total	Total Incl O&P
0010	**CONCRETE IN PLACE** R033105-20									
0020	Including forms (4 uses), Grade 60 rebar, concrete (Portland cement R033105-70									
0050	Type I), placement and finishing unless otherwise indicated R033105-80									
0300	Beams (3500 psi), 5 kip per L.F., 10′ span	C-14A	15.62	12.804	C.Y.	320	605	48	973	1,400
0350	25′ span	″	18.55	10.782		335	510	40.50	885.50	1,250
0500	Chimney foundations (5000 psi), over 5 C.Y.	C-14C	32.22	3.476		150	157	.97	307.97	420
0510	(3500 psi), under 5 C.Y.	″	23.71	4.724		177	213	1.32	391.32	545
0700	Columns, square (4000 psi), 12″ x 12″, less than 2% reinforcing	C-14A	11.96	16.722		365	790	63	1,218	1,775
0720	2% to 3% reinforcing		10.13	19.743		565	935	74	1,574	2,225
0740	Over 3% reinforcing		9.03	22.148		840	1,050	83.50	1,973.50	2,700
0800	16″ x 16″, less than 2% reinforcing		16.22	12.330		286	585	46.50	917.50	1,325
0820	2% to 3% reinforcing		12.57	15.911		480	750	60	1,290	1,825
0840	Over 3% reinforcing		10.25	19.512		735	920	73.50	1,728.50	2,375
0900	24″ x 24″, less than 2% reinforcing		23.66	8.453		241	400	32	673	950
0920	2% to 3% reinforcing		17.71	11.293		425	535	42.50	1,002.50	1,375
0940	Over 3% reinforcing		14.15	14.134		670	670	53	1,393	1,900
1000	36″ x 36″, less than 2% reinforcing		33.69	5.936		212	281	22.50	515.50	720
1020	2% to 3% reinforcing		23.32	8.576		370	405	32.50	807.50	1,100
1040	Over 3% reinforcing		17.82	11.223		625	530	42	1,197	1,600
1100	Columns, round (4000 psi), tied, 12″ diameter, less than 2% reinforcing		20.97	9.537		315	450	36	801	1,125
1120	2% to 3% reinforcing		15.27	13.098		515	620	49.50	1,184.50	1,625
1140	Over 3% reinforcing		12.11	16.515		780	780	62	1,622	2,200
1200	16″ diameter, less than 2% reinforcing		31.49	6.351		289	300	24	613	830
1220	2% to 3% reinforcing		19.12	10.460		490	495	39.50	1,024.50	1,400
1240	Over 3% reinforcing		13.77	14.524		735	685	54.50	1,474.50	2,000
1300	20″ diameter, less than 2% reinforcing		41.04	4.873		289	230	18.30	537.30	710
1320	2% to 3% reinforcing		24.05	8.316		475	395	31.50	901.50	1,200
1340	Over 3% reinforcing		17.01	11.758		735	555	44	1,334	1,750
1400	24″ diameter, less than 2% reinforcing		51.85	3.857		269	182	14.50	465.50	610
1420	2% to 3% reinforcing		27.06	7.391		470	350	28	848	1,125
1440	Over 3% reinforcing		18.29	10.935		715	515	41	1,271	1,675
1500	36″ diameter, less than 2% reinforcing		75.04	2.665		266	126	10	402	510
1520	2% to 3% reinforcing		37.49	5.335		445	252	20	717	920
1540	Over 3% reinforcing		22.84	8.757		695	415	33	1,143	1,475
1900	Elevated slab (4000 psi), flat slab with drops, 125 psf Sup. Load, 20′ span	C-14B	38.45	5.410		261	255	19.55	535.55	725
1950	30′ span		50.99	4.079		276	192	14.75	482.75	635
2100	Flat plate, 125 psf Sup. Load, 15′ span		30.24	6.878		240	325	25	590	820
2150	25′ span		49.60	4.194		249	198	15.15	462.15	615
2300	Waffle const., 30″ domes, 125 psf Sup. Load, 20′ span		37.07	5.611		259	265	20.50	544.50	740
2350	30′ span		44.07	4.720		241	223	17.05	481.05	650
2500	One way joists, 30″ pans, 125 psf Sup. Load, 15′ span		27.38	7.597		310	360	27.50	697.50	955

03 30 53.40 Concrete In Place		Crew	Daily Output	Labor-Hours	Unit	Material	2015 Bare Costs Labor	Equipment	Total	Total Incl O&P
2550	25′ span	C-14B	31.15	6.677	C.Y.	292	315	24	631	860
2700	One way beam & slab, 125 psf Sup. Load, 15′ span		20.59	10.102		259	475	36.50	770.50	1,100
2750	25′ span		28.36	7.334		245	345	26.50	616.50	865
2900	Two way beam & slab, 125 psf Sup. Load, 15′ span		24.04	8.652		250	410	31	691	975
2950	25′ span		35.87	5.799		216	273	21	510	705
3100	Elevated slabs, flat plate, including finish, not									
3110	including forms or reinforcing									
3150	Regular concrete (4000 psi), 4″ slab	C-8	2613	.021	S.F.	1.43	.90	.28	2.61	3.33
3200	6″ slab		2585	.022		2.09	.91	.28	3.28	4.07
3250	2-1/2″ thick floor fill		2685	.021		.94	.87	.27	2.08	2.74
3300	Lightweight, 110# per C.F., 2-1/2″ thick floor fill		2585	.022		1.46	.91	.28	2.65	3.38
3400	Cellular concrete, 1-5/8″ fill, under 5000 S.F.		2000	.028		.99	1.17	.36	2.52	3.38
3450	Over 10,000 S.F.		2200	.025		.94	1.07	.33	2.34	3.12
3500	Add per floor for 3 to 6 stories high		31800	.002			.07	.02	.09	.14
3520	For 7 to 20 stories high		21200	.003			.11	.03	.14	.22
3540	Equipment pad (3000 psi), 3′ x 3′ x 6″ thick	C-14H	45	1.067	Ea.	47	49.50	.69	97.19	133
3550	4′ x 4′ x 6″ thick		30	1.600		69.50	74	1.04	144.54	198
3560	5′ x 5′ x 8″ thick		18	2.667		122	124	1.73	247.73	335
3570	6′ x 6′ x 8″ thick		14	3.429		164	159	2.23	325.23	440
3580	8′ x 8′ x 10″ thick		8	6		350	278	3.90	631.90	845
3590	10′ x 10′ x 12″ thick		5	9.600		595	445	6.25	1,046.25	1,375
3800	Footings (3000 psi), spread under 1 C.Y.	C-14C	28	4	C.Y.	166	180	1.12	347.12	480
3825	1 C.Y. to 5 C.Y.		43	2.605		201	117	.73	318.73	415
3850	Over 5 C.Y.		75	1.493		185	67.50	.42	252.92	315
3900	Footings, strip (3000 psi), 18″ x 9″, unreinforced	C-14L	40	2.400		125	105	.79	230.79	310
3920	18″ x 9″, reinforced	C-14C	35	3.200		148	144	.90	292.90	400
3925	20″ x 10″, unreinforced	C-14L	45	2.133		122	93.50	.70	216.20	288
3930	20″ x 10″, reinforced	C-14C	40	2.800		140	126	.78	266.78	360
3935	24″ x 12″, unreinforced	C-14L	55	1.745		120	76.50	.58	197.08	258
3940	24″ x 12″, reinforced	C-14C	48	2.333		139	105	.65	244.65	325
3945	36″ x 12″, unreinforced	C-14L	70	1.371		116	60	.45	176.45	226
3950	36″ x 12″, reinforced	C-14C	60	1.867		133	84	.52	217.52	285
4000	Foundation mat (3000 psi), under 10 C.Y.		38.67	2.896		204	131	.81	335.81	440
4050	Over 20 C.Y.		56.40	1.986		178	89.50	.56	268.06	345
4200	Wall, free-standing (3000 psi), 8″ thick, 8′ high	C-14D	45.83	4.364		160	204	16.40	380.40	530
4250	14′ high		27.26	7.337		190	345	27.50	562.50	800
4260	12″ thick, 8′ high		64.32	3.109		145	146	11.70	302.70	410
4270	14′ high		40.01	4.999		154	234	18.80	406.80	570
4300	15″ thick, 8′ high		80.02	2.499		139	117	9.40	265.40	355
4350	12′ high		51.26	3.902		139	183	14.65	336.65	465
4500	18′ high		48.85	4.094		157	192	15.40	364.40	505
4520	Handicap access ramp (4000 psi), railing both sides, 3′ wide	C-14H	14.58	3.292	L.F.	320	153	2.14	475.14	600
4525	5′ wide		12.22	3.928		330	182	2.55	514.55	665
4530	With 6″ curb and rails both sides, 3′ wide		8.55	5.614		330	260	3.65	593.65	790
4535	5′ wide		7.31	6.566		335	305	4.27	644.27	870
4650	Slab on grade (3500 psi), not including finish, 4″ thick	C-14E	60.75	1.449	C.Y.	124	67.50	.51	192.01	248
4700	6″ thick	"	92	.957	"	119	44.50	.33	163.83	204
4701	Thickened slab edge (3500 psi), for slab on grade poured									
4702	monolithically with slab; depth is in addition to slab thickness;									
4703	formed vertical outside edge, earthen bottom and inside slope									
4705	8″ deep x 8″ wide bottom, unreinforced	C-14L	2190	.044	L.F.	3.47	1.92	.01	5.40	7
4710	8″ x 8″, reinforced	C-14C	1670	.067		5.75	3.02	.02	8.79	11.25
4715	12″ deep x 12″ wide bottom, unreinforced	C-14L	1800	.053		7.05	2.34	.02	9.41	11.60

03 30 53.40 Concrete In Place		Crew	Daily Output	Labor-Hours	Unit	Material	2015 Bare Costs Labor	Equipment	Total	Total Incl O&P
4720	12" x 12", reinforced	C-14C	1310	.086	L.F.	11.20	3.85	.02	15.07	18.65
4725	16" deep x 16" wide bottom, unreinforced	C-14L	1440	.067		11.85	2.92	.02	14.79	17.80
4730	16" x 16", reinforced	C-14C	1120	.100		16.80	4.51	.03	21.34	26
4735	20" deep x 20" wide bottom, unreinforced	C-14L	1150	.083		17.90	3.66	.03	21.59	25.50
4740	20" x 20", reinforced	C-14C	920	.122		24	5.50	.03	29.53	35.50
4745	24" deep x 24" wide bottom, unreinforced	C-14L	930	.103		25	4.53	.03	29.56	35.50
4750	24" x 24", reinforced	C-14C	740	.151		33.50	6.80	.04	40.34	47.50
4751	Slab on grade (3500 psi), incl. troweled finish, not incl. forms									
4760	or reinforcing, over 10,000 S.F., 4" thick	C-14F	3425	.021	S.F.	1.35	.90	.01	2.26	2.94
4820	6" thick		3350	.021		1.98	.92	.01	2.91	3.66
4840	8" thick		3184	.023		2.71	.97	.01	3.69	4.54
4900	12" thick		2734	.026		4.06	1.13	.01	5.20	6.30
4950	15" thick		2505	.029		5.10	1.23	.01	6.34	7.55
5000	Slab on grade (3000 psi), incl. broom finish, not incl. forms									
5001	or reinforcing, 4" thick	C-14G	2873	.019	S.F.	1.32	.82	.01	2.15	2.78
5010	6" thick		2590	.022		2.07	.91	.01	2.99	3.74
5020	8" thick		2320	.024		2.70	1.02	.01	3.73	4.60
5200	Lift slab in place above the foundation, incl. forms, reinforcing,									
5210	concrete (4000 psi) and columns, over 20,000 S.F. per floor	C-14B	2113	.098	S.F.	6.90	4.64	.36	11.90	15.55
5250	10,000 S.F. to 20,000 S.F. per floor		1650	.126		7.55	5.95	.46	13.96	18.50
5300	Under 10,000 S.F. per floor		1500	.139		8.20	6.55	.50	15.25	20.50
5500	Lightweight, ready mix, including screed finish only,									
5510	not including forms or reinforcing									
5550	1:4 (2500 psi) for structural roof decks	C-14B	260	.800	C.Y.	166	37.50	2.89	206.39	248
5600	1:6 (3000 psi) for ground slab with radiant heat	C-14F	92	.783		168	33.50	.34	201.84	239
5650	1:3:2 (2000 psi) with sand aggregate, roof deck	C-14B	260	.800		164	37.50	2.89	204.39	246
5700	Ground slab (2000 psi)	C-14F	107	.673		164	29	.29	193.29	227
5900	Pile caps (3000 psi), incl. forms and reinf., sq. or rect., under 10 C.Y.	C-14C	54.14	2.069		168	93.50	.58	262.08	340
5950	Over 10 C.Y.		75	1.493		157	67.50	.42	224.92	283
6000	Triangular or hexagonal, under 10 C.Y.		53	2.113		123	95.50	.59	219.09	292
6050	Over 10 C.Y.		85	1.318		138	59.50	.37	197.87	249
6200	Retaining walls (3000 psi), gravity, 4' high see Section 32 32	C-14D	66.20	3.021		140	141	11.35	292.35	400
6250	10' high		125	1.600		134	75	6	215	276
6300	Cantilever, level backfill loading, 8' high		70	2.857		150	134	10.75	294.75	395
6350	16' high		91	2.198		145	103	8.25	256.25	335
6800	Stairs (3500 psi), not including safety treads, free standing, 3'-6" wide	C-14H	83	.578	LF Nose	5.60	27	.38	32.98	50
6850	Cast on ground		125	.384	"	4.63	17.80	.25	22.68	34.50
7000	Stair landings, free standing		200	.240	S.F.	4.52	11.10	.16	15.78	23
7050	Cast on ground		475	.101	"	3.52	4.68	.07	8.27	11.55
9000	Minimum labor/equipment charge	2 Carp	1	16	Job		750		750	1,225

03 31 13 – Heavyweight Structural Concrete

03 31 13.25 Concrete, Hand Mix

	03 31 13.25 Concrete, Hand Mix	Crew	Daily Output	Labor-Hours	Unit	Material	2015 Bare Costs Labor	Equipment	Total	Total Incl O&P
0010	**CONCRETE, HAND MIX** for small quantities or remote areas									
0050	Includes bulk local aggregate, bulk sand, bagged Portland									
0060	cement (Type I) and water, using gas powered cement mixer									
0125	2500 psi	C-30	135	.059	C.F.	3.29	2.23	1.28	6.80	8.65
0130	3000 psi		135	.059		3.52	2.23	1.28	7.03	8.95
0135	3500 psi		135	.059		3.66	2.23	1.28	7.17	9.10
0140	4000 psi		135	.059		3.82	2.23	1.28	7.33	9.25
0145	4500 psi		135	.059		4	2.23	1.28	7.51	9.45
0150	5000 psi		135	.059		4.26	2.23	1.28	7.77	9.75
0300	Using pre-bagged dry mix and wheelbarrow (80-lb. bag = 0.6 C.F.)									
0340	4000 psi	1 Clab	48	.167	C.F.	6.25	6.25		12.50	17.20

03 31 13.30 Concrete, Volumetric Site-Mixed

	03 31 13.30 Concrete, Volumetric Site-Mixed	Crew	Daily Output	Labor-Hours	Unit	Material	Labor	Equipment	Total	Total Incl O&P
0010	**CONCRETE, VOLUMETRIC SITE-MIXED**									
0015	Mixed on-site in volumetric truck									
0020	Includes local aggregate, sand, Portland cement (Type I) and water									
0025	Excludes all additives and treatments									
0100	3000 psi, 1 C.Y. mixed and discharged				C.Y.	177			177	194
0110	2 C.Y.					138			138	151
0120	3 C.Y.					124			124	136
0130	4 C.Y.					113			113	124
0140	5 C.Y.					106			106	117
0200	For truck holding/waiting time past first 2 on-site hours, add				Hr.	78			78	86
0210	For trip charge beyond first 20 miles, each way, add				Mile	3.50			3.50	3.85
0220	For each additional increase of 500 psi, add				Ea.	4.17			4.17	4.59

03 31 13.35 Heavyweight Concrete, Ready Mix

	03 31 13.35 Heavyweight Concrete, Ready Mix	Crew	Daily Output	Labor-Hours	Unit	Material	Labor	Equipment	Total	Total Incl O&P
0010	**HEAVYWEIGHT CONCRETE, READY MIX**, delivered									
0012	Includes local aggregate, sand, Portland cement (Type I) and water									
0015	Excludes all additives and treatments									
0020	2000 psi				C.Y.	97			97	107
0100	2500 psi					99.50			99.50	109
0150	3000 psi					102			102	112
0200	3500 psi					104			104	115
0300	4000 psi					107			107	118
0350	4500 psi					110			110	121
0400	5000 psi					113			113	125
0411	6000 psi					116			116	128
0412	8000 psi					123			123	135
0413	10,000 psi					129			129	142
0414	12,000 psi					135			135	149
1000	For high early strength (Portland cement Type III), add					10%				
1010	For structural lightweight with regular sand, add					25%				
1300	For winter concrete (hot water), add					4.05			4.05	4.46
1410	For mid-range water reducer, add					3.31			3.31	3.64
1420	For high-range water reducer/superplasticizer, add					5.65			5.65	6.20
1430	For retarder, add					3.23			3.23	3.55
1440	For non-Chloride accelerator, add					5.50			5.50	6.05
1450	For Chloride accelerator, per 1%, add					2.90			2.90	3.19
1460	For fiber reinforcing, synthetic (1 lb./C.Y.), add					6.65			6.65	7.30
1500	For Saturday delivery, add					10.80			10.80	11.90
1510	For truck holding/waiting time past 1st hour per load, add				Hr.	94			94	103
1520	For short load (less than 4 C.Y.), add per load				Ea.	60.50			60.50	66.50
2000	For all lightweight aggregate, add				C.Y.	45%				

03 31 13 – Heavyweight Structural Concrete

03 31 13.35 Heavyweight Concrete, Ready Mix		Crew	Daily Output	Labor-Hours	Unit	Material	2015 Bare Costs Labor	Equipment	Total	Total Incl O&P
4000	Flowable fill: ash, cement, aggregate, water									
4100	40 – 80 psi				C.Y.	77			77	85
4150	Structural: ash, cement, aggregate, water & sand									
4200	50 psi				C.Y.	77			77	85
4250	140 psi					78			78	85.50
4300	500 psi					80			80	88
4350	1000 psi				↓	83.50			83.50	91.50

03 31 13.70 Placing Concrete		Crew	Daily Output	Labor-Hours	Unit	Material	Labor	Equipment	Total	Total Incl O&P
0010	**PLACING CONCRETE** R033105-70									
0020	Includes labor and equipment to place, level (strike off) and consolidate									
0050	Beams, elevated, small beams, pumped	C-20	60	1.067	C.Y.		43	13.05	56.05	84.50
0100	With crane and bucket	C-7	45	1.600			65.50	27	92.50	136
0200	Large beams, pumped	C-20	90	.711			28.50	8.70	37.20	56
0250	With crane and bucket	C-7	65	1.108			45.50	18.70	64.20	94
0400	Columns, square or round, 12" thick, pumped	C-20	60	1.067			43	13.05	56.05	84.50
0450	With crane and bucket	C-7	40	1.800			73.50	30.50	104	153
0600	18" thick, pumped	C-20	90	.711			28.50	8.70	37.20	56
0650	With crane and bucket	C-7	55	1.309			53.50	22	75.50	111
0800	24" thick, pumped	C-20	92	.696			28	8.50	36.50	55
0850	With crane and bucket	C-7	70	1.029			42	17.35	59.35	87
1000	36" thick, pumped	C-20	140	.457			18.45	5.60	24.05	36
1050	With crane and bucket	C-7	100	.720			29.50	12.15	41.65	61
1400	Elevated slabs, less than 6" thick, pumped	C-20	140	.457			18.45	5.60	24.05	36
1450	With crane and bucket	C-7	95	.758			31	12.80	43.80	64
1500	6" to 10" thick, pumped	C-20	160	.400			16.15	4.89	21.04	31.50
1550	With crane and bucket	C-7	110	.655			27	11.05	38.05	55.50
1600	Slabs over 10" thick, pumped	C-20	180	.356			14.35	4.35	18.70	28.50
1650	With crane and bucket	C-7	130	.554			22.50	9.35	31.85	47
1900	Footings, continuous, shallow, direct chute	C-6	120	.400			15.65	.52	16.17	26
1950	Pumped	C-20	150	.427			17.25	5.20	22.45	34
2000	With crane and bucket	C-7	90	.800			32.50	13.50	46	68
2100	Footings, continuous, deep, direct chute	C-6	140	.343			13.45	.45	13.90	22.50
2150	Pumped	C-20	160	.400			16.15	4.89	21.04	31.50
2200	With crane and bucket	C-7	110	.655			27	11.05	38.05	55.50
2400	Footings, spread, under 1 C.Y., direct chute	C-6	55	.873			34	1.13	35.13	57
2450	Pumped	C-20	65	.985			40	12.05	52.05	78
2500	With crane and bucket	C-7	45	1.600			65.50	27	92.50	136
2600	Over 5 C.Y., direct chute	C-6	120	.400			15.65	.52	16.17	26
2650	Pumped	C-20	150	.427			17.25	5.20	22.45	34
2700	With crane and bucket	C-7	100	.720			29.50	12.15	41.65	61
2900	Foundation mats, over 20 C.Y., direct chute	C-6	350	.137			5.35	.18	5.53	8.95
2950	Pumped	C-20	400	.160			6.45	1.96	8.41	12.65
3000	With crane and bucket	C-7	300	.240			9.80	4.05	13.85	20.50
3200	Grade beams, direct chute	C-6	150	.320			12.55	.42	12.97	21
3250	Pumped	C-20	180	.356			14.35	4.35	18.70	28.50
3300	With crane and bucket	C-7	120	.600			24.50	10.10	34.60	51
3500	High rise, for more than 5 stories, pumped, add per story	C-20	2100	.030			1.23	.37	1.60	2.41
3510	With crane and bucket, add per story	C-7	2100	.034			1.40	.58	1.98	2.91
3700	Pile caps, under 5 C.Y., direct chute	C-6	90	.533			21	.69	21.69	35
3750	Pumped	C-20	110	.582			23.50	7.10	30.60	46
3800	With crane and bucket	C-7	80	.900			37	15.15	52.15	76
3850	Pile cap, 5 C.Y. to 10 C.Y., direct chute	C-6	175	.274	↓		10.75	.36	11.11	17.90

03 31 13.70 Placing Concrete		Crew	Daily Output	Labor-Hours	Unit	Material	2015 Bare Costs Labor	Equipment	Total	Total Incl O&P
3900	Pumped	C-20	200	.320	C.Y.		12.95	3.91	16.86	25.50
3950	With crane and bucket	C-7	150	.480			19.65	8.10	27.75	41
4000	Over 10 C.Y., direct chute	C-6	215	.223			8.75	.29	9.04	14.55
4050	Pumped	C-20	240	.267			10.75	3.26	14.01	21
4100	With crane and bucket	C-7	185	.389			15.95	6.55	22.50	33
4300	Slab on grade, up to 6" thick, direct chute	C-6	110	.436			17.10	.57	17.67	28.50
4350	Pumped	C-20	130	.492			19.90	6	25.90	39
4400	With crane and bucket	C-7	110	.655			27	11.05	38.05	55.50
4600	Over 6" thick, direct chute	C-6	165	.291			11.40	.38	11.78	18.95
4650	Pumped	C-20	185	.346			14	4.23	18.23	27
4700	With crane and bucket	C-7	145	.497			20.50	8.35	28.85	42
4900	Walls, 8" thick, direct chute	C-6	90	.533			21	.69	21.69	35
4950	Pumped	C-20	100	.640			26	7.85	33.85	50.50
5000	With crane and bucket	C-7	80	.900			37	15.15	52.15	76
5050	12" thick, direct chute	C-6	100	.480			18.80	.62	19.42	31
5100	Pumped	C-20	110	.582			23.50	7.10	30.60	46
5200	With crane and bucket	C-7	90	.800			32.50	13.50	46	68
5300	15" thick, direct chute	C-6	105	.457			17.90	.59	18.49	29.50
5350	Pumped	C-20	120	.533			21.50	6.50	28	42
5400	With crane and bucket	C-7	95	.758			31	12.80	43.80	64
5600	Wheeled concrete dumping, add to placing costs above									
5610	Walking cart, 50' haul, add	C-18	32	.281	C.Y.		10.65	1.88	12.53	19.50
5620	150' haul, add		24	.375			14.20	2.50	16.70	26.50
5700	250' haul, add		18	.500			18.90	3.34	22.24	34.50
5800	Riding cart, 50' haul, add	C-19	80	.113			4.25	1.25	5.50	8.35
5810	150' haul, add		60	.150			5.65	1.66	7.31	11.15
5900	250' haul, add		45	.200			7.55	2.22	9.77	14.85
6000	Concrete in-fill for pan-type metal stairs and landings. Manual placement									
6010	includes up to 50' horizontal haul from point of concrete discharge.									
6100	Stair pan treads, 2" deep									
6110	Flights in 1st floor level up/down from discharge point	C-8A	3200	.015	S.F.		.61		.61	.98
6120	2nd floor level		2500	.019			.78		.78	1.25
6130	3rd floor level		2000	.024			.97		.97	1.57
6140	4th floor level		1800	.027			1.08		1.08	1.74
6200	Intermediate stair landings, pan-type 4" deep									
6210	Flights in 1st floor level up/down from discharge point	C-8A	2000	.024	S.F.		.97		.97	1.57
6220	2nd floor level		1500	.032			1.29		1.29	2.09
6230	3rd floor level		1200	.040			1.62		1.62	2.61
6240	4th floor level		1000	.048			1.94		1.94	3.14
9000	Minimum labor/equipment charge	C-6	2	24	Job		940	31	971	1,550

03 35 Concrete Finishing

03 35 13 – High-Tolerance Concrete Floor Finishing

03 35 13.30 Finishing Floors, High Tolerance		Crew	Daily Output	Labor-Hours	Unit	Material	2015 Bare Costs Labor	Equipment	Total	Total Incl O&P
0010	**FINISHING FLOORS, HIGH TOLERANCE**									
0012	Finishing of fresh concrete flatwork requires that concrete									
0013	first be placed, struck off & consolidated									
0015	Basic finishing for various unspecified flatwork									
0100	Bull float only	C-10	4000	.006	S.F.		.26		.26	.41
0125	Bull float & manual float		2000	.012			.51		.51	.82
0150	Bull float, manual float, & broom finish, w/edging & joints		1850	.013			.55		.55	.88
0200	Bull float, manual float & manual steel trowel		1265	.019			.81		.81	1.29
0210	For specified Random Access Floors in ACI Classes 1, 2, 3 and 4 to achieve									
0215	Composite Overall Floor Flatness and Levelness values up to FF35/FL25									
0250	Bull float, machine float & machine trowel (walk-behind)	C-10C	1715	.014	S.F.		.60	.03	.63	.98
0300	Power screed, bull float, machine float & trowel (walk-behind)	C-10D	2400	.010			.43	.05	.48	.73
0350	Power screed, bull float, machine float & trowel (ride-on)	C-10E	4000	.006			.26	.06	.32	.48
0352	For specified Random Access Floors in ACI Classes 5, 6, 7 and 8 to achieve									
0354	Composite Overall Floor Flatness and Levelness values up to FF50/FL50									
0356	Add for two-dimensional restraightening after power float	C-10	6000	.004	S.F.		.17		.17	.27
0358	For specified Random or Defined Access Floors in ACI Class 9 to achieve									
0360	Composite Overall Floor Flatness and Levelness values up to FF100/FL100									
0362	Add for two-dimensional restraightening after bull float & power float	C-10	3000	.008	S.F.		.34		.34	.54
0364	For specified Superflat Defined Access Floors in ACI Class 9 to achieve									
0366	Minimum Floor Flatness and Levelness values of FF100/FL100									
0368	Add for 2-dim'l restraightening after bull float, power float, power trowel	C-10	2000	.012	S.F.		.51		.51	.82
9100	Minimum labor/equipment charge	″	2	12	Job		510		510	815

03 35 16 – Heavy-Duty Concrete Floor Finishing

03 35 16.30 Finishing Floors, Heavy-Duty		Crew	Daily Output	Labor-Hours	Unit	Material	Labor	Equipment	Total	Total Incl O&P
0010	**FINISHING FLOORS, HEAVY-DUTY**									
1800	Floor abrasives, dry shake on fresh concrete, .25 psf, aluminum oxide	1 Cefi	850	.009	S.F.	.53	.42		.95	1.25
1850	Silicon carbide		850	.009		.79	.42		1.21	1.54
2000	Floor hardeners, dry shake, metallic, light service, .50 psf		850	.009		.54	.42		.96	1.26
2050	Medium service, .75 psf		750	.011		.80	.48		1.28	1.64
2100	Heavy service, 1.0 psf		650	.012		1.07	.55		1.62	2.06
2150	Extra heavy, 1.5 psf		575	.014		1.61	.63		2.24	2.76
2300	Non-metallic, light service, .50 psf		850	.009		.21	.42		.63	.91
2350	Medium service, .75 psf		750	.011		.32	.48		.80	1.11
2400	Heavy service, 1.00 psf		650	.012		.43	.55		.98	1.35
2450	Extra heavy, 1.50 psf		575	.014		.64	.63		1.27	1.70
2800	Trap rock wearing surface, dry shake, for monolithic floors									
2810	2.0 psf	C-10B	1250	.032	S.F.	.02	1.30	.22	1.54	2.35
3800	Dustproofing, liquid, for cured concrete, solvent-based, 1 coat	1 Cefi	1900	.004		.18	.19		.37	.49
3850	2 coats		1300	.006		.63	.28		.91	1.14
4000	Epoxy-based, 1 coat		1500	.005		.14	.24		.38	.53
4050	2 coats		1500	.005		.28	.24		.52	.69

03 35 19 – Colored Concrete Finishing

03 35 19.30 Finishing Floors, Colored		Crew	Daily Output	Labor-Hours	Unit	Material	Labor	Equipment	Total	Total Incl O&P
0010	**FINISHING FLOORS, COLORED**									
3000	Floor coloring, dry shake on fresh concrete (0.6 psf)	1 Cefi	1300	.006	S.F.	.44	.28		.72	.92
3050	(1.0 psf)	″	625	.013	″	.73	.58		1.31	1.72
3100	Colored dry shake powder only				Lb.	.73			.73	.81
3600	1/2″ topping using 0.6 psf dry shake powdered color	C-10B	590	.068	S.F.	4.96	2.75	.46	8.17	10.40
3650	1.0 psf dry shake powdered color	″	590	.068	″	5.25	2.75	.46	8.46	10.75

03 35 Concrete Finishing

03 35 23 – Exposed Aggregate Concrete Finishing

03 35 23.30 Finishing Floors, Exposed Aggregate		Crew	Daily Output	Labor-Hours	Unit	Material	2015 Bare Costs Labor	Equipment	Total	Total Incl O&P
0010	**FINISHING FLOORS, EXPOSED AGGREGATE**									
1600	Exposed local aggregate finish, seeded on fresh concrete, 3 lb. per S.F.	1 Cefi	625	.013	S.F.	.47	.58		1.05	1.42
1650	4 lb. per S.F.	"	465	.017	"	.87	.77		1.64	2.18

03 35 29 – Tooled Concrete Finishing

03 35 29.30 Finishing Floors, Tooled

Line	Description	Crew	Daily Output	Labor-Hours	Unit	Material	Labor	Equipment	Total	Total Incl O&P
0010	**FINISHING FLOORS, TOOLED**									
4400	Stair finish, fresh concrete, float finish	1 Cefi	275	.029	S.F.		1.31		1.31	2.07
4500	Steel trowel finish		200	.040			1.80		1.80	2.85
4600	Silicon carbide finish, dry shake on fresh concrete, .25 psf		150	.053		.53	2.40		2.93	4.37

03 35 29.60 Finishing Walls

Line	Description	Crew	Daily Output	Labor-Hours	Unit	Material	Labor	Equipment	Total	Total Incl O&P
0010	**FINISHING WALLS**									
0020	Break ties and patch voids	1 Cefi	540	.015	S.F.	.04	.67		.71	1.09
0050	Burlap rub with grout		450	.018		.04	.80		.84	1.31
0100	Carborundum rub, dry		270	.030			1.33		1.33	2.11
0150	Wet rub		175	.046			2.06		2.06	3.25
0300	Bush hammer, green concrete	B-39	1000	.048			1.91	.23	2.14	3.37
0350	Cured concrete	"	650	.074			2.94	.36	3.30	5.15
0500	Acid etch	1 Cefi	575	.014		.14	.63		.77	1.14
0600	Float finish, 1/16" thick	"	300	.027		.36	1.20		1.56	2.30
0700	Sandblast, light penetration	E-11	1100	.029		.46	1.23	.21	1.90	2.88
0750	Heavy penetration	"	375	.085		.92	3.61	.63	5.16	7.95
0850	Grind form fins flush	1 Clab	700	.011	L.F.		.43		.43	.70
9000	Minimum labor/equipment charge	C-10	2	12	Job		510		510	815

03 35 33 – Stamped Concrete Finishing

03 35 33.50 Slab Texture Stamping

Line	Description	Crew	Daily Output	Labor-Hours	Unit	Material	Labor	Equipment	Total	Total Incl O&P
0010	**SLAB TEXTURE STAMPING**									
0050	Stamping requires that concrete first be placed, struck off, consolidated,									
0060	bull floated and free of bleed water. Decorative stamping tasks include:									
0100	Step 1 - first application of dry shake colored hardener	1 Cefi	6400	.001	S.F.	.43	.06		.49	.56
0110	Step 2 - bull float		6400	.001			.06		.06	.09
0130	Step 3 - second application of dry shake colored hardener		6400	.001		.21	.06		.27	.32
0140	Step 4 - bull float, manual float & steel trowel	3 Cefi	1280	.019			.84		.84	1.33
0150	Step 5 - application of dry shake colored release agent	1 Cefi	6400	.001		.10	.06		.16	.20
0160	Step 6 - place, tamp & remove mats	3 Cefi	2400	.010		1.43	.45		1.88	2.29
0170	Step 7 - touch up edges, mat joints & simulated grout lines	1 Cefi	1280	.006			.28		.28	.44
0300	Alternate stamping estimating method includes all tasks above	4 Cefi	800	.040		2.17	1.80		3.97	5.25
0400	Step 8 - pressure wash @ 3000 psi after 24 hours	1 Cefi	1600	.005			.23		.23	.36
0500	Step 9 - roll 2 coats cure/seal compound when dry	"	800	.010		.62	.45		1.07	1.39

03 35 43 – Polished Concrete Finishing

03 35 43.10 Polished Concrete Floors

Line	Description	Crew	Daily Output	Labor-Hours	Unit	Material	Labor	Equipment	Total	Total Incl O&P
0010	**POLISHED CONCRETE FLOORS** R033543-10									
0015	Processing of cured concrete to include grinding, honing,									
0020	and polishing of interior floors with 22" segmented diamond									
0025	planetary floor grinder (2 passes in different directions per grit)									
0100	Removal of pre-existing coatings, dry, with carbide discs using									
0105	dry vacuum pick-up system, final hand sweeping									
0110	Glue, adhesive or tar	J-4	1.60	15	M.S.F.	19.75	640	135	794.75	1,200
0120	Paint, epoxy, 1 coat		3.60	6.667		19.75	284	60	363.75	545
0130	2 coats		1.80	13.333		19.75	565	120	704.75	1,050
0200	Grinding and edging, wet, including wet vac pick-up and auto									

03 35 Concrete Finishing

03 35 43 – Polished Concrete Finishing

03 35 43.10 Polished Concrete Floors		Crew	Daily Output	Labor-Hours	Unit	2015 Bare Costs Material	Labor	Equipment	Total	Total Incl O&P
0205	scrubbing between grit changes									
0210	40-grit diamond/metal matrix	J-4A	1.60	20	M.S.F.	31.50	825	281	1,137.50	1,675
0220	80-grit diamond/metal matrix		2	16		31.50	660	225	916.50	1,325
0230	120-grit diamond/metal matrix		2.40	13.333		31.50	550	187	768.50	1,125
0240	200-grit diamond/metal matrix		2.80	11.429		31.50	470	161	662.50	970
0300	Spray on dye or stain (1 coat)	1 Cefi	16	.500		216	22.50		238.50	274
0400	Spray on densifier/hardener (2 coats)	"	8	1		350	45		395	455
0410	Auto scrubbing after 2nd coat, when dry	J-4B	16	.500			18.80	14.60	33.40	47
0500	Honing and edging, wet, including wet vac pick-up and auto									
0505	scrubbing between grit changes									
0510	100-grit diamond/resin matrix	J-4A	2.80	11.429	M.S.F.	31.50	470	161	662.50	970
0520	200-grit diamond/resin matrix	"	2.80	11.429	"	31.50	470	161	662.50	970
0530	Dry, including dry vacuum pick-up system, final hand sweeping									
0540	400-grit diamond/resin matrix	J-4A	2.80	11.429	M.S.F.	31.50	470	161	662.50	970
0600	Polishing and edging, dry, including dry vac pick-up and hand									
0605	sweeping between grit changes									
0610	800-grit diamond/resin matrix	J-4A	2.80	11.429	M.S.F.	31.50	470	161	662.50	970
0620	1500-grit diamond/resin matrix		2.80	11.429		31.50	470	161	662.50	970
0630	3000-grit diamond/resin matrix		2.80	11.429		31.50	470	161	662.50	970
0700	Auto scrubbing after final polishing step	J-4B	16	.500			18.80	14.60	33.40	47

03 37 Specialty Placed Concrete

03 37 13 – Shotcrete

03 37 13.30 Gunite (Dry-Mix)

		Crew	Daily Output	Labor-Hours	Unit	Material	Labor	Equipment	Total	Total Incl O&P
0010	**GUNITE (DRY-MIX)**									
0020	Typical in place, 1" layers, no mesh included	C-16	2000	.028	S.F.	.35	1.17	.20	1.72	2.49
0100	Mesh for gunite 2 x 2, #12	2 Rodm	800	.020		.63	1.05		1.68	2.41
0150	#4 reinforcing bars @ 6" each way	"	500	.032		1.61	1.68		3.29	4.52
0300	Typical in place, including mesh, 2" thick, flat surfaces	C-16	1000	.056		1.32	2.34	.39	4.05	5.65
0350	Curved surfaces		500	.112		1.32	4.69	.79	6.80	9.85
0500	4" thick, flat surfaces		750	.075		2.02	3.13	.52	5.67	7.85
0550	Curved surfaces		350	.160		2.02	6.70	1.12	9.84	14.25
0900	Prepare old walls, no scaffolding, good condition	C-10	1000	.024			1.02		1.02	1.63
0950	Poor condition	"	275	.087			3.71		3.71	5.95
1100	For high finish requirement or close tolerance, add						50%			
1150	Very high						110%			
9000	Minimum labor/equipment charge	C-10	1	24	Job		1,025		1,025	1,625

03 37 13.60 Shotcrete (Wet-Mix)

		Crew	Daily Output	Labor-Hours	Unit	Material	Labor	Equipment	Total	Total Incl O&P
0010	**SHOTCRETE (WET-MIX)**									
0020	Wet mix, placed @ up to 12 C.Y. per hour, 3000 psi	C-8C	80	.600	C.Y.	113	25	5.70	143.70	170
0100	Up to 35 C.Y. per hour	C-8E	240	.200	"	101	8.20	2.21	111.41	128
1010	Fiber reinforced, 1" thick	C-8C	1740	.028	S.F.	.83	1.14	.26	2.23	3.05
1020	2" thick		900	.053		1.66	2.20	.51	4.37	5.95
1030	3" thick		825	.058		2.48	2.40	.55	5.43	7.25
1040	4" thick		750	.064		3.31	2.65	.61	6.57	8.60

03 39 Concrete Curing

03 39 13 – Water Concrete Curing

03 39 13.50 Water Curing		Crew	Daily Output	Labor-Hours	Unit	Material	Labor	Equipment	Total	Total Incl O&P
						2015 Bare Costs				
0010	**WATER CURING**									
0015	With burlap, 4 uses assumed, 7.5 oz.	2 Clab	55	.291	C.S.F.	13.05	10.95		24	32.50
0100	10 oz.	″	55	.291	″	23.50	10.95		34.45	44
0400	Curing blankets, 1″ to 2″ thick, buy				S.F.	.19			.19	.21
9000	Minimum labor/equipment charge	1 Clab	5	1.600	Job		60		60	98.50

03 39 23 – Membrane Concrete Curing

03 39 23.13 Chemical Compound Membrane Concrete Curing		Crew	Daily Output	Labor-Hours	Unit	Material	Labor	Equipment	Total	Total Incl O&P
0010	**CHEMICAL COMPOUND MEMBRANE CONCRETE CURING**									
0300	Sprayed membrane curing compound	2 Clab	95	.168	C.S.F.	11.05	6.35		17.40	22.50
0700	Curing compound, solvent based, 400 S.F./gal., 55 gallon lots				Gal.	24.50			24.50	27
0720	5 gallon lots					30.50			30.50	34
0800	Curing compound, water based, 250 S.F./gal., 55 gallon lots					22.50			22.50	25
0820	5 gallon lots				↓	25.50			25.50	28.50

03 39 23.23 Sheet Membrane Concrete Curing		Crew	Daily Output	Labor-Hours	Unit	Material	Labor	Equipment	Total	Total Incl O&P
0010	**SHEET MEMBRANE CONCRETE CURING**									
0200	Curing blanket, burlap/poly, 2-ply	2 Clab	70	.229	C.S.F.	23	8.60		31.60	39.50

03 41 Precast Structural Concrete

03 41 13 – Precast Concrete Hollow Core Planks

03 41 13.50 Precast Slab Planks		Crew	Daily Output	Labor-Hours	Unit	Material	Labor	Equipment	Total	Total Incl O&P
0010	**PRECAST SLAB PLANKS**									
0020	Prestressed roof/floor members, grouted, solid, 4″ thick	C-11	2400	.030	S.F.	5.85	1.56	.76	8.17	10
0050	6″ thick		2800	.026		6.65	1.34	.65	8.64	10.40
0100	Hollow, 8″ thick		3200	.023		7.10	1.17	.57	8.84	10.50
0150	10″ thick		3600	.020		7.75	1.04	.50	9.29	10.90
0200	12″ thick	↓	4000	.018	↓	7.90	.93	.45	9.28	10.80

03 41 16 – Precast Concrete Slabs

03 41 16.20 Precast Concrete Channel Slabs		Crew	Daily Output	Labor-Hours	Unit	Material	Labor	Equipment	Total	Total Incl O&P
0010	**PRECAST CONCRETE CHANNEL SLABS**									
0335	Lightweight concrete channel slab, long runs, 2-3/4″ thick	C-12	1575	.030	S.F.	8.50	1.42	.42	10.34	12.10
0375	3-3/4″ thick		1550	.031		8.75	1.44	.42	10.61	12.45
0475	4-3/4″ thick		1525	.031		9.75	1.46	.43	11.64	13.60
1275	Short pieces, 2-3/4″ thick		785	.061		12.75	2.84	.83	16.42	19.60
1375	3-3/4″ thick		770	.062		13.15	2.90	.85	16.90	20
1475	4-3/4″ thick	↓	762	.063	↓	14.65	2.93	.86	18.44	22

03 41 16.50 Precast Lightweight Concrete Plank		Crew	Daily Output	Labor-Hours	Unit	Material	Labor	Equipment	Total	Total Incl O&P
0010	**PRECAST LIGHTWEIGHT CONCRETE PLANK**									
0015	Lightweight plank, nailable, T&G, 2″ thick	C-12	1800	.027	S.F.	7.65	1.24	.36	9.25	10.80
0150	For premium ceiling finish, add				″	10%				
0200	For sloping roofs, slope over 4 in 12, add						25%			
0250	Slope over 6 in 12, add						150%			

03 41 23 – Precast Concrete Stairs

03 41 23.50 Precast Stairs		Crew	Daily Output	Labor-Hours	Unit	Material	Labor	Equipment	Total	Total Incl O&P
0010	**PRECAST STAIRS**									
0020	Precast concrete treads on steel stringers, 3′ wide	C-12	75	.640	Riser	135	30	8.70	173.70	206
0300	Front entrance, 5′ wide with 48″ platform, 2 risers		16	3	Flight	485	140	41	666	800
0350	5 risers		12	4		765	186	54.50	1,005.50	1,200
0500	6′ wide, 2 risers		15	3.200		535	149	43.50	727.50	880
0550	5 risers	↓	11	4.364	↓	845	203	59.50	1,107.50	1,325

03 41 23 – Precast Concrete Stairs

	03 41 23.50 Precast Stairs	Crew	Daily Output	Labor-Hours	Unit	2015 Bare Costs Material	Labor	Equipment	Total	Total Incl O&P
0700	7′ wide, 2 risers	C-12	14	3.429	Flight	685	160	46.50	891.50	1,050
0750	5 risers		10	4.800		1,125	223	65.50	1,413.50	1,675
1200	Basement entrance stairwell, 6 steps, incl. steel bulkhead door	B-51	22	2.182		1,450	83.50	11.15	1,544.65	1,725
1250	14 steps	″	11	4.364		2,400	167	22.50	2,589.50	2,925

03 41 33 – Precast Structural Pretensioned Concrete

03 41 33.10 Precast Beams

		Crew	Daily Output	Labor-Hours	Unit	Material	Labor	Equipment	Total	Total Incl O&P
0010	**PRECAST BEAMS**									
0011	L-shaped, 20′ span, 12″ x 20″	C-11	32	2.250	Ea.	3,000	117	56.50	3,173.50	3,575
0060	18″ x 36″		24	3		4,125	156	75.50	4,356.50	4,875
0100	24″ x 44″		22	3.273		4,925	170	82.50	5,177.50	5,825
0150	30′ span, 12″ x 36″		24	3		5,550	156	75.50	5,781.50	6,475
0200	18″ x 44″		20	3.600		6,775	187	90.50	7,052.50	7,875
0250	24″ x 52″		16	4.500		8,125	234	113	8,472	9,450
0400	40′ span, 12″ x 52″		20	3.600		8,625	187	90.50	8,902.50	9,925
0450	18″ x 52″		16	4.500		9,625	234	113	9,972	11,100
0500	24″ x 52″		12	6		10,800	310	151	11,261	12,600
1200	Rectangular, 20′ span, 12″ x 20″		32	2.250		2,925	117	56.50	3,098.50	3,475
1250	18″ x 36″		24	3		3,600	156	75.50	3,831.50	4,325
1300	24″ x 44″		22	3.273		4,325	170	82.50	4,577.50	5,150
1400	30′ span, 12″ x 36″		24	3		4,875	156	75.50	5,106.50	5,700
1450	18″ x 44″		20	3.600		5,850	187	90.50	6,127.50	6,875
1500	24″ x 52″		16	4.500		7,075	234	113	7,422	8,300
1600	40′ span, 12″ x 52″		20	3.600		7,225	187	90.50	7,502.50	8,375
1650	18″ x 52″		16	4.500		8,225	234	113	8,572	9,575
1700	24″ x 52″		12	6		9,425	310	151	9,886	11,100
2000	"T" shaped, 20′ span, 12″ x 20″		32	2.250		3,425	117	56.50	3,598.50	4,050
2050	18″ x 36″		24	3		4,600	156	75.50	4,831.50	5,425
2100	24″ x 44″		22	3.273		5,525	170	82.50	5,777.50	6,475
2200	30′ span, 12″ x 36″		24	3		6,325	156	75.50	6,556.50	7,300
2250	18″ x 44″		20	3.600		7,675	187	90.50	7,952.50	8,850
2300	24″ x 52″		16	4.500		9,175	234	113	9,522	10,600
2500	40′ span, 12″ x 52″		20	3.600		10,400	187	90.50	10,677.50	11,900
2550	18″ x 52″		16	4.500		11,000	234	113	11,347	12,700
2600	24″ x 52″		12	6		12,200	310	151	12,661	14,200

03 41 33.15 Precast Columns

		Crew	Daily Output	Labor-Hours	Unit	Material	Labor	Equipment	Total	Total Incl O&P
0010	**PRECAST COLUMNS**									
0020	Rectangular to 12′ high, 16″ x 16″	C-11	120	.600	L.F.	172	31	15.10	218.10	261
0050	24″ x 24″		96	.750		234	39	18.90	291.90	345
0300	24′ high, 28″ x 28″		192	.375		271	19.50	9.45	299.95	345
0350	36″ x 36″		144	.500		365	26	12.60	403.60	460
0700	24′ high, 1 haunch, 12″ x 12″		32	2.250	Ea.	3,625	117	56.50	3,798.50	4,275
0800	20″ x 20″		28	2.571	″	4,975	134	65	5,174	5,775

03 41 33.25 Precast Joists

		Crew	Daily Output	Labor-Hours	Unit	Material	Labor	Equipment	Total	Total Incl O&P
0010	**PRECAST JOISTS**									
0015	40 psf L.L., 6″ deep for 12′ spans	C-12	600	.080	L.F.	24	3.72	1.09	28.81	34
0050	8″ deep for 16′ spans		575	.083		40	3.88	1.14	45.02	51.50
0100	10″ deep for 20′ spans		550	.087		70	4.06	1.19	75.25	85
0150	12″ deep for 24′ spans		525	.091		96	4.25	1.25	101.50	114

03 41 Precast Structural Concrete

03 41 33 – Precast Structural Pretensioned Concrete

03 41 33.60 Precast Tees		Crew	Daily Output	Labor-Hours	Unit	Material	Labor	Equipment	Total	Total Incl O&P
						2015 Bare Costs				
0010	**PRECAST TEES**									
0020	Quad tee, short spans, roof	C-11	7200	.010	S.F.	7.65	.52	.25	8.42	9.60
0050	Floor		7200	.010		7.65	.52	.25	8.42	9.60
0200	Double tee, floor members, 60′ span		8400	.009		9.10	.45	.22	9.77	11
0250	80′ span		8000	.009		11.80	.47	.23	12.50	14.05
0300	Roof members, 30′ span		4800	.015		7.65	.78	.38	8.81	10.20
0350	50′ span		6400	.011		8.40	.58	.28	9.26	10.60
0400	Wall members, up to 55′ high		3600	.020		11.35	1.04	.50	12.89	14.90
0500	Single tee roof members, 40′ span		3200	.023		11.30	1.17	.57	13.04	15.10
0550	80′ span		5120	.014		11.70	.73	.35	12.78	14.55
0600	100′ span		6000	.012		17.15	.62	.30	18.07	20.50
0650	120′ span	↓	6000	.012	↓	18.80	.62	.30	19.72	22
1000	Double tees, floor members									
1100	Lightweight, 20″ x 8′ wide, 45′ span	C-11	20	3.600	Ea.	3,025	187	90.50	3,302.50	3,750
1150	24″ x 8′ wide, 50′ span		18	4		3,350	208	101	3,659	4,175
1200	32″ x 10′ wide, 60′ span		16	4.500		5,050	234	113	5,397	6,075
1250	Standard weight, 12″ x 8′ wide, 20′ span		22	3.273		1,225	170	82.50	1,477.50	1,750
1300	16″ x 8′ wide, 25′ span		20	3.600		1,525	187	90.50	1,802.50	2,100
1350	18″ x 8′ wide, 30′ span		20	3.600		1,825	187	90.50	2,102.50	2,450
1400	20″ x 8′ wide, 45′ span		18	4		2,750	208	101	3,059	3,500
1450	24″ x 8′ wide, 50′ span		16	4.500		3,050	234	113	3,397	3,875
1500	32″ x 10′ wide, 60′ span	↓	14	5.143	↓	4,575	267	130	4,972	5,650
2000	Roof members									
2050	Lightweight, 20″ x 8′ wide, 40′ span	C-11	20	3.600	Ea.	2,700	187	90.50	2,977.50	3,375
2100	24″ x 8′ wide, 50′ span		18	4		3,350	208	101	3,659	4,175
2150	32″ x 10′ wide, 60′ span		16	4.500		5,050	234	113	5,397	6,075
2200	Standard weight, 12″ x 8′ wide, 30′ span		22	3.273		1,825	170	82.50	2,077.50	2,425
2250	16″ x 8′ wide, 30′ span		20	3.600		1,925	187	90.50	2,202.50	2,550
2300	18″ x 8′ wide, 30′ span		20	3.600		2,025	187	90.50	2,302.50	2,650
2350	20″ x 8′ wide, 40′ span		18	4		2,450	208	101	2,759	3,175
2400	24″ x 8′ wide, 50′ span		16	4.500		3,050	234	113	3,397	3,875
2450	32″ x 10′ wide, 60′ span	↓	14	5.143	↓	4,575	267	130	4,972	5,650

03 45 Precast Architectural Concrete

03 45 13 – Faced Architectural Precast Concrete

03 45 13.50 Precast Wall Panels		Crew	Daily Output	Labor-Hours	Unit	Material	Labor	Equipment	Total	Total Incl O&P
0010	**PRECAST WALL PANELS**									
0050	Uninsulated, smooth gray									
0150	Low rise, 4′ x 8′ x 4″ thick	C-11	320	.225	S.F.	20.50	11.70	5.65	37.85	50
0210	8′ x 8′, 4″ thick		576	.125		20.50	6.50	3.15	30.15	37.50
0250	8′ x 16′ x 4″ thick		1024	.070		20.50	3.65	1.77	25.92	31
0600	High rise, 4′ x 8′ x 4″ thick		288	.250		20.50	13	6.30	39.80	53
0650	8′ x 8′ x 4″ thick		512	.141		20.50	7.30	3.54	31.34	39.50
0700	8′ x 16′ x 4″ thick		768	.094		20.50	4.87	2.36	27.73	33.50
0750	10′ x 20′, 6″ thick	↓	1400	.051		34.50	2.67	1.30	38.47	44
0800	Insulated panel, 2″ polystyrene, add					1.04			1.04	1.14
0850	2″ urethane, add					.80			.80	.88
1200	Finishes, white, add					2.93			2.93	3.22
1250	Exposed aggregate, add					2.17			2.17	2.39
1300	Granite faced, domestic, add				↓	29.50			29.50	32.50

03 45 Precast Architectural Concrete

03 45 13 – Faced Architectural Precast Concrete

03 45 13.50 Precast Wall Panels		Crew	Daily Output	Labor-Hours	Unit	Material	Labor	Equipment	Total	Total Incl O&P
						2015 Bare Costs				
1350	Brick faced, modular, red, add				S.F.	9			9	9.90
2200	Fiberglass reinforced cement with urethane core									
2210	R20, 8′ x 8′, 5″ plain finish	E-2	750	.075	S.F.	21	3.86	2.01	26.87	32.50
2220	Exposed aggregate or brick finish	″	600	.093	″	32	4.83	2.52	39.35	46

03 47 Site-Cast Concrete

03 47 13 – Tilt-Up Concrete

03 47 13.50 Tilt-Up Wall Panels

Line	Description	Crew	Daily Output	Labor-Hours	Unit	Material	Labor	Equipment	Total	Total Incl O&P
0010	**TILT-UP WALL PANELS**									
0015	Wall panel construction, walls only, 5-1/2″ thick	C-14	1600	.090	S.F.	5.55	4.16	1.02	10.73	13.95
0100	7-1/2″ thick		1550	.093		6.90	4.29	1.05	12.24	15.65
0500	Walls and columns, 5-1/2″ thick walls, 12″ x 12″ columns		1565	.092		8.30	4.25	1.04	13.59	17.15
0550	7-1/2″ thick wall, 12″ x 12″ columns		1370	.105		10.10	4.85	1.19	16.14	20.50
0800	Columns only, site precast, 12″ x 12″		200	.720	L.F.	20	33.50	8.10	61.60	85
0850	16″ x 16″		105	1.371	″	29	63.50	15.45	107.95	152

03 48 Precast Concrete Specialties

03 48 43 – Precast Concrete Trim

03 48 43.40 Precast Lintels

Line	Description	Crew	Daily Output	Labor-Hours	Unit	Material	Labor	Equipment	Total	Total Incl O&P
0010	**PRECAST LINTELS**, smooth gray, prestressed, stock units only									
0800	4″ wide, 8″ high, x 4′ long	D-10	28	1.143	Ea.	25	52.50	16.90	94.40	131
0850	8′ long		24	1.333		60.50	61.50	19.70	141.70	187
1000	6″ wide, 8″ high, x 4′ long		26	1.231		35.50	56.50	18.20	110.20	151
1050	10′ long		22	1.455		91	67	21.50	179.50	232
1200	8″ wide, 8″ high, x 4′ long		24	1.333		45	61.50	19.70	126.20	170
1250	12′ long		20	1.600		146	73.50	23.50	243	305
1275	For custom sizes, types, colors, or finishes of precast lintels, add					150%				

03 48 43.90 Precast Window Sills

Line	Description	Crew	Daily Output	Labor-Hours	Unit	Material	Labor	Equipment	Total	Total Incl O&P
0010	**PRECAST WINDOW SILLS**									
0600	Precast concrete, 4″ tapers to 3″, 9″ wide	D-1	70	.229	L.F.	11.90	9.60		21.50	29
0650	11″ wide		60	.267		15.75	11.25		27	35.50
0700	13″ wide, 3 1/2″ tapers to 2 1/2″, 12″ wall		50	.320		15.50	13.45		28.95	39

03 51 Cast Roof Decks

03 51 13 – Cementitious Wood Fiber Decks

03 51 13.50 Cementitious/Wood Fiber Planks

Line	Description	Crew	Daily Output	Labor-Hours	Unit	Material	Labor	Equipment	Total	Total Incl O&P
0010	**CEMENTITIOUS/WOOD FIBER PLANKS** R051223-50									
0050	Plank, beveled edge, 1″ thick	2 Carp	1000	.016	S.F.	2.50	.75		3.25	3.98
0100	1-1/2″ thick		975	.016		3.25	.77		4.02	4.84
0150	T & G, 2″ thick		950	.017		2.75	.79		3.54	4.33
0200	2-1/2″ thick		925	.017		3	.81		3.81	4.63
0250	3″ thick		900	.018		3.40	.83		4.23	5.10
1000	Bulb tee, sub-purlin and grout, 6′ span, add	E-1	5000	.005		2.16	.25	.03	2.44	2.83
1100	8′ span	″	4200	.006		2.16	.30	.03	2.49	2.93

03 51 Cast Roof Decks

03 51 16 – Gypsum Concrete Roof Decks

03 51 16.50 Gypsum Roof Deck		Crew	Daily Output	Labor-Hours	Unit	Material	2015 Bare Costs Labor	Equipment	Total	Total Incl O&P
0010	**GYPSUM ROOF DECK**									
1000	Poured gypsum, 2″ thick	C-8	6000	.009	S.F.	1.45	.39	.12	1.96	2.35
1100	3″ thick	″	4800	.012	″	2.17	.49	.15	2.81	3.35

03 52 Lightweight Concrete Roof Insulation

03 52 16 – Lightweight Insulating Concrete

03 52 16.13 Lightweight Cellular Insulating Concrete

Line	Description	Crew	Daily Output	Labor-Hours	Unit	Material	Labor	Equipment	Total	Total Incl O&P
0010	**LIGHTWEIGHT CELLULAR INSULATING CONCRETE**									
0020	Portland cement and foaming agent G	C-8	50	1.120	C.Y.	123	47	14.40	184.40	226

03 52 16.16 Lightweight Aggregate Insulating Concrete

Line	Description	Crew	Daily Output	Labor-Hours	Unit	Material	Labor	Equipment	Total	Total Incl O&P
0010	**LIGHTWEIGHT AGGREGATE INSULATING CONCRETE**									
0100	Poured vermiculite or perlite, field mix,									
0110	1:6 field mix G	C-8	50	1.120	C.Y.	252	47	14.40	313.40	370
0200	Ready mix, 1:6 mix, roof fill, 2″ thick G		10000	.006	S.F.	1.40	.23	.07	1.70	2
0250	3″ thick G		7700	.007		2.10	.30	.09	2.49	2.90
0400	Expanded volcanic glass rock, 1″ thick G	2 Carp	1500	.011		.52	.50		1.02	1.39
0450	3″ thick G	″	1200	.013		1.56	.63		2.19	2.75

03 53 Concrete Topping

03 53 16 – Iron-Aggregate Concrete Topping

03 53 16.50 Floor Topping

Line	Description	Crew	Daily Output	Labor-Hours	Unit	Material	Labor	Equipment	Total	Total Incl O&P
0010	**FLOOR TOPPING**									
0400	Integral topping/finish, on fresh concrete, using 1:1:2 mix, 3/16″ thick	C-10B	1000	.040	S.F.	.10	1.62	.27	1.99	3.03
0450	1/2″ thick		950	.042		.28	1.71	.29	2.28	3.39
0500	3/4″ thick		850	.047		.42	1.91	.32	2.65	3.89
0600	1″ thick		750	.053		.56	2.16	.36	3.08	4.50
0800	Granolithic topping, on fresh or cured concrete, 1:1:1-1/2 mix, 1/2″ thick		590	.068		.31	2.75	.46	3.52	5.30
0820	3/4″ thick		580	.069		.46	2.80	.47	3.73	5.55
0850	1″ thick		575	.070		.62	2.82	.47	3.91	5.75
0950	2″ thick		500	.080		1.23	3.24	.55	5.02	7.20
1200	Heavy duty, 1:1:2, 3/4″ thick, preshrunk, gray, 20 M.S.F.		320	.125		.80	5.05	.85	6.70	10
1300	100 M.S.F.		380	.105		.42	4.27	.72	5.41	8.15

03 54 Cast Underlayment

03 54 13 – Gypsum Cement Underlayment

03 54 13.50 Poured Gypsum Underlayment

Line	Description	Crew	Daily Output	Labor-Hours	Unit	Material	Labor	Equipment	Total	Total Incl O&P
0010	**POURED GYPSUM UNDERLAYMENT**									
0400	Underlayment, gypsum based, self-leveling 2500 psi, pumped, 1/2″ thick	C-8	24000	.002	S.F.	.36	.10	.03	.49	.59
0500	3/4″ thick		20000	.003		.54	.12	.04	.70	.83
0600	1″ thick		16000	.004		.72	.15	.05	.92	1.09
1400	Hand placed, 1/2″ thick	C-18	450	.020		.36	.76	.13	1.25	1.79
1500	3/4″ thick	″	300	.030		.54	1.13	.20	1.87	2.68

03 54 Cast Underlayment

03 54 16 – Hydraulic Cement Underlayment

03 54 16.50 Cement Underlayment		Crew	Daily Output	Labor-Hours	Unit	Material	Labor	Equipment	Total	Total Incl O&P
						2015 Bare Costs				
0010	**CEMENT UNDERLAYMENT**									
2510	Underlayment, P.C based, self-leveling, 4100 psi, pumped, 1/4″ thick	C-8	20000	.003	S.F.	1.55	.12	.04	1.71	1.93
2520	1/2″ thick		19000	.003		3.09	.12	.04	3.25	3.64
2530	3/4″ thick		18000	.003		4.64	.13	.04	4.81	5.35
2540	1″ thick		17000	.003		6.20	.14	.04	6.38	7.05
2550	1-1/2″ thick		15000	.004		9.30	.16	.05	9.51	10.50
2560	Hand placed, 1/2″ thick	C-18	450	.020		3.09	.76	.13	3.98	4.79
2610	Topping, P.C based, self-leveling, 6100 psi, pumped, 1/4″ thick	C-8	20000	.003		2.26	.12	.04	2.42	2.72
2620	1/2″ thick		19000	.003		4.52	.12	.04	4.68	5.20
2630	3/4″ thick		18000	.003		6.80	.13	.04	6.97	7.70
2660	1″ thick		17000	.003		9.05	.14	.04	9.23	10.20
2670	1-1/2″ thick		15000	.004		13.55	.16	.05	13.76	15.20
2680	Hand placed, 1/2″ thick	C-18	450	.020		4.52	.76	.13	5.41	6.35

03 62 Non-Shrink Grouting

03 62 13 – Non-Metallic Non-Shrink Grouting

03 62 13.50 Grout, Non-Metallic Non-shrink

		Crew	Daily Output	Labor-Hours	Unit	Material	Labor	Equipment	Total	Total Incl O&P
0010	**GROUT, NON-METALLIC NON-SHRINK**									
0300	Non-shrink, non-metallic, 1″ deep	1 Cefi	35	.229	S.F.	6.50	10.30		16.80	23.50
0350	2″ deep	″	25	.320	″	13	14.40		27.40	37.50

03 62 16 – Metallic Non-Shrink Grouting

03 62 16.50 Grout, Metallic Non-Shrink

		Crew	Daily Output	Labor-Hours	Unit	Material	Labor	Equipment	Total	Total Incl O&P
0010	**GROUT, METALLIC NON-SHRINK**									
0020	Column & machine bases, non-shrink, metallic, 1″ deep	1 Cefi	35	.229	S.F.	10.25	10.30		20.55	27.50
0050	2″ deep	″	25	.320	″	20.50	14.40		34.90	45.50

03 63 Epoxy Grouting

03 63 05 – Grouting of Dowels and Fasteners

03 63 05.10 Epoxy Only

		Crew	Daily Output	Labor-Hours	Unit	Material	Labor	Equipment	Total	Total Incl O&P
0010	**EPOXY ONLY**									
1500	Chemical anchoring, epoxy cartridge, excludes layout, drilling, fastener									
1530	For fastener 3/4″ diam. x 6″ embedment	2 Skwk	72	.222	Ea.	5.65	10.80		16.45	24
1535	1″ diam. x 8″ embedment		66	.242		8.45	11.80		20.25	28.50
1540	1-1/4″ diam. x 10″ embedment		60	.267		16.90	12.95		29.85	39.50
1545	1-3/4″ diam. x 12″ embedment		54	.296		28	14.40		42.40	54.50
1550	14″ embedment		48	.333		34	16.20		50.20	63.50
1555	2″ diam. x 12″ embedment		42	.381		45	18.55		63.55	79.50
1560	18″ embedment		32	.500		56.50	24.50		81	102

03 81 Concrete Cutting

03 81 13 – Flat Concrete Sawing

03 81 13.50 Concrete Floor/Slab Cutting		Crew	Daily Output	Labor-Hours	Unit	Material	2015 Bare Costs Labor	Equipment	Total	Total Incl O&P
0010	**CONCRETE FLOOR/SLAB CUTTING**									
0050	Includes blade cost, layout and set-up time									
0300	Saw cut concrete slabs, plain, up to 3" deep	B-89	1060	.015	L.F.	.14	.66	.46	1.26	1.72
0320	Each additional inch of depth		3180	.005		.05	.22	.15	.42	.57
0400	Mesh reinforced, up to 3" deep		980	.016		.16	.72	.50	1.38	1.86
0420	Each additional inch of depth		2940	.005		.05	.24	.17	.46	.62
0500	Rod reinforced, up to 3" deep		800	.020		.19	.88	.61	1.68	2.28
0520	Each additional inch of depth		2400	.007		.06	.29	.20	.55	.76
0590	Minimum labor/equipment charge		2	8	Job		350	244	594	830

03 81 13.75 Concrete Saw Blades		Crew	Daily Output	Labor-Hours	Unit	Material	Labor	Equipment	Total	Total Incl O&P
0010	**CONCRETE SAW BLADES**									
3000	Blades for saw cutting, included in cutting line items									
3020	Diamond, 12" diameter				Ea.	241			241	265
3040	18" diameter					465			465	510
3080	24" diameter					770			770	845
3120	30" diameter					1,100			1,100	1,200
3160	36" diameter					1,425			1,425	1,550
3200	42" diameter					2,625			2,625	2,875

03 81 16 – Track Mounted Concrete Wall Sawing

03 81 16.50 Concrete Wall Cutting		Crew	Daily Output	Labor-Hours	Unit	Material	Labor	Equipment	Total	Total Incl O&P
0010	**CONCRETE WALL CUTTING**									
0750	Includes blade cost, layout and set-up time									
0800	Concrete walls, hydraulic saw, plain, per inch of depth	B-89B	250	.064	L.F.	.05	2.81	3.43	6.29	8.30
0820	Rod reinforcing, per inch of depth		150	.107	"	.06	4.68	5.70	10.44	13.85
0890	Minimum labor/equipment charge		2	8	Job		350	430	780	1,025

03 82 Concrete Boring

03 82 13 – Concrete Core Drilling

03 82 13.10 Core Drilling		Crew	Daily Output	Labor-Hours	Unit	Material	Labor	Equipment	Total	Total Incl O&P
0010	**CORE DRILLING**									
0015	Includes bit cost, layout and set-up time									
0020	Reinforced concrete slab, up to 6" thick									
0100	1" diameter core	B-89A	17	.941	Ea.	.18	40.50	6.80	47.48	73.50
0150	For each additional inch of slab thickness in same hole, add		1440	.011		.03	.48	.08	.59	.90
0200	2" diameter core		16.50	.970		.28	42	7	49.28	76
0250	For each additional inch of slab thickness in same hole, add		1080	.015		.05	.64	.11	.80	1.21
0300	3" diameter core		16	1		.38	43	7.25	50.63	79
0350	For each additional inch of slab thickness in same hole, add		720	.022		.06	.96	.16	1.18	1.81
0500	4" diameter core		15	1.067		.49	46	7.70	54.19	84
0550	For each additional inch of slab thickness in same hole, add		480	.033		.08	1.44	.24	1.76	2.70
0700	6" diameter core		14	1.143		.76	49.50	8.25	58.51	90.50
0750	For each additional inch of slab thickness in same hole, add		360	.044		.13	1.92	.32	2.37	3.62
0900	8" diameter core		13	1.231		1.05	53	8.90	62.95	97.50
0950	For each additional inch of slab thickness in same hole, add		288	.056		.17	2.40	.40	2.97	4.54
1100	10" diameter core		12	1.333		1.48	57.50	9.65	68.63	106
1150	For each additional inch of slab thickness in same hole, add		240	.067		.25	2.88	.48	3.61	5.50
1300	12" diameter core		11	1.455		1.78	62.50	10.50	74.78	116
1350	For each additional inch of slab thickness in same hole, add		206	.078		.30	3.35	.56	4.21	6.40
1500	14" diameter core		10	1.600		2.08	69	11.55	82.63	128
1550	For each additional inch of slab thickness in same hole, add		180	.089		.35	3.83	.64	4.82	7.35

03 82 Concrete Boring

03 82 13 – Concrete Core Drilling

03 82 13.10 Core Drilling		Crew	Daily Output	Labor-Hours	Unit	Material	2015 Bare Costs Labor	Equipment	Total	Total Incl O&P
1700	18" diameter core	B-89A	9	1.778	Ea.	2.82	76.50	12.85	92.17	142
1750	For each additional inch of slab thickness in same hole, add		144	.111		.47	4.79	.80	6.06	9.20
1754	24" diameter core		8	2		4.02	86.50	14.45	104.97	161
1756	For each additional inch of slab thickness in same hole, add		120	.133		.67	5.75	.96	7.38	11.20
1760	For horizontal holes, add to above						20%	20%		
1770	Prestressed hollow core plank, 8" thick									
1780	1" diameter core	B-89A	17.50	.914	Ea.	.24	39.50	6.60	46.34	72
1790	For each additional inch of plank thickness in same hole, add		3840	.004		.03	.18	.03	.24	.35
1794	2" diameter core		17.25	.928		.38	40	6.70	47.08	73.50
1796	For each additional inch of plank thickness in same hole, add		2880	.006		.05	.24	.04	.33	.48
1800	3" diameter core		17	.941		.51	40.50	6.80	47.81	74
1810	For each additional inch of plank thickness in same hole, add		1920	.008		.06	.36	.06	.48	.73
1820	4" diameter core		16.50	.970		.66	42	7	49.66	76.50
1830	For each additional inch of plank thickness in same hole, add		1280	.013		.08	.54	.09	.71	1.07
1840	6" diameter core		15.50	1.032		1.01	44.50	7.45	52.96	82
1850	For each additional inch of plank thickness in same hole, add		960	.017		.13	.72	.12	.97	1.44
1860	8" diameter core		15	1.067		1.40	46	7.70	55.10	85
1870	For each additional inch of plank thickness in same hole, add		768	.021		.17	.90	.15	1.22	1.83
1880	10" diameter core		14	1.143		1.98	49.50	8.25	59.73	92
1890	For each additional inch of plank thickness in same hole, add		640	.025		.25	1.08	.18	1.51	2.23
1900	12" diameter core		13.50	1.185		2.37	51	8.55	61.92	95.50
1910	For each additional inch of plank thickness in same hole, add		548	.029		.30	1.26	.21	1.77	2.61
1999	Drilling, core, minimum labor/equipment charge		2	8	Job		345	58	403	630
3000	Bits for core drilling, included in drilling line items									
3010	Diamond, premium, 1" diameter				Ea.	70.50			70.50	78
3020	2" diameter					113			113	124
3030	3" diameter					152			152	168
3040	4" diameter					197			197	217
3060	6" diameter					305			305	335
3080	8" diameter					420			420	460
3110	10" diameter					595			595	650
3120	12" diameter					710			710	780
3140	14" diameter					835			835	915
3180	18" diameter					1,125			1,125	1,250
3240	24" diameter					1,600			1,600	1,775

03 82 16 – Concrete Drilling

03 82 16.10 Concrete Impact Drilling		Crew	Daily Output	Labor-Hours	Unit	Material	Labor	Equipment	Total	Total Incl O&P
0010	CONCRETE IMPACT DRILLING									
0020	Includes bit cost, layout and set-up time, no anchors									
0050	Up to 4" deep in concrete/brick floors/walls									
0100	Holes, 1/4" diameter	1 Carp	75	.107	Ea.	.07	5		5.07	8.25
0150	For each additional inch of depth in same hole, add		430	.019		.02	.87		.89	1.45
0200	3/8" diameter		63	.127		.06	5.95		6.01	9.80
0250	For each additional inch of depth in same hole, add		340	.024		.01	1.10		1.11	1.83
0300	1/2" diameter		50	.160		.06	7.50		7.56	12.35
0350	For each additional inch of depth in same hole, add		250	.032		.01	1.50		1.51	2.48
0400	5/8" diameter		48	.167		.09	7.85		7.94	12.95
0450	For each additional inch of depth in same hole, add		240	.033		.02	1.56		1.58	2.58
0500	3/4" diameter		45	.178		.12	8.35		8.47	13.85
0550	For each additional inch of depth in same hole, add		220	.036		.03	1.71		1.74	2.83
0600	7/8" diameter		43	.186		.16	8.75		8.91	14.45
0650	For each additional inch of depth in same hole, add		210	.038		.04	1.79		1.83	2.97

03 82 Concrete Boring

03 82 16 – Concrete Drilling

03 82 16.10 Concrete Impact Drilling		Crew	Daily Output	Labor-Hours	Unit	Material	Labor	Equipment	Total	Total Incl O&P
						2015 Bare Costs				
0700	1" diameter	1 Carp	40	.200	Ea.	.17	9.40		9.57	15.60
0750	For each additional inch of depth in same hole, add		190	.042		.04	1.98		2.02	3.29
0800	1-1/4" diameter		38	.211		.26	9.90		10.16	16.50
0850	For each additional inch of depth in same hole, add		180	.044		.06	2.09		2.15	3.49
0900	1-1/2" diameter		35	.229		.39	10.75		11.14	18.05
0950	For each additional inch of depth in same hole, add	↓	165	.048	↓	.10	2.28		2.38	3.84
1000	For ceiling installations, add						40%			

Estimating Tips

04 05 00 Common Work Results for Masonry

- The terms mortar and grout are often used interchangeably, and incorrectly. Mortar is used to bed masonry units, seal the entry of air and moisture, provide architectural appearance, and allow for size variations in the units. Grout is used primarily in reinforced masonry construction and is used to bond the masonry to the reinforcing steel. Common mortar types are M(2500 psi), S(1800 psi), N(750 psi), and O(350 psi), and conform to ASTM C270. Grout is either fine or coarse and conforms to ASTM C476, and in-place strengths generally exceed 2500 psi. Mortar and grout are different components of masonry construction and are placed by entirely different methods. An estimator should be aware of their unique uses and costs.
- Mortar is included in all assembled masonry line items. The mortar cost, part of the assembled masonry material cost, includes all ingredients, all labor, and all equipment required. Please see reference number R040513-10.
- Waste, specifically the loss/droppings of mortar and the breakage of brick and block, is included in all masonry assemblies in this division. A factor of 25% is added for mortar and 3% for brick and concrete masonry units.
- Scaffolding or staging is not included in any of the Division 4 costs. Refer to Subdivision 01 54 23 for scaffolding and staging costs.

04 20 00 Unit Masonry

- The most common types of unit masonry are brick and concrete masonry. The major classifications of brick are building brick (ASTM C62), facing brick (ASTM C216), glazed brick, fire brick, and pavers. Many varieties of texture and appearance can exist within these classifications, and the estimator would be wise to check local custom and availability within the project area. For repair and remodeling jobs, matching the existing brick may be the most important criteria.
- Brick and concrete block are priced by the piece and then converted into a price per square foot of wall. Openings less than two square feet are generally ignored by the estimator because any savings in units used is offset by the cutting and trimming required.
- It is often difficult and expensive to find and purchase small lots of historic brick. Costs can vary widely. Many design issues affect costs, selection of mortar mix, and repairs or replacement of masonry materials. Cleaning techniques must be reflected in the estimate.
- All masonry walls, whether interior or exterior, require bracing. The cost of bracing walls during construction should be included by the estimator, and this bracing must remain in place until permanent bracing is complete. Permanent bracing of masonry walls is accomplished by masonry itself, in the form of pilasters or abutting wall corners, or by anchoring the walls to the structural frame. Accessories in the form of anchors, anchor slots, and ties are used, but their supply and installation can be by different trades. For instance, anchor slots on spandrel beams and columns are supplied and welded in place by the steel fabricator, but the ties from the slots into the masonry are installed by the bricklayer. Regardless of the installation method, the estimator must be certain that these accessories are accounted for in pricing.

Reference Numbers

Reference numbers are shown in shaded boxes at the beginning of some major classifications. These numbers refer to related items in the Reference Section. The reference information may be an estimating procedure, an alternate pricing method, or technical information.

Note: Not all subdivisions listed here necessarily appear in this publication. ■

04 01 Maintenance of Masonry

04 01 20 – Maintenance of Unit Masonry

04 01 20.20 Pointing Masonry

		Crew	Daily Output	Labor-Hours	Unit	Material	2015 Bare Costs Labor	Equipment	Total	Total Incl O&P
0010	**POINTING MASONRY**									
0300	Cut and repoint brick, hard mortar, running bond	1 Bric	80	.100	S.F.	.56	4.62		5.18	8.10
0320	Common bond		77	.104		.56	4.80		5.36	8.40
0360	Flemish bond		70	.114		.59	5.25		5.84	9.25
0400	English bond		65	.123		.59	5.70		6.29	9.90
0600	Soft old mortar, running bond		100	.080		.56	3.69		4.25	6.60
0620	Common bond		96	.083		.56	3.85		4.41	6.85
0640	Flemish bond		90	.089		.59	4.10		4.69	7.35
0680	English bond		82	.098		.59	4.50		5.09	8
0700	Stonework, hard mortar		140	.057	L.F.	.74	2.64		3.38	5.10
0720	Soft old mortar		160	.050	"	.74	2.31		3.05	4.58
1000	Repoint, mask and grout method, running bond		95	.084	S.F.	.74	3.89		4.63	7.10
1020	Common bond		90	.089		.74	4.10		4.84	7.50
1040	Flemish bond		86	.093		.78	4.29		5.07	7.85
1060	English bond		77	.104		.78	4.80		5.58	8.65
2000	Scrub coat, sand grout on walls, thin mix, brushed		120	.067		2.96	3.08		6.04	8.25
2020	Troweled		98	.082		4.12	3.77		7.89	10.70
9000	Minimum labor/equipment charge		3	2.667	Job		123		123	200

04 01 20.30 Pointing CMU

		Crew	Daily Output	Labor-Hours	Unit	Material	Labor	Equipment	Total	Total Incl O&P
0010	**POINTING CMU**									
0300	Cut and repoint block, hard mortar, running bond	1 Bric	190	.042	S.F.	.23	1.94		2.17	3.41
0310	Stacked bond		200	.040		.23	1.85		2.08	3.25
0600	Soft old mortar, running bond		230	.035		.23	1.61		1.84	2.86
0610	Stacked bond		245	.033		.23	1.51		1.74	2.70

04 01 20.40 Sawing Masonry

		Crew	Daily Output	Labor-Hours	Unit	Material	Labor	Equipment	Total	Total Incl O&P
0010	**SAWING MASONRY**									
0050	Brick or block by hand, per inch depth	A-1	125	.064	L.F.	.05	2.41	.65	3.11	4.71

04 01 20.41 Unit Masonry Stabilization

		Crew	Daily Output	Labor-Hours	Unit	Material	Labor	Equipment	Total	Total Incl O&P
0010	**UNIT MASONRY STABILIZATION**									
0100	Structural repointing method									
0110	Cut / grind mortar joint	1 Bric	240	.033	L.F.		1.54		1.54	2.50
0120	Clean and mask joint		2500	.003		.11	.15		.26	.36
0130	Epoxy paste and 1/4" FRP rod		240	.033		1.67	1.54		3.21	4.34
0132	3/8" FRP rod		160	.050		2.48	2.31		4.79	6.50
0134	1/4" CFRP rod		240	.033		14.50	1.54		16.04	18.45
0136	3/8" CFRP rod		160	.050		18.35	2.31		20.66	24
0140	Remove masking		14400	.001			.03		.03	.04
0300	Structural fabric method									
0310	Primer	1 Bric	600	.013	S.F.	.90	.62		1.52	1.99
0320	Apply filling/leveling paste		720	.011		.72	.51		1.23	1.62
0330	Epoxy, glass fiber fabric		720	.011		8.10	.51		8.61	9.75
0340	Carbon fiber fabric		720	.011		18.90	.51		19.41	22

04 01 20.50 Toothing Masonry

		Crew	Daily Output	Labor-Hours	Unit	Material	Labor	Equipment	Total	Total Incl O&P
0010	**TOOTHING MASONRY**									
0500	Brickwork, soft old mortar	1 Clab	40	.200	V.L.F.		7.50		7.50	12.35
0520	Hard mortar		30	.267			10.05		10.05	16.45
0700	Blockwork, soft old mortar		70	.114			4.30		4.30	7.05
0720	Hard mortar		50	.160			6		6	9.85
9000	Minimum labor/equipment charge		4	2	Job		75		75	123

04 01 Maintenance of Masonry

04 01 30 – Unit Masonry Cleaning

04 01 30.20 Cleaning Masonry		Crew	Daily Output	Labor-Hours	Unit	Material	2015 Bare Costs Labor	Equipment	Total	Total Incl O&P
0010	**CLEANING MASONRY**									
0200	By chemical, brush and rinse, new work, light construction dust	D-1	1000	.016	S.F.	.06	.67		.73	1.16
0220	Medium construction dust		800	.020		.09	.84		.93	1.46
0240	Heavy construction dust, drips or stains		600	.027		.11	1.12		1.23	1.96
0260	Low pressure wash and rinse, light restoration, light soil		800	.020		.13	.84		.97	1.51
0270	Average soil, biological staining		400	.040		.19	1.68		1.87	2.95
0280	Heavy soil, biological and mineral staining, paint		330	.048		.26	2.04		2.30	3.61
0300	High pressure wash and rinse, heavy restoration, light soil		600	.027		.12	1.12		1.24	1.96
0310	Average soil, biological staining		400	.040		.17	1.68		1.85	2.93
0320	Heavy soil, biological and mineral staining, paint		250	.064		.23	2.69		2.92	4.63
0400	High pressure wash, water only, light soil	C-29	500	.016			.60	.14	.74	1.14
0420	Average soil, biological staining		375	.021			.80	.19	.99	1.53
0440	Heavy soil, biological and mineral staining, paint		250	.032			1.20	.28	1.48	2.28
0800	High pressure water and chemical, light soil		450	.018		.16	.67	.16	.99	1.44
0820	Average soil, biological staining		300	.027		.23	1	.23	1.46	2.16
0840	Heavy soil, biological and mineral staining, paint		200	.040		.31	1.50	.35	2.16	3.20
1200	Sandblast, wet system, light soil	J-6	1750	.018		.31	.76	.13	1.20	1.72
1220	Average soil, biological staining		1100	.029		.46	1.21	.21	1.88	2.70
1240	Heavy soil, biological and mineral staining, paint		700	.046		.62	1.91	.34	2.87	4.12
1400	Dry system, light soil		2500	.013		.31	.53	.09	.93	1.30
1420	Average soil, biological staining		1750	.018		.46	.76	.13	1.35	1.89
1440	Heavy soil, biological and mineral staining, paint		1000	.032		.62	1.34	.24	2.20	3.09
1800	For walnut shells, add					.69			.69	.76
1820	For corn chips, add					.69			.69	.76
2000	Steam cleaning, light soil	A-1H	750	.011			.40	.10	.50	.77
2020	Average soil, biological staining		625	.013			.48	.12	.60	.92
2040	Heavy soil, biological and mineral staining		375	.021			.80	.20	1	1.54
4000	Add for masking doors and windows	1 Clab	800	.010		.07	.38		.45	.70
4200	Add for pedestrian protection				Job				10%	10%
9000	Minimum labor/equipment charge	D-4	2	16	"		685	66.50	751.50	1,175

04 01 30.60 Brick Washing		Crew	Daily Output	Labor-Hours	Unit	Material	Labor	Equipment	Total	Total Incl O&P
0010	**BRICK WASHING**									
0012	Acid cleanser, smooth brick surface	1 Bric	560	.014	S.F.	.04	.66		.70	1.12
0050	Rough brick		400	.020		.06	.92		.98	1.56
0060	Stone, acid wash		600	.013		.07	.62		.69	1.08
1000	Muriatic acid, price per gallon in 5 gallon lots				Gal.	8.70			8.70	9.55

04 05 Common Work Results for Masonry

04 05 05 – Selective Demolition for Masonry

04 05 05.10 Selective Demolition		Crew	Daily Output	Labor-Hours	Unit	Material	Labor	Equipment	Total	Total Incl O&P
0010	**SELECTIVE DEMOLITION** R024119-10									
0200	Bond beams, 8" block with #4 bar	2 Clab	32	.500	L.F.		18.80		18.80	31
0300	Concrete block walls, unreinforced, 2" thick		1200	.013	S.F.		.50		.50	.82
0310	4" thick		1150	.014			.52		.52	.86
0320	6" thick		1100	.015			.55		.55	.90
0330	8" thick		1050	.015			.57		.57	.94
0340	10" thick		1000	.016			.60		.60	.99
0360	12" thick		950	.017			.63		.63	1.04
0380	Reinforced alternate courses, 2" thick		1130	.014			.53		.53	.87
0390	4" thick		1080	.015			.56		.56	.91

04 05 05 – Selective Demolition for Masonry

04 05 05.10 Selective Demolition		Crew	Daily Output	Labor-Hours	Unit	Material	2015 Bare Costs Labor	Equipment	Total	Total Incl O&P
0400	6" thick	2 Clab	1035	.015	S.F.		.58		.58	.95
0410	8" thick		990	.016			.61		.61	1
0420	10" thick		940	.017			.64		.64	1.05
0430	12" thick		890	.018			.68		.68	1.11
0440	Reinforced alternate courses & vertically 48" OC, 4" thick		900	.018			.67		.67	1.10
0450	6" thick		850	.019			.71		.71	1.16
0460	8" thick		800	.020			.75		.75	1.23
0480	10" thick		750	.021			.80		.80	1.32
0490	12" thick		700	.023			.86		.86	1.41
1000	Chimney, 16" x 16", soft old mortar	1 Clab	55	.145	C.F.		5.45		5.45	8.95
1020	Hard mortar		40	.200			7.50		7.50	12.35
1030	16" x 20", soft old mortar		55	.145			5.45		5.45	8.95
1040	Hard mortar		40	.200			7.50		7.50	12.35
1050	16" x 24", soft old mortar		55	.145			5.45		5.45	8.95
1060	Hard mortar		40	.200			7.50		7.50	12.35
1080	20" x 20", soft old mortar		55	.145			5.45		5.45	8.95
1100	Hard mortar		40	.200			7.50		7.50	12.35
1110	20" x 24", soft old mortar		55	.145			5.45		5.45	8.95
1120	Hard mortar		40	.200			7.50		7.50	12.35
1140	20" x 32", soft old mortar		55	.145			5.45		5.45	8.95
1160	Hard mortar		40	.200			7.50		7.50	12.35
1200	48" x 48", soft old mortar		55	.145			5.45		5.45	8.95
1220	Hard mortar		40	.200			7.50		7.50	12.35
1250	Metal, high temp steel jacket, 24" diameter	E-2	130	.431	V.L.F.		22.50	11.60	34.10	52
1260	60" diameter	"	60	.933			48.50	25	73.50	112
1280	Flue lining, up to 12" x 12"	1 Clab	200	.040			1.50		1.50	2.47
1282	Up to 24" x 24"		150	.053			2.01		2.01	3.29
2000	Columns, 8" x 8", soft old mortar		48	.167			6.25		6.25	10.30
2020	Hard mortar		40	.200			7.50		7.50	12.35
2060	16" x 16", soft old mortar		16	.500			18.80		18.80	31
2100	Hard mortar		14	.571			21.50		21.50	35
2140	24" x 24", soft old mortar		8	1			37.50		37.50	61.50
2160	Hard mortar		6	1.333			50		50	82
2200	36" x 36", soft old mortar		4	2			75		75	123
2220	Hard mortar		3	2.667			100		100	164
2230	Alternate pricing method, soft old mortar		30	.267	C.F.		10.05		10.05	16.45
2240	Hard mortar		23	.348	"		13.10		13.10	21.50
3000	Copings, precast or masonry, to 8" wide									
3020	Soft old mortar	1 Clab	180	.044	L.F.		1.67		1.67	2.74
3040	Hard mortar	"	160	.050	"		1.88		1.88	3.08
3100	To 12" wide									
3120	Soft old mortar	1 Clab	160	.050	L.F.		1.88		1.88	3.08
3140	Hard mortar	"	140	.057	"		2.15		2.15	3.52
4000	Fireplace, brick, 30" x 24" opening									
4020	Soft old mortar	1 Clab	2	4	Ea.		150		150	247
4040	Hard mortar		1.25	6.400			241		241	395
4100	Stone, soft old mortar		1.50	5.333			201		201	330
4120	Hard mortar		1	8			300		300	495
5000	Veneers, brick, soft old mortar		140	.057	S.F.		2.15		2.15	3.52
5020	Hard mortar		125	.064			2.41		2.41	3.95
5050	Glass block, up to 4" thick		500	.016			.60		.60	.99
5100	Granite and marble, 2" thick		180	.044			1.67		1.67	2.74
5120	4" thick		170	.047			1.77		1.77	2.90

04 05 Common Work Results for Masonry

04 05 05 – Selective Demolition for Masonry

04 05 05.10 Selective Demolition		Crew	Daily Output	Labor-Hours	Unit	Material	2015 Bare Costs Labor	Equipment	Total	Total Incl O&P
5140	Stone, 4" thick	1 Clab	180	.044	S.F.		1.67		1.67	2.74
5160	8" thick		175	.046			1.72		1.72	2.82
5400	Alternate pricing method, stone, 4" thick		60	.133	C.F.		5		5	8.20
5420	8" thick		85	.094	"		3.54		3.54	5.80
9000	Minimum labor/equipment charge		2	4	Job		150		150	247

04 05 13 – Masonry Mortaring

04 05 13.10 Cement

Code	Description	Crew	Daily Output	Labor-Hours	Unit	Material	Labor	Equipment	Total	Total Incl O&P
0010	**CEMENT**									
0100	Masonry, 70 lb. bag, T.L. lots				Bag	12.15			12.15	13.35
0150	L.T.L. lots					12.85			12.85	14.15
0200	White, 70 lb. bag, T.L. lots					15.85			15.85	17.45
0250	L.T.L. lots					16.70			16.70	18.40

04 05 13.20 Lime

Code	Description	Crew	Daily Output	Labor-Hours	Unit	Material	Labor	Equipment	Total	Total Incl O&P
0010	**LIME**									
0020	Masons, hydrated, 50 lb. bag, T.L. lots				Bag	10.10			10.10	11.10
0050	L.T.L. lots					11.10			11.10	12.20
0200	Finish, double hydrated, 50 lb. bag, T.L. lots					8.65			8.65	9.55
0250	L.T.L. lots					9.55			9.55	10.50

04 05 13.23 Surface Bonding Masonry Mortaring

Code	Description	Crew	Daily Output	Labor-Hours	Unit	Material	Labor	Equipment	Total	Total Incl O&P
0010	**SURFACE BONDING MASONRY MORTARING**									
0020	Gray or white colors, not incl. block work	1 Bric	540	.015	S.F.	.13	.68		.81	1.26

04 05 13.30 Mortar

Code	Description	Crew	Daily Output	Labor-Hours	Unit	Material	Labor	Equipment	Total	Total Incl O&P
0010	**MORTAR** R042110-50									
0020	With masonry cement									
0100	Type M, 1:1:6 mix	1 Brhe	143	.056	C.F.	5.25	2.13		7.38	9.20
0200	Type N, 1:3 mix		143	.056		5.40	2.13		7.53	9.40
0300	Type O, 1:3 mix		143	.056		4.18	2.13		6.31	8.05
0400	Type PM, 1:1:6 mix, 2500 psi		143	.056		6	2.13		8.13	10.05
0500	Type S, 1/2:1:4 mix		143	.056		5.25	2.13		7.38	9.25
2000	With portland cement and lime									
2100	Type M, 1:1/4:3 mix	1 Brhe	143	.056	C.F.	8.95	2.13		11.08	13.25
2200	Type N, 1:1:6 mix, 750 psi		143	.056		7.15	2.13		9.28	11.35
2300	Type O, 1:2:9 mix (Pointing Mortar)		143	.056		8.35	2.13		10.48	12.60
2400	Type PL, 1:1/2:4 mix, 2500 psi		143	.056		6	2.13		8.13	10.05
2600	Type S, 1:1/2:4 mix, 1800 psi		143	.056		8.20	2.13		10.33	12.50
2650	Pre-mixed, type S or N					5.05			5.05	5.55
2700	Mortar for glass block	1 Brhe	143	.056		11.05	2.13		13.18	15.60
2900	Mortar for fire brick, dry mix, 10 lb. pail				Ea.	27			27	29.50

04 05 13.91 Masonry Restoration Mortaring

Code	Description	Crew	Daily Output	Labor-Hours	Unit	Material	Labor	Equipment	Total	Total Incl O&P
0010	**MASONRY RESTORATION MORTARING**									
0020	Masonry restoration mix				Lb.	.29			.29	.32
0050	White				"	.29			.29	.32

04 05 13.93 Mortar Pigments

Code	Description	Crew	Daily Output	Labor-Hours	Unit	Material	Labor	Equipment	Total	Total Incl O&P
0010	**MORTAR PIGMENTS**, 50 lb. bags (2 bags per M bricks)									
0020	Color admixture, range 2 to 10 lb. per bag of cement, light colors				Lb.	4.37			4.37	4.81
0050	Medium colors					6.40			6.40	7
0100	Dark colors					13.25			13.25	14.60

04 05 Common Work Results for Masonry

04 05 13 – Masonry Mortaring

04 05 13.95 Sand		Crew	Daily Output	Labor-Hours	Unit	2015 Bare Costs Material	Labor	Equipment	Total	Total Incl O&P
0010	**SAND**, screened and washed at pit									
0020	For mortar, per ton				Ton	19.35			19.35	21.50
0050	With 10 mile haul					33.50			33.50	37
0100	With 30 mile haul					65.50			65.50	72
0200	Screened and washed, at the pit				C.Y.	27			27	29.50
0250	With 10 mile haul					47			47	51.50
0300	With 30 mile haul					91			91	100

04 05 13.98 Mortar Admixtures

		Crew	Daily Output	Labor-Hours	Unit	Material	Labor	Equipment	Total	Total Incl O&P
0010	**MORTAR ADMIXTURES**									
0020	Waterproofing admixture, per quart (1 qt. to 2 bags of masonry cement)				Qt.	3.22			3.22	3.54

04 05 16 – Masonry Grouting

04 05 16.30 Grouting

		Crew	Daily Output	Labor-Hours	Unit	Material	Labor	Equipment	Total	Total Incl O&P
0010	**GROUTING**									
0011	Bond beams & lintels, 8″ deep, 6″ thick, 0.15 C.F. per L.F.	D-4	1480	.022	L.F.	.66	.92	.09	1.67	2.32
0020	8″ thick, 0.2 C.F. per L.F.		1400	.023		1.06	.98	.10	2.14	2.84
0050	10″ thick, 0.25 C.F. per L.F.		1200	.027		1.11	1.14	.11	2.36	3.18
0060	12″ thick, 0.3 C.F. per L.F.		1040	.031		1.33	1.31	.13	2.77	3.72
0200	Concrete block cores, solid, 4″ thk., by hand, 0.067 C.F./S.F. of wall	D-8	1100	.036	S.F.	.30	1.56		1.86	2.87
0210	6″ thick, pumped, 0.175 C.F. per S.F.	D-4	720	.044		.77	1.90	.19	2.86	4.12
0250	8″ thick, pumped, 0.258 C.F. per S.F.		680	.047		1.14	2.01	.20	3.35	4.72
0300	10″ thick, pumped, 0.340 C.F. per S.F.		660	.048		1.50	2.07	.20	3.77	5.20
0350	12″ thick, pumped, 0.422 C.F. per S.F.		640	.050		1.87	2.14	.21	4.22	5.75
0500	Cavity walls, 2″ space, pumped, 0.167 C.F./S.F. of wall		1700	.019		.74	.80	.08	1.62	2.20
0550	3″ space, 0.250 C.F./S.F.		1200	.027		1.11	1.14	.11	2.36	3.18
0600	4″ space, 0.333 C.F. per S.F.		1150	.028		1.47	1.19	.12	2.78	3.67
0700	6″ space, 0.500 C.F. per S.F.		800	.040		2.21	1.71	.17	4.09	5.35
0800	Door frames, 3′ x 7′ opening, 2.5 C.F. per opening		60	.533	Opng.	11.05	23	2.22	36.27	51.50
0850	6′ x 7′ opening, 3.5 C.F. per opening		45	.711	″	15.45	30.50	2.97	48.92	69.50
2000	Grout, C476, for bond beams, lintels and CMU cores		350	.091	C.F.	4.42	3.91	.38	8.71	11.60
9000	Minimum labor/equipment charge	1 Bric	2	4	Job		185		185	300

04 05 19 – Masonry Anchorage and Reinforcing

04 05 19.05 Anchor Bolts

		Crew	Daily Output	Labor-Hours	Unit	Material	Labor	Equipment	Total	Total Incl O&P
0010	**ANCHOR BOLTS**									
0015	Installed in fresh grout in CMU bond beams or filled cores, no templates									
0020	Hooked, with nut and washer, 1/2″ diam., 8″ long	1 Bric	132	.061	Ea.	1.33	2.80		4.13	6
0030	12″ long		131	.061		1.48	2.82		4.30	6.20
0040	5/8″ diameter, 8″ long		129	.062		2.98	2.86		5.84	7.95
0050	12″ long		127	.063		3.67	2.91		6.58	8.75
0060	3/4″ diameter, 8″ long		127	.063		3.67	2.91		6.58	8.75
0070	12″ long		125	.064		4.59	2.95		7.54	9.85

04 05 19.16 Masonry Anchors

		Crew	Daily Output	Labor-Hours	Unit	Material	Labor	Equipment	Total	Total Incl O&P
0010	**MASONRY ANCHORS**									
0020	For brick veneer, galv., corrugated, 7/8″ x 7″, 22 Ga.	1 Bric	10.50	.762	C	14.35	35		49.35	73
0100	24 Ga.		10.50	.762		9.85	35		44.85	68
0150	16 Ga.		10.50	.762		28	35		63	87.50
0200	Buck anchors, galv., corrugated, 16 ga., 2″ bend, 8″ x 2″		10.50	.762		56.50	35		91.50	119
0250	8″ x 3″		10.50	.762		58.50	35		93.50	122
0300	Adjustable, rectangular, 4-1/8″ wide									
0350	Anchor and tie, 3/16″ wire, mill galv.									
0400	2-3/4″ eye, 3-1/4″ tie	1 Bric	1.05	7.619	M	450	350		800	1,075

04 05 19 – Masonry Anchorage and Reinforcing

04 05 19.16 Masonry Anchors		Crew	Daily Output	Labor-Hours	Unit	Material	Labor	Equipment	Total	Total Incl O&P
						2015 Bare Costs				
0500	4-3/4" tie	1 Bric	1.05	7.619	M	485	350		835	1,100
0520	5-1/2" tie		1.05	7.619		525	350		875	1,150
0550	4-3/4" eye, 3-1/4" tie		1.05	7.619		495	350		845	1,125
0570	4-3/4" tie		1.05	7.619		535	350		885	1,150
0580	5-1/2" tie		1.05	7.619		585	350		935	1,225
0660	Cavity wall, Z-type, galvanized, 6" long, 1/8" diam.		10.50	.762	C	25	35		60	84.50
0670	3/16" diameter		10.50	.762		29	35		64	89
0680	1/4" diameter		10.50	.762		40.50	35		75.50	102
0850	8" long, 3/16" diameter		10.50	.762		25	35		60	84.50
0855	1/4" diameter		10.50	.762		49	35		84	111
1000	Rectangular type, galvanized, 1/4" diameter, 2" x 6"		10.50	.762		72.50	35		107.50	137
1050	4" x 6"		10.50	.762		87.50	35		122.50	154
1100	3/16" diameter, 2" x 6"		10.50	.762		45	35		80	107
1150	4" x 6"		10.50	.762		51.50	35		86.50	114
1200	Mesh wall tie, 1/2" mesh, hot dip galvanized									
1400	16 ga., 12" long, 3" wide	1 Bric	9	.889	C	88.50	41		129.50	164
1420	6" wide		9	.889		130	41		171	210
1440	12" wide		8.50	.941		208	43.50		251.50	300
1500	Rigid partition anchors, plain, 8" long, 1" x 1/8"		10.50	.762		235	35		270	315
1550	1" x 1/4"		10.50	.762		277	35		312	360
1580	1-1/2" x 1/8"		10.50	.762		259	35		294	340
1600	1-1/2" x 1/4"		10.50	.762		325	35		360	410
1650	2" x 1/8"		10.50	.762		305	35		340	390
1700	2" x 1/4"		10.50	.762		405	35		440	500
2000	Column flange ties, wire, galvanized									
2300	3/16" diameter, up to 3" wide	1 Bric	10.50	.762	C	87.50	35		122.50	153
2350	To 5" wide		10.50	.762		95	35		130	161
2400	To 7" wide		10.50	.762		102	35		137	170
2600	To 9" wide		10.50	.762		109	35		144	177
2650	1/4" diameter, up to 3" wide		10.50	.762		110	35		145	178
2700	To 5" wide		10.50	.762		119	35		154	188
2800	To 7" wide		10.50	.762		133	35		168	204
2850	To 9" wide		10.50	.762		144	35		179	215
2900	For hot dip galvanized, add					35%				
4000	Channel slots, 1-3/8" x 1/2" x 8"									
4100	12 ga., plain	1 Bric	10.50	.762	C	213	35		248	291
4150	16 ga., galvanized	"	10.50	.762	"	146	35		181	217
4200	Channel slot anchors									
4300	16 ga., galvanized, 1-1/4" x 3-1/2"				C	56.50			56.50	62.50
4350	1-1/4" x 5-1/2"					66			66	73
4400	1-1/4" x 7-1/2"					75			75	82.50
4500	1/8" plain, 1-1/4" x 3-1/2"					143			143	157
4550	1-1/4" x 5-1/2"					152			152	167
4600	1-1/4" x 7-1/2"					166			166	183
4700	For corrugation, add					76.50			76.50	84.50
4750	For hot dip galvanized, add					35%				
5000	Dowels									
5100	Plain, 1/4" diameter, 3" long				C	55.50			55.50	61
5150	4" long					61.50			61.50	67.50
5200	6" long					75			75	82.50
5300	3/8" diameter, 3" long					73			73	80.50
5350	4" long					91			91	100
5400	6" long					105			105	115

04 05 19.16 Masonry Anchors		Crew	Daily Output	Labor-Hours	Unit	2015 Bare Costs Material	Labor	Equipment	Total	Total Incl O&P
5500	1/2" diameter, 3" long				C	107			107	117
5550	4" long					127			127	139
5600	6" long					164			164	180
5700	5/8" diameter, 3" long					146			146	161
5750	4" long					180			180	198
5800	6" long					247			247	272
6000	3/4" diameter, 3" long					182			182	200
6100	4" long					231			231	254
6150	6" long					330			330	365
6300	For hot dip galvanized, add					35%				
04 05 19.26 Masonry Reinforcing Bars										
0010	**MASONRY REINFORCING BARS** R040519-50									
0015	Steel bars A615, placed horiz., #3 & #4 bars	1 Bric	450	.018	Lb.	.48	.82		1.30	1.87
0020	#5 & #6 bars		800	.010		.48	.46		.94	1.28
0050	Placed vertical, #3 & #4 bars		350	.023		.48	1.06		1.54	2.25
0060	#5 & #6 bars		650	.012		.48	.57		1.05	1.45
0200	Joint reinforcing, regular truss, to 6" wide, mill std galvanized		30	.267	C.L.F.	22.50	12.30		34.80	44.50
0250	12" wide		20	.400		26	18.45		44.45	58.50
0400	Cavity truss with drip section, to 6" wide		30	.267		22.50	12.30		34.80	44.50
0450	12" wide		20	.400		26	18.45		44.45	58.50
0500	Joint reinforcing, ladder type, mill std galvanized									
0600	9 ga. sides, 9 ga. ties, 4" wall	1 Bric	30	.267	C.L.F.	19.85	12.30		32.15	42
0650	6" wall		30	.267		18.40	12.30		30.70	40.50
0700	8" wall		25	.320		19.70	14.75		34.45	45.50
0750	10" wall		20	.400		22	18.45		40.45	54
0800	12" wall		20	.400		23	18.45		41.45	55.50
1000	Truss type									
1100	9 ga. sides, 9 ga. ties, 4" wall	1 Bric	30	.267	C.L.F.	21.50	12.30		33.80	43.50
1150	6" wall		30	.267		22	12.30		34.30	44
1200	8" wall		25	.320		26	14.75		40.75	53
1250	10" wall		20	.400		21.50	18.45		39.95	53.50
1300	12" wall		20	.400		22.50	18.45		40.95	54.50
1500	3/16" sides, 9 ga. ties, 4" wall		30	.267		25	12.30		37.30	47.50
1550	6" wall		30	.267		31.50	12.30		43.80	54.50
1600	8" wall		25	.320		33	14.75		47.75	60
1650	10" wall		20	.400		33.50	18.45		51.95	67
1700	12" wall		20	.400		33.50	18.45		51.95	67
2000	3/16" sides, 3/16" ties, 4" wall		30	.267		35.50	12.30		47.80	59
2050	6" wall		30	.267		36.50	12.30		48.80	60
2100	8" wall		25	.320		37.50	14.75		52.25	65.50
2150	10" wall		20	.400		39.50	18.45		57.95	73.50
2200	12" wall		20	.400		41.50	18.45		59.95	76
2500	Cavity truss type, galvanized									
2600	9 ga. sides, 9 ga. ties, 4" wall	1 Bric	25	.320	C.L.F.	37	14.75		51.75	65
2650	6" wall		25	.320		35	14.75		49.75	62.50
2700	8" wall		20	.400		39	18.45		57.45	73
2750	10" wall		15	.533		41	24.50		65.50	85
2800	12" wall		15	.533		37.50	24.50		62	81.50
3000	3/16" sides, 9 ga. ties, 4" wall		25	.320		35.50	14.75		50.25	63
3050	6" wall		25	.320		32.50	14.75		47.25	60
3100	8" wall		20	.400		47	18.45		65.45	81.50
3150	10" wall		15	.533		33.50	24.50		58	77

04 05 Common Work Results for Masonry

04 05 19 – Masonry Anchorage and Reinforcing

04 05 19.26 Masonry Reinforcing Bars		Crew	Daily Output	Labor-Hours	Unit	Material	2015 Bare Costs Labor	Equipment	Total	Total Incl O&P
3200	12" wall	1 Bric	15	.533	C.L.F.	37.50	24.50		62	81
3500	For hot dip galvanizing, add				Ton	460			460	505

04 05 23 – Masonry Accessories

04 05 23.13 Masonry Control and Expansion Joints

		Crew	Daily Output	Labor-Hours	Unit	Material	Labor	Equipment	Total	Total Incl O&P
0010	**MASONRY CONTROL AND EXPANSION JOINTS**									
0020	Rubber, for double wythe 8" minimum wall (Brick/CMU)	1 Bric	400	.020	L.F.	2.07	.92		2.99	3.78
0025	"T" shaped		320	.025		1.23	1.15		2.38	3.23
0030	Cross-shaped for CMU units		280	.029		1.45	1.32		2.77	3.75
0050	PVC, for double wythe 8" minimum wall (Brick/CMU)		400	.020		1.47	.92		2.39	3.12
0120	"T" shaped		320	.025		.75	1.15		1.90	2.71
0160	Cross-shaped for CMU units		280	.029		.92	1.32		2.24	3.16

04 05 23.19 Masonry Cavity Drainage, Weepholes, and Vents

		Crew	Daily Output	Labor-Hours	Unit	Material	Labor	Equipment	Total	Total Incl O&P
0010	**MASONRY CAVITY DRAINAGE, WEEPHOLES, AND VENTS**									
0020	Extruded aluminum, 4" deep, 2-3/8" x 8-1/8"	1 Bric	30	.267	Ea.	34	12.30		46.30	57.50
0050	5" x 8-1/8"		25	.320		45	14.75		59.75	73.50
0100	2-1/4" x 25"		25	.320		78	14.75		92.75	110
0150	5" x 16-1/2"		22	.364		62.50	16.80		79.30	96
0175	5" x 24"		22	.364		84	16.80		100.80	120
0200	6" x 16-1/2"		22	.364		90.50	16.80		107.30	127
0250	7-3/4" x 16-1/2"		20	.400		77	18.45		95.45	115
0400	For baked enamel finish, add					35%				
0500	For cast aluminum, painted, add					60%				
1000	Stainless steel ventilators, 6" x 6"	1 Bric	25	.320		210	14.75		224.75	256
1050	8" x 8"		24	.333		232	15.40		247.40	281
1100	12" x 12"		23	.348		266	16.05		282.05	320
1150	12" x 6"		24	.333		254	15.40		269.40	305
1200	Foundation block vent, galv., 1-1/4" thk, 8" high, 16" long, no damper		30	.267		16.20	12.30		28.50	38
1250	For damper, add					3.11			3.11	3.42

04 05 23.95 Wall Plugs

		Crew	Daily Output	Labor-Hours	Unit	Material	Labor	Equipment	Total	Total Incl O&P
0010	**WALL PLUGS** (for nailing to brickwork)									
0020	25 ga., galvanized, plain	1 Bric	10.50	.762	C	26.50	35		61.50	86
0050	Wood filled	"	10.50	.762	"	68	35		103	132

04 21 Clay Unit Masonry

04 21 13 – Brick Masonry

04 21 13.13 Brick Veneer Masonry

		Crew	Daily Output	Labor-Hours	Unit	Material	Labor	Equipment	Total	Total Incl O&P
0010	**BRICK VENEER MASONRY**, T.L. lots, excl. scaff., grout & reinforcing									
0015	Material costs incl. 3% brick and 25% mortar waste									
0020	Standard, select common, 4" x 2-2/3" x 8" (6.75/S.F.)	D-8	1.50	26.667	M	615	1,150		1,765	2,525
0050	Red, 4" x 2-2/3" x 8", running bond		1.50	26.667		550	1,150		1,700	2,450
0100	Full header every 6th course (7.88/S.F.)		1.45	27.586		550	1,175		1,725	2,525
0150	English, full header every 2nd course (10.13/S.F.)		1.40	28.571		550	1,225		1,775	2,600
0200	Flemish, alternate header every course (9.00/S.F.)		1.40	28.571		550	1,225		1,775	2,600
0250	Flemish, alt. header every 6th course (7.13/S.F.)		1.45	27.586		550	1,175		1,725	2,525
0300	Full headers throughout (13.50/S.F.)		1.40	28.571		545	1,225		1,770	2,600
0350	Rowlock course (13.50/S.F.)		1.35	29.630		545	1,275		1,820	2,675
0400	Rowlock stretcher (4.50/S.F.)		1.40	28.571		555	1,225		1,780	2,625
0450	Soldier course (6.75/S.F.)		1.40	28.571		550	1,225		1,775	2,600
0500	Sailor course (4.50/S.F.)		1.30	30.769		555	1,325		1,880	2,775
0601	Buff or gray face, running bond, (6.75/S.F.)		1.50	26.667		550	1,150		1,700	2,450

04 21 13.13 Brick Veneer Masonry		Crew	Daily Output	Labor-Hours	Unit	Material	2015 Bare Costs Labor	Equipment	Total	Total Incl O&P
0700	Glazed face, 4" x 2-2/3" x 8", running bond	D-8	1.40	28.571	M	1,825	1,225		3,050	4,025
0750	Full header every 6th course (7.88/S.F.)		1.35	29.630		1,750	1,275		3,025	4,000
1000	Jumbo, 6" x 4" x 12",(3.00/S.F.)		1.30	30.769		1,725	1,325		3,050	4,050
1051	Norman, 4" x 2-2/3" x 12" (4.50/S.F.)		1.45	27.586		1,175	1,175		2,350	3,225
1100	Norwegian, 4" x 3-1/5" x 12" (3.75/S.F.)		1.40	28.571		1,400	1,225		2,625	3,550
1150	Economy, 4" x 4" x 8" (4.50 per S.F.)		1.40	28.571		915	1,225		2,140	3,000
1201	Engineer, 4" x 3-1/5" x 8", (5.63/S.F.)		1.45	27.586		645	1,175		1,820	2,625
1251	Roman, 4" x 2" x 12", (6.00/S.F.)		1.50	26.667		1,175	1,150		2,325	3,150
1300	S.C.R. 6" x 2-2/3" x 12" (4.50/S.F.)		1.40	28.571		1,350	1,225		2,575	3,500
1350	Utility, 4" x 4" x 12" (3.00/S.F.)	↓	1.08	37.037		1,600	1,600		3,200	4,325
1360	For less than truck load lots, add				↓	15%				
1400	For battered walls, add						30%			
1450	For corbels, add						75%			
1500	For curved walls, add						30%			
1550	For pits and trenches, deduct						20%			
1999	Alternate method of figuring by square foot									
2000	Standard, sel. common, 4" x 2-2/3" x 8", (6.75/S.F.)	D-8	230	.174	S.F.	4.16	7.45		11.61	16.75
2020	Red, 4" x 2-2/3" x 8", running bond		220	.182		3.72	7.80		11.52	16.80
2050	Full header every 6th course (7.88/S.F.)		185	.216		4.34	9.30		13.64	19.85
2100	English, full header every 2nd course (10.13/S.F.)		140	.286		5.55	12.25		17.80	26
2150	Flemish, alternate header every course (9.00/S.F.)		150	.267		4.94	11.45		16.39	24
2200	Flemish, alt. header every 6th course (7.13/S.F.)		205	.195		3.93	8.35		12.28	17.90
2250	Full headers throughout (13.50/S.F.)		105	.381		7.40	16.35		23.75	34.50
2300	Rowlock course (13.50/S.F.)		100	.400		7.40	17.15		24.55	36
2350	Rowlock stretcher (4.50/S.F.)		310	.129		2.51	5.55		8.06	11.75
2400	Soldier course (6.75/S.F.)		200	.200		3.72	8.60		12.32	18.05
2450	Sailor course (4.50/S.F.)		290	.138		2.51	5.90		8.41	12.40
2600	Buff or gray face, running bond, (6.75/S.F.)		220	.182		3.93	7.80		11.73	17
2700	Glazed face brick, running bond		210	.190		11.75	8.15		19.90	26.50
2750	Full header every 6th course (7.88/S.F.)		170	.235		13.70	10.10		23.80	31.50
3000	Jumbo, 6" x 4" x 12" running bond (3.00/S.F.)		435	.092		4.67	3.95		8.62	11.55
3050	Norman, 4" x 2-2/3" x 12" running bond, (4.5/S.F.)		320	.125		5.95	5.35		11.30	15.25
3100	Norwegian, 4" x 3-1/5" x 12" (3.75/S.F.)		375	.107		5.15	4.58		9.73	13.10
3150	Economy, 4" x 4" x 8" (4.50/S.F.)		310	.129		4.08	5.55		9.63	13.50
3200	Engineer, 4" x 3-1/5" x 8" (5.63/S.F.)		260	.154		3.61	6.60		10.21	14.70
3250	Roman, 4" x 2" x 12" (6.00/S.F.)		250	.160		6.95	6.85		13.80	18.75
3300	SCR, 6" x 2-2/3" x 12" (4.50/S.F.)		310	.129		6	5.55		11.55	15.60
3350	Utility, 4" x 4" x 12" (3.00/S.F.)	↓	360	.111	↓	4.65	4.77		9.42	12.85
3360	For less than truck load lots, add				M	15%				
3370	For battered walls, add						30%			
3380	For corbels, add						75%			
3400	For cavity wall construction, add						15%			
3450	For stacked bond, add						10%			
3500	For interior veneer construction, add						15%			
3510	For pits and trenches, deduct						20%			
3550	For curved walls, add						30%			
9000	Minimum labor/equipment charge	D-1	2	8	Job		335		335	550

04 21 13.14 Thin Brick Veneer		Crew	Daily Output	Labor-Hours	Unit	Material	Labor	Equipment	Total	Total Incl O&P
0010	**THIN BRICK VENEER**									
0015	Material costs incl. 3% brick and 25% mortar waste									
0020	On & incl. metal panel support sys, modular, 2-2/3" x 5/8" x 8", red	D-7	92	.174	S.F.	9	6.65		15.65	20.50
0100	Closure, 4" x 5/8" x 8"	↓	110	.145	↓	8.80	5.55		14.35	18.40

04 21 13 – Brick Masonry

04 21 13.14 Thin Brick Veneer		Crew	Daily Output	Labor-Hours	Unit	Material	Labor	Equipment	Total	Total Incl O&P
						2015 Bare Costs				
0110	Norman, 2-2/3" x 5/8" x 12"	D-7	110	.145	S.F.	8.85	5.55		14.40	18.45
0120	Utility, 4" x 5/8" x 12"		125	.128		8.55	4.89		13.44	17.10
0130	Emperor, 4" x 3/4" x 16"		175	.091		9.65	3.49		13.14	16.15
0140	Super emperor, 8" x 3/4" x 16"		195	.082		9.95	3.13		13.08	15.90
0150	For L shaped corners with 4" return, add				L.F.	9.25			9.25	10.20
0200	On masonry/plaster back-up, modular, 2-2/3" x 5/8" x 8", red	D-7	137	.117	S.F.	4.21	4.46		8.67	11.70
0210	Closure, 4" x 5/8" x 8"		165	.097		3.98	3.70		7.68	10.25
0220	Norman, 2-2/3" x 5/8" x 12"		165	.097		4.02	3.70		7.72	10.25
0230	Utility, 4" x 5/8" x 12"		185	.086		3.75	3.30		7.05	9.35
0240	Emperor, 4" x 3/4" x 16"		260	.062		4.85	2.35		7.20	9.05
0250	Super emperor, 8" x 3/4" x 16"		285	.056		5.15	2.14		7.29	9.05
0260	For L shaped corners with 4" return, add				L.F.	9.25			9.25	10.20
0270	For embedment into pre-cast concrete panels, add				S.F.	14.40			14.40	15.85

04 21 13.15 Chimney		Crew	Daily Output	Labor-Hours	Unit	Material	Labor	Equipment	Total	Total Incl O&P
0010	**CHIMNEY**, excludes foundation, scaffolding, grout and reinforcing									
0100	Brick, 16" x 16", 8" flue	D-1	18.20	.879	V.L.F.	24	37		61	86.50
0150	16" x 20" with one 8" x 12" flue		16	1		37.50	42		79.50	110
0200	16" x 24" with two 8" x 8" flues		14	1.143		55	48		103	139
0250	20" x 20" with one 12" x 12" flue		13.70	1.168		45.50	49		94.50	130
0300	20" x 24" with two 8" x 12" flues		12	1.333		62.50	56		118.50	160
0350	20" x 32" with two 12" x 12" flues		10	1.600		80	67.50		147.50	198

04 21 13.18 Columns		Crew	Daily Output	Labor-Hours	Unit	Material	Labor	Equipment	Total	Total Incl O&P
0010	**COLUMNS**, solid, excludes scaffolding, grout and reinforcing									
0050	Brick, 8" x 8", 9 brick per V.L.F.	D-1	56	.286	V.L.F.	4.76	12.05		16.81	25
0100	12" x 8", 13.5 brick per V.L.F.		37	.432		7.15	18.20		25.35	37.50
0200	12" x 12", 20 brick per V.L.F.		25	.640		10.60	27		37.60	55.50
0300	16" x 12", 27 brick per V.L.F.		19	.842		14.30	35.50		49.80	73
0400	16" x 16", 36 brick per V.L.F.		14	1.143		19.05	48		67.05	99.50
0500	20" x 16", 45 brick per V.L.F.		11	1.455		24	61		85	126
0600	20" x 20", 56 brick per V.L.F.		9	1.778		29.50	75		104.50	155
0700	24" x 20", 68 brick per V.L.F.		7	2.286		36	96		132	197
0800	24" x 24", 81 brick per V.L.F.		6	2.667		43	112		155	230
1000	36" x 36", 182 brick per V.L.F.		3	5.333		96.50	225		321.50	470
9000	Minimum labor/equipment charge		2	8	Job		335		335	550

04 21 13.30 Oversized Brick		Crew	Daily Output	Labor-Hours	Unit	Material	Labor	Equipment	Total	Total Incl O&P
0010	**OVERSIZED BRICK**, excludes scaffolding, grout and reinforcing									
0100	Veneer, 4" x 2.25" x 16"	D-8	387	.103	S.F.	5.10	4.44		9.54	12.85
0102	8" x 2.25" x 16", multicell		265	.151		16.10	6.50		22.60	28.50
0105	4" x 2.75" x 16"		412	.097		5.30	4.17		9.47	12.65
0107	8" x 2.75" x 16", multicell		295	.136		16.10	5.80		21.90	27
0110	4" x 4" x 16"		460	.087		3.48	3.73		7.21	9.90
0120	4" x 8" x 16"		533	.075		4.08	3.22		7.30	9.75
0122	4" x 8" x 16" multi cell		327	.122		15.10	5.25		20.35	25
0125	Loadbearing, 6" x 4" x 16", grouted and reinforced		387	.103		10.30	4.44		14.74	18.55
0130	8" x 4" x 16", grouted and reinforced		327	.122		11.30	5.25		16.55	21
0132	10" x 4" x 16", grouted and reinforced		327	.122		23	5.25		28.25	33.50
0135	6" x 8" x 16", grouted and reinforced		440	.091		13.40	3.90		17.30	21
0140	8" x 8" x 16", grouted and reinforced		400	.100		14.30	4.29		18.59	22.50
0145	Curtainwall/reinforced veneer, 6" x 4" x 16"		387	.103		14.65	4.44		19.09	23.50
0150	8" x 4" x 16"		327	.122		17.80	5.25		23.05	28
0152	10" x 4" x 16"		327	.122		25	5.25		30.25	36
0155	6" x 8" x 16"		440	.091		18.20	3.90		22.10	26.50

04 21 Clay Unit Masonry

04 21 13 – Brick Masonry

04 21 13.30 Oversized Brick		Crew	Daily Output	Labor-Hours	Unit	Material	2015 Bare Costs Labor	Equipment	Total	Total Incl O&P
0160	8" x 8" x 16"	D-8	400	.100	S.F.	25.50	4.29		29.79	35
0200	For 1 to 3 slots in face, add					15%				
0210	For 4 to 7 slots in face, add					25%				
0220	For bond beams, add					20%				
0230	For bullnose shapes, add					20%				
0240	For open end knockout, add					10%				
0250	For white or gray color group, add					10%				
0260	For 135 degree corner, add					250%				

04 21 13.35 Common Building Brick

		Crew	Daily Output	Labor-Hours	Unit	Material	Labor	Equipment	Total	Total Incl O&P
0010	**COMMON BUILDING BRICK**, C62, TL lots, material only R042110-20									
0020	Standard				M	490			490	540
0050	Select				"	500			500	550

04 21 13.40 Structural Brick

		Crew	Daily Output	Labor-Hours	Unit	Material	Labor	Equipment	Total	Total Incl O&P
0010	**STRUCTURAL BRICK** C652, Grade SW, incl. mortar, scaffolding not incl.									
0100	Standard unit, 4-5/8" x 2-3/4" x 9-5/8"	D-8	245	.163	S.F.	4.08	7		11.08	15.90
0120	Bond beam		225	.178		4.08	7.65		11.73	16.90
0140	V cut bond beam		225	.178		4.08	7.65		11.73	16.90
0160	Stretcher quoin, 5-5/8" x 2-3/4" x 9-5/8"		245	.163		7.55	7		14.55	19.70
0180	Corner quoin		245	.163		7.55	7		14.55	19.70
0200	Corner, 45 deg, 4-5/8" x 2-3/4" x 10-7/16"		235	.170		7.55	7.30		14.85	20

04 21 13.45 Face Brick

		Crew	Daily Output	Labor-Hours	Unit	Material	Labor	Equipment	Total	Total Incl O&P
0010	**FACE BRICK** Material Only, C216, TL lots R042110-20									
0300	Standard modular, 4" x 2-2/3" x 8"				M	435			435	480
0450	Economy, 4" x 4" x 8"					775			775	855
0510	Economy, 4" x 4" x 12"					1,225			1,225	1,350
0550	Jumbo, 6" x 4" x 12"					1,400			1,400	1,550
0610	Jumbo, 8" x 4" x 12"					1,400			1,400	1,550
0650	Norwegian, 4" x 3-1/5" x 12"					1,225			1,225	1,350
0710	Norwegian, 6" x 3-1/5" x 12"					1,525			1,525	1,675
0850	Standard glazed, plain colors, 4" x 2-2/3" x 8"					1,600			1,600	1,750
1000	Deep trim shades, 4" x 2-2/3" x 8"					2,025			2,025	2,225
1080	Jumbo utility, 4" x 4" x 12"					1,400			1,400	1,525
1120	4" x 8" x 8"					1,775			1,775	1,950
1140	4" x 8" x 16"					5,225			5,225	5,750
1260	Engineer, 4" x 3-1/5" x 8"					525			525	575
1350	King, 4" x 2-3/4" x 10"					485			485	530
1770	Standard modular, double glazed, 4" x 2-2/3" x 8"					2,400			2,400	2,650
1850	Jumbo, colored glazed ceramic, 6" x 4" x 12"					2,525			2,525	2,775
2050	Jumbo utility, glazed, 4" x 4" x 12"					4,750			4,750	5,225
2100	4" x 8" x 8"					5,600			5,600	6,150
2150	4" x 16" x 8"					6,550			6,550	7,200
2170	For less than truck load lots, add					15			15	16.50
2180	For buff or gray brick, add					16			16	17.60
3050	Used brick					405			405	445
3150	Add for brick to match existing work, minimum					5%				
3200	Maximum					50%				

04 21 26 – Glazed Structural Clay Tile Masonry

04 21 26.10 Structural Facing Tile

		Crew	Daily Output	Labor-Hours	Unit	Material	Labor	Equipment	Total	Total Incl O&P
0010	**STRUCTURAL FACING TILE**, std. colors, excl. scaffolding, grout, reinforcing									
0020	6T series, 5-1/3" x 12", 2.3 pieces per S.F., glazed 1 side, 2" thick	D-8	225	.178	S.F.	8.95	7.65		16.60	22.50
0100	4" thick		220	.182		12.20	7.80		20	26

04 21 26 – Glazed Structural Clay Tile Masonry

04 21 26.10 Structural Facing Tile		Crew	Daily Output	Labor-Hours	Unit	2015 Bare Costs Material	Labor	Equipment	Total	Total Incl O&P
0150	Glazed 2 sides	D-8	195	.205	S.F.	16	8.80		24.80	32
0250	6" thick		210	.190		18.20	8.15		26.35	33.50
0300	Glazed 2 sides		185	.216		21.50	9.30		30.80	39
0400	8" thick		180	.222		24	9.55		33.55	42
0500	Special shapes, group 1		400	.100	Ea.	7.70	4.29		11.99	15.45
0550	Group 2		375	.107		12.50	4.58		17.08	21
0600	Group 3		350	.114		16.05	4.90		20.95	25.50
0650	Group 4		325	.123		33.50	5.30		38.80	45
0700	Group 5		300	.133		39.50	5.70		45.20	53
0750	Group 6		275	.145		54	6.25		60.25	69.50
1000	Fire rated, 4" thick, 1 hr. rating		210	.190	S.F.	17.45	8.15		25.60	32.50
1300	Acoustic, 4" thick		210	.190	"	34	8.15		42.15	51
2000	8W series, 8" x 16", 1.125 pieces per S.F.									
2050	2" thick, glazed 1 side	D-8	360	.111	S.F.	10.45	4.77		15.22	19.25
2100	4" thick, glazed 1 side		345	.116		14.35	4.98		19.33	24
2150	Glazed 2 sides		325	.123		17.15	5.30		22.45	27.50
2200	6" thick, glazed 1 side		330	.121		24.50	5.20		29.70	35.50
2250	8" thick, glazed 1 side		310	.129		25	5.55		30.55	36
2500	Special shapes, group 1		300	.133	Ea.	14.85	5.70		20.55	25.50
2550	Group 2		280	.143		19.85	6.15		26	32
2600	Group 3		260	.154		21	6.60		27.60	34
2650	Group 4		250	.160		43.50	6.85		50.35	59
2700	Group 5		240	.167		39.50	7.15		46.65	55
2750	Group 6		230	.174		85.50	7.45		92.95	106
3000	4" thick, glazed 1 side		345	.116	S.F.	13.95	4.98		18.93	23.50
3100	Acoustic, 4" thick		345	.116	"	18.55	4.98		23.53	28.50
3120	4W series, 8" x 8", 2.25 pieces per S.F.									
3125	2" thick, glazed 1 side	D-8	360	.111	S.F.	9.60	4.77		14.37	18.30
3130	4" thick, glazed 1 side		345	.116		11.30	4.98		16.28	20.50
3135	Glazed 2 sides		325	.123		15.55	5.30		20.85	25.50
3140	6" thick, glazed 1 side		330	.121		16.05	5.20		21.25	26
3150	8" thick, glazed 1 side		310	.129		23.50	5.55		29.05	34.50
3155	Special shapes, group I		300	.133	Ea.	7.50	5.70		13.20	17.55
3160	Group II		280	.143	"	8.65	6.15		14.80	19.45
3200	For designer colors, add					25%				
3300	For epoxy mortar joints, add				S.F.	1.74			1.74	1.91
9000	Minimum labor/equipment charge	D-1	2	8	Job		335		335	550

04 21 29 – Terra Cotta Masonry

04 21 29.10 Terra Cotta Masonry Components		Crew	Daily Output	Labor-Hours	Unit	Material	Labor	Equipment	Total	Total Incl O&P
0010	**TERRA COTTA MASONRY COMPONENTS**									
0020	Coping, split type, not glazed, 9" wide	D-1	90	.178	L.F.	13.15	7.50		20.65	26.50
0100	13" wide		80	.200		18.35	8.40		26.75	33.50
0200	Coping, split type, glazed, 9" wide		90	.178		11.95	7.50		19.45	25.50
0250	13" wide		80	.200		15.75	8.40		24.15	31
0500	Partition or back-up blocks, scored, in C.L. lots									
0700	Non-load bearing 12" x 12", 3" thick, special order	D-8	550	.073	S.F.	16.85	3.12		19.97	23.50
0750	4" thick, standard		500	.080		5.35	3.43		8.78	11.45
0800	6" thick		450	.089		7.20	3.81		11.01	14.10
0850	8" thick		400	.100		9.05	4.29		13.34	16.95
1000	Load bearing, 12" x 12", 4" thick, in walls		500	.080		5.45	3.43		8.88	11.60
1050	In floors		750	.053		5.45	2.29		7.74	9.70
1200	6" thick, in walls		450	.089		7.35	3.81		11.16	14.30

04 21 Clay Unit Masonry

04 21 29 – Terra Cotta Masonry

04 21 29.10 Terra Cotta Masonry Components		Crew	Daily Output	Labor-Hours	Unit	2015 Bare Costs Material	Labor	Equipment	Total	Total Incl O&P
1250	In floors	D-8	675	.059	S.F.	7.35	2.54		9.89	12.25
1400	8" thick, in walls		400	.100		9.15	4.29		13.44	17.05
1450	In floors		575	.070		9.15	2.99		12.14	14.90
1600	10" thick, in walls, special order		350	.114		24.50	4.90		29.40	35
1650	In floors, special order		500	.080		24.50	3.43		27.93	32.50
1800	12" thick, in walls, special order		300	.133		23	5.70		28.70	35
1850	In floors, special order		450	.089		23	3.81		26.81	31.50
2000	For reinforcing with steel rods, add to above					15%	5%			
2100	For smooth tile instead of scored, add					3.66			3.66	4.03
2200	For L.C.L. quantities, add					10%	10%			
9000	Minimum labor/equipment charge	D-1	2	8	Job		335		335	550

04 21 29.20 Terra Cotta Tile		Crew	Daily Output	Labor-Hours	Unit	Material	Labor	Equipment	Total	Total Incl O&P
0010	**TERRA COTTA TILE**, on walls, dry set, 1/2" thick									
0100	Square, hexagonal or lattice shapes, unglazed	1 Tilf	135	.059	S.F.	4.62	2.54		7.16	9.10
0300	Glazed, plain colors		130	.062		7.10	2.63		9.73	11.95
0400	Intense colors		125	.064		8.30	2.74		11.04	13.50

04 22 Concrete Unit Masonry

04 22 10 – Concrete Masonry Units

04 22 10.10 Concrete Block		Crew	Daily Output	Labor-Hours	Unit	Material	Labor	Equipment	Total	Total Incl O&P
0010	**CONCRETE BLOCK** Material Only R042210-20									
0020	2" x 8" x 16" solid, normal-weight, 2,000 psi				Ea.	1.04			1.04	1.14
0050	3,500 psi R042110-50					1.11			1.11	1.22
0100	5,000 psi					1.27			1.27	1.40
0150	Lightweight, std.					1.22			1.22	1.34
0300	3" x 8" x 16" solid, normal-weight, 2000 psi					.92			.92	1.01
0350	3,500 psi					1.27			1.27	1.40
0400	5,000 psi					1.44			1.44	1.58
0450	Lightweight, std.					1.18			1.18	1.30
0600	4" x 8" x 16" hollow, normal-weight, 2000 psi					1.19			1.19	1.31
0650	3,500 psi					1.31			1.31	1.44
0700	5000 psi					1.54			1.54	1.69
0750	Lightweight, std.					1.28			1.28	1.41
1300	Solid, normal-weight, 2,000 psi					1.42			1.42	1.56
1350	3,500 psi					1.58			1.58	1.74
1400	5,000 psi					1.34			1.34	1.47
1450	Lightweight, std.					1.13			1.13	1.24
1600	6" x 8" x 16" hollow, normal-weight, 2,000 psi					1.54			1.54	1.69
1650	3,500 psi					1.76			1.76	1.94
1700	5,000 psi					1.89			1.89	2.08
1750	Lightweight, std.					1.84			1.84	2.02
2300	Solid, normal-weight, 2,000 psi					1.69			1.69	1.86
2350	3,500 psi					1.50			1.50	1.65
2400	5,000 psi					2			2	2.20
2450	Lightweight, std.					2.19			2.19	2.41
2600	8" x 8" x 16" hollow, normal-weight, 2000 psi					1.88			1.88	2.07
2650	3500 psi					1.68			1.68	1.85
2700	5,000 psi					2.30			2.30	2.53
2750	Lightweight, std.					2.22			2.22	2.44
3200	Solid, normal-weight, 2,000 psi					2.50			2.50	2.75
3250	3,500 psi					2.86			2.86	3.15

04 22 Concrete Unit Masonry

04 22 10 – Concrete Masonry Units

04 22 10.10 Concrete Block		Crew	Daily Output	Labor-Hours	Unit	2015 Bare Costs Material	Labor	Equipment	Total	Total Incl O&P
3300	5,000 psi				Ea.	2.94			2.94	3.23
3350	Lightweight, std.					2.81			2.81	3.09
3400	10" x 8" x 16" hollow, normal-weight, 2000 psi					1.88			1.88	2.07
3410	3500 psi					1.68			1.68	1.85
3420	5,000 psi					2.30			2.30	2.53
3430	Lightweight, std.					2.22			2.22	2.44
3480	Solid, normal-weight, 2,000 psi					2.50			2.50	2.75
3490	3,500 psi					2.86			2.86	3.15
3500	5,000 psi					2.94			2.94	3.23
3510	Lightweight, std.					2.81			2.81	3.09
3600	12" x 8" x 16" hollow, normal-weight, 2000 psi					2.72			2.72	2.99
3650	3,500 psi					2.76			2.76	3.04
3700	5,000 psi					3.08			3.08	3.39
3750	Lightweight, std.					2.77			2.77	3.05
4300	Solid, normal-weight, 2,000 psi					3.85			3.85	4.24
4350	3,500 psi					3.35			3.35	3.69
4400	5,000 psi					3.25			3.25	3.58
4500	Lightweight, std.					2.71			2.71	2.98
04 22 10.11 Autoclave Aerated Concrete Block										
0010	**AUTOCLAVE AERATED CONCRETE BLOCK**, excl. scaffolding, grout & reinforcing									
0050	Solid, 4" x 8" x 24", incl. mortar G	D-8	600	.067	S.F.	1.48	2.86		4.34	6.30
0060	6" x 8" x 24" R042110-50 G		600	.067		2.24	2.86		5.10	7.10
0070	8" x 8" x 24" G		575	.070		2.98	2.99		5.97	8.15
0080	10" x 8" x 24" G		575	.070		3.64	2.99		6.63	8.85
0090	12" x 8" x 24" G		550	.073		4.47	3.12		7.59	10
04 22 10.12 Chimney Block										
0010	**CHIMNEY BLOCK**, excludes scaffolding, grout and reinforcing									
0220	1 piece, with 8" x 8" flue, 16" x 16"	D-1	28	.571	V.L.F.	18.50	24		42.50	59.50
0230	2 piece, 16" x 16"		26	.615		23	26		49	67
0240	2 piece, with 8" x 12" flue, 16" x 20"		24	.667		31.50	28		59.50	80.50
04 22 10.14 Concrete Block, Back-Up										
0010	**CONCRETE BLOCK, BACK-UP**, C90, 2000 psi R042210-20									
0020	Normal weight, 8" x 16" units, tooled joint 1 side									
0050	Not-reinforced, 2000 psi, 2" thick R042110-50	D-8	475	.084	S.F.	1.50	3.61		5.11	7.55
0200	4" thick		460	.087		1.79	3.73		5.52	8
0300	6" thick		440	.091		2.31	3.90		6.21	8.90
0350	8" thick		400	.100		2.84	4.29		7.13	10.10
0400	10" thick		330	.121		2.96	5.20		8.16	11.70
0450	12" thick	D-9	310	.155		4.05	6.50		10.55	15.05
1000	Reinforced, alternate courses, 4" thick	D-8	450	.089		1.95	3.81		5.76	8.35
1100	6" thick		430	.093		2.48	3.99		6.47	9.25
1150	8" thick		395	.101		3.03	4.35		7.38	10.40
1200	10" thick		320	.125		3.12	5.35		8.47	12.20
1250	12" thick	D-9	300	.160		4.21	6.75		10.96	15.60
2000	Lightweight, not reinforced, 4" thick	D-8	460	.087		1.89	3.73		5.62	8.15
2100	6" thick		445	.090		2.65	3.86		6.51	9.20
2150	8" thick		435	.092		3.22	3.95		7.17	9.95
2200	10" thick		410	.098		3.88	4.19		8.07	11.05
2250	12" thick	D-9	390	.123		4.10	5.20		9.30	12.95
3000	Reinforced, alternate courses, 4" thick	D-8	450	.089		2.05	3.81		5.86	8.45
3100	6" thick		430	.093		2.82	3.99		6.81	9.60
3150	8" thick		420	.095		3.42	4.09		7.51	10.40

04 22 10 – Concrete Masonry Units

04 22 10.14 Concrete Block, Back-Up		Crew	Daily Output	Labor-Hours	Unit	Material (2015 Bare Costs)	Labor (2015 Bare Costs)	Equipment (2015 Bare Costs)	Total (2015 Bare Costs)	Total Incl O&P
3200	10" thick	D-8	400	.100	S.F.	4.04	4.29		8.33	11.45
3250	12" thick	D-9	380	.126		4.27	5.30		9.57	13.35
9000	Minimum labor/equipment charge	D-1	2	8	Job		335		335	550

04 22 10.16 Concrete Block, Bond Beam		Crew	Daily Output	Labor-Hours	Unit	Material	Labor	Equipment	Total	Total Incl O&P
0010	**CONCRETE BLOCK, BOND BEAM**, C90, 2000 psi									
0020	Not including grout or reinforcing									
0125	Regular block, 6" thick	D-8	584	.068	L.F.	2.65	2.94		5.59	7.70
0130	8" high, 8" thick	"	565	.071		2.71	3.04		5.75	7.90
0150	12" thick	D-9	510	.094		4.02	3.96		7.98	10.85
0525	Lightweight, 6" thick	D-8	592	.068		2.73	2.90		5.63	7.70
0530	8" high, 8" thick	"	575	.070		3.29	2.99		6.28	8.50
0550	12" thick	D-9	520	.092		4.43	3.89		8.32	11.15
2000	Including grout and 2 #5 bars									
2100	Regular block, 8" high, 8" thick	D-8	300	.133	L.F.	4.78	5.70		10.48	14.55
2150	12" thick	D-9	250	.192		6.65	8.10		14.75	20.50
2500	Lightweight, 8" high, 8" thick	D-8	305	.131		5.35	5.65		11	15.05
2550	12" thick	D-9	255	.188		7.05	7.90		14.95	20.50
9000	Minimum labor/equipment charge	D-1	2	8	Job		335		335	550

04 22 10.18 Concrete Block, Column		Crew	Daily Output	Labor-Hours	Unit	Material	Labor	Equipment	Total	Total Incl O&P
0010	**CONCRETE BLOCK, COLUMN** or pilaster									
0050	Including vertical reinforcing (4-#4 bars) and grout									
0160	1 piece unit, 16" x 16"	D-1	26	.615	V.L.F.	15.90	26		41.90	59.50
0170	2 piece units, 16" x 20"		24	.667		20.50	28		48.50	68
0180	20" x 20"		22	.727		26	30.50		56.50	78.50
0190	22" x 24"		18	.889		39	37.50		76.50	104
0200	20" x 32"		14	1.143		45	48		93	128

04 22 10.19 Concrete Block, Insulation Inserts		Crew	Daily Output	Labor-Hours	Unit	Material	Labor	Equipment	Total	Total Incl O&P
0010	**CONCRETE BLOCK, INSULATION INSERTS**									
0100	Styrofoam, plant installed, add to block prices									
0200	8" x 16" units, 6" thick				S.F.	1.20			1.20	1.32
0250	8" thick					1.35			1.35	1.49
0300	10" thick					1.40			1.40	1.54
0350	12" thick					1.55			1.55	1.71
0500	8" x 8" units, 8" thick					1.20			1.20	1.32
0550	12" thick					1.40			1.40	1.54

04 22 10.23 Concrete Block, Decorative		Crew	Daily Output	Labor-Hours	Unit	Material	Labor	Equipment	Total	Total Incl O&P
0010	**CONCRETE BLOCK, DECORATIVE**, C90, 2000 psi									
0020	Embossed, simulated brick face									
0100	8" x 16" units, 4" thick	D-8	400	.100	S.F.	2.70	4.29		6.99	9.95
0200	8" thick		340	.118		2.95	5.05		8	11.45
0250	12" thick		300	.133		4.98	5.70		10.68	14.75
0400	Embossed both sides									
0500	8" thick	D-8	300	.133	S.F.	3.97	5.70		9.67	13.65
0550	12" thick	"	275	.145	"	5.20	6.25		11.45	15.90
1000	Fluted high strength									
1100	8" x 16" x 4" thick, flutes 1 side,	D-8	345	.116	S.F.	3.83	4.98		8.81	12.30
1150	Flutes 2 sides		335	.119		4.40	5.10		9.50	13.20
1200	8" thick		300	.133		5.15	5.70		10.85	15
1250	For special colors, add					.62			.62	.68
1400	Deep grooved, smooth face									
1450	8" x 16" x 4" thick	D-8	345	.116	S.F.	2.46	4.98		7.44	10.80
1500	8" thick	"	300	.133	"	3.89	5.70		9.59	13.60

04 22 10.23 Concrete Block, Decorative		Crew	Daily Output	Labor-Hours	Unit	Material	2015 Bare Costs Labor	Equipment	Total	Total Incl O&P
2000	Formblock, incl. inserts & reinforcing									
2100	8" x 16" x 8" thick	D-8	345	.116	S.F.	3.44	4.98		8.42	11.90
2150	12" thick	"	310	.129	"	4.55	5.55		10.10	14
2500	Ground face									
2600	8" x 16" x 4" thick	D-8	345	.116	S.F.	3.58	4.98		8.56	12.05
2650	6" thick		325	.123		3.98	5.30		9.28	13
2700	8" thick		300	.133		4.39	5.70		10.09	14.10
2750	12" thick	D-9	265	.181		5.60	7.65		13.25	18.55
2900	For special colors, add, minimum					15%				
2950	For special colors, add, maximum					45%				
4000	Slump block									
4100	4" face height x 16" x 4" thick	D-1	165	.097	S.F.	3.89	4.08		7.97	10.95
4150	6" thick		160	.100		5.65	4.21		9.86	13.10
4200	8" thick		155	.103		5.60	4.35		9.95	13.20
4250	10" thick		140	.114		10.95	4.81		15.76	19.90
4300	12" thick		130	.123		11.85	5.20		17.05	21.50
4400	6" face height x 16" x 6" thick		155	.103		5.25	4.35		9.60	12.85
4450	8" thick		150	.107		7.95	4.49		12.44	16.05
4500	10" thick		130	.123		12.80	5.20		18	22.50
4550	12" thick		120	.133		13.15	5.60		18.75	23.50
5000	Split rib profile units, 1" deep ribs, 8 ribs									
5100	8" x 16" x 4" thick	D-8	345	.116	S.F.	3.84	4.98		8.82	12.35
5150	6" thick		325	.123		4.37	5.30		9.67	13.40
5200	8" thick		300	.133		4.94	5.70		10.64	14.75
5250	12" thick	D-9	275	.175		5.80	7.35		13.15	18.30
5400	For special deeper colors, 4" thick, add					1.24			1.24	1.37
5450	12" thick, add					1.28			1.28	1.40
5600	For white, 4" thick, add					1.24			1.24	1.37
5650	6" thick, add					1.27			1.27	1.39
5700	8" thick, add					1.29			1.29	1.42
5750	12" thick, add					1.33			1.33	1.47
6000	Split face									
6100	8" x 16" x 4" thick	D-8	350	.114	S.F.	3.48	4.90		8.38	11.85
6150	6" thick		325	.123		3.98	5.30		9.28	12.95
6200	8" thick		300	.133		4.47	5.70		10.17	14.20
6250	12" thick	D-9	270	.178		5.40	7.50		12.90	18.15
6300	For scored, add					.37			.37	.41
6400	For special deeper colors, 4" thick, add					.60			.60	.66
6450	6" thick, add					.71			.71	.78
6500	8" thick, add					.73			.73	.81
6550	12" thick, add					.76			.76	.83
6650	For white, 4" thick, add					1.23			1.23	1.35
6700	6" thick, add					1.24			1.24	1.37
6750	8" thick, add					1.25			1.25	1.38
6800	12" thick, add					1.28			1.28	1.40
7000	Scored ground face, 2 to 5 scores									
7100	8" x 16" x 4" thick	D-8	340	.118	S.F.	7.55	5.05		12.60	16.50
7150	6" thick		310	.129		8.40	5.55		13.95	18.20
7200	8" thick		290	.138		9.45	5.90		15.35	20
7250	12" thick	D-9	265	.181		12.70	7.65		20.35	26.50
8000	Hexagonal face profile units, 8" x 16" units									
8100	4" thick, hollow	D-8	340	.118	S.F.	3.63	5.05		8.68	12.20
8200	Solid		340	.118		4.65	5.05		9.70	13.30

04 22 10.23 Concrete Block, Decorative		Crew	Daily Output	Labor-Hours	Unit	Material	Labor	Equipment	Total	Total Incl O&P
						2015 Bare Costs				
8300	6″ thick, hollow	D-8	310	.129	S.F.	3.82	5.55		9.37	13.20
8350	8″ thick, hollow	↓	290	.138	↓	4.36	5.90		10.26	14.45
8500	For stacked bond, add						26%			
8550	For high rise construction, add per story	D-8	67.80	.590	M.S.F.		25.50		25.50	41
8600	For scored block, add					10%				
8650	For honed or ground face, per face, add				Ea.	1.13			1.13	1.24
8700	For honed or ground end, per end, add				″	1.13			1.13	1.24
8750	For bullnose block, add					10%				
8800	For special color, add					13%				
9000	Minimum labor/equipment charge	D-1	2	8	Job		335		335	550

04 22 10.24 Concrete Block, Exterior		Crew	Daily Output	Labor-Hours	Unit	Material	Labor	Equipment	Total	Total Incl O&P
0010	**CONCRETE BLOCK, EXTERIOR**, C90, 2000 psi									
0020	Reinforced alt courses, tooled joints 2 sides									
0100	Normal weight, 8″ x 16″ x 6″ thick	D-8	395	.101	S.F.	2.32	4.35		6.67	9.60
0200	8″ thick		360	.111		4	4.77		8.77	12.15
0250	10″ thick	↓	290	.138		4.24	5.90		10.14	14.30
0300	12″ thick	D-9	250	.192		4.84	8.10		12.94	18.45
0500	Lightweight, 8″ x 16″ x 6″ thick	D-8	450	.089		3.10	3.81		6.91	9.60
0600	8″ thick		430	.093		3.94	3.99		7.93	10.85
0650	10″ thick	↓	395	.101		4.08	4.35		8.43	11.55
0700	12″ thick	D-9	350	.137	↓	4.12	5.75		9.87	13.95
9000	Minimum labor/equipment charge	D-1	2	8	Job		335		335	550

04 22 10.26 Concrete Block Foundation Wall		Crew	Daily Output	Labor-Hours	Unit	Material	Labor	Equipment	Total	Total Incl O&P
0010	**CONCRETE BLOCK FOUNDATION WALL**, C90/C145									
0050	Normal-weight, cut joints, horiz joint reinf, no vert reinf.									
0200	Hollow, 8″ x 16″ x 6″ thick	D-8	455	.088	S.F.	2.89	3.77		6.66	9.30
0250	8″ thick		425	.094		3.45	4.04		7.49	10.35
0300	10″ thick	↓	350	.114		3.57	4.90		8.47	11.90
0350	12″ thick	D-9	300	.160		4.68	6.75		11.43	16.10
0500	Solid, 8″ x 16″ block, 6″ thick	D-8	440	.091		3.05	3.90		6.95	9.70
0550	8″ thick	″	415	.096		4.15	4.14		8.29	11.30
0600	12″ thick	D-9	350	.137	↓	5.95	5.75		11.70	15.95
1000	Reinforced, #4 vert @ 48″									
1100	Hollow, 8″ x 16″ block, 4″ thick	D-8	455	.088	S.F.	2.88	3.77		6.65	9.30
1125	6″ thick		445	.090		3.76	3.86		7.62	10.45
1150	8″ thick		415	.096		4.69	4.14		8.83	11.90
1200	10″ thick	↓	340	.118		5.15	5.05		10.20	13.90
1250	12″ thick	D-9	290	.166		6.65	6.95		13.60	18.65
1500	Solid, 8″ x 16″ block, 6″ thick	D-8	430	.093		3.08	3.99		7.07	9.90
1600	8″ thick	″	405	.099		4.15	4.24		8.39	11.45
1650	12″ thick	D-9	340	.141	↓	5.95	5.95		11.90	16.20
9000	Minimum labor/equipment charge	D-1	2	8	Job		335		335	550

04 22 10.28 Concrete Block, High Strength		Crew	Daily Output	Labor-Hours	Unit	Material	Labor	Equipment	Total	Total Incl O&P
0010	**CONCRETE BLOCK, HIGH STRENGTH** R042210-20									
0050	Hollow, reinforced alternate courses, 8″ x 16″ units									
0200	3500 psi, 4″ thick	D-8	440	.091	S.F.	2.24	3.90		6.14	8.80
0250	6″ thick		395	.101		2.23	4.35		6.58	9.50
0300	8″ thick	↓	360	.111		3.92	4.77		8.69	12.05
0350	12″ thick	D-9	250	.192		4.72	8.10		12.82	18.35
0500	5000 psi, 4″ thick	D-8	440	.091		2.02	3.90		5.92	8.55
0550	6″ thick		395	.101		2.88	4.35		7.23	10.20
0600	8″ thick	↓	360	.111	↓	4.09	4.77		8.86	12.25

04 22 Concrete Unit Masonry

04 22 10 – Concrete Masonry Units

04 22 10.28 Concrete Block, High Strength		Crew	Daily Output	Labor-Hours	Unit	Material	Labor	Equipment	Total	Total Incl O&P
						2015 Bare Costs				
0650	12″ thick	D-9	300	.160	S.F.	4.73	6.75		11.48	16.15
1000	For 75% solid block, add					30%				
1050	For 100% solid block, add					50%				

04 22 10.30 Concrete Block, Interlocking

		Crew	Daily Output	Labor-Hours	Unit	Material	Labor	Equipment	Total	Total Incl O&P
0010	**CONCRETE BLOCK, INTERLOCKING**									
0100	Not including grout or reinforcing									
0200	8″ x 16″ units, 2,000 psi, 8″ thick	D-1	245	.065	S.F.	2.90	2.75		5.65	7.65
0300	12″ thick		220	.073		4.30	3.06		7.36	9.70
0350	16″ thick		185	.086		6.45	3.64		10.09	13
0400	Including grout & reinforcing, 8″ thick	D-4	245	.131		7.75	5.60	.54	13.89	18.10
0450	12″ thick		220	.145		9.35	6.20	.61	16.16	21
0500	16″ thick		185	.173		11.70	7.40	.72	19.82	25.50
9000	Minimum labor/equipment charge	D-1	2	8	Job		335		335	550

04 22 10.32 Concrete Block, Lintels

		Crew	Daily Output	Labor-Hours	Unit	Material	Labor	Equipment	Total	Total Incl O&P
0010	**CONCRETE BLOCK, LINTELS**, C90, normal weight									
0100	Including grout and horizontal reinforcing									
0200	8″ x 8″ x 8″, 1 #4 bar	D-4	300	.107	L.F.	3.74	4.56	.44	8.74	11.95
0250	2 #4 bars		295	.108		3.96	4.63	.45	9.04	12.35
0400	8″ x 16″ x 8″, 1 #4 bar		275	.116		3.83	4.97	.49	9.29	12.80
0450	2 #4 bars		270	.119		4.05	5.05	.49	9.59	13.20
1000	12″ x 8″ x 8″, 1 #4 bar		275	.116		5.20	4.97	.49	10.66	14.30
1100	2 #4 bars		270	.119		5.40	5.05	.49	10.94	14.70
1150	2 #5 bars		270	.119		5.65	5.05	.49	11.19	14.95
1200	2 #6 bars		265	.121		5.95	5.15	.50	11.60	15.45
1500	12″ x 16″ x 8″, 1 #4 bar		250	.128		5.80	5.45	.53	11.78	15.85
1600	2 #3 bars		245	.131		5.80	5.60	.54	11.94	16
1650	2 #4 bars		245	.131		6	5.60	.54	12.14	16.20
1700	2 #5 bars		240	.133		6.25	5.70	.56	12.51	16.70
9000	Minimum labor/equipment charge	D-1	2	8	Job		335		335	550

04 22 10.33 Lintel Block

		Crew	Daily Output	Labor-Hours	Unit	Material	Labor	Equipment	Total	Total Incl O&P
0010	**LINTEL BLOCK**									
3481	Lintel block 6″ x 8″ x 8″	D-1	300	.053	Ea.	1.28	2.25		3.53	5.05
3501	6″ x 16″ x 8″		275	.058		2	2.45		4.45	6.20
3521	8″ x 8″ x 8″		275	.058		1.15	2.45		3.60	5.25
3561	8″ x 16″ x 8″		250	.064		2.02	2.69		4.71	6.60

04 22 10.34 Concrete Block, Partitions

		Crew	Daily Output	Labor-Hours	Unit	Material	Labor	Equipment	Total	Total Incl O&P
0010	**CONCRETE BLOCK, PARTITIONS**, excludes scaffolding R042210-20									
0100	Acoustical slotted block									
0200	4″ thick, type A-1	D-8	315	.127	S.F.	4.06	5.45		9.51	13.30
0210	8″ thick		275	.145		5.20	6.25		11.45	15.85
0250	8″ thick, type Q		275	.145		9.05	6.25		15.30	20
0260	4″ thick, type RSC		315	.127		6.30	5.45		11.75	15.80
0270	6″ thick		295	.136		6.30	5.80		12.10	16.40
0280	8″ thick		275	.145		6.30	6.25		12.55	17.10
0290	12″ thick		250	.160		6.30	6.85		13.15	18.10
0300	8″ thick, type RSR		275	.145		6.30	6.25		12.55	17.10
0400	8″ thick, type RSC/RF		275	.145		7.10	6.25		13.35	18
0410	10″ thick		260	.154		7.65	6.60		14.25	19.15
0420	12″ thick		250	.160		8.25	6.85		15.10	20
0430	12″ thick, type RSC/RF-4		250	.160		9.80	6.85		16.65	22
0500	NRC .60 type R, 8″ thick		265	.151		8.45	6.50		14.95	19.85
0600	NRC .65 type RR, 8″ thick		265	.151		7.90	6.50		14.40	19.20

04 22 Concrete Unit Masonry

04 22 10 – Concrete Masonry Units

04 22 10.34 Concrete Block, Partitions		Crew	Daily Output	Labor-Hours	Unit	2015 Bare Costs Material	Labor	Equipment	Total	Total Incl O&P
0700	NRC .65 type 4R-RF, 8" thick	D-8	265	.151	S.F.	8.70	6.50		15.20	20
0710	NRC .70 type R, 12" thick		245	.163		9.15	7		16.15	21.50
1000	Lightweight block, tooled joints, 2 sides, hollow									
1100	Not reinforced, 8" x 16" x 4" thick	D-8	440	.091	S.F.	1.80	3.90		5.70	8.35
1150	6" thick		410	.098		2.56	4.19		6.75	9.60
1200	8" thick		385	.104		3.13	4.46		7.59	10.70
1250	10" thick		370	.108		3.79	4.64		8.43	11.70
1300	12" thick	D-9	350	.137		4.01	5.75		9.76	13.80
1500	Reinforced alternate courses, 4" thick	D-8	435	.092		1.95	3.95		5.90	8.55
1600	6" thick		405	.099		2.70	4.24		6.94	9.85
1650	8" thick		380	.105		3.28	4.52		7.80	10.95
1700	10" thick		365	.110		3.95	4.70		8.65	12
1750	12" thick	D-9	345	.139		4.18	5.85		10.03	14.15
2000	Not reinforced, 8" x 24" x 4" thick, hollow		460	.104		1.31	4.39		5.70	8.60
2100	6" thick		440	.109		1.84	4.59		6.43	9.45
2150	8" thick		415	.116		2.29	4.87		7.16	10.40
2200	10" thick		385	.125		2.77	5.25		8.02	11.60
2250	12" thick		365	.132		2.96	5.55		8.51	12.25
2400	Reinforced alternate courses, 4" thick		455	.105		1.46	4.44		5.90	8.85
2500	6" thick		435	.110		2	4.65		6.65	9.75
2550	8" thick		410	.117		2.43	4.93		7.36	10.70
2600	10" thick		380	.126		2.93	5.30		8.23	11.85
2650	12" thick		360	.133		3.13	5.60		8.73	12.60
2800	Solid, not reinforced, 8" x 16" x 2" thick	D-8	440	.091		1.61	3.90		5.51	8.10
2900	4" thick		420	.095		1.63	4.09		5.72	8.45
2950	6" thick		390	.103		2.96	4.40		7.36	10.40
3000	8" thick		365	.110		3.80	4.70		8.50	11.85
3050	10" thick		350	.114		3.91	4.90		8.81	12.30
3100	12" thick	D-9	330	.145		3.94	6.10		10.04	14.30
3300	Solid, reinforced alternate courses, 4" thick	D-8	415	.096		1.78	4.14		5.92	8.70
3400	6" thick		385	.104		3.28	4.46		7.74	10.85
3450	8" thick		360	.111		3.94	4.77		8.71	12.10
3500	10" thick		345	.116		4.08	4.98		9.06	12.60
3550	12" thick	D-9	325	.148		4.12	6.20		10.32	14.65
4000	Regular block, tooled joints, 2 sides, hollow									
4100	Not reinforced, 8" x 16" x 4" thick	D-8	430	.093	S.F.	1.70	3.99		5.69	8.35
4150	6" thick		400	.100		2.22	4.29		6.51	9.45
4200	8" thick		375	.107		2.74	4.58		7.32	10.45
4250	10" thick		360	.111		2.86	4.77		7.63	10.90
4300	12" thick	D-9	340	.141		3.95	5.95		9.90	14
4500	Reinforced alternate courses, 8" x 16" x 4" thick	D-8	425	.094		1.86	4.04		5.90	8.60
4550	6" thick		395	.101		2.39	4.35		6.74	9.70
4600	8" thick		370	.108		2.94	4.64		7.58	10.80
4650	10" thick		355	.113		3.98	4.84		8.82	12.25
4700	12" thick	D-9	335	.143		4.12	6.05		10.17	14.35
4900	Solid, not reinforced, 2" thick	D-8	435	.092		1.44	3.95		5.39	8
5000	3" thick		430	.093		1.35	3.99		5.34	8
5050	4" thick		415	.096		1.96	4.14		6.10	8.90
5100	6" thick		385	.104		2.39	4.46		6.85	9.90
5150	8" thick		360	.111		3.45	4.77		8.22	11.55
5200	12" thick	D-9	325	.148		5.25	6.20		11.45	15.85
5500	Solid, reinforced alternate courses, 4" thick	D-8	420	.095		2.11	4.09		6.20	8.95
5550	6" thick		380	.105		2.53	4.52		7.05	10.15

04 22 Concrete Unit Masonry

04 22 10 – Concrete Masonry Units

04 22 10.34 Concrete Block, Partitions		Crew	Daily Output	Labor-Hours	Unit	2015 Bare Costs Material	Labor	Equipment	Total	Total Incl O&P
5600	8" thick	D-8	355	.113	S.F.	3.59	4.84		8.43	11.80
5650	12" thick	D-9	320	.150		4.38	6.30		10.68	15.10
9000	Minimum labor/equipment charge	D-1	2	8	Job		335		335	550

04 22 10.38 Concrete Brick

No.	Description	Crew	Daily Output	Labor-Hours	Unit	Material	Labor	Equipment	Total	Total Incl O&P
0010	**CONCRETE BRICK**, C55, grade N, type 1									
0100	Regular, 4 x 2-1/4 x 8	D-8	660	.061	Ea.	.54	2.60		3.14	4.82
0125	Rusticated, 4 x 2-1/4 x 8		660	.061		.60	2.60		3.20	4.90
0150	Frog, 4 x 2-1/4 x 8		660	.061		.58	2.60		3.18	4.87
0200	Double, 4 x 4-7/8 x 8		535	.075		.95	3.21		4.16	6.25

04 22 10.42 Concrete Block, Screen Block

No.	Description	Crew	Daily Output	Labor-Hours	Unit	Material	Labor	Equipment	Total	Total Incl O&P
0010	**CONCRETE BLOCK, SCREEN BLOCK**									
0200	8" x 16", 4" thick	D-8	330	.121	S.F.	2.34	5.20		7.54	11
0300	8" thick		270	.148		3.71	6.35		10.06	14.45
0350	12" x 12", 4" thick		290	.138		4.27	5.90		10.17	14.35
0500	8" thick		250	.160		7.15	6.85		14	19

04 22 10.44 Glazed Concrete Block

No.	Description	Crew	Daily Output	Labor-Hours	Unit	Material	Labor	Equipment	Total	Total Incl O&P
0010	**GLAZED CONCRETE BLOCK** C744									
0100	Single face, 8" x 16" units, 2" thick	D-8	360	.111	S.F.	9.60	4.77		14.37	18.35
0200	4" thick		345	.116		9.90	4.98		14.88	19
0250	6" thick		330	.121		10.70	5.20		15.90	20
0300	8" thick		310	.129		11.45	5.55		17	21.50
0350	10" thick		295	.136		13.05	5.80		18.85	24
0400	12" thick	D-9	280	.171		13.95	7.20		21.15	27
0700	Double face, 8" x 16" units, 4" thick	D-8	340	.118		13.85	5.05		18.90	23.50
0750	6" thick		320	.125		17	5.35		22.35	27.50
0800	8" thick		300	.133		17.80	5.70		23.50	29
1000	Jambs, bullnose or square, single face, 8" x 16", 2" thick		315	.127	Ea.	18.35	5.45		23.80	29
1050	4" thick		285	.140	"	19.30	6		25.30	31
1200	Caps, bullnose or square, 8" x 16", 2" thick		420	.095	L.F.	17.05	4.09		21.14	25.50
1250	4" thick		380	.105	"	18.50	4.52		23.02	28
1256	Corner, bullnose or square, 2" thick		280	.143	Ea.	19.95	6.15		26.10	32
1258	4" thick		270	.148		21.50	6.35		27.85	34
1260	6" thick		260	.154		27	6.60		33.60	40.50
1270	8" thick		250	.160		32.50	6.85		39.35	47
1280	10" thick		240	.167		33	7.15		40.15	48
1290	12" thick		230	.174		35	7.45		42.45	50.50
1500	Cove base, 8" x 16", 2" thick		315	.127	L.F.	8.55	5.45		14	18.30
1550	4" thick		285	.140		8.65	6		14.65	19.35
1600	6" thick		265	.151		9.30	6.50		15.80	21
1650	8" thick		245	.163		9.75	7		16.75	22
9000	Minimum labor/equipment charge	D-1	2	8	Job		335		335	550

04 23 Glass Unit Masonry

04 23 13 – Vertical Glass Unit Masonry

04 23 13.10 Glass Block		Crew	Daily Output	Labor-Hours	Unit	2015 Bare Costs Material	Labor	Equipment	Total	Total Incl O&P
0010	**GLASS BLOCK**									
0100	Plain, 4" thick, under 1,000 S.F., 6" x 6"	D-8	115	.348	S.F.	26.50	14.95		41.45	53.50
0150	8" x 8"		160	.250		15.35	10.75		26.10	34.50
0160	end block		160	.250		59	10.75		69.75	82.50
0170	90 deg corner		160	.250		53.50	10.75		64.25	76.50
0180	45 deg corner		160	.250		44	10.75		54.75	66
0200	12" x 12"		175	.229		22	9.80		31.80	40.50
0210	4" x 8"		160	.250		28	10.75		38.75	48.50
0220	6" x 8"		160	.250		19.90	10.75		30.65	39.50
0300	1,000 to 5,000 S.F., 6" x 6"		135	.296		26	12.70		38.70	49
0350	8" x 8"		190	.211		15.05	9.05		24.10	31.50
0400	12" x 12"		215	.186		21.50	8		29.50	37
0410	4" x 8"		215	.186		27.50	8		35.50	43
0420	6" x 8"		215	.186		19.50	8		27.50	34.50
0500	Over 5,000 S.F., 6" x 6"		145	.276		25	11.85		36.85	47
0550	8" x 8"		215	.186		14.60	8		22.60	29
0600	12" x 12"		240	.167		21	7.15		28.15	34.50
0610	4" x 8"		240	.167		26.50	7.15		33.65	41
0620	6" x 8"		240	.167		18.90	7.15		26.05	32.50
0700	For solar reflective blocks, add					100%				
1000	Thinline, plain, 3-1/8" thick, under 1,000 S.F., 6" x 6"	D-8	115	.348	S.F.	21	14.95		35.95	47.50
1050	8" x 8"		160	.250		11.60	10.75		22.35	30
1200	Over 5,000 S.F., 6" x 6"		145	.276		19.80	11.85		31.65	41.50
1250	8" x 8"		215	.186		11.05	8		19.05	25
1400	For cleaning block after installation (both sides), add		1000	.040		.16	1.72		1.88	2.97
4000	Accessories									
4100	Anchors, 20 ga. galv., 1-3/4" wide x 24" long				Ea.	4.96			4.96	5.45
4200	Emulsion asphalt				Gal.	10.65			10.65	11.70
4300	Expansion joint, fiberglass				L.F.	.61			.61	.67
4400	Steel mesh, double galvanized				"	.96			.96	1.06
9000	Minimum labor/equipment charge	D-1	2	8	Job		335		335	550

04 24 Adobe Unit Masonry

04 24 16 – Manufactured Adobe Unit Masonry

04 24 16.06 Adobe Brick		Crew	Daily Output	Labor-Hours	Unit	Material	Labor	Equipment	Total	Total Incl O&P
0010	**ADOBE BRICK**, Semi-stabilized, with cement mortar									
0060	Brick, 10" x 4" x 14", 2.6/S.F. G	D-8	560	.071	S.F.	4.50	3.07		7.57	9.95
0080	12" x 4" x 16", 2.3/S.F. G		580	.069		6.85	2.96		9.81	12.30
0100	10" x 4" x 16", 2.3/S.F. G		590	.068		6.40	2.91		9.31	11.80
0120	8" x 4" x 16", 2.3/S.F. G		560	.071		4.89	3.07		7.96	10.40
0140	4" x 4" x 16", 2.3/S.F. G		540	.074		4.79	3.18		7.97	10.40
0160	6" x 4" x 16", 2.3/S.F. G		540	.074		4.44	3.18		7.62	10.05
0180	4" x 4" x 12", 3.0/S.F. G		520	.077		5.15	3.30		8.45	11.05
0200	8" x 4" x 12", 3.0/S.F. G		520	.077		4.16	3.30		7.46	9.95

04 25 Unit Masonry Panels

04 25 20 – Pre-Fabricated Masonry Panels

04 25 20.10 Brick and Epoxy Mortar Panels		Crew	Daily Output	Labor-Hours	Unit	Material	2015 Bare Costs Labor	Equipment	Total	Total Incl O&P
0010	**BRICK AND EPOXY MORTAR PANELS**									
0020	Prefabricated brick & epoxy mortar, 4" thick, minimum	C-11	775	.093	S.F.	8	4.83	2.34	15.17	19.85
0100	Maximum	"	500	.144		10	7.50	3.63	21.13	28
0200	For 2" concrete back-up, add					50%				
0300	For 1" urethane & 3" concrete back-up, add				↓	70%				

04 27 Multiple-Wythe Unit Masonry

04 27 10 – Multiple-Wythe Masonry

04 27 10.10 Cornices

		Crew	Daily Output	Labor-Hours	Unit	Material	Labor	Equipment	Total	Total Incl O&P
0010	**CORNICES**									
0110	Face bricks, 12 brick/S.F.	D-1	30	.533	SF Face	5.65	22.50		28.15	42.50
0150	15 brick/S.F.		23	.696	"	6.75	29.50		36.25	55
9000	Minimum labor/equipment charge	↓	1.50	10.667	Job		450		450	730

04 27 10.30 Brick Walls

		Crew	Daily Output	Labor-Hours	Unit	Material	Labor	Equipment	Total	Total Incl O&P
0010	**BRICK WALLS**, including mortar, excludes scaffolding R042110-20									
0020	Estimating by number of brick									
0140	Face brick, 4" thick wall, 6.75 brick/S.F.	D-8	1.45	27.586	M	540	1,175		1,715	2,525
0150	Common brick, 4" thick wall, 6.75 brick/S.F. R042110-50		1.60	25		595	1,075		1,670	2,400
0204	8" thick, 13.50 bricks per S.F.		1.80	22.222		615	955		1,570	2,225
0250	12" thick, 20.25 bricks per S.F.		1.90	21.053		620	905		1,525	2,150
0304	16" thick, 27.00 bricks per S.F.		2	20		630	860		1,490	2,100
0500	Reinforced, face brick, 4" thick wall, 6.75 brick/S.F.		1.40	28.571		565	1,225		1,790	2,625
0520	Common brick, 4" thick wall, 6.75 brick/S.F.		1.55	25.806		620	1,100		1,720	2,475
0550	8" thick, 13.50 bricks per S.F.		1.75	22.857		640	980		1,620	2,300
0600	12" thick, 20.25 bricks per S.F.		1.85	21.622		645	930		1,575	2,200
0650	16" thick, 27.00 bricks per S.F.	↓	1.95	20.513	↓	655	880		1,535	2,150
0790	Alternate method of figuring by square foot									
0800	Face brick, 4" thick wall, 6.75 brick/S.F.	D-8	215	.186	S.F.	3.65	8		11.65	17
0850	Common brick, 4" thick wall, 6.75 brick/S.F.		240	.167		4.03	7.15		11.18	16.10
0900	8" thick, 13.50 bricks per S.F.		135	.296		8.30	12.70		21	29.50
1000	12" thick, 20.25 bricks per S.F.		95	.421		12.55	18.05		30.60	43.50
1050	16" thick, 27.00 bricks per S.F.		75	.533		17	23		40	55.50
1200	Reinforced, face brick, 4" thick wall, 6.75 brick/S.F.		210	.190		3.82	8.15		11.97	17.50
1220	Common brick, 4" thick wall, 6.75 brick/S.F.		235	.170		4.20	7.30		11.50	16.50
1250	8" thick, 13.50 bricks per S.F.		130	.308		8.65	13.20		21.85	31
1300	12" thick, 20.25 bricks per S.F.		90	.444		13.05	19.05		32.10	45.50
1350	16" thick, 27.00 bricks per S.F.	↓	70	.571	↓	17.65	24.50		42.15	59.50
9000	Minimum labor/equipment charge	D-1	2	8	Job		335		335	550

04 27 10.40 Steps

		Crew	Daily Output	Labor-Hours	Unit	Material	Labor	Equipment	Total	Total Incl O&P
0010	**STEPS**									
0012	Entry steps, select common brick	D-1	.30	53.333	M	500	2,250		2,750	4,200

04 41 Dry-Placed Stone

04 41 10 – Dry Placed Stone

04 41 10.10 Rough Stone Wall		Crew	Daily Output	Labor-Hours	Unit	Material	2015 Bare Costs Labor	Equipment	Total	Total Incl O&P
0011	**ROUGH STONE WALL**, Dry									
0012	Dry laid (no mortar), under 18" thick G	D-1	60	.267	C.F.	12.90	11.25		24.15	32.50
0100	Random fieldstone, under 18" thick G	D-12	60	.533		12.90	22.50		35.40	51
0150	Over 18" thick G	"	63	.508		15.50	21.50		37	52
0500	Field stone veneer G	D-8	120	.333	S.F.	12.20	14.30		26.50	37
0510	Valley stone veneer G		120	.333		12.20	14.30		26.50	37
0520	River stone veneer G		120	.333		12.20	14.30		26.50	37
0600	Rubble stone walls, in mortar bed, up to 18" thick G	D-11	75	.320	C.F.	15.55	14.10		29.65	40
9000	Minimum labor/equipment charge	D-1	2	8	Job		335		335	550

04 43 Stone Masonry

04 43 10 – Masonry with Natural and Processed Stone

04 43 10.05 Ashlar Veneer		Crew	Daily Output	Labor-Hours	Unit	Material	Labor	Equipment	Total	Total Incl O&P
0011	**ASHLAR VENEER** 4" + or - thk, random or random rectangular									
0150	Sawn face, split joints, low priced stone	D-8	140	.286	S.F.	11.60	12.25		23.85	32.50
0200	Medium priced stone		130	.308		13.35	13.20		26.55	36
0300	High priced stone		120	.333		17.90	14.30		32.20	43
0600	Seam face, split joints, medium price stone		125	.320		19	13.75		32.75	43.50
0700	High price stone		120	.333		18.75	14.30		33.05	44
1000	Split or rock face, split joints, medium price stone		125	.320		11.40	13.75		25.15	35
1100	High price stone		120	.333		17.25	14.30		31.55	42.50

04 43 10.10 Bluestone		Crew	Daily Output	Labor-Hours	Unit	Material	Labor	Equipment	Total	Total Incl O&P
0010	**BLUESTONE**, cut to size									
0500	Sills, natural cleft, 10" wide to 6' long, 1-1/2" thick	D-11	70	.343	L.F.	13.35	15.15		28.50	39
0550	2" thick		63	.381		14.60	16.80		31.40	43.50
0600	Smooth finish, 1-1/2" thick		70	.343		12	15.15		27.15	37.50
0650	2" thick		63	.381		13	16.80		29.80	42
0800	Thermal finish, 1-1/2" thick		70	.343		12	15.15		27.15	37.50
0850	2" thick		63	.381		13	16.80		29.80	42
1000	Stair treads, natural cleft, 12" wide, 6' long, 1-1/2" thick	D-10	115	.278		12	12.80	4.12	28.92	38
1050	2" thick		105	.305		13	14	4.51	31.51	42
1100	Smooth finish, 1-1/2" thick		115	.278		12	12.80	4.12	28.92	38
1150	2" thick		105	.305		13	14	4.51	31.51	42
1300	Thermal finish, 1-1/2" thick		115	.278		12	12.80	4.12	28.92	38
1350	2" thick		105	.305		13	14	4.51	31.51	42
2000	Coping, finished top & 2 sides, 12" to 6'									
2100	Natural cleft, 1-1/2" thick	D-10	115	.278	L.F.	12.50	12.80	4.12	29.42	39
2150	2" thick		105	.305		13.50	14	4.51	32.01	42.50
2200	Smooth finish, 1-1/2" thick		115	.278		12.50	12.80	4.12	29.42	39
2250	2" thick		105	.305		13.50	14	4.51	32.01	42.50
2300	Thermal finish, 1-1/2" thick		115	.278		12.50	12.80	4.12	29.42	39
2350	2" thick		105	.305		13.50	14	4.51	32.01	42.50
9000	Minimum labor/equipment charge	D-1	2.50	6.400	Job		269		269	440

04 43 10.45 Granite		Crew	Daily Output	Labor-Hours	Unit	Material	Labor	Equipment	Total	Total Incl O&P
0010	**GRANITE**, cut to size									
0050	Veneer, polished face, 3/4" to 1-1/2" thick									
0150	Low price, gray, light gray, etc.	D-10	130	.246	S.F.	26.50	11.35	3.64	41.49	52
0180	Medium price, pink, brown, etc.		130	.246		29.50	11.35	3.64	44.49	55
0220	High price, red, black, etc.		130	.246		42	11.35	3.64	56.99	69
0300	1-1/2" to 2-1/2" thick, veneer									

04 43 Stone Masonry

04 43 10 – Masonry with Natural and Processed Stone

04 43 10.45 Granite		Crew	Daily Output	Labor-Hours	Unit	Material	2015 Bare Costs Labor	Equipment	Total	Total Incl O&P
0350	Low price, gray, light gray, etc.	D-10	130	.246	S.F.	28.50	11.35	3.64	43.49	54
0500	Medium price, pink, brown, etc.		130	.246		33.50	11.35	3.64	48.49	59.50
0550	High price, red, black, etc.		130	.246		52.50	11.35	3.64	67.49	80
0700	2-1/2" to 4" thick, veneer									
0750	Low price, gray, light gray, etc.	D-10	110	.291	S.F.	38.50	13.40	4.30	56.20	68.50
0850	Medium price, pink, brown, etc.		110	.291		44	13.40	4.30	61.70	74.50
0950	High price, red, black, etc.		110	.291		63	13.40	4.30	80.70	95.50
1000	For bush hammered finish, deduct					5%				
1050	Coarse rubbed finish, deduct					10%				
1100	Honed finish, deduct					5%				
1150	Thermal finish, deduct					18%				
1800	Carving or bas-relief, from templates or plaster molds									
1850	Low price, gray, light gray, etc.	D-10	80	.400	C.F.	178	18.40	5.90	202.30	232
1875	Medium price, pink, brown, etc.		80	.400		350	18.40	5.90	374.30	420
1900	High price, red, black, etc.		80	.400		520	18.40	5.90	544.30	605
2000	Intricate or hand finished pieces									
2010	Mouldings, radius cuts, bullnose edges, etc.									
2050	Add for low price gray, light gray, etc.					30%				
2075	Add for medium price, pink, brown, etc.					30%				
2100	Add for high price red, black, etc.					300%				
2450	For radius under 5', add				L.F.	100%				
2500	Steps, copings, etc., finished on more than one surface									
2550	Low price, gray, light gray, etc.	D-10	50	.640	C.F.	91	29.50	9.45	129.95	158
2575	Medium price, pink, brown, etc.		50	.640		118	29.50	9.45	156.95	188
2600	High price, red, black, etc.		50	.640		146	29.50	9.45	184.95	218
2700	Pavers, Belgian block, 8"-13" long, 4"-6" wide, 4"-6" deep	D-11	120	.200	S.F.	27	8.80		35.80	44
2800	Pavers, 4" x 4" x 4" blocks, split face and joints									
2850	Low price, gray, light gray, etc.	D-11	80	.300	S.F.	13.10	13.25		26.35	36
2875	Medium price, pinks, browns, etc.		80	.300		21	13.25		34.25	44.50
2900	High price, red, black, etc.		80	.300		29	13.25		42.25	53.50
3000	Pavers, 4" x 4" x 4", thermal face, sawn joints									
3050	Low price, gray, light gray, etc.	D-11	65	.369	S.F.	24.50	16.30		40.80	53
3075	Medium price, pink, brown, etc.		65	.369		28	16.30		44.30	57.50
3100	High price, red, black, etc.		65	.369		32	16.30		48.30	62
4000	Soffits, 2" thick, low price, gray, light gray	D-13	35	1.371		37.50	61.50	13.50	112.50	156
4050	Medium price, pink, brown, etc.		35	1.371		64.50	61.50	13.50	139.50	185
4100	High price, red, black, etc.		35	1.371		91.50	61.50	13.50	166.50	214
4200	Low price, gray, light gray, etc.		35	1.371		63	61.50	13.50	138	184
4250	Medium price, pink, brown, etc.		35	1.371		91	61.50	13.50	166	214
4300	High price, red, black, etc.		35	1.371		119	61.50	13.50	194	245
9000	Minimum labor/equipment charge	D-1	2	8	Job		335		335	550

04 43 10.50 Lightweight Natural Stone		Crew	Daily Output	Labor-Hours	Unit	Material	Labor	Equipment	Total	Total Incl O&P
0011	**LIGHTWEIGHT NATURAL STONE** Lava type									
0100	Veneer, rubble face, sawed back, irregular shapes G	D-10	130	.246	S.F.	8.50	11.35	3.64	23.49	31.50
0200	Sawed face and back, irregular shapes G	"	130	.246	"	8.50	11.35	3.64	23.49	31.50

04 43 10.55 Limestone		Crew	Daily Output	Labor-Hours	Unit	Material	Labor	Equipment	Total	Total Incl O&P
0010	**LIMESTONE**, cut to size									
0020	Veneer facing panels									
0500	Texture finish, light stick, 4-1/2" thick, 5' x 12'	D-4	300	.107	S.F.	19.50	4.56	.44	24.50	29.50
0750	5" thick, 5' x 14' panels	D-10	275	.116		20.50	5.35	1.72	27.57	33
1000	Sugarcube finish, 2" Thick, 3' x 5' panels		275	.116		28.50	5.35	1.72	35.57	42
1050	3" Thick, 4' x 9' panels		275	.116		22.50	5.35	1.72	29.57	35.50

04 43 10 – Masonry with Natural and Processed Stone

04 43 10.55 Limestone		Crew	Daily Output	Labor-Hours	Unit	2015 Bare Costs Material	Labor	Equipment	Total	Total Incl O&P
1200	4" Thick, 5' x 11' panels	D-10	275	.116	S.F.	30	5.35	1.72	37.07	43.50
1400	Sugarcube, textured finish, 4-1/2" thick, 5' x 12'		275	.116		31	5.35	1.72	38.07	44.50
1450	5" thick, 5' x 14' panels		275	.116		32	5.35	1.72	39.07	46
2000	Coping, sugarcube finish, top & 2 sides		30	1.067	C.F.	67.50	49	15.80	132.30	171
2100	Sills, lintels, jambs, trim, stops, sugarcube finish, simple		20	1.600		67.50	73.50	23.50	164.50	219
2150	Detailed		20	1.600		67.50	73.50	23.50	164.50	219
2300	Steps, extra hard, 14" wide, 6" rise		50	.640	L.F.	25	29.50	9.45	63.95	85.50
3000	Quoins, plain finish, 6" x 12" x 12"	D-12	25	1.280	Ea.	37.50	54.50		92	130
3050	6" x 16" x 24"	"	25	1.280	"	50	54.50		104.50	144
9000	Minimum labor/equipment charge	D-1	2	8	Job		335		335	550

04 43 10.60 Marble		Crew	Daily Output	Labor-Hours	Unit	Material	Labor	Equipment	Total	Total Incl O&P
0011	**MARBLE**, ashlar, split face, 4" + or - thick, random									
0040	Lengths 1' to 4' & heights 2" to 7-1/2", average	D-8	175	.229	S.F.	17.45	9.80		27.25	35
0100	Base, polished, 3/4" or 7/8" thick, polished, 6" high	D-10	65	.492	L.F.	12	22.50	7.30	41.80	57.50
0300	Carvings or bas relief, from templates, simple design		80	.400	S.F.	144	18.40	5.90	168.30	194
0350	Intricate design		80	.400	"	335	18.40	5.90	359.30	405
0600	Columns, cornices, mouldings, etc.									
0650	Hand or special machine cut, simple design	D-10	35	.914	C.F.	52.50	42	13.50	108	140
0700	Intricate design	"	35	.914	"	278	42	13.50	333.50	390
1000	Facing, polished finish, cut to size, 3/4" to 7/8" thick									
1050	Carrara or equal	D-10	130	.246	S.F.	19	11.35	3.64	33.99	43.50
1100	Arabescato or equal		130	.246		39	11.35	3.64	53.99	65.50
1300	1-1/4" thick, Botticino Classico or equal		125	.256		24	11.80	3.79	39.59	49.50
1350	Statuarietto or equal		125	.256		41.50	11.80	3.79	57.09	68.50
1500	2" thick, Crema Marfil or equal		120	.267		42.50	12.25	3.94	58.69	71
1550	Cafe Pinta or equal		120	.267		70	12.25	3.94	86.19	101
1700	Rubbed finish, cut to size, 4" thick									
1740	Average	D-10	100	.320	S.F.	40	14.70	4.73	59.43	73
1780	Maximum	"	100	.320	"	69	14.70	4.73	88.43	105
2200	Window sills, 6" x 3/4" thick	D-1	85	.188	L.F.	10.90	7.90		18.80	25
2500	Flooring, polished tiles, 12" x 12" x 3/8" thick									
2510	Thin set, Giallo Solare or equal	D-11	90	.267	S.F.	14.75	11.75		26.50	35.50
2600	Sky Blue or equal		90	.267		16	11.75		27.75	37
2700	Mortar bed, Giallo Solare or equal		65	.369		14.75	16.30		31.05	43
2740	Sky Blue or equal		65	.369		16	16.30		32.30	44
2780	Travertine, 3/8" thick, Sierra or equal	D-10	130	.246		9.25	11.35	3.64	24.24	32.50
2790	Silver or equal	"	130	.246		25.50	11.35	3.64	40.49	50.50
2800	Patio tile, non-slip, 1/2" thick, flame finish	D-11	75	.320		11.90	14.10		26	36
2900	Shower or toilet partitions, 7/8" thick partitions									
3050	3/4" or 1-1/4" thick stiles, polished 2 sides, average	D-11	75	.320	S.F.	45.50	14.10		59.60	73
3201	Soffits, add to above prices				"	20%	100%			
3210	Stairs, risers, 7/8" thick x 6" high	D-10	115	.278	L.F.	15.25	12.80	4.12	32.17	42
3360	Treads, 12" wide x 1-1/4" thick	"	115	.278	"	43.50	12.80	4.12	60.42	73
3500	Thresholds, 3' long, 7/8" thick, 4" to 5" wide, plain	D-12	24	1.333	Ea.	35.50	57		92.50	132
3550	Beveled		24	1.333	"	70.50	57		127.50	170
3700	Window stools, polished, 7/8" thick, 5" wide		85	.376	L.F.	21.50	16.05		37.55	49.50
9000	Minimum labor/equipment charge	D-1	2	8	Job		335		335	550

04 43 10.75 Sandstone or Brownstone		Crew	Daily Output	Labor-Hours	Unit	Material	Labor	Equipment	Total	Total Incl O&P
0011	**SANDSTONE OR BROWNSTONE**									
0100	Sawed face veneer, 2-1/2" thick, to 2' x 4' panels	D-10	130	.246	S.F.	19	11.35	3.64	33.99	43.50
0150	4" thick, to 3'-6" x 8'panels		100	.320		19	14.70	4.73	38.43	50
0300	Split face, random sizes		100	.320		13.65	14.70	4.73	33.08	44

04 43 10.75 Sandstone or Brownstone

		Crew	Daily Output	Labor-Hours	Unit	Material	Labor	Equipment	Total	Total Incl O&P
0350	Cut stone trim (limestone)									
0360	Ribbon stone, 4" thick, 5' pieces	D-8	120	.333	Ea.	155	14.30		169.30	194
0370	Cove stone, 4" thick, 5' pieces		105	.381		156	16.35		172.35	198
0380	Cornice stone, 10" to 12" wide		90	.444		192	19.05		211.05	242
0390	Band stone, 4" thick, 5' pieces		145	.276		99.50	11.85		111.35	128
0410	Window and door trim, 3" to 4" wide		160	.250		84.50	10.75		95.25	110
0420	Key stone, 18" long		60	.667		88.50	28.50		117	144
9000	Minimum labor/equipment charge	D-1	2.50	6.400	Job		269		269	440

04 43 10.80 Slate

		Crew	Daily Output	Labor-Hours	Unit	Material	Labor	Equipment	Total	Total Incl O&P
0010	**SLATE**									
0040	Pennsylvania - blue gray to black									
0050	Vermont - unfading green, mottled green & purple, gray & purple									
0100	Virginia - blue black									
0200	Exterior paving, natural cleft, 1" thick									
0250	6" x 6" Pennsylvania	D-12	100	.320	S.F.	7.15	13.65		20.80	30
0300	Vermont		100	.320		11.25	13.65		24.90	34.50
0350	Virginia		100	.320		14.95	13.65		28.60	38.50
0500	24" x 24", Pennsylvania		120	.267		13.80	11.35		25.15	33.50
0550	Vermont		120	.267		28	11.35		39.35	49
0600	Virginia		120	.267		21.50	11.35		32.85	42.50
0700	18" x 30" Pennsylvania		120	.267		15.65	11.35		27	35.50
0750	Vermont		120	.267		28	11.35		39.35	49
0800	Virginia		120	.267		19.30	11.35		30.65	39.50
1000	Interior flooring, natural cleft, 1/2" thick									
1100	6" x 6" Pennsylvania	D-12	100	.320	S.F.	4.24	13.65		17.89	26.50
1150	Vermont		100	.320		9.90	13.65		23.55	33
1200	Virginia		100	.320		11.80	13.65		25.45	35
1300	24" x 24" Pennsylvania		120	.267		8.20	11.35		19.55	27.50
1350	Vermont		120	.267		22.50	11.35		33.85	43.50
1400	Virginia		120	.267		15.60	11.35		26.95	35.50
1500	18" x 24" Pennsylvania		120	.267		8.20	11.35		19.55	27.50
1550	Vermont		120	.267		18	11.35		29.35	38.50
1600	Virginia		120	.267		15.85	11.35		27.20	36
2000	Facing panels, 1-1/4" thick, to 4' x 4' panels									
2100	Natural cleft finish, Pennsylvania	D-10	180	.178	S.F.	36	8.20	2.63	46.83	55.50
2110	Vermont		180	.178		29	8.20	2.63	39.83	47.50
2120	Virginia		180	.178		35	8.20	2.63	45.83	55
2150	Sand rubbed finish, surface, add					10.80			10.80	11.85
2200	Honed finish, add					7.80			7.80	8.55
2500	Ribbon, natural cleft finish, 1" thick, to 9 S.F.	D-10	80	.400		13.90	18.40	5.90	38.20	51.50
2550	Sand rubbed finish		80	.400		18.80	18.40	5.90	43.10	56.50
2600	Honed finish		80	.400		17.50	18.40	5.90	41.80	55.50
2700	1-1/2" thick		78	.410		18.05	18.90	6.05	43	57
2750	Sand rubbed finish		78	.410		24	18.90	6.05	48.95	63
2800	Honed finish		78	.410		22.50	18.90	6.05	47.45	62
2850	2" thick		76	.421		21.50	19.35	6.25	47.10	62.50
2900	Sand rubbed finish		76	.421		30	19.35	6.25	55.60	71.50
2950	Honed finish		76	.421		27.50	19.35	6.25	53.10	69
3100	Stair landings, 1" thick, black, clear	D-1	65	.246		21	10.35		31.35	40.50
3200	Ribbon	"	65	.246		23.50	10.35		33.85	42.50
3500	Stair treads, sand finish, 1" thick x 12" wide									
3550	Under 3 L.F.	D-10	85	.376	L.F.	23.50	17.30	5.55	46.35	59.50

04 43 Stone Masonry

04 43 10 – Masonry with Natural and Processed Stone

04 43 10.80 Slate		Crew	Daily Output	Labor-Hours	Unit	Material	2015 Bare Costs Labor	Equipment	Total	Total Incl O&P
3600	3 L.F. to 6 L.F.	D-10	120	.267	L.F.	25	12.25	3.94	41.19	51.50
3700	Ribbon, sand finish, 1" thick x 12" wide									
3750	To 6 L.F.	D-10	120	.267	L.F.	21	12.25	3.94	37.19	47.50
4000	Stools or sills, sand finish, 1" thick, 6" wide	D-12	160	.200		12.20	8.50		20.70	27.50
4100	Honed finish		160	.200		11.65	8.50		20.15	26.50
4200	10" wide		90	.356		18.80	15.15		33.95	45
4250	Honed finish		90	.356		17.50	15.15		32.65	44
4400	2" thick, 6" wide		140	.229		19.60	9.75		29.35	37.50
4450	Honed finish		140	.229		18.65	9.75		28.40	36.50
4600	10" wide		90	.356		30.50	15.15		45.65	58.50
4650	Honed finish		90	.356		29	15.15		44.15	56.50
4800	For lengths over 3', add					25%				
9000	Minimum labor/equipment charge	D-1	2.50	6.400	Job		269		269	440

04 43 10.85 Window Sill		Crew	Daily Output	Labor-Hours	Unit	Material	Labor	Equipment	Total	Total Incl O&P
0010	**WINDOW SILL**									
0020	Bluestone, thermal top, 10" wide, 1-1/2" thick	D-1	85	.188	S.F.	10	7.90		17.90	24
0050	2" thick		75	.213	"	10	9		19	25.50
0100	Cut stone, 5" x 8" plain		48	.333	L.F.	12.10	14.05		26.15	36.50
0200	Face brick on edge, brick, 8" wide		80	.200		5.05	8.40		13.45	19.25
0400	Marble, 9" wide, 1" thick		85	.188		8.65	7.90		16.55	22.50
0900	Slate, colored, unfading, honed, 12" wide, 1" thick		85	.188		8.50	7.90		16.40	22.50
0950	2" thick		70	.229		8.50	9.60		18.10	25
9000	Minimum labor/equipment charge	1 Bric	2	4	Job		185		185	300

04 51 Flue Liner Masonry

04 51 10 – Clay Flue Lining

04 51 10.10 Flue Lining		Crew	Daily Output	Labor-Hours	Unit	Material	Labor	Equipment	Total	Total Incl O&P
0010	**FLUE LINING**, including mortar									
0020	Clay, 8" x 8"	D-1	125	.128	V.L.F.	5.50	5.40		10.90	14.80
0100	8" x 12"		103	.155		8.15	6.55		14.70	19.60
0200	12" x 12"		93	.172		10.75	7.25		18	23.50
0300	12" x 18"		84	.190		21.50	8		29.50	36.50
0400	18" x 18"		75	.213		28	9		37	45
0500	20" x 20"		66	.242		40	10.20		50.20	60.50
0600	24" x 24"		56	.286		51.50	12.05		63.55	76.50
1000	Round, 18" diameter		66	.242		37	10.20		47.20	57.50
1100	24" diameter		47	.340		72.50	14.35		86.85	103

04 54 Refractory Brick Masonry

04 54 10 – Refractory Brick Work

04 54 10.10 Fire Brick		Crew	Daily Output	Labor-Hours	Unit	Material	Labor	Equipment	Total	Total Incl O&P
0010	**FIRE BRICK**									
0012	Low duty, 2000°F, 9" x 2-1/2" x 4-1/2"	D-1	.60	26.667	M	1,425	1,125		2,550	3,400
0050	High duty, 3000°F	"	.60	26.667	"	2,600	1,125		3,725	4,675

04 54 10.20 Fire Clay		Crew	Daily Output	Labor-Hours	Unit	Material	Labor	Equipment	Total	Total Incl O&P
0010	**FIRE CLAY**									
0020	Gray, high duty, 100 lb. bag				Bag	33			33	36.50
0050	100 lb. drum, premixed (400 brick per drum)				Drum	41			41	45

04 57 Masonry Fireplaces

04 57 10 – Brick or Stone Fireplaces

04 57 10.10 Fireplace

	04 57 10.10 Fireplace	Crew	Daily Output	Labor-Hours	Unit	Material	2015 Bare Costs Labor	Equipment	Total	Total Incl O&P
0010	**FIREPLACE**									
0100	Brick fireplace, not incl. foundations or chimneys									
0110	30" x 29" opening, incl. chamber, plain brickwork	D-1	.40	40	Ea.	550	1,675		2,225	3,350
0200	Fireplace box only (110 brick)	"	2	8	"	157	335		492	720
0300	For elaborate brickwork and details, add					35%	35%			
0400	For hearth, brick & stone, add	D-1	2	8	Ea.	205	335		540	775
0410	For steel, damper, cleanouts, add		4	4		19.70	168		187.70	296
0600	Plain brickwork, incl. metal circulator		.50	32		995	1,350		2,345	3,300
0800	Face brick only, standard size, 8" x 2-2/3" x 4"		.30	53.333	M	550	2,250		2,800	4,250
0900	Stone fireplace, fieldstone, add				SF Face	8.50			8.50	9.35
1000	Cut stone, add				"	8			8	8.80
9000	Minimum labor/equipment charge	D-1	2	8	Job		335		335	550

04 71 Manufactured Brick Masonry

04 71 10 – Simulated or Manufactured Brick

04 71 10.10 Simulated Brick

	04 71 10.10 Simulated Brick	Crew	Daily Output	Labor-Hours	Unit	Material	Labor	Equipment	Total	Total Incl O&P
0010	**SIMULATED BRICK**									
0020	Aluminum, baked on colors	1 Carp	200	.040	S.F.	4.31	1.88		6.19	7.80
0050	Fiberglass panels		200	.040		8.05	1.88		9.93	11.95
0100	Urethane pieces cemented in mastic		150	.053		8.25	2.50		10.75	13.15
0150	Vinyl siding panels		200	.040		10	1.88		11.88	14.10
0160	Cement base, brick, incl. mastic	D-1	100	.160		9.15	6.75		15.90	21
0170	Corner		50	.320	V.L.F.	23	13.45		36.45	47.50
0180	Stone face, incl. mastic		100	.160	S.F.	9.85	6.75		16.60	22
0190	Corner		50	.320	V.L.F.	28	13.45		41.45	53

04 72 Cast Stone Masonry

04 72 10 – Cast Stone Masonry Features

04 72 10.10 Coping

	04 72 10.10 Coping	Crew	Daily Output	Labor-Hours	Unit	Material	Labor	Equipment	Total	Total Incl O&P
0010	**COPING**, stock units									
0050	Precast concrete, 10" wide, 4" tapers to 3-1/2", 8" wall	D-1	75	.213	L.F.	17.10	9		26.10	33.50
0100	12" wide, 3-1/2" tapers to 3", 10" wall		70	.229		18.45	9.60		28.05	36
0110	14" wide, 4" tapers to 3-1/2", 12" wall		65	.246		20.50	10.35		30.85	39.50
0150	16" wide, 4" tapers to 3-1/2", 14" wall		60	.267		22	11.25		33.25	43
0250	Precast concrete corners		40	.400	Ea.	29	16.85		45.85	59
0300	Limestone for 12" wall, 4" thick		90	.178	L.F.	14.70	7.50		22.20	28.50
0350	6" thick		80	.200		22	8.40		30.40	38
0500	Marble, to 4" thick, no wash, 9" wide		90	.178		12.05	7.50		19.55	25.50
0550	12" wide		80	.200		16.15	8.40		24.55	31.50
0700	Terra cotta, 9" wide		90	.178		6.35	7.50		13.85	19.15
0750	12" wide		80	.200		8.65	8.40		17.05	23
0800	Aluminum, for 12" wall		80	.200		8.90	8.40		17.30	23.50
9000	Minimum labor/equipment charge		2	8	Job		335		335	550

04 72 20 – Cultured Stone Veneer

04 72 20.10 Cultured Stone Veneer Components

	04 72 20.10 Cultured Stone Veneer Components	Crew	Daily Output	Labor-Hours	Unit	Material	Labor	Equipment	Total	Total Incl O&P
0010	**CULTURED STONE VENEER COMPONENTS**									
0110	On wood frame and sheathing substrate, random sized cobbles, corner stones	D-8	70	.571	V.L.F.	10.15	24.50		34.65	51
0120	Field stones		140	.286	S.F.	7.40	12.25		19.65	28
0130	Random sized flats, corner stones		70	.571	V.L.F.	10.05	24.50		34.55	51

04 72 Cast Stone Masonry

04 72 20 – Cultured Stone Veneer

04 72 20.10 Cultured Stone Veneer Components		Crew	Daily Output	Labor-Hours	Unit	2015 Bare Costs Material	Labor	Equipment	Total	Total Incl O&P
0140	Field stones	D-8	140	.286	S.F.	8.70	12.25		20.95	29.50
0150	Horizontal lined ledgestones, corner stones		75	.533	V.L.F.	10.15	23		33.15	48
0160	Field stones		150	.267	S.F.	7.40	11.45		18.85	26.50
0170	Random shaped flats, corner stones		65	.615	V.L.F.	10.15	26.50		36.65	54
0180	Field stones		150	.267	S.F.	7.40	11.45		18.85	26.50
0190	Random shaped/textured face, corner stones		65	.615	V.L.F.	10.15	26.50		36.65	54
0200	Field stones		130	.308	S.F.	7.40	13.20		20.60	29.50
0210	Random shaped river rock, corner stones		65	.615	V.L.F.	10.15	26.50		36.65	54
0220	Field stones		130	.308	S.F.	7.40	13.20		20.60	29.50
0240	On concrete or CMU substrate, random sized cobbles, corner stones		70	.571	V.L.F.	9.55	24.50		34.05	50.50
0250	Field stones		140	.286	S.F.	7.10	12.25		19.35	28
0260	Random sized flats, corner stones		70	.571	V.L.F.	9.45	24.50		33.95	50.50
0270	Field stones		140	.286	S.F.	8.40	12.25		20.65	29
0280	Horizontal lined ledgestones, corner stones		75	.533	V.L.F.	9.55	23		32.55	47.50
0290	Field stones		150	.267	S.F.	7.10	11.45		18.55	26.50
0300	Random shaped flats, corner stones		70	.571	V.L.F.	9.55	24.50		34.05	50.50
0310	Field stones		140	.286	S.F.	7.10	12.25		19.35	28
0320	Random shaped/textured face, corner stones		65	.615	V.L.F.	9.55	26.50		36.05	53.50
0330	Field stones		130	.308	S.F.	7.10	13.20		20.30	29.50
0340	Random shaped river rock, corner stones		65	.615	V.L.F.	9.55	26.50		36.05	53.50
0350	Field stones		130	.308	S.F.	7.10	13.20		20.30	29.50
0360	Cultured stone veneer, #15 felt weather resistant barrier	1 Clab	3700	.002	Sq.	5.40	.08		5.48	6.10
0370	Expanded metal lath, diamond, 2.5 lb./S.Y., galvanized	1 Lath	85	.094	S.Y.	2.72	4.05		6.77	9.35
0390	Water table or window sill, 18" long	1 Bric	80	.100	Ea.	10.30	4.62		14.92	18.85

04 73 Manufactured Stone Masonry

04 73 20 – Simulated or Manufactured Stone

04 73 20.10 Simulated Stone		Crew	Daily Output	Labor-Hours	Unit	Material	Labor	Equipment	Total	Total Incl O&P
0010	**SIMULATED STONE**									
0100	Insulated fiberglass panels, 5/8" ply backer	L-4	200	.120	S.F.	10.80	5.30		16.10	20.50

Estimating Tips

05 05 00 Common Work Results for Metals

- Nuts, bolts, washers, connection angles, and plates can add a significant amount to both the tonnage of a structural steel job and the estimated cost. As a rule of thumb, add 10% to the total weight to account for these accessories.
- Type 2 steel construction, commonly referred to as "simple construction," consists generally of field-bolted connections with lateral bracing supplied by other elements of the building, such as masonry walls or x-bracing. The estimator should be aware, however, that shop connections may be accomplished by welding or bolting. The method may be particular to the fabrication shop and may have an impact on the estimated cost.

05 10 00 Structural Steel

- Steel items can be obtained from two sources: a fabrication shop or a metals service center. Fabrication shops can fabricate items under more controlled conditions than can crews in the field. They are also more efficient and can produce items more economically. Metal service centers serve as a source of long mill shapes to both fabrication shops and contractors.
- Most line items in this structural steel subdivision, and most items in 05 50 00 Metal Fabrications, are indicated as being shop fabricated. The bare material cost for these shop fabricated items is the "Invoice Cost" from the shop and includes the mill base price of steel plus mill extras, transportation to the shop, shop drawings and detailing where warranted, shop fabrication and handling, sandblasting and a shop coat of primer paint, all necessary structural bolts, and delivery to the job site. The bare labor cost and bare equipment cost for these shop fabricated items is for field installation or erection.
- Line items in Subdivision 05 12 23.40 Lightweight Framing, and other items scattered in Division 5, are indicated as being field fabricated. The bare material cost for these field fabricated items is the "Invoice Cost" from the metals service center and includes the mill base price of steel plus mill extras, transportation to the metals service center, material handling, and delivery of long lengths of mill shapes to the job site. Material costs for structural bolts and welding rods should be added to the estimate. The bare labor cost and bare equipment cost for these items is for both field fabrication and field installation or erection, and include time for cutting, welding and drilling in the fabricated metal items. Drilling into concrete and fasteners to fasten field fabricated items to other work are not included and should be added to the estimate.

05 20 00 Steel Joist Framing

- In any given project the total weight of open web steel joists is determined by the loads to be supported and the design. However, economies can be realized in minimizing the amount of labor used to place the joists. This is done by maximizing the joist spacing, and therefore minimizing the number of joists required to be installed on the job. Certain spacings and locations may be required by the design, but in other cases maximizing the spacing and keeping it as uniform as possible will keep the costs down.

05 30 00 Steel Decking

- The takeoff and estimating of metal deck involves more than simply the area of the floor or roof and the type of deck specified or shown on the drawings. Many different sizes and types of openings may exist. Small openings for individual pipes or conduits may be drilled after the floor/roof is installed, but larger openings may require special deck lengths as well as reinforcing or structural support. The estimator should determine who will be supplying this reinforcing. Additionally, some deck terminations are part of the deck package, such as screed angles and pour stops, and others will be part of the steel contract, such as angles attached to structural members and cast-in-place angles and plates. The estimator must ensure that all pieces are accounted for in the complete estimate.

05 50 00 Metal Fabrications

- The most economical steel stairs are those that use common materials, standard details, and most importantly, a uniform and relatively simple method of field assembly. Commonly available A36 channels and plates are very good choices for the main stringers of the stairs, as are angles and tees for the carrier members. Risers and treads are usually made by specialty shops, and it is most economical to use a typical detail in as many places as possible. The stairs should be pre-assembled and shipped directly to the site. The field connections should be simple and straightforward to be accomplished efficiently, and with minimum equipment and labor.

Reference Numbers

Reference numbers are shown in shaded boxes at the beginning of some major classifications. These numbers refer to related items in the Reference Section. The reference information may be an estimating procedure, an alternate pricing method, or technical information.

Note: Not all subdivisions listed here necessarily appear in this publication. ■

05 01 Maintenance of Metals

05 01 10 – Maintenance of Structural Metal Framing

05 01 10.51 Cleaning of Structural Metal Framing		Crew	Daily Output	Labor-Hours	Unit	2015 Bare Costs Material	Labor	Equipment	Total	Total Incl O&P
0010	**CLEANING OF STRUCTURAL METAL FRAMING**									
6125	Steel surface treatments, PDCA guidelines									
6170	Wire brush, hand (SSPC-SP2)	1 Psst	400	.020	S.F.	.02	.83		.85	1.57
6180	Power tool (SSPC-SP3)	″	700	.011		.09	.47		.56	.97
6215	Pressure washing, up to 5000 psi, 5000-15,000 S.F./day	1 Pord	10000	.001			.03		.03	.05
6220	Steam cleaning, 600 psi @ 300 F, 1250 – 2500 S.F./day		2000	.004			.16		.16	.26
6225	Water blasting, up to 25,000 psi, 1750 – 3500 S.F./day		2500	.003			.13		.13	.21
6230	Brush-off blast (SSPC-SP7)	E-11	1750	.018		.15	.77	.13	1.05	1.66
6235	Com'l blast (SSPC-SP6), loose scale, fine pwder rust, 2.0#/S.F. sand		1200	.027		.31	1.13	.20	1.64	2.51
6240	Tight mill scale, little/no rust, 3.0#/S.F. sand		1000	.032		.46	1.35	.24	2.05	3.11
6245	Exist coat blistered/pitted, 4.0#/S.F. sand		875	.037		.62	1.55	.27	2.44	3.66
6250	Exist coat badly pitted/nodules, 6.7#/S.F. sand		825	.039		1.03	1.64	.29	2.96	4.29
6255	Near white blast (SSPC-SP10), loose scale, fine rust, 5.6#/S.F. sand		450	.071		.86	3.01	.52	4.39	6.75
6260	Tight mill scale, little/no rust, 6.9#/S.F. sand		325	.098		1.06	4.16	.73	5.95	9.15
6265	Exist coat blistered/pitted, 9.0#/S.F. sand		225	.142		1.39	6	1.05	8.44	13.10
6270	Exist coat badly pitted/nodules, 11.3#/S.F. sand		150	.213		1.74	9	1.57	12.31	19.30

05 05 Common Work Results for Metals

05 05 05 – Selective Demolition for Metals

05 05 05.10 Selective Demolition, Metals

Line	Description	Crew	Daily Output	Labor-Hours	Unit	Material	Labor	Equipment	Total	Total Incl O&P
0010	**SELECTIVE DEMOLITION, METALS** R024119-10									
0015	Excludes shores, bracing, cutting, loading, hauling, dumping									
0020	Remove nuts only up to 3/4″ diameter	1 Sswk	480	.017	Ea.		.88		.88	1.59
0030	7/8″ to 1-1/4″ diameter		240	.033			1.75		1.75	3.17
0040	1-3/8″ to 2″ diameter		160	.050			2.63		2.63	4.76
0060	Unbolt and remove structural bolts up to 3/4″ diameter		240	.033			1.75		1.75	3.17
0070	7/8″ to 2″ diameter		160	.050			2.63		2.63	4.76
0140	Light weight framing members, remove whole or cut up, up to 20 lb.		240	.033			1.75		1.75	3.17
0150	21 – 40 lb.	2 Sswk	210	.076			4.01		4.01	7.25
0160	41 – 80 lb.	3 Sswk	180	.133			7		7	12.70
0170	81 – 120 lb.	4 Sswk	150	.213			11.25		11.25	20.50
0230	Structural members, remove whole or cut up, up to 500 lb.	E-19	48	.500			26	19.75	45.75	67
0240	1/4 – 2 tons	E-18	36	1.111			58.50	26.50	85	132
0250	2 – 5 tons	E-24	30	1.067			55.50	24.50	80	125
0260	5 – 10 tons	E-20	24	2.667			138	48.50	186.50	297
0270	10 – 15 tons	E-2	18	3.111			161	84	245	375
0340	Fabricated item, remove whole or cut up, up to 20 lb.	1 Sswk	96	.083			4.39		4.39	7.95
0350	21 – 40 lb.	2 Sswk	84	.190			10.05		10.05	18.10
0360	41 – 80 lb.	3 Sswk	72	.333			17.55		17.55	31.50
0370	81 – 120 lb.	4 Sswk	60	.533			28		28	51
0380	121 – 500 lb.	E-19	48	.500			26	19.75	45.75	67
0390	501 – 1000 lb.	″	36	.667			34.50	26.50	61	89
0500	Steel roof decking, uncovered, bare	B-2	5000	.008	S.F.		.30		.30	.50
2290	Tube steel column, 8′ high	B-1J	8	2	Ea.		75.50		75.50	124
2700	Floor grating, supports excluded	B-68D	800	.030	S.F.		1.24	.41	1.65	2.47
2705	Deckplate, supports excluded	″	800	.030			1.24	.41	1.65	2.47
2800	Structural steel platform, interior	B-68E	96	.417			22	1.93	23.93	41.50
2805	Exterior		48	.833			44	3.86	47.86	83.50
2810	Demolish, catwalk		32	1.250	L.F.		66	5.80	71.80	125
2815	Ladder with cage		100	.400	V.L.F.		21	1.85	22.85	40
2820	without cage		128	.313	″		16.50	1.45	17.95	31.50

05 05 05 – Selective Demolition for Metals

05 05 05.10 Selective Demolition, Metals		Crew	Daily Output	Labor-Hours	Unit	Material	2015 Bare Costs Labor	Equipment	Total	Total Incl O&P
2825	Hand rail	B-68E	520	.077	L.F.		4.06	.36	4.42	7.75
2830	Guard rail		160	.250	"		13.20	1.16	14.36	25.50
2835	Guard post	2 Sswk	8	2	Ea.		105		105	190
2840	Steel bar grating	B-68E	600	.067	S.F.		3.52	.31	3.83	6.70
2845	Duct supports	"	80	.500	L.F.		26.50	2.32	28.82	50
2950	Minimum labor/equipment charge	E-19	2	12	Job		625	475	1,100	1,600

05 05 13 – Shop-Applied Coatings for Metal

05 05 13.50 Paints and Protective Coatings

		Crew	Daily Output	Labor-Hours	Unit	Material	Labor	Equipment	Total	Total Incl O&P
0010	PAINTS AND PROTECTIVE COATINGS									
5900	Galvanizing structural steel in shop, under 1 ton R050516-30				Ton	550			550	605
5950	1 ton to 20 tons					505			505	555
6000	Over 20 tons					460			460	505

05 05 19 – Post-Installed Concrete Anchors

05 05 19.10 Chemical Anchors

		Crew	Daily Output	Labor-Hours	Unit	Material	Labor	Equipment	Total	Total Incl O&P
0010	CHEMICAL ANCHORS									
0020	Includes layout & drilling									
1430	Chemical anchor, w/rod & epoxy cartridge, 3/4" diam. x 9-1/2" long	B-89A	27	.593	Ea.	9.80	25.50	4.28	39.58	57
1435	1" diameter x 11-3/4" long		24	.667		17.55	29	4.82	51.37	71.50
1440	1-1/4" diameter x 14" long		21	.762		36.50	33	5.50	75	99.50
1445	1-3/4" diameter x 15" long		20	.800		65	34.50	5.80	105.30	134
1450	18" long		17	.941		78	40.50	6.80	125.30	160
1455	2" diameter x 18" long		16	1		103	43	7.25	153.25	191
1460	24" long		15	1.067		134	46	7.70	187.70	231

05 05 19.20 Expansion Anchors

		Crew	Daily Output	Labor-Hours	Unit	Material	Labor	Equipment	Total	Total Incl O&P
0010	EXPANSION ANCHORS									
0100	Anchors for concrete, brick or stone, no layout and drilling									
0200	Expansion shields, zinc, 1/4" diameter, 1-5/16" long, single G	1 Carp	90	.089	Ea.	.43	4.17		4.60	7.30
0300	1-3/8" long, double G		85	.094		.54	4.42		4.96	7.85
0400	3/8" diameter, 1-1/2" long, single G		85	.094		.64	4.42		5.06	7.95
0500	2" long, double G		80	.100		1.18	4.70		5.88	9
0600	1/2" diameter, 2-1/16" long, single G		80	.100		1.18	4.70		5.88	9
0700	2-1/2" long, double G		75	.107		1.91	5		6.91	10.30
0800	5/8" diameter, 2-5/8" long, single G		75	.107		2.05	5		7.05	10.45
0900	2-3/4" long, double G		70	.114		2.70	5.35		8.05	11.75
1000	3/4" diameter, 2-3/4" long, single G		70	.114		3.09	5.35		8.44	12.20
1100	3-15/16" long, double G		65	.123		5	5.80		10.80	14.95
2100	Hollow wall anchors for gypsum wall board, plaster or tile									
2300	1/8" diameter, short G	1 Carp	160	.050	Ea.	.23	2.35		2.58	4.10
2400	Long G		150	.053		.23	2.50		2.73	4.35
2500	3/16" diameter, short G		150	.053		.40	2.50		2.90	4.54
2600	Long G		140	.057		.59	2.68		3.27	5.05
2700	1/4" diameter, short G		140	.057		.59	2.68		3.27	5.05
2800	Long G		130	.062		.65	2.89		3.54	5.45
3000	Toggle bolts, bright steel, 1/8" diameter, 2" long G		85	.094		.19	4.42		4.61	7.45
3100	4" long G		80	.100		.25	4.70		4.95	8
3200	3/16" diameter, 3" long G		80	.100		.26	4.70		4.96	8
3300	6" long G		75	.107		.36	5		5.36	8.60
3400	1/4" diameter, 3" long G		75	.107		.34	5		5.34	8.55
3500	6" long G		70	.114		.51	5.35		5.86	9.35
3600	3/8" diameter, 3" long G		70	.114		.84	5.35		6.19	9.70
3700	6" long G		60	.133		1.36	6.25		7.61	11.75

05 05 19.20 Expansion Anchors			Crew	Daily Output	Labor-Hours	Unit	2015 Bare Costs Material	Labor	Equipment	Total	Total Incl O&P
3800	1/2" diameter, 4" long	G	1 Carp	60	.133	Ea.	1.88	6.25		8.13	12.30
3900	6" long	G		50	.160		2.45	7.50		9.95	15
4000	Nailing anchors										
4100	Nylon nailing anchor, 1/4" diameter, 1" long		1 Carp	3.20	2.500	C	21	117		138	215
4200	1-1/2" long			2.80	2.857		26	134		160	249
4300	2" long			2.40	3.333		30.50	157		187.50	291
4400	Metal nailing anchor, 1/4" diameter, 1" long	G		3.20	2.500		17.85	117		134.85	212
4500	1-1/2" long	G		2.80	2.857		23.50	134		157.50	246
4600	2" long	G		2.40	3.333		27.50	157		184.50	288
5000	Screw anchors for concrete, masonry,										
5100	stone & tile, no layout or drilling included										
5700	Lag screw shields, 1/4" diameter, short	G	1 Carp	90	.089	Ea.	.34	4.17		4.51	7.20
5800	Long	G		85	.094		.38	4.42		4.80	7.65
5900	3/8" diameter, short	G		85	.094		.67	4.42		5.09	8
6000	Long	G		80	.100		.87	4.70		5.57	8.65
6100	1/2" diameter, short	G		80	.100		.95	4.70		5.65	8.75
6200	Long	G		75	.107		1.22	5		6.22	9.55
6300	5/8" diameter, short	G		70	.114		1.45	5.35		6.80	10.40
6400	Long	G		65	.123		1.86	5.80		7.66	11.50
6600	Lead, #6 & #8, 3/4" long	G		260	.031		.18	1.44		1.62	2.57
6700	#10 - #14, 1-1/2" long	G		200	.040		.37	1.88		2.25	3.49
6800	#16 & #18, 1-1/2" long	G		160	.050		.41	2.35		2.76	4.30
6900	Plastic, #6 & #8, 3/4" long			260	.031		.04	1.44		1.48	2.41
7000	#8 & #10, 7/8" long			240	.033		.04	1.56		1.60	2.60
7100	#10 & #12, 1" long			220	.036		.05	1.71		1.76	2.86
7200	#14 & #16, 1-1/2" long			160	.050		.07	2.35		2.42	3.93
8000	Wedge anchors, not including layout or drilling										
8050	Carbon steel, 1/4" diameter, 1-3/4" long	G	1 Carp	150	.053	Ea.	.41	2.50		2.91	4.55
8100	3-1/4" long	G		140	.057		.54	2.68		3.22	4.99
8150	3/8" diameter, 2-1/4" long	G		145	.055		.50	2.59		3.09	4.80
8200	5" long	G		140	.057		.88	2.68		3.56	5.35
8250	1/2" diameter, 2-3/4" long	G		140	.057		.99	2.68		3.67	5.50
8300	7" long	G		125	.064		1.70	3		4.70	6.80
8350	5/8" diameter, 3-1/2" long	G		130	.062		1.86	2.89		4.75	6.80
8400	8-1/2" long	G		115	.070		3.95	3.27		7.22	9.70
8450	3/4" diameter, 4-1/4" long	G		115	.070		2.90	3.27		6.17	8.55
8500	10" long	G		95	.084		6.60	3.95		10.55	13.75
8550	1" diameter, 6" long	G		100	.080		9.30	3.76		13.06	16.40
8575	9" long	G		85	.094		12.10	4.42		16.52	20.50
8600	12" long	G		75	.107		13.10	5		18.10	22.50
8650	1-1/4" diameter, 9" long	G		70	.114		24.50	5.35		29.85	36
8700	12" long	G		60	.133		31.50	6.25		37.75	45
8750	For type 303 stainless steel, add						350%				
8800	For type 316 stainless steel, add						450%				
8950	Self-drilling concrete screw, hex washer head, 3/16" diam. x 1-3/4" long	G	1 Carp	300	.027	Ea.	.20	1.25		1.45	2.27
8960	2-1/4" long	G		250	.032		.24	1.50		1.74	2.72
8970	Phillips flat head, 3/16" diam. x 1-3/4" long	G		300	.027		.21	1.25		1.46	2.28
8980	2-1/4" long	G		250	.032		.23	1.50		1.73	2.71
9000	Minimum labor/equipment charge			4	2	Job		94		94	154

05 05 21.10 Cutting Steel

	05 05 21.10 Cutting Steel	Crew	Daily Output	Labor-Hours	Unit	Material	2015 Bare Costs Labor	Equipment	Total	Total Incl O&P
0010	**CUTTING STEEL**									
0020	Hand burning, incl. preparation, torch cutting & grinding, no staging									
0050	Steel to 1/4″ thick	E-25	400	.020	L.F.	.19	1.09	.03	1.31	2.21
0100	1/2″ thick		320	.025		.35	1.37	.04	1.76	2.90
0150	3/4″ thick		260	.031		.58	1.68	.04	2.30	3.73
0200	1″ thick		200	.040		.84	2.19	.06	3.09	4.93
9000	Minimum labor/equipment charge		2	4	Job		219	5.70	224.70	400

05 05 21.15 Drilling Steel

	05 05 21.15 Drilling Steel	Crew	Daily Output	Labor-Hours	Unit	Material	Labor	Equipment	Total	Total Incl O&P
0010	**DRILLING STEEL**									
1910	Drilling & layout for steel, up to 1/4″ deep, no anchor									
1920	Holes, 1/4″ diameter	1 Sswk	112	.071	Ea.	.09	3.76		3.85	6.90
1925	For each additional 1/4″ depth, add		336	.024		.09	1.25		1.34	2.37
1930	3/8″ diameter		104	.077		.09	4.05		4.14	7.40
1935	For each additional 1/4″ depth, add		312	.026		.09	1.35		1.44	2.54
1940	1/2″ diameter		96	.083		.10	4.39		4.49	8.05
1945	For each additional 1/4″ depth, add		288	.028		.10	1.46		1.56	2.75
1950	5/8″ diameter		88	.091		.14	4.79		4.93	8.80
1955	For each additional 1/4″ depth, add		264	.030		.14	1.60		1.74	3.04
1960	3/4″ diameter		80	.100		.18	5.25		5.43	9.70
1965	For each additional 1/4″ depth, add		240	.033		.18	1.75		1.93	3.37
1970	7/8″ diameter		72	.111		.23	5.85		6.08	10.80
1975	For each additional 1/4″ depth, add		216	.037		.23	1.95		2.18	3.78
1980	1″ diameter		64	.125		.24	6.60		6.84	12.15
1985	For each additional 1/4″ depth, add		192	.042		.24	2.19		2.43	4.22
1990	For drilling up, add						40%			
2000	Minimum labor/equipment charge	1 Carp	4	2	Job		94		94	154

05 05 21.90 Welding Steel

	05 05 21.90 Welding Steel	Crew	Daily Output	Labor-Hours	Unit	Material	Labor	Equipment	Total	Total Incl O&P
0010	**WELDING STEEL**, Structural R050521-20									
0020	Field welding, 1/8″ E6011, cost per welder, no operating engineer	E-14	8	1	Hr.	4.33	54.50	18.25	77.08	124
0200	With 1/2 operating engineer	E-13	8	1.500		4.33	79	18.25	101.58	162
0300	With 1 operating engineer	E-12	8	2		4.33	103	18.25	125.58	201
0500	With no operating engineer, 2# weld rod per ton	E-14	8	1	Ton	4.33	54.50	18.25	77.08	124
0600	8# E6011 per ton	″	2	4		17.30	219	73	309.30	495
0800	With one operating engineer per welder, 2# E6011 per ton	E-12	8	2		4.33	103	18.25	125.58	201
0900	8# E6011 per ton	″	2	8		17.30	415	73	505.30	805
1200	Continuous fillet, down welding									
1300	Single pass, 1/8″ thick, 0.1#/L.F.	E-14	150	.053	L.F.	.22	2.91	.97	4.10	6.55
1400	3/16″ thick, 0.2#/L.F.		75	.107		.43	5.85	1.94	8.22	13.15
1500	1/4″ thick, 0.3#/L.F.		50	.160		.65	8.75	2.92	12.32	19.70
1610	5/16″ thick, 0.4#/L.F.		38	.211		.87	11.50	3.84	16.21	26
1800	3 passes, 3/8″ thick, 0.5#/L.F.		30	.267		1.08	14.55	4.86	20.49	33
2010	4 passes, 1/2″ thick, 0.7#/L.F.		22	.364		1.52	19.85	6.65	28.02	45
2200	5 to 6 passes, 3/4″ thick, 1.3#/L.F.		12	.667		2.81	36.50	12.15	51.46	82.50
2400	8 to 11 passes, 1″ thick, 2.4#/L.F.		6	1.333		5.20	73	24.50	102.70	164
2600	For vertical joint welding, add						20%			
2700	Overhead joint welding, add						300%			
2900	For semi-automatic welding, obstructed joints, deduct						5%			
3000	Exposed joints, deduct						15%			
4000	Cleaning and welding plates, bars, or rods									
4010	to existing beams, columns, or trusses	E-14	12	.667	L.F.	1.08	36.50	12.15	49.73	80.50
9000	Minimum labor/equipment charge	″	4	2	Job		109	36.50	145.50	238

05 05 23.10 Bolts and Hex Nuts

	05 05 23.10 Bolts and Hex Nuts		Crew	Daily Output	Labor-Hours	Unit	Material	2015 Bare Costs Labor	Equipment	Total	Total Incl O&P
0010	**BOLTS & HEX NUTS**, Steel, A307										
0100	1/4" diameter, 1/2" long	G	1 Sswk	140	.057	Ea.	.06	3.01		3.07	5.50
0200	1" long	G		140	.057		.07	3.01		3.08	5.55
0300	2" long	G		130	.062		.10	3.24		3.34	5.95
0400	3" long	G		130	.062		.15	3.24		3.39	6
0500	4" long	G		120	.067		.17	3.51		3.68	6.55
0600	3/8" diameter, 1" long	G		130	.062		.14	3.24		3.38	6
0700	2" long	G		130	.062		.18	3.24		3.42	6.05
0800	3" long	G		120	.067		.24	3.51		3.75	6.60
0900	4" long	G		120	.067		.30	3.51		3.81	6.70
1000	5" long	G		115	.070		.38	3.66		4.04	7
1100	1/2" diameter, 1-1/2" long	G		120	.067		.40	3.51		3.91	6.80
1200	2" long	G		120	.067		.46	3.51		3.97	6.85
1300	4" long	G		115	.070		.75	3.66		4.41	7.45
1400	6" long	G		110	.073		1.05	3.83		4.88	8.05
1500	8" long	G		105	.076		1.38	4.01		5.39	8.75
1600	5/8" diameter, 1-1/2" long	G		120	.067		.98	3.51		4.49	7.45
1700	2" long	G		120	.067		1.09	3.51		4.60	7.55
1800	4" long	G		115	.070		1.59	3.66		5.25	8.35
1900	6" long	G		110	.073		2.05	3.83		5.88	9.15
2000	8" long	G		105	.076		3.06	4.01		7.07	10.60
2100	10" long	G		100	.080		3.87	4.21		8.08	11.85
2200	3/4" diameter, 2" long	G		120	.067		1.15	3.51		4.66	7.60
2300	4" long	G		110	.073		1.65	3.83		5.48	8.70
2400	6" long	G		105	.076		2.12	4.01		6.13	9.60
2500	8" long	G		95	.084		3.20	4.43		7.63	11.50
2600	10" long	G		85	.094		4.20	4.96		9.16	13.55
2700	12" long	G		80	.100		4.92	5.25		10.17	14.90
2800	1" diameter, 3" long	G		105	.076		2.69	4.01		6.70	10.20
2900	6" long	G		90	.089		3.94	4.68		8.62	12.80
3000	12" long	G		75	.107		7.10	5.60		12.70	17.95
3100	For galvanized, add						75%				
3200	For stainless, add						350%				

05 05 23.25 High Strength Bolts

	05 05 23.25 High Strength Bolts		Crew	Daily Output	Labor-Hours	Unit	Material	Labor	Equipment	Total	Total Incl O&P
0010	**HIGH STRENGTH BOLTS** R050523-10										
0020	A325 Type 1, structural steel, bolt-nut-washer set										
0100	1/2" diameter x 1-1/2" long	G	1 Sswk	130	.062	Ea.	.97	3.24		4.21	6.90
0120	2" long	G		125	.064		1.05	3.37		4.42	7.25
0150	3" long	G		120	.067		1.46	3.51		4.97	7.95
0170	5/8" diameter x 1-1/2" long	G		125	.064		1.64	3.37		5.01	7.90
0180	2" long	G		120	.067		1.77	3.51		5.28	8.30
0190	3" long	G		115	.070		2.19	3.66		5.85	9
0200	3/4" diameter x 2" long	G		120	.067		2.65	3.51		6.16	9.25
0220	3" long	G		115	.070		3.19	3.66		6.85	10.10
0250	4" long	G		110	.073		3.92	3.83		7.75	11.20
0300	6" long	G		105	.076		5.10	4.01		9.11	12.85
0350	8" long	G		95	.084		10.15	4.43		14.58	19.15
0360	7/8" diameter x 2" long	G		115	.070		3.66	3.66		7.32	10.65
0365	3" long	G		110	.073		4.33	3.83		8.16	11.65
0370	4" long	G		105	.076		5.25	4.01		9.26	13
0380	6" long	G		100	.080		6.65	4.21		10.86	14.95
0390	8" long	G		90	.089		10.60	4.68		15.28	20

05 05 23 – Metal Fastenings

05 05 23.25 High Strength Bolts			Crew	Daily Output	Labor-Hours	Unit	2015 Bare Costs Material	Labor	Equipment	Total	Total Incl O&P
0400	1" diameter x 2" long	G	1 Sswk	105	.076	Ea.	4.58	4.01		8.59	12.30
0420	3" long	G		100	.080		5.20	4.21		9.41	13.30
0450	4" long	G		95	.084		5.85	4.43		10.28	14.45
0500	6" long	G		90	.089		7.80	4.68		12.48	17.05
0550	8" long	G		85	.094		13.55	4.96		18.51	24
0600	1-1/4" diameter x 3" long	G		85	.094		10.70	4.96		15.66	21
0650	4" long	G		80	.100		11.65	5.25		16.90	22.50
0700	6" long	G		75	.107		15.20	5.60		20.80	27
0750	8" long	G		70	.114		19.35	6		25.35	32.50
1020	A490, bolt-nut-washer set										
1170	5/8" diameter x 1-1/2" long	G	1 Sswk	125	.064	Ea.	4.48	3.37		7.85	11.05
1180	2" long	G		120	.067		5.35	3.51		8.86	12.20
1190	3" long	G		115	.070		6.55	3.66		10.21	13.80
1200	3/4" diameter x 2" long	G		120	.067		3.92	3.51		7.43	10.65
1220	3" long	G		115	.070		4.64	3.66		8.30	11.70
1250	4" long	G		110	.073		5.40	3.83		9.23	12.85
1300	6" long	G		105	.076		7.95	4.01		11.96	16
1350	8" long	G		95	.084		13.50	4.43		17.93	23
1360	7/8" diameter x 2" long	G		115	.070		5.90	3.66		9.56	13.10
1365	3" long	G		110	.073		6.95	3.83		10.78	14.55
1370	4" long	G		105	.076		8.65	4.01		12.66	16.75
1380	6" long	G		100	.080		12.15	4.21		16.36	21
1390	8" long	G		90	.089		17.70	4.68		22.38	28
1400	1" diameter x 2" long	G		105	.076		7.80	4.01		11.81	15.85
1420	3" long	G		100	.080		9.45	4.21		13.66	18
1450	4" long	G		95	.084		10.95	4.43		15.38	20
1500	6" long	G		90	.089		14.65	4.68		19.33	24.50
1550	8" long	G		85	.094		23.50	4.96		28.46	34.50
1600	1-1/4" diameter x 3" long	G		85	.094		37.50	4.96		42.46	50
1650	4" long	G		80	.100		43	5.25		48.25	57
1700	6" long	G		75	.107		60.50	5.60		66.10	76.50
1750	8" long	G		70	.114		79	6		85	98

05 05 23.30 Lag Screws			Crew	Daily Output	Labor-Hours	Unit	Material	Labor	Equipment	Total	Total Incl O&P
0010	**LAG SCREWS**										
0020	Steel, 1/4" diameter, 2" long	G	1 Carp	200	.040	Ea.	.09	1.88		1.97	3.18
0100	3/8" diameter, 3" long	G		150	.053		.30	2.50		2.80	4.43
0200	1/2" diameter, 3" long	G		130	.062		.64	2.89		3.53	5.45
0300	5/8" diameter, 3" long	G		120	.067		1.15	3.13		4.28	6.40

05 05 23.35 Machine Screws			Crew	Daily Output	Labor-Hours	Unit	Material	Labor	Equipment	Total	Total Incl O&P
0010	**MACHINE SCREWS**										
0020	Steel, round head, #8 x 1" long	G	1 Carp	4.80	1.667	C	3.93	78.50		82.43	132
0110	#8 x 2" long	G		2.40	3.333		6.20	157		163.20	264
0200	#10 x 1" long	G		4	2		5.20	94		99.20	160
0300	#10 x 2" long	G		2	4		8.65	188		196.65	320

05 05 23.50 Powder Actuated Tools and Fasteners			Crew	Daily Output	Labor-Hours	Unit	Material	Labor	Equipment	Total	Total Incl O&P
0010	**POWDER ACTUATED TOOLS & FASTENERS**										
0020	Stud driver, .22 caliber, single shot					Ea.	148			148	163
0100	.27 caliber, semi automatic, strip					"	440			440	485
0300	Powder load, single shot, .22 cal, power level 2, brown					C	5.35			5.35	5.90
0400	Strip, .27 cal, power level 4, red						7.70			7.70	8.50
0600	Drive pin, .300 x 3/4" long	G	1 Carp	4.80	1.667		4.11	78.50		82.61	133
0700	.300 x 3" long with washer	G	"	4	2		12.65	94		106.65	168

05 05 23 – Metal Fastenings

05 05 23.55 Rivets			Crew	Daily Output	Labor-Hours	Unit	Material	2015 Bare Costs Labor	Equipment	Total	Total Incl O&P
0010	RIVETS										
0100	Aluminum rivet & mandrel, 1/2" grip length x 1/8" diameter	G	1 Carp	4.80	1.667	C	7.65	78.50		86.15	136
0200	3/16" diameter	G		4	2		11.40	94		105.40	167
0300	Aluminum rivet, steel mandrel, 1/8" diameter	G		4.80	1.667		10.25	78.50		88.75	139
0400	3/16" diameter	G		4	2		16.45	94		110.45	172
0500	Copper rivet, steel mandrel, 1/8" diameter	G		4.80	1.667		9.50	78.50		88	138
0800	Stainless rivet & mandrel, 1/8" diameter	G		4.80	1.667		24.50	78.50		103	155
0900	3/16" diameter	G		4	2		39	94		133	197
1000	Stainless rivet, steel mandrel, 1/8" diameter	G		4.80	1.667		15.70	78.50		94.20	145
1100	3/16" diameter	G		4	2		25.50	94		119.50	182
1200	Steel rivet and mandrel, 1/8" diameter	G		4.80	1.667		7.40	78.50		85.90	136
1300	3/16" diameter	G		4	2		12	94		106	167
1400	Hand riveting tool, standard					Ea.	70.50			70.50	77.50
1500	Deluxe						365			365	400
1600	Power riveting tool, standard						500			500	555
1700	Deluxe						1,650			1,650	1,825

05 05 23.70 Structural Blind Bolts			Crew	Daily Output	Labor-Hours	Unit	Material	Labor	Equipment	Total	Total Incl O&P
0010	STRUCTURAL BLIND BOLTS										
0100	1/4" diameter x 1/4" grip	G	1 Sswk	240	.033	Ea.	1.24	1.75		2.99	4.53
0150	1/2" grip	G		216	.037		1.33	1.95		3.28	4.98
0200	3/8" diameter x 1/2" grip	G		232	.034		1.75	1.82		3.57	5.20
0250	3/4" grip	G		208	.038		1.84	2.02		3.86	5.70
0300	1/2" diameter x 1/2" grip	G		224	.036		3.99	1.88		5.87	7.80
0350	3/4" grip	G		200	.040		5.60	2.11		7.71	9.95
0400	5/8" diameter x 3/4" grip	G		216	.037		8.25	1.95		10.20	12.60
0450	1" grip	G		192	.042		9.50	2.19		11.69	14.40

05 05 23.80 Vibration and Bearing Pads			Crew	Daily Output	Labor-Hours	Unit	Material	Labor	Equipment	Total	Total Incl O&P
0010	VIBRATION & BEARING PADS										
0300	Laminated synthetic rubber impregnated cotton duck, 1/2" thick		2 Sswk	24	.667	S.F.	69	35		104	140
0400	1" thick			20	.800		135	42		177	225
0600	Neoprene bearing pads, 1/2" thick			24	.667		26.50	35		61.50	92.50
0700	1" thick			20	.800		52.50	42		94.50	134
0900	Fabric reinforced neoprene, 5000 psi, 1/2" thick			24	.667		11.50	35		46.50	76
1000	1" thick			20	.800		23	42		65	102
1200	Felt surfaced vinyl pads, cork and sisal, 5/8" thick			24	.667		29	35		64	95.50
1300	1" thick			20	.800		52.50	42		94.50	134
1500	Teflon bonded to 10 ga. carbon steel, 1/32" layer			24	.667		51.50	35		86.50	120
1600	3/32" layer			24	.667		77	35		112	148
1800	Bonded to 10 ga. stainless steel, 1/32" layer			24	.667		91	35		126	164
1900	3/32" layer			24	.667		120	35		155	196
2100	Circular machine leveling pad & stud					Kip	6.45			6.45	7.10

05 05 23.85 Weld Shear Connectors			Crew	Daily Output	Labor-Hours	Unit	Material	Labor	Equipment	Total	Total Incl O&P
0010	WELD SHEAR CONNECTORS										
0020	3/4" diameter, 3-3/16" long	G	E-10	960	.017	Ea.	.53	.89	.47	1.89	2.72
0030	3-3/8" long	G		950	.017		.56	.90	.47	1.93	2.76
0200	3-7/8" long	G		945	.017		.60	.91	.48	1.99	2.82
0300	4-3/16" long	G		935	.017		.63	.92	.48	2.03	2.88
0500	4-7/8" long	G		930	.017		.70	.92	.48	2.10	2.97
0600	5-3/16" long	G		920	.017		.73	.93	.49	2.15	3.03
0800	5-3/8" long	G		910	.018		.74	.94	.49	2.17	3.05
0900	6-3/16" long	G		905	.018		.81	.95	.50	2.26	3.15

05 05 23 – Metal Fastenings

05 05 23.85 Weld Shear Connectors			Crew	Daily Output	Labor-Hours	Unit	Material	2015 Bare Costs Labor	Equipment	Total	Total Incl O&P
1000	7-3/16" long	G	E-10	895	.018	Ea.	1	.96	.50	2.46	3.38
1100	8-3/16" long	G		890	.018		1.10	.96	.50	2.56	3.51
1500	7/8" diameter, 3-11/16" long	G		920	.017		.86	.93	.49	2.28	3.18
1600	4-3/16" long	G		910	.018		.93	.94	.49	2.36	3.26
1700	5-3/16" long	G		905	.018		1.05	.95	.50	2.50	3.42
1800	6-3/16" long	G		895	.018		1.17	.96	.50	2.63	3.57
1900	7-3/16" long	G		890	.018		1.30	.96	.50	2.76	3.73
2000	8-3/16" long	G		880	.018		1.42	.98	.51	2.91	3.89
9000	Minimum labor/equipment charge		1 Sswk	2	4	Job		211		211	380

05 05 23.87 Weld Studs			Crew	Daily Output	Labor-Hours	Unit	Material	Labor	Equipment	Total	Total Incl O&P
0010	**WELD STUDS**										
0020	1/4" diameter, 2-11/16" long	G	E-10	1120	.014	Ea.	.35	.77	.40	1.52	2.22
0100	4-1/8" long	G		1080	.015		.33	.79	.42	1.54	2.26
0200	3/8" diameter, 4-1/8" long	G		1080	.015		.38	.79	.42	1.59	2.32
0300	6-1/8" long	G		1040	.015		.49	.83	.43	1.75	2.50
0400	1/2" diameter, 2-1/8" long	G		1040	.015		.35	.83	.43	1.61	2.35
0500	3-1/8" long	G		1025	.016		.43	.84	.44	1.71	2.46
0600	4-1/8" long	G		1010	.016		.50	.85	.44	1.79	2.58
0700	5-5/16" long	G		990	.016		.62	.87	.45	1.94	2.75
0800	6-1/8" long	G		975	.016		.67	.88	.46	2.01	2.84
0900	8-1/8" long	G		960	.017		.95	.89	.47	2.31	3.17
1000	5/8" diameter, 2-11/16" long	G		1000	.016		.61	.86	.45	1.92	2.72
1010	4-3/16" long	G		990	.016		.76	.87	.45	2.08	2.91
1100	6-9/16" long	G		975	.016		.99	.88	.46	2.33	3.19
1200	8-3/16" long	G		960	.017		1.33	.89	.47	2.69	3.59
9000	Minimum labor/equipment charge		1 Sswk	2	4	Job		211		211	380

05 05 23.90 Welding Rod			Crew	Daily Output	Labor-Hours	Unit	Material	Labor	Equipment	Total	Total Incl O&P
0010	**WELDING ROD**										
0020	Steel, type 6011, 1/8" diam., less than 500#					Lb.	2.16			2.16	2.38
0100	500# to 2,000#						1.95			1.95	2.15
0200	2,000# to 5,000#						1.83			1.83	2.02
0300	5/32" diameter, less than 500#						2.20			2.20	2.42
0310	500# to 2,000#						1.98			1.98	2.18
0320	2,000# to 5,000#						1.86			1.86	2.05
0400	3/16" diam., less than 500#						2.46			2.46	2.71
0500	500# to 2,000#						2.22			2.22	2.44
0600	2,000# to 5,000#						2.09			2.09	2.30
0620	Steel, type 6010, 1/8" diam., less than 500#						2.23			2.23	2.45
0630	500# to 2,000#						2.01			2.01	2.21
0640	2,000# to 5,000#						1.89			1.89	2.08
0650	Steel, type 7018 Low Hydrogen, 1/8" diam., less than 500#						2.16			2.16	2.38
0660	500# to 2,000#						1.95			1.95	2.15
0670	2,000# to 5,000#						1.83			1.83	2.02
0700	Steel, type 7024 Jet Weld, 1/8" diam., less than 500#						2.50			2.50	2.75
0710	500# to 2,000#						2.25			2.25	2.48
0720	2,000# to 5,000#						2.12			2.12	2.33
1550	Aluminum, type 4043 TIG, 1/8" diam., less than 10#						5.10			5.10	5.65
1560	10# to 60#						4.61			4.61	5.05
1570	Over 60#						4.33			4.33	4.77
1600	Aluminum, type 5356 TIG, 1/8" diam., less than 10#						5.45			5.45	6
1610	10# to 60#						4.90			4.90	5.40
1620	Over 60#						4.61			4.61	5.05

05 05 Common Work Results for Metals

05 05 23 – Metal Fastenings

05 05 23.90 Welding Rod		Crew	Daily Output	Labor-Hours	Unit	Material	2015 Bare Costs Labor	Equipment	Total	Total Incl O&P
1900	Cast iron, type 8 Nickel, 1/8″ diam., less than 500#				Lb.	22			22	24
1910	500# to 1,000#					19.75			19.75	21.50
1920	Over 1,000#					18.55			18.55	20.50
2000	Stainless steel, type 316/316L, 1/8″ diam., less than 500#					7.10			7.10	7.85
2100	500# to 1000#					6.40			6.40	7.05
2220	Over 1000#					6.05			6.05	6.65

05 12 Structural Steel Framing

05 12 23 – Structural Steel for Buildings

05 12 23.05 Canopy Framing

Code	Description		Crew	Daily Output	Labor-Hours	Unit	Material	Labor	Equipment	Total	Total Incl O&P
0010	**CANOPY FRAMING**										
0020	6″ and 8″ members, shop fabricated	G	E-4	3000	.011	Lb.	1.59	.57	.05	2.21	2.82
9000	Minimum labor/equipment charge		1 Sswk	1	8	Job		420		420	760

05 12 23.10 Ceiling Supports

Code	Description		Crew	Daily Output	Labor-Hours	Unit	Material	Labor	Equipment	Total	Total Incl O&P
0010	**CEILING SUPPORTS**										
1000	Entrance door/folding partition supports, shop fabricated	G	E-4	60	.533	L.F.	26.50	28.50	2.43	57.43	82.50
1100	Linear accelerator door supports	G		14	2.286		121	121	10.40	252.40	365
1200	Lintels or shelf angles, hung, exterior hot dipped galv.	G		267	.120		18.10	6.35	.55	25	32
1250	Two coats primer paint instead of galv.	G		267	.120		15.65	6.35	.55	22.55	29.50
1400	Monitor support, ceiling hung, expansion bolted	G		4	8	Ea.	420	425	36.50	881.50	1,275
1450	Hung from pre-set inserts	G		6	5.333		450	283	24.50	757.50	1,025
1600	Motor supports for overhead doors	G		4	8		214	425	36.50	675.50	1,050
1700	Partition support for heavy folding partitions, without pocket	G		24	1.333	L.F.	60.50	71	6.10	137.60	201
1750	Supports at pocket only	G		12	2.667		121	142	12.15	275.15	400
2000	Rolling grilles & fire door supports	G		34	.941		51.50	50	4.29	105.79	152
2100	Spider-leg light supports, expansion bolted to ceiling slab	G		8	4	Ea.	172	213	18.25	403.25	595
2150	Hung from pre-set inserts	G		12	2.667	″	186	142	12.15	340.15	475
2400	Toilet partition support	G		36	.889	L.F.	60.50	47	4.05	111.55	156
2500	X-ray travel gantry support	G		12	2.667	″	207	142	12.15	361.15	495

05 12 23.15 Columns, Lightweight

Code	Description		Crew	Daily Output	Labor-Hours	Unit	Material	Labor	Equipment	Total	Total Incl O&P
0010	**COLUMNS, LIGHTWEIGHT**										
1000	Lightweight units (lally), 3-1/2″ diameter		E-2	780	.072	L.F.	8.25	3.71	1.94	13.90	17.75
1050	4″ diameter		″	900	.062	″	10.15	3.22	1.68	15.05	18.65
5800	Adjustable jack post, 8′ maximum height, 2-3/4″ diameter	G				Ea.	52.50			52.50	58
5850	4″ diameter	G				″	84			84	92.50

05 12 23.17 Columns, Structural

Code	Description		Crew	Daily Output	Labor-Hours	Unit	Material	Labor	Equipment	Total	Total Incl O&P
0010	**COLUMNS, STRUCTURAL**	R051223-20									
0015	Made from recycled materials										
0020	Shop fab'd for 100-ton, 1-2 story project, bolted connections										
0800	Steel, concrete filled, extra strong pipe, 3-1/2″ diameter		E-2	660	.085	L.F.	43.50	4.39	2.29	50.18	58
0830	4″ diameter			780	.072		48.50	3.71	1.94	54.15	62
0890	5″ diameter			1020	.055		58	2.84	1.48	62.32	70.50
0930	6″ diameter			1200	.047		77	2.41	1.26	80.67	90
0940	8″ diameter			1100	.051		77	2.63	1.37	81	90.50
1100	For galvanizing, add					Lb.	.25			.25	.28
1300	For web ties, angles, etc., add per added lb.		1 Sswk	945	.008		1.33	.45		1.78	2.27
1500	Steel pipe, extra strong, no concrete, 3″ to 5″ diameter	G	E-2	16000	.004		1.33	.18	.09	1.60	1.88
1600	6″ to 12″ diameter	G		14000	.004		1.33	.21	.11	1.65	1.94
1700	Steel pipe, extra strong, no concrete, 3″ diameter x 12′-0″	G		60	.933	Ea.	163	48.50	25	236.50	291
1750	4″ diameter x 12′-0″	G		58	.966		238	50	26	314	380

05 12 Structural Steel Framing

05 12 23 – Structural Steel for Buildings

05 12 23.17 Columns, Structural			Crew	Daily Output	Labor-Hours	Unit	Material	2015 Bare Costs Labor	Equipment	Total	Total Incl O&P
1800	6" diameter x 12'-0"	G	E-2	54	1.037	Ea.	455	53.50	28	536.50	625
1850	8" diameter x 14'-0"	G		50	1.120		805	58	30	893	1,025
1900	10" diameter x 16'-0"	G		48	1.167		1,150	60.50	31.50	1,242	1,425
1950	12" diameter x 18'-0"	G		45	1.244		1,550	64.50	33.50	1,648	1,875
3300	Structural tubing, square, A500GrB, 4" to 6" square, light section	G		11270	.005	Lb.	1.33	.26	.13	1.72	2.06
3600	Heavy section	G		32000	.002	"	1.33	.09	.05	1.47	1.67
4000	Concrete filled, add					L.F.	4.20			4.20	4.63
4500	Structural tubing, square, 4" x 4" x 1/4" x 12'-0"	G	E-2	58	.966	Ea.	219	50	26	295	355
4550	6" x 6" x 1/4" x 12'-0"	G		54	1.037		360	53.50	28	441.50	520
4600	8" x 8" x 3/8" x 14'-0"	G		50	1.120		775	58	30	863	990
4650	10" x 10" x 1/2" x 16'-0"	G		48	1.167		1,450	60.50	31.50	1,542	1,725
5100	Structural tubing, rect., 5" to 6" wide, light section	G		8000	.007	Lb.	1.33	.36	.19	1.88	2.30
5200	Heavy section	G		12000	.005		1.33	.24	.13	1.70	2.02
5300	7" to 10" wide, light section	G		15000	.004		1.33	.19	.10	1.62	1.91
5400	Heavy section	G		18000	.003		1.33	.16	.08	1.57	1.83
5500	Structural tubing, rect., 5" x 3" x 1/4" x 12'-0"	G		58	.966	Ea.	212	50	26	288	350
5550	6" x 4" x 5/16" x 12'-0"	G		54	1.037		330	53.50	28	411.50	490
5600	8" x 4" x 3/8" x 12'-0"	G		54	1.037		485	53.50	28	566.50	655
5650	10" x 6" x 3/8" x 14'-0"	G		50	1.120		775	58	30	863	990
5700	12" x 8" x 1/2" x 16'-0"	G		48	1.167		1,425	60.50	31.50	1,517	1,725
6800	W Shape, A992 steel, 2 tier, W8 x 24	G		1080	.052	L.F.	35	2.68	1.40	39.08	44.50
6850	W8 x 31	G		1080	.052		45	2.68	1.40	49.08	55.50
6900	W8 x 48	G		1032	.054		70	2.81	1.46	74.27	83.50
6950	W8 x 67	G		984	.057		97.50	2.94	1.54	101.98	114
7000	W10 x 45	G		1032	.054		65.50	2.81	1.46	69.77	78.50
7050	W10 x 68	G		984	.057		99	2.94	1.54	103.48	116
7100	W10 x 112	G		960	.058		163	3.02	1.57	167.59	187
7150	W12 x 50	G		1032	.054		73	2.81	1.46	77.27	86.50
7200	W12 x 87	G		984	.057		127	2.94	1.54	131.48	146
7250	W12 x 120	G		960	.058		175	3.02	1.57	179.59	199
7300	W12 x 190	G		912	.061		277	3.18	1.66	281.84	310
7350	W14 x 74	G		984	.057		108	2.94	1.54	112.48	126
7400	W14 x 120	G		960	.058		175	3.02	1.57	179.59	199
7450	W14 x 176	G		912	.061		257	3.18	1.66	261.84	289
8090	For projects 75 to 99 tons, add					All	10%				
8092	50 to 74 tons, add						20%				
8094	25 to 49 tons, add						30%	10%			
8096	10 to 24 tons, add						50%	25%			
8098	2 to 9 tons, add						75%	50%			
8099	Less than 2 tons, add						100%	100%			
9000	Minimum labor/equipment charge		1 Sswk	1	8	Job		420		420	760

05 12 23.20 Curb Edging			Crew	Daily Output	Labor-Hours	Unit	Material	Labor	Equipment	Total	Total Incl O&P
0010	**CURB EDGING**										
0020	Steel angle w/anchors, shop fabricated, on forms, 1" x 1", 0.8#/L.F.	G	E-4	350	.091	L.F.	1.67	4.86	.42	6.95	11.10
0100	2" x 2" angles, 3.92#/L.F.	G		330	.097		6.65	5.15	.44	12.24	17.10
0200	3" x 3" angles, 6.1#/L.F.	G		300	.107		10.50	5.65	.49	16.64	22.50
0300	4" x 4" angles, 8.2#/L.F.	G		275	.116		13.85	6.20	.53	20.58	27
1000	6" x 4" angles, 12.3#/L.F.	G		250	.128		20.50	6.80	.58	27.88	35.50
1050	Steel channels with anchors, on forms, 3" channel, 5#/L.F.	G		290	.110		8.35	5.85	.50	14.70	20.50
1100	4" channel, 5.4#/L.F.	G		270	.119		9	6.30	.54	15.84	22
1200	6" channel, 8.2#/L.F.	G		255	.125		13.85	6.65	.57	21.07	28
1300	8" channel, 11.5#/L.F.	G		225	.142		19.10	7.55	.65	27.30	35.50

05 12 23.20 Curb Edging			Crew	Daily Output	Labor-Hours	Unit	2015 Bare Costs Material	Labor	Equipment	Total	Total Incl O&P
1400	10" channel, 15.3#/L.F.	G	E-4	180	.178	L.F.	25	9.45	.81	35.26	45.50
1500	12" channel, 20.7#/L.F.	G		140	.229		33.50	12.15	1.04	46.69	60
2000	For curved edging, add						35%	10%			
9000	Minimum labor/equipment charge		E-4	4	8	Job		425	36.50	461.50	810

05 12 23.40 Lightweight Framing			Crew	Daily Output	Labor-Hours	Unit	Material	Labor	Equipment	Total	Total Incl O&P
0010	**LIGHTWEIGHT FRAMING** R051223-35										
0015	Made from recycled materials										
0200	For load-bearing steel studs see Section 05 41 13.30										
0400	Angle framing, field fabricated, 4" and larger R051223-45	G	E-3	440	.055	Lb.	.77	2.91	.33	4.01	6.45
0450	Less than 4" angles	G		265	.091	"	.80	4.83	.55	6.18	10.25
0460	1/2" x 1/2" x 1/8"	G		200	.120	L.F.	.16	6.40	.73	7.29	12.50
0462	3/4" x 3/4" x 1/8"	G		160	.150		.45	8	.91	9.36	15.95
0464	1" x 1" x 1/8"	G		135	.178		.64	9.50	1.08	11.22	19.05
0466	1-1/4" x 1-1/4" x 3/16"	G		115	.209		1.18	11.15	1.27	13.60	22.50
0468	1-1/2" x 1-1/2" x 3/16"	G		100	.240		1.43	12.80	1.46	15.69	26
0470	2" x 2" x 1/4"	G		90	.267		2.54	14.20	1.62	18.36	30
0472	2-1/2" x 2-1/2" x 1/4"	G		72	.333		3.26	17.75	2.03	23.04	38
0474	3" x 2" x 3/8"	G		65	.369		4.69	19.70	2.24	26.63	43
0476	3" x 3" x 3/8"	G		57	.421		5.70	22.50	2.56	30.76	49.50
0600	Channel framing, field fabricated, 8" and larger	G		500	.048	Lb.	.80	2.56	.29	3.65	5.80
0650	Less than 8" channels	G		335	.072	"	.80	3.82	.44	5.06	8.25
0660	C2 x 1.78	G		115	.209	L.F.	1.42	11.15	1.27	13.84	23
0662	C3 x 4.1	G		80	.300		3.26	16	1.82	21.08	34.50
0664	C4 x 5.4	G		66	.364		4.29	19.40	2.21	25.90	42
0666	C5 x 6.7	G		57	.421		5.35	22.50	2.56	30.41	49
0668	C6 x 8.2	G		55	.436		6.30	23.50	2.65	32.45	52
0670	C7 x 9.8	G		40	.600		7.80	32	3.65	43.45	70.50
0672	C8 x 11.5	G		36	.667		9.15	35.50	4.05	48.70	78.50
0710	Structural bar tee, field fabricated, 3/4" x 3/4" x 1/8"	G		160	.150		.45	8	.91	9.36	15.95
0712	1" x 1" x 1/8"	G		135	.178		.64	9.50	1.08	11.22	19.05
0714	1-1/2" x 1-1/2" x 1/4"	G		114	.211		1.86	11.25	1.28	14.39	24
0716	2" x 2" x 1/4"	G		89	.270		2.54	14.40	1.64	18.58	30.50
0718	2-1/2" x 2-1/2" x 3/8"	G		72	.333		4.69	17.75	2.03	24.47	39.50
0720	3" x 3" x 3/8"	G		57	.421		5.70	22.50	2.56	30.76	49.50
0730	Structural zee, field fabricated, 1-1/4" x 1-3/4" x 1-3/4"	G		114	.211		.60	11.25	1.28	13.13	22.50
0732	2-11/16" x 3" x 2-11/16"	G		114	.211		1.42	11.25	1.28	13.95	23.50
0734	3-1/16" x 4" x 3-1/16"	G		133	.180		2.14	9.60	1.10	12.84	21
0736	3-1/4" x 5" x 3-1/4"	G		133	.180		2.92	9.60	1.10	13.62	22
0738	3-1/2" x 6" x 3-1/2"	G		160	.150		4.40	8	.91	13.31	20.50
0740	Junior beam, field fabricated, 3"	G		80	.300		4.53	16	1.82	22.35	36
0742	4"	G		72	.333		6.10	17.75	2.03	25.88	41
0744	5"	G		67	.358		7.95	19.10	2.18	29.23	45.50
0746	6"	G		62	.387		9.95	20.50	2.35	32.80	51
0748	7"	G		57	.421		12.15	22.50	2.56	37.21	56.50
0750	8"	G		53	.453		14.65	24	2.75	41.40	62.50
1000	Continuous slotted channel framing system, shop fab, simple framing	G	2 Sswk	2400	.007	Lb.	4.11	.35		4.46	5.15
1200	Complex framing	G	"	1600	.010		4.64	.53		5.17	6.05
1250	Plate & bar stock for reinforcing beams and trusses	G					1.46			1.46	1.60
1300	Cross bracing, rods, shop fabricated, 3/4" diameter	G	E-3	700	.034		1.59	1.83	.21	3.63	5.30
1310	7/8" diameter	G		850	.028		1.59	1.51	.17	3.27	4.66
1320	1" diameter	G		1000	.024		1.59	1.28	.15	3.02	4.22
1330	Angle, 5" x 5" x 3/8"	G		2800	.009		1.59	.46	.05	2.10	2.64

05 12 23.40 Lightweight Framing

	05 12 23.40 Lightweight Framing		Crew	Daily Output	Labor-Hours	Unit	2015 Bare Costs Material	Labor	Equipment	Total	Total Incl O&P
1350	Hanging lintels, shop fabricated	G	E-3	850	.028	Lb.	1.59	1.51	.17	3.27	4.66
1380	Roof frames, shop fabricated, 3′-0″ square, 5′ span	G	E-2	4200	.013		1.59	.69	.36	2.64	3.36
1400	Tie rod, not upset, 1-1/2″ to 4″ diameter, with turnbuckle	G	2 Sswk	800	.020		1.72	1.05		2.77	3.79
1420	No turnbuckle	G		700	.023		1.66	1.20		2.86	4
1500	Upset, 1-3/4″ to 4″ diameter, with turnbuckle	G		800	.020		1.72	1.05		2.77	3.79
1520	No turnbuckle	G		700	.023		1.66	1.20		2.86	4
9000	Minimum labor/equipment charge			2	8	Job		420		420	760

05 12 23.45 Lintels

	05 12 23.45 Lintels		Crew	Daily Output	Labor-Hours	Unit	Material	Labor	Equipment	Total	Total Incl O&P
0010	**LINTELS**										
0015	Made from recycled materials										
0020	Plain steel angles, shop fabricated, under 500 lb.	G	1 Bric	550	.015	Lb.	1.02	.67		1.69	2.21
0100	500 to 1000 lb.	G		640	.013		.99	.58		1.57	2.03
0200	1,000 to 2,000 lb.	G		640	.013		.97	.58		1.55	2
0300	2,000 to 4,000 lb.	G		640	.013		.94	.58		1.52	1.97
0500	For built-up angles and plates, add to above	G					1.33			1.33	1.46
0700	For engineering, add to above						.13			.13	.15
0900	For galvanizing, add to above, under 500 lb.						.30			.30	.33
0950	500 to 2,000 lb.						.28			.28	.30
1000	Over 2,000 lb.						.25			.25	.28
2000	Steel angles, 3-1/2″ x 3″, 1/4″ thick, 2′-6″ long	G	1 Bric	47	.170	Ea.	14.30	7.85		22.15	28.50
2100	4′-6″ long	G		26	.308		26	14.20		40.20	51.50
2500	3-1/2″ x 3-1/2″ x 5/16″, 5′-0″ long	G		18	.444		38	20.50		58.50	75.50
2600	4″ x 3-1/2″, 1/4″ thick, 5′-0″ long	G		21	.381		33	17.60		50.60	64.50
2700	9′-0″ long	G		12	.667		59	31		90	115
2800	4″ x 3-1/2″ x 5/16″, 7′-0″ long	G		12	.667		57	31		88	113
2900	5″ x 3-1/2″ x 5/16″, 10′-0″ long	G		8	1		92	46		138	176
9000	Minimum labor/equipment charge			4	2	Job		92.50		92.50	150

05 12 23.60 Pipe Support Framing

	05 12 23.60 Pipe Support Framing		Crew	Daily Output	Labor-Hours	Unit	Material	Labor	Equipment	Total	Total Incl O&P
0010	**PIPE SUPPORT FRAMING**										
0020	Under 10#/L.F., shop fabricated	G	E-4	3900	.008	Lb.	1.78	.44	.04	2.26	2.78
0200	10.1 to 15#/L.F.	G		4300	.007		1.75	.40	.03	2.18	2.67
0400	15.1 to 20#/L.F.	G		4800	.007		1.72	.35	.03	2.10	2.56
0600	Over 20#/L.F.	G		5400	.006		1.70	.32	.03	2.05	2.47

05 12 23.65 Plates

	05 12 23.65 Plates		Crew	Daily Output	Labor-Hours	Unit	Material	Labor	Equipment	Total	Total Incl O&P
0010	**PLATES**	R051223-80									
0015	Made from recycled materials										
0020	For connections & stiffener plates, shop fabricated										
0050	1/8″ thick (5.1 lb./S.F.)	G				S.F.	6.75			6.75	7.45
0100	1/4″ thick (10.2 lb./S.F.)	G					13.50			13.50	14.85
0300	3/8″ thick (15.3 lb./S.F.)	G					20.50			20.50	22.50
0400	1/2″ thick (20.4 lb./S.F.)	G					27			27	29.50
0450	3/4″ thick (30.6 lb./S.F.)	G					40.50			40.50	44.50
0500	1″ thick (40.8 lb./S.F.)	G					54			54	59.50
2000	Steel plate, warehouse prices, no shop fabrication										
2100	1/4″ thick (10.2 lb./S.F.)	G				S.F.	7.30			7.30	8

05 12 23.70 Stressed Skin Steel Roof and Ceiling System

	05 12 23.70 Stressed Skin Steel Roof and Ceiling System		Crew	Daily Output	Labor-Hours	Unit	Material	Labor	Equipment	Total	Total Incl O&P
0010	**STRESSED SKIN STEEL ROOF & CEILING SYSTEM**										
0020	Double panel flat roof, spans to 100′	G	E-2	1150	.049	S.F.	10.60	2.52	1.31	14.43	17.50
0100	Double panel convex roof, spans to 200′	G		960	.058		17.25	3.02	1.57	21.84	26
0200	Double panel arched roof, spans to 300′	G		760	.074		26.50	3.81	1.99	32.30	38

05 12 Structural Steel Framing

05 12 23 – Structural Steel for Buildings

05 12 23.75 Structural Steel Members			Crew	Daily Output	Labor-Hours	Unit	Material	Labor	Equipment	Total	Total Incl O&P
							2015 Bare Costs				
0010	**STRUCTURAL STEEL MEMBERS**	R051223-10									
0015	Made from recycled materials	R051223-15									
0020	Shop fab'd for 100-ton, 1-2 story project, bolted connections										
0102	Beam or girder, W 6 x 9	G	E-2	600	.093	L.F.	13.10	4.83	2.52	20.45	25.50
0302	W 8 x 10	G		600	.093		14.60	4.83	2.52	21.95	27.50
0502	x 31	G		550	.102		45	5.25	2.75	53	61.50
0702	W 10 x 22	G		600	.093		32	4.83	2.52	39.35	46.50
0902	x 49	G		550	.102		71.50	5.25	2.75	79.50	90.50
1102	W 12 x 16	G		880	.064		23.50	3.29	1.72	28.51	33
1302	x 22	G		880	.064		32	3.29	1.72	37.01	43
1502	x 26	G		880	.064		38	3.29	1.72	43.01	49
1702	x 72	G		640	.088		105	4.53	2.36	111.89	126
1902	W 14 x 26	G		990	.057		38	2.93	1.53	42.46	48.50
2102	x 30	G		900	.062		43.50	3.22	1.68	48.40	55.50
2302	x 34	G		810	.069		49.50	3.58	1.87	54.95	63
2502	x 120	G		720	.078		175	4.02	2.10	181.12	201
2702	W 16 x 26	G		1000	.056		38	2.90	1.51	42.41	48
2902	x 31	G		900	.062		45	3.22	1.68	49.90	57
3102	x 40	G		800	.070		58.50	3.62	1.89	64.01	72.50
3302	W 18 x 35	G	E-5	960	.083		51	4.35	1.73	57.08	65.50
3502	x 40	G		960	.083		58.50	4.35	1.73	64.58	73.50
3702	x 50	G		912	.088		73	4.58	1.82	79.40	90
3902	x 55	G		912	.088		80	4.58	1.82	86.40	98
4102	W 21 x 44	G		1064	.075		64	3.93	1.56	69.49	79
4302	x 50	G		1064	.075		73	3.93	1.56	78.49	88.50
4502	x 62	G		1036	.077		90.50	4.03	1.60	96.13	108
4702	x 68	G		1036	.077		99	4.03	1.60	104.63	118
4902	W 24 x 55	G		1110	.072		80	3.76	1.49	85.25	96.50
5102	x 62	G		1110	.072		90.50	3.76	1.49	95.75	108
5302	x 68	G		1110	.072		99	3.76	1.49	104.25	117
5502	x 76	G		1110	.072		111	3.76	1.49	116.25	130
5702	x 84	G		1080	.074		122	3.87	1.53	127.40	144
5902	W 27 x 94	G		1190	.067		137	3.51	1.39	141.90	159
6102	W 30 x 99	G		1200	.067		144	3.48	1.38	148.86	167
6302	x 108	G		1200	.067		157	3.48	1.38	161.86	181
6502	x 116	G		1160	.069		169	3.60	1.43	174.03	194
6702	W 33 x 118	G		1176	.068		172	3.55	1.41	176.96	197
6902	x 130	G		1134	.071		189	3.68	1.46	194.14	216
7102	x 141	G		1134	.071		206	3.68	1.46	211.14	234
7302	W 36 x 135	G		1170	.068		197	3.57	1.42	201.99	224
7502	x 150	G		1170	.068		219	3.57	1.42	223.99	248
7702	x 194	G		1125	.071		283	3.71	1.47	288.18	320
7902	x 231	G		1125	.071		335	3.71	1.47	340.18	380
8102	x 302	G		1035	.077		440	4.04	1.60	445.64	495
8490	For projects 75 to 99 tons, add						10%				
8492	50 to 74 tons, add						20%				
8494	25 to 49 tons, add						30%	10%			
8496	10 to 24 tons, add						50%	25%			
8498	2 to 9 tons, add						75%	50%			
8499	Less than 2 tons, add						100%	100%			
9000	Minimum labor/equipment charge		E-2	2	28	Job		1,450	755	2,205	3,350

05 12 Structural Steel Framing

05 12 23 – Structural Steel for Buildings

05 12 23.77 Structural Steel Projects			Crew	Daily Output	Labor-Hours	Unit	2015 Bare Costs Material	Labor	Equipment	Total	Total Incl O&P
0010	**STRUCTURAL STEEL PROJECTS** R050516-30										
0015	Made from recycled materials										
0020	Shop fab'd for 100-ton, 1-2 story project, bolted connections										
0200	Apartments, nursing homes, etc., 1 to 2 stories R050523-10	G	E-5	10.30	7.767	Ton	2,650	405	161	3,216	3,825
0300	3 to 6 stories	G	″	10.10	7.921		2,700	415	164	3,279	3,875
0400	7 to 15 stories R051223-10	G	E-6	14.20	9.014		2,750	470	130	3,350	4,000
0500	Over 15 stories	G	″	13.90	9.209		2,850	480	133	3,463	4,150
0700	Offices, hospitals, etc., steel bearing, 1 to 2 stories R051223-15	G	E-5	10.30	7.767		2,650	405	161	3,216	3,825
0800	3 to 6 stories	G	E-6	14.40	8.889		2,700	465	128	3,293	3,925
0900	7 to 15 stories R051223-20	G		14.20	9.014		2,750	470	130	3,350	4,000
1000	Over 15 stories	G		13.90	9.209		2,850	480	133	3,463	4,150
1100	For multi-story masonry wall bearing construction, add R051223-25	G						30%			
1300	Industrial bldgs., 1 story, beams & girders, steel bearing	G	E-5	12.90	6.202		2,650	325	128	3,103	3,625
1400	Masonry bearing	G	″	10	8		2,650	420	166	3,236	3,850
1500	Industrial bldgs., 1 story, under 10 tons,										
1510	steel from warehouse, trucked	G	E-2	7.50	7.467	Ton	3,175	385	201	3,761	4,400
1600	1 story with roof trusses, steel bearing	G	E-5	10.60	7.547		3,125	395	156	3,676	4,325
1700	Masonry bearing	G	″	8.30	9.639		3,125	505	200	3,830	4,550
1900	Monumental structures, banks, stores, etc., simple connections	G	E-6	13	9.846		2,650	515	142	3,307	4,000
2000	Moment/composite connections	G	″	9	14.222		4,400	745	206	5,351	6,400
2200	Churches, simple connections	G	E-5	11.60	6.897		2,475	360	143	2,978	3,500
2300	Moment/composite connections	G	″	5.20	15.385		3,275	805	320	4,400	5,400
2800	Power stations, fossil fuels, simple connections	G	E-6	11	11.636		2,650	610	168	3,428	4,175
2900	Moment/composite connections	G		5.70	22.456		3,975	1,175	325	5,475	6,800
2950	Nuclear fuels, non-safety steel, simple connections	G		7	18.286		2,650	955	264	3,869	4,925
3000	Moment/composite connections	G		5.50	23.273		3,975	1,225	335	5,535	6,900
3040	Safety steel, simple connections	G		2.50	51.200		3,875	2,675	740	7,290	9,800
3070	Moment/composite connections	G		1.50	85.333		5,100	4,450	1,225	10,775	14,800
3100	Roof trusses, simple connections	G	E-5	13	6.154		3,700	320	127	4,147	4,775
3200	Moment/composite connections	G		8.30	9.639		4,500	505	200	5,205	6,050
3210	Schools, simple connections	G		14.50	5.517		2,650	288	114	3,052	3,550
3220	Moment/composite connections	G		8.30	9.639		3,875	505	200	4,580	5,350
3400	Welded construction, simple commercial bldgs., 1 to 2 stories	G	E-7	7.60	10.526		2,700	550	237	3,487	4,200
3500	7 to 15 stories	G	E-9	8.30	15.422		3,125	805	270	4,200	5,175
3700	Welded rigid frame, 1 story, simple connections	G	E-7	15.80	5.063		2,750	264	114	3,128	3,625
3800	Moment/composite connections	G	″	5.50	14.545		3,575	760	330	4,665	5,625
3810	Fabrication shop costs (incl in project bare material cost, above)										
3820	Mini mill base price, Grade A992	G				Ton	800			800	880
3830	Mill extras plus delivery to warehouse						275			275	305
3835	Delivery from warehouse to fabrication shop						85			85	93.50
3840	Shop extra for shop drawings and detailing						295			295	325
3850	Shop fabricating and handling						920			920	1,000
3860	Shop sandblasting and primer coat of paint						155			155	171
3870	Shop delivery to the job site						120			120	132
3880	Total material cost, shop fabricated, primed, delivered						2,650			2,650	2,925
3900	High strength steel mill spec extras:										
3950	A529, A572 (50 ksi) and A36: same as A992 steel (no extra)										
4000	Add to A992 price for A572 (60, 65 ksi)	G				Ton	80			80	88
4100	A242 and A588 Weathering	G				″	80			80	88
4200	Mill size extras for W-Shapes: 0 to 30 plf: no extra charge										
4210	Member sizes 31 to 65 plf, deduct	G				Ton	.01			.01	.01
4220	Member sizes 66 to 100 plf, deduct	G					5			5	5.50

05 12 23.77 Structural Steel Projects			Crew	Daily Output	Labor-Hours	Unit	Material	2015 Bare Costs Labor	Equipment	Total	Total Incl O&P
4230	Member sizes 101 to 387 plf, add	G				Ton	55.50			55.50	61
4300	Column base plates, light, up to 150 lb.	G	2 Sswk	2000	.008	Lb.	1.46	.42		1.88	2.36
4400	Heavy, over 150 lb.	G	E-2	7500	.007	"	1.52	.39	.20	2.11	2.58
4600	Castellated beams, light sections, to 50#/L.F., simple connections	G		10.70	5.234	Ton	2,775	271	141	3,187	3,675
4700	Moment/composite connections	G		7	8		3,050	415	216	3,681	4,300
4900	Heavy sections, over 50 plf, simple connections	G		11.70	4.786		2,925	248	129	3,302	3,775
5000	Moment/composite connections	G		7.80	7.179		3,175	370	194	3,739	4,375
5390	For projects 75 to 99 tons, add						10%				
5392	50 to 74 tons, add						20%				
5394	25 to 49 tons, add						30%	10%			
5396	10 to 24 tons, add						50%	25%			
5398	2 to 9 tons, add						75%	50%			
5399	Less than 2 tons, add						100%	100%			

05 12 23.78 Structural Steel Secondary Members

Line	Description	Crew	Daily Output	Labor-Hours	Unit	Material	Labor	Equipment	Total	Total Incl O&P
0010	**STRUCTURAL STEEL SECONDARY MEMBERS**									
0015	Made from recycled materials									
0020	Shop fabricated for 20-ton girt/purlin framing package, materials only									
0100	Girts/purlins, C/Z-shapes, includes clips and bolts									
0110	6" x 2-1/2" x 2-1/2", 16 ga., 3.0 lb./L.F.				L.F.	3.58			3.58	3.94
0115	14 ga., 3.5 lb./L.F.					4.17			4.17	4.59
0120	8" x 2-3/4" x 2-3/4", 16 ga., 3.4 lb./L.F.					4.05			4.05	4.46
0125	14 ga., 4.1 lb./L.F.					4.89			4.89	5.40
0130	12 ga., 5.6 lb./L.F.					6.70			6.70	7.35
0135	10" x 3-1/2" x 3-1/2", 14 ga., 4.7 lb./L.F.					5.60			5.60	6.15
0140	12 ga., 6.7 lb./L.F.					8			8	8.80
0145	12" x 3-1/2" x 3-1/2", 14 ga., 5.3 lb./L.F.					6.30			6.30	6.95
0150	12 ga., 7.4 lb./L.F.					8.80			8.80	9.70
0200	Eave struts, C-shape, includes clips and bolts									
0210	6" x 4" x 3", 16 ga., 3.1 lb./L.F.				L.F.	3.70			3.70	4.07
0215	14 ga., 3.9 lb./L.F.					4.65			4.65	5.10
0220	8" x 4" x 3", 16 ga., 3.5 lb./L.F.					4.17			4.17	4.59
0225	14 ga., 4.4 lb./L.F.					5.25			5.25	5.75
0230	12 ga., 6.2 lb./L.F.					7.40			7.40	8.15
0235	10" x 5" x 3", 14 ga., 5.2 lb./L.F.					6.20			6.20	6.80
0240	12 ga., 7.3 lb./L.F.					8.70			8.70	9.60
0245	12" x 5" x 4", 14 ga., 6.0 lb./L.F.					7.15			7.15	7.85
0250	12 ga., 8.4 lb./L.F.					10			10	11
0300	Rake/base angle, excludes concrete drilling and expansion anchors									
0310	2" x 2", 14 ga., 1.0 lb./L.F.	2 Sswk	640	.025	L.F.	1.19	1.32		2.51	3.69
0315	3" x 2", 14 ga., 1.3 lb./L.F.		535	.030		1.55	1.57		3.12	4.56
0320	3" x 3", 14 ga., 1.6 lb./L.F.		500	.032		1.91	1.68		3.59	5.15
0325	4" x 3", 14 ga., 1.8 lb./L.F.		480	.033		2.15	1.75		3.90	5.55
0600	Installation of secondary members, erection only									
0610	Girts, purlins, eave struts, 16 ga., 6" deep	E-18	100	.400	Ea.		21	9.50	30.50	47.50
0615	8" deep		80	.500			26.50	11.85	38.35	59.50
0620	14 ga., 6" deep		80	.500			26.50	11.85	38.35	59.50
0625	8" deep		65	.615			32.50	14.60	47.10	73
0630	10" deep		55	.727			38.50	17.25	55.75	86.50
0635	12" deep		50	.800			42	19	61	95.50
0640	12 ga., 8" deep		50	.800			42	19	61	95.50
0645	10" deep		45	.889			47	21	68	106
0650	12" deep		40	1			52.50	23.50	76	119

05 12 Structural Steel Framing

05 12 23 – Structural Steel for Buildings

05 12 23.78 Structural Steel Secondary Members		Crew	Daily Output	Labor-Hours	Unit	2015 Bare Costs Material	Labor	Equipment	Total	Total Incl O&P
0900	For less than 20-ton job lots									
0905	For 15 to 19 tons, add					10%				
0910	For 10 to 14 tons, add					25%				
0915	For 5 to 9 tons, add					50%	50%	50%		
0920	For 1 to 4 tons, add					75%	75%	75%		
0925	For less than 1 ton, add					100%	100%	100%		
0990	Minimum labor/equipment charge	E-18	1	40	Job		2,100	950	3,050	4,775

05 12 23.80 Subpurlins		Crew	Daily Output	Labor-Hours	Unit	Material	Labor	Equipment	Total	Total Incl O&P
0010	**SUBPURLINS** R051223-50									
0015	Made from recycled materials									
0020	Bulb tees, shop fabricated, painted, 32-5/8" O.C., 40 psf L.L.									
0200	Type 218, max 10'-2" span, 3.19 plf, 2-1/8" high x 2-1/8" wide G	E-1	3100	.008	S.F.	1.66	.40	.05	2.11	2.57
1420	For 24-5/8" spacing, add					33%	33%			
1430	For 48-5/8" spacing, deduct				↓	33%	33%			

05 14 Structural Aluminum Framing

05 14 23 – Non-Exposed Structural Aluminum Framing

05 14 23.05 Aluminum Shapes		Crew	Daily Output	Labor-Hours	Unit	Material	Labor	Equipment	Total	Total Incl O&P
0010	**ALUMINUM SHAPES**									
0015	Made from recycled materials									
0020	Structural shapes, 1" to 10" members, under 1 ton G	E-2	4000	.014	Lb.	3.85	.72	.38	4.95	5.95
0050	1 to 5 tons G		4300	.013		3.52	.67	.35	4.54	5.45
0100	Over 5 tons G		4600	.012		3.34	.63	.33	4.30	5.15
0300	Extrusions, over 5 tons, stock shapes G		1330	.042		3.28	2.18	1.14	6.60	8.65
0400	Custom shapes G	↓	1330	.042	↓	3.41	2.18	1.14	6.73	8.80

05 15 Wire Rope Assemblies

05 15 16 – Steel Wire Rope Assemblies

05 15 16.05 Accessories for Steel Wire Rope		Crew	Daily Output	Labor-Hours	Unit	Material	Labor	Equipment	Total	Total Incl O&P
0010	**ACCESSORIES FOR STEEL WIRE ROPE**									
0015	Made from recycled materials									
1500	Thimbles, heavy duty, 1/4" G	E-17	160	.100	Ea.	.52	5.35		5.87	10.25
1510	1/2" G		160	.100		2.29	5.35		7.64	12.20
1520	3/4" G		105	.152		5.20	8.20		13.40	20.50
1530	1" G		52	.308		10.40	16.50		26.90	41.50
1540	1-1/4" G		38	.421		16	22.50		38.50	58.50
1550	1-1/2" G		13	1.231		45	66		111	169
1560	1-3/4" G		8	2		93	107		200	296
1570	2" G		6	2.667		135	143		278	410
1580	2-1/4" G		4	4		183	215		398	590
1600	Clips, 1/4" diameter G		160	.100		2.74	5.35		8.09	12.70
1610	3/8" diameter G		160	.100		3.01	5.35		8.36	13
1620	1/2" diameter G		160	.100		4.84	5.35		10.19	15
1630	3/4" diameter G		102	.157		7.85	8.40		16.25	24
1640	1" diameter G		64	.250		13.05	13.40		26.45	38.50
1650	1-1/4" diameter G		35	.457		21.50	24.50		46	68
1670	1-1/2" diameter G		26	.615		29	33		62	91.50
1680	1-3/4" diameter G		16	1		67.50	53.50		121	171
1690	2" diameter G	↓	12	1.333	↓	75	71.50		146.50	212

05 15 16.05 Accessories for Steel Wire Rope			Crew	Daily Output	Labor-Hours	Unit	Material	2015 Bare Costs Labor	Equipment	Total	Total Incl O&P
1700	2-1/4" diameter	G	E-17	10	1.600	Ea.	110	86		196	276
1800	Sockets, open swage, 1/4" diameter	G		160	.100		47	5.35		52.35	61.50
1810	1/2" diameter	G		77	.208		68	11.15		79.15	94.50
1820	3/4" diameter	G		19	.842		105	45		150	198
1830	1" diameter	G		9	1.778		188	95.50		283.50	380
1840	1-1/4" diameter	G		5	3.200		262	172		434	600
1850	1-1/2" diameter	G		3	5.333		575	286		861	1,150
1860	1-3/4" diameter	G		3	5.333		1,025	286		1,311	1,650
1870	2" diameter	G		1.50	10.667		1,550	570		2,120	2,725
1900	Closed swage, 1/4" diameter	G		160	.100		28	5.35		33.35	40
1910	1/2" diameter	G		104	.154		48	8.25		56.25	68
1920	3/4" diameter	G		32	.500		72	27		99	128
1930	1" diameter	G		15	1.067		126	57		183	242
1940	1-1/4" diameter	G		7	2.286		189	123		312	430
1950	1-1/2" diameter	G		4	4		345	215		560	770
1960	1-3/4" diameter	G		3	5.333		505	286		791	1,075
1970	2" diameter	G		2	8		985	430		1,415	1,850
2000	Open spelter, galv., 1/4" diameter	G		160	.100		59.50	5.35		64.85	75
2010	1/2" diameter	G		70	.229		62	12.25		74.25	90
2020	3/4" diameter	G		26	.615		93	33		126	162
2030	1" diameter	G		10	1.600		258	86		344	440
2040	1-1/4" diameter	G		5	3.200		370	172		542	715
2050	1-1/2" diameter	G		4	4		785	215		1,000	1,250
2060	1-3/4" diameter	G		2	8		1,375	430		1,805	2,275
2070	2" diameter	G		1.20	13.333		1,575	715		2,290	3,025
2080	2-1/2" diameter	G		1	16		2,900	860		3,760	4,725
2100	Closed spelter, galv., 1/4" diameter	G		160	.100		49.50	5.35		54.85	64
2110	1/2" diameter	G		88	.182		53	9.75		62.75	76
2120	3/4" diameter	G		30	.533		80.50	28.50		109	140
2130	1" diameter	G		13	1.231		171	66		237	305
2140	1-1/4" diameter	G		7	2.286		274	123		397	520
2150	1-1/2" diameter	G		6	2.667		590	143		733	910
2160	1-3/4" diameter	G		2.80	5.714		785	305		1,090	1,425
2170	2" diameter	G		2	8		970	430		1,400	1,850
2200	Jaw & jaw turnbuckles, 1/4" x 4"	G		160	.100		16.25	5.35		21.60	27.50
2250	1/2" x 6"	G		96	.167		20.50	8.95		29.45	38.50
2260	1/2" x 9"	G		77	.208		27.50	11.15		38.65	50
2270	1/2" x 12"	G		66	.242		31	13		44	57.50
2300	3/4" x 6"	G		38	.421		40	22.50		62.50	85.50
2310	3/4" x 9"	G		30	.533		44.50	28.50		73	101
2320	3/4" x 12"	G		28	.571		57.50	30.50		88	119
2330	3/4" x 18"	G		23	.696		68.50	37.50		106	143
2350	1" x 6"	G		17	.941		78	50.50		128.50	177
2360	1" x 12"	G		13	1.231		85.50	66		151.50	213
2370	1" x 18"	G		10	1.600		128	86		214	296
2380	1" x 24"	G		9	1.778		141	95.50		236.50	325
2400	1-1/4" x 12"	G		7	2.286		144	123		267	380
2410	1-1/4" x 18"	G		6.50	2.462		178	132		310	435
2420	1-1/4" x 24"	G		5.60	2.857		240	153		393	540
2450	1-1/2" x 12"	G		5.20	3.077		465	165		630	810
2460	1-1/2" x 18"	G		4	4		495	215		710	935
2470	1-1/2" x 24"	G		3.20	5		665	268		933	1,225
2500	1-3/4" x 18"	G		3.20	5		1,000	268		1,268	1,575

05 15 16 – Steel Wire Rope Assemblies

05 15 16.05 Accessories for Steel Wire Rope		Crew	Daily Output	Labor-Hours	Unit	Material	2015 Bare Costs Labor	Equipment	Total	Total Incl O&P
2510	1-3/4" x 24"	G	E-17	2.80	5.714	Ea.	1,150	305		1,455	1,800
2550	2" x 24"	G	↓	1.60	10	↓	1,550	535		2,085	2,675

05 15 16.50 Steel Wire Rope

Code	Description		Crew	Daily Output	Labor-Hours	Unit	Material	Labor	Equipment	Total	Total Incl O&P
0010	STEEL WIRE ROPE										
0015	Made from recycled materials										
0020	6 x 19, bright, fiber core, 5000' rolls, 1/2" diameter	G				L.F.	.85			.85	.93
0050	Steel core	G					1.12			1.12	1.23
0100	Fiber core, 1" diameter	G					2.86			2.86	3.15
0150	Steel core	G					3.26			3.26	3.59
0300	6 x 19, galvanized, fiber core, 1/2" diameter	G					1.25			1.25	1.38
0350	Steel core	G					1.43			1.43	1.57
0400	Fiber core, 1" diameter	G					3.67			3.67	4.03
0450	Steel core	G					3.84			3.84	4.23
0500	6 x 7, bright, IPS, fiber core, <500 L.F. w/acc., 1/4" diameter	G	E-17	6400	.003		1.51	.13		1.64	1.90
0510	1/2" diameter	G		2100	.008		3.68	.41		4.09	4.79
0520	3/4" diameter	G		960	.017		6.65	.89		7.54	8.95
0550	6 x 19, bright, IPS, IWRC, <500 L.F. w/acc., 1/4" diameter	G		5760	.003		.94	.15		1.09	1.30
0560	1/2" diameter	G		1730	.009		1.52	.50		2.02	2.57
0570	3/4" diameter	G		770	.021		2.64	1.11		3.75	4.91
0580	1" diameter	G		420	.038		4.47	2.04		6.51	8.60
0590	1-1/4" diameter	G		290	.055		7.40	2.96		10.36	13.50
0600	1-1/2" diameter	G	↓	192	.083		9.10	4.47		13.57	18.15
0610	1-3/4" diameter	G	E-18	240	.167		14.55	8.75	3.95	27.25	36
0620	2" diameter	G		160	.250		18.65	13.15	5.95	37.75	50
0630	2-1/4" diameter	G	↓	160	.250		25	13.15	5.95	44.10	57
0650	6 x 37, bright, IPS, IWRC, <500 L.F. w/acc., 1/4" diameter	G	E-17	6400	.003		1.11	.13		1.24	1.46
0660	1/2" diameter	G		1730	.009		1.88	.50		2.38	2.97
0670	3/4" diameter	G		770	.021		3.04	1.11		4.15	5.35
0680	1" diameter	G		430	.037		4.83	2		6.83	8.90
0690	1-1/4" diameter	G		290	.055		7.30	2.96		10.26	13.35
0700	1-1/2" diameter	G	↓	190	.084		10.45	4.52		14.97	19.65
0710	1-3/4" diameter	G	E-18	260	.154		16.55	8.10	3.65	28.30	36.50
0720	2" diameter	G		200	.200		21.50	10.55	4.74	36.79	47.50
0730	2-1/4" diameter	G	↓	160	.250		28.50	13.15	5.95	47.60	60.50
0800	6 x 19 & 6 x 37, swaged, 1/2" diameter	G	E-17	1220	.013		2.47	.70		3.17	3.99
0810	9/16" diameter	G		1120	.014		2.87	.77		3.64	4.55
0820	5/8" diameter	G		930	.017		3.41	.92		4.33	5.40
0830	3/4" diameter	G		640	.025		4.34	1.34		5.68	7.20
0840	7/8" diameter	G		480	.033		5.50	1.79		7.29	9.30
0850	1" diameter	G		350	.046		6.70	2.45		9.15	11.80
0860	1-1/8" diameter	G		288	.056		8.20	2.98		11.18	14.45
0870	1-1/4" diameter	G		230	.070		9.95	3.73		13.68	17.70
0880	1-3/8" diameter	G	↓	192	.083		11.50	4.47		15.97	21
0890	1-1/2" diameter	G	E-18	300	.133	↓	13.95	7	3.16	24.11	31

05 15 16.60 Galvanized Steel Wire Rope and Accessories

Code	Description		Crew	Daily Output	Labor-Hours	Unit	Material	Labor	Equipment	Total	Total Incl O&P
0010	GALVANIZED STEEL WIRE ROPE & ACCESSORIES										
0015	Made from recycled materials										
3000	Aircraft cable, galvanized, 7 x 7 x 1/8"	G	E-17	5000	.003	L.F.	.20	.17		.37	.53
3100	Clamps, 1/8"	G	"	125	.128	Ea.	1.96	6.85		8.81	14.55

05 15 Wire Rope Assemblies

05 15 16 – Steel Wire Rope Assemblies

05 15 16.70 Temporary Cable Safety Railing		Crew	Daily Output	Labor-Hours	Unit	Material	Labor	Equipment	Total	Total Incl O&P
						2015 Bare Costs				
0010	**TEMPORARY CABLE SAFETY RAILING**, Each 100′ strand incl.									
0020	2 eyebolts, 1 turnbuckle, 100′ cable, 2 thimbles, 6 clips									
0025	Made from recycled materials									
0100	One strand using 1/4″ cable & accessories G	2 Sswk	4	4	C.L.F.	209	211		420	610
0200	1/2″ cable & accessories G	″	2	8	″	440	420		860	1,250

05 21 Steel Joist Framing

05 21 13 – Deep Longspan Steel Joist Framing

05 21 13.50 Deep Longspan Joists

		Crew	Daily Output	Labor-Hours	Unit	Material	Labor	Equipment	Total	Total Incl O&P
0010	**DEEP LONGSPAN JOISTS**									
3010	DLH series, 40-ton job lots, bolted cross bridging, shop primer									
3015	Made from recycled materials									
3040	Spans to 144′ (shipped in 2 pieces) G	E-7	13	6.154	Ton	1,925	320	139	2,384	2,850
3200	52DLH11, 26 lb./L.F. G		2000	.040	L.F.	24	2.09	.90	26.99	31
3220	52DLH16, 45 lb./L.F. G		2000	.040		43.50	2.09	.90	46.49	52
3240	56DLH11, 26 lb./L.F. G		2000	.040		25	2.09	.90	27.99	32
3260	56DLH16, 46 lb./L.F. G		2000	.040		44.50	2.09	.90	47.49	53.50
3280	60DLH12, 29 lb./L.F. G		2000	.040		28	2.09	.90	30.99	35.50
3300	60DLH17, 52 lb./L.F. G		2000	.040		50	2.09	.90	52.99	59.50
3320	64DLH12, 31 lb./L.F. G		2200	.036		30	1.90	.82	32.72	37.50
3340	64DLH17, 52 lb./L.F. G		2200	.036		50	1.90	.82	52.72	59.50
3360	68DLH13, 37 lb./L.F. G		2200	.036		35.50	1.90	.82	38.22	44
3380	68DLH18, 61 lb./L.F. G		2200	.036		59	1.90	.82	61.72	69
3400	72DLH14, 41 lb./L.F. G		2200	.036		39.50	1.90	.82	42.22	48
3420	72DLH19, 70 lb./L.F. G	↓	2200	.036	↓	67.50	1.90	.82	70.22	79
3500	For less than 40-ton job lots									
3502	For 30 to 39 tons, add					10%				
3504	20 to 29 tons, add					20%				
3506	10 to 19 tons, add					30%				
3507	5 to 9 tons, add					50%	25%			
3508	1 to 4 tons, add					75%	50%			
3509	Less than 1 ton, add					100%	100%			
4010	SLH series, 40-ton job lots, bolted cross bridging, shop primer									
4040	Spans to 200′ (shipped in 3 pieces) G	E-7	13	6.154	Ton	1,975	320	139	2,434	2,900
4200	80SLH15, 40 lb./L.F. G		1500	.053	L.F.	39.50	2.78	1.20	43.48	49.50
4220	80SLH20, 75 lb./L.F. G		1500	.053		74.50	2.78	1.20	78.48	88
4240	88SLH16, 46 lb./L.F. G		1500	.053		45.50	2.78	1.20	49.48	56
4260	88SLH21, 89 lb./L.F. G		1500	.053		88.50	2.78	1.20	92.48	103
4280	96SLH17, 52 lb./L.F. G		1500	.053		51.50	2.78	1.20	55.48	62.50
4300	96SLH22, 102 lb./L.F. G		1500	.053		101	2.78	1.20	104.98	117
4320	104SLH18, 59 lb./L.F. G		1800	.044		58.50	2.32	1	61.82	69.50
4340	104SLH23, 109 lb./L.F. G		1800	.044		108	2.32	1	111.32	124
4360	112SLH19, 67 lb./L.F. G		1800	.044		66.50	2.32	1	69.82	78
4380	112SLH24, 131 lb./L.F. G		1800	.044		130	2.32	1	133.32	148
4400	120SLH20, 77 lb./L.F. G		1800	.044		76.50	2.32	1	79.82	89
4420	120SLH25, 152 lb./L.F. G	↓	1800	.044	↓	151	2.32	1	154.32	171
6100	For less than 40-ton job lots									
6102	For 30 to 39 tons, add					10%				
6104	20 to 29 tons, add					20%				
6106	10 to 19 tons, add					30%				

05 21 Steel Joist Framing

05 21 13 – Deep Longspan Steel Joist Framing

05 21 13.50 Deep Longspan Joists			Crew	Daily Output	Labor-Hours	Unit	Material (2015 Bare Costs)	Labor (2015 Bare Costs)	Equipment (2015 Bare Costs)	Total (2015 Bare Costs)	Total Incl O&P
6107	5 to 9 tons, add						50%	25%			
6108	1 to 4 tons, add						75%	50%			
6109	Less than 1 ton, add						100%	100%			

05 21 16 – Longspan Steel Joist Framing

05 21 16.50 Longspan Joists			Crew	Daily Output	Labor-Hours	Unit	Material	Labor	Equipment	Total	Total Incl O&P
0010	**LONGSPAN JOISTS**										
2000	LH series, 40-ton job lots, bolted cross bridging, shop primer										
2015	Made from recycled materials										
2040	Longspan joists, LH series, up to 96′	G	E-7	13	6.154	Ton	1,825	320	139	2,284	2,750
2200	18LH04, 12 lb./L.F.	G		1400	.057	L.F.	11	2.98	1.29	15.27	18.75
2220	18LH08, 19 lb./L.F.	G		1400	.057		17.40	2.98	1.29	21.67	26
2240	20LH04, 12 lb./L.F.	G		1400	.057		11	2.98	1.29	15.27	18.75
2260	20LH08, 19 lb./L.F.	G		1400	.057		17.40	2.98	1.29	21.67	26
2280	24LH05, 13 lb./L.F.	G		1400	.057		11.90	2.98	1.29	16.17	19.75
2300	24LH10, 23 lb./L.F.	G		1400	.057		21	2.98	1.29	25.27	29.50
2320	28LH06, 16 lb./L.F.	G		1800	.044		14.65	2.32	1	17.97	21.50
2340	28LH11, 25 lb./L.F.	G		1800	.044		23	2.32	1	26.32	30
2360	32LH08, 17 lb./L.F.	G		1800	.044		15.60	2.32	1	18.92	22.50
2380	32LH13, 30 lb./L.F.	G		1800	.044		27.50	2.32	1	30.82	35.50
2400	36LH09, 21 lb./L.F.	G		1800	.044		19.25	2.32	1	22.57	26
2420	36LH14, 36 lb./L.F.	G		1800	.044		33	2.32	1	36.32	41.50
2440	40LH10, 21 lb./L.F.	G		2200	.036		19.25	1.90	.82	21.97	25.50
2460	40LH15, 36 lb./L.F.	G		2200	.036		33	1.90	.82	35.72	41
2480	44LH11, 22 lb./L.F.	G		2200	.036		20	1.90	.82	22.72	26.50
2500	44LH16, 42 lb./L.F.	G		2200	.036		38.50	1.90	.82	41.22	47
2520	48LH11, 22 lb./L.F.	G		2200	.036		20	1.90	.82	22.72	26.50
2540	48LH16, 42 lb./L.F.	G		2200	.036		38.50	1.90	.82	41.22	47
2600	For less than 40-ton job lots										
2602	For 30 to 39 tons, add						10%				
2604	20 to 29 tons, add						20%				
2606	10 to 19 tons, add						30%				
2607	5 to 9 tons, add						50%	25%			
2608	1 to 4 tons, add						75%	50%			
2609	Less than 1 ton, add						100%	100%			
6000	For welded cross bridging, add							30%			

05 21 19 – Open Web Steel Joist Framing

05 21 19.10 Open Web Joists			Crew	Daily Output	Labor-Hours	Unit	Material	Labor	Equipment	Total	Total Incl O&P
0010	**OPEN WEB JOISTS**										
0015	Made from recycled materials										
0050	K series, 40-ton lots, horiz. bridging, spans to 30′, shop primer	G	E-7	12	6.667	Ton	1,650	350	150	2,150	2,600
0130	8K1, 5.1 lb./L.F.	G		1200	.067	L.F.	4.22	3.48	1.50	9.20	12.45
0140	10K1, 5.0 lb./L.F.	G		1200	.067		4.14	3.48	1.50	9.12	12.35
0160	12K3, 5.7 lb./L.F.	G		1500	.053		4.72	2.78	1.20	8.70	11.45
0180	14K3, 6.0 lb./L.F.	G		1500	.053		4.97	2.78	1.20	8.95	11.70
0200	16K3, 6.3 lb./L.F.	G		1800	.044		5.20	2.32	1	8.52	10.95
0220	16K6, 8.1 lb./L.F.	G		1800	.044		6.70	2.32	1	10.02	12.60
0240	18K5, 7.7 lb./L.F.	G		2000	.040		6.40	2.09	.90	9.39	11.70
0260	18K9, 10.2 lb./L.F.	G		2000	.040		8.45	2.09	.90	11.44	14
0440	K series, 30′ to 50′ spans	G		17	4.706	Ton	1,625	246	106	1,977	2,350
0500	20K5, 8.2 lb./L.F.	G		2000	.040	L.F.	6.65	2.09	.90	9.64	12.05
0520	20K9, 10.8 lb./L.F.	G		2000	.040		8.80	2.09	.90	11.79	14.35
0540	22K5, 8.8 lb./L.F.	G		2000	.040		7.15	2.09	.90	10.14	12.55

05 21 Steel Joist Framing

05 21 19 – Open Web Steel Joist Framing

05 21 19.10 Open Web Joists			Crew	Daily Output	Labor-Hours	Unit	Material	2015 Bare Costs Labor	Equipment	Total	Total Incl O&P
0560	22K9, 11.3 lb./L.F.	G	E-7	2000	.040	L.F.	9.20	2.09	.90	12.19	14.80
0580	24K6, 9.7 lb./L.F.	G		2200	.036		7.90	1.90	.82	10.62	12.90
0600	24K10, 13.1 lb./L.F.	G		2200	.036		10.65	1.90	.82	13.37	15.95
0620	26K6, 10.6 lb./L.F.	G		2200	.036		8.60	1.90	.82	11.32	13.70
0640	26K10, 13.8 lb./L.F.	G		2200	.036		11.20	1.90	.82	13.92	16.60
0660	28K8, 12.7 lb./L.F.	G		2400	.033		10.30	1.74	.75	12.79	15.25
0680	28K12, 17.1 lb./L.F.	G		2400	.033		13.90	1.74	.75	16.39	19.20
0700	30K8, 13.2 lb./L.F.	G		2400	.033		10.75	1.74	.75	13.24	15.70
0720	30K12, 17.6 lb./L.F.	G		2400	.033		14.30	1.74	.75	16.79	19.65
0800	For less than 40-ton job lots										
0802	For 30 to 39 tons, add						10%				
0804	20 to 29 tons, add						20%				
0806	10 to 19 tons, add						30%				
0807	5 to 9 tons, add						50%	25%			
0808	1 to 4 tons, add						75%	50%			
0809	Less than 1 ton, add						100%	100%			
1010	CS series, 40-ton job lots, horizontal bridging, shop primer										
1040	Spans to 30′	G	E-7	12	6.667	Ton	1,700	350	150	2,200	2,650
1100	10CS2, 7.5 lb./L.F.	G		1200	.067	L.F.	6.40	3.48	1.50	11.38	14.80
1120	12CS2, 8.0 lb./L.F.	G		1500	.053		6.80	2.78	1.20	10.78	13.75
1140	14CS2, 8.0 lb./L.F.	G		1500	.053		6.80	2.78	1.20	10.78	13.75
1160	16CS2, 8.5 lb./L.F.	G		1800	.044		7.25	2.32	1	10.57	13.15
1180	16CS4, 14.5 lb./L.F.	G		1800	.044		12.35	2.32	1	15.67	18.75
1200	18CS2, 9.0 lb./L.F.	G		2000	.040		7.65	2.09	.90	10.64	13.10
1220	18CS4, 15.0 lb./L.F.	G		2000	.040		12.75	2.09	.90	15.74	18.75
1240	20CS2, 9.5 lb./L.F.	G		2000	.040		8.10	2.09	.90	11.09	13.60
1260	20CS4, 16.5 lb./L.F.	G		2000	.040		14.05	2.09	.90	17.04	20
1280	22CS2, 10.0 lb./L.F.	G		2000	.040		8.50	2.09	.90	11.49	14.05
1300	22CS4, 16.5 lb./L.F.	G		2000	.040		14.05	2.09	.90	17.04	20
1320	24CS2, 10.0 lb./L.F.	G		2200	.036		8.50	1.90	.82	11.22	13.60
1340	24CS4, 16.5 lb./L.F.	G		2200	.036		14.05	1.90	.82	16.77	19.70
1360	26CS2, 10.0 lb./L.F.	G		2200	.036		8.50	1.90	.82	11.22	13.60
1380	26CS4, 16.5 lb./L.F.	G		2200	.036		14.05	1.90	.82	16.77	19.70
1400	28CS2, 10.5 lb./L.F.	G		2400	.033		8.95	1.74	.75	11.44	13.70
1420	28CS4, 16.5 lb./L.F.	G		2400	.033		14.05	1.74	.75	16.54	19.35
1440	30CS2, 11.0 lb./L.F.	G		2400	.033		9.35	1.74	.75	11.84	14.20
1460	30CS4, 16.5 lb./L.F.	G		2400	.033		14.05	1.74	.75	16.54	19.35
1500	For less than 40-ton job lots										
1502	For 30 to 39 tons, add						10%				
1504	20 to 29 tons, add						20%				
1506	10 to 19 tons, add						30%				
1507	5 to 9 tons, add						50%	25%			
1508	1 to 4 tons, add						75%	50%			
1509	Less than 1 ton, add						100%	100%			
6200	For shop prime paint other than mfrs. standard, add						20%				
6300	For bottom chord extensions, add per chord	G				Ea.	36			36	39.50
6400	Individual steel bearing plate, 6″ x 6″ x 1/4″ with J-hook	G	1 Bric	160	.050	″	7.95	2.31		10.26	12.50
9000	Minimum labor and equipment, bar joists		F-6	1	40	Job		1,775	655	2,430	3,600

05 21 Steel Joist Framing

05 21 23 – Steel Joist Girder Framing

05 21 23.50 Joist Girders			Crew	Daily Output	Labor-Hours	Unit	Material	Labor	Equipment	Total	Total Incl O&P
							2015 Bare Costs				
0010	**JOIST GIRDERS**										
0015	Made from recycled materials										
7020	Joist girders, 40-ton job lots, shop primer	G	E-5	13	6.154	Ton	1,650	320	127	2,097	2,525
7100	For less than 40-ton job lots										
7102	For 30 to 39 tons, add						10%				
7104	20 to 29 tons, add						20%				
7106	10 to 19 tons, add						30%				
7107	5 to 9 tons, add						50%	25%			
7108	1 to 4 tons, add						75%	50%			
7109	Less than 1 ton, add						100%	100%			
8000	Trusses, 40-ton job lots, shop fabricated WT chords, shop primer	G	E-5	11	7.273	Ton	5,425	380	151	5,956	6,775
8100	For less than 40-ton job lots										
8102	For 30 to 39 tons, add						10%				
8104	20 to 29 tons, add						20%				
8106	10 to 19 tons, add						30%				
8107	5 to 9 tons, add						50%	25%			
8108	1 to 4 tons, add						75%	50%			
8109	Less than 1 ton, add						100%	100%			

05 31 Steel Decking

05 31 13 – Steel Floor Decking

05 31 13.50 Floor Decking

Line	Description		Crew	Daily Output	Labor-Hours	Unit	Material	Labor	Equipment	Total	Total Incl O&P
0010	**FLOOR DECKING** R053100-10										
0015	Made from recycled materials										
5100	Non-cellular composite decking, galvanized, 1-1/2" deep, 16 ga.	G	E-4	3500	.009	S.F.	3.44	.49	.04	3.97	4.72
5120	18 ga.	G		3650	.009		2.78	.47	.04	3.29	3.94
5140	20 ga.	G		3800	.008		2.22	.45	.04	2.71	3.29
5200	2" deep, 22 ga.	G		3860	.008		1.93	.44	.04	2.41	2.96
5300	20 ga.	G		3600	.009		2.13	.47	.04	2.64	3.23
5400	18 ga.	G		3380	.009		2.73	.50	.04	3.27	3.96
5500	16 ga.	G		3200	.010		3.41	.53	.05	3.99	4.76
5700	3" deep, 22 ga.	G		3200	.010		2.10	.53	.05	2.68	3.32
5800	20 ga.	G		3000	.011		2.34	.57	.05	2.96	3.65
5900	18 ga.	G		2850	.011		2.90	.60	.05	3.55	4.33
6000	16 ga.	G		2700	.012		3.87	.63	.05	4.55	5.45
9000	Minimum labor/equipment charge		1 Sswk	1	8	Job		420		420	760

05 31 23 – Steel Roof Decking

05 31 23.50 Roof Decking

Line	Description		Crew	Daily Output	Labor-Hours	Unit	Material	Labor	Equipment	Total	Total Incl O&P
0010	**ROOF DECKING**										
0015	Made from recycled materials										
2100	Open type, 1-1/2" deep, Type B, wide rib, galv., 22 ga., under 50 sq.	G	E-4	4500	.007	S.F.	2.05	.38	.03	2.46	2.97
2200	50-500 squares	G		4900	.007		1.59	.35	.03	1.97	2.41
2400	Over 500 squares	G		5100	.006		1.47	.33	.03	1.83	2.25
2600	20 ga., under 50 squares	G		3865	.008		2.39	.44	.04	2.87	3.47
2650	50-500 squares	G		4170	.008		1.92	.41	.04	2.37	2.89
2700	Over 500 squares	G		4300	.007		1.72	.40	.03	2.15	2.65
2900	18 ga., under 50 squares	G		3800	.008		3.09	.45	.04	3.58	4.25
2950	50-500 squares	G		4100	.008		2.47	.41	.04	2.92	3.51
3000	Over 500 squares	G		4300	.007		2.22	.40	.03	2.65	3.20
3050	16 ga., under 50 squares	G		3700	.009		4.17	.46	.04	4.67	5.45

05 31 23 – Steel Roof Decking

05 31 23.50 Roof Decking			Crew	Daily Output	Labor-Hours	Unit	Material	2015 Bare Costs Labor	Equipment	Total	Total Incl O&P
3060	50-500 squares	G	E-4	4000	.008	S.F.	3.34	.43	.04	3.81	4.48
3100	Over 500 squares	G		4200	.008		3	.41	.03	3.44	4.07
3200	3" deep, Type N, 22 ga., under 50 squares	G		3600	.009		3.01	.47	.04	3.52	4.20
3250	50-500 squares	G		3800	.008		2.41	.45	.04	2.90	3.50
3260	over 500 squares	G		4000	.008		2.17	.43	.04	2.64	3.20
3300	20 ga., under 50 squares	G		3400	.009		3.26	.50	.04	3.80	4.53
3350	50-500 squares	G		3600	.009		2.61	.47	.04	3.12	3.76
3360	over 500 squares	G		3800	.008		2.35	.45	.04	2.84	3.43
3400	18 ga., under 50 squares	G		3200	.010		4.21	.53	.05	4.79	5.65
3450	50-500 squares	G		3400	.009		3.37	.50	.04	3.91	4.65
3460	over 500 squares	G		3600	.009		3.03	.47	.04	3.54	4.22
3500	16 ga., under 50 squares	G		3000	.011		5.55	.57	.05	6.17	7.15
3550	50-500 squares	G		3200	.010		4.45	.53	.05	5.03	5.90
3560	over 500 squares	G		3400	.009		4	.50	.04	4.54	5.35
3700	4-1/2" deep, Type J, 20 ga., over 50 squares	G		2700	.012		3.64	.63	.05	4.32	5.20
3800	18 ga.	G		2460	.013		4.80	.69	.06	5.55	6.60
3900	16 ga.	G		2350	.014		6.25	.72	.06	7.03	8.30
4100	6" deep, Type H, 18 ga., over 50 squares	G		2000	.016		5.75	.85	.07	6.67	7.95
4200	16 ga.	G		1930	.017		7.20	.88	.08	8.16	9.55
4300	14 ga.	G		1860	.017		9.25	.91	.08	10.24	11.95
4500	7-1/2" deep, Type H, 18 ga., over 50 squares	G		1690	.019		6.85	1.01	.09	7.95	9.40
4600	16 ga.	G		1590	.020		8.50	1.07	.09	9.66	11.40
4700	14 ga.	G	↓	1490	.021	↓	10.60	1.14	.10	11.84	13.85
4800	For painted instead of galvanized, deduct						5%				
5000	For acoustical perforated with fiberglass insulation, add					S.F.	25%				
5100	For type F intermediate rib instead of type B wide rib, add	G					25%				
5150	For type A narrow rib instead of type B wide rib, add	G				↓	25%				

05 31 33 – Steel Form Decking

05 31 33.50 Form Decking			Crew	Daily Output	Labor-Hours	Unit	Material	2015 Bare Costs Labor	Equipment	Total	Total Incl O&P
0010	**FORM DECKING**										
0015	Made from recycled materials										
6100	Slab form, steel, 28 ga., 9/16" deep, Type UFS, uncoated	G	E-4	4000	.008	S.F.	1.55	.43	.04	2.02	2.51
6200	Galvanized	G		4000	.008		1.37	.43	.04	1.84	2.32
6220	24 ga., 1" deep, Type UF1X, uncoated	G		3900	.008		1.49	.44	.04	1.97	2.47
6240	Galvanized	G		3900	.008		1.75	.44	.04	2.23	2.76
6300	24 ga., 1-5/16" deep, Type UFX, uncoated	G		3800	.008		1.58	.45	.04	2.07	2.59
6400	Galvanized	G		3800	.008		1.86	.45	.04	2.35	2.90
6500	22 ga., 1-5/16" deep, uncoated	G		3700	.009		2	.46	.04	2.50	3.07
6600	Galvanized	G		3700	.009		2.04	.46	.04	2.54	3.11
6700	22 ga., 2" deep, uncoated	G		3600	.009		2.60	.47	.04	3.11	3.75
6800	Galvanized	G	↓	3600	.009	↓	2.55	.47	.04	3.06	3.70
7000	Sheet metal edge closure form, 12" wide with 2 bends, galvanized										
7100	18 ga.	G	E-14	360	.022	L.F.	4.20	1.21	.41	5.82	7.25
7200	16 ga.	G	"	360	.022	"	5.70	1.21	.41	7.32	8.90

05 35 Raceway Decking Assemblies

05 35 13 – Steel Cellular Decking

05 35 13.50 Cellular Decking			Crew	Daily Output	Labor-Hours	Unit	2015 Bare Costs Material	Labor	Equipment	Total	Total Incl O&P
0010	**CELLULAR DECKING**										
0015	Made from recycled materials										
0200	Cellular units, galv, 1-1/2" deep, Type BC, 20-20 ga., over 15 squares	G	E-4	1460	.022	S.F.	8.10	1.17	.10	9.37	11.10
0250	18-20 ga.	G		1420	.023		9.20	1.20	.10	10.50	12.40
0300	18-18 ga.	G		1390	.023		9.40	1.22	.11	10.73	12.70
0320	16-18 ga.	G		1360	.024		11.25	1.25	.11	12.61	14.75
0340	16-16 ga.	G		1330	.024		12.50	1.28	.11	13.89	16.20
0400	3" deep, Type NC, galvanized, 20-20 ga.	G		1375	.023		8.90	1.24	.11	10.25	12.15
0500	18-20 ga.	G		1350	.024		10.75	1.26	.11	12.12	14.20
0600	18-18 ga.	G		1290	.025		10.70	1.32	.11	12.13	14.25
0700	16-18 ga.	G		1230	.026		12.05	1.38	.12	13.55	15.95
0800	16-16 ga.	G		1150	.028		13.15	1.48	.13	14.76	17.25
1000	4-1/2" deep, Type JC, galvanized, 18-20 ga.	G		1100	.029		12.40	1.55	.13	14.08	16.60
1100	18-18 ga.	G		1040	.031		12.30	1.64	.14	14.08	16.65
1200	16-18 ga.	G		980	.033		13.90	1.74	.15	15.79	18.55
1300	16-16 ga.	G	↓	935	.034	↓	15.10	1.82	.16	17.08	20
1500	For acoustical deck, add						15%				
1700	For cells used for ventilation, add						15%				
1900	For multi-story or congested site, add							50%			
8000	Metal deck and trench, 2" thick, 20 ga., combination										
8010	60% cellular, 40% non-cellular, inserts and trench	G	R-4	1100	.036	S.F.	16.20	1.94	.13	18.27	21.50

05 41 Structural Metal Stud Framing

05 41 13 – Load-Bearing Metal Stud Framing

05 41 13.05 Bracing			Crew	Daily Output	Labor-Hours	Unit	Material	Labor	Equipment	Total	Total Incl O&P
0010	**BRACING**, shear wall X-bracing, per 10' x 10' bay, one face										
0015	Made of recycled materials										
0120	Metal strap, 20 ga. x 4" wide	G	2 Carp	18	.889	Ea.	17.55	41.50		59.05	88
0130	6" wide	G		18	.889		29.50	41.50		71	101
0160	18 ga. x 4" wide	G		16	1		30.50	47		77.50	111
0170	6" wide	G	↓	16	1	↓	45	47		92	127
0410	Continuous strap bracing, per horizontal row on both faces										
0420	Metal strap, 20 ga. x 2" wide, studs 12" O.C.	G	1 Carp	7	1.143	C.L.F.	53	53.50		106.50	146
0430	16" O.C.	G		8	1		53	47		100	135
0440	24" O.C.	G		10	.800		53	37.50		90.50	120
0450	18 ga. x 2" wide, studs 12" O.C.	G		6	1.333		75	62.50		137.50	186
0460	16" O.C.	G		7	1.143		75	53.50		128.50	171
0470	24" O.C.	G	↓	8	1	↓	75	47		122	160

05 41 13.10 Bridging			Crew	Daily Output	Labor-Hours	Unit	Material	Labor	Equipment	Total	Total Incl O&P
0010	**BRIDGING**, solid between studs w/1-1/4" leg track, per stud bay										
0015	Made from recycled materials										
0200	Studs 12" O.C., 18 ga. x 2-1/2" wide	G	1 Carp	125	.064	Ea.	.89	3		3.89	5.90
0210	3-5/8" wide	G		120	.067		1.07	3.13		4.20	6.35
0220	4" wide	G		120	.067		1.13	3.13		4.26	6.40
0230	6" wide	G		115	.070		1.48	3.27		4.75	7
0240	8" wide	G		110	.073		1.84	3.41		5.25	7.60
0300	16 ga. x 2-1/2" wide	G		115	.070		1.13	3.27		4.40	6.60
0310	3-5/8" wide	G		110	.073		1.38	3.41		4.79	7.10
0320	4" wide	G		110	.073		1.47	3.41		4.88	7.20
0330	6" wide	G		105	.076		1.87	3.58		5.45	7.90
0340	8" wide	G	↓	100	.080	↓	2.34	3.76		6.10	8.75

	05 41 13.10 Bridging		Crew	Daily Output	Labor-Hours	Unit	Material	2015 Bare Costs Labor	Equipment	Total	Total Incl O&P
1200	Studs 16" O.C., 18 ga. x 2-1/2" wide	G	1 Carp	125	.064	Ea.	1.14	3		4.14	6.15
1210	3-5/8" wide	G		120	.067		1.37	3.13		4.50	6.65
1220	4" wide	G		120	.067		1.45	3.13		4.58	6.75
1230	6" wide	G		115	.070		1.90	3.27		5.17	7.45
1240	8" wide	G		110	.073		2.36	3.41		5.77	8.20
1300	16 ga. x 2-1/2" wide	G		115	.070		1.45	3.27		4.72	6.95
1310	3-5/8" wide	G		110	.073		1.77	3.41		5.18	7.55
1320	4" wide	G		110	.073		1.88	3.41		5.29	7.65
1330	6" wide	G		105	.076		2.39	3.58		5.97	8.50
1340	8" wide	G		100	.080		3	3.76		6.76	9.45
2200	Studs 24" O.C., 18 ga. x 2-1/2" wide	G		125	.064		1.65	3		4.65	6.75
2210	3-5/8" wide	G		120	.067		1.98	3.13		5.11	7.35
2220	4" wide	G		120	.067		2.10	3.13		5.23	7.45
2230	6" wide	G		115	.070		2.75	3.27		6.02	8.35
2240	8" wide	G		110	.073		3.41	3.41		6.82	9.35
2300	16 ga. x 2-1/2" wide	G		115	.070		2.10	3.27		5.37	7.65
2310	3-5/8" wide	G		110	.073		2.55	3.41		5.96	8.40
2320	4" wide	G		110	.073		2.72	3.41		6.13	8.60
2330	6" wide	G		105	.076		3.46	3.58		7.04	9.65
2340	8" wide	G		100	.080		4.34	3.76		8.10	10.95
3000	Continuous bridging, per row										
3100	16 ga. x 1-1/2" channel thru studs 12" O.C.	G	1 Carp	6	1.333	C.L.F.	48	62.50		110.50	156
3110	16" O.C.	G		7	1.143		48	53.50		101.50	141
3120	24" O.C.	G		8.80	.909		48	42.50		90.50	123
4100	2" x 2" angle x 18 ga., studs 12" O.C.	G		7	1.143		75	53.50		128.50	171
4110	16" O.C.	G		9	.889		75	41.50		116.50	151
4120	24" O.C.	G		12	.667		75	31.50		106.50	134
4200	16 ga., studs 12" O.C.	G		5	1.600		94.50	75		169.50	227
4210	16" O.C.	G		7	1.143		94.50	53.50		148	192
4220	24" O.C.	G		10	.800		94.50	37.50		132	166

	05 41 13.25 Framing, Boxed Headers/Beams		Crew	Daily Output	Labor-Hours	Unit	Material	Labor	Equipment	Total	Total Incl O&P
0010	**FRAMING, BOXED HEADERS/BEAMS**										
0015	Made from recycled materials										
0200	Double, 18 ga. x 6" deep	G	2 Carp	220	.073	L.F.	5.10	3.41		8.51	11.20
0210	8" deep	G		210	.076		5.65	3.58		9.23	12.05
0220	10" deep	G		200	.080		6.90	3.76		10.66	13.75
0230	12" deep	G		190	.084		7.55	3.95		11.50	14.80
0300	16 ga. x 8" deep	G		180	.089		6.50	4.17		10.67	14
0310	10" deep	G		170	.094		7.90	4.42		12.32	15.90
0320	12" deep	G		160	.100		8.60	4.70		13.30	17.15
0400	14 ga. x 10" deep	G		140	.114		9.10	5.35		14.45	18.80
0410	12" deep	G		130	.123		10	5.80		15.80	20.50
1210	Triple, 18 ga. x 8" deep	G		170	.094		8.20	4.42		12.62	16.25
1220	10" deep	G		165	.097		9.85	4.55		14.40	18.30
1230	12" deep	G		160	.100		10.85	4.70		15.55	19.60
1300	16 ga. x 8" deep	G		145	.110		9.50	5.20		14.70	18.90
1310	10" deep	G		140	.114		11.35	5.35		16.70	21.50
1320	12" deep	G		135	.119		12.40	5.55		17.95	22.50
1400	14 ga. x 10" deep	G		115	.139		12.40	6.55		18.95	24.50
1410	12" deep	G		110	.145		13.70	6.85		20.55	26.50

05 41 13 – Load-Bearing Metal Stud Framing

05 41 13.30 Framing, Stud Walls		Crew	Daily Output	Labor-Hours	Unit	Material	2015 Bare Costs Labor	Equipment	Total	Total Incl O&P
0010	**FRAMING, STUD WALLS** w/top & bottom track, no openings,									
0020	Headers, beams, bridging or bracing									
0025	Made from recycled materials									
4100	8' high walls, 18 ga. x 2-1/2" wide, studs 12" O.C. G	2 Carp	54	.296	L.F.	8.50	13.90		22.40	32.50
4110	16" O.C. G		77	.208		6.80	9.75		16.55	23.50
4120	24" O.C. G		107	.150		5.10	7		12.10	17.10
4130	3-5/8" wide, studs 12" O.C. G		53	.302		10.05	14.15		24.20	34
4140	16" O.C. G		76	.211		8.05	9.90		17.95	25
4150	24" O.C. G		105	.152		6.05	7.15		13.20	18.40
4160	4" wide, studs 12" O.C. G		52	.308		10.55	14.45		25	35
4170	16" O.C. G		74	.216		8.45	10.15		18.60	26
4180	24" O.C. G		103	.155		6.35	7.30		13.65	18.95
4190	6" wide, studs 12" O.C. G		51	.314		13.40	14.75		28.15	38.50
4200	16" O.C. G		73	.219		10.75	10.30		21.05	28.50
4210	24" O.C. G		101	.158		8.10	7.45		15.55	21
4220	8" wide, studs 12" O.C. G		50	.320		16.30	15		31.30	42.50
4230	16" O.C. G		72	.222		13.10	10.45		23.55	31.50
4240	24" O.C. G		100	.160		9.90	7.50		17.40	23
4300	16 ga. x 2-1/2" wide, studs 12" O.C. G		47	.340		10.10	16		26.10	37
4310	16" O.C. G		68	.235		8	11.05		19.05	27
4320	24" O.C. G		94	.170		5.90	8		13.90	19.60
4330	3-5/8" wide, studs 12" O.C. G		46	.348		12.05	16.35		28.40	40.50
4340	16" O.C. G		66	.242		9.55	11.40		20.95	29
4350	24" O.C. G		92	.174		7.05	8.15		15.20	21
4360	4" wide, studs 12" O.C. G		45	.356		12.65	16.70		29.35	41.50
4370	16" O.C. G		65	.246		10	11.55		21.55	30
4380	24" O.C. G		90	.178		7.40	8.35		15.75	22
4390	6" wide, studs 12" O.C. G		44	.364		15.80	17.05		32.85	45.50
4400	16" O.C. G		64	.250		12.55	11.75		24.30	33
4410	24" O.C. G		88	.182		9.30	8.55		17.85	24.50
4420	8" wide, studs 12" O.C. G		43	.372		19.50	17.45		36.95	50
4430	16" O.C. G		63	.254		15.50	11.90		27.40	36.50
4440	24" O.C. G		86	.186		11.50	8.75		20.25	27
5100	10' high walls, 18 ga. x 2-1/2" wide, studs 12" O.C. G		54	.296		10.20	13.90		24.10	34
5110	16" O.C. G		77	.208		8.05	9.75		17.80	25
5120	24" O.C. G		107	.150		5.95	7		12.95	18.05
5130	3-5/8" wide, studs 12" O.C. G		53	.302		12.05	14.15		26.20	36.50
5140	16" O.C. G		76	.211		9.55	9.90		19.45	26.50
5150	24" O.C. G		105	.152		7.05	7.15		14.20	19.50
5160	4" wide, studs 12" O.C. G		52	.308		12.65	14.45		27.10	37.50
5170	16" O.C. G		74	.216		10.05	10.15		20.20	27.50
5180	24" O.C. G		103	.155		7.40	7.30		14.70	20
5190	6" wide, studs 12" O.C. G		51	.314		16	14.75		30.75	41.50
5200	16" O.C. G		73	.219		12.70	10.30		23	31
5210	24" O.C. G		101	.158		9.40	7.45		16.85	22.50
5220	8" wide, studs 12" O.C. G		50	.320		19.50	15		34.50	46
5230	16" O.C. G		72	.222		15.50	10.45		25.95	34
5240	24" O.C. G		100	.160		11.50	7.50		19	25
5300	16 ga. x 2-1/2" wide, studs 12" O.C. G		47	.340		12.20	16		28.20	39.50
5310	16" O.C. G		68	.235		9.55	11.05		20.60	28.50
5320	24" O.C. G		94	.170		6.95	8		14.95	21
5330	3-5/8" wide, studs 12" O.C. G		46	.348		14.55	16.35		30.90	43

05 41 13.30 Framing, Stud Walls			Crew	Daily Output	Labor-Hours	Unit	Material	2015 Bare Costs Labor	Equipment	Total	Total Incl O&P
5340	16″ O.C.	G	2 Carp	66	.242	L.F.	11.40	11.40		22.80	31
5350	24″ O.C.	G		92	.174		8.30	8.15		16.45	22.50
5360	4″ wide, studs 12″ O.C.	G		45	.356		15.25	16.70		31.95	44.50
5370	16″ O.C.	G		65	.246		12	11.55		23.55	32
5380	24″ O.C.	G		90	.178		8.70	8.35		17.05	23.50
5390	6″ wide, studs 12″ O.C.	G		44	.364		19	17.05		36.05	49
5400	16″ O.C.	G		64	.250		14.95	11.75		26.70	35.50
5410	24″ O.C.	G		88	.182		10.90	8.55		19.45	26
5420	8″ wide, studs 12″ O.C.	G		43	.372		23.50	17.45		40.95	54.50
5430	16″ O.C.	G		63	.254		18.50	11.90		30.40	40
5440	24″ O.C.	G		86	.186		13.50	8.75		22.25	29
6190	12′ high walls, 18 ga. x 6″ wide, studs 12″ O.C.	G		41	.390		18.65	18.30		36.95	50.50
6200	16″ O.C.	G		58	.276		14.70	12.95		27.65	37
6210	24″ O.C.	G		81	.198		10.75	9.25		20	27
6220	8″ wide, studs 12″ O.C.	G		40	.400		22.50	18.80		41.30	56
6230	16″ O.C.	G		57	.281		17.90	13.20		31.10	41
6240	24″ O.C.	G		80	.200		13.10	9.40		22.50	30
6390	16 ga. x 6″ wide, studs 12″ O.C.	G		35	.457		22.50	21.50		44	59.50
6400	16″ O.C.	G		51	.314		17.40	14.75		32.15	43
6410	24″ O.C.	G		70	.229		12.55	10.75		23.30	31.50
6420	8″ wide, studs 12″ O.C.	G		34	.471		27.50	22		49.50	66.50
6430	16″ O.C.	G		50	.320		21.50	15		36.50	48
6440	24″ O.C.	G		69	.232		15.50	10.90		26.40	35
6530	14 ga. x 3-5/8″ wide, studs 12″ O.C.	G		34	.471		21	22		43	59.50
6540	16″ O.C.	G		48	.333		16.55	15.65		32.20	44
6550	24″ O.C.	G		65	.246		11.90	11.55		23.45	32
6560	4″ wide, studs 12″ O.C.	G		33	.485		22.50	23		45.50	62
6570	16″ O.C.	G		47	.340		17.55	16		33.55	45.50
6580	24″ O.C.	G		64	.250		12.65	11.75		24.40	33
6730	12 ga. x 3-5/8″ wide, studs 12″ O.C.	G		31	.516		29.50	24		53.50	72
6740	16″ O.C.	G		43	.372		23	17.45		40.45	53.50
6750	24″ O.C.	G		59	.271		16.05	12.75		28.80	38.50
6760	4″ wide, studs 12″ O.C.	G		30	.533		31.50	25		56.50	75.50
6770	16″ O.C.	G		42	.381		24.50	17.90		42.40	56
6780	24″ O.C.	G		58	.276		17.15	12.95		30.10	40
7390	16′ high walls, 16 ga. x 6″ wide, studs 12″ O.C.	G		33	.485		28.50	23		51.50	69
7400	16″ O.C.	G		48	.333		22.50	15.65		38.15	50
7410	24″ O.C.	G		67	.239		15.80	11.20		27	36
7420	8″ wide, studs 12″ O.C.	G		32	.500		35.50	23.50		59	77.50
7430	16″ O.C.	G		47	.340		27.50	16		43.50	56.50
7440	24″ O.C.	G		66	.242		19.50	11.40		30.90	40
7560	14 ga. x 4″ wide, studs 12″ O.C.	G		31	.516		29	24		53	71.50
7570	16″ O.C.	G		45	.356		22.50	16.70		39.20	52
7580	24″ O.C.	G		61	.262		15.90	12.30		28.20	37.50
7590	6″ wide, studs 12″ O.C.	G		30	.533		36.50	25		61.50	81
7600	16″ O.C.	G		44	.364		28.50	17.05		45.55	59
7610	24″ O.C.	G		60	.267		20	12.50		32.50	42.50
7760	12 ga. x 4″ wide, studs 12″ O.C.	G		29	.552		41	26		67	87.50
7770	16″ O.C.	G		40	.400		31.50	18.80		50.30	65.50
7780	24″ O.C.	G		55	.291		22	13.65		35.65	46.50
7790	6″ wide, studs 12″ O.C.	G		28	.571		51.50	27		78.50	101
7800	16″ O.C.	G		39	.410		39.50	19.25		58.75	75
7810	24″ O.C.	G		54	.296		27.50	13.90		41.40	53.50

05 41 Structural Metal Stud Framing

05 41 13 – Load-Bearing Metal Stud Framing

05 41 13.30 Framing, Stud Walls			Crew	Daily Output	Labor-Hours	Unit	Material	2015 Bare Costs Labor	Equipment	Total	Total Incl O&P
8590	20′ high walls, 14 ga. x 6″ wide, studs 12″ O.C.	G	2 Carp	29	.552	L.F.	45	26		71	91.50
8600	16″ O.C.	G		42	.381		34.50	17.90		52.40	67.50
8610	24″ O.C.	G		57	.281		24	13.20		37.20	48
8620	8″ wide, studs 12″ O.C.	G		28	.571		48.50	27		75.50	97.50
8630	16″ O.C.	G		41	.390		37.50	18.30		55.80	71
8640	24″ O.C.	G		56	.286		26.50	13.40		39.90	51
8790	12 ga. x 6″ wide, studs 12″ O.C.	G		27	.593		64	28		92	116
8800	16″ O.C.	G		37	.432		48.50	20.50		69	87
8810	24″ O.C.	G		51	.314		33.50	14.75		48.25	61
8820	8″ wide, studs 12″ O.C.	G		26	.615		77.50	29		106.50	133
8830	16″ O.C.	G		36	.444		59	21		80	99
8840	24″ O.C.	G		50	.320		41	15		56	69.50
9000	Minimum labor/equipment charge			4	4	Job		188		188	310

05 42 Cold-Formed Metal Joist Framing

05 42 13 – Cold-Formed Metal Floor Joist Framing

05 42 13.05 Bracing

			Crew	Daily Output	Labor-Hours	Unit	Material	Labor	Equipment	Total	Total Incl O&P
0010	**BRACING**, continuous, per row, top & bottom										
0015	Made from recycled materials										
0120	Flat strap, 20 ga. x 2″ wide, joists at 12″ O.C.	G	1 Carp	4.67	1.713	C.L.F.	55	80.50		135.50	193
0130	16″ O.C.	G		5.33	1.501		53.50	70.50		124	175
0140	24″ O.C.	G		6.66	1.201		51.50	56.50		108	149
0150	18 ga. x 2″ wide, joists at 12″ O.C.	G		4	2		74	94		168	236
0160	16″ O.C.	G		4.67	1.713		73	80.50		153.50	212
0170	24″ O.C.	G		5.33	1.501		71.50	70.50		142	195

05 42 13.10 Bridging

			Crew	Daily Output	Labor-Hours	Unit	Material	Labor	Equipment	Total	Total Incl O&P
0010	**BRIDGING**, solid between joists w/1-1/4″ leg track, per joist bay										
0015	Made from recycled materials										
0230	Joists 12″ O.C., 18 ga. track x 6″ wide	G	1 Carp	80	.100	Ea.	1.48	4.70		6.18	9.35
0240	8″ wide	G		75	.107		1.84	5		6.84	10.20
0250	10″ wide	G		70	.114		2.29	5.35		7.64	11.30
0260	12″ wide	G		65	.123		2.60	5.80		8.40	12.30
0330	16 ga. track x 6″ wide	G		70	.114		1.87	5.35		7.22	10.85
0340	8″ wide	G		65	.123		2.34	5.80		8.14	12.05
0350	10″ wide	G		60	.133		2.91	6.25		9.16	13.45
0360	12″ wide	G		55	.145		3.35	6.85		10.20	14.90
0440	14 ga. track x 8″ wide	G		60	.133		2.93	6.25		9.18	13.50
0450	10″ wide	G		55	.145		3.64	6.85		10.49	15.20
0460	12″ wide	G		50	.160		4.20	7.50		11.70	16.90
0550	12 ga. track x 10″ wide	G		45	.178		5.35	8.35		13.70	19.55
0560	12″ wide	G		40	.200		5.45	9.40		14.85	21.50
1230	16″ O.C., 18 ga. track x 6″ wide	G		80	.100		1.90	4.70		6.60	9.80
1240	8″ wide	G		75	.107		2.36	5		7.36	10.80
1250	10″ wide	G		70	.114		2.94	5.35		8.29	12.05
1260	12″ wide	G		65	.123		3.33	5.80		9.13	13.10
1330	16 ga. track x 6″ wide	G		70	.114		2.39	5.35		7.74	11.45
1340	8″ wide	G		65	.123		3	5.80		8.80	12.75
1350	10″ wide	G		60	.133		3.73	6.25		9.98	14.35
1360	12″ wide	G		55	.145		4.29	6.85		11.14	15.90
1440	14 ga. track x 8″ wide	G		60	.133		3.76	6.25		10.01	14.40
1450	10″ wide	G		55	.145		4.67	6.85		11.52	16.35

05 42 13.10 Bridging			Crew	Daily Output	Labor-Hours	Unit	2015 Bare Costs Material	Labor	Equipment	Total	Total Incl O&P
1460	12" wide	G	1 Carp	50	.160	Ea.	5.40	7.50		12.90	18.20
1550	12 ga. track x 10" wide	G		45	.178		6.85	8.35		15.20	21
1560	12" wide	G		40	.200		7	9.40		16.40	23
2230	24" O.C., 18 ga. track x 6" wide	G		80	.100		2.75	4.70		7.45	10.70
2240	8" wide	G		75	.107		3.41	5		8.41	11.95
2250	10" wide	G		70	.114		4.25	5.35		9.60	13.45
2260	12" wide	G		65	.123		4.82	5.80		10.62	14.75
2330	16 ga. track x 6" wide	G		70	.114		3.46	5.35		8.81	12.60
2340	8" wide	G		65	.123		4.34	5.80		10.14	14.25
2350	10" wide	G		60	.133		5.40	6.25		11.65	16.20
2360	12" wide	G		55	.145		6.20	6.85		13.05	18.05
2440	14 ga. track x 8" wide	G		60	.133		5.45	6.25		11.70	16.25
2450	10" wide	G		55	.145		6.75	6.85		13.60	18.65
2460	12" wide	G		50	.160		7.80	7.50		15.30	21
2550	12 ga. track x 10" wide	G		45	.178		9.90	8.35		18.25	24.50
2560	12" wide	G	↓	40	.200	↓	10.10	9.40		19.50	26.50

05 42 13.25 Framing, Band Joist

			Crew	Daily Output	Labor-Hours	Unit	Material	Labor	Equipment	Total	Total Incl O&P
0010	**FRAMING, BAND JOIST** (track) fastened to bearing wall										
0015	Made from recycled materials										
0220	18 ga. track x 6" deep	G	2 Carp	1000	.016	L.F.	1.21	.75		1.96	2.56
0230	8" deep	G		920	.017		1.50	.82		2.32	2.99
0240	10" deep	G		860	.019		1.87	.87		2.74	3.49
0320	16 ga. track x 6" deep	G		900	.018		1.52	.83		2.35	3.04
0330	8" deep	G		840	.019		1.91	.89		2.80	3.57
0340	10" deep	G		780	.021		2.37	.96		3.33	4.19
0350	12" deep	G		740	.022		2.73	1.02		3.75	4.66
0430	14 ga. track x 8" deep	G		750	.021		2.39	1		3.39	4.27
0440	10" deep	G		720	.022		2.97	1.04		4.01	4.98
0450	12" deep	G		700	.023		3.42	1.07		4.49	5.55
0540	12 ga. track x 10" deep	G		670	.024		4.35	1.12		5.47	6.60
0550	12" deep	G	↓	650	.025	↓	4.45	1.16		5.61	6.80

05 42 13.30 Framing, Boxed Headers/Beams

			Crew	Daily Output	Labor-Hours	Unit	Material	Labor	Equipment	Total	Total Incl O&P
0010	**FRAMING, BOXED HEADERS/BEAMS**										
0015	Made from recycled materials										
0200	Double, 18 ga. x 6" deep	G	2 Carp	220	.073	L.F.	5.10	3.41		8.51	11.20
0210	8" deep	G		210	.076		5.65	3.58		9.23	12.05
0220	10" deep	G		200	.080		6.90	3.76		10.66	13.75
0230	12" deep	G		190	.084		7.55	3.95		11.50	14.80
0300	16 ga. x 8" deep	G		180	.089		6.50	4.17		10.67	14
0310	10" deep	G		170	.094		7.90	4.42		12.32	15.90
0320	12" deep	G		160	.100		8.60	4.70		13.30	17.15
0400	14 ga. x 10" deep	G		140	.114		9.10	5.35		14.45	18.80
0410	12" deep	G		130	.123		10	5.80		15.80	20.50
0500	12 ga. x 10" deep	G		110	.145		12	6.85		18.85	24.50
0510	12" deep	G		100	.160		13.25	7.50		20.75	27
1210	Triple, 18 ga. x 8" deep	G		170	.094		8.20	4.42		12.62	16.25
1220	10" deep	G		165	.097		9.85	4.55		14.40	18.30
1230	12" deep	G		160	.100		10.85	4.70		15.55	19.60
1300	16 ga. x 8" deep	G		145	.110		9.50	5.20		14.70	18.90
1310	10" deep	G		140	.114		11.35	5.35		16.70	21.50
1320	12" deep	G		135	.119		12.40	5.55		17.95	22.50
1400	14 ga. x 10" deep	G	↓	115	.139	↓	13.15	6.55		19.70	25

05 42 Cold-Formed Metal Joist Framing

05 42 13 – Cold-Formed Metal Floor Joist Framing

05 42 13.30 Framing, Boxed Headers/Beams

	05 42 13.30 Framing, Boxed Headers/Beams		Crew	Daily Output	Labor-Hours	Unit	2015 Bare Costs Material	Labor	Equipment	Total	Total Incl O&P
1410	12" deep	G	2 Carp	110	.145	L.F.	14.50	6.85		21.35	27
1500	12 ga. x 10" deep	G		90	.178		17.50	8.35		25.85	33
1510	12" deep	G	↓	85	.188	↓	19.40	8.85		28.25	36

05 42 13.40 Framing, Joists

	05 42 13.40 Framing, Joists		Crew	Daily Output	Labor-Hours	Unit	Material	Labor	Equipment	Total	Total Incl O&P
0010	**FRAMING, JOISTS**, no band joists (track), web stiffeners, headers,										
0020	Beams, bridging or bracing										
0025	Made from recycled materials										
0030	Joists (2" flange) and fasteners, materials only										
0220	18 ga. x 6" deep	G				L.F.	1.58			1.58	1.73
0230	8" deep	G					1.86			1.86	2.04
0240	10" deep	G					2.18			2.18	2.40
0320	16 ga. x 6" deep	G					1.93			1.93	2.13
0330	8" deep	G					2.31			2.31	2.54
0340	10" deep	G					2.70			2.70	2.97
0350	12" deep	G					3.07			3.07	3.37
0430	14 ga. x 8" deep	G					2.90			2.90	3.19
0440	10" deep	G					3.34			3.34	3.67
0450	12" deep	G					3.80			3.80	4.18
0540	12 ga. x 10" deep	G					4.86			4.86	5.35
0550	12" deep	G				↓	5.50			5.50	6.10
1010	Installation of joists to band joists, beams & headers, labor only										
1220	18 ga. x 6" deep		2 Carp	110	.145	Ea.		6.85		6.85	11.20
1230	8" deep			90	.178			8.35		8.35	13.70
1240	10" deep			80	.200			9.40		9.40	15.40
1320	16 ga. x 6" deep			95	.168			7.90		7.90	12.95
1330	8" deep			70	.229			10.75		10.75	17.60
1340	10" deep			60	.267			12.50		12.50	20.50
1350	12" deep			55	.291			13.65		13.65	22.50
1430	14 ga. x 8" deep			65	.246			11.55		11.55	18.95
1440	10" deep			45	.356			16.70		16.70	27.50
1450	12" deep			35	.457			21.50		21.50	35
1540	12 ga. x 10" deep			40	.400			18.80		18.80	31
1550	12" deep			30	.533	↓		25		25	41
9000	Minimum labor/equipment charge		↓	4	4	Job		188		188	310

05 42 13.45 Framing, Web Stiffeners

	05 42 13.45 Framing, Web Stiffeners		Crew	Daily Output	Labor-Hours	Unit	Material	Labor	Equipment	Total	Total Incl O&P
0010	**FRAMING, WEB STIFFENERS** at joist bearing, fabricated from										
0020	Stud piece (1-5/8" flange) to stiffen joist (2" flange)										
0025	Made from recycled materials										
2120	For 6" deep joist, with 18 ga. x 2-1/2" stud	G	1 Carp	120	.067	Ea.	.85	3.13		3.98	6.10
2130	3-5/8" stud	G		110	.073		1	3.41		4.41	6.70
2140	4" stud	G		105	.076		1.05	3.58		4.63	7
2150	6" stud	G		100	.080		1.32	3.76		5.08	7.60
2160	8" stud	G		95	.084		1.60	3.95		5.55	8.25
2220	8" deep joist, with 2-1/2" stud	G		120	.067		1.14	3.13		4.27	6.40
2230	3-5/8" stud	G		110	.073		1.34	3.41		4.75	7.05
2240	4" stud	G		105	.076		1.41	3.58		4.99	7.40
2250	6" stud	G		100	.080		1.77	3.76		5.53	8.10
2260	8" stud	G		95	.084		2.14	3.95		6.09	8.85
2320	10" deep joist, with 2-1/2" stud	G		110	.073		1.41	3.41		4.82	7.15
2330	3-5/8" stud	G		100	.080		1.66	3.76		5.42	8
2340	4" stud	G		95	.084		1.74	3.95		5.69	8.40
2350	6" stud	G	↓	90	.089	↓	2.19	4.17		6.36	9.25

05 42 13 – Cold-Formed Metal Floor Joist Framing

05 42 13.45 Framing, Web Stiffeners			Crew	Daily Output	Labor-Hours	Unit	2015 Bare Costs Material	Labor	Equipment	Total	Total Incl O&P
2360	8″ stud	G	1 Carp	85	.094	Ea.	2.66	4.42		7.08	10.15
2420	12″ deep joist, with 2-1/2″ stud	G		110	.073		1.70	3.41		5.11	7.45
2430	3-5/8″ stud	G		100	.080		2	3.76		5.76	8.35
2440	4″ stud	G		95	.084		2.10	3.95		6.05	8.80
2450	6″ stud	G		90	.089		2.64	4.17		6.81	9.75
2460	8″ stud	G		85	.094		3.20	4.42		7.62	10.75
3130	For 6″ deep joist, with 16 ga. x 3-5/8″ stud	G		100	.080		1.25	3.76		5.01	7.55
3140	4″ stud	G		95	.084		1.31	3.95		5.26	7.95
3150	6″ stud	G		90	.089		1.62	4.17		5.79	8.65
3160	8″ stud	G		85	.094		2	4.42		6.42	9.45
3230	8″ deep joist, with 3-5/8″ stud	G		100	.080		1.68	3.76		5.44	8
3240	4″ stud	G		95	.084		1.76	3.95		5.71	8.45
3250	6″ stud	G		90	.089		2.17	4.17		6.34	9.25
3260	8″ stud	G		85	.094		2.68	4.42		7.10	10.20
3330	10″ deep joist, with 3-5/8″ stud	G		85	.094		2.08	4.42		6.50	9.55
3340	4″ stud	G		80	.100		2.17	4.70		6.87	10.10
3350	6″ stud	G		75	.107		2.69	5		7.69	11.15
3360	8″ stud	G		70	.114		3.32	5.35		8.67	12.45
3430	12″ deep joist, with 3-5/8″ stud	G		85	.094		2.50	4.42		6.92	10
3440	4″ stud	G		80	.100		2.62	4.70		7.32	10.60
3450	6″ stud	G		75	.107		3.24	5		8.24	11.75
3460	8″ stud	G		70	.114		4	5.35		9.35	13.20
4230	For 8″ deep joist, with 14 ga. x 3-5/8″ stud	G		90	.089		2.08	4.17		6.25	9.15
4240	4″ stud	G		85	.094		2.20	4.42		6.62	9.65
4250	6″ stud	G		80	.100		2.76	4.70		7.46	10.75
4260	8″ stud	G		75	.107		2.95	5		7.95	11.45
4330	10″ deep joist, with 3-5/8″ stud	G		75	.107		2.57	5		7.57	11.05
4340	4″ stud	G		70	.114		2.72	5.35		8.07	11.80
4350	6″ stud	G		65	.123		3.42	5.80		9.22	13.20
4360	8″ stud	G		60	.133		3.65	6.25		9.90	14.25
4430	12″ deep joist, with 3-5/8″ stud	G		75	.107		3.10	5		8.10	11.60
4440	4″ stud	G		70	.114		3.28	5.35		8.63	12.40
4450	6″ stud	G		65	.123		4.12	5.80		9.92	14
4460	8″ stud	G		60	.133		4.40	6.25		10.65	15.10
5330	For 10″ deep joist, with 12 ga. x 3-5/8″ stud	G		65	.123		3.72	5.80		9.52	13.55
5340	4″ stud	G		60	.133		3.97	6.25		10.22	14.60
5350	6″ stud	G		55	.145		5	6.85		11.85	16.70
5360	8″ stud	G		50	.160		6.05	7.50		13.55	18.95
5430	12″ deep joist, with 3-5/8″ stud	G		65	.123		4.48	5.80		10.28	14.40
5440	4″ stud	G		60	.133		4.78	6.25		11.03	15.50
5450	6″ stud	G		55	.145		6	6.85		12.85	17.80
5460	8″ stud	G	↓	50	.160	↓	7.30	7.50		14.80	20.50

05 42 23 – Cold-Formed Metal Roof Joist Framing

05 42 23.05 Framing, Bracing			Crew	Daily Output	Labor-Hours	Unit	Material	Labor	Equipment	Total	Total Incl O&P
0010	**FRAMING, BRACING**										
0015	Made from recycled materials										
0020	Continuous bracing, per row										
0100	16 ga. x 1-1/2″ channel thru rafters/trusses @ 16″ O.C.	G	1 Carp	4.50	1.778	C.L.F.	48	83.50		131.50	190
0120	24″ O.C.	G		6	1.333		48	62.50		110.50	156
0300	2″ x 2″ angle x 18 ga., rafters/trusses @ 16″ O.C.	G		6	1.333		75	62.50		137.50	186
0320	24″ O.C.	G		8	1		75	47		122	160
0400	16 ga., rafters/trusses @ 16″ O.C.	G	↓	4.50	1.778	↓	94.50	83.50		178	241

05 42 Cold-Formed Metal Joist Framing

05 42 23 – Cold-Formed Metal Roof Joist Framing

05 42 23.05 Framing, Bracing

			Crew	Daily Output	Labor-Hours	Unit	2015 Bare Costs Material	Labor	Equipment	Total	Total Incl O&P
0420	24″ O.C.	G	1 Carp	6.50	1.231	C.L.F.	94.50	58		152.50	199

05 42 23.10 Framing, Bridging

			Crew	Daily Output	Labor-Hours	Unit	Material	Labor	Equipment	Total	Total Incl O&P
0010	**FRAMING, BRIDGING**										
0015	Made from recycled materials										
0020	Solid, between rafters w/1-1/4″ leg track, per rafter bay										
1200	Rafters 16″ O.C., 18 ga. x 4″ deep	G	1 Carp	60	.133	Ea.	1.45	6.25		7.70	11.85
1210	6″ deep	G		57	.140		1.90	6.60		8.50	12.90
1220	8″ deep	G		55	.145		2.36	6.85		9.21	13.80
1230	10″ deep	G		52	.154		2.94	7.20		10.14	15.10
1240	12″ deep	G		50	.160		3.33	7.50		10.83	15.95
2200	24″ O.C., 18 ga. x 4″ deep	G		60	.133		2.10	6.25		8.35	12.55
2210	6″ deep	G		57	.140		2.75	6.60		9.35	13.80
2220	8″ deep	G		55	.145		3.41	6.85		10.26	14.95
2230	10″ deep	G		52	.154		4.25	7.20		11.45	16.50
2240	12″ deep	G		50	.160		4.82	7.50		12.32	17.60

05 42 23.50 Framing, Parapets

			Crew	Daily Output	Labor-Hours	Unit	Material	Labor	Equipment	Total	Total Incl O&P
0010	**FRAMING, PARAPETS**										
0015	Made from recycled materials										
0100	3′ high installed on 1st story, 18 ga. x 4″ wide studs, 12″ O.C.	G	2 Carp	100	.160	L.F.	5.30	7.50		12.80	18.15
0110	16″ O.C.	G		150	.107		4.52	5		9.52	13.15
0120	24″ O.C.	G		200	.080		3.73	3.76		7.49	10.25
0200	6″ wide studs, 12″ O.C.	G		100	.160		6.80	7.50		14.30	19.75
0210	16″ O.C.	G		150	.107		5.80	5		10.80	14.55
0220	24″ O.C.	G		200	.080		4.80	3.76		8.56	11.45
1100	Installed on 2nd story, 18 ga. x 4″ wide studs, 12″ O.C.	G		95	.168		5.30	7.90		13.20	18.80
1110	16″ O.C.	G		145	.110		4.52	5.20		9.72	13.45
1120	24″ O.C.	G		190	.084		3.73	3.95		7.68	10.60
1200	6″ wide studs, 12″ O.C.	G		95	.168		6.80	7.90		14.70	20.50
1210	16″ O.C.	G		145	.110		5.80	5.20		11	14.85
1220	24″ O.C.	G		190	.084		4.80	3.95		8.75	11.80
2100	Installed on gable, 18 ga. x 4″ wide studs, 12″ O.C.	G		85	.188		5.30	8.85		14.15	20.50
2110	16″ O.C.	G		130	.123		4.52	5.80		10.32	14.40
2120	24″ O.C.	G		170	.094		3.73	4.42		8.15	11.35
2200	6″ wide studs, 12″ O.C.	G		85	.188		6.80	8.85		15.65	22
2210	16″ O.C.	G		130	.123		5.80	5.80		11.60	15.80
2220	24″ O.C.	G		170	.094		4.80	4.42		9.22	12.55

05 42 23.60 Framing, Roof Rafters

			Crew	Daily Output	Labor-Hours	Unit	Material	Labor	Equipment	Total	Total Incl O&P
0010	**FRAMING, ROOF RAFTERS**										
0015	Made from recycled materials										
0100	Boxed ridge beam, double, 18 ga. x 6″ deep	G	2 Carp	160	.100	L.F.	5.10	4.70		9.80	13.30
0110	8″ deep	G		150	.107		5.65	5		10.65	14.40
0120	10″ deep	G		140	.114		6.90	5.35		12.25	16.40
0130	12″ deep	G		130	.123		7.55	5.80		13.35	17.75
0200	16 ga. x 6″ deep	G		150	.107		5.80	5		10.80	14.55
0210	8″ deep	G		140	.114		6.50	5.35		11.85	15.95
0220	10″ deep	G		130	.123		7.90	5.80		13.70	18.10
0230	12″ deep	G		120	.133		8.60	6.25		14.85	19.70
1100	Rafters, 2″ flange, material only, 18 ga. x 6″ deep	G					1.58			1.58	1.73
1110	8″ deep	G					1.86			1.86	2.04
1120	10″ deep	G					2.18			2.18	2.40
1130	12″ deep	G					2.52			2.52	2.77
1200	16 ga. x 6″ deep	G					1.93			1.93	2.13

05 42 23 – Cold-Formed Metal Roof Joist Framing

05 42 23.60 Framing, Roof Rafters			Crew	Daily Output	Labor-Hours	Unit	Material	2015 Bare Costs Labor	Equipment	Total	Total Incl O&P
1210	8″ deep	G				L.F.	2.31			2.31	2.54
1220	10″ deep	G					2.70			2.70	2.97
1230	12″ deep	G				↓	3.07			3.07	3.37
2100	Installation only, ordinary rafter to 4:12 pitch, 18 ga. x 6″ deep		2 Carp	35	.457	Ea.		21.50		21.50	35
2110	8″ deep			30	.533			25		25	41
2120	10″ deep			25	.640			30		30	49.50
2130	12″ deep			20	.800			37.50		37.50	61.50
2200	16 ga. x 6″ deep			30	.533			25		25	41
2210	8″ deep			25	.640			30		30	49.50
2220	10″ deep			20	.800			37.50		37.50	61.50
2230	12″ deep		↓	15	1.067	↓		50		50	82
8100	Add to labor, ordinary rafters on steep roofs							25%			
8110	Dormers & complex roofs							50%			
8200	Hip & valley rafters to 4:12 pitch							25%			
8210	Steep roofs							50%			
8220	Dormers & complex roofs							75%			
8300	Hip & valley jack rafters to 4:12 pitch							50%			
8310	Steep roofs							75%			
8320	Dormers & complex roofs							100%			
9000	Minimum labor/equipment charge		2 Carp	4	4	Job		188		188	310

05 42 23.70 Framing, Soffits and Canopies			Crew	Daily Output	Labor-Hours	Unit	Material	Labor	Equipment	Total	Total Incl O&P
0010	**FRAMING, SOFFITS & CANOPIES**										
0015	Made from recycled materials										
0130	Continuous ledger track @ wall, studs @ 16″ O.C., 18 ga. x 4″ wide	G	2 Carp	535	.030	L.F.	.97	1.40		2.37	3.36
0140	6″ wide	G		500	.032		1.27	1.50		2.77	3.85
0150	8″ wide	G		465	.034		1.57	1.62		3.19	4.38
0160	10″ wide	G		430	.037		1.96	1.75		3.71	5
0230	Studs @ 24″ O.C., 18 ga. x 4″ wide	G		800	.020		.92	.94		1.86	2.56
0240	6″ wide	G		750	.021		1.21	1		2.21	2.97
0250	8″ wide	G		700	.023		1.50	1.07		2.57	3.41
0260	10″ wide	G	↓	650	.025	↓	1.87	1.16		3.03	3.95
1000	Horizontal soffit and canopy members, material only										
1030	1-5/8″ flange studs, 18 ga. x 4″ deep	G				L.F.	1.26			1.26	1.39
1040	6″ deep	G					1.58			1.58	1.74
1050	8″ deep	G					1.92			1.92	2.11
1140	2″ flange joists, 18 ga. x 6″ deep	G					1.80			1.80	1.98
1150	8″ deep	G					2.12			2.12	2.34
1160	10″ deep	G				↓	2.50			2.50	2.75
4030	Installation only, 18 ga., 1-5/8″ flange x 4″ deep		2 Carp	130	.123	Ea.		5.80		5.80	9.45
4040	6″ deep			110	.145			6.85		6.85	11.20
4050	8″ deep			90	.178			8.35		8.35	13.70
4140	2″ flange, 18 ga. x 6″ deep			110	.145			6.85		6.85	11.20
4150	8″ deep			90	.178			8.35		8.35	13.70
4160	10″ deep		↓	80	.200			9.40		9.40	15.40
6010	Clips to attach facia to rafter tails, 2″ x 2″ x 18 ga. angle	G	1 Carp	120	.067		.88	3.13		4.01	6.10
6020	16 ga. angle	G	″	100	.080	↓	1.12	3.76		4.88	7.40
9000	Minimum labor/equipment charge		2 Carp	4	4	Job		188		188	310

05 44 Cold-Formed Metal Trusses

05 44 13 – Cold-Formed Metal Roof Trusses

05 44 13.60 Framing, Roof Trusses		Crew	Daily Output	Labor-Hours	Unit	Material	Labor	Equipment	Total	Total Incl O&P
						2015 Bare Costs				
0010	FRAMING, ROOF TRUSSES									
0015	Made from recycled materials									
0020	Fabrication of trusses on ground, Fink (W) or King Post, to 4:12 pitch									
0120	18 ga. x 4" chords, 16' span G	2 Carp	12	1.333	Ea.	59	62.50		121.50	168
0130	20' span G		11	1.455		73.50	68.50		142	193
0140	24' span G		11	1.455		88	68.50		156.50	209
0150	28' span G		10	1.600		103	75		178	236
0160	32' span G		10	1.600		118	75		193	252
0250	6" chords, 28' span G		9	1.778		129	83.50		212.50	279
0260	32' span G		9	1.778		148	83.50		231.50	300
0270	36' span G		8	2		166	94		260	335
0280	40' span G		8	2		185	94		279	355
1120	5:12 to 8:12 pitch, 18 ga. x 4" chords, 16' span G		10	1.600		67	75		142	197
1130	20' span G		9	1.778		84	83.50		167.50	230
1140	24' span G		9	1.778		101	83.50		184.50	248
1150	28' span G		8	2		118	94		212	283
1160	32' span G		8	2		134	94		228	300
1250	6" chords, 28' span G		7	2.286		148	107		255	340
1260	32' span G		7	2.286		169	107		276	360
1270	36' span G		6	2.667		190	125		315	415
1280	40' span G		6	2.667		211	125		336	435
2120	9:12 to 12:12 pitch, 18 ga. x 4" chords, 16' span G		8	2		84	94		178	247
2130	20' span G		7	2.286		105	107		212	292
2140	24' span G		7	2.286		126	107		233	315
2150	28' span G		6	2.667		147	125		272	365
2160	32' span G		6	2.667		168	125		293	390
2250	6" chords, 28' span G		5	3.200		185	150		335	450
2260	32' span G		5	3.200		211	150		361	480
2270	36' span G		4	4		238	188		426	570
2280	40' span G		4	4	↓	264	188		452	600
4900	Minimum labor/equipment charge	↓	4	4	Job		188		188	310
5120	Erection only of roof trusses, to 4:12 pitch, 16' span	F-6	48	.833	Ea.		37	13.60	50.60	75
5130	20' span		46	.870			38.50	14.20	52.70	78
5140	24' span		44	.909			40	14.85	54.85	82
5150	28' span		42	.952			42	15.55	57.55	85.50
5160	32' span		40	1			44	16.35	60.35	90
5170	36' span		38	1.053			46.50	17.20	63.70	94.50
5180	40' span		36	1.111			49	18.15	67.15	100
5220	5:12 to 8:12 pitch, 16' span		42	.952			42	15.55	57.55	85.50
5230	20' span		40	1			44	16.35	60.35	90
5240	24' span		38	1.053			46.50	17.20	63.70	94.50
5250	28' span		36	1.111			49	18.15	67.15	100
5260	32' span		34	1.176			52	19.25	71.25	106
5270	36' span		32	1.250			55	20.50	75.50	113
5280	40' span		30	1.333			59	22	81	120
5320	9:12 to 12:12 pitch, 16' span		36	1.111			49	18.15	67.15	100
5330	20' span		34	1.176			52	19.25	71.25	106
5340	24' span		32	1.250			55	20.50	75.50	113
5350	28' span		30	1.333			59	22	81	120
5360	32' span		28	1.429			63	23.50	86.50	129
5370	36' span		26	1.538			68	25	93	139
5380	40' span		24	1.667	↓		73.50	27.50	101	150
9000	Minimum labor/equipment charge	↓	2	20	Job		885	325	1,210	1,775

05 51 Metal Stairs

05 51 13 – Metal Pan Stairs

05 51 13.50 Pan Stairs			Crew	Daily Output	Labor-Hours	Unit	2015 Bare Costs Material	Labor	Equipment	Total	Total Incl O&P
0010	**PAN STAIRS**, shop fabricated, steel stringers										
0015	Made from recycled materials										
0200	Cement fill metal pan, picket rail, 3'-6" wide	G	E-4	35	.914	Riser	500	48.50	4.17	552.67	645
0300	4'-0" wide	G		30	1.067		560	56.50	4.86	621.36	720
0350	Wall rail, both sides, 3'-6" wide	G		53	.604		380	32	2.75	414.75	480
1500	Landing, steel pan, conventional	G		160	.200	S.F.	66	10.65	.91	77.56	93
1600	Pre-erected	G		255	.125	"	118	6.65	.57	125.22	143
1700	Pre-erected, steel pan tread, 3'-6" wide, 2 line pipe rail	G	E-2	87	.644	Riser	550	33.50	17.35	600.85	680

05 51 16 – Metal Floor Plate Stairs

05 51 16.50 Floor Plate Stairs			Crew	Daily Output	Labor-Hours	Unit	Material	Labor	Equipment	Total	Total Incl O&P
0010	**FLOOR PLATE STAIRS**, shop fabricated, steel stringers										
0015	Made from recycled materials										
0400	Cast iron tread and pipe rail, 3'-6" wide	G	E-4	35	.914	Riser	535	48.50	4.17	587.67	680
0500	Checkered plate tread, industrial, 3'-6" wide	G		28	1.143		330	60.50	5.20	395.70	475
0550	Circular, for tanks, 3'-0" wide	G		33	.970		370	51.50	4.42	425.92	505
0600	For isolated stairs, add							100%			
0800	Custom steel stairs, 3'-6" wide, economy	G	E-4	35	.914		500	48.50	4.17	552.67	645
0810	Medium priced	G		30	1.067		660	56.50	4.86	721.36	835
0900	Deluxe	G		20	1.600		825	85	7.30	917.30	1,075
1100	For 4' wide stairs, add						5%	5%			
1300	For 5' wide stairs, add						10%	10%			

05 51 19 – Metal Grating Stairs

05 51 19.50 Grating Stairs			Crew	Daily Output	Labor-Hours	Unit	Material	Labor	Equipment	Total	Total Incl O&P
0010	**GRATING STAIRS**, shop fabricated, steel stringers, safety nosing on treads										
0015	Made from recycled materials										
0020	Grating tread and pipe railing, 3'-6" wide	G	E-4	35	.914	Riser	330	48.50	4.17	382.67	455
0100	4'-0" wide	G		30	1.067	"	430	56.50	4.86	491.36	580
9000	Minimum labor/equipment charge			2	16	Job		850	73	923	1,600

05 51 23 – Metal Fire Escapes

05 51 23.25 Fire Escapes			Crew	Daily Output	Labor-Hours	Unit	Material	Labor	Equipment	Total	Total Incl O&P
0010	**FIRE ESCAPES**, shop fabricated										
0200	2' wide balcony, 1" x 1/4" bars 1-1/2" O.C., with railing	G	2 Sswk	10	1.600	L.F.	58.50	84		142.50	217
0400	1st story cantilevered stair, standard, with railing	G		.50	32	Ea.	2,425	1,675		4,100	5,725
0500	Cable counterweighted, with railing	G		.40	40	"	2,250	2,100		4,350	6,275
0700	36" x 40" platform & fixed stair, with railing	G		.40	40	Flight	1,075	2,100		3,175	4,975
0900	For 3'-6" wide escapes, add to above						100%	150%			

05 51 23.50 Fire Escape Stairs			Crew	Daily Output	Labor-Hours	Unit	Material	Labor	Equipment	Total	Total Incl O&P
0010	**FIRE ESCAPE STAIRS**, portable										
0100	Portable ladder					Ea.	112			112	123

05 51 33 – Metal Ladders

05 51 33.13 Vertical Metal Ladders			Crew	Daily Output	Labor-Hours	Unit	Material	Labor	Equipment	Total	Total Incl O&P
0010	**VERTICAL METAL LADDERS**, shop fabricated										
0015	Made from recycled materials										
0020	Steel, 20" wide, bolted to concrete, with cage	G	E-4	50	.640	V.L.F.	64	34	2.92	100.92	135
0100	Without cage	G		85	.376		38	20	1.72	59.72	80
0300	Aluminum, bolted to concrete, with cage	G		50	.640		120	34	2.92	156.92	197
0400	Without cage	G		85	.376		51	20	1.72	72.72	94
9000	Minimum labor/equipment charge			2	16	Job		850	73	923	1,600

05 51 Metal Stairs

05 51 33 – Metal Ladders

05 51 33.16 Inclined Metal Ladders		Crew	Daily Output	Labor-Hours	Unit	Material	2015 Bare Costs Labor	Equipment	Total	Total Incl O&P
0010	**INCLINED METAL LADDERS**, shop fabricated									
0015	Made from recycled materials									
3900	Industrial ships ladder, steel, 24" W, grating treads, 2 line pipe rail G	E-4	30	1.067	Riser	193	56.50	4.86	254.36	320
4000	Aluminum G	"	30	1.067	"	270	56.50	4.86	331.36	405

05 51 33.23 Alternating Tread Ladders		Crew	Daily Output	Labor-Hours	Unit	Material	Labor	Equipment	Total	Total Incl O&P
0010	**ALTERNATING TREAD LADDERS**, shop fabricated									
0015	Made from recycled materials									
0800	Alternating tread ladders, 68-degree angle of incline									
0810	8 foot vertical rise, steel, 149 lb., standard paint color	B-68G	3	5.333	Ea.	2,275	281	111	2,667	3,125
0820	Non-standard paint color		3	5.333		2,625	281	111	3,017	3,525
0830	Galvanized		3	5.333		2,650	281	111	3,042	3,525
0840	Stainless		3	5.333		3,850	281	111	4,242	4,850
0850	Aluminum, 87 lb.		3	5.333		2,800	281	111	3,192	3,700
1010	10 foot vertical rise, steel, 181 lb., standard paint color		2.75	5.818		2,750	305	121	3,176	3,725
1020	Non-standard paint color		2.75	5.818		3,150	305	121	3,576	4,175
1030	Galvanized		2.75	5.818		3,200	305	121	3,626	4,225
1040	Stainless		2.75	5.818		4,625	305	121	5,051	5,775
1050	Aluminum, 103 lb.		2.75	5.818		3,375	305	121	3,801	4,425
1210	12 foot vertical rise, steel, 245 lb., standard paint color		2.50	6.400		3,225	335	133	3,693	4,300
1220	Non-standard paint color		2.50	6.400		3,650	335	133	4,118	4,775
1230	Galvanized		2.50	6.400		3,750	335	133	4,218	4,875
1240	Stainless		2.50	6.400		5,400	335	133	5,868	6,700
1250	Aluminum, 103 lb.		2.50	6.400		5,200	335	133	5,668	6,475
1410	14 foot vertical rise, steel, 281 lb., standard paint color		2.25	7.111		3,700	375	147	4,222	4,900
1420	Non-standard paint color		2.25	7.111		4,175	375	147	4,697	5,425
1430	Galvanized		2.25	7.111		4,325	375	147	4,847	5,575
1440	Stainless		2.25	7.111		6,200	375	147	6,722	7,625
1450	Aluminum, 136 lb.		2.25	7.111		4,525	375	147	5,047	5,800
1610	16 foot vertical rise, steel, 317 lb., standard paint color		2	8		4,175	420	166	4,761	5,550
1620	Non-standard paint color		2	8		4,700	420	166	5,286	6,100
1630	Galvanized		2	8		4,875	420	166	5,461	6,300
1640	Stainless		2	8		6,975	420	166	7,561	8,625
1650	Aluminum, 153 lb.	↓	2	8	↓	5,100	420	166	5,686	6,575

05 52 Metal Railings

05 52 13 – Pipe and Tube Railings

05 52 13.50 Railings, Pipe		Crew	Daily Output	Labor-Hours	Unit	Material	Labor	Equipment	Total	Total Incl O&P
0010	**RAILINGS, PIPE**, shop fab'd, 3'-6" high, posts @ 5' O.C.									
0015	Made from recycled materials									
0020	Aluminum, 2 rail, satin finish, 1-1/4" diameter G	E-4	160	.200	L.F.	36	10.65	.91	47.56	59.50
0030	Clear anodized G		160	.200		44	10.65	.91	55.56	68.50
0040	Dark anodized G		160	.200		49	10.65	.91	60.56	74
0080	1-1/2" diameter, satin finish G		160	.200		42.50	10.65	.91	54.06	66.50
0090	Clear anodized G		160	.200		47.50	10.65	.91	59.06	72
0100	Dark anodized G		160	.200		52.50	10.65	.91	64.06	77.50
0140	Aluminum, 3 rail, 1-1/4" diam., satin finish G		137	.234		54.50	12.40	1.07	67.97	83
0150	Clear anodized G		137	.234		67.50	12.40	1.07	80.97	98
0160	Dark anodized G		137	.234		75	12.40	1.07	88.47	106
0200	1-1/2" diameter, satin finish G		137	.234		64.50	12.40	1.07	77.97	94.50
0210	Clear anodized G	↓	137	.234	↓	74	12.40	1.07	87.47	105

05 52 Metal Railings

05 52 13 – Pipe and Tube Railings

05 52 13.50 Railings, Pipe			Crew	Daily Output	Labor-Hours	Unit	2015 Bare Costs Material	Labor	Equipment	Total	Total Incl O&P
0220	Dark anodized	G	E-4	137	.234	L.F.	80.50	12.40	1.07	93.97	112
0500	Steel, 2 rail, on stairs, primed, 1-1/4" diameter	G		160	.200		25	10.65	.91	36.56	47.50
0520	1-1/2" diameter	G		160	.200		27	10.65	.91	38.56	50
0540	Galvanized, 1-1/4" diameter	G		160	.200		34	10.65	.91	45.56	57.50
0560	1-1/2" diameter	G		160	.200		38.50	10.65	.91	50.06	62
0580	Steel, 3 rail, primed, 1-1/4" diameter	G		137	.234		37.50	12.40	1.07	50.97	64.50
0600	1-1/2" diameter	G		137	.234		39	12.40	1.07	52.47	66.50
0620	Galvanized, 1-1/4" diameter	G		137	.234		52.50	12.40	1.07	65.97	81
0640	1-1/2" diameter	G		137	.234		60.50	12.40	1.07	73.97	90
0700	Stainless steel, 2 rail, 1-1/4" diam. #4 finish	G		137	.234		110	12.40	1.07	123.47	145
0720	High polish	G		137	.234		178	12.40	1.07	191.47	220
0740	Mirror polish	G		137	.234		223	12.40	1.07	236.47	269
0760	Stainless steel, 3 rail, 1-1/2" diam., #4 finish	G		120	.267		166	14.15	1.22	181.37	210
0770	High polish	G		120	.267		275	14.15	1.22	290.37	330
0780	Mirror finish	G		120	.267		335	14.15	1.22	350.37	395
0900	Wall rail, alum. pipe, 1-1/4" diam., satin finish	G		213	.150		20	8	.69	28.69	37
0905	Clear anodized	G		213	.150		25	8	.69	33.69	42.50
0910	Dark anodized	G		213	.150		29.50	8	.69	38.19	47.50
0915	1-1/2" diameter, satin finish	G		213	.150		22.50	8	.69	31.19	40
0920	Clear anodized	G		213	.150		28	8	.69	36.69	46
0925	Dark anodized	G		213	.150		35	8	.69	43.69	53.50
0930	Steel pipe, 1-1/4" diameter, primed	G		213	.150		15	8	.69	23.69	31.50
0935	Galvanized	G		213	.150		22	8	.69	30.69	39
0940	1-1/2" diameter	G		176	.182		15.50	9.65	.83	25.98	35.50
0945	Galvanized	G		213	.150		22	8	.69	30.69	39
0955	Stainless steel pipe, 1-1/2" diam., #4 finish	G		107	.299		88	15.90	1.36	105.26	127
0960	High polish	G		107	.299		179	15.90	1.36	196.26	227
0965	Mirror polish	G		107	.299		212	15.90	1.36	229.26	263
2000	2-line pipe rail (1-1/2" T&B) with 1/2" pickets @ 4-1/2" O.C.,										
2005	attached handrail on brackets										
2010	42" high aluminum, satin finish, straight & level	G	E-4	120	.267	L.F.	198	14.15	1.22	213.37	245
2050	42" high steel, primed, straight & level	G	"	120	.267		127	14.15	1.22	142.37	167
4000	For curved and level rails, add						10%	10%			
4100	For sloped rails for stairs, add						30%	30%			
9000	Minimum labor/equipment charge		1 Sswk	2	4	Job		211		211	380

05 52 16 – Industrial Railings

05 52 16.50 Railings, Industrial			Crew	Daily Output	Labor-Hours	Unit	Material	Labor	Equipment	Total	Total Incl O&P
0010	**RAILINGS, INDUSTRIAL**, shop fab'd, 3'-6" high, posts @ 5' O.C.										
0020	2 rail, 3'-6" high, 1-1/2" pipe	G	E-4	255	.125	L.F.	27	6.65	.57	34.22	42.50
0100	2" angle rail	G	"	255	.125		25	6.65	.57	32.22	40
0200	For 4" high kick plate, 10 ga., add	G					5.65			5.65	6.20
0300	1/4" thick, add	G					7.40			7.40	8.15
0500	For curved level rails, add						10%	10%			
0550	For sloped rails for stairs, add						30%	30%			
9000	Minimum labor/equipment charge		1 Sswk	2	4	Job		211		211	380

05 53 13.10 Floor Grating, Aluminum

Line	Description		Crew	Daily Output	Labor-Hours	Unit	Material (2015 Bare Costs)	Labor (2015 Bare Costs)	Equipment (2015 Bare Costs)	Total (2015 Bare Costs)	Total Incl O&P
0010	**FLOOR GRATING, ALUMINUM**, field fabricated from panels										
0015	Made from recycled materials										
0110	Bearing bars @ 1-3/16" O.C., cross bars @ 4" O.C.,										
0111	Up to 300 S.F., 1" x 1/8" bar	G	E-4	900	.036	S.F.	17.35	1.89	.16	19.40	22.50
0112	Over 300 S.F.	G		850	.038		15.80	2	.17	17.97	21
0113	1-1/4" x 1/8" bar, up to 300 S.F.	G		800	.040		18.05	2.13	.18	20.36	24
0114	Over 300 S.F.	G		1000	.032		16.40	1.70	.15	18.25	21.50
0122	1-1/4" x 3/16" bar, up to 300 S.F.	G		750	.043		27.50	2.27	.19	29.96	35
0124	Over 300 S.F.	G		1000	.032		25	1.70	.15	26.85	30.50
0132	1-1/2" x 3/16" bar, up to 300 S.F.	G		700	.046		32	2.43	.21	34.64	39.50
0134	Over 300 S.F.	G		1000	.032		29	1.70	.15	30.85	35
0136	1-3/4" x 3/16" bar, up to 300 S.F.	G		500	.064		35.50	3.40	.29	39.19	45.50
0138	Over 300 S.F.	G		1000	.032		32.50	1.70	.15	34.35	38.50
0146	2-1/4" x 3/16" bar, up to 300 S.F.	G		600	.053		45	2.83	.24	48.07	55
0148	Over 300 S.F.	G		1000	.032		41	1.70	.15	42.85	48
0162	Cross bars @ 2" O.C., 1" x 1/8", up to 300 S.F.	G		600	.053		30.50	2.83	.24	33.57	39
0164	Over 300 S.F.	G		1000	.032		27.50	1.70	.15	29.35	33.50
0172	1-1/4" x 3/16" bar, up to 300 S.F.	G		600	.053		49.50	2.83	.24	52.57	60
0174	Over 300 S.F.	G		1000	.032		45	1.70	.15	46.85	52.50
0182	1-1/2" x 3/16" bar, up to 300 S.F.	G		600	.053		57	2.83	.24	60.07	68.50
0184	Over 300 S.F.	G		1000	.032		52	1.70	.15	53.85	60
0186	1-3/4" x 3/16" bar, up to 300 S.F.	G		600	.053		62	2.83	.24	65.07	73.50
0188	Over 300 S.F.	G		1000	.032		56	1.70	.15	57.85	65
0200	For straight cuts, add					L.F.	4.30			4.30	4.73
0212	Bearing bars @ 15/16" O.C., 1" x 1/8", up to 300 S.F.	G	E-4	520	.062	S.F.	29.50	3.27	.28	33.05	38.50
0214	Over 300 S.F.	G		920	.035		27	1.85	.16	29.01	33
0222	1-1/4" x 3/16", up to 300 S.F.	G		520	.062		49	3.27	.28	52.55	59.50
0224	Over 300 S.F.	G		920	.035		44.50	1.85	.16	46.51	52.50
0232	1-1/2" x 3/16", up to 300 S.F.	G		520	.062		56.50	3.27	.28	60.05	68
0234	Over 300 S.F.	G		920	.035		51	1.85	.16	53.01	60
0300	For curved cuts, add					L.F.	5.30			5.30	5.85
0400	For straight banding, add	G					5.50			5.50	6.05
0500	For curved banding, add	G					6.60			6.60	7.25
0600	For aluminum checkered plate nosings, add	G					7.15			7.15	7.85
0700	For straight toe plate, add	G					10.80			10.80	11.90
0800	For curved toe plate, add	G					12.60			12.60	13.85
1000	For cast aluminum abrasive nosings, add	G					10.50			10.50	11.55
1200	Expanded aluminum, .65# per S.F.	G	E-4	1050	.030	S.F.	12.45	1.62	.14	14.21	16.80
1400	Extruded I bars are 10% less than 3/16" bars										
1600	Heavy duty, all extruded plank, 3/4" deep, 1.8# per S.F.	G	E-4	1100	.029	S.F.	26.50	1.55	.13	28.18	32
1700	1-1/4" deep, 2.9# per S.F.	G		1000	.032		29	1.70	.15	30.85	35
1800	1-3/4" deep, 4.2# per S.F.	G		925	.035		40.50	1.84	.16	42.50	48
1900	2-1/4" deep, 5.0# per S.F.	G		875	.037		60.50	1.94	.17	62.61	70
2100	For safety serrated surface, add						15%				

05 53 13.70 Floor Grating, Steel

Line	Description		Crew	Daily Output	Labor-Hours	Unit	Material (2015 Bare Costs)	Labor (2015 Bare Costs)	Equipment (2015 Bare Costs)	Total (2015 Bare Costs)	Total Incl O&P
0010	**FLOOR GRATING, STEEL**, field fabricated from panels										
0015	Made from recycled materials										
0300	Platforms, to 12' high, rectangular	G	E-4	3150	.010	Lb.	3.21	.54	.05	3.80	4.56
0400	Circular	G	"	2300	.014	"	4.01	.74	.06	4.81	5.80
0410	Painted bearing bars @ 1-3/16"										
0412	Cross bars @ 4" O.C., 3/4" x 1/8" bar, up to 300 S.F.	G	E-2	500	.112	S.F.	8.25	5.80	3.02	17.07	22.50
0414	Over 300 S.F.	G		750	.075		7.50	3.86	2.01	13.37	17.20

05 53 13 – Bar Gratings

05 53 13.70 Floor Grating, Steel			Crew	Daily Output	Labor-Hours	Unit	Material	Labor	Equipment	Total	Total Incl O&P
							2015 Bare Costs				
0422	1-1/4" x 3/16", up to 300 S.F.	G	E-2	400	.140	S.F.	12.75	7.25	3.78	23.78	31
0424	Over 300 S.F.	G		600	.093		11.60	4.83	2.52	18.95	24
0432	1-1/2" x 3/16", up to 300 S.F.	G		400	.140		14.85	7.25	3.78	25.88	33
0434	Over 300 S.F.	G		600	.093		13.50	4.83	2.52	20.85	26
0436	1-3/4" x 3/16", up to 300 S.F.	G		400	.140		18.50	7.25	3.78	29.53	37.50
0438	Over 300 S.F.	G		600	.093		16.80	4.83	2.52	24.15	29.50
0452	2-1/4" x 3/16", up to 300 S.F.	G		300	.187		23	9.65	5.05	37.70	48
0454	Over 300 S.F.	G		450	.124		21	6.45	3.36	30.81	38
0462	Cross bars @ 2" O.C., 3/4" x 1/8", up to 300 S.F.	G		500	.112		15.15	5.80	3.02	23.97	30
0464	Over 300 S.F.	G		750	.075		12.60	3.86	2.01	18.47	23
0472	1-1/4" x 3/16", up to 300 S.F.	G		400	.140		19.80	7.25	3.78	30.83	39
0474	Over 300 S.F.	G		600	.093		16.50	4.83	2.52	23.85	29.50
0482	1-1/2" x 3/16", up to 300 S.F.	G		400	.140		22.50	7.25	3.78	33.53	41.50
0484	Over 300 S.F.	G		600	.093		18.60	4.83	2.52	25.95	31.50
0486	1-3/4" x 3/16", up to 300 S.F.	G		400	.140		33.50	7.25	3.78	44.53	54
0488	Over 300 S.F.	G		600	.093		28	4.83	2.52	35.35	42
0502	2-1/4" x 3/16", up to 300 S.F.	G		300	.187		33	9.65	5.05	47.70	58.50
0504	Over 300 S.F.	G		450	.124		27.50	6.45	3.36	37.31	45
0601	Painted bearing bars @ 15/16" O.C., cross bars @ 4" O.C.,										
0612	Up to 300 S.F., 3/4" x 3/16" bars	G	E-4	850	.038	S.F.	12.20	2	.17	14.37	17.20
0622	1-1/4" x 3/16" bars	G		600	.053		16	2.83	.24	19.07	23
0632	1-1/2" x 3/16" bars	G		550	.058		17.95	3.09	.27	21.31	25.50
0636	1-3/4" x 3/16" bars	G		450	.071		24.50	3.78	.32	28.60	34
0652	2-1/4" x 3/16" bars	G	E-2	300	.187		27.50	9.65	5.05	42.20	52.50
0662	Cross bars @ 2" O.C., up to 300 S.F., 3/4" x 3/16"	G		500	.112		15.80	5.80	3.02	24.62	31
0672	1-1/4" x 3/16" bars	G		400	.140		20	7.25	3.78	31.03	39
0682	1-1/2" x 3/16" bars	G		400	.140		23	7.25	3.78	34.03	42
0686	1-3/4" x 3/16" bars	G		300	.187		29	9.65	5.05	43.70	54
0690	For galvanized grating, add						25%				
0800	For straight cuts, add					L.F.	6.15			6.15	6.75
0900	For curved cuts, add						7.80			7.80	8.60
1000	For straight banding, add	G					6.60			6.60	7.25
1100	For curved banding, add	G					8.70			8.70	9.55
1200	For checkered plate nosings, add	G					7.65			7.65	8.40
1300	For straight toe or kick plate, add	G					13.70			13.70	15.05
1400	For curved toe or kick plate, add	G					15.50			15.50	17.05
1500	For abrasive nosings, add	G					10.15			10.15	11.15
1510	For stair treads, see Section 05 55 13.50										
1600	For safety serrated surface, bearing bars @ 1-3/16" O.C., add						15%				
1700	Bearing bars @ 15/16" O.C., add						25%				
2000	Stainless steel gratings, close spaced, 1" x 1/8" bars, up to 300 S.F.	G	E-4	450	.071	S.F.	73	3.78	.32	77.10	87.50
2100	Standard spacing, 3/4" x 1/8" bars	G		500	.064		84	3.40	.29	87.69	98.50
2200	1-1/4" x 3/16" bars	G		400	.080		80	4.25	.36	84.61	96
2400	Expanded steel grating, at ground, 3.0# per S.F.	G		900	.036		8.05	1.89	.16	10.10	12.50
2500	3.14# per S.F.	G		900	.036		7.15	1.89	.16	9.20	11.50
2600	4.0# per S.F.	G		850	.038		8.60	2	.17	10.77	13.30
2650	4.27# per S.F.	G		850	.038		9.30	2	.17	11.47	14.05
2700	5.0# per S.F.	G		800	.040		13.80	2.13	.18	16.11	19.20
2800	6.25# per S.F.	G		750	.043		17.80	2.27	.19	20.26	24
2900	7.0# per S.F.	G		700	.046		19.65	2.43	.21	22.29	26
3100	For flattened expanded steel grating, add						8%				
3300	For elevated installation above 15', add							15%			

05 53 Metal Gratings

05 53 16 – Plank Gratings

05 53 16.50 Grating Planks			Crew	Daily Output	Labor-Hours	Unit	Material	2015 Bare Costs Labor	Equipment	Total	Total Incl O&P
0010	**GRATING PLANKS**, field fabricated from planks										
0020	Aluminum, 9-1/2" wide, 14 ga., 2" rib	G	E-4	950	.034	L.F.	25.50	1.79	.15	27.44	31.50
0200	Galvanized steel, 9-1/2" wide, 14 ga., 2-1/2" rib	G		950	.034		16.80	1.79	.15	18.74	22
0300	4" rib	G		950	.034		18.50	1.79	.15	20.44	24
0500	12 ga., 2-1/2" rib	G		950	.034		16.90	1.79	.15	18.84	22
0600	3" rib	G		950	.034		22.50	1.79	.15	24.44	28
0800	Stainless steel, type 304, 16 ga., 2" rib	G		950	.034		37	1.79	.15	38.94	44
0900	Type 316	G		950	.034		57	1.79	.15	58.94	66.50

05 53 19 – Floor Grating Frame

05 53 19.30 Grating Frame			Crew	Daily Output	Labor-Hours	Unit	Material	Labor	Equipment	Total	Total Incl O&P
0010	**GRATING FRAME**, field fabricated										
0020	Aluminum, for gratings 1" to 1-1/2" deep	G	1 Sswk	70	.114	L.F.	3.78	6		9.78	15
0100	For each corner, add	G				Ea.	5.65			5.65	6.25

05 54 Metal Floor Plates

05 54 13 – Floor Plates

05 54 13.20 Checkered Plates			Crew	Daily Output	Labor-Hours	Unit	Material	Labor	Equipment	Total	Total Incl O&P
0010	**CHECKERED PLATES**, steel, field fabricated										
0015	Made from recycled materials										
0020	1/4" & 3/8", 2000 to 5000 S.F., bolted	G	E-4	2900	.011	Lb.	.88	.59	.05	1.52	2.09
0100	Welded	G		4400	.007	"	.84	.39	.03	1.26	1.66
0300	Pit or trench cover and frame, 1/4" plate, 2' to 3' wide	G		100	.320	S.F.	11.35	17	1.46	29.81	44.50
0400	For galvanizing, add	G				Lb.	.29			.29	.32
0500	Platforms, 1/4" plate, no handrails included, rectangular	G	E-4	4200	.008		3.21	.41	.03	3.65	4.30
0600	Circular	G	"	2500	.013		4.01	.68	.06	4.75	5.70

05 54 13.70 Trench Covers			Crew	Daily Output	Labor-Hours	Unit	Material	Labor	Equipment	Total	Total Incl O&P
0010	**TRENCH COVERS**, field fabricated										
0020	Cast iron grating with bar stops and angle frame, to 18" wide	G	1 Sswk	20	.400	L.F.	213	21		234	272
0100	Frame only (both sides of trench), 1" grating	G		45	.178		1.37	9.35		10.72	18.40
0150	2" grating	G		35	.229		3.23	12.05		15.28	25.50
0200	Aluminum, stock units, including frames and										
0210	3/8" plain cover plate, 4" opening	G	E-4	205	.156	L.F.	21	8.30	.71	30.01	39
0300	6" opening	G		185	.173		25.50	9.20	.79	35.49	45.50
0400	10" opening	G		170	.188		34.50	10	.86	45.36	57
0500	16" opening	G		155	.206		48	10.95	.94	59.89	74
0700	Add per inch for additional widths to 24"	G					2			2	2.20
0900	For custom fabrication, add						50%				
1100	For 1/4" plain cover plate, deduct						12%				
1500	For cover recessed for tile, 1/4" thick, deduct						12%				
1600	3/8" thick, add						5%				
1800	For checkered plate cover, 1/4" thick, deduct						12%				
1900	3/8" thick, add						2%				
2100	For slotted or round holes in cover, 1/4" thick, add						3%				
2200	3/8" thick, add						4%				
2300	For abrasive cover, add						12%				

05 55 Metal Stair Treads and Nosings

05 55 13 – Metal Stair Treads

05 55 13.50 Stair Treads			Crew	Daily Output	Labor-Hours	Unit	2015 Bare Costs Material	Labor	Equipment	Total	Total Incl O&P
0010	**STAIR TREADS**, stringers and bolts not included										
3000	Diamond plate treads, steel, 1/8" thick										
3005	Open riser, black enamel										
3010	9" deep x 36" long	G	2 Sswk	48	.333	Ea.	91.50	17.55		109.05	132
3020	42" long	G		48	.333		96.50	17.55		114.05	138
3030	48" long	G		48	.333		101	17.55		118.55	143
3040	11" deep x 36" long	G		44	.364		96.50	19.15		115.65	141
3050	42" long	G		44	.364		102	19.15		121.15	147
3060	48" long	G		44	.364		108	19.15		127.15	154
3110	Galvanized, 9" deep x 36" long	G		48	.333		144	17.55		161.55	191
3120	42" long	G		48	.333		154	17.55		171.55	201
3130	48" long	G		48	.333		163	17.55		180.55	211
3140	11" deep x 36" long	G		44	.364		149	19.15		168.15	199
3150	42" long	G		44	.364		163	19.15		182.15	214
3160	48" long	G		44	.364		169	19.15		188.15	221
3200	Closed riser, black enamel										
3210	12" deep x 36" long	G	2 Sswk	40	.400	Ea.	110	21		131	159
3220	42" long	G		40	.400		119	21		140	169
3230	48" long	G		40	.400		126	21		147	176
3240	Galvanized, 12" deep x 36" long	G		40	.400		173	21		194	229
3250	42" long	G		40	.400		189	21		210	246
3260	48" long	G		40	.400		198	21		219	256
4000	Bar grating treads										
4005	Steel, 1-1/4" x 3/16" bars, anti-skid nosing, black enamel										
4010	8-5/8" deep x 30" long	G	2 Sswk	48	.333	Ea.	54	17.55		71.55	91
4020	36" long	G		48	.333		63.50	17.55		81.05	101
4030	48" long	G		48	.333		97.50	17.55		115.05	139
4040	10-15/16" deep x 36" long	G		44	.364		70	19.15		89.15	112
4050	48" long	G		44	.364		100	19.15		119.15	145
4060	Galvanized, 8-5/8" deep x 30" long	G		48	.333		62	17.55		79.55	100
4070	36" long	G		48	.333		73.50	17.55		91.05	113
4080	48" long	G		48	.333		108	17.55		125.55	151
4090	10-15/16" deep x 36" long	G		44	.364		86.50	19.15		105.65	130
4100	48" long	G		44	.364		112	19.15		131.15	158
4200	Aluminum, 1-1/4" x 3/16" bars, serrated, with nosing										
4210	7-5/8" deep x 18" long	G	2 Sswk	52	.308	Ea.	50	16.20		66.20	84.50
4220	24" long	G		52	.308		59	16.20		75.20	94.50
4230	30" long	G		52	.308		68.50	16.20		84.70	105
4240	36" long	G		52	.308		177	16.20		193.20	225
4250	8-13/16" deep x 18" long	G		48	.333		67	17.55		84.55	106
4260	24" long	G		48	.333		99.50	17.55		117.05	141
4270	30" long	G		48	.333		115	17.55		132.55	158
4280	36" long	G		48	.333		194	17.55		211.55	246
4290	10" deep x 18" long	G		44	.364		130	19.15		149.15	178
4300	30" long	G		44	.364		173	19.15		192.15	225
4310	36" long	G		44	.364		210	19.15		229.15	266
5000	Channel grating treads										
5005	Steel, 14 ga., 2-12" thick, galvanized										
5010	9" deep x 36" long	G	2 Sswk	48	.333	Ea.	103	17.55		120.55	145
5020	48" long	G		48	.333	"	139	17.55		156.55	184
9000	Minimum labor/equipment charge			4	4	Job		211		211	380

05 55 Metal Stair Treads and Nosings

05 55 19 – Metal Stair Tread Covers

05 55 19.50 Stair Tread Covers for Renovation		Crew	Daily Output	Labor-Hours	Unit	Material	2015 Bare Costs Labor	Equipment	Total	Total Incl O&P
0010	**STAIR TREAD COVERS FOR RENOVATION**									
0205	Extruded tread cover with nosing, pre-drilled, includes screws									
0210	Aluminum with black abrasive strips, 9″ wide x 3′ long	1 Carp	24	.333	Ea.	107	15.65		122.65	144
0220	4′ long		22	.364		142	17.05		159.05	184
0230	5′ long		20	.400		177	18.80		195.80	225
0240	11″ wide x 3′ long		24	.333		139	15.65		154.65	179
0250	4′ long		22	.364		179	17.05		196.05	224
0260	5′ long		20	.400		230	18.80		248.80	283
0305	Black abrasive strips with yellow front strips									
0310	Aluminum, 9″ wide x 3′ long	1 Carp	24	.333	Ea.	116	15.65		131.65	153
0320	4′ long		22	.364		154	17.05		171.05	198
0330	5′ long		20	.400		196	18.80		214.80	247
0340	11″ wide x 3′ long		24	.333		149	15.65		164.65	190
0350	4′ long		22	.364		192	17.05		209.05	239
0360	5′ long		20	.400		247	18.80		265.80	305
0405	Black abrasive strips with photoluminescent front strips									
0410	Aluminum, 9″ wide x 3′ long	1 Carp	24	.333	Ea.	156	15.65		171.65	197
0420	4′ long		22	.364		172	17.05		189.05	217
0430	5′ long		20	.400		215	18.80		233.80	267
0440	11″ wide x 3′ long		24	.333		153	15.65		168.65	194
0450	4′ long		22	.364		204	17.05		221.05	252
0460	5′ long		20	.400		255	18.80		273.80	310
9000	Minimum labor/equipment charge		4	2	Job		94		94	154

05 56 Metal Castings

05 56 13 – Metal Construction Castings

05 56 13.50 Construction Castings		Crew	Daily Output	Labor-Hours	Unit	Material	Labor	Equipment	Total	Total Incl O&P
0010	**CONSTRUCTION CASTINGS**									
0020	Manhole covers and frames, see Section 33 44 13.13									
0100	Column bases, cast iron, 16″ x 16″, approx. 65 lb. G	E-4	46	.696	Ea.	142	37	3.17	182.17	226
0200	32″ x 32″, approx. 256 lb. G	″	23	1.391		520	74	6.35	600.35	715
0400	Cast aluminum for wood columns, 8″ x 8″ G	1 Carp	32	.250		41	11.75		52.75	65
0500	12″ x 12″ G	″	32	.250		68.50	11.75		80.25	95
0600	Miscellaneous C.I. castings, light sections, less than 150 lb. G	E-4	3200	.010	Lb.	8.95	.53	.05	9.53	10.85
1100	Heavy sections, more than 150 lb. G		4200	.008		4.66	.41	.03	5.10	5.90
1300	Special low volume items G		3200	.010		11.25	.53	.05	11.83	13.35
1500	For ductile iron, add					100%				

05 58 Formed Metal Fabrications

05 58 13 – Column Covers

05 58 13.05 Column Covers		Crew	Daily Output	Labor-Hours	Unit	Material	Labor	Equipment	Total	Total Incl O&P
0010	**COLUMN COVERS**									
0015	Made from recycled materials									
0020	Excludes structural steel, light ga. metal framing, misc. metals, sealants									
0100	Round covers, 2 halves with 2 vertical joints for backer rod and sealant									
0110	Up to 12′ high, no horizontal joints									
0120	12″ diameter, 0.125″ aluminum, anodized/painted finish G	2 Sswk	32	.500	V.L.F.	30.50	26.50		57	81
0130	Type 304 stainless steel, 16 gauge, #4 brushed finish G		32	.500		46.50	26.50		73	98.50
0140	Type 316 stainless steel, 16 gauge, #4 brushed finish G		32	.500		54	26.50		80.50	107

05 58 Formed Metal Fabrications

05 58 13 – Column Covers

05 58 13.05 Column Covers			Crew	Daily Output	Labor-Hours	Unit	Material	2015 Bare Costs Labor	Equipment	Total	Total Incl O&P
0150	18″ diameter, aluminum	G	2 Sswk	32	.500	V.L.F.	46	26.50		72.50	98
0160	Type 304 stainless steel	G		32	.500		69.50	26.50		96	124
0170	Type 316 stainless steel	G		32	.500		81	26.50		107.50	137
0180	24″ diameter, aluminum	G		32	.500		61	26.50		87.50	115
0190	Type 304 stainless steel	G		32	.500		93	26.50		119.50	150
0200	Type 316 stainless steel	G		32	.500		108	26.50		134.50	167
0210	30″ diameter, aluminum	G		30	.533		76.50	28		104.50	135
0220	Type 304 stainless steel	G		30	.533		116	28		144	179
0230	Type 316 stainless steel	G		30	.533		135	28		163	199
0240	36″ diameter, aluminum			30	.533		91.50	28		119.50	152
0250	Type 304 stainless steel	G		30	.533		139	28		167	204
0260	Type 316 stainless steel	G		30	.533		162	28		190	229
0400	Up to 24′ high, 2 stacked sections with 1 horizontal joint										
0410	18″ diameter, aluminum	G	2 Sswk	28	.571	V.L.F.	48	30		78	108
0450	Type 304 stainless steel	G		28	.571		73	30		103	135
0460	Type 316 stainless steel	G		28	.571		85	30		115	148
0470	24″ diameter, aluminum	G		28	.571		64	30		94	125
0480	Type 304 stainless steel	G		28	.571		97.50	30		127.50	162
0490	Type 316 stainless steel	G		28	.571		113	30		143	180
0500	30″ diameter, aluminum	G		24	.667		80	35		115	152
0510	Type 304 stainless steel	G		24	.667		122	35		157	198
0520	Type 316 stainless steel	G		24	.667		142	35		177	220
0530	36″ diameter, aluminum	G		24	.667		96.50	35		131.50	170
0540	Type 304 stainless steel	G		24	.667		146	35		181	225
0550	Type 316 stainless steel	G		24	.667		170	35		205	251

05 58 21 – Formed Chain

05 58 21.05 Alloy Steel Chain			Crew	Daily Output	Labor-Hours	Unit	Material	Labor	Equipment	Total	Total Incl O&P
0010	**ALLOY STEEL CHAIN**, Grade 80, for lifting										
0015	Self-colored, cut lengths, 1/4″	G	E-17	4	4	C.L.F.	765	215		980	1,225
0020	3/8″	G		2	8		980	430		1,410	1,850
0030	1/2″	G		1.20	13.333		1,550	715		2,265	3,025
0040	5/8″	G		.72	22.222		2,450	1,200		3,650	4,825
0050	3/4″	G	E-18	.48	83.333		3,300	4,375	1,975	9,650	13,600
0060	7/8″	G		.40	100		5,725	5,275	2,375	13,375	18,200
0070	1″	G		.35	114		7,975	6,025	2,700	16,700	22,400
0080	1-1/4″	G		.24	166		13,400	8,775	3,950	26,125	34,600
0110	Hook, Grade 80, Clevis slip, 1/4″	G				Ea.	26			26	28.50
0120	3/8″	G					30.50			30.50	33.50
0130	1/2″	G					49.50			49.50	54.50
0140	5/8″	G					73			73	80
0150	3/4″	G					101			101	111
0160	Hook, Grade 80, eye/sling w/hammerlock coupling, 15 Ton	G					340			340	375
0170	22 Ton	G					925			925	1,025
0180	37 Ton	G					2,825			2,825	3,100

05 58 23 – Formed Metal Guards

05 58 23.90 Window Guards			Crew	Daily Output	Labor-Hours	Unit	Material	Labor	Equipment	Total	Total Incl O&P
0010	**WINDOW GUARDS**, shop fabricated										
0015	Expanded metal, steel angle frame, permanent	G	E-4	350	.091	S.F.	23	4.86	.42	28.28	34.50
0025	Steel bars, 1/2″ x 1/2″, spaced 5″ O.C.	G	″	290	.110	″	15.90	5.85	.50	22.25	28.50
0030	Hinge mounted, add	G				Opng.	46			46	50.50
0040	Removable type, add	G				″	29			29	32
0050	For galvanized guards, add					S.F.	35%				

05 58 Formed Metal Fabrications

05 58 23 – Formed Metal Guards

05 58 23.90 Window Guards			Crew	Daily Output	Labor-Hours	Unit	2015 Bare Costs Material	Labor	Equipment	Total	Total Incl O&P
0070	For pivoted or projected type, add					S.F.	105%	40%			
0100	Mild steel, stock units, economy	G	E-4	405	.079		6.25	4.20	.36	10.81	14.90
0200	Deluxe	G		405	.079		12.70	4.20	.36	17.26	22
0400	Woven wire, stock units, 3/8″ channel frame, 3′ x 5′ opening	G		40	.800	Opng.	169	42.50	3.65	215.15	267
0500	4′ x 6′ opening	G		38	.842		270	45	3.84	318.84	380
0800	Basket guards for above, add	G					233			233	256
1000	Swinging guards for above, add	G					79.50			79.50	87.50
9000	Minimum labor/equipment charge		1 Sswk	2	4	Job		211		211	380

05 58 25 – Formed Lamp Posts

05 58 25.40 Lamp Posts			Crew	Daily Output	Labor-Hours	Unit	Material	Labor	Equipment	Total	Total Incl O&P
0010	**LAMP POSTS**										
0020	Aluminum, 7′ high, stock units, post only	G	1 Carp	16	.500	Ea.	82	23.50		105.50	129
0100	Mild steel, plain	G		16	.500	"	73	23.50		96.50	119
9000	Minimum labor/equipment charge			4	2	Job		94		94	154

05 71 Decorative Metal Stairs

05 71 13 – Fabricated Metal Spiral Stairs

05 71 13.50 Spiral Stairs			Crew	Daily Output	Labor-Hours	Unit	Material	Labor	Equipment	Total	Total Incl O&P
0010	**SPIRAL STAIRS**										
1805	Shop fabricated, custom ordered										
1810	Aluminum, 5′-0″ diameter, plain units	G	E-4	45	.711	Riser	575	38	3.24	616.24	705
1820	Fancy units	G		45	.711		1,100	38	3.24	1,141.24	1,275
1900	Cast iron, 4′-0″ diameter, plain units	G		45	.711		540	38	3.24	581.24	665
1920	Fancy Units	G		25	1.280		735	68	5.85	808.85	940
2000	Steel, industrial checkered plate, 4′ diameter	G		45	.711		540	38	3.24	581.24	665
2200	6′ diameter	G		40	.800		650	42.50	3.65	696.15	795
3100	Spiral stair kits, 12 stacking risers to fit exact floor height										
3110	Steel, flat metal treads, primed, 3′-6″ diameter	G	2 Carp	1.60	10	Flight	1,275	470		1,745	2,175
3120	4′-0″ diameter	G		1.45	11.034		1,450	520		1,970	2,450
3130	4′-6″ diameter	G		1.35	11.852		1,600	555		2,155	2,675
3140	5′-0″ diameter	G		1.25	12.800		1,750	600		2,350	2,900
3210	Galvanized, 3′-6″ diameter	G		1.60	10		1,525	470		1,995	2,450
3220	4′-0″ diameter	G		1.45	11.034		1,750	520		2,270	2,775
3230	4′-6″ diameter	G		1.35	11.852		1,925	555		2,480	3,025
3240	5′-0″ diameter	G		1.25	12.800		2,100	600		2,700	3,275
3310	Checkered plate tread, primed, 3′-6″ diameter	G		1.45	11.034		1,500	520		2,020	2,500
3320	4′-0″ diameter	G		1.35	11.852		1,700	555		2,255	2,775
3330	4′-6″ diameter	G		1.25	12.800		1,850	600		2,450	3,025
3340	5′-0″ diameter	G		1.15	13.913		2,025	655		2,680	3,300
3410	Galvanized, 3′-6″ diameter	G		1.45	11.034		1,800	520		2,320	2,850
3420	4′-0″ diameter	G		1.35	11.852		2,050	555		2,605	3,150
3430	4′-6″ diameter	G		1.25	12.800		2,225	600		2,825	3,425
3440	5′-0″ diameter	G		1.15	13.913		2,425	655		3,080	3,750
3510	Red oak covers on flat metal treads, 3′-6″ diameter			1.35	11.852		2,600	555		3,155	3,775
3520	4′-0″ diameter			1.25	12.800		2,875	600		3,475	4,150
3530	4′-6″ diameter			1.15	13.913		3,100	655		3,755	4,475
3540	5′-0″ diameter			1.05	15.238		3,375	715		4,090	4,900

05 73 Decorative Metal Railings

05 73 16 – Wire Rope Decorative Metal Railings

05 73 16.10 Cable Railings			Crew	Daily Output	Labor-Hours	Unit	Material	2015 Bare Costs Labor	Equipment	Total	Total Incl O&P
0010	**CABLE RAILINGS**, with 316 stainless steel 1 x 19 cable, 3/16" diameter										
0015	Made from recycled materials										
0100	1-3/4" diameter stainless steel posts x 42" high, cables 4" OC	G	2 Sswk	25	.640	L.F.	33	33.50		66.50	97.50

05 73 23 – Ornamental Railings

05 73 23.50 Railings, Ornamental			Crew	Daily Output	Labor-Hours	Unit	Material	Labor	Equipment	Total	Total Incl O&P
0010	**RAILINGS, ORNAMENTAL**, 3'-6" high, posts @ 6' O.C.										
0020	Bronze or stainless, hand forged, plain	G	2 Sswk	24	.667	L.F.	122	35		157	198
0100	Fancy	G		18	.889		223	47		270	330
0200	Aluminum, panelized, plain	G		24	.667		13.60	35		48.60	78.50
0300	Fancy	G		18	.889		30	47		77	118
0400	Wrought iron, hand forged, plain	G		24	.667		81	35		116	153
0500	Fancy	G		18	.889		159	47		206	260
0550	Steel, panelized, plain	G		24	.667		20.50	35		55.50	86
0560	Fancy	G		18	.889		32	47		79	120
0600	Composite metal/wood/glass, plain			18	.889		119	47		166	216
0700	Fancy			12	1.333		238	70		308	390
9000	Minimum labor/equipment charge		1 Sswk	2	4	Job		211		211	380

05 75 Decorative Formed Metal

05 75 13 – Columns

05 75 13.10 Aluminum Columns			Crew	Daily Output	Labor-Hours	Unit	Material	Labor	Equipment	Total	Total Incl O&P
0010	**ALUMINUM COLUMNS**										
0015	Made from recycled materials										
0020	Aluminum, extruded, stock units, no cap or base, 6" diameter	G	E-4	240	.133	L.F.	10.50	7.10	.61	18.21	25
0100	8" diameter	G		170	.188		13.85	10	.86	24.71	34
0200	10" diameter	G		150	.213		18.95	11.35	.97	31.27	42.50
0300	12" diameter	G		140	.229		32.50	12.15	1.04	45.69	59
0400	15" diameter	G		120	.267		48	14.15	1.22	63.37	79.50
0410	Caps and bases, plain, 6" diameter	G				Set	23.50			23.50	26
0420	8" diameter	G					30			30	33
0430	10" diameter	G					42			42	46.50
0440	12" diameter	G					68.50			68.50	75
0450	15" diameter	G					110			110	121
0460	Caps, ornamental, plain	G					340			340	370
0470	Fancy	G					1,675			1,675	1,850
0500	For square columns, add to column prices above					L.F.	50%				
0700	Residential, flat, 8' high, plain	G	E-4	20	1.600	Ea.	97.50	85	7.30	189.80	269
0720	Fancy	G		20	1.600		190	85	7.30	282.30	370
0740	Corner type, plain	G		20	1.600		168	85	7.30	260.30	345
0760	Fancy	G		20	1.600		330	85	7.30	422.30	525

05 75 13.20 Columns, Ornamental			Crew	Daily Output	Labor-Hours	Unit	Material	Labor	Equipment	Total	Total Incl O&P
0010	**COLUMNS, ORNAMENTAL**, shop fabricated										
6400	Mild steel, flat, 9" wide, stock units, painted, plain	G	E-4	160	.200	V.L.F.	9.15	10.65	.91	20.71	30.50
6450	Fancy	G		160	.200		17.75	10.65	.91	29.31	40
6500	Corner columns, painted, plain	G		160	.200		15.75	10.65	.91	27.31	37.50
6550	Fancy	G		160	.200		31	10.65	.91	42.56	54.50

Estimating Tips

06 05 00 Common Work Results for Wood, Plastics, and Composites

- Common to any wood-framed structure are the accessory connector items such as screws, nails, adhesives, hangers, connector plates, straps, angles, and hold-downs. For typical wood-framed buildings, such as residential projects, the aggregate total for these items can be significant, especially in areas where seismic loading is a concern. For floor and wall framing, the material cost is based on 10 to 25 lbs. per MBF. Hold-downs, hangers, and other connectors should be taken off by the piece.

  Included with material costs are fasteners for a normal installation. RSMeans engineers use manufacturer's recommendations, written specifications, and/or standard construction practice for size and spacing of fasteners. Prices for various fasteners are shown for informational purposes only. Adjustments should be made if unusual fastening conditions exist.

06 10 00 Carpentry

- Lumber is a traded commodity and therefore sensitive to supply and demand in the marketplace. Even in "budgetary" estimating of wood-framed projects, it is advisable to call local suppliers for the latest market pricing.
- Common quantity units for wood-framed projects are "thousand board feet" (MBF). A board foot is a volume of wood, 1" x 1' x 1', or 144 cubic inches. Board-foot quantities are generally calculated using nominal material dimensions— dressed sizes are ignored. Board foot per lineal foot of any stick of lumber can be calculated by dividing the nominal cross-sectional area by 12. As an example, 2,000 lineal feet of 2 x 12 equates to 4 MBF by dividing the nominal area, 2 x 12, by 12, which equals 2, and multiplying by 2,000 to give 4,000 board feet. This simple rule applies to all nominal dimensioned lumber.
- Waste is an issue of concern at the quantity takeoff for any area of construction. Framing lumber is sold in even foot lengths, i.e., 10', 12', 14', 16' and, depending on spans, wall heights, and the grade of lumber, waste is inevitable. A rule of thumb for lumber waste is 5%–10% depending on material quality and the complexity of the framing.
- Wood in various forms and shapes is used in many projects, even where the main structural framing is steel, concrete, or masonry. Plywood as a back-up partition material and 2x boards used as blocking and cant strips around roof edges are two common examples. The estimator should ensure that the costs of all wood materials are included in the final estimate.

06 20 00 Finish Carpentry

- It is necessary to consider the grade of workmanship when estimating labor costs for erecting millwork and interior finish. In practice, there are three grades: premium, custom, and economy. The RSMeans daily output for base and case moldings is in the range of 200 to 250 L.F. per carpenter per day. This is appropriate for most average custom-grade projects. For premium projects, an adjustment to productivity of 25%–50% should be made, depending on the complexity of the job.

Reference Numbers

Reference numbers are shown in shaded boxes at the beginning of some major classifications. These numbers refer to related items in the Reference Section. The reference information may be an estimating procedure, an alternate pricing method, or technical information.

Note: Not all subdivisions listed here necessarily appear in this publication. ■

06 05 05 – Selective Demolition for Wood, Plastics, and Composites

06 05 05.10 Selective Demolition Wood Framing		Crew	Daily Output	Labor-Hours	Unit	2015 Bare Costs Material	Labor	Equipment	Total	Total Incl O&P
0010	**SELECTIVE DEMOLITION WOOD FRAMING** R024119-10									
0100	Timber connector, nailed, small	1 Clab	96	.083	Ea.		3.13		3.13	5.15
0110	Medium		60	.133			5		5	8.20
0120	Large		48	.167			6.25		6.25	10.30
0130	Bolted, small		48	.167			6.25		6.25	10.30
0140	Medium		32	.250			9.40		9.40	15.40
0150	Large		24	.333			12.55		12.55	20.50
2958	Beams, 2" x 6"	2 Clab	1100	.015	L.F.		.55		.55	.90
2960	2" x 8"		825	.019			.73		.73	1.20
2965	2" x 10"		665	.024			.90		.90	1.48
2970	2" x 12"		550	.029			1.09		1.09	1.79
2972	2" x 14"		470	.034			1.28		1.28	2.10
2975	4" x 8"	B-1	413	.058			2.22		2.22	3.65
2980	4" x 10"		330	.073			2.78		2.78	4.56
2985	4" x 12"		275	.087			3.34		3.34	5.45
3000	6" x 8"		275	.087			3.34		3.34	5.45
3040	6" x 10"		220	.109			4.17		4.17	6.85
3080	6" x 12"		185	.130			4.96		4.96	8.15
3120	8" x 12"		140	.171			6.55		6.55	10.75
3160	10" x 12"		110	.218			8.35		8.35	13.70
3162	Alternate pricing method		1.10	21.818	M.B.F.		835		835	1,375
3170	Blocking, in 16" OC wall framing, 2" x 4"	1 Clab	600	.013	L.F.		.50		.50	.82
3172	2" x 6"		400	.020			.75		.75	1.23
3174	In 24" OC wall framing, 2" x 4"		600	.013			.50		.50	.82
3176	2" x 6"		400	.020			.75		.75	1.23
3178	Alt method, wood blocking removal from wood framing		.40	20	M.B.F.		750		750	1,225
3179	Wood blocking removal from steel framing		.36	22.222	"		835		835	1,375
3180	Bracing, let in, 1" x 3", studs 16" OC		1050	.008	L.F.		.29		.29	.47
3181	Studs 24" OC		1080	.007			.28		.28	.46
3182	1" x 4", studs 16" OC		1050	.008			.29		.29	.47
3183	Studs 24" OC		1080	.007			.28		.28	.46
3184	1" x 6", studs 16" OC		1050	.008			.29		.29	.47
3185	Studs 24" OC		1080	.007			.28		.28	.46
3186	2" x 3", studs 16" OC		800	.010			.38		.38	.62
3187	Studs 24" OC		830	.010			.36		.36	.59
3188	2" x 4", studs 16" OC		800	.010			.38		.38	.62
3189	Studs 24" OC		830	.010			.36		.36	.59
3190	2" x 6", studs 16" OC		800	.010			.38		.38	.62
3191	Studs 24" OC		830	.010			.36		.36	.59
3192	2" x 8", studs 16" OC		800	.010			.38		.38	.62
3193	Studs 24" OC		830	.010			.36		.36	.59
3194	"T" shaped metal bracing, studs at 16" OC		1060	.008			.28		.28	.47
3195	Studs at 24" OC		1200	.007			.25		.25	.41
3196	Metal straps, studs at 16" OC		1200	.007			.25		.25	.41
3197	Studs at 24" OC		1240	.006			.24		.24	.40
3200	Columns, round, 8' to 14' tall		40	.200	Ea.		7.50		7.50	12.35
3202	Dimensional lumber sizes	2 Clab	1.10	14.545	M.B.F.		545		545	895
3250	Blocking, between joists	1 Clab	320	.025	Ea.		.94		.94	1.54
3252	Bridging, metal strap, between joists		320	.025	Pr.		.94		.94	1.54
3254	Wood, between joists		320	.025	"		.94		.94	1.54
3260	Door buck, studs, header & access., 8' high 2" x 4" wall, 3' wide		32	.250	Ea.		9.40		9.40	15.40
3261	4' wide		32	.250			9.40		9.40	15.40
3262	5' wide		32	.250			9.40		9.40	15.40

06 05 05 – Selective Demolition for Wood, Plastics, and Composites

06 05 05.10 Selective Demolition Wood Framing		Crew	Daily Output	Labor-Hours	Unit	2015 Bare Costs Material	Labor	Equipment	Total	Total Incl O&P
3263	6′ wide	1 Clab	32	.250	Ea.		9.40		9.40	15.40
3264	8′ wide		30	.267			10.05		10.05	16.45
3265	10′ wide		30	.267			10.05		10.05	16.45
3266	12′ wide		30	.267			10.05		10.05	16.45
3267	2″ x 6″ wall, 3′ wide		32	.250			9.40		9.40	15.40
3268	4′ wide		32	.250			9.40		9.40	15.40
3269	5′ wide		32	.250			9.40		9.40	15.40
3270	6′ wide		32	.250			9.40		9.40	15.40
3271	8′ wide		30	.267			10.05		10.05	16.45
3272	10′ wide		30	.267			10.05		10.05	16.45
3273	12′ wide		30	.267			10.05		10.05	16.45
3274	Window buck, studs, header & access, 8′ high 2″ x 4″ wall, 2′ wide		24	.333			12.55		12.55	20.50
3275	3′ wide		24	.333			12.55		12.55	20.50
3276	4′ wide		24	.333			12.55		12.55	20.50
3277	5′ wide		24	.333			12.55		12.55	20.50
3278	6′ wide		24	.333			12.55		12.55	20.50
3279	7′ wide		24	.333			12.55		12.55	20.50
3280	8′ wide		22	.364			13.65		13.65	22.50
3281	10′ wide		22	.364			13.65		13.65	22.50
3282	12′ wide		22	.364			13.65		13.65	22.50
3283	2″ x 6″ wall, 2′ wide		24	.333			12.55		12.55	20.50
3284	3′ wide		24	.333			12.55		12.55	20.50
3285	4′ wide		24	.333			12.55		12.55	20.50
3286	5′ wide		24	.333			12.55		12.55	20.50
3287	6′ wide		24	.333			12.55		12.55	20.50
3288	7′ wide		24	.333			12.55		12.55	20.50
3289	8′ wide		22	.364			13.65		13.65	22.50
3290	10′ wide		22	.364			13.65		13.65	22.50
3291	12′ wide		22	.364			13.65		13.65	22.50
3360	Deck or porch decking		825	.010	L.F.		.36		.36	.60
3400	Fascia boards, 1″ x 6″		500	.016			.60		.60	.99
3440	1″ x 8″		450	.018			.67		.67	1.10
3480	1″ x 10″		400	.020			.75		.75	1.23
3490	2″ x 6″		450	.018			.67		.67	1.10
3500	2″ x 8″		400	.020			.75		.75	1.23
3510	2″ x 10″		350	.023			.86		.86	1.41
3610	Furring, on wood walls or ceiling		4000	.002	S.F.		.08		.08	.12
3620	On masonry or concrete walls or ceiling		1200	.007	″		.25		.25	.41
3800	Headers over openings, 2 @ 2″ x 6″		110	.073	L.F.		2.73		2.73	4.48
3840	2 @ 2″ x 8″		100	.080			3.01		3.01	4.93
3880	2 @ 2″ x 10″		90	.089			3.34		3.34	5.50
3885	Alternate pricing method		.26	30.651	M.B.F.		1,150		1,150	1,900
3920	Joists, 1″ x 4″		1250	.006	L.F.		.24		.24	.39
3930	1″ x 6″		1135	.007			.27		.27	.43
3940	1″ x 8″		1000	.008			.30		.30	.49
3950	1″ x 10″		895	.009			.34		.34	.55
3960	1″ x 12″		765	.010			.39		.39	.64
4200	2″ x 4″	2 Clab	1000	.016			.60		.60	.99
4230	2″ x 6″		970	.016			.62		.62	1.02
4240	2″ x 8″		940	.017			.64		.64	1.05
4250	2″ x 10″		910	.018			.66		.66	1.08
4280	2″ x 12″		880	.018			.68		.68	1.12
4281	2″ x 14″		850	.019			.71		.71	1.16

06 05 05 – Selective Demolition for Wood, Plastics, and Composites

06 05 05.10 Selective Demolition Wood Framing		Crew	Daily Output	Labor-Hours	Unit	Material	2015 Bare Costs Labor	Equipment	Total	Total Incl O&P
4282	Composite joists, 9-1/2"	2 Clab	960	.017	L.F.		.63		.63	1.03
4283	11-7/8"		930	.017			.65		.65	1.06
4284	14"		897	.018			.67		.67	1.10
4285	16"		865	.019			.70		.70	1.14
4290	Wood joists, alternate pricing method		1.50	10.667	M.B.F.		400		400	660
4500	Open web joist, 12" deep		500	.032	L.F.		1.20		1.20	1.97
4505	14" deep		475	.034			1.27		1.27	2.08
4510	16" deep		450	.036			1.34		1.34	2.19
4520	18" deep		425	.038			1.42		1.42	2.32
4530	24" deep		400	.040			1.50		1.50	2.47
4550	Ledger strips, 1" x 2"	1 Clab	1200	.007			.25		.25	.41
4560	1" x 3"		1200	.007			.25		.25	.41
4570	1" x 4"		1200	.007			.25		.25	.41
4580	2" x 2"		1100	.007			.27		.27	.45
4590	2" x 4"		1000	.008			.30		.30	.49
4600	2" x 6"		1000	.008			.30		.30	.49
4601	2" x 8 or 2" x 10"		800	.010			.38		.38	.62
4602	4" x 6"		600	.013			.50		.50	.82
4604	4" x 8"		450	.018			.67		.67	1.10
5400	Posts, 4" x 4"	2 Clab	800	.020			.75		.75	1.23
5405	4" x 6"		550	.029			1.09		1.09	1.79
5410	4" x 8"		440	.036			1.37		1.37	2.24
5425	4" x 10"		390	.041			1.54		1.54	2.53
5430	4" x 12"		350	.046			1.72		1.72	2.82
5440	6" x 6"		400	.040			1.50		1.50	2.47
5445	6" x 8"		350	.046			1.72		1.72	2.82
5450	6" x 10"		320	.050			1.88		1.88	3.08
5455	6" x 12"		290	.055			2.07		2.07	3.40
5480	8" x 8"		300	.053			2.01		2.01	3.29
5500	10" x 10"		240	.067			2.51		2.51	4.11
5660	Tongue and groove floor planks		2	8	M.B.F.		300		300	495
5682	Rafters, ordinary, 16" OC, 2" x 4"		880	.018	S.F.		.68		.68	1.12
5683	2" x 6"		840	.019			.72		.72	1.17
5684	2" x 8"		820	.020			.73		.73	1.20
5685	2" x 10"		820	.020			.73		.73	1.20
5686	2" x 12"		810	.020			.74		.74	1.22
5687	24" OC, 2" x 4"		1170	.014			.51		.51	.84
5688	2" x 6"		1117	.014			.54		.54	.88
5689	2" x 8"		1091	.015			.55		.55	.90
5690	2" x 10"		1091	.015			.55		.55	.90
5691	2" x 12"		1077	.015			.56		.56	.92
5795	Rafters, ordinary, 2" x 4" (alternate method)		862	.019	L.F.		.70		.70	1.14
5800	2" x 6" (alternate method)		850	.019			.71		.71	1.16
5840	2" x 8" (alternate method)		837	.019			.72		.72	1.18
5855	2" x 10" (alternate method)		825	.019			.73		.73	1.20
5865	2" x 12" (alternate method)		812	.020			.74		.74	1.21
5870	Sill plate, 2" x 4"	1 Clab	1170	.007			.26		.26	.42
5871	2" x 6"		780	.010			.39		.39	.63
5872	2" x 8"		586	.014			.51		.51	.84
5873	Alternate pricing method		.78	10.256	M.B.F.		385		385	630
5885	Ridge board, 1" x 4"	2 Clab	900	.018	L.F.		.67		.67	1.10
5886	1" x 6"		875	.018			.69		.69	1.13
5887	1" x 8"		850	.019			.71		.71	1.16

06 05 05 – Selective Demolition for Wood, Plastics, and Composites

06 05 05.10 Selective Demolition Wood Framing		Crew	Daily Output	Labor-Hours	Unit	2015 Bare Costs Material	Labor	Equipment	Total	Total Incl O&P
5888	1″ x 10″	2 Clab	825	.019	L.F.		.73		.73	1.20
5889	1″ x 12″		800	.020			.75		.75	1.23
5890	2″ x 4″		900	.018			.67		.67	1.10
5892	2″ x 6″		875	.018			.69		.69	1.13
5894	2″ x 8″		850	.019			.71		.71	1.16
5896	2″ x 10″		825	.019			.73		.73	1.20
5898	2″ x 12″		800	.020			.75		.75	1.23
5900	Hip & valley rafters, 2″ x 6″		500	.032			1.20		1.20	1.97
5940	2″ x 8″		420	.038			1.43		1.43	2.35
6050	Rafter tie, 1″ x 4″		1250	.013			.48		.48	.79
6052	1″ x 6″		1135	.014			.53		.53	.87
6054	2″ x 4″		1000	.016			.60		.60	.99
6056	2″ x 6″		970	.016			.62		.62	1.02
6070	Sleepers, on concrete, 1″ x 2″	1 Clab	4700	.002			.06		.06	.10
6075	1″ x 3″		4000	.002			.08		.08	.12
6080	2″ x 4″		3000	.003			.10		.10	.16
6085	2″ x 6″		2600	.003			.12		.12	.19
6086	Sheathing from roof, 5/16″	2 Clab	1600	.010	S.F.		.38		.38	.62
6088	3/8″		1525	.010			.39		.39	.65
6090	1/2″		1400	.011			.43		.43	.70
6092	5/8″		1300	.012			.46		.46	.76
6094	3/4″		1200	.013			.50		.50	.82
6096	Board sheathing from roof		1400	.011			.43		.43	.70
6100	Sheathing, from walls, 1/4″		1200	.013			.50		.50	.82
6110	5/16″		1175	.014			.51		.51	.84
6120	3/8″		1150	.014			.52		.52	.86
6130	1/2″		1125	.014			.53		.53	.88
6140	5/8″		1100	.015			.55		.55	.90
6150	3/4″		1075	.015			.56		.56	.92
6152	Board sheathing from walls		1500	.011			.40		.40	.66
6158	Subfloor/roof deck, with boards		2200	.007			.27		.27	.45
6159	Subfloor/roof deck, with tongue & groove boards		2000	.008			.30		.30	.49
6160	Plywood, 1/2″ thick		768	.021			.78		.78	1.28
6162	5/8″ thick		760	.021			.79		.79	1.30
6164	3/4″ thick		750	.021			.80		.80	1.32
6165	1-1/8″ thick		720	.022			.84		.84	1.37
6166	Underlayment, particle board, 3/8″ thick	1 Clab	780	.010			.39		.39	.63
6168	1/2″ thick		768	.010			.39		.39	.64
6170	5/8″ thick		760	.011			.40		.40	.65
6172	3/4″ thick		750	.011			.40		.40	.66
6200	Stairs and stringers, straight run	2 Clab	40	.400	Riser		15.05		15.05	24.50
6240	With platforms, winders or curves	″	26	.615	″		23		23	38
6300	Components, tread	1 Clab	110	.073	Ea.		2.73		2.73	4.48
6320	Riser		80	.100	″		3.76		3.76	6.15
6390	Stringer, 2″ x 10″		260	.031	L.F.		1.16		1.16	1.90
6400	2″ x 12″		260	.031			1.16		1.16	1.90
6410	3″ x 10″		250	.032			1.20		1.20	1.97
6420	3″ x 12″		250	.032			1.20		1.20	1.97
6590	Wood studs, 2″ x 3″	2 Clab	3076	.005			.20		.20	.32
6600	2″ x 4″		2000	.008			.30		.30	.49
6640	2″ x 6″		1600	.010			.38		.38	.62
6720	Wall framing, including studs, plates and blocking, 2″ x 4″	1 Clab	600	.013	S.F.		.50		.50	.82
6740	2″ x 6″		480	.017	″		.63		.63	1.03

06 05 Common Work Results for Wood, Plastics, and Composites

06 05 05 – Selective Demolition for Wood, Plastics, and Composites

06 05 05.10 Selective Demolition Wood Framing		Crew	Daily Output	Labor-Hours	Unit	2015 Bare Costs Material	Labor	Equipment	Total	Total Incl O&P
6750	Headers, 2" x 4"	1 Clab	1125	.007	L.F.		.27		.27	.44
6755	2" x 6"		1125	.007			.27		.27	.44
6760	2" x 8"		1050	.008			.29		.29	.47
6765	2" x 10"		1050	.008			.29		.29	.47
6770	2" x 12"		1000	.008			.30		.30	.49
6780	4" x 10"		525	.015			.57		.57	.94
6785	4" x 12"		500	.016			.60		.60	.99
6790	6" x 8"		560	.014			.54		.54	.88
6795	6" x 10"		525	.015			.57		.57	.94
6797	6" x 12"		500	.016			.60		.60	.99
7000	Trusses									
7050	12' span	2 Clab	74	.216	Ea.		8.15		8.15	13.35
7150	24' span	F-3	66	.606			29	9.90	38.90	58.50
7200	26' span		64	.625			30	10.20	40.20	60
7250	28' span		62	.645			31	10.55	41.55	62
7300	30' span		58	.690			33	11.30	44.30	66.50
7350	32' span		56	.714			34	11.70	45.70	68.50
7400	34' span		54	.741			35.50	12.10	47.60	71.50
7450	36' span		52	.769			37	12.60	49.60	74
8000	Soffit, T & G wood	1 Clab	520	.015	S.F.		.58		.58	.95
8010	Hardboard, vinyl or aluminum	"	640	.013			.47		.47	.77
8030	Plywood	2 Carp	315	.051			2.38		2.38	3.91
9000	Minimum labor/equipment charge	1 Clab	4	2	Job		75		75	123
9500	See Section 02 41 19.19 for rubbish handling									

06 05 05.20 Selective Demolition Millwork and Trim

		Crew	Daily Output	Labor-Hours	Unit	Material	Labor	Equipment	Total	Total Incl O&P
0010	**SELECTIVE DEMOLITION MILLWORK AND TRIM** R024119-10									
1000	Cabinets, wood, base cabinets, per L.F.	2 Clab	80	.200	L.F.		7.50		7.50	12.35
1020	Wall cabinets, per L.F.	"	80	.200	"		7.50		7.50	12.35
1060	Remove and reset, base cabinets	2 Carp	18	.889	Ea.		41.50		41.50	68.50
1070	Wall cabinets		20	.800			37.50		37.50	61.50
1072	Oven cabinet, 7' high		11	1.455			68.50		68.50	112
1074	Cabinet door, up to 2' high	1 Clab	66	.121			4.56		4.56	7.45
1076	2' - 4' high	"	46	.174			6.55		6.55	10.70
1100	Steel, painted, base cabinets	2 Clab	60	.267	L.F.		10.05		10.05	16.45
1120	Wall cabinets		60	.267	"		10.05		10.05	16.45
1200	Casework, large area		320	.050	S.F.		1.88		1.88	3.08
1220	Selective		200	.080	"		3.01		3.01	4.93
1500	Counter top, straight runs		200	.080	L.F.		3.01		3.01	4.93
1510	L, U or C shapes		120	.133			5		5	8.20
1550	Remove and reset, straight runs	2 Carp	50	.320			15		15	24.50
1560	L, U or C shape	"	40	.400			18.80		18.80	31
2000	Paneling, 4' x 8' sheets	2 Clab	2000	.008	S.F.		.30		.30	.49
2100	Boards, 1" x 4"		700	.023			.86		.86	1.41
2120	1" x 6"		750	.021			.80		.80	1.32
2140	1" x 8"		800	.020			.75		.75	1.23
3000	Trim, baseboard, to 6" wide		1200	.013	L.F.		.50		.50	.82
3040	Greater than 6" and up to 12" wide		1000	.016			.60		.60	.99
3080	Remove and reset, minimum	2 Carp	400	.040			1.88		1.88	3.08
3090	Maximum	"	300	.053			2.50		2.50	4.10
3100	Ceiling trim	2 Clab	1000	.016			.60		.60	.99
3120	Chair rail		1200	.013			.50		.50	.82
3140	Railings with balusters		240	.067			2.51		2.51	4.11

06 05 05 – Selective Demolition for Wood, Plastics, and Composites

06 05 05.20 Selective Demolition Millwork and Trim		Crew	Daily Output	Labor-Hours	Unit	2015 Bare Costs Material	Labor	Equipment	Total	Total Incl O&P
3160	Wainscoting	2 Clab	700	.023	S.F.		.86		.86	1.41
9000	Minimum labor/equipment charge	1 Clab	4	2	Job		75		75	123

06 05 23 – Wood, Plastic, and Composite Fastenings

06 05 23.10 Nails

Line	Description	Crew	Daily Output	Labor-Hours	Unit	Material	Labor	Equipment	Total	Total Incl O&P
0010	**NAILS**, material only, based upon 50# box purchase									
0020	Copper nails, plain				Lb.	9.85			9.85	10.80
0400	Stainless steel, plain					9.20			9.20	10.15
0500	Box, 3d to 20d, bright					1.37			1.37	1.51
0520	Galvanized					1.68			1.68	1.85
0600	Common, 3d to 60d, plain					1.30			1.30	1.43
0700	Galvanized					1.97			1.97	2.17
0800	Aluminum					10.80			10.80	11.90
1000	Annular or spiral thread, 4d to 60d, plain					2.63			2.63	2.89
1200	Galvanized					3.15			3.15	3.47
1400	Drywall nails, plain					1.50			1.50	1.65
1600	Galvanized					2.20			2.20	2.42
1800	Finish nails, 4d to 10d, plain					1.25			1.25	1.38
2000	Galvanized					1.86			1.86	2.05
2100	Aluminum					5.85			5.85	6.45
2300	Flooring nails, hardened steel, 2d to 10d, plain					3.38			3.38	3.72
2400	Galvanized					4.04			4.04	4.44
2500	Gypsum lath nails, 1-1/8", 13 ga. flathead, blued					2			2	2.20
2600	Masonry nails, hardened steel, 3/4" to 3" long, plain					1.71			1.71	1.88
2700	Galvanized					3.27			3.27	3.60
2900	Roofing nails, threaded, galvanized					1.76			1.76	1.94
3100	Aluminum					4.85			4.85	5.35
3300	Compressed lead head, threaded, galvanized					2.49			2.49	2.74
3600	Siding nails, plain shank, galvanized					2.02			2.02	2.22
3800	Aluminum					5.85			5.85	6.45
5000	Add to prices above for cement coating					.11			.11	.12
5200	Zinc or tin plating					.14			.14	.15
5500	Vinyl coated sinkers, 8d to 16d					2.50			2.50	2.75

06 05 23.40 Sheet Metal Screws

Line	Description	Crew	Daily Output	Labor-Hours	Unit	Material	Labor	Equipment	Total	Total Incl O&P
0010	**SHEET METAL SCREWS**									
0020	Steel, standard, #8 x 3/4", plain				C	3.20			3.20	3.52
0100	Galvanized					3.20			3.20	3.52
0300	#10 x 1", plain					4.40			4.40	4.84
0400	Galvanized					4.40			4.40	4.84
0600	With washers, #14 x 1", plain					9.65			9.65	10.60
0700	Galvanized					9.65			9.65	10.60
0900	#14 x 2", plain					15			15	16.50
1000	Galvanized					15			15	16.50
1500	Self-drilling, with washers, (pinch point) #8 x 3/4", plain					7.30			7.30	8
1600	Galvanized					7.30			7.30	8
1800	#10 x 3/4", plain					7.80			7.80	8.55
1900	Galvanized					7.80			7.80	8.55
3000	Stainless steel w/aluminum or neoprene washers, #14 x 1", plain					8.10			8.10	8.95
3100	#14 x 2", plain					10			10	11

06 05 23.50 Wood Screws

Line	Description	Crew	Daily Output	Labor-Hours	Unit	Material	Labor	Equipment	Total	Total Incl O&P
0010	**WOOD SCREWS**									
0020	#8 x 1" long, steel				C	2.70			2.70	2.97
0100	Brass					12.20			12.20	13.40

06 05 23 – Wood, Plastic, and Composite Fastenings

06 05 23.50 Wood Screws		Crew	Daily Output	Labor-Hours	Unit	2015 Bare Costs Material	Labor	Equipment	Total	Total Incl O&P
0200	#8, 2″ long, steel				C	4.16			4.16	4.58
0300	Brass					22			22	24
0400	#10, 1″ long, steel					3.30			3.30	3.63
0500	Brass					15.60			15.60	17.15
0600	#10, 2″ long, steel					5.30			5.30	5.85
0700	Brass					27			27	29.50
0800	#10, 3″ long, steel					8.85			8.85	9.75
1000	#12, 2″ long, steel					7.30			7.30	8.05
1100	Brass					34.50			34.50	38
1500	#12, 3″ long, steel					11.15			11.15	12.30
2000	#12, 4″ long, steel					18.75			18.75	20.50

06 05 23.60 Timber Connectors		Crew	Daily Output	Labor-Hours	Unit	Material	Labor	Equipment	Total	Total Incl O&P
0010	**TIMBER CONNECTORS**									
0020	Add up cost of each part for total cost of connection									
0100	Connector plates, steel, with bolts, straight	2 Carp	75	.213	Ea.	30.50	10		40.50	50
0110	Tee, 7 ga.		50	.320		35	15		50	63
0120	T- Strap, 14 ga., 12″ x 8″ x 2″		50	.320		35	15		50	63
0150	Anchor plates, 7 ga., 9″ x 7″		75	.213		30.50	10		40.50	50
0200	Bolts, machine, sq. hd. with nut & washer, 1/2″ diameter, 4″ long	1 Carp	140	.057		.75	2.68		3.43	5.25
0300	7-1/2″ long		130	.062		1.37	2.89		4.26	6.25
0500	3/4″ diameter, 7-1/2″ long		130	.062		3.20	2.89		6.09	8.25
0610	Machine bolts, w/nut, washer, 3/4″ diam., 15″ L, HD's & beam hangers		95	.084		5.95	3.95		9.90	13.05
0800	Drilling bolt holes in timber, 1/2″ diameter		450	.018	Inch		.83		.83	1.37
0900	1″ diameter		350	.023	″		1.07		1.07	1.76
1100	Framing anchor, angle, 3″ x 3″ x 1-1/2″, 12 ga		175	.046	Ea.	2.61	2.15		4.76	6.40
1150	Framing anchors, 18 ga., 4-1/2″ x 2-3/4″		175	.046		2.61	2.15		4.76	6.40
1160	Framing anchors, 18 ga., 4-1/2″ x 3″		175	.046		2.61	2.15		4.76	6.40
1170	Clip anchors plates, 18 ga., 12″ x 1-1/8″		175	.046		2.61	2.15		4.76	6.40
1250	Holdowns, 3 ga. base, 10 ga. body		8	1		25	47		72	105
1260	Holdowns, 7 ga. 11-1/16″ x 3-1/4″		8	1		25	47		72	105
1270	Holdowns, 7 ga. 14-3/8″ x 3-1/8″		8	1		25	47		72	105
1275	Holdowns, 12 ga. 8″ x 2-1/2″		8	1		25	47		72	105
1300	Joist and beam hangers, 18 ga. galv., for 2″ x 4″ joist		175	.046		.73	2.15		2.88	4.32
1400	2″ x 6″ to 2″ x 10″ joist		165	.048		1.39	2.28		3.67	5.25
1600	16 ga. galv., 3″ x 6″ to 3″ x 10″ joist		160	.050		2.96	2.35		5.31	7.10
1700	3″ x 10″ to 3″ x 14″ joist		160	.050		4.92	2.35		7.27	9.25
1800	4″ x 6″ to 4″ x 10″ joist		155	.052		3.07	2.42		5.49	7.35
1900	4″ x 10″ to 4″ x 14″ joist		155	.052		5	2.42		7.42	9.45
2000	Two-2″ x 6″ to two-2″ x 10″ joists		150	.053		4.19	2.50		6.69	8.70
2100	Two-2″ x 10″ to two-2″ x 14″ joists		150	.053		4.68	2.50		7.18	9.25
2300	3/16″ thick, 6″ x 8″ joist		145	.055		65	2.59		67.59	76
2400	6″ x 10″ joist		140	.057		67.50	2.68		70.18	79
2500	6″ x 12″ joist		135	.059		70.50	2.78		73.28	82
2700	1/4″ thick, 6″ x 14″ joist		130	.062		73	2.89		75.89	85
2900	Plywood clips, extruded aluminum H clip, for 3/4″ panels					.24			.24	.26
3000	Galvanized 18 ga. back-up clip					.18			.18	.20
3200	Post framing, 16 ga. galv. for 4″ x 4″ base, 2 piece	1 Carp	130	.062		16.40	2.89		19.29	23
3300	Cap		130	.062		23	2.89		25.89	29.50
3500	Rafter anchors, 18 ga. galv., 1-1/2″ wide, 5-1/4″ long		145	.055		.49	2.59		3.08	4.79
3600	10-3/4″ long		145	.055		1.46	2.59		4.05	5.85
3800	Shear plates, 2-5/8″ diameter		120	.067		2.38	3.13		5.51	7.75
3900	4″ diameter		115	.070		5.65	3.27		8.92	11.55

06 05 Common Work Results for Wood, Plastics, and Composites

06 05 23 – Wood, Plastic, and Composite Fastenings

06 05 23.60 Timber Connectors		Crew	Daily Output	Labor-Hours	Unit	2015 Bare Costs Material	Labor	Equipment	Total	Total Incl O&P
4000	Sill anchors, embedded in concrete or block, 25-1/2" long	1 Carp	115	.070	Ea.	12.70	3.27		15.97	19.30
4100	Spike grids, 3" x 6"		120	.067		.96	3.13		4.09	6.20
4400	Split rings, 2-1/2" diameter		120	.067		1.96	3.13		5.09	7.30
4500	4" diameter		110	.073		2.97	3.41		6.38	8.85
4550	Tie plate, 20 ga., 7" x 3 1/8"		110	.073		2.97	3.41		6.38	8.85
4560	Tie plate, 20 ga., 5" x 4 1/8"		110	.073		2.97	3.41		6.38	8.85
4575	Twist straps, 18 ga., 12" x 1 1/4"		110	.073		2.97	3.41		6.38	8.85
4580	Twist straps, 18 ga., 16" x 1 1/4"		110	.073		2.97	3.41		6.38	8.85
4600	Strap ties, 20 ga., 2 -1/16" wide, 12 13/16" long		180	.044		.94	2.09		3.03	4.45
4700	Strap ties, 16 ga., 1-3/8" wide, 12" long		180	.044		.94	2.09		3.03	4.45
4800	21-5/8" x 1-1/4"		160	.050		2.96	2.35		5.31	7.10
5000	Toothed rings, 2-5/8" or 4" diameter		90	.089		1.78	4.17		5.95	8.80
5200	Truss plates, nailed, 20 ga., up to 32' span		17	.471	Truss	12.90	22		34.90	50
5400	Washers, 2" x 2" x 1/8"				Ea.	.40			.40	.44
5500	3" x 3" x 3/16"				"	1.06			1.06	1.17
9000	Minimum labor/equipment charge	1 Carp	4	2	Job		94		94	154

06 05 23.80 Metal Bracing		Crew	Daily Output	Labor-Hours	Unit	Material	Labor	Equipment	Total	Total Incl O&P
0010	**METAL BRACING**									
0302	Let-in, "T" shaped, 22 ga. galv. steel, studs at 16" O.C.	1 Carp	580	.014	L.F.	.81	.65		1.46	1.95
0402	Studs at 24" O.C.		600	.013		.81	.63		1.44	1.92
0502	Steel straps, 16 ga. galv. steel, studs at 16" O.C.		600	.013		1.05	.63		1.68	2.19
0602	Studs at 24" O.C.		620	.013		1.05	.61		1.66	2.15

06 11 Wood Framing

06 11 10 – Framing with Dimensional, Engineered or Composite Lumber

06 11 10.01 Forest Stewardship Council Certification		Crew	Daily Output	Labor-Hours	Unit	Material	Labor	Equipment	Total	Total Incl O&P
0010	**FOREST STEWARDSHIP COUNCIL CERTIFICATION**									
0020	For Forest Stewardship Council (FSC) cert dimension lumber, add [G]					65%				

06 11 10.02 Blocking		Crew	Daily Output	Labor-Hours	Unit	Material	Labor	Equipment	Total	Total Incl O&P
0010	**BLOCKING**									
2600	Miscellaneous, to wood construction									
2620	2" x 4"	1 Carp	.17	47.059	M.B.F.	635	2,200		2,835	4,325
2625	Pneumatic nailed		.21	38.095		640	1,800		2,440	3,625
2660	2" x 8"		.27	29.630		690	1,400		2,090	3,025
2665	Pneumatic nailed		.33	24.242		695	1,150		1,845	2,650
2720	To steel construction									
2740	2" x 4"	1 Carp	.14	57.143	M.B.F.	635	2,675		3,310	5,100
2780	2" x 8"		.21	38.095	"	690	1,800		2,490	3,675
9000	Minimum labor/equipment charge		4	2	Job		94		94	154

06 11 10.04 Wood Bracing		Crew	Daily Output	Labor-Hours	Unit	Material	Labor	Equipment	Total	Total Incl O&P
0010	**WOOD BRACING**									
0012	Let-in, with 1" x 6" boards, studs @ 16" O.C.	1 Carp	150	.053	L.F.	.73	2.50		3.23	4.90
0202	Studs @ 24" O.C.	"	230	.035	"	.73	1.63		2.36	3.48

06 11 10.06 Bridging		Crew	Daily Output	Labor-Hours	Unit	Material	Labor	Equipment	Total	Total Incl O&P
0010	**BRIDGING**									
0012	Wood, for joists 16" O.C., 1" x 3"	1 Carp	130	.062	Pr.	.63	2.89		3.52	5.45
0017	Pneumatic nailed		170	.047		.68	2.21		2.89	4.37
0102	2" x 3" bridging		130	.062		.69	2.89		3.58	5.50
0107	Pneumatic nailed		170	.047		.69	2.21		2.90	4.38
0302	Steel, galvanized, 18 ga., for 2" x 10" joists at 12" O.C.		130	.062		.92	2.89		3.81	5.75

06 11 10 – Framing with Dimensional, Engineered or Composite Lumber

06 11 10.06 Bridging		Crew	Daily Output	Labor-Hours	Unit	2015 Bare Costs Material	Labor	Equipment	Total	Total Incl O&P
0352	16" O.C.	1 Carp	135	.059	Pr.	.93	2.78		3.71	5.60
0402	24" O.C.		140	.057		1.90	2.68		4.58	6.50
0602	For 2" x 14" joists at 16" O.C.		130	.062		1.40	2.89		4.29	6.30
0902	Compression type, 16" O.C., 2" x 8" joists		200	.040		.93	1.88		2.81	4.10
1002	2" x 12" joists		200	.040		.93	1.88		2.81	4.10

06 11 10.10 Beam and Girder Framing

Code	Description	Crew	Daily Output	Labor-Hours	Unit	Material	Labor	Equipment	Total	Total Incl O&P
0010	**BEAM AND GIRDER FRAMING** R061110-30									
1000	Single, 2" x 6"	2 Carp	700	.023	L.F.	.67	1.07		1.74	2.49
1005	Pneumatic nailed		812	.020		.67	.92		1.59	2.26
1020	2" x 8"		650	.025		.92	1.16		2.08	2.90
1025	Pneumatic nailed		754	.021		.93	1		1.93	2.65
1040	2" x 10"		600	.027		1.34	1.25		2.59	3.52
1045	Pneumatic nailed		696	.023		1.35	1.08		2.43	3.25
1060	2" x 12"		550	.029		1.72	1.37		3.09	4.14
1065	Pneumatic nailed		638	.025		1.73	1.18		2.91	3.84
1080	2" x 14"		500	.032		2.01	1.50		3.51	4.67
1085	Pneumatic nailed		580	.028		2.02	1.30		3.32	4.35
1100	3" x 8"		550	.029		2.55	1.37		3.92	5.05
1120	3" x 10"		500	.032		3.22	1.50		4.72	6
1140	3" x 12"		450	.036		3.86	1.67		5.53	7
1160	3" x 14"		400	.040		4.51	1.88		6.39	8.05
1170	4" x 6"	F-3	1100	.036		2.84	1.74	.59	5.17	6.60
1180	4" x 8"		1000	.040		2.89	1.92	.65	5.46	7
1200	4" x 10"		950	.042		4.28	2.02	.69	6.99	8.75
1220	4" x 12"		900	.044		5.15	2.13	.73	8.01	9.90
1240	4" x 14"		850	.047		6	2.25	.77	9.02	11.10
1250	6" x 8"		525	.076		7.40	3.65	1.25	12.30	15.45
1260	6" x 10"		500	.080		6.45	3.83	1.31	11.59	14.80
1290	8" x 12"		300	.133		15.20	6.40	2.18	23.78	29.50
2000	Double, 2" x 6"	2 Carp	625	.026		1.33	1.20		2.53	3.44
2005	Pneumatic nailed		725	.022		1.34	1.04		2.38	3.18
2020	2" x 8"		575	.028		1.84	1.31		3.15	4.17
2025	Pneumatic nailed		667	.024		1.86	1.13		2.99	3.89
2040	2" x 10"		550	.029		2.68	1.37		4.05	5.20
2045	Pneumatic nailed		638	.025		2.70	1.18		3.88	4.90
2060	2" x 12"		525	.030		3.45	1.43		4.88	6.15
2065	Pneumatic nailed		610	.026		3.47	1.23		4.70	5.85
2080	2" x 14"		475	.034		4.02	1.58		5.60	7
2085	Pneumatic nailed		551	.029		4.05	1.36		5.41	6.70
3000	Triple, 2" x 6"		550	.029		2	1.37		3.37	4.44
3005	Pneumatic nailed		638	.025		2.01	1.18		3.19	4.14
3020	2" x 8"		525	.030		2.77	1.43		4.20	5.40
3025	Pneumatic nailed		609	.026		2.78	1.23		4.01	5.10
3040	2" x 10"		500	.032		4.02	1.50		5.52	6.90
3045	Pneumatic nailed		580	.028		4.04	1.30		5.34	6.55
3060	2" x 12"		475	.034		5.15	1.58		6.73	8.30
3065	Pneumatic nailed		551	.029		5.20	1.36		6.56	7.95
3080	2" x 14"		450	.036		6.70	1.67		8.37	10.10
3085	Pneumatic nailed		522	.031		6.05	1.44		7.49	9.05
9000	Minimum labor/equipment charge	1 Carp	2	4	Job		188		188	310

06 11 Wood Framing

06 11 10 – Framing with Dimensional, Engineered or Composite Lumber

06 11 10.12 Ceiling Framing		Crew	Daily Output	Labor-Hours	Unit	2015 Bare Costs Material	Labor	Equipment	Total	Total Incl O&P
0010	**CEILING FRAMING**									
6000	Suspended, 2" x 3"	2 Carp	1000	.016	L.F.	.39	.75		1.14	1.66
6050	2" x 4"		900	.018		.42	.83		1.25	1.84
6100	2" x 6"		800	.020		.67	.94		1.61	2.27
6150	2" x 8"		650	.025		.92	1.16		2.08	2.90
9000	Minimum labor/equipment charge	1 Carp	4	2	Job		94		94	154

06 11 10.14 Posts and Columns

Line	Description	Crew	Daily Output	Labor-Hours	Unit	Material	Labor	Equipment	Total	Total Incl O&P
0010	**POSTS AND COLUMNS**									
0100	4" x 4"	2 Carp	390	.041	L.F.	1.73	1.93		3.66	5.05
0150	4" x 6"		275	.058		2.84	2.73		5.57	7.60
0200	4" x 8"		220	.073		2.89	3.41		6.30	8.80
0250	6" x 6"		215	.074		5.05	3.49		8.54	11.30
0300	6" x 8"		175	.091		7.40	4.29		11.69	15.20
0350	6" x 10"		150	.107		6.45	5		11.45	15.30
9000	Minimum labor/equipment charge	1 Carp	2	4	Job		188		188	310

06 11 10.18 Joist Framing

Line	Description	Crew	Daily Output	Labor-Hours	Unit	Material	Labor	Equipment	Total	Total Incl O&P
0010	**JOIST FRAMING** R061110-30									
2000	Joists, 2" x 4"	2 Carp	1250	.013	L.F.	.42	.60		1.02	1.46
2005	Pneumatic nailed		1438	.011		.43	.52		.95	1.33
2100	2" x 6"		1250	.013		.67	.60		1.27	1.72
2105	Pneumatic nailed		1438	.011		.67	.52		1.19	1.60
2150	2" x 8"		1100	.015		.92	.68		1.60	2.13
2155	Pneumatic nailed		1265	.013		.93	.59		1.52	1.99
2200	2" x 10"		900	.018		1.34	.83		2.17	2.84
2205	Pneumatic nailed		1035	.015		1.35	.73		2.08	2.67
2250	2" x 12"		875	.018		1.72	.86		2.58	3.31
2255	Pneumatic nailed		1006	.016		1.73	.75		2.48	3.13
2300	2" x 14"		770	.021		2.01	.98		2.99	3.81
2305	Pneumatic nailed		886	.018		2.02	.85		2.87	3.62
2350	3" x 6"		925	.017		1.90	.81		2.71	3.41
2400	3" x 10"		780	.021		3.22	.96		4.18	5.10
2450	3" x 12"		600	.027		3.86	1.25		5.11	6.30
2500	4" x 6"		800	.020		2.84	.94		3.78	4.66
2550	4" x 10"		600	.027		4.28	1.25		5.53	6.75
2600	4" x 12"		450	.036		5.15	1.67		6.82	8.40
2605	Sister joist, 2" x 6"		800	.020		.67	.94		1.61	2.27
2606	Pneumatic nailed		960	.017		.67	.78		1.45	2.02
2610	2" x 8"		640	.025		.92	1.17		2.09	2.93
2611	Pneumatic nailed		768	.021		.93	.98		1.91	2.62
2615	2" x 10"		535	.030		1.34	1.40		2.74	3.77
2616	Pneumatic nailed		642	.025		1.35	1.17		2.52	3.40
2620	2" x 12"		455	.035		1.72	1.65		3.37	4.61
2625	Pneumatic nailed		546	.029		1.73	1.38		3.11	4.16
3000	Composite wood joist 9-1/2" deep		.90	17.778	M.L.F.	1,800	835		2,635	3,350
3010	11-1/2" deep		.88	18.182		1,975	855		2,830	3,575
3020	14" deep		.82	19.512		2,100	915		3,015	3,800
3030	16" deep		.78	20.513		3,475	965		4,440	5,400
4000	Open web joist 12" deep		.88	18.182		3,525	855		4,380	5,275
4002	Per linear foot		880	.018	L.F.	3.52	.85		4.37	5.25
4004	Treated, per linear foot		880	.018	"	4.39	.85		5.24	6.25
4010	14" deep		.82	19.512	M.L.F.	3,725	915		4,640	5,600
4012	Per linear foot		820	.020	L.F.	3.72	.92		4.64	5.60

06 11 10 – Framing with Dimensional, Engineered or Composite Lumber

06 11 10.18 Joist Framing		Crew	Daily Output	Labor-Hours	Unit	2015 Bare Costs Material	Labor	Equipment	Total	Total Incl O&P
4014	Treated, per linear foot	2 Carp	820	.020	L.F.	4.74	.92		5.66	6.70
4020	16″ deep		.78	20.513	M.L.F.	3,825	965		4,790	5,775
4022	Per linear foot		780	.021	L.F.	3.83	.96		4.79	5.80
4024	Treated, per linear foot		780	.021	″	4.99	.96		5.95	7.10
4030	18″ deep		.74	21.622	M.L.F.	4,000	1,025		5,025	6,075
4032	Per linear foot		740	.022	L.F.	4	1.02		5.02	6.05
4034	Treated, per linear foot		740	.022	″	5.30	1.02		6.32	7.50
6000	Composite rim joist, 1-1/4″ x 9-1/2″		.90	17.778	M.L.F.	1,775	835		2,610	3,325
6010	1-1/4″ x 11-1/2″		.88	18.182		2,050	855		2,905	3,650
6020	1-1/4″ x 14-1/2″		.82	19.512		2,750	915		3,665	4,525
6030	1-1/4″ x 16-1/2″		.78	20.513		3,375	965		4,340	5,300
9000	Minimum labor/equipment charge	1 Carp	4	2	Job		94		94	154

06 11 10.24 Miscellaneous Framing		Crew	Daily Output	Labor-Hours	Unit	Material	Labor	Equipment	Total	Total Incl O&P
0010	**MISCELLANEOUS FRAMING**									
2000	Firestops, 2″ x 4″	2 Carp	780	.021	L.F.	.42	.96		1.38	2.05
2005	Pneumatic nailed		952	.017		.43	.79		1.22	1.76
2100	2″ x 6″		600	.027		.67	1.25		1.92	2.78
2105	Pneumatic nailed		732	.022		.67	1.03		1.70	2.42
5000	Nailers, treated, wood construction, 2″ x 4″		800	.020		.51	.94		1.45	2.10
5005	Pneumatic nailed		960	.017		.51	.78		1.29	1.84
5100	2″ x 6″		750	.021		.78	1		1.78	2.50
5105	Pneumatic nailed		900	.018		.78	.83		1.61	2.23
5120	2″ x 8″		700	.023		1.05	1.07		2.12	2.92
5125	Pneumatic nailed		840	.019		1.06	.89		1.95	2.63
5200	Steel construction, 2″ x 4″		750	.021		.51	1		1.51	2.20
5220	2″ x 6″		700	.023		.78	1.07		1.85	2.62
5240	2″ x 8″		650	.025		1.05	1.16		2.21	3.05
7000	Rough bucks, treated, for doors or windows, 2″ x 6″		400	.040		.78	1.88		2.66	3.94
7005	Pneumatic nailed		480	.033		.78	1.56		2.34	3.42
7100	2″ x 8″		380	.042		1.05	1.98		3.03	4.40
7105	Pneumatic nailed		456	.035		1.06	1.65		2.71	3.86
8000	Stair stringers, 2″ x 10″		130	.123		1.34	5.80		7.14	10.90
8100	2″ x 12″		130	.123		1.72	5.80		7.52	11.35
8150	3″ x 10″		125	.128		3.22	6		9.22	13.40
8200	3″ x 12″		125	.128		3.86	6		9.86	14.10
9000	Minimum labor/equipment charge	1 Carp	4	2	Job		94		94	154

06 11 10.26 Partitions		Crew	Daily Output	Labor-Hours	Unit	Material	Labor	Equipment	Total	Total Incl O&P
0010	**PARTITIONS**									
0020	Single bottom and double top plate, no waste, std. & better lumber									
0180	2″ x 4″ studs, 8′ high, studs 12″ O.C.	2 Carp	80	.200	L.F.	4.65	9.40		14.05	20.50
0185	12″ O.C., pneumatic nailed		96	.167		4.69	7.85		12.54	18
0200	16″ O.C.		100	.160		3.81	7.50		11.31	16.50
0205	16″ O.C., pneumatic nailed		120	.133		3.84	6.25		10.09	14.45
0300	24″ O.C.		125	.128		2.96	6		8.96	13.10
0305	24″ O.C., pneumatic nailed		150	.107		2.98	5		7.98	11.50
0380	10′ high, studs 12″ O.C.		80	.200		5.50	9.40		14.90	21.50
0385	12″ O.C., pneumatic nailed		96	.167		5.55	7.85		13.40	18.95
0400	16″ O.C.		100	.160		4.44	7.50		11.94	17.20
0405	16″ O.C., pneumatic nailed		120	.133		4.47	6.25		10.72	15.15
0500	24″ O.C.		125	.128		3.38	6		9.38	13.55
0505	24″ O.C., pneumatic nailed		150	.107		3.41	5		8.41	11.95
0580	12′ high, studs 12″ O.C.		65	.246		6.35	11.55		17.90	26

06 11 Wood Framing

06 11 10 – Framing with Dimensional, Engineered or Composite Lumber

06 11 10.26 Partitions		Crew	Daily Output	Labor-Hours	Unit	Material	2015 Bare Costs Labor	Equipment	Total	Total Incl O&P
0585	12" O.C., pneumatic nailed	2 Carp	78	.205	L.F.	6.40	9.65		16.05	23
0600	16" O.C.		80	.200		5.10	9.40		14.50	21
0605	16" O.C., pneumatic nailed		96	.167		5.10	7.85		12.95	18.45
0700	24" O.C.		100	.160		3.81	7.50		11.31	16.50
0705	24" O.C., pneumatic nailed		120	.133		3.84	6.25		10.09	14.45
0780	2" x 6" studs, 8' high, studs 12" O.C.		70	.229		7.35	10.75		18.10	25.50
0785	12" O.C., pneumatic nailed		84	.190		7.40	8.95		16.35	23
0800	16" O.C.		90	.178		6	8.35		14.35	20.50
0805	16" O.C., pneumatic nailed		108	.148		6.05	6.95		13	18.05
0900	24" O.C.		115	.139		4.66	6.55		11.21	15.85
0905	24" O.C., pneumatic nailed		138	.116		4.70	5.45		10.15	14.05
0980	10' high, studs 12" O.C.		70	.229		8.65	10.75		19.40	27
0985	12" O.C., pneumatic nailed		84	.190		8.75	8.95		17.70	24.50
1000	16" O.C.		90	.178		7	8.35		15.35	21.50
1005	16" O.C., pneumatic nailed		108	.148		7.05	6.95		14	19.15
1100	24" O.C.		115	.139		5.35	6.55		11.90	16.55
1105	24" O.C., pneumatic nailed		138	.116		5.35	5.45		10.80	14.80
1180	12' high, studs 12" O.C.		55	.291		10	13.65		23.65	33.50
1185	12" O.C., pneumatic nailed		66	.242		10.05	11.40		21.45	29.50
1200	16" O.C.		70	.229		8	10.75		18.75	26.50
1205	16" O.C., pneumatic nailed		84	.190		8.05	8.95		17	23.50
1300	24" O.C.		90	.178		6	8.35		14.35	20.50
1305	24" O.C., pneumatic nailed		108	.148		6.05	6.95		13	18.05
1400	For horizontal blocking, 2" x 4", add		600	.027		.42	1.25		1.67	2.52
1500	2" x 6", add		600	.027		.67	1.25		1.92	2.78
1600	For openings, add		250	.064			3		3	4.92
1702	Headers for above openings, material only, add				B.F.	.76			.76	.84
9000	Minimum labor/equipment charge	1 Carp	4	2	Job		94		94	154

06 11 10.28 Porch or Deck Framing		Crew	Daily Output	Labor-Hours	Unit	Material	Labor	Equipment	Total	Total Incl O&P
0010	**PORCH OR DECK FRAMING**									
0100	Treated lumber, posts or columns, 4" x 4"	2 Carp	390	.041	L.F.	1.34	1.93		3.27	4.63
0110	4" x 6"		275	.058		1.91	2.73		4.64	6.60
0120	4" x 8"		220	.073		3.81	3.41		7.22	9.80
0130	Girder, single, 4" x 4"		675	.024		1.34	1.11		2.45	3.29
0140	4" x 6"		600	.027		1.91	1.25		3.16	4.15
0150	4" x 8"		525	.030		3.81	1.43		5.24	6.55
0160	Double, 2" x 4"		625	.026		1.05	1.20		2.25	3.12
0170	2" x 6"		600	.027		1.60	1.25		2.85	3.81
0180	2" x 8"		575	.028		2.15	1.31		3.46	4.51
0190	2" x 10"		550	.029		2.86	1.37		4.23	5.40
0200	2" x 12"		525	.030		3.95	1.43		5.38	6.70
0210	Triple, 2" x 4"		575	.028		1.57	1.31		2.88	3.86
0220	2" x 6"		550	.029		2.40	1.37		3.77	4.88
0230	2" x 8"		525	.030		3.23	1.43		4.66	5.90
0240	2" x 10"		500	.032		4.29	1.50		5.79	7.20
0250	2" x 12"		475	.034		5.95	1.58		7.53	9.10
0260	Ledger, bolted 4' O.C., 2" x 4"		400	.040		.66	1.88		2.54	3.81
0270	2" x 6"		395	.041		.93	1.90		2.83	4.14
0280	2" x 8"		390	.041		1.20	1.93		3.13	4.48
0290	2" x 10"		385	.042		1.54	1.95		3.49	4.90
0300	2" x 12"		380	.042		2.08	1.98		4.06	5.55
0310	Joists, 2" x 4"		1250	.013		.52	.60		1.12	1.56

06 11 10 – Framing with Dimensional, Engineered or Composite Lumber

06 11 10.28 Porch or Deck Framing		Crew	Daily Output	Labor-Hours	Unit	2015 Bare Costs Material	Labor	Equipment	Total	Total Incl O&P
0320	2" x 6"	2 Carp	1250	.013	L.F.	.79	.60		1.39	1.86
0330	2" x 8"		1100	.015		1.07	.68		1.75	2.30
0340	2" x 10"		900	.018		1.42	.83		2.25	2.94
0350	2" x 12"		875	.018		1.71	.86		2.57	3.29
0360	Railings and trim , 1" x 4"	1 Carp	300	.027		.47	1.25		1.72	2.57
0370	2" x 2"		300	.027		.33	1.25		1.58	2.42
0380	2" x 4"		300	.027		.51	1.25		1.76	2.61
0390	2" x 6"		300	.027		.79	1.25		2.04	2.91
0400	Decking, 1" x 4"		275	.029	S.F.	2.16	1.37		3.53	4.62
0410	2" x 4"		300	.027		1.73	1.25		2.98	3.95
0420	2" x 6"		320	.025		1.69	1.17		2.86	3.78
0430	5/4" x 6"		320	.025		2.25	1.17		3.42	4.40
0440	Balusters, square, 2" x 2"	2 Carp	660	.024	L.F.	.34	1.14		1.48	2.24
0450	Turned, 2" x 2"		420	.038		.45	1.79		2.24	3.42
0460	Stair stringer, 2" x 10"		130	.123		1.42	5.80		7.22	11
0470	2" x 12"		130	.123		1.71	5.80		7.51	11.35
0480	Stair treads, 1" x 4"		140	.114		2.17	5.35		7.52	11.20
0490	2" x 4"		140	.114		.52	5.35		5.87	9.35
0500	2" x 6"		160	.100		.78	4.70		5.48	8.55
0510	5/4" x 6"		160	.100		1.05	4.70		5.75	8.85
0520	Turned handrail post, 4" x 4"		64	.250	Ea.	31	11.75		42.75	53.50
0530	Lattice panel, 4' x 8', 1/2"		1600	.010	S.F.	.81	.47		1.28	1.66
0535	3/4"		1600	.010	"	1.21	.47		1.68	2.10
0540	Cedar, posts or columns, 4" x 4"		390	.041	L.F.	3.93	1.93		5.86	7.50
0550	4" x 6"		275	.058		7.35	2.73		10.08	12.60
0560	4" x 8"		220	.073		10	3.41		13.41	16.60
0800	Decking, 1" x 4"		550	.029		1.81	1.37		3.18	4.23
0810	2" x 4"		600	.027		3.62	1.25		4.87	6.05
0820	2" x 6"		640	.025		6.60	1.17		7.77	9.15
0830	5/4" x 6"		640	.025		4.45	1.17		5.62	6.80
0840	Railings and trim, 1" x 4"		600	.027		1.81	1.25		3.06	4.04
0860	2" x 4"		600	.027		3.62	1.25		4.87	6.05
0870	2" x 6"		600	.027		6.60	1.25		7.85	9.30
0920	Stair treads, 1" x 4"		140	.114		1.81	5.35		7.16	10.80
0930	2" x 4"		140	.114		3.62	5.35		8.97	12.80
0940	2" x 6"		160	.100		6.60	4.70		11.30	14.95
0950	5/4" x 6"		160	.100		4.45	4.70		9.15	12.60
0980	Redwood, posts or columns, 4" x 4"		390	.041		6.45	1.93		8.38	10.20
0990	4" x 6"		275	.058		12.60	2.73		15.33	18.35
1000	4" x 8"		220	.073		23.50	3.41		26.91	31.50
1240	Decking, 1" x 4"	1 Carp	275	.029	S.F.	3.97	1.37		5.34	6.60
1260	2" x 6"		340	.024		7.50	1.10		8.60	10.05
1270	5/4" x 6"		320	.025		4.77	1.17		5.94	7.15
1280	Railings and trim, 1" x 4"	2 Carp	600	.027	L.F.	1.17	1.25		2.42	3.34
1310	2" x 6"		600	.027		7.50	1.25		8.75	10.30
1420	Alternative decking, wood/plastic composite, 5/4" x 6" G		640	.025		3.14	1.17		4.31	5.35
1440	1" x 4" square edge fir		550	.029		2.18	1.37		3.55	4.64
1450	1" x 4" tongue and groove fir		450	.036		1.46	1.67		3.13	4.34
1460	1" x 4" mahogany		550	.029		2.01	1.37		3.38	4.45
1462	5/4" x 6" PVC		550	.029		3.73	1.37		5.10	6.35
1465	Framing, porch or deck, alt deck fastening, screws, add	1 Carp	240	.033	S.F.		1.56		1.56	2.56
1470	Accessories, joist hangers, 2" x 4"		160	.050	Ea.	.73	2.35		3.08	4.65
1480	2" x 6" through 2" x 12"		150	.053		1.39	2.50		3.89	5.65

06 11 10 – Framing with Dimensional, Engineered or Composite Lumber

06 11 10.28 Porch or Deck Framing		Crew	Daily Output	Labor-Hours	Unit	Material	2015 Bare Costs Labor	Equipment	Total	Total Incl O&P
1530	Post footing, incl excav, backfill, tube form & concrete, 4′ deep, 8″ dia	F-7	12	2.667	Ea.	12.15	113		125.15	198
1540	10″ diameter		11	2.909		17.75	123		140.75	222
1550	12″ diameter		10	3.200		23.50	135		158.50	248

06 11 10.30 Roof Framing

Code	Description	Crew	Daily Output	Labor-Hours	Unit	Material	Labor	Equipment	Total	Total Incl O&P
0010	**ROOF FRAMING**									
1900	Rough fascia, 2″ x 6″	2 Carp	250	.064	L.F.	.67	3		3.67	5.65
2000	2″ x 8″		225	.071		.92	3.34		4.26	6.45
2100	2″ x 10″		180	.089		1.34	4.17		5.51	8.30
5000	Rafters, to 4 in 12 pitch, 2″ x 6″, ordinary		1000	.016		.67	.75		1.42	1.96
5060	2″ x 8″, ordinary		950	.017		.92	.79		1.71	2.31
5120	2″ x 10″, ordinary		630	.025		1.34	1.19		2.53	3.42
5180	2″ x 12″, ordinary		575	.028		1.72	1.31		3.03	4.04
5300	Hip and valley rafters, 2″ x 6″, ordinary		760	.021		.67	.99		1.66	2.35
5360	2″ x 8″, ordinary		720	.022		.92	1.04		1.96	2.72
5420	2″ x 10″, ordinary		570	.028		1.34	1.32		2.66	3.63
5480	2″ x 12″, ordinary		525	.030		1.72	1.43		3.15	4.25
5540	Hip and valley jacks, 2″ x 6″, ordinary		600	.027		.67	1.25		1.92	2.78
5600	2″ x 8″, ordinary		490	.033		.92	1.53		2.45	3.52
5660	2″ x 10″, ordinary		450	.036		1.34	1.67		3.01	4.21
5720	2″ x 12″, ordinary		375	.043		1.72	2		3.72	5.20
5761	For slopes steeper than 4 in 12, add						30%			
5780	Rafter tie, 1″ x 4″, #3	2 Carp	800	.020	L.F.	.47	.94		1.41	2.06
5790	2″ x 4″, #3		800	.020		.42	.94		1.36	2.01
5800	Ridge board, #2 or better, 1″ x 6″		600	.027		.73	1.25		1.98	2.85
5820	1″ x 8″		550	.029		1.20	1.37		2.57	3.56
5840	1″ x 10″		500	.032		1.56	1.50		3.06	4.17
5860	2″ x 6″		500	.032		.67	1.50		2.17	3.19
5880	2″ x 8″		450	.036		.92	1.67		2.59	3.75
5900	2″ x 10″		400	.040		1.34	1.88		3.22	4.55
5920	Roof cants, split, 4″ x 4″		650	.025		1.73	1.16		2.89	3.79
5940	6″ x 6″		600	.027		5.05	1.25		6.30	7.60
5960	Roof curbs, untreated, 2″ x 6″		520	.031		.67	1.44		2.11	3.10
5980	2″ x 12″		400	.040		1.72	1.88		3.60	4.98
6000	Sister rafters, 2″ x 6″		800	.020		.67	.94		1.61	2.27
6020	2″ x 8″		640	.025		.92	1.17		2.09	2.93
6040	2″ x 10″		535	.030		1.34	1.40		2.74	3.77
6060	2″ x 12″		455	.035		1.72	1.65		3.37	4.61
9000	Minimum labor/equipment charge	1 Carp	4	2	Job		94		94	154

06 11 10.32 Sill and Ledger Framing

Code	Description	Crew	Daily Output	Labor-Hours	Unit	Material	Labor	Equipment	Total	Total Incl O&P
0010	**SILL AND LEDGER FRAMING**									
2002	Ledgers, nailed, 2″ x 4″	2 Carp	755	.021	L.F.	.42	.99		1.41	2.10
2052	2″ x 6″		600	.027		.67	1.25		1.92	2.78
2102	Bolted, not including bolts, 3″ x 6″		325	.049		1.88	2.31		4.19	5.85
2152	3″ x 12″		233	.069		3.83	3.22		7.05	9.50
2602	Mud sills, redwood, construction grade, 2″ x 4″		895	.018		2.25	.84		3.09	3.85
2622	2″ x 6″		780	.021		3.40	.96		4.36	5.30
4002	Sills, 2″ x 4″		600	.027		.42	1.25		1.67	2.51
4052	2″ x 6″		550	.029		.66	1.37		2.03	2.96
4082	2″ x 8″		500	.032		.91	1.50		2.41	3.46
4101	2″ x 10″		450	.036		1.32	1.67		2.99	4.19
4121	2″ x 12″		400	.040		1.70	1.88		3.58	4.95
4202	Treated, 2″ x 4″		550	.029		.50	1.37		1.87	2.79

06 11 10 – Framing with Dimensional, Engineered or Composite Lumber

06 11 10.32 Sill and Ledger Framing		Crew	Daily Output	Labor-Hours	Unit	2015 Bare Costs Material	Labor	Equipment	Total	Total Incl O&P
4222	2″ x 6″	2 Carp	500	.032	L.F.	.77	1.50		2.27	3.31
4242	2″ x 8″		450	.036		1.04	1.67		2.71	3.88
4261	2″ x 10″		400	.040		1.38	1.88		3.26	4.60
4281	2″ x 12″		350	.046		1.92	2.15		4.07	5.65
4402	4″ x 4″		450	.036		1.30	1.67		2.97	4.17
4422	4″ x 6″		350	.046		1.85	2.15		4	5.55
4462	4″ x 8″		300	.053		3.73	2.50		6.23	8.20
4481	4″ x 10″		260	.062		4.67	2.89		7.56	9.90
9000	Minimum labor/equipment charge	1 Carp	4	2	Job		94		94	154

06 11 10.34 Sleepers

		Crew	Daily Output	Labor-Hours	Unit	Material	Labor	Equipment	Total	Total Incl O&P
0010	**SLEEPERS**									
0100	On concrete, treated, 1″ x 2″	2 Carp	2350	.007	L.F.	.27	.32		.59	.82
0150	1″ x 3″		2000	.008		.45	.38		.83	1.12
0200	2″ x 4″		1500	.011		.55	.50		1.05	1.42
0250	2″ x 6″		1300	.012		.86	.58		1.44	1.90
9000	Minimum labor/equipment charge	1 Carp	4	2	Job		94		94	154

06 11 10.36 Soffit and Canopy Framing

		Crew	Daily Output	Labor-Hours	Unit	Material	Labor	Equipment	Total	Total Incl O&P
0010	**SOFFIT AND CANOPY FRAMING**									
1002	Canopy or soffit framing , 1″ x 4″	2 Carp	900	.018	L.F.	.47	.83		1.30	1.89
1021	1″ x 6″		850	.019		.73	.88		1.61	2.25
1042	1″ x 8″		750	.021		1.20	1		2.20	2.96
1102	2″ x 4″		620	.026		.42	1.21		1.63	2.46
1121	2″ x 6″		560	.029		.67	1.34		2.01	2.93
1142	2″ x 8″		500	.032		.92	1.50		2.42	3.47
1202	3″ x 4″		500	.032		1.07	1.50		2.57	3.64
1221	3″ x 6″		400	.040		1.90	1.88		3.78	5.15
1242	3″ x 10″		300	.053		3.22	2.50		5.72	7.65
9000	Minimum labor/equipment charge	1 Carp	4	2	Job		94		94	154

06 11 10.38 Treated Lumber Framing Material

		Crew	Daily Output	Labor-Hours	Unit	Material	Labor	Equipment	Total	Total Incl O&P
0010	**TREATED LUMBER FRAMING MATERIAL**									
0100	2″ x 4″				M.B.F.	755			755	830
0110	2″ x 6″					770			770	845
0120	2″ x 8″					780			780	855
0130	2″ x 10″					830			830	910
0140	2″ x 12″					960			960	1,050
0200	4″ x 4″					975			975	1,075
0210	4″ x 6″					925			925	1,025
0220	4″ x 8″					1,400			1,400	1,550

06 11 10.40 Wall Framing

		Crew	Daily Output	Labor-Hours	Unit	Material	Labor	Equipment	Total	Total Incl O&P
0010	**WALL FRAMING** R061110-30									
0100	Door buck, studs, header, access, 8′ H, 2″ x 4″ wall, 3′ W	1 Carp	32	.250	Ea.	17.85	11.75		29.60	39
0110	4′ wide		32	.250		19.15	11.75		30.90	40.50
0120	5′ wide		32	.250		23	11.75		34.75	45
0130	6′ wide		32	.250		25	11.75		36.75	47
0140	8′ wide		30	.267		35.50	12.50		48	59.50
0150	10′ wide		30	.267		48.50	12.50		61	74
0160	12′ wide		30	.267		62.50	12.50		75	89
0170	2″ x 6″ wall, 3′ wide		32	.250		25.50	11.75		37.25	47.50
0180	4′ wide		32	.250		27	11.75		38.75	49
0190	5′ wide		32	.250		31	11.75		42.75	53.50
0200	6′ wide		32	.250		33	11.75		44.75	55.50
0210	8′ wide		30	.267		43.50	12.50		56	68

06 11 10 – Framing with Dimensional, Engineered or Composite Lumber

06 11 10.40 Wall Framing		Crew	Daily Output	Labor-Hours	Unit	2015 Bare Costs Material	Labor	Equipment	Total	Total Incl O&P
0220	10′ wide	1 Carp	30	.267	Ea.	56.50	12.50		69	82.50
0230	12′ wide		30	.267		70.50	12.50		83	98
0240	Window buck, studs, header & access, 8′ high 2″ x 4″ wall, 2′ wide		24	.333		18.80	15.65		34.45	46
0250	3′ wide		24	.333		22	15.65		37.65	50
0260	4′ wide		24	.333		24	15.65		39.65	52
0270	5′ wide		24	.333		28	15.65		43.65	56.50
0280	6′ wide		24	.333		31.50	15.65		47.15	60
0290	7′ wide		24	.333		40	15.65		55.65	69.50
0300	8′ wide		22	.364		44.50	17.05		61.55	76.50
0310	10′ wide		22	.364		59	17.05		76.05	92.50
0320	12′ wide		22	.364		75	17.05		92.05	111
0330	2″ x 6″ wall, 2′ wide		24	.333		28.50	15.65		44.15	57
0340	3′ wide		24	.333		32	15.65		47.65	60.50
0350	4′ wide		24	.333		34.50	15.65		50.15	63.50
0360	5′ wide		24	.333		39	15.65		54.65	68.50
0370	6′ wide		24	.333		43	15.65		58.65	72.50
0380	7′ wide		24	.333		52.50	15.65		68.15	83.50
0390	8′ wide		22	.364		57.50	17.05		74.55	91
0400	10′ wide		22	.364		72.50	17.05		89.55	108
0410	12′ wide		22	.364		90.50	17.05		107.55	128
2002	Headers over openings, 2″ x 6″	2 Carp	360	.044	L.F.	.67	2.09		2.76	4.15
2007	2″ x 6″, pneumatic nailed		432	.037		.67	1.74		2.41	3.59
2052	2″ x 8″		340	.047		.92	2.21		3.13	4.63
2057	2″ x 8″, pneumatic nailed		408	.039		.93	1.84		2.77	4.04
2101	2″ x 10″		320	.050		1.34	2.35		3.69	5.30
2106	2″ x 10″, pneumatic nailed		384	.042		1.35	1.96		3.31	4.69
2152	2″ x 12″		300	.053		1.72	2.50		4.22	6
2157	2″ x 12″, pneumatic nailed		360	.044		1.73	2.09		3.82	5.35
2180	4″ x 8″		260	.062		2.89	2.89		5.78	7.90
2185	4″ x 8″, pneumatic nailed		312	.051		2.90	2.41		5.31	7.15
2191	4″ x 10″		240	.067		4.28	3.13		7.41	9.85
2196	4″ x 10″, pneumatic nailed		288	.056		4.30	2.61		6.91	9
2202	4″ x 12″		190	.084		5.15	3.95		9.10	12.15
2207	4″ x 12″, pneumatic nailed		228	.070		5.15	3.30		8.45	11.10
2241	6″ x 10″		165	.097		6.45	4.55		11	14.55
2246	6″ x 10″, pneumatic nailed		198	.081		6.50	3.79		10.29	13.35
2251	6″ x 12″		140	.114		8.20	5.35		13.55	17.85
2256	6″ x 12″, pneumatic nailed		168	.095		8.25	4.47		12.72	16.40
5002	Plates, untreated, 2″ x 3″		850	.019		.39	.88		1.27	1.88
5007	2″ x 3″, pneumatic nailed		1020	.016		.40	.74		1.14	1.64
5022	2″ x 4″		800	.020		.42	.94		1.36	2.01
5027	2″ x 4″, pneumatic nailed		960	.017		.43	.78		1.21	1.75
5041	2″ x 6″		750	.021		.67	1		1.67	2.37
5046	2″ x 6″, pneumatic nailed		900	.018		.67	.83		1.50	2.11
5122	Studs, 8′ high wall, 2″ x 3″		1200	.013		.39	.63		1.02	1.46
5127	2″ x 3″, pneumatic nailed		1440	.011		.40	.52		.92	1.28
5142	2″ x 4″		1100	.015		.42	.68		1.10	1.58
5147	2″ x 4″, pneumatic nailed		1320	.012		.43	.57		1	1.40
5162	2″ x 6″		1000	.016		.67	.75		1.42	1.96
5167	2″ x 6″, pneumatic nailed		1200	.013		.67	.63		1.30	1.77
5182	3″ x 4″		800	.020		1.07	.94		2.01	2.72
5187	3″ x 4″, pneumatic nailed		960	.017		1.08	.78		1.86	2.46
8200	For 12′ high walls, deduct						5%			

06 11 Wood Framing

06 11 10 – Framing with Dimensional, Engineered or Composite Lumber

06 11 10.40 Wall Framing		Crew	Daily Output	Labor-Hours	Unit	2015 Bare Costs Material	Labor	Equipment	Total	Total Incl O&P
8220	For stub wall, 6′ high, add						20%			
8240	3′ high, add						40%			
8250	For second story & above, add						5%			
8300	For dormer & gable, add						15%			
9000	Minimum labor/equipment charge	1 Carp	4	2	Job		94		94	154

06 11 10.42 Furring

Code	Description	Crew	Daily Output	Labor-Hours	Unit	Material	Labor	Equipment	Total	Total Incl O&P
0010	**FURRING**									
0012	Wood strips, 1″ x 2″, on walls, on wood	1 Carp	550	.015	L.F.	.24	.68		.92	1.38
0015	On wood, pneumatic nailed		710	.011		.24	.53		.77	1.13
0300	On masonry		495	.016		.26	.76		1.02	1.53
0400	On concrete		260	.031		.26	1.44		1.70	2.66
0600	1″ x 3″, on walls, on wood		550	.015		.39	.68		1.07	1.54
0605	On wood, pneumatic nailed		710	.011		.39	.53		.92	1.29
0700	On masonry		495	.016		.42	.76		1.18	1.70
0800	On concrete		260	.031		.42	1.44		1.86	2.83
0850	On ceilings, on wood		350	.023		.39	1.07		1.46	2.18
0855	On wood, pneumatic nailed		450	.018		.39	.83		1.22	1.79
0900	On masonry		320	.025		.42	1.17		1.59	2.38
0950	On concrete		210	.038		.42	1.79		2.21	3.39
9000	Minimum labor/equipment charge		4	2	Job		94		94	154

06 11 10.44 Grounds

Code	Description	Crew	Daily Output	Labor-Hours	Unit	Material	Labor	Equipment	Total	Total Incl O&P
0010	**GROUNDS**									
0020	For casework, 1″ x 2″ wood strips, on wood	1 Carp	330	.024	L.F.	.24	1.14		1.38	2.13
0100	On masonry		285	.028		.26	1.32		1.58	2.45
0200	On concrete		250	.032		.26	1.50		1.76	2.75
0400	For plaster, 3/4″ deep, on wood		450	.018		.24	.83		1.07	1.63
0500	On masonry		225	.036		.26	1.67		1.93	3.03
0600	On concrete		175	.046		.26	2.15		2.41	3.81
0700	On metal lath		200	.040		.26	1.88		2.14	3.37
9000	Minimum labor/equipment charge		4	2	Job		94		94	154

06 12 Structural Panels

06 12 10 – Structural Insulated Panels

06 12 10.10 OSB Faced Panels

Code	Description		Crew	Daily Output	Labor-Hours	Unit	Material	Labor	Equipment	Total	Total Incl O&P
0010	**OSB FACED PANELS**										
0100	Structural insul. panels, 7/16″ OSB both faces, EPS insul, 3-5/8″ T	G	F-3	2075	.019	S.F.	3.60	.92	.32	4.84	5.80
0110	5-5/8″ thick	G		1725	.023		4.05	1.11	.38	5.54	6.70
0120	7-3/8″ thick	G		1425	.028		4.40	1.34	.46	6.20	7.55
0130	9-3/8″ thick	G		1125	.036		4.70	1.70	.58	6.98	8.55
0140	7/16″ OSB one face, EPS insul, 3-5/8″ thick	G		2175	.018		3.70	.88	.30	4.88	5.85
0150	5-5/8″ thick	G		1825	.022		4.30	1.05	.36	5.71	6.85
0160	7-3/8″ thick	G		1525	.026		4.80	1.26	.43	6.49	7.80
0170	9-3/8″ thick	G		1225	.033		5.30	1.56	.53	7.39	9
0190	7/16″ OSB - 1/2″ GWB faces , EPS insul, 3-5/8″ T	G		2075	.019		3.29	.92	.32	4.53	5.45
0200	5-5/8″ thick	G		1725	.023		3.90	1.11	.38	5.39	6.50
0210	7-3/8″ thick	G		1425	.028		4.42	1.34	.46	6.22	7.55
0220	9-3/8″ thick	G		1125	.036		5	1.70	.58	7.28	8.90
0240	7/16″ OSB - 1/2″ MRGWB faces , EPS insul, 3-5/8″ T	G		2075	.019		3.39	.92	.32	4.63	5.60
0250	5-5/8″ thick	G		1725	.023		4	1.11	.38	5.49	6.65
0260	7-3/8″ thick	G		1425	.028		4.52	1.34	.46	6.32	7.65

06 12 Structural Panels

06 12 10 – Structural Insulated Panels

06 12 10.10 OSB Faced Panels			Crew	Daily Output	Labor-Hours	Unit	2015 Bare Costs Material	Labor	Equipment	Total	Total Incl O&P
0270	9-3/8" thick	G	F-3	1125	.036	S.F.	5.10	1.70	.58	7.38	9.05
0300	For 1/2" GWB added to OSB skin, add	G					1.28			1.28	1.41
0310	For 1/2" MRGWB added to OSB skin, add	G					1.28			1.28	1.41
0320	For one T1-11 skin, add to OSB-OSB	G					1.90			1.90	2.09
0330	For one 19/32" CDX skin, add to OSB-OSB	G					1.46			1.46	1.61
0500	Structural insulated panel, 7/16" OSB both sides, straw core										
0510	4-3/8" T, walls (w/sill, splines, plates)	G	F-6	2400	.017	S.F.	7.40	.74	.27	8.41	9.60
0520	Floors (w/splines)	G		2400	.017		7.40	.74	.27	8.41	9.60
0530	Roof (w/splines)	G		2400	.017		7.40	.74	.27	8.41	9.60
0550	7-7/8" T, walls (w/sill, splines, plates)	G		2400	.017		11.20	.74	.27	12.21	13.80
0560	Floors (w/splines)	G		2400	.017		11.20	.74	.27	12.21	13.80
0570	Roof (w/splines)	G		2400	.017		11.20	.74	.27	12.21	13.80

06 12 19 – Composite Shearwall Panels

06 12 19.10 Steel and Wood Composite Shearwall Panels		Crew	Daily Output	Labor-Hours	Unit	Material	Labor	Equipment	Total	Total Incl O&P
0010	STEEL & WOOD COMPOSITE SHEARWALL PANELS									
0020	Anchor bolts, 36" long (must be placed in wet concrete)	1 Carp	150	.053	Ea.	32	2.50		34.50	39.50
0030	On concrete, 2 x 4 & 2 x 6 walls, 7' - 10' high, 360 lb. shear, 12" wide	2 Carp	8	2		420	94		514	615
0040	715 lb. shear, 15" wide		8	2		500	94		594	705
0050	1860 lb. shear, 18" wide		8	2		520	94		614	725
0060	2780 lb. shear, 21" wide		8	2		620	94		714	835
0070	3790 lb. shear, 24" wide		8	2		710	94		804	935
0080	2 x 6 walls, 11' to 13' high, 1180 lb. shear, 18" wide		6	2.667		635	125		760	905
0090	1555 lb. shear, 21" wide		6	2.667		800	125		925	1,075
0100	2280 lb. shear, 24" wide		6	2.667		895	125		1,020	1,200
0110	For installing above on wood floor frame, add									
0120	Coupler nuts, threaded rods, bolts, shear transfer plate kit	1 Carp	16	.500	Ea.	63	23.50		86.50	108
0130	Framing anchors, angle (2 required)	"	96	.083	"	2.61	3.91		6.52	9.25
0140	For blocking see Section 06 11 10.02									
0150	For installing above, first floor to second floor, wood floor frame, add									
0160	Add stack option to first floor wall panel				Ea.	69.50			69.50	76
0170	Threaded rods, bolts, shear transfer plate kit	1 Carp	16	.500		76	23.50		99.50	122
0180	Framing anchors, angle (2 required)	"	96	.083		2.61	3.91		6.52	9.25
0190	For blocking see section 06 11 10.02									
0200	For installing stacked panels, balloon framing									
0210	Add stack option to first floor wall panel				Ea.	69.50			69.50	76
0220	Threaded rods, bolts kit	1 Carp	16	.500	"	44	23.50		67.50	87

06 13 Heavy Timber Construction

06 13 23 – Heavy Timber Framing

06 13 23.10 Heavy Framing		Crew	Daily Output	Labor-Hours	Unit	Material	Labor	Equipment	Total	Total Incl O&P
0010	HEAVY FRAMING									
0020	Beams, single 6" x 10"	2 Carp	1.10	14.545	M.B.F.	1,600	685		2,285	2,875
0100	Single 8" x 16"		1.20	13.333	"	2,000	625		2,625	3,225
0202	Built from 2" lumber, multiple 2" x 14"		900	.018	B.F.	.85	.83		1.68	2.31
0212	Built from 3" lumber, multiple 3" x 6"		700	.023		1.25	1.07		2.32	3.14
0222	Multiple 3" x 8"		800	.020		1.27	.94		2.21	2.93
0232	Multiple 3" x 10"		900	.018		1.28	.83		2.11	2.77
0242	Multiple 3" x 12"		1000	.016		1.28	.75		2.03	2.63
0252	Built from 4" lumber, multiple 4" x 6"		800	.020		1.41	.94		2.35	3.09
0262	Multiple 4" x 8"		900	.018		1.07	.83		1.90	2.55

06 13 Heavy Timber Construction

06 13 23 – Heavy Timber Framing

06 13 23.10 Heavy Framing		Crew	Daily Output	Labor-Hours	Unit	Material	2015 Bare Costs Labor	Equipment	Total	Total Incl O&P
0272	Multiple 4″ x 10″	2 Carp	1000	.016	B.F.	1.28	.75		2.03	2.63
0282	Multiple 4″ x 12″		1100	.015		1.28	.68		1.96	2.52
0292	Columns, structural grade, 1500f, 4″ x 4″		450	.036	L.F.	1.70	1.67		3.37	4.61
0302	6″ x 6″		225	.071		4.21	3.34		7.55	10.10
0402	8″ x 8″		240	.067		7.20	3.13		10.33	13.05
0502	10″ x 10″		90	.178		12.10	8.35		20.45	27
0602	12″ x 12″		70	.229		18.35	10.75		29.10	37.50
0802	Floor planks, 2″ thick, T & G, 2″ x 6″		1050	.015	B.F.	1.59	.72		2.31	2.92
0902	2″ x 10″		1100	.015		1.59	.68		2.27	2.87
1102	3″ thick, 3″ x 6″		1050	.015		1.59	.72		2.31	2.92
1202	3″ x 10″		1100	.015		1.59	.68		2.27	2.87
1402	Girders, structural grade, 12″ x 12″		800	.020		1.53	.94		2.47	3.22
1502	10″ x 16″		1000	.016		2.40	.75		3.15	3.87
2302	Roof purlins, 4″ thick, structural grade		1050	.015		1.07	.72		1.79	2.35
9000	Minimum labor/equipment charge	1 Carp	2	4	Job		188		188	310

06 15 Wood Decking

06 15 16 – Wood Roof Decking

06 15 16.10 Solid Wood Roof Decking		Crew	Daily Output	Labor-Hours	Unit	Material	Labor	Equipment	Total	Total Incl O&P
0010	**SOLID WOOD ROOF DECKING**									
0350	Cedar planks, 2″ thick	2 Carp	350	.046	S.F.	6.25	2.15		8.40	10.35
0400	3″ thick		320	.050		9.35	2.35		11.70	14.15
0500	4″ thick		250	.064		12.45	3		15.45	18.60
0550	6″ thick		200	.080		18.70	3.76		22.46	26.50
0650	Douglas fir, 2″ thick		350	.046		2.69	2.15		4.84	6.50
0700	3″ thick		320	.050		4.04	2.35		6.39	8.30
0800	4″ thick		250	.064		5.40	3		8.40	10.80
0850	6″ thick		200	.080		8.05	3.76		11.81	15.05
0950	Hemlock, 2″ thick		350	.046		2.74	2.15		4.89	6.55
1000	3″ thick		320	.050		4.11	2.35		6.46	8.35
1100	4″ thick		250	.064		5.50	3		8.50	10.90
1150	6″ thick		200	.080		8.20	3.76		11.96	15.20
1250	Western white spruce, 2″ thick		350	.046		1.75	2.15		3.90	5.45
1300	3″ thick		320	.050		2.62	2.35		4.97	6.75
1400	4″ thick		250	.064		3.50	3		6.50	8.75
1450	6″ thick		200	.080		5.25	3.76		9.01	11.90
9000	Minimum labor/equipment charge	1 Carp	2	4	Job		188		188	310

06 15 23 – Laminated Wood Decking

06 15 23.10 Laminated Roof Deck		Crew	Daily Output	Labor-Hours	Unit	Material	Labor	Equipment	Total	Total Incl O&P
0010	**LAMINATED ROOF DECK**									
0020	Pine or hemlock, 3″ thick	2 Carp	425	.038	S.F.	6.20	1.77		7.97	9.70
0100	4″ thick		325	.049		8.10	2.31		10.41	12.70
0300	Cedar, 3″ thick		425	.038		7	1.77		8.77	10.60
0400	4″ thick		325	.049		9.40	2.31		11.71	14.15
0600	Fir, 3″ thick		425	.038		5.40	1.77		7.17	8.85
0700	4″ thick		325	.049		7.40	2.31		9.71	11.95
9000	Minimum labor/equipment charge	1 Carp	3	2.667	Job		125		125	205

06 16 Sheathing

06 16 13 – Insulating Sheathing

06 16 13.10 Insulating Sheathing		Crew	Daily Output	Labor-Hours	Unit	Material	Labor	Equipment	Total	Total Incl O&P
						2015 Bare Costs				
0010	**INSULATING SHEATHING**									
0020	Expanded polystyrene, 1#/C.F. density, 3/4" thick R2.89 G	2 Carp	1400	.011	S.F.	.33	.54		.87	1.24
0030	1" thick R3.85 G		1300	.012		.39	.58		.97	1.38
0040	2" thick R7.69 G		1200	.013		.64	.63		1.27	1.74
0050	Extruded polystyrene, 15 PSI compressive strength, 1" thick, R5 G		1300	.012		.69	.58		1.27	1.71
0060	2" thick, R10 G		1200	.013		.86	.63		1.49	1.97
0070	Polyisocyanurate, 2#/C.F. density, 3/4" thick G		1400	.011		.60	.54		1.14	1.54
0080	1" thick G		1300	.012		.62	.58		1.20	1.64
0090	1-1/2" thick G		1250	.013		.78	.60		1.38	1.85
0100	2" thick G	↓	1200	.013	↓	.97	.63		1.60	2.10

06 16 23 – Subflooring

06 16 23.10 Subfloor		Crew	Daily Output	Labor-Hours	Unit	Material	Labor	Equipment	Total	Total Incl O&P
0010	**SUBFLOOR**									
0011	Plywood, CDX, 1/2" thick	2 Carp	1500	.011	SF Flr.	.66	.50		1.16	1.54
0017	Pneumatic nailed		1860	.009		.66	.40		1.06	1.38
0102	5/8" thick		1350	.012		.78	.56		1.34	1.77
0107	Pneumatic nailed		1674	.010		.78	.45		1.23	1.60
0202	3/4" thick		1250	.013		.93	.60		1.53	2.01
0207	Pneumatic nailed		1550	.010		.93	.48		1.41	1.81
0302	1-1/8" thick, 2-4-1 including underlayment		1050	.015		2.03	.72		2.75	3.40
0440	With boards, 1" x 6", S4S, laid regular		900	.018		1.57	.83		2.40	3.10
0452	1" x 8", laid regular		1000	.016		1.90	.75		2.65	3.32
0462	Laid diagonal		850	.019		1.90	.88		2.78	3.54
0502	1" x 10", laid regular		1100	.015		1.95	.68		2.63	3.26
0602	Laid diagonal	↓	900	.018	↓	1.95	.83		2.78	3.51
8990	Subfloor adhesive, 3/8" bead	1 Carp	2300	.003	L.F.	.12	.16		.28	.40
9000	Minimum labor/equipment charge	"	4	2	Job		94		94	154

06 16 26 – Underlayment

06 16 26.10 Wood Product Underlayment		Crew	Daily Output	Labor-Hours	Unit	Material	Labor	Equipment	Total	Total Incl O&P
0010	**WOOD PRODUCT UNDERLAYMENT** R061636-20									
0015	Plywood, underlayment grade, 1/4" thick	2 Carp	1500	.011	S.F.	.83	.50		1.33	1.73
0018	Pneumatic nailed		1860	.009		.83	.40		1.23	1.57
0030	3/8" thick		1500	.011		.92	.50		1.42	1.83
0070	Pneumatic nailed		1860	.009		.92	.40		1.32	1.67
0102	1/2" thick		1450	.011		1.10	.52		1.62	2.06
0107	Pneumatic nailed		1798	.009		1.10	.42		1.52	1.89
0202	5/8" thick		1400	.011		1.20	.54		1.74	2.20
0207	Pneumatic nailed		1736	.009		1.20	.43		1.63	2.03
0302	3/4" thick		1300	.012		1.43	.58		2.01	2.52
0306	Pneumatic nailed		1612	.010		1.43	.47		1.90	2.33
0502	Particle board, 3/8" thick		1500	.011		.40	.50		.90	1.26
0507	Pneumatic nailed		1860	.009		.40	.40		.80	1.10
0602	1/2" thick		1450	.011		.43	.52		.95	1.32
0607	Pneumatic nailed		1798	.009		.43	.42		.85	1.15
0802	5/8" thick		1400	.011		.52	.54		1.06	1.45
0807	Pneumatic nailed		1736	.009		.52	.43		.95	1.28
0902	3/4" thick		1300	.012		.63	.58		1.21	1.64
0907	Pneumatic nailed		1612	.010		.63	.47		1.10	1.45
1100	Hardboard, underlayment grade, 4' x 4', .215" thick G	↓	1500	.011	↓	.58	.50		1.08	1.46
9000	Minimum labor/equipment charge	1 Carp	4	2	Job		94		94	154

06 16 Sheathing

06 16 33 – Wood Board Sheathing

06 16 33.10 Board Sheathing		Crew	Daily Output	Labor-Hours	Unit	Material	2015 Bare Costs Labor	Equipment	Total	Total Incl O&P
0009	**BOARD SHEATHING**									
0010	Roof, 1" x 6" boards, laid horizontal	2 Carp	725	.022	S.F.	1.57	1.04		2.61	3.43
0020	On steep roof		520	.031		1.57	1.44		3.01	4.10
0040	On dormers, hips, & valleys		480	.033		1.57	1.56		3.13	4.29
0050	Laid diagonal		650	.025		1.57	1.16		2.73	3.62
0070	1" x 8" boards, laid horizontal		875	.018		1.90	.86		2.76	3.50
0080	On steep roof		635	.025		1.90	1.18		3.08	4.03
0090	On dormers, hips, & valleys		580	.028		1.95	1.30		3.25	4.26
0100	Laid diagonal		725	.022		1.90	1.04		2.94	3.79
0110	Skip sheathing, 1" x 4", 7" OC	1 Carp	1200	.007		.58	.31		.89	1.15
0120	1" x 6", 9" OC		1450	.006		.72	.26		.98	1.21
0180	Tongue and groove sheathing/decking, 1" x 6"		1000	.008		1.80	.38		2.18	2.60
0190	2" x 6"		1000	.008		3.81	.38		4.19	4.81
0200	Walls, 1" x 6" boards, laid regular	2 Carp	650	.025		1.57	1.16		2.73	3.62
0210	Laid diagonal		585	.027		1.57	1.28		2.85	3.83
0220	1" x 8" boards, laid regular		765	.021		1.90	.98		2.88	3.70
0230	Laid diagonal		650	.025		1.90	1.16		3.06	3.98

06 16 36 – Wood Panel Product Sheathing

06 16 36.10 Sheathing

		Crew	Daily Output	Labor-Hours	Unit	Material	Labor	Equipment	Total	Total Incl O&P
0010	**SHEATHING** R061636-20									
0012	Plywood on roofs, CDX									
0032	5/16" thick	2 Carp	1600	.010	S.F.	.56	.47		1.03	1.38
0037	Pneumatic nailed R061110-30		1952	.008		.56	.39		.95	1.24
0052	3/8" thick		1525	.010		.60	.49		1.09	1.47
0057	Pneumatic nailed		1860	.009		.60	.40		1	1.32
0102	1/2" thick		1400	.011		.66	.54		1.20	1.60
0103	Pneumatic nailed		1708	.009		.66	.44		1.10	1.44
0202	5/8" thick		1300	.012		.78	.58		1.36	1.81
0207	Pneumatic nailed		1586	.010		.78	.47		1.25	1.64
0302	3/4" thick		1200	.013		.93	.63		1.56	2.05
0307	Pneumatic nailed		1464	.011		.93	.51		1.44	1.86
0502	Plywood on walls, with exterior CDX, 3/8" thick		1200	.013		.60	.63		1.23	1.69
0507	Pneumatic nailed		1488	.011		.60	.50		1.10	1.49
0602	1/2" thick		1125	.014		.66	.67		1.33	1.81
0607	Pneumatic nailed		1395	.011		.66	.54		1.20	1.60
0702	5/8" thick		1050	.015		.78	.72		1.50	2.03
0707	Pneumatic nailed		1302	.012		.78	.58		1.36	1.81
0802	3/4" thick		975	.016		.93	.77		1.70	2.28
0807	Pneumatic nailed		1209	.013		.93	.62		1.55	2.04
1000	For shear wall construction, add						20%			
1200	For structural 1 exterior plywood, add				S.F.	10%				
3000	Wood fiber, regular, no vapor barrier, 1/2" thick	2 Carp	1200	.013		.61	.63		1.24	1.70
3100	5/8" thick		1200	.013		.77	.63		1.40	1.88
3300	No vapor barrier, in colors, 1/2" thick		1200	.013		.74	.63		1.37	1.84
3400	5/8" thick		1200	.013		.78	.63		1.41	1.89
3600	With vapor barrier one side, white, 1/2" thick		1200	.013		.60	.63		1.23	1.69
3700	Vapor barrier 2 sides, 1/2" thick		1200	.013		.83	.63		1.46	1.94
3800	Asphalt impregnated, 25/32" thick		1200	.013		.33	.63		.96	1.39
3850	Intermediate, 1/2" thick		1200	.013		.25	.63		.88	1.31
4500	Oriented strand board, on roof, 7/16" thick G		1460	.011		.50	.51		1.01	1.39
4505	Pneumatic nailed G		1780	.009		.50	.42		.92	1.24
4550	1/2" thick G		1400	.011		.50	.54		1.04	1.43

06 16 Sheathing

06 16 36 – Wood Panel Product Sheathing

06 16 36.10 Sheathing		Crew	Daily Output	Labor-Hours	Unit	Material	Labor	Equipment	Total	Total Incl O&P
						2015 Bare Costs				
4555	Pneumatic nailed G	2 Carp	1736	.009	S.F.	.50	.43		.93	1.26
4600	5/8" thick G		1300	.012		.69	.58		1.27	1.71
4605	Pneumatic nailed G		1586	.010		.69	.47		1.16	1.54
4610	On walls, 7/16" thick		1200	.013		.50	.63		1.13	1.58
4615	Pneumatic nailed		1488	.011		.50	.50		1	1.38
4620	1/2" thick		1195	.013		.50	.63		1.13	1.58
4625	Pneumatic nailed		1325	.012		.50	.57		1.07	1.48
4630	5/8" thick		1050	.015		.69	.72		1.41	1.93
4635	Pneumatic nailed		1302	.012		.69	.58		1.27	1.71
4700	Oriented strand board, factory laminated W.R. barrier, on roof, 1/2" thick G		1400	.011		.78	.54		1.32	1.74
4705	Pneumatic nailed G		1736	.009		.78	.43		1.21	1.57
4720	5/8" thick G		1300	.012		.93	.58		1.51	1.97
4725	Pneumatic nailed G		1586	.010		.93	.47		1.40	1.80
4730	5/8" thick, T&G G		1150	.014		.93	.65		1.58	2.09
4735	Pneumatic nailed, T&G G		1400	.011		.93	.54		1.47	1.90
4740	On walls, 7/16" thick G		1200	.013		.66	.63		1.29	1.76
4745	Pneumatic nailed G		1488	.011		.66	.50		1.16	1.56
4750	1/2" thick G		1195	.013		.78	.63		1.41	1.89
4755	Pneumatic nailed G		1325	.012	↓	.78	.57		1.35	1.79
4800	Joint sealant tape, 3-1/2"		7600	.002	L.F.	.31	.10		.41	.50
4810	Joint sealant tape, 6"	↓	7600	.002	"	.44	.10		.54	.64
9000	Minimum labor/equipment charge	1 Carp	2	4	Job		188		188	310

06 16 43 – Gypsum Sheathing

06 16 43.10 Gypsum Sheathing		Crew	Daily Output	Labor-Hours	Unit	Material	Labor	Equipment	Total	Total Incl O&P
0010	**GYPSUM SHEATHING**									
0020	Gypsum, weatherproof, 1/2" thick	2 Carp	1125	.014	S.F.	.45	.67		1.12	1.59
0040	With embedded glass mats	"	1100	.015	"	.72	.68		1.40	1.91

06 17 Shop-Fabricated Structural Wood

06 17 33 – Wood I-Joists

06 17 33.10 Wood and Composite I-Joists		Crew	Daily Output	Labor-Hours	Unit	Material	Labor	Equipment	Total	Total Incl O&P
0010	**WOOD AND COMPOSITE I-JOISTS**									
0100	Plywood webs, incl. bridging & blocking, panels 24" O.C.									
1200	15' to 24' span, 50 psf live load	F-5	2400	.013	SF Flr.	2.02	.63		2.65	3.27
1300	55 psf live load		2250	.014		2.23	.67		2.90	3.56
1400	24' to 30' span, 45 psf live load		2600	.012		2.36	.58		2.94	3.55
1500	55 psf live load	↓	2400	.013	↓	3.93	.63		4.56	5.35

06 17 53 – Shop-Fabricated Wood Trusses

06 17 53.10 Roof Trusses		Crew	Daily Output	Labor-Hours	Unit	Material	Labor	Equipment	Total	Total Incl O&P
0010	**ROOF TRUSSES**									
0100	Fink (W) or King post type, 2'-0" O.C.									
0200	Metal plate connected, 4 in 12 slope									
0210	24' to 29' span	F-3	3000	.013	SF Flr.	1.70	.64	.22	2.56	3.15
0300	30' to 43' span		3000	.013		2.16	.64	.22	3.02	3.65
0400	44' to 60' span	↓	3000	.013		2.29	.64	.22	3.15	3.80
0700	Glued and nailed, add				↓	50%				

06 18 Glued-Laminated Construction

06 18 13 – Glued-Laminated Beams

	06 18 13.20 Laminated Framing	Crew	Daily Output	Labor-Hours	Unit	2015 Bare Costs Material	Labor	Equipment	Total	Total Incl O&P
0010	**LAMINATED FRAMING**									
0020	30 lb., short term live load, 15 lb. dead load									
0200	Straight roof beams, 20' clear span, beams 8' O.C.	F-3	2560	.016	SF Flr.	2.04	.75	.26	3.05	3.74
0300	Beams 16' O.C.		3200	.013		1.48	.60	.20	2.28	2.82
0500	40' clear span, beams 8' O.C.		3200	.013		3.90	.60	.20	4.70	5.50
0600	Beams 16' O.C.		3840	.010		3.20	.50	.17	3.87	4.52
0800	60' clear span, beams 8' O.C.	F-4	2880	.017		6.70	.79	.39	7.88	9.10
0900	Beams 16' O.C.	"	3840	.013		5	.59	.29	5.88	6.80
1100	Tudor arches, 30' to 40' clear span, frames 8' O.C.	F-3	1680	.024		8.75	1.14	.39	10.28	11.95
1200	Frames 16' O.C.	"	2240	.018		6.85	.86	.29	8	9.25
1400	50' to 60' clear span, frames 8' O.C.	F-4	2200	.022		9.45	1.04	.51	11	12.65
1500	Frames 16' O.C.		2640	.018		8.05	.86	.43	9.34	10.70
1700	Radial arches, 60' clear span, frames 8' O.C.		1920	.025		8.85	1.19	.59	10.63	12.25
1800	Frames 16' O.C.		2880	.017		6.80	.79	.39	7.98	9.15
2000	100' clear span, frames 8' O.C.		1600	.030		9.15	1.42	.71	11.28	13.15
2100	Frames 16' O.C.		2400	.020		8.05	.95	.47	9.47	10.90
2300	120' clear span, frames 8' O.C.		1440	.033		12.15	1.58	.78	14.51	16.75
2400	Frames 16' O.C.		1920	.025		11.10	1.19	.59	12.88	14.75
2600	Bowstring trusses, 20' O.C., 40' clear span	F-3	2400	.017		5.50	.80	.27	6.57	7.65
2700	60' clear span	F-4	3600	.013		4.92	.63	.31	5.86	6.75
2800	100' clear span		4000	.012		6.95	.57	.28	7.80	8.90
2900	120' clear span		3600	.013		7.45	.63	.31	8.39	9.55
3000	For less than 1000 B.F., add					20%				
3100	For premium appearance, add to S.F. prices					5%				
3300	For industrial type, deduct					15%				
3500	For stain and varnish, add					5%				
3900	For 3/4" laminations, add to straight					25%				
4100	Add to curved					15%				
4300	Alternate pricing method: (use nominal footage of									
4310	components). Straight beams, camber less than 6"	F-3	3.50	11.429	M.B.F.	2,925	545	187	3,657	4,300
4400	Columns, including hardware		2	20		3,125	960	325	4,410	5,350
4600	Curved members, radius over 32'		2.50	16		3,200	765	262	4,227	5,075
4700	Radius 10' to 32'		3	13.333		3,175	640	218	4,033	4,800
4900	For complicated shapes, add maximum					100%				
5100	For pressure treating, add to straight					35%				
5200	Add to curved					45%				
6000	Laminated veneer members, southern pine or western species									
6050	1-3/4" wide x 5-1/2" deep	2 Carp	480	.033	L.F.	3.44	1.56		5	6.35
6100	9-1/2" deep		480	.033		4.74	1.56		6.30	7.75
6150	14" deep		450	.036		7.85	1.67		9.52	11.40
6200	18" deep		450	.036		10.65	1.67		12.32	14.50
6300	Parallel strand members, southern pine or western species									
6350	1-3/4" wide x 9-1/4" deep	2 Carp	480	.033	L.F.	5.10	1.56		6.66	8.15
6400	11-1/4" deep		450	.036		5.75	1.67		7.42	9.10
6450	14" deep		400	.040		7.85	1.88		9.73	11.75
6500	3-1/2" wide x 9-1/4" deep		480	.033		16.30	1.56		17.86	20.50
6550	11-1/4" deep		450	.036		20.50	1.67		22.17	25
6600	14" deep		400	.040		24	1.88		25.88	29.50
6650	7" wide x 9-1/4" deep		450	.036		34.50	1.67		36.17	40.50
6700	11-1/4" deep		420	.038		43.50	1.79		45.29	50.50
6750	14" deep		400	.040		52	1.88		53.88	60
9000	Minimum labor/equipment charge	F-3	2.50	16	Job		765	262	1,027	1,550

06 22 13 – Standard Pattern Wood Trim

06 22 13.15 Moldings, Base		Crew	Daily Output	Labor-Hours	Unit	Material	Labor	Equipment	Total	Total Incl O&P
						2015 Bare Costs				
0010	MOLDINGS, BASE									
5100	Classic profile, 5/8" x 5-1/2", finger jointed and primed	1 Carp	250	.032	L.F.	1.46	1.50		2.96	4.06
5105	Poplar		240	.033		1.54	1.56		3.10	4.25
5110	Red oak		220	.036		2.36	1.71		4.07	5.40
5115	Maple		220	.036		3.57	1.71		5.28	6.75
5120	Cherry		220	.036		4.24	1.71		5.95	7.45
5125	3/4 x 7-1/2", finger jointed and primed		250	.032		1.85	1.50		3.35	4.49
5130	Poplar		240	.033		2.46	1.56		4.02	5.25
5135	Red oak		220	.036		3.43	1.71		5.14	6.55
5140	Maple		220	.036		4.75	1.71		6.46	8
5145	Cherry		220	.036		5.65	1.71		7.36	9
5150	Modern profile, 5/8" x 3-1/2", finger jointed and primed		250	.032		.89	1.50		2.39	3.44
5155	Poplar		240	.033		.98	1.56		2.54	3.64
5160	Red oak		220	.036		1.62	1.71		3.33	4.58
5165	Maple		220	.036		2.50	1.71		4.21	5.55
5170	Cherry		220	.036		2.70	1.71		4.41	5.75
5175	Ogee profile, 7/16" x 3", finger jointed and primed		250	.032		.59	1.50		2.09	3.11
5180	Poplar		240	.033		.64	1.56		2.20	3.26
5185	Red oak		220	.036		.90	1.71		2.61	3.79
5200	9/16" x 3-1/2", finger jointed and primed		250	.032		.66	1.50		2.16	3.18
5205	Pine		240	.033		1.13	1.56		2.69	3.80
5210	Red oak		220	.036		2.38	1.71		4.09	5.40
5215	9/16" x 4-1/2", red oak		220	.036		3.56	1.71		5.27	6.70
5220	5/8" x 3-1/2", finger jointed and primed		250	.032		.89	1.50		2.39	3.44
5225	Poplar		240	.033		.98	1.56		2.54	3.64
5230	Red oak		220	.036		1.62	1.71		3.33	4.58
5235	Maple		220	.036		2.50	1.71		4.21	5.55
5240	Cherry		220	.036		2.70	1.71		4.41	5.75
5245	5/8" x 4", finger jointed and primed		250	.032		1.15	1.50		2.65	3.72
5250	Poplar		240	.033		1.26	1.56		2.82	3.94
5255	Red oak		220	.036		1.81	1.71		3.52	4.79
5260	Maple		220	.036		2.79	1.71		4.50	5.85
5265	Cherry		220	.036		2.93	1.71		4.64	6
5270	Rectangular profile, oak, 3/8" x 1-1/4"		260	.031		1.25	1.44		2.69	3.75
5275	1/2" x 2-1/2"		255	.031		1.94	1.47		3.41	4.54
5280	1/2" x 3-1/2"		250	.032		2.36	1.50		3.86	5.05
5285	1" x 6"		240	.033		4.09	1.56		5.65	7.05
5290	1" x 8"		240	.033		5.70	1.56		7.26	8.80
5295	Pine, 3/8" x 1-3/4"		260	.031		.46	1.44		1.90	2.87
5300	7/16" x 2-1/2"		255	.031		.74	1.47		2.21	3.22
5305	1" x 6"		240	.033		.83	1.56		2.39	3.47
5310	1" x 8"		240	.033		1.06	1.56		2.62	3.72
5315	Shoe, 1/2" x 3/4", primed		260	.031		.48	1.44		1.92	2.90
5320	Pine		240	.033		.30	1.56		1.86	2.89
5325	Poplar		240	.033		.38	1.56		1.94	2.98
5330	Red oak		220	.036		.52	1.71		2.23	3.37
5335	Maple		220	.036		.69	1.71		2.40	3.56
5340	Cherry		220	.036		.78	1.71		2.49	3.66
5345	11/16" x 1-1/2", pine		240	.033		.70	1.56		2.26	3.33
5350	Caps, 11/16" x 1-3/8", pine		240	.033		.55	1.56		2.11	3.16
5355	3/4" x 1-3/4", finger jointed and primed		260	.031		.70	1.44		2.14	3.14
5360	Poplar		240	.033		.99	1.56		2.55	3.65
5365	Red oak	↓	220	.036	↓	1.11	1.71		2.82	4.02

06 22 13 – Standard Pattern Wood Trim

06 22 13.15 Moldings, Base		Crew	Daily Output	Labor-Hours	Unit	2015 Bare Costs Material	Labor	Equipment	Total	Total Incl O&P
5370	Maple	1 Carp	220	.036	L.F.	1.55	1.71		3.26	4.50
5375	Cherry		220	.036		3.21	1.71		4.92	6.35
5380	Combination base & shoe, 9/16" x 3-1/2" & 1/2" x 3/4", pine		125	.064		1.43	3		4.43	6.50
5385	Three piece oak, 6" high		80	.100		5.70	4.70		10.40	14
5390	Including 3/4" x 1" base shoe		70	.114		5.90	5.35		11.25	15.30
5395	Flooring cant strip, 3/4" x 3/4", pre-finished pine		260	.031		.51	1.44		1.95	2.93
5400	For pre-finished, stain and clear coat, add					.51			.51	.56
5405	Clear coat only, add					.42			.42	.46
9000	Minimum labor/equipment charge	1 Carp	4	2	Job		94		94	154

06 22 13.30 Moldings, Casings		Crew	Daily Output	Labor-Hours	Unit	Material	Labor	Equipment	Total	Total Incl O&P
0010	**MOLDINGS, CASINGS**									
0085	Apron, 9/16" x 2-1/2", pine	1 Carp	250	.032	L.F.	1.63	1.50		3.13	4.25
0090	5/8" x 2-1/2", pine		250	.032		1.63	1.50		3.13	4.25
0110	5/8" x 3-1/2", pine		220	.036		1.86	1.71		3.57	4.84
0300	Band, 11/16" x 1-1/8", pine		270	.030		.75	1.39		2.14	3.10
0310	11/16" x 1-1/2", finger jointed and primed		270	.030		.76	1.39		2.15	3.11
0320	Pine		270	.030		.96	1.39		2.35	3.33
0330	11/16" x 1-3/4", finger jointed and primed		270	.030		.95	1.39		2.34	3.32
0350	Pine		270	.030		1.16	1.39		2.55	3.55
0355	Beaded, 3/4" x 3-1/2", finger jointed and primed		220	.036		1.09	1.71		2.80	4
0360	Poplar		220	.036		1.09	1.71		2.80	4
0365	Red oak		220	.036		1.62	1.71		3.33	4.58
0370	Maple		220	.036		2.50	1.71		4.21	5.55
0375	Cherry		220	.036		3.02	1.71		4.73	6.10
0380	3/4" x 4", finger jointed and primed		220	.036		1.15	1.71		2.86	4.06
0385	Poplar		220	.036		1.41	1.71		3.12	4.35
0390	Red oak		220	.036		1.98	1.71		3.69	4.98
0395	Maple		220	.036		2.60	1.71		4.31	5.65
0400	Cherry		220	.036		3.29	1.71		5	6.40
0405	3/4" x 5-1/2", finger jointed and primed		200	.040		1.15	1.88		3.03	4.34
0410	Poplar		200	.040		1.89	1.88		3.77	5.15
0415	Red oak		200	.040		2.78	1.88		4.66	6.15
0420	Maple		200	.040		3.66	1.88		5.54	7.10
0425	Cherry		200	.040		4.22	1.88		6.10	7.70
0430	Classic profile, 3/4" x 2-3/4", finger jointed and primed		250	.032		.77	1.50		2.27	3.31
0435	Poplar		250	.032		.98	1.50		2.48	3.54
0440	Red oak		250	.032		1.43	1.50		2.93	4.03
0445	Maple		250	.032		2.05	1.50		3.55	4.71
0450	Cherry		250	.032		2.32	1.50		3.82	5
0455	Fluted, 3/4" x 3-1/2", poplar		220	.036		1.09	1.71		2.80	4
0460	Red oak		220	.036		1.62	1.71		3.33	4.58
0465	Maple		220	.036		2.50	1.71		4.21	5.55
0470	Cherry		220	.036		3.02	1.71		4.73	6.10
0475	3/4" x 4", poplar		220	.036		1.41	1.71		3.12	4.35
0480	Red oak		220	.036		1.98	1.71		3.69	4.98
0485	Maple		220	.036		2.60	1.71		4.31	5.65
0490	Cherry		220	.036		3.29	1.71		5	6.40
0495	3/4" x 5-1/2", poplar		200	.040		1.89	1.88		3.77	5.15
0500	Red oak		200	.040		2.78	1.88		4.66	6.15
0505	Maple		200	.040		3.66	1.88		5.54	7.10
0510	Cherry		200	.040		4.22	1.88		6.10	7.70
0515	3/4" x 7-1/2", poplar		190	.042		1.27	1.98		3.25	4.64

06 22 13 – Standard Pattern Wood Trim

06 22 13.30 Moldings, Casings		Crew	Daily Output	Labor-Hours	Unit	Material	2015 Bare Costs Labor	Equipment	Total	Total Incl O&P
0520	Red oak	1 Carp	190	.042	L.F.	3.85	1.98		5.83	7.45
0525	Maple		190	.042		5.55	1.98		7.53	9.35
0530	Cherry		190	.042		6.65	1.98		8.63	10.55
0535	3/4" x 9-1/2", poplar		180	.044		4.04	2.09		6.13	7.85
0540	Red oak		180	.044		6.45	2.09		8.54	10.45
0545	Maple		180	.044		8.80	2.09		10.89	13.10
0550	Cherry		180	.044		9.60	2.09		11.69	14
0555	Modern profile, 9/16" x 2-1/4", poplar		250	.032		.56	1.50		2.06	3.07
0560	Red oak		250	.032		.76	1.50		2.26	3.29
0565	11/16" x 2-1/2", finger jointed & primed		250	.032		.84	1.50		2.34	3.38
0570	Pine		250	.032		1.28	1.50		2.78	3.87
0575	3/4" x 2-1/2", poplar		250	.032		.86	1.50		2.36	3.40
0580	Red oak		250	.032		1.17	1.50		2.67	3.75
0585	Maple		250	.032		1.76	1.50		3.26	4.39
0590	Cherry		250	.032		2.23	1.50		3.73	4.91
0595	Mullion, 5/16" x 2", pine		270	.030		.86	1.39		2.25	3.22
0600	9/16" x 2-1/2", fingerjointed and primed		250	.032		.98	1.50		2.48	3.54
0605	Pine		250	.032		1.28	1.50		2.78	3.87
0610	Red oak		250	.032		2.86	1.50		4.36	5.60
0615	1-1/16" x 3-3/4", red oak		220	.036		6.95	1.71		8.66	10.40
0620	Ogee, 7/16" x 2-1/2", poplar		250	.032		.56	1.50		2.06	3.07
0625	Red oak		250	.032		.77	1.50		2.27	3.31
0630	9/16" x 2-1/4", finger jointed and primed		250	.032		.49	1.50		1.99	3
0635	Poplar		250	.032		.56	1.50		2.06	3.07
0640	Red oak		250	.032		.76	1.50		2.26	3.29
0645	11/16" x 2-1/2", finger jointed and primed		250	.032		.72	1.50		2.22	3.25
0700	Pine		250	.032		1.41	1.50		2.91	4.01
0701	Red oak		250	.032		2.70	1.50		4.20	5.45
0730	11/16" x 3-1/2", finger jointed and primed		220	.036		1.11	1.71		2.82	4.02
0750	Pine		220	.036		1.72	1.71		3.43	4.69
0755	3/4" x 2-1/2", finger jointed and primed		250	.032		.89	1.50		2.39	3.44
0760	Poplar		250	.032		.86	1.50		2.36	3.40
0765	Red oak		250	.032		1.17	1.50		2.67	3.75
0770	Maple		250	.032		1.76	1.50		3.26	4.39
0775	Cherry		250	.032		2.23	1.50		3.73	4.91
0780	3/4" x 3-1/2", finger jointed and primed		220	.036		.89	1.71		2.60	3.78
0785	Poplar		220	.036		1.09	1.71		2.80	4
0790	Red oak		220	.036		1.62	1.71		3.33	4.58
0795	Maple		220	.036		2.50	1.71		4.21	5.55
0800	Cherry		220	.036		3.02	1.71		4.73	6.10
4700	Square profile, 1" x 1", teak		215	.037		2.26	1.75		4.01	5.35
4800	Rectangular profile, 1" x 3", teak		200	.040		6.20	1.88		8.08	9.95
9000	Minimum labor/equipment charge		4	2	Job		94		94	154

06 22 13.35 Moldings, Ceilings		Crew	Daily Output	Labor-Hours	Unit	Material	Labor	Equipment	Total	Total Incl O&P
0010	**MOLDINGS, CEILINGS**									
0600	Bed, 9/16" x 1-3/4", pine	1 Carp	270	.030	L.F.	1.06	1.39		2.45	3.44
0650	9/16" x 2", pine		270	.030		1.14	1.39		2.53	3.53
0710	9/16" x 1-3/4", oak		270	.030		1.86	1.39		3.25	4.32
1200	Cornice, 9/16" x 1-3/4", pine		270	.030		.96	1.39		2.35	3.33
1300	9/16" x 2-1/4", pine		265	.030		1.27	1.42		2.69	3.71
1350	Cove, 1/2" x 2-1/4", poplar		265	.030		1	1.42		2.42	3.42
1360	Red oak		265	.030		1.33	1.42		2.75	3.78

06 22 13 – Standard Pattern Wood Trim

06 22 13.35 Moldings, Ceilings		Crew	Daily Output	Labor-Hours	Unit	Material	2015 Bare Costs Labor	Equipment	Total	Total Incl O&P
1370	Hard maple	1 Carp	265	.030	L.F.	1.83	1.42		3.25	4.33
1380	Cherry		265	.030		2.24	1.42		3.66	4.78
2400	9/16" x 1-3/4", pine		270	.030		.97	1.39		2.36	3.35
2500	11/16" x 2-3/4", pine		265	.030		1.80	1.42		3.22	4.30
2510	Crown, 5/8" x 5/8", poplar		300	.027		.43	1.25		1.68	2.52
2520	Red oak		300	.027		.51	1.25		1.76	2.61
2530	Hard maple		300	.027		.69	1.25		1.94	2.81
2540	Cherry		300	.027		.73	1.25		1.98	2.85
2600	9/16" x 3-5/8", pine		250	.032		2.02	1.50		3.52	4.68
2700	11/16" x 4-1/4", pine		250	.032		2.90	1.50		4.40	5.65
2705	Oak		250	.032		6.40	1.50		7.90	9.50
2710	3/4" x 1-3/4", poplar		270	.030		.70	1.39		2.09	3.05
2720	Red oak		270	.030		1.03	1.39		2.42	3.41
2730	Hard maple		270	.030		1.36	1.39		2.75	3.77
2740	Cherry		270	.030		1.68	1.39		3.07	4.13
2750	3/4" x 2", poplar		270	.030		.92	1.39		2.31	3.29
2760	Red oak		270	.030		1.24	1.39		2.63	3.64
2770	Hard maple		270	.030		1.65	1.39		3.04	4.09
2780	Cherry		270	.030		1.84	1.39		3.23	4.30
2790	3/4" x 2-3/4", poplar		265	.030		1	1.42		2.42	3.42
2800	Red oak		265	.030		1.54	1.42		2.96	4.01
2810	Hard maple		265	.030		2.51	1.42		3.93	5.10
2820	Cherry		265	.030		2.35	1.42		3.77	4.90
2830	3/4" x 3-1/2", poplar		250	.032		1.27	1.50		2.77	3.85
2840	Red oak		250	.032		1.92	1.50		3.42	4.57
2850	Hard maple		250	.032		2.51	1.50		4.01	5.20
2860	Cherry		250	.032		2.94	1.50		4.44	5.70
2870	FJP poplar		250	.032		.90	1.50		2.40	3.45
2880	3/4" x 5", poplar		245	.033		1.89	1.53		3.42	4.59
2890	Red oak		245	.033		2.79	1.53		4.32	5.60
2900	Hard maple		245	.033		3.67	1.53		5.20	6.55
2910	Cherry		245	.033		4.24	1.53		5.77	7.15
2920	FJP poplar		245	.033		1.29	1.53		2.82	3.93
2930	3/4" x 6-1/4", poplar		240	.033		2.26	1.56		3.82	5.05
2940	Red oak		240	.033		3.40	1.56		4.96	6.30
2950	Hard maple		240	.033		4.44	1.56		6	7.45
2960	Cherry		240	.033		5.25	1.56		6.81	8.35
2970	7/8" x 8-3/4", poplar		220	.036		3.51	1.71		5.22	6.65
2980	Red oak		220	.036		5.35	1.71		7.06	8.65
2990	Hard maple		220	.036		7.35	1.71		9.06	10.90
3000	Cherry		220	.036		8.40	1.71		10.11	12.05
3010	1" x 7-1/4", poplar		220	.036		3.71	1.71		5.42	6.90
3020	Red oak		220	.036		5.75	1.71		7.46	9.15
3030	Hard maple		220	.036		8	1.71		9.71	11.55
3040	Cherry		220	.036		9.30	1.71		11.01	13.05
3050	1-1/16" x 4-1/4", poplar		250	.032		2.25	1.50		3.75	4.93
3060	Red oak		250	.032		2.64	1.50		4.14	5.35
3070	Hard maple		250	.032		3.49	1.50		4.99	6.30
3080	Cherry		250	.032		4.98	1.50		6.48	7.90
3090	Dentil crown, 3/4" x 5", poplar		250	.032		1.89	1.50		3.39	4.54
3100	Red oak		250	.032		2.79	1.50		4.29	5.55
3110	Hard maple		250	.032		3.67	1.50		5.17	6.50
3120	Cherry	↓	250	.032	↓	4.24	1.50		5.74	7.10

06 22 13 – Standard Pattern Wood Trim

06 22 13.35 Moldings, Ceilings		Crew	Daily Output	Labor-Hours	Unit	Material	2015 Bare Costs Labor	Equipment	Total	Total Incl O&P
3130	Dentil piece for above, 1/2" x 1/2", poplar	1 Carp	300	.027	L.F.	2.68	1.25		3.93	4.99
3140	Red oak		300	.027		3.08	1.25		4.33	5.45
3150	Hard maple		300	.027		3.50	1.25		4.75	5.90
3160	Cherry		300	.027		3.81	1.25		5.06	6.25
9000	Minimum labor/equipment charge		4	2	Job		94		94	154

06 22 13.40 Moldings, Exterior

		Crew	Daily Output	Labor-Hours	Unit	Material	Labor	Equipment	Total	Total Incl O&P
0010	**MOLDINGS, EXTERIOR**									
0100	Band board, cedar, rough sawn, 1" x 2"	1 Carp	300	.027	L.F.	.54	1.25		1.79	2.65
0110	1" x 3"		300	.027		.81	1.25		2.06	2.94
0120	1" x 4"		250	.032		1.07	1.50		2.57	3.64
0130	1" x 6"		250	.032		1.61	1.50		3.11	4.23
0140	1" x 8"		225	.036		2.15	1.67		3.82	5.10
0150	1" x 10"		225	.036		2.67	1.67		4.34	5.70
0160	1" x 12"		200	.040		3.21	1.88		5.09	6.60
0240	STK, 1" x 2"		300	.027		.51	1.25		1.76	2.61
0250	1" x 3"		300	.027		.56	1.25		1.81	2.67
0260	1" x 4"		250	.032		.65	1.50		2.15	3.17
0270	1" x 6"		250	.032		1.03	1.50		2.53	3.59
0280	1" x 8"		225	.036		1.60	1.67		3.27	4.50
0290	1" x 10"		225	.036		2.14	1.67		3.81	5.10
0300	1" x 12"		200	.040		3.54	1.88		5.42	6.95
0310	Pine, #2, 1" x 2"		300	.027		.25	1.25		1.50	2.33
0320	1" x 3"		300	.027		.40	1.25		1.65	2.49
0330	1" x 4"		250	.032		.49	1.50		1.99	3
0340	1" x 6"		250	.032		.75	1.50		2.25	3.28
0350	1" x 8"		225	.036		1.22	1.67		2.89	4.08
0360	1" x 10"		225	.036		1.59	1.67		3.26	4.49
0370	1" x 12"		200	.040		2.05	1.88		3.93	5.35
0380	D & better, 1" x 2"		300	.027		.41	1.25		1.66	2.50
0390	1" x 3"		300	.027		.61	1.25		1.86	2.72
0400	1" x 4"		250	.032		.82	1.50		2.32	3.36
0410	1" x 6"		250	.032		1.12	1.50		2.62	3.69
0420	1" x 8"		225	.036		1.62	1.67		3.29	4.52
0430	1" x 10"		225	.036		2.03	1.67		3.70	4.97
0440	1" x 12"		200	.040		2.63	1.88		4.51	5.95
0450	Redwood, clear all heart, 1" x 2"		300	.027		.63	1.25		1.88	2.75
0460	1" x 3"		300	.027		.94	1.25		2.19	3.09
0470	1" x 4"		250	.032		1.19	1.50		2.69	3.77
0480	1" x 6"		252	.032		1.78	1.49		3.27	4.39
0490	1" x 8"		225	.036		2.36	1.67		4.03	5.35
0500	1" x 10"		225	.036		3.96	1.67		5.63	7.10
0510	1" x 12"		200	.040		4.75	1.88		6.63	8.35
0530	Corner board, cedar, rough sawn, 1" x 2"		225	.036		.54	1.67		2.21	3.34
0540	1" x 3"		225	.036		.81	1.67		2.48	3.63
0550	1" x 4"		200	.040		1.07	1.88		2.95	4.26
0560	1" x 6"		200	.040		1.61	1.88		3.49	4.85
0570	1" x 8"		200	.040		2.15	1.88		4.03	5.45
0580	1" x 10"		175	.046		2.67	2.15		4.82	6.45
0590	1" x 12"		175	.046		3.21	2.15		5.36	7.05
0670	STK, 1" x 2"		225	.036		.51	1.67		2.18	3.30
0680	1" x 3"		225	.036		.56	1.67		2.23	3.36
0690	1" x 4"		200	.040		.63	1.88		2.51	3.78

06 22 13 – Standard Pattern Wood Trim

06 22 13.40 Moldings, Exterior		Crew	Daily Output	Labor-Hours	Unit	2015 Bare Costs Material	Labor	Equipment	Total	Total Incl O&P
0700	1" x 6"	1 Carp	200	.040	L.F.	1.03	1.88		2.91	4.21
0710	1" x 8"		200	.040		1.60	1.88		3.48	4.84
0720	1" x 10"		175	.046		2.14	2.15		4.29	5.85
0730	1" x 12"		175	.046		3.54	2.15		5.69	7.40
0740	Pine, #2, 1" x 2"		225	.036		.25	1.67		1.92	3.02
0750	1" x 3"		225	.036		.40	1.67		2.07	3.18
0760	1" x 4"		200	.040		.49	1.88		2.37	3.62
0770	1" x 6"		200	.040		.75	1.88		2.63	3.90
0780	1" x 8"		200	.040		1.22	1.88		3.10	4.42
0790	1" x 10"		175	.046		1.59	2.15		3.74	5.25
0800	1" x 12"		175	.046		2.05	2.15		4.20	5.75
0810	D & better, 1" x 2"		225	.036		.41	1.67		2.08	3.19
0820	1" x 3"		225	.036		.61	1.67		2.28	3.41
0830	1" x 4"		200	.040		.82	1.88		2.70	3.98
0840	1" x 6"		200	.040		1.12	1.88		3	4.31
0850	1" x 8"		200	.040		1.62	1.88		3.50	4.86
0860	1" x 10"		175	.046		2.03	2.15		4.18	5.75
0870	1" x 12"		175	.046		2.63	2.15		4.78	6.40
0880	Redwood, clear all heart, 1" x 2"		225	.036		.63	1.67		2.30	3.44
0890	1" x 3"		225	.036		.94	1.67		2.61	3.78
0900	1" x 4"		200	.040		1.19	1.88		3.07	4.39
0910	1" x 6"		200	.040		1.78	1.88		3.66	5.05
0920	1" x 8"		200	.040		2.36	1.88		4.24	5.65
0930	1" x 10"		175	.046		3.96	2.15		6.11	7.90
0940	1" x 12"		175	.046		4.75	2.15		6.90	8.75
0950	Cornice board, cedar, rough sawn, 1" x 2"		330	.024		.54	1.14		1.68	2.47
0960	1" x 3"		290	.028		.81	1.30		2.11	3.01
0970	1" x 4"		250	.032		1.07	1.50		2.57	3.64
0980	1" x 6"		250	.032		1.61	1.50		3.11	4.23
0990	1" x 8"		200	.040		2.15	1.88		4.03	5.45
1000	1" x 10"		180	.044		2.67	2.09		4.76	6.35
1010	1" x 12"		180	.044		3.21	2.09		5.30	6.95
1020	STK, 1" x 2"		330	.024		.51	1.14		1.65	2.43
1030	1" x 3"		290	.028		.56	1.30		1.86	2.74
1040	1" x 4"		250	.032		.65	1.50		2.15	3.17
1050	1" x 6"		250	.032		1.03	1.50		2.53	3.59
1060	1" x 8"		200	.040		1.60	1.88		3.48	4.84
1070	1" x 10"		180	.044		2.14	2.09		4.23	5.75
1080	1" x 12"		180	.044		3.54	2.09		5.63	7.30
1500	Pine, #2, 1" x 2"		330	.024		.25	1.14		1.39	2.15
1510	1" x 3"		290	.028		.27	1.30		1.57	2.42
1600	1" x 4"		250	.032		.49	1.50		1.99	3
1700	1" x 6"		250	.032		.75	1.50		2.25	3.28
1800	1" x 8"		200	.040		1.22	1.88		3.10	4.42
1900	1" x 10"		180	.044		1.59	2.09		3.68	5.15
2000	1" x 12"		180	.044		2.05	2.09		4.14	5.65
2020	D & better, 1" x 2"		330	.024		.41	1.14		1.55	2.32
2030	1" x 3"		290	.028		.61	1.30		1.91	2.79
2040	1" x 4"		250	.032		.82	1.50		2.32	3.36
2050	1" x 6"		250	.032		1.12	1.50		2.62	3.69
2060	1" x 8"		200	.040		1.62	1.88		3.50	4.86
2070	1" x 10"		180	.044		2.03	2.09		4.12	5.65
2080	1" x 12"		180	.044		2.63	2.09		4.72	6.30

06 22 13.40 Moldings, Exterior		Crew	Daily Output	Labor-Hours	Unit	Material	Labor	Equipment	Total	Total Incl O&P
						2015 Bare Costs				
2090	Redwood, clear all heart, 1″ x 2″	1 Carp	330	.024	L.F.	.63	1.14		1.77	2.57
2100	1″ x 3″		290	.028		.94	1.30		2.24	3.16
2110	1″ x 4″		250	.032		1.19	1.50		2.69	3.77
2120	1″ x 6″		250	.032		1.78	1.50		3.28	4.41
2130	1″ x 8″		200	.040		2.36	1.88		4.24	5.65
2140	1″ x 10″		180	.044		3.96	2.09		6.05	7.80
2150	1″ x 12″		180	.044		4.75	2.09		6.84	8.65
2160	3 piece, 1″ x 2″, 1″ x 4″, 1″ x 6″, rough sawn cedar		80	.100		3.24	4.70		7.94	11.25
2180	STK cedar		80	.100		2.19	4.70		6.89	10.10
2200	#2 pine		80	.100		1.49	4.70		6.19	9.35
2210	D & better pine		80	.100		2.35	4.70		7.05	10.30
2220	Clear all heart redwood		80	.100		3.60	4.70		8.30	11.65
2230	1″ x 8″, 1″ x 10″, 1″ x 12″, rough sawn cedar		65	.123		8	5.80		13.80	18.25
2240	STK cedar		65	.123		7.25	5.80		13.05	17.45
2300	#2 pine		65	.123		4.87	5.80		10.62	14.75
2320	D & better pine		65	.123		6.25	5.80		12.05	16.35
2330	Clear all heart redwood		65	.123		11.05	5.80		16.85	21.50
2340	Door/window casing, cedar, rough sawn, 1″ x 2″		275	.029		.54	1.37		1.91	2.84
2350	1″ x 3″		275	.029		.81	1.37		2.18	3.13
2360	1″ x 4″		250	.032		1.07	1.50		2.57	3.64
2370	1″ x 6″		250	.032		1.61	1.50		3.11	4.23
2380	1″ x 8″		230	.035		2.15	1.63		3.78	5.05
2390	1″ x 10″		230	.035		2.67	1.63		4.30	5.60
2395	1″ x 12″		210	.038		3.21	1.79		5	6.45
2410	STK, 1″ x 2″		275	.029		.51	1.37		1.88	2.80
2420	1″ x 3″		275	.029		.56	1.37		1.93	2.86
2430	1″ x 4″		250	.032		.65	1.50		2.15	3.17
2440	1″ x 6″		250	.032		1.03	1.50		2.53	3.59
2450	1″ x 8″		230	.035		1.60	1.63		3.23	4.44
2460	1″ x 10″		230	.035		2.14	1.63		3.77	5.05
2470	1″ x 12″		210	.038		3.54	1.79		5.33	6.80
2550	Pine, #2, 1″ x 2″		275	.029		.25	1.37		1.62	2.52
2560	1″ x 3″		275	.029		.40	1.37		1.77	2.68
2570	1″ x 4″		250	.032		.49	1.50		1.99	3
2580	1″ x 6″		250	.032		.75	1.50		2.25	3.28
2590	1″ x 8″		230	.035		1.22	1.63		2.85	4.02
2600	1″ x 10″		230	.035		1.59	1.63		3.22	4.43
2610	1″ x 12″		210	.038		2.05	1.79		3.84	5.20
2620	Pine, D & better, 1″ x 2″		275	.029		.41	1.37		1.78	2.69
2630	1″ x 3″		275	.029		.61	1.37		1.98	2.91
2640	1″ x 4″		250	.032		.82	1.50		2.32	3.36
2650	1″ x 6″		250	.032		1.12	1.50		2.62	3.69
2660	1″ x 8″		230	.035		1.62	1.63		3.25	4.46
2670	1″ x 10″		230	.035		2.03	1.63		3.66	4.91
2680	1″ x 12″		210	.038		2.63	1.79		4.42	5.80
2690	Redwood, clear all heart, 1″ x 2″		275	.029		.63	1.37		2	2.94
2695	1″ x 3″		275	.029		.94	1.37		2.31	3.28
2710	1″ x 4″		250	.032		1.19	1.50		2.69	3.77
2715	1″ x 6″		250	.032		1.78	1.50		3.28	4.41
2730	1″ x 8″		230	.035		2.36	1.63		3.99	5.25
2740	1″ x 10″		230	.035		3.96	1.63		5.59	7.05
2750	1″ x 12″		210	.038		4.75	1.79		6.54	8.20
3500	Bellyband, pine, 11/16″ x 4-1/4″	↓	250	.032	↓	3.23	1.50		4.73	6

06 22 13 – Standard Pattern Wood Trim

06 22 13.40 Moldings, Exterior		Crew	Daily Output	Labor-Hours	Unit	Material	2015 Bare Costs Labor	Equipment	Total	Total Incl O&P
3610	Brickmold, pine, 1-1/4" x 2"	1 Carp	200	.040	L.F.	2.43	1.88		4.31	5.75
3620	FJP, 1-1/4" x 2"		200	.040		1.09	1.88		2.97	4.28
5100	Fascia, cedar, rough sawn, 1" x 2"		275	.029		.54	1.37		1.91	2.84
5110	1" x 3"		275	.029		.81	1.37		2.18	3.13
5120	1" x 4"		250	.032		1.07	1.50		2.57	3.64
5200	1" x 6"		250	.032		1.61	1.50		3.11	4.23
5300	1" x 8"		230	.035		2.15	1.63		3.78	5.05
5310	1" x 10"		230	.035		2.67	1.63		4.30	5.60
5320	1" x 12"		210	.038		3.21	1.79		5	6.45
5400	2" x 4"		220	.036		1.05	1.71		2.76	3.96
5500	2" x 6"		220	.036		1.58	1.71		3.29	4.54
5600	2" x 8"		200	.040		2.11	1.88		3.99	5.40
5700	2" x 10"		180	.044		2.62	2.09		4.71	6.30
5800	2" x 12"		170	.047		6.25	2.21		8.46	10.50
6120	STK, 1" x 2"		275	.029		.51	1.37		1.88	2.80
6130	1" x 3"		275	.029		.56	1.37		1.93	2.86
6140	1" x 4"		250	.032		.65	1.50		2.15	3.17
6150	1" x 6"		250	.032		1.03	1.50		2.53	3.59
6160	1" x 8"		230	.035		1.60	1.63		3.23	4.44
6170	1" x 10"		230	.035		2.14	1.63		3.77	5.05
6180	1" x 12"		210	.038		3.54	1.79		5.33	6.80
6185	2" x 2"		260	.031		.63	1.44		2.07	3.07
6190	Pine, #2, 1" x 2"		275	.029		.25	1.37		1.62	2.52
6200	1" x 3"		275	.029		.40	1.37		1.77	2.68
6210	1" x 4"		250	.032		.49	1.50		1.99	3
6220	1" x 6"		250	.032		.75	1.50		2.25	3.28
6230	1" x 8"		230	.035		1.22	1.63		2.85	4.02
6240	1" x 10"		230	.035		1.59	1.63		3.22	4.43
6250	1" x 12"		210	.038		2.05	1.79		3.84	5.20
6260	D & better, 1" x 2"		275	.029		.41	1.37		1.78	2.69
6270	1" x 3"		275	.029		.61	1.37		1.98	2.91
6280	1" x 4"		250	.032		.82	1.50		2.32	3.36
6290	1" x 6"		250	.032		1.12	1.50		2.62	3.69
6300	1" x 8"		230	.035		1.62	1.63		3.25	4.46
6310	1" x 10"		230	.035		2.03	1.63		3.66	4.91
6312	1" x 12"		210	.038		2.63	1.79		4.42	5.80
6330	Southern yellow, 1-1/4" x 5"		240	.033		2.27	1.56		3.83	5.05
6340	1-1/4" x 6"		240	.033		2.52	1.56		4.08	5.35
6350	1-1/4" x 8"		215	.037		3.43	1.75		5.18	6.65
6360	1-1/4" x 12"		190	.042		5.05	1.98		7.03	8.80
6370	Redwood, clear all heart, 1" x 2"		275	.029		.63	1.37		2	2.94
6380	1" x 3"		275	.029		1.19	1.37		2.56	3.55
6390	1" x 4"		250	.032		1.19	1.50		2.69	3.77
6400	1" x 6"		250	.032		1.78	1.50		3.28	4.41
6410	1" x 8"		230	.035		2.36	1.63		3.99	5.25
6420	1" x 10"		230	.035		3.96	1.63		5.59	7.05
6430	1" x 12"		210	.038		4.75	1.79		6.54	8.20
6440	1-1/4" x 5"		240	.033		1.86	1.56		3.42	4.60
6450	1-1/4" x 6"		240	.033		2.22	1.56		3.78	5
6460	1-1/4" x 8"		215	.037		3.53	1.75		5.28	6.75
6470	1-1/4" x 12"		190	.042		7.10	1.98		9.08	11.10
6580	Frieze, cedar, rough sawn, 1" x 2"		275	.029		.54	1.37		1.91	2.84
6590	1" x 3"	↓	275	.029	↓	.81	1.37		2.18	3.13

06 22 13.40 Moldings, Exterior		Crew	Daily Output	Labor-Hours	Unit	Material	Labor	Equipment	Total	Total Incl O&P
6600	1" x 4"	1 Carp	250	.032	L.F.	1.07	1.50		2.57	3.64
6610	1" x 6"		250	.032		1.61	1.50		3.11	4.23
6620	1" x 8"		250	.032		2.15	1.50		3.65	4.83
6630	1" x 10"		225	.036		2.67	1.67		4.34	5.70
6640	1" x 12"		200	.040		3.18	1.88		5.06	6.60
6650	STK, 1" x 2"		275	.029		.51	1.37		1.88	2.80
6660	1" x 3"		275	.029		.56	1.37		1.93	2.86
6670	1" x 4"		250	.032		.65	1.50		2.15	3.17
6680	1" x 6"		250	.032		1.03	1.50		2.53	3.59
6690	1" x 8"		250	.032		1.60	1.50		3.10	4.22
6700	1" x 10"		225	.036		2.14	1.67		3.81	5.10
6710	1" x 12"		200	.040		3.54	1.88		5.42	6.95
6790	Pine, #2, 1" x 2"		275	.029		.25	1.37		1.62	2.52
6800	1" x 3"		275	.029		.40	1.37		1.77	2.68
6810	1" x 4"		250	.032		.49	1.50		1.99	3
6820	1" x 6"		250	.032		.75	1.50		2.25	3.28
6830	1" x 8"		250	.032		1.22	1.50		2.72	3.80
6840	1" x 10"		225	.036		1.59	1.67		3.26	4.49
6850	1" x 12"		200	.040		2.05	1.88		3.93	5.35
6860	D & better, 1" x 2"		275	.029		.41	1.37		1.78	2.69
6870	1" x 3"		275	.029		.61	1.37		1.98	2.91
6880	1" x 4"		250	.032		.82	1.50		2.32	3.36
6890	1" x 6"		250	.032		1.12	1.50		2.62	3.69
6900	1" x 8"		250	.032		1.62	1.50		3.12	4.24
6910	1" x 10"		225	.036		2.03	1.67		3.70	4.97
6920	1" x 12"		200	.040		2.63	1.88		4.51	5.95
6930	Redwood, clear all heart, 1" x 2"		275	.029		.63	1.37		2	2.94
6940	1" x 3"		275	.029		.94	1.37		2.31	3.28
6950	1" x 4"		250	.032		1.19	1.50		2.69	3.77
6960	1" x 6"		250	.032		1.78	1.50		3.28	4.41
6970	1" x 8"		250	.032		2.36	1.50		3.86	5.05
6980	1" x 10"		225	.036		3.96	1.67		5.63	7.10
6990	1" x 12"		200	.040		4.75	1.88		6.63	8.35
7000	Grounds, 1" x 1", cedar, rough sawn		300	.027		.28	1.25		1.53	2.35
7010	STK		300	.027		.31	1.25		1.56	2.39
7020	Pine, #2		300	.027		.16	1.25		1.41	2.22
7030	D & better		300	.027		.25	1.25		1.50	2.33
7050	Redwood		300	.027		.39	1.25		1.64	2.47
7060	Rake/verge board, cedar, rough sawn, 1" x 2"		225	.036		.54	1.67		2.21	3.34
7070	1" x 3"		225	.036		.81	1.67		2.48	3.63
7080	1" x 4"		200	.040		1.07	1.88		2.95	4.26
7090	1" x 6"		200	.040		1.61	1.88		3.49	4.85
7100	1" x 8"		190	.042		2.15	1.98		4.13	5.60
7110	1" x 10"		190	.042		2.67	1.98		4.65	6.20
7120	1" x 12"		180	.044		3.21	2.09		5.30	6.95
7130	STK, 1" x 2"		225	.036		.51	1.67		2.18	3.30
7140	1" x 3"		225	.036		.56	1.67		2.23	3.36
7150	1" x 4"		200	.040		.65	1.88		2.53	3.79
7160	1" x 6"		200	.040		1.03	1.88		2.91	4.21
7170	1" x 8"		190	.042		1.60	1.98		3.58	5
7180	1" x 10"		190	.042		2.14	1.98		4.12	5.60
7190	1" x 12"		180	.044		3.54	2.09		5.63	7.30
7200	Pine, #2, 1" x 2"		225	.036		.25	1.67		1.92	3.02

06 22 13 – Standard Pattern Wood Trim

06 22 13.40 Moldings, Exterior		Crew	Daily Output	Labor-Hours	Unit	2015 Bare Costs Material	Labor	Equipment	Total	Total Incl O&P
7210	1" x 3"	1 Carp	225	.036	L.F.	.40	1.67		2.07	3.18
7220	1" x 4"		200	.040		.49	1.88		2.37	3.62
7230	1" x 6"		200	.040		.75	1.88		2.63	3.90
7240	1" x 8"		190	.042		1.22	1.98		3.20	4.58
7250	1" x 10"		190	.042		1.59	1.98		3.57	4.99
7260	1" x 12"		180	.044		2.05	2.09		4.14	5.65
7340	D & better, 1" x 2"		225	.036		.41	1.67		2.08	3.19
7350	1" x 3"		225	.036		.61	1.67		2.28	3.41
7360	1" x 4"		200	.040		.82	1.88		2.70	3.98
7370	1" x 6"		200	.040		1.12	1.88		3	4.31
7380	1" x 8"		190	.042		1.62	1.98		3.60	5
7390	1" x 10"		190	.042		2.03	1.98		4.01	5.45
7400	1" x 12"		180	.044		2.63	2.09		4.72	6.30
7410	Redwood, clear all heart, 1" x 2"		225	.036		.63	1.67		2.30	3.44
7420	1" x 3"		225	.036		.94	1.67		2.61	3.78
7430	1" x 4"		200	.040		1.19	1.88		3.07	4.39
7440	1" x 6"		200	.040		1.78	1.88		3.66	5.05
7450	1" x 8"		190	.042		2.36	1.98		4.34	5.85
7460	1" x 10"		190	.042		3.96	1.98		5.94	7.60
7470	1" x 12"		180	.044		4.75	2.09		6.84	8.65
7480	2" x 4"		200	.040		2.28	1.88		4.16	5.60
7490	2" x 6"		182	.044		3.43	2.06		5.49	7.15
7500	2" x 8"		165	.048		4.56	2.28		6.84	8.75
7630	Soffit, cedar, rough sawn, 1" x 2"	2 Carp	440	.036		.54	1.71		2.25	3.40
7640	1" x 3"		440	.036		.81	1.71		2.52	3.69
7650	1" x 4"		420	.038		1.07	1.79		2.86	4.11
7660	1" x 6"		420	.038		1.61	1.79		3.40	4.70
7670	1" x 8"		420	.038		2.15	1.79		3.94	5.30
7680	1" x 10"		400	.040		2.67	1.88		4.55	6
7690	1" x 12"		400	.040		3.21	1.88		5.09	6.60
7700	STK, 1" x 2"		440	.036		.51	1.71		2.22	3.36
7710	1" x 3"		440	.036		.56	1.71		2.27	3.42
7720	1" x 4"		420	.038		.65	1.79		2.44	3.64
7730	1" x 6"		420	.038		1.03	1.79		2.82	4.06
7740	1" x 8"		420	.038		1.60	1.79		3.39	4.69
7750	1" x 10"		400	.040		2.14	1.88		4.02	5.45
7760	1" x 12"		400	.040		3.54	1.88		5.42	6.95
7770	Pine, #2, 1" x 2"		440	.036		.25	1.71		1.96	3.08
7780	1" x 3"		440	.036		.40	1.71		2.11	3.24
7790	1" x 4"		420	.038		.49	1.79		2.28	3.47
7800	1" x 6"		420	.038		.75	1.79		2.54	3.75
7810	1" x 8"		420	.038		1.22	1.79		3.01	4.27
7820	1" x 10"		400	.040		1.59	1.88		3.47	4.83
7830	1" x 12"		400	.040		2.05	1.88		3.93	5.35
7840	D & better, 1" x 2"		440	.036		.41	1.71		2.12	3.25
7850	1" x 3"		440	.036		.61	1.71		2.32	3.47
7860	1" x 4"		420	.038		.82	1.79		2.61	3.83
7870	1" x 6"		420	.038		1.12	1.79		2.91	4.16
7880	1" x 8"		420	.038		1.62	1.79		3.41	4.71
7890	1" x 10"		400	.040		2.03	1.88		3.91	5.30
7900	1" x 12"		400	.040		2.63	1.88		4.51	5.95
7910	Redwood, clear all heart, 1" x 2"		440	.036		.63	1.71		2.34	3.50
7920	1" x 3"		440	.036		.94	1.71		2.65	3.84

06 22 Millwork

06 22 13 – Standard Pattern Wood Trim

06 22 13.40 Moldings, Exterior

	06 22 13.40 Moldings, Exterior	Crew	Daily Output	Labor-Hours	Unit	Material	Labor	Equipment	Total	Total Incl O&P
						2015 Bare Costs				
7930	1" x 4"	2 Carp	420	.038	L.F.	1.19	1.79		2.98	4.24
7940	1" x 6"		420	.038		1.78	1.79		3.57	4.88
7950	1" x 8"		420	.038		2.36	1.79		4.15	5.50
7960	1" x 10"		400	.040		3.96	1.88		5.84	7.45
7970	1" x 12"		400	.040		4.75	1.88		6.63	8.35
8050	Trim, crown molding, pine, 11/16" x 4-1/4"	1 Carp	250	.032		4.54	1.50		6.04	7.45
8060	Back band, 11/16" x 1-1/16"		250	.032		.99	1.50		2.49	3.55
8070	Insect screen frame stock, 1-1/16" x 1-3/4"		395	.020		2.39	.95		3.34	4.19
8080	Dentils, 2-1/2" x 2-1/2" x 4", 6" O.C.		30	.267		1.22	12.50		13.72	22
8100	Fluted, 5-1/2"		165	.048		5.05	2.28		7.33	9.30
8110	Stucco bead, 1-3/8" x 1-5/8"		250	.032		2.50	1.50		4	5.20
9000	Minimum labor/equipment charge		4	2	Job		94		94	154

06 22 13.45 Moldings, Trim

	06 22 13.45 Moldings, Trim	Crew	Daily Output	Labor-Hours	Unit	Material	Labor	Equipment	Total	Total Incl O&P
0010	**MOLDINGS, TRIM**									
0200	Astragal, stock pine, 11/16" x 1-3/4"	1 Carp	255	.031	L.F.	1.42	1.47		2.89	3.97
0250	1-5/16" x 2-3/16"		240	.033		2.10	1.56		3.66	4.87
0800	Chair rail, stock pine, 5/8" x 2-1/2"		270	.030		1.58	1.39		2.97	4.02
0900	5/8" x 3-1/2"		240	.033		2.40	1.56		3.96	5.20
1000	Closet pole, stock pine, 1-1/8" diameter		200	.040		1.16	1.88		3.04	4.36
1100	Fir, 1-5/8" diameter		200	.040		2.20	1.88		4.08	5.50
3300	Half round, stock pine, 1/4" x 1/2"		270	.030		.24	1.39		1.63	2.54
3350	1/2" x 1"		255	.031		.73	1.47		2.20	3.21
3400	Handrail, fir, single piece, stock, hardware not included									
3450	1-1/2" x 1-3/4"	1 Carp	80	.100	L.F.	2.56	4.70		7.26	10.50
3470	Pine, 1-1/2" x 1-3/4"		80	.100		2.40	4.70		7.10	10.35
3500	1-1/2" x 2-1/2"		76	.105		2.46	4.94		7.40	10.80
3600	Lattice, stock pine, 1/4" x 1-1/8"		270	.030		.35	1.39		1.74	2.67
3700	1/4" x 1-3/4"		250	.032		.92	1.50		2.42	3.47
3800	Miscellaneous, custom, pine, 1" x 1"		270	.030		.44	1.39		1.83	2.76
3850	1" x 2"		265	.030		.87	1.42		2.29	3.28
3900	1" x 3"		240	.033		1.31	1.56		2.87	4
4100	Birch or oak, nominal 1" x 1"		240	.033		.42	1.56		1.98	3.02
4200	Nominal 1" x 3"		215	.037		1.26	1.75		3.01	4.24
4400	Walnut, nominal 1" x 1"		215	.037		.66	1.75		2.41	3.59
4500	Nominal 1" x 3"		200	.040		1.99	1.88		3.87	5.25
4700	Teak, nominal 1" x 1"		215	.037		2.80	1.75		4.55	5.95
4800	Nominal 1" x 3"		200	.040		8.40	1.88		10.28	12.35
4900	Quarter round, stock pine, 1/4" x 1/4"		275	.029		.24	1.37		1.61	2.50
4950	3/4" x 3/4"		255	.031		.50	1.47		1.97	2.96
5600	Wainscot moldings, 1-1/8" x 9/16", 2' high, minimum		76	.105	S.F.	11.65	4.94		16.59	21
5700	Maximum		65	.123	"	15.75	5.80		21.55	27
9000	Minimum labor/equipment charge		4	2	Job		94		94	154

06 22 13.50 Moldings, Window and Door

	06 22 13.50 Moldings, Window and Door	Crew	Daily Output	Labor-Hours	Unit	Material	Labor	Equipment	Total	Total Incl O&P
0010	**MOLDINGS, WINDOW AND DOOR**									
2800	Door moldings, stock, decorative, 1-1/8" wide, plain	1 Carp	17	.471	Set	47.50	22		69.50	88.50
2900	Detailed		17	.471	"	91.50	22		113.50	137
2960	Clear pine door jamb, no stops, 11/16" x 4-9/16"		240	.033	L.F.	5.20	1.56		6.76	8.25
3150	Door trim set, 1 head and 2 sides, pine, 2-1/2 wide		12	.667	Opng.	24	31.50		55.50	77.50
3170	3-1/2" wide		11	.727	"	29	34		63	88
3250	Glass beads, stock pine, 3/8" x 1/2"		275	.029	L.F.	.33	1.37		1.70	2.60
3270	3/8" x 7/8"		270	.030		.42	1.39		1.81	2.74
4850	Parting bead, stock pine, 3/8" x 3/4"		275	.029		.44	1.37		1.81	2.72

06 22 Millwork

06 22 13 – Standard Pattern Wood Trim

06 22 13.50 Moldings, Window and Door		Crew	Daily Output	Labor-Hours	Unit	Material	Labor	Equipment	Total	Total Incl O&P
						2015 Bare Costs				
4870	1/2" x 3/4"	1 Carp	255	.031	L.F.	.42	1.47		1.89	2.87
5000	Stool caps, stock pine, 11/16" x 3-1/2"		200	.040		2.07	1.88		3.95	5.35
5100	1-1/16" x 3-1/4"		150	.053		3.24	2.50		5.74	7.65
5300	Threshold, oak, 3' long, inside, 5/8" x 3-5/8"		32	.250	Ea.	6.50	11.75		18.25	26.50
5400	Outside, 1-1/2" x 7-5/8"		16	.500	"	45	23.50		68.50	88
5900	Window trim sets, including casings, header, stops,									
5910	stool and apron, 2-1/2" wide, FJP	1 Carp	13	.615	Opng.	31.50	29		60.50	82
5950	Pine		10	.800		37	37.50		74.50	103
6000	Oak		6	1.333		64.50	62.50		127	174
9000	Minimum labor/equipment charge		4	2	Job		94		94	154

06 22 13.60 Moldings, Soffits		Crew	Daily Output	Labor-Hours	Unit	Material	Labor	Equipment	Total	Total Incl O&P
0010	**MOLDINGS, SOFFITS**									
0200	Soffits, pine, 1" x 4"	2 Carp	420	.038	L.F.	.47	1.79		2.26	3.44
0210	1" x 6"		420	.038		.72	1.79		2.51	3.72
0220	1" x 8"		420	.038		1.19	1.79		2.98	4.24
0230	1" x 10"		400	.040		1.55	1.88		3.43	4.78
0240	1" x 12"		400	.040		2.01	1.88		3.89	5.30
0250	STK cedar, 1" x 4"		420	.038		.62	1.79		2.41	3.61
0260	1" x 6"		420	.038		1	1.79		2.79	4.03
0270	1" x 8"		420	.038		1.57	1.79		3.36	4.66
0280	1" x 10"		400	.040		2.10	1.88		3.98	5.40
0290	1" x 12"		400	.040		3.50	1.88		5.38	6.95
1000	Exterior AC plywood, 1/4" thick		400	.040	S.F.	.88	1.88		2.76	4.05
1050	3/8" thick		400	.040		.92	1.88		2.80	4.09
1100	1/2" thick		400	.040		1.10	1.88		2.98	4.29
1150	Polyvinyl chloride, white, solid	1 Carp	230	.035		2.10	1.63		3.73	4.99
1160	Perforated	"	230	.035		2.10	1.63		3.73	4.99
1170	Accessories, "J" channel 5/8"	2 Carp	700	.023	L.F.	.45	1.07		1.52	2.26
9000	Minimum labor/equipment charge	"	5	3.200	Job		150		150	246

06 25 Prefinished Paneling

06 25 13 – Prefinished Hardboard Paneling

06 25 13.10 Paneling, Hardboard			Crew	Daily Output	Labor-Hours	Unit	Material	Labor	Equipment	Total	Total Incl O&P
0010	**PANELING, HARDBOARD**										
0050	Not incl. furring or trim, hardboard, tempered, 1/8" thick	G	2 Carp	500	.032	S.F.	.41	1.50		1.91	2.91
0100	1/4" thick	G		500	.032		.64	1.50		2.14	3.16
0300	Tempered pegboard, 1/8" thick	G		500	.032		.40	1.50		1.90	2.90
0400	1/4" thick	G		500	.032		.68	1.50		2.18	3.21
0600	Untempered hardboard, natural finish, 1/8" thick	G		500	.032		.42	1.50		1.92	2.92
0700	1/4" thick	G		500	.032		.51	1.50		2.01	3.02
0900	Untempered pegboard, 1/8" thick	G		500	.032		.44	1.50		1.94	2.94
1000	1/4" thick	G		500	.032		.48	1.50		1.98	2.99
1200	Plastic faced hardboard, 1/8" thick	G		500	.032		.66	1.50		2.16	3.19
1300	1/4" thick	G		500	.032		.89	1.50		2.39	3.44
1500	Plastic faced pegboard, 1/8" thick	G		500	.032		.67	1.50		2.17	3.20
1600	1/4" thick	G		500	.032		.85	1.50		2.35	3.40
1800	Wood grained, plain or grooved, 1/8" thick	G		500	.032		.69	1.50		2.19	3.22
1900	1/4" thick	G		425	.038		1.40	1.77		3.17	4.44
2100	Moldings, wood grained MDF			500	.032	L.F.	.41	1.50		1.91	2.91
2200	Pine			425	.038	"	1.40	1.77		3.17	4.44
9000	Minimum labor/equipment charge		1 Carp	2	4	Job		188		188	310

06 25 Prefinished Paneling

06 25 16 – Prefinished Plywood Paneling

06 25 16.10 Paneling, Plywood		Crew	Daily Output	Labor-Hours	Unit	Material	2015 Bare Costs Labor	Equipment	Total	Total Incl O&P
0010	**PANELING, PLYWOOD**									
2400	Plywood, prefinished, 1/4" thick, 4' x 8' sheets									
2410	with vertical grooves. Birch faced, economy	2 Carp	500	.032	S.F.	1.40	1.50		2.90	4
2420	Average		420	.038		1.20	1.79		2.99	4.25
2430	Custom		350	.046		1.25	2.15		3.40	4.90
2600	Mahogany, African		400	.040		2.75	1.88		4.63	6.10
2700	Philippine (Lauan)		500	.032		.65	1.50		2.15	3.18
2900	Oak		500	.032		1.40	1.50		2.90	4
3000	Cherry		400	.040		2.05	1.88		3.93	5.35
3200	Rosewood		320	.050		2.90	2.35		5.25	7.05
3400	Teak		400	.040		2.90	1.88		4.78	6.25
3600	Chestnut		375	.043		4.80	2		6.80	8.60
3800	Pecan		400	.040		2.50	1.88		4.38	5.85
3900	Walnut, average		500	.032		2.40	1.50		3.90	5.10
3950	Custom		400	.040		5.25	1.88		7.13	8.90
4000	Plywood, prefinished, 3/4" thick, stock grades, economy		320	.050		1.56	2.35		3.91	5.55
4100	Average		224	.071		4.68	3.35		8.03	10.65
4300	Architectural grade, custom		224	.071		5.20	3.35		8.55	11.20
4400	Luxury		160	.100		5.20	4.70		9.90	13.40
4600	Plywood, "A" face, birch, VC, 1/2" thick, natural		450	.036		2.05	1.67		3.72	5
4700	Select		450	.036		2.15	1.67		3.82	5.10
4900	Veneer core, 3/4" thick, natural		320	.050		2.24	2.35		4.59	6.30
5000	Select		320	.050		2.44	2.35		4.79	6.55
5200	Lumber core, 3/4" thick, natural		320	.050		3.05	2.35		5.40	7.20
5500	Plywood, knotty pine, 1/4" thick, A2 grade		450	.036		1.70	1.67		3.37	4.61
5600	A3 grade		450	.036		2.10	1.67		3.77	5.05
5800	3/4" thick, veneer core, A2 grade		320	.050		2.15	2.35		4.50	6.20
5900	A3 grade		320	.050		2.42	2.35		4.77	6.50
6100	Aromatic cedar, 1/4" thick, plywood		400	.040		2.20	1.88		4.08	5.50
6200	1/4" thick, particle board	↓	400	.040	↓	1.05	1.88		2.93	4.24
9000	Minimum labor/equipment charge	1 Carp	2	4	Job		188		188	310

06 25 26 – Panel System

06 25 26.10 Panel Systems		Crew	Daily Output	Labor-Hours	Unit	Material	Labor	Equipment	Total	Total Incl O&P
0010	**PANEL SYSTEMS**									
0100	Raised panel, eng. wood core w/wood veneer, std., paint grade	2 Carp	300	.053	S.F.	11.20	2.50		13.70	16.40
0110	Oak veneer		300	.053		25.50	2.50		28	32
0120	Maple veneer		300	.053		32.50	2.50		35	40
0130	Cherry veneer		300	.053		37	2.50		39.50	45
0300	Class I fire rated, paint grade		300	.053		13	2.50		15.50	18.40
0310	Oak veneer		300	.053		30	2.50		32.50	37
0320	Maple veneer		300	.053		40.50	2.50		43	48.50
0330	Cherry veneer		300	.053		49	2.50		51.50	57.50
0510	Beadboard, 5/8" MDF, standard, primed		300	.053		8.45	2.50		10.95	13.40
0520	Oak veneer, unfinished		300	.053		13.35	2.50		15.85	18.80
0530	Maple veneer, unfinished		300	.053		14.65	2.50		17.15	20
0610	Rustic paneling, 5/8" MDF, standard, maple veneer, unfinished	↓	300	.053	↓	18.15	2.50		20.65	24

06 26 Board Paneling

06 26 13 – Profile Board Paneling

06 26 13.10 Paneling, Boards		Crew	Daily Output	Labor-Hours	Unit	Material	2015 Bare Costs Labor	Equipment	Total	Total Incl O&P
0010	**PANELING, BOARDS**									
6400	Wood board paneling, 3/4″ thick, knotty pine	2 Carp	300	.053	S.F.	1.91	2.50		4.41	6.20
6500	Rough sawn cedar		300	.053		3.20	2.50		5.70	7.60
6700	Redwood, clear, 1″ x 4″ boards		300	.053		4.96	2.50		7.46	9.55
6900	Aromatic cedar, closet lining, boards		275	.058		2.32	2.73		5.05	7.05
9000	Minimum labor/equipment charge	1 Carp	2	4	Job		188		188	310

06 43 Wood Stairs and Railings

06 43 13 – Wood Stairs

06 43 13.20 Prefabricated Wood Stairs

Code	Description	Crew	Daily Output	Labor-Hours	Unit	Material	Labor	Equipment	Total	Total Incl O&P
0010	**PREFABRICATED WOOD STAIRS**									
0100	Box stairs, prefabricated, 3′-0″ wide									
0110	Oak treads, up to 14 risers	2 Carp	39	.410	Riser	92.50	19.25		111.75	134
0600	With pine treads for carpet, up to 14 risers	″	39	.410	″	59.50	19.25		78.75	97
1100	For 4′ wide stairs, add				Flight	25%				
1550	Stairs, prefabricated stair handrail with balusters	1 Carp	30	.267	L.F.	78.50	12.50		91	107
1700	Basement stairs, prefabricated, pine treads									
1710	Pine risers, 3′ wide, up to 14 risers	2 Carp	52	.308	Riser	59.50	14.45		73.95	89
4000	Residential, wood, oak treads, prefabricated		1.50	10.667	Flight	1,200	500		1,700	2,150
4200	Built in place		.44	36.364	″	2,175	1,700		3,875	5,175
4400	Spiral, oak, 4′-6″ diameter, unfinished, prefabricated,									
4500	incl. railing, 9′ high	2 Carp	1.50	10.667	Flight	3,425	500		3,925	4,600
9000	Minimum labor/equipment charge	″	3	5.333	Job		250		250	410

06 43 13.40 Wood Stair Parts

Code	Description	Crew	Daily Output	Labor-Hours	Unit	Material	Labor	Equipment	Total	Total Incl O&P
0010	**WOOD STAIR PARTS**									
0020	Pin top balusters, 1-1/4″, oak, 34″	1 Carp	96	.083	Ea.	5.10	3.91		9.01	12
0030	38″		96	.083		5.85	3.91		9.76	12.85
0040	42″		96	.083		6.35	3.91		10.26	13.40
0050	Poplar, 34″		96	.083		3.05	3.91		6.96	9.75
0060	38″		96	.083		3.88	3.91		7.79	10.65
0070	42″		96	.083		8.85	3.91		12.76	16.10
0080	Maple, 34″		96	.083		4.90	3.91		8.81	11.80
0090	38″		96	.083		5.60	3.91		9.51	12.60
0100	42″		96	.083		6.50	3.91		10.41	13.55
0130	Primed, 34″		96	.083		3	3.91		6.91	9.70
0140	38″		96	.083		3.62	3.91		7.53	10.40
0150	42″		96	.083		4.42	3.91		8.33	11.25
0180	Box top balusters, 1-1/4″, oak, 34″		60	.133		8.95	6.25		15.20	20
0190	38″		60	.133		9.95	6.25		16.20	21
0200	42″		60	.133		10.95	6.25		17.20	22.50
0210	Poplar, 34″		60	.133		6.25	6.25		12.50	17.15
0220	38″		60	.133		6.95	6.25		13.20	17.90
0230	42″		60	.133		7.50	6.25		13.75	18.50
0240	Maple, 34″		60	.133		8.25	6.25		14.50	19.35
0250	38″		60	.133		9	6.25		15.25	20
0260	42″		60	.133		10	6.25		16.25	21.50
0290	Primed, 34″		60	.133		7	6.25		13.25	17.95
0300	38″		60	.133		8	6.25		14.25	19.05
0310	42″		60	.133		8.35	6.25		14.60	19.45
0340	Square balusters, cut from lineal stock, pine, 1-1/16″ x 1-1/16″		180	.044	L.F.	1.50	2.09		3.59	5.05
0350	1-5/16″ x 1-5/16″		180	.044		2.10	2.09		4.19	5.75

06 43 13 – Wood Stairs

06 43 13.40 Wood Stair Parts		Crew	Daily Output	Labor-Hours	Unit	Material	Labor	Equipment	Total	Total Incl O&P
						2015 Bare Costs				
0360	1-5/8" x 1-5/8"	1 Carp	180	.044	L.F.	3.30	2.09		5.39	7.05
0370	Turned newel, oak, 3-1/2" square, 48" high		8	1	Ea.	92	47		139	178
0380	62" high		8	1		92	47		139	178
0390	Poplar, 3-1/2" square, 48" high		8	1		54	47		101	137
0400	62" high		8	1		68	47		115	152
0410	Maple, 3-1/2" square, 48" high		8	1		72	47		119	156
0420	62" high		8	1		92	47		139	178
0430	Square newel, oak, 3-1/2" square, 48" high		8	1		54	47		101	137
0440	58" high		8	1		68	47		115	152
0450	Poplar, 3-1/2" square, 48" high		8	1		35	47		82	116
0460	58" high		8	1		42	47		89	123
0470	Maple, 3" square, 48" high		8	1		52	47		99	134
0480	58" high		8	1		64	47		111	148
0490	Railings, oak, economy		96	.083	L.F.	8.65	3.91		12.56	15.90
0500	Average		96	.083		13	3.91		16.91	20.50
0510	Custom		96	.083		16.50	3.91		20.41	24.50
0520	Maple, economy		96	.083		11	3.91		14.91	18.50
0530	Average		96	.083		13.50	3.91		17.41	21.50
0540	Custom		96	.083		16.95	3.91		20.86	25
0550	Oak, for bending rail, economy		48	.167		23.50	7.85		31.35	39
0560	Average		48	.167		26	7.85		33.85	41.50
0570	Custom		48	.167		29.50	7.85		37.35	45
0580	Maple, for bending rail, economy		48	.167		32	7.85		39.85	48
0590	Average		48	.167		32	7.85		39.85	48
0600	Custom		48	.167		32	7.85		39.85	48
0610	Risers, oak, 3/4" x 8", 36" long		80	.100	Ea.	13	4.70		17.70	22
0620	42" long		70	.114		15.15	5.35		20.50	25.50
0630	48" long		63	.127		17.30	5.95		23.25	29
0640	54" long		56	.143		19.50	6.70		26.20	32.50
0650	60" long		50	.160		21.50	7.50		29	36.50
0660	72" long		42	.190		26	8.95		34.95	43
0670	Poplar, 3/4" x 8", 36" long		80	.100		12.50	4.70		17.20	21.50
0680	42" long		71	.113		14.55	5.30		19.85	24.50
0690	48" long		63	.127		16.65	5.95		22.60	28
0700	54" long		56	.143		18.70	6.70		25.40	31.50
0710	60" long		50	.160		21	7.50		28.50	35.50
0720	72" long		42	.190		25	8.95		33.95	42
0730	Pine, 1" x 8", 36" long		80	.100		3.57	4.70		8.27	11.65
0740	42" long		70	.114		4.17	5.35		9.52	13.40
0750	48" long		63	.127		4.76	5.95		10.71	15
0760	54" long		56	.143		5.35	6.70		12.05	16.90
0770	60" long		50	.160		5.95	7.50		13.45	18.85
0780	72" long		42	.190		7.15	8.95		16.10	22.50
0790	Treads, oak, no returns, 1-1/32" x 11-1/2" x 36" long		32	.250		27	11.75		38.75	49
0800	42" long		32	.250		31.50	11.75		43.25	54
0810	48" long		32	.250		36	11.75		47.75	59
0820	54" long		32	.250		40.50	11.75		52.25	64
0830	60" long		32	.250		45	11.75		56.75	69
0840	72" long		32	.250		54	11.75		65.75	79
0850	Mitred return one end, 1-1/32" x 11-1/2" x 36" long		24	.333		36	15.65		51.65	65
0860	42" long		24	.333		42	15.65		57.65	71.50
0870	48" long		24	.333		48	15.65		63.65	78.50
0880	54" long		24	.333		54	15.65		69.65	85

06 43 Wood Stairs and Railings

06 43 13 – Wood Stairs

06 43 13.40 Wood Stair Parts		Crew	Daily Output	Labor-Hours	Unit	Material	Labor	Equipment	Total	Total Incl O&P
						2015 Bare Costs				
0890	60″ long	1 Carp	24	.333	Ea.	60	15.65		75.65	91.50
0900	72″ long		24	.333		72	15.65		87.65	105
0910	Mitred return two ends, 1-1/32″ x 11-1/2″ x 36″ long		12	.667		46	31.50		77.50	102
0920	42″ long		12	.667		53.50	31.50		85	111
0930	48″ long		12	.667		61.50	31.50		93	119
0940	54″ long		12	.667		69	31.50		100.50	128
0950	60″ long		12	.667		76.50	31.50		108	136
0960	72″ long		12	.667		92	31.50		123.50	153
0970	Starting step, oak, 48″, bullnose		8	1		172	47		219	266
0980	Double end bullnose		8	1		254	47		301	355
1030	Skirt board, pine, 1″ x 10″		55	.145	L.F.	1.55	6.85		8.40	12.90
1040	1″ x 12″		52	.154	″	2.01	7.20		9.21	14.05
1050	Oak landing tread, 1-1/16″ thick		54	.148	S.F.	9	6.95		15.95	21.50
1060	Oak cove molding		96	.083	L.F.	1	3.91		4.91	7.50
1070	Oak stringer molding		96	.083	″	4	3.91		7.91	10.80
1090	Rail bolt, 5/16″ x 3-1/2″		48	.167	Ea.	2.75	7.85		10.60	15.90
1100	5/16″ x 4-1/2″		48	.167		2.75	7.85		10.60	15.90
1120	Newel post anchor		16	.500		13	23.50		36.50	53
1130	Tapered plug, 1/2″		240	.033		1	1.56		2.56	3.66
1140	1″		240	.033		.99	1.56		2.55	3.65
9000	Minimum labor/equipment charge		3	2.667	Job		125		125	205

06 43 16 – Wood Railings

06 43 16.10 Wood Handrails and Railings		Crew	Daily Output	Labor-Hours	Unit	Material	Labor	Equipment	Total	Total Incl O&P
0010	**WOOD HANDRAILS AND RAILINGS**									
0020	Custom design, architectural grade, hardwood, plain	1 Carp	38	.211	L.F.	12.05	9.90		21.95	29.50
0100	Shaped		30	.267		62.50	12.50		75	89.50
0300	Stock interior railing with spindles 4″ O.C., 4′ long		40	.200		38.50	9.40		47.90	58
0400	8′ long		48	.167		38.50	7.85		46.35	55.50
9000	Minimum labor/equipment charge		3	2.667	Job		125		125	205

06 44 Ornamental Woodwork

06 44 19 – Wood Grilles

06 44 19.10 Grilles		Crew	Daily Output	Labor-Hours	Unit	Material	Labor	Equipment	Total	Total Incl O&P
0010	**GRILLES** and panels, hardwood, sanded									
0020	2′ x 4′ to 4′ x 8′, custom designs, unfinished, economy	1 Carp	38	.211	S.F.	62	9.90		71.90	84.50
0050	Average		30	.267		69	12.50		81.50	96.50
0100	Custom		19	.421		72	19.75		91.75	112
9000	Minimum labor/equipment charge		2	4	Job		188		188	310

06 44 33 – Wood Mantels

06 44 33.10 Fireplace Mantels		Crew	Daily Output	Labor-Hours	Unit	Material	Labor	Equipment	Total	Total Incl O&P
0010	**FIREPLACE MANTELS**									
0015	6″ molding, 6′ x 3′-6″ opening, plain, paint grade	1 Carp	5	1.600	Opng.	440	75		515	605
0100	Ornate, oak		5	1.600		590	75		665	775
0300	Prefabricated pine, colonial type, stock, deluxe		2	4		1,500	188		1,688	1,950
0400	Economy		3	2.667		690	125		815	965
9000	Minimum labor/equipment charge		3	2.667	Job		125		125	205

06 44 33.20 Fireplace Mantel Beam		Crew	Daily Output	Labor-Hours	Unit	Material	Labor	Equipment	Total	Total Incl O&P
0010	**FIREPLACE MANTEL BEAM**									
0020	Rough texture wood, 4″ x 8″	1 Carp	36	.222	L.F.	8.30	10.45		18.75	26
0100	4″ x 10″		35	.229	″	10.90	10.75		21.65	29.50

06 44 33 – Wood Mantels

06 44 33.20 Fireplace Mantel Beam		Crew	Daily Output	Labor-Hours	Unit	Material	2015 Bare Costs Labor	Equipment	Total	Total Incl O&P
0300	Laminated hardwood, 2-1/4" x 10-1/2" wide, 6' long	1 Carp	5	1.600	Ea.	110	75		185	244
0400	8' long		5	1.600	"	150	75		225	288
0600	Brackets for above, rough sawn		12	.667	Pr.	10	31.50		41.50	62.50
0700	Laminated		12	.667	"	15	31.50		46.50	68
9000	Minimum labor/equipment charge	↓	4	2	Job		94		94	154

06 44 39 – Wood Posts and Columns

06 44 39.10 Decorative Beams

		Crew	Daily Output	Labor-Hours	Unit	Material	Labor	Equipment	Total	Total Incl O&P
0010	**DECORATIVE BEAMS**									
0020	Rough sawn cedar, non-load bearing, 4" x 4"	2 Carp	180	.089	L.F.	1.19	4.17		5.36	8.15
0100	4" x 6"		170	.094		1.79	4.42		6.21	9.20
0200	4" x 8"		160	.100		2.39	4.70		7.09	10.35
0300	4" x 10"		150	.107		3.63	5		8.63	12.20
0400	4" x 12"		140	.114		4.61	5.35		9.96	13.85
0500	8" x 8"	↓	130	.123	↓	4.78	5.80		10.58	14.70
9000	Minimum labor/equipment charge	1 Carp	3	2.667	Job		125		125	205

06 44 39.20 Columns

		Crew	Daily Output	Labor-Hours	Unit	Material	Labor	Equipment	Total	Total Incl O&P
0010	**COLUMNS**									
0050	Aluminum, round colonial, 6" diameter	2 Carp	80	.200	V.L.F.	19	9.40		28.40	36.50
0100	8" diameter		62.25	.257		22	12.05		34.05	44
0200	10" diameter		55	.291		22.50	13.65		36.15	47.50
0250	Fir, stock units, hollow round, 6" diameter		80	.200		29.50	9.40		38.90	48
0300	8" diameter		80	.200		35.50	9.40		44.90	54.50
0350	10" diameter		70	.229		44.50	10.75		55.25	66.50
0360	12" diameter		65	.246		54.50	11.55		66.05	79
0400	Solid turned, to 8' high, 3-1/2" diameter		80	.200		9.60	9.40		19	26
0500	4-1/2" diameter		75	.213		11.90	10		21.90	29.50
0600	5-1/2" diameter		70	.229		16	10.75		26.75	35
0800	Square columns, built-up, 5" x 5"		65	.246		14.20	11.55		25.75	34.50
0900	Solid, 3-1/2" x 3-1/2"		130	.123		9.60	5.80		15.40	20
1600	Hemlock, tapered, T & G, 12" diam., 10' high		100	.160		39.50	7.50		47	56
1700	16' high		65	.246		73	11.55		84.55	99
1900	14" diameter, 10' high		100	.160		113	7.50		120.50	137
2000	18' high		65	.246		103	11.55		114.55	132
2200	18" diameter, 12' high		65	.246		165	11.55		176.55	201
2300	20' high		50	.320		118	15		133	155
2500	20" diameter, 14' high		40	.400		180	18.80		198.80	229
2600	20' high	↓	35	.457		170	21.50		191.50	222
2800	For flat pilasters, deduct				↓	33%				
3000	For splitting into halves, add				Ea.	106			106	117
4000	Rough sawn cedar posts, 4" x 4"	2 Carp	250	.064	V.L.F.	3.90	3		6.90	9.20
4100	4" x 6"		235	.068		6.80	3.20		10	12.75
4200	6" x 6"		220	.073		9.90	3.41		13.31	16.50
4300	8" x 8"	↓	200	.080	↓	19.20	3.76		22.96	27
9000	Minimum labor/equipment charge	1 Carp	3	2.667	Job		125		125	205

06 48 Wood Frames

06 48 13 – Exterior Wood Door Frames

06 48 13.10 Exterior Wood Door Frames and Accessories		Crew	Daily Output	Labor-Hours	Unit	2015 Bare Costs Material	Labor	Equipment	Total	Total Incl O&P
0010	**EXTERIOR WOOD DOOR FRAMES AND ACCESSORIES**									
0400	Exterior frame, incl. ext. trim, pine, 5/4 x 4-9/16" deep	2 Carp	375	.043	L.F.	6.60	2		8.60	10.60
0420	5-3/16" deep		375	.043		7.90	2		9.90	11.95
0440	6-9/16" deep		375	.043		8.80	2		10.80	12.95
0600	Oak, 5/4 x 4-9/16" deep		350	.046		19.75	2.15		21.90	25
0620	5-3/16" deep		350	.046		21.50	2.15		23.65	27.50
0640	6-9/16" deep		350	.046		19.50	2.15		21.65	25
1000	Sills, 8/4 x 8" deep, oak, no horns		100	.160		6.65	7.50		14.15	19.60
1020	2" horns		100	.160		20.50	7.50		28	35
1040	3" horns		100	.160		20.50	7.50		28	35
1100	8/4 x 10" deep, oak, no horns		90	.178		6.40	8.35		14.75	21
1120	2" horns		90	.178		26.50	8.35		34.85	42.50
1140	3" horns		90	.178		26.50	8.35		34.85	42.50
2000	Wood frame & trim, ext, colonial, 3' opng, fluted pilasters, flat head		22	.727	Ea.	505	34		539	610
2010	Dentil head		21	.762		580	36		616	695
2020	Ram's head		20	.800		695	37.50		732.50	820
2100	5'-4" opening, in-swing, fluted pilasters, flat head		17	.941		440	44		484	560
2120	Ram's head		15	1.067		1,400	50		1,450	1,600
2140	Out swing, fluted pilasters, flat head		17	.941		520	44		564	645
2160	Ram's head		15	1.067		1,475	50		1,525	1,700
2400	6'-0" opening, in-swing, fluted pilasters, flat head		16	1		520	47		567	645
2420	Ram's head		10	1.600		1,475	75		1,550	1,750
2460	Out-swing, fluted pilasters, flat head		16	1		520	47		567	645
2480	Ram's head		10	1.600		1,475	75		1,550	1,750
2600	For two sidelights, flat head, add		30	.533	Opng.	226	25		251	290
2620	Ram's head, add		20	.800	"	840	37.50		877.50	985
2700	Custom birch frame, 3'-0" opening		16	1	Ea.	240	47		287	340
2750	6'-0" opening		16	1		360	47		407	470
2900	Exterior, modern, plain trim, 3' opng., in-swing, FJP		26	.615		46.50	29		75.50	98.50
2920	Fir		24	.667		55	31.50		86.50	112
2940	Oak		22	.727		63	34		97	126

06 48 16 – Interior Wood Door Frames

06 48 16.10 Interior Wood Door Jamb and Frames		Crew	Daily Output	Labor-Hours	Unit	Material	Labor	Equipment	Total	Total Incl O&P
0010	**INTERIOR WOOD DOOR JAMB AND FRAMES**									
3000	Interior frame, pine, 11/16" x 3-5/8" deep	2 Carp	375	.043	L.F.	4.44	2		6.44	8.15
3020	4-9/16" deep		375	.043		4.92	2		6.92	8.70
3200	Oak, 11/16" x 3-5/8" deep		350	.046		9.85	2.15		12	14.30
3220	4-9/16" deep		350	.046		9.95	2.15		12.10	14.45
3240	5-3/16" deep		350	.046		13.90	2.15		16.05	18.75
3400	Walnut, 11/16" x 3-5/8" deep		350	.046		9	2.15		11.15	13.40
3420	4-9/16" deep		350	.046		9.45	2.15		11.60	13.90
3440	5-3/16" deep		350	.046		9.55	2.15		11.70	14
3600	Pocket door frame		16	1	Ea.	81	47		128	166
3800	Threshold, oak, 5/8" x 3-5/8" deep		200	.080	L.F.	3.56	3.76		7.32	10.05
3820	4-5/8" deep		190	.084		4.13	3.95		8.08	11.05
3840	5-5/8" deep		180	.089		6.50	4.17		10.67	14
9000	Minimum labor/equipment charge	1 Carp	4	2	Job		94		94	154

06 49 Wood Screens and Exterior Wood Shutters

06 49 19 – Exterior Wood Shutters

06 49 19.10 Shutters, Exterior		Crew	Daily Output	Labor-Hours	Unit	Material	Labor	Equipment	Total	Total Incl O&P
						2015 Bare Costs				
0010	SHUTTERS, EXTERIOR									
0012	Aluminum, louvered, 1'-4" wide, 3'-0" long	1 Carp	10	.800	Pr.	200	37.50		237.50	282
0200	4'-0" long		10	.800		240	37.50		277.50	325
0300	5'-4" long		10	.800		281	37.50		318.50	370
0400	6'-8" long		9	.889		355	41.50		396.50	460
1000	Pine, louvered, primed, each 1'-2" wide, 3'-3" long		10	.800		216	37.50		253.50	300
1100	4'-7" long		10	.800		270	37.50		307.50	360
1250	Each 1'-4" wide, 3'-0" long		10	.800		223	37.50		260.50	305
1350	5'-3" long		10	.800		330	37.50		367.50	420
1500	Each 1'-6" wide, 3'-3" long		10	.800		244	37.50		281.50	330
1600	4'-7" long		10	.800		325	37.50		362.50	415
1620	Cedar, louvered, 1'-2" wide, 5'-7" long		10	.800		315	37.50		352.50	405
1630	Each 1'-4" wide, 2'-2" long		10	.800		175	37.50		212.50	255
1640	3'-0" long		10	.800		217	37.50		254.50	300
1650	3'-3" long		10	.800		228	37.50		265.50	315
1660	3'-11" long		10	.800		265	37.50		302.50	355
1670	4'-3" long		10	.800		275	37.50		312.50	365
1680	5'-3" long		10	.800		290	37.50		327.50	380
1690	5'-11" long		10	.800		350	37.50		387.50	445
1700	Door blinds, 6'-9" long, each 1'-3" wide		9	.889		380	41.50		421.50	485
1710	1'-6" wide		9	.889		435	41.50		476.50	550
1720	Cedar, solid raised panel, each 1'-4" wide, 3'-3" long		10	.800		315	37.50		352.50	405
1730	3'-11" long		10	.800		360	37.50		397.50	455
1740	4'-3" long		10	.800		365	37.50		402.50	460
1750	4'-7" long		10	.800		390	37.50		427.50	490
1760	4'-11" long		10	.800		415	37.50		452.50	520
1770	5'-11" long		10	.800		520	37.50		557.50	630
1800	Door blinds, 6'-9" long, each 1'-3" wide		9	.889		535	41.50		576.50	660
1900	1'-6" wide		9	.889		630	41.50		671.50	765
2500	Polystyrene, solid raised panel, each 1'-4" wide, 3'-3" long		10	.800		71.50	37.50		109	140
2600	3'-11" long		10	.800		93	37.50		130.50	164
2700	4'-7" long		10	.800		105	37.50		142.50	178
2800	5'-3" long		10	.800		120	37.50		157.50	194
2900	6'-8" long		9	.889		152	41.50		193.50	236
4500	Polystyrene, louvered, each 1'-2" wide, 3'-3" long		10	.800		36	37.50		73.50	101
4600	4'-7" long		10	.800		46	37.50		83.50	112
4750	5'-3" long		10	.800		59	37.50		96.50	127
4850	6'-8" long		9	.889		68	41.50		109.50	144
6000	Vinyl, louvered, each 1'-2" x 4'-7" long		10	.800		64	37.50		101.50	132
6200	Each 1'-4" x 6'-8" long		9	.889		76	41.50		117.50	152
9000	Minimum labor/equipment charge		4	2	Job		94		94	154

06 51 13.10 Recycled Plastic Lumber			Crew	Daily Output	Labor-Hours	Unit	Material	2015 Bare Costs Labor	Equipment	Total	Total Incl O&P
0010	**RECYCLED PLASTIC LUMBER**										
4000	Sheeting, recycled plastic, black or white, 4′ x 8′ x 1/8″	G	2 Carp	1100	.015	S.F.	1.26	.68		1.94	2.51
4010	4′ x 8′ x 3/16″	G		1100	.015		1.90	.68		2.58	3.21
4020	4′ x 8′ x 1/4″	G		950	.017		2.25	.79		3.04	3.78
4030	4′ x 8′ x 3/8″	G		950	.017		3.80	.79		4.59	5.50
4040	4′ x 8′ x 1/2″	G		900	.018		5	.83		5.83	6.85
4050	4′ x 8′ x 5/8″	G		900	.018		7.50	.83		8.33	9.60
4060	4′ x 8′ x 3/4″	G		850	.019		9.40	.88		10.28	11.80
4070	Add for colors	G				Ea.	5%				
8500	100% recycled plastic, var colors, NLB, 2″ x 2″	G				L.F.	1.76			1.76	1.94
8510	2″ x 4″	G					3.65			3.65	4.02
8520	2″ x 6″	G					5.75			5.75	6.35
8530	2″ x 8″	G					7.90			7.90	8.70
8540	2″ x 10″	G					11.50			11.50	12.65
8550	5/4″ x 4″	G					4.45			4.45	4.90
8560	5/4″ x 6″	G					5			5	5.50
8570	1″ x 6″	G					2.87			2.87	3.16
8580	1/2″ x 8″	G					3.05			3.05	3.36
8590	2″ x 10″ T & G	G					11.50			11.50	12.65
8600	3″ x 10″ T & G	G					15.30			15.30	16.85
8610	Add for premium colors	G					20%				

06 51 13.12 Structural Plastic Lumber		Crew	Daily Output	Labor-Hours	Unit	Material	Labor	Equipment	Total	Total Incl O&P
0010	**STRUCTURAL PLASTIC LUMBER**									
1320	Plastic lumber, posts or columns, 4″ x 4″	2 Carp	390	.041	L.F.	9	1.93		10.93	13.05
1325	4″ x 6″		275	.058		13.15	2.73		15.88	19
1330	4″ x 8″		220	.073		19.20	3.41		22.61	26.50
1340	Girder, single, 4″ x 4″		675	.024		9	1.11		10.11	11.70
1345	4″ x 6″		600	.027		13.15	1.25		14.40	16.55
1350	4″ x 8″		525	.030		19.20	1.43		20.63	23.50
1352	Double, 2″ x 4″		625	.026		8.25	1.20		9.45	11
1354	2″ x 6″		600	.027		12.65	1.25		13.90	16
1356	2″ x 8″		575	.028		16.10	1.31		17.41	19.85
1358	2″ x 10″		550	.029		20	1.37		21.37	24
1360	2″ x 12″		525	.030		25	1.43		26.43	30
1362	Triple, 2″ x 4″		575	.028		12.35	1.31		13.66	15.75
1364	2″ x 6″		550	.029		19	1.37		20.37	23
1366	2″ x 8″		525	.030		24	1.43		25.43	29
1368	2″ x 10″		500	.032		30	1.50		31.50	35.50
1370	2″ x 12″		475	.034		37	1.58		38.58	43.50
1372	Ledger, bolted 4′ O.C., 2″ x 4″		400	.040		4.26	1.88		6.14	7.75
1374	2″ x 6″		550	.029		6.40	1.37		7.77	9.30
1376	2″ x 8″		390	.041		8.15	1.93		10.08	12.15
1378	2″ x 10″		385	.042		10.10	1.95		12.05	14.30
1380	2″ x 12″		380	.042		12.50	1.98		14.48	17
1382	Joists, 2″ x 4″		1250	.013		4.12	.60		4.72	5.50
1384	2″ x 6″		1250	.013		6.35	.60		6.95	7.95
1386	2″ x 8″		1100	.015		8.05	.68		8.73	9.95
1388	2″ x 10″		500	.032		10.10	1.50		11.60	13.55
1390	2″ x 12″		875	.018		12.40	.86		13.26	15.05
1392	Railings and trim , 5/4″ x 4″	1 Carp	300	.027		4.25	1.25		5.50	6.75
1394	2″ x 2″		300	.027		2.35	1.25		3.60	4.64
1396	2″ x 4″		300	.027		4.10	1.25		5.35	6.55

06 51 Structural Plastic Shapes and Plates

06 51 13 – Plastic Lumber

06 51 13.12 Structural Plastic Lumber		Crew	Daily Output	Labor-Hours	Unit	2015 Bare Costs Material	Labor	Equipment	Total	Total Incl O&P
1398	2" x 6"	1 Carp	300	.027	L.F.	6.30	1.25		7.55	9

06 52 Plastic Structural Assemblies

06 52 10 – Fiberglass Structural Assemblies

06 52 10.20 Fiberglass Stair Treads

		Crew	Daily Output	Labor-Hours	Unit	Material	Labor	Equipment	Total	Total Incl O&P
0010	**FIBERGLASS STAIR TREADS**									
0100	24" wide	2 Sswk	52	.308	Ea.	31.50	16.20		47.70	64.50
0140	30" wide		52	.308		39.50	16.20		55.70	73
0180	36" wide		52	.308		47.50	16.20		63.70	81.50
0220	42" wide		52	.308		55.50	16.20		71.70	90.50

06 52 10.30 Fiberglass Grating

		Crew	Daily Output	Labor-Hours	Unit	Material	Labor	Equipment	Total	Total Incl O&P
0010	**FIBERGLASS GRATING**									
0100	Molded, green (for mod. corrosive environment)									
0140	1" x 4" mesh, 1" thick	2 Sswk	400	.040	S.F.	8.50	2.11		10.61	13.15
0180	1-1/2" square mesh, 1" thick		400	.040		18.70	2.11		20.81	24.50
0220	1-1/4" thick		400	.040		9.35	2.11		11.46	14.05
0260	1-1/2" thick		400	.040		23	2.11		25.11	29
0300	2" square mesh, 2" thick		320	.050		26	2.63		28.63	34
1000	Orange (for highly corrosive environment)									
1040	1" x 4" mesh, 1" thick	2 Sswk	400	.040	S.F.	16.50	2.11		18.61	22
1080	1-1/2" square mesh, 1" thick		400	.040		19.55	2.11		21.66	25.50
1120	1-1/4" thick		400	.040		20	2.11		22.11	26
1160	1-1/2" thick		400	.040		20.50	2.11		22.61	26.50
1200	2" square mesh, 2" thick		320	.050		20.50	2.63		23.13	27.50
3000	Pultruded, green (for mod. corrosive environment)									
3040	1" O.C. bar spacing, 1" thick	2 Sswk	400	.040	S.F.	14.85	2.11		16.96	20
3080	1-1/2" thick		320	.050		15.30	2.63		17.93	21.50
3120	1-1/2" O.C. bar spacing, 1" thick		400	.040		13.70	2.11		15.81	18.85
3160	1-1/2" thick		400	.040		18.05	2.11		20.16	23.50
4000	Grating support legs, fixed height, no base				Ea.	46.50			46.50	51.50
4040	With base					41			41	45
4080	Adjustable to 60"					61.50			61.50	67.50

06 52 10.40 Fiberglass Floor Grating

		Crew	Daily Output	Labor-Hours	Unit	Material	Labor	Equipment	Total	Total Incl O&P
0010	**FIBERGLASS FLOOR GRATING**									
0100	Reinforced polyester, fire retardant, 1" x 4" grid, 1" thick	E-4	510	.063	S.F.	14	3.34	.29	17.63	22
0200	1-1/2" x 6" mesh, 1-1/2" thick		500	.064		16.50	3.40	.29	20.19	24.50
0300	With grit surface, 1-1/2" x 6" grid, 1-1/2" thick		500	.064		16.80	3.40	.29	20.49	25

06 63 Plastic Railings

06 63 10 – Plastic (PVC) Railings

06 63 10.10 Plastic Railings

		Crew	Daily Output	Labor-Hours	Unit	Material	Labor	Equipment	Total	Total Incl O&P
0010	**PLASTIC RAILINGS**									
0100	Horizontal PVC handrail with balusters, 3-1/2" wide, 36" high	1 Carp	96	.083	L.F.	24.50	3.91		28.41	33.50
0150	42" high		96	.083		28	3.91		31.91	37.50
0200	Angled PVC handrail with balusters, 3-1/2" wide, 36" high		72	.111		28.50	5.20		33.70	39.50
0250	42" high		72	.111		31.50	5.20		36.70	43
0300	Post sleeve for 4 x 4 post		96	.083		12.80	3.91		16.71	20.50
0400	Post cap for 4 x 4 post, flat profile		48	.167	Ea.	12.80	7.85		20.65	27
0450	Newel post style profile		48	.167		23.50	7.85		31.35	39

06 63 Plastic Railings

06 63 10 – Plastic (PVC) Railings

06 63 10.10 Plastic Railings		Crew	Daily Output	Labor-Hours	Unit	Material	Labor	Equipment	Total	Total Incl O&P
						2015 Bare Costs				
0500	Raised corbeled profile	1 Carp	48	.167	Ea.	35.50	7.85		43.35	52
0550	Post base trim for 4 x 4 post	↓	96	.083	↓	18.25	3.91		22.16	26.50

06 65 Plastic Trim

06 65 10 – PVC Trim

06 65 10.10 PVC Trim, Exterior

06 65 10.10 PVC Trim, Exterior		Crew	Daily Output	Labor-Hours	Unit	Material	Labor	Equipment	Total	Total Incl O&P
0010	**PVC TRIM, EXTERIOR**									
0100	Cornerboards, 5/4" x 6" x 6"	1 Carp	240	.033	L.F.	8.45	1.56		10.01	11.85
0110	Door/window casing, 1" x 4"		200	.040		1.39	1.88		3.27	4.61
0120	1" x 6"		200	.040		2.11	1.88		3.99	5.40
0130	1" x 8"		195	.041		2.78	1.93		4.71	6.20
0140	1" x 10"		195	.041		3.50	1.93		5.43	7
0150	1" x 12"		190	.042		4.33	1.98		6.31	8
0160	5/4" x 4"		195	.041		1.69	1.93		3.62	5
0170	5/4" x 6"		195	.041		2.72	1.93		4.65	6.15
0180	5/4" x 8"		190	.042		3.56	1.98		5.54	7.15
0190	5/4" x 10"		190	.042		4.56	1.98		6.54	8.25
0200	5/4" x 12"		185	.043		5.25	2.03		7.28	9.15
0210	Fascia, 1" x 4"		250	.032		1.39	1.50		2.89	3.99
0220	1" x 6"		250	.032		2.11	1.50		3.61	4.78
0230	1" x 8"		225	.036		2.78	1.67		4.45	5.80
0240	1" x 10"		225	.036		3.50	1.67		5.17	6.60
0250	1" x 12"		200	.040		4.33	1.88		6.21	7.85
0260	5/4" x 4"		240	.033		1.69	1.56		3.25	4.42
0270	5/4" x 6"		240	.033		2.72	1.56		4.28	5.55
0280	5/4" x 8"		215	.037		3.56	1.75		5.31	6.80
0290	5/4" x 10"		215	.037		4.56	1.75		6.31	7.85
0300	5/4" x 12"		190	.042		5.25	1.98		7.23	9.05
0310	Frieze, 1" x 4"		250	.032		1.39	1.50		2.89	3.99
0320	1" x 6"		250	.032		2.11	1.50		3.61	4.78
0330	1" x 8"		225	.036		2.78	1.67		4.45	5.80
0340	1" x 10"		225	.036		3.50	1.67		5.17	6.60
0350	1" x 12"		200	.040		4.33	1.88		6.21	7.85
0360	5/4" x 4"		240	.033		1.69	1.56		3.25	4.42
0370	5/4" x 6"		240	.033		2.72	1.56		4.28	5.55
0380	5/4" x 8"		215	.037		3.56	1.75		5.31	6.80
0390	5/4" x 10"		215	.037		4.56	1.75		6.31	7.85
0400	5/4" x 12"		190	.042		5.25	1.98		7.23	9.05
0410	Rake, 1" x 4"		200	.040		1.39	1.88		3.27	4.61
0420	1" x 6"		200	.040		2.11	1.88		3.99	5.40
0430	1" x 8"		190	.042		2.78	1.98		4.76	6.30
0440	1" x 10"		190	.042		3.50	1.98		5.48	7.10
0450	1" x 12"		180	.044		4.33	2.09		6.42	8.20
0460	5/4" x 4"		195	.041		1.69	1.93		3.62	5
0470	5/4" x 6"		195	.041		2.72	1.93		4.65	6.15
0480	5/4" x 8"		185	.043		3.56	2.03		5.59	7.25
0490	5/4" x 10"		185	.043		4.56	2.03		6.59	8.35
0500	5/4" x 12"		175	.046		5.25	2.15		7.40	9.30
0510	Rake trim, 1" x 4"		225	.036		1.39	1.67		3.06	4.27
0520	1" x 6"		225	.036		2.11	1.67		3.78	5.05
0560	5/4" x 4"	↓	220	.036	↓	1.69	1.71		3.40	4.66

06 65 Plastic Trim

06 65 10 – PVC Trim

06 65 10.10 PVC Trim, Exterior		Crew	Daily Output	Labor-Hours	Unit	Material	Labor	Equipment	Total	Total Incl O&P
						2015 Bare Costs				
0570	5/4" x 6"	1 Carp	220	.036	L.F.	2.72	1.71		4.43	5.80
0610	Soffit, 1" x 4"	2 Carp	420	.038		1.39	1.79		3.18	4.46
0620	1" x 6"		420	.038		2.11	1.79		3.90	5.25
0630	1" x 8"		420	.038		2.78	1.79		4.57	6
0640	1" x 10"		400	.040		3.50	1.88		5.38	6.95
0650	1" x 12"		400	.040		4.33	1.88		6.21	7.85
0660	5/4" x 4"		410	.039		1.69	1.83		3.52	4.86
0670	5/4" x 6"		410	.039		2.72	1.83		4.55	6
0680	5/4" x 8"		410	.039		3.56	1.83		5.39	6.90
0690	5/4" x 10"		390	.041		4.56	1.93		6.49	8.15
0700	5/4" x 12"	↓	390	.041	↓	5.25	1.93		7.18	8.95

06 80 Composite Fabrications

06 80 10 – Composite Decking

06 80 10.10 Woodgrained Composite Decking

		Crew	Daily Output	Labor-Hours	Unit	Material	Labor	Equipment	Total	Total Incl O&P
0010	**WOODGRAINED COMPOSITE DECKING**									
0100	Woodgrained composite decking, 1" x 6"	2 Carp	640	.025	L.F.	3.54	1.17		4.71	5.80
0110	Grooved edge		660	.024		3.68	1.14		4.82	5.90
0120	2" x 6"		640	.025		3.67	1.17		4.84	5.95
0130	Encased, 1" x 6"		640	.025		3.67	1.17		4.84	5.95
0140	Grooved edge		660	.024		3.81	1.14		4.95	6.05
0150	2" x 6"	↓	640	.025	↓	4.98	1.17		6.15	7.40

06 81 Composite Railings

06 81 10 – Encased Railings

06 81 10.10 Encased Composite Railings

		Crew	Daily Output	Labor-Hours	Unit	Material	Labor	Equipment	Total	Total Incl O&P
0010	**ENCASED COMPOSITE RAILINGS**									
0100	Encased composite railing, 6' long, 36" high, incl. balusters	1 Carp	16	.500	Ea.	156	23.50		179.50	210
0110	42" high, incl. balusters		16	.500		172	23.50		195.50	229
0120	8' long, 36" high, incl. balusters		12	.667		156	31.50		187.50	224
0130	42" high, incl. balusters		12	.667		172	31.50		203.50	242
0140	Accessories, post sleeve, 4" x 4", 39" long		32	.250		23	11.75		34.75	45
0150	96" long		24	.333		86.50	15.65		102.15	121
0160	6" x 6", 39" long		32	.250		47	11.75		58.75	71
0170	96" long		24	.333		137	15.65		152.65	176
0180	Accessories, post skirt, 4" x 4"		96	.083		4	3.91		7.91	10.80
0190	6" x 6"		96	.083		4.85	3.91		8.76	11.75
0200	Post cap, 4" x 4", flat		48	.167		6	7.85		13.85	19.45
0210	Pyramid		48	.167		6	7.85		13.85	19.45
0220	Post cap, 6" x 6", flat		48	.167		9.60	7.85		17.45	23.50
0230	Pyramid	↓	48	.167	↓	9.60	7.85		17.45	23.50

Division Notes

		CREW	DAILY OUTPUT	LABOR-HOURS	UNIT	BARE COSTS MAT.	BARE COSTS LABOR	BARE COSTS EQUIP.	BARE COSTS TOTAL	TOTAL INCL O&P

Division 7 – Thermal & Moisture Protection

Estimating Tips

07 10 00 Dampproofing and Waterproofing

- Be sure of the job specifications before pricing this subdivision. The difference in cost between waterproofing and dampproofing can be great. Waterproofing will hold back standing water. Dampproofing prevents the transmission of water vapor. Also included in this section are vapor retarding membranes.

07 20 00 Thermal Protection

- Insulation and fireproofing products are measured by area, thickness, volume or R-value. Specifications may give only what the specific R-value should be in a certain situation. The estimator may need to choose the type of insulation to meet that R-value.

07 30 00 Steep Slope Roofing
07 40 00 Roofing and Siding Panels

- Many roofing and siding products are bought and sold by the square. One square is equal to an area that measures 100 square feet.

  This simple change in unit of measure could create a large error if the estimator is not observant. Accessories necessary for a complete installation must be figured into any calculations for both material and labor.

07 50 00 Membrane Roofing
07 60 00 Flashing and Sheet Metal
07 70 00 Roofing and Wall Specialties and Accessories

- The items in these subdivisions compose a roofing system. No one component completes the installation, and all must be estimated. Built-up or single-ply membrane roofing systems are made up of many products and installation trades. Wood blocking at roof perimeters or penetrations, parapet coverings, reglets, roof drains, gutters, downspouts, sheet metal flashing, skylights, smoke vents, and roof hatches all need to be considered along with the roofing material. Several different installation trades will need to work together on the roofing system. Inherent difficulties in the scheduling and coordination of various trades must be accounted for when estimating labor costs.

07 90 00 Joint Protection

- To complete the weather-tight shell, the sealants and caulkings must be estimated. Where different materials meet—at expansion joints, at flashing penetrations, and at hundreds of other locations throughout a construction project—they provide another line of defense against water penetration. Often, an entire system is based on the proper location and placement of caulking or sealants. The detailed drawings that are included as part of a set of architectural plans show typical locations for these materials. When caulking or sealants are shown at typical locations, this means the estimator must include them for all the locations where this detail is applicable. Be careful to keep different types of sealants separate, and remember to consider backer rods and primers if necessary.

Reference Numbers

Reference numbers are shown in shaded boxes at the beginning of some major classifications. These numbers refer to related items in the Reference Section. The reference information may be an estimating procedure, an alternate pricing method, or technical information.

Note: Not all subdivisions listed here necessarily appear in this publication. ■

07 01 50 – Maintenance of Membrane Roofing

07 01 50.10 Roof Coatings		Crew	Daily Output	Labor-Hours	Unit	2015 Bare Costs Material	Labor	Equipment	Total	Total Incl O&P
0010	**ROOF COATINGS**									
0012	Asphalt, brush grade, material only				Gal.	8.95			8.95	9.80
0200	Asphalt base, fibered aluminum coating G					8.25			8.25	9.10
0300	Asphalt primer, 5 gallon					7.50			7.50	8.25
0600	Coal tar pitch, 200 lb. barrels				Ton	1,525			1,525	1,675
0700	Tar roof cement, 5 gal. lots				Gal.	14.05			14.05	15.45
0800	Glass fibered roof & patching cement, 5 gallon				"	8.25			8.25	9.10
0900	Reinforcing glass membrane, 450 S.F./roll				Ea.	59.50			59.50	65.50
1000	Neoprene roof coating, 5 gal., 2 gal./sq.				Gal.	30.50			30.50	33.50
1100	Roof patch & flashing cement, 5 gallon					8.70			8.70	9.55
1200	Roof resaturant, glass fibered, 3 gal./sq.					8.85			8.85	9.75
1600	Reflective roof coating, white, elastomeric, approx. 50 S.F. per gal. G					18.90			18.90	21

07 01 90 – Maintenance of Joint Protection

07 01 90.81 Joint Sealant Replacement		Crew	Daily Output	Labor-Hours	Unit	Material	Labor	Equipment	Total	Total Incl O&P
0010	**JOINT SEALANT REPLACEMENT**									
0050	Control joints in concrete floors/slabs									
0100	Option 1 for joints with hard dry sealant									
0110	Step 1: Sawcut to remove 95% of old sealant									
0112	1/4" wide x 1/2" deep, with single saw blade	C-27	4800	.003	L.F.	.02	.15	.03	.20	.30
0114	3/8" wide x 3/4" deep, with single saw blade		4000	.004		.03	.18	.04	.25	.37
0116	1/2" wide x 1" deep, with double saw blades		3600	.004		.06	.20	.05	.31	.44
0118	3/4" wide x 1-1/2" deep, with double saw blades		3200	.005		.13	.23	.05	.41	.56
0120	Step 2: Water blast joint faces and edges	C-29	2500	.003			.12	.03	.15	.23
0130	Step 3: Air blast joint faces and edges	C-28	2000	.004			.18	.01	.19	.29
0140	Step 4: Sand blast joint faces and edges	E-11	2000	.016			.68	.12	.80	1.30
0150	Step 5: Air blast joint faces and edges	C-28	2000	.004			.18	.01	.19	.29
0200	Option 2 for joints with soft pliable sealant									
0210	Step 1: Plow joint with rectangular blade	B-62	2600	.009	L.F.		.38	.07	.45	.69
0220	Step 2: Sawcut to re-face joint faces									
0222	1/4" wide x 1/2" deep, with single saw blade	C-27	2400	.007	L.F.	.02	.30	.07	.39	.58
0224	3/8" wide x 3/4" deep, with single saw blade		2000	.008		.04	.36	.08	.48	.71
0226	1/2" wide x 1" deep, with double saw blades		1800	.009		.09	.40	.09	.58	.82
0228	3/4" wide x 1-1/2" deep, with double saw blades		1600	.010		.17	.45	.10	.72	1.02
0230	Step 3: Water blast joint faces and edges	C-29	2500	.003			.12	.03	.15	.23
0240	Step 4: Air blast joint faces and edges	C-28	2000	.004			.18	.01	.19	.29
0250	Step 5: Sand blast joint faces and edges	E-11	2000	.016			.68	.12	.80	1.30
0260	Step 6: Air blast joint faces and edges	C-28	2000	.004			.18	.01	.19	.29
0290	For saw cutting new control joints, see Section 03 15 16.20									
8910	For backer rod, see Section 07 91 23.10									
8920	For joint sealant, see Sections 03 15 16.30 or 07 92 13.20									

07 05 Common Work Results for Thermal and Moisture Protection

07 05 05 – Selective Demolition for Thermal and Moisture Protection

07 05 05.10 Selective Demo., Thermal and Moist. Protection		Crew	Daily Output	Labor-Hours	Unit	Material	2015 Bare Costs Labor	Equipment	Total	Total Incl O&P
0010	SELECTIVE DEMO., THERMAL AND MOISTURE PROTECTION									
0020	Caulking/sealant, to 1" x 1" joint R024119-10	1 Clab	600	.013	L.F.		.50		.50	.82
0120	Downspouts, including hangers		350	.023	"		.86		.86	1.41
0220	Flashing, sheet metal		290	.028	S.F.		1.04		1.04	1.70
0420	Gutters, aluminum or wood, edge hung		240	.033	L.F.		1.25		1.25	2.05
0520	Built-in		100	.080	"		3.01		3.01	4.93
0620	Insulation, air/vapor barrier		3500	.002	S.F.		.09		.09	.14
0670	Batts or blankets		1400	.006	C.F.		.21		.21	.35
0720	Foamed or sprayed in place	2 Clab	1000	.016	B.F.		.60		.60	.99
0770	Loose fitting	1 Clab	3000	.003	C.F.		.10		.10	.16
0870	Rigid board		3450	.002	B.F.		.09		.09	.14
1120	Roll roofing, cold adhesive		12	.667	Sq.		25		25	41
1170	Roof accessories, adjustable metal chimney flashing		9	.889	Ea.		33.50		33.50	55
1325	Plumbing vent flashing		32	.250	"		9.40		9.40	15.40
1375	Ridge vent strip, aluminum		310	.026	L.F.		.97		.97	1.59
1620	Skylight to 10 S.F.		8	1	Ea.		37.50		37.50	61.50
2120	Roof edge, aluminum soffit and fascia		570	.014	L.F.		.53		.53	.87
2170	Concrete coping, up to 12" wide	2 Clab	160	.100			3.76		3.76	6.15
2220	Drip edge	1 Clab	1000	.008			.30		.30	.49
2270	Gravel stop		950	.008			.32		.32	.52
2370	Sheet metal coping, up to 12" wide		240	.033			1.25		1.25	2.05
2470	Roof insulation board, over 2" thick	B-2	7800	.005	B.F.		.19		.19	.32
2520	Up to 2" thick	"	3900	.010	S.F.		.39		.39	.64
2620	Roof ventilation, louvered gable vent	1 Clab	16	.500	Ea.		18.80		18.80	31
2670	Remove, roof hatch	G-3	15	2.133			100		100	160
2675	Rafter vents	1 Clab	960	.008			.31		.31	.51
2720	Soffit vent and/or fascia vent		575	.014	L.F.		.52		.52	.86
2775	Soffit vent strip, aluminum, 3" to 4" wide		160	.050			1.88		1.88	3.08
2820	Roofing accessories, shingle moulding, to 1" x 4"		1600	.005			.19		.19	.31
2870	Cant strip	B-2	2000	.020			.76		.76	1.25
2920	Concrete block walkway	1 Clab	230	.035			1.31		1.31	2.14
3070	Roofing, felt paper, 15#		70	.114	Sq.		4.30		4.30	7.05
3125	#30 felt		30	.267	"		10.05		10.05	16.45
3170	Asphalt shingles, 1 layer	B-2	3500	.011	S.F.		.43		.43	.71
3180	2 layers		1750	.023	"		.87		.87	1.42
3370	Modified bitumen		26	1.538	Sq.		58.50		58.50	96
3420	Built-up, no gravel, 3 ply		25	1.600			61		61	99.50
3470	4 ply		21	1.905			72.50		72.50	119
3620	5 ply		1600	.025	S.F.		.95		.95	1.56
3720	5 ply, with gravel		890	.045			1.71		1.71	2.80
3725	Loose gravel removal		5000	.008			.30		.30	.50
3730	Embedded gravel removal		2000	.020			.76		.76	1.25
3870	Fiberglass sheet		1200	.033			1.27		1.27	2.08
4120	Slate shingles		1900	.021			.80		.80	1.31
4170	Ridge shingles, clay or slate		2000	.020	L.F.		.76		.76	1.25
4320	Single ply membrane, attached at seams		52	.769	Sq.		29		29	48
4370	Ballasted		75	.533			20.50		20.50	33
4420	Fully adhered		39	1.026			39		39	64
4550	Roof hatch, 2'-6" x 3'-0"	1 Clab	10	.800	Ea.		30		30	49.50
4670	Wood shingles	B-2	2200	.018	S.F.		.69		.69	1.13
4820	Sheet metal roofing	"	2150	.019			.71		.71	1.16
4970	Siding, horizontal wood clapboards	1 Clab	380	.021			.79		.79	1.30
5025	Exterior insulation finish system	"	120	.067			2.51		2.51	4.11

07 05 Common Work Results for Thermal and Moisture Protection

07 05 05 – Selective Demolition for Thermal and Moisture Protection

07 05 05.10 Selective Demo., Thermal and Moist. Protection		Crew	Daily Output	Labor-Hours	Unit	Material	2015 Bare Costs Labor	Equipment	Total	Total Incl O&P
5070	Tempered hardboard, remove and reset	1 Carp	380	.021	S.F.		.99		.99	1.62
5120	Tempered hardboard sheet siding	″	375	.021			1		1	1.64
5170	Metal, corner strips	1 Clab	850	.009	L.F.		.35		.35	.58
5225	Horizontal strips		444	.018	S.F.		.68		.68	1.11
5320	Vertical strips		400	.020			.75		.75	1.23
5520	Wood shingles		350	.023			.86		.86	1.41
5620	Stucco siding		360	.022			.84		.84	1.37
5670	Textured plywood		725	.011			.41		.41	.68
5720	Vinyl siding		510	.016			.59		.59	.97
5770	Corner strips		900	.009	L.F.		.33		.33	.55
5870	Wood, boards, vertical		400	.020	S.F.		.75		.75	1.23
5920	Waterproofing, protection/drain board	2 Clab	3900	.004	B.F.		.15		.15	.25
5970	Over 1/2″ thick		1750	.009	S.F.		.34		.34	.56
6020	To 1/2″ thick		2000	.008	″		.30		.30	.49
9000	Minimum labor/equipment charge	1 Clab	2	4	Job		150		150	247

07 11 Dampproofing

07 11 13 – Bituminous Dampproofing

07 11 13.10 Bituminous Asphalt Coating

Line	Description	Crew	Daily Output	Labor-Hours	Unit	Material	Labor	Equipment	Total	Total Incl O&P
0010	**BITUMINOUS ASPHALT COATING**									
0030	Brushed on, below grade, 1 coat	1 Rofc	665	.012	S.F.	.22	.48		.70	1.12
0100	2 coat		500	.016		.45	.64		1.09	1.65
0300	Sprayed on, below grade, 1 coat		830	.010		.22	.39		.61	.95
0400	2 coat		500	.016		.44	.64		1.08	1.64
0500	Asphalt coating, with fibers				Gal.	8.25			8.25	9.10
0600	Troweled on, asphalt with fibers, 1/16″ thick	1 Rofc	500	.016	S.F.	.36	.64		1	1.56
0700	1/8″ thick		400	.020		.64	.80		1.44	2.15
1000	1/2″ thick		350	.023		2.07	.92		2.99	3.93
9000	Minimum labor/equipment charge		3	2.667	Job		107		107	193

07 11 16 – Cementitious Dampproofing

07 11 16.20 Cementitious Parging

Line	Description	Crew	Daily Output	Labor-Hours	Unit	Material	Labor	Equipment	Total	Total Incl O&P
0010	**CEMENTITIOUS PARGING**									
0020	Portland cement, 2 coats, 1/2″ thick	D-1	250	.064	S.F.	.33	2.69		3.02	4.74
0100	Waterproofed Portland cement, 1/2″ thick, 2 coats	″	250	.064	″	3.93	2.69		6.62	8.70

07 12 Built-up Bituminous Waterproofing

07 12 13 – Built-Up Asphalt Waterproofing

07 12 13.20 Membrane Waterproofing

Line	Description	Crew	Daily Output	Labor-Hours	Unit	Material	Labor	Equipment	Total	Total Incl O&P
0010	**MEMBRANE WATERPROOFING**									
0012	On slabs, 1 ply, felt, mopped	G-1	3000	.019	S.F.	.42	.70	.19	1.31	1.94
0015	On walls, 1 ply, felt, mopped		3000	.019		.42	.70	.19	1.31	1.94
0100	On slabs, 1 ply, glass fiber fabric, mopped		2100	.027		.46	1	.27	1.73	2.61
0105	On walls, 1 ply, glass fiber fabric, mopped		2100	.027		.46	1	.27	1.73	2.61
0300	On slabs, 2 ply, felt, mopped		2500	.022		.83	.84	.23	1.90	2.68
0305	On walls, 2 ply, felt, mopped		2500	.022		.83	.84	.23	1.90	2.68
0400	On slabs, 2 ply, glass fiber fabric, mopped		1650	.034		1.01	1.27	.34	2.62	3.79
0405	On walls, 2 ply, glass fiber fabric, mopped		1650	.034		1.01	1.27	.34	2.62	3.79
0600	On slabs, 3 ply, felt, mopped		2100	.027		1.25	1	.27	2.52	3.48
0605	On walls, 3 ply, felt, mopped		2100	.027		1.25	1	.27	2.52	3.48

07 12 Built-up Bituminous Waterproofing

07 12 13 – Built-Up Asphalt Waterproofing

07 12 13.20 Membrane Waterproofing		Crew	Daily Output	Labor-Hours	Unit	2015 Bare Costs Material	Labor	Equipment	Total	Total Incl O&P
0700	On slabs, 3 ply, glass fiber fabric, mopped	G-1	1550	.036	S.F.	1.37	1.36	.37	3.10	4.36
0705	On walls, 3 ply, glass fiber fabric, mopped		1550	.036		1.37	1.36	.37	3.10	4.36
0710	Asphaltic hardboard protection board, 1/8" thick	2 Rofc	500	.032		.61	1.28		1.89	2.99
0715	1/4" thick		450	.036		1.70	1.43		3.13	4.45
1000	EPS membrane protection board, 1/4"		3500	.005		.33	.18		.51	.70
1050	3/8" thick		3500	.005		.36	.18		.54	.73
1060	1/2" thick		3500	.005		.40	.18		.58	.76
1070	Fiberglass fabric, black, 20/10 mesh		116	.138	Sq.	15.40	5.55		20.95	27
9000	Minimum labor/equipment charge		2	8	Job		320		320	580

07 13 Sheet Waterproofing

07 13 53 – Elastomeric Sheet Waterproofing

07 13 53.10 Elastomeric Sheet Waterproofing and Access.		Crew	Daily Output	Labor-Hours	Unit	Material	Labor	Equipment	Total	Total Incl O&P
0010	**ELASTOMERIC SHEET WATERPROOFING AND ACCESS.**									
0090	EPDM, plain, 45 mils thick	2 Rofc	580	.028	S.F.	1.41	1.11		2.52	3.55
0100	60 mils thick		570	.028		1.47	1.13		2.60	3.65
0300	Nylon reinforced sheets, 45 mils thick		580	.028		1.55	1.11		2.66	3.70
0400	60 mils thick		570	.028		1.64	1.13		2.77	3.84
0600	Vulcanizing splicing tape for above, 2" wide				C.L.F.	60.50			60.50	66.50
0700	4" wide				"	121			121	133
0900	Adhesive, bonding, 60 S.F. per gal.				Gal.	27			27	29.50
1000	Splicing, 75 S.F. per gal.				"	41			41	45
1200	Neoprene sheets, plain, 45 mils thick	2 Rofc	580	.028	S.F.	1.70	1.11		2.81	3.87
1300	60 mils thick		570	.028		2.17	1.13		3.30	4.42
1500	Nylon reinforced, 45 mils thick		580	.028		1.96	1.11		3.07	4.16
1600	60 mils thick		570	.028		3.05	1.13		4.18	5.40
1800	120 mils thick		500	.032		6.05	1.28		7.33	8.95
1900	Adhesive, splicing, 150 S.F. per gal. per coat				Gal.	41			41	45
2100	Fiberglass reinforced, fluid applied, 1/8" thick	2 Rofc	500	.032	S.F.	1.63	1.28		2.91	4.11
2200	Polyethylene and rubberized asphalt sheets, 60 mils thick		550	.029		.92	1.17		2.09	3.12
2400	Polyvinyl chloride sheets, plain, 10 mils thick		580	.028		.15	1.11		1.26	2.17
2500	20 mils thick		570	.028		.19	1.13		1.32	2.24
2700	30 mils thick		560	.029		.24	1.15		1.39	2.33
3000	Adhesives, trowel grade, 40-100 S.F. per gal.				Gal.	24			24	26.50
3100	Brush grade, 100-250 S.F. per gal.				"	21.50			21.50	24
3300	Bitumen modified polyurethane, fluid applied, 55 mils thick	2 Rofc	665	.024	S.F.	.93	.96		1.89	2.76
9000	Minimum labor/equipment charge	"	2	8	Job		320		320	580

07 16 Cementitious and Reactive Waterproofing

07 16 16 – Crystalline Waterproofing

07 16 16.20 Cementitious Waterproofing		Crew	Daily Output	Labor-Hours	Unit	Material	Labor	Equipment	Total	Total Incl O&P
0010	**CEMENTITIOUS WATERPROOFING**									
0020	1/8" application, sprayed on	G-2A	1000	.024	S.F.	.73	.86	.72	2.31	3.10
0050	4 coat cementitious metallic slurry	1 Cefi	1.20	6.667	C.S.F.	32	300		332	510

07 17 Bentonite Waterproofing

07 17 13 – Bentonite Panel Waterproofing

07 17 13.10 Bentonite		Crew	Daily Output	Labor-Hours	Unit	Material	Labor	Equipment	Total	Total Incl O&P
						2015 Bare Costs				
0010	**BENTONITE**									
0020	Panels, 4′ x 4′, 3/16″ thick	1 Rofc	625	.013	S.F.	1.57	.51		2.08	2.66
0100	Rolls, 3/8″ thick, with geotextile fabric both sides	″	550	.015	″	1.47	.58		2.05	2.67
0300	Granular bentonite, 50 lb. bags (.625 C.F.)				Bag	20.50			20.50	22.50
0400	3/8″ thick, troweled on	1 Rofc	475	.017	S.F.	1.02	.68		1.70	2.34
0500	Drain board, expanded polystyrene, 1-1/2″ thick	1 Rohe	1600	.005		.38	.15		.53	.68
0510	2″ thick		1600	.005		.50	.15		.65	.82
0520	3″ thick		1600	.005		.75	.15		.90	1.10
0530	4″ thick		1600	.005		1	.15		1.15	1.37
0600	With filter fabric, 1-1/2″ thick		1600	.005		.44	.15		.59	.75
0625	2″ thick		1600	.005		.56	.15		.71	.89
0650	3″ thick		1600	.005		.81	.15		.96	1.17
0675	4″ thick		1600	.005		1.06	.15		1.21	1.44

07 19 Water Repellents

07 19 19 – Silicone Water Repellents

07 19 19.10 Silicone Based Water Repellents		Crew	Daily Output	Labor-Hours	Unit	Material	Labor	Equipment	Total	Total Incl O&P
0010	**SILICONE BASED WATER REPELLENTS**									
0020	Water base liquid, roller applied	2 Rofc	7000	.002	S.F.	.53	.09		.62	.75
0200	Silicone or stearate, sprayed on CMU, 1 coat	1 Rofc	4000	.002		.34	.08		.42	.51
0300	2 coats		3000	.003		.68	.11		.79	.94
9000	Minimum labor/equipment charge		3	2.667	Job		107		107	193

07 21 Thermal Insulation

07 21 13 – Board Insulation

07 21 13.10 Rigid Insulation			Crew	Daily Output	Labor-Hours	Unit	Material	Labor	Equipment	Total	Total Incl O&P
0010	**RIGID INSULATION**, for walls										
0040	Fiberglass, 1.5#/C.F., unfaced, 1″ thick, R4.1	G	1 Carp	1000	.008	S.F.	.27	.38		.65	.92
0060	1-1/2″ thick, R6.2	G		1000	.008		.40	.38		.78	1.06
0080	2″ thick, R8.3	G		1000	.008		.45	.38		.83	1.12
0120	3″ thick, R12.4	G		800	.010		.56	.47		1.03	1.39
0370	3#/C.F., unfaced, 1″ thick, R4.3	G		1000	.008		.52	.38		.90	1.19
0390	1-1/2″ thick, R6.5	G		1000	.008		.78	.38		1.16	1.48
0400	2″ thick, R8.7	G		890	.009		1.05	.42		1.47	1.85
0420	2-1/2″ thick, R10.9	G		800	.010		1.10	.47		1.57	1.98
0440	3″ thick, R13	G		800	.010		1.59	.47		2.06	2.52
0520	Foil faced, 1″ thick, R4.3	G		1000	.008		.90	.38		1.28	1.61
0540	1-1/2″ thick, R6.5	G		1000	.008		1.35	.38		1.73	2.11
0560	2″ thick, R8.7	G		890	.009		1.69	.42		2.11	2.55
0580	2-1/2″ thick, R10.9	G		800	.010		1.98	.47		2.45	2.95
0600	3″ thick, R13	G		800	.010		2.18	.47		2.65	3.17
1600	Isocyanurate, 4′ x 8′ sheet, foil faced, both sides										
1610	1/2″ thick	G	1 Carp	800	.010	S.F.	.31	.47		.78	1.11
1620	5/8″ thick	G		800	.010		.33	.47		.80	1.13
1630	3/4″ thick	G		800	.010		.36	.47		.83	1.17
1640	1″ thick	G		800	.010		.52	.47		.99	1.34
1650	1-1/2″ thick	G		730	.011		.63	.51		1.14	1.53
1660	2″ thick	G		730	.011		.80	.51		1.31	1.72
1670	3″ thick	G		730	.011		1.80	.51		2.31	2.82

07 21 Thermal Insulation

07 21 13 – Board Insulation

07 21 13.10 Rigid Insulation			Crew	Daily Output	Labor-Hours	Unit	Material	2015 Bare Costs Labor	Equipment	Total	Total Incl O&P
1680	4" thick	G	1 Carp	730	.011	S.F.	2.05	.51		2.56	3.10
1700	Perlite, 1" thick, R2.77	G		800	.010		.42	.47		.89	1.23
1750	2" thick, R5.55	G		730	.011		.75	.51		1.26	1.67
1900	Extruded polystyrene, 25 PSI compressive strength, 1" thick, R5	G		800	.010		.53	.47		1	1.35
1940	2" thick R10	G		730	.011		1.04	.51		1.55	1.98
1960	3" thick, R15	G		730	.011		1.50	.51		2.01	2.49
2100	Expanded polystyrene, 1" thick, R3.85	G		800	.010		.25	.47		.72	1.05
2120	2" thick, R7.69	G		730	.011		.50	.51		1.01	1.39
2140	3" thick, R11.49	G		730	.011		.75	.51		1.26	1.67
9000	Minimum labor/equipment charge			4	2	Job		94		94	154

07 21 13.13 Foam Board Insulation			Crew	Daily Output	Labor-Hours	Unit	Material	Labor	Equipment	Total	Total Incl O&P
0010	**FOAM BOARD INSULATION**										
0600	Polystyrene, expanded, 1" thick, R4	G	1 Carp	680	.012	S.F.	.25	.55		.80	1.18
0700	2" thick, R8	G		675	.012	"	.50	.56		1.06	1.46
9000	Minimum labor/equipment charge			4	2	Job		94		94	154

07 21 16 – Blanket Insulation

07 21 16.10 Blanket Insulation for Floors/Ceilings			Crew	Daily Output	Labor-Hours	Unit	Material	Labor	Equipment	Total	Total Incl O&P
0010	**BLANKET INSULATION FOR FLOORS/CEILINGS**										
0020	Including spring type wire fasteners										
2000	Fiberglass, blankets or batts, paper or foil backing										
2100	3-1/2" thick, R13	G	1 Carp	700	.011	S.F.	.37	.54		.91	1.29
2150	6-1/4" thick, R19	G		600	.013		.49	.63		1.12	1.57
2210	9-1/2" thick, R30	G		500	.016		.71	.75		1.46	2.01
2220	12" thick, R38	G		475	.017		.96	.79		1.75	2.36
3000	Unfaced, 3-1/2" thick, R13	G		600	.013		.31	.63		.94	1.37
3010	6-1/4" thick, R19	G		500	.016		.36	.75		1.11	1.63
3020	9-1/2" thick, R30	G		450	.018		.58	.83		1.41	2.01
3030	12" thick, R38	G		425	.019		.74	.88		1.62	2.26
9000	Minimum labor/equipment charge			4	2	Job		94		94	154

07 21 16.20 Blanket Insulation for Walls			Crew	Daily Output	Labor-Hours	Unit	Material	Labor	Equipment	Total	Total Incl O&P
0010	**BLANKET INSULATION FOR WALLS**										
0020	Kraft faced fiberglass, 3-1/2" thick, R11, 15" wide	G	1 Carp	1350	.006	S.F.	.27	.28		.55	.76
0030	23" wide	G		1600	.005		.27	.23		.50	.68
0060	R13, 11" wide	G		1150	.007		.30	.33		.63	.87
0080	15" wide	G		1350	.006		.30	.28		.58	.79
0100	23" wide	G		1600	.005		.30	.23		.53	.71
0110	R15, 11" wide	G		1150	.007		.45	.33		.78	1.04
0120	15" wide	G		1350	.006		.45	.28		.73	.96
0130	23" wide	G		1600	.005		.45	.23		.68	.88
0140	6" thick, R19, 11" wide	G		1150	.007		.42	.33		.75	1
0160	15" wide	G		1350	.006		.42	.28		.70	.92
0180	23" wide	G		1600	.005		.42	.23		.65	.84
0182	R21, 11" wide	G		1150	.007		.61	.33		.94	1.21
0184	15" wide	G		1350	.006		.61	.28		.89	1.13
0186	23" wide	G		1600	.005		.61	.23		.84	1.05
0188	9" thick, R30, 11" wide	G		985	.008		.71	.38		1.09	1.40
0200	15" wide	G		1150	.007		.71	.33		1.04	1.32
0220	23" wide	G		1350	.006		.71	.28		.99	1.24
0230	12" thick, R38, 11" wide	G		985	.008		.96	.38		1.34	1.68
0240	15" wide	G		1150	.007		.96	.33		1.29	1.60
0260	23" wide	G		1350	.006		.96	.28		1.24	1.52
0410	Foil faced fiberglass, 3-1/2" thick, R13, 11" wide	G		1150	.007		.45	.33		.78	1.04

07 21 Thermal Insulation

07 21 16 – Blanket Insulation

07 21 16.20 Blanket Insulation for Walls			Crew	Daily Output	Labor-Hours	Unit	2015 Bare Costs Material	Labor	Equipment	Total	Total Incl O&P
0420	15" wide	G	1 Carp	1350	.006	S.F.	.45	.28		.73	.96
0440	23" wide	G		1600	.005		.45	.23		.68	.88
0442	R15, 11" wide	G		1150	.007		.47	.33		.80	1.06
0444	15" wide	G		1350	.006		.47	.28		.75	.98
0446	23" wide	G		1600	.005		.47	.23		.70	.90
0448	6" thick, R19, 11" wide	G		1150	.007		.60	.33		.93	1.20
0460	15" wide	G		1350	.006		.60	.28		.88	1.12
0480	23" wide	G		1600	.005		.60	.23		.83	1.04
0482	R21, 11" wide	G		1150	.007		.62	.33		.95	1.22
0484	15" wide	G		1350	.006		.62	.28		.90	1.14
0486	23" wide	G		1600	.005		.62	.23		.85	1.06
0488	9" thick, R30, 11" wide	G		985	.008		.90	.38		1.28	1.61
0500	15" wide	G		1150	.007		.90	.33		1.23	1.53
0550	23" wide	G		1350	.006		.90	.28		1.18	1.45
0560	12" thick, R38, 11" wide	G		985	.008		1.05	.38		1.43	1.78
0570	15" wide	G		1150	.007		1.05	.33		1.38	1.70
0580	23" wide	G		1350	.006		1.05	.28		1.33	1.62
0620	Unfaced fiberglass, 3-1/2" thick, R13, 11" wide	G		1150	.007		.31	.33		.64	.88
0820	15" wide	G		1350	.006		.31	.28		.59	.80
0830	23" wide	G		1600	.005		.31	.23		.54	.72
0832	R15, 11" wide	G		1150	.007		.42	.33		.75	1
0836	23" wide	G		1600	.005		.42	.23		.65	.84
0838	6" thick, R19, 11" wide	G		1150	.007		.36	.33		.69	.94
0860	15" wide	G		1150	.007		.36	.33		.69	.94
0880	23" wide	G		1350	.006		.36	.28		.64	.86
0882	R21, 11" wide	G		1150	.007		.54	.33		.87	1.13
0886	15" wide	G		1350	.006		.54	.28		.82	1.05
0888	23" wide	G		1600	.005		.54	.23		.77	.97
0890	9" thick, R30, 11" wide	G		985	.008		.58	.38		.96	1.26
0900	15" wide	G		1150	.007		.58	.33		.91	1.18
0920	23" wide	G		1350	.006		.58	.28		.86	1.10
0930	12" thick, R38, 11" wide	G		985	.008		.74	.38		1.12	1.43
0940	15" wide	G		1000	.008		.74	.38		1.12	1.43
0960	23" wide	G		1150	.007		.74	.33		1.07	1.35
1300	Wall or ceiling insulation, mineral wool batts										
1320	3-1/2" thick, R15	G	1 Carp	1600	.005	S.F.	.60	.23		.83	1.04
1340	5-1/2" thick, R23	G		1600	.005		.94	.23		1.17	1.42
1380	7-1/4" thick, R30	G		1350	.006		1.24	.28		1.52	1.83
1700	Non-rigid insul, recycled blue cotton fiber, unfaced batts, R13, 16" wide	G		1600	.005		1.01	.23		1.24	1.49
1710	R19, 16" wide	G		1600	.005		1.38	.23		1.61	1.90
1850	Friction fit wire insulation supports, 16" O.C.			960	.008	Ea.	.07	.39		.46	.72
9000	Minimum labor/equipment charge			4	2	Job		94		94	154

07 21 19 – Foamed In Place Insulation

07 21 19.10 Masonry Foamed In Place Insulation			Crew	Daily Output	Labor-Hours	Unit	Material	Labor	Equipment	Total	Total Incl O&P
0010	**MASONRY FOAMED IN PLACE INSULATION**										
0100	Amino-plast foam, injected into block core, 6" block	G	G-2A	6000	.004	Ea.	.15	.14	.12	.41	.55
0110	8" block	G		5000	.005		.19	.17	.14	.50	.66
0120	10" block	G		4000	.006		.23	.22	.18	.63	.84
0130	12" block	G		3000	.008		.31	.29	.24	.84	1.10
0140	Injected into cavity wall	G		13000	.002	B.F.	.05	.07	.06	.18	.24
0150	Preparation, drill holes into mortar joint every 4 VLF, 5/8" dia		1 Clab	960	.008	Ea.		.31		.31	.51
0160	7/8" dia			680	.012			.44		.44	.73

07 21 Thermal Insulation

07 21 19 – Foamed In Place Insulation

07 21 19.10 Masonry Foamed In Place Insulation			Crew	Daily Output	Labor-Hours	Unit	Material (2015 Bare Costs)	Labor (2015 Bare Costs)	Equipment (2015 Bare Costs)	Total (2015 Bare Costs)	Total Incl O&P
0170	Patch drilled holes, 5/8" diameter		1 Clab	1800	.004	Ea.	.03	.17		.20	.31
0180	7/8" diameter			1200	.007		.05	.25		.30	.46
9000	Minimum labor/equipment charge		G-2A	2	12	Job		430	360	790	1,150
9010	Minimum labor/equipment charge		1 Clab	4	2	"		75		75	123

07 21 23 – Loose-Fill Insulation

07 21 23.10 Poured Loose-Fill Insulation

Line	Description		Crew	Daily Output	Labor-Hours	Unit	Material	Labor	Equipment	Total	Total Incl O&P
0010	**POURED LOOSE-FILL INSULATION**										
0020	Cellulose fiber, R3.8 per inch	G	1 Carp	200	.040	C.F.	.69	1.88		2.57	3.84
0021	4" thick	G		1000	.008	S.F.	.17	.38		.55	.80
0022	6" thick	G		800	.010	"	.28	.47		.75	1.08
0080	Fiberglass wool, R4 per inch	G		200	.040	C.F.	.55	1.88		2.43	3.68
0081	4" thick	G		600	.013	S.F.	.19	.63		.82	1.24
0082	6" thick	G		400	.020	"	.26	.94		1.20	1.83
0100	Mineral wool, R3 per inch	G		200	.040	C.F.	.41	1.88		2.29	3.53
0101	4" thick	G		600	.013	S.F.	.14	.63		.77	1.18
0102	6" thick	G		400	.020	"	.21	.94		1.15	1.77
0300	Polystyrene, R4 per inch	G		200	.040	C.F.	1.60	1.88		3.48	4.84
0301	4" thick	G		600	.013	S.F.	.53	.63		1.16	1.61
0302	6" thick	G		400	.020	"	.80	.94		1.74	2.42
0400	Perlite, R2.78 per inch	G		200	.040	C.F.	5.20	1.88		7.08	8.85
0401	4" thick	G		1000	.008	S.F.	1.73	.38		2.11	2.53
0402	6" thick	G		800	.010	"	2.61	.47		3.08	3.64
9000	Minimum labor/equipment charge			4	2	Job		94		94	154

07 21 23.20 Masonry Loose-Fill Insulation

Line	Description		Crew	Daily Output	Labor-Hours	Unit	Material	Labor	Equipment	Total	Total Incl O&P
0010	**MASONRY LOOSE-FILL INSULATION**, vermiculite or perlite										
0100	In cores of concrete block, 4" thick wall, .115 C.F./S.F.	G	D-1	4800	.003	S.F.	.60	.14		.74	.89
0200	6" thick wall, .175 C.F./S.F.	G		3000	.005		.91	.22		1.13	1.37
0300	8" thick wall, .258 C.F./S.F.	G		2400	.007		1.34	.28		1.62	1.94
0400	10" thick wall, .340 C.F./S.F.	G		1850	.009		1.77	.36		2.13	2.54
0500	12" thick wall, .422 C.F./S.F.	G		1200	.013		2.20	.56		2.76	3.33
0600	Poured cavity wall, vermiculite or perlite, water repellant	G		250	.064	C.F.	5.20	2.69		7.89	10.15
0700	Foamed in place, urethane in 2-5/8" cavity	G	G-2A	1035	.023	S.F.	1.38	.83	.69	2.90	3.74
0800	For each 1" added thickness, add	G	"	2372	.010	"	.53	.36	.30	1.19	1.55

07 21 26 – Blown Insulation

07 21 26.10 Blown Insulation

Line	Description		Crew	Daily Output	Labor-Hours	Unit	Material	Labor	Equipment	Total	Total Incl O&P
0010	**BLOWN INSULATION** Ceilings, with open access										
0020	Cellulose, 3-1/2" thick, R13	G	G-4	5000	.005	S.F.	.24	.18	.08	.50	.65
0030	5-3/16" thick, R19	G		3800	.006		.35	.24	.11	.70	.91
0050	6-1/2" thick, R22	G		3000	.008		.45	.31	.13	.89	1.15
0100	8-11/16" thick, R30	G		2600	.009		.61	.35	.15	1.11	1.42
0120	10-7/8" thick, R38	G		1800	.013		.78	.51	.22	1.51	1.94
1000	Fiberglass, 5.5" thick, R11	G		3800	.006		.19	.24	.11	.54	.72
1050	6" thick, R12	G		3000	.008		.26	.31	.13	.70	.94
1100	8.8" thick, R19	G		2200	.011		.33	.42	.18	.93	1.24
1200	10" thick, R22	G		1800	.013		.38	.51	.22	1.11	1.51
1300	11.5" thick, R26	G		1500	.016		.46	.61	.27	1.34	1.79
1350	13" thick, R30	G		1400	.017		.53	.66	.29	1.48	1.98
1450	16" thick, R38	G		1145	.021		.68	.80	.35	1.83	2.44
1500	20" thick, R49	G		920	.026		.89	1	.44	2.33	3.10
9000	Minimum labor/equipment charge			4	6	Job		230	100	330	485

07 21 Thermal Insulation

07 21 27 – Reflective Insulation

07 21 27.10 Reflective Insulation Options			Crew	Daily Output	Labor-Hours	Unit	Material	2015 Bare Costs Labor	Equipment	Total	Total Incl O&P
0010	**REFLECTIVE INSULATION OPTIONS**										
0020	Aluminum foil on reinforced scrim	G	1 Carp	19	.421	C.S.F.	14.20	19.75		33.95	48
0100	Reinforced with woven polyolefin	G		19	.421		22	19.75		41.75	56.50
0500	With single bubble air space, R8.8	G		15	.533		28	25		53	72
0600	With double bubble air space, R9.8	G		15	.533		32	25		57	76
9000	Minimum labor/equipment charge			4	2	Job		94		94	154

07 21 29 – Sprayed Insulation

07 21 29.10 Sprayed-On Insulation			Crew	Daily Output	Labor-Hours	Unit	Material	Labor	Equipment	Total	Total Incl O&P
0010	**SPRAYED-ON INSULATION**										
0020	Fibrous/cementitious, finished wall, 1" thick, R3.7	G	G-2	2050	.012	S.F.	.31	.46	.07	.84	1.16
0100	Attic, 5.2" thick, R19	G		1550	.015	"	.42	.61	.09	1.12	1.54
0200	Fiberglass, R4 per inch, vertical	G		1600	.015	B.F.	.18	.59	.08	.85	1.25
0210	Horizontal	G		1200	.020	"	.18	.79	.11	1.08	1.60
0300	Closed cell, spray polyurethane foam, 2 pounds per cubic foot density										
0310	1" thick	G	G-2A	6000	.004	S.F.	.53	.14	.12	.79	.96
0320	2" thick	G		3000	.008		1.05	.29	.24	1.58	1.92
0330	3" thick	G		2000	.012		1.58	.43	.36	2.37	2.87
0335	3-1/2" thick	G		1715	.014		1.84	.50	.42	2.76	3.36
0340	4" thick	G		1500	.016		2.10	.57	.48	3.15	3.84
0350	5" thick	G		1200	.020		2.63	.72	.60	3.95	4.81
0355	5-1/2" thick	G		1090	.022		2.89	.79	.66	4.34	5.30
0360	6" thick	G		1000	.024		3.15	.86	.72	4.73	5.75
9000	Minimum labor/equipment charge		G-2	2	12	Job		475	66.50	541.50	840

07 22 Roof and Deck Insulation

07 22 16 – Roof Board Insulation

07 22 16.10 Roof Deck Insulation			Crew	Daily Output	Labor-Hours	Unit	Material	Labor	Equipment	Total	Total Incl O&P
0010	**ROOF DECK INSULATION**, fastening excluded										
0016	Asphaltic cover board, fiberglass lined, 1/8" thick		1 Rofc	1400	.006	S.F.	.47	.23		.70	.93
0018	1/4" thick			1400	.006		.94	.23		1.17	1.44
0020	Fiberboard low density, 1/2" thick R1.39	G		1300	.006		.30	.25		.55	.78
0030	1" thick R2.78	G		1040	.008		.52	.31		.83	1.13
0080	1-1/2" thick R4.17	G		1040	.008		.80	.31		1.11	1.44
0100	2" thick R5.56	G		1040	.008		1.06	.31		1.37	1.73
0110	Fiberboard high density, 1/2" thick R1.3	G		1300	.006		.30	.25		.55	.78
0120	1" thick R2.5	G		1040	.008		.58	.31		.89	1.20
0130	1-1/2" thick R3.8	G		1040	.008		.88	.31		1.19	1.53
0200	Fiberglass, 3/4" thick R2.78	G		1300	.006		.61	.25		.86	1.12
0400	15/16" thick R3.70	G		1300	.006		.81	.25		1.06	1.34
0460	1-1/16" thick R4.17	G		1300	.006		1.01	.25		1.26	1.56
0600	1-5/16" thick R5.26	G		1300	.006		1.37	.25		1.62	1.96
0650	2-1/16" thick R8.33	G		1040	.008		1.45	.31		1.76	2.16
0700	2-7/16" thick R10	G		1040	.008		1.68	.31		1.99	2.41
0800	Gypsum cover board, fiberglass mat facer, 1/4" thick			1400	.006		.47	.23		.70	.93
0810	1/2" thick			1300	.006		.57	.25		.82	1.08
0820	5/8" thick			1200	.007		.61	.27		.88	1.15
0830	Primed fiberglass mat facer, 1/4" thick			1400	.006		.51	.23		.74	.97
0840	1/2" thick			1300	.006		.62	.25		.87	1.13
0850	5/8" thick			1200	.007		.65	.27		.92	1.20
1650	Perlite, 1/2" thick R1.32	G		1365	.006		.28	.24		.52	.73

07 22 Roof and Deck Insulation

07 22 16 – Roof Board Insulation

07 22 16.10 Roof Deck Insulation			Crew	Daily Output	Labor-Hours	Unit	2015 Bare Costs Material	Labor	Equipment	Total	Total Incl O&P
1655	3/4" thick R2.08	G	1 Rofc	1040	.008	S.F.	.34	.31		.65	.93
1660	1" thick R2.78	G		1040	.008		.50	.31		.81	1.11
1670	1-1/2" thick R4.17	G		1040	.008		.73	.31		1.04	1.36
1680	2" thick R5.56	G		910	.009		1	.35		1.35	1.74
1685	2-1/2" thick R6.67	G		910	.009		1.30	.35		1.65	2.07
1690	Tapered for drainage	G		1040	.008	B.F.	1.01	.31		1.32	1.67
1700	Polyisocyanurate, 2#/C.F. density, 3/4" thick	G		1950	.004	S.F.	.46	.16		.62	.81
1705	1" thick	G		1820	.004		.48	.18		.66	.85
1715	1-1/2" thick	G		1625	.005		.64	.20		.84	1.06
1725	2" thick	G		1430	.006		.83	.22		1.05	1.32
1735	2-1/2" thick	G		1365	.006		1.06	.24		1.30	1.59
1745	3" thick	G		1300	.006		1.27	.25		1.52	1.85
1755	3-1/2" thick	G		1300	.006		1.95	.25		2.20	2.60
1765	Tapered for drainage	G		1820	.004	B.F.	.64	.18		.82	1.02
1900	Extruded Polystyrene										
1910	15 PSI compressive strength, 1" thick, R5	G	1 Rofc	1950	.004	S.F.	.55	.16		.71	.91
1920	2" thick, R10	G		1625	.005		.72	.20		.92	1.15
1930	3" thick, R15	G		1300	.006		1.43	.25		1.68	2.02
1932	4" thick, R20	G		1300	.006		1.93	.25		2.18	2.57
1934	Tapered for drainage	G		1950	.004	B.F.	.59	.16		.75	.95
1940	25 PSI compressive strength, 1" thick, R5	G		1950	.004	S.F.	.69	.16		.85	1.06
1942	2" thick, R10	G		1625	.005		1.31	.20		1.51	1.80
1944	3" thick, R15	G		1300	.006		2	.25		2.25	2.65
1946	4" thick, R20	G		1300	.006		2.76	.25		3.01	3.49
1948	Tapered for drainage	G		1950	.004	B.F.	.61	.16		.77	.97
1950	40 psi compressive strength, 1" thick, R5	G		1950	.004	S.F.	.53	.16		.69	.88
1952	2" thick, R10	G		1625	.005		1.01	.20		1.21	1.47
1954	3" thick, R15	G		1300	.006		1.46	.25		1.71	2.05
1956	4" thick, R20	G		1300	.006		1.91	.25		2.16	2.55
1958	Tapered for drainage	G		1820	.004	B.F.	.76	.18		.94	1.16
1960	60 PSI compressive strength, 1" thick, R5	G		1885	.004	S.F.	.74	.17		.91	1.12
1962	2" thick, R10	G		1560	.005		1.41	.21		1.62	1.92
1964	3" thick, R15	G		1270	.006		2.29	.25		2.54	2.98
1966	4" thick, R20	G		1235	.006		2.85	.26		3.11	3.60
1968	Tapered for drainage	G		1820	.004	B.F.	.96	.18		1.14	1.38
2010	Expanded polystyrene, 1#/C.F. density, 3/4" thick, R2.89	G		1950	.004	S.F.	.19	.16		.35	.51
2020	1" thick, R3.85	G		1950	.004		.25	.16		.41	.58
2100	2" thick, R7.69	G		1625	.005		.50	.20		.70	.91
2110	3" thick, R11.49	G		1625	.005		.75	.20		.95	1.19
2120	4" thick, R15.38	G		1625	.005		1	.20		1.20	1.46
2130	5" thick, R19.23	G		1495	.005		1.25	.21		1.46	1.77
2140	6" thick, R23.26	G		1495	.005		1.50	.21		1.71	2.04
2150	Tapered for drainage	G		1950	.004	B.F.	.53	.16		.69	.88
2400	Composites with 2" EPS										
2410	1" fiberboard	G	1 Rofc	1325	.006	S.F.	1.40	.24		1.64	1.98
2420	7/16" oriented strand board	G		1040	.008		1.13	.31		1.44	1.80
2430	1/2" plywood	G		1040	.008		1.39	.31		1.70	2.09
2440	1" perlite	G		1040	.008		1.14	.31		1.45	1.81
2450	Composites with 1-1/2" polyisocyanurate										
2460	1" fiberboard	G	1 Rofc	1040	.008	S.F.	1.19	.31		1.50	1.87
2470	1" perlite	G		1105	.007		1.06	.29		1.35	1.69
2480	7/16" oriented strand board	G		1040	.008		.93	.31		1.24	1.58
3000	Fastening alternatives, coated screws, 2" long			3744	.002	Ea.	.05	.09		.14	.22

07 22 Roof and Deck Insulation

07 22 16 – Roof Board Insulation

07 22 16.10 Roof Deck Insulation		Crew	Daily Output	Labor-Hours	Unit	Material	2015 Bare Costs Labor	Equipment	Total	Total Incl O&P
3010	4" long	1 Rofc	3120	.003	Ea.	.10	.10		.20	.30
3020	6" long		2675	.003		.17	.12		.29	.41
3030	8" long		2340	.003		.25	.14		.39	.53
3040	10" long		1872	.004		.43	.17		.60	.78
3050	Pre-drill and drive wedge spike, 2-1/2"		1248	.006		.37	.26		.63	.87
3060	3-1/2"		1101	.007		.48	.29		.77	1.06
3070	4-1/2"		936	.009		.60	.34		.94	1.28
3075	3" galvanized deck plates		7488	.001		.07	.04		.11	.16
3080	Spot mop asphalt	G-1	295	.190	Sq.	5.55	7.10	1.93	14.58	21
3090	Full mop asphalt	"	192	.292		11.10	10.95	2.96	25.01	35
3110	Low-rise polyurethane adhesive, from 5 gallon kit, 12" OC beads	1 Rofc	45	.178		33	7.15		40.15	49
3120	6" OC beads		32	.250		65.50	10.05		75.55	90
3130	4" OC beads		30	.267		98.50	10.70		109.20	127
9000	Minimum labor/equipment charge		3.25	2.462	Job		98.50		98.50	178

07 24 Exterior Insulation and Finish Systems

07 24 13 – Polymer-Based Exterior Insulation and Finish System

07 24 13.10 Exterior Insulation and Finish Systems		Crew	Daily Output	Labor-Hours	Unit	Material	Labor	Equipment	Total	Total Incl O&P
0010	**EXTERIOR INSULATION AND FINISH SYSTEMS**									
0095	Field applied, 1" EPS insulation G	J-1	390	.103	S.F.	1.91	4.21	.36	6.48	9.25
0100	With 1/2" cement board sheathing G		268	.149		2.65	6.10	.52	9.27	13.30
0105	2" EPS insulation G		390	.103		2.16	4.21	.36	6.73	9.55
0110	With 1/2" cement board sheathing G		268	.149		2.90	6.10	.52	9.52	13.55
0115	3" EPS insulation G		390	.103		2.41	4.21	.36	6.98	9.80
0120	With 1/2" cement board sheathing G		268	.149		3.15	6.10	.52	9.77	13.85
0125	4" EPS insulation G		390	.103		2.66	4.21	.36	7.23	10.10
0130	With 1/2" cement board sheathing G		268	.149		4.14	6.10	.52	10.76	14.95
0140	Premium finish add		1265	.032		.33	1.30	.11	1.74	2.55
0150	Heavy duty reinforcement add		914	.044		.83	1.79	.15	2.77	3.95
0160	2.5#/S.Y. metal lath substrate add	1 Lath	75	.107	S.Y.	2.72	4.59		7.31	10.20
0170	3.4#/S.Y. metal lath substrate add	"	75	.107	"	4.13	4.59		8.72	11.75
0180	Color or texture change,	J-1	1265	.032	S.F.	.77	1.30	.11	2.18	3.04
0190	With substrate leveling base coat	1 Plas	530	.015		.77	.65		1.42	1.89
0210	With substrate sealing base coat	1 Pord	1224	.007		.10	.26		.36	.53
0370	V groove shape in panel face				L.F.	.62			.62	.68
0380	U groove shape in panel face				"	.80			.80	.88
0433	Crack repair, acrylic rubber, fluid applied, 20 mils thick	1 Plas	350	.023	S.F.	1.45	.98		2.43	3.17
0437	50 mils thick, reinforced	"	200	.040	"	2.65	1.72		4.37	5.65
0440	For higher than one story, add						25%			

07 25 Weather Barriers

07 25 10 – Weather Barriers or Wraps

07 25 10.10 Weather Barriers		Crew	Daily Output	Labor-Hours	Unit	2015 Bare Costs Material	Labor	Equipment	Total	Total Incl O&P
0010	WEATHER BARRIERS									
0400	Asphalt felt paper, 15#	1 Carp	37	.216	Sq.	5.40	10.15		15.55	22.50
0401	Per square foot	"	3700	.002	S.F.	.05	.10		.15	.23
0450	Housewrap, exterior, spun bonded polypropylene									
0470	Small roll	1 Carp	3800	.002	S.F.	.15	.10		.25	.32
0480	Large roll	"	4000	.002	"	.14	.09		.23	.30
2100	Asphalt felt roof deck vapor barrier, class 1 metal decks	1 Rofc	37	.216	Sq.	22	8.65		30.65	39.50
2200	For all other decks	"	37	.216		16.50	8.65		25.15	34
2800	Asphalt felt, 50% recycled content, 15 lb., 4 sq. per roll	1 Carp	36	.222		5.40	10.45		15.85	23
2810	30 lb., 2 sq. per roll	"	36	.222		10.80	10.45		21.25	29
3000	Building wrap, spunbonded polyethylene	2 Carp	8000	.002	S.F.	.15	.09		.24	.32
9960	Minimum labor/equipment charge	1 Carp	2	4	Job		188		188	310

07 26 Vapor Retarders

07 26 10 – Above-Grade Vapor Retarders

07 26 10.10 Vapor Retarders		Crew	Daily Output	Labor-Hours	Unit	Material	Labor	Equipment	Total	Total Incl O&P
0010	VAPOR RETARDERS									
0020	Aluminum and kraft laminated, foil 1 side G	1 Carp	37	.216	Sq.	12.50	10.15		22.65	30.50
0100	Foil 2 sides G		37	.216		14	10.15		24.15	32
0600	Polyethylene vapor barrier, standard, 2 mil G		37	.216		1.50	10.15		11.65	18.30
0700	4 mil G		37	.216		2.90	10.15		13.05	19.85
0900	6 mil G		37	.216		4.02	10.15		14.17	21
1200	10 mil G		37	.216		8.85	10.15		19	26.50
1300	Clear reinforced, fire retardant, 8 mil G		37	.216		10.85	10.15		21	28.50
1350	Cross laminated type, 3 mil G		37	.216		7.60	10.15		17.75	25
1400	4 mil G		37	.216		7.95	10.15		18.10	25.50
1800	Reinf. waterproof, 2 mil polyethylene backing, 1 side		37	.216		6.05	10.15		16.20	23.50
1900	2 sides		37	.216		7.95	10.15		18.10	25.50
2400	Waterproofed kraft with sisal or fiberglass fibers		37	.216		12.75	10.15		22.90	30.50
9950	Minimum labor/equipment charge		4	2	Job		94		94	154

07 27 Air Barriers

07 27 26 – Fluid-Applied Membrane Air Barriers

07 27 26.10 Fluid Applied Membrane Air Barrier		Crew	Daily Output	Labor-Hours	Unit	Material	Labor	Equipment	Total	Total Incl O&P
0010	FLUID APPLIED MEMBRANE AIR BARRIER									
0100	Spray applied vapor barrier, 25 S.F./gallon	1 Pord	1375	.006	S.F.	.01	.23		.24	.40

07 31 Shingles and Shakes

07 31 13 – Asphalt Shingles

07 31 13.10 Asphalt Roof Shingles		Crew	Daily Output	Labor-Hours	Unit	Material	Labor	Equipment	Total	Total Incl O&P
0010	ASPHALT ROOF SHINGLES									
0100	Standard strip shingles									
0150	Inorganic, class A, 25 year	1 Rofc	5.50	1.455	Sq.	79.50	58.50		138	193
0155	Pneumatic nailed		7	1.143		79.50	46		125.50	171
0200	30 year		5	1.600		94	64		158	219
0205	Pneumatic nailed		6.25	1.280		94	51.50		145.50	196
0250	Standard laminated multi-layered shingles									

07 31 13 – Asphalt Shingles

07 31 13.10 Asphalt Roof Shingles		Crew	Daily Output	Labor-Hours	Unit	Material	2015 Bare Costs Labor	Equipment	Total	Total Incl O&P
0300	Class A, 240-260 lb./square	1 Rofc	4.50	1.778	Sq.	110	71.50		181.50	250
0305	Pneumatic nailed		5.63	1.422		110	57		167	224
0350	Class A, 250-270 lb./square		4	2		110	80		190	266
0355	Pneumatic nailed		5	1.600		110	64		174	237
0400	Premium, laminated multi-layered shingles									
0450	Class A, 260-300 lb./square	1 Rofc	3.50	2.286	Sq.	153	91.50		244.50	335
0455	Pneumatic nailed		4.37	1.831		153	73.50		226.50	300
0500	Class A, 300-385 lb./square		3	2.667		230	107		337	445
0505	Pneumatic nailed		3.75	2.133		230	85.50		315.50	410
0800	#15 felt underlayment		64	.125		5.40	5		10.40	15
0825	#30 felt underlayment		58	.138		10.60	5.55		16.15	21.50
0850	Self adhering polyethylene and rubberized asphalt underlayment		22	.364		75	14.60		89.60	109
0900	Ridge shingles		330	.024	L.F.	2.10	.97		3.07	4.07
0905	Pneumatic nailed		412.50	.019	"	2.10	.78		2.88	3.71
1000	For steep roofs (7 to 12 pitch or greater), add						50%			
9000	Minimum labor/equipment charge	1 Rofc	3	2.667	Job		107		107	193

07 31 16 – Metal Shingles

07 31 16.10 Aluminum Shingles		Crew	Daily Output	Labor-Hours	Unit	Material	Labor	Equipment	Total	Total Incl O&P
0010	**ALUMINUM SHINGLES**									
0020	Mill finish, .019 thick	1 Carp	5	1.600	Sq.	221	75		296	365
0100	.020" thick	"	5	1.600		224	75		299	370
0300	For colors, add					21			21	23
0600	Ridge cap, .024" thick	1 Carp	170	.047	L.F.	3.63	2.21		5.84	7.60
0700	End wall flashing, .024" thick		170	.047		2.10	2.21		4.31	5.95
0900	Valley section, .024" thick		170	.047		3.57	2.21		5.78	7.55
1000	Starter strip, .024" thick		400	.020		1.67	.94		2.61	3.38
1200	Side wall flashing, .024" thick		170	.047		2.03	2.21		4.24	5.85
1500	Gable flashing, .024" thick		400	.020		1.63	.94		2.57	3.33
9000	Minimum labor/equipment charge		3	2.667	Job		125		125	205

07 31 16.20 Steel Shingles		Crew	Daily Output	Labor-Hours	Unit	Material	Labor	Equipment	Total	Total Incl O&P
0010	**STEEL SHINGLES**									
0012	Galvanized, 26 ga.	1 Rots	2.20	3.636	Sq.	335	146		481	635
0200	24 ga.	"	2.20	3.636		335	146		481	635
0300	For colored galvanized shingles, add					56			56	61.50
9000	Minimum labor/equipment charge	1 Rots	3	2.667	Job		107		107	193

07 31 26 – Slate Shingles

07 31 26.10 Slate Roof Shingles		Crew	Daily Output	Labor-Hours	Unit	Material	Labor	Equipment	Total	Total Incl O&P
0010	**SLATE ROOF SHINGLES**									
0100	Buckingham Virginia black, 3/16" - 1/4" thick G	1 Rots	1.75	4.571	Sq.	470	183		653	845
0200	1/4" thick G		1.75	4.571		470	183		653	845
0900	Pennsylvania black, Bangor, #1 clear G		1.75	4.571		490	183		673	870
1200	Vermont, unfading, green, mottled green G		1.75	4.571		475	183		658	855
1300	Semi-weathering green & gray G		1.75	4.571		345	183		528	710
1400	Purple G		1.75	4.571		425	183		608	795
1500	Black or gray G		1.75	4.571		455	183		638	830
2500	Slate roof repair, extensive replacement		1	8		600	320		920	1,250
2600	Repair individual pieces, scattered		19	.421	Ea.	7	16.90		23.90	38
2700	Ridge shingles, slate		200	.040	L.F.	10	1.60		11.60	13.90
9000	Minimum labor/equipment charge		3	2.667	Job		107		107	193

07 31 29 – Wood Shingles and Shakes

07 31 29.13 Wood Shingles

	07 31 29.13 Wood Shingles	Crew	Daily Output	Labor-Hours	Unit	Material	2015 Bare Costs Labor	Equipment	Total	Total Incl O&P
0010	**WOOD SHINGLES**									
0012	16" No. 1 red cedar shingles, 5" exposure, on roof	1 Carp	2.50	3.200	Sq.	287	150		437	560
0015	Pneumatic nailed		3.25	2.462		287	116		403	505
0200	7-1/2" exposure, on walls		2.05	3.902		191	183		374	510
0205	Pneumatic nailed		2.67	2.996		191	141		332	440
0300	18" No. 1 red cedar perfections, 5-1/2" exposure, on roof		2.75	2.909		246	137		383	495
0305	Pneumatic nailed		3.57	2.241		246	105		351	445
0500	7-1/2" exposure, on walls		2.25	3.556		181	167		348	475
0505	Pneumatic nailed		2.92	2.740		181	129		310	410
0600	Resquared, and rebutted, 5-1/2" exposure, on roof		3	2.667		279	125		404	510
0605	Pneumatic nailed		3.90	2.051		279	96.50		375.50	465
0900	7-1/2" exposure, on walls		2.45	3.265		205	153		358	475
0905	Pneumatic nailed		3.18	2.516		205	118		323	420
1000	Add to above for fire retardant shingles					56			56	61.50
1060	Preformed ridge shingles	1 Carp	400	.020	L.F.	5	.94		5.94	7.05
2000	White cedar shingles, 16" long, extras, 5" exposure, on roof		2.40	3.333	Sq.	192	157		349	470
2005	Pneumatic nailed		3.12	2.564		192	120		312	410
2050	5" exposure on walls		2	4		192	188		380	520
2055	Pneumatic nailed		2.60	3.077		192	144		336	450
2100	7-1/2" exposure, on walls		2	4		137	188		325	460
2105	Pneumatic nailed		2.60	3.077		137	144		281	390
2150	"B" grade, 5" exposure on walls		2	4		165	188		353	490
2155	Pneumatic nailed		2.60	3.077		165	144		309	420
2300	For 15# organic felt underlayment on roof, 1 layer, add		64	.125		5.40	5.85		11.25	15.55
2400	2 layers, add		32	.250		10.80	11.75		22.55	31
2600	For steep roofs (7/12 pitch or greater), add to above						50%			
2700	Panelized systems, No.1 cedar shingles on 5/16" CDX plywood									
2800	On walls, 8' strips, 7" or 14" exposure	2 Carp	700	.023	S.F.	6	1.07		7.07	8.35
3500	On roofs, 8' strips, 7" or 14" exposure	1 Carp	3	2.667	Sq.	600	125		725	865
3505	Pneumatic nailed		4	2	"	600	94		694	815
9000	Minimum labor/equipment charge		3	2.667	Job		125		125	205

07 31 29.16 Wood Shakes

	07 31 29.16 Wood Shakes	Crew	Daily Output	Labor-Hours	Unit	Material	2015 Bare Costs Labor	Equipment	Total	Total Incl O&P
0010	**WOOD SHAKES**									
1100	Hand-split red cedar shakes, 1/2" thick x 24" long, 10" exp. on roof	1 Carp	2.50	3.200	Sq.	272	150		422	545
1105	Pneumatic nailed		3.25	2.462		272	116		388	490
1110	3/4" thick x 24" long, 10" exp. on roof		2.25	3.556		272	167		439	575
1115	Pneumatic nailed		2.92	2.740		272	129		401	510
1200	1/2" thick, 18" long, 8-1/2" exp. on roof		2	4		180	188		368	510
1205	Pneumatic nailed		2.60	3.077		180	144		324	435
1210	3/4" thick x 18" long, 8 1/2" exp. on roof		1.80	4.444		180	209		389	540
1215	Pneumatic nailed		2.34	3.419		180	161		341	460
1255	10" exp. on walls		2	4		174	188		362	500
1260	10" exposure on walls, pneumatic nailed		2.60	3.077		174	144		318	430
1700	Add to above for fire retardant shakes, 24" long					56			56	61.50
1800	18" long					56			56	61.50
1810	Ridge shakes	1 Carp	350	.023	L.F.	5	1.07		6.07	7.25

07 32 Roof Tiles

07 32 13 – Clay Roof Tiles

07 32 13.10 Clay Tiles		Crew	Daily Output	Labor-Hours	Unit	Material	Labor (2015 Bare Costs)	Equipment	Total	Total Incl O&P
0010	**CLAY TILES**, including accessories									
0300	Flat shingle, interlocking, 15", 166 pcs/sq, fireflashed blend	3 Rots	6	4	Sq.	505	160		665	845
0500	Terra cotta red		6	4		510	160		670	850
0600	Roman pan and top, 18", 102 pcs/sq, fireflashed blend		5.50	4.364		555	175		730	925
0640	Terra cotta red	1 Rots	2.40	3.333		555	134		689	850
1100	Barrel mission tile, 18", 166 pcs/sq, fireflashed blend	3 Rots	5.50	4.364		410	175		585	765
1140	Terra cotta red		5.50	4.364		410	175		585	765
1700	Scalloped edge flat shingle, 14", 145 pcs/sq, fireflashed blend		6	4		1,125	160		1,285	1,550
1800	Terra cotta red		6	4		1,100	160		1,260	1,525
3010	#15 felt underlayment	1 Rofc	64	.125		5.40	5		10.40	15
3020	#30 felt underlayment		58	.138		10.60	5.55		16.15	21.50
3040	Polyethylene and rubberized asph. underlayment		22	.364		75	14.60		89.60	109
9010	Minimum labor/equipment charge	1 Rots	2	4	Job		160		160	290

07 32 16 – Concrete Roof Tiles

07 32 16.10 Concrete Tiles		Crew	Daily Output	Labor-Hours	Unit	Material	Labor	Equipment	Total	Total Incl O&P
0010	**CONCRETE TILES**									
0020	Corrugated, 13" x 16-1/2", 90 per sq., 950 lb. per sq.									
0050	Earthtone colors, nailed to wood deck	1 Rots	1.35	5.926	Sq.	106	238		344	545
0150	Blues		1.35	5.926		106	238		344	545
0200	Greens		1.35	5.926		106	238		344	545
0250	Premium colors		1.35	5.926		106	238		344	545
0500	Shakes, 13" x 16-1/2", 90 per sq., 950 lb. per sq.									
0600	All colors, nailed to wood deck	1 Rots	1.50	5.333	Sq.	108	214		322	505
1500	Accessory pieces, ridge & hip, 10" x 16-1/2", 8 lb. each	"	120	.067	Ea.	3.20	2.67		5.87	8.35
1700	Rake, 6-1/2" x 16-3/4", 9 lb. each					3.20			3.20	3.52
1800	Mansard hip, 10" x 16-1/2", 9.2 lb. each					3.20			3.20	3.52
1900	Hip starter, 10" x 16-1/2", 10.5 lb. each					9.90			9.90	10.90
2000	3 or 4 way apex, 10" each side, 11.5 lb. each					11.40			11.40	12.55
9000	Minimum labor/equipment charge	1 Rots	3	2.667	Job		107		107	193

07 32 19 – Metal Roof Tiles

07 32 19.10 Metal Roof Tiles		Crew	Daily Output	Labor-Hours	Unit	Material	Labor	Equipment	Total	Total Incl O&P
0010	**METAL ROOF TILES**									
0020	Accessories included, .032" thick aluminum, mission tile	1 Carp	2.50	3.200	Sq.	825	150		975	1,150
0200	Spanish tiles		3	2.667	"	570	125		695	830
9000	Minimum labor/equipment charge		3	2.667	Job		125		125	205

07 33 Natural Roof Coverings

07 33 63 – Vegetated Roofing

07 33 63.10 Green Roof Systems		Crew	Daily Output	Labor-Hours	Unit	Material	Labor	Equipment	Total	Total Incl O&P
0010	**GREEN ROOF SYSTEMS**									
0020	Soil mixture for green roof 30% sand, 55% gravel, 15% soil									
0100	Hoist and spread soil mixture 4 inch depth up to five stories tall roof G	B-13B	4000	.014	S.F.	.25	.57	.28	1.10	1.52
0150	6 inch depth G		2667	.021		.38	.86	.42	1.66	2.29
0200	8 inch depth G		2000	.028		.50	1.15	.56	2.21	3.03
0250	10 inch depth G		1600	.035		.63	1.43	.70	2.76	3.80
0300	12 inch depth G		1335	.042		.76	1.72	.84	3.32	4.55
0310	Alt. man-made soil mix, hoist & spread, 4" deep up to 5 stories tall roof G		4000	.014		1.86	.57	.28	2.71	3.29
0350	Mobilization 55 ton crane to site G	1 Eqhv	3.60	2.222	Ea.		115		115	182
0355	Hoisting cost to five stories per day (Avg. 28 picks per day) G	B-13B	1	56	Day		2,300	1,125	3,425	4,975
0360	Mobilization or demobilization, 100 ton crane to site driver & escort G	A-3E	2.50	6.400	Ea.		294	62.50	356.50	540

07 33 Natural Roof Coverings

07 33 63 – Vegetated Roofing

07 33 63.10 Green Roof Systems			Crew	Daily Output	Labor-Hours	Unit	2015 Bare Costs Material	Labor	Equipment	Total	Total Incl O&P
0365	Hoisting cost six to ten stories per day (Avg. 21 picks per day)	G	B-13C	1	56	Day		2,300	1,675	3,975	5,550
0370	Hoist and spread soil mixture 4 inch depth six to ten stories tall roof	G		4000	.014	S.F.	.25	.57	.42	1.24	1.67
0375	6 inch depth	G		2667	.021		.38	.86	.63	1.87	2.51
0380	8 inch depth	G		2000	.028		.50	1.15	.83	2.48	3.33
0385	10 inch depth	G		1600	.035		.63	1.43	1.04	3.10	4.17
0390	12 inch depth	G		1335	.042		.76	1.72	1.25	3.73	4.99
0400	Green roof edging treated lumber 4" x 4" no hoisting included	G	2 Carp	400	.040	L.F.	1.34	1.88		3.22	4.55
0410	4" x 6"	G		400	.040		1.91	1.88		3.79	5.20
0420	4" x 8"	G		360	.044		3.81	2.09		5.90	7.60
0430	4" x 6" double stacked	G		300	.053		3.83	2.50		6.33	8.30
0500	Green roof edging redwood lumber 4" x 4" no hoisting included	G		400	.040		6.45	1.88		8.33	10.15
0510	4" x 6"	G		400	.040		12.60	1.88		14.48	16.95
0520	4" x 8"	G		360	.044		23.50	2.09		25.59	29.50
0530	4" x 6" double stacked	G		300	.053		25	2.50		27.50	31.50
0550	Components, not including membrane or insulation:										
0560	Fluid applied rubber membrane, reinforced, 215 mil thick	G	G-5	350	.114	S.F.	.28	4.17	.55	5	8.45
0570	Root barrier	G	2 Rofc	775	.021		.60	.83		1.43	2.16
0580	Moisture retention barrier and reservoir	G	"	900	.018		2.45	.71		3.16	3.99
0600	Planting sedum, light soil, potted, 2-1/4" diameter, two per S.F.	G	1 Clab	420	.019		4.90	.72		5.62	6.55
0610	one per S.F.	G	"	840	.010		2.45	.36		2.81	3.29
0630	Planting sedum mat per S.F. including shipping (4000 S.F. min)	G	4 Clab	4000	.008		5.90	.30		6.20	6.95
0640	Installation sedum mat system (no soil required) per S.F. (4000 S.F. min)	G	"	4000	.008		8.20	.30		8.50	9.50
0645	Note: pricing of sedum mats shipped in full truck loads (4000-5000 S.F.)										

07 41 Roof Panels

07 41 13 – Metal Roof Panels

07 41 13.10 Aluminum Roof Panels

		Crew	Daily Output	Labor-Hours	Unit	Material	Labor	Equipment	Total	Total Incl O&P
0010	**ALUMINUM ROOF PANELS**									
0020	Corrugated or ribbed, .0155" thick, natural	G-3	1200	.027	S.F.	.99	1.25		2.24	3.09
0300	Painted		1200	.027		1.44	1.25		2.69	3.58
0400	Corrugated, .018" thick, on steel frame, natural finish		1200	.027		1.30	1.25		2.55	3.43
0600	Painted		1200	.027		1.60	1.25		2.85	3.76
0700	Corrugated, on steel frame, natural, .024" thick		1200	.027		1.85	1.25		3.10	4.04
0800	Painted		1200	.027		2.25	1.25		3.50	4.48
0900	.032" thick, natural		1200	.027		2.39	1.25		3.64	4.63
1200	Painted		1200	.027		3.06	1.25		4.31	5.35
1300	V-Beam, on steel frame construction, .032" thick, natural		1200	.027		2.53	1.25		3.78	4.78
1500	Painted		1200	.027		3.10	1.25		4.35	5.40
1600	.040" thick, natural		1200	.027		3.07	1.25		4.32	5.40
1800	Painted		1200	.027		3.70	1.25		4.95	6.05
1900	.050" thick, natural		1200	.027		3.69	1.25		4.94	6.05
2100	Painted		1200	.027		4.45	1.25		5.70	6.90
2200	For roofing on wood frame, deduct		4600	.007		.08	.33		.41	.61
2400	Ridge cap, .032" thick, natural		800	.040	L.F.	2.91	1.87		4.78	6.20
9000	Minimum labor/equipment charge	1 Rofc	3	2.667	Job		107		107	193

07 41 13.20 Steel Roofing Panels

		Crew	Daily Output	Labor-Hours	Unit	Material	Labor	Equipment	Total	Total Incl O&P
0010	**STEEL ROOFING PANELS**									
0012	Corrugated or ribbed, on steel framing, 30 ga. galv	G-3	1100	.029	S.F.	1.60	1.36		2.96	3.94
0100	28 ga.		1050	.030		1.65	1.43		3.08	4.11
0300	26 ga.		1000	.032		1.68	1.50		3.18	4.25
0400	24 ga.		950	.034		2.90	1.58		4.48	5.70

07 41 Roof Panels

07 41 13 – Metal Roof Panels

07 41 13.20 Steel Roofing Panels		Crew	Daily Output	Labor-Hours	Unit	Material	2015 Bare Costs Labor	Equipment	Total	Total Incl O&P
0600	Colored, 28 ga.	G-3	1050	.030	S.F.	1.67	1.43		3.10	4.13
0700	26 ga.		1000	.032		1.98	1.50		3.48	4.58
0710	Flat profile, 1-3/4" standing seams, 10" wide, standard finish, 26 ga.		1000	.032		3.75	1.50		5.25	6.55
0715	24 ga.		950	.034		4.35	1.58		5.93	7.30
0720	22 ga.		900	.036		5.40	1.66		7.06	8.60
0725	Zinc aluminum alloy finish, 26 ga.		1000	.032		3	1.50		4.50	5.70
0730	24 ga.		950	.034		3.55	1.58		5.13	6.45
0735	22 ga.		900	.036		4.05	1.66		5.71	7.15
0740	12" wide, standard finish, 26 ga.		1000	.032		3.75	1.50		5.25	6.55
0745	24 ga.		950	.034		4.90	1.58		6.48	7.95
0750	Zinc aluminum alloy finish, 26 ga.		1000	.032		4.26	1.50		5.76	7.10
0755	24 ga.		950	.034		3.50	1.58		5.08	6.40
0840	Flat profile, 1" x 3/8" batten, 12" wide, standard finish, 26 ga.		1000	.032		3.30	1.50		4.80	6.05
0845	24 ga.		950	.034		3.90	1.58		5.48	6.80
0850	22 ga.		900	.036		4.65	1.66		6.31	7.75
0855	Zinc aluminum alloy finish, 26 ga.		1000	.032		3.20	1.50		4.70	5.90
0860	24 ga.		950	.034		3.60	1.58		5.18	6.50
0865	22 ga.		900	.036		4.15	1.66		5.81	7.25
0870	16-1/2" wide, standard finish, 24 ga.		950	.034		3.85	1.58		5.43	6.75
0875	22 ga.		900	.036		4.30	1.66		5.96	7.40
0880	Zinc aluminum alloy finish, 24 ga.		950	.034		3.35	1.58		4.93	6.20
0885	22 ga.		900	.036		3.75	1.66		5.41	6.80
0890	Flat profile, 2" x 2" batten, 12" wide, standard finish, 26 ga.		1000	.032		3.85	1.50		5.35	6.65
0895	24 ga.		950	.034		4.55	1.58		6.13	7.55
0900	22 ga.		900	.036		5.55	1.66		7.21	8.75
0905	Zinc aluminum alloy finish, 26 ga.		1000	.032		3.55	1.50		5.05	6.30
0910	24 ga.		950	.034		4.05	1.58		5.63	7
0915	22 ga.		900	.036		4.70	1.66		6.36	7.80
0920	16-1/2" wide, standard finish, 24 ga.		950	.034		4.20	1.58		5.78	7.15
0925	22 ga.		900	.036		4.90	1.66		6.56	8.05
0930	Zinc aluminum alloy finish, 24 ga.		950	.034		3.80	1.58		5.38	6.70
0935	22 ga.		900	.036		4.30	1.66		5.96	7.40
1200	Ridge, galvanized, 10" wide		800	.040	L.F.	3.10	1.87		4.97	6.40
1210	20" wide		750	.043	"	4.10	2		6.10	7.70
9000	Minimum labor/equipment charge	1 Rofc	2	4	Job		160		160	290

07 41 33 – Plastic Roof Panels

07 41 33.10 Fiberglass Panels		Crew	Daily Output	Labor-Hours	Unit	Material	Labor	Equipment	Total	Total Incl O&P
0010	**FIBERGLASS PANELS**									
0012	Corrugated panels, roofing, 8 oz. per S.F.	G-3	1000	.032	S.F.	2.10	1.50		3.60	4.71
0100	12 oz. per S.F.		1000	.032		4.18	1.50		5.68	7
0300	Corrugated siding, 6 oz. per S.F.		880	.036		2.10	1.70		3.80	5.05
0400	8 oz. per S.F.		880	.036		2.10	1.70		3.80	5.05
0500	Fire retardant		880	.036		3.80	1.70		5.50	6.90
0600	12 oz. siding, textured		880	.036		3.75	1.70		5.45	6.85
0700	Fire retardant		880	.036		4.35	1.70		6.05	7.50
0900	Flat panels, 6 oz. per S.F., clear or colors		880	.036		2.40	1.70		4.10	5.35
1100	Fire retardant, class A		880	.036		3.45	1.70		5.15	6.55
1300	8 oz. per S.F., clear or colors		880	.036		2.52	1.70		4.22	5.50
1700	Sandwich panels, fiberglass, 1-9/16" thick, panels to 20 S.F.		180	.178		36	8.30		44.30	53
1900	As above, but 2-3/4" thick, panels to 100 S.F.		265	.121		26	5.65		31.65	37.50
9000	Minimum labor/equipment charge	1 Rofc	2	4	Job		160		160	290

07 42 13 – Metal Wall Panels

07 42 13.10 Mansard Panels

	07 42 13.10 Mansard Panels	Crew	Daily Output	Labor-Hours	Unit	Material	Labor	Equipment	Total	Total Incl O&P
						2015 Bare Costs				
0010	**MANSARD PANELS**									
0600	Aluminum, stock units, straight surfaces	1 Shee	115	.070	S.F.	3.90	3.89		7.79	10.45
0700	Concave or convex surfaces		75	.107	"	2.06	5.95		8.01	11.65
0800	For framing, to 5′ high, add		115	.070	L.F.	3.60	3.89		7.49	10.10
0900	Soffits, to 1′ wide		125	.064	S.F.	2.20	3.58		5.78	8.05
9000	Minimum labor/equipment charge		2.50	3.200	Job		179		179	283

07 42 13.20 Aluminum Siding Panels

	07 42 13.20 Aluminum Siding Panels	Crew	Daily Output	Labor-Hours	Unit	Material	Labor	Equipment	Total	Total Incl O&P
0010	**ALUMINUM SIDING PANELS**									
0012	Corrugated, on steel framing, .019 thick, natural finish	G-3	775	.041	S.F.	1.52	1.93		3.45	4.77
0100	Painted		775	.041		1.66	1.93		3.59	4.93
0400	Farm type, .021″ thick on steel frame, natural		775	.041		1.55	1.93		3.48	4.81
0600	Painted		775	.041		1.65	1.93		3.58	4.92
0700	Industrial type, corrugated, on steel, .024″ thick, mill		775	.041		2.15	1.93		4.08	5.45
0900	Painted		775	.041		2.31	1.93		4.24	5.65
1000	.032″ thick, mill		775	.041		2.45	1.93		4.38	5.80
1200	Painted		775	.041		3	1.93		4.93	6.40
1300	V-Beam, on steel frame, .032″ thick, mill		775	.041		2.80	1.93		4.73	6.20
1500	Painted		775	.041		3.20	1.93		5.13	6.60
1600	.040″ thick, mill		775	.041		3.38	1.93		5.31	6.80
1800	Painted		775	.041		3.94	1.93		5.87	7.45
1900	.050″ thick, mill		775	.041		3.98	1.93		5.91	7.50
2100	Painted		775	.041		4.73	1.93		6.66	8.30
2200	Ribbed, 3″ profile, on steel frame, .032″ thick, natural		775	.041		2.40	1.93		4.33	5.75
2400	Painted		775	.041		3.05	1.93		4.98	6.45
2500	.040″ thick, natural		775	.041		2.75	1.93		4.68	6.15
2700	Painted		775	.041		3.20	1.93		5.13	6.60
2750	.050″ thick, natural		775	.041		3.15	1.93		5.08	6.55
2760	Painted		775	.041		3.63	1.93		5.56	7.10
3300	For siding on wood frame, deduct from above		2800	.011		.09	.53		.62	.96
3400	Screw fasteners, aluminum, self tapping, neoprene washer, 1″				M	210			210	231
3600	Stitch screws, self tapping, with neoprene washer, 5/8″				"	158			158	174
3630	Flashing, sidewall, .032″ thick	G-3	800	.040	L.F.	2.96	1.87		4.83	6.25
3650	End wall, .040″ thick		800	.040		3.46	1.87		5.33	6.80
3670	Closure strips, corrugated, .032″ thick		800	.040		.88	1.87		2.75	3.97
3680	Ribbed, 4″ or 8″, .032″ thick		800	.040		.88	1.87		2.75	3.97
3690	V-beam, .040″ thick		800	.040		1.18	1.87		3.05	4.30
3800	Horizontal, colored clapboard, 8″ wide, plain	2 Carp	515	.031	S.F.	2.45	1.46		3.91	5.10
3810	Insulated		515	.031		2.67	1.46		4.13	5.35
3830	8″ embossed, painted		515	.031		2.65	1.46		4.11	5.30
3840	Insulated		515	.031		2.84	1.46		4.30	5.50
3860	12″ painted, smooth		600	.027		2.50	1.25		3.75	4.80
3870	Insulated		600	.027		2.67	1.25		3.92	4.99
3890	12″ embossed, painted		600	.027		2.65	1.25		3.90	4.97
3900	Insulated		515	.031		2.55	1.46		4.01	5.20
4000	Vertical board & batten, colored, non-insulated		515	.031		1.90	1.46		3.36	4.48
4200	For simulated wood design, add					.12			.12	.13
4300	Corners for above, outside	2 Carp	515	.031	V.L.F.	3.36	1.46		4.82	6.10
4500	Inside corners	"	515	.031	"	1.61	1.46		3.07	4.16
9000	Minimum labor/equipment charge	1 Carp	3	2.667	Job		125		125	205

07 42 Wall Panels

07 42 13 – Metal Wall Panels

07 42 13.30 Steel Siding		Crew	Daily Output	Labor-Hours	Unit	Material	2015 Bare Costs Labor	Equipment	Total	Total Incl O&P
0010	**STEEL SIDING**									
0020	Beveled, vinyl coated, 8″ wide	1 Carp	265	.030	S.F.	1.85	1.42		3.27	4.36
0050	10″ wide	″	275	.029		1.88	1.37		3.25	4.31
0080	Galv, corrugated or ribbed, on steel frame, 30 ga.	G-3	800	.040		1.20	1.87		3.07	4.32
0100	28 ga.		795	.040		1.28	1.88		3.16	4.43
0300	26 ga.		790	.041		1.80	1.90		3.70	5
0400	24 ga.		785	.041		1.85	1.91		3.76	5.10
0600	22 ga.		770	.042		2.15	1.94		4.09	5.50
0700	Colored, corrugated/ribbed, on steel frame, 10 yr. finish, 28 ga.		800	.040		1.95	1.87		3.82	5.15
0900	26 ga.		795	.040		2.09	1.88		3.97	5.30
1000	24 ga.		790	.041		2.37	1.90		4.27	5.65
1020	20 ga.		785	.041		3	1.91		4.91	6.35
1200	Factory sandwich panel, 26 ga., 1″ insulation, galvanized		380	.084		5.35	3.94		9.29	12.15
1300	Colored 1 side		380	.084		6.60	3.94		10.54	13.55
1500	Galvanized 2 sides		380	.084		7.95	3.94		11.89	15.05
1600	Colored 2 sides		380	.084		8.25	3.94		12.19	15.35
1800	Acrylic paint face, regular paint liner		380	.084		6.20	3.94		10.14	13.10
1900	For 2″ thick polystyrene, add					1			1	1.10
2000	22 ga., galv, 2″ insulation, baked enamel exterior	G-3	360	.089		12.15	4.16		16.31	20
2100	Polyvinylidene exterior finish	″	360	.089		12.80	4.16		16.96	21
9000	Minimum labor/equipment charge	1 Carp	3	2.667	Job		125		125	205

07 44 Faced Panels

07 44 73 – Metal Faced Panels

07 44 73.10 Metal Faced Panels and Accessories		Crew	Daily Output	Labor-Hours	Unit	Material	Labor	Equipment	Total	Total Incl O&P
0010	**METAL FACED PANELS AND ACCESSORIES**									
0400	Textured aluminum, 4′ x 8′ x 5/16″ plywood backing, single face	2 Shee	375	.043	S.F.	4.07	2.39		6.46	8.25
0600	Double face		375	.043		5.25	2.39		7.64	9.55
0700	4′ x 10′ x 5/16″ plywood backing, single face		375	.043		4.26	2.39		6.65	8.45
0900	Double face		375	.043		5.80	2.39		8.19	10.15
1000	4′ x 12′ x 5/16″ plywood backing, single face		375	.043		4.26	2.39		6.65	8.45
1300	Smooth aluminum, 1/4″ plywood panel, fluoropolymer finish, double face		375	.043		6.05	2.39		8.44	10.40
1350	Clear anodized finish, double face		375	.043		10.05	2.39		12.44	14.80
1400	Double face textured aluminum, structural panel, 1″ EPS insulation		375	.043		5.70	2.39		8.09	10
1500	Accessories, outside corner	1 Shee	175	.046	L.F.	1.90	2.56		4.46	6.15
1600	Inside corner		175	.046		1.34	2.56		3.90	5.50
1800	Batten mounting clip		200	.040		.49	2.24		2.73	4.07
1900	Low profile batten		480	.017		.61	.93		1.54	2.14
2100	High profile batten		480	.017		1.41	.93		2.34	3.02
2200	Water table		200	.040		2.12	2.24		4.36	5.85
2400	Horizontal joint connector		200	.040		1.67	2.24		3.91	5.35
2500	Corner cap		200	.040		1.84	2.24		4.08	5.55
2700	H - moulding		480	.017		1.25	.93		2.18	2.85

07 46 23 – Wood Siding

07 46 23.10 Wood Board Siding		Crew	Daily Output	Labor-Hours	Unit	Material	2015 Bare Costs Labor	Equipment	Total	Total Incl O&P
0010	WOOD BOARD SIDING									
3200	Wood, cedar bevel, A grade, 1/2" x 6"	1 Carp	295	.027	S.F.	4.14	1.27		5.41	6.65
3300	1/2" x 8"		330	.024		5.90	1.14		7.04	8.35
3500	3/4" x 10", clear grade		375	.021		6.30	1		7.30	8.55
3600	"B" grade		375	.021		3.79	1		4.79	5.80
3800	Cedar, rough sawn, 1" x 4", A grade, natural		220	.036		6.55	1.71		8.26	10.05
3900	Stained		220	.036		6.70	1.71		8.41	10.15
4100	1" x 12", board & batten, #3 & Btr., natural		420	.019		4.48	.89		5.37	6.40
4200	Stained		420	.019		4.81	.89		5.70	6.75
4400	1" x 8" channel siding, #3 & Btr., natural		330	.024		4.57	1.14		5.71	6.85
4500	Stained		330	.024		4.92	1.14		6.06	7.25
4700	Redwood, clear, beveled, vertical grain, 1/2" x 4"		220	.036		4.11	1.71		5.82	7.30
4750	1/2" x 6"		295	.027		4.49	1.27		5.76	7.05
4800	1/2" x 8"		330	.024		4.89	1.14		6.03	7.25
5000	3/4" x 10"		375	.021		4.50	1		5.50	6.60
5200	Channel siding, 1" x 10", B grade		375	.021		4.12	1		5.12	6.15
5250	Redwood, T&G boards, B grade, 1" x 4"		220	.036		2.54	1.71		4.25	5.60
5270	1" x 8"		330	.024		4.10	1.14		5.24	6.40
5400	White pine, rough sawn, 1" x 8", natural		330	.024		2.19	1.14		3.33	4.28
5500	Stained		330	.024		2.19	1.14		3.33	4.28
9000	Minimum labor/equipment charge		2	4	Job		188		188	310

07 46 29 – Plywood Siding

07 46 29.10 Plywood Siding Options		Crew	Daily Output	Labor-Hours	Unit	Material	Labor	Equipment	Total	Total Incl O&P
0010	PLYWOOD SIDING OPTIONS									
0900	Plywood, medium density overlaid, 3/8" thick	2 Carp	750	.021	S.F.	1.25	1		2.25	3.02
1000	1/2" thick		700	.023		1.44	1.07		2.51	3.34
1100	3/4" thick		650	.025		1.85	1.16		3.01	3.93
1600	Texture 1-11, cedar, 5/8" thick, natural		675	.024		2.57	1.11		3.68	4.65
1700	Factory stained		675	.024		2.80	1.11		3.91	4.90
1900	Texture 1-11, fir, 5/8" thick, natural		675	.024		1.57	1.11		2.68	3.55
2000	Factory stained		675	.024		1.80	1.11		2.91	3.80
2050	Texture 1-11, S.Y.P., 5/8" thick, natural		675	.024		1.37	1.11		2.48	3.33
2100	Factory stained		675	.024		1.44	1.11		2.55	3.40
2200	Rough sawn cedar, 3/8" thick, natural		675	.024		1.25	1.11		2.36	3.20
2300	Factory stained		675	.024		1.50	1.11		2.61	3.47
2500	Rough sawn fir, 3/8" thick, natural		675	.024		.84	1.11		1.95	2.74
2600	Factory stained		675	.024		1.04	1.11		2.15	2.96
2800	Redwood, textured siding, 5/8" thick		675	.024		1.92	1.11		3.03	3.93
3000	Polyvinyl chloride coated, 3/8" thick		750	.021		1.11	1		2.11	2.86
9000	Minimum labor/equipment charge	1 Carp	2	4	Job		188		188	310

07 46 33 – Plastic Siding

07 46 33.10 Vinyl Siding		Crew	Daily Output	Labor-Hours	Unit	Material	Labor	Equipment	Total	Total Incl O&P
0010	VINYL SIDING									
3995	Clapboard profile, woodgrain texture, .048 thick, double 4	2 Carp	495	.032	S.F.	1.04	1.52		2.56	3.64
4000	Double 5		550	.029		1.04	1.37		2.41	3.39
4005	Single 8		495	.032		1.26	1.52		2.78	3.88
4010	Single 10		550	.029		1.51	1.37		2.88	3.91
4015	.044 thick, double 4		495	.032		.99	1.52		2.51	3.58
4020	Double 5		550	.029		.99	1.37		2.36	3.33
4025	.042 thick, double 4		495	.032		.92	1.52		2.44	3.51
4030	Double 5		550	.029		.92	1.37		2.29	3.26
4035	Cross sawn texture, .040 thick, double 4		495	.032		.67	1.52		2.19	3.23

07 46 33 – Plastic Siding

07 46 33.10 Vinyl Siding		Crew	Daily Output	Labor-Hours	Unit	2015 Bare Costs Material	Labor	Equipment	Total	Total Incl O&P
4040	Double 5	2 Carp	550	.029	S.F.	.67	1.37		2.04	2.98
4045	Smooth texture, .042 thick, double 4		495	.032		.75	1.52		2.27	3.32
4050	Double 5		550	.029		.75	1.37		2.12	3.07
4055	Single 8		495	.032		.75	1.52		2.27	3.32
4060	Cedar texture, .044 thick, double 4		495	.032		1.01	1.52		2.53	3.61
4065	Double 6		600	.027		1.01	1.25		2.26	3.17
4070	Dutch lap profile, woodgrain texture, .048 thick, double 5		550	.029		1.04	1.37		2.41	3.39
4075	.044 thick, double 4.5		525	.030		1.01	1.43		2.44	3.47
4080	.042 thick, double 4.5		525	.030		.84	1.43		2.27	3.28
4085	.040 thick, double 4.5		525	.030		.67	1.43		2.10	3.09
4100	Shake profile, 10" wide		400	.040		3.45	1.88		5.33	6.90
4105	Vertical pattern, .046 thick, double 5		550	.029		1.42	1.37		2.79	3.81
4110	.044 thick, triple 3		550	.029		1.56	1.37		2.93	3.96
4115	.040 thick, triple 4		550	.029		1.50	1.37		2.87	3.90
4120	.040 thick, triple 2.66		550	.029		1.81	1.37		3.18	4.24
4125	Insulation, fan folded extruded polystyrene, 1/4"		2000	.008		.28	.38		.66	.93
4130	3/8"		2000	.008		.31	.38		.69	.96
4135	Accessories, J channel, 5/8" pocket		700	.023	L.F.	.46	1.07		1.53	2.26
4140	3/4" pocket		695	.023		.50	1.08		1.58	2.32
4145	1-1/4" pocket		680	.024		.77	1.10		1.87	2.65
4150	Flexible, 3/4" pocket		600	.027		2.33	1.25		3.58	4.61
4155	Under sill finish trim		500	.032		.50	1.50		2	3.01
4160	Vinyl starter strip		700	.023		.58	1.07		1.65	2.40
4165	Aluminum starter strip		700	.023		.28	1.07		1.35	2.07
4170	Window casing, 2-1/2" wide, 3/4" pocket		510	.031		1.57	1.47		3.04	4.13
4175	Outside corner, woodgrain finish, 4" face, 3/4" pocket		700	.023		1.92	1.07		2.99	3.88
4180	5/8" pocket		700	.023		1.92	1.07		2.99	3.88
4185	Smooth finish, 4" face, 3/4" pocket		700	.023		1.92	1.07		2.99	3.88
4190	7/8" pocket		690	.023		1.89	1.09		2.98	3.86
4195	1-1/4" pocket		700	.023		1.33	1.07		2.40	3.23
4200	Soffit and fascia, 1' overhang, solid		120	.133		4.29	6.25		10.54	14.95
4205	Vented		120	.133		4.29	6.25		10.54	14.95
4207	18" overhang, solid		110	.145		5	6.85		11.85	16.70
4208	Vented		110	.145		5	6.85		11.85	16.70
4210	2' overhang, solid		100	.160		5.70	7.50		13.20	18.60
4215	Vented		100	.160		5.70	7.50		13.20	18.60
4217	3' overhang, solid		100	.160		7.10	7.50		14.60	20
4218	Vented		100	.160		7.10	7.50		14.60	20
4220	Colors for siding and soffits, add				S.F.	.14			.14	.15
4225	Colors for accessories and trim, add				L.F.	.30			.30	.33
9000	Minimum labor/equipment charge	1 Carp	3	2.667	Job		125		125	205

07 46 33.20 Polypropylene Siding

		Crew	Daily Output	Labor-Hours	Unit	Material	Labor	Equipment	Total	Total Incl O&P
0010	**POLYPROPYLENE SIDING**									
4090	Shingle profile, random grooves, double 7	2 Carp	400	.040	S.F.	2.99	1.88		4.87	6.35
4092	Cornerpost for above	1 Carp	365	.022	L.F.	12	1.03		13.03	14.90
4095	Triple 5	2 Carp	400	.040	S.F.	2.99	1.88		4.87	6.35
4097	Cornerpost for above	1 Carp	365	.022	L.F.	11.05	1.03		12.08	13.85
5000	Staggered butt, double 7"	2 Carp	400	.040	S.F.	3.45	1.88		5.33	6.90
5002	Cornerpost for above	1 Carp	365	.022	L.F.	12	1.03		13.03	14.90
5010	Half round, double 6-1/4"	2 Carp	360	.044	S.F.	3.40	2.09		5.49	7.15
5020	Shake profile, staggered butt, double 9"	"	510	.031	"	3.45	1.47		4.92	6.20
5022	Cornerpost for above	1 Carp	365	.022	L.F.	9	1.03		10.03	11.60

07 46 Siding

07 46 33 – Plastic Siding

07 46 33.20 Polypropylene Siding		Crew	Daily Output	Labor-Hours	Unit	Material	Labor	Equipment	Total	Total Incl O&P
						2015 Bare Costs				
5030	Straight butt, double 7"	2 Carp	400	.040	S.F.	3.40	1.88		5.28	6.85
5032	Cornerpost for above	1 Carp	365	.022	L.F.	11.90	1.03		12.93	14.75
6000	Accessories, J channel, 5/8" pocket	2 Carp	700	.023		.46	1.07		1.53	2.26
6010	3/4" pocket		695	.023		.50	1.08		1.58	2.32
6020	1-1/4" pocket		680	.024		.77	1.10		1.87	2.65
6030	Aluminum starter strip		700	.023		.28	1.07		1.35	2.07

07 46 46 – Fiber Cement Siding

07 46 46.10 Fiber Cement Siding

Code	Description	Crew	Daily Output	Labor-Hours	Unit	Material	Labor	Equipment	Total	Total Incl O&P
0010	**FIBER CEMENT SIDING**									
0020	Lap siding, 5/16" thick, 6" wide, 4-3/4" exposure, smooth texture	2 Carp	415	.039	S.F.	1.18	1.81		2.99	4.27
0025	Woodgrain texture		415	.039		1.18	1.81		2.99	4.27
0030	7-1/2" wide, 6-1/4" exposure, smooth texture		425	.038		1.46	1.77		3.23	4.51
0035	Woodgrain texture		425	.038		1.46	1.77		3.23	4.51
0040	8" wide, 6-3/4" exposure, smooth texture		425	.038		1.42	1.77		3.19	4.46
0045	Roughsawn texture		425	.038		1.42	1.77		3.19	4.46
0050	9-1/2" wide, 8-1/4" exposure, smooth texture		440	.036		1.22	1.71		2.93	4.14
0055	Woodgrain texture		440	.036		1.22	1.71		2.93	4.14
0060	12" wide, 10-3/8" exposure, smooth texture		455	.035		2.01	1.65		3.66	4.92
0065	Woodgrain texture		455	.035		2.01	1.65		3.66	4.92
0070	Panel siding, 5/16" thick, smooth texture		750	.021		1.25	1		2.25	3.02
0075	Stucco texture		750	.021		1.25	1		2.25	3.02
0080	Grooved woodgrain texture		750	.021		1.25	1		2.25	3.02
0085	V - grooved woodgrain texture		750	.021		1.25	1		2.25	3.02
0088	Shingle siding, 48" x 15-1/4" panels, 7" exposure		700	.023		3.94	1.07		5.01	6.10
0090	Wood starter strip		400	.040	L.F.	.42	1.88		2.30	3.54

07 46 73 – Soffit

07 46 73.10 Soffit Options

Code	Description	Crew	Daily Output	Labor-Hours	Unit	Material	Labor	Equipment	Total	Total Incl O&P
0010	**SOFFIT OPTIONS**									
0012	Aluminum, residential, .020" thick	1 Carp	210	.038	S.F.	2	1.79		3.79	5.15
0100	Baked enamel on steel, 16 or 18 ga.		105	.076		5.90	3.58		9.48	12.35
0300	Polyvinyl chloride, white, solid		230	.035		2.10	1.63		3.73	4.99
0400	Perforated		230	.035		2.10	1.63		3.73	4.99
0500	For colors, add					.15			.15	.17
9000	Minimum labor/equipment charge	1 Carp	3	2.667	Job		125		125	205

07 51 Built-Up Bituminous Roofing

07 51 13 – Built-Up Asphalt Roofing

07 51 13.10 Built-Up Roofing Components

Code	Description	Crew	Daily Output	Labor-Hours	Unit	Material	Labor	Equipment	Total	Total Incl O&P
0010	**BUILT-UP ROOFING COMPONENTS**									
0012	Asphalt saturated felt, #30, 2 square per roll	1 Rofc	58	.138	Sq.	10.60	5.55		16.15	21.50
0200	#15, 4 sq. per roll, plain or perforated, not mopped		58	.138		5.40	5.55		10.95	15.95
0300	Roll roofing, smooth, #65		15	.533		9.65	21.50		31.15	49
0500	#90		12	.667		37	26.50		63.50	89
0520	Mineralized		12	.667		35	26.50		61.50	87
0540	D.C. (Double coverage), 19" selvage edge		10	.800		52.50	32		84.50	116
0580	Adhesive (lap cement)				Gal.	8.85			8.85	9.75
0800	Steep, flat or dead level asphalt, 10 ton lots, packaged				Ton	925			925	1,025
9000	Minimum labor/equipment charge	1 Rofc	4	2	Job		80		80	145

07 51 13 – Built-Up Asphalt Roofing

07 51 13.13 Cold-Applied Built-Up Asphalt Roofing		Crew	Daily Output	Labor-Hours	Unit	2015 Bare Costs Material	Labor	Equipment	Total	Total Incl O&P
0010	COLD-APPLIED BUILT-UP ASPHALT ROOFING									
0020	3 ply system, installation only (components listed below)	G-5	50	.800	Sq.		29	3.86	32.86	57
0100	Spunbond poly. fabric, 1.35 oz./S.Y., 36"W, 10.8 sq./roll				Ea.	135			135	149
0500	Base & finish coat, 3 gal./sq., 5 gal./can				Gal.	7.75			7.75	8.55
0600	Coating, ceramic granules, 1/2 sq./bag				Ea.	21.50			21.50	23.50
0700	Aluminum, 2 gal./sq.				Gal.	13.50			13.50	14.85
0800	Emulsion, fibered or non-fibered, 4 gal./sq.				"	6.50			6.50	7.15

07 51 13.20 Built-Up Roofing Systems

Line	Description	Crew	Daily Output	Labor-Hours	Unit	Material	Labor	Equipment	Total	Total Incl O&P
0010	BUILT-UP ROOFING SYSTEMS									
0120	Asphalt flood coat with gravel/slag surfacing, not including									
0140	Insulation, flashing or wood nailers									
0200	Asphalt base sheet, 3 plies #15 asphalt felt, mopped	G-1	22	2.545	Sq.	100	95.50	26	221.50	310
0350	On nailable decks		21	2.667		102	100	27	229	325
0500	4 plies #15 asphalt felt, mopped		20	2.800		136	105	28.50	269.50	370
0550	On nailable decks		19	2.947		120	111	30	261	365
0700	Coated glass base sheet, 2 plies glass (type IV), mopped		22	2.545		107	95.50	26	228.50	320
0850	3 plies glass, mopped		20	2.800		130	105	28.50	263.50	365
0950	On nailable decks		19	2.947		122	111	30	263	365
1100	4 plies glass fiber felt (type IV), mopped		20	2.800		160	105	28.50	293.50	400
1150	On nailable decks		19	2.947		145	111	30	286	390
1200	Coated & saturated base sheet, 3 plies #15 asph. felt, mopped		20	2.800		112	105	28.50	245.50	345
1250	On nailable decks		19	2.947		104	111	30	245	345
1300	4 plies #15 asphalt felt, mopped		22	2.545		130	95.50	26	251.50	345
2000	Asphalt flood coat, smooth surface									
2200	Asphalt base sheet & 3 plies #15 asphalt felt, mopped	G-1	24	2.333	Sq.	105	87.50	23.50	216	299
2400	On nailable decks		23	2.435		96.50	91.50	24.50	212.50	298
2600	4 plies #15 asphalt felt, mopped		24	2.333		123	87.50	23.50	234	320
2700	On nailable decks		23	2.435		115	91.50	24.50	231	320
2900	Coated glass fiber base sheet, mopped, and 2 plies of									
2910	glass fiber felt (type IV)	G-1	25	2.240	Sq.	102	84	23	209	290
3100	On nailable decks		24	2.333		96.50	87.50	23.50	207.50	290
3200	3 plies, mopped		23	2.435		125	91.50	24.50	241	330
3300	On nailable decks		22	2.545		117	95.50	26	238.50	330
3800	4 plies glass fiber felt (type IV), mopped		23	2.435		147	91.50	24.50	263	355
3900	On nailable decks		22	2.545		140	95.50	26	261.50	355
4000	Coated & saturated base sheet, 3 plies #15 asph. felt, mopped		24	2.333		107	87.50	23.50	218	300
4200	On nailable decks		23	2.435		99	91.50	24.50	215	300
4300	4 plies #15 organic felt, mopped		22	2.545		125	95.50	26	246.50	340
4500	Coal tar pitch with gravel/slag surfacing									
4600	4 plies #15 tarred felt, mopped	G-1	21	2.667	Sq.	233	100	27	360	470
4800	3 plies glass fiber felt (type IV), mopped	"	19	2.947	"	193	111	30	334	445
5000	Coated glass fiber base sheet, and 2 plies of									
5010	glass fiber felt, (type IV), mopped	G-1	19	2.947	Sq.	196	111	30	337	450
5300	On nailable decks		18	3.111		170	117	31.50	318.50	435
5600	4 plies glass fiber felt (type IV), mopped		21	2.667		264	100	27	391	500
5800	On nailable decks		20	2.800		237	105	28.50	370.50	485

07 51 13.30 Cants

Line	Description	Crew	Daily Output	Labor-Hours	Unit	Material	Labor	Equipment	Total	Total Incl O&P
0010	CANTS									
0012	Lumber, treated, 4" x 4" cut diagonally	1 Rofc	325	.025	L.F.	1.75	.99		2.74	3.71
0300	Mineral or fiber, trapezoidal, 1" x 4" x 48"		325	.025		.25	.99		1.24	2.06
0400	1-1/2" x 5-5/8" x 48"		325	.025		.44	.99		1.43	2.26
9000	Minimum labor/equipment charge		4	2	Job		80		80	145

07 51 Built-Up Bituminous Roofing

07 51 13 – Built-Up Asphalt Roofing

07 51 13.40 Felts

	07 51 13.40 Felts	Crew	Daily Output	Labor-Hours	Unit	2015 Bare Costs Material	Labor	Equipment	Total	Total Incl O&P
0010	**FELTS**									
0012	Glass fibered roofing felt, #15, not mopped	1 Rofc	58	.138	Sq.	9.65	5.55		15.20	20.50
0300	Base sheet, #80, channel vented		58	.138		42.50	5.55		48.05	57
0400	#70, coated		58	.138		17.20	5.55		22.75	29
0500	Cap, #87, mineral surfaced		58	.138		80	5.55		85.55	98
0600	Flashing membrane, #65		16	.500		9.65	20		29.65	46.50
0800	Coal tar fibered, #15, no mopping		58	.138		18.10	5.55		23.65	30
0900	Asphalt felt, #15, 4 sq. per roll, no mopping		58	.138		5.40	5.55		10.95	15.95
1100	#30, 2 sq. per roll		58	.138		10.60	5.55		16.15	21.50
1200	Double coated, #33		58	.138		11	5.55		16.55	22
1400	#40, base sheet		58	.138		9.65	5.55		15.20	20.50
1450	Coated and saturated		58	.138		12	5.55		17.55	23
1500	Tarred felt, organic, #15, 4 sq. rolls		58	.138		13.95	5.55		19.50	25.50
1550	#30, 2 sq. roll		58	.138		28	5.55		33.55	40.50
1700	Add for mopping above felts, per ply, asphalt, 24 lb. per sq.	G-1	192	.292		11.10	10.95	2.96	25.01	35
1800	Coal tar mopping, 30 lb. per sq.		186	.301		23	11.30	3.06	37.36	49
1900	Flood coat, with asphalt, 60 lb. per sq.		60	.933		27.50	35	9.50	72	104
2000	With coal tar, 75 lb. per sq.		56	1		57	37.50	10.15	104.65	142
9000	Minimum labor/equipment charge	1 Rofc	4	2	Job		80		80	145

07 51 13.50 Walkways for Built-Up Roofs

	07 51 13.50 Walkways for Built-Up Roofs	Crew	Daily Output	Labor-Hours	Unit	Material	Labor	Equipment	Total	Total Incl O&P
0010	**WALKWAYS FOR BUILT-UP ROOFS**									
0020	Asphalt impregnated, 3′ x 6′ x 1/2″ thick	1 Rofc	400	.020	S.F.	1.56	.80		2.36	3.16
0100	3′ x 3′ x 3/4″ thick	"	400	.020		4.48	.80		5.28	6.40
0300	Concrete patio blocks, 2″ thick, natural	1 Clab	115	.070		3.37	2.62		5.99	8
0400	Colors	"	115	.070		3.73	2.62		6.35	8.40
0600	100% recycled rubber, 3′ x 4′ x 3/8″ G	1 Rofc	400	.020	L.F.	6.35	.80		7.15	8.40
0610	3′ x 4′ x 1/2″ G		400	.020		6.80	.80		7.60	8.90
0620	3′ x 4′ x 3/4″ G		400	.020		8.45	.80		9.25	10.75
9000	Minimum labor/equipment charge		2.75	2.909	Job		117		117	211

07 52 Modified Bituminous Membrane Roofing

07 52 13 – Atactic-Polypropylene-Modified Bituminous Membrane Roofing

07 52 13.10 APP Modified Bituminous Membrane

	07 52 13.10 APP Modified Bituminous Membrane	Crew	Daily Output	Labor-Hours	Unit	Material	Labor	Equipment	Total	Total Incl O&P
0010	**APP MODIFIED BITUMINOUS MEMBRANE** R075213-30									
0020	Base sheet, #15 glass fiber felt, nailed to deck	1 Rofc	58	.138	Sq.	10.65	5.55		16.20	21.50
0030	Spot mopped to deck	G-1	295	.190		15.20	7.10	1.93	24.23	31.50
0040	Fully mopped to deck	"	192	.292		20.50	10.95	2.96	34.41	46
0050	#15 organic felt, nailed to deck	1 Rofc	58	.138		6.40	5.55		11.95	17.05
0060	Spot mopped to deck	G-1	295	.190		10.95	7.10	1.93	19.98	27
0070	Fully mopped to deck	"	192	.292		16.50	10.95	2.96	30.41	41
2100	APP mod., smooth surf. cap sheet, poly. reinf., torched, 160 mils	G-5	2100	.019	S.F.	.68	.69	.09	1.46	2.11
2150	170 mils		2100	.019		.70	.69	.09	1.48	2.13
2200	Granule surface cap sheet, poly. reinf., torched, 180 mils		2000	.020		.89	.73	.10	1.72	2.41
2250	Smooth surface flashing, torched, 160 mils		1260	.032		.68	1.16	.15	1.99	3.01
2300	170 mils		1260	.032		.70	1.16	.15	2.01	3.03
2350	Granule surface flashing, torched, 180 mils		1260	.032		.89	1.16	.15	2.20	3.24
2400	Fibrated aluminum coating	1 Rofc	3800	.002		.08	.08		.16	.24
2450	Seam heat welding	"	205	.039	L.F.	.08	1.56		1.64	2.92

07 52 Modified Bituminous Membrane Roofing

07 52 16 – Styrene-Butadiene-Styrene Modified Bituminous Membrane Roofing

07 52 16.10 SBS Modified Bituminous Membrane		Crew	Daily Output	Labor-Hours	Unit	2015 Bare Costs Material	Labor	Equipment	Total	Total Incl O&P
0010	**SBS MODIFIED BITUMINOUS MEMBRANE**									
0080	Mod bit rfng, SBS mod, gran surf cap sheet, poly reinf									
0650	120 to 149 mils thick	G-1	2000	.028	S.F.	1.25	1.05	.28	2.58	3.59
0750	150 to 160 mils	"	2000	.028		1.72	1.05	.28	3.05	4.10
1150	For reflective granules, add					.71	1.02	.28	2.01	2.93
1600	Smooth surface cap sheet, mopped, 145 mils	G-1	2100	.027		.78	1	.27	2.05	2.97
1620	Lightweight base sheet, fiberglass reinforced, 35 to 47 mil		2100	.027		.27	1	.27	1.54	2.41
1625	Heavyweight base/ply sheet, reinforced, 87 to 120 mil thick	↓	2100	.027		.89	1	.27	2.16	3.09
1650	Granulated walkpad, 180 – 220 mils	1 Rofc	400	.020		1.82	.80		2.62	3.45
1700	Smooth surface flashing, 145 mils	G-1	1260	.044		.78	1.67	.45	2.90	4.37
1800	150 mils		1260	.044		.49	1.67	.45	2.61	4.05
1900	Granular surface flashing, 150 mils		1260	.044		.65	1.67	.45	2.77	4.23
2000	160 mils	↓	1260	.044		.71	1.67	.45	2.83	4.29
2010	Elastomeric asphalt primer	1 Rofc	2600	.003		.17	.12		.29	.41
2015	Roofing asphalt, 30 lb. per square	G-1	19000	.003		.14	.11	.03	.28	.38
2020	Cold process adhesive, 20 – 30 mils thick	1 Rofc	750	.011		.24	.43		.67	1.03
2025	Self adhering vapor retarder, 30 to 45 mils thick	G-5	2150	.019	↓	1.02	.68	.09	1.79	2.45
2050	Seam heat welding	1 Rofc	205	.039	L.F.	.08	1.56		1.64	2.92

07 53 Elastomeric Membrane Roofing

07 53 16 – Chlorosulfonate-Polyethylene Roofing

07 53 16.10 Chlorosulfonated Polyethylene Roofing

		Crew	Daily Output	Labor-Hours	Unit	Material	Labor	Equipment	Total	Total Incl O&P
0010	**CHLOROSULFONATED POLYETHYLENE ROOFING**									
0800	Chlorosulfonated polyethylene (CSPE)									
0900	45 mils, heat welded seams, plate attachment	G-5	35	1.143	Sq.	310	41.50	5.50	357	420
1100	Heat welded seams, plate attachment and ballasted		26	1.538		320	56	7.45	383.45	465
1200	60 mils, heat welded seams, plate attachment		35	1.143		410	41.50	5.50	457	530
1300	Heat welded seams, plate attachment and ballasted	↓	26	1.538	↓	420	56	7.45	483.45	570

07 53 23 – Ethylene-Propylene-Diene-Monomer Roofing

07 53 23.20 Ethylene-Propylene-Diene-Monomer Roofing

		Crew	Daily Output	Labor-Hours	Unit	Material	Labor	Equipment	Total	Total Incl O&P
0010	**ETHYLENE-PROPYLENE-DIENE-MONOMER ROOFING (EPDM)**									
3500	Ethylene-propylene-diene-monomer (EPDM), 45 mils, 0.28 psf									
3600	Loose-laid & ballasted with stone (10 psf)	G-5	51	.784	Sq.	85	28.50	3.79	117.29	149
3700	Mechanically attached		35	1.143		79	41.50	5.50	126	168
3800	Fully adhered with adhesive	↓	26	1.538	↓	111	56	7.45	174.45	231
4500	60 mils, 0.40 psf									
4600	Loose-laid & ballasted with stone (10 psf)	G-5	51	.784	Sq.	100	28.50	3.79	132.29	166
4700	Mechanically attached		35	1.143		93	41.50	5.50	140	184
4800	Fully adhered with adhesive	↓	26	1.538		125	56	7.45	188.45	247
4810	45 mil, .28 psf, membrane only					50.50			50.50	55.50
4820	60 mil, .40 psf, membrane only				↓	63			63	69.50
4850	Seam tape for membrane, 3" x 100' roll				Ea.	40			40	43.50
4900	Batten strips, 10' sections					3.43			3.43	3.77
4910	Cover tape for batten strips, 6" x 100' roll				↓	183			183	202
4930	Plate anchors				M	78			78	86
4970	Adhesive for fully adhered systems, 60 S.F./gal.				Gal.	19.65			19.65	21.50

07 53 Elastomeric Membrane Roofing

07 53 29 – Polyisobutylene Roofing

07 53 29.10 Polyisobutylene Roofing		Crew	Daily Output	Labor-Hours	Unit	Material	2015 Bare Costs Labor	Equipment	Total	Total Incl O&P
0010	**POLYISOBUTYLENE ROOFING**									
7500	Polyisobutylene (PIB), 100 mils, 0.57 psf									
7600	Loose-laid & ballasted with stone/gravel (10 psf)	G-5	51	.784	Sq.	200	28.50	3.79	232.29	276
7700	Partially adhered with adhesive		35	1.143		240	41.50	5.50	287	345
7800	Hot asphalt attachment		35	1.143		230	41.50	5.50	277	335
7900	Fully adhered with contact cement	↓	26	1.538	↓	250	56	7.45	313.45	385

07 54 Thermoplastic Membrane Roofing

07 54 19 – Polyvinyl-Chloride Roofing

07 54 19.10 Polyvinyl-Chloride Roofing (PVC)

		Crew	Daily Output	Labor-Hours	Unit	Material	Labor	Equipment	Total	Total Incl O&P
0010	**POLYVINYL-CHLORIDE ROOFING (PVC)**									
8200	Heat welded seams									
8700	Reinforced, 48 mils, 0.33 psf									
8750	Loose-laid & ballasted with stone/gravel (12 psf)	G-5	51	.784	Sq.	118	28.50	3.79	150.29	186
8800	Mechanically attached		35	1.143		111	41.50	5.50	158	204
8850	Fully adhered with adhesive	↓	26	1.538	↓	153	56	7.45	216.45	277
8860	Reinforced, 60 mils, .40 psf									
8870	Loose-laid & ballasted with stone/gravel (12 psf)	G-5	51	.784	Sq.	120	28.50	3.79	152.29	188
8880	Mechanically attached		35	1.143		112	41.50	5.50	159	206
8890	Fully adhered with adhesive	↓	26	1.538	↓	154	56	7.45	217.45	279

07 54 23 – Thermoplastic-Polyolefin Roofing

07 54 23.10 Thermoplastic Polyolefin Roofing (T.P.O)

		Crew	Daily Output	Labor-Hours	Unit	Material	Labor	Equipment	Total	Total Incl O&P
0010	**THERMOPLASTIC POLYOLEFIN ROOFING (T.P.O.)**									
0100	45 mil, loose laid & ballasted with stone (1/2 ton/sq.)	G-5	51	.784	Sq.	87.50	28.50	3.79	119.79	152
0120	Fully adhered		25	1.600		78	58.50	7.75	144.25	199
0140	Mechanically attached		34	1.176		77	43	5.70	125.70	169
0160	Self adhered		35	1.143		87	41.50	5.50	134	178
0180	60 mil membrane, heat welded seams, ballasted		50	.800		100	29	3.86	132.86	167
0200	Fully adhered		25	1.600		90	58.50	7.75	156.25	213
0220	Mechanically attached		34	1.176		93.50	43	5.70	142.20	187
0240	Self adhered	↓	35	1.143	↓	107	41.50	5.50	154	200

07 54 30 – Ketone Ethylene Ester Roofing

07 54 30.10 Ketone Ethylene Ester Roofing

		Crew	Daily Output	Labor-Hours	Unit	Material	Labor	Equipment	Total	Total Incl O&P
0010	**KETONE ETHYLENE ESTER ROOFING**									
0100	Ketone ethylene ester roofing, 50 mil, fully adhered	G-5	26	1.538	Sq.	215	56	7.45	278.45	345
0120	Mechanically attached		35	1.143		141	41.50	5.50	188	237
0140	Ballasted with stone	↓	51	.784		149	28.50	3.79	181.29	219
0160	50 mil, fleece backed, adhered w/hot asphalt	G-1	26	2.154	↓	174	81	22	277	360
0180	Accessories, pipe boot	1 Rofc	32	.250	Ea.	24.50	10.05		34.55	44.50
0200	Pre-formed corners		32	.250	"	8.95	10.05		19	28
0220	Ketone clad metal, including up to 4 bends	↓	330	.024	S.F.	3.95	.97		4.92	6.10
0240	Walkway pad	2 Rofc	800	.020	"	4.27	.80		5.07	6.15
0260	Stripping material	1 Rofc	310	.026	L.F.	.95	1.04		1.99	2.92

07 55 Protected Membrane Roofing

07 55 10 – Protected Membrane Roofing Components

07 55 10.10 Protected Membrane Roofing Components		Crew	Daily Output	Labor-Hours	Unit	Material	2015 Bare Costs Labor	Equipment	Total	Total Incl O&P
0010	**PROTECTED MEMBRANE ROOFING COMPONENTS**									
0100	Choose roofing membrane from 07 50									
0120	Then choose roof deck insulation from 07 22									
0130	Filter fabric	2 Rofc	10000	.002	S.F.	.12	.06		.18	.26
0140	Ballast, 3/8" - 1/2" in place	G-1	36	1.556	Ton	20	58.50	15.80	94.30	144
0150	3/4" - 1-1/2" in place	"	36	1.556	"	20	58.50	15.80	94.30	144
0200	2" concrete blocks, natural	1 Clab	115	.070	S.F.	3.37	2.62		5.99	8
0210	Colors	"	115	.070	"	3.73	2.62		6.35	8.40

07 56 Fluid-Applied Roofing

07 56 10 – Fluid-Applied Roofing Elastomers

07 56 10.10 Elastomeric Roofing

Line	Description	Crew	Daily Output	Labor-Hours	Unit	Material	Labor	Equipment	Total	Total Incl O&P
0010	**ELASTOMERIC ROOFING**									
0020	Acrylic, 44% solids, 2 coats, on corrugated metal	2 Rofc	2400	.007	S.F.	.58	.27		.85	1.11
0025	On smooth metal		3000	.005		.46	.21		.67	.90
0030	On foam or modified bitumen		1500	.011		.92	.43		1.35	1.78
0035	On concrete		1500	.011		.92	.43		1.35	1.78
0040	On tar and gravel		1500	.011		.92	.43		1.35	1.78
0045	36% solids, 2 coats, on corrugated metal		2400	.007		.56	.27		.83	1.10
0050	On smooth metal		3000	.005		.45	.21		.66	.88
0055	On foam or modified bitumen		1500	.011		.90	.43		1.33	1.76
0060	On concrete		1500	.011		.90	.43		1.33	1.76
0065	On tar and gravel		1500	.011		.90	.43		1.33	1.76
0070	Primer if required, 2 coats on corrugated metal		2400	.007		.52	.27		.79	1.05
0075	On smooth metal		3000	.005		.52	.21		.73	.96
0080	On foam or modified bitumen		1500	.011		.75	.43		1.18	1.60
0085	On concrete		1500	.011		.56	.43		.99	1.39
0090	On tar & gravel/rolled roof		1500	.011		.75	.43		1.18	1.60
0110	Acrylic rubber, fluid applied, 20 mils thick	G-5	2000	.020		2	.73	.10	2.83	3.63
0120	50 mils, reinforced		1200	.033		3	1.22	.16	4.38	5.70
0130	For walking surface, add		900	.044		1.05	1.62	.21	2.88	4.33
0300	Neoprene, fluid applied, 20 mil thick, not-reinforced	G-1	1135	.049		1.32	1.85	.50	3.67	5.35
0600	Non-woven polyester, reinforced		960	.058		1.45	2.19	.59	4.23	6.20
0700	5 coat neoprene deck, 60 mil thick, under 10,000 S.F.		325	.172		4.40	6.45	1.75	12.60	18.45
0900	Over 10,000 S.F.		625	.090		4.40	3.36	.91	8.67	11.90
9000	Minimum labor/equipment charge	1 Rofc	2	4	Job		160		160	290

07 57 Coated Foamed Roofing

07 57 13 – Sprayed Polyurethane Foam Roofing

07 57 13.10 Sprayed Polyurethane Foam Roofing (S.P.F.)

Line	Description	Crew	Daily Output	Labor-Hours	Unit	Material	Labor	Equipment	Total	Total Incl O&P
0010	**SPRAYED POLYURETHANE FOAM ROOFING (S.P.F.)**									
0100	Primer for metal substrate (when required)	G-2A	3000	.008	S.F.	.44	.29	.24	.97	1.24
0200	Primer for non-metal substrate (when required)		3000	.008		.17	.29	.24	.70	.95
0300	Closed cell spray, polyurethane foam, 3 lb. per C.F. density, 1", R6.7		15000	.002		.66	.06	.05	.77	.88
0400	2", R13.4		13125	.002		1.32	.07	.05	1.44	1.62
0500	3", R18.6		11485	.002		1.98	.08	.06	2.12	2.38
0550	4", R24.8		10080	.002		2.64	.09	.07	2.80	3.14
0700	Spray-on silicone coating		2500	.010		1.13	.34	.29	1.76	2.16
0800	Warranty 5-20 year manufacturer's								.15	.15

07 57 Coated Foamed Roofing

07 57 13 – Sprayed Polyurethane Foam Roofing

07 57 13.10 Sprayed Polyurethane Foam Roofing (S.P.F.)		Crew	Daily Output	Labor-Hours	Unit	2015 Bare Costs Material	Labor	Equipment	Total	Total Incl O&P
0900	Warranty 20 year, no dollar limit				S.F.				.20	.20

07 58 Roll Roofing

07 58 10 – Asphalt Roll Roofing

07 58 10.10 Roll Roofing

Line	Description	Crew	Daily Output	Labor-Hours	Unit	Material	Labor	Equipment	Total	Total Incl O&P
0010	**ROLL ROOFING**									
0100	Asphalt, mineral surface									
0200	1 ply #15 organic felt, 1 ply mineral surfaced									
0300	Selvage roofing, lap 19", nailed & mopped	G-1	27	2.074	Sq.	74.50	78	21	173.50	246
0400	3 plies glass fiber felt (type IV), 1 ply mineral surfaced									
0500	Selvage roofing, lapped 19", mopped	G-1	25	2.240	Sq.	126	84	23	233	315
0600	Coated glass fiber base sheet, 2 plies of glass fiber									
0700	Felt (type IV), 1 ply mineral surfaced selvage									
0800	Roofing, lapped 19", mopped	G-1	25	2.240	Sq.	133	84	23	240	325
0900	On nailable decks	"	24	2.333	"	122	87.50	23.50	233	320
1000	3 plies glass fiber felt (type III), 1 ply mineral surfaced									
1100	Selvage roofing, lapped 19", mopped	G-1	25	2.240	Sq.	126	84	23	233	315

07 61 Sheet Metal Roofing

07 61 13 – Standing Seam Sheet Metal Roofing

07 61 13.10 Standing Seam Sheet Metal Roofing, Field Fab.

Line	Description	Crew	Daily Output	Labor-Hours	Unit	Material	Labor	Equipment	Total	Total Incl O&P
0010	**STANDING SEAM SHEET METAL ROOFING, FIELD FABRICATED**									
0400	Copper standing seam roofing, over 10 squares, 16 oz., 125 lb. per sq.	1 Shee	1.30	6.154	Sq.	960	345		1,305	1,600
0600	18 oz., 140 lb. per sq.		1.20	6.667		1,075	375		1,450	1,775
0700	20 oz., 150 lb. per sq.		1.10	7.273		1,175	405		1,580	1,950
1200	For abnormal conditions or small areas, add					25%	100%			
1300	For lead-coated copper, add					25%				
9000	Minimum labor/equipment charge	1 Shee	2	4	Job		224		224	355

07 61 16 – Batten Seam Sheet Metal Roofing

07 61 16.10 Batten Seam Sheet Metal Roofing, Field Fabricated

Line	Description	Crew	Daily Output	Labor-Hours	Unit	Material	Labor	Equipment	Total	Total Incl O&P
0010	**BATTEN SEAM SHEET METAL ROOFING, FIELD FABRICATED**									
0012	Copper batten seam roofing, over 10 sq., 16 oz., 130 lb. per sq.	1 Shee	1.10	7.273	Sq.	1,225	405		1,630	1,975
0020	Lead batten seam roofing, 5 lb. per S.F.		1.20	6.667		1,350	375		1,725	2,100
0100	Zinc/copper alloy batten seam roofing, .020 thick		1.20	6.667		1,225	375		1,600	1,950
0200	Copper roofing, batten seam, over 10 sq., 18 oz., 145 lb. per sq.		1	8		1,350	450		1,800	2,200
0300	20 oz., 160 lb. per sq.		1	8		1,500	450		1,950	2,325
0500	Stainless steel batten seam roofing, type 304, 28 ga.		1.20	6.667		585	375		960	1,225
0600	26 ga.		1.15	6.957		545	390		935	1,225
0800	Zinc, copper alloy roofing, batten seam, .027" thick		1.15	6.957		1,550	390		1,940	2,325
0900	.032" thick		1.10	7.273		1,975	405		2,380	2,825
1000	.040" thick		1.05	7.619		2,475	425		2,900	3,400
9000	Minimum labor/equipment charge		2	4	Job		224		224	355

07 61 19 – Flat Seam Sheet Metal Roofing

07 61 19.10 Flat Seam Sheet Metal Roofing, Field Fabricated

Line	Description	Crew	Daily Output	Labor-Hours	Unit	Material	Labor	Equipment	Total	Total Incl O&P
0010	**FLAT SEAM SHEET METAL ROOFING, FIELD FABRICATED**									
0900	Copper flat seam roofing, over 10 squares, 16 oz., 115 lb./sq.	1 Shee	1.20	6.667	Sq.	890	375		1,265	1,575
0950	18 oz., 130 lb./sq.		1.15	6.957		995	390		1,385	1,725
1000	20 oz., 145 lb./sq.		1.10	7.273		1,100	405		1,505	1,850

07 61 Sheet Metal Roofing

07 61 19 – Flat Seam Sheet Metal Roofing

07 61 19.10 Flat Seam Sheet Metal Roofing, Field Fabricated		Crew	Daily Output	Labor-Hours	Unit	2015 Bare Costs Material	Labor	Equipment	Total	Total Incl O&P
1008	Zinc flat seam roofing, .020" thick	1 Shee	1.20	6.667	Sq.	1,050	375		1,425	1,750
1010	.027" thick		1.15	6.957		1,325	390		1,715	2,075
1020	.032" thick		1.12	7.143		1,675	400		2,075	2,475
1030	.040" thick		1.05	7.619		2,100	425		2,525	3,000
1100	Lead flat seam roofing, 5 lb. per S.F.		1.30	6.154		1,150	345		1,495	1,825
9000	Minimum labor/equipment charge		2.75	2.909	Job		163		163	257

07 62 Sheet Metal Flashing and Trim

07 62 10 – Sheet Metal Trim

07 62 10.10 Sheet Metal Cladding

Code	Description	Crew	Daily Output	Labor-Hours	Unit	Material	Labor	Equipment	Total	Total Incl O&P
0010	**SHEET METAL CLADDING**									
0100	Aluminum, up to 6 bends, .032" thick, window casing	1 Carp	180	.044	S.F.	1.40	2.09		3.49	4.96
0200	Window sill		72	.111	L.F.	1.40	5.20		6.60	10.10
0300	Door casing		180	.044	S.F.	1.40	2.09		3.49	4.96
0400	Fascia		250	.032		1.40	1.50		2.90	4
0500	Rake trim		225	.036		1.40	1.67		3.07	4.28
0700	.024" thick, window casing		180	.044		1.22	2.09		3.31	4.76
0800	Window sill		72	.111	L.F.	1.22	5.20		6.42	9.90
0900	Door casing		180	.044	S.F.	1.22	2.09		3.31	4.76
1000	Fascia		250	.032		1.22	1.50		2.72	3.80
1100	Rake trim		225	.036		1.22	1.67		2.89	4.08
1200	Vinyl coated aluminum, up to 6 bends, window casing		180	.044		1.35	2.09		3.44	4.91
1300	Window sill		72	.111	L.F.	1.35	5.20		6.55	10.05
1400	Door casing		180	.044	S.F.	1.35	2.09		3.44	4.91
1500	Fascia		250	.032		1.35	1.50		2.85	3.95
1600	Rake trim		225	.036		1.35	1.67		3.02	4.23

07 65 Flexible Flashing

07 65 10 – Sheet Metal Flashing

07 65 10.10 Sheet Metal Flashing and Counter Flashing

Code	Description	Crew	Daily Output	Labor-Hours	Unit	Material	Labor	Equipment	Total	Total Incl O&P
0010	**SHEET METAL FLASHING AND COUNTER FLASHING**									
0011	Including up to 4 bends									
0020	Aluminum, mill finish, .013" thick	1 Rofc	145	.055	S.F.	.79	2.21		3	4.87
0030	.016" thick		145	.055		.91	2.21		3.12	5
0060	.019" thick		145	.055		1.15	2.21		3.36	5.25
0100	.032" thick		145	.055		1.35	2.21		3.56	5.50
0200	.040" thick		145	.055		2.31	2.21		4.52	6.55
0300	.050" thick		145	.055		2.43	2.21		4.64	6.65
0325	Mill finish 5" x 7" step flashing, .016" thick		1920	.004	Ea.	.15	.17		.32	.47
0350	Mill finish 12" x 12" step flashing, .016" thick		1600	.005	"	.50	.20		.70	.91
0400	Painted finish, add				S.F.	.29			.29	.32
1000	Mastic-coated 2 sides, .005" thick	1 Rofc	330	.024		1.82	.97		2.79	3.76
1100	.016" thick		330	.024		2.01	.97		2.98	3.97
1600	Copper, 16 oz., sheets, under 1000 lb.		115	.070		7.65	2.79		10.44	13.50
1700	Over 4000 lb.		155	.052		7.65	2.07		9.72	12.20
1900	20 oz. sheets, under 1000 lb.		110	.073		9.40	2.92		12.32	15.60
2000	Over 4000 lb.		145	.055		8.95	2.21		11.16	13.80
2200	24 oz. sheets, under 1000 lb.		105	.076		13.30	3.06		16.36	20
2300	Over 4000 lb.		135	.059		12.65	2.38		15.03	18.20

07 65 Flexible Flashing

07 65 10 – Sheet Metal Flashing

	07 65 10.10 Sheet Metal Flashing and Counter Flashing	Crew	Daily Output	Labor-Hours	Unit	Material	Labor	Equipment	Total	Total Incl O&P
						2015 Bare Costs				
2500	32 oz. sheets, under 1000 lb.	1 Rofc	100	.080	S.F.	17.80	3.21		21.01	25.50
2600	Over 4000 lb.		130	.062		16.90	2.47		19.37	23
2700	W shape for valleys, 16 oz., 24" wide		100	.080	L.F.	15.40	3.21		18.61	23
5800	Lead, 2.5 lb. per S.F., up to 12" wide		135	.059	S.F.	4.99	2.38		7.37	9.80
5900	Over 12" wide		135	.059		3.99	2.38		6.37	8.70
8900	Stainless steel sheets, 32 ga.,		155	.052		3	2.07		5.07	7.05
9000	28 ga.		155	.052		3.99	2.07		6.06	8.15
9100	26 ga.		155	.052		4.25	2.07		6.32	8.40
9200	24 ga.		155	.052		4.75	2.07		6.82	9
9290	For mechanically keyed flashing, add					40%				
9320	Steel sheets, galvanized, 20 ga.	1 Rofc	130	.062	S.F.	1.25	2.47		3.72	5.85
9322	22 ga.		135	.059		1.21	2.38		3.59	5.60
9324	24 ga.		140	.057		.92	2.29		3.21	5.15
9326	26 ga.		148	.054		.80	2.17		2.97	4.80
9328	28 ga.		155	.052		.69	2.07		2.76	4.50
9340	30 ga.		160	.050		.58	2.01		2.59	4.26
9400	Terne coated stainless steel, .015" thick, 28 ga		155	.052		7.35	2.07		9.42	11.85
9500	.018" thick, 26 ga		155	.052		8.30	2.07		10.37	12.85
9600	Zinc and copper alloy (brass), .020" thick		155	.052		9.30	2.07		11.37	14
9700	.027" thick		155	.052		11	2.07		13.07	15.85
9800	.032" thick		155	.052		14	2.07		16.07	19.15
9900	.040" thick		155	.052		18.50	2.07		20.57	24
9950	Minimum labor/equipment charge		3	2.667	Job		107		107	193

07 65 12 – Fabric and Mastic Flashings

	07 65 12.10 Fabric and Mastic Flashing and Counter Flashing	Crew	Daily Output	Labor-Hours	Unit	Material	Labor	Equipment	Total	Total Incl O&P
0010	**FABRIC AND MASTIC FLASHING AND COUNTER FLASHING**									
1300	Asphalt flashing cement, 5 gallon				Gal.	9.20			9.20	10.10
4900	Fabric, asphalt-saturated cotton, specification grade	1 Rofc	35	.229	S.Y.	2.96	9.15		12.11	19.80
5000	Utility grade		35	.229		1.42	9.15		10.57	18.10
5300	Close-mesh fabric, saturated, 17 oz. per S.Y.		35	.229		2.15	9.15		11.30	18.90
5500	Fiberglass, resin-coated		35	.229		.97	9.15		10.12	17.60
8500	Shower pan, bituminous membrane, 7 oz.		155	.052	S.F.	1.62	2.07		3.69	5.50

07 65 13 – Laminated Sheet Flashing

	07 65 13.10 Laminated Sheet Flashing	Crew	Daily Output	Labor-Hours	Unit	Material	Labor	Equipment	Total	Total Incl O&P
0010	**LAMINATED SHEET FLASHING**, Including up to 4 bends									
0500	Aluminum, fabric-backed 2 sides, mill finish, .004" thick	1 Rofc	330	.024	S.F.	1.54	.97		2.51	3.45
0700	.005" thick		330	.024		1.82	.97		2.79	3.76
0750	Mastic-backed, self adhesive		460	.017		3.51	.70		4.21	5.10
0800	Mastic-coated 2 sides, .004" thick		330	.024		1.54	.97		2.51	3.45
2800	Copper, paperbacked 1 side, 2 oz.		330	.024		1.84	.97		2.81	3.78
2900	3 oz.		330	.024		2.15	.97		3.12	4.13
3100	Paperbacked 2 sides, 2 oz.		330	.024		1.97	.97		2.94	3.93
3150	3 oz.		330	.024		2.25	.97		3.22	4.24
3200	5 oz.		330	.024		3.03	.97		4	5.10
3250	7 oz.		330	.024		5.30	.97		6.27	7.60
3400	Mastic-backed 2 sides, copper, 2 oz.		330	.024		1.82	.97		2.79	3.76
3500	3 oz.		330	.024		2.20	.97		3.17	4.18
3700	5 oz.		330	.024		3.30	.97		4.27	5.40
3800	Fabric-backed 2 sides, copper, 2 oz.		330	.024		2.10	.97		3.07	4.07
4000	3 oz.		330	.024		2.40	.97		3.37	4.40
4100	5 oz.		330	.024		3.40	.97		4.37	5.50
4300	Copper-clad stainless steel, .015" thick, under 500 lb.		115	.070		6.40	2.79		9.19	12.10

07 65 Flexible Flashing

07 65 13 – Laminated Sheet Flashing

07 65 13.10 Laminated Sheet Flashing		Crew	Daily Output	Labor-Hours	Unit	Material	Labor	Equipment	Total	Total Incl O&P
						2015 Bare Costs				
4400	Over 2000 lb.	1 Rofc	155	.052	S.F.	6.10	2.07		8.17	10.45
4600	.018" thick, under 500 lb.		100	.080		7.20	3.21		10.41	13.70
4700	Over 2000 lb.		145	.055		6.80	2.21		9.01	11.50
8550	Shower pan, 3 ply copper and fabric, 3 oz.		155	.052		3.50	2.07		5.57	7.60
8600	7 oz.		155	.052		4.20	2.07		6.27	8.35
9300	Stainless steel, paperbacked 2 sides, .005" thick	↓	330	.024	↓	3.62	.97		4.59	5.75

07 65 19 – Plastic Sheet Flashing

07 65 19.10 Plastic Sheet Flashing and Counter Flashing		Crew	Daily Output	Labor-Hours	Unit	Material	Labor	Equipment	Total	Total Incl O&P
0010	**PLASTIC SHEET FLASHING AND COUNTER FLASHING**									
7300	Polyvinyl chloride, black, 10 mil	1 Rofc	285	.028	S.F.	.21	1.13		1.34	2.26
7400	20 mil		285	.028		.28	1.13		1.41	2.34
7600	30 mil		285	.028		.35	1.13		1.48	2.42
7700	60 mil		285	.028		.85	1.13		1.98	2.97
7900	Black or white for exposed roofs, 60 mil	↓	285	.028	↓	1.80	1.13		2.93	4.01
8060	PVC tape, 5" x 45 mils, for joint covers, 100 L.F./roll				Ea.	170			170	187
8850	Polyvinyl chloride, 30 mil	1 Rofc	160	.050	S.F.	1.30	2.01		3.31	5.05

07 65 23 – Rubber Sheet Flashing

07 65 23.10 Rubber Sheet Flashing and Counterflashing		Crew	Daily Output	Labor-Hours	Unit	Material	Labor	Equipment	Total	Total Incl O&P
0010	**RUBBER SHEET FLASHING AND COUNTERFLASHING**									
4810	EPDM 90 mils, 1" diameter pipe flashing	1 Rofc	32	.250	Ea.	19.35	10.05		29.40	39.50
4820	2" diameter		30	.267		19.25	10.70		29.95	40.50
4830	3" diameter		28	.286		20.50	11.45		31.95	43.50
4840	4" diameter		24	.333		23.50	13.35		36.85	50
4850	6" diameter		22	.364	↓	23.50	14.60		38.10	52.50
8100	Rubber, butyl, 1/32" thick		285	.028	S.F.	1.92	1.13		3.05	4.14
8200	1/16" thick		285	.028		2.60	1.13		3.73	4.89
8300	Neoprene, cured, 1/16" thick		285	.028		2.23	1.13		3.36	4.48
8400	1/8" thick	↓	285	.028	↓	6.05	1.13		7.18	8.70

07 71 Roof Specialties

07 71 16 – Manufactured Counterflashing Systems

07 71 16.10 Roof Drain Boot		Crew	Daily Output	Labor-Hours	Unit	Material	Labor	Equipment	Total	Total Incl O&P
0010	**ROOF DRAIN BOOT**									
0100	Cast iron, 4" diameter	1 Shee	125	.064	L.F.	50	3.58		53.58	61
0300	4" x 3"		125	.064		57.50	3.58		61.08	69
0400	5" x 4"	↓	125	.064	↓	117	3.58		120.58	135

07 71 19 – Manufactured Gravel Stops and Fasciae

07 71 19.10 Gravel Stop		Crew	Daily Output	Labor-Hours	Unit	Material	Labor	Equipment	Total	Total Incl O&P
0010	**GRAVEL STOP**									
0020	Aluminum, .050" thick, 4" face height, mill finish	1 Shee	145	.055	L.F.	6.30	3.09		9.39	11.75
0080	Duranodic finish		145	.055		6.75	3.09		9.84	12.30
0100	Painted		145	.055		6.95	3.09		10.04	12.50
0300	6" face height		135	.059		6.45	3.32		9.77	12.35
0350	Duranodic finish		135	.059		7.25	3.32		10.57	13.25
0400	Painted		135	.059		8.20	3.32		11.52	14.25
0600	8" face height		125	.064		7.35	3.58		10.93	13.75
0650	Duranodic finish		125	.064		8.40	3.58		11.98	14.90
0700	Painted		125	.064		8.75	3.58		12.33	15.30
0900	12" face height, .080 thick, 2 piece	↓	100	.080	↓	9.95	4.48		14.43	17.95

07 71 Roof Specialties

07 71 19 – Manufactured Gravel Stops and Fasciae

07 71 19.10 Gravel Stop

	07 71 19.10 Gravel Stop	Crew	Daily Output	Labor-Hours	Unit	2015 Bare Costs Material	Labor	Equipment	Total	Total Incl O&P
0950	Duranodic finish	1 Shee	100	.080	L.F.	10.25	4.48		14.73	18.35
1000	Painted		100	.080		12	4.48		16.48	20.50
1200	Copper, 16 oz., 3" face height		145	.055		24	3.09		27.09	31.50
1300	6" face height		135	.059		33	3.32		36.32	42
1350	Galv steel, 24 ga., 4" leg, plain, with continuous cleat, 4" face		145	.055		6.25	3.09		9.34	11.75
1360	6" face height		145	.055		6.50	3.09		9.59	12
1500	Polyvinyl chloride, 6" face height		135	.059		5.20	3.32		8.52	10.95
1600	9" face height		125	.064		6	3.58		9.58	12.25
1800	Stainless steel, 24 ga., 6" face height		135	.059		15.30	3.32		18.62	22
1900	12" face height		100	.080		22.50	4.48		26.98	32
2100	20 ga., 6" face height		135	.059		17.80	3.32		21.12	25
2200	12" face height		100	.080		27	4.48		31.48	36.50
9000	Minimum labor/equipment charge		3.50	2.286	Job		128		128	202

07 71 19.30 Fascia

	07 71 19.30 Fascia	Crew	Daily Output	Labor-Hours	Unit	Material	Labor	Equipment	Total	Total Incl O&P
0010	**FASCIA**									
0100	Aluminum, reverse board and batten, .032" thick, colored, no furring incl	1 Shee	145	.055	S.F.	6.75	3.09		9.84	12.30
0300	Steel, galv and enameled, stock, no furring, long panels		145	.055		4.25	3.09		7.34	9.55
0600	Short panels		115	.070		5.80	3.89		9.69	12.55
9000	Minimum labor/equipment charge		4	2	Job		112		112	177

07 71 23 – Manufactured Gutters and Downspouts

07 71 23.10 Downspouts

	07 71 23.10 Downspouts	Crew	Daily Output	Labor-Hours	Unit	Material	Labor	Equipment	Total	Total Incl O&P
0010	**DOWNSPOUTS**									
0020	Aluminum, embossed, .020" thick, 2" x 3"	1 Shee	190	.042	L.F.	.98	2.36		3.34	4.80
0100	Enameled		190	.042		1.37	2.36		3.73	5.25
0300	.024" thick, 2' x 3"		180	.044		2.05	2.49		4.54	6.20
0400	3" x 4"		140	.057		2.33	3.20		5.53	7.60
0600	Round, corrugated aluminum, 3" diameter, .020" thick		190	.042		1.95	2.36		4.31	5.85
0700	4" diameter, .025" thick		140	.057		2.53	3.20		5.73	7.85
0900	Wire strainer, round, 2" diameter		155	.052	Ea.	2.67	2.89		5.56	7.50
1000	4" diameter		155	.052		3.67	2.89		6.56	8.60
1200	Rectangular, perforated, 2" x 3"		145	.055		2.30	3.09		5.39	7.40
1300	3" x 4"		145	.055		3.32	3.09		6.41	8.50
1500	Copper, round, 16 oz., stock, 2" diameter		190	.042	L.F.	7.60	2.36		9.96	12.05
1600	3" diameter		190	.042		7.50	2.36		9.86	11.95
1800	4" diameter		145	.055		9.70	3.09		12.79	15.50
1900	5" diameter		130	.062		17.05	3.44		20.49	24.50
2100	Rectangular, corrugated copper, stock, 2" x 3"		190	.042		8.55	2.36		10.91	13.10
2200	3" x 4"		145	.055		10.40	3.09		13.49	16.30
2400	Rectangular, plain copper, stock, 2" x 3"		190	.042		11.05	2.36		13.41	15.85
2500	3" x 4"		145	.055		13.95	3.09		17.04	20
2700	Wire strainers, rectangular, 2" x 3"		145	.055	Ea.	17.05	3.09		20.14	23.50
2800	3" x 4"		145	.055		17.80	3.09		20.89	24.50
3000	Round, 2" diameter		145	.055		6.40	3.09		9.49	11.90
3100	3" diameter		145	.055		7.10	3.09		10.19	12.65
3300	4" diameter		145	.055		12.10	3.09		15.19	18.15
3400	5" diameter		115	.070		22.50	3.89		26.39	30.50
3600	Lead-coated copper, round, stock, 2" diameter		190	.042	L.F.	22	2.36		24.36	28
3700	3" diameter		190	.042		22	2.36		24.36	28
3900	4" diameter		145	.055		23	3.09		26.09	30.50
4000	5" diameter, corrugated		130	.062		22	3.44		25.44	30
4200	6" diameter, corrugated		105	.076		30	4.26		34.26	40
4300	Rectangular, corrugated, stock, 2" x 3"		190	.042		15.40	2.36		17.76	20.50

07 71 Roof Specialties

07 71 23 – Manufactured Gutters and Downspouts

07 71 23.10 Downspouts		Crew	Daily Output	Labor-Hours	Unit	2015 Bare Costs Material	Labor	Equipment	Total	Total Incl O&P
4500	Plain, stock, 2″ x 3″	1 Shee	190	.042	L.F.	24	2.36		26.36	30
4600	3″ x 4″		145	.055		33	3.09		36.09	41.50
4800	Steel, galvanized, round, corrugated, 2″ or 3″ diameter, 28 ga.		190	.042		2.08	2.36		4.44	6
4900	4″ diameter, 28 ga.		145	.055		2.52	3.09		5.61	7.65
5100	5″ diameter, 26 ga.		130	.062		4	3.44		7.44	9.85
5400	6″ diameter, 28 ga.		105	.076		4.20	4.26		8.46	11.35
5500	26 ga.		105	.076		4	4.26		8.26	11.15
5700	Rectangular, corrugated, 28 ga., 2″ x 3″		190	.042		1.94	2.36		4.30	5.85
5800	3″ x 4″		145	.055		1.82	3.09		4.91	6.85
6000	Rectangular, plain, 28 ga., galvanized, 2″ x 3″		190	.042		3.65	2.36		6.01	7.75
6100	3″ x 4″		145	.055		4.06	3.09		7.15	9.35
6300	Epoxy painted, 24 ga., corrugated, 2″ x 3″		190	.042		2.25	2.36		4.61	6.20
6400	3″ x 4″		145	.055		2.80	3.09		5.89	7.95
6600	Wire strainers, rectangular, 2″ x 3″		145	.055	Ea.	17.05	3.09		20.14	23.50
6700	3″ x 4″		145	.055		17.80	3.09		20.89	24.50
6900	Round strainers, 2″ or 3″ diameter		145	.055		3.86	3.09		6.95	9.10
7000	4″ diameter		145	.055		5.80	3.09		8.89	11.20
7800	Stainless steel tubing, schedule 5, 2″ x 3″ or 3″ diameter		190	.042	L.F.	38.50	2.36		40.86	46
7900	3″ x 4″ or 4″ diameter		145	.055		49	3.09		52.09	59
8100	4″ x 5″ or 5″ diameter		135	.059		101	3.32		104.32	116
8200	Vinyl, rectangular, 2″ x 3″		210	.038		2.08	2.13		4.21	5.65
8300	Round, 2-1/2″		220	.036		1.32	2.03		3.35	4.66
9000	Minimum labor/equipment charge		4	2	Job		112		112	177

07 71 23.20 Downspout Elbows		Crew	Daily Output	Labor-Hours	Unit	Material	Labor	Equipment	Total	Total Incl O&P
0010	**DOWNSPOUT ELBOWS**									
0020	Aluminum, embossed, 2″ x 3″, .020″ thick	1 Shee	100	.080	Ea.	.95	4.48		5.43	8.10
0100	Enameled		100	.080		1.75	4.48		6.23	9
0200	Embossed, 3″ x 4″, .025″ thick		100	.080		4.51	4.48		8.99	12
0300	Enameled		100	.080		3.75	4.48		8.23	11.20
0400	Embossed, corrugated, 3″ diameter, .020″ thick		100	.080		2.74	4.48		7.22	10.05
0500	4″ diameter, .025″ thick		100	.080		5.80	4.48		10.28	13.40
0600	Copper, 16 oz., 2″ diameter		100	.080		9.45	4.48		13.93	17.45
0700	3″ diameter		100	.080		9	4.48		13.48	16.95
0800	4″ diameter		100	.080		13.90	4.48		18.38	22.50
1000	2″ x 3″ corrugated		100	.080		9.35	4.48		13.83	17.35
1100	3″ x 4″ corrugated		100	.080		13.25	4.48		17.73	21.50
1300	Vinyl, 2-1/2″ diameter, 45 or 75 degree bend		100	.080		3.88	4.48		8.36	11.30
1400	Tee Y junction		75	.107		12.90	5.95		18.85	23.50
9000	Minimum labor/equipment charge		4	2	Job		112		112	177

07 71 23.30 Gutters		Crew	Daily Output	Labor-Hours	Unit	Material	Labor	Equipment	Total	Total Incl O&P
0010	**GUTTERS**									
0012	Aluminum, stock units, 5″ K type, .027″ thick, plain	1 Shee	125	.064	L.F.	2.72	3.58		6.30	8.65
0100	Enameled		125	.064		2.66	3.58		6.24	8.55
0300	5″ K type type, .032″ thick, plain		125	.064		3.23	3.58		6.81	9.20
0400	Enameled		125	.064		3.02	3.58		6.60	8.95
0700	Copper, half round, 16 oz., stock units, 4″ wide		125	.064		8.90	3.58		12.48	15.45
0900	5″ wide		125	.064		7.25	3.58		10.83	13.60
1000	6″ wide		118	.068		11.85	3.79		15.64	19
1200	K type, 16 oz., stock, 5″ wide		125	.064		8.05	3.58		11.63	14.50
1300	6″ wide		125	.064		8.50	3.58		12.08	15
1500	Lead coated copper, 16 oz., half round, stock, 4″ wide		125	.064		13.80	3.58		17.38	21
1600	6″ wide		118	.068		21.50	3.79		25.29	30

07 71 23 – Manufactured Gutters and Downspouts

07 71 23.30 Gutters		Crew	Daily Output	Labor-Hours	Unit	Material	Labor	Equipment	Total	Total Incl O&P
						2015 Bare Costs				
1800	K type, stock, 5″ wide	1 Shee	125	.064	L.F.	18.35	3.58		21.93	25.50
1900	6″ wide		125	.064		17.90	3.58		21.48	25.50
2100	Copper clad stainless steel, K type, 5″ wide		125	.064		7.50	3.58		11.08	13.90
2200	6″ wide		125	.064		9.45	3.58		13.03	16.05
2400	Steel, galv, half round or box, 28 ga., 5″ wide, plain		125	.064		2.10	3.58		5.68	7.95
2500	Enameled		125	.064		2.10	3.58		5.68	7.95
2700	26 ga., stock, 5″ wide		125	.064		2.18	3.58		5.76	8.05
2800	6″ wide		125	.064		2.71	3.58		6.29	8.65
3000	Vinyl, O.G., 4″ wide	1 Carp	115	.070		1.25	3.27		4.52	6.75
3100	5″ wide		115	.070		1.55	3.27		4.82	7.05
3200	4″ half round, stock units		115	.070		1.30	3.27		4.57	6.80
3250	Joint connectors				Ea.	2.90			2.90	3.19
3300	Wood, clear treated cedar, fir or hemlock, 3″ x 4″	1 Carp	100	.080	L.F.	10	3.76		13.76	17.15
3400	4″ x 5″	″	100	.080	″	16.75	3.76		20.51	24.50
5000	Accessories, end cap, K type, aluminum 5″	1 Shee	625	.013	Ea.	.66	.72		1.38	1.86
5010	6″		625	.013		1.44	.72		2.16	2.71
5020	Copper, 5″		625	.013		3.22	.72		3.94	4.67
5030	6″		625	.013		3.54	.72		4.26	5
5040	Lead coated copper, 5″		625	.013		12.50	.72		13.22	14.90
5050	6″		625	.013		13.35	.72		14.07	15.85
5060	Copper clad stainless steel, 5″		625	.013		3	.72		3.72	4.43
5070	6″		625	.013		3.60	.72		4.32	5.10
5080	Galvanized steel, 5″		625	.013		1.28	.72		2	2.54
5090	6″		625	.013		2.18	.72		2.90	3.53
5100	Vinyl, 4″	1 Carp	625	.013		13.15	.60		13.75	15.50
5110	5″	″	625	.013		13.15	.60		13.75	15.50
5120	Half round, copper, 4″	1 Shee	625	.013		4.53	.72		5.25	6.10
5130	5″		625	.013		4.53	.72		5.25	6.10
5140	6″		625	.013		7.65	.72		8.37	9.60
5150	Lead coated copper, 5″		625	.013		14.15	.72		14.87	16.70
5160	6″		625	.013		21	.72		21.72	24.50
5170	Copper clad stainless steel, 5″		625	.013		4.50	.72		5.22	6.10
5180	6″		625	.013		4.50	.72		5.22	6.10
5190	Galvanized steel, 5″		625	.013		2.25	.72		2.97	3.61
5200	6″		625	.013		2.80	.72		3.52	4.21
5210	Outlet, aluminum, 2″ x 3″		420	.019		.62	1.07		1.69	2.36
5220	3″ x 4″		420	.019		1.05	1.07		2.12	2.84
5230	2-3/8″ round		420	.019		.56	1.07		1.63	2.30
5240	Copper, 2″ x 3″		420	.019		6.50	1.07		7.57	8.85
5250	3″ x 4″		420	.019		7.85	1.07		8.92	10.35
5260	2-3/8″ round		420	.019		4.55	1.07		5.62	6.70
5270	Lead coated copper, 2″ x 3″		420	.019		25.50	1.07		26.57	29.50
5280	3″ x 4″		420	.019		28.50	1.07		29.57	33
5290	2-3/8″ round		420	.019		25.50	1.07		26.57	29.50
5300	Copper clad stainless steel, 2″ x 3″		420	.019		6.50	1.07		7.57	8.85
5310	3″ x 4″		420	.019		7.85	1.07		8.92	10.35
5320	2-3/8″ round		420	.019		4.55	1.07		5.62	6.70
5330	Galvanized steel, 2″ x 3″		420	.019		3.18	1.07		4.25	5.20
5340	3″ x 4″		420	.019		4.90	1.07		5.97	7.10
5350	2-3/8″ round		420	.019		3.98	1.07		5.05	6.05
5360	K type mitres, aluminum		65	.123		3	6.90		9.90	14.15
5370	Copper		65	.123		17.40	6.90		24.30	30
5380	Lead coated copper		65	.123		54.50	6.90		61.40	71

07 71 23 – Manufactured Gutters and Downspouts

07 71 23.30 Gutters		Crew	Daily Output	Labor-Hours	Unit	Material	Labor	Equipment	Total	Total Incl O&P
5390	Copper clad stainless steel	1 Shee	65	.123	Ea.	27.50	6.90		34.40	41.50
5400	Galvanized steel		65	.123		22.50	6.90		29.40	36
5420	Half round mitres, copper		65	.123		63.50	6.90		70.40	81
5430	Lead coated copper		65	.123		89.50	6.90		96.40	109
5440	Copper clad stainless steel		65	.123		56	6.90		62.90	73
5450	Galvanized steel		65	.123		25.50	6.90		32.40	39
5460	Vinyl mitres and outlets		65	.123		9.75	6.90		16.65	21.50
5470	Sealant		940	.009	L.F.	.01	.48		.49	.76
5480	Soldering		96	.083	"	.19	4.66		4.85	7.55
9000	Minimum labor/equipment charge		3.75	2.133	Job		119		119	188

07 71 23.35 Gutter Guard		Crew	Daily Output	Labor-Hours	Unit	Material	Labor	Equipment	Total	Total Incl O&P
0010	**GUTTER GUARD**									
0020	6" wide strip, aluminum mesh	1 Carp	500	.016	L.F.	2.42	.75		3.17	3.89
0100	Vinyl mesh		500	.016	"	2.61	.75		3.36	4.10
9000	Minimum labor/equipment charge		4	2	Job		94		94	154

07 71 26 – Reglets

07 71 26.10 Reglets and Accessories		Crew	Daily Output	Labor-Hours	Unit	Material	Labor	Equipment	Total	Total Incl O&P
0010	**REGLETS AND ACCESSORIES**									
0020	Reglet, aluminum, .025" thick, in parapet	1 Carp	225	.036	L.F.	1.93	1.67		3.60	4.86
0300	16 oz. copper		225	.036		6.40	1.67		8.07	9.80
0400	Galvanized steel, 24 ga.		225	.036		1.34	1.67		3.01	4.21
0600	Stainless steel, .020" thick		225	.036		3.75	1.67		5.42	6.85
0900	Counter flashing for above, 12" wide, .032" aluminum	1 Shee	150	.053		2.40	2.98		5.38	7.35
1200	16 oz. copper		150	.053		6.10	2.98		9.08	11.40
1300	Galvanized steel, 26 ga.		150	.053		1.42	2.98		4.40	6.25
1500	Stainless steel, .020" thick		150	.053		6.15	2.98		9.13	11.45
9000	Minimum labor/equipment charge	1 Carp	3	2.667	Job		125		125	205

07 71 29 – Manufactured Roof Expansion Joints

07 71 29.10 Expansion Joints		Crew	Daily Output	Labor-Hours	Unit	Material	Labor	Equipment	Total	Total Incl O&P
0010	**EXPANSION JOINTS**									
0300	Butyl or neoprene center with foam insulation, metal flanges									
0400	Aluminum, .032" thick for openings to 2-1/2"	1 Rofc	165	.048	L.F.	11.30	1.94		13.24	15.95
0600	For joint openings to 3-1/2"		165	.048		11.30	1.94		13.24	15.95
0610	For joint openings to 5"		165	.048		13.10	1.94		15.04	17.90
0620	For joint openings to 8"		165	.048		15.80	1.94		17.74	21
0700	Copper, 16 oz. for openings to 2-1/2"		165	.048		16.70	1.94		18.64	22
0900	For joint openings to 3-1/2"		165	.048		16.70	1.94		18.64	22
0910	For joint openings to 5"		165	.048		18.90	1.94		20.84	24.50
0920	For joint openings to 8"		165	.048		22	1.94		23.94	27.50
1000	Galvanized steel, 26 ga. for openings to 2-1/2"		165	.048		9.80	1.94		11.74	14.30
1200	For joint openings to 3-1/2"		165	.048		9.80	1.94		11.74	14.30
1210	For joint openings to 5"		165	.048		11.30	1.94		13.24	15.95
1220	For joint openings to 8"		165	.048		14.80	1.94		16.74	19.80
1300	Lead-coated copper, 16 oz. for openings to 2-1/2"		165	.048		32	1.94		33.94	38.50
1500	For joint openings to 3-1/2"		165	.048		32	1.94		33.94	38.50
1600	Stainless steel, .018", for openings to 2-1/2"		165	.048		15.40	1.94		17.34	20.50
1800	For joint openings to 3-1/2"		165	.048		15.40	1.94		17.34	20.50
1810	For joint openings to 5"		165	.048		17.20	1.94		19.14	22.50
1820	For joint openings to 8"		165	.048		21	1.94		22.94	26.50
1900	Neoprene, double-seal type with thick center, 4-1/2" wide		125	.064		14.50	2.57		17.07	20.50
1950	Polyethylene bellows, with galv steel flat flanges		100	.080		6.20	3.21		9.41	12.60

07 71 Roof Specialties

07 71 29 – Manufactured Roof Expansion Joints

07 71 29.10 Expansion Joints		Crew	Daily Output	Labor-Hours	Unit	2015 Bare Costs Material	Labor	Equipment	Total	Total Incl O&P
1960	With galvanized angle flanges	1 Rofc	100	.080	L.F.	6.40	3.21		9.61	12.85
2000	Roof joint with extruded aluminum cover, 2″	1 Shee	115	.070		28	3.89		31.89	37
2100	Roof joint, plastic curbs, foam center, standard	1 Rofc	100	.080		13	3.21		16.21	20
2200	Large	″	100	.080		17	3.21		20.21	24.50
2500	Roof to wall joint with extruded aluminum cover	1 Shee	115	.070		28	3.89		31.89	37
2700	Wall joint, closed cell foam on PVC cover, 9″ wide	1 Rofc	125	.064		5	2.57		7.57	10.15
2800	12″ wide	″	115	.070		6	2.79		8.79	11.65
9000	Minimum labor/equipment charge	1 Shee	3	2.667	Job		149		149	235

07 71 43 – Drip Edge

07 71 43.10 Drip Edge, Rake Edge, Ice Belts		Crew	Daily Output	Labor-Hours	Unit	Material	Labor	Equipment	Total	Total Incl O&P
0010	**DRIP EDGE, RAKE EDGE, ICE BELTS**									
0020	Aluminum, .016″ thick, 5″ wide, mill finish	1 Carp	400	.020	L.F.	.55	.94		1.49	2.15
0100	White finish		400	.020		.63	.94		1.57	2.23
0200	8″ wide, mill finish		400	.020		1.45	.94		2.39	3.14
0300	Ice belt, 28″ wide, mill finish		100	.080		7.65	3.76		11.41	14.60
0310	Vented, mill finish		400	.020		1.99	.94		2.93	3.73
0320	Painted finish		400	.020		2.24	.94		3.18	4
0400	Galvanized, 5″ wide		400	.020		.53	.94		1.47	2.12
0500	8″ wide, mill finish		400	.020		.80	.94		1.74	2.42
0510	Rake edge, aluminum, 1-1/2″ x 1-1/2″		400	.020		.30	.94		1.24	1.87
0520	3-1/2″ x 1-1/2″		400	.020		.42	.94		1.36	2
9000	Minimum labor/equipment charge		4	2	Job		94		94	154

07 72 Roof Accessories

07 72 23 – Relief Vents

07 72 23.10 Roof Vents		Crew	Daily Output	Labor-Hours	Unit	Material	Labor	Equipment	Total	Total Incl O&P
0010	**ROOF VENTS**									
0020	Mushroom shape, for built-up roofs, aluminum	1 Rofc	30	.267	Ea.	65	10.70		75.70	91
0100	PVC, 6″ high		30	.267	″	21	10.70		31.70	42.50
9000	Minimum labor/equipment charge		2.75	2.909	Job		117		117	211

07 72 23.20 Vents		Crew	Daily Output	Labor-Hours	Unit	Material	Labor	Equipment	Total	Total Incl O&P
0010	**VENTS**									
0100	Soffit or eave, aluminum, mill finish, strips, 2-1/2″ wide	1 Carp	200	.040	L.F.	.43	1.88		2.31	3.55
0200	3″ wide		200	.040		.41	1.88		2.29	3.53
0300	Enamel finish, 3″ wide		200	.040		.50	1.88		2.38	3.63
0400	Mill finish, rectangular, 4″ x 16″		72	.111	Ea.	1.45	5.20		6.65	10.15
0500	8″ x 16″		72	.111		2.26	5.20		7.46	11.05
2420	Roof ventilator	Q-9	16	1		52	50.50		102.50	137
2500	Vent, roof vent	1 Rofc	24	.333		26	13.35		39.35	52.50

07 72 26 – Ridge Vents

07 72 26.10 Ridge Vents and Accessories		Crew	Daily Output	Labor-Hours	Unit	Material	Labor	Equipment	Total	Total Incl O&P
0010	**RIDGE VENTS AND ACCESSORIES**									
0100	Aluminum strips, mill finish	1 Rofc	160	.050	L.F.	2.75	2.01		4.76	6.65
0150	Painted finish		160	.050	″	4.05	2.01		6.06	8.10
0200	Connectors		48	.167	Ea.	4.77	6.70		11.47	17.35
0300	End caps		48	.167	″	2.03	6.70		8.73	14.35
0400	Galvanized strips		160	.050	L.F.	3.58	2.01		5.59	7.55
0430	Molded polyethylene, shingles not included		160	.050	″	2.74	2.01		4.75	6.65
0440	End plugs		48	.167	Ea.	2.03	6.70		8.73	14.35
0450	Flexible roll, shingles not included		160	.050	L.F.	2.38	2.01		4.39	6.25

07 72 Roof Accessories

07 72 26 – Ridge Vents

07 72 26.10 Ridge Vents and Accessories		Crew	Daily Output	Labor-Hours	Unit	2015 Bare Costs Material	Labor	Equipment	Total	Total Incl O&P
2300	Ridge vent strip, mill finish	1 Shee	155	.052	L.F.	3.85	2.89		6.74	8.80

07 72 33 – Roof Hatches

07 72 33.10 Roof Hatch Options

		Crew	Daily Output	Labor-Hours	Unit	Material	Labor	Equipment	Total	Total Incl O&P
0010	**ROOF HATCH OPTIONS**									
0500	2′-6″ x 3′, aluminum curb and cover	G-3	10	3.200	Ea.	1,025	150		1,175	1,375
0520	Galvanized steel curb and aluminum cover		10	3.200		640	150		790	945
0540	Galvanized steel curb and cover		10	3.200		715	150		865	1,025
0600	2′-6″ x 4′-6″, aluminum curb and cover		9	3.556		1,375	166		1,541	1,800
0800	Galvanized steel curb and aluminum cover		9	3.556		885	166		1,051	1,250
0900	Galvanized steel curb and cover		9	3.556		965	166		1,131	1,325
1100	4′ x 4′ aluminum curb and cover		8	4		1,675	187		1,862	2,150
1120	Galvanized steel curb and aluminum cover		8	4		1,725	187		1,912	2,200
1140	Galvanized steel curb and cover		8	4		1,200	187		1,387	1,600
1200	2′-6″ x 8′-0″, aluminum curb and cover		6.60	4.848		1,900	227		2,127	2,475
1400	Galvanized steel curb and aluminum cover		6.60	4.848		1,800	227		2,027	2,350
1500	Galvanized steel curb and cover		6.60	4.848		1,525	227		1,752	2,050
1800	For plexiglass panels, 2′-6″ x 3′-0″, add to above					440			440	485
9000	Minimum labor/equipment charge	2 Carp	2	8	Job		375		375	615

07 72 36 – Smoke Vents

07 72 36.10 Smoke Hatches

		Crew	Daily Output	Labor-Hours	Unit	Material	Labor	Equipment	Total	Total Incl O&P
0010	**SMOKE HATCHES**									
0200	For 3′-0″ long, add to roof hatches from Section 07 72 33.10				Ea.	25%	5%			
0250	For 4′-0″ long, add to roof hatches from Section 07 72 33.10					20%	5%			
0300	For 8′-0″ long, add to roof hatches from Section 07 72 33.10					10%	5%			

07 72 36.20 Smoke Vent Options

		Crew	Daily Output	Labor-Hours	Unit	Material	Labor	Equipment	Total	Total Incl O&P
0010	**SMOKE VENT OPTIONS**									
0100	4′ x 4′ aluminum cover and frame	G-3	13	2.462	Ea.	2,050	115		2,165	2,425
0200	Galvanized steel cover and frame		13	2.462		1,800	115		1,915	2,150
0300	4′ x 8′ aluminum cover and frame		8	4		2,800	187		2,987	3,375
0400	Galvanized steel cover and frame		8	4		2,375	187		2,562	2,925
9000	Minimum labor/equipment charge	2 Carp	2	8	Job		375		375	615

07 72 53 – Snow Guards

07 72 53.10 Snow Guard Options

		Crew	Daily Output	Labor-Hours	Unit	Material	Labor	Equipment	Total	Total Incl O&P
0010	**SNOW GUARD OPTIONS**									
0100	Slate & asphalt shingle roofs, fastened with nails	1 Rofc	160	.050	Ea.	12.55	2.01		14.56	17.40
0200	Standing seam metal roofs, fastened with set screws		48	.167		16.95	6.70		23.65	31
0300	Surface mount for metal roofs, fastened with solder		48	.167		6.05	6.70		12.75	18.75
0400	Double rail pipe type, including pipe		130	.062	L.F.	32	2.47		34.47	39.50

07 72 73 – Pitch Pockets

07 72 73.10 Pitch Pockets, Variable Sizes

		Crew	Daily Output	Labor-Hours	Unit	Material	Labor	Equipment	Total	Total Incl O&P
0010	**PITCH POCKETS, VARIABLE SIZES**									
0100	Adjustable, 4″ to 7″, welded corners, 4″ deep	1 Rofc	48	.167	Ea.	16.25	6.70		22.95	30
0200	Side extenders, 6″	″	240	.033	″	2.50	1.34		3.84	5.15

07 72 80 – Vents

07 72 80.30 Vent Options

		Crew	Daily Output	Labor-Hours	Unit	Material	Labor	Equipment	Total	Total Incl O&P
0010	**VENT OPTIONS**									
0020	Plastic, for insulated decks, 1 per M.S.F.	1 Rofc	40	.200	Ea.	21	8		29	37.50
0100	Heavy duty		20	.400		38	16.05		54.05	71
0300	Aluminum		30	.267		22	10.70		32.70	43.50
0800	Polystyrene baffles, 12″ wide for 16″ O.C. rafter spacing	1 Carp	90	.089		.42	4.17		4.59	7.30

07 72 Roof Accessories

07 72 80 – Vents

07 72 80.30 Vent Options		Crew	Daily Output	Labor-Hours	Unit	Material	2015 Bare Costs Labor	Equipment	Total	Total Incl O&P
0900	For 24" O.C. rafter spacing	1 Carp	110	.073	Ea.	.77	3.41		4.18	6.45
9000	Minimum labor/equipment charge	↓	3	2.667	Job		125		125	205

07 76 Roof Pavers

07 76 16 – Roof Decking Pavers

07 76 16.10 Roof Pavers and Supports		Crew	Daily Output	Labor-Hours	Unit	Material	Labor	Equipment	Total	Total Incl O&P
0010	ROOF PAVERS AND SUPPORTS									
1000	Roof decking pavers, concrete blocks, 2" thick, natural	1 Clab	115	.070	S.F.	3.37	2.62		5.99	8
1100	Colors		115	.070	"	3.73	2.62		6.35	8.40
1200	Support pedestal, bottom cap		960	.008	Ea.	3	.31		3.31	3.81
1300	Top cap		960	.008		4.80	.31		5.11	5.80
1400	Leveling shims, 1/16"		1920	.004		1.20	.16		1.36	1.58
1500	1/8"		1920	.004		1.20	.16		1.36	1.58
1600	Buffer pad		960	.008	↓	2.50	.31		2.81	3.26
1700	PVC legs (4" SDR 35)		2880	.003	Inch	.12	.10		.22	.30
2000	Alternate pricing method, system in place	↓	101	.079	S.F.	7	2.98		9.98	12.60

07 81 Applied Fireproofing

07 81 16 – Cementitious Fireproofing

07 81 16.10 Sprayed Cementitious Fireproofing		Crew	Daily Output	Labor-Hours	Unit	Material	Labor	Equipment	Total	Total Incl O&P
0010	SPRAYED CEMENTITIOUS FIREPROOFING									
0050	Not including canvas protection, normal density									
0100	Per 1" thick, on flat plate steel	G-2	3000	.008	S.F.	.53	.32	.04	.89	1.14
0200	Flat decking		2400	.010		.53	.40	.06	.99	1.28
0400	Beams		1500	.016		.53	.63	.09	1.25	1.70
0500	Corrugated or fluted decks		1250	.019		.79	.76	.11	1.66	2.21
0700	Columns, 1-1/8" thick		1100	.022		.59	.86	.12	1.57	2.17
0800	2-3/16" thick		700	.034		1.25	1.36	.19	2.80	3.78
0900	For canvas protection, add	↓	5000	.005	↓	.08	.19	.03	.30	.43
1000	Not including canvas protection, high density									
1100	Per 1" thick, on flat plate steel	G-2	3000	.008	S.F.	1.85	.32	.04	2.21	2.60
1110	On flat decking		2400	.010		1.85	.40	.06	2.31	2.74
1120	On beams		1500	.016		1.85	.63	.09	2.57	3.16
1130	Corrugated or fluted decks		1250	.019		1.85	.76	.11	2.72	3.38
1140	Columns, 1-1/8" thick		1100	.022		2.08	.86	.12	3.06	3.81
1150	2-3/16" thick		1100	.022		4.16	.86	.12	5.14	6.10
1170	For canvas protection, add	↓	5000	.005	↓	.08	.19	.03	.30	.43
1200	Not including canvas protection, retrofitting									
1210	Per 1" thick, on flat plate steel	G-2	1500	.016	S.F.	.38	.63	.09	1.10	1.54
1220	On flat decking		1200	.020		.38	.79	.11	1.28	1.82
1230	On beams		750	.032		.38	1.27	.18	1.83	2.66
1240	Corrugated or fluted decks		625	.038		.58	1.52	.21	2.31	3.31
1250	Columns, 1-1/8" thick		550	.044		.43	1.73	.24	2.40	3.53
1260	2-3/16" thick		500	.048		.86	1.90	.27	3.03	4.30
1400	Accessories, preliminary spattered texture coat	↓	4500	.005		.02	.21	.03	.26	.39
1410	Bonding agent	1 Plas	1000	.008	↓	.09	.34		.43	.65
1500	Intumescent epoxy fireproofing on wire mesh, 3/16" thick									
1550	1 hour rating, exterior use	G-2	136	.176	S.F.	7.35	7	.98	15.33	20.50
1600	Magnesium oxychloride, 35# to 40# density, 1/4" thick	↓	3000	.008	↓	1.55	.32	.04	1.91	2.27

07 81 Applied Fireproofing

07 81 16 – Cementitious Fireproofing

07 81 16.10 Sprayed Cementitious Fireproofing		Crew	Daily Output	Labor-Hours	Unit	2015 Bare Costs Material	Labor	Equipment	Total	Total Incl O&P
1650	1/2″ thick	G-2	2000	.012	S.F.	3.10	.47	.07	3.64	4.25
1700	60# to 70# density, 1/4″ thick		3000	.008		2.05	.32	.04	2.41	2.82
1750	1/2″ thick		2000	.012		4.15	.47	.07	4.69	5.40
2000	Vermiculite cement, troweled or sprayed, 1/4″ thick		3000	.008		1.40	.32	.04	1.76	2.10
2050	1/2″ thick		2000	.012		2.75	.47	.07	3.29	3.87
9000	Minimum labor/equipment charge		3	8	Job		315	44.50	359.50	560

07 84 Firestopping

07 84 13 – Penetration Firestopping

07 84 13.10 Firestopping

07 84 13.10 Firestopping		Crew	Daily Output	Labor-Hours	Unit	Material	Labor	Equipment	Total	Total Incl O&P
0010	**FIRESTOPPING** R078413-30									
0100	Metallic piping, non insulated									
0110	Through walls, 2″ diameter	1 Carp	16	.500	Ea.	17.20	23.50		40.70	57.50
0120	4″ diameter		14	.571		26	27		53	73
0130	6″ diameter		12	.667		35.50	31.50		67	90.50
0140	12″ diameter		10	.800		62.50	37.50		100	130
0150	Through floors, 2″ diameter		32	.250		9.80	11.75		21.55	30
0160	4″ diameter		28	.286		14.20	13.40		27.60	37.50
0170	6″ diameter		24	.333		18.50	15.65		34.15	46
0180	12″ diameter		20	.400		31.50	18.80		50.30	65.50
0190	Metallic piping, insulated									
0200	Through walls, 2″ diameter	1 Carp	16	.500	Ea.	24	23.50		47.50	65
0210	4″ diameter		14	.571		33	27		60	80
0220	6″ diameter		12	.667		42	31.50		73.50	97.50
0230	12″ diameter		10	.800		68.50	37.50		106	137
0240	Through floors, 2″ diameter		32	.250		16.60	11.75		28.35	37.50
0250	4″ diameter		28	.286		21	13.40		34.40	45
0260	6″ diameter		24	.333		25.50	15.65		41.15	53.50
0270	12″ diameter		20	.400		31.50	18.80		50.30	65.50
0280	Non metallic piping, non insulated									
0290	Through walls, 2″ diameter	1 Carp	12	.667	Ea.	68.50	31.50		100	127
0300	4″ diameter		10	.800		86	37.50		123.50	157
0310	6″ diameter		8	1		120	47		167	209
0330	Through floors, 2″ diameter		16	.500		54	23.50		77.50	98
0340	4″ diameter		6	1.333		66.50	62.50		129	177
0350	6″ diameter		6	1.333		80.50	62.50		143	192
0370	Ductwork, insulated & non insulated, round									
0380	Through walls, 6″ diameter	1 Carp	12	.667	Ea.	35	31.50		66.50	90
0390	12″ diameter		10	.800		69.50	37.50		107	138
0400	18″ diameter		8	1		113	47		160	202
0410	Through floors, 6″ diameter		16	.500		18.60	23.50		42.10	59
0420	12″ diameter		14	.571		34	27		61	81.50
0430	18″ diameter		12	.667		59	31.50		90.50	117
0440	Ductwork, insulated & non insulated, rectangular									
0450	With stiffener/closure angle, through walls, 6″ x 12″	1 Carp	8	1	Ea.	28	47		75	108
0460	12″ x 24″		6	1.333		37.50	62.50		100	144
0470	24″ x 48″		4	2		107	94		201	271
0480	With stiffener/closure angle, through floors, 6″ x 12″		10	.800		15.40	37.50		52.90	78.50
0490	12″ x 24″		8	1		28	47		75	108
0500	24″ x 48″		6	1.333		54	62.50		116.50	163
0510	Multi trade openings									

07 84 Firestopping

07 84 13 – Penetration Firestopping

07 84 13.10 Firestopping		Crew	Daily Output	Labor-Hours	Unit	Material	2015 Bare Costs Labor	Equipment	Total	Total Incl O&P
0520	Through walls, 6″ x 12″	1 Carp	2	4	Ea.	59	188		247	375
0530	12″ x 24″	″	1	8		238	375		613	875
0540	24″ x 48″	2 Carp	1	16		950	750		1,700	2,275
0550	48″ x 96″	″	.75	21.333		3,700	1,000		4,700	5,725
0560	Through floors, 6″ x 12″	1 Carp	2	4		39	188		227	355
0570	12″ x 24″	″	1	8		157	375		532	790
0580	24″ x 48″	2 Carp	.75	21.333		625	1,000		1,625	2,350
0590	48″ x 96″	″	.50	32	↓	2,450	1,500		3,950	5,125
0600	Structural penetrations, through walls									
0610	Steel beams, W8 x 10	1 Carp	8	1	Ea.	37.50	47		84.50	118
0620	W12 x 14		6	1.333		59	62.50		121.50	168
0630	W21 x 44		5	1.600		118	75		193	253
0640	W36 x 135		3	2.667		287	125		412	520
0650	Bar joists, 18″ deep		6	1.333		54.50	62.50		117	163
0660	24″ deep		6	1.333		67.50	62.50		130	177
0670	36″ deep		5	1.600		101	75		176	235
0680	48″ deep	↓	4	2	↓	118	94		212	284
0690	Construction joints, floor slab at exterior wall									
0700	Precast, brick, block or drywall exterior									
0710	2″ wide joint	1 Carp	125	.064	L.F.	8.50	3		11.50	14.25
0720	4″ wide joint	″	75	.107	″	17	5		22	27
0730	Metal panel, glass or curtain wall exterior									
0740	2″ wide joint	1 Carp	40	.200	L.F.	20	9.40		29.40	37.50
0750	4″ wide joint	″	25	.320	″	28	15		43	55
0760	Floor slab to drywall partition									
0770	Flat joint	1 Carp	100	.080	L.F.	8.40	3.76		12.16	15.40
0780	Fluted joint		50	.160		17.40	7.50		24.90	31.50
0790	Etched fluted joint	↓	75	.107	↓	11.10	5		16.10	20.50
0800	Floor slab to concrete/masonry partition									
0810	Flat joint	1 Carp	75	.107	L.F.	18.75	5		23.75	28.50
0820	Fluted joint	″	50	.160	″	22.50	7.50		30	37
0830	Concrete/CMU wall joints									
0840	1″ wide	1 Carp	100	.080	L.F.	10.25	3.76		14.01	17.45
0850	2″ wide		75	.107		18.75	5		23.75	28.50
0860	4″ wide	↓	50	.160	↓	38	7.50		45.50	54
0870	Concrete/CMU floor joints									
0880	1″ wide	1 Carp	200	.040	L.F.	5.15	1.88		7.03	8.75
0890	2″ wide		150	.053		9.40	2.50		11.90	14.45
0900	4″ wide	↓	100	.080	↓	17.90	3.76		21.66	26

07 91 Preformed Joint Seals

07 91 13 – Compression Seals

07 91 13.10 Compression Seals		Crew	Daily Output	Labor-Hours	Unit	Material	Labor	Equipment	Total	Total Incl O&P
0010	**COMPRESSION SEALS**									
4900	O-ring type cord, 1/4″	1 Bric	472	.017	L.F.	.37	.78		1.15	1.68
4910	1/2″		440	.018		.95	.84		1.79	2.42
4920	3/4″		424	.019		1.80	.87		2.67	3.40
4930	1″		408	.020		3.40	.91		4.31	5.20
4940	1-1/4″		384	.021		6.30	.96		7.26	8.50
4950	1-1/2″		368	.022		7.95	1		8.95	10.40
4960	1-3/4″	↓	352	.023	↓	13.25	1.05		14.30	16.30

07 91 Preformed Joint Seals

07 91 13 – Compression Seals

07 91 13.10 Compression Seals		Crew	Daily Output	Labor-Hours	Unit	2015 Bare Costs Material	Labor	Equipment	Total	Total Incl O&P
4970	2″	1 Bric	344	.023	L.F.	18.50	1.07		19.57	22.50

07 91 16 – Joint Gaskets

07 91 16.10 Joint Gaskets

Line	Description	Crew	Daily Output	Labor-Hours	Unit	Material	Labor	Equipment	Total	Total Incl O&P
0010	**JOINT GASKETS**									
4400	Joint gaskets, neoprene, closed cell w/adh, 1/8″ x 3/8″	1 Bric	240	.033	L.F.	.28	1.54		1.82	2.81
4500	1/4″ x 3/4″		215	.037		.59	1.72		2.31	3.44
4700	1/2″ x 1″		200	.040		1.30	1.85		3.15	4.43
4800	3/4″ x 1-1/2″		165	.048		1.45	2.24		3.69	5.25

07 91 23 – Backer Rods

07 91 23.10 Backer Rods

Line	Description	Crew	Daily Output	Labor-Hours	Unit	Material	Labor	Equipment	Total	Total Incl O&P
0010	**BACKER RODS**									
0032	Backer rod, polyethylene, 1/4″ diameter	1 Bric	460	.017	L.F.	.02	.80		.82	1.33
0052	1/2″ diameter		460	.017		.03	.80		.83	1.35
0072	3/4″ diameter		460	.017		.06	.80		.86	1.37
0092	1″ diameter		460	.017		.09	.80		.89	1.41

07 91 26 – Joint Fillers

07 91 26.10 Joint Fillers

Line	Description	Crew	Daily Output	Labor-Hours	Unit	Material	Labor	Equipment	Total	Total Incl O&P
0010	**JOINT FILLERS**									
4360	Butyl rubber filler, 1/4″ x 1/4″	1 Bric	290	.028	L.F.	.22	1.27		1.49	2.32
4365	1/2″ x 1/2″		250	.032		.89	1.48		2.37	3.38
4370	1/2″ x 3/4″		210	.038		1.34	1.76		3.10	4.33
4375	3/4″ x 3/4″		230	.035		2.01	1.61		3.62	4.82
4380	1″ x 1″		180	.044		2.68	2.05		4.73	6.30
4390	For coloring, add					12%				
4980	Polyethylene joint backing, 1/4″ x 2″	1 Bric	2.08	3.846	C.L.F.	12	178		190	300
4990	1/4″ x 6″		1.28	6.250	″	28	288		316	500
5600	Silicone, room temp vulcanizing foam seal, 1/4″ x 1/2″		1312	.006	L.F.	.33	.28		.61	.83
5610	1/2″ x 1/2″		656	.012		.67	.56		1.23	1.65
5620	1/2″ x 3/4″		442	.018		1	.84		1.84	2.46
5630	3/4″ x 3/4″		328	.024		1.50	1.13		2.63	3.48
5640	1/8″ x 1″		1312	.006		.33	.28		.61	.83
5650	1/8″ x 3″		442	.018		1	.84		1.84	2.46
5670	1/4″ x 3″		295	.027		2	1.25		3.25	4.24
5680	1/4″ x 6″		148	.054		3.99	2.49		6.48	8.45
5690	1/2″ x 6″		82	.098		8	4.50		12.50	16.15
5700	1/2″ x 9″		52.50	.152		12	7.05		19.05	24.50
5710	1/2″ x 12″		33	.242		15.95	11.20		27.15	36

07 92 Joint Sealants

07 92 13 – Elastomeric Joint Sealants

07 92 13.20 Caulking and Sealant Options

Line	Description	Crew	Daily Output	Labor-Hours	Unit	Material	Labor	Equipment	Total	Total Incl O&P
0010	**CAULKING AND SEALANT OPTIONS**									
0050	Latex acrylic based, bulk				Gal.	26.50			26.50	29.50
0055	Bulk in place 1/4″ x 1/4″ bead	1 Bric	300	.027	L.F.	.08	1.23		1.31	2.09
0060	1/4″ x 3/8″		294	.027		.14	1.26		1.40	2.19
0065	1/4″ x 1/2″		288	.028		.19	1.28		1.47	2.30
0075	3/8″ x 3/8″		284	.028		.21	1.30		1.51	2.35
0080	3/8″ x 1/2″		280	.029		.28	1.32		1.60	2.46
0085	3/8″ x 5/8″		276	.029		.35	1.34		1.69	2.57

07 92 Joint Sealants

07 92 13 – Elastomeric Joint Sealants

07 92 13.20 Caulking and Sealant Options

07 92 13.20	Caulking and Sealant Options	Crew	Daily Output	Labor-Hours	Unit	2015 Bare Costs Material	Labor	Equipment	Total	Total Incl O&P
0095	3/8″ x 3/4″	1 Bric	272	.029	L.F.	.42	1.36		1.78	2.67
0100	1/2″ x 1/2″		275	.029		.37	1.34		1.71	2.59
0105	1/2″ x 5/8″		269	.030		.47	1.37		1.84	2.74
0110	1/2″ x 3/4″		263	.030		.56	1.40		1.96	2.90
0115	1/2″ x 7/8″		256	.031		.65	1.44		2.09	3.07
0120	1/2″ x 1″		250	.032		.75	1.48		2.23	3.22
0125	3/4″ x 3/4″		244	.033		.84	1.51		2.35	3.39
0130	3/4″ x 1″		225	.036		1.12	1.64		2.76	3.90
0135	1″ x 1″		200	.040		1.50	1.85		3.35	4.65
0190	Cartridges				Gal.	32			32	35.50
0200	11 fl. oz. cartridge				Ea.	2.76			2.76	3.04
0500	1/4″ x 1/2″	1 Bric	288	.028	L.F.	.23	1.28		1.51	2.34
0600	1/2″ x 1/2″		275	.029		.45	1.34		1.79	2.68
0800	3/4″ x 3/4″		244	.033		1.01	1.51		2.52	3.58
0900	3/4″ x 1″		225	.036		1.35	1.64		2.99	4.16
1000	1″ x 1″		200	.040		1.69	1.85		3.54	4.86
1400	Butyl based, bulk				Gal.	35			35	38.50
1500	Cartridges				″	38.50			38.50	42
1700	1/4″ x 1/2″, 154 L.F./gal.	1 Bric	288	.028	L.F.	.23	1.28		1.51	2.34
1800	1/2″ x 1/2″, 77 L.F./gal.	″	275	.029	″	.45	1.34		1.79	2.68
2300	Polysulfide compounds, 1 component, bulk				Gal.	72			72	79
2600	1 or 2 component, in place, 1/4″ x 1/4″, 308 L.F./gal.	1 Bric	300	.027	L.F.	.23	1.23		1.46	2.26
2700	1/2″ x 1/4″, 154 L.F./gal.		288	.028		.47	1.28		1.75	2.60
2900	3/4″ x 3/8″, 68 L.F./gal.		272	.029		1.06	1.36		2.42	3.37
3000	1″ x 1/2″, 38 L.F./gal.		250	.032		1.89	1.48		3.37	4.48
3200	Polyurethane, 1 or 2 component				Gal.	49			49	54
3500	Bulk, in place, 1/4″ x 1/4″	1 Bric	300	.027	L.F.	.16	1.23		1.39	2.18
3655	1/2″ x 1/4″		288	.028		.32	1.28		1.60	2.44
3800	3/4″ x 3/8″		272	.029		.72	1.36		2.08	3.01
3900	1″ x 1/2″		250	.032		1.28	1.48		2.76	3.81
4100	Silicone rubber, bulk				Gal.	49			49	54
4200	Cartridges				″	51			51	56

07 92 16 – Rigid Joint Sealants

07 92 16.10 Rigid Joint Sealants

Line	Description	Crew	Daily Output	Labor-Hours	Unit	Material	Labor	Equipment	Total	Total Incl O&P
0010	**RIGID JOINT SEALANTS**									
5802	Tapes, sealant, PVC foam adhesive, 1/16″ x 1/4″				L.F.	.10			.10	.11
5902	1/16″ x 1/2″					.10			.10	.11
5952	1/16″ x 1″					.16			.16	.18
6002	1/8″ x 1/2″					.10			.10	.10

07 92 19 – Acoustical Joint Sealants

07 92 19.10 Acoustical Sealant

Line	Description	Crew	Daily Output	Labor-Hours	Unit	Material	Labor	Equipment	Total	Total Incl O&P
0010	**ACOUSTICAL SEALANT**									
0020	Acoustical sealant, elastomeric, cartridges				Ea.	8.50			8.50	9.35
0025	In place, 1/4″ x 1/4″	1 Bric	300	.027	L.F.	.35	1.23		1.58	2.38
0030	1/4″ x 1/2″		288	.028		.69	1.28		1.97	2.85
0035	1/2″ x 1/2″		275	.029		1.39	1.34		2.73	3.70
0040	1/2″ x 3/4″		263	.030		2.08	1.40		3.48	4.57
0045	3/4″ x 3/4″		244	.033		3.12	1.51		4.63	5.90
0050	1″ x 1″		200	.040		5.55	1.85		7.40	9.10

07 95 Expansion Control

07 95 13 – Expansion Joint Cover Assemblies

07 95 13.50 Expansion Joint Assemblies		Crew	Daily Output	Labor-Hours	Unit	2015 Bare Costs Material	Labor	Equipment	Total	Total Incl O&P
0010	**EXPANSION JOINT ASSEMBLIES**									
0200	Floor cover assemblies, 1" space, aluminum	1 Sswk	38	.211	L.F.	16.85	11.10		27.95	38.50
0300	Bronze		38	.211		54.50	11.10		65.60	80
0500	2" space, aluminum		38	.211		16.85	11.10		27.95	38.50
0600	Bronze		38	.211		54.50	11.10		65.60	80
0800	Wall and ceiling assemblies, 1" space, aluminum		38	.211		14.85	11.10		25.95	36.50
0900	Bronze		38	.211		48.50	11.10		59.60	73.50
1100	2" space, aluminum		38	.211		15.15	11.10		26.25	36.50
1200	Bronze		38	.211		48.50	11.10		59.60	73.50
1400	Floor to wall assemblies, 1" space, aluminum		38	.211		18	11.10		29.10	40
1500	Bronze or stainless		38	.211		59	11.10		70.10	85
1700	Gym floor angle covers, aluminum, 3" x 3" angle		46	.174		18	9.15		27.15	36.50
1800	3" x 4" angle		46	.174		21	9.15		30.15	39.50
2000	Roof closures, aluminum, flat roof, low profile, 1" space		57	.140		22.50	7.40		29.90	38.50
2100	High profile		57	.140		24	7.40		31.40	40
2300	Roof to wall, low profile, 1" space		57	.140		21.50	7.40		28.90	37
2400	High profile		57	.140		24	7.40		31.40	40
9000	Minimum labor/equipment charge		2	4	Job		211		211	380

Estimating Tips

08 10 00 Doors and Frames

All exterior doors should be addressed for their energy conservation (insulation and seals).

- Most metal doors and frames look alike, but there may be significant differences among them. When estimating these items, be sure to choose the line item that most closely compares to the specification or door schedule requirements regarding:
  - type of metal
  - metal gauge
  - door core material
  - fire rating
  - finish
- Wood and plastic doors vary considerably in price. The primary determinant is the veneer material. Lauan, birch, and oak are the most common veneers. Other variables include the following:
  - hollow or solid core
  - fire rating
  - flush or raised panel
  - finish
- Door pricing includes bore for cylindrical lockset and mortise for hinges.

08 30 00 Specialty Doors and Frames

- There are many varieties of special doors, and they are usually priced per each. Add frames, hardware, or operators required for a complete installation.

08 40 00 Entrances, Storefronts, and Curtain Walls

- Glazed curtain walls consist of the metal tube framing and the glazing material. The cost data in this subdivision is presented for the metal tube framing alone or the composite wall. If your estimate requires a detailed takeoff of the framing, be sure to add the glazing cost and any tints.

08 50 00 Windows

- Most metal windows are delivered preglazed. However, some metal windows are priced without glass. Refer to 08 80 00 Glazing for glass pricing. The grade C indicates commercial grade windows, usually ASTM C-35.
- All wood windows and vinyl are priced preglazed. The glazing is insulating glass. Add the cost of screens and grills if required, and not already included.

08 70 00 Hardware

- Hardware costs add considerably to the cost of a door. The most efficient method to determine the hardware requirements for a project is to review the door and hardware schedule together. One type of door may have different hardware, depending on the door usage.
- Door hinges are priced by the pair, with most doors requiring 1-1/2 pairs per door. The hinge prices do not include installation labor, because it is included in door installation. Hinges are classified according to the frequency of use, base material, and finish.

08 80 00 Glazing

- Different openings require different types of glass. The most common types are:
  - float
  - tempered
  - insulating
  - impact-resistant
  - ballistic-resistant
- Most exterior windows are glazed with insulating glass. Entrance doors and window walls, where the glass is less than 18" from the floor, are generally glazed with tempered glass. Interior windows and some residential windows are glazed with float glass.
- Coastal communities require the use of impact-resistant glass, dependant on wind speed.
- The insulation or 'u' value is a strong consideration, along with solar heat gain, to determine total energy efficiency.

Reference Numbers

Reference numbers are shown in shaded boxes at the beginning of some major classifications. These numbers refer to related items in the Reference Section. The reference information may be an estimating procedure, an alternate pricing method, or technical information.

Note: Not all subdivisions listed here necessarily appear in this publication. ■

08 01 Operation and Maintenance of Openings

08 01 11 – Operation and Maintenance of Metal Doors and Frames

08 01 11.10 Door and Window Maintenance		Crew	Daily Output	Labor-Hours	Unit	Material	2015 Bare Costs Labor	Equipment	Total	Total Incl O&P
0010	**DOOR & WINDOW MAINTENANCE**									
0012	Remove weather stripping from door or window	1 Carp	45	.178	Ea.		8.35		8.35	13.70
0060	Remove lockset		20	.400			18.80		18.80	31
0200	Remove damaged louver		58.46	.137			6.45		6.45	10.55
1130	Install door	2 Carp	13.07	1.224			57.50		57.50	94
1140	Remove deadbolt	1 Carp	13.85	.578			27		27	44.50
1150	Remove panic bar		7.69	1.040			49		49	80
1420	Remove door, plane to fit, rail, reinstall door		9.60	.833			39		39	64

08 01 14 – Operation and Maintenance of Wood Doors

08 01 14.15 Door and Window Maintenance		Crew	Daily Output	Labor-Hours	Unit	Material	Labor	Equipment	Total	Total Incl O&P
0010	**DOOR & WINDOW MAINTENANCE**									
0050	Rehang single door	1 Carp	32	.250	Ea.		11.75		11.75	19.25
0100	Rehang, double door		16	.500	Pr.		23.50		23.50	38.50
0350	Install window sash cord		32	.250	Ea.	3.80	11.75		15.55	23.50

08 01 53 – Operation and Maintenance of Plastic Windows

08 01 53.81 Solid Vinyl Replacement Windows		Crew	Daily Output	Labor-Hours	Unit	Material	Labor	Equipment	Total	Total Incl O&P
0010	**SOLID VINYL REPLACEMENT WINDOWS**									
0020	Double hung, insulated glass, up to 83 united inches G	2 Carp	8	2	Ea.	340	94		434	530
0040	84 to 93 G		8	2		380	94		474	570
0060	94 to 101 G		6	2.667		380	125		505	620
0080	102 to 111 G		6	2.667		395	125		520	640
0100	112 to 120 G		6	2.667		430	125		555	680
0120	For each united inch over 120 , add G		800	.020	Inch	5.50	.94		6.44	7.60
0140	Casement windows, one operating sash , 42 to 60 united inches G		8	2	Ea.	230	94		324	405
0160	61 to 70 G		8	2		260	94		354	440
0180	71 to 80 G		8	2		280	94		374	465
0200	81 to 96 G		8	2		300	94		394	485
0220	Two operating sash, 58 to 78 united inches G		8	2		455	94		549	655
0240	79 to 88 G		8	2		485	94		579	690
0260	89 to 98 G		8	2		525	94		619	735
0280	99 to 108 G		6	2.667		550	125		675	810
0300	109 to 121 G		6	2.667		590	125		715	855
0320	Two operating.one fixed sash, 73 to 108 united inches G		8	2		715	94		809	945
0340	109 to 118 G		8	2		755	94		849	985
0360	119 to 128 G		6	2.667		775	125		900	1,050
0380	129 to 138 G		6	2.667		870	125		995	1,150
0400	139 to 156 G		6	2.667		945	125		1,070	1,250
0420	Four operating sash, 98 to 118 united inches G		8	2		1,025	94		1,119	1,275
0440	119 to 128 G		8	2		1,100	94		1,194	1,350
0460	129 to 138 G		6	2.667		1,150	125		1,275	1,475
0480	139 to 148 G		6	2.667		1,225	125		1,350	1,525
0500	149 to 168 G		6	2.667		1,300	125		1,425	1,625
0520	169 to 178 G		6	2.667		1,400	125		1,525	1,725
0560	Fixed picture window, up to 63 united inches G		8	2		180	94		274	350
0580	64 to 83 G		8	2		206	94		300	380
0600	84 to 101 G		8	2		257	94		351	435
0620	For each united inch over 101, add G		900	.018	Inch	3.10	.83		3.93	4.78
0800	Cellulose fiber insulation, poured into sash balance cavity G	1 Carp	36	.222	C.F.	.69	10.45		11.14	17.85
0820	Silicone caulking at perimeter G	"	800	.010	L.F.	.17	.47		.64	.95

08 05 05 – Selective Demolition for Openings

08 05 05.10 Selective Demolition Doors		Crew	Daily Output	Labor-Hours	Unit	Material	2015 Bare Costs Labor	Equipment	Total	Total Incl O&P
0010	**SELECTIVE DEMOLITION DOORS** R024119-10									
0200	Doors, exterior, 1-3/4" thick, single, 3' x 7' high	1 Clab	16	.500	Ea.		18.80		18.80	31
0202	Doors, exterior, 3' - 6' wide x 7' high		12	.667			25		25	41
0210	3' x 8' high		10	.800			30		30	49.50
0215	Double, 3' x 8' high		6	1.333			50		50	82
0220	Double, 6' x 7' high		12	.667			25		25	41
0500	Interior, 1-3/8" thick, single, 3' x 7' high		20	.400			15.05		15.05	24.50
0520	Double, 6' x 7' high		16	.500			18.80		18.80	31
0700	Bi-folding, 3' x 6'-8" high		20	.400			15.05		15.05	24.50
0720	6' x 6'-8" high		18	.444			16.70		16.70	27.50
0900	Bi-passing, 3' x 6'-8" high		16	.500			18.80		18.80	31
0940	6' x 6'-8" high		14	.571			21.50		21.50	35
1500	Remove and reset, hollow core	1 Carp	8	1			47		47	77
1520	Solid		6	1.333			62.50		62.50	103
2000	Frames, including trim, metal		8	1			47		47	77
2200	Wood	2 Carp	32	.500			23.50		23.50	38.50
2201	Alternate pricing method	1 Carp	200	.040	L.F.		1.88		1.88	3.08
2205	Remove door hardware	"	45.70	.175	Ea.		8.20		8.20	13.45
3000	Special doors, counter doors	2 Carp	6	2.667			125		125	205
3001	Special doors, counter doors	2 Clab	6	2.667			100		100	164
3100	Double acting	2 Carp	10	1.600			75		75	123
3101	Double acting	2 Clab	10	1.600			60		60	98.50
3200	Floor door (trap type), or access type	2 Carp	8	2			94		94	154
3300	Glass, sliding, including frames		12	1.333			62.50		62.50	103
3400	Overhead, commercial, 12' x 12' high		4	4			188		188	310
3440	up to 20' x 16' high		3	5.333			250		250	410
3445	up to 35' x 30' high		1	16			750		750	1,225
3500	Residential, 9' x 7' high		8	2			94		94	154
3540	16' x 7' high		7	2.286			107		107	176
3600	Remove and reset, small		4	4			188		188	310
3620	Large		2.50	6.400			300		300	490
3700	Roll-up grille		5	3.200			150		150	246
3800	Revolving door		2	8			375		375	615
3900	Storefront swing door		3	5.333			250		250	410
3902	Cafe/bar swing door	2 Clab	8	2			75		75	123
4000	Residential lockset, exterior	1 Carp	28	.286			13.40		13.40	22
4224	Pocket door		8	1			47		47	77
4300	Remove and reset deadbolt		8	1			47		47	77
5585	Remove panic device	1 Clab	10	.800			30		30	49.50
5590	Remove mail slot		45	.178			6.70		6.70	10.95
5595	Remove peep hole		45	.178			6.70		6.70	10.95
5600	Remove door sidelight	1 Carp	6	1.333			62.50		62.50	103
5604	Remove 9' x 7' garage door	2 Carp	9.50	1.684			79		79	130
5624	16' x 7' garage door		8	2			94		94	154
5626	9' x 7' swing up garage door		7	2.286			107		107	176
5628	16' x 7' swing up garage door		7	2.286			107		107	176
5630	Garage door demolition, remove garage door track	1 Carp	6	1.333			62.50		62.50	103
5644	Remove overhead door opener	1 Clab	5	1.600			60		60	98.50
5774	Remove heavy gage sectional door, 20 ga., 8' x 8'	2 Carp	7	2.286			107		107	176
5784	10' x 10'		6	2.667			125		125	205
5794	12' x 12'		5	3.200			150		150	246
5805	14' x 14'		4	4			188		188	310
5850	8' x 8' sliding door & frame	B-68D	1.60	15			620	207	827	1,225

08 05 Common Work Results for Openings

08 05 05 – Selective Demolition for Openings

08 05 05.10 Selective Demolition Doors		Crew	Daily Output	Labor-Hours	Unit	2015 Bare Costs Material	Labor	Equipment	Total	Total Incl O&P
5860	12′ x 22′ wand fire door and frame	B-68D	.64	37.500	Ea.		1,550	520	2,070	3,100
6334	Remove shower door unit	1 Clab	16	.500			18.80		18.80	31
6384	Remove French door unit	"	8	1			37.50		37.50	61.50
6600	Demo flexible transparent strip entrance	3 Shee	115	.209	SF Surf		11.70		11.70	18.45
7100	Remove double swing pneumatic doors, openers and sensors	2 Skwk	.50	32	Opng.		1,550		1,550	2,525
7110	Remove automatic operators, industrial, sliding doors, to 12′ wide	"	.40	40	"		1,950		1,950	3,150
7550	Hangar door demo	2 Sswk	330	.048	S.F.		2.55		2.55	4.61
7570	Remove shock absorbing door	"	1.90	8.421	Opng.		445		445	800
9000	Minimum labor/equipment charge	1 Carp	4	2	Job		94		94	154

08 05 05.20 Selective Demolition of Windows		Crew	Daily Output	Labor-Hours	Unit	Material	Labor	Equipment	Total	Total Incl O&P
0010	**SELECTIVE DEMOLITION OF WINDOWS** R024119-10									
0200	Aluminum, including trim, to 12 S.F.	1 Clab	16	.500	Ea.		18.80		18.80	31
0240	To 25 S.F.		11	.727			27.50		27.50	45
0280	To 50 S.F.		5	1.600			60		60	98.50
0320	Storm windows/screens, to 12 S.F.		27	.296			11.15		11.15	18.25
0360	To 25 S.F.		21	.381			14.30		14.30	23.50
0400	To 50 S.F.		16	.500			18.80		18.80	31
0600	Glass, up to 10 SF per window		200	.040	S.F.		1.50		1.50	2.47
0620	Over 10 SF per window		150	.053	"		2.01		2.01	3.29
1000	Steel, including trim, to 12 S.F.		13	.615	Ea.		23		23	38
1020	To 25 S.F.		9	.889			33.50		33.50	55
1040	To 50 S.F.		4	2			75		75	123
2000	Wood, including trim, to 12 S.F.		22	.364			13.65		13.65	22.50
2020	To 25 S.F.		18	.444			16.70		16.70	27.50
2060	To 50 S.F.		13	.615			23		23	38
2065	To 180 S.F.		8	1			37.50		37.50	61.50
4300	Remove bay/bow window	2 Carp	6	2.667			125		125	205
4400	Remove skylight, prefabricated glass block with metal frame	G-3	180	.178	S.F.		8.30		8.30	13.35
4410	Remove skylight, plstc domes, flush/curb mtd	"	395	.081	"		3.79		3.79	6.05
5020	Remove and reset window, up to a 2′x2′ widow	1 Carp	6	1.333	Ea.		62.50		62.50	103
5040	Up to a 3′x3′ window		4	2			94		94	154
5080	Up to a 4′x5′ window		2	4			188		188	310
6000	Screening only	1 Clab	4000	.002	S.F.		.08		.08	.12
9000	Minimum labor/equipment charge	"	4	2	Job		75		75	123

08 11 Metal Doors and Frames

08 11 16 – Aluminum Doors and Frames

08 11 16.10 Entrance Doors		Crew	Daily Output	Labor-Hours	Unit	Material	Labor	Equipment	Total	Total Incl O&P
0010	**ENTRANCE DOORS** and frame, Aluminum, narrow stile									
0011	Including standard hardware, clear finish, no glass									
0012	Top and bottom offset pivots, 1/4″ beveled glass stops, threshold									
0013	Dead bolt lock with inside thumb screw, standard push pull									
0020	3′-0″ x 7′-0″ opening	2 Sswk	2	8	Ea.	915	420		1,335	1,750
0025	Anodizing aluminum entr. door & frame, add					104			104	114
0030	3′-6″ x 7′-0″ opening	2 Sswk	2	8		840	420		1,260	1,675
0100	3′-0″ x 10′-0″ opening, 3′ high transom		1.80	8.889		1,300	470		1,770	2,275
0200	3′-6″ x 10′-0″ opening, 3′ high transom		1.80	8.889		1,350	470		1,820	2,350
0280	5′-0″ x 7′-0″ opening		2	8		1,425	420		1,845	2,325
0300	6′-0″ x 7′-0″ opening		1.30	12.308		1,200	650		1,850	2,500
0301	6′-0″ x 7′-0″ opening		1.30	12.308	Pr.	1,200	650		1,850	2,500
0400	6′-0″ x 10′-0″ opening, 3′ high transom		1.10	14.545	"	1,675	765		2,440	3,200

08 11 Metal Doors and Frames

08 11 16 – Aluminum Doors and Frames

08 11 16.10 Entrance Doors		Crew	Daily Output	Labor-Hours	Unit	Material	2015 Bare Costs Labor	Equipment	Total	Total Incl O&P
0520	3′-0″ x 7′-0″ opening, wide stile	2 Sswk	2	8	Ea.	1,000	420		1,420	1,850
0540	3′-6″ x 7′-0″ opening		2	8		1,200	420		1,620	2,075
0560	5′-0″ x 7′-0″ opening		2	8		1,525	420		1,945	2,425
0580	6′-0″ x 7′-0″ opening		1.30	12.308	Pr.	1,575	650		2,225	2,925
0600	7′-0″ x 7′-0″ opening		1	16	″	1,725	840		2,565	3,400
1200	For non-standard size, add				Leaf	80%				
1250	For installation of non-standard size, add						20%			
1300	Light bronze finish, add				Leaf	36%				
1400	Dark bronze finish, add					25%				
1500	For black finish, add					40%				
1600	Concealed panic device, add					940			940	1,025
1700	Electric striker release, add				Opng.	280			280	310
1800	Floor check, add				Leaf	650			650	710
1900	Concealed closer, add				″	530			530	580
9000	Minimum labor/equipment charge	2 Carp	4	4	Job		188		188	310

08 11 63 – Metal Screen and Storm Doors and Frames

08 11 63.23 Aluminum Screen and Storm Doors and Frames		Crew	Daily Output	Labor-Hours	Unit	Material	2015 Bare Costs Labor	Equipment	Total	Total Incl O&P
0010	**ALUMINUM SCREEN AND STORM DOORS AND FRAMES**									
0020	Combination storm and screen									
0420	Clear anodic coating, 2′-8″ wide	2 Carp	14	1.143	Ea.	205	53.50		258.50	315
0440	3′-0″ wide	″	14	1.143	″	178	53.50		231.50	283
0500	For 7′ door height, add					8%				
1020	Mill finish, 2′-8″ wide	2 Carp	14	1.143	Ea.	235	53.50		288.50	345
1040	3′-0″ wide	″	14	1.143		258	53.50		311.50	370
1100	For 7′-0″ door, add					8%				
1520	White painted, 2′-8″ wide	2 Carp	14	1.143		286	53.50		339.50	405
1540	3′-0″ wide	″	14	1.143		310	53.50		363.50	430
1600	For 7′-0″ door, add					8%				
2000	Wood door & screen, see Section 08 14 33.20									
9000	Minimum labor/equipment charge	1 Carp	4	2	Job		94		94	154

08 12 Metal Frames

08 12 13 – Hollow Metal Frames

08 12 13.13 Standard Hollow Metal Frames		Crew	Daily Output	Labor-Hours	Unit	Material	2015 Bare Costs Labor	Equipment	Total	Total Incl O&P
0010	**STANDARD HOLLOW METAL FRAMES**									
0020	16 ga., up to 5-3/4″ jamb depth									
0025	3′-0″ x 6′-8″ single G	2 Carp	16	1	Ea.	146	47		193	237
0028	3′-6″ wide, single G		16	1		153	47		200	246
0030	4′-0″ wide, single G		16	1		152	47		199	244
0040	6′-0″ wide, double G		14	1.143		204	53.50		257.50	310
0045	8′-0″ wide, double G		14	1.143		213	53.50		266.50	320
0100	3′-0″ x 7′-0″ single		16	1		151	47		198	243
0110	3′-6″ wide, single		16	1		159	47		206	251
0112	4′-0″ wide, single		16	1		159	47		206	251
0140	6′-0″ wide, double		14	1.143		194	53.50		247.50	300
0145	8′-0″ wide, double		14	1.143		229	53.50		282.50	340
1000	16 ga., up to 4-7/8″ deep, 3′-0″ x 7′-0″ single G		16	1		166	47		213	260
1140	6′-0″ wide, double G		14	1.143		188	53.50		241.50	295
1200	16 ga., 8-3/4″ deep, 3′-0″ x 7′-0″ single G		16	1		199	47		246	296
1240	6′-0″ wide, double G		14	1.143		234	53.50		287.50	345

08 12 13.13 Standard Hollow Metal Frames			Crew	Daily Output	Labor-Hours	Unit	2015 Bare Costs Material	Labor	Equipment	Total	Total Incl O&P
2800	14 ga., up to 3-7/8" deep, 3'-0" x 7'-0" single	G	2 Carp	16	1	Ea.	181	47		228	276
2840	6'-0" wide, double	G		14	1.143		217	53.50		270.50	325
2900	14 ga., 4-3/4" deep, 2'-0" x 7'-0" high, single	G		15	1.067		189	50		239	290
2910	2'-4" wide	G		15	1.067		195	50		245	297
2920	2'-6" wide	G		15	1.067		195	50		245	297
2930	2'-8" wide	G		15	1.067		196	50		246	298
2940	3'-0" wide	G		15	1.067		198	50		248	299
2950	3'-4" wide	G		15	1.067		201	50		251	305
2960	6'-0" wide, double	G		12	1.333		240	62.50		302.50	365
3000	14 ga., up to 5-3/4" deep, 3'-0" x 6'-8" single	G		16	1		155	47		202	248
3002	3'-6" wide, single	G		16	1		198	47		245	294
3005	4'-0" wide, single	G		16	1		203	47		250	300
3600	up to 5-3/4" jamb depth, 4'-0" x 7'-0" single			15	1.067		185	50		235	285
3620	6'-0" wide, double			12	1.333		235	62.50		297.50	360
3640	8'-0" wide, double			12	1.333		246	62.50		308.50	375
3700	8'-0" high, 4'-0" wide, single			15	1.067		235	50		285	340
3740	8'-0" wide, double			12	1.333		290	62.50		352.50	425
4000	6-3/4" deep, 4'-0" x 7'-0" single	G		15	1.067		222	50		272	325
4020	6'-0" wide, double	G		12	1.333		276	62.50		338.50	410
4040	8'-0", wide double	G		12	1.333		284	62.50		346.50	415
4100	8'-0" high, 4'-0" wide, single	G		15	1.067		276	50		326	385
4140	8'-0" wide, double	G		12	1.333		320	62.50		382.50	460
4400	8-3/4" deep, 4'-0" x 7'-0" , single			15	1.067		252	50		302	360
4440	8'-0" wide, double			12	1.333		315	62.50		377.50	450
4500	4'-0" x 8'-0" , single			15	1.067		283	50		333	390
4540	8'-0" wide, double			12	1.333		350	62.50		412.50	490
4900	For welded frames, add						63			63	69.50
5380	Steel frames, KD, 14 ga., "B" label, to 5-3/4" throat, to 3'-0" x 7'-0"		2 Carp	15	1.067		215	50		265	320
5400	14 ga., "B" label, up to 5-3/4" deep, 4'-0" x 7'-0" single	G		15	1.067		210	50		260	315
5440	8'-0" wide, double	G		12	1.333		272	62.50		334.50	400
5800	6-3/4" deep, 7'-0" high, 4'-0" wide, single	G		15	1.067		218	50		268	320
5840	8'-0" wide, double	G		12	1.333		385	62.50		447.50	530
6200	8-3/4" deep, 4'-0" x 7'-0" single	G		15	1.067		291	50		341	400
6240	8'-0" wide, double	G		12	1.333		380	62.50		442.50	520
6300	For "A" label use same price as "B" label										
6400	For baked enamel finish, add						30%	15%			
6500	For galvanizing, add						20%				
6600	For hospital stop, add					Ea.	295			295	325
6620	For hospital stop, stainless steel add					"	380			380	420
7900	Transom lite frames, fixed, add		2 Carp	155	.103	S.F.	54	4.85		58.85	67.50
8000	Movable, add		"	130	.123	"	68	5.80		73.80	84.50
9000	Minimum labor/equipment charge		1 Carp	4	2	Job		94		94	154

08 12 13.18 Transoms and Sidelights		Crew	Daily Output	Labor-Hours	Unit	Material	Labor	Equipment	Total	Total Incl O&P
0010	**TRANSOMS & SIDELIGHTS**									
0160	Transom sash, 6-3/4" thick, 3'-0" x 1'-4"	2 Carp	11	1.455	Ea.	181	68.50		249.50	310
0180	3'-4" x 1'-4"		11	1.455		181	68.50		249.50	310
0200	6'-0" x 1'-4"		6	2.667		210	125		335	435
0220	4-3/4" thick, 3'-0" x 1'-4"		11	1.455		175	68.50		243.50	305
0240	3'-4" x 1'-4"		11	1.455		175	68.50		243.50	305
0260	6'-0" x 1'-4"		6	2.667		208	125		333	435

08 12 Metal Frames

08 12 13 – Hollow Metal Frames

08 12 13.20 Wrap Around Drywall Frames		Crew	Daily Output	Labor-Hours	Unit	Material	Labor	Equipment	Total	Total Incl O&P
						2015 Bare Costs				
0010	**WRAP AROUND DRYWALL FRAMES**									
0400	Wrap-around drywall frame, 16 ga., 6-1/4" x 7'-0" x 2'-6" wide	2 Carp	16	1	Ea.	169	47		216	263
0430	2'-8" wide		16	1		169	47		216	263
0460	3'-0" wide		15	1.067		169	50		219	268
0490	3'-6" wide		15	1.067		177	50		227	276
0520	5'-0" wide		14	1.143		207	53.50		260.50	315
0580	6'-0" wide		14	1.143		210	53.50		263.50	320
0610	7'-0" wide		13	1.231		212	58		270	330
2000	18 ga., 4-3/4" x 7'-0 x 2'-6" wide		16	1		159	47		206	252
2030	2'-8" wide		16	1		159	47		206	252
2060	3'-0" wide		15	1.067		159	50		209	257
2090	3'-6" wide		15	1.067		163	50		213	261
2120	5'-0" wide		14	1.143		170	53.50		223.50	275
2150	5'-4" wide		14	1.143		194	53.50		247.50	300
2200	6'-0" wide		14	1.143		194	53.50		247.50	300
2250	7'-0" wide		13	1.231		199	58		257	315
3700	Fire door frames, 6-1/4" x 7'-0" x 2'-6" wide		16	1		202	47		249	300
3730	2'-8" wide		16	1		202	47		249	300
3760	3'-0" wide		15	1.067		203	50		253	305
3820	5'-0" wide		14	1.143		234	53.50		287.50	345
3850	5'-4" wide		14	1.143		238	53.50		291.50	350
3880	6'-0" wide		14	1.143		238	53.50		291.50	350
9000	Minimum labor/equipment charge	1 Carp	2	4	Job		188		188	310

08 12 13.25 Channel Metal Frames		Crew	Daily Output	Labor-Hours	Unit	Material	Labor	Equipment	Total	Total Incl O&P
0010	**CHANNEL METAL FRAMES**									
0020	Steel channels with anchors and bar stops									
0100	6" channel @ 8.2#/L.F., 3' x 7' door, weighs 150# G	E-4	13	2.462	Ea.	239	131	11.20	381.20	510
0200	8" channel @ 11.5#/L.F., 6' x 8' door, weighs 275# G		9	3.556		435	189	16.20	640.20	840
0300	8' x 12' door, weighs 400# G		6.50	4.923		635	262	22.50	919.50	1,200
0400	10" channel @ 15.3#/L.F., 10' x 10' door, weighs 500# G		6	5.333		795	283	24.50	1,102.50	1,400
0500	12' x 12' door, weighs 600# G		5.50	5.818		955	310	26.50	1,291.50	1,650
0600	12" channel @ 20.7#/L.F., 12' x 12' door, weighs 825# G		4.50	7.111		1,300	380	32.50	1,712.50	2,175
0700	12' x 16' door, weighs 1000# G		4	8		1,600	425	36.50	2,061.50	2,550
0800	For frames without bar stops, light sections, deduct					15%				
0900	Heavy sections, deduct					10%				
9000	Minimum labor/equipment charge	E-4	4	8	Job		425	36.50	461.50	810

08 12 13.53 Borrowed Lites		Crew	Daily Output	Labor-Hours	Unit	Material	Labor	Equipment	Total	Total Incl O&P
0010	**BORROWED LITES**									
0100	Hollow metal section 20 ga., 3-1/2" with glass stop	2 Carp	100	.160	L.F.	8.15	7.50		15.65	21.50
0110	3-3/4" with glass stop		100	.160		8.30	7.50		15.80	21.50
0120	4" with glass stop		100	.160		8.45	7.50		15.95	21.50
0130	4-5/8" with glass stop		100	.160		8.75	7.50		16.25	22
0140	4-7/8" with glass stop		100	.160		8.90	7.50		16.40	22
0150	5" with glass stop		100	.160		9	7.50		16.50	22
0160	5-3/8" with glass stop		100	.160		9.25	7.50		16.75	22.50
0300	Hollow metal section 18 ga., 3-1/2" with glass stop		80	.200		9	9.40		18.40	25.50
0310	3-3/4" with glass stop		80	.200		9.20	9.40		18.60	25.50
0320	4" with glass stop		80	.200		9.30	9.40		18.70	25.50
0330	4-5/8" with glass stop		80	.200		9.75	9.40		19.15	26
0340	4-7/8" with glass stop		80	.200		10	9.40		19.40	26.50
0350	5" with glass stop		80	.200		10.10	9.40		19.50	26.50
0360	5-3/8" with glass stop		80	.200		10.35	9.40		19.75	27

08 12 13 – Hollow Metal Frames

08 12 13.53 Borrowed Lites		Crew	Daily Output	Labor-Hours	Unit	Material	2015 Bare Costs Labor	Equipment	Total	Total Incl O&P
0370	6-5/8" with glass stop	2 Carp	80	.200	L.F.	11.25	9.40		20.65	28
0380	6-7/8" with glass stop		80	.200		11.40	9.40		20.80	28
0390	7-1/4" with glass stop		80	.200		11.90	9.40		21.30	28.50
0500	Mullion section 18 ga., 3-1/2" with double stop		50	.320		21	15		36	47.50
0510	3-3/4" with double stop		50	.320		21.50	15		36.50	48
0530	4-5/8" with double stop		50	.320		23	15		38	49.50
0540	4-7/8" with double stop		50	.320		23.50	15		38.50	50
0550	5" with double stop		50	.320		23.50	15		38.50	50
0560	5-3/8" with double stop		50	.320		24	15		39	51
0570	6-5/8" with double stop		50	.320		26	15		41	53.50
0580	6-7/8" with double stop		50	.320		26.50	15		41.50	54
0590	7-1/4" with double stop		50	.320		27	15		42	54.50
1000	Assembled frame 2'-0" x 2'-0" 16 ga., 3-3/4" with glass stop		10	1.600	Ea.	180	75		255	320
1010	3'-0" x 2'-0"		10	1.600		180	75		255	320
1020	4'-0" x 2'-0"		10	1.600		180	75		255	320
1030	5'-0" x 2'-0"		10	1.600		180	75		255	320
1040	6'-0" x 2'-0"		8	2		270	94		364	450
1050	7'-0" x 2'-0"		8	2		270	94		364	450
1060	8'-0" x 2'-0"		8	2		270	94		364	450
1100	2'-0" x 7'-0"		8	2		270	94		364	450
1110	3'-0" x 7'-0"		8	2		270	94		364	450
1120	4'-0" x 7'-0"		8	2		270	94		364	450
1130	5'-0" x 7'-0"		8	2		270	94		364	450
1140	6'-0" x 7'-0"		6	2.667		295	125		420	530
1150	7'-0" x 7'-0"		6	2.667		295	125		420	530
1160	8'-0" x 7'-0"		6	2.667		295	125		420	530
1200	Assembled frame 2'-0" x 2'-0" 16 ga., 4-7/8" with glass stop		10	1.600		180	75		255	320
1210	3'-0" x 2'-0"		10	1.600		180	75		255	320
1220	4'-0" x 2'-0"		8	2		180	94		274	350
1230	5'-0" x 2'-0"		8	2		180	94		274	350
1240	6'-0" x 2'-0"		8	2		270	94		364	450
1250	7'-0" x 2'-0"		8	2		270	94		364	450
1260	8'-0" x 2'-0"		8	2		270	94		364	450
1300	2'-0" x 7'-0"		8	2		270	94		364	450
1310	3'-0" x 7'-0"		8	2		270	94		364	450
1320	4'-0" x 7'-0"		8	2		270	94		364	450
1330	5'-0" x 7'-0"		6	2.667		270	125		395	500
1340	6'-0" x 7'-0"		6	2.667		295	125		420	530
1350	7'-0" x 7'-0"		6	2.667		270	125		395	500
1360	8'-0" x 7'-0"		6	2.667		295	125		420	530
1400	Assembled frame 2'-0" x 3'-0" 16 ga., 6-1/8" with glass stop		6	2.667		180	125		305	405
1410	3'-0" x 3'-0"		8	2		180	94		274	350
1420	4'-0" x 3'-0"		8	2		180	94		274	350
1430	5'-0" x 3'-0"		6	2.667		180	125		305	405
1440	6'-0" x 3'-0"		6	2.667		270	125		395	500
1450	7'-0" x 3'-0"		6	2.667		270	125		395	500
1460	8'-0" x 3'-0"	↓	10	1.600	↓	270	75		345	420

08 13 Metal Doors

08 13 13 – Hollow Metal Doors

08 13 13.13 Standard Hollow Metal Doors			Crew	Daily Output	Labor-Hours	Unit	2015 Bare Costs Material	Labor	Equipment	Total	Total Incl O&P
0010	**STANDARD HOLLOW METAL DOORS** R081313-20										
0015	Flush, full panel, hollow core										
0017	When noted doors are prepared but do not include glass or louvers										
0020	1-3/8″ thick, 20 ga., 2′-0″ x 6′-8″	G	2 Carp	20	.800	Ea.	335	37.50		372.50	430
0040	2′-8″ x 6′-8″	G		18	.889		350	41.50		391.50	455
0060	3′-0″ x 6′-8″	G		17	.941		350	44		394	460
0100	3′-0″ x 7′-0″			17	.941		360	44		404	470
0120	For vision lite, add						94.50			94.50	104
0140	For narrow lite, add						102			102	113
0320	Half glass, 20 ga., 2′-0″ x 6′-8″	G	2 Carp	20	.800		490	37.50		527.50	600
0340	2′-8″ x 6′-8″	G		18	.889		515	41.50		556.50	635
0360	3′-0″ x 6′-8″			17	.941		510	44		554	635
0400	3′-0″ x 7′-0″	G		17	.941		625	44		669	765
0410	1-3/8″ thick, 18 ga., 2′-0″ x 6′-8″	G		20	.800		400	37.50		437.50	500
0420	3′-0″ x 6′-8″	G		17	.941		405	44		449	520
0425	3′-0″ x 7′-0″			17	.941		415	44		459	535
0450	For vision lite, add						94.50			94.50	104
0452	For narrow lite, add						102			102	113
0460	Half glass, 18 ga., 2′-0″ x 6′-8″	G	2 Carp	20	.800		555	37.50		592.50	675
0465	2′-8″ x 6′-8″	G		18	.889		575	41.50		616.50	705
0470	3′-0″ x 6′-8″	G		17	.941		565	44		609	695
0475	3′-0″ x 7′-0″	G		17	.941		565	44		609	700
0500	Hollow core, 1-3/4″ thick, full panel, 20 ga., 2′-8″ x 6′-8″	G		18	.889		420	41.50		461.50	530
0520	3′-0″ x 6′-8″	G		17	.941		420	44		464	535
0640	3′-0″ x 7′-0″	G		17	.941		435	44		479	555
0680	4′-0″ x 7′-0″	G		15	1.067		630	50		680	775
0700	4′-0″ x 8′-0″	G		13	1.231		735	58		793	905
1000	18 ga., 2′-8″ x 6′-8″	G		17	.941		490	44		534	615
1020	3′-0″ x 6′-8″	G		16	1		475	47		522	600
1120	3′-0″ x 7′-0″	G		17	.941		515	44		559	640
1150	3′-6″ x 7′-0″		1 Carp	14	.571		570	27		597	675
1180	4′-0″ x 7′-0″	G	2 Carp	14	1.143		630	53.50		683.50	785
1200	4′-0″ x 8′-0″	G	″	17	.941		735	44		779	885
1212	For vision lite, add						94.50			94.50	104
1214	For narrow lite, add						102			102	113
1230	Half glass, 20 ga., 2′-8″ x 6′-8″	G	2 Carp	20	.800		575	37.50		612.50	695
1240	3′-0″ x 6′-8″	G		18	.889		575	41.50		616.50	705
1260	3′-0″ x 7′-0″	G		18	.889		595	41.50		636.50	720
1280	Embossed panel, 1-3/4″ thick, poly core, 20 ga., 3′-0″ x 7′-0″	G		18	.889		415	41.50		456.50	525
1290	Half glass, 1-3/4″ thick, poly core, 20 ga., 3′-0″ x 7′-0″	G		18	.889		610	41.50		651.50	740
1320	18 ga., 2′-8″ x 6′-8″	G		18	.889		640	41.50		681.50	775
1340	3′-0″ x 6′-8″	G		17	.941		635	44		679	770
1360	3′-0″ x 7′-0″	G		17	.941		645	44		689	785
1380	4′-0″ x 7′-0″			15	1.067		785	50		835	945
1400	4′-0″ x 8′-0″	G		14	1.143		890	53.50		943.50	1,075
1500	Flush full panel, 16 ga., steel hollow core										
1520	2′-0″ x 6′-8″	G	2 Carp	20	.800	Ea.	555	37.50		592.50	670
1530	2′-8″ x 6′-8″	G		20	.800		555	37.50		592.50	675
1540	3′-0″ x 6′-8″	G		20	.800		550	37.50		587.50	665
1560	2′-8″ x 7′-0″	G		18	.889		575	41.50		616.50	700
1570	3′-0″ x 7′-0″	G		18	.889		560	41.50		601.50	685
1580	3′-6″ x 7′-0″	G		18	.889		655	41.50		696.50	790
1590	4′-0″ x 7′-0″	G		18	.889		725	41.50		766.50	865

08 13 13.13 Standard Hollow Metal Doors			Crew	Daily Output	Labor-Hours	Unit	2015 Bare Costs Material	Labor	Equipment	Total	Total Incl O&P
1600	2'-8" x 8'-0"	G	2 Carp	18	.889	Ea.	700	41.50		741.50	840
1620	3'-0" x 8'-0"	G		18	.889		720	41.50		761.50	860
1630	3'-6" x 8'-0"	G		18	.889		800	41.50		841.50	950
1640	4'-0" x 8'-0"	G		18	.889		850	41.50		891.50	1,000
1650	1-13/16", 14 Ga., 2'-8" x 7'-0"	G		10	1.600		1,125	75		1,200	1,375
1670	3'-0" x 7'-0"	G		10	1.600		1,075	75		1,150	1,325
1690	3'-6" x 7'-0"	G		10	1.600		1,200	75		1,275	1,425
1700	4'-0" x 7'-0"	G		10	1.600		1,250	75		1,325	1,500
1720	Insulated, 1-3/4" thick, full panel, 18 ga., 3'-0" x 6'-8"	G		15	1.067		475	50		525	605
1740	2'-8" x 7'-0"	G		16	1		505	47		552	630
1760	3'-0" x 7'-0"	G		15	1.067		490	50		540	620
1800	4'-0" x 8'-0"	G		13	1.231		740	58		798	905
1820	Half glass, 18 ga., 3'-0" x 6'-8"	G		16	1		635	47		682	770
1840	2'-8" x 7'-0"	G		17	.941		660	44		704	800
1860	3'-0" x 7'-0"	G		16	1		685	47		732	830
1900	4'-0" x 8'-0"	G		14	1.143		670	53.50		723.50	825
2000	For vision lite, add						94.50			94.50	104
2010	For narrow lite, add						102			102	113
8100	For bottom louver, add						282			282	310
8110	For baked enamel finish, add						30%	15%			
8120	For galvanizing, add						20%				
8170	For soundproofing STC 40, add					Ea.	1,725			1,725	1,900
8190	For 3 hour door, add						370			370	405
8270	For dutch door with shelf, add to standard door						300			300	330
9000	Minimum labor/equipment charge		1 Carp	4	2	Job		94		94	154

08 13 13.15 Metal Fire Doors			Crew	Daily Output	Labor-Hours	Unit	Material	Labor	Equipment	Total	Total Incl O&P
0010	**METAL FIRE DOORS**	R081313-20									
0015	Steel, flush, "B" label, 90 minute										
0020	Full panel, 20 ga., 2'-0" x 6'-8"		2 Carp	20	.800	Ea.	420	37.50		457.50	520
0040	2'-8" x 6'-8"			18	.889		435	41.50		476.50	550
0060	3'-0" x 6'-8"			17	.941		435	44		479	555
0080	3'-0" x 7'-0"			17	.941		455	44		499	575
0140	18 ga., 3'-0" x 6'-8"			16	1		495	47		542	620
0160	2'-8" x 7'-0"			17	.941		520	44		564	645
0180	3'-0" x 7'-0"			16	1		505	47		552	630
0200	4'-0" x 7'-0"			15	1.067		650	50		700	795
0220	For "A" label, 3 hour, 18 ga., use same price as "B" label										
0240	For vision lite, add					Ea.	156			156	172
0300	Full panel, 16 ga., 2'-0" x 6'-8"		2 Carp	20	.800		545	37.50		582.50	660
0310	2'-8" x 6'-8"			18	.889		545	41.50		586.50	670
0320	3'-0" x 6'-8"			17	.941		540	44		584	670
0350	2'-8" x 7'-0"			17	.941		565	44		609	695
0360	3'-0" x 7'-0"			16	1		545	47		592	675
0370	4'-0" x 7'-0"			15	1.067		705	50		755	860
0520	Flush, "B" label 90 min., egress core, 20 ga., 2'-0" x 6'-8"			18	.889		655	41.50		696.50	790
0540	2'-8" x 6'-8"			17	.941		665	44		709	805
0560	3'-0" x 6'-8"			16	1		665	47		712	805
0580	3'-0" x 7'-0"			16	1		685	47		732	825
0640	Flush, "A" label 3 hour, egress core, 18 ga., 3'-0" x 6'-8"			15	1.067		720	50		770	875
0660	2'-8" x 7'-0"			16	1		750	47		797	895
0680	3'-0" x 7'-0"			15	1.067		740	50		790	895
0700	4'-0" x 7'-0"			14	1.143		885	53.50		938.50	1,050

08 13 Metal Doors

08 13 13 – Hollow Metal Doors

08 13 13.15 Metal Fire Doors		Crew	Daily Output	Labor-Hours	Unit	Material	2015 Bare Costs Labor	Equipment	Total	Total Incl O&P
9000	Minimum labor/equipment charge	1 Carp	4	2	Job		94		94	154

08 13 13.20 Residential Steel Doors			Crew	Daily Output	Labor-Hours	Unit	Material	Labor	Equipment	Total	Total Incl O&P
0010	RESIDENTIAL STEEL DOORS										
0020	Prehung, insulated, exterior										
0030	Embossed, full panel, 2'-8" x 6'-8"	G	2 Carp	17	.941	Ea.	315	44		359	420
0040	3'-0" x 6'-8"	G		15	1.067		270	50		320	380
0070	5'-4" x 6'-8", double	G		8	2		630	94		724	850
0220	Half glass, 2'-8" x 6'-8"	G		17	.941		320	44		364	425
0240	3'-0" x 6'-8"	G		16	1		320	47		367	425
0260	3'-0" x 7'-0"	G		16	1		370	47		417	480
0270	5'-4" x 6'-8", double	G		8	2		650	94		744	870
1320	Flush face, full panel, 2'-8" x 6'-8"	G		16	1		264	47		311	365
1340	3'-0" x 6'-8"	G		15	1.067		264	50		314	370
1360	3'-0" x 7'-0"	G		15	1.067		296	50		346	405
1380	5'-4" x 6'-8", double	G		8	2		560	94		654	770
1420	Half glass, 2'-8" x 6'-8"	G		17	.941		325	44		369	430
1440	3'-0" x 6'-8"	G		16	1		325	47		372	430
1460	3'-0" x 7'-0"	G		16	1		360	47		407	470
1480	5'-4" x 6'-8", double	G		8	2		640	94		734	860
1500	Sidelight, full lite, 1'-0" x 6'-8" with grille	G					252			252	277
1510	1'-0" x 6'-8", low e	G					268			268	295
1520	1'-0" x 6'-8", half lite	G					280			280	310
1530	1'-0" x 6'-8", half lite, low e	G					284			284	310
2300	Interior, residential, closet, bi-fold, 2'-0" x 6'-8"	G	2 Carp	16	1		161	47		208	254
2330	3'-0" wide	G		16	1		200	47		247	297
2360	4'-0" wide	G		15	1.067		260	50		310	370
2400	5'-0" wide	G		14	1.143		315	53.50		368.50	440
2420	6'-0" wide	G		13	1.231		335	58		393	465
9000	Minimum labor/equipment charge		1 Carp	4	2	Job		94		94	154

08 13 13.25 Doors Hollow Metal			Crew	Daily Output	Labor-Hours	Unit	Material	Labor	Equipment	Total	Total Incl O&P
0010	DOORS HOLLOW METAL										
0500	Exterior, commercial, flush, 20 ga., 1-3/4" x 7'-0" x 2'-6" wide	G	2 Carp	15	1.067	Ea.	395	50		445	510
0530	2'-8" wide	G		15	1.067		435	50		485	555
0560	3'-0" wide	G		14	1.143		435	53.50		488.50	565
0590	3'-6" wide	G		14	1.143		440	53.50		493.50	575
1000	18 ga., 1-3/4" x 7'-0" x 2'-6" wide	G		15	1.067		495	50		545	625
1030	2'-8" wide	G		15	1.067		500	50		550	630
1060	3'-0" wide	G		14	1.143		485	53.50		538.50	625
1500	16 ga., 1-3/4 x 7'-0" x 2'-6" wide	G		15	1.067		560	50		610	695
1530	2'-8" wide	G		15	1.067		565	50		615	705
1560	3'-0" wide	G		14	1.143		555	53.50		608.50	700
1590	3'-6" wide	G		14	1.143		645	53.50		698.50	800
2900	Fire door, "A" label, 18 gauge, 1-3/4" x 2'-6" x 7'-0"	G		15	1.067		645	50		695	790
2930	2'-8" wide	G		15	1.067		665	50		715	810
2960	3'-0" wide	G		14	1.143		640	53.50		693.50	795
2990	3'-6" wide	G		14	1.143		735	53.50		788.50	895
3100	"B" label, 2'-6" wide	G		15	1.067		585	50		635	725
3130	2'-8" wide	G		15	1.067		590	50		640	730
3160	3'-0" wide	G		14	1.143		580	53.50		633.50	730

08 13 Metal Doors

08 13 16 – Aluminum Doors

08 13 16.10 Commercial Aluminum Doors		Crew	Daily Output	Labor-Hours	Unit	2015 Bare Costs Material	Labor	Equipment	Total	Total Incl O&P
0010	**COMMERCIAL ALUMINUM DOORS**, flush, no glazing									
5000	Flush panel doors, pair of 2'-6" x 7'-0"	2 Sswk	2	8	Pr.	1,475	420		1,895	2,350
5050	3'-0" x 7'-0", single		2.50	6.400	Ea.	825	335		1,160	1,525
5100	Pair of 3'-0" x 7'-0"		2	8	Pr.	1,550	420		1,970	2,475
5150	3'-6" x 7'-0", single		2.50	6.400	Ea.	985	335		1,320	1,675

08 14 Wood Doors

08 14 13 – Carved Wood Doors

08 14 13.10 Types of Wood Doors, Carved

		Crew	Daily Output	Labor-Hours	Unit	Material	Labor	Equipment	Total	Total Incl O&P
0010	**TYPES OF WOOD DOORS, CARVED**									
3000	Solid wood, 1-3/4" thick stile and rail									
3020	Mahogany, 3'-0" x 7'-0", six panel	2 Carp	14	1.143	Ea.	1,125	53.50		1,178.50	1,350
3030	With two lites		10	1.600		1,850	75		1,925	2,150
3040	3'-6" x 8'-0", six panel		10	1.600		1,500	75		1,575	1,775
3050	With two lites		8	2		2,500	94		2,594	2,900
3100	Pine, 3'-0" x 7'-0", six panel		14	1.143		525	53.50		578.50	670
3110	With two lites		10	1.600		825	75		900	1,025
3120	3'-6" x 8'-0", six panel		10	1.600		920	75		995	1,125
3130	With two lites		8	2		1,825	94		1,919	2,175
3200	Red oak, 3'-0" x 7'-0", six panel		14	1.143		1,750	53.50		1,803.50	2,000
3210	With two lites		10	1.600		2,300	75		2,375	2,650
3220	3'-6" x 8'-0", six panel		10	1.600		2,600	75		2,675	2,975
3230	With two lites		8	2		3,300	94		3,394	3,775
4000	Hand carved door, mahogany									
4020	3'-0" x 7'-0", simple design	2 Carp	14	1.143	Ea.	1,750	53.50		1,803.50	2,025
4030	Intricate design		11	1.455		3,700	68.50		3,768.50	4,175
4040	3'-6" x 8'-0", simple design		10	1.600		3,000	75		3,075	3,425
4050	Intricate design		8	2		3,700	94		3,794	4,225
4400	For custom finish, add					475			475	525
4600	Side light, mahogany, 7'-0" x 1'-6" wide, 4 lites	2 Carp	18	.889		1,100	41.50		1,141.50	1,275
4610	6 lites		14	1.143		2,625	53.50		2,678.50	3,000
4620	8'-0" x 1'-6" wide, 4 lites		14	1.143		1,800	53.50		1,853.50	2,075
4630	6 lites		10	1.600		2,100	75		2,175	2,425
4640	Side light, oak, 7'-0" x 1'-6" wide, 4 lites		18	.889		1,200	41.50		1,241.50	1,400
4650	6 lites		14	1.143		2,100	53.50		2,153.50	2,400
4660	8'-0" x 1'-6" wide, 4 lites		14	1.143		1,100	53.50		1,153.50	1,300
4670	6 lites		10	1.600		2,100	75		2,175	2,425

08 14 16 – Flush Wood Doors

08 14 16.09 Smooth Wood Doors

		Crew	Daily Output	Labor-Hours	Unit	Material	Labor	Equipment	Total	Total Incl O&P
0010	**SMOOTH WOOD DOORS**									
0015	Flush, interior, hollow core									
0025	Lauan face, 1-3/8", 3'-0" x 6'-8"	2 Carp	17	.941	Ea.	54.50	44		98.50	133
0030	4'-0" x 6'-8"		16	1		126	47		173	216
0080	1-3/4", 2'-0" x 6'-8"		17	.941		39	44		83	116
0085	3'-0" x 6'-8"	1 Carp	16	.500		54.50	23.50		78	98.50
0108	3'-0" x 7'-0"	2 Carp	16	1		145	47		192	237
0112	Pair of 3'-0" x 7'-0"		9	1.778	Pr.	104	83.50		187.50	252
0140	Birch face, 1-3/8", 2'-6" x 6'-8"		17	.941	Ea.	85	44		129	166
0180	3'-0" x 6'-8"		17	.941		95.50	44		139.50	178
0200	4'-0" x 6'-8"		16	1		158	47		205	251

08 14 Wood Doors

08 14 16 – Flush Wood Doors

08 14 16.09 Smooth Wood Doors		Crew	Daily Output	Labor-Hours	Unit	2015 Bare Costs Material	Labor	Equipment	Total	Total Incl O&P
0202	1-3/4", 2'-0" x 6'-8"	2 Carp	17	.941	Ea.	50.50	44		94.50	128
0204	2'-4" x 7'-0"		16	1		96.50	47		143.50	183
0206	2'-6" x 7'-0"		16	1		100	47		147	187
0210	3'-0" x 7'-0"		16	1		110	47		157	198
0214	Pair of 3'-0" x 7'-0"		9	1.778	Pr.	213	83.50		296.50	370
0218	Birch face, 1-3/4", 3'-0" x 8'-0"		7	2.286	Ea.	320	107		427	530
0220	Oak face, 1-3/8", 2'-0" x 6'-8"		17	.941		104	44		148	187
0280	3'-0" x 6'-8"		17	.941		115	44		159	200
0300	4'-0" x 6'-8"		16	1		139	47		186	230
0305	1-3/4", 2'-6" x 6'-8"		17	.941		110	44		154	194
0310	3'-0" x 7'-0"		16	1		211	47		258	310
0320	Walnut face, 1-3/8", 2'-0" x 6'-8"		17	.941		182	44		226	274
0340	2'-6" x 6'-8"		17	.941		189	44		233	281
0380	3'-0" x 6'-8"		17	.941		197	44		241	289
0400	4'-0" x 6'-8"		16	1		216	47		263	315
0430	For 7'-0" high, add					26			26	28.50
0440	For 8'-0" high, add					36			36	39.50
0480	For prefinishing, clear, add					46			46	50.50
0500	For prefinishing, stain, add					57			57	62.50
1320	M.D. overlay on hardboard, 1-3/8", 2'-0" x 6'-8"	2 Carp	17	.941		115	44		159	200
1340	2'-6" x 6'-8"		17	.941		115	44		159	200
1380	3'-0" x 6'-8"		17	.941		126	44		170	212
1400	4'-0" x 6'-8"		16	1		178	47		225	273
1420	For 7'-0" high, add					15.75			15.75	17.35
1440	For 8'-0" high, add					31.50			31.50	34.50
1720	H.P. plastic laminate, 1-3/8", 2'-0" x 6'-8"	2 Carp	16	1		260	47		307	365
1740	2'-6" x 6'-8"		16	1		260	47		307	365
1780	3'-0" x 6'-8"		15	1.067		290	50		340	400
1800	4'-0" x 6'-8"		14	1.143		385	53.50		438.50	510
1820	For 7'-0" high, add					15.75			15.75	17.35
1840	For 8'-0" high, add					31.50			31.50	34.50
2020	Particle core, lauan face, 1-3/8", 2'-6" x 6'-8"	2 Carp	15	1.067		92	50		142	183
2040	3'-0" x 6'-8"		14	1.143		94	53.50		147.50	191
2080	3'-0" x 7'-0"		13	1.231		101	58		159	206
2085	4'-0" x 7'-0"		12	1.333		123	62.50		185.50	238
2110	1-3/4", 3'-0" x 7'-0"		13	1.231		145	58		203	255
2120	Birch face, 1-3/8", 2'-6" x 6'-8"		15	1.067		103	50		153	195
2140	3'-0" x 6'-8"		14	1.143		113	53.50		166.50	212
2180	3'-0" x 7'-0"		13	1.231		123	58		181	230
2200	4'-0" x 7'-0"		12	1.333		139	62.50		201.50	256
2205	1-3/4", 3'-0" x 7'-0"		13	1.231		123	58		181	230
2220	Oak face, 1-3/8", 2'-6" x 6'-8"		15	1.067		116	50		166	210
2240	3'-0" x 6'-8"		14	1.143		128	53.50		181.50	229
2280	3'-0" x 7'-0"		13	1.231		133	58		191	241
2300	4'-0" x 7'-0"		12	1.333		156	62.50		218.50	275
2305	1-3/4" 3'-0" x 7'-0"		13	1.231		190	58		248	305
2320	Walnut face, 1-3/8", 2'-0" x 6'-8"		15	1.067		123	50		173	217
2340	2'-6" x 6'-8"		14	1.143		139	53.50		192.50	241
2380	3'-0" x 6'-8"		13	1.231		156	58		214	267
2400	4'-0" x 6'-8"		12	1.333		206	62.50		268.50	330
2440	For 8'-0" high, add					41			41	45
2460	For 8'-0" high walnut, add					36			36	39.50
2720	For prefinishing, clear, add					36			36	39.50

08 14 Wood Doors

08 14 16 – Flush Wood Doors

08 14 16.09 Smooth Wood Doors		Crew	Daily Output	Labor-Hours	Unit	2015 Bare Costs Material	Labor	Equipment	Total	Total Incl O&P
2740	For prefinishing, stain, add				Ea.	53			53	58.50
3320	M.D. overlay on hardboard, 1-3/8", 2'-6" x 6'-8"	2 Carp	14	1.143		106	53.50		159.50	205
3340	3'-0" x 6'-8"		13	1.231		115	58		173	222
3380	3'-0" x 7'-0"		12	1.333		117	62.50		179.50	232
3400	4'-0" x 7'-0"		10	1.600		155	75		230	294
3440	For 8'-0" height, add					37			37	40.50
3460	For solid wood core, add					42			42	46
3720	H.P. plastic laminate, 1-3/8", 2'-6" x 6'-8"	2 Carp	13	1.231		160	58		218	271
3740	3'-0" x 6'-8"		12	1.333		185	62.50		247.50	305
3780	3'-0" x 7'-0"		11	1.455		190	68.50		258.50	320
3800	4'-0" x 7'-0"		8	2		225	94		319	400
3840	For 8'-0" height, add					37			37	40.50
3860	For solid wood core, add					42			42	46
4000	Exterior, flush, solid core, birch, 1-3/4" x 2'-6" x 7'-0"	2 Carp	15	1.067		173	50		223	272
4020	2'-8" wide		15	1.067		159	50		209	257
4040	3'-0" wide		14	1.143		209	53.50		262.50	320
4045	3'-0" x 8'-0"	1 Carp	8	1		430	47		477	550
4100	Oak faced 1-3/4" x 2'-6" x 7'-0"	2 Carp	15	1.067		215	50		265	320
4120	2'-8" wide		15	1.067		225	50		275	330
4140	3'-0" wide		14	1.143		230	53.50		283.50	340
4200	Walnut faced, 1-3/4" x 2'-6" x 7'-0"		15	1.067		310	50		360	420
4220	2'-8" wide		15	1.067		315	50		365	425
4240	3'-0" wide		14	1.143		320	53.50		373.50	440
4300	For 6'-8" high door, deduct from 7'-0" door					16.80			16.80	18.50
5000	Wood doors, for vision lite, add					94.50			94.50	104
5010	Wood doors, for narrow lite, add					102			102	113
5015	Wood doors, for bottom (or top) louver, add					282			282	310
9000	Minimum labor/equipment charge	1 Carp	4	2	Job		94		94	154

08 14 16.10 Wood Doors Decorator		Crew	Daily Output	Labor-Hours	Unit	Material	Labor	Equipment	Total	Total Incl O&P
0010	**WOOD DOORS DECORATOR**									
1800	Exterior, flush, solid wood core, birch 1-3/4" x 2'-6" x 7'-0"	2 Carp	15	1.067	Ea.	320	50		370	430
1820	2'-8" wide		15	1.067		325	50		375	440
1840	3'-0" wide		14	1.143		335	53.50		388.50	460
1900	Oak faced, 1-3/4" x 2'-6" x 7'-0"		15	1.067		320	50		370	430
1920	2'-8" wide		15	1.067		330	50		380	445
1940	3'-0" wide		14	1.143		340	53.50		393.50	465
2100	Walnut faced, 1-3/4" x 2'-6"x 7'-0"		15	1.067		385	50		435	505
2120	2'-8" wide		15	1.067		400	50		450	520
2140	3'-0" wide		14	1.143		430	53.50		483.50	565

08 14 16.20 Wood Fire Doors		Crew	Daily Output	Labor-Hours	Unit	Material	Labor	Equipment	Total	Total Incl O&P
0010	**WOOD FIRE DOORS**									
0020	Particle core, 7 face plys, "B" label,									
0040	1 hour, birch face, 1-3/4" x 2'-6" x 6'-8"	2 Carp	14	1.143	Ea.	420	53.50		473.50	550
0080	3'-0" x 6'-8"		13	1.231		400	58		458	535
0090	3'-0" x 7'-0"		12	1.333		435	62.50		497.50	585
0100	4'-0" x 7'-0"		12	1.333		540	62.50		602.50	700
0140	Oak face, 2'-6" x 6'-8"		14	1.143		450	53.50		503.50	585
0180	3'-0" x 6'-8"		13	1.231		455	58		513	595
0190	3'-0" x 7'-0"		12	1.333		460	62.50		522.50	610
0200	4'-0" x 7'-0"		12	1.333		590	62.50		652.50	755
0240	Walnut face, 2'-6" x 6'-8"		14	1.143		470	53.50		523.50	605
0280	3'-0" x 6'-8"		13	1.231		490	58		548	635

08 14 Wood Doors

08 14 16 – Flush Wood Doors

08 14 16.20 Wood Fire Doors		Crew	Daily Output	Labor-Hours	Unit	Material	Labor (2015 Bare Costs)	Equipment	Total	Total Incl O&P
0290	3'-0" x 7'-0"	2 Carp	12	1.333	Ea.	510	62.50		572.50	665
0300	4'-0" x 7'-0"		12	1.333		630	62.50		692.50	800
0440	M.D. overlay on hardboard, 2'-6" x 6'-8"		15	1.067		315	50		365	425
0480	3'-0" x 6'-8"		14	1.143		375	53.50		428.50	505
0490	3'-0" x 7'-0"		13	1.231		395	58		453	530
0500	4'-0" x 7'-0"		12	1.333		415	62.50		477.50	560
0740	90 minutes, birch face, 1-3/4" x 2'-6" x 6'-8"		14	1.143		325	53.50		378.50	445
0780	3'-0" x 6'-8"		13	1.231		320	58		378	445
0790	3'-0" x 7'-0"		12	1.333		390	62.50		452.50	535
0800	4'-0" x 7'-0"		12	1.333		545	62.50		607.50	700
0840	Oak face, 2'-6" x 6'-8"		14	1.143		430	53.50		483.50	565
0880	3'-0" x 6'-8"		13	1.231		440	58		498	580
0890	3'-0" x 7'-0"		12	1.333		455	62.50		517.50	605
0900	4'-0" x 7'-0"		12	1.333		590	62.50		652.50	755
0940	Walnut face, 2'-6" x 6'-8"		14	1.143		400	53.50		453.50	530
0980	3'-0" x 6'-8"		13	1.231		410	58		468	545
0990	3'-0" x 7'-0"		12	1.333		470	62.50		532.50	620
1000	4'-0" x 7'-0"		12	1.333		620	62.50		682.50	785
1140	M.D. overlay on hardboard, 2'-6" x 6'-8"		15	1.067		355	50		405	470
1180	3'-0" x 6'-8"		14	1.143		375	53.50		428.50	505
1190	3'-0" x 7'-0"		13	1.231		405	58		463	540
1200	4'-0" x 7'-0"		12	1.333		455	62.50		517.50	605
1240	For 8'-0" height, add					75			75	82.50
1260	For 8'-0" height walnut, add					90			90	99
2200	Custom architectural "B" label, flush, 1-3/4" thick, birch,									
2210	Solid core									
2220	2'-6" x 7'-0"	2 Carp	15	1.067	Ea.	305	50		355	415
2260	3'-0" x 7'-0"		14	1.143		315	53.50		368.50	435
2300	4'-0" x 7'-0"		13	1.231		400	58		458	535
2420	4'-0" x 8'-0"		11	1.455		430	68.50		498.50	585
2480	For oak veneer, add					50%				
2500	For walnut veneer, add					75%				
9000	Minimum labor/equipment charge	1 Carp	4	2	Job		94		94	154

08 14 33 – Stile and Rail Wood Doors

08 14 33.10 Wood Doors Paneled		Crew	Daily Output	Labor-Hours	Unit	Material	Labor	Equipment	Total	Total Incl O&P
0010	**WOOD DOORS PANELED**									
0020	Interior, six panel, hollow core, 1-3/8" thick									
0040	Molded hardboard, 2'-0" x 6'-8"	2 Carp	17	.941	Ea.	62	44		106	141
0060	2'-6" x 6'-8"		17	.941		64	44		108	143
0070	2'-8" x 6'-8"		17	.941		67	44		111	146
0080	3'-0" x 6'-8"		17	.941		72	44		116	152
0140	Embossed print, molded hardboard, 2'-0" x 6'-8"		17	.941		64	44		108	143
0160	2'-6" x 6'-8"		17	.941		64	44		108	143
0180	3'-0" x 6'-8"		17	.941		72	44		116	152
0540	Six panel, solid, 1-3/8" thick, pine, 2'-0" x 6'-8"		15	1.067		155	50		205	253
0560	2'-6" x 6'-8"		14	1.143		170	53.50		223.50	275
0580	3'-0" x 6'-8"		13	1.231		145	58		203	255
1020	Two panel, bored rail, solid, 1-3/8" thick, pine, 1'-6" x 6'-8"		16	1		270	47		317	375
1040	2'-0" x 6'-8"		15	1.067		355	50		405	470
1060	2'-6" x 6'-8"		14	1.143		400	53.50		453.50	530
1340	Two panel, solid, 1-3/8" thick, fir, 2'-0" x 6'-8"		15	1.067		160	50		210	258
1360	2'-6" x 6'-8"		14	1.143		210	53.50		263.50	320

08 14 Wood Doors

08 14 33 – Stile and Rail Wood Doors

08 14 33.10 Wood Doors Paneled		Crew	Daily Output	Labor-Hours	Unit	Material	Labor	Equipment	Total	Total Incl O&P
						2015 Bare Costs				
1380	3′-0″ x 6′-8″	2 Carp	13	1.231	Ea.	415	58		473	550
1740	Five panel, solid, 1-3/8″ thick, fir, 2′-0″ x 6′-8″		15	1.067		280	50		330	390
1760	2′-6″ x 6′-8″		14	1.143		420	53.50		473.50	550
1780	3′-0″ x 6′-8″		13	1.231		420	58		478	555
4000	Ext., hollow core, 1-3/4″ thick, hdbd., with 10″ x 10″ wdw., 2′-8″ x 6′-8″		16	1		168	47		215	261
4050	3′-0″ x 6′-8″		16	1		186	47		233	282
4100	Solid core, 2′-8″ x 6′-8″		16	1		193	47		240	289
4150	3′-0″ x 6′-8″		16	1		193	47		240	289
4190	Exterior, Knotty pine, paneled, 1-3/4″, 3′-0″ x 6′-8″		16	1		780	47		827	935
4195	Double 1-3/4″, 3′-0″ x 6′-8″		16	1		1,550	47		1,597	1,800
4200	Ash, paneled, 1-3/4″, 3′-0″ x 6′-8″		16	1		890	47		937	1,050
4205	Double 1-3/4″, 3′-0″ x 6′-8″		16	1		1,775	47		1,822	2,025
4210	Cherry, paneled, 1-3/4″, 3′-0″ x 6′-8″		16	1		890	47		937	1,050
4215	Double 1-3/4″, 3′-0″ x 6′-8″		16	1		1,775	47		1,822	2,025
4230	Ash, paneled, 1-3/4″, 3′-0″ x 8′-0″		16	1		950	47		997	1,125
4235	Double 1-3/4″, 3′-0″ x 8′-0″		16	1		1,900	47		1,947	2,175
4240	Hard Maple, paneled, 1-3/4″, 3′-0″ x 8′-0″		16	1		950	47		997	1,125
4245	Double 1-3/4″, 3′-0″ x 8′-0″		16	1		1,900	47		1,947	2,175
4250	Cherry, paneled, 1-3/4″, 3′-0″ x 8′-0″		16	1		1,000	47		1,047	1,175
4255	Double 1-3/4″, 3′-0″ x 8′-0″		16	1		2,000	47		2,047	2,275
9000	Minimum labor/equipment charge	1 Carp	4	2	Job		94		94	154

08 14 33.20 Wood Doors Residential		Crew	Daily Output	Labor-Hours	Unit	Material	Labor	Equipment	Total	Total Incl O&P
0010	**WOOD DOORS RESIDENTIAL**									
0200	Exterior, combination storm & screen, pine									
0220	Cross buck, 6′-9″ x 2′-6″ wide	2 Carp	11	1.455	Ea.	340	68.50		408.50	485
0260	2′-8″ wide		10	1.600		300	75		375	455
0280	3′-0″ wide		9	1.778		310	83.50		393.50	475
0300	7′-1″ x 3′-0″ wide		9	1.778		335	83.50		418.50	505
0400	Full lite, 6′-9″ x 2′-6″ wide		11	1.455		320	68.50		388.50	460
0420	2′-8″ wide		10	1.600		320	75		395	475
0440	3′-0″ wide		9	1.778		325	83.50		408.50	490
0500	7′-1″ x 3′-0″ wide		9	1.778		355	83.50		438.50	525
0700	Dutch door, pine, 1-3/4″ x 2′-8″ x 6′-8″ , 6 panel		12	1.333		790	62.50		852.50	970
0720	Half glass		10	1.600		900	75		975	1,125
0800	3′-0″ wide, 6 panel		12	1.333		790	62.50		852.50	970
0820	Half glass		10	1.600		900	75		975	1,125
1000	Entrance door, colonial, 1-3/4″ x 6′-8″ x 2′-8″ wide		16	1		500	47		547	625
1020	6 panel pine, 3′-0″ wide		15	1.067		445	50		495	570
1100	8 panel pine, 2′-8″ wide		16	1		580	47		627	710
1120	3′-0″ wide		15	1.067		570	50		620	705
1200	For tempered safety glass lites, (min of 2) add					79			79	87
1300	Flush, birch, solid core, 1-3/4″ x 6′-8″ x 2′-8″ wide	2 Carp	16	1		113	47		160	202
1320	3′-0″ wide		15	1.067		119	50		169	212
1350	7′-0″ x 2′-8″ wide		16	1		122	47		169	211
1360	3′-0″ wide		15	1.067		143	50		193	239
1380	For tempered safety glass lites, add					105			105	116
1420	6′-8″ x 3′-0″ wide, fir	2 Carp	16	1		470	47		517	590
1720	Carved mahogany 3′-0″ x 6′-8″		15	1.067		1,375	50		1,425	1,600
1800	Lauan, solid core, 1-3/4″ x 7′-0″ x 2′-4″ wide		16	1		110	47		157	199
1810	2′-6″ wide		15	1.067		114	50		164	207
1820	2′-8″ wide		9	1.778		118	83.50		201.50	267
1830	3′-0″ wide		16	1		130	47		177	220

08 14 Wood Doors

08 14 33 – Stile and Rail Wood Doors

08 14 33.20 Wood Doors Residential		Crew	Daily Output	Labor-Hours	Unit	Material	2015 Bare Costs Labor	Equipment	Total	Total Incl O&P
1840	3′-4″ wide	2 Carp	16	1	Ea.	205	47		252	305
1850	Pair of 3′-0″ wide		15	1.067	Pr.	259	50		309	365
2700	Interior, closet, bi-fold, w/hardware, no frame or trim incl.									
2720	Flush, birch, 2′-6″ x 6′-8″	2 Carp	13	1.231	Ea.	67.50	58		125.50	169
2740	3′-0″ wide		13	1.231		71.50	58		129.50	173
2760	4′-0″ wide		12	1.333		109	62.50		171.50	223
2780	5′-0″ wide		11	1.455		105	68.50		173.50	228
2800	6′-0″ wide		10	1.600		128	75		203	264
2804	Flush lauan 2′-0″ x 6′-8″		14	1.143		52.50	53.50		106	146
2810	8′-0″ wide		9	1.778		195	83.50		278.50	350
2817	6′-0″ wide		9	1.778		113	83.50		196.50	261
2820	Flush, hardboard, primed, 6′-8″ x 2′-6″ wide		13	1.231		56.50	58		114.50	157
2840	3′-0″ wide		13	1.231		64	58		122	165
2860	4′-0″ wide		12	1.333		94	62.50		156.50	206
2880	5′-0″ wide		11	1.455		101	68.50		169.50	223
2900	6′-0″ wide		10	1.600		139	75		214	276
3000	Raised panel pine, 6′-6″ or 6′-8″ x 2′-6″ wide		13	1.231		198	58		256	315
3020	3′-0″ wide		13	1.231		278	58		336	400
3040	4′-0″ wide		12	1.333		305	62.50		367.50	440
3060	5′-0″ wide		11	1.455		365	68.50		433.50	510
3080	6′-0″ wide		10	1.600		400	75		475	565
3200	Louvered, pine 6′-6″ or 6′-8″ x 2′-6″ wide		13	1.231		144	58		202	253
3220	3′-0″ wide		13	1.231		208	58		266	325
3240	4′-0″ wide		12	1.333		235	62.50		297.50	360
3260	5′-0″ wide		11	1.455		262	68.50		330.50	400
3280	6′-0″ wide		10	1.600		289	75		364	445
4400	Bi-passing closet, incl. hardware and frame, no trim incl.									
4420	Flush, lauan, 6′-8″ x 4′-0″ wide	2 Carp	12	1.333	Opng.	176	62.50		238.50	297
4440	5′-0″ wide		11	1.455		187	68.50		255.50	315
4460	6′-0″ wide		10	1.600		209	75		284	355
4600	Flush, birch, 6′-8″ x 4′-0″ wide		12	1.333		223	62.50		285.50	350
4620	5′-0″ wide		11	1.455		212	68.50		280.50	345
4640	6′-0″ wide		10	1.600		260	75		335	410
4800	Louvered, pine, 6′-8″ x 4′-0″ wide		12	1.333		410	62.50		472.50	560
4820	5′-0″ wide		11	1.455		395	68.50		463.50	545
4840	6′-0″ wide		10	1.600		535	75		610	710
5000	Paneled, pine, 6′-8″ x 4′-0″ wide		12	1.333		515	62.50		577.50	670
5020	5′-0″ wide		11	1.455		410	68.50		478.50	560
5040	6′-0″ wide		10	1.600		610	75		685	795
5042	8′-0″ wide		12	1.333		1,025	62.50		1,087.50	1,225
6100	Folding accordion, closet, including track and frame									
6120	Vinyl, 2 layer, stock	2 Carp	10	1.600	Ea.	66	75		141	196
6140	Woven mahogany and vinyl, stock		10	1.600		57.50	75		132.50	186
6160	Wood slats with vinyl overlay, stock		10	1.600		150	75		225	288
6180	Economy vinyl, stock		10	1.600		41.50	75		116.50	169
6200	Rigid PVC		10	1.600		55.50	75		130.50	184
7310	Passage doors, flush, no frame included									
7320	Hardboard, hollow core, 1-3/8″ x 6′-8″ x 1′-6″ wide	2 Carp	18	.889	Ea.	40.50	41.50		82	113
7330	2′-0″ wide		18	.889		43	41.50		84.50	116
7340	2′-6″ wide		18	.889		47.50	41.50		89	121
7350	2′-8″ wide		18	.889		49	41.50		90.50	123
7360	3′-0″ wide		17	.941		52	44		96	130
7420	Lauan, hollow core, 1-3/8″ x 6′-8″ x 1′-6″ wide		18	.889		35	41.50		76.50	107

08 14 33 – Stile and Rail Wood Doors

08 14 33.20 Wood Doors Residential		Crew	Daily Output	Labor-Hours	Unit	Material	Labor (2015 Bare Costs)	Equipment	Total	Total Incl O&P
7440	2'-0" wide	2 Carp	18	.889	Ea.	35	41.50		76.50	107
7450	2'-4" wide		18	.889		38.50	41.50		80	111
7460	2'-6" wide		18	.889		38.50	41.50		80	111
7480	2'-8" wide		18	.889		40	41.50		81.50	113
7500	3'-0" wide		17	.941		42.50	44		86.50	119
7700	Birch, hollow core, 1-3/8" x 6'-8" x 1'-6" wide		18	.889		41	41.50		82.50	114
7720	2'-0" wide		18	.889		42.50	41.50		84	115
7740	2'-6" wide		18	.889		50	41.50		91.50	124
7760	2'-8" wide		18	.889		50.50	41.50		92	124
7780	3'-0" wide		17	.941		52	44		96	130
8000	Pine louvered, 1-3/8" x 6'-8" x 1'-6" wide		19	.842		120	39.50		159.50	197
8020	2'-0" wide		18	.889		131	41.50		172.50	213
8040	2'-6" wide		18	.889		150	41.50		191.50	234
8060	2'-8" wide		18	.889		161	41.50		202.50	246
8080	3'-0" wide		17	.941		175	44		219	265
8300	Pine paneled, 1-3/8" x 6'-8" x 1'-6" wide		19	.842		128	39.50		167.50	206
8320	2'-0" wide		18	.889		153	41.50		194.50	237
8330	2'-4" wide		18	.889		182	41.50		223.50	269
8340	2'-6" wide		18	.889		182	41.50		223.50	269
8360	2'-8" wide		18	.889		185	41.50		226.50	273
8380	3'-0" wide		17	.941		204	44		248	297
9000	Passage doors, flush, no frame, birch, solid core, 1-3/8" x 2'-4" x 7'-0"		16	1		118	47		165	207
9020	2'-8" wide		16	1		125	47		172	215
9040	3'-0" wide		16	1		135	47		182	226
9060	3'-4" wide		15	1.067		156	50		206	254
9080	Pair of 3'-0" wide		9	1.778	Pr.	267	83.50		350.50	430
9100	Lauan, solid core, 1-3/8" x 7'-0" x 2'-4" wide		16	1	Ea.	108	47		155	196
9120	2'-8" wide		16	1		116	47		163	204
9140	3'-0" wide		16	1		123	47		170	212
9160	3'-4" wide		15	1.067		129	50		179	224
9180	Pair of 3'-0" wide		9	1.778	Pr.	223	83.50		306.50	380
9200	Hardboard, solid core, 1-3/8" x 7'-0" x 2'-4" wide		16	1	Ea.	134	47		181	225
9220	2'-8" wide		16	1		139	47		186	230
9240	3'-0" wide		16	1		143	47		190	234
9260	3'-4" wide		15	1.067		158	50		208	256
9900	Minimum labor/equipment charge	1 Carp	4	2	Job		94		94	154

08 14 35 – Torrified Doors

08 14 35.10 Torrified Exterior Doors		Crew	Daily Output	Labor-Hours	Unit	Material	Labor	Equipment	Total	Total Incl O&P
0010	**TORRIFIED EXTERIOR DOORS**									
0020	Wood doors made from torrified wood, exterior.									
0030	All doors require a finish be applied, all glass is insulated									
0040	All doors require pilot holes for all fasteners									
0100	6 panel, Paint grade poplar, 1-3/4" x 3'-0"x 6'-8"	2 Carp	12	1.333	Ea.	1,075	62.50		1,137.50	1,275
0120	Half glass 3'-0"x 6'-8"	"	12	1.333		1,175	62.50		1,237.50	1,400
0200	Side lite, full glass, 1-3/4" x 1'-2" x 6'-8"					905			905	995
0220	Side lite, half glass, 1-3/4" x 1'-2" x 6'-8"					905			905	995
0300	Raised Face, 2 Panel, Paint grade poplar, 1-3/4" x 3'-0"x 7'-0"	2 Carp	12	1.333		1,275	62.50		1,337.50	1,500
0320	Side lite, raised face, half glass, 1-3/4" x 1'-2" x 7'-0"					1,050			1,050	1,175
0500	6 panel, Fir, 1-3/4" x 3'-0"x 6'-8"	2 Carp	12	1.333		1,550	62.50		1,612.50	1,800
0520	Half glass 3'-0"x 6'-8"	"	12	1.333		1,650	62.50		1,712.50	1,925
0600	Side lite, full glass, 1-3/4" x 1'-2" x 6'-8"					1,150			1,150	1,275
0620	Side lite, half glass, 1-3/4" x 1'-2" x 6'-8"					1,450			1,450	1,600

08 14 Wood Doors

08 14 35 – Torrified Doors

08 14 35.10 Torrified Exterior Doors		Crew	Daily Output	Labor-Hours	Unit	2015 Bare Costs Material	Labor	Equipment	Total	Total Incl O&P
0700	6 panel, Mahogany, 1-3/4″ x 3′-0″x 6′-8″	2 Carp	12	1.333	Ea.	1,625	62.50		1,687.50	1,875
0800	Side lite, full glass, 1-3/4″ x 1′-2″ x 6′-8″					1,225			1,225	1,350
0820	Side lite, half glass, 1-3/4″ x 1′-2″ x 6′-8″					1,200			1,200	1,325

08 14 40 – Interior Cafe Doors

08 14 40.10 Cafe Style Doors		Crew	Daily Output	Labor-Hours	Unit	Material	Labor	Equipment	Total	Total Incl O&P
0010	**CAFE STYLE DOORS**									
6520	Interior cafe doors, 2′-6″ opening, stock, panel pine	2 Carp	16	1	Ea.	218	47		265	315
6540	3′-0″ opening	″	16	1	″	240	47		287	340
6550	Louvered pine									
6560	2′-6″ opening	2 Carp	16	1	Ea.	180	47		227	275
8000	3′-0″ opening		16	1		193	47		240	289
8010	2′-6″ opening, hardwood		16	1		264	47		311	365
8020	3′-0″ opening		16	1		281	47		328	385
9000	Minimum labor/equipment charge	1 Carp	4	2	Job		94		94	154

08 16 Composite Doors

08 16 13 – Fiberglass Doors

08 16 13.10 Entrance Doors, Fiberous Glass			Crew	Daily Output	Labor-Hours	Unit	Material	Labor	Equipment	Total	Total Incl O&P
0010	**ENTRANCE DOORS, FIBEROUS GLASS**										
0020	Exterior, fiberglass, door, 2′-8″ wide x 6′-8″ high	G	2 Carp	15	1.067	Ea.	270	50		320	380
0040	3′-0″ wide x 6′-8″ high	G		15	1.067		270	50		320	380
0060	3′-0″ wide x 7′-0″ high	G		15	1.067		460	50		510	585
0080	3′-0″ wide x 6′-8″ high, with two lites	G		15	1.067		315	50		365	425
0100	3′-0″ wide x 8′-0″ high, with two lites	G		15	1.067		525	50		575	660
0110	Half glass, 3′-0″ wide x 6′-8″ high	G		15	1.067		435	50		485	560
0120	3′-0″ wide x 6′-8″ high, low e	G		15	1.067		465	50		515	590
0130	3′-0″ wide x 8′-0″ high	G		15	1.067		585	50		635	725
0140	3′-0″ wide x 8′-0″ high, low e	G		15	1.067		655	50		705	800
0150	Side lights, 1′-0″ wide x 6′-8″ high	G					266			266	293
0160	1′-0″ wide x 6′-8″ high, low e	G					281			281	310
0180	1′-0″ wide x 6′-8″ high, full glass	G					310			310	340
0190	1′-0″ wide x 6′-8″ high, low e	G					340			340	370

08 16 14 – French Doors

08 16 14.10 Exterior Doors With Glass Lites		Crew	Daily Output	Labor-Hours	Unit	Material	Labor	Equipment	Total	Total Incl O&P
0010	**EXTERIOR DOORS WITH GLASS LITES**									
0020	French, Fir, 1-3/4″, 3′-0″wide x 6′-8″ high	2 Carp	12	1.333	Ea.	600	62.50		662.50	765
0025	Double		12	1.333		1,200	62.50		1,262.50	1,425
0030	Maple, 1-3/4″, 3′-0″wide x 6′-8″ high		12	1.333		675	62.50		737.50	850
0035	Double		12	1.333		1,350	62.50		1,412.50	1,575
0040	Cherry, 1-3/4″, 3′-0″wide x 6′-8″ high		12	1.333		790	62.50		852.50	970
0045	Double		12	1.333		1,575	62.50		1,637.50	1,825
0100	Mahogany, 1-3/4″, 3′-0″wide x 8′-0″ high		10	1.600		800	75		875	1,000
0105	Double		10	1.600		1,600	75		1,675	1,875
0110	Fir, 1-3/4″, 3′-0″wide x 8′-0″ high		10	1.600		1,200	75		1,275	1,450
0115	Double		10	1.600		2,400	75		2,475	2,775
0120	Oak, 1-3/4″, 3′-0″wide x 8′-0″ high		10	1.600		1,825	75		1,900	2,125
0125	Double		10	1.600		3,650	75		3,725	4,125

08 17 Integrated Door Opening Assemblies

08 17 13 – Integrated Metal Door Opening Assemblies

08 17 13.10 Hollow Metal Doors and Frames

	08 17 13.10 Hollow Metal Doors and Frames	Crew	Daily Output	Labor-Hours	Unit	2015 Bare Costs Material	Labor	Equipment	Total	Total Incl O&P
0010	**HOLLOW METAL DOORS AND FRAMES**									
0100	Prehung, flush, 18 ga., 1-3/4" x 6'-8" x 2'-8" wide	1 Carp	16	.500	Ea.	665	23.50		688.50	770
0120	3'-0" wide		15	.533		620	25		645	725
0130	3'-6" wide		13	.615		710	29		739	830
0140	4'-0" wide		10	.800		770	37.50		807.50	905
0300	Double, 1-3/4" x 3'-0" x 6'-8"		5	1.600		700	75		775	895
0350	3'-0" x 8'-0"		4	2		1,550	94		1,644	1,850

08 17 13.20 Stainless Steel Doors and Frames

	08 17 13.20 Stainless Steel Doors and Frames		Crew	Daily Output	Labor-Hours	Unit	Material	Labor	Equipment	Total	Total Incl O&P
0010	**STAINLESS STEEL DOORS AND FRAMES**										
0020	Stainless Steel(304) prehung 24 g 2'-6" x 6'-8" door w/16 g frame	G	2 Carp	6	2.667	Ea.	1,800	125		1,925	2,175
0025	2'-8" x 6'-8"	G		6	2.667		1,825	125		1,950	2,225
0030	3'-0" x 6'-8"	G		6	2.667		1,825	125		1,950	2,200
0040	3'-0" x 7'-0"	G		6	2.667		1,850	125		1,975	2,225
0050	4'-0" x 7'-0"	G		5	3.200		2,400	150		2,550	2,900
0100	Stainless Steel(316) prehung 24 g 2'-6" x 6'-8" door w/16 g frame	G		6	2.667		2,700	125		2,825	3,175
0110	2'-8" x 6'-8"	G		6	2.667		2,725	125		2,850	3,175
0120	3'-0" x 6'-8"	G		6	2.667		2,500	125		2,625	2,950
0150	3'-0" x 7'-0"	G		6	2.667		2,675	125		2,800	3,150
0160	4'-0" x 7'-0"	G		6	2.667		3,300	125		3,425	3,825
0300	Stainless Steel(304) prehung 18 g 2'-6" x 6'-8" door w/16 g frame	G		6	2.667		3,075	125		3,200	3,575
0310	2'-8" x 6'-8"	G		6	2.667		3,100	125		3,225	3,600
0320	3'-0" x 6'-8"	G		6	2.667		3,150	125		3,275	3,675
0350	3'-0" x 7'-0"	G		6	2.667		3,200	125		3,325	3,725
0360	4'-0" x 7'-0"	G		5	3.200		3,400	150		3,550	3,975
0500	Stainless steel, prehung door, foam core, 14 ga, 3'-0" x 7'-0"	G		5	3.200		3,400	150		3,550	4,000
0600	Stainless steel, prehung double door, foam core, 14 ga, 3'-0" x 7'-0"	G		4	4		6,700	188		6,888	7,675

08 17 23 – Integrated Wood Door Opening Assemblies

08 17 23.10 Pre-Hung Doors

	08 17 23.10 Pre-Hung Doors	Crew	Daily Output	Labor-Hours	Unit	Material	Labor	Equipment	Total	Total Incl O&P
0010	**PRE-HUNG DOORS**									
0300	Exterior, wood, comb. storm & screen, 6'-9" x 2'-6" wide	2 Carp	15	1.067	Ea.	296	50		346	405
0320	2'-8" wide		15	1.067		296	50		346	405
0340	3'-0" wide		15	1.067		305	50		355	415
0360	For 7'-0" high door, add					30			30	33
1600	Entrance door, flush, birch, solid core									
1620	4-5/8" solid jamb, 1-3/4" x 6'-8" x 2'-8" wide	2 Carp	16	1	Ea.	289	47		336	395
1640	3'-0" wide	"	16	1		375	47		422	490
1680	For 7'-0" high door, add					25			25	27.50
2000	Entrance door, colonial, 6 panel pine									
2020	4-5/8" solid jamb, 1-3/4" x 6'-8" x 2'-8" wide	2 Carp	16	1	Ea.	640	47		687	780
2040	3'-0" wide	"	16	1		675	47		722	815
2060	For 7'-0" high door, add					54			54	59
2200	For 5-5/8" solid jamb, add					41.50			41.50	46
2230	French style, exterior, 1 lite, 1-3/4" x 3'-0" x 6'-8"	1 Carp	14	.571		590	27		617	695
2235	9 lites	"	14	.571		660	27		687	770
2245	15 lites	2 Carp	14	1.143		640	53.50		693.50	795
2250	Double, 15 lites, 2'-0" x 6'-8", 4'-0" opening		7	2.286	Pr.	1,200	107		1,307	1,500
2260	2'-6" x 6'-8", 5' -0" opening		7	2.286		1,325	107		1,432	1,625
2280	3'-0" x 6'-8", 6'-0" opening		7	2.286		1,350	107		1,457	1,650
2430	3'-0" x 7'-0", 15 lites		14	1.143	Ea.	950	53.50		1,003.50	1,150
2432	Two 3'-0" x 7'-0"		7	2.286	Pr.	1,975	107		2,082	2,325
2435	3'-0" x 8'-0"		14	1.143	Ea.	1,000	53.50		1,053.50	1,200
2437	Two, 3'-0" x 8'-0"		7	2.286	Pr.	2,075	107		2,182	2,475

08 17 Integrated Door Opening Assemblies

08 17 23 – Integrated Wood Door Opening Assemblies

08 17 23.10 Pre-Hung Doors		Crew	Daily Output	Labor-Hours	Unit	Material	2015 Bare Costs Labor	Equipment	Total	Total Incl O&P
4000	Interior, passage door, 4-5/8" solid jamb									
4350	Paneled, primed, hollow core, 2'-8" wide	2 Carp	17	.941	Ea.	116	44		160	201
4360	3'-0" wide		17	.941		180	44		224	271
4370	Pine, louvered, 2'-8"x 6'-8"		17	.941		193	44		237	285
4380	3'-0"		17	.941		203	44		247	296
4400	Lauan, flush, solid core, 1-3/8" x 6'-8" x 2'-6" wide		17	.941		186	44		230	278
4420	2'-8" wide		17	.941		186	44		230	278
4440	3'-0" wide		16	1		201	47		248	298
4600	Hollow core, 1-3/8" x 6'-8" x 2'-6" wide		17	.941		125	44		169	211
4620	2'-8" wide		17	.941		125	44		169	210
4640	3'-0" wide		16	1		140	47		187	231
4700	For 7'-0" high door, add					35.50			35.50	39
5000	Birch, flush, solid core, 1-3/8" x 6'-8" x 2'-6" wide	2 Carp	17	.941		275	44		319	380
5020	2'-8" wide		17	.941		201	44		245	294
5040	3'-0" wide		16	1		300	47		347	410
5200	Hollow core, 1-3/8" x 6'-8" x 2'-6" wide		17	.941		222	44		266	315
5220	2'-8" wide		17	.941		266	44		310	365
5240	3'-0" wide		16	1		230	47		277	330
5280	For 7'-0" high door, add					30.50			30.50	33.50
5500	Hardboard paneled, 1-3/8" x 6'-8" x 2'-6" wide	2 Carp	17	.941		145	44		189	232
5520	2'-8" wide		17	.941		153	44		197	241
5540	3'-0" wide		16	1		150	47		197	242
6000	Pine paneled, 1-3/8" x 6'-8" x 2'-6" wide		17	.941		255	44		299	355
6020	2'-8" wide		17	.941		274	44		318	375
6040	3'-0" wide		16	1		282	47		329	385
7600	Oak, 6 panel, 1-3/4" x 6'-8" x 3'-0"	1 Carp	17	.471		895	22		917	1,025
8200	Birch, flush, solid core, 1-3/4" x 6'-8" x 2'-4" wide		17	.471		229	22		251	288
8220	2'-6" wide		17	.471		227	22		249	286
8240	2'-8" wide		17	.471		221	22		243	279
8260	3'-0" wide		16	.500		229	23.50		252.50	291
8280	3'-6" wide		15	.533		360	25		385	435
8500	Pocket door frame with lauan, flush, hollow core , 1-3/8" x 3'-0" x 6'-8"		17	.471		223	22		245	282
9000	Minimum labor/equipment charge		4	2	Job		94		94	154

08 31 Access Doors and Panels

08 31 13 – Access Doors and Frames

08 31 13.10 Types of Framed Access Doors		Crew	Daily Output	Labor-Hours	Unit	Material	Labor	Equipment	Total	Total Incl O&P
0010	**TYPES OF FRAMED ACCESS DOORS**									
1000	Fire rated door with lock									
1100	Metal, 12" x 12"	1 Carp	10	.800	Ea.	165	37.50		202.50	244
1150	18" x 18"		9	.889		220	41.50		261.50	310
1200	24" x 24"		9	.889		325	41.50		366.50	430
1250	24" x 36"		8	1		335	47		382	440
1300	24" x 48"		8	1		430	47		477	550
1350	36" x 36"		7.50	1.067		495	50		545	625
1400	48" x 48"		7.50	1.067		635	50		685	775
1600	Stainless steel, 12" x 12"		10	.800		282	37.50		319.50	370
1650	18" x 18"		9	.889		410	41.50		451.50	520
1700	24" x 24"		9	.889		500	41.50		541.50	620
1750	24" x 36"		8	1		640	47		687	780
2000	Flush door for finishing									

08 31 13.10 Types of Framed Access Doors

	08 31 13.10 Types of Framed Access Doors	Crew	Daily Output	Labor-Hours	Unit	Material	2015 Bare Costs Labor	Equipment	Total	Total Incl O&P
2100	Metal 8" x 8"	1 Carp	10	.800	Ea.	38	37.50		75.50	104
2150	12" x 12"	"	10	.800	"	45	37.50		82.50	111
3000	Recessed door for acoustic tile									
3100	Metal, 12" x 12"	1 Carp	4.50	1.778	Ea.	82	83.50		165.50	227
3150	12" x 24"		4.50	1.778		100	83.50		183.50	247
3200	24" x 24"		4	2		130	94		224	297
3250	24" x 36"		4	2		180	94		274	350
4000	Recessed door for drywall									
4100	Metal 12" x 12"	1 Carp	6	1.333	Ea.	80	62.50		142.50	191
4150	12" x 24"		5.50	1.455		114	68.50		182.50	237
4200	24" x 36"		5	1.600		182	75		257	325
6000	Standard door									
6100	Metal, 8" x 8"	1 Carp	10	.800	Ea.	45	37.50		82.50	111
6150	12" x 12"		10	.800		50	37.50		87.50	117
6200	18" x 18"		9	.889		70	41.50		111.50	146
6250	24" x 24"		9	.889		85	41.50		126.50	162
6300	24" x 36"		8	1		125	47		172	215
6350	36" x 36"		8	1		145	47		192	237
6500	Stainless steel, 8" x 8"		10	.800		85	37.50		122.50	155
6550	12" x 12"		10	.800		110	37.50		147.50	183
6600	18" x 18"		9	.889		205	41.50		246.50	295
6650	24" x 24"		9	.889		265	41.50		306.50	360
9000	Minimum labor/equipment charge		4	2	Job		94		94	154

08 31 13.20 Bulkhead/Cellar Doors

	08 31 13.20 Bulkhead/Cellar Doors	Crew	Daily Output	Labor-Hours	Unit	Material	Labor	Equipment	Total	Total Incl O&P
0010	**BULKHEAD/CELLAR DOORS**									
0020	Steel, not incl. sides, 44" x 62"	1 Carp	5.50	1.455	Ea.	560	68.50		628.50	725
0100	52" x 73"		5.10	1.569		790	73.50		863.50	990
0500	With sides and foundation plates, 57" x 45" x 24"		4.70	1.702		845	80		925	1,050
0600	42" x 49" x 51"		4.30	1.860		915	87.50		1,002.50	1,150
9000	Minimum labor/equipment charge		2	4	Job		188		188	310

08 31 13.30 Commercial Floor Doors

	08 31 13.30 Commercial Floor Doors	Crew	Daily Output	Labor-Hours	Unit	Material	Labor	Equipment	Total	Total Incl O&P
0010	**COMMERCIAL FLOOR DOORS**									
0020	Aluminum tile, steel frame, one leaf, 2' x 2' opng.	2 Sswk	3.50	4.571	Opng.	880	241		1,121	1,400
0021	Aluminum tile, steel frame, one leaf, 2' x 2' opng.	L-4	3.50	6.857		880	305		1,185	1,475
0050	3'-6" x 3'-6" opening	2 Sswk	3.50	4.571		1,600	241		1,841	2,175
0051	3'-6" x 3'-6" opening	L-4	3.50	6.857		1,600	305		1,905	2,250
0500	Double leaf, 4' x 4' opening	2 Sswk	3	5.333		1,750	281		2,031	2,425
0501	Double leaf, 4' x 4' opening	L-4	3	8		1,750	355		2,105	2,500
0550	5' x 5' opening	2 Sswk	3	5.333		3,100	281		3,381	3,900
0551	5' x 5' opening	L-4	3	8		3,100	355		3,455	3,975
9000	Minimum labor/equipment charge	2 Sswk	2	8	Job		420		420	760

08 31 13.35 Industrial Floor Doors

	08 31 13.35 Industrial Floor Doors	Crew	Daily Output	Labor-Hours	Unit	Material	Labor	Equipment	Total	Total Incl O&P
0010	**INDUSTRIAL FLOOR DOORS**									
0020	Steel 300 psf L.L., single leaf, 2' x 2', 175#	2 Sswk	6	2.667	Opng.	760	140		900	1,100
0050	3' x 3' opening, 300#		5.50	2.909		1,075	153		1,228	1,450
0300	Double leaf, 4' x 4' opening, 455#		5	3.200		2,275	168		2,443	2,800
0350	5' x 5' opening, 645#		4.50	3.556		3,300	187		3,487	3,975
1000	Aluminum, 300 psf L.L., single leaf, 2' x 2', 60#		6	2.667		800	140		940	1,125
1050	3' x 3' opening, 100#		5.50	2.909		1,300	153		1,453	1,700
1500	Double leaf, 4' x 4' opening, 160#		5	3.200		2,100	168		2,268	2,600
1550	5' x 5' opening, 235#		4.50	3.556		2,800	187		2,987	3,425
2000	Aluminum, 150 psf L.L., single leaf, 2' x 2', 60#		6	2.667		720	140		860	1,050

08 31 Access Doors and Panels

08 31 13 – Access Doors and Frames

08 31 13.35 Industrial Floor Doors		Crew	Daily Output	Labor-Hours	Unit	Material	2015 Bare Costs Labor	Equipment	Total	Total Incl O&P
2050	3′ x 3′ opening, 95#	2 Sswk	5.50	2.909	Opng.	1,200	153		1,353	1,600
2500	Double leaf, 4′ x 4′ opening, 150#		5	3.200		1,450	168		1,618	1,900
2550	5′ x 5′ opening, 230#		4.50	3.556		1,950	187		2,137	2,475
9000	Minimum labor/equipment charge		2	8	Job		420		420	760

08 31 13.40 Kennel Doors

Code	Description	Crew	Daily Output	Labor-Hours	Unit	Material	Labor	Equipment	Total	Total Incl O&P
0010	**KENNEL DOORS**									
0020	2 way, swinging type, 13″ x 19″ opening	2 Carp	11	1.455	Opng.	90	68.50		158.50	211
0100	17″ x 29″ opening		11	1.455		110	68.50		178.50	233
0200	9″ x 9″ opening, electronic with accessories		11	1.455		144	68.50		212.50	270

08 32 Sliding Glass Doors

08 32 13 – Sliding Aluminum-Framed Glass Doors

08 32 13.10 Sliding Aluminum Doors

Code	Description	Crew	Daily Output	Labor-Hours	Unit	Material	Labor	Equipment	Total	Total Incl O&P
0010	**SLIDING ALUMINUM DOORS**									
0350	Aluminum, 5/8″ tempered insulated glass, 6′ wide									
0400	Premium	2 Carp	4	4	Ea.	1,575	188		1,763	2,025
0450	Economy		4	4		815	188		1,003	1,200
0500	8′ wide, premium		3	5.333		1,650	250		1,900	2,225
0550	Economy		3	5.333		1,425	250		1,675	1,950
0600	12′ wide, premium		2.50	6.400		2,975	300		3,275	3,775
0650	Economy		2.50	6.400		1,550	300		1,850	2,200
4000	Aluminum, baked on enamel, temp glass, 6′-8″ x 10′-0″ wide		4	4		1,075	188		1,263	1,475
4020	Insulating glass, 6′-8″ x 6′-0″ wide		4	4		935	188		1,123	1,325
4040	8′-0″ wide		3	5.333		1,100	250		1,350	1,600
4060	10′-0″ wide		2	8		1,350	375		1,725	2,100
4080	Anodized, temp glass, 6′-8″ x 6′-0″ wide		4	4		455	188		643	810
4100	8′-0″ wide		3	5.333		575	250		825	1,050
4120	10′-0″ wide		2	8		645	375		1,020	1,325
5000	Aluminum sliding glass door system									
5010	Sliding door 4′ wide opening single side	2 Carp	2	8	Ea.	5,400	375		5,775	6,575
5015	8′ wide opening single side		2	8		8,100	375		8,475	9,525
5020	Telescoping glass door system, 4′ wide opening biparting		2	8		4,500	375		4,875	5,575
5025	8′ wide opening biparting		2	8		5,400	375		5,775	6,575
5030	Folding glass door, 4′ wide opening biparting		2	8		7,200	375		7,575	8,550
5035	8′ wide opening biparting		2	8		9,000	375		9,375	10,500
5040	ICU-CCU Sliding telescoping glass door, 4′ x 7′, single side opening		2	8		2,650	375		3,025	3,550
5045	8′ x 7′, single side opening		2	8		4,125	375		4,500	5,150
7000	Electric swing door operator and control, single door w/sensors	1 Carp	4	2		2,525	94		2,619	2,925
7005	Double door w/sensors		2	4		4,950	188		5,138	5,750
7010	Electric folding door operator and control, single door w/sensors		4	2		5,850	94		5,944	6,575
7015	Bi-folding door		4	2		7,025	94		7,119	7,875
7020	Electric swing door operator and control, single door		4	2		2,700	94		2,794	3,125

08 32 19 – Sliding Wood-Framed Glass Doors

08 32 19.15 Sliding Glass Vinyl-Clad Wood Doors

Code	Description		Crew	Daily Output	Labor-Hours	Unit	Material	Labor	Equipment	Total	Total Incl O&P
0010	**SLIDING GLASS VINYL-CLAD WOOD DOORS**										
0020	Glass, sliding vinyl clad, insul. glass, 6′-0″ x 6′-8″	G	2 Carp	4	4	Opng.	1,500	188		1,688	1,975
0025	6′-0″ x 6′-10″ high	G		4	4		1,650	188		1,838	2,125
0030	6′-0″ x 8′-0″ high	G		4	4		1,975	188		2,163	2,475
0050	5′-0″ x 6′-8″ high	G		4	4		1,500	188		1,688	1,950
0100	8′-0″ x 6′-10″ high	G		4	4		2,025	188		2,213	2,525

08 32 Sliding Glass Doors

08 32 19 – Sliding Wood-Framed Glass Doors

08 32 19.15 Sliding Glass Vinyl-Clad Wood Doors		Crew	Daily Output	Labor-Hours	Unit	Material	Labor	Equipment	Total	Total Incl O&P
						2015 Bare Costs				
0150	8'-0" x 8'-0" high G	2 Carp	4	4	Opng.	2,225	188		2,413	2,750
0500	4 leaf, 9'-0" x 6'-10" high G		3	5.333		3,250	250		3,500	3,975
0550	9'-0" x 8'-0" high G		3	5.333		3,775	250		4,025	4,550
0600	12'-0" x 6'-10" high G		3	5.333		3,875	250		4,125	4,675
0650	12'-0" x 8'-0" high G		3	5.333		3,875	250		4,125	4,675
9000	Minimum labor/equipment charge	1 Carp	4	2	Job		94		94	154

08 33 Coiling Doors and Grilles

08 33 13 – Coiling Counter Doors

08 33 13.10 Counter Doors, Coiling Type		Crew	Daily Output	Labor-Hours	Unit	Material	Labor	Equipment	Total	Total Incl O&P
0010	**COUNTER DOORS, COILING TYPE**									
0020	Manual, incl. frame and hardware, galv. stl., 4' roll-up, 6' long	2 Carp	2	8	Opng.	1,225	375		1,600	1,975
0300	Galvanized steel, UL label		1.80	8.889		1,225	415		1,640	2,025
0600	Stainless steel, 4' high roll-up, 6' long		2	8		2,150	375		2,525	2,975
0700	10' long		1.80	8.889		2,500	415		2,915	3,425
2000	Aluminum, 4' high, 4' long		2.20	7.273		1,525	340		1,865	2,225
2020	6' long		2	8		1,800	375		2,175	2,625
2040	8' long		1.90	8.421		2,050	395		2,445	2,900
2060	10' long		1.80	8.889		2,100	415		2,515	2,975
2080	14' long		1.40	11.429		2,675	535		3,210	3,825
2100	6' high, 4' long		2	8		1,850	375		2,225	2,650
2120	6' long		1.60	10		1,675	470		2,145	2,600
2140	10' long		1.40	11.429		2,200	535		2,735	3,300
9000	Minimum labor/equipment charge	1 Carp	2	4	Job		188		188	310

08 33 16 – Coiling Counter Grilles

08 33 16.10 Coiling Grilles		Crew	Daily Output	Labor-Hours	Unit	Material	Labor	Equipment	Total	Total Incl O&P
0010	**COILING GRILLES**									
0015	Aluminum, manual, incl. frame, mill finish									
0020	Top coiling, 4' high, 4' long	2 Sswk	3.20	5	Opng.	1,575	263		1,838	2,200
0030	6' long		3.20	5		1,825	263		2,088	2,500
0040	8' long		2.40	6.667		2,050	350		2,400	2,875
0050	12' long		2.40	6.667		2,475	350		2,825	3,350
0060	16' long		1.60	10		2,650	525		3,175	3,875
0070	6' high, 4' long		3.20	5		1,825	263		2,088	2,500
0080	6' long		3.20	5		1,800	263		2,063	2,450
0090	8' long		2.40	6.667		1,800	350		2,150	2,600
0100	12' long		1.60	10		2,375	525		2,900	3,575
0110	16' long		1.20	13.333		2,825	700		3,525	4,400
0200	Side coiling, 8' high, 12' long		.60	26.667		2,350	1,400		3,750	5,100
0220	18' long		.50	32		4,450	1,675		6,125	7,950
0240	24' long		.40	40		3,775	2,100		5,875	7,950
0260	12' high, 12' long		.50	32		3,400	1,675		5,075	6,775
0280	18' long		.40	40		5,300	2,100		7,400	9,625
0300	24' long		.28	57.143		6,725	3,000		9,725	12,800
9000	Minimum labor/equipment charge		1	16	Job		840		840	1,525

08 33 23 – Overhead Coiling Doors

08 33 23.10 Coiling Service Doors		Crew	Daily Output	Labor-Hours	Unit	Material	Labor	Equipment	Total	Total Incl O&P
0010	**COILING SERVICE DOORS** Steel, manual, 20 ga., incl. hardware									
0050	8' x 8' high	2 Sswk	1.60	10	Ea.	1,150	525		1,675	2,225
0100	10' x 10' high	"	1.40	11.429		1,925	600		2,525	3,175

08 33 Coiling Doors and Grilles

08 33 23 - Overhead Coiling Doors

08 33 23.10 Coiling Service Doors		Crew	Daily Output	Labor-Hours	Unit	Material	2015 Bare Costs Labor	Equipment	Total	Total Incl O&P
0101	10′ x 10′ high	L-4	1.40	17.143	Ea.	1,925	760		2,685	3,350
0130	12′ x 12′ high, standard	2 Sswk	1.20	13.333		2,175	700		2,875	3,675
0160	10′ x 20′ high, standard		.50	32		2,625	1,675		4,300	5,950
0200	20′ x 10′ high		1	16		3,150	840		3,990	5,000
0300	12′ x 12′ high		1.20	13.333		1,950	700		2,650	3,400
0303	Doors, rolling service, steel, manual, 20ga., 12′ x 12′, incl. hardware		.92	17.391		1,950	915		2,865	3,775
0400	20′ x 12′ high		.90	17.778		2,175	935		3,110	4,100
0420	14′ x 12′ high		.80	20		2,975	1,050		4,025	5,175
0430	14′ x 13′ high		.80	20		3,075	1,050		4,125	5,300
0500	14′ x 14′ high		.80	20		3,100	1,050		4,150	5,325
0520	14′ x 16′ high		.80	20	Opng.	2,850	1,050		3,900	5,050
0540	14′ x 20′ high		.56	28.571		4,050	1,500		5,550	7,175
0550	18′ x 18′ high		.60	26.667		2,525	1,400		3,925	5,300
0600	20′ x 16′ high		.60	26.667	Ea.	3,725	1,400		5,125	6,625
0700	10′ x 20′ high		.50	32	"	2,625	1,675		4,300	5,950
0750	16′ x 24′ high		.48	33.333	Opng.	6,400	1,750		8,150	10,200
1000	12′ x 12′, crank operated, crank on door side		.80	20	Ea.	1,775	1,050		2,825	3,875
1100	Crank thru wall		.70	22.857		2,050	1,200		3,250	4,450
1300	For vision panel, add					360			360	400
1600	3′ x 7′ pass door within rolling steel door, new construction					1,925			1,925	2,125
1700	Existing construction	2 Sswk	2	8		2,050	420		2,470	3,000
2000	Class A fire doors, manual, 20 ga., 8′ x 8′ high		1.40	11.429		1,600	600		2,200	2,825
2100	10′ x 10′ high		1.10	14.545		2,175	765		2,940	3,775
2200	20′ x 10′ high		.80	20		4,450	1,050		5,500	6,800
2300	12′ x 12′ high		1	16		3,425	840		4,265	5,300
2304	Overhead door, roll up, fire rated 12′ x 14′		.90	17.778		3,900	935		4,835	6,000
2400	20′ x 12′ high		.80	20		4,825	1,050		5,875	7,225
2500	14′ x 14′ high		.60	26.667		3,750	1,400		5,150	6,650
2600	20′ x 16′ high		.50	32		5,900	1,675		7,575	9,550
2700	10′ x 20′ high		.40	40		4,675	2,100		6,775	8,925
2730	Manual steel rollup fire doors		160	.100	S.F.	34.50	5.25		39.75	47.50
2740	For motor operated, add		1	16	Ea.	1,200	840		2,040	2,850
3000	For 18 ga. doors, add				S.F.	1.65			1.65	1.82
3300	For enamel finish, add				"	1.90			1.90	2.09
3600	For safety edge bottom bar, pneumatic, add				L.F.	23			23	25
3700	Electric, add					43			43	47.50
4000	For weatherstripping, extruded rubber, jambs, add					14.40			14.40	15.85
4100	Hood, add					8.20			8.20	9
4200	Sill, add					5.15			5.15	5.65
4500	Motor operators, to 14′ x 14′ opening	2 Sswk	5	3.200	Ea.	1,150	168		1,318	1,575
4600	Over 14′ x 14′, jack shaft type	"	5	3.200		1,075	168		1,243	1,475
4700	For fire door, additional fusible link, add					28			28	31
5100	Radio control operator to 12′ x 12′ door	2 Sswk	3	5.333		880	281		1,161	1,475
5120	Receiver to 12′ x 12′ door		13	1.231		195	65		260	330
5140	Transmitter to 12′ x 12′ door		49	.327		56.50	17.20		73.70	93
9000	Minimum labor/equipment charge		1	16	Job		840		840	1,525

08 34 Special Function Doors

08 34 13 – Cold Storage Doors

08 34 13.10 Doors for Cold Area Storage		Crew	Daily Output	Labor-Hours	Unit	2015 Bare Costs Material	Labor	Equipment	Total	Total Incl O&P
0010	**DOORS FOR COLD AREA STORAGE**									
0020	Single, 20 ga. galvanized steel									
0300	Horizontal sliding, 5′ x 7′, manual operation, 3.5″ thick	2 Carp	2	8	Ea.	3,250	375		3,625	4,200
0400	4″ thick		2	8		3,250	375		3,625	4,200
0500	6″ thick		2	8		3,425	375		3,800	4,400
0800	5′ x 7′, power operation, 2″ thick		1.90	8.421		5,600	395		5,995	6,800
0900	4″ thick		1.90	8.421		5,700	395		6,095	6,925
1000	6″ thick		1.90	8.421		6,500	395		6,895	7,800
1300	9′ x 10′, manual operation, 2″ insulation		1.70	9.412		4,550	440		4,990	5,725
1400	4″ insulation		1.70	9.412		4,675	440		5,115	5,875
1500	6″ insulation		1.70	9.412		5,575	440		6,015	6,875
1800	Power operation, 2″ insulation		1.60	10		7,675	470		8,145	9,225
1900	4″ insulation		1.60	10		7,900	470		8,370	9,450
2000	6″ insulation		1.70	9.412		8,850	440		9,290	10,500
2300	For stainless steel face, add					25%				
3000	Hinged, lightweight, 3′ x 7′-0″, 2″ thick	2 Carp	2	8	Ea.	1,425	375		1,800	2,175
3050	4″ thick		1.90	8.421		1,775	395		2,170	2,600
3300	Polymer doors, 3′ x 7′-0″		1.90	8.421		1,350	395		1,745	2,125
3350	6″ thick		1.40	11.429		2,350	535		2,885	3,450
3600	Stainless steel, 3′ x 7′-0″, 4″ thick		1.90	8.421		1,725	395		2,120	2,550
3650	6″ thick		1.40	11.429		2,950	535		3,485	4,100
3900	Painted, 3′ x 7′-0″, 4″ thick		1.90	8.421		1,275	395		1,670	2,050
3950	6″ thick		1.40	11.429		2,350	535		2,885	3,450
5000	Bi-parting, electric operated									
5010	6′ x 8′ opening, galv. faces, 4″ thick for cooler	2 Carp	.80	20	Opng.	7,225	940		8,165	9,475
5050	For freezer, 4″ thick		.80	20		7,950	940		8,890	10,300
5300	For door buck framing and door protection, add		2.50	6.400		640	300		940	1,200
6000	Galvanized batten door, galvanized hinges, 4′ x 7′		2	8		1,825	375		2,200	2,650
6050	6′ x 8′		1.80	8.889		2,475	415		2,890	3,400
6500	Fire door, 3 hr., 6′ x 8′, single slide		.80	20		8,275	940		9,215	10,700
6550	Double, bi-parting		.70	22.857		12,200	1,075		13,275	15,200
9000	Minimum labor/equipment charge	1 Carp	2	4	Job		188		188	310

08 34 16 – Hangar Doors

08 34 16.10 Aircraft Hangar Doors		Crew	Daily Output	Labor-Hours	Unit	Material	Labor	Equipment	Total	Total Incl O&P
0010	**AIRCRAFT HANGAR DOORS**									
0020	Bi-fold, ovhd., 20 psf wind load, incl. elec. oper.									
0100	12′ high x 40′	2 Sswk	240	.067	S.F.	16.45	3.51		19.96	24.50
0200	16′ high x 60′		230	.070		19.80	3.66		23.46	28.50
0300	20′ high x 80′		220	.073		21.50	3.83		25.33	30.50

08 34 36 – Darkroom Doors

08 34 36.10 Various Types of Darkroom Doors		Crew	Daily Output	Labor-Hours	Unit	Material	Labor	Equipment	Total	Total Incl O&P
0010	**VARIOUS TYPES OF DARKROOM DOORS**									
0015	Revolving, standard, 2 way, 36″ diameter	2 Carp	3.10	5.161	Opng.	2,900	242		3,142	3,575
0020	41″ diameter		3.10	5.161		3,000	242		3,242	3,725
0050	3 way, 51″ diameter		1.40	11.429		3,825	535		4,360	5,075
1000	4 way, 49″ diameter		1.40	11.429		3,950	535		4,485	5,225
2000	Hinged safety, 2 way, 41″ diameter		2.30	6.957		3,750	325		4,075	4,625
2500	3 way, 51″ diameter		1.40	11.429		4,050	535		4,585	5,350
3000	Pop out safety, 2 way, 41″ diameter		3.10	5.161		4,025	242		4,267	4,825
4000	3 way, 51″ diameter		1.40	11.429		4,875	535		5,410	6,250
5000	Wheelchair-type, pop out, 51″ diameter		1.40	11.429		5,500	535		6,035	6,925
5020	72″ diameter		.90	17.778		8,950	835		9,785	11,200

08 34 Special Function Doors

08 34 36 – Darkroom Doors

08 34 36.10 Various Types of Darkroom Doors		Crew	Daily Output	Labor-Hours	Unit	Material	2015 Bare Costs Labor	Equipment	Total	Total Incl O&P
9300	For complete darkrooms, see Section 13 21 53.50									

08 34 53 – Security Doors and Frames

08 34 53.20 Steel Door

		Crew	Daily Output	Labor-Hours	Unit	Material	Labor	Equipment	Total	Total Incl O&P
0010	**STEEL DOOR** with ballistic core and welded frame both 14 ga.									
0050	Flush, UL 752 Level 3, 1-3/4", 3'-0" x 6'-8"	2 Carp	1.50	10.667	Opng.	2,325	500		2,825	3,400
0055	1-3/4", 3'-6" x 6'-8"		1.50	10.667		2,425	500		2,925	3,500
0060	1-3/4", 4'-0" x 6'-8"		1.20	13.333		2,650	625		3,275	3,950
0100	UL 752 Level 8, 1-3/4", 3'-0" x 6'-8"		1.50	10.667		10,200	500		10,700	12,000
0105	1-3/4", 3'-6" x 6'-8"		1.50	10.667		10,300	500		10,800	12,100
0110	1-3/4", 4'-0" x 6'-8"		1.20	13.333		11,700	625		12,325	13,900
0120	UL 752 Level 3, 1-3/4", 3'-0" x 7'-0"		1.50	10.667		2,350	500		2,850	3,425
0125	1-3/4", 3'-6" x 7'-0"		1.50	10.667		2,475	500		2,975	3,550
0130	1-3/4", 4'-0" x 7'-0"		1.20	13.333		2,675	625		3,300	3,975
0150	UL 752 Level 8, 1-3/4", 3'-0" x 7'-0"		1.50	10.667		10,200	500		10,700	12,000
0155	1-3/4", 3'-6" x 7'-0"		1.50	10.667		10,300	500		10,800	12,100
0160	1-3/4", 4'-0" x 7'-0"		1.20	13.333		11,700	625		12,325	13,900
1000	Safe Room sliding door and hardware, 1-3/4", 3'-0" x 7'-0" UL 752 Level 3		.50	32		23,500	1,500		25,000	28,400
1050	Safe Room swinging door and hardware, 1-3/4", 3'-0" x 7'-0" UL 752 Level 3		.50	32		28,100	1,500		29,600	33,400

08 34 53.30 Wood Ballistic Doors

		Crew	Daily Output	Labor-Hours	Unit	Material	Labor	Equipment	Total	Total Incl O&P
0010	**WOOD BALLISTIC DOORS** with frames and hardware									
0050	Wood, 1-3/4", 3'-0" x 7'-0" UL 752 Level 3	2 Carp	1.50	10.667	Opng.	2,250	500		2,750	3,300

08 34 56 – Security Gates

08 34 56.10 Gates

		Crew	Daily Output	Labor-Hours	Unit	Material	Labor	Equipment	Total	Total Incl O&P
0010	**GATES**									
0015	Driveway Gates include mounting hardware									
0500	Wood, security gate, driveway, dual, 10' wide	H-4	.80	25	Opng.	3,900	1,100		5,000	6,075
0505	12' wide		.80	25		4,300	1,100		5,400	6,500
0510	15' wide		.80	25		5,100	1,100		6,200	7,375
0600	Steel, security gate, driveway, single, 10' wide		.80	25		2,150	1,100		3,250	4,125
0605	12' wide		.80	25		2,225	1,100		3,325	4,200
0620	Steel, security gate, driveway, dual, 12' wide		.80	25		2,050	1,100		3,150	4,050
0625	14' wide		.80	25		2,200	1,100		3,300	4,200
0630	16' wide		.80	25		2,525	1,100		3,625	4,550
0700	Aluminum, security gate, driveway, dual, 10' wide		.80	25		3,000	1,100		4,100	5,075
0705	12' wide		.80	25		3,200	1,100		4,300	5,300
0710	16' wide		.80	25		3,700	1,100		4,800	5,850
1000	Security gate, driveway, opener 12 VDC				Ea.	800			800	880
1010	Wireless					1,800			1,800	1,975
1020	Security gate, driveway, opener 24 VDC					1,300			1,300	1,425
1030	Wireless					1,450			1,450	1,600
1040	Security gate, driveway, opener 12 VDC, solar panel 10 watt	1 Elec	2	4		300	219		519	670
1050	20 watt	"	2	4		450	219		669	835

08 34 59 – Vault Doors and Day Gates

08 34 59.10 Secure Storage Doors

		Crew	Daily Output	Labor-Hours	Unit	Material	Labor	Equipment	Total	Total Incl O&P
0010	**SECURE STORAGE DOORS**									
0020	Door and frame, 32" x 78", clear opening									
0100	1 hour test, 32" door, weighs 750 lb.	2 Sswk	1.50	10.667	Opng.	6,625	560		7,185	8,325
0200	2 hour test, 32" door, weighs 950 lb.		1.30	12.308		8,200	650		8,850	10,200
0250	40" door, weighs 1130 lb.		1	16		9,250	840		10,090	11,700
0300	4 hour test, 32" door, weighs 1025 lb.		1.20	13.333		8,650	700		9,350	10,800
0350	40" door, weighs 1140 lb.		.90	17.778		9,975	935		10,910	12,700

08 34 Special Function Doors

08 34 59 – Vault Doors and Day Gates

08 34 59.10 Secure Storage Doors		Crew	Daily Output	Labor-Hours	Unit	Material	Labor	Equipment	Total	Total Incl O&P
						2015 Bare Costs				
0600	For time lock, two movement, add	1 Elec	2	4	Ea.	1,800	219		2,019	2,325
0800	Day gate, painted, steel, 32" wide	2 Sswk	1.50	10.667		2,000	560		2,560	3,225
0850	40" wide		1.40	11.429		2,100	600		2,700	3,375
0900	Aluminum, 32" wide		1.50	10.667		3,200	560		3,760	4,550
0950	40" wide		1.40	11.429		3,400	600		4,000	4,825
2050	Security vault door, class I, 3' wide, 3 1/2" thick	E-24	.19	166	Opng.	15,700	8,700	3,825	28,225	36,700
2100	Class II, 3' wide, 7" thick		.19	166		18,600	8,700	3,825	31,125	39,900
2150	Class III, 9R, 3' wide, 10" thick		.13	250		23,500	13,000	5,750	42,250	55,000
2160	Class V, type 1, 40" door		2.48	12.903	Ea.	7,125	675	297	8,097	9,325
2170	Class V, type 2, 40" door		2.48	12.903		6,850	675	297	7,822	9,025
2180	Day gate for class V vault	2 Sswk	2	8		2,025	420		2,445	2,975

08 34 63 – Detention Doors and Frames

08 34 63.13 Steel Detention Doors and Frames		Crew	Daily Output	Labor-Hours	Unit	Material	Labor	Equipment	Total	Total Incl O&P
0010	**STEEL DETENTION DOORS AND FRAMES**									
0500	Rolling cell door, bar front, 7/8" bars, 4" O.C., 7' H, 5' W, with hardware G	E-4	2	16	Ea.	5,325	850	73	6,248	7,475
0550	Actuator for rolling cell door, bar front	2 Skwk	2	8		4,250	390		4,640	5,300
1000	Doors & frames, 3' x 7', complete, with hardware, single plate	E-4	4	8		4,600	425	36.50	5,061.50	5,850
1650	Double plate	"	4	8		5,600	425	36.50	6,061.50	6,950

08 34 73 – Sound Control Door Assemblies

08 34 73.10 Acoustical Doors		Crew	Daily Output	Labor-Hours	Unit	Material	Labor	Equipment	Total	Total Incl O&P
0010	**ACOUSTICAL DOORS**									
0020	Including framed seals, 3' x 7', wood, 40 STC rating	2 Carp	1.50	10.667	Ea.	1,300	500		1,800	2,250
0100	Steel, 41 STC rating		1.50	10.667		3,275	500		3,775	4,425
0200	45 STC rating		1.50	10.667		3,675	500		4,175	4,875
0300	48 STC rating		1.50	10.667		4,275	500		4,775	5,550
0400	52 STC rating		1.50	10.667		4,900	500		5,400	6,225
9000	Minimum labor/equipment charge	1 Carp	4	2	Job		94		94	154

08 36 Panel Doors

08 36 13 – Sectional Doors

08 36 13.10 Overhead Commercial Doors		Crew	Daily Output	Labor-Hours	Unit	Material	Labor	Equipment	Total	Total Incl O&P
0010	**OVERHEAD COMMERCIAL DOORS**									
1000	Stock, sectional, heavy duty, wood, 1-3/4" thick, 8' x 8' high	2 Carp	2	8	Ea.	1,050	375		1,425	1,775
1100	10' x 10' high		1.80	8.889		1,500	415		1,915	2,325
1200	12' x 12' high		1.50	10.667		2,050	500		2,550	3,075
1300	Chain hoist, 14' x 14' high		1.30	12.308		3,300	580		3,880	4,575
1400	12' x 16' high		1	16		3,275	750		4,025	4,825
1500	20' x 8' high		1.30	12.270		2,675	575		3,250	3,900
1600	20' x 16' high		.65	24.615		5,475	1,150		6,625	7,925
1800	Center mullion openings, 8' high		4	4		1,225	188		1,413	1,650
1900	20' high		2	8		2,050	375		2,425	2,875
2100	For medium duty custom door, deduct					5%	5%			
2150	For medium duty stock doors, deduct					10%	5%			
2300	Fiberglass and aluminum, heavy duty, sectional, 12' x 12' high	2 Carp	1.50	10.667	Ea.	2,700	500		3,200	3,800
2450	Chain hoist, 20' x 20' high		.50	32		6,675	1,500		8,175	9,800
2600	Steel, 24 ga. sectional, manual, 8' x 8' high		2	8		820	375		1,195	1,525
2650	10' x 10' high		1.80	8.889		1,100	415		1,515	1,900
2700	12' x 12' high		1.50	10.667		1,350	500		1,850	2,300
2800	Chain hoist, 20' x 14' high		.70	22.857		3,725	1,075		4,800	5,850
2850	For 1-1/4" rigid insulation and 26 ga. galv.									

08 36 13.10 Overhead Commercial Doors		Crew	Daily Output	Labor-Hours	Unit	Material	2015 Bare Costs Labor	Equipment	Total	Total Incl O&P
2860	back panel, add				S.F.	4.75			4.75	5.25
2900	For electric trolley operator, 1/3 H.P., to 12′ x 12′, add	1 Carp	2	4	Ea.	950	188		1,138	1,350
2950	Over 12′ x 12′, 1/2 H.P., add		1	8		1,125	375		1,500	1,875
2980	Overhead, for row of clear lites add		1	8		110	375		485	735
9000	Minimum labor/equipment charge	2 Carp	1.50	10.667	Job		500		500	820

08 36 13.20 Residential Garage Doors		Crew	Daily Output	Labor-Hours	Unit	Material	Labor	Equipment	Total	Total Incl O&P
0010	**RESIDENTIAL GARAGE DOORS**									
0050	Hinged, wood, custom, double door, 9′ x 7′	2 Carp	4	4	Ea.	800	188		988	1,200
0070	16′ x 7′		3	5.333		1,200	250		1,450	1,725
0200	Overhead, sectional, incl. hardware, fiberglass, 9′ x 7′, standard		5.28	3.030		945	142		1,087	1,275
0220	Deluxe		5.28	3.030		1,150	142		1,292	1,475
0300	16′ x 7′, standard		6	2.667		1,575	125		1,700	1,925
0320	Deluxe		6	2.667		2,150	125		2,275	2,550
0500	Hardboard, 9′ x 7′, standard		8	2		625	94		719	845
0520	Deluxe		8	2		815	94		909	1,050
0600	16′ x 7′, standard		6	2.667		1,225	125		1,350	1,550
0620	Deluxe		6	2.667		1,425	125		1,550	1,775
0700	Metal, 9′ x 7′, standard		5.28	3.030		740	142		882	1,050
0720	Deluxe		8	2		915	94		1,009	1,150
0800	16′ x 7′, standard		3	5.333		940	250		1,190	1,425
0820	Deluxe		6	2.667		1,400	125		1,525	1,750
0900	Wood, 9′ x 7′, standard		8	2		970	94		1,064	1,225
0920	Deluxe		8	2		2,150	94		2,244	2,500
1000	16′ x 7′, standard		6	2.667		1,625	125		1,750	2,000
1020	Deluxe		6	2.667		3,025	125		3,150	3,525
1800	Door hardware, sectional	1 Carp	4	2		350	94		444	540
1810	Door tracks only		4	2		163	94		257	335
1820	One side only		7	1.143		120	53.50		173.50	220
3000	Swing-up, including hardware, fiberglass, 9′ x 7′, standard	2 Carp	8	2		1,000	94		1,094	1,250
3020	Deluxe		8	2		1,100	94		1,194	1,350
3100	16′ x 7′, standard		6	2.667		1,250	125		1,375	1,575
3120	Deluxe		6	2.667		1,600	125		1,725	1,950
3200	Hardboard, 9′ x 7′, standard		8	2		550	94		644	760
3220	Deluxe		8	2		650	94		744	870
3300	16′ x 7′, standard		6	2.667		670	125		795	940
3320	Deluxe		6	2.667		850	125		975	1,150
3400	Metal, 9′ x 7′, standard		8	2		600	94		694	815
3420	Deluxe		8	2		965	94		1,059	1,200
3500	16′ x 7′, standard		6	2.667		800	125		925	1,075
3520	Deluxe		6	2.667		1,100	125		1,225	1,400
3600	Wood, 9′ x 7′, standard		8	2		700	94		794	925
3620	Deluxe		8	2		1,125	94		1,219	1,375
3700	16′ x 7′, standard		6	2.667		900	125		1,025	1,200
3720	Deluxe		6	2.667		2,100	125		2,225	2,500
3900	Door hardware only, swing up	1 Carp	4	2		168	94		262	340
3920	One side only		7	1.143		90	53.50		143.50	187
4000	For electric operator, economy, add		8	1		425	47		472	545
4100	Deluxe, including remote control		8	1		610	47		657	750
4500	For transmitter/receiver control , add to operator				Total	110			110	121
4600	Transmitters, additional				"	60			60	66
7010	Garage doors, row of lites				Ea.	115			115	127
9000	Minimum labor/equipment charge	1 Carp	2.50	3.200	Job		150		150	246

08 36 Panel Doors

08 36 19 – Multi-Leaf Vertical Lift Doors

08 36 19.10 Sectional Vertical Lift Doors		Crew	Daily Output	Labor-Hours	Unit	2015 Bare Costs Material	Labor	Equipment	Total	Total Incl O&P
0010	SECTIONAL VERTICAL LIFT DOORS									
0020	Motorized, 14 ga. steel, incl. frame and control panel									
0050	16′ x 16′ high	L-10	.50	48	Ea.	21,500	2,550	1,300	25,350	29,600
0100	10′ x 20′ high		1.30	18.462		34,800	980	505	36,285	40,600
0120	15′ x 20′ high		1.30	18.462		42,600	980	505	44,085	49,200
0140	20′ x 20′ high		1	24		49,900	1,275	655	51,830	58,000
0160	25′ x 20′ high		1	24		56,000	1,275	655	57,930	64,500
0170	32′ x 24′ high		.75	32		48,700	1,700	870	51,270	57,500
0180	20′ x 25′ high		1	24		57,500	1,275	655	59,430	66,000
0200	25′ x 25′ high		.70	34.286		66,000	1,825	935	68,760	77,000
0220	25′ x 30′ high		.70	34.286		71,000	1,825	935	73,760	82,500
0240	30′ x 30′ high		.70	34.286		82,500	1,825	935	85,260	94,500
0260	35′ x 30′ high		.70	34.286		92,500	1,825	935	95,260	106,000

08 38 Traffic Doors

08 38 13 – Flexible Strip Doors

08 38 13.10 Flexible Transparent Strip Doors

Line	Description	Crew	Daily Output	Labor-Hours	Unit	Material	Labor	Equipment	Total	Total Incl O&P
0010	FLEXIBLE TRANSPARENT STRIP DOORS									
0100	12″ strip width, 2/3 overlap	3 Shee	135	.178	SF Surf	7.60	9.95		17.55	24
0200	Full overlap		115	.209		9.60	11.70		21.30	29
0220	8″ strip width, 1/2 overlap		140	.171		6.20	9.60		15.80	22
0240	Full overlap		120	.200		7.80	11.20		19	26.50
0300	Add for suspension system, header mount				L.F.	9.15			9.15	10.05
0400	Wall mount				″	9.45			9.45	10.40

08 38 16 – Flexible Traffic Doors

08 38 16.10 Rubber Doors

Line	Description	Crew	Daily Output	Labor-Hours	Unit	Material	Labor	Equipment	Total	Total Incl O&P
0010	RUBBER DOORS									
0015	Incl. frame and hardware, 14″ x 16″ vision panel,									
0020	48″ wear panel, pair of 2′-6″ x 7′	2 Sswk	1.60	10	Ea.	2,225	525		2,750	3,400
0040	Pair of 3′ x 7′		1.50	10.667		2,550	560		3,110	3,825
0060	Pair of 4′ x 7′		1.40	11.429		3,050	600		3,650	4,450

08 38 19 – Rigid Traffic Doors

08 38 19.20 Double Acting Swing Doors

Line	Description	Crew	Daily Output	Labor-Hours	Unit	Material	Labor	Equipment	Total	Total Incl O&P
0010	DOUBLE ACTING SWING DOORS									
0020	Including frame, closer, hardware and vision panel									
1000	Polymer, 7′-0″ high, 4′-0″ wide	2 Carp	4.20	3.810	Pr.	2,100	179		2,279	2,600
1025	6′-0″ wide		4	4		2,200	188		2,388	2,725
1050	6′-8″ wide		4	4		2,400	188		2,588	2,950
2000	3/4″ thick, stainless steel									
2010	Stainless steel, 7′ high opening, 4′ wide	2 Carp	4	4	Pr.	2,600	188		2,788	3,150
2050	7′ wide		3.80	4.211	″	2,800	198		2,998	3,400
9000	Minimum labor/equipment charge		2	8	Job		375		375	615

08 38 19.30 Shock Absorbing Doors

Line	Description	Crew	Daily Output	Labor-Hours	Unit	Material	Labor	Equipment	Total	Total Incl O&P
0010	SHOCK ABSORBING DOORS									
0020	Rigid, no frame, 1-1/2″ thick, 5′ x 7′	2 Sswk	1.90	8.421	Opng.	1,525	445		1,970	2,475
0100	8′ x 8′		1.80	8.889		2,000	470		2,470	3,050
0500	Flexible, no frame, insulated, .16″ thick, economy, 5′ x 7′		2	8		1,750	420		2,170	2,675
0600	Deluxe		1.90	8.421		2,625	445		3,070	3,675
1000	8′ x 8′ opening, economy		2	8		2,750	420		3,170	3,775

08 38 Traffic Doors

08 38 19 – Rigid Traffic Doors

08 38 19.30 Shock Absorbing Doors		Crew	Daily Output	Labor-Hours	Unit	2015 Bare Costs Material	Labor	Equipment	Total	Total Incl O&P
1100	Deluxe	2 Sswk	1.90	8.421	Opng.	3,500	445		3,945	4,650
9000	Minimum labor/equipment charge		2	8	Job		420		420	760

08 41 Entrances and Storefronts

08 41 13 – Aluminum-Framed Entrances and Storefronts

08 41 13.20 Tube Framing

Line	Description	Crew	Daily Output	Labor-Hours	Unit	Material	Labor	Equipment	Total	Total Incl O&P
0010	**TUBE FRAMING**, For window walls and store fronts, aluminum stock									
0050	Plain tube frame, mill finish, 1-3/4″ x 1-3/4″	2 Glaz	103	.155	L.F.	9.90	7		16.90	22.50
0150	1-3/4″ x 4″		98	.163		13.35	7.35		20.70	26.50
0200	1-3/4″ x 4-1/2″		95	.168		16	7.60		23.60	30
0250	2″ x 6″		89	.180		23.50	8.10		31.60	39
0350	4″ x 4″		87	.184		26.50	8.30		34.80	43
0400	4-1/2″ x 4-1/2″		85	.188		28	8.50		36.50	45
0450	Glass bead		240	.067		3.07	3.01		6.08	8.25
1000	Flush tube frame, mill finish, 1/4″ glass, 1-3/4″ x 4″, open header		80	.200		13.25	9		22.25	29.50
1050	Open sill		82	.195		10.85	8.80		19.65	26
1100	Closed back header		83	.193		19	8.70		27.70	35
1150	Closed back sill		85	.188		18.15	8.50		26.65	34
1160	Tube fmg., spandrel cover both sides, alum 1″ wide	1 Sswk	85	.094	S.F.	98	4.96		102.96	117
1170	Tube fmg., spandrel cover both sides, alum 2″ wide	″	85	.094	″	38.50	4.96		43.46	51
1200	Vertical mullion, one piece	2 Glaz	75	.213	L.F.	19.90	9.60		29.50	37.50
1250	Two piece		73	.219		21	9.90		30.90	39.50
1300	90° or 180° vertical corner post		75	.213		32.50	9.60		42.10	51
1400	1-3/4″ x 4-1/2″, open header		80	.200		16	9		25	32.50
1450	Open sill		82	.195		13.55	8.80		22.35	29
1500	Closed back header		83	.193		19.10	8.70		27.80	35
1550	Closed back sill		85	.188		18.95	8.50		27.45	35
1600	Vertical mullion, one piece		75	.213		21.50	9.60		31.10	39
1650	Two piece		73	.219		22.50	9.90		32.40	40.50
1700	90° or 180° vertical corner post		75	.213		23	9.60		32.60	41
2000	Flush tube frame, mil fin.,ins. glass w/thml brk, 2″ x 4-1/2″, open header		75	.213		16.30	9.60		25.90	33.50
2050	Open sill		77	.208		13.70	9.35		23.05	30.50
2100	Closed back header		78	.205		15.50	9.25		24.75	32
2150	Closed back sill		80	.200		14.95	9		23.95	31
2200	Vertical mullion, one piece		70	.229		17.20	10.30		27.50	35.50
2250	Two piece		68	.235		18.60	10.60		29.20	37.50
2300	90° or 180° vertical corner post		70	.229		17.70	10.30		28	36.50
5000	Flush tube frame, mill fin., thermal brk., 2-1/4″ x 4-1/2″, open header		74	.216		17.15	9.75		26.90	34.50
5050	Open sill		75	.213		15.10	9.60		24.70	32
5100	Vertical mullion, one piece		69	.232		17.75	10.45		28.20	36.50
5150	Two piece		67	.239		21.50	10.75		32.25	41
5200	90° or 180° vertical corner post		69	.232		19.10	10.45		29.55	38
5295	Flush tube frame, mill finish, 4-1/2″ x 4-1/2″		85	.188		28	8.50		36.50	45
5300	Plain tube frame, mill finish, 2″ x 3″ jamb		115	.139		10.90	6.25		17.15	22
5310	2″ x 4″		115	.139		16.60	6.25		22.85	28.50
5320	3″ x 5-1/2″		107	.150		26.50	6.75		33.25	40.50
5330	Head, 2″ x 3″		200	.080		19.95	3.61		23.56	28
5340	Mullion, 2″ x 3″		200	.080		20.50	3.61		24.11	28.50
5350	2″ x 4″		115	.139		24	6.25		30.25	36.50
5360	3″ x 5-1/2″		107	.150		31.50	6.75		38.25	46
5370	4″ corner mullion		90	.178		29.50	8		37.50	45.50

08 41 Entrances and Storefronts

08 41 13 – Aluminum-Framed Entrances and Storefronts

08 41 13.20 Tube Framing		Crew	Daily Output	Labor-Hours	Unit	Material	2015 Bare Costs Labor	Equipment	Total	Total Incl O&P
5380	Horizontal, 2″ x 3″	2 Glaz	115	.139	L.F.	23.50	6.25		29.75	36
5390	3″ x 5-1/2″		107	.150		29	6.75		35.75	43
5430	Sill section, 1/8″ x 6″		133	.120		36	5.45		41.45	48.50
5440	1/8″ x 7″		133	.120		43	5.45		48.45	56.50
5450	1/8″ x 8-1/2″		133	.120		50.50	5.45		55.95	64.50
5460	Column covers, aluminum, 1/8″ x 26″		57	.281		65	12.65		77.65	92
5470	1/8″ x 34″		53	.302		83.50	13.60		97.10	114
5480	1/8″ x 38″		53	.302		93	13.60		106.60	124
5700	Vertical mullions, clear finish, 1/4″ thick glass		528	.030		12.45	1.37		13.82	15.90
5720	3/8″ thick		445	.036		12.45	1.62		14.07	16.35
5740	1/2″ thick		400	.040		12.90	1.80		14.70	17.10
5760	3/4″ thick		344	.047		13.35	2.10		15.45	18.10
5780	1″ thick		304	.053		13.50	2.37		15.87	18.70
6980	Door stop (snap in)		380	.042		3.40	1.90		5.30	6.80
7000	For joints, 90°, clip type, add				Ea.	25.50			25.50	28.50
7050	Screw spline joint, add					24			24	26
7100	For joint other than 90°, add					49.50			49.50	54.50
8000	For bronze anodized aluminum, add					15%				
8020	For black finish, add					30%				
8050	For stainless steel materials, add					350%				
8100	For monumental grade, add					53%				
8150	For steel stiffener, add	2 Glaz	200	.080	L.F.	11.40	3.61		15.01	18.40
8200	For 2 to 5 stories, add per story				Story		8%			
9000	Minimum labor/equipment charge	2 Glaz	2	8	Job		360		360	585

08 41 19 – Stainless-Steel-Framed Entrances and Storefronts

08 41 19.10 Stainless-Steel and Glass Entrance Unit		Crew	Daily Output	Labor-Hours	Unit	Material	Labor	Equipment	Total	Total Incl O&P
0010	**STAINLESS-STEEL AND GLASS ENTRANCE UNIT**, narrow stiles									
0020	3′ x 7′ opening, including hardware, minimum	2 Sswk	1.60	10	Opng.	6,800	525		7,325	8,425
0050	Average		1.40	11.429		7,250	600		7,850	9,050
0100	Maximum		1.20	13.333		7,825	700		8,525	9,900
1000	For solid bronze entrance units, statuary finish, add					64%				
1100	Without statuary finish, add					45%				
2000	Balanced doors, 3′ x 7′, economy	2 Sswk	.90	17.778	Ea.	9,200	935		10,135	11,800
2100	Premium		.70	22.857	″	15,500	1,200		16,700	19,300
9000	Minimum labor/equipment charge		2	8	Job		420		420	760

08 41 26 – All-Glass Entrances and Storefronts

08 41 26.10 Window Walls Aluminum, Stock		Crew	Daily Output	Labor-Hours	Unit	Material	Labor	Equipment	Total	Total Incl O&P
0010	**WINDOW WALLS ALUMINUM, STOCK**, including glazing									
0020	Minimum	H-2	160	.150	S.F.	46.50	6.40		52.90	62
0050	Average		140	.171		64	7.30		71.30	82.50
0100	Maximum		110	.218		172	9.30		181.30	204
0500	For translucent sandwich wall systems, see Section 07 41 33.10									
0850	Cost of the above walls depends on material,									
0860	finish, repetition, and size of units.									
0870	The larger the opening, the lower the S.F. cost									
1200	Double glazed acoustical window wall for airports,									

08 41 26.20 All-Glass Entrance Doors		Crew	Daily Output	Labor-Hours	Unit	Material	Labor	Equipment	Total	Total Incl O&P
0010	**ALL-GLASS ENTRANCE DOORS**									
0015	Hardware and sst trim, temp glass, 1/2″ thk, 3′ x 7′, single	2 Sswk	1.85	8.649	Ea.	3,150	455		3,605	4,300
0020	Pair of 3′ x 7′		1.05	15.238	Pr.	5,000	800		5,800	6,950
0040	3/4″ thick, 3′ x 7′, single		1.85	8.649	Ea.	5,775	455		6,230	7,175

08 41 Entrances and Storefronts

08 41 26 – All-Glass Entrances and Storefronts

08 41 26.20 All-Glass Entrance Doors		Crew	Daily Output	Labor-Hours	Unit	Material	2015 Bare Costs Labor	Equipment	Total	Total Incl O&P
0060	Pair of 3′ x 7′	2 Sswk	1.05	15.238	Pr.	11,500	800		12,300	14,200

08 42 Entrances

08 42 26 – All-Glass Entrances

08 42 26.10 Swinging Glass Doors

		Crew	Daily Output	Labor-Hours	Unit	Material	Labor	Equipment	Total	Total Incl O&P
0010	**SWINGING GLASS DOORS**									
0020	Including hardware, 1/2″ thick, tempered, 3′ x 7′ opening	2 Glaz	2	8	Opng.	2,200	360		2,560	3,000
0100	6′ x 7′ opening		1.40	11.429	″	4,400	515		4,915	5,650
9000	Minimum labor/equipment charge		2	8	Job		360		360	585

08 42 33 – Revolving Door Entrances

08 42 33.10 Circular Rotating Entrance Doors

		Crew	Daily Output	Labor-Hours	Unit	Material	Labor	Equipment	Total	Total Incl O&P
0010	**CIRCULAR ROTATING ENTRANCE DOORS**, Aluminum									
0020	6′-10″ to 7′ high, stock units, minimum	4 Sswk	.75	42.667	Opng.	22,300	2,250		24,550	28,600
0050	Average		.60	53.333		26,300	2,800		29,100	34,000
0100	Maximum		.45	71.111		32,700	3,750		36,450	42,800
1000	Stainless steel		.30	105		41,000	5,575		46,575	55,000
1100	Solid bronze		.15	213		48,000	11,200		59,200	73,500
1500	For automatic controls, add	2 Elec	2	8		15,100	440		15,540	17,300

08 42 36 – Balanced Door Entrances

08 42 36.10 Balanced Entrance Doors

		Crew	Daily Output	Labor-Hours	Unit	Material	Labor	Equipment	Total	Total Incl O&P
0010	**BALANCED ENTRANCE DOORS**									
0020	Hardware & frame, alum. & glass, 3′ x 7′, econ.	2 Sswk	.90	17.778	Ea.	6,600	935		7,535	8,975
0150	Premium		.70	22.857	″	7,925	1,200		9,125	10,900
9000	Minimum labor/equipment charge		1	16	Job		840		840	1,525

08 43 Storefronts

08 43 13 – Aluminum-Framed Storefronts

08 43 13.10 Aluminum-Framed Entrance Doors and Frames

		Crew	Daily Output	Labor-Hours	Unit	Material	Labor	Equipment	Total	Total Incl O&P
0010	**ALUMINUM-FRAMED ENTRANCE DOORS AND FRAMES**									
0015	Standard hardware and glass stops but no glass									
0020	Entrance door, 3′ x 7′ opening, clear anodized finish	2 Sswk	7	2.286	Opng.	525	120		645	795
0040	Bronze finish		7	2.286		530	120		650	800
0060	Black finish		7	2.286		575	120		695	845
0200	3′-6″ x 7′-0″, mill finish		7	2.286		635	120		755	910
0220	Bronze finish		7	2.286		655	120		775	935
0240	Black finish		7	2.286		750	120		870	1,050
0500	6′ x 7′ opening, clear finish		6	2.667		840	140		980	1,175
0520	Bronze finish		6	2.667		910	140		1,050	1,250
0600	Door Frame for above doors 3′-0″ x 7′-0″, mill finish		6	2.667		405	140		545	700
0620	Bronze finish		6	2.667		465	140		605	765
0640	Black finish		6	2.667		505	140		645	810
0700	3′-6″ x 7′-0″, mill finish		6	2.667		330	140		470	620
0720	Bronze finish		6	2.667		330	140		470	620
0740	Black finish		6	2.667		330	140		470	620
0800	6′-0″ x 7′-0″, mill finish		6	2.667		330	140		470	620
0820	Bronze finish		6	2.667		335	140		475	625
0840	Black finish		6	2.667		360	140		500	650
1000	With 3′ high transom above, 3′ x 7′ opening, clear finish		5.50	2.909		490	153		643	815

08 43 Storefronts

08 43 13 – Aluminum-Framed Storefronts

08 43 13.10 Aluminum-Framed Entrance Doors and Frames		Crew	Daily Output	Labor-Hours	Unit	Material	2015 Bare Costs Labor	Equipment	Total	Total Incl O&P
1050	Bronze finish	2 Sswk	5.50	2.909	Opng.	510	153		663	835
1100	Black finish		5.50	2.909		535	153		688	860
1300	3'-6" x 7'-0" opening, clear finish		5.50	2.909		340	153		493	650
1320	Bronze finish		5.50	2.909		355	153		508	670
1340	Black finish		5.50	2.909		365	153		518	675
1500	6' x 7' opening, clear finish		5.50	2.909		595	153		748	930
1550	Bronze finish		5.50	2.909		615	153		768	950
1600	Black finish		5.50	2.909		675	153		828	1,025
8000	For 8' high doors add				Ea.	200			200	220
9000	Minimum labor/equipment charge	2 Sswk	4	4	Job		211		211	380

08 43 13.20 Storefront Systems		Crew	Daily Output	Labor-Hours	Unit	Material	Labor	Equipment	Total	Total Incl O&P
0010	**STOREFRONT SYSTEMS**, aluminum frame clear 3/8" plate glass									
0020	incl. 3' x 7' door with hardware (400 sq. ft. max. wall)									
0500	Wall height to 12' high, commercial grade	2 Glaz	150	.107	S.F.	22.50	4.81		27.31	33
0600	Institutional grade		130	.123		28	5.55		33.55	40
0700	Monumental grade		115	.139		40.50	6.25		46.75	54.50
1000	6' x 7' door with hardware, commercial grade		135	.119		30	5.35		35.35	41.50
1100	Institutional grade		115	.139		28.50	6.25		34.75	41.50
1200	Monumental grade		100	.160		54.50	7.20		61.70	71.50
1500	For bronze anodized finish, add					15%				
1600	For black anodized finish, add					36%				
1700	For stainless steel framing, add to monumental					78%				
9000	Minimum labor/equipment charge	2 Glaz	1	16	Job		720		720	1,175

08 43 29 – Sliding Storefronts

08 43 29.10 Sliding Panels		Crew	Daily Output	Labor-Hours	Unit	Material	Labor	Equipment	Total	Total Incl O&P
0010	**SLIDING PANELS**									
0020	Mall fronts, aluminum & glass, 15' x 9' high	2 Glaz	1.30	12.308	Opng.	3,475	555		4,030	4,725
0100	24' x 9' high		.70	22.857		4,925	1,025		5,950	7,100
0200	48' x 9' high, with fixed panels		.90	17.778		9,100	800		9,900	11,300
0500	For bronze finish, add					17%				
9000	Minimum labor/equipment charge	2 Glaz	1	16	Job		720		720	1,175

08 44 Curtain Wall and Glazed Assemblies

08 44 13 – Glazed Aluminum Curtain Walls

08 44 13.10 Glazed Curtain Walls		Crew	Daily Output	Labor-Hours	Unit	Material	Labor	Equipment	Total	Total Incl O&P
0010	**GLAZED CURTAIN WALLS**, aluminum, stock, including glazing									
0020	Minimum	H-1	205	.156	S.F.	37.50	7.65		45.15	54.50
0050	Average, single glazed		195	.164		53.50	8		61.50	72.50
0150	Average, double glazed		180	.178		69	8.70		77.70	91
0200	Maximum		160	.200		181	9.80		190.80	216

08 45 Translucent Wall and Roof Assemblies

08 45 10 – Translucent Roof Assemblies

08 45 10.10 Skyroofs		Crew	Daily Output	Labor-Hours	Unit	2015 Bare Costs Material	Labor	Equipment	Total	Total Incl O&P
0010	**SKYROOFS,**									
1200	Skylights, circular, clear, double glazed acrylic									
1230	30″ diameter	2 Carp	3	5.333	Ea.	3,000	250		3,250	3,700
1250	60″ diameter		3	5.333		4,000	250		4,250	4,800
1290	96″ diameter		2	8		5,000	375		5,375	6,125
1300	Skylight Barrel Vault, clear, double glazed, acrylic									
1330	3′-0″ X 12′-0″	G-3	3	10.667	Ea.	5,000	500		5,500	6,300
1350	4′-0″ X 12′-0″		3	10.667		5,500	500		6,000	6,850
1390	5′-0″ X 12′-0″		2	16		6,000	750		6,750	7,800
1400	Skylight Pyramid, Aluminum frame, clear low-E laminated glass									
1410	The glass is installed in the frame except where noted									
1430	Square, 3′ X 3′	G-3	3	10.667	Ea.	6,000	500		6,500	7,400
1440	4′ X 4′		3	10.667		7,000	500		7,500	8,500
1450	5′ X 5′, glass must be field installed		3	10.667		8,000	500		8,500	9,600
1460	6′ X 6′, glass must be field installed		2	16		10,000	750		10,750	12,200
1550	Install pre-cut laminated glass in aluminum frame on a flat roof	2 Glaz	55	.291	SF Surf		13.10		13.10	21.50
1560	Install pre-cut laminated glass in aluminum frame on a sloped roof	″	40	.400	″		18.05		18.05	29.50
9000	Minimum labor/equipment charge	2 Carp	8	2	Job		94		94	154

08 51 Metal Windows

08 51 13 – Aluminum Windows

08 51 13.10 Aluminum Sash

Code	Description	Crew	Daily Output	Labor-Hours	Unit	Material	Labor	Equipment	Total	Total Incl O&P
0010	**ALUMINUM SASH**									
0020	Stock, grade C, glaze & trim not incl., casement	2 Sswk	200	.080	S.F.	39	4.21		43.21	50.50
0050	Double hung		200	.080		39.50	4.21		43.71	51
0100	Fixed casement		200	.080		17.35	4.21		21.56	26.50
0150	Picture window		200	.080		18.50	4.21		22.71	28
0200	Projected window		200	.080		35.50	4.21		39.71	46.50
0250	Single hung		200	.080		16.55	4.21		20.76	26
0300	Sliding		200	.080		21.50	4.21		25.71	31
1000	Mullions for above, tubular		240	.067	L.F.	6.15	3.51		9.66	13.15
2000	Custom aluminum sash, grade HC, glazing not included		140	.114	S.F.	40	6		46	55
9000	Minimum labor/equipment charge	1 Sswk	2	4	Job		211		211	380

08 51 13.20 Aluminum Windows

Code	Description	Crew	Daily Output	Labor-Hours	Unit	Material	Labor	Equipment	Total	Total Incl O&P
0010	**ALUMINUM WINDOWS**, incl. frame and glazing, commercial grade									
1000	Stock units, casement, 3′-1″ x 3′-2″ opening	2 Sswk	10	1.600	Ea.	375	84		459	560
1050	Add for storms					120			120	132
1600	Projected, with screen, 3′-1″ x 3′-2″ opening	2 Sswk	10	1.600		355	84		439	540
1700	Add for storms					117			117	129
2000	4′-5″ x 5′-3″ opening	2 Sswk	8	2		400	105		505	630
2100	Add for storms					126			126	139
2500	Enamel finish windows, 3′-1″ x 3′-2″	2 Sswk	10	1.600		360	84		444	545
2600	4′-5″ x 5′-3″		8	2		405	105		510	635
3000	Single hung, 2′ x 3′ opening, enameled, standard glazed		10	1.600		208	84		292	380
3100	Insulating glass		10	1.600		252	84		336	430
3300	2′-8″ x 6′-8″ opening, standard glazed		8	2		365	105		470	590
3400	Insulating glass		8	2		475	105		580	710
3700	3′-4″ x 5′-0″ opening, standard glazed		9	1.778		300	93.50		393.50	500
3800	Insulating glass		9	1.778		335	93.50		428.50	535
3890	Awning type, 3′ x 3′ opening standard glass		14	1.143		425	60		485	580
3900	Insulating glass		14	1.143		450	60		510	605

08 51 Metal Windows

08 51 13 – Aluminum Windows

08 51 13.20 Aluminum Windows		Crew	Daily Output	Labor-Hours	Unit	Material	Labor	Equipment	Total	Total Incl O&P
						2015 Bare Costs				
3910	3′ x 4′ opening, standard glass	2 Sswk	10	1.600	Ea.	490	84		574	690
3920	Insulating glass		10	1.600		565	84		649	770
3930	3′ x 5′-4″ opening, standard glass		10	1.600		590	84		674	800
3940	Insulating glass		10	1.600		695	84		779	915
3950	4′ x 5′-4″ opening, standard glass		9	1.778		650	93.50		743.50	885
3960	Insulating glass		9	1.778		775	93.50		868.50	1,025
4000	Sliding aluminum, 3′ x 2′ opening, standard glazed		10	1.600		217	84		301	390
4100	Insulating glass		10	1.600		232	84		316	405
4300	5′ x 3′ opening, standard glazed		9	1.778		330	93.50		423.50	535
4400	Insulating glass		9	1.778		385	93.50		478.50	595
4600	8′ x 4′ opening, standard glazed		6	2.667		350	140		490	640
4700	Insulating glass		6	2.667		565	140		705	875
5000	9′ x 5′ opening, standard glazed		4	4		530	211		741	965
5100	Insulating glass		4	4		850	211		1,061	1,325
5500	Sliding, with thermal barrier and screen, 6′ x 4′, 2 track		8	2		725	105		830	985
5700	4 track		8	2		910	105		1,015	1,200
6000	For above units with bronze finish, add					15%				
6200	For installation in concrete openings, add					8%				
7000	Double hung, insulating glass, 2′-0″ x 2′-0″	2 Sswk	11	1.455		115	76.50		191.50	265
7020	2′-0″ x 2′-6″		11	1.455		124	76.50		200.50	274
7040	3′-0″ x 1′-6″		10	1.600		270	84		354	450
7060	3′-0″ x 2′-0″		10	1.600		315	84		399	495
7080	3′-0″ x 2′-6″		10	1.600		380	84		464	570
7100	3′-0″ x 3′-0″		10	1.600		425	84		509	615
7120	3′-0″ x 3′-6″		10	1.600		445	84		529	640
7140	3′-0″ x 4′-0″		10	1.600		480	84		564	680
7160	3′-0″ x 5′-0″		10	1.600		530	84		614	730
7180	3′-0″ x 6′-0″		9	1.778		560	93.50		653.50	785

08 51 23 – Steel Windows

08 51 23.10 Steel Sash

Line	Description	Crew	Daily Output	Labor-Hours	Unit	Material	Labor	Equipment	Total	Total Incl O&P
0010	**STEEL SASH** Custom units, glazing and trim not included									
0100	Casement, 100% vented	2 Sswk	200	.080	S.F.	66.50	4.21		70.71	80.50
0200	50% vented		200	.080		54.50	4.21		58.71	67
0300	Fixed		200	.080		29	4.21		33.21	39.50
1000	Projected, commercial, 40% vented		200	.080		51.50	4.21		55.71	64
1100	Intermediate, 50% vented		200	.080		58.50	4.21		62.71	71.50
1500	Industrial, horizontally pivoted		200	.080		53	4.21		57.21	65.50
1600	Fixed		200	.080		31	4.21		35.21	41.50
2000	Industrial security sash, 50% vented		200	.080		57.50	4.21		61.71	70.50
2100	Fixed		200	.080		47	4.21		51.21	59
2500	Picture window		200	.080		30	4.21		34.21	40.50
3000	Double hung		200	.080		59.50	4.21		63.71	73
5000	Mullions for above, open interior face		240	.067	L.F.	10.45	3.51		13.96	17.85
5100	With interior cover		240	.067	″	17.30	3.51		20.81	25.50
6100	Triple glazing for above, add	2 Glaz	85	.188	S.F.	12.50	8.50		21	27.50
9000	Minimum labor/equipment charge	1 Sswk	2	4	Job		211		211	380

08 51 23.20 Steel Windows

Line	Description	Crew	Daily Output	Labor-Hours	Unit	Material	Labor	Equipment	Total	Total Incl O&P
0010	**STEEL WINDOWS** Stock, including frame, trim and insul. glass									
0020	See Section 13 34 19.50									
1000	Custom units, double hung, 2′-8″ x 4′-6″ opening	2 Sswk	12	1.333	Ea.	705	70		775	905
1100	2′-4″ x 3′-9″ opening		12	1.333		585	70		655	765
1500	Commercial projected, 3′-9″ x 5′-5″ opening		10	1.600		1,225	84		1,309	1,500

08 51 Metal Windows

08 51 23 – Steel Windows

08 51 23.20 Steel Windows		Crew	Daily Output	Labor-Hours	Unit	2015 Bare Costs Material	Labor	Equipment	Total	Total Incl O&P
1600	6′-9″ x 4′-1″ opening	2 Sswk	7	2.286	Ea.	1,625	120		1,745	2,000
2000	Intermediate projected, 2′-9″ x 4′-1″ opening		12	1.333		685	70		755	880
2100	4′-1″ x 5′-5″ opening		10	1.600		1,400	84		1,484	1,700
9000	Minimum labor/equipment charge	1 Sswk	3	2.667	Job		140		140	254

08 51 23.40 Basement Utility Windows

		Crew	Daily Output	Labor-Hours	Unit	Material	Labor	Equipment	Total	Total Incl O&P
0010	**BASEMENT UTILITY WINDOWS**									
0015	1′-3″ x 2′-8″	1 Carp	16	.500	Ea.	128	23.50		151.50	179
1100	1′-7″ x 2′-8″	″	16	.500	″	140	23.50		163.50	193

08 51 66 – Metal Window Screens

08 51 66.10 Screens

		Crew	Daily Output	Labor-Hours	Unit	Material	Labor	Equipment	Total	Total Incl O&P
0010	**SCREENS**									
0020	For metal sash, aluminum or bronze mesh, flat screen	2 Sswk	1200	.013	S.F.	4.30	.70		5	6
0500	Wicket screen, inside window		1000	.016		6.65	.84		7.49	8.80
0800	Security screen, aluminum frame with stainless steel cloth		1200	.013		24	.70		24.70	27.50
0900	Steel grate, painted, on steel frame		1600	.010		13.10	.53		13.63	15.35
1000	Screens for solar louvers		160	.100		24.50	5.25		29.75	36.50
4000	See Section 05 58 23.90									

08 52 Wood Windows

08 52 10 – Plain Wood Windows

08 52 10.20 Awning Window

		Crew	Daily Output	Labor-Hours	Unit	Material	Labor	Equipment	Total	Total Incl O&P
0010	**AWNING WINDOW**, Including frame, screens and grilles									
0100	34″ x 22″, insulated glass	1 Carp	10	.800	Ea.	270	37.50		307.50	360
0200	Low E glass		10	.800		271	37.50		308.50	360
0300	40″ x 28″, insulated glass		9	.889		315	41.50		356.50	415
0400	Low E Glass		9	.889		340	41.50		381.50	445
0500	48″ x 36″, insulated glass		8	1		465	47		512	590
0600	Low E glass		8	1		490	47		537	615
4000	Impact windows, minimum, add					60%				
4010	Impact windows, maximum, add					160%				
9000	Minimum labor/equipment charge	1 Carp	4	2	Job		94		94	154

08 52 10.30 Wood Windows

		Crew	Daily Output	Labor-Hours	Unit	Material	Labor	Equipment	Total	Total Incl O&P
0010	**WOOD WINDOWS**, double hung									
0020	Including frame, double insulated glass, screens and grilles									
0040	Double hung, 2′-2″ x 3′-4″ high	2 Carp	15	1.067	Ea.	210	50		260	315
0060	2′-2″ x 4′-4″		14	1.143		229	53.50		282.50	340
0080	2′-6″ x 3′-4″		13	1.231		219	58		277	335
0100	2′-6″ x 4′-0″		12	1.333		229	62.50		291.50	355
0120	2′-6″ x 4′-8″		12	1.333		250	62.50		312.50	380
0140	2′-10″ x 3′-4″		10	1.600		224	75		299	370
0160	2′-10″ x 4′-0″		10	1.600		247	75		322	395
0180	3′-7″ x 3′-4″		9	1.778		254	83.50		337.50	415
0200	3′-7″ x 5′-4″		9	1.778		284	83.50		367.50	445
0220	3′-10″ x 5′-4″		8	2		510	94		604	715
3800	Triple glazing for above, add					25%				

08 52 10.40 Casement Window

		Crew	Daily Output	Labor-Hours	Unit	Material	Labor	Equipment	Total	Total Incl O&P
0010	**CASEMENT WINDOW**, including frame, screen and grilles									
0100	2′-0″ x 3′-0″ H, dbl. insulated glass G	1 Carp	10	.800	Ea.	268	37.50		305.50	355
0150	Low E glass G		10	.800		259	37.50		296.50	345
0200	2′-0″ x 4′-6″ high, double insulated glass G		9	.889		360	41.50		401.50	465

08 52 Wood Windows

08 52 10 – Plain Wood Windows

08 52 10.40 Casement Window

	08 52 10.40 Casement Window		Crew	Daily Output	Labor-Hours	Unit	2015 Bare Costs Material	Labor	Equipment	Total	Total Incl O&P
0250	Low E glass	G	1 Carp	9	.889	Ea.	370	41.50		411.50	475
0260	Casement 4'-2" x 4'-2" double insulated glass	G		11	.727		875	34		909	1,025
0270	4'-0" x 4'-0" Low E glass	G		11	.727		535	34		569	645
0290	6'-4" x 5'-7" Low E glass	G		9	.889		1,125	41.50		1,166.50	1,300
0300	2'-4" x 6'-0" high, double insulated glass	G		8	1		440	47		487	560
0350	Low E glass	G		8	1		475	47		522	600
0522	Vinyl clad, premium, double insulated glass, 2'-0" x 3'-0"	G		10	.800		271	37.50		308.50	360
0524	2'-0" x 4'-0"	G		9	.889		315	41.50		356.50	420
0525	2'-0" x 5'-0"	G		8	1		360	47		407	475
0528	2'-0" x 6'-0"	G		8	1		380	47		427	495
0600	3'-0" x 5'-0"	G		8	1		665	47		712	805
0700	4'-0" x 3'-0"	G		8	1		730	47		777	880
0710	4'-0" x 4'-0"	G		8	1		625	47		672	760
0720	4'-8" x 4'-0"	G		8	1		690	47		737	835
0730	4'-8" x 5'-0"	G		6	1.333		790	62.50		852.50	970
0740	4'-8" x 6'-0"	G		6	1.333		880	62.50		942.50	1,075
0750	6'-0" x 4'-0"	G		6	1.333		805	62.50		867.50	990
0800	6'-0" x 5'-0"	G		6	1.333		895	62.50		957.50	1,100
0900	5'-6" x 5'-6"	G	2 Carp	15	1.067		1,425	50		1,475	1,650
2000	Bay, casement units, 8' x 5', w/screens, dbl. insul. glass			2.50	6.400	Opng.	1,600	300		1,900	2,250
2100	Low E glass			2.50	6.400	"	1,675	300		1,975	2,350
8190	For installation, add per leaf					Ea.		15%			
8200	For multiple leaf units, deduct for stationary sash										
8220	2' high					Ea.	23			23	25.50
8240	4'-6" high						26			26	28.50
8260	6' high						34.50			34.50	38
8300	Impact windows, minimum, add						60%				
8310	Impact windows, maximum, add						160%				
9000	Minimum labor/equipment charge		1 Carp	3	2.667	Job		125		125	205

08 52 10.50 Double Hung

	08 52 10.50 Double Hung		Crew	Daily Output	Labor-Hours	Unit	Material	Labor	Equipment	Total	Total Incl O&P
0010	**DOUBLE HUNG**, Including frame, screens and grilles										
0100	2'-0" x 3'-0" high, low E insul. glass	G	1 Carp	10	.800	Ea.	210	37.50		247.50	293
0200	3'-0" x 4'-0" high, double insulated glass	G		9	.889		279	41.50		320.50	375
0300	4'-0" x 4'-6" high, low E insulated glass	G		8	1		320	47		367	430
8000	Impact windows, minimum, add						60%				
8010	Impact windows, maximum, add						160%				
9000	Minimum labor/equipment charge		1 Carp	3	2.667	Job		125		125	205

08 52 10.55 Picture Window

	08 52 10.55 Picture Window	Crew	Daily Output	Labor-Hours	Unit	Material	Labor	Equipment	Total	Total Incl O&P
0010	**PICTURE WINDOW**, Including frame and grilles									
0100	3'-6" x 4'-0" high, dbl. insulated glass	2 Carp	12	1.333	Ea.	420	62.50		482.50	570
0150	Low E glass		12	1.333		435	62.50		497.50	580
0200	4'-0" x 4'-6" high, double insulated glass		11	1.455		550	68.50		618.50	715
0250	Low E glass		11	1.455		530	68.50		598.50	695
0300	5'-0" x 4'-0" high, double insulated glass		11	1.455		580	68.50		648.50	750
0350	Low E glass		11	1.455		605	68.50		673.50	775
0400	6'-0" x 4'-6" high, double insulated glass		10	1.600		625	75		700	815
0450	Low E glass		10	1.600		635	75		710	825

08 52 10.65 Wood Sash

	08 52 10.65 Wood Sash	Crew	Daily Output	Labor-Hours	Unit	Material	Labor	Equipment	Total	Total Incl O&P
0010	**WOOD SASH**, Including glazing but not trim									
0050	Custom, 5'-0" x 4'-0", 1" dbl. glazed, 3/16" thick lites	2 Carp	3.20	5	Ea.	230	235		465	640
0100	1/4" thick lites		5	3.200		245	150		395	515
0200	1" thick, triple glazed		5	3.200		415	150		565	700

08 52 Wood Windows

08 52 10 – Plain Wood Windows

08 52 10.65 Wood Sash			Crew	Daily Output	Labor-Hours	Unit	Material	2015 Bare Costs Labor	Equipment	Total	Total Incl O&P
0300	7'-0" x 4'-6" high, 1" double glazed, 3/16" thick lites		2 Carp	4.30	3.721	Ea.	420	175		595	745
0400	1/4" thick lites			4.30	3.721		475	175		650	805
0500	1" thick, triple glazed			4.30	3.721		540	175		715	880
0600	8'-6" x 5'-0" high, 1" double glazed, 3/16" thick lites			3.50	4.571		565	215		780	975
0700	1/4" thick lites			3.50	4.571		620	215		835	1,025
0800	1" thick, triple glazed			3.50	4.571		625	215		840	1,025
0900	Window frames only, based on perimeter length					L.F.	4.02			4.02	4.42
1200	Window sill, stock, per lineal foot						8.50			8.50	9.35
1250	Casing, stock						3.30			3.30	3.63
08 52 10.70 Sliding Windows											
0010	**SLIDING WINDOWS**										
0100	3'-0" x 3'-0" high, double insulated	G	1 Carp	10	.800	Ea.	284	37.50		321.50	370
0120	Low E glass	G		10	.800		310	37.50		347.50	400
0200	4'-0" x 3'-6" high, double insulated	G		9	.889		355	41.50		396.50	460
0220	Low E glass	G		9	.889		360	41.50		401.50	470
0300	6'-0" x 5'-0" high, double insulated	G		8	1		490	47		537	610
0320	Low E glass	G		8	1		530	47		577	655
9000	Minimum labor/equipment charge			3	2.667	Job		125		125	205

08 52 13 – Metal-Clad Wood Windows

08 52 13.10 Awning Windows, Metal-Clad			Crew	Daily Output	Labor-Hours	Unit	Material	Labor	Equipment	Total	Total Incl O&P
0010	**AWNING WINDOWS, METAL-CLAD**										
2000	Metal clad, awning deluxe, double insulated glass, 34" x 22"		1 Carp	9	.889	Ea.	247	41.50		288.50	340
2050	36" x 25"			9	.889		272	41.50		313.50	370
2100	40" x 22"			9	.889		291	41.50		332.50	390
2150	40" x 30"			9	.889		340	41.50		381.50	445
2200	48" x 28"			8	1		345	47		392	455
2250	60" x 36"			8	1		370	47		417	485
08 52 13.20 Casement Windows, Metal-Clad											
0010	**CASEMENT WINDOWS, METAL-CLAD**										
0100	Metal clad, deluxe, dbl. insul. glass, 2'-0" x 3'-0" high	G	1 Carp	10	.800	Ea.	279	37.50		316.50	365
0120	2'-0" x 4'-0" high	G		9	.889		315	41.50		356.50	415
0130	2'-0" x 5'-0" high	G		8	1		325	47		372	435
0140	2'-0" x 6'-0" high	G		8	1		365	47		412	475
0300	Metal clad, casement, bldrs mdl, 6'-0" x 4'-0", dbl. insltd gls, 3 panels		2 Carp	10	1.600		1,200	75		1,275	1,450
0310	9'-0" x 4'-0" , 4 panels			8	2		1,550	94		1,644	1,850
0320	10'-0" x 5'-0", 5 panels			7	2.286		2,100	107		2,207	2,500
0330	12'-0" x 6'-0", 6 panels			6	2.667		2,700	125		2,825	3,150
08 52 13.30 Double-Hung Windows, Metal-clad											
0010	**DOUBLE-HUNG WINDOWS, METAL-CLAD**										
0100	Metal clad, deluxe, dbl. insul. glass, 2'-6" x 3'-0" high	G	1 Carp	10	.800	Ea.	272	37.50		309.50	360
0120	3'-0" x 3'-6" high	G		10	.800		315	37.50		352.50	405
0140	3'-0" x 4'-0" high	G		9	.889		325	41.50		366.50	430
0160	3'-0" x 4'-6" high	G		9	.889		345	41.50		386.50	450
0180	3'-0" x 5'-0" high	G		8	1		370	47		417	485
0200	3'-6" x 6'-0" high	G		8	1		450	47		497	570
08 52 13.35 Picture and Sliding Windows Metal-Clad											
0010	**PICTURE AND SLIDING WINDOWS METAL-CLAD**										
2000	Metal clad, dlx picture, dbl. insul. glass, 4'-0" x 4'-0" high		2 Carp	12	1.333	Ea.	375	62.50		437.50	515
2100	4'-0" x 6'-0" high			11	1.455		545	68.50		613.50	710
2200	5'-0" x 6'-0" high			10	1.600		610	75		685	795
2300	6'-0" x 6'-0" high			10	1.600		695	75		770	890

08 52 13 – Metal-Clad Wood Windows

08 52 13.35 Picture and Sliding Windows Metal-Clad		Crew	Daily Output	Labor-Hours	Unit	Material	2015 Bare Costs Labor	Equipment	Total	Total Incl O&P
2400	Metal clad, dlx sliding, double insulated glass, 3'-0" x 3'-0" high G	1 Carp	10	.800	Ea.	330	37.50		367.50	420
2420	4'-0" x 3'-6" high G		9	.889		400	41.50		441.50	510
2440	5'-0" x 4'-0" high G		9	.889		480	41.50		521.50	595
2460	6'-0" x 5'-0" high G		8	1		730	47		777	875
9000	Minimum labor/equipment charge	2 Carp	2.75	5.818	Job		273		273	450

08 52 13.40 Bow and Bay Windows, Metal-Clad		Crew	Daily Output	Labor-Hours	Unit	Material	Labor	Equipment	Total	Total Incl O&P
0010	**BOW AND BAY WINDOWS, METAL-CLAD**									
0100	Metal clad, deluxe, dbl. insul. glass, 8'-0" x 5'-0" high, 4 panels	2 Carp	10	1.600	Ea.	1,675	75		1,750	1,950
0120	10'-0" x 5'-0" high, 5 panels		8	2		1,800	94		1,894	2,125
0140	10'-0" x 6'-0" high, 5 panels		7	2.286		2,100	107		2,207	2,500
0160	12'-0" x 6'-0" high, 6 panels		6	2.667		2,925	125		3,050	3,425
0400	Double hung, bldrs. model, bay, 8' x 4' high, dbl. insulated glass		10	1.600		1,325	75		1,400	1,575
0440	Low E glass		10	1.600		1,425	75		1,500	1,700
0460	9'-0" x 5'-0" high, double insulated glass		6	2.667		1,425	125		1,550	1,775
0480	Low E glass		6	2.667		1,500	125		1,625	1,850
0500	Metal clad, deluxe, dbl. insul. glass, 7'-0" x 4'-0" high		10	1.600		1,275	75		1,350	1,525
0520	8'-0" x 4'-0" high		8	2		1,300	94		1,394	1,600
0540	8'-0" x 5'-0" high		7	2.286		1,350	107		1,457	1,675
0560	9'-0" x 5'-0" high		6	2.667		1,450	125		1,575	1,775

08 52 16 – Plastic-Clad Wood Windows

08 52 16.10 Bow Window		Crew	Daily Output	Labor-Hours	Unit	Material	Labor	Equipment	Total	Total Incl O&P
0010	**BOW WINDOW** Including frames, screens, and grilles									
0020	End panels operable									
1000	Bow type, casement, wood, bldrs. mdl., 8' x 5' dbl. insltd glass, 4 panel	2 Carp	10	1.600	Ea.	1,500	75		1,575	1,775
1050	Low E glass		10	1.600		1,325	75		1,400	1,575
1100	10'-0" x 5'-0", double insulated glass, 6 panels		6	2.667		1,350	125		1,475	1,700
1200	Low E glass, 6 panels		6	2.667		1,450	125		1,575	1,800
1300	Vinyl clad, bldrs. model, double insulated glass, 6'-0" x 4'-0", 3 panel		10	1.600		1,025	75		1,100	1,250
1340	9'-0" x 4'-0", 4 panel		8	2		1,350	94		1,444	1,650
1380	10'-0" x 6'-0", 5 panels		7	2.286		2,250	107		2,357	2,650
1420	12'-0" x 6'-0", 6 panels		6	2.667		2,925	125		3,050	3,425
2000	Bay window, 8' x 5', dbl. insul glass		10	1.600		1,875	75		1,950	2,200
2050	Low E glass		10	1.600		2,275	75		2,350	2,625
2100	12'-0" x 6'-0", double insulated glass, 6 panels		6	2.667		2,350	125		2,475	2,775
2200	Low E glass		6	2.667		2,375	125		2,500	2,825
2280	6'-0" x 4'-0"		11	1.455		1,250	68.50		1,318.50	1,475
2300	Vinyl clad, premium, double insulated glass, 8'-0" x 5'-0"		10	1.600		1,750	75		1,825	2,050
2340	10'-0" x 5'-0"		8	2		2,300	94		2,394	2,675
2380	10'-0" x 6'-0"		7	2.286		2,625	107		2,732	3,050
2420	12'-0" x 6'-0"		6	2.667		3,200	125		3,325	3,725
3300	Vinyl clad, premium, double insulated glass, 7'-0" x 4'-6"		10	1.600		1,375	75		1,450	1,625
3340	8'-0" x 4'-6"		8	2		1,400	94		1,494	1,675
3380	8'-0" x 5'-0"		7	2.286		1,450	107		1,557	1,775
3420	9'-0" x 5'-0"		6	2.667		1,500	125		1,625	1,850
9000	Minimum labor/equipment charge		2.50	6.400	Job		300		300	490

08 52 16.15 Awning Window Vinyl-Clad		Crew	Daily Output	Labor-Hours	Unit	Material	Labor	Equipment	Total	Total Incl O&P
0010	**AWNING WINDOW VINYL-CLAD** Including frames, screens, and grilles									
0240	Vinyl clad, 34" x 22"	1 Carp	10	.800	Ea.	264	37.50		301.50	350
0280	36" x 28"		9	.889		305	41.50		346.50	405
0300	36" x 36"		9	.889		340	41.50		381.50	445
0340	40" x 22"		10	.800		288	37.50		325.50	375
0360	48" x 28"		8	1		370	47		417	480

08 52 Wood Windows

08 52 16 – Plastic-Clad Wood Windows

08 52 16.15 Awning Window Vinyl-Clad			Crew	Daily Output	Labor-Hours	Unit	Material	2015 Bare Costs Labor	Equipment	Total	Total Incl O&P
0380	60" x 36"		1 Carp	8	1	Ea.	500	47		547	625

08 52 16.30 Palladian Windows

			Crew	Daily Output	Labor-Hours	Unit	Material	Labor	Equipment	Total	Total Incl O&P
0010	**PALLADIAN WINDOWS**										
0020	Vinyl clad, double insulated glass, including frame and grilles										
0040	3'-2" x 2'-6" high		2 Carp	11	1.455	Ea.	1,250	68.50		1,318.50	1,475
0060	3'-2" x 4'-10"			11	1.455		1,750	68.50		1,818.50	2,000
0080	3'-2" x 6'-4"			10	1.600		1,700	75		1,775	2,000
0100	4'-0" x 4'-0"			10	1.600		1,525	75		1,600	1,800
0120	4'-0" x 5'-4"		3 Carp	10	2.400		1,875	113		1,988	2,225
0140	4'-0" x 6'-0"			9	2.667		1,950	125		2,075	2,350
0160	4'-0" x 7'-4"			9	2.667		2,125	125		2,250	2,525
0180	5'-5" x 4'-10"			9	2.667		2,275	125		2,400	2,700
0200	5'-5" x 6'-10"			9	2.667		2,575	125		2,700	3,050
0220	5'-5" x 7'-9"			9	2.667		2,800	125		2,925	3,275
0240	6'-0" x 7'-11"			8	3		3,475	141		3,616	4,050
0260	8'-0" x 6'-0"			8	3		3,075	141		3,216	3,600

08 52 16.35 Double-Hung Window

			Crew	Daily Output	Labor-Hours	Unit	Material	Labor	Equipment	Total	Total Incl O&P
0010	**DOUBLE-HUNG WINDOW** Including frames, screens, and grilles										
0300	Vinyl clad, premium, double insulated glass, 2'-6" x 3'-0"	G	1 Carp	10	.800	Ea.	310	37.50		347.50	400
0305	2'-6" x 4'-0"	G		10	.800		360	37.50		397.50	455
0400	3'-0" x 3'-6"	G		10	.800		335	37.50		372.50	430
0500	3'-0" x 4'-0"	G		9	.889		395	41.50		436.50	505
0600	3'-0" x 4'-6"	G		9	.889		410	41.50		451.50	520
0700	3'-0" x 5'-0"	G		8	1		445	47		492	560
0790	3'-4" x 5'-0"	G		8	1		455	47		502	575
0800	3'-6" x 6'-0"	G		8	1		490	47		537	615
0820	4'-0" x 5'-0"	G		7	1.143		560	53.50		613.50	705
0830	4'-0" x 6'-0"	G		7	1.143		700	53.50		753.50	855

08 52 16.40 Transom Windows

			Crew	Daily Output	Labor-Hours	Unit	Material	Labor	Equipment	Total	Total Incl O&P
0010	**TRANSOM WINDOWS**										
0050	Vinyl clad, premium, double insulated glass, 32" x 8"		1 Carp	16	.500	Ea.	188	23.50		211.50	245
0100	36" x 8"			16	.500		199	23.50		222.50	258
0110	36" x 12"			16	.500		212	23.50		235.50	272
0200	44" x 48"			12	.667		580	31.50		611.50	690
1000	Vinyl clad, premium, dbl. insul. glass, 4'-0" x 4'-0"		2 Carp	12	1.333		510	62.50		572.50	665
1100	4'-0" x 6'-0"			11	1.455		935	68.50		1,003.50	1,125
1200	5'-0" x 6'-0"			10	1.600		1,050	75		1,125	1,275
1300	6'-0" x 6'-0"			10	1.600		1,050	75		1,125	1,275

08 52 16.70 Vinyl Clad, Premium, Dbl. Insulated Glass

			Crew	Daily Output	Labor-Hours	Unit	Material	Labor	Equipment	Total	Total Incl O&P
0010	**VINYL CLAD, PREMIUM, DBL. INSULATED GLASS**										
1000	Sliding, 3'-0" x 3'-0"	G	1 Carp	10	.800	Ea.	605	37.50		642.50	725
1050	4'-0" x 3'-6"	G		9	.889		685	41.50		726.50	825
1100	5'-0" x 4'-0"	G		9	.889		900	41.50		941.50	1,050
1150	6'-0" x 5'-0"	G		8	1		1,125	47		1,172	1,325

08 52 50 – Window Accessories

08 52 50.10 Window Grille or Muntin

			Crew	Daily Output	Labor-Hours	Unit	Material	Labor	Equipment	Total	Total Incl O&P
0010	**WINDOW GRILLE OR MUNTIN**, snap in type										
0020	Standard pattern interior grilles										
2000	Wood, awning window, glass size 28" x 16" high		1 Carp	30	.267	Ea.	28	12.50		40.50	51.50
2060	44" x 24" high			32	.250		40	11.75		51.75	63.50
2100	Casement, glass size, 20" x 36" high			30	.267		32	12.50		44.50	55.50

08 52 Wood Windows

08 52 50 – Window Accessories

08 52 50.10 Window Grille or Muntin		Crew	Daily Output	Labor-Hours	Unit	Material	2015 Bare Costs Labor	Equipment	Total	Total Incl O&P
2180	20" x 56" high	1 Carp	32	.250	Ea.	43	11.75		54.75	67
2200	Double hung, glass size, 16" x 24" high		24	.333	Set	51	15.65		66.65	82
2280	32" x 32" high		34	.235	"	131	11.05		142.05	162
2500	Picture, glass size, 48" x 48" high		30	.267	Ea.	120	12.50		132.50	153
2580	60" x 68" high		28	.286	"	183	13.40		196.40	224
2600	Sliding, glass size, 14" x 36" high		24	.333	Set	35.50	15.65		51.15	64.50
2680	36" x 36" high		22	.364	"	43.50	17.05		60.55	75.50
9000	Minimum labor/equipment charge	↓	5	1.600	Job		75		75	123

08 52 66 – Wood Window Screens

08 52 66.10 Wood Screens		Crew	Daily Output	Labor-Hours	Unit	Material	Labor	Equipment	Total	Total Incl O&P
0010	**WOOD SCREENS**									
0020	Over 3 S.F., 3/4" frames	2 Carp	375	.043	S.F.	4.89	2		6.89	8.70
0100	1-1/8" frames	"	375	.043	"	8.20	2		10.20	12.30
9000	Minimum labor/equipment charge	1 Carp	4	2	Job		94		94	154

08 52 69 – Wood Storm Windows

08 52 69.10 Storm Windows		Crew	Daily Output	Labor-Hours	Unit	Material	Labor	Equipment	Total	Total Incl O&P
0010	**STORM WINDOWS**, aluminum residential									
0300	Basement, mill finish, incl. fiberglass screen									
0320	1'-10" x 1'-0" high G	2 Carp	30	.533	Ea.	35	25		60	79.50
0340	2'-9" x 1'-6" high G		30	.533		38	25		63	83
0360	3'-4" x 2'-0" high G	↓	30	.533	↓	45	25		70	90.50
1600	Double-hung, combination, storm & screen									
2000	Clear anodic coating, 2'-0" x 3'-5" high G	2 Carp	30	.533	Ea.	95	25		120	146
2020	2'-6" x 5'-0" high G		28	.571		117	27		144	172
2040	4'-0" x 6'-0" high G		25	.640		130	30		160	193
2400	White painted, 2'-0" x 3'-5" high G		30	.533		90	25		115	140
2420	2'-6" x 5'-0" high G		28	.571		95	27		122	149
2440	4'-0" x 6'-0" high G		25	.640		110	30		140	171
2600	Mill finish, 2'-0" x 3'-5" high G		30	.533		85	25		110	135
2620	2'-6" x 5'-0" high G		28	.571		90	27		117	143
2640	4'-0" x 6'-8" high G	↓	25	.640	↓	110	30		140	171
9410	Minimum labor/equipment charge	1 Carp	4	2	Job		94		94	154

08 53 Plastic Windows

08 53 13 – Vinyl Windows

08 53 13.20 Vinyl Single Hung Windows		Crew	Daily Output	Labor-Hours	Unit	Material	Labor	Equipment	Total	Total Incl O&P
0010	**VINYL SINGLE HUNG WINDOWS**, insulated glass									
0100	Grids, low E, J fin, ext. jambs, 21" x 53" G	2 Carp	18	.889	Ea.	198	41.50		239.50	287
0110	21" x 57" G		17	.941		200	44		244	293
0120	21" x 65" G		16	1		205	47		252	305
0130	25" x 41" G		20	.800		190	37.50		227.50	271
0140	25" x 49" G		18	.889		200	41.50		241.50	289
0150	25" x 57" G		17	.941		205	44		249	299
0160	25" x 65" G		16	1		240	47		287	340
0170	29" x 41" G		18	.889		195	41.50		236.50	284
0180	29" x 53" G		18	.889		205	41.50		246.50	295
0190	29" x 57" G		17	.941		210	44		254	305
0200	29" x 65" G		16	1		215	47		262	315
0210	33" x 41" G		20	.800		200	37.50		237.50	282
0220	33" x 53" G	↓	18	.889	↓	215	41.50		256.50	305

08 53 Plastic Windows

08 53 13 – Vinyl Windows

08 53 13.20 Vinyl Single Hung Windows			Crew	Daily Output	Labor-Hours	Unit	Material	2015 Bare Costs Labor	Equipment	Total	Total Incl O&P
0230	33″ x 57″	G	2 Carp	17	.941	Ea.	215	44		259	310
0240	33″ x 65″	G		16	1		220	47		267	320
0250	37″ x 41″	G		20	.800		225	37.50		262.50	310
0260	37″ x 53″	G		18	.889		235	41.50		276.50	330
0270	37″ x 57″	G		17	.941		240	44		284	335
0280	37″ x 65″	G		16	1		256	47		303	360

08 53 13.30 Vinyl Double Hung Windows			Crew	Daily Output	Labor-Hours	Unit	Material	Labor	Equipment	Total	Total Incl O&P
0010	**VINYL DOUBLE HUNG WINDOWS**, insulated glass										
0100	Grids, low E, J fin, ext. jambs, 21″ x 53″	G	2 Carp	18	.889	Ea.	215	41.50		256.50	305
0102	21″ x 37″	G		18	.889		223	41.50		264.50	315
0104	21″ x 41″	G		18	.889		232	41.50		273.50	325
0106	21″ x 49″	G		18	.889		247	41.50		288.50	340
0110	21″ x 57″	G		17	.941		263	44		307	360
0120	21″ x 65″	G		16	1		279	47		326	380
0128	25″ x 37″	G		20	.800		234	37.50		271.50	320
0130	25″ x 41″	G		20	.800		242	37.50		279.50	330
0140	25″ x 49″	G		18	.889		258	41.50		299.50	355
0145	25″ x 53″	G		18	.889		265	41.50		306.50	360
0150	25″ x 57″	G		17	.941		274	44		318	375
0160	25″ x 65″	G		16	1		290	47		337	395
0162	25″ x 69″	G		16	1		297	47		344	400
0164	25″ x 77″	G		16	1		325	47		372	430
0168	29″ x 37″	G		18	.889		242	41.50		283.50	335
0170	29″ x 41″	G		18	.889		250	41.50		291.50	345
0172	29″ x 49″	G		18	.889		267	41.50		308.50	365
0180	29″ x 53″	G		18	.889		271	41.50		312.50	365
0190	29″ x 57″	G		17	.941		280	44		324	385
0200	29″ x 65″	G		16	1		300	47		347	405
0202	29″ x 69″	G		16	1		310	47		357	415
0205	29″ x 77″	G		16	1		335	47		382	440
0208	33″ x 37″	G		20	.800		255	37.50		292.50	345
0210	33″ x 41″	G		20	.800		263	37.50		300.50	350
0215	33″ x 49″	G		20	.800		280	37.50		317.50	370
0220	33″ x 53″	G		18	.889		285	41.50		326.50	385
0230	33″ x 57″	G		17	.941		297	44		341	400
0240	33″ x 65″	G		16	1		315	47		362	420
0242	33″ x 69″	G		16	1		330	47		377	435
0246	33″ x 77″	G		16	1		350	47		397	460
0250	37″ x 41″	G		20	.800		289	37.50		326.50	380
0255	37″ x 49″	G		20	.800		282	37.50		319.50	370
0260	37″ x 53″	G		18	.889		315	41.50		356.50	415
0270	37″ x 57″	G		17	.941		325	44		369	435
0280	37″ x 65″	G		16	1		350	47		397	460
0282	37″ x 69″	G		16	1		360	47		407	470
0286	37″ x 77″	G		16	1		380	47		427	490
0300	Solid vinyl, average quality, double insulated glass, 2′-0″ x 3′-0″	G	1 Carp	10	.800		291	37.50		328.50	380
0310	3′-0″ x 4′-0″	G		9	.889		207	41.50		248.50	296
0320	4′-0″ x 4′-6″	G		8	1		335	47		382	440
0330	Premium, double insulated glass, 2′-6″ x 3′-0″	G		10	.800		271	37.50		308.50	360
0340	3′-0″ x 3′-6″	G		9	.889		292	41.50		333.50	390
0350	3′-0″ x 4″-0″	G		9	.889		315	41.50		356.50	420
0360	3′-0″ x 4′-6″	G		9	.889		330	41.50		371.50	435

08 53 Plastic Windows

08 53 13 – Vinyl Windows

08 53 13.30 Vinyl Double Hung Windows		Crew	Daily Output	Labor-Hours	Unit	Material	Labor	Equipment	Total	Total Incl O&P
						2015 Bare Costs				
0370	3'-0" x 5'-0" G	1 Carp	8	1	Ea.	355	47		402	465
0380	3'-6" x 6"-0" G	↓	8	1	↓	380	47		427	495
08 53 13.40 Vinyl Casement Windows										
0010	**VINYL CASEMENT WINDOWS**, insulated glass									
0015	Grids, low E, J fin, extension jambs, screens									
0100	One lite, 21" x 41" G	2 Carp	20	.800	Ea.	295	37.50		332.50	385
0110	21" x 47" G		20	.800		320	37.50		357.50	415
0120	21" x 53" G		20	.800		345	37.50		382.50	440
0128	24" x 35" G		19	.842		283	39.50		322.50	375
0130	24" x 41" G		19	.842		310	39.50		349.50	405
0140	24" x 47" G		19	.842		335	39.50		374.50	430
0150	24" x 53" G		19	.842		360	39.50		399.50	460
0158	28" x 35" G		19	.842		300	39.50		339.50	395
0160	28" x 41" G		19	.842		330	39.50		369.50	430
0170	28" x 47" G		19	.842		350	39.50		389.50	450
0180	28" x 53" G		19	.842		385	39.50		424.50	490
0184	28" x 59" G		19	.842		395	39.50		434.50	500
0188	Two lites, 33" x 35" G		18	.889		470	41.50		511.50	585
0190	33" x 41" G		18	.889		500	41.50		541.50	620
0200	33" x 47" G		18	.889		540	41.50		581.50	660
0210	33" x 53" G		18	.889		575	41.50		616.50	700
0212	33" x 59" G		18	.889		610	41.50		651.50	740
0215	33" x 72" G		18	.889		635	41.50		676.50	765
0220	41" x 41" G		18	.889		545	41.50		586.50	670
0230	41" x 47" G		18	.889		585	41.50		626.50	710
0240	41" x 53" G		17	.941		620	44		664	755
0242	41" x 59" G		17	.941		650	44		694	790
0246	41" x 72" G		17	.941		685	44		729	830
0250	47" x 41" G		17	.941		555	44		599	685
0260	47" x 47" G		17	.941		590	44		634	725
0270	47" x 53" G		17	.941		625	44		669	765
0272	47" x 59" G		17	.941		680	44		724	825
0280	56" x 41" G		15	1.067		595	50		645	735
0290	56" x 47" G		15	1.067		625	50		675	770
0300	56" x 53" G		15	1.067		680	50		730	830
0302	56" x 59" G		15	1.067		710	50		760	860
0310	56" x 72" G	↓	15	1.067		770	50		820	930
0340	Solid vinyl, premium, double insulated glass, 2'-0" x 3'-0" high G	1 Carp	10	.800		270	37.50		307.50	360
0360	2'-0" x 4'-0" high G		9	.889		299	41.50		340.50	400
0380	2'-0" x 5'-0" high G	↓	8	1	↓	335	47		382	445
08 53 13.50 Vinyl Picture Windows										
0010	**VINYL PICTURE WINDOWS**, insulated glass									
0100	Grids, low E, J fin, ext. jambs, 33" x 47"	2 Carp	12	1.333	Ea.	280	62.50		342.50	415
0110	35" x 71"		12	1.333		375	62.50		437.50	520
0120	47" x 35"		12	1.333		300	62.50		362.50	435
0130	47" x 41"		12	1.333		390	62.50		452.50	535
0140	47" x 47"		12	1.333		345	62.50		407.50	480
0150	47" x 53"		11	1.455		370	68.50		438.50	515
0160	71" x 35"		11	1.455		390	68.50		458.50	540
0170	71" x 41"		11	1.455		410	68.50		478.50	560
0180	71" x 47"	↓	11	1.455	↓	440	68.50		508.50	595

08 54 Composite Windows

08 54 13 – Fiberglass Windows

08 54 13.10 Fiberglass Single Hung Windows			Crew	Daily Output	Labor-Hours	Unit	Material	2015 Bare Costs Labor	Equipment	Total	Total Incl O&P
0010	FIBERGLASS SINGLE HUNG WINDOWS										
0100	Grids, low E, 18″ x 24″	G	2 Carp	18	.889	Ea.	335	41.50		376.50	440
0110	18″ x 40″	G		17	.941		340	44		384	450
0130	24″ x 40″	G		20	.800		360	37.50		397.50	455
0230	36″ x 36″	G		17	.941		370	44		414	480
0250	36″ x 48″	G		20	.800		405	37.50		442.50	505
0260	36″ x 60″	G		18	.889		445	41.50		486.50	560
0280	36″ x 72″	G		16	1		470	47		517	590
0290	48″ x 40″	G		16	1		470	47		517	590

08 54 13.30 Fiberglass Slider Windows

			Crew	Daily Output	Labor-Hours	Unit	Material	Labor	Equipment	Total	Total Incl O&P
0010	FIBERGLASS SLIDER WINDOWS										
0100	Grids, low E, 36″ x 24″	G	2 Carp	20	.800	Ea.	340	37.50		377.50	435
0110	36″ x 36″	G	″	20	.800	″	390	37.50		427.50	490

08 54 13.50 Fiberglass Bay Windows

			Crew	Daily Output	Labor-Hours	Unit	Material	Labor	Equipment	Total	Total Incl O&P
0010	FIBERGLASS BAY WINDOWS										
0150	48″ x 36″	G	2 Carp	11	1.455	Ea.	1,100	68.50		1,168.50	1,300

08 56 Special Function Windows

08 56 46 – Radio-Frequency-Interference Shielding Windows

08 56 46.10 Radio-Frequency-interference Mesh

		Crew	Daily Output	Labor-Hours	Unit	Material	Labor	Equipment	Total	Total Incl O&P
0010	RADIO-FREQUENCY-INTERFERENCE MESH									
0100	16 mesh copper 0.011″ wire				S.F.	4.79			4.79	5.25
0150	22 mesh copper 0.015″ wire					6.10			6.10	6.70
0200	100 mesh copper 0.022″ wire					8.65			8.65	9.50
0250	100 mesh stainless steel 0.0012″ wire					14			14	15.40

08 56 63 – Detention Windows

08 56 63.13 Visitor Cubicle Windows

		Crew	Daily Output	Labor-Hours	Unit	Material	Labor	Equipment	Total	Total Incl O&P
0010	VISITOR CUBICLE WINDOWS									
4000	Visitor cubicle, vision panel, no intercom	E-4	2	16	Ea.	3,200	850	73	4,123	5,125

08 61 Roof Windows

08 61 13 – Metal Roof Windows

08 61 13.10 Roof Windows

		Crew	Daily Output	Labor-Hours	Unit	Material	Labor	Equipment	Total	Total Incl O&P
0010	**ROOF WINDOWS**, fixed high perf tmpd glazing, metallic framed									
0020	46″ x 21-1/2″, Flashed for shingled roof	1 Carp	8	1	Ea.	270	47		317	375
0100	46″ x 28″		8	1		300	47		347	405
0125	57″ x 44″		6	1.333		370	62.50		432.50	510
0130	72″ x 28″		7	1.143		370	53.50		423.50	495
0150	Fixed, laminated tempered glazing, 46″ x 21-1/2″		8	1		455	47		502	575
0175	46″ x 28″		8	1		505	47		552	630
0200	57″ x 44″		6	1.333		475	62.50		537.50	625
0500	Vented flashing set for shingled roof, 46″ x 21-1/2″		7	1.143		455	53.50		508.50	590
0525	46″ x 28″		6	1.333		505	62.50		567.50	660
0550	57″ x 44″		5	1.600		635	75		710	820
0560	72″ x 28″		5	1.600		635	75		710	820
0575	Flashing set for low pitched roof, 46″ x 21-1/2″		7	1.143		525	53.50		578.50	670
0600	46″ x 28″		7	1.143		575	53.50		628.50	725
0625	57″ x 44″		5	1.600		715	75		790	910

08 61 Roof Windows

08 61 13 – Metal Roof Windows

08 61 13.10 Roof Windows		Crew	Daily Output	Labor-Hours	Unit	Material	2015 Bare Costs Labor	Equipment	Total	Total Incl O&P
0650	Flashing set for curb 46" x 21-1/2"	1 Carp	7	1.143	Ea.	605	53.50		658.50	755
0675	46" x 28"		7	1.143		660	53.50		713.50	815
0700	57" x 44"	↓	5	1.600	↓	810	75		885	1,025

08 62 Unit Skylights

08 62 13 – Domed Unit Skylights

08 62 13.10 Domed Skylights			Crew	Daily Output	Labor-Hours	Unit	Material	Labor	Equipment	Total	Total Incl O&P
0010	**DOMED SKYLIGHTS**										
0020	Skylight, fixed dome type, 22" x 22"	G	G-3	12	2.667	Ea.	216	125		341	440
0030	22" x 46"	G		10	3.200		269	150		419	535
0040	30" x 30"	G		12	2.667		286	125		411	515
0050	30" x 46"	G		10	3.200		380	150		530	660
0110	Fixed, double glazed, 22" x 27"	G		12	2.667		275	125		400	505
0120	22" x 46"	G		10	3.200		293	150		443	560
0130	44" x 46"	G		10	3.200		430	150		580	715
0210	Operable, double glazed, 22" x 27"	G		12	2.667		405	125		530	645
0220	22" x 46"	G		10	3.200		465	150		615	755
0230	44" x 46"	G		10	3.200	↓	875	150		1,025	1,200
9000	Minimum labor/equipment charge		↓	2	16	Job		750		750	1,200

08 62 13.20 Skylights			Crew	Daily Output	Labor-Hours	Unit	Material	Labor	Equipment	Total	Total Incl O&P
0010	**SKYLIGHTS**, flush or curb mounted										
2120	Ventilating insulated plexiglass dome with										
2130	curb mounting, 36" x 36"	G	G-3	12	2.667	Ea.	480	125		605	730
2150	52" x 52"	G		12	2.667		670	125		795	935
2160	28" x 52"	G		10	3.200		490	150		640	780
2170	36" x 52"	G	↓	10	3.200		545	150		695	835
2180	For electric opening system, add	G					315			315	345
2210	Operating skylight, with thermopane glass, 24" x 48"	G	G-3	10	3.200		585	150		735	885
2220	32" x 48"	G		9	3.556	↓	615	166		781	940
2300	Insulated safety glass with aluminum frame	G		160	.200	S.F.	94	9.35		103.35	118
2386	Skylight,non venting,non insul plexiglass dome with curb mount46"x46"		↓	13.91	2.301	Ea.	325	108		433	530
4000	Skylight, solar tube kit, incl dome, flashing, diffuser, 1 pipe, 10" dia.	G	1 Carp	2	4		235	188		423	570
4010	14" diam.	G		2	4		325	188		513	670
4020	21" diam.	G		2	4		410	188		598	760
4030	Accessories for, 1' long x 9" dia pipe	G		24	.333		47	15.65		62.65	77
4040	2' long x 9" dia pipe	G		24	.333		37	15.65		52.65	66
4050	4' long x 9" dia pipe	G		20	.400		70	18.80		88.80	108
4060	1' long x 13" dia pipe	G		24	.333		65	15.65		80.65	97
4070	2' long x 13" dia pipe	G		24	.333		51	15.65		66.65	81.50
4080	4' long x 13" dia pipe	G		20	.400		97	18.80		115.80	138
4090	6.5" turret ext for 21" dia pipe	G		16	.500		101	23.50		124.50	150
4100	12' long x 21" dia flexible pipe	G		12	.667		82	31.50		113.50	142
4110	45 degree elbow, 10"	G		16	.500		144	23.50		167.50	197
4120	14"	G		16	.500		76	23.50		99.50	122
4130	Interior decorative ring, 9"	G		20	.400		35	18.80		53.80	69.50
4140	13"	G	↓	20	.400	↓	55	18.80		73.80	91.50

08 63 Metal-Framed Skylights

08 63 13 – Domed Metal-Framed Skylights

08 63 13.20 Skylight Rigid Metal-Framed		Crew	Daily Output	Labor-Hours	Unit	2015 Bare Costs Material	Labor	Equipment	Total	Total Incl O&P
0010	**SKYLIGHT RIGID METAL-FRAMED** Skylght framing is aluminum									
0050	Fixed acrlic double domes, curb mount, 25-1/2" x 25-1/2"	G-3	10	3.200	Ea.	200	150		350	460
0060	25-1/2" x 33-1/2"		10	3.200		230	150		380	495
0070	25-1/2" x 49-1/2"		6	5.333		265	249		514	690
0080	33-1/2" x 33-1/2"		8	4		290	187		477	620
0090	37-1/2" x 25-1/2"		8	4		240	187		427	565
0100	37-1/2" x 37-1/2"		8	4		235	187		422	560
0110	37-1/2" x 49-1/2"		6	5.333		400	249		649	840
0120	49-1/2" x 33-1/2"		6	5.333		355	249		604	790
0130	49-1/2" x 49-1/2"		6	5.333		445	249		694	890
1000	Fixed tempered glass, curb mount, 17-1/2" x 33-1/2"		10	3.200		170	150		320	425
1020	17-1/2" x 49-1/2"		6	5.333		190	249		439	610
1030	25-1/2" x 25-1/2"		10	3.200		170	150		320	425
1040	25-1/2" x 33-1/2"		10	3.200		200	150		350	460
1050	25-1/2" x 37-1/2"		8	4		210	187		397	530
1060	25-1/2" x 49-1/2"		6	5.333		222	249		471	645
1070	25-1/2" x 73-1/2"		6	5.333		375	249		624	810
1080	33-1/2" x 33-1/2"		8	4		235	187		422	560
2000	Manual vent tempered glass & screen, curb, 25-1/2" x 25-1/2"		10	3.200		410	150		560	690
2020	25-1/2" x 37-1/2"		10	3.200		465	150		615	750
2030	25-1/2" x 49-1/2"		8	4		505	187		692	855
2040	33-1/2" x 33-1/2"		8	4		535	187		722	890
2050	33-1/2" x 49-1/2"		6	5.333		675	249		924	1,150
2060	37-1/2" x 37-1/2"		6	5.333		615	249		864	1,075
2070	49-1/2" x 49-1/2"		6	5.333		790	249		1,039	1,275
3000	Electric vent tempered glass , curb mount, 25-1/2" x 25-1/2"		10	3.200		960	150		1,110	1,300
3020	25-1/2" x 37-1/2"		10	3.200		1,050	150		1,200	1,400
3030	25-1/2" x 49-1/2"		8	4		1,125	187		1,312	1,525
3040	33-1/2" x 33-1/2"		8	4		1,125	187		1,312	1,550
3050	33-1/2" x 49-1/2"		6	5.333		1,225	249		1,474	1,725
3060	37-1/2" x 37-1/2"		6	5.333		1,200	249		1,449	1,700
3070	49-1/2" x 49-1/2"	↓	6	5.333	↓	1,325	249		1,574	1,850

08 71 Door Hardware

08 71 13 – Automatic Door Operators

08 71 13.10 Automatic Openers Commercial		Crew	Daily Output	Labor-Hours	Unit	Material	Labor	Equipment	Total	Total Incl O&P
0010	**AUTOMATIC OPENERS COMMERCIAL**									
0020	Pneumatic, incl opener, motion sens, control box, tubing, compressor									
0050	For single swing door, per opening	2 Skwk	.80	20	Ea.	4,550	975		5,525	6,575
0100	Pair, per opening		.50	32	Opng.	7,375	1,550		8,925	10,700
1000	For single sliding door, per opening		.60	26.667		4,975	1,300		6,275	7,575
1300	Bi-parting pair	↓	.50	32	↓	7,475	1,550		9,025	10,800
1420	Electronic door opener incl motion sens, 12 V control box, motor									
1450	For single swing door, per opening	2 Skwk	.80	20	Opng.	3,600	975		4,575	5,550
1500	Pair, per opening		.50	32		6,650	1,550		8,200	9,850
1600	For single sliding door, per opening		.60	26.667		4,400	1,300		5,700	6,950
1700	Bi-parting pair	↓	.50	32	↓	5,350	1,550		6,900	8,425
1750	Handicap actuator buttons, 2, including 12 V DC wiring, add	1 Carp	1.50	5.333	Pr.	470	250		720	925
2000	Electric Panic Button for door	2 Skwk	.50	32	Opng.	213	1,550		1,763	2,750

08 71 Door Hardware

08 71 13 – Automatic Door Operators

08 71 13.20 Automatic Openers Industrial		Crew	Daily Output	Labor-Hours	Unit	2015 Bare Costs Material	Labor	Equipment	Total	Total Incl O&P
0010	**AUTOMATIC OPENERS INDUSTRIAL**									
0015	Sliding doors up to 6′ wide	2 Skwk	.60	26.667	Opng.	5,900	1,300		7,200	8,600
0200	To 12′ wide	"	.40	40	"	7,075	1,950		9,025	10,900
0400	Over 12′ wide, add per L.F. of excess				L.F.	800			800	880
1000	Swing doors, to 5′ wide	2 Skwk	.80	20	Ea.	3,500	975		4,475	5,425
1860	Add for controls, wall pushbutton, 3 button		4	4		240	195		435	580
1870	Control pull cord		4.30	3.721		195	181		376	510

08 71 20 – Hardware

08 71 20.10 Bolts, Flush

		Crew	Daily Output	Labor-Hours	Unit	Material	Labor	Equipment	Total	Total Incl O&P
0010	**BOLTS, FLUSH**									
0020	Standard, concealed	1 Carp	7	1.143	Ea.	23.50	53.50		77	114
0800	Automatic fire exit	"	5	1.600		278	75		353	430
1600	Electrified dead bolt	1 Elec	3	2.667		140	146		286	380
3000	Barrel, brass, 2″ long	1 Carp	40	.200		7.80	9.40		17.20	24
3020	4″ long		40	.200		13	9.40		22.40	29.50
3060	6″ long		40	.200		26	9.40		35.40	44

08 71 20.15 Hardware

		Crew	Daily Output	Labor-Hours	Unit	Material	Labor	Equipment	Total	Total Incl O&P
0010	**HARDWARE**									
0020	Average percentage for hardware, total job cost									
0500	Total hardware for building, average distribution				Job	85%	15%			
1000	Door hardware, apartment, interior	1 Carp	4	2	Door	455	94		549	655
1300	Average, door hardware, motel/hotel interior, with access card		4	2		580	94		674	795
1500	Hospital bedroom, average quality		4	2		640	94		734	860
2000	High quality		3	2.667		740	125		865	1,025
2100	Pocket door		6	1.333	Ea.	100	62.50		162.50	213
2250	School, single exterior, incl. lever, incl. panic device		3	2.667	Door	1,300	125		1,425	1,650
2500	Single interior, regular use, lever included		3	2.667		635	125		760	905
2550	Average, door hdwe., school, classroom, ANSI F84, lever handl		3	2.667		850	125		975	1,150
2600	Average, door hdwe.set, school, classroom, ANSI F88, incl. lever		3	2.667		905	125		1,030	1,200
2850	Stairway, single interior		3	2.667		590	125		715	850
3100	Double exterior, with panic device		2	4	Pr.	2,425	188		2,613	2,975
6020	Add for fire alarm door holder, electro-magnetic	1 Elec	4	2	Ea.	103	109		212	282

08 71 20.20 Door Protectors

		Crew	Daily Output	Labor-Hours	Unit	Material	Labor	Equipment	Total	Total Incl O&P
0010	**DOOR PROTECTORS**									
0020	1-3/4″ x 3/4″ U channel	2 Carp	80	.200	L.F.	28	9.40		37.40	46.50
0021	1-3/4″ x 1-1/4″ U channel		80	.200	"	30	9.40		39.40	48.50
1000	Tear drop, spring-stl, 8″ high x 19″ long		15	1.067	Ea.	120	50		170	214
1010	8″ high x 32″ long		15	1.067		180	50		230	280
1100	Tear drop, stainless stl., 8″ high x 19″ long		15	1.067		315	50		365	425
1200	8″ high x 32″ long		15	1.067		400	50		450	520

08 71 20.30 Door Closers

		Crew	Daily Output	Labor-Hours	Unit	Material	Labor	Equipment	Total	Total Incl O&P
0010	**DOOR CLOSERS** Adjustible backcheck, multiple mounting									
0015	and rack and pinion									
0020	Standard Regular Arm	1 Carp	6	1.333	Ea.	192	62.50		254.50	315
0040	Hold open arm		6	1.333		200	62.50		262.50	325
0100	Fusible link		6.50	1.231		165	58		223	277
0210	Light duty, regular arm		6	1.333		109	62.50		171.50	223
0220	Parallel arm		6	1.333		133	62.50		195.50	249
0230	Hold open arm		6	1.333		117	62.50		179.50	232
0240	Fusible link arm		6	1.333		148	62.50		210.50	266
0250	Medium duty, regular arm		6	1.333		117	62.50		179.50	232

08 71 Door Hardware

08 71 20 – Hardware

08 71 20.30 Door Closers		Crew	Daily Output	Labor-Hours	Unit	Material	2015 Bare Costs Labor	Equipment	Total	Total Incl O&P
0500	Surface mount regular arm	1 Carp	6.50	1.231	Ea.	153	58		211	263
0550	Fusible link		6.50	1.231		145	58		203	255
1500	Concealed closers, normal use, head, pivot hung, interior		5.50	1.455		345	68.50		413.50	490
1510	Exterior		5.50	1.455		400	68.50		468.50	550
1520	Overhead concealed, all sizes, regular arm		5.50	1.455		200	68.50		268.50	330
1525	Concealed arm		5	1.600		315	75		390	470
1530	Concealed in door, all sizes, regular arm		5.50	1.455		325	68.50		393.50	470
1535	Concealed arm		5	1.600		255	75		330	405
1560	Floor concealed, all sizes, single acting		2.20	3.636		500	171		671	830
1565	Double acting		2.20	3.636		475	171		646	805
1570	Interior, floor, offset pivot, single acting		3.50	2.286		830	107		937	1,100
1590	Exterior		3.50	2.286		905	107		1,012	1,175
1610	Hold open arm		6	1.333		420	62.50		482.50	565
1620	Double acting, standard arm		6	1.333		760	62.50		822.50	940
1630	Hold open arm		6	1.333		765	62.50		827.50	945
1640	Floor, center hung, single acting, bottom arm		6	1.333		410	62.50		472.50	555
1650	Double acting		6	1.333		465	62.50		527.50	615
1660	Offset hung, single acting, bottom arm		6	1.333		550	62.50		612.50	710
2000	Backcheck and adjustable power, hinge face mount									
5000	For cast aluminum cylinder, deduct				Ea.	35			35	38.50
5040	For delayed action, add					46			46	50.50
5080	For fusible link arm, add					35			35	38.50
5120	For shock absorbing arm, add					50			50	55
5160	For spring power adjustment, add					40			40	44
6000	Closer-holder, hinge face mount, all sizes, exposed arm	1 Carp	6.50	1.231		205	58		263	320
6500	Electro magnetic closer/holder									
6510	Single point, no detector	1 Carp	4	2	Ea.	505	94		599	715
6515	Including detector		4	2		665	94		759	890
6520	Multi-point, no detector		4	2		885	94		979	1,125
6524	Including detector		4	2		1,325	94		1,419	1,600
6550	Electric automatic operators									
6555	Operator	1 Carp	4	2	Ea.	1,975	94		2,069	2,325
6570	Wall plate actuator		4	2		217	94		311	395
7000	Electronic closer-holder, hinge facemount, concealed arm		5	1.600		420	75		495	585
7400	With built-in detector		5	1.600		605	75		680	790
8000	Surface mounted, stand. duty, parallel arm, primed, traditional		6	1.333		199	62.50		261.50	320
8030	Light duty		6	1.333		143	62.50		205.50	261
8050	Heavy duty		6	1.333		234	62.50		296.50	360
8100	Standard duty, parallel arm, modern		6	1.333		243	62.50		305.50	370
8150	Heavy duty		6	1.333		274	62.50		336.50	405
9000	Minimum labor/equipment charge		4	2	Job		94		94	154

08 71 20.35 Panic Devices		Crew	Daily Output	Labor-Hours	Unit	Material	Labor	Equipment	Total	Total Incl O&P
0010	**PANIC DEVICES**									
0015	For rim locks, single door exit only	1 Carp	6	1.333	Ea.	460	62.50		522.50	610
0016	For rim locks, double door exit only		6	1.333		275	62.50		337.50	410
0020	Outside key and pull		5	1.600		570	75		645	755
0200	Bar and vertical rod, exit only		5	1.600		865	75		940	1,075
0203	Install panic bar		3.85	2.080		865	97.50		962.50	1,125
0205	Panic device for exit door					865			865	955
0210	Outside key and pull	1 Carp	4	2		975	94		1,069	1,225
0220	Surface vertical rod w/thumb piece, brass, US26D		4	2		1,075	94		1,169	1,325
0400	Bar and concealed rod		4	2		745	94		839	975

08 71 20.35 Panic Devices		Crew	Daily Output	Labor-Hours	Unit	Material	2015 Bare Costs Labor	Equipment	Total	Total Incl O&P
0500	Concealed vertical rod w/lever handle, brass, US26D	1 Carp	4	2	Ea.	1,125	94		1,219	1,400
0600	Touch bar, exit only		6	1.333		555	62.50		617.50	715
0610	Outside key and pull		5	1.600		685	75		760	875
0700	Touch bar and vertical rod, exit only		5	1.600		845	75		920	1,050
0710	Outside key and pull		4	2		970	94		1,064	1,225
0800	Touch bar, low profile, exit only		6	1.333		345	62.50		407.50	480
0810	Outside key and pull		5	1.600		410	75		485	575
0900	Touch bar and vertical rod, low profile, exit only		5	1.600		640	75		715	830
0910	Outside key and pull		4	2		740	94		834	970
1000	Mortise, bar, exit only		4	2		725	94		819	955
1100	Mortise bar, with thumb piece, brass, US26D		4	2		1,125	94		1,219	1,400
1600	Touch bar, exit only		4	2		760	94		854	990
2000	Narrow stile, rim mounted, bar, exit only		6	1.333		610	62.50		672.50	775
2010	Outside key and pull		5	1.600		890	75		965	1,100
2020	Rim type with thumb piece, brass, US26D		5	1.600		1,000	75		1,075	1,225
2200	Bar and vertical rod, exit only		5	1.600		920	75		995	1,125
2210	Outside key and pull		4	2		920	94		1,014	1,150
2400	Bar and concealed rod, exit only		3	2.667		1,075	125		1,200	1,375
2420	Bar and concealed rods, exit only, fire rated		3	2.667		900	125		1,025	1,200
3000	Mortise, bar, exit only		4	2		600	94		694	815
3600	Touch bar, exit only		4	2		770	94		864	1,000
4000	Double doors, exit only		2	4	Pr.	1,425	188		1,613	1,850
4500	Exit & entrance		2	4	"	1,975	188		2,163	2,475
6000	Trim, rim mounted, cylinder and pull		25	.320	Ea.	177	15		192	219
6100	Cylinder, pull and thumb piece		20	.400		195	18.80		213.80	246
6200	Pull only		30	.267		96.50	12.50		109	127
6400	Mortise, cylinder and pull		25	.320		160	15		175	201
6500	Cylinder, pull and thumb piece		20	.400		190	18.80		208.80	240
6600	Pull only		30	.267		147	12.50		159.50	183
9000	Minimum labor/equipment charge		2.50	3.200	Job		150		150	246

08 71 20.40 Lockset		Crew	Daily Output	Labor-Hours	Unit	Material	Labor	Equipment	Total	Total Incl O&P
0010	**LOCKSET**, Standard duty									
0020	Non-keyed, passage, w/sect.trim	1 Carp	12	.667	Ea.	70	31.50		101.50	129
0100	Privacy		12	.667		75	31.50		106.50	134
0400	Keyed, single cylinder function		10	.800		108	37.50		145.50	180
0420	Hotel (see also Section 08 71 20.15)		8	1		200	47		247	297
0500	Lever handled, keyed, single cylinder function		10	.800		128	37.50		165.50	203
1000	Heavy duty with sectional trim, non-keyed, passages		12	.667		125	31.50		156.50	190
1100	Privacy		12	.667		155	31.50		186.50	223
1400	Keyed, single cylinder function		10	.800		185	37.50		222.50	266
1420	Hotel		8	1		500	47		547	625
1600	Communicating		10	.800		300	37.50		337.50	390
1690	For re-core cylinder, add					50			50	55
1800	Average quality	1 Carp	14	.571		50	27		77	99
1820	Heavy duty		8	1		220	47		267	320
3800	Cipher lockset w/key pad (security item)		13	.615		920	29		949	1,075
3900	Cipher lockset with dial for swinging doors (security item)		13	.615		1,800	29		1,829	2,050
3920	with dial for swinging doors & drill resistant plate (security item)		12	.667		2,200	31.50		2,231.50	2,475
3950	Cipher lockset with dial for safe/vault door (security item)		12	.667		1,500	31.50		1,531.50	1,700
3980	Keyless, pushbutton type									
4000	Residential/light commercial, deadbolt, standard	1 Carp	9	.889	Ea.	140	41.50		181.50	223
4010	Heavy duty		9	.889		230	41.50		271.50	320

08 71 Door Hardware

08 71 20 – Hardware

08 71 20.40 Lockset		Crew	Daily Output	Labor-Hours	Unit	Material	2015 Bare Costs Labor	Equipment	Total	Total Incl O&P
4020	Industrial, heavy duty, with deadbolt	1 Carp	9	.889	Ea.	380	41.50		421.50	485
4030	Key override		9	.889		390	41.50		431.50	500
4040	Lever activated handle		9	.889		415	41.50		456.50	525
4050	Key override		9	.889		440	41.50		481.50	555
4060	Double sided pushbutton type		8	1		755	47		802	905
4070	Key override		8	1		790	47		837	945
9000	Minimum labor/equipment charge		6	1.333	Job		62.50		62.50	103

08 71 20.41 Dead Locks

Code	Description	Crew	Daily Output	Labor-Hours	Unit	Material	Labor	Equipment	Total	Total Incl O&P
0010	**DEAD LOCKS**									
0011	Mortise heavy duty outside key (security item)	1 Carp	9	.889	Ea.	175	41.50		216.50	262
0020	Double cylinder		9	.889		175	41.50		216.50	262
0100	Medium duty, outside key		10	.800		110	37.50		147.50	183
0110	Double cylinder		10	.800		120	37.50		157.50	194
1000	Tubular, standard duty, outside key		10	.800		45	37.50		82.50	111
1010	Double cylinder		10	.800		60	37.50		97.50	128
1200	Night latch, outside key		10	.800		45	37.50		82.50	111
1203	Deadlock night latch		7.70	1.039		45	49		94	130
1420	Deadbolt lock, single cylinder		10	.800		48.50	37.50		86	115
1440	Double cylinder		10	.800		63	37.50		100.50	131

08 71 20.42 Mortise Locksets

Code	Description	Crew	Daily Output	Labor-Hours	Unit	Material	Labor	Equipment	Total	Total Incl O&P
0010	**MORTISE LOCKSETS**, Comm., wrought knobs & full escutcheon trim									
0015	Assumes mortise is cut									
0020	Non-keyed, passage, grade 3	1 Carp	9	.889	Ea.	149	41.50		190.50	233
0030	Grade 1		8	1		405	47		452	520
0040	Privacy set Grade 3		9	.889		169	41.50		210.50	255
0050	Grade 1		8	1		455	47		502	575
0100	Keyed, office/entrance/apartment, Grade 2		8	1		197	47		244	294
0110	Grade 1		7	1.143		515	53.50		568.50	660
0120	Single cylinder, typical, Grade 3		8	1		189	47		236	285
0130	Grade 1		7	1.143		500	53.50		553.50	640
0200	Hotel, room, Grade 3		7	1.143		190	53.50		243.50	297
0210	Grade 1 (see also Section 08 71 20.15)		6	1.333		510	62.50		572.50	665
0300	Double cylinder, Grade 3		8	1		225	47		272	325
0310	Grade 1		7	1.143		515	53.50		568.50	660
1000	Wrought knobs and sectional trim, non-keyed, passage, Grade 3		10	.800		130	37.50		167.50	205
1010	Grade 1		9	.889		405	41.50		446.50	515
1040	Privacy, Grade 3		10	.800		145	37.50		182.50	222
1050	Grade 1		9	.889		455	41.50		496.50	570
1100	Keyed, entrance, office/apartment, Grade 3		9	.889		220	41.50		261.50	310
1103	Install lockset		6.92	1.156		220	54.50		274.50	330
1110	Grade 1		8	1		520	47		567	645
1120	Single cylinder, Grade 3		9	.889		225	41.50		266.50	315
1130	Grade 1		8	1		500	47		547	625
2000	Cast knobs and full escutcheon trim									
2010	Non-keyed, passage, Grade 3	1 Carp	9	.889	Ea.	275	41.50		316.50	375
2020	Grade 1		8	1		380	47		427	495
2040	Privacy, Grade 3		9	.889		320	41.50		361.50	420
2050	Grade 1		8	1		440	47		487	560
2120	Keyed, single cylinder, Grade 3		8	1		330	47		377	440
2123	Mortise lock		6.15	1.301		330	61		391	465
2130	Grade 1		7	1.143		525	53.50		578.50	670
3000	Cast knob and sectional trim, non-keyed, passage, Grade 3		10	.800		210	37.50		247.50	293

08 71 20.42 Mortise Locksets

	08 71 20.42 Mortise Locksets	Crew	Daily Output	Labor-Hours	Unit	Material	Labor (2015 Bare Costs)	Equipment	Total	Total Incl O&P
3010	Grade 1	1 Carp	10	.800	Ea.	380	37.50		417.50	480
3040	Privacy, Grade 3		10	.800		225	37.50		262.50	310
3050	Grade 1		10	.800		440	37.50		477.50	545
3100	Keyed, office/entrance/apartment, Grade 3		9	.889		255	41.50		296.50	350
3110	Grade 1		9	.889		580	41.50		621.50	710
3120	Single cylinder, Grade 3		9	.889		260	41.50		301.50	355
3130	Grade 1		9	.889		525	41.50		566.50	650
3190	For re-core cylinder, add					75			75	82.50
5000	Wrought steel case, brass base, knob US26D									
5020	Closet, nonkeyed passage	1 Carp	8	1	Ea.	250	47		297	350
5040	Bath/bedroom, keyed		8	1		277	47		324	380
5060	Entrance, keyed		8	1		360	47		407	470
5080	Classroom, outside keyed		8	1		385	47		432	495
5100	Storeroom, keyed		8	1		375	47		422	490
5120	Front door, keyed		8	1		375	47		422	490
5140	Dormitory/exit, keyed		8	1		375	47		422	490

08 71 20.45 Peepholes

	08 71 20.45 Peepholes	Crew	Daily Output	Labor-Hours	Unit	Material	Labor	Equipment	Total	Total Incl O&P
0010	**PEEPHOLES**									
2010	Peephole	1 Carp	32	.250	Ea.	15.80	11.75		27.55	36.50
2020	Peephole, wide view	"	32	.250	"	16.50	11.75		28.25	37.50

08 71 20.50 Door Stops

	08 71 20.50 Door Stops	Crew	Daily Output	Labor-Hours	Unit	Material	Labor	Equipment	Total	Total Incl O&P
0010	**DOOR STOPS**									
0020	Holder & bumper, floor or wall	1 Carp	32	.250	Ea.	34.50	11.75		46.25	57
1300	Wall bumper, 4" diameter, with rubber pad, aluminum		32	.250		11.70	11.75		23.45	32
1600	Door bumper, floor type, aluminum		32	.250		8.10	11.75		19.85	28
1620	Brass		32	.250		10	11.75		21.75	30.50
1630	Bronze		32	.250		18.20	11.75		29.95	39.50
1900	Plunger type, door mounted		32	.250		28	11.75		39.75	50
2500	Holder, floor type, aluminum		32	.250		33.50	11.75		45.25	56.50
2520	Wall type, aluminum		32	.250		35	11.75		46.75	58
2530	Overhead type, bronze		32	.250		102	11.75		113.75	132
2540	Plunger type, aluminum		32	.250		28	11.75		39.75	50
2560	Brass		32	.250		43	11.75		54.75	66.50
3000	Electro-magnetic, wall mounted, US3		3	2.667		258	125		383	490
3020	Floor mounted, US3		3	2.667		315	125		440	555
4000	Doorstop, ceiling mounted		3	2.667		20	125		145	227
4030	Doorstop, header		3	2.667		45	125		170	255
9000	Minimum labor/equipment charge		6	1.333	Job		62.50		62.50	103

08 71 20.55 Push-Pull Plates

	08 71 20.55 Push-Pull Plates	Crew	Daily Output	Labor-Hours	Unit	Material	Labor	Equipment	Total	Total Incl O&P
0010	**PUSH-PULL PLATES**									
0090	Push plate, 0.050 thick, 3" x 12", aluminum	1 Carp	12	.667	Ea.	6.25	31.50		37.75	58.50
0100	4" x 16"		12	.667		12.80	31.50		44.30	65.50
0110	6" x 16"		12	.667		8.50	31.50		40	61
0120	8" x 16"		12	.667		9.90	31.50		41.40	62.50
0200	Push plate, 0.050 thick, 3" x 12", brass		12	.667		14	31.50		45.50	67
0210	4" x 16"		12	.667		17.50	31.50		49	71
0220	6" x 16"		12	.667		27	31.50		58.50	81
0230	8" x 16"		12	.667		35	31.50		66.50	90
0250	Push plate, 0.050 thick, 3" x 12", satin brass		12	.667		14.25	31.50		45.75	67
0260	4" x 16"		12	.667		17.70	31.50		49.20	71
0270	6" x 16"		12	.667		27	31.50		58.50	81
0280	8" x 16"		12	.667		35	31.50		66.50	90

08 71 20.55 Push-Pull Plates		Crew	Daily Output	Labor-Hours	Unit	Material	Labor	Equipment	Total	Total Incl O&P
						2015 Bare Costs				
0490	Push plate, 0.050 thick, 3" x 12", bronze	1 Carp	12	.667	Ea.	17	31.50		48.50	70
0500	4" x 16"		12	.667		25	31.50		56.50	79
0510	6" x 16"		12	.667		32	31.50		63.50	86.50
0520	8" x 16"		12	.667		38.50	31.50		70	94
0600	Push plate, antimicrobial copper alloy finish, 3.5" x 15"		13	.615		19.80	29		48.80	69.50
0610	4" x 16"		13	.615		23.50	29		52.50	73
0620	6" x 16"		13	.615		26	29		55	76
0630	6" x 20"		13	.615		26	29		55	76
0740	Push plate, 0.050 thick, 3" x 12", stainless steel		12	.667		12	31.50		43.50	64.50
0750	4" x 16"		12	.667		27.50	31.50		59	81.50
0760	6" x 16"		12	.667		18.65	31.50		50.15	72
0780	8" x 16"		12	.667		23.50	31.50		55	77.50
0790	Push plate, 0.050 thick, 3" x 12", satin stainless steel		12	.667		7	31.50		38.50	59
0810	4" x 16"		12	.667		8.50	31.50		40	61
0820	6" x 16"		12	.667		11.70	31.50		43.20	64.50
0830	8" x 16"		12	.667		16	31.50		47.50	69
0980	Pull plate, 0.050 thick, 3" x 12", aluminum		12	.667		25.50	31.50		57	79.50
1000	4" x 16"		12	.667		25.50	31.50		57	79.50
1050	Pull plate, 0.050 thick, 3" x 12", brass		12	.667		39.50	31.50		71	95
1060	4" x 16"		12	.667		34.50	31.50		66	89.50
1080	Pull plate, 0.050 thick, 3" x 12", bronze		12	.667		50	31.50		81.50	107
1100	4" x 16"		12	.667		56	31.50		87.50	113
1180	Pull plate, 0.050 thick, 3" x 12", stainless steel		12	.667		45.50	31.50		77	102
1200	4" x 16"		12	.667		58.50	31.50		90	116
1250	Pull plate, 0.050 thick, 3" x 12", chrome		12	.667		44.50	31.50		76	101
1270	4" x 16"		12	.667		46.50	31.50		78	103
1500	Pull handle and push bar, aluminum		11	.727		118	34		152	186
2000	Bronze		10	.800		159	37.50		196.50	237
9800	Minimum labor/equipment charge		5	1.600	Job		75		75	123

08 71 20.60 Entrance Locks		Crew	Daily Output	Labor-Hours	Unit	Material	Labor	Equipment	Total	Total Incl O&P
0010	**ENTRANCE LOCKS**									
0015	Cylinder, grip handle deadlocking latch	1 Carp	9	.889	Ea.	175	41.50		216.50	262
0020	Deadbolt		8	1		175	47		222	270
0100	Push and pull plate, dead bolt		8	1		225	47		272	325
0200	Push bar and pull, dead bolt, bronze		7	1.143		270	53.50		323.50	385
0240	Push bar and pull bar, dead bolt, bronze		7	1.143		400	53.50		453.50	530
0900	For handicapped lever, add					150			150	165

08 71 20.65 Thresholds		Crew	Daily Output	Labor-Hours	Unit	Material	Labor	Equipment	Total	Total Incl O&P
0010	**THRESHOLDS**									
0011	Threshold 3' long saddles aluminum	1 Carp	48	.167	L.F.	9.40	7.85		17.25	23
0100	Aluminum, 8" wide, 1/2" thick		12	.667	Ea.	49	31.50		80.50	105
0500	Bronze		60	.133	L.F.	42	6.25		48.25	57
0600	Bronze, panic threshold, 5" wide, 1/2" thick		12	.667	Ea.	158	31.50		189.50	225
0700	Rubber, 1/2" thick, 5-1/2" wide		20	.400		40	18.80		58.80	75
0800	2-3/4" wide		20	.400		45	18.80		63.80	80.50
1950	ADA Compliant Thresholds									
2000	Threshold, wood oak 3-1/2" wide x 24" long	1 Carp	12	.667	Ea.	9	31.50		40.50	61.50
2010	3-1/2" wide x 36" long		12	.667		13.50	31.50		45	66.50
2020	3-1/2" wide x 48" long		12	.667		18	31.50		49.50	71.50
2030	4-1/2" wide x 24" long		12	.667		12	31.50		43.50	64.50
2040	4-1/2" wide x 36" long		12	.667		17	31.50		48.50	70
2050	4-1/2" wide x 48" long		12	.667		23	31.50		54.50	77

08 71 20 – Hardware

08 71 20.65 Thresholds		Crew	Daily Output	Labor-Hours	Unit	Material	Labor	Equipment	Total	Total Incl O&P
						2015 Bare Costs				
2060	6-1/2" wide x 24" long	1 Carp	12	.667	Ea.	16.50	31.50		48	69.50
2070	6-1/2" wide x 36" long		12	.667		25	31.50		56.50	79
2080	6-1/2" wide x 48" long		12	.667		33	31.50		64.50	88
2090	Threshold, wood cherry 3-1/2" wide x 24" long		12	.667		13.50	31.50		45	66.50
2100	3-1/2" wide x 36" long		12	.667		21	31.50		52.50	74.50
2110	3-1/2" wide x 48" long		12	.667		27	31.50		58.50	81
2120	4-1/2" wide x 24" long		12	.667		17.50	31.50		49	71
2130	4-1/2" wide x 36" long		12	.667		25.50	31.50		57	79.50
2140	4-1/2" wide x 48" long		12	.667		34	31.50		65.50	89
2150	6-1/2" wide x 24" long		12	.667		25.50	31.50		57	79.50
2160	6-1/2" wide x 36" long		12	.667		36.50	31.50		68	91.50
2170	6-1/2" wide x 48" long		12	.667		49	31.50		80.50	106
2180	Threshold, wood walnut 3-1/2" wide x 24" long		12	.667		15.50	31.50		47	68.50
2190	3-1/2" wide x 36" long		12	.667		23	31.50		54.50	77
2200	3-1/2" wide x 48" long		12	.667		30	31.50		61.50	84.50
2210	4-1/2" wide x 24" long		12	.667		19.50	31.50		51	73
2220	4-1/2" wide x 36" long		12	.667		28.50	31.50		60	83
2230	4-1/2" wide x 48" long		12	.667		38	31.50		69.50	93.50
2240	6-1/2" wide x 24" long		12	.667		28	31.50		59.50	82.50
2250	6-1/2" wide x 36" long		12	.667		41.50	31.50		73	97
2260	6-1/2" wide x 48" long		12	.667		55	31.50		86.50	112
2300	Threshold, aluminum 4" wide x 36" long		12	.667		33	31.50		64.50	88
2310	4" wide x 48" long		12	.667		41	31.50		72.50	96.50
2320	4" wide x 72" long		12	.667		66	31.50		97.50	124
2330	5" wide x 36" long		12	.667		45	31.50		76.50	101
2340	5" wide x 48" long		12	.667		56	31.50		87.50	113
2350	5" wide x 72" long		12	.667		89	31.50		120.50	150
2360	6" wide x 36" long		12	.667		53	31.50		84.50	110
2370	6" wide x 48" long		12	.667		68	31.50		99.50	127
2380	6" wide x 72" long		12	.667		106	31.50		137.50	169
2390	7" wide x 36" long		12	.667		69	31.50		100.50	128
2400	7" wide x 48" long		12	.667		90	31.50		121.50	151
2410	7" wide x 72" long		12	.667		137	31.50		168.50	203
2500	Threshold, ramp, aluminum or rubber 24" x 24"		12	.667		190	31.50		221.50	261
9000	Minimum labor/equipment charge		4	2	Job		94		94	154

08 71 20.70 Floor Checks

		Crew	Daily Output	Labor-Hours	Unit	Material	Labor	Equipment	Total	Total Incl O&P
0010	**FLOOR CHECKS**									
0020	For over 3' wide doors single acting	1 Carp	2.50	3.200	Ea.	745	150		895	1,075
0500	Double acting	"	2.50	3.200	"	860	150		1,010	1,200

08 71 20.75 Door Hardware Accessories

		Crew	Daily Output	Labor-Hours	Unit	Material	Labor	Equipment	Total	Total Incl O&P
0010	**DOOR HARDWARE ACCESSORIES**									
0050	Door closing coordinator, 36" (for paired openings up to 56")	1 Carp	8	1	Ea.	98	47		145	185
0060	48" (for paired openings up to 84")		8	1		105	47		152	193
0070	56" (for paired openings up to 96")		8	1		116	47		163	204

08 71 20.80 Hasps

		Crew	Daily Output	Labor-Hours	Unit	Material	Labor	Equipment	Total	Total Incl O&P
0010	**HASPS**, steel assembly									
0015	3"	1 Carp	26	.308	Ea.	4.90	14.45		19.35	29
0020	4-1/2"		13	.615		6.60	29		35.60	55
0040	6"		12.50	.640		9.35	30		39.35	60

08 71 Door Hardware

08 71 20 – Hardware

08 71 20.90 Hinges		Crew	Daily Output	Labor-Hours	Unit	2015 Bare Costs Material	Labor	Equipment	Total	Total Incl O&P
0010	**HINGES**									
0012	Full mortise, avg. freq., steel base, USP, 4-1/2" x 4-1/2"				Pr.	36			36	40
0040	US26D					65.50			65.50	72
0080	US10A					49.50			49.50	54.50
0100	5" x 5", USP					57.50			57.50	63
0200	6" x 6", USP					119			119	131
0400	Brass base, 4-1/2" x 4-1/2", US10					58.50			58.50	64.50
0440	US26D					62.50			62.50	68.50
0480	US10B					63			63	69.50
0500	5" x 5", US10					86			86	94.50
0600	6" x 6", US10					162			162	178
0800	Stainless steel base, 4-1/2" x 4-1/2", US32					74			74	81.50
0900	For non removable pin, add (security item)				Ea.	4.88			4.88	5.35
0910	For floating pin, driven tips, add					3.30			3.30	3.63
0930	For hospital type tip on pin, add					13.95			13.95	15.35
0940	For steeple type tip on pin, add					18.95			18.95	21
0950	Full mortise, high frequency, steel base, 3-1/2" x 3-1/2", US26D				Pr.	30			30	33
1000	4-1/2" x 4-1/2", USP					63			63	69
1040	US26D					56.50			56.50	62.50
1080	US26					68			68	75
1100	5" x 5", USP					50			50	55
1200	6" x 6", USP					134			134	148
1300	8" x 8", USP					293			293	320
1400	Brass base, 3-1/2" x 3-1/2", US4					51			51	56
1430	4-1/2" x 4-1/2", US10					73.50			73.50	81
1440	US26D					102			102	112
1480	US10B					118			118	130
1500	5" x 5", US10					122			122	134
1600	6" x 6", US10					165			165	181
1700	8" x 8", US10					385			385	425
1800	Stainless steel base, 4-1/2" x 4-1/2", US32					103			103	114
1810	5" x 4-1/2", US32					137			137	151
1930	For hospital type tip on pin, add				Ea.	13.15			13.15	14.45
1950	Full mortise, low frequency, steel base, 3-1/2" x 3-1/2", US26D				Pr.	23.50			23.50	26
2000	4-1/2" x 4-1/2", USP					22			22	24.50
2040	US26D					23.50			23.50	25.50
2080	US10A					48.50			48.50	53
2100	5" x 5", USP					47.50			47.50	52
2200	6" x 6", USP					89			89	98
2300	4-1/2" x 4-1/2", US3					16.85			16.85	18.50
2310	5" x 5", US3					41			41	45
2400	Brass bass, 4-1/2" x 4-1/2", US10					53			53	58.50
2440	US26D					61			61	67
2480	US10A					55.50			55.50	61
2500	5" x 5", US10					76.50			76.50	84
2800	Stainless steel base, 4-1/2" x 4-1/2", US32					74.50			74.50	82
3000	Half surface, half mortise, or full surface, average frequency									
3010	Steel base, 4-1/2" x 4-1/2", USP				Pr.	44.50			44.50	49
3040	US26D					69			69	76
3080	US10A					88.50			88.50	97
3100	5" x 5", USP					81.50			81.50	90
3400	Brass base, 4-1/2" x 4-1/2", US10					164			164	181

08 71 20 – Hardware

08 71 20.90 Hinges		Crew	Daily Output	Labor-Hours	Unit	2015 Bare Costs Material	Labor	Equipment	Total	Total Incl O&P
3440	US26D				Pr.	172			172	189
3480	US10B					185			185	204
3500	5" x 5", US10					178			178	196
3800	Stainless steel base, 4-1/2" x 4-1/2", US32				↓	226			226	249
4000	Half surface, half mortise or full surface, high frequency									
4010	Steel base, 4-1/2" x 4-1/2", USP				Pr.	99			99	109
4040	US26D					133			133	146
4080	US10A					153			153	168
4100	5" x 5", USP					142			142	156
4400	Brass base, 4-1/2" x 4-1/2", US10					169			169	186
4440	US26D					175			175	193
4480	US10B					214			214	235
4500	5" x 5", US10					216			216	237
4800	Stainless steel base, 4-1/2" x 4-1/2", US32				↓	222			222	245
5000	Half surface, half mortise, or full surface, low frequency									
5010	Steel base, 4-1/2" x 4-1/2", USP				Pr.	46			46	50.50
5040	US26D					54.50			54.50	60
5080	US10A					78.50			78.50	86.50
8000	Install hinge	1 Carp	34	.235	↓		11.05		11.05	18.10

08 71 20.91 Special Hinges		Crew	Daily Output	Labor-Hours	Unit	Material	Labor	Equipment	Total	Total Incl O&P
0010	**SPECIAL HINGES**									
0015	Paumelle, high frequency									
0020	Steel base, 6" x 4-1/2", US10				Pr.	145			145	160
0040	US26D					181			181	199
0080	US10A				↓	190			190	209
0100	Brass base, 5" x 4-1/2", US10				Ea.	248			248	273
0140	US26D					253			253	278
0180	US3				↓	259			259	284
0200	Paumelle, average frequency, steel base, 4-1/2" x 3-1/2", US10				Pr.	98.50			98.50	108
0240	US26D					103			103	113
0280	US10A				↓	106			106	116
0400	Olive knuckle, low frequency, brass base, 6" x 4-1/2", US10				Ea.	144			144	159
0440	US26					156			156	171
0480	US3				↓	150			150	165
0800	Emergency door pivot, average frequency									
0810	Brass base, 4-7/8" jamb plate, USP				Pr.	68			68	75
0840	US26D					75			75	82.50
0880	For emergency door stop and hold back, add				↓	64.50			64.50	71
1000	Electric hinge with concealed conductor, average frequency									
1010	Steel base, 4-1/2" x 4-1/2", US26D				Pr.	310			310	340
1100	Bronze base, 4-1/2" x 4-1/2", US26D				"	320			320	350
1200	Electric hinge with concealed conductor, high frequency									
1210	Steel base, 4-1/2" x 4-1/2", US26D				Pr.	262			262	288
1400	Non template, full mortise, low frequency									
1410	Steel base, 4" x 4", USP				Pr.	23.50			23.50	26
1440	US26D					34			34	37
1480	US10A				↓	23			23	25.50
1500	Non template, full mortise, average frequency									
1510	Steel base, 4" x 4", USP				Pr.	36			36	39.50
1540	US26D					34.50			34.50	38
1580	US10A					45			45	49.50
1600	Double weight, 800 lb., steel base, removable pin, 5" x 6", USP				↓	410			410	450

08 71 20.91 Special Hinges		Crew	Daily Output	Labor-Hours	Unit	Material	Labor	Equipment	Total	Total Incl O&P
						2015 Bare Costs				
1700	Steel base-welded pin, 5" x 6", USP				Pr.	166			166	183
1800	Triple weight, 2000 lb., steel base, welded pin, 5" x 6", USP					535			535	590
2000	Pivot reinf., high frequency, steel base, 7-3/4" door plate, USP					149			149	164
2040	US26D					240			240	263
2080	US10A					236			236	259
2200	Bronze base, 7-3/4" door plate, US10					232			232	255
2240	US26D					300			300	330
2280	US10A					360			360	395
3000	Swing clear, full mortise, full or half surface, high frequency,									
3010	Steel base, 5" high, USP				Pr.	143			143	157
3040	US26D					150			150	165
3080	US10A					190			190	209
3200	Swing clear, full mortise, average frequency									
3210	Steel base, 4-1/2" high, USP				Pr.	128			128	141
3280	US10A				"	144			144	158
3400	Swing clear, half mortise, high frequency									
3410	Steel base, 4-1/2" high, USP				Pr.	139			139	153
3440	US26D					149			149	163
3480	US10A					214			214	235
4000	Wide throw, average frequency, steel base, 4-1/2" x 6", USP					94.50			94.50	104
4040	US26D					104			104	114
4080	US10A					108			108	119
4100	5" x 7", USP					110			110	121
4200	High frequency, steel base, 4-1/2" x 6", USP					112			112	124
4240	US26D					155			155	171
4280	US10A					175			175	193
4300	5" x 7", USP					121			121	134
4400	Wide throw, low frequency, steel base, 4-1/2" x 6", USP					77			77	84.50
4440	US26D					89.50			89.50	98
4480	US10A					90			90	98.50
4500	5" x 7", USP					95			95	105
4600	Spring hinge, single acting, 6" flange, steel				Ea.	50			50	55
4700	Brass					94.50			94.50	104
4900	Double acting, 6" flange, steel					80			80	88
4950	Brass					133			133	146
8000	Continuous hinges									
8010	Steel, piano, 2" x 72"	1 Carp	20	.400	Ea.	22	18.80		40.80	55
8020	Brass, piano, 1-1/16" x 30"		30	.267		8	12.50		20.50	29.50
8030	Acrylic, piano, 1-3/4" x 12"		40	.200		15	9.40		24.40	32
8040	Aluminum, door, standard duty, 7'		3	2.667		135	125		260	355
8050	Heavy duty, 7'		3	2.667		142	125		267	360
8060	8'		3	2.667		160	125		285	380
8070	Steel, door, heavy duty, 7'		3	2.667		200	125		325	425
8080	8'		3	2.667		250	125		375	480
8090	Stainless steel, door, heavy duty, 7'		3	2.667		260	125		385	490
8100	8'		3	2.667		280	125		405	515
9000	Continuous hinge, steel, full mortise, heavy duty, 96 inch		2	4		460	188		648	815
9200	Continuous geared hinge, aluminum, full mortise, standard duty, 83 inch		3	2.667		120	125		245	335
9250	Continuous geared hinge, aluminum, full mortise, heavy duty, 83 inch		3	2.667		200	125		325	425

08 71 20 – Hardware

08 71 20.92 Mortised Hinges		Crew	Daily Output	Labor-Hours	Unit	Material	2015 Bare Costs Labor	Equipment	Total	Total Incl O&P
0010	**MORTISED HINGES**									
0200	Average frequency, steel plated, ball bearing, 3-1/2" x 3-1/2"				Pr.	27.50			27.50	30.50
0300	Bronze, ball bearing					32.50			32.50	35.50
0900	High frequency, steel plated, ball bearing					73.50			73.50	81
1100	Bronze, ball bearing					73.50			73.50	81
1300	Average frequency, steel plated, ball bearing, 4-1/2" x 4-1/2"					35			35	38.50
1500	Bronze, ball bearing, to 36" wide					36.50			36.50	40.50
1700	Low frequency, steel, plated, plain bearing					19.40			19.40	21.50
1900	Bronze, plain bearing					27			27	30

08 71 20.95 Kick Plates		Crew	Daily Output	Labor-Hours	Unit	Material	Labor	Equipment	Total	Total Incl O&P
0010	**KICK PLATES**									
0020	Stainless steel, .050, 16 ga., 8" x 28", US32	1 Carp	15	.533	Ea.	38	25		63	83
0030	8" x 30"		15	.533		41	25		66	86
0040	8" x 34"		15	.533		46	25		71	91.50
0050	10" x 28"		15	.533		76	25		101	125
0060	10" x 30"		15	.533		82	25		107	131
0070	10" x 34"		15	.533		92	25		117	142
0080	Mop/Kick, 4" x 28"		15	.533		34	25		59	78.50
0090	4" x 30"		15	.533		36	25		61	80.50
0100	4" x 34"		15	.533		41	25		66	86
0110	6" x 28"		15	.533		43	25		68	88.50
0120	6" x 30"		15	.533		47	25		72	92.50
0130	6" x 34"		15	.533		53	25		78	99.50
0500	Bronze, .050", 8" x 28"		15	.533		66.50	25		91.50	115
0510	8" x 30"		15	.533		65	25		90	113
0520	8" x 34"		15	.533		73	25		98	122
0530	10" x 28"		15	.533		75	25		100	124
0540	10" x 30"		15	.533		80	25		105	129
0550	10" x 34"		15	.533		91	25		116	141
0560	Mop/Kick, 4" x 28"		15	.533		33	25		58	77.50
0570	4" x 30"		15	.533		36	25		61	80.50
0580	4" x 34"		15	.533		37	25		62	81.50
0590	6" x 28"		15	.533		46	25		71	91.50
0600	6" x 30"		15	.533		52	25		77	98
0610	6" x 34"		15	.533		56	25		81	103
1000	Acrylic, .125", 8" x 26"		15	.533		28	25		53	72
1010	8" x 36"		15	.533		38	25		63	83
1020	8" x 42"		15	.533		45	25		70	90.50
1030	10" x 26"		15	.533		35	25		60	79.50
1040	10" x 36"		15	.533		48	25		73	94
1050	10" x 42"		15	.533		68.50	25		93.50	117
1060	Mop/Kick, 4" x 26"		15	.533		17	25		42	59.50
1070	4" x 36"		15	.533		24	25		49	67.50
1080	4" x 42"		15	.533		27	25		52	70.50
1090	6" x 26"		15	.533		23	25		48	66.50
1100	6" x 36"		15	.533		34	25		59	78.50
1110	6" x 42"		15	.533		39	25		64	84
1220	Brass, .050", 8" x 26"		15	.533		56	25		81	103
1230	8" x 36"		15	.533		75	25		100	124
1240	8" x 42"		15	.533		86	25		111	136
1250	10" x 26"		15	.533		71	25		96	119
1260	10" x 36"		15	.533		91	25		116	141

08 71 20 – Hardware

08 71 20.95 Kick Plates		Crew	Daily Output	Labor-Hours	Unit	Material	2015 Bare Costs Labor	Equipment	Total	Total Incl O&P
1270	10" x 42"	1 Carp	15	.533	Ea.	105	25		130	157
1320	Mop/Kick, 4" x 26"		15	.533		28	25		53	72
1330	4" x 36"		15	.533		39	25		64	84
1340	4" x 42"		15	.533		44	25		69	89.50
1350	6" x 26"		15	.533		38	25		63	83
1360	6" x 36"		15	.533		48	25		73	94
1370	6" x 42"		15	.533		55	25		80	102
1800	Aluminum, .050", 8" x 26"		15	.533		35	25		60	79.50
1810	8" x 36"		15	.533		40	25		65	85
1820	8" x 42"		15	.533		47	25		72	92.50
1830	10" x 26"		15	.533		36	25		61	80.50
1840	10" x 36"		15	.533		50	25		75	96
1850	10" x 42"		15	.533		59	25		84	106
1860	Mop/Kick, 4" x 26"		15	.533		15	25		40	57.50
1870	4" x 36"		15	.533		20	25		45	63
1880	4" x 42"		15	.533		24	25		49	67.50
1890	6" x 26"		15	.533		22	25		47	65
1900	6" x 36"		15	.533		30	25		55	74
1910	6" x 42"		15	.533		35	25		60	79.50
9000	Minimum labor/equipment charge		6	1.333	Job		62.50		62.50	103

08 71 21 – Astragals

08 71 21.10 Exterior Mouldings, Astragals		Crew	Daily Output	Labor-Hours	Unit	Material	Labor	Equipment	Total	Total Incl O&P
0010	EXTERIOR MOULDINGS, ASTRAGALS									
0400	One piece, overlapping cadmium plated steel, flat, 3/16" x 2"	1 Carp	90	.089	L.F.	4	4.17		8.17	11.25
0600	Prime coated steel, flat, 1/8" x 3"		90	.089		5.75	4.17		9.92	13.20
0800	Stainless steel, flat, 3/32" x 1-5/8"		90	.089		16	4.17		20.17	24.50
1000	Aluminum, flat, 1/8" x 2"		90	.089		4.10	4.17		8.27	11.35
1200	Nail on, "T" extrusion		120	.067		1.90	3.13		5.03	7.25
1300	Vinyl bulb insert		105	.076		2.50	3.58		6.08	8.60
1600	Screw on, "T" extrusion		90	.089		3.75	4.17		7.92	11
1700	Vinyl insert		75	.107		4.50	5		9.50	13.15
2000	"L" extrusion, neoprene bulbs		75	.107		4.10	5		9.10	12.70
2100	Neoprene sponge insert		75	.107		6.90	5		11.90	15.80
2200	Magnetic		75	.107		10.60	5		15.60	19.85
2400	Spring hinged security seal, with cam		75	.107		6.80	5		11.80	15.70
2600	Spring loaded locking bolt, vinyl insert		45	.178		9.20	8.35		17.55	24
2800	Neoprene sponge strip, "Z" shaped, aluminum		60	.133		8.30	6.25		14.55	19.40
2900	Solid neoprene strip, nail on aluminum strip		90	.089		4.05	4.17		8.22	11.30
3000	One piece stile protection									
3020	Neoprene fabric loop, nail on aluminum strips	1 Carp	60	.133	L.F.	1.10	6.25		7.35	11.45
3110	Flush mounted aluminum extrusion, 1/2" x 1-1/4"		60	.133		6.90	6.25		13.15	17.85
3140	3/4" x 1-3/8"		60	.133		4.10	6.25		10.35	14.75
3160	1-1/8" x 1-3/4"		60	.133		4.70	6.25		10.95	15.40
3300	Mortise, 9/16" x 3/4"		60	.133		4.10	6.25		10.35	14.75
3320	13/16" x 1-3/8"		60	.133		4.30	6.25		10.55	15
3600	Spring bronze strip, nail on type		105	.076		1.85	3.58		5.43	7.90
3620	Screw on, with retainer		75	.107		2.70	5		7.70	11.15
3800	Flexible stainless steel housing, pile insert, 1/2" door		105	.076		7.25	3.58		10.83	13.85
3820	3/4" door		105	.076		8.10	3.58		11.68	14.75
4000	Extruded aluminum retainer, flush mount, pile insert		105	.076		2.25	3.58		5.83	8.35
4080	Mortise, felt insert		90	.089		4.55	4.17		8.72	11.85
4160	Mortise with spring, pile insert		90	.089		3.40	4.17		7.57	10.60

08 71 Door Hardware

08 71 21 – Astragals

08 71 21.10 Exterior Mouldings, Astragals		Crew	Daily Output	Labor-Hours	Unit	Material	Labor	Equipment	Total	Total Incl O&P
						2015 Bare Costs				
4400	Rigid vinyl retainer, mortise, pile insert	1 Carp	105	.076	L.F.	2.70	3.58		6.28	8.80
4600	Wool pile filler strip, aluminum backing	↓	105	.076	↓	2.70	3.58		6.28	8.80
5000	Two piece overlapping astragal, extruded aluminum retainer									
5010	Pile insert	1 Carp	60	.133	L.F.	3.35	6.25		9.60	13.95
5020	Vinyl bulb insert		60	.133		1.85	6.25		8.10	12.30
5040	Vinyl flap insert		60	.133		3.65	6.25		9.90	14.25
5060	Solid neoprene flap insert		60	.133		6.55	6.25		12.80	17.45
5080	Hypalon rubber flap insert		60	.133		6.65	6.25		12.90	17.55
5090	Snap on cover, pile insert		60	.133		9.45	6.25		15.70	20.50
5400	Magnetic aluminum, surface mounted		60	.133		23	6.25		29.25	35.50
5500	Interlocking aluminum, 5/8" x 1" neoprene bulb insert		45	.178		5.70	8.35		14.05	19.95
5600	Adjustable aluminum, 9/16" x 21/32", pile insert		45	.178		17.55	8.35		25.90	33
5800	Magnetic, adjustable, 9/16" x 21/32"	↓	45	.178	↓	23	8.35		31.35	38.50
6000	Two piece stile protection									
6010	Cloth backed rubber loop, 1" gap, nail on aluminum strips	1 Carp	45	.178	L.F.	4.35	8.35		12.70	18.50
6040	Screw on aluminum strips		45	.178		6.55	8.35		14.90	21
6100	1-1/2" gap, screw on aluminum extrusion		45	.178		5.85	8.35		14.20	20
6240	Vinyl fabric loop, slotted aluminum extrusion, 1" gap		45	.178		2.20	8.35		10.55	16.10
6300	1-1/4" gap	↓	45	.178	↓	6.20	8.35		14.55	20.50

08 71 25 – Weatherstripping

08 71 25.10 Mechanical Seals, Weatherstripping

Line	Description	Crew	Daily Output	Labor-Hours	Unit	Material	Labor	Equipment	Total	Total Incl O&P
0010	**MECHANICAL SEALS, WEATHERSTRIPPING**									
1000	Doors, wood frame, interlocking, for 3' x 7' door, zinc	1 Carp	3	2.667	Opng.	44	125		169	254
1100	Bronze		3	2.667		56	125		181	267
1300	6' x 7' opening, zinc		2	4		54	188		242	370
1400	Bronze	↓	2	4	↓	65	188		253	380
1700	Wood frame, spring type, bronze									
1800	3' x 7' door	1 Carp	7.60	1.053	Opng.	23	49.50		72.50	107
1900	6' x 7' door	"	7	1.143	"	29.50	53.50		83	121
2200	Metal frame, spring type, bronze									
2300	3' x 7' door	1 Carp	3	2.667	Opng.	46.50	125		171.50	256
2400	6' x 7' door	"	2.50	3.200	"	52	150		202	305
2500	For stainless steel, spring type, add					133%				
2700	Metal frame, extruded sections, 3' x 7' door, aluminum	1 Carp	3	2.667	Opng.	28	125		153	236
2800	Bronze		3	2.667		82	125		207	295
3100	6' x 7' door, aluminum		1.50	5.333		35	250		285	450
3200	Bronze	↓	1.50	5.333	↓	137	250		387	560
3500	Threshold weatherstripping									
3650	Door sweep, flush mounted, aluminum	1 Carp	25	.320	Ea.	19	15		34	45.50
3700	Vinyl		25	.320		18	15		33	44.50
5000	Garage door bottom weatherstrip, 12' aluminum, clear		14	.571		25	27		52	71.50
5010	Bronze		14	.571		90	27		117	143
5050	Bottom protection, Rubber		14	.571		37	27		64	84.50
5100	Threshold		14	.571	↓	72	27		99	123
9000	Minimum labor/equipment charge	↓	3	2.667	Job		125		125	205

08 71 63 – Detention Door Hardware

08 71 63.10 Detention Locks and Keys

Line	Description	Crew	Daily Output	Labor-Hours	Unit	Material	Labor	Equipment	Total	Total Incl O&P
0010	**DETENTION LOCKS AND KEYS**									
0050	Jail cell, mechanical deadlock, with paracentric key	1 Carp	8	1	Ea.	385	47		432	500
0060	with mogul key		8	1		710	47		757	855
0070	Mechanical snap lock, with paracentric key		8	1		385	47		432	500
0080	with mogul key	↓	8	1	↓	700	47		747	845

08 71 Door Hardware

08 71 63 – Detention Door Hardware

08 71 63.10 Detention Locks and Keys		Crew	Daily Output	Labor-Hours	Unit	2015 Bare Costs Material	Labor	Equipment	Total	Total Incl O&P
0090	Solenoid operated electro-mechanical deadlock	1 Carp	8	1	Ea.	2,375	47		2,422	2,700
0100	Solenoid operated electro-mechanical deadlatch		8	1		2,050	47		2,097	2,350
0110	Electric lock cylinders	↓	8	1		350	47		397	455
0120	Mogul keys					45			45	49.50
0130	Paracentric keys				↓	134			134	147

08 74 Access Control Hardware

08 74 13 – Card Key Access Control Hardware

08 74 13.50 Card Key Access

		Crew	Daily Output	Labor-Hours	Unit	Material	Labor	Equipment	Total	Total Incl O&P
0010	**CARD KEY ACCESS**									
0020	Computerized system , processor, proximity reader and cards									
0030	Does not inculde door hardware, lockset or wiring									
0040	Card key system for 1 door				Ea.	1,225			1,225	1,350
0060	Card key system for 2 doors					2,125			2,125	2,350
0080	Card key system for 4 doors					2,650			2,650	2,900
0100	Processor for card key access system					850			850	935
0160	Magnetic lock for electric access, 600 Pound holding force					190			190	209
0170	Magnetic lock for electric access, 1200 Pound holding force					190			190	209
0200	Proximity card reader				↓	130			130	143

08 74 13.60 Entrance Card Systems

		Crew	Daily Output	Labor-Hours	Unit	Material	Labor	Equipment	Total	Total Incl O&P
0010	**ENTRANCE CARD SYSTEMS**									
0100	Entrance card, barium ferrite				Ea.	4.50			4.50	4.95
0120	Credential					6			6	6.60
0140	Proximity					8.50			8.50	9.35
0160	Weigand					9.75			9.75	10.75
0500	Entrance card reader, barium ferrite	R-19	4	5		295	274		569	750
0520	Credential		4	5		235	274		509	685
0540	Proximity		4	5		315	274		589	775
0560	Weigand		4	5		415	274		689	880
0600	Local processor for card system		4	5		1,750	274		2,024	2,350
0650	Scanner, eye retina	↓	4	5		6,000	274		6,274	7,025
0700	Gate opener, cantilever	R-18	3	8.667		2,350	370		2,720	3,175
0750	Switch, tamper	R-19	6	3.333		22	183		205	305
0900	Accessories, electric door strike/bolt		5	4		77.50	219		296.50	425
0920	Electromagnetic lock		5	4		169	219		388	525
0940	Keypad for card reader	↓	4	5	↓	580	274		854	1,075

08 74 16 – Keypad Access Control Hardware

08 74 16.50 Keypad Access

		Crew	Daily Output	Labor-Hours	Unit	Material	Labor	Equipment	Total	Total Incl O&P
0010	**KEYPAD ACCESS**									
0340	Digital keypad, int/ext, basic, excl. striker/power/wiring	1 Elec	3	2.667	Ea.	109	146		255	345
0350	Lockset, mechanical push-button type, complete, incl hardware	1 Carp	4	2	"	380	94		474	570

08 74 19 – Biometric Identity Access Control Hardware

08 74 19.50 Biometric Identity Access

		Crew	Daily Output	Labor-Hours	Unit	Material	Labor	Equipment	Total	Total Incl O&P
0010	**BIOMETRIC IDENTITY ACCESS**									
0200	Fingerprint scanner unit, excl striker/power supply	1 Elec	4	2	Ea.	1,400	109		1,509	1,725
0210	Fingerprint scanner unit, for computer keyboard access		8	1		1,100	54.50		1,154.50	1,275
0220	Hand geometry scanner, mem of 512 users, excl striker/power		3	2.667		2,100	146		2,246	2,525
0230	Memory upgrade for, adds 9,700 user profiles		8	1		300	54.50		354.50	415
0240	Adds 32,500 user profiles		8	1		600	54.50		654.50	745
0250	Prison type, memory of 256 users, excl striker, power	↓	3	2.667	↓	2,600	146		2,746	3,075

08 74 Access Control Hardware

08 74 19 – Biometric Identity Access Control Hardware

08 74 19.50 Biometric Identity Access		Crew	Daily Output	Labor-Hours	Unit	Material	Labor	Equipment	Total	Total Incl O&P
						2015 Bare Costs				
0260	Memory upgrade for, adds 3,300 user profiles	1 Elec	8	1	Ea.	250	54.50		304.50	360
0270	Adds 9,700 user profiles		8	1		460	54.50		514.50	590
0280	Adds 27,900 user profiles		8	1		610	54.50		664.50	755
0290	All weather, mem of 512 users, excl striker/power		3	2.667		3,900	146		4,046	4,525
0300	Facial & fingerprint scanner, combination unit, excl striker/power		3	2.667		4,300	146		4,446	4,950
0310	Access for, for initial setup, excl striker/power		3	2.667		1,100	146		1,246	1,425

08 74 23 – Access Control Accessories

08 74 23.50 Security Access Control Accessories		Crew	Daily Output	Labor-Hours	Unit	Material	Labor	Equipment	Total	Total Incl O&P
0010	**SECURITY ACCESS CONTROL ACCESSORIES**									
0360	Scanner/reader access, power supply/transf, 110 V to 12/24 V	1 Elec	4	2	Ea.	250	109		359	445
0370	Elec/mag strikers, 12/24V		4	2		85	109		194	263
0380	1 hour battery backup power supply		4	2		50	109		159	224
0390	Deadbolt, digital, batt-operated, indoor/outdoor, complete, incl hardware	1 Carp	4	2		250	94		344	430

08 75 Window Hardware

08 75 30 – Weatherstripping

08 75 30.10 Mechanical Weather Seals		Crew	Daily Output	Labor-Hours	Unit	Material	Labor	Equipment	Total	Total Incl O&P
0010	**MECHANICAL WEATHER SEALS**, Window, double hung, 3' X 5'									
0020	Zinc	1 Carp	7.20	1.111	Opng.	20	52		72	108
0100	Bronze		7.20	1.111		40	52		92	130
0500	As above but heavy duty, zinc		4.60	1.739		20	81.50		101.50	156
0600	Bronze		4.60	1.739		70	81.50		151.50	211
9000	Minimum labor/equipment charge	1 Clab	4.60	1.739	Job		65.50		65.50	107

08 79 Hardware Accessories

08 79 13 – Key Storage Equipment

08 79 13.10 Key Cabinets		Crew	Daily Output	Labor-Hours	Unit	Material	Labor	Equipment	Total	Total Incl O&P
0010	**KEY CABINETS**									
0020	Wall mounted, 60 key capacity	1 Carp	20	.400	Ea.	93.50	18.80		112.30	134
0200	Drawer type, 600 key capacity	1 Clab	15	.533		800	20		820	915
0300	2,400 key capacity		20	.400		4,200	15.05		4,215.05	4,650
0400	Tray type, 20 key capacity		50	.160		67.50	6		73.50	84.50
0500	50 key capacity		40	.200		105	7.50		112.50	128

08 79 20 – Door Accessories

08 79 20.10 Door Hardware Accessories		Crew	Daily Output	Labor-Hours	Unit	Material	Labor	Equipment	Total	Total Incl O&P
0010	**DOOR HARDWARE ACCESSORIES**									
0140	Door bolt, surface, 4"	1 Carp	32	.250	Ea.	11.70	11.75		23.45	32
0160	Door latch	"	12	.667	"	8.55	31.50		40.05	61
0200	Sliding closet door									
0220	Track and hanger, single	1 Carp	10	.800	Ea.	58.50	37.50		96	126
0240	Double		8	1		80	47		127	165
0260	Door guide, single		48	.167		30	7.85		37.85	46
0280	Double		48	.167		40	7.85		47.85	57
0600	Deadbolt and lock cover plate, brass or stainless steel		30	.267		28	12.50		40.50	51.50
0620	Hole cover plate, brass or chrome		35	.229		8	10.75		18.75	26.50
2240	Mortise lockset, passage, lever handle		9	.889		160	41.50		201.50	245
4000	Security chain, standard		18	.444		10	21		31	45

08 81 Glass Glazing

08 81 10 – Float Glass

08 81 10.10 Various Types and Thickness of Float Glass		Crew	Daily Output	Labor-Hours	Unit	Material	Labor	Equipment	Total	Total Incl O&P
						2015 Bare Costs				
0010	**VARIOUS TYPES AND THICKNESS OF FLOAT GLASS**									
0020	3/16" Plain	2 Glaz	130	.123	S.F.	5.05	5.55		10.60	14.55
0200	Tempered, clear		130	.123		6.95	5.55		12.50	16.65
0300	Tinted		130	.123		8	5.55		13.55	17.80
0600	1/4" thick, clear, plain		120	.133		5.95	6		11.95	16.30
0700	Tinted		120	.133		9.10	6		15.10	19.75
0800	Tempered, clear		120	.133		8.85	6		14.85	19.50
0900	Tinted		120	.133		10.90	6		16.90	22
1600	3/8" thick, clear, plain		75	.213		10.20	9.60		19.80	27
1700	Tinted		75	.213		15.70	9.60		25.30	33
1800	Tempered, clear		75	.213		16.75	9.60		26.35	34
1900	Tinted		75	.213		18.85	9.60		28.45	36
2200	1/2" thick, clear, plain		55	.291		17.30	13.10		30.40	40.50
2300	Tinted		55	.291		27.50	13.10		40.60	51.50
2400	Tempered, clear		55	.291		25	13.10		38.10	49
2500	Tinted		55	.291		26	13.10		39.10	50
2800	5/8" thick, clear, plain		45	.356		27.50	16.05		43.55	56
2900	Tempered, clear		45	.356		31.50	16.05		47.55	60.50
3200	3/4" thick, clear, plain		35	.457		35.50	20.50		56	72.50
3300	Tempered, clear		35	.457		41	20.50		61.50	79
3600	1" thick, clear, plain	↓	30	.533		59	24		83	104
8900	For low emissivity coating for 3/16" & 1/4" only, add to above				↓	18%				
9000	Minimum labor/equipment charge	1 Glaz	2	4	Job		180		180	293

08 81 13 – Decorative Glass Glazing

08 81 13.10 Beveled Glass

		Crew	Daily Output	Labor-Hours	Unit	Material	Labor	Equipment	Total	Total Incl O&P
0010	**BEVELED GLASS**, with design patterns									
0020	Simple pattern	2 Glaz	150	.107	S.F.	60.50	4.81		65.31	74.50
0050	Intricate pattern	"	125	.128	"	133	5.75		138.75	155

08 81 13.30 Sandblasted Glass

		Crew	Daily Output	Labor-Hours	Unit	Material	Labor	Equipment	Total	Total Incl O&P
0010	**SANDBLASTED GLASS**, float glass									
0020	1/8" thick	2 Glaz	160	.100	S.F.	11	4.51		15.51	19.40
0100	3/16" thick		130	.123		12.15	5.55		17.70	22.50
0500	1/4" thick		120	.133		12.65	6		18.65	23.50
0600	3/8" thick	↓	75	.213	↓	13.60	9.60		23.20	30.50

08 81 17 – Fire Glass

08 81 17.10 Fire Resistant Glass

		Crew	Daily Output	Labor-Hours	Unit	Material	Labor	Equipment	Total	Total Incl O&P
0010	**FIRE RESISTANT GLASS**									
0020	Fire Glass Minimum	2 Glaz	40	.400	S.F.	35	18.05		53.05	68
0030	Mid Range		40	.400		76	18.05		94.05	113
0050	High End	↓	40	.400	↓	340	18.05		358.05	405

08 81 20 – Vision Panels

08 81 20.10 Full Vision

		Crew	Daily Output	Labor-Hours	Unit	Material	Labor	Equipment	Total	Total Incl O&P
0010	**FULL VISION**, window system with 3/4" glass mullions									
0020	Up to 10' high	H-2	130	.185	S.F.	63.50	7.85		71.35	82.50
0100	10' to 20' high, minimum		110	.218		67.50	9.30		76.80	89.50
0150	Average		100	.240		73	10.20		83.20	96.50
0200	Maximum	↓	80	.300	↓	81.50	12.80		94.30	111
9000	Minimum labor/equipment charge	1 Glaz	2	4	Job		180		180	293

08 81 Glass Glazing

08 81 25 – Glazing Variables

08 81 25.10 Applications of Glazing			Crew	Daily Output	Labor-Hours	Unit	2015 Bare Costs Material	Labor	Equipment	Total	Total Incl O&P
0010	**APPLICATIONS OF GLAZING**										
0600	For glass replacement, add					S.F.		100%			
0700	For gasket settings, add					L.F.	5.75			5.75	6.35
0900	For sloped glazing, add					S.F.		26%			
2000	Fabrication, polished edges, 1/4" thick					Inch	.55			.55	.61
2100	1/2" thick						1.30			1.30	1.43
2500	Mitered edges, 1/4" thick						1.30			1.30	1.43
2600	1/2" thick						2.15			2.15	2.37

08 81 30 – Insulating Glass

08 81 30.10 Reduce Heat Transfer Glass			Crew	Daily Output	Labor-Hours	Unit	Material	Labor	Equipment	Total	Total Incl O&P
0010	**REDUCE HEAT TRANSFER GLASS**										
0015	2 lites 1/8" float, 1/2" thk under 15 S.F.										
0020	Clear	G	2 Glaz	95	.168	S.F.	10.05	7.60		17.65	23.50
0100	Tinted	G		95	.168		14.05	7.60		21.65	28
0200	2 lites 3/16" float, for 5/8" thk unit, 15 to 30 S.F., clear	G		90	.178		13.70	8		21.70	28
0300	Tinted	G		90	.178		13.75	8		21.75	28
0400	1" thk, dbl. glazed, 1/4" float, 30-70 S.F., clear	G		75	.213		16.80	9.60		26.40	34
0500	Tinted	G		75	.213		23.50	9.60		33.10	41.50
0600	1" thick double glazed, 1/4" float, 1/4" wire			75	.213		23.50	9.60		33.10	41.50
0700	1/4" float, 1/4" tempered			75	.213		31	9.60		40.60	49.50
0800	1/4" wire, 1/4" tempered			75	.213		29.50	9.60		39.10	48
2000	Both lites, light & heat reflective	G		85	.188		31.50	8.50		40	48.50
2500	Heat reflective, film inside, 1" thick unit, clear	G		85	.188		27.50	8.50		36	44.50
2600	Tinted	G		85	.188		28.50	8.50		37	45.50
3000	Film on weatherside, clear, 1/2" thick unit	G		95	.168		19.60	7.60		27.20	34
3100	5/8" thick unit	G		90	.178		20	8		28	35
3200	1" thick unit	G		85	.188		27	8.50		35.50	44
3350	Clear heat reflective film on inside	G	1 Glaz	50	.160		12.90	7.20		20.10	26
3360	Heat reflective film on inside, trinted	G		25	.320		13.70	14.45		28.15	38.50
3370	Heat reflective film on the inside metalized	G		20	.400		14	18.05		32.05	45
5000	Spectrally selective film, on ext, blocks solar gain/allows 70% of light	G	2 Glaz	95	.168		14.25	7.60		21.85	28
9000	Minimum labor/equipment charge		1 Glaz	2	4	Job		180		180	293

08 81 35 – Translucent Glass

08 81 35.10 Obscure Glass		Crew	Daily Output	Labor-Hours	Unit	Material	Labor	Equipment	Total	Total Incl O&P
0010	**OBSCURE GLASS**									
0020	1/8" thick, textured	2 Glaz	140	.114	S.F.	11.60	5.15		16.75	21
0100	Color		125	.128		13.70	5.75		19.45	24.50
0300	7/32" thick, textured		120	.133		12.70	6		18.70	23.50
0400	Color		105	.152		15.95	6.85		22.80	28.50

08 81 35.20 Patterned Glass		Crew	Daily Output	Labor-Hours	Unit	Material	Labor	Equipment	Total	Total Incl O&P
0010	**PATTERNED GLASS**, colored									
0020	1/8" thick	2 Glaz	140	.114	S.F.	9.35	5.15		14.50	18.65
0300	7/32" thick	"	120	.133	"	11.70	6		17.70	22.50

08 81 45 – Sheet Glass

08 81 45.10 Window Glass, Sheet		Crew	Daily Output	Labor-Hours	Unit	Material	Labor	Equipment	Total	Total Incl O&P
0010	**WINDOW GLASS, SHEET** gray									
0020	1/8" thick	2 Glaz	160	.100	S.F.	5.80	4.51		10.31	13.70
0200	1/4" thick	"	130	.123	"	7.10	5.55		12.65	16.80

08 81 Glass Glazing

08 81 50 – Spandrel Glass

08 81 50.10 Glass for Non Vision Areas		Crew	Daily Output	Labor-Hours	Unit	2015 Bare Costs Material	Labor	Equipment	Total	Total Incl O&P
0010	**GLASS FOR NON VISION AREAS**, 1/4" thick standard colors									
0020	Up to 1000 S.F.	2 Glaz	110	.145	S.F.	16.95	6.55		23.50	29.50
0200	1,000 to 2,000 S.F.	"	120	.133	"	15.70	6		21.70	27
0300	For custom colors, add				Total	10%				
0500	For 3/8" thick, add				S.F.	11.90			11.90	13.10
1000	For double coated, 1/4" thick, add					4.25			4.25	4.68
1200	For insulation on panels, add					6.95			6.95	7.65
2000	Panels, insulated, with aluminum backed fiberglass, 1" thick	2 Glaz	120	.133		16.85	6		22.85	28.50
2100	2" thick	"	120	.133		20	6		26	32

08 81 55 – Window Glass

08 81 55.10 Sheet Glass		Crew	Daily Output	Labor-Hours	Unit	Material	Labor	Equipment	Total	Total Incl O&P
0010	**SHEET GLASS**(window), clear float, stops, putty bed									
0015	1/8" thick, clear float	2 Glaz	480	.033	S.F.	3.65	1.50		5.15	6.45
0500	3/16" thick, clear		480	.033		5.90	1.50		7.40	8.90
0600	Tinted		480	.033		7.45	1.50		8.95	10.65
0700	Tempered		480	.033		9.20	1.50		10.70	12.55
9000	Minimum labor/equipment charge		5	3.200	Job		144		144	234

08 81 65 – Wire Glass

08 81 65.10 Glass Reinforced With Wire		Crew	Daily Output	Labor-Hours	Unit	Material	Labor	Equipment	Total	Total Incl O&P
0010	**GLASS REINFORCED WITH WIRE**									
0012	1/4" thick rough obscure	2 Glaz	135	.119	S.F.	23.50	5.35		28.85	34.50
1000	Polished wire, 1/4" thick, diamond, clear		135	.119		28	5.35		33.35	39
1500	Pinstripe, obscure		135	.119		41	5.35		46.35	53.50

08 83 Mirrors

08 83 13 – Mirrored Glass Glazing

08 83 13.10 Mirrors		Crew	Daily Output	Labor-Hours	Unit	Material	Labor	Equipment	Total	Total Incl O&P
0010	**MIRRORS**, No frames, wall type, 1/4" plate glass, polished edge									
0100	Up to 5 S.F.	2 Glaz	125	.128	S.F.	9.50	5.75		15.25	19.80
0200	Over 5 S.F.		160	.100		9.25	4.51		13.76	17.50
0500	Door type, 1/4" plate glass, up to 12 S.F.		160	.100		8.70	4.51		13.21	16.90
1000	Float glass, up to 10 S.F., 1/8" thick		160	.100		5.90	4.51		10.41	13.80
1100	3/16" thick		150	.107		7.30	4.81		12.11	15.85
1500	12" x 12" wall tiles, square edge, clear		195	.082		2.22	3.70		5.92	8.45
1600	Veined		195	.082		5.85	3.70		9.55	12.40
2000	1/4" thick, stock sizes, one way transparent		125	.128		19.60	5.75		25.35	31
2010	Bathroom, unframed, laminated		160	.100		13.90	4.51		18.41	22.50
2500	Tempered		160	.100		17.80	4.51		22.31	27

08 83 13.15 Reflective Glass		Crew	Daily Output	Labor-Hours	Unit	Material	Labor	Equipment	Total	Total Incl O&P
0010	**REFLECTIVE GLASS**									
0100	1/4" float with fused metallic oxide fixed [G]	2 Glaz	115	.139	S.F.	16.95	6.25		23.20	29
0500	1/4" float glass with reflective applied coating [G]	"	115	.139	"	13.70	6.25		19.95	25.50

08 84 Plastic Glazing

08 84 10 – Plexiglass Glazing

08 84 10.10 Plexiglass Acrylic		Crew	Daily Output	Labor-Hours	Unit	Material	2015 Bare Costs Labor	Equipment	Total	Total Incl O&P
0010	**PLEXIGLASS ACRYLIC**, clear, masked,									
0020	1/8" thick, cut sheets	2 Glaz	170	.094	S.F.	12	4.24		16.24	20
0200	Full sheets		195	.082		5	3.70		8.70	11.50
0500	1/4" thick, cut sheets		165	.097		14	4.37		18.37	22.50
0600	Full sheets		185	.086		9	3.90		12.90	16.25
0900	3/8" thick, cut sheets		155	.103		20	4.66		24.66	29.50
1000	Full sheets		180	.089		15	4.01		19.01	23
1300	1/2" thick, cut sheets		135	.119		28	5.35		33.35	39.50
1400	Full sheets		150	.107		20	4.81		24.81	30
1700	3/4" thick, cut sheets		115	.139		69	6.25		75.25	86
1800	Full sheets		130	.123		40	5.55		45.55	53
2100	1" thick, cut sheets		105	.152		77.50	6.85		84.35	96.50
2200	Full sheets		125	.128		48	5.75		53.75	62.50
3000	Colored, 1/8" thick, cut sheets		170	.094		18	4.24		22.24	26.50
3200	Full sheets		195	.082		11	3.70		14.70	18.10
3500	1/4" thick, cut sheets		165	.097		20	4.37		24.37	29
3600	Full sheets		185	.086		14	3.90		17.90	22
4000	Mirrors, untinted, cut sheets, 1/8" thick		185	.086		12	3.90		15.90	19.55
4200	1/4" thick		180	.089		16	4.01		20.01	24

08 84 20 – Polycarbonate

08 84 20.10 Thermoplastic		Crew	Daily Output	Labor-Hours	Unit	Material	Labor	Equipment	Total	Total Incl O&P
0010	**THERMOPLASTIC**, clear, masked, cut sheets									
0020	1/8" thick	2 Glaz	170	.094	S.F.	14	4.24		18.24	22.50
0500	3/16" thick		165	.097		16	4.37		20.37	24.50
1000	1/4" thick		155	.103		17	4.66		21.66	26.50
1500	3/8" thick		150	.107		26	4.81		30.81	36.50
9000	Minimum labor/equipment charge	1 Glaz	2	4	Job		180		180	293

08 87 Glazing Surface Films

08 87 13 – Solar Control Films

08 87 13.10 Solar Films On Glass		Crew	Daily Output	Labor-Hours	Unit	Material	Labor	Equipment	Total	Total Incl O&P
0010	**SOLAR FILMS ON GLASS** (glass not included)									
2000	Minimum G	2 Glaz	180	.089	S.F.	6.80	4.01		10.81	14
2050	Maximum G	"	225	.071	"	15.30	3.21		18.51	22

08 87 23 – Safety and Security Films

08 87 23.13 Safety Films		Crew	Daily Output	Labor-Hours	Unit	Material	Labor	Equipment	Total	Total Incl O&P
0010	**SAFETY FILMS**									
2010	Window protection film, controls blast damage	2 Glaz	80	.200	S.F.	6.80	9		15.80	22

08 87 23.16 Security Films		Crew	Daily Output	Labor-Hours	Unit	Material	Labor	Equipment	Total	Total Incl O&P
0010	**SECURITY FILMS**, clear, 32000 psi tensile strength, adhered to glass									
0100	.002" thick, daylight installation	H-2	950	.025	S.F.	2.70	1.08		3.78	4.72
0150	.004" thick, daylight installation		800	.030		3.35	1.28		4.63	5.75
0200	.006" thick, daylight installation		700	.034		3.60	1.46		5.06	6.35
0210	Install for anchorage		600	.040		4	1.70		5.70	7.15
0400	.007" thick, daylight installation		600	.040		4.45	1.70		6.15	7.65
0410	Install for anchorage		500	.048		4.94	2.04		6.98	8.80
0500	.008" thick, daylight installation		500	.048		5	2.04		7.04	8.85
0510	Install for anchorage		500	.048		5.55	2.04		7.59	9.45
0600	.015" thick, daylight installation		400	.060		8.80	2.56		11.36	13.85
0610	Install for anchorage		400	.060		5.55	2.56		8.11	10.25

08 87 Glazing Surface Films

08 87 23 – Safety and Security Films

08 87 23.16 Security Films		Crew	Daily Output	Labor-Hours	Unit	Material	2015 Bare Costs Labor	Equipment	Total	Total Incl O&P
0900	Security film anchorage, mechanical attachment and cover plate	H-3	370	.043	L.F.	9.90	1.74		11.64	13.75
0950	Security film anchorage, wet glaze structural caulking	1 Glaz	225	.036	"	.99	1.60		2.59	3.69
1000	Adhered security film removal	1 Clab	275	.029	S.F.		1.09		1.09	1.79

08 88 Special Function Glazing

08 88 40 – Acoustical Glass Units

08 88 40.10 Sound Reduction Units

		Crew	Daily Output	Labor-Hours	Unit	Material	Labor	Equipment	Total	Total Incl O&P
0010	**SOUND REDUCTION UNITS**, 1lite at 3/8", 1 lite at 3/16"									
0020	For 1" thick	2 Glaz	100	.160	S.F.	34	7.20		41.20	49
0100	For 4" thick	"	80	.200	"	58.50	9		67.50	78.50

08 88 52 – Prefabricated Glass Block Windows

08 88 52.10 Prefabricated Glass Block Windows

		Crew	Daily Output	Labor-Hours	Unit	Material	Labor	Equipment	Total	Total Incl O&P
0010	**PREFABRICATED GLASS BLOCK WINDOWS**									
0015	Includes frame and silicone seal									
0020	Glass Block Window 16" x 8"	2 Glaz	6	2.667	Ea.	185	120		305	400
0025	16" x 16"		6	2.667		223	120		343	440
0030	16" x 24"		6	2.667		265	120		385	485
0050	16" x 48"		4	4		380	180		560	715
0100	Glass Block Window 24" x 8"		6	2.667		211	120		331	425
0110	24" x 16"		6	2.667		265	120		385	485
0120	24" x 24"		6	2.667		315	120		435	540
0150	24" x 48"		6	2.667		470	120		590	710
0200	Glass Block Window 32" x 8"		6	2.667		211	120		331	425
0210	32" x 16"		6	2.667		300	120		420	525
0220	32" x 24"		4	4		365	180		545	695
0250	32" x 48"		4	4		555	180		735	905
0300	Glass Block Window 40" x 8"		6	2.667		271	120		391	495
0310	40" x 16"		6	2.667		340	120		460	570
0320	40" x 24"		6	2.667		410	120		530	650
0350	40" x 48"		6	2.667		645	120		765	905
0400	Glass Block Window 48" x 8"		6	2.667		300	120		420	525
0410	48" x 16"		6	2.667		380	120		500	610
0420	48" x 24"		6	2.667		470	120		590	710
0450	48" x 48"		6	2.667		730	120		850	1,000
0500	Glass Block Window 56" x 8"		6	2.667		330	120		450	560
0510	56" x 16"		4	4		415	180		595	755
0520	56" x 24"		4	4		520	180		700	865
0550	56" x 48"		4	4		825	180		1,005	1,200

08 88 52.20 Prefabricated Glass Block Windows Hurricane Resistant

		Crew	Daily Output	Labor-Hours	Unit	Material	Labor	Equipment	Total	Total Incl O&P
0010	**PREFABRICATED GLASS BLOCK WINDOWS HURRICANE RESISTANT**									
0015	Includes frame and silicone seal									
3020	Glass Block Window 16" x 16"	2 Glaz	6	2.667	Ea.	500	120		620	745
3030	16" x 24"		6	2.667		580	120		700	835
3040	16" x 32"		4	4		640	180		820	1,000
3050	16" x 40"		4	4		690	180		870	1,050
3060	16" x 48"		4	4		700	180		880	1,075
3070	16" x 56"		4	4		880	180		1,060	1,275
3080	Glass Block Window 24" x 24"		6	2.667		640	120		760	900
3090	24" x 32"		4	4		700	180		880	1,075
3100	24" x 40"		4	4		880	180		1,060	1,275

08 88 Special Function Glazing

08 88 52 – Prefabricated Glass Block Windows

08 88 52.20 Prefabricated Glass Block Windows Hurricane Resistant		Crew	Daily Output	Labor-Hours	Unit	Material	2015 Bare Costs Labor	Equipment	Total	Total Incl O&P
3110	24" x 48"	2 Glaz	4	4	Ea.	910	180		1,090	1,300
3120	24" x 56"		4	4		950	180		1,130	1,350
3130	Glass Block Window 32" x 32"		4	4		800	180		980	1,175
3140	32" x 40"		4	4		900	180		1,080	1,275
3150	32" x 48"		4	4		1,050	180		1,230	1,450
3160	32" x 56"		4	4		1,200	180		1,380	1,625
3200	Glass Block Window 40" x 40"		4	4		1,025	180		1,205	1,425
3210	40" x 48"		3	5.333		1,250	241		1,491	1,775
3220	40" x 56"		3	5.333		1,400	241		1,641	1,950
3260	Glass Block Window 48" x 48"		3	5.333		1,450	241		1,691	2,000
3270	48" x 56"		3	5.333		1,625	241		1,866	2,175
3310	Glass Block Window 56" x 56"	↓	3	5.333	↓	1,775	241		2,016	2,350

08 88 56 – Ballistics-Resistant Glazing

08 88 56.10 Laminated Glass		Crew	Daily Output	Labor-Hours	Unit	Material	Labor	Equipment	Total	Total Incl O&P
0010	**LAMINATED GLASS**									
0020	Clear float .03" vinyl 1/4"	2 Glaz	90	.178	S.F.	12.40	8		20.40	26.50
0100	3/8" thick		78	.205		22.50	9.25		31.75	39.50
0200	.06" vinyl, 1/2" thick		65	.246		25.50	11.10		36.60	46
1000	5/8" thick		90	.178		29.50	8		37.50	45.50
2000	Bullet-resisting, 1-3/16" thick, to 15 S.F.		16	1		105	45		150	189
2100	Over 15 S.F.		16	1		119	45		164	204
2200	2" thick, to 15 S.F.		10	1.600		140	72		212	271
2300	Over 15 S.F.		10	1.600		120	72		192	249
2500	2-1/4" thick, to 15 S.F.		12	1.333		179	60		239	295
2600	Over 15 S.F.		12	1.333		168	60		228	283
2700	Level 2 (.357 magnum), NIJ and UL		12	1.333		77	60		137	182
2750	Level 3A (.44 magnum) NIJ, UL 3		12	1.333		82	60		142	188
2800	Level 4 (AK-47) NIJ, UL 7 & 8		12	1.333		113	60		173	222
2850	Level 5 (M-16) UL		12	1.333		116	60		176	226
2900	Level 3 (7.62 Armor Piercing) NIJ, UL 4 & 5	↓	12	1.333	↓	137	60		197	249

08 91 Louvers

08 91 16 – Operable Louvers

08 91 16.10 Movable Blade Louvers		Crew	Daily Output	Labor-Hours	Unit	Material	Labor	Equipment	Total	Total Incl O&P
0010	**MOVABLE BLADE LOUVERS**									
0100	PVC, commercial grade, 12" x 12"	1 Shee	20	.400	Ea.	82	22.50		104.50	126
0110	16" x 16"		14	.571		96	32		128	157
0120	18" x 18"		14	.571		99	32		131	160
0130	24" x 24"		14	.571		140	32		172	205
0140	30" x 30"		12	.667		168	37.50		205.50	244
0150	36" x 36"		12	.667		197	37.50		234.50	276
0300	Stainless steel, commercial grade, 12" x 12"		20	.400		219	22.50		241.50	277
0310	16" x 16"		14	.571		265	32		297	345
0320	18" x 18"		14	.571		282	32		314	360
0330	20" x 20"		14	.571		325	32		357	405
0340	24" x 24"		14	.571		385	32		417	475
0350	30" x 30"		12	.667		430	37.50		467.50	535
0360	36" x 36"	↓	12	.667	↓	520	37.50		557.50	635

08 91 Louvers

08 91 19 – Fixed Louvers

08 91 19.10 Aluminum Louvers

	08 91 19.10 Aluminum Louvers	Crew	Daily Output	Labor-Hours	Unit	2015 Bare Costs Material	Labor	Equipment	Total	Total Incl O&P
0010	**ALUMINUM LOUVERS**									
0020	Aluminum with screen, residential, 8" x 8"	1 Carp	38	.211	Ea.	20	9.90		29.90	38
0100	12" x 12"		38	.211		16	9.90		25.90	34
0200	12" x 18"		35	.229		20	10.75		30.75	39.50
0250	14" x 24"		30	.267		29	12.50		41.50	52.50
0300	18" x 24"		27	.296		32	13.90		45.90	58
0500	24" x 30"		24	.333		60	15.65		75.65	91.50
0700	Triangle, adjustable, small		20	.400		55	18.80		73.80	91.50
0800	Large		15	.533		76	25		101	125
1200	Extruded aluminum, see Section 23 37 15.40									
2100	Midget, aluminum, 3/4" deep, 1" diameter	1 Carp	85	.094	Ea.	.75	4.42		5.17	8.10
2150	3" diameter		60	.133		2.47	6.25		8.72	12.95
2200	4" diameter		50	.160		4.95	7.50		12.45	17.75
2250	6" diameter		30	.267		4.10	12.50		16.60	25
3000	PVC, commercial grade, 12" x 12"	1 Shee	20	.400		97	22.50		119.50	143
3010	12" x 18"		20	.400		109	22.50		131.50	156
3020	12" x 24"		20	.400		146	22.50		168.50	197
3030	14" x 24"		20	.400		152	22.50		174.50	203
3100	Aluminum, commercial grade, 12" x 12"		20	.400		171	22.50		193.50	224
3110	24" x 24"		16	.500		249	28		277	320
3120	24" x 36"		16	.500		300	28		328	375
3130	36" x 24"		16	.500		310	28		338	385
3140	36" x 36"		14	.571		390	32		422	480
3150	36" x 48"		10	.800		400	45		445	510
3160	48" x 36"		10	.800		390	45		435	500
3170	48" x 48"		10	.800		435	45		480	545
3180	60" x 48"		8	1		485	56		541	625
3190	60" x 60"		8	1		600	56		656	750

08 91 19.20 Steel Louvers

	08 91 19.20 Steel Louvers	Crew	Daily Output	Labor-Hours	Unit	Material	Labor	Equipment	Total	Total Incl O&P
0010	**STEEL LOUVERS**									
3300	Galvanized Steel, fixed blades, commercial grade, 18" x 18"	1 Shee	20	.400	Ea.	197	22.50		219.50	253
3310	24" x 24"		20	.400		232	22.50		254.50	291
3320	24" x 36"		16	.500		300	28		328	375
3330	36" x 24"		16	.500		298	28		326	375
3340	36" x 36"		14	.571		380	32		412	470
3350	36" x 48"		10	.800		385	45		430	495
3360	48" x 36"		10	.800		385	45		430	495
3370	48" x 48"		10	.800		425	45		470	535
3380	60" x 48"		10	.800		490	45		535	610
3390	60" x 60"		10	.800		590	45		635	715

08 91 26 – Door Louvers

08 91 26.10 Steel Louvers, 18 Gauge, Fixed Blade

	08 91 26.10 Steel Louvers, 18 Gauge, Fixed Blade	Crew	Daily Output	Labor-Hours	Unit	Material	Labor	Equipment	Total	Total Incl O&P
0010	**STEEL LOUVERS, 18 GAUGE, FIXED BLADE**									
0050	12" x 12", with enamel or powder coat	1 Carp	20	.400	Ea.	82	18.80		100.80	121
0055	18" x 12"		20	.400		87.50	18.80		106.30	127
0060	18" x 18"		20	.400		103	18.80		121.80	145
0065	24" x 12"		20	.400		108	18.80		126.80	150
0070	24" x 18"		20	.400		117	18.80		135.80	160
0075	24" x 24"		20	.400		144	18.80		162.80	189
0100	12" x 12", galvanized		20	.400		72.50	18.80		91.30	111
0105	18" x 12"		20	.400		85.50	18.80		104.30	125

08 91 Louvers

08 91 26 – Door Louvers

08 91 26.10 Steel Louvers, 18 Gauge, Fixed Blade		Crew	Daily Output	Labor-Hours	Unit	Material	Labor	Equipment	Total	Total Incl O&P
0115	24" x 12"	1 Carp	20	.400	Ea.	102	18.80		120.80	143
0125	24" x 24"	↓	20	.400	↓	143	18.80		161.80	188

08 95 Vents

08 95 13 – Soffit Vents

08 95 13.10 Wall Louvers

		Crew	Daily Output	Labor-Hours	Unit	Material	Labor	Equipment	Total	Total Incl O&P
0010	**WALL LOUVERS**									
2330	Soffit vent, continuous, 3" wide, aluminum, mill finish	1 Carp	200	.040	L.F.	.67	1.88		2.55	3.82
2340	Baked enamel finish		200	.040	"	5.50	1.88		7.38	9.15
2400	Under eaves vent, aluminum, mill finish, 16" x 4"		48	.167	Ea.	1.90	7.85		9.75	14.95
2500	16" x 8"	↓	48	.167	"	2.18	7.85		10.03	15.25

08 95 16 – Wall Vents

08 95 16.10 Louvers

		Crew	Daily Output	Labor-Hours	Unit	Material	Labor	Equipment	Total	Total Incl O&P
0010	**LOUVERS**									
0020	Redwood, 2'-0" diameter, full circle	1 Carp	16	.500	Ea.	190	23.50		213.50	248
0100	Half circle		16	.500		180	23.50		203.50	237
0200	Octagonal		16	.500		142	23.50		165.50	195
0300	Triangular, 5/12 pitch, 5'-0" at base		16	.500		200	23.50		223.50	259
1000	Rectangular, 1'-4" x 1'-3"		16	.500		19.50	23.50		43	60
1100	Rectangular, 1'-4" x 1'-8"		16	.500		28	23.50		51.50	69.50
1200	1'-4" x 2'-2"		15	.533		31	25		56	75.50
1300	1'-9" x 2'-2"		15	.533		37.50	25		62.50	82.50
1400	2'-3" x 2'-2"		14	.571		49.50	27		76.50	98.50
1700	2'-4" x 2'-11"		13	.615		49.50	29		78.50	102
2000	Aluminum, 12" x 16"		25	.320		20	15		35	46.50
2010	16" x 20"		25	.320		27.50	15		42.50	55
2020	24" x 30"		25	.320		59	15		74	89.50
2100	6' triangle		12	.667		168	31.50		199.50	237
3100	Round, 2'-2" diameter		16	.500		124	23.50		147.50	175
7000	Vinyl gable vent, 8" x 8"		38	.211		14	9.90		23.90	31.50
7020	12" x 12"		38	.211		27	9.90		36.90	45.50
7080	12" x 18"		35	.229		35	10.75		45.75	56
7200	18" x 24"		30	.267	↓	45	12.50		57.50	70
9000	Minimum labor/equipment charge	↓	3.50	2.286	Job		107		107	176

Estimating Tips

General

- Room Finish Schedule: A complete set of plans should contain a room finish schedule. If one is not available, it would be well worth the time and effort to obtain one.

09 20 00 Plaster and Gypsum Board

- Lath is estimated by the square yard plus a 5% allowance for waste. Furring, channels, and accessories are measured by the linear foot. An extra foot should be allowed for each accessory miter or stop.
- Plaster is also estimated by the square yard. Deductions for openings vary by preference, from zero deduction to 50% of all openings over 2 feet in width. The estimator should allow one extra square foot for each linear foot of horizontal interior or exterior angle located below the ceiling level. Also, double the areas of small radius work.
- Drywall accessories, studs, track, and acoustical caulking are all measured by the linear foot. Drywall taping is figured by the square foot. Gypsum wallboard is estimated by the square foot. No material deductions should be made for door or window openings under 32 S.F.

09 60 00 Flooring

- Tile and terrazzo areas are taken off on a square foot basis. Trim and base materials are measured by the linear foot. Accent tiles are listed per each. Two basic methods of installation are used. Mud set is approximately 30% more expensive than thin set. In terrazzo work, be sure to include the linear footage of embedded decorative strips, grounds, machine rubbing, and power cleanup.
- Wood flooring is available in strip, parquet, or block configuration. The latter two types are set in adhesives with quantities estimated by the square foot. The laying pattern will influence labor costs and material waste. In addition to the material and labor for laying wood floors, the estimator must make allowances for sanding and finishing these areas, unless the flooring is prefinished.
- Sheet flooring is measured by the square yard. Roll widths vary, so consideration should be given to use the most economical width, as waste must be figured into the total quantity. Consider also the installation methods available, direct glue down or stretched.

09 70 00 Wall Finishes

- Wall coverings are estimated by the square foot. The area to be covered is measured, length by height of wall above baseboards, to calculate the square footage of each wall. This figure is divided by the number of square feet in the single roll which is being used. Deduct, in full, the areas of openings such as doors and windows. Where a pattern match is required allow 25%–30% waste.

09 80 00 Acoustic Treatment

- Acoustical systems fall into several categories. The takeoff of these materials should be by the square foot of area with a 5% allowance for waste. Do not forget about scaffolding, if applicable, when estimating these systems.

09 90 00 Painting and Coating

- A major portion of the work in painting involves surface preparation. Be sure to include cleaning, sanding, filling, and masking costs in the estimate.
- Protection of adjacent surfaces is not included in painting costs. When considering the method of paint application, an important factor is the amount of protection and masking required. These must be estimated separately and may be the determining factor in choosing the method of application.

Reference Numbers

Reference numbers are shown in shaded boxes at the beginning of some major classifications. These numbers refer to related items in the Reference Section. The reference information may be an estimating procedure, an alternate pricing method, or technical information.

Note: Not all subdivisions listed here necessarily appear in this publication. ■

09 01 Maintenance of Finishes

09 01 60 – Maintenance of Flooring

09 01 60.10 Carpet Maintenance		Crew	Daily Output	Labor-Hours	Unit	Material	2015 Bare Costs Labor	Equipment	Total	Total Incl O&P
0011	**CARPET MAINTENANCE** See Section 01 93 13.09									

09 01 70 – Maintenance of Wall Finishes

09 01 70.10 Gypsum Wallboard Repairs

Code	Description	Crew	Daily Output	Labor-Hours	Unit	Material	Labor	Equipment	Total	Total Incl O&P
0010	**GYPSUM WALLBOARD REPAIRS**									
0100	Fill and sand, pin/nail holes	1 Carp	960	.008	Ea.		.39		.39	.64
0110	Screw head pops		480	.017			.78		.78	1.28
0120	Dents, up to 2" square		48	.167		.01	7.85		7.86	12.85
0130	2" to 4" square		24	.333		.03	15.65		15.68	25.50
0140	Cut square, patch, sand and finish, holes, up to 2" square		12	.667		.03	31.50		31.53	51.50
0150	2" to 4" square		11	.727		.09	34		34.09	56
0160	4" to 8" square		10	.800		.23	37.50		37.73	62
0170	8" to 12" square		8	1		.46	47		47.46	77.50
0180	12" to 32" square		6	1.333		1.55	62.50		64.05	105
0210	16" by 48"		5	1.600		2.65	75		77.65	126
0220	32" by 48"		4	2		4.09	94		98.09	159
0230	48" square		3.50	2.286		5.75	107		112.75	182
0240	60" square		3.20	2.500		9.50	117		126.50	202
0500	Skim coat surface with joint compound		1600	.005	S.F.	.03	.23		.26	.42
0510	Prepare, retape and refinish joints		60	.133	L.F.	.64	6.25		6.89	10.95
9000	Minimum labor/equipment charge		2	4	Job		188		188	310

09 05 Common Work Results for Finishes

09 05 05 – Selective Demolition for Finishes

09 05 05.10 Selective Demolition, Ceilings

Code	Description	Crew	Daily Output	Labor-Hours	Unit	Material	Labor	Equipment	Total	Total Incl O&P
0010	**SELECTIVE DEMOLITION, CEILINGS** R024119-10									
0200	Ceiling, drywall, furred and nailed or screwed	2 Clab	800	.020	S.F.		.75		.75	1.23
0220	On metal frame		760	.021			.79		.79	1.30
0240	On suspension system, including system		720	.022			.84		.84	1.37
1000	Plaster, lime and horse hair, on wood lath, incl. lath		700	.023			.86		.86	1.41
1020	On metal lath		570	.028			1.06		1.06	1.73
1100	Gypsum, on gypsum lath		720	.022			.84		.84	1.37
1120	On metal lath		500	.032			1.20		1.20	1.97
1200	Suspended ceiling, mineral fiber, 2' x 2' or 2' x 4'		1500	.011			.40		.40	.66
1250	On suspension system, incl. system		1200	.013			.50		.50	.82
1500	Tile, wood fiber, 12" x 12", glued		900	.018			.67		.67	1.10
1540	Stapled		1500	.011			.40		.40	.66
1580	On suspension system, incl. system		760	.021			.79		.79	1.30
2000	Wood, tongue and groove, 1" x 4"		1000	.016			.60		.60	.99
2040	1" x 8"		1100	.015			.55		.55	.90
2400	Plywood or wood fiberboard, 4' x 8' sheets		1200	.013			.50		.50	.82
9000	Minimum labor/equipment charge	1 Clab	2	4	Job		150		150	247

09 05 05.20 Selective Demolition, Flooring

Code	Description	Crew	Daily Output	Labor-Hours	Unit	Material	Labor	Equipment	Total	Total Incl O&P
0010	**SELECTIVE DEMOLITION, FLOORING** R024119-10									
0200	Brick with mortar	2 Clab	475	.034	S.F.		1.27		1.27	2.08
0400	Carpet, bonded, including surface scraping		2000	.008			.30		.30	.49
0480	Tackless		9000	.002			.07		.07	.11
0550	Carpet tile, releasable adhesive		5000	.003			.12		.12	.20
0560	Permanent adhesive		1850	.009			.33		.33	.53
0600	Composition, acrylic or epoxy		400	.040			1.50		1.50	2.47
0700	Concrete, scarify skin	A-1A	225	.036			1.73	.95	2.68	3.86

09 05 Common Work Results for Finishes

09 05 05 – Selective Demolition for Finishes

09 05 05.20 Selective Demolition, Flooring		Crew	Daily Output	Labor-Hours	Unit	Material	2015 Bare Costs Labor	Equipment	Total	Total Incl O&P
0800	Resilient, sheet goods	2 Clab	1400	.011	S.F.		.43		.43	.70
0820	For gym floors	"	900	.018			.67		.67	1.10
0850	Vinyl or rubber cove base	1 Clab	1000	.008	L.F.		.30		.30	.49
0860	Vinyl or rubber cove base, molded corner	"	1000	.008	Ea.		.30		.30	.49
0870	For glued and caulked installation, add to labor						50%			
0900	Vinyl composition tile, 12" x 12"	2 Clab	1000	.016	S.F.		.60		.60	.99
2000	Tile, ceramic, thin set		675	.024			.89		.89	1.46
2020	Mud set		625	.026			.96		.96	1.58
2200	Marble, slate, thin set		675	.024			.89		.89	1.46
2220	Mud set		625	.026			.96		.96	1.58
2600	Terrazzo, thin set		450	.036			1.34		1.34	2.19
2620	Mud set		425	.038			1.42		1.42	2.32
2640	Terrazzo, cast in place		300	.053			2.01		2.01	3.29
3000	Wood, block, on end	1 Carp	400	.020			.94		.94	1.54
3010	Wood blk floor, includes shot blasting	B-63B	103.28	.310			12.55	3.09	15.64	24
3200	Parquet	1 Carp	450	.018			.83		.83	1.37
3400	Strip flooring, interior, 2-1/4" x 25/32" thick		325	.025			1.16		1.16	1.89
3500	Exterior, porch flooring, 1" x 4"		220	.036			1.71		1.71	2.80
3800	Subfloor, tongue and groove, 1" x 6"		325	.025			1.16		1.16	1.89
3820	1" x 8"		430	.019			.87		.87	1.43
3840	1" x 10"		520	.015			.72		.72	1.18
4000	Plywood, nailed		600	.013			.63		.63	1.03
4100	Glued and nailed		400	.020			.94		.94	1.54
4200	Hardboard, 1/4" thick		760	.011			.49		.49	.81
8000	Remove flooring, bead blast, simple floor plan	A-1A	1000	.008			.39	.21	.60	.87
8100	complex floor plan		400	.020			.97	.54	1.51	2.17
8150	Mastic only		1500	.005			.26	.14	.40	.58
8200	Floor demolition, raised access floor	B-1J	400	.040			1.51		1.51	2.48
9000	Minimum labor/equipment charge	1 Clab	4	2	Job		75		75	123

09 05 05.30 Selective Demolition, Walls and Partitions

		Crew	Daily Output	Labor-Hours	Unit	Material	Labor	Equipment	Total	Total Incl O&P
0010	**SELECTIVE DEMOLITION, WALLS AND PARTITIONS** R024119-10									
0020	Walls, concrete, reinforced	B-39	120	.400	C.F.		15.90	1.94	17.84	28
0025	Plain	"	160	.300			11.95	1.46	13.41	21
0100	Brick, 4" to 12" thick	B-9	220	.182			6.90	1.06	7.96	12.50
0200	Concrete block, 4" thick		1150	.035	S.F.		1.32	.20	1.52	2.39
0280	8" thick		1050	.038			1.45	.22	1.67	2.61
0300	Exterior stucco 1" thick over mesh		3200	.013			.48	.07	.55	.86
1000	Drywall, nailed or screwed	1 Clab	1000	.008			.30		.30	.49
1010	2 layers		400	.020			.75		.75	1.23
1020	Glued and nailed		900	.009			.33		.33	.55
1500	Fiberboard, nailed		900	.009			.33		.33	.55
1520	Glued and nailed		800	.010			.38		.38	.62
1568	Plenum barrier, sheet lead		300	.027			1		1	1.64
1600	Glass block		65	.123			4.63		4.63	7.60
2000	Movable walls, metal, 5' high		300	.027			1		1	1.64
2020	8' high		400	.020			.75		.75	1.23
2200	Metal or wood studs, finish 2 sides, fiberboard	B-1	520	.046			1.77		1.77	2.90
2250	Lath and plaster		260	.092			3.53		3.53	5.80
2300	Plasterboard (drywall)		520	.046			1.77		1.77	2.90
2350	Plywood		450	.053			2.04		2.04	3.35
2800	Paneling, 4' x 8' sheets	1 Clab	475	.017			.63		.63	1.04
3000	Plaster, lime and horsehair, on wood lath		400	.020			.75		.75	1.23

09 05 Common Work Results for Finishes

09 05 05 – Selective Demolition for Finishes

09 05 05.30 Selective Demolition, Walls and Partitions		Crew	Daily Output	Labor-Hours	Unit	Material	2015 Bare Costs Labor	Equipment	Total	Total Incl O&P
3020	On metal lath	1 Clab	335	.024	S.F.		.90		.90	1.47
3400	Gypsum or perlite, on gypsum lath		410	.020			.73		.73	1.20
3420	On metal lath		300	.027			1		1	1.64
3450	Plaster, interior gypsum, acoustic, or cement		60	.133	S.Y.		5		5	8.20
3500	Stucco, on masonry		145	.055			2.07		2.07	3.40
3510	Commercial 3-coat		80	.100			3.76		3.76	6.15
3520	Interior stucco		25	.320			12.05		12.05	19.75
3750	Terra cotta block and plaster, to 6" thick	B-1	175	.137	S.F.		5.25		5.25	8.60
3753	Remove damaged glass block	"	134.62	.178			6.80		6.80	11.20
3760	Tile, ceramic, on walls, thin set	1 Clab	300	.027			1		1	1.64
3765	Mud set		250	.032			1.20		1.20	1.97
3800	Toilet partitions, slate or marble		5	1.600	Ea.		60		60	98.50
3820	Metal or plastic		8	1	"		37.50		37.50	61.50
3950	Demolish coreboard partitions	B-1J	1548.64	.010	S.F.		.39		.39	.64
3955	w/ lift included	B-68D	900	.027			1.10	.37	1.47	2.20
3960	Demolish metal wall panels	B-1J	774.32	.021			.78		.78	1.28
3965	w/ lift included	B-1K	400	.040			1.89		1.89	3.09
3970	Demolish weld curtain with frame		600	.027			1.26		1.26	2.06
3975	Demolish wire mesh panels		400	.040			1.89		1.89	3.09
3980	Demolish gyp & mtl stud wall, forktruck included	B-68D	520	.046			1.91	.64	2.55	3.80
3990	Demolish dust curtain, lift included (curtain furn by ASI)	B-1K	2000	.008			.38		.38	.62
3995	Demolish dust curtain, lift included	"	2000	.008			.38		.38	.62
5000	Wallcovering, vinyl	1 Pape	700	.011			.47		.47	.75
5010	With release agent		1500	.005			.22		.22	.35
5025	Wallpaper, 2 layers or less, by hand		250	.032			1.31		1.31	2.10
5035	3 layers or more		165	.048			1.98		1.98	3.19
5040	Designer		480	.017			.68		.68	1.10
9000	Minimum labor/equipment charge	1 Clab	4	2	Job		75		75	123

09 05 71 – Acoustic Underlayment

09 05 71.10 Acoustical Underlayment		Crew	Daily Output	Labor-Hours	Unit	Material	Labor	Equipment	Total	Total Incl O&P
0010	ACOUSTICAL UNDERLAYMENT									
4000	Nylon matting 0.4" thick, with carbon black spinerette									
4010	plus polyester fabric, on floor	D-7	1600	.010	S.F.	1.36	.38		1.74	2.10
4200	Fiberglass reinf. backer board underlayment, 7/16" thick, on floor	"	1500	.011	"	2.77	.41		3.18	3.69

09 21 Plaster and Gypsum Board Assemblies

09 21 13 – Plaster Assemblies

09 21 13.10 Plaster Partition Wall		Crew	Daily Output	Labor-Hours	Unit	Material	Labor	Equipment	Total	Total Incl O&P
0010	PLASTER PARTITION WALL									
0400	Stud walls, 3.4 lb. metal lath, 3 coat gypsum plaster, 2 sides									
0600	2" x 4" wood studs, 16" O.C.	J-2	315	.152	S.F.	3.39	6.30	.44	10.13	14.25
0700	2-1/2" metal studs, 25 ga., 12" O.C.		325	.148		3.14	6.10	.43	9.67	13.65
0800	3-5/8" metal studs, 25 ga., 16" O.C.		320	.150		3.16	6.20	.44	9.80	13.85
0900	Gypsum lath, 2 coat vermiculite plaster, 2 sides									
1000	2" x 4" wood studs, 16" O.C.	J-2	355	.135	S.F.	3.77	5.60	.39	9.76	13.50
1200	2-1/2" metal studs, 25 ga., 12" O.C.		365	.132		3.35	5.45	.38	9.18	12.75
1300	3-5/8" metal studs, 25 ga., 16" O.C.		360	.133		3.44	5.50	.39	9.33	13

09 21 16.23 Gypsum Board Shaft Wall Assemblies

	09 21 16.23 Gypsum Board Shaft Wall Assemblies	Crew	Daily Output	Labor-Hours	Unit	2015 Bare Costs Material	Labor	Equipment	Total	Total Incl O&P
0010	**GYPSUM BOARD SHAFT WALL ASSEMBLIES**									
0020	Cavity type on 25 ga. J-track & C-H studs, 24" O.C.									
0030	1" thick coreboard wall liner on shaft side									
0040	2-hour assembly with double layer									
0060	5/8" fire rated gypsum board on room side	2 Carp	220	.073	S.F.	2.10	3.41		5.51	7.90
0100	3-hour assembly with triple layer									
0300	5/8" fire rated gypsum board on room side	2 Carp	180	.089	S.F.	1.77	4.17		5.94	8.80
0400	4-hour assembly, 1" coreboard, 5/8" fire rated gypsum board									
0600	and 3/4" galv. metal furring channels, 24" O.C., with									
0700	Double layer 5/8" fire rated gypsum board on room side	2 Carp	110	.145	S.F.	1.67	6.85		8.52	13.05
0900	For taping & finishing, add per side	1 Carp	1050	.008	"	.05	.36		.41	.64
1000	For insulation, see Section 07 21									
5200	For work over 8' high, add	2 Carp	3060	.005	S.F.		.25		.25	.40
5300	For distribution cost over 3 stories high, add per story	"	6100	.003	"		.12		.12	.20

09 21 16.33 Partition Wall

	09 21 16.33 Partition Wall	Crew	Daily Output	Labor-Hours	Unit	Material	Labor	Equipment	Total	Total Incl O&P
0010	**PARTITION WALL** Stud wall, 8' to 12' high									
0050	1/2", interior, gypsum board, std, tape & finish 2 sides									
0500	Installed on and incl., 2" x 4" wood studs, 16" O.C.	2 Carp	310	.052	S.F.	1.16	2.42		3.58	5.25
1000	Metal studs, NLB, 25 ga., 16" O.C., 3-5/8" wide		350	.046		1.06	2.15		3.21	4.69
1200	6" wide		330	.048		1.19	2.28		3.47	5.05
1400	Water resistant, on 2" x 4" wood studs, 16" O.C.		310	.052		1.34	2.42		3.76	5.45
1600	Metal studs, NLB, 25 ga., 16" O.C., 3-5/8" wide		350	.046		1.24	2.15		3.39	4.89
1800	6" wide		330	.048		1.37	2.28		3.65	5.25
2000	Fire res., 2 layers, 1-1/2 hr., on 2" x 4" wood studs, 16" O.C.		210	.076		1.96	3.58		5.54	8
2200	Metal studs, NLB, 25 ga., 16" O.C., 3-5/8" wide		250	.064		1.86	3		4.86	6.95
2400	6" wide		230	.070		1.99	3.27		5.26	7.55
2600	Fire & water res., 2 layers, 1-1/2 hr., 2" x 4" studs, 16" O.C.		210	.076		1.96	3.58		5.54	8
2800	Metal studs, NLB, 25 ga., 16" O.C., 3-5/8" wide		250	.064		1.86	3		4.86	6.95
3000	6" wide		230	.070		1.99	3.27		5.26	7.55
3200	5/8", interior, gypsum board, standard, tape & finish 2 sides									
3400	Installed on and including 2" x 4" wood studs, 16" O.C.	2 Carp	300	.053	S.F.	1.22	2.50		3.72	5.45
3600	24" O.C.		330	.048		1.12	2.28		3.40	4.96
3800	Metal studs, NLB, 25 ga., 16" O.C., 3-5/8" wide		340	.047		1.12	2.21		3.33	4.85
4000	6" wide		320	.050		1.25	2.35		3.60	5.20
4200	24" O.C., 3-5/8" wide		360	.044		1.02	2.09		3.11	4.55
4400	6" wide		340	.047		1.12	2.21		3.33	4.85
4800	Water resistant, on 2" x 4" wood studs, 16" O.C.		300	.053		1.40	2.50		3.90	5.65
5000	24" O.C.		330	.048		1.30	2.28		3.58	5.15
5200	Metal studs, NLB, 25 ga. 16" O.C., 3-5/8" wide		340	.047		1.30	2.21		3.51	5.05
5400	6" wide		320	.050		1.43	2.35		3.78	5.40
5600	24" O.C., 3-5/8" wide		360	.044		1.20	2.09		3.29	4.75
5800	6" wide		340	.047		1.30	2.21		3.51	5.05
6000	Fire resistant, 2 layers, 2 hr., on 2" x 4" wood studs, 16" O.C.		205	.078		1.82	3.66		5.48	8
6200	24" O.C.		235	.068		1.82	3.20		5.02	7.25
6400	Metal studs, NLB, 25 ga., 16" O.C., 3-5/8" wide		245	.065		1.85	3.07		4.92	7.10
6600	6" wide		225	.071		1.95	3.34		5.29	7.60
6800	24" O.C., 3-5/8" wide		265	.060		1.72	2.83		4.55	6.55
7000	6" wide		245	.065		1.82	3.07		4.89	7.05
7200	Fire & water resistant, 2 layers, 2 hr., 2" x 4" studs, 16" O.C.		205	.078		1.92	3.66		5.58	8.10
7400	24" O.C.		235	.068		1.82	3.20		5.02	7.25
7600	Metal studs, NLB, 25 ga., 16" O.C., 3-5/8" wide		245	.065		1.82	3.07		4.89	7.05
7800	6" wide		225	.071		1.95	3.34		5.29	7.60

09 21 Plaster and Gypsum Board Assemblies

09 21 16 – Gypsum Board Assemblies

09 21 16.33 Partition Wall		Crew	Daily Output	Labor-Hours	Unit	Material	2015 Bare Costs Labor	Equipment	Total	Total Incl O&P
8000	24" O.C., 3-5/8" wide	2 Carp	265	.060	S.F.	1.72	2.83		4.55	6.55
8200	6" wide		245	.065		1.82	3.07		4.89	7.05
8600	1/2" blueboard, mesh tape both sides									
8620	Installed on and including 2" x 4" wood studs, 16" O.C.	2 Carp	300	.053	S.F.	1.22	2.50		3.72	5.45
8640	Metal studs, NLB, 25 ga., 16" O.C., 3-5/8" wide		340	.047		1.12	2.21		3.33	4.85
8660	6" wide		320	.050		1.25	2.35		3.60	5.20
8800	Hospital security partition, 5/8" fiber reinf. high abuse gyp. bd.									
8810	Mtl. studs, NLB, 20 ga., 16" O.C., 3-5/8" wide, w/sec. mesh, gyp. bd.	2 Carp	208	.077	S.F.	4.18	3.61		7.79	10.50
9000	Exterior, 1/2" gypsum sheathing, 1/2" gypsum finished, interior,									
9100	including foil faced insulation, metal studs, 20 ga.									
9200	16" O.C., 3-5/8" wide	2 Carp	290	.055	S.F.	1.70	2.59		4.29	6.10
9400	6" wide		270	.059		1.88	2.78		4.66	6.60
9600	Partitions, for work over 8' high, add		1530	.010			.49		.49	.80

09 22 Supports for Plaster and Gypsum Board

09 22 03 – Fastening Methods for Finishes

09 22 03.20 Drilling Plaster/Drywall		Crew	Daily Output	Labor-Hours	Unit	Material	Labor	Equipment	Total	Total Incl O&P
0010	**DRILLING PLASTER/DRYWALL**									
1100	Drilling & layout for drywall/plaster walls, up to 1" deep, no anchor									
1200	Holes, 1/4" diameter	1 Carp	150	.053	Ea.	.01	2.50		2.51	4.11
1300	3/8" diameter		140	.057		.01	2.68		2.69	4.41
1400	1/2" diameter		130	.062		.01	2.89		2.90	4.75
1500	3/4" diameter		120	.067		.01	3.13		3.14	5.15
1600	1" diameter		110	.073		.02	3.41		3.43	5.60
1700	1-1/4" diameter		100	.080		.03	3.76		3.79	6.20
1800	1-1/2" diameter		90	.089		.05	4.17		4.22	6.90
1900	For ceiling installations, add						40%			

09 22 13 – Metal Furring

09 22 13.13 Metal Channel Furring		Crew	Daily Output	Labor-Hours	Unit	Material	Labor	Equipment	Total	Total Incl O&P
0010	**METAL CHANNEL FURRING**									
0030	Beams and columns, 7/8" channels, galvanized, 12" O.C.	1 Lath	155	.052	S.F.	.40	2.22		2.62	3.94
0050	16" O.C.		170	.047		.33	2.02		2.35	3.55
0070	24" O.C.		185	.043		.22	1.86		2.08	3.17
0100	Ceilings, on steel, 7/8" channels, galvanized, 12" O.C.		210	.038		.37	1.64		2.01	2.98
0300	16" O.C.		290	.028		.33	1.19		1.52	2.23
0400	24" O.C.		420	.019		.22	.82		1.04	1.53
0600	1-5/8" channels, galvanized, 12" O.C.		190	.042		.49	1.81		2.30	3.39
0700	16" O.C.		260	.031		.44	1.32		1.76	2.56
0900	24" O.C.		390	.021		.29	.88		1.17	1.71
0930	7/8" channels with sound isolation clips, 12" O.C.		120	.067		1.74	2.87		4.61	6.45
0940	16" O.C.		100	.080		1.31	3.44		4.75	6.85
0950	24" O.C.		165	.048		.87	2.08		2.95	4.24
0960	1-5/8" channels, galvanized, 12" O.C.		110	.073		1.86	3.13		4.99	6.95
0970	16" O.C.		100	.080		1.40	3.44		4.84	6.95
0980	24" O.C.		155	.052		.93	2.22		3.15	4.52
1000	Walls, 7/8" channels, galvanized, 12" O.C.		235	.034		.37	1.46		1.83	2.70
1200	16" O.C.		265	.030		.33	1.30		1.63	2.40
1300	24" O.C.		350	.023		.22	.98		1.20	1.79
1500	1-5/8" channels, galvanized, 12" O.C.		210	.038		.49	1.64		2.13	3.12
1600	16" O.C.		240	.033		.44	1.43		1.87	2.74

09 22 13 – Metal Furring

09 22 13.13 Metal Channel Furring		Crew	Daily Output	Labor-Hours	Unit	2015 Bare Costs Material	Labor	Equipment	Total	Total Incl O&P
1800	24" O.C.	1 Lath	305	.026	S.F.	.29	1.13		1.42	2.10
1920	7/8" channels with sound isolation clips, 12" O.C.		125	.064		1.74	2.75		4.49	6.25
1940	16" O.C.		100	.080		1.31	3.44		4.75	6.85
1950	24" O.C.		150	.053		.87	2.29		3.16	4.57
1960	1-5/8" channels, galvanized, 12" O.C.		115	.070		1.86	2.99		4.85	6.75
1970	16" O.C.		95	.084		1.40	3.62		5.02	7.25
1980	24" O.C.		140	.057		.93	2.46		3.39	4.90
9000	Minimum labor/equipment charge		4	2	Job		86		86	135

09 22 16 – Non-Structural Metal Framing

09 22 16.13 Non-Structural Metal Stud Framing		Crew	Daily Output	Labor-Hours	Unit	Material	Labor	Equipment	Total	Total Incl O&P
0010	**NON-STRUCTURAL METAL STUD FRAMING**									
1600	Non-load bearing, galv., 8' high, 25 ga. 1-5/8" wide, 16" O.C.	1 Carp	619	.013	S.F.	.27	.61		.88	1.28
1610	24" O.C.		950	.008		.20	.40		.60	.87
1620	2-1/2" wide, 16" O.C.		613	.013		.33	.61		.94	1.36
1630	24" O.C.		938	.009		.24	.40		.64	.93
1640	3-5/8" wide, 16" O.C.		600	.013		.39	.63		1.02	1.46
1650	24" O.C.		925	.009		.29	.41		.70	.99
1660	4" wide, 16" O.C.		594	.013		.43	.63		1.06	1.51
1670	24" O.C.		925	.009		.32	.41		.73	1.03
1680	6" wide, 16" O.C.		588	.014		.52	.64		1.16	1.62
1690	24" O.C.		906	.009		.39	.41		.80	1.11
1700	20 ga. studs, 1-5/8" wide, 16" O.C.		494	.016		.34	.76		1.10	1.62
1710	24" O.C.		763	.010		.25	.49		.74	1.09
1720	2-1/2" wide, 16" O.C.		488	.016		.42	.77		1.19	1.72
1730	24" O.C.		750	.011		.31	.50		.81	1.17
1740	3-5/8" wide, 16" O.C.		481	.017		.48	.78		1.26	1.81
1750	24" O.C.		738	.011		.36	.51		.87	1.23
1760	4" wide, 16" O.C.		475	.017		.57	.79		1.36	1.93
1770	24" O.C.		738	.011		.43	.51		.94	1.30
1780	6" wide, 16" O.C.		469	.017		.66	.80		1.46	2.04
1790	24" O.C.		725	.011		.50	.52		1.02	1.40
2000	Non-load bearing, galv., 10' high, 25 ga. 1-5/8" wide, 16" O.C.		495	.016		.25	.76		1.01	1.52
2100	24" O.C.		760	.011		.19	.49		.68	1.01
2200	2-1/2" wide, 16" O.C.		490	.016		.31	.77		1.08	1.60
2250	24" O.C.		750	.011		.23	.50		.73	1.07
2300	3-5/8" wide, 16" O.C.		480	.017		.37	.78		1.15	1.68
2350	24" O.C.		740	.011		.27	.51		.78	1.13
2400	4" wide, 16" O.C.		475	.017		.41	.79		1.20	1.75
2450	24" O.C.		740	.011		.30	.51		.81	1.16
2500	6" wide, 16" O.C.		470	.017		.49	.80		1.29	1.85
2550	24" O.C.		725	.011		.36	.52		.88	1.25
2600	20 ga. studs, 1-5/8" wide, 16" O.C.		395	.020		.32	.95		1.27	1.91
2650	24" O.C.		610	.013		.23	.62		.85	1.27
2700	2-1/2" wide, 16" O.C.		390	.021		.39	.96		1.35	2.01
2750	24" O.C.		600	.013		.29	.63		.92	1.35
2800	3-5/8" wide, 16" OC		385	.021		.46	.98		1.44	2.10
2850	24" O.C.		590	.014		.34	.64		.98	1.41
2900	4" wide, 16" O.C.		380	.021		.54	.99		1.53	2.21
2950	24" O.C.		590	.014		.40	.64		1.04	1.48
3000	6" wide, 16" O.C.		375	.021		.63	1		1.63	2.33
3050	24" O.C.		580	.014		.46	.65		1.11	1.57
3060	Non-load bearing, galv., 12' high, 25 ga. 1-5/8" wide, 16" O.C.		413	.019		.24	.91		1.15	1.76

09 22 16 – Non-Structural Metal Framing

09 22 16.13 Non-Structural Metal Stud Framing		Crew	Daily Output	Labor-Hours	Unit	Material	2015 Bare Costs Labor	Equipment	Total	Total Incl O&P
3070	24" O.C.	1 Carp	633	.013	S.F.	.18	.59		.77	1.16
3080	2-1/2" wide, 16" O.C.		408	.020		.30	.92		1.22	1.84
3090	24" O.C.		625	.013		.22	.60		.82	1.23
3100	3-5/8" wide, 16" O.C.		400	.020		.35	.94		1.29	1.93
3110	24" O.C.		617	.013		.26	.61		.87	1.28
3120	4" wide, 16" O.C.		396	.020		.39	.95		1.34	1.98
3130	24" O.C.		617	.013		.28	.61		.89	1.31
3140	6" wide, 16" O.C.		392	.020		.47	.96		1.43	2.09
3150	24" O.C.		604	.013		.34	.62		.96	1.40
3160	20 ga. studs, 1-5/8" wide, 16" O.C.		329	.024		.30	1.14		1.44	2.20
3170	24" O.C.		508	.016		.22	.74		.96	1.45
3180	2-1/2" wide, 16" O.C.		325	.025		.38	1.16		1.54	2.30
3190	24" O.C.		500	.016		.27	.75		1.02	1.53
3200	3-5/8" wide, 16" O.C.		321	.025		.44	1.17		1.61	2.40
3210	24" O.C.		492	.016		.32	.76		1.08	1.60
3220	4" wide, 16" O.C.		317	.025		.52	1.19		1.71	2.51
3230	24" O.C.		492	.016		.38	.76		1.14	1.66
3240	6" wide, 16" O.C.		313	.026		.60	1.20		1.80	2.63
3250	24" O.C.		483	.017		.44	.78		1.22	1.75
5000	Load bearing studs, see Section 05 41 13.30									
9000	Minimum labor/equipment charge	1 Carp	4	2	Job		94		94	154

09 22 26 – Suspension Systems

09 22 26.13 Ceiling Suspension Systems		Crew	Daily Output	Labor-Hours	Unit	Material	Labor	Equipment	Total	Total Incl O&P
0010	**CEILING SUSPENSION SYSTEMS** for gypsum board or plaster									
8000	Suspended ceilings, including carriers									
8200	1-1/2" carriers, 24" O.C. with:									
8300	7/8" channels, 16" O.C.	1 Lath	275	.029	S.F.	.54	1.25		1.79	2.56
8320	24" O.C.		310	.026		.42	1.11		1.53	2.22
8400	1-5/8" channels, 16" O.C.		205	.039		.64	1.68		2.32	3.35
8420	24" O.C.		250	.032		.50	1.38		1.88	2.72
8600	2" carriers, 24" O.C. with:									
8700	7/8" channels, 16" O.C.	1 Lath	250	.032	S.F.	.59	1.38		1.97	2.82
8720	24" O.C.		285	.028		.48	1.21		1.69	2.43
8800	1-5/8" channels, 16" O.C.		190	.042		.70	1.81		2.51	3.62
8820	24" O.C.		225	.036		.55	1.53		2.08	3.02

09 22 36 – Lath

09 22 36.13 Gypsum Lath		Crew	Daily Output	Labor-Hours	Unit	Material	Labor	Equipment	Total	Total Incl O&P
0010	**GYPSUM LATH**									
0020	Plain or perforated, nailed, 3/8" thick	1 Lath	85	.094	S.Y.	3.06	4.05		7.11	9.70
0100	1/2" thick		80	.100		2.43	4.30		6.73	9.40
0300	Clipped to steel studs, 3/8" thick		75	.107		3.06	4.59		7.65	10.55
0400	1/2" thick		70	.114		2.43	4.91		7.34	10.40
1500	For ceiling installations, add		216	.037			1.59		1.59	2.51
1600	For columns and beams, add		170	.047			2.02		2.02	3.19
9000	Minimum labor/equipment charge		4.25	1.882	Job		81		81	127

09 22 36.23 Metal Lath		Crew	Daily Output	Labor-Hours	Unit	Material	Labor	Equipment	Total	Total Incl O&P
0010	**METAL LATH**									
0020	Diamond, expanded, 2.5 lb. per S.Y., painted				S.Y.	3.61			3.61	3.97
0100	Galvanized					2.72			2.72	2.99
0300	3.4 lb. per S.Y., painted					4.08			4.08	4.49
0400	Galvanized					4.13			4.13	4.54

09 22 36 – Lath

09 22 36.23 Metal Lath		Crew	Daily Output	Labor-Hours	Unit	Material	2015 Bare Costs Labor	Equipment	Total	Total Incl O&P
0600	For 15# asphalt sheathing paper, add				S.Y.	.49			.49	.53
0900	Flat rib, 1/8" high, 2.75 lb., painted					3.40			3.40	3.74
1000	Foil backed					3.58			3.58	3.94
1200	3.4 lb. per S.Y., painted					4.24			4.24	4.66
1300	Galvanized					4.35			4.35	4.79
1500	For 15# asphalt sheathing paper, add					.49			.49	.53
1800	High rib, 3/8" high, 3.4 lb. per S.Y., painted					4.12			4.12	4.53
1900	Galvanized				↓	3.70			3.70	4.07
2400	3/4" high, painted, .60 lb. per S.F.				S.F.	.62			.62	.68
2500	.75 lb. per S.F.				"	1.33			1.33	1.46
2800	Stucco mesh, painted, 3.6 lb.				S.Y.	3.69			3.69	4.06
3000	K-lath, perforated, absorbent paper, regular					4.42			4.42	4.86
3100	Heavy duty					5.20			5.20	5.75
3300	Waterproof, heavy duty, grade B backing					5.10			5.10	5.60
3400	Fire resistant backing					5.65			5.65	6.20
3600	2.5 lb. diamond painted, on wood framing, on walls	1 Lath	85	.094		3.61	4.05		7.66	10.30
3700	On ceilings		75	.107		3.61	4.59		8.20	11.15
3900	3.4 lb. diamond painted, on wood framing, on walls		80	.100		4.24	4.30		8.54	11.40
4000	On ceilings		70	.114		4.24	4.91		9.15	12.40
4200	3.4 lb. diamond painted, wired to steel framing		75	.107		4.24	4.59		8.83	11.85
4300	On ceilings		60	.133		4.24	5.75		9.99	13.70
4500	Columns and beams, wired to steel		40	.200		4.24	8.60		12.84	18.20
4600	Cornices, wired to steel		35	.229		4.24	9.85		14.09	20
4800	Screwed to steel studs, 2.5 lb.		80	.100		3.61	4.30		7.91	10.70
4900	3.4 lb.		75	.107		4.08	4.59		8.67	11.70
5100	Rib lath, painted, wired to steel, on walls, 2.5 lb.		75	.107		3.40	4.59		7.99	10.95
5200	3.4 lb.		70	.114		4.12	4.91		9.03	12.30
5400	4.0 lb.	↓	65	.123		5.65	5.30		10.95	14.55
5500	For self-furring lath, add					.11			.11	.12
5700	Suspended ceiling system, incl. 3.4 lb. diamond lath, painted	1 Lath	15	.533		4.31	23		27.31	40.50
5800	Galvanized	"	15	.533	↓	4.24	23		27.24	40.50
6000	Hollow metal stud partitions, 3.4 lb. painted lath both sides									
6010	Non-load bearing, 25 ga., w/rib lath 2-1/2" studs, 12" O.C.	1 Lath	20.30	.394	S.Y.	11.75	16.95		28.70	39.50
6300	16" O.C.		21.10	.379		11	16.30		27.30	37.50
6350	24" O.C.		22.70	.352		10.30	15.15		25.45	35.50
6400	3-5/8" studs, 16" O.C.		19.50	.410		11.55	17.65		29.20	40.50
6600	24" O.C.		20.40	.392		10.65	16.85		27.50	38.50
6700	4" studs, 16" O.C.		20.40	.392		11.90	16.85		28.75	39.50
6900	24" O.C.		21.60	.370		10.95	15.95		26.90	37
7000	6" studs, 16" O.C.		19.50	.410		12.65	17.65		30.30	42
7100	24" O.C.		21.10	.379		11.50	16.30		27.80	38
7200	L.B. partitions, 16 ga., w/rib lath, 2-1/2" studs, 16" O.C.		20	.400		12.45	17.20		29.65	40.50
7300	3-5/8" studs, 16 ga.		19.70	.406		14.15	17.45		31.60	43
7500	4" studs, 16 ga.		19.50	.410		14.65	17.65		32.30	44
7600	6" studs, 16 ga.		18.70	.428	↓	17.30	18.40		35.70	48
9000	Minimum labor/equipment charge	↓	4.25	1.882	Job		81		81	127

09 22 36.43 Security Mesh		Crew	Daily Output	Labor-Hours	Unit	Material	Labor	Equipment	Total	Total Incl O&P
0010	**SECURITY MESH**, expanded metal, flat, screwed to framing									
0100	On walls, 3/4", 1.76 lb./S.F.	2 Carp	1500	.011	S.F.	1.84	.50		2.34	2.84
0110	1-1/2", 1.14 lb./S.F.		1600	.010		1.41	.47		1.88	2.32
0200	On ceilings, 3/4", 1.76 lb./S.F.		1350	.012		1.84	.56		2.40	2.93
0210	1-1/2", 1.14 lb./S.F.	↓	1450	.011	↓	1.41	.52		1.93	2.40

09 22 Supports for Plaster and Gypsum Board

09 22 36 – Lath

09 22 36.83 Accessories, Plaster		Crew	Daily Output	Labor-Hours	Unit	2015 Bare Costs Material	Labor	Equipment	Total	Total Incl O&P
0010	**ACCESSORIES, PLASTER**									
0020	Casing bead, expanded flange, galvanized	1 Lath	2.70	2.963	C.L.F.	55	127		182	262
0200	Foundation weep screed, galvanized	"	2.70	2.963		52	127		179	258
0900	Channels, cold rolled, 16 ga., 3/4" deep, galvanized					37			37	40.50
1200	1-1/2" deep, 16 ga., galvanized					49			49	54
1620	Corner bead, expanded bullnose, 3/4" radius, #10, galvanized	1 Lath	2.60	3.077		24.50	132		156.50	235
1650	#1, galvanized		2.55	3.137		48.50	135		183.50	265
1670	Expanded wing, 2-3/4" wide, #1, galvanized		2.65	3.019		37	130		167	245
1700	Inside corner (corner rite), 3" x 3", painted		2.60	3.077		20.50	132		152.50	231
1750	Strip-ex, 4" wide, painted		2.55	3.137		24	135		159	239
1800	Expansion joint, 3/4" grounds, limited expansion, galv., 1 piece		2.70	2.963		75	127		202	284
2100	Extreme expansion, galvanized, 2 piece		2.60	3.077		140	132		272	360

09 23 Gypsum Plastering

09 23 13 – Acoustical Gypsum Plastering

09 23 13.10 Perlite or Vermiculite Plaster		Crew	Daily Output	Labor-Hours	Unit	Material	Labor	Equipment	Total	Total Incl O&P
0010	**PERLITE OR VERMICULITE PLASTER**									
0020	In 100 lb. bags, under 200 bags				Bag	17.55			17.55	19.35
0100	Over 200 bags				"	16.80			16.80	18.45
0300	2 coats, no lath included, on walls	J-1	92	.435	S.Y.	5.85	17.85	1.53	25.23	36.50
0400	On ceilings	"	79	.506		5.85	21	1.78	28.63	41.50
0600	On and incl. 3/8" gypsum lath, on metal studs	J-2	84	.571		9.55	23.50	1.67	34.72	50
0700	On ceilings	"	70	.686		9.55	28.50	2	40.05	57.50
0900	3 coats, no lath included, on walls	J-1	74	.541		6.40	22	1.90	30.30	44.50
1000	On ceilings	"	63	.635		6.40	26	2.23	34.63	51
1200	On and incl. painted metal lath, on metal studs	J-2	72	.667		10.50	27.50	1.95	39.95	57.50
1300	On ceilings		61	.787		10.50	32.50	2.30	45.30	66
1500	On and incl. suspended metal lath ceiling		37	1.297		10.70	53.50	3.79	67.99	101
1700	For irregular or curved surfaces, add to above						30%			
1800	For columns and beams, add to above						50%			
1900	For soffits, add to ceiling prices						40%			
9000	Minimum labor/equipment charge	1 Plas	1	8	Job		345		345	550

09 23 20 – Gypsum Plaster

09 23 20.10 Gypsum Plaster On Walls and Ceilings		Crew	Daily Output	Labor-Hours	Unit	Material	Labor	Equipment	Total	Total Incl O&P
0010	**GYPSUM PLASTER ON WALLS AND CEILINGS**									
0020	80# bag, less than 1 ton				Bag	15.95			15.95	17.55
0100	Over 1 ton				"	13.95			13.95	15.35
0300	2 coats, no lath included, on walls	J-1	105	.381	S.Y.	3.61	15.60	1.34	20.55	30.50
0400	On ceilings	"	92	.435		3.61	17.85	1.53	22.99	34
0600	On and incl. 3/8" gypsum lath on steel, on walls	J-2	97	.495		6.65	20.50	1.45	28.60	41.50
0700	On ceilings	"	83	.578		6.65	24	1.69	32.34	47
0900	3 coats, no lath included, on walls	J-1	87	.460		5.20	18.85	1.61	25.66	37.50
1000	On ceilings	"	78	.513		5.20	21	1.80	28	41
1200	On and including painted metal lath, on wood studs	J-2	86	.558		10.20	23	1.63	34.83	50
1300	On ceilings	"	76.50	.627		10.20	26	1.83	38.03	55
1600	For irregular or curved surfaces, add						30%			
1800	For columns & beams, add						50%			
9000	Minimum labor/equipment charge	1 Plas	1	8	Job		345		345	550

09 23 Gypsum Plastering

09 23 20 – Gypsum Plaster

09 23 20.20 Gauging Plaster		Crew	Daily Output	Labor-Hours	Unit	Material	Labor	Equipment	Total	Total Incl O&P
0010	**GAUGING PLASTER**									
0020	100 lb. bags, less than 1 ton				Bag	19.35			19.35	21.50
0100	Over 1 ton				"	18.35			18.35	20
09 23 20.30 Keenes Cement										
0010	**KEENES CEMENT**									
0020	In 100 lb. bags, less than 1 ton				Bag	22			22	24
0100	Over 1 ton				"	20			20	22.50
0300	Finish only, add to plaster prices, standard	J-1	215	.186	S.Y.	1.95	7.65	.65	10.25	15.05
0400	High quality	"	144	.278	"	1.97	11.40	.98	14.35	21.50

(2015 Bare Costs: Material, Labor, Equipment, Total)

09 24 Cement Plastering

09 24 23 – Cement Stucco

09 24 23.40 Stucco		Crew	Daily Output	Labor-Hours	Unit	Material	Labor	Equipment	Total	Total Incl O&P
0010	**STUCCO**									
0015	3 coats 1" thick, float finish, with mesh, on wood frame	J-2	63	.762	S.Y.	6.25	31.50	2.22	39.97	59.50
0100	On masonry construction, no mesh incl.	J-1	67	.597		2.55	24.50	2.10	29.15	44
0300	For trowel finish, add	1 Plas	170	.047			2.02		2.02	3.23
0400	For 3/4" thick, on masonry, deduct	J-1	880	.045		.63	1.86	.16	2.65	3.85
0600	For coloring add		685	.058		.40	2.39	.20	2.99	4.50
0700	For special texture add		200	.200		1.41	8.20	.70	10.31	15.40
0900	For soffits, add	J-2	155	.310		2.18	12.80	.90	15.88	24
1000	Exterior stucco, with bonding agent, 3 coats, on walls, no mesh incl.	J-1	200	.200		3.65	8.20	.70	12.55	17.90
1200	Ceilings		180	.222		3.65	9.10	.78	13.53	19.45
1300	Beams		80	.500		3.65	20.50	1.76	25.91	39
1500	Columns		100	.400		3.65	16.40	1.40	21.45	31.50
1550	Minimum labor/equipment charge	1 Plas	1	8	Job		345		345	550
1600	Mesh, painted, nailed to wood, 1.8 lb.	1 Lath	60	.133	S.Y.	6.20	5.75		11.95	15.85
1800	3.6 lb.		55	.145		3.69	6.25		9.94	13.90
1900	Wired to steel, painted, 1.8 lb.		53	.151		6.20	6.50		12.70	17
2100	3.6 lb.		50	.160		3.69	6.90		10.59	14.90
9000	Minimum labor/equipment charge		4	2	Job		86		86	135

09 25 Other Plastering

09 25 23 – Lime Based Plastering

09 25 23.10 Venetian Plaster		Crew	Daily Output	Labor-Hours	Unit	Material	Labor	Equipment	Total	Total Incl O&P
0010	**VENETIAN PLASTER**									
0100	Walls, 1 coat primer, roller applied	1 Plas	950	.008	S.F.	.16	.36		.52	.76
0200	Plaster, 3 coats, incl. sanding	2 Plas	700	.023	"	.46	.98		1.44	2.07
0210	For pigment, light colors add per ea.				Ea.	3			3	3.30
0220	For pigment, dark colors add per ea.				"	9			9	9.90
0300	For sealer/wax coat incl. burnishing, add	1 Plas	300	.027	S.F.	.41	1.15		1.56	2.28

09 26 Veneer Plastering

09 26 13 – Gypsum Veneer Plastering

09 26 13.20 Blueboard		Crew	Daily Output	Labor-Hours	Unit	2015 Bare Costs Material	Labor	Equipment	Total	Total Incl O&P
0010	**BLUEBOARD** For use with thin coat									
0100	plaster application see Section 09 26 13.80									
1000	3/8" thick, on walls or ceilings, standard, no finish included	2 Carp	1900	.008	S.F.	.34	.40		.74	1.02
1100	With thin coat plaster finish		875	.018		.45	.86		1.31	1.91
1400	On beams, columns, or soffits, standard, no finish included		675	.024		.39	1.11		1.50	2.25
1450	With thin coat plaster finish		475	.034		.50	1.58		2.08	3.14
3000	1/2" thick, on walls or ceilings, standard, no finish included		1900	.008		.33	.40		.73	1.01
3100	With thin coat plaster finish		875	.018		.44	.86		1.30	1.90
3300	Fire resistant, no finish included		1900	.008		.33	.40		.73	1.01
3400	With thin coat plaster finish		875	.018		.44	.86		1.30	1.90
3450	On beams, columns, or soffits, standard, no finish included		675	.024		.38	1.11		1.49	2.24
3500	With thin coat plaster finish		475	.034		.49	1.58		2.07	3.13
3700	Fire resistant, no finish included		675	.024		.38	1.11		1.49	2.24
3800	With thin coat plaster finish		475	.034		.49	1.58		2.07	3.13
5000	5/8" thick, on walls or ceilings, fire resistant, no finish included		1900	.008		.34	.40		.74	1.02
5100	With thin coat plaster finish		875	.018		.45	.86		1.31	1.91
5500	On beams, columns, or soffits, no finish included		675	.024		.39	1.11		1.50	2.25
5600	With thin coat plaster finish		475	.034		.50	1.58		2.08	3.14
6000	For high ceilings, over 8' high, add		3060	.005			.25		.25	.40
6500	For over 3 stories high, add per story		6100	.003			.12		.12	.20
9000	Minimum labor/equipment charge	1 Carp	2	4	Job		188		188	310

09 26 13.80 Thin Coat Plaster		Crew	Daily Output	Labor-Hours	Unit	Material	Labor	Equipment	Total	Total Incl O&P
0010	**THIN COAT PLASTER**									
0012	1 coat veneer, not incl. lath	J-1	3600	.011	S.F.	.11	.46	.04	.61	.89
1000	In 50 lb. bags				Bag	15.25			15.25	16.80

09 28 Backing Boards and Underlayments

09 28 13 – Cementitious Backing Boards

09 28 13.10 Cementitious Backerboard		Crew	Daily Output	Labor-Hours	Unit	Material	Labor	Equipment	Total	Total Incl O&P
0010	**CEMENTITIOUS BACKERBOARD**									
0070	Cementitious backerboard, on floor, 3' x 4' x 1/2" sheets	2 Carp	525	.030	S.F.	.78	1.43		2.21	3.21
0080	3' x 5' x 1/2" sheets		525	.030		.76	1.43		2.19	3.19
0090	3' x 6' x 1/2" sheets		525	.030		.74	1.43		2.17	3.16
0100	3' x 4' x 5/8" sheets		525	.030		.99	1.43		2.42	3.44
0110	3' x 5' x 5/8" sheets		525	.030		.99	1.43		2.42	3.44
0120	3' x 6' x 5/8" sheets		525	.030		.96	1.43		2.39	3.41
0150	On wall, 3' x 4' x 1/2" sheets		350	.046		.78	2.15		2.93	4.38
0160	3' x 5' x 1/2" sheets		350	.046		.76	2.15		2.91	4.36
0170	3' x 6' x 1/2" sheets		350	.046		.74	2.15		2.89	4.33
0180	3' x 4' x 5/8" sheets		350	.046		.99	2.15		3.14	4.61
0190	3' x 5' x 5/8" sheets		350	.046		.99	2.15		3.14	4.61
0200	3' x 6' x 5/8" sheets		350	.046		.96	2.15		3.11	4.58
0250	On counter, 3' x 4' x 1/2" sheets		180	.089		.78	4.17		4.95	7.70
0260	3' x 5' x 1/2" sheets		180	.089		.76	4.17		4.93	7.70
0270	3' x 6' x 1/2" sheets		180	.089		.74	4.17		4.91	7.65
0300	3' x 4' x 5/8" sheets		180	.089		.99	4.17		5.16	7.95
0310	3' x 5' x 5/8" sheets		180	.089		.99	4.17		5.16	7.95
0320	3' x 6' x 5/8" sheets		180	.089		.96	4.17		5.13	7.90

09 29 10 – Gypsum Board Panels

09 29 10.30 Gypsum Board		Crew	Daily Output	Labor-Hours	Unit	2015 Bare Costs Material	Labor	Equipment	Total	Total Incl O&P
0010	**GYPSUM BOARD** on walls & ceilings R092910-10									
0100	Nailed or screwed to studs unless otherwise noted									
0110	1/4" thick, on walls or ceilings, standard, no finish included	2 Carp	1330	.012	S.F.	.35	.56		.91	1.32
0115	1/4" thick, on walls or ceilings, flexible, no finish included		1050	.015		.50	.72		1.22	1.72
0117	1/4" thick, on columns or soffits, flexible, no finish included		1050	.015		.50	.72		1.22	1.72
0130	1/4" thick, standard, no finish included, less than 800 S.F.		510	.031		.35	1.47		1.82	2.80
0150	3/8" thick, on walls, standard, no finish included		2000	.008		.34	.38		.72	.99
0200	On ceilings, standard, no finish included		1800	.009		.34	.42		.76	1.05
0250	On beams, columns, or soffits, no finish included		675	.024		.34	1.11		1.45	2.19
0300	1/2" thick, on walls, standard, no finish included		2000	.008		.30	.38		.68	.95
0350	Taped and finished (level 4 finish)		965	.017		.35	.78		1.13	1.66
0390	With compound skim coat (level 5 finish)		775	.021		.40	.97		1.37	2.03
0400	Fire resistant, no finish included		2000	.008		.35	.38		.73	1.01
0450	Taped and finished (level 4 finish)		965	.017		.40	.78		1.18	1.72
0490	With compound skim coat (level 5 finish)		775	.021		.45	.97		1.42	2.09
0500	Water resistant, no finish included		2000	.008		.39	.38		.77	1.05
0550	Taped and finished (level 4 finish)		965	.017		.44	.78		1.22	1.76
0590	With compound skim coat (level 5 finish)		775	.021		.49	.97		1.46	2.13
0600	Prefinished, vinyl, clipped to studs		900	.018		.48	.83		1.31	1.90
0700	Mold resistant, no finish included		2000	.008		.44	.38		.82	1.10
0710	Taped and finished (level 4 finish)		965	.017		.49	.78		1.27	1.82
0720	With compound skim coat (level 5 finish)		775	.021		.54	.97		1.51	2.18
1000	On ceilings, standard, no finish included		1800	.009		.30	.42		.72	1.01
1050	Taped and finished (level 4 finish)		765	.021		.35	.98		1.33	1.99
1090	With compound skim coat (level 5 finish)		610	.026		.40	1.23		1.63	2.46
1100	Fire resistant, no finish included		1800	.009		.35	.42		.77	1.07
1150	Taped and finished (level 4 finish)		765	.021		.40	.98		1.38	2.05
1195	With compound skim coat (level 5 finish)		610	.026		.45	1.23		1.68	2.52
1200	Water resistant, no finish included		1800	.009		.39	.42		.81	1.11
1250	Taped and finished (level 4 finish)		765	.021		.44	.98		1.42	2.09
1290	With compound skim coat (level 5 finish)		610	.026		.49	1.23		1.72	2.56
1310	Mold resistant, no finish included		1800	.009		.44	.42		.86	1.16
1320	Taped and finished (level 4 finish)		765	.021		.49	.98		1.47	2.15
1330	With compound skim coat (level 5 finish)		610	.026		.54	1.23		1.77	2.61
1350	Sag resistant, no finish included		1600	.010		.34	.47		.81	1.14
1360	Taped and finished (level 4 finish)		765	.021		.39	.98		1.37	2.04
1370	With compound skim coat (level 5 finish)		610	.026		.44	1.23		1.67	2.50
1500	On beams, columns, or soffits, standard, no finish included		675	.024		.35	1.11		1.46	2.20
1550	Taped and finished (level 4 finish)		540	.030		.35	1.39		1.74	2.66
1590	With compound skim coat (level 5 finish)		475	.034		.40	1.58		1.98	3.03
1600	Fire resistant, no finish included		675	.024		.35	1.11		1.46	2.21
1650	Taped and finished (level 4 finish)		540	.030		.40	1.39		1.79	2.72
1690	With compound skim coat (level 5 finish)		475	.034		.45	1.58		2.03	3.09
1700	Water resistant, no finish included		675	.024		.45	1.11		1.56	2.31
1750	Taped and finished (level 4 finish)		540	.030		.44	1.39		1.83	2.76
1790	With compound skim coat (level 5 finish)		475	.034		.49	1.58		2.07	3.13
1800	Mold resistant, no finish included		675	.024		.51	1.11		1.62	2.38
1810	Taped and finished (level 4 finish)		540	.030		.49	1.39		1.88	2.82
1820	With compound skim coat (level 5 finish)		475	.034		.54	1.58		2.12	3.18
1850	Sag resistant, no finish included		675	.024		.39	1.11		1.50	2.25
1860	Taped and finished (level 4 finish)		540	.030		.39	1.39		1.78	2.71
1870	With compound skim coat (level 5 finish)		475	.034		.44	1.58		2.02	3.07
2000	5/8" thick, on walls, standard, no finish included		2000	.008		.33	.38		.71	.98

09 29 Gypsum Board

09 29 10 – Gypsum Board Panels

09 29 10.30 Gypsum Board		Crew	Daily Output	Labor-Hours	Unit	Material	2015 Bare Costs Labor	Equipment	Total	Total Incl O&P
2050	Taped and finished (level 4 finish)	2 Carp	965	.017	S.F.	.38	.78		1.16	1.69
2090	With compound skim coat (level 5 finish)		775	.021		.43	.97		1.40	2.06
2100	Fire resistant, no finish included		2000	.008		.34	.38		.72	.99
2150	Taped and finished (level 4 finish)		965	.017		.39	.78		1.17	1.71
2195	With compound skim coat (level 5 finish)		775	.021		.44	.97		1.41	2.07
2200	Water resistant, no finish included		2000	.008		.42	.38		.80	1.08
2250	Taped and finished (level 4 finish)		965	.017		.47	.78		1.25	1.79
2290	With compound skim coat (level 5 finish)		775	.021		.52	.97		1.49	2.16
2300	Prefinished, vinyl, clipped to studs		900	.018		.76	.83		1.59	2.21
2510	Mold resistant, no finish included		2000	.008		.46	.38		.84	1.13
2520	Taped and finished (level 4 finish)		965	.017		.51	.78		1.29	1.84
2530	With compound skim coat (level 5 finish)		775	.021		.56	.97		1.53	2.21
3000	On ceilings, standard, no finish included		1800	.009		.33	.42		.75	1.04
3050	Taped and finished (level 4 finish)		765	.021		.38	.98		1.36	2.02
3090	With compound skim coat (level 5 finish)		615	.026		.43	1.22		1.65	2.47
3100	Fire resistant, no finish included		1800	.009		.34	.42		.76	1.05
3150	Taped and finished (level 4 finish)		765	.021		.39	.98		1.37	2.04
3190	With compound skim coat (level 5 finish)		615	.026		.44	1.22		1.66	2.48
3200	Water resistant, no finish included		1800	.009		.42	.42		.84	1.14
3250	Taped and finished (level 4 finish)		765	.021		.47	.98		1.45	2.12
3290	With compound skim coat (level 5 finish)		615	.026		.52	1.22		1.74	2.57
3300	Mold resistant, no finish included		1800	.009		.46	.42		.88	1.19
3310	Taped and finished (level 4 finish)		765	.021		.51	.98		1.49	2.17
3320	With compound skim coat (level 5 finish)		615	.026		.56	1.22		1.78	2.62
3500	On beams, columns, or soffits, no finish included		675	.024		.38	1.11		1.49	2.24
3550	Taped and finished (level 4 finish)		475	.034		.43	1.58		2.01	3.07
3590	With compound skim coat (level 5 finish)		380	.042		.49	1.98		2.47	3.78
3600	Fire resistant, no finish included		675	.024		.39	1.11		1.50	2.25
3650	Taped and finished (level 4 finish)		475	.034		.45	1.58		2.03	3.08
3690	With compound skim coat (level 5 finish)		380	.042		.44	1.98		2.42	3.72
3700	Water resistant, no finish included		675	.024		.48	1.11		1.59	2.35
3750	Taped and finished (level 4 finish)		475	.034		.52	1.58		2.10	3.16
3790	With compound skim coat (level 5 finish)		380	.042		.54	1.98		2.52	3.83
3800	Mold resistant, no finish included		675	.024		.53	1.11		1.64	2.40
3810	Taped and finished (level 4 finish)		475	.034		.56	1.58		2.14	3.21
3820	With compound skim coat (level 5 finish)		380	.042		.58	1.98		2.56	3.88
4000	Fireproofing, beams or columns, 2 layers, 1/2" thick, incl finish		330	.048		.79	2.28		3.07	4.60
4010	Mold resistant		330	.048		.97	2.28		3.25	4.80
4050	5/8" thick		300	.053		.77	2.50		3.27	4.95
4060	Mold resistant		300	.053		1.01	2.50		3.51	5.20
4100	3 layers, 1/2" thick		225	.071		1.19	3.34		4.53	6.75
4110	Mold resistant		225	.071		1.46	3.34		4.80	7.05
4150	5/8" thick		210	.076		1.16	3.58		4.74	7.15
4160	Mold resistant		210	.076		1.52	3.58		5.10	7.50
5050	For 1" thick coreboard on columns		480	.033		.78	1.56		2.34	3.42
5100	For foil-backed board, add					.15			.15	.17
5200	For work over 8' high, add	2 Carp	3060	.005			.25		.25	.40
5270	For textured spray, add	2 Lath	1600	.010		.04	.43		.47	.72
5300	For distribution cost over 3 stories high, add per story	2 Carp	6100	.003			.12		.12	.20
5350	For finishing inner corners, add		950	.017	L.F.	.10	.79		.89	1.41
5355	For finishing outer corners, add		1250	.013		.22	.60		.82	1.24
5500	For acoustical sealant, add per bead	1 Carp	500	.016		.04	.75		.79	1.28
5550	Sealant, 1 quart tube				Ea.	7.05			7.05	7.80

09 29 Gypsum Board

09 29 10 – Gypsum Board Panels

09 29 10.30 Gypsum Board		Crew	Daily Output	Labor-Hours	Unit	2015 Bare Costs Material	Labor	Equipment	Total	Total Incl O&P
6000	Gypsum sound dampening panels									
6010	1/2" thick on walls, multi-layer, light weight, no finish included	2 Carp	1500	.011	S.F.	1.87	.50		2.37	2.88
6015	Taped and finished (level 4 finish)		725	.022		1.92	1.04		2.96	3.81
6020	With compound skim coat (level 5 finish)		580	.028		1.97	1.30		3.27	4.29
6025	5/8" thick on walls, for wood studs, no finish included		1500	.011		2.17	.50		2.67	3.21
6030	Taped and finished (level 4 finish)		725	.022		2.22	1.04		3.26	4.14
6035	With compound skim coat (level 5 finish)		580	.028		2.27	1.30		3.57	4.62
6040	For metal stud, no finish included		1500	.011		2.06	.50		2.56	3.09
6045	Taped and finished (level 4 finish)		725	.022		2.11	1.04		3.15	4.02
6050	With compound skim coat (level 5 finish)		580	.028		2.16	1.30		3.46	4.50
6055	Abuse resist, no finish included		1500	.011		3.75	.50		4.25	4.95
6060	Taped and finished (level 4 finish)		725	.022		3.80	1.04		4.84	5.90
6065	With compound skim coat (level 5 finish)		580	.028		3.85	1.30		5.15	6.35
6070	Shear rated, no finish included		1500	.011		4.30	.50		4.80	5.55
6075	Taped and finished (level 4 finish)		725	.022		4.35	1.04		5.39	6.50
6080	With compound skim coat (level 5 finish)		580	.028		4.40	1.30		5.70	6.95
6085	For SCIF applications, no finish included		1500	.011		4.72	.50		5.22	6
6090	Taped and finished (level 4 finish)		725	.022		4.77	1.04		5.81	6.95
6095	With compound skim coat (level 5 finish)		580	.028		4.82	1.30		6.12	7.40
6100	1-3/8" thick on walls, THX Certified, no finish included		1500	.011		8.40	.50		8.90	10.05
6105	Taped and finished (level 4 finish)		725	.022		8.45	1.04		9.49	11
6110	With compound skim coat (level 5 finish)		580	.028		8.50	1.30		9.80	11.45
6115	5/8" thick on walls, score & snap installation, no finish included		2000	.008		1.69	.38		2.07	2.48
6120	Taped and finished (level 4 finish)		965	.017		1.74	.78		2.52	3.19
6125	With compound skim coat (level 5 finish)		775	.021		1.79	.97		2.76	3.56
7020	5/8" thick on ceilings, for wood joists, no finish included		1200	.013		2.17	.63		2.80	3.42
7025	Taped and finished (level 4 finish)		510	.031		2.22	1.47		3.69	4.85
7030	With compound skim coat (level 5 finish)		410	.039		2.27	1.83		4.10	5.50
7035	For metal joists, no finish included		1200	.013		2.06	.63		2.69	3.30
7040	Taped and finished (level 4 finish)		510	.031		2.11	1.47		3.58	4.73
7045	With compound skim coat (level 5 finish)		410	.039		2.16	1.83		3.99	5.40
7050	Abuse resist, no finish included		1200	.013		3.75	.63		4.38	5.15
7055	Taped and finished (level 4 finish)		510	.031		3.80	1.47		5.27	6.60
7060	With compound skim coat (level 5 finish)		410	.039		3.85	1.83		5.68	7.25
7065	Shear rated, no finish included		1200	.013		4.30	.63		4.93	5.75
7070	Taped and finished (level 4 finish)		510	.031		4.35	1.47		5.82	7.20
7075	With compound skim coat (level 5 finish)		410	.039		4.40	1.83		6.23	7.85
7080	For SCIF applications, no finish included		1200	.013		4.72	.63		5.35	6.25
7085	Taped and finished (level 4 finish)		510	.031		4.77	1.47		6.24	7.65
7090	With compound skim coat (level 5 finish)		410	.039		4.82	1.83		6.65	8.30
8010	5/8" thick on ceilings, score & snap installation, no finish included		1600	.010		1.69	.47		2.16	2.63
8015	Taped and finished (level 4 finish)		680	.024		1.74	1.10		2.84	3.72
8020	With compound skim coat (level 5 finish)		545	.029		1.79	1.38		3.17	4.23
9000	Minimum labor/equipment charge	1 Carp	2	4	Job		188		188	310

09 29 10.50 High Abuse Gypsum Board		Crew	Daily Output	Labor-Hours	Unit	Material	Labor	Equipment	Total	Total Incl O&P
0010	**HIGH ABUSE GYPSUM BOARD**, fiber reinforced, nailed or									
0100	screwed to studs unless otherwise noted									
0110	1/2" thick, on walls, no finish included	2 Carp	1800	.009	S.F.	.70	.42		1.12	1.45
0120	Taped and finished (level 4 finish)		870	.018		.75	.86		1.61	2.24
0130	With compound skim coat (level 5 finish)		700	.023		.80	1.07		1.87	2.64
0150	On ceilings, no finish included		1620	.010		.70	.46		1.16	1.53
0160	Taped and finished (level 4 finish)		690	.023		.75	1.09		1.84	2.60

09 29 Gypsum Board

09 29 10 – Gypsum Board Panels

09 29 10.50 High Abuse Gypsum Board		Crew	Daily Output	Labor-Hours	Unit	Material	Labor	Equipment	Total	Total Incl O&P
						2015 Bare Costs				
0170	With compound skim coat (level 5 finish)	2 Carp	550	.029	S.F.	.80	1.37		2.17	3.12
0210	5/8" thick, on walls, no finish included		1800	.009		.85	.42		1.27	1.62
0220	Taped and finished (level 4 finish)		870	.018		.90	.86		1.76	2.41
0230	With compound skim coat (level 5 finish)		700	.023		.95	1.07		2.02	2.81
0250	On ceilings, no finish included		1620	.010		.85	.46		1.31	1.70
0260	Taped and finished (level 4 finish)		690	.023		.90	1.09		1.99	2.77
0270	With compound skim coat (level 5 finish)		550	.029		.95	1.37		2.32	3.29
0310	5/8" thick, on walls, very high impact, no finish included		1800	.009		.98	.42		1.40	1.76
0320	Taped and finished (level 4 finish)		870	.018		1.03	.86		1.89	2.55
0330	With compound skim coat (level 5 finish)		700	.023		1.08	1.07		2.15	2.95
0350	On ceilings, no finish included		1620	.010		.98	.46		1.44	1.84
0360	Taped and finished (level 4 finish)		690	.023		1.03	1.09		2.12	2.91
0370	With compound skim coat (level 5 finish)		550	.029		1.08	1.37		2.45	3.43
0400	High abuse, gypsum core, paper face									
0410	1/2" thick, on walls, no finish included	2 Carp	1800	.009	S.F.	.62	.42		1.04	1.36
0420	Taped and finished (level 4 finish)		870	.018		.67	.86		1.53	2.15
0430	With compound skim coat (level 5 finish)		700	.023		.72	1.07		1.79	2.55
0450	On ceilings, no finish included		1620	.010		.62	.46		1.08	1.44
0460	Taped and finished (level 4 finish)		690	.023		.67	1.09		1.76	2.51
0470	With compound skim coat (level 5 finish)		550	.029		.72	1.37		2.09	3.03
0510	5/8" thick, on walls, no finish included		1800	.009		.66	.42		1.08	1.41
0520	Taped and finished (level 4 finish)		870	.018		.71	.86		1.57	2.20
0530	With compound skim coat (level 5 finish)		700	.023		.76	1.07		1.83	2.60
0550	On ceilings, no finish included		1620	.010		.66	.46		1.12	1.49
0560	Taped and finished (level 4 finish)		690	.023		.71	1.09		1.80	2.56
0570	With compound skim coat (level 5 finish)		550	.029		.76	1.37		2.13	3.08
1000	For high ceilings, over 8' high, add		2750	.006			.27		.27	.45
1010	For distribution cost over 3 stories high, add per story		5500	.003			.14		.14	.22

09 29 15 – Gypsum Board Accessories

09 29 15.10 Accessories, Gypsum Board		Crew	Daily Output	Labor-Hours	Unit	Material	Labor	Equipment	Total	Total Incl O&P
0010	**ACCESSORIES, GYPSUM BOARD**									
0020	Casing bead, galvanized steel	1 Carp	2.90	2.759	C.L.F.	24	130		154	239
0100	Vinyl		3	2.667		22	125		147	230
0300	Corner bead, galvanized steel, 1" x 1"		4	2		14.70	94		108.70	170
0400	1-1/4" x 1-1/4"		3.50	2.286		16.15	107		123.15	194
0600	Vinyl		4	2		20	94		114	176
0900	Furring channel, galv. steel, 7/8" deep, standard		2.60	3.077		33.50	144		177.50	274
1000	Resilient		2.55	3.137		25.50	147		172.50	269
1100	J trim, galvanized steel, 1/2" wide		3	2.667		22	125		147	229
1120	5/8" wide		2.95	2.712		31	127		158	243
1140	L trim, galvanized		3	2.667		19.30	125		144.30	226
1150	U trim, galvanized		2.95	2.712		22.50	127		149.50	234
1160	Screws #6 x 1" A				M	10.05			10.05	11.05
1170	#6 x 1-5/8" A				"	15.10			15.10	16.60
1200	For stud partitions, see Section 05 41 13.30 and 09 22 16.13									
1500	Z stud, galvanized steel, 1-1/2" wide	1 Carp	2.60	3.077	C.L.F.	38.50	144		182.50	280
1600	2" wide		2.55	3.137	"	62	147		209	310
9000	Minimum labor/equipment charge		3	2.667	Job		125		125	205

09 30 13 – Ceramic Tiling

09 30 13.10 Ceramic Tile		Crew	Daily Output	Labor-Hours	Unit	Material	Labor	Equipment	Total	Total Incl O&P
						2015 Bare Costs				
0010	**CERAMIC TILE**									
0020	Backsplash, thinset, average grade tiles	1 Tilf	50	.160	S.F.	2.46	6.85		9.31	13.50
0022	Custom grade tiles		50	.160		4.93	6.85		11.78	16.20
0024	Luxury grade tiles		50	.160		9.85	6.85		16.70	21.50
0026	Economy grade tiles		50	.160		2.25	6.85		9.10	13.30
0050	Base, using 1′ x 4″ high pc. with 1″ x 1″ tiles, mud set	D-7	82	.195	L.F.	5.20	7.45		12.65	17.45
0100	Thin set	"	128	.125		4.84	4.77		9.61	12.85
0300	For 6″ high base, 1″ x 1″ tile face, add					.77			.77	.85
0400	For 2″ x 2″ tile face, add to above					.42			.42	.46
0600	Cove base, 4-1/4″ x 4-1/4″ high, mud set	D-7	91	.176		3.99	6.70		10.69	15
0700	Thin set		128	.125		3.87	4.77		8.64	11.80
0900	6″ x 4-1/4″ high, mud set		100	.160		4.54	6.10		10.64	14.65
1000	Thin set		137	.117		4.42	4.46		8.88	11.90
1200	Sanitary cove base, 6″ x 4-1/4″ high, mud set		93	.172		4.36	6.55		10.91	15.15
1300	Thin set		124	.129		4.24	4.93		9.17	12.45
1500	6″ x 6″ high, mud set		84	.190		5.35	7.25		12.60	17.40
1600	Thin set		117	.137		5.20	5.20		10.40	14
1800	Bathroom accessories, average (soap dish, tooth brush holder)		82	.195	Ea.	12	7.45		19.45	25
1900	Bathtub, 5′, rec. 4-1/4″ x 4-1/4″ tile wainscot, adhesive set 6′ high		2.90	5.517		156	211		367	505
2100	7′ high wainscot		2.50	6.400		179	244		423	580
2200	8′ high wainscot		2.20	7.273		190	278		468	650
2400	Bullnose trim, 4-1/4″ x 4-1/4″, mud set		82	.195	L.F.	3.92	7.45		11.37	16.05
2500	Thin set		128	.125		3.84	4.77		8.61	11.75
2700	2″ x 6″ bullnose trim, mud set		84	.190		4.05	7.25		11.30	15.95
2800	Thin set		124	.129		3.99	4.93		8.92	12.20
3000	Floors, natural clay, random or uniform, thin set, color group 1		183	.087	S.F.	4.15	3.34		7.49	9.80
3100	Color group 2		183	.087		5.85	3.34		9.19	11.70
3255	Floors, glazed, thin set, 6″ x 6″, color group 1		300	.053		4.45	2.04		6.49	8.10
3260	8″ x 8″ tile		300	.053		4.45	2.04		6.49	8.10
3270	12″ x 12″ tile		290	.055		6.25	2.11		8.36	10.25
3280	16″ x 16″ tile		280	.057		6.70	2.18		8.88	10.85
3281	18″ x 18″ tile		270	.059		8.65	2.26		10.91	13.05
3282	20″ x 20″ tile		260	.062		9.90	2.35		12.25	14.60
3283	24″ x 24″ tile		250	.064		11.25	2.44		13.69	16.20
3285	Border, 6″ x 12″ tile		200	.080		12.55	3.05		15.60	18.60
3290	3″ x 12″ tile		200	.080		40	3.05		43.05	49
3300	Porcelain type, 1 color, color group 2, 1″ x 1″		183	.087		5.20	3.34		8.54	10.95
3310	2″ x 2″ or 2″ x 1″, thin set		190	.084		6.10	3.22		9.32	11.80
3350	For random blend, 2 colors, add					1			1	1.10
3360	4 colors, add					1.50			1.50	1.65
3370	For color group 3, add					.65			.65	.72
3380	For abrasive non-slip tile, add					.44			.44	.48
4300	Specialty tile, 4-1/4″ x 4-1/4″ x 1/2″, decorator finish	D-7	183	.087		10.40	3.34		13.74	16.70
4500	Add for epoxy grout, 1/16″ joint, 1″ x 1″ tile		800	.020		.67	.76		1.43	1.95
4600	2″ x 2″ tile		820	.020		.62	.74		1.36	1.86
4610	Add for epoxy grout, 1/8″ joint, 8″ x 8″ x 3/8″ tile, add		900	.018		1.44	.68		2.12	2.66
4800	Pregrouted sheets, walls, 4-1/4″ x 4-1/4″, 6″ x 4-1/4″									
4810	and 8-1/2″ x 4-1/4″, 4 S.F. sheets, silicone grout	D-7	240	.067	S.F.	5.10	2.55		7.65	9.65
5100	Floors, unglazed, 2 S.F. sheets,									
5110	Urethane adhesive	D-7	180	.089	S.F.	5.10	3.39		8.49	10.95
5400	Walls, interior, thin set, 4-1/4″ x 4-1/4″ tile		190	.084		2.26	3.22		5.48	7.60
5500	6″ x 4-1/4″ tile		190	.084		2.92	3.22		6.14	8.30
5700	8-1/2″ x 4-1/4″ tile		190	.084		4.86	3.22		8.08	10.45

09 30 13 – Ceramic Tiling

09 30 13.10 Ceramic Tile		Crew	Daily Output	Labor-Hours	Unit	Material	Labor (2015 Bare Costs)	Equipment	Total	Total Incl O&P
5800	6" x 6" tile	D-7	175	.091	S.F.	3.28	3.49		6.77	9.10
5810	8" x 8" tile		170	.094		4.44	3.59		8.03	10.60
5820	12" x 12" tile		160	.100		4.35	3.82		8.17	10.85
5830	16" x 16" tile		150	.107		4.77	4.07		8.84	11.70
6000	Decorated wall tile, 4-1/4" x 4-1/4", color group 1		270	.059		3.18	2.26		5.44	7.05
6100	Color group 4		180	.089		49.50	3.39		52.89	60
6300	Exterior walls, frostproof, mud set, 4-1/4" x 4-1/4"		102	.157		7.15	6		13.15	17.30
6400	1-3/8" x 1-3/8"		93	.172		6.10	6.55		12.65	17.05
6600	Crystalline glazed, 4-1/4" x 4-1/4", mud set, plain		100	.160		4.36	6.10		10.46	14.45
6700	4-1/4" x 4-1/4", scored tile		100	.160		5.80	6.10		11.90	16.05
6900	6" x 6" plain		93	.172		6.70	6.55		13.25	17.70
7000	For epoxy grout, 1/16" joints, 4-1/4" tile, add		800	.020		.41	.76		1.17	1.66
7200	For tile set in dry mortar, add		1735	.009			.35		.35	.56
7300	For tile set in Portland cement mortar, add		290	.055		.16	2.11		2.27	3.51
9300	Ceramic tiles, recycled glass, standard colors, 2" x 2" thru 6" x 6" G		190	.084		21	3.22		24.22	28.50
9310	6" x 6" G		175	.091		21.50	3.49		24.99	29
9320	8" x 8" G		170	.094		22.50	3.59		26.09	30.50
9330	12" x 12" G		160	.100		22.50	3.82		26.32	31
9340	Earthtones, 2" x 2" to 4" x 8" G		190	.084		25	3.22		28.22	32.50
9350	6" x 6" G		175	.091		25	3.49		28.49	33
9360	8" x 8" G		170	.094		26	3.59		29.59	34
9370	12" x 12" G		160	.100		26	3.82		29.82	34.50
9380	Deep colors, 2" x 2" to 4" x 8" G		190	.084		29.50	3.22		32.72	37.50
9390	6" x 6" G		175	.091		29.50	3.49		32.99	38
9400	8" x 8" G		170	.094		31	3.59		34.59	39.50
9410	12" x 12" G		160	.100		31	3.82		34.82	40
9500	Minimum labor/equipment charge		3.25	4.923	Job		188		188	297

09 30 13.20 Ceramic Tile Repairs		Crew	Daily Output	Labor-Hours	Unit	Material	Labor	Equipment	Total	Total Incl O&P
0010	CERAMIC TILE REPAIRS									
1000	Grout removal, carbide tipped, rotary grinder	1 Clab	240	.033	L.F.		1.25		1.25	2.05
1100	Regrout tile 4-1/2 x 4-1/2, or larger, wall	1 Tilf	100	.080	S.F.	.14	3.42		3.56	5.55
1150	Floor		125	.064		.15	2.74		2.89	4.50
1200	Seal tile and grout		360	.022			.95		.95	1.50

09 30 13.45 Ceramic Tile Accessories		Crew	Daily Output	Labor-Hours	Unit	Material	Labor	Equipment	Total	Total Incl O&P
0010	CERAMIC TILE ACCESSORIES									
0100	Spacers, 1/8"				C	1.98			1.98	2.18
1310	Sealer for natural stone tile, installed	1 Tilf	650	.012	S.F.	.05	.53		.58	.89

09 30 16 – Quarry Tiling

09 30 16.10 Quarry Tile		Crew	Daily Output	Labor-Hours	Unit	Material	Labor	Equipment	Total	Total Incl O&P
0010	QUARRY TILE									
0100	Base, cove or sanitary, mud set, to 5" high, 1/2" thick	D-7	110	.145	L.F.	5.35	5.55		10.90	14.65
0300	Bullnose trim, red, mud set, 6" x 6" x 1/2" thick		120	.133		4.39	5.10		9.49	12.90
0400	4" x 4" x 1/2" thick		110	.145		4.50	5.55		10.05	13.70
0600	4" x 8" x 1/2" thick, using 8" as edge		130	.123		4.50	4.70		9.20	12.35
0700	Floors, mud set, 1,000 S.F. lots, red, 4" x 4" x 1/2" thick		120	.133	S.F.	8.05	5.10		13.15	16.90
0900	6" x 6" x 1/2" thick		140	.114		7.55	4.36		11.91	15.20
1000	4" x 8" x 1/2" thick		130	.123		5.55	4.70		10.25	13.50
1300	For waxed coating, add					.75			.75	.83
1500	For non-standard colors, add					.46			.46	.51
1600	For abrasive surface, add					.52			.52	.57
1800	Brown tile, imported, 6" x 6" x 3/4"	D-7	120	.133		7.95	5.10		13.05	16.80
1900	8" x 8" x 1"		110	.145		8.70	5.55		14.25	18.30

09 30 Tiling

09 30 16 – Quarry Tiling

09 30 16.10 Quarry Tile		Crew	Daily Output	Labor-Hours	Unit	2015 Bare Costs Material	Labor	Equipment	Total	Total Incl O&P
2100	For thin set mortar application, deduct	D-7	700	.023	S.F.		.87		.87	1.38
2200	For epoxy grout & mortar, 6" x 6" x 1/2", add		350	.046		2.04	1.75		3.79	5
2700	Stair tread, 6" x 6" x 3/4", plain		50	.320		7.20	12.20		19.40	27
2800	Abrasive		47	.340		6.05	13		19.05	27
3000	Wainscot, 6" x 6" x 1/2", thin set, red		105	.152		4.68	5.80		10.48	14.35
3100	Non-standard colors		105	.152		5.20	5.80		11	14.90
3300	Window sill, 6" wide, 3/4" thick		90	.178	L.F.	9.20	6.80		16	21
3400	Corners		80	.200	Ea.	6.65	7.65		14.30	19.35
9000	Minimum labor/equipment charge		3.25	4.923	Job		188		188	297

09 30 23 – Glass Mosaic Tiling

09 30 23.10 Glass Mosaics		Crew	Daily Output	Labor-Hours	Unit	Material	Labor	Equipment	Total	Total Incl O&P
0010	**GLASS MOSAICS** 3/4" tile on 12" sheets, standard grout									
0300	Color group 1 & 2	D-7	73	.219	S.F.	17.20	8.35		25.55	32
0350	Color group 3		73	.219		20.50	8.35		28.85	35.50
0400	Color group 4		73	.219		26.50	8.35		34.85	42.50
0450	Color group 5		73	.219		29	8.35		37.35	45
0500	Color group 6		73	.219		40	8.35		48.35	57
0600	Color group 7		73	.219		40.50	8.35		48.85	57.50
0700	Color group 8, golds, silvers & specialties		64	.250		41	9.55		50.55	60.50
1020	1" tile on 12" sheets, opalescent finish		73	.219		16.85	8.35		25.20	32
1040	1" x 2" tile on 12" sheet, blend		73	.219		18.15	8.35		26.50	33
1060	2" tile on 12" sheet, blend		73	.219		16.50	8.35		24.85	31.50
1080	5/8" x random tile, linear, on 12" sheet, blend		73	.219		26	8.35		34.35	41.50
1600	Dots on 12" sheet		73	.219		26	8.35		34.35	41.50
1700	For glass mosaic tiles set in dry mortar, add		290	.055		.45	2.11		2.56	3.83
1720	For glass mosaic tile set in Portland cement mortar, add		290	.055		.01	2.11		2.12	3.34
1730	For polyblend sanded tile grout		96.15	.166	Lb.	2.19	6.35		8.54	12.45

09 30 29 – Metal Tiling

09 30 29.10 Metal Tile		Crew	Daily Output	Labor-Hours	Unit	Material	Labor	Equipment	Total	Total Incl O&P
0010	**METAL TILE** 4' x 4' sheet, 24 ga., tile pattern, nailed									
0200	Stainless steel	2 Carp	512	.031	S.F.	28	1.47		29.47	33.50
0400	Aluminized steel	"	512	.031	"	15.10	1.47		16.57	19
9000	Minimum labor/equipment charge	1 Carp	4	2	Job		94		94	154

09 34 Waterproofing-Membrane Tiling

09 34 13 – Waterproofing-Membrane Ceramic Tiling

09 34 13.10 Ceramic Tile Waterproofing Membrane		Crew	Daily Output	Labor-Hours	Unit	Material	Labor	Equipment	Total	Total Incl O&P
0010	**CERAMIC TILE WATERPROOFING MEMBRANE**									
0020	On floors, including thinset									
0030	Fleece laminated polyethylene grid, 1/8" thick	D-7	250	.064	S.F.	2.26	2.44		4.70	6.35
0040	5/16" thick	"	250	.064	"	2.58	2.44		5.02	6.70
0050	On walls, including thinset									
0060	Fleece laminated polyethylene sheet, 8 mil thick	D-7	480	.033	S.F.	2.26	1.27		3.53	4.50
0070	Accessories, including thinset									
0080	Joint and corner sheet, 4 mils thick, 5" wide	1 Tilf	240	.033	L.F.	1.33	1.43		2.76	3.72
0090	7-1/4" wide		180	.044		1.69	1.90		3.59	4.86
0100	10" wide		120	.067		2.06	2.85		4.91	6.80
0110	Pre-formed corners, inside		32	.250	Ea.	6.90	10.70		17.60	24.50
0120	Outside		32	.250		7.65	10.70		18.35	25.50
0130	2" flanged floor drain with 6" stainless steel grate		16	.500		370	21.50		391.50	445

09 34 Waterproofing-Membrane Tiling

09 34 13 – Waterproofing-Membrane Ceramic Tiling

09 34 13.10 Ceramic Tile Waterproofing Membrane		Crew	Daily Output	Labor-Hours	Unit	2015 Bare Costs Material	Labor	Equipment	Total	Total Incl O&P
0140	EPS, sloped shower floor	1 Tilf	480	.017	S.F.	4.95	.71		5.66	6.60
0150	Curb		32	.250	L.F.	14	10.70		24.70	32.50

09 51 Acoustical Ceilings

09 51 23 – Acoustical Tile Ceilings

09 51 23.10 Suspended Acoustic Ceiling Tiles

		Crew	Daily Output	Labor-Hours	Unit	Material	Labor	Equipment	Total	Total Incl O&P
0010	**SUSPENDED ACOUSTIC CEILING TILES**, not including									
0100	suspension system									
0300	Fiberglass boards, film faced, 2′ x 2′ or 2′ x 4′, 5/8″ thick	1 Carp	625	.013	S.F.	1.24	.60		1.84	2.35
0400	3/4″ thick		600	.013		2.63	.63		3.26	3.92
0500	3″ thick, thermal, R11		450	.018		2.42	.83		3.25	4.03
0600	Glass cloth faced fiberglass, 3/4″ thick		500	.016		2.65	.75		3.40	4.15
0700	1″ thick		485	.016		3.13	.77		3.90	4.71
0820	1-1/2″ thick, nubby face		475	.017		2.53	.79		3.32	4.08
1110	Mineral fiber tile, lay-in, 2′ x 2′ or 2′ x 4′, 5/8″ thick, fine texture		625	.013		.91	.60		1.51	1.99
1115	Rough textured		625	.013		.85	.60		1.45	1.93
1125	3/4″ thick, fine textured		600	.013		1.94	.63		2.57	3.16
1130	Rough textured		600	.013		1.56	.63		2.19	2.75
1135	Fissured		600	.013		1.96	.63		2.59	3.19
1150	Tegular, 5/8″ thick, fine textured		470	.017		1.03	.80		1.83	2.44
1155	Rough textured		470	.017		1.14	.80		1.94	2.56
1165	3/4″ thick, fine textured		450	.018		2.14	.83		2.97	3.72
1170	Rough textured		450	.018		1.43	.83		2.26	2.94
1175	Fissured		450	.018		2.16	.83		2.99	3.75
1185	For plastic film face, add					.75			.75	.83
1190	For fire rating, add					.44			.44	.48
1300	Metal panel, lay-in, 2′ x 2′, sq. edge	1 Carp	500	.016		9.55	.75		10.30	11.75
1350	Tegular edge		500	.016		13.20	.75		13.95	15.75
1400	2′ x 4′, sq. edge		500	.016		12.90	.75		13.65	15.45
1450	Tegular edge		500	.016		13.20	.75		13.95	15.75
1500	Perforated alum. clip-in, 2′ x 2′		500	.016		13.45	.75		14.20	16
1550	2′ x 4′		500	.016		10.85	.75		11.60	13.20
1600	Solid alum. planks, 3-1/4″x12′, open reveal		500	.016		2.35	.75		3.10	3.82
1650	Closed reveal		500	.016		3	.75		3.75	4.53
1700	7-1/4″x12′, open reveal		500	.016		4	.75		4.75	5.65
1750	Closed reveal		500	.016		5.10	.75		5.85	6.85
1775	Metal, open cell, 2′x2′, 6″ cell		500	.016		8	.75		8.75	10.05
1800	8″ cell		500	.016		8.85	.75		9.60	11
1825	2′x4′, 6″ cell		500	.016		5.10	.75		5.85	6.85
1850	8″ cell		500	.016		5.10	.75		5.85	6.85
1870	Translucent lay-in panels, 2′x2′		500	.016		23	.75		23.75	26
1890	2′x6′		500	.016		17.20	.75		17.95	20
3720	Mineral fiber, 24″ x 24″ or 48″, reveal edge, painted, 5/8″ thick		600	.013		1.15	.63		1.78	2.30
3740	3/4″ thick		575	.014		1.52	.65		2.17	2.74
5020	66 – 78% recycled content, 3/4″ thick G		600	.013		1.93	.63		2.56	3.15
5040	Mylar, 42% recycled content, 3/4″ thick G		600	.013		4.54	.63		5.17	6
6000	Remove and replace ceiling tiles, min fiber, 2x2 or 2x4, 5/8″thk.		335	.024		.91	1.12		2.03	2.84
9000	Minimum labor/equipment charge		4	2	Job		94		94	154

09 51 Acoustical Ceilings

09 51 23 – Acoustical Tile Ceilings

	09 51 23.30 Suspended Ceilings, Complete	Crew	Daily Output	Labor-Hours	Unit	2015 Bare Costs Material	Labor	Equipment	Total	Total Incl O&P
0010	**SUSPENDED CEILINGS, COMPLETE**, including standard									
0100	suspension system but not incl. 1-1/2" carrier channels									
0600	Fiberglass ceiling board, 2' x 4' x 5/8", plain faced	1 Carp	500	.016	S.F.	1.97	.75		2.72	3.40
0700	Offices, 2' x 4' x 3/4"		380	.021		3.36	.99		4.35	5.30
0800	Mineral fiber, on 15/16" T bar susp. 2' x 2' x 3/4" lay-in board		345	.023		2.89	1.09		3.98	4.96
0810	2' x 4' x 5/8" tile		380	.021		2.31	.99		3.30	4.17
0820	Tegular, 2' x 2' x 5/8" tile on 9/16" grid		250	.032		2.40	1.50		3.90	5.10
0830	2' x 4' x 3/4" tile		275	.029		2.60	1.37		3.97	5.10
0900	Luminous panels, prismatic, acrylic		255	.031		3.37	1.47		4.84	6.10
1200	Metal pan with acoustic pad, steel		75	.107		4.51	5		9.51	13.15
1300	Painted aluminum		75	.107		3.07	5		8.07	11.60
1500	Aluminum, degreased finish		75	.107		5.20	5		10.20	13.90
1600	Stainless steel		75	.107		9.75	5		14.75	18.90
1800	Tile, Z bar suspension, 5/8" mineral fiber tile		150	.053		2.32	2.50		4.82	6.65
1900	3/4" mineral fiber tile		150	.053		2.48	2.50		4.98	6.85
2402	For strip lighting, see Section 26 51 13.50									
2500	For rooms under 500 S.F., add				S.F.		25%			
9000	Minimum labor/equipment charge	1 Carp	2	4	Job		188		188	310

09 51 53 – Direct-Applied Acoustical Ceilings

	09 51 53.10 Ceiling Tile	Crew	Daily Output	Labor-Hours	Unit	Material	Labor	Equipment	Total	Total Incl O&P
0010	**CEILING TILE**, stapled or cemented									
0100	12" x 12" or 12" x 24", not including furring									
0600	Mineral fiber, vinyl coated, 5/8" thick	1 Carp	300	.027	S.F.	2.18	1.25		3.43	4.45
0700	3/4" thick		300	.027		2.40	1.25		3.65	4.69
0900	Fire rated, 3/4" thick, plain faced		300	.027		1.27	1.25		2.52	3.45
1000	Plastic coated face		300	.027		1.84	1.25		3.09	4.07
1200	Aluminum faced, 5/8" thick, plain		300	.027		1.66	1.25		2.91	3.88
3700	Wall application of above, add		1000	.008			.38		.38	.62
3900	For ceiling primer, add					.13			.13	.14
4000	For ceiling cement, add					.39			.39	.43
9000	Minimum labor/equipment charge	1 Carp	4	2	Job		94		94	154

09 53 Acoustical Ceiling Suspension Assemblies

09 53 23 – Metal Acoustical Ceiling Suspension Assemblies

	09 53 23.30 Ceiling Suspension Systems	Crew	Daily Output	Labor-Hours	Unit	Material	Labor	Equipment	Total	Total Incl O&P
0010	**CEILING SUSPENSION SYSTEMS** for boards and tile									
0050	Class A suspension system, 15/16" T bar, 2' x 4' grid	1 Carp	800	.010	S.F.	.73	.47		1.20	1.58
0300	2' x 2' grid		650	.012		.95	.58		1.53	1.99
0310	25% recycled steel, 2' x 4' grid G		800	.010		.77	.47		1.24	1.62
0320	2' x 2' grid G		650	.012		.96	.58		1.54	2.01
0350	For 9/16" grid, add					.16			.16	.18
0360	For fire rated grid, add					.09			.09	.10
0370	For colored grid, add					.21			.21	.23
0400	Concealed Z bar suspension system, 12" module	1 Carp	520	.015		.84	.72		1.56	2.10
0600	1-1/2" carrier channels, 4' O.C., add	"	470	.017		.11	.80		.91	1.43
0700	Carrier channels for ceilings with									
0900	recessed lighting fixtures, add	1 Carp	460	.017	S.F.	.20	.82		1.02	1.56
1040	Hanging wire, 12 ga., 4' long		65	.123	C.S.F.	.37	5.80		6.17	9.85
1080	8' long		65	.123	"	.74	5.80		6.54	10.25
3000	Seismic ceiling bracing, IBC Site Class D, Occupancy Category II									

09 53 Acoustical Ceiling Suspension Assemblies

09 53 23 – Metal Acoustical Ceiling Suspension Assemblies

09 53 23.30 Ceiling Suspension Systems		Crew	Daily Output	Labor-Hours	Unit	2015 Bare Costs Material	Labor	Equipment	Total	Total Incl O&P
3050	For ceilings less than 2500 S.F.									
3060	Seismic clips at attached walls	1 Carp	180	.044	Ea.	1.12	2.09		3.21	4.65
3100	For ceilings greater than 2500 S.F., add									
3120	Seismic clips, joints at cross tees	1 Carp	120	.067	Ea.	3.29	3.13		6.42	8.75
3140	At cross tees and mains, mains field cut	"	60	.133	"	3.29	6.25		9.54	13.85
3200	Compression posts, telescopic, attached to structure above									
3210	To 30" high	1 Carp	26	.308	Ea.	39	14.45		53.45	66.50
3220	30" to 48" high		25.50	.314		43.50	14.75		58.25	72
3230	48" to 84" high		25	.320		52.50	15		67.50	82
3240	84" to 102" high		24.50	.327		60	15.35		75.35	91
3250	102" to 120" high		24	.333		85.50	15.65		101.15	120
3260	120" to 144" high		24	.333		95	15.65		110.65	131
3300	Stabilizer bars									
3310	12" long	1 Carp	240	.033	Ea.	.97	1.56		2.53	3.63
3320	24" long		235	.034		.92	1.60		2.52	3.63
3330	36" long		230	.035		.89	1.63		2.52	3.66
3340	48" long		220	.036		.73	1.71		2.44	3.61
3400	Wire support for light fixtures, per L.F. height to structure above									
3410	Less than 10 lb.	1 Carp	400	.020	L.F.	.28	.94		1.22	1.85
3420	10 lb. to 56 lb.	"	240	.033	"	.56	1.56		2.12	3.17
3500	Retrofit existing suspended ceiling grid to current code									
3510	Less than 2500 S.F. (using clips @ perimeter)	1 Carp	1455	.006	S.F.	.07	.26		.33	.49
3520	Greater than 2500 S.F. (using compression posts and clips)		550	.015		.44	.68		1.12	1.61
4000	Remove and replace ceiling grid, class A, 15/16" T bar, 2' x 2' grid		435	.018		.95	.86		1.81	2.46
4010	2' x 4' grid		535	.015		.73	.70		1.43	1.96

09 54 Specialty Ceilings

09 54 26 – Suspended Wood Ceilings

09 54 26.10 Wood Ceilings		Crew	Daily Output	Labor-Hours	Unit	Material	Labor	Equipment	Total	Total Incl O&P
0010	**WOOD CEILINGS**									
1000	4" - 6" wood slats on heavy duty 15/16" T-bar grid	2 Carp	250	.064	S.F.	24	3		27	31.50

09 54 33 – Decorative Panel Ceilings

09 54 33.20 Metal Panel Ceilings		Crew	Daily Output	Labor-Hours	Unit	Material	Labor	Equipment	Total	Total Incl O&P
0010	**METAL PANEL CEILINGS**									
0020	Lay-in or screwed to furring, not including grid									
0100	Tin ceilings, 2' x 2' or 2' x 4', bare steel finish	2 Carp	300	.053	S.F.	2.46	2.50		4.96	6.80
0120	Painted white finish		300	.053	"	3.71	2.50		6.21	8.20
0140	Copper, chrome or brass finish		300	.053	L.F.	6.55	2.50		9.05	11.30
0200	Cornice molding, 2-1/2" to 3-1/2" wide, 4' long, bare steel finish		200	.080	S.F.	2.21	3.76		5.97	8.60
0220	Painted white finish		200	.080		2.75	3.76		6.51	9.15
0240	Copper, chrome or brass finish		200	.080		3.91	3.76		7.67	10.45
0320	5" to 6-1/2" wide, 4' long, bare steel finish		150	.107		3.19	5		8.19	11.70
0340	Painted white finish		150	.107		4.05	5		9.05	12.65
0360	Copper, chrome or brass finish		150	.107		6.45	5		11.45	15.30
0420	Flat molding, 3-1/2" to 5" wide, 4' long, bare steel finish		250	.064		3.71	3		6.71	9
0440	Painted white finish		250	.064		3.96	3		6.96	9.30
0460	Copper, chrome or brass finish		250	.064		7.50	3		10.50	13.15

09 61 Flooring Treatment

09 61 19 – Concrete Floor Staining

09 61 19.40 Floors, Interior		Crew	Daily Output	Labor-Hours	Unit	Material (2015 Bare Costs)	Labor (2015 Bare Costs)	Equipment (2015 Bare Costs)	Total (2015 Bare Costs)	Total Incl O&P
0010	**FLOORS, INTERIOR**									
0300	Acid stain and sealer									
0310	Stain, one coat	1 Pord	650	.012	S.F.	.12	.50		.62	.93
0320	Two coats		570	.014		.23	.57		.80	1.17
0330	Acrylic sealer, one coat		2600	.003		.23	.12		.35	.45
0340	Two coats		1400	.006		.46	.23		.69	.88

09 62 Specialty Flooring

09 62 19 – Laminate Flooring

09 62 19.10 Floating Floor		Crew	Daily Output	Labor-Hours	Unit	Material	Labor	Equipment	Total	Total Incl O&P
0010	**FLOATING FLOOR**									
8300	Floating floor, laminate, wood pattern strip, complete	1 Clab	133	.060	S.F.	4.35	2.26		6.61	8.50
8310	Components, T & G wood composite strips					3.91			3.91	4.30
8320	Film					.14			.14	.15
8330	Foam					.25			.25	.28
8340	Adhesive					.65			.65	.72
8350	Installation kit					.17			.17	.19
8360	Trim, 2″ wide x 3′ long				L.F.	4.30			4.30	4.73
8370	Reducer moulding				″	5.70			5.70	6.25

09 62 23 – Bamboo Flooring

09 62 23.10 Flooring, Bamboo		Crew	Daily Output	Labor-Hours	Unit	Material	Labor	Equipment	Total	Total Incl O&P
0010	**FLOORING, BAMBOO**									
8600	Flooring, wood, bamboo strips, unfinished, 5/8″ x 4″ x 3′ G	1 Carp	255	.031	S.F.	4.60	1.47		6.07	7.45
8610	5/8″ x 4″ x 4′ G		275	.029		4.77	1.37		6.14	7.50
8620	5/8″ x 4″ x 6′ G		295	.027		5.25	1.27		6.52	7.85
8630	Finished, 5/8″ x 4″ x 3′ G		255	.031		5.05	1.47		6.52	7.95
8640	5/8″ x 4″ x 4′ G		275	.029		5.30	1.37		6.67	8.10
8650	5/8″ x 4″ x 6′ G		295	.027		4.61	1.27		5.88	7.15
8660	Stair treads, unfinished, 1-1/16″ x 11-1/2″ x 4′ G		18	.444	Ea.	44	21		65	82.50
8670	Finished, 1-1/16″ x 11-1/2″ x 4′ G		18	.444		78	21		99	120
8680	Stair risers, unfinished, 5/8″ x 7-1/2″ x 4′ G		18	.444		16.30	21		37.30	52
8690	Finished, 5/8″ x 7-1/2″ x 4′ G		18	.444		31	21		52	68
8700	Stair nosing, unfinished, 6′ long G		16	.500		36	23.50		59.50	78
8710	Finished, 6′ long G		16	.500		42.50	23.50		66	85.50

09 63 Masonry Flooring

09 63 13 – Brick Flooring

09 63 13.10 Miscellaneous Brick Flooring		Crew	Daily Output	Labor-Hours	Unit	Material	Labor	Equipment	Total	Total Incl O&P
0010	**MISCELLANEOUS BRICK FLOORING**									
0020	Acid-proof shales, red, 8″ x 3-3/4″ x 1-1/4″ thick	D-7	.43	37.209	M	695	1,425		2,120	3,025
0050	2-1/4″ thick	D-1	.40	40		965	1,675		2,640	3,800
0200	Acid-proof clay brick, 8″ x 3-3/4″ x 2-1/4″ thick G		.40	40		935	1,675		2,610	3,775
0250	9″ x 4-1/2″ x 3″ G		95	.168	S.F.	4.14	7.10		11.24	16.10
0260	Cast ceramic, pressed, 4″ x 8″ x 1/2″, unglazed	D-7	100	.160		6.35	6.10		12.45	16.60
0270	Glazed		100	.160		8.45	6.10		14.55	18.95
0280	Hand molded flooring, 4″ x 8″ x 3/4″, unglazed		95	.168		8.35	6.45		14.80	19.35
0290	Glazed		95	.168		10.50	6.45		16.95	21.50
0300	8″ hexagonal, 3/4″ thick, unglazed		85	.188		9.20	7.20		16.40	21.50
0310	Glazed		85	.188		16.60	7.20		23.80	29.50

09 63 Masonry Flooring

09 63 13 – Brick Flooring

09 63 13.10 Miscellaneous Brick Flooring		Crew	Daily Output	Labor-Hours	Unit	Material	Labor	Equipment	Total	Total Incl O&P
						2015 Bare Costs				
0400	Heavy duty industrial, cement mortar bed, 2" thick, not incl. brick	D-1	80	.200	S.F.	1.06	8.40		9.46	14.85
0450	Acid-proof joints, 1/4" wide	"	65	.246		1.46	10.35		11.81	18.45
0500	Pavers, 8" x 4", 1" to 1-1/4" thick, red	D-7	95	.168		3.69	6.45		10.14	14.20
0510	Ironspot	"	95	.168		5.20	6.45		11.65	15.90
0540	1-3/8" to 1-3/4" thick, red	D-1	95	.168		3.56	7.10		10.66	15.45
0560	Ironspot		95	.168		5.15	7.10		12.25	17.25
0580	2-1/4" thick, red		90	.178		3.62	7.50		11.12	16.20
0590	Ironspot		90	.178		5.60	7.50		13.10	18.40
0700	Paver, adobe brick, 6" x 12", 1/2" joint G		42	.381		1.30	16.05		17.35	27.50
0710	Mexican red, 12" x 12" G	1 Tilf	48	.167		1.66	7.15		8.81	13.10
0720	Saltillo, 12" x 12" G	"	48	.167		1.40	7.15		8.55	12.80
0800	For sidewalks and patios with pavers, see Section 32 14 16.10									
0870	For epoxy joints, add	D-1	600	.027	S.F.	2.81	1.12		3.93	4.92
0880	For Furan underlayment, add	"	600	.027		2.33	1.12		3.45	4.39
0890	For waxed surface, steam cleaned, add	A-1H	1000	.008		.20	.30	.08	.58	.79
9000	Minimum labor/equipment charge	1 Bric	2	4	Job		185		185	300

09 63 40 – Stone Flooring

09 63 40.10 Marble		Crew	Daily Output	Labor-Hours	Unit	Material	Labor	Equipment	Total	Total Incl O&P
0010	**MARBLE**									
0020	Thin gauge tile, 12" x 6", 3/8", white Carara	D-7	60	.267	S.F.	14.40	10.20		24.60	32
0100	Travertine		60	.267		11.60	10.20		21.80	29
0200	12" x 12" x 3/8", thin set, floors		60	.267		10.15	10.20		20.35	27.50
0300	On walls		52	.308		9.85	11.75		21.60	29.50
1000	Marble threshold, 4" wide x 36" long x 5/8" thick, white		60	.267	Ea.	10.15	10.20		20.35	27.50
9000	Minimum labor/equipment charge		3	5.333	Job		204		204	320
09 63 40.20 Slate Tile										
0010	**SLATE TILE**									
0020	Vermont, 6" x 6" x 1/4" thick, thin set	D-7	180	.089	S.F.	7.50	3.39		10.89	13.60
0200	See also Section 32 14 40.10									
9000	Minimum labor/equipment charge	D-7	3	5.333	Job		204		204	320
09 63 40.95 Concrete Floors										
0010	**CONCRETE FLOORS** And toppings, see Section 03 35 29.30									

09 64 Wood Flooring

09 64 16 – Wood Block Flooring

09 64 16.10 End Grain Block Flooring		Crew	Daily Output	Labor-Hours	Unit	Material	Labor	Equipment	Total	Total Incl O&P
0010	**END GRAIN BLOCK FLOORING**									
0020	End grain flooring, coated, 2" thick	1 Carp	295	.027	S.F.	3.57	1.27		4.84	6
0400	Natural finish, 1" thick, fir		125	.064		3.69	3		6.69	9
0600	1-1/2" thick, pine		125	.064		3.62	3		6.62	8.90
0700	2" thick, pine		125	.064		4.44	3		7.44	9.80
9000	Minimum labor/equipment charge		2	4	Job		188		188	310

09 64 19 – Wood Composition Flooring

09 64 19.10 Wood Composition		Crew	Daily Output	Labor-Hours	Unit	Material	Labor	Equipment	Total	Total Incl O&P
0010	**WOOD COMPOSITION** Gym floors									
0100	2-1/4" x 6-7/8" x 3/8", on 2" grout setting bed	D-7	150	.107	S.F.	6.15	4.07		10.22	13.20
0200	Thin set, on concrete	"	250	.064		5.60	2.44		8.04	10
0300	Sanding and finishing, add	1 Carp	200	.040		.84	1.88		2.72	4

09 64 Wood Flooring

09 64 23 – Wood Parquet Flooring

09 64 23.10 Wood Parquet		Crew	Daily Output	Labor-Hours	Unit	Material	Labor	Equipment	Total	Total Incl O&P
						2015 Bare Costs				
0010	**WOOD PARQUET** flooring									
5200	Parquetry, 5/16" thk, no finish, oak, plain pattern	1 Carp	160	.050	S.F.	5.25	2.35		7.60	9.60
5300	Intricate pattern		100	.080		9.60	3.76		13.36	16.70
5500	Teak, plain pattern		160	.050		5.85	2.35		8.20	10.30
5600	Intricate pattern		100	.080		10.05	3.76		13.81	17.20
5650	13/16" thick, select grade oak, plain pattern		160	.050		9.75	2.35		12.10	14.55
5700	Intricate pattern		100	.080		16.40	3.76		20.16	24
5800	Custom parquetry, including finish, plain pattern		100	.080		16.80	3.76		20.56	24.50
5900	Intricate pattern		50	.160		24	7.50		31.50	38.50
6700	Parquetry, prefinished white oak, 5/16" thick, plain pattern		160	.050		7.90	2.35		10.25	12.55
6800	Intricate pattern		100	.080		8.45	3.76		12.21	15.40
7000	Walnut or teak, parquetry, plain pattern		160	.050		8	2.35		10.35	12.65
7100	Intricate pattern		100	.080		11.50	3.76		15.26	18.80
7200	Acrylic wood parquet blocks, 12" x 12" x 5/16",									
7210	Irradiated, set in epoxy	1 Carp	160	.050	S.F.	10	2.35		12.35	14.85

09 64 29 – Wood Strip and Plank Flooring

09 64 29.10 Wood		Crew	Daily Output	Labor-Hours	Unit	Material	Labor	Equipment	Total	Total Incl O&P
0010	**WOOD**									
0020	Fir, vertical grain, 1" x 4", not incl. finish, grade B & better	1 Carp	255	.031	S.F.	2.79	1.47		4.26	5.50
0100	C grade & better		255	.031		2.63	1.47		4.10	5.30
4000	Maple, strip, 25/32" x 2-1/4", not incl. finish, select		170	.047		4.95	2.21		7.16	9.05
4100	#2 & better		170	.047		4.29	2.21		6.50	8.35
4300	33/32" x 3-1/4", not incl. finish, #1 grade		170	.047		4.56	2.21		6.77	8.60
4400	#2 & better		170	.047		4.06	2.21		6.27	8.10
4600	Oak, white or red, 25/32" x 2-1/4", not incl. finish									
4700	#1 common	1 Carp	170	.047	S.F.	3.19	2.21		5.40	7.15
4900	Select quartered, 2-1/4" wide		170	.047		3.89	2.21		6.10	7.90
5000	Clear		170	.047		4.01	2.21		6.22	8.05
6100	Prefinished, white oak, prime grade, 2-1/4" wide		170	.047		4.69	2.21		6.90	8.75
6200	3-1/4" wide		185	.043		5.10	2.03		7.13	8.95
6400	Ranch plank		145	.055		7.15	2.59		9.74	12.10
6500	Hardwood blocks, 9" x 9", 25/32" thick		160	.050		6	2.35		8.35	10.45
7400	Yellow pine, 3/4" x 3-1/8", T & G, C & better, not incl. finish		200	.040		1.49	1.88		3.37	4.72
7550	Refinish old floors, see Section 01 93 13.09									
7800	Sanding and finishing, 2 coats polyurethane	1 Clab	295	.027	S.F.	.90	1.02		1.92	2.66
7900	Subfloor and underlayment, see Section 06 16									
8015	Transition molding, 2 1/4" wide, 5' long	1 Carp	19.20	.417	Ea.	10.85	19.55		30.40	44
9000	Minimum labor/equipment charge	"	2	4	Job		188		188	310

09 64 66 – Wood Athletic Flooring

09 64 66.10 Gymnasium Flooring		Crew	Daily Output	Labor-Hours	Unit	Material	Labor	Equipment	Total	Total Incl O&P
0010	**GYMNASIUM FLOORING**									
0600	Gym floor, in mastic, over 2 ply felt, #2 & better									
0700	25/32" thick maple	1 Carp	100	.080	S.F.	3.99	3.76		7.75	10.55
0900	33/32" thick maple		98	.082		4.99	3.83		8.82	11.80
1000	For 1/2" corkboard underlayment, add		750	.011		.97	.50		1.47	1.89
1300	For #1 grade maple, add					.51			.51	.56
1600	Maple flooring, over sleepers, #2 & better									
1700	25/32" thick	1 Carp	85	.094	S.F.	4.70	4.42		9.12	12.40
1900	33/32" thick	"	83	.096		5.45	4.53		9.98	13.40
2000	For #1 grade, add					.55			.55	.61
2200	For 3/4" subfloor, add	1 Carp	350	.023		1.22	1.07		2.29	3.10

09 64 Wood Flooring

09 64 66 – Wood Athletic Flooring

09 64 66.10 Gymnasium Flooring		Crew	Daily Output	Labor-Hours	Unit	2015 Bare Costs Material	Labor	Equipment	Total	Total Incl O&P
2300	With two 1/2" subfloors, 25/32" thick	1 Carp	69	.116	S.F.	5.90	5.45		11.35	15.35
2500	Maple, incl. finish, #2 & btr., 25/32" thick, on rubber									
2600	Sleepers, with two 1/2" subfloors	1 Carp	76	.105	S.F.	6.30	4.94		11.24	15.05
2800	With steel spline, double connection to channels	"	73	.110		6.75	5.15		11.90	15.85
2900	For 33/32" maple, add					.72			.72	.79
3100	For #1 grade maple, add					.55			.55	.61
3500	For termite proofing all of the above, add					.29			.29	.32
3700	Portable hardwood, prefinished panels	1 Carp	83	.096		8.40	4.53		12.93	16.65
3720	Insulated with polystyrene, 1" thick, add		165	.048		.72	2.28		3	4.52
3750	Running tracks, Sitka spruce surface, 25/32" x 2-1/4"		62	.129		15.40	6.05		21.45	27
3770	3/4" plywood surface, finished	↓	100	.080	↓	3.65	3.76		7.41	10.15

09 65 Resilient Flooring

09 65 10 – Resilient Tile Underlayment

09 65 10.10 Latex Underlayment

		Crew	Daily Output	Labor-Hours	Unit	Material	Labor	Equipment	Total	Total Incl O&P
0010	LATEX UNDERLAYMENT									
3600	Latex underlayment, 1/8" thk., cementitious for resilient flooring	1 Tilf	160	.050	S.F.	1.19	2.14		3.33	4.69
4000	Liquid, fortified				Gal.	33			33	36

09 65 13 – Resilient Base and Accessories

09 65 13.13 Resilient Base

		Crew	Daily Output	Labor-Hours	Unit	Material	Labor	Equipment	Total	Total Incl O&P
0010	RESILIENT BASE									
0690	1/8" vinyl base, 2 1/2" H, straight or cove, standard colors	1 Tilf	315	.025	L.F.	.67	1.09		1.76	2.46
0700	4" high		315	.025		1.32	1.09		2.41	3.17
0710	6" high		315	.025	↓	1.40	1.09		2.49	3.26
0720	Corners, 2 1/2" high		315	.025	Ea.	2.02	1.09		3.11	3.94
0730	4" high		315	.025		2.14	1.09		3.23	4.07
0740	6" high		315	.025	↓	2.45	1.09		3.54	4.42
0800	1/8" rubber base, 2 1/2" H, straight or cove, standard colors		315	.025	L.F.	1.13	1.09		2.22	2.96
1100	4" high		315	.025		1.02	1.09		2.11	2.84
1110	6" high		315	.025	↓	1.77	1.09		2.86	3.67
1150	Corners, 2 1/2" high		315	.025	Ea.	2	1.09		3.09	3.92
1153	4" high		315	.025		2.05	1.09		3.14	3.98
1155	6" high	↓	315	.025	↓	2.51	1.09		3.60	4.48
1450	For premium color/finish add					50%				
1500	Millwork profile	1 Tilf	315	.025	L.F.	5.85	1.09		6.94	8.15

09 65 13.23 Resilient Stair Treads and Risers

		Crew	Daily Output	Labor-Hours	Unit	Material	Labor	Equipment	Total	Total Incl O&P
0010	RESILIENT STAIR TREADS AND RISERS									
0300	Rubber, molded tread, 12" wide, 5/16" thick, black	1 Tilf	115	.070	L.F.	14.75	2.98		17.73	21
0400	Colors		115	.070		15.35	2.98		18.33	21.50
0600	1/4" thick, black		115	.070		13.35	2.98		16.33	19.40
0700	Colors		115	.070		14.95	2.98		17.93	21
0900	Grip strip safety tread, colors, 5/16" thick		115	.070		20.50	2.98		23.48	27
1000	3/16" thick		120	.067	↓	15.25	2.85		18.10	21.50
1200	Landings, smooth sheet rubber, 1/8" thick		120	.067	S.F.	7.80	2.85		10.65	13.10
1300	3/16" thick		120	.067	"	8.30	2.85		11.15	13.65
1500	Nosings, 3" wide, 3/16" thick, black		140	.057	L.F.	4.23	2.45		6.68	8.50
1600	Colors		140	.057		4.88	2.45		7.33	9.20
1800	Risers, 7" high, 1/8" thick, flat		250	.032		7.70	1.37		9.07	10.60
1900	Coved		250	.032		8.55	1.37		9.92	11.55
2100	Vinyl, molded tread, 12" wide, colors, 1/8" thick	↓	115	.070	↓	5.35	2.98		8.33	10.55

09 65 Resilient Flooring

09 65 13 – Resilient Base and Accessories

09 65 13.23 Resilient Stair Treads and Risers		Crew	Daily Output	Labor-Hours	Unit	2015 Bare Costs Material	Labor	Equipment	Total	Total Incl O&P
2200	1/4" thick	1 Tilf	115	.070	L.F.	6.70	2.98		9.68	12.10
2300	Landing material, 1/8" thick		200	.040	S.F.	6.05	1.71		7.76	9.35
2400	Riser, 7" high, 1/8" thick, coved		175	.046	L.F.	2.70	1.96		4.66	6.05
2500	Tread and riser combined, 1/8" thick		80	.100	"	9.95	4.28		14.23	17.65
9000	Minimum labor/equipment charge		3	2.667	Job		114		114	180

09 65 16 – Resilient Sheet Flooring

09 65 16.10 Rubber and Vinyl Sheet Flooring

		Crew	Daily Output	Labor-Hours	Unit	Material	Labor	Equipment	Total	Total Incl O&P
0010	**RUBBER AND VINYL SHEET FLOORING**									
5500	Linoleum, sheet goods G	1 Tilf	360	.022	S.F.	3.59	.95		4.54	5.45
5900	Rubber, sheet goods, 36" wide, 1/8" thick		120	.067		7.75	2.85		10.60	13.05
5950	3/16" thick		100	.080		10.50	3.42		13.92	16.95
6000	1/4" thick		90	.089		12.45	3.80		16.25	19.65
8000	Vinyl sheet goods, backed, .065" thick, plain pattern/colors		250	.032		4.09	1.37		5.46	6.65
8050	Intricate pattern/ colors		200	.040		4.57	1.71		6.28	7.75
8100	.080" thick, plain pattern/colors		230	.035		4.10	1.49		5.59	6.85
8150	Intricate pattern/colors		200	.040		5.90	1.71		7.61	9.20
8200	.125" thick, plain pattern/colors		230	.035		4.25	1.49		5.74	7.05
8250	intricate pattern/colors		200	.040		7.50	1.71		9.21	10.95
8400	For welding seams, add		100	.080	L.F.	.25	3.42		3.67	5.70
8450	For integral cove base, add		175	.046	"	.75	1.96		2.71	3.92
8700	Adhesive cement, 1 gallon per 200 to 300 S.F.				Gal.	27			27	29.50
8800	Asphalt primer, 1 gallon per 300 S.F.					14			14	15.40
8900	Emulsion, 1 gallon per 140 S.F.					18			18	19.80

09 65 19 – Resilient Tile Flooring

09 65 19.10 Miscellaneous Resilient Tile Flooring

		Crew	Daily Output	Labor-Hours	Unit	Material	Labor	Equipment	Total	Total Incl O&P
0010	**MISCELLANEOUS RESILIENT TILE FLOORING**									
2200	Cork tile, standard finish, 1/8" thick G	1 Tilf	315	.025	S.F.	6.95	1.09		8.04	9.35
2250	3/16" thick G		315	.025		6.65	1.09		7.74	9
2300	5/16" thick G		315	.025		8.30	1.09		9.39	10.80
2350	1/2" thick G		315	.025		10.95	1.09		12.04	13.70
2500	Urethane finish, 1/8" thick G		315	.025		8.10	1.09		9.19	10.60
2550	3/16" thick G		315	.025		8.25	1.09		9.34	10.75
2600	5/16" thick G		315	.025		8.85	1.09		9.94	11.45
2650	1/2" thick G		315	.025		12.15	1.09		13.24	15.05
6700	Synthetic turf, 3/8" thick		90	.089		4.54	3.80		8.34	11
6750	Interlocking 2' x 2' squares, 1/2" thick, not									
6810	cemented, for playgrounds, 3/8" thick	1 Tilf	210	.038	S.F.	4.69	1.63		6.32	7.75
6850	1/2" thick	"	190	.042	"	5.10	1.80		6.90	8.45

09 65 19.19 Vinyl Composition Tile Flooring

		Crew	Daily Output	Labor-Hours	Unit	Material	Labor	Equipment	Total	Total Incl O&P
0010	**VINYL COMPOSITION TILE FLOORING**									
7000	Vinyl composition tile, 12" x 12", 1/16" thick	1 Tilf	500	.016	S.F.	1.18	.68		1.86	2.38
7050	Embossed		500	.016		2.21	.68		2.89	3.51
7100	Marbleized		500	.016		2.21	.68		2.89	3.51
7150	Solid		500	.016		2.85	.68		3.53	4.22
7200	3/32" thick, embossed		500	.016		1.51	.68		2.19	2.74
7250	Marbleized		500	.016		2.54	.68		3.22	3.87
7300	Solid		500	.016		2.36	.68		3.04	3.68
7350	1/8" thick, marbleized		500	.016		2.40	.68		3.08	3.72
7400	Solid		500	.016		1.53	.68		2.21	2.76
7450	Conductive		500	.016		6.05	.68		6.73	7.75

09 65 Resilient Flooring

09 65 19 – Resilient Tile Flooring

09 65 19.23 Vinyl Tile Flooring

	09 65 19.23 Vinyl Tile Flooring	Crew	Daily Output	Labor-Hours	Unit	2015 Bare Costs Material	Labor	Equipment	Total	Total Incl O&P
0010	**VINYL TILE FLOORING**									
7500	Vinyl tile, 12" x 12", 3/32" thick,	1 Tilf	500	.016	S.F.	3.59	.68		4.27	5.05
7550	3/32" thick, premium colors/patterns		500	.016		7.25	.68		7.93	9.10
7600	1/8" thick, standard colors/patterns		500	.016		5.65	.68		6.33	7.30
7650	Solid colors		500	.016		3.25	.68		3.93	4.66
7700	Marbleized or Travertine pattern		500	.016		5.90	.68		6.58	7.60
7750	Florentine pattern		500	.016		6.30	.68		6.98	8.05
7800	Premium colors/patterns		500	.016		6.20	.68		6.88	7.95
9500	Minimum labor/equipment charge		4	2	Job		85.50		85.50	135

09 65 19.33 Rubber Tile Flooring

	09 65 19.33 Rubber Tile Flooring	Crew	Daily Output	Labor-Hours	Unit	Material	Labor	Equipment	Total	Total Incl O&P
0010	**RUBBER TILE FLOORING**									
6050	Rubber tile, marbleized colors, 12" x 12", 1/8" thick	1 Tilf	400	.020	S.F.	5.70	.86		6.56	7.60
6100	3/16" thick		400	.020		9.25	.86		10.11	11.55
6300	Special tile, plain colors, 1/8" thick		400	.020		7.70	.86		8.56	9.85
6350	3/16" thick		400	.020		9.05	.86		9.91	11.30
6410	Raised, radial or square, .5 mm black		400	.020		8.25	.86		9.11	10.40
6430	.5 mm colored		400	.020		9.65	.86		10.51	11.95
6450	For golf course, skating rink, etc., 1/4" thick		275	.029		9.95	1.25		11.20	12.90

09 65 33 – Conductive Resilient Flooring

09 65 33.10 Conductive Rubber and Vinyl Flooring

	09 65 33.10 Conductive Rubber and Vinyl Flooring	Crew	Daily Output	Labor-Hours	Unit	Material	Labor	Equipment	Total	Total Incl O&P
0010	**CONDUCTIVE RUBBER AND VINYL FLOORING**									
1700	Conductive flooring, rubber tile, 1/8" thick	1 Tilf	315	.025	S.F.	6.95	1.09		8.04	9.35
1800	Homogeneous vinyl tile, 1/8" thick	"	315	.025	"	6.40	1.09		7.49	8.75

09 66 Terrazzo Flooring

09 66 13 – Portland Cement Terrazzo Flooring

09 66 13.10 Portland Cement Terrazzo

	09 66 13.10 Portland Cement Terrazzo	Crew	Daily Output	Labor-Hours	Unit	Material	Labor	Equipment	Total	Total Incl O&P
0010	**PORTLAND CEMENT TERRAZZO**, cast-in-place									
0020	Cove base, 6" high, 16 ga. zinc	1 Mstz	20	.400	L.F.	3.28	17.20		20.48	30.50
0100	Curb, 6" high and 6" wide		6	1.333		5.65	57.50		63.15	96.50
0300	Divider strip for floors, 14 ga., 1-1/4" deep, zinc		375	.021		1.42	.92		2.34	3.01
0400	Brass		375	.021		2.46	.92		3.38	4.16
0600	Heavy top strip 1/4" thick, 1-1/4" deep, zinc		300	.027		2.17	1.15		3.32	4.20
0900	Galv. bottoms, brass		300	.027		3.25	1.15		4.40	5.40
1200	For thin set floors, 16 ga., 1/2" x 1/2", zinc		350	.023		1.12	.98		2.10	2.78
1300	Brass		350	.023		2.32	.98		3.30	4.10
1500	Floor, bonded to concrete, 1-3/4" thick, gray cement	J-3	75	.213	S.F.	3.35	8.35	4.05	15.75	21.50
1600	White cement, mud set		75	.213		3.73	8.35	4.05	16.13	22
1800	Not bonded, 3" total thickness, gray cement		70	.229		4.17	8.95	4.34	17.46	23.50
1900	White cement, mud set		70	.229		4.86	8.95	4.34	18.15	24.50
2100	For Venetian terrazzo, 1" topping, add					50%	50%			
2200	For heavy duty abrasive terrazzo, add					50%	50%			
2700	Monolithic terrazzo, 1/2" thick									
2710	10' panels	J-3	125	.128	S.F.	3.14	5	2.43	10.57	14.05
3000	Stairs, cast in place, pan filled treads		30	.533	L.F.	3.57	21	10.10	34.67	48
3100	Treads and risers		14	1.143	"	6.15	45	21.50	72.65	102
3300	For stair landings, add to floor prices						50%			
3400	Stair stringers and fascia	J-3	30	.533	S.F.	5.25	21	10.10	36.35	50
3600	For abrasive metal nosings on stairs, add		150	.107	L.F.	9.15	4.18	2.02	15.35	18.95
3700	For abrasive surface finish, add		600	.027	S.F.	1.54	1.05	.51	3.10	3.90

09 66 Terrazzo Flooring

09 66 13 – Portland Cement Terrazzo Flooring

09 66 13.10 Portland Cement Terrazzo		Crew	Daily Output	Labor-Hours	Unit	Material	Labor	Equipment	Total	Total Incl O&P
						2015 Bare Costs				
3900	For raised abrasive strips, add	J-3	150	.107	L.F.	1.34	4.18	2.02	7.54	10.30
4000	Wainscot, bonded, 1-1/2" thick		30	.533	S.F.	3.93	21	10.10	35.03	48.50
4200	1/4" thick	↓	40	.400	"	5.55	15.70	7.60	28.85	39.50
4300	Stone chips, onyx gemstone, per 50 lb. bag				Bag	16.80			16.80	18.50
9000	Minimum labor/equipment charge	1 Mstz	1	8	Job		345		345	540

09 66 16 – Terrazzo Floor Tile

09 66 16.10 Tile or Terrazzo Base		Crew	Daily Output	Labor-Hours	Unit	Material	Labor	Equipment	Total	Total Incl O&P
0010	TILE OR TERRAZZO BASE									
0020	Scratch coat only	1 Mstz	150	.053	S.F.	.43	2.29		2.72	4.09
0500	Scratch and brown coat only		75	.107	"	.82	4.58		5.40	8.15
9000	Minimum labor/equipment charge	↓	1	8	Job		345		345	540
09 66 16.13 Portland Cement Terrazzo Floor Tile										
0010	PORTLAND CEMENT TERRAZZO FLOOR TILE									
1200	Floor tiles, non-slip, 1" thick, 12" x 12"	D-1	60	.267	S.F.	20	11.25		31.25	40.50
1300	1-1/4" thick, 12" x 12"		60	.267		20.50	11.25		31.75	41.50
1500	16" x 16"		50	.320		22.50	13.45		35.95	46.50
1600	1-1/2" thick, 16" x 16"	↓	45	.356		20.50	14.95		35.45	47
1800	For Venetian terrazzo, add					6.15			6.15	6.80
1900	For white cement, add				↓	.58			.58	.64
09 66 16.16 Plastic Matrix Terrazzo Floor Tile										
0010	PLASTIC MATRIX TERRAZZO FLOOR TILE									
0100	12" x 12", 3/16" thick, Floor tiles w/marble chips	1 Tilf	500	.016	S.F.	7.30	.68		7.98	9.15
0200	12" x 12", 3/16" thick, Floor tiles w/glass chips		500	.016		7.90	.68		8.58	9.75
0300	12" x 12", 3/16" thick, Floor tiles w/recycled content	↓	500	.016	↓	6.20	.68		6.88	7.95
09 66 16.30 Terrazzo, Precast										
0010	TERRAZZO, PRECAST									
0020	Base, 6" high, straight	1 Mstz	70	.114	L.F.	11.95	4.91		16.86	21
0100	Cove		60	.133		12.80	5.75		18.55	23
0300	8" high, straight		60	.133		11.45	5.75		17.20	21.50
0400	Cove	↓	50	.160		16.85	6.85		23.70	29.50
0600	For white cement, add					.45			.45	.50
0700	For 16 ga. zinc toe strip, add					1.72			1.72	1.89
0900	Curbs, 4" x 4" high	1 Mstz	40	.200		32.50	8.60		41.10	49
1000	8" x 8" high	"	30	.267		38	11.45		49.45	60
2400	Stair treads, 1-1/2" thick, non-slip, three line pattern	2 Mstz	70	.229		41	9.80		50.80	60.50
2500	Nosing and two lines		70	.229		41	9.80		50.80	60.50
2700	2" thick treads, straight		60	.267		45.50	11.45		56.95	68
2800	Curved		50	.320		59	13.75		72.75	86.50
3000	Stair risers, 1" thick, to 6" high, straight sections		60	.267		10.80	11.45		22.25	30
3100	Cove		50	.320		15.65	13.75		29.40	38.50
3300	Curved, 1" thick, to 6" high, vertical		48	.333		21.50	14.30		35.80	46
3400	Cove		38	.421		40.50	18.10		58.60	73
3600	Stair tread and riser, single piece, straight, smooth surface		60	.267		52.50	11.45		63.95	76
3700	Non skid surface		40	.400		68	17.20		85.20	102
3900	Curved tread and riser, smooth surface		40	.400		74	17.20		91.20	108
4000	Non skid surface		32	.500		93	21.50		114.50	136
4200	Stair stringers, notched, 1" thick		25	.640		30	27.50		57.50	76.50
4300	2" thick		22	.727	↓	35.50	31		66.50	88.50
4500	Stair landings, structural, non-slip, 1-1/2" thick		85	.188	S.F.	33	8.10		41.10	49
4600	3" thick	↓	75	.213		46.50	9.15		55.65	66
4800	Wainscot, 12" x 12" x 1" tiles	1 Mstz	12	.667	↓	6.85	28.50		35.35	52.50

09 66 Terrazzo Flooring

09 66 16 – Terrazzo Floor Tile

09 66 16.30 Terrazzo, Precast		Crew	Daily Output	Labor-Hours	Unit	Material	Labor	Equipment	Total	Total Incl O&P
						2015 Bare Costs				
4900	16" x 16" x 1-1/2" tiles	1 Mstz	8	1	S.F.	14.25	43		57.25	83.50
9500	Minimum labor/equipment charge	1 Tilf	2	4	Job		171		171	270

09 66 23 – Resinous Matrix Terrazzo Flooring

09 66 23.13 Polyacrylate Modified Cementitious Terrazzo Flooring

		Crew	Daily Output	Labor-Hours	Unit	Material	Labor	Equipment	Total	Total Incl O&P
0010	**POLYACRYLATE MODIFIED CEMENTITIOUS TERRAZZO FLOORING**									
3150	Polyacrylate, 1/4" thick, granite chips	C-6	735	.065	S.F.	3.50	2.56	.08	6.14	8.10
3170	Recycled porcelain		480	.100		4.51	3.92	.13	8.56	11.50
3200	3/8" thick, granite chips		620	.077		4.58	3.03	.10	7.71	10.10
3220	Recycled porcelain	↓	480	.100	↓	6.40	3.92	.13	10.45	13.60

09 66 23.16 Epoxy-Resin Terrazzo Flooring

		Crew	Daily Output	Labor-Hours	Unit	Material	Labor	Equipment	Total	Total Incl O&P
0010	**EPOXY-RESIN TERRAZZO FLOORING**									
1800	Epoxy terrazzo, 1/4" thick, chemical resistant, granite chips	J-3	200	.080	S.F.	5.95	3.14	1.52	10.61	13.15
1900	Recycled porcelain		150	.107		9.05	4.18	2.02	15.25	18.80
2500	Epoxy terrazzo, 1/4" thick, granite chips		200	.080		5.15	3.14	1.52	9.81	12.30
2550	Average		175	.091		5.45	3.59	1.74	10.78	13.55
2600	Recycled porcelain	↓	150	.107	↓	6.35	4.18	2.02	12.55	15.85

09 66 33 – Conductive Terrazzo Flooring

09 66 33.10 Conductive Terrazzo

		Crew	Daily Output	Labor-Hours	Unit	Material	Labor	Equipment	Total	Total Incl O&P
0010	**CONDUCTIVE TERRAZZO**									
2400	Bonded conductive floor for hospitals	J-3	90	.178	S.F.	4.84	6.95	3.37	15.16	20

09 66 33.13 Conductive Epoxy-Resin Terrazzo

		Crew	Daily Output	Labor-Hours	Unit	Material	Labor	Equipment	Total	Total Incl O&P
0010	**CONDUCTIVE EPOXY-RESIN TERRAZZO**									
2100	Epoxy terrazzo, 1/4" thick, conductive, granite chips	J-3	100	.160	S.F.	8	6.30	3.04	17.34	22
2200	Recycled porcelain	"	90	.178	"	10.40	6.95	3.37	20.72	26

09 66 33.19 Conductive Plastic-Matrix Terrazzo Flooring

		Crew	Daily Output	Labor-Hours	Unit	Material	Labor	Equipment	Total	Total Incl O&P
0010	**CONDUCTIVE PLASTIC-MATRIX TERRAZZO FLOORING**									
3300	Conductive, 1/4" thick, granite chips	C-6	450	.107	S.F.	7.45	4.18	.14	11.77	15.15
3330	Recycled porcelain		305	.157		10	6.15	.20	16.35	21.50
3350	3/8" thick, granite chips		365	.132		9.80	5.15	.17	15.12	19.40
3370	Recycled porcelain		255	.188		12.75	7.35	.24	20.34	26.50
3450	Granite, conductive, 1/4" thick, 20% chip		695	.069		9.25	2.71	.09	12.05	14.65
3470	50% chip		420	.114		11.90	4.48	.15	16.53	20.50
3500	3/8" thick, 20% chip		695	.069		13.50	2.71	.09	16.30	19.35
3520	50% chip	↓	380	.126	↓	16.30	4.95	.16	21.41	26

09 67 Fluid-Applied Flooring

09 67 13 – Elastomeric Liquid Flooring

09 67 13.13 Elastomeric Liquid Flooring

		Crew	Daily Output	Labor-Hours	Unit	Material	Labor	Equipment	Total	Total Incl O&P
0010	**ELASTOMERIC LIQUID FLOORING**									
0020	Cementitious acrylic, 1/4" thick	C-6	520	.092	S.F.	1.66	3.62	.12	5.40	7.85
0100	3/8" thick	"	450	.107		2.10	4.18	.14	6.42	9.25
0200	Methyl methachrylate, 1/4" thick	C-8A	3000	.016		6.50	.65		7.15	8.20
0210	1/8" thick	"	3000	.016		5.50	.65		6.15	7.10
0300	Cupric oxychloride, on bond coat, simple configs and patterns	C-6	480	.100		3.58	3.92	.13	7.63	10.50
0400	Complex configurations and patterns		420	.114		6	4.48	.15	10.63	14.05
2400	Mastic, hot laid, 2 coat, 1-1/2" thick, standard, simple configurations and		690	.070		4.15	2.73	.09	6.97	9.10
2500	Maximum		520	.092		5.35	3.62	.12	9.09	11.90
2700	Acid-proof, minimum		605	.079		5.35	3.11	.10	8.56	11
2800	Maximum	↓	350	.137	↓	7.40	5.35	.18	12.93	17.05

09 67 Fluid-Applied Flooring

09 67 13 – Elastomeric Liquid Flooring

09 67 13.13 Elastomeric Liquid Flooring		Crew	Daily Output	Labor-Hours	Unit	2015 Bare Costs Material	Labor	Equipment	Total	Total Incl O&P
3000	Neoprene, troweled on, 1/4" thick, minimum	C-6	545	.088	S.F.	4.09	3.45	.11	7.65	10.25
3100	Maximum		430	.112		5.55	4.37	.15	10.07	13.40
4300	Polyurethane, with suspended vinyl chips, clear		1065	.045		7.60	1.77	.06	9.43	11.30
4500	Pigmented		860	.056		11.05	2.19	.07	13.31	15.80

09 67 26 – Quartz Flooring

09 67 26.26 Quartz Flooring		Crew	Daily Output	Labor-Hours	Unit	Material	Labor	Equipment	Total	Total Incl O&P
0010	**QUARTZ FLOORING**									
0600	Epoxy, with colored quartz chips, broadcast, 3/8" thick	C-6	675	.071	S.F.	2.81	2.79	.09	5.69	7.75
0700	1/2" thick		490	.098		4.05	3.84	.13	8.02	10.85
0900	Troweled, minimum		560	.086		3.57	3.36	.11	7.04	9.50
1000	Maximum		480	.100		5.40	3.92	.13	9.45	12.45
1200	Heavy duty epoxy topping, 1/4" thick,									
1300	500 to 1,000 S.F.	C-6	420	.114	S.F.	5.55	4.48	.15	10.18	13.60
1500	1,000 to 2,000 S.F.		450	.107		5.05	4.18	.14	9.37	12.50
1600	Over 10,000 S.F.		480	.100		4.66	3.92	.13	8.71	11.70
3600	Polyester, with colored quartz chips, 1/16" thick, minimum		1065	.045		3.16	1.77	.06	4.99	6.40
3700	Maximum		560	.086		4.19	3.36	.11	7.66	10.20
3900	1/8" thick, minimum		810	.059		3.67	2.32	.08	6.07	7.90
4000	Maximum		675	.071		4.73	2.79	.09	7.61	9.85
4200	Polyester, heavy duty, compared to epoxy, add		2590	.019		1.51	.73	.02	2.26	2.87

09 67 66 – Fluid-Applied Athletic Flooring

09 67 66.10 Polyurethane		Crew	Daily Output	Labor-Hours	Unit	Material	Labor	Equipment	Total	Total Incl O&P
0010	**POLYURETHANE**									
4400	Thermoset, prefabricated in place, indoor									
4500	3/8" thick for basketball, gyms, etc.	1 Tilf	100	.080	S.F.	5.40	3.42		8.82	11.35
4600	1/2" thick for professional sports		95	.084		7.25	3.60		10.85	13.65
4700	Outdoor, 1/4" thick, smooth, for tennis		100	.080		5.60	3.42		9.02	11.60
5000	Poured in place, indoor, with finish, 1/4" thick		80	.100		3.99	4.28		8.27	11.15
5050	3/8" thick		65	.123		4.84	5.25		10.09	13.60
5100	1/2" thick		50	.160		6.75	6.85		13.60	18.25

09 68 Carpeting

09 68 05 – Carpet Accessories

09 68 05.11 Flooring Transition Strip		Crew	Daily Output	Labor-Hours	Unit	Material	Labor	Equipment	Total	Total Incl O&P
0010	**FLOORING TRANSITION STRIP**									
0107	Clamp down brass divider, 12' strip, vinyl to carpet	1 Tilf	31.25	.256	Ea.	13.30	10.95		24.25	32
0117	Vinyl to hard surface	"	31.25	.256	"	13.30	10.95		24.25	32

09 68 10 – Carpet Pad

09 68 10.10 Commercial Grade Carpet Pad		Crew	Daily Output	Labor-Hours	Unit	Material	Labor	Equipment	Total	Total Incl O&P
0010	**COMMERCIAL GRADE CARPET PAD**									
9000	Sponge rubber pad, 20 oz./sq. yd.	1 Tilf	150	.053	S.Y.	4.51	2.28		6.79	8.55
9100	40-62 oz./sq. yd.		150	.053		8.75	2.28		11.03	13.25
9200	Felt pad, 20 oz./sq. yd.		150	.053		5.30	2.28		7.58	9.45
9300	32 to 56 oz./sq. yd.		150	.053		8.95	2.28		11.23	13.45
9400	Bonded urethane pad, 2.7 density		150	.053		5.95	2.28		8.23	10.15
9500	13.0 Density		150	.053		8	2.28		10.28	12.40
9600	Prime urethane pad, 2.7 density		150	.053		3.15	2.28		5.43	7.10
9700	13.0 density		150	.053		4.95	2.28		7.23	9.05
9800	Carpet pad, for 'double stick' installation, add					.90	4.10		5	7

09 68 Carpeting

09 68 13 – Tile Carpeting

09 68 13.10 Carpet Tile		Crew	Daily Output	Labor-Hours	Unit	Material	2015 Bare Costs Labor	Equipment	Total	Total Incl O&P
0010	**CARPET TILE**									
0100	Tufted nylon, 18" x 18", hard back, 20 oz.	1 Tilf	80	.100	S.Y.	24	4.28		28.28	33.50
0110	26 oz.		80	.100		32	4.28		36.28	42
0200	Cushion back, 20 oz.		80	.100		28	4.28		32.28	37.50
0210	26 oz.		80	.100		43	4.28		47.28	54
1100	Tufted, 24" x 24", hard back, 24 oz. nylon		80	.100		30	4.28		34.28	40
1180	35 oz.		80	.100		35	4.28		39.28	45.50
5060	42 oz.		80	.100		46	4.28		50.28	58

09 68 16 – Sheet Carpeting

09 68 16.10 Sheet Carpet		Crew	Daily Output	Labor-Hours	Unit	Material	Labor	Equipment	Total	Total Incl O&P
0010	**SHEET CARPET**									
0700	Nylon, level loop, 26 oz., light to medium traffic	1 Tilf	75	.107	S.Y.	25	4.57		29.57	34.50
0720	28 oz., light to medium traffic		75	.107		32	4.57		36.57	42
0900	32 oz., medium traffic		75	.107		42.50	4.57		47.07	54
1100	40 oz., medium to heavy traffic		75	.107		59.50	4.57		64.07	72.50
2920	Nylon plush, 30 oz., medium traffic		57	.140		29.50	6		35.50	42
3000	36 oz., medium traffic		75	.107		34.50	4.57		39.07	45
3100	42 oz., medium to heavy traffic		70	.114		45	4.89		49.89	57.50
3200	46 oz., medium to heavy traffic		70	.114		49	4.89		53.89	62
3300	54 oz., heavy traffic		70	.114		55	4.89		59.89	68.50
3340	60 oz., heavy traffic		70	.114		62.50	4.89		67.39	77
3665	Olefin, 24 oz., light to medium traffic		75	.107		18.65	4.57		23.22	27.50
3670	26 oz., medium traffic		75	.107		13.10	4.57		17.67	21.50
3680	28 oz., medium to heavy traffic		75	.107		23.50	4.57		28.07	33
3700	32 oz., medium to heavy traffic		75	.107		27	4.57		31.57	36.50
3730	42 oz., heavy traffic		70	.114		28.50	4.89		33.39	39
4110	Wool, level loop, 40 oz., medium traffic		75	.107		106	4.57		110.57	124
4500	50 oz., medium to heavy traffic		75	.107		100	4.57		104.57	117
4700	Patterned, 32 oz., medium to heavy traffic		70	.114		91	4.89		95.89	108
4900	48 oz., heavy traffic		70	.114		100	4.89		104.89	118
5000	For less than full roll (approx. 1500 S.F.), add					25%				
5100	For small rooms, less than 12' wide, add						25%			
5200	For large open areas (no cuts), deduct						25%			
5600	For bound carpet baseboard, add	1 Tilf	300	.027	L.F.	3	1.14		4.14	5.10
5610	For stairs, not incl. price of carpet, add	"	30	.267	Riser		11.40		11.40	18.05
5620	For borders and patterns, add to labor						18%			
8950	For tackless, stretched installation, add padding from 09 68 10.10 to above									
9850	For brand-named specific fiber, add				S.Y.	25%				
9910	Minimum labor/equipment charge	1 Tilf	3	2.667	Job		114		114	180

09 68 20 – Athletic Carpet

09 68 20.10 Indoor Athletic Carpet		Crew	Daily Output	Labor-Hours	Unit	Material	Labor	Equipment	Total	Total Incl O&P
0010	**INDOOR ATHLETIC CARPET**									
3700	Polyethylene, in rolls, no base incl., landscape surfaces	1 Tilf	275	.029	S.F.	4.09	1.25		5.34	6.45
3800	Nylon action surface, 1/8" thick		275	.029		3.76	1.25		5.01	6.10
3900	1/4" thick		275	.029		5.45	1.25		6.70	7.90
4000	3/8" thick		275	.029		6.80	1.25		8.05	9.45
4100	Golf tee surface with foam back		235	.034		6.75	1.46		8.21	9.75
4200	Practice putting, knitted nylon surface		235	.034		5.70	1.46		7.16	8.60
5500	Polyvinyl chloride, sheet goods for gyms, 1/4" thick		80	.100		7.90	4.28		12.18	15.45
5600	3/8" thick		60	.133		11.80	5.70		17.50	22

09 69 Access Flooring

09 69 13 – Rigid-Grid Access Flooring

09 69 13.10 Access Floors		Crew	Daily Output	Labor-Hours	Unit	2015 Bare Costs Material	Labor	Equipment	Total	Total Incl O&P
0010	**ACCESS FLOORS**									
0015	Access floor package including panel, pedestal, stringers & laminate cover									
0100	Computer room, greater than 6,000 S.F.	4 Carp	750	.043	S.F.	7.50	2		9.50	11.55
0110	Less than 6,000 S.F.	2 Carp	375	.043		8.15	2		10.15	12.30
0120	Office, greater than 6,000 S.F.	4 Carp	1050	.030		5.35	1.43		6.78	8.25
0250	Panels, particle board or steel, 1250# load, no covering, under 6,000 S.F.	2 Carp	600	.027		3.75	1.25		5	6.20
0300	Over 6,000 S.F.		640	.025		3.21	1.17		4.38	5.45
0400	Aluminum, 24″ panels		500	.032		33	1.50		34.50	38.50
0600	For carpet covering, add					8.75			8.75	9.65
0700	For vinyl floor covering, add					9.05			9.05	9.95
0900	For high pressure laminate covering, add					7.50			7.50	8.25
0910	For snap on stringer system, add	2 Carp	1000	.016		1.56	.75		2.31	2.95
0950	Office applications, steel or concrete panels,									
0960	no covering, over 6,000 S.F.	2 Carp	960	.017	S.F.	10.55	.78		11.33	12.90
1000	Machine cutouts after initial installation	1 Carp	50	.160	Ea.	20	7.50		27.50	34.50
1050	Pedestals, 6″ to 12″	2 Carp	85	.188		8.40	8.85		17.25	24
1100	Air conditioning grilles, 4″ x 12″	1 Carp	17	.471		68.50	22		90.50	111
1150	4″ x 18″	″	14	.571		93.50	27		120.50	147
1200	Approach ramps, steel	2 Carp	60	.267	S.F.	24.50	12.50		37	47.50
1300	Aluminum	″	40	.400	″	34	18.80		52.80	68.50
1500	Handrail, 2 rail, aluminum	1 Carp	15	.533	L.F.	109	25		134	160

09 72 Wall Coverings

09 72 19 – Textile Wall Covering

09 72 19.10 Textile Wall Covering

		Crew	Daily Output	Labor-Hours	Unit	Material	Labor	Equipment	Total	Total Incl O&P
0010	**TEXTILE WALL COVERING**, including sizing; add 10-30% waste @ takeoff									
0020	Silk	1 Pape	640	.013	S.F.	4.15	.51		4.66	5.40
0030	Cotton		640	.013		6.65	.51		7.16	8.10
0040	Linen		640	.013		1.85	.51		2.36	2.86
0050	Blend		640	.013		2.97	.51		3.48	4.09

09 72 20 – Natural Fiber Wall Covering

09 72 20.10 Natural Fiber Wall Covering

		Crew	Daily Output	Labor-Hours	Unit	Material	Labor	Equipment	Total	Total Incl O&P
0010	**NATURAL FIBER WALL COVERING**, including sizing; add 10-30% waste @ takeoff									
0015	Bamboo	1 Pape	640	.013	S.F.	2.26	.51		2.77	3.31
0030	Burlap		640	.013		2.03	.51		2.54	3.05
0045	Jute		640	.013		1.29	.51		1.80	2.24
0060	Sisal		640	.013		1.46	.51		1.97	2.43

09 72 23 – Wallpapering

09 72 23.10 Wallpaper

		Crew	Daily Output	Labor-Hours	Unit	Material	Labor	Equipment	Total	Total Incl O&P
0010	**WALLPAPER** including sizing; add 10-30 percent waste @ takeoff									
0050	Aluminum foil	1 Pape	275	.029	S.F.	1.04	1.19		2.23	3.05
0100	Copper sheets, .025″ thick, vinyl backing		240	.033		5.55	1.36		6.91	8.30
0300	Phenolic backing		240	.033		7.20	1.36		8.56	10.15
0600	Cork tiles, light or dark, 12″ x 12″ x 3/16″		240	.033		4.51	1.36		5.87	7.15
0700	5/16″ thick		235	.034		3.13	1.39		4.52	5.70
0900	1/4″ basketweave		240	.033		3.50	1.36		4.86	6.05
1000	1/2″ natural, non-directional pattern		240	.033		6.90	1.36		8.26	9.80
1100	3/4″ natural, non-directional pattern		240	.033		11.35	1.36		12.71	14.70
1200	Granular surface, 12″ x 36″, 1/2″ thick		385	.021		1.31	.85		2.16	2.81
1300	1″ thick		370	.022		1.68	.88		2.56	3.27

09 72 Wall Coverings

09 72 23 – Wallpapering

09 72 23.10 Wallpaper		Crew	Daily Output	Labor-Hours	Unit	2015 Bare Costs Material	Labor	Equipment	Total	Total Incl O&P
1500	Polyurethane coated, 12" x 12" x 3/16" thick	1 Pape	240	.033	S.F.	4.08	1.36		5.44	6.70
1600	5/16" thick		235	.034		6.60	1.39		7.99	9.50
1800	Cork wallpaper, paperbacked, natural		480	.017		2.06	.68		2.74	3.37
1900	Colors		480	.017		2.87	.68		3.55	4.26
2100	Flexible wood veneer, 1/32" thick, plain woods		100	.080		2.44	3.27		5.71	7.95
2200	Exotic woods		95	.084		3.69	3.44		7.13	9.60
2400	Gypsum-based, fabric-backed, fire resistant									
2500	for masonry walls, 21 oz./S.Y.	1 Pape	800	.010	S.F.	.85	.41		1.26	1.60
2600	Average		720	.011		1.27	.45		1.72	2.13
2700	Small quantities		640	.013		.76	.51		1.27	1.66
2750	Acrylic, modified, semi-rigid PVC, .028" thick	2 Carp	330	.048		1.37	2.28		3.65	5.25
2800	.040" thick	"	320	.050		1.81	2.35		4.16	5.85
3000	Vinyl wall covering, fabric-backed, lightweight, type 1 (12-15 oz./S.Y.)	1 Pape	640	.013		.97	.51		1.48	1.89
3300	Medium weight, type 2 (20-24 oz./S.Y.)		480	.017		.90	.68		1.58	2.09
3400	Heavy weight, type 3 (28 oz./S.Y.)		435	.018		1.37	.75		2.12	2.72
3600	Adhesive, 5 gal. lots (18 S.Y./gal.)				Gal.	14.20			14.20	15.60
3700	Wallpaper, average workmanship, solid pattern, low cost paper	1 Pape	640	.013	S.F.	.61	.51		1.12	1.49
3900	basic patterns (matching required), avg. cost paper		535	.015		1.11	.61		1.72	2.20
4000	Paper at $85 per double roll, quality workmanship		435	.018		2.20	.75		2.95	3.63
4100	Linen wall covering, paper backed									
4150	Flame treatment				S.F.	1.01			1.01	1.11
4190	Grass cloth, natural fabric G	1 Pape	400	.020		2.01	.82		2.83	3.53
4200	Grass cloths with lining paper G		400	.020		.91	.82		1.73	2.32
4300	Premium texture/color G		350	.023		2.91	.94		3.85	4.70
5990	Wallpaper removal, 1 layer		800	.010		.03	.41		.44	.69
6000	Wallpaper removal, 3 layer		400	.020		.08	.82		.90	1.41
9000	Minimum labor/equipment charge		2	4	Job		164		164	263

09 77 Special Wall Surfacing

09 77 30 – Fiberglass Reinforced Panels

09 77 30.10 Fiberglass Reinforced Plastic Panels		Crew	Daily Output	Labor-Hours	Unit	Material	Labor	Equipment	Total	Total Incl O&P
0010	**FIBERGLASS REINFORCED PLASTIC PANELS**, .090" thick									
0020	On walls, adhesive mounted, embossed surface	2 Carp	640	.025	S.F.	1.11	1.17		2.28	3.14
0030	Smooth surface		640	.025		1.37	1.17		2.54	3.43
0040	Fire rated, embossed surface		640	.025		1.99	1.17		3.16	4.11
0050	Nylon rivet mounted, on drywall, embossed surface		480	.033		1.11	1.56		2.67	3.78
0060	Smooth surface		480	.033		1.37	1.56		2.93	4.07
0070	Fire rated, embossed surface		480	.033		1.99	1.56		3.55	4.75
0080	On masonry, embossed surface		320	.050		1.11	2.35		3.46	5.05
0090	Smooth surface		320	.050		1.37	2.35		3.72	5.35
0100	Fire rated, embossed surface		320	.050		1.99	2.35		4.34	6.05
0110	Nylon rivet and adhesive mounted, on drywall, embossed surface		240	.067		1.26	3.13		4.39	6.55
0120	Smooth surface		240	.067		1.26	3.13		4.39	6.55
0130	Fire rated, embossed surface		240	.067		2.19	3.13		5.32	7.55
0140	On masonry, embossed surface		190	.084		1.26	3.95		5.21	7.90
0150	Smooth surface		190	.084		1.26	3.95		5.21	7.90
0160	Fire rated, embossed surface		190	.084		2.19	3.95		6.14	8.90
0170	For moldings add	1 Carp	250	.032	L.F.	.26	1.50		1.76	2.75
0180	On ceilings, for lay in grid system, embossed surface		400	.020	S.F.	1.11	.94		2.05	2.76
0190	Smooth surface		400	.020		1.37	.94		2.31	3.05
0200	Fire rated, embossed surface		400	.020		1.99	.94		2.93	3.73

09 77 Special Wall Surfacing

09 77 43 – Panel Systems

09 77 43.20 Slatwall Panels and Accessories		Crew	Daily Output	Labor-Hours	Unit	Material	Labor	Equipment	Total	Total Incl O&P
						2015 Bare Costs				
0010	SLATWALL PANELS AND ACCESSORIES									
0100	Slatwall panel, 4′ x 8′ x 3/4″ T, MDF, paint grade	1 Carp	500	.016	S.F.	1.41	.75		2.16	2.78
0110	Melamine finish		500	.016		2.12	.75		2.87	3.56
0120	High pressure plastic laminate finish		500	.016		3.72	.75		4.47	5.30
0125	Wood veneer		500	.016		3.41	.75		4.16	4.98
0130	Aluminum channel inserts, add					2.43			2.43	2.67
0200	Accessories, corner forms, 8′ L				L.F.	5.80			5.80	6.40
0210	T-connector, 8′ L					8.40			8.40	9.25
0220	J-mold, 8′ L					1.26			1.26	1.39
0230	Edge cap, 8′ L					1.33			1.33	1.46
0240	Finish end cap, 8′ L					3.39			3.39	3.73
0300	Display hook, metal, 4″ L				Ea.	.43			.43	.47
0310	6″ L					.48			.48	.53
0320	8″ L					.52			.52	.57
0330	10″ L					.59			.59	.65
0340	12″ L					.64			.64	.70
0350	Acrylic, 4″ L					.92			.92	1.01
0360	6″ L					1.06			1.06	1.17
0370	8″ L					1.11			1.11	1.22
0380	10″ L					1.25			1.25	1.38
0400	Waterfall hanger, metal, 12″ - 16″					3.84			3.84	4.22
0410	Acrylic					11.30			11.30	12.40
0500	Shelf bracket, metal, 8″					1.79			1.79	1.97
0510	10″					1.97			1.97	2.17
0520	12″					2.17			2.17	2.39
0530	14″					2.39			2.39	2.63
0540	16″					2.81			2.81	3.09
0550	Acrylic, 8″					3.79			3.79	4.17
0560	10″					4.03			4.03	4.43
0570	12″					4.26			4.26	4.69
0580	14″					4.48			4.48	4.93
0600	Shelf, acrylic, 12″ x 16″ x 1/4″					17.30			17.30	19
0610	12″ x 24″ x 1/4″					24			24	26

09 81 Acoustic Insulation

09 81 16 – Acoustic Blanket Insulation

09 81 16.10 Sound Attenuation Blanket		Crew	Daily Output	Labor-Hours	Unit	Material	Labor	Equipment	Total	Total Incl O&P
0010	SOUND ATTENUATION BLANKET									
0020	Blanket, 1″ thick	1 Carp	925	.009	S.F.	.25	.41		.66	.95
0500	1-1/2″ thick		920	.009		.25	.41		.66	.95
1000	2″ thick		915	.009		.36	.41		.77	1.07
1500	3″ thick		910	.009		.52	.41		.93	1.25
2000	Wall hung, STC 18 – 21, 1″ thick, 4′ x 20′	2 Carp	22	.727	Ea.	385	34		419	475
2010	10′ x 20′	″	19	.842		960	39.50		999.50	1,125
2020	Wall hung, STC 27 – 28, 3″ thick, 4′ x 20′	3 Carp	12	2		545	94		639	755
2030	10′ x 20′	″	9	2.667		1,375	125		1,500	1,700
3000	Thermal or acoustical batt above ceiling, 2″ thick	1 Carp	900	.009	S.F.	.50	.42		.92	1.23
3100	3″ thick		900	.009		.75	.42		1.17	1.51
3200	4″ thick		900	.009		.94	.42		1.36	1.71
3400	Urethane plastic foam, open cell, on wall, 2″ thick	2 Carp	2050	.008		3.11	.37		3.48	4.02

09 81 Acoustic Insulation

09 81 16 – Acoustic Blanket Insulation

09 81 16.10 Sound Attenuation Blanket		Crew	Daily Output	Labor-Hours	Unit	2015 Bare Costs Material	Labor	Equipment	Total	Total Incl O&P
3500	3" thick	2 Carp	1550	.010	S.F.	4.13	.48		4.61	5.35
3600	4" thick		1050	.015		5.80	.72		6.52	7.50
3700	On ceiling, 2" thick		1700	.009		3.10	.44		3.54	4.13
3800	3" thick		1300	.012		4.13	.58		4.71	5.50
3900	4" thick		900	.018		5.80	.83		6.63	7.70
9000	Minimum labor/equipment charge	1 Carp	5	1.600	Job		75		75	123

09 84 Acoustic Room Components

09 84 13 – Fixed Sound-Absorptive Panels

09 84 13.10 Fixed Panels

		Crew	Daily Output	Labor-Hours	Unit	Material	Labor	Equipment	Total	Total Incl O&P
0010	**FIXED PANELS** Perforated steel facing, painted with									
0100	Fiberglass or mineral filler, no backs, 2-1/4" thick, modular									
0200	space units, ceiling or wall hung, white or colored	1 Carp	100	.080	S.F.	8.80	3.76		12.56	15.80
0300	Fiberboard sound deadening panels, 1/2" thick	"	600	.013	"	.33	.63		.96	1.39
0500	Fiberglass panels, 4' x 8' x 1" thick, with									
0600	glass cloth face for walls, cemented	1 Carp	155	.052	S.F.	8.35	2.42		10.77	13.15
0700	1-1/2" thick, dacron covered, inner aluminum frame,									
0710	wall mounted	1 Carp	300	.027	S.F.	8.90	1.25		10.15	11.85
0900	Mineral fiberboard panels, fabric covered, 30" x 108",									
1000	3/4" thick, concealed spline, wall mounted	1 Carp	150	.053	S.F.	6.40	2.50		8.90	11.15
9000	Minimum labor/equipment charge	"	4	2	Job		94		94	154

09 84 36 – Sound-Absorbing Ceiling Units

09 84 36.10 Barriers

		Crew	Daily Output	Labor-Hours	Unit	Material	Labor	Equipment	Total	Total Incl O&P
0010	**BARRIERS** Plenum									
0600	Aluminum foil, fiberglass reinf., parallel with joists	1 Carp	275	.029	S.F.	1.08	1.37		2.45	3.43
0700	Perpendicular to joists		180	.044		1.08	2.09		3.17	4.61
0900	Aluminum mesh, kraft paperbacked		275	.029		.79	1.37		2.16	3.11
0970	Fiberglass batts, kraft faced, 3-1/2" thick		1400	.006		.37	.27		.64	.85
0980	6" thick		1300	.006		.66	.29		.95	1.20
1000	Sheet lead, 1 lb., 1/64" thick, perpendicular to joists		150	.053		6.40	2.50		8.90	11.15
1100	Vinyl foam reinforced, 1/8" thick, 1.0 lb. per S.F.		150	.053		4.49	2.50		6.99	9.05

09 91 Painting

09 91 03 – Paint Restoration

09 91 03.20 Sanding

		Crew	Daily Output	Labor-Hours	Unit	Material	Labor	Equipment	Total	Total Incl O&P
0010	**SANDING** and puttying interior trim, compared to R099100-10									
0100	Painting 1 coat, on quality work				L.F.		100%			
0300	Medium work						50%			
0400	Industrial grade						25%			
0500	Surface protection, placement and removal									
0510	Basic drop cloths	1 Pord	6400	.001	S.F.		.05		.05	.08
0520	Masking with paper		800	.010		.07	.40		.47	.73
0530	Volume cover up (using plastic sheathing, or building paper)		16000	.001			.02		.02	.03

09 91 Painting

09 91 03 – Paint Restoration

09 91 03.30 Exterior Surface Preparation		Crew	Daily Output	Labor-Hours	Unit	Material	2015 Bare Costs Labor	Equipment	Total	Total Incl O&P
0010	**EXTERIOR SURFACE PREPARATION** R099100-10									
0015	Doors, per side, not incl. frames or trim									
0020	Scrape & sand									
0030	Wood, flush	1 Pord	616	.013	S.F.		.52		.52	.84
0040	Wood, detail		496	.016			.65		.65	1.05
0050	Wood, louvered		280	.029			1.15		1.15	1.85
0060	Wood, overhead		616	.013			.52		.52	.84
0070	Wire brush									
0080	Metal, flush	1 Pord	640	.013	S.F.		.50		.50	.81
0090	Metal, detail		520	.015			.62		.62	1
0100	Metal, louvered		360	.022			.90		.90	1.44
0110	Metal or fibr., overhead		640	.013			.50		.50	.81
0120	Metal, roll up		560	.014			.58		.58	.93
0130	Metal, bulkhead		640	.013			.50		.50	.81
0140	Power wash, based on 2500 lb. operating pressure									
0150	Metal, flush	A-1H	2240	.004	S.F.		.13	.03	.16	.26
0160	Metal, detail		2120	.004			.14	.04	.18	.27
0170	Metal, louvered		2000	.004			.15	.04	.19	.29
0180	Metal or fibr., overhead		2400	.003			.13	.03	.16	.24
0190	Metal, roll up		2400	.003			.13	.03	.16	.24
0200	Metal, bulkhead		2200	.004			.14	.03	.17	.26
0400	Windows, per side, not incl. trim									
0410	Scrape & sand									
0420	Wood, 1-2 lite	1 Pord	320	.025	S.F.		1.01		1.01	1.62
0430	Wood, 3-6 lite		280	.029			1.15		1.15	1.85
0440	Wood, 7-10 lite		240	.033			1.34		1.34	2.16
0450	Wood, 12 lite		200	.040			1.61		1.61	2.59
0460	Wood, Bay/Bow		320	.025			1.01		1.01	1.62
0470	Wire brush									
0480	Metal, 1-2 lite	1 Pord	480	.017	S.F.		.67		.67	1.08
0490	Metal, 3-6 lite		400	.020			.81		.81	1.30
0500	Metal, Bay/Bow		480	.017			.67		.67	1.08
0510	Power wash, based on 2500 lb. operating pressure									
0520	1-2 lite	A-1H	4400	.002	S.F.		.07	.02	.09	.13
0530	3-6 lite		4320	.002			.07	.02	.09	.13
0540	7-10 lite		4240	.002			.07	.02	.09	.14
0550	12 lite		4160	.002			.07	.02	.09	.14
0560	Bay/Bow		4400	.002			.07	.02	.09	.13
0600	Siding, scrape and sand, light=10-30%, med.=30-70%									
0610	Heavy=70-100% of surface to sand									
0650	Texture 1-11, light	1 Pord	480	.017	S.F.		.67		.67	1.08
0660	Med.		440	.018			.73		.73	1.18
0670	Heavy		360	.022			.90		.90	1.44
0680	Wood shingles, shakes, light		440	.018			.73		.73	1.18
0690	Med.		360	.022			.90		.90	1.44
0700	Heavy		280	.029			1.15		1.15	1.85
0710	Clapboard, light		520	.015			.62		.62	1
0720	Med.		480	.017			.67		.67	1.08
0730	Heavy		400	.020			.81		.81	1.30
0740	Wire brush									
0750	Aluminum, light	1 Pord	600	.013	S.F.		.54		.54	.86
0760	Med.		520	.015			.62		.62	1

09 91 03.30 Exterior Surface Preparation

	09 91 03.30 Exterior Surface Preparation	Crew	Daily Output	Labor-Hours	Unit	Material	2015 Bare Costs Labor	Equipment	Total	Total Incl O&P
0770	Heavy	1 Pord	440	.018	S.F.		.73		.73	1.18
0780	Pressure wash, based on 2500 lb. operating pressure									
0790	Stucco	A-1H	3080	.003	S.F.		.10	.02	.12	.19
0800	Aluminum or vinyl		3200	.003			.09	.02	.11	.18
0810	Siding, masonry, brick & block		2400	.003			.13	.03	.16	.24
1300	Miscellaneous, wire brush									
1310	Metal, pedestrian gate	1 Pord	100	.080	S.F.		3.23		3.23	5.20
1320	Aluminum chain link, both sides		250	.032			1.29		1.29	2.08
1400	Existing galvanized surface, clean and prime, prep for painting		380	.021		.11	.85		.96	1.50
8000	For chemical washing, see Section 04 01 30									
8010	For steam cleaning, see Section 04 01 30.20									
8020	For sand blasting, see Section 03 35 29.60 and 05 01 10.51									

09 91 03.40 Interior Surface Preparation

	09 91 03.40 Interior Surface Preparation	Crew	Daily Output	Labor-Hours	Unit	Material	Labor	Equipment	Total	Total Incl O&P
0010	**INTERIOR SURFACE PREPARATION**									
0020	Doors, per side, not incl. frames or trim									
0030	Scrape & sand									
0040	Wood, flush	1 Pord	616	.013	S.F.		.52		.52	.84
0050	Wood, detail		496	.016			.65		.65	1.05
0060	Wood, louvered		280	.029			1.15		1.15	1.85
0070	Wire brush									
0080	Metal, flush	1 Pord	640	.013	S.F.		.50		.50	.81
0090	Metal, detail		520	.015			.62		.62	1
0100	Metal, louvered		360	.022			.90		.90	1.44
0110	Hand wash									
0120	Wood, flush	1 Pord	2160	.004	S.F.		.15		.15	.24
0130	Wood, detailed		2000	.004			.16		.16	.26
0140	Wood, louvered		1360	.006			.24		.24	.38
0150	Metal, flush		2160	.004			.15		.15	.24
0160	Metal, detail		2000	.004			.16		.16	.26
0170	Metal, louvered		1360	.006			.24		.24	.38
0400	Windows, per side, not incl. trim									
0410	Scrape & sand									
0420	Wood, 1-2 lite	1 Pord	360	.022	S.F.		.90		.90	1.44
0430	Wood, 3-6 lite		320	.025			1.01		1.01	1.62
0440	Wood, 7-10 lite		280	.029			1.15		1.15	1.85
0450	Wood, 12 lite		240	.033			1.34		1.34	2.16
0460	Wood, Bay/Bow		360	.022			.90		.90	1.44
0470	Wire brush									
0480	Metal, 1-2 lite	1 Pord	520	.015	S.F.		.62		.62	1
0490	Metal, 3-6 lite		440	.018			.73		.73	1.18
0500	Metal, Bay/Bow		520	.015			.62		.62	1
0600	Walls, sanding, light=10-30%, medium - 30-70%,									
0610	heavy=70-100% of surface to sand									
0650	Walls, sand									
0660	Gypsum board or plaster, light	1 Pord	3077	.003	S.F.		.10		.10	.17
0670	Gypsum board or plaster, medium		2160	.004			.15		.15	.24
0680	Gypsum board or plaster, heavy		923	.009			.35		.35	.56
0690	Wood, T&G, light		2400	.003			.13		.13	.22
0700	Wood, T&G, med.		1600	.005			.20		.20	.32
0710	Wood, T&G, heavy		800	.010			.40		.40	.65
0720	Walls, wash									
0730	Gypsum board or plaster	1 Pord	3200	.003	S.F.		.10		.10	.16

09 91 03 – Paint Restoration

09 91 03.40 Interior Surface Preparation		Crew	Daily Output	Labor-Hours	Unit	Material	2015 Bare Costs Labor	Equipment	Total	Total Incl O&P
0740	Wood, T&G	1 Pord	3200	.003	S.F.		.10		.10	.16
0750	Masonry, brick & block, smooth		2800	.003			.12		.12	.19
0760	Masonry, brick & block, coarse		2000	.004			.16		.16	.26
8000	For chemical washing, see Section 04 01 30									
8010	For steam cleaning, see Section 04 01 30.20									
8020	For sand blasting, see Section 03 35 29.60 and 05 01 10.51									
9010	Minimum labor/equipment charge	1 Pord	3	2.667	Job		108		108	173

09 91 03.41 Scrape After Fire Damage

		Crew	Daily Output	Labor-Hours	Unit	Material	Labor	Equipment	Total	Total Incl O&P
0010	**SCRAPE AFTER FIRE DAMAGE**									
0050	Boards, 1" x 4"	1 Pord	336	.024	L.F.		.96		.96	1.54
0060	1" x 6"		260	.031			1.24		1.24	2
0070	1" x 8"		207	.039			1.56		1.56	2.51
0080	1" x 10"		174	.046			1.86		1.86	2.98
0500	Framing, 2" x 4"		265	.030			1.22		1.22	1.96
0510	2" x 6"		221	.036			1.46		1.46	2.35
0520	2" x 8"		190	.042			1.70		1.70	2.73
0530	2" x 10"		165	.048			1.96		1.96	3.14
0540	2" x 12"		144	.056			2.24		2.24	3.60
1000	Heavy framing, 3" x 4"		226	.035			1.43		1.43	2.30
1010	4" x 4"		210	.038			1.54		1.54	2.47
1020	4" x 6"		191	.042			1.69		1.69	2.72
1030	4" x 8"		165	.048			1.96		1.96	3.14
1040	4" x 10"		144	.056			2.24		2.24	3.60
1060	4" x 12"		131	.061			2.46		2.46	3.96
2900	For sealing, light damage		825	.010	S.F.	.14	.39		.53	.78
2920	Heavy damage		460	.017	"	.31	.70		1.01	1.47
9000	Minimum labor/equipment charge		3	2.667	Job		108		108	173

09 91 13 – Exterior Painting

09 91 13.30 Fences

		Crew	Daily Output	Labor-Hours	Unit	Material	Labor	Equipment	Total	Total Incl O&P
0010	**FENCES** R099100-20									
0100	Chain link or wire metal, one side, water base									
0110	Roll & brush, first coat	1 Pord	960	.008	S.F.	.08	.34		.42	.63
0120	Second coat		1280	.006		.07	.25		.32	.49
0130	Spray, first coat		2275	.004		.08	.14		.22	.32
0140	Second coat		2600	.003		.08	.12		.20	.29
0150	Picket, water base									
0160	Roll & brush, first coat	1 Pord	865	.009	S.F.	.08	.37		.45	.69
0170	Second coat		1050	.008		.08	.31		.39	.58
0180	Spray, first coat		2275	.004		.08	.14		.22	.32
0190	Second coat		2600	.003		.08	.12		.20	.29
0200	Stockade, water base									
0210	Roll & brush, first coat	1 Pord	1040	.008	S.F.	.08	.31		.39	.59
0220	Second coat		1200	.007		.08	.27		.35	.52
0230	Spray, first coat		2275	.004		.08	.14		.22	.32
0240	Second coat		2600	.003		.08	.12		.20	.29
9000	Minimum labor/equipment charge		2	4	Job		161		161	259

09 91 13.42 Miscellaneous, Exterior

		Crew	Daily Output	Labor-Hours	Unit	Material	Labor	Equipment	Total	Total Incl O&P
0010	**MISCELLANEOUS, EXTERIOR** R099100-20									
0015	For painting metals, see Section 09 97 13.23									
0100	Railing, ext., decorative wood, incl. cap & baluster									
0110	Newels & spindles @ 12" O.C.									
0120	Brushwork, stain, sand, seal & varnish									

09 91 13.42 Miscellaneous, Exterior		Crew	Daily Output	Labor-Hours	Unit	2015 Bare Costs Material	Labor	Equipment	Total	Total Incl O&P
0130	First coat	1 Pord	90	.089	L.F.	.81	3.59		4.40	6.65
0140	Second coat	"	120	.067	"	.81	2.69		3.50	5.20
0150	Rough sawn wood, 42" high, 2" x 2" verticals, 6" O.C.									
0160	Brushwork, stain, each coat	1 Pord	90	.089	L.F.	.26	3.59		3.85	6.05
0170	Wrought iron, 1" rail, 1/2" sq. verticals									
0180	Brushwork, zinc chromate, 60" high, bars 6" O.C.									
0190	Primer	1 Pord	130	.062	L.F.	.86	2.48		3.34	4.94
0200	Finish coat		130	.062		1.13	2.48		3.61	5.25
0210	Additional coat	↓	190	.042	↓	1.32	1.70		3.02	4.18
0220	Shutters or blinds, single panel, 2' x 4', paint all sides									
0230	Brushwork, primer	1 Pord	20	.400	Ea.	.66	16.15		16.81	26.50
0240	Finish coat, exterior latex		20	.400		.62	16.15		16.77	26.50
0250	Primer & 1 coat, exterior latex		13	.615		1.14	25		26.14	41.50
0260	Spray, primer		35	.229		.96	9.20		10.16	15.85
0270	Finish coat, exterior latex		35	.229		1.33	9.20		10.53	16.25
0280	Primer & 1 coat, exterior latex	↓	20	.400	↓	1.04	16.15		17.19	27
0290	For louvered shutters, add				S.F.	10%				
0300	Stair stringers, exterior, metal									
0310	Roll & brush, zinc chromate, to 14", each coat	1 Pord	320	.025	L.F.	.38	1.01		1.39	2.03
0320	Rough sawn wood, 4" x 12"									
0330	Roll & brush, exterior latex, each coat	1 Pord	215	.037	L.F.	.09	1.50		1.59	2.51
0340	Trellis/lattice, 2" x 2" @ 3" O.C. with 2" x 8" supports									
0350	Spray, latex, per side, each coat	1 Pord	475	.017	S.F.	.09	.68		.77	1.19
0450	Decking, ext., sealer, alkyd, brushwork, sealer coat		1140	.007		.10	.28		.38	.57
0460	1st coat		1140	.007		.10	.28		.38	.57
0470	2nd coat		1300	.006		.07	.25		.32	.48
0500	Paint, alkyd, brushwork, primer coat		1140	.007		.11	.28		.39	.59
0510	1st coat		1140	.007		.13	.28		.41	.60
0520	2nd coat		1300	.006		.09	.25		.34	.50
0600	Sand paint, alkyd, brushwork, 1 coat		150	.053	↓	.14	2.15		2.29	3.61
9000	Minimum labor/equipment charge	↓	2	4	Job		161		161	259

09 91 13.60 Siding Exterior		Crew	Daily Output	Labor-Hours	Unit	Material	Labor	Equipment	Total	Total Incl O&P
0010	**SIDING EXTERIOR**, Alkyd (oil base) R099100-10									
0450	Steel siding, oil base, paint 1 coat, brushwork	2 Pord	2015	.008	S.F.	.10	.32		.42	.62
0500	Spray R099100-20		4550	.004		.15	.14		.29	.40
0800	Paint 2 coats, brushwork		1300	.012		.20	.50		.70	1.02
1000	Spray		2750	.006		.17	.23		.40	.57
1200	Stucco, rough, oil base, paint 2 coats, brushwork		1300	.012		.20	.50		.70	1.02
1400	Roller		1625	.010		.21	.40		.61	.87
1600	Spray		2925	.005		.22	.22		.44	.60
1800	Texture 1-11 or clapboard, oil base, primer coat, brushwork		1300	.012		.15	.50		.65	.96
2000	Spray		4550	.004		.15	.14		.29	.39
2100	Paint 1 coat, brushwork		1300	.012		.15	.50		.65	.96
2200	Spray		4550	.004		.15	.14		.29	.39
2400	Paint 2 coats, brushwork		810	.020		.30	.80		1.10	1.60
2600	Spray		2600	.006		.33	.25		.58	.76
3000	Stain 1 coat, brushwork		1520	.011		.09	.42		.51	.77
3200	Spray		5320	.003		.10	.12		.22	.31
3400	Stain 2 coats, brushwork		950	.017		.17	.68		.85	1.28
4000	Spray		3050	.005		.19	.21		.40	.55
4200	Wood shingles, oil base primer coat, brushwork		1300	.012		.14	.50		.64	.95
4400	Spray	↓	3900	.004	↓	.13	.17		.30	.41

09 91 13 – Exterior Painting

09 91 13.60 Siding Exterior		Crew	Daily Output	Labor-Hours	Unit	Material	2015 Bare Costs Labor	Equipment	Total	Total Incl O&P
4600	Paint 1 coat, brushwork	2 Pord	1300	.012	S.F.	.12	.50		.62	.94
4800	Spray		3900	.004		.15	.17		.32	.44
5000	Paint 2 coats, brushwork		810	.020		.25	.80		1.05	1.55
5200	Spray		2275	.007		.23	.28		.51	.72
5800	Stain 1 coat, brushwork		1500	.011		.09	.43		.52	.78
6000	Spray		3900	.004		.09	.17		.26	.36
6500	Stain 2 coats, brushwork		950	.017		.17	.68		.85	1.28
7000	Spray		2660	.006		.24	.24		.48	.65
8000	For latex paint, deduct					10%				
8100	For work over 12' H, from pipe scaffolding, add						15%			
8200	For work over 12' H, from extension ladder, add						25%			
8300	For work over 12' H, from swing staging, add						35%			
9000	Minimum labor/equipment charge	1 Pord	2	4	Job		161		161	259

09 91 13.62 Siding, Misc.		Crew	Daily Output	Labor-Hours	Unit	Material	Labor	Equipment	Total	Total Incl O&P
0010	**SIDING, MISC.**, latex paint R099100-10									
0100	Aluminum siding									
0110	Brushwork, primer R099100-20	2 Pord	2275	.007	S.F.	.06	.28		.34	.53
0120	Finish coat, exterior latex		2275	.007		.06	.28		.34	.52
0130	Primer & 1 coat exterior latex		1300	.012		.13	.50		.63	.94
0140	Primer & 2 coats exterior latex		975	.016		.19	.66		.85	1.26
0150	Mineral fiber shingles									
0160	Brushwork, primer	2 Pord	1495	.011	S.F.	.15	.43		.58	.85
0170	Finish coat, industrial enamel		1495	.011		.18	.43		.61	.89
0180	Primer & 1 coat enamel		810	.020		.33	.80		1.13	1.64
0190	Primer & 2 coats enamel		540	.030		.51	1.20		1.71	2.49
0200	Roll, primer		1625	.010		.17	.40		.57	.82
0210	Finish coat, industrial enamel		1625	.010		.20	.40		.60	.86
0220	Primer & 1 coat enamel		975	.016		.36	.66		1.02	1.46
0230	Primer & 2 coats enamel		650	.025		.56	.99		1.55	2.22
0240	Spray, primer		3900	.004		.13	.17		.30	.41
0250	Finish coat, industrial enamel		3900	.004		.16	.17		.33	.45
0260	Primer & 1 coat enamel		2275	.007		.29	.28		.57	.78
0270	Primer & 2 coats enamel		1625	.010		.46	.40		.86	1.14
0280	Waterproof sealer, first coat		4485	.004		.09	.14		.23	.32
0290	Second coat		5235	.003		.08	.12		.20	.29
0300	Rough wood incl. shingles, shakes or rough sawn siding									
0310	Brushwork, primer	2 Pord	1280	.013	S.F.	.13	.50		.63	.96
0320	Finish coat, exterior latex		1280	.013		.10	.50		.60	.92
0330	Primer & 1 coat exterior latex		960	.017		.24	.67		.91	1.34
0340	Primer & 2 coats exterior latex		700	.023		.34	.92		1.26	1.85
0350	Roll, primer		2925	.005		.18	.22		.40	.55
0360	Finish coat, exterior latex		2925	.005		.12	.22		.34	.48
0370	Primer & 1 coat exterior latex		1790	.009		.30	.36		.66	.91
0380	Primer & 2 coats exterior latex		1300	.012		.42	.50		.92	1.26
0390	Spray, primer		3900	.004		.15	.17		.32	.43
0400	Finish coat, exterior latex		3900	.004		.09	.17		.26	.37
0410	Primer & 1 coat exterior latex		2600	.006		.24	.25		.49	.67
0420	Primer & 2 coats exterior latex		2080	.008		.34	.31		.65	.87
0430	Waterproof sealer, first coat		4485	.004		.16	.14		.30	.40
0440	Second coat		4485	.004		.09	.14		.23	.32
0450	Smooth wood incl. butt, T&G, beveled, drop or B&B siding									
0460	Brushwork, primer	2 Pord	2325	.007	S.F.	.10	.28		.38	.56

09 91 Painting

09 91 13 – Exterior Painting

09 91 13.62 Siding, Misc.		Crew	Daily Output	Labor-Hours	Unit	Material	2015 Bare Costs Labor	Equipment	Total	Total Incl O&P
0470	Finish coat, exterior latex	2 Pord	1280	.013	S.F.	.10	.50		.60	.92
0480	Primer & 1 coat exterior latex		800	.020		.20	.81		1.01	1.52
0490	Primer & 2 coats exterior latex		630	.025		.30	1.02		1.32	1.98
0500	Roll, primer		2275	.007		.11	.28		.39	.58
0510	Finish coat, exterior latex		2275	.007		.11	.28		.39	.58
0520	Primer & 1 coat exterior latex		1300	.012		.22	.50		.72	1.04
0530	Primer & 2 coats exterior latex		975	.016		.33	.66		.99	1.42
0540	Spray, primer		4550	.004		.08	.14		.22	.32
0550	Finish coat, exterior latex		4550	.004		.09	.14		.23	.33
0560	Primer & 1 coat exterior latex		2600	.006		.18	.25		.43	.60
0570	Primer & 2 coats exterior latex		1950	.008		.27	.33		.60	.83
0580	Waterproof sealer, first coat		5230	.003		.09	.12		.21	.29
0590	Second coat		5980	.003		.09	.11		.20	.26
0600	For oil base paint, add					10%				
9000	Minimum labor/equipment charge	1 Pord	2	4	Job		161		161	259

09 91 13.70 Doors and Windows, Exterior		Crew	Daily Output	Labor-Hours	Unit	Material	Labor	Equipment	Total	Total Incl O&P
0010	**DOORS AND WINDOWS, EXTERIOR** R099100-10									
0100	Door frames & trim, only									
0110	Brushwork, primer R099100-20	1 Pord	512	.016	L.F.	.06	.63		.69	1.08
0120	Finish coat, exterior latex		512	.016		.08	.63		.71	1.10
0130	Primer & 1 coat, exterior latex		300	.027		.14	1.08		1.22	1.88
0135	2 coats, exterior latex, both sides		15	.533	Ea.	6.95	21.50		28.45	42
0140	Primer & 2 coats, exterior latex		265	.030	L.F.	.22	1.22		1.44	2.20
0150	Doors, flush, both sides, incl. frame & trim									
0160	Roll & brush, primer	1 Pord	10	.800	Ea.	4.55	32.50		37.05	57
0170	Finish coat, exterior latex		10	.800		5.95	32.50		38.45	58.50
0180	Primer & 1 coat, exterior latex		7	1.143		10.50	46		56.50	85.50
0190	Primer & 2 coats, exterior latex		5	1.600		16.40	64.50		80.90	122
0200	Brushwork, stain, sealer & 2 coats polyurethane		4	2		28.50	80.50		109	162
0210	Doors, French, both sides, 10-15 lite, incl. frame & trim									
0220	Brushwork, primer	1 Pord	6	1.333	Ea.	2.27	54		56.27	89
0230	Finish coat, exterior latex		6	1.333		2.97	54		56.97	90
0240	Primer & 1 coat, exterior latex		3	2.667		5.25	108		113.25	179
0250	Primer & 2 coats, exterior latex		2	4		8.05	161		169.05	268
0260	Brushwork, stain, sealer & 2 coats polyurethane		2.50	3.200		10.30	129		139.30	219
0270	Doors, louvered, both sides, incl. frame & trim									
0280	Brushwork, primer	1 Pord	7	1.143	Ea.	4.55	46		50.55	79
0290	Finish coat, exterior latex		7	1.143		5.95	46		51.95	80.50
0300	Primer & 1 coat, exterior latex		4	2		10.50	80.50		91	142
0310	Primer & 2 coats, exterior latex		3	2.667		16.10	108		124.10	191
0320	Brushwork, stain, sealer & 2 coats polyurethane		4.50	1.778		28.50	71.50		100	147
0330	Doors, panel, both sides, incl. frame & trim									
0340	Roll & brush, primer	1 Pord	6	1.333	Ea.	4.55	54		58.55	91.50
0350	Finish coat, exterior latex		6	1.333		5.95	54		59.95	93
0360	Primer & 1 coat, exterior latex		3	2.667		10.50	108		118.50	185
0370	Primer & 2 coats, exterior latex		2.50	3.200		16.10	129		145.10	226
0380	Brushwork, stain, sealer & 2 coats polyurethane		3	2.667		28.50	108		136.50	205
0400	Windows, per ext. side, based on 15 S.F.									
0410	1 to 6 lite									
0420	Brushwork, primer	1 Pord	13	.615	Ea.	.90	25		25.90	41
0430	Finish coat, exterior latex		13	.615		1.17	25		26.17	41.50
0440	Primer & 1 coat, exterior latex		8	1		2.07	40.50		42.57	67.50

09 91 13.70 Doors and Windows, Exterior		Crew	Daily Output	Labor-Hours	Unit	Material	2015 Bare Costs Labor	Equipment	Total	Total Incl O&P
0450	Primer & 2 coats, exterior latex	1 Pord	6	1.333	Ea.	3.17	54		57.17	90
0460	Stain, sealer & 1 coat varnish	↓	7	1.143	↓	4.06	46		50.06	78.50
0470	7 to 10 lite									
0480	Brushwork, primer	1 Pord	11	.727	Ea.	.90	29.50		30.40	48
0490	Finish coat, exterior latex		11	.727		1.17	29.50		30.67	48.50
0500	Primer & 1 coat, exterior latex		7	1.143		2.07	46		48.07	76.50
0510	Primer & 2 coats, exterior latex		5	1.600		3.17	64.50		67.67	107
0520	Stain, sealer & 1 coat varnish	↓	6	1.333	↓	4.06	54		58.06	91
0530	12 lite									
0540	Brushwork, primer	1 Pord	10	.800	Ea.	.90	32.50		33.40	53
0550	Finish coat, exterior latex		10	.800		1.17	32.50		33.67	53.50
0560	Primer & 1 coat, exterior latex		6	1.333		2.07	54		56.07	89
0570	Primer & 2 coats, exterior latex		5	1.600		3.17	64.50		67.67	107
0580	Stain, sealer & 1 coat varnish	↓	6	1.333		4.15	54		58.15	91
0590	For oil base paint, add				↓	10%				
9000	Minimum labor/equipment charge	1 Pord	2	4	Job		161		161	259

09 91 13.80 Trim, Exterior		Crew	Daily Output	Labor-Hours	Unit	Material	Labor	Equipment	Total	Total Incl O&P
0010	**TRIM, EXTERIOR** R099100-10									
0100	Door frames & trim (see Doors, interior or exterior)									
0110	Fascia, latex paint, one coat coverage R099100-20									
0120	1" x 4", brushwork	1 Pord	640	.013	L.F.	.02	.50		.52	.84
0130	Roll		1280	.006		.03	.25		.28	.44
0140	Spray		2080	.004		.02	.16		.18	.27
0150	1" x 6" to 1" x 10", brushwork		640	.013		.08	.50		.58	.90
0160	Roll		1230	.007		.09	.26		.35	.52
0170	Spray		2100	.004		.07	.15		.22	.32
0180	1" x 12", brushwork		640	.013		.08	.50		.58	.90
0190	Roll		1050	.008		.09	.31		.40	.59
0200	Spray	↓	2200	.004	↓	.07	.15		.22	.31
0210	Gutters & downspouts, metal, zinc chromate paint									
0220	Brushwork, gutters, 5", first coat	1 Pord	640	.013	L.F.	.40	.50		.90	1.25
0230	Second coat		960	.008		.38	.34		.72	.95
0240	Third coat		1280	.006		.30	.25		.55	.75
0250	Downspouts, 4", first coat		640	.013		.40	.50		.90	1.25
0260	Second coat		960	.008		.38	.34		.72	.95
0270	Third coat	↓	1280	.006	↓	.30	.25		.55	.75
0280	Gutters & downspouts, wood									
0290	Brushwork, gutters, 5", primer	1 Pord	640	.013	L.F.	.06	.50		.56	.88
0300	Finish coat, exterior latex		640	.013		.07	.50		.57	.89
0310	Primer & 1 coat exterior latex		400	.020		.14	.81		.95	1.45
0320	Primer & 2 coats exterior latex		325	.025		.22	.99		1.21	1.84
0330	Downspouts, 4", primer		640	.013		.06	.50		.56	.88
0340	Finish coat, exterior latex		640	.013		.07	.50		.57	.89
0350	Primer & 1 coat exterior latex		400	.020		.14	.81		.95	1.45
0360	Primer & 2 coats exterior latex	↓	325	.025	↓	.11	.99		1.10	1.72
0370	Molding, exterior, up to 14" wide									
0380	Brushwork, primer	1 Pord	640	.013	L.F.	.07	.50		.57	.89
0390	Finish coat, exterior latex		640	.013		.08	.50		.58	.90
0400	Primer & 1 coat exterior latex		400	.020		.17	.81		.98	1.48
0410	Primer & 2 coats exterior latex		315	.025		.17	1.02		1.19	1.83
0420	Stain & fill		1050	.008		.10	.31		.41	.60
0430	Shellac	↓	1850	.004	↓	.13	.17		.30	.42

09 91 Painting

09 91 13 – Exterior Painting

09 91 13.80 Trim, Exterior		Crew	Daily Output	Labor-Hours	Unit	Material	Labor	Equipment	Total	Total Incl O&P
0440	Varnish	1 Pord	1275	.006	L.F.	.11	.25		.36	.53
9000	Minimum labor/equipment charge	↓	2	4	Job		161		161	259

(Material, Labor, Equipment and Total are 2015 Bare Costs.)

09 91 13.90 Walls, Masonry (CMU), Exterior		Crew	Daily Output	Labor-Hours	Unit	Material	Labor	Equipment	Total	Total Incl O&P
0010	**WALLS, MASONRY (CMU), EXTERIOR**									
0360	Concrete masonry units (CMU), smooth surface									
0370	Brushwork, latex, first coat	1 Pord	640	.013	S.F.	.07	.50		.57	.89
0380	Second coat		960	.008		.06	.34		.40	.60
0390	Waterproof sealer, first coat		736	.011		.25	.44		.69	.97
0400	Second coat		1104	.007		.25	.29		.54	.74
0410	Roll, latex, paint, first coat		1465	.005		.09	.22		.31	.44
0420	Second coat		1790	.004		.07	.18		.25	.36
0430	Waterproof sealer, first coat		1680	.005		.25	.19		.44	.58
0440	Second coat		2060	.004		.25	.16		.41	.52
0450	Spray, latex, paint, first coat		1950	.004		.07	.17		.24	.34
0460	Second coat		2600	.003		.05	.12		.17	.26
0470	Waterproof sealer, first coat		2245	.004		.25	.14		.39	.50
0480	Second coat	↓	2990	.003	↓	.25	.11		.36	.44
0490	Concrete masonry unit (CMU), porous									
0500	Brushwork, latex, first coat	1 Pord	640	.013	S.F.	.14	.50		.64	.97
0510	Second coat		960	.008		.07	.34		.41	.62
0520	Waterproof sealer, first coat		736	.011		.25	.44		.69	.97
0530	Second coat		1104	.007		.25	.29		.54	.74
0540	Roll latex, first coat		1465	.005		.11	.22		.33	.47
0550	Second coat		1790	.004		.07	.18		.25	.37
0560	Waterproof sealer, first coat		1680	.005		.25	.19		.44	.58
0570	Second coat		2060	.004		.25	.16		.41	.52
0580	Spray latex, first coat		1950	.004		.08	.17		.25	.36
0590	Second coat		2600	.003		.05	.12		.17	.26
0600	Waterproof sealer, first coat		2245	.004		.25	.14		.39	.50
0610	Second coat		2990	.003	↓	.25	.11		.36	.44
9000	Minimum labor/equipment charge	↓	2	4	Job		161		161	259

09 91 23 – Interior Painting

09 91 23.20 Cabinets and Casework		Crew	Daily Output	Labor-Hours	Unit	Material	Labor	Equipment	Total	Total Incl O&P
0010	**CABINETS AND CASEWORK** R099100-10									
1000	Primer coat, oil base, brushwork	1 Pord	650	.012	S.F.	.07	.50		.57	.88
2000	Paint, oil base, brushwork, 1 coat R099100-20		650	.012		.11	.50		.61	.92
2500	2 coats		400	.020		.21	.81		1.02	1.53
3000	Stain, brushwork, wipe off		650	.012		.09	.50		.59	.89
4000	Shellac, 1 coat, brushwork		650	.012		.11	.50		.61	.92
4500	Varnish, 3 coats, brushwork, sand after 1st coat	↓	325	.025		.28	.99		1.27	1.91
5000	For latex paint, deduct				↓	10%				
6300	Strip, prep and refinish wood furniture									
6310	Remove paint using chemicals, wood furniture	1 Pord	28	.286	S.F.	1.58	11.55		13.13	20.50
6320	Prep for painting, sanding		75	.107		.23	4.30		4.53	7.15
6350	Stain and wipe, brushwork		600	.013		.09	.54		.63	.95
6355	Spray applied		900	.009		.08	.36		.44	.66
6360	Sealer or varnish, brushwork		1080	.007		.09	.30		.39	.58
6365	Spray applied		2100	.004		.09	.15		.24	.35
6370	Paint, primer, brushwork		720	.011		.07	.45		.52	.80
6375	Spray applied		2100	.004		.07	.15		.22	.32
6380	Finish coat, brushwork		810	.010		.11	.40		.51	.76
6385	Spray applied	↓	2100	.004	↓	.10	.15		.25	.36

09 91 Painting

09 91 23 – Interior Painting

09 91 23.20 Cabinets and Casework		Crew	Daily Output	Labor-Hours	Unit	Material	2015 Bare Costs Labor	Equipment	Total	Total Incl O&P
9010	Minimum labor/equipment charge	1 Pord	2	4	Job		161		161	259

09 91 23.33 Doors and Windows, Interior Alkyd (Oil Base)

		Crew	Daily Output	Labor-Hours	Unit	Material	Labor	Equipment	Total	Total Incl O&P
0010	**DOORS AND WINDOWS, INTERIOR ALKYD (OIL BASE)** R099100-10									
0500	Flush door & frame, 3′ x 7′, oil, primer, brushwork	1 Pord	10	.800	Ea.	3.45	32.50		35.95	56
1000	Paint, 1 coat R099100-20		10	.800		4.14	32.50		36.64	56.50
1200	2 coats		6	1.333		5	54		59	92
1220	3 coats		5	1.600		15.10	64.50		79.60	121
1400	Stain, brushwork, wipe off		18	.444		1.81	17.95		19.76	31
1600	Shellac, 1 coat, brushwork		25	.320		2.25	12.90		15.15	23.50
1800	Varnish, 3 coats, brushwork, sand after 1st coat		9	.889		5.85	36		41.85	64
2000	Panel door & frame, 3′ x 7′, oil, primer, brushwork		6	1.333		2.66	54		56.66	89.50
2200	Paint, 1 coat		6	1.333		4.14	54		58.14	91
2400	2 coats		3	2.667		10.95	108		118.95	185
2420	3 coats		2	4		14.60	161		175.60	275
2600	Stain, brushwork, panel door, 3′ x 7′, not incl. frame		16	.500		1.81	20		21.81	34.50
2800	Shellac, 1 coat, brushwork		22	.364		2.25	14.65		16.90	26
3000	Varnish, 3 coats, brushwork, sand after 1st coat	↓	7.50	1.067	↓	5.85	43		48.85	75.50
4400	Windows, including frame and trim, per side									
4600	Colonial type, 6/6 lites, 2′ x 3′, oil, primer, brushwork	1 Pord	14	.571	Ea.	.42	23		23.42	37.50
5800	Paint, 1 coat		14	.571		.65	23		23.65	37.50
6000	2 coats		9	.889		1.27	36		37.27	59
6010	3 coats		7	1.143		1.89	46		47.89	76
6200	3′ x 5′ opening, 6/6 lites, primer coat, brushwork		12	.667		1.05	27		28.05	44
6400	Paint, 1 coat		12	.667		1.64	27		28.64	45
6600	2 coats		7	1.143		3.18	46		49.18	77.50
6610	3 coats		6	1.333		4.72	54		58.72	91.50
6800	4′ x 8′ opening, 6/6 lites, primer coat, brushwork		8	1		2.24	40.50		42.74	67.50
7000	Paint, 1 coat		8	1		3.49	40.50		43.99	69
7200	2 coats		5	1.600		6.80	64.50		71.30	111
7210	3 coats		4	2		10.05	80.50		90.55	141
7500	Standard, 6/6 lites, 2′ x 3′, primer coat, brushwork		14	.571		.42	23		23.42	37.50
7520	Paint 1 coat		14	.571		.65	23		23.65	37.50
7540	2 coats		9	.889		1.27	36		37.27	59
7560	3 coats		7	1.143		1.89	46		47.89	76
7580	3′ x 5′, 6/6 lites, primer coat, brushwork		12	.667		1.05	27		28.05	44
7600	Paint 1 coat		12	.667		1.64	27		28.64	45
7620	2 coats		7	1.143		3.18	46		49.18	77.50
7640	3 coats		6	1.333		4.72	54		58.72	91.50
7660	4′ x 8′, 6/6 lites, primer coat, brushwork		8	1		2.24	40.50		42.74	67.50
7680	Paint 1 coat		8	1		3.49	40.50		43.99	69
7700	2 coats		5	1.600		6.80	64.50		71.30	111
7720	3 coats		4	2		10.05	80.50		90.55	141
8000	Single lite type, 2′ x 3′, oil base, primer coat, brushwork		33	.242		.42	9.80		10.22	16.15
8200	Paint, 1 coat		33	.242		.65	9.80		10.45	16.40
8400	2 coats		20	.400		1.27	16.15		17.42	27.50
8410	3 coats		16	.500		1.89	20		21.89	34.50
8600	3′ x 5′ opening, primer coat, brushwork		20	.400		1.05	16.15		17.20	27
8800	Paint, 1 coat		20	.400		1.64	16.15		17.79	28
8900	2 coats		13	.615		3.18	25		28.18	43.50
9010	3 coats		10	.800		4.72	32.50		37.22	57
9200	4′ x 8′ opening, primer coat, brushwork		14	.571		2.24	23		25.24	39.50
9400	Paint, 1 coat	↓	14	.571	↓	3.49	23		26.49	41

09 91 23.33 Doors and Windows, Interior Alkyd (Oil Base)		Crew	Daily Output	Labor-Hours	Unit	2015 Bare Costs Material	Labor	Equipment	Total	Total Incl O&P
9600	2 coats	1 Pord	8	1	Ea.	6.80	40.50		47.30	72.50
9610	3 coats		7	1.143		10.05	46		56.05	85
9900	Minimum labor/equipment charge		2	4	Job		161		161	259

09 91 23.35 Doors and Windows, Interior Latex		Crew	Daily Output	Labor-Hours	Unit	2015 Bare Costs Material	Labor	Equipment	Total	Total Incl O&P
0010	**DOORS & WINDOWS, INTERIOR LATEX** R099100-10									
0100	Doors, flush, both sides, incl. frame & trim									
0110	Roll & brush, primer	1 Pord	10	.800	Ea.	4.13	32.50		36.63	56.50
0120	Finish coat, latex		10	.800		5.30	32.50		37.80	58
0130	Primer & 1 coat latex		7	1.143		9.40	46		55.40	84.50
0140	Primer & 2 coats latex		5	1.600		14.40	64.50		78.90	120
0160	Spray, both sides, primer		20	.400		4.35	16.15		20.50	31
0170	Finish coat, latex		20	.400		5.55	16.15		21.70	32
0180	Primer & 1 coat latex		11	727		9.95	29.50		39.45	58
0190	Primer & 2 coats latex		8	1		15.25	40.50		55.75	82
0200	Doors, French, both sides, 10-15 lite, incl. frame & trim									
0210	Roll & brush, primer	1 Pord	6	1.333	Ea.	2.06	54		56.06	89
0220	Finish coat, latex		6	1.333		2.65	54		56.65	89.50
0230	Primer & 1 coat latex		3	2.667		4.71	108		112.71	178
0240	Primer & 2 coats latex		2	4		7.20	161		168.20	267
0260	Doors, louvered, both sides, incl. frame & trim									
0270	Roll & brush, primer	1 Pord	7	1.143	Ea.	4.13	46		50.13	78.50
0280	Finish coat, latex		7	1.143		5.30	46		51.30	80
0290	Primer & 1 coat, latex		4	2		9.20	80.50		89.70	140
0300	Primer & 2 coats, latex		3	2.667		14.70	108		122.70	189
0320	Spray, both sides, primer		20	.400		4.35	16.15		20.50	31
0330	Finish coat, latex		20	.400		5.55	16.15		21.70	32
0340	Primer & 1 coat, latex		11	.727		9.95	29.50		39.45	58
0350	Primer & 2 coats, latex		8	1		15.60	40.50		56.10	82
0360	Doors, panel, both sides, incl. frame & trim									
0370	Roll & brush, primer	1 Pord	6	1.333	Ea.	4.35	54		58.35	91.50
0380	Finish coat, latex		6	1.333		5.30	54		59.30	92.50
0390	Primer & 1 coat, latex		3	2.667		9.40	108		117.40	183
0400	Primer & 2 coats, latex		2.50	3.200		14.70	129		143.70	224
0420	Spray, both sides, primer		10	.800		4.35	32.50		36.85	57
0430	Finish coat, latex		10	.800		5.55	32.50		38.05	58
0440	Primer & 1 coat, latex		5	1.600		9.95	64.50		74.45	115
0450	Primer & 2 coats, latex		4	2		15.60	80.50		96.10	147
0460	Windows, per interior side, based on 15 S.F.									
0470	1 to 6 lite									
0480	Brushwork, primer	1 Pord	13	.615	Ea.	.81	25		25.81	41
0490	Finish coat, enamel		13	.615		1.04	25		26.04	41
0500	Primer & 1 coat enamel		8	1		1.86	40.50		42.36	67
0510	Primer & 2 coats enamel		6	1.333		2.90	54		56.90	89.50
0530	7 to 10 lite									
0540	Brushwork, primer	1 Pord	11	.727	Ea.	.81	29.50		30.31	48
0550	Finish coat, enamel		11	.727		1.04	29.50		30.54	48
0560	Primer & 1 coat enamel		7	1.143		1.86	46		47.86	76
0570	Primer & 2 coats enamel		5	1.600		2.90	64.50		67.40	107
0590	12 lite									
0600	Brushwork, primer	1 Pord	10	.800	Ea.	.81	32.50		33.31	53
0610	Finish coat, enamel		10	.800		1.04	32.50		33.54	53
0620	Primer & 1 coat enamel		6	1.333		1.86	54		55.86	88.50

09 91 23.35 Doors and Windows, Interior Latex		Crew	Daily Output	Labor-Hours	Unit	Material	Labor	Equipment	Total	Total Incl O&P
						2015 Bare Costs				
0630	Primer & 2 coats enamel	1 Pord	5	1.600	Ea.	2.90	64.50		67.40	107
0650	For oil base paint, add					10%				
9000	Minimum labor/equipment charge	1 Pord	2	4	Job		161		161	259

09 91 23.39 Doors and Windows, Interior Latex, Zero Voc

		Crew	Daily Output	Labor-Hours	Unit	Material	Labor	Equipment	Total	Total Incl O&P
0010	**DOORS & WINDOWS, INTERIOR LATEX, ZERO VOC**									
0100	Doors flush, both sides, incl. frame & trim									
0110	Roll & brush, primer G	1 Pord	10	.800	Ea.	4.90	32.50		37.40	57.50
0120	Finish coat, latex G		10	.800		5.70	32.50		38.20	58.50
0130	Primer & 1 coat latex G		7	1.143		10.60	46		56.60	85.50
0140	Primer & 2 coats latex G		5	1.600		16	64.50		80.50	122
0160	Spray, both sides, primer G		20	.400		5.15	16.15		21.30	31.50
0170	Finish coat, latex G		20	.400		6	16.15		22.15	32.50
0180	Primer & 1 coat latex G		11	.727		11.25	29.50		40.75	59.50
0190	Primer & 2 coats latex G		8	1		16.95	40.50		57.45	83.50
0200	Doors, French, both sides, 10-15 lite, incl. frame & trim									
0210	Roll & brush, primer G	1 Pord	6	1.333	Ea.	2.45	54		56.45	89
0220	Finish coat, latex G		6	1.333		2.86	54		56.86	89.50
0230	Primer & 1 coat latex G		3	2.667		5.30	108		113.30	179
0240	Primer & 2 coats latex G		2	4		8	161		169	268
0360	Doors, panel, both sides, incl. frame & trim									
0370	Roll & brush, primer G	1 Pord	6	1.333	Ea.	5.15	54		59.15	92
0380	Finish coat, latex G		6	1.333		5.70	54		59.70	93
0390	Primer & 1 coat, latex G		3	2.667		10.60	108		118.60	185
0400	Primer & 2 coats, latex G		2.50	3.200		16.35	129		145.35	226
0420	Spray, both sides, primer G		10	.800		5.15	32.50		37.65	57.50
0430	Finish coat, latex G		10	.800		6	32.50		38.50	58.50
0440	Primer & 1 coat, latex G		5	1.600		11.25	64.50		75.75	116
0450	Primer & 2 coats, latex G		4	2		17.30	80.50		97.80	149
0460	Windows, per interior side, based on 15 S.F.									
0470	1 to 6 lite									
0480	Brushwork, primer G	1 Pord	13	.615	Ea.	.97	25		25.97	41
0490	Finish coat, enamel G		13	.615		1.13	25		26.13	41
0500	Primer & 1 coat enamel G		8	1		2.09	40.50		42.59	67.50
0510	Primer & 2 coats enamel G		6	1.333		3.22	54		57.22	90
9000	Minimum labor/equipment charge		2	4	Job		161		161	259

09 91 23.40 Floors, Interior

		Crew	Daily Output	Labor-Hours	Unit	Material	Labor	Equipment	Total	Total Incl O&P
0010	**FLOORS, INTERIOR** R099100-10									
0100	Concrete paint, latex									
0110	Brushwork									
0120	1st coat	1 Pord	975	.008	S.F.	.15	.33		.48	.69
0130	2nd coat		1150	.007		.10	.28		.38	.56
0140	3rd coat		1300	.006		.08	.25		.33	.49
0150	Roll									
0160	1st coat	1 Pord	2600	.003	S.F.	.20	.12		.32	.42
0170	2nd coat		3250	.002		.12	.10		.22	.29
0180	3rd coat		3900	.002		.09	.08		.17	.23
0190	Spray									
0200	1st coat	1 Pord	2600	.003	S.F.	.17	.12		.29	.39
0210	2nd coat		3250	.002		.09	.10		.19	.26
0220	3rd coat		3900	.002		.07	.08		.15	.21

	09 91 23.44 Anti-Slip Floor Treatments	Crew	Daily Output	Labor-Hours	Unit	2015 Bare Costs Material	Labor	Equipment	Total	Total Incl O&P
0010	**ANTI-SLIP FLOOR TREATMENTS**									
1000	Walking surface treatment, ADA compliant, mop on and rinse									
1100	For tile, terrazzo, stone or smooth concrete	1 Pord	4000	.002	S.F.	.32	.08		.40	.49
1110	For marble		4000	.002		.21	.08		.29	.36
1120	For wood		4000	.002		.20	.08		.28	.35
1130	For baths and showers		500	.016		.37	.65		1.02	1.45
2000	Granular additive for paint or sealer, add to paint cost					.02			.02	.02

09 91 23.52 Miscellaneous, Interior

	09 91 23.52 Miscellaneous, Interior	Crew	Daily Output	Labor-Hours	Unit	Material	Labor	Equipment	Total	Total Incl O&P
0010	**MISCELLANEOUS, INTERIOR** R099100-10									
2400	Floors, conc./wood, oil base, primer/sealer coat, brushwork	2 Pord	1950	.008	S.F.	.09	.33		.42	.63
2450	Roller		5200	.003		.10	.12		.22	.31
2600	Spray		6000	.003		.10	.11		.21	.28
2650	Paint 1 coat, brushwork		1950	.008		.10	.33		.43	.64
2800	Roller		5200	.003		.10	.12		.22	.31
2850	Spray		6000	.003		.11	.11		.22	.29
3000	Stain, wood floor, brushwork, 1 coat		4550	.004		.09	.14		.23	.32
3200	Roller		5200	.003		.09	.12		.21	.30
3250	Spray		6000	.003		.09	.11		.20	.27
3400	Varnish, wood floor, brushwork		4550	.004		.09	.14		.23	.33
3450	Roller		5200	.003		.10	.12		.22	.31
3600	Spray		6000	.003		.10	.11		.21	.28
3650	For dust proofing or anti skid, see Section 03 35 29.30									
3800	Grilles, per side, oil base, primer coat, brushwork	1 Pord	520	.015	S.F.	.14	.62		.76	1.15
3850	Spray		1140	.007		.15	.28		.43	.62
3880	Paint 1 coat, brushwork		520	.015		.22	.62		.84	1.24
3900	Spray		1140	.007		.24	.28		.52	.73
3920	Paint 2 coats, brushwork		325	.025		.42	.99		1.41	2.07
3940	Spray		650	.012		.48	.50		.98	1.33
3950	Prime & paint 1 coat		325	.025		.36	.99		1.35	1.99
3960	Prime & paint 2 coats		270	.030		.35	1.20		1.55	2.31
4500	Louvers, one side, primer, brushwork		524	.015		.09	.62		.71	1.09
4520	Paint one coat, brushwork		520	.015		.10	.62		.72	1.10
4530	Spray		1140	.007		.11	.28		.39	.58
4540	Paint two coats, brushwork		325	.025		.18	.99		1.17	1.80
4550	Spray		650	.012		.20	.50		.70	1.02
4560	Paint three coats, brushwork		270	.030		.27	1.20		1.47	2.22
4570	Spray		500	.016		.30	.65		.95	1.37
4600	Miscellaneous surfaces, metallic paint, spray applied									
4610	Water based, non-tintable, warm silver	1 Pord	1140	.007	S.F.	.75	.28		1.03	1.29
4620	Rusted iron		1140	.007		.97	.28		1.25	1.53
4630	Low VOC, tintable		1140	.007		.33	.28		.61	.82
5000	Pipe, 1" - 4" diameter, primer or sealer coat, oil base, brushwork	2 Pord	1250	.013	L.F.	.10	.52		.62	.94
5100	Spray		2165	.007		.09	.30		.39	.58
5200	Paint 1 coat, brushwork		1250	.013		.11	.52		.63	.95
5300	Spray		2165	.007		.10	.30		.40	.59
5350	Paint 2 coats, brushwork		775	.021		.19	.83		1.02	1.55
5400	Spray		1240	.013		.22	.52		.74	1.08
5420	Paint 3 coats, brushwork		775	.021		.29	.83		1.12	1.66
5450	5" - 8" diameter, primer or sealer coat, brushwork		620	.026		.20	1.04		1.24	1.89
5500	Spray		1085	.015		.32	.60		.92	1.31
5550	Paint 1 coat, brushwork		620	.026		.29	1.04		1.33	1.99
5600	Spray		1085	.015		.33	.60		.93	1.32

09 91 Painting

09 91 23 - Interior Painting

09 91 23.52 Miscellaneous, Interior		Crew	Daily Output	Labor-Hours	Unit	Material	Labor	Equipment	Total	Total Incl O&P
						2015 Bare Costs				
5650	Paint 2 coats, brushwork	2 Pord	385	.042	L.F.	.39	1.68		2.07	3.13
5700	Spray		620	.026		.43	1.04		1.47	2.14
5720	Paint 3 coats, brushwork		385	.042		.58	1.68		2.26	3.33
5750	9" - 12" diameter, primer or sealer coat, brushwork		415	.039		.29	1.56		1.85	2.82
5800	Spray		725	.022		.35	.89		1.24	1.82
5850	Paint 1 coat, brushwork		415	.039		.30	1.56		1.86	2.83
6000	Spray		725	.022		.33	.89		1.22	1.79
6200	Paint 2 coats, brushwork		260	.062		.58	2.48		3.06	4.63
6250	Spray		415	.039		.64	1.56		2.20	3.21
6270	Paint 3 coats, brushwork		260	.062		.86	2.48		3.34	4.94
6300	13" - 16" diameter, primer or sealer coat, brushwork		310	.052		.39	2.08		2.47	3.78
6350	Spray		540	.030		.44	1.20		1.64	2.40
6400	Paint 1 coat, brushwork		310	.052		.40	2.08		2.48	3.79
6450	Spray		540	.030		.44	1.20		1.64	2.41
6500	Paint 2 coats, brushwork		195	.082		.78	3.31		4.09	6.15
6550	Spray		310	.052		.86	2.08		2.94	4.30
6600	Radiators, per side, primer, brushwork	1 Pord	520	.015	S.F.	.09	.62		.71	1.10
6620	Paint, one coat		520	.015		.09	.62		.71	1.10
6640	Two coats		340	.024		.18	.95		1.13	1.73
6660	Three coats		283	.028		.27	1.14		1.41	2.13
7000	Trim, wood, incl. puttying, under 6" wide									
7200	Primer coat, oil base, brushwork	1 Pord	650	.012	L.F.	.03	.50		.53	.84
7250	Paint, 1 coat, brushwork		650	.012		.05	.50		.55	.86
7400	2 coats		400	.020		.11	.81		.92	1.42
7450	3 coats		325	.025		.16	.99		1.15	1.77
7500	Over 6" wide, primer coat, brushwork		650	.012		.07	.50		.57	.88
7550	Paint, 1 coat, brushwork		650	.012		.11	.50		.61	.92
7600	2 coats		400	.020		.21	.81		1.02	1.53
7650	3 coats		325	.025		.31	.99		1.30	1.95
8000	Cornice, simple design, primer coat, oil base, brushwork		650	.012	S.F.	.07	.50		.57	.88
8250	Paint, 1 coat		650	.012		.11	.50		.61	.92
8300	2 coats		400	.020		.21	.81		1.02	1.53
8350	Ornate design, primer coat		350	.023		.07	.92		.99	1.56
8400	Paint, 1 coat		350	.023		.11	.92		1.03	1.60
8450	2 coats		400	.020		.21	.81		1.02	1.53
8600	Balustrades, primer coat, oil base, brushwork		520	.015		.07	.62		.69	1.08
8650	Paint, 1 coat		520	.015		.11	.62		.73	1.12
8700	2 coats		325	.025		.21	.99		1.20	1.83
8900	Trusses and wood frames, primer coat, oil base, brushwork		800	.010		.07	.40		.47	.73
8950	Spray		1200	.007		.07	.27		.34	.51
9000	Paint 1 coat, brushwork		750	.011		.11	.43		.54	.81
9200	Spray		1200	.007		.12	.27		.39	.56
9220	Paint 2 coats, brushwork		500	.016		.21	.65		.86	1.27
9240	Spray		600	.013		.24	.54		.78	1.12
9260	Stain, brushwork, wipe off		600	.013		.09	.54		.63	.95
9280	Varnish, 3 coats, brushwork		275	.029		.28	1.17		1.45	2.20
9350	For latex paint, deduct					10%				
9900	Minimum labor/equipment charge	1 Pord	2	4	Job		161		161	259

09 91 23.62 Electrostatic Painting		Crew	Daily Output	Labor-Hours	Unit	Material	Labor	Equipment	Total	Total Incl O&P
						2015 Bare Costs				
0010	**ELECTROSTATIC PAINTING**									
0100	In shop									
0200	Flat surfaces (lockers, casework, elevator doors. etc.)									
0300	One coat	1 Pord	200	.040	S.F.	.53	1.61		2.14	3.17
0400	Two coats	"	120	.067	"	.77	2.69		3.46	5.15
0500	Irregular surfaces (furniture, door frames, etc.)									
0600	One coat	1 Pord	150	.053	S.F.	.53	2.15		2.68	4.04
0700	Two coats	"	100	.080	"	.77	3.23		4	6.05
0800	On site									
0900	Flat surfaces (lockers, casework, elevator doors, etc.)									
1000	One coat	1 Pord	150	.053	S.F.	.53	2.15		2.68	4.04
1100	Two coats	"	100	.080	"	.77	3.23		4	6.05
1200	Irregular surfaces (furniture, door frames, etc)									
1300	One coat	1 Pord	115	.070	S.F.	.53	2.81		3.34	5.10
1400	Two coats		70	.114		.77	4.61		5.38	8.25
2000	Anti-microbial coating, hospital application		150	.053		.06	2.15		2.21	3.53

09 91 23.72 Walls and Ceilings, Interior		Crew	Daily Output	Labor-Hours	Unit	Material	Labor	Equipment	Total	Total Incl O&P
0010	**WALLS AND CEILINGS, INTERIOR** R099100-10									
0100	Concrete, drywall or plaster, latex , primer or sealer coat R099100-20									
0200	Smooth finish, brushwork	1 Pord	1150	.007	S.F.	.06	.28		.34	.52
0240	Roller		1350	.006		.06	.24		.30	.45
0280	Spray		2750	.003		.05	.12		.17	.25
0300	Sand finish, brushwork		975	.008		.06	.33		.39	.60
0340	Roller		1150	.007		.06	.28		.34	.52
0380	Spray		2275	.004		.05	.14		.19	.29
0400	Paint 1 coat, smooth finish, brushwork		1200	.007		.07	.27		.34	.50
0440	Roller		1300	.006		.07	.25		.32	.47
0480	Spray		2275	.004		.05	.14		.19	.29
0500	Sand finish, brushwork		1050	.008		.06	.31		.37	.56
0540	Roller		1600	.005		.07	.20		.27	.39
0580	Spray		2100	.004		.02	.15		.17	.27
0800	Paint 2 coats, smooth finish, brushwork		680	.012		.13	.47		.60	.91
0840	Roller		800	.010		.13	.40		.53	.80
0880	Spray		1625	.005		.12	.20		.32	.46
0900	Sand finish, brushwork		605	.013		.13	.53		.66	1.01
0940	Roller		1020	.008		.13	.32		.45	.66
0980	Spray		1700	.005		.12	.19		.31	.45
1200	Paint 3 coats, smooth finish, brushwork		510	.016		.20	.63		.83	1.24
1240	Roller		650	.012		.20	.50		.70	1.02
1280	Spray		1625	.005		.19	.20		.39	.52
1300	Sand finish, brushwork		454	.018		.33	.71		1.04	1.50
1340	Roller		680	.012		.35	.47		.82	1.14
1380	Spray		1133	.007		.30	.28		.58	.79
1600	Glaze coating, 2 coats, spray, clear		1200	.007		.49	.27		.76	.97
1640	Multicolor		1200	.007		.99	.27		1.26	1.52
1660	Painting walls, complete, including surface prep, primer &									
1670	2 coats finish, on drywall or plaster, with roller	1 Pord	325	.025	S.F.	.20	.99		1.19	1.82
1700	For oil base paint, add					10%				
1800	For ceiling installations, add						25%			
2000	Masonry or concrete block, primer/sealer, latex paint									
2100	Primer, smooth finish, brushwork	1 Pord	1000	.008	S.F.	.11	.32		.43	.64
2110	Roller		1150	.007		.11	.28		.39	.57

09 91 23.72 Walls and Ceilings, Interior		Crew	Daily Output	Labor-Hours	Unit	2015 Bare Costs Material	Labor	Equipment	Total	Total Incl O&P
2180	Spray	1 Pord	2400	.003	S.F.	.10	.13		.23	.33
2200	Sand finish, brushwork		850	.009		.11	.38		.49	.73
2210	Roller		975	.008		.11	.33		.44	.65
2280	Spray		2050	.004		.10	.16		.26	.36
2400	Finish coat, smooth finish, brush		1100	.007		.08	.29		.37	.56
2410	Roller		1300	.006		.08	.25		.33	.49
2480	Spray		2400	.003		.07	.13		.20	.30
2500	Sand finish, brushwork		950	.008		.08	.34		.42	.64
2510	Roller		1090	.007		.08	.30		.38	.57
2580	Spray		2040	.004		.07	.16		.23	.33
2800	Primer plus one finish coat, smooth brush		525	.015		.30	.61		.91	1.32
2810	Roller		615	.013		.19	.53		.72	1.05
2880	Spray		1200	.007		.17	.27		.44	.61
2900	Sand finish, brushwork		450	.018		.19	.72		.91	1.36
2910	Roller		515	.016		.19	.63		.82	1.22
2980	Spray		1025	.008		.17	.31		.48	.69
3200	Primer plus 2 finish coats, smooth, brush		355	.023		.27	.91		1.18	1.75
3210	Roller		415	.019		.27	.78		1.05	1.54
3280	Spray		800	.010		.23	.40		.63	.91
3300	Sand finish, brushwork		305	.026		.27	1.06		1.33	1.99
3310	Roller		350	.023		.27	.92		1.19	1.77
3380	Spray		675	.012		.23	.48		.71	1.03
3600	Glaze coating, 3 coats, spray, clear		900	.009		.70	.36		1.06	1.35
3620	Multicolor		900	.009		1.15	.36		1.51	1.85
4000	Block filler, 1 coat, brushwork		425	.019		.12	.76		.88	1.35
4100	Silicone, water repellent, 2 coats, spray		2000	.004		.33	.16		.49	.63
4120	For oil base paint, add					10%				
8200	For work 8' - 15' H, add						10%			
8300	For work over 15' H, add						20%			
8400	For light textured surfaces, add						10%			
8410	Heavy textured, add						25%			
9900	Minimum labor/equipment charge	1 Pord	2	4	Job		161		161	259

09 91 23.74 Walls and Ceilings, Interior, Zero VOC Latex			Crew	Daily Output	Labor-Hours	Unit	Material	Labor	Equipment	Total	Total Incl O&P
0010	**WALLS AND CEILINGS, INTERIOR, ZERO VOC LATEX**										
0100	Concrete, dry wall or plaster, latex, primer or sealer coat										
0200	Smooth finish, brushwork	G	1 Pord	1150	.007	S.F.	.06	.28		.34	.52
0240	Roller	G		1350	.006		.06	.24		.30	.45
0280	Spray	G		2750	.003		.05	.12		.17	.24
0300	Sand finish, brushwork	G		975	.008		.06	.33		.39	.60
0340	Roller	G		1150	.007		.07	.28		.35	.52
0380	Spray	G		2275	.004		.05	.14		.19	.29
0400	Paint 1 coat, smooth finish, brushwork	G		1200	.007		.08	.27		.35	.52
0440	Roller	G		1300	.006		.08	.25		.33	.49
0480	Spray	G		2275	.004		.07	.14		.21	.30
0500	Sand finish, brushwork	G		1050	.008		.07	.31		.38	.57
0540	Roller	G		1600	.005		.08	.20		.28	.41
0580	Spray	G		2100	.004		.07	.15		.22	.32
0800	Paint 2 coats, smooth finish, brushwork	G		680	.012		.15	.47		.62	.93
0840	Roller	G		800	.010		.16	.40		.56	.82
0880	Spray	G		1625	.005		.13	.20		.33	.47
0900	Sand finish, brushwork	G		605	.013		.15	.53		.68	1.02
0940	Roller	G		1020	.008		.16	.32		.48	.68

09 91 Painting

09 91 23 – Interior Painting

09 91 23.74 Walls and Ceilings, Interior, Zero VOC Latex		Crew	Daily Output	Labor-Hours	Unit	Material	2015 Bare Costs Labor	Equipment	Total	Total Incl O&P
0980	Spray G	1 Pord	1700	.005	S.F.	.13	.19		.32	.46
1200	Paint 3 coats, smooth finish, brushwork G		510	.016		.22	.63		.85	1.26
1240	Roller G		650	.012		.23	.50		.73	1.06
1280	Spray G		1625	.005		.20	.20		.40	.54
1800	For ceiling installations, add G						25%			
8200	For work 8' - 15' H, add						10%			
8300	For work over 15' H, add						20%			
9900	Minimum labor/equipment charge	1 Pord	2	4	Job		161		161	259

09 91 23.75 Dry Fall Painting		Crew	Daily Output	Labor-Hours	Unit	Material	Labor	Equipment	Total	Total Incl O&P
0010	DRY FALL PAINTING R099100-10									
0100	Sprayed on walls, gypsum board or plaster									
0220	One coat R099100-20	1 Pord	2600	.003	S.F.	.06	.12		.18	.26
0250	Two coats		1560	.005		.11	.21		.32	.45
0280	Concrete or textured plaster, one coat		1560	.005		.06	.21		.27	.39
0310	Two coats		1300	.006		.11	.25		.36	.52
0340	Concrete block, one coat		1560	.005		.06	.21		.27	.39
0370	Two coats		1300	.006		.11	.25		.36	.52
0400	Wood, one coat		877	.009		.06	.37		.43	.65
0430	Two coats		650	.012		.11	.50		.61	.92
0440	On ceilings, gypsum board or plaster									
0470	One coat	1 Pord	1560	.005	S.F.	.06	.21		.27	.39
0500	Two coats		1300	.006		.11	.25		.36	.52
0530	Concrete or textured plaster, one coat		1560	.005		.06	.21		.27	.39
0560	Two coats		1300	.006		.11	.25		.36	.52
0570	Structural steel, bar joists or metal deck, one coat		1560	.005		.06	.21		.27	.39
0580	Two coats		1040	.008		.11	.31		.42	.62
9900	Minimum labor/equipment charge		2	4	Job		161		161	259

09 93 Staining and Transparent Finishing

09 93 23 – Interior Staining and Finishing

09 93 23.10 Varnish		Crew	Daily Output	Labor-Hours	Unit	Material	Labor	Equipment	Total	Total Incl O&P
0010	VARNISH									
0012	1 coat + sealer, on wood trim, brush, no sanding included	1 Pord	400	.020	S.F.	.07	.81		.88	1.38
0020	1 coat + sealer, on wood trim, brush, no sanding included, no VOC		400	.020		.21	.81		1.02	1.53
0100	Hardwood floors, 2 coats, no sanding included, roller		1890	.004		.15	.17		.32	.44
9000	Minimum labor/equipment charge		4	2	Job		80.50		80.50	130

09 96 High-Performance Coatings

09 96 23 – Graffiti-Resistant Coatings

09 96 23.10 Graffiti Resistant Treatments		Crew	Daily Output	Labor-Hours	Unit	Material	Labor	Equipment	Total	Total Incl O&P
0010	GRAFFITI RESISTANT TREATMENTS, sprayed on walls									
0100	Non-sacrificial, permanent non-stick coating, clear, on metals	1 Pord	2000	.004	S.F.	2.06	.16		2.22	2.53
0200	Concrete		2000	.004		2.35	.16		2.51	2.84
0300	Concrete block		2000	.004		3.03	.16		3.19	3.60
0400	Brick		2000	.004		3.44	.16		3.60	4.04
0500	Stone		2000	.004		3.44	.16		3.60	4.04
0600	Unpainted wood		2000	.004		3.97	.16		4.13	4.63
2000	Semi-permanent cross linking polymer primer, on metals		2000	.004		.70	.16		.86	1.03
2100	Concrete		2000	.004		.84	.16		1	1.18

09 96 23 – Graffiti-Resistant Coatings

09 96 23.10 Graffiti Resistant Treatments		Crew	Daily Output	Labor-Hours	Unit	Material	Labor	Equipment	Total	Total Incl O&P
						2015 Bare Costs				
2200	Concrete block	1 Pord	2000	.004	S.F.	1.05	.16		1.21	1.41
2300	Brick		2000	.004		.84	.16		1	1.18
2400	Stone		2000	.004		.84	.16		1	1.18
2500	Unpainted wood		2000	.004		1.16	.16		1.32	1.54
3000	Top coat, on metals		2000	.004		.55	.16		.71	.86
3100	Concrete		2000	.004		.62	.16		.78	.95
3200	Concrete block		2000	.004		.87	.16		1.03	1.22
3300	Brick		2000	.004		.73	.16		.89	1.06
3400	Stone		2000	.004		.73	.16		.89	1.06
3500	Unpainted wood		2000	.004		.87	.16		1.03	1.22
5000	Sacrificial, water based, on metal		2000	.004		.32	.16		.48	.62
5100	Concrete		2000	.004		.32	.16		.48	.62
5200	Concrete block		2000	.004		.32	.16		.48	.62
5300	Brick		2000	.004		.32	.16		.48	.62
5400	Stone		2000	.004		.32	.16		.48	.62
5500	Unpainted wood	↓	2000	.004	↓	.32	.16		.48	.62
8000	Cleaner for use after treatment									
8100	Towels or wipes, per package of 30				Ea.	.63			.63	.70
8200	Aerosol spray, 24 oz. can				"	18			18	19.80

09 96 46 – Intumescent Coatings

09 96 46.10 Coatings, Intumescent		Crew	Daily Output	Labor-Hours	Unit	Material	Labor	Equipment	Total	Total Incl O&P
0010	**COATINGS, INTUMESCENT**, spray applied									
0100	On exterior structural steel, 0.25" d.f.t.	1 Pord	475	.017	S.F.	.41	.68		1.09	1.54
0150	0.51" d.f.t.		350	.023		.41	.92		1.33	1.93
0200	0.98" d.f.t.		280	.029		.41	1.15		1.56	2.30
0300	On interior structural steel, 0.108" d.f.t.		300	.027		.41	1.08		1.49	2.18
0350	0.310" d.f.t.		150	.053		.41	2.15		2.56	3.91
0400	0.670" d.f.t.	↓	100	.080	↓	.41	3.23		3.64	5.65

09 96 53 – Elastomeric Coatings

09 96 53.10 Coatings, Elastomeric		Crew	Daily Output	Labor-Hours	Unit	Material	Labor	Equipment	Total	Total Incl O&P
0010	**COATINGS, ELASTOMERIC**									
0020	High build, water proof, one coat system									
0100	Concrete, brush	1 Pord	650	.012	S.F.	.27	.50		.77	1.10

09 96 56 – Epoxy Coatings

09 96 56.20 Wall Coatings		Crew	Daily Output	Labor-Hours	Unit	Material	Labor	Equipment	Total	Total Incl O&P
0010	**WALL COATINGS**									
0100	Acrylic glazed coatings, matte	1 Pord	525	.015	S.F.	.31	.61		.92	1.33
0200	Gloss		305	.026		.65	1.06		1.71	2.42
0300	Epoxy coatings, solvent based		525	.015		.40	.61		1.01	1.43
0400	Water based		170	.047		1.20	1.90		3.10	4.37
0600	Exposed aggregate, troweled on, 1/16" to 1/4", solvent based		235	.034		.62	1.37		1.99	2.89
0700	Water based (epoxy or polyacrylate)		130	.062		1.33	2.48		3.81	5.45
0900	1/2" to 5/8" aggregate, solvent based		130	.062		1.20	2.48		3.68	5.30
1000	Water based		80	.100		2.08	4.04		6.12	8.80
1200	1" aggregate size, solvent based		90	.089		2.12	3.59		5.71	8.10
1300	Water based		55	.145		3.23	5.85		9.08	13
1500	Exposed aggregate, sprayed on, 1/8" aggregate, solvent based		295	.027		.57	1.09		1.66	2.39
1600	Water based		145	.055		1.05	2.23		3.28	4.74
1800	High build epoxy, 50 mil, solvent based		390	.021		.68	.83		1.51	2.08
1900	Water based		95	.084		1.15	3.40		4.55	6.70
2100	Laminated epoxy with fiberglass, solvent based	↓	295	.027	↓	.73	1.09		1.82	2.56

09 96 High-Performance Coatings

09 96 56 – Epoxy Coatings

09 96 56.20 Wall Coatings		Crew	Daily Output	Labor-Hours	Unit	Material	Labor	Equipment	Total	Total Incl O&P
						2015 Bare Costs				
2200	Water based	1 Pord	145	.055	S.F.	1.32	2.23		3.55	5.05
2400	Sprayed perlite or vermiculite, 1/16" thick, solvent based		2935	.003		.27	.11		.38	.48
2500	Water based		640	.013		.74	.50		1.24	1.62
2700	Vinyl plastic wall coating, solvent based		735	.011		.33	.44		.77	1.07
2800	Water based		240	.033		.82	1.34		2.16	3.06
3000	Urethane on smooth surface, 2 coats, solvent based		1135	.007		.27	.28		.55	.76
3100	Water based		665	.012		.59	.49		1.08	1.43
3300	3 coat, solvent based		840	.010		.35	.38		.73	1.01
3400	Water based		470	.017		.79	.69		1.48	1.97
3600	Ceramic-like glazed coating, cementitious, solvent based		440	.018		.48	.73		1.21	1.71
3700	Water based		345	.023		.81	.94		1.75	2.39
3900	Resin base, solvent based		640	.013		.33	.50		.83	1.17
4000	Water based	↓	330	.024	↓	.54	.98		1.52	2.16

09 97 Special Coatings

09 97 10 – Coatings and Paints

09 97 10.10 Coatings and Paints

		Crew	Daily Output	Labor-Hours	Unit	Material	Labor	Equipment	Total	Total Incl O&P
0010	**COATINGS & PAINTS** in 5 gallon lots R099100-20									
0050	For 100 gallons or more, deduct					10%				
0100	Paint, Exterior alkyd (oil base)									
0200	Flat				Gal.	44.50			44.50	48.50
0300	Gloss					40.50			40.50	44.50
0400	Primer				↓	37.50			37.50	41
0500	Latex (water base)									
0600	Acrylic stain				Gal.	35.50			35.50	39.50
0700	Gloss enamel					33			33	36.50
0800	Flat					27.50			27.50	30.50
0900	Primer					24			24	26.50
1000	Semi-gloss				↓	30.50			30.50	33.50
1054	Low VOC									
1055	Primer				Gal.	48			48	53
1060	Satin					15			15	16.50
1065	Door and trim				↓	59			59	65
1100	Interior, alkyd (oil base)									
1200	Enamel undercoat				Gal.	36.50			36.50	40
1300	Flat					49			49	54
1400	Gloss					46.50			46.50	51
1500	Primer sealer					28			28	31
1600	Semi-gloss				↓	46.50			46.50	51
1700	Latex (water base)									
1800	Enamel undercoat				Gal.	24.50			24.50	26.50
1900	Flat					21.50			21.50	24
2000	Floor and deck					27			27	29.50
2100	Gloss					30.50			30.50	33.50
2200	Primer sealer					21.50			21.50	24
2300	Semi-gloss				↓	29.50			29.50	32.50
2320	Low VOC									
2330	Wallboard primer				Gal.	40			40	44
2335	Flat					43.50			43.50	47.50
2340	Semi-gloss					50.50			50.50	55.50
2345	Eggshell				↓	49.50			49.50	54.50

09 97 10.10 Coatings and Paints		Crew	Daily Output	Labor-Hours	Unit	Material	Labor	Equipment	Total	Total Incl O&P
						2015 Bare Costs				
2350	Stain				Gal.	75			75	82.50
2400	Masonry, Exterior									
2500	Alkali resistant primer				Gal.	28			28	30.50
2600	Block filler, epoxy					29.50			29.50	32.50
2700	Latex					21.50			21.50	23.50
2800	Latex, flat					21.50			21.50	24
2900	Semi-gloss				↓	34			34	37
3000	Masonry, Interior									
3100	Alkali resistant primer				Gal.	31.50			31.50	34.50
3200	Block filler, epoxy					24.50			24.50	27
3300	Latex					20.50			20.50	22.50
3400	Floor, alkyd					41.50			41.50	46
3500	Latex					29.50			29.50	32.50
3600	Latex, flat acrylic					22			22	24
3700	Flat emulsion					21			21	23
3800	Sealer					21			21	23
3900	Semi-gloss				↓	27.50			27.50	30
4000	Metal									
4100	Galvanizing paint				Gal.	93.50			93.50	103
4200	High heat					62.50			62.50	69
4300	Heat resistant					38.50			38.50	42.50
4400	Machinery enamel, alkyd					49.50			49.50	54
4500	Metal pretreatment (polyvinyl butyral)					112			112	123
4600	Rust inhibitor, ferrous metal					43			43	47.50
4700	Zinc chromate					151			151	166
4800	Zinc rich primer				↓	173			173	190
4900	Varnish and Stain									
5000	Alkyd, clear				Gal.	37.50			37.50	41
5100	Polyurethane, clear					41.50			41.50	45.50
5200	Primer sealer					32.50			32.50	36
5300	Stain, semi-transparent					34.50			34.50	38
5400	Solid color				↓	37			37	40.50
5500	Coatings									
5600	Heavy duty									
5700	Acrylic urethane				Gal.	56.50			56.50	62
5800	Chlorinated rubber					65			65	71.50
5900	Coal tar epoxy					76			76	83.50
6000	Polyamide epoxy, finish					65.50			65.50	72
6100	Primer					58.50			58.50	64.50
6200	Silicone alkyd					52.50			52.50	57.50
6300	2 component solvent based acrylic epoxy					83			83	91.50
6400	Polyester epoxy					83.50			83.50	92
6500	Vinyl				↓	37.50			37.50	41
6600	Special/Miscellaneous									
6700	Aluminum				Gal.	45			45	49.50
6900	Dry fall out, flat					16.90			16.90	18.60
7000	Fire retardant, intumescent					50.50			50.50	56
7100	Linseed oil					21.50			21.50	23.50
7200	Shellac					43			43	47
7300	Swimming pool, epoxy or urethane base					53			53	58
7400	Rubber base					65			65	71.50
7500	Texture paint					21.50			21.50	24
7600	Turpentine				↓	29			29	32

09 97 Special Coatings

09 97 10 – Coatings and Paints

09 97 10.10 Coatings and Paints		Crew	Daily Output	Labor-Hours	Unit	Material	Labor	Equipment	Total	Total Incl O&P
						2015 Bare Costs				
7700	Water repellent, 5% silicone				Gal.	25			25	27.50
7800	Insulating additive				↓	25.50			25.50	28

09 97 13 – Steel Coatings

09 97 13.23 Exterior Steel Coatings		Crew	Daily Output	Labor-Hours	Unit	Material	Labor	Equipment	Total	Total Incl O&P
						2015 Bare Costs				
0010	**EXTERIOR STEEL COATINGS** R050516-30									
6100	Cold galvanizing, brush in field	1 Psst	1100	.007	S.F.	.23	.30		.53	.82
6510	Paints & protective coatings, sprayed in field									
6520	Alkyds, primer	2 Psst	3600	.004	S.F.	.09	.18		.27	.44
6540	Gloss topcoats		3200	.005		.08	.21		.29	.48
6560	Silicone alkyd		3200	.005		.15	.21		.36	.55
6610	Epoxy, primer		3000	.005		.29	.22		.51	.73
6630	Intermediate or topcoat		2800	.006		.26	.24		.50	.73
6650	Enamel coat		2800	.006		.33	.24		.57	.80
6700	Epoxy ester, primer		2800	.006		.42	.24		.66	.90
6720	Topcoats		2800	.006		.22	.24		.46	.68
6810	Latex primer		3600	.004		.06	.18		.24	.41
6830	Topcoats		3200	.005		.07	.21		.28	.47
6910	Universal primers, one part, phenolic, modified alkyd		2000	.008		.37	.33		.70	1.03
6940	Two part, epoxy spray		2000	.008		.33	.33		.66	.99
7000	Zinc rich primers, self cure, spray, inorganic		1800	.009		.86	.37		1.23	1.64
7010	Epoxy, spray, organic	↓	1800	.009	↓	.26	.37		.63	.98
7020	Above one story, spray painting simple structures, add						25%			
7030	Intricate structures, add						50%			

Estimating Tips

General

- The items in this division are usually priced per square foot or each.
- Many items in Division 10 require some type of support system or special anchors that are not usually furnished with the item. The required anchors must be added to the estimate in the appropriate division.
- Some items in Division 10, such as lockers, may require assembly before installation. Verify the amount of assembly required. Assembly can often exceed installation time.

10 20 00 Interior Specialties

- Support angles and blocking are not included in the installation of toilet compartments, shower/dressing compartments, or cubicles. Appropriate line items from Divisions 5 or 6 may need to be added to support the installations.
- Toilet partitions are priced by the stall. A stall consists of a side wall, pilaster, and door with hardware. Toilet tissue holders and grab bars are extra.
- The required acoustical rating of a folding partition can have a significant impact on costs. Verify the sound transmission coefficient rating of the panel priced to the specification requirements.
- Grab bar installation does not include supplemental blocking or backing to support the required load. When grab bars are installed at an existing facility, provisions must be made to attach the grab bars to solid structure.

Reference Numbers

Reference numbers are shown in shaded boxes at the beginning of some major classifications. These numbers refer to related items in the Reference Section. The reference information may be an estimating procedure, an alternate pricing method, or technical information.

Note: Not all subdivisions listed here necessarily appear in this publication. ■

10 05 Common Work Results for Specialties

10 05 05 – Selective Demolition for Specialties

10 05 05.10 Selective Demolition, Specialties		Crew	Daily Output	Labor-Hours	Unit	Material	2015 Bare Costs Labor	Equipment	Total	Total Incl O&P
0010	**SELECTIVE DEMOLITION, SPECIALTIES** R024119-10									
1100	Boards and panels, wall mounted	2 Clab	15	1.067	Ea.		40		40	66
1200	Cases, for directory and/or bulletin boards, including doors		24	.667			25		25	41
1850	Shower partitions, cabinet or stall, including base and door		8	2			75		75	123
1855	Shower receptor, terrazzo or concrete	1 Clab	14	.571			21.50		21.50	35
1900	Curtain track or rod, hospital type, ceiling mounted or suspended	″	220	.036	L.F.		1.37		1.37	2.24
1910	Toilet cubicles, remove	2 Clab	8	2	Ea.		75		75	123
1930	Urinal screen, remove	1 Clab	12	.667	″		25		25	41
2650	Wall guard, misc. wall or corner protection	″	320	.025	L.F.		.94		.94	1.54
2750	Access floor, metal panel system, including pedestals, covering	2 Clab	850	.019	S.F.		.71		.71	1.16
3050	Fireplace, prefab, freestanding or wall hung, including hood and screen	1 Clab	2	4	Ea.		150		150	247
3054	Chimney top, simulated brick, 4′ high	″	15	.533			20		20	33
3200	Stove, woodburning, cast iron	2 Clab	2	8			300		300	495
3440	Weathervane, residential	1 Clab	12	.667			25		25	41
3500	Flagpole, groundset, to 70′ high, excluding base/foundation	K-1	1	16			690	305	995	1,450
3555	To 30′ high	″	2.50	6.400			275	121	396	585
4000	Removal of traffic signs, including supports									
4020	To 10 S.F.	B-80B	16	2	Ea.		80.50	15.25	95.75	148
4030	11 S.F. to 20 S.F.	″	5	6.400			258	49	307	475
4040	21 S.F. to 40 S.F.	B-14	1.80	26.667			1,050	202	1,252	1,950
4050	41 S.F. to 100 S.F.	B-13	1.30	43.077			1,775	565	2,340	3,500
4070	Remove traffic posts to 12′-0″ high	B-6	100	.240			9.90	3.64	13.54	20
4300	Letter, signs or plaques, exterior on wall	1 Clab	20	.400			15.05		15.05	24.50
4310	Signs, street, reflective aluminum, including post and bracket		60	.133			5		5	8.20
4320	Door signs interior on door 6″ x 6″, selective demolition		20	.400			15.05		15.05	24.50
4550	Turnstiles, manual or electric	2 Clab	2	8			300		300	495
5050	Lockers	1 Clab	15	.533	Opng.		20		20	33
5250	Cabinets, recessed	Q-12	12	1.333	Ea.		67.50		67.50	105
5260	Mail boxes, Horiz., Key Lock, front loading, Remove	1 Carp	34	.235	″		11.05		11.05	18.10
5350	Awning, fabric, including frame	2 Clab	100	.160	S.F.		6		6	9.85
6050	Partition, woven wire		1400	.011			.43		.43	.70
6100	Folding gate, security, door or window		500	.032			1.20		1.20	1.97
6580	Acoustic air wall		650	.025			.93		.93	1.52
7550	Telephone enclosure, exterior, post mounted		3	5.333	Ea.		201		201	330
8850	Scale, platform, excludes foundation or pit		.25	64	″		2,400		2,400	3,950

10 11 Visual Display Units

10 11 13 – Chalkboards

10 11 13.13 Fixed Chalkboards		Crew	Daily Output	Labor-Hours	Unit	Material	Labor	Equipment	Total	Total Incl O&P
0010	**FIXED CHALKBOARDS** Porcelain enamel steel									
3900	Wall hung									
4000	Aluminum frame and chalktrough									
4200	3′ x 4′	2 Carp	16	1	Ea.	248	47		295	350
4300	3′ x 5′		15	1.067		320	50		370	430
4500	4′ x 8′		14	1.143		430	53.50		483.50	565
4600	4′ x 12′		13	1.231		600	58		658	755
4700	Wood frame and chalktrough									
4800	3′ x 4′	2 Carp	16	1	Ea.	194	47		241	291
5000	3′ x 5′		15	1.067		243	50		293	350
5100	4′ x 5′		14	1.143		255	53.50		308.50	370
5300	4′ x 8′		13	1.231		345	58		403	475

10 11 Visual Display Units

10 11 13 – Chalkboards

10 11 13.13 Fixed Chalkboards		Crew	Daily Output	Labor-Hours	Unit	Material	2015 Bare Costs Labor	Equipment	Total	Total Incl O&P
5400	Liquid chalk, white porcelain enamel, wall hung									
5420	Deluxe units, aluminum trim and chalktrough									
5450	4′ x 4′	2 Carp	16	1	Ea.	253	47		300	355
5500	4′ x 8′		14	1.143		390	53.50		443.50	520
5550	4′ x 12′		12	1.333		535	62.50		597.50	695
5700	Wood trim and chalktrough									
5900	4′ x 4′	2 Carp	16	1	Ea.	715	47		762	860
6000	4′ x 6′		15	1.067		810	50		860	970
6200	4′ x 8′		14	1.143		970	53.50		1,023.50	1,175
6250	Economy dry erase board, melamine, 36″ high		104	.154	S.F.	15.60	7.20		22.80	29
6300	Liquid chalk, felt tip markers				Ea.	2.12			2.12	2.33
6500	Erasers					1.93			1.93	2.12
6600	Board cleaner, 8 oz. bottle					6.20			6.20	6.80
9000	Minimum labor/equipment charge	2 Carp	3	5.333	Job		250		250	410

10 11 13.23 Modular-Support-Mounted Chalkboards		Crew	Daily Output	Labor-Hours	Unit	Material	Labor	Equipment	Total	Total Incl O&P
0010	**MODULAR-SUPPORT-MOUNTED CHALKBOARDS**									
0400	Sliding chalkboards									
0450	Vertical, one sliding board with back panel, wall mounted									
0500	8′ x 4′	2 Carp	8	2	Ea.	2,225	94		2,319	2,600
0520	8′ x 8′		7.50	2.133		3,225	100		3,325	3,725
0540	8′ x 12′		7	2.286		4,175	107		4,282	4,775
0600	Two sliding boards, with back panel									
0620	8′ x 4′	2 Carp	8	2	Ea.	3,425	94		3,519	3,925
0640	8′ x 8′		7.50	2.133		5,025	100		5,125	5,725
0660	8′ x 12′		7	2.286		8,275	107		8,382	9,275
0700	Horizontal, two track									
0800	4′ x 8′, 2 sliding panels	2 Carp	8	2	Ea.	1,950	94		2,044	2,300
0820	4′ x 12′, 2 sliding panels		7.50	2.133		2,550	100		2,650	2,975
0840	4′ x 16′, 4 sliding panels		7	2.286		3,425	107		3,532	3,950
0900	Four track, four sliding panels									
0920	4′ x 8′	2 Carp	8	2	Ea.	3,150	94		3,244	3,600
0940	4′ x 12′		7.50	2.133		4,100	100		4,200	4,700
0960	4′ x 16′		7	2.286		5,350	107		5,457	6,050
1200	Vertical, motor operated									
1400	One sliding panel with back panel									
1450	10′ x 4′	2 Carp	4	4	Ea.	5,325	188		5,513	6,175
1500	10′ x 10′		3.75	4.267		6,425	200		6,625	7,400
1550	10′ x 16′		3.50	4.571		7,575	215		7,790	8,700
1700	Two sliding panels with back panel									
1750	10′ x 4′	2 Carp	4	4	Ea.	9,475	188		9,663	10,700
1800	10′ x 10′		3.75	4.267		10,600	200		10,800	12,000
1850	10′ x 16′		3.50	4.571		12,600	215		12,815	14,300
2000	Three sliding panels with back panel									
2100	10′ x 4′	2 Carp	4	4	Ea.	13,200	188		13,388	14,800
2150	10′ x 10′		3.75	4.267		14,700	200		14,900	16,400
2200	10′ x 16′		3.50	4.571		17,500	215		17,715	19,600
2400	For projection screen, glass beaded, add				S.F.	4.57			4.57	5.05
2500	For remote control, 1 panel control, add				Ea.	360			360	395
2600	2 panel control, add				″	620			620	680
2800	For units without back panels, deduct				S.F.	4.84			4.84	5.30
2850	For liquid chalk porcelain panels, add				″	5.15			5.15	5.70
3000	Swing leaf, any comb. of chalkboard & cork, aluminum frame									

10 11 Visual Display Units

10 11 13 – Chalkboards

10 11 13.23 Modular-Support-Mounted Chalkboards		Crew	Daily Output	Labor-Hours	Unit	2015 Bare Costs Material	Labor	Equipment	Total	Total Incl O&P
3100	Floor style, 6 panels									
3150	30" x 40" panels				Ea.	1,550			1,550	1,700
3200	48" x 40" panels				"	2,600			2,600	2,875
3300	Wall mounted, 6 panels									
3400	30" x 40" panels	2 Carp	16	1	Ea.	1,475	47		1,522	1,700
3450	48" x 40" panels	"	16	1	"	1,800	47		1,847	2,075
3600	Extra panels for swing leaf units									
3700	30" x 40" panels				Ea.	294			294	325
3750	48" x 40" panels				"	365			365	400

10 11 13.43 Portable Chalkboards		Crew	Daily Output	Labor-Hours	Unit	Material	Labor	Equipment	Total	Total Incl O&P
0010	**PORTABLE CHALKBOARDS**									
0100	Freestanding, reversible									
0120	Economy, wood frame, 4' x 6'									
0140	Chalkboard both sides				Ea.	610			610	670
0160	Chalkboard one side, cork other side				"	575			575	630
0200	Standard, lightweight satin finished aluminum, 4' x 6'									
0220	Chalkboard both sides				Ea.	635			635	695
0240	Chalkboard one side, cork other side				"	640			640	705
0300	Deluxe, heavy duty extruded aluminum, 4' x 6'									
0320	Chalkboard both sides				Ea.	1,050			1,050	1,150
0340	Chalkboard one side, cork other side				"	970			970	1,075

10 11 16 – Markerboards

10 11 16.53 Electronic Markerboards		Crew	Daily Output	Labor-Hours	Unit	Material	Labor	Equipment	Total	Total Incl O&P
0010	**ELECTRONIC MARKERBOARDS**									
0100	Wall hung or free standing, 3' x 4' to 4' x 6'	2 Carp	8	2	S.F.	87.50	94		181.50	250
0150	5' x 6' to 4' x 8'		8	2	"	61	94		155	221
0500	Interactive projection module for existing whiteboards	↓	8	2	Ea.	1,300	94		1,394	1,575

10 11 23 – Tackboards

10 11 23.10 Fixed Tackboards		Crew	Daily Output	Labor-Hours	Unit	Material	Labor	Equipment	Total	Total Incl O&P
0010	**FIXED TACKBOARDS**									
0020	Cork sheets, unbacked, no frame, 1/4" thick	2 Carp	290	.055	S.F.	1.54	2.59		4.13	5.95
0100	1/2" thick		290	.055		4.14	2.59		6.73	8.80
0300	Fabric-face, no frame, on 7/32" cork underlay		290	.055		6.85	2.59		9.44	11.80
0400	On 1/4" cork on 1/4" hardboard		290	.055		8.25	2.59		10.84	13.35
0600	With edges wrapped		290	.055		9.90	2.59		12.49	15.10
0700	On 7/16" fire retardant core		290	.055		6.50	2.59		9.09	11.40
0900	With edges wrapped	↓	290	.055		8.20	2.59		10.79	13.30
1000	Designer fabric only, cut to size					2.70			2.70	2.97
1200	1/4" vinyl cork, on 1/4" hardboard, no frame	2 Carp	290	.055		8.45	2.59		11.04	13.50
1300	On 1/4" coreboard		290	.055	↓	5.45	2.59		8.04	10.20
2000	For map and display rail, economy, add		385	.042	L.F.	3.06	1.95		5.01	6.55
2100	Deluxe, add		350	.046	"	4.70	2.15		6.85	8.65
2120	Prefabricated, 1/4" cork, 3' x 5' with aluminum frame		16	1	Ea.	132	47		179	222
2140	Wood frame		16	1		156	47		203	248
2160	4' x 4' with aluminum frame		16	1		135	47		182	225
2180	Wood frame		16	1		177	47		224	272
2200	4' x 8' with aluminum frame		14	1.143		270	53.50		323.50	385
2210	With wood frame		14	1.143		241	53.50		294.50	355
2220	4' x 12' with aluminum frame	↓	12	1.333	↓	400	62.50		462.50	545
2230	Bulletin board case, single glass door, with lock									
2240	36" x 24", economy	2 Carp	12	1.333	Ea.	315	62.50		377.50	450

10 11 Visual Display Units

10 11 23 – Tackboards

10 11 23.10 Fixed Tackboards		Crew	Daily Output	Labor-Hours	Unit	2015 Bare Costs Material	Labor	Equipment	Total	Total Incl O&P
2250	Deluxe	2 Carp	12	1.333	Ea.	370	62.50		432.50	510
2260	42″ x 30″, economy		12	1.333		380	62.50		442.50	525
2270	Deluxe		12	1.333		510	62.50		572.50	665
2300	Glass enclosed cabinets, alum., cork panel, hinged doors									
2400	3′ x 3′, 1 door	2 Carp	12	1.333	Ea.	605	62.50		667.50	770
2500	4′ x 4′, 2 door		11	1.455		1,000	68.50		1,068.50	1,200
2600	4′ x 7′, 3 door		10	1.600		1,775	75		1,850	2,075
2800	4′ x 10′, 4 door		8	2		2,350	94		2,444	2,750
2900	For lights, add per door opening	1 Elec	13	.615		165	33.50		198.50	234
3100	Horizontal sliding units, 4 doors, 4′ x 8′, 8′ x 4′	2 Carp	9	1.778		1,950	83.50		2,033.50	2,275
3200	4′ x 12′		7	2.286		2,550	107		2,657	2,975
3400	8 doors, 4′ x 16′		5	3.200		3,450	150		3,600	4,025
3500	4′ x 24′		4	4		4,650	188		4,838	5,400
9000	Minimum labor/equipment charge		4	4	Job		188		188	310

10 11 23.20 Control Boards

		Crew	Daily Output	Labor-Hours	Unit	Material	Labor	Equipment	Total	Total Incl O&P
0010	**CONTROL BOARDS**									
0020	Magnetic, porcelain finish, 18″ x 24″, framed	2 Carp	8	2	Ea.	194	94		288	365
0100	24″ x 36″		7.50	2.133		268	100		368	460
0200	36″ x 48″		7	2.286		370	107		477	580
0300	48″ x 72″		6	2.667		650	125		775	920
0400	48″ x 96″		5	3.200		1,075	150		1,225	1,425
1000	Hospital patient display board, 4-color custom design									
1010	Porcelain steel dry erase board, 36″ x 24″	2 Carp	7.50	2.133	Ea.	246	100		346	435

10 13 Directories

10 13 10 – Building Directories

10 13 10.10 Directory Boards

		Crew	Daily Output	Labor-Hours	Unit	Material	Labor	Equipment	Total	Total Incl O&P
0010	**DIRECTORY BOARDS**									
0050	Plastic, glass covered, 30″ x 20″	2 Carp	3	5.333	Ea.	198	250		448	625
0100	36″ x 48″		2	8		845	375		1,220	1,550
0300	Grooved cork, 30″ x 20″		3	5.333		405	250		655	855
0400	36″ x 48″		2	8		555	375		930	1,225
0600	Black felt, 30″ x 20″		3	5.333		239	250		489	675
0700	36″ x 48″		2	8		465	375		840	1,125
0900	Outdoor, weatherproof, black plastic, 36″ x 24″		2	8		760	375		1,135	1,450
1000	36″ x 36″		1.50	10.667		880	500		1,380	1,800
1800	Indoor, economy, open face, 18″ x 24″		7	2.286		163	107		270	355
1900	24″ x 36″		7	2.286		154	107		261	345
2000	36″ x 24″		6	2.667		154	125		279	375
2100	36″ x 48″		6	2.667		251	125		376	480
2400	Building directory, alum., black felt panels, 1 door, 24″ x 18″		4	4		315	188		503	655
2500	36″ x 24″		3.50	4.571		385	215		600	775
2600	48″ x 32″		3	5.333		610	250		860	1,075
2700	2 door, 36″ x 48″		2.50	6.400		680	300		980	1,225
2800	36″ x 60″		2	8		860	375		1,235	1,550
2900	48″ x 60″		1	16		970	750		1,720	2,300
3100	For bronze enamel finish, add					15%				
3200	For bronze anodized finish, add					25%				
3400	For illuminated directory, single door unit, add					138			138	151
3500	For 6″ header panel, 6 letters per foot, add				L.F.	21.50			21.50	23.50
5000	Building directory, illuminated, 7″ sections, bronze finish									

10 13 Directories

10 13 10 – Building Directories

10 13 10.10 Directory Boards		Crew	Daily Output	Labor-Hours	Unit	Material	Labor	Equipment	Total	Total Incl O&P
						2015 Bare Costs				
5100	19" x 36", 2 sections, 80 name capacity				Ea.	2,675			2,675	2,925
5200	26-1/4" x 36", 3 sections, 120 name capacity					3,175			3,175	3,500
5300	33-5/8" x 36", 4 sections, 160 name capacity					3,625			3,625	4,000
5400	48-3/4" x 36", 6 sections, 240 name capacity					4,700			4,700	5,175
5500	63-1/2" x 36", 8 sections, 320 name capacity					5,700			5,700	6,275
5600	Engraving charge per namestrip				↓	14.20			14.20	15.60
6050	Building directory, electronic display, alum. frame, wall mounted	2 Carp	32	.500	S.F.	2,625	23.50		2,648.50	2,925
6100	Free standing	"	60	.267	"	3,700	12.50		3,712.50	4,100
9000	Minimum labor/equipment charge	1 Carp	1	8	Job		375		375	615

10 14 Signage

10 14 19 – Dimensional Letter Signage

10 14 19.10 Exterior Signs

		Crew	Daily Output	Labor-Hours	Unit	Material	Labor	Equipment	Total	Total Incl O&P
0010	**EXTERIOR SIGNS**									
0020	Letters, 2" high, 3/8" deep, cast bronze	1 Carp	24	.333	Ea.	25	15.65		40.65	53
0140	1/2" deep, cast aluminum		18	.444		25	21		46	61.50
0160	Cast bronze		32	.250		30	11.75		41.75	52.50
0300	6" high, 5/8" deep, cast aluminum		24	.333		29	15.65		44.65	57.50
0400	Cast bronze		24	.333		62.50	15.65		78.15	94.50
0600	8" high, 3/4" deep, cast aluminum		14	.571		36	27		63	83.50
0700	Cast bronze		20	.400		88	18.80		106.80	128
0900	10" high, 1" deep, cast aluminum		18	.444		53	21		74	92
1000	Bronze		18	.444		104	21		125	148
1200	12" high, 1-1/4" deep, cast aluminum		12	.667		53.50	31.50		85	111
1500	Cast bronze		18	.444		127	21		148	173
1600	14" high, 2-5/16" deep, cast aluminum		12	.667		101	31.50		132.50	163
1800	Fabricated stainless steel, 6" high, 2" deep		20	.400		41.50	18.80		60.30	76.50
1900	12" high, 3" deep		18	.444		67	21		88	108
2100	18" high, 3" deep		12	.667		109	31.50		140.50	171
2200	24" high, 4" deep		10	.800		212	37.50		249.50	295
2700	Acrylic, on high density foam, 12" high, 2" deep		20	.400		19.80	18.80		38.60	53
2800	18" high, 2" deep	↓	18	.444		37.50	21		58.50	75
3900	Plaques, custom, 20" x 30", for up to 450 letters, cast aluminum	2 Carp	4	4		1,850	188		2,038	2,325
4000	Cast bronze		4	4		1,750	188		1,938	2,200
4200	30" x 36", up to 900 letters cast aluminum		3	5.333		2,625	250		2,875	3,300
4300	Cast bronze		3	5.333		4,025	250		4,275	4,825
4500	36" x 48", for up to 1300 letters, cast bronze		2	8		4,650	375		5,025	5,750
4800	Signs, reflective alum. directional signs, dbl. face, 2-way, w/bracket		30	.533		144	25		169	199
4900	4-way	↓	30	.533		231	25		256	295
5100	Exit signs, 24 ga. alum., 14" x 12" surface mounted	1 Carp	30	.267		47.50	12.50		60	73
5200	10" x 7"		20	.400		25.50	18.80		44.30	59
5400	Bracket mounted, double face, 12" x 10"	↓	30	.267		56	12.50		68.50	82.50
5500	Sticky back, stock decals, 14" x 10"	1 Clab	50	.160		26.50	6		32.50	39
6000	Interior elec., wall mount, fiberglass panels, 2 lamps, 6"	1 Elec	8	1		92	54.50		146.50	186
6100	8"	"	8	1		114	54.50		168.50	211
6400	Replacement sign faces, 6" or 8"	1 Clab	50	.160	↓	62.50	6		68.50	78.50
9000	Minimum labor/equipment charge	1 Carp	4	2	Job		94		94	154

10 14 Signage

10 14 23 – Panel Signage

10 14 23.13 Engraved Panel Signage		Crew	Daily Output	Labor-Hours	Unit	Material	2015 Bare Costs Labor	Equipment	Total	Total Incl O&P
0010	**ENGRAVED PANEL SIGNAGE**, interior									
1010	Flexible door sign, adhesive back, w/Braille, 5/8" letters, 4" x 4"	1 Clab	32	.250	Ea.	33	9.40		42.40	52
1050	6" x 6"		32	.250		48.50	9.40		57.90	69
1100	8" x 2"		32	.250		33	9.40		42.40	52
1150	8" x 4"		32	.250		43.50	9.40		52.90	63
1200	8" x 8"		32	.250		53	9.40		62.40	74
1250	12" x 2"		32	.250		36	9.40		45.40	55
1300	12" x 6"		32	.250		39	9.40		48.40	58.50
1350	12" x 12"		32	.250		150	9.40		159.40	180
1500	Graphic symbols, 2" x 2"		32	.250		12	9.40		21.40	28.50
1550	6" x 6"		32	.250		31	9.40		40.40	49.50
1600	8" x 8"		32	.250		39	9.40		48.40	58
2010	Corridor, stock acrylic, 2-sided, with mounting bracket, 2" x 8"	1 Carp	24	.333		24.50	15.65		40.15	52.50
2020	2" x 10"		24	.333		35.50	15.65		51.15	64.50
2050	3" x 8"		24	.333		28.50	15.65		44.15	57
2060	3" x 10"		24	.333		40	15.65		55.65	70
2070	3" x 12"		24	.333		37	15.65		52.65	66
2100	4" x 8"		24	.333		21	15.65		36.65	48.50
2110	4" x 10"		24	.333		38	15.65		53.65	67.50
2120	4" x 12"		24	.333		51	15.65		66.65	82

10 14 53 – Traffic Signage

10 14 53.20 Traffic Signs		Crew	Daily Output	Labor-Hours	Unit	Material	Labor	Equipment	Total	Total Incl O&P
0010	**TRAFFIC SIGNS**									
0012	Stock, 24" x 24", no posts, .080" alum. reflectorized	B-80	70	.457	Ea.	85	18.85	10.30	114.15	135
0100	High intensity		70	.457		97.50	18.85	10.30	126.65	150
0300	30" x 30", reflectorized		70	.457		123	18.85	10.30	152.15	177
0400	High intensity		70	.457		135	18.85	10.30	164.15	190
0600	Guide and directional signs, 12" x 18", reflectorized		70	.457		34.50	18.85	10.30	63.65	79.50
0700	High intensity		70	.457		52	18.85	10.30	81.15	99
0900	18" x 24", stock signs, reflectorized		70	.457		47	18.85	10.30	76.15	93.50
1000	High intensity		70	.457		52	18.85	10.30	81.15	99
1200	24" x 24", stock signs, reflectorized		70	.457		57	18.85	10.30	86.15	104
1300	High intensity		70	.457		62	18.85	10.30	91.15	110
1500	Add to above for steel posts, galvanized, 10'-0" upright, bolted		200	.160		32.50	6.60	3.60	42.70	50.50
1600	12'-0" upright, bolted		140	.229		39	9.40	5.15	53.55	64
1800	Highway road signs, aluminum, over 20 S.F., reflectorized		350	.091	S.F.	33.50	3.77	2.06	39.33	45.50
2000	High intensity		350	.091		33.50	3.77	2.06	39.33	45.50
2200	Highway, suspended over road, 80 S.F. min., reflectorized		165	.194		32	8	4.37	44.37	53.50
2300	High intensity		165	.194		30.50	8	4.37	42.87	52

10 17 Telephone Specialties

10 17 16 – Telephone Enclosures

10 17 16.10 Commercial Telephone Enclosures		Crew	Daily Output	Labor-Hours	Unit	Material	Labor	Equipment	Total	Total Incl O&P
0010	**COMMERCIAL TELEPHONE ENCLOSURES**									
0300	Shelf type, wall hung, recessed	2 Carp	5	3.200	Ea.	745	150		895	1,075
0400	Surface mount		5	3.200	"	1,625	150		1,775	2,025
9000	Minimum labor/equipment charge		4	4	Job		188		188	310

10 21 Compartments and Cubicles

10 21 13 – Toilet Compartments

10 21 13.13 Metal Toilet Compartments		Crew	Daily Output	Labor-Hours	Unit	2015 Bare Costs Material	Labor	Equipment	Total	Total Incl O&P
0010	**METAL TOILET COMPARTMENTS**									
0110	Cubicles, ceiling hung									
0200	Powder coated steel	2 Carp	4	4	Ea.	525	188		713	885
0500	Stainless steel	"	4	4		1,075	188		1,263	1,475
0600	For handicap units, incl. 52" grab bars, add					450			450	495
0900	Floor and ceiling anchored									
1000	Powder coated steel	2 Carp	5	3.200	Ea.	590	150		740	890
1300	Stainless steel	"	5	3.200		1,275	150		1,425	1,650
1400	For handicap units, incl. 52" grab bars, add					315			315	345
1610	Floor anchored									
1700	Powder coated steel	2 Carp	7	2.286	Ea.	590	107		697	825
2000	Stainless steel	"	7	2.286		1,400	107		1,507	1,725
2100	For handicap units, incl. 52" grab bars, add					310			310	345
2200	For juvenile units, deduct					41.50			41.50	45.50
2450	Floor anchored, headrail braced									
2500	Powder coated steel	2 Carp	6	2.667	Ea.	390	125		515	635
2804	Stainless steel	"	4.60	3.478		1,025	163		1,188	1,400
2900	For handicap units, incl. 52" grab bars, add					370			370	410
3000	Wall hung partitions, powder coated steel	2 Carp	7	2.286		635	107		742	875
3300	Stainless steel	"	7	2.286		1,650	107		1,757	2,000
3400	For handicap units, incl. 52" grab bars, add					370			370	410
4000	Screens, entrance, floor mounted, 58" high, 48" wide									
4200	Powder coated steel	2 Carp	15	1.067	Ea.	242	50		292	350
4500	Stainless steel	"	15	1.067	"	910	50		960	1,075
4650	Urinal screen, 18" wide									
4704	Powder coated steel	2 Carp	6.15	2.602	Ea.	211	122		333	430
5004	Stainless steel	"	6.15	2.602	"	580	122		702	840
5100	Floor mounted, head rail braced									
5300	Powder coated steel	2 Carp	8	2	Ea.	230	94		324	405
5600	Stainless steel	"	8	2	"	570	94		664	780
5750	Pilaster, flush									
5800	Powder coated steel	2 Carp	10	1.600	Ea.	278	75		353	430
6100	Stainless steel		10	1.600		625	75		700	815
6300	Post braced, powder coated steel		10	1.600		163	75		238	300
6600	Stainless steel		10	1.600		450	75		525	620
6700	Wall hung, bracket supported									
6800	Powder coated steel	2 Carp	10	1.600	Ea.	163	75		238	300
7100	Stainless steel		10	1.600		278	75		353	430
7400	Flange supported, powder coated steel		10	1.600		106	75		181	239
7700	Stainless steel		10	1.600		310	75		385	465
7800	Wedge type, powder coated steel		10	1.600		134	75		209	271
8100	Stainless steel		10	1.600		575	75		650	760
9000	Minimum labor/equipment charge	1 Carp	2.50	3.200	Job		150		150	246

10 21 13.14 Metal Toilet Compartment Components		Crew	Daily Output	Labor-Hours	Unit	Material	Labor	Equipment	Total	Total Incl O&P
0010	**METAL TOILET COMPARTMENT COMPONENTS**									
0100	Pilasters									
0110	Overhead braced, powder coated steel, 7" wide x 82" high	2 Carp	22.20	.721	Ea.	73.50	34		107.50	137
0120	Stainless steel		22.20	.721		125	34		159	193
0130	Floor braced, powder coated steel, 7" wide x 70" high		23.30	.687		128	32		160	194
0140	Stainless steel		23.30	.687		245	32		277	325
0150	Ceiling hung, powder coated steel, 7" wide x 83" high		13.30	1.203		136	56.50		192.50	242
0160	Stainless steel		13.30	1.203		274	56.50		330.50	395

10 21 Compartments and Cubicles

10 21 13 – Toilet Compartments

10 21 13.14 Metal Toilet Compartment Components		Crew	Daily Output	Labor-Hours	Unit	2015 Bare Costs Material	Labor	Equipment	Total	Total Incl O&P
0170	Wall hung, powder coated steel, 3" wide x 58" high	2 Carp	18.90	.847	Ea.	133	40		173	211
0180	Stainless steel		18.90	.847		193	40		233	278
0200	Panels									
0210	Powder coated steel, 31" wide x 58" high	2 Carp	18.90	.847	Ea.	137	40		177	215
0220	Stainless steel		18.90	.847		365	40		405	465
0230	Powder coated steel, 53" wide x 58" high		18.90	.847		169	40		209	251
0240	Stainless steel		18.90	.847		475	40		515	585
0250	Powder coated steel, 63" wide x 58" high		18.90	.847		205	40		245	291
0260	Stainless steel		18.90	.847		520	40		560	640
0300	Doors									
0310	Powder coated steel, 24" wide x 58" high	2 Carp	14.10	1.135	Ea.	139	53.50		192.50	241
0320	Stainless steel		14.10	1.135		290	53.50		343.50	410
0330	Powder coated steel, 26" wide x 58" high		14.10	1.135		141	53.50		194.50	244
0340	Stainless steel		14.10	1.135		300	53.50		353.50	420
0350	Powder coated steel, 28" wide x 58" high		14.10	1.135		162	53.50		215.50	266
0360	Stainless steel		14.10	1.135		335	53.50		388.50	455
0370	Powder coated steel, 36" wide x 58" high		14.10	1.135		174	53.50		227.50	280
0380	Stainless steel		14.10	1.135		375	53.50		428.50	500
0400	Headrails									
0410	For powder coated steel, 62" long	2 Carp	65	.246	Ea.	22	11.55		33.55	43.50
0420	Stainless steel		65	.246		22	11.55		33.55	43.50
0430	For powder coated steel, 84" long		50	.320		31.50	15		46.50	59.50
0440	Stainless steel		50	.320		31.50	15		46.50	59.50
0450	For powder coated steel, 120" long		30	.533		43	25		68	88.50
0460	Stainless steel		30	.533		42.50	25		67.50	88
9000	Minimum labor/equipment charge	1 Carp	4	2	Job		94		94	154

10 21 13.16 Plastic-Laminate-Clad Toilet Compartments		Crew	Daily Output	Labor-Hours	Unit	Material	Labor	Equipment	Total	Total Incl O&P
0010	**PLASTIC-LAMINATE-CLAD TOILET COMPARTMENTS**									
0110	Cubicles, ceiling hung									
0300	Plastic laminate on particle board	2 Carp	4	4	Ea.	520	188		708	880
0600	For handicap units, incl. 52" grab bars, add				"	450			450	495
0900	Floor and ceiling anchored									
1100	Plastic laminate on particle board	2 Carp	5	3.200	Ea.	790	150		940	1,100
1400	For handicap units, incl. 52" grab bars, add				"	315			315	345
1610	Floor mounted									
1800	Plastic laminate on particle board	2 Carp	7	2.286	Ea.	535	107		642	760
2450	Floor mounted, headrail braced									
2600	Plastic laminate on particle board	2 Carp	6	2.667	Ea.	750	125		875	1,025
3400	For handicap units, incl. 52" grab bars, add					370			370	410
4300	Entrance screen, floor mtd., plas. lam., 58" high, 48" wide	2 Carp	15	1.067		610	50		660	750
4800	Urinal screen, 18" wide, ceiling braced, plastic laminate		8	2		194	94		288	365
5400	Floor mounted, headrail braced		8	2		200	94		294	375
5900	Pilaster, flush, plastic laminate		10	1.600		505	75		580	685
6400	Post braced, plastic laminate		10	1.600		305	75		380	460
6700	Wall hung, bracket supported									
6900	Plastic laminate on particle board	2 Carp	10	1.600	Ea.	94.50	75		169.50	227
7450	Flange supported									
7500	Plastic laminate on particle board	2 Carp	10	1.600	Ea.	230	75		305	375
9000	Minimum labor/equipment charge	1 Carp	2.50	3.200	Job		150		150	246

10 21 13 – Toilet Compartments

10 21 13.17 Plastic-Lam. Clad Toilet Compart. Components		Crew	Daily Output	Labor-Hours	Unit	2015 Bare Costs Material	Labor	Equipment	Total	Total Incl O&P
0010	**PLASTIC-LAMINATE CLAD TOILET COMPARTMENT COMPONENTS**									
0100	Pilasters									
0110	Overhead braced, 7" wide x 82" high	2 Carp	22.20	.721	Ea.	99.50	34		133.50	165
0130	Floor anchored, 7" wide x 70" high		23.30	.687		99.50	32		131.50	162
0150	Ceiling hung, 7" wide x 83" high		13.30	1.203		104	56.50		160.50	207
0180	Wall hung, 3" wide x 58" high		18.90	.847		93	40		133	167
0200	Panels									
0210	31" wide x 58" high	2 Carp	18.90	.847	Ea.	148	40		188	227
0230	51" wide x 58" high		18.90	.847		202	40		242	287
0250	63" wide x 58" high		18.90	.847		235	40		275	325
0300	Doors									
0310	24" wide x 58" high	2 Carp	14.10	1.135	Ea.	142	53.50		195.50	244
0330	26" wide x 58" high		14.10	1.135		147	53.50		200.50	250
0350	28" wide x 58" high		14.10	1.135		152	53.50		205.50	255
0370	36" wide x 58" high		14.10	1.135		182	53.50		235.50	288
0400	Headrails									
0410	62" long	2 Carp	65	.246	Ea.	22.50	11.55		34.05	44
0430	84" long		60	.267		31	12.50		43.50	54.50
0450	120" long		30	.533		42	25		67	87
9000	Minimum labor/equipment charge	1 Carp	4	2	Job		94		94	154

10 21 13.19 Plastic Toilet Compartments		Crew	Daily Output	Labor-Hours	Unit	Material	Labor	Equipment	Total	Total Incl O&P
0010	**PLASTIC TOILET COMPARTMENTS**									
0110	Cubicles, ceiling hung									
0250	Phenolic	2 Carp	4	4	Ea.	865	188		1,053	1,250
0600	For handicap units, incl. 52" grab bars, add				"	450			450	495
0900	Floor and ceiling anchored									
1050	Phenolic	2 Carp	5	3.200	Ea.	810	150		960	1,125
1400	For handicap units, incl. 52" grab bars, add				"	315			315	345
1610	Floor mounted									
1750	Phenolic	2 Carp	7	2.286	Ea.	750	107		857	1,000
2100	For handicap units, incl. 52" grab bars, add					310			310	345
2200	For juvenile units, deduct					41.50			41.50	45.50
2450	Floor mounted, headrail braced									
2550	Phenolic	2 Carp	6	2.667	Ea.	750	125		875	1,025
9000	Minimum labor/equipment charge	1 Carp	2.50	3.200	Job		150		150	246

10 21 13.20 Plastic Toilet Compartment Components		Crew	Daily Output	Labor-Hours	Unit	Material	Labor	Equipment	Total	Total Incl O&P
0010	**PLASTIC TOILET COMPARTMENT COMPONENTS**									
0100	Pilasters									
0110	Overhead braced, polymer plastic, 7" wide x 82" high	2 Carp	22.20	.721	Ea.	121	34		155	189
0120	Phenolic		22.20	.721		151	34		185	222
0130	Floor braced, polymer plastic, 7" wide x 70" high		23.30	.687		171	32		203	242
0140	Phenolic		23.30	.687		142	32		174	209
0150	Ceiling hung, polymer plastic, 7" wide x 83" high		13.30	1.203		171	56.50		227.50	281
0160	Phenolic		13.30	1.203		161	56.50		217.50	270
0180	Wall hung, phenolic, 3" wide x 58" high		18.90	.847		96	40		136	170
0200	Panels									
0203	Polymer plastic, 18" high x 55" high	2 Carp	18.90	.847	Ea.	252	40		292	340
0206	Phenolic, 18" wide x 58" high		18.90	.847		222	40		262	310
0210	Polymer plastic, 31" high x 55" high		18.90	.847		305	40		345	400
0220	Phenolic, 31" wide x 58" high		18.90	.847		263	40		303	355
0223	Polymer plastic, 48" high x 55" high		18.90	.847		470	40		510	580
0226	Phenolic, 48" wide x 58" high		18.90	.847		430	40		470	535

10 21 Compartments and Cubicles

10 21 13 – Toilet Compartments

10 21 13.20 Plastic Toilet Compartment Components		Crew	Daily Output	Labor-Hours	Unit	Material	Labor	Equipment	Total	Total Incl O&P
						2015 Bare Costs				
0230	Polymer plastic, 51" wide x 55" high	2 Carp	18.90	.847	Ea.	410	40		450	515
0240	Phenolic, 51" wide x 58" high		18.90	.847		400	40		440	505
0250	Polymer plastic, 63" wide x 55" high		18.90	.847		560	40		600	680
0260	Phenolic, 63" wide x 58" high		18.90	.847		420	40		460	525
0300	Doors									
0310	Polymer plastic, 24" wide x 55" high	2 Carp	14.10	1.135	Ea.	211	53.50		264.50	320
0320	Phenolic, 24" wide x 58" high		14.10	1.135		296	53.50		349.50	415
0330	Polymer plastic, 26" high x 55" high		14.10	1.135		225	53.50		278.50	335
0340	Phenolic, 26" wide x 58" high		14.10	1.135		315	53.50		368.50	435
0350	Polymer plastic, 28" wide x 55" high		14.10	1.135		250	53.50		303.50	365
0360	Phenolic, 28" wide x 58" high		14.10	1.135		330	53.50		383.50	455
0370	Polymer plastic, 36" wide x 55" high		14.10	1.135		291	53.50		344.50	410
0380	Phenolic, 36" wide x 58" high		14.10	1.135		415	53.50		468.50	545
0400	Headrails									
0410	For polymer plastic, 62" long	2 Carp	65	.246	Ea.	22.50	11.55		34.05	44
0420	Phenolic		65	.246		22.50	11.55		34.05	44
0430	For polymer plastic, 84" long		50	.320		32	15		47	60
0440	Phenolic		50	.320		32	15		47	60
0450	For polymer plastic, 120" long		30	.533		43	25		68	88.50
0460	Phenolic		30	.533		43	25		68	88.50
9000	Minimum labor/equipment charge	1 Carp	4	2	Job		94		94	154
10 21 13.40 Stone Toilet Compartments										
0010	**STONE TOILET COMPARTMENTS**									
0100	Cubicles, ceiling hung, marble	2 Marb	2	8	Ea.	1,800	350		2,150	2,550
0600	For handicap units, incl. 52" grab bars, add					450			450	495
0800	Floor & ceiling anchored, marble	2 Marb	2.50	6.400		1,975	278		2,253	2,625
1400	For handicap units, incl. 52" grab bars, add					315			315	345
1600	Floor mounted, marble	2 Marb	3	5.333		1,225	232		1,457	1,725
2400	Floor mounted, headrail braced, marble	"	3	5.333		1,150	232		1,382	1,650
2900	For handicap units, incl. 52" grab bars, add					370			370	410
4100	Entrance screen, floor mounted marble, 58" high, 48" wide	2 Marb	9	1.778		795	77.50		872.50	995
4600	Urinal screen, 18" wide, ceiling braced, marble	D-1	6	2.667		755	112		867	1,025
5100	Floor mounted, head rail braced									
5200	Marble	D-1	6	2.667	Ea.	645	112		757	895
5700	Pilaster, flush, marble		9	1.778		840	75		915	1,050
6200	Post braced, marble		9	1.778		825	75		900	1,025
9000	Minimum labor/equipment charge	1 Carp	2.50	3.200	Job		150		150	246

10 21 23 – Cubicle Curtains and Track

10 21 23.16 Cubicle Track and Hardware		Crew	Daily Output	Labor-Hours	Unit	Material	Labor	Equipment	Total	Total Incl O&P
0010	**CUBICLE TRACK AND HARDWARE**									
0020	Curtain track, box channel, ceiling mounted	1 Carp	135	.059	L.F.	6.25	2.78		9.03	11.40
0100	Suspended	"	100	.080	"	8.30	3.76		12.06	15.25
0300	Curtains, nylon mesh tops, fire resistant, 11 oz. per lineal yard									
0310	Polyester oxford cloth, 9' ceiling height	1 Carp	425	.019	L.F.	16.15	.88		17.03	19.20
0500	8' ceiling height		425	.019		7.35	.88		8.23	9.55
0550	Polyester, antimicrobial, 9' ceiling height		425	.019		19.65	.88		20.53	23
0560	8' ceiling height		425	.019		17.50	.88		18.38	20.50
0700	Designer oxford cloth		425	.019		7.35	.88		8.23	9.55
0800	I.V. track systems									
0820	I.V. track, oval	1 Carp	135	.059	L.F.	7.80	2.78		10.58	13.10
0830	I.V. trolley		32	.250	Ea.	41	11.75		52.75	64.50
0840	I.V. pendent, (tree, 5 hook)		32	.250	"	171	11.75		182.75	207

10 22 Partitions

10 22 13 – Wire Mesh Partitions

10 22 13.10 Partitions, Woven Wire

	10 22 13.10 Partitions, Woven Wire	Crew	Daily Output	Labor-Hours	Unit	2015 Bare Costs Material	Labor	Equipment	Total	Total Incl O&P
0010	**PARTITIONS, WOVEN WIRE** for tool or stockroom enclosures									
0100	Channel frame, 1-1/2" diamond mesh, 10 ga. wire, painted									
0300	Wall panels, 4'-0" wide, 7' high	2 Carp	25	.640	Ea.	144	30		174	208
0400	8' high		23	.696		162	32.50		194.50	233
0600	10' high		18	.889		190	41.50		231.50	278
0700	For 5' wide panels, add					5%				
0900	Ceiling panels, 10' long, 2' wide	2 Carp	25	.640		130	30		160	193
1000	4' wide		15	1.067		197	50		247	299
1200	Panel with service window & shelf, 5' wide, 7' high		20	.800		375	37.50		412.50	470
1300	8' high		15	1.067		450	50		500	575
1500	Sliding doors, full height, 3' wide, 7' high		6	2.667		490	125		615	740
1600	10' high		5	3.200		525	150		675	820
1800	6' wide sliding door, 7' full height		5	3.200		670	150		820	985
1900	10' high		4	4		830	188		1,018	1,225
2100	Swinging doors, 3' wide, 7' high, no transom		6	2.667		310	125		435	550
2200	7' high, 3' transom		5	3.200		375	150		525	655

10 22 16 – Folding Gates

10 22 16.10 Security Gates

	10 22 16.10 Security Gates	Crew	Daily Output	Labor-Hours	Unit	Material	Labor	Equipment	Total	Total Incl O&P
0010	**SECURITY GATES** for roll up type, see Section 08 33 13.10									
0300	Scissors type folding gate, ptd. steel, single, 6-1/2' high, 5-1/2' wide	2 Sswk	4	4	Opng.	226	211		437	630
0350	6-1/2' wide		4	4		246	211		457	650
0400	7-1/2' wide		4	4		237	211		448	640
0600	Double gate, 8' high, 8' wide		2.50	6.400		390	335		725	1,050
0650	10' wide		2.50	6.400		425	335		760	1,075
0700	12' wide		2	8		620	420		1,040	1,450
0750	14' wide		2	8		630	420		1,050	1,450
0900	Door gate, folding steel, 4' wide, 61" high		4	4		139	211		350	535
1000	71" high		4	4		169	211		380	565
1200	81" high		4	4		195	211		406	595
1300	Window gates, 2' to 4' wide, 31" high		4	4		80	211		291	470
1500	55" high		3.75	4.267		122	225		347	540
1600	79" high		3.50	4.571		144	241		385	595

10 22 19 – Demountable Partitions

10 22 19.43 Demountable Composite Partitions

	10 22 19.43 Demountable Composite Partitions	Crew	Daily Output	Labor-Hours	Unit	Material	Labor	Equipment	Total	Total Incl O&P
0010	**DEMOUNTABLE COMPOSITE PARTITIONS**, add for doors									
0100	Do not deduct door openings from total L.F.									
0900	Demountable gypsum system on 2" to 2-1/2"									
1000	steel studs, 9' high, 3" to 3-3/4" thick									
1200	Vinyl clad gypsum	2 Carp	48	.333	L.F.	60	15.65		75.65	91.50
1300	Fabric clad gypsum		44	.364		150	17.05		167.05	193
1500	Steel clad gypsum		40	.400		167	18.80		185.80	215
1600	1.75 system, aluminum framing, vinyl clad hardboard,									
1800	paper honeycomb core panel, 1-3/4" to 2-1/2" thick									
1900	9' high	2 Carp	48	.333	L.F.	101	15.65		116.65	137
2100	7' high		60	.267		90.50	12.50		103	120
2200	5' high		80	.200		76.50	9.40		85.90	99.50
2250	Unitized gypsum system									
2300	Unitized panel, 9' high, 2" to 2-1/2" thick									
2350	Vinyl clad gypsum	2 Carp	48	.333	L.F.	130	15.65		145.65	169
2400	Fabric clad gypsum	"	44	.364	"	214	17.05		231.05	263
2500	Unitized mineral fiber system									
2510	Unitized panel, 9' high, 2-1/4" thick, aluminum frame									

10 22 19 – Demountable Partitions

10 22 19.43 Demountable Composite Partitions		Crew	Daily Output	Labor-Hours	Unit	Material	Labor	Equipment	Total	Total Incl O&P
						2015 Bare Costs				
2550	Vinyl clad mineral fiber	2 Carp	48	.333	L.F.	129	15.65		144.65	168
2600	Fabric clad mineral fiber	"	44	.364	"	193	17.05		210.05	240
2800	Movable steel walls, modular system									
2900	Unitized panels, 9′ high, 48″ wide									
3100	Baked enamel, pre-finished	2 Carp	60	.267	L.F.	146	12.50		158.50	182
3200	Fabric clad steel		56	.286	"	212	13.40		225.40	255
5310	Trackless wall, cork finish, semi-acoustic, 1-5/8″ thick, unsealed		325	.049	S.F.	38.50	2.31		40.81	46.50
5320	Sealed		190	.084		42.50	3.95		46.45	53.50
5330	Acoustic, 2″ thick, unsealed		305	.052		36.50	2.46		38.96	44
5340	Sealed		225	.071		56	3.34		59.34	67
5500	For acoustical partitions, add, unsealed					2.36			2.36	2.60
5550	Sealed					11			11	12.10
5700	For doors, see Sections 08 11 & 08 16									
5800	For door hardware, see Section 08 71									
6100	In-plant modular office system, w/prehung hollow core door									
6200	3″ thick polystyrene core panels									
6250	12′ x 12′, 2 wall	2 Clab	3.80	4.211	Ea.	4,025	158		4,183	4,700
6300	4 wall		1.90	8.421		6,150	315		6,465	7,275
6350	16′ x 16′, 2 wall		3.60	4.444		6,125	167		6,292	7,025
6400	4 wall		1.80	8.889		8,325	335		8,660	9,700
9000	Minimum labor/equipment charge	2 Carp	3	5.333	Job		250		250	410

10 22 23 – Portable Partitions, Screens, and Panels

10 22 23.13 Wall Screens		Crew	Daily Output	Labor-Hours	Unit	Material	Labor	Equipment	Total	Total Incl O&P
0010	**WALL SCREENS**, divider panels, free standing, fiber core									
0020	Fabric face straight									
0100	3′-0″ long, 4′-0″ high	2 Carp	100	.160	L.F.	123	7.50		130.50	147
0200	5′-0″ high		90	.178		103	8.35		111.35	127
0500	6′-0″ high		75	.213		104	10		114	131
0900	5′-0″ long, 4′-0″ high		175	.091		67.50	4.29		71.79	81.50
1000	5′-0″ high		150	.107		75.50	5		80.50	91
1500	6″-0″ high		125	.128		90.50	6		96.50	109
1600	6′-0″ long, 5′-0″ high		162	.099		75.50	4.64		80.14	90.50
3200	Economical panels, fabric face, 4′-0″ long, 5′-0″ high		132	.121		50	5.70		55.70	64.50
3250	6′-0″ high		112	.143		55.50	6.70		62.20	72
3300	5′-0″ long, 5′-0″ high		150	.107		54.50	5		59.50	67.50
3350	6′-0″ high		125	.128		50	6		56	65
3450	Acoustical panels, 60 to 90 NRC, 3′-0″ long, 5′-0″ high		90	.178		76.50	8.35		84.85	97.50
3550	6′-0″ high		75	.213		89.50	10		99.50	115
3600	5′-0″ long, 5′-0″ high		150	.107		61	5		66	75
3650	6′-0″ high		125	.128		69	6		75	85.50
3700	6′-0″ long, 5′-0″ high		162	.099		53	4.64		57.64	66
3750	6′-0″ high		138	.116		82	5.45		87.45	99
3800	Economy acoustical panels, 40 N.R.C., 4′-0″ long, 5′-0″ high		132	.121		50	5.70		55.70	64.50
3850	6′-0″ high		112	.143		55.50	6.70		62.20	72
3900	5′-0″ long, 6′-0″ high		125	.128		50	6		56	65
3950	6′-0″ long, 5′-0″ high		162	.099		47	4.64		51.64	59.50
4000	Metal chalkboard, 6′-6″ high, chalkboard, 1 side		125	.128		120	6		126	142
4100	Metal chalkboard, 2 sides		120	.133		137	6.25		143.25	161
4300	Tackboard, both sides		123	.130		109	6.10		115.10	129
9000	Minimum labor/equipment charge		3	5.333	Job		250		250	410

10 22 Partitions

10 22 23 – Portable Partitions, Screens, and Panels

	10 22 23.23 Movable Panel Systems	Crew	Daily Output	Labor-Hours	Unit	2015 Bare Costs Material	Labor	Equipment	Total	Total Incl O&P
0010	**MOVABLE PANEL SYSTEMS**									
0030	Fabric panel, class A fire rated									
0040	Minimum N.R.C. 0.95, STC 28, fabric edged									
0200	42" high, 24" wide	2 Clab	30	.533	Ea.	125	20		145	170
0220	36" wide		28	.571		178	21.50		199.50	230
0240	48" wide		26	.615		231	23		254	292
0260	60" wide		25	.640		315	24		339	390
0400	54" high, 24" wide		30	.533		163	20		183	212
0420	36" wide		28	.571		176	21.50		197.50	229
0440	48" wide		26	.615		204	23		227	262
0460	60" wide		25	.640		340	24		364	415
0600	60" high, 24" wide		29	.552		151	20.50		171.50	200
0620	36" wide		27	.593		226	22.50		248.50	285
0640	48" wide		25	.640		300	24		324	370
0660	60" wide		24	.667		375	25		400	455
0800	72" high, 24" wide		29	.552		181	20.50		201.50	233
1000	36" wide		27	.593		271	22.50		293.50	335
1200	48" wide		25	.640		360	24		384	435
1400	60" wide		24	.667		450	25		475	540
2000	Hardwood edged, 42" high, 24" wide		30	.533		106	20		126	149
2100	36" wide		28	.571		158	21.50		179.50	209
2120	48" wide		26	.615		211	23		234	270
2180	60" wide		25	.640		264	24		288	330
2200	54" high, 24" wide		30	.533		136	20		156	182
2220	36" wide		28	.571		251	21.50		272.50	310
2240	48" wide		26	.615		271	23		294	335
2300	60" wide		25	.640		340	24		364	415
2400	60" high, 24" wide		29	.552		151	20.50		171.50	200
2420	36" wide		27	.593		226	22.50		248.50	286
2440	48" wide		25	.640		300	24		324	370
2460	60" wide		24	.667		375	25		400	455
2800	72" high, 24" wide		29	.552		181	20.50		201.50	233
2820	36" wide		27	.593		325	22.50		347.50	395
2840	48" wide		25	.640		460	24		484	545
2860	60" wide		24	.667		450	25		475	540
3000	Fabric panel, straight, N.R.C. <= .50									
3110	60" high, 24" wide	2 Clab	29	.552	Ea.	164	20.50		184.50	214
3120	30" wide		28	.571		184	21.50		205.50	237
3130	36" wide		27	.593		187	22.50		209.50	243
3140	48" wide		26	.615		201	23		224	259
3150	60" wide		25	.640		271	24		295	340
3160	72" wide		24	.667		283	25		308	350
3210	72" high, 24" wide		29	.552		174	20.50		194.50	226
3220	30" wide		28	.571		179	21.50		200.50	231
3230	36" wide		27	.593		219	22.50		241.50	278
3240	48" wide		26	.615		222	23		245	282
3250	60" wide		25	.640		249	24		273	315
3260	72" wide		24	.667		330	25		355	400
3310	N.R.C. => .75, 42" high, 24" wide		30	.533		189	20		209	241
3320	30" wide		29	.552		192	20.50		212.50	245
3330	36" wide		28	.571		210	21.50		231.50	265
3340	48" wide		27	.593		230	22.50		252.50	289

10 22 23 – Portable Partitions, Screens, and Panels

10 22 23.23 Movable Panel Systems		Crew	Daily Output	Labor-Hours	Unit	2015 Bare Costs Material	Labor	Equipment	Total	Total Incl O&P
3350	60" wide	2 Clab	26	.615	Ea.	250	23		273	310
3360	72" wide		25	.640		287	24		311	355
3410	60" high, 24" wide		29	.552		229	20.50		249.50	286
3420	30" wide		28	.571		246	21.50		267.50	305
3430	36" wide		27	.593		259	22.50		281.50	320
3440	48" wide		26	.615		260	23		283	325
3450	60" wide		25	.640		305	24		329	375
3460	72" wide		24	.667		320	25		345	390
3510	72" high, 24" wide		29	.552		278	20.50		298.50	340
3520	30" wide		28	.571		289	21.50		310.50	355
3530	36" wide		27	.593		299	22.50		321.50	365
3540	48" wide		26	.615		310	23		333	380
3550	60" wide		25	.640		345	24		369	420
3560	72" wide		24	.667		380	25		405	455
3910	Connector kit, straight					35			35	38.50
3920	Corner, 2-way					28.50			28.50	31
3930	3-way					16.40			16.40	18.05
3940	4-way					15.05			15.05	16.60
3950	T-leg					23.50			23.50	26

10 22 33 – Accordion Folding Partitions

10 22 33.10 Partitions, Accordion Folding		Crew	Daily Output	Labor-Hours	Unit	Material	Labor	Equipment	Total	Total Incl O&P
0010	**PARTITIONS, ACCORDION FOLDING**									
0100	Vinyl covered, over 150 S.F., frame not included									
0300	Residential, 1.25 lb. per S.F., 8' maximum height	2 Carp	300	.053	S.F.	25	2.50		27.50	31.50
0400	Commercial, 1.75 lb. per S.F., 8' maximum height		225	.071		28.50	3.34		31.84	37
0600	2 lb. per S.F., 17' maximum height		150	.107		29.50	5		34.50	40.50
0700	Industrial, 4 lb. per S.F., 20' maximum height		75	.213		44.50	10		54.50	65.50
0900	Acoustical, 3 lb. per S.F., 17' maximum height		100	.160		31.50	7.50		39	47.50
1200	5 lb. per S.F., 20' maximum height		95	.168		44	7.90		51.90	61.50
1300	5.5 lb. per S.F., 17' maximum height		90	.178		51.50	8.35		59.85	70
1400	Fire rated, 4.5 psf, 20' maximum height		160	.100		51.50	4.70		56.20	64
1500	Vinyl clad wood or steel, electric operation, 5.0 psf		160	.100		63	4.70		67.70	76.50
1900	Wood, non-acoustic, birch or mahogany, to 10' high		300	.053		34	2.50		36.50	41
9000	Minimum labor/equipment charge		4	4	Job		188		188	310

10 22 39 – Folding Panel Partitions

10 22 39.10 Partitions, Folding Panel		Crew	Daily Output	Labor-Hours	Unit	Material	Labor	Equipment	Total	Total Incl O&P
0010	**PARTITIONS, FOLDING PANEL**, acoustic, wood									
0100	Vinyl faced, to 18' high, 6 psf, economy trim	2 Carp	60	.267	S.F.	57	12.50		69.50	83
0150	Standard trim		45	.356		68	16.70		84.70	102
0200	Premium trim		30	.533		87.50	25		112.50	138
0400	Plastic laminate or hardwood finish, standard trim		60	.267		58.50	12.50		71	85
0500	Premium trim		30	.533		62.50	25		87.50	110
0600	Wood, low acoustical type, 4.5 psf, to 14' high		50	.320		42.50	15		57.50	71.50
1100	Steel, acoustical, 9 to 12 lb. per S.F., vinyl faced, standard trim		60	.267		60.50	12.50		73	87
1200	Premium trim		30	.533		74	25		99	123
1700	Aluminum framed, acoustical, to 12' high, 5.5 psf, standard trim		60	.267		41	12.50		53.50	65.50
1800	Premium trim		30	.533		49.50	25		74.50	95
2000	6.5 lb. per S.F., standard trim		60	.267		43	12.50		55.50	67.50
2100	Premium trim		30	.533		53	25		78	99.50
9000	Minimum labor/equipment charge		4	4	Job		188		188	310

10 22 Partitions

10 22 43 – Sliding Partitions

10 22 43.10 Partitions, Sliding		Crew	Daily Output	Labor-Hours	Unit	Material	2015 Bare Costs Labor	Equipment	Total	Total Incl O&P
0010	**PARTITIONS, SLIDING**									
0020	Acoustic air wall, 1-5/8″ thick, standard trim	2 Carp	375	.043	S.F.	33	2		35	40
0100	Premium trim		365	.044		56.50	2.06		58.56	65.50
0300	2-1/4″ thick, standard trim		360	.044		37	2.09		39.09	44.50
0400	Premium trim		330	.048		65	2.28		67.28	75
0600	For track type, add to above				L.F.	121			121	133
0700	Overhead track type, acoustical, 3″ thick, 11 psf, standard trim	2 Carp	350	.046	S.F.	83.50	2.15		85.65	95
0800	Premium trim	″	300	.053	″	100	2.50		102.50	114

10 26 Wall and Door Protection

10 26 13 – Corner Guards

10 26 13.10 Metal Corner Guards		Crew	Daily Output	Labor-Hours	Unit	Material	Labor	Equipment	Total	Total Incl O&P
0010	**METAL CORNER GUARDS**									
0020	Steel angle w/anchors, 1″ x 1″ x 1/4″, 1.5#/L.F.	2 Carp	160	.100	L.F.	7.10	4.70		11.80	15.50
0100	2″ x 2″ x 1/4″ angles, 3.2#/L.F.		150	.107		10.20	5		15.20	19.40
0200	3″ x 3″ x 5/16″ angles, 6.1#/L.F.		140	.114		14.75	5.35		20.10	25
0300	4″ x 4″ x 5/16″ angles, 8.2#/L.F.		120	.133		19.95	6.25		26.20	32.50
0350	For angles drilled and anchored to masonry, add					15%	120%			
0370	Drilled and anchored to concrete, add					20%	170%			
0400	For galvanized angles, add					35%				
0450	For stainless steel angles, add					100%				
0500	Steel door track/wheel guards, 4′ - 0″ high	E-4	22	1.455	Ea.	109	77.50	6.65	193.15	267
0800	Pipe bumper for truck doors, 8′ long, 6″ diameter, filled		20	1.600		625	85	7.30	717.30	850
0900	8″ diameter		20	1.600		725	85	7.30	817.30	960
1000	Wall protection, stainless steel, 16 ga, 48″ x 36″ tall, screwed to studs	2 Skwk	500	.032	S.F.	8.25	1.56		9.81	11.65
1050	Wall end guard, stainless steel, 16 ga, 36″ tall, screwed to studs	1 Skwk	30	.267	Ea.	25	12.95		37.95	48.50
9000	Minimum labor/equipment charge	1 Carp	2	4	Job		188		188	310

10 26 13.20 Corner Protection		Crew	Daily Output	Labor-Hours	Unit	Material	Labor	Equipment	Total	Total Incl O&P
0010	**CORNER PROTECTION**									
0100	Stainless steel, 16 ga., adhesive mount, 3-1/2″ leg	1 Carp	80	.100	L.F.	24	4.70		28.70	34
0200	12 ga. stainless, adhesive mount	″	80	.100		26.50	4.70		31.20	36.50
0300	For screw mount, add						10%			
0500	Vinyl acrylic, adhesive mount, 3″ leg	1 Carp	128	.063		9.25	2.93		12.18	15
0550	1-1/2″ leg		160	.050		4.83	2.35		7.18	9.15
0600	Screw mounted, 3″ leg		80	.100		10.10	4.70		14.80	18.80
0650	1-1/2″ leg		100	.080		4.57	3.76		8.33	11.20
0700	Clear plastic, screw mounted, 2-1/2″		60	.133		4.37	6.25		10.62	15.05
1000	Vinyl cover, alum. retainer, surface mount, 3″ x 3″		48	.167		10.45	7.85		18.30	24.50
1050	2″ x 2″		48	.167		9.45	7.85		17.30	23.50
1100	Flush mounted, 3″ x 3″		32	.250		20.50	11.75		32.25	42
1150	2″ x 2″		32	.250		16.70	11.75		28.45	37.50

10 26 16 – Bumper Guards

10 26 16.10 Wallguard		Crew	Daily Output	Labor-Hours	Unit	Material	Labor	Equipment	Total	Total Incl O&P
0010	**WALLGUARD**									
0400	Rub rail, vinyl, adhesive mounted	1 Carp	185	.043	L.F.	8.65	2.03		10.68	12.90
0500	Neoprene, aluminum backing, 1-1/2″ x 2″		110	.073		8.80	3.41		12.21	15.30
1000	Trolley rail, PVC, clipped to wall, 5″ high		185	.043		8.80	2.03		10.83	13.05
1050	8″ high		180	.044		14.70	2.09		16.79	19.55
1200	Bed bumper, vinyl acrylic, alum. retainer, 21″ long		10	.800	Ea.	40.50	37.50		78	106
1300	53″ long with aligner		9	.889	″	102	41.50		143.50	181

10 26 Wall and Door Protection

10 26 16 – Bumper Guards

10 26 16.10 Wallguard		Crew	Daily Output	Labor-Hours	Unit	Material	2015 Bare Costs Labor	Equipment	Total	Total Incl O&P
1400	Bumper, vinyl cover, alum. retain., cush. mnt., 1-1/2″ x 2-3/4″	1 Carp	80	.100	L.F.	14.20	4.70		18.90	23.50
1500	2″ x 4-1/4″		80	.100		20.50	4.70		25.20	30.50
1600	Surface mounted, 1-3/4″ x 3-5/8″		80	.100		11.95	4.70		16.65	21
1700	Bumper rail, stainless steel, flat bar on brackets, 4″ x 1/4″	2 Skwk	120	.133		42	6.50		48.50	57
1750	Wallguard stainless steel baseboard, 12″ tall, adhesive applied	″	260	.062		33	2.99		35.99	41.50
2000	Crash rail, vinyl cover, alum. retainer, 1″ x 4″	1 Carp	110	.073		11.05	3.41		14.46	17.75
2100	1″ x 8″		90	.089		18.15	4.17		22.32	27
2150	Vinyl inserts, aluminum plate, 1″ x 2-1/2″		110	.073		14.90	3.41		18.31	22
2200	1″ x 5″		90	.089		23	4.17		27.17	32.50
3000	Handrail/bumper, vinyl cover, alum. retainer									
3010	Bracket mounted, flat rail, 5-1/2″	1 Carp	80	.100	L.F.	18.30	4.70		23	27.50
3100	6-1/2″		80	.100		23	4.70		27.70	32.50
3200	Bronze bracket, 1-3/4″ diam. rail		80	.100		16.50	4.70		21.20	26
4000	Handrail, with antimicrobial copper alloy, #6 finish, 1-1/2″ OD		80	.100		7.50	4.70		12.20	15.95

10 28 Toilet, Bath, and Laundry Accessories

10 28 13 – Toilet Accessories

10 28 13.13 Commercial Toilet Accessories

		Crew	Daily Output	Labor-Hours	Unit	Material	Labor	Equipment	Total	Total Incl O&P
0010	**COMMERCIAL TOILET ACCESSORIES**									
0200	Curtain rod, stainless steel, 5′ long, 1″ diameter	1 Carp	13	.615	Ea.	28.50	29		57.50	78.50
0300	1-1/4″ diameter		13	.615		29	29		58	79
0350	Chrome, 1″ diameter		13	.615		32.50	29		61.50	83
0360	For vinyl curtain, add		1950	.004	S.F.	.91	.19		1.10	1.32
0400	Diaper changing station, horizontal, wall mounted, plastic		10	.800	Ea.	229	37.50		266.50	315
0420	Vertical		10	.800		229	37.50		266.50	315
0430	Oval shaped		10	.800		225	37.50		262.50	310
0440	Recessed, with stainless steel flange		6	1.333		565	62.50		627.50	730
0500	Dispenser units, combined soap & towel dispensers,									
0510	mirror and shelf, flush mounted	1 Carp	10	.800	Ea.	310	37.50		347.50	400
0600	Towel dispenser and waste receptacle,									
0610	18 gallon capacity	1 Carp	10	.800	Ea.	300	37.50		337.50	390
0800	Grab bar, straight, 1-1/4″ diameter, stainless steel, 18″ long		24	.333		29	15.65		44.65	57.50
0900	24″ long		23	.348		29	16.35		45.35	58.50
1000	30″ long		22	.364		31.50	17.05		48.55	62.50
1100	36″ long		20	.400		38.50	18.80		57.30	73.50
1105	42″ long		20	.400		46	18.80		64.80	81.50
1120	Corner, 36″ long		20	.400		85.50	18.80		104.30	125
1200	1-1/2″ diameter, 24″ long		23	.348		31	16.35		47.35	61
1300	36″ long		20	.400		33.50	18.80		52.30	67.50
1310	42″ long		18	.444		38	21		59	75.50
1500	Tub bar, 1-1/4″ diameter, 24″ x 36″		14	.571		92.50	27		119.50	146
1600	Plus vertical arm		12	.667		97.50	31.50		129	159
1900	End tub bar, 1″ diameter, 90° angle, 16″ x 32″		12	.667		109	31.50		140.50	172
2010	Tub/shower/toilet, 2-wall, 36″ x 24″		12	.667		91	31.50		122.50	152
2110	Antimicrobial copper alloy finish, straight, 18″ long		24	.333		68.50	15.65		84.15	101
2120	24″ long		23	.348		75.50	16.35		91.85	110
2130	36″ long		20	.400		90	18.80		108.80	130
2140	48″ long		19	.421		103	19.75		122.75	146
2300	Hand dryer, surface mounted, electric, 115 volt, 20 amp		4	2		445	94		539	640
2400	230 volt, 10 amp		4	2		745	94		839	975
2450	Hand dryer, touch free, 1400 watt, 81,000 rpm		4	2		1,025	94		1,119	1,275

10 28 13.13 Commercial Toilet Accessories		Crew	Daily Output	Labor-Hours	Unit	Material	Labor	Equipment	Total	Total Incl O&P
						2015 Bare Costs				
2600	Hat and coat strip, stainless steel, 4 hook, 36" long	1 Carp	24	.333	Ea.	68	15.65		83.65	101
2700	6 hook, 60" long		20	.400		124	18.80		142.80	168
3000	Mirror, with stainless steel 3/4" square frame, 18" x 24"		20	.400		49	18.80		67.80	85
3100	36" x 24"		15	.533		109	25		134	161
3200	48" x 24"		10	.800		150	37.50		187.50	227
3300	72" x 24"		6	1.333		289	62.50		351.50	425
3500	With 5" stainless steel shelf, 18" x 24"		20	.400		194	18.80		212.80	244
3600	36" x 24"		15	.533		236	25		261	300
3700	48" x 24"		10	.800		245	37.50		282.50	330
3800	72" x 24"		6	1.333		300	62.50		362.50	435
4100	Mop holder strip, stainless steel, 5 holders, 48" long		20	.400		89	18.80		107.80	129
4200	Napkin/tampon dispenser, recessed		15	.533		590	25		615	690
4220	Semi-recessed		6.50	1.231		310	58		368	435
4250	Napkin receptacle, recessed		6.50	1.231		165	58		223	277
4300	Robe hook, single, regular		36	.222		18.95	10.45		29.40	38
4400	Heavy duty, concealed mounting		36	.222		19.60	10.45		30.05	38.50
4600	Soap dispenser, chrome, surface mounted, liquid		20	.400		46.50	18.80		65.30	82
4700	Powder		20	.400		56.50	18.80		75.30	93.50
5000	Recessed stainless steel, liquid		10	.800		154	37.50		191.50	231
5600	Shelf, stainless steel, 5" wide, 18 ga., 24" long		24	.333		80.50	15.65		96.15	114
5700	48" long		16	.500		157	23.50		180.50	211
5800	8" wide shelf, 18 ga., 24" long		22	.364		69	17.05		86.05	104
5900	48" long		14	.571		121	27		148	177
6000	Toilet seat cover dispenser, stainless steel, recessed		20	.400		167	18.80		185.80	214
6050	Surface mounted		15	.533		34.50	25		59.50	79
6100	Toilet tissue dispenser, surface mounted, SS, single roll		30	.267		17.80	12.50		30.30	40
6200	Double roll		24	.333		23.50	15.65		39.15	51.50
6240	Plastic, twin/jumbo dbl. roll		24	.333		28.50	15.65		44.15	56.50
6400	Towel bar, stainless steel, 18" long		23	.348		41.50	16.35		57.85	73
6500	30" long		21	.381		111	17.90		128.90	153
6610	Antimicrobial copper alloy finish, 3/4" round, straight, w/o mounting				L.F.	10.95			10.95	12.05
6620	Antimicrobial copper alloy finish, 1" round, straight, w/o mounting				"	13.15			13.15	14.50
6630	24" long, including mounting	1 Carp	23	.348	Ea.	93.50	16.35		109.85	130
6700	Towel dispenser, stainless steel, surface mounted		16	.500		44.50	23.50		68	87.50
6800	Flush mounted, recessed		10	.800		257	37.50		294.50	345
6900	Plastic, touchless, battery operated		16	.500		88	23.50		111.50	136
7000	Towel holder, hotel type, 2 guest size		20	.400		52	18.80		70.80	88.50
7200	Towel shelf, stainless steel, 24" long, 8" wide		20	.400		60	18.80		78.80	97
7400	Tumbler holder, for tumbler only		30	.267		17.80	12.50		30.30	40
7410	Tumbler holder, recessed		20	.400		9.80	18.80		28.60	42
7500	Soap, tumbler & toothbrush		30	.267		19.60	12.50		32.10	42
7510	Tumbler & toothbrush holder		20	.400		13.60	18.80		32.40	46
7590	Air freshner, stainless steel, recessed		16	.500		42	23.50		65.50	84.50
7700	Wall urn ash receiver, surface mount, 11" long		12	.667		95	31.50		126.50	156
7800	7-1/2", long		18	.444		102	21		123	146
8000	Waste receptacles, stainless steel, with top, 13 gallon		10	.800		295	37.50		332.50	385
8100	36 gallon		8	1		405	47		452	520
8200	Shower seat with adjustable back		30	.267		315	12.50		327.50	370
9000	Minimum labor/equipment charge		5	1.600	Job		75		75	123

10 28 Toilet, Bath, and Laundry Accessories

10 28 16 – Bath Accessories

10 28 16.20 Medicine Cabinets		Crew	Daily Output	Labor-Hours	Unit	Material	Labor	Equipment	Total	Total Incl O&P
						2015 Bare Costs				
0010	**MEDICINE CABINETS**									
0020	With mirror, sst frame, 16" x 22", unlighted	1 Carp	14	.571	Ea.	98	27		125	152
0100	Wood frame		14	.571		128	27		155	185
0300	Sliding mirror doors, 20" x 16" x 4-3/4", unlighted		7	1.143		124	53.50		177.50	224
0400	24" x 19" x 8-1/2", lighted		5	1.600		179	75		254	320
0600	Triple door, 30" x 32", unlighted, plywood body		7	1.143		325	53.50		378.50	450
0700	Steel body		7	1.143		375	53.50		428.50	500
0900	Oak door, wood body, beveled mirror, single door		7	1.143		199	53.50		252.50	305
1000	Double door		6	1.333		380	62.50		442.50	525
1200	Hotel cabinets, stainless, with lower shelf, unlighted		10	.800		200	37.50		237.50	281
1300	Lighted		5	1.600		305	75		380	460
9000	Minimum labor/equipment charge		4	2	Job		94		94	154

10 28 19 – Tub and Shower Enclosures

10 28 19.10 Partitions, Shower		Crew	Daily Output	Labor-Hours	Unit	Material	Labor	Equipment	Total	Total Incl O&P
0010	**PARTITIONS, SHOWER** floor mounted, no plumbing									
0400	Cabinet, one piece, fiberglass, 32" x 32"	2 Carp	5	3.200	Ea.	530	150		680	830
0420	36" x 36"		5	3.200		570	150		720	870
0440	36" x 48"		5	3.200		1,375	150		1,525	1,750
0460	Acrylic, 32" x 32"		5	3.200		310	150		460	585
0480	36" x 36"		5	3.200		1,025	150		1,175	1,375
0500	36" x 48"		5	3.200		1,500	150		1,650	1,925
0520	Shower door for above, clear plastic, 24" wide	1 Carp	8	1		186	47		233	281
0540	28" wide		8	1		206	47		253	305
0560	Tempered glass, 24" wide		8	1		200	47		247	297
0580	28" wide		8	1		226	47		273	325
2400	Glass stalls, with doors, no receptors, chrome on brass	2 Shee	3	5.333		1,650	298		1,948	2,300
2700	Anodized aluminum	"	4	4		1,150	224		1,374	1,625
2900	Marble shower stall, stock design, with shower door	2 Marb	1.20	13.333		2,475	580		3,055	3,675
3000	With curtain		1.30	12.308		2,200	535		2,735	3,275
3200	Receptors, precast terrazzo, 32" x 32"		14	1.143		360	49.50		409.50	475
3300	48" x 34"		9.50	1.684		465	73.50		538.50	630
3500	Plastic, simulated terrazzo receptor, 32" x 32"		14	1.143		148	49.50		197.50	244
3600	32" x 48"		12	1.333		215	58		273	330
3800	Precast concrete, colors, 32" x 32"		14	1.143		186	49.50		235.50	286
3900	48" x 48"		8	2		251	87		338	420
4100	Shower doors, economy plastic, 24" wide	1 Shee	9	.889		144	49.50		193.50	237
4200	Tempered glass door, economy		8	1		258	56		314	375
4400	Folding, tempered glass, aluminum frame		6	1.333		390	74.50		464.50	550
4500	Sliding, tempered glass, 48" opening		6	1.333		540	74.50		614.50	710
4700	Deluxe, tempered glass, chrome on brass frame, 42" to 44"		8	1		385	56		441	510
4800	39" to 48" wide		1	8		625	450		1,075	1,400
4850	On anodized aluminum frame, obscure glass		2	4		540	224		764	945
4900	Clear glass		1	8		625	450		1,075	1,400
5100	Shower enclosure, tempered glass, anodized alum. frame									
5120	2 panel & door, corner unit, 32" x 32"	1 Shee	2	4	Ea.	1,025	224		1,249	1,500
5140	Neo-angle corner unit, 16" x 24" x 16"	"	2	4		1,075	224		1,299	1,550
5200	Shower surround, 3 wall, polypropylene, 32" x 32"	1 Carp	4	2		455	94		549	655
5220	PVC, 32" x 32"		4	2		380	94		474	575
5240	Fiberglass		4	2		420	94		514	615
5250	2 wall, polypropylene, 32" x 32"		4	2		315	94		409	505
5270	PVC		4	2		395	94		489	590
5290	Fiberglass		4	2		400	94		494	595

10 28 Toilet, Bath, and Laundry Accessories

10 28 19 – Tub and Shower Enclosures

10 28 19.10 Partitions, Shower		Crew	Daily Output	Labor-Hours	Unit	Material	Labor	Equipment	Total	Total Incl O&P
						2015 Bare Costs				
5300	Tub doors, tempered glass & frame, obscure glass	1 Shee	8	1	Ea.	229	56		285	340
5400	Clear glass		6	1.333		535	74.50		609.50	710
5600	Chrome plated, brass frame, obscure glass		8	1		305	56		361	425
5700	Clear glass		6	1.333		745	74.50		819.50	935
5900	Tub/shower enclosure, temp. glass, alum. frame, obscure glass		2	4		410	224		634	805
6200	Clear glass		1.50	5.333		855	298		1,153	1,400
6500	On chrome-plated brass frame, obscure glass		2	4		565	224		789	980
6600	Clear glass		1.50	5.333		1,200	298		1,498	1,800
6800	Tub surround, 3 wall, polypropylene	1 Carp	4	2		256	94		350	435
6900	PVC		4	2		390	94		484	585
7000	Fiberglass, obscure glass		4	2		400	94		494	595
7100	Clear glass		3	2.667		680	125		805	955
9990	Minimum labor/equipment charge	1 Shee	2.50	3.200	Job		179		179	283

10 28 23 – Laundry Accessories

10 28 23.13 Built-In Ironing Boards		Crew	Daily Output	Labor-Hours	Unit	Material	Labor	Equipment	Total	Total Incl O&P
0010	BUILT-IN IRONING BOARDS									
0020	Including cabinet, board & light, 42"	1 Carp	2	4	Ea.	395	188		583	745
0100	46"	"	1.50	5.333	"	515	250		765	975

10 31 Manufactured Fireplaces

10 31 13 – Manufactured Fireplace Chimneys

10 31 13.10 Fireplace Chimneys		Crew	Daily Output	Labor-Hours	Unit	Material	Labor	Equipment	Total	Total Incl O&P
0010	FIREPLACE CHIMNEYS									
0500	Chimney dbl. wall, all stainless, over 8'-6", 7" diam., add to fireplace	1 Carp	33	.242	V.L.F.	83.50	11.40		94.90	110
0600	10" diameter, add to fireplace		32	.250		132	11.75		143.75	164
0700	12" diameter, add to fireplace		31	.258		158	12.10		170.10	194
0800	14" diameter, add to fireplace		30	.267		227	12.50		239.50	271
1000	Simulated brick chimney top, 4' high, 16" x 16"		10	.800	Ea.	425	37.50		462.50	530
1100	24" x 24"		7	1.143	"	540	53.50		593.50	685

10 31 13.20 Chimney Accessories		Crew	Daily Output	Labor-Hours	Unit	Material	Labor	Equipment	Total	Total Incl O&P
0010	CHIMNEY ACCESSORIES									
0020	Chimney screens, galv., 13" x 13" flue	1 Bric	8	1	Ea.	58.50	46		104.50	139
0050	24" x 24" flue		5	1.600		124	74		198	257
0200	Stainless steel, 13" x 13" flue		8	1		97.50	46		143.50	182
0250	20" x 20" flue		5	1.600		152	74		226	287
2400	Squirrel and bird screens, galvanized, 8" x 8" flue		16	.500		53	23		76	96
2450	13" x 13" flue		12	.667		55	31		86	111
9000	Minimum labor/equipment charge		3.50	2.286	Job		105		105	172

10 31 16 – Manufactured Fireplace Forms

10 31 16.10 Fireplace Forms		Crew	Daily Output	Labor-Hours	Unit	Material	Labor	Equipment	Total	Total Incl O&P
0010	FIREPLACE FORMS									
1800	Fireplace forms, no accessories, 32" opening	1 Bric	3	2.667	Ea.	670	123		793	940
1900	36" opening		2.50	3.200		855	148		1,003	1,175
2000	40" opening		2	4		1,125	185		1,310	1,550
2100	78" opening		1.50	5.333		1,650	246		1,896	2,225

10 31 Manufactured Fireplaces

10 31 23 – Prefabricated Fireplaces

10 31 23.10 Fireplace, Prefabricated		Crew	Daily Output	Labor-Hours	Unit	Material	Labor	Equipment	Total	Total Incl O&P
						2015 Bare Costs				
0010	**FIREPLACE, PREFABRICATED**, free standing or wall hung									
0100	With hood & screen, painted	1 Carp	1.30	6.154	Ea.	1,375	289		1,664	2,000
0150	Average		1	8		1,625	375		2,000	2,400
0200	Stainless steel		.90	8.889		3,050	415		3,465	4,050
1500	Simulated logs, gas fired, 40,000 BTU, 2′ long, manual safety pilot		7	1.143	Set	425	53.50		478.50	555
1600	Adjustable flame remote pilot		6	1.333		1,100	62.50		1,162.50	1,325
1700	Electric, 1,500 BTU, 1′-6″ long, incandescent flame		7	1.143		197	53.50		250.50	305
1800	1,500 BTU, LED flame		6	1.333		300	62.50		362.50	440

10 32 Fireplace Specialties

10 32 13 – Fireplace Dampers

10 32 13.10 Dampers

		Crew	Daily Output	Labor-Hours	Unit	Material	Labor	Equipment	Total	Total Incl O&P
0010	**DAMPERS**									
0800	Damper, rotary control, steel, 30″ opening	1 Bric	6	1.333	Ea.	119	61.50		180.50	231
0850	Cast iron, 30″ opening		6	1.333		125	61.50		186.50	237
0880	36″ opening		6	1.333		127	61.50		188.50	240
0900	48″ opening		6	1.333		167	61.50		228.50	284
0920	60″ opening		6	1.333		355	61.50		416.50	490
0950	72″ opening		5	1.600		425	74		499	585
1000	84″ opening, special order		5	1.600		910	74		984	1,125
1050	96″ opening, special order		4	2		925	92.50		1,017.50	1,175
1200	Steel plate, poker control, 60″ opening		8	1		320	46		366	430
1250	84″ opening, special order		5	1.600		585	74		659	765
1400	"Universal" type, chain operated, 32″ x 20″ opening		8	1		250	46		296	350
1450	48″ x 24″ opening		5	1.600		375	74		449	530

10 32 23 – Fireplace Doors

10 32 23.10 Doors

		Crew	Daily Output	Labor-Hours	Unit	Material	Labor	Equipment	Total	Total Incl O&P
0010	**DOORS**									
0400	Cleanout doors and frames, cast iron, 8″ x 8″	1 Bric	12	.667	Ea.	41	31		72	95
0450	12″ x 12″		10	.800		108	37		145	179
0500	18″ x 24″		8	1		150	46		196	240
0550	Cast iron frame, steel door, 24″ x 30″		5	1.600		315	74		389	465
1600	Dutch Oven door and frame, cast iron, 12″ x 15″ opening		13	.615		131	28.50		159.50	191
1650	Copper plated, 12″ x 15″ opening		13	.615		257	28.50		285.50	330

10 35 Stoves

10 35 13 – Heating Stoves

10 35 13.10 Woodburning Stoves

		Crew	Daily Output	Labor-Hours	Unit	Material	Labor	Equipment	Total	Total Incl O&P
0010	**WOODBURNING STOVES**									
0015	Cast iron, °1500sf	2 Carp	1.30	12.308	Ea.	1,175	580		1,755	2,250
0020	1500-2000sf		1	16		2,025	750		2,775	3,450
0030	2,000 sf		.80	20		2,775	940		3,715	4,600
0050	For gas log lighter, add					45			45	49.50

10 43 Emergency Aid Specialties

10 43 13 – Defibrillator Cabinets

10 43 13.05 Defibrillator Cabinets		Crew	Daily Output	Labor-Hours	Unit	Material	2015 Bare Costs Labor	Equipment	Total	Total Incl O&P
0010	**DEFIBRILLATOR CABINETS**, not equipped, stainless steel									
0050	Defibrillator cabinet, stainless steel with strobe & alarm 12" x 27"	1 Carp	10	.800	Ea.	430	37.50		467.50	530
0100	Automatic External Defibrillator	"	30	.267	"	1,350	12.50		1,362.50	1,500

10 44 Fire Protection Specialties

10 44 13 – Fire Protection Cabinets

10 44 13.53 Fire Equipment Cabinets

		Crew	Daily Output	Labor-Hours	Unit	Material	Labor	Equipment	Total	Total Incl O&P
0010	**FIRE EQUIPMENT CABINETS**, not equipped, 20 ga. steel box									
0040	recessed, D.S. glass in door, box size given									
1000	Portable extinguisher, single, 8" x 12" x 27", alum. door & frame	Q-12	8	2	Ea.	155	101		256	330
1100	Steel door and frame		8	2		116	101		217	286
1200	Stainless steel door and frame		8	2		205	101		306	385
2000	Portable extinguisher, large, 8" x 12" x 36", alum. door & frame		8	2		258	101		359	440
2100	Steel door and frame		8	2		163	101		264	335
2200	Stainless steel door and frame		8	2		270	101		371	455
2500	8" x 16" x 38", aluminum door & frame		8	2		236	101		337	415
2600	Steel door and frame		8	2		224	101		325	405
2700	Fire blanket & extinguisher cab, inc blanket, rec stl., 14" x 40" x 8"		7	2.286		188	116		304	390
2800	Fire blanket cab, inc blanket, surf mtd, stl, 15"x10"x5", w/pwdr coat fin		8	2		94	101		195	262
3000	Hose rack assy., 1-1/2" valve & 100' hose, 24" x 40" x 5-1/2"									
3100	Aluminum door and frame	Q-12	6	2.667	Ea.	370	135		505	615
3200	Steel door and frame		6	2.667		246	135		381	480
3300	Stainless steel door and frame		6	2.667		455	135		590	715
4000	Hose rack assy., 2-1/2" x 1-1/2" valve, 100' hose, 24" x 40" x 8"									
4100	Aluminum door and frame	Q-12	6	2.667	Ea.	375	135		510	620
4200	Steel door and frame		6	2.667		253	135		388	490
4300	Stainless steel door and frame		6	2.667		495	135		630	755
5000	Hose rack assy., 2-1/2" x 1-1/2" valve, 100' hose									
5010	and extinguisher, 30" x 40" x 8"									
5100	Aluminum door and frame	Q-12	5	3.200	Ea.	475	162		637	780
5200	Steel door and frame		5	3.200		264	162		426	545
5300	Stainless steel door and frame		5	3.200		535	162		697	845
6000	Hose rack assy., 1-1/2" valve, 100' hose									
6010	and 2-1/2" FD valve, 24" x 44" x 8"									
6100	Aluminum door and frame	Q-12	5	3.200	Ea.	405	162		567	700
6200	Steel door and frame		5	3.200		258	162		420	535
6300	Stainless steel door and frame		5	3.200		560	162		722	875
7000	Hose rack assy., 1-1/2" valve & 100' hose, 2-1/2" FD valve									
7010	and extinguisher, 30" x 44" x 8"									
7100	Aluminum door and frame	Q-12	5	3.200	Ea.	500	162		662	805
7200	Steel door and frame		5	3.200		315	162		477	600
7300	Stainless steel door and frame		5	3.200		665	162		827	990
8000	Valve cabinet for 2-1/2" FD angle valve, 18" x 18" x 8"									
8100	Aluminum door and frame	Q-12	12	1.333	Ea.	164	67.50		231.50	285
8200	Steel door and frame		12	1.333		136	67.50		203.50	254
8300	Stainless steel door and frame		12	1.333		221	67.50		288.50	350

10 44 Fire Protection Specialties

10 44 16 – Fire Extinguishers

10 44 16.13 Portable Fire Extinguishers		Crew	Daily Output	Labor-Hours	Unit	Material	2015 Bare Costs Labor	Equipment	Total	Total Incl O&P
0010	**PORTABLE FIRE EXTINGUISHERS**									
0140	CO_2, with hose and "H" horn, 10 lb.				Ea.	275			275	305
0160	15 lb.					355			355	390
0180	20 lb.					395			395	435
1000	Dry chemical, pressurized									
1040	Standard type, portable, painted, 2-1/2 lb.				Ea.	37.50			37.50	41
1060	5 lb.					51			51	56.50
1080	10 lb.					81			81	89.50
1100	20 lb.					136			136	150
1120	30 lb.					425			425	470
1300	Standard type, wheeled, 150 lb.					2,425			2,425	2,650
2000	ABC all purpose type, portable, 2-1/2 lb.					21			21	23.50
2060	5 lb.					26.50			26.50	29
2080	9-1/2 lb.					44			44	48
2100	20 lb.					79.50			79.50	87.50
3500	Halotron 1, 2-1/2 lb.					126			126	139
3600	5 lb.					198			198	218
3700	11 lb.					390			390	430
5000	Pressurized water, 2-1/2 gallon, stainless steel					102			102	112
5060	With anti-freeze					106			106	116
9400	Installation of extinguishers, 12 or more, on nailable surface	1 Carp	30	.267			12.50		12.50	20.50
9420	On masonry or concrete	"	15	.533			25		25	41

10 44 16.16 Wheeled Fire Extinguisher Units		Crew	Daily Output	Labor-Hours	Unit	Material	Labor	Equipment	Total	Total Incl O&P
0010	**WHEELED FIRE EXTINGUISHER UNITS**									
0350	CO_2, portable, with swivel horn									
0360	Wheeled type, cart mounted, 50 lb.				Ea.	1,125			1,125	1,250
0400	100 lb.				"	3,850			3,850	4,250
2200	ABC all purpose type									
2300	Wheeled, 45 lb.				Ea.	735			735	810
2360	150 lb.				"	1,850			1,850	2,025

10 51 Lockers

10 51 13 – Metal Lockers

10 51 13.10 Lockers		Crew	Daily Output	Labor-Hours	Unit	Material	Labor	Equipment	Total	Total Incl O&P
0011	**LOCKERS** steel, baked enamel, pre-assembled									
0110	Single tier box locker, 12" x 15" x 72"	1 Shee	20	.400	Ea.	223	22.50		245.50	282
0120	18" x 15" x 72"		20	.400		240	22.50		262.50	300
0130	12" x 18" x 72"		20	.400		228	22.50		250.50	287
0140	18" x 18" x 72"		20	.400		283	22.50		305.50	345
0410	Double tier, 12" x 15" x 36"		30	.267		232	14.90		246.90	279
0420	18" x 15" x 36"		30	.267		235	14.90		249.90	283
0430	12" x 18" x 36"		30	.267		264	14.90		278.90	315
0440	18" x 18" x 36"		30	.267		241	14.90		255.90	289
0500	Two person, 18" x 15" x 72"		20	.400		291	22.50		313.50	355
0510	18" x 18" x 72"		20	.400		325	22.50		347.50	390
0520	Duplex, 15" x 15" x 72"		20	.400		325	22.50		347.50	390
0530	15" x 21" x 72"		20	.400		365	22.50		387.50	435
0600	5 tier box lockers, unassembled		30	.267	Opng.	49	14.90		63.90	77.50
0700	Set up		24	.333		52.50	18.65		71.15	87.50
0900	6 tier box lockers, unassembled		36	.222		37.50	12.45		49.95	60.50

10 51 Lockers

10 51 13 – Metal Lockers

10 51 13.10 Lockers		Crew	Daily Output	Labor-Hours	Unit	2015 Bare Costs Material	Labor	Equipment	Total	Total Incl O&P
1000	Set up	1 Shee	30	.267	Opng.	46	14.90		60.90	74
1100	Wire meshed wardrobe, floor. mtd., open front varsity type		7.50	1.067	Ea.	272	59.50		331.50	395
2400	16-person locker unit with clothing rack									
2500	72 wide x 15″ deep x 72″ high	1 Shee	15	.533	Ea.	455	30		485	545
2550	18″ deep	″	15	.533	″	605	30		635	710
3000	Wall mounted lockers, 4 person, with coat bar									
3100	48″ wide x 18″ deep x 12″ high	1 Shee	20	.400	Ea.	325	22.50		347.50	390
3250	Rack w/24 wire mesh baskets		1.50	5.333	Set	400	298		698	910
3260	30 baskets		1.25	6.400		350	360		710	950
3270	36 baskets		.95	8.421		510	470		980	1,300
3280	42 baskets		.80	10		555	560		1,115	1,500
3300	For built-in lock with 2 keys, add				Ea.	13.20			13.20	14.50
3600	For hanger rods, add					1.90			1.90	2.09
3650	For number plate kit, 100 plates #1 - #100, add	1 Shee	4	2		82	112		194	267
3700	For locker base, closed front panel		90	.089		7.25	4.97		12.22	15.85
3710	End panel, bolted		36	.222		9	12.45		21.45	29.50
3800	For sloping top, 12″ wide		24	.333		31.50	18.65		50.15	64
3810	15″ wide		24	.333		34	18.65		52.65	67
3820	18″ wide		24	.333		35.50	18.65		54.15	69
3850	Sloping top end panel, 12″ deep		72	.111		12.20	6.20		18.40	23
3860	15″ deep		72	.111		12.20	6.20		18.40	23
3870	18″ deep		72	.111		12.20	6.20		18.40	23
3900	For finish end panels, steel, 60″ high, 15″ deep		12	.667		36	37.50		73.50	98.50
3910	72″ high, 12″ deep		12	.667		29	37.50		66.50	91
3920	18″ deep		12	.667		43	37.50		80.50	106
5000	For 'ready to assemble' lockers,									
5010	Add to labor						75%			
5020	Deduct from material					20%				
6000	Heavy duty for detention facility, tamper proof, 14 ga. welded steel, solid									
6100	24″ W x 24″ D x 74″ H, single tier	1 Shee	18	.444	Ea.	515	25		540	610
6110	Double tier		18	.444		505	25		530	595
6120	Triple tier		18	.444		520	25		545	615
9000	Minimum labor/equipment charge		2.50	3.200	Job		179		179	283

10 51 26 – Plastic Lockers

10 51 26.13 Recycled Plastic Lockers			Crew	Daily Output	Labor-Hours	Unit	Material	Labor	Equipment	Total	Total Incl O&P
0011	**RECYCLED PLASTIC LOCKERS**, 30% recycled										
0110	Single tier box locker, 12″ x 12″ x 72″	G	1 Shee	8	1	Ea.	450	56		506	585
0120	12″ x 15″ x 72″	G		8	1		470	56		526	610
0130	12″ x 18″ x 72″	G		8	1		460	56		516	595
0410	Double tier, 12″ x 12″ x 72″	G		21	.381		470	21.50		491.50	550
0420	12″ x 15″ x 72″	G		21	.381		495	21.50		516.50	580
0430	12″ x 18″ x 72″	G		21	.381		485	21.50		506.50	565

10 51 53 – Locker Room Benches

10 51 53.10 Benches		Crew	Daily Output	Labor-Hours	Unit	Material	Labor	Equipment	Total	Total Incl O&P
0010	**BENCHES**									
2100	Locker bench, laminated maple, top only	1 Shee	100	.080	L.F.	24.50	4.48		28.98	34
2200	Pedestals, steel pipe		25	.320	Ea.	44	17.90		61.90	77
2250	Plastic, 9.5″ top with PVC pedestals		80	.100	L.F.	59.50	5.60		65.10	74

10 55 Postal Specialties

10 55 23 – Mail Boxes

10 55 23.10 Commercial Mail Boxes		Crew	Daily Output	Labor-Hours	Unit	Material	2015 Bare Costs Labor	Equipment	Total	Total Incl O&P
0010	COMMERCIAL MAIL BOXES									
0020	Horiz., key lock, 5"H x 6"W x 15"D, alum., rear load	1 Carp	34	.235	Ea.	40	11.05		51.05	62
0100	Front loading		34	.235		40	11.05		51.05	62
0200	Double, 5"H x 12"W x 15"D, rear loading		26	.308		66.50	14.45		80.95	97
0300	Front loading		26	.308		70	14.45		84.45	101
0500	Quadruple, 10"H x 12"W x 15"D, rear loading		20	.400		106	18.80		124.80	148
0600	Front loading		20	.400		88.50	18.80		107.30	128
0800	Vertical, front load, 15"H x 5"W x 6"D, alum., per compartment		34	.235		40	11.05		51.05	62
0900	Bronze, duranodic finish		34	.235		46.50	11.05		57.55	69
1000	Steel, enameled		34	.235		40	11.05		51.05	62
1700	Alphabetical directories, 120 names		10	.800		125	37.50		162.50	200
1800	Letter collection box		6	1.333		715	62.50		777.50	890
1830	Lobby collection boxes, aluminum	2 Shee	5	3.200		1,750	179		1,929	2,200
1840	Bronze or stainless	"	4.50	3.556		1,800	199		1,999	2,325
1900	Letter slot, residential	1 Carp	20	.400		80	18.80		98.80	119
2000	Post office type		8	1		104	47		151	191
2250	Key keeper, single key, aluminum		26	.308		39.50	14.45		53.95	67
2300	Steel, enameled		26	.308		75	14.45		89.45	106
9000	Minimum labor/equipment charge		5	1.600	Job		75		75	123

10 56 Storage Assemblies

10 56 13 – Metal Storage Shelving

10 56 13.10 Shelving		Crew	Daily Output	Labor-Hours	Unit	Material	Labor	Equipment	Total	Total Incl O&P
0010	SHELVING									
0020	Metal, industrial, cross-braced, 3' wide, 12" deep	1 Sswk	175	.046	SF Shlf	7.65	2.41		10.06	12.75
0100	24" deep		330	.024		5.05	1.28		6.33	7.85
0300	4' wide, 12" deep		185	.043		6.50	2.28		8.78	11.25
0400	24" deep		380	.021		4.62	1.11		5.73	7.10
1200	Enclosed sides, cross-braced back, 3' wide, 12" deep		175	.046		11.85	2.41		14.26	17.40
1300	24" deep		290	.028		8.20	1.45		9.65	11.70
1500	Fully enclosed, sides and back, 3' wide, 12" deep		150	.053		15.60	2.81		18.41	22.50
1600	24" deep		255	.031		10.40	1.65		12.05	14.45
1800	4' wide, 12" deep		150	.053		9.95	2.81		12.76	16
1900	24" deep		290	.028		8.30	1.45		9.75	11.80
2200	Wide span, 1600 lb. capacity per shelf, 6' wide, 24" deep		380	.021		7	1.11		8.11	9.70
2400	36" deep		440	.018		6.85	.96		7.81	9.25
2600	8' wide, 24" deep		440	.018		6.45	.96		7.41	8.85
2800	36" deep		520	.015		6.20	.81		7.01	8.25
4000	Pallet racks, steel frame 5,000 lb. capacity, 8' long, 36" deep	2 Sswk	450	.036		9.25	1.87		11.12	13.60
4200	42" deep		500	.032		8.10	1.68		9.78	12
4400	48" deep		520	.031		7.55	1.62		9.17	11.25
9000	Minimum labor/equipment charge	1 Carp	4	2	Job		94		94	154

10 56 13.20 Parts Bins		Crew	Daily Output	Labor-Hours	Unit	Material	Labor	Equipment	Total	Total Incl O&P
0010	**PARTS BINS** metal, gray baked enamel finish									
0100	6'-3" high, 3' wide									
0300	12 bins, 18" wide x 12" high, 12" deep	2 Clab	10	1.600	Ea.	320	60		380	450
0400	24" deep		10	1.600		405	60		465	545
0600	72 bins, 6" wide x 6" high, 12" deep		8	2		540	75		615	720
0700	18" deep		8	2		850	75		925	1,050
1000	7'-3" high, 3' wide									
1200	14 bins, 18" wide x 12" high, 12" deep	2 Clab	10	1.600	Ea.	350	60		410	485

10 56 Storage Assemblies

10 56 13 – Metal Storage Shelving

10 56 13.20 Parts Bins		Crew	Daily Output	Labor-Hours	Unit	Material	Labor	Equipment	Total	Total Incl O&P
						2015 Bare Costs				
1300	24" deep	2 Clab	10	1.600	Ea.	440	60		500	580
1500	84 bins, 6" wide x 6" high, 12" deep		8	2		925	75		1,000	1,150
1600	24" deep		8	2		1,125	75		1,200	1,350

10 57 Wardrobe and Closet Specialties

10 57 13 – Hat and Coat Racks

10 57 13.10 Coat Racks and Wardrobes		Crew	Daily Output	Labor-Hours	Unit	Material	Labor	Equipment	Total	Total Incl O&P
0010	**COAT RACKS AND WARDROBES**									
0020	Hat & coat rack, floor model, 6 hangers									
0050	Standing, beech wood, 21" x 21" x 72", chrome				Ea.	237			237	260
0100	18 ga. tubular steel, 21" x 21" x 69", wood walnut				"	299			299	330
0500	16 ga. steel frame, 22 ga. steel shelves									
0650	Single pedestal, 30" x 18" x 63"				Ea.	262			262	289
0800	Single face rack, 29" x 18-1/2" x 62"					315			315	345
0900	51" x 18-1/2" x 70"					410			410	450
0910	Double face rack, 39" x 26" x 70"					390			390	430
0920	63" x 26" x 70"					465			465	515
0940	For 2" ball casters, add				Set	88			88	97
1400	Utility hook strips, 3/8" x 2-1/2" x 18", 6 hooks	1 Carp	48	.167	Ea.	62	7.85		69.85	81
1500	34" long, 12 hooks	"	48	.167	"	66	7.85		73.85	85.50
1650	Wall mounted racks, 16 ga. steel frame, 22 ga. steel shelves									
1850	12" x 15" x 26", 6 hangers	1 Carp	32	.250	Ea.	155	11.75		166.75	189
2000	12" x 15" x 50", 12 hangers	"	32	.250	"	181	11.75		192.75	218
2150	Wardrobe cabinet, steel, baked enamel finish									
2300	36" x 21" x 78", incl. top shelf & hanger rod				Ea.	315			315	350
2400	Wardrobe, 24" x 24" x 76", KD, w/door, hospital, baked enamel steel	1 Carp	2	4		635	188		823	1,000
2500	Hardwood	"	2	4		1,150	188		1,338	1,575

10 57 23 – Closet and Utility Shelving

10 57 23.19 Wood Closet and Utility Shelving		Crew	Daily Output	Labor-Hours	Unit	Material	Labor	Equipment	Total	Total Incl O&P
0010	**WOOD CLOSET AND UTILITY SHELVING**									
0020	Pine, clear grade, no edge band, 1" x 8"	1 Carp	115	.070	L.F.	3.49	3.27		6.76	9.20
0100	1" x 10"		110	.073		4.34	3.41		7.75	10.35
0200	1" x 12"		105	.076		5.25	3.58		8.83	11.60
0600	Plywood, 3/4" thick with lumber edge, 12" wide		75	.107		1.87	5		6.87	10.25
0700	24" wide		70	.114		3.30	5.35		8.65	12.45
0900	Bookcase, clear grade pine, shelves 12" O.C., 8" deep, per S.F. shelf		70	.114	S.F.	11.35	5.35		16.70	21.50
1000	12" deep shelves		65	.123	"	17	5.80		22.80	28
1200	Adjustable closet rod and shelf, 12" wide, 3' long		20	.400	Ea.	11.15	18.80		29.95	43.50
1300	8' long		15	.533	"	21.50	25		46.50	64.50
1500	Prefinished shelves with supports, stock, 8" wide		75	.107	L.F.	4.58	5		9.58	13.25
1600	10" wide		70	.114	"	6.50	5.35		11.85	15.95
9000	Minimum labor/equipment charge		4	2	Job		94		94	154

10 73 Protective Covers

10 73 13 – Awnings

10 73 13.10 Awnings, Fabric		Crew	Daily Output	Labor-Hours	Unit	Material	2015 Bare Costs Labor	Equipment	Total	Total Incl O&P
0010	**AWNINGS, FABRIC**									
0020	Including acrylic canvas and frame, standard design									
0100	Door and window, slope, 3′ high, 4′ wide	1 Carp	4.50	1.778	Ea.	720	83.50		803.50	925
0110	6′ wide		3.50	2.286		925	107		1,032	1,200
0120	8′ wide		3	2.667		1,125	125		1,250	1,450
0200	Quarter round convex, 4′ wide		3	2.667		1,125	125		1,250	1,425
0210	6′ wide		2.25	3.556		1,450	167		1,617	1,875
0220	8′ wide		1.80	4.444		1,775	209		1,984	2,300
0300	Dome, 4′ wide		7.50	1.067		430	50		480	555
0310	6′ wide		3.50	2.286		970	107		1,077	1,250
0320	8′ wide		2	4		1,725	188		1,913	2,200
0350	Elongated dome, 4′ wide		1.33	6.015		1,625	282		1,907	2,250
0360	6′ wide		1.11	7.207		1,925	340		2,265	2,675
0370	8′ wide		1	8		2,275	375		2,650	3,125
1000	Entry or walkway, peak, 12′ long, 4′ wide	2 Carp	.90	17.778		5,225	835		6,060	7,125
1010	6′ wide		.60	26.667		8,050	1,250		9,300	10,900
1020	8′ wide		.40	40		11,100	1,875		12,975	15,300
1100	Radius with dome end, 4′ wide		1.10	14.545		3,950	685		4,635	5,475
1110	6′ wide		.70	22.857		6,350	1,075		7,425	8,750
1120	8′ wide		.50	32		9,050	1,500		10,550	12,400
2000	Retractable lateral arm awning, manual									
2010	To 12′ wide, 8′ - 6″ projection	2 Carp	1.70	9.412	Ea.	1,175	440		1,615	2,025
2020	To 14′ wide, 8′ - 6″ projection		1.10	14.545		1,375	685		2,060	2,650
2030	To 19′ wide, 8′ - 6″ projection		.85	18.824		1,875	885		2,760	3,500
2040	To 24′ wide, 8′ - 6″ projection		.67	23.881		2,350	1,125		3,475	4,450
2050	Motor for above, add	1 Carp	2.67	3		1,000	141		1,141	1,350
3000	Patio/deck canopy with frame									
3010	12′ wide, 12′ projection	2 Carp	2	8	Ea.	1,675	375		2,050	2,450
3020	16′ wide, 14′ projection	″	1.20	13.333		2,600	625		3,225	3,875
9000	For fire retardant canvas, add					7%				
9010	For lettering or graphics, add					35%				
9020	For painted or coated acrylic canvas, deduct					8%				
9030	For translucent or opaque vinyl canvas, add					10%				
9040	For 6 or more units, deduct					20%	15%			

10 73 16 – Canopies

10 73 16.20 Metal Canopies		Crew	Daily Output	Labor-Hours	Unit	Material	Labor	Equipment	Total	Total Incl O&P
0010	**METAL CANOPIES**									
0020	Wall hung, .032″, aluminum, prefinished, 8′ x 10′	K-2	1.30	18.462	Ea.	2,150	900	233	3,283	4,200
0300	8′ x 20′		1.10	21.818		3,650	1,075	276	5,001	6,175
0500	10′ x 10′		1.30	18.462		2,825	900	233	3,958	4,925
0700	10′ x 20′		1.10	21.818		4,600	1,075	276	5,951	7,225
1000	12′ x 20′		1	24		5,250	1,175	305	6,730	8,150
1360	12′ x 30′		.80	30		7,875	1,475	380	9,730	11,700
1700	12′ x 40′		.60	40		10,500	1,950	505	12,955	15,600
1900	For free standing units, add					20%	10%			
2300	Aluminum entrance canopies, flat soffit, .032″									
2500	3′-6″ x 4′-0″, clear anodized	2 Carp	4	4	Ea.	915	188		1,103	1,300
2700	Bronze anodized		4	4		1,625	188		1,813	2,075
3000	Polyurethane painted		4	4		1,300	188		1,488	1,750
3300	4′-6″ x 10′-0″, clear anodized		2	8		2,525	375		2,900	3,400
3500	Bronze anodized		2	8		3,225	375		3,600	4,175
3700	Polyurethane painted		2	8		2,700	375		3,075	3,600

10 73 Protective Covers

10 73 16 – Canopies

10 73 16.20 Metal Canopies		Crew	Daily Output	Labor-Hours	Unit	Material	2015 Bare Costs Labor	Equipment	Total	Total Incl O&P
4000	Wall downspout, 10 L.F., clear anodized	1 Carp	7	1.143	Ea.	157	53.50		210.50	261
4300	Bronze anodized		7	1.143		274	53.50		327.50	390
4500	Polyurethane painted		7	1.143		236	53.50		289.50	345
7000	Carport, baked vinyl finish, .032", 20' x 10', no foundations, flat panel	K-2	4	6	Car	4,000	293	76	4,369	5,000
7250	Insulated flat panel		2	12	"	6,300	585	152	7,037	8,125
7500	Walkway cover, to 12' wide, stl., vinyl finish, .032",no fndtns., flat		250	.096	S.F.	22.50	4.68	1.21	28.39	34
7750	Arched		200	.120	"	55	5.85	1.52	62.37	72.50
9000	Minimum labor/equipment charge	2 Carp	2	8	Job		375		375	615

10 74 Manufactured Exterior Specialties

10 74 29 – Steeples

10 74 29.10 Prefabricated Steeples

		Crew	Daily Output	Labor-Hours	Unit	Material	Labor	Equipment	Total	Total Incl O&P
0010	**PREFABRICATED STEEPLES**									
4000	Steeples, translucent fiberglass, 30" square, 15' high	F-3	2	20	Ea.	8,375	960	325	9,660	11,100
4150	25' high		1.80	22.222		9,750	1,075	365	11,190	12,800
4350	Opaque fiberglass, 24" square, 14' high		2	20		6,975	960	325	8,260	9,550
4500	28' high		1.80	22.222		6,375	1,075	365	7,815	9,125
4600	Aluminum, baked finish, 16" square, 14' high					5,700			5,700	6,275
4620	20' high, 3'-6" base					9,550			9,550	10,500
4640	35' high, 8' base					36,700			36,700	40,400
4660	60' high, 14' base					77,500			77,500	85,000
4680	152' high, custom					632,000			632,000	695,000
4700	Porcelain enamel steeples, custom, 40' high	F-3	.50	80		14,300	3,825	1,300	19,425	23,400
4800	60' high	"	.30	133		24,800	6,375	2,175	33,350	40,100

10 74 33 – Weathervanes

10 74 33.10 Residential Weathervanes

		Crew	Daily Output	Labor-Hours	Unit	Material	Labor	Equipment	Total	Total Incl O&P
0010	**RESIDENTIAL WEATHERVANES**									
0020	Residential types, 18" to 24"	1 Carp	8	1	Ea.	151	47		198	243
0100	24" to 48"		2	4	"	1,700	188		1,888	2,175
9000	Minimum labor/equipment charge		4	2	Job		94		94	154

10 74 46 – Window Wells

10 74 46.10 Area Window Wells

		Crew	Daily Output	Labor-Hours	Unit	Material	Labor	Equipment	Total	Total Incl O&P
0010	**AREA WINDOW WELLS**, Galvanized steel									
0020	20 ga., 3'-2" wide, 1' deep	1 Sswk	29	.276	Ea.	17.45	14.50		31.95	45.50
0100	2' deep		23	.348		31	18.30		49.30	67.50
0300	16 ga., 3'-2" wide, 1' deep		29	.276		23	14.50		37.50	52
0400	3' deep		23	.348		47	18.30		65.30	84.50
0600	Welded grating for above, 15 lb., painted		45	.178		87.50	9.35		96.85	113
0700	Galvanized		45	.178		118	9.35		127.35	147
0900	Translucent plastic cap for above		60	.133		19	7		26	33.50

10 75 Flagpoles

10 75 16 – Ground-Set Flagpoles

10 75 16.10 Flagpoles		Crew	Daily Output	Labor-Hours	Unit	Material	2015 Bare Costs Labor	Equipment	Total	Total Incl O&P
0010	**FLAGPOLES**, ground set									
0050	Not including base or foundation									
0100	Aluminum, tapered, ground set 20′ high	K-1	2	8	Ea.	1,050	345	152	1,547	1,875
0200	25′ high		1.70	9.412		1,100	405	178	1,683	2,075
0300	30′ high		1.50	10.667		1,300	460	202	1,962	2,400
0400	35′ high		1.40	11.429		1,800	490	217	2,507	3,050
0500	40′ high		1.20	13.333		2,775	575	253	3,603	4,275
0600	50′ high		1	16		3,175	690	305	4,170	4,950
0700	60′ high		.90	17.778		4,950	765	335	6,050	7,050
0800	70′ high		.80	20		8,375	860	380	9,615	11,000
1100	Counterbalanced, internal halyard, 20′ high		1.80	8.889		2,450	380	168	2,998	3,500
1200	30′ high		1.50	10.667		2,675	460	202	3,337	3,925
1300	40′ high		1.30	12.308		6,475	530	233	7,238	8,250
1400	50′ high		1	16		9,025	690	305	10,020	11,400
2820	Aluminum, electronically operated, 30′ high		1.40	11.429		4,150	490	217	4,857	5,625
2840	35′ high		1.30	12.308		4,950	530	233	5,713	6,550
2860	39′ high		1.10	14.545		6,025	625	276	6,926	7,950
2880	45′ high		1	16		6,350	690	305	7,345	8,450
2900	50′ high		.90	17.778		8,175	765	335	9,275	10,600
3000	Fiberglass, tapered, ground set, 23′ high		2	8		580	345	152	1,077	1,350
3100	29′-7″ high		1.50	10.667		1,525	460	202	2,187	2,650
3200	36′-1″ high		1.40	11.429		2,000	490	217	2,707	3,250
3300	39′-5″ high		1.20	13.333		2,100	575	253	2,928	3,525
3400	49′-2″ high		1	16		3,900	690	305	4,895	5,750
3500	59′ high	↓	.90	17.778	↓	4,825	765	335	5,925	6,925
4300	Steel, direct imbedded installation									
4400	Internal halyard, 20′ high	K-1	2.50	6.400	Ea.	1,375	275	121	1,771	2,075
4500	25′ high		2.50	6.400		2,050	275	121	2,446	2,825
4600	30′ high		2.30	6.957		2,450	299	132	2,881	3,300
4700	40′ high		2.10	7.619		3,725	330	144	4,199	4,800
4800	50′ high		1.90	8.421		4,325	360	160	4,845	5,525
5000	60′ high		1.80	8.889		7,200	380	168	7,748	8,725
5100	70′ high		1.60	10		7,850	430	190	8,470	9,525
5200	80′ high		1.40	11.429		10,100	490	217	10,807	12,100
5300	90′ high		1.20	13.333		15,500	575	253	16,328	18,200
5500	100′ high	↓	1	16	↓	17,600	690	305	18,595	20,900
6400	Wood poles, tapered, clear vertical grain fir with tilting									
6410	base, not incl. foundation, 4″ butt, 25′ high	K-1	1.90	8.421	Ea.	1,400	360	160	1,920	2,300
6800	6″ butt, 30′ high	″	1.30	12.308	″	2,600	530	233	3,363	3,975
7300	Foundations for flagpoles, including									
7400	excavation and concrete, to 35′ high poles	C-1	10	3.200	Ea.	685	143		828	990
7600	40′ to 50′ high		3.50	9.143		1,275	410		1,685	2,075
7700	Over 60′ high	↓	2	16	↓	1,575	715		2,290	2,900

10 75 23 – Wall-Mounted Flagpoles

10 75 23.10 Flagpoles		Crew	Daily Output	Labor-Hours	Unit	Material	Labor	Equipment	Total	Total Incl O&P
0010	**FLAGPOLES**, structure mounted									
0100	Fiberglass, vertical wall set, 19′-8″ long	K-1	1.50	10.667	Ea.	1,150	460	202	1,812	2,250
0200	23′ long		1.40	11.429		1,425	490	217	2,132	2,600
0300	26′-3″ long		1.30	12.308		2,050	530	233	2,813	3,375
0800	19′-8″ long outrigger		1.30	12.308		1,325	530	233	2,088	2,575
1300	Aluminum, vertical wall set, tapered, with base, 20′ high		1.20	13.333		1,075	575	253	1,903	2,400
1400	29′-6″ high	↓	1	16	↓	2,650	690	305	3,645	4,375

10 75 Flagpoles

10 75 23 – Wall-Mounted Flagpoles

10 75 23.10 Flagpoles		Crew	Daily Output	Labor-Hours	Unit	Material	2015 Bare Costs Labor	Equipment	Total	Total Incl O&P
2400	Outrigger poles with base, 12′ long	K-1	1.30	12.308	Ea.	1,100	530	233	1,863	2,325
2500	14′ long	↓	1	16	↓	1,425	690	305	2,420	3,025

10 86 Security Mirrors and Domes

10 86 10 – Security Mirrors

10 86 10.10 Exterior Traffic Control Mirrors

		Crew	Daily Output	Labor-Hours	Unit	Material	Labor	Equipment	Total	Total Incl O&P
0010	**EXTERIOR TRAFFIC CONTROL MIRRORS**									
0100	Convex, stainless steel, 20 ga., 26″ diameter	1 Carp	12	.667	Ea.	213	31.50		244.50	286

10 86 20 – Security Domes

10 86 20.10 Domes

		Crew	Daily Output	Labor-Hours	Unit	Material	Labor	Equipment	Total	Total Incl O&P
0010	**DOMES** for security cameras (CCTV)									
0100	Ceiling mounted, 10″ diameter	1 Carp	30	.267	Ea.	11.60	12.50		24.10	33.50
0110	12″ diameter	″	30	.267	″	7.55	12.50		20.05	29

10 88 Scales

10 88 05 – Commercial Scales

10 88 05.10 Scales

		Crew	Daily Output	Labor-Hours	Unit	Material	Labor	Equipment	Total	Total Incl O&P
0010	**SCALES**									
0700	Truck scales, incl. steel weigh bridge,									
0800	not including foundation, pits									
1550	Digital, electronic, 100 ton capacity, steel deck 12′ x 10′ platform	3 Carp	.20	120	Ea.	14,100	5,625		19,725	24,700
1600	40′ x 10′ platform		.14	171		28,100	8,050		36,150	44,100
1640	60′ x 10′ platform		.13	184		37,100	8,675		45,775	55,000
1680	70′ x 10′ platform	↓	.12	200		39,400	9,400		48,800	59,000
2000	For standard automatic printing device, add					1,250			1,250	1,375
2100	For remote reading electronic system, add					2,725			2,725	3,000
2300	Concrete foundation pits for above, 8′ x 6′, 5 C.Y. required	C-1	.50	64		1,075	2,850		3,925	5,850
2400	14′ x 6′ platform, 10 C.Y. required		.35	91.429		1,575	4,075		5,650	8,400
2600	50′ x 10′ platform, 30 C.Y. required		.25	128		2,125	5,700		7,825	11,700
2700	70′ x 10′ platform, 40 C.Y. required	↓	.15	213		4,625	9,525		14,150	20,700
2750	Crane scales, dial, 1 ton capacity					1,150			1,150	1,275
2780	5 ton capacity					1,550			1,550	1,700
2800	Digital, 1 ton capacity					1,900			1,900	2,075
2850	10 ton capacity				↓	4,800			4,800	5,300
2900	Low profile electronic warehouse scale,									
3000	not incl. printer, 4′ x 4′ platform, 10,000 lb. capacity	2 Carp	.30	53.333	Ea.	1,425	2,500		3,925	5,650
3300	5′ x 7′ platform, 10,000 lb. capacity		.25	64		4,825	3,000		7,825	10,200
3400	20,000 lb. capacity	↓	.20	80		5,925	3,750		9,675	12,700
3500	For printers, incl. time, date & numbering, add					890			890	980
3800	Portable, beam type, capacity 1000#, platform 18″ x 24″					785			785	860
3900	Dial type, capacity 2000#, platform 24″ x 24″					1,425			1,425	1,550
4000	Digital type, capacity 1000#, platform 24″ x 30″					2,275			2,275	2,500
4100	Portable contractor truck scales, 50 ton cap., 40′ x 10′ platform					33,600			33,600	37,000
4200	60′ x 10′ platform				↓	31,600			31,600	34,800

Estimating Tips

General

- The items in this division are usually priced per square foot or each. Many of these items are purchased by the owner for installation by the contractor. Check the specifications for responsibilities and include time for receiving, storage, installation, and mechanical and electrical hookups in the appropriate divisions.
- Many items in Division 11 require some type of support system that is not usually furnished with the item. Examples of these systems include blocking for the attachment of casework and support angles for ceiling-hung projection screens. The required blocking or supports must be added to the estimate in the appropriate division.
- Some items in Division 11 may require assembly or electrical hookups. Verify the amount of assembly required or the need for a hard electrical connection and add the appropriate costs.

Reference Numbers

Reference numbers are shown in shaded boxes at the beginning of some major classifications. These numbers refer to related items in the Reference Section. The reference information may be an estimating procedure, an alternate pricing method, or technical information.

Note: Not all subdivisions listed here necessarily appear in this publication. ■

11 05 Common Work Results for Equipment

11 05 05 – Selective Demolition for Equipment

11 05 05.10 Selective Demolition		Crew	Daily Output	Labor-Hours	Unit	Material	2015 Bare Costs Labor	2015 Bare Costs Equipment	2015 Bare Costs Total	Total Incl O&P
0010	**SELECTIVE DEMOLITION** R024119-10									
0130	Central vacuum, motor unit, residential or commercial	1 Clab	2	4	Ea.		150		150	247
0210	Vault door and frame	2 Skwk	2	8			390		390	630
0215	Day gate, for vault	"	3	5.333			259		259	420
0380	Bank equipment, teller window, bullet resistant	1 Clab	1.20	6.667			251		251	410
0381	Counter	2 Clab	1.50	10.667	Station		400		400	660
0382	Drive-up window, including drawer and glass		1.50	10.667	"		400		400	660
0383	Thru-wall boxes and chests, selective demolition		2.50	6.400	Ea.		241		241	395
0384	Bullet resistant partitions		20	.800	L.F.		30		30	49.50
0385	Pneumatic tube system, 2 lane drive-up	L-3	.45	35.556	Ea.		1,825		1,825	2,900
0386	Safety deposit box	1 Clab	50	.160	Opng.		6		6	9.85
0387	Surveillance system, video, complete	2 Elec	2	8	Ea.		440		440	680
0410	Church equipment, misc moveable fixtures	2 Clab	1	16			600		600	985
0412	Steeple, to 28′ high	F-3	3	13.333			640	218	858	1,300
0414	40′ to 60′ high	"	.80	50			2,400	820	3,220	4,800
0510	Library equipment, bookshelves, wood, to 90″ high	1 Clab	20	.400	L.F.		15.05		15.05	24.50
0515	Carrels, hardwood, 36″ x 24″	"	9	.889	Ea.		33.50		33.50	55
0630	Stage equipment, light control panel	1 Elec	1	8	"		440		440	680
0632	Border lights		40	.200	L.F.		10.95		10.95	16.95
0634	Spotlights		8	1	Ea.		54.50		54.50	84.50
0636	Telescoping platforms and risers	2 Clab	175	.091	SF Stg.		3.44		3.44	5.65
1020	Barber equipment, hydraulic chair	1 Clab	40	.200	Ea.		7.50		7.50	12.35
1030	Checkout counter, supermarket or warehouse conveyor	2 Clab	18	.889			33.50		33.50	55
1040	Food cases, refrigerated or frozen	Q-5	6	2.667			143		143	224
1190	Laundry equipment, commercial	L-6	3	4			229		229	355
1360	Movie equipment, lamphouse, to 4000 watt, incl rectifier	1 Elec	4	2			109		109	169
1365	Sound system, incl amplifier	"	1.25	6.400			350		350	540
1410	Air compressor, to 5 H.P.	2 Clab	2.50	6.400			241		241	395
1412	Lubrication equipment, automotive, 3 reel type, incl pump, excl piping	L-4	1	24	Set		1,050		1,050	1,725
1414	Booth, spray paint, complete, to 26′ long	"	.80	30	"		1,325		1,325	2,175
1560	Parking equipment, cashier booth	B-22	2	15	Ea.		660	103	763	1,200
1600	Loading dock equipment, dock bumpers, rubber	1 Clab	50	.160	"		6		6	9.85
1610	Door seal for door perimeter	"	50	.160	L.F.		6		6	9.85
1620	Platform lifter, fixed, 6′ x 8′, 5000 lb. capacity	E-16	1.50	10.667	Ea.		570	97.50	667.50	1,125
1630	Dock leveller	"	2	8			430	73	503	855
1640	Lights, single or double arm	1 Elec	8	1			54.50		54.50	84.50
1650	Shelter, fabric, truck or train	1 Clab	1.50	5.333			201		201	330
1790	Waste handling equipment, commercial compactor	L-4	2	12			530		530	865
1792	Commercial or municipal incinerator, gas	"	2	12			530		530	865
1795	Crematory, excluding building	Q-3	.25	128			7,150		7,150	11,200
1910	Detection equipment, cell bar front	E-4	4	8			425	36.50	461.50	810
1912	Cell door and frame		8	4			213	18.25	231.25	405
1914	Prefab cell, 4′ to 5′ wide, 7′ to 8′ high, 7′ deep		8	4			213	18.25	231.25	405
1916	Cot, bolted, single		40	.800			42.50	3.65	46.15	81
1918	Visitor cubicle		4	8			425	36.50	461.50	810
2850	Hydraulic gates, canal, flap, knife, slide or sluice, to 18″ diameter	L-5A	8	4			212	75	287	455
2852	19″ to 36″ diameter		6	5.333			282	100	382	605
2854	37″ to 48″ diameter		2	16			845	300	1,145	1,800
2856	49″ to 60″ diameter		1	32			1,700	600	2,300	3,625
2858	Over 60″ diameter		.30	106			5,650	2,000	7,650	12,100
3100	Sewage pumping system, prefabricated, to 1000 GPM	C-17D	.20	420			20,700	4,050	24,750	38,000
3110	Sewage treatment, holding tank for recirc chemical water closet	1 Plum	8	1			58.50		58.50	91.50
3900	Wastewater treatment system, to 1500 gallons	B-21	2	14			605	69	674	1,050

11 05 Common Work Results for Equipment

11 05 05 – Selective Demolition for Equipment

11 05 05.10 Selective Demolition		Crew	Daily Output	Labor-Hours	Unit	Material	2015 Bare Costs Labor	Equipment	Total	Total Incl O&P
4050	Food storage equipment, walk-in refrigerator/freezer	2 Clab	64	.250	S.F.		9.40		9.40	15.40
4052	Shelving, stainless steel, 4 tier or dunnage rack	1 Clab	12	.667	Ea.		25		25	41
4100	Food preparation equipment, small countertop		18	.444			16.70		16.70	27.50
4150	Food delivery carts, heated cabinets		18	.444			16.70		16.70	27.50
4200	Cooking equipment, commercial range	Q-1	12	1.333			70.50		70.50	110
4250	Hood and ventilation equipment, kitchen exhaust hood, excl fire prot	1 Clab	3	2.667			100		100	164
4255	Fire protection system	Q-1	3	5.333			282		282	440
4300	Food dispensing equipment, countertop items	1 Clab	15	.533			20		20	33
4310	Serving counter	"	65	.123	L.F.		4.63		4.63	7.60
4350	Ice machine, ice cube maker, flakers and storage bins, to 2000 lb./day	Q-1	1.60	10	Ea.		530		530	825
4400	Cleaning and disposal, commercial dishwasher, to 50 racks per hour	L-6	1	12			690		690	1,075
4405	To 275 racks per hour	L-4	1	24			1,050		1,050	1,725
4410	Dishwasher hood	2 Clab	5	3.200			120		120	197
4420	Garbage disposal, commercial, to 5 H.P.	L-1	8	2			113		113	176
4540	Water heater, residential, to 80 gal/day	"	5	3.200			181		181	282
4542	Water softener, automatic	2 Plum	10	1.600			94		94	146
4544	Disappearing stairway, to 15′ floor height	2 Clab	6	2.667			100		100	164
4710	Darkroom equipment, light	L-7	10	2.800			127		127	206
4712	Heavy	"	1.50	18.667			845		845	1,375
4720	Doors	2 Clab	3.50	4.571	Opng.		172		172	282
4830	Bowling alley, complete, incl pinsetter, scorer, counters, misc supplies	4 Clab	.40	80	Lane		3,000		3,000	4,925
4840	Health club equipment, circuit training apparatus	2 Clab	2	8	Set		300		300	495
4842	Squat racks	"	10	1.600	Ea.		60		60	98.50
4860	School equipment, basketball backstop	L-2	2	8			330		330	540
4862	Table and benches, folding, in wall, 14′ long	L-4	4	6			266		266	435
4864	Bleachers, telescoping, to 30 tier	F-5	120	.267	Seat		12.65		12.65	20.50
4866	Boxing ring, elevated	L-4	.20	120	Ea.		5,300		5,300	8,675
4867	Boxing ring, floor level	"	2	12			530		530	865
4868	Exercise equipment	1 Clab	6	1.333			50		50	82
4870	Gym divider	L-4	1000	.024	S.F.		1.06		1.06	1.73
4875	Scoreboard	R-3	2	10	Ea.		545	69	614	920
4880	Shooting range, incl bullet traps, targets, excl structure	L-9	1	36	Point		1,550		1,550	2,575
5200	Vocational shop equipment	2 Clab	8	2	Ea.		75		75	123
6200	Fume hood, incl countertop, excl HVAC	"	6	2.667	L.F.		100		100	164
7100	Medical sterilizing, distiller, water, steam heated, 50 gal. capacity	1 Plum	2.80	2.857	Ea.		168		168	262
7200	Medical equipment, surgery table, minor	1 Clab	1	8			300		300	495
7210	Surgical lights, doctors office, single or double arm	2 Elec	3	5.333			292		292	450
7300	Physical therapy, table	2 Clab	4	4			150		150	247
7310	Whirlpool bath, fixed, incl mixing valves	1 Plum	4	2			117		117	183
7400	Dental equipment, chair, electric or hydraulic	1 Clab	.75	10.667			400		400	660
7410	Central suction system	1 Plum	2	4			235		235	365
7420	Drill console with accessories	1 Clab	3.20	2.500			94		94	154
7430	X-ray unit	"	4	2			75		75	123
7440	X-ray developer	1 Plum	10	.800			47		47	73

11 05 10 – Equipment Installation

11 05 10.10 Industrial Equipment Installation		Crew	Daily Output	Labor-Hours	Unit	Material	Labor	Equipment	Total	Total Incl O&P
0010	**INDUSTRIAL EQUIPMENT INSTALLATION**									
0020	Industrial equipment, minimum	E-2	12	4.667	Ton		241	126	367	560
0200	Maximum	"	2	28	"		1,450	755	2,205	3,350

11 11 Vehicle Service Equipment

11 11 13 – Compressed-Air Vehicle Service Equipment

11 11 13.10 Compressed Air Equipment		Crew	Daily Output	Labor-Hours	Unit	Material	Labor	Equipment	Total	Total Incl O&P
						2015 Bare Costs				
0010	**COMPRESSED AIR EQUIPMENT**									
0030	Compressors, electric, 1-1/2 H.P., standard controls	L-4	1.50	16	Ea.	455	710		1,165	1,650
0550	Dual controls		1.50	16		810	710		1,520	2,050
0600	5 H.P., 115/230 volt, standard controls		1	24		2,575	1,050		3,625	4,550
0650	Dual controls	↓	1	24	↓	3,400	1,050		4,450	5,450

11 11 19 – Vehicle Lubrication Equipment

11 11 19.10 Lubrication Equipment		Crew	Daily Output	Labor-Hours	Unit	Material	Labor	Equipment	Total	Total Incl O&P
0010	**LUBRICATION EQUIPMENT**									
3000	Lube equipment, 3 reel type, with pumps, not including piping	L-4	.50	48	Set	8,800	2,125		10,925	13,200
3100	Hose reel, including hose, oil/lube, 1000 PSI	2 Sswk	2	8	Ea.	740	420		1,160	1,575
3200	Grease, 5000 PSI		2	8		800	420		1,220	1,650
3300	Air, 50 feet, 160 PSI		2	8		875	420		1,295	1,725
3350	25 feet, 160 PSI	↓	2	8	↓	525	420		945	1,325

11 11 33 – Vehicle Spray Painting Equipment

11 11 33.10 Spray Painting Equipment		Crew	Daily Output	Labor-Hours	Unit	Material	Labor	Equipment	Total	Total Incl O&P
0010	**SPRAY PAINTING EQUIPMENT**									
4000	Spray painting booth, 26′ long, complete	L-4	.40	60	Ea.	16,400	2,650		19,050	22,300

11 12 Parking Control Equipment

11 12 13 – Parking Key and Card Control Units

11 12 13.10 Parking Control Units		Crew	Daily Output	Labor-Hours	Unit	Material	Labor	Equipment	Total	Total Incl O&P
0010	**PARKING CONTROL UNITS**									
5100	Card reader	1 Elec	2	4	Ea.	1,950	219		2,169	2,500
5120	Proximity with customer display	2 Elec	1	16		5,575	875		6,450	7,475
6000	Parking control software, basic functionality	1 Elec	.50	16		23,900	875		24,775	27,600
6020	multi-function	"	.20	40	↓	104,500	2,200		106,700	118,500

11 12 16 – Parking Ticket Dispensers

11 12 16.10 Ticket Dispensers		Crew	Daily Output	Labor-Hours	Unit	Material	Labor	Equipment	Total	Total Incl O&P
0010	**TICKET DISPENSERS**									
5900	Ticket spitter with time/date stamp, standard	2 Elec	2	8	Ea.	6,200	440		6,640	7,500
5920	Mag stripe encoding	"	2	8	"	18,800	440		19,240	21,400

11 12 26 – Parking Fee Collection Equipment

11 12 26.13 Parking Fee Coin Collection Equipment		Crew	Daily Output	Labor-Hours	Unit	Material	Labor	Equipment	Total	Total Incl O&P
0010	**PARKING FEE COIN COLLECTION EQUIPMENT**									
5200	Cashier booth, average	B-22	1	30	Ea.	10,400	1,325	207	11,932	13,900
5300	Collector station, pay on foot	2 Elec	.20	80		111,000	4,375		115,375	129,000
5320	Credit card only	"	.50	32	↓	20,500	1,750		22,250	25,200

11 12 26.23 Fee Equipment		Crew	Daily Output	Labor-Hours	Unit	Material	Labor	Equipment	Total	Total Incl O&P
0010	**FEE EQUIPMENT**									
5600	Fee computer	1 Elec	1.50	5.333	Ea.	14,200	292		14,492	16,100

11 12 33 – Parking Gates

11 12 33.13 Lift Arm Parking Gates		Crew	Daily Output	Labor-Hours	Unit	Material	Labor	Equipment	Total	Total Incl O&P
0010	**LIFT ARM PARKING GATES**									
5000	Barrier gate with programmable controller	2 Elec	3	5.333	Ea.	3,400	292		3,692	4,175
5020	Industrial		3	5.333		5,050	292		5,342	6,000
5050	Non-programmable, with reader and 12′ arm		3	5.333		1,875	292		2,167	2,500
5500	Exit verifier	↓	1	16		17,600	875		18,475	20,800
5700	Full sign, 4″ letters	1 Elec	2	4	↓	1,225	219		1,444	1,700

11 12 Parking Control Equipment

11 12 33 – Parking Gates

11 12 33.13 Lift Arm Parking Gates		Crew	Daily Output	Labor-Hours	Unit	Material	2015 Bare Costs Labor	Equipment	Total	Total Incl O&P
5800	Inductive loop	2 Elec	4	4	Ea.	171	219		390	530
5950	Vehicle detector, microprocessor based	1 Elec	3	2.667		425	146		571	695
7100	Traffic spike unit, flush mount, spring loaded, 72" L	B-89	4	4		1,250	175	122	1,547	1,800
7200	Surface mount, 72" L	2 Skwk	10	1.600		2,275	78		2,353	2,625

11 13 Loading Dock Equipment

11 13 13 – Loading Dock Bumpers

11 13 13.10 Dock Bumpers

Line	Description	Crew	Daily Output	Labor-Hours	Unit	Material	Labor	Equipment	Total	Total Incl O&P
0010	**DOCK BUMPERS** Bolts not included									
0012	2" x 6" to 4" x 8", average	1 Carp	300	.027	B.F.	.70	1.25		1.95	2.82
0050	Bumpers, rubber blocks 4-1/2" thick, 10" high, 14" long		26	.308	Ea.	64	14.45		78.45	93.50
0200	24" long		22	.364		96	17.05		113.05	134
0300	36" long		17	.471		78	22		100	122
0500	12" high, 14" long		25	.320		101	15		116	137
0550	24" long		20	.400		112	18.80		130.80	154
0600	36" long		15	.533		125	25		150	178
0800	Rubber blocks 6" thick, 10" high, 14" long		22	.364		104	17.05		121.05	142
0850	24" long		18	.444		123	21		144	169
0900	36" long		13	.615		188	29		217	255
0910	20" high, 11" long		13	.615		148	29		177	211
0920	Extruded rubber bumpers, T section, 22" x 22" x 3" thick		41	.195		65.50	9.15		74.65	87
0940	Molded rubber bumpers, 24" x 12" x 3" thick		20	.400		57.50	18.80		76.30	94.50
1000	Welded installation of above bumpers	E-14	8	1		3.78	54.50	18.25	76.53	123
1100	For drilled anchors, add per anchor	1 Carp	36	.222		6.80	10.45		17.25	24.50
1300	Steel bumpers, see Section 10 26 13.10									

11 13 16 – Loading Dock Seals and Shelters

11 13 16.10 Dock Seals and Shelters

Line	Description	Crew	Daily Output	Labor-Hours	Unit	Material	Labor	Equipment	Total	Total Incl O&P
0010	**DOCK SEALS AND SHELTERS**									
3600	Door seal for door perimeter, 12" x 12", vinyl covered	1 Carp	26	.308	L.F.	32.50	14.45		46.95	59
3900	Folding gates, see Section 10 22 16.10									
6200	Shelters, fabric, for truck or train, scissor arms, minimum	1 Carp	1	8	Ea.	1,775	375		2,150	2,575
6300	Maximum	"	.50	16	"	2,425	750		3,175	3,900

11 13 19 – Stationary Loading Dock Equipment

11 13 19.10 Dock Equipment

Line	Description	Crew	Daily Output	Labor-Hours	Unit	Material	Labor	Equipment	Total	Total Incl O&P
0010	**DOCK EQUIPMENT**									
2200	Dock boards, heavy duty, 60" x 60", aluminum, 5,000 lb. capacity				Ea.	1,450			1,450	1,600
2700	9,000 lb. capacity					1,625			1,625	1,775
3200	15,000 lb. capacity					1,750			1,750	1,925
4200	Platform lifter, 6' x 6', portable, 3,000 lb. capacity					9,250			9,250	10,200
4250	4,000 lb. capacity					11,400			11,400	12,500
4400	Fixed, 6' x 8', 5,000 lb. capacity	E-16	.70	22.857		9,975	1,225	208	11,408	13,500
4500	Levelers, hinged for trucks, 10 ton capacity, 6' x 8'		1.08	14.815		5,025	795	135	5,955	7,125
4650	7' x 8'		1.08	14.815		5,975	795	135	6,905	8,150
4670	Air bag power operated, 10 ton cap., 6' x 8'		1.08	14.815		5,850	795	135	6,780	8,000
4680	7' x 8'		1.08	14.815		5,875	795	135	6,805	8,025
4700	Hydraulic, 10 ton capacity, 6' x 8'		1.08	14.815		8,750	795	135	9,680	11,200
4800	7' x 8'		1.08	14.815		9,425	795	135	10,355	12,000
6000	Dock leveler, 15 ton capacity									
6100	I beam construction, mechanical, 6' x 8'	E-16	.50	32	Ea.	4,400	1,725	292	6,417	8,250
6150	Hydraulic		.50	32		5,225	1,725	292	7,242	9,175

11 13 Loading Dock Equipment

11 13 19 – Stationary Loading Dock Equipment

11 13 19.10 Dock Equipment		Crew	Daily Output	Labor-Hours	Unit	2015 Bare Costs Material	Labor	Equipment	Total	Total Incl O&P
6200	Formed beam deck construction, mechanical, 6′ x 8′	E-16	.50	32	Ea.	3,250	1,725	292	5,267	7,000
6250	Hydraulic		.50	32		4,450	1,725	292	6,467	8,325
6300	Edge of dock leveler, mechanical, 15 ton capacity		2	8		970	430	73	1,473	1,925
7000	22.5 ton capacity									
7100	Vertical storing dock leveler, hydraulic, 6′ x 6′	E-16	.40	40	Ea.	7,200	2,150	365	9,715	12,200

11 13 26 – Loading Dock Lights

11 13 26.10 Dock Lights

		Crew	Daily Output	Labor-Hours	Unit	Material	Labor	Equipment	Total	Total Incl O&P
0010	**DOCK LIGHTS**									
5000	Lights for loading docks, single arm, 24″ long	1 Elec	3.80	2.105	Ea.	138	115		253	330
5700	Double arm, 60″ long	"	3.80	2.105	"	211	115		326	410

11 14 Pedestrian Control Equipment

11 14 13 – Pedestrian Gates

11 14 13.13 Portable Posts and Railings

		Crew	Daily Output	Labor-Hours	Unit	Material	Labor	Equipment	Total	Total Incl O&P
0010	**PORTABLE POSTS AND RAILINGS**									
0020	Portable for pedestrian traffic control, standard				Ea.	134			134	147
0300	Deluxe posts				"	210			210	231
0600	Ropes for above posts, plastic covered, 1-1/2″ diameter				L.F.	17.25			17.25	19
0700	Chain core				"	11.65			11.65	12.80
1500	Portable security or safety barrier, black with 7′ yellow strap				Ea.	213			213	234
1510	12′ yellow strap					238			238	262
1550	Sign holder, standard design					76			76	83.50

11 14 13.19 Turnstiles

		Crew	Daily Output	Labor-Hours	Unit	Material	Labor	Equipment	Total	Total Incl O&P
0010	**TURNSTILES**									
0020	One way, 4 arm, 46″ diameter, economy, manual	2 Carp	5	3.200	Ea.	1,700	150		1,850	2,125
0100	Electric		1.20	13.333		2,050	625		2,675	3,300
0300	High security, galv., 5′-5″ diameter, 7′ high, manual		1	16		5,875	750		6,625	7,700
0350	Electric		.60	26.667		8,225	1,250		9,475	11,100
0420	Three arm, 24″ opening, light duty, manual		2	8		3,500	375		3,875	4,475
0450	Heavy duty		1.50	10.667		5,000	500		5,500	6,325
0460	Manual, with registering & controls, light duty		2	8		3,750	375		4,125	4,775
0470	Heavy duty		1.50	10.667		4,100	500		4,600	5,325
0480	Electric, heavy duty		1.10	14.545		4,775	685		5,460	6,375
0500	For coin or token operating, add					710			710	780
1200	One way gate with horizontal bars, 5′-5″ diameter									
1300	7′ high, recreation or transit type	2 Carp	.80	20	Ea.	5,350	940		6,290	7,450
1500	For electronic counter, add				"	211			211	232

11 21 Retail and Service Equipment

11 21 13 – Cash Registers and Checking Equipment

11 21 13.10 Checkout Counter

		Crew	Daily Output	Labor-Hours	Unit	Material	Labor	Equipment	Total	Total Incl O&P
0010	**CHECKOUT COUNTER**									
0020	Supermarket conveyor, single belt	2 Clab	10	1.600	Ea.	3,175	60		3,235	3,600
0100	Double belt, power take-away		9	1.778		4,575	67		4,642	5,125
0400	Double belt, power take-away, incl. side scanning		7	2.286		5,375	86		5,461	6,050
0800	Warehouse or bulk type		6	2.667		6,350	100		6,450	7,150
1000	Scanning system, 2 lanes, w/registers, scan gun & memory				System	16,600			16,600	18,300
1100	10 lanes, single processor, full scan, with scales				"	158,000			158,000	173,500

11 21 13 – Cash Registers and Checking Equipment

11 21 13.10 Checkout Counter		Crew	Daily Output	Labor-Hours	Unit	Material	2015 Bare Costs Labor	Equipment	Total	Total Incl O&P
2000	Register, restaurant, minimum				Ea.	740			740	815
2100	Maximum					3,125			3,125	3,425
2150	Store, minimum					740			740	815
2200	Maximum				↓	3,125			3,125	3,425

11 21 33 – Checkroom Equipment

11 21 33.10 Clothes Check Equipment		Crew	Daily Output	Labor-Hours	Unit	Material	Labor	Equipment	Total	Total Incl O&P
0010	CLOTHES CHECK EQUIPMENT									
0030	Clothes check rack, free standing, st. stl., 2-tier, 90 bag capacity				Ea.	1,525			1,525	1,700
0050	Wall mounted, 45 bag capacity	L-2	8	2		750	82.50		832.50	960
0100	Garment checking bag, green mesh fabric, 21" H x 17" W with 4.5" hook				↓	19.10			19.10	21

11 21 53 – Barber and Beauty Shop Equipment

11 21 53.10 Barber Equipment		Crew	Daily Output	Labor-Hours	Unit	Material	Labor	Equipment	Total	Total Incl O&P
0010	BARBER EQUIPMENT									
0020	Chair, hydraulic, movable, minimum	1 Carp	24	.333	Ea.	560	15.65		575.65	640
0050	Maximum	"	16	.500		3,475	23.50		3,498.50	3,875
0200	Wall hung styling station with mirrors, minimum	L-2	8	2		460	82.50		542.50	640
0300	Maximum	"	4	4		2,325	165		2,490	2,825
0500	Sink, hair washing basin, rough plumbing not incl.	1 Plum	8	1		495	58.50		553.50	635
1000	Sterilizer, liquid solution for tools					161			161	177
1100	Total equipment, rule of thumb, per chair, minimum	L-8	1	20		1,925	985		2,910	3,725
1150	Maximum	"	1	20	↓	5,150	985		6,135	7,275

11 21 73 – Commercial Laundry and Dry Cleaning Equipment

11 21 73.13 Dry Cleaning Equipment		Crew	Daily Output	Labor-Hours	Unit	Material	Labor	Equipment	Total	Total Incl O&P
0010	DRY CLEANING EQUIPMENT									
2000	Dry cleaners, electric, 20 lb. capacity, not incl. rough-in	L-1	.20	80	Ea.	33,700	4,525		38,225	44,100
2050	25 lb. capacity		.17	94.118		48,600	5,325		53,925	62,000
2100	30 lb. capacity		.15	106		51,000	6,050		57,050	66,000
2150	60 lb. capacity	↓	.09	177	↓	79,000	10,100		89,100	102,500

11 21 73.16 Drying and Conditioning Equipment		Crew	Daily Output	Labor-Hours	Unit	Material	Labor	Equipment	Total	Total Incl O&P
0010	DRYING AND CONDITIONING EQUIPMENT									
0100	Dryers, Not including rough-in									
1500	Industrial, 30 lb. capacity	1 Plum	2	4	Ea.	3,175	235		3,410	3,850
1600	50 lb. capacity	"	1.70	4.706		3,400	276		3,676	4,175
4700	Lint collector, ductwork not included, 8,000 to 10,000 CFM	Q-10	.30	80	↓	9,200	4,175		13,375	16,700

11 21 73.19 Finishing Equipment		Crew	Daily Output	Labor-Hours	Unit	Material	Labor	Equipment	Total	Total Incl O&P
0010	FINISHING EQUIPMENT									
3500	Folders, blankets & sheets, minimum	1 Elec	.17	47.059	Ea.	33,100	2,575		35,675	40,400
3700	King size with automatic stacker		.10	80		60,000	4,375		64,375	73,000
3800	For conveyor delivery, add	↓	.45	17.778		14,500	970		15,470	17,400
4900	Spreader feeders, 240V, 2 station	L-6	.70	17.143		56,000	985		56,985	63,500
4920	4 station	"	.35	34.286	↓	68,000	1,975		69,975	78,000

11 21 73.23 Commercial Ironing Equipment		Crew	Daily Output	Labor-Hours	Unit	Material	Labor	Equipment	Total	Total Incl O&P
0010	COMMERCIAL IRONING EQUIPMENT									
4500	Ironers, institutional, 110", single roll	1 Elec	.20	40	Ea.	32,100	2,200		34,300	38,700
4800	Pressers, low capacity air operated	L-6	1.75	6.857		9,650	395		10,045	11,200
4820	Hand operated		1.75	6.857		8,900	395		9,295	10,400
4840	Ironer 48", 240V		3.50	3.429		115,000	197		115,197	127,000
6600	Hand operated presser		.70	17.143		6,175	985		7,160	8,325
6620	Mushroom press 115 V	↓	.70	17.143	↓	7,700	985		8,685	10,000

11 21 Retail and Service Equipment

11 21 73 – Commercial Laundry and Dry Cleaning Equipment

11 21 73.26 Commercial Washers and Extractors		Crew	Daily Output	Labor-Hours	Unit	2015 Bare Costs Material	Labor	Equipment	Total	Total Incl O&P
0010	**COMMERCIAL WASHERS AND EXTRACTORS**, not including rough-in									
6000	Combination washer/extractor, 20 lb. capacity	L-6	1.50	8	Ea.	5,950	460		6,410	7,250
6100	30 lb. capacity		.80	15		9,275	860		10,135	11,600
6200	50 lb. capacity		.68	17.647		10,900	1,000		11,900	13,600
6300	75 lb. capacity		.30	40		20,400	2,300		22,700	26,100
6350	125 lb. capacity		.16	75		27,700	4,300		32,000	37,100
6380	Washer extractor/dryer, 110 lb., 240V		1	12		8,900	690		9,590	10,900
6400	Washer extractor, 135 lb., 240V		1	12		25,900	690		26,590	29,600
6450	Pass through		1	12		69,500	690		70,190	77,500
6500	200 lb. washer extractor		1	12		72,000	690		72,690	80,000
6550	Pass through		1	12		75,500	690		76,190	84,000
6600	Extractor, low capacity		1.75	6.857		7,375	395		7,770	8,700

11 21 73.33 Coin-Operated Laundry Equipment		Crew	Daily Output	Labor-Hours	Unit	Material	Labor	Equipment	Total	Total Incl O&P
0010	**COIN-OPERATED LAUNDRY EQUIPMENT**									
0990	Dryer, gas fired									
1000	Commercial, 30 lb. capacity, coin operated, single	1 Plum	3	2.667	Ea.	3,300	157		3,457	3,875
1100	Double stacked	"	2	4		7,275	235		7,510	8,375
4860	Coin dry cleaner 20 lb.	L-6	1.75	6.857		28,800	395		29,195	32,300
5290	Clothes washer									
5300	Commercial, coin operated, average	1 Plum	3	2.667	Ea.	1,250	157		1,407	1,625

11 21 83 – Photo Processing Equipment

11 21 83.13 Darkroom Equipment		Crew	Daily Output	Labor-Hours	Unit	Material	Labor	Equipment	Total	Total Incl O&P
0010	**DARKROOM EQUIPMENT**									
0020	Developing sink, 5" deep, 24" x 48"	Q-1	2	8	Ea.	4,150	425		4,575	5,225
0050	48" x 52"		1.70	9.412		4,425	495		4,920	5,650
0200	10" deep, 24" x 48"		1.70	9.412		1,550	495		2,045	2,475
0250	24" x 108"		1.50	10.667		3,650	565		4,215	4,900
0500	Dryers, dehumidified filtered air, 36" x 25" x 68" high	L-7	6	4.667		4,300	212		4,512	5,075
0550	48" x 25" x 68" high		5	5.600		10,100	254		10,354	11,500
2000	Processors, automatic, color print, minimum		4	7		15,900	320		16,220	18,000
2050	Maximum		.60	46.667		25,000	2,125		27,125	31,000
2300	Black and white print, minimum		2	14		12,100	635		12,735	14,300
2350	Maximum		.80	35		62,000	1,600		63,600	70,500
2600	Manual processor, 16" x 20" maximum print size		2	14		9,575	635		10,210	11,500
2650	20" x 24" maximum print size		1	28		9,100	1,275		10,375	12,100
3000	Viewing lights, 20" x 24"		6	4.667		288	212		500	660
3100	20" x 24" with color correction		6	4.667		345	212		557	725
3500	Washers, round, minimum sheet 11" x 14"	Q-1	2	8		3,275	425		3,700	4,250
3550	Maximum sheet 20" x 24"		1	16		3,600	845		4,445	5,300
3800	Square, minimum sheet 20" x 24"		1	16		3,150	845		3,995	4,800
3900	Maximum sheet 50" x 56"		.80	20		4,900	1,050		5,950	7,050
4500	Combination tank sink, tray sink, washers, with									
4510	Dry side tables, average	Q-1	.45	35.556	Ea.	10,800	1,875		12,675	14,700

11 22 Banking Equipment

11 22 13 – Vault Equipment

11 22 13.16 Safes		Crew	Daily Output	Labor-Hours	Unit	2015 Bare Costs Material	Labor	Equipment	Total	Total Incl O&P
0010	**SAFES**									
0200	Office, 1 hr. rating, 30" x 18" x 18"				Ea.	2,275			2,275	2,500
0250	40" x 18" x 18"					4,875			4,875	5,350
0300	60" x 36" x 18", double door					7,375			7,375	8,125
0600	Data, 1 hr. rating, 27" x 19" x 16"					4,550			4,550	5,025
0700	63" x 34" x 16"					14,500			14,500	16,000
0750	Diskette, 1 hr., 14" x 12" x 11", inside					4,050			4,050	4,450
0800	Money, "B" label, 9" x 14" x 14"					540			540	595
0900	Tool resistive, 24" x 24" x 20"					4,150			4,150	4,550
1050	Tool and torch resistive, 24" x 24" x 20"					7,875			7,875	8,675
1150	Jewelers, 23" x 20" x 18"					8,775			8,775	9,650
1200	63" x 25" x 18"					14,300			14,300	15,800
1300	For handling into building, add, minimum	A-2	8.50	2.824			108	29	137	207
1400	Maximum	"	.78	30.769			1,175	315	1,490	2,250

11 22 16 – Teller and Service Equipment

11 22 16.13 Teller Equipment Systems

11 22 16.13 Teller Equipment Systems		Crew	Daily Output	Labor-Hours	Unit	Material	Labor	Equipment	Total	Total Incl O&P
0010	**TELLER EQUIPMENT SYSTEMS**									
0020	Alarm system, police	2 Elec	1.60	10	Ea.	4,975	545		5,520	6,325
0100	With vault alarm	"	.40	40		19,700	2,200		21,900	25,100
0400	Bullet resistant teller window, 44" x 60"	1 Glaz	.60	13.333		3,850	600		4,450	5,225
0500	48" x 60"	"	.60	13.333		5,600	600		6,200	7,125
3000	Counters for banks, frontal only	2 Carp	1	16	Station	1,850	750		2,600	3,250
3100	Complete with steel undercounter	"	.50	32	"	3,625	1,500		5,125	6,425
4600	Door and frame, bullet-resistant, with vision panel, minimum	2 Sswk	1.10	14.545	Ea.	5,575	765		6,340	7,500
4700	Maximum		1.10	14.545		7,550	765		8,315	9,675
4800	Drive-up window, drawer & mike, not incl. glass, minimum		1	16		6,900	840		7,740	9,125
4900	Maximum		.50	32		9,225	1,675		10,900	13,200
5000	Night depository, with chest, minimum		1	16		7,550	840		8,390	9,825
5100	Maximum		.50	32		10,700	1,675		12,375	14,900
5200	Package receiver, painted		3.20	5		1,375	263		1,638	2,000
5300	Stainless steel		3.20	5		2,350	263		2,613	3,050
5400	Partitions, bullet-resistant, 1-3/16" glass, 8' high	2 Carp	10	1.600	L.F.	201	75		276	345
5450	Acrylic	"	10	1.600	"	380	75		455	545
5500	Pneumatic tube systems, 2 lane drive-up, complete	L-3	.25	64	Total	26,100	3,275		29,375	34,000
5550	With T.V. viewer	"	.20	80	"	50,500	4,100		54,600	62,000
5570	Safety deposit boxes, minimum	1 Sswk	44	.182	Opng.	58	9.55		67.55	81.50
5580	Maximum, 10" x 15" opening		19	.421		123	22		145	175
5590	Teller locker, average		15	.533		1,600	28		1,628	1,800
5600	Pass thru, bullet-res. window, painted steel, 24" x 36"	2 Sswk	1.60	10	Ea.	2,650	525		3,175	3,875
5700	48" x 48"		1.20	13.333		2,700	700		3,400	4,250
5800	72" x 40"		.80	20		4,250	1,050		5,300	6,575
5900	For stainless steel frames, add					20%				
6100	Surveillance system, video camera, complete	2 Elec	1	16	Ea.	9,675	875		10,550	12,100
6110	For each additional camera, add				"	1,000			1,000	1,100
6120	CCTV system, see Section 27 41 33.10									
6200	Twenty-four hour teller, single unit,									
6300	automated deposit, cash and memo	L-3	.25	64	Ea.	45,200	3,275		48,475	55,000
7000	Vault front, see Section 08 34 59.10									

11 30 13.15 Cooking Equipment

	11 30 13.15 Cooking Equipment	Crew	Daily Output	Labor-Hours	Unit	Material	Labor (2015 Bare Costs)	Equipment	Total	Total Incl O&P
0010	**COOKING EQUIPMENT**									
0020	Cooking range, 30″ free standing, 1 oven, minimum	2 Clab	10	1.600	Ea.	440	60		500	585
0050	Maximum		4	4		2,025	150		2,175	2,475
0150	2 oven, minimum		10	1.600		1,225	60		1,285	1,425
0200	Maximum		10	1.600		2,550	60		2,610	2,900
0350	Built-in, 30″ wide, 1 oven, minimum	1 Elec	6	1.333		680	73		753	865
0400	Maximum	2 Carp	2	8		1,350	375		1,725	2,125
0500	2 oven, conventional, minimum		4	4		1,325	188		1,513	1,750
0550	1 conventional, 1 microwave, maximum		2	8		1,800	375		2,175	2,600
0700	Free-standing, 1 oven, 21″ wide range, minimum	2 Clab	10	1.600		455	60		515	600
0750	21″ wide, maximum	″	4	4		450	150		600	740
0900	Countertop cooktops, 4 burner, standard, minimum	1 Elec	6	1.333		296	73		369	440
0950	Maximum		3	2.667		1,300	146		1,446	1,675
1050	As above, but with grill and griddle attachment, minimum		6	1.333		1,250	73		1,323	1,500
1100	Maximum		3	2.667		3,700	146		3,846	4,300
1200	Induction cooktop, 30″ wide		3	2.667		1,200	146		1,346	1,550
1250	Microwave oven, minimum		4	2		120	109		229	300
1300	Maximum		2	4		465	219		684	850

11 30 13.16 Refrigeration Equipment

	11 30 13.16 Refrigeration Equipment	Crew	Daily Output	Labor-Hours	Unit	Material	Labor	Equipment	Total	Total Incl O&P
0010	**REFRIGERATION EQUIPMENT**									
2000	Deep freeze, 15 to 23 C.F., minimum	2 Clab	10	1.600	Ea.	610	60		670	770
2050	Maximum		5	3.200		820	120		940	1,100
2200	30 C.F., minimum		8	2		705	75		780	900
2250	Maximum		3	5.333		910	201		1,111	1,325
5200	Icemaker, automatic, 20 lb. per day	1 Plum	7	1.143		895	67		962	1,100
5350	51 lb. per day	″	2	4		1,575	235		1,810	2,125
5450	Refrigerator, no frost, 6 C.F.	2 Clab	15	1.067		244	40		284	335
5500	Refrigerator, no frost, 10 C.F. to 12 C.F., minimum		10	1.600		425	60		485	565
5600	Maximum		6	2.667		450	100		550	660
5750	14 C.F. to 16 C.F., minimum		9	1.778		505	67		572	665
5800	Maximum		5	3.200		620	120		740	880
5950	18 C.F. to 20 C.F., minimum		8	2		630	75		705	815
6000	Maximum		4	4		1,425	150		1,575	1,825
6150	21 C.F. to 29 C.F., minimum		7	2.286		965	86		1,051	1,225
6200	Maximum		3	5.333		3,375	201		3,576	4,025
6790	Energy-star qualified, 18 C.F., minimum G	2 Carp	4	4		510	188		698	870
6795	Maximum G		2	8		1,175	375		1,550	1,900
6797	21.7 C.F., minimum G		4	4		1,025	188		1,213	1,425
6799	Maximum G		4	4		2,100	188		2,288	2,625

11 30 13.17 Kitchen Cleaning Equipment

	11 30 13.17 Kitchen Cleaning Equipment	Crew	Daily Output	Labor-Hours	Unit	Material	Labor	Equipment	Total	Total Incl O&P
0010	**KITCHEN CLEANING EQUIPMENT**									
2750	Dishwasher, built-in, 2 cycles, minimum	L-1	4	4	Ea.	238	227		465	615
2800	Maximum		2	8		435	455		890	1,175
2950	4 or more cycles, minimum		4	4		375	227		602	770
2960	Average		4	4		500	227		727	905
3000	Maximum		2	8		1,100	455		1,555	1,900
3100	Energy-star qualified, minimum G		4	4		370	227		597	760
3110	Maximum G		2	8		1,500	455		1,955	2,350

11 30 13.18 Waste Disposal Equipment

	11 30 13.18 Waste Disposal Equipment	Crew	Daily Output	Labor-Hours	Unit	Material	Labor	Equipment	Total	Total Incl O&P
0010	**WASTE DISPOSAL EQUIPMENT**									
1750	Compactor, residential size, 4 to 1 compaction, minimum	1 Carp	5	1.600	Ea.	645	75		720	835
1800	Maximum	″	3	2.667		1,025	125		1,150	1,325

11 30 13 – Residential Appliances

11 30 13.18 Waste Disposal Equipment		Crew	Daily Output	Labor-Hours	Unit	2015 Bare Costs Material	Labor	Equipment	Total	Total Incl O&P
3300	Garbage disposal, sink type, minimum	L-1	10	1.600	Ea.	87.50	90.50		178	237
3350	Maximum	"	10	1.600	↓	220	90.50		310.50	385
11 30 13.19 Kitchen Ventilation Equipment										
0010	**KITCHEN VENTILATION EQUIPMENT**									
4150	Hood for range, 2 speed, vented, 30" wide, minimum	L-3	5	3.200	Ea.	70.50	164		234.50	340
4200	Maximum		3	5.333		780	273		1,053	1,300
4300	42" wide, minimum		5	3.200		168	164		332	445
4330	Custom		5	3.200		1,650	164		1,814	2,075
4350	Maximum	↓	3	5.333		2,025	273		2,298	2,650
4500	For ventless hood, 2 speed, add					18.65			18.65	20.50
4650	For vented 1 speed, deduct from maximum				↓	50			50	55
11 30 13.24 Washers										
0010	**WASHERS**									
5000	Residential, 4 cycle, average	1 Plum	3	2.667	Ea.	875	157		1,032	1,200
6650	Washing machine, automatic, minimum		3	2.667		485	157		642	775
6700	Maximum		1	8		1,525	470		1,995	2,400
6750	Energy star qualified, front loading, minimum G		3	2.667		655	157		812	965
6760	Maximum G		1	8		1,600	470		2,070	2,475
6764	Top loading, minimum G		3	2.667		450	157		607	740
6766	Maximum G	↓	3	2.667	↓	1,250	157		1,407	1,625
11 30 13.25 Dryers										
0010	**DRYERS**									
0500	Gas fired residential, 16 lb. capacity, average	1 Plum	3	2.667	Ea.	675	157		832	985
6770	Electric, front loading, energy-star qualified, minimum G	L-2	3	5.333		385	220		605	785
6780	Maximum G	"	2	8		1,925	330		2,255	2,650
7450	Vent kits for dryers	1 Carp	10	.800	↓	39	37.50		76.50	104

11 30 15 – Miscellaneous Residential Appliances

11 30 15.13 Sump Pumps		Crew	Daily Output	Labor-Hours	Unit	Material	Labor	Equipment	Total	Total Incl O&P
0010	**SUMP PUMPS**									
6400	Cellar drainer, pedestal, 1/3 H.P., molded PVC base	1 Plum	3	2.667	Ea.	135	157		292	395
6450	Solid brass	"	2	4	"	289	235		524	685
6460	Sump pump, see also Section 22 14 29.16									
11 30 15.23 Water Heaters										
0010	**WATER HEATERS**									
6900	Electric, glass lined, 30 gallon, minimum	L-1	5	3.200	Ea.	430	181		611	750
6950	Maximum		3	5.333		595	300		895	1,125
7100	80 gallon, minimum		2	8		1,225	455		1,680	2,050
7150	Maximum	↓	1	16		1,700	905		2,605	3,250
7180	Gas, glass lined, 30 gallon, minimum	2 Plum	5	3.200		805	188		993	1,175
7220	Maximum		3	5.333		1,125	315		1,440	1,725
7260	50 gallon, minimum		2.50	6.400		845	375		1,220	1,500
7300	Maximum	↓	1.50	10.667	↓	1,175	625		1,800	2,250
7310	Water heater, see also Section 22 33 30.13									
11 30 15.43 Air Quality										
0010	**AIR QUALITY**									
2450	Dehumidifier, portable, automatic, 15 pint	1 Elec	4	2	Ea.	152	109		261	335
2550	40 pint		3.75	2.133		209	117		326	410
3550	Heater, electric, built-in, 1250 watt, ceiling type, minimum		4	2		107	109		216	287
3600	Maximum		3	2.667		175	146		321	420
3700	Wall type, minimum		4	2		172	109		281	360
3750	Maximum	↓	3	2.667	↓	185	146		331	430

11 30 Residential Equipment

11 30 15 – Miscellaneous Residential Appliances

11 30 15.43 Air Quality		Crew	Daily Output	Labor-Hours	Unit	Material	2015 Bare Costs Labor	Equipment	Total	Total Incl O&P
3900	1500 watt wall type, with blower	1 Elec	4	2	Ea.	172	109		281	360
3950	3000 watt		3	2.667		350	146		496	610
4850	Humidifier, portable, 8 gallons per day					133			133	146
5000	15 gallons per day					211			211	232

11 30 33 – Retractable Stairs

11 30 33.10 Disappearing Stairway

Code	Description	Crew	Daily Output	Labor-Hours	Unit	Material	Labor	Equipment	Total	Total Incl O&P
0010	**DISAPPEARING STAIRWAY** No trim included									
0100	Custom grade, pine, 8′-6″ ceiling, minimum	1 Carp	4	2	Ea.	177	94		271	350
0150	Average		3.50	2.286		253	107		360	455
0200	Maximum		3	2.667		325	125		450	565
0500	Heavy duty, pivoted, from 7′-7″ to 12′-10″ floor to floor		3	2.667		740	125		865	1,025
0600	16′-0″ ceiling		2	4		1,525	188		1,713	1,975
0800	Economy folding, pine, 8′-6″ ceiling		4	2		176	94		270	350
0900	9′-6″ ceiling		4	2		196	94		290	370
1100	Automatic electric, aluminum, floor to floor height, 8′ to 9′	2 Carp	1	16		8,550	750		9,300	10,600
1400	11′ to 12′		.90	17.778		9,075	835		9,910	11,400
1700	14′ to 15′		.70	22.857		9,775	1,075		10,850	12,600
9000	Minimum labor/equipment charge	1 Carp	2	4	Job		188		188	310

11 32 Unit Kitchens

11 32 13 – Metal Unit Kitchens

11 32 13.10 Commercial Unit Kitchens

Code	Description	Crew	Daily Output	Labor-Hours	Unit	Material	Labor	Equipment	Total	Total Incl O&P
0010	**COMMERCIAL UNIT KITCHENS**									
1500	Combination range, refrigerator and sink, 30″ wide, minimum	L-1	2	8	Ea.	1,100	455		1,555	1,925
1550	Maximum		1	16		1,475	905		2,380	3,025
1570	60″ wide, average		1.40	11.429		1,525	650		2,175	2,700
1590	72″ wide, average		1.20	13.333		1,625	755		2,380	2,950
1600	Office model, 48″ wide		2	8		2,025	455		2,480	2,950
1620	Refrigerator and sink only		2.40	6.667		2,525	380		2,905	3,375
1640	Combination range, refrigerator, sink, microwave									
1660	Oven and ice maker	L-1	.80	20	Ea.	4,550	1,125		5,675	6,775

11 41 Foodservice Storage Equipment

11 41 13 – Refrigerated Food Storage Cases

11 41 13.10 Refrigerated Food Cases

Code	Description	Crew	Daily Output	Labor-Hours	Unit	Material	Labor	Equipment	Total	Total Incl O&P
0010	**REFRIGERATED FOOD CASES**									
0030	Dairy, multi-deck, 12′ long	Q-5	3	5.333	Ea.	11,300	287		11,587	12,800
0100	For rear sliding doors, add					1,800			1,800	2,000
0200	Delicatessen case, service deli, 12′ long, single deck	Q-5	3.90	4.103		7,750	221		7,971	8,875
0300	Multi-deck, 18 S.F. shelf display		3	5.333		7,150	287		7,437	8,325
0400	Freezer, self-contained, chest-type, 30 C.F.		3.90	4.103		8,525	221		8,746	9,750
0500	Glass door, upright, 78 C.F.		3.30	4.848		10,200	261		10,461	11,600
0600	Frozen food, chest type, 12′ long		3.30	4.848		8,225	261		8,486	9,450
0700	Glass door, reach-in, 5 door		3	5.333		14,100	287		14,387	15,900
0800	Island case, 12′ long, single deck		3.30	4.848		7,275	261		7,536	8,425
0900	Multi-deck		3	5.333		8,500	287		8,787	9,800
1000	Meat case, 12′ long, single deck		3.30	4.848		7,275	261		7,536	8,425
1050	Multi-deck		3.10	5.161		10,400	278		10,678	11,800
1100	Produce, 12′ long, single deck		3.30	4.848		6,875	261		7,136	7,950

11 41 13 – Refrigerated Food Storage Cases

11 41 13.10 Refrigerated Food Cases		Crew	Daily Output	Labor-Hours	Unit	Material	Labor	Equipment	Total	Total Incl O&P
1200	Multi-deck	Q-5	3.10	5.161	Ea.	8,725	278		9,003	10,000

11 41 13.20 Refrigerated Food Storage Equipment

		Crew	Daily Output	Labor-Hours	Unit	Material	Labor	Equipment	Total	Total Incl O&P
0010	**REFRIGERATED FOOD STORAGE EQUIPMENT**									
2350	Cooler, reach-in, beverage, 6′ long	Q-1	6	2.667	Ea.	3,600	141		3,741	4,175
4300	Freezers, reach-in, 44 C.F.		4	4		4,425	211		4,636	5,175
4500	68 C.F.		3	5.333		4,825	282		5,107	5,750
4600	Freezer, pre-fab, 8′ x 8′ w/refrigeration	2 Carp	.45	35.556		11,100	1,675		12,775	14,900
4620	8′ x 12′		.35	45.714		11,200	2,150		13,350	15,800
4640	8′ x 16′		.25	64		14,300	3,000		17,300	20,600
4660	8′ x 20′		.17	94.118		19,400	4,425		23,825	28,600
4680	Reach-in, 1 compartment	Q-1	4	4		2,500	211		2,711	3,100
4685	Energy star rated G	R-18	7.80	3.333		2,575	143		2,718	3,050
4700	2 compartment	Q-1	3	5.333		4,100	282		4,382	4,975
4705	Energy star rated G	R-18	6.20	4.194		3,100	180		3,280	3,700
4710	3 compartment	Q-1	3	5.333		5,300	282		5,582	6,275
4715	Energy star rated G	R-18	5.60	4.643		4,175	199		4,374	4,925
8320	Refrigerator, reach-in, 1 compartment		7.80	3.333		2,425	143		2,568	2,900
8325	Energy star rated G		7.80	3.333		2,575	143		2,718	3,050
8330	2 compartment		6.20	4.194		3,800	180		3,980	4,475
8335	Energy star rated G		6.20	4.194		3,100	180		3,280	3,700
8340	3 compartment		5.60	4.643		4,775	199		4,974	5,575
8345	Energy star rated G		5.60	4.643		4,175	199		4,374	4,925
8350	Pre-fab, with refrigeration, 8′ x 8′	2 Carp	.45	35.556		7,050	1,675		8,725	10,500
8360	8′ x 12′		.35	45.714		8,000	2,150		10,150	12,300
8370	8′ x 16′		.25	64		12,300	3,000		15,300	18,500
8380	8′ x 20′		.17	94.118		15,700	4,425		20,125	24,500
8390	Pass-thru/roll-in, 1 compartment	R-18	7.80	3.333		4,500	143		4,643	5,175
8400	2 compartment		6.24	4.167		6,325	179		6,504	7,225
8410	3 compartment		5.60	4.643		8,625	199		8,824	9,825
8420	Walk-in, alum, door & floor only, no refrig, 6′ x 6′ x 7′-6″	2 Carp	1.40	11.429		8,775	535		9,310	10,500
8430	10′ x 6′ x 7′-6″		.55	29.091		12,700	1,375		14,075	16,300
8440	12′ x 14′ x 7′-6″		.25	64		17,500	3,000		20,500	24,100
8450	12′ x 20′ x 7′-6″		.17	94.118		19,300	4,425		23,725	28,500
8460	Refrigerated cabinets, mobile					3,925			3,925	4,325
8470	Refrigerator/freezer, reach-in, 1 compartment	R-18	5.60	4.643		5,800	199		5,999	6,700
8480	2 compartment	″	4.80	5.417		7,575	232		7,807	8,700

11 41 13.30 Wine Cellar

		Crew	Daily Output	Labor-Hours	Unit	Material	Labor	Equipment	Total	Total Incl O&P
0010	**WINE CELLAR**, refrigerated, Redwood interior, carpeted, walk-in type									
0020	6′-8″ high, including racks									
0200	80″ W x 48″ D for 900 bottles	2 Carp	1.50	10.667	Ea.	4,300	500		4,800	5,550
0250	80″ W x 72″ D for 1300 bottles		1.33	12.030		5,225	565		5,790	6,675
0300	80″ W x 94″ D for 1900 bottles		1.17	13.675		6,350	640		6,990	8,025
0400	80″ W x 124″ D for 2500 bottles		1	16		7,450	750		8,200	9,425
0600	Portable cabinets, red oak, reach-in temp.& humidity controlled									
0650	26-5/8″W x 26-1/2″D x 68″H for 235 bottles				Ea.	3,575			3,575	3,925
0660	32″W x 21-1/2″D x 73-1/2″H for 144 bottles					2,900			2,900	3,200
0670	32″W x 29-1/2″D x 73-1/2″H for 288 bottles					3,875			3,875	4,275
0680	39-1/2″W x 29-1/2″D x 86-1/2″H for 440 bottles					4,100			4,100	4,500
0690	52-1/2″W x 29-1/2″D x 73-1/2″H for 468 bottles					4,375			4,375	4,825
0700	52-1/2″W x 29-1/2″D x 86-1/2″H for 572 bottles					4,475			4,475	4,925
0730	Portable, red oak, can be built-in with glass door									
0750	23-7/8″W x 24″D x 34-1/2″H for 50 bottles				Ea.	940			940	1,025

11 41 Foodservice Storage Equipment

11 41 33 – Foodservice Shelving

11 41 33.20 Metal Food Storage Shelving		Crew	Daily Output	Labor-Hours	Unit	2015 Bare Costs Material	Labor	Equipment	Total	Total Incl O&P
0010	**METAL FOOD STORAGE SHELVING**									
8600	Stainless steel shelving, louvered 4-tier, 20" x 3'	1 Clab	6	1.333	Ea.	1,400	50		1,450	1,625
8605	20" x 4'		6	1.333		1,550	50		1,600	1,775
8610	20" x 6'		6	1.333		2,200	50		2,250	2,500
8615	24" x 3'		6	1.333		1,975	50		2,025	2,250
8620	24" x 4'		6	1.333		2,350	50		2,400	2,675
8625	24" x 6'		6	1.333		3,275	50		3,325	3,675
8630	Flat 4-tier, 20" x 3'		6	1.333		1,150	50		1,200	1,350
8635	20" x 4'		6	1.333		1,400	50		1,450	1,625
8640	20" x 5'		6	1.333		1,625	50		1,675	1,850
8645	24" x 3'		6	1.333		1,275	50		1,325	1,475
8650	24" x 4'		6	1.333		2,225	50		2,275	2,525
8655	24" x 6'		6	1.333		2,675	50		2,725	3,000
8700	Galvanized shelving, louvered 4-tier, 20" x 3'		6	1.333		760	50		810	915
8705	20" x 4'		6	1.333		860	50		910	1,025
8710	20" x 6'		6	1.333		1,000	50		1,050	1,175
8715	24" x 3'		6	1.333		705	50		755	860
8720	24" x 4'		6	1.333		995	50		1,045	1,175
8725	24" x 6'		6	1.333		1,325	50		1,375	1,525
8730	Flat 4-tier, 20" x 3'		6	1.333		700	50		750	850
8735	20" x 4'		6	1.333		695	50		745	845
8740	20" x 6'		6	1.333		905	50		955	1,075
8745	24" x 3'		6	1.333		670	50		720	815
8750	24" x 4'		6	1.333		755	50		805	910
8755	24" x 6'		6	1.333		950	50		1,000	1,125
8760	Stainless steel dunnage rack, 24" x 3'		8	1		330	37.50		367.50	425
8765	24" x 4'		8	1		425	37.50		462.50	525
8770	Galvanized dunnage rack, 24" x 3'		8	1		168	37.50		205.50	247
8775	24" x 4'		8	1		190	37.50		227.50	271

11 42 Food Preparation Equipment

11 42 10 – Commercial Food Preparation Equipment

11 42 10.10 Choppers, Mixers and Misc. Equipment		Crew	Daily Output	Labor-Hours	Unit	2015 Bare Costs Material	Labor	Equipment	Total	Total Incl O&P
0010	**CHOPPERS, MIXERS AND MISC. EQUIPMENT**									
1700	Choppers, 5 pounds	R-18	7	3.714	Ea.	2,150	159		2,309	2,625
1720	16 pounds		5	5.200		2,200	223		2,423	2,775
1740	35 to 40 pounds		4	6.500		3,550	279		3,829	4,350
1840	Coffee brewer, 5 burners	1 Plum	3	2.667		1,325	157		1,482	1,700
1850	Coffee urn, twin 6 gallon urns		2	4		2,400	235		2,635	3,025
1860	Single, 3 gallon		3	2.667		1,825	157		1,982	2,250
3000	Fast food equipment, total package, minimum	6 Skwk	.08	600		203,500	29,200		232,700	271,500
3100	Maximum	"	.07	685		277,500	33,400		310,900	359,000
3800	Food mixers, bench type, 20 quarts	L-7	7	4		2,775	182		2,957	3,350
3850	40 quarts		5.40	5.185		6,725	235		6,960	7,775
3900	60 quarts		5	5.600		11,300	254		11,554	12,900
4040	80 quarts		3.90	7.179		12,700	325		13,025	14,500
4100	Floor type, 20 quarts		15	1.867		3,225	84.50		3,309.50	3,700
4120	60 quarts		14	2		9,950	91		10,041	11,000
4140	80 quarts		12	2.333		15,400	106		15,506	17,200
4160	140 quarts		8.60	3.256		25,600	148		25,748	28,400

11 42 Food Preparation Equipment

11 42 10 – Commercial Food Preparation Equipment

11 42 10.10 Choppers, Mixers and Misc. Equipment		Crew	Daily Output	Labor-Hours	Unit	Material	2015 Bare Costs Labor	Equipment	Total	Total Incl O&P
6700	Peelers, small	R-18	8	3.250	Ea.	1,900	139		2,039	2,300
6720	Large	"	6	4.333		4,825	186		5,011	5,600
6800	Pulper/extractor, close coupled, 5 HP	1 Plum	1.90	4.211		3,525	247		3,772	4,250
8580	Slicer with table	R-18	9	2.889	↓	4,675	124		4,799	5,350

11 43 Food Delivery Carts and Conveyors

11 43 13 – Food Delivery Carts

11 43 13.10 Mobile Carts, Racks and Trays

Line	Description	Crew	Daily Output	Labor-Hours	Unit	Material	Labor	Equipment	Total	Total Incl O&P
0010	**MOBILE CARTS, RACKS AND TRAYS**									
1650	Cabinet, heated, 1 compartment, reach-in	R-18	5.60	4.643	Ea.	3,300	199		3,499	3,950
1655	Pass-thru roll-in		5.60	4.643		3,825	199		4,024	4,525
1660	2 compartment, reach-in	↓	4.80	5.417		9,050	232		9,282	10,300
1670	Mobile					3,525			3,525	3,875
2000	Hospital food cart, hot and cold service, 20 tray capacity					15,300			15,300	16,800
6850	Mobile rack w/pan slide					1,400			1,400	1,525
9180	Tray and silver dispenser, mobile	1 Clab	16	.500	↓	915	18.80		933.80	1,025

11 44 Food Cooking Equipment

11 44 13 – Commercial Ranges

11 44 13.10 Cooking Equipment

Line	Description	Crew	Daily Output	Labor-Hours	Unit	Material	Labor	Equipment	Total	Total Incl O&P
0010	**COOKING EQUIPMENT**									
0020	Bake oven, gas, one section	Q-1	8	2	Ea.	5,450	106		5,556	6,175
0300	Two sections		7	2.286		9,075	121		9,196	10,200
0600	Three sections	↓	6	2.667		11,200	141		11,341	12,500
0900	Electric convection, single deck	L-7	4	7		6,225	320		6,545	7,375
1300	Broiler, without oven, standard	Q-1	8	2		3,550	106		3,656	4,075
1550	Infrared	L-7	4	7		7,425	320		7,745	8,675
4750	Fryer, with twin baskets, modular model	Q-1	7	2.286		1,250	121		1,371	1,575
5000	Floor model, on 6" legs	"	5	3.200		2,425	169		2,594	2,950
5100	Extra single basket, large					100			100	110
5170	Energy star rated, 50 lb. capacity [G]	R-18	4	6.500		4,650	279		4,929	5,550
5175	85 lb. capacity [G]	"	4	6.500		8,525	279		8,804	9,825
5300	Griddle, SS, 24" plate, w/4" legs, elec, 208 V, 3 phase, 3' long	Q-1	7	2.286		1,400	121		1,521	1,750
5550	4' long	"	6	2.667		2,250	141		2,391	2,700
6200	Iced tea brewer	1 Plum	3.44	2.326		750	137		887	1,050
6350	Kettle, w/steam jacket, tilting, w/positive lock, SS, 20 gallons	L-7	7	4		8,225	182		8,407	9,350
6600	60 gallons	"	6	4.667		11,000	212		11,212	12,400
6900	Range, restaurant type, 6 burners and 1 standard oven, 36" wide	Q-1	7	2.286		2,500	121		2,621	2,950
6950	Convection		7	2.286		4,450	121		4,571	5,100
7150	2 standard ovens, 24" griddle, 60" wide		6	2.667		4,575	141		4,716	5,250
7200	1 standard, 1 convection oven		6	2.667		9,475	141		9,616	10,600
7450	Heavy duty, single 34" standard oven, open top		5	3.200		5,050	169		5,219	5,825
7500	Convection oven		5	3.200		5,450	169		5,619	6,275
7700	Griddle top		6	2.667		2,850	141		2,991	3,375
7750	Convection oven	↓	6	2.667		7,775	141		7,916	8,775
7760	Induction cooker, electric	L-7	7	4		1,850	182		2,032	2,350
8850	Steamer, electric 27 KW		7	4		10,700	182		10,882	12,100
9100	Electric, 10 KW or gas 100,000 BTU	↓	5	5.600		6,325	254		6,579	7,375
9150	Toaster, conveyor type, 16-22 slices per minute				↓	1,075			1,075	1,200

11 44 Food Cooking Equipment

11 44 13 – Commercial Ranges

11 44 13.10 Cooking Equipment		Crew	Daily Output	Labor-Hours	Unit	Material	Labor	Equipment	Total	Total Incl O&P
						2015 Bare Costs				
9160	Pop-up, 2 slot				Ea.	615			615	680
9200	For deluxe models of above equipment, add					75%				
9400	Rule of thumb: Equipment cost based									
9410	on kitchen work area									
9420	Office buildings, minimum	L-7	77	.364	S.F.	90.50	16.50		107	127
9450	Maximum		58	.483		153	22		175	204
9550	Public eating facilities, minimum		77	.364		119	16.50		135.50	158
9600	Maximum		46	.609		193	27.50		220.50	257
9750	Hospitals, minimum		58	.483		122	22		144	170
9800	Maximum	↓	39	.718	↓	225	32.50		257.50	300

11 46 Food Dispensing Equipment

11 46 13 – Bar Equipment

11 46 13.10 Bar Equipment		Crew	Daily Output	Labor-Hours	Unit	Material	Labor	Equipment	Total	Total Incl O&P
0010	**BAR EQUIPMENT**									
0100	Bar die, flat wall type 41" high	Q-1	15	1.067	L.F.	265	56.50		321.50	380
0150	Blender station with sink, 14"		6	2.667	Ea.	595	141		736	875
0200	Cocktail station 36"		6	2.667		1,425	141		1,566	1,800
0250	Drainboard with glass rack	↓	6	2.667		365	141		506	625
0300	Glass storage cabinet	1 Clab	6	1.333		620	50		670	760
0350	Glass and plate chiller, 4.4 C.F.	Q-1	6	2.667	↓	2,150	141		2,291	2,600

11 46 16 – Service Line Equipment

11 46 16.10 Commercial Food Dispensing Equipment		Crew	Daily Output	Labor-Hours	Unit	Material	Labor	Equipment	Total	Total Incl O&P
0010	**COMMERCIAL FOOD DISPENSING EQUIPMENT**									
1050	Butter pat dispenser	1 Clab	13	.615	Ea.	1,000	23		1,023	1,150
1100	Bread dispenser, counter top		13	.615		890	23		913	1,025
1900	Cup and glass dispenser, drop in		4	2		1,075	75		1,150	1,300
1920	Disposable cup, drop in		16	.500		495	18.80		513.80	575
2650	Dish dispenser, drop in, 12"		11	.727		2,250	27.50		2,277.50	2,525
2660	Mobile	↓	10	.800		2,625	30		2,655	2,950
3300	Food warmer, counter, 1.2 KW					665			665	735
3550	1.6 KW					2,150			2,150	2,350
3600	Well, hot food, built-in, rectangular, 12" x 20"	R-30	10	2.600		815	115		930	1,075
3610	Circular, 7 qt.		10	2.600		420	115		535	645
3620	Refrigerated, 2 compartments		10	2.600		3,075	115		3,190	3,550
3630	3 compartments		9	2.889		3,700	128		3,828	4,275
3640	4 compartments		8	3.250		4,350	144		4,494	5,025
4720	Frost cold plate	↓	9	2.889		18,400	128		18,528	20,500
5700	Hot chocolate dispenser	1 Plum	4	2		1,175	117		1,292	1,450
5750	Ice dispenser 567 pound	Q-1	6	2.667		5,200	141		5,341	5,950
6250	Jet spray dispenser	R-18	4.50	5.778		3,175	248		3,423	3,900
6300	Juice dispenser, concentrate	"	4.50	5.778		1,900	248		2,148	2,475
6690	Milk dispenser, bulk, 2 flavor	R-30	8	3.250		1,850	144		1,994	2,250
6695	3 flavor	"	8	3.250	↓	2,475	144		2,619	2,950
8800	Serving counter, straight	1 Carp	40	.200	L.F.	925	9.40		934.40	1,050
8820	Curved section	"	30	.267	"	1,125	12.50		1,137.50	1,275
8825	Solid surface, see Section 12 36 61.16									
8860	Sneeze guard with lights, 60" L	1 Clab	16	.500	Ea.	395	18.80		413.80	465
8900	Sneeze guard, stainless steel and glass, single sided									
8910	Portable, 48" W				Ea.	305			305	335

11 46 Food Dispensing Equipment

11 46 16 – Service Line Equipment

11 46 16.10 Commercial Food Dispensing Equipment		Crew	Daily Output	Labor-Hours	Unit	Material	Labor	Equipment	Total	Total Incl O&P
						2015 Bare Costs				
8920	Portable, 72" W				Ea.	325			325	360
8930	Adjustable, 36" W	1 Carp	24	.333		253	15.65		268.65	305
8940	Adjustable, 48" W	"	20	.400		320	18.80		338.80	385
9100	Soft serve ice cream machine, medium	R-18	11	2.364		12,100	101		12,201	13,500
9110	Large	"	9	2.889	↓	21,800	124		21,924	24,100

11 46 83 – Ice Machines

11 46 83.10 Commercial Ice Equipment

		Crew	Daily Output	Labor-Hours	Unit	Material	Labor	Equipment	Total	Total Incl O&P
0010	**COMMERCIAL ICE EQUIPMENT**									
5800	Ice cube maker, 50 pounds per day	Q-1	6	2.667	Ea.	1,600	141		1,741	2,000
5810	65 pounds per day, energy star rated		6	2.667		1,475	141		1,616	1,850
5900	250 pounds per day		1.20	13.333		2,675	705		3,380	4,050
5950	300 pounds per day, remote condensing		1.20	13.333		2,350	705		3,055	3,700
6050	500 pounds per day		4	4		2,775	211		2,986	3,375
6060	With bin		1.20	13.333		3,400	705		4,105	4,825
6070	Modular, with bin and condenser		1.20	13.333		3,875	705		4,580	5,375
6090	1000 pounds per day, with bin		1	16		4,950	845		5,795	6,750
6100	Ice flakers, 300 pounds per day		1.60	10		2,800	530		3,330	3,900
6120	600 pounds per day		.95	16.842		3,950	890		4,840	5,725
6130	1000 pounds per day		.75	21.333		4,750	1,125		5,875	6,975
6140	2000 pounds per day	↓	.65	24.615		21,600	1,300		22,900	25,700
6160	Ice storage bin, 500 pound capacity	Q-5	1	16		1,050	860		1,910	2,500
6180	1000 pound	"	.56	28.571	↓	2,525	1,525		4,050	5,175

11 48 Foodservice Cleaning and Disposal Equipment

11 48 13 – Commercial Dishwashers

11 48 13.10 Dishwashers

		Crew	Daily Output	Labor-Hours	Unit	Material	Labor	Equipment	Total	Total Incl O&P
0010	**DISHWASHERS**									
2700	Dishwasher, commercial, rack type									
2720	10 to 12 racks per hour	Q-1	3.20	5	Ea.	3,525	264		3,789	4,275
2730	Energy star rated, 35 to 40 racks/hour G		1.30	12.308		4,700	650		5,350	6,200
2740	50 to 60 racks/hour G	↓	1.30	12.308		10,200	650		10,850	12,200
2800	Automatic, 190 to 230 racks per hour	L-6	.35	34.286		13,900	1,975		15,875	18,400
2820	235 to 275 racks per hour		.25	48		31,500	2,750		34,250	38,900
2840	8,750 to 12,500 dishes per hour	↓	.10	120	↓	55,500	6,875		62,375	71,500
2950	Dishwasher hood, canopy type	L-3A	10	1.200	L.F.	865	61.50		926.50	1,050
2960	Pant leg type	"	2.50	4.800	Ea.	8,700	246		8,946	9,975
5200	Garbage disposal 1.5 HP, 100 GPH	L-1	4.80	3.333		2,175	189		2,364	2,700
5210	3 HP, 120 GPH		4.60	3.478		2,625	197		2,822	3,200
5220	5 HP, 250 GPH	↓	4.50	3.556	↓	3,600	202		3,802	4,300
6750	Pot sink, 3 compartment	1 Plum	7.25	1.103	L.F.	980	65		1,045	1,175
6760	Pot washer, low temp wash/rinse		1.60	5	Ea.	4,050	294		4,344	4,900
6770	High pressure wash, high temperature rinse	↓	1.20	6.667		36,900	390		37,290	41,200
9170	Trash compactor, small, up to 125 lb. compacted weight	L-4	4	6		24,200	266		24,466	27,000
9175	Large, up to 175 lb. compacted weight	"	3	8	↓	29,500	355		29,855	33,100

11 52 Audio-Visual Equipment

11 52 13 – Projection Screens

11 52 13.10 Projection Screens, Wall or Ceiling Hung		Crew	Daily Output	Labor-Hours	Unit	2015 Bare Costs Material	Labor	Equipment	Total	Total Incl O&P
0010	**PROJECTION SCREENS, WALL OR CEILING HUNG**, matte white									
0100	Manually operated, economy	2 Carp	500	.032	S.F.	5.90	1.50		7.40	8.95
0300	Intermediate		450	.036		6.90	1.67		8.57	10.35
0400	Deluxe		400	.040		9.55	1.88		11.43	13.60
0600	Electric operated, matte white, 25 S.F., economy		5	3.200	Ea.	865	150		1,015	1,200
0700	Deluxe		4	4		1,800	188		1,988	2,275
0900	50 S.F., economy		3	5.333		705	250		955	1,175
1000	Deluxe		2	8		2,000	375		2,375	2,825
1200	Heavy duty, electric operated, 200 S.F.		1.50	10.667		3,950	500		4,450	5,175
1300	400 S.F.		1	16		4,875	750		5,625	6,575
1500	Rigid acrylic in wall, for rear projection, 1/4" thick	2 Glaz	30	.533	S.F.	47.50	24		71.50	91.50
1600	1/2" thick (maximum size 10' x 20')	"	25	.640	"	84.50	29		113.50	140
9000	Minimum labor/equipment charge	2 Carp	3	5.333	Job		250		250	410

11 52 16 – Projectors

11 52 16.10 Movie Equipment		Crew	Daily Output	Labor-Hours	Unit	Material	Labor	Equipment	Total	Total Incl O&P
0010	**MOVIE EQUIPMENT**									
0020	Changeover, minimum				Ea.	470			470	520
0100	Maximum					915			915	1,000
0400	Film transport, incl. platters and autowind, minimum					5,075			5,075	5,600
0500	Maximum					14,400			14,400	15,900
0800	Lamphouses, incl. rectifiers, xenon, 1,000 watt	1 Elec	2	4		6,725	219		6,944	7,750
0900	1,600 watt		2	4		7,175	219		7,394	8,250
1000	2,000 watt		1.50	5.333		7,700	292		7,992	8,925
1100	4,000 watt		1.50	5.333		9,550	292		9,842	11,000
1400	Lenses, anamorphic, minimum					1,300			1,300	1,425
1500	Maximum					2,900			2,900	3,200
1800	Flat 35 mm, minimum					1,125			1,125	1,225
1900	Maximum					1,750			1,750	1,925
2200	Pedestals, for projectors					1,500			1,500	1,650
2300	Console type					10,800			10,800	11,900
2600	Projector mechanisms, incl. soundhead, 35 mm, minimum					11,200			11,200	12,300
2700	Maximum					15,400			15,400	16,900
3000	Projection screens, rigid, in wall, acrylic, 1/4" thick	2 Glaz	195	.082	S.F.	42.50	3.70		46.20	52.50
3100	1/2" thick	"	130	.123	"	49	5.55		54.55	63
3300	Electric operated, heavy duty, 400 S.F.	2 Carp	1	16	Ea.	3,000	750		3,750	4,500
3320	Theater projection screens, matte white, including frames	"	200	.080	S.F.	6.70	3.76		10.46	13.55
3400	Also see Section 11 52 13.10									
3700	Sound systems, incl. amplifier, mono, minimum	1 Elec	.90	8.889	Ea.	3,350	485		3,835	4,425
3800	Dolby/Super Sound, maximum		.40	20		18,300	1,100		19,400	21,800
4100	Dual system, 2 channel, front surround, minimum		.70	11.429		4,675	625		5,300	6,125
4200	Dolby/Super Sound, 4 channel, maximum		.40	20		16,700	1,100		17,800	20,100
4500	Sound heads, 35 mm					5,350			5,350	5,900
4900	Splicer, tape system, minimum					750			750	825
5000	Tape type, maximum					1,350			1,350	1,475
5300	Speakers, recessed behind screen, minimum	1 Elec	2	4		1,075	219		1,294	1,525
5400	Maximum	"	1	8		3,125	440		3,565	4,125
5700	Seating, painted steel, upholstered, minimum	2 Carp	35	.457		133	21.50		154.50	182
5800	Maximum	"	28	.571		425	27		452	515
6100	Rewind tables, minimum					2,675			2,675	2,950
6200	Maximum					4,775			4,775	5,250
7000	For automation, varying sophistication, minimum	1 Elec	1	8	System	2,425	440		2,865	3,325
7100	Maximum	2 Elec	.30	53.333	"	5,625	2,925		8,550	10,700

11 52 Audio-Visual Equipment

11 52 16 – Projectors

11 52 16.20 Movie Equipment- Digital		Crew	Daily Output	Labor-Hours	Unit	Material	2015 Bare Costs Labor	Equipment	Total	Total Incl O&P
0010	MOVIE EQUIPMENT- DIGITAL									
1000	Digital 2K projection system, 98" DMD	1 Elec	2	4	Ea.	44,500	219		44,719	49,300
1100	OEM lens		2	4		5,425	219		5,644	6,325
2000	Pedestal with power distribution		2	4		2,075	219		2,294	2,625
3000	Software	↓	2	4	↓	1,750	219		1,969	2,275

11 53 Laboratory Equipment

11 53 03 – Laboratory Test Equipment

11 53 03.13 Test Equipment

Code	Description	Crew	Daily Output	Labor-Hours	Unit	Material	Labor	Equipment	Total	Total Incl O&P
0010	TEST EQUIPMENT									
1700	Thermometer, electric, portable				Ea.	500			500	550
1800	Titration unit, four 2000 ml reservoirs				"	5,550			5,550	6,125

11 53 13 – Laboratory Fume Hoods

11 53 13.13 Recirculating Laboratory Fume Hoods

Code	Description	Crew	Daily Output	Labor-Hours	Unit	Material	Labor	Equipment	Total	Total Incl O&P
0010	RECIRCULATING LABORATORY FUME HOODS									
0600	Fume hood, with countertop & base, not including HVAC									
0610	Simple, minimum	2 Carp	5.40	2.963	L.F.	500	139		639	780
0620	Complex, including fixtures		2.40	6.667		805	315		1,120	1,400
0630	Special, maximum	↓	1.70	9.412	↓	840	440		1,280	1,650
0670	Service fixtures, average				Ea.	240			240	264
0680	For sink assembly with hot and cold water, add	1 Plum	1.40	5.714		735	335		1,070	1,325
0750	Glove box, fiberglass, bacteriological					17,100			17,100	18,800
0760	Controlled atmosphere					19,600			19,600	21,600
0770	Radioisotope					17,100			17,100	18,800
0780	Carcinogenic				↓	17,100			17,100	18,800

11 53 13.23 Exhaust Hoods

Code	Description	Crew	Daily Output	Labor-Hours	Unit	Material	Labor	Equipment	Total	Total Incl O&P
0010	EXHAUST HOODS									
0650	Ductwork, minimum	2 Shee	1	16	Hood	4,075	895		4,970	5,900
0660	Maximum	"	.50	32	"	6,275	1,800		8,075	9,725

11 53 16 – Laboratory Incubators

11 53 16.13 Incubators

Code	Description	Crew	Daily Output	Labor-Hours	Unit	Material	Labor	Equipment	Total	Total Incl O&P
0010	INCUBATORS									
1000	Incubators, minimum				Ea.	3,050			3,050	3,350
1010	Maximum				"	11,800			11,800	13,000

11 53 19 – Laboratory Sterilizers

11 53 19.13 Sterilizers

Code	Description	Crew	Daily Output	Labor-Hours	Unit	Material	Labor	Equipment	Total	Total Incl O&P
0010	STERILIZERS									
0700	Glassware washer, undercounter, minimum	L-1	1.80	8.889	Ea.	6,325	505		6,830	7,725
0710	Maximum	"	1	16		13,300	905		14,205	16,000
1850	Utensil washer-sanitizer	1 Plum	2	4	↓	11,300	235		11,535	12,800

11 53 23 – Laboratory Refrigerators

11 53 23.13 Refrigerators

Code	Description	Crew	Daily Output	Labor-Hours	Unit	Material	Labor	Equipment	Total	Total Incl O&P
0010	REFRIGERATORS									
1200	Blood bank, 28.6 C.F. emergency signal				Ea.	9,650			9,650	10,600
1210	Reach-in, 16.9 C.F.				"	8,525			8,525	9,375

11 53 Laboratory Equipment

11 53 33 – Emergency Safety Appliances

11 53 33.13 Emergency Equipment		Crew	Daily Output	Labor-Hours	Unit	Material	Labor	Equipment	Total	Total Incl O&P
						2015 Bare Costs				
0010	EMERGENCY EQUIPMENT									
1400	Safety equipment, eye wash, hand held				Ea.	410			410	450
1450	Deluge shower				"	770			770	850

11 53 43 – Service Fittings and Accessories

11 53 43.13 Fittings

		Crew	Daily Output	Labor-Hours	Unit	Material	Labor	Equipment	Total	Total Incl O&P
0010	FITTINGS									
1600	Sink, one piece plastic, flask wash, hose, free standing	1 Plum	1.60	5	Ea.	1,950	294		2,244	2,600
1610	Epoxy resin sink, 25" x 16" x 10"	"	2	4	"	221	235		456	610
1950	Utility table, acid resistant top with drawers	2 Carp	30	.533	L.F.	164	25		189	221
8000	Alternate pricing method: as percent of lab furniture									
8050	Installation, not incl. plumbing & duct work				% Furn.				22%	22%
8100	Plumbing, final connections, simple system								10%	10%
8110	Moderately complex system								15%	15%
8120	Complex system								20%	20%
8150	Electrical, simple system								10%	10%
8160	Moderately complex system								20%	20%
8170	Complex system								35%	35%

11 53 53 – Biological Safety Cabinets

11 53 53.10 Pharmacy Cabinets

		Crew	Daily Output	Labor-Hours	Unit	Material	Labor	Equipment	Total	Total Incl O&P
0010	PHARMACY CABINETS, vertical flow									
0100	Class II, type B2, 6' L	2 Carp	1.50	10.667	Ea.	13,200	500		13,700	15,300

11 57 Vocational Shop Equipment

11 57 10 – Shop Equipment

11 57 10.10 Vocational School Shop Equipment

		Crew	Daily Output	Labor-Hours	Unit	Material	Labor	Equipment	Total	Total Incl O&P
0010	VOCATIONAL SCHOOL SHOP EQUIPMENT									
0020	Benches, work, wood, average	2 Carp	5	3.200	Ea.	635	150		785	945
0100	Metal, average		5	3.200		550	150		700	850
0400	Combination belt & disc sander, 6"		4	4		1,650	188		1,838	2,125
0700	Drill press, floor mounted, 12", 1/2 H.P.		4	4		415	188		603	765
0800	Dust collector, not incl. ductwork, 6" diameter	1 Shee	1.10	7.273		4,650	405		5,055	5,775
0810	Dust collector bag, 20" diameter	"	5	1.600		440	89.50		529.50	625
1000	Grinders, double wheel, 1/2 H.P.	2 Carp	5	3.200		217	150		367	485
1300	Jointer, 4", 3/4 H.P.		4	4		1,375	188		1,563	1,800
1600	Kilns, 16 C.F., to 2000°		4	4		1,475	188		1,663	1,925
1900	Lathe, woodworking, 10", 1/2 H.P.		4	4		545	188		733	910
2200	Planer, 13" x 6"		4	4		1,100	188		1,288	1,500
2500	Potter's wheel, motorized		4	4		1,125	188		1,313	1,550
2800	Saws, band, 14", 3/4 H.P.		4	4		930	188		1,118	1,325
3100	Metal cutting band saw, 14"		4	4		2,525	188		2,713	3,075
3400	Radial arm saw, 10", 2 H.P.		4	4		1,375	188		1,563	1,800
3700	Scroll saw, 24"		4	4		585	188		773	955
4000	Table saw, 10", 3 H.P.		4	4		2,725	188		2,913	3,300
4300	Welder AC arc, 30 amp capacity		4	4		3,175	188		3,363	3,800

11 61 Broadcast, Theater, and Stage Equipment

11 61 23 – Folding and Portable Stages

11 61 23.10 Portable Stages		Crew	Daily Output	Labor-Hours	Unit	Material	2015 Bare Costs Labor	Equipment	Total	Total Incl O&P
0010	**PORTABLE STAGES**									
1500	Flooring, portable oak parquet, 3′ x 3′ sections				S.F.	13.55			13.55	14.90
1600	Cart to carry 225 S.F. of flooring				Ea.	405			405	445
5000	Stages, portable with steps, folding legs, stock, 8″ high				SF Stg.	32.50			32.50	36
5100	16″ high					49.50			49.50	54.50
5200	32″ high					53.50			53.50	59
5300	40″ high					60			60	66
6000	Telescoping platforms, extruded alum., straight, minimum	4 Carp	157	.204		34.50	9.55		44.05	53
6100	Maximum		77	.416		48	19.50		67.50	84.50
6500	Pie-shaped, minimum		150	.213		74	10		84	98
6600	Maximum		70	.457		83	21.50		104.50	126
6800	For 3/4″ plywood covered deck, deduct					4.21			4.21	4.63
7000	Band risers, steel frame, plywood deck, minimum	4 Carp	275	.116		31	5.45		36.45	43
7100	Maximum	"	138	.232		69.50	10.90		80.40	94.50
7500	Chairs for above, self-storing, minimum	2 Carp	43	.372	Ea.	110	17.45		127.45	150
7600	Maximum	"	40	.400	"	194	18.80		212.80	245

11 61 33 – Rigging Systems and Controls

11 61 33.10 Controls		Crew	Daily Output	Labor-Hours	Unit	Material	Labor	Equipment	Total	Total Incl O&P
0010	**CONTROLS**									
0050	Control boards with dimmers and breakers, minimum	1 Elec	1	8	Ea.	12,600	440		13,040	14,600
0100	Average		.50	16		39,700	875		40,575	45,100
0150	Maximum		.20	40		129,000	2,200		131,200	145,500
8000	Rule of thumb: total stage equipment, minimum	4 Carp	100	.320	SF Stg.	98.50	15		113.50	133
8100	Maximum	"	25	1.280	"	555	60		615	710

11 61 43 – Stage Curtains

11 61 43.10 Curtains		Crew	Daily Output	Labor-Hours	Unit	Material	Labor	Equipment	Total	Total Incl O&P
0010	**CURTAINS**									
0500	Curtain track, straight, light duty	2 Carp	20	.800	L.F.	27.50	37.50		65	92
0600	Heavy duty		18	.889		62	41.50		103.50	137
0700	Curved sections		12	1.333		177	62.50		239.50	298
1000	Curtains, velour, medium weight		600	.027	S.F.	8	1.25		9.25	10.85
1150	Silica based yarn, inherently fire retardant		50	.320	"	15.40	15		30.40	41.50

11 62 Musical Equipment

11 62 16 – Carillons

11 62 16.10 Bell Tower Equipment		Crew	Daily Output	Labor-Hours	Unit	Material	Labor	Equipment	Total	Total Incl O&P
0010	**BELL TOWER EQUIPMENT**									
0300	Carillon, 4 octave (48 bells), with keyboard				System	996,000			996,000	1,095,500
0320	2 octave (24 bells)					468,500			468,500	515,500
0340	3 to 4 bell peal, minimum					117,000			117,000	129,000
0360	Maximum					703,000			703,000	773,000
0380	Cast bronze bell, average				Ea.	105,500			105,500	116,000
0400	Electronic, digital, minimum					17,600			17,600	19,300
0410	With keyboard, maximum					88,000			88,000	96,500

11 66 13 – Exercise Equipment

11 66 13.10 Physical Training Equipment

	11 66 13.10 Physical Training Equipment	Crew	Daily Output	Labor-Hours	Unit	Material	2015 Bare Costs Labor	Equipment	Total	Total Incl O&P
0010	**PHYSICAL TRAINING EQUIPMENT**									
0020	Abdominal rack, 2 board capacity				Ea.	490			490	535
0050	Abdominal board, upholstered					665			665	730
0200	Bicycle trainer, minimum					485			485	535
0300	Deluxe, electric					4,350			4,350	4,775
0400	Barbell set, chrome plated steel, 25 lb.					252			252	277
0420	100 lb.					465			465	510
0450	200 lb.					705			705	780
0500	Weight plates, cast iron, per lb.				Lb.	5.25			5.25	5.80
0520	Storage rack, 10 station				Ea.	910			910	1,000
0600	Circuit training apparatus, 12 machines minimum	2 Clab	1.25	12.800	Set	29,000	480		29,480	32,700
0700	Average		1	16		35,600	600		36,200	40,200
0800	Maximum		.75	21.333		42,200	800		43,000	47,700
0820	Dumbbell set, cast iron, with rack and 5 pair					625			625	690
0900	Squat racks	2 Clab	5	3.200	Ea.	900	120		1,020	1,175
1200	Multi-station gym machine, 5 station					5,050			5,050	5,575
1250	9 station					11,700			11,700	12,900
1280	Rowing machine, hydraulic					1,775			1,775	1,950
1300	Treadmill, manual					1,100			1,100	1,225
1320	Motorized					3,550			3,550	3,900
1340	Electronic					3,750			3,750	4,125
1360	Cardio-testing					4,600			4,600	5,050
1400	Treatment/massage tables, minimum					595			595	655
1420	Deluxe, with accessories					730			730	800
4150	Exercise equipment, bicycle trainer					760			760	835
4180	Chinning bar, adjustable, wall mounted	1 Carp	5	1.600		212	75		287	355
4200	Exercise ladder, 16' x 1'-7", suspended	L-2	3	5.333		1,375	220		1,595	1,850
4210	High bar, floor plate attached	1 Carp	4	2		2,250	94		2,344	2,625
4240	Parallel bars, adjustable		4	2		1,700	94		1,794	2,025
4270	Uneven parallel bars, adjustable		4	2		3,225	94		3,319	3,700
4280	Wall mounted, adjustable	L-2	1.50	10.667	Set	865	440		1,305	1,675
4300	Rope, ceiling mounted, 18' long	1 Carp	3.66	2.186	Ea.	190	103		293	375
4330	Side horse, vaulting		5	1.600		1,375	75		1,450	1,625
4360	Treadmill, motorized, deluxe, training type		5	1.600		3,850	75		3,925	4,350
4390	Weight lifting multi-station, minimum	2 Clab	1	16		335	600		935	1,350
4450	Maximum	"	.50	32		14,900	1,200		16,100	18,400

11 66 23 – Gymnasium Equipment

11 66 23.13 Basketball Equipment

	11 66 23.13 Basketball Equipment	Crew	Daily Output	Labor-Hours	Unit	Material	Labor	Equipment	Total	Total Incl O&P
0010	**BASKETBALL EQUIPMENT**									
1000	Backstops, wall mtd., 6' extended, fixed, minimum	L-2	1	16	Ea.	1,375	660		2,035	2,600
1100	Maximum		1	16		1,900	660		2,560	3,175
1200	Swing up, minimum		1	16		1,450	660		2,110	2,650
1250	Maximum		1	16		2,850	660		3,510	4,200
1300	Portable, manual, heavy duty, spring operated		1.90	8.421		12,700	345		13,045	14,600
1400	Ceiling suspended, stationary, minimum		.78	20.513		4,050	845		4,895	5,850
1450	Fold up, with accessories, maximum		.40	40		6,075	1,650		7,725	9,375
1600	For electrically operated, add	1 Elec	1	8		2,350	440		2,790	3,250
5800	Wall pads, 1-1/2" thick, standard (not fire rated)	2 Carp	640	.025	S.F.	6.25	1.17		7.42	8.80

11 66 23.19 Boxing Ring

	11 66 23.19 Boxing Ring	Crew	Daily Output	Labor-Hours	Unit	Material	Labor	Equipment	Total	Total Incl O&P
0010	**BOXING RING**									
4100	Elevated, 22' x 22'	L-4	.10	240	Ea.	7,075	10,600		17,675	25,100
4110	For cellular plastic foam padding, add		.10	240		1,025	10,600		11,625	18,500

11 66 Athletic Equipment

11 66 23 – Gymnasium Equipment

11 66 23.19 Boxing Ring		Crew	Daily Output	Labor-Hours	Unit	2015 Bare Costs Material	Labor	Equipment	Total	Total Incl O&P
4120	Floor level, including posts and ropes only, 20′ x 20′	L-4	.80	30	Ea.	4,550	1,325		5,875	7,175
4130	Canvas, 30′ x 30′		5	4.800		1,275	212		1,487	1,750

11 66 23.47 Gym Mats

		Crew	Daily Output	Labor-Hours	Unit	Material	Labor	Equipment	Total	Total Incl O&P
0010	**GYM MATS**									
5500	2″ thick, naugahyde covered				S.F.	3.71			3.71	4.08
5600	Vinyl/nylon covered					8.05			8.05	8.85
6000	Wrestling mats, 1″ thick, heavy duty					5.60			5.60	6.15

11 66 43 – Interior Scoreboards

11 66 43.10 Scoreboards

		Crew	Daily Output	Labor-Hours	Unit	Material	Labor	Equipment	Total	Total Incl O&P
0010	**SCOREBOARDS**									
7000	Baseball, minimum	R-3	1.30	15.385	Ea.	4,125	835	106	5,066	5,975
7200	Maximum		.05	400		18,100	21,700	2,750	42,550	56,500
7300	Football, minimum		.86	23.256		5,325	1,275	160	6,760	8,025
7400	Maximum		.20	100		15,900	5,425	690	22,015	26,700
7500	Basketball (one side), minimum		2.07	9.662		2,400	525	66.50	2,991.50	3,550
7600	Maximum		.30	66.667		3,525	3,625	460	7,610	10,000
7700	Hockey-basketball (four sides), minimum		.25	80		5,675	4,350	550	10,575	13,600
7800	Maximum		.15	133		5,750	7,250	920	13,920	18,600

11 66 53 – Gymnasium Dividers

11 66 53.10 Divider Curtains

		Crew	Daily Output	Labor-Hours	Unit	Material	Labor	Equipment	Total	Total Incl O&P
0010	**DIVIDER CURTAINS**									
4500	Gym divider curtain, mesh top, vinyl bottom, manual	L-4	500	.048	S.F.	9	2.12		11.12	13.35
4700	Electric roll up	L-7	400	.070	″	12.05	3.18		15.23	18.40

11 67 Recreational Equipment

11 67 13 – Bowling Alley Equipment

11 67 13.10 Bowling Alleys

		Crew	Daily Output	Labor-Hours	Unit	Material	Labor	Equipment	Total	Total Incl O&P
0010	**BOWLING ALLEYS** Including alley, pinsetter, scorer,									
0020	Counters and misc. supplies, minimum	4 Carp	.20	160	Lane	42,200	7,500		49,700	58,500
0150	Average		.19	168		46,400	7,900		54,300	64,000
0300	Maximum		.18	177		53,500	8,350		61,850	72,000
0400	Combo table ball rack, add					1,150			1,150	1,275
0600	For automatic scorer, add, minimum					8,325			8,325	9,150
0700	Maximum					10,000			10,000	11,000

11 67 23 – Shooting Range Equipment

11 67 23.10 Shooting Range

		Crew	Daily Output	Labor-Hours	Unit	Material	Labor	Equipment	Total	Total Incl O&P
0010	**SHOOTING RANGE** Incl. bullet traps, target provisions, controls,									
0100	Separators, ceiling system, etc. Not incl. structural shell									
0200	Commercial	L-9	.64	56.250	Point	28,300	2,425		30,725	35,300
0300	Law enforcement		.28	128		38,500	5,525		44,025	51,500
0400	National Guard armories		.71	50.704		20,700	2,175		22,875	26,500
0500	Reserve training centers		.71	50.704		15,900	2,175		18,075	21,200
0600	Schools and colleges		.32	112		35,400	4,825		40,225	47,100
0700	Major academies		.19	189		52,500	8,150		60,650	71,500
0800	For acoustical treatment, add					10%	10%			
0900	For lighting, add					28%	25%			
1000	For plumbing, add					5%	5%			
1100	For ventilating system, add, minimum					40%	40%			
1200	Add, average					25%	25%			

11 67 Recreational Equipment

11 67 23 – Shooting Range Equipment

11 67 23.10 Shooting Range		Crew	Daily Output	Labor-Hours	Unit	Material	Labor	Equipment	Total	Total Incl O&P
						2015 Bare Costs				
1300	Add, maximum				Point	35%	35%			

11 68 Play Field Equipment and Structures

11 68 13 – Playground Equipment

11 68 13.10 Free-Standing Playground Equipment

		Crew	Daily Output	Labor-Hours	Unit	Material	Labor	Equipment	Total	Total Incl O&P
0010	**FREE-STANDING PLAYGROUND EQUIPMENT** See also individual items									
0200	Bike rack, 10′ long, permanent G	B-1	12	2	Ea.	425	76.50		501.50	595
0240	Climber, arch, 6′ high, 12′ long, 5′ wide		4	6		775	230		1,005	1,225
0260	Fitness trail, with signs, 9 to 10 stations, treated pine, minimum		.25	96		7,200	3,675		10,875	14,000
0270	Maximum		.17	141		16,900	5,400		22,300	27,500
0280	Metal, minimum		.25	96		12,000	3,675		15,675	19,200
0285	Maximum		.17	141		30,000	5,400		35,400	41,900
0300	Redwood, minimum		.25	96		10,900	3,675		14,575	18,000
0310	Maximum		.17	141		15,500	5,400		20,900	25,900
0320	16 to 20 station, treated pine, minimum		.17	141		11,400	5,400		16,800	21,500
0330	Maximum		.13	184		12,500	7,075		19,575	25,300
0340	Metal, minimum		.17	141		12,700	5,400		18,100	22,800
0350	Maximum		.13	184		16,100	7,075		23,175	29,300
0360	Redwood, minimum		.17	141		16,600	5,400		22,000	27,200
0370	Maximum		.13	184		25,500	7,075		32,575	39,600
0392	Upper body warm-up station		2.60	9.231		2,025	355		2,380	2,800
0394	Bench stepper station		2.60	9.231		2,875	355		3,230	3,725
0396	Standing push up station		2.60	9.231		1,075	355		1,430	1,775
0398	Upper body stretch station		2.60	9.231		1,700	355		2,055	2,450
0400	Horizontal monkey ladder, 14′ long, 6′ high		4	6		905	230		1,135	1,375
0590	Parallel bars, 10′ long		4	6		320	230		550	725
0600	Posts, tether ball set, 2-3/8″ O.D.		12	2		455	76.50		531.50	625
0800	Poles, multiple purpose, 10′-6″ long		12	2	Pr.	179	76.50		255.50	320
1000	Ground socket for movable posts, 2-3/8″ post		10	2.400		102	92		194	263
1100	3-1/2″ post		10	2.400		167	92		259	335
1300	See-saw, spring, steel, 2 units		6	4	Ea.	740	153		893	1,075
1400	4 units		4	6		1,275	230		1,505	1,800
1500	6 units		3	8		1,675	305		1,980	2,350
1700	Shelter, fiberglass golf tee, 3 person		4.60	5.217		4,375	200		4,575	5,125
1900	Slides, stainless steel bed, 12′ long, 6′ high		3	8		3,325	305		3,630	4,150
2000	20′ long, 10′ high		2	12		3,675	460		4,135	4,800
2200	Swings, plain seats, 8′ high, 4 seats		2	12		1,150	460		1,610	2,025
2300	8 seats		1.30	18.462		2,175	705		2,880	3,525
2500	12′ high, 4 seats		2	12		2,150	460		2,610	3,125
2600	8 seats		1.30	18.462		3,525	705		4,230	5,025
2800	Whirlers, 8′ diameter		3	8		2,775	305		3,080	3,550
2900	10′ diameter		3	8		6,275	305		6,580	7,400

11 68 13.20 Modular Playground

		Crew	Daily Output	Labor-Hours	Unit	Material	Labor	Equipment	Total	Total Incl O&P
0010	**MODULAR PLAYGROUND** Basic components									
0100	Deck, square, steel, 48″ x 48″	B-1	1	24	Ea.	520	920		1,440	2,075
0110	Recycled polyurethane		1	24		515	920		1,435	2,075
0120	Triangular, steel, 48″ side		1	24		680	920		1,600	2,250
0130	Post, steel, 5″ square		18	1.333	L.F.	40.50	51		91.50	128
0140	Aluminum, 2-3/8″ square		20	1.200		39.50	46		85.50	119
0150	5″ square		18	1.333		40.50	51		91.50	128
0160	Roof, square poly, 54″ side		18	1.333	Ea.	1,400	51		1,451	1,625

11 68 Play Field Equipment and Structures

11 68 13 – Playground Equipment

11 68 13.20 Modular Playground		Crew	Daily Output	Labor-Hours	Unit	2015 Bare Costs Material	Labor	Equipment	Total	Total Incl O&P
0170	Wheelchair transfer module, for 3′ high deck	B-1	3	8	Ea.	2,950	305		3,255	3,750
0180	Guardrail, pipe, 36″ high		60	.400	L.F.	198	15.30		213.30	243
0190	Steps, deck-to-deck, 3 – 8″ steps		8	3	Ea.	1,125	115		1,240	1,425
0200	Activity panel, crawl through panel		2	12		500	460		960	1,300
0210	Alphabet/spelling panel		2	12		555	460		1,015	1,375
0360	With guardrails		3	8		1,850	305		2,155	2,525
0370	Crawl tunnel, straight, 56″ long		4	6		1,200	230		1,430	1,700
0380	90°, 4′ long		4	6		1,400	230		1,630	1,925
1200	Slide, tunnel, for 56″ high deck		8	3		1,750	115		1,865	2,125
1210	Straight, poly		8	3		390	115		505	620
1220	Stainless steel, 54″ high deck		6	4		690	153		843	1,000
1230	Curved, poly, 40″ high deck		6	4		865	153		1,018	1,200
1240	Spiral slide, 56″ - 72″ high		5	4.800		4,475	184		4,659	5,225
1300	Ladder, vertical, for 24″ - 72″ high deck		5	4.800		555	184		739	910
1310	Horizontal, 8′ long		5	4.800		695	184		879	1,075
1320	Corkscrew climber, 6′ high		3	8		1,125	305		1,430	1,750
1330	Fire pole for 72″ high deck		6	4		655	153		808	970
1340	Bridge, ring climber, 8′ long		4	6		1,900	230		2,130	2,450
1350	Suspension		4	6	L.F.	360	230		590	770

11 68 16 – Play Structures

11 68 16.10 Handball/Squash Court		Crew	Daily Output	Labor-Hours	Unit	Material	Labor	Equipment	Total	Total Incl O&P
0010	**HANDBALL/SQUASH COURT**, outdoor									
0900	Handball or squash court, outdoor, wood	2 Carp	.50	32	Ea.	5,050	1,500		6,550	8,025
1000	Masonry handball/squash court	D-1	.30	53.333	"	25,100	2,250		27,350	31,300

11 68 16.30 Platform/Paddle Tennis Court		Crew	Daily Output	Labor-Hours	Unit	Material	Labor	Equipment	Total	Total Incl O&P
0010	**PLATFORM/PADDLE TENNIS COURT** Complete with lighting, etc.									
0100	Aluminum slat deck with aluminum frame	B-1	.08	300	Court	60,500	11,500		72,000	86,000
0500	Aluminum slat deck with wood frame	C-1	.12	266		63,500	11,900		75,400	89,000
0800	Aluminum deck heater, add	B-1	1.18	20.339		2,575	780		3,355	4,125
0900	Douglas fir planking with wood frame 2″ x 6″ x 30′	C-1	.12	266		59,500	11,900		71,400	85,000
1000	Plywood deck with steel frame		.12	266		59,500	11,900		71,400	85,000
1100	Steel slat deck with wood frame		.12	266		39,300	11,900		51,200	63,000

11 68 33 – Athletic Field Equipment

11 68 33.13 Football Field Equipment		Crew	Daily Output	Labor-Hours	Unit	Material	Labor	Equipment	Total	Total Incl O&P
0010	**FOOTBALL FIELD EQUIPMENT**									
0020	Goal posts, steel, football, double post	B-1	1.50	16	Pr.	3,800	610		4,410	5,175
0100	Deluxe, single post		1.50	16		2,625	610		3,235	3,900
0300	Football, convertible to soccer		1.50	16		2,825	610		3,435	4,125
0500	Soccer, regulation		2	12		1,425	460		1,885	2,300

11 71 Medical Sterilizing Equipment

11 71 10 – Medical Sterilizers & Distillers

11 71 10.10 Sterilizers and Distillers		Crew	Daily Output	Labor-Hours	Unit	2015 Bare Costs Material	Labor	Equipment	Total	Total Incl O&P
0010	**STERILIZERS AND DISTILLERS**									
0700	Distiller, water, steam heated, 50 gal. capacity	1 Plum	1.40	5.714	Ea.	19,200	335		19,535	21,700
3010	Portable, top loading, 105 – 135 degree C, 3 to 30 psi, 50 L chamber					10,000			10,000	11,000
3020	Stainless steel basket, 10.7" diam. x 11.8" H					223			223	245
3025	Stainless steel pail, 10.7" diam. x 10.7" H					243			243	267
3050	85 L chamber					14,300			14,300	15,700
3060	Stainless steel basket, 15.3" diam. x 11.5" H					305			305	335
3065	Stainless steel pail, 15.3" diam. x 11" H					425			425	470
5600	Sterilizers, floor loading, 26" x 62" x 42", single door, steam					122,000			122,000	134,500
5650	Double door, steam					206,000			206,000	227,000
5800	General purpose, 20" x 20" x 38", single door					13,100			13,100	14,400
6000	Portable, counter top, steam, minimum					3,550			3,550	3,900
6020	Maximum					4,325			4,325	4,750
6050	Portable, counter top, gas, 17" x 15" x 32-1/2"					39,800			39,800	43,700
6150	Manual washer/sterilizer, 16" x 16" x 26"	1 Plum	2	4		54,500	235		54,735	60,500
6200	Steam generators, electric 10 kW to 180 kW, freestanding									
6250	Minimum	1 Elec	3	2.667	Ea.	8,775	146		8,921	9,875
6300	Maximum	"	.70	11.429		29,200	625		29,825	33,100
8200	Bed pan washer-sanitizer	1 Plum	2	4		7,500	235		7,735	8,625

11 72 Examination and Treatment Equipment

11 72 13 – Examination Equipment

11 72 13.13 Examination Equipment		Crew	Daily Output	Labor-Hours	Unit	Material	Labor	Equipment	Total	Total Incl O&P
0010	**EXAMINATION EQUIPMENT**									
0300	Blood pressure unit, mercurial, wall				Ea.	152			152	167
0400	Diagnostic set, wall					780			780	860
4400	Scale, physician's, with height rod					320			320	350

11 72 53 – Treatment Equipment

11 72 53.13 Medical Treatment Equipment		Crew	Daily Output	Labor-Hours	Unit	Material	Labor	Equipment	Total	Total Incl O&P
0010	**MEDICAL TREATMENT EQUIPMENT**									
6300	Exam light, portable, 14" flexible arm				Ea.	197			197	217
6500	Surgery table, minor minimum	1 Sswk	.70	11.429		12,100	600		12,700	14,500
6520	Maximum	"	.50	16		20,200	840		21,040	23,800
6700	Surgical lights, doctor's office, single arm	2 Elec	2	8		2,475	440		2,915	3,400
6750	Dual arm	"	1	16		4,575	875		5,450	6,375

11 73 Patient Care Equipment

11 73 10 – Patient Treatment Equipment

11 73 10.10 Treatment Equipment		Crew	Daily Output	Labor-Hours	Unit	Material	Labor	Equipment	Total	Total Incl O&P
0010	**TREATMENT EQUIPMENT**									
0750	Exam room furnishings, average per room				Ea.	7,050			7,050	7,750
1800	Heat therapy unit, humidified, 26" x 78" x 28"				"	3,600			3,600	3,975
2100	Hubbard tank with accessories, stainless steel,									
2110	125 GPM at 45 psi water pressure				Ea.	27,200			27,200	29,900
2150	For electric overhead hoist, add					2,975			2,975	3,250
2900	K-Module for heat therapy, 20 oz. capacity, 75°F to 110°F					425			425	465
3600	Paraffin bath, 126°F, auto controlled					1,100			1,100	1,200
3900	Parallel bars for walking training, 12'-0"					1,450			1,450	1,600
4600	Station, dietary, medium, with ice					16,700			16,700	18,400

11 73 Patient Care Equipment

11 73 10 – Patient Treatment Equipment

11 73 10.10 Treatment Equipment		Crew	Daily Output	Labor-Hours	Unit	2015 Bare Costs Material	Labor	Equipment	Total	Total Incl O&P
4700	Medicine				Ea.	7,550			7,550	8,300
7000	Tables, physical therapy, walk off, electric	2 Carp	3	5.333		3,375	250		3,625	4,100
7150	Standard, vinyl top with base cabinets, minimum		3	5.333		1,025	250		1,275	1,525
7200	Maximum		2	8		5,625	375		6,000	6,800
7250	Table, hospital, adjustable height					1,200			1,200	1,325
8400	Whirlpool bath, mobile, sst, 18" x 24" x 60"					4,900			4,900	5,400
8450	Fixed, incl. mixing valves	1 Plum	2	4		9,700	235		9,935	11,100

11 73 10.20 Bariatric Equipment		Crew	Daily Output	Labor-Hours	Unit	Material	Labor	Equipment	Total	Total Incl O&P
0010	**BARIATRIC EQUIPMENT**									
5000	Patient lift, electric operated, arm style									
5110	400 lb. capacity				Ea.	2,000			2,000	2,200
5120	450 lb. capacity					3,050			3,050	3,350
5130	600 lb. capacity					3,200			3,200	3,525
5140	700 lb. capacity					5,050			5,050	5,550
5150	1,000 lb. capacity					10,800			10,800	11,900
5200	Overhead, 4-post, 1,000 lb. capacity					10,400			10,400	11,500
5300	Overhead, track type, 450 lb. capacity, not including track					3,000			3,000	3,275
5500	For fabric sling, add					315			315	345
5550	For digital scale, add					780			780	855

11 74 Dental Equipment

11 74 10 – Dental Office Equipment

11 74 10.10 Diagnostic and Treatment Equipment		Crew	Daily Output	Labor-Hours	Unit	Material	Labor	Equipment	Total	Total Incl O&P
0010	**DIAGNOSTIC AND TREATMENT EQUIPMENT**									
0020	Central suction system, minimum	1 Plum	1.20	6.667	Ea.	1,575	390		1,965	2,325
0100	Maximum	"	.90	8.889		4,450	520		4,970	5,725
0300	Air compressor, minimum	1 Skwk	.80	10		3,025	485		3,510	4,125
0400	Maximum		.50	16		8,950	780		9,730	11,100
0600	Chair, electric or hydraulic, minimum		.50	16		2,250	780		3,030	3,750
0700	Maximum		.25	32		7,825	1,550		9,375	11,100
0800	Doctor's/assistant's stool, minimum					254			254	279
0850	Maximum					715			715	785
1000	Drill console with accessories, minimum	1 Skwk	1.60	5		2,100	243		2,343	2,700
1100	Maximum		1.60	5		4,900	243		5,143	5,800
2000	Light, ceiling mounted, minimum		8	1		1,175	48.50		1,223.50	1,350
2100	Maximum		8	1		2,025	48.50		2,073.50	2,300
2200	Unit light, minimum	2 Skwk	5.33	3.002		735	146		881	1,050
2210	Maximum		5.33	3.002		1,575	146		1,721	1,950
2220	Track light, minimum		3.20	5		1,575	243		1,818	2,125
2230	Maximum		3.20	5		2,650	243		2,893	3,325
2300	Sterilizers, steam portable, minimum					1,300			1,300	1,425
2350	Maximum					10,500			10,500	11,600
2600	Steam, institutional					3,275			3,275	3,600
2650	Dry heat, electric, portable, 3 trays					1,225			1,225	1,350
2700	Ultra-sonic cleaner, portable, minimum					445			445	490
2750	Maximum (institutional)					1,325			1,325	1,475
3000	X-ray unit, wall, minimum	1 Skwk	4	2		2,350	97.50		2,447.50	2,725
3010	Maximum		4	2		4,050	97.50		4,147.50	4,600
3100	Panoramic unit		.60	13.333		15,700	650		16,350	18,400
3105	Deluxe, minimum	2 Skwk	1.60	10		16,700	485		17,185	19,100
3110	Maximum	"	1.60	10		41,000	485		41,485	45,900

11 74 Dental Equipment

11 74 10 – Dental Office Equipment

11 74 10.10 Diagnostic and Treatment Equipment		Crew	Daily Output	Labor-Hours	Unit	Material	2015 Bare Costs Labor	Equipment	Total	Total Incl O&P
3500	Developers, X-ray, average	1 Plum	5.33	1.501	Ea.	5,000	88		5,088	5,650
3600	Maximum	"	5.33	1.501	↓	8,200	88		8,288	9,125

11 76 Operating Room Equipment

11 76 10 – Operating Room Equipment

11 76 10.10 Surgical Equipment

		Crew	Daily Output	Labor-Hours	Unit	Material	Labor	Equipment	Total	Total Incl O&P
0010	**SURGICAL EQUIPMENT**									
5000	Scrub, surgical, stainless steel, single station, minimum	1 Plum	3	2.667	Ea.	4,300	157		4,457	4,975
5100	Maximum					6,975			6,975	7,675
6550	Major surgery table, minimum	1 Sswk	.50	16		26,300	840		27,140	30,400
6570	Maximum		.50	16		28,300	840		29,140	32,700
6600	Hydraulic, hand-held control, general surgery		.60	13.333		30,400	700		31,100	34,700
6650	Stationary, universal	↓	.50	16		39,200	840		40,040	44,600
6800	Surgical lights, major operating room, dual head, minimum	2 Elec	1	16		4,375	875		5,250	6,150
6850	Maximum		1	16		30,000	875		30,875	34,500
6900	Ceiling mount articulation, single arm	↓	1	16	↓	3,800	875		4,675	5,525

11 77 Radiology Equipment

11 77 10 – Radiology Equipment

11 77 10.10 X-Ray Equipment

		Crew	Daily Output	Labor-Hours	Unit	Material	Labor	Equipment	Total	Total Incl O&P
0010	**X-RAY EQUIPMENT**									
8700	X-ray, mobile, minimum				Ea.	16,800			16,800	18,500
8750	Maximum					77,500			77,500	85,500
8900	Stationary, minimum					43,200			43,200	47,500
8950	Maximum					224,500			224,500	247,000
9150	Developing processors, minimum					4,750			4,750	5,225
9200	Maximum				↓	12,900			12,900	14,100

11 78 Mortuary Equipment

11 78 13 – Mortuary Refrigerators

11 78 13.10 Mortuary and Autopsy Equipment

		Crew	Daily Output	Labor-Hours	Unit	Material	Labor	Equipment	Total	Total Incl O&P
0010	**MORTUARY AND AUTOPSY EQUIPMENT**									
0015	Autopsy table, standard	1 Plum	1	8	Ea.	9,750	470		10,220	11,400
0020	Deluxe	"	.60	13.333		15,100	785		15,885	17,800
3200	Mortuary refrigerator, end operated, 2 capacity					12,300			12,300	13,500
3300	6 capacity				↓	22,200			22,200	24,400

11 78 16 – Crematorium Equipment

11 78 16.10 Crematory

		Crew	Daily Output	Labor-Hours	Unit	Material	Labor	Equipment	Total	Total Incl O&P
0010	**CREMATORY**									
1500	Crematory, not including building, 1 place	Q-3	.20	160	Ea.	72,500	8,950		81,450	93,500
1750	2 place	"	.10	320	"	103,500	17,900		121,400	142,000

11 81 Facility Maintenance Equipment

11 81 19 – Vacuum Cleaning Systems

11 81 19.10 Vacuum Cleaning		Crew	Daily Output	Labor-Hours	Unit	Material	2015 Bare Costs Labor	Equipment	Total	Total Incl O&P
0010	VACUUM CLEANING									
0020	Central, 3 inlet, residential	1 Skwk	.90	8.889	Total	1,075	430		1,505	1,875
0200	Commercial		.70	11.429		1,225	555		1,780	2,250
0400	5 inlet system, residential		.50	16		1,500	780		2,280	2,925
0600	7 inlet system, commercial		.40	20		1,700	975		2,675	3,450
0800	9 inlet system, residential		.30	26.667		3,750	1,300		5,050	6,225
4010	Rule of thumb: First 1200 S.F., installed								1,425	1,575
4020	For each additional S.F., add				S.F.				.26	.26

11 82 Facility Solid Waste Handling Equipment

11 82 19 – Packaged Incinerators

11 82 19.10 Packaged Gas Fired Incinerators		Crew	Daily Output	Labor-Hours	Unit	Material	Labor	Equipment	Total	Total Incl O&P
0010	PACKAGED GAS FIRED INCINERATORS									
4400	Incinerator, gas, not incl. chimney, elec. or pipe, 50#/hr., minimum	Q-3	.80	40	Ea.	38,500	2,225		40,725	45,800
4420	Maximum		.70	45.714		40,500	2,550		43,050	48,500
4440	200 lb. per hr., minimum (batch type)		.60	53.333		68,000	2,975		70,975	79,000
4460	Maximum (with feeder)		.50	64		76,000	3,575		79,575	89,000
4480	400 lb. per hr., minimum (batch type)		.30	106		79,000	5,950		84,950	96,500
4500	Maximum (with feeder)		.25	128		101,000	7,150		108,150	122,500
4520	800 lb. per hr., with feeder, minimum		.20	160		121,500	8,950		130,450	147,500
4540	Maximum		.17	188		182,000	10,500		192,500	217,000
4560	1,200 lb. per hr., with feeder, minimum		.15	213		154,000	11,900		165,900	187,500
4580	Maximum		.11	290		200,000	16,300		216,300	245,500
4600	2,000 lb. per hr., with feeder, minimum		.10	320		405,000	17,900		422,900	473,500
4620	Maximum		.05	640		607,000	35,800		642,800	724,000
4700	For heat recovery system, add, minimum		.25	128		81,000	7,150		88,150	100,000
4710	Add, maximum		.11	290		253,000	16,300		269,300	304,000
4720	For automatic ash conveyer, add		.50	64		33,700	3,575		37,275	42,700
4750	Large municipal incinerators, incl. stack, minimum		.25	128	Ton/day	20,500	7,150		27,650	33,700
4850	Maximum		.10	320	"	54,500	17,900		72,400	88,000

11 82 26 – Facility Waste Compactors

11 82 26.10 Compactors		Crew	Daily Output	Labor-Hours	Unit	Material	Labor	Equipment	Total	Total Incl O&P
0010	COMPACTORS									
0020	Compactors, 115 volt, 250#/hr., chute fed	L-4	1	24	Ea.	12,100	1,050		13,150	15,000
0100	Hand fed		2.40	10		15,100	445		15,545	17,300
0300	Multi-bag, 230 volt, 600#/hr., chute fed		1	24		15,100	1,050		16,150	18,300
0400	Hand fed		1	24		15,300	1,050		16,350	18,500
0500	Containerized, hand fed, 2 to 6 C.Y. containers, 250#/hr.		1	24		15,100	1,050		16,150	18,300
0550	For chute fed, add per floor		1	24		1,400	1,050		2,450	3,250
1000	Heavy duty industrial compactor, 0.5 C.Y. capacity		1	24		10,100	1,050		11,150	12,800
1050	1.0 C.Y. capacity		1	24		15,200	1,050		16,250	18,400
1100	3.0 C.Y. capacity		.50	48		25,900	2,125		28,025	32,000
1150	5.0 C.Y. capacity		.50	48		32,600	2,125		34,725	39,300
1200	Combination shredder/compactor (5,000 lb./hr.)		.50	48		63,500	2,125		65,625	73,000
1400	For handling hazardous waste materials, 55 gallon drum packer, std.					19,700			19,700	21,700
1410	55 gallon drum packer w/HEPA filter					24,600			24,600	27,100
1420	55 gallon drum packer w/charcoal & HEPA filter					32,800			32,800	36,100
1430	All of the above made explosion proof, add					1,450			1,450	1,575
5500	Shredder, municipal use, 35 ton per hour					304,500			304,500	335,000
5600	60 ton per hour					648,500			648,500	713,500

11 82 Facility Solid Waste Handling Equipment

11 82 26 – Facility Waste Compactors

11 82 26.10 Compactors		Crew	Daily Output	Labor-Hours	Unit	Material	Labor	Equipment	Total	Total Incl O&P
						2015 Bare Costs				
5750	Shredder & baler, 50 ton per day				Ea.	608,000			608,000	669,000
5800	Shredder, industrial, minimum					24,000			24,000	26,400
5850	Maximum					128,500			128,500	141,500
5900	Baler, industrial, minimum					9,625			9,625	10,600
5950	Maximum				↓	560,500			560,500	616,500
6000	Transfer station compactor, with power unit									
6050	and pedestal, not including pit, 50 ton per hour				Ea.	192,500			192,500	211,500

11 82 39 – Medical Waste Disposal Systems

11 82 39.10 Off-Site Disposal		Crew	Daily Output	Labor-Hours	Unit	Material	Labor	Equipment	Total	Total Incl O&P
0010	**OFF-SITE DISPOSAL**									
0100	Medical waste disposal, Red Bag system, pick up & treat, 200 lb. per week				Week	177			177	195
0110	Per month				Month	700			700	770
0150	Red bags, 7-10 gal., 1.2 mil, pkg of 500				Ea.	65.50			65.50	72.50
0200	15 gal., package of 250					61.50			61.50	68
0250	33 gal., package of 250					64			64	70
0300	45 gal., package of 100				↓	54.50			54.50	60

11 82 39.20 Disposal Carts		Crew	Daily Output	Labor-Hours	Unit	Material	Labor	Equipment	Total	Total Incl O&P
0010	**DISPOSAL CARTS**									
2010	Medical waste disposal cart, HDPE, w/lid, 28 gal. capacity				Ea.	261			261	287
2020	96 gal. capacity					305			305	335
2030	150 gal. capacity, low profile					470			470	515
2040	200 gal. capacity				↓	910			910	1,000

11 82 39.30 Medical Waste Sanitizers		Crew	Daily Output	Labor-Hours	Unit	Material	Labor	Equipment	Total	Total Incl O&P
0010	**MEDICAL WASTE SANITIZERS**									
2010	Small, hand loaded, 1.5 C.Y., 225 lb. capacity				Ea.	78,000			78,000	86,000
2020	Medium, cart loaded, 6.25 C.Y., 938 lb. capacity					104,000			104,000	114,500
2030	Large, cart loaded, 15 C.Y., 2250 lb. capacity					130,500			130,500	143,500
3010	Cart, aluminum, 75 lb. capacity					2,150			2,150	2,375
3020	95 lb. capacity					2,350			2,350	2,575
4010	Stainless steel, 173 lb. capacity					3,075			3,075	3,375
4020	232 lb. capacity					3,450			3,450	3,775
4030	Cart lift, hydraulic scissor type					6,100			6,100	6,700
4040	Portable aluminum ramp					1,725			1,725	1,900
4050	Fold-down steel tracks					1,375			1,375	1,525
4060	Pull-out drawer, small					6,575			6,575	7,225
4070	Medium					9,475			9,475	10,400
4080	Large				↓	13,400			13,400	14,800
5000	Medical waste treatment, sanitize, on-site									
5010	Less than 15,000 lb. per month				Lb.	.20			.20	.22
5020	Over 15,000 lb. per month				"	.16			.16	.18

11 91 Religious Equipment

11 91 13 – Baptisteries

11 91 13.10 Baptistry		Crew	Daily Output	Labor-Hours	Unit	Material	Labor	Equipment	Total	Total Incl O&P
						2015 Bare Costs				
0010	BAPTISTRY									
0150	Fiberglass, 3'-6" deep, x 13'-7" long,									
0160	steps at both ends, incl. plumbing, minimum	L-8	1	20	Ea.	5,725	985		6,710	7,900
0200	Maximum	"	.70	28.571		9,325	1,400		10,725	12,600
0250	Add for filter, heater and lights					1,850			1,850	2,050

11 91 23 – Sanctuary Equipment

11 91 23.10 Sanctuary Furnishings		Crew	Daily Output	Labor-Hours	Unit	Material	Labor	Equipment	Total	Total Incl O&P
0010	SANCTUARY FURNISHINGS									
0020	Altar, wood, custom design, plain	1 Carp	1.40	5.714	Ea.	2,550	268		2,818	3,275
0050	Deluxe	"	.20	40		12,400	1,875		14,275	16,700
0070	Granite or marble, average	2 Marb	.50	32		13,300	1,400		14,700	17,000
0090	Deluxe	"	.20	80		38,000	3,475		41,475	47,500
0100	Arks, prefabricated, plain	2 Carp	.80	20		9,750	940		10,690	12,300
0130	Deluxe, maximum	"	.20	80		138,500	3,750		142,250	158,500
0500	Reconciliation room, wood, prefabricated, single, plain	1 Carp	.60	13.333		3,175	625		3,800	4,525
0550	Deluxe		.40	20		8,750	940		9,690	11,200
0650	Double, plain		.40	20		6,375	940		7,315	8,550
0700	Deluxe		.20	40		19,000	1,875		20,875	24,000
1000	Lecterns, wood, plain		5	1.600		845	75		920	1,050
1100	Deluxe		2	4		6,125	188		6,313	7,050
2000	Pulpits, hardwood, prefabricated, plain		2	4		1,475	188		1,663	1,925
2100	Deluxe		1.60	5		10,100	235		10,335	11,500
2500	Railing, hardwood, average		25	.320	L.F.	208	15		223	254
3000	Seating, individual, oak, contour, laminated		21	.381	Person	178	17.90		195.90	226
3100	Cushion seat		21	.381		162	17.90		179.90	208
3200	Fully upholstered		21	.381		158	17.90		175.90	204
3300	Combination, self-rising		21	.381		305	17.90		322.90	365
3500	For cherry, add					30%				
5000	Wall cross, aluminum, extruded, 2" x 2" section	1 Carp	34	.235	L.F.	218	11.05		229.05	258
5150	4" x 4" section		29	.276		315	12.95		327.95	365
5300	Bronze, extruded, 1" x 2" section		31	.258		430	12.10		442.10	490
5350	2-1/2" x 2-1/2" section		34	.235		650	11.05		661.05	735
5450	Solid bar stock, 1/2" x 3" section		29	.276		855	12.95		867.95	960
5600	Fiberglass, stock		34	.235		147	11.05		158.05	179
5700	Stainless steel, 4" deep, channel section		29	.276		690	12.95		702.95	780
5800	4" deep box section		29	.276		940	12.95		952.95	1,050

11 97 Security Equipment

11 97 30 – Security Drawers

11 97 30.10 Pass Through Drawer		Crew	Daily Output	Labor-Hours	Unit	Material	Labor	Equipment	Total	Total Incl O&P
0010	PASS THROUGH DRAWER									
0100	Pass-thru drawer for personal items, 18" x 15" x 24"	1 Skwk	2	4	Ea.	2,800	195		2,995	3,400
0110	Including speakers	"	1.50	5.333	"	3,150	259		3,409	3,900

11 98 Detention Equipment

11 98 30 – Detention Cell Equipment

11 98 30.10 Cell Equipment		Crew	Daily Output	Labor-Hours	Unit	2015 Bare Costs Material	Labor	Equipment	Total	Total Incl O&P
0010	**CELL EQUIPMENT**									
3000	Toilet apparatus including wash basin, average	L-8	1.50	13.333	Ea.	3,400	655		4,055	4,825

Estimating Tips

General

- The items in this division are usually priced per square foot or each. Most of these items are purchased by the owner and installed by the contractor. Do not assume the items in Division 12 will be purchased and installed by the contractor. Check the specifications for responsibilities and include receiving, storage, installation, and mechanical and electrical hookups in the appropriate divisions.
- Some items in this division require some type of support system that is not usually furnished with the item. Examples of these systems include blocking for the attachment of casework and heavy drapery rods. The required blocking must be added to the estimate in the appropriate division.

Reference Numbers

Reference numbers are shown in shaded boxes at the beginning of some major classifications. These numbers refer to related items in the Reference Section. The reference information may be an estimating procedure, an alternate pricing method, or technical information.

Note: Not all subdivisions listed here necessarily appear in this publication. ■

12 05 Common Work Results for Furnishings

12 05 05 – Selective Demolition for Furnishings

12 05 05.10 Selective Demolition, Interiors		Crew	Daily Output	Labor-Hours	Unit	2015 Bare Costs Material	Labor	Equipment	Total	Total Incl O&P
0010	SELECTIVE DEMOLITION, INTERIORS R024119-10									
3100	Casework, metal base cabinets	2 Clab	20	.800	L.F.		30		30	49.50
3110	Cabinet base trim		400	.040			1.50		1.50	2.47
3120	Countertop, stainless steel		80	.200			7.50		7.50	12.35
3130	Wall cabinets, wood, 84" high		30	.533			20		20	33
3500	Laboratory casework, tall storage cabinets, 84" high		40	.400			15.05		15.05	24.50
3510	Wall cabinets, metal		40	.400			15.05		15.05	24.50
4830	Floor mats, recessed or link mats	1 Clab	300	.027	S.F.		1		1	1.64
4832	Skate lock tile		200	.040			1.50		1.50	2.47
4834	Duckboard		300	.027			1		1	1.64
4920	Blinds, interior, horizontal or vertical		150	.053	L.F.		2.01		2.01	3.29
4922	Wood folding panels		35	.229	Pr.		8.60		8.60	14.10
4924	Shades, interior		700	.011	S.F.		.43		.43	.70
4930	Drapery hardware, traverse rods		35	.229	Ea.		8.60		8.60	14.10
4950	Blast curtains, including hardware		25	.320			12.05		12.05	19.75
5200	Fixed seating, per seat	2 Carp	44	.364			17.05		17.05	28
6400	Booth, restaurant	2 Clab	80	.200	L.F.		7.50		7.50	12.35
7400	Office systems furniture, cubicle	"	3000	.005	S.F.		.20		.20	.33

12 05 13 – Fabrics

12 05 13.10 Upholstery Materials		Crew	Daily Output	Labor-Hours	Unit	Material	Labor	Equipment	Total	Total Incl O&P
0010	UPHOLSTERY MATERIALS									
1000	Fabrics, fabric blends, minimum				S.Y.	39			39	42.50
1020	Maximum					100			100	110
1200	Nylons, minimum					36			36	39.50
1220	Maximum					58			58	63.50
1400	Polyester, minimum					28			28	31
1420	Maximum					71.50			71.50	78.50
1600	Silk, minimum					88			88	96.50
1620	Maximum					152			152	167
1800	Wool, minimum					62.50			62.50	69
1820	Maximum					122			122	135
5000	Leather, minimum				S.F.	8.15			8.15	8.95
5020	Maximum				"	9.40			9.40	10.35
6000	Vinyl, minimum				S.Y.	26			26	28.50
6020	Maximum				"	42.50			42.50	47

12 12 Wall Decorations

12 12 19 – Framed Prints

12 12 19.10 Art Work		Crew	Daily Output	Labor-Hours	Unit	Material	Labor	Equipment	Total	Total Incl O&P
0010	ART WORK framed									
1000	Photography, minimum	1 Carp	36	.222	Ea.	84	10.45		94.45	110
1050	Maximum		30	.267		525	12.50		537.50	600
2000	Posters, minimum		36	.222		47	10.45		57.45	68.50
2050	Maximum		30	.267		690	12.50		702.50	780
3000	Reproductions, minimum		36	.222		89.50	10.45		99.95	116
3050	Maximum		30	.267		830	12.50		842.50	930

12 21 Window Blinds

12 21 13 – Horizontal Louver Blinds

12 21 13.13 Metal Horizontal Louver Blinds		Crew	Daily Output	Labor-Hours	Unit	Material	2015 Bare Costs Labor	Equipment	Total	Total Incl O&P
0010	METAL HORIZONTAL LOUVER BLINDS									
0020	Horizontal, 1" aluminum slats, solid color, stock	1 Carp	590	.014	S.F.	4.90	.64		5.54	6.45
0070	Horizontal, 1" aluminum slats, custom color		590	.014		5.40	.64		6.04	7
0250	2" aluminum slats, solid color, stock		590	.014		5.40	.64		6.04	7
0275	2" aluminum slats, custom color		590	.014		5.70	.64		6.34	7.35
1000	Alternate method of figuring:									
1300	1" aluminum slats, 48" wide, 48" high	1 Carp	30	.267	Ea.	53	12.50		65.50	78.50
1320	72" high		29	.276		66	12.95		78.95	93.50
1340	96" high		28	.286		159	13.40		172.40	197
1400	72" wide, 72" high		25	.320		64	15		79	94.50
1420	96" high		23	.348		99	16.35		115.35	136
1480	96" wide, 96" high		20	.400		110	18.80		128.80	152
1490	For special colors, add					12%				

12 21 13.33 Vinyl Horizontal Louver Blinds		Crew	Daily Output	Labor-Hours	Unit	Material	Labor	Equipment	Total	Total Incl O&P
0010	VINYL HORIZONTAL LOUVER BLINDS									
0100	2" composite, 48" wide, 48" high	1 Carp	30	.267	Ea.	92	12.50		104.50	122
0120	72" high		29	.276		131	12.95		143.95	165
0140	96" high		28	.286		180	13.40		193.40	220
0200	60" wide, 60" high		27	.296		99.50	13.90		113.40	132
0220	72" high		25	.320		114	15		129	151
0240	96" high		24	.333		182	15.65		197.65	226
0300	72" wide, 72" high		25	.320		194	15		209	239
0320	96" high		23	.348		271	16.35		287.35	325
0400	96" wide, 96" high		20	.400		315	18.80		333.80	375
1000	2" faux wood, 48" wide, 48" high		30	.267		59	12.50		71.50	85.50
1020	72" high		29	.276		81	12.95		93.95	110
1040	96" high		28	.286		100	13.40		113.40	132
1300	72" wide, 72" high		25	.320		125	15		140	162
1320	96" high		23	.348		196	16.35		212.35	242
1400	96" wide, 96" high		20	.400		217	18.80		235.80	269

12 21 16 – Vertical Louver Blinds

12 21 16.13 Metal Vertical Louver Blinds		Crew	Daily Output	Labor-Hours	Unit	Material	Labor	Equipment	Total	Total Incl O&P
0010	METAL VERTICAL LOUVER BLINDS									
1500	Vertical, 3" PVC strips, minimum	1 Carp	460	.017	S.F.	7.95	.82		8.77	10.10
1600	Maximum		400	.020		23.50	.94		24.44	27.50
1800	4" aluminum slats, minimum		460	.017		8.10	.82		8.92	10.25
1900	Maximum		400	.020		14.95	.94		15.89	18
1990	Alternate method of figuring:									
2000	2" aluminum slats, 48" wide, 48" high	1 Carp	30	.267	Ea.	76	12.50		88.50	104
2050	72" high		29	.276		104	12.95		116.95	136
2100	96" high		28	.286		112	13.40		125.40	146
2200	72" wide, 72" high		25	.320		147	15		162	187
2250	96" high		23	.348		176	16.35		192.35	221
2300	96" wide, 96" high		20	.400		269	18.80		287.80	325

12 22 Curtains and Drapes

12 22 13 – Draperies

12 22 13.10 Custom Draperies		Crew	Daily Output	Labor-Hours	Unit	Material	2015 Bare Costs Labor	Equipment	Total	Total Incl O&P
0010	**CUSTOM DRAPERIES** (Material only)									
0050	Lined, minimum				S.F.	6.25			6.25	6.85
0100	Maximum					15.30			15.30	16.85
0200	Unlined, minimum					3.13			3.13	3.44
0300	Maximum					9.65			9.65	10.65
0400	Lightproof type, add, minimum					1.50			1.50	1.65
0500	Maximum				▼	9			9	9.90
0800	Valances, pleated, 10″ to 18″ depth, minimum				L.F.	1.73			1.73	1.90
0900	Maximum				″	10			10	11
2000	Alternate method, lined overlaps and returns, average									
2020	32″ thru 48″ wide, 26″ to 39″ long				Ea.	192			192	211
2040	40″ to 63″ long					194			194	213
2060	64″ to 72″ long					199			199	219
2080	73″ to 81″ long					226			226	248
2100	82″ to 90″ long					232			232	255
2200	91″ to 99″ long					249			249	274
2400	100″ to 108″ long					260			260	286
2600	109″ to 120″ long					274			274	300
2800	121″ to 130″ long					280			280	310
3000	48″ thru 72″ wide, 26″ to 39″ long					279			279	305
3050	40″ to 63″ long					298			298	330
3100	64″ to 72″ long					310			310	340
3150	73″ to 81″ long					325			325	360
3200	82″ to 90″ long					355			355	395
3250	91″ to 99″ long					370			370	410
3300	100″ to 108″ long					385			385	425
3350	109″ to 120″ long					410			410	450
3400	121″ to 130″ long					445			445	490
3450	64″ thru 96″ wide, 26″ to 39″ long					370			370	410
3500	40″ to 63″ long					400			400	440
3550	64″ to 72″ long					415			415	455
3600	73″ to 81″ long					445			445	485
3650	82″ to 90″ long					470			470	515
3700	91″ to 99″ long					500			500	550
3750	100″ to 108″ long					515			515	565
3800	109″ to 120″ long					545			545	600
3820	121″ to 130″ long					585			585	645
3840	80″ thru 120″ wide, 26″ to 39″ long					470			470	520
3860	40″ to 63″ long					500			500	550
3880	64″to 72″ long					515			515	565
4000	73″ to 81″ long					555			555	610
4020	82″ to 90″ long					585			585	640
4080	91″ to 99″ long					615			615	680
4100	100″ to 108″ long					655			655	720
4150	109″ to 120″ long					680			680	745
4200	121″ to 130″ long					710			710	780
4250	96″ thru 144″ wide, 26″ to 39″ long					555			555	610
4300	40″ to 63″ long					600			600	660
4400	64″ to 72″ long					610			610	675
4500	73″ to 81″ long					645			645	710
4600	82″ to 90″ long					700			700	770
4700	91″ to 99″ long					745			745	820
4800	100″ to 108″ long				▼	780			780	860

12 22 Curtains and Drapes

12 22 13 – Draperies

12 22 13.10 Custom Draperies		Crew	Daily Output	Labor-Hours	Unit	Material	2015 Bare Costs Labor	Equipment	Total	Total Incl O&P
4900	109" to 120" long				Ea.	825			825	905
5000	121" to 130" long					885			885	975
5100	112" thru 168" wide, 26" to 39" long					655			655	720
5200	40" to 63" long					700			700	775
5300	64" to 72" long					730			730	800
5400	73" to 81" long					770			770	845
5500	82" to 90" long					815			815	895
5600	91" to 99" long					865			865	955
5700	100" to 108" long					915			915	1,000
5800	109" to 120" long					1,025			1,025	1,125
6000	121" to 130" long					1,100			1,100	1,200
6100	128" thru 192" wide, 26" to 39" long					740			740	815
6200	40" to 63" long					800			800	880
6300	64" to 72" long					830			830	910
6400	73" to 81" long					885			885	975
6450	82" to 90" long					925			925	1,025
6500	91" to 99" long					985			985	1,075
6550	100" to 108" long					1,050			1,050	1,150
6600	109" to 120" long					1,100			1,100	1,225
6650	121" to 130" long					1,175			1,175	1,300
6700	144" thru 216" wide, 26" to 39" long					845			845	925
6750	40" to 63" long					925			925	1,025
6800	64" to 72" long					985			985	1,075
6850	73" to 81" long					1,000			1,000	1,100
6880	82" to 90" long					1,050			1,050	1,150
6900	91" to 99" long					1,100			1,100	1,225
6920	100" to 108" long					1,175			1,175	1,300
6980	109" to 120" long					1,250			1,250	1,350
7000	121" to 130" long					1,325			1,325	1,475
7100	160" thru 240" wide, 26" to 39" long					930			930	1,025
7150	40" to 63" long					1,000			1,000	1,100
7200	64" to 72" long					1,025			1,025	1,125
7250	73" to 81" long					1,100			1,100	1,200
7300	82" to 90" long					1,175			1,175	1,275
7350	91" to 99" long					1,225			1,225	1,350
7400	100" to 108" long					1,300			1,300	1,425
7500	109" to 120" long					1,375			1,375	1,500
7600	121" to 130" long				↓	1,475			1,475	1,625
8800	Drapery installation, hardware & drapes,									
9000	Labor cost only, minimum	1 Clab	75	.107	L.F.		4.01		4.01	6.60
9100	Maximum	"	20	.400	"		15.05		15.05	24.50

12 22 16 – Drapery Track and Accessories

12 22 16.10 Drapery Hardware		Crew	Daily Output	Labor-Hours	Unit	Material	Labor	Equipment	Total	Total Incl O&P
0010	**DRAPERY HARDWARE**									
0030	Standard traverse, per foot, minimum	1 Carp	59	.136	L.F.	7	6.35		13.35	18.15
0100	Maximum		51	.157	"	10.20	7.35		17.55	23.50
0200	Decorative traverse, 28"-48", minimum		22	.364	Ea.	24.50	17.05		41.55	55
0220	Maximum		21	.381		50	17.90		67.90	84.50
0300	48"-84", minimum		20	.400		22.50	18.80		41.30	55.50
0320	Maximum		19	.421		64	19.75		83.75	103
0400	66"-120", minimum		18	.444		43	21		64	81
0420	Maximum	↓	17	.471	↓	97.50	22		119.50	143

12 22 Curtains and Drapes

12 22 16 – Drapery Track and Accessories

12 22 16.10 Drapery Hardware		Crew	Daily Output	Labor-Hours	Unit	Material	2015 Bare Costs Labor	Equipment	Total	Total Incl O&P
0500	84"-156", minimum	1 Carp	16	.500	Ea.	62.50	23.50		86	108
0520	Maximum		15	.533		133	25		158	187
0600	130"-240", minimum		14	.571		33.50	27		60.50	81
0620	Maximum		13	.615		184	29		213	250
0700	Slide rings, each, minimum					1.06			1.06	1.17
0720	Maximum					2.05			2.05	2.26
4000	Traverse rods, adjustable, 28" to 48"	1 Carp	22	.364		22.50	17.05		39.55	53
4020	48" to 84"		20	.400		29	18.80		47.80	62.50
4040	66" to 120"		18	.444		35	21		56	72.50
4060	84" to 156"		16	.500		40	23.50		63.50	82
4080	100" to 180"		14	.571		46.50	27		73.50	95
4090	156" to 228"		13	.615		55.50	29		84.50	109
4100	228" to 312"		13	.615		64.50	29		93.50	118
4200	Double rods, adjustable, 30" to 48"		9	.889		42.50	41.50		84	115
4220	48" to 86"		9	.889		59	41.50		100.50	133
4240	86" to 150"		8	1		65	47		112	149
4260	100" to 180"		7	1.143		68.50	53.50		122	164
4300	Curtain rod & brackets, adjustable, 30" to 48"		9	.889		29.50	41.50		71	101
4320	48" to 86"		9	.889		42	41.50		83.50	115
4340	86" to 150"		8	1		52	47		99	134
4360	100" to 180"		7	1.143		64.50	53.50		118	159
5000	Stationary rods, first 2'					8.25			8.25	9.10
5020	Each additional foot, add				L.F.	3.90			3.90	4.29

12 22 16.20 Blast Curtains		Crew	Daily Output	Labor-Hours	Unit	Material	Labor	Equipment	Total	Total Incl O&P
0010	**BLAST CURTAINS** per L.F. horizontal opening width, off-white or gray fabric									
0100	Blast curtains, drapery system, complete, including hardware, minimum	1 Carp	10.25	.780	L.F.	189	36.50		225.50	268
0120	Average		10.25	.780		204	36.50		240.50	284
0140	Maximum		10.25	.780		235	36.50		271.50	320

12 23 Interior Shutters

12 23 10 – Wood Interior Shutters

12 23 10.10 Wood Interior Shutters		Crew	Daily Output	Labor-Hours	Unit	Material	Labor	Equipment	Total	Total Incl O&P
0010	**WOOD INTERIOR SHUTTERS**, louvered									
0200	Two panel, 27" wide, 36" high	1 Carp	5	1.600	Set	150	75		225	288
0300	33" wide, 36" high		5	1.600		194	75		269	335
0500	47" wide, 36" high		5	1.600		260	75		335	410
1000	Four panel, 27" wide, 36" high		5	1.600		220	75		295	365
1100	33" wide, 36" high		5	1.600		282	75		357	435
1300	47" wide, 36" high		5	1.600		375	75		450	540

12 23 10.13 Wood Panels		Crew	Daily Output	Labor-Hours	Unit	Material	Labor	Equipment	Total	Total Incl O&P
0010	**WOOD PANELS**									
3000	Wood folding panels with movable louvers, 7" x 20" each	1 Carp	17	.471	Pr.	79.50	22		101.50	124
3300	8" x 28" each		17	.471		79.50	22		101.50	124
3450	9" x 36" each		17	.471		91.50	22		113.50	137
3600	10" x 40" each		17	.471		100	22		122	146
4000	Fixed louver type, stock units, 8" x 20" each		17	.471		94	22		116	139
4150	10" x 28" each		17	.471		79.50	22		101.50	124
4300	12" x 36" each		17	.471		94	22		116	139
4450	18" x 40" each		17	.471		134	22		156	184
5000	Insert panel type, stock, 7" x 20" each		17	.471		21	22		43	59

12 23 Interior Shutters

12 23 10 – Wood Interior Shutters

12 23 10.13 Wood Panels		Crew	Daily Output	Labor-Hours	Unit	Material	2015 Bare Costs Labor	Equipment	Total	Total Incl O&P
5150	8" x 28" each	1 Carp	17	.471	Pr.	38	22		60	78
5300	9" x 36" each		17	.471		48.50	22		70.50	89
5450	10" x 40" each		17	.471		52	22		74	93
5600	Raised panel type, stock, 10" x 24" each		17	.471		247	22		269	310
5650	12" x 26" each		17	.471		247	22		269	310
5700	14" x 30" each		17	.471		273	22		295	335
5750	16" x 36" each		17	.471		300	22		322	370
6000	For custom built pine, add					22%				
6500	For custom built hardwood blinds, add					42%				

12 24 Window Shades

12 24 13 – Roller Window Shades

12 24 13.10 Shades		Crew	Daily Output	Labor-Hours	Unit	Material	Labor	Equipment	Total	Total Incl O&P
0010	**SHADES**									
0020	Basswood, roll-up, stain finish, 3/8" slats	1 Carp	300	.027	S.F.	14.95	1.25		16.20	18.45
0200	7/8" slats		300	.027		14.10	1.25		15.35	17.55
0300	Vertical side slide, stain finish, 3/8" slats		300	.027		19.30	1.25		20.55	23.50
0400	7/8" slats		300	.027		19.30	1.25		20.55	23.50
0500	For fire retardant finishes, add					16%				
0600	For "B" rated finishes, add					20%				
0900	Mylar, single layer, non-heat reflective	1 Carp	685	.012		5.05	.55		5.60	6.45
0910	Mylar, single layer, heat reflective		685	.012		5.45	.55		6	6.90
1000	Double layered, heat reflective		685	.012		5.85	.55		6.40	7.35
1100	Triple layered, heat reflective		685	.012		6.40	.55		6.95	7.90
1200	For metal roller instead of wood, add per				Shade	4.51			4.51	4.96
1300	Vinyl coated cotton, standard	1 Carp	685	.012	S.F.	2.90	.55		3.45	4.09
1400	Lightproof decorator shades		685	.012		3	.55		3.55	4.20
1500	Vinyl, lightweight, 4 ga.		685	.012		.61	.55		1.16	1.57
1600	Heavyweight, 6 ga.		685	.012		1.87	.55		2.42	2.96
1700	Vinyl laminated fiberglass, 6 ga., translucent		685	.012		2.60	.55		3.15	3.76
1800	Lightproof		685	.012		4.27	.55		4.82	5.60
2000	Polyester, room darkening, with continuous cord, GEI									
2010	36" x 72"	1 Carp	38	.211	Ea.	224	9.90		233.90	263
2020	48" x 72"		28	.286		293	13.40		306.40	340
2030	60" x 72"		23	.348		345	16.35		361.35	405
2040	72" x 72"		19	.421		395	19.75		414.75	470
5011	Insulative shades G		125	.064	S.F.	11.80	3		14.80	17.85
6011	Solar screening, fiberglass G		85	.094	"	6.75	4.42		11.17	14.65
8011	Interior insulative shutter									
8111	Stock unit, 15" x 60" G	1 Carp	17	.471	Pr.	11.90	22		33.90	49

12 32 16 – Manufactured Plastic-Laminate-Clad Casework

12 32 16.20 Plastic Laminate Casework Doors		Crew	Daily Output	Labor-Hours	Unit	Material	2015 Bare Costs Labor	Equipment	Total	Total Incl O&P
0010	PLASTIC LAMINATE CASEWORK DOORS									
1000	For casework frames, see Section 12 32 23.15									
1100	For casework hardware, see Section 12 32 23.35									
6000	Plastic laminate on particle board									
6100	12" wide, 18" high	1 Carp	25	.320	Ea.	24	15		39	51
6120	24" high		24	.333		32	15.65		47.65	60.50
6140	30" high		23	.348		40	16.35		56.35	71
6160	36" high		21	.381		48	17.90		65.90	82
6200	48" high		16	.500		64	23.50		87.50	109
6250	60" high		13	.615		79.50	29		108.50	135
6300	72" high		12	.667		95.50	31.50		127	157
6320	15" wide, 18" high		24.50	.327		32	15.35		47.35	60
6340	24" high		23.50	.340		42.50	16		58.50	73
6360	30" high		22.50	.356		53	16.70		69.70	86
6380	36" high		20.50	.390		64	18.30		82.30	100
6400	48" high		15.50	.516		85	24		109	133
6450	60" high		12.50	.640		106	30		136	167
6480	72" high		11.50	.696		128	32.50		160.50	194
6500	18" wide, 18" high		24	.333		36	15.65		51.65	65
6550	24" high		23	.348		48	16.35		64.35	79.50
6600	30" high		22	.364		60	17.05		77.05	94
6650	36" high		20	.400		71.50	18.80		90.30	110
6700	48" high		15	.533		95.50	25		120.50	146
6750	60" high		12	.667		120	31.50		151.50	184
6800	72" high		11	.727		143	34		177	214

12 32 16.25 Plastic Laminate Drawer Fronts		Crew	Daily Output	Labor-Hours	Unit	Material	Labor	Equipment	Total	Total Incl O&P
0010	PLASTIC LAMINATE DRAWER FRONTS									
2800	Plastic laminate on particle board front									
3000	4" high, 12" wide	1 Carp	17	.471	Ea.	4.51	22		26.51	41
3200	18" wide		16	.500		6.75	23.50		30.25	46
3600	24" wide		15	.533		9.05	25		34.05	51
3800	6" high, 12" wide		16	.500		6.85	23.50		30.35	46
4000	18" wide		15	.533		10.85	25		35.85	53
4500	24" wide		14	.571		13.70	27		40.70	59
4800	9" high, 12" wide		15	.533		10.25	25		35.25	52.50
5000	18" wide		14	.571		15.40	27		42.40	61
5200	24" wide		13	.615		20.50	29		49.50	70

12 32 23 – Hardwood Casework

12 32 23.10 Manufactured Wood Casework, Stock Units		Crew	Daily Output	Labor-Hours	Unit	Material	Labor	Equipment	Total	Total Incl O&P
0010	MANUFACTURED WOOD CASEWORK, STOCK UNITS									
0700	Kitchen base cabinets, hardwood, not incl. counter tops,									
0710	24" deep, 35" high, prefinished									
0800	One top drawer, one door below, 12" wide	2 Carp	24.80	.645	Ea.	265	30.50		295.50	340
0820	15" wide		24	.667		276	31.50		307.50	355
0840	18" wide		23.30	.687		300	32		332	385
0860	21" wide		22.70	.705		305	33		338	395
0880	24" wide		22.30	.717		365	33.50		398.50	455
1000	Four drawers, 12" wide		24.80	.645		279	30.50		309.50	355
1020	15" wide		24	.667		283	31.50		314.50	360
1040	18" wide		23.30	.687		310	32		342	400
1060	24" wide		22.30	.717		345	33.50		378.50	435
1200	Two top drawers, two doors below, 27" wide		22	.727		390	34		424	485

12 32 Manufactured Wood Casework

12 32 23 – Hardwood Casework

12 32 23.10 Manufactured Wood Casework, Stock Units		Crew	Daily Output	Labor-Hours	Unit	Material	2015 Bare Costs Labor	Equipment	Total	Total Incl O&P
1220	30″ wide	2 Carp	21.40	.748	Ea.	425	35		460	530
1240	33″ wide		20.90	.766		440	36		476	545
1260	36″ wide		20.30	.788		455	37		492	560
1280	42″ wide		19.80	.808		480	38		518	590
1300	48″ wide		18.90	.847		515	40		555	635
1500	Range or sink base, two doors below, 30″ wide		21.40	.748		350	35		385	445
1520	33″ wide		20.90	.766		375	36		411	475
1540	36″ wide		20.30	.788		395	37		432	495
1560	42″ wide		19.80	.808		415	38		453	515
1580	48″ wide		18.90	.847		435	40		475	545
1800	For sink front units, deduct					161			161	177
2000	Corner base cabinets, 36″ wide, standard	2 Carp	18	.889		625	41.50		666.50	760
2100	Lazy Susan with revolving door	″	16.50	.970		840	45.50		885.50	1,000
4000	Kitchen wall cabinets, hardwood, 12″ deep with two doors									
4050	12″ high, 30″ wide	2 Carp	24.80	.645	Ea.	237	30.50		267.50	310
4100	36″ wide		24	.667		282	31.50		313.50	360
4400	15″ high, 30″ wide		24	.667		241	31.50		272.50	315
4420	33″ wide		23.30	.687		298	32		330	385
4440	36″ wide		22.70	.705		290	33		323	375
4450	42″ wide		22.70	.705		325	33		358	415
4700	24″ high, 30″ wide		23.30	.687		325	32		357	410
4720	36″ wide		22.70	.705		355	33		388	445
4740	42″ wide		22.30	.717		405	33.50		438.50	505
5000	30″ high, one door, 12″ wide		22	.727		216	34		250	294
5020	15″ wide		21.40	.748		241	35		276	325
5040	18″ wide		20.90	.766		265	36		301	350
5060	24″ wide		20.30	.788		310	37		347	400
5300	Two doors, 27″ wide		19.80	.808		340	38		378	435
5320	30″ wide		19.30	.829		355	39		394	460
5340	36″ wide		18.80	.851		405	40		445	515
5360	42″ wide		18.50	.865		445	40.50		485.50	555
5380	48″ wide		18.40	.870		500	41		541	615
6000	Corner wall, 30″ high, 24″ wide		18	.889		355	41.50		396.50	460
6050	30″ wide		17.20	.930		380	43.50		423.50	485
6100	36″ wide		16.50	.970		430	45.50		475.50	545
6500	Revolving Lazy Susan		15.20	1.053		480	49.50		529.50	605
7000	Broom cabinet, 84″ high, 24″ deep, 18″ wide		10	1.600		650	75		725	840
7500	Oven cabinets, 84″ high, 24″ deep, 27″ wide		8	2		1,000	94		1,094	1,250
7750	Valance board trim		396	.040	L.F.	13	1.90		14.90	17.40
7780	Toe kick trim	1 Carp	256	.031	″	2.79	1.47		4.26	5.45
7790	Base cabinet corner filler		16	.500	Ea.	41.50	23.50		65	84
7800	Cabinet filler, 3″ x 24″		20	.400		18.05	18.80		36.85	51
7810	3″ x 30″		20	.400		22.50	18.80		41.30	56
7820	3″ x 42″		18	.444		31.50	21		52.50	68.50
7830	3″ x 80″		16	.500		60	23.50		83.50	105
7850	Cabinet panel		50	.160	S.F.	8.65	7.50		16.15	22
9000	For deluxe models of all cabinets, add					40%				
9500	For custom built in place, add					25%	10%			
9558	Rule of thumb, kitchen cabinets not including									
9560	appliances & counter top, minimum	2 Carp	30	.533	L.F.	176	25		201	234
9600	Maximum	″	25	.640	″	395	30		425	485
9610	For metal cabinets, see Section 12 35 70.13									
9700	Minimum labor/equipment charge	1 Carp	3	2.667	Job		125		125	205

12 32 Manufactured Wood Casework

12 32 23 – Hardwood Casework

12 32 23.15 Manufactured Wood Casework Frames		Crew	Daily Output	Labor-Hours	Unit	Material	Labor	Equipment	Total	Total Incl O&P
						2015 Bare Costs				
0010	**MANUFACTURED WOOD CASEWORK FRAMES**									
0050	Base cabinets, counter storage, 36" high									
0100	One bay, 18" wide	1 Carp	2.70	2.963	Ea.	171	139		310	415
0200	24" wide		2.50	3.200		201	150		351	465
0300	36" wide		2.30	3.478		239	163		402	530
0400	Two bay, 36" wide		2.20	3.636		261	171		432	565
1000	48" wide		2	4		282	188		470	620
1050	72" wide		1.80	4.444		330	209		539	705
1100	Three bay, 54" wide		1.50	5.333		310	250		560	750
1200	72" wide		1.30	6.154		385	289		674	895
1800	108" wide		1.10	7.273		445	340		785	1,050
2000	Four bay, 72" wide		1.10	7.273		435	340		775	1,025
2100	96" wide		1	8		485	375		860	1,150
2500	144" wide		.90	8.889		585	415		1,000	1,325
2800	Bookcases, one bay, 7' high, 18" wide		2.40	3.333		201	157		358	480
3000	24" wide		2.30	3.478		231	163		394	520
3100	36" wide		2.20	3.636		282	171		453	590
3500	Two bay, 36" wide		1.60	5		292	235		527	705
3800	48" wide		1.50	5.333		365	250		615	810
4000	72" wide		1.40	5.714		445	268		713	930
4100	Three bay, 54" wide		1.20	6.667		485	315		800	1,050
4200	72" wide		1.10	7.273		535	340		875	1,150
4500	108" wide		1	8		685	375		1,060	1,375
4600	Four bay, 72" wide		.95	8.421		545	395		940	1,250
4700	96" wide		.90	8.889		660	415		1,075	1,425
5000	144" wide		.85	9.412		905	440		1,345	1,725
5100	Coat racks, one bay, 7' high, 24" wide		4.50	1.778		201	83.50		284.50	360
5200	36" wide		4.30	1.860		219	87.50		306.50	385
5300	Two bay, 48" wide		2.75	2.909		279	137		416	530
5600	72" wide		2.50	3.200		320	150		470	600
5800	Three bay, 72" wide		2.10	3.810		410	179		589	745
6000	108" wide		1.90	4.211		445	198		643	810
6100	Wall mounted cabinet, one bay, 24" high, 18" wide		3.60	2.222		110	104		214	292
6400	24" wide		3.50	2.286		120	107		227	310
6600	36" wide		3.40	2.353		151	110		261	345
6800	Two bay, 36" wide		2.20	3.636		161	171		332	455
7000	48" wide		2.10	3.810		181	179		360	490
7200	72" wide		2	4		231	188		419	565
7400	Three bay, 54" wide		1.70	4.706		201	221		422	580
7600	72" wide		1.60	5		241	235		476	650
7800	108" wide		1.50	5.333		300	250		550	740
8000	Four bay, 72" wide		1.40	5.714		240	268		508	705
8100	96" wide		1.30	6.154		281	289		570	785
8200	144" wide		1.20	6.667		380	315		695	930
8400	30" high, one bay, 18" wide		3.60	2.222		120	104		224	305
8600	24" wide		3.40	2.353		139	110		249	335
8800	36" wide		3.20	2.500		159	117		276	365
9000	Two bay, 36" wide		2.15	3.721		160	175		335	460
9100	48" wide		2	4		190	188		378	520
9200	72" wide		1.85	4.324		229	203		432	585
9400	Three bay, 54" wide		1.60	5		199	235		434	605
9600	72" wide		1.50	5.333		239	250		489	675

12 32 Manufactured Wood Casework

12 32 23 – Hardwood Casework

12 32 23.15 Manufactured Wood Casework Frames		Crew	Daily Output	Labor-Hours	Unit	2015 Bare Costs Material	Labor	Equipment	Total	Total Incl O&P
9650	108″ wide	1 Carp	1.40	5.714	Ea.	530	268		798	1,025
9700	Four bay, 72″ wide		1.30	6.154		249	289		538	750
9750	96″ wide		1.22	6.557		299	310		609	835
9780	144″ wide		1.15	6.957		405	325		730	980
9800	Wardrobe, 7′ high, single, 24″ wide		2.70	2.963		222	139		361	470
9850	36″ wide		2.50	3.200		335	150		485	610
9880	Partition & adjustable shelves, 48″ wide		1.70	4.706		282	221		503	670
9890	72″ wide		1.55	5.161		455	242		697	895
9950	Partition, adjustable shelves & drawers, 48″ wide		1.40	5.714		425	268		693	905
9960	72″ wide		1.25	6.400		555	300		855	1,100
9970	Minimum labor/equipment charge		4	2	Job		94		94	154

12 32 23.20 Manufactured Hardwood Casework Doors		Crew	Daily Output	Labor-Hours	Unit	Material	Labor	Equipment	Total	Total Incl O&P
0010	**MANUFACTURED HARDWOOD CASEWORK DOORS**									
2000	Glass panel, hardwood frame									
2200	12″ wide, 18″ high	1 Carp	34	.235	Ea.	27	11.05		38.05	47.50
2400	24″ high		33	.242		36	11.40		47.40	58
2600	30″ high		32	.250		45	11.75		56.75	69
2800	36″ high		30	.267		54	12.50		66.50	80
3000	48″ high		23	.348		72	16.35		88.35	106
3200	60″ high		17	.471		90	22		112	135
3400	72″ high		15	.533		108	25		133	160
3600	15″ wide, 18″ high		33	.242		34	11.40		45.40	55.50
3800	24″ high		32	.250		45	11.75		56.75	69
4000	30″ high		30	.267		56.50	12.50		69	82.50
4250	36″ high		28	.286		67.50	13.40		80.90	96.50
4300	48″ high		22	.364		90	17.05		107.05	127
4350	60″ high		16	.500		113	23.50		136.50	163
4400	72″ high		14	.571		135	27		162	193
4450	18″ wide, 18″ high		32	.250		40.50	11.75		52.25	64
4500	24″ high		30	.267		54	12.50		66.50	80
4550	30″ high		29	.276		67.50	12.95		80.45	95.50
4600	36″ high		27	.296		81	13.90		94.90	112
4650	48″ high		21	.381		108	17.90		125.90	149
4700	60″ high		15	.533		135	25		160	190
4750	72″ high		13	.615		162	29		191	226
5000	Hardwood, raised panel									
5100	12″ wide, 18″ high	1 Carp	16	.500	Ea.	28.50	23.50		52	70
5150	24″ high		15.50	.516		38	24		62	81.50
5200	30″ high		15	.533		47.50	25		72.50	93.50
5250	36″ high		14	.571		57	27		84	107
5300	48″ high		11	.727		76	34		110	140
5320	60″ high		8	1		95	47		142	182
5340	72″ high		7	1.143		114	53.50		167.50	213
5360	15″ wide, 18″ high		15.50	.516		35.50	24		59.50	78.50
5380	24″ high		15	.533		47.50	25		72.50	93.50
5400	30″ high		14.50	.552		59.50	26		85.50	108
5420	36″ high		13.50	.593		71.50	28		99.50	124
5440	48″ high		10.50	.762		95	36		131	164
5460	60″ high		7.50	1.067		119	50		169	213
5480	72″ high		6.50	1.231		143	58		201	252
5500	18″ wide, 18″ high		15	.533		43	25		68	88
5550	24″ high		14.50	.552		57	26		83	105

12 32 23 – Hardwood Casework

12 32 23.20 Manufactured Hardwood Casework Doors		Crew	Daily Output	Labor-Hours	Unit	Material	Labor	Equipment	Total	Total Incl O&P
						2015 Bare Costs				
5600	30″ high	1 Carp	14	.571	Ea.	71.50	27		98.50	123
5650	36″ high		13	.615		85.50	29		114.50	142
5700	48″ high		10	.800		114	37.50		151.50	187
5750	60″ high		7	1.143		143	53.50		196.50	245
5800	72″ high		6	1.333		171	62.50		233.50	291
9000	Minimum labor/equipment charge		4	2	Job		94		94	154

12 32 23.25 Manufactured Wood Casework Drawer Fronts		Crew	Daily Output	Labor-Hours	Unit	Material	Labor	Equipment	Total	Total Incl O&P
0010	**MANUFACTURED WOOD CASEWORK DRAWER FRONTS**									
0100	Solid hardwood front									
1000	4″ high, 12″ wide	1 Carp	17	.471	Ea.	4.33	22		26.33	41
1200	18″ wide		16	.500		6.50	23.50		30	45.50
1400	24″ wide		15	.533		8.65	25		33.65	50.50
1600	6″ high, 12″ wide		16	.500		6.50	23.50		30	45.50
1800	18″ wide		15	.533		9.75	25		34.75	52
2000	24″ wide		14	.571		13	27		40	58.50
2200	9″ high, 12″ wide		15	.533		9.75	25		34.75	52
2400	18″ wide		14	.571		14.60	27		41.60	60
2600	24″ wide		13	.615		19.50	29		48.50	69
9000	Minimum labor/equipment charge		4	2	Job		94		94	154

12 32 23.30 Manufactured Wood Casework Vanities		Crew	Daily Output	Labor-Hours	Unit	Material	Labor	Equipment	Total	Total Incl O&P
0010	**MANUFACTURED WOOD CASEWORK VANITIES**									
8000	Vanity bases, 2 doors, 30″ high, 21″ deep, 24″ wide	2 Carp	20	.800	Ea.	310	37.50		347.50	400
8050	30″ wide		16	1		370	47		417	480
8100	36″ wide		13.33	1.200		360	56.50		416.50	490
8150	48″ wide		11.43	1.400		470	65.50		535.50	630
9000	For deluxe models of all vanities, add to above					40%				
9500	For custom built in place, add to above					25%	10%			

12 32 23.35 Manufactured Wood Casework Hardware		Crew	Daily Output	Labor-Hours	Unit	Material	Labor	Equipment	Total	Total Incl O&P
0010	**MANUFACTURED WOOD CASEWORK HARDWARE**									
1000	Catches, minimum	1 Carp	235	.034	Ea.	1.22	1.60		2.82	3.96
1020	Average		119.40	.067		3.96	3.15		7.11	9.50
1040	Maximum		80	.100		7.55	4.70		12.25	16
2000	Door/drawer pulls, handles									
2200	Handles and pulls, projecting, metal, minimum	1 Carp	48	.167	Ea.	5	7.85		12.85	18.35
2220	Average		42	.190		7.75	8.95		16.70	23
2240	Maximum		36	.222		10.60	10.45		21.05	29
2300	Wood, minimum		48	.167		5.25	7.85		13.10	18.65
2320	Average		42	.190		7.05	8.95		16	22.50
2340	Maximum		36	.222		9.65	10.45		20.10	28
2400	Drawer pulls, antimicrobial copper alloy finish		50	.160		18.75	7.50		26.25	33
2600	Flush, metal, minimum		48	.167		5.25	7.85		13.10	18.65
2620	Average		42	.190		7.05	8.95		16	22.50
2640	Maximum		36	.222		9.65	10.45		20.10	28
2900	Drawer knobs, antimicrobial copper alloy finish		50	.160		12.50	7.50		20	26
3000	Drawer tracks/glides, minimum		48	.167	Pr.	8.95	7.85		16.80	22.50
3020	Average		32	.250		15.15	11.75		26.90	36
3040	Maximum		24	.333		26	15.65		41.65	54
4000	Cabinet hinges, minimum		160	.050		3.02	2.35		5.37	7.15
4020	Average		95.24	.084		6.95	3.94		10.89	14.10
4040	Maximum		68	.118		11.45	5.50		16.95	21.50
5000	Cabinet locks, minimum		47.90	.167	Ea.	5.70	7.85		13.55	19.10
5020	Average		23.95	.334		8.95	15.70		24.65	35.50

12 32 Manufactured Wood Casework

12 32 23 – Hardwood Casework

12 32 23.35 Manufactured Wood Casework Hardware		Crew	Daily Output	Labor-Hours	Unit	2015 Bare Costs Material	Labor	Equipment	Total	Total Incl O&P
5040	Maximum	1 Carp	16	.500	Ea.	21.50	23.50		45	62.50
7000	Appliance pulls, antimicrobial copper alloy finish	↓	50	.160	L.F.	62.50	7.50		70	81.50

12 35 Specialty Casework

12 35 50 – Educational/Library Casework

12 35 50.13 Educational Casework		Crew	Daily Output	Labor-Hours	Unit	Material	Labor	Equipment	Total	Total Incl O&P
0010	**EDUCATIONAL CASEWORK**									
5000	School, 24" deep, metal, 84" high units	2 Carp	15	1.067	L.F.	430	50		480	550
5150	Counter height units		20	.800		288	37.50		325.50	375
5450	Wood, custom fabricated, 32" high counter		20	.800		240	37.50		277.50	325
5600	Add for counter top		56	.286		25.50	13.40		38.90	50
5800	84" high wall units	↓	15	1.067	↓	465	50		515	590
6000	Laminated plastic finish is same price as wood									

12 35 53 – Laboratory Casework

12 35 53.13 Metal Laboratory Casework		Crew	Daily Output	Labor-Hours	Unit	Material	Labor	Equipment	Total	Total Incl O&P
0010	**METAL LABORATORY CASEWORK**									
0020	Cabinets, base, door units, metal	2 Carp	18	.889	L.F.	231	41.50		272.50	325
0300	Drawer units		18	.889		515	41.50		556.50	635
0700	Tall storage cabinets, open, 7' high		20	.800		495	37.50		532.50	605
0900	With glazed doors		20	.800		740	37.50		777.50	875
1300	Wall cabinets, metal, 12-1/2" deep, open		20	.800		166	37.50		203.50	245
1500	With doors	↓	20	.800	↓	345	37.50		382.50	440
6300	Rule of thumb: lab furniture including installation & connection									
6320	High school				S.F.				35	39
6340	College								52	57
6360	Clinical, health care								45	49.50
6380	Industrial				↓				72.50	79.50

12 35 59 – Display Casework

12 35 59.10 Display Cases		Crew	Daily Output	Labor-Hours	Unit	Material	Labor	Equipment	Total	Total Incl O&P
0010	**DISPLAY CASES** Free standing, all glass									
0020	Aluminum frame, 42" high x 36" x 12" deep	2 Carp	8	2	Ea.	1,225	94		1,319	1,500
0100	70" high x 48" x 18" deep	"	6	2.667		3,775	125		3,900	4,350
0500	For wood bases, add					9%				
0600	For hardwood frames, deduct					8%				
0700	For bronze, baked enamel finish, add				↓	10%				
2000	Wall mounted, glass front, aluminum frame									
2010	Non-illuminated, one section 3' x 4' x 1'-4"	2 Carp	5	3.200	Ea.	2,175	150		2,325	2,650
2100	5' x 4' x 1'-4"		5	3.200		2,525	150		2,675	3,025
2200	6' x 4' x 1'-4"		4	4		3,050	188		3,238	3,650
2500	Two sections, 8' x 4' x 1'-4"		2	8		2,225	375		2,600	3,075
2600	10' x 4' x 1'-4"		2	8		2,725	375		3,100	3,625
3000	Three sections, 16' x 4' x 1'-4"	↓	1.50	10.667	↓	4,125	500		4,625	5,375
3500	For fluorescent lights, add				Section	330			330	365
4000	Table exhibit cases, 2' wide, 3' high, 4' long, flat top	2 Carp	5	3.200	Ea.	1,375	150		1,525	1,750
4100	3' wide, 3' high, 4' long, sloping top	"	3	5.333	"	825	250		1,075	1,325

12 35 Specialty Casework

12 35 70 – Healthcare Casework

12 35 70.13 Hospital Casework		Crew	Daily Output	Labor-Hours	Unit	Material	Labor	Equipment	Total	Total Incl O&P
						2015 Bare Costs				
0010	**HOSPITAL CASEWORK**									
0500	Base cabinets, laminated plastic	2 Carp	10	1.600	L.F.	275	75		350	425
0700	Enameled steel		10	1.600		253	75		328	400
1000	Stainless steel		10	1.600		505	75		580	680
1200	For all drawers, add					28.50			28.50	31.50
1300	Cabinet base trim, 4" high, enameled steel	2 Carp	200	.080		46	3.76		49.76	56.50
1400	Stainless steel		200	.080		92	3.76		95.76	107
1450	Countertop, laminated plastic, no backsplash		40	.400		47.50	18.80		66.30	83
1650	With backsplash		40	.400		59	18.80		77.80	96
1800	For sink cutout, add		12.20	1.311	Ea.		61.50		61.50	101
1900	Stainless steel counter top		40	.400	L.F.	153	18.80		171.80	200
2000	For drop-in stainless 43" x 21" sink, add				Ea.	1,000			1,000	1,100
2050	Laminate with antimicrobial finish #4	2 Carp	40	.400	L.F.	32	18.80		50.80	66
2500	Wall cabinets, laminated plastic		15	1.067		206	50		256	310
2600	Enameled steel		15	1.067		253	50		303	360
2700	Stainless steel		15	1.067		505	50		555	635
3000	Hospital cabinets, stainless steel with glass door(s), lockable									
3010	One door, 24" W x 18" D x 60" H	2 Clab	18	.889	Ea.	2,500	33.50		2,533.50	2,775
3020	Two doors, 36" W x 18" D x 60" H		15	1.067		2,825	40		2,865	3,175
3030	36" W x 24" D x 67" H		15	1.067		4,250	40		4,290	4,750
3040	48" W x 24" D x 66" H		12	1.333		4,000	50		4,050	4,475
3050	48" W x 24" D x 72" H		12	1.333		5,050	50		5,100	5,625
3060	60" W x 24" D x 72" H		9	1.778		5,500	67		5,567	6,150

12 35 70.16 Nurse Station Casework		Crew	Daily Output	Labor-Hours	Unit	Material	Labor	Equipment	Total	Total Incl O&P
0010	**NURSE STATION CASEWORK**									
2100	Door type, laminated plastic	2 Carp	10	1.600	L.F.	320	75		395	475
2200	Enameled steel		10	1.600		305	75		380	460
2300	Stainless steel		10	1.600		610	75		685	795
2400	For drawer type, add					258			258	284

12 35 80 – Commercial Kitchen Casework

12 35 80.13 Metal Kitchen Casework		Crew	Daily Output	Labor-Hours	Unit	Material	Labor	Equipment	Total	Total Incl O&P
0010	**METAL KITCHEN CASEWORK**									
3500	Base cabinets, metal, minimum	2 Carp	30	.533	L.F.	74	25		99	123
3600	Maximum		25	.640		188	30		218	257
3700	Wall cabinets, metal, minimum		30	.533		74	25		99	123
3800	Maximum		25	.640		170	30		200	237

12 36 Countertops

12 36 16 – Metal Countertops

12 36 16.10 Stainless Steel Countertops		Crew	Daily Output	Labor-Hours	Unit	Material	Labor	Equipment	Total	Total Incl O&P
0010	**STAINLESS STEEL COUNTERTOPS**									
3200	Stainless steel, custom	1 Carp	24	.333	S.F.	153	15.65		168.65	194

12 36 23 – Plastic Countertops

12 36 23.13 Plastic-Laminate-Clad Countertops		Crew	Daily Output	Labor-Hours	Unit	Material	Labor	Equipment	Total	Total Incl O&P
0010	**PLASTIC-LAMINATE-CLAD COUNTERTOPS**									
0020	Stock, 24" wide w/backsplash, minimum	1 Carp	30	.267	L.F.	17	12.50		29.50	39
0100	Maximum	"	25	.320	"	34.50	15		49.50	62.50
1700	For end splash, add				Ea.	18.35			18.35	20
9000	Minimum labor/equipment charge	1 Carp	3.75	2.133	Job		100		100	164

12 36 Countertops

12 36 23 – Plastic Countertops

12 36 23.30 Plastic Laminate Countertop Components		Crew	Daily Output	Labor-Hours	Unit	Material	2015 Bare Costs Labor	Equipment	Total	Total Incl O&P
0010	PLASTIC LAMINATE COUNTERTOP COMPONENTS									
1000	Edging, 24" wide, 1-1/2" thick									
1500	Plastic laminate, minimum	1 Carp	25	.320	L.F.	3.94	15		18.94	29
1520	Average		24	.333		7.20	15.65		22.85	33.50
1540	Maximum		22	.364		10.85	17.05		27.90	40
2500	Hardwood, minimum		20	.400		6.10	18.80		24.90	38
2520	Average		18	.444		12.25	21		33.25	47.50
2540	Maximum		16	.500		18.35	23.50		41.85	58.50
2600	Backsplash, add to above, minimum		36	.222		1.57	10.45		12.02	18.85
2620	Average		35	.229		2.13	10.75		12.88	19.95
2640	Maximum		34	.235		4.24	11.05		15.29	23
2700	For metal cove, add					1.02			1.02	1.12
2900	Postformed backsplash, add to above									
2920	Minimum	1 Carp	96	.083	L.F.	7.45	3.91		11.36	14.60
2940	Average		96	.083		8.50	3.91		12.41	15.75
2960	Maximum		96	.083		9.50	3.91		13.41	16.85
3500	Well openings (for computers etc.)		2.50	3.200	Ea.		150		150	246
3900	Cutouts for sinks, lavatories		12	.667	"		31.50		31.50	51.50

12 36 40 – Stone Countertops

12 36 40.10 Natural Stone Countertops		Crew	Daily Output	Labor-Hours	Unit	Material	Labor	Equipment	Total	Total Incl O&P
0010	NATURAL STONE COUNTERTOPS									
2500	Marble, stock, with splash, 1/2" thick, minimum	1 Bric	17	.471	L.F.	42	21.50		63.50	81.50
2700	3/4" thick, maximum		13	.615		105	28.50		133.50	162
2720	Marble, 24" wide, no splash		10	.800		46	37		83	111
2740	4" backsplash		10	.800		53	37		90	119
2800	Granite, average, 1-1/4" thick, 24" wide, no splash		13.01	.615		138	28.50		166.50	197

12 36 53 – Laboratory Countertops

12 36 53.10 Laboratory Countertops and Sinks		Crew	Daily Output	Labor-Hours	Unit	Material	Labor	Equipment	Total	Total Incl O&P
0010	LABORATORY COUNTERTOPS AND SINKS									
0020	Countertops, epoxy resin, not incl. base cabinets, acid-proof, minimum	2 Carp	82	.195	S.F.	40.50	9.15		49.65	59.50
0030	Maximum		70	.229		50	10.75		60.75	72.50
0040	Stainless steel		82	.195		131	9.15		140.15	159

12 36 61 – Simulated Stone Countertops

12 36 61.16 Solid Surface Countertops		Crew	Daily Output	Labor-Hours	Unit	Material	Labor	Equipment	Total	Total Incl O&P
0010	**SOLID SURFACE COUNTERTOPS**, Acrylic polymer									
0020	Pricing for orders of 100 L.F. or greater									
0100	25" wide, solid colors	2 Carp	28	.571	L.F.	54.50	27		81.50	104
0200	Patterned colors		28	.571		69	27		96	120
0300	Premium patterned colors		28	.571		86.50	27		113.50	139
0400	With silicone attached 4" backsplash, solid colors		27	.593		60	28		88	112
0500	Patterned colors		27	.593		76	28		104	129
0600	Premium patterned colors		27	.593		94.50	28		122.50	150
0700	With hard seam attached 4" backsplash, solid colors		23	.696		60	32.50		92.50	120
0800	Patterned colors		23	.696		76	32.50		108.50	137
0900	Premium patterned colors		23	.696		94.50	32.50		127	158
1000	Pricing for order of 51 – 99 L.F.									
1100	25" wide, solid colors	2 Carp	24	.667	L.F.	63	31.50		94.50	121
1200	Patterned colors		24	.667		79.50	31.50		111	139
1300	Premium patterned colors		24	.667		99.50	31.50		131	161
1400	With silicone attached 4" backsplash, solid colors		23	.696		69	32.50		101.50	130

12 36 61 – Simulated Stone Countertops

12 36 61.16 Solid Surface Countertops		Crew	Daily Output	Labor-Hours	Unit	Material	2015 Bare Costs Labor	Equipment	Total	Total Incl O&P
1500	Patterned colors	2 Carp	23	.696	L.F.	87.50	32.50		120	150
1600	Premium patterned colors		23	.696		109	32.50		141.50	174
1700	With hard seam attached 4" backsplash, solid colors		20	.800		69	37.50		106.50	138
1800	Patterned colors		20	.800		87.50	37.50		125	158
1900	Premium patterned colors		20	.800		109	37.50		146.50	182
2000	Pricing for order of 1 – 50 L.F.									
2100	25" wide, solid colors	2 Carp	20	.800	L.F.	73.50	37.50		111	143
2200	Patterned colors		20	.800		93.50	37.50		131	165
2300	Premium patterned colors		20	.800		117	37.50		154.50	191
2400	With silicone attached 4" backsplash, solid colors		19	.842		81	39.50		120.50	154
2500	Patterned colors		19	.842		102	39.50		141.50	178
2600	Premium patterned colors		19	.842		128	39.50		167.50	205
2700	With hard seam attached 4" backsplash, solid colors		15	1.067		81	50		131	171
2800	Patterned colors		15	1.067		102	50		152	195
2900	Premium patterned colors		15	1.067		128	50		178	222
3000	Sinks, pricing for order of 100 or greater units									
3100	Single bowl, hard seamed, solid colors, 13" x 17"	1 Carp	3	2.667	Ea.	370	125		495	610
3200	10" x 15"		7	1.143		170	53.50		223.50	275
3300	Cutouts for sinks		8	1			47		47	77
3400	Sinks, pricing for order of 51 – 99 units									
3500	Single bowl, hard seamed, solid colors, 13" x 17"	1 Carp	2.55	3.137	Ea.	425	147		572	705
3600	10" x 15"		6	1.333		196	62.50		258.50	320
3700	Cutouts for sinks		7	1.143			53.50		53.50	88
3800	Sinks, pricing for order of 1 – 50 units									
3900	Single bowl, hard seamed, solid colors, 13" x 17"	1 Carp	2	4	Ea.	500	188		688	860
4000	10" x 15"		4.55	1.758		230	82.50		312.50	390
4100	Cutouts for sinks		5.25	1.524			71.50		71.50	117
4200	Cooktop cutouts, pricing for 100 or greater units		4	2		27	94		121	184
4300	51 – 99 units		3.40	2.353		31.50	110		141.50	216
4400	1 – 50 units		3	2.667		36.50	125		161.50	246

12 36 61.17 Solid Surface Vanity Tops		Crew	Daily Output	Labor-Hours	Unit	Material	Labor	Equipment	Total	Total Incl O&P
0010	**SOLID SURFACE VANITY TOPS**									
0015	Solid surface, center bowl, 17" x 19"	1 Carp	12	.667	Ea.	190	31.50		221.50	261
0020	19" x 25"		12	.667		194	31.50		225.50	266
0030	19" x 31"		12	.667		227	31.50		258.50	300
0040	19" x 37"		12	.667		264	31.50		295.50	345
0050	22" x 25"		10	.800		345	37.50		382.50	440
0060	22" x 31"		10	.800		405	37.50		442.50	505
0070	22" x 37"		10	.800		470	37.50		507.50	580
0080	22" x 43"		10	.800		535	37.50		572.50	650
0090	22" x 49"		10	.800		595	37.50		632.50	715
0110	22" x 55"		8	1		675	47		722	820
0120	22" x 61"		8	1		770	47		817	925
0220	Double bowl, 22" x 61"		8	1		870	47		917	1,025
0230	Double bowl, 22" x 73"		8	1		950	47		997	1,125
0240	For aggregate colors, add					35%				
0250	For faucets and fittings, see Section 22 41 39.10									

12 36 61.19 Quartz Agglomerate Countertops		Crew	Daily Output	Labor-Hours	Unit	Material	Labor	Equipment	Total	Total Incl O&P
0010	**QUARTZ AGGLOMERATE COUNTERTOPS**									
0100	25" wide, 4" backsplash, color group A, minimum	2 Carp	15	1.067	L.F.	64.50	50		114.50	153
0110	Maximum		15	1.067		90	50		140	181
0120	Color group B, minimum		15	1.067		66.50	50		116.50	156

12 36 Countertops

12 36 61 – Simulated Stone Countertops

12 36 61.19 Quartz Agglomerate Countertops		Crew	Daily Output	Labor-Hours	Unit	Material	Labor	Equipment	Total	Total Incl O&P
						2015 Bare Costs				
0130	Maximum	2 Carp	15	1.067	L.F.	94.50	50		144.50	186
0140	Color group C, minimum		15	1.067		78	50		128	168
0150	Maximum		15	1.067		107	50		157	199
0160	Color group D, minimum		15	1.067		84.50	50		134.50	175
0170	Maximum		15	1.067		115	50		165	208

12 43 Portable Lamps

12 43 13 – Lamps

12 43 13.23 Miscellaneous Lamps		Crew	Daily Output	Labor-Hours	Unit	Material	Labor	Equipment	Total	Total Incl O&P
0010	**MISCELLANEOUS LAMPS**									
1000	Ceramic, desk, minimum				Ea.	62			62	68
1020	Maximum					141			141	155
1200	End table, minimum					58			58	64
1220	Maximum					118			118	130
1300	Night stand, minimum					44			44	48.50
1320	Maximum					100			100	110
1400	Wall, minimum					71.50			71.50	78.50
1420	Maximum					243			243	267
2000	Glass, desk, minimum					153			153	168
2020	Maximum					510			510	560
2100	End table, minimum					184			184	202
2120	Maximum					415			415	455
2200	Floor, minimum					440			440	485
2250	Maximum					1,425			1,425	1,575
2280	Night stand, minimum					173			173	191
2300	Maximum					305			305	335
2350	Pendant, minimum					86.50			86.50	95.50
2400	Maximum					910			910	1,000
3000	Metal, desk, minimum					50			50	55
3100	Maximum					121			121	134
3200	End table, minimum					65			65	71.50
3250	Maximum					90			90	99
3300	Wall, minimum					68			68	75
3350	Maximum					140			140	154
3400	Floor, minimum					78			78	86
3450	Maximum					460			460	505
3500	Night stand, minimum					44			44	48.50
3550	Maximum					115			115	126
3600	Pendant, minimum					45			45	49.50
3650	Maximum					665			665	730
4000	Stone, desk, minimum					71.50			71.50	78.50
4100	Maximum					142			142	156
4200	End table, minimum					77.50			77.50	85
4300	Maximum					132			132	145
4400	Night stand, minimum					49			49	54
4500	Maximum					94			94	104
5000	Wall, minimum					71.50			71.50	78.50
6000	Maximum					114			114	126
8000	Replacement shades, lamp, desk, minimum					12.25			12.25	13.45
8020	Maximum					43			43	47
8040	End table, minimum					18.85			18.85	21

12 43 Portable Lamps

12 43 13 – Lamps

12 43 13.23 Miscellaneous Lamps		Crew	Daily Output	Labor-Hours	Unit	Material	2015 Bare Costs Labor	Equipment	Total	Total Incl O&P
8060	Maximum				Ea.	37.50			37.50	41.50
8100	Night stand minimum					10.20			10.20	11.20
8120	Maximum					41			41	45
8140	Wall, minimum					10.20			10.20	11.20
8160	Maximum				↓	18.30			18.30	20

12 45 Bedroom Furnishings

12 45 13 – Bed Linens

12 45 13.13 Blankets

		Crew	Daily Output	Labor-Hours	Unit	Material	Labor	Equipment	Total	Total Incl O&P
0010	**BLANKETS**									
7700	Bedspreads, unquilted, twin, minimum				Ea.	67			67	74
7800	Maximum					112			112	123
7850	Full, minimum					80.50			80.50	88.50
7900	Maximum					163			163	180
8000	Queen, minimum					93			93	102
8100	Maximum					184			184	202
8200	Quilted, twin, minimum					80			80	88
8300	Maximum					97			97	107
8400	Full, minimum					91.50			91.50	100
8500	Maximum					159			159	174
8600	Queen, minimum					104			104	115
8700	Maximum				↓	184			184	203

12 46 Furnishing Accessories

12 46 13 – Ash Receptacles

12 46 13.10 Ash/Trash Receivers

		Crew	Daily Output	Labor-Hours	Unit	Material	Labor	Equipment	Total	Total Incl O&P
0010	**ASH/TRASH RECEIVERS**									
1000	Ash urn, cylindrical metal									
1020	8″ diameter, 20″ high	1 Clab	60	.133	Ea.	158	5		163	182
1040	8″ diam., 25″ high		60	.133		231	5		236	262
1060	10″ diameter, 26″ high		60	.133		126	5		131	147
1080	12″ diam., 30″ high	↓	60	.133	↓	179	5		184	205
2000	Combination ash/trash urn, metal									
2020	8″ diameter, 20″ high	1 Clab	60	.133	Ea.	158	5		163	182
2040	8″ diam., 25″ high		60	.133		190	5		195	217
2050	10″ diameter, 26″ high		60	.133		126	5		131	147
2060	12″ diam., 30″ high	↓	60	.133	↓	305	5		310	345

12 46 19 – Clocks

12 46 19.50 Wall Clocks

		Crew	Daily Output	Labor-Hours	Unit	Material	Labor	Equipment	Total	Total Incl O&P
0010	**WALL CLOCKS**									
0080	12″ diameter, single face	1 Elec	8	1	Ea.	130	54.50		184.50	228
0100	Double face	″	6.20	1.290	″	300	70.50		370.50	440

12 46 33 – Waste Receptacles

12 46 33.13 Trash Receptacles

		Crew	Daily Output	Labor-Hours	Unit	Material	Labor	Equipment	Total	Total Incl O&P
0010	**TRASH RECEPTACLES**									
4000	Trash receptacle, metal									
4020	8″ diameter, 15″ high	1 Clab	60	.133	Ea.	73.50	5		78.50	89
4040	10″ diameter, 18″ high	↓	60	.133	↓	137	5		142	159

12 46 Furnishing Accessories

12 46 33 – Waste Receptacles

12 46 33.13 Trash Receptacles		Crew	Daily Output	Labor-Hours	Unit	2015 Bare Costs Material	Labor	Equipment	Total	Total Incl O&P
4060	16″ diam., 16″ high	1 Clab	60	.133	Ea.	147	5		152	170
4100	18″ diam., 32″ high		60	.133		206	5		211	234
5000	Plastic, fire resistant									
5020	Rectangular 11″ x 8″ x 12″ high	1 Clab	60	.133	Ea.	22	5		27	32
5040	16″ x 8″ x 14″ high	″	60	.133	″	31	5		36	42
5500	Plastic, with lid									
5520	35 gallon	1 Clab	60	.133	Ea.	145	5		150	167
5540	45 gallon		60	.133		233	5		238	265
5550	Plastic recycling barrel, w/lid & wheels, 32 gal. G		60	.133		84.50	5		89.50	101
5560	65 gal. G		60	.133		545	5		550	610
5570	95 gal. G		60	.133		1,025	5		1,030	1,125

12 46 36 – Desk Accessories

12 46 36.10 Commercial Desk Accessories		Crew	Daily Output	Labor-Hours	Unit	Material	Labor	Equipment	Total	Total Incl O&P
0010	**COMMERCIAL DESK ACCESSORIES**									
0300	Bookends, minimum				Ea.	12			12	13.20
0320	Maximum					22.50			22.50	25
1000	Calendar with pad, minimum					20.50			20.50	22.50
1020	Maximum					28.50			28.50	31.50
1200	Carafe, tray, minimum					53			53	58.50
1220	Maximum					81.50			81.50	89.50
1300	Desk pad, minimum					24			24	26.50
1320	Maximum					92			92	101
1400	Double pen set with pens, minimum					60			60	66
1420	Maximum					83.50			83.50	92
1500	Letter tray, minimum					16.25			16.25	17.90
1520	Maximum					48.50			48.50	53.50
1600	Memo box, minimum					5.80			5.80	6.35
1620	Maximum					28.50			28.50	31.50

12 48 Rugs and Mats

12 48 13 – Entrance Floor Mats and Frames

12 48 13.13 Entrance Floor Mats		Crew	Daily Output	Labor-Hours	Unit	Material	Labor	Equipment	Total	Total Incl O&P
0010	**ENTRANCE FLOOR MATS**									
0020	Recessed, black rubber, 3/8″ thick, solid	1 Clab	155	.052	S.F.	26	1.94		27.94	31.50
0050	Perforated		155	.052		16.15	1.94		18.09	21
0100	1/2″ thick, solid		155	.052		19.40	1.94		21.34	24.50
0150	Perforated		155	.052		23.50	1.94		25.44	28.50
0200	In colors, 3/8″ thick, solid		155	.052		21	1.94		22.94	26
0250	Perforated		155	.052		21.50	1.94		23.44	26.50
0300	1/2″ thick, solid		155	.052		27	1.94		28.94	32.50
0350	Perforated		155	.052		27.50	1.94		29.44	33.50
1225	Recessed, alum. rail, hinged mat, 7/16″ thk									
1250	Carpet insert	1 Clab	360	.022	S.F.	49.50	.84		50.34	56
1275	Vinyl insert		360	.022		49.50	.84		50.34	56
1300	Abrasive insert		360	.022		49.50	.84		50.34	56
1325	Recessed, vinyl rail, hinged mat, 7/16″ thk									
1350	Carpet insert	1 Clab	360	.022	S.F.	55	.84		55.84	62
1375	Vinyl insert		360	.022		55	.84		55.84	62
1400	Abrasive insert		360	.022		55	.84		55.84	62
2000	Recycled rubber tire tile, 12″ x 12″ x 3/8″ thick G		125	.064		10.10	2.41		12.51	15.05

12 48 Rugs and Mats

12 48 13 – Entrance Floor Mats and Frames

12 48 13.13 Entrance Floor Mats		Crew	Daily Output	Labor-Hours	Unit	Material	2015 Bare Costs Labor	Equipment	Total	Total Incl O&P
2510	Natural cocoa fiber, 1/2" thick G	1 Clab	125	.064	S.F.	8.35	2.41		10.76	13.15
2520	3/4" thick G		125	.064		6.90	2.41		9.31	11.55
2530	1" thick G		125	.064		9.60	2.41		12.01	14.50
3000	Hospital tacky mats, package of 30 with frame				Ea.	56			56	61.50
3010	4 packages of 30				"	86.50			86.50	95.50

12 51 Office Furniture

12 51 16 – Case Goods

12 51 16.13 Metal Case Goods

		Crew	Daily Output	Labor-Hours	Unit	Material	Labor	Equipment	Total	Total Incl O&P
0010	**METAL CASE GOODS**									
0020	Desks, 29" high, double pedestal, 30" x 60", metal, minimum				Ea.	595			595	655
0030	Maximum					1,550			1,550	1,700
0400	36" x 72", metal, minimum					800			800	880
0420	Maximum					1,950			1,950	2,150
0600	Desks, single pedestal, 30" x 60", metal, minimum					540			540	595
0620	Maximum					1,250			1,250	1,400
0720	Desks, secretarial, 30" x 60", metal, minimum					485			485	535
0730	Maximum					860			860	945
0740	Return, 20" x 42", minimum					360			360	395
0750	Maximum					555			555	610
0940	59" x 12" x 23" high, steel, minimum					305			305	340
0960	Maximum					390			390	430
0970	Keyboard shelf, standard					101			101	111
0980	Articulating					530			530	585
0990	Center drawer, 18" wide					140			140	154
1020	Credenza, 18" to 22" x 60" to 72" metal, minimum					510			510	560
1040	Maximum					1,550			1,550	1,700
1240	Bookcase, 36" x 12" x 29" high					173			173	190
1260	52" high					265			265	291
1320	Computer stand, mobile, 25" x 24" x 38" high					300			300	330
1370	Computer desk with hutch, 47" x 24" x 46" high					435			435	475
1400	Printer stand, 22" x 24" x 34" high					232			232	255

12 51 16.16 Wood Case Goods

		Crew	Daily Output	Labor-Hours	Unit	Material	Labor	Equipment	Total	Total Incl O&P
0010	**WOOD CASE GOODS**									
0150	Desk, 29" high, double pedestal, 30" x 60"									
0160	Wood, minimum				Ea.	740			740	815
0180	Maximum				"	3,025			3,025	3,325
0550	Desk, 29" high, double pedestal, 36" x 72"									
0560	Wood, minimum				Ea.	650			650	715
0580	Maximum				"	4,050			4,050	4,450
0630	Single pedestal, 30" x 60"									
0640	Wood, minimum				Ea.	600			600	660
0650	Maximum					945			945	1,050
0670	Executive return, 24" x 42", with box, file, wood, minimum					390			390	430
0680	Maximum					945			945	1,050
0790	Desk, 29" high, secretarial, 30" x 60"									
0800	Wood, minimum				Ea.	510			510	560
0810	Maximum					3,075			3,075	3,375
0820	Return, 20" x 42", minimum					320			320	350
0830	Maximum					1,150			1,150	1,250
0900	Desktop organizer, 72" x 14" x 36" high, wood, minimum					180			180	198

12 51 16 – Case Goods

12 51 16.16 Wood Case Goods		Crew	Daily Output	Labor-Hours	Unit	Material	2015 Bare Costs Labor	Equipment	Total	Total Incl O&P
0920	Maximum				Ea.	465			465	510
1110	Furniture, credenza, 29″ high, 18″ to 22″ x 60″ to 72″									
1120	Wood, minimum				Ea.	725			725	795
1140	Maximum					2,675			2,675	2,950
1150	Hutch/bookcase, minimum					570			570	625
1160	Maximum					2,225			2,225	2,450
1200	Bookcase, 36″ x 12″ x 30″ high, wood					194			194	213
1220	48″ high					286			286	315
1300	Computer stand, mobile, wood, 25″ x 24″ x 38″ high					530			530	585
1360	Computer desk with hutch, wood, 47″ x 24″ x 46″ high					585			585	645

12 51 16.26 Resinite Case Goods

		Crew	Daily Output	Labor-Hours	Unit	Material	Labor	Equipment	Total	Total Incl O&P
0010	**RESINITE CASE GOODS**									
1340	Computer stand, resinite				Ea.	325			325	355
1380	Desk with hutch				″	490			490	540

12 51 19 – Filing Cabinets

12 51 19.13 Lateral Filing Cabinets

		Crew	Daily Output	Labor-Hours	Unit	Material	Labor	Equipment	Total	Total Incl O&P
0010	**LATERAL FILING CABINETS**, metal, baked enamel finish									
1060	Lateral, 36″ wide, minimum				Ea.	350			350	385
1080	Maximum					475			475	520
1160	Lateral, 36″ wide, minimum					475			475	520
1180	Maximum					890			890	980
1200	Wood, 2 drawer, lateral, minimum					495			495	545
1210	Maximum					1,000			1,000	1,100

12 51 19.16 Vertical Filing Cabinets

		Crew	Daily Output	Labor-Hours	Unit	Material	Labor	Equipment	Total	Total Incl O&P
0010	**VERTICAL FILING CABINETS**, metal, baked enamel finish									
1000	2 drawer, vertical, minimum				Ea.	164			164	181
1020	Maximum					395			395	435
1100	4 drawer, vertical, minimum					310			310	340
1120	Maximum					530			530	580

12 51 19.23 Flat Files

		Crew	Daily Output	Labor-Hours	Unit	Material	Labor	Equipment	Total	Total Incl O&P
0010	**FLAT FILES**, metal, baked enamel finish									
2010	Steel, 5 drawer, 40″ wide x 27″ deep				Ea.	935			935	1,025
2020	46″ wide x 33″ deep					1,075			1,075	1,175
2030	8 drawer, 40″ wide x 27″ deep					1,625			1,625	1,775
2040	46″ wide x 33″ deep					1,600			1,600	1,750
2050	Base, 40″ wide x 27″ deep x 5″ high					263			263	290
2060	46″ wide x 33″ deep					288			288	315

12 51 19.26 Hanging Files

		Crew	Daily Output	Labor-Hours	Unit	Material	Labor	Equipment	Total	Total Incl O&P
0010	**HANGING FILES**, metal, baked enamel finish									
2100	File stand				Ea.	325			325	355
2110	Clamps, package of 6				″	129			129	142

12 51 23 – Office Tables

12 51 23.13 Wood Tables

		Crew	Daily Output	Labor-Hours	Unit	Material	Labor	Equipment	Total	Total Incl O&P
0010	**WOOD TABLES**									
5000	Sled base, laminate top, coffee, minimum				Ea.	202			202	222
5100	Maximum					520			520	570
5150	End, minimum					155			155	170
5200	Maximum					289			289	320
5250	Wood cube, coffee					740			740	815
5350	End					620			620	685

12 51 Office Furniture

12 51 23 – Office Tables

12 51 23.13 Wood Tables		Crew	Daily Output	Labor-Hours	Unit	Material	2015 Bare Costs Labor	Equipment	Total	Total Incl O&P
5700	Designer table, Mies Barcelona, 40" sq., glass top, s.s. legs				Ea.	1,325			1,325	1,450
5750	Saarinen, 42" round, plastic top, metal pedestal base					1,650			1,650	1,825
5800	Aulenti coffee table, 45" square, marble top and leg base					10,300			10,300	11,300
5840	Bruer, Laccio, 21-1/2" x 19", plastic top, tubular legs					550			550	605
5860	F. knoll, 24" square, glass top, solid steel leg base, chrome				↓	1,975			1,975	2,150
12 51 23.23 Metal Tables										
0010	**METAL TABLES**									
5400	All metal drum, 14" diameter, 14" high				Ea.	395			395	435
5420	21" high					465			465	510
5480	18" diameter, 18" high					575			575	635
5500	22" diameter, 15" high					750			750	825
5550	20" high				↓	825			825	905
5600	For 3/4" glass top, add				S.F.	73			73	80.50
5650	For 1" marble top, add				"	73			73	80.50
7500	Table, training, with modesty panel, 30" x 60"				Ea.	570			570	630
8000	Tables, modular									
8010	Rectangular, 24" x 48"				Ea.	289			289	320
8020	24" x 60"					310			310	340
8030	24" x 72"					400			400	440
8040	30" x 60"					1,050			1,050	1,175
8050	30" x 72"					1,075			1,075	1,175
8100	Corner triangle, 24" deep					360			360	395
8110	30" deep					405			405	445
8200	Half round, 30" x 60"					865			865	955
8300	Trapezoid, 30" x 60"					475			475	520
8400	Crescent, 30" x 60"				↓	1,075			1,075	1,175
12 51 23.33 Conference Tables										
0010	**CONFERENCE TABLES**									
6010	Segmented, oval, 72" x 144"				Ea.	4,875			4,875	5,350
6050	Boat, 96" x 42", minimum					800			800	875
6150	Maximum					3,675			3,675	4,050
6200	120" x 48", minimum					1,425			1,425	1,550
6250	Maximum					4,500			4,500	4,950
6300	144" x 48", minimum					1,600			1,600	1,775
6350	Maximum					5,725			5,725	6,300
6400	168" x 60", minimum					2,750			2,750	3,025
6450	Maximum					8,200			8,200	9,025
6500	192" x 60", minimum					3,200			3,200	3,525
6550	Maximum					9,100			9,100	10,000
6680	240" x 60", minimum					4,650			4,650	5,125
6700	Maximum					16,200			16,200	17,800
6720	Rectangle, 96" x 42", minimum					1,325			1,325	1,450
6740	Maximum					3,675			3,675	4,050
6760	120" x 48", minimum					2,750			2,750	3,025
6780	Maximum					4,725			4,725	5,200
6800	144" x 60", minimum					4,200			4,200	4,600
6820	Maximum					6,825			6,825	7,500
6840	168" x 60", minimum					5,100			5,100	5,625
6860	Maximum					8,200			8,200	9,025
6880	192" x 60", minimum					5,950			5,950	6,550
6900	Maximum					9,100			9,100	10,000
6960	240" x 60", minimum				↓	7,075			7,075	7,775

12 51 Office Furniture

12 51 23 – Office Tables

12 51 23.33 Conference Tables		Crew	Daily Output	Labor-Hours	Unit	2015 Bare Costs Material	Labor	Equipment	Total	Total Incl O&P
6980	Maximum				Ea.	14,800			14,800	16,300

12 52 Seating

12 52 13 – Chairs

12 52 13.10 Chairs, Folding and Stack

Code	Description	Crew	Daily Output	Labor-Hours	Unit	Material	Labor	Equipment	Total	Total Incl O&P
0010	**CHAIRS, FOLDING & STACK**									
2000	Folding, all steel, baked enamel finish,									
2100	Form fitting seat and backrests				Ea.	37.50			37.50	41.50
2200	Upholstered seat and back					63.50			63.50	69.50
2300	Polypropylene seat and back w/frame					60.50			60.50	66.50
2500	Chair caddy for above					325			325	360
5000	Stack chair									
5300	Hardwood frame, seat and back									
5320	Minimum				Ea.	65.50			65.50	72
5340	Maximum				"	237			237	260
5400	Upholstered seat and back									
5420	Minimum				Ea.	65			65	71.50
5440	Maximum				"	360			360	395
5600	Metal frame, upholstered seat and back									
5620	Minimum				Ea.	53.50			53.50	59
5640	Maximum				"	370			370	410
5700	Plastic shell, metal legs									
5720	Minimum				Ea.	48			48	53
5740	Maximum					143			143	157
5800	Chair caddy for above					267			267	294
5900	Tablet arms, minimum					167			167	184
5920	Maximum					206			206	226

12 52 19 – Upholstered Seating

12 52 19.13 Upholstered Office Seating

Code	Description	Crew	Daily Output	Labor-Hours	Unit	Material	Labor	Equipment	Total	Total Incl O&P
0010	**UPHOLSTERED OFFICE SEATING**									
3800	Lounge chair, upholstered, minimum				Ea.	400			400	440
3850	Maximum					2,425			2,425	2,650
4000	Sofa, two seat, upholstered, minimum					865			865	950
4100	Maximum					3,525			3,525	3,875
4650	Three seat, minimum					1,250			1,250	1,375
4680	Maximum					4,950			4,950	5,450
4700	Modular seating, lounge chair unit, upholstered, minimum					1,975			1,975	2,175
4750	Maximum					2,875			2,875	3,175
4780	Corner unit, minimum					1,350			1,350	1,500
4800	Maximum					3,175			3,175	3,500

12 52 23 – Office Seating

12 52 23.13 Office Chairs

Code	Description	Crew	Daily Output	Labor-Hours	Unit	Material	Labor	Equipment	Total	Total Incl O&P
0010	**OFFICE CHAIRS**									
2000	Standard office chair, executive, minimum				Ea.	305			305	335
2150	Maximum					2,075			2,075	2,300
2200	Management, minimum					231			231	254
2250	Maximum					2,225			2,225	2,450
2280	Task, minimum					169			169	186
2290	Maximum					560			560	615
2300	Arm kit, minimum					77			77	85

12 52 Seating

12 52 23 – Office Seating

12 52 23.13 Office Chairs		Crew	Daily Output	Labor-Hours	Unit	Material	2015 Bare Costs Labor	Equipment	Total	Total Incl O&P
2320	Maximum				Ea.	117			117	129
2340	Ergonomic, executive, minimum					540			540	595
2380	Maximum					855			855	940
2390	Management, minimum					460			460	505
2400	Maximum					1,650			1,650	1,825
2450	Task, minimum					152			152	167
2500	Maximum					640			640	705
2550	Arm kit, minimum					45.50			45.50	50
2600	Maximum					112			112	124
3000	Side/guest chairs, upholstered, sled base, metal, min.					155			155	170
3100	Maximum					565			565	625
3200	Wood, minimum					220			220	242
3250	Maximum					385			385	420
3400	Traditional, wood leg, minimum					175			175	192
3500	Maximum					1,175			1,175	1,275
3600	Conference chair, upholstered, metal leg, minimum					197			197	216
3700	Maximum				↓	955			955	1,050

12 52 23.23 Multiple Office Seating Units

		Crew	Daily Output	Labor-Hours	Unit	Material	Labor	Equipment	Total	Total Incl O&P
0010	**MULTIPLE OFFICE SEATING UNITS**									
1000	Area seating, full upholstered, 3 seat straight unit, minimum				Ea.	1,225			1,225	1,350
1020	Maximum					2,600			2,600	2,850
1100	Four seat with corner table, minimum					3,175			3,175	3,500
1120	Maximum					2,725			2,725	3,000
1200	2 seat with in-line table, minimum					1,575			1,575	1,750
1220	Maximum					1,400			1,400	1,550
2000	Individual seat with table, minimum					805			805	885
2020	Maximum				↓	990			990	1,100

12 54 Hospitality Furniture

12 54 13 – Hotel and Motel Furniture

12 54 13.10 Hotel Furniture

		Crew	Daily Output	Labor-Hours	Unit	Material	Labor	Equipment	Total	Total Incl O&P
0010	**HOTEL FURNITURE**									
0020	Standard quality set, minimum				Room	2,400			2,400	2,650
0200	Maximum				"	8,600			8,600	9,450
0300	Bed frame				Ea.	80			80	88
0400	Bench, upholstered, 42" x 18" x 18"					240			240	264
0420	18" x 18" x 23"					365			365	400
0500	Desk section, one drawer, 34" x 20" x 30"					269			269	296
0600	Free standing, 42" x 22" x 30"					305			305	335
0700	Desk chair, upholstered, minimum					110			110	121
0720	Maximum					180			180	198
1000	Dressers, uniplex, 2 drawer					400			400	440
1100	3 drawer					450			450	495
2000	Guest tables, 30" diameter					140			140	154
2100	34" diameter					230			230	253
3000	Headboards, free standing, twin					146			146	160
3050	Full					209			209	230
3100	Queen					214			214	236
3200	Wall mounted, twin					110			110	121
3250	Full					199			199	219
3300	Queen				↓	214			214	236

12 54 Hospitality Furniture

12 54 13 – Hotel and Motel Furniture

12 54 13.10 Hotel Furniture		Crew	Daily Output	Labor-Hours	Unit	Material	2015 Bare Costs Labor	Equipment	Total	Total Incl O&P
4000	Lounge chair, full upholstered, minimum				Ea.	370			370	405
4050	Maximum					450			450	495
4200	Open arms, minimum					97			97	107
4250	Maximum					390			390	430
5000	Mattress/box springs, twin					315			315	350
5050	Full					490			490	540
5100	Queen					550			550	605
5150	Mirror, framed, 29" x 45"					173			173	191
6000	Sleep sofas, twin, minimum					615			615	680
6050	Maximum					780			780	860
6100	Full, minimum					750			750	825
6150	Maximum					995			995	1,100
6200	Queen, minimum					760			760	835
6250	Maximum					1,075			1,075	1,200
7000	Table, wood top, cocktail, 54" x 24" x 16" high					370			370	405
7020	Corner, 30" x 30" x 21" high					340			340	375
7040	End, 22" x 28" x 21" high					340			340	375

12 54 13.20 Mattress and Box Springs		Crew	Daily Output	Labor-Hours	Unit	Material	Labor	Equipment	Total	Total Incl O&P
0010	**MATTRESS & BOX SPRINGS** per set									
1000	Hospital, 34" x 84"				Ea.	580			580	640
2000	Hotel/motel, twin, minimum					380			380	420
2020	Maximum					645			645	705
2200	Full, minimum					460			460	505
2220	Maximum					770			770	850
2400	Queen, minimum					580			580	640
2420	Maximum					880			880	970
3000	Stow away bed with head board, twin size					700			700	770

12 54 16 – Restaurant Furniture

12 54 16.10 Tables, Folding		Crew	Daily Output	Labor-Hours	Unit	Material	Labor	Equipment	Total	Total Incl O&P
0010	**TABLES, FOLDING** Laminated plastic tops									
1000	Tubular steel legs with glides									
1020	18" x 60", minimum				Ea.	272			272	299
1040	Maximum					1,550			1,550	1,700
1100	18" x 72", minimum					305			305	335
1120	Maximum					1,625			1,625	1,800
1200	18" x 96", minimum					335			335	370
1220	Maximum					1,975			1,975	2,175
1400	30" x 48", minimum					213			213	235
1420	Maximum					298			298	330
1500	30" x 60", minimum					265			265	292
1520	Maximum					1,825			1,825	2,000
1600	30" x 72", minimum					370			370	405
1620	Maximum					2,125			2,125	2,325
1700	30" x 96", minimum					293			293	325
1720	Maximum					2,675			2,675	2,950
1800	36" x 72", minimum					325			325	355
1820	Maximum					2,375			2,375	2,625
1840	36" x 96", minimum					335			335	370
1860	Maximum					3,125			3,125	3,425
2000	Round, wood stained, plywood top, 60" diameter, minimum					213			213	234
2020	Maximum					295			295	325
2200	72" diameter, minimum					355			355	390

12 54 16 – Restaurant Furniture

12 54 16.10 Tables, Folding		Crew	Daily Output	Labor-Hours	Unit	Material	2015 Bare Costs Labor	Equipment	Total	Total Incl O&P
2220	Maximum				Ea.	440			440	485
4000	Mobile storage carts									
4020	For 72" tables, flat, maximum 10 tables				Ea.	840			840	925
4040	96" tables, flat, maximum 10 tables					850			850	935
4060	Rounds, on edge, maximum 10 tables					990			990	1,100

12 54 16.20 Furniture, Restaurant

		Crew	Daily Output	Labor-Hours	Unit	Material	Labor	Equipment	Total	Total Incl O&P
0010	**FURNITURE, RESTAURANT**									
0020	Bars, built-in, front bar	1 Carp	5	1.600	L.F.	280	75		355	435
0200	Back bar	"	5	1.600	"	203	75		278	345
0300	Booth seating, see Section 12 54 16.70									
2000	Chair, bentwood side chair, metal, minimum				Ea.	100			100	110
2020	Maximum					117			117	129
2100	Wood, minimum					209			209	230
2120	Maximum					244			244	269
2400	Bruer Cesca, cane seat & back, arms, minimum					575			575	635
2420	Maximum					675			675	745
2500	Side, minimum					575			575	635
2520	Maximum					675			675	745
2600	Upholstered seat & back, arms, minimum					163			163	179
2620	Maximum					460			460	510
2640	Side, minimum					450			450	495
2660	Maximum					530			530	580
2700	Corbusier, arm chair, cane seat and back, minimum					224			224	247
2720	Maximum					244			244	269
2740	Fledermaus, fabric seat, ash frame, minimum					207			207	228
2760	Maximum					254			254	279
2780	Hoffman arm, fabric seat, cane back, minimum					236			236	260
2800	Maximum					258			258	284
2820	Side, minimum					185			185	203
2840	Maximum					199			199	218
2860	Lombard, minimum					127			127	140
2880	Maximum					165			165	182
2900	Mies, side chair, leather seat and back, minimum					545			545	600
2920	Maximum					805			805	885
2940	Napoleon, upholstered seat, wood back, minimum					91			91	100
2960	Maximum					116			116	128
3000	Prague, arm, upholstered, minimum					169			169	186
3020	Maximum					365			365	400
3040	Side, upholstered, minimum					168			168	185
3060	Maximum					207			207	227
3080	Contemporary leather seat & back, chrome frame, min.					340			340	370
3100	Maximum					495			495	540
3120	Foam padded seat & back, s. steel barstock, min.					905			905	995
3140	Maximum					960			960	1,050
3160	Chrome tube cantilever frame, padded arms, min.					139			139	153
3180	Maximum					355			355	395
4200	"Tub" style, minimum					238			238	262
4220	Maximum					540			540	595
4280	Bent wood arms and legs with back bow, minimum					66.50			66.50	73
4300	Maximum					440			440	485
4320	Sled base, minimum					139			139	153
4340	Maximum					440			440	485

12 54 16 – Restaurant Furniture

12 54 16.20 Furniture, Restaurant		Crew	Daily Output	Labor-Hours	Unit	2015 Bare Costs Material	Labor	Equipment	Total	Total Incl O&P
4360	Armchair foam padded seat & back,carved wood frame min.				Ea.	405			405	445
4380	Maximum					770			770	850
4400	Ornate wood frame and legs, minimum					430			430	475
4420	Maximum					510			510	560
4440	Heavy wood frame and legs, tailored back, minimum					685			685	755
4460	Maximum					795			795	875
4480	Button tufted back, minimum					445			445	490
4500	Maximum					1,025			1,025	1,125
4520	"Dining" wood frame and legs, minimum					330			330	365
4550	Maximum					385			385	425
4600	"Queen Ann" wood frame and legs, minimum					450			450	495
4620	Maximum					500			500	550
4660	Upholstered arms and wood legs, minimum					635			635	700
4680	Maximum					745			745	820
4700	Wicker, foam padded seat and back, barrel design, minimum					560			560	615
4720	Maximum					630			630	695
4740	Couch design, minimum					415			415	455
4760	Maximum					505			505	555
4780	Chrome tube round cantilever base, minimum					239			239	263
4800	Maximum					525			525	575
4820	Chrome tube square cantilever base, minimum					229			229	252
4840	Maximum					510			510	565
4860	Misc. chair, upholst. seat, wood arms, bow back & legs, min.					186			186	205
4880	Maximum					259			259	285
4900	Open curved wood back, minimum					420			420	465
4950	Maximum					470			470	515
5000	Upholstered seat and open back, wood frame arms, min.					385			385	425
5100	Maximum					450			450	495
5150	Wood frame sled base, minimum					216			216	238
5200	Maximum					360			360	395
5250	Wood saddle seat, "Windsor", wood frame & legs, min.					221			221	243
5300	Maximum					258			258	284
5350	Brass nail trim leather, minimum					260			260	285
5400	Maximum					296			296	325
5450	Wood frame, fabric, minimum					222			222	244
5500	Maximum					480			480	525
6000	Stools, upholst. seat & back, chrome tube cantilever frame, min.					350			350	385
6150	Swivel on chrome post mount with spread base, minimum					204			204	224
6250	"Vienna" wood back and legs, minimum					273			273	300
6350	"Napoleon" wood back and legs, minimum					93.50			93.50	103
6450	Wood swivel seat and curved spindle back, leg base					162			162	178
6600	Upholstered seat no back, bent wood legs					325			325	360
7000	With back & straight wood frame and legs					325			325	355
7200	Wood swivel seat, "Captain" wood back and legs					206			206	227
7400	Veneer seat and curved wood back with arms, leg base					196			196	216
7600	Upholstered swivel seat no back, wood legs					78			78	85.50
8000	With back & wood legs					570			570	625
8200	Upholstered seat & wood back, wood legs					540			540	595
8400	Upholstered swivel seat and back, wood legs					220			220	242
8600	Square tube frame and legs					56			56	61.50
8800	Metal frame and legs				↓	56			56	61.50

12 54 Hospitality Furniture

12 54 16 – Restaurant Furniture

12 54 16.30 Table Bases

12 54 16.30 Table Bases		Crew	Daily Output	Labor-Hours	Unit	Material	2015 Bare Costs Labor	Equipment	Total	Total Incl O&P
0010	**TABLE BASES**									
0040	Dining height									
1000	Metal disk design, minimum				Ea.	109			109	120
1200	Maximum					266			266	292
1400	Wood, minimum					435			435	480
1600	Maximum					480			480	525
2000	Fluted wheel, minimum					115			115	127
2200	Maximum					198			198	218
2400	Heavy cast iron, minimum					87.50			87.50	96
2600	Maximum					256			256	281
2800	Hobnail design, minimum					106			106	117
3000	Maximum					325			325	360
3200	Manhole design, minimum					122			122	134
3400	Maximum					325			325	360
3600	Ring design, minimum					170			170	187
3800	Maximum					280			280	310
4000	Trumpet design, minimum					305			305	335
4200	Maximum					335			335	370
4400	Tubular, round or rectangular shape, minimum					121			121	133
4600	Maximum					254			254	280
5000	Cocktail height bases, minimum					141			141	155
5200	Maximum					155			155	171
5400	For foot ring, add, minimum					37			37	41
5600	Maximum				↓	45.50			45.50	50

12 54 16.40 Table Tops

12 54 16.40 Table Tops		Crew	Daily Output	Labor-Hours	Unit	Material	Labor	Equipment	Total	Total Incl O&P
0010	**TABLE TOPS** Laminated plastic top and edge									
0040	24" wide, 24" long, minimum				Ea.	79.50			79.50	87.50
1000	Maximum					273			273	300
1200	30" long, minimum					92			92	101
1400	Maximum					400			400	445
1600	36" long, minimum					117			117	129
1800	Maximum					425			425	465
2000	48" long, minimum					141			141	155
2200	Maximum					500			500	550
2400	60" long, minimum					227			227	250
2600	Maximum					780			780	860
2800	72" long, minimum					277			277	305
3000	Maximum					825			825	905
3050	30" wide, 30" long, minimum					97			97	107
3150	Maximum					425			425	465
3200	36" long, minimum					122			122	134
3250	Maximum					500			500	550
3400	42" long, minimum					126			126	139
3600	Maximum					330			330	365
3800	48" long, minimum					158			158	174
4000	Maximum					550			550	605
4200	60" long, minimum					224			224	246
4400	Maximum					710			710	780
4600	72" long, minimum					275			275	300
4800	Maximum					835			835	920
5000	36" wide, 36" long, minimum					129			129	142
5200	Maximum				↓	500			500	550

12 54 16 – Restaurant Furniture

12 54 16.40 Table Tops

12 54 16.40 Table Tops		Crew	Daily Output	Labor-Hours	Unit	Material	2015 Bare Costs Labor	Equipment	Total	Total Incl O&P
5400	48" long, minimum				Ea.	188			188	206
5600	Maximum					650			650	715
5800	60" long, minimum					267			267	293
6000	Maximum					835			835	920
6200	72" long, minimum					305			305	335
6400	Maximum					985			985	1,075
6500	42" wide, 42" long, minimum					232			232	255
6600	Maximum					815			815	895
6700	Round, 24" diameter, minimum					87.50			87.50	96.50
6800	Maximum					335			335	370
6900	30" diameter, minimum					115			115	127
7000	Maximum					500			500	550
7200	36" diameter, minimum					148			148	163
7400	Maximum					575			575	630
7600	42" diameter, minimum					237			237	261
8000	Maximum					875			875	965
8200	48" diameter, minimum					267			267	293
8400	Maximum					930			930	1,025
8800	54" diameter, minimum					425			425	470
9000	Maximum					1,200			1,200	1,325
9200	60" diameter, minimum					455			455	500
9400	Maximum				▼	1,500			1,500	1,650

12 54 16.50 Table Tops

12 54 16.50 Table Tops		Crew	Daily Output	Labor-Hours	Unit	Material	Labor	Equipment	Total	Total Incl O&P
0010	**TABLE TOPS** laminate top and hardwood edge									
0040	24" wide, 24" long, minimum				Ea.	253			253	278
1000	Maximum					710			710	780
1200	30" long, minimum					280			280	310
1400	Maximum					760			760	835
1600	36" long, minimum					254			254	279
1800	Maximum					795			795	875
2000	48" long, minimum					370			370	405
2100	Maximum					875			875	965
2200	60" long, minimum					525			525	580
2400	Maximum					960			960	1,050
2600	72" long, minimum					595			595	655
2800	Maximum					1,075			1,075	1,175
3000	30" wide, 30" long, minimum					305			305	335
3100	Maximum					810			810	890
3200	36" long, minimum					340			340	370
3400	Maximum					865			865	955
3600	42" long, minimum					360			360	395
3800	Maximum					1,150			1,150	1,275
4000	48" long, minimum					400			400	440
4100	Maximum					935			935	1,025
4200	60" long, minimum					540			540	595
4300	Maximum					1,000			1,000	1,100
4400	72" long, minimum					435			435	475
4600	Maximum					1,050			1,050	1,150
4800	36" wide, 36" long, minimum					370			370	410
5000	Maximum					890			890	975
5200	48" long, minimum					435			435	480
5400	Maximum				▼	1,025			1,025	1,125

12 54 16 – Restaurant Furniture

12 54 16.50 Table Tops	Crew	Daily Output	Labor-Hours	Unit	Material	Labor	Equipment	2015 Bare Costs Total	Total Incl O&P
5600 60″ long, minimum				Ea.	620			620	680
5800 Maximum					1,125			1,125	1,225
6000 72″ long, minimum					675			675	740
6200 Maximum					1,225			1,225	1,350
6400 42″ wide, 42″ long, minimum					450			450	495
6600 Maximum					1,075			1,075	1,175
6800 Round, 24″ diameter, minimum					460			460	505
6900 Maximum					750			750	830
7000 30″ diameter, minimum					505			505	560
7200 Maximum					850			850	930
7400 36″ diameter, minimum					560			560	615
7600 Maximum					1,000			1,000	1,100
7800 42″ diameter, minimum					650			650	715
7900 Maximum					1,225			1,225	1,350
8000 48″ diameter, minimum					695			695	765
8150 Maximum					1,300			1,300	1,450
8200 54″ diameter, minimum					1,050			1,050	1,150
8250 Maximum					1,700			1,700	1,875
8300 60″ diameter, minimum					1,050			1,050	1,150
8350 Maximum				↓	1,700			1,700	1,875

12 54 16.60 Table Tops	Crew	Daily Output	Labor-Hours	Unit	Material	Labor	Equipment	Total	Total Incl O&P
0010 **TABLE TOPS** Polyester resin top & edge									
0050 Add inlay top material to cost									
0100 24″ wide, 24″ long				Ea.	450			450	495
1200 30″ long					450			450	495
1600 36″ long					470			470	515
2000 48″ long					485			485	530
2400 60″ long					570			570	630
2800 72″ long					670			670	740
3200 30″ wide, 30″ long					475			475	520
3600 36″ long					560			560	620
4000 42″ long					630			630	690
4400 48″ long					620			620	685
4800 60″ long					705			705	780
5200 72″ long					860			860	945
5600 36″ wide, 36″ long					560			560	620
6000 48″ long					680			680	745
6400 60″ long					825			825	910
6800 72″ long					995			995	1,100
7200 42″ wide, 42″ long					660			660	725
7600 Round, 24″ diameter					630			630	695
8000 30″ diameter					655			655	720
8400 36″ diameter					775			775	855
8800 42″ diameter					1,050			1,050	1,150
9150 48″ diameter					1,150			1,150	1,275
9250 54″ diameter					1,550			1,550	1,725
9350 60″ diameter				↓	1,625			1,625	1,775

12 54 16.70 Booths	Crew	Daily Output	Labor-Hours	Unit	Material	Labor	Equipment	Total	Total Incl O&P
0010 **BOOTHS**									
1000 Banquet, upholstered seat and back, custom									
1500 Straight, minimum	2 Carp	40	.400	L.F.	201	18.80		219.80	252
1520 Maximum	↓	36	.444	↓	390	21		411	465

12 54 Hospitality Furniture

12 54 16 – Restaurant Furniture

12 54 16.70 Booths		Crew	Daily Output	Labor-Hours	Unit	2015 Bare Costs Material	Labor	Equipment	Total	Total Incl O&P
1600	"L" or "U" shape, minimum	2 Carp	35	.457	L.F.	205	21.50		226.50	261
1620	Maximum	↓	30	.533	↓	365	25		390	440
1800	Upholstered outside finished backs for									
1810	single booths and custom banquets									
1820	Minimum	2 Carp	44	.364	L.F.	23	17.05		40.05	53.50
1840	Maximum	"	40	.400	"	69.50	18.80		88.30	108
3000	Fixed seating, one piece plastic chair and									
3010	plastic laminate table top									
3100	Two seat, 24" x 24" table, minimum	F-7	30	1.067	Ea.	810	45		855	965
3120	Maximum		26	1.231		1,150	52		1,202	1,350
3200	Four seat, 24" x 48" table, minimum		28	1.143		805	48.50		853.50	965
3220	Maximum		24	1.333		1,375	56.50		1,431.50	1,600
3300	Six seat, 24" x 76" table, minimum		26	1.231		1,450	52		1,502	1,675
3320	Maximum		22	1.455		2,000	61.50		2,061.50	2,300
3400	Eight seat, 24" x 102" table, minimum		20	1.600		1,950	67.50		2,017.50	2,225
3420	Maximum	↓	18	1.778	↓	2,525	75		2,600	2,900
4000	Free standing, wood fiber core with									
4010	plastic laminate face, single booth									
4100	24" wide	2 Carp	38	.421	Ea.	385	19.75		404.75	455
4150	48" wide		34	.471		465	22		487	545
4200	60" wide		30	.533		570	25		595	665
4300	Double booth, 24" wide		32	.500		570	23.50		593.50	665
4350	48" wide		28	.571		760	27		787	880
4400	60" wide	↓	26	.615	↓	955	29		984	1,100
4600	Upholstered seat and back									
4650	Foursome, single booth, minimum	2 Carp	38	.421	Ea.	435	19.75		454.75	510
4700	Maximum		30	.533		1,650	25		1,675	1,850
4800	Double booth, minimum		32	.500		955	23.50		978.50	1,100
4850	Maximum	↓	26	.615	↓	2,550	29		2,579	2,850
5000	Mount in floor, wood fiber core with									
5010	plastic laminate face, single booth									
5050	24" wide	F-7	30	1.067	Ea.	310	45		355	415
5100	48" wide		28	1.143		395	48.50		443.50	515
5150	60" wide		26	1.231		570	52		622	710
5200	Double booth, 24" wide		26	1.231		500	52		552	635
5250	48" wide		24	1.333		640	56.50		696.50	800
5300	60" wide	↓	22	1.455	↓	910	61.50		971.50	1,100

12 55 Detention Furniture

12 55 13 – Detention Bunks

12 55 13.13 Cots

		Crew	Daily Output	Labor-Hours	Unit	Material	Labor	Equipment	Total	Total Incl O&P
0010	**COTS**									
2500	Bolted, single, painted steel	E-4	20	1.600	Ea.	335	85	7.30	427.30	530
2700	Stainless steel	"	20	1.600	"	975	85	7.30	1,067.30	1,225

12 56 Institutional Furniture

12 56 33 – Classroom Furniture

12 56 33.10 Furniture, School		Crew	Daily Output	Labor-Hours	Unit	Material	2015 Bare Costs Labor	Equipment	Total	Total Incl O&P
0010	**FURNITURE, SCHOOL**									
0500	Classroom, movable chair & desk type, minimum				Set				73.50	81
0600	Maximum				"				155	171
1000	Chair, molded plastic									
1100	Integral tablet arm, minimum				Ea.	100			100	110
1150	Maximum					189			189	207
2000	Desk, single pedestal, top book compartment, minimum					91.50			91.50	101
2020	Maximum					188			188	206
2100	Side book compartment, minimum					131			131	144
2120	Maximum					173			173	190
2200	Flip top, minimum					228			228	251
2220	Maximum					277			277	305
3000	Preschool, moulded plastic chairs, minimum					22.50			22.50	25
3020	Maximum					59.50			59.50	65.50
3800	Tables, plastic laminate top, 24″ wide, 36″ long					66			66	72.50
3820	48″ long					68.50			68.50	75
3840	60″ long					134			134	148
3900	30″ wide, 48″ long					98.50			98.50	109
3920	60″ long					147			147	162
3940	72″ long					173			173	190
4000	36″ wide, 36″ long					126			126	139
4020	60″ long					187			187	206
4040	72″ long					232			232	255
4200	Round, 36″ diam.					79			79	87
4220	42″ diam.					100			100	110
4230	48″ diam.					105			105	115
4240	60″ diam.				↓	186			186	205

12 56 43 – Dormitory Furniture

12 56 43.10 Dormitory Furnishings		Crew	Daily Output	Labor-Hours	Unit	Material	Labor	Equipment	Total	Total Incl O&P
0010	**DORMITORY FURNISHINGS**									
0200	Bookcase, two shelf				Ea.	185			185	203
0300	Bunkable bed, twin, minimum					375			375	415
0320	Maximum					550			550	605
1000	Chest, four drawer, minimum					360			360	395
1020	Maximum				↓	690			690	760
1050	Built-in, minimum	2 Carp	13	1.231	L.F.	120	58		178	227
1150	Maximum		10	1.600		221	75		296	365
1200	Desk top, built-in, laminated plastic, 24″ deep, minimum		50	.320		44	15		59	73
1300	Maximum		40	.400		132	18.80		150.80	177
1450	30″ deep, minimum		50	.320		56.50	15		71.50	86.50
1550	Maximum		40	.400		247	18.80		265.80	305
1750	Dressing unit, built-in, minimum		12	1.333		179	62.50		241.50	300
1850	Maximum	↓	8	2	↓	540	94		634	745
2000	Desk, single pedestal				Ea.	530			530	580
2100	Hutch/bookcase, with light					265			265	292
3000	Ladder					131			131	144
4000	Mirror, with frame					120			120	132
5000	Nightstand					295			295	325
6000	Wall unit, open shelving					480			480	525
7000	Wardrobe				↓	545			545	595
8000	Rule of thumb: total cost for furniture, minimum				Student				2,525	2,800
8050	Maximum				"				4,850	5,350

12 56 51 – Library Furniture

12 56 51.10 Library Furnishings		Crew	Daily Output	Labor-Hours	Unit	Material	2015 Bare Costs Labor	Equipment	Total	Total Incl O&P
0010	**LIBRARY FURNISHINGS**									
0100	Attendant desk, 36" x 62" x 29" high	1 Carp	16	.500	Ea.	1,900	23.50		1,923.50	2,150
0200	Book display, "A" frame display, both sides, 42" x 42" x 60" high		16	.500		1,250	23.50		1,273.50	1,425
0220	Table with bulletin board, 42" x 24" x 49" high		16	.500		740	23.50		763.50	855
0300	Book trucks, descending platform,									
0320	Small, 14" x 30" x 35" high	1 Carp	16	.500	Ea.	750	23.50		773.50	865
0340	Large, 14" x 40" x 42" high		16	.500		790	23.50		813.50	910
0800	Card catalogue, 30 tray unit		16	.500		3,350	23.50		3,373.50	3,725
0840	60 tray unit		16	.500		6,675	23.50		6,698.50	7,400
0880	72 tray unit	2 Carp	16	1		8,700	47		8,747	9,625
0960	120 tray unit	"	16	1		12,100	47		12,147	13,400
1000	Carrels, single face, initial unit	1 Carp	16	.500		820	23.50		843.50	940
1050	Additional unit	"	16	.500		775	23.50		798.50	895
1500	Double face, initial unit	2 Carp	16	1		1,300	47		1,347	1,500
1550	Additional unit		16	1		730	47		777	880
1600	Cloverleaf		11	1.455		2,400	68.50		2,468.50	2,750
1710	Carrels, hardwood, 36" x 24", minimum	1 Carp	5	1.600		785	75		860	985
1720	Maximum		4	2		2,000	94		2,094	2,350
2000	Chairs, sled base, arms, minimum		24	.333		204	15.65		219.65	250
2050	Maximum		16	.500		186	23.50		209.50	244
2100	No arms, minimum		24	.333		133	15.65		148.65	172
2150	Maximum		16	.500		160	23.50		183.50	214
2500	Standard leg base, arms, minimum		24	.333		198	15.65		213.65	243
2520	Maximum		16	.500		435	23.50		458.50	515
2600	No arms, minimum		24	.333		170	15.65		185.65	213
2620	Maximum		16	.500		290	23.50		313.50	360
2700	Card catalog file, 60 trays, complete					7,925			7,925	8,700
2720	Alternate method: each tray					132			132	145
3000	Charge desk, modular unit, 35" x 27" x 39" high									
3020	Wood front and edges, plastic laminate tops									
3100	Book return	1 Carp	16	.500	Ea.	780	23.50		803.50	895
3150	Book truck port		16	.500		730	23.50		753.50	845
3400	Corner		16	.500		1,175	23.50		1,198.50	1,350
3450	Cupboard		16	.500		1,425	23.50		1,448.50	1,625
3500	Detachable end panel		16	.500		350	23.50		373.50	425
3550	Gate		16	.500		500	23.50		523.50	585
3600	Knee space		16	.500		1,250	23.50		1,273.50	1,425
3650	Open storage		16	.500		1,200	23.50		1,223.50	1,375
3700	Station charge		16	.500		1,475	23.50		1,498.50	1,675
3750	Work station		16	.500		1,325	23.50		1,348.50	1,525
3800	Charging desk, built-in, with counter, plastic laminated top		7	1.143	L.F.	305	53.50		358.50	425
4000	Dictionary stand, stationary		16	.500	Ea.	690	23.50		713.50	795
4020	Revolving		16	.500		214	23.50		237.50	275
4200	Exhibit case, table style, 60" x 28" x 36"		11	.727		2,625	34		2,659	2,925
4500	Globe stand		16	.500		815	23.50		838.50	935
4800	Magazine rack		16	.500		1,000	23.50		1,023.50	1,150
5000	Newspaper rack		16	.500		810	23.50		833.50	935
6010	Bookshelf, metal, 90" high, 10" shelf, double face		11.50	.696	L.F.	150	32.50		182.50	219
6020	Single face		12	.667	"	124	31.50		155.50	189
6050	For 8" shelving, subtract from above					10%				
6060	For 12" shelving, add to above					10%				
6070	For 42" high with countertop, subtract from above					20%				

12 56 Institutional Furniture

12 56 51 – Library Furniture

12 56 51.10 Library Furnishings		Crew	Daily Output	Labor-Hours	Unit	Material	Labor	Equipment	Total	Total Incl O&P
						2015 Bare Costs				
6100	Mobile compacted shelving, hand crank, 9'-0" high									
6110	Double face, including track, 3' section				Ea.	1,175			1,175	1,275
6150	For electrical operation, add					25%				
6200	Magazine shelving, 82" high, 12" deep, single face	1 Carp	11.50	.696	L.F.	151	32.50		183.50	220
6210	Double face		11.50	.696	"	248	32.50		280.50	325
7100	Index, single tier, 48" x 72"		16	.500	Ea.	2,125	23.50		2,148.50	2,375
7150	Double tier, 48" x 72"		16	.500		2,425	23.50		2,448.50	2,725
7200	Reading table, laminated top, 60" x 36"					705			705	775
8000	Parsons table, 29" high, plastic lam. top, wood legs & edges									
8010	36" x 36"	2 Carp	16	1	Ea.	475	47		522	600
8020	36" x 60"		16	1		600	47		647	735
8030	36" x 72"		16	1		750	47		797	900
8040	36" x 84"		16	1		1,100	47		1,147	1,300
8050	42" x 90"		16	1		1,225	47		1,272	1,425
8060	48" x 72"		16	1		1,200	47		1,247	1,400
8070	48" x 120"		16	1		2,200	47		2,247	2,500
8110	42" diameter		16	1		605	47		652	740
8120	48" diameter		16	1		700	47		747	845
8130	60" diameter		16	1		920	47		967	1,100
8500	Study, panel ends, plastic laminate surfaces 29" high, 36" x 60"		16	1		545	47		592	670
8510	36" x 72"		16	1		515	47		562	640
8515	36" x 90"		16	1		590	47		637	725
8525	48" x 72"		16	1		1,125	47		1,172	1,300

12 56 70 – Healthcare Furniture

12 56 70.10 Furniture, Hospital		Crew	Daily Output	Labor-Hours	Unit	Material	Labor	Equipment	Total	Total Incl O&P
0010	**FURNITURE, HOSPITAL**									
0020	Beds, manual, minimum				Ea.	800			800	880
0100	Maximum					2,550			2,550	2,800
0600	All electric hospital beds, minimum					1,625			1,625	1,800
0700	Maximum					4,500			4,500	4,925
0900	Manual, nursing home beds, minimum					770			770	850
1000	Maximum					2,200			2,200	2,400
1020	Overbed table, laminated top, minimum					460			460	505
1040	Maximum					885			885	975
1100	Patient wall systems, not incl. plumbing, minimum				Room	1,425			1,425	1,575
1200	Maximum				"	1,925			1,925	2,125
2000	Geriatric chairs, minimum				Ea.	440			440	480
2020	Maximum				"	765			765	840

12 59 Systems Furniture

12 59 13 – Panel-Hung Component System Furniture

12 59 13.10 Furniture, Office Systems		Crew	Daily Output	Labor-Hours	Unit	Material	Labor	Equipment	Total	Total Incl O&P
0010	**FURNITURE, OFFICE SYSTEMS**, Panel hung									
0100	Acoustic panel, 43" high, 24" wide, NRC .85				Ea.	320			320	350
0110	30" wide					355			355	395
0120	36" wide					395			395	435
0130	48" wide					470			470	520
0200	64" high, 30" wide					425			425	465
0210	36" wide					455			455	500
0220	42" wide					525			525	580

12 59 Systems Furniture

12 59 13 – Panel-Hung Component System Furniture

12 59 13.10 Furniture, Office Systems		Crew	Daily Output	Labor-Hours	Unit	Material	2015 Bare Costs Labor	Equipment	Total	Total Incl O&P
0230	48" wide				Ea.	560			560	615
0240	60" wide					615			615	675
0400	Bookshelf, 36" wide					148			148	163
0410	42" wide					155			155	171
0420	48" wide				↓	160			160	176
1000	Connectors, brackets and supports									
1200	Bracket, cantilever, 20" deep, 24" deep				Ea.	50			50	55
1220	Countertop, per pair					21			21	23
1240	Flat, 20" deep, 24" wide					25.50			25.50	28.50
1260	Worksurface kit, per pair					21			21	23
1300	Connector kit, ell, 43" high, 90" deep					89			89	98
1320	64" high					52.50			52.50	58
1340	Straight, 43" high					36.50			36.50	40.50
1360	64" high					36.50			36.50	40.50
1400	Support column for peninsula					128			128	141
2000	Countertop, 15" deep, 36" wide					169			169	186
2020	48" wide					188			188	207
3000	Duplex receptacle circuit 1					18.35			18.35	20
3010	Circuit 2					18.35			18.35	20
3100	Electric power harness, 36" wide					120			120	132
3120	42" wide					252			252	277
3140	60" wide					126			126	139
4000	End cover, panel, 43" high					36.50			36.50	40.50
4020	64" high					36.50			36.50	40.50
4100	Finish, variable height, 2-way					55			55	60.50
4600	Overhead cabinet with door, 42" wide					345			345	380
4620	60" wide					525			525	580
5000	Pedestal spacer, 22" deep, 15" wide					60			60	66
5100	28" deep, 15" wide					72			72	79.50
5200	Box, box, file, 22" deep, 26" high					420			420	460
5300	28" deep					440			440	480
5400	File, file, 22" deep, 26" high					390			390	430
5600	Lateral file, 2 drawer, 30" wide					555			555	610
6000	Power, base in-feed cable					133			133	147
7400	Task light, recessed, 30" - 36" wide					163			163	179
7420	42" - 48" wide					176			176	194
7600	60" wide					191			191	210
8000	Worksurface, radius edge, 24" deep, 36" wide					175			175	192
8020	42" wide					215			215	237
8040	48" wide					230			230	253
8060	60" wide					288			288	315
8080	72" wide					325			325	360
8200	30" deep, 42" wide					276			276	305
8220	60" wide					320			320	355
8240	72" wide					365			365	405
8400	Corner, 24" deep, 36" wide					420			420	465
8420	42" wide					480			480	525
8500	Peninsula, 36" wide, 66" long				↓	550			550	605

12 59 Systems Furniture

12 59 13 – Panel-Hung Component System Furniture

12 59 13.10 Furniture, Office Systems		Crew	Daily Output	Labor-Hours	Unit	2015 Bare Costs Material	Labor	Equipment	Total	Total Incl O&P
9000	For installation of systems furniture, add 5%									

12 59 16 – Free-Standing Component System Furniture

12 59 16.10 Furniture, Office Systems		Crew	Daily Output	Labor-Hours	Unit	Material	Labor	Equipment	Total	Total Incl O&P
0010	**FURNITURE, OFFICE SYSTEMS**, Freestanding									
0100	Desk table, 24" deep x 48" wide				Ea.	251			251	276
0110	30" deep x 48" wide					375			375	415
0120	30" deep x 60" wide					495			495	545
0130	30" deep x 72" wide					715			715	790
0200	File, mobile, 2 drawer, 22" deep					480			480	525
0210	Suspended, 2 file					500			500	550
0220	Undercounter					580			580	635
0230	2 box, 1 file					590			590	645
0400	Keyboard platform with articulating arm					191			191	210
0500	End support leg					167			167	183
0800	Modesty panel, 48" wide					101			101	111
0810	Return					110			110	121
0820	Pencil/utility drawer					142			142	156
0830	Privacy screen, 48" wide x 17-20" high					294			294	325
0840	72" wide					455			455	500
0910	Printer stand, 30" deep x 36" wide					625			625	690
0920	Universal					660			660	725
0930	Paper basket					73.50			73.50	81
1010	Storage unit with doors, 60" wide					770			770	845
1020	72" wide					1,075			1,075	1,175
1210	Table, peninsula, 30" x 72"					915			915	1,000
1220	Return, 24" x 48"					750			750	825
1230	Round conference return, 42" diameter					550			550	605
1510	Work surface, 24" deep x 48" wide					340			340	375
1520	30" deep x 30" wide					320			320	355
1530	60" wide					440			440	485
1540	72" wide					435			435	480
1550	Corner bridge, 30" x 42"					194			194	213
1560	Peninsula, 30" x 60"					550			550	605
1570	Connector plate					18.35			18.35	20

12 59 23 – Desk System Furniture

12 59 23.10 Work Stations		Crew	Daily Output	Labor-Hours	Unit	Material	Labor	Equipment	Total	Total Incl O&P
0010	**WORK STATIONS**									
1000	Secretarial work station, minimum				Ea.	4,125			4,125	4,550
1020	Maximum					14,100			14,100	15,500
1400	Management work station, minimum					4,700			4,700	5,175
1420	Maximum					15,200			15,200	16,700
1800	Executive work station, minimum					10,400			10,400	11,500
1820	Maximum					26,800			26,800	29,500
1840	For installation of systems furniture, add 15%									

12 61 Fixed Audience Seating

12 61 13 – Upholstered Audience Seating

12 61 13.13 Auditorium Chairs		Crew	Daily Output	Labor-Hours	Unit	Material	2015 Bare Costs Labor	Equipment	Total	Total Incl O&P
0010	**AUDITORIUM CHAIRS**									
2000	All veneer construction	2 Carp	22	.727	Ea.	232	34		266	310
2200	Veneer back, padded seat		22	.727		242	34		276	320
2350	Fully upholstered, spring seat		22	.727		242	34		276	320
2450	For tablet arms, add					68			68	74.50
2500	For fire retardancy, CATB-133, add					30			30	33

12 61 13.23 Lecture Hall Seating		Crew	Daily Output	Labor-Hours	Unit	Material	Labor	Equipment	Total	Total Incl O&P
0010	**LECTURE HALL SEATING**									
1000	Pedestal type, minimum	2 Carp	22	.727	Ea.	191	34		225	266
1200	Maximum	"	14.50	1.103	"	490	52		542	625

12 63 Stadium and Arena Seating

12 63 13 – Stadium and Arena Bench Seating

12 63 13.13 Bleachers		Crew	Daily Output	Labor-Hours	Unit	Material	Labor	Equipment	Total	Total Incl O&P
0010	**BLEACHERS**									
3000	Telescoping, manual to 15 tier, minimum	F-5	65	.492	Seat	91.50	23.50		115	140
3100	Maximum		60	.533		137	25.50		162.50	193
3300	16 to 20 tier, minimum		60	.533		220	25.50		245.50	284
3400	Maximum		55	.582		275	27.50		302.50	345
3600	21 to 30 tier, minimum		50	.640		229	30.50		259.50	300
3700	Maximum		40	.800		300	38		338	390
3900	For integral power operation, add, minimum	2 Elec	300	.053		46	2.92		48.92	55
4000	Maximum	"	250	.064		73.50	3.50		77	86
5000	Benches, folding, in wall, 14′ table, 2 benches	L-4	2	12	Set	775	530		1,305	1,725

12 67 Pews and Benches

12 67 13 – Pews

12 67 13.13 Sanctuary Pews		Crew	Daily Output	Labor-Hours	Unit	Material	Labor	Equipment	Total	Total Incl O&P
0010	**SANCTUARY PEWS**									
1500	Bench type, hardwood, minimum	1 Carp	20	.400	L.F.	94.50	18.80		113.30	135
1550	Maximum	"	15	.533		187	25		212	247
1570	For kneeler, add					22.50			22.50	24.50

12 92 Interior Planters and Artificial Plants

12 92 13 – Interior Artificial Plants

12 92 13.10 Plants		Crew	Daily Output	Labor-Hours	Unit	Material	Labor	Equipment	Total	Total Incl O&P
0010	**PLANTS** Permanent only, weighted									
0020	Preserved or polyester leaf, natural wood									
0030	Or molded trunk. For pots see Section 12 92 33.10									
0040	For fill see Section 31 23 16.00									
0100	Plants, acuba, 5′ high				Ea.	350			350	385
0200	Apidistra, 4′ high					310			310	345
0300	Beech, variegated, 5′ high					305			305	335
0400	Birds nest fern, 6′ high					550			550	605
0500	Croton, 4′ high					134			134	147
0600	Diffenbachia, 3′ high					123			123	136
0700	Ficus Benjamina, 3′ high					146			146	161

12 92 Interior Planters and Artificial Plants

12 92 13 – Interior Artificial Plants

12 92 13.10 Plants		Crew	Daily Output	Labor-Hours	Unit	Material	Labor	Equipment	Total	Total Incl O&P
						2015 Bare Costs				
0720	6′ high				Ea.	370			370	405
0760	Nitida, 3′ high					330			330	360
0780	6′ high					575			575	635
0800	Helicona					305			305	335
1200	Palm, green date fan, 5′ high					315			315	350
1220	7′ high					400			400	440
1280	Green chamdora date, 5′ high					288			288	315
1300	7′ high					440			440	485
1400	Green giant, 7′ high					685			685	750
1600	Rubber plant, 5′ high					232			232	255
1800	Schefflera, 3′ high					159			159	175
1820	4′ high					190			190	209
1840	5′ high					300			300	330
1860	6′ high					415			415	455
1880	7′ high					620			620	680
1900	Spathiphyllum, 3′ high					220			220	242
4000	Trees, polyester or preserved, with									
4020	Natural trunks									
4100	Acuba, 10′ high				Ea.	1,725			1,725	1,900
4120	12′ high					2,425			2,425	2,675
4140	14′ high					3,375			3,375	3,725
4400	Bamboo, 10′ high					755			755	830
4420	12′ high					875			875	960
4800	Beech, 10′ high					2,325			2,325	2,550
4820	12′ high					2,675			2,675	2,925
4840	14′ high					3,600			3,600	3,950
5000	Birch, 10′ high					2,325			2,325	2,550
5020	12′ high					2,675			2,675	2,925
5040	14′ high					2,900			2,900	3,200
5060	16′ high					3,600			3,600	3,950
5080	18′ high					4,050			4,050	4,475
5500	Ficus Benjamina, 10′ high					1,700			1,700	1,875
5520	12′ high					2,175			2,175	2,400
5540	14′ high					2,450			2,450	2,700
5560	16′ high					3,425			3,425	3,750
5580	18′ high					5,125			5,125	5,650
6000	Magnolia, 10′ high					2,500			2,500	2,750
6020	12′ high					3,550			3,550	3,900
6040	14′ high					4,850			4,850	5,350
6500	Maple, 10′ high					267			267	294
6520	12′ high					400			400	440
6540	14′ high					495			495	545
7000	Palms, chamadora, 10′ high					490			490	535
7020	12′ high					620			620	685
7040	Date, 10′ high					670			670	740
7060	12′ high					770			770	845

12 92 33 – Interior Planters

12 92 33.10 Planters		Crew	Daily Output	Labor-Hours	Unit	Material	Labor	Equipment	Total	Total Incl O&P
0010	**PLANTERS**									
1000	Fiberglass, hanging, 12″ diameter, 7″ high				Ea.	118			118	129
1100	15″ diameter, 7″ high					171			171	188
1200	36″ diameter, 8″ high					235			235	259

12 92 Interior Planters and Artificial Plants

12 92 33 – Interior Planters

12 92 33.10 Planters		Crew	Daily Output	Labor-Hours	Unit	Material	Labor	Equipment	Total	Total Incl O&P
						2015 Bare Costs				
1500	Rectangular, 48" long, 16" high x 15" wide				Ea.	665			665	735
1550	16" high x 24" wide					725			725	800
1600	24" high x 24" wide					1,075			1,075	1,175
1650	60" long, 30" high, 28" wide					1,025			1,025	1,125
1700	72" long, 16" high, 15" wide					850			850	935
1750	21" high, 24" wide					1,025			1,025	1,125
1800	30" high, 24" wide					1,175			1,175	1,300
2000	Round, 12" diameter, 13" high					154			154	170
2050	25" high					203			203	223
2150	14" diameter, 15" high					167			167	183
2200	16" diameter, 16" high					188			188	207
2250	18" diameter, 19" high					227			227	250
2300	23" high					295			295	325
2350	20" diameter, 16" high					205			205	226
2400	18" high					226			226	248
2450	21" high					291			291	320
2500	22" diameter, 10" high					239			239	263
2550	24" diameter, 16" high					285			285	315
2600	19" high					340			340	375
2650	25" high					395			395	430
2700	36" high					610			610	675
2750	48" high					855			855	940
2800	30" diameter, 16" high					325			325	355
2850	18" high					325			325	360
2900	21" high					365			365	400
3000	24" high					375			375	415
3350	27" high					415			415	455
3400	36" diameter, 16" high					410			410	450
3450	18" high					425			425	470
3500	21" high					460			460	505
3550	24" high					490			490	540
3600	27" high					530			530	585
3650	30" high					575			575	635
3700	48" diameter, 16" high					675			675	745
3750	21" high					730			730	800
3800	24" high					815			815	900
3850	27" high					860			860	945
3900	30" high					930			930	1,025
3950	36" high					980			980	1,075
4000	60" diameter, 16" high					965			965	1,050
4100	21" high					1,100			1,100	1,200
4150	27" high					1,225			1,225	1,350
4200	30" high					1,350			1,350	1,475
4250	33" high					1,550			1,550	1,700
4300	36" high					1,875			1,875	2,075
4400	39" high					2,025			2,025	2,225
5000	Square, 10" side, 20" high					193			193	212
5100	14" side, 15" high					234			234	257
5200	18" side, 19" high					230			230	253
5300	20" side, 16" high					282			282	310
5320	18" high					475			475	520
5340	21" high					375			375	415
5400	24" side, 16" high					350			350	385

12 92 Interior Planters and Artificial Plants

12 92 33 – Interior Planters

12 92 33.10 Planters		Crew	Daily Output	Labor-Hours	Unit	Material	2015 Bare Costs Labor	Equipment	Total	Total Incl O&P
5420	21" high				Ea.	540			540	590
5440	25" high					515			515	570
5460	30" side, 16" high					555			555	610
5480	24" high					665			665	730
5490	27" high					885			885	975
5500	Round, 36" diameter, 16" high					500			500	550
5510	18" high					695			695	765
5520	21" high					765			765	845
5530	24" high					850			850	935
5540	27" high					965			965	1,075
5550	30" high					1,325			1,325	1,475
5800	48" diameter, 16" high					670			670	740
5820	21" high					750			750	825
5840	24" high					935			935	1,025
5860	27" high					1,025			1,025	1,125
5880	30" high					305			305	335
5900	60" diameter, 16" high					665			665	730
5920	21" high					770			770	845
5940	27" high					840			840	925
5960	30" high					950			950	1,050
5980	36" high					1,025			1,025	1,125
6000	Metal bowl, 32" diameter, 8" high, minimum					555			555	610
6050	Maximum					755			755	830
6100	Rectangle, 30" long x 12" wide, 6" high, minimum					430			430	475
6200	Maximum					600			600	660
6300	36" long 12" wide, 6" high, minimum					850			850	935
6400	Maximum					475			475	520
6500	Square, 15" side, minimum					605			605	665
6600	Maximum					935			935	1,025
6700	20" side, minimum					1,350			1,350	1,475
6800	Maximum					1,025			1,025	1,125
6900	Round, 6" diameter x 6" high, minimum					62.50			62.50	69
7000	Maximum					74.50			74.50	82
7100	8" diameter x 8" high, minimum					83.50			83.50	92
7200	Maximum					89.50			89.50	98.50
7300	10" diameter x 11" high, minimum					107			107	118
7400	Maximum					159			159	174
7420	12" diameter x 13" high, minimum					99			99	109
7440	Maximum					180			180	198
7500	14" diameter x 15" high, minimum					134			134	147
7550	Maximum					235			235	258
7580	16" diameter x 17" high, minimum					144			144	159
7600	Maximum					257			257	282
7620	18" diameter x 19" high, minimum					175			175	192
7640	Maximum					315			315	345
7680	22" diameter x 20" high, minimum					220			220	242
7700	Maximum					400			400	440
7750	24" diameter x 21" high, minimum					285			285	315
7800	Maximum					535			535	590
7850	31" diameter x 18" high, minimum					620			620	685
7900	Maximum					1,575			1,575	1,750
7950	38" diameter x 24" high, minimum					1,050			1,050	1,150
8000	Maximum					2,675			2,675	2,925

12 92 Interior Planters and Artificial Plants

12 92 33 – Interior Planters

12 92 33.10 Planters		Crew	Daily Output	Labor-Hours	Unit	2015 Bare Costs Material	Labor	Equipment	Total	Total Incl O&P
8050	48" diameter x 24" high, minimum				Ea.	1,400			1,400	1,525
8150	Maximum					2,750			2,750	3,025
8500	Plastic laminate faced, fiberglass liner, square									
8520	14" sq., 15" high				Ea.	630			630	690
8540	24" sq., 16" high					805			805	885
8580	36" sq., 21" high					1,100			1,100	1,200
8600	Rectangle 36" long, 12" wide, 10" high					680			680	750
8650	48" long, 12" wide, 10" high					755			755	830
8700	48" long, 12" wide, 24" high					1,125			1,125	1,225
8750	Wood, fiberglass liner, square									
8780	14" square, 15" high, minimum				Ea.	455			455	500
8800	Maximum					560			560	615
8820	24" sq., 16" high, minimum					560			560	615
8840	Maximum					740			740	815
8860	36" sq., 21" high, minimum					720			720	790
8880	Maximum					1,075			1,075	1,200
9000	Rectangle, 36" long x 12" wide, 10" high, minimum					500			500	550
9050	Maximum					650			650	715
9100	48" long x 12" wide, 10" high, minimum					540			540	595
9120	Maximum					680			680	750
9200	48" long x 12" wide, 24" high, minimum					650			650	715
9300	Maximum					950			950	1,050
9400	Plastic cylinder, molded, 10" diameter, 10" high					18.45			18.45	20.50
9500	11" diameter, 11" high					35			35	38.50
9600	13" diameter, 12" high					57			57	63
9700	16" diameter, 14" high					66.50			66.50	73

12 93 Interior Public Space Furnishings

12 93 23 – Trash and Litter Receptacles

12 93 23.10 Trash Receptacles

Line	Description	Crew	Daily Output	Labor-Hours	Unit	Material	Labor	Equipment	Total	Total Incl O&P
0010	**TRASH RECEPTACLES**									
0020	Fiberglass, 2' square, 18" high	2 Clab	30	.533	Ea.	560	20		580	650
0100	2' square, 2'-6" high		30	.533		795	20		815	910
0300	Circular , 2' diameter, 18" high		30	.533		475	20		495	555
0400	2' diameter, 2'-6" high		30	.533		530	20		550	620
0500	Recycled plastic, var colors, round, 32 gal., 28" x 38" H [G]		5	3.200		510	120		630	760
0510	32 gal., 31" x 32" H [G]		5	3.200		585	120		705	840
1000	Alum. frame, hardboard panels, steel drum base,									
1020	30 gal. capacity, silk screen on plastic finish	2 Clab	25	.640	Ea.	460	24		484	545
1040	Aggregate finish		25	.640		590	24		614	690
1100	50 gal. capacity, silk screen on plastic finish		20	.800		695	30		725	815
1140	Aggregate finish		20	.800		750	30		780	875
1200	Formed plastic liner, 14 gal., silk screen on plastic finish		40	.400		390	15.05		405.05	455
1240	Aggregate finish		40	.400		430	15.05		445.05	495
1300	30 gal. capacity, silk screen on plastic finish		35	.457		510	17.20		527.20	590
1340	Aggregate finish		35	.457		560	17.20		577.20	645
1400	Redwood slats, plastic liner, leg base, 14 gal. capacity,									
1420	Varnish w/routed message	2 Clab	40	.400	Ea.	415	15.05		430.05	480
2000	Concrete, precast, 2' to 2-1/2' wide, 3' high, sandblasted	"	15	1.067	"	730	40		770	870
3000	Galv. steel frame and panels, leg base, poly bag retainer,									
3020	40 gal. capacity, silk screen on enamel finish	2 Clab	25	.640	Ea.	430	24		454	510

12 93 Interior Public Space Furnishings

12 93 23 – Trash and Litter Receptacles

12 93 23.10 Trash Receptacles		Crew	Daily Output	Labor-Hours	Unit	Material	2015 Bare Costs Labor	Equipment	Total	Total Incl O&P
3040	Aggregate finish	2 Clab	25	.640	Ea.	470	24		494	555
3200	Formed plastic liner, 50 gal., silk screen on enamel finish		20	.800		865	30		895	1,000
3240	Aggregate finish		20	.800		910	30		940	1,050
4000	Perforated steel, pole mounted, 12" diam., 10 gal., painted		25	.640		120	24		144	172
4040	Redwood slats		25	.640		360	24		384	435
4100	22 gal. capacity, painted		25	.640		160	24		184	216
4140	Redwood slats		25	.640		420	24		444	500
4500	Galvanized steel street basket, 52 gal. capacity, unpainted		40	.400		395	15.05		410.05	460
9110	Plastic, with dome lid, 32 gal. capacity		35	.457		58	17.20		75.20	92
9120	Recycled plastic slats, plastic dome lid, 32 gal. capacity	↓	35	.457	↓	284	17.20		301.20	340

12 93 23.20 Trash Closure		Crew	Daily Output	Labor-Hours	Unit	Material	Labor	Equipment	Total	Total Incl O&P
0010	**TRASH CLOSURE**									
0020	Steel with pullover cover, 2'-3" wide, 4'-7" high, 6'-2" long	2 Clab	5	3.200	Ea.	1,950	120		2,070	2,350
0100	10'-1" long		4	4		2,425	150		2,575	2,925
0300	Wood, 10' wide, 6' high, 10' long	↓	1.20	13.333	↓	1,750	500		2,250	2,750

Division 13 – Special Construction

Estimating Tips

General

- The items and systems in this division are usually estimated, purchased, supplied, and installed as a unit by one or more subcontractors. The estimator must ensure that all parties are operating from the same set of specifications and assumptions, and that all necessary items are estimated and will be provided. Many times the complex items and systems are covered, but the more common ones, such as excavation or a crane, are overlooked for the very reason that everyone assumes nobody could miss them. The estimator should be the central focus and be able to ensure that all systems are complete.
- Another area where problems can develop in this division is at the interface between systems. The estimator must ensure, for instance, that anchor bolts, nuts, and washers are estimated and included for the air-supported structures and pre-engineered buildings to be bolted to their foundations. Utility supply is a common area where essential items or pieces of equipment can be missed or overlooked, because each subcontractor may feel it is another's responsibility. The estimator should also be aware of certain items which may be supplied as part of a package but installed by others, and ensure that the installing contractor's estimate includes the cost of installation. Conversely, the estimator must also ensure that items are not costed by two different subcontractors, resulting in an inflated overall estimate.

13 30 00 Special Structures

- The foundations and floor slab, as well as rough mechanical and electrical, should be estimated, as this work is required for the assembly and erection of the structure. Generally, as noted in the book, the pre-engineered building comes as a shell. Pricing is based on the size and structural design parameters stated in the reference section. Additional features, such as windows and doors with their related structural framing, must also be included by the estimator. Here again, the estimator must have a clear understanding of the scope of each portion of the work and all the necessary interfaces.

Reference Numbers

Reference numbers are shown in shaded boxes at the beginning of some major classifications. These numbers refer to related items in the Reference Section. The reference information may be an estimating procedure, an alternate pricing method, or technical information.

Note: Not all subdivisions listed here necessarily appear in this publication. ■

13 05 Common Work Results for Special Construction

13 05 05 – Selective Demolition for Special Construction

13 05 05.10 Selective Demolition, Air Supported Structures

	13 05 05.10 Selective Demolition, Air Supported Structures		Crew	Daily Output	Labor-Hours	Unit	Material	2015 Bare Costs Labor	Equipment	Total	Total Incl O&P
0010	SELECTIVE DEMOLITION, AIR SUPPORTED STRUCTURES										
0020	Tank covers, scrim, dbl. layer, vinyl poly w/hdwe., blower & controls										
0050	Round and rectangular	R024119-10	B-2	9000	.004	S.F.		.17		.17	.28
0100	Warehouse structures										
0120	Poly/vinyl fabric, 28 oz., incl. tension cables & inflation system		4 Clab	9000	.004	SF Flr.		.13		.13	.22
0150	Reinforced vinyl, 12 oz., 3000 S.F.		″	5000	.006			.24		.24	.39
0200	12,000 to 24,000 S.F.		8 Clab	20000	.003			.12		.12	.20
0250	Tedlar vinyl fabric, 28 oz. w/liner, to 3000 S.F.		4 Clab	5000	.006			.24		.24	.39
0300	12,000 to 24,000 S.F.		8 Clab	20000	.003			.12		.12	.20
0350	Greenhouse/shelter, woven polyethylene with liner										
0400	3000 S.F.		4 Clab	5000	.006	SF Flr.		.24		.24	.39
0450	12,000 to 24,000 S.F.		8 Clab	20000	.003			.12		.12	.20
0500	Tennis/gymnasium, poly/vinyl fabric, 28 oz., incl. thermal liner		4 Clab	9000	.004			.13		.13	.22
0600	Stadium/convention center, teflon coated fiberglass, incl. thermal liner		9 Clab	40000	.002			.07		.07	.11
0700	Doors, air lock, 15′ long, 10′ x 10′		2 Carp	1.50	10.667	Ea.		500		500	820
0720	15′ x 15′			.80	20			940		940	1,550
0750	Revolving personnel door, 6′ diam. x 6′-6″ high			1.50	10.667			500		500	820

13 05 05.20 Selective Demolition, Garden Houses

	13 05 05.20 Selective Demolition, Garden Houses		Crew	Daily Output	Labor-Hours	Unit	Material	Labor	Equipment	Total	Total Incl O&P
0010	SELECTIVE DEMOLITION, GARDEN HOUSES	R024119-10									
0020	Prefab, wood, excl foundation, average		2 Clab	400	.040	SF Flr.		1.50		1.50	2.47

13 05 05.25 Selective Demolition, Geodesic Domes

	13 05 05.25 Selective Demolition, Geodesic Domes	Crew	Daily Output	Labor-Hours	Unit	Material	Labor	Equipment	Total	Total Incl O&P
0010	SELECTIVE DEMOLITION, GEODESIC DOMES									
0050	Shell only, interlocking plywood panels, 30′ diameter	F-5	3.20	10	Ea.		475		475	780
0060	34′ diameter		2.30	13.913			660		660	1,075
0070	39′ diameter		2	16			760		760	1,250
0080	45′ diameter	F-3	2.20	18.182			870	297	1,167	1,750
0090	55′ diameter		2	20			960	325	1,285	1,900
0100	60′ diameter		2	20			960	325	1,285	1,900
0110	65′ diameter		1.60	25			1,200	410	1,610	2,400

13 05 05.30 Selective Demolition, Greenhouses

	13 05 05.30 Selective Demolition, Greenhouses		Crew	Daily Output	Labor-Hours	Unit	Material	Labor	Equipment	Total	Total Incl O&P
0010	SELECTIVE DEMOLITION, GREENHOUSES	R024119-10									
0020	Resi-type, free standing, excl. foundations, 9′ long x 8′ wide		2 Clab	160	.100	SF Flr.		3.76		3.76	6.15
0030	9′ long x 11′ wide			170	.094			3.54		3.54	5.80
0040	9′ long x 14′ wide			220	.073			2.73		2.73	4.48
0050	9′ long x 17′ wide			320	.050			1.88		1.88	3.08
0060	Lean-to type, 4′ wide			64	.250			9.40		9.40	15.40
0070	7′ wide			120	.133			5		5	8.20
0080	Geodesic hemisphere, 1/8″ plexiglass glazing, 8′ diam.			4	4	Ea.		150		150	247
0090	24′ diam.			.80	20			750		750	1,225
0100	48′ diam.			.40	40			1,500		1,500	2,475

13 05 05.35 Selective Demolition, Hangars

	13 05 05.35 Selective Demolition, Hangars	Crew	Daily Output	Labor-Hours	Unit	Material	Labor	Equipment	Total	Total Incl O&P
0010	SELECTIVE DEMOLITION, HANGARS									
0020	T type hangars, prefab, steel , galv roof & walls, incl doors, excl fndtn	E-2	2550	.022	SF Flr.		1.14	.59	1.73	2.64
0030	Circular type, prefab, steel frame, plastic skin, incl foundation, 80′ diam	″	.50	112	Total		5,800	3,025	8,825	13,400

13 05 05.45 Selective Demolition, Lightning Protection

	13 05 05.45 Selective Demolition, Lightning Protection	Crew	Daily Output	Labor-Hours	Unit	Material	Labor	Equipment	Total	Total Incl O&P
0010	SELECTIVE DEMOLITION, LIGHTNING PROTECTION									
0020	Air terminal & base, copper, 3/8″ diam. x 10″, to 75′ h	1 Clab	16	.500	Ea.		18.80		18.80	31
0030	1/2″ diam. x 12″, over 75′ h		16	.500			18.80		18.80	31
0050	Aluminum, 1/2″ diam. x 12″, to 75′ h		16	.500			18.80		18.80	31
0060	5/8″ diam. x 12″, over 75′ h		16	.500			18.80		18.80	31
0070	Cable, copper, 220 lb. per thousand feet, to 75′ high		640	.013	L.F.		.47		.47	.77

13 05 05 – Selective Demolition for Special Construction

13 05 05.45 Selective Demolition, Lightning Protection		Crew	Daily Output	Labor-Hours	Unit	Material	2015 Bare Costs Labor	Equipment	Total	Total Incl O&P
0080	375 lb. per thousand feet, over 75′ high	1 Clab	460	.017	L.F.		.65		.65	1.07
0090	Aluminum, 101 lb. per thousand feet, to 75′ high		560	.014			.54		.54	.88
0100	199 lb. per thousand feet, over 75′ high		480	.017			.63		.63	1.03
0110	Arrester, 175 V AC, to ground		16	.500	Ea.		18.80		18.80	31
0120	650 V AC, to ground		13	.615	"		23		23	38

13 05 05.50 Selective Demolition, Pre-Engineered Steel Buildings		Crew	Daily Output	Labor-Hours	Unit	Material	Labor	Equipment	Total	Total Incl O&P
0010	**SELECTIVE DEMOLITION, PRE-ENGINEERED STEEL BUILDINGS**									
0500	Pre-engd. steel bldgs., rigid frame, clear span & multi post, excl. salvage									
0550	3,500 to 7,500 S.F.	L-10	1000	.024	SF Flr.		1.27	.65	1.92	2.93
0600	7,501 to 12,500 S.F.		1500	.016			.85	.44	1.29	1.95
0650	12,500 S.F. or greater		1650	.015			.77	.40	1.17	1.78
0700	Pre-engd. steel building components									
0710	Entrance canopy, including frame 4′ x 4′	E-24	8	4	Ea.		209	92	301	465
0720	4′ x 8′	"	7	4.571			238	105	343	535
0730	HM doors, self framing, single leaf	2 Skwk	8	2			97.50		97.50	158
0740	Double leaf		5	3.200			156		156	253
0760	Gutter, eave type		600	.027	L.F.		1.30		1.30	2.11
0770	Sash, single slide, double slide or fixed		24	.667	Ea.		32.50		32.50	52.50
0780	Skylight, fiberglass, to 30 S.F.		16	1			48.50		48.50	79
0785	Roof vents, circular, 12″ to 24″ diameter		12	1.333			65		65	105
0790	Continuous, 10′ long		8	2			97.50		97.50	158
0900	Shelters, aluminum frame									
0910	Acrylic glazing, 3′ x 9′ x 8′ high	2 Skwk	2	8	Ea.		390		390	630
0920	9′ x 12′ x 8′ high	"	1.50	10.667	"		520		520	845

13 05 05.55 Selective Demolition, Residential Garages		Crew	Daily Output	Labor-Hours	Unit	Material	Labor	Equipment	Total	Total Incl O&P
0010	**SELECTIVE DEMOLITION, RESIDENTIAL GARAGES** R024119-10									
0020	Garage, residential, prefab shell, stock, wood, single car	2 Clab	1.50	10.667	Ea.		400		400	660
0030	Two car	"	1.10	14.545	"		545		545	895

13 05 05.60 Selective Demolition, Silos		Crew	Daily Output	Labor-Hours	Unit	Material	Labor	Equipment	Total	Total Incl O&P
0010	**SELECTIVE DEMOLITION, SILOS**									
0020	Conc stave, indstrl, conical/sloping bott, excl fndtn, 12′ diam., 35′ h	E-24	.18	177	Ea.		9,275	4,100	13,375	20,800
0030	16′ diam., 45′ h		.12	266			13,900	6,150	20,050	31,200
0040	25′ diam., 75′ h		.08	400			20,900	9,200	30,100	46,700
0050	Steel, factory fabricated, 30,000 gal. cap, painted or epoxy lined	L-5	2	28			1,475	370	1,845	3,025

13 05 05.65 Selective Demolition, Sound Control		Crew	Daily Output	Labor-Hours	Unit	Material	Labor	Equipment	Total	Total Incl O&P
0010	**SELECTIVE DEMOLITION, SOUND CONTROL** R024119-10									
0120	Acoustical enclosure, 4″ thick walls & ceiling panels, 8 lb./S.F.	3 Carp	144	.167	SF Surf		7.85		7.85	12.85
0130	10.5 lb./S.F.		128	.188			8.80		8.80	14.45
0140	Reverb chamber, parallel walls, 4″ thick		120	.200			9.40		9.40	15.40
0150	Skewed walls, parallel roof, 4″ thick		110	.218			10.25		10.25	16.80
0160	Skewed walls/roof, 4″ layer/air space		96	.250			11.75		11.75	19.25
0170	Sound-absorbing panels, painted metal, 2′-6″ x 8′, under 1,000 S.F.		430	.056			2.62		2.62	4.29
0180	Over 1,000 S.F.		480	.050			2.35		2.35	3.85
0190	Flexible transparent curtain, clear	3 Shee	430	.056			3.12		3.12	4.93
0192	50% clear, 50% foam		430	.056			3.12		3.12	4.93
0194	25% clear, 75% foam		430	.056			3.12		3.12	4.93
0196	100% foam		430	.056			3.12		3.12	4.93
0200	Audio-masking sys., incl. speakers, amplfr., signal gnrtr.									
0205	Ceiling mounted, 5,000 S.F.	2 Elec	4800	.003	S.F.		.18		.18	.28
0210	10,000 S.F.		5600	.003			.16		.16	.24
0220	Plenum mounted, 5,000 S.F.		7600	.002			.12		.12	.18
0230	10,000 S.F.		8800	.002			.10		.10	.15

13 05 Common Work Results for Special Construction

13 05 05 – Selective Demolition for Special Construction

13 05 05.70 Selective Demolition, Special Purpose Rooms		Crew	Daily Output	Labor-Hours	Unit	Material	2015 Bare Costs Labor	Equipment	Total	Total Incl O&P
0010	SELECTIVE DEMOLITION, SPECIAL PURPOSE ROOMS R024119-10									
0100	Audiometric rooms, under 500 S.F. surface	4 Carp	200	.160	SF Surf		7.50		7.50	12.30
0110	Over 500 S.F. surface	"	240	.133	"		6.25		6.25	10.25
0200	Clean rooms, 12′ x 12′ soft wall, class 100	1 Carp	.30	26.667	Ea.		1,250		1,250	2,050
0210	Class 1000		.30	26.667			1,250		1,250	2,050
0220	Class 10,000		.35	22.857			1,075		1,075	1,750
0230	Class 100,000		.35	22.857			1,075		1,075	1,750
0300	Darkrooms, shell complete, 8′ high	2 Carp	220	.073	SF Flr.		3.41		3.41	5.60
0310	12′ high		110	.145	"		6.85		6.85	11.20
0350	Darkrooms doors, mini-cylindrical, revolving		4	4	Ea.		188		188	310
0400	Music room, practice modular		140	.114	SF Surf		5.35		5.35	8.80
0500	Refrigeration structures and finishes									
0510	Wall finish, 2 coat portland cement plaster, 1/2″ thick	1 Clab	200	.040	S.F.		1.50		1.50	2.47
0520	Fiberglass panels, 1/8″ thick		400	.020			.75		.75	1.23
0530	Ceiling finish, polystyrene plastic, 1″ to 2″ thick		500	.016			.60		.60	.99
0540	4″ thick		450	.018			.67		.67	1.10
0550	Refrigerator, prefab aluminum walk-in, 7′-6″ high, 6′ x 6′ OD	2 Carp	100	.160	SF Flr.		7.50		7.50	12.30
0560	10′ x 10′ OD		160	.100			4.70		4.70	7.70
0570	Over 150 S.F.		200	.080			3.76		3.76	6.15
0600	Sauna, prefabricated, including heater & controls, 7′ high, to 30 S.F.		120	.133			6.25		6.25	10.25
0610	To 40 S.F.		140	.114			5.35		5.35	8.80
0620	To 60 S.F.		175	.091			4.29		4.29	7.05
0630	To 100 S.F.		220	.073			3.41		3.41	5.60
0640	To 130 S.F.		250	.064			3		3	4.92
0650	Steam bath, heater, timer, head, single, to 140 C.F.	1 Plum	2.20	3.636	Ea.		213		213	335
0660	To 300 C.F.		2.20	3.636			213		213	335
0670	Steam bath, comm. size, w/blow-down assembly, to 800 C.F.		1.80	4.444			261		261	405
0680	To 2500 C.F.		1.60	5			294		294	460
0690	Steam bath, comm. size, multiple, for motels, apts, 500 C.F., 2 baths		2	4			235		235	365
0700	1,000 C.F., 4 baths		1.40	5.714			335		335	525

13 05 05.75 Selective Demolition, Storage Tanks		Crew	Daily Output	Labor-Hours	Unit	Material	Labor	Equipment	Total	Total Incl O&P
0010	SELECTIVE DEMOLITION, STORAGE TANKS									
0500	Steel tank, single wall, above ground, not incl. fdn., pumps or piping									
0510	Single wall, 275 gallon R024119-10	Q-1	3	5.333	Ea.		282		282	440
0520	550 thru 2,000 gallon	B-34P	2	12			600	335	935	1,300
0530	5,000 thru 10,000 gallon	B-34Q	2	12			600	630	1,230	1,650
0540	15,000 thru 30,000 gallon	B-34S	2	16			845	1,775	2,620	3,275
0600	Steel tank, double wall, above ground not incl. fdn., pumps & piping									
0620	500 thru 2,000 gallon	B-34P	2	12	Ea.		600	335	935	1,300

13 05 05.85 Selective Demolition, Swimming Pool Equip		Crew	Daily Output	Labor-Hours	Unit	Material	Labor	Equipment	Total	Total Incl O&P
0010	SELECTIVE DEMOLITION, SWIMMING POOL EQUIP									
0020	Diving stand, stainless steel, 3 meter	2 Clab	3	5.333	Ea.		201		201	330
0030	1 meter		5	3.200			120		120	197
0040	Diving board, 16′ long, aluminum		5.40	2.963			111		111	183
0050	Fiberglass		5.40	2.963			111		111	183
0070	Ladders, heavy duty, stainless steel, 2 tread		14	1.143			43		43	70.50
0080	4 tread		12	1.333			50		50	82
0090	Lifeguard chair, stainless steel, fixed		5	3.200			120		120	197
0100	Slide, tubular, fiberglass, aluminum handrails & ladder, 5′, straight		4	4			150		150	247
0110	8′, curved		6	2.667			100		100	164
0120	10′, curved		3	5.333			201		201	330
0130	12′ straight, with platform		2.50	6.400			241		241	395

13 05 Common Work Results for Special Construction

13 05 05 – Selective Demolition for Special Construction

13 05 05.85 Selective Demolition, Swimming Pool Equip		Crew	Daily Output	Labor-Hours	Unit	Material	2015 Bare Costs Labor	Equipment	Total	Total Incl O&P
0140	Removable access ramp, stainless steel	2 Clab	4	4	Ea.		150		150	247
0150	Removable stairs, stainless steel, collapsible		4	4			150		150	247

13 05 05.90 Selective Demolition, Tension Structures

		Crew	Daily Output	Labor-Hours	Unit	Material	Labor	Equipment	Total	Total Incl O&P
0010	**SELECTIVE DEMOLITION, TENSION STRUCTURES**									
0020	Steel/alum. frame, fabric shell, 60′ clear span, 6,000 S.F.	B-41	2000	.022	SF Flr.		.86	.14	1	1.56
0030	12,000 S.F.		2200	.020			.78	.13	.91	1.41
0040	80′ clear span, 20,800 S.F.		2440	.018			.70	.12	.82	1.28
0050	100′ clear span, 10,000 S.F.	L-5	4350	.013			.68	.17	.85	1.40
0060	26,000 S.F.		4600	.012			.64	.16	.80	1.32
0070	36,000 S.F.		5000	.011			.59	.15	.74	1.21
0080	120′ clear span, 24,000 S.F.		6000	.009			.49	.12	.61	1.02
0090	150′ clear span, 30,000 S.F.		7000	.008			.42	.11	.53	.87
0100	200′ clear span, 40,000 S.F.	E-6	16000	.008			.42	.12	.54	.87
0110	For roll-up door, 12′ x 14′	L-2	2	8	Ea.		330		330	540

13 05 05.95 Selective Demo, X-Ray/Radio Freq Protection

		Crew	Daily Output	Labor-Hours	Unit	Material	Labor	Equipment	Total	Total Incl O&P
0010	**SELECTIVE DEMO, X-RAY/RADIO FREQ PROTECTION**									
0020	Shielding lead, lined door frame, excl. hdwe., 1/16″ thick	1 Clab	4.80	1.667	Ea.		62.50		62.50	103
0030	Lead sheets, 1/16″ thick R024119-10	2 Clab	270	.059	S.F.		2.23		2.23	3.65
0040	1/8″ thick		240	.067			2.51		2.51	4.11
0050	Lead shielding, 1/4″ thick		270	.059			2.23		2.23	3.65
0060	1/2″ thick		240	.067			2.51		2.51	4.11
0070	Lead glass, 1/4″ thick, 2.0 mm LE, 12″ x 16″	2 Glaz	16	1	Ea.		45		45	73
0080	24″ x 36″		8	2			90		90	146
0090	36″ x 60″		4	4			180		180	293
0100	Lead glass window frame, with 1/16″ lead & voice passage, 36″ x 60″		4	4			180		180	293
0110	Lead glass window frame, 24″ x 36″		8	2			90		90	146
0120	Lead gypsum board, 5/8″ thick with 1/16″ lead	2 Clab	320	.050	S.F.		1.88		1.88	3.08
0130	1/8″ lead		280	.057			2.15		2.15	3.52
0140	1/32″ lead		400	.040			1.50		1.50	2.47
0150	Butt joints, 1/8″ lead or thicker, 2″ x 7′ long batten strip		480	.033	Ea.		1.25		1.25	2.05
0160	X-ray protection, average radiography room, up to 300 S.F., 1/16″ lead, min		.50	32	Total		1,200		1,200	1,975
0170	Maximum		.30	53.333			2,000		2,000	3,300
0180	Deep therapy X-ray room, 250 kV cap, up to 300 S.F., 1/4″ lead, min		.20	80			3,000		3,000	4,925
0190	Maximum		.12	133			5,025		5,025	8,225
0880	Radio frequency shielding, prefab or screen-type copper or steel, minimum		360	.044	SF Surf		1.67		1.67	2.74
0890	Average		310	.052			1.94		1.94	3.18
0895	Maximum		290	.055			2.07		2.07	3.40

13 11 Swimming Pools

13 11 13 – Below-Grade Swimming Pools

13 11 13.50 Swimming Pools

		Crew	Daily Output	Labor-Hours	Unit	Material	Labor	Equipment	Total	Total Incl O&P
0010	**SWIMMING POOLS** Residential in-ground, vinyl lined, concrete									
0020	Swimming pools,resi in-ground,vyl lined,conc sides,W/ equip ,sand bot	B-52	300	.187	SF Surf	23.50	8.15	1.98	33.63	41
0100	Metal or polystyrene sides	B-14	410	.117		19.55	4.66	.89	25.10	30
0200	Add for vermiculite bottom					1.49			1.49	1.64
0500	Gunite bottom and sides, white plaster finish									
0600	12′ x 30′ pool	B-52	145	.386	SF Surf	43.50	16.85	4.10	64.45	80
0720	16′ x 32′ pool		155	.361		39	15.75	3.83	58.58	72.50
0750	20′ x 40′ pool		250	.224		35	9.75	2.38	47.13	57
0810	Concrete bottom and sides, tile finish									

13 11 Swimming Pools

13 11 13 – Below-Grade Swimming Pools

13 11 13.50 Swimming Pools		Crew	Daily Output	Labor-Hours	Unit	Material	2015 Bare Costs Labor	Equipment	Total	Total Incl O&P
0820	12' x 30' pool	B-52	80	.700	SF Surf	44	30.50	7.45	81.95	106
0830	16' x 32' pool		95	.589		36.50	25.50	6.25	68.25	89
0840	20' x 40' pool		130	.431		29	18.80	4.57	52.37	67.50
1100	Motel, gunite with plaster finish, incl. medium									
1150	capacity filtration & chlorination	B-52	115	.487	SF Surf	53.50	21	5.15	79.65	99
1200	Municipal, gunite with plaster finish, incl. high									
1250	capacity filtration & chlorination	B-52	100	.560	SF Surf	69.50	24.50	5.95	99.95	122
1350	Add for formed gutters				L.F.	102			102	112
1360	Add for stainless steel gutters				"	300			300	330
1600	For water heating system, see Section 23 52 28.10									
1700	Filtration and deck equipment only, as % of total				Total				20%	20%
1800	Deck equipment, rule of thumb, 20' x 40' pool				SF Pool				1.18	1.30
1900	5000 S.F. pool				"				1.73	1.90
3000	Painting pools, preparation + 3 coats, 20' x 40' pool, epoxy	2 Pord	.33	48.485	Total	1,775	1,950		3,725	5,100
3100	Rubber base paint, 18 gallons	"	.33	48.485		1,225	1,950		3,175	4,500
3500	42' x 82' pool, 75 gallons, epoxy paint	3 Pord	.14	171		7,500	6,925		14,425	19,400
3600	Rubber base paint	"	.14	171		5,075	6,925		12,000	16,700

13 11 46 – Swimming Pool Accessories

13 11 46.50 Swimming Pool Equipment		Crew	Daily Output	Labor-Hours	Unit	Material	Labor	Equipment	Total	Total Incl O&P
0010	**SWIMMING POOL EQUIPMENT**									
0020	Diving stand, stainless steel, 3 meter	2 Carp	.40	40	Ea.	15,100	1,875		16,975	19,700
0300	1 meter		2.70	5.926		9,175	278		9,453	10,600
0600	Diving boards, 16' long, aluminum		2.70	5.926		3,950	278		4,228	4,800
0700	Fiberglass		2.70	5.926		3,250	278		3,528	4,025
0800	14' long, aluminum		2.70	5.926		3,575	278		3,853	4,375
0850	Fiberglass		2.70	5.926		3,225	278		3,503	3,975
1200	Ladders, heavy duty, stainless steel, 2 tread		7	2.286		810	107		917	1,075
1500	4 tread		6	2.667		1,075	125		1,200	1,375
1800	Lifeguard chair, stainless steel, fixed		2.70	5.926		3,225	278		3,503	4,000
1900	Portable					2,775			2,775	3,050
2100	Lights, underwater, 12 volt, with transformer, 300 watt	1 Elec	1	8		330	440		770	1,050
2200	110 volt, 500 watt, standard		1	8		294	440		734	1,000
2400	Low water cutoff type		1	8		300	440		740	1,000
2800	Heaters, see Section 23 52 28.10									
3000	Pool covers, reinforced vinyl	3 Clab	1800	.013	S.F.	1.13	.50		1.63	2.06
3050	Automatic, electric								8.75	9.65
3100	Vinyl, for winter, 400 SF max pool surface	3 Clab	3200	.008		.26	.28		.54	.75
3200	With water tubes, 400SF max pool surface	"	3000	.008		.29	.30		.59	.81
3250	Sealed air bubble polyethylene solar blanket, 16 mils					.30			.30	.33
3300	Slides, tubular, fiberglass, aluminum handrails & ladder, 5'-0", straight	2 Carp	1.60	10	Ea.	3,575	470		4,045	4,725
3320	8'-0", curved		3	5.333		7,225	250		7,475	8,350
3400	10'-0", curved		1	16		22,000	750		22,750	25,400
3420	12'-0", straight with platform		1.20	13.333		13,800	625		14,425	16,200
4500	Hydraulic lift, movable pool bottom, single ram									
4520	Under 1,000 S.F. area	L-9	72	.500	S.F.	160	21.50		181.50	212
4600	Four ram lift, over 1,000 S.F.	"	109	.330	"	130	14.20		144.20	167
5000	Removable access ramp, stainless steel	2 Clab	2	8	Ea.	5,800	300		6,100	6,875

13 17 Tubs and Pools

13 17 13 – Hot Tubs

13 17 13.10 Redwood Hot Tub System		Crew	Daily Output	Labor-Hours	Unit	2015 Bare Costs Material	Labor	Equipment	Total	Total Incl O&P
0010	**REDWOOD HOT TUB SYSTEM**									
7050	4′ diameter x 4′ deep	Q-1	1	16	Ea.	3,225	845		4,070	4,875
7100	5′ diameter x 4′ deep		1	16		4,100	845		4,945	5,825
7150	6′ diameter x 4′ deep		.80	20		4,950	1,050		6,000	7,075
7200	8′ diameter x 4′ deep	↓	.80	20	↓	7,250	1,050		8,300	9,625

13 17 33 – Whirlpool Tubs

13 17 33.10 Whirlpool Bath		Crew	Daily Output	Labor-Hours	Unit	Material	Labor	Equipment	Total	Total Incl O&P
0010	**WHIRLPOOL BATH**									
6000	Whirlpool, bath with vented overflow, molded fiberglass									
6100	66″ x 36″ x 24″	Q-1	1	16	Ea.	3,475	845		4,320	5,150
6400	72″ x 36″ x 21″		1	16		1,800	845		2,645	3,325
6500	60″ x 34″ x 21″		1	16		1,800	845		2,645	3,300
6600	72″ x 42″ x 23″	↓	1	16	↓	2,175	845		3,020	3,725

13 18 Ice Rinks

13 18 13 – Ice Rink Floor Systems

13 18 13.50 Ice Skating		Crew	Daily Output	Labor-Hours	Unit	Material	Labor	Equipment	Total	Total Incl O&P
0010	**ICE SKATING** Equipment incl. refrigeration, plumbing & cooling									
0020	coils & concrete slab, 85′ x 200′ rink									
0300	55° system, 5 mos., 100 ton				Total	575,000			575,000	632,500
0700	90° system, 12 mos., 135 ton				″	650,000			650,000	715,000
1200	Subsoil heating system (recycled from compressor), 85′ x 200′	Q-7	.27	118	Ea.	40,000	6,750		46,750	54,500
1300	Subsoil insulation, 2 lb. polystyrene with vapor barrier, 85′ x 200′	2 Carp	.14	114	″	30,000	5,375		35,375	41,800

13 18 16 – Ice Rink Dasher Boards

13 18 16.50 Ice Rink Dasher Boards		Crew	Daily Output	Labor-Hours	Unit	Material	Labor	Equipment	Total	Total Incl O&P
0010	**ICE RINK DASHER BOARDS**									
1000	Dasher boards, 1/2″ H.D. polyethylene faced steel frame, 3′ acrylic									
1020	screen at sides, 5′ acrylic ends, 85′ x 200′	F-5	.06	533	Ea.	135,000	25,300		160,300	190,000
1100	Fiberglass & aluminum construction, same sides and ends	″	.06	533	″	155,000	25,300		180,300	212,000

13 21 Controlled Environment Rooms

13 21 13 – Clean Rooms

13 21 13.50 Clean Room Components		Crew	Daily Output	Labor-Hours	Unit	Material	Labor	Equipment	Total	Total Incl O&P
0010	**CLEAN ROOM COMPONENTS**									
1100	Clean room, soft wall, 12′ x 12′, Class 100	1 Carp	.18	44.444	Ea.	18,600	2,075		20,675	23,900
1110	Class 1,000		.18	44.444		15,400	2,075		17,475	20,400
1120	Class 10,000		.21	38.095		13,000	1,800		14,800	17,200
1130	Class 100,000	↓	.21	38.095	↓	12,000	1,800		13,800	16,100
2800	Ceiling grid support, slotted channel struts 4′-0″ O.C., ea. way				S.F.				5.90	6.50
3000	Ceiling panel, vinyl coated foil on mineral substrate									
3020	Sealed, non-perforated				S.F.				1.27	1.40
4000	Ceiling panel seal, silicone sealant, 150 L.F./gal.	1 Carp	150	.053	L.F.	.34	2.50		2.84	4.47
4100	Two sided adhesive tape	″	240	.033	″	.12	1.56		1.68	2.69
4200	Clips, one per panel				Ea.	.99			.99	1.09
6000	HEPA filter, 2′ x 4′, 99.97% eff., 3″ dp beveled frame (silicone seal)					470			470	515
6040	6″ deep skirted frame (channel seal)					440			440	485
6100	99.99% efficient, 3″ deep beveled frame (silicone seal)					525			525	580
6140	6″ deep skirted frame (channel seal)				↓	455			455	500

13 21 Controlled Environment Rooms

13 21 13 – Clean Rooms

13 21 13.50 Clean Room Components		Crew	Daily Output	Labor-Hours	Unit	2015 Bare Costs Material	Labor	Equipment	Total	Total Incl O&P
6200	99.999% efficient, 3" deep beveled frame (silicone seal)				Ea.	605			605	665
6240	6" deep skirted frame (channel seal)					485			485	530
7000	Wall panel systems, including channel strut framing									
7020	Polyester coated aluminum, particle board				S.F.				18.20	20
7100	Porcelain coated aluminum, particle board								32	35
7400	Wall panel support, slotted channel struts, to 12' high								16.35	18

13 21 26 – Cold Storage Rooms

13 21 26.50 Refrigeration		Crew	Daily Output	Labor-Hours	Unit	Material	Labor	Equipment	Total	Total Incl O&P
0010	**REFRIGERATION**									
0020	Curbs, 12" high, 4" thick, concrete	2 Carp	58	.276	L.F.	5.15	12.95		18.10	26.50
1000	Doors, see Section 08 34 13.10									
2400	Finishes, 2 coat portland cement plaster, 1/2" thick	1 Plas	48	.167	S.F.	1.41	7.15		8.56	13
2500	For galvanized reinforcing mesh, add	1 Lath	335	.024		.94	1.03		1.97	2.65
2700	3/16" thick latex cement	1 Plas	88	.091		2.40	3.90		6.30	8.90
2900	For glass cloth reinforced ceilings, add	"	450	.018		.57	.76		1.33	1.85
3100	Fiberglass panels, 1/8" thick	1 Carp	149.45	.054		3.18	2.51		5.69	7.60
3200	Polystyrene, plastic finish ceiling, 1" thick		274	.029		2.94	1.37		4.31	5.50
3400	2" thick		274	.029		3.36	1.37		4.73	5.95
3500	4" thick		219	.037		3.71	1.72		5.43	6.90
3800	Floors, concrete, 4" thick	1 Cefi	93	.086		1.32	3.87		5.19	7.55
3900	6" thick	"	85	.094		2.07	4.24		6.31	8.95
4000	Insulation, 1" to 6" thick, cork				B.F.	1.35			1.35	1.49
4100	Urethane					.52			.52	.57
4300	Polystyrene, regular					.53			.53	.58
4400	Bead board					.25			.25	.28
4600	Installation of above, add per layer	2 Carp	657.60	.024	S.F.	.45	1.14		1.59	2.37
4700	Wall and ceiling juncture		298.90	.054	L.F.	2.19	2.51		4.70	6.55
4900	Partitions, galvanized sandwich panels, 4" thick, stock		219.20	.073	S.F.	9.20	3.43		12.63	15.70
5000	Aluminum or fiberglass		219.20	.073	"	10.05	3.43		13.48	16.70
5200	Prefab walk-in, 7'-6" high, aluminum, incl. refrigeration, door & floor									
5210	not incl. partitions, 6' x 6'	2 Carp	54.80	.292	SF Flr.	164	13.70		177.70	204
5500	10' x 10'		82.20	.195		132	9.15		141.15	160
5700	12' x 14'		109.60	.146		119	6.85		125.85	142
5800	12' x 20'		109.60	.146		103	6.85		109.85	125
6100	For 8'-6" high, add					5%				
6300	Rule of thumb for complete units, w/o doors & refrigeration, cooler	2 Carp	146	.110		149	5.15		154.15	172
6400	Freezer		109.60	.146		176	6.85		182.85	205
6600	Shelving, plated or galvanized, steel wire type		360	.044	SF Hor.	13.05	2.09		15.14	17.75
6700	Slat shelf type		375	.043		16.10	2		18.10	21
6900	For stainless steel shelving, add					300%				
7000	Vapor barrier, on wood walls	2 Carp	1644	.010	S.F.	.20	.46		.66	.97
7200	On masonry walls	"	1315	.012	"	.49	.57		1.06	1.48
7500	For air curtain doors, see Section 23 34 33.10									

13 21 48 – Sound-Conditioned Rooms

13 21 48.10 Anechoic Chambers		Crew	Daily Output	Labor-Hours	Unit	Material	Labor	Equipment	Total	Total Incl O&P
0010	**ANECHOIC CHAMBERS** Standard units, 7' ceiling heights									
0100	Area for pricing is net inside dimensions									
0300	200 cycles per second cutoff, 25 S.F. floor area				SF Flr.	1,625			1,625	1,800
0400	50 S.F.								1,050	1,150
0600	75 S.F.								1,000	1,100
0700	100 S.F.					1,225			1,225	1,350
0900	For 150 cycles per second cutoff, add to 100 S.F. room								30%	30%

13 21 Controlled Environment Rooms

13 21 48 – Sound-Conditioned Rooms

13 21 48.10 Anechoic Chambers		Crew	Daily Output	Labor-Hours	Unit	2015 Bare Costs Material	Labor	Equipment	Total	Total Incl O&P
1000	For 100 cycles per second cutoff, add to 100 S.F. room				SF Flr.				45%	45%
13 21 48.15 Audiometric Rooms										
0010	**AUDIOMETRIC ROOMS**									
0020	Under 500 S.F. surface	4 Carp	98	.327	SF Surf	52.50	15.35		67.85	82.50
0100	Over 500 S.F. surface	"	120	.267	"	50	12.50		62.50	75.50

13 21 53 – Darkrooms

13 21 53.50 Darkrooms		Crew	Daily Output	Labor-Hours	Unit	Material	Labor	Equipment	Total	Total Incl O&P
0010	**DARKROOMS**									
0020	Shell, complete except for door, 64 S.F., 8′ high	2 Carp	128	.125	SF Flr.	51	5.85		56.85	65.50
0100	12′ high		64	.250		66.50	11.75		78.25	92.50
0500	120 S.F. floor, 8′ high		120	.133		37	6.25		43.25	51.50
0600	12′ high		60	.267		50.50	12.50		63	76
0800	240 S.F. floor, 8′ high		120	.133		27	6.25		33.25	40
0900	12′ high		60	.267	▼	37	12.50		49.50	61
1200	Mini-cylindrical, revolving, unlined, 4′ diameter		3.50	4.571	Ea.	2,725	215		2,940	3,350
1400	5′-6″ diameter	▼	2.50	6.400		5,675	300		5,975	6,725
1600	Add for lead lining, inner cylinder, 1/32″ thick					1,650			1,650	1,825
1700	1/16″ thick					4,450			4,450	4,875
1800	Add for lead lining, inner and outer cylinder, 1/32″ thick					3,050			3,050	3,375
1900	1/16″ thick				▼	6,825			6,825	7,500
2000	For darkroom door, see Section 08 34 36.10									

13 21 56 – Music Rooms

13 21 56.50 Music Rooms		Crew	Daily Output	Labor-Hours	Unit	Material	Labor	Equipment	Total	Total Incl O&P
0010	**MUSIC ROOMS**									
0020	Practice room, modular, perforated steel, under 500 S.F.	2 Carp	70	.229	SF Surf	32	10.75		42.75	53
0100	Over 500 S.F.	"	80	.200	"	27	9.40		36.40	45.50

13 24 Special Activity Rooms

13 24 16 – Saunas

13 24 16.50 Saunas and Heaters		Crew	Daily Output	Labor-Hours	Unit	Material	Labor	Equipment	Total	Total Incl O&P
0010	**SAUNAS AND HEATERS**									
0020	Prefabricated, incl. heater & controls, 7′ high, 6′ x 4′, C/C	L-7	2.20	12.727	Ea.	5,175	580		5,755	6,650
0050	6′ x 4′, C/P		2	14		4,700	635		5,335	6,175
0400	6′ x 5′, C/C		2	14		5,875	635		6,510	7,500
0450	6′ x 5′, C/P		2	14		5,325	635		5,960	6,875
0600	6′ x 6′, C/C		1.80	15.556		6,225	705		6,930	8,000
0650	6′ x 6′, C/P		1.80	15.556		5,675	705		6,380	7,375
0800	6′ x 9′, C/C		1.60	17.500		7,975	795		8,770	10,100
0850	6′ x 9′, C/P		1.60	17.500		7,225	795		8,020	9,225
1000	8′ x 12′, C/C		1.10	25.455		11,700	1,150		12,850	14,800
1050	8′ x 12′, C/P		1.10	25.455		10,500	1,150		11,650	13,500
1200	8′ x 8′, C/C		1.40	20		9,175	910		10,085	11,600
1250	8′ x 8′, C/P		1.40	20		8,450	910		9,360	10,800
1400	8′ x 10′, C/C		1.20	23.333		10,200	1,050		11,250	12,900
1450	8′ x 10′, C/P		1.20	23.333		9,250	1,050		10,300	11,900
1600	10′ x 12′, C/C		1	28		12,200	1,275		13,475	15,600
1650	10′ x 12′, C/P	▼	1	28		11,000	1,275		12,275	14,200
1700	Door only, cedar, 2′x6′, with 1′x4′ tempered insulated glass window	2 Carp	3.40	4.706		750	221		971	1,175
1800	Prehung, incl. jambs, pulls & hardware	"	12	1.333		745	62.50		807.50	925
2500	Heaters only (incl. above), wall mounted, to 200 C.F.				▼	685			685	755

13 24 Special Activity Rooms

13 24 16 – Saunas

13 24 16.50 Saunas and Heaters		Crew	Daily Output	Labor-Hours	Unit	Material	2015 Bare Costs Labor	Equipment	Total	Total Incl O&P
2750	To 300 C.F.				Ea.	930			930	1,025
3000	Floor standing, to 720 C.F., 10,000 watts, w/controls	1 Elec	3	2.667		2,950	146		3,096	3,475
3250	To 1,000 C.F., 16,000 watts	"	3	2.667	↓	3,825	146		3,971	4,425

13 24 26 – Steam Baths

13 24 26.50 Steam Baths and Components		Crew	Daily Output	Labor-Hours	Unit	Material	Labor	Equipment	Total	Total Incl O&P
0010	**STEAM BATHS AND COMPONENTS**									
0020	Heater, timer & head, single, to 140 C.F.	1 Plum	1.20	6.667	Ea.	2,200	390		2,590	3,000
0500	To 300 C.F.		1.10	7.273		2,425	425		2,850	3,350
1000	Commercial size, with blow-down assembly, to 800 C.F.		.90	8.889		5,725	520		6,245	7,125
1500	To 2500 C.F.	↓	.80	10		7,650	585		8,235	9,350
2000	Multiple, motels, apts., 2 baths, w/blow-down assm., 500 C.F.	Q-1	1.30	12.308		6,325	650		6,975	8,000
2500	4 baths	"	.70	22.857		10,200	1,200		11,400	13,100
2700	Conversion unit for residential tub, including door				↓	3,550			3,550	3,925

13 28 Athletic and Recreational Special Construction

13 28 33 – Athletic and Recreational Court Walls

13 28 33.50 Sport Court		Crew	Daily Output	Labor-Hours	Unit	Material	Labor	Equipment	Total	Total Incl O&P
0010	**SPORT COURT**									
0020	Floors, No. 2 & better maple, 25/32" thick				SF Flr.				6.05	6.65
0100	Walls, laminated plastic bonded to galv. steel studs				SF Wall				7.70	8.45
0300	Squash, regulation court in existing building, minimum				Court	36,800			36,800	40,400
0400	Maximum				"	41,000			41,000	45,000
0450	Rule of thumb for components:									
0470	Walls	3 Carp	.15	160	Court	11,000	7,500		18,500	24,400
0500	Floor	"	.25	96		8,725	4,500		13,225	17,000
0550	Lighting	2 Elec	.60	26.667		2,100	1,450		3,550	4,550
0600	Handball, racquetball court in existing building, minimum	C-1	.20	160		39,800	7,150		46,950	55,500
0800	Maximum	"	.10	320		43,100	14,300		57,400	71,000
0900	Rule of thumb for components: walls	3 Carp	.12	200		12,600	9,400		22,000	29,300
1000	Floor		.25	96		8,725	4,500		13,225	17,000
1100	Ceiling	↓	.33	72.727		4,200	3,425		7,625	10,200
1200	Lighting	2 Elec	.60	26.667	↓	2,200	1,450		3,650	4,675

13 31 Fabric Structures

13 31 13 – Air-Supported Fabric Structures

13 31 13.09 Air Supported Tank Covers		Crew	Daily Output	Labor-Hours	Unit	Material	Labor	Equipment	Total	Total Incl O&P
0010	**AIR SUPPORTED TANK COVERS**, vinyl polyester									
0100	Scrim, double layer, with hardware, blower, standby & controls									
0200	Round, 75' diameter	B-2	4500	.009	S.F.	11.75	.34		12.09	13.50
0300	100' diameter		5000	.008		10.70	.30		11	12.25
0400	150' diameter		5000	.008		8.45	.30		8.75	9.80
0500	Rectangular, 20' x 20'		4500	.009		23	.34		23.34	26
0600	30' x 40'		4500	.009		23	.34		23.34	26
0700	50' x 60'	↓	4500	.009		23	.34		23.34	26
0800	For single wall construction, deduct, minimum					.79			.79	.87
0900	Maximum					2.33			2.33	2.56
1000	For maximum resistance to atmosphere or cold, add				↓	1.14			1.14	1.25
1100	For average shipping charges, add				Total	1,975			1,975	2,175

13 31 13 – Air-Supported Fabric Structures

13 31 13.13 Single-Walled Air-Supported Structures		Crew	Daily Output	Labor-Hours	Unit	2015 Bare Costs Material	Labor	Equipment	Total	Total Incl O&P
0010	**SINGLE-WALLED AIR-SUPPORTED STRUCTURES**									
0020	Site preparation, incl. anchor placement and utilities	B-11B	1000	.016	SF Flr.	1.16	.69	.30	2.15	2.72
0030	For concrete, see Section 03 30 53.40									
0050	Warehouse, polyester/vinyl fabric, 28 oz., over 10 yr. life, welded									
0060	Seams, tension cables, primary & auxiliary inflation system,									
0070	airlock, personnel doors and liner									
0100	5,000 S.F.	4 Clab	5000	.006	SF Flr.	26.50	.24		26.74	29.50
0250	12,000 S.F.	"	6000	.005		18.85	.20		19.05	21.50
0400	24,000 S.F.	8 Clab	12000	.005		13.30	.20		13.50	14.95
0500	50,000 S.F.	"	12500	.005	↓	12.30	.19		12.49	13.85
0700	12 oz. reinforced vinyl fabric, 5 yr. life, sewn seams,									
0710	accordion door, including liner									
0750	3000 S.F.	4 Clab	3000	.011	SF Flr.	13.20	.40		13.60	15.20
0800	12,000 S.F.	"	6000	.005		11.25	.20		11.45	12.70
0850	24,000 S.F.	8 Clab	12000	.005		9.50	.20		9.70	10.80
0950	Deduct for single layer					1.03			1.03	1.13
1000	Add for welded seams					1.50			1.50	1.65
1050	Add for double layer, welded seams included				↓	3			3	3.30
1250	Tedlar/vinyl fabric, 28 oz., with liner, over 10 yr. life,									
1260	incl. overhead and personnel doors									
1300	3000 S.F.	4 Clab	3000	.011	SF Flr.	24.50	.40		24.90	27.50
1450	12,000 S.F.	"	6000	.005		17.20	.20		17.40	19.30
1550	24,000 S.F.	8 Clab	12000	.005		13.30	.20		13.50	15
1700	Deduct for single layer				↓	2			2	2.20
2250	Greenhouse/shelter, woven polyethylene with liner, 2 yr. life,									
2260	sewn seams, including doors									
2300	3000 S.F.	4 Clab	3000	.011	SF Flr.	16	.40		16.40	18.25
2350	12,000 S.F.	"	6000	.005		14	.20		14.20	15.75
2450	24,000 S.F.	8 Clab	12000	.005		12	.20		12.20	13.55
2550	Deduct for single layer				↓	.98			.98	1.08
2600	Tennis/gymnasium, polyester/vinyl fabric, 28 oz., over 10 yr. life,									
2610	including thermal liner, heat and lights									
2650	7,200 S.F.	4 Clab	6000	.005	SF Flr.	23.50	.20		23.70	26.50
2750	13,000 S.F.	"	6500	.005		18	.19		18.19	20
2850	Over 24,000 S.F.	8 Clab	12000	.005		16.45	.20		16.65	18.40
2860	For low temperature conditions, add				↓	1.14			1.14	1.25
2870	For average shipping charges, add				Total	5,600			5,600	6,150
2900	Thermal liner, translucent reinforced vinyl				SF Flr.	1.14			1.14	1.25
2950	Metalized mylar fabric and mesh, double liner				"	2.33			2.33	2.56
3050	Stadium/convention center, teflon coated fiberglass, heavy weight,									
3060	over 20 yr. life, incl. thermal liner and heating system									
3100	Minimum	9 Clab	26000	.003	SF Flr.	57.50	.10		57.60	63
3110	Maximum	"	19000	.004	"	68	.14		68.14	75
3400	Doors, air lock, 15' long, 10' x 10'	2 Carp	.80	20	Ea.	20,400	940		21,340	24,100
3600	15' x 15'	"	.50	32		30,300	1,500		31,800	35,900
3700	For each added 5' length, add					5,425			5,425	5,950
3900	Revolving personnel door, 6' diameter, 6'-6" high	2 Carp	.80	20	↓	15,200	940		16,140	18,300

13 31 Fabric Structures

13 31 23 – Tensioned Fabric Structures

13 31 23.50 Tension Structures		Crew	Daily Output	Labor-Hours	Unit	Material	2015 Bare Costs Labor	Equipment	Total	Total Incl O&P
0010	**TENSION STRUCTURES** Rigid steel/alum. frame, vinyl coated poly									
0100	Fabric shell, 60′ clear span, not incl. foundations or floors									
0200	6,000 S.F.	B-41	1000	.044	SF Flr.	13.80	1.71	.29	15.80	18.30
0300	12,000 S.F.		1100	.040		13.20	1.56	.26	15.02	17.40
0400	80′ to 99′ clear span, 20,800 S.F.		1220	.036		13	1.40	.24	14.64	16.85
0410	100′ to 119′ clear span, 10,000 S.F.	L-5	2175	.026		13.70	1.36	.34	15.40	17.85
0430	26,000 S.F.		2300	.024		12.75	1.29	.32	14.36	16.70
0450	36,000 S.F.		2500	.022		12.60	1.18	.29	14.07	16.25
0460	120′ to 149′ clear span, 24,000 S.F.		3000	.019		13.90	.99	.25	15.14	17.25
0470	150′ to 199′ clear span, 30,000 S.F.		6000	.009		14.45	.49	.12	15.06	16.85
0480	200′ clear span, 40,000 S.F.	E-6	8000	.016		17.75	.84	.23	18.82	21.50
0500	For roll-up door, 12′ x 14′, add	L-2	1	16	Ea.	5,500	660		6,160	7,125

13 34 Fabricated Engineered Structures

13 34 13 – Glazed Structures

13 34 13.13 Greenhouses		Crew	Daily Output	Labor-Hours	Unit	Material	Labor	Equipment	Total	Total Incl O&P
0010	**GREENHOUSES**, Shell only, stock units, not incl. 2′ stub walls,									
0020	foundation, floors, heat or compartments									
0300	Residential type, free standing, 8′-6″ long x 7′-6″ wide	2 Carp	59	.271	SF Flr.	20	12.75		32.75	43
0400	10′-6″ wide		85	.188		37	8.85		45.85	55
0600	13′-6″ wide		108	.148		39	6.95		45.95	54
0700	17′-0″ wide		160	.100		43.50	4.70		48.20	55.50
0900	Lean-to type, 3′-10″ wide		34	.471		41.50	22		63.50	82
1000	6′-10″ wide		58	.276		50	12.95		62.95	76
1500	Commercial, custom, truss frame, incl. equip., plumbing, elec.,									
1550	benches and controls, under 2,000 S.F.				SF Flr.	13			13	14.30
1700	Over 5,000 S.F.				″	11.95			11.95	13.15
2000	Institutional, custom, rigid frame, including compartments and									
2050	multi-controls, under 500 S.F.				SF Flr.	25.50			25.50	28
2150	Over 2,000 S.F.				″	10.70			10.70	11.75
3700	For 1/4″ tempered glass, add				SF Surf	1.34			1.34	1.47
3900	Cooling, 1200 CFM exhaust fan, add				Ea.	310			310	340
4000	7850 CFM					1,050			1,050	1,150
4200	For heaters, 10 MBH, add					215			215	237
4300	60 MBH, add					780			780	855
4500	For benches, 2′ x 8′, add					160			160	176
4600	4′ x 10′, add					195			195	214
4800	For ventilation & humidity control w/ 4 integrated outlets, add				Total	240			240	264
4900	For environmental controls and automation, 8 outputs, 9 stages, add				″	765			765	840
5100	For humidification equipment, add				Ea.	299			299	330
5200	For vinyl shading, add				S.F.	.24			.24	.26
6000	Geodesic hemisphere, 1/8″ plexiglass glazing									
6050	8′ diameter	2 Carp	2	8	Ea.	6,250	375		6,625	7,500
6150	24′ diameter		.35	45.714		14,000	2,150		16,150	18,800
6250	48′ diameter		.20	80		33,000	3,750		36,750	42,400

13 34 13.19 Swimming Pool Enclosures		Crew	Daily Output	Labor-Hours	Unit	Material	Labor	Equipment	Total	Total Incl O&P
0010	**SWIMMING POOL ENCLOSURES** Translucent, free standing									
0020	not including foundations, heat or light									
0200	Economy	2 Carp	200	.080	SF Hor.	37	3.76		40.76	47
0600	Deluxe	″	70	.229		93	10.75		103.75	120

13 34 Fabricated Engineered Structures

13 34 13 – Glazed Structures

13 34 13.19 Swimming Pool Enclosures		Crew	Daily Output	Labor-Hours	Unit	Material	2015 Bare Costs Labor	Equipment	Total	Total Incl O&P
0700	For motorized roof, 40% opening, solid roof, add				SF Hor.	21			21	23
0800	Skylight type roof, add				↓	13.50			13.50	14.85

13 34 16 – Grandstands and Bleachers

13 34 16.13 Grandstands

		Crew	Daily Output	Labor-Hours	Unit	Material	Labor	Equipment	Total	Total Incl O&P
0010	**GRANDSTANDS** Permanent, municipal, including foundation									
0300	Steel, economy				Seat	19.90			19.90	22
0400	Steel, deluxe					22.50			22.50	24.50
0900	Composite, steel, wood and plastic, stock design, economy					38.50			38.50	42
1000	Deluxe				↓	68.50			68.50	75

13 34 16.53 Bleachers

		Crew	Daily Output	Labor-Hours	Unit	Material	Labor	Equipment	Total	Total Incl O&P
0010	**BLEACHERS**									
0020	Bleachers, outdoor, portable, 5 tiers, 42 seats	2 Sswk	120	.133	Seat	90	7		97	112
0100	5 tiers, 54 seats		80	.200		81.50	10.55		92.05	109
0200	10 tiers, 104 seats		120	.133		92.50	7		99.50	115
0300	10 tiers, 144 seats	↓	80	.200	↓	83	10.55		93.55	110
0500	Permanent bleachers, aluminum seat, steel frame, 24" row									
0600	8 tiers, 80 seats	2 Sswk	60	.267	Seat	67	14.05		81.05	99.50
0700	8 tiers, 160 seats		48	.333		58	17.55		75.55	95.50
0925	15 tiers, 154 to 165 seats		60	.267		94	14.05		108.05	129
0975	15 tiers, 214 to 225 seats		60	.267		84.50	14.05		98.55	119
1050	15 tiers, 274 to 285 seats		60	.267		77	14.05		91.05	110
1200	Seat backs only, 30" row, fiberglass		160	.100		23	5.25		28.25	35
1300	Steel and wood	↓	160	.100	↓	22	5.25		27.25	34
1400	NOTE: average seating is 1.5' in width									

13 34 19 – Metal Building Systems

13 34 19.50 Pre-Engineered Steel Buildings

		Crew	Daily Output	Labor-Hours	Unit	Material	Labor	Equipment	Total	Total Incl O&P
0010	**PRE-ENGINEERED STEEL BUILDINGS** R133419-10									
0100	Clear span rigid frame, 26 ga. colored roofing and siding									
0150	20' to 29' wide, 10' eave height	E-2	425	.132	SF Flr.	8.80	6.80	3.55	19.15	25.50
0160	14' eave height		350	.160		9.50	8.30	4.32	22.12	29.50
0170	16' eave height		320	.175		10.20	9.05	4.72	23.97	32.50
0180	20' eave height		275	.204		11.15	10.55	5.50	27.20	36.50
0190	24' eave height		240	.233		12.30	12.05	6.30	30.65	41.50
0200	30' to 49' wide, 10' eave height		535	.105		6.75	5.40	2.82	14.97	20
0300	14' eave height		450	.124		7.30	6.45	3.36	17.11	23
0400	16' eave height		415	.135		7.80	7	3.64	18.44	25
0500	20' eave height		360	.156		8.45	8.05	4.20	20.70	28
0600	24' eave height		320	.175		9.30	9.05	4.72	23.07	31.50
0700	50' to 100' wide, 10' eave height		770	.073		5.75	3.76	1.96	11.47	15.05
0900	16' eave height		600	.093		6.60	4.83	2.52	13.95	18.45
1000	20' eave height		490	.114		7.15	5.90	3.08	16.13	21.50
1100	24' eave height	↓	435	.129	↓	7.90	6.65	3.47	18.02	24
1200	Clear span tapered beam frame, 26 ga. colored roofing/siding									
1300	30' to 39' wide, 10' eave height	E-2	535	.105	SF Flr.	7.65	5.40	2.82	15.87	21
1400	14' eave height		450	.124		8.45	6.45	3.36	18.26	24
1500	16' eave height		415	.135		8.90	7	3.64	19.54	26
1600	20' eave height		360	.156		9.80	8.05	4.20	22.05	29.50
1700	40' wide, 10' eave height		600	.093		6.80	4.83	2.52	14.15	18.70
1800	14' eave height		510	.110		7.55	5.70	2.96	16.21	21.50
1900	16' eave height		475	.118		7.90	6.10	3.18	17.18	23
2000	20' eave height	↓	415	.135	↓	8.70	7	3.64	19.34	26

13 34 19.50 Pre-Engineered Steel Buildings		Crew	Daily Output	Labor-Hours	Unit	2015 Bare Costs Material	Labor	Equipment	Total	Total Incl O&P
2100	50′ to 79′ wide, 10′ eave height	E-2	770	.073	SF Flr.	6.40	3.76	1.96	12.12	15.75
2200	14′ eave height		675	.083		6.95	4.29	2.24	13.48	17.55
2300	16′ eave height		635	.088		7.20	4.56	2.38	14.14	18.55
2400	20′ eave height		490	.114		8	5.90	3.08	16.98	22.50
2410	80′ to 100′ wide, 10′ eave height		935	.060		5.65	3.10	1.62	10.37	13.45
2420	14′ eave height		750	.075		6.25	3.86	2.01	12.12	15.80
2430	16′ eave height		685	.082		6.50	4.23	2.21	12.94	17
2440	20′ eave height		560	.100		6.95	5.15	2.70	14.80	19.65
2460	101′ to 120′ wide, 10′ eave height		950	.059		5.20	3.05	1.59	9.84	12.80
2470	14′ eave height		770	.073		5.75	3.76	1.96	11.47	15.10
2480	16′ eave height		675	.083		6.15	4.29	2.24	12.68	16.70
2490	20′ eave height		560	.100		6.55	5.15	2.70	14.40	19.25
2500	Single post 2-span frame, 26 ga. colored roofing and siding									
2600	80′ wide, 14′ eave height	E-2	740	.076	SF Flr.	5.75	3.92	2.04	11.71	15.40
2700	16′ eave height		695	.081		6.10	4.17	2.17	12.44	16.40
2800	20′ eave height		625	.090		6.60	4.64	2.42	13.66	18
2900	24′ eave height		570	.098		7.25	5.10	2.65	15	19.75
3000	100′ wide, 14′ eave height		835	.067		5.60	3.47	1.81	10.88	14.20
3100	16′ eave height		795	.070		5.20	3.64	1.90	10.74	14.20
3200	20′ eave height		730	.077		6.30	3.97	2.07	12.34	16.20
3300	24′ eave height		670	.084		7	4.32	2.26	13.58	17.75
3400	120′ wide, 14′ eave height		870	.064		6.45	3.33	1.74	11.52	14.80
3500	16′ eave height		830	.067		5.80	3.49	1.82	11.11	14.45
3600	20′ eave height		765	.073		6.30	3.79	1.97	12.06	15.70
3700	24′ eave height		705	.079		6.90	4.11	2.14	13.15	17.15
3800	Double post 3-span frame, 26 ga. colored roofing and siding									
3900	150′ wide, 14′ eave height	E-2	925	.061	SF Flr.	4.57	3.13	1.63	9.33	12.35
4000	16′ eave height		890	.063		4.77	3.26	1.70	9.73	12.80
4100	20′ eave height		820	.068		5.20	3.53	1.84	10.57	13.95
4200	24′ eave height		765	.073		5.75	3.79	1.97	11.51	15.10
4300	Triple post 4-span frame, 26 ga. colored roofing and siding									
4400	160′ wide, 14′ eave height	E-2	970	.058	SF Flr.	4.50	2.99	1.56	9.05	11.85
4500	16′ eave height		930	.060		4.69	3.12	1.62	9.43	12.40
4600	20′ eave height		870	.064		4.62	3.33	1.74	9.69	12.80
4700	24′ eave height		815	.069		5.25	3.56	1.85	10.66	14.05
4800	200′ wide, 14′ eave height		1030	.054		4.13	2.81	1.47	8.41	11.05
4900	16′ eave height		995	.056		4.28	2.91	1.52	8.71	11.50
5000	20′ eave height		935	.060		4.72	3.10	1.62	9.44	12.40
5100	24′ eave height		885	.063		5.30	3.27	1.71	10.28	13.45
5200	Accessory items: add to the basic building cost above									
5250	Eave overhang, 2′ wide, 26 ga., with soffit	E-2	360	.156	L.F.	31.50	8.05	4.20	43.75	53
5300	4′ wide, without soffit		300	.187		27.50	9.65	5.05	42.20	52.50
5350	With soffit		250	.224		40	11.60	6.05	57.65	71
5400	6′ wide, without soffit		250	.224		35.50	11.60	6.05	53.15	66
5450	With soffit		200	.280		48	14.50	7.55	70.05	86.50
5500	Entrance canopy, incl. frame, 4′ x 4′		25	2.240	Ea.	475	116	60.50	651.50	790
5550	4′ x 8′		19	2.947	"	550	153	79.50	782.50	960
5600	End wall roof overhang, 4′ wide, without soffit		850	.066	L.F.	17.60	3.41	1.78	22.79	27.50
5650	With soffit		500	.112	"	28	5.80	3.02	36.82	44.50
5700	Doors, HM self-framing, incl. butts, lockset and trim									
5750	Single leaf, 3070 (3′ x 7′), economy	2 Sswk	5	3.200	Opng.	585	168		753	950
5800	Deluxe		4	4		640	211		851	1,075
5825	Glazed		4	4		745	211		956	1,200

13 34 19.50 Pre-Engineered Steel Buildings		Crew	Daily Output	Labor-Hours	Unit	Material	Labor	Equipment	Total	Total Incl O&P
						2015 Bare Costs				
5850	3670 (3'-6" x 7')	2 Sswk	4	4	Opng.	815	211		1,026	1,275
5900	4070 (4' x 7')		3	5.333		885	281		1,166	1,475
5950	Double leaf, 6070 (6' x 7')		2	8		1,100	420		1,520	1,950
6000	Glazed		2	8		1,400	420		1,820	2,300
6050	Framing only, for openings, 3' x 7'		4	4		185	211		396	585
6100	10' x 10'		3	5.333		610	281		891	1,175
6150	For windows below, 2020 (2' x 2')		6	2.667		196	140		336	470
6200	4030 (4' x 3')		5	3.200		239	168		407	570
6250	Flashings, 26 ga., corner or eave, painted		240	.067	L.F.	4.47	3.51		7.98	11.25
6300	Galvanized		240	.067		4.10	3.51		7.61	10.85
6350	Rake flashing, painted		240	.067		4.83	3.51		8.34	11.65
6400	Galvanized		240	.067		4.40	3.51		7.91	11.20
6450	Ridge flashing, 18" wide, painted		240	.067		6.45	3.51		9.96	13.45
6500	Galvanized		240	.067		7.10	3.51		10.61	14.15
6550	Gutter, eave type, 26 ga., painted		320	.050		6.95	2.63		9.58	12.40
6650	Valley type, between buildings, painted		120	.133		12.85	7		19.85	27
6710	Insulation, rated .6 lb. density, unfaced 4" thick, R13	2 Carp	2300	.007	S.F.	.42	.33		.75	1
6720	6" thick, R19		2300	.007		.58	.33		.91	1.18
6730	10" thick, R30		2300	.007		1.10	.33		1.43	1.75
6750	Insulation, rated .6 lb. density, poly/scrim/foil (PSF) faced									
6760	4" thick R13	2 Carp	2300	.007	S.F.	.63	.33		.96	1.23
6770	6" thick, R19		2300	.007		.89	.33		1.22	1.52
6780	9 1/2" thick, R30		2300	.007		.98	.33		1.31	1.62
6800	Insulation, rated .6 lb. density, vinyl faced 1-1/2" thick, R5		2300	.007		.31	.33		.64	.88
6850	3" thick, R10		2300	.007		.32	.33		.65	.89
6900	4" thick, R13		2300	.007		.42	.33		.75	1
6920	6" thick, R19		2300	.007		.55	.33		.88	1.15
6930	10" thick, R30		2300	.007		1.44	.33		1.77	2.12
6950	Foil/scrim/kraft (FSK) faced, 1-1/2" thick, R5		2300	.007		.33	.33		.66	.90
7000	2" thick, R6		2300	.007		.43	.33		.76	1.01
7050	3" thick, R10		2300	.007		.43	.33		.76	1.01
7100	4" thick, R13		2300	.007		.45	.33		.78	1.04
7110	6" thick, R19		2300	.007		.66	.33		.99	1.27
7120	10" thick, R30		2300	.007		.92	.33		1.25	1.55
7150	Metalized polyester/scrim/kraft (PSK) facing,1-1/2" thk, R5		2300	.007		.53	.33		.86	1.12
7200	2" thick, R6		2300	.007		.63	.33		.96	1.23
7250	3" thick, R11		2300	.007		.72	.33		1.05	1.33
7300	4" thick, R13		2300	.007		.81	.33		1.14	1.43
7310	6" thick, R19		2300	.007		1.06	.33		1.39	1.71
7320	10" thick, R30		2300	.007		1.18	.33		1.51	1.84
7350	Vinyl/scrim/foil (VSF), 1-1/2" thick, R5		2300	.007		.47	.33		.80	1.06
7400	2" thick, R6		2300	.007		.60	.33		.93	1.20
7450	3" thick, R10		2300	.007		.65	.33		.98	1.26
7500	4" thick, R13		2300	.007		.79	.33		1.12	1.41
7510	Vinyl/scrim/vinyl (VSV) 4" thick, R13		2300	.007		.53	.33		.86	1.12
7520	6" thick, R19		2300	.007		.68	.33		1.01	1.29
7530	9 1/2" thick, R30		2300	.007		.92	.33		1.25	1.55
7580	10" thick, R19		2300	.007		1.41	.33		1.74	2.09
7585	Vinyl/scrim/polyester (VSP), 4" thick, R13		2300	.007		.57	.33		.90	1.17
7590	6" thick, R19		2300	.007		.70	.33		1.03	1.31
7600	10" thick, R30		2300	.007		1.09	.33		1.42	1.74
7635	Insulation installation, over the purlin, second layer, up to 4" thick, add						90%			
7640	Insulation installation, between the purlins, up to 4" thick, add						100%			

13 34 Fabricated Engineered Structures

13 34 19 – Metal Building Systems

13 34 19.50 Pre-Engineered Steel Buildings		Crew	Daily Output	Labor-Hours	Unit	Material	2015 Bare Costs Labor	Equipment	Total	Total Incl O&P
7650	Sash, single slide, glazed, with screens, 2020 (2′ x 2′)	E-1	22	1.091	Opng.	127	56.50	6.65	190.15	246
7700	3030 (3′ x 3′)		14	1.714		285	89	10.40	384.40	480
7750	4030 (4′ x 3′)		13	1.846		380	96	11.20	487.20	600
7800	6040 (6′ x 4′)		12	2		760	104	12.15	876.15	1,025
7850	Double slide sash, 3030 (3′ x 3′)		14	1.714		224	89	10.40	323.40	410
7900	6040 (6′ x 4′)		12	2		595	104	12.15	711.15	850
7950	Fixed glass, no screens, 3030 (3′ x 3′)		14	1.714		220	89	10.40	319.40	410
8000	6040 (6′ x 4′)		12	2		585	104	12.15	701.15	840
8050	Prefinished storm sash, 3030 (3′ x 3′)		70	.343		80	17.80	2.08	99.88	121
8100	Siding and roofing, see Sections 07 41 13.00 & 07 42 13.00									
8200	Skylight, fiberglass panels, to 30 S.F.	E-1	10	2.400	Ea.	125	125	14.60	264.60	370
8250	Larger sizes, add for excess over 30 S.F.	"	300	.080	S.F.	4.18	4.16	.49	8.83	12.40
8300	Roof vents, turbine ventilator, wind driven									
8350	No damper, includes base, galvanized									
8400	12″ diameter	Q-9	10	1.600	Ea.	91	80.50		171.50	227
8450	20″ diameter		8	2		256	101		357	440
8500	24″ diameter		8	2		390	101		491	585
8600	Continuous, 26 ga., 10′ long, 9″ wide	2 Sswk	4	4		36	211		247	420
8650	12″ wide	"	4	4		36	211		247	420

13 34 23 – Fabricated Structures

13 34 23.10 Comfort Stations

		Crew	Daily Output	Labor-Hours	Unit	Material	Labor	Equipment	Total	Total Incl O&P
0010	**COMFORT STATIONS** Prefab., stock, w/doors, windows & fixt.									
0100	Not incl. interior finish or electrical									
0300	Mobile, on steel frame, 2 unit				S.F.	163			163	180
0350	7 unit					271			271	298
0400	Permanent, including concrete slab, 2 unit	B-12J	50	.320		249	14.30	17.65	280.95	315
0500	6 unit	"	43	.372		191	16.60	20.50	228.10	260
0600	Alternate pricing method, mobile, 2 fixture				Fixture	5,650			5,650	6,225
0650	7 fixture					9,750			9,750	10,700
0700	Permanent, 2 unit	B-12J	.70	22.857		20,500	1,025	1,250	22,775	25,600
0750	6 unit	"	.50	32		17,400	1,425	1,775	20,600	23,500

13 34 23.15 Domes

		Crew	Daily Output	Labor-Hours	Unit	Material	Labor	Equipment	Total	Total Incl O&P
0010	**DOMES**									
0020	Domes, rev. alum., elec. drive, for astronomy obsv. shell only, stock units									
0600	10′-6″ diameter	2 Carp	.25	64	Ea.	38,000	3,000		41,000	46,700
0900	18′-6″ diameter		.17	94.118		78,000	4,425		82,425	93,500
1200	24′-6″ diameter		.08	200		103,000	9,400		112,400	129,000
1500	Domes, bulk storage, shell only, dual radius hemisphere, arch, steel									
1600	framing, corrugated steel covering, 150′ diameter	E-2	550	.102	SF Flr.	35	5.25	2.75	43	50
1700	400′ diameter	"	720	.078		28.50	4.02	2.10	34.62	40.50
1800	Wood framing, wood decking, to 400′ diameter	F-4	400	.120		37.50	5.70	2.82	46.02	54
1900	Radial framed wood (2″ x 6″), 1/2″ thick									
2000	plywood, asphalt shingles, 50′ diameter	F-3	2000	.020	SF Flr.	72.50	.96	.33	73.79	81.50
2100	60′ diameter		1900	.021		62	1.01	.34	63.35	70
2200	72′ diameter		1800	.022		51.50	1.06	.36	52.92	58.50
2300	116′ diameter		1730	.023		35	1.11	.38	36.49	40.50
2400	150′ diameter		1500	.027		37.50	1.28	.44	39.22	44

13 34 23.16 Fabricated Control Booths

		Crew	Daily Output	Labor-Hours	Unit	Material	Labor	Equipment	Total	Total Incl O&P
0010	**FABRICATED CONTROL BOOTHS**									
0100	Guard House, prefab conc. w/bullet resistant doors & windows, roof & wiring									
0110	8′ x 8′, Level III	L-10	1	24	Ea.	44,200	1,275	655	46,130	51,500
0120	8′ x 8′, Level IV	"	1	24	"	50,500	1,275	655	52,430	58,500

13 34 23 – Fabricated Structures

13 34 23.25 Garage Costs

	13 34 23.25 Garage Costs	Crew	Daily Output	Labor-Hours	Unit	Material	Labor (2015 Bare Costs)	Equipment	Total	Total Incl O&P
0010	**GARAGE COSTS**									
0020	Public parking, average				Car				18,400	20,200
0100	See also Square Foot Costs in Reference Section									
0300	Residential, wood, 12′ x 20′, one car prefab shell, stock, economy	2 Carp	1	16	Total	5,375	750		6,125	7,150
0350	Custom		.67	23.881		5,975	1,125		7,100	8,425
0400	Two car, 24′ x 20′, economy		.67	23.881		10,100	1,125		11,225	13,000
0450	Custom		.50	32		12,100	1,500		13,600	15,800

13 34 23.30 Garden House

	13 34 23.30 Garden House	Crew	Daily Output	Labor-Hours	Unit	Material	Labor	Equipment	Total	Total Incl O&P
0010	**GARDEN HOUSE** Prefab wood, no floors or foundations									
0100	6′ x 6′	2 Carp	200	.080	SF Flr.	51.50	3.76		55.26	62.50
0300	8′ x 12′	"	48	.333	"	38	15.65		53.65	67.50

13 34 23.35 Geodesic Domes

	13 34 23.35 Geodesic Domes	Crew	Daily Output	Labor-Hours	Unit	Material	Labor	Equipment	Total	Total Incl O&P
0010	**GEODESIC DOMES** Shell only, interlocking plywood panels									
0400	30′ diameter	F-5	1.60	20	Ea.	23,000	950		23,950	26,900
0500	33′ diameter		1.14	28.070		24,300	1,325		25,625	28,900
0600	40′ diameter		1	32		28,200	1,525		29,725	33,500
0700	45′ diameter	F-3	1.13	35.556		29,500	1,700	580	31,780	35,900
0750	56′ diameter		1	40		53,000	1,925	655	55,580	62,500
0800	60′ diameter		1	40		60,000	1,925	655	62,580	70,000
0850	67′ diameter		.80	50		84,500	2,400	820	87,720	98,000
1100	Aluminum panel, with 6″ insulation									
1200	100′ diameter				SF Flr.	30.50			30.50	33.50
1300	500′ diameter				"	29.50			29.50	32.50
1600	Aluminum framed, plexiglass closure panels									
1700	40′ diameter				SF Flr.	75.50			75.50	83
1800	200′ diameter				"	69.50			69.50	76.50
2100	Aluminum framed, aluminum closure panels									
2200	40′ diameter				SF Flr.	24.50			24.50	27
2300	100′ diameter					23.50			23.50	26
2400	200′ diameter					23.50			23.50	26
2500	For VRP faced bonded fiberglass insulation, add								10	10
2700	Aluminum framed, fiberglass sandwich panel closure									
2800	6′ diameter	2 Carp	150	.107	SF Flr.	33	5		38	44.50
2900	28′ diameter	"	350	.046	"	30	2.15		32.15	36.50

13 34 23.45 Kiosks

	13 34 23.45 Kiosks	Crew	Daily Output	Labor-Hours	Unit	Material	Labor	Equipment	Total	Total Incl O&P
0010	**KIOSKS**									
0020	Round, advertising type, 5′ diameter, 7′ high, aluminum wall, illuminated				Ea.	23,500			23,500	25,800
0100	Aluminum wall, non-illuminated					22,500			22,500	24,700
0500	Rectangular, 5′ x 9′, 7′-6″ high, aluminum wall, illuminated					25,500			25,500	28,000
0600	Aluminum wall, non-illuminated					24,000			24,000	26,400

13 34 23.60 Portable Booths

	13 34 23.60 Portable Booths	Crew	Daily Output	Labor-Hours	Unit	Material	Labor	Equipment	Total	Total Incl O&P
0010	**PORTABLE BOOTHS** Prefab. aluminum with doors, windows, ext. roof									
0100	lights wiring & insulation, 15 S.F. building, O.D., painted				S.F.	266			266	293
0300	30 S.F. building					214			214	235
0400	50 S.F. building					169			169	186
0600	80 S.F. building					147			147	161
0700	100 S.F. building					123			123	135
0900	Acoustical booth, 27 Db @ 1,000 Hz, 15 S.F. floor				Ea.	3,550			3,550	3,925
1000	7′ x 7′-6″, including light & ventilation					7,325			7,325	8,050
1200	Ticket booth, galv. steel, not incl. foundations., 4′ x 4′					4,625			4,625	5,100
1300	4′ x 6′					6,825			6,825	7,500

13 34 Fabricated Engineered Structures

13 34 23 – Fabricated Structures

13 34 23.70 Shelters		Crew	Daily Output	Labor-Hours	Unit	Material	2015 Bare Costs Labor	Equipment	Total	Total Incl O&P
0010	**SHELTERS**									
0020	Aluminum frame, acrylic glazing, 3' x 9' x 8' high	2 Sswk	1.14	14.035	Ea.	3,075	740		3,815	4,700
0100	9' x 12' x 8' high	"	.73	21.918	"	7,300	1,150		8,450	10,100

13 34 43 – Aircraft Hangars

13 34 43.50 Hangars		Crew	Daily Output	Labor-Hours	Unit	Material	Labor	Equipment	Total	Total Incl O&P
0010	**HANGARS** Prefabricated steel T hangars, Galv. steel roof &									
0100	walls, incl. electric bi-folding doors									
0110	not including floors or foundations, 4 unit	E-2	1275	.044	SF Flr.	12.75	2.27	1.19	16.21	19.30
0130	8 unit		1063	.053		11.60	2.73	1.42	15.75	19.05
0900	With bottom rolling doors, 4 unit		1386	.040		11.75	2.09	1.09	14.93	17.75
1000	8 unit	↓	966	.058	↓	10.55	3	1.56	15.11	18.55
1200	Alternate pricing method:									
1300	Galv. roof and walls, electric bi-folding doors, 4 plane	E-2	1.06	52.830	Plane	16,900	2,725	1,425	21,050	25,000
1500	8 plane		.91	61.538		13,800	3,175	1,650	18,625	22,600
1600	With bottom rolling doors, 4 plane		1.25	44.800		15,600	2,325	1,200	19,125	22,500
1800	8 plane	↓	.97	57.732	↓	12,600	2,975	1,550	17,125	20,900
2000	Circular type, prefab., steel frame, plastic skin, electric									
2010	door, including foundations, 80' diameter,									

13 34 53 – Agricultural Structures

13 34 53.50 Silos		Crew	Daily Output	Labor-Hours	Unit	Material	Labor	Equipment	Total	Total Incl O&P
0010	**SILOS**									
0500	Steel, factory fab., 30,000 gallon cap., painted, economy	L-5	1	56	Ea.	22,100	2,950	735	25,785	30,400
0700	Deluxe		.50	112		35,100	5,925	1,475	42,500	50,500
0800	Epoxy lined, economy		1	56		36,100	2,950	735	39,785	45,800
1000	Deluxe	↓	.50	112	↓	45,700	5,925	1,475	53,100	62,500

13 34 63 – Natural Fiber Construction

13 34 63.50 Straw Bale Construction		Crew	Daily Output	Labor-Hours	Unit	Material	Labor	Equipment	Total	Total Incl O&P
0010	**STRAW BALE CONSTRUCTION**									
2020	Straw bales in walls w/modified post and beam frame G	2 Carp	320	.050	S.F.	6.15	2.35		8.50	10.60

13 36 Towers

13 36 13 – Metal Towers

13 36 13.50 Control Towers		Crew	Daily Output	Labor-Hours	Unit	Material	Labor	Equipment	Total	Total Incl O&P
0010	**CONTROL TOWERS**									
0020	Modular 12' x 10', incl. instruments				Ea.	757,000			757,000	832,500
0500	With standard 40' tower				"	1,191,000			1,191,000	1,310,000
1000	Temporary portable control towers, 8' x 12',									
1010	complete with one position communications				Ea.				266,000	293,000

13 42 Building Modules

13 42 63 – Detention Cell Modules

13 42 63.16 Steel Detention Cell Modules		Crew	Daily Output	Labor-Hours	Unit	2015 Bare Costs Material	Labor	Equipment	Total	Total Incl O&P
0010	**STEEL DETENTION CELL MODULES**									
2000	Cells, prefab., 5′ to 6′ wide, 7′ to 8′ high, 7′ to 8′ deep,									
2010	bar front, cot, not incl. plumbing	E-4	1.50	21.333	Ea.	9,775	1,125	97.50	10,997.50	12,900

13 47 Facility Protection

13 47 13 – Cathodic Protection

13 47 13.16 Cathodic Prot. for Underground Storage Tanks

Line	Description	Crew	Daily Output	Labor-Hours	Unit	Material	Labor	Equipment	Total	Total Incl O&P
0010	**CATHODIC PROTECTION FOR UNDERGROUND STORAGE TANKS**									
1000	Anodes, magnesium type, 9 #	R-15	18.50	2.595	Ea.	41.50	140	16.30	197.80	281
1010	17 #		13	3.692		71.50	199	23	293.50	415
1020	32 #		10	4.800		124	258	30	412	570
1030	48 #		7.20	6.667		164	360	42	566	780
1100	Graphite type w/epoxy cap, 3″ x 60″ (32 #)	R-22	8.40	4.438		128	206		334	465
1110	4″ x 80″ (68 #)		6	6.213		241	289		530	720
1120	6″ x 72″ (80 #)		5.20	7.169		1,500	335		1,835	2,175
1130	6″ x 36″ (45 #)		9.60	3.883		760	181		941	1,125
2000	Rectifiers, silicon type, air cooled, 28 V/10 A	R-19	3.50	5.714		2,300	315		2,615	3,025
2010	20 V/20 A		3.50	5.714		2,375	315		2,690	3,075
2100	Oil immersed, 28 V/10 A		3	6.667		3,175	365		3,540	4,050
2110	20 V/20 A		3	6.667		3,175	365		3,540	4,075
3000	Anode backfill, coke breeze	R-22	3850	.010	Lb.	.22	.45		.67	.95
4000	Cable, HMWPE, No. 8		2.40	15.533	M.L.F.	500	720		1,220	1,700
4010	No. 6		2.40	15.533		745	720		1,465	1,975
4020	No. 4		2.40	15.533		1,125	720		1,845	2,400
4030	No. 2		2.40	15.533		1,750	720		2,470	3,075
4040	No. 1		2.20	16.945		2,400	790		3,190	3,900
4050	No. 1/0		2.20	16.945		2,975	790		3,765	4,525
4060	No. 2/0		2.20	16.945		4,625	790		5,415	6,350
4070	No. 4/0		2	18.640		6,000	865		6,865	7,975
5000	Test station, 7 terminal box, flush curb type w/lockable cover	R-19	12	1.667	Ea.	71.50	91.50		163	220
5010	Reference cell, 2″ dia PVC conduit, cplg., plug, set flush	″	4.80	4.167	″	151	228		379	520

13 48 Sound, Vibration, and Seismic Control

13 48 13 – Manufactured Sound and Vibration Control Components

13 48 13.50 Audio Masking

Line	Description	Crew	Daily Output	Labor-Hours	Unit	Material	Labor	Equipment	Total	Total Incl O&P
0010	**AUDIO MASKING**, acoustical enclosure, 4″ thick wall and ceiling									
0020	8# per S.F., up to 12′ span	3 Carp	72	.333	SF Surf	33	15.65		48.65	61.50
0300	Better quality panels, 10.5# per S.F.		64	.375		37	17.60		54.60	70
0400	Reverb-chamber, 4″ thick, parallel walls		60	.400		46.50	18.80		65.30	82
0600	Skewed wall, parallel roof, 4″ thick panels		55	.436		53	20.50		73.50	92
0700	Skewed walls, skewed roof, 4″ layers, 4″ air space		48	.500		59.50	23.50		83	104
0900	Sound-absorbing panels, pntd. mtl., 2′-6″ x 8′, under 1,000 S.F.		215	.112		12.30	5.25		17.55	22
1100	Over 1000 S.F.		240	.100		11.85	4.70		16.55	21
1200	Fabric faced		240	.100		9.60	4.70		14.30	18.30
1500	Flexible transparent curtain, clear	3 Shee	215	.112		7.45	6.25		13.70	18.05
1600	50% foam		215	.112		10.40	6.25		16.65	21.50
1700	75% foam		215	.112		10.40	6.25		16.65	21.50
1800	100% foam		215	.112		10.40	6.25		16.65	21.50
3100	Audio masking system, including speakers, amplification									

13 48 Sound, Vibration, and Seismic Control

13 48 13 – Manufactured Sound and Vibration Control Components

13 48 13.50 Audio Masking		Crew	Daily Output	Labor-Hours	Unit	2015 Bare Costs Material	Labor	Equipment	Total	Total Incl O&P
3110	and signal generator									
3200	Ceiling mounted, 5,000 S.F.	2 Elec	2400	.007	S.F.	1.26	.36		1.62	1.95
3300	10,000 S.F.		2800	.006		1.02	.31		1.33	1.60
3400	Plenum mounted, 5,000 S.F.		3800	.004		1.08	.23		1.31	1.55
3500	10,000 S.F.		4400	.004		.73	.20		.93	1.11

13 49 Radiation Protection

13 49 13 – Integrated X-Ray Shielding Assemblies

13 49 13.50 Lead Sheets

		Crew	Daily Output	Labor-Hours	Unit	Material	Labor	Equipment	Total	Total Incl O&P
0010	**LEAD SHEETS**									
0300	Lead sheets, 1/16″ thick	2 Lath	135	.119	S.F.	10.50	5.10		15.60	19.55
0400	1/8″ thick		120	.133		31	5.75		36.75	43
0500	Lead shielding, 1/4″ thick		135	.119		41.50	5.10		46.60	54
0550	1/2″ thick		120	.133		73.50	5.75		79.25	90
0950	Lead headed nails (average 1 lb. per sheet)				Lb.	8			8	8.80
1000	Butt joints in 1/8″ lead or thicker, 2″ batten strip x 7′ long	2 Lath	240	.067	Ea.	28	2.87		30.87	35.50
1200	X-ray protection, average radiography or fluoroscopy									
1210	room, up to 300 S.F. floor, 1/16″ lead, economy	2 Lath	.25	64	Total	10,100	2,750		12,850	15,400
1500	7′-0″ walls, deluxe	″	.15	106	″	12,200	4,575		16,775	20,600
1600	Deep therapy X-ray room, 250 kV capacity,									
1800	up to 300 S.F. floor, 1/4″ lead, economy	2 Lath	.08	200	Total	28,300	8,600		36,900	44,600
1900	7′-0″ walls, deluxe		.06	266	″	34,900	11,500		46,400	56,500
1999	Minimum labor/equipment charge		4.50	3.556	Job		153		153	241

13 49 19 – Lead-Lined Materials

13 49 19.50 Shielding Lead

		Crew	Daily Output	Labor-Hours	Unit	Material	Labor	Equipment	Total	Total Incl O&P
0010	**SHIELDING LEAD**									
0100	Laminated lead in wood doors, 1/16″ thick, no hardware				S.F.	52.50			52.50	57.50
0200	Lead lined door frame, not incl. hardware,									
0210	1/16″ thick lead, butt prepared for hardware	1 Lath	2.40	3.333	Ea.	810	143		953	1,125
0850	Window frame with 1/16″ lead and voice passage, 36″ x 60″	2 Glaz	2	8		4,200	360		4,560	5,175
0870	24″ x 36″ frame		4	4		2,175	180		2,355	2,700
0900	Lead gypsum board, 5/8″ thick with 1/16″ lead		160	.100	S.F.	10.95	4.51		15.46	19.35
0910	1/8″ lead		140	.114		23	5.15		28.15	34
0930	1/32″ lead	2 Lath	200	.080		7.95	3.44		11.39	14.15

13 49 21 – Lead Glazing

13 49 21.50 Lead Glazing

		Crew	Daily Output	Labor-Hours	Unit	Material	Labor	Equipment	Total	Total Incl O&P
0010	**LEAD GLAZING**									
0600	Lead glass, 1/4″ thick, 2.0 mm LE, 12″ x 16″	2 Glaz	13	1.231	Ea.	380	55.50		435.50	505
0700	24″ x 36″		8	2		1,325	90		1,415	1,600
0800	36″ x 60″		2	8		3,675	360		4,035	4,600
2000	X-ray viewing panels, clear lead plastic									
2010	7 mm thick, 0.3 mm LE, 2.3 lb./S.F.	H-3	139	.115	S.F.	241	4.64		245.64	273
2020	12 mm thick, 0.5 mm LE, 3.9 lb./S.F.		82	.195		355	7.85		362.85	405
2030	18 mm thick, 0.8 mm LE, 5.9 lb./S.F.		54	.296		405	11.95		416.95	465
2040	22 mm thick, 1.0 mm LE, 7.2 lb./S.F.		44	.364		530	14.65		544.65	610
2050	35 mm thick, 1.5 mm LE, 11.5 lb./S.F.		28	.571		815	23		838	935
2060	46 mm thick, 2.0 mm LE, 15.0 lb./S.F.		21	.762		1,050	30.50		1,080.50	1,225
2090	For panels 12 S.F. to 48 S.F., add crating charge				Ea.	50			50	50

13 49 23 – Integrated RFI/EMI Shielding Assemblies

13 49 23.50 Modular Shielding Partitions		Crew	Daily Output	Labor-Hours	Unit	Material	2015 Bare Costs Labor	Equipment	Total	Total Incl O&P
0010	**MODULAR SHIELDING PARTITIONS**									
4000	X-ray barriers, modular, panels mounted within framework for									
4002	attaching to floor, wall or ceiling, upper portion is clear lead									
4005	plastic window panels 48"H, lower portion is opaque leaded									
4008	steel panels 36"H, structural supports not incl.									
4010	1-section barrier, 36"W x 84"H overall									
4020	0.5 mm LE panels	H-3	6.40	2.500	Ea.	8,100	101		8,201	9,075
4030	0.8 mm LE panels		6.40	2.500		8,725	101		8,826	9,775
4040	1.0 mm LE panels		5.33	3.002		10,200	121		10,321	11,400
4050	1.5 mm LE panels		5.33	3.002		13,600	121		13,721	15,200
4060	2-section barrier, 72"W x 84"H overall									
4070	0.5 mm LE panels	H-3	4	4	Ea.	11,800	161		11,961	13,300
4080	0.8 mm LE panels		4	4		13,100	161		13,261	14,700
4090	1.0 mm LE panels		3.56	4.494		16,000	181		16,181	17,900
5000	1.5 mm LE panels		3.20	5		22,800	201		23,001	25,400
5010	3-section barrier, 108"W x 84"H overall									
5020	0.5 mm LE panels	H-3	3.20	5	Ea.	17,700	201		17,901	19,800
5030	0.8 mm LE panels		3.20	5		19,600	201		19,801	21,800
5040	1.0 mm LE panels		2.67	5.993		24,000	241		24,241	26,800
5050	1.5 mm LE panels		2.46	6.504		34,200	262		34,462	38,000
7000	X-ray barriers, mobile, mounted within framework w/casters on									
7005	bottom, clear lead plastic window panels on upper portion,									
7010	opaque on lower, 30"W x 75"H overall, incl. framework									
7020	24"H upper w/0.5 mm LE, 48"H lower w/0.8 mm LE	1 Carp	16	.500	Ea.	3,800	23.50		3,823.50	4,225
7030	48"W x 75"H overall, incl. framework									
7040	36"H upper w/0.5 mm LE, 36"H lower w/0.8 mm LE	1 Carp	16	.500	Ea.	6,175	23.50		6,198.50	6,850
7050	36"H upper w/1.0 mm LE, 36"H lower w/1.5 mm LE	"	16	.500	"	7,300	23.50		7,323.50	8,100
7060	72"W x 75"H overall, incl. framework									
7070	36"H upper w/0.5 mm LE, 36"H lower w/0.8 mm LE	1 Carp	16	.500	Ea.	7,300	23.50		7,323.50	8,100
7080	36"H upper w/1.0 mm LE, 36"H lower w/1.5 mm LE	"	16	.500	"	9,150	23.50		9,173.50	10,100

13 49 33 – Radio Frequency Shielding

13 49 33.50 Shielding, Radio Frequency		Crew	Daily Output	Labor-Hours	Unit	Material	2015 Bare Costs Labor	Equipment	Total	Total Incl O&P
0010	**SHIELDING, RADIO FREQUENCY**									
0020	Prefabricated, galvanized steel	2 Carp	375	.043	SF Surf	4.46	2		6.46	8.20
0040	5 oz., copper floor panel		480	.033		3.68	1.56		5.24	6.60
0050	5 oz., copper wall/ceiling panel		155	.103		3.68	4.85		8.53	12
0100	12 oz., copper floor panel		470	.034		7.85	1.60		9.45	11.25
0110	12 oz., copper wall/ceiling panel		140	.114		7.85	5.35		13.20	17.45
0150	Door, copper/wood laminate, 4' x 7'		1.50	10.667	Ea.	7,400	500		7,900	8,950
0200	RF modular shielding panels, walls & ceilings	E-1	180	.133	S.F.	18.15	6.95	.81	25.91	33
0210	Floor liner, steel sheet, 14 ga.		430	.056		2.48	2.90	.34	5.72	8.15
0215	11 ga.		140	.171		3.59	8.90	1.04	13.53	20.50
0220	Wall liner, steel sheet, 14 ga.		180	.133		2.62	6.95	.81	10.38	15.80
0225	11 ga.		140	.171		3.78	8.90	1.04	13.72	21
0230	Steel plate, 1/4"		90	.267		6.95	13.85	1.62	22.42	33.50
0235	Ceiling liner, steel sheet, 14 ga.		180	.133		2.48	6.95	.81	10.24	15.65
0250	Ceiling hangers		45	.533	Ea.	30.50	27.50	3.24	61.24	85
0275	Wall supports		45	.533	"	30.50	27.50	3.24	61.24	85
0300	Shielding transition, 11 ga. preformed angles		1365	.018	L.F.	30.50	.91	.11	31.52	35
0500	Protection, door, 3' W x 7' H, to 120 Db elec/plane wave	Q-11	9	3.556	Ea.	9,200	189		9,389	10,400
0550	Double door, 6' W x 7' H		6	5.333		14,800	284		15,084	16,800
0600	Wave guide vents, 2" diameter		67	.478		199	25.50		224.50	259

13 49 Radiation Protection

13 49 33 – Radio Frequency Shielding

13 49 33.50 Shielding, Radio Frequency		Crew	Daily Output	Labor-Hours	Unit	2015 Bare Costs Material	Labor	Equipment	Total	Total Incl O&P
0610	12″ diameter	Q-11	17	1.882	Ea.	995	100		1,095	1,250
0620	12″ x 6″		17	1.882		325	100		425	520
0630	12″ x 12″		17	1.882		430	100		530	630
0640	15″ x 15″		17	1.882		490	100		590	700
0650	30″ x 14″	↓	11	2.909	↓	920	155		1,075	1,275

13 53 Meteorological Instrumentation

13 53 09 – Weather Instrumentation

13 53 09.50 Weather Station

		Crew	Daily Output	Labor-Hours	Unit	Material	Labor	Equipment	Total	Total Incl O&P
0010	**WEATHER STATION**									
0020	Remote recording, solar powered, with rain gauge & display, 400 ft range				Ea.	775			775	850
0100	1 mile range				″	1,900			1,900	2,075

Estimating Tips

General

- Many products in Division 14 will require some type of support or blocking for installation not included with the item itself. Examples are supports for conveyors or tube systems, attachment points for lifts, and footings for hoists or cranes. Add these supports in the appropriate division.

14 10 00 Dumbwaiters
14 20 00 Elevators

- Dumbwaiters and elevators are estimated and purchased in a method similar to buying a car. The manufacturer has a base unit with standard features. Added to this base unit price will be whatever options the owner or specifications require. Increased load capacity, additional vertical travel, additional stops, higher speed, and cab finish options are items to be considered. When developing an estimate for dumbwaiters and elevators, remember that some items needed by the installers may have to be included as part of the general contract.

  Examples are:
  - shaftway
  - rail support brackets
  - machine room
  - electrical supply
  - sill angles
  - electrical connections
  - pits
  - roof penthouses
  - pit ladders

  Check the job specifications and drawings before pricing.

- Installation of elevators and handicapped lifts in historic structures can require significant additional costs. The associated structural requirements may involve cutting into and repairing finishes, moldings, flooring, etc. The estimator must account for these special conditions.

14 30 00 Escalators and Moving Walks

- Escalators and moving walks are specialty items installed by specialty contractors. There are numerous options associated with these items. For specific options, contact a manufacturer or contractor. In a method similar to estimating dumbwaiters and elevators, you should verify the extent of general contract work and add items as necessary.

14 40 00 Lifts
14 90 00 Other Conveying Equipment

- Products such as correspondence lifts, chutes, and pneumatic tube systems, as well as other items specified in this subdivision, may require trained installers. The general contractor might not have any choice as to who will perform the installation, or when it will be performed. Long lead times are often required for these products, making early decisions in scheduling necessary.

Reference Numbers

Reference numbers are shown in shaded boxes at the beginning of some major classifications. These numbers refer to related items in the Reference Section. The reference information may be an estimating procedure, an alternate pricing method, or technical information.

Note: Not all subdivisions listed here necessarily appear in this publication. ■

14 05 Common Work Results for Conveying Equipment

14 05 05 – Selective Demolition for Conveying Equipment

14 05 05.10 Dumbwaiter Removal		Crew	Daily Output	Labor-Hours	Unit	Material	2015 Bare Costs Labor	Equipment	Total	Total Incl O&P
0010	DUMBWAITER REMOVAL									
0050	Dumbwaiter removal, cab, track and equipment	2 Elev	4	4	Stop		305		305	470
14 05 05.20 Elevator Removal										
0010	ELEVATOR REMOVAL									
0050	Elevator removal, cab, track and equipment	2 Elev	2	8	Stop		610		610	945
14 05 05.30 Escalator/Moving Walk Removal										
0010	ESCALATOR/MOVING WALK REMOVAL									
0050	Escalator removal, 10′ floor to floor	M-1	.14	228	Ea.		16,600	330	16,930	26,000
0100	Escalator removal, 15′ floor to floor		.12	266			19,400	385	19,785	30,300
0200	Escalator removal, 20′ floor to floor		.10	320			23,300	460	23,760	36,400
0300	Escalator removal, 25′ floor to floor		.08	400			29,100	575	29,675	45,500
0400	Moving ramp/walk removal		10	3.200	L.F.		233	4.61	237.61	365
14 05 05.40 Lift Removal										
0010	LIFT REMOVAL									
0050	Chair lift removal, minimum	2 Elev	2	8	Ea.		610		610	945
0060	Maximum	"	1	16			1,225		1,225	1,900
0100	Single post lift removal	L-4	7.50	3.200			142		142	231
0200	Double post lift removal		5	4.800			212		212	345
0300	Four post lift removal		3	8			355		355	580
14 05 05.50 Distribution System Removal										
0010	DISTRIBUTION SYSTEM REMOVAL									
0050	Distribution system removal, motorized cart	4 Mill	.30	106	Station		5,250		5,250	8,250
0100	Distribution system removal, conveyor belt	2 Mill	80	.200	L.F.		9.85		9.85	15.45
0200	Distribution system removal, overhead monorail	1 Elev	50	.160			12.25		12.25	18.90
0300	Distribution system removal, pneumatic tube	2 Stpi	150	.107			6.35		6.35	9.95
0400	Remove linen/trash chute	2 Shee	8	2	Floor		112		112	177
14 05 05.60 Crane Removal										
0010	CRANE REMOVAL									
0050	Remove crane rail	E-4	300	.107	L.F.		5.65	.49	6.14	10.80
0100	Remove overhead bridge crane	M-4	1	36	Ea.		2,050	176	2,226	3,400

14 11 Manual Dumbwaiters

14 11 10 – Hand Operated Dumbwaiters

14 11 10.20 Manual Dumbwaiters		Crew	Daily Output	Labor-Hours	Unit	Material	2015 Bare Costs Labor	Equipment	Total	Total Incl O&P
0010	MANUAL DUMBWAITERS									
0020	2 stop, hand powered, up to 75 lb. capacity	2 Elev	.75	21.333	Ea.	3,000	1,625		4,625	5,825
0100	76 lb capacity and up		.50	32	"	6,700	2,450		9,150	11,200
0300	For each additional stop, add		.75	21.333	Stop	1,100	1,625		2,725	3,725

14 12 Electric Dumbwaiters

14 12 10 – Dumbwaiters

14 12 10.10 Electric Dumbwaiters		Crew	Daily Output	Labor-Hours	Unit	Material	2015 Bare Costs Labor	Equipment	Total	Total Incl O&P
0010	ELECTRIC DUMBWAITERS									
0020	2 stop, up to 75 lb capacity	2 Elev	.13	123	Ea.	7,400	9,425		16,825	22,700
0100	76 lb capacity and up		.11	145	"	22,300	11,100		33,400	41,700
0600	For each additional stop, add		.54	29.630	Stop	3,300	2,275		5,575	7,125

14 21 Electric Traction Elevators

14 21 13 – Electric Traction Freight Elevators

14 21 13.10 Electric Traction Freight Elevators and Options

		Crew	Daily Output	Labor-Hours	Unit	Material	Labor	Equipment	Total	Total Incl O&P
0010	ELECTRIC TRACTION FREIGHT ELEVATORS AND OPTIONS R142000-10									
0425	Electric freight, base unit, 4000 lb., 200 fpm, 4 stop, std. fin.	2 Elev	.05	320	Ea.	110,000	24,500		134,500	159,500
0450	For 5000 lb. capacity, add R142000-40					5,850			5,850	6,425
0500	For 6000 lb. capacity, add					14,500			14,500	15,900
0525	For 7000 lb. capacity, add					18,100			18,100	19,900
0550	For 8000 lb. capacity, add					22,300			22,300	24,500
0575	For 10000 lb. capacity, add					30,500			30,500	33,600
0600	For 12000 lb. capacity, add					37,500			37,500	41,300
0625	For 16000 lb. capacity, add					45,000			45,000	49,500
0650	For 20000 lb. capacity, add					50,000			50,000	55,000
0675	For increased speed, 250 fpm, add					17,000			17,000	18,700
0700	300 fpm, geared electric, add					20,600			20,600	22,700
0725	350 fpm, geared electric, add					25,100			25,100	27,600
0750	400 fpm, geared electric, add					29,400			29,400	32,300
0775	500 fpm, gearless electric, add					36,400			36,400	40,100
0800	600 fpm, gearless electric, add					43,600			43,600	47,900
0825	700 fpm, gearless electric, add					52,000			52,000	57,000
0850	800 fpm, gearless electric, add					59,500			59,500	65,500
0875	For class "B" loading, add					4,800			4,800	5,275
0900	For class "C-1" loading, add					6,950			6,950	7,650
0925	For class "C-2" loading, add					8,000			8,000	8,800
0950	For class "C-3" loading, add					10,700			10,700	11,700
0975	For travel over 40 V.L.F., add	2 Elev	7.25	2.207	V.L.F.	645	169		814	970
1000	For number of stops over 4, add	"	.27	59.259	Stop	4,225	4,525		8,750	11,700

14 21 23 – Electric Traction Passenger Elevators

14 21 23.10 Electric Traction Passenger Elevators and Options

		Crew	Daily Output	Labor-Hours	Unit	Material	Labor	Equipment	Total	Total Incl O&P
0010	ELECTRIC TRACTION PASSENGER ELEVATORS AND OPTIONS									
1625	Electric pass., base unit, 2000 lb., 200 fpm, 4 stop, std. fin.	2 Elev	.05	320	Ea.	92,000	24,500		116,500	139,000
1650	For 2500 lb. capacity, add					3,800			3,800	4,175
1675	For 3000 lb. capacity, add					4,400			4,400	4,825
1700	For 3500 lb. capacity, add					5,675			5,675	6,250
1725	For 4000 lb. capacity, add					6,825			6,825	7,500
1750	For 4500 lb. capacity, add					9,150			9,150	10,100
1775	For 5000 lb. capacity, add					11,400			11,400	12,600
1800	For increased speed, 250 fpm, geared electric, add					4,800			4,800	5,275
1825	300 fpm, geared electric, add					6,575			6,575	7,225
1850	350 fpm, geared electric, add					8,775			8,775	9,650
1875	400 fpm, geared electric, add					11,500			11,500	12,600
1900	500 fpm, gearless electric, add					28,000			28,000	30,800
1925	600 fpm, gearless electric, add					43,700			43,700	48,000
1950	700 fpm, gearless electric, add					50,000			50,000	55,500
1975	800 fpm, gearless electric, add					55,500			55,500	61,000

14 21 Electric Traction Elevators

14 21 23 – Electric Traction Passenger Elevators

14 21 23.10 Electric Traction Passenger Elevators and Options		Crew	Daily Output	Labor-Hours	Unit	Material	2015 Bare Costs Labor	Equipment	Total	Total Incl O&P
2000	For travel over 40 V.L.F., add	2 Elev	7.25	2.207	V.L.F.	695	169		864	1,025
2025	For number of stops over 4, add		.27	59.259	Stop	3,075	4,525		7,600	10,400
2400	Electric hospital, base unit, 4000 lb., 200 fpm, 4 stop, std fin.	↓	.05	320	Ea.	79,500	24,500		104,000	125,500
2425	For 4500 lb. capacity, add					5,475			5,475	6,025
2450	For 5000 lb. capacity, add					7,175			7,175	7,900
2475	For increased speed, 250 fpm, geared electric, add					4,925			4,925	5,425
2500	300 fpm, geared electric, add					7,600			7,600	8,375
2525	350 fpm, geared electric, add					8,775			8,775	9,650
2550	400 fpm, geared electric, add					11,500			11,500	12,700
2575	500 fpm, gearless electric, add					32,600			32,600	35,900
2600	600 fpm, gearless electric, add					47,600			47,600	52,500
2625	700 fpm, gearless electric, add					52,500			52,500	58,000
2650	800 fpm, gearless electric, add				↓	59,000			59,000	65,000
2675	For travel over 40 V.L.F., add	2 Elev	7.25	2.207	V.L.F.	435	169		604	740
2700	For number of stops over 4, add	"	.27	59.259	Stop	4,325	4,525		8,850	11,800

14 21 33 – Electric Traction Residential Elevators

14 21 33.20 Residential Elevators		Crew	Daily Output	Labor-Hours	Unit	Material	Labor	Equipment	Total	Total Incl O&P
0010	**RESIDENTIAL ELEVATORS**									
7000	Residential, cab type, 1 floor, 2 stop, economy model	2 Elev	.20	80	Ea.	11,400	6,125		17,525	22,100
7100	Custom model		.10	160		19,300	12,200		31,500	40,100
7200	2 floor, 3 stop, economy model		.12	133		17,000	10,200		27,200	34,400
7300	Custom model	↓	.06	266	↓	27,700	20,400		48,100	62,000

14 24 Hydraulic Elevators

14 24 13 – Hydraulic Freight Elevators

14 24 13.10 Hydraulic Freight Elevators and Options		Crew	Daily Output	Labor-Hours	Unit	Material	Labor	Equipment	Total	Total Incl O&P
0010	**HYDRAULIC FREIGHT ELEVATORS AND OPTIONS**									
1025	Hydraulic freight, base unit, 2000 lb., 50 fpm, 2 stop, std. fin.	2 Elev	.10	160	Ea.	79,500	12,200		91,700	106,500
1050	For 2500 lb. capacity, add					3,875			3,875	4,250
1075	For 3000 lb. capacity, add					5,250			5,250	5,775
1100	For 3500 lb. capacity, add					8,350			8,350	9,200
1125	For 4000 lb. capacity, add					9,750			9,750	10,700
1150	For 4500 lb. capacity, add					14,300			14,300	15,700
1175	For 5000 lb. capacity, add					15,000			15,000	16,500
1200	For 6000 lb. capacity, add					15,800			15,800	17,400
1225	For 7000 lb. capacity, add					23,500			23,500	25,800
1250	For 8000 lb. capacity, add					30,900			30,900	33,900
1275	For 10000 lb. capacity, add					37,800			37,800	41,600
1300	For 12000 lb. capacity, add					49,100			49,100	54,000
1325	For 16000 lb. capacity, add					70,000			70,000	77,000
1350	For 20000 lb. capacity, add					80,500			80,500	89,000
1375	For increased speed, 100 fpm, add					1,550			1,550	1,700
1400	125 fpm, add					3,425			3,425	3,775
1425	150 fpm, add					4,675			4,675	5,150
1450	175 fpm, add					7,150			7,150	7,875
1475	For class "B" loading, add					4,525			4,525	4,975
1500	For class "C-1" loading, add					6,900			6,900	7,600
1525	For class "C-2" loading, add					7,950			7,950	8,750
1550	For class "C-3" loading, add				↓	10,700			10,700	11,800
1575	For travel over 20 V.L.F., add	2 Elev	7.25	2.207	V.L.F.	845	169		1,014	1,200

14 24 Hydraulic Elevators

14 24 13 – Hydraulic Freight Elevators

14 24 13.10 Hydraulic Freight Elevators and Options		Crew	Daily Output	Labor-Hours	Unit	2015 Bare Costs Material	Labor	Equipment	Total	Total Incl O&P
1600	For number of stops over 2, add	2 Elev	.27	59.259	Stop	2,150	4,525		6,675	9,350

14 24 23 – Hydraulic Passenger Elevators

14 24 23.10 Hydraulic Passenger Elevators and Options		Crew	Daily Output	Labor-Hours	Unit	Material	Labor	Equipment	Total	Total Incl O&P
0010	HYDRAULIC PASSENGER ELEVATORS AND OPTIONS									
2050	Hyd. pass., base unit, 1500 lb., 100 fpm, 2 stop, std. fin.	2 Elev	.10	160	Ea.	37,700	12,200		49,900	60,500
2075	For 2000 lb. capacity, add					810			810	890
2100	For 2500 lb. capacity, add					2,800			2,800	3,100
2125	For 3000 lb. capacity, add					3,975			3,975	4,375
2150	For 3500 lb. capacity, add					6,875			6,875	7,550
2175	For 4000 lb. capacity, add					8,225			8,225	9,050
2200	For 4500 lb. capacity, add					10,800			10,800	11,900
2225	For 5000 lb. capacity, add					15,200			15,200	16,700
2250	For increased speed, 125 fpm, add					1,975			1,975	2,175
2275	150 fpm, add					2,550			2,550	2,825
2300	175 fpm, add					4,850			4,850	5,325
2325	200 fpm, add					9,275			9,275	10,200
2350	For travel over 12 V.L.F., add	2 Elev	7.25	2.207	V.L.F.	690	169		859	1,025
2375	For number of stops over 2, add		.27	59.259	Stop	960	4,525		5,485	8,050
2725	Hydraulic hospital, base unit, 4000 lb., 100 fpm, 2 stop, std. fin.		.10	160	Ea.	59,500	12,200		71,700	84,500
2775	For 4500 lb. capacity, add					6,700			6,700	7,350
2800	For 5000 lb. capacity, add					9,800			9,800	10,800
2825	For increased speed, 125 fpm, add					2,450			2,450	2,700
2850	150 fpm, add					3,450			3,450	3,800
2875	175 fpm, add					5,775			5,775	6,350
2900	200 fpm, add					8,450			8,450	9,300
2925	For travel over 12 V.L.F., add	2 Elev	7.25	2.207	V.L.F.	530	169		699	840
2950	For number of stops over 2, add	"	.27	59.259	Stop	4,175	4,525		8,700	11,600

14 27 Custom Elevator Cabs and Doors

14 27 13 – Custom Elevator Cab Finishes

14 27 13.10 Cab Finishes		Crew	Daily Output	Labor-Hours	Unit	Material	Labor	Equipment	Total	Total Incl O&P
0010	CAB FINISHES									
3325	Passenger elevator cab finishes (based on 3500 lb. cab size)									
3350	Acrylic panel ceiling				Ea.	755			755	830
3375	Aluminum eggcrate ceiling					865			865	950
3400	Stainless steel doors					3,950			3,950	4,350
3425	Carpet flooring					610			610	670
3450	Epoxy flooring					465			465	510
3475	Quarry tile flooring					875			875	960
3500	Slate flooring					1,600			1,600	1,750
3525	Textured rubber flooring					650			650	715
3550	Stainless steel walls					4,100			4,100	4,500
3575	Stainless steel returns at door					1,175			1,175	1,275
4450	Hospital elevator cab finishes (based on 3500 lb. cab size)									
4475	Aluminum eggcrate ceiling				Ea.	900			900	990
4500	Stainless steel doors					3,950			3,950	4,350
4525	Epoxy flooring					465			465	510
4550	Quarry tile flooring					875			875	960
4575	Textured rubber flooring					650			650	715
4600	Stainless steel walls					4,550			4,550	5,000

14 27 Custom Elevator Cabs and Doors

14 27 13 – Custom Elevator Cab Finishes

14 27 13.10 Cab Finishes		Crew	Daily Output	Labor-Hours	Unit	Material	2015 Bare Costs Labor	Equipment	Total	Total Incl O&P
4625	Stainless steel returns at door				Ea.	960			960	1,050

14 28 Elevator Equipment and Controls

14 28 10 – Elevator Equipment and Control Options

14 28 10.10 Elevator Controls and Doors

		Crew	Daily Output	Labor-Hours	Unit	Material	Labor	Equipment	Total	Total Incl O&P
0010	**ELEVATOR CONTROLS AND DOORS**									
2975	Passenger elevator options									
3000	2 car group automatic controls	2 Elev	.66	24.242	Ea.	4,625	1,850		6,475	7,925
3025	3 car group automatic controls		.44	36.364		8,825	2,775		11,600	14,000
3050	4 car group automatic controls		.33	48.485		17,600	3,700		21,300	25,100
3075	5 car group automatic controls		.26	61.538		31,800	4,700		36,500	42,200
3100	6 car group automatic controls		.22	72.727		64,500	5,575		70,075	79,000
3125	Intercom service		3	5.333		955	410		1,365	1,675
3150	Duplex car selective collective		.66	24.242		8,100	1,850		9,950	11,800
3175	Center opening 1 speed doors		2	8		1,950	610		2,560	3,100
3200	Center opening 2 speed doors		2	8		2,750	610		3,360	3,975
3225	Rear opening doors (opposite front)		2	8		4,225	610		4,835	5,600
3250	Side opening 2 speed doors		2	8		4,350	610		4,960	5,725
3275	Automatic emergency power switching		.66	24.242		1,200	1,850		3,050	4,150
3300	Manual emergency power switching		8	2		515	153		668	800
3625	Hall finishes, stainless steel doors					1,425			1,425	1,575
3650	Stainless steel frames					1,450			1,450	1,600
3675	12 month maintenance contract								3,600	3,950
3700	Signal devices, hall lanterns	2 Elev	8	2		520	153		673	805
3725	Position indicators, up to 3		9.40	1.702		475	130		605	725
3750	Position indicators, per each over 3		32	.500		420	38.50		458.50	520
3775	High speed heavy duty door opener					3,250			3,250	3,575
3800	Variable voltage, O.H. gearless machine, min.	2 Elev	.16	100		34,000	7,650		41,650	49,200
3815	Maximum		.07	228		81,000	17,500		98,500	116,000
3825	Basement installed geared machine		.33	48.485		49,200	3,700		52,900	59,500
3850	Freight elevator options									
3875	Doors, bi-parting	2 Elev	.66	24.242	Ea.	7,550	1,850		9,400	11,200
3900	Power operated door and gate	"	.66	24.242		24,300	1,850		26,150	29,600
3925	Finishes, steel plate floor					1,825			1,825	2,000
3950	14 ga. 1/4" x 4' steel plate walls					2,000			2,000	2,200
3975	12 month maintenance contract								3,600	3,950
4000	Signal devices, hall lanterns	2 Elev	8	2		505	153		658	790
4025	Position indicators, up to 3		9.40	1.702		490	130		620	735
4050	Position indicators, per each over 3		32	.500		425	38.50		463.50	530
4075	Variable voltage basement installed geared machine		.66	24.242		20,000	1,850		21,850	24,900
4100	Hospital elevator options									
4125	2 car group automatic controls	2 Elev	.66	24.242	Ea.	4,850	1,850		6,700	8,200
4150	3 car group automatic controls		.44	36.364		9,275	2,775		12,050	14,500
4175	4 car group automatic controls		.33	48.485		12,400	3,700		16,100	19,400
4200	5 car group automatic controls		.26	61.538		33,500	4,700		38,200	44,200
4225	6 car group automatic controls		.22	72.727		67,500	5,575		73,075	83,000
4250	Intercom service		3	5.333		1,000	410		1,410	1,725
4275	Duplex car selective collective		.66	24.242		8,100	1,850		9,950	11,800
4300	Center opening 1 speed doors		2	8		1,950	610		2,560	3,100
4325	Center opening 2 speed doors		2	8		2,575	610		3,185	3,800
4350	Rear opening doors (opposite front)		2	8		4,250	610		4,860	5,625

14 28 Elevator Equipment and Controls

14 28 10 – Elevator Equipment and Control Options

14 28 10.10 Elevator Controls and Doors		Crew	Daily Output	Labor-Hours	Unit	2015 Bare Costs Material	Labor	Equipment	Total	Total Incl O&P
4375	Side opening 2 speed doors	2 Elev	2	8	Ea.	6,350	610		6,960	7,925
4400	Automatic emergency power switching		.66	24.242		1,175	1,850		3,025	4,125
4425	Manual emergency power switching		8	2		505	153		658	790
4675	Hall finishes, stainless steel doors					1,600			1,600	1,750
4700	Stainless steel frames					1,475			1,475	1,625
4725	12 month maintenance contract								3,600	3,950
4750	Signal devices, hall lanterns	2 Elev	8	2		490	153		643	770
4775	Position indicators, up to 3		9.40	1.702		475	130		605	720
4800	Position indicators, per each over 3		32	.500		415	38.50		453.50	520
4825	High speed heavy duty door opener					3,250			3,250	3,575
4850	Variable voltage, O.H. gearless machine, min.	2 Elev	.16	100		51,500	7,650		59,150	68,500
4865	Maximum		.07	228		79,500	17,500		97,000	114,500
4875	Basement installed geared machine		.33	48.485		20,300	3,700		24,000	28,100
5000	Drilling for piston, casing included, 18″ diameter	B-48	80	.700	V.L.F.	56.50	30	36.50	123	150

14 31 Escalators

14 31 10 – Glass and Steel Escalators

14 31 10.10 Escalators

		Crew	Daily Output	Labor-Hours	Unit	Material	Labor	Equipment	Total	Total Incl O&P
0010	**ESCALATORS**									
1000	Glass, 32″ wide x 10′ floor to floor height	M-1	.07	457	Ea.	86,500	33,200	660	120,360	147,000
1010	48″ wide x 10′ floor to floor height		.07	457		93,500	33,200	660	127,360	154,500
1020	32″ wide x 15′ floor to floor height		.06	533		91,000	38,800	770	130,570	161,000
1030	48″ wide x 15′ floor to floor height		.06	533		96,500	38,800	770	136,070	167,000
1040	32″ wide x 20′ floor to floor height		.05	653		96,500	47,500	940	144,940	180,500
1050	48″ wide x 20′ floor to floor height		.05	653		105,000	47,500	940	153,440	189,500
1060	32″ wide x 25′ floor to floor height		.04	800		105,500	58,000	1,150	164,650	207,000
1070	48″ wide x 25′ floor to floor height		.04	800		122,000	58,000	1,150	181,150	225,000
1080	Enameled steel, 32″ wide x 10′ floor to floor height		.07	457		93,500	33,200	660	127,360	155,000
1090	48″ wide x 10′ floor to floor height		.07	457		101,500	33,200	660	135,360	163,500
1110	32″ wide x 15′ floor to floor height		.06	533		98,500	38,800	770	138,070	169,500
1120	48″ wide x 15′ floor to floor height		.06	533		104,500	38,800	770	144,070	176,000
1130	32″ wide x 20′ floor to floor height		.05	653		105,000	47,500	940	153,440	189,500
1140	48″ wide x 20′ floor to floor height		.05	653		113,500	47,500	940	161,940	199,000
1150	32″ wide x 25′ floor to floor height		.04	800		114,000	58,000	1,150	173,150	216,500
1160	48″ wide x 25′ floor to floor height		.04	800		131,500	58,000	1,150	190,650	235,500
1170	Stainless steel, 32″ wide x 10′ floor to floor height		.07	457		99,000	33,200	660	132,860	160,500
1180	48″ wide x 10′ floor to floor height		.07	457		106,500	33,200	660	140,360	169,000
1500	32″ wide x 15′ floor to floor height		.06	533		104,000	38,800	770	143,570	175,500
1700	48″ wide x 15′ floor to floor height		.06	533		110,000	38,800	770	149,570	182,000
1750	32″ wide x 18′ floor to floor height		.05	615		102,500	44,700	885	148,085	182,500
1775	48″ wide x 18′ floor to floor height		.05	615		111,500	44,700	885	157,085	192,500
2300	32″ wide x 25′ floor to floor height		.04	800		119,500	58,000	1,150	178,650	222,500
2500	48″ wide x 25′ floor to floor height		.04	800		138,000	58,000	1,150	197,150	242,500

14 32 Moving Walks

14 32 10 – Moving Walkways

14 32 10.10 Moving Walks		Crew	Daily Output	Labor-Hours	Unit	Material	2015 Bare Costs Labor	Equipment	Total	Total Incl O&P
0010	MOVING WALKS R143210-20									
0020	Walk, 27" tread width, minimum	M-1	6.50	4.923	L.F.	870	360	7.10	1,237.10	1,525
0100	300' to 500', maximum		4.43	7.223		1,200	525	10.40	1,735.40	2,150
0300	48" tread width walk, minimum		4.43	7.223		1,950	525	10.40	2,485.40	2,975
0400	100' to 350', maximum		3.82	8.377		2,300	610	12.05	2,922.05	3,475
0600	Ramp, 12° incline, 36" tread width, minimum		5.27	6.072		1,600	440	8.75	2,048.75	2,475
0700	70' to 90' maximum		3.82	8.377		2,300	610	12.05	2,922.05	3,475
0900	48" tread width, minimum		3.57	8.964		2,350	650	12.90	3,012.90	3,600
1000	40' to 70', maximum	↓	2.91	10.997	↓	2,950	800	15.85	3,765.85	4,500

14 42 Wheelchair Lifts

14 42 13 – Inclined Wheelchair Lifts

14 42 13.10 Inclined Wheelchair Lifts and Stairclimbers

		Crew	Daily Output	Labor-Hours	Unit	Material	Labor	Equipment	Total	Total Incl O&P
0010	INCLINED WHEELCHAIR LIFTS AND STAIRCLIMBERS									
7700	Stair climber (chair lift), single seat, minimum	2 Elev	1	16	Ea.	5,325	1,225		6,550	7,775
7800	Maximum		.20	80		7,350	6,125		13,475	17,500
8700	Stair lift, minimum		1	16		14,500	1,225		15,725	17,800
8900	Maximum	↓	.20	80	↓	22,900	6,125		29,025	34,700

14 42 16 – Vertical Wheelchair Lifts

14 42 16.10 Wheelchair Lifts

		Crew	Daily Output	Labor-Hours	Unit	Material	Labor	Equipment	Total	Total Incl O&P
0010	WHEELCHAIR LIFTS									
8000	Wheelchair lift, minimum	2 Elev	1	16	Ea.	7,325	1,225		8,550	9,950
8500	Maximum	"	.50	32	"	17,300	2,450		19,750	22,800

14 45 Vehicle Lifts

14 45 10 – Hydraulic Vehicle Lifts

14 45 10.10 Hydraulic Lifts

		Crew	Daily Output	Labor-Hours	Unit	Material	Labor	Equipment	Total	Total Incl O&P
0010	HYDRAULIC LIFTS									
2200	Single post, 8000 lb. capacity	L-4	.40	60	Ea.	5,550	2,650		8,200	10,500
2810	Double post, 6000 lb. capacity		2.67	8.989		7,825	400		8,225	9,275
2815	9000 lb. capacity		2.29	10.480		18,600	465		19,065	21,300
2820	15,000 lb. capacity		2	12		21,000	530		21,530	24,000
2822	Four post, 26,000 lb. capacity		1.80	13.333		14,200	590		14,790	16,600
2825	30,000 lb. capacity		1.60	15		46,400	665		47,065	52,000
2830	Ramp style, 4 post, 25,000 lb. capacity		2	12		18,200	530		18,730	20,900
2835	35,000 lb. capacity		1	24		84,500	1,050		85,550	94,500
2840	50,000 lb. capacity		1	24		94,500	1,050		95,550	105,500
2845	75,000 lb. capacity	↓	1	24		110,000	1,050		111,050	122,500
2850	For drive thru tracks, add, minimum					1,175			1,175	1,275
2855	Maximum					2,000			2,000	2,200
2860	Ramp extensions, 3' (set of 2)					960			960	1,050
2865	Rolling jack platform					3,325			3,325	3,675
2870	Electric/hydraulic jacking beam					8,925			8,925	9,825
2880	Scissor lift, portable, 6000 lb. capacity				↓	8,750			8,750	9,625

14 91 Facility Chutes

14 91 33 – Laundry and Linen Chutes

14 91 33.10 Chutes		Crew	Daily Output	Labor-Hours	Unit	2015 Bare Costs Material	Labor	Equipment	Total	Total Incl O&P
0011	**CHUTES**, linen, trash or refuse									
0050	Aluminized steel, 16 ga., 18″ diameter	2 Shee	3.50	4.571	Floor	1,700	256		1,956	2,275
0100	24″ diameter		3.20	5		1,775	280		2,055	2,400
0200	30″ diameter		3	5.333		2,125	298		2,423	2,800
0300	36″ diameter		2.80	5.714		2,625	320		2,945	3,400
0400	Galvanized steel, 16 ga., 18″ diameter		3.50	4.571		1,000	256		1,256	1,500
0500	24″ diameter		3.20	5		1,125	280		1,405	1,700
0600	30″ diameter		3	5.333		1,275	298		1,573	1,875
0700	36″ diameter		2.80	5.714		1,500	320		1,820	2,150
0800	Stainless steel, 18″ diameter		3.50	4.571		3,000	256		3,256	3,700
0900	24″ diameter		3.20	5		3,150	280		3,430	3,900
1000	30″ diameter		3	5.333		3,750	298		4,048	4,600
1005	36″ diameter		2.80	5.714		3,950	320		4,270	4,825
1200	Linen chute bottom collector, aluminized steel		4	4	Ea.	1,400	224		1,624	1,900
1300	Stainless steel		4	4		1,800	224		2,024	2,325
1500	Refuse, bottom hopper, aluminized steel, 18″ diameter		3	5.333		1,025	298		1,323	1,600
1600	24″ diameter		3	5.333		1,250	298		1,548	1,850
1800	36″ diameter		3	5.333		2,500	298		2,798	3,225

14 91 82 – Trash Chutes

14 91 82.10 Trash Chutes and Accessories

		Crew	Daily Output	Labor-Hours	Unit	Material	Labor	Equipment	Total	Total Incl O&P
0010	**TRASH CHUTES AND ACCESSORIES**									
2900	Package chutes, spiral type, minimum	2 Shee	4.50	3.556	Floor	2,425	199		2,624	3,000
3000	Maximum	"	1.50	10.667	"	6,325	595		6,920	7,925
9000	Minimum labor/equipment charge	1 Shee	1	8	Job		450		450	705

14 92 Pneumatic Tube Systems

14 92 10 – Conventional, Automatic and Computer Controlled Pneumatic Tube Systems

14 92 10.10 Pneumatic Tube Systems

		Crew	Daily Output	Labor-Hours	Unit	Material	Labor	Equipment	Total	Total Incl O&P
0010	**PNEUMATIC TUBE SYSTEMS**									
0020	100′ long, single tube, 2 stations, stock									
0100	3″ diameter	2 Stpi	.12	133	Total	3,375	7,975		11,350	16,100
0300	4″ diameter	"	.09	177	"	4,275	10,600		14,875	21,300
0400	Twin tube, two stations or more, conventional system									
0600	2-1/2″ round	2 Stpi	62.50	.256	L.F.	38	15.30		53.30	66
0700	3″ round		46	.348		38	21		59	74.50
0900	4″ round		49.60	.323		48	19.25		67.25	83
1000	4″ x 7″ oval		37.60	.426		89	25.50		114.50	138
1050	Add for blower		2	8	System	5,200	480		5,680	6,475
1110	Plus for each round station, add		7.50	2.133	Ea.	1,350	127		1,477	1,700
1150	Plus for each oval station, add		7.50	2.133	"	1,350	127		1,477	1,700
1200	Alternate pricing method: base cost, economy model		.75	21.333	Total	5,750	1,275		7,025	8,325
1300	Custom model		.25	64	"	11,500	3,825		15,325	18,700
1500	Plus total system length, add, for economy model		93.40	.171	L.F.	8.25	10.25		18.50	25
1600	For custom model		37.60	.426	"	25	25.50		50.50	67
1800	Completely automatic system, 4″ round, 15 to 50 stations		.29	55.172	Station	20,100	3,300		23,400	27,300
2200	51 to 144 stations		.32	50		15,600	3,000		18,600	21,900
2400	6″ round or 4″ x 7″ oval, 15 to 50 stations		.24	66.667		25,200	3,975		29,175	33,900
2800	51 to 144 stations		.23	69.565		21,100	4,150		25,250	29,700

Division Notes

I	CREW	DAILY OUTPUT	LABOR-HOURS	UNIT	BARE COSTS MAT.	BARE COSTS LABOR	BARE COSTS EQUIP.	BARE COSTS TOTAL	TOTAL INCL O&P

Estimating Tips

Pipe for fire protection and all uses is located in Subdivisions 21 11 13 and 22 11 13.

The labor adjustment factors listed in Subdivision 22 01 02.20 also apply to Division 21.

Many, but not all, areas in the U.S. require backflow protection in the fire system. It is advisable to check local building codes for specific requirements.

For your reference, the following is a list of the most applicable Fire Codes and Standards which may be purchased from the NFPA, 1 Batterymarch Park, Quincy, MA 02169-7471.

- NFPA 1: Uniform Fire Code
- NFPA 10: Portable Fire Extinguishers
- NFPA 11: Low-, Medium-, and High-Expansion Foam
- NFPA 12: Carbon Dioxide Extinguishing Systems (Also companion 12A)
- NFPA 13: Installation of Sprinkler Systems (Also companion 13D, 13E, and 13R)
- NFPA 14: Installation of Standpipe and Hose Systems
- NFPA 15: Water Spray Fixed Systems for Fire Protection
- NFPA 16: Installation of Foam-Water Sprinkler and Foam-Water Spray Systems
- NFPA 17: Dry Chemical Extinguishing Systems (Also companion 17A)
- NFPA 18: Wetting Agents
- NFPA 20: Installation of Stationary Pumps for Fire Protection
- NFPA 22: Water Tanks for Private Fire Protection
- NFPA 24: Installation of Private Fire Service Mains and their Appurtenances
- NFPA 25: Inspection, Testing and Maintenance of Water-Based Fire Protection

Reference Numbers

Reference numbers are shown in shaded boxes at the beginning of some major classifications. These numbers refer to related items in the Reference Section. The reference information may be an estimating procedure, an alternate pricing method, or technical information.

Note: Not all subdivisions listed here necessarily appear in this publication. ■

21 05 Common Work Results for Fire Suppression

21 05 23 – General-Duty Valves for Water-Based Fire-Suppression Piping

21 05 23.50 General-Duty Valves		Crew	Daily Output	Labor-Hours	Unit	2015 Bare Costs Material	Labor	Equipment	Total	Total Incl O&P
0010	**GENERAL-DUTY VALVES**, for water-based fire suppression									
6200	Valves and components									
6210	Alarm, includes									
6220	retard chamber, trim, gauges, alarm line strainer									
6260	3″ size	Q-12	3	5.333	Ea.	1,625	269		1,894	2,225
6280	4″ size	″	2	8		1,700	405		2,105	2,500
6300	6″ size	Q-13	4	8		1,900	430		2,330	2,775
6320	8″ size	″	3	10.667		2,125	570		2,695	3,250
6500	Check, swing, C.I. body, brass fittings, auto. ball drip									
6520	4″ size	Q-12	3	5.333	Ea.	350	269		619	805
6540	6″ size	Q-13	4	8		650	430		1,080	1,375
6580	8″ size	″	3	10.667		1,225	570		1,795	2,250
6800	Check, wafer, butterfly type, C.I. body, bronze fittings									
6820	4″ size	Q-12	4	4	Ea.	1,000	202		1,202	1,425
6840	6″ size	Q-13	5.50	5.818		1,625	310		1,935	2,250
6860	8″ size		5	6.400		1,850	340		2,190	2,575
6880	10″ size		4.50	7.111		2,250	380		2,630	3,075
8800	Flow control valve, includes trim and gauges, 2″ size	Q-12	2	8		4,700	405		5,105	5,800
8820	3″ size	″	1.50	10.667		5,175	540		5,715	6,525
8840	4″ size	Q-13	2.80	11.429		5,775	610		6,385	7,300
8860	6″ size	″	2	16		6,675	855		7,530	8,675
9200	Pressure operated relief valve, brass body	1 Spri	18	.444		580	25		605	675
9600	Waterflow indicator, with recycling retard and									
9610	two single pole retard switches, 2″ thru 6″ pipe size	1 Spri	8	1	Ea.	137	56		193	239
9990	Minimum labor/equipment charge	″	3	2.667	Job		150		150	234

21 11 Facility Fire-Suppression Water-Service Piping

21 11 13 – Facility Water Distribution Piping

21 11 13.16 Pipe, Plastic

		Crew	Daily Output	Labor-Hours	Unit	Material	Labor	Equipment	Total	Total Incl O&P
0010	**PIPE, PLASTIC**									
0020	CPVC, fire suppression, (C-UL-S. FM, NFPA 13, 13D & 13R)									
0030	Socket joint, no couplings or hangers									
0100	SDR 13.5, (ASTM F442)									
0120	3/4″ diameter	Q-12	420	.038	L.F.	1.67	1.93		3.60	4.85
0130	1″ diameter		340	.047		2.58	2.38		4.96	6.55
0140	1-1/4″ diameter		260	.062		4.10	3.11		7.21	9.35
0150	1-1/2″ diameter		190	.084		5.65	4.26		9.91	12.90
0160	2″ diameter		140	.114		9	5.80		14.80	18.95
0170	2-1/2″ diameter		130	.123		16.65	6.20		22.85	28
0180	3″ diameter		120	.133		25.50	6.75		32.25	38.50

21 11 13.18 Pipe Fittings, Plastic

		Crew	Daily Output	Labor-Hours	Unit	Material	Labor	Equipment	Total	Total Incl O&P
0010	**PIPE FITTINGS, PLASTIC**									
0020	CPVC, fire suppression, (C-UL-S. FM, NFPA 13, 13D & 13R)									
0030	Socket joint									
0100	90° Elbow									
0120	3/4″	1 Plum	26	.308	Ea.	2.10	18.05		20.15	30.50
0130	1″		22.70	.352		4.62	20.50		25.12	37.50
0140	1-1/4″		20.20	.396		5.85	23.50		29.35	43
0150	1-1/2″		18.20	.440		8.30	26		34.30	49
0160	2″	Q-1	33.10	.483		10.30	25.50		35.80	51.50
0170	2-1/2″		24.20	.661		19.80	35		54.80	76.50

21 11 13.18 Pipe Fittings, Plastic		Crew	Daily Output	Labor-Hours	Unit	Material	Labor	Equipment	Total	Total Incl O&P
						2015 Bare Costs				
0180	3"	Q-1	20.80	.769	Ea.	27	40.50		67.50	93
0200	45° Elbow									
0210	3/4"	1 Plum	26	.308	Ea.	2.89	18.05		20.94	31
0220	1"		22.70	.352		3.39	20.50		23.89	36
0230	1-1/4"		20.20	.396		4.90	23.50		28.40	42
0240	1-1/2"		18.20	.440		6.85	26		32.85	47.50
0250	2"	Q-1	33.10	.483		8.50	25.50		34	49.50
0260	2-1/2"		24.20	.661		15.30	35		50.30	71.50
0270	3"		20.80	.769		22	40.50		62.50	87.50
0300	Tee									
0310	3/4"	1 Plum	17.30	.462	Ea.	2.89	27		29.89	45.50
0320	1"		15.20	.526		5.70	31		36.70	54.50
0330	1-1/4"		13.50	.593		8.55	35		43.55	64
0340	1-1/2"		12.10	.661		12.60	39		51.60	74.50
0350	2"	Q-1	20	.800		18.65	42.50		61.15	86.50
0360	2-1/2"		16.20	.988		30.50	52		82.50	115
0370	3"		13.90	1.151		47.50	61		108.50	147
0400	Tee, reducing x any size									
0420	1"	1 Plum	15.20	.526	Ea.	4.84	31		35.84	53.50
0430	1-1/4"		13.50	.593		8.85	35		43.85	64.50
0440	1-1/2"		12.10	.661		10.75	39		49.75	72.50
0450	2"	Q-1	20	.800		19.85	42.50		62.35	88
0460	2-1/2"		16.20	.988		23.50	52		75.50	108
0470	3"		13.90	1.151		27	61		88	125
0500	Coupling									
0510	3/4"	1 Plum	26	.308	Ea.	2.02	18.05		20.07	30
0520	1"		22.70	.352		2.67	20.50		23.17	35.50
0530	1-1/4"		20.20	.396		3.89	23.50		27.39	41
0540	1-1/2"		18.20	.440		5.55	26		31.55	46
0550	2"	Q-1	33.10	.483		7.50	25.50		33	48.50
0560	2-1/2"		24.20	.661		11.45	35		46.45	67
0570	3"		20.80	.769		14.90	40.50		55.40	80
0600	Coupling, reducing									
0610	1" x 3/4"	1 Plum	22.70	.352	Ea.	2.67	20.50		23.17	35.50
0620	1-1/4" x 1"		20.20	.396		4.04	23.50		27.54	41
0630	1-1/2" x 3/4"		18.20	.440		6.05	26		32.05	46.50
0640	1-1/2" x 1"		18.20	.440		5.85	26		31.85	46.50
0650	1-1/2" x 1-1/4"		18.20	.440		5.55	26		31.55	46
0660	2" x 1"	Q-1	33.10	.483		7.80	25.50		33.30	48.50
0670	2" x 1-1/2"	"	33.10	.483		7.50	25.50		33	48.50
0700	Cross									
0720	3/4"	1 Plum	13	.615	Ea.	4.54	36		40.54	61.50
0730	1"		11.30	.708		5.70	41.50		47.20	71.50
0740	1-1/4"		10.10	.792		7.85	46.50		54.35	81
0750	1-1/2"		9.10	.879		10.90	51.50		62.40	92.50
0760	2"	Q-1	16.60	.964		17.75	51		68.75	99
0770	2-1/2"	"	12.10	1.322		39	70		109	152
0800	Cap									
0820	3/4"	1 Plum	52	.154	Ea.	1.12	9.05		10.17	15.35
0830	1"		45	.178		1.74	10.45		12.19	18.20
0840	1-1/4"		40	.200		2.81	11.75		14.56	21.50
0850	1-1/2"		36.40	.220		3.89	12.90		16.79	24.50
0860	2"	Q-1	66	.242		5.85	12.80		18.65	26.50

21 11 Facility Fire-Suppression Water-Service Piping

21 11 13 – Facility Water Distribution Piping

21 11 13.18 Pipe Fittings, Plastic		Crew	Daily Output	Labor-Hours	Unit	Material	Labor	Equipment	Total	Total Incl O&P
						2015 Bare Costs				
0870	2-1/2"	Q-1	48.40	.331	Ea.	8.45	17.45		25.90	36.50
0880	3"	↓	41.60	.385	↓	13.60	20.50		34.10	46.50
0900	Adapter, sprinkler head, female w/metal thd. insert, (s x FNPT)									
0920	3/4" x 1/2"	1 Plum	52	.154	Ea.	5.45	9.05		14.50	20
0930	1" x 1/2"		45	.178		5.75	10.45		16.20	22.50
0940	1" x 3/4"	↓	45	.178	↓	9.05	10.45		19.50	26.50

21 11 16 – Facility Fire Hydrants

21 11 16.50 Fire Hydrants for Buildings		Crew	Daily Output	Labor-Hours	Unit	Material	Labor	Equipment	Total	Total Incl O&P
0010	**FIRE HYDRANTS FOR BUILDINGS**									
3750	Hydrants, wall, w/caps, single, flush, polished brass									
3800	2-1/2" x 2-1/2"	Q-12	5	3.200	Ea.	213	162		375	490
3840	2-1/2" x 3"		5	3.200		430	162		592	725
3860	3" x 3"	↓	4.80	3.333		350	168		518	650
3900	For polished chrome, add				↓	20%				
3950	Double, flush, polished brass									
4000	2-1/2" x 2-1/2" x 4"	Q-12	5	3.200	Ea.	570	162		732	885
4040	2-1/2" x 2-1/2" x 6"		4.60	3.478		825	176		1,001	1,175
4080	3" x 3" x 4"		4.90	3.265		1,225	165		1,390	1,600
4120	3" x 3" x 6"	↓	4.50	3.556		1,250	180		1,430	1,650
4200	For polished chrome, add				↓	10%				
4350	Double, projecting, polished brass									
4400	2-1/2" x 2-1/2" x 4"	Q-12	5	3.200	Ea.	254	162		416	530
4450	2-1/2" x 2-1/2" x 6"	"	4.60	3.478	"	520	176		696	845
4460	Valve control, dbl. flush/projecting hydrant, cap &									
4470	chain, extension rod & cplg., escutcheon, polished brass	Q-12	8	2	Ea.	300	101		401	490
4480	Four-way square, flush, polished brass									
4540	2-1/2"(4) x 6"	Q-12	3.60	4.444	Ea.	3,075	225		3,300	3,725

21 11 19 – Fire-Department Connections

21 11 19.50 Connections for the Fire-Department		Crew	Daily Output	Labor-Hours	Unit	Material	Labor	Equipment	Total	Total Incl O&P
0010	**CONNECTIONS FOR THE FIRE-DEPARTMENT**									
6000	Roof manifold, horiz., brass, without valves & caps									
6040	2-1/2" x 2-1/2" x 4"	Q-12	4.80	3.333	Ea.	168	168		336	450
6060	2-1/2" x 2-1/2" x 6"		4.60	3.478		184	176		360	480
6080	2-1/2" x 2-1/2" x 2-1/2" x 4"		4.60	3.478		271	176		447	575
6090	2-1/2" x 2-1/2" x 2-1/2" x 6"	↓	4.60	3.478		280	176		456	585
7000	Sprinkler line tester, cast brass				↓	24			24	26.50
7140	Standpipe connections, wall, w/plugs & chains									
7160	Single, flush, brass, 2-1/2" x 2-1/2"	Q-12	5	3.200	Ea.	159	162		321	430
7180	2-1/2" x 3"	"	5	3.200	"	164	162		326	435
7240	For polished chrome, add					15%				
7280	Double, flush, polished brass									
7300	2-1/2" x 2-1/2" x 4"	Q-12	5	3.200	Ea.	520	162		682	825
7330	2-1/2" x 2-1/2" x 6"		4.60	3.478		725	176		901	1,075
7340	3" x 3" x 4"		4.90	3.265		945	165		1,110	1,300
7370	3" x 3" x 6"	↓	4.50	3.556	↓	1,125	180		1,305	1,525
7400	For polished chrome, add					15%				
7440	For sill cock combination, add				Ea.	90.50			90.50	99.50
7580	Double projecting, polished brass									
7600	2-1/2" x 2-1/2" x 4"	Q-12	5	3.200	Ea.	475	162		637	775
7630	2-1/2" x 2-1/2" x 6"	"	4.60	3.478	"	800	176		976	1,150
7680	For polished chrome, add					15%				
7900	Three way, flush, polished brass									

21 11 Facility Fire-Suppression Water-Service Piping

21 11 19 – Fire-Department Connections

21 11 19.50 Connections for the Fire-Department		Crew	Daily Output	Labor-Hours	Unit	2015 Bare Costs Material	Labor	Equipment	Total	Total Incl O&P
7920	2-1/2″ (3) x 4″	Q-12	4.80	3.333	Ea.	1,650	168		1,818	2,100
7930	2-1/2″ (3) x 6″	″	4.80	3.333		1,650	168		1,818	2,100
8000	For polished chrome, add					9%				
8020	Three way, projecting, polished brass									
8040	2-1/2″(3) x 4″	Q-12	4.80	3.333	Ea.	1,575	168		1,743	2,000
8070	2-1/2″ (3) x 6″	″	4.60	3.478		1,575	176		1,751	2,000
8100	For polished chrome, add					12%				
8200	Four way, square, flush, polished brass,									
8240	2-1/2″(4) x 6″	Q-12	3.60	4.444	Ea.	1,250	225		1,475	1,725
8300	For polished chrome, add				″	10%				
8550	Wall, vertical, flush, cast brass									
8600	Two way, 2-1/2″ x 2-1/2″ x 4″	Q-12	5	3.200	Ea.	395	162		557	690
8660	Four way, 2-1/2″(4) x 6″		3.80	4.211		1,300	213		1,513	1,750
8680	Six way, 2-1/2″(6) x 6″		3.40	4.706		1,550	238		1,788	2,100
8700	For polished chrome, add					10%				
8800	Sidewalk siamese unit, polished brass, two way									
8820	2-1/2″ x 2-1/2″ x 4″	Q-12	2.50	6.400	Ea.	635	325		960	1,200
8850	2-1/2″ x 2-1/2″ x 6″		2	8		780	405		1,185	1,500
8860	3″ x 3″ x 4″		2.50	6.400		900	325		1,225	1,500
8890	3″ x 3″ x 6″		2	8		1,225	405		1,630	1,975
8940	For polished chrome, add					12%				
9100	Sidewalk siamese unit, polished brass, three way									
9120	2-1/2″ x 2-1/2″ x 2-1/2″ x 6″	Q-12	2	8	Ea.	1,000	405		1,405	1,725
9160	For polished chrome, add				″	15%				
9990	Minimum labor/equipment charge	1 Spri	4	2	Job		112		112	176

21 12 Fire-Suppression Standpipes

21 12 13 – Fire-Suppression Hoses and Nozzles

21 12 13.50 Fire Hoses and Nozzles		Crew	Daily Output	Labor-Hours	Unit	Material	Labor	Equipment	Total	Total Incl O&P
0010	**FIRE HOSES AND NOZZLES** R211226-10									
0200	Adapters, rough brass, straight hose threads									
0220	One piece, female to male, rocker lugs R211226-20									
0240	1″ x 1″				Ea.	48			48	53
0260	1-1/2″ x 1″					48			48	52.50
0280	1-1/2″ x 1-1/2″					12.90			12.90	14.20
0300	2″ x 1-1/2″					66			66	72.50
0320	2″ x 2″					42			42	46
0340	2-1/2″ x 1-1/2″					14.70			14.70	16.15
0360	2-1/2″ x 2″					38			38	41.50
0380	2-1/2″ x 2-1/2″					19			19	21
0400	3″ x 2-1/2″					125			125	137
0420	3″ x 3″					83.50			83.50	92
0500	For polished brass, add					50%				
0520	For polished chrome, add					75%				
0700	One piece, female to male, hexagon									
0740	1-1/2″ x 3/4″				Ea.	40.50			40.50	44.50
0760	2″ x 1-1/2″					84			84	92.50
0780	2-1/2″ x 1″					161			161	178
0800	2-1/2″ x 1-1/2″					55.50			55.50	61
0820	2-1/2″ x 2″					38			38	41.50
0840	3″ x 2-1/2″					70			70	77

21 12 13 – Fire-Suppression Hoses and Nozzles

21 12 13.50 Fire Hoses and Nozzles		Crew	Daily Output	Labor-Hours	Unit	Material	2015 Bare Costs Labor	Equipment	Total	Total Incl O&P
0900	For polished chrome, add				Ea.	75%				
1100	Swivel, female to female, pin lugs									
1120	1-1/2" x 1-1/2"				Ea.	61.50			61.50	67.50
1200	2-1/2" x 2-1/2"					120			120	132
1260	For polished brass, add					50%				
1280	For polished chrome, add					75%				
1400	Couplings, sngl. & dbl. jacket, pin lug or rocker lug, cast brass									
1410	1-1/2"				Ea.	53.50			53.50	58.50
1420	2-1/2"				"	70			70	77
1500	For polished brass, add					20%				
1520	For polished chrome, add					40%				
1580	Reducing, F x M, interior installation, cast brass									
1590	2" x 1-1/2"				Ea.	66			66	72.50
1600	2-1/2" x 1-1/2"					14.70			14.70	16.15
1680	For polished brass, add					50%				
1720	For polished chrome, add					75%				
2200	Hose, less couplings									
2260	Synthetic jacket, lined, 300 lb. test, 1-1/2" diameter	Q-12	2600	.006	L.F.	3.24	.31		3.55	4.05
2280	2-1/2" diameter		2200	.007		5.60	.37		5.97	6.70
2360	High strength, 500 lb. test, 1-1/2" diameter		2600	.006		3.35	.31		3.66	4.18
2380	2-1/2" diameter		2200	.007		5.90	.37		6.27	7.05
5000	Nipples, straight hose to tapered iron pipe, brass									
5060	Female to female, 1-1/2" x 1-1/2"				Ea.	19.85			19.85	22
5100	2-1/2" x 2-1/2"					37			37	40.50
5190	For polished chrome, add					75%				
5200	Double male or male to female, 1" x 1"					42			42	46
5220	1-1/2" x 1"					62			62	68
5230	1-1/2" x 1-1/2"					14.35			14.35	15.75
5260	2" x 1-1/2"					67.50			67.50	74
5270	2" x 2"					84			84	92.50
5280	2-1/2" x 1-1/2"					55.50			55.50	61
5300	2-1/2" x 2"					45.50			45.50	50
5310	2-1/2" x 2-1/2"					25.50			25.50	28.50
5340	For polished chrome, add					75%				
5600	Nozzles, brass									
5620	Adjustable fog, 3/4" booster line				Ea.	111			111	122
5630	1" booster line					137			137	150
5640	1-1/2" leader line					101			101	112
5660	2-1/2" direct connection					151			151	166
5680	2-1/2" playpipe nozzle					223			223	246
5780	For chrome plated, add					8%				
5850	Electrical fire, adjustable fog, no shock									
5900	1-1/2"				Ea.	415			415	455
5920	2-1/2"					560			560	615
5980	For polished chrome, add					6%				
6200	Heavy duty, comb. adj. fog and str. stream, with handle									
6210	1" booster line				Ea.	375			375	410
6240	1-1/2"					450			450	495
6260	2-1/2", for playpipe					750			750	825
6280	2-1/2" direct connection					560			560	615
6300	2-1/2" playpipe combination					560			560	615
6480	For polished chrome, add					7%				
6500	Plain fog, polished brass, 1-1/2"					118			118	129

21 12 13 – Fire-Suppression Hoses and Nozzles

21 12 13.50 Fire Hoses and Nozzles		Crew	Daily Output	Labor-Hours	Unit	Material	2015 Bare Costs Labor	Equipment	Total	Total Incl O&P
6540	Chrome plated, 1-1/2"				Ea.	117			117	128
6700	Plain stream, polished brass, 1-1/2" x 10"					48			48	52.50
6760	2-1/2" x 15" x 7/8" or 1-1/2"				↓	100			100	110
6860	For polished chrome, add					20%				
7000	Underwriters playpipe, 2-1/2" x 30" with 1-1/8" tip				Ea.	239			239	263
9200	Storage house, hose only, primed steel					855			855	940
9220	Aluminum					1,875			1,875	2,050
9280	Hose and hydrant house, primed steel					1,150			1,150	1,275
9300	Aluminum					1,675			1,675	1,825
9340	Tools, crowbar and brackets	1 Carp	12	.667		70	31.50		101.50	129
9360	Combination hydrant wrench and spanner				↓	31.50			31.50	35
9380	Fire axe and brackets									
9400	6 lb.	1 Carp	12	.667	Ea.	116	31.50		147.50	179
9900	Minimum labor/equipment charge	1 Plum	2	4	Job		235		235	365

21 12 16 – Fire-Suppression Hose Reels

21 12 16.50 Fire-Suppression Hose Reels

		Crew	Daily Output	Labor-Hours	Unit	Material	Labor	Equipment	Total	Total Incl O&P
0010	**FIRE-SUPPRESSION HOSE REELS**									
2990	Hose reel, swinging, for 1-1/2" polyester neoprene lined hose									
3000	50' long	Q-12	14	1.143	Ea.	131	58		189	235
3020	100' long		14	1.143		184	58		242	293
3060	For 2-1/2" cotton rubber hose, 75' long		14	1.143		212	58		270	325
3100	150' long	↓	14	1.143	↓	248	58		306	365

21 12 19 – Fire-Suppression Hose Racks

21 12 19.50 Fire Hose Racks

		Crew	Daily Output	Labor-Hours	Unit	Material	Labor	Equipment	Total	Total Incl O&P
0010	**FIRE HOSE RACKS**									
2600	Hose rack, swinging, for 1-1/2" diameter hose,									
2620	Enameled steel, 50' & 75' lengths of hose	Q-12	20	.800	Ea.	61	40.50		101.50	130
2640	100' and 125' lengths of hose		20	.800		61	40.50		101.50	130
2680	Chrome plated, 50' and 75' lengths of hose		20	.800		98.50	40.50		139	171
2700	100' and 125' lengths of hose	↓	20	.800		104	40.50		144.50	178
2780	For hose rack nipple, 1-1/2" polished brass, add					27			27	29.50
2820	2-1/2" polished brass, add					49			49	54
2840	1-1/2" polished chrome, add					39.50			39.50	43.50
2860	2-1/2" polished chrome, add				↓	52			52	57.50

21 12 23 – Fire-Suppression Hose Valves

21 12 23.70 Fire Hose Valves

		Crew	Daily Output	Labor-Hours	Unit	Material	Labor	Equipment	Total	Total Incl O&P
0010	**FIRE HOSE VALVES**									
0020	Angle, combination pressure adjust/restricting, rough brass									
0030	1-1/2" R211226-20	1 Spri	12	.667	Ea.	85.50	37.50		123	153
0040	2-1/2"	"	7	1.143	"	175	64		239	292
0042	Nonpressure adjustable/restricting, rough brass									
0044	1-1/2"	1 Spri	12	.667	Ea.	53.50	37.50		91	117
0046	2-1/2"	"	7	1.143	"	90	64		154	199
0050	For polished brass, add					30%				
0060	For polished chrome, add					40%				
0080	Wheel handle, 300 lb., 1-1/2"	1 Spri	12	.667	Ea.	95	37.50		132.50	163
0090	2-1/2"	"	7	1.143	"	175	64		239	292
0100	For polished brass, add					35%				
0110	For polished chrome, add					50%				
1000	Ball drip, automatic, rough brass, 1/2"	1 Spri	20	.400	Ea.	16.15	22.50		38.65	53
1010	3/4"	"	20	.400	"	18	22.50		40.50	55

21 12 Fire-Suppression Standpipes

21 12 23 – Fire-Suppression Hose Valves

21 12 23.70 Fire Hose Valves		Crew	Daily Output	Labor-Hours	Unit	2015 Bare Costs Material	Labor	Equipment	Total	Total Incl O&P
1100	Ball, 175 lb., sprinkler system, FM/UL, threaded, bronze									
1120	Slow close									
1150	1" size	1 Spri	19	.421	Ea.	241	23.50		264.50	300
1160	1-1/4" size		15	.533		260	30		290	335
1170	1-1/2" size		13	.615		440	34.50		474.50	540
1180	2" size		11	.727		415	41		456	525
1190	2-1/2" size	Q-12	15	1.067		560	54		614	705
1230	For supervisory switch kit, all sizes									
1240	One circuit, add	1 Spri	48	.167	Ea.	165	9.35		174.35	197
1280	Quarter turn for trim									
1300	1/2" size	1 Spri	22	.364	Ea.	32	20.50		52.50	67
1310	3/4" size		20	.400		34.50	22.50		57	72.50
1320	1" size		19	.421		38	23.50		61.50	79
1330	1-1/4" size		15	.533		62.50	30		92.50	116
1340	1-1/2" size		13	.615		78.50	34.50		113	140
1350	2" size		11	.727		93.50	41		134.50	167
1400	Caps, polished brass with chain, 3/4"					42			42	46
1420	1"					53.50			53.50	59
1440	1-1/2"					14.35			14.35	15.75
1460	2-1/2"					21.50			21.50	24
1480	3"					33.50			33.50	37
1900	Escutcheon plate, for angle valves, polished brass, 1-1/2"					15.20			15.20	16.70
1920	2-1/2"					24.50			24.50	27
1940	3"					31			31	34
1980	For polished chrome, add					15%				
2000	Foam, control valve, 3"	1 Spri	6	1.333		1,825	75		1,900	2,150
2020	Supply valve, 2-1/2"		7	1.143		154	64		218	269
2040	Proportioner, 8"		2	4		3,000	225		3,225	3,650
2060	Oscillating foam monitor with electric remote control	Q-12	5.33	3.002		11,300	152		11,452	12,600
3000	Gate, hose, wheel handle, N.R.S., rough brass, 1-1/2"	1 Spri	12	.667		135	37.50		172.50	207
3040	2-1/2", 300 lb.	"	7	1.143		189	64		253	310
3080	For polished brass, add					40%				
3090	For polished chrome, add					50%				
3800	Hydrant, screw type, crank handle, brass									
3840	2-1/2" size	Q-12	11	1.455	Ea.	345	73.50		418.50	495
3880	For chrome, same price									
4200	Hydrolator, vent and draining, rough brass, 1-1/2"	1 Spri	12	.667	Ea.	84	37.50		121.50	151
4280	For polished brass, add					50%				
4290	For polished chrome, add					90%				
5000	Pressure restricting, adjustable rough brass, 1-1/2"	1 Spri	12	.667		107	37.50		144.50	176
5020	2-1/2"	"	7	1.143		160	64		224	276
5080	For polished brass, add					30%				
5090	For polished chrome, add					45%				
8000	Wye, leader line, ball type, swivel female x male x male									
8040	2-1/2" x 1-1/2" x 1-1/2" polished brass				Ea.	262			262	288
8060	2-1/2" x 1-1/2" x 1-1/2" polished chrome				"	245			245	270

21 13 13.50 Wet-Pipe Sprinkler System Components		Crew	Daily Output	Labor-Hours	Unit	2015 Bare Costs Material	Labor	Equipment	Total	Total Incl O&P
0010	**WET-PIPE SPRINKLER SYSTEM COMPONENTS**									
1100	Alarm, electric pressure switch (circuit closer)	1 Spri	26	.308	Ea.	88.50	17.30		105.80	125
1140	For explosion proof, max 20 PSI, contacts close or open		26	.308		535	17.30		552.30	610
1220	Water motor, complete with gong		4	2		415	112		527	630
1800	Firecycle system, controls, includes panel,									
1820	batteries, solenoid valves and pressure switches	Q-13	1	32	Ea.	20,900	1,700		22,600	25,700
1860	Detector	1 Spri	16	.500	″	715	28		743	830
1900	Flexible sprinkler head connectors									
1910	Braided stainless steel hose with mounting bracket									
1920	1/2″ and 3/4″ outlet size									
1940	40″ length	1 Spri	30	.267	Ea.	70.50	14.95		85.45	101
1960	60″ length	″	22	.364	″	81	20.50		101.50	121
1982	May replace hard-pipe armovers									
1984	For wet, pre-action, deluge or dry pipe systems									
2000	Release, emergency, manual, for hydraulic or pneumatic system	1 Spri	12	.667	Ea.	218	37.50		255.50	298
2060	Release, thermostatic, for hydraulic or pneumatic release line		20	.400		675	22.50		697.50	775
2200	Sprinkler cabinets, 6 head capacity		16	.500		79	28		107	131
2260	12 head capacity		16	.500		83.50	28		111.50	136
2340	Sprinkler head escutcheons, standard, brass tone, 1″ size		40	.200		2.85	11.25		14.10	20.50
2360	Chrome, 1″ size		40	.200		3.04	11.25		14.29	21
2400	Recessed type, bright brass		40	.200		10.10	11.25		21.35	28.50
2440	Chrome or white enamel		40	.200		3.35	11.25		14.60	21
2600	Sprinkler heads, not including supply piping									
3700	Standard spray, pendent or upright, brass, 135°F to 286°F									
3720	1/2″ NPT, 3/8″ orifice	1 Spri	16	.500	Ea.	15.40	28		43.40	61
3730	1/2″ NPT, 7/16″ orifice		16	.500		15.20	28		43.20	61
3740	1/2″ NPT, 1/2″ orifice		16	.500		9.90	28		37.90	55
3760	1/2″ NPT, 17/32″ orifice		16	.500		12.90	28		40.90	58
3780	3/4″ NPT, 17/32″ orifice		16	.500		11.90	28		39.90	57
3800	For open sprinklers, deduct					15%				
3840	For chrome, add				Ea.	3.38			3.38	3.72
3860	For wax and lead coating, add					35			35	38
3880	For wax coating, add					21			21	23
3900	For lead coating, add					22.50			22.50	25
3920	For 360°F, same cost									
3930	For 400°F				Ea.	93			93	102
3940	For 500°F				″	93			93	102
4200	Sidewall, vertical brass, 135°F to 286°F									
4240	1/2″ NPT, 1/2″ orifice	1 Spri	16	.500	Ea.	25	28		53	71.50
4280	3/4″ NPT, 17/32″ orifice	″	16	.500		70	28		98	121
4360	For satin chrome, add					2.75			2.75	3.03
4400	For 360°F, same cost									
4500	Sidewall, horizontal, brass, 135°F to 286°F									
4520	1/2″ NPT, 1/2″ orifice	1 Spri	16	.500	Ea.	25	28		53	71.50
4540	For 360°F, same cost									
4800	Recessed pendent, brass, 135°F to 286°F									
4820	1/2″ NPT, 3/8″ orifice	1 Spri	10	.800	Ea.	42.50	45		87.50	117
4830	1/2″ NPT, 7/16″ orifice		10	.800		18.40	45		63.40	90
4840	1/2″ NPT, 1/2″ orifice		10	.800		14.25	45		59.25	85.50
4860	1/2″ NPT, 17/32″ orifice		10	.800		42.50	45		87.50	117
4900	For satin chrome, add					2.75			2.75	3.03
5000	Recessed-vertical sidewall, brass, 135°F to 286°F									
5020	1/2″ NPT, 3/8″ orifice	1 Spri	10	.800	Ea.	31	45		76	104

21 13 13 – Wet-Pipe Sprinkler Systems

21 13 13.50 Wet-Pipe Sprinkler System Components		Crew	Daily Output	Labor-Hours	Unit	Material	Labor	Equipment	Total	Total Incl O&P
						2015 Bare Costs				
5030	1/2" NPT, 7/16" orifice	1 Spri	10	.800	Ea.	31	45		76	104
5040	1/2" NPT, 1/2" orifice	↓	10	.800	↓	31	45		76	104
5100	For bright nickel, same cost									
5600	Concealed, complete with cover plate									
5620	1/2" NPT, 1/2" orifice, 135°F to 212°F	1 Spri	9	.889	Ea.	24	50		74	104
5800	Window, brass, 1/2" NPT, 1/4" orifice		16	.500		34	28		62	81
5810	1/2" NPT, 5/16" orifice		16	.500		34	28		62	81
5820	1/2" NPT, 3/8" orifice		16	.500		34	28		62	81
5830	1/2" NPT, 7/16" orifice		16	.500		34	28		62	81
5840	1/2" NPT, 1/2" orifice	↓	16	.500		36	28		64	83.50
5860	For polished chrome, add					4.55			4.55	5
5880	3/4" NPT, 5/8" orifice	1 Spri	16	.500		37.50	28		65.50	85
5890	3/4 NPT, 3/4" orifice	"	16	.500		37.50	28		65.50	85
6000	Sprinkler head guards, bright zinc, 1/2" NPT					6			6	6.60
6020	Bright zinc, 3/4" NPT					6			6	6.60
6100	Sprinkler head wrenches, standard head					24.50			24.50	26.50
6120	Recessed head					37.50			37.50	41
6160	Tamper switch, (valve supervisory switch)	1 Spri	16	.500	↓	102	28		130	156

21 13 16 – Dry-Pipe Sprinkler Systems

21 13 16.50 Dry-Pipe Sprinkler System Components		Crew	Daily Output	Labor-Hours	Unit	Material	Labor	Equipment	Total	Total Incl O&P
0010	**DRY-PIPE SPRINKLER SYSTEM COMPONENTS**									
0600	Accelerator	1 Spri	8	1	Ea.	755	56		811	920
0800	Air compressor for dry pipe system, automatic, complete R211313-20									
0820	30 gal. system capacity, 3/4 HP	1 Spri	1.30	6.154	Ea.	765	345		1,110	1,375
0860	30 gal. system capacity, 1 HP		1.30	6.154		790	345		1,135	1,400
0910	30 gal. system capacity, 1-1/2 HP		1.30	6.154		825	345		1,170	1,450
0920	30 gal. system capacity, 2 HP		1.30	6.154		875	345		1,220	1,500
0960	Air pressure maintenance control		24	.333		345	18.70		363.70	410
1600	Dehydrator package, incl. valves and nipples	↓	12	.667	↓	710	37.50		747.50	845
2600	Sprinkler heads, not including supply piping									
2640	Dry, pendent, 1/2" orifice, 3/4" or 1" NPT									
2660	1/2" to 6" length	1 Spri	14	.571	Ea.	124	32		156	186
2670	6-1/4" to 8" length		14	.571		129	32		161	192
2680	8-1/4" to 12" length		14	.571		135	32		167	198
2690	12-1/4" to 15" length		14	.571		140	32		172	204
2700	15-1/4" to 18" length		14	.571		145	32		177	210
2710	18-1/4" to 21" length		13	.615		151	34.50		185.50	220
2720	21-1/4" to 24" length		13	.615		156	34.50		190.50	226
2730	24-1/4" to 27" length		13	.615		162	34.50		196.50	232
2740	27-1/4" to 30" length		13	.615		168	34.50		202.50	238
2750	30-1/4" to 33" length		13	.615		173	34.50		207.50	245
2760	33-1/4" to 36" length		13	.615		179	34.50		213.50	251
2780	36-1/4" to 39" length		12	.667		185	37.50		222.50	262
2790	39-1/4" to 42" length	↓	12	.667		190	37.50		227.50	267
2800	For each inch or fraction, add				↓	2.84			2.84	3.12
6330	Valves and components									
6340	Alarm test/shut off valve, 1/2"	1 Spri	20	.400	Ea.	21.50	22.50		44	59
8000	Dry pipe air check valve, 3" size	Q-12	2	8		1,775	405		2,180	2,575
8200	Dry pipe valve, incl. trim and gauges, 3" size		2	8		2,575	405		2,980	3,475
8220	4" size	↓	1	16		2,850	810		3,660	4,425
8240	6" size	Q-13	2	16		3,350	855		4,205	5,000
8280	For accelerator trim with gauges, add	1 Spri	8	1	↓	239	56		295	350

21 13 Fire-Suppression Sprinkler Systems

21 13 20 – Fire-Cycle Sprinkler Systems

21 13 20.50 Firecycle Fire-Suppression Sprinkler Systems		Crew	Daily Output	Labor-Hours	Unit	2015 Bare Costs Material	Labor	Equipment	Total	Total Incl O&P
0010	**FIRECYCLE FIRE-SUPPRESSION SPRINKLER SYSTEMS**									
8400	Firecycle package, includes swing check									
8420	and flow control valves with required trim									
8440	2" size	Q-12	2	8	Ea.	4,625	405		5,030	5,700
8460	3" size		1.50	10.667		5,075	540		5,615	6,425
8480	4" size		1	16		5,650	810		6,460	7,500
8500	6" size	Q-13	1.40	22.857		6,525	1,225		7,750	9,075

21 13 26 – Deluge Fire-Suppression Sprinkler Systems

21 13 26.50 Deluge Fire-Suppression Sprinkler Sys. Comp.		Crew	Daily Output	Labor-Hours	Unit	Material	Labor	Equipment	Total	Total Incl O&P
0010	**DELUGE FIRE-SUPPRESSION SPRINKLER SYSTEM COMPONENTS**									
1400	Deluge system, monitoring panel w/deluge valve & trim	1 Spri	18	.444	Ea.	10,800	25		10,825	11,900
6200	Valves and components									
7000	Deluge, assembly, incl. trim, pressure									
7020	operated relief, emergency release, gauges									
7040	2" size	Q-12	2	8	Ea.	3,675	405		4,080	4,650
7060	3" size		1.50	10.667		4,100	540		4,640	5,375
7080	4" size		1	16		4,700	810		5,510	6,425
7100	6" size	Q-13	1.80	17.778		5,525	950		6,475	7,550
7800	Pneumatic actuator, bronze, required on all									
7820	pneumatic release systems, any size deluge	1 Spri	18	.444	Ea.	415	25		440	495

21 13 39 – Foam-Water Systems

21 13 39.50 Foam-Water System Components		Crew	Daily Output	Labor-Hours	Unit	Material	Labor	Equipment	Total	Total Incl O&P
0010	**FOAM-WATER SYSTEM COMPONENTS**									
2600	Sprinkler heads, not including supply piping									
3600	Foam-water, pendent or upright, 1/2" NPT	1 Spri	12	.667	Ea.	210	37.50		247.50	289

21 21 Carbon-Dioxide Fire-Extinguishing Systems

21 21 16 – Carbon-Dioxide Fire-Extinguishing Equipment

21 21 16.50 CO2 Fire Extinguishing System		Crew	Daily Output	Labor-Hours	Unit	Material	Labor	Equipment	Total	Total Incl O&P
0010	**CO_2 FIRE EXTINGUISHING SYSTEM**									
0042	For detectors and control stations, see Section 28 31 23.50									
0100	Control panel, single zone with batteries (2 zones det., 1 suppr.)	1 Elec	1	8	Ea.	1,725	440		2,165	2,575
0150	Multizone (4) with batteries (8 zones det., 4 suppr.)	"	.50	16		3,275	875		4,150	4,950
1000	Dispersion nozzle, CO_2, 3" x 5"	1 Plum	18	.444		67	26		93	114
2000	Extinguisher, CO_2 system, high pressure, 75 lb. cylinder	Q-1	6	2.667		1,275	141		1,416	1,625
2100	100 lb. cylinder	"	5	3.200		1,300	169		1,469	1,725
3000	Electro/mechanical release	L-1	4	4		167	227		394	540
3400	Manual pull station	1 Plum	6	1.333		60.50	78.50		139	189
4000	Pneumatic damper release	"	8	1		223	58.50		281.50	340

21 22 Clean-Agent Fire-Extinguishing Systems

21 22 16 – Clean-Agent Fire-Extinguishing Equipment

21 22 16.50 FM200 Fire Extinguishing System		Crew	Daily Output	Labor-Hours	Unit	Material	2015 Bare Costs Labor	Equipment	Total	Total Incl O&P
0010	FM200 FIRE EXTINGUISHING SYSTEM									
1100	Dispersion nozzle FM200, 1-1/2"	1 Plum	14	.571	Ea.	67	33.50		100.50	126
2400	Extinguisher, FM200 system, filled, with mounting bracket									
2460	26 lb. container	Q-1	8	2	Ea.	2,300	106		2,406	2,700
2480	44 lb. container		7	2.286		3,050	121		3,171	3,575
2500	63 lb. container		6	2.667		3,575	141		3,716	4,150
2520	101 lb. container		5	3.200		4,775	169		4,944	5,525
2540	196 lb. container	↓	4	4	↓	7,775	211		7,986	8,875
6000	FM200 system, simple nozzle layout, with broad dispersion				C.F.	1.76			1.76	1.94
6020	Complex nozzle layout and/or including underfloor dispersion				"	3.50			3.50	3.85

21 31 Centrifugal Fire Pumps

21 31 13 – Electric-Drive, Centrifugal Fire Pumps

21 31 13.50 Electric-Drive Fire Pumps		Crew	Daily Output	Labor-Hours	Unit	Material	Labor	Equipment	Total	Total Incl O&P
0010	ELECTRIC-DRIVE FIRE PUMPS Including controller, fittings and relief valve									
3100	250 GPM, 55 psi, 15 HP, 3550 RPM, 2" pump	Q-13	.70	45.714	Ea.	14,500	2,450		16,950	19,700
3200	500 GPM, 50 psi, 27 HP, 1770 RPM, 4" pump		.68	47.059		14,900	2,525		17,425	20,300
3250	500 GPM, 100 psi, 47 HP, 3550 RPM, 3" pump		.66	48.485		16,100	2,600		18,700	21,800
3300	500 GPM, 125 psi, 64 HP, 3550 RPM, 3" pump		.62	51.613		17,700	2,750		20,450	23,700
3350	750 GPM, 50 psi, 44 HP, 1770 RPM, 5" pump		.64	50		15,500	2,675		18,175	21,300
3400	750 GPM, 100 psi, 66 HP, 3550 RPM, 4" pump		.58	55.172		17,900	2,950		20,850	24,300
3450	750 GPM, 165 psi, 120 HP, 3550 RPM, 4" pump		.56	57.143		24,400	3,050		27,450	31,600
3500	1000 GPM, 50 psi, 48 HP 1770 RPM, 5" pump		.60	53.333		16,600	2,850		19,450	22,700
3550	1000 GPM, 100 psi, 86 HP, 3550 RPM, 5" pump		.54	59.259		22,600	3,175		25,775	29,800
3600	1000 GPM, 150 psi, 142 HP, 3550 RPM, 5" pump		.50	64		26,400	3,425		29,825	34,500
3650	1000 GPM, 200 psi, 245 HP, 1770 RPM, 6" pump		.36	88.889		39,500	4,750		44,250	51,000
3660	1250 GPM, 75 psi, 75 HP, 1770 RPM, 5" pump		.55	58.182		21,500	3,100		24,600	28,500
3700	1500 GPM, 50 psi, 66 HP, 1770 RPM, 6" pump		.50	64		20,900	3,425		24,325	28,300
3750	1500 GPM, 100 psi, 139 HP, 1770 RPM, 6" pump		.46	69.565		26,100	3,725		29,825	34,500
3800	1500 GPM, 150 psi, 200 HP, 1770 RPM, 6" pump		.36	88.889		41,800	4,750		46,550	53,500
3850	1500 GPM, 200 psi, 279 HP, 1770 RPM, 6" pump		.32	100		44,700	5,350		50,050	57,500
3900	2000 GPM, 100 psi, 167 HP, 1770 RPM, 6" pump		.34	94.118		28,800	5,025		33,825	39,600
3950	2000 GPM, 150 psi, 292 HP, 1770 RPM, 6" pump		.28	114		42,300	6,100		48,400	56,000
4000	2500 GPM, 100 psi, 213 HP, 1770 RPM, 8" pump		.30	106		35,800	5,700		41,500	48,300
4040	2500 GPM, 135 psi, 339 HP, 1770 RPM, 8" pump		.26	123		50,500	6,575		57,075	66,000
4100	3000 GPM, 100 psi, 250 HP, 1770 RPM, 8" pump		.28	114		54,000	6,100		60,100	69,000
4150	3000 GPM, 140 psi, 428 HP, 1770 RPM, 10" pump		.24	133		59,500	7,125		66,625	76,500
4200	3500 GPM, 100 psi, 300 HP, 1770 RPM, 10" pump		.26	123		54,500	6,575		61,075	70,000
4250	3500 GPM, 140 psi, 450 HP, 1770 RPM, 10" pump	↓	.24	133		69,000	7,125		76,125	87,000
5000	For jockey pump 1", 3 HP, with control, add	Q-12	2	8	↓	2,600	405		3,005	3,475

21 31 16 – Diesel-Drive, Centrifugal Fire Pumps

21 31 16.50 Diesel-Drive Fire Pumps		Crew	Daily Output	Labor-Hours	Unit	Material	Labor	Equipment	Total	Total Incl O&P
0010	DIESEL-DRIVE FIRE PUMPS Including controller, fittings and relief valve									
0050	500 GPM, 50 psi, 27 HP, 4" pump	Q-13	.64	50	Ea.	33,800	2,675		36,475	41,300
0100	500 GPM, 100 psi, 62 HP, 4" pump		.60	53.333		39,600	2,850		42,450	48,100
0150	500 GPM, 125 psi, 78 HP, 4" pump		.56	57.143		42,100	3,050		45,150	51,000
0200	750 GPM, 50 psi, 44 HP, 5" pump		.60	53.333		34,800	2,850		37,650	42,700
0250	750 GPM, 100 psi, 80 HP, 4" pump		.56	57.143		39,100	3,050		42,150	47,800
0300	750 GPM, 165 psi, 203 HP, 5" pump		.52	61.538		49,700	3,300		53,000	59,500
0350	1000 GPM, 50 psi, 48 HP, 5" pump	↓	.58	55.172	↓	36,800	2,950		39,750	45,100

21 31 16 – Diesel-Drive, Centrifugal Fire Pumps

21 31 16.50 Diesel-Drive Fire Pumps		Crew	Daily Output	Labor-Hours	Unit	2015 Bare Costs Material	Labor	Equipment	Total	Total Incl O&P
0400	1000 GPM, 100 psi, 89 HP, 4" pump	Q-13	.56	57.143	Ea.	39,400	3,050		42,450	48,100
0450	1000 GPM, 150 psi, 148 HP, 4" pump		.48	66.667		51,000	3,575		54,575	61,500
0470	1000 GPM, 200 psi, 280 HP, 5" pump		.40	80		59,500	4,275		63,775	72,000
0480	1250 GPM, 75 psi, 75 HP, 5" pump		.54	59.259		39,900	3,175		43,075	48,900
0500	1500 GPM, 50 psi, 66 HP, 6" pump		.50	64		39,100	3,425		42,525	48,400
0550	1500 GPM, 100 psi, 140 HP, 6" pump		.46	69.565		50,000	3,725		53,725	61,000
0600	1500 GPM, 150 psi, 228 HP, 6" pump		.42	76.190		56,000	4,075		60,075	68,500
0650	1500 GPM, 200 psi, 279 HP, 6" pump		.38	84.211		70,500	4,500		75,000	84,500
0700	2000 GPM, 100 psi, 167 HP, 6" pump		.34	94.118		49,900	5,025		54,925	63,000
0750	2000 GPM, 150 psi, 284 HP, 6" pump		.30	106		64,000	5,700		69,700	79,500
0800	2500 GPM, 100 psi, 213 HP, 8" pump		.32	100		54,500	5,350		59,850	68,000
0820	2500 GPM, 150 psi, 365 HP, 8" pump		.26	123		69,500	6,575		76,075	87,000
0850	3000 GPM, 100 psi, 250 HP, 8" pump		.28	114		71,000	6,100		77,100	88,000
0900	3000 GPM, 150 psi, 384 HP, 10" pump		.20	160		92,500	8,550		101,050	115,500
0950	3500 GPM, 100 psi, 300 HP, 10" pump		.24	133		71,000	7,125		78,125	89,000
1000	3500 GPM, 150 psi, 518 HP, 10" pump	↓	.20	160	↓	100,000	8,550		108,550	123,500

Division Notes

	CREW	DAILY OUTPUT	LABOR-HOURS	UNIT	BARE COSTS MAT.	BARE COSTS LABOR	BARE COSTS EQUIP.	BARE COSTS TOTAL	TOTAL INCL O&P

Estimating Tips

22 10 00 Plumbing Piping and Pumps

This subdivision is primarily basic pipe and related materials. The pipe may be used by any of the mechanical disciplines, i.e., plumbing, fire protection, heating, and air conditioning.

Note: CPVC plastic piping approved for fire protection is located in 21 11 13.

- The labor adjustment factors listed in Subdivision 22 01 02.20 apply throughout Divisions 21, 22, and 23. CAUTION: the correct percentage may vary for the same items. For example, the percentage add for the basic pipe installation should be based on the maximum height that the craftsman must install for that particular section. If the pipe is to be located 14' above the floor but it is suspended on threaded rod from beams, the bottom flange of which is 18' high (4' rods), then the height is actually 18' and the add is 20%. The pipe coverer, however, does not have to go above the 14', and so the add should be 10%.
- Most pipe is priced first as straight pipe with a joint (coupling, weld, etc.) every 10' and a hanger usually every 10'. There are exceptions with hanger spacing such as for cast iron pipe (5') and plastic pipe (3 per 10'). Following each type of pipe there are several lines listing sizes and the amount to be subtracted to delete couplings and hangers. This is for pipe that is to be buried or supported together on trapeze hangers. The reason that the couplings are deleted is that these runs are usually long, and frequently longer lengths of pipe are used. By deleting the couplings, the estimator is expected to look up and add back the correct reduced number of couplings.
- When preparing an estimate, it may be necessary to approximate the fittings. Fittings usually run between 25% and 50% of the cost of the pipe. The lower percentage is for simpler runs, and the higher number is for complex areas, such as mechanical rooms.
- For historic restoration projects, the systems must be as invisible as possible, and pathways must be sought for pipes, conduit, and ductwork. While installations in accessible spaces (such as basements and attics) are relatively straightforward to estimate, labor costs may be more difficult to determine when delivery systems must be concealed.

22 40 00 Plumbing Fixtures

- Plumbing fixture costs usually require two lines: the fixture itself and its "rough-in, supply, and waste."
- In the Assemblies Section (Plumbing D2010) for the desired fixture, the System Components Group at the center of the page shows the fixture on the first line. The rest of the list (fittings, pipe, tubing, etc.) will total up to what we refer to in the Unit Price section as "Rough-in, supply, waste, and vent." Note that for most fixtures we allow a nominal 5' of tubing to reach from the fixture to a main or riser.
- Remember that gas- and oil-fired units need venting.

Reference Numbers

Reference numbers are shown in shaded boxes at the beginning of some major classifications. These numbers refer to related items in the Reference Section. The reference information may be an estimating procedure, an alternate pricing method, or technical information.

Note: Not all subdivisions listed here necessarily appear in this publication. ■

22 01 02 – Labor Adjustments

22 01 02.10 Boilers, General		Crew	Daily Output	Labor-Hours	Unit	Material	2015 Bare Costs Labor	Equipment	Total	Total Incl O&P
0010	**BOILERS, GENERAL**, Prices do not include flue piping, elec. wiring,									
0020	gas or oil piping, boiler base, pad, or tankless unless noted									
0100	Boiler H.P.: 10 KW = 34 lb./steam/hr. = 33,475 BTU/hr.									
0150	To convert SFR to BTU rating: Hot water, 150 x SFR;									
0160	Forced hot water, 180 x SFR; steam, 240 x SFR									

22 01 02.20 Labor Adjustment Factors		Crew	Daily Output	Labor-Hours	Unit	Material	2015 Bare Costs Labor	Equipment	Total	Total Incl O&P
0010	**LABOR ADJUSTMENT FACTORS**, (For Div. 21, 22 and 23) R220102-20									
0100	Labor factors, The below are reasonable suggestions, however									
0110	each project must be evaluated for its own peculiarities, and									
0120	the adjustments be increased or decreased depending on the									
0130	severity of the special conditions.									
1000	Add to labor for elevated installation (Above floor level)									
1080	10′ to 14.5′ high						10%			
1100	15′ to 19.5′ high						20%			
1120	20′ to 24.5′ high						25%			
1140	25′ to 29.5′ high						35%			
1160	30′ to 34.5′ high						40%			
1180	35′ to 39.5′ high						50%			
1200	40′ and higher						55%			
2000	Add to labor for crawl space									
2100	3′ high						40%			
2140	4′ high						30%			
3000	Add to labor for multi-story building									
3100	Add per floor for floors 3 thru 19						2%			
3140	Add per floor for floors 20 and up						4%			
4000	Add to labor for working in existing occupied buildings									
4100	Hospital						35%			
4140	Office building						25%			
4180	School						20%			
4220	Factory or warehouse						15%			
4260	Multi dwelling						15%			
5000	Add to labor, miscellaneous									
5100	Cramped shaft						35%			
5140	Congested area						15%			
5180	Excessive heat or cold						30%			
9000	Labor factors, The above are reasonable suggestions, however									
9010	each project should be evaluated for its own peculiarities.									
9100	Other factors to be considered are:									
9140	Movement of material and equipment through finished areas									
9180	Equipment room									
9220	Attic space									
9260	No service road									
9300	Poor unloading/storage area									
9340	Congested site area/heavy traffic									

22 05 05 – Selective Demolition for Plumbing

22 05 05.10 Plumbing Demolition		Crew	Daily Output	Labor-Hours	Unit	Material	2015 Bare Costs Labor	Equipment	Total	Total Incl O&P
0010	**PLUMBING DEMOLITION** R220105-10									
0400	Air compressor, up thru 2 H.P.	Q-1	10	1.600	Ea.		84.50		84.50	132
0410	3 H.P. thru 7-1/2 H.P. R024119-10		5.60	2.857			151		151	235
0420	10 H.P. thru 15 H.P.		1.40	11.429			605		605	940
0430	20 H.P. thru 30 H.P.	Q-2	1.30	18.462			1,000		1,000	1,575
0500	Backflow preventer, up thru 2" diameter	1 Plum	17	.471			27.50		27.50	43
0510	2-1/2" thru 3" diameter	Q-1	10	1.600			84.50		84.50	132
0520	4" thru 6" diameter	"	5	3.200			169		169	264
0530	8" thru 10" diameter	Q-2	3	8			440		440	685
0700	Carriers and supports									
0710	Fountains, sinks, lavatories and urinals	1 Plum	14	.571	Ea.		33.50		33.50	52.50
0720	Water closets	"	12	.667	"		39		39	61
0730	Grinder pump or sewage ejector system									
0732	Simplex	Q-1	7	2.286	Ea.		121		121	188
0734	Duplex	"	2.80	5.714			300		300	470
0738	Hot water dispenser	1 Plum	36	.222			13.05		13.05	20.50
0740	Hydrant, wall		26	.308			18.05		18.05	28
0744	Ground		12	.667			39		39	61
0760	Cleanouts and drains, up thru 4" pipe diameter		10	.800			47		47	73
0764	5" thru 8" pipe diameter	Q-1	10	1.600			84.50		84.50	132
0780	Industrial safety fixtures	1 Plum	8	1			58.50		58.50	91.50
1020	Fixtures, including 10' piping									
1100	Bathtubs, cast iron	1 Plum	4	2	Ea.		117		117	183
1120	Fiberglass		6	1.333			78.50		78.50	122
1140	Steel		5	1.600			94		94	146
1150	Bidet	Q-1	7	2.286			121		121	188
1200	Lavatory, wall hung	1 Plum	10	.800			47		47	73
1220	Counter top		8	1			58.50		58.50	91.50
1300	Sink, single compartment		8	1			58.50		58.50	91.50
1320	Double compartment		7	1.143			67		67	105
1340	Shower, stall and receptor	Q-1	6	2.667			141		141	220
1350	Group	"	7	2.286			121		121	188
1400	Water closet, floor mounted	1 Plum	8	1			58.50		58.50	91.50
1420	Wall mounted	"	7	1.143			67		67	105
1440	Wash fountain, 36" diameter	Q-2	8	3			164		164	256
1442	54" diameter	"	7	3.429			188		188	293
1500	Urinal, floor mounted	1 Plum	4	2			117		117	183
1520	Wall mounted	"	7	1.143			67		67	105
1590	Whirl pool or hot tub	Q-1	2.60	6.154			325		325	505
1600	Water fountains, free standing	1 Plum	8	1			58.50		58.50	91.50
1620	Wall or deck mounted		6	1.333			78.50		78.50	122
1800	Medical gas specialties		8	1			58.50		58.50	91.50
1900	Piping fittings, single connection, up thru 1-1/2" diameter		30	.267			15.65		15.65	24.50
1910	2" thru 4" diameter		14	.571			33.50		33.50	52.50
1980	Pipe hanger/support removal		80	.100			5.85		5.85	9.15
1990	Glass pipe with fittings, 1" thru 3" diameter		200	.040	L.F.		2.35		2.35	3.66
1992	4" thru 6" diameter		150	.053			3.13		3.13	4.88
2000	Piping, metal, up thru 1-1/2" diameter		200	.040			2.35		2.35	3.66
2050	2" thru 3-1/2" diameter		150	.053			3.13		3.13	4.88
2100	4" thru 6" diameter	2 Plum	100	.160			9.40		9.40	14.65
2150	8" thru 14" diameter	"	60	.267			15.65		15.65	24.50
2153	16" thru 20" diameter	Q-18	70	.343			19.10	.82	19.92	31
2155	24" thru 26" diameter		55	.436			24.50	1.05	25.55	39

22 05 05 – Selective Demolition for Plumbing

22 05 05.10 Plumbing Demolition		Crew	Daily Output	Labor-Hours	Unit	Material	2015 Bare Costs Labor	Equipment	Total	Total Incl O&P
2156	30″ thru 36″ diameter	Q-18	40	.600	L.F.		33.50	1.44	34.94	53.50
2160	Plastic pipe with fittings, up thru 1-1/2″ diameter	1 Plum	250	.032			1.88		1.88	2.93
2162	2″ thru 3″ diameter	″	200	.040			2.35		2.35	3.66
2164	4″ thru 6″ diameter	Q-1	200	.080			4.23		4.23	6.60
2166	8″ thru 14″ diameter		150	.107			5.65		5.65	8.80
2168	16″ diameter		100	.160			8.45		8.45	13.20
2170	Prison fixtures, lavatory or sink		18	.889	Ea.		47		47	73
2172	Shower		5.60	2.857			151		151	235
2174	Urinal or water closet		13	1.231			65		65	101
2180	Pumps, all fractional horse-power		12	1.333			70.50		70.50	110
2184	1 H.P. thru 5 H.P.		6	2.667			141		141	220
2186	7-1/2 H.P. thru 15 H.P.		2.50	6.400			340		340	525
2188	20 H.P. thru 25 H.P.	Q-2	4	6			330		330	515
2190	30 H.P. thru 60 H.P.		.80	30			1,650		1,650	2,575
2192	75 H.P. thru 100 H.P.		.60	40			2,200		2,200	3,425
2194	150 H.P.		.50	48			2,625		2,625	4,100
2198	Pump, sump or submersible	1 Plum	12	.667			39		39	61
2200	Receptors and interceptors, up thru 20 GPM	″	8	1			58.50		58.50	91.50
2204	25 thru 100 GPM	Q-1	6	2.667			141		141	220
2208	125 thru 300 GPM	″	2.40	6.667			350		350	550
2211	325 thru 500 GPM	Q-2	2.60	9.231			505		505	790
2212	Deduct for salvage, aluminum scrap				Ton				700	770
2214	Brass scrap								2,450	2,675
2216	Copper scrap								3,200	3,525
2218	Lead scrap								520	570
2220	Steel scrap								180	200
2230	Temperature maintenance cable	1 Plum	1200	.007	L.F.		.39		.39	.61
2250	Water heater, 40 gal.	″	6	1.333	Ea.		78.50		78.50	122
3100	Tanks, water heaters and liquid containers									
3110	Up thru 45 gallons	Q-1	22	.727	Ea.		38.50		38.50	60
3120	50 thru 120 gallons		14	1.143			60.50		60.50	94
3130	130 thru 240 gallons		7.60	2.105			111		111	173
3140	250 thru 500 gallons		5.40	2.963			157		157	244
3150	600 thru 1000 gallons	Q-2	1.60	15			820		820	1,275
3160	1100 thru 2000 gallons		.70	34.286			1,875		1,875	2,925
3170	2100 thru 4000 gallons		.50	48			2,625		2,625	4,100
6000	Remove and reset fixtures, easy access	1 Plum	6	1.333			78.50		78.50	122
6100	Difficult access		4	2			117		117	183
9000	Minimum labor/equipment charge		2	4	Job		235		235	365
9100	Valve, metal valves or strainers and similar, up thru 1-1/2″ diameter	1 Stpi	28	.286	Ea.		17.05		17.05	26.50
9110	2″ thru 3″ diameter	Q-1	11	1.455			77		77	120
9120	4″ thru 6″ diameter	″	8	2			106		106	165
9130	8″ thru 14″ diameter	Q-2	8	3			164		164	256
9140	16″ thru 20″ diameter		2	12			655		655	1,025
9150	24″ diameter		1.20	20			1,100		1,100	1,700
9200	Valve, plastic, up thru 1-1/2″ diameter	1 Plum	42	.190			11.20		11.20	17.45
9210	2″ thru 3″ diameter		15	.533			31.50		31.50	49
9220	4″ thru 6″ diameter		12	.667			39		39	61
9300	Vent flashing and caps		55	.145			8.55		8.55	13.30
9350	Water filter, commercial, 1″ thru 1-1/2″	Q-1	2	8			425		425	660
9360	2″ thru 2-1/2″	″	1.60	10			530		530	825
9400	Water heaters									
9410	Up thru 245 GPH	Q-1	2.40	6.667	Ea.		350		350	550

22 05 05 – Selective Demolition for Plumbing

22 05 05.10 Plumbing Demolition		Crew	Daily Output	Labor-Hours	Unit	2015 Bare Costs Material	Labor	Equipment	Total	Total Incl O&P
9420	250 thru 756 GPH	Q-1	1.60	10	Ea.		530		530	825
9430	775 thru 1640 GPH	↓	.80	20			1,050		1,050	1,650
9440	1650 thru 4000 GPH	Q-2	.50	48			2,625		2,625	4,100
9470	Water softener	Q-1	2	8	↓		425		425	660

22 05 23 – General-Duty Valves for Plumbing Piping

22 05 23.10 Valves, Brass

		Crew	Daily Output	Labor-Hours	Unit	Material	Labor	Equipment	Total	Total Incl O&P
0010	**VALVES, BRASS**									
0032	For motorized valves, see Section 23 09 53.10									
0500	Gas cocks, threaded									
0510	1/4"	1 Plum	26	.308	Ea.	14.70	18.05		32.75	44
0520	3/8"		24	.333		14.70	19.55		34.25	46.50
0530	1/2"		24	.333		13.15	19.55		32.70	45
0540	3/4"		22	.364		15.50	21.50		37	50.50
0550	1"		19	.421		31	24.50		55.50	72.50
0560	1-1/4"		15	.533		44.50	31.50		76	98
0570	1-1/2"		13	.615		61	36		97	124
0580	2"	↓	11	.727	↓	103	42.50		145.50	181
0672	For larger sizes use lubricated plug valve, Section 23 05 23.70									

22 05 23.20 Valves, Bronze

		Crew	Daily Output	Labor-Hours	Unit	Material	Labor	Equipment	Total	Total Incl O&P
0010	**VALVES, BRONZE** R220523-90									
1020	Angle, 150 lb., rising stem, threaded									
1030	1/8"	1 Plum	24	.333	Ea.	128	19.55		147.55	171
1040	1/4"		24	.333		128	19.55		147.55	171
1050	3/8"		24	.333		130	19.55		149.55	174
1060	1/2"		22	.364		142	21.50		163.50	190
1070	3/4"		20	.400		193	23.50		216.50	250
1080	1"		19	.421		279	24.50		303.50	345
1090	1-1/4"		15	.533		360	31.50		391.50	445
1100	1-1/2"		13	.615		470	36		506	570
1110	2"	↓	11	.727	↓	755	42.50		797.50	895
1300	Ball									
1304	Soldered									
1312	3/8"	1 Plum	21	.381	Ea.	14.95	22.50		37.45	51.50
1316	1/2"		18	.444		14.95	26		40.95	57
1320	3/4"		17	.471		25	27.50		52.50	70
1324	1"		15	.533		31	31.50		62.50	83.50
1328	1-1/4"		13	.615		52	36		88	114
1332	1-1/2"		11	.727		66.50	42.50		109	140
1336	2"		9	.889		83.50	52		135.50	174
1340	2-1/2"		7	1.143		395	67		462	540
1344	3"	↓	5	1.600	↓	455	94		549	645
1350	Single union end									
1358	3/8"	1 Plum	21	.381	Ea.	21.50	22.50		44	58.50
1362	1/2"		18	.444		22.50	26		48.50	65.50
1366	3/4"		17	.471		43	27.50		70.50	90.50
1370	1"		15	.533		52	31.50		83.50	106
1374	1-1/4"		13	.615		88.50	36		124.50	154
1378	1-1/2"		11	.727		111	42.50		153.50	189
1382	2"	↓	9	.889	↓	173	52		225	272
1398	Threaded, 150 psi									
1400	1/4"	1 Plum	24	.333	Ea.	14.15	19.55		33.70	46
1430	3/8"	↓	24	.333	↓	14.15	19.55		33.70	46

22 05 23.20 Valves, Bronze		Crew	Daily Output	Labor-Hours	Unit	Material	Labor	Equipment	Total	Total Incl O&P
						2015 Bare Costs				
1450	1/2"	1 Plum	22	.364	Ea.	14.15	21.50		35.65	49
1460	3/4"		20	.400		23.50	23.50		47	62
1470	1"		19	.421		33.50	24.50		58	75.50
1480	1-1/4"		15	.533		58.50	31.50		90	114
1490	1-1/2"		13	.615		76.50	36		112.50	141
1500	2"		11	.727		93	42.50		135.50	169
1510	2-1/2"		9	.889		310	52		362	420
1520	3"		8	1		470	58.50		528.50	610
1600	Butterfly, 175 psi, full port, solder or threaded ends									
1610	Stainless steel disc and stem									
1620	1/4"	1 Plum	24	.333	Ea.	17.35	19.55		36.90	49.50
1630	3/8"		24	.333		17.35	19.55		36.90	49.50
1640	1/2"		22	.364		18.65	21.50		40.15	54
1650	3/4"		20	.400		29.50	23.50		53	69
1660	1"		19	.421		36	24.50		60.50	78
1670	1-1/4"		15	.533		58	31.50		89.50	113
1680	1-1/2"		13	.615		75	36		111	139
1690	2"		11	.727		94	42.50		136.50	170
1750	Check, swing, class 150, regrinding disc, threaded									
1800	1/8"	1 Plum	24	.333	Ea.	66.50	19.55		86.05	104
1830	1/4"		24	.333		66.50	19.55		86.05	104
1840	3/8"		24	.333		70.50	19.55		90.05	108
1850	1/2"		24	.333		75.50	19.55		95.05	114
1860	3/4"		20	.400		100	23.50		123.50	147
1870	1"		19	.421		144	24.50		168.50	197
1880	1-1/4"		15	.533		208	31.50		239.50	278
1890	1-1/2"		13	.615		242	36		278	325
1900	2"		11	.727		355	42.50		397.50	455
1910	2-1/2"	Q-1	15	1.067		800	56.50		856.50	970
1920	3"	"	13	1.231		1,075	65		1,140	1,275
2000	For 200 lb., add					5%	10%			
2040	For 300 lb., add					15%	15%			
2060	Check swing, 300#, sweat, 3/8" size	1 Plum	24	.333	Ea.	63	19.55		82.55	99.50
2070	1/2"		24	.333		63	19.55		82.55	99.50
2080	3/4"		20	.400		79.50	23.50		103	124
2090	1"		19	.421		114	24.50		138.50	164
2100	1-1/4"		15	.533		159	31.50		190.50	224
2110	1-1/2"		13	.615		194	36		230	270
2120	2"		11	.727		296	42.50		338.50	390
2130	2-1/2"	Q-1	15	1.067		650	56.50		706.50	805
2140	3"	"	13	1.231		870	65		935	1,050
2350	Check, lift, class 150, horizontal composition disc, threaded									
2430	1/4"	1 Plum	24	.333	Ea.	153	19.55		172.55	199
2440	3/8"		24	.333		194	19.55		213.55	244
2450	1/2"		24	.333		169	19.55		188.55	216
2460	3/4"		20	.400		206	23.50		229.50	263
2470	1"		19	.421		292	24.50		316.50	360
2480	1-1/4"		15	.533		395	31.50		426.50	485
2490	1-1/2"		13	.615		470	36		506	570
2500	2"		11	.727		780	42.50		822.50	925
2850	Gate, N.R.S., soldered, 125 psi									
2900	3/8"	1 Plum	24	.333	Ea.	61	19.55		80.55	97.50
2920	1/2"		24	.333		54	19.55		73.55	90

22 05 23.20 Valves, Bronze		Crew	Daily Output	Labor-Hours	Unit	Material	Labor	Equipment	Total	Total Incl O&P
						2015 Bare Costs				
2940	3/4"	1 Plum	20	.400	Ea.	63.50	23.50		87	107
2950	1"		19	.421		76.50	24.50		101	123
2960	1-1/4"		15	.533		126	31.50		157.50	188
2970	1-1/2"		13	.615		142	36		178	214
2980	2"		11	.727		185	42.50		227.50	271
2990	2-1/2"	Q-1	15	1.067		450	56.50		506.50	585
3000	3"	"	13	1.231		560	65		625	715
3350	Threaded, class 150									
3410	1/4"	1 Plum	24	.333	Ea.	75.50	19.55		95.05	114
3420	3/8"		24	.333		75.50	19.55		95.05	114
3430	1/2"		24	.333		71	19.55		90.55	109
3440	3/4"		20	.400		81.50	23.50		105	126
3450	1"		19	.421		102	24.50		126.50	151
3460	1-1/4" size		15	.533		138	31.50		169.50	201
3470	1-1/2"		13	.615		198	36		234	275
3480	2"		11	.727		237	42.50		279.50	325
3490	2-1/2"	Q-1	15	1.067		605	56.50		661.50	755
3500	3" size	"	13	1.231		855	65		920	1,050
3850	Rising stem, soldered, 300 psi									
3900	3/8"	1 Plum	24	.333	Ea.	141	19.55		160.55	186
3920	1/2"		24	.333		116	19.55		135.55	158
3940	3/4"		20	.400		132	23.50		155.50	182
3950	1"		19	.421		179	24.50		203.50	236
3960	1-1/4"		15	.533		247	31.50		278.50	320
3970	1-1/2"		13	.615		300	36		336	385
3980	2"		11	.727		480	42.50		522.50	595
3990	2-1/2"	Q-1	15	1.067		1,025	56.50		1,081.50	1,225
4000	3"	"	13	1.231		1,575	65		1,640	1,850
4250	Threaded, class 150									
4310	1/4"	1 Plum	24	.333	Ea.	68.50	19.55		88.05	106
4320	3/8"		24	.333		68.50	19.55		88.05	106
4330	1/2"		24	.333		63	19.55		82.55	99.50
4340	3/4"		20	.400		73.50	23.50		97	117
4350	1"		19	.421		98.50	24.50		123	147
4360	1-1/4"		15	.533		134	31.50		165.50	196
4370	1-1/2"		13	.615		169	36		205	242
4380	2"		11	.727		227	42.50		269.50	315
4390	2-1/2"	Q-1	15	1.067		530	56.50		586.50	675
4400	3"	"	13	1.231		740	65		805	910
4500	For 300 psi, threaded, add					100%	15%			
4540	For chain operated type, add					15%				
4850	Globe, class 150, rising stem, threaded									
4920	1/4"	1 Plum	24	.333	Ea.	97	19.55		116.55	137
4940	3/8"		24	.333		95.50	19.55		115.05	136
4950	1/2"		24	.333		95.50	19.55		115.05	136
4960	3/4"		20	.400		99	23.50		122.50	146
4970	1"		19	.421		154	24.50		178.50	208
4980	1-1/4"		15	.533		245	31.50		276.50	320
4990	1-1/2"		13	.615		320	36		356	410
5000	2"		11	.727		465	42.50		507.50	580
5010	2-1/2"	Q-1	15	1.067		1,175	56.50		1,231.50	1,375
5020	3"	"	13	1.231		1,675	65		1,740	1,925
5120	For 300 lb. threaded, add					50%	15%			

22 05 23 – General-Duty Valves for Plumbing Piping

22 05 23.20 Valves, Bronze		Crew	Daily Output	Labor-Hours	Unit	2015 Bare Costs Material	Labor	Equipment	Total	Total Incl O&P
5130	Globe, 300 lb., sweat, 3/8" size	1 Plum	24	.333	Ea.	121	19.55		140.55	164
5140	1/2"		24	.333		124	19.55		143.55	168
5150	3/4"		20	.400		169	23.50		192.50	222
5160	1"		19	.421		234	24.50		258.50	296
5170	1-1/4"		15	.533		370	31.50		401.50	455
5180	1-1/2"		13	.615		445	36		481	545
5190	2"		11	.727		660	42.50		702.50	790
5200	2-1/2"	Q-1	15	1.067		1,275	56.50		1,331.50	1,525
5210	3"	"	13	1.231		1,650	65		1,715	1,925
5600	Relief, pressure & temperature, self-closing, ASME, threaded									
5640	3/4"	1 Plum	28	.286	Ea.	253	16.75		269.75	305
5650	1"		24	.333		405	19.55		424.55	475
5660	1-1/4"		20	.400		895	23.50		918.50	1,025
5670	1-1/2"		18	.444		1,250	26		1,276	1,425
5680	2"		16	.500		1,350	29.50		1,379.50	1,550
5950	Pressure, poppet type, threaded									
6000	1/2"	1 Plum	30	.267	Ea.	78.50	15.65		94.15	111
6040	3/4"	"	28	.286	"	73.50	16.75		90.25	107
6400	Pressure, water, ASME, threaded									
6440	3/4"	1 Plum	28	.286	Ea.	116	16.75		132.75	153
6450	1"		24	.333		260	19.55		279.55	315
6460	1-1/4"		20	.400		390	23.50		413.50	465
6470	1-1/2"		18	.444		570	26		596	665
6480	2"		16	.500		820	29.50		849.50	950
6490	2-1/2"		15	.533		3,150	31.50		3,181.50	3,525
6900	Reducing, water pressure									
6920	300 psi to 25-75 psi, threaded or sweat									
6940	1/2"	1 Plum	24	.333	Ea.	395	19.55		414.55	465
6950	3/4"		20	.400		405	23.50		428.50	485
6960	1"		19	.421		630	24.50		654.50	730
6970	1-1/4"		15	.533		1,100	31.50		1,131.50	1,250
6980	1-1/2"		13	.615		1,650	36		1,686	1,875
6990	2"		11	.727		2,450	42.50		2,492.50	2,775
7100	For built-in by-pass or 10-35 psi, add					39			39	42.50
7700	High capacity, 250 psi to 25-75 psi, threaded									
7740	1/2"	1 Plum	24	.333	Ea.	540	19.55		559.55	625
7780	3/4"		20	.400		540	23.50		563.50	630
7790	1"		19	.421		775	24.50		799.50	895
7800	1-1/4"		15	.533		1,350	31.50		1,381.50	1,525
7810	1-1/2"		13	.615		1,975	36		2,011	2,225
7820	2"		11	.727		2,875	42.50		2,917.50	3,225
7830	2-1/2"		9	.889		4,100	52		4,152	4,600
7840	3"		8	1		4,925	58.50		4,983.50	5,500
7850	3" flanged (iron body)	Q-1	10	1.600		4,850	84.50		4,934.50	5,450
7860	4" flanged (iron body)	"	8	2		6,300	106		6,406	7,100
7920	For higher pressure, add					25%				
8350	Tempering, water, sweat connections									
8400	1/2"	1 Plum	24	.333	Ea.	98	19.55		117.55	139
8440	3/4"	"	20	.400	"	126	23.50		149.50	176
8650	Threaded connections									
8700	1/2"	1 Plum	24	.333	Ea.	126	19.55		145.55	170
8740	3/4"		20	.400		770	23.50		793.50	880
8750	1"		19	.421		865	24.50		889.50	995

22 05 23 – General-Duty Valves for Plumbing Piping

22 05 23.20 Valves, Bronze		Crew	Daily Output	Labor-Hours	Unit	Material	2015 Bare Costs Labor	Equipment	Total	Total Incl O&P
8760	1-1/4"	1 Plum	15	.533	Ea.	1,350	31.50		1,381.50	1,525
8770	1-1/2"		13	.615		1,475	36		1,511	1,650
8780	2"		11	.727		2,200	42.50		2,242.50	2,500
9000	Minimum labor/equipment charge		4	2	Job		117		117	183

22 05 23.40 Valves, Lined, Corrosion Resistant/High Purity

		Crew	Daily Output	Labor-Hours	Unit	Material	Labor	Equipment	Total	Total Incl O&P
0010	**VALVES, LINED, CORROSION RESISTANT/HIGH PURITY** R220523-90									
3500	Check lift, 125 lb., cast iron flanged									
3510	Horizontal PPL or SL lined									
3530	1"	1 Plum	14	.571	Ea.	635	33.50		668.50	755
3540	1-1/2"		11	.727		770	42.50		812.50	910
3550	2"		8	1		895	58.50		953.50	1,075
3560	2-1/2"	Q-1	5	3.200		1,175	169		1,344	1,575
3570	3"		4.50	3.556		1,450	188		1,638	1,900
3590	4"		3	5.333		1,925	282		2,207	2,575
3610	6"	Q-2	3	8		3,275	440		3,715	4,275
3620	8"	"	2.50	9.600		7,225	525		7,750	8,775
4250	Vertical PPL or SL lined									
4270	1"	1 Plum	14	.571	Ea.	570	33.50		603.50	685
4290	1-1/2"		11	.727		705	42.50		747.50	840
4300	2"		8	1		820	58.50		878.50	995
4310	2-1/2"	Q-1	5	3.200		1,250	169		1,419	1,650
4320	3"		4.50	3.556		1,250	188		1,438	1,650
4340	4"		3	5.333		1,625	282		1,907	2,250
4360	6"	Q-2	3	8		2,750	440		3,190	3,700
4370	8"	"	2.50	9.600		4,975	525		5,500	6,275

22 05 23.60 Valves, Plastic

		Crew	Daily Output	Labor-Hours	Unit	Material	Labor	Equipment	Total	Total Incl O&P
0010	**VALVES, PLASTIC** R220523-90									
1100	Angle, PVC, threaded									
1110	1/4"	1 Plum	26	.308	Ea.	39.50	18.05		57.55	71.50
1120	1/2"		26	.308		56.50	18.05		74.55	90
1130	3/4"		25	.320		67	18.80		85.80	104
1140	1"		23	.348		81.50	20.50		102	122
1150	Ball, PVC, socket or threaded, true union									
1230	1/2"	1 Plum	26	.308	Ea.	40.50	18.05		58.55	72.50
1240	3/4"		25	.320		40.50	18.80		59.30	74
1250	1"		23	.348		48	20.50		68.50	85
1260	1-1/4"		21	.381		83.50	22.50		106	127
1270	1-1/2"		20	.400		83.50	23.50		107	129
1280	2"		17	.471		110	27.50		137.50	164
1290	2-1/2"	Q-1	26	.615		188	32.50		220.50	258
1300	3"		24	.667		282	35		317	365
1310	4"		20	.800		450	42.50		492.50	560
1360	For PVC, flanged, add					100%	15%			
1450	Double union 1/2"	1 Plum	26	.308		32.50	18.05		50.55	64
1460	3/4"		25	.320		42.50	18.80		61.30	76
1470	1"		23	.348		50.50	20.50		71	87.50
1480	1-1/4"		21	.381		69.50	22.50		92	112
1490	1-1/2"		20	.400		84	23.50		107.50	129
1500	2"		17	.471		111	27.50		138.50	165
1650	CPVC, socket or threaded, single union									
1700	1/2"	1 Plum	26	.308	Ea.	58	18.05		76.05	91.50
1720	3/4"		25	.320		77	18.80		95.80	114

22 05 23.60 Valves, Plastic		Crew	Daily Output	Labor-Hours	Unit	2015 Bare Costs Material	Labor	Equipment	Total	Total Incl O&P
1730	1"	1 Plum	23	.348	Ea.	87.50	20.50		108	128
1750	1-1/4"		21	.381		140	22.50		162.50	189
1760	1-1/2"		20	.400		140	23.50		163.50	191
1770	2"		17	.471		192	27.50		219.50	255
1780	3"	Q-1	24	.667		760	35		795	890
1840	For CPVC, flanged, add					65%	15%			
1880	For true union, socket or threaded, add					50%	5%			
2050	Polypropylene, threaded									
2100	1/4"	1 Plum	26	.308	Ea.	45	18.05		63.05	77.50
2120	3/8"		26	.308		45	18.05		63.05	77.50
2130	1/2"		26	.308		45	18.05		63.05	77.50
2140	3/4"		25	.320		54.50	18.80		73.30	89.50
2150	1"		23	.348		62.50	20.50		83	101
2160	1-1/4"		21	.381		84	22.50		106.50	128
2170	1-1/2"		20	.400		103	23.50		126.50	151
2180	2"		17	.471		138	27.50		165.50	195
2190	3"	Q-1	24	.667		315	35		350	400
2200	4"	"	20	.800		550	42.50		592.50	670
2550	PVC, three way, socket or threaded									
2600	1/2"	1 Plum	26	.308	Ea.	62.50	18.05		80.55	96.50
2640	3/4"		25	.320		71	18.80		89.80	108
2650	1"		23	.348		76.50	20.50		97	116
2660	1-1/2"		20	.400		181	23.50		204.50	236
2670	2"		17	.471		206	27.50		233.50	270
2680	3"	Q-1	24	.667		430	35		465	530
2740	For flanged, add					60%	15%			
3150	Ball check, PVC, socket or threaded									
3200	1/4"	1 Plum	26	.308	Ea.	45.50	18.05		63.55	78.50
3220	3/8"		26	.308		45.50	18.05		63.55	78.50
3240	1/2"		26	.308		45.50	18.05		63.55	78.50
3250	3/4"		25	.320		51	18.80		69.80	85.50
3260	1"		23	.348		64	20.50		84.50	103
3270	1-1/4"		21	.381		107	22.50		129.50	153
3280	1-1/2"		20	.400		107	23.50		130.50	155
3290	2"		17	.471		146	27.50		173.50	204
3310	3"	Q-1	24	.667		405	35		440	500
3320	4"	"	20	.800		575	42.50		617.50	700
3360	For PVC, flanged, add					50%	15%			
3750	CPVC, socket or threaded									
3800	1/2"	1 Plum	26	.308	Ea.	70.50	18.05		88.55	106
3840	3/4"		25	.320		83.50	18.80		102.30	122
3850	1"		23	.348		99.50	20.50		120	141
3860	1-1/2"		20	.400		171	23.50		194.50	225
3870	2"		17	.471		230	27.50		257.50	296
3880	3"	Q-1	24	.667		620	35		655	735
3920	4"	"	20	.800		840	42.50		882.50	990
3930	For CPVC, flanged, add					40%	15%			
4340	Polypropylene, threaded									
4360	1/2"	1 Plum	26	.308	Ea.	44	18.05		62.05	76.50
4400	3/4"		25	.320		61	18.80		79.80	96.50
4440	1"		23	.348		65	20.50		85.50	104
4450	1-1/2"		20	.400		125	23.50		148.50	174
4460	2"		17	.471		158	27.50		185.50	217

22 05 23.60 Valves, Plastic		Crew	Daily Output	Labor-Hours	Unit	Material	2015 Bare Costs Labor	Equipment	Total	Total Incl O&P
4500	For polypropylene flanged, add				Ea.	200%	15%			
4850	Foot valve, PVC, socket or threaded									
4900	1/2"	1 Plum	34	.235	Ea.	73	13.80		86.80	102
4930	3/4"		32	.250		83	14.70		97.70	115
4940	1"		28	.286		107	16.75		123.75	144
4950	1-1/4"		27	.296		206	17.40		223.40	254
4960	1-1/2"		26	.308		206	18.05		224.05	255
4970	2"		24	.333		365	19.55		384.55	430
4980	3"		20	.400		835	23.50		858.50	955
4990	4"		18	.444		1,275	26		1,301	1,475
5000	For flanged, add					25%	10%			
5050	CPVC, socket or threaded									
5060	1/2"	1 Plum	34	.235	Ea.	72	13.80		85.80	101
5070	3/4"		32	.250		83	14.70		97.70	115
5080	1"		28	.286		102	16.75		118.75	139
5090	1-1/4"		27	.296		164	17.40		181.40	207
5100	1-1/2"		26	.308		164	18.05		182.05	208
5110	2"		24	.333		211	19.55		230.55	263
5120	3"		20	.400		430	23.50		453.50	505
5130	4"		18	.444		780	26		806	895
5140	For flanged, add					25%	10%			
5280	Needle valve, PVC, threaded									
5300	1/4"	1 Plum	26	.308	Ea.	55.50	18.05		73.55	89
5340	3/8"		26	.308		64	18.05		82.05	98.50
5360	1/2"		26	.308		64	18.05		82.05	98.50
5380	For polypropylene, add					10%				
5800	Y check, PVC, socket or threaded									
5820	1/2"	1 Plum	26	.308	Ea.	55	18.05		73.05	88.50
5840	3/4"		25	.320		89	18.80		107.80	127
5850	1"		23	.348		96.50	20.50		117	138
5860	1-1/4"		21	.381		151	22.50		173.50	202
5870	1-1/2"		20	.400		165	23.50		188.50	219
5880	2"		17	.471		205	27.50		232.50	269
5890	2-1/2"		15	.533		435	31.50		466.50	525
5900	3"	Q-1	24	.667		435	35		470	530
5910	4"	"	20	.800		710	42.50		752.50	845
5960	For PVC flanged, add					45%	15%			
6350	Y sediment strainer, PVC, socket or threaded									
6400	1/2"	1 Plum	26	.308	Ea.	55	18.05		73.05	88.50
6440	3/4"		24	.333		58.50	19.55		78.05	94.50
6450	1"		23	.348		69.50	20.50		90	109
6460	1-1/4"		21	.381		117	22.50		139.50	163
6470	1-1/2"		20	.400		117	23.50		140.50	165
6480	2"		17	.471		144	27.50		171.50	201
6490	2-1/2"		15	.533		350	31.50		381.50	435
6500	3"	Q-1	24	.667		350	35		385	440
6510	4"	"	20	.800		585	42.50		627.50	705
6560	For PVC, flanged, add					55%	15%			
9000	Minimum labor/equipment charge	1 Plum	3.50	2.286	Job		134		134	209

22 05 29.10 Hangers & Supp. for Plumb'g/HVAC Pipe/Equip.		Crew	Daily Output	Labor-Hours	Unit	2015 Bare Costs Material	Labor	Equipment	Total	Total Incl O&P
0010	**HANGERS AND SUPPORTS FOR PLUMB'G/HVAC PIPE/EQUIP.**									
0011	TYPE numbers per MSS-SP58									
0050	Brackets									
0060	Beam side or wall, malleable iron, TYPE 34									
0070	3/8" threaded rod size	1 Plum	48	.167	Ea.	3.11	9.80		12.91	18.65
0080	1/2" threaded rod size		48	.167		4.39	9.80		14.19	20
0090	5/8" threaded rod size		48	.167		8.20	9.80		18	24.50
0100	3/4" threaded rod size		48	.167		11.30	9.80		21.10	27.50
0110	7/8" threaded rod size		48	.167		11.70	9.80		21.50	28
0120	For concrete installation, add						30%			
0150	Wall, welded steel, medium, TYPE 32									
0160	0 size, 12" wide, 18" deep	1 Plum	34	.235	Ea.	172	13.80		185.80	211
0170	1 size, 18" wide, 24" deep		34	.235		211	13.80		224.80	254
0180	2 size, 24" wide, 30" deep		34	.235		290	13.80		303.80	340
0300	Clamps									
0310	C-clamp, for mounting on steel beam flange, w/locknut, TYPE 23									
0320	3/8" threaded rod size	1 Plum	160	.050	Ea.	2.05	2.94		4.99	6.85
0330	1/2" threaded rod size		160	.050		2.20	2.94		5.14	7
0340	5/8" threaded rod size		160	.050		3.70	2.94		6.64	8.65
0350	3/4" threaded rod size		160	.050		4.55	2.94		7.49	9.60
0400	High temperature to 1050°F, alloy steel									
0410	4" pipe size	Q-1	106	.151	Ea.	26.50	7.95		34.45	42
0420	6" pipe size		106	.151		45.50	7.95		53.45	62.50
0430	8" pipe size		97	.165		50.50	8.70		59.20	69
0440	10" pipe size		84	.190		74	10.05		84.05	97
0450	12" pipe size		72	.222		83	11.75		94.75	110
0460	14" pipe size		64	.250		209	13.20		222.20	251
0470	16" pipe size		56	.286		230	15.10		245.10	276
0500	I-beam, for mounting on bottom flange, strap iron, TYPE 21									
0510	2" flange size	1 Plum	96	.083	Ea.	17.65	4.89		22.54	27
0520	3" flange size		95	.084		21	4.94		25.94	30.50
0530	4" flange size		93	.086		5.75	5.05		10.80	14.20
0540	5" flange size		92	.087		6.30	5.10		11.40	14.90
0550	6" flange size		90	.089		7.15	5.20		12.35	16
0560	7" flange size		88	.091		7.95	5.35		13.30	17.05
0570	8" flange size		86	.093		8.55	5.45		14	17.90
0600	One hole, vertical mounting, malleable iron									
0610	1/2" pipe size	1 Plum	160	.050	Ea.	1.08	2.94		4.02	5.75
0620	3/4" pipe size		145	.055		1.17	3.24		4.41	6.35
0630	1" pipe size		136	.059		1.24	3.45		4.69	6.75
0640	1-1/4" pipe size		128	.063		2.19	3.67		5.86	8.10
0650	1-1/2" pipe size		120	.067		2.52	3.91		6.43	8.85
0660	2" pipe size		112	.071		3.64	4.19		7.83	10.55
0670	2-1/2" pipe size		104	.077		7.05	4.52		11.57	14.85
0680	3" pipe size		96	.083		8.55	4.89		13.44	17.05
0690	3-1/2" pipe size		90	.089		10.45	5.20		15.65	19.65
0700	4" pipe size		84	.095		11.90	5.60		17.50	22
0750	Riser or extension pipe, carbon steel, TYPE 8									
0760	3/4" pipe size	1 Plum	48	.167	Ea.	2.52	9.80		12.32	18
0770	1" pipe size		47	.170		2.64	10		12.64	18.50
0780	1-1/4" pipe size		46	.174		3.21	10.20		13.41	19.45
0790	1-1/2" pipe size		45	.178		3.42	10.45		13.87	20

22 05 29 – Hangers and Supports for Plumbing Piping and Equipment

22 05 29.10 Hangers & Supp. for Plumb'g/HVAC Pipe/Equip.		Crew	Daily Output	Labor-Hours	Unit	Material	Labor	Equipment	Total	Total Incl O&P
						2015 Bare Costs				
0800	2" pipe size	1 Plum	43	.186	Ea.	3.54	10.90		14.44	21
0810	2-1/2" pipe size		41	.195		3.68	11.45		15.13	22
0820	3" pipe size		40	.200		4	11.75		15.75	22.50
0830	3-1/2" pipe size		39	.205		4.50	12.05		16.55	24
0840	4" pipe size		38	.211		5.45	12.35		17.80	25.50
0850	5" pipe size		37	.216		6.80	12.70		19.50	27.50
0860	6" pipe size		36	.222		8.35	13.05		21.40	29.50
0870	8" pipe size		34	.235		14.60	13.80		28.40	37.50
0880	10" pipe size		32	.250		22.50	14.70		37.20	48
0890	12" pipe size		28	.286		30	16.75		46.75	59
0900	For plastic coating 3/4" to 4", add					190%				
0910	For copper plating 3/4" to 4", add					58%				
0950	Two piece, complete, carbon steel, medium weight, TYPE 4									
0960	1/2" pipe size	Q-1	137	.117	Ea.	2.29	6.15		8.44	12.10
0970	3/4" pipe size		134	.119		2.29	6.30		8.59	12.35
0980	1" pipe size		132	.121		2.59	6.40		8.99	12.85
0990	1-1/4" pipe size		130	.123		2.80	6.50		9.30	13.25
1000	1-1/2" pipe size		126	.127		3.64	6.70		10.34	14.45
1010	2" pipe size		124	.129		4.48	6.80		11.28	15.60
1020	2-1/2" pipe size		120	.133		4.84	7.05		11.89	16.30
1030	3" pipe size		117	.137		5.35	7.20		12.55	17.15
1040	3-1/2" pipe size		114	.140		10	7.40		17.40	22.50
1050	4" pipe size		110	.145		7.10	7.70		14.80	19.80
1060	5" pipe size		106	.151		17.40	7.95		25.35	31.50
1070	6" pipe size		104	.154		19.20	8.15		27.35	33.50
1080	8" pipe size		100	.160		23.50	8.45		31.95	38.50
1090	10" pipe size		96	.167		46	8.80		54.80	64.50
1100	12" pipe size		89	.180		55.50	9.50		65	76.50
1110	14" pipe size		82	.195		73.50	10.30		83.80	97
1120	16" pipe size		68	.235		95	12.45		107.45	124
1130	For galvanized, add					45%				
1150	Insert, concrete									
1160	Wedge type, carbon steel body, malleable iron nut, galvanized									
1170	1/4" threaded rod size	1 Plum	96	.083	Ea.	9.55	4.89		14.44	18.15
1180	3/8" threaded rod size		96	.083		10.30	4.89		15.19	19
1190	1/2" threaded rod size		96	.083		10.65	4.89		15.54	19.40
1200	5/8" threaded rod size		96	.083		11.25	4.89		16.14	20
1210	3/4" threaded rod size		96	.083		11.95	4.89		16.84	21
1220	7/8" threaded rod size		96	.083		12.35	4.89		17.24	21.50
1250	Pipe guide sized for insulation									
1260	No. 1, 1" pipe size, 1" thick insulation	1 Stpi	26	.308	Ea.	124	18.40		142.40	165
1270	No. 2, 1-1/4"-2" pipe size, 1" thick insulation		23	.348		151	21		172	199
1280	No. 3, 1-1/4"-2" pipe size, 1-1/2" thick insulation		21	.381		151	23		174	202
1290	No. 4, 2-1/2"-3-1/2" pipe size, 1-1/2" thick insulation		18	.444		151	26.50		177.50	208
1300	No. 5, 4"-5" pipe size, 1-1/2" thick insulation		16	.500		171	30		201	236
1310	No. 6, 5"-6" pipe size, 2" thick insulation	Q-5	21	.762		199	41		240	283
1320	No. 7, 8" pipe size, 2" thick insulation		16	1		251	54		305	360
1330	No. 8, 10" pipe size, 2" thick insulation		12	1.333		370	71.50		441.50	520
1340	No. 9, 12" pipe size, 2" thick insulation	Q-6	17	1.412		370	78.50		448.50	535
1350	No. 10, 12"-14" pipe size, 2-1/2" thick insulation		16	1.500		435	83.50		518.50	610
1360	No. 11, 16" pipe size, 2-1/2" thick insulation		10.50	2.286		435	127		562	680
1370	No. 12, 16"-18" pipe size, 3" thick insulation		9	2.667		640	149		789	935
1380	No. 13, 20" pipe size, 3" thick insulation		7.50	3.200		640	178		818	985

22 05 Common Work Results for Plumbing

22 05 29 – Hangers and Supports for Plumbing Piping and Equipment

22 05 29.10 Hangers & Supp. for Plumb'g/HVAC Pipe/Equip.		Crew	Daily Output	Labor-Hours	Unit	Material	2015 Bare Costs Labor	Equipment	Total	Total Incl O&P
1390	No. 14, 24" pipe size, 3" thick insulation	Q-6	7	3.429	Ea.	845	191		1,036	1,225
1400	Bands									
1410	Adjustable band, carbon steel, for non-insulated pipe, TYPE 7									
1420	1/2" pipe size	Q-1	142	.113	Ea.	.30	5.95		6.25	9.65
1430	3/4" pipe size		140	.114		.30	6.05		6.35	9.75
1440	1" pipe size		137	.117		.30	6.15		6.45	9.95
1450	1-1/4" pipe size		134	.119		.33	6.30		6.63	10.20
1460	1-1/2" pipe size		131	.122		.33	6.45		6.78	10.40
1470	2" pipe size		129	.124		.33	6.55		6.88	10.55
1480	2-1/2" pipe size		125	.128		.60	6.75		7.35	11.20
1490	3" pipe size		122	.131		.64	6.95		7.59	11.50
1500	3-1/2" pipe size		119	.134		1.04	7.10		8.14	12.25
1510	4" pipe size		114	.140		1.04	7.40		8.44	12.70
1520	5" pipe size		110	.145		1.66	7.70		9.36	13.85
1530	6" pipe size		108	.148		1.94	7.85		9.79	14.35
1540	8" pipe size		104	.154		2.93	8.15		11.08	15.90
1550	For copper plated, add					50%				
1560	For galvanized, add					30%				
1570	For plastic coating, add					30%				
1600	Adjusting nut malleable iron, steel band, TYPE 9									
1610	1/2" pipe size, galvanized band	Q-1	137	.117	Ea.	4.88	6.15		11.03	14.95
1620	3/4" pipe size, galvanized band		135	.119		4.88	6.25		11.13	15.10
1630	1" pipe size, galvanized band		132	.121		5.05	6.40		11.45	15.55
1640	1-1/4" pipe size, galvanized band		129	.124		5.05	6.55		11.60	15.75
1650	1-1/2" pipe size, galvanized band		126	.127		5.20	6.70		11.90	16.20
1660	2" pipe size, galvanized band		124	.129		5.45	6.80		12.25	16.65
1670	2-1/2" pipe size, galvanized band		120	.133		8.60	7.05		15.65	20.50
1680	3" pipe size, galvanized band		117	.137		9	7.20		16.20	21
1690	3-1/2" pipe size, galvanized band		114	.140		35.50	7.40		42.90	50.50
1700	4" pipe size, cadmium plated band		110	.145		23	7.70		30.70	37
1740	For plastic coated band, add					35%				
1750	For completely copper coated, add					45%				
1800	Clevis, adjustable, carbon steel, for non-insulated pipe, TYPE 1									
1810	1/2" pipe size	Q-1	137	.117	Ea.	.93	6.15		7.08	10.60
1820	3/4" pipe size		135	.119		.93	6.25		7.18	10.75
1830	1" pipe size		132	.121		.98	6.40		7.38	11.10
1840	1-1/4" pipe size		129	.124		1.11	6.55		7.66	11.40
1850	1-1/2" pipe size		126	.127		1.22	6.70		7.92	11.80
1860	2" pipe size		124	.129		1.43	6.80		8.23	12.20
1870	2-1/2" pipe size		120	.133		2.39	7.05		9.44	13.65
1880	3" pipe size		117	.137		2.58	7.20		9.78	14.10
1890	3-1/2" pipe size		114	.140		3	7.40		10.40	14.85
1900	4" pipe size		110	.145		3.24	7.70		10.94	15.55
1910	5" pipe size		106	.151		4.48	7.95		12.43	17.40
1920	6" pipe size		104	.154		5.45	8.15		13.60	18.65
1930	8" pipe size		100	.160		7.95	8.45		16.40	22
1940	10" pipe size		96	.167		14.20	8.80		23	29.50
1950	12" pipe size		89	.180		18.40	9.50		27.90	35
1960	14" pipe size		82	.195		31	10.30		41.30	50
1970	16" pipe size		68	.235		49.50	12.45		61.95	74
1980	For galvanized, add					66%				
1990	For copper plated 1/2" to 4", add					77%				
2000	For light weight 1/2" to 4", deduct					13%				

22 05 29 – Hangers and Supports for Plumbing Piping and Equipment

22 05 29.10 Hangers & Supp. for Plumb'g/HVAC Pipe/Equip.		Crew	Daily Output	Labor-Hours	Unit	Material	2015 Bare Costs Labor	Equipment	Total	Total Incl O&P
2010	Insulated pipe type, 3/4" to 12" pipe, add					180%				
2020	Insulated pipe type, chrome-moly U-strap, add					530%				
2250	Split ring, malleable iron, for non-insulated pipe, TYPE 11									
2260	1/2" pipe size	Q-1	137	.117	Ea.	4.39	6.15		10.54	14.45
2270	3/4" pipe size		135	.119		4.43	6.25		10.68	14.60
2280	1" pipe size		132	.121		4.70	6.40		11.10	15.15
2290	1-1/4" pipe size		129	.124		5.80	6.55		12.35	16.60
2300	1-1/2" pipe size		126	.127		7.05	6.70		13.75	18.25
2310	2" pipe size		124	.129		8	6.80		14.80	19.45
2320	2-1/2" pipe size		120	.133		11.55	7.05		18.60	23.50
2330	3" pipe size		117	.137		14.45	7.20		21.65	27
2340	3-1/2" pipe size		114	.140		15.05	7.40		22.45	28
2350	4" pipe size		110	.145		15.05	7.70		22.75	28.50
2360	5" pipe size		106	.151		20	7.95		27.95	34.50
2370	6" pipe size		104	.154		44.50	8.15		52.65	61.50
2380	8" pipe size	↓	100	.160	↓	65.50	8.45		73.95	85
2390	For copper plated, add					8%				
2532	Turnbuckle, TYPE 13									
2534	3/8"	1 Plum	80	.100	Ea.	4.82	5.85		10.67	14.45
2535	1/2"		72	.111		5.45	6.50		11.95	16.15
2536	5/8"		64	.125		9.45	7.35		16.80	22
2537	3/4"		56	.143		12.40	8.40		20.80	26.50
2538	7/8"		48	.167		29.50	9.80		39.30	48
2539	1"		40	.200		42	11.75		53.75	64.50
2540	1-1/4"	↓	32	.250	↓	78	14.70		92.70	109
2650	Rods, carbon steel									
2660	Continuous thread									
2670	1/4" thread size	1 Plum	144	.056	L.F.	1.70	3.26		4.96	6.95
2680	3/8" thread size		144	.056		1.81	3.26		5.07	7.10
2690	1/2" thread size		144	.056		2.86	3.26		6.12	8.25
2700	5/8" thread size		144	.056		4.05	3.26		7.31	9.55
2710	3/4" thread size		144	.056		7.15	3.26		10.41	12.95
2720	7/8" thread size		144	.056		8.95	3.26		12.21	14.95
2725	1/4" thread size, bright finish		144	.056		1.24	3.26		4.50	6.45
2726	1/2" thread size, bright finish	↓	144	.056	↓	4.23	3.26		7.49	9.75
2730	For galvanized, add					40%				
2860	Pipe hanger assy, adj. clevis, saddle, rod, clamp, insul. allowance									
2864	1/2" pipe size	Q-5	35	.457	Ea.	30	24.50		54.50	71.50
2866	3/4" pipe size		34.80	.460		31	24.50		55.50	73
2868	1" pipe size		34.60	.462		18.50	25		43.50	59.50
2869	1-1/4" pipe size		34.30	.466		18.90	25		43.90	60
2870	1-1/2" pipe size		33.90	.472		19.20	25.50		44.70	60.50
2872	2" pipe size		33.30	.480		21.50	26		47.50	64
2874	2-1/2" pipe size		32.30	.495		25	26.50		51.50	69
2876	3" pipe size		31.20	.513		30	27.50		57.50	76
2880	4" pipe size		30.70	.521		32.50	28		60.50	79
2884	6" pipe size		29.80	.537		41.50	29		70.50	90.50
2888	8" pipe size		28	.571		52.50	30.50		83	106
2892	10" pipe size		25.20	.635		59.50	34		93.50	119
2896	12" pipe size	↓	23.20	.690	↓	88.50	37		125.50	155
2900	Rolls									
2910	Adjustable yoke, carbon steel with CI roll, TYPE 43									
2920	2-1/2" pipe size	Q-1	137	.117	Ea.	10.60	6.15		16.75	21.50

22 05 29 – Hangers and Supports for Plumbing Piping and Equipment

22 05 29.10 Hangers & Supp. for Plumb'g/HVAC Pipe/Equip.		Crew	Daily Output	Labor-Hours	Unit	2015 Bare Costs Material	Labor	Equipment	Total	Total Incl O&P
2930	3" pipe size	Q-1	131	.122	Ea.	11.25	6.45		17.70	22.50
2940	3-1/2" pipe size		124	.129		15.80	6.80		22.60	28
2950	4" pipe size		117	.137		15.80	7.20		23	28.50
2960	5" pipe size		110	.145		23	7.70		30.70	37
2970	6" pipe size		104	.154		30.50	8.15		38.65	46
2980	8" pipe size		96	.167		43.50	8.80		52.30	62
2990	10" pipe size		80	.200		53.50	10.55		64.05	75.50
3000	12" pipe size		68	.235		85.50	12.45		97.95	113
3010	14" pipe size		56	.286		162	15.10		177.10	202
3020	16" pipe size	↓	48	.333	↓	195	17.60		212.60	242
3050	Chair, carbon steel with CI roll									
3060	2" pipe size	1 Plum	68	.118	Ea.	11.80	6.90		18.70	23.50
3070	2-1/2" pipe size		65	.123		12.80	7.20		20	25.50
3080	3" pipe size		62	.129		13.80	7.55		21.35	27
3090	3-1/2" pipe size		60	.133		17.80	7.85		25.65	32
3100	4" pipe size		58	.138		16.25	8.10		24.35	30.50
3110	5" pipe size		56	.143		19.10	8.40		27.50	34
3120	6" pipe size		53	.151		25.50	8.85		34.35	42
3130	8" pipe size		50	.160		34.50	9.40		43.90	52.50
3140	10" pipe size		48	.167		47	9.80		56.80	67.50
3150	12" pipe size	↓	46	.174		72	10.20		82.20	95.50
3170	Single pipe roll, (see line 2650 for rods), TYPE 41, 1" pipe size	Q-1	137	.117		7.90	6.15		14.05	18.30
3180	1-1/4" pipe size		131	.122		8.15	6.45		14.60	19.05
3190	1-1/2" pipe size		129	.124		8.45	6.55		15	19.50
3200	2" pipe size		124	.129		8.75	6.80		15.55	20.50
3210	2-1/2" pipe size		118	.136		9.10	7.15		16.25	21
3220	3" pipe size		115	.139		9.40	7.35		16.75	22
3230	3-1/2" pipe size		113	.142		10.30	7.50		17.80	23
3240	4" pipe size		112	.143		10.60	7.55		18.15	23.50
3250	5" pipe size		110	.145		12.10	7.70		19.80	25.50
3260	6" pipe size		101	.158		26	8.35		34.35	41.50
3270	8" pipe size		90	.178		26.50	9.40		35.90	44
3280	10" pipe size		80	.200		35	10.55		45.55	55
3290	12" pipe size		68	.235		57.50	12.45		69.95	82.50
3291	14" pipe size		56	.286		87	15.10		102.10	120
3292	16" pipe size		48	.333		112	17.60		129.60	151
3293	18" pipe size		40	.400		132	21		153	178
3294	20" pipe size		35	.457		156	24		180	210
3296	24" pipe size		30	.533		251	28		279	320
3297	30" pipe size		25	.640		475	34		509	575
3298	36" pipe size	↓	20	.800	↓	580	42.50		622.50	705
3300	Saddles (add vertical pipe riser, usually 3" diameter)									
3310	Pipe support, complete, adjust., CI saddle, TYPE 36									
3320	2-1/2" pipe size	1 Plum	96	.083	Ea.	92.50	4.89		97.39	110
3330	3" pipe size		88	.091		94	5.35		99.35	112
3340	3-1/2" pipe size		79	.101		95.50	5.95		101.45	114
3350	4" pipe size		68	.118		139	6.90		145.90	164
3360	5" pipe size		64	.125		142	7.35		149.35	167
3370	6" pipe size		59	.136		142	7.95		149.95	168
3380	8" pipe size		53	.151		147	8.85		155.85	176
3390	10" pipe size		50	.160		171	9.40		180.40	203
3400	12" pipe size	↓	48	.167	↓	181	9.80		190.80	214
3450	For standard pipe support, one piece, CI deduct					34%				

22 05 29 – Hangers and Supports for Plumbing Piping and Equipment

22 05 29.10 Hangers & Supp. for Plumb'g/HVAC Pipe/Equip.		Crew	Daily Output	Labor-Hours	Unit	Material	2015 Bare Costs Labor	Equipment	Total	Total Incl O&P
3460	For stanchion support, CI with steel yoke, deduct					60%				
3550	Insulation shield 1" thick, 1/2" pipe size, TYPE 40	1 Asbe	100	.080	Ea.	2.74	4.19		6.93	9.75
3560	3/4" pipe size		100	.080		3.06	4.19		7.25	10.10
3570	1" pipe size		98	.082		3.30	4.27		7.57	10.50
3580	1-1/4" pipe size		98	.082		3.52	4.27		7.79	10.70
3590	1-1/2" pipe size		96	.083		3.45	4.36		7.81	10.80
3600	2" pipe size		96	.083		3.65	4.36		8.01	11
3610	2-1/2" pipe size		94	.085		3.77	4.46		8.23	11.30
3620	3" pipe size		94	.085		4.53	4.46		8.99	12.15
3630	2" thick, 3-1/2" pipe size		92	.087		4.84	4.55		9.39	12.60
3640	4" pipe size		92	.087		6.75	4.55		11.30	14.75
3650	5" pipe size		90	.089		9.15	4.65		13.80	17.55
3660	6" pipe size		90	.089		10.45	4.65		15.10	18.95
3670	8" pipe size		88	.091		17	4.76		21.76	26.50
3680	10" pipe size		88	.091		21.50	4.76		26.26	31
3690	12" pipe size		86	.093		24.50	4.87		29.37	35
3700	14" pipe size		86	.093		40.50	4.87		45.37	52.50
3710	16" pipe size		84	.095		43.50	4.99		48.49	55.50
3720	18" pipe size		84	.095		48	4.99		52.99	61
3730	20" pipe size		82	.098		53	5.10		58.10	66
3732	24" pipe size	↓	80	.100	↓	62	5.25		67.25	77
3750	Covering protection saddle, TYPE 39									
3760	1" covering size									
3770	3/4" pipe size	1 Plum	68	.118	Ea.	4.61	6.90		11.51	15.80
3780	1" pipe size		68	.118		4.61	6.90		11.51	15.80
3790	1-1/4" pipe size		68	.118		4.61	6.90		11.51	15.80
3800	1-1/2" pipe size		66	.121		4.92	7.10		12.02	16.50
3810	2" pipe size		66	.121		4.92	7.10		12.02	16.50
3820	2-1/2" pipe size		64	.125		4.92	7.35		12.27	16.85
3830	3" pipe size		64	.125		6.60	7.35		13.95	18.70
3840	3-1/2" pipe size		62	.129		7.30	7.55		14.85	19.85
3850	4" pipe size		62	.129		7.30	7.55		14.85	19.85
3860	5" pipe size		60	.133		7.30	7.85		15.15	20.50
3870	6" pipe size	↓	60	.133	↓	8.75	7.85		16.60	22
3900	1-1/2" covering size									
3910	3/4" pipe size	1 Plum	68	.118	Ea.	5.50	6.90		12.40	16.80
3920	1" pipe size		68	.118		5.50	6.90		12.40	16.80
3930	1-1/4" pipe size		68	.118		5.50	6.90		12.40	16.80
3940	1-1/2" pipe size		66	.121		5.50	7.10		12.60	17.15
3950	2" pipe size		66	.121		5.95	7.10		13.05	17.65
3960	2-1/2" pipe size		64	.125		7.20	7.35		14.55	19.35
3970	3" pipe size		64	.125		7.20	7.35		14.55	19.35
3980	3-1/2" pipe size		62	.129		7.60	7.55		15.15	20
3990	4" pipe size		62	.129		7.60	7.55		15.15	20
4000	5" pipe size		60	.133		7.60	7.85		15.45	20.50
4010	6" pipe size		60	.133		10.25	7.85		18.10	23.50
4020	8" pipe size	↓	58	.138	↓	12.85	8.10		20.95	27
4028	2" covering size									
4029	2-1/2" pipe size	1 Plum	62	.129	Ea.	7.25	7.55		14.80	19.80
4032	3" pipe size		60	.133		8.55	7.85		16.40	21.50
4033	4" pipe size		58	.138		8.55	8.10		16.65	22
4034	6" pipe size		56	.143		12.25	8.40		20.65	26.50
4035	8" pipe size	↓	54	.148	↓	14.40	8.70		23.10	29.50

22 05 29 – Hangers and Supports for Plumbing Piping and Equipment

22 05 29.10 Hangers & Supp. for Plumb'g/HVAC Pipe/Equip.		Crew	Daily Output	Labor-Hours	Unit	2015 Bare Costs Material	Labor	Equipment	Total	Total Incl O&P
4080	10" pipe size	1 Plum	58	.138	Ea.	15.35	8.10		23.45	29.50
4090	12" pipe size		56	.143		35	8.40		43.40	51.50
4100	14" pipe size		56	.143		35	8.40		43.40	51.50
4110	16" pipe size		54	.148		45.50	8.70		54.20	63.50
4120	18" pipe size		54	.148		45.50	8.70		54.20	63.50
4130	20" pipe size		52	.154		50.50	9.05		59.55	70
4150	24" pipe size		50	.160		59.50	9.40		68.90	79.50
4160	30" pipe size		48	.167		65	9.80		74.80	87
4180	36" pipe size		45	.178		75.50	10.45		85.95	99.50
4200	Sockets									
4210	Rod end, malleable iron, TYPE 16									
4220	1/4" thread size	1 Plum	240	.033	Ea.	1.42	1.96		3.38	4.61
4230	3/8" thread size		240	.033		1.42	1.96		3.38	4.61
4240	1/2" thread size		230	.035		1.87	2.04		3.91	5.25
4250	5/8" thread size		225	.036		3.50	2.09		5.59	7.10
4260	3/4" thread size		220	.036		4.91	2.13		7.04	8.75
4270	7/8" thread size		210	.038		6.90	2.24		9.14	11.10
4290	Strap, 1/2" pipe size, TYPE 26	Q-1	142	.113		1.54	5.95		7.49	11
4300	3/4" pipe size		140	.114		1.61	6.05		7.66	11.15
4310	1" pipe size		137	.117		2.37	6.15		8.52	12.20
4320	1-1/4" pipe size		134	.119		2.41	6.30		8.71	12.50
4330	1-1/2" pipe size		131	.122		2.57	6.45		9.02	12.90
4340	2" pipe size		129	.124		2.59	6.55		9.14	13.05
4350	2-1/2" pipe size		125	.128		3.97	6.75		10.72	14.90
4360	3" pipe size		122	.131		5.20	6.95		12.15	16.55
4370	3-1/2" pipe size		119	.134		5.40	7.10		12.50	17.05
4380	4" pipe size		114	.140		5.70	7.40		13.10	17.80
4400	U-bolt, carbon steel									
4410	Standard, with nuts, TYPE 42									
4420	1/2" pipe size	1 Plum	160	.050	Ea.	.98	2.94		3.92	5.65
4430	3/4" pipe size		158	.051		1.03	2.97		4	5.75
4450	1" pipe size		152	.053		1.07	3.09		4.16	6
4460	1-1/4" pipe size		148	.054		1.35	3.17		4.52	6.45
4470	1-1/2" pipe size		143	.056		1.43	3.28		4.71	6.65
4480	2" pipe size		139	.058		1.55	3.38		4.93	6.95
4490	2-1/2" pipe size		134	.060		2.53	3.50		6.03	8.25
4500	3" pipe size		128	.063		2.86	3.67		6.53	8.85
4510	3-1/2" pipe size		122	.066		3.08	3.85		6.93	9.40
4520	4" pipe size		117	.068		3.12	4.01		7.13	9.70
4530	5" pipe size		114	.070		3.66	4.12		7.78	10.50
4540	6" pipe size		111	.072		6.95	4.23		11.18	14.25
4550	8" pipe size		109	.073		7.60	4.31		11.91	15.05
4560	10" pipe size		107	.075		13.20	4.39		17.59	21.50
4570	12" pipe size		104	.077		19.15	4.52		23.67	28
4580	For plastic coating on 1/2" thru 6" size, add					150%				
4700	U-hook, carbon steel, requires mounting screws or bolts									
4710	3/4" thru 2" pipe size									
4720	6" long	1 Plum	96	.083	Ea.	.73	4.89		5.62	8.45
4730	8" long		96	.083		1.06	4.89		5.95	8.80
4740	10" long		96	.083		1.64	4.89		6.53	9.45
4750	12" long		96	.083		2.24	4.89		7.13	10.10
4760	For copper plated, add					50%				
7000	Roof supports									

22 05 29 – Hangers and Supports for Plumbing Piping and Equipment

22 05 29.10 Hangers & Supp. for Plumb'g/HVAC Pipe/Equip.		Crew	Daily Output	Labor-Hours	Unit	2015 Bare Costs Material	Labor	Equipment	Total	Total Incl O&P
7006	Duct									
7010	Rectangular, open, 12" off roof									
7020	To 18" wide	Q-9	26	.615	Ea.	150	31		181	214
7030	To 24" wide	"	22	.727		175	36.50		211.50	251
7040	To 36" wide	Q-10	30	.800		200	42		242	286
7050	To 48" wide		28	.857		225	45		270	320
7060	To 60" wide		24	1		250	52		302	360
7100	Equipment									
7120	Equipment support	Q-5	20	.800	Ea.	65	43		108	139
7300	Pipe									
7310	Roller type									
7320	Up to 2-1/2" diam. pipe									
7324	3-1/2" off roof	Q-5	24	.667	Ea.	16	36		52	73.50
7326	Up to 10" off roof	"	20	.800	"	21	43		64	90
7340	2-1/2" to 3-1/2" diam. pipe									
7342	Up to 16" off roof	Q-5	18	.889	Ea.	45.50	48		93.50	125
7360	4" to 5" diam. pipe									
7362	Up to 12" off roof	Q-5	16	1	Ea.	65	54		119	156
7400	Strut/channel type									
7410	Up to 2-1/2" diam. pipe									
7424	3-1/2" off roof	Q-5	24	.667	Ea.	14	36		50	71.50
7426	Up to 10" off roof	"	20	.800	"	19	43		62	88
7440	Strut and roller type									
7452	2-1/2" to 3-1/2" diam. pipe									
7454	Up to 16" off roof	Q-5	18	.889	Ea.	65.50	48		113.50	147
7460	Strut and hanger type									
7470	Up to 3" diam. pipe									
7474	Up to 8" off roof	Q-5	19	.842	Ea.	41.50	45.50		87	117
8000	Pipe clamp, plastic, 1/2" CTS	1 Plum	80	.100		.23	5.85		6.08	9.40
8010	3/4" CTS		73	.110		.24	6.45		6.69	10.30
8020	1" CTS		68	.118		.54	6.90		7.44	11.35
8080	Economy clamp, 1/4" CTS		175	.046		.05	2.68		2.73	4.24
8090	3/8" CTS		168	.048		.07	2.80		2.87	4.44
8100	1/2" CTS		160	.050		.07	2.94		3.01	4.66
8110	3/4" CTS		145	.055		.15	3.24		3.39	5.20
8200	Half clamp, 1/2" CTS		80	.100		.08	5.85		5.93	9.25
8210	3/4" CTS		73	.110		.12	6.45		6.57	10.20
8300	Suspension clamp, 1/2" CTS		80	.100		.23	5.85		6.08	9.40
8310	3/4" CTS		73	.110		.24	6.45		6.69	10.30
8320	1" CTS		68	.118		.54	6.90		7.44	11.35
8400	Insulator, 1/2" CTS		80	.100		.37	5.85		6.22	9.55
8410	3/4" CTS		73	.110		.39	6.45		6.84	10.50
8420	1" CTS		68	.118		.40	6.90		7.30	11.20
8800	Wire cable support system									
8810	Cable with hook terminal and locking device									
8830	2 mm, (.079") dia cable, (100 lb. cap.)									
8840	1 m, (3.3') length, with hook	1 Shee	96	.083	Ea.	3.03	4.66		7.69	10.70
8850	2 m, (6.6') length, with hook		84	.095		3.50	5.35		8.85	12.25
8860	3 m, (9.9') length, with hook		72	.111		3.97	6.20		10.17	14.15
8870	5 m, (16.4') length, with hook	Q-9	60	.267		4.97	13.45		18.42	26.50
8880	10 m, (32.8') length, with hook	"	30	.533		7.20	27		34.20	50.50
8900	3mm, (.118") dia cable, (200 lb. cap.)									
8910	1 m, (3.3') length, with hook	1 Shee	96	.083	Ea.	3.88	4.66		8.54	11.60

22 05 29 – Hangers and Supports for Plumbing Piping and Equipment

22 05 29.10 Hangers & Supp. for Plumb'g/HVAC Pipe/Equip.		Crew	Daily Output	Labor-Hours	Unit	2015 Bare Costs Material	Labor	Equipment	Total	Total Incl O&P
8920	2 m, (6.6') length, with hook	1 Shee	84	.095	Ea.	4.36	5.35		9.71	13.20
8930	3 m, (9.9') length, with hook		72	.111		4.81	6.20		11.01	15.10
8940	5 m, (16.4') length, with hook	Q-9	60	.267		5.95	13.45		19.40	27.50
8950	10 m, (32.8') length, with hook	"	30	.533		8.55	27		35.55	52
9000	Cable system accessories									
9010	Anchor bolt, 3/8", with nut	1 Shee	140	.057	Ea.	1.16	3.20		4.36	6.35
9020	Air duct corner protector		160	.050		.52	2.80		3.32	4.99
9030	Air duct support attachment		140	.057		.99	3.20		4.19	6.15
9040	Flange clip, hammer-on style									
9044	For flange thickness 3/32" - 9/64", 160 lb. cap.	1 Shee	180	.044	Ea.	.29	2.49		2.78	4.24
9048	For flange thickness 1/8" - 1/4", 200 lb. cap.		160	.050		.29	2.80		3.09	4.74
9052	For flange thickness 5/16" - 1/2", 200 lb. cap.		150	.053		.56	2.98		3.54	5.35
9056	For flange thickness 9/16" - 3/4", 200 lb. cap.		140	.057		.75	3.20		3.95	5.90
9060	Wire insulation protection tube		180	.044	L.F.	.34	2.49		2.83	4.29
9070	Wire cutter				Ea.	41.50			41.50	45.50

22 05 33 – Heat Tracing for Plumbing Piping

22 05 33.20 Temperature Maintenance Cable		Crew	Daily Output	Labor-Hours	Unit	Material	Labor	Equipment	Total	Total Incl O&P
0010	**TEMPERATURE MAINTENANCE CABLE**									
0040	Components									
0080	Heating cable									
0100	208 V									
0150	140°F	Q-1	1060.80	.015	L.F.	9.20	.80		10	11.40
0200	120 V									
0220	125°F	Q-1	1060.80	.015	L.F.	8.10	.80		8.90	10.15
0300	Power kit w/1 end seal	1 Elec	48.80	.164	Ea.	76.50	8.95		85.45	98
0310	Splice kit		35.50	.225		90.50	12.35		102.85	119
0320	End seal		160	.050		8.80	2.74		11.54	13.95
0330	Tee kit w/1 end seal		26.80	.299		96	16.35		112.35	131
0340	Powered splice w/2 end seals		20	.400		102	22		124	146
0350	Powered tee kit w/3 end seals		18	.444		131	24.50		155.50	182
0360	Cross kit w/2 end seals		18.60	.430		135	23.50		158.50	186
0500	Recommended thickness of fiberglass insulation									
0510	Pipe size									
0520	1/2" - 1" use 1" insulation									
0530	1-1/4" - 2" use 1-1/2" insulation									
0540	2-1/2" - 6" use 2" insulation									
0560	NOTE: For pipe sizes 1-1/4" and smaller use 1/4" larger diameter									
0570	insulation to allow room for installation over cable.									

22 05 48 – Vibration and Seismic Controls for Plumbing Piping and Equipment

22 05 48.40 Vibration Absorbers		Crew	Daily Output	Labor-Hours	Unit	Material	Labor	Equipment	Total	Total Incl O&P
0010	**VIBRATION ABSORBERS**									
0100	Hangers, neoprene flex									
0200	10 – 120 lb. capacity				Ea.	26			26	28.50
0220	75 – 550 lb. capacity					37			37	40.50
0240	250 – 1100 lb. capacity					72			72	79
0260	1000 – 4000 lb. capacity					134			134	148
0500	Spring flex, 60 lb. capacity					36			36	39.50
0520	450 lb. capacity					80.50			80.50	88.50
0540	900 lb. capacity					72			72	79
0560	1100 – 1300 lb. capacity					72			72	79
0600	Rubber in shear									
0610	45 – 340 lb., up to 1/2" rod size	1 Stpi	22	.364	Ea.	14.30	21.50		35.80	50

22 05 48 – Vibration and Seismic Controls for Plumbing Piping and Equipment

22 05 48.40 Vibration Absorbers		Crew	Daily Output	Labor-Hours	Unit	2015 Bare Costs Material	Labor	Equipment	Total	Total Incl O&P
0620	130 – 700 lb., up to 3/4" rod size	1 Stpi	20	.400	Ea.	31	24		55	71.50
0630	50 – 1000 lb., up to 3/4" rod size	↓	18	.444		36.50	26.50		63	81.50
1000	Mounts, neoprene, 45 – 380 lb. capacity					17.25			17.25	19
1020	250 – 1100 lb. capacity					49.50			49.50	54
1040	1000 – 4000 lb. capacity					111			111	122
1100	Spring flex, 60 lb. capacity					70.50			70.50	78
1120	165 lb. capacity					72			72	79
1140	260 lb. capacity					74.50			74.50	82
1160	450 lb. capacity					96.50			96.50	106
1180	600 lb. capacity					99.50			99.50	110
1200	750 lb. capacity					110			110	121
1220	900 lb. capacity					108			108	119
1240	1100 lb. capacity					117			117	129
1260	1300 lb. capacity					117			117	129
1280	1500 lb. capacity					176			176	194
1300	1800 lb. capacity					189			189	208
1320	2200 lb. capacity					189			189	208
1340	2600 lb. capacity					197			197	216
2000	Pads, cork rib, 18" x 18" x 1", 10-50 psi					179			179	197
2020	18" x 36" x 1", 10-50 psi					350			350	385
2100	Shear flexible pads, 18" x 18" x 3/8", 20-70 psi					78			78	86
2120	18" x 36" x 3/8", 20-70 psi				↓	179			179	197
3000	Note overlap in capacities due to deflections									

22 05 53 – Identification for Plumbing Piping and Equipment

22 05 53.10 Piping System Identification Labels		Crew	Daily Output	Labor-Hours	Unit	Material	Labor	Equipment	Total	Total Incl O&P
0010	**PIPING SYSTEM IDENTIFICATION LABELS**									
0100	Indicate contents and flow direction									
0106	Pipe markers									
0110	Plastic snap around									
0114	1/2" pipe	1 Plum	80	.100	Ea.	6.50	5.85		12.35	16.30
0116	3/4" pipe		80	.100		6.50	5.85		12.35	16.30
0118	1" pipe		80	.100		6.50	5.85		12.35	16.30
0120	2" pipe		75	.107		8.10	6.25		14.35	18.65
0122	3" pipe		70	.114		12.40	6.70		19.10	24
0124	4" pipe		60	.133		12.40	7.85		20.25	26
0126	6" pipe		60	.133		13.20	7.85		21.05	26.50
0128	8" pipe		56	.143		17.40	8.40		25.80	32.50
0130	10" pipe		56	.143		17.40	8.40		25.80	32.50
0200	Over 10" pipe size	↓	50	.160	↓	21	9.40		30.40	37.50
1110	Self adhesive									
1114	1" pipe	1 Plum	80	.100	Ea.	3.20	5.85		9.05	12.65
1116	2" pipe		75	.107		3.20	6.25		9.45	13.25
1118	3" pipe		70	.114		5.10	6.70		11.80	16.05
1120	4" pipe		60	.133		5	7.85		12.85	17.70
1122	6" pipe		60	.133		5	7.85		12.85	17.70
1124	8" pipe		56	.143		7.20	8.40		15.60	21
1126	10" pipe		56	.143		7.20	8.40		15.60	21
1200	Over 10" pipe size	↓	50	.160	↓	10.20	9.40		19.60	26
2000	Valve tags									
2010	Numbered plus identifying legend									
2100	Brass, 2" diameter	1 Plum	40	.200	Ea.	3	11.75		14.75	21.50
2200	Plastic, 1-1/2" diam.	"	40	.200	"	2.70	11.75		14.45	21.50

22 05 76 – Facility Drainage Piping Cleanouts

22 05 76.10 Cleanouts		Crew	Daily Output	Labor-Hours	Unit	2015 Bare Costs Material	Labor	Equipment	Total	Total Incl O&P
0010	**CLEANOUTS**									
0060	Floor type									
0080	Round or square, scoriated nickel bronze top									
0100	2″ pipe size	1 Plum	10	.800	Ea.	197	47		244	290
0120	3″ pipe size		8	1		295	58.50		353.50	415
0140	4″ pipe size		6	1.333		295	78.50		373.50	445
0160	5″ pipe size		4	2		370	117		487	595
0180	6″ pipe size	Q-1	6	2.667		375	141		516	635
0200	8″ pipe size	″	4	4		660	211		871	1,050
0340	Recessed for tile, same price									
0980	Round top, recessed for terrazzo									
1000	2″ pipe size	1 Plum	9	.889	Ea.	197	52		249	299
1080	3″ pipe size		6	1.333		295	78.50		373.50	445
1100	4″ pipe size		4	2		295	117		412	510
1120	5″ pipe size	Q-1	6	2.667		375	141		516	635
1140	6″ pipe size		5	3.200		375	169		544	680
1160	8″ pipe size		4	4		660	211		871	1,050
2000	Round scoriated nickel bronze top, extra heavy duty									
2060	2″ pipe size	1 Plum	9	.889	Ea.	270	52		322	380
2080	3″ pipe size		6	1.333		370	78.50		448.50	525
2100	4″ pipe size		4	2		370	117		487	590
2120	5″ pipe size	Q-1	6	2.667		440	141		581	705
2140	6″ pipe size		5	3.200		440	169		609	750
2160	8″ pipe size		4	4		735	211		946	1,150
4000	Wall type, square smooth cover, over wall frame									
4060	2″ pipe size	1 Plum	14	.571	Ea.	320	33.50		353.50	410
4080	3″ pipe size		12	.667		345	39		384	440
4100	4″ pipe size		10	.800		370	47		417	485
4120	5″ pipe size		9	.889		540	52		592	675
4140	6″ pipe size		8	1		595	58.50		653.50	745
4160	8″ pipe size	Q-1	11	1.455		795	77		872	995
5000	Extension, C.I.; bronze countersunk plug, 8″ long									
5040	2″ pipe size	1 Plum	16	.500	Ea.	169	29.50		198.50	232
5060	3″ pipe size		14	.571		196	33.50		229.50	269
5080	4″ pipe size		13	.615		200	36		236	276
5100	5″ pipe size		12	.667		310	39		349	405
5120	6″ pipe size		11	.727		380	42.50		422.50	480
9000	Minimum labor/equipment charge		3	2.667	Job		157		157	244

22 05 76.20 Cleanout Tees

Line	Description	Crew	Daily Output	Labor-Hours	Unit	Material	Labor	Equipment	Total	Total Incl O&P
0010	**CLEANOUT TEES**									
0100	Cast iron, B&S, with countersunk plug									
0200	2″ pipe size	1 Plum	4	2	Ea.	267	117		384	475
0220	3″ pipe size		3.60	2.222		292	130		422	525
0240	4″ pipe size		3.30	2.424		365	142		507	620
0260	5″ pipe size	Q-1	5.50	2.909		790	154		944	1,100
0280	6″ pipe size	″	5	3.200		980	169		1,149	1,350
0300	8″ pipe size	Q-3	5	6.400		1,125	360		1,485	1,775
0500	For round smooth access cover, same price									
0600	For round scoriated access cover, same price									
0700	For square smooth access cover, add				Ea.	60%				
4000	Plastic, tees and adapters. Add plugs									
4010	ABS, DWV									

22 05 Common Work Results for Plumbing

22 05 76 – Facility Drainage Piping Cleanouts

22 05 76.20 Cleanout Tees		Crew	Daily Output	Labor-Hours	Unit	Material	2015 Bare Costs Labor	Equipment	Total	Total Incl O&P
4020	Cleanout tee, 1-1/2" pipe size	1 Plum	15	.533	Ea.	23	31.50		54.50	74
4030	2" pipe size	Q-1	27	.593		25	31.50		56.50	76.50
4040	3" pipe size		21	.762		47	40.50		87.50	115
4050	4" pipe size		16	1		101	53		154	194
4100	Cleanout plug, 1-1/2" pipe size	1 Plum	32	.250		3.94	14.70		18.64	27.50
4110	2" pipe size	Q-1	56	.286		5.20	15.10		20.30	29
4120	3" pipe size		36	.444		8.35	23.50		31.85	45.50
4130	4" pipe size		30	.533		14.65	28		42.65	60
4180	Cleanout adapter fitting, 1-1/2" pipe size	1 Plum	32	.250		6.25	14.70		20.95	30
4190	2" pipe size	Q-1	56	.286		9.60	15.10		24.70	34
4200	3" pipe size		36	.444		24.50	23.50		48	63.50
4210	4" pipe size		30	.533		45	28		73	93.50
5000	PVC, DWV									
5010	Cleanout tee, 1-1/2" pipe size	1 Plum	15	.533	Ea.	17.10	31.50		48.60	68
5020	2" pipe size	Q-1	27	.593		19.95	31.50		51.45	71
5030	3" pipe size		21	.762		35.50	40.50		76	102
5040	4" pipe size		16	1		69.50	53		122.50	159
5090	Cleanout plug, 1-1/2" pipe size	1 Plum	32	.250		4.08	14.70		18.78	27.50
5100	2" pipe size	Q-1	56	.286		4.53	15.10		19.63	28.50
5110	3" pipe size		36	.444		8.10	23.50		31.60	45.50
5120	4" pipe size		30	.533		11.95	28		39.95	57
5130	6" pipe size		24	.667		39	35		74	98
5170	Cleanout adapter fitting, 1-1/2" pipe size	1 Plum	32	.250		5.45	14.70		20.15	29
5180	2" pipe size	Q-1	56	.286		7	15.10		22.10	31
5190	3" pipe size		36	.444		19.75	23.50		43.25	58
5200	4" pipe size		30	.533		32.50	28		60.50	79.50
5210	6" pipe size		24	.667		94.50	35		129.50	159
9000	Minimum labor/equipment charge	1 Plum	2.75	2.909	Job		171		171	266

22 07 Plumbing Insulation

22 07 16 – Plumbing Equipment Insulation

22 07 16.10 Insulation for Plumbing Equipment

		Crew	Daily Output	Labor-Hours	Unit	Material	Labor	Equipment	Total	Total Incl O&P
0010	**INSULATION FOR PLUMBING EQUIPMENT**									
2900	Domestic water heater wrap kit									
2920	1-1/2" with vinyl jacket, 20-60 gal. G	1 Plum	8	1	Ea.	18.60	58.50		77.10	112
2925	50 to 80 gallons G	"	8	1	"	23.50	58.50		82	118
9000	Minimum labor/equipment charge	1 Stpi	4	2	Job		120		120	186

22 07 19 – Plumbing Piping Insulation

22 07 19.10 Piping Insulation

		Crew	Daily Output	Labor-Hours	Unit	Material	Labor	Equipment	Total	Total Incl O&P
0010	**PIPING INSULATION**									
0100	Rule of thumb, as a percentage of total mechanical costs				Job				10%	10%
0110	Insulation req'd. is based on the surface size/area to be covered									
0230	Insulated protectors, (ADA)									
0235	For exposed piping under sinks or lavatories									
0240	Vinyl coated foam, velcro tabs									
0245	P Trap, 1-1/4" or 1-1/2"	1 Plum	32	.250	Ea.	22.50	14.70		37.20	47.50
0260	Valve and supply cover									
0265	1/2", 3/8", and 7/16" pipe size	1 Plum	32	.250	Ea.	22	14.70		36.70	47
0280	Tailpiece offset (wheelchair)									
0285	1-1/4" pipe size	1 Plum	32	.250	Ea.	13.65	14.70		28.35	38

22 07 Plumbing Insulation

22 07 19 – Plumbing Piping Insulation

22 07 19.10 Piping Insulation			Crew	Daily Output	Labor-Hours	Unit	Material	Labor	Equipment	Total	Total Incl O&P
							2015 Bare Costs				
0600	Pipe covering (price copper tube one size less than IPS)										
4280	Cellular glass, closed cell foam, all service jacket, sealant,										
4281	working temp. (-450°F to +900°F), 0 water vapor transmission										
4284	1" wall										
4286	1/2" iron pipe size	G	Q-14	120	.133	L.F.	6.80	6.30		13.10	17.55
4300	1-1/2" wall,										
4301	1" iron pipe size	G	Q-14	105	.152	L.F.	10.40	7.20		17.60	23
4304	2-1/2" iron pipe size	G		90	.178		14.20	8.40		22.60	29
4306	3" iron pipe size	G		85	.188		17.60	8.85		26.45	33.50
4308	4" iron pipe size	G		70	.229		21.50	10.75		32.25	41
4310	5" iron pipe size	G		65	.246		22	11.60		33.60	42.50
4320	2" wall,										
4322	1" iron pipe size	G	Q-14	100	.160	L.F.	13.85	7.55		21.40	27.50
4324	2-1/2" iron pipe size	G		85	.188		21.50	8.85		30.35	38
4326	3" iron pipe size	G		80	.200		21.50	9.45		30.95	38.50
4328	4" iron pipe size	G		65	.246		26	11.60		37.60	47
4330	5" iron pipe size	G		60	.267		27	12.55		39.55	49.50
4332	6" iron pipe size	G		50	.320		31	15.10		46.10	58.50
4336	8" iron pipe size	G		40	.400		36.50	18.85		55.35	71
4338	10" iron pipe size	G		35	.457		45.50	21.50		67	84.50
4350	2-1/2" wall,										
4360	12" iron pipe size	G	Q-14	32	.500	L.F.	60.50	23.50		84	105
4362	14" iron pipe size	G	"	28	.571	"	67	27		94	117
4370	3" wall,										
4378	6" iron pipe size	G	Q-14	48	.333	L.F.	41.50	15.70		57.20	71
4380	8" iron pipe size	G		38	.421		52	19.85		71.85	89
4382	10" iron pipe size	G		33	.485		60	23		83	103
4384	16" iron pipe size	G		25	.640		84.50	30		114.50	142
4386	18" iron pipe size	G		22	.727		108	34.50		142.50	173
4388	20" iron pipe size	G		20	.800		117	37.50		154.50	190
4400	3-1/2" wall,										
4412	12" iron pipe size	G	Q-14	27	.593	L.F.	71.50	28		99.50	124
4414	14" iron pipe size	G	"	25	.640	"	74	30		104	130
4430	4" wall,										
4446	16" iron pipe size	G	Q-14	22	.727	L.F.	96	34.50		130.50	161
4448	18" iron pipe size	G		20	.800		105	37.50		142.50	176
4450	20" iron pipe size	G		18	.889		118	42		160	198
4480	Fittings, average with fabric and mastic										
4484	1" wall,										
4486	1/2" iron pipe size	G	1 Asbe	40	.200	Ea.	5.85	10.45		16.30	23.50
4500	1-1/2" wall,										
4502	1" iron pipe size	G	1 Asbe	38	.211	Ea.	7.80	11		18.80	26.50
4504	2-1/2" iron pipe size	G		32	.250		14.15	13.10		27.25	36.50
4506	3" iron pipe size	G		30	.267		14.15	13.95		28.10	38
4508	4" iron pipe size	G		28	.286		17.85	14.95		32.80	43.50
4510	5" iron pipe size	G		24	.333		22	17.45		39.45	52
4520	2" wall,										
4522	1" iron pipe size	G	1 Asbe	36	.222	Ea.	11.30	11.65		22.95	31
4524	2-1/2" iron pipe size	G		30	.267		17.85	13.95		31.80	42
4526	3" iron pipe size	G		28	.286		17.85	14.95		32.80	43.50
4528	4" iron pipe size	G		24	.333		22	17.45		39.45	52
4530	5" iron pipe size	G		22	.364		32.50	19.05		51.55	66
4532	6" iron pipe size	G		20	.400		45.50	21		66.50	83.50

22 07 19 – Plumbing Piping Insulation

22 07 19.10 Piping Insulation		Crew	Daily Output	Labor-Hours	Unit	Material	Labor	Equipment	Total	Total Incl O&P
						2015 Bare Costs				
4536	8" iron pipe size G	1 Asbe	12	.667	Ea.	76	35		111	140
4538	10" iron pipe size G	↓	8	1	↓	88.50	52.50		141	182
4550	2-1/2" wall,									
4560	12" iron pipe size G	1 Asbe	6	1.333	Ea.	162	70		232	290
4562	14" iron pipe size G	"	4	2	"	214	105		319	405
4570	3" wall,									
4578	6" iron pipe size G	1 Asbe	16	.500	Ea.	65.50	26		91.50	115
4580	8" iron pipe size G		10	.800		98	42		140	176
4582	10" iron pipe size G	↓	6	1.333	↓	133	70		203	258
4900	Calcium silicate, with 8 oz. canvas cover									
5100	1" wall, 1/2" iron pipe size G	Q-14	170	.094	L.F.	4.07	4.44		8.51	11.65
5130	3/4" iron pipe size G		170	.094		4.10	4.44		8.54	11.65
5140	1" iron pipe size G		170	.094		3.98	4.44		8.42	11.55
5150	1-1/4" iron pipe size G		165	.097		4.07	4.57		8.64	11.85
5160	1-1/2" iron pipe size G		165	.097		4.13	4.57		8.70	11.90
5170	2" iron pipe size G		160	.100		4.74	4.71		9.45	12.80
5180	2-1/2" iron pipe size G		160	.100		5.05	4.71		9.76	13.15
5190	3" iron pipe size G		150	.107		5.45	5.05		10.50	14.10
5200	4" iron pipe size G		140	.114		6.75	5.40		12.15	16.10
5210	5" iron pipe size G		135	.119		7.10	5.60		12.70	16.85
5220	6" iron pipe size G		130	.123		7.55	5.80		13.35	17.65
5280	1-1/2" wall, 1/2" iron pipe size G		150	.107		4.40	5.05		9.45	12.95
5310	3/4" iron pipe size G		150	.107		4.48	5.05		9.53	13.05
5320	1" iron pipe size G		150	.107		4.88	5.05		9.93	13.45
5330	1-1/4" iron pipe size G		145	.110		5.25	5.20		10.45	14.15
5340	1-1/2" iron pipe size G		145	.110		5.65	5.20		10.85	14.55
5350	2" iron pipe size G		140	.114		6.20	5.40		11.60	15.45
5360	2-1/2" iron pipe size G		140	.114		6.75	5.40		12.15	16.10
5370	3" iron pipe size G		135	.119		7.10	5.60		12.70	16.80
5380	4" iron pipe size G		125	.128		8.20	6.05		14.25	18.75
5390	5" iron pipe size G		120	.133		9.25	6.30		15.55	20.50
5400	6" iron pipe size G		110	.145		9.55	6.85		16.40	21.50
5460	2" wall, 1/2" iron pipe size G		135	.119		6.80	5.60		12.40	16.45
5490	3/4" iron pipe size G		135	.119		7.15	5.60		12.75	16.85
5500	1" iron pipe size G		135	.119		7.50	5.60		13.10	17.25
5510	1-1/4" iron pipe size G		130	.123		8.05	5.80		13.85	18.15
5520	1-1/2" iron pipe size G		130	.123		8.40	5.80		14.20	18.55
5530	2" iron pipe size G		125	.128		8.90	6.05		14.95	19.50
5540	2-1/2" iron pipe size G		125	.128		10.55	6.05		16.60	21.50
5550	3" iron pipe size G		120	.133		10.65	6.30		16.95	22
5560	4" iron pipe size G		115	.139		12.30	6.55		18.85	24
5570	5" iron pipe size G		110	.145		14	6.85		20.85	26.50
5580	6" iron pipe size G	↓	105	.152	↓	15.30	7.20		22.50	28.50
5600	Calcium silicate, no cover									
5720	1" wall, 1/2" iron pipe size G	Q-14	180	.089	L.F.	3.70	4.19		7.89	10.80
5740	3/4" iron pipe size G		180	.089		3.70	4.19		7.89	10.80
5750	1" iron pipe size G		180	.089		3.55	4.19		7.74	10.65
5760	1-1/4" iron pipe size G		175	.091		3.61	4.31		7.92	10.90
5770	1-1/2" iron pipe size G		175	.091		3.64	4.31		7.95	10.95
5780	2" iron pipe size G		170	.094		4.20	4.44		8.64	11.75
5790	2-1/2" iron pipe size G		170	.094		4.44	4.44		8.88	12.05
5800	3" iron pipe size G		160	.100		4.77	4.71		9.48	12.85
5810	4" iron pipe size G	↓	150	.107	↓	5.95	5.05		11	14.65

22 07 Plumbing Insulation

22 07 19 – Plumbing Piping Insulation

22 07 19.10 Piping Insulation			Crew	Daily Output	Labor-Hours	Unit	Material	2015 Bare Costs Labor	Equipment	Total	Total Incl O&P
5820	5″ iron pipe size	G	Q-14	145	.110	L.F.	6.20	5.20		11.40	15.15
5830	6″ iron pipe size	G		140	.114		6.50	5.40		11.90	15.80
5900	1-1/2″ wall, 1/2″ iron pipe size	G		160	.100		3.91	4.71		8.62	11.90
5920	3/4″ iron pipe size	G		160	.100		3.98	4.71		8.69	12
5930	1″ iron pipe size	G		160	.100		4.34	4.71		9.05	12.35
5940	1-1/4″ iron pipe size	G		155	.103		4.67	4.87		9.54	12.95
5950	1-1/2″ iron pipe size	G		155	.103		5	4.87		9.87	13.30
5960	2″ iron pipe size	G		150	.107		5.55	5.05		10.60	14.20
5970	2-1/2″ iron pipe size	G		150	.107		6.05	5.05		11.10	14.75
5980	3″ iron pipe size	G		145	.110		6.30	5.20		11.50	15.30
5990	4″ iron pipe size	G		135	.119		7.30	5.60		12.90	17.05
6000	5″ iron pipe size	G		130	.123		8.25	5.80		14.05	18.35
6010	6″ iron pipe size	G		120	.133		8.40	6.30		14.70	19.35
6020	7″ iron pipe size	G		115	.139		9.85	6.55		16.40	21.50
6030	8″ iron pipe size	G		105	.152		11.15	7.20		18.35	24
6040	9″ iron pipe size	G		100	.160		13.35	7.55		20.90	27
6050	10″ iron pipe size	G		95	.168		14.75	7.95		22.70	29
6060	12″ iron pipe size	G		90	.178		17.50	8.40		25.90	32.50
6070	14″ iron pipe size	G		85	.188		19.80	8.85		28.65	36.50
6080	16″ iron pipe size	G		80	.200		22	9.45		31.45	39.50
6090	18″ iron pipe size	G		75	.213		24.50	10.05		34.55	43
6120	2″ wall, 1/2″ iron pipe size	G		145	.110		6.20	5.20		11.40	15.15
6140	3/4″ iron pipe size	G		145	.110		6.50	5.20		11.70	15.50
6150	1″ iron pipe size	G		145	.110		6.85	5.20		12.05	15.90
6160	1-1/4″ iron pipe size	G		140	.114		7.35	5.40		12.75	16.75
6170	1-1/2″ iron pipe size	G		140	.114		7.70	5.40		13.10	17.10
6180	2″ iron pipe size	G		135	.119		8.05	5.60		13.65	17.85
6190	2-1/2″ iron pipe size	G		135	.119		9.70	5.60		15.30	19.65
6200	3″ iron pipe size	G		130	.123		9.75	5.80		15.55	20
6210	4″ iron pipe size	G		125	.128		11.25	6.05		17.30	22
6220	5″ iron pipe size	G		120	.133		12.85	6.30		19.15	24.50
6230	6″ iron pipe size	G		115	.139		14.05	6.55		20.60	26
6240	7″ iron pipe size	G		110	.145		15.20	6.85		22.05	28
6250	8″ iron pipe size	G		105	.152		16.75	7.20		23.95	30
6260	9″ iron pipe size	G		100	.160		18.90	7.55		26.45	33
6270	10″ iron pipe size	G		95	.168		21	7.95		28.95	36
6280	12″ iron pipe size	G		90	.178		23.50	8.40		31.90	39
6290	14″ iron pipe size	G		85	.188		25.50	8.85		34.35	42.50
6300	16″ iron pipe size	G		80	.200		28.50	9.45		37.95	46
6310	18″ iron pipe size	G		75	.213		31	10.05		41.05	50
6320	20″ iron pipe size	G		65	.246		38	11.60		49.60	60.50
6330	22″ iron pipe size	G		60	.267		42	12.55		54.55	66.50
6340	24″ iron pipe size	G		55	.291		43	13.70		56.70	69.50
6360	3″ wall, 1/2″ iron pipe size	G		115	.139		11.20	6.55		17.75	23
6380	3/4″ iron pipe size	G		115	.139		11.30	6.55		17.85	23
6390	1″ iron pipe size	G		115	.139		11.35	6.55		17.90	23
6400	1-1/4″ iron pipe size	G		110	.145		11.55	6.85		18.40	23.50
6410	1-1/2″ iron pipe size	G		110	.145		11.60	6.85		18.45	24
6420	2″ iron pipe size	G		105	.152		12	7.20		19.20	25
6430	2-1/2″ iron pipe size	G		105	.152		14.10	7.20		21.30	27
6440	3″ iron pipe size	G		100	.160		14.20	7.55		21.75	28
6450	4″ iron pipe size	G		95	.168		18.15	7.95		26.10	33
6460	5″ iron pipe size	G		90	.178		19.70	8.40		28.10	35

22 07 19 – Plumbing Piping Insulation

22 07 19.10 Piping Insulation		Crew	Daily Output	Labor-Hours	Unit	Material	2015 Bare Costs Labor	Equipment	Total	Total Incl O&P
6470	6" iron pipe size	G	Q-14	90	.178	L.F.	22	8.40		30.40	38
6480	7" iron pipe size	G		85	.188		24.50	8.85		33.35	41.50
6490	8" iron pipe size	G		85	.188		26.50	8.85		35.35	43.50
6500	9" iron pipe size	G		80	.200		30	9.45		39.45	48
6510	10" iron pipe size	G		75	.213		32	10.05		42.05	51
6520	12" iron pipe size	G		70	.229		35	10.75		45.75	56.50
6530	14" iron pipe size	G		65	.246		39.50	11.60		51.10	62
6540	16" iron pipe size	G		60	.267		43.50	12.55		56.05	68
6550	18" iron pipe size	G		55	.291		48	13.70		61.70	74.50
6560	20" iron pipe size	G		50	.320		57	15.10		72.10	86.50
6570	22" iron pipe size	G		45	.356		62	16.75		78.75	95
6580	24" iron pipe size	G	↓	40	.400	↓	66.50	18.85		85.35	104
6600	Fiberglass, with all service jacket										
6640	1/2" wall, 1/2" iron pipe size	G	Q-14	250	.064	L.F.	.68	3.02		3.70	5.60
6660	3/4" iron pipe size	G		240	.067		.77	3.14		3.91	5.90
6670	1" iron pipe size	G		230	.070		.80	3.28		4.08	6.15
6680	1-1/4" iron pipe size	G		220	.073		.85	3.43		4.28	6.45
6690	1-1/2" iron pipe size	G		220	.073		.98	3.43		4.41	6.60
6700	2" iron pipe size	G		210	.076		1.05	3.59		4.64	6.90
6710	2-1/2" iron pipe size	G		200	.080		1.09	3.77		4.86	7.25
6840	1" wall, 1/2" iron pipe size	G		240	.067		.83	3.14		3.97	5.95
6860	3/4" iron pipe size	G		230	.070		.90	3.28		4.18	6.25
6870	1" iron pipe size	G		220	.073		.97	3.43		4.40	6.55
6880	1-1/4" iron pipe size	G		210	.076		1.05	3.59		4.64	6.90
6890	1-1/2" iron pipe size	G		210	.076		1.13	3.59		4.72	7
6900	2" iron pipe size	G		200	.080		1.22	3.77		4.99	7.40
6910	2-1/2" iron pipe size	G		190	.084		1.39	3.97		5.36	7.95
6920	3" iron pipe size	G		180	.089		1.49	4.19		5.68	8.40
6930	3-1/2" iron pipe size	G		170	.094		1.61	4.44		6.05	8.90
6940	4" iron pipe size	G		150	.107		1.97	5.05		7.02	10.25
6950	5" iron pipe size	G		140	.114		2.22	5.40		7.62	11.10
6960	6" iron pipe size	G		120	.133		2.36	6.30		8.66	12.70
6970	7" iron pipe size	G		110	.145		2.81	6.85		9.66	14.10
6980	8" iron pipe size	G		100	.160		3.85	7.55		11.40	16.35
6990	9" iron pipe size	G		90	.178		4.06	8.40		12.46	17.90
7000	10" iron pipe size	G		90	.178		4.09	8.40		12.49	17.95
7010	12" iron pipe size	G		80	.200		4.45	9.45		13.90	20
7020	14" iron pipe size	G		80	.200		5.35	9.45		14.80	21
7030	16" iron pipe size	G		70	.229		6.80	10.75		17.55	25
7040	18" iron pipe size	G		70	.229		7.55	10.75		18.30	25.50
7050	20" iron pipe size	G		60	.267		8.40	12.55		20.95	29.50
7060	24" iron pipe size	G		60	.267		10.25	12.55		22.80	31.50
7080	1-1/2" wall, 1/2" iron pipe size	G		230	.070		1.57	3.28		4.85	7
7100	3/4" iron pipe size	G		220	.073		1.57	3.43		5	7.25
7110	1" iron pipe size	G		210	.076		1.68	3.59		5.27	7.60
7120	1-1/4" iron pipe size	G		200	.080		1.83	3.77		5.60	8.05
7130	1-1/2" iron pipe size	G		200	.080		1.92	3.77		5.69	8.15
7140	2" iron pipe size	G		190	.084		2.11	3.97		6.08	8.70
7150	2-1/2" iron pipe size	G		180	.089		2.27	4.19		6.46	9.25
7160	3" iron pipe size	G		170	.094		2.37	4.44		6.81	9.75
7170	3-1/2" iron pipe size	G		160	.100		2.60	4.71		7.31	10.45
7180	4" iron pipe size	G		140	.114		2.70	5.40		8.10	11.60
7190	5" iron pipe size	G	↓	130	.123	↓	3.02	5.80		8.82	12.60

22 07 19 – Plumbing Piping Insulation

22 07 19.10 Piping Insulation			Crew	Daily Output	Labor-Hours	Unit	Material	Labor	Equipment	Total	Total Incl O&P
							2015 Bare Costs				
7200	6" iron pipe size	G	Q-14	110	.145	L.F.	3.19	6.85		10.04	14.50
7210	7" iron pipe size	G		100	.160		3.54	7.55		11.09	16
7220	8" iron pipe size	G		90	.178		4.47	8.40		12.87	18.35
7230	9" iron pipe size	G		85	.188		4.66	8.85		13.51	19.40
7240	10" iron pipe size	G		80	.200		4.82	9.45		14.27	20.50
7250	12" iron pipe size	G		75	.213		5.45	10.05		15.50	22
7260	14" iron pipe size	G		70	.229		6.55	10.75		17.30	24.50
7270	16" iron pipe size	G		65	.246		8.50	11.60		20.10	28
7280	18" iron pipe size	G		60	.267		9.55	12.55		22.10	30.50
7290	20" iron pipe size	G		55	.291		9.75	13.70		23.45	33
7300	24" iron pipe size	G		50	.320		11.95	15.10		27.05	37
7320	2" wall, 1/2" iron pipe size	G		220	.073		2.43	3.43		5.86	8.15
7340	3/4" iron pipe size	G		210	.076		2.51	3.59		6.10	8.50
7350	1" iron pipe size	G		200	.080		2.67	3.77		6.44	9
7360	1-1/4" iron pipe size	G		190	.084		2.82	3.97		6.79	9.50
7370	1-1/2" iron pipe size	G		190	.084		2.94	3.97		6.91	9.65
7380	2" iron pipe size	G		180	.089		3.10	4.19		7.29	10.15
7390	2-1/2" iron pipe size	G		170	.094		3.34	4.44		7.78	10.80
7400	3" iron pipe size	G		160	.100		3.54	4.71		8.25	11.50
7410	3-1/2" iron pipe size	G		150	.107		3.84	5.05		8.89	12.30
7420	4" iron pipe size	G		130	.123		4.13	5.80		9.93	13.85
7430	5" iron pipe size	G		120	.133		4.73	6.30		11.03	15.30
7440	6" iron pipe size	G		100	.160		4.88	7.55		12.43	17.45
7450	7" iron pipe size	G		90	.178		5.55	8.40		13.95	19.55
7460	8" iron pipe size	G		80	.200		5.95	9.45		15.40	21.50
7470	9" iron pipe size	G		75	.213		6.50	10.05		16.55	23.50
7480	10" iron pipe size	G		70	.229		7.10	10.75		17.85	25
7490	12" iron pipe size	G		65	.246		7.95	11.60		19.55	27.50
7500	14" iron pipe size	G		60	.267		10.25	12.55		22.80	31.50
7510	16" iron pipe size	G		55	.291		11.25	13.70		24.95	34.50
7520	18" iron pipe size	G		50	.320		12.55	15.10		27.65	38
7530	20" iron pipe size	G		45	.356		14.25	16.75		31	42.50
7540	24" iron pipe size	G		40	.400		15.30	18.85		34.15	47.50
7560	2-1/2" wall, 1/2" iron pipe size	G		210	.076		2.88	3.59		6.47	8.90
7562	3/4" iron pipe size	G		200	.080		3.01	3.77		6.78	9.35
7564	1" iron pipe size	G		190	.084		3.14	3.97		7.11	9.85
7566	1-1/4" iron pipe size	G		185	.086		3.25	4.08		7.33	10.15
7568	1-1/2" iron pipe size	G		180	.089		3.42	4.19		7.61	10.50
7570	2" iron pipe size	G		170	.094		3.59	4.44		8.03	11.10
7572	2-1/2" iron pipe size	G		160	.100		4.13	4.71		8.84	12.15
7574	3" iron pipe size	G		150	.107		4.35	5.05		9.40	12.90
7576	3-1/2" iron pipe size	G		140	.114		4.73	5.40		10.13	13.85
7578	4" iron pipe size	G		120	.133		4.99	6.30		11.29	15.60
7580	5" iron pipe size	G		110	.145		5.95	6.85		12.80	17.55
7582	6" iron pipe size	G		90	.178		7.15	8.40		15.55	21.50
7584	7" iron pipe size	G		80	.200		7.20	9.45		16.65	23
7586	8" iron pipe size	G		70	.229		7.65	10.75		18.40	25.50
7588	9" iron pipe size	G		65	.246		8.35	11.60		19.95	28
7590	10" iron pipe size	G		60	.267		9.10	12.55		21.65	30
7592	12" iron pipe size	G		55	.291		11.05	13.70		24.75	34
7594	14" iron pipe size	G		50	.320		12.90	15.10		28	38
7596	16" iron pipe size	G		45	.356		14.80	16.75		31.55	43.50
7598	18" iron pipe size	G		40	.400		16.05	18.85		34.90	48

22 07 19.10 Piping Insulation		Crew	Daily Output	Labor-Hours	Unit	2015 Bare Costs Material	Labor	Equipment	Total	Total Incl O&P
7602	24" iron pipe size	G	Q-14	30	.533	L.F.	21	25		46	64
7620	3" wall, 1/2" iron pipe size	G		200	.080		3.69	3.77		7.46	10.10
7622	3/4" iron pipe size	G		190	.084		3.90	3.97		7.87	10.70
7624	1" iron pipe size	G		180	.089		4.13	4.19		8.32	11.30
7626	1-1/4" iron pipe size	G		175	.091		4.21	4.31		8.52	11.60
7628	1-1/2" iron pipe size	G		170	.094		4.40	4.44		8.84	12
7630	2" iron pipe size	G		160	.100		4.76	4.71		9.47	12.85
7632	2-1/2" iron pipe size	G		150	.107		4.95	5.05		10	13.55
7634	3" iron pipe size	G		140	.114		5.25	5.40		10.65	14.45
7636	3-1/2" iron pipe size	G		130	.123		5.80	5.80		11.60	15.70
7638	4" iron pipe size	G		110	.145		6.25	6.85		13.10	17.90
7640	5" iron pipe size	G		100	.160		7.05	7.55		14.60	19.90
7642	6" iron pipe size	G		80	.200		7.55	9.45		17	23.50
7644	7" iron pipe size	G		70	.229		8.65	10.75		19.40	27
7646	8" iron pipe size	G		60	.267		9.40	12.55		21.95	30.50
7648	9" iron pipe size	G		55	.291		10.15	13.70		23.85	33
7650	10" iron pipe size	G		50	.320		10.90	15.10		26	36
7652	12" iron pipe size	G		45	.356		13.55	16.75		30.30	42
7654	14" iron pipe size	G		40	.400		15.65	18.85		34.50	47.50
7656	16" iron pipe size	G		35	.457		17.75	21.50		39.25	54
7658	18" iron pipe size	G		32	.500		18.95	23.50		42.45	59
7660	20" iron pipe size	G		30	.533		20.50	25		45.50	63
7662	24" iron pipe size	G		28	.571		26	27		53	72.50
7664	26" iron pipe size	G		26	.615		29	29		58	78
7666	30" iron pipe size	G	↓	24	.667	↓	34	31.50		65.50	88
7800	For fiberglass with standard canvas jacket, deduct						5%				
7802	For fittings, add 3 L.F. for each fitting										
7804	plus 4 L.F. for each flange of the fitting										
7810	Finishes										
7814	For single layer of felt, add						10%	10%			
7816	For roofing paper, 45 lb. to 55 lb., add						25%	10%			
7879	Rubber tubing, flexible closed cell foam										
7880	3/8" wall, 1/4" iron pipe size	G	1 Asbe	120	.067	L.F.	.33	3.49		3.82	5.95
7900	3/8" iron pipe size	G		120	.067		.36	3.49		3.85	6
7910	1/2" iron pipe size	G		115	.070		.41	3.64		4.05	6.30
7920	3/4" iron pipe size	G		115	.070		.46	3.64		4.10	6.35
7930	1" iron pipe size	G		110	.073		.52	3.81		4.33	6.65
7940	1-1/4" iron pipe size	G		110	.073		.59	3.81		4.40	6.75
7950	1-1/2" iron pipe size	G		110	.073		.73	3.81		4.54	6.90
8100	1/2" wall, 1/4" iron pipe size	G		90	.089		.54	4.65		5.19	8.10
8120	3/8" iron pipe size	G		90	.089		.60	4.65		5.25	8.15
8130	1/2" iron pipe size	G		89	.090		.67	4.71		5.38	8.30
8140	3/4" iron pipe size	G		89	.090		.75	4.71		5.46	8.40
8150	1" iron pipe size	G		88	.091		.82	4.76		5.58	8.55
8160	1-1/4" iron pipe size	G		87	.092		.95	4.81		5.76	8.80
8170	1-1/2" iron pipe size	G		87	.092		1.15	4.81		5.96	9
8180	2" iron pipe size	G		86	.093		1.47	4.87		6.34	9.45
8190	2-1/2" iron pipe size	G		86	.093		1.85	4.87		6.72	9.90
8200	3" iron pipe size	G		85	.094		2.06	4.93		6.99	10.15
8210	3-1/2" iron pipe size	G		85	.094		2.86	4.93		7.79	11.05
8220	4" iron pipe size	G		80	.100		3.06	5.25		8.31	11.75
8230	5" iron pipe size	G		80	.100		4.17	5.25		9.42	13
8240	6" iron pipe size	G	↓	75	.107	↓	4.17	5.60		9.77	13.60

22 07 19 – Plumbing Piping Insulation

22 07 19.10 Piping Insulation		Crew	Daily Output	Labor-Hours	Unit	Material	2015 Bare Costs Labor	Equipment	Total	Total Incl O&P
8300	3/4" wall, 1/4" iron pipe size	G	1 Asbe	90	.089	L.F.	.85	4.65		5.50	8.45
8320	3/8" iron pipe size	G		90	.089		.92	4.65		5.57	8.50
8330	1/2" iron pipe size	G		89	.090		1.10	4.71		5.81	8.75
8340	3/4" iron pipe size	G		89	.090		1.35	4.71		6.06	9.05
8350	1" iron pipe size	G		88	.091		1.54	4.76		6.30	9.35
8360	1-1/4" iron pipe size	G		87	.092		2.05	4.81		6.86	10
8370	1-1/2" iron pipe size	G		87	.092		2.32	4.81		7.13	10.30
8380	2" iron pipe size	G		86	.093		2.68	4.87		7.55	10.80
8390	2-1/2" iron pipe size	G		86	.093		3.58	4.87		8.45	11.80
8400	3" iron pipe size	G		85	.094		4.08	4.93		9.01	12.40
8410	3-1/2" iron pipe size	G		85	.094		4.81	4.93		9.74	13.20
8420	4" iron pipe size	G		80	.100		5.10	5.25		10.35	14.05
8430	5" iron pipe size	G		80	.100		6.15	5.25		11.40	15.20
8440	6" iron pipe size	G		80	.100		7.45	5.25		12.70	16.60
8444	1" wall, 1/2" iron pipe size	G		86	.093		2.05	4.87		6.92	10.10
8445	3/4" iron pipe size	G		84	.095		2.48	4.99		7.47	10.75
8446	1" iron pipe size	G		84	.095		2.89	4.99		7.88	11.20
8447	1-1/4" iron pipe size	G		82	.098		3.23	5.10		8.33	11.75
8448	1-1/2" iron pipe size	G		82	.098		3.76	5.10		8.86	12.35
8449	2" iron pipe size	G		80	.100		4.95	5.25		10.20	13.85
8450	2-1/2" iron pipe size	G	↓	80	.100	↓	6.45	5.25		11.70	15.50
8456	Rubber insulation tape, 1/8" x 2" x 30'	G				Ea.	12.05			12.05	13.25
9600	Minimum labor/equipment charge		1 Plum	4	2	Job		117		117	183

22 07 19.30 Piping Insulation Protective Jacketing, PVC		Crew	Daily Output	Labor-Hours	Unit	Material	Labor	Equipment	Total	Total Incl O&P
0010	**PIPING INSULATION PROTECTIVE JACKETING, PVC**									
0100	PVC, white, 48" lengths cut from roll goods									
0120	20 mil thick									
0140	Size based on OD of insulation									
0150	1-1/2" ID	Q-14	270	.059	L.F.	.20	2.79		2.99	4.71
0152	2" ID		260	.062		.27	2.90		3.17	4.96
0154	2-1/2" ID		250	.064		.32	3.02		3.34	5.20
0156	3" ID		240	.067		.39	3.14		3.53	5.50
0158	3-1/2" ID		230	.070		.44	3.28		3.72	5.75
0160	4" ID		220	.073		.50	3.43		3.93	6.05
0162	4-1/2" ID		210	.076		.56	3.59		4.15	6.35
0164	5" ID		200	.080		.61	3.77		4.38	6.70
0166	5-1/2" ID		190	.084		.67	3.97		4.64	7.15
0168	6" ID		180	.089		.73	4.19		4.92	7.55
0170	6-1/2" ID		175	.091		.79	4.31		5.10	7.80
0172	7" ID		170	.094		.84	4.44		5.28	8.05
0174	7-1/2" ID		164	.098		.91	4.60		5.51	8.40
0176	8" ID		161	.099		.96	4.68		5.64	8.60
0178	8-1/2" ID		158	.101		1.02	4.77		5.79	8.75
0180	9" ID		155	.103		1.08	4.87		5.95	9
0182	9-1/2" ID		152	.105		1.14	4.96		6.10	9.20
0184	10" ID		149	.107		1.19	5.05		6.24	9.45
0186	10-1/2" ID		146	.110		1.25	5.15		6.40	9.70
0188	11" ID		143	.112		1.31	5.25		6.56	9.95
0190	11-1/2" ID		140	.114		1.37	5.40		6.77	10.15
0192	12" ID		137	.117		1.42	5.50		6.92	10.40
0194	12-1/2" ID		134	.119		1.49	5.65		7.14	10.70
0195	13" ID	↓	132	.121	↓	1.54	5.70		7.24	10.90

22 07 19.30 Piping Insulation Protective Jacketing, PVC		Crew	Daily Output	Labor-Hours	Unit	Material	2015 Bare Costs Labor	Equipment	Total	Total Incl O&P
0196	13-1/2" ID	Q-14	132	.121	L.F.	1.60	5.70		7.30	10.95
0198	14" ID		130	.123		1.66	5.80		7.46	11.15
0200	15" ID		128	.125		1.78	5.90		7.68	11.40
0202	16" ID		126	.127		1.89	6		7.89	11.70
0204	17" ID		124	.129		2.01	6.10		8.11	11.95
0206	18" ID		122	.131		2.13	6.20		8.33	12.30
0208	19" ID		120	.133		2.24	6.30		8.54	12.55
0210	20" ID		118	.136		2.35	6.40		8.75	12.85
0212	21" ID		116	.138		2.47	6.50		8.97	13.15
0214	22" ID		114	.140		2.58	6.60		9.18	13.50
0216	23" ID		112	.143		2.69	6.75		9.44	13.75
0218	24" ID		110	.145		2.81	6.85		9.66	14.10
0220	25" ID		108	.148		2.93	7		9.93	14.40
0222	26" ID		106	.151		3.04	7.10		10.14	14.80
0224	27" ID		104	.154		3.16	7.25		10.41	15.15
0226	28" ID		102	.157		3.27	7.40		10.67	15.50
0228	29" ID		100	.160		3.39	7.55		10.94	15.85
0230	30" ID		98	.163		3.51	7.70		11.21	16.20
0300	For colors, add				Ea.	10%				
1000	30 mil thick									
1010	Size based on OD of insulation									
1020	2" ID	Q-14	260	.062	L.F.	.40	2.90		3.30	5.10
1022	2-1/2" ID		250	.064		.48	3.02		3.50	5.40
1024	3" ID		240	.067		.58	3.14		3.72	5.70
1026	3-1/2" ID		230	.070		.65	3.28		3.93	5.95
1028	4" ID		220	.073		.74	3.43		4.17	6.30
1030	4-1/2" ID		210	.076		.82	3.59		4.41	6.65
1032	5" ID		200	.080		.91	3.77		4.68	7.05
1034	5-1/2" ID		190	.084		.99	3.97		4.96	7.50
1036	6" ID		180	.089		1.08	4.19		5.27	7.95
1038	6-1/2" ID		175	.091		1.16	4.31		5.47	8.25
1040	7" ID		170	.094		1.25	4.44		5.69	8.55
1042	7-1/2" ID		164	.098		1.34	4.60		5.94	8.85
1044	8" ID		161	.099		1.43	4.68		6.11	9.10
1046	8-1/2" ID		158	.101		1.51	4.77		6.28	9.30
1048	9" ID		155	.103		1.60	4.87		6.47	9.55
1050	9-1/2" ID		152	.105		1.68	4.96		6.64	9.80
1052	10" ID		149	.107		1.77	5.05		6.82	10.10
1054	10-1/2" ID		146	.110		1.85	5.15		7	10.35
1056	11" ID		143	.112		1.94	5.25		7.19	10.65
1058	11-1/2" ID		140	.114		2.02	5.40		7.42	10.85
1060	12" ID		137	.117		2.11	5.50		7.61	11.15
1062	12-1/2" ID		134	.119		2.20	5.65		7.85	11.45
1063	13" ID		132	.121		2.29	5.70		7.99	11.70
1064	13-1/2" ID		132	.121		2.37	5.70		8.07	11.80
1066	14" ID		130	.123		2.46	5.80		8.26	12
1068	15" ID		128	.125		2.63	5.90		8.53	12.35
1070	16" ID		126	.127		2.80	6		8.80	12.70
1072	17" ID		124	.129		2.97	6.10		9.07	13
1074	18" ID		122	.131		3.15	6.20		9.35	13.40
1076	19" ID		120	.133		3.32	6.30		9.62	13.75
1078	20" ID		118	.136		3.49	6.40		9.89	14.10
1080	21" ID		116	.138		3.66	6.50		10.16	14.50

22 07 19.30 Piping Insulation Protective Jacketing, PVC		Crew	Daily Output	Labor-Hours	Unit	Material	2015 Bare Costs Labor	Equipment	Total	Total Incl O&P
1082	22" ID	Q-14	114	.140	L.F.	3.83	6.60		10.43	14.85
1084	23" ID		112	.143		4	6.75		10.75	15.20
1086	24" ID		110	.145		4.17	6.85		11.02	15.60
1088	25" ID		108	.148		4.34	7		11.34	15.95
1090	26" ID		106	.151		4.51	7.10		11.61	16.40
1092	27" ID		104	.154		4.68	7.25		11.93	16.80
1094	28" ID		102	.157		4.85	7.40		12.25	17.25
1096	29" ID		100	.160		5	7.55		12.55	17.60
1098	30" ID		98	.163		5.20	7.70		12.90	18.05
1300	For colors, add				Ea.	10%				
2000	PVC, white, fitting covers									
2020	Fiberglass insulation inserts included with sizes 1-3/4" thru 9-3/4"									
2030	Size is based on OD of insulation									
2040	90° Elbow fitting									
2060	1-3/4"	Q-14	135	.119	Ea.	.47	5.60		6.07	9.50
2062	2"		130	.123		.58	5.80		6.38	9.95
2064	2-1/4"		128	.125		.65	5.90		6.55	10.15
2068	2-1/2"		126	.127		.71	6		6.71	10.40
2070	2-3/4"		123	.130		.77	6.15		6.92	10.70
2072	3"		120	.133		.82	6.30		7.12	11
2074	3-3/8"		116	.138		.93	6.50		7.43	11.45
2076	3-3/4"		113	.142		1.04	6.65		7.69	11.90
2078	4-1/8"		110	.145		1.31	6.85		8.16	12.45
2080	4-3/4"		105	.152		1.58	7.20		8.78	13.30
2082	5-1/4"		100	.160		1.84	7.55		9.39	14.10
2084	5-3/4"		95	.168		2.43	7.95		10.38	15.40
2086	6-1/4"		90	.178		4.05	8.40		12.45	17.90
2088	6-3/4"		87	.184		4.27	8.65		12.92	18.65
2090	7-1/4"		85	.188		5.35	8.85		14.20	20
2092	7-3/4"		83	.193		5.60	9.10		14.70	21
2094	8-3/4"		80	.200		7.20	9.45		16.65	23
2096	9-3/4"		77	.208		9.65	9.80		19.45	26.50
2098	10-7/8"		74	.216		10.75	10.20		20.95	28.50
2100	11-7/8"		71	.225		12.35	10.60		22.95	30.50
2102	12-7/8"		68	.235		15.55	11.10		26.65	35
2104	14-1/8"		66	.242		17.80	11.45		29.25	38
2106	15-1/8"		64	.250		19.15	11.80		30.95	40
2108	16-1/8"		63	.254		21	11.95		32.95	42.50
2110	17-1/8"		62	.258		23	12.15		35.15	45
2112	18-1/8"		61	.262		30	12.35		42.35	53
2114	19-1/8"		60	.267		40.50	12.55		53.05	64.50
2116	20-1/8"		59	.271		52	12.80		64.80	78
2200	45° Elbow fitting									
2220	1-3/4" thru 9-3/4" same price as 90° Elbow fitting									
2320	10-7/8"	Q-14	74	.216	Ea.	10.75	10.20		20.95	28.50
2322	11-7/8"		71	.225		11.85	10.60		22.45	30
2324	12-7/8"		68	.235		13.25	11.10		24.35	32.50
2326	14-1/8"		66	.242		15.10	11.45		26.55	35
2328	15-1/8"		64	.250		16.25	11.80		28.05	37
2330	16-1/8"		63	.254		18.55	11.95		30.50	40
2332	17-1/8"		62	.258		21	12.15		33.15	42.50
2334	18-1/8"		61	.262		25.50	12.35		37.85	48
2336	19-1/8"		60	.267		35	12.55		47.55	58.50

22 07 19.30 Piping Insulation Protective Jacketing, PVC		Crew	Daily Output	Labor-Hours	Unit	Material	Labor	Equipment	Total	Total Incl O&P
						2015 Bare Costs				
2338	20-1/8"	Q-14	59	.271	Ea.	39.50	12.80		52.30	64
2400	Tee fitting									
2410	1-3/4"	Q-14	96	.167	Ea.	.88	7.85		8.73	13.60
2412	2"		94	.170		1	8		9	14
2414	2-1/4"		91	.176		1.09	8.30		9.39	14.50
2416	2-1/2"		88	.182		1.19	8.55		9.74	15.05
2418	2-3/4"		85	.188		1.31	8.85		10.16	15.70
2420	3"		82	.195		1.42	9.20		10.62	16.35
2422	3-3/8"		79	.203		1.63	9.55		11.18	17.15
2424	3-3/4"		76	.211		1.85	9.90		11.75	18
2426	4-1/8"		73	.219		2.19	10.35		12.54	19
2428	4-3/4"		70	.229		2.73	10.75		13.48	20.50
2430	5-1/4"		67	.239		3.28	11.25		14.53	21.50
2432	5-3/4"		63	.254		4.36	11.95		16.31	24
2434	6-1/4"		60	.267		5.75	12.55		18.30	26.50
2436	6-3/4"		59	.271		7.10	12.80		19.90	28.50
2438	7-1/4"		57	.281		11.45	13.25		24.70	34
2440	7-3/4"		54	.296		12.55	13.95		26.50	36.50
2442	8-3/4"		52	.308		15.30	14.50		29.80	40.50
2444	9-3/4"		50	.320		18	15.10		33.10	44
2446	10-7/8"		48	.333		18.25	15.70		33.95	45.50
2448	11-7/8"		47	.340		20.50	16.05		36.55	48.50
2450	12-7/8"		46	.348		22.50	16.40		38.90	51
2452	14-1/8"		45	.356		24.50	16.75		41.25	54
2454	15-1/8"		44	.364		26.50	17.15		43.65	56.50
2456	16-1/8"		43	.372		28.50	17.55		46.05	59
2458	17-1/8"		42	.381		30.50	17.95		48.45	62.50
2460	18-1/8"		41	.390		33.50	18.40		51.90	66.50
2462	19-1/8"		40	.400		36.50	18.85		55.35	70.50
2464	20-1/8"		39	.410		40	19.35		59.35	75
4000	Mechanical grooved fitting cover, including insert									
4020	90° Elbow fitting									
4030	3/4" & 1"	Q-14	140	.114	Ea.	3.70	5.40		9.10	12.70
4040	1-1/4" & 1-1/2"		135	.119		4.68	5.60		10.28	14.15
4042	2"		130	.123		7.40	5.80		13.20	17.45
4044	2-1/2"		125	.128		8.20	6.05		14.25	18.70
4046	3"		120	.133		9.20	6.30		15.50	20
4048	3-1/2"		115	.139		10.65	6.55		17.20	22.50
4050	4"		110	.145		11.85	6.85		18.70	24
4052	5"		100	.160		14.75	7.55		22.30	28.50
4054	6"		90	.178		22	8.40		30.40	37.50
4056	8"		80	.200		23.50	9.45		32.95	41
4058	10"		75	.213		30.50	10.05		40.55	49.50
4060	12"		68	.235		43.50	11.10		54.60	66
4062	14"		65	.246		43.50	11.60		55.10	66.50
4064	16"		63	.254		59.50	11.95		71.45	85
4066	18"		61	.262		81.50	12.35		93.85	110
4100	45° Elbow fitting									
4120	3/4" & 1"	Q-14	140	.114	Ea.	3.29	5.40		8.69	12.25
4130	1-1/4" & 1-1/2"		135	.119		4.19	5.60		9.79	13.60
4140	2"		130	.123		6.60	5.80		12.40	16.55
4142	2-1/2"		125	.128		7.30	6.05		13.35	17.75
4144	3"		120	.133		8.20	6.30		14.50	19.10

22 07 Plumbing Insulation

22 07 19 – Plumbing Piping Insulation

22 07 19.30 Piping Insulation Protective Jacketing, PVC		Crew	Daily Output	Labor-Hours	Unit	Material	2015 Bare Costs Labor	Equipment	Total	Total Incl O&P
4146	3-1/2"	Q-14	115	.139	Ea.	8.65	6.55		15.20	20
4148	4"		110	.145		10.55	6.85		17.40	22.50
4150	5"		100	.160		13.10	7.55		20.65	26.50
4152	6"		90	.178		19.35	8.40		27.75	35
4154	8"		80	.200		21	9.45		30.45	38
4156	10"		75	.213		27	10.05		37.05	45.50
4158	12"		68	.235		39	11.10		50.10	60.50
4160	14"		65	.246		50.50	11.60		62.10	74
4162	16"		63	.254		61	11.95		72.95	87
4164	18"	↓	61	.262	↓	72	12.35		84.35	99.50
4200	Tee fitting									
4220	3/4" & 1"	Q-14	93	.172	Ea.	4.82	8.10		12.92	18.35
4230	1-1/4" & 1-1/2"		90	.178		6.10	8.40		14.50	20
4240	2"		87	.184		9.60	8.65		18.25	24.50
4242	2-1/2"		84	.190		10.70	9		19.70	26
4244	3"		80	.200		11.95	9.45		21.40	28.50
4246	3-1/2"		77	.208		12.65	9.80		22.45	29.50
4248	4"		73	.219		15.45	10.35		25.80	33.50
4250	5"		67	.239		17.05	11.25		28.30	37
4252	6"		60	.267		25	12.55		37.55	47.50
4254	8"		54	.296		27.50	13.95		41.45	52.50
4256	10"		50	.320		39.50	15.10		54.60	67.50
4258	12"		46	.348		56.50	16.40		72.90	89
4260	14"		43	.372		68.50	17.55		86.05	104
4262	16"		42	.381		82	17.95		99.95	120
4264	18"	↓	41	.390	↓	90.50	18.40		108.90	129

22 07 19.40 Pipe Insulation Protective Jacketing, Aluminum		Crew	Daily Output	Labor-Hours	Unit	Material	Labor	Equipment	Total	Total Incl O&P
0010	**PIPE INSULATION PROTECTIVE JACKETING, ALUMINUM**									
0100	Metal roll jacketing									
0120	Aluminum with polykraft moisture barrier									
0140	Smooth, based on OD of insulation, .016" thick									
0180	1/2" ID	Q-14	220	.073	L.F.	.20	3.43		3.63	5.70
0190	3/4" ID		215	.074		.26	3.51		3.77	5.95
0200	1" ID		210	.076		.32	3.59		3.91	6.10
0210	1-1/4" ID		205	.078		.38	3.68		4.06	6.30
0220	1-1/2" ID		202	.079		.44	3.73		4.17	6.50
0230	1-3/4" ID		199	.080		.50	3.79		4.29	6.65
0240	2" ID		195	.082		.56	3.87		4.43	6.80
0250	2-1/4" ID		191	.084		.62	3.95		4.57	7.05
0260	2-1/2" ID		187	.086		.69	4.03		4.72	7.25
0270	2-3/4" ID		184	.087		.75	4.10		4.85	7.45
0280	3" ID		180	.089		.82	4.19		5.01	7.65
0290	3-1/4" ID		176	.091		.88	4.28		5.16	7.85
0300	3-1/2" ID		172	.093		.94	4.38		5.32	8.10
0310	3-3/4" ID		169	.095		1	4.46		5.46	8.25
0320	4" ID		165	.097		1.06	4.57		5.63	8.50
0330	4-1/4" ID		161	.099		1.12	4.68		5.80	8.80
0340	4-1/2" ID		157	.102		1.19	4.80		5.99	9
0350	4-3/4" ID		154	.104		1.25	4.90		6.15	9.25
0360	5" ID		150	.107		1.31	5.05		6.36	9.55
0370	5-1/4" ID		146	.110		1.37	5.15		6.52	9.80
0380	5-1/2" ID	↓	143	.112	↓	1.43	5.25		6.68	10.05

22 07 19.40 Pipe Insulation Protective Jacketing, Aluminum		Crew	Daily Output	Labor-Hours	Unit	Material	2015 Bare Costs Labor	Equipment	Total	Total Incl O&P
0390	5-3/4" ID	Q-14	139	.115	L.F.	1.49	5.45		6.94	10.35
0400	6" ID		135	.119		1.55	5.60		7.15	10.70
0410	6-1/4" ID		133	.120		1.61	5.65		7.26	10.85
0420	6-1/2" ID		131	.122		1.67	5.75		7.42	11.10
0430	7" ID		128	.125		1.79	5.90		7.69	11.40
0440	7-1/4" ID		125	.128		1.85	6.05		7.90	11.75
0450	7-1/2" ID		123	.130		1.91	6.15		8.06	11.95
0460	8" ID		121	.132		2.03	6.25		8.28	12.25
0470	8-1/2" ID		119	.134		2.15	6.35		8.50	12.55
0480	9" ID		116	.138		2.27	6.50		8.77	12.95
0490	9-1/2" ID		114	.140		2.39	6.60		8.99	13.30
0500	10" ID		112	.143		2.51	6.75		9.26	13.55
0510	10-1/2" ID		110	.145		2.63	6.85		9.48	13.90
0520	11" ID		107	.150		2.75	7.05		9.80	14.40
0530	11-1/2" ID		105	.152		2.88	7.20		10.08	14.70
0540	12" ID		103	.155		3	7.30		10.30	15.05
0550	12-1/2" ID		100	.160		3.12	7.55		10.67	15.55
0560	13" ID		99	.162		3.24	7.60		10.84	15.80
0570	14" ID		98	.163		3.49	7.70		11.19	16.20
0580	15" ID		96	.167		3.73	7.85		11.58	16.75
0590	16" ID		95	.168		3.98	7.95		11.93	17.15
0600	17" ID		93	.172		4.22	8.10		12.32	17.70
0610	18" ID		92	.174		4.38	8.20		12.58	17.95
0620	19" ID		90	.178		4.70	8.40		13.10	18.60
0630	20" ID		89	.180		4.95	8.45		13.40	19.05
0640	21" ID		87	.184		5.20	8.65		13.85	19.65
0650	22" ID		86	.186		5.45	8.75		14.20	20
0660	23" ID		84	.190		5.70	9		14.70	20.50
0670	24" ID		83	.193		5.90	9.10		15	21
0710	For smooth .020" thick, add					27%	10%			
0720	For smooth .024" thick, add					52%	20%			
0730	For smooth .032" thick, add					104%	33%			
0800	For stucco embossed, add					1%				
0820	For corrugated, add					2.50%				
0900	White aluminum with polysurlyn moisture barrier									
0910	Smooth, % is an add to polykraft lines of same thickness									
0940	For smooth .016" thick, add				L.F.	35%				
0960	For smooth .024" thick, add				"	22%				
1000	Aluminum fitting covers									
1010	Size is based on OD of insulation									
1020	90° LR elbow, 2 piece									
1100	1-1/2"	Q-14	140	.114	Ea.	3.45	5.40		8.85	12.45
1110	1-3/4"		135	.119		3.45	5.60		9.05	12.80
1120	2"		130	.123		4.34	5.80		10.14	14.05
1130	2-1/4"		128	.125		4.34	5.90		10.24	14.20
1140	2-1/2"		126	.127		4.34	6		10.34	14.35
1150	2-3/4"		123	.130		4.34	6.15		10.49	14.60
1160	3"		120	.133		4.71	6.30		11.01	15.30
1170	3-1/4"		117	.137		4.71	6.45		11.16	15.55
1180	3-1/2"		115	.139		5.55	6.55		12.10	16.70
1190	3-3/4"		113	.142		5.70	6.65		12.35	17.05
1200	4"		110	.145		6	6.85		12.85	17.60
1210	4-1/4"		108	.148		6.15	7		13.15	17.95

22 07 19.40 Pipe Insulation Protective Jacketing, Aluminum		Crew	Daily Output	Labor-Hours	Unit	Material	2015 Bare Costs Labor	Equipment	Total	Total Incl O&P
1220	4-1/2"	Q-14	106	.151	Ea.	6.15	7.10		13.25	18.20
1230	4-3/4"		104	.154		6.15	7.25		13.40	18.40
1240	5"		102	.157		6.95	7.40		14.35	19.50
1250	5-1/4"		100	.160		8	7.55		15.55	21
1260	5-1/2"		97	.165		9.95	7.75		17.70	23.50
1270	5-3/4"		95	.168		9.95	7.95		17.90	23.50
1280	6"		92	.174		8.50	8.20		16.70	22.50
1290	6-1/4"		90	.178		8.50	8.40		16.90	23
1300	6-1/2"		87	.184		10.25	8.65		18.90	25
1310	7"		85	.188		12.40	8.85		21.25	28
1320	7-1/4"		84	.190		12.40	9		21.40	28
1330	7-1/2"		83	.193		18.25	9.10		27.35	34.50
1340	8"		82	.195		13.60	9.20		22.80	30
1350	8-1/2"		80	.200		25	9.45		34.45	42.50
1360	9"		78	.205		25	9.65		34.65	43
1370	9-1/2"		77	.208		18.50	9.80		28.30	36.50
1380	10"		76	.211		18.50	9.90		28.40	36.50
1390	10-1/2"		75	.213		18.90	10.05		28.95	37
1400	11"		74	.216		18.90	10.20		29.10	37.50
1410	11-1/2"		72	.222		21.50	10.45		31.95	40.50
1420	12"		71	.225		21.50	10.60		32.10	40.50
1430	12-1/2"		69	.232		38	10.95		48.95	59.50
1440	13"		68	.235		38	11.10		49.10	60
1450	14"		66	.242		51.50	11.45		62.95	75
1460	15"		64	.250		53.50	11.80		65.30	78
1470	16"		63	.254		58.50	11.95		70.45	84
2000	45° Elbow, 2 piece									
2010	2-1/2"	Q-14	126	.127	Ea.	3.59	6		9.59	13.55
2020	2-3/4"		123	.130		3.59	6.15		9.74	13.80
2030	3"		120	.133		4.05	6.30		10.35	14.55
2040	3-1/4"		117	.137		4.05	6.45		10.50	14.80
2050	3-1/2"		115	.139		4.62	6.55		11.17	15.65
2060	3-3/4"		113	.142		4.62	6.65		11.27	15.85
2070	4"		110	.145		4.70	6.85		11.55	16.15
2080	4-1/4"		108	.148		5.35	7		12.35	17.10
2090	4-1/2"		106	.151		5.35	7.10		12.45	17.35
2100	4-3/4"		104	.154		5.35	7.25		12.60	17.55
2110	5"		102	.157		6.30	7.40		13.70	18.85
2120	5-1/4"		100	.160		6.30	7.55		13.85	19.05
2130	5-1/2"		97	.165		6.60	7.75		14.35	19.75
2140	6"		92	.174		6.60	8.20		14.80	20.50
2150	6-1/2"		87	.184		9.10	8.65		17.75	24
2160	7"		85	.188		9.10	8.85		17.95	24.50
2170	7-1/2"		83	.193		9.20	9.10		18.30	25
2180	8"		82	.195		9.20	9.20		18.40	25
2190	8-1/2"		80	.200		11.45	9.45		20.90	28
2200	9"		78	.205		11.45	9.65		21.10	28
2210	9-1/2"		77	.208		15.90	9.80		25.70	33.50
2220	10"		76	.211		15.90	9.90		25.80	33.50
2230	10-1/2"		75	.213		14.85	10.05		24.90	32.50
2240	11"		74	.216		14.85	10.20		25.05	33
2250	11-1/2"		72	.222		17.85	10.45		28.30	36.50
2260	12"		71	.225		17.85	10.60		28.45	36.50

22 07 19 – Plumbing Piping Insulation

22 07 19.40 Pipe Insulation Protective Jacketing, Aluminum		Crew	Daily Output	Labor-Hours	Unit	2015 Bare Costs Material	Labor	Equipment	Total	Total Incl O&P
2270	13"	Q-14	68	.235	Ea.	21	11.10		32.10	41
2280	14"		66	.242		25.50	11.45		36.95	47
2290	15"		64	.250		39.50	11.80		51.30	62.50
2300	16"		63	.254		43	11.95		54.95	67
2310	17"		62	.258		42	12.15		54.15	65.50
2320	18"		61	.262		48	12.35		60.35	72.50
2330	19"		60	.267		58.50	12.55		71.05	84.50
2340	20"		59	.271		56.50	12.80		69.30	82.50
2350	21"		58	.276		61	13		74	88
3000	Tee, 4 piece									
3010	2-1/2"	Q-14	88	.182	Ea.	29	8.55		37.55	46
3020	2-3/4"		86	.186		29	8.75		37.75	46
3030	3"		84	.190		32.50	9		41.50	50.50
3040	3-1/4"		82	.195		32.50	9.20		41.70	51
3050	3-1/2"		80	.200		34.50	9.45		43.95	52.50
3060	4"		78	.205		34.50	9.65		44.15	53.50
3070	4-1/4"		76	.211		36.50	9.90		46.40	56
3080	4-1/2"		74	.216		36.50	10.20		46.70	56.50
3090	4-3/4"		72	.222		36.50	10.45		46.95	57
3100	5"		70	.229		37.50	10.75		48.25	59
3110	5-1/4"		68	.235		37.50	11.10		48.60	59.50
3120	5-1/2"		66	.242		39.50	11.45		50.95	62
3130	6"		64	.250		39.50	11.80		51.30	62.50
3140	6-1/2"		60	.267		43.50	12.55		56.05	67.50
3150	7"		58	.276		43.50	13		56.50	68.50
3160	7-1/2"		56	.286		48	13.45		61.45	74.50
3170	8"		54	.296		48	13.95		61.95	75.50
3180	8-1/2"		52	.308		49.50	14.50		64	78
3190	9"		50	.320		49.50	15.10		64.60	78.50
3200	9-1/2"		49	.327		36.50	15.40		51.90	65
3210	10"		48	.333		36.50	15.70		52.20	66
3220	10-1/2"		47	.340		39	16.05		55.05	69
3230	11"		46	.348		39	16.40		55.40	69.50
3240	11-1/2"		45	.356		41	16.75		57.75	72.50
3250	12"		44	.364		41	17.15		58.15	73
3260	13"		43	.372		43	17.55		60.55	75.50
3270	14"		42	.381		45.50	17.95		63.45	79
3280	15"		41	.390		49	18.40		67.40	83.50
3290	16"		40	.400		50.50	18.85		69.35	86
3300	17"		39	.410		58	19.35		77.35	94.50
3310	18"		38	.421		60	19.85		79.85	98
3320	19"		37	.432		66.50	20.50		87	106
3330	20"		36	.444		68	21		89	108
3340	22"		35	.457		86.50	21.50		108	130
3350	23"		34	.471		89	22		111	133
3360	24"		31	.516		91	24.50		115.50	139

22 07 19 – Plumbing Piping Insulation

22 07 19.50 Pipe Insulation Protective Jacketing, St. Stl.		Crew	Daily Output	Labor-Hours	Unit	Material	2015 Bare Costs Labor	Equipment	Total	Total Incl O&P
0010	**PIPE INSULATION PROTECTIVE JACKETING, STAINLESS STEEL**									
0100	Metal roll jacketing									
0120	Type 304 with moisture barrier									
0140	Smooth, based on OD of insulation, .010" thick									
0260	2-1/2" ID	Q-14	250	.064	L.F.	2.12	3.02		5.14	7.20
0270	2-3/4" ID		245	.065		2.31	3.08		5.39	7.50
0280	3" ID		240	.067		2.50	3.14		5.64	7.80
0290	3-1/4" ID		235	.068		2.69	3.21		5.90	8.10
0300	3-1/2" ID		230	.070		2.88	3.28		6.16	8.40
0310	3-3/4" ID		225	.071		3.07	3.35		6.42	8.80
0320	4" ID		220	.073		3.26	3.43		6.69	9.10
0330	4-1/4" ID		215	.074		3.44	3.51		6.95	9.45
0340	4-1/2" ID		210	.076		3.63	3.59		7.22	9.75
0350	5" ID		200	.080		4.01	3.77		7.78	10.45
0360	5-1/2" ID		190	.084		4.39	3.97		8.36	11.25
0370	6" ID		180	.089		4.76	4.19		8.95	12
0380	6-1/2" ID		175	.091		5.15	4.31		9.46	12.60
0390	7" ID		170	.094		5.50	4.44		9.94	13.20
0400	7-1/2" ID		164	.098		5.90	4.60		10.50	13.90
0410	8" ID		161	.099		6.25	4.68		10.93	14.45
0420	8-1/2" ID		158	.101		6.65	4.77		11.42	14.95
0430	9" ID		155	.103		7.05	4.87		11.92	15.55
0440	9-1/2" ID		152	.105		7.40	4.96		12.36	16.10
0450	10" ID		149	.107		7.80	5.05		12.85	16.70
0460	10-1/2" ID		146	.110		8.15	5.15		13.30	17.30
0470	11" ID		143	.112		8.55	5.25		13.80	17.90
0480	12" ID		137	.117		9.30	5.50		14.80	19.05
0490	13" ID		132	.121		10.05	5.70		15.75	20.50
0500	14" ID		130	.123		10.80	5.80		16.60	21
0700	For smooth .016" thick, add					45%	33%			
1000	Stainless steel, Type 316, fitting covers									
1010	Size is based on OD of insulation									
1020	90° LR Elbow, 2 piece									
1100	1-1/2"	Q-14	126	.127	Ea.	12.50	6		18.50	23.50
1110	2-3/4"		123	.130		12.50	6.15		18.65	23.50
1120	3"		120	.133		13.05	6.30		19.35	24.50
1130	3-1/4"		117	.137		13.05	6.45		19.50	25
1140	3-1/2"		115	.139		13.75	6.55		20.30	25.50
1150	3-3/4"		113	.142		14.50	6.65		21.15	26.50
1160	4"		110	.145		15.95	6.85		22.80	28.50
1170	4-1/4"		108	.148		21	7		28	34.50
1180	4-1/2"		106	.151		21	7.10		28.10	35
1190	5"		102	.157		21.50	7.40		28.90	36
1200	5-1/2"		97	.165		32.50	7.75		40.25	48
1210	6"		92	.174		35.50	8.20		43.70	52
1220	6-1/2"		87	.184		49	8.65		57.65	67.50
1230	7"		85	.188		49	8.85		57.85	68
1240	7-1/2"		83	.193		57	9.10		66.10	77
1250	8"		80	.200		57	9.45		66.45	77.50
1260	8-1/2"		80	.200		59	9.45		68.45	80
1270	9"		78	.205		89.50	9.65		99.15	114
1280	9-1/2"		77	.208		88	9.80		97.80	113

22 07 Plumbing Insulation

22 07 19 – Plumbing Piping Insulation

22 07 19.50 Pipe Insulation Protective Jacketing, St. Stl.		Crew	Daily Output	Labor-Hours	Unit	Material	2015 Bare Costs Labor	Equipment	Total	Total Incl O&P
1290	10"	Q-14	76	.211	Ea.	88	9.90		97.90	113
1300	10-1/2"		75	.213		105	10.05		115.05	132
1310	11"		74	.216		101	10.20		111.20	127
1320	12"		71	.225		114	10.60		124.60	143
1330	13"		68	.235		158	11.10		169.10	192
1340	14"		66	.242		159	11.45		170.45	193
2000	45° Elbow, 2 piece									
2010	2-1/2"	Q-14	126	.127	Ea.	10.40	6		16.40	21
2020	2-3/4"		123	.130		10.40	6.15		16.55	21.50
2030	3"		120	.133		11.20	6.30		17.50	22.50
2040	3-1/4"		117	.137		11.20	6.45		17.65	22.50
2050	3-1/2"		115	.139		11.35	6.55		17.90	23
2060	3-3/4"		113	.142		11.35	6.65		18	23.50
2070	4"		110	.145		14.60	6.85		21.45	27
2080	4-1/4"		108	.148		20.50	7		27.50	33.50
2090	4-1/2"		106	.151		20.50	7.10		27.60	34
2100	4-3/4"		104	.154		20.50	7.25		27.75	34
2110	5"		102	.157		21	7.40		28.40	35
2120	5-1/2"		97	.165		21	7.75		28.75	35.50
2130	6"		92	.174		24.50	8.20		32.70	40
2140	6-1/2"		87	.184		24.50	8.65		33.15	41
2150	7"		85	.188		42	8.85		50.85	60.50
2160	7-1/2"		83	.193		42.50	9.10		51.60	61.50
2170	8"		82	.195		42.50	9.20		51.70	62
2180	8-1/2"		80	.200		50	9.45		59.45	70
2190	9"		78	.205		50	9.65		59.65	70.50
2200	9-1/2"		77	.208		58.50	9.80		68.30	80.50
2210	10"		76	.211		58.50	9.90		68.40	80.50
2220	10-1/2"		75	.213		70.50	10.05		80.55	93.50
2230	11"		74	.216		70.50	10.20		80.70	94
2240	12"		71	.225		76.50	10.60		87.10	101
2250	13"		68	.235		90	11.10		101.10	117

22 11 Facility Water Distribution

22 11 13 – Facility Water Distribution Piping

22 11 13.23 Pipe/Tube, Copper

		Crew	Daily Output	Labor-Hours	Unit	Material	Labor	Equipment	Total	Total Incl O&P
0010	**PIPE/TUBE, COPPER**, Solder joints R221113-50									
1000	Type K tubing, couplings & clevis hanger assemblies 10' O.C.									
1100	1/4" diameter	1 Plum	84	.095	L.F.	3.71	5.60		9.31	12.80
1120	3/8" diameter		82	.098		4.31	5.75		10.06	13.70
1140	1/2" diameter		78	.103		4.78	6		10.78	14.65
1160	5/8" diameter		77	.104		5.90	6.10		12	15.95
1180	3/4" diameter		74	.108		7.80	6.35		14.15	18.45
1200	1" diameter		66	.121		10.35	7.10		17.45	22.50
1220	1-1/4" diameter		56	.143		12.20	8.40		20.60	26.50
1240	1-1/2" diameter		50	.160		15.60	9.40		25	32
1260	2" diameter		40	.200		23.50	11.75		35.25	44
1280	2-1/2" diameter	Q-1	60	.267		35	14.10		49.10	60.50
1300	3" diameter		54	.296		48	15.65		63.65	77
1320	3-1/2" diameter		42	.381		64	20		84	102
1330	4" diameter		38	.421		80	22		102	123

22 11 13.23 Pipe/Tube, Copper		Crew	Daily Output	Labor-Hours	Unit	2015 Bare Costs Material	Labor	Equipment	Total	Total Incl O&P
1340	5″ diameter	Q-1	32	.500	L.F.	152	26.50		178.50	208
1360	6″ diameter	Q-2	38	.632		222	34.50		256.50	298
1380	8″ diameter	"	34	.706		385	38.50		423.50	485
1390	For other than full hard temper, add					13%				
1440	For silver solder, add						15%			
1800	For medical clean, (oxygen class), add					12%				
1950	To delete cplgs. & hngrs., 1/4″-1″ pipe, subtract					27%	60%			
1960	1-1/4″ -3″ pipe, subtract					14%	52%			
1970	3-1/2″-5″ pipe, subtract					10%	60%			
1980	6″-8″ pipe, subtract					19%	53%			
2000	Type L tubing, couplings & clevis hanger assemblies 10′ O.C.									
2100	1/4″ diameter	1 Plum	88	.091	L.F.	2.77	5.35		8.12	11.35
2120	3/8″ diameter		84	.095		3.50	5.60		9.10	12.55
2140	1/2″ diameter		81	.099		3.70	5.80		9.50	13.10
2160	5/8″ diameter		79	.101		5.35	5.95		11.30	15.10
2180	3/4″ diameter		76	.105		5.20	6.20		11.40	15.35
2200	1″ diameter		68	.118		7.85	6.90		14.75	19.35
2220	1-1/4″ diameter		58	.138		10.35	8.10		18.45	24
2240	1-1/2″ diameter		52	.154		12.95	9.05		22	28.50
2260	2″ diameter		42	.190		18.80	11.20		30	38
2280	2-1/2″ diameter	Q-1	62	.258		29.50	13.65		43.15	54
2300	3″ diameter		56	.286		37	15.10		52.10	64
2320	3-1/2″ diameter		43	.372		53	19.65		72.65	88.50
2340	4″ diameter		39	.410		65.50	21.50		87	106
2360	5″ diameter		34	.471		120	25		145	171
2380	6″ diameter	Q-2	40	.600		166	33		199	234
2400	8″ diameter	"	36	.667		285	36.50		321.50	370
2410	For other than full hard temper, add					21%				
2590	For silver solder, add						15%			
2900	For medical clean, (oxygen class), add					12%				
2940	To delete cplgs. & hngrs., 1/4″-1″ pipe, subtract					37%	63%			
2960	1-1/4″-3″ pipe, subtract					12%	53%			
2970	3-1/2″-5″ pipe, subtract					12%	63%			
2980	6″-8″ pipe, subtract					24%	55%			
3000	Type M tubing, couplings & clevis hanger assemblies 10′ O.C.									
3100	1/4″ diameter	1 Plum	90	.089	L.F.	3	5.20		8.20	11.45
3120	3/8″ diameter		87	.092		3.05	5.40		8.45	11.75
3140	1/2″ diameter		84	.095		3.05	5.60		8.65	12.05
3160	5/8″ diameter		81	.099		4.49	5.80		10.29	14
3180	3/4″ diameter		78	.103		4.13	6		10.13	13.95
3200	1″ diameter		70	.114		6.60	6.70		13.30	17.70
3220	1-1/4″ diameter		60	.133		8.80	7.85		16.65	22
3240	1-1/2″ diameter		54	.148		11.55	8.70		20.25	26.50
3260	2″ diameter		44	.182		17.35	10.65		28	36
3280	2-1/2″ diameter	Q-1	64	.250		25.50	13.20		38.70	48.50
3300	3″ diameter		58	.276		34	14.55		48.55	60
3320	3-1/2″ diameter		45	.356		47.50	18.80		66.30	81.50
3340	4″ diameter		40	.400		60	21		81	99
3360	5″ diameter		36	.444		120	23.50		143.50	169
3370	6″ diameter	Q-2	42	.571		166	31.50		197.50	231
3380	8″ diameter	"	38	.632		285	34.50		319.50	370
3440	For silver solder, add						15%			
3960	To delete cplgs. & hngrs., 1/4″-1″ pipe, subtract					35%	65%			

22 11 13.23 Pipe/Tube, Copper		Crew	Daily Output	Labor-Hours	Unit	Material	2015 Bare Costs Labor	Equipment	Total	Total Incl O&P
3970	1-1/4″-3″ pipe, subtract					19%	56%			
3980	3-1/2″-5″ pipe, subtract					13%	65%			
3990	6″-8″ pipe, subtract					28%	58%			
4000	Type DWV tubing, couplings & clevis hanger assemblies 10′ O.C.									
4100	1-1/4″ diameter	1 Plum	60	.133	L.F.	9.15	7.85		17	22.50
4120	1-1/2″ diameter		54	.148		11.15	8.70		19.85	26
4140	2″ diameter		44	.182		14.60	10.65		25.25	33
4160	3″ diameter	Q-1	58	.276		26.50	14.55		41.05	51.50
4180	4″ diameter		40	.400		44	21		65	81
4200	5″ diameter		36	.444		107	23.50		130.50	155
4220	6″ diameter	Q-2	42	.571		153	31.50		184.50	217
4240	8″ diameter	″	38	.632		310	34.50		344.50	395
4730	To delete cplgs. & hngrs., 1-1/4″-2″ pipe, subtract					16%	53%			
4740	3″-4″ pipe, subtract					13%	60%			
4750	5″ 8″ pipe, subtract					23%	58%			
5200	ACR tubing, type L, hard temper, cleaned and									
5220	capped, no couplings or hangers									
5240	3/8″ OD				L.F.	1.51			1.51	1.66
5250	1/2″ OD					2.30			2.30	2.53
5260	5/8″ OD					2.81			2.81	3.09
5270	3/4″ OD					3.97			3.97	4.37
5280	7/8″ OD					4.41			4.41	4.85
5290	1-1/8″ OD					6.40			6.40	7.05
5300	1-3/8″ OD					8.60			8.60	9.45
5310	1-5/8″ OD					11			11	12.10
5320	2-1/8″ OD					17.30			17.30	19.05
5330	2-5/8″ OD					26			26	28.50
5340	3-1/8″ OD					34.50			34.50	38
5350	3-5/8″ OD					45			45	49.50
5360	4-1/8″ OD					57.50			57.50	63
5380	ACR tubing, type L, hard, cleaned and capped									
5381	No couplings or hangers									
5384	3/8″	1 Stpi	160	.050	L.F.	1.51	2.99		4.50	6.30
5385	1/2″		160	.050		2.30	2.99		5.29	7.20
5386	5/8″		160	.050		2.81	2.99		5.80	7.75
5387	3/4″		130	.062		3.97	3.68		7.65	10.10
5388	7/8″		130	.062		4.41	3.68		8.09	10.60
5389	1-1/8″		115	.070		6.40	4.16		10.56	13.55
5390	1-3/8″		100	.080		8.60	4.78		13.38	16.90
5391	1-5/8″		90	.089		11	5.30		16.30	20.50
5392	2-1/8″		80	.100		17.30	6		23.30	28.50
5393	2-5/8″	Q-5	125	.128		26	6.90		32.90	39.50
5394	3-1/8″		105	.152		34.50	8.20		42.70	51
5395	4-1/8″		95	.168		57.50	9.05		66.55	77
5800	Refrigeration tubing, dryseal, 50′ coils									
5840	1/8″ OD				Coil	32			32	35.50
5850	3/16″ OD					37.50			37.50	41.50
5860	1/4″ OD					43.50			43.50	47.50
5870	5/16″ OD					58			58	64
5880	3/8″ OD					60			60	66
5890	1/2″ OD					87			87	96
5900	5/8″ OD					112			112	124
5910	3/4″ OD					133			133	147

22 11 13.23 Pipe/Tube, Copper		Crew	Daily Output	Labor-Hours	Unit	Material	Labor (2015 Bare Costs)	Equipment	Total	Total Incl O&P
5920	7/8″ OD				Coil	199			199	219
5930	1-1/8″ OD					300			300	330
5940	1-3/8″ OD					520			520	575
5950	1-5/8″ OD					660			660	730
9000	Minimum labor/equipment charge	1 Plum	4	2	Job		117		117	183
9400	Sub assemblies used in assembly systems									
9410	Chilled water unit, coil connections per unit under 10 ton	Q-5	.80	20	System	1,175	1,075		2,250	2,975
9420	Chilled water unit, coil connections per unit 10 ton and up		1	16		1,900	860		2,760	3,450
9430	Chilled water dist. piping per ton, less than 61 ton systems		26	.615		21	33		54	74.50
9440	Chilled water dist. piping per ton, 61 through 120 ton systems	Q-6	31	.774		48.50	43		91.50	121
9450	Chilled water dist. piping/ton, 135 ton systems and up	Q-8	25.40	1.260		64.50	71.50	2.27	138.27	185
9510	Refrigerant piping/ton of cooling for remote condensers	Q-5	2	8		375	430		805	1,075
9520	Refrigerant piping per ton up to 10 ton w/remote condensing unit		2.40	6.667		176	360		536	755
9530	Refrigerant piping per ton, 20 ton w/remote condensing unit		2	8		259	430		689	955
9540	Refrigerant piping per ton, 40 ton w/remote condensing unit		1.90	8.421		330	455		785	1,075
9550	Refrigerant piping per ton, 75-80 ton w/remote condensing unit	Q-6	2.40	10		485	560		1,045	1,400
9560	Refrigerant piping per ton, 100 ton w/remote condensing unit	″	2.20	10.909		610	610		1,220	1,625

22 11 13.25 Pipe/Tube Fittings, Copper

		Crew	Daily Output	Labor-Hours	Unit	Material	Labor	Equipment	Total	Total Incl O&P
0010	**PIPE/TUBE FITTINGS, COPPER**, Wrought unless otherwise noted									
0020	For silver solder, add						15%			
0040	Solder joints, copper x copper									
0070	90° elbow, 1/4″	1 Plum	22	.364	Ea.	3.28	21.50		24.78	37
0090	3/8″		22	.364		3.11	21.50		24.61	37
0100	1/2″		20	.400		1.10	23.50		24.60	37.50
0110	5/8″		19	.421		2.46	24.50		26.96	41
0120	3/4″		19	.421		2.46	24.50		26.96	41
0130	1″		16	.500		6.05	29.50		35.55	52.50
0140	1-1/4″		15	.533		9	31.50		40.50	59
0150	1-1/2″		13	.615		14.05	36		50.05	72
0160	2″		11	.727		25.50	42.50		68	94.50
0170	2-1/2″	Q-1	13	1.231		51.50	65		116.50	158
0180	3″		11	1.455		64.50	77		141.50	191
0190	3-1/2″		10	1.600		225	84.50		309.50	380
0200	4″		9	1.778		152	94		246	315
0210	5″		6	2.667		645	141		786	930
0220	6″	Q-2	9	2.667		865	146		1,011	1,175
0230	8″	″	8	3		3,475	164		3,639	4,075
0250	45° elbow, 1/4″	1 Plum	22	.364		5.90	21.50		27.40	40
0270	3/8″		22	.364		5	21.50		26.50	39
0280	1/2″		20	.400		2.01	23.50		25.51	38.50
0290	5/8″		19	.421		9.55	24.50		34.05	49
0300	3/4″		19	.421		3.52	24.50		28.02	42.50
0310	1″		16	.500		8.85	29.50		38.35	55.50
0320	1-1/4″		15	.533		11.85	31.50		43.35	62
0330	1-1/2″		13	.615		14.25	36		50.25	72
0340	2″		11	.727		24	42.50		66.50	92.50
0350	2-1/2″	Q-1	13	1.231		50.50	65		115.50	157
0360	3″		13	1.231		70.50	65		135.50	179
0370	3-1/2″		10	1.600		125	84.50		209.50	270
0380	4″		9	1.778		151	94		245	310
0390	5″		6	2.667		545	141		686	820
0400	6″	Q-2	9	2.667		860	146		1,006	1,175

22 11 13.25 Pipe/Tube Fittings, Copper		Crew	Daily Output	Labor-Hours	Unit	Material	2015 Bare Costs Labor	Equipment	Total	Total Incl O&P
0410	8"	Q-2	8	3	Ea.	3,400	164		3,564	3,975
0450	Tee, 1/4"	1 Plum	14	.571		6.55	33.50		40.05	59.50
0470	3/8"		14	.571		5.25	33.50		38.75	58.50
0480	1/2"		13	.615		1.87	36		37.87	58.50
0490	5/8"		12	.667		12.05	39		51.05	74.50
0500	3/4"		12	.667		4.51	39		43.51	66
0510	1"		10	.800		13.95	47		60.95	88.50
0520	1-1/4"		9	.889		18.85	52		70.85	102
0530	1-1/2"		8	1		29	58.50		87.50	124
0540	2"		7	1.143		45.50	67		112.50	155
0550	2-1/2"	Q-1	8	2		91.50	106		197.50	266
0560	3"		7	2.286		131	121		252	330
0570	3-1/2"		6	2.667		380	141		521	640
0580	4"		5	3.200		320	169		489	615
0590	5"		4	4		975	211		1,186	1,400
0600	6"	Q-2	6	4		1,325	219		1,544	1,825
0610	8"	"	5	4.800		5,125	263		5,388	6,050
0612	Tee, reducing on the outlet, 1/4"	1 Plum	15	.533		16.45	31.50		47.95	67
0613	3/8"		15	.533		15.40	31.50		46.90	66
0614	1/2"		14	.571		10.05	33.50		43.55	63.50
0615	5/8"		13	.615		32	36		68	92
0616	3/4"		12	.667		6.45	39		45.45	68
0617	1"		11	.727		18.50	42.50		61	87
0618	1-1/4"		10	.800		20.50	47		67.50	95.50
0619	1-1/2"		9	.889		21.50	52		73.50	106
0620	2"		8	1		39	58.50		97.50	135
0621	2-1/2"	Q-1	9	1.778		107	94		201	263
0622	3"		8	2		110	106		216	286
0623	4"		6	2.667		203	141		344	445
0624	5"		5	3.200		1,125	169		1,294	1,525
0625	6"	Q-2	7	3.429		1,550	188		1,738	2,000
0626	8"	"	6	4		5,950	219		6,169	6,900
0630	Tee, reducing on the run, 1/4"	1 Plum	15	.533		13.90	31.50		45.40	64.50
0631	3/8"		15	.533		19.60	31.50		51.10	70.50
0632	1/2"		14	.571		11.35	33.50		44.85	65
0633	5/8"		13	.615		19.40	36		55.40	78
0634	3/4"		12	.667		15.15	39		54.15	77.50
0635	1"		11	.727		16.65	42.50		59.15	85
0636	1-1/4"		10	.800		26	47		73	102
0637	1-1/2"		9	.889		46	52		98	133
0638	2"		8	1		59	58.50		117.50	157
0639	2-1/2"	Q-1	9	1.778		152	94		246	315
0640	3"		8	2		202	106		308	385
0641	4"		6	2.667		425	141		566	685
0642	5"		5	3.200		1,075	169		1,244	1,450
0643	6"	Q-2	7	3.429		1,625	188		1,813	2,100
0644	8"	"	6	4		5,950	219		6,169	6,900
0650	Coupling, 1/4"	1 Plum	24	.333		.75	19.55		20.30	31.50
0670	3/8"		24	.333		1	19.55		20.55	31.50
0680	1/2"		22	.364		.83	21.50		22.33	34.50
0690	5/8"		21	.381		2.50	22.50		25	38
0700	3/4"		21	.381		1.67	22.50		24.17	37
0710	1"		18	.444		5.55	26		31.55	46.50

22 11 13.25 Pipe/Tube Fittings, Copper		Crew	Daily Output	Labor-Hours	Unit	2015 Bare Costs Material	Labor	Equipment	Total	Total Incl O&P
0715	1-1/4"	1 Plum	17	.471	Ea.	5.85	27.50		33.35	49.50
0716	1-1/2"		15	.533		7.70	31.50		39.20	57.50
0718	2"		13	.615		12.90	36		48.90	70.50
0721	2-1/2"	Q-1	15	1.067		27.50	56.50		84	118
0722	3"		13	1.231		38.50	65		103.50	144
0724	3-1/2"		8	2		74	106		180	247
0726	4"		7	2.286		81.50	121		202.50	278
0728	5"		6	2.667		187	141		328	425
0731	6"	Q-2	8	3		310	164		474	595
0732	8"	"	7	3.429		1,000	188		1,188	1,400
0741	Coupling, reducing, concentric									
0743	1/2"	1 Plum	23	.348	Ea.	2.16	20.50		22.66	34.50
0745	3/4"		21.50	.372		4.70	22		26.70	39
0747	1"		19.50	.410		5.95	24		29.95	44
0748	1-1/4"		18	.444		8.30	26		34.30	49.50
0749	1-1/2"		16	.500		10.80	29.50		40.30	58
0751	2"		14	.571		23	33.50		56.50	77.50
0752	2-1/2"		13	.615		47	36		83	109
0753	3"	Q-1	14	1.143		53.50	60.50		114	153
0755	4"	"	8	2		113	106		219	289
0757	5"	Q-2	7.50	3.200		585	175		760	915
0759	6"		7	3.429		1,075	188		1,263	1,500
0761	8"		6.50	3.692		2,150	202		2,352	2,675
0771	Cap, sweat									
0773	1/2"	1 Plum	40	.200	Ea.	.80	11.75		12.55	19.20
0775	3/4"		38	.211		1.48	12.35		13.83	21
0777	1"		32	.250		3.52	14.70		18.22	27
0778	1-1/4"		29	.276		4.66	16.20		20.86	30.50
0779	1-1/2"		26	.308		6.80	18.05		24.85	35.50
0781	2"		22	.364		12.30	21.50		33.80	47
0791	Flange, sweat									
0793	3"	Q-1	22	.727	Ea.	174	38.50		212.50	251
0795	4"		18	.889		243	47		290	340
0797	5"		12	1.333		455	70.50		525.50	610
0799	6"	Q-2	18	1.333		475	73		548	635
0801	8"	"	16	1.500		1,000	82		1,082	1,225
0850	Unions, 1/4"	1 Plum	21	.381		26	22.50		48.50	64
0870	3/8"		21	.381		26.50	22.50		49	64
0880	1/2"		19	.421		14.05	24.50		38.55	54
0890	5/8"		18	.444		60	26		86	107
0900	3/4"		18	.444		17.60	26		43.60	60
0910	1"		15	.533		30.50	31.50		62	82.50
0920	1-1/4"		14	.571		50	33.50		83.50	107
0930	1-1/2"		12	.667		65.50	39		104.50	133
0940	2"		10	.800		113	47		160	197
0950	2-1/2"	Q-1	12	1.333		245	70.50		315.50	380
0960	3"	"	10	1.600		635	84.50		719.50	825
0980	Adapter, copper x male IPS, 1/4"	1 Plum	20	.400		10.85	23.50		34.35	48.50
0990	3/8"		20	.400		5.40	23.50		28.90	42.50
1000	1/2"		18	.444		2.29	26		28.29	43
1010	3/4"		17	.471		3.84	27.50		31.34	47
1020	1"		15	.533		9.95	31.50		41.45	60
1030	1-1/4"		13	.615		14.55	36		50.55	72.50

22 11 13 – Facility Water Distribution Piping

22 11 13.25 Pipe/Tube Fittings, Copper		Crew	Daily Output	Labor-Hours	Unit	Material	Labor	Equipment	Total	Total Incl O&P
						2015 Bare Costs				
1040	1-1/2"	1 Plum	12	.667	Ea.	16.70	39		55.70	79.50
1050	2"		11	.727		28.50	42.50		71	97.50
1060	2-1/2"	Q-1	10.50	1.524		108	80.50		188.50	245
1070	3"		10	1.600		135	84.50		219.50	280
1080	3-1/2"		9	1.778		161	94		255	325
1090	4"		8	2		182	106		288	365
1200	5", cast		6	2.667		4,600	141		4,741	5,275
1210	6", cast	Q-2	8.50	2.824		5,125	155		5,280	5,875
1250	Cross, 1/2"	1 Plum	10	.800		18	47		65	93
1260	3/4"		9.50	.842		30.50	49.50		80	111
1270	1"		8	1		38	58.50		96.50	134
1280	1-1/4"		7.50	1.067		85.50	62.50		148	192
1290	1-1/2"		6.50	1.231		122	72.50		194.50	247
1300	2"		5.50	1.455		237	85.50		322.50	395
1310	2-1/2"	Q-1	6.50	2.462		530	130		660	790
1320	3"	"	5.50	2.909		620	154		774	925
1500	Tee fitting, mechanically formed, (Type 1, 'branch sizes up to 2 in.')									
1520	1/2" run size, 3/8" to 1/2" branch size	1 Plum	80	.100	Ea.		5.85		5.85	9.15
1530	3/4" run size, 3/8" to 3/4" branch size		60	.133			7.85		7.85	12.20
1540	1" run size, 3/8" to 1" branch size		54	.148			8.70		8.70	13.55
1550	1-1/4" run size, 3/8" to 1-1/4" branch size		48	.167			9.80		9.80	15.25
1560	1-1/2" run size, 3/8" to 1-1/2" branch size		40	.200			11.75		11.75	18.30
1570	2" run size, 3/8" to 2" branch size		35	.229			13.40		13.40	21
1580	2-1/2" run size, 1/2" to 2" branch size		32	.250			14.70		14.70	23
1590	3" run size, 1" to 2" branch size		26	.308			18.05		18.05	28
1600	4" run size, 1" to 2" branch size		24	.333			19.55		19.55	30.50
1640	Tee fitting, mechanically formed, (Type 2, branches 2-1/2" thru 4")									
1650	2-1/2" run size, 2-1/2" branch size	1 Plum	12.50	.640	Ea.		37.50		37.50	58.50
1660	3" run size, 2-1/2" to 3" branch size		12	.667			39		39	61
1670	3-1/2" run size, 2-1/2" to 3-1/2" branch size		11	.727			42.50		42.50	66.50
1680	4" run size, 2-1/2" to 4" branch size		10.50	.762			44.50		44.50	70
1698	5" run size, 2" to 4" branch size		9.50	.842			49.50		49.50	77
1700	6" run size, 2" to 4" branch size		8.50	.941			55.50		55.50	86
1710	8" run size, 2" to 4" branch size		7	1.143			67		67	105
1800	ACR fittings, OD size									
1802	Tee, straight									
1808	5/8"	1 Stpi	12	.667	Ea.	1.87	40		41.87	64
1810	3/4"		12	.667		12.05	40		52.05	75.50
1812	7/8"		10	.800		4.51	48		52.51	79.50
1813	1"		10	.800		33	48		81	111
1814	1-1/8"		10	.800		13.95	48		61.95	90
1816	1-3/8"		9	.889		18.85	53		71.85	104
1818	1-5/8"		8	1		29	60		89	125
1820	2-1/8"		7	1.143		45.50	68.50		114	157
1822	2-5/8"	Q-5	8	2		91.50	108		199.50	269
1824	3-1/8"		7	2.286		131	123		254	335
1826	4-1/8"		5	3.200		320	172		492	620
1830	90° elbow									
1836	5/8"	1 Stpi	19	.421	Ea.	3.67	25		28.67	43
1838	3/4"		19	.421		6.65	25		31.65	46.50
1840	7/8"		16	.500		6.60	30		36.60	54
1842	1-1/8"		16	.500		8.85	30		38.85	56.50
1844	1-3/8"		15	.533		9	32		41	59.50

22 11 13.25 Pipe/Tube Fittings, Copper		Crew	Daily Output	Labor-Hours	Unit	Material	Labor	Equipment	Total	Total Incl O&P
						2015 Bare Costs				
1846	1-5/8"	1 Stpi	13	.615	Ea.	14.05	37		51.05	73
1848	2-1/8"	↓	11	.727		25.50	43.50		69	96
1850	2-5/8"	Q-5	13	1.231		51.50	66		117.50	160
1852	3-1/8"		11	1.455		64.50	78		142.50	193
1854	4-1/8"	↓	9	1.778	↓	170	95.50		265.50	335
1860	Coupling									
1866	5/8"	1 Stpi	21	.381	Ea.	1.15	23		24.15	37
1868	3/4"		21	.381		2.19	23		25.19	38
1870	7/8"		18	.444		1.67	26.50		28.17	43.50
1871	1"		18	.444		5.70	26.50		32.20	48
1872	1-1/8"		18	.444		3.33	26.50		29.83	45
1874	1-3/8"		17	.471		5.85	28		33.85	50.50
1876	1-5/8"		15	.533		7.70	32		39.70	58
1878	2 1/8"	↓	13	.615		12.90	37		49.90	71.50
1880	2-5/8"	Q-5	15	1.067		29	57.50		86.50	122
1882	3-1/8"		13	1.231		39.50	66		105.50	147
1884	4-1/8"	↓	7	2.286	↓	90	123		213	291
2000	DWV, solder joints, copper x copper									
2030	90° Elbow, 1-1/4"	1 Plum	13	.615	Ea.	10.45	36		46.45	68
2050	1-1/2"		12	.667		14.55	39		53.55	77
2070	2"	↓	10	.800		28	47		75	104
2090	3"	Q-1	10	1.600		56.50	84.50		141	195
2100	4"	"	9	1.778		335	94		429	510
2150	45° Elbow, 1-1/4"	1 Plum	13	.615		9.65	36		45.65	67
2170	1-1/2"		12	.667		8	39		47	70
2180	2"	↓	10	.800		18.35	47		65.35	93
2190	3"	Q-1	10	1.600		36.50	84.50		121	173
2200	4"	"	9	1.778		184	94		278	350
2250	Tee, Sanitary, 1-1/4"	1 Plum	9	.889		20.50	52		72.50	105
2270	1-1/2"		8	1		26	58.50		84.50	120
2290	2"	↓	7	1.143		30	67		97	138
2310	3"	Q-1	7	2.286		138	121		259	340
2330	4"	"	6	2.667		330	141		471	580
2400	Coupling, 1-1/4"	1 Plum	14	.571		4.91	33.50		38.41	58
2420	1-1/2"		13	.615		6.10	36		42.10	63
2440	2"	↓	11	.727		8.45	42.50		50.95	76
2460	3"	Q-1	11	1.455		15.40	77		92.40	137
2480	4"	"	10	1.600	↓	49	84.50		133.50	186
2602	Traps, see Section 22 13 16.60									
6992	Tube connector fittings, See Section 22 11 13.76 for plastic ftng.									
7000	Insert type Brass/copper, 100 psi @ 180°F, CTS									
7010	Adapter MPT 3/8" x 3/8" CTS	1 Plum	29	.276	Ea.	2.93	16.20		19.13	28.50
7020	1/2" x 1/2"		26	.308		2.95	18.05		21	31.50
7030	3/4" x 1/2"		26	.308		3.77	18.05		21.82	32
7040	3/4" x 3/4"		25	.320		4.35	18.80		23.15	34.50
7050	Adapter CTS 1/2" x 1/2" sweat		24	.333		3.96	19.55		23.51	35
7060	3/4" x 3/4" sweat		22	.364		1.39	21.50		22.89	35
7070	Coupler center set 3/8" CTS		25	.320		1.44	18.80		20.24	31
7080	1/2" CTS		23	.348		3.96	20.50		24.46	36.50
7090	3/4" CTS		22	.364		1.39	21.50		22.89	35
7100	Elbow 90°, copper 3/8"		25	.320		3.16	18.80		21.96	33
7110	1/2" CTS		23	.348		2.02	20.50		22.52	34
7120	3/4" CTS	↓	22	.364	↓	2.46	21.50		23.96	36

22 11 13.25 Pipe/Tube Fittings, Copper		Crew	Daily Output	Labor-Hours	Unit	Material	2015 Bare Costs Labor	Equipment	Total	Total Incl O&P
7130	Tee copper 3/8" CTS	1 Plum	17	.471	Ea.	3.75	27.50		31.25	47
7140	1/2" CTS		15	.533		2.58	31.50		34.08	52
7150	3/4" CTS		14	.571		3.96	33.50		37.46	57
7160	3/8" x 3/8" x 1/2"		16	.500		2.98	29.50		32.48	49.50
7170	1/2" x 3/8" x 1/2"		15	.533		1.93	31.50		33.43	51
7180	3/4" x 1/2" x 3/4"		14	.571		2.98	33.50		36.48	56
9000	Minimum labor/equipment charge		4	2	Job		117		117	183

22 11 13.29 Pipe, Fittings and Valves, Copper, Pressed-Joint		Crew	Daily Output	Labor-Hours	Unit	Material	Labor	Equipment	Total	Total Incl O&P
0010	**PIPE, FITTINGS AND VALVES, COPPER, PRESSED-JOINT**									
0040	Pipe/tube includes coupling & clevis type hanger assy's, 10' O.C.									
0120	Type K									
0130	1/2" diameter	1 Plum	78	.103	L.F.	4.98	6		10.98	14.90
0134	3/4" diameter		74	.108		8.05	6.35		14.40	18.75
0138	1" diameter		66	.121		10.65	7.10		17.75	23
0142	1-1/4" diameter		56	.143		12.70	8.40		21.10	27
0146	1-1/2" diameter		50	.160		16.80	9.40		26.20	33
0150	2" diameter		40	.200		24.50	11.75		36.25	45.50
0154	2-1/2" diameter	Q-1	60	.267		40	14.10		54.10	66
0158	3" diameter		54	.296		54	15.65		69.65	84
0162	4" diameter		38	.421		86	22		108	129
0180	To delete cplgs. & hngrs., 1/2" pipe, subtract					19%	48%			
0184	3/4" -2" pipe, subtract					14%	46%			
0186	2-1/2"-4" pipe, subtract					24%	34%			
0220	Type L									
0230	1/2" diameter	1 Plum	81	.099	L.F.	3.90	5.80		9.70	13.35
0234	3/4" diameter		76	.105		5.45	6.20		11.65	15.65
0238	1" diameter		68	.118		8.10	6.90		15	19.70
0242	1-1/4" diameter		58	.138		10.80	8.10		18.90	24.50
0246	1-1/2" diameter		52	.154		14.15	9.05		23.20	29.50
0250	2" diameter		42	.190		20	11.20		31.20	39.50
0254	2-1/2" diameter	Q-1	62	.258		35	13.65		48.65	59.50
0258	3" diameter		56	.286		43.50	15.10		58.60	71
0262	4" diameter		39	.410		71.50	21.50		93	113
0280	To delete cplgs. & hngrs., 1/2" - 1"pipe, subtract					21%	52%			
0284	1-1/4"-2" pipe, subtract					17%	46%			
0286	2-1/2"-4" pipe, subtract					23%	35%			
0320	Type M									
0330	1/2" diameter	1 Plum	84	.095	L.F.	3.25	5.60		8.85	12.30
0334	3/4" diameter		78	.103		4.38	6		10.38	14.20
0338	1" diameter		70	.114		6.85	6.70		13.55	18
0342	1-1/4" diameter		60	.133		9.25	7.85		17.10	22.50
0346	1-1/2" diameter		54	.148		12.75	8.70		21.45	27.50
0350	2" diameter		44	.182		18.55	10.65		29.20	37
0354	2-1/2" diameter	Q-1	64	.250		31	13.20		44.20	54.50
0358	3" diameter		58	.276		40.50	14.55		55.05	67
0362	4" diameter		40	.400		66	21		87	106
0380	To delete cplgs. & hngrs., 1/2" pipe, subtract					32%	49%			
0384	3/4"-2" pipe, subtract					21%	46%			
0386	2-1/2"-4" pipe, subtract					25%	36%			
1600	Fittings									
1610	Press joints, copper x copper									
1620	Note: Reducing fittings show most expensive size combination.									

22 11 13.29 Pipe, Fittings and Valves, Copper, Pressed-Joint		Crew	Daily Output	Labor-Hours	Unit	2015 Bare Costs Material	Labor	Equipment	Total	Total Incl O&P
1800	90° elbow, 1/2"	1 Plum	36.60	.219	Ea.	3.11	12.85		15.96	23.50
1810	3/4"		27.50	.291		5.25	17.10		22.35	32.50
1820	1"		25.90	.309		10.50	18.15		28.65	40
1830	1-1/4"		20.90	.383		21	22.50		43.50	58.50
1840	1-1/2"		18.30	.437		40	25.50		65.50	84
1850	2"		15.70	.510		55.50	30		85.50	108
1860	2-1/2"	Q-1	25.90	.618		161	32.50		193.50	229
1870	3"		22	.727		203	38.50		241.50	283
1880	4"		16.30	.982		251	52		303	355
2000	45° elbow, 1/2"	1 Plum	36.60	.219		3.61	12.85		16.46	24
2010	3/4"		27.50	.291		4.44	17.10		21.54	31.50
2020	1"		25.90	.309		14.20	18.15		32.35	44
2030	1-1/4"		20.90	.383		20.50	22.50		43	57.50
2040	1-1/2"		18.30	.437		33	25.50		58.50	76
2050	2"		15.70	.510		46	30		76	97
2060	2-1/2"	Q-1	25.90	.618		107	32.50		139.50	168
2070	3"		22	.727		152	38.50		190.50	228
2080	4"		16.30	.982		212	52		264	315
2200	Tee, 1/2"	1 Plum	27.50	.291		4.74	17.10		21.84	31.50
2210	3/4"		20.70	.386		8.35	22.50		30.85	44.50
2220	1"		19.40	.412		15	24		39	54.50
2230	1-1/4"		15.70	.510		24	30		54	73
2240	1-1/2"		13.80	.580		49	34		83	107
2250	2"		11.80	.678		61	40		101	129
2260	2-1/2"	Q-1	19.40	.825		199	43.50		242.50	287
2270	3"		16.50	.970		249	51		300	355
2280	4"		12.20	1.311		350	69.50		419.50	495
2400	Tee, reducing on the outlet									
2410	3/4"	1 Plum	20.70	.386	Ea.	7	22.50		29.50	43
2420	1"		19.40	.412		17.15	24		41.15	57
2430	1-1/4"		15.70	.510		24	30		54	73
2440	1-1/2"		13.80	.580		52.50	34		86.50	111
2450	2"		11.80	.678		84	40		124	155
2460	2-1/2"	Q-1	19.40	.825		278	43.50		321.50	375
2470	3"		16.50	.970		305	51		356	415
2480	4"		12.20	1.311		420	69.50		489.50	570
2600	Tee, reducing on the run									
2610	3/4"	1 Plum	20.70	.386	Ea.	13.35	22.50		35.85	50
2620	1"		19.40	.412		26	24		50	66.50
2630	1-1/4"		15.70	.510		50	30		80	102
2640	1-1/2"		13.80	.580		76.50	34		110.50	137
2650	2"		11.80	.678		84	40		124	155
2660	2-1/2"	Q-1	19.40	.825		284	43.50		327.50	385
2670	3"		16.50	.970		340	51		391	455
2680	4"		12.20	1.311		485	69.50		554.50	645
2800	Coupling, 1/2"	1 Plum	36.60	.219		2.80	12.85		15.65	23
2810	3/4"		27.50	.291		4.23	17.10		21.33	31
2820	1"		25.90	.309		8.55	18.15		26.70	38
2830	1-1/4"		20.90	.383		10.60	22.50		33.10	46.50
2840	1-1/2"		18.30	.437		19.70	25.50		45.20	61.50
2850	2"		15.70	.510		25	30		55	74
2860	2-1/2"	Q-1	25.90	.618		80.50	32.50		113	140
2870	3"		22	.727		102	38.50		140.50	173

22 11 13.29 Pipe, Fittings and Valves, Copper, Pressed-Joint		Crew	Daily Output	Labor-Hours	Unit	Material	2015 Bare Costs Labor	Equipment	Total	Total Incl O&P
2880	4"	Q-1	16.30	.982	Ea.	142	52		194	237
3000	Union, 1/2"	1 Plum	36.60	.219		23	12.85		35.85	45.50
3010	3/4"		27.50	.291		30.50	17.10		47.60	60
3020	1"		25.90	.309		48.50	18.15		66.65	82
3030	1-1/4"		20.90	.383		69.50	22.50		92	112
3040	1-1/2"		18.30	.437		94.50	25.50		120	144
3050	2"		15.70	.510		154	30		184	216
3200	Adapter, tube to MPT									
3210	1/2"	1 Plum	15.10	.530	Ea.	3.65	31		34.65	52.50
3220	3/4"		13.60	.588		6.60	34.50		41.10	61.50
3230	1"		11.60	.690		12.05	40.50		52.55	76.50
3240	1-1/4"		9.90	.808		26.50	47.50		74	103
3250	1-1/2"		9	.889		37	52		89	123
3260	2"		7.90	1.013		71.50	59.50		131	171
3270	2-1/2"	Q-1	13.40	1.194		161	63		224	277
3280	3"		10.60	1.509		203	79.50		282.50	345
3290	4"		7.70	2.078		235	110		345	430
3400	Adapter, tube to FPT									
3410	1/2"	1 Plum	15.10	.530	Ea.	4.40	31		35.40	53.50
3420	3/4"		13.60	.588		7.15	34.50		41.65	62
3430	1"		11.60	.690		13.15	40.50		53.65	77.50
3440	1-1/4"		9.90	.808		29.50	47.50		77	107
3450	1-1/2"		9	.889		43	52		95	129
3460	2"		7.90	1.013		73.50	59.50		133	173
3470	2-1/2"	Q-1	13.40	1.194		216	63		279	335
3480	3"		10.60	1.509		283	79.50		362.50	435
3490	4"		7.70	2.078		365	110		475	570
3600	Flange									
3620	1"	1 Plum	36.20	.221	Ea.	123	12.95		135.95	155
3630	1-1/4"		29.30	.273		178	16.05		194.05	221
3640	1-1/2"		25.60	.313		197	18.35		215.35	246
3650	2"		22	.364		221	21.50		242.50	277
3660	2-1/2"	Q-1	36.20	.442		247	23.50		270.50	310
3670	3"		30.80	.519		299	27.50		326.50	375
3680	4"		22.80	.702		335	37		372	430
3800	Cap, 1/2"	1 Plum	53.10	.151		5.95	8.85		14.80	20.50
3810	3/4"		39.80	.201		10.05	11.80		21.85	29.50
3820	1"		37.50	.213		15.50	12.50		28	36.50
3830	1-1/4"		30.40	.263		17.55	15.45		33	43.50
3840	1-1/2"		26.60	.301		27	17.65		44.65	57.50
3850	2"		22.80	.351		33	20.50		53.50	68.50
3860	2-1/2"	Q-1	37.50	.427		106	22.50		128.50	151
3870	3"		31.90	.502		134	26.50		160.50	189
3880	4"		23.60	.678		178	36		214	252
4000	Reducer									
4010	3/4"	1 Plum	27.50	.291	Ea.	12.65	17.10		29.75	40.50
4020	1"		25.90	.309		22	18.15		40.15	53
4030	1-1/4"		20.90	.383		32.50	22.50		55	71
4040	1-1/2"		18.30	.437		39	25.50		64.50	83
4050	2"		15.70	.510		53	30		83	105
4060	2-1/2"	Q-1	25.90	.618		159	32.50		191.50	226
4070	3"		22	.727		206	38.50		244.50	286
4080	4"		16.30	.982		265	52		317	375

22 11 13.29 Pipe, Fittings and Valves, Copper, Pressed-Joint		Crew	Daily Output	Labor-Hours	Unit	Material	2015 Bare Costs Labor	Equipment	Total	Total Incl O&P
4100	Stub out, 1/2"	1 Plum	50	.160	Ea.	7.85	9.40		17.25	23.50
4110	3/4"		35	.229		13.90	13.40		27.30	36.50
4120	1"		30	.267		19.05	15.65		34.70	45.50
6000	Valves									
6200	Ball valve									
6210	1/2"	1 Plum	25.60	.313	Ea.	43.50	18.35		61.85	76
6220	3/4"		19.20	.417		61	24.50		85.50	105
6230	1"		18.10	.442		75	26		101	123
6240	1-1/4"		14.70	.544		122	32		154	184
6250	1-1/2"		12.80	.625		169	36.50		205.50	242
6260	2"		11	.727		269	42.50		311.50	365
6400	Check valve									
6410	1/2"	1 Plum	30	.267	Ea.	32.50	15.65		48.15	60.50
6420	3/4"		22.50	.356		41.50	21		62.50	78.50
6430	1"		21.20	.377		48	22		70	87.50
6440	1-1/4"		17.20	.465		69.50	27.50		97	119
6450	1-1/2"		15	.533		98.50	31.50		130	158
6460	2"		12.90	.620		180	36.50		216.50	255
6600	Butterfly valve, lug type									
6660	2-1/2"	Q-1	9	1.778	Ea.	154	94		248	315
6670	3"		8	2		188	106		294	370
6680	4"		5	3.200		235	169		404	520

22 11 13.44 Pipe, Steel		Crew	Daily Output	Labor-Hours	Unit	Material	Labor	Equipment	Total	Total Incl O&P
0010	**PIPE, STEEL** R221113-50									
0020	All pipe sizes are to Spec. A-53 unless noted otherwise									
0032	Schedule 10, see Line 22 11 13.48 0500									
0050	Schedule 40, threaded, with couplings, and clevis hanger									
0060	assemblies sized for covering, 10' O.C.									
0540	Black, 1/4" diameter	1 Plum	66	.121	L.F.	6.55	7.10		13.65	18.30
0550	3/8" diameter		65	.123		7.30	7.20		14.50	19.30
0560	1/2" diameter		63	.127		3.32	7.45		10.77	15.30
0570	3/4" diameter		61	.131		3.91	7.70		11.61	16.30
0580	1" diameter		53	.151		5.05	8.85		13.90	19.35
0590	1-1/4" diameter	Q-1	89	.180		6.20	9.50		15.70	21.50
0600	1-1/2" diameter		80	.200		7.10	10.55		17.65	24.50
0610	2" diameter		64	.250		9	13.20		22.20	30.50
0620	2-1/2" diameter		50	.320		14	16.90		30.90	42
0630	3" diameter		43	.372		17.80	19.65		37.45	50
0640	3-1/2" diameter		40	.400		24.50	21		45.50	60
0650	4" diameter		36	.444		27	23.50		50.50	66
0809	A-106, gr. A/B, seamless w/cplgs. & clevis hanger assemblies									
0811	1/4" diameter	1 Plum	66	.121	L.F.	8.60	7.10		15.70	20.50
0812	3/8" diameter		65	.123		8.55	7.20		15.75	20.50
0813	1/2" diameter		63	.127		9.55	7.45		17	22
0814	3/4" diameter		61	.131		10.95	7.70		18.65	24
0815	1" diameter		53	.151		12.60	8.85		21.45	27.50
0816	1-1/4" diameter	Q-1	89	.180		15.10	9.50		24.60	31.50
0817	1-1/2" diameter		80	.200		21.50	10.55		32.05	40.50
0819	2" diameter		64	.250		19.05	13.20		32.25	41.50
0821	2-1/2" diameter		50	.320		23	16.90		39.90	52
0822	3" diameter		43	.372		30	19.65		49.65	63.50
0823	4" diameter		36	.444		46.50	23.50		70	87.50

22 11 13.44 Pipe, Steel		Crew	Daily Output	Labor-Hours	Unit	Material	2015 Bare Costs Labor	Equipment	Total	Total Incl O&P
1220	To delete coupling & hanger, subtract									
1230	1/4" diam. to 3/4" diam.					31%	56%			
1240	1" diam. to 1-1/2" diam.					23%	51%			
1250	2" diam. to 4" diam.					23%	41%			
1280	All pipe sizes are to Spec. A-53 unless noted otherwise									
1281	Schedule 40, threaded, with couplings and clevis hanger									
1282	assemblies sized for covering, 10' O. C.									
1290	Galvanized, 1/4" diameter	1 Plum	66	.121	L.F.	9.10	7.10		16.20	21
1300	3/8" diameter		65	.123		9.95	7.20		17.15	22
1310	1/2" diameter		63	.127		3.69	7.45		11.14	15.70
1320	3/4" diameter		61	.131		4.24	7.70		11.94	16.65
1330	1" diameter		53	.151		5.75	8.85		14.60	20
1340	1-1/4" diameter	Q-1	89	.180		7	9.50		16.50	22.50
1350	1-1/2" diameter		80	.200		8.05	10.55		18.60	25.50
1360	2" diameter		64	.250		10.35	13.20		23.55	32
1370	2-1/2" diameter		50	.320		16.30	16.90		33.20	44.50
1380	3" diameter		43	.372		20.50	19.65		40.15	53.50
1390	3-1/2" diameter		40	.400		26	21		47	62
1400	4" diameter		36	.444		30	23.50		53.50	69.50
1750	To delete coupling & hanger, subtract									
1760	1/4" diam. to 3/4" diam.					31%	56%			
1770	1" diam. to 1-1/2" diam.					23%	51%			
1780	2" diam. to 4" diam.					23%	41%			
2000	Welded, sch. 40, on yoke & roll hanger assy's, sized for covering, 10' O.C.									
2040	Black, 1" diameter	Q-15	93	.172	L.F.	4.72	9.10	.62	14.44	20
2050	1-1/4" diameter		84	.190		5.90	10.05	.69	16.64	23
2060	1-1/2" diameter		76	.211		6.45	11.10	.76	18.31	25.50
2070	2" diameter		61	.262		7.85	13.85	.95	22.65	31
2080	2-1/2" diameter		47	.340		12.75	18	1.23	31.98	43.50
2090	3" diameter		43	.372		15.45	19.65	1.34	36.44	49
2100	3-1/2" diameter		39	.410		18.55	21.50	1.48	41.53	56
2110	4" diameter		37	.432		21.50	23	1.56	46.06	61
2120	5" diameter		32	.500		34.50	26.50	1.81	62.81	81
2130	6" diameter	Q-16	36	.667		43	36.50	1.60	81.10	106
2140	8" diameter		29	.828		68.50	45.50	1.99	115.99	148
2150	10" diameter		24	1		89.50	55	2.40	146.90	186
2160	12" diameter		19	1.263		105	69	3.03	177.03	227
2170	14" diameter, (two rod roll type hanger for 14" diam. and up)		15	1.600		98.50	87.50	3.84	189.84	249
2180	16" diameter, (two rod roll type hanger)		13	1.846		148	101	4.43	253.43	325
2190	18" diameter, (two rod roll type hanger)		11	2.182		141	120	5.25	266.25	345
2200	20" diameter, (two rod roll type hanger)		9	2.667		143	146	6.40	295.40	390
2220	24" diameter, (two rod roll type hanger)		8	3		183	164	7.20	354.20	465
2345	Sch. 40, A-53, gr. A/B, ERW, welded w/hngrs.									
2346	2" diameter	Q-15	61	.262	L.F.	7.70	13.85	.95	22.50	31
2347	2-1/2" diameter		47	.340		10.10	18	1.23	29.33	40.50
2348	3" diameter		43	.372		12.60	19.65	1.34	33.59	46
2349	4" diameter		38	.421		17.15	22	1.52	40.67	55
2350	6" diameter	Q-16	37	.649		28	35.50	1.56	65.06	88
2351	8" diameter		29	.828		67.50	45.50	1.99	114.99	147
2352	10" diameter		24	1		88	55	2.40	145.40	185
2560	To delete hanger, subtract									
2570	1" diam. to 1-1/2" diam.					15%	34%			
2580	2" diam. to 3-1/2" diam.					9%	21%			

22 11 13 – Facility Water Distribution Piping

22 11 13.44 Pipe, Steel		Crew	Daily Output	Labor-Hours	Unit	Material	2015 Bare Costs Labor	Equipment	Total	Total Incl O&P
2590	4″ diam. to 12″ diam.					5%	12%			
2596	14″ diam. to 24″ diam.					3%	10%			
3250	Flanged, 150 lb. weld neck, on yoke & roll hangers									
3260	sized for covering, 10′ O.C.									
3290	Black, 1″ diameter	Q-15	70	.229	L.F.	9.90	12.10	.83	22.83	30.50
3300	1-1/4″ diameter		64	.250		11.10	13.20	.90	25.20	33.50
3310	1-1/2″ diameter		58	.276		11.65	14.55	1	27.20	36.50
3320	2″ diameter		45	.356		13.90	18.80	1.28	33.98	46
3330	2-1/2″ diameter		36	.444		19.50	23.50	1.60	44.60	60
3340	3″ diameter		32	.500		23	26.50	1.81	51.31	68
3350	3-1/2″ diameter		29	.552		28	29	1.99	58.99	78.50
3360	4″ diameter		26	.615		31	32.50	2.22	65.72	87
3370	5″ diameter		21	.762		49.50	40.50	2.75	92.75	121
3380	6″ diameter	Q-16	25	.960		59	52.50	2.30	113.80	150
3390	8″ diameter		19	1.263		96.50	69	3.03	168.53	217
3400	10″ diameter		16	1.500		138	82	3.60	223.60	283
3410	12″ diameter		14	1.714		165	94	4.11	263.11	335
3470	For 300 lb. flanges, add					63%				
3480	For 600 lb. flanges, add					310%				
3960	To delete flanges & hanger, subtract									
3970	1″ diam. to 2″ diam.					76%	65%			
3980	2-1/2″ diam. to 4″ diam.					62%	59%			
3990	5″ diam. to 12″ diam.					60%	46%			
4750	Schedule 80, threaded, with couplings, and clevis hanger assemblies									
4760	sized for covering, 10′ O.C.									
4790	Black, 1/4″ diameter	1 Plum	54	.148	L.F.	10.95	8.70		19.65	25.50
4800	3/8″ diameter		53	.151		13.90	8.85		22.75	29
4810	1/2″ diameter		52	.154		4.77	9.05		13.82	19.35
4820	3/4″ diameter		50	.160		5.45	9.40		14.85	20.50
4830	1″ diameter		45	.178		7.20	10.45		17.65	24.50
4840	1-1/4″ diameter	Q-1	75	.213		9.45	11.25		20.70	28
4850	1-1/2″ diameter		69	.232		10.65	12.25		22.90	31
4860	2″ diameter		56	.286		14.10	15.10		29.20	39
4870	2-1/2″ diameter		44	.364		21	19.20		40.20	53
4880	3″ diameter		38	.421		27	22		49	64
4890	3-1/2″ diameter		35	.457		33	24		57	73.50
4900	4″ diameter		32	.500		38.50	26.50		65	83.50
5061	A-106, gr. A/B seamless with cplgs. & clevis hanger assemblies, 1/4″ diam.	1 Plum	63	.127		10.55	7.45		18	23.50
5062	3/8″ diameter		62	.129		12.25	7.55		19.80	25.50
5063	1/2″ diameter		61	.131		10.15	7.70		17.85	23
5064	3/4″ diameter		57	.140		11.75	8.25		20	26
5065	1″ diameter		51	.157		13.75	9.20		22.95	29.50
5066	1-1/4″ diameter	Q-1	85	.188		18.15	9.95		28.10	35.50
5067	1-1/2″ diameter		77	.208		26.50	11		37.50	46
5071	2″ diameter		61	.262		32.50	13.85		46.35	57
5072	2-1/2″ diameter		48	.333		24.50	17.60		42.10	54.50
5073	3″ diameter		41	.390		32	20.50		52.50	67
5074	4″ diameter		35	.457		46.50	24		70.50	88.50
5430	To delete coupling & hanger, subtract									
5440	1/4″ diam. to 1/2″ diam.					31%	54%			
5450	3/4″ diam. to 1-1/2″ diam.					28%	49%			
5460	2″ diam. to 4″ diam.					21%	40%			
5510	Galvanized, 1/4″ diameter	1 Plum	54	.148	L.F.	9.90	8.70		18.60	24.50

22 11 13.44 Pipe, Steel		Crew	Daily Output	Labor-Hours	Unit	2015 Bare Costs Material	Labor	Equipment	Total	Total Incl O&P
5520	3/8″ diameter	1 Plum	53	.151	L.F.	10.85	8.85		19.70	26
5530	1/2″ diameter		52	.154		5.30	9.05		14.35	19.95
5540	3/4″ diameter		50	.160		6.40	9.40		15.80	21.50
5550	1″ diameter		45	.178		8.50	10.45		18.95	25.50
5560	1-1/4″ diameter	Q-1	75	.213		11.35	11.25		22.60	30
5570	1-1/2″ diameter		69	.232		13.15	12.25		25.40	33.50
5580	2″ diameter		56	.286		17.45	15.10		32.55	42.50
5590	2-1/2″ diameter		44	.364		26	19.20		45.20	58.50
5600	3″ diameter		38	.421		34	22		56	71.50
5610	3-1/2″ diameter		35	.457		53	24		77	96
5620	4″ diameter		32	.500		40	26.50		66.50	85
5930	To delete coupling & hanger, subtract									
5940	1/4″ diam. to 1/2″ diam.					31%	54%			
5950	3/4″ diam. to 1-1/2″ diam.					28%	49%			
5960	2″ diam. to 4″ diam.					21%	40%			
6000	Welded, on yoke & roller hangers									
6010	sized for covering, 10′ O.C.									
6040	Black, 1″ diameter	Q-15	85	.188	L.F.	7.35	9.95	.68	17.98	24.50
6050	1-1/4″ diameter		79	.203		9.50	10.70	.73	20.93	28
6060	1-1/2″ diameter		72	.222		10.75	11.75	.80	23.30	31
6070	2″ diameter		57	.281		13.55	14.85	1.01	29.41	39
6080	2-1/2″ diameter		44	.364		20	19.20	1.31	40.51	53.50
6090	3″ diameter		40	.400		25.50	21	1.44	47.94	62.50
6100	3-1/2″ diameter		34	.471		31	25	1.70	57.70	75
6110	4″ diameter		33	.485		36	25.50	1.75	63.25	81.50
6120	5″ diameter, A-106B		26	.615		51.50	32.50	2.22	86.22	109
6130	6″ diameter, A-106B	Q-16	30	.800		70.50	44	1.92	116.42	149
6140	8″ diameter, A-106B		25	.960		106	52.50	2.30	160.80	202
6150	10″ diameter, A-106B		20	1.200		172	65.50	2.88	240.38	295
6160	12″ diameter, A-106B		15	1.600		385	87.50	3.84	476.34	565
6540	To delete hanger, subtract									
6550	1″ diam. to 1-1/2″ diam.					30%	14%			
6560	2″ diam. to 3″ diam.					23%	9%			
6570	3-1/2″ diam. to 5″ diam.					12%	6%			
6580	6″ diam. to 12″ diam.					10%	4%			
7250	Flanged, 300 lb. weld neck, on yoke & roll hangers									
7260	sized for covering, 10′ O.C.									
7290	Black, 1″ diameter	Q-15	66	.242	L.F.	13.75	12.80	.88	27.43	36
7300	1-1/4″ diameter		61	.262		15.95	13.85	.95	30.75	40
7310	1-1/2″ diameter		54	.296		17.20	15.65	1.07	33.92	44.50
7320	2″ diameter		42	.381		22	20	1.38	43.38	57
7330	2-1/2″ diameter		33	.485		29.50	25.50	1.75	56.75	74.50
7340	3″ diameter		29	.552		35	29	1.99	65.99	86
7350	3-1/2″ diameter		24	.667		47	35	2.41	84.41	109
7360	4″ diameter		23	.696		52	37	2.51	91.51	117
7370	5″ diameter		19	.842		75.50	44.50	3.04	123.04	156
7380	6″ diameter	Q-16	23	1.043		95	57	2.50	154.50	196
7390	8″ diameter		17	1.412		147	77.50	3.39	227.89	287
7400	10″ diameter		14	1.714		247	94	4.11	345.11	425
7410	12″ diameter		12	2		480	110	4.80	594.80	705
7470	For 600 lb. flanges, add					100%				
7940	To delete flanges & hanger, subtract									
7950	1″ diam. to 1-1/2″ diam.					75%	66%			

22 11 13.44 Pipe, Steel		Crew	Daily Output	Labor-Hours	Unit	Material	Labor	Equipment	Total	Total Incl O&P
						2015 Bare Costs				
7960	2" diam. to 3" diam.					62%	60%			
7970	3-1/2" diam. to 5" diam.					54%	66%			
7980	6" diam. to 12" diam.					55%	62%			
8040	Galvanized, 1" diameter	Q-15	66	.242	L.F.	15.05	12.80	.88	28.73	37.50
8050	1-1/4" diameter		61	.262		17.85	13.85	.95	32.65	42
8060	1-1/2" diameter		54	.296		19.65	15.65	1.07	36.37	47
8070	2" diameter		42	.381		25	20	1.38	46.38	60.50
8080	2-1/2" diameter		33	.485		34.50	25.50	1.75	61.75	80
8090	3" diameter		29	.552		42	29	1.99	72.99	93.50
8100	3-1/2" diameter		24	.667		67	35	2.41	104.41	132
8110	4" diameter		23	.696		53.50	37	2.51	93.01	119
8120	5" diameter, A-106B		19	.842		78	44.50	3.04	125.54	159
8130	6" diameter, A-106B	Q-16	23	1.043		98	57	2.50	157.50	200
8140	8" diameter, A-106B		17	1.412		180	77.50	3.39	260.89	325
8150	10" diameter, A-106B		14	1.714		330	94	4.11	428.11	515
8160	12" diameter, A-106B		12	2		610	110	4.80	724.80	845
8240	For 600 lb. flanges, add					100%				
8300	To delete flanges & hangers, subtract									
8310	1" diam. to 1-1/2" diam.					72%	66%			
8320	2" diam. to 3" diam.					59%	60%			
8330	3-1/2" diam. to 5" diam.					51%	66%			
8340	6" diam. to 12" diam.					49%	62%			
9000	Threading pipe labor, one end, all schedules through 80									
9010	1/4" through 3/4" pipe size	1 Plum	80	.100	Ea.		5.85		5.85	9.15
9020	1" through 2" pipe size		73	.110			6.45		6.45	10.05
9030	2-1/2" pipe size		53	.151			8.85		8.85	13.80
9040	3" pipe size		50	.160			9.40		9.40	14.65
9050	3-1/2" pipe size	Q-1	89	.180			9.50		9.50	14.80
9060	4" pipe size		73	.219			11.60		11.60	18.05
9070	5" pipe size		53	.302			15.95		15.95	25
9080	6" pipe size		46	.348			18.40		18.40	28.50
9090	8" pipe size		29	.552			29		29	45.50
9100	10" pipe size		21	.762			40.50		40.50	63
9110	12" pipe size		13	1.231			65		65	101
9120	Cutting pipe labor, one cut									
9124	Shop fabrication, machine cut									
9126	Schedule 40, straight pipe									
9128	2" pipe size or less	1 Stpi	62	.129	Ea.		7.70		7.70	12.05
9130	2-1/2" pipe size		56	.143			8.55		8.55	13.30
9132	3" pipe size		42	.190			11.40		11.40	17.75
9134	4" pipe size		31	.258			15.40		15.40	24
9136	5" pipe size		26	.308			18.40		18.40	28.50
9138	6" pipe size		19	.421			25		25	39
9140	8" pipe size		14	.571			34		34	53.50
9142	10" pipe size		10	.800			48		48	74.50
9144	12" pipe size		7	1.143			68.50		68.50	107
9146	14" pipe size	Q-5	10.50	1.524			82		82	128
9148	16" pipe size		8.60	1.860			100		100	156
9150	18" pipe size		7	2.286			123		123	192
9152	20" pipe size		5.80	2.759			148		148	231
9154	24" pipe size		4	4			215		215	335
9160	Schedule 80, straight pipe									
9164	2" pipe size or less	1 Stpi	42	.190	Ea.		11.40		11.40	17.75

22 11 13 – Facility Water Distribution Piping

22 11 13.44 Pipe, Steel		Crew	Daily Output	Labor-Hours	Unit	Material	2015 Bare Costs Labor	Equipment	Total	Total Incl O&P
9166	2-1/2" pipe size	1 Stpi	37	.216	Ea.		12.90		12.90	20
9168	3" pipe size		31	.258			15.40		15.40	24
9170	4" pipe size		23	.348			21		21	32.50
9172	5" pipe size		18	.444			26.50		26.50	41.50
9174	6" pipe size		14.60	.548			32.50		32.50	51
9176	8" pipe size		10	.800			48		48	74.50
9178	10" pipe size	Q-5	14	1.143			61.50		61.50	96
9180	12" pipe size	"	10	1.600			86		86	134
9200	Welding labor per joint									
9210	Schedule 40,									
9230	1/2" pipe size	Q-15	32	.500	Ea.		26.50	1.81	28.31	43
9240	3/4" pipe size		27	.593			31.50	2.14	33.64	51.50
9250	1" pipe size		23	.696			37	2.51	39.51	60.50
9260	1-1/4" pipe size		20	.800			42.50	2.89	45.39	69
9270	1-1/2" pipe size		19	.842			44.50	3.04	47.54	73
9280	2" pipe size		16	1			53	3.61	56.61	86.50
9290	2-1/2" pipe size		13	1.231			65	4.44	69.44	106
9300	3" pipe size		12	1.333			70.50	4.81	75.31	115
9310	4" pipe size		10	1.600			84.50	5.80	90.30	138
9320	5" pipe size		9	1.778			94	6.40	100.40	153
9330	6" pipe size		8	2			106	7.20	113.20	173
9340	8" pipe size		5	3.200			169	11.55	180.55	277
9350	10" pipe size		4	4			211	14.45	225.45	345
9360	12" pipe size		3	5.333			282	19.25	301.25	460
9370	14" pipe size		2.60	6.154			325	22	347	530
9380	16" pipe size		2.20	7.273			385	26.50	411.50	630
9390	18" pipe size		2	8			425	29	454	690
9400	20" pipe size		1.80	8.889			470	32	502	765
9410	22" pipe size		1.70	9.412			495	34	529	815
9420	24" pipe size		1.50	10.667			565	38.50	603.50	925
9450	Schedule 80,									
9460	1/2" pipe size	Q-15	27	.593	Ea.		31.50	2.14	33.64	51.50
9470	3/4" pipe size		23	.696			37	2.51	39.51	60.50
9480	1" pipe size		20	.800			42.50	2.89	45.39	69
9490	1-1/4" pipe size		19	.842			44.50	3.04	47.54	73
9500	1-1/2" pipe size		18	.889			47	3.21	50.21	76.50
9510	2" pipe size		15	1.067			56.50	3.85	60.35	92
9520	2-1/2" pipe size		12	1.333			70.50	4.81	75.31	115
9530	3" pipe size		11	1.455			77	5.25	82.25	126
9540	4" pipe size		8	2			106	7.20	113.20	173
9550	5" pipe size		6	2.667			141	9.65	150.65	231
9560	6" pipe size		5	3.200			169	11.55	180.55	277
9570	8" pipe size		4	4			211	14.45	225.45	345
9580	10" pipe size		3	5.333			282	19.25	301.25	460
9590	12" pipe size		2	8			425	29	454	690
9600	14" pipe size	Q-16	2.60	9.231			505	22	527	815
9610	16" pipe size		2.30	10.435			570	25	595	920
9620	18" pipe size		2	12			655	29	684	1,050
9630	20" pipe size		1.80	13.333			730	32	762	1,175
9640	22" pipe size		1.60	15			820	36	856	1,325
9650	24" pipe size		1.50	16			875	38.50	913.50	1,425
9990	Minimum labor/equipment charge	1 Plum	3	2.667	Job		157		157	244

22 11 13 – Facility Water Distribution Piping

22 11 13.45 Pipe Fittings, Steel, Threaded	Crew	Daily Output	Labor-Hours	Unit	Material	Labor	Equipment	Total	Total Incl O&P
					2015 Bare Costs				
0010 **PIPE FITTINGS, STEEL, THREADED** R221113-50									
0020 Cast Iron									
0040 Standard weight, black									
0060 90° Elbow, straight									
0070 1/4"	1 Plum	16	.500	Ea.	9.70	29.50		39.20	56.50
0080 3/8"		16	.500		14.05	29.50		43.55	61.50
0090 1/2"		15	.533		6.15	31.50		37.65	56
0100 3/4"		14	.571		6.40	33.50		39.90	59.50
0110 1"		13	.615		7.60	36		43.60	65
0120 1-1/4"	Q-1	22	.727		10.75	38.50		49.25	72
0130 1-1/2"		20	.800		14.90	42.50		57.40	82.50
0140 2"		18	.889		23.50	47		70.50	98.50
0150 2-1/2"		14	1.143		56	60.50		116.50	156
0160 3"		10	1.600		91.50	84.50		176	233
0170 3-1/2"		8	2		248	106		354	440
0180 4"		6	2.667		169	141		310	405
0250 45° Elbow, straight									
0260 1/4"	1 Plum	16	.500	Ea.	11.95	29.50		41.45	59
0270 3/8"		16	.500		12.85	29.50		42.35	60
0280 1/2"		15	.533		9.35	31.50		40.85	59.50
0300 3/4"		14	.571		9.50	33.50		43	63
0320 1"		13	.615		11.10	36		47.10	68.50
0330 1-1/4"	Q-1	22	.727		14.95	38.50		53.45	76.50
0340 1-1/2"		20	.800		24.50	42.50		67	93
0350 2"		18	.889		28.50	47		75.50	105
0360 2-1/2"		14	1.143		74	60.50		134.50	176
0370 3"		10	1.600		118	84.50		202.50	261
0380 3-1/2"		8	2		271	106		377	465
0400 4"		6	2.667		244	141		385	490
0500 Tee, straight									
0510 1/4"	1 Plum	10	.800	Ea.	15.20	47		62.20	89.50
0520 3/8"		10	.800		14.75	47		61.75	89.50
0530 1/2"		9	.889		9.60	52		61.60	92
0540 3/4"		9	.889		11.20	52		63.20	94
0550 1"		8	1		9.95	58.50		68.45	102
0560 1-1/4"	Q-1	14	1.143		18.15	60.50		78.65	114
0570 1-1/2"		13	1.231		23.50	65		88.50	127
0580 2"		11	1.455		33	77		110	156
0590 2-1/2"		9	1.778		85.50	94		179.50	240
0600 3"		6	2.667		131	141		272	365
0610 3-1/2"		5	3.200		264	169		433	555
0620 4"		4	4		264	211		475	620
0700 Standard weight, galvanized cast iron									
0720 90° Elbow, straight									
0730 1/4"	1 Plum	16	.500	Ea.	15.95	29.50		45.45	63.50
0740 3/8"		16	.500		15.95	29.50		45.45	63.50
0750 1/2"		15	.533		18.20	31.50		49.70	69
0760 3/4"		14	.571		17.85	33.50		51.35	72
0770 1"		13	.615		20.50	36		56.50	79
0780 1-1/4"	Q-1	22	.727		32	38.50		70.50	95.50
0790 1-1/2"		20	.800		44	42.50		86.50	115
0800 2"		18	.889		65	47		112	145

22 11 13 – Facility Water Distribution Piping

22 11 13.45 Pipe Fittings, Steel, Threaded		Crew	Daily Output	Labor-Hours	Unit	Material	Labor (2015 Bare Costs)	Equipment	Total	Total Incl O&P
0810	2-1/2"	Q-1	14	1.143	Ea.	134	60.50		194.50	241
0820	3"		10	1.600		203	84.50		287.50	355
0830	3-1/2"		8	2		325	106		431	520
0840	4"	↓	6	2.667	↓	370	141		511	630
0900	45° Elbow, straight									
0910	1/4"	1 Plum	16	.500	Ea.	18.15	29.50		47.65	66
0920	3/8"		16	.500		18.15	29.50		47.65	66
0930	1/2"		15	.533		18.15	31.50		49.65	69
0940	3/4"		14	.571		21	33.50		54.50	75.50
0950	1"	↓	13	.615		25.50	36		61.50	84.50
0960	1-1/4"	Q-1	22	.727		38	38.50		76.50	102
0970	1-1/2"		20	.800		54	42.50		96.50	126
0980	2"		18	.889		76	47		123	157
0990	2-1/2"		14	1.143		122	60.50		182.50	228
1000	3"		10	1.600		238	84.50		322.50	395
1010	3-1/2"		8	2		435	106		541	645
1020	4"	↓	6	2.667	↓	435	141		576	700
1100	Tee, straight									
1110	1/4"	1 Plum	10	.800	Ea.	16.75	47		63.75	91.50
1120	3/8"		10	.800		19.30	47		66.30	94
1130	1/2"		9	.889		18.60	52		70.60	102
1140	3/4"		9	.889		25.50	52		77.50	110
1150	1"	↓	8	1		28	58.50		86.50	122
1160	1-1/4"	Q-1	14	1.143		49	60.50		109.50	148
1170	1-1/2"		13	1.231		64.50	65		129.50	172
1180	2"		11	1.455		80.50	77		157.50	209
1190	2-1/2"		9	1.778		164	94		258	325
1200	3"		6	2.667		365	141		506	620
1210	3-1/2"		5	3.200		460	169		629	770
1220	4"	↓	4	4	↓	515	211		726	900
1300	Extra heavy weight, black									
1310	Couplings, steel straight									
1320	1/4"	1 Plum	19	.421	Ea.	3.77	24.50		28.27	42.50
1330	3/8"		19	.421		4.10	24.50		28.60	43
1340	1/2"		19	.421		5.55	24.50		30.05	44.50
1350	3/4"		18	.444		5.90	26		31.90	47
1360	1"	↓	15	.533		7.55	31.50		39.05	57.50
1370	1-1/4"	Q-1	26	.615		12.10	32.50		44.60	64
1380	1-1/2"		24	.667		12.10	35		47.10	68.50
1390	2"		21	.762		18.45	40.50		58.95	83.50
1400	2-1/2"		18	.889		27.50	47		74.50	103
1410	3"		14	1.143		32.50	60.50		93	130
1420	3-1/2"		12	1.333		43.50	70.50		114	158
1430	4"	↓	10	1.600	↓	51.50	84.50		136	189
1510	90° Elbow, straight									
1520	1/2"	1 Plum	15	.533	Ea.	31.50	31.50		63	84
1530	3/4"		14	.571		32	33.50		65.50	88
1540	1"	↓	13	.615		39	36		75	99.50
1550	1-1/4"	Q-1	22	.727		58	38.50		96.50	124
1560	1-1/2"		20	.800		72	42.50		114.50	145
1580	2"		18	.889		89	47		136	171
1590	2-1/2"		14	1.143		216	60.50		276.50	330
1600	3"	↓	10	1.600	↓	285	84.50		369.50	445

22 11 13 – Facility Water Distribution Piping

22 11 13.45 Pipe Fittings, Steel, Threaded	Crew	Daily Output	Labor-Hours	Unit	Material	Labor	Equipment	Total	Total Incl O&P
					2015 Bare Costs				
1610 4"	Q-1	6	2.667	Ea.	585	141		726	860
1650 45° Elbow, straight									
1660 1/2"	1 Plum	15	.533	Ea.	45	31.50		76.50	98.50
1670 3/4"		14	.571		43.50	33.50		77	101
1680 1"		13	.615		52.50	36		88.50	114
1690 1-1/4"	Q-1	22	.727		86	38.50		124.50	155
1700 1-1/2"		20	.800		95	42.50		137.50	170
1710 2"		18	.889		135	47		182	221
1720 2-1/2"		14	1.143		233	60.50		293.50	350
1800 Tee, straight									
1810 1/2"	1 Plum	9	.889	Ea.	50	52		102	137
1820 3/4"		9	.889		49.50	52		101.50	136
1830 1"		8	1		59.50	58.50		118	157
1840 1-1/4"	Q-1	14	1.143		89	60.50		149.50	192
1850 1-1/2"		13	1.231		115	65		180	227
1860 2"		11	1.455		142	77		219	276
1870 2-1/2"		9	1.778		305	94		399	480
1880 3"		6	2.667		415	141		556	680
1890 4"		4	4		810	211		1,021	1,225
4000 Standard weight, black									
4010 Couplings, steel straight, merchants									
4030 1/4"	1 Plum	19	.421	Ea.	1.01	24.50		25.51	39.50
4040 3/8"		19	.421		1.22	24.50		25.72	40
4050 1/2"		19	.421		1.30	24.50		25.80	40
4060 3/4"		18	.444		1.64	26		27.64	42.50
4070 1"		15	.533		2.31	31.50		33.81	51.50
4080 1-1/4"	Q-1	26	.615		2.95	32.50		35.45	54
4090 1-1/2"		24	.667		3.73	35		38.73	59
4100 2"		21	.762		5.35	40.50		45.85	69
4110 2-1/2"		18	.889		16.75	47		63.75	91.50
4120 3"		14	1.143		23.50	60.50		84	120
4130 3-1/2"		12	1.333		41.50	70.50		112	156
4140 4"		10	1.600		41.50	84.50		126	178
4166 Plug, 1/4"	1 Plum	38	.211		2.73	12.35		15.08	22.50
4167 3/8"		38	.211		2.73	12.35		15.08	22.50
4168 1/2"		38	.211		2.73	12.35		15.08	22.50
4169 3/4"		32	.250		8.25	14.70		22.95	32
4170 1"		30	.267		8.80	15.65		24.45	34
4171 1-1/4"	Q-1	52	.308		10.10	16.25		26.35	36.50
4172 1-1/2"		48	.333		14.35	17.60		31.95	43.50
4173 2"		42	.381		18.55	20		38.55	52
4176 2-1/2"		36	.444		27.50	23.50		51	66.50
4180 4"		20	.800		47.50	42.50		90	118
5000 Malleable iron, 150 lb.									
5020 Black									
5040 90° elbow, straight									
5060 1/4"	1 Plum	16	.500	Ea.	4.78	29.50		34.28	51.50
5070 3/8"		16	.500		4.78	29.50		34.28	51.50
5080 1/2"		15	.533		3.30	31.50		34.80	52.50
5090 3/4"		14	.571		3.99	33.50		37.49	57
5100 1"		13	.615		6.95	36		42.95	64
5110 1-1/4"	Q-1	22	.727		11.45	38.50		49.95	72.50
5120 1-1/2"		20	.800		15.05	42.50		57.55	82.50

22 11 13 – Facility Water Distribution Piping

22 11 13.45 Pipe Fittings, Steel, Threaded	Crew	Daily Output	Labor-Hours	Unit	Material	2015 Bare Costs Labor	Equipment	Total	Total Incl O&P
5130 2"	Q-1	18	.889	Ea.	26	47		73	102
5140 2-1/2"		14	1.143		58	60.50		118.50	158
5150 3"		10	1.600		84.50	84.50		169	225
5160 3-1/2"		8	2		232	106		338	420
5170 4"		6	2.667		182	141		323	420
5250 45° elbow, straight									
5270 1/4"	1 Plum	16	.500	Ea.	7.20	29.50		36.70	54
5280 3/8"		16	.500		7.20	29.50		36.70	54
5290 1/2"		15	.533		5.45	31.50		36.95	55
5300 3/4"		14	.571		6.75	33.50		40.25	60
5310 1"		13	.615		8.50	36		44.50	66
5320 1-1/4"	Q-1	22	.727		15.05	38.50		53.55	76.50
5330 1-1/2"		20	.800		18.60	42.50		61.10	86.50
5340 2"		18	.889		28	47		75	104
5350 2-1/2"		14	1.143		81.50	60.50		142	184
5360 3"		10	1.600		188	84.50		272.50	340
5370 3-1/2"		8	2		208	106		314	395
5380 4"		6	2.667		208	141		349	450
5450 Tee, straight									
5470 1/4"	1 Plum	10	.800	Ea.	6.95	47		53.95	80.50
5480 3/8"		10	.800		6.95	47		53.95	80.50
5490 1/2"		9	.889		4.42	52		56.42	86.50
5500 3/4"		9	.889		6.35	52		58.35	88.50
5510 1"		8	1		10.85	58.50		69.35	103
5520 1-1/4"	Q-1	14	1.143		17.60	60.50		78.10	113
5530 1-1/2"		13	1.231		22	65		87	125
5540 2"		11	1.455		37.50	77		114.50	161
5550 2-1/2"		9	1.778		80.50	94		174.50	235
5560 3"		6	2.667		119	141		260	350
5570 3-1/2"		5	3.200		275	169		444	565
5580 4"		4	4		286	211		497	645
5650 Coupling									
5670 1/4"	1 Plum	19	.421	Ea.	5.95	24.50		30.45	45
5680 3/8"		19	.421		5.95	24.50		30.45	45
5690 1/2"		19	.421		4.58	24.50		29.08	43.50
5700 3/4"		18	.444		5.35	26		31.35	46.50
5710 1"		15	.533		8	31.50		39.50	58
5720 1-1/4"	Q-1	26	.615		10.40	32.50		42.90	62
5730 1-1/2"		24	.667		14.05	35		49.05	70.50
5740 2"		21	.762		21	40.50		61.50	86
5750 2-1/2"		18	.889		57.50	47		104.50	136
5760 3"		14	1.143		77.50	60.50		138	180
5770 3-1/2"		12	1.333		156	70.50		226.50	282
5780 4"		10	1.600		156	84.50		240.50	305
5840 Reducer, concentric, 1/4"	1 Plum	19	.421		6.25	24.50		30.75	45.50
5850 3/8"		19	.421		8	24.50		32.50	47.50
5860 1/2"		19	.421		6.30	24.50		30.80	45.50
5870 3/4"		16	.500		7.50	29.50		37	54.50
5880 1"		15	.533		12.60	31.50		44.10	63
5890 1-1/4"	Q-1	26	.615		14.10	32.50		46.60	66
5900 1-1/2"		24	.667		20	35		55	77.50
5910 2"		21	.762		29	40.50		69.50	95
5911 2-1/2"		18	.889		65	47		112	145

22 11 13 – Facility Water Distribution Piping

22 11 13.45 Pipe Fittings, Steel, Threaded		Crew	Daily Output	Labor-Hours	Unit	Material	Labor	Equipment	Total	Total Incl O&P
						2015 Bare Costs				
5981	Bushing, 1/4"	1 Plum	19	.421	Ea.	1.15	24.50		25.65	40
5982	3/8"		19	.421		6.10	24.50		30.60	45
5983	1/2"		19	.421		6.10	24.50		30.60	45
5984	3/4"		16	.500		6.15	29.50		35.65	53
5985	1"		15	.533		8.70	31.50		40.20	58.50
5986	1-1/4"	Q-1	26	.615		10.85	32.50		43.35	62.50
5987	1-1/2"		24	.667		9.35	35		44.35	65.50
5988	2"		21	.762		11.65	40.50		52.15	76
5989	Cap, 1/4"	1 Plum	38	.211		4.79	12.35		17.14	24.50
5991	3/8"		38	.211		5.85	12.35		18.20	25.50
5992	1/2"		38	.211		3.27	12.35		15.62	23
5993	3/4"		32	.250		4.53	14.70		19.23	28
5994	1"		30	.267		5.50	15.65		21.15	30.50
5995	1-1/4"	Q-1	52	.308		7.15	16.25		23.40	33.50
5996	1-1/2"		48	.333		9.95	17.60		27.55	38.50
5997	2"		42	.381		14.50	20		34.50	47.50
6000	For galvanized elbows, tees, and couplings add					20%				
6058	For galvanized reducers, caps and bushings add					20%				
7000	Union, with brass seat									
7010	1/4"	1 Plum	15	.533	Ea.	25.50	31.50		57	77
7020	3/8"		15	.533		16.15	31.50		47.65	67
7030	1/2"		14	.571		14.60	33.50		48.10	68.50
7040	3/4"		13	.615		16.80	36		52.80	75
7050	1"		12	.667		22	39		61	85
7060	1-1/4"	Q-1	21	.762		31.50	40.50		72	97.50
7070	1-1/2"		19	.842		39	44.50		83.50	113
7080	2"		17	.941		45.50	49.50		95	128
7090	2-1/2"		13	1.231		136	65		201	250
7100	3"		9	1.778		163	94		257	325
7250	For galvanized unions, add					15%				
9757	Forged steel, 3000 lb.									
9758	Black									
9760	90° Elbow, 1/4"	1 Plum	16	.500	Ea.	20.50	29.50		50	68.50
9761	3/8"		16	.500		20.50	29.50		50	68.50
9762	1/2"		15	.533		15.70	31.50		47.20	66.50
9763	3/4"		14	.571		19.70	33.50		53.20	74
9764	1"		13	.615		29.50	36		65.50	88.50
9765	1-1/4"	Q-1	22	.727		55.50	38.50		94	121
9766	1-1/2"		20	.800		72	42.50		114.50	145
9767	2"		18	.889		87	47		134	169
9780	45° Elbow 1/4"	1 Plum	16	.500		26	29.50		55.50	74.50
9781	3/8"		16	.500		26	29.50		55.50	74.50
9782	1/2"		15	.533		25.50	31.50		57	77
9783	3/4"		14	.571		29.50	33.50		63	85
9784	1"		13	.615		40.50	36		76.50	101
9785	1-1/4"	Q-1	22	.727		56	38.50		94.50	122
9786	1-1/2"		20	.800		80	42.50		122.50	154
9787	2"		18	.889		110	47		157	195
9800	Tee, 1/4"	1 Plum	10	.800		25	47		72	101
9801	3/8"		10	.800		25	47		72	101
9802	1/2"		9	.889		22.50	52		74.50	106
9803	3/4"		9	.889		30.50	52		82.50	115
9804	1"		8	1		40	58.50		98.50	136

22 11 Facility Water Distribution

22 11 13 – Facility Water Distribution Piping

22 11 13.45 Pipe Fittings, Steel, Threaded		Crew	Daily Output	Labor-Hours	Unit	Material	2015 Bare Costs Labor	Equipment	Total	Total Incl O&P
9805	1-1/4"	Q-1	14	1.143	Ea.	77.50	60.50		138	179
9806	1-1/2"		13	1.231		90.50	65		155.50	201
9807	2"		11	1.455		116	77		193	248
9820	Reducer, concentric, 1/4"	1 Plum	19	.421		12.55	24.50		37.05	52.50
9821	3/8"		19	.421		13.05	24.50		37.55	53
9822	1/2"		17	.471		13.05	27.50		40.55	57.50
9823	3/4"		16	.500		15.50	29.50		45	63
9824	1"		15	.533		20	31.50		51.50	71
9825	1-1/4"	Q-1	26	.615		35.50	32.50		68	89.50
9826	1-1/2"		24	.667		38.50	35		73.50	97
9827	2"		21	.762		55.50	40.50		96	124
9840	Cap, 1/4"	1 Plum	38	.211		7.30	12.35		19.65	27.50
9841	3/8"		38	.211		7.70	12.35		20.05	27.50
9842	1/2"		34	.235		7.50	13.80		21.30	30
9843	3/4"		32	.250		10.45	14.70		25.15	34.50
9844	1"		30	.267		16	15.65		31.65	42
9845	1-1/4"	Q-1	52	.308		26	16.25		42.25	54
9846	1-1/2"		48	.333		31	17.60		48.60	61.50
9847	2"		42	.381		44.50	20		64.50	80.50
9860	Plug, 1/4"	1 Plum	38	.211		3.47	12.35		15.82	23
9861	3/8"		38	.211		3.64	12.35		15.99	23.50
9862	1/2"		34	.235		3.82	13.80		17.62	25.50
9863	3/4"		32	.250		4.76	14.70		19.46	28.50
9864	1"		30	.267		7.30	15.65		22.95	32.50
9865	1-1/4"	Q-1	52	.308		15	16.25		31.25	42
9866	1-1/2"		48	.333		17.15	17.60		34.75	46.50
9867	2"		42	.381		27	20		47	61.50
9880	Union, bronze seat, 1/4"	1 Plum	15	.533		57	31.50		88.50	112
9881	3/8"		15	.533		57	31.50		88.50	112
9882	1/2"		14	.571		54.50	33.50		88	113
9883	3/4"		13	.615		73	36		109	137
9884	1"		12	.667		82.50	39		121.50	152
9885	1-1/4"	Q-1	21	.762		160	40.50		200.50	239
9886	1-1/2"		19	.842		165	44.50		209.50	251
9887	2"		17	.941		195	49.50		244.50	293
9900	Coupling, 1/4"	1 Plum	19	.421		8	24.50		32.50	47.50
9901	3/8"		19	.421		8	24.50		32.50	47.50
9902	1/2"		17	.471		6.55	27.50		34.05	50
9903	3/4"		16	.500		8.55	29.50		38.05	55.50
9904	1"		15	.533		14.85	31.50		46.35	65.50
9905	1-1/4"	Q-1	26	.615		25	32.50		57.50	78
9906	1-1/2"		24	.667		32	35		67	90
9907	2"		21	.762		40	40.50		80.50	107
9990	Minimum labor/equipment charge	1 Plum	4	2	Job		117		117	183

22 11 13.47 Pipe Fittings, Steel		Crew	Daily Output	Labor-Hours	Unit	Material	Labor	Equipment	Total	Total Incl O&P
0010	**PIPE FITTINGS, STEEL**, Flanged, Welded & Special									
0020	Flanged joints, C.I., standard weight, black. One gasket & bolt									
0040	set, mat'l only, required at each joint, not included (see line 0620)									
0060	90° Elbow, straight, 1-1/2" pipe size	Q-1	14	1.143	Ea.	550	60.50		610.50	700
0080	2" pipe size		13	1.231		325	65		390	455
0090	2-1/2" pipe size		12	1.333		350	70.50		420.50	495
0100	3" pipe size		11	1.455		292	77		369	440

22 11 13 – Facility Water Distribution Piping

22 11 13.47 Pipe Fittings, Steel		Crew	Daily Output	Labor-Hours	Unit	2015 Bare Costs Material	Labor	Equipment	Total	Total Incl O&P
0110	4" pipe size	Q-1	8	2	Ea.	355	106		461	555
0120	5" pipe size		7	2.286		855	121		976	1,125
0130	6" pipe size	Q-2	9	2.667		565	146		711	850
0140	8" pipe size		8	3		975	164		1,139	1,325
0150	10" pipe size		7	3.429		2,150	188		2,338	2,650
0160	12" pipe size		6	4		4,300	219		4,519	5,100
0200	45° Elbow, straight, 1-1/2" pipe size	Q-1	14	1.143		665	60.50		725.50	825
0220	2" pipe size		13	1.231		470	65		535	615
0230	2-1/2" pipe size		12	1.333		500	70.50		570.50	660
0240	3" pipe size		11	1.455		485	77		562	655
0250	4" pipe size		8	2		545	106		651	765
0260	5" pipe size		7	2.286		1,300	121		1,421	1,650
0270	6" pipe size	Q-2	9	2.667		895	146		1,041	1,225
0280	8" pipe size		8	3		1,300	164		1,464	1,700
0290	10" pipe size		7	3.429		2,775	188		2,963	3,350
0300	12" pipe size		6	4		4,225	219		4,444	5,000
0350	Tee, straight, 1-1/2" pipe size	Q-1	10	1.600		635	84.50		719.50	825
0370	2" pipe size		9	1.778		355	94		449	535
0380	2-1/2" pipe size		8	2		515	106		621	730
0390	3" pipe size		7	2.286		360	121		481	590
0400	4" pipe size		5	3.200		550	169		719	870
0410	5" pipe size		4	4		1,475	211		1,686	1,950
0420	6" pipe size	Q-2	6	4		800	219		1,019	1,225
0430	8" pipe size		5	4.800		1,375	263		1,638	1,900
0440	10" pipe size		4	6		3,675	330		4,005	4,550
0450	12" pipe size		3	8		5,775	440		6,215	7,025
0500	For galvanized elbows and tees, add					100%				
0520	For extra heavy weight elbows and tees, add					140%				
0620	Gasket and bolt set, 150#, 1/2" pipe size	1 Plum	20	.400		2.75	23.50		26.25	39.50
0622	3/4" pipe size		19	.421		2.90	24.50		27.40	41.50
0624	1" pipe size		18	.444		2.96	26		28.96	44
0626	1-1/4" pipe size		17	.471		4.02	27.50		31.52	47.50
0628	1-1/2" pipe size		15	.533		4.16	31.50		35.66	53.50
0630	2" pipe size		13	.615		5.95	36		41.95	63
0640	2-1/2" pipe size		12	.667		6.15	39		45.15	68
0650	3" pipe size		11	.727		6.35	42.50		48.85	73.50
0660	3-1/2" pipe size		9	.889		13.05	52		65.05	96
0670	4" pipe size		8	1		14.50	58.50		73	107
0680	5" pipe size		7	1.143		18.85	67		85.85	126
0690	6" pipe size		6	1.333		21.50	78.50		100	146
0700	8" pipe size		5	1.600		21.50	94		115.50	170
0710	10" pipe size		4.50	1.778		45	104		149	213
0720	12" pipe size		4.20	1.905		41.50	112		153.50	220
0730	14" pipe size		4	2		40.50	117		157.50	228
0740	16" pipe size		3	2.667		46	157		203	295
0750	18" pipe size		2.70	2.963		88.50	174		262.50	370
0760	20" pipe size		2.30	3.478		144	204		348	480
0780	24" pipe size		1.90	4.211		180	247		427	585
0790	26" pipe size		1.60	5		244	294		538	730
0810	30" pipe size		1.40	5.714		475	335		810	1,050
0830	36" pipe size		1.10	7.273		890	425		1,315	1,650
0850	For 300 lb. gasket set, add					40%				
2000	Flanged unions, 125 lb., black, 1/2" pipe size	1 Plum	17	.471	Ea.	82.50	27.50		110	134

22 11 13 – Facility Water Distribution Piping

22 11 13.47 Pipe Fittings, Steel		Crew	Daily Output	Labor-Hours	Unit	Material	2015 Bare Costs Labor	Equipment	Total	Total Incl O&P
2040	3/4" pipe size	1 Plum	17	.471	Ea.	109	27.50		136.50	163
2050	1" pipe size	↓	16	.500		105	29.50		134.50	162
2060	1-1/4" pipe size	Q-1	28	.571		125	30		155	184
2070	1-1/2" pipe size		27	.593		115	31.50		146.50	176
2080	2" pipe size		26	.615		135	32.50		167.50	200
2090	2-1/2" pipe size		24	.667		183	35		218	257
2100	3" pipe size		22	.727		207	38.50		245.50	288
2110	3-1/2" pipe size		18	.889		350	47		397	460
2120	4" pipe size		16	1		282	53		335	395
2130	5" pipe size	↓	14	1.143		660	60.50		720.50	820
2140	6" pipe size	Q-2	19	1.263		615	69		684	790
2150	8" pipe size	"	16	1.500		1,425	82		1,507	1,700
2200	For galvanized unions, add				↓	150%				
2290	Threaded flange									
2300	Cast iron									
2310	Black, 125 lb., per flange									
2320	1" pipe size	1 Plum	27	.296	Ea.	47	17.40		64.40	79
2330	1-1/4" pipe size	Q-1	44	.364		56.50	19.20		75.70	92
2340	1-1/2" pipe size		40	.400		52	21		73	90.50
2350	2" pipe size		36	.444		52	23.50		75.50	94
2360	2-1/2" pipe size		28	.571		61	30		91	114
2370	3" pipe size		20	.800		78.50	42.50		121	153
2380	3-1/2" pipe size		16	1		110	53		163	204
2390	4" pipe size		12	1.333		106	70.50		176.50	226
2400	5" pipe size	↓	10	1.600		149	84.50		233.50	296
2410	6" pipe size	Q-2	14	1.714		169	94		263	330
2420	8" pipe size		12	2		267	110		377	465
2430	10" pipe size		10	2.400		475	131		606	725
2440	12" pipe size	↓	8	3	↓	1,050	164		1,214	1,400
2460	For galvanized flanges, add					95%				
2490	Blind flange									
2492	Cast iron									
2494	Black, 125 lb., per flange									
2496	1" pipe size	1 Plum	27	.296	Ea.	75	17.40		92.40	110
2500	1-1/2" pipe size	Q-1	40	.400		84.50	21		105.50	126
2502	2" pipe size		36	.444		95	23.50		118.50	142
2504	2-1/2" pipe size		28	.571		102	30		132	159
2506	3" pipe size		20	.800		125	42.50		167.50	204
2508	4" pipe size		12	1.333		161	70.50		231.50	287
2510	5" pipe size	↓	10	1.600		261	84.50		345.50	420
2512	6" pipe size	Q-2	14	1.714		284	94		378	455
2514	8" pipe size		12	2		445	110		555	660
2516	10" pipe size		10	2.400		660	131		791	930
2518	12" pipe size	↓	8	3	↓	1,250	164		1,414	1,625
2520	For galvanized flanges, add					80%				
2570	Threaded flange									
2580	Forged steel,									
2590	Black 150 lb., per flange									
2600	1/2" pipe size	1 Plum	30	.267	Ea.	26.50	15.65		42.15	53.50
2610	3/4" pipe size		28	.286		26.50	16.75		43.25	55
2620	1" pipe size	↓	27	.296		26.50	17.40		43.90	56
2630	1-1/4" pipe size	Q-1	44	.364		26.50	19.20		45.70	59
2640	1-1/2" pipe size	↓	40	.400	↓	26.50	21		47.50	62

22 11 13 – Facility Water Distribution Piping

22 11 13.47 Pipe Fittings, Steel		Crew	Daily Output	Labor-Hours	Unit	Material	Labor	Equipment	Total	Total Incl O&P
						2015 Bare Costs				
2650	2" pipe size	Q-1	36	.444	Ea.	29.50	23.50		53	69
2660	2-1/2" pipe size		28	.571		36	30		66	87
2670	3" pipe size		20	.800		36.50	42.50		79	106
2690	4" pipe size		12	1.333		42.50	70.50		113	157
2700	5" pipe size		10	1.600		66.50	84.50		151	206
2710	6" pipe size	Q-2	14	1.714		73	94		167	227
2720	8" pipe size		12	2		126	110		236	310
2730	10" pipe size		10	2.400		228	131		359	455
2860	Black 300 lb., per flange									
2870	1/2" pipe size	1 Plum	30	.267	Ea.	29	15.65		44.65	56.50
2880	3/4" pipe size		28	.286		29	16.75		45.75	58
2890	1" pipe size		27	.296		29	17.40		46.40	59
2900	1-1/4" pipe size	Q-1	44	.364		29	19.20		48.20	62
2910	1-1/2" pipe size		40	.400		29	21		50	65
2920	2" pipe size		36	.444		34	23.50		57.50	74
2930	2-1/2" pipe size		28	.571		47.50	30		77.50	99
2940	3" pipe size		20	.800		50.50	42.50		93	122
2960	4" pipe size		12	1.333		73	70.50		143.50	190
2970	6" pipe size	Q-2	14	1.714		136	94		230	296
3000	Weld joint, butt, carbon steel, standard weight									
3040	90° elbow, long radius									
3050	1/2" pipe size	Q-15	16	1	Ea.	45	53	3.61	101.61	136
3060	3/4" pipe size		16	1		45	53	3.61	101.61	136
3070	1" pipe size		16	1		21	53	3.61	77.61	109
3080	1-1/4" pipe size		14	1.143		21	60.50	4.13	85.63	122
3090	1-1/2" pipe size		13	1.231		21	65	4.44	90.44	129
3100	2" pipe size		10	1.600		22.50	84.50	5.80	112.80	163
3110	2-1/2" pipe size		8	2		27.50	106	7.20	140.70	203
3120	3" pipe size		7	2.286		29	121	8.25	158.25	229
3130	4" pipe size		5	3.200		48	169	11.55	228.55	330
3136	5" pipe size		4	4		103	211	14.45	328.45	460
3140	6" pipe size	Q-16	5	4.800		106	263	11.50	380.50	540
3150	8" pipe size		3.75	6.400		200	350	15.35	565.35	780
3160	10" pipe size		3	8		400	440	19.20	859.20	1,150
3170	12" pipe size		2.50	9.600		590	525	23	1,138	1,500
3180	14" pipe size		2	12		970	655	29	1,654	2,125
3190	16" pipe size		1.50	16		1,325	875	38.50	2,238.50	2,900
3191	18" pipe size		1.25	19.200		1,525	1,050	46	2,621	3,375
3192	20" pipe size		1.15	20.870		2,450	1,150	50	3,650	4,525
3194	24" pipe size		1.02	23.529		3,450	1,300	56.50	4,806.50	5,850
3200	45° Elbow, long									
3210	1/2" pipe size	Q-15	16	1	Ea.	62.50	53	3.61	119.11	155
3220	3/4" pipe size		16	1		62.50	53	3.61	119.11	155
3230	1" pipe size		16	1		22	53	3.61	78.61	111
3240	1-1/4" pipe size		14	1.143		22	60.50	4.13	86.63	123
3250	1-1/2" pipe size		13	1.231		22	65	4.44	91.44	130
3260	2" pipe size		10	1.600		22	84.50	5.80	112.30	163
3270	2-1/2" pipe size		8	2		26.50	106	7.20	139.70	202
3280	3" pipe size		7	2.286		27.50	121	8.25	156.75	228
3290	4" pipe size		5	3.200		49.50	169	11.55	230.05	330
3296	5" pipe size		4	4		75	211	14.45	300.45	430
3300	6" pipe size	Q-16	5	4.800		97.50	263	11.50	372	530
3310	8" pipe size		3.75	6.400		163	350	15.35	528.35	740

22 11 13.47 Pipe Fittings, Steel		Crew	Daily Output	Labor-Hours	Unit	Material	2015 Bare Costs Labor	Equipment	Total	Total Incl O&P
3320	10" pipe size	Q-16	3	8	Ea.	325	440	19.20	784.20	1,050
3330	12" pipe size		2.50	9.600		460	525	23	1,008	1,350
3340	14" pipe size		2	12		615	655	29	1,299	1,725
3341	16" pipe size		1.50	16		1,100	875	38.50	2,013.50	2,650
3342	18" pipe size		1.25	19.200		1,575	1,050	46	2,671	3,425
3343	20" pipe size		1.15	20.870		1,625	1,150	50	2,825	3,600
3345	24" pipe size		1.05	22.857		2,450	1,250	55	3,755	4,700
3346	26" pipe size		.85	28.235		2,700	1,550	68	4,318	5,475
3347	30" pipe size		.45	53.333		2,975	2,925	128	6,028	7,975
3349	36" pipe size		.38	63.158		3,275	3,450	152	6,877	9,200
3350	Tee, straight									
3360	1/2" pipe size	Q-15	10	1.600	Ea.	110	84.50	5.80	200.30	259
3370	3/4" pipe size		10	1.600		110	84.50	5.80	200.30	259
3380	1" pipe size		10	1.600		54.50	84.50	5.80	144.80	198
3390	1-1/4" pipe size		9	1.778		68.50	94	6.40	168.90	229
3400	1-1/2" pipe size		8	2		68.50	106	7.20	181.70	248
3410	2" pipe size		6	2.667		54.50	141	9.65	205.15	291
3420	2-1/2" pipe size		5	3.200		75.50	169	11.55	256.05	360
3430	3" pipe size		4	4		84	211	14.45	309.45	440
3440	4" pipe size		3	5.333		118	282	19.25	419.25	590
3446	5" pipe size		2.50	6.400		195	340	23	558	765
3450	6" pipe size	Q-16	3	8		203	440	19.20	662.20	930
3460	8" pipe size		2.50	9.600		355	525	23	903	1,225
3470	10" pipe size		2	12		695	655	29	1,379	1,825
3480	12" pipe size		1.60	15		975	820	36	1,831	2,400
3481	14" pipe size		1.30	18.462		1,700	1,000	44.50	2,744.50	3,500
3482	16" pipe size		1	24		1,900	1,325	57.50	3,282.50	4,225
3483	18" pipe size		.80	30		3,025	1,650	72	4,747	5,975
3484	20" pipe size		.75	32		4,750	1,750	77	6,577	8,025
3486	24" pipe size		.70	34.286		6,150	1,875	82.50	8,107.50	9,775
3487	26" pipe size		.55	43.636		6,775	2,400	105	9,280	11,300
3488	30" pipe size		.30	80		7,425	4,375	192	11,992	15,200
3490	36" pipe size		.25	96		8,200	5,250	230	13,680	17,500
3491	Eccentric reducer, 1-1/2" pipe size	Q-15	14	1.143		63.50	60.50	4.13	128.13	169
3492	2" pipe size		11	1.455		43	77	5.25	125.25	173
3493	2-1/2" pipe size		9	1.778		50	94	6.40	150.40	208
3494	3" pipe size		8	2		59	106	7.20	172.20	238
3495	4" pipe size		6	2.667		75.50	141	9.65	226.15	315
3496	6" pipe size	Q-16	5	4.800		232	263	11.50	506.50	680
3497	8" pipe size		4	6		345	330	14.40	689.40	910
3498	10" pipe size		3	8		435	440	19.20	894.20	1,175
3499	12" pipe size		2.50	9.600		605	525	23	1,153	1,500
3501	Cap, 1-1/2" pipe size	Q-15	28	.571		24.50	30	2.06	56.56	76.50
3502	2" pipe size		22	.727		22	38.50	2.63	63.13	87
3503	2-1/2" pipe size		18	.889		30	47	3.21	80.21	109
3504	3" pipe size		16	1		33.50	53	3.61	90.11	123
3505	4" pipe size		12	1.333		45	70.50	4.81	120.31	165
3506	6" pipe size	Q-16	10	2.400		92	131	5.75	228.75	310
3507	8" pipe size		8	3		138	164	7.20	309.20	415
3508	10" pipe size		6	4		212	219	9.60	440.60	585
3509	12" pipe size		5	4.800		283	263	11.50	557.50	735
3511	14" pipe size		4	6		293	330	14.40	637.40	855
3512	16" pipe size		4	6		455	330	14.40	799.40	1,025

22 11 13.47 Pipe Fittings, Steel		Crew	Daily Output	Labor-Hours	Unit	Material	2015 Bare Costs Labor	Equipment	Total	Total Incl O&P
3513	18" pipe size	Q-16	3	8	Ea.	665	440	19.20	1,124.20	1,450
3517	Weld joint, butt, carbon steel, extra strong									
3519	90° elbow, long									
3520	1/2" pipe size	Q-15	13	1.231	Ea.	55.50	65	4.44	124.94	167
3530	3/4" pipe size		12	1.333		55.50	70.50	4.81	130.81	177
3540	1" pipe size		11	1.455		27	77	5.25	109.25	156
3550	1-1/4" pipe size		10	1.600		27	84.50	5.80	117.30	168
3560	1-1/2" pipe size		9	1.778		27	94	6.40	127.40	183
3570	2" pipe size		8	2		27.50	106	7.20	140.70	203
3580	2-1/2" pipe size		7	2.286		38.50	121	8.25	167.75	240
3590	3" pipe size		6	2.667		49.50	141	9.65	200.15	285
3600	4" pipe size		4	4		81.50	211	14.45	306.95	435
3606	5" pipe size		3.50	4.571		195	242	16.50	453.50	605
3610	6" pipe size	Q-16	4.50	5.333		206	292	12.80	510.80	695
3620	8" pipe size		3.50	6.857		390	375	16.45	781.45	1,025
3630	10" pipe size		2.50	9.600		830	525	23	1,378	1,750
3640	12" pipe size		2.25	10.667		1,025	585	25.50	1,635.50	2,075
3650	45° Elbow, long									
3660	1/2" pipe size	Q-15	13	1.231	Ea.	62	65	4.44	131.44	174
3670	3/4" pipe size		12	1.333		62	70.50	4.81	137.31	183
3680	1" pipe size		11	1.455		29	77	5.25	111.25	157
3690	1-1/4" pipe size		10	1.600		29	84.50	5.80	119.30	170
3700	1-1/2" pipe size		9	1.778		29	94	6.40	129.40	185
3710	2" pipe size		8	2		29	106	7.20	142.20	204
3720	2-1/2" pipe size		7	2.286		64	121	8.25	193.25	268
3730	3" pipe size		6	2.667		37	141	9.65	187.65	272
3740	4" pipe size		4	4		58.50	211	14.45	283.95	410
3746	5" pipe size		3.50	4.571		139	242	16.50	397.50	545
3750	6" pipe size	Q-16	4.50	5.333		160	292	12.80	464.80	645
3760	8" pipe size		3.50	6.857		277	375	16.45	668.45	910
3770	10" pipe size		2.50	9.600		535	525	23	1,083	1,425
3780	12" pipe size		2.25	10.667		790	585	25.50	1,400.50	1,800
3800	Tee, straight									
3810	1/2" pipe size	Q-15	9	1.778	Ea.	136	94	6.40	236.40	300
3820	3/4" pipe size		8.50	1.882		131	99.50	6.80	237.30	305
3830	1" pipe size		8	2		52	106	7.20	165.20	230
3840	1-1/4" pipe size		7	2.286		52	121	8.25	181.25	255
3850	1-1/2" pipe size		6	2.667		52	141	9.65	202.65	288
3860	2" pipe size		5	3.200		59.50	169	11.55	240.05	340
3870	2-1/2" pipe size		4	4		99	211	14.45	324.45	455
3880	3" pipe size		3.50	4.571		124	242	16.50	382.50	530
3890	4" pipe size		2.50	6.400		149	340	23	512	715
3896	5" pipe size		2.25	7.111		375	375	25.50	775.50	1,025
3900	6" pipe size	Q-16	2.25	10.667		286	585	25.50	896.50	1,250
3910	8" pipe size		2	12		545	655	29	1,229	1,650
3920	10" pipe size		1.75	13.714		845	750	33	1,628	2,150
3930	12" pipe size		1.50	16		1,225	875	38.50	2,138.50	2,775
4000	Eccentric reducer, 1-1/2" pipe size	Q-15	10	1.600		17.30	84.50	5.80	107.60	157
4010	2" pipe size		9	1.778		47.50	94	6.40	147.90	206
4020	2-1/2" pipe size		8	2		71.50	106	7.20	184.70	251
4030	3" pipe size		7	2.286		59	121	8.25	188.25	262
4040	4" pipe size		5	3.200		95.50	169	11.55	276.05	380
4046	5" pipe size		4.70	3.404		266	180	12.30	458.30	585

	22 11 13.47 Pipe Fittings, Steel	Crew	Daily Output	Labor-Hours	Unit	Material	2015 Bare Costs Labor	Equipment	Total	Total Incl O&P
4050	6" pipe size	Q-16	4.50	5.333	Ea.	259	292	12.80	563.80	755
4060	8" pipe size		3.50	6.857		390	375	16.45	781.45	1,025
4070	10" pipe size		2.50	9.600		665	525	23	1,213	1,575
4080	12" pipe size		2.25	10.667		845	585	25.50	1,455.50	1,875
4090	14" pipe size		2.10	11.429		1,575	625	27.50	2,227.50	2,725
4100	16" pipe size		1.90	12.632		2,025	690	30.50	2,745.50	3,325
4151	Cap, 1-1/2" pipe size	Q-15	24	.667		24.50	35	2.41	61.91	84.50
4152	2" pipe size		18	.889		22	47	3.21	72.21	101
4153	2-1/2" pipe size		16	1		30	53	3.61	86.61	119
4154	3" pipe size		14	1.143		33.50	60.50	4.13	98.13	135
4155	4" pipe size		10	1.600		45	84.50	5.80	135.30	188
4156	6" pipe size	Q-16	9	2.667		92	146	6.40	244.40	335
4157	8" pipe size		7	3.429		138	188	8.25	334.25	455
4158	10" pipe size		5	4.800		212	263	11.50	486.50	655
4159	12" pipe size		4	6		283	330	14.40	627.40	840
4190	Weld fittings, reducing, standard weight									
4200	Welding ring w/spacer pins, 2" pipe size				Ea.	2.40			2.40	2.64
4210	2-1/2" pipe size					2.45			2.45	2.70
4220	3" pipe size					2.50			2.50	2.75
4230	4" pipe size					2.65			2.65	2.92
4236	5" pipe size					3.35			3.35	3.69
4240	6" pipe size					3.60			3.60	3.96
4250	8" pipe size					4.20			4.20	4.62
4260	10" pipe size					4.70			4.70	5.15
4270	12" pipe size					5.40			5.40	5.95
4280	14" pipe size					6.45			6.45	7.10
4290	16" pipe size					7.60			7.60	8.35
4300	18" pipe size					8.60			8.60	9.45
4310	20" pipe size					8.95			8.95	9.85
4330	24" pipe size					11.10			11.10	12.20
4340	26" pipe size					13.70			13.70	15.05
4350	30" pipe size					16.75			16.75	18.45
4370	36" pipe size					20.50			20.50	22.50
5000	Weld joint, socket, forged steel, 3000 lb., schedule 40 pipe									
5010	90° elbow, straight									
5020	1/4" pipe size	Q-15	22	.727	Ea.	25	38.50	2.63	66.13	90.50
5030	3/8" pipe size		22	.727		25	38.50	2.63	66.13	90.50
5040	1/2" pipe size		20	.800		13.95	42.50	2.89	59.34	84.50
5050	3/4" pipe size		20	.800		14.55	42.50	2.89	59.94	85
5060	1" pipe size		20	.800		18.70	42.50	2.89	64.09	89.50
5070	1-1/4" pipe size		18	.889		36	47	3.21	86.21	116
5080	1-1/2" pipe size		16	1		41.50	53	3.61	98.11	132
5090	2" pipe size		12	1.333		61.50	70.50	4.81	136.81	183
5100	2-1/2" pipe size		10	1.600		174	84.50	5.80	264.30	330
5110	3" pipe size		8	2		300	106	7.20	413.20	505
5120	4" pipe size		6	2.667		795	141	9.65	945.65	1,100
5130	45° Elbow, straight									
5134	1/4" pipe size	Q-15	22	.727	Ea.	25	38.50	2.63	66.13	90.50
5135	3/8" pipe size		22	.727		25	38.50	2.63	66.13	90.50
5136	1/2" pipe size		20	.800		18.60	42.50	2.89	63.99	89.50
5137	3/4" pipe size		20	.800		21.50	42.50	2.89	66.89	93
5140	1" pipe size		20	.800		28.50	42.50	2.89	73.89	100
5150	1-1/4" pipe size		18	.889		39	47	3.21	89.21	119

22 11 13 – Facility Water Distribution Piping

22 11 13.47 Pipe Fittings, Steel		Crew	Daily Output	Labor-Hours	Unit	Material	Labor	Equipment	Total	Total Incl O&P
						2015 Bare Costs				
5160	1-1/2" pipe size	Q-15	16	1	Ea.	47.50	53	3.61	104.11	139
5170	2" pipe size		12	1.333		76.50	70.50	4.81	151.81	199
5180	2-1/2" pipe size		10	1.600		201	84.50	5.80	291.30	360
5190	3" pipe size		8	2		335	106	7.20	448.20	540
5200	4" pipe size		6	2.667		660	141	9.65	810.65	955
5250	Tee, straight									
5254	1/4" pipe size	Q-15	15	1.067	Ea.	27.50	56.50	3.85	87.85	123
5255	3/8" pipe size		15	1.067		27.50	56.50	3.85	87.85	123
5256	1/2" pipe size		13	1.231		17.30	65	4.44	86.74	125
5257	3/4" pipe size		13	1.231		21	65	4.44	90.44	129
5260	1" pipe size		13	1.231		29	65	4.44	98.44	137
5270	1-1/4" pipe size		12	1.333		44	70.50	4.81	119.31	163
5280	1-1/2" pipe size		11	1.455		58.50	77	5.25	140.75	190
5290	2" pipe size		8	2		84.50	106	7.20	197.70	266
5300	2-1/2" pipe size		6	2.667		243	141	9.65	393.65	500
5310	3" pipe size		5	3.200		580	169	11.55	760.55	915
5320	4" pipe size		4	4		935	211	14.45	1,160.45	1,375
5350	For reducing sizes, add					60%				
5450	Couplings									
5451	1/4" pipe size	Q-15	23	.696	Ea.	15.90	37	2.51	55.41	78
5452	3/8" pipe size		23	.696		15.90	37	2.51	55.41	78
5453	1/2" pipe size		21	.762		7.35	40.50	2.75	50.60	74
5454	3/4" pipe size		21	.762		9.50	40.50	2.75	52.75	76.50
5460	1" pipe size		20	.800		10.50	42.50	2.89	55.89	80.50
5470	1-1/4" pipe size		20	.800		18.60	42.50	2.89	63.99	89.50
5480	1-1/2" pipe size		18	.889		20.50	47	3.21	70.71	99
5490	2" pipe size		14	1.143		33	60.50	4.13	97.63	135
5500	2-1/2" pipe size		12	1.333		73.50	70.50	4.81	148.81	196
5510	3" pipe size		9	1.778		165	94	6.40	265.40	335
5520	4" pipe size		7	2.286		248	121	8.25	377.25	470
5570	Union, 1/4" pipe size		21	.762		34	40.50	2.75	77.25	104
5571	3/8" pipe size		21	.762		34	40.50	2.75	77.25	104
5572	1/2" pipe size		19	.842		28.50	44.50	3.04	76.04	104
5573	3/4" pipe size		19	.842		33.50	44.50	3.04	81.04	109
5574	1" pipe size		19	.842		43	44.50	3.04	90.54	120
5575	1-1/4" pipe size		17	.941		69.50	49.50	3.40	122.40	158
5576	1-1/2" pipe size		15	1.067		75	56.50	3.85	135.35	175
5577	2" pipe size		11	1.455		108	77	5.25	190.25	245
5600	Reducer, 1/4" pipe size		23	.696		40.50	37	2.51	80.01	105
5601	3/8" pipe size		23	.696		43	37	2.51	82.51	107
5602	1/2" pipe size		21	.762		26	40.50	2.75	69.25	94.50
5603	3/4" pipe size		21	.762		26	40.50	2.75	69.25	94.50
5604	1" pipe size		21	.762		36.50	40.50	2.75	79.75	106
5605	1-1/4" pipe size		19	.842		46.50	44.50	3.04	94.04	124
5607	1-1/2" pipe size		17	.941		50	49.50	3.40	102.90	136
5608	2" pipe size		13	1.231		55	65	4.44	124.44	166
5612	Cap, 1/4" pipe size		46	.348		15.15	18.40	1.26	34.81	46.50
5613	3/8" pipe size		46	.348		15.15	18.40	1.26	34.81	46.50
5614	1/2" pipe size		42	.381		9.25	20	1.38	30.63	43
5615	3/4" pipe size		42	.381		10.80	20	1.38	32.18	45
5616	1" pipe size		42	.381		16.35	20	1.38	37.73	51
5617	1-1/4" pipe size		38	.421		19.30	22	1.52	42.82	57
5618	1-1/2" pipe size		34	.471		27	25	1.70	53.70	70.50

22 11 13.47 Pipe Fittings, Steel		Crew	Daily Output	Labor-Hours	Unit	Material	Labor	Equipment	Total	Total Incl O&P
						2015 Bare Costs				
5619	2" pipe size	Q-15	26	.615	Ea.	41.50	32.50	2.22	76.22	98.50
5630	T-O-L, 1/4" pipe size, nozzle		23	.696		7.10	37	2.51	46.61	68
5631	3/8" pipe size, nozzle		23	.696		7.20	37	2.51	46.71	68
5632	1/2" pipe size, nozzle		22	.727		7.10	38.50	2.63	48.23	70.50
5633	3/4" pipe size, nozzle		21	.762		8.20	40.50	2.75	51.45	75
5634	1" pipe size, nozzle		20	.800		9.55	42.50	2.89	54.94	79.50
5635	1-1/4" pipe size, nozzle		18	.889		15.10	47	3.21	65.31	93
5636	1-1/2" pipe size, nozzle		16	1		15.10	53	3.61	71.71	103
5637	2" pipe size, nozzle		12	1.333		17.20	70.50	4.81	92.51	134
5638	2-1/2" pipe size, nozzle		10	1.600		57.50	84.50	5.80	147.80	202
5639	4" pipe size, nozzle		6	2.667		131	141	9.65	281.65	375
5640	W-O-L, 1/4" pipe size, nozzle		23	.696		16.80	37	2.51	56.31	78.50
5641	3/8" pipe size, nozzle		23	.696		15.85	37	2.51	55.36	77.50
5642	1/2" pipe size, nozzle		22	.727		14.70	38.50	2.63	55.83	79
5643	3/4" pipe size, nozzle		21	.762		15.50	40.50	2.75	58.75	83
5644	1" pipe size, nozzle		20	.800		16.20	42.50	2.89	61.59	87
5645	1-1/4" pipe size, nozzle		18	.889		19.25	47	3.21	69.46	97.50
5646	1-1/2" pipe size, nozzle		16	1		19.25	53	3.61	75.86	107
5647	2" pipe size, nozzle		12	1.333		19.45	70.50	4.81	94.76	137
5648	2-1/2" pipe size, nozzle		10	1.600		44.50	84.50	5.80	134.80	187
5649	3" pipe size, nozzle		8	2		48.50	106	7.20	161.70	226
5650	4" pipe size, nozzle		6	2.667		61.50	141	9.65	212.15	298
5651	5" pipe size, nozzle		5	3.200		152	169	11.55	332.55	445
5652	6" pipe size, nozzle		4	4		177	211	14.45	402.45	540
5653	8" pipe size, nozzle		3	5.333		340	282	19.25	641.25	835
5654	10" pipe size, nozzle		2.60	6.154		490	325	22	837	1,075
5655	12" pipe size, nozzle		2.20	7.273		930	385	26.50	1,341.50	1,650
5674	S-O-L, 1/4" pipe size, outlet		23	.696		9.45	37	2.51	48.96	70.50
5675	3/8" pipe size, outlet		23	.696		9.45	37	2.51	48.96	70.50
5676	1/2" pipe size, outlet		22	.727		8.80	38.50	2.63	49.93	72.50
5677	3/4" pipe size, outlet		21	.762		8.95	40.50	2.75	52.20	76
5678	1" pipe size, outlet		20	.800		9.90	42.50	2.89	55.29	80
5679	1-1/4" pipe size, outlet		18	.889		16.55	47	3.21	66.76	94.50
5680	1-1/2" pipe size, outlet		16	1		16.55	53	3.61	73.16	105
5681	2" pipe size, outlet	↓	12	1.333	↓	18.80	70.50	4.81	94.11	136
6000	Weld-on flange, forged steel									
6020	Slip-on, 150 lb. flange (welded front and back)									
6050	1/2" pipe size	Q-15	18	.889	Ea.	17.50	47	3.21	67.71	96
6060	3/4" pipe size		18	.889		17.50	47	3.21	67.71	96
6070	1" pipe size		17	.941		17.50	49.50	3.40	70.40	100
6080	1-1/4" pipe size		16	1		17.50	53	3.61	74.11	106
6090	1-1/2" pipe size		15	1.067		17.50	56.50	3.85	77.85	111
6100	2" pipe size		12	1.333		19.45	70.50	4.81	94.76	137
6110	2-1/2" pipe size		10	1.600		25	84.50	5.80	115.30	166
6120	3" pipe size		9	1.778		26.50	94	6.40	126.90	182
6130	3-1/2" pipe size		7	2.286		33	121	8.25	162.25	234
6140	4" pipe size		6	2.667		33	141	9.65	183.65	267
6150	5" pipe size	↓	5	3.200		57.50	169	11.55	238.05	340
6160	6" pipe size	Q-16	6	4		54.50	219	9.60	283.10	410
6170	8" pipe size		5	4.800		82.50	263	11.50	357	515
6180	10" pipe size		4	6		143	330	14.40	487.40	690
6190	12" pipe size		3	8		212	440	19.20	671.20	940
6191	14" pipe size	↓	2.50	9.600	↓	281	525	23	829	1,150

22 11 13 – Facility Water Distribution Piping

22 11 13.47 Pipe Fittings, Steel		Crew	Daily Output	Labor-Hours	Unit	Material	2015 Bare Costs Labor	Equipment	Total	Total Incl O&P
6192	16" pipe size	Q-16	1.80	13.333	Ea.	440	730	32	1,202	1,675
6200	300 lb. flange									
6210	1/2" pipe size	Q-15	17	.941	Ea.	23.50	49.50	3.40	76.40	107
6220	3/4" pipe size		17	.941		23.50	49.50	3.40	76.40	107
6230	1" pipe size		16	1		23.50	53	3.61	80.11	112
6240	1-1/4" pipe size		13	1.231		23.50	65	4.44	92.94	131
6250	1-1/2" pipe size		12	1.333		23.50	70.50	4.81	98.81	141
6260	2" pipe size		11	1.455		31	77	5.25	113.25	160
6270	2-1/2" pipe size		9	1.778		35	94	6.40	135.40	192
6280	3" pipe size		7	2.286		38	121	8.25	167.25	239
6290	4" pipe size		6	2.667		55.50	141	9.65	206.15	292
6300	5" pipe size		4	4		95	211	14.45	320.45	450
6310	6" pipe size	Q-16	5	4.800		95.50	263	11.50	370	530
6320	8" pipe size		4	6		163	330	14.40	507.40	710
6330	10" pipe size		3.40	7.059		278	385	16.95	679.95	930
6340	12" pipe size		2.80	8.571		340	470	20.50	830.50	1,125
6400	Welding neck, 150 lb. flange									
6410	1/2" pipe size	Q-15	40	.400	Ea.	25.50	21	1.44	47.94	62.50
6420	3/4" pipe size		36	.444		25.50	23.50	1.60	50.60	66.50
6430	1" pipe size		32	.500		25.50	26.50	1.81	53.81	71
6440	1-1/4" pipe size		29	.552		25.50	29	1.99	56.49	75.50
6450	1-1/2" pipe size		26	.615		25.50	32.50	2.22	60.22	81
6460	2" pipe size		20	.800		29.50	42.50	2.89	74.89	101
6470	2-1/2" pipe size		16	1		32	53	3.61	88.61	121
6480	3" pipe size		14	1.143		35	60.50	4.13	99.63	137
6500	4" pipe size		10	1.600		42.50	84.50	5.80	132.80	185
6510	5" pipe size		8	2		66.50	106	7.20	179.70	246
6520	6" pipe size	Q-16	10	2.400		64	131	5.75	200.75	281
6530	8" pipe size		7	3.429		113	188	8.25	309.25	425
6540	10" pipe size		6	4		180	219	9.60	408.60	550
6550	12" pipe size		5	4.800		262	263	11.50	536.50	710
6551	14" pipe size		4.50	5.333		380	292	12.80	684.80	885
6552	16" pipe size		3	8		570	440	19.20	1,029.20	1,325
6553	18" pipe size		2.50	9.600		800	525	23	1,348	1,725
6554	20" pipe size		2.30	10.435		975	570	25	1,570	2,000
6556	24" pipe size		2	12		1,300	655	29	1,984	2,500
6557	26" pipe size		1.70	14.118		1,425	775	34	2,234	2,825
6558	30" pipe size		.90	26.667		1,650	1,450	64	3,164	4,150
6559	36" pipe size		.75	32		1,900	1,750	77	3,727	4,875
6560	300 lb. flange									
6570	1/2" pipe size	Q-15	36	.444	Ea.	31	23.50	1.60	56.10	72.50
6580	3/4" pipe size		34	.471		31	25	1.70	57.70	75
6590	1" pipe size		30	.533		31	28	1.93	60.93	80
6600	1-1/4" pipe size		28	.571		31	30	2.06	63.06	83.50
6610	1-1/2" pipe size		24	.667		31	35	2.41	68.41	91.50
6620	2" pipe size		18	.889		38.50	47	3.21	88.71	119
6630	2-1/2" pipe size		14	1.143		44	60.50	4.13	108.63	147
6640	3" pipe size		12	1.333		44.50	70.50	4.81	119.81	164
6650	4" pipe size		8	2		73	106	7.20	186.20	253
6660	5" pipe size		7	2.286		111	121	8.25	240.25	320
6670	6" pipe size	Q-16	9	2.667		113	146	6.40	265.40	360
6680	8" pipe size		6	4		195	219	9.60	423.60	565
6690	10" pipe size		5	4.800		355	263	11.50	629.50	815

22 11 13.47 Pipe Fittings, Steel		Crew	Daily Output	Labor-Hours	Unit	Material	2015 Bare Costs Labor	Equipment	Total	Total Incl O&P
6700	12" pipe size	Q-16	4	6	Ea.	455	330	14.40	799.40	1,025
6710	14" pipe size		3.50	6.857		980	375	16.45	1,371.45	1,675
6720	16" pipe size		2	12		1,150	655	29	1,834	2,325
7740	Plain ends for plain end pipe, mechanically coupled									
7750	Cplg. & labor required at joints not included, add 1 per									
7760	joint for installed price, see line 9180									
7770	Malleable iron, painted, unless noted otherwise									
7800	90° Elbow 1"				Ea.	112			112	124
7810	1-1/2"					133			133	146
7820	2"					198			198	218
7830	2-1/2"					232			232	255
7840	3"					240			240	264
7860	4"					272			272	300
7870	5" welded steel					315			315	345
7880	6"					395			395	435
7890	8" welded steel					745			745	815
7900	10" welded steel					945			945	1,050
7910	12" welded steel					1,050			1,050	1,150
7970	45° Elbow 1"					67			67	73.50
7980	1-1/2"					100			100	110
7990	2"					210			210	231
8000	2-1/2"					210			210	231
8010	3"					240			240	264
8030	4"					251			251	276
8040	5" welded steel					315			315	345
8050	6"					360			360	395
8060	8"					420			420	460
8070	10" welded steel					435			435	480
8080	12" welded steel					785			785	865
8140	Tee, straight 1"					126			126	138
8150	1-1/2"					162			162	178
8160	2"					162			162	178
8170	2-1/2"					211			211	232
8180	3"					325			325	355
8200	4"					465			465	510
8210	5" welded steel					635			635	695
8220	6"					550			550	605
8230	8" welded steel					795			795	875
8240	10" welded steel					1,250			1,250	1,375
8250	12" welded steel					1,450			1,450	1,600
8340	Segmentally welded steel, painted									
8390	Wye 2"				Ea.	201			201	221
8400	2-1/2"					201			201	221
8410	3"					225			225	247
8430	4"					335			335	370
8440	5"					410			410	450
8450	6"					560			560	620
8460	8"					730			730	800
8470	10"					1,075			1,075	1,175
8480	12"					1,675			1,675	1,825
8540	Wye, lateral 2"					218			218	240
8550	2-1/2"					251			251	276
8560	3"					298			298	330

22 11 13 – Facility Water Distribution Piping

22 11 13.47	Pipe Fittings, Steel	Crew	Daily Output	Labor-Hours	Unit	2015 Bare Costs Material	Labor	Equipment	Total	Total Incl O&P
8580	4"				Ea.	410			410	450
8590	5"					695			695	765
8600	6"					720			720	790
8610	8"					1,200			1,200	1,325
8620	10"					1,200			1,200	1,325
8630	12"					2,300			2,300	2,525
8690	Cross, 2"					212			212	234
8700	2-1/2"					212			212	234
8710	3"					253			253	279
8730	4"					345			345	380
8740	5"					490			490	540
8750	6"					650			650	715
8760	8"					830			830	915
8770	10"					1,225			1,225	1,325
8780	12"					1,750			1,750	1,925
8800	Tees, reducing 2" x 1"					138			138	152
8810	2" x 1-1/2"					138			138	152
8820	3" x 1"					138			138	152
8830	3" x 1-1/2"					138			138	152
8840	3" x 2"					139			139	153
8850	4" x 1"					203			203	223
8860	4" x 1-1/2"					203			203	223
8870	4" x 2"					200			200	220
8880	4" x 2-1/2"					206			206	227
8890	4" x 3"					206			206	227
8900	6" x 2"					300			300	330
8910	6" x 3"					325			325	355
8920	6" x 4"					355			355	390
8930	8" x 2"					365			365	405
8940	8" x 3"					375			375	410
8950	8" x 4"					400			400	440
8960	8" x 5"					430			430	475
8970	8" x 6"					465			465	515
8980	10" x 4"					620			620	680
8990	10" x 6"					620			620	680
9000	10" x 8"					655			655	720
9010	12" x 6"					955			955	1,050
9020	12" x 8"					955			955	1,050
9030	12" x 10"				↓	995			995	1,100
9080	Adapter nipples 3" long									
9090	1"				Ea.	19.40			19.40	21.50
9100	1-1/2"					19.40			19.40	21.50
9110	2"					19.40			19.40	21.50
9120	2-1/2"					22.50			22.50	24.50
9130	3"					27			27	30
9140	4"					44.50			44.50	49
9150	6"				↓	117			117	129
9180	Coupling, mechanical, plain end pipe to plain end pipe or fitting									
9190	1"	Q-1	29	.552	Ea.	68	29		97	120
9200	1-1/2"		28	.571		68	30		98	122
9210	2"		27	.593		68	31.50		99.50	124
9220	2-1/2"		26	.615		68	32.50		100.50	125
9230	3"	↓	25	.640	↓	99	34		133	162

22 11 13 – Facility Water Distribution Piping

22 11 13.47 Pipe Fittings, Steel		Crew	Daily Output	Labor-Hours	Unit	Material	2015 Bare Costs Labor	Equipment	Total	Total Incl O&P
9240	3-1/2"	Q-1	24	.667	Ea.	114	35		149	181
9250	4"	↓	22	.727		114	38.50		152.50	186
9260	5"	Q-2	28	.857		163	47		210	252
9270	6"		24	1		199	55		254	305
9280	8"		19	1.263		350	69		419	495
9290	10"		16	1.500		455	82		537	630
9300	12"	↓	12	2	↓	570	110		680	800
9940	For galvanized fittings for plain end pipe, add					20%				
9990	Minimum labor/equipment charge	Q-15	3	5.333	Job		282	19.25	301.25	460

22 11 13.48 Pipe, Fittings and Valves, Steel, Grooved-Joint		Crew	Daily Output	Labor-Hours	Unit	Material	Labor	Equipment	Total	Total Incl O&P
0010	**PIPE, FITTINGS AND VALVES, STEEL, GROOVED-JOINT**									
0012	Fittings are ductile iron. Steel fittings noted.									
0020	Pipe includes coupling & clevis type hanger assemblies, 10' O.C.									
0500	Schedule 10, black									
0550	2" diameter	1 Plum	43	.186	L.F.	7.15	10.90		18.05	25
0560	2-1/2" diameter	Q-1	61	.262		8.85	13.85		22.70	31.50
0570	3" diameter		55	.291		10.15	15.35		25.50	35
0580	3-1/2" diameter		53	.302		12.80	15.95		28.75	39
0590	4" diameter		49	.327		13.40	17.25		30.65	42
0600	5" diameter	↓	40	.400		16.15	21		37.15	51
0610	6" diameter	Q-2	46	.522		20.50	28.50		49	67
0620	8" diameter	"	41	.585	↓	28	32		60	81
0700	To delete couplings & hangers, subtract									
0710	2" diam. to 5" diam.					25%	20%			
0720	6" diam. to 8" diam.					27%	15%			
1000	Schedule 40, black									
1040	3/4" diameter	1 Plum	71	.113	L.F.	5.50	6.60		12.10	16.30
1050	1" diameter		63	.127		5.30	7.45		12.75	17.45
1060	1-1/4" diameter		58	.138		6.60	8.10		14.70	19.90
1070	1-1/2" diameter		51	.157		7.35	9.20		16.55	22.50
1080	2" diameter	↓	40	.200		8.70	11.75		20.45	28
1090	2-1/2" diameter	Q-1	57	.281		13.65	14.85		28.50	38
1100	3" diameter		50	.320		16.75	16.90		33.65	45
1110	4" diameter		45	.356		23.50	18.80		42.30	55.50
1120	5" diameter	↓	37	.432		38	23		61	77.50
1130	6" diameter	Q-2	42	.571		48.50	31.50		80	102
1140	8" diameter		37	.649		78.50	35.50		114	142
1150	10" diameter		31	.774		106	42.50		148.50	183
1160	12" diameter		27	.889		118	48.50		166.50	206
1170	14" diameter		20	1.200		122	65.50		187.50	238
1180	16" diameter		17	1.412		183	77.50		260.50	325
1190	18" diameter		14	1.714		183	94		277	345
1200	20" diameter		12	2		214	110		324	405
1210	24" diameter	↓	10	2.400	↓	241	131		372	470
1740	To delete coupling & hanger, subtract									
1750	3/4" diam. to 2" diam.					65%	27%			
1760	2-1/2" diam. to 5" diam.					41%	18%			
1770	6" diam. to 12" diam.					31%	13%			
1780	14" diam. to 24" diam.					35%	10%			
1800	Galvanized									
1840	3/4" diameter	1 Plum	71	.113	L.F.	5.80	6.60		12.40	16.65
1850	1" diameter	↓	63	.127	↓	7	7.45		14.45	19.35

22 11 13 – Facility Water Distribution Piping

22 11 13.48 Pipe, Fittings and Valves, Steel, Grooved-Joint		Crew	Daily Output	Labor-Hours	Unit	Material	2015 Bare Costs Labor	Equipment	Total	Total Incl O&P
1860	1-1/4" diameter	1 Plum	58	.138	L.F.	8.95	8.10		17.05	22.50
1870	1-1/2" diameter		51	.157		10.10	9.20		19.30	25.50
1880	2" diameter		40	.200		12.50	11.75		24.25	32
1890	2-1/2" diameter	Q-1	57	.281		17.60	14.85		32.45	42.50
1900	3" diameter		50	.320		22.50	16.90		39.40	51
1910	4" diameter		45	.356		32	18.80		50.80	64.50
1920	5" diameter		37	.432		60	23		83	101
1930	6" diameter	Q-2	42	.571		64.50	31.50		96	120
1940	8" diameter		37	.649		76	35.50		111.50	139
1950	10" diameter		31	.774		122	42.50		164.50	200
1960	12" diameter		27	.889		148	48.50		196.50	239
2540	To delete coupling & hanger, subtract									
2550	3/4" diam. to 2" diam.					36%	27%			
2560	2-1/2" diam. to 5" diam.					19%	18%			
2570	6" diam. to 12" diam.					14%	13%			
2600	Schedule 80, black									
2610	3/4" diameter	1 Plum	65	.123	L.F.	6.60	7.20		13.80	18.50
2650	1" diameter		61	.131		8.10	7.70		15.80	21
2660	1-1/4" diameter		55	.145		10.40	8.55		18.95	24.50
2670	1-1/2" diameter		49	.163		11.80	9.60		21.40	28
2680	2" diameter		38	.211		14.80	12.35		27.15	35.50
2690	2-1/2" diameter	Q-1	54	.296		21	15.65		36.65	48
2700	3" diameter		48	.333		27	17.60		44.60	57
2710	4" diameter		44	.364		38	19.20		57.20	72
2720	5" diameter		35	.457		55.50	24		79.50	98.50
2730	6" diameter	Q-2	40	.600		74.50	33		107.50	133
2740	8" diameter		35	.686		113	37.50		150.50	183
2750	10" diameter		29	.828		181	45.50		226.50	270
2760	12" diameter		24	1		395	55		450	520
3240	To delete coupling & hanger, subtract									
3250	3/4" diam. to 2" diam.					30%	25%			
3260	2-1/2" diam. to 5" diam.					14%	17%			
3270	6" diam. to 12" diam.					12%	12%			
3300	Galvanized									
3310	3/4" diameter	1 Plum	65	.123	L.F.	7.55	7.20		14.75	19.55
3350	1" diameter		61	.131		9.35	7.70		17.05	22.50
3360	1-1/4" diameter		55	.145		12.30	8.55		20.85	27
3370	1-1/2" diameter		46	.174		14.30	10.20		24.50	31.50
3380	2" diameter		38	.211		18.60	12.35		30.95	40
3390	2-1/2" diameter	Q-1	54	.296		26.50	15.65		42.15	54
3400	3" diameter		48	.333		34.50	17.60		52.10	65.50
3410	4" diameter		44	.364		40.50	19.20		59.70	74.50
3420	5" diameter		35	.457		57.50	24		81.50	101
3430	6" diameter	Q-2	40	.600		78.50	33		111.50	138
3440	8" diameter		35	.686		147	37.50		184.50	220
3450	10" diameter		29	.828		271	45.50		316.50	370
3460	12" diameter		24	1		530	55		585	670
3920	To delete coupling & hanger, subtract									
3930	3/4" diam. to 2" diam.					30%	25%			
3940	2-1/2" diam. to 5" diam.					15%	17%			
3950	6" diam. to 12" diam.					11%	12%			
3990	Fittings: coupling material required at joints not incl. in fitting price.									
3994	Add 1 selected coupling, material only, per joint for installed price.									

22 11 13 – Facility Water Distribution Piping

22 11 13.48 Pipe, Fittings and Valves, Steel, Grooved-Joint		Crew	Daily Output	Labor-Hours	Unit	Material	2015 Bare Costs Labor	Equipment	Total	Total Incl O&P
4000	Elbow, 90° or 45°, painted									
4030	3/4" diameter	1 Plum	50	.160	Ea.	60	9.40		69.40	80.50
4040	1" diameter		50	.160		32	9.40		41.40	49.50
4050	1-1/4" diameter		40	.200		32	11.75		43.75	53.50
4060	1-1/2" diameter		33	.242		32	14.25		46.25	57
4070	2" diameter		25	.320		32	18.80		50.80	64.50
4080	2-1/2" diameter	Q-1	40	.400		32	21		53	68
4090	3" diameter		33	.485		56.50	25.50		82	103
4100	4" diameter		25	.640		61.50	34		95.50	120
4110	5" diameter		20	.800		146	42.50		188.50	227
4120	6" diameter	Q-2	25	.960		172	52.50		224.50	271
4130	8" diameter		21	1.143		360	62.50		422.50	495
4140	10" diameter		18	1.333		625	73		698	805
4150	12" diameter		15	1.600		1,000	87.50		1,087.50	1,225
4170	14" diameter		12	2		860	110		970	1,125
4180	16" diameter		11	2.182		1,125	120		1,245	1,400
4190	18" diameter	Q-3	15	2.133		1,425	119		1,544	1,725
4200	20" diameter		13	2.462		1,875	138		2,013	2,275
4210	24" diameter		11	2.909		2,700	163		2,863	3,225
4250	For galvanized elbows, add					26%				
4690	Tee, painted									
4700	3/4" diameter	1 Plum	38	.211	Ea.	64.50	12.35		76.85	90.50
4740	1" diameter		33	.242		50	14.25		64.25	77
4750	1-1/4" diameter		27	.296		50	17.40		67.40	82
4760	1-1/2" diameter		22	.364		50	21.50		71.50	88.50
4770	2" diameter		17	.471		50	27.50		77.50	98
4780	2-1/2" diameter	Q-1	27	.593		50	31.50		81.50	104
4790	3" diameter		22	.727		68	38.50		106.50	135
4800	4" diameter		17	.941		103	49.50		152.50	192
4810	5" diameter		13	1.231		241	65		306	365
4820	6" diameter	Q-2	17	1.412		278	77.50		355.50	425
4830	8" diameter		14	1.714		610	94		704	815
4840	10" diameter		12	2		1,225	110		1,335	1,525
4850	12" diameter		10	2.400		1,700	131		1,831	2,075
4851	14" diameter		9	2.667		1,150	146		1,296	1,500
4852	16" diameter		8	3		1,300	164		1,464	1,675
4853	18" diameter	Q-3	11	2.909		1,625	163		1,788	2,050
4854	20" diameter		10	3.200		2,325	179		2,504	2,850
4855	24" diameter		8	4		3,550	224		3,774	4,250
4900	For galvanized tees, add					24%				
4939	Couplings									
4940	Flexible, standard, painted									
4950	3/4" diameter	1 Plum	100	.080	Ea.	17.80	4.70		22.50	27
4960	1" diameter		100	.080		17.80	4.70		22.50	27
4970	1-1/4" diameter		80	.100		23.50	5.85		29.35	34.50
4980	1-1/2" diameter		67	.119		25.50	7		32.50	39
4990	2" diameter		50	.160		27	9.40		36.40	44.50
5000	2-1/2" diameter	Q-1	80	.200		31.50	10.55		42.05	51.50
5010	3" diameter		67	.239		35	12.60		47.60	58
5020	3-1/2" diameter		57	.281		50	14.85		64.85	78
5030	4" diameter		50	.320		50.50	16.90		67.40	82
5040	5" diameter		40	.400		76.50	21		97.50	118
5050	6" diameter	Q-2	50	.480		90.50	26.50		117	141

22 11 13.48 Pipe, Fittings and Valves, Steel, Grooved-Joint		Crew	Daily Output	Labor-Hours	Unit	Material	2015 Bare Costs Labor	Equipment	Total	Total Incl O&P
5070	8″ diameter	Q-2	42	.571	Ea.	147	31.50		178.50	211
5090	10″ diameter		35	.686		240	37.50		277.50	325
5110	12″ diameter		32	.750		272	41		313	365
5120	14″ diameter		24	1		390	55		445	515
5130	16″ diameter		20	1.200		510	65.50		575.50	665
5140	18″ diameter		18	1.333		595	73		668	770
5150	20″ diameter		16	1.500		940	82		1,022	1,150
5160	24″ diameter		13	1.846		1,025	101		1,126	1,275
5200	For galvanized couplings, add					33%				
5750	Flange, w/groove gasket, black steel									
5754	See Line 22 11 13.47 0620 for gasket & bolt set									
5760	ANSI class 125 and 150, painted									
5780	2″ pipe size	1 Plum	23	.348	Ea.	106	20.50		126.50	149
5790	2-1/2″ pipe size	Q-1	37	.432		132	23		155	181
5800	3″ pipe size		31	.516		142	27.50		169.50	199
5820	4″ pipe size		23	.696		189	37		226	266
5830	5″ pipe size		19	.842		220	44.50		264.50	310
5840	6″ pipe size	Q-2	23	1.043		240	57		297	355
5850	8″ pipe size		17	1.412		271	77.50		348.50	420
5860	10″ pipe size		14	1.714		430	94		524	615
5870	12″ pipe size		12	2		560	110		670	785
5880	14″ pipe size		10	2.400		1,100	131		1,231	1,400
5890	16″ pipe size		9	2.667		1,275	146		1,421	1,625
5900	18″ pipe size		6	4		1,575	219		1,794	2,075
5910	20″ pipe size		5	4.800		1,875	263		2,138	2,475
5920	24″ pipe size		4.50	5.333		2,400	292		2,692	3,100
8000	Butterfly valve, 2 position handle, with standard trim									
8010	1-1/2″ pipe size	1 Plum	50	.160	Ea.	257	9.40		266.40	298
8020	2″ pipe size	"	38	.211		257	12.35		269.35	300
8030	3″ pipe size	Q-1	50	.320		370	16.90		386.90	430
8050	4″ pipe size	"	38	.421		405	22		427	480
8070	6″ pipe size	Q-2	38	.632		820	34.50		854.50	955
8080	8″ pipe size		27	.889		1,100	48.50		1,148.50	1,300
8090	10″ pipe size		20	1.200		1,800	65.50		1,865.50	2,100
8200	With stainless steel trim									
8240	1-1/2″ pipe size	1 Plum	50	.160	Ea.	325	9.40		334.40	375
8250	2″ pipe size	"	38	.211		325	12.35		337.35	380
8270	3″ pipe size	Q-1	50	.320		440	16.90		456.90	505
8280	4″ pipe size	"	38	.421		475	22		497	560
8300	6″ pipe size	Q-2	38	.632		890	34.50		924.50	1,025
8310	8″ pipe size		27	.889		2,200	48.50		2,248.50	2,500
8320	10″ pipe size		20	1.200		3,675	65.50		3,740.50	4,125
9000	Cut one groove, labor									
9010	3/4″ pipe size	Q-1	152	.105	Ea.		5.55		5.55	8.65
9020	1″ pipe size		140	.114			6.05		6.05	9.40
9030	1-1/4″ pipe size		124	.129			6.80		6.80	10.65
9040	1-1/2″ pipe size		114	.140			7.40		7.40	11.55
9050	2″ pipe size		104	.154			8.15		8.15	12.70
9060	2-1/2″ pipe size		96	.167			8.80		8.80	13.75
9070	3″ pipe size		88	.182			9.60		9.60	15
9080	3-1/2″ pipe size		83	.193			10.20		10.20	15.90
9090	4″ pipe size		78	.205			10.85		10.85	16.90
9100	5″ pipe size		72	.222			11.75		11.75	18.30

22 11 13 – Facility Water Distribution Piping

22 11 13.48 Pipe, Fittings and Valves, Steel, Grooved-Joint		Crew	Daily Output	Labor-Hours	Unit	Material	2015 Bare Costs Labor	Equipment	Total	Total Incl O&P
9110	6" pipe size	Q-1	70	.229	Ea.		12.10		12.10	18.85
9120	8" pipe size		54	.296			15.65		15.65	24.50
9130	10" pipe size		38	.421			22		22	34.50
9140	12" pipe size		30	.533			28		28	44
9150	14" pipe size		20	.800			42.50		42.50	66
9160	16" pipe size		19	.842			44.50		44.50	69.50
9170	18" pipe size		18	.889			47		47	73
9180	20" pipe size		17	.941			49.50		49.50	77.50
9190	24" pipe size		15	1.067			56.50		56.50	88
9210	Roll one groove									
9220	3/4" pipe size	Q-1	266	.060	Ea.		3.18		3.18	4.96
9230	1" pipe size		228	.070			3.71		3.71	5.80
9240	1-1/4" pipe size		200	.080			4.23		4.23	6.60
9250	1-1/2" pipe size		178	.090			4.75		4.75	7.40
9260	2" pipe size		116	.138			7.30		7.30	11.35
9270	2-1/2" pipe size		110	.145			7.70		7.70	12
9280	3" pipe size		100	.160			8.45		8.45	13.20
9290	3-1/2" pipe size		94	.170			9		9	14.05
9300	4" pipe size		86	.186			9.85		9.85	15.35
9310	5" pipe size		84	.190			10.05		10.05	15.70
9320	6" pipe size		80	.200			10.55		10.55	16.50
9330	8" pipe size		66	.242			12.80		12.80	20
9340	10" pipe size		58	.276			14.55		14.55	22.50
9350	12" pipe size		46	.348			18.40		18.40	28.50
9360	14" pipe size		30	.533			28		28	44
9370	16" pipe size		28	.571			30		30	47
9380	18" pipe size		27	.593			31.50		31.50	49
9390	20" pipe size		25	.640			34		34	52.50
9400	24" pipe size		23	.696			37		37	57.50
9990	Minimum labor/equipment charge	1 Plum	4	2	Job		117		117	183

22 11 13.60 Tubing, Stainless Steel		Crew	Daily Output	Labor-Hours	Unit	Material	Labor	Equipment	Total	Total Incl O&P
0010	**TUBING, STAINLESS STEEL**									
5000	Tubing									
5010	Type 304, no joints, no hangers									
5020	.035 wall									
5021	1/4"	1 Plum	160	.050	L.F.	3.12	2.94		6.06	8
5022	3/8"		160	.050		3.94	2.94		6.88	8.90
5023	1/2"		160	.050		5.05	2.94		7.99	10.20
5024	5/8"		160	.050		6.20	2.94		9.14	11.45
5025	3/4"		133	.060		7.50	3.53		11.03	13.75
5026	7/8"		133	.060		7.90	3.53		11.43	14.20
5027	1"		114	.070		8.70	4.12		12.82	16
5040	.049 wall									
5041	1/4"	1 Plum	160	.050	L.F.	3.67	2.94		6.61	8.60
5042	3/8"		160	.050		5.05	2.94		7.99	10.20
5043	1/2"		160	.050		5.60	2.94		8.54	10.80
5044	5/8"		160	.050		7.60	2.94		10.54	12.95
5045	3/4"		133	.060		8.10	3.53		11.63	14.40
5046	7/8"		133	.060		8.45	3.53		11.98	14.75
5047	1"		114	.070		8.95	4.12		13.07	16.30
5060	.065 wall									
5061	1/4"	1 Plum	160	.050	L.F.	4.11	2.94		7.05	9.10

22 11 13 – Facility Water Distribution Piping

22 11 13.60 Tubing, Stainless Steel		Crew	Daily Output	Labor-Hours	Unit	Material	2015 Bare Costs Labor	Equipment	Total	Total Incl O&P
5062	3/8"	1 Plum	160	.050	L.F.	5.65	2.94		8.59	10.85
5063	1/2"		160	.050		6.85	2.94		9.79	12.15
5064	5/8"		160	.050		8.30	2.94		11.24	13.70
5065	3/4"		133	.060		8.85	3.53		12.38	15.20
5066	7/8"		133	.060		10.15	3.53		13.68	16.65
5067	1"		114	.070		9.95	4.12		14.07	17.35
5210	Type 316									
5220	.035 wall									
5221	1/4"	1 Plum	160	.050	L.F.	3.96	2.94		6.90	8.95
5222	3/8"		160	.050		5	2.94		7.94	10.10
5223	1/2"		160	.050		7	2.94		9.94	12.30
5224	5/8"		160	.050		9.10	2.94		12.04	14.60
5225	3/4"		133	.060		10.50	3.53		14.03	17.05
5226	7/8"		133	.060		15	3.53		18.53	22
5227	1"		114	.070		11.90	4.12		16.02	19.55
5240	.049 wall									
5241	1/4"	1 Plum	160	.050	L.F.	5.45	2.94		8.39	10.60
5242	3/8"		160	.050		7.15	2.94		10.09	12.45
5243	1/2"		160	.050		7.85	2.94		10.79	13.25
5244	5/8"		160	.050		10.90	2.94		13.84	16.60
5245	3/4"		133	.060		10.50	3.53		14.03	17.05
5246	7/8"		133	.060		13.50	3.53		17.03	20.50
5247	1"		114	.070		12.30	4.12		16.42	20
5260	.065 wall									
5261	1/4"	1 Plum	160	.050	L.F.	5.20	2.94		8.14	10.30
5262	3/8"		160	.050		8.30	2.94		11.24	13.70
5263	1/2"		160	.050		8.90	2.94		11.84	14.40
5264	5/8"		160	.050		12.40	2.94		15.34	18.25
5265	3/4"		133	.060		13.35	3.53		16.88	20
5266	7/8"		133	.060		15.80	3.53		19.33	23
5267	1"		114	.070		15.05	4.12		19.17	23

22 11 13.61 Tubing Fittings, Stainless Steel		Crew	Daily Output	Labor-Hours	Unit	Material	Labor	Equipment	Total	Total Incl O&P
0010	**TUBING FITTINGS, STAINLESS STEEL**									
8200	Tube fittings, compression type									
8202	Type 316									
8204	90° elbow									
8206	1/4"	1 Plum	24	.333	Ea.	17.60	19.55		37.15	50
8207	3/8"		22	.364		21.50	21.50		43	57.50
8208	1/2"		22	.364		35.50	21.50		57	72.50
8209	5/8"		21	.381		40	22.50		62.50	79
8210	3/4"		21	.381		63	22.50		85.50	104
8211	7/8"		20	.400		96.50	23.50		120	143
8212	1"		20	.400		120	23.50		143.50	169
8220	Union tee									
8222	1/4"	1 Plum	15	.533	Ea.	25	31.50		56.50	76.50
8224	3/8"		15	.533		32	31.50		63.50	84.50
8225	1/2"		15	.533		49.50	31.50		81	104
8226	5/8"		14	.571		55.50	33.50		89	114
8227	3/4"		14	.571		74	33.50		107.50	134
8228	7/8"		13	.615		148	36		184	220
8229	1"		13	.615		159	36		195	232
8234	Union									

	22 11 13.61 Tubing Fittings, Stainless Steel	Crew	Daily Output	Labor-Hours	Unit	Material	2015 Bare Costs Labor	Equipment	Total	Total Incl O&P
8236	1/4"	1 Plum	24	.333	Ea.	12.20	19.55		31.75	44
8237	3/8"		22	.364		17.40	21.50		38.90	52.50
8238	1/2"		22	.364		26	21.50		47.50	62
8239	5/8"		21	.381		34	22.50		56.50	72
8240	3/4"		21	.381		42	22.50		64.50	81
8241	7/8"		20	.400		68.50	23.50		92	112
8242	1"		20	.400		72.50	23.50		96	117
8250	Male connector									
8252	1/4" x 1/4"	1 Plum	24	.333	Ea.	7.80	19.55		27.35	39
8253	3/8" x 3/8"		22	.364		12.20	21.50		33.70	47
8254	1/2" x 1/2"		22	.364		18.05	21.50		39.55	53.50
8256	3/4" x 3/4"		21	.381		27.50	22.50		50	65.50
8258	1" x 1"		20	.400		48.50	23.50		72	89.50

	22 11 13.64 Pipe, Stainless Steel	Crew	Daily Output	Labor-Hours	Unit	Material	Labor	Equipment	Total	Total Incl O&P
0010	**PIPE, STAINLESS STEEL**									
0020	Welded, with clevis type hanger assemblies, 10' O.C.									
0500	Schedule 5, type 304									
0540	1/2" diameter	Q-15	128	.125	L.F.	8.25	6.60	.45	15.30	19.90
0550	3/4" diameter		116	.138		10.25	7.30	.50	18.05	23
0560	1" diameter		103	.155		11.35	8.20	.56	20.11	26
0570	1-1/4" diameter		93	.172		15.45	9.10	.62	25.17	32
0580	1-1/2" diameter		85	.188		18.60	9.95	.68	29.23	37
0590	2" diameter		69	.232		31	12.25	.84	44.09	54
0600	2-1/2" diameter		53	.302		33.50	15.95	1.09	50.54	63
0610	3" diameter		48	.333		40	17.60	1.20	58.80	73
0620	4" diameter		44	.364		51	19.20	1.31	71.51	87.50
0630	5" diameter		36	.444		103	23.50	1.60	128.10	152
0640	6" diameter	Q-16	42	.571		95.50	31.50	1.37	128.37	156
0650	8" diameter		34	.706		146	38.50	1.69	186.19	223
0660	10" diameter		26	.923		202	50.50	2.22	254.72	305
0670	12" diameter		21	1.143		265	62.50	2.74	330.24	395
0700	To delete hangers, subtract									
0710	1/2" diam. to 1-1/2" diam.					8%	19%			
0720	2" diam. to 5" diam.					4%	9%			
0730	6" diam. to 12" diam.					3%	4%			
0750	For small quantities, add				L.F.	10%				
1250	Schedule 5, type 316									
1290	1/2" diameter	Q-15	128	.125	L.F.	10.30	6.60	.45	17.35	22
1300	3/4" diameter		116	.138		14.75	7.30	.50	22.55	28
1310	1" diameter		103	.155		16.45	8.20	.56	25.21	31.50
1320	1-1/4" diameter		93	.172		20	9.10	.62	29.72	37.50
1330	1-1/2" diameter		85	.188		27.50	9.95	.68	38.13	47
1340	2" diameter		69	.232		42.50	12.25	.84	55.59	66.50
1350	2-1/2" diameter		53	.302		54.50	15.95	1.09	71.54	86
1360	3" diameter		48	.333		69	17.60	1.20	87.80	105
1370	4" diameter		44	.364		85.50	19.20	1.31	106.01	125
1380	5" diameter		36	.444		163	23.50	1.60	188.10	217
1390	6" diameter	Q-16	42	.571		159	31.50	1.37	191.87	226
1400	8" diameter		34	.706		239	38.50	1.69	279.19	325
1410	10" diameter		26	.923		340	50.50	2.22	392.72	455
1420	12" diameter		21	1.143		440	62.50	2.74	505.24	585
1490	For small quantities, add					10%				

22 11 13.64 Pipe, Stainless Steel		Crew	Daily Output	Labor-Hours	Unit	Material	Labor	Equipment	Total	Total Incl O&P
						2015 Bare Costs				
1940	To delete hanger, subtract									
1950	1/2" diam. to 1-1/2" diam.					5%	19%			
1960	2" diam. to 5" diam.					3%	9%			
1970	6" diam. to 12" diam.					2%	4%			
2000	Schedule 10, type 304									
2040	1/4" diameter	Q-15	131	.122	L.F.	5.40	6.45	.44	12.29	16.50
2050	3/8" diameter		128	.125		6.45	6.60	.45	13.50	17.90
2060	1/2" diameter		125	.128		6.75	6.75	.46	13.96	18.50
2070	3/4" diameter		113	.142		7.80	7.50	.51	15.81	21
2080	1" diameter		100	.160		11.15	8.45	.58	20.18	26
2090	1-1/4" diameter		91	.176		12.70	9.30	.63	22.63	29
2100	1-1/2" diameter		83	.193		14.10	10.20	.70	25	32
2110	2" diameter		67	.239		17.10	12.60	.86	30.56	39.50
2120	2-1/2" diameter		51	.314		22	16.55	1.13	39.68	51.50
2130	3" diameter		46	.348		26.50	18.40	1.26	46.16	59
2140	4" diameter		42	.381		32.50	20	1.38	53.88	68.50
2150	5" diameter		35	.457		43	24	1.65	68.65	87
2160	6" diameter	Q-16	40	.600		51	33	1.44	85.44	109
2170	8" diameter		33	.727		88.50	40	1.75	130.25	161
2180	10" diameter		25	.960		122	52.50	2.30	176.80	219
2190	12" diameter		21	1.143		152	62.50	2.74	217.24	268
2250	For small quantities, add					10%				
2650	To delete hanger, subtract									
2660	1/4" diam. to 3/4" diam.					9%	22%			
2670	1" diam. to 2" diam.					4%	15%			
2680	2-1/2" diam. to 5" diam.					3%	8%			
2690	6" diam. to 12" diam.					3%	4%			
2750	Schedule 10, type 316									
2790	1/4" diameter	Q-15	131	.122	L.F.	6.50	6.45	.44	13.39	17.70
2800	3/8" diameter		128	.125		7.60	6.60	.45	14.65	19.15
2810	1/2" diameter		125	.128		8.45	6.75	.46	15.66	20.50
2820	3/4" diameter		113	.142		9.95	7.50	.51	17.96	23
2830	1" diameter		100	.160		14.10	8.45	.58	23.13	29.50
2840	1-1/4" diameter		91	.176		17.90	9.30	.63	27.83	35
2850	1-1/2" diameter		83	.193		19.75	10.20	.70	30.65	38.50
2860	2" diameter		67	.239		24	12.60	.86	37.46	47
2870	2-1/2" diameter		51	.314		31	16.55	1.13	48.68	61.50
2880	3" diameter		46	.348		36.50	18.40	1.26	56.16	70
2890	4" diameter		42	.381		46.50	20	1.38	67.88	84
2900	5" diameter		35	.457		56.50	24	1.65	82.15	101
2910	6" diameter	Q-16	40	.600		68.50	33	1.44	102.94	129
2920	8" diameter		33	.727		130	40	1.75	171.75	207
2930	10" diameter		25	.960		171	52.50	2.30	225.80	273
2940	12" diameter		21	1.143		214	62.50	2.74	279.24	335
2990	For small quantities, add					10%				
3430	To delete hanger, subtract									
3440	1/4" diam. to 3/4" diam.					6%	22%			
3450	1" diam. to 2" diam.					3%	15%			
3460	2-1/2" diam. to 5" diam.					2%	8%			
3470	6" diam. to 12" diam.					2%	4%			
3500	Threaded, couplings and clevis hanger assemblies, 10' O.C.									
3520	Schedule 40, type 304									
3540	1/4" diameter	1 Plum	54	.148	L.F.	10.85	8.70		19.55	25.50

22 11 13.64 Pipe, Stainless Steel		Crew	Daily Output	Labor-Hours	Unit	Material	Labor	Equipment	Total	Total Incl O&P
						2015 Bare Costs				
3550	3/8" diameter	1 Plum	53	.151	L.F.	11.05	8.85		19.90	26
3560	1/2" diameter		52	.154		12.85	9.05		21.90	28.50
3570	3/4" diameter		51	.157		13.35	9.20		22.55	29
3580	1" diameter		45	.178		19.30	10.45		29.75	37.50
3590	1-1/4" diameter	Q-1	76	.211		27	11.10		38.10	47
3600	1-1/2" diameter		69	.232		29.50	12.25		41.75	51
3610	2" diameter		57	.281		45	14.85		59.85	72.50
3620	2-1/2" diameter		44	.364		77	19.20		96.20	115
3630	3" diameter		38	.421		100	22		122	145
3640	4" diameter	Q-2	51	.471		138	26		164	191
3740	For small quantities, add					10%				
4200	To delete couplings & hangers, subtract									
4210	1/4" diam. to 3/4" diam.					15%	56%			
4220	1" diam. to 2" diam.					18%	49%			
4230	2-1/2" diam. to 4" diam.					34%	40%			
4250	Schedule 40, type 316									
4290	1/4" diameter	1 Plum	54	.148	L.F.	11.65	8.70		20.35	26.50
4300	3/8" diameter		53	.151		12.70	8.85		21.55	28
4310	1/2" diameter		52	.154		15.50	9.05		24.55	31
4320	3/4" diameter		51	.157		17.95	9.20		27.15	34
4330	1" diameter		45	.178		25	10.45		35.45	44
4340	1-1/4" diameter	Q-1	76	.211		34	11.10		45.10	54.50
4350	1-1/2" diameter		69	.232		38.50	12.25		50.75	61.50
4360	2" diameter		57	.281		53	14.85		67.85	81.50
4370	2-1/2" diameter		44	.364		90.50	19.20		109.70	130
4380	3" diameter		38	.421		116	22		138	163
4390	4" diameter	Q-2	51	.471		153	26		179	208
4490	For small quantities, add					10%				
4900	To delete couplings & hangers, subtract									
4910	1/4" diam. to 3/4" diam.					12%	56%			
4920	1" diam. to 2" diam.					14%	49%			
4930	2-1/2" diam. to 4" diam.					27%	40%			
5000	Schedule 80, type 304									
5040	1/4" diameter	1 Plum	53	.151	L.F.	18.10	8.85		26.95	34
5050	3/8" diameter		52	.154		20.50	9.05		29.55	37
5060	1/2" diameter		51	.157		25	9.20		34.20	42
5070	3/4" diameter		48	.167		26.50	9.80		36.30	44.50
5080	1" diameter		43	.186		33	10.90		43.90	53.50
5090	1-1/4" diameter	Q-1	73	.219		44.50	11.60		56.10	67
5100	1-1/2" diameter		67	.239		56	12.60		68.60	81.50
5110	2" diameter		54	.296		69.50	15.65		85.15	101
5190	For small quantities, add					10%				
5700	To delete couplings & hangers, subtract									
5710	1/4" diam. to 3/4" diam.					10%	53%			
5720	1" diam. to 2" diam.					14%	47%			
5750	Schedule 80, type 316									
5790	1/4" diameter	1 Plum	53	.151	L.F.	21.50	8.85		30.35	37.50
5800	3/8" diameter		52	.154		26.50	9.05		35.55	43
5810	1/2" diameter		51	.157		31	9.20		40.20	49
5820	3/4" diameter		48	.167		34	9.80		43.80	53
5830	1" diameter		43	.186		44.50	10.90		55.40	66
5840	1-1/4" diameter	Q-1	73	.219		64	11.60		75.60	88.50
5850	1-1/2" diameter		67	.239		68.50	12.60		81.10	95

22 11 13.64 Pipe, Stainless Steel		Crew	Daily Output	Labor-Hours	Unit	2015 Bare Costs Material	Labor	Equipment	Total	Total Incl O&P
5860	2" diameter	Q-1	54	.296	L.F.	86	15.65		101.65	120
5950	For small quantities, add					10%				
7000	To delete couplings & hangers, subtract									
7010	1/4" diam. to 3/4" diam.					9%	53%			
7020	1" diam. to 2" diam.					14%	47%			
8000	Weld joints with clevis type hanger assemblies, 10' O.C.									
8010	Schedule 40, type 304									
8050	1/8" pipe size	Q-15	126	.127	L.F.	8.70	6.70	.46	15.86	20.50
8060	1/4" pipe size		125	.128		9.60	6.75	.46	16.81	21.50
8070	3/8" pipe size		122	.131		9.60	6.95	.47	17.02	22
8080	1/2" pipe size		118	.136		10.95	7.15	.49	18.59	23.50
8090	3/4" pipe size		109	.147		10.80	7.75	.53	19.08	24.50
8100	1" pipe size		95	.168		15.15	8.90	.61	24.66	31.50
8110	1-1/4" pipe size		86	.186		20.50	9.85	.67	31.02	38.50
8120	1-1/2" pipe size		78	.205		22	10.85	.74	33.59	41.50
8130	2" pipe size		62	.258		32.50	13.65	.93	47.08	58.50
8140	2-1/2" pipe size		49	.327		48.50	17.25	1.18	66.93	81.50
8150	3" pipe size		44	.364		61	19.20	1.31	81.51	99
8160	3-1/2" pipe size		44	.364		72.50	19.20	1.31	93.01	111
8170	4" pipe size		39	.410		83	21.50	1.48	105.98	127
8180	5" pipe size		32	.500		110	26.50	1.81	138.31	164
8190	6" pipe size	Q-16	37	.649		128	35.50	1.56	165.06	198
8200	8" pipe size		29	.828		179	45.50	1.99	226.49	270
8210	10" pipe size		24	1		305	55	2.40	362.40	425
8220	12" pipe size		20	1.200		490	65.50	2.88	558.38	645
8300	Schedule 40, type 316									
8310	1/8" pipe size	Q-15	126	.127	L.F.	9.25	6.70	.46	16.41	21
8320	1/4" pipe size		125	.128		10.15	6.75	.46	17.36	22.50
8330	3/8" pipe size		122	.131		10.95	6.95	.47	18.37	23.50
8340	1/2" pipe size		118	.136		13.15	7.15	.49	20.79	26
8350	3/4" pipe size		109	.147		14.85	7.75	.53	23.13	29
8360	1" pipe size		95	.168		20	8.90	.61	29.51	37
8370	1-1/4" pipe size		86	.186		26	9.85	.67	36.52	44.50
8380	1-1/2" pipe size		78	.205		29.50	10.85	.74	41.09	50
8390	2" pipe size		62	.258		38.50	13.65	.93	53.08	64.50
8400	2-1/2" pipe size		49	.327		56	17.25	1.18	74.43	90
8410	3" pipe size		44	.364		69	19.20	1.31	89.51	107
8420	3-1/2" pipe size		44	.364		80.50	19.20	1.31	101.01	120
8430	4" pipe size		39	.410		87	21.50	1.48	109.98	131
8440	5" pipe size		32	.500		120	26.50	1.81	148.31	175
8450	6" pipe size	Q-16	37	.649		138	35.50	1.56	175.06	209
8460	8" pipe size		29	.828		184	45.50	1.99	231.49	275
8470	10" pipe size		24	1		245	55	2.40	302.40	360
8480	12" pipe size		20	1.200		330	65.50	2.88	398.38	470
8500	Schedule 80, type 304									
8510	1/4" pipe size	Q-15	110	.145	L.F.	16.75	7.70	.53	24.98	31
8520	3/8" pipe size		109	.147		19.05	7.75	.53	27.33	33.50
8530	1/2" pipe size		106	.151		23	7.95	.54	31.49	38.50
8540	3/4" pipe size		96	.167		24	8.80	.60	33.40	41
8550	1" pipe size		87	.184		28.50	9.70	.66	38.86	47.50
8560	1-1/4" pipe size		81	.198		34	10.45	.71	45.16	54
8570	1-1/2" pipe size		74	.216		43.50	11.40	.78	55.68	66.50
8580	2" pipe size		58	.276		53	14.55	1	68.55	82

22 11 13.64 Pipe, Stainless Steel		Crew	Daily Output	Labor-Hours	Unit	Material	2015 Bare Costs Labor	Equipment	Total	Total Incl O&P
8590	2-1/2" pipe size	Q-15	46	.348	L.F.	68	18.40	1.26	87.66	105
8600	3" pipe size		41	.390		78.50	20.50	1.41	100.41	120
8610	4" pipe size		33	.485		96.50	25.50	1.75	123.75	148
8630	6" pipe size	Q-16	30	.800		171	44	1.92	216.92	259
8640	Schedule 80, type 316									
8650	1/4" pipe size	Q-15	110	.145	L.F.	19.95	7.70	.53	28.18	34.50
8660	3/8" pipe size		109	.147		24.50	7.75	.53	32.78	39.50
8670	1/2" pipe size		106	.151		29	7.95	.54	37.49	44.50
8680	3/4" pipe size		96	.167		30.50	8.80	.60	39.90	48
8690	1" pipe size		87	.184		38.50	9.70	.66	48.86	58.50
8700	1-1/4" pipe size		81	.198		50.50	10.45	.71	61.66	72.50
8710	1-1/2" pipe size		74	.216		53	11.40	.78	65.18	76.50
8720	2" pipe size		58	.276		64	14.55	1	79.55	94
8730	2-1/2" pipe size		46	.348		80	18.40	1.26	99.66	118
8740	3" pipe size		41	.390		92.50	20.50	1.41	114.41	136
8760	4" pipe size		33	.485		122	25.50	1.75	149.25	176
8770	6" pipe size	Q-16	30	.800		219	44	1.92	264.92	310
9100	Threading pipe labor, sst, one end, schedules 40 & 80									
9110	1/4" through 3/4" pipe size	1 Plum	61.50	.130	Ea.		7.65		7.65	11.90
9120	1" through 2" pipe size		55.90	.143			8.40		8.40	13.10
9130	2-1/2" pipe size		41.50	.193			11.30		11.30	17.65
9140	3" pipe size		38.50	.208			12.20		12.20	19
9150	3-1/2" pipe size	Q-1	68.40	.234			12.35		12.35	19.30
9160	4" pipe size		73	.219			11.60		11.60	18.05
9170	5" pipe size		40.70	.393			21		21	32.50
9180	6" pipe size		35.40	.452			24		24	37
9190	8" pipe size		22.30	.717			38		38	59
9200	10" pipe size		16.10	.994			52.50		52.50	82
9210	12" pipe size		12.30	1.301			68.50		68.50	107
9250	Welding labor per joint for stainless steel									
9260	Schedule 5 and 10									
9270	1/4" pipe size	Q-15	36	.444	Ea.		23.50	1.60	25.10	38.50
9280	3/8" pipe size		35	.457			24	1.65	25.65	39.50
9290	1/2" pipe size		35	.457			24	1.65	25.65	39.50
9300	3/4" pipe size		28	.571			30	2.06	32.06	49.50
9310	1" pipe size		25	.640			34	2.31	36.31	55
9320	1-1/4" pipe size		22	.727			38.50	2.63	41.13	63
9330	1-1/2" pipe size		21	.762			40.50	2.75	43.25	66
9340	2" pipe size		18	.889			47	3.21	50.21	76.50
9350	2-1/2" pipe size		12	1.333			70.50	4.81	75.31	115
9360	3" pipe size		9.73	1.644			87	5.95	92.95	143
9370	4" pipe size		7.37	2.171			115	7.85	122.85	188
9380	5" pipe size		6.15	2.602			137	9.40	146.40	224
9390	6" pipe size		5.71	2.802			148	10.10	158.10	242
9400	8" pipe size		3.69	4.336			229	15.65	244.65	370
9410	10" pipe size		2.91	5.498			290	19.85	309.85	475
9420	12" pipe size		2.31	6.926			365	25	390	600
9500	Schedule 40									
9510	1/4" pipe size	Q-15	28	.571	Ea.		30	2.06	32.06	49.50
9520	3/8" pipe size		27	.593			31.50	2.14	33.64	51.50
9530	1/2" pipe size		25.40	.630			33.50	2.27	35.77	54.50
9540	3/4" pipe size		22.22	.720			38	2.60	40.60	62.50
9550	1" pipe size		20.25	.790			41.50	2.85	44.35	68

22 11 13.64 Pipe, Stainless Steel		Crew	Daily Output	Labor-Hours	Unit	Material	2015 Bare Costs Labor	Equipment	Total	Total Incl O&P
9560	1-1/4" pipe size	Q-15	18.82	.850	Ea.		45	3.07	48.07	73.50
9570	1-1/2" pipe size		17.78	.900			47.50	3.25	50.75	77.50
9580	2" pipe size		15.09	1.060			56	3.83	59.83	91.50
9590	2-1/2" pipe size		7.96	2.010			106	7.25	113.25	174
9600	3" pipe size		6.43	2.488			131	9	140	215
9610	4" pipe size		4.88	3.279			173	11.85	184.85	283
9620	5" pipe size		4.26	3.756			198	13.55	211.55	325
9630	6" pipe size		3.77	4.244			224	15.30	239.30	365
9640	8" pipe size		2.44	6.557			345	23.50	368.50	565
9650	10" pipe size		1.92	8.333			440	30	470	720
9660	12" pipe size		1.52	10.526			555	38	593	905
9750	Schedule 80									
9760	1/4" pipe size	Q-15	21.55	.742	Ea.		39	2.68	41.68	64
9770	3/8" pipe size		20.75	.771			40.50	2.78	43.28	66.50
9780	1/2" pipe size		19.54	.819			43.50	2.96	46.46	71
9790	3/4" pipe size		17.09	.936			49.50	3.38	52.88	80.50
9800	1" pipe size		15.58	1.027			54.50	3.71	58.21	88.50
9810	1-1/4" pipe size		14.48	1.105			58.50	3.99	62.49	95.50
9820	1-1/2" pipe size		13.68	1.170			62	4.22	66.22	101
9830	2" pipe size		11.61	1.378			73	4.98	77.98	119
9840	2-1/2" pipe size		6.12	2.614			138	9.45	147.45	225
9850	3" pipe size		4.94	3.239			171	11.70	182.70	280
9860	4" pipe size		3.75	4.267			225	15.40	240.40	365
9870	5" pipe size		3.27	4.893			259	17.65	276.65	425
9880	6" pipe size		2.90	5.517			291	19.90	310.90	475
9890	8" pipe size		1.87	8.556			450	31	481	740
9900	10" pipe size		1.48	10.811			570	39	609	935
9910	12" pipe size		1.17	13.675			720	49.50	769.50	1,175
9920	Schedule 160, 1/2" pipe size		17	.941			49.50	3.40	52.90	81
9930	3/4" pipe size		14.81	1.080			57	3.90	60.90	93.50
9940	1" pipe size		13.50	1.185			62.50	4.28	66.78	102
9950	1-1/4" pipe size		12.55	1.275			67.50	4.60	72.10	110
9960	1-1/2" pipe size		11.85	1.350			71.50	4.87	76.37	116
9970	2" pipe size		10	1.600			84.50	5.80	90.30	138
9980	3" pipe size		4.28	3.738			198	13.50	211.50	325
9990	4" pipe size		3.25	4.923			260	17.75	277.75	425

22 11 13.66 Pipe Fittings, Stainless Steel		Crew	Daily Output	Labor-Hours	Unit	Material	Labor	Equipment	Total	Total Incl O&P
0010	**PIPE FITTINGS, STAINLESS STEEL**									
0100	Butt weld joint, schedule 5, type 304									
0120	90° Elbow, long									
0140	1/2"	Q-15	17.50	.914	Ea.	19.80	48.50	3.30	71.60	101
0150	3/4"		14	1.143		19.80	60.50	4.13	84.43	121
0160	1"		12.50	1.280		21	67.50	4.62	93.12	133
0170	1-1/4"		11	1.455		28.50	77	5.25	110.75	157
0180	1-1/2"		10.50	1.524		24	80.50	5.50	110	159
0190	2"		9	1.778		28.50	94	6.40	128.90	185
0200	2-1/2"		6	2.667		66	141	9.65	216.65	305
0210	3"		4.86	3.292		72.50	174	11.90	258.40	365
0220	3-1/2"		4.27	3.747		182	198	13.55	393.55	525
0230	4"		3.69	4.336		99	229	15.65	343.65	480
0240	5"		3.08	5.195		385	274	18.75	677.75	875
0250	6"	Q-16	4.29	5.594		297	305	13.45	615.45	820

22 11 13.66 Pipe Fittings, Stainless Steel		Crew	Daily Output	Labor-Hours	Unit	Material	2015 Bare Costs Labor	Equipment	Total	Total Incl O&P
0260	8″	Q-16	2.76	8.696	Ea.	625	475	21	1,121	1,450
0270	10″		2.18	11.009		970	605	26.50	1,601.50	2,050
0280	12″		1.73	13.873		1,375	760	33.50	2,168.50	2,725
0320	For schedule 5, type 316, add					30%				
0600	45° Elbow, long									
0620	1/2″	Q-15	17.50	.914	Ea.	19.80	48.50	3.30	71.60	101
0630	3/4″		14	1.143		19.80	60.50	4.13	84.43	121
0640	1″		12.50	1.280		21	67.50	4.62	93.12	133
0650	1-1/4″		11	1.455		28.50	77	5.25	110.75	157
0660	1-1/2″		10.50	1.524		24	80.50	5.50	110	159
0670	2″		9	1.778		28.50	94	6.40	128.90	185
0680	2-1/2″		6	2.667		66	141	9.65	216.65	305
0690	3″		4.86	3.292		72.50	174	11.90	258.40	365
0700	3-1/2″		4.27	3.747		182	198	13.55	393.55	525
0710	4″		3.69	4.336		83.50	229	15.65	328.15	465
0720	5″		3.08	5.195		310	274	18.75	602.75	790
0730	6″	Q-16	4.29	5.594		209	305	13.45	527.45	725
0740	8″		2.76	8.696		440	475	21	936	1,250
0750	10″		2.18	11.009		770	605	26.50	1,401.50	1,825
0760	12″		1.73	13.873		965	760	33.50	1,758.50	2,250
0800	For schedule 5, type 316, add					25%				
1100	Tee, straight									
1130	1/2″	Q-15	11.66	1.372	Ea.	59.50	72.50	4.95	136.95	184
1140	3/4″		9.33	1.715		59.50	90.50	6.20	156.20	213
1150	1″		8.33	1.921		62.50	101	6.95	170.45	235
1160	1-1/4″		7.33	2.183		50.50	115	7.90	173.40	244
1170	1-1/2″		7	2.286		49.50	121	8.25	178.75	252
1180	2″		6	2.667		51.50	141	9.65	202.15	288
1190	2-1/2″		4	4		124	211	14.45	349.45	485
1200	3″		3.24	4.938		198	261	17.85	476.85	645
1210	3-1/2″		2.85	5.614		260	297	20.50	577.50	775
1220	4″		2.46	6.504		141	345	23.50	509.50	715
1230	5″		2	8		445	425	29	899	1,175
1240	6″	Q-16	2.85	8.421		350	460	20	830	1,125
1250	8″		1.84	13.043		750	715	31.50	1,496.50	1,975
1260	10″		1.45	16.552		1,200	905	39.50	2,144.50	2,800
1270	12″		1.15	20.870		1,675	1,150	50	2,875	3,675
1320	For schedule 5, type 316, add					25%				
2000	Butt weld joint, schedule 10, type 304									
2020	90° elbow, long									
2040	1/2″	Q-15	17	.941	Ea.	13.50	49.50	3.40	66.40	96
2050	3/4″		14	1.143		13.50	60.50	4.13	78.13	113
2060	1″		12.50	1.280		14.25	67.50	4.62	86.37	126
2070	1-1/4″		11	1.455		19.50	77	5.25	101.75	147
2080	1-1/2″		10.50	1.524		16.50	80.50	5.50	102.50	150
2090	2″		9	1.778		19.50	94	6.40	119.90	175
2100	2-1/2″		6	2.667		45	141	9.65	195.65	280
2110	3″		4.86	3.292		42	174	11.90	227.90	330
2120	3-1/2″		4.27	3.747		124	198	13.55	335.55	460
2130	4″		3.69	4.336		70.50	229	15.65	315.15	450
2140	5″		3.08	5.195		263	274	18.75	555.75	740
2150	6″	Q-16	4.29	5.594		203	305	13.45	521.45	720
2160	8″		2.76	8.696		430	475	21	926	1,250

22 11 13 – Facility Water Distribution Piping

22 11 13.66 Pipe Fittings, Stainless Steel		Crew	Daily Output	Labor-Hours	Unit	Material	Labor	Equipment	Total	Total Incl O&P
						2015 Bare Costs				
2170	10"	Q-16	2.18	11.009	Ea.	660	605	26.50	1,291.50	1,700
2180	12"	↓	1.73	13.873	↓	940	760	33.50	1,733.50	2,225
2500	45° elbow, long									
2520	1/2"	Q-15	17.50	.914	Ea.	13.50	48.50	3.30	65.30	94
2530	3/4"		14	1.143		13.50	60.50	4.13	78.13	113
2540	1"		12.50	1.280		14.25	67.50	4.62	86.37	126
2550	1-1/4"		11	1.455		19.50	77	5.25	101.75	147
2560	1-1/2"		10.50	1.524		16.50	80.50	5.50	102.50	150
2570	2"		9	1.778		19.50	94	6.40	119.90	175
2580	2-1/2"		6	2.667		45	141	9.65	195.65	280
2590	3"		4.86	3.292		34	174	11.90	219.90	320
2600	3-1/2"		4.27	3.747		124	198	13.55	335.55	460
2610	4"		3.69	4.336		57	229	15.65	301.65	435
2620	5"	↓	3.08	5.195		210	274	18.75	502.75	680
2630	6"	Q-16	4.29	5.594		143	305	13.45	461.45	650
2640	8"		2.76	8.696		300	475	21	796	1,100
2650	10"		2.18	11.009		525	605	26.50	1,156.50	1,550
2660	12"	↓	1.73	13.873	↓	655	760	33.50	1,448.50	1,925
3000	Tee, straight									
3030	1/2"	Q-15	11.66	1.372	Ea.	40.50	72.50	4.95	117.95	163
3040	3/4"		9.33	1.715		40.50	90.50	6.20	137.20	192
3050	1"		8.33	1.921		43	101	6.95	150.95	213
3060	1-1/4"		7.33	2.183		49	115	7.90	171.90	242
3070	1-1/2"		7	2.286		49	121	8.25	178.25	251
3080	2"		6	2.667		51	141	9.65	201.65	287
3090	2-1/2"		4	4		35.50	211	14.45	260.95	385
3100	3"		3.24	4.938		67	261	17.85	345.85	500
3110	3-1/2"		2.85	5.614		98.50	297	20.50	416	595
3120	4"		2.46	6.504		96	345	23.50	464.50	665
3130	5"	↓	2	8		305	425	29	759	1,025
3140	6"	Q-16	2.85	8.421		240	460	20	720	1,000
3150	8"		1.84	13.043		510	715	31.50	1,256.50	1,725
3151	10"		1.45	16.552		825	905	39.50	1,769.50	2,375
3152	12"	↓	1.15	20.870		1,150	1,150	50	2,350	3,075
3154	For schedule 10, type 316, add				↓	25%				
3281	Butt weld joint, schedule 40, type 304									
3284	90° Elbow, long, 1/2"	Q-15	12.70	1.260	Ea.	15.75	66.50	4.55	86.80	126
3288	3/4"		11.10	1.441		15.75	76	5.20	96.95	142
3289	1"		10.13	1.579		16.50	83.50	5.70	105.70	154
3290	1-1/4"		9.40	1.702		22	90	6.15	118.15	171
3300	1-1/2"		8.89	1.800		17.25	95	6.50	118.75	174
3310	2"		7.55	2.119		25	112	7.65	144.65	210
3320	2-1/2"		3.98	4.020		47.50	212	14.50	274	400
3330	3"		3.21	4.984		61	263	18	342	495
3340	3-1/2"		2.83	5.654		225	299	20.50	544.50	735
3350	4"		2.44	6.557		105	345	23.50	473.50	680
3360	5"	↓	2.13	7.512		355	395	27	777	1,050
3370	6"	Q-16	2.83	8.481		310	465	20.50	795.50	1,100
3380	8"		1.83	13.115		615	720	31.50	1,366.50	1,825
3390	10"		1.44	16.667		1,300	915	40	2,255	2,900
3400	12"	↓	1.14	21.053	↓	1,650	1,150	50.50	2,850.50	3,675
3410	For schedule 40, type 316, add					25%				
3460	45° Elbow, long, 1/2"	Q-15	12.70	1.260	Ea.	15.75	66.50	4.55	86.80	126

22 11 13.66 Pipe Fittings, Stainless Steel		Crew	Daily Output	Labor-Hours	Unit	Material	2015 Bare Costs Labor	Equipment	Total	Total Incl O&P
3470	3/4"	Q-15	11.10	1.441	Ea.	15.75	76	5.20	96.95	142
3480	1"		10.13	1.579		16.50	83.50	5.70	105.70	154
3490	1-1/4"		9.40	1.702		22	90	6.15	118.15	171
3500	1-1/2"		8.89	1.800		17.25	95	6.50	118.75	174
3510	2"		7.55	2.119		25	112	7.65	144.65	210
3520	2-1/2"		3.98	4.020		47.50	212	14.50	274	400
3530	3"		3.21	4.984		48	263	18	329	485
3540	3-1/2"		2.83	5.654		225	299	20.50	544.50	735
3550	4"		2.44	6.557		75	345	23.50	443.50	650
3560	5"		2.13	7.512		248	395	27	670	920
3570	6"	Q-16	2.83	8.481		218	465	20.50	703.50	985
3580	8"		1.83	13.115		430	720	31.50	1,181.50	1,625
3590	10"		1.44	16.667		910	915	40	1,865	2,475
3600	12"		1.14	21.053		1,150	1,150	50.50	2,350.50	3,125
3610	For schedule 40, type 316, add					25%				
3660	Tee, straight 1/2"	Q-15	8.46	1.891	Ea.	40.50	100	6.85	147.35	208
3670	3/4"		7.40	2.162		40.50	114	7.80	162.30	231
3680	1"		6.74	2.374		43	125	8.55	176.55	252
3690	1-1/4"		6.27	2.552		100	135	9.20	244.20	330
3700	1-1/2"		5.92	2.703		46	143	9.75	198.75	284
3710	2"		5.03	3.181		68.50	168	11.50	248	350
3720	2-1/2"		2.65	6.038		86.50	320	22	428.50	620
3730	3"		2.14	7.477		87	395	27	509	740
3740	3-1/2"		1.88	8.511		164	450	30.50	644.50	915
3750	4"		1.62	9.877		164	520	35.50	719.50	1,025
3760	5"		1.42	11.268		380	595	40.50	1,015.50	1,400
3770	6"	Q-16	1.88	12.766		365	700	30.50	1,095.50	1,525
3780	8"		1.22	19.672		735	1,075	47	1,857	2,525
3790	10"		.96	25		1,425	1,375	60	2,860	3,775
3800	12"		.76	31.579		1,900	1,725	76	3,701	4,875
3810	For schedule 40, type 316, add					25%				
3820	Tee, reducing on outlet, 3/4" x 1/2"	Q-15	7.73	2.070	Ea.	59.50	109	7.45	175.95	245
3822	1" x 1/2"		7.24	2.210		58.50	117	8	183.50	255
3824	1" x 3/4"		6.96	2.299		54	121	8.30	183.30	258
3826	1-1/4" x 1"		6.43	2.488		132	131	9	272	360
3828	1-1/2" x 1/2"		6.58	2.432		79.50	128	8.80	216.30	297
3830	1-1/2" x 3/4"		6.35	2.520		74.50	133	9.10	216.60	300
3832	1-1/2" x 1"		6.18	2.589		57	137	9.35	203.35	286
3834	2" x 1"		5.50	2.909		102	154	10.50	266.50	365
3836	2" x 1-1/2"		5.30	3.019		85.50	159	10.90	255.40	355
3838	2-1/2" x 2"		3.15	5.079		129	268	18.35	415.35	580
3840	3" x 1-1/2"		2.72	5.882		131	310	21	462	655
3842	3" x 2"		2.65	6.038		109	320	22	451	645
3844	4" x 2"		2.10	7.619		272	405	27.50	704.50	960
3846	4" x 3"		1.77	9.040		197	480	32.50	709.50	1,000
3848	5" x 4"		1.48	10.811		455	570	39	1,064	1,425
3850	6" x 3"	Q-16	2.19	10.959		505	600	26.50	1,131.50	1,525
3852	6" x 4"		2.04	11.765		435	645	28	1,108	1,500
3854	8" x 4"		1.46	16.438		1,025	900	39.50	1,964.50	2,575
3856	10" x 8"		.69	34.783		1,725	1,900	83.50	3,708.50	4,950
3858	12" x 10"		.55	43.636		2,275	2,400	105	4,780	6,375
3950	Reducer, concentric, 3/4" x 1/2"	Q-15	11.85	1.350		30	71.50	4.87	106.37	149
3952	1" x 3/4"		10.60	1.509		31.50	79.50	5.45	116.45	165

22 11 13.66 Pipe Fittings, Stainless Steel		Crew	Daily Output	Labor-Hours	Unit	Material	Labor	Equipment	Total	Total Incl O&P
						2015 Bare Costs				
3954	1-1/4" x 3/4"	Q-15	10.19	1.570	Ea.	77.50	83	5.65	166.15	220
3956	1-1/4" x 1"		9.76	1.639		39	86.50	5.90	131.40	185
3958	1-1/2" x 3/4"		9.88	1.619		65.50	85.50	5.85	156.85	211
3960	1-1/2" x 1"		9.47	1.690		52.50	89.50	6.10	148.10	204
3962	2" x 1"		8.65	1.850		30	97.50	6.70	134.20	192
3964	2" x 1-1/2"		8.16	1.961		26.50	104	7.10	137.60	199
3966	2-1/2" x 1"		5.71	2.802		113	148	10.10	271.10	365
3968	2-1/2" x 2"		5.21	3.071		58	162	11.10	231.10	330
3970	3" x 1"		4.88	3.279		78	173	11.85	262.85	370
3972	3" x 1-1/2"		4.72	3.390		40	179	12.25	231.25	335
3974	3" x 2"		4.51	3.548		37.50	187	12.80	237.30	350
3976	4" x 2"		3.69	4.336		53.50	229	15.65	298.15	430
3978	4" x 3"		2.77	5.776		40	305	21	366	540
3980	5" x 3"		2.56	6.250		237	330	22.50	589.50	800
3982	5" x 4"	↓	2.27	7.048		191	370	25.50	586.50	820
3984	6" x 3"	Q-16	3.57	6.723		122	370	16.15	508.15	725
3986	6" x 4"		3.19	7.524		101	410	18.05	529.05	775
3988	8" x 4"		2.44	9.836		315	540	23.50	878.50	1,200
3990	8" x 6"		2.22	10.811		236	590	26	852	1,225
3992	10" x 6"		1.91	12.565		440	690	30	1,160	1,600
3994	10" x 8"		1.61	14.907		365	815	36	1,216	1,725
3995	12" x 6"		1.63	14.724		675	805	35.50	1,515.50	2,025
3996	12" x 8"		1.41	17.021		620	930	41	1,591	2,175
3997	12" x 10"	↓	1.27	18.898	↓	435	1,025	45.50	1,505.50	2,150
4000	Socket weld joint, 3000 lb., type 304									
4100	90° Elbow									
4140	1/4"	Q-15	13.47	1.188	Ea.	44	63	4.29	111.29	151
4150	3/8"		12.97	1.234		57	65	4.45	126.45	170
4160	1/2"		12.21	1.310		62.50	69	4.73	136.23	182
4170	3/4"		10.68	1.498		72	79	5.40	156.40	208
4180	1"		9.74	1.643		108	87	5.95	200.95	261
4190	1-1/4"		9.05	1.768		189	93.50	6.40	288.90	360
4200	1-1/2"		8.55	1.871		229	99	6.75	334.75	415
4210	2"	↓	7.26	2.204	↓	370	116	7.95	493.95	595
4300	45° Elbow									
4340	1/4"	Q-15	13.47	1.188	Ea.	83	63	4.29	150.29	194
4350	3/8"		12.97	1.234		83	65	4.45	152.45	198
4360	1/2"		12.21	1.310		83	69	4.73	156.73	204
4370	3/4"		10.68	1.498		94	79	5.40	178.40	232
4380	1"		9.74	1.643		136	87	5.95	228.95	292
4390	1-1/4"		9.05	1.768		224	93.50	6.40	323.90	400
4400	1-1/2"		8.55	1.871		225	99	6.75	330.75	410
4410	2"	↓	7.26	2.204	↓	410	116	7.95	533.95	640
4500	Tee									
4540	1/4"	Q-15	8.97	1.784	Ea.	58	94	6.45	158.45	218
4550	3/8"		8.64	1.852		69	98	6.70	173.70	236
4560	1/2"		8.13	1.968		85	104	7.10	196.10	263
4570	3/4"		7.12	2.247		98	119	8.10	225.10	300
4580	1"		6.48	2.469		132	130	8.90	270.90	360
4590	1-1/4"		6.03	2.653		234	140	9.60	383.60	490
4600	1-1/2"		5.69	2.812		335	149	10.15	494.15	615
4610	2"	↓	4.83	3.313	↓	510	175	11.95	696.95	845
5000	Socket weld joint, 3000 lb., type 316									

22 11 13 – Facility Water Distribution Piping

22 11 13.66 Pipe Fittings, Stainless Steel		Crew	Daily Output	Labor-Hours	Unit	Material	2015 Bare Costs Labor	Equipment	Total	Total Incl O&P
5100	90° Elbow									
5140	1/4"	Q-15	13.47	1.188	Ea.	54	63	4.29	121.29	162
5150	3/8"		12.97	1.234		63	65	4.45	132.45	176
5160	1/2"		12.21	1.310		76	69	4.73	149.73	197
5170	3/4"		10.68	1.498		100	79	5.40	184.40	239
5180	1"		9.74	1.643		142	87	5.95	234.95	298
5190	1-1/4"		9.05	1.768		253	93.50	6.40	352.90	430
5200	1-1/2"		8.55	1.871		286	99	6.75	391.75	475
5210	2"		7.26	2.204		485	116	7.95	608.95	725
5300	45° Elbow									
5340	1/4"	Q-15	13.47	1.188	Ea.	104	63	4.29	171.29	217
5350	3/8"		12.97	1.234		104	65	4.45	173.45	221
5360	1/2"		12.21	1.310		104	69	4.73	177.73	227
5370	3/4"		10.68	1.498		121	79	5.40	205.40	262
5380	1"		9.74	1.643		182	87	5.95	274.95	340
5390	1-1/4"		9.05	1.768		258	93.50	6.40	357.90	435
5400	1-1/2"		8.55	1.871		292	99	6.75	397.75	480
5410	2"		7.26	2.204		435	116	7.95	558.95	670
5500	Tee									
5540	1/4"	Q-15	8.97	1.784	Ea.	73	94	6.45	173.45	235
5550	3/8"		8.64	1.852		89.50	98	6.70	194.20	258
5560	1/2"		8.13	1.968		99.50	104	7.10	210.60	279
5570	3/4"		7.12	2.247		123	119	8.10	250.10	330
5580	1"		6.48	2.469		188	130	8.90	326.90	420
5590	1-1/4"		6.03	2.653		299	140	9.60	448.60	560
5600	1-1/2"		5.69	2.812		420	149	10.15	579.15	710
5610	2"		4.83	3.313		670	175	11.95	856.95	1,025
5700	For socket weld joint, 6000 lb., type 304 and 316, add					100%				
6000	Threaded companion flange									
6010	Stainless steel, 150 lb., type 304									
6020	1/2" diam.	1 Plum	30	.267	Ea.	45	15.65		60.65	74
6030	3/4" diam.		28	.286		50	16.75		66.75	81
6040	1" diam.		27	.296		55	17.40		72.40	87.50
6050	1-1/4" diam.	Q-1	44	.364		71	19.20		90.20	108
6060	1-1/2" diam.		40	.400		71	21		92	111
6070	2" diam.		36	.444		93	23.50		116.50	139
6080	2-1/2" diam.		28	.571		130	30		160	190
6090	3" diam.		20	.800		136	42.50		178.50	216
6110	4" diam.		12	1.333		186	70.50		256.50	315
6130	6" diam.	Q-2	14	1.714		335	94		429	510
6140	8" diam.	"	12	2		645	110		755	880
6150	For type 316 add					40%				
6260	Weld flanges, stainless steel, type 304									
6270	Slip on, 150 lb. (welded, front and back)									
6280	1/2" diam.	Q-15	12.70	1.260	Ea.	40	66.50	4.55	111.05	153
6290	3/4" diam.		11.11	1.440		41	76	5.20	122.20	170
6300	1" diam.		10.13	1.579		45	83.50	5.70	134.20	186
6310	1-1/4" diam.		9.41	1.700		61	90	6.15	157.15	214
6320	1-1/2" diam.		8.89	1.800		61	95	6.50	162.50	222
6330	2" diam.		7.55	2.119		79	112	7.65	198.65	270
6340	2-1/2" diam.		3.98	4.020		111	212	14.50	337.50	470
6350	3" diam.		3.21	4.984		119	263	18	400	560
6370	4" diam.		2.44	6.557		162	345	23.50	530.50	745

22 11 13.66 Pipe Fittings, Stainless Steel		Crew	Daily Output	Labor-Hours	Unit	2015 Bare Costs Material	Labor	Equipment	Total	Total Incl O&P
6390	6″ diam.	Q-16	1.89	12.698	Ea.	247	695	30.50	972.50	1,375
6400	8″ diam.	″	1.22	19.672		465	1,075	47	1,587	2,250
6410	For type 316, add					40%				
6530	Weld neck 150 lb.									
6540	1/2″ diam.	Q-15	25.40	.630	Ea.	20.50	33.50	2.27	56.27	77
6550	3/4″ diam.		22.22	.720		25	38	2.60	65.60	89.50
6560	1″ diam.		20.25	.790		27	41.50	2.85	71.35	97.50
6570	1-1/4″ diam.		18.82	.850		43.50	45	3.07	91.57	121
6580	1-1/2″ diam.		17.78	.900		38	47.50	3.25	88.75	119
6590	2″ diam.		15.09	1.060		40.50	56	3.83	100.33	136
6600	2-1/2″ diam.		7.96	2.010		65	106	7.25	178.25	246
6610	3″ diam.		6.43	2.488		66.50	131	9	206.50	288
6630	4″ diam.		4.88	3.279		101	173	11.85	285.85	395
6640	5″ diam.		4.26	3.756		132	198	13.55	343.55	470
6650	6″ diam.	Q-16	5.66	4.240		145	232	10.20	387.20	530
6652	8″ diam.		3.65	6.575		261	360	15.80	636.80	865
6654	10″ diam.		2.88	8.333		365	455	20	840	1,125
6656	12″ diam.		2.28	10.526		705	575	25.50	1,305.50	1,700
6670	For type 316 add					23%				
7000	Threaded joint, 150 lb., type 304									
7030	90° elbow									
7040	1/8″	1 Plum	13	.615	Ea.	25.50	36		61.50	84.50
7050	1/4″		13	.615		25.50	36		61.50	84.50
7070	3/8″		13	.615		29.50	36		65.50	89
7080	1/2″		12	.667		26.50	39		65.50	90
7090	3/4″		11	.727		31.50	42.50		74	101
7100	1″		10	.800		43	47		90	121
7110	1-1/4″	Q-1	17	.941		66.50	49.50		116	151
7120	1-1/2″		16	1		77	53		130	168
7130	2″		14	1.143		111	60.50		171.50	217
7140	2-1/2″		11	1.455		271	77		348	420
7150	3″		8	2		395	106		501	600
7160	4″	Q-2	11	2.182		670	120		790	920
7180	45° elbow									
7190	1/8″	1 Plum	13	.615	Ea.	37	36		73	97.50
7200	1/4″		13	.615		37	36		73	97.50
7210	3/8″		13	.615		37.50	36		73.50	97.50
7220	1/2″		12	.667		38	39		77	103
7230	3/4″		11	.727		41	42.50		83.50	112
7240	1″		10	.800		47.50	47		94.50	125
7250	1-1/4″	Q-1	17	.941		66	49.50		115.50	150
7260	1-1/2″		16	1		85.50	53		138.50	177
7270	2″		14	1.143		121	60.50		181.50	227
7280	2-1/2″		11	1.455		380	77		457	540
7290	3″		8	2		560	106		666	780
7300	4″	Q-2	11	2.182		1,000	120		1,120	1,275
7320	Tee, straight									
7330	1/8″	1 Plum	9	.889	Ea.	39	52		91	125
7340	1/4″		9	.889		39	52		91	125
7350	3/8″		9	.889		41.50	52		93.50	128
7360	1/2″		8	1		39.50	58.50		98	135
7370	3/4″		7	1.143		42	67		109	152
7380	1″		6.50	1.231		54	72.50		126.50	173

22 11 13.66 Pipe Fittings, Stainless Steel		Crew	Daily Output	Labor-Hours	Unit	Material	Labor	Equipment	Total	Total Incl O&P
						2015 Bare Costs				
7390	1-1/4"	Q-1	11	1.455	Ea.	91	77		168	220
7400	1-1/2"		10	1.600		116	84.50		200.50	260
7410	2"		9	1.778		145	94		239	305
7420	2-1/2"		7	2.286		390	121		511	620
7430	3"		5	3.200		590	169		759	915
7440	4"	Q-2	7	3.429		1,475	188		1,663	1,900
7460	Coupling, straight									
7470	1/8"	1 Plum	19	.421	Ea.	10.25	24.50		34.75	50
7480	1/4"		19	.421		12.10	24.50		36.60	52
7490	3/8"		19	.421		14.50	24.50		39	54.50
7500	1/2"		19	.421		19.30	24.50		43.80	59.50
7510	3/4"		18	.444		25.50	26		51.50	69
7520	1"		15	.533		41.50	31.50		73	94.50
7530	1-1/4"	Q-1	26	.615		66.50	32.50		99	124
7540	1-1/2"		24	.667		74.50	35		109.50	137
7550	2"		21	.762		123	40.50		163.50	199
7560	2-1/2"		18	.889		287	47		334	390
7570	3"		14	1.143		390	60.50		450.50	525
7580	4"	Q-2	16	1.500		550	82		632	735
7600	Reducer, concentric, 1/2"	1 Plum	12	.667		20.50	39		59.50	84
7610	3/4"		11	.727		26.50	42.50		69	96
7612	1"		10	.800		44.50	47		91.50	122
7614	1-1/4"	Q-1	17	.941		94.50	49.50		144	182
7616	1-1/2"		16	1		103	53		156	196
7618	2"		14	1.143		161	60.50		221.50	271
7620	2-1/2"		11	1.455		400	77		477	560
7622	3"		8	2		445	106		551	655
7624	4"	Q-2	11	2.182		735	120		855	995
7710	Union									
7720	1/8"	1 Plum	12	.667	Ea.	47.50	39		86.50	114
7730	1/4"		12	.667		47.50	39		86.50	114
7740	3/8"		12	.667		55.50	39		94.50	122
7750	1/2"		11	.727		69.50	42.50		112	143
7760	3/4"		10	.800		96	47		143	178
7770	1"		9	.889		138	52		190	234
7780	1-1/4"	Q-1	16	1		340	53		393	460
7790	1-1/2"		15	1.067		370	56.50		426.50	495
7800	2"		13	1.231		465	65		530	615
7810	2-1/2"		10	1.600		880	84.50		964.50	1,100
7820	3"		7	2.286		1,175	121		1,296	1,500
7830	4"	Q-2	10	2.400		1,625	131		1,756	1,975
7850	For 150 lb., type 316, add					25%				
8750	Threaded joint, 2000 lb., type 304									
8770	90° Elbow									
8780	1/8"	1 Plum	13	.615	Ea.	32.50	36		68.50	92
8790	1/4"		13	.615		32.50	36		68.50	92
8800	3/8"		13	.615		40	36		76	101
8810	1/2"		12	.667		54.50	39		93.50	121
8820	3/4"		11	.727		63	42.50		105.50	136
8830	1"		10	.800		83.50	47		130.50	165
8840	1-1/4"	Q-1	17	.941		136	49.50		185.50	228
8850	1-1/2"		16	1		218	53		271	325
8860	2"		14	1.143		305	60.50		365.50	430

22 11 13.66 Pipe Fittings, Stainless Steel		Crew	Daily Output	Labor-Hours	Unit	2015 Bare Costs Material	Labor	Equipment	Total	Total Incl O&P
8880	45° Elbow									
8890	1/8"	1 Plum	13	.615	Ea.	67.50	36		103.50	131
8900	1/4"		13	.615		67.50	36		103.50	131
8910	3/8"		13	.615		84.50	36		120.50	150
8920	1/2"		12	.667		84.50	39		123.50	154
8930	3/4"		11	.727		93	42.50		135.50	169
8940	1"		10	.800		111	47		158	195
8950	1-1/4"	Q-1	17	.941		194	49.50		243.50	291
8960	1-1/2"		16	1		260	53		313	370
8970	2"		14	1.143		299	60.50		359.50	425
8990	Tee, straight									
9000	1/8"	1 Plum	9	.889	Ea.	42	52		94	128
9010	1/4"		9	.889		42	52		94	128
9020	3/8"		9	.889		54	52		106	141
9030	1/2"		8	1		69	58.50		127.50	168
9040	3/4"		7	1.143		78.50	67		145.50	191
9050	1"		6.50	1.231		108	72.50		180.50	232
9060	1-1/4"	Q-1	11	1.455		182	77		259	320
9070	1-1/2"		10	1.600		305	84.50		389.50	465
9080	2"		9	1.778		420	94		514	610
9100	For couplings and unions use 3000 lb., type 304									
9120	2000 lb., type 316									
9130	90° Elbow									
9140	1/8"	1 Plum	13	.615	Ea.	40.50	36		76.50	101
9150	1/4"		13	.615		40.50	36		76.50	101
9160	3/8"		13	.615		47	36		83	109
9170	1/2"		12	.667		61	39		100	129
9180	3/4"		11	.727		75	42.50		117.50	149
9190	1"		10	.800		111	47		158	196
9200	1-1/4"	Q-1	17	.941		202	49.50		251.50	300
9210	1-1/2"		16	1		233	53		286	340
9220	2"		14	1.143		395	60.50		455.50	530
9240	45° Elbow									
9250	1/8"	1 Plum	13	.615	Ea.	85.50	36		121.50	151
9260	1/4"		13	.615		85.50	36		121.50	151
9270	3/8"		13	.615		85.50	36		121.50	151
9280	1/2"		12	.667		85.50	39		124.50	155
9300	3/4"		11	.727		96.50	42.50		139	173
9310	1"		10	.800		146	47		193	233
9320	1-1/4"	Q-1	17	.941		208	49.50		257.50	305
9330	1-1/2"		16	1		285	53		338	400
9340	2"		14	1.143		350	60.50		410.50	480
9360	Tee, straight									
9370	1/8"	1 Plum	9	.889	Ea.	51.50	52		103.50	138
9380	1/4"		9	.889		51.50	52		103.50	138
9390	3/8"		9	.889		63	52		115	151
9400	1/2"		8	1		78	58.50		136.50	178
9410	3/4"		7	1.143		96.50	67		163.50	211
9420	1"		6.50	1.231		147	72.50		219.50	275
9430	1-1/4"	Q-1	11	1.455		239	77		316	385
9440	1-1/2"		10	1.600		345	84.50		429.50	505
9450	2"		9	1.778		545	94		639	745
9470	For couplings and unions use 3000 lb., type 316									

22 11 13 – Facility Water Distribution Piping

22 11 13.66 Pipe Fittings, Stainless Steel		Crew	Daily Output	Labor-Hours	Unit	Material	2015 Bare Costs Labor	Equipment	Total	Total Incl O&P
9490	3000 lb., type 304									
9510	Coupling									
9520	1/8"	1 Plum	19	.421	Ea.	12.80	24.50		37.30	52.50
9530	1/4"		19	.421		13.90	24.50		38.40	54
9540	3/8"		19	.421		16.20	24.50		40.70	56.50
9550	1/2"		19	.421		19.50	24.50		44	60
9560	3/4"		18	.444		26	26		52	69
9570	1"		15	.533		44	31.50		75.50	97.50
9580	1-1/4"	Q-1	26	.615		107	32.50		139.50	168
9590	1-1/2"		24	.667		125	35		160	193
9600	2"		21	.762		165	40.50		205.50	244
9620	Union									
9630	1/8"	1 Plum	12	.667	Ea.	98	39		137	169
9640	1/4"		12	.667		98	39		137	169
9650	3/8"		12	.667		105	39		144	176
9660	1/2"		11	.727		105	42.50		147.50	182
9670	3/4"		10	.800		128	47		175	214
9680	1"		9	.889		198	52		250	299
9690	1-1/4"	Q-1	16	1		350	53		403	470
9700	1-1/2"		15	1.067		410	56.50		466.50	545
9710	2"		13	1.231		550	65		615	705
9730	3000 lb., type 316									
9750	Coupling									
9770	1/8"	1 Plum	19	.421	Ea.	14.40	24.50		38.90	54.50
9780	1/4"		19	.421		16.25	24.50		40.75	56.50
9790	3/8"		19	.421		17.40	24.50		41.90	57.50
9800	1/2"		19	.421		24.50	24.50		49	65.50
9810	3/4"		18	.444		36.50	26		62.50	80.50
9820	1"		15	.533		60.50	31.50		92	116
9830	1-1/4"	Q-1	26	.615		136	32.50		168.50	201
9840	1-1/2"		24	.667		159	35		194	230
9850	2"		21	.762		219	40.50		259.50	305
9870	Union									
9880	1/8"	1 Plum	12	.667	Ea.	108	39		147	179
9890	1/4"		12	.667		108	39		147	179
9900	3/8"		12	.667		125	39		164	198
9910	1/2"		11	.727		125	42.50		167.50	204
9920	3/4"		10	.800		163	47		210	252
9930	1"		9	.889		257	52		309	365
9940	1-1/4"	Q-1	16	1		450	53		503	580
9950	1-1/2"		15	1.067		555	56.50		611.50	700
9960	2"		13	1.231		705	65		770	875

22 11 13.74 Pipe, Plastic		Crew	Daily Output	Labor-Hours	Unit	Material	Labor	Equipment	Total	Total Incl O&P
0010	**PIPE, PLASTIC**									
0020	Fiberglass reinforced, couplings 10' O.C., clevis hanger assy's, 3 per 10'									
0080	General service									
0120	2" diameter	Q-1	59	.271	L.F.	12.55	14.35		26.90	36.50
0140	3" diameter		52	.308		16.10	16.25		32.35	43.50
0150	4" diameter		48	.333		21	17.60		38.60	50.50
0160	6" diameter		39	.410		35.50	21.50		57	73.50
0170	8" diameter	Q-2	49	.490		55	27		82	103
0180	10" diameter		41	.585		80	32		112	138

22 11 13.74 Pipe, Plastic		Crew	Daily Output	Labor-Hours	Unit	Material	Labor	Equipment	Total	Total Incl O&P
						2015 Bare Costs				
0190	12″ diameter	Q-2	36	.667	L.F.	103	36.50		139.50	170
0600	PVC, high impact/pressure, cplgs. 10′ O.C., clevis hanger assy's, 3 per 10′									
1020	Schedule 80									
1070	1/2″ diameter	1 Plum	50	.160	L.F.	6.75	9.40		16.15	22
1080	3/4″ diameter		47	.170		7.80	10		17.80	24
1090	1″ diameter		43	.186		9.65	10.90		20.55	27.50
1100	1-1/4″ diameter		39	.205		11.85	12.05		23.90	32
1110	1-1/2″ diameter		34	.235		13.40	13.80		27.20	36.50
1120	2″ diameter	Q-1	55	.291		16.65	15.35		32	42.50
1140	3″ diameter		50	.320		31	16.90		47.90	61
1150	4″ diameter		46	.348		43.50	18.40		61.90	76.50
1170	6″ diameter		38	.421		94.50	22		116.50	139
1730	To delete coupling & hangers, subtract									
1740	1/2″ diam.					62%	80%			
1750	3/4″ diam. to 1-1/4″ diam.					58%	73%			
1760	1-1/2″ diam. to 6″ diam.					40%	57%			
1800	PVC, couplings 10′ O.C., clevis hanger assemblies, 3 per 10′									
1820	Schedule 40									
1860	1/2″ diameter	1 Plum	54	.148	L.F.	4.88	8.70		13.58	18.90
1870	3/4″ diameter		51	.157		5.20	9.20		14.40	20
1880	1″ diameter		46	.174		5.85	10.20		16.05	22.50
1890	1-1/4″ diameter		42	.190		6.55	11.20		17.75	24.50
1900	1-1/2″ diameter		36	.222		6.85	13.05		19.90	28
1910	2″ diameter	Q-1	59	.271		8.05	14.35		22.40	31.50
1920	2-1/2″ diameter		56	.286		10.35	15.10		25.45	35
1930	3″ diameter		53	.302		12.60	15.95		28.55	39
1940	4″ diameter		48	.333		16.10	17.60		33.70	45
1950	5″ diameter		43	.372		26.50	19.65		46.15	59.50
1960	6″ diameter		39	.410		27.50	21.50		49	64
1970	8″ diameter	Q-2	48	.500		36.50	27.50		64	82.50
1980	10″ diameter		43	.558		75	30.50		105.50	130
1990	12″ diameter		42	.571		90	31.50		121.50	148
2000	14″ diameter		31	.774		143	42.50		185.50	224
2010	16″ diameter		23	1.043		205	57		262	315
2340	To delete coupling & hangers, subtract									
2360	1/2″ diam. to 1-1/4″ diam.					65%	74%			
2370	1-1/2″ diam. to 6″ diam.					44%	57%			
2380	8″ diam. to 12″ diam.					41%	53%			
2390	14″ diam. to 16″ diam.					48%	45%			
2420	Schedule 80									
2440	1/4″ diameter	1 Plum	58	.138	L.F.	4.35	8.10		12.45	17.45
2450	3/8″ diameter		55	.145		4.35	8.55		12.90	18.10
2460	1/2″ diameter		50	.160		5	9.40		14.40	20
2470	3/4″ diameter		47	.170		5.35	10		15.35	21.50
2480	1″ diameter		43	.186		6.10	10.90		17	24
2490	1-1/4″ diameter		39	.205		6.85	12.05		18.90	26.50
2500	1-1/2″ diameter		34	.235		7.30	13.80		21.10	29.50
2510	2″ diameter	Q-1	55	.291		8.20	15.35		23.55	33
2520	2-1/2″ diameter		52	.308		9.85	16.25		26.10	36.50
2530	3″ diameter		50	.320		13.95	16.90		30.85	42
2540	4″ diameter		46	.348		19.55	18.40		37.95	50
2550	5″ diameter		42	.381		29.50	20		49.50	63.50
2560	6″ diameter		38	.421		35.50	22		57.50	74

22 11 13.74 Pipe, Plastic		Crew	Daily Output	Labor-Hours	Unit	Material	Labor	Equipment	Total	Total Incl O&P
						2015 Bare Costs				
2570	8″ diameter	Q-2	47	.511	L.F.	45	28		73	93
2580	10″ diameter		42	.571		106	31.50		137.50	165
2590	12″ diameter		38	.632		128	34.50		162.50	195
2830	To delete coupling & hangers, subtract									
2840	1/4″ diam. to 1/2″ diam.					66%	80%			
2850	3/4″ diam. to 1-1/4″ diam.					61%	73%			
2860	1-1/2″ diam. to 6″ diam.					41%	57%			
2870	8″ diam. to 12″ diam.					31%	50%			
2900	Schedule 120									
2910	1/2″ diameter	1 Plum	50	.160	L.F.	5.50	9.40		14.90	20.50
2950	3/4″ diameter		47	.170		6	10		16	22
2960	1″ diameter		43	.186		7.05	10.90		17.95	25
2970	1-1/4″ diameter		39	.205		8.30	12.05		20.35	28
2980	1-1/2″ diameter		33	.242		9.05	14.25		23.30	32
2990	2″ diameter	Q-1	54	.296		10.75	15.65		26.40	36.50
3000	2-1/2″ diameter		52	.308		15.85	16.25		32.10	43
3010	3″ diameter		49	.327		19.10	17.25		36.35	48
3020	4″ diameter		45	.356		27.50	18.80		46.30	59.50
3030	6″ diameter		37	.432		49.50	23		72.50	90
3240	To delete coupling & hangers, subtract									
3250	1/2″ diam. to 1-1/4″ diam.					52%	74%			
3260	1-1/2″ diam. to 4″ diam.					30%	57%			
3270	6″ diam.					17%	50%			
3300	PVC, pressure, couplings 10′ O.C., clevis hanger assy's, 3 per 10′									
3310	SDR 26, 160 psi									
3350	1-1/4″ diameter	1 Plum	42	.190	L.F.	6.20	11.20		17.40	24.50
3360	1-1/2″ diameter	"	36	.222		6.40	13.05		19.45	27.50
3370	2″ diameter	Q-1	59	.271		7.30	14.35		21.65	30.50
3380	2-1/2″ diameter		56	.286		9.80	15.10		24.90	34.50
3390	3″ diameter		53	.302		12.30	15.95		28.25	38.50
3400	4″ diameter		48	.333		16.50	17.60		34.10	45.50
3420	6″ diameter		39	.410		30.50	21.50		52	67.50
3430	8″ diameter	Q-2	48	.500		45	27.50		72.50	92
3660	To delete coupling & clevis hanger assy's, subtract									
3670	1-1/4″ diam.					63%	68%			
3680	1-1/2″ diam. to 4″ diam.					48%	57%			
3690	6″ diam. to 8″ diam.					60%	54%			
3720	SDR 21, 200 psi, 1/2″ diameter	1 Plum	54	.148	L.F.	4.77	8.70		13.47	18.80
3740	3/4″ diameter		51	.157		5.10	9.20		14.30	19.95
3750	1″ diameter		46	.174		5.60	10.20		15.80	22
3760	1-1/4″ diameter		42	.190		6.40	11.20		17.60	24.50
3770	1-1/2″ diameter		36	.222		6.75	13.05		19.80	28
3780	2″ diameter	Q-1	59	.271		7.65	14.35		22	31
3790	2-1/2″ diameter		56	.286		11.90	15.10		27	36.50
3800	3″ diameter		53	.302		13.10	15.95		29.05	39.50
3810	4″ diameter		48	.333		20	17.60		37.60	50
3830	6″ diameter		39	.410		34	21.50		55.50	71
3840	8″ diameter	Q-2	48	.500		50	27.50		77.50	97.50
4000	To delete coupling & hangers, subtract									
4010	1/2″ diam. to 3/4″ diam.					71%	77%			
4020	1″ diam. to 1-1/4″ diam.					63%	70%			
4030	1-1/2″ diam. to 6″ diam.					44%	57%			
4040	8″ diam.					46%	54%			

22 11 13 – Facility Water Distribution Piping

22 11 13.74 Pipe, Plastic		Crew	Daily Output	Labor-Hours	Unit	Material	Labor	Equipment	Total	Total Incl O&P
						2015 Bare Costs				
4100	DWV type, schedule 40, couplings 10′ O.C., clevis hanger assy's, 3 per 10′									
4210	ABS, schedule 40, foam core type									
4212	Plain end black									
4214	1-1/2″ diameter	1 Plum	39	.205	L.F.	5.20	12.05		17.25	24.50
4216	2″ diameter	Q-1	62	.258		5.60	13.65		19.25	27.50
4218	3″ diameter		56	.286		8	15.10		23.10	32.50
4220	4″ diameter		51	.314		10.20	16.55		26.75	37
4222	6″ diameter		42	.381		18	20		38	51.50
4240	To delete coupling & hangers, subtract									
4244	1-1/2″ diam. to 6″ diam.					43%	48%			
4400	PVC									
4410	1-1/4″ diameter	1 Plum	42	.190	L.F.	5.35	11.20		16.55	23.50
4420	1-1/2″ diameter	″	36	.222		5.30	13.05		18.35	26.50
4460	2″ diameter	Q-1	59	.271		5.65	14.35		20	28.50
4470	3″ diameter		53	.302		8.05	15.95		24	34
4480	4″ diameter		48	.333		10	17.60		27.60	38.50
4490	6″ diameter		39	.410		16.40	21.50		37.90	52
4500	8″ diameter	Q-2	48	.500		21.50	27.50		49	66.50
4510	To delete coupling & hangers, subtract									
4520	1-1/4″ diam. to 1-1/2″ diam.					48%	60%			
4530	2″ diam. to 8″ diam.					42%	54%			
4550	PVC, schedule 40, foam core type									
4552	Plain end, white									
4554	1-1/2″ diameter	1 Plum	39	.205	L.F.	5	12.05		17.05	24.50
4556	2″ diameter	Q-1	62	.258		5.35	13.65		19	27.50
4558	3″ diameter		56	.286		7.45	15.10		22.55	31.50
4560	4″ diameter		51	.314		9.10	16.55		25.65	36
4562	6″ diameter		42	.381		14.55	20		34.55	47.50
4564	8″ diameter	Q-2	51	.471		18.75	26		44.75	60.50
4568	10″ diameter		48	.500		23.50	27.50		51	68
4570	12″ diameter		46	.522		28	28.50		56.50	75.50
4580	To delete coupling & hangers, subtract									
4582	1-1/2″ dia to 2″ dia					58%	54%			
4584	3″ dia to 12″ dia					46%	42%			
4800	PVC, clear pipe, cplgs. 10′ O.C., clevis hanger assy's 3 per 10′, Sched. 40									
4840	1/4″ diameter	1 Plum	59	.136	L.F.	4.91	7.95		12.86	17.80
4850	3/8″ diameter		56	.143		5.25	8.40		13.65	18.85
4860	1/2″ diameter		54	.148		5.90	8.70		14.60	20
4870	3/4″ diameter		51	.157		6.60	9.20		15.80	21.50
4880	1″ diameter		46	.174		8.15	10.20		18.35	25
4890	1-1/4″ diameter		42	.190		9.60	11.20		20.80	28
4900	1-1/2″ diameter		36	.222		10.70	13.05		23.75	32.50
4910	2″ diameter	Q-1	59	.271		13	14.35		27.35	37
4920	2-1/2″ diameter		56	.286		19	15.10		34.10	44.50
4930	3″ diameter		53	.302		23.50	15.95		39.45	50.50
4940	3-1/2″ diameter		50	.320		32.50	16.90		49.40	62
4950	4″ diameter		48	.333		33.50	17.60		51.10	64.50
5250	To delete coupling & hangers, subtract									
5260	1/4″ diam. to 3/8″ diam.					60%	81%			
5270	1/2″ diam. to 3/4″ diam.					41%	77%			
5280	1″ diam. to 1-1/2″ diam.					26%	67%			
5290	2″ diam. to 4″ diam.					16%	58%			
5300	CPVC, socket joint, couplings 10′ O.C., clevis hanger assemblies, 3 per 10′									

22 11 13.74 Pipe, Plastic		Crew	Daily Output	Labor-Hours	Unit	Material	Labor	Equipment	Total	Total Incl O&P
						2015 Bare Costs				
5302	Schedule 40									
5304	1/2" diameter	1 Plum	54	.148	L.F.	5.90	8.70		14.60	20
5305	3/4" diameter		51	.157		6.75	9.20		15.95	22
5306	1" diameter		46	.174		8.10	10.20		18.30	25
5307	1-1/4" diameter		42	.190		9.65	11.20		20.85	28
5308	1-1/2" diameter		36	.222		10.85	13.05		23.90	32.50
5309	2" diameter	Q-1	59	.271		12.85	14.35		27.20	36.50
5310	2-1/2" diameter		56	.286		19.90	15.10		35	45.50
5311	3" diameter		53	.302		23.50	15.95		39.45	51
5312	4" diameter		48	.333		31.50	17.60		49.10	62.50
5314	6" diameter		43	.372		57.50	19.65		77.15	93.50
5318	To delete coupling & hangers, subtract									
5319	1/2" diam. to 3/4" diam.					37%	77%			
5320	1" diam. to 1-1/4" diam.					27%	70%			
5321	1-1/2" diam. to 3" diam.					21%	57%			
5322	4" diam. to 6" diam.					16%	57%			
5324	Schedule 80									
5325	1/2" diameter	1 Plum	50	.160	L.F.	6.60	9.40		16	22
5326	3/4" diameter		47	.170		7	10		17	23.50
5327	1" diameter		43	.186		8.70	10.90		19.60	26.50
5328	1-1/4" diameter		39	.205		10.60	12.05		22.65	30.50
5329	1-1/2" diameter		34	.235		12.05	13.80		25.85	35
5330	2" diameter	Q-1	55	.291		14.70	15.35		30.05	40
5331	2-1/2" diameter		52	.308		22.50	16.25		38.75	50
5332	3" diameter		50	.320		26	16.90		42.90	55
5333	4" diameter		46	.348		35.50	18.40		53.90	67.50
5334	6" diameter		38	.421		67.50	22		89.50	109
5335	8" diameter	Q-2	47	.511		170	28		198	231
5339	To delete couplings & hangers, subtract									
5340	1/2" diam. to 3/4" diam.					44%	77%			
5341	1" diam. to 1-1/4" diam.					32%	71%			
5342	1-1/2" diam. to 4" diam.					25%	58%			
5343	6" diam. to 8" diam.					20%	53%			
5360	CPVC, threaded, couplings 10' O.C., clevis hanger assemblies, 3 per 10'									
5380	Schedule 40									
5460	1/2" diameter	1 Plum	54	.148	L.F.	6.70	8.70		15.40	21
5470	3/4" diameter		51	.157		8.15	9.20		17.35	23.50
5480	1" diameter		46	.174		9.55	10.20		19.75	26.50
5490	1-1/4" diameter		42	.190		10.80	11.20		22	29.50
5500	1-1/2" diameter		36	.222		11.80	13.05		24.85	33.50
5510	2" diameter	Q-1	59	.271		14	14.35		28.35	38
5520	2-1/2" diameter		56	.286		21	15.10		36.10	46.50
5530	3" diameter		53	.302		25.50	15.95		41.45	53
5540	4" diameter		48	.333		38.50	17.60		56.10	70
5550	6" diameter		43	.372		60.50	19.65		80.15	97.50
5730	To delete coupling & hangers, subtract									
5740	1/2" diam. to 3/4" diam.					37%	77%			
5750	1" diam. to 1-1/4" diam.					27%	70%			
5760	1-1/2" diam. to 3" diam.					21%	57%			
5770	4" diam. to 6" diam.					16%	57%			
5800	Schedule 80									
5860	1/2" diameter	1 Plum	50	.160	L.F.	7.40	9.40		16.80	23
5870	3/4" diameter		47	.170		8.40	10		18.40	25

22 11 13 – Facility Water Distribution Piping

22 11 13.74 Pipe, Plastic		Crew	Daily Output	Labor-Hours	Unit	Material	Labor	Equipment	Total	Total Incl O&P
						2015 Bare Costs				
5880	1" diameter	1 Plum	43	.186	L.F.	10.15	10.90		21.05	28
5890	1-1/4" diameter		39	.205		11.70	12.05		23.75	31.50
5900	1-1/2" diameter		34	.235		13	13.80		26.80	36
5910	2" diameter	Q-1	55	.291		15.85	15.35		31.20	41.50
5920	2-1/2" diameter		52	.308		23.50	16.25		39.75	51.50
5930	3" diameter		50	.320		27.50	16.90		44.40	57
5940	4" diameter		46	.348		42.50	18.40		60.90	75.50
5950	6" diameter		38	.421		71	22		93	113
5960	8" diameter	Q-2	47	.511		169	28		197	230
6060	To delete couplings & hangers, subtract									
6070	1/2" diam. to 3/4" diam.					44%	77%			
6080	1" diam. to 1-1/4" diam.					32%	71%			
6090	1-1/2" diam. to 4" diam.					25%	58%			
6100	6" diam. to 8" diam.					20%	53%			
6240	CTS, 1/2" diameter	1 Plum	54	.148	L.F.	5.50	8.70		14.20	19.60
6250	3/4" diameter		51	.157		6.75	9.20		15.95	22
6260	1" diameter		46	.174		10.10	10.20		20.30	27
6270	1 1/4"		42	.190		14.75	11.20		25.95	33.50
6280	1 1/2" diameter		36	.222		18.35	13.05		31.40	40.50
6290	2" diameter	Q-1	59	.271		28.50	14.35		42.85	53.50
6370	To delete coupling & hangers, subtract									
6380	1/2" diam.					51%	79%			
6390	3/4" diam.					40%	76%			
6392	1" thru 2" diam.					72%	68%			
6500	Residential installation, plastic pipe									
6510	Couplings 10' O.C., strap hangers 3 per 10'									
6520	PVC, Schedule 40									
6530	1/2" diameter	1 Plum	138	.058	L.F.	1.02	3.40		4.42	6.45
6540	3/4" diameter		128	.063		1.23	3.67		4.90	7.05
6550	1" diameter		119	.067		1.81	3.95		5.76	8.15
6560	1-1/4" diameter		111	.072		2.20	4.23		6.43	9
6570	1-1/2" diameter		104	.077		2.44	4.52		6.96	9.75
6580	2" diameter	Q-1	197	.081		3.10	4.29		7.39	10.10
6590	2-1/2" diameter		162	.099		5.10	5.20		10.30	13.75
6600	4" diameter		123	.130		8.60	6.85		15.45	20
6700	PVC, DWV, Schedule 40									
6720	1-1/4" diameter	1 Plum	100	.080	L.F.	1.84	4.70		6.54	9.30
6730	1-1/2" diameter	"	94	.085		1.77	5		6.77	9.75
6740	2" diameter	Q-1	178	.090		2.09	4.75		6.84	9.70
6760	4" diameter	"	110	.145		5.70	7.70		13.40	18.30
7280	PEX, flexible, no couplings or hangers									
7282	Note: For labor costs add 25% to the couplings and fittings labor total.									
7285	For fittings see section 23 83 16.10 7000									
7300	Non-barrier type, hot/cold tubing rolls									
7310	1/4" diameter x 100'				L.F.	.49			.49	.54
7350	3/8" diameter x 100'					.55			.55	.61
7360	1/2" diameter x 100'					.61			.61	.67
7370	1/2" diameter x 500'					.61			.61	.67
7380	1/2" diameter x 1000'					.61			.61	.67
7400	3/4" diameter x 100'					1.11			1.11	1.22
7410	3/4" diameter x 500'					1.11			1.11	1.22
7420	3/4" diameter x 1000'					1.11			1.11	1.22
7460	1" diameter x 100'					1.90			1.90	2.09

22 11 13 – Facility Water Distribution Piping

22 11 13.74 Pipe, Plastic		Crew	Daily Output	Labor-Hours	Unit	Material	2015 Bare Costs Labor	Equipment	Total	Total Incl O&P
7470	1" diameter x 300'				L.F.	1.90			1.90	2.09
7480	1" diameter x 500'					1.90			1.90	2.09
7500	1-1/4" diameter x 100'					3.23			3.23	3.55
7510	1-1/4" diameter x 300'					3.23			3.23	3.55
7540	1-1/2" diameter x 100'					4.39			4.39	4.83
7550	1-1/2" diameter x 300'				↓	4.39			4.39	4.83
7596	Most sizes available in red or blue									
7700	Non-barrier type, hot/cold tubing straight lengths									
7710	1/2" diameter x 20'				L.F.	.60			.60	.66
7750	3/4" diameter x 20'					1.10			1.10	1.21
7760	1" diameter x 20'					1.90			1.90	2.09
7770	1-1/4" diameter x 20'					3.23			3.23	3.55
7780	1-1/2" diameter x 20'					4.39			4.39	4.83
7790	2" diameter				↓	8.60			8.60	9.45
7796	Most sizes available in red or blue									
9000	Polypropylene pipe									
9002	For fusion weld fittings and accessories see line 22 11 13.76 9400									
9004	Note: sizes 1/2" thru 4" use socket fusion									
9005	Sizes 6" thru 10" use butt fusion									
9010	SDR 7.4, (domestic hot water piping)									
9011	Enhanced to minimize thermal expansion and high temperature life									
9016	13' lengths, size is ID, includes joints 13' O.C. and hangers 3 per 10'									
9020	3/8" diameter	1 Plum	53	.151	L.F.	2.35	8.85		11.20	16.40
9022	1/2" diameter		52	.154		2.80	9.05		11.85	17.20
9024	3/4" diameter		50	.160		3.31	9.40		12.71	18.30
9026	1" diameter		45	.178		4.34	10.45		14.79	21
9028	1-1/4" diameter		40	.200		6.20	11.75		17.95	25
9030	1-1/2" diameter	↓	35	.229		8.75	13.40		22.15	30.50
9032	2" diameter	Q-1	58	.276		12.40	14.55		26.95	36
9034	2-1/2" diameter		55	.291		17	15.35		32.35	42.50
9036	3" diameter		52	.308		22.50	16.25		38.75	50
9038	3-1/2" diameter		49	.327		31	17.25		48.25	61.50
9040	4" diameter		46	.348		34	18.40		52.40	66
9042	6" diameter	↓	39	.410		46	21.50		67.50	84.50
9044	8" diameter	Q-2	48	.500		69.50	27.50		97	119
9046	10" diameter	"	43	.558	↓	110	30.50		140.50	169
9050	To delete joint & hangers, subtract									
9052	3/8" diam. to 1" diam.					45%	65%			
9054	1-1/4" diam. to 4" diam.					15%	45%			
9056	6" diam. to 10" diam.					5%	24%			
9060	SDR 11, (domestic cold water piping)									
9062	13' lengths, size is ID, includes joints 13' O.C. and hangers 3 per 10'									
9064	1/2" diameter	1 Plum	57	.140	L.F.	2.48	8.25		10.73	15.60
9066	3/4" diameter		54	.148		2.80	8.70		11.50	16.65
9068	1" diameter		49	.163		3.41	9.60		13.01	18.70
9070	1-1/4" diameter		45	.178		4.83	10.45		15.28	21.50
9072	1-1/2" diameter	↓	40	.200		6.40	11.75		18.15	25.50
9074	2" diameter	Q-1	62	.258		9	13.65		22.65	31.50
9076	2-1/2" diameter		59	.271		11.85	14.35		26.20	35.50
9078	3" diameter		56	.286		15.90	15.10		31	41
9080	3-1/2" diameter		53	.302		23	15.95		38.95	50.50
9082	4" diameter		50	.320		26	16.90		42.90	55
9084	6" diameter	↓	47	.340	↓	31.50	18		49.50	62.50

22 11 Facility Water Distribution

22 11 13 – Facility Water Distribution Piping

22 11 13.74 Pipe, Plastic		Crew	Daily Output	Labor-Hours	Unit	Material	Labor (2015 Bare Costs)	Equipment	Total	Total Incl O&P
9086	8″ diameter	Q-2	51	.471	L.F.	47.50	26		73.50	92.50
9088	10″ diameter	″	46	.522		72.50	28.50		101	124
9090	To delete joint & hangers, subtract									
9092	1/2″ diam. to 1″ diam.					45%	65%			
9094	1-1/4″ diam. to 4″ diam.					15%	45%			
9096	6″ diam. to 10″ diam.					5%	24%			
9900	Minimum labor/equipment charge	1 Plum	4	2	Job		117		117	183
22 11 13.76 Pipe Fittings, Plastic										
0010	**PIPE FITTINGS, PLASTIC**									
0030	Epoxy resin, fiberglass reinforced, general service									
0100	3″	Q-1	20.80	.769	Ea.	113	40.50		153.50	188
0110	4″		16.50	.970		115	51		166	206
0120	6″		10.10	1.584		225	83.50		308.50	380
0130	8″	Q-2	9.30	2.581		415	141		556	675
0140	10″		8.50	2.824		520	155		675	815
0150	12″		7.60	3.158		750	173		923	1,100
0380	Couplings									
0410	2″	Q-1	33.10	.483	Ea.	22.50	25.50		48	65
0420	3″		20.80	.769		24.50	40.50		65	90.50
0430	4″		16.50	.970		33	51		84	117
0440	6″		10.10	1.584		78.50	83.50		162	217
0450	8″	Q-2	9.30	2.581		133	141		274	365
0460	10″		8.50	2.824		195	155		350	455
0470	12″		7.60	3.158		262	173		435	560
0473	High corrosion resistant couplings, add					30%				
2100	PVC schedule 80, socket joint									
2110	90° elbow, 1/2″	1 Plum	30.30	.264	Ea.	2.48	15.50		17.98	26.50
2130	3/4″		26	.308		3.17	18.05		21.22	31.50
2140	1″		22.70	.352		4.95	20.50		25.45	38
2150	1-1/4″		20.20	.396		6.85	23.50		30.35	44
2160	1-1/2″		18.20	.440		7.30	26		33.30	48
2170	2″	Q-1	33.10	.483		8.85	25.50		34.35	49.50
2180	3″		20.80	.769		23.50	40.50		64	89
2190	4″		16.50	.970		35.50	51		86.50	119
2200	6″		10.10	1.584		101	83.50		184.50	242
2210	8″	Q-2	9.30	2.581		277	141		418	525
2250	45° elbow, 1/2″	1 Plum	30.30	.264		4.67	15.50		20.17	29
2270	3/4″		26	.308		7.15	18.05		25.20	36
2280	1″		22.70	.352		10.70	20.50		31.20	44.50
2290	1-1/4″		20.20	.396		13.60	23.50		37.10	51.50
2300	1-1/2″		18.20	.440		16.10	26		42.10	57.50
2310	2″	Q-1	33.10	.483		21	25.50		46.50	63
2320	3″		20.80	.769		53.50	40.50		94	122
2330	4″		16.50	.970		96	51		147	186
2340	6″		10.10	1.584		121	83.50		204.50	264
2350	8″	Q-2	9.30	2.581		262	141		403	510
2400	Tee, 1/2″	1 Plum	20.20	.396		7	23.50		30.50	44
2420	3/4″		17.30	.462		7.35	27		34.35	50.50
2430	1″		15.20	.526		9.15	31		40.15	58
2440	1-1/4″		13.50	.593		25.50	35		60.50	82.50
2450	1-1/2″		12.10	.661		25.50	39		64.50	88.50
2460	2″	Q-1	20	.800		31.50	42.50		74	101

22 11 13.76 Pipe Fittings, Plastic		Crew	Daily Output	Labor-Hours	Unit	Material	Labor	Equipment	Total	Total Incl O&P
						2015 Bare Costs				
2470	3"	Q-1	13.90	1.151	Ea.	43	61		104	142
2480	4"		11	1.455		49.50	77		126.50	175
2490	6"		6.70	2.388		169	126		295	385
2500	8"	Q-2	6.20	3.871		395	212		607	760
2510	Flange, socket, 150 lb., 1/2"	1 Plum	55.60	.144		13.60	8.45		22.05	28
2514	3/4"		47.60	.168		14.50	9.85		24.35	31.50
2518	1"		41.70	.192		16.15	11.25		27.40	35.50
2522	1-1/2"		33.30	.240		16.65	14.10		30.75	40.50
2526	2"	Q-1	60.60	.264		22.50	13.95		36.45	47
2530	4"		30.30	.528		49	28		77	97.50
2534	6"		18.50	.865		77	45.50		122.50	156
2538	8"	Q-2	17.10	1.404		138	77		215	271
2550	Coupling, 1/2"	1 Plum	30.30	.264		4.49	15.50		19.99	29
2570	3/4"		26	.308		6.05	18.05		24.10	34.50
2580	1"		22.70	.352		6.25	20.50		26.75	39.50
2590	1-1/4"		20.20	.396		9.50	23.50		33	47
2600	1-1/2"		18.20	.440		10.25	26		36.25	51.50
2610	2"	Q-1	33.10	.483		11	25.50		36.50	52
2620	3"		20.80	.769		31	40.50		71.50	97.50
2630	4"		16.50	.970		39	51		90	123
2640	6"		10.10	1.584		84	83.50		167.50	223
2650	8"	Q-2	9.30	2.581		114	141		255	345
2660	10"		8.50	2.824		400	155		555	680
2670	12"		7.60	3.158		450	173		623	765
2700	PVC (white), schedule 40, socket joints									
2760	90° elbow, 1/2"	1 Plum	33.30	.240	Ea.	.48	14.10		14.58	22.50
2770	3/4"		28.60	.280		.54	16.40		16.94	26
2780	1"		25	.320		.97	18.80		19.77	30.50
2790	1-1/4"		22.20	.360		1.71	21		22.71	35
2800	1-1/2"		20	.400		1.85	23.50		25.35	38.50
2810	2"	Q-1	36.40	.440		2.89	23		25.89	39
2820	2-1/2"		26.70	.599		8.80	31.50		40.30	59
2830	3"		22.90	.699		10.50	37		47.50	69
2840	4"		18.20	.879		18.80	46.50		65.30	93
2850	5"		12.10	1.322		48.50	70		118.50	163
2860	6"		11.10	1.441		60	76		136	185
2870	8"	Q-2	10.30	2.330		154	128		282	370
2980	45° elbow, 1/2"	1 Plum	33.30	.240		.80	14.10		14.90	23
2990	3/4"		28.60	.280		1.23	16.40		17.63	27
3000	1"		25	.320		1.47	18.80		20.27	31
3010	1-1/4"		22.20	.360		2.03	21		23.03	35
3020	1-1/2"		20	.400		2.60	23.50		26.10	39.50
3030	2"	Q-1	36.40	.440		3.38	23		26.38	39.50
3040	2-1/2"		26.70	.599		8.80	31.50		40.30	59
3050	3"		22.90	.699		13.65	37		50.65	72.50
3060	4"		18.20	.879		24.50	46.50		71	99.50
3070	5"		12.10	1.322		48.50	70		118.50	163
3080	6"		11.10	1.441		60.50	76		136.50	186
3090	8"	Q-2	10.30	2.330		146	128		274	360
3180	Tee, 1/2"	1 Plum	22.20	.360		.60	21		21.60	33.50
3190	3/4"		19	.421		.69	24.50		25.19	39.50
3200	1"		16.70	.479		1.29	28		29.29	45.50
3210	1-1/4"		14.80	.541		2.01	31.50		33.51	51.50

22 11 13 – Facility Water Distribution Piping

22 11 13.76 Pipe Fittings, Plastic		Crew	Daily Output	Labor-Hours	Unit	2015 Bare Costs Material	Labor	Equipment	Total	Total Incl O&P
3220	1-1/2"	1 Plum	13.30	.602	Ea.	2.45	35.50		37.95	57.50
3230	2"	Q-1	24.20	.661		3.56	35		38.56	58.50
3240	2-1/2"		17.80	.899		11.75	47.50		59.25	87
3250	3"		15.20	1.053		15.45	55.50		70.95	104
3260	4"		12.10	1.322		28	70		98	140
3270	5"		8.10	1.975		67.50	104		171.50	237
3280	6"		7.40	2.162		94	114		208	281
3290	8"	Q-2	6.80	3.529		218	193		411	540
3380	Coupling, 1/2"	1 Plum	33.30	.240		.32	14.10		14.42	22.50
3390	3/4"		28.60	.280		.44	16.40		16.84	26
3400	1"		25	.320		.77	18.80		19.57	30.50
3410	1-1/4"		22.20	.360		1.06	21		22.06	34
3420	1-1/2"		20	.400		1.13	23.50		24.63	37.50
3430	2"	Q-1	36.40	.440		1.72	23		24.72	38
3440	2-1/2"		26.70	.599		3.81	31.50		35.31	53.50
3450	3"		22.90	.699		5.95	37		42.95	64
3460	4"		18.20	.879		8.65	46.50		55.15	82
3470	5"		12.10	1.322		15.80	70		85.80	126
3480	6"		11.10	1.441		27.50	76		103.50	149
3490	8"	Q-2	10.30	2.330		51	128		179	255
4500	DWV, ABS, non pressure, socket joints									
4540	1/4 Bend, 1-1/4"	1 Plum	20.20	.396	Ea.	2.57	23.50		26.07	39.50
4560	1-1/2"	"	18.20	.440		1.92	26		27.92	42
4570	2"	Q-1	33.10	.483		3.05	25.50		28.55	43.50
4580	3"		20.80	.769		7.70	40.50		48.20	72
4590	4"		16.50	.970		15.95	51		66.95	97.50
4600	6"		10.10	1.584		65	83.50		148.50	203
4650	1/8 Bend, same as 1/4 Bend									
4800	Tee, sanitary									
4820	1-1/4"	1 Plum	13.50	.593	Ea.	3.42	35		38.42	58.50
4830	1-1/2"	"	12.10	.661		2.92	39		41.92	63.50
4840	2"	Q-1	20	.800		4.52	42.50		47.02	71
4850	3"		13.90	1.151		12.40	61		73.40	109
4860	4"		11	1.455		22	77		99	144
4862	Tee, sanitary, reducing, 2" x 1-1/2"		22	.727		3.94	38.50		42.44	64.50
4864	3" x 2"		15.30	1.046		9.05	55.50		64.55	96
4868	4" x 3"		12.10	1.322		20	70		90	131
4870	Combination Y and 1/8 bend									
4872	1-1/2"	1 Plum	12.10	.661	Ea.	7	39		46	68
4874	2"	Q-1	20	.800		7.85	42.50		50.35	74.50
4876	3"		13.90	1.151		18.50	61		79.50	116
4878	4"		11	1.455		35.50	77		112.50	159
4880	3" x 1-1/2"		15.50	1.032		18.05	54.50		72.55	105
4882	4" x 3"		12.10	1.322		28	70		98	140
4900	Wye, 1-1/4"	1 Plum	13.50	.593		3.94	35		38.94	59
4902	1-1/2"	"	12.10	.661		4.49	39		43.49	65.50
4904	2"	Q-1	20	.800		5.85	42.50		48.35	72.50
4906	3"		13.90	1.151		13.80	61		74.80	110
4908	4"		11	1.455		28	77		105	151
4910	6"		6.70	2.388		81	126		207	286
4918	3" x 1-1/2"		15.50	1.032		11.05	54.50		65.55	97
4920	4" x 3"		12.10	1.322		22	70		92	134
4922	6" x 4"		6.90	2.319		66.50	123		189.50	265

22 11 13.76 Pipe Fittings, Plastic		Crew	Daily Output	Labor-Hours	Unit	Material	2015 Bare Costs Labor	Equipment	Total	Total Incl O&P
4930	Double Wye, 1-1/2"	1 Plum	9.10	.879	Ea.	13.95	51.50		65.45	96
4932	2"	Q-1	16.60	.964		16.65	51		67.65	98
4934	3"		10.40	1.538		38	81.50		119.50	169
4936	4"		8.25	1.939		74	102		176	242
4940	2" x 1-1/2"		16.80	.952		14.55	50.50		65.05	94.50
4942	3" x 2"		10.60	1.509		28	79.50		107.50	155
4944	4" x 3"		8.45	1.893		58.50	100		158.50	221
4946	6" x 4"		7.25	2.207		90	117		207	281
4950	Reducer bushing, 2" x 1-1/2"		36.40	.440		1.56	23		24.56	37.50
4952	3" x 1-1/2"		27.30	.586		6.65	31		37.65	56
4954	4" x 2"		18.20	.879		12.70	46.50		59.20	86.50
4956	6" x 4"		11.10	1.441		34.50	76		110.50	157
4960	Couplings, 1-1/2"	1 Plum	18.20	.440		.95	26		26.95	41
4962	2"	Q-1	33.10	.483		1.28	25.50		26.78	41.50
4963	3"		20.80	.769		3.58	40.50		44.08	67.50
4964	4"		16.50	.970		6.45	51		57.45	87
4966	6"		10.10	1.584		27	83.50		110.50	161
4970	2" x 1-1/2"		33.30	.480		2.75	25.50		28.25	42.50
4972	3" x 1-1/2"		21	.762		7.65	40.50		48.15	71.50
4974	4" x 3"		16.70	.958		12.15	50.50		62.65	92.50
4978	Closet flange, 4"	1 Plum	32	.250		6.65	14.70		21.35	30.50
4980	4" x 3"	"	34	.235		7.55	13.80		21.35	30
5000	DWV, PVC, schedule 40, socket joints									
5040	1/4 bend, 1-1/4"	1 Plum	20.20	.396	Ea.	5.20	23.50		28.70	42
5060	1-1/2"	"	18.20	.440		1.48	26		27.48	41.50
5070	2"	Q-1	33.10	.483		2.33	25.50		27.83	42.50
5080	3"		20.80	.769		6.85	40.50		47.35	71
5090	4"		16.50	.970		13.05	51		64.05	94.50
5100	6"		10.10	1.584		46	83.50		129.50	182
5105	8"	Q-2	9.30	2.581		65	141		206	293
5110	1/4 bend, long sweep, 1-1/2"	1 Plum	18.20	.440		3.42	26		29.42	44
5112	2"	Q-1	33.10	.483		3.83	25.50		29.33	44
5114	3"		20.80	.769		8.75	40.50		49.25	73
5116	4"		16.50	.970		16.65	51		67.65	98.50
5150	1/8 bend, 1-1/4"	1 Plum	20.20	.396		3.40	23.50		26.90	40
5170	1-1/2"	"	18.20	.440		1.44	26		27.44	41.50
5180	2"	Q-1	33.10	.483		2.08	25.50		27.58	42.50
5190	3"		20.80	.769		6.15	40.50		46.65	70.50
5200	4"		16.50	.970		10.40	51		61.40	91.50
5210	6"		10.10	1.584		42	83.50		125.50	177
5215	8"	Q-2	9.30	2.581		158	141		299	395
5250	Tee, sanitary 1-1/4"	1 Plum	13.50	.593		5.20	35		40.20	60.50
5254	1-1/2"	"	12.10	.661		2.58	39		41.58	63.50
5255	2"	Q-1	20	.800		3.80	42.50		46.30	70
5256	3"		13.90	1.151		10	61		71	106
5257	4"		11	1.455		17.75	77		94.75	140
5259	6"		6.70	2.388		74	126		200	279
5261	8"	Q-2	6.20	3.871		173	212		385	520
5264	2" x 1-1/2"	Q-1	22	.727		3.35	38.50		41.85	63.50
5266	3" x 1-1/2"		15.50	1.032		7.05	54.50		61.55	93
5268	4" x 3"		12.10	1.322		21.50	70		91.50	133
5271	6" x 4"		6.90	2.319		71.50	123		194.50	270
5314	Combination Y & 1/8 bend, 1-1/2"	1 Plum	12.10	.661		6.25	39		45.25	67.50

22 11 13.76 Pipe Fittings, Plastic		Crew	Daily Output	Labor-Hours	Unit	Material	2015 Bare Costs Labor	Equipment	Total	Total Incl O&P
5315	2"	Q-1	20	.800	Ea.	6.70	42.50		49.20	73.50
5317	3"		13.90	1.151		16.60	61		77.60	113
5318	4"		11	1.455		33	77		110	156
5324	Combination Y & 1/8 bend, reducing									
5325	2" x 2" x 1-1/2"	Q-1	22	.727	Ea.	8.80	38.50		47.30	69.50
5327	3" x 3" x 1-1/2"		15.50	1.032		15.65	54.50		70.15	102
5328	3" x 3" x 2"		15.30	1.046		11.35	55.50		66.85	98.50
5329	4" x 4" x 2"		12.20	1.311		17.85	69.50		87.35	128
5331	Wye, 1-1/4"	1 Plum	13.50	.593		6.65	35		41.65	62
5332	1-1/2"	"	12.10	.661		4.87	39		43.87	66
5333	2"	Q-1	20	.800		4.63	42.50		47.13	71
5334	3"		13.90	1.151		12.50	61		73.50	109
5335	4"		11	1.455		22.50	77		99.50	145
5336	6"		6.70	2.388		68.50	126		194.50	273
5337	8"	Q-2	6.20	3.871		124	212		336	465
5341	2" x 1-1/2"	Q-1	22	.727		5.70	38.50		44.20	66.50
5342	3" x 1-1/2"		15.50	1.032		8.20	54.50		62.70	94
5343	4" x 3"		12.10	1.322		18.40	70		88.40	129
5344	6" x 4"		6.90	2.319		50.50	123		173.50	247
5345	8" x 6"	Q-2	6.40	3.750		93.50	205		298.50	425
5347	Double wye, 1-1/2"	1 Plum	9.10	.879		10.55	51.50		62.05	92
5348	2"	Q-1	16.60	.964		12.20	51		63.20	93
5349	3"		10.40	1.538		25	81.50		106.50	155
5350	4"		8.25	1.939		50.50	102		152.50	216
5353	Double wye, reducing									
5354	2" x 2" x 1-1/2" x 1-1/2"	Q-1	16.80	.952	Ea.	10.75	50.50		61.25	90.50
5355	3" x 3" x 2" x 2"		10.60	1.509		18.45	79.50		97.95	145
5356	4" x 4" x 3" x 3"		8.45	1.893		40	100		140	200
5357	6" x 6" x 4" x 4"		7.25	2.207		138	117		255	335
5374	Coupling, 1-1/4"	1 Plum	20.20	.396		3.15	23.50		26.65	40
5376	1-1/2"	"	18.20	.440		.72	26		26.72	41
5378	2"	Q-1	33.10	.483		.95	25.50		26.45	41
5380	3"		20.80	.769		3.29	40.50		43.79	67
5390	4"		16.50	.970		5.65	51		56.65	86
5400	6"		10.10	1.584		18.40	83.50		101.90	151
5402	8"	Q-2	9.30	2.581		31.50	141		172.50	256
5404	2" x 1-1/2"	Q-1	33.30	.480		2.13	25.50		27.63	42
5406	3" x 1-1/2"		21	.762		6.25	40.50		46.75	70
5408	4" x 3"		16.70	.958		10.15	50.50		60.65	90
5410	Reducer bushing, 2" x 1-1/4"		36.50	.438		2.82	23		25.82	39
5412	3" x 1-1/2"		27.30	.586		5.85	31		36.85	55
5414	4" x 2"		18.20	.879		10.20	46.50		56.70	83.50
5416	6" x 4"		11.10	1.441		27	76		103	149
5418	8" x 6"	Q-2	10.20	2.353		46	129		175	252
5425	Closet flange 4"	Q-1	32	.500		7.20	26.50		33.70	49
5426	4" x 3"	"	34	.471		5.80	25		30.80	45.50
5450	Solvent cement for PVC, industrial grade, per quart				Qt.	27.50			27.50	30.50
5500	CPVC, Schedule 80, threaded joints									
5540	90° Elbow, 1/4"	1 Plum	32	.250	Ea.	12.15	14.70		26.85	36.50
5560	1/2"		30.30	.264		7.05	15.50		22.55	32
5570	3/4"		26	.308		10.55	18.05		28.60	39.50
5580	1"		22.70	.352		14.80	20.50		35.30	49
5590	1-1/4"		20.20	.396		28.50	23.50		52	68

22 11 13.76 Pipe Fittings, Plastic		Crew	Daily Output	Labor-Hours	Unit	Material	Labor	Equipment	Total	Total Incl O&P
						2015 Bare Costs				
5600	1-1/2"	1 Plum	18.20	.440	Ea.	30.50	26		56.50	74
5610	2"	Q-1	33.10	.483		41	25.50		66.50	85
5620	2-1/2"		24.20	.661		128	35		163	195
5630	3"		20.80	.769		137	40.50		177.50	214
5640	4"		16.50	.970		214	51		265	315
5650	6"		10.10	1.584		252	83.50		335.50	410
5660	45° Elbow same as 90° Elbow									
5700	Tee, 1/4"	1 Plum	22	.364	Ea.	23.50	21.50		45	59.50
5702	1/2"		20.20	.396		23.50	23.50		47	62.50
5704	3/4"		17.30	.462		34	27		61	80
5706	1"		15.20	.526		36.50	31		67.50	88.50
5708	1-1/4"		13.50	.593		37	35		72	95
5710	1-1/2"		12.10	.661		38.50	39		77.50	103
5712	2"	Q-1	20	.800		43	42.50		85.50	114
5714	2-1/2"		16.20	.988		211	52		263	315
5716	3"		13.90	1.151		245	61		306	365
5718	4"		11	1.455		580	77		657	755
5720	6"		6.70	2.388		670	126		796	930
5730	Coupling, 1/4"	1 Plum	32	.250		15.45	14.70		30.15	40
5732	1/2"		30.30	.264		12.75	15.50		28.25	38
5734	3/4"		26	.308		20.50	18.05		38.55	50.50
5736	1"		22.70	.352		23.50	20.50		44	58.50
5738	1-1/4"		20.20	.396		25	23.50		48.50	63.50
5740	1-1/2"		18.20	.440		26.50	26		52.50	69.50
5742	2"	Q-1	33.10	.483		31.50	25.50		57	74.50
5744	2-1/2"		24.20	.661		56.50	35		91.50	117
5746	3"		20.80	.769		65.50	40.50		106	136
5748	4"		16.50	.970		133	51		184	227
5750	6"		10.10	1.584		182	83.50		265.50	330
5752	8"	Q-2	9.30	2.581		390	141		531	645
5900	CPVC, Schedule 80, socket joints									
5904	90° Elbow, 1/4"	1 Plum	32	.250	Ea.	11.55	14.70		26.25	35.50
5906	1/2"		30.30	.264		4.53	15.50		20.03	29
5908	3/4"		26	.308		5.80	18.05		23.85	34.50
5910	1"		22.70	.352		9.20	20.50		29.70	42.50
5912	1-1/4"		20.20	.396		19.85	23.50		43.35	58.50
5914	1-1/2"		18.20	.440		22	26		48	64.50
5916	2"	Q-1	33.10	.483		26.50	25.50		52	69.50
5918	2-1/2"		24.20	.661		61.50	35		96.50	122
5920	3"		20.80	.769		69.50	40.50		110	140
5922	4"		16.50	.970		126	51		177	218
5924	6"		10.10	1.584		252	83.50		335.50	410
5926	8"		9.30	1.720		620	91		711	820
5930	45° Elbow, 1/4"	1 Plum	32	.250		17.20	14.70		31.90	42
5932	1/2"		30.30	.264		5.55	15.50		21.05	30
5934	3/4"		26	.308		7.95	18.05		26	37
5936	1"		22.70	.352		12.75	20.50		33.25	46.50
5938	1-1/4"		20.20	.396		25	23.50		48.50	64
5940	1-1/2"		18.20	.440		25.50	26		51.50	68
5942	2"	Q-1	33.10	.483		29	25.50		54.50	72
5944	2-1/2"		24.20	.661		59	35		94	119
5946	3"		20.80	.769		75.50	40.50		116	147
5948	4"		16.50	.970		104	51		155	194

22 11 13.76 Pipe Fittings, Plastic		Crew	Daily Output	Labor-Hours	Unit	Material	2015 Bare Costs Labor	Equipment	Total	Total Incl O&P
5950	6"	Q-1	10.10	1.584	Ea.	305	83.50		388.50	465
5952	8"		9.30	1.720		665	91		756	870
5960	Tee, 1/4"	1 Plum	22	.364		10.60	21.50		32.10	45
5962	1/2"		20.20	.396		10.60	23.50		34.10	48
5964	3/4"		17.30	.462		10.80	27		37.80	54.50
5966	1"		15.20	.526		13.25	31		44.25	62.50
5968	1-1/4"		13.50	.593		28	35		63	85
5970	1-1/2"		12.10	.661		32	39		71	95.50
5972	2"	Q-1	20	.800		36	42.50		78.50	106
5974	2-1/2"		16.20	.988		90.50	52		142.50	182
5976	3"		13.90	1.151		90.50	61		151.50	195
5978	4"		11	1.455		121	77		198	253
5980	6"		6.70	2.388		315	126		441	540
5982	8"	Q-2	6.20	3.871		900	212		1,112	1,325
5990	Coupling, 1/4"	1 Plum	32	.250		12.30	14.70		27	36.50
5992	1/2"		30.30	.264		4.79	15.50		20.29	29.50
5994	3/4"		26	.308		6.70	18.05		24.75	35.50
5996	1"		22.70	.352		9	20.50		29.50	42.50
5998	1-1/4"		20.20	.396		13.50	23.50		37	51.50
6000	1-1/2"		18.20	.440		17	26		43	58.50
6002	2"	Q-1	33.10	.483		19.75	25.50		45.25	61.50
6004	2-1/2"		24.20	.661		44	35		79	103
6006	3"		20.80	.769		47.50	40.50		88	116
6008	4"		16.50	.970		62.50	51		113.50	149
6010	6"		10.10	1.584		147	83.50		230.50	293
6012	8"	Q-2	9.30	2.581		400	141		541	660
6200	CTS, 100 psi at 180°F, hot and cold water									
6230	90° Elbow, 1/2"	1 Plum	20	.400	Ea.	.48	23.50		23.98	37
6250	3/4"		19	.421		.80	24.50		25.30	39.50
6251	1"		16	.500		2.19	29.50		31.69	48.50
6252	1-1/4"		15	.533		4.65	31.50		36.15	54
6253	1-1/2"		14	.571		8.40	33.50		41.90	62
6254	2"	Q-1	23	.696		16.10	37		53.10	75
6260	45° Elbow, 1/2"	1 Plum	20	.400		.62	23.50		24.12	37
6280	3/4"		19	.421		1.06	24.50		25.56	39.50
6281	1"		16	.500		2.71	29.50		32.21	49
6282	1-1/4"		15	.533		5.75	31.50		37.25	55.50
6283	1-1/2"		14	.571		8.30	33.50		41.80	61.50
6284	2"	Q-1	23	.696		17.40	37		54.40	76.50
6290	Tee, 1/2"	1 Plum	13	.615		.62	36		36.62	57
6310	3/4"		12	.667		1.17	39		40.17	62.50
6311	1"		11	.727		6.10	42.50		48.60	73.50
6312	1-1/4"		10	.800		9.40	47		56.40	83.50
6313	1-1/2"		10	.800		12.25	47		59.25	86.50
6314	2"	Q-1	17	.941		19.80	49.50		69.30	99.50
6320	Coupling, 1/2"	1 Plum	22	.364		.40	21.50		21.90	34
6340	3/4"		21	.381		.53	22.50		23.03	35.50
6341	1"		18	.444		2.38	26		28.38	43
6342	1-1/4"		17	.471		3.07	27.50		30.57	46.50
6343	1-1/2"		16	.500		4.34	29.50		33.84	51
6344	2"	Q-1	28	.571		8.25	30		38.25	56
6360	Solvent cement for CPVC, commercial grade, per quart				Qt.	45			45	49.50
7340	PVC flange, slip-on, Sch 80 std., 1/2"	1 Plum	22	.364	Ea.	13.60	21.50		35.10	48.50

22 11 13 – Facility Water Distribution Piping

22 11 13.76 Pipe Fittings, Plastic		Crew	Daily Output	Labor-Hours	Unit	Material	2015 Bare Costs Labor	Equipment	Total	Total Incl O&P
7350	3/4"	1 Plum	21	.381	Ea.	14.50	22.50		37	51
7360	1"		18	.444		16.15	26		42.15	58.50
7370	1-1/4"		17	.471		16.65	27.50		44.15	61.50
7380	1-1/2"		16	.500		17	29.50		46.50	64.50
7390	2"	Q-1	26	.615		22.50	32.50		55	75.50
7400	2-1/2"		24	.667		35	35		70	93.50
7410	3"		18	.889		38.50	47		85.50	116
7420	4"		15	1.067		49	56.50		105.50	142
7430	6"		10	1.600		77	84.50		161.50	217
7440	8"	Q-2	11	2.182		138	120		258	335
7550	Union, schedule 40, socket joints, 1/2"	1 Plum	19	.421		5.20	24.50		29.70	44.50
7560	3/4"		18	.444		5.35	26		31.35	46.50
7570	1"		15	.533		5.55	31.50		37.05	55
7580	1-1/4"		14	.571		16.65	33.50		50.15	71
7590	1-1/2"		13	.615		18.60	36		54.60	77
7600	2"	Q-1	20	.800		25	42.50		67.50	93.50
7992	Polybutyl/polyethyl pipe, for copper fittings see Line 22 11 13.25 7000									
8000	Compression type, PVC, 160 psi cold water									
8010	Coupling, 3/4" CTS	1 Plum	21	.381	Ea.	3.48	22.50		25.98	39
8020	1" CTS		18	.444		4.32	26		30.32	45.50
8030	1-1/4" CTS		17	.471		6.10	27.50		33.60	49.50
8040	1-1/2" CTS		16	.500		8.30	29.50		37.80	55
8050	2" CTS		15	.533		11.65	31.50		43.15	62
8060	Female adapter, 3/4" FPT x 3/4" CTS		23	.348		5.70	20.50		26.20	38.50
8070	3/4" FPT x 1" CTS		21	.381		6.60	22.50		29.10	42.50
8080	1" FPT x 1" CTS		20	.400		6.60	23.50		30.10	44
8090	1-1/4" FPT x 1-1/4" CTS		18	.444		8.65	26		34.65	50
8100	1-1/2" FPT x 1-1/2" CTS		16	.500		9.90	29.50		39.40	57
8110	2" FPT x 2" CTS		13	.615		14.35	36		50.35	72.50
8130	Male adapter, 3/4" MPT x 3/4" CTS		23	.348		4.74	20.50		25.24	37
8140	3/4" MPT x 1" CTS		21	.381		5.65	22.50		28.15	41
8150	1" MPT x 1" CTS		20	.400		5.65	23.50		29.15	42.50
8160	1-1/4" MPT x 1-1/4" CTS		18	.444		7.60	26		33.60	49
8170	1-1/2" MPT x 1-1/2" CTS		16	.500		9.15	29.50		38.65	56
8180	2" MPT x 2" CTS		13	.615		11.85	36		47.85	69.50
8200	Spigot adapter, 3/4" IPS x 3/4" CTS		23	.348		2.70	20.50		23.20	35
8210	3/4" IPS x 1" CTS		21	.381		3.30	22.50		25.80	38.50
8220	1" IPS x 1" CTS		20	.400		3.30	23.50		26.80	40
8230	1-1/4" IPS x 1-1/4" CTS		18	.444		4.99	26		30.99	46
8240	1-1/2" IPS x 1-1/2" CTS		16	.500		5.25	29.50		34.75	52
8250	2" IPS x 2" CTS		13	.615		6.45	36		42.45	63.50
8270	Price includes insert stiffeners									
8280	250 psi is same price as 160 psi									
8300	Insert type, nylon, 160 & 250 psi, cold water									
8310	Clamp ring stainless steel, 3/4" IPS	1 Plum	115	.070	Ea.	2.62	4.08		6.70	9.25
8320	1" IPS		107	.075		2.67	4.39		7.06	9.80
8330	1-1/4" IPS		101	.079		2.69	4.65		7.34	10.20
8340	1-1/2" IPS		95	.084		3.61	4.94		8.55	11.65
8350	2" IPS		85	.094		4.15	5.50		9.65	13.15
8370	Coupling, 3/4" IPS		22	.364		.89	21.50		22.39	34.50
8380	1" IPS		19	.421		.92	24.50		25.42	39.50
8390	1-1/4" IPS		18	.444		1.37	26		27.37	42
8400	1-1/2" IPS		17	.471		1.62	27.50		29.12	45

22 11 13 – Facility Water Distribution Piping

22 11 13.76 Pipe Fittings, Plastic		Crew	Daily Output	Labor-Hours	Unit	Material	2015 Bare Costs Labor	Equipment	Total	Total Incl O&P
8410	2" IPS	1 Plum	16	.500	Ea.	3.60	29.50		33.10	50
8430	Elbow, 90°, 3/4" IPS		22	.364		1.77	21.50		23.27	35.50
8440	1" IPS		19	.421		1.96	24.50		26.46	40.50
8450	1-1/4" IPS		18	.444		2.19	26		28.19	43
8460	1-1/2" IPS		17	.471		2.58	27.50		30.08	46
8470	2" IPS		16	.500		3.60	29.50		33.10	50
8490	Male adapter, 3/4" IPS x 3/4" MPT		25	.320		.89	18.80		19.69	30.50
8500	1" IPS x 1" MPT		21	.381		.92	22.50		23.42	36
8510	1-1/4" IPS x 1-1/4" MPT		20	.400		1.45	23.50		24.95	38
8520	1-1/2" IPS x 1-1/2" MPT		18	.444		1.62	26		27.62	42.50
8530	2" IPS x 2" MPT		15	.533		3.11	31.50		34.61	52.50
8550	Tee, 3/4" IPS		14	.571		1.71	33.50		35.21	54.50
8560	1" IPS		13	.615		2.23	36		38.23	59
8570	1-1/4" IPS		12	.667		3.48	39		42.48	65
8580	1-1/2" IPS		11	.727		3.95	42.50		46.45	71
8590	2" IPS	↓	10	.800	↓	7.80	47		54.80	81.50
8610	Insert type, PVC, 100 psi @ 180°F, hot & cold water									
8620	Coupler, male, 3/8" CTS x 3/8" MPT	1 Plum	29	.276	Ea.	.72	16.20		16.92	26.50
8630	3/8" CTS x 1/2" MPT		28	.286		.72	16.75		17.47	27
8640	1/2" CTS x 1/2" MPT		27	.296		.72	17.40		18.12	28
8650	1/2" CTS x 3/4" MPT		26	.308		2.29	18.05		20.34	30.50
8660	3/4" CTS x 1/2" MPT		25	.320		2.07	18.80		20.87	32
8670	3/4" CTS x 3/4" MPT		25	.320		2.46	18.80		21.26	32
8700	Coupling, 3/8" CTS x 1/2" CTS		25	.320		4.58	18.80		23.38	34.50
8710	1/2" CTS		23	.348		5.50	20.50		26	38
8730	3/4" CTS		22	.364		11.60	21.50		33.10	46.50
8750	Elbow 90°, 3/8" CTS		25	.320		4	18.80		22.80	34
8760	1/2" CTS		23	.348		4.73	20.50		25.23	37
8770	3/4" CTS		22	.364		7.25	21.50		28.75	41.50
8800	Rings, crimp, copper, 3/8" CTS		120	.067		.22	3.91		4.13	6.35
8810	1/2" CTS		117	.068		.22	4.01		4.23	6.50
8820	3/4" CTS		115	.070		.29	4.08		4.37	6.65
8850	Reducer tee, bronze, 3/8" x 3/8" x 1/2" CTS		17	.471		4.35	27.50		31.85	48
8860	1/2" x 1/2" x 3/4" CTS		15	.533		5.55	31.50		37.05	55
8870	3/4" x 1/2" x 1/2" CTS		14	.571		5.55	33.50		39.05	58.50
8890	3/4" x 3/4" x 1/2" CTS		14	.571		5.65	33.50		39.15	59
8900	1" x 1/2" x 1/2" CTS		14	.571		9.35	33.50		42.85	63
8930	Tee, 3/8" CTS		17	.471		3.31	27.50		30.81	46.50
8940	1/2" CTS		15	.533		2.27	31.50		33.77	51.50
8950	3/4" CTS	↓	14	.571	↓	3.50	33.50		37	56.50
8960	Copper rings included in fitting price									
9000	Flare type, assembled, acetal, hot & cold water									
9010	Coupling, 1/4" & 3/8" CTS	1 Plum	24	.333	Ea.	3.36	19.55		22.91	34
9020	1/2" CTS		22	.364		3.84	21.50		25.34	37.50
9030	3/4" CTS		21	.381		5.70	22.50		28.20	41.50
9040	1" CTS		18	.444		7.25	26		33.25	48.50
9050	Elbow 90°, 1/4" CTS		26	.308		3.73	18.05		21.78	32
9060	3/8" CTS		24	.333		4	19.55		23.55	35
9070	1/2" CTS		22	.364		4.73	21.50		26.23	38.50
9080	3/4" CTS		21	.381		7.25	22.50		29.75	43
9090	1" CTS		18	.444		9.15	26		35.15	50.50
9110	Tee, 1/4" CTS		16	.500		4.09	29.50		33.59	50.50
9114	3/8" CTS	↓	15	.533	↓	4.13	31.50		35.63	53.50

22 11 13.76 Pipe Fittings, Plastic		Crew	Daily Output	Labor-Hours	Unit	Material	2015 Bare Costs Labor	Equipment	Total	Total Incl O&P
9120	1/2" CTS	1 Plum	14	.571	Ea.	5.25	33.50		38.75	58.50
9130	3/4" CTS		13	.615		8.25	36		44.25	65.50
9140	1" CTS		12	.667		11.05	39		50.05	73
9400	Polypropylene, fittings and accessories									
9404	Fittings fusion welded, sizes are I.D.									
9408	Note: sizes 1/2" thru 4" use socket fusion									
9410	Sizes 6" thru 10" use butt fusion									
9416	Coupling									
9420	3/8"	1 Plum	39	.205	Ea.	.88	12.05		12.93	19.75
9422	1/2"		37.40	.214		1.15	12.55		13.70	21
9424	3/4"		35.40	.226		1.28	13.25		14.53	22
9426	1"		29.70	.269		1.69	15.80		17.49	26.50
9428	1-1/4"		27.60	.290		2.02	17		19.02	28.50
9430	1-1/2"		24.80	.323		4.25	18.95		23.20	34
9432	2"	Q-1	43	.372		8.50	19.65		28.15	40
9434	2-1/2"		35.60	.449		9.50	23.50		33	47.50
9436	3"		30.90	.518		21	27.50		48.50	65.50
9438	3-1/2"		27.80	.576		34	30.50		64.50	85
9440	4"		25	.640		44.50	34		78.50	102
9442	Reducing coupling, female to female									
9446	2" to 1-1/2"	Q-1	49	.327	Ea.	12.25	17.25		29.50	40.50
9448	2-1/2" to 2"		41.20	.388		13.45	20.50		33.95	47
9450	3" to 2-1/2"		33.10	.483		18.15	25.50		43.65	60
9470	Reducing bushing, female to female									
9472	1/2" to 3/8"	1 Plum	38.20	.209	Ea.	1.15	12.30		13.45	20.50
9474	3/4" to 3/8" or 1/2"		36.50	.219		1.28	12.85		14.13	21.50
9476	1" to 3/4" or 1/2"		33.10	.242		1.71	14.20		15.91	24
9478	1-1/4" to 3/4" or 1"		28.70	.279		2.63	16.35		18.98	28.50
9480	1-1/2" to 1/2" thru 1-1/4"		26.20	.305		4.36	17.90		22.26	33
9482	2" to 1/2" thru 1-1/2"	Q-1	43	.372		8.70	19.65		28.35	40
9484	2-1/2" to 1/2" thru 2"		41.20	.388		9.75	20.50		30.25	42.50
9486	3" to 1-1/2" thru 2-1/2"		33.10	.483		21.50	25.50		47	64
9488	3-1/2" to 2" thru 3"		29.20	.548		34.50	29		63.50	83
9490	4" to 2-1/2" thru 3-1/2"		26.20	.611		54	32.50		86.50	110
9491	6" to 4" SDR 7.4		16.50	.970		65	51		116	152
9492	6" to 4" SDR 11		16.50	.970		65	51		116	152
9493	8" to 6" SDR 7.4		10.10	1.584		95.50	83.50		179	236
9494	8" to 6" SDR 11		10.10	1.584		87.50	83.50		171	228
9495	10" to 8" SDR 7.4		7.80	2.051		131	108		239	315
9496	10" to 8" SDR 11		7.80	2.051		114	108		222	295
9500	90° Elbow									
9504	3/8"	1 Plum	39	.205	Ea.	.97	12.05		13.02	19.85
9506	1/2"		37.40	.214		1.23	12.55		13.78	21
9508	3/4"		35.40	.226		1.58	13.25		14.83	22
9510	1"		29.70	.269		2.28	15.80		18.08	27
9512	1-1/4"		27.60	.290		3.51	17		20.51	30.50
9514	1-1/2"		24.80	.323		7.55	18.95		26.50	38
9516	2"	Q-1	43	.372		11.60	19.65		31.25	43.50
9518	2-1/2"		35.60	.449		26	23.50		49.50	65.50
9520	3"		30.90	.518		43	27.50		70.50	89.50
9522	3-1/2"		27.80	.576		61	30.50		91.50	115
9524	4"		25	.640		93.50	34		127.50	156
9526	6" SDR 7.4		5.55	2.883		107	152		259	355

22 11 13 – Facility Water Distribution Piping

22 11 13.76 Pipe Fittings, Plastic		Crew	Daily Output	Labor-Hours	Unit	Material	Labor	Equipment	Total	Total Incl O&P
						2015 Bare Costs				
9528	6" SDR 11	Q-1	5.55	2.883	Ea.	83.50	152		235.50	330
9530	8" SDR 7.4	Q-2	8.10	2.963		415	162		577	715
9532	8" SDR 11		8.10	2.963		355	162		517	645
9534	10" SDR 7.4		7.50	3.200		635	175		810	975
9536	10" SDR 11		7.50	3.200		575	175		750	910
9551	45° Elbow									
9554	3/8"	1 Plum	39	.205	Ea.	.94	12.05		12.99	19.85
9556	1/2"		37.40	.214		1.23	12.55		13.78	21
9558	3/4"		35.40	.226		1.58	13.25		14.83	22
9564	1"		29.70	.269		2.28	15.80		18.08	27
9566	1-1/4"		27.60	.290		3.51	17		20.51	30.50
9568	1-1/2"		24.80	.323		7.55	18.95		26.50	38
9570	2"	Q-1	43	.372		11.45	19.65		31.10	43
9572	2-1/2"		35.60	.449		25.50	23.50		49	65
9574	3"		30.90	.518		47	27.50		74.50	94
9576	3-1/2"		27.80	.576		67	30.50		97.50	121
9578	4"		25	.640		103	34		137	166
9580	6" SDR 7.4		5.55	2.883		119	152		271	370
9582	6" SDR 11		5.55	2.883		98	152		250	345
9584	8" SDR 7.4	Q-2	8.10	2.963		325	162		487	615
9586	8" SDR 11		8.10	2.963		284	162		446	565
9588	10" SDR 7.4		7.50	3.200		525	175		700	850
9590	10" SDR 11		7.50	3.200		435	175		610	750
9600	Tee									
9604	3/8"	1 Plum	26	.308	Ea.	1.22	18.05		19.27	29.50
9606	1/2"		24.90	.321		1.66	18.85		20.51	31.50
9608	3/4"		23.70	.338		2.28	19.80		22.08	33.50
9610	1"		19.90	.402		2.90	23.50		26.40	40
9612	1-1/4"		18.50	.432		4.43	25.50		29.93	44.50
9614	1-1/2"		16.60	.482		12.65	28.50		41.15	58
9616	2"	Q-1	26.80	.597		18.10	31.50		49.60	69
9618	2-1/2"		23.80	.672		30.50	35.50		66	89
9620	3"		20.60	.777		50	41		91	119
9622	3-1/2"		18.40	.870		78	46		124	158
9624	4"		16.70	.958		104	50.50		154.50	193
9626	6" SDR 7.4		3.70	4.324		129	228		357	495
9628	6" SDR 11		3.70	4.324		98	228		326	465
9630	8" SDR 7.4	Q-2	5.40	4.444		355	243		598	770
9632	8" SDR 11		5.40	4.444		310	243		553	725
9634	10" SDR 7.4		5	4.800		610	263		873	1,075
9636	10" SDR 11		5	4.800		530	263		793	990
9638	For reducing tee use same tee price									
9660	End cap									
9662	3/8"	1 Plum	78	.103	Ea.	1.22	6		7.22	10.75
9664	1/2"		74.60	.107		1.80	6.30		8.10	11.80
9666	3/4"		71.40	.112		2.28	6.60		8.88	12.75
9668	1"		59.50	.134		2.76	7.90		10.66	15.35
9670	1-1/4"		55.60	.144		4.36	8.45		12.81	17.95
9672	1-1/2"		49.50	.162		6	9.50		15.50	21.50
9674	2"	Q-1	80	.200		10.05	10.55		20.60	27.50
9676	2-1/2"		71.40	.224		14.55	11.85		26.40	34.50
9678	3"		61.70	.259		33	13.70		46.70	57.50
9680	3-1/2"		55.20	.290		39.50	15.30		54.80	67.50

22 11 13 – Facility Water Distribution Piping

22 11 13.76 Pipe Fittings, Plastic		Crew	Daily Output	Labor-Hours	Unit	Material	Labor	Equipment	Total	Total Incl O&P
						2015 Bare Costs				
9682	4"	Q-1	50	.320	Ea.	60.50	16.90		77.40	93
9684	6" SDR 7.4		26.50	.604		83	32		115	141
9686	6" SDR 11		26.50	.604		83	32		115	141
9688	8" SDR 7.4	Q-2	16.30	1.472		83	80.50		163.50	218
9690	8" SDR 11		16.30	1.472		72	80.50		152.50	205
9692	10" SDR 7.4		14.90	1.611		124	88		212	275
9694	10" SDR 11		14.90	1.611		106	88		194	254
9800	Accessories and tools									
9802	Pipe clamps for suspension, not including rod or beam clamp									
9804	3/8"	1 Plum	74	.108	Ea.	2.13	6.35		8.48	12.25
9805	1/2"		70	.114		2.67	6.70		9.37	13.40
9806	3/4"		68	.118		3.07	6.90		9.97	14.15
9807	1"		66	.121		3.35	7.10		10.45	14.80
9808	1-1/4"		64	.125		3.44	7.35		10.79	15.25
9809	1-1/2"		62	.129		3.76	7.55		11.31	15.95
9810	2"	Q-1	110	.145		4.66	7.70		12.36	17.15
9811	2-1/2"		104	.154		5.95	8.15		14.10	19.25
9812	3"		98	.163		6.45	8.65		15.10	20.50
9813	3-1/2"		92	.174		7.05	9.20		16.25	22
9814	4"		86	.186		7.90	9.85		17.75	24
9815	6"		70	.229		9.65	12.10		21.75	29.50
9816	8"	Q-2	100	.240		34.50	13.15		47.65	58.50
9817	10"	"	94	.255		39.50	14		53.50	65.50
9820	Pipe cutter									
9822	For 3/8" thru 1-1/4"				Ea.	118			118	129
9824	For 1-1/2" thru 4"				"	310			310	340
9826	Note: Pipes may be cut with standard									
9827	iron saw with blades for plastic.									
9982	For plastic hangers see Line 22 05 29.10 8000									
9986	For copper/brass fittings see Line 22 11 13.25 7000									
9990	Minimum labor/equipment charge	1 Plum	4	2	Job		117		117	183

22 11 19 – Domestic Water Piping Specialties

22 11 19.10 Flexible Connectors		Crew	Daily Output	Labor-Hours	Unit	Material	Labor	Equipment	Total	Total Incl O&P
0010	**FLEXIBLE CONNECTORS**, Corrugated, 7/8" O.D., 1/2" I.D.									
0050	Gas, seamless brass, steel fittings									
0200	12" long	1 Plum	36	.222	Ea.	17.40	13.05		30.45	39.50
0220	18" long		36	.222		21.50	13.05		34.55	44.50
0240	24" long		34	.235		25.50	13.80		39.30	49.50
0260	30" long		34	.235		27.50	13.80		41.30	52
0280	36" long		32	.250		30.50	14.70		45.20	56.50
0320	48" long		30	.267		38.50	15.65		54.15	67
0340	60" long		30	.267		46	15.65		61.65	75
0360	72" long		30	.267		53	15.65		68.65	83
2000	Water, copper tubing, dielectric separators									
2100	12" long	1 Plum	36	.222	Ea.	17.35	13.05		30.40	39.50
2220	15" long		36	.222		19.30	13.05		32.35	41.50
2240	18" long		36	.222		21	13.05		34.05	43.50
2260	24" long		34	.235		26	13.80		39.80	50
9000	Minimum labor/equipment charge		4	2	Job		117		117	183

22 11 19 – Domestic Water Piping Specialties

22 11 19.14 Flexible Metal Hose		Crew	Daily Output	Labor-Hours	Unit	Material	2015 Bare Costs Labor	Equipment	Total	Total Incl O&P
0010	**FLEXIBLE METAL HOSE**, Connectors, standard lengths									
0100	Bronze braided, bronze ends									
0120	3/8" diameter x 12"	1 Stpi	26	.308	Ea.	21	18.40		39.40	51.50
0140	1/2" diameter x 12"		24	.333		22	19.90		41.90	55
0160	3/4" diameter x 12"		20	.400		32	24		56	72.50
0180	1" diameter x 18"		19	.421		38	25		63	81
0200	1-1/2" diameter x 18"		13	.615		52	37		89	115
0220	2" diameter x 18"		11	.727		68	43.50		111.50	143
1000	Carbon steel ends									
1020	1/4" diameter x 12"	1 Stpi	28	.286	Ea.	15.50	17.05		32.55	43.50
1040	3/8" diameter x 12"		26	.308		15.90	18.40		34.30	46
1060	1/2" diameter x 12"		24	.333		17.55	19.90		37.45	50.50
1080	1/2" diameter x 24"		24	.333		47.50	19.90		67.40	83
1120	3/4" diameter x 12"		20	.400		28.50	24		52.50	69
1140	3/4" diameter x 24"		20	.400		58.50	24		82.50	102
1160	3/4" diameter x 36"		20	.400		69.50	24		93.50	114
1180	1" diameter x 18"		19	.421		35.50	25		60.50	78
1200	1" diameter x 30"		19	.421		67.50	25		92.50	113
1220	1" diameter x 36"		19	.421		87.50	25		112.50	136
1240	1-1/4" diameter x 18"		15	.533		49.50	32		81.50	104
1260	1-1/4" diameter x 36"		15	.533		109	32		141	170
1280	1-1/2" diameter x 18"		13	.615		69.50	37		106.50	134
1300	1-1/2" diameter x 36"		13	.615		118	37		155	188
1320	2" diameter x 24"		11	.727		96.50	43.50		140	174
1340	2" diameter x 36"		11	.727		146	43.50		189.50	229
1360	2-1/2" diameter x 24"		9	.889		221	53		274	325
1380	2-1/2" diameter x 36"		9	.889		560	53		613	700
1400	3" diameter x 24"		7	1.143		305	68.50		373.50	440
1420	3" diameter x 36"		7	1.143		670	68.50		738.50	840
2000	Carbon steel braid, carbon steel solid ends									
2100	1/2" diameter x 12"	1 Stpi	24	.333	Ea.	42.50	19.90		62.40	78
2120	3/4" diameter x 12"		20	.400		64.50	24		88.50	109
2140	1" diameter x 12"		19	.421		92	25		117	140
2160	1-1/4" diameter x 12"		15	.533		28	32		60	80.50
2180	1-1/2" diameter x 12"		13	.615		30	37		67	90.50
3000	Stainless steel braid, welded on carbon steel ends									
3100	1/2" diameter x 12"	1 Stpi	24	.333	Ea.	59	19.90		78.90	96
3120	3/4" diameter x 12"		20	.400		71	24		95	116
3140	3/4" diameter x 24"		20	.400		82	24		106	128
3160	3/4" diameter x 36"		20	.400		94	24		118	142
3180	1" diameter x 12"		19	.421		84.50	25		109.50	132
3200	1" diameter x 24"		19	.421		102	25		127	151
3220	1" diameter x 36"		19	.421		117	25		142	168
3240	1-1/4" diameter x 12"		15	.533		128	32		160	190
3260	1-1/4" diameter x 24"		15	.533		139	32		171	203
3280	1-1/4" diameter x 36"		15	.533		158	32		190	224
3300	1-1/2" diameter x 12"		13	.615		44	37		81	106
3320	1-1/2" diameter x 24"		13	.615		152	37		189	225
3340	1-1/2" diameter x 36"		13	.615		178	37		215	254
3400	Metal stainless steel braid, over corrugated stainless steel, flanged ends									
3410	150 PSI									
3420	1/2" diameter x 12"	1 Stpi	24	.333	Ea.	120	19.90		139.90	163

22 11 19 – Domestic Water Piping Specialties

22 11 19.14 Flexible Metal Hose		Crew	Daily Output	Labor-Hours	Unit	Material	Labor	Equipment	Total	Total Incl O&P
						2015 Bare Costs				
3430	1" diameter x 12"	1 Stpi	20	.400	Ea.	180	24		204	236
3440	1-1/2" diameter x 12"		15	.533		204	32		236	274
3450	2-1/2" diameter x 9"		12	.667		108	40		148	181
3460	3" diameter x 9"		9	.889		93	53		146	185
3470	4" diameter x 9"		7	1.143		113	68.50		181.50	231
3480	4" diameter x 30"		5	1.600		560	95.50		655.50	765
3490	4" diameter x 36"		4.80	1.667		580	99.50		679.50	795
3500	6" diameter x 11"		5	1.600		188	95.50		283.50	355
3510	6" diameter x 36"		3.80	2.105		685	126		811	945
3520	8" diameter x 12"		4	2		410	120		530	635
3530	10" diameter x 13"		3	2.667		625	159		784	940
3540	12" diameter x 14"	Q-5	4	4		950	215		1,165	1,375

22 11 19.18 Mixing Valve		Crew	Daily Output	Labor-Hours	Unit	Material	Labor	Equipment	Total	Total Incl O&P
0010	**MIXING VALVE**, Automatic, water tempering.									
0040	1/2" size	1 Stpi	19	.421	Ea.	555	25		580	650
0050	3/4" size		18	.444		555	26.50		581.50	650
0100	1" size		16	.500		845	30		875	975
0120	1-1/4" size		13	.615		1,150	37		1,187	1,325
0140	1-1/2" size		10	.800		1,400	48		1,448	1,600
0160	2" size		8	1		1,750	60		1,810	2,025
0170	2-1/2" size		6	1.333		1,750	79.50		1,829.50	2,050
0180	3" size		4	2		4,125	120		4,245	4,725
0190	4" size		3	2.667		4,125	159		4,284	4,800
9000	Minimum labor/equipment charge		5	1.600	Job		95.50		95.50	149

22 11 19.22 Pressure Reducing Valve		Crew	Daily Output	Labor-Hours	Unit	Material	Labor	Equipment	Total	Total Incl O&P
0010	**PRESSURE REDUCING VALVE**, Steam, pilot operated.									
0100	Threaded, iron body									
0200	1-1/2" size	1 Stpi	8	1	Ea.	1,825	60		1,885	2,100
0220	2" size	"	5	1.600	"	2,100	95.50		2,195.50	2,450
1000	Flanged, iron body, 125 lb. flanges									
1020	2" size	1 Stpi	8	1	Ea.	2,100	60		2,160	2,425
1040	2-1/2" size	"	4	2		2,525	120		2,645	2,975
1060	3" size	Q-5	4.50	3.556		3,075	191		3,266	3,675
1080	4" size	"	3	5.333		4,425	287		4,712	5,325
1500	For 250 lb. flanges, add					5%				

22 11 19.26 Pressure Regulators		Crew	Daily Output	Labor-Hours	Unit	Material	Labor	Equipment	Total	Total Incl O&P
0010	**PRESSURE REGULATORS**									
0200	Oil, light, hot water, ordinary steam, threaded									
0220	Bronze body, 1/4" size	1 Stpi	24	.333	Ea.	154	19.90		173.90	200
0230	3/8" size		24	.333		166	19.90		185.90	214
0240	1/2" size		24	.333		206	19.90		225.90	258
0250	3/4" size		20	.400		246	24		270	310
0260	1" size		19	.421		375	25		400	450
0270	1-1/4" size		15	.533		505	32		537	605
0280	1-1/2" size		13	.615		540	37		577	655
0290	2" size		11	.727		985	43.50		1,028.50	1,150
0320	Iron body, 1/4" size		24	.333		120	19.90		139.90	163
0330	3/8" size		24	.333		141	19.90		160.90	186
0340	1/2" size		24	.333		150	19.90		169.90	196
0350	3/4" size		20	.400		182	24		206	238
0360	1" size		19	.421		237	25		262	300
0370	1-1/4" size		15	.533		330	32		362	410

22 11 19 – Domestic Water Piping Specialties

22 11 19.26 Pressure Regulators		Crew	Daily Output	Labor-Hours	Unit	Material	Labor	Equipment	Total	Total Incl O&P
						2015 Bare Costs				
0380	1-1/2" size	1 Stpi	13	.615	Ea.	365	37		402	465
0390	2" size	↓	11	.727	↓	530	43.50		573.50	655
0500	Oil, heavy, viscous fluids, threaded									
0520	Bronze body, 3/8" size	1 Stpi	24	.333	Ea.	264	19.90		283.90	320
0530	1/2" size		24	.333		340	19.90		359.90	405
0540	3/4" size		20	.400		380	24		404	455
0550	1" size		19	.421		470	25		495	560
0560	1-1/4" size		15	.533		665	32		697	785
0570	1-1/2" size		13	.615		760	37		797	900
0600	Iron body, 3/8" size		24	.333		210	19.90		229.90	262
0620	1/2" size		24	.333		255	19.90		274.90	310
0630	3/4" size		20	.400		289	24		313	360
0640	1" size		19	.421		340	25		365	415
0650	1-1/4" size		15	.533		495	32		527	595
0660	1-1/2" size	↓	13	.615	↓	530	37		567	640
0800	Process steam, wet or super heated, monel trim, threaded									
0820	Bronze body, 1/4" size	1 Stpi	24	.333	Ea.	600	19.90		619.90	690
0830	3/8" size		24	.333		600	19.90		619.90	690
0840	1/2" size		24	.333		690	19.90		709.90	790
0850	3/4" size		20	.400		810	24		834	935
0860	1" size		19	.421		1,025	25		1,050	1,175
0870	1-1/4" size		15	.533		1,275	32		1,307	1,450
0880	1-1/2" size		13	.615		1,525	37		1,562	1,725
0920	Iron body, max 125 PSIG press out, 1/4" size		24	.333		795	19.90		814.90	905
0930	3/8" size		24	.333		795	19.90		814.90	905
0940	1/2" size		24	.333		990	19.90		1,009.90	1,125
0950	3/4" size		20	.400		1,200	24		1,224	1,375
0960	1" size		19	.421		1,525	25		1,550	1,725
0970	1-1/4" size		15	.533		1,850	32		1,882	2,075
0980	1-1/2" size	↓	13	.615	↓	2,225	37		2,262	2,475
3000	Steam, high capacity, bronze body, stainless steel trim									
3020	Threaded, 1/2" diameter	1 Stpi	24	.333	Ea.	1,975	19.90		1,994.90	2,200
3030	3/4" diameter		24	.333		2,150	19.90		2,169.90	2,375
3040	1" diameter		19	.421		2,400	25		2,425	2,700
3060	1-1/4" diameter		15	.533		2,500	32		2,532	2,800
3080	1-1/2" diameter		13	.615		3,025	37		3,062	3,375
3100	2" diameter	↓	11	.727		3,725	43.50		3,768.50	4,150
3120	2-1/2" diameter	Q-5	12	1.333		4,650	71.50		4,721.50	5,225
3140	3" diameter	"	11	1.455	↓	5,075	78		5,153	5,700
3500	Flanged connection, iron body, 125 lb. W.S.P.									
3520	3" diameter	Q-5	11	1.455	Ea.	5,800	78		5,878	6,525
3540	4" diameter	"	5	3.200	"	7,325	172		7,497	8,325
9002	For water pressure regulators, see Section 22 05 23.20									

22 11 19.30 Pressure and Temperature Safety Plug		Crew	Daily Output	Labor-Hours	Unit	Material	Labor	Equipment	Total	Total Incl O&P
0010	PRESSURE & TEMPERATURE SAFETY PLUG									
1000	3/4" external thread, 3/8" diam. element									
1020	Carbon steel									
1050	7-1/2" insertion	1 Stpi	32	.250	Ea.	45	14.95		59.95	73
1120	304 stainless steel									
1150	7-1/2" insertion	1 Stpi	32	.250	Ea.	50.50	14.95		65.45	79
1220	316 Stainless steel									
1250	7-1/2" insertion	1 Stpi	32	.250	Ea.	69.50	14.95		84.45	100

22 11 19 – Domestic Water Piping Specialties

22 11 19.32 Pressure and Temperature Measurement Plug		Crew	Daily Output	Labor-Hours	Unit	Material	2015 Bare Costs Labor	Equipment	Total	Total Incl O&P
0010	**PRESSURE & TEMPERATURE MEASUREMENT PLUG**									
0020	A permanent access port for insertion of									
0030	a pressure or temperature measuring probe									
0100	Plug, brass									
0110	1/4" MNPT, 1-1/2" long	1 Stpi	32	.250	Ea.	5.40	14.95		20.35	29.50
0120	3" long		31	.258		10.45	15.40		25.85	35.50
0140	1/2" MNPT, 1-1/2" long		30	.267		8.65	15.95		24.60	34.50
0150	3" long		29	.276		12.90	16.50		29.40	39.50
0200	Pressure gauge probe adapter									
0210	1/8" diameter, 1-1/2" probe				Ea.	26			26	28.50
0220	3" probe				"	53			53	58
0300	Temperature gauge, 5" stem									
0310	Analog				Ea.	69.50			69.50	76.50
0330	Digital				"	75.50			75.50	83
0400	Pressure gauge, compound									
0410	1/4" MNPT				Ea.	30.50			30.50	33.50
0500	Pressure and temperature test kit									
0510	Contains 2 thermometers and 2 pressure gauges									
0520	Kit				Ea.	355			355	390

22 11 19.34 Sleeves and Escutcheons

		Crew	Daily Output	Labor-Hours	Unit	Material	Labor	Equipment	Total	Total Incl O&P
0010	**SLEEVES & ESCUTCHEONS**									
0100	Pipe sleeve									
0110	Steel, w/water stop, 12" long, with link seal									
0120	2" diam. for 1/2" carrier pipe	1 Plum	8.40	.952	Ea.	59.50	56		115.50	153
0130	2-1/2" diam. for 3/4" carrier pipe		8	1		67	58.50		125.50	166
0140	2-1/2" diam. for 1" carrier pipe		8	1		63.50	58.50		122	162
0150	3" diam. for 1-1/4" carrier pipe		7.20	1.111		81.50	65		146.50	192
0160	3-1/2" diam. for 1-1/2" carrier pipe		6.80	1.176		82	69		151	198
0170	4" diam. for 2" carrier pipe		6	1.333		88.50	78.50		167	220
0180	4" diam. for 2-1/2" carrier pipe		6	1.333		89.50	78.50		168	221
0190	5" diam. for 3" carrier pipe		5.40	1.481		102	87		189	248
0200	6" diam. for 4" carrier pipe		4.80	1.667		122	98		220	288
0210	10" diam. for 6" carrier pipe	Q-1	8	2		244	106		350	435
0220	12" diam. for 8" carrier pipe		7.20	2.222		325	117		442	545
0230	14" diam. for 10" carrier pipe		6.40	2.500		345	132		477	580
0240	16" diam. for 12" carrier pipe		5.80	2.759		380	146		526	645
0250	18" diam. for 14" carrier pipe		5.20	3.077		705	163		868	1,025
0260	24" diam. for 18" carrier pipe		4	4		1,050	211		1,261	1,475
0270	24" diam. for 20" carrier pipe		4	4		885	211		1,096	1,300
0280	30" diam. for 24" carrier pipe		3.20	5		1,300	264		1,564	1,825
0500	Wall sleeve									
0510	Ductile iron with rubber gasket seal									
0520	3"	1 Plum	8.40	.952	Ea.	700	56		756	855
0530	4"		7.20	1.111		750	65		815	925
0540	6"		6	1.333		920	78.50		998.50	1,150
0550	8"		4	2		1,125	117		1,242	1,425
0560	10"		3	2.667		1,375	157		1,532	1,775
0570	12"		2.40	3.333		1,625	196		1,821	2,100
5000	Escutcheon									
5100	Split ring, pipe									
5110	Chrome plated									
5120	1/2"	1 Plum	160	.050	Ea.	1.09	2.94		4.03	5.80

22 11 19 – Domestic Water Piping Specialties

22 11 19.34 Sleeves and Escutcheons		Crew	Daily Output	Labor-Hours	Unit	2015 Bare Costs Material	Labor	Equipment	Total	Total Incl O&P
5130	3/4"	1 Plum	160	.050	Ea.	1.21	2.94		4.15	5.90
5140	1"		135	.059		1.30	3.48		4.78	6.90
5150	1-1/2"		115	.070		1.83	4.08		5.91	8.35
5160	2"		100	.080		2.03	4.70		6.73	9.55
5170	4"		80	.100		4.50	5.85		10.35	14.10
5180	6"		68	.118		7.50	6.90		14.40	19
5400	Shallow flange type									
5410	Chrome plated steel									
5420	1/2" CTS	1 Plum	180	.044	Ea.	.32	2.61		2.93	4.42
5430	3/4" CTS		180	.044		.34	2.61		2.95	4.44
5440	1/2" IPS		180	.044		.36	2.61		2.97	4.47
5450	3/4" IPS		180	.044		.40	2.61		3.01	4.51
5460	1" IPS		175	.046		.46	2.68		3.14	4.69
5470	1-1/2" IPS		170	.047		.76	2.76		3.52	5.15
5480	2" IPS		160	.050		.91	2.94		3.85	5.60

22 11 19.38 Water Supply Meters

		Crew	Daily Output	Labor-Hours	Unit	Material	Labor	Equipment	Total	Total Incl O&P
0010	**WATER SUPPLY METERS**									
1000	Detector, serves dual systems such as fire and domestic or									
1020	process water, wide range cap., UL and FM approved									
1100	3" mainline x 2" by-pass, 400 GPM	Q-1	3.60	4.444	Ea.	7,450	235		7,685	8,550
1140	4" mainline x 2" by-pass, 700 GPM	"	2.50	6.400		7,450	340		7,790	8,700
1180	6" mainline x 3" by-pass, 1600 GPM	Q-2	2.60	9.231		11,400	505		11,905	13,300
1220	8" mainline x 4" by-pass, 2800 GPM		2.10	11.429		16,900	625		17,525	19,500
1260	10" mainline x 6" by-pass, 4400 GPM		2	12		24,100	655		24,755	27,500
1300	10" x 12" mainlines x 6" by-pass, 5400 GPM		1.70	14.118		32,700	775		33,475	37,200
2000	Domestic/commercial, bronze									
2020	Threaded									
2060	5/8" diameter, to 20 GPM	1 Plum	16	.500	Ea.	50	29.50		79.50	101
2080	3/4" diameter, to 30 GPM		14	.571		91	33.50		124.50	153
2100	1" diameter, to 50 GPM		12	.667		138	39		177	213
2300	Threaded/flanged									
2340	1-1/2" diameter, to 100 GPM	1 Plum	8	1	Ea.	340	58.50		398.50	460
2360	2" diameter, to 160 GPM	"	6	1.333	"	460	78.50		538.50	625
2600	Flanged, compound									
2640	3" diameter, 320 GPM	Q-1	3	5.333	Ea.	3,125	282		3,407	3,875
2660	4" diameter, to 500 GPM		1.50	10.667		5,000	565		5,565	6,375
2680	6" diameter, to 1,000 GPM		1	16		7,975	845		8,820	10,100
2700	8" diameter, to 1,800 GPM		.80	20		12,500	1,050		13,550	15,400
7000	Turbine									
7260	Flanged									
7300	2" diameter, to 160 GPM	1 Plum	7	1.143	Ea.	610	67		677	775
7320	3" diameter, to 450 GPM	Q-1	3.60	4.444		1,275	235		1,510	1,775
7340	4" diameter, to 650 GPM	"	2.50	6.400		2,100	340		2,440	2,850
7360	6" diameter, to 1800 GPM	Q-2	2.60	9.231		3,775	505		4,280	4,950
7380	8" diameter, to 2500 GPM		2.10	11.429		5,975	625		6,600	7,550
7400	10" diameter, to 5500 GPM		1.70	14.118		8,075	775		8,850	10,100
9000	Minimum labor/equipment charge	1 Plum	3.25	2.462	Job		144		144	225

22 11 19 – Domestic Water Piping Specialties

22 11 19.42 Backflow Preventers		Crew	Daily Output	Labor-Hours	Unit	Material	2015 Bare Costs Labor	Equipment	Total	Total Incl O&P
0010	**BACKFLOW PREVENTERS**, Includes valves									
0020	and four test cocks, corrosion resistant, automatic operation									
1000	Double check principle									
1010	Threaded, with ball valves									
1020	3/4" pipe size	1 Plum	16	.500	Ea.	231	29.50		260.50	300
1030	1" pipe size		14	.571		264	33.50		297.50	345
1040	1-1/2" pipe size		10	.800		560	47		607	695
1050	2" pipe size		7	1.143		655	67		722	825
1080	Threaded, with gate valves									
1100	3/4" pipe size	1 Plum	16	.500	Ea.	1,075	29.50		1,104.50	1,250
1120	1" pipe size		14	.571		1,100	33.50		1,133.50	1,250
1140	1-1/2" pipe size		10	.800		1,400	47		1,447	1,625
1160	2" pipe size		7	1.143		1,725	67		1,792	2,000
1300	Flanged, valves are OS&Y									
1380	3" pipe size	Q-1	4.50	3.556	Ea.	2,875	188		3,063	3,450
1400	4" pipe size	"	3	5.333		3,475	282		3,757	4,275
1420	6" pipe size	Q-2	3	8		5,450	440		5,890	6,675
4000	Reduced pressure principle									
4100	Threaded, bronze, valves are ball									
4120	3/4" pipe size	1 Plum	16	.500	Ea.	445	29.50		474.50	535
4140	1" pipe size		14	.571		480	33.50		513.50	580
4150	1-1/4" pipe size		12	.667		845	39		884	990
4160	1-1/2" pipe size		10	.800		960	47		1,007	1,125
4180	2" pipe size		7	1.143		1,075	67		1,142	1,300
5000	Flanged, bronze, valves are OS&Y									
5060	2-1/2" pipe size	Q-1	5	3.200	Ea.	4,025	169		4,194	4,700
5080	3" pipe size		4.50	3.556		4,650	188		4,838	5,400
5100	4" pipe size		3	5.333		5,500	282		5,782	6,500
5120	6" pipe size	Q-2	3	8		8,750	440		9,190	10,300
5600	Flanged, iron, valves are OS&Y									
5660	2-1/2" pipe size	Q-1	5	3.200	Ea.	3,025	169		3,194	3,600
5680	3" pipe size		4.50	3.556		3,175	188		3,363	3,800
5700	4" pipe size		3	5.333		3,975	282		4,257	4,825
5720	6" pipe size	Q-2	3	8		5,775	440		6,215	7,025
5740	8" pipe size		2	12		10,100	655		10,755	12,100
5760	10" pipe size		1	24		13,600	1,325		14,925	17,000
9010	Minimum labor/equipment charge	1 Plum	2	4	Job		235		235	365

22 11 19.50 Vacuum Breakers		Crew	Daily Output	Labor-Hours	Unit	Material	Labor	Equipment	Total	Total Incl O&P
0010	**VACUUM BREAKERS**									
0013	See also backflow preventers Section 22 11 19.42									
1000	Anti-siphon continuous pressure type									
1010	Max. 150 PSI - 210°F									
1020	Bronze body									
1030	1/2" size	1 Stpi	24	.333	Ea.	187	19.90		206.90	237
1040	3/4" size		20	.400		187	24		211	244
1050	1" size		19	.421		194	25		219	252
1060	1-1/4" size		15	.533		380	32		412	470
1070	1-1/2" size		13	.615		470	37		507	575
1080	2" size		11	.727		485	43.50		528.50	600
1200	Max. 125 PSI with atmospheric vent									
1210	Brass, in-line construction									
1220	1/4" size	1 Stpi	24	.333	Ea.	117	19.90		136.90	160

22 11 19 – Domestic Water Piping Specialties

22 11 19.50 Vacuum Breakers

22 11 19.50 Vacuum Breakers		Crew	Daily Output	Labor-Hours	Unit	2015 Bare Costs Material	Labor	Equipment	Total	Total Incl O&P
1230	3/8″ size	1 Stpi	24	.333	Ea.	117	19.90		136.90	160
1260	For polished chrome finish, add					13%				
2000	Anti-siphon, non-continuous pressure type									
2010	Hot or cold water 125 PSI - 210°F									
2020	Bronze body									
2030	1/4″ size	1 Stpi	24	.333	Ea.	64	19.90		83.90	102
2040	3/8″ size		24	.333		64	19.90		83.90	102
2050	1/2″ size		24	.333		72.50	19.90		92.40	111
2060	3/4″ size		20	.400		86.50	24		110.50	133
2070	1″ size		19	.421		134	25		159	186
2080	1-1/4″ size		15	.533		235	32		267	310
2090	1-1/2″ size		13	.615		276	37		313	365
2100	2″ size		11	.727		430	43.50		473.50	540
2110	2 1/2″ size		8	1		1,225	60		1,285	1,450
2120	3″ size		6	1.333		1,625	79.50		1,704.50	1,925
2150	For polished chrome finish, add					50%				
3000	Air gap fitting									
3020	1/2″ NPT size	1 Plum	19	.421	Ea.	50	24.50		74.50	93.50
3030	1″ NPT size	″	15	.533		57.50	31.50		89	113
3040	2″ NPT size	Q-1	21	.762		113	40.50		153.50	187
3050	3″ NPT size		14	1.143		223	60.50		283.50	340
3060	4″ NPT size		10	1.600		223	84.50		307.50	375

22 11 19.54 Water Hammer Arresters/Shock Absorbers

Line	Description	Crew	Daily Output	Labor-Hours	Unit	Material	Labor	Equipment	Total	Total Incl O&P
0010	**WATER HAMMER ARRESTERS/SHOCK ABSORBERS**									
0490	Copper									
0500	3/4″ male I.P.S. For 1 to 11 fixtures	1 Plum	12	.667	Ea.	28	39		67	92
0600	1″ male I.P.S. For 12 to 32 fixtures		8	1		45.50	58.50		104	142
0700	1-1/4″ male I.P.S. For 33 to 60 fixtures		8	1		47	58.50		105.50	143
0800	1-1/2″ male I.P.S. For 61 to 113 fixtures		8	1		67.50	58.50		126	166
0900	2″ male I.P.S. For 114 to 154 fixtures		8	1		98.50	58.50		157	200
1000	2-1/2″ male I.P.S. For 155 to 330 fixtures		4	2		305	117		422	520
9000	Minimum labor/equipment charge		3.50	2.286	Job		134		134	209

22 11 19.64 Hydrants

Line	Description	Crew	Daily Output	Labor-Hours	Unit	Material	Labor	Equipment	Total	Total Incl O&P
0010	**HYDRANTS**									
0050	Wall type, moderate climate, bronze, encased									
0200	3/4″ IPS connection	1 Plum	16	.500	Ea.	745	29.50		774.50	860
0300	1″ IPS connection		14	.571		850	33.50		883.50	990
0500	Anti-siphon type, 3/4″ connection		16	.500		640	29.50		669.50	750
1000	Non-freeze, bronze, exposed									
1100	3/4″ IPS connection, 4″ to 9″ thick wall	1 Plum	14	.571	Ea.	500	33.50		533.50	605
1120	10″ to 14″ thick wall		12	.667		545	39		584	660
1140	15″ to 19″ thick wall		12	.667		605	39		644	725
1160	20″ to 24″ thick wall		10	.800		640	47		687	780
1200	For 1″ IPS connection, add					15%	10%			
1240	For 3/4″ adapter type vacuum breaker, add				Ea.	63			63	69.50
1280	For anti-siphon type, add				″	132			132	145
2000	Non-freeze bronze, encased, anti-siphon type									
2100	3/4″ IPS connection, 5″ to 9″ thick wall	1 Plum	14	.571	Ea.	1,275	33.50		1,308.50	1,450
2120	10″ to 14″ thick wall		12	.667		1,300	39		1,339	1,500
2140	15″ to 19″ thick wall		12	.667		1,375	39		1,414	1,550
2160	20″ to 24″ thick wall		10	.800		1,425	47		1,472	1,650
2200	For 1″ IPS connection, add					10%	10%			

22 11 19 – Domestic Water Piping Specialties

22 11 19.64 Hydrants		Crew	Daily Output	Labor-Hours	Unit	Material	2015 Bare Costs Labor	Equipment	Total	Total Incl O&P
3000	Ground box type, bronze frame, 3/4" IPS connection									
3080	Non-freeze, all bronze, polished face, set flush									
3100	2 feet depth of bury	1 Plum	8	1	Ea.	950	58.50		1,008.50	1,150
3120	3 feet depth of bury		8	1		1,025	58.50		1,083.50	1,225
3140	4 feet depth of bury		8	1		1,075	58.50		1,133.50	1,300
3160	5 feet depth of bury		7	1.143		1,175	67		1,242	1,375
3180	6 feet depth of bury		7	1.143		1,250	67		1,317	1,475
3200	7 feet depth of bury		6	1.333		1,300	78.50		1,378.50	1,575
3220	8 feet depth of bury		5	1.600		1,375	94		1,469	1,675
3240	9 feet depth of bury		4	2		1,450	117		1,567	1,775
3260	10 feet depth of bury		4	2		1,525	117		1,642	1,850
3400	For 1" IPS connection, add					15%	10%			
3450	For 1-1/4" IPS connection, add					325%	14%			
3500	For 1-1/2" connection, add					370%	18%			
3550	For 2" connection, add					445%	24%			
3600	For tapped drain port in box, add					86			86	94.50
4000	Non-freeze, CI body, bronze frame & scoriated cover									
4010	with hose storage									
4100	2 feet depth of bury	1 Plum	7	1.143	Ea.	1,750	67		1,817	2,025
4120	3 feet depth of bury		7	1.143		1,825	67		1,892	2,100
4140	4 feet depth of bury		7	1.143		1,875	67		1,942	2,175
4160	5 feet depth of bury		6.50	1.231		1,925	72.50		1,997.50	2,225
4180	6 feet depth of bury		6	1.333		1,950	78.50		2,028.50	2,275
4200	7 feet depth of bury		5.50	1.455		2,025	85.50		2,110.50	2,350
4220	8 feet depth of bury		5	1.600		2,100	94		2,194	2,450
4240	9 feet depth of bury		4.50	1.778		2,150	104		2,254	2,550
4260	10 feet depth of bury		4	2		2,225	117		2,342	2,625
4280	For 1" IPS connection, add					390			390	425
4300	For tapped drain port in box, add					86			86	94.50
5000	Moderate climate, all bronze, polished face									
5020	and scoriated cover, set flush									
5100	3/4" IPS connection	1 Plum	16	.500	Ea.	655	29.50		684.50	770
5120	1" IPS connection	"	14	.571		810	33.50		843.50	945
5200	For tapped drain port in box, add					86			86	94.50
6000	Ground post type, all non-freeze, all bronze, aluminum casing									
6010	guard, exposed head, 3/4" IPS connection									
6100	2 feet depth of bury	1 Plum	8	1	Ea.	925	58.50		983.50	1,125
6120	3 feet depth of bury		8	1		1,000	58.50		1,058.50	1,200
6140	4 feet depth of bury		8	1		1,075	58.50		1,133.50	1,275
6160	5 feet depth of bury		7	1.143		1,150	67		1,217	1,375
6180	6 feet depth of bury		7	1.143		1,225	67		1,292	1,450
6200	7 feet depth of bury		6	1.333		1,325	78.50		1,403.50	1,575
6220	8 feet depth of bury		5	1.600		1,400	94		1,494	1,700
6240	9 feet depth of bury		4	2		1,500	117		1,617	1,825
6260	10 feet depth of bury		4	2		1,575	117		1,692	1,900
6300	For 1" IPS connection, add					40%	10%			
6350	For 1-1/4" IPS connection, add					140%	14%			
6400	For 1-1/2" IPS connection, add					225%	18%			
6450	For 2" IPS connection, add					315%	24%			
9000	Minimum labor/equipment charge	1 Plum	3	2.667	Job		157		157	244

22 11 Facility Water Distribution

22 11 23 – Domestic Water Pumps

22 11 23.10 General Utility Pumps		Crew	Daily Output	Labor-Hours	Unit	Material	Labor	Equipment	Total	Total Incl O&P
						2015 Bare Costs				
0010	**GENERAL UTILITY PUMPS**									
2000	Single stage									
3000	Double suction,									
3140	50 HP, 5"D. x 6"S.	Q-2	.33	72.727	Ea.	9,175	3,975		13,150	16,300
3180	60 HP, 6"D. x 8"S.	Q-3	.30	106		13,400	5,950		19,350	24,000
3190	75 HP, to 2500 GPM		.28	114		20,100	6,375		26,475	32,100
3220	100 HP, to 3000 GPM		.26	123		25,500	6,875		32,375	38,800
3240	150 HP, to 4000 GPM		.24	133		36,000	7,450		43,450	51,000

22 11 23.13 Domestic-Water Packaged Booster Pumps

		Crew	Daily Output	Labor-Hours	Unit	Material	Labor	Equipment	Total	Total Incl O&P
0010	**DOMESTIC-WATER PACKAGED BOOSTER PUMPS**									
0200	Pump system, with diaphragm tank, control, press. switch									
0300	1 HP pump	Q-1	1.30	12.308	Ea.	6,500	650		7,150	8,175
0400	1-1/2 HP pump		1.25	12.800		6,550	675		7,225	8,275
0420	2 HP pump		1.20	13.333		6,725	705		7,430	8,500
0440	3 HP pump		1.10	14.545		6,825	770		7,595	8,700
0460	5 HP pump	Q-2	1.50	16		7,550	875		8,425	9,675
0480	7-1/2 HP pump		1.42	16.901		8,400	925		9,325	10,700
0500	10 HP pump		1.34	17.910		8,800	980		9,780	11,200
2000	Pump system, variable speed, base, controls, starter									
2010	Duplex, 100' head									
2020	400 GPM, 7-1/2HP, 4" discharge	Q-2	.70	34.286	Ea.	38,900	1,875		40,775	45,600
2025	Triplex, 100' head									
2030	1000 GPM, 15HP, 6" discharge	Q-2	.50	48	Ea.	56,500	2,625		59,125	66,500
2040	1700 GPM, 30HP, 6" discharge	"	.30	80	"	68,500	4,375		72,875	82,000

22 12 Facility Potable-Water Storage Tanks

22 12 21 – Facility Underground Potable-Water Storage Tanks

22 12 21.13 Fiberglass, Underground Potable-Water Storage Tanks

		Crew	Daily Output	Labor-Hours	Unit	Material	Labor	Equipment	Total	Total Incl O&P
0010	**FIBERGLASS, UNDERGROUND POTABLE-WATER STORAGE TANKS**									
0020	Excludes excavation, backfill, & piping									
0030	Single wall									
2000	600 gallon capacity	B-21B	3.75	10.667	Ea.	3,300	435	174	3,909	4,525
2010	1,000 gallon capacity		3.50	11.429		4,250	465	187	4,902	5,650
2020	2,000 gallon capacity		3.25	12.308		6,200	500	201	6,901	7,825
2030	4,000 gallon capacity		3	13.333		8,575	545	218	9,338	10,600
2040	6,000 gallon capacity		2.65	15.094		9,625	615	247	10,487	11,900
2050	8,000 gallon capacity		2.30	17.391		11,500	710	284	12,494	14,200
2060	10,000 gallon capacity		2	20		13,200	815	325	14,340	16,300
2070	12,000 gallon capacity		1.50	26.667		18,300	1,100	435	19,835	22,400
2080	15,000 gallon capacity		1	40		21,200	1,625	655	23,480	26,700
2090	20,000 gallon capacity		.75	53.333		27,300	2,175	870	30,345	34,500
2100	25,000 gallon capacity		.50	80		40,800	3,275	1,300	45,375	51,500
2110	30,000 gallon capacity		.35	114		49,400	4,675	1,875	55,950	64,000
2120	40,000 gallon capacity		.30	133		71,500	5,450	2,175	79,125	90,000

22 12 23 – Facility Indoor Potable-Water Storage Tanks

22 12 23.13 Facility Steel, Indoor Pot.-Water Storage Tanks

		Crew	Daily Output	Labor-Hours	Unit	Material	Labor	Equipment	Total	Total Incl O&P
0010	**FACILITY STEEL, INDOOR POT.-WATER STORAGE TANKS**									
2000	Galvanized steel, 15 gal., 14" diam., 26" LOA	1 Plum	12	.667	Ea.	1,425	39		1,464	1,625
2060	30 gal., 14" diam. x 49" LOA		11	.727		1,700	42.50		1,742.50	1,925
2080	80 gal., 20" diam. x 64" LOA		9	.889		2,625	52		2,677	2,975

22 12 Facility Potable-Water Storage Tanks

22 12 23 – Facility Indoor Potable-Water Storage Tanks

22 12 23.13 Facility Steel, Indoor Pot.-Water Storage Tanks		Crew	Daily Output	Labor-Hours	Unit	Material	2015 Bare Costs Labor	Equipment	Total	Total Incl O&P
2100	135 gal., 24" diam. x 75" LOA	1 Plum	6	1.333	Ea.	3,825	78.50		3,903.50	4,325
2120	240 gal., 30" diam. x 86" LOA		4	2		7,475	117		7,592	8,400
2140	300 gal., 36" diam. x 76" LOA		3	2.667		10,600	157		10,757	11,900
2160	400 gal., 36" diam. x 100" LOA	Q-1	4	4		13,100	211		13,311	14,700
2180	500 gal., 36" diam., x 126" LOA	"	3	5.333		16,200	282		16,482	18,200
3000	Glass lined, P.E., 80 gal., 20" diam. x 60" LOA	1 Plum	9	.889		2,800	52		2,852	3,150
3060	140 gal., 24" diam. x 72" LOA		6	1.333		3,825	78.50		3,903.50	4,350
3080	200 gal., 30" diam. x 71" LOA		4	2		5,225	117		5,342	5,900
3100	350 gal., 36" diam. x 86" LOA		3	2.667		6,300	157		6,457	7,175
3120	450 gal., 42" diam. x 78" LOA	Q-1	4	4		8,575	211		8,786	9,750
3140	600 gal., 48" diam. x 81" LOA		3	5.333		10,500	282		10,782	11,900
3160	750 gal., 48" diam. x 105" LOA		3	5.333		11,700	282		11,982	13,300
3180	900 gal.,54" diam. x 95" LOA		2.50	6.400		16,400	340		16,740	18,500
3200	1250 gal., 54" diam. x 129" LOA		2	8		21,000	425		21,425	23,800
3220	1800 gal., 54" diam. x 181" LOA		1.50	10.667		23,500	565		24,065	26,800
3240	2000 gal., 60" diam. x 165" LOA		1	16		27,900	845		28,745	32,000
3260	3500 gal., 72" diam. x 201" LOA	Q-2	1.50	16		34,100	875		34,975	39,000

22 13 Facility Sanitary Sewerage

22 13 16 – Sanitary Waste and Vent Piping

22 13 16.20 Pipe, Cast Iron

		Crew	Daily Output	Labor-Hours	Unit	Material	Labor	Equipment	Total	Total Incl O&P
0010	**PIPE, CAST IRON**, Soil, on clevis hanger assemblies, 5' O.C. R221113-50									
0020	Single hub, service wt., lead & oakum joints 10' O.C.									
2120	2" diameter	Q-1	63	.254	L.F.	11.35	13.40		24.75	33.50
2140	3" diameter		60	.267		15	14.10		29.10	38.50
2160	4" diameter		55	.291		18.55	15.35		33.90	44.50
2180	5" diameter	Q-2	76	.316		25.50	17.30		42.80	55
2200	6" diameter	"	73	.329		31	18		49	62.50
2220	8" diameter	Q-3	59	.542		47	30.50		77.50	99.50
2240	10" diameter		54	.593		74.50	33		107.50	134
2260	12" diameter		48	.667		105	37.50		142.50	174
2261	15" diameter		40	.800		156	44.50		200.50	242
2320	For service weight, double hub, add					10%				
2340	For extra heavy, single hub, add					48%	4%			
2360	For extra heavy, double hub, add					71%	4%			
2400	Lead for caulking, (1#/diam. in.)	Q-1	160	.100	Lb.	1.04	5.30		6.34	9.40
2420	Oakum for caulking, (1/8#/diam. in.)	"	40	.400	"	3.60	21		24.60	37
2960	To delete hangers, subtract									
2970	2" diam. to 4" diam.					16%	19%			
2980	5" diam. to 8" diam.					14%	14%			
2990	10" diam. to 15" diam.					13%	19%			
3000	Single hub, service wt., push-on gasket joints 10' O.C.									
3010	2" diameter	Q-1	66	.242	L.F.	12.30	12.80		25.10	33.50
3020	3" diameter		63	.254		16.20	13.40		29.60	39
3030	4" diameter		57	.281		20	14.85		34.85	45
3040	5" diameter	Q-2	79	.304		28	16.65		44.65	56.50
3050	6" diameter	"	75	.320		33.50	17.55		51.05	64.50
3060	8" diameter	Q-3	62	.516		52.50	29		81.50	103
3070	10" diameter		56	.571		84	32		116	143
3080	12" diameter		49	.653		118	36.50		154.50	186
3082	15" diameter		40	.800		171	44.50		215.50	258

22 13 16 – Sanitary Waste and Vent Piping

22 13 16.20 Pipe, Cast Iron		Crew	Daily Output	Labor-Hours	Unit	Material	Labor	Equipment	Total	Total Incl O&P
						2015 Bare Costs				
3100	For service weight, double hub, add					65%				
3110	For extra heavy, single hub, add					48%	4%			
3120	For extra heavy, double hub, add					29%	4%			
3130	To delete hangers, subtract									
3140	2" diam. to 4" diam.					12%	21%			
3150	5" diam. to 8" diam.					10%	16%			
3160	10" diam. to 15" diam.					9%	21%			
4000	No hub, couplings 10' O.C.									
4100	1-1/2" diameter	Q-1	71	.225	L.F.	11.05	11.90		22.95	30.50
4120	2" diameter		67	.239		11.25	12.60		23.85	32
4140	3" diameter		64	.250		15.20	13.20		28.40	37
4160	4" diameter	↓	58	.276		19.10	14.55		33.65	43.50
4180	5" diameter	Q-2	83	.289		25.50	15.85		41.35	52.50
4200	6" diameter	"	79	.304		32	16.65		48.65	61.50
4220	8" diameter	Q-3	69	.464		54.50	26		80.50	101
4240	10" diameter		61	.525		88.50	29.50		118	143
4244	12" diameter		58	.552		110	31		141	169
4248	15" diameter	↓	52	.615	↓	163	34.50		197.50	233
4280	To delete hangers, subtract									
4290	1-1/2" diam. to 6" diam.					22%	47%			
4300	8" diam. to 10" diam.					21%	44%			
4310	12" diam. to 15" diam.					19%	40%			
9000	Minimum labor/equipment charge	1 Plum	4	2	Job		117		117	183

22 13 16.30 Pipe Fittings, Cast Iron		Crew	Daily Output	Labor-Hours	Unit	Material	Labor	Equipment	Total	Total Incl O&P
0010	**PIPE FITTINGS, CAST IRON**, Soil									
0040	Hub and spigot, service weight, lead & oakum joints									
0080	1/4 bend, 2"	Q-1	16	1	Ea.	19.90	53		72.90	105
0120	3"		14	1.143		26.50	60.50		87	123
0140	4"	↓	13	1.231		41.50	65		106.50	147
0160	5"	Q-2	18	1.333		58	73		131	178
0180	6"	"	17	1.412		72	77.50		149.50	201
0200	8"	Q-3	11	2.909		217	163		380	495
0220	10"		10	3.200		320	179		499	630
0224	12"	↓	9	3.556		430	199		629	785
0266	Closet bend, 3" diameter with flange 10" x 16"	Q-1	14	1.143		127	60.50		187.50	234
0268	16" x 16"		12	1.333		141	70.50		211.50	266
0270	Closet bend, 4" diameter, 2-1/2" x 4" ring, 6" x 16"		13	1.231		118	65		183	231
0280	8" x 16"		13	1.231		106	65		171	218
0290	10" x 12"		12	1.333		102	70.50		172.50	222
0300	10" x 18"		11	1.455		136	77		213	270
0310	12" x 16"		11	1.455		119	77		196	251
0330	16" x 16"		10	1.600		148	84.50		232.50	295
0340	1/8 bend, 2"		16	1		14.15	53		67.15	98
0350	3"		14	1.143		22	60.50		82.50	119
0360	4"	↓	13	1.231		33.50	65		98.50	138
0380	5"	Q-2	18	1.333		45.50	73		118.50	164
0400	6"	"	17	1.412		54.50	77.50		132	181
0420	8"	Q-3	11	2.909		164	163		327	435
0440	10"		10	3.200		236	179		415	540
0460	12"	↓	9	3.556		445	199		644	800
0500	Sanitary tee, 2"	Q-1	10	1.600		27.50	84.50		112	163
0540	3"	↓	9	1.778	↓	45	94		139	196

22 13 16.30 Pipe Fittings, Cast Iron		Crew	Daily Output	Labor-Hours	Unit	Material	Labor	Equipment	Total	Total Incl O&P
						2015 Bare Costs				
0620	4"	Q-1	8	2	Ea.	55	106		161	226
0700	5"	Q-2	12	2		109	110		219	291
0800	6"	"	11	2.182		126	120		246	325
0880	8"	Q-3	7	4.571		330	256		586	760
1000	Tee, 2"	Q-1	10	1.600		40	84.50		124.50	176
1060	3"		9	1.778		59.50	94		153.50	212
1120	4"		8	2		76.50	106		182.50	249
1200	5"	Q-2	12	2		162	110		272	350
1300	6"	"	11	2.182		160	120		280	360
1380	8"	Q-3	7	4.571		293	256		549	725
1400	Combination Y and 1/8 bend									
1420	2"	Q-1	10	1.600	Ea.	34.50	84.50		119	170
1460	3"		9	1.778		52.50	94		146.50	204
1520	4"		8	2		72.50	106		178.50	245
1540	5"	Q-2	12	2		138	110		248	325
1560	6"		11	2.182		174	120		294	380
1580	8"		7	3.429		430	188		618	770
1582	12"	Q-3	6	5.333		880	298		1,178	1,425
1600	Double Y, 2"	Q-1	8	2		61.50	106		167.50	233
1610	3"		7	2.286		76.50	121		197.50	272
1620	4"		6.50	2.462		100	130		230	315
1630	5"	Q-2	9	2.667		183	146		329	430
1640	6"	"	8	3		262	164		426	545
1650	8"	Q-3	5.50	5.818		635	325		960	1,200
1660	10"		5	6.400		1,425	360		1,785	2,100
1670	12"		4.50	7.111		1,625	395		2,020	2,425
1740	Reducer, 3" x 2"	Q-1	15	1.067		19.45	56.50		75.95	110
1750	4" x 2"		14.50	1.103		22	58.50		80.50	116
1760	4" x 3"		14	1.143		25.50	60.50		86	122
1770	5" x 2"		14	1.143		53.50	60.50		114	153
1780	5" x 3"		13.50	1.185		56.50	62.50		119	160
1790	5" x 4"		13	1.231		32.50	65		97.50	137
1800	6" x 2"		13.50	1.185		51	62.50		113.50	154
1810	6" x 3"		13	1.231		51.50	65		116.50	158
1830	6" x 4"		12.50	1.280		51	67.50		118.50	161
1840	6" x 5"		11	1.455		55	77		132	181
1880	8" x 3"	Q-2	13.50	1.778		99	97.50		196.50	261
1900	8" x 4"		13	1.846		85	101		186	252
1920	8" x 5"		12	2		89.50	110		199.50	270
1940	8" x 6"		12	2		87.50	110		197.50	267
1960	Increaser, 2" x 3"	Q-1	15	1.067		46.50	56.50		103	139
1980	2" x 4"		14	1.143		46.50	60.50		107	145
2000	2" x 5"		13	1.231		57	65		122	164
2020	3" x 4"		13	1.231		51	65		116	157
2040	3" x 5"		13	1.231		57	65		122	164
2060	3" x 6"		12	1.333		68.50	70.50		139	186
2070	4" x 5"		13	1.231		60.50	65		125.50	168
2080	4" x 6"		12	1.333		69	70.50		139.50	186
2090	4" x 8"	Q-2	13	1.846		143	101		244	315
2100	5" x 6"	Q-1	11	1.455		103	77		180	233
2110	5" x 8"	Q-2	12	2		166	110		276	355
2120	6" x 8"		12	2		166	110		276	355
2130	6" x 10"		8	3		299	164		463	585

22 13 16 – Sanitary Waste and Vent Piping

22 13 16.30 Pipe Fittings, Cast Iron		Crew	Daily Output	Labor-Hours	Unit	Material	Labor	Equipment	Total	Total Incl O&P
						2015 Bare Costs				
2140	8″ x 10″	Q-2	6.50	3.692	Ea.	305	202		507	650
2150	10″ x 12″		5.50	4.364		540	239		779	970
2500	Y, 2″	Q-1	10	1.600		25.50	84.50		110	160
2510	3″		9	1.778		47	94		141	198
2520	4″		8	2		62.50	106		168.50	234
2530	5″	Q-2	12	2		111	110		221	293
2540	6″	″	11	2.182		144	120		264	345
2550	8″	Q-3	7	4.571		350	256		606	785
2560	10″		6	5.333		570	298		868	1,100
2570	12″		5	6.400		1,300	360		1,660	2,000
2580	15″		4	8		2,875	445		3,320	3,850
3000	For extra heavy, add					44%	4%			
3600	Hub and spigot, service weight gasket joint									
3605	Note: gaskets and joint labor have									
3606	been included with all listed fittings.									
3610	1/4 bend, 2″	Q-1	20	.800	Ea.	29.50	42.50		72	98
3620	3″		17	.941		38.50	49.50		88	120
3630	4″		15	1.067		56.50	56.50		113	151
3640	5″	Q-2	21	1.143		82	62.50		144.50	188
3650	6″	″	19	1.263		97	69		166	215
3660	8″	Q-3	12	2.667		272	149		421	530
3670	10″		11	2.909		415	163		578	710
3680	12″		10	3.200		550	179		729	890
3700	Closet bend, 3″ diameter with ring 10″ x 16″	Q-1	17	.941		139	49.50		188.50	231
3710	16″ x 16″		15	1.067		154	56.50		210.50	257
3730	Closet bend, 4″ diameter, 1″ x 4″ ring, 6″ x 16″		15	1.067		133	56.50		189.50	235
3740	8″ x 16″		15	1.067		121	56.50		177.50	221
3750	10″ x 12″		14	1.143		117	60.50		177.50	223
3760	10″ x 18″		13	1.231		152	65		217	268
3770	12″ x 16″		13	1.231		134	65		199	249
3780	16″ x 16″		12	1.333		163	70.50		233.50	290
3800	1/8 bend, 2″		20	.800		23.50	42.50		66	92
3810	3″		17	.941		34.50	49.50		84	116
3820	4″		15	1.067		48.50	56.50		105	142
3830	5″	Q-2	21	1.143		69.50	62.50		132	174
3840	6″	″	19	1.263		79.50	69		148.50	196
3850	8″	Q-3	12	2.667		219	149		368	475
3860	10″		11	2.909		330	163		493	620
3870	12″		10	3.200		570	179		749	905
3900	Sanitary Tee, 2″	Q-1	12	1.333		46.50	70.50		117	161
3910	3″		10	1.600		69	84.50		153.50	208
3920	4″		9	1.778		85.50	94		179.50	240
3930	5″	Q-2	13	1.846		157	101		258	330
3940	6″	″	11	2.182		176	120		296	380
3950	8″	Q-3	8.50	3.765		440	210		650	810
3980	Tee, 2″	Q-1	12	1.333		59	70.50		129.50	175
3990	3″		10	1.600		84	84.50		168.50	225
4000	4″		9	1.778		107	94		201	264
4010	5″	Q-2	13	1.846		210	101		311	390
4020	6″	″	11	2.182		210	120		330	415
4030	8″	Q-3	8	4		405	224		629	795
4060	Combination Y and 1/8 bend									
4070	2″	Q-1	12	1.333	Ea.	53.50	70.50		124	169

22 13 16.30 Pipe Fittings, Cast Iron		Crew	Daily Output	Labor-Hours	Unit	Material	2015 Bare Costs Labor	Equipment	Total	Total Incl O&P
4080	3"	Q-1	10	1.600	Ea.	77	84.50		161.50	217
4090	4"		9	1.778		103	94		197	260
4100	5"	Q-2	13	1.846		186	101		287	360
4110	6"	"	11	2.182		224	120		344	430
4120	8"	Q-3	8	4		540	224		764	945
4121	12"	"	7	4.571		1,125	256		1,381	1,650
4160	Double Y, 2"	Q-1	10	1.600		90	84.50		174.50	231
4170	3"		8	2		113	106		219	289
4180	4"		7	2.286		146	121		267	350
4190	5"	Q-2	10	2.400		255	131		386	485
4200	6"	"	9	2.667		335	146		481	600
4210	8"	Q-3	6	5.333		800	298		1,098	1,350
4220	10"		5	6.400		1,700	360		2,060	2,425
4230	12"		4.50	7.111		2,000	395		2,395	2,825
4260	Reducer, 3" x 2"	Q-1	17	.941		41	49.50		90.50	123
4270	4" x 2"		16.50	.970		47	51		98	132
4280	4" x 3"		16	1		53	53		106	141
4290	5" x 2"		16	1		87	53		140	178
4300	5" x 3"		15.50	1.032		93	54.50		147.50	187
4310	5" x 4"		15	1.067		72	56.50		128.50	167
4320	6" x 2"		15.50	1.032		85	54.50		139.50	179
4330	6" x 3"		15	1.067		88.50	56.50		145	186
4336	6" x 4"		14	1.143		91	60.50		151.50	194
4340	6" x 5"		13	1.231		104	65		169	215
4360	8" x 3"	Q-2	15	1.600		166	87.50		253.50	320
4370	8" x 4"		15	1.600		155	87.50		242.50	310
4380	8" x 5"		14	1.714		169	94		263	330
4390	8" x 6"		14	1.714		167	94		261	330
4430	Increaser, 2" x 3"	Q-1	17	.941		58.50	49.50		108	142
4440	2" x 4"		16	1		62	53		115	151
4450	2" x 5"		15	1.067		81	56.50		137.50	178
4460	3" x 4"		15	1.067		66.50	56.50		123	161
4470	3" x 5"		15	1.067		81	56.50		137.50	178
4480	3" x 6"		14	1.143		93.50	60.50		154	197
4490	4" x 5"		15	1.067		84.50	56.50		141	181
4500	4" x 6"		14	1.143		94	60.50		154.50	197
4510	4" x 8"	Q-2	15	1.600		198	87.50		285.50	355
4520	5" x 6"	Q-1	13	1.231		127	65		192	241
4530	5" x 8"	Q-2	14	1.714		221	94		315	390
4540	6" x 8"		14	1.714		221	94		315	390
4550	6" x 10"		10	2.400		395	131		526	640
4560	8" x 10"		8.50	2.824		400	155		555	680
4570	10" x 12"		7.50	3.200		665	175		840	1,000
4600	Y, 2"	Q-1	12	1.333		44	70.50		114.50	159
4610	3"		10	1.600		71	84.50		155.50	211
4620	4"		9	1.778		93.50	94		187.50	249
4630	5"	Q-2	13	1.846		159	101		260	335
4640	6"	"	11	2.182		194	120		314	400
4650	8"	Q-3	8	4		460	224		684	860
4660	10"		7	4.571		760	256		1,016	1,225
4670	12"		6	5.333		1,550	298		1,848	2,175
4672	15"		5	6.400		3,150	360		3,510	4,025
4900	For extra heavy, add					44%	4%			

22 13 16.30 Pipe Fittings, Cast Iron		Crew	Daily Output	Labor-Hours	Unit	Material	2015 Bare Costs Labor	Equipment	Total	Total Incl O&P
4940	Gasket and making push-on joint									
4950	2"	Q-1	40	.400	Ea.	9.40	21		30.40	43.50
4960	3"		35	.457		12.20	24		36.20	51
4970	4"		32	.500		15.30	26.50		41.80	58
4980	5"	Q-2	43	.558		24	30.50		54.50	74
4990	6"	"	40	.600		25	33		58	79
5000	8"	Q-3	32	1		55	56		111	148
5010	10"		29	1.103		95.50	61.50		157	201
5020	12"		25	1.280		122	71.50		193.50	246
5022	15"		21	1.524		146	85		231	294
5030	Note: gaskets and joint labor have									
5040	Been included with all listed fittings.									
5990	No hub									
6000	Cplg. & labor required at joints not incl. in fitting									
6010	price. Add 1 coupling per joint for installed price									
6020	1/4 Bend, 1-1/2"				Ea.	10.20			10.20	11.25
6060	2"					11.15			11.15	12.25
6080	3"					16.25			16.25	17.85
6120	4"					23			23	25.50
6140	5"					55.50			55.50	61
6160	6"					55.50			55.50	61
6180	8"					156			156	172
6184	1/4 Bend, long sweep, 1-1/2"					25.50			25.50	28
6186	2"					24.50			24.50	26.50
6188	3"					29.50			29.50	32.50
6189	4"					47			47	51.50
6190	5"					90.50			90.50	99.50
6191	6"					103			103	114
6192	8"					282			282	310
6193	10"					570			570	625
6200	1/8 Bend, 1-1/2"					8.60			8.60	9.45
6210	2"					9.60			9.60	10.55
6212	3"					12.80			12.80	14.10
6214	4"					16.80			16.80	18.50
6216	5"					35			35	38.50
6218	6"					37.50			37.50	41
6220	8"					108			108	118
6222	10"					204			204	225
6380	Sanitary Tee, tapped, 1-1/2"					20.50			20.50	22.50
6382	2" x 1-1/2"					17.95			17.95	19.70
6384	2"					19.25			19.25	21
6386	3" x 2"					28.50			28.50	31.50
6388	3"					49.50			49.50	54.50
6390	4" x 1-1/2"					25.50			25.50	28
6392	4" x 2"					29			29	32
6393	4"					29			29	32
6394	6" x 1-1/2"					66			66	73
6396	6" x 2"					67.50			67.50	74
6459	Sanitary Tee, 1-1/2"					14.35			14.35	15.80
6460	2"					15.35			15.35	16.85
6470	3"					18.90			18.90	21
6472	4"					36			36	39.50
6474	5"					83.50			83.50	92

22 13 16.30 Pipe Fittings, Cast Iron		Crew	Daily Output	Labor-Hours	Unit	Material	2015 Bare Costs Labor	Equipment	Total	Total Incl O&P
6476	6"				Ea.	85.50			85.50	94
6478	8"					345			345	380
6730	Y, 1-1/2"					14.50			14.50	15.95
6740	2"					14.20			14.20	15.65
6750	3"					20.50			20.50	22.50
6760	4"					33			33	36.50
6762	5"					78.50			78.50	86
6764	6"					88			88	97
6768	8"					207			207	228
6769	10"					460			460	505
6770	12"					905			905	995
6771	15"					2,025			2,025	2,225
6791	Y, reducing, 3" x 2"					15.35			15.35	16.85
6792	4" x 2"					22			22	24.50
6793	5" x 2"					48.50			48.50	53.50
6794	6" x 2"					54			54	59.50
6795	6" x 4"					70			70	77
6796	8" x 4"					121			121	133
6797	8" x 6"					149			149	163
6798	10" x 6"					335			335	370
6799	10" x 8"					400			400	440
6800	Double Y, 2"					22.50			22.50	25
6920	3"					41.50			41.50	45.50
7000	4"					84.50			84.50	92.50
7100	6"					149			149	164
7120	8"					425			425	470
7200	Combination Y and 1/8 Bend									
7220	1-1/2"				Ea.	15.45			15.45	17
7260	2"					16.15			16.15	17.80
7320	3"					25.50			25.50	28
7400	4"					49			49	54
7480	5"					101			101	111
7500	6"					135			135	148
7520	8"					315			315	345
7800	Reducer, 3" x 2"					7.85			7.85	8.60
7820	4" x 2"					12.10			12.10	13.30
7840	4" x 3"					12.10			12.10	13.30
7842	6" x 3"					32.50			32.50	36
7844	6" x 4"					29.50			29.50	32.50
7846	6" x 5"					33.50			33.50	36.50
7848	8" x 2"					51.50			51.50	57
7850	8" x 3"					47.50			47.50	52.50
7852	8" x 4"					50.50			50.50	55.50
7854	8" x 5"					57			57	62.50
7856	8" x 6"					56			56	61.50
7858	10" x 4"					99			99	109
7860	10" x 6"					104			104	115
7862	10" x 8"					122			122	134
7864	12" x 4"					204			204	225
7866	12" x 6"					219			219	241
7868	12" x 8"					225			225	248
7870	12" x 10"					225			225	248
7872	15" x 4"					425			425	470

22 13 Facility Sanitary Sewerage

22 13 16 – Sanitary Waste and Vent Piping

22 13 16.30 Pipe Fittings, Cast Iron		Crew	Daily Output	Labor-Hours	Unit	Material	Labor	Equipment	Total	Total Incl O&P
						2015 Bare Costs				
7874	15″ x 6″				Ea.	400			400	440
7876	15″ x 8″					465			465	510
7878	15″ x 10″					480			480	530
7880	15″ x 12″					485			485	535
8000	Coupling, standard (by CISPI Mfrs.)									
8020	1-1/2″	Q-1	48	.333	Ea.	13.95	17.60		31.55	43
8040	2″		44	.364		15.25	19.20		34.45	47
8080	3″		38	.421		17.10	22		39.10	53.50
8120	4″		33	.485		20	25.50		45.50	62
8160	5″	Q-2	44	.545		41	30		71	91.50
8180	6″	″	40	.600		46.50	33		79.50	103
8200	8″	Q-3	33	.970		97	54		151	192
8220	10″	″	26	1.231		124	69		193	243
8300	Coupling, cast iron clamp & neoprene gasket (by MG)									
8310	1-1/2″	Q-1	48	.333	Ea.	6.40	17.60		24	34.50
8320	2″		44	.364		8.35	19.20		27.55	39
8330	3″		38	.421		12.65	22		34.65	48.50
8340	4″		33	.485		14.60	25.50		40.10	56
8350	5″	Q-2	44	.545		20.50	30		50.50	69
8360	6″	″	40	.600		32.50	33		65.50	87.50
8380	8″	Q-3	33	.970		116	54		170	213
8400	10″	″	26	1.231		189	69		258	315
8600	Coupling, Stainless steel, heavy duty									
8620	1-1/2″	Q-1	48	.333	Ea.	10	17.60		27.60	38.50
8630	2″		44	.364		10.40	19.20		29.60	41.50
8640	2″ x 1-1/2″		44	.364		12.20	19.20		31.40	43.50
8650	3″		38	.421		11.30	22		33.30	47
8660	4″		33	.485		12.75	25.50		38.25	54
8670	4″ x 3″		33	.485		18.40	25.50		43.90	60
8680	5″	Q-2	44	.545		24.50	30		54.50	73.50
8690	6″	″	40	.600		30.50	33		63.50	85.50
8700	8″	Q-3	33	.970		52	54		106	142
8710	10″		26	1.231		65.50	69		134.50	179
8712	12″		22	1.455		102	81.50		183.50	239
8715	15″		18	1.778		120	99.50		219.50	287
9000	Minimum labor/equipment charge	1 Plum	4	2	Job		117		117	183

22 13 16.40 Pipe Fittings, Cast Iron for Drainage		Crew	Daily Output	Labor-Hours	Unit	Material	Labor	Equipment	Total	Total Incl O&P
0010	**PIPE FITTINGS, CAST IRON FOR DRAINAGE**, Special									
1000	Drip pan elbow, (safety valve discharge elbow)									
1010	Cast iron, threaded inlet									
1014	2-1/2″	Q-1	8	2	Ea.	1,300	106		1,406	1,600
1015	3″		6.40	2.500		1,400	132		1,532	1,750
1017	4″		4.80	3.333		1,900	176		2,076	2,350
1018	Cast iron, flanged inlet									
1019	6″	Q-2	3.60	6.667	Ea.	1,950	365		2,315	2,725
1020	8″	″	2.60	9.231	″	2,050	505		2,555	3,050

22 13 16.50 Shower Drains		Crew	Daily Output	Labor-Hours	Unit	Material	Labor	Equipment	Total	Total Incl O&P
0010	**SHOWER DRAINS**									
2780	Shower, with strainer, uniform diam. trap, bronze top									
2800	2″ and 3″ pipe size	Q-1	8	2	Ea.	480	106		586	695
2820	4″ pipe size	″	7	2.286		485	121		606	720
2840	For galvanized body, add					189			189	208

22 13 16 – Sanitary Waste and Vent Piping

22 13 16.50 Shower Drains		Crew	Daily Output	Labor-Hours	Unit	Material	Labor	Equipment	Total	Total Incl O&P
						2015 Bare Costs				
2860	With strainer, backwater valve, drum trap									
2880	1-1/2", 2" & 3" pipe size	Q-1	8	2	Ea.	385	106		491	590
2890	4" pipe size	"	7	2.286		530	121		651	775
2900	For galvanized body, add					193			193	213

22 13 16.60 Traps		Crew	Daily Output	Labor-Hours	Unit	Material	Labor	Equipment	Total	Total Incl O&P
0010	**TRAPS**									
0030	Cast iron, service weight									
0050	Running P trap, without vent									
1100	2"	Q-1	16	1	Ea.	143	53		196	240
1140	3"		14	1.143		143	60.50		203.50	251
1150	4"		13	1.231		143	65		208	258
1160	6"	Q-2	17	1.412		635	77.50		712.50	815
1180	Running trap, single hub, with vent									
2080	3" pipe size, 3" vent	Q-1	14	1.143	Ea.	118	60.50		178.50	223
2120	4" pipe size, 4" vent	"	13	1.231		154	65		219	270
2140	5" pipe size, 4" vent	Q-2	11	2.182		245	120		365	455
2160	6" pipe size, 4" vent		10	2.400		660	131		791	930
2180	6" pipe size, 6" vent		8	3		715	164		879	1,050
2200	8" pipe size, 4" vent	Q-3	10	3.200		3,050	179		3,229	3,625
2220	8" pipe size, 6" vent	"	8	4		2,350	224		2,574	2,950
2300	For double hub, vent, add					10%	20%			
2800	S trap,									
2850	4" pipe size	Q-1	13	1.231	Ea.	68.50	65		133.50	177
3000	P trap, B&S, 2" pipe size		16	1		34	53		87	120
3040	3" pipe size		14	1.143		50.50	60.50		111	150
3060	4" pipe size		13	1.231		73	65		138	181
3080	5" pipe size	Q-2	18	1.333		162	73		235	292
3100	6" pipe size	"	17	1.412		225	77.50		302.50	370
3120	8" pipe size	Q-3	11	2.909		680	163		843	1,000
3130	10" pipe size	"	10	3.200		1,275	179		1,454	1,675
3150	P trap, no hub, 1-1/2" pipe size	Q-1	17	.941		18.40	49.50		67.90	97.50
3160	2" pipe size		16	1		17.35	53		70.35	102
3170	3" pipe size		14	1.143		38	60.50		98.50	136
3180	4" pipe size		13	1.231		67.50	65		132.50	175
3190	6" pipe size	Q-2	17	1.412		164	77.50		241.50	300
3350	Deep seal trap, B&S									
3400	1-1/4" pipe size	Q-1	14	1.143	Ea.	53	60.50		113.50	152
3410	1-1/2" pipe size		14	1.143		53	60.50		113.50	152
3420	2" pipe size		14	1.143		49	60.50		109.50	148
3440	3" pipe size		12	1.333		62.50	70.50		133	179
3460	4" pipe size		11	1.455		99	77		176	229
3500	For trap primer connection, add					125			125	137
3540	For trap with floor cleanout, add					70%	5%			
3580	For trap with adjustable cleanout, add	Q-1	10	1.600	Ea.	284	84.50		368.50	440
4700	Copper, drainage, drum trap									
4800	3" x 5" solid, 1-1/2" pipe size	1 Plum	16	.500	Ea.	106	29.50		135.50	163
4840	3" x 6" swivel, 1-1/2" pipe size	"	16	.500	"	160	29.50		189.50	222
5100	P trap, standard pattern									
5200	1-1/4" pipe size	1 Plum	18	.444	Ea.	78.50	26		104.50	127
5240	1-1/2" pipe size		17	.471		72	27.50		99.50	122
5260	2" pipe size		15	.533		111	31.50		142.50	171
5280	3" pipe size		11	.727		281	42.50		323.50	375

22 13 16.60 Traps		Crew	Daily Output	Labor-Hours	Unit	2015 Bare Costs Material	Labor	Equipment	Total	Total Incl O&P
5340	With cleanout, swivel joint and slip joint									
5360	1-1/4" pipe size	1 Plum	18	.444	Ea.	99.50	26		125.50	150
5400	1-1/2" pipe size		17	.471		106	27.50		133.50	160
5420	2" pipe size		15	.533		177	31.50		208.50	243
5750	Chromed brass, tubular, P trap, without cleanout, 20 Ga.									
5800	1-1/4" pipe size	1 Plum	18	.444	Ea.	19.25	26		45.25	61.50
5840	1-1/2" pipe size	"	17	.471	"	22	27.50		49.50	67
5900	With cleanout, 20 Ga.									
5940	1-1/4" pipe size	1 Plum	18	.444	Ea.	31	26		57	74.50
6000	1-1/2" pipe size	"	17	.471	"	33.50	27.50		61	80
6350	S trap, without cleanout, 20 Ga.									
6400	1-1/4" pipe size	1 Plum	18	.444	Ea.	50	26		76	95.50
6440	1-1/2" pipe size	"	17	.471	"	57.50	27.50		85	106
6550	With cleanout, 20 Ga.									
6600	1-1/4" pipe size	1 Plum	18	.444	Ea.	61.50	26		87.50	108
6640	1-1/2" pipe size	"	17	.471		64	27.50		91.50	113
6660	Corrosion resistant, glass, P trap, 1-1/2" pipe size	Q-1	17	.941		72	49.50		121.50	157
6670	2" pipe size		16	1		94	53		147	186
6680	3" pipe size		14	1.143		193	60.50		253.50	305
6690	4" pipe size		13	1.231		287	65		352	415
6700	6" pipe size	Q-2	17	1.412		1,100	77.50		1,177.50	1,350
6710	ABS DWV P trap, solvent weld joint									
6720	1-1/2" pipe size	1 Plum	18	.444	Ea.	6.30	26		32.30	47.50
6722	2" pipe size		17	.471		8.30	27.50		35.80	52
6724	3" pipe size		15	.533		32.50	31.50		64	85
6726	4" pipe size		14	.571		65	33.50		98.50	124
6732	PVC DWV P trap, solvent weld joint									
6733	1-1/2" pipe size	1 Plum	18	.444	Ea.	4.95	26		30.95	46
6734	2" pipe size		17	.471		6.70	27.50		34.20	50.50
6735	3" pipe size		15	.533		22.50	31.50		54	74
6736	4" pipe size		14	.571		51.50	33.50		85	110
6760	PP DWV, dilution trap, 1-1/2" pipe size		16	.500		226	29.50		255.50	295
6770	P trap, 1-1/2" pipe size		17	.471		56.50	27.50		84	105
6780	2" pipe size		16	.500		83	29.50		112.50	138
6790	3" pipe size		14	.571		153	33.50		186.50	221
6800	4" pipe size		13	.615		257	36		293	340
6830	S trap, 1-1/2" pipe size		16	.500		46.50	29.50		76	97
6840	2" pipe size		15	.533		69.50	31.50		101	126
6850	Universal trap, 1-1/2" pipe size		14	.571		92	33.50		125.50	154
6860	PVC DWV hub x hub, basin trap, 1-1/4" pipe size		18	.444		40	26		66	84.50
6870	Sink P trap, 1-1/2" pipe size		18	.444		9.65	26		35.65	51
6880	Tubular S trap, 1-1/2" pipe size		17	.471		23.50	27.50		51	68.50
6890	PVC sch. 40 DWV, drum trap									
6900	1-1/2" pipe size	1 Plum	16	.500	Ea.	23.50	29.50		53	72
6910	P trap, 1-1/2" pipe size		18	.444		5.85	26		31.85	47
6920	2" pipe size		17	.471		7.90	27.50		35.40	51.50
6930	3" pipe size		15	.533		27	31.50		58.50	78.50
6940	4" pipe size		14	.571		61	33.50		94.50	120
6950	P trap w/clean out, 1-1/2" pipe size		18	.444		9.65	26		35.65	51
6960	2" pipe size		17	.471		16.40	27.50		43.90	61
6970	P trap adjustable, 1-1/2" pipe size		17	.471		7.75	27.50		35.25	51.50
6980	P trap adj. w/union & cleanout, 1-1/2" pipe size		16	.500		30.50	29.50		60	79.50
7000	Trap primer, flow through type, 1/2" diameter		24	.333		42.50	19.55		62.05	77

22 13 16 – Sanitary Waste and Vent Piping

22 13 16.60 Traps

	22 13 16.60 Traps	Crew	Daily Output	Labor-Hours	Unit	Material	Labor	Equipment	Total	Total Incl O&P
						2015 Bare Costs				
7100	With sediment strainer	1 Plum	22	.364	Ea.	46.50	21.50		68	84.50
7450	Trap primer distribution unit									
7500	2 openings	1 Plum	18	.444	Ea.	28	26		54	71.50
7540	3 openings		17	.471		30	27.50		57.50	76
7560	4 openings		16	.500		35.50	29.50		65	85
7850	Trap primer manifold									
7900	2 outlet	1 Plum	18	.444	Ea.	58.50	26		84.50	105
7940	4 outlet		16	.500		94.50	29.50		124	150
7960	6 outlet		15	.533		131	31.50		162.50	193
7980	8 outlet		13	.615		166	36		202	240
9000	Minimum labor/equipment charge		3	2.667	Job		157		157	244

22 13 16.80 Vent Flashing and Caps

	22 13 16.80 Vent Flashing and Caps	Crew	Daily Output	Labor-Hours	Unit	Material	Labor	Equipment	Total	Total Incl O&P
0010	**VENT FLASHING AND CAPS**									
0120	Vent caps									
0140	Cast iron									
0180	2-1/2" - 3-5/8" pipe	1 Plum	21	.381	Ea.	45	22.50		67.50	84.50
0190	4" - 4-1/8" pipe	"	19	.421	"	55	24.50		79.50	99
0900	Vent flashing									
1000	Aluminum with lead ring									
1020	1-1/4" pipe	1 Plum	20	.400	Ea.	8	23.50		31.50	45.50
1030	1-1/2" pipe		20	.400		8.45	23.50		31.95	46
1040	2" pipe		18	.444		8.60	26		34.60	50
1050	3" pipe		17	.471		9.55	27.50		37.05	53.50
1060	4" pipe		16	.500		11.50	29.50		41	58.50
1350	Copper with neoprene ring									
1400	1-1/4" pipe	1 Plum	20	.400	Ea.	61.50	23.50		85	104
1430	1-1/2" pipe		20	.400		61.50	23.50		85	104
1440	2" pipe		18	.444		61.50	26		87.50	108
1450	3" pipe		17	.471		74.50	27.50		102	125
1460	4" pipe		16	.500		74.50	29.50		104	128
2000	Galvanized with neoprene ring									
2020	1-1/4" pipe	1 Plum	20	.400	Ea.	16.95	23.50		40.45	55
2030	1-1/2" pipe		20	.400		16.95	23.50		40.45	55
2040	2" pipe		18	.444		17.55	26		43.55	60
2050	3" pipe		17	.471		18.55	27.50		46.05	63.50
2060	4" pipe		16	.500		21.50	29.50		51	69.50
2980	Neoprene, one piece									
3000	1-1/4" pipe	1 Plum	24	.333	Ea.	5.05	19.55		24.60	36
3030	1-1/2" pipe		24	.333		5.10	19.55		24.65	36
3040	2" pipe		23	.348		5.10	20.50		25.60	37.50
3050	3" pipe		21	.381		5.85	22.50		28.35	41.50
3060	4" pipe		20	.400		8.65	23.50		32.15	46
4000	Lead, 4#, 8" skirt, vent through roof									
4100	2" pipe	1 Plum	18	.444	Ea.	41.50	26		67.50	86.50
4110	3" pipe		17	.471		47.50	27.50		75	95
4120	4" pipe		16	.500		57	29.50		86.50	109
4130	6" pipe		14	.571		81	33.50		114.50	142
9000	Minimum labor/equipment charge		4	2	Job		117		117	183

22 13 19 – Sanitary Waste Piping Specialties

22 13 19.13 Sanitary Drains		Crew	Daily Output	Labor-Hours	Unit	Material	Labor	Equipment	Total	Total Incl O&P
						2015 Bare Costs				
0010	**SANITARY DRAINS**									
0400	Deck, auto park, C.I., 13" top									
0440	3", 4", 5", and 6" pipe size	Q-1	8	2	Ea.	1,450	106		1,556	1,750
0480	For galvanized body, add				"	780			780	855
0800	Promenade, heelproof grate, C.I., 14" top									
0840	2", 3", and 4" pipe size	Q-1	10	1.600	Ea.	550	84.50		634.50	735
0860	5" and 6" pipe size		9	1.778		685	94		779	900
0880	8" pipe size	↓	8	2		810	106		916	1,050
0940	For galvanized body, add					400			400	440
0960	With polished bronze top, 2"-3"-4" diam.				↓	930			930	1,025
1200	Promenade, heelproof grate, C.I., lateral, 14" top									
1240	2", 3" and 4" pipe size	Q-1	10	1.600	Ea.	720	84.50		804.50	920
1260	5" and 6" pipe size		9	1.778		855	94		949	1,075
1280	8" pipe size	↓	8	2		980	106		1,086	1,250
1340	For galvanized body, add					410			410	450
1360	For polished bronze top, add				↓	375			375	415
1500	Promenade, slotted grate, C.I., 11" top									
1540	2", 3", 4", 5", and 6" pipe size	Q-1	12	1.333	Ea.	495	70.50		565.50	655
1600	For galvanized body, add					230			230	253
1640	With polished bronze top				↓	790			790	870
2000	Floor, medium duty, C.I., deep flange, 7" diam. top									
2040	2" and 3" pipe size	Q-1	12	1.333	Ea.	208	70.50		278.50	340
2080	For galvanized body, add					96.50			96.50	106
2120	With polished bronze top				↓	315			315	345
2160	Heavy duty, C.I., 12" dia anti-tilt grate									
2180	2", 3", 4", 5" and 6" pipe size	Q-1	10	1.600	Ea.	585	84.50		669.50	770
2220	For galvanized body, add					296			296	325
2240	With polished bronze top				↓	875			875	965
2300	X-Heavy duty, C.I., 15" antitilt grate									
2320	4", 5", 6", and 8" pipe size	Q-1	8	2	Ea.	1,250	106		1,356	1,550
2360	For galvanized body, add					480			480	530
2380	With polished bronze top				↓	1,750			1,750	1,925
2400	Heavy duty, with sediment bucket, C.I., 12" diam. loose grate									
2420	2", 3", 4", 5", and 6" pipe size	Q-1	9	1.778	Ea.	690	94		784	905
2440	For galvanized body, add					440			440	485
2460	With polished bronze top				↓	975			975	1,075
2500	Heavy duty, cleanout & trap w/bucket, C.I., 15" top									
2540	2", 3", and 4" pipe size	Q-1	6	2.667	Ea.	6,575	141		6,716	7,450
2560	For galvanized body, add					1,675			1,675	1,850
2580	With polished bronze top				↓	7,300			7,300	8,025
2600	Medium duty, with perforated SS basket, C.I., body,									
2610	18" top for refuse container washing area									
2620	2" thru 6" pipe size	Q-1	4	4	Ea.	4,475	211		4,686	5,250
2630	Acid resistant									
2638	PVC									
2640	2", 3" and 4" pipe size	Q-1	16	1	Ea.	320	53		373	435
2644	Cast iron, epoxy coated									
2646	2", 3" and 4" pipe size	Q-1	14	1.143	Ea.	730	60.50		790.50	900
2650	PVC or ABS thermoplastic									
2660	3" and 4" pipe size	Q-1	16	1	Ea.	350	53		403	470
2680	Extra heavy duty, oil intercepting, gas seal cone,									
2690	with cleanout, loose grate, C.I., body 16" top									

22 13 19 – Sanitary Waste Piping Specialties

22 13 19.13 Sanitary Drains		Crew	Daily Output	Labor-Hours	Unit	Material	Labor	Equipment	Total	Total Incl O&P
						2015 Bare Costs				
2700	3" and 4" diameter outlet, 4" slab depth	Q-1	4	4	Ea.	7,100	211		7,311	8,150
2720	4" diameter outlet, 8" slab depth		3	5.333		7,100	282		7,382	8,275
2740	4" diam. outlet, 10"-12" slab depth, 16"top		2	8		8,000	425		8,425	9,450
2910	Prison cell, vandal-proof, 1-1/2", and 2" diam. pipe		12	1.333		445	70.50		515.50	600
2920	3" pipe size	↓	10	1.600	↓	500	84.50		584.50	680
2930	Trap drain, light duty, backwater valve C.I. top									
2950	8" diameter top, 2" pipe size	Q-1	12	1.333	Ea.	390	70.50		460.50	535
2960	10" diameter top, 3" pipe size		10	1.600		540	84.50		624.50	720
2970	12" diameter top, 4" pipe size	↓	8	2	↓	745	106		851	985

22 13 19.14 Floor Receptors		Crew	Daily Output	Labor-Hours	Unit	Material	Labor	Equipment	Total	Total Incl O&P
0010	**FLOOR RECEPTORS**, For connection to 2", 3" & 4" diameter pipe									
0200	12-1/2" square top, 25 sq. in. open area	Q-1	10	1.600	Ea.	970	84.50		1,054.50	1,200
0300	For grate with 4" diam. x 3-3/4" high funnel, add					174			174	191
0400	For grate with 6" diameter x 6" high funnel, add					223			223	246
0500	For full hinged grate with open center, add					81			81	89.50
0600	For aluminum bucket, add					146			146	160
0700	For acid-resisting bucket, add					274			274	300
0900	For stainless steel mesh bucket liner, add					216			216	238
1000	For bronze antisplash dome strainer, add					104			104	114
1100	For partial solid cover, add					52.50			52.50	58
1200	For trap primer connection, add					88			88	97
2000	12-5/8" diameter top, 40 sq. in. open area	Q-1	10	1.600	↓	770	84.50		854.50	975
2100	For options, add same prices as square top									
3000	8" x 4" rectangular top, 7.5 sq. in. open area	Q-1	14	1.143	Ea.	745	60.50		805.50	915
3100	For trap primer connections, add					88			88	97
4000	24" x 16" rectangular top, 70 sq. in. open area	Q-1	4	4		4,350	211		4,561	5,125
4100	For trap primer connection, add				↓	174			174	191
9000	Minimum labor/equipment charge	Q-1	3	5.333	Job		282		282	440

22 13 19.15 Sink Waste Treatment		Crew	Daily Output	Labor-Hours	Unit	Material	Labor	Equipment	Total	Total Incl O&P
0010	**SINK WASTE TREATMENT**, System for commercial kitchens									
0100	includes clock timer, & fittings									
0200	System less chemical, wall mounted cabinet	1 Plum	16	.500	Ea.	550	29.50		579.50	650
2000	Chemical, 1 gallon, add					68.50			68.50	75.50
2100	6 gallons, add					310			310	340
2200	15 gallons, add					840			840	925
2300	30 gallons, add					1,575			1,575	1,725
2400	55 gallons, add				↓	2,675			2,675	2,950

22 13 19.39 Floor Drain Trap Seal		Crew	Daily Output	Labor-Hours	Unit	Material	Labor	Equipment	Total	Total Incl O&P
0010	**FLOOR DRAIN TRAP SEAL**									
0100	Inline									
0110	2"	1 Plum	20	.400	Ea.	45	23.50		68.50	86
0120	3"		16	.500		49	29.50		78.50	100
0130	3.5"		14	.571		52	33.50		85.50	110
0140	4"	↓	10	.800	↓	56	47		103	135

22 13 23 – Sanitary Waste Interceptors

22 13 23.10 Interceptors		Crew	Daily Output	Labor-Hours	Unit	Material	Labor	Equipment	Total	Total Incl O&P
0010	**INTERCEPTORS**									
0150	Grease, fabricated steel, 4 GPM, 8 lb. fat capacity	1 Plum	4	2	Ea.	1,150	117		1,267	1,450
0200	7 GPM, 14 lb. fat capacity		4	2		1,600	117		1,717	1,950
1000	10 GPM, 20 lb. fat capacity		4	2		1,875	117		1,992	2,250
1040	15 GPM, 30 lb. fat capacity	↓	4	2	↓	2,800	117		2,917	3,250

22 13 23 – Sanitary Waste Interceptors

22 13 23.10 Interceptors		Crew	Daily Output	Labor-Hours	Unit	2015 Bare Costs Material	Labor	Equipment	Total	Total Incl O&P
1060	20 GPM, 40 lb. fat capacity	1 Plum	3	2.667	Ea.	3,425	157		3,582	4,000
1080	25 GPM, 50 lb. fat capacity	Q-1	3.50	4.571		3,825	242		4,067	4,600
1100	35 GPM, 70 lb. fat capacity		3	5.333		4,750	282		5,032	5,675
1120	50 GPM, 100 lb. fat capacity		2	8		6,300	425		6,725	7,575
1140	75 GPM, 150 lb. fat capacity		2	8		14,300	425		14,725	16,400
1160	100 GPM, 200 lb. fat capacity		2	8		15,100	425		15,525	17,300
1180	150 GPM, 300 lb. fat capacity		2	8		15,800	425		16,225	18,100
1200	200 GPM, 400 lb. fat capacity		1.50	10.667		23,000	565		23,565	26,200
1220	250 GPM, 500 lb. fat capacity		1.30	12.308		26,700	650		27,350	30,400
1240	300 GPM, 600 lb. fat capacity		1	16		31,500	845		32,345	36,000
1260	400 GPM, 800 lb. fat capacity	Q-2	1.20	20		38,600	1,100		39,700	44,100
1280	500 GPM, 1000 lb. fat capacity	″	1	24		45,700	1,325		47,025	52,000
1580	For seepage pan, add					7%				
3000	Hair, cast iron, 1-1/4″ and 1-1/2″ pipe connection	1 Plum	8	1	Ea.	435	58.50		493.50	565
3100	For chrome-plated cast iron, add					266			266	293
3200	For polished bronze, add					266			266	293
3400	Lint interceptor, fabricated steel									
3410	Size based on 10 GPM per machine									
3420	30 GPM, 2″ pipe size	Q-1	3	5.333	Ea.	5,075	282		5,357	6,050
3430	70 GPM, 3″ pipe size		2.50	6.400		5,925	340		6,265	7,025
3440	100 GPM, 4″ pipe size		2	8		6,975	425		7,400	8,300
3450	200 GPM, 4″ pipe size		1.50	10.667		8,425	565		8,990	10,200
3460	300 GPM, 6″ pipe size		1	16		9,775	845		10,620	12,100
3470	400 GPM, 6″ pipe size	Q-2	1.20	20		11,200	1,100		12,300	14,000
3480	500 GPM, 6″ pipe size	″	1	24		12,500	1,325		13,825	15,900
4000	Oil, fabricated steel, 10 GPM, 2″ pipe size	1 Plum	4	2		2,575	117		2,692	3,000
4100	15 GPM, 2″ or 3″ pipe size		4	2		3,525	117		3,642	4,075
4120	20 GPM, 2″ or 3″ pipe size		3	2.667		4,625	157		4,782	5,325
4140	25 GPM, 2″ or 3″ pipe size	Q-1	3.50	4.571		4,650	242		4,892	5,475
4160	35 GPM, 2″, 3″, or 4″ pipe size		3	5.333		5,650	282		5,932	6,650
4180	50 GPM, 2″, 3″, or 4″ pipe size		2	8		7,600	425		8,025	9,000
4200	75 GPM, 3″ pipe size		2	8		13,200	425		13,625	15,200
4220	100 GPM, 3″ pipe size		2	8		14,300	425		14,725	16,400
4240	150 GPM, 4″ pipe size		2	8		17,700	425		18,125	20,100
4260	200 GPM, 4″ pipe size		1.50	10.667		24,900	565		25,465	28,300
4280	250 GPM, 5″ pipe size		1.30	12.308		28,900	650		29,550	32,800
4300	300 GPM, 5″ pipe size		1	16		32,700	845		33,545	37,300
4320	400 GPM, 6″ pipe size	Q-2	1.20	20		42,100	1,100		43,200	48,000
4340	500 GPM, 6″ pipe size	″	1	24		53,000	1,325		54,325	60,500
5000	Sand interceptor, fabricated steel									
5020	20 GPM, 4″ pipe size	Q-1	3	5.333	Ea.	6,350	282		6,632	7,450
5030	50 GPM, 4″ pipe size		2.50	6.400		10,800	340		11,140	12,400
5040	150 GPM, 4″ pipe size		2	8		32,600	425		33,025	36,600
5050	250 GPM, 6″ pipe size		1.30	12.308		33,900	650		34,550	38,300
5060	500 GPM, 6″ pipe size	Q-2	1	24		43,500	1,325		44,825	49,900
6000	Solids, precious metals recovery, C.I., 1-1/4″ to 2″ pipe	1 Plum	4	2		655	117		772	905
6100	Dental Lab., large, C.I., 1-1/2″ to 2″ pipe		3	2.667		2,275	157		2,432	2,775
9000	Minimum labor/equipment charge		3	2.667	Job		157		157	244

22 13 Facility Sanitary Sewerage

22 13 26 – Sanitary Waste Separators

22 13 26.10 Separators

	22 13 26.10 Separators	Crew	Daily Output	Labor-Hours	Unit	Material	2015 Bare Costs Labor	Equipment	Total	Total Incl O&P
0010	**SEPARATORS**, Entrainment eliminator, steel body, 150 PSIG									
0100	1/4" size	1 Stpi	24	.333	Ea.	259	19.90		278.90	315
0120	1/2" size		24	.333		268	19.90		287.90	325
0140	3/4" size		20	.400		280	24		304	350
0160	1" size		19	.421		289	25		314	360
0180	1-1/4" size		15	.533		305	32		337	390
0200	1-1/2" size		13	.615		340	37		377	430
0220	2" size		11	.727		375	43.50		418.50	485
0240	2-1/2" size	Q-5	15	1.067		1,600	57.50		1,657.50	1,875
0260	3" size		13	1.231		1,800	66		1,866	2,100
0280	4" size		10	1.600		2,050	86		2,136	2,375
0300	5" size		6	2.667		2,425	143		2,568	2,900
0320	6" size		3	5.333		2,650	287		2,937	3,350
0340	8" size	Q-6	4.40	5.455		3,350	305		3,655	4,175
0360	10" size	"	4	6		4,750	335		5,085	5,750
1000	For 300 PSIG, add					15%				

22 13 29 – Sanitary Sewerage Pumps

22 13 29.13 Wet-Pit-Mounted, Vertical Sewerage Pumps

	22 13 29.13 Wet-Pit-Mounted, Vertical Sewerage Pumps	Crew	Daily Output	Labor-Hours	Unit	Material	Labor	Equipment	Total	Total Incl O&P
0010	**WET-PIT-MOUNTED, VERTICAL SEWERAGE PUMPS**									
0020	Controls incl. alarm/disconnect panel w/wire. Excavation not included									
0260	Simplex, 9 GPM at 60 PSIG, 91 gal. tank				Ea.	3,325			3,325	3,675
0300	Unit with manway, 26" I.D., 18" high					3,700			3,700	4,075
0340	26" I.D., 36" high					3,750			3,750	4,125
0380	43" I.D., 4' high					3,975			3,975	4,375
0600	Simplex, 9 GPM at 60 PSIG, 150 gal. tank, indoor					3,525			3,525	3,900
0700	Unit with manway, 26" I.D., 36" high					4,225			4,225	4,650
0740	26" I.D., 4' high					4,450			4,450	4,900
2000	Duplex, 18 GPM at 60 PSIG, 150 gal. tank, indoor					7,050			7,050	7,750
2060	Unit with manway, 43" I.D., 4' high					8,100			8,100	8,925
2400	For core only					1,875			1,875	2,050
3000	Indoor residential type installation									
3020	Simplex, 9 GPM at 60 PSIG, 91 gal. HDPE tank				Ea.	3,350			3,350	3,675

22 13 29.14 Sewage Ejector Pumps

	22 13 29.14 Sewage Ejector Pumps	Crew	Daily Output	Labor-Hours	Unit	Material	Labor	Equipment	Total	Total Incl O&P
0010	**SEWAGE EJECTOR PUMPS**, With operating and level controls									
0100	Simplex system incl. tank, cover, pump 15' head									
0500	37 gal. PE tank, 12 GPM, 1/2 HP, 2" discharge	Q-1	3.20	5	Ea.	480	264		744	935
0510	3" discharge		3.10	5.161		520	273		793	995
0530	87 GPM, .7 HP, 2" discharge		3.20	5		735	264		999	1,225
0540	3" discharge		3.10	5.161		795	273		1,068	1,300
0600	45 gal. coated stl. tank, 12 GPM, 1/2 HP, 2" discharge		3	5.333		855	282		1,137	1,375
0610	3" discharge		2.90	5.517		890	291		1,181	1,425
0630	87 GPM, .7 HP, 2" discharge		3	5.333		1,100	282		1,382	1,650
0640	3" discharge		2.90	5.517		1,150	291		1,441	1,725
0660	134 GPM, 1 HP, 2" discharge		2.80	5.714		1,175	300		1,475	1,775
0680	3" discharge		2.70	5.926		1,250	315		1,565	1,875
0700	70 gal. PE tank, 12 GPM, 1/2 HP, 2" discharge		2.60	6.154		920	325		1,245	1,525
0710	3" discharge		2.40	6.667		980	350		1,330	1,625
0730	87 GPM, 0.7 HP, 2" discharge		2.50	6.400		1,200	340		1,540	1,825
0740	3" discharge		2.30	6.957		1,275	370		1,645	1,975
0760	134 GPM, 1 HP, 2" discharge		2.20	7.273		1,300	385		1,685	2,025
0770	3" discharge		2	8		1,375	425		1,800	2,175

22 13 Facility Sanitary Sewerage

22 13 29 – Sanitary Sewerage Pumps

22 13 29.14 Sewage Ejector Pumps		Crew	Daily Output	Labor-Hours	Unit	Material	2015 Bare Costs Labor	Equipment	Total	Total Incl O&P
0800	75 gal. coated stl. tank, 12 GPM, 1/2 HP, 2" discharge	Q-1	2.40	6.667	Ea.	1,025	350		1,375	1,675
0810	3" discharge		2.20	7.273		1,075	385		1,460	1,775
0830	87 GPM, .7 HP, 2" discharge		2.30	6.957		1,300	370		1,670	2,000
0840	3" discharge		2.10	7.619		1,350	405		1,755	2,125
0860	134 GPM, 1 HP, 2" discharge		2	8		1,375	425		1,800	2,175
0880	3" discharge	↓	1.80	8.889	↓	1,450	470		1,920	2,325
1040	Duplex system incl. tank, covers, pumps									
1060	110 gal. fiberglass tank, 24 GPM, 1/2 HP, 2" discharge	Q-1	1.60	10	Ea.	1,850	530		2,380	2,850
1080	3" discharge		1.40	11.429		1,950	605		2,555	3,075
1100	174 GPM, .7 HP, 2" discharge		1.50	10.667		2,400	565		2,965	3,500
1120	3" discharge		1.30	12.308		2,475	650		3,125	3,750
1140	268 GPM, 1 HP, 2" discharge		1.20	13.333		2,600	705		3,305	3,950
1160	3" discharge	↓	1	16		2,700	845		3,545	4,275
1260	135 gal. coated stl. tank, 24 GPM, 1/2 HP, 2" discharge	Q-2	1.70	14.118		1,900	775		2,675	3,300
2000	3" discharge		1.60	15		2,025	820		2,845	3,500
2640	174 GPM, .7 HP, 2" discharge		1.60	15		2,500	820		3,320	4,025
2660	3" discharge		1.50	16		2,625	875		3,500	4,275
2700	268 GPM, 1 HP, 2" discharge		1.30	18.462		2,700	1,000		3,700	4,550
3040	3" discharge		1.10	21.818		2,875	1,200		4,075	5,050
3060	275 gal. coated stl. tank, 24 GPM, 1/2 HP, 2" discharge		1.50	16		2,375	875		3,250	4,000
3080	3" discharge		1.40	17.143		2,425	940		3,365	4,125
3100	174 GPM, .7 HP, 2" discharge		1.40	17.143		3,075	940		4,015	4,850
3120	3" discharge		1.30	18.462		3,250	1,000		4,250	5,150
3140	268 GPM, 1 HP, 2" discharge		1.10	21.818		3,375	1,200		4,575	5,575
3160	3" discharge	↓	.90	26.667	↓	3,550	1,450		5,000	6,175
3260	Pump system accessories, add									
3300	Alarm horn and lights, 115 V mercury switch	Q-1	8	2	Ea.	97	106		203	271
3340	Switch, mag. contactor, alarm bell, light, 3 level control		5	3.200		475	169		644	790
3380	Alternator, mercury switch activated		4	4	↓	855	211		1,066	1,275
9000	Minimum labor/equipment charge	↓	2.50	6.400	Job		340		340	525

22 14 Facility Storm Drainage

22 14 23 – Storm Drainage Piping Specialties

22 14 23.33 Backwater Valves		Crew	Daily Output	Labor-Hours	Unit	Material	2015 Bare Costs Labor	Equipment	Total	Total Incl O&P
0010	**BACKWATER VALVES**, C.I. Body									
6980	Bronze gate and automatic flapper valves									
7000	3" and 4" pipe size	Q-1	13	1.231	Ea.	2,050	65		2,115	2,350
7100	5" and 6" pipe size	"	13	1.231	"	3,125	65		3,190	3,550
7240	Bronze flapper valve, bolted cover									
7260	2" pipe size	Q-1	16	1	Ea.	595	53		648	740
7280	3" pipe size		14.50	1.103		935	58.50		993.50	1,125
7300	4" pipe size	↓	13	1.231		1,150	65		1,215	1,375
7320	5" pipe size	Q-2	18	1.333		1,650	73		1,723	1,950
7340	6" pipe size	"	17	1.412		1,650	77.50		1,727.50	1,950
7360	8" pipe size	Q-3	10	3.200		2,250	179		2,429	2,750
7380	10" pipe size	"	9	3.556	↓	3,700	199		3,899	4,350
7500	For threaded cover, same cost									
7540	Revolving disk type, same cost as flapper type									

22 14 Facility Storm Drainage

22 14 26 – Facility Storm Drains

22 14 26.13 Roof Drains		Crew	Daily Output	Labor-Hours	Unit	Material	2015 Bare Costs Labor	Equipment	Total	Total Incl O&P
0010	**ROOF DRAINS**									
0140	Cornice, C.I., 45° or 90° outlet									
0200	3″ and 4″ pipe size	Q-1	12	1.333	Ea.	330	70.50		400.50	475
0260	For galvanized body, add					75.50			75.50	83
0280	For polished bronze dome, add					85.50			85.50	94
3860	Roof, flat metal deck, C.I. body, 12″ C.I. dome									
3880	2″ pipe size	Q-1	15	1.067	Ea.	287	56.50		343.50	405
3890	3″ pipe size		14	1.143		395	60.50		455.50	530
3900	4″ pipe size		13	1.231		395	65		460	535
3910	5″ pipe size		12	1.333		535	70.50		605.50	700
3920	6″ pipe size		10	1.600		680	84.50		764.50	880
4280	Integral expansion joint, C.I. body, 12″ C.I. dome									
4300	2″ pipe size	Q-1	8	2	Ea.	590	106		696	815
4320	3″ pipe size		7	2.286		610	121		731	860
4340	4″ pipe size		6	2.667		655	141		796	940
4360	5″ pipe size		4	4		820	211		1,031	1,225
4380	6″ pipe size		3	5.333		875	282		1,157	1,400
4400	8″ pipe size		3	5.333		1,375	282		1,657	1,950
4440	For galvanized body, add					345			345	380
4620	Main, all aluminum, 12″ low profile dome									
4640	2″, 3″ and 4″ pipe size	Q-1	14	1.143	Ea.	435	60.50		495.50	575
4660	5″ and 6″ pipe size		13	1.231		575	65		640	735
4680	8″ pipe size		10	1.600		715	84.50		799.50	920
4690	Main, CI body, 12″ poly. dome, 2″, 3″, & 4″ pipe		8	2		325	106		431	520
4710	5″ and 6″ pipe size		6	2.667		465	141		606	730
4720	8″ pipe size		4	4		605	211		816	995
4730	For underdeck clamp, add		22	.727		224	38.50		262.50	305
4740	For vandalproof dome, add					56.50			56.50	62.50
4750	For galvanized body, add					440			440	485
4760	Main, ABS body and dome, 2″ pipe size	Q-1	14	1.143		131	60.50		191.50	238
4780	3″ pipe size		14	1.143		131	60.50		191.50	238
4800	4″ pipe size		14	1.143		131	60.50		191.50	238
4820	For underdeck clamp, add		24	.667		27	35		62	85
4900	Terrace planting area, with perforated overflow, C.I.									
4920	2″, 3″ and 4″ pipe size	Q-1	8	2	Ea.	590	106		696	815
9000	Minimum labor/equipment charge	1 Plum	4	2	Job		117		117	183

22 14 26.16 Facility Area Drains		Crew	Daily Output	Labor-Hours	Unit	Material	Labor	Equipment	Total	Total Incl O&P
0010	**FACILITY AREA DRAINS**									
4980	Scupper floor, oblique strainer, C.I.									
5000	6″ x 7″ top, 2″, 3″ and 4″ pipe size	Q-1	16	1	Ea.	283	53		336	395
5100	8″ x 12″ top, 5″ and 6″ pipe size	″	14	1.143		550	60.50		610.50	700
5160	For galvanized body, add					40%				
5200	For polished bronze strainer, add					85%				

22 14 26.19 Facility Trench Drains		Crew	Daily Output	Labor-Hours	Unit	Material	Labor	Equipment	Total	Total Incl O&P
0010	**FACILITY TRENCH DRAINS**									
5980	Trench, floor, heavy duty, modular, C.I., 12″ x 12″ top									
6000	2″, 3″, 4″, 5″, & 6″ pipe size	Q-1	8	2	Ea.	895	106		1,001	1,150
6100	For unit with polished bronze top		8	2		1,325	106		1,431	1,650
6200	For 12″ extension section, C.I. top		8	2		895	106		1,001	1,150
6240	For 12″ extension section, polished bronze top		8	2		1,400	106		1,506	1,725
6600	Trench, floor, for cement concrete encasement									
6610	Not including trenching or concrete									

22 14 Facility Storm Drainage

22 14 26 – Facility Storm Drains

22 14 26.19 Facility Trench Drains		Crew	Daily Output	Labor-Hours	Unit	2015 Bare Costs Material	Labor	Equipment	Total	Total Incl O&P
6640	Polyester polymer concrete									
6650	4″ internal width, with grate									
6660	Light duty steel grate	Q-1	120	.133	L.F.	35	7.05		42.05	49.50
6670	Medium duty steel grate		115	.139		40.50	7.35		47.85	56
6680	Heavy duty iron grate		110	.145		62	7.70		69.70	80
6700	12″ internal width, with grate									
6770	Heavy duty galvanized grate	Q-1	80	.200	L.F.	169	10.55		179.55	202
6800	Fiberglass									
6810	8″ internal width, with grate									
6820	Medium duty galvanized grate	Q-1	115	.139	L.F.	111	7.35		118.35	133
6830	Heavy duty iron grate	″	110	.145	″	106	7.70		113.70	128

22 14 29 – Sump Pumps

22 14 29.13 Wet-Pit-Mounted, Vertical Sump Pumps		Crew	Daily Output	Labor-Hours	Unit	Material	Labor	Equipment	Total	Total Incl O&P
0010	**WET-PIT-MOUNTED, VERTICAL SUMP PUMPS**									
0400	Molded PVC base, 21 GPM at 15′ head, 1/3 HP	1 Plum	5	1.600	Ea.	135	94		229	295
0800	Iron base, 21 GPM at 15′ head, 1/3 HP		5	1.600		164	94		258	325
1200	Solid brass, 21 GPM at 15′ head, 1/3 HP		5	1.600		289	94		383	465
2000	Sump pump, single stage									
2010	25 GPM, 1 HP, 1-1/2″ discharge	Q-1	1.80	8.889	Ea.	3,825	470		4,295	4,950
2020	75 GPM, 1-1/2 HP, 2″ discharge		1.50	10.667		4,050	565		4,615	5,325
2030	100 GPM, 2 HP, 2-1/2″ discharge		1.30	12.308		4,125	650		4,775	5,575
2040	150 GPM, 3 HP, 3″ discharge		1.10	14.545		4,125	770		4,895	5,750
2050	200 GPM, 3 HP, 3″ discharge		1	16		4,375	845		5,220	6,150
2060	300 GPM, 10 HP, 4″ discharge	Q-2	1.20	20		4,725	1,100		5,825	6,900
2070	500 GPM, 15 HP, 5″ discharge		1.10	21.818		5,375	1,200		6,575	7,775
2080	800 GPM, 20 HP, 6″ discharge		1	24		6,350	1,325		7,675	9,025
2090	1000 GPM, 30 HP, 6″ discharge		.85	28.235		6,975	1,550		8,525	10,100
2100	1600 GPM, 50 HP, 8″ discharge		.72	33.333		10,900	1,825		12,725	14,900
2110	2000 GPM, 60 HP, 8″ discharge	Q-3	.85	37.647		11,100	2,100		13,200	15,500
2202	For general purpose float switch, copper coated float, add	Q-1	5	3.200		108	169		277	385

22 14 29.16 Submersible Sump Pumps		Crew	Daily Output	Labor-Hours	Unit	Material	Labor	Equipment	Total	Total Incl O&P
0010	**SUBMERSIBLE SUMP PUMPS**									
1000	Elevator sump pumps, automatic									
1010	Complete systems, pump, oil detector, controls and alarm									
1020	1-1/2″discharge, does not include the sump pit/tank.									
1040	1/3 HP, 115 V	1 Plum	4.40	1.818	Ea.	1,875	107		1,982	2,225
1050	1/2 HP, 115 V		4	2		1,925	117		2,042	2,300
1060	1/2 HP, 230 V		4	2		1,975	117		2,092	2,350
1070	3/4 HP, 115 V		3.60	2.222		2,025	130		2,155	2,425
1080	3/4 HP, 230 V		3.60	2.222		2,075	130		2,205	2,475
1100	Sump pump only									
1110	1/3 HP, 115 V	1 Plum	6.40	1.250	Ea.	170	73.50		243.50	300
1120	1/2 HP, 115 V		5.80	1.379		243	81		324	395
1130	1/2 HP, 230 V		5.80	1.379		287	81		368	440
1140	3/4 HP, 115 V		5.40	1.481		325	87		412	490
1150	3/4 HP, 230 V		5.40	1.481		370	87		457	545
1200	Oil detector, control and alarm only									
1210	115 V	1 Plum	8	1	Ea.	1,700	58.50		1,758.50	1,975
1220	230 V	″	8	1	″	1,700	58.50		1,758.50	1,975
7000	Sump pump, automatic									
7100	Plastic, 1-1/4″ discharge, 1/4 HP	1 Plum	6.40	1.250	Ea.	138	73.50		211.50	266
7140	1/3 HP		6	1.333		200	78.50		278.50	340

22 14 Facility Storm Drainage

22 14 29 – Sump Pumps

22 14 29.16 Submersible Sump Pumps		Crew	Daily Output	Labor-Hours	Unit	Material	Labor	Equipment	Total	Total Incl O&P
						2015 Bare Costs				
7160	1/2 HP	1 Plum	5.40	1.481	Ea.	246	87		333	405
7180	1-1/2" discharge, 1/2 HP		5.20	1.538		281	90.50		371.50	450
7500	Cast iron, 1-1/4" discharge, 1/4 HP		6	1.333		194	78.50		272.50	335
7540	1/3 HP		6	1.333		229	78.50		307.50	375
7560	1/2 HP		5	1.600		277	94		371	450
9000	Minimum labor/equipment charge		4	2	Job		117		117	183

22 14 53 – Rainwater Storage Tanks

22 14 53.13 Fiberglass, Rainwater Storage Tank

		Crew	Daily Output	Labor-Hours	Unit	Material	Labor	Equipment	Total	Total Incl O&P
0010	**FIBERGLASS, RAINWATER STORAGE TANK**									
2000	600 gallon	B-21B	3.75	10.667	Ea.	3,300	435	174	3,909	4,525
2010	1,000 gallon		3.50	11.429		4,250	465	187	4,902	5,650
2020	2,000 gallon		3.25	12.308		6,200	500	201	6,901	7,825
2030	4,000 gallon		3	13.333		8,575	545	218	9,338	10,600
2040	6,000 gallon		2.65	15.094		9,625	615	247	10,487	11,900
2050	8,000 gallon		2.30	17.391		11,500	710	284	12,494	14,200
2060	10,000 gallon		2	20		13,200	815	325	14,340	16,300
2070	12,000 gallon		1.50	26.667		18,300	1,100	435	19,835	22,400
2080	15,000 gallon		1	40		21,200	1,625	655	23,480	26,700
2090	20,000 gallon		.75	53.333		27,300	2,175	870	30,345	34,500
2100	25,000 gallon		.50	80		40,800	3,275	1,300	45,375	51,500
2110	30,000 gallon		.35	114		49,400	4,675	1,875	55,950	64,000
2120	40,000 gallon		.30	133		71,500	5,450	2,175	79,125	90,000

22 15 General Service Compressed-Air Systems

22 15 13 – General Service Compressed-Air Piping

22 15 13.10 Compressor Accessories

		Crew	Daily Output	Labor-Hours	Unit	Material	Labor	Equipment	Total	Total Incl O&P
0010	**COMPRESSOR ACCESSORIES**									
1700	Refrigerated air dryers with ambient air filters									
1710	10 CFM	Q-5	8	2	Ea.	960	108		1,068	1,225
1720	25 CFM		6.60	2.424		1,075	130		1,205	1,375
1730	50 CFM		6.20	2.581		1,950	139		2,089	2,375
1740	75 CFM		5.80	2.759		2,525	148		2,673	3,000
1750	100 CFM		5.60	2.857		2,675	154		2,829	3,200
3460	Air filter, regulator, lubricator combination									
3470	Flush mount									
3480	Adjustable range 0-140 psi									
3500	1/8" NPT, 34 SCFM	1 Stpi	17	.471	Ea.	108	28		136	163
3510	1/4" NPT, 61 SCFM		17	.471		179	28		207	241
3520	3/8" NPT, 85 SCFM		16	.500		184	30		214	250
3530	1/2" NPT, 150 SCFM		15	.533		179	32		211	247
3540	3/4" NPT, 171 SCFM		14	.571		237	34		271	315
3550	1" NPT, 150 SCFM		13	.615		370	37		407	465
4000	Couplers, air line, sleeve type									
4010	Female, connection size NPT									
4020	1/4"	1 Stpi	38	.211	Ea.	9.95	12.60		22.55	30.50
4030	3/8"		36	.222		8.55	13.30		21.85	30
4040	1/2"		35	.229		21.50	13.65		35.15	45.50
4050	3/4"		34	.235		21.50	14.05		35.55	46
4100	Male									
4110	1/4"	1 Stpi	38	.211	Ea.	9	12.60		21.60	29.50

22 15 General Service Compressed-Air Systems

22 15 13 – General Service Compressed-Air Piping

22 15 13.10 Compressor Accessories		Crew	Daily Output	Labor-Hours	Unit	Material	Labor	Equipment	Total	Total Incl O&P
						2015 Bare Costs				
4120	3/8"	1 Stpi	36	.222	Ea.	10.60	13.30		23.90	32
4130	1/2"		35	.229		21	13.65		34.65	45
4140	3/4"		34	.235		21.50	14.05		35.55	46
4150	Coupler, combined male and female halves									
4160	1/2"	1 Stpi	17	.471	Ea.	43	28		71	91
4170	3/4"	"	15	.533	"	43.50	32		75.50	97

22 15 19 – General Service Packaged Air Compressors and Receivers

22 15 19.10 Air Compressors

Line	Description	Crew	Daily Output	Labor-Hours	Unit	Material	Labor	Equipment	Total	Total Incl O&P
0010	**AIR COMPRESSORS**									
5250	Air, reciprocating air cooled, splash lubricated, tank mounted									
5300	Single stage, 1 phase, 140 psi									
5303	1/2 HP, 30 gal. tank	1 Stpi	3	2.667	Ea.	1,950	159		2,109	2,400
5305	3/4 HP, 30 gal. tank		2.60	3.077		1,950	184		2,134	2,425
5307	1 HP, 30 gal. tank		2.20	3.636		2,450	217		2,667	3,050
5309	2 HP, 30 gal. tank	Q-5	4	4		2,625	215		2,840	3,225
5310	3 HP, 30 gal. tank		3.60	4.444		3,350	239		3,589	4,075
5314	3 HP, 60 gal. tank		3.50	4.571		3,550	246		3,796	4,275
5320	5 HP, 60 gal. tank		3.20	5		3,725	269		3,994	4,525
5330	5 HP, 80 gal. tank		3	5.333		4,100	287		4,387	4,950
5340	7.5 HP, 80 gal. tank		2.60	6.154		5,300	330		5,630	6,350
5600	2 stage pkg., 3 phase									
5650	6 CFM at 125 psi 1-1/2 HP, 60 gal. tank	Q-5	3	5.333	Ea.	3,375	287		3,662	4,150
5670	10.9 CFM at 125 psi, 3 HP, 80 gal. tank		1.50	10.667		3,800	575		4,375	5,075
5680	38.7 CFM at 125 psi, 10 HP, 120 gal. tank		.60	26.667		6,825	1,425		8,250	9,725
5690	105 CFM at 125 psi, 25 HP, 250 gal. tank	Q-6	.60	40		14,100	2,225		16,325	19,000
5800	With single stage pump									
5850	8.3 CFM at 125 psi, 2 HP, 80 gal. tank	Q-6	3.50	6.857	Ea.	3,775	380		4,155	4,750
5860	38.7 CFM at 125 psi, 10 HP, 120 gal. tank	"	.90	26.667	"	6,625	1,475		8,100	9,625
6000	Reciprocating, 2 stage, tank mtd, 3 Ph., Cap. rated @175 PSIG									
6050	Pressure lubricated, hvy. duty, 9.7 CFM, 3 HP, 120 gal. tank	Q-5	1.30	12.308	Ea.	5,775	660		6,435	7,375
6054	5 CFM, 1-1/2 HP, 80 gal. tank		2.80	5.714		3,575	305		3,880	4,425
6056	6.4 CFM, 2 HP, 80 gal. tank		2	8		3,775	430		4,205	4,825
6058	8.1 CFM, 3 HP, 80 gal. tank		1.70	9.412		3,800	505		4,305	4,975
6059	14.8 CFM, 5 HP, 80 gal. tank		1	16		4,100	860		4,960	5,850
6060	16.5 CFM, 5 HP, 120 gal. tank		1	16		5,900	860		6,760	7,825
6063	13 CFM, 6 HP, 80 gal. tank		.90	17.778		5,875	955		6,830	7,975
6066	19.8 CFM, 7.5 HP, 80 gal. tank		.80	20		5,875	1,075		6,950	8,150
6070	25.8 CFM, 7-1/2 HP, 120 gal. tank		.80	20		7,975	1,075		9,050	10,400
6078	34.8 CFM, 10 HP, 80 gal. tank		.70	22.857		8,050	1,225		9,275	10,800
6080	34.8 CFM, 10 HP, 120 gal. tank		.60	26.667		8,500	1,425		9,925	11,600
6090	53.7 CFM, 15 HP, 120 gal. tank	Q-6	.80	30		10,300	1,675		11,975	13,900
6100	76.7 CFM, 20 HP, 120 gal. tank		.70	34.286		13,500	1,900		15,400	17,800
6104	76.7 CFM, 20 HP, 240 gal. tank		.68	35.294		14,800	1,975		16,775	19,400
6110	90.1 CFM, 25 HP, 120 gal. tank		.63	38.095		14,100	2,125		16,225	18,800
6120	101 CFM, 30 HP, 120 gal. tank		.57	42.105		15,200	2,350		17,550	20,400
6130	101 CFM, 30 HP, 250 gal. tank		.52	46.154		16,400	2,575		18,975	22,100
6200	Oil-less, 13.6 CFM, 5 HP, 120 gal. tank	Q-5	.88	18.182		16,900	980		17,880	20,100
6210	13.6 CFM, 5 HP, 250 gal. tank		.80	20		18,200	1,075		19,275	21,700
6220	18.2 CFM, 7.5 HP, 120 gal. tank		.73	21.918		16,900	1,175		18,075	20,500
6230	18.2 CFM, 7.5 HP, 250 gal. tank		.67	23.881		18,200	1,275		19,475	22,000
6250	30.5 CFM, 10 HP, 120 gal. tank		.57	28.070		19,900	1,500		21,400	24,300
6260	30.5 CFM, 10 HP, 250 gal. tank		.53	30.189		21,200	1,625		22,825	25,800

22 15 General Service Compressed-Air Systems

22 15 19 – General Service Packaged Air Compressors and Receivers

22 15 19.10 Air Compressors		Crew	Daily Output	Labor-Hours	Unit	Material	Labor	Equipment	Total	Total Incl O&P
						2015 Bare Costs				
6270	41.3 CFM, 15 HP, 120 gal. tank	Q-6	.70	34.286	Ea.	21,700	1,900		23,600	26,900
6280	41.3 CFM, 15 HP, 250 gal. tank	"	.67	35.821	↓	22,900	2,000		24,900	28,300

22 31 Domestic Water Softeners

22 31 13 – Residential Domestic Water Softeners

22 31 13.10 Residential Water Softeners

		Crew	Daily Output	Labor-Hours	Unit	Material	Labor	Equipment	Total	Total Incl O&P
0010	**RESIDENTIAL WATER SOFTENERS**									
7350	Water softener, automatic, to 30 grains per gallon	2 Plum	5	3.200	Ea.	405	188		593	740
7400	To 100 grains per gallon	"	4	4	"	660	235		895	1,100

22 31 16 – Commercial Domestic Water Softeners

22 31 16.10 Water Softeners

		Crew	Daily Output	Labor-Hours	Unit	Material	Labor	Equipment	Total	Total Incl O&P
0010	**WATER SOFTENERS**									
5800	Softener systems, automatic, intermediate sizes									
5820	available, may be used in multiples.									
6000	Hardness capacity between regenerations and flow									
6100	150,000 grains, 37 GPM cont., 51 GPM peak	Q-1	1.20	13.333	Ea.	6,075	705		6,780	7,775
6200	300,000 grains, 81 GPM cont., 113 GPM peak		1	16		9,850	845		10,695	12,100
6300	750,000 grains, 160 GPM cont., 230 GPM peak		.80	20		12,800	1,050		13,850	15,800
6400	900,000 grains, 185 GPM cont., 270 GPM peak	↓	.70	22.857	↓	20,700	1,200		21,900	24,600

22 32 Domestic Water Filtration Equipment

22 32 19 – Domestic-Water Off-Floor Cartridge Filters

22 32 19.10 Water Filters

		Crew	Daily Output	Labor-Hours	Unit	Material	Labor	Equipment	Total	Total Incl O&P
0010	**WATER FILTERS**, Purification and treatment.									
1000	Cartridge style, dirt and rust type	1 Plum	12	.667	Ea.	200	39		239	281
1200	Replacement cartridge		32	.250		19.60	14.70		34.30	44.50
1600	Taste and odor type		12	.667		246	39		285	330
1700	Replacement cartridge		32	.250		40.50	14.70		55.20	67.50
3000	Central unit, dirt/rust/odor/taste/scale		4	2		545	117		662	780
3100	Replacement cartridge, standard		20	.400		75.50	23.50		99	120
3600	Replacement cartridge, heavy duty	↓	20	.400	↓	95	23.50		118.50	141
8000	Commercial, fully automatic or push button automatic									
8200	Iron removal, 660 GPH, 1" pipe size	Q-1	1.50	10.667	Ea.	3,075	565		3,640	4,250
8240	1500 GPH, 1-1/4" pipe size		1	16		5,175	845		6,020	7,025
8280	2340 GPH, 1-1/2" pipe size		.80	20		5,675	1,050		6,725	7,900
8320	3420 GPH, 2" pipe size		.60	26.667		10,500	1,400		11,900	13,700
8360	4620 GPH, 2-1/2" pipe size		.50	32		16,600	1,700		18,300	20,900
8500	Neutralizer for acid water, 780 GPH, 1" pipe size		1.50	10.667		2,975	565		3,540	4,125
8540	1140 GPH, 1-1/4" pipe size		1	16		3,350	845		4,195	5,000
8580	1740 GPH, 1-1/2" pipe size		.80	20		4,850	1,050		5,900	6,975
8620	2520 GPH, 2" pipe size		.60	26.667		6,300	1,400		7,700	9,125
8660	3480 GPH, 2-1/2" pipe size		.50	32		10,500	1,700		12,200	14,100
8800	Sediment removal, 780 GPH, 1" pipe size		1.50	10.667		2,825	565		3,390	4,000
8840	1140 GPH, 1-1/4" pipe size		1	16		3,400	845		4,245	5,050
8880	1740 GPH, 1-1/2" pipe size		.80	20		4,500	1,050		5,550	6,600
8920	2520 GPH, 2" pipe size		.60	26.667		6,500	1,400		7,900	9,350
8960	3480 GPH, 2-1/2" pipe size		.50	32		10,200	1,700		11,900	13,800
9200	Taste and odor removal, 660 GPH, 1" pipe size		1.50	10.667		3,950	565		4,515	5,225
9240	1500 GPH, 1-1/4" pipe size	↓	1	16	↓	6,800	845		7,645	8,800

22 32 Domestic Water Filtration Equipment

22 32 19 – Domestic-Water Off-Floor Cartridge Filters

22 32 19.10 Water Filters		Crew	Daily Output	Labor-Hours	Unit	Material	Labor (2015 Bare Costs)	Equipment	Total	Total Incl O&P
9280	2340 GPH, 1-1/2" pipe size	Q-1	.80	20	Ea.	7,725	1,050		8,775	10,200
9320	3420 GPH, 2" pipe size		.60	26.667		11,900	1,400		13,300	15,300
9360	4620 GPH, 2-1/2" pipe size		.50	32		18,300	1,700		20,000	22,800

22 33 Electric Domestic Water Heaters

22 33 13 – Instantaneous Electric Domestic Water Heaters

22 33 13.10 Hot Water Dispensers

		Crew	Daily Output	Labor-Hours	Unit	Material	Labor	Equipment	Total	Total Incl O&P
0010	**HOT WATER DISPENSERS**									
0160	Commercial, 100 cup, 11.3 amp	1 Plum	14	.571	Ea.	510	33.50		543.50	620
3180	Household, 60 cup	"	14	.571	"	269	33.50		302.50	350

22 33 13.20 Instantaneous Electric Point-Of-Use Water Heaters

		Crew	Daily Output	Labor-Hours	Unit	Material	Labor	Equipment	Total	Total Incl O&P
0010	**INSTANTANEOUS ELECTRIC POINT-OF-USE WATER HEATERS**									
8965	Point of use, electric, glass lined									
8969	Energy saver									
8970	2.5 gal. single element G	1 Plum	2.80	2.857	Ea.	245	168		413	530
8971	4 gal. single element G		2.80	2.857		253	168		421	540
8974	6 gal. single element G		2.50	3.200		261	188		449	580
8975	10 gal. single element G		2.50	3.200		330	188		518	655
8976	15 gal. single element G		2.40	3.333		370	196		566	710
8977	20 gal. single element G		2.40	3.333		410	196		606	755
8978	30 gal. single element G		2.30	3.478		480	204		684	850
8979	40 gal. single element G		2.20	3.636		800	213		1,013	1,225
8988	Commercial (ASHRAE energy std. 90)									
8989	6 gallon G	1 Plum	2.50	3.200	Ea.	665	188		853	1,025
8990	10 gallon G		2.50	3.200		710	188		898	1,075
8991	15 gallon G		2.40	3.333		750	196		946	1,125
8992	20 gallon G		2.40	3.333		785	196		981	1,175
8993	30 gallon G		2.30	3.478		1,700	204		1,904	2,200
8995	Under the sink, copper, w/bracket									
8996	2.5 gallon G	1 Plum	4	2	Ea.	485	117		602	720
9000	Minimum labor/equipment charge	"	1.75	4.571	Job		268		268	420

22 33 30 – Residential, Electric Domestic Water Heaters

22 33 30.13 Residential, Small-Capacity Elec. Water Heaters

		Crew	Daily Output	Labor-Hours	Unit	Material	Labor	Equipment	Total	Total Incl O&P
0010	**RESIDENTIAL, SMALL-CAPACITY ELECTRIC DOMESTIC WATER HEATERS**									
1000	Residential, electric, glass lined tank, 5 yr., 10 gal., single element	1 Plum	2.30	3.478	Ea.	330	204		534	680
1040	20 gallon, single element		2.20	3.636		410	213		623	785
1060	30 gallon, double element		2.20	3.636		475	213		688	860
1080	40 gallon, double element		2	4		800	235		1,035	1,250
1100	52 gallon, double element		2	4		895	235		1,130	1,350
1120	66 gallon, double element		1.80	4.444		1,200	261		1,461	1,725
1140	80 gallon, double element		1.60	5		1,350	294		1,644	1,925
1180	120 gallon, double element		1.40	5.714		1,900	335		2,235	2,600

22 33 33 – Light-Commercial Electric Domestic Water Heaters

22 33 33.10 Commercial Electric Water Heaters

		Crew	Daily Output	Labor-Hours	Unit	Material	Labor	Equipment	Total	Total Incl O&P
0010	**COMMERCIAL ELECTRIC WATER HEATERS**									
4000	Commercial, 100° rise. NOTE: for each size tank, a range of									
4010	heaters between the ones shown are available									
4020	Electric									
4100	5 gal., 3 kW, 12 GPH, 208 volt	1 Plum	2	4	Ea.	2,850	235		3,085	3,525
4120	10 gal., 6 kW, 25 GPH, 208 volt		2	4		3,175	235		3,410	3,850

22 33 Electric Domestic Water Heaters

22 33 33 – Light-Commercial Electric Domestic Water Heaters

22 33 33.10 Commercial Electric Water Heaters		Crew	Daily Output	Labor-Hours	Unit	Material	Labor	Equipment	Total	Total Incl O&P
						2015 Bare Costs				
4130	30 gal., 24 kW, 98 GPH, 208 volt	1 Plum	1.92	4.167	Ea.	5,200	245		5,445	6,100
4136	40 gal., 36 kW, 148 GPH, 208 volt		1.88	4.255		6,225	250		6,475	7,250
4140	50 gal., 9 kW, 37 GPH, 208 volt		1.80	4.444		4,350	261		4,611	5,175
4160	50 gal., 36 kW, 148 GPH, 208 volt		1.80	4.444		6,650	261		6,911	7,700
4180	80 gal., 12 kW, 49 GPH, 208 volt		1.50	5.333		5,350	315		5,665	6,400
4200	80 gal., 36 kW, 148 GPH, 208 volt		1.50	5.333		7,400	315		7,715	8,650
4220	100 gal., 36 kW, 148 GPH, 208 volt		1.20	6.667		7,725	390		8,115	9,100
4240	120 gal., 36 kW, 148 GPH, 208 volt		1.20	6.667		8,050	390		8,440	9,450
4260	150 gal., 15 kW , 61 GPH, 480 volt		1	8		18,900	470		19,370	21,500
4280	150 gal., 120 kW, 490 GPH, 480 volt		1	8		26,600	470		27,070	29,900
4300	200 gal., 15 kW, 61 GPH, 480 volt	Q-1	1.70	9.412		20,500	495		20,995	23,300
4320	200 gal., 120 kW , 490 GPH, 480 volt		1.70	9.412		28,000	495		28,495	31,600
4340	250 gal., 15 kW, 61 GPH, 480 volt		1.50	10.667		21,100	565		21,665	24,100
4360	250 gal., 150 kW, 615 GPH, 480 volt		1.50	10.667		30,800	565		31,365	34,800
4380	300 gal., 30 kW, 123 GPH, 480 volt		1.30	12.308		23,300	650		23,950	26,600
4400	300 gal., 180 kW, 738 GPH, 480 volt		1.30	12.308		41,700	650		42,350	46,900
4420	350 gal., 30 kW, 123 GPH, 480 volt		1.10	14.545		24,600	770		25,370	28,200
4440	350 gal., 180 kW, 738 GPH, 480 volt		1.10	14.545		34,700	770		35,470	39,300
4460	400 gal., 30 kW, 123 GPH, 480 volt		1	16		27,800	845		28,645	31,900
4480	400 gal., 210 kW, 860 GPH, 480 volt		1	16		40,000	845		40,845	45,300
4500	500 gal., 30 kW, 123 GPH, 480 volt		.80	20		32,500	1,050		33,550	37,400
4520	500 gal., 240 kW, 984 GPH, 480 volt		.80	20		47,800	1,050		48,850	54,000
4540	600 gal., 30 kW, 123 GPH, 480 volt	Q-2	1.20	20		24,700	1,100		25,800	28,900
4560	600 gal., 300 kW, 1230 GPH, 480 volt		1.20	20		37,600	1,100		38,700	43,000
4580	700 gal., 30 kW, 123 GPH, 480 volt		1	24		26,000	1,325		27,325	30,700
4600	700 gal., 300 kW, 1230 GPH, 480 volt		1	24		39,100	1,325		40,425	45,100
4620	800 gal., 60 kW, 245 GPH, 480 volt		.90	26.667		34,800	1,450		36,250	40,600
4640	800 gal., 300 kW, 1230 GPH, 480 volt		.90	26.667		40,000	1,450		41,450	46,300
4660	1000 gal., 60 kW, 245 GPH, 480 volt		.70	34.286		30,800	1,875		32,675	36,800
4680	1000 gal., 480 kW, 1970 GPH, 480 volt		.70	34.286		51,000	1,875		52,875	59,500
4700	1200 gal., 60 kW, 245 GPH, 480 volt		.60	40		52,500	2,200		54,700	61,000
4720	1200 gal., 480 kW, 1970 GPH, 480 volt		.60	40		79,500	2,200		81,700	91,000
4740	1500 gal., 60 kW, 245 GPH, 480 volt		.50	48		69,000	2,625		71,625	80,000
4760	1500 gal., 480 kW, 1970 GPH, 480 volt		.50	48		95,500	2,625		98,125	109,500
5400	Modulating step control for under 90 kW, 2-5 steps	1 Elec	5.30	1.509		810	82.50		892.50	1,025
5440	1 through 5 steps beyond standard		3.20	2.500		221	137		358	455
5460	6 through 10 steps beyond standard		2.70	2.963		455	162		617	750
5480	11 through 18 steps beyond standard		1.60	5		680	274		954	1,175

22 34 Fuel-Fired Domestic Water Heaters

22 34 13 – Instantaneous, Tankless, Gas Domestic Water Heaters

22 34 13.10 Instantaneous, Tankless, Gas Water Heaters		Crew	Daily Output	Labor-Hours	Unit	Material	Labor	Equipment	Total	Total Incl O&P
0010	**INSTANTANEOUS, TANKLESS, GAS WATER HEATERS**									
9410	Natural gas/propane, 3.2 GPM G	1 Plum	2	4	Ea.	370	235		605	770
9420	6.4 GPM G		1.90	4.211		625	247		872	1,075
9430	8.4 GPM G		1.80	4.444		730	261		991	1,200
9440	9.5 GPM G		1.60	5		930	294		1,224	1,475

22 34 Fuel-Fired Domestic Water Heaters

22 34 30 – Residential Gas Domestic Water Heaters

22 34 30.13 Residential, Atmos, Gas Domestic Wtr Heaters		Crew	Daily Output	Labor-Hours	Unit	Material	2015 Bare Costs Labor	Equipment	Total	Total Incl O&P
0010	RESIDENTIAL, ATMOSPHERIC, GAS DOMESTIC WATER HEATERS									
2000	Gas fired, foam lined tank, 10 yr., vent not incl.									
2040	30 gallon	1 Plum	2	4	Ea.	895	235		1,130	1,350
2060	40 gallon		1.90	4.211		895	247		1,142	1,375
2080	50 gallon		1.80	4.444		935	261		1,196	1,425
2090	60 gallon		1.70	4.706		1,325	276		1,601	1,875
2100	75 gallon		1.50	5.333		1,350	315		1,665	2,000
2120	100 gallon		1.30	6.154		1,600	360		1,960	2,350
2900	Water heater, safety-drain pan, 26" round		20	.400		37	23.50		60.50	77.50

22 34 36 – Commercial Gas Domestic Water Heaters

22 34 36.13 Commercial, Atmos., Gas Domestic Water Htrs.		Crew	Daily Output	Labor-Hours	Unit	Material	Labor	Equipment	Total	Total Incl O&P
0010	COMMERCIAL, ATMOSPHERIC, GAS DOMESTIC WATER HEATERS									
6000	Gas fired, flush jacket, std. controls, vent not incl.									
6040	75 MBH input, 73 GPH	1 Plum	1.40	5.714	Ea.	3,500	335		3,835	4,375
6060	98 MBH input, 95 GPH		1.40	5.714		5,300	335		5,635	6,350
6080	120 MBH input, 110 GPH		1.20	6.667		5,500	390		5,890	6,625
6100	120 MBH input, 115 GPH		1.10	7.273		6,625	425		7,050	7,975
6120	140 MBH input, 130 GPH		1	8		7,225	470		7,695	8,675
6140	155 MBH input, 150 GPH		.80	10		8,200	585		8,785	9,950
6160	180 MBH input, 170 GPH		.70	11.429		8,675	670		9,345	10,600
6180	200 MBH input, 192 GPH		.60	13.333		8,925	785		9,710	11,100
6200	250 MBH input, 245 GPH		.50	16		9,325	940		10,265	11,800
6220	260 MBH input, 250 GPH	Q-1	.80	20		9,900	1,050		10,950	12,600
6240	360 MBH input, 360 GPH		.80	20		11,700	1,050		12,750	14,600
6260	500 MBH input, 480 GPH		.70	22.857		16,400	1,200		17,600	20,000
6280	725 MBH input, 690 GPH		.60	26.667		19,300	1,400		20,700	23,400
6900	For low water cutoff, add	1 Plum	8	1		350	58.50		408.50	475
6960	For bronze body hot water circulator, add	"	4	2		1,925	117		2,042	2,300

22 34 46 – Oil-Fired Domestic Water Heaters

22 34 46.10 Residential Oil-Fired Water Heaters		Crew	Daily Output	Labor-Hours	Unit	Material	Labor	Equipment	Total	Total Incl O&P
0010	RESIDENTIAL OIL-FIRED WATER HEATERS									
3000	Oil fired, glass lined tank, 5 yr., vent not included, 30 gallon	1 Plum	2	4	Ea.	1,175	235		1,410	1,650
3040	50 gallon		1.80	4.444		1,375	261		1,636	1,925
3060	70 gallon		1.50	5.333		1,975	315		2,290	2,675

22 34 46.20 Commercial Oil-Fired Water Heaters		Crew	Daily Output	Labor-Hours	Unit	Material	Labor	Equipment	Total	Total Incl O&P
0010	COMMERCIAL OIL-FIRED WATER HEATERS									
8000	Oil fired, glass lined, UL listed, std. controls, vent not incl.									
8060	140 gal., 140 MBH input, 134 GPH	Q-1	2.13	7.512	Ea.	19,900	395		20,295	22,500
8080	140 gal., 199 MBH input, 191 GPH		2	8		20,600	425		21,025	23,400
8100	140 gal., 255 MBH input, 247 GPH		1.60	10		21,200	530		21,730	24,100
8120	140 gal., 270 MBH input, 259 GPH		1.20	13.333		26,200	705		26,905	29,900
8140	140 gal., 400 MBH input, 384 GPH		1	16		27,100	845		27,945	31,100
8160	140 gal., 540 MBH input, 519 GPH		.96	16.667		28,100	880		28,980	32,300
8180	140 gal., 720 MBH input, 691 GPH		.92	17.391		28,600	920		29,520	32,900
8200	221 gal., 300 MBH input, 288 GPH		.88	18.182		37,800	960		38,760	43,100
8220	221 gal., 600 MBH input, 576 GPH		.86	18.605		42,200	985		43,185	47,900
8240	221 gal., 800 MBH input, 768 GPH		.82	19.512		42,500	1,025		43,525	48,300
8260	201 gal., 1000 MBH input, 960 GPH	Q-2	1.26	19.048		43,500	1,050		44,550	49,400
8280	201 gal., 1250 MBH input, 1200 GPH		1.22	19.672		43,900	1,075		44,975	50,000
8300	201 gal., 1500 MBH input, 1441 GPH		1.16	20.690		47,800	1,125		48,925	54,500
8320	411 gal., 600 MBH input, 576 GPH		1.12	21.429		48,200	1,175		49,375	55,000

22 34 Fuel-Fired Domestic Water Heaters

22 34 46 – Oil-Fired Domestic Water Heaters

22 34 46.20 Commercial Oil-Fired Water Heaters		Crew	Daily Output	Labor-Hours	Unit	Material	2015 Bare Costs Labor	Equipment	Total	Total Incl O&P
8340	411 gal., 800 MBH input, 768 GPH	Q-2	1.08	22.222	Ea.	48,300	1,225		49,525	55,000
8360	411 gal., 1000 MBH input, 960 GPH		1.04	23.077		51,000	1,275		52,275	58,000
8380	411 gal., 1250 MBH input, 1200 GPH		.98	24.490		52,000	1,350		53,350	59,500
8400	397 gal., 1500 MBH input, 1441 GPH		.92	26.087		55,500	1,425		56,925	63,000
8420	397 gal., 1750 MBH input, 1681 GPH		.86	27.907		57,000	1,525		58,525	65,500
8430	397 gal., 2000 MBH input, 1921 GPH		.82	29.268		62,000	1,600		63,600	70,500
8440	375 gal., 2250 MBH input, 2161 GPH		.76	31.579		63,500	1,725		65,225	72,500
8450	375 gal., 2500 MBH input, 2401 GPH		.82	29.268		66,000	1,600		67,600	75,000
8900	For low water cutoff, add	1 Plum	8	1		350	58.50		408.50	475
8960	For bronze body hot water circulator, add	"	4	2		710	117		827	965

22 35 Domestic Water Heat Exchangers

22 35 30 – Water Heating by Steam

22 35 30.10 Water Heating Transfer Package		Crew	Daily Output	Labor-Hours	Unit	Material	Labor	Equipment	Total	Total Incl O&P
0010	**WATER HEATING TRANSFER PACKAGE**, Complete controls,									
0020	expansion tank, converter, air separator									
1000	Hot water, 180°F enter, 200°F leaving, 15# steam									
1010	One pump system, 28 GPM	Q-6	.75	32	Ea.	20,000	1,775		21,775	24,900
1020	35 GPM		.70	34.286		21,800	1,900		23,700	27,000
1040	55 GPM		.65	36.923		25,800	2,050		27,850	31,500
1060	130 GPM		.55	43.636		32,500	2,425		34,925	39,600
1080	255 GPM		.40	60		42,900	3,350		46,250	52,500
1100	550 GPM		.30	80		58,000	4,450		62,450	71,000
1120	800 GPM		.25	96		69,000	5,350		74,350	84,500
1220	Two pump system, 28 GPM		.70	34.286		27,300	1,900		29,200	33,000
1240	35 GPM		.65	36.923		31,900	2,050		33,950	38,300
1260	55 GPM		.60	40		34,400	2,225		36,625	41,400
1280	130 GPM		.50	48		45,200	2,675		47,875	54,000
1300	255 GPM		.35	68.571		58,000	3,825		61,825	70,000
1320	550 GPM		.25	96		73,000	5,350		78,350	88,500
1340	800 GPM		.20	120		94,000	6,700		100,700	114,000

22 41 Residential Plumbing Fixtures

22 41 06 – Plumbing Fixtures General

22 41 06.10 Plumbing Fixture Notes		Crew	Daily Output	Labor-Hours	Unit	Material	Labor	Equipment	Total	Total Incl O&P
0010	**PLUMBING FIXTURE NOTES**, Incl. trim fittings unless otherwise noted									
0080	For rough-in, supply, waste, and vent, see add for each type									
0122	For electric water coolers, see Section 22 47 16.10									
0160	For color, unless otherwise noted, add				Ea.	20%				

22 41 13 – Residential Water Closets, Urinals, and Bidets

22 41 13.13 Water Closets		Crew	Daily Output	Labor-Hours	Unit	Material	Labor	Equipment	Total	Total Incl O&P
0010	**WATER CLOSETS** R224000-30									
0022	For seats, see Section 22 41 13.44									
0032	For automatic flush, see Line 22 42 39.10 0972									
0150	Tank type, vitreous china, incl. seat, supply pipe w/stop, 1.6 gpf or noted									
0200	Wall hung									
0400	Two piece, close coupled	Q-1	5.30	3.019	Ea.	630	159		789	945
0960	For rough-in, supply, waste, vent and carrier	"	2.73	5.861	"	1,025	310		1,335	1,600
0999	Floor mounted									

22 41 13 – Residential Water Closets, Urinals, and Bidets

22 41 13.13 Water Closets		Crew	Daily Output	Labor-Hours	Unit	Material	Labor	Equipment	Total	Total Incl O&P
						2015 Bare Costs				
1020	One piece, low profile	Q-1	5.30	3.019	Ea.	1,125	159		1,284	1,475
1050	One piece		5.30	3.019		1,100	159		1,259	1,450
1100	Two piece, close coupled		5.30	3.019		237	159		396	510
1102	Economy		5.30	3.019		132	159		291	395
1110	Two piece, close coupled, dual flush		5.30	3.019		310	159		469	595
1140	Two piece, close coupled, 1.28 gpf, ADA G	↓	5.30	3.019	↓	310	159		469	595
1960	For color, add					30%				
1980	For rough-in, supply, waste and vent	Q-1	3.05	5.246	Ea.	330	277		607	795

22 41 13.19 Bidets		Crew	Daily Output	Labor-Hours	Unit	Material	Labor	Equipment	Total	Total Incl O&P
0010	**BIDETS**									
0180	Vitreous china, with trim on fixture	Q-1	5	3.200	Ea.	595	169		764	920
0200	With trim for wall mounting		5	3.200		710	169		879	1,050
9600	For rough-in, supply, waste and vent, add	↓	1.78	8.989	↓	325	475		800	1,100

22 41 13.44 Toilet Seats		Crew	Daily Output	Labor-Hours	Unit	Material	Labor	Equipment	Total	Total Incl O&P
0010	**TOILET SEATS**									
0100	Molded composition, white									
0150	Industrial, w/o cover, open front, regular bowl	1 Plum	24	.333	Ea.	21	19.55		40.55	53.50
0200	With self-sustaining hinge		24	.333		22.50	19.55		42.05	55
0220	With self-sustaining check hinge		24	.333		22.50	19.55		42.05	55
0240	Extra heavy, with check hinge	↓	24	.333	↓	27.50	19.55		47.05	61
0260	Elongated bowl, same price									
0300	Junior size, w/o cover, open front	1 Plum	24	.333	Ea.	40.50	19.55		60.05	75
0320	Regular primary bowl, open front		24	.333		40.50	19.55		60.05	75
0340	Regular baby bowl, open front, check hinge		24	.333		37	19.55		56.55	71
0380	Open back & front, w/o cover, reg. or elongated bowl	↓	24	.333	↓	29	19.55		48.55	62.50
0400	Residential									
0420	Regular bowl, w/cover, closed front	1 Plum	24	.333	Ea.	29	19.55		48.55	62
0440	Open front	"	24	.333	"	26	19.55		45.55	59.50
0460	Elongated bowl, add					25%				
0500	Self-raising hinge, w/o cover, open front									
0520	Regular bowl	1 Plum	24	.333	Ea.	101	19.55		120.55	142
0540	Elongated bowl	"	24	.333	"	23	19.55		42.55	56
0700	Molded wood, white, with cover									
0720	Closed front, regular bowl, square back	1 Plum	24	.333	Ea.	10.90	19.55		30.45	42.50
0740	Extended back		24	.333		13.80	19.55		33.35	45.50
0780	Elongated bowl, square back		24	.333		14.10	19.55		33.65	46
0800	Open front	↓	24	.333	↓	15.05	19.55		34.60	47
0850	Decorator styles									
0890	Vinyl top, patterned	1 Plum	24	.333	Ea.	22	19.55		41.55	54.50
0900	Vinyl padded, plain colors, regular bowl		24	.333		22	19.55		41.55	54.50
0930	Elongated bowl	↓	24	.333	↓	24	19.55		43.55	57
1000	Solid plastic, white									
1030	Industrial, w/o cover, open front, regular bowl	1 Plum	24	.333	Ea.	26	19.55		45.55	59.50
1080	Extra heavy, concealed check hinge		24	.333		18.75	19.55		38.30	51
1100	Self-sustaining hinge		24	.333		23	19.55		42.55	56
1150	Elongated bowl		24	.333		29	19.55		48.55	62.50
1170	Concealed check		24	.333		17.25	19.55		36.80	49.50
1190	Self-sustaining hinge, concealed check		24	.333		51.50	19.55		71.05	87
1220	Residential, with cover, closed front, regular bowl		24	.333		45	19.55		64.55	80
1240	Elongated bowl		24	.333		55	19.55		74.55	90.50
1260	Open front, regular bowl		24	.333		40	19.55		59.55	74.50
1280	Elongated bowl	↓	24	.333	↓	48.50	19.55		68.05	84

22 41 Residential Plumbing Fixtures

22 41 16 – Residential Lavatories and Sinks

22 41 16.13 Lavatories		Crew	Daily Output	Labor-Hours	Unit	Material	2015 Bare Costs Labor	Equipment	Total	Total Incl O&P
0010	**LAVATORIES**, With trim, white unless noted otherwise R224000-30									
0500	Vanity top, porcelain enamel on cast iron									
0600	20" x 18"	Q-1	6.40	2.500	Ea.	335	132		467	575
0640	33" x 19" oval		6.40	2.500		475	132		607	730
0680	20" x 17" oval		6.40	2.500		172	132		304	395
0720	19" round		6.40	2.500		435	132		567	685
0760	20" x 12" triangular bowl	↓	6.40	2.500	↓	273	132		405	505
0860	For color, add					25%				
1000	Cultured marble, 19" x 17", single bowl	Q-1	6.40	2.500	Ea.	175	132		307	400
1040	25" x 19", single bowl		6.40	2.500		206	132		338	435
1080	31" x 19", single bowl		6.40	2.500		216	132		348	445
1120	25" x 22", single bowl		6.40	2.500		215	132		347	440
1160	37" x 22", single bowl		6.40	2.500		245	132		377	475
1200	49" x 22", single bowl	↓	6.40	2.500	↓	292	132		424	525
1580	For color, same price									
1900	Stainless steel, self-rimming, 25" x 22", single bowl, ledge	Q-1	6.40	2.500	Ea.	365	132		497	610
1960	17" x 22", single bowl		6.40	2.500		355	132		487	595
2040	18-3/4" round		6.40	2.500		825	132		957	1,100
2600	Steel, enameled, 20" x 17", single bowl		5.80	2.759		161	146		307	405
2660	19" round		5.80	2.759		171	146		317	415
2720	18" round	↓	5.80	2.759	↓	138	146		284	380
2860	For color, add					10%				
2900	Vitreous china, 20" x 16", single bowl	Q-1	5.40	2.963	Ea.	260	157		417	530
2960	20" x 17", single bowl		5.40	2.963		176	157		333	440
3020	19" round, single bowl		5.40	2.963		174	157		331	435
3080	19" x 16", single bowl		5.40	2.963		267	157		424	540
3140	17" x 14", single bowl		5.40	2.963		214	157		371	480
3200	22" x 13", single bowl	↓	5.40	2.963	↓	267	157		424	540
3560	For color, add					50%				
3580	Rough-in, supply, waste and vent for all above lavatories	Q-1	2.30	6.957	Ea.	231	370		601	830
4000	Wall hung									
4040	Porcelain enamel on cast iron, 16" x 14", single bowl	Q-1	8	2	Ea.	520	106		626	735
4060	18" x 15" single bowl		8	2		405	106		511	610
4120	19" x 17", single bowl		8	2		425	106		531	635
4180	20" x 18", single bowl		8	2		277	106		383	470
4240	22" x 19", single bowl	↓	8	2	↓	700	106		806	935
4580	For color, add					30%				
6000	Vitreous china, 18" x 15", single bowl with backsplash	Q-1	7	2.286	Ea.	231	121		352	440
6060	19" x 17", single bowl		7	2.286		207	121		328	415
6120	20" x 18", single bowl		7	2.286		284	121		405	500
6210	27" x 20", wheelchair type	↓	7	2.286	↓	815	121		936	1,075
6500	For color, add					30%				
6960	Rough-in, supply, waste and vent for above lavatories	Q-1	1.66	9.639	Ea.	455	510		965	1,300
7000	Pedestal type									
7600	Vitreous china, 27" x 21", white	Q-1	6.60	2.424	Ea.	700	128		828	970
7610	27" x 21", colored		6.60	2.424		880	128		1,008	1,175
7620	27" x 21", premium color		6.60	2.424		995	128		1,123	1,300
7660	26" x 20", white		6.60	2.424		795	128		923	1,075
7670	26" x 20", colored		6.60	2.424		1,000	128		1,128	1,300
7680	26" x 20", premium color		6.60	2.424		1,125	128		1,253	1,450
7700	24" x 20", white		6.60	2.424		405	128		533	645
7710	24" x 20", colored	↓	6.60	2.424	↓	475	128		603	725

22 41 Residential Plumbing Fixtures

22 41 16 – Residential Lavatories and Sinks

22 41 16.13 Lavatories		Crew	Daily Output	Labor-Hours	Unit	Material	2015 Bare Costs Labor	Equipment	Total	Total Incl O&P
7720	24" x 20", premium color	Q-1	6.60	2.424	Ea.	495	128		623	745
7760	21" x 18", white		6.60	2.424		291	128		419	520
7770	21" x 18", colored		6.60	2.424		291	128		419	520
7990	Rough-in, supply, waste and vent for pedestal lavatories	↓	1.66	9.639	↓	455	510		965	1,300
9000	Minimum labor/equipment charge	1 Plum	3	2.667	Job		157		157	244

22 41 16.16 Sinks

Code	Description	Crew	Daily Output	Labor-Hours	Unit	Material	Labor	Equipment	Total	Total Incl O&P
0010	**SINKS**, With faucets and drain R224000-30									
2000	Kitchen, counter top style, P.E. on C.I., 24" x 21" single bowl	Q-1	5.60	2.857	Ea.	285	151		436	550
2100	31" x 22" single bowl		5.60	2.857		615	151		766	915
2200	32" x 21" double bowl		4.80	3.333		355	176		531	665
3000	Stainless steel, self rimming, 19" x 18" single bowl		5.60	2.857		590	151		741	885
3100	25" x 22" single bowl		5.60	2.857		660	151		811	960
3200	33" x 22" double bowl		4.80	3.333		960	176		1,136	1,325
3300	43" x 22" double bowl		4.80	3.333		1,125	176		1,301	1,500
3400	22" x 43" triple bowl		4.40	3.636		1,350	192		1,542	1,775
3500	Corner double bowl each 14" x 16"		4.80	3.333		830	176		1,006	1,200
4000	Steel, enameled, with ledge, 24" x 21" single bowl		5.60	2.857		510	151		661	795
4100	32" x 21" double bowl	↓	4.80	3.333		495	176		671	820
4960	For color sinks except stainless steel, add					10%				
4980	For rough-in, supply, waste and vent, counter top sinks	Q-1	2.14	7.477	↓	260	395		655	900
5000	Kitchen, raised deck, P.E. on C.I.									
5100	32" x 21", dual level, double bowl	Q-1	2.60	6.154	Ea.	420	325		745	965
5200	42" x 21", double bowl & disposer well	"	2.20	7.273		1,175	385		1,560	1,900
5700	For color, add					20%				
5790	For rough-in, supply, waste & vent, sinks	Q-1	1.85	8.649	↓	260	455		715	1,000

22 41 19 – Residential Bathtubs

22 41 19.10 Baths

Code	Description	Crew	Daily Output	Labor-Hours	Unit	Material	Labor	Equipment	Total	Total Incl O&P
0010	**BATHS** R224000-30									
0100	Tubs, recessed porcelain enamel on cast iron, with trim									
0180	48" x 42"	Q-1	4	4	Ea.	2,625	211		2,836	3,225
0220	72" x 36"	"	3	5.333	"	2,725	282		3,007	3,450
0300	Mat bottom									
0340	4'-6" long	Q-1	5	3.200	Ea.	1,350	169		1,519	1,775
0380	5' long		4.40	3.636		1,125	192		1,317	1,550
0420	5'-6" long		4	4		1,775	211		1,986	2,275
0480	Above floor drain, 5' long		4	4		785	211		996	1,200
0560	Corner 48" x 44"	↓	4.40	3.636	↓	2,625	192		2,817	3,200
0750	For color, add					30%				
2000	Enameled formed steel, 4'-6" long	Q-1	5.80	2.759	Ea.	495	146		641	770
2300	Above floor drain, 5' long	"	5.50	2.909	"	535	154		689	830
2350	For color, add					10%				
4000	Soaking, acrylic, w/pop-up drain 66" x 36" x 20" deep	Q-1	5.50	2.909	Ea.	2,300	154		2,454	2,775
4100	60" x 42" x 20" deep		5	3.200		1,175	169		1,344	1,575
4200	72" x 42" x 23" deep	↓	4.80	3.333	↓	1,750	176		1,926	2,200
4600	Module tub & showerwall surround, molded fiberglass									
4610	5' long x 34" wide x 76" high ♿	Q-1	4	4	Ea.	840	211		1,051	1,250
4750	Handicap with 1-1/2" OD grab bar, antiskid bottom									
4760	60" x 32-3/4" x 72" high	Q-1	4	4	Ea.	840	211		1,051	1,250
4770	60" x 30" x 71" high with molded seat		3.50	4.571		875	242		1,117	1,350
9600	Rough-in, supply, waste and vent, for all above tubs, add		2.07	7.729	↓	345	410		755	1,025
9900	Minimum labor/equipment charge	↓	3	5.333	Job		282		282	440

22 41 Residential Plumbing Fixtures

22 41 23 – Residential Showers

22 41 23.20 Showers		Crew	Daily Output	Labor-Hours	Unit	Material	Labor (2015 Bare Costs)	Equipment	Total	Total Incl O&P
0010	**SHOWERS** R224000-30									
1500	Stall, with drain only. Add for valve and door/curtain									
1510	Baked enamel, molded stone receptor, 30″ square	Q-1	5.20	3.077	Ea.	1,150	163		1,313	1,525
1520	32″ square		5	3.200		1,175	169		1,344	1,550
1530	36″ square		4.80	3.333		2,925	176		3,101	3,475
1540	Terrazzo receptor, 32″ square		5	3.200		1,350	169		1,519	1,750
1560	36″ square		4.80	3.333		1,475	176		1,651	1,900
1580	36″ corner angle		4.80	3.333		1,725	176		1,901	2,175
1600	For color, add					10%				
3000	Fiberglass, one piece, with 3 walls, 32″ x 32″ square	Q-1	5.50	2.909	Ea.	490	154		644	780
3100	36″ x 36″ square	″	5.50	2.909	″	505	154		659	795
3200	Handicap, 1-1/2″ O.D. grab bars, nonskid floor									
3210	48″ x 34-1/2″ x 72″ corner seat	Q-1	5	3.200	Ea.	720	169		889	1,050
3220	60″ x 34-1/2″ x 72″ corner seat		4	4		750	211		961	1,150
3230	48″ x 34-1/2″ x 72″ fold up seat		5	3.200		2,225	169		2,394	2,725
3250	64″ x 65-3/4″ x 81-1/2″ fold. seat, whlchr.		3.80	4.211		2,325	222		2,547	2,900
4000	Polypropylene, stall only, w/molded-stone floor, 30″ x 30″		2	8		635	425		1,060	1,350
4100	32″ x 32″		2	8		650	425		1,075	1,375
4200	Rough-in, supply, waste and vent for above showers		2.05	7.805		345	410		755	1,025

22 41 23.40 Shower System Components		Crew	Daily Output	Labor-Hours	Unit	Material	Labor	Equipment	Total	Total Incl O&P
0010	**SHOWER SYSTEM COMPONENTS**									
4500	Receptor only									
4510	For tile, 36″ x 36″	1 Plum	4	2	Ea.	365	117		482	590
4520	Fiberglass receptor only, 32″ x 32″		8	1		114	58.50		172.50	217
4530	34″ x 34″		7.80	1.026		128	60		188	234
4540	36″ x 36″		7.60	1.053		133	62		195	243
4600	Rectangular									
4620	32″ x 48″	1 Plum	7.40	1.081	Ea.	158	63.50		221.50	273
4630	34″ x 54″		7.20	1.111		190	65		255	310
4640	34″ x 60″		7	1.143		201	67		268	325
5000	Built-in, head, arm, 2.5 GPM valve		4	2		80	117		197	271
5200	Head, arm, by-pass, integral stops, handles		3.60	2.222		255	130		385	485
5500	Head, water economizer, 1.6 GPM G		24	.333		45	19.55		64.55	80
5800	Mixing valve, built-in		6	1.333		153	78.50		231.50	291
5900	Exposed		6	1.333		660	78.50		738.50	850

22 41 36 – Residential Laundry Trays

22 41 36.10 Laundry Sinks		Crew	Daily Output	Labor-Hours	Unit	Material	Labor	Equipment	Total	Total Incl O&P
0010	**LAUNDRY SINKS**, With trim									
0020	Porcelain enamel on cast iron, black iron frame									
0050	24″ x 21″, single compartment	Q-1	6	2.667	Ea.	580	141		721	855
0100	26″ x 21″, single compartment	″	6	2.667	″	610	141		751	890
2000	Molded stone, on wall hanger or legs									
2020	22″ x 23″, single compartment	Q-1	6	2.667	Ea.	167	141		308	405
2100	45″ x 21″, double compartment	″	5	3.200	″	330	169		499	630
3000	Plastic, on wall hanger or legs									
3020	18″ x 23″, single compartment	Q-1	6.50	2.462	Ea.	135	130		265	350
3100	20″ x 24″, single compartment		6.50	2.462		154	130		284	375
3200	36″ x 23″, double compartment		5.50	2.909		185	154		339	445
3300	40″ x 24″, double compartment		5.50	2.909		278	154		432	545
5000	Stainless steel, counter top, 22″ x 17″ single compartment		6	2.667		64	141		205	291
5200	33″ x 22″, double compartment		5	3.200		79	169		248	350

22 41 36 – Residential Laundry Trays

22 41 36.10 Laundry Sinks		Crew	Daily Output	Labor-Hours	Unit	Material	2015 Bare Costs Labor	Equipment	Total	Total Incl O&P
9600	Rough-in, supply, waste and vent, for all laundry sinks	Q-1	2.14	7.477	Ea.	260	395		655	900
9810	Minimum labor/equipment charge	1 Plum	3	2.667	Job		157		157	244

22 41 39 – Residential Faucets, Supplies and Trim

22 41 39.10 Faucets and Fittings		Crew	Daily Output	Labor-Hours	Unit	Material	2015 Bare Costs Labor	Equipment	Total	Total Incl O&P
0010	**FAUCETS AND FITTINGS**									
0150	Bath, faucets, diverter spout combination, sweat	1 Plum	8	1	Ea.	86.50	58.50		145	187
0200	For integral stops, IPS unions, add					109			109	120
0300	Three valve combinations, spout, head, arm, flange, sweat	1 Plum	6	1.333		88	78.50		166.50	219
0400	For integral stops, IPS unions, add				Pr.	64.50			64.50	71
0420	Bath, press-bal mix valve w/diverter, spout, shower head, arm/flange	1 Plum	8	1	Ea.	168	58.50		226.50	277
0500	Drain, central lift, 1-1/2" IPS male		20	.400		71	23.50		94.50	115
0600	Trip lever, 1-1/2" IPS male		20	.400		45	23.50		68.50	86
0700	Pop up, 1-1/2" IPS male		18	.444		53	26		79	99
0800	Chain and stopper, 1-1/2" IPS male		24	.333		32.50	19.55		52.05	66
0810	Bidet									
0812	Fitting, over the rim, swivel spray/pop-up drain	1 Plum	8	1	Ea.	206	58.50		264.50	320
1000	Kitchen sink faucets, top mount, cast spout		10	.800		61.50	47		108.50	141
1100	For spray, add		24	.333		16.15	19.55		35.70	48.50
1110	For basket strainer w/tail piece, add		24	.333		14.85	19.55		34.40	47
1200	Wall type, swing tube spout		10	.800		74.50	47		121.50	155
1240	For soap dish, add					3.60			3.60	3.96
1250	For basket strainer w/tail piece, add					51			51	56
1300	Single control lever handle									
1310	With pull out spray									
1320	Polished chrome	1 Plum	10	.800	Ea.	196	47		243	288
2000	Laundry faucets, shelf type, IPS or copper unions		12	.667		49.50	39		88.50	116
2100	Lavatory faucet, centerset, without drain		10	.800		44.50	47		91.50	122
2120	With pop-up drain		6.66	1.201		62.50	70.50		133	179
2130	For acrylic handles, add					5.15			5.15	5.65
2150	Concealed, 12" centers	1 Plum	10	.800		101	47		148	184
2160	With pop-up drain	"	6.66	1.201		118	70.50		188.50	240
2210	Porcelain cross handles and pop-up drain									
2220	Polished chrome	1 Plum	6.66	1.201	Ea.	189	70.50		259.50	320
2230	Polished brass	"	6.66	1.201	"	298	70.50		368.50	440
2260	Single lever handle and pop-up drain									
2280	Satin nickel	1 Plum	6.66	1.201	Ea.	277	70.50		347.50	415
2290	Polished chrome		6.66	1.201		198	70.50		268.50	330
2600	Shelfback, 4" to 6" centers, 17 Ga. tailpiece		10	.800		78	47		125	159
2650	With pop-up drain		6.66	1.201		95	70.50		165.50	214
2700	Shampoo faucet with supply tube		24	.333		48.50	19.55		68.05	84
2800	Self-closing, center set		10	.800		131	47		178	217
2810	Automatic sensor and operator, with faucet head G		6.15	1.301		450	76.50		526.50	615
4000	Shower by-pass valve with union		18	.444		68.50	26		94.50	116
4100	Shower arm with flange and head		22	.364		19.25	21.50		40.75	54.50
4140	Shower, hand held, pin mount, massage action, chrome		22	.364		76	21.50		97.50	117
4142	Polished brass		22	.364		142	21.50		163.50	190
4144	Shower, hand held, wall mtd, adj. spray, 2 wall mounts, chrome		20	.400		102	23.50		125.50	149
4146	Polished brass		20	.400		192	23.50		215.50	249
4148	Shower, hand held head, bar mounted 24", adj. spray, chrome		20	.400		164	23.50		187.50	217
4150	Polished brass		20	.400		340	23.50		363.50	405
4200	Shower thermostatic mixing valve, concealed, with shower head trim kit		8	1		345	58.50		403.50	470
4220	Shower pressure balancing mixing valve,									

22 41 39 – Residential Faucets, Supplies and Trim

22 41 39.10 Faucets and Fittings

	22 41 39.10 Faucets and Fittings	Crew	Daily Output	Labor-Hours	Unit	Material	2015 Bare Costs Labor	Equipment	Total	Total Incl O&P
4230	With shower head, arm, flange and diverter tub spout									
4240	Chrome	1 Plum	6.14	1.303	Ea.	360	76.50		436.50	515
4250	Satin nickel		6.14	1.303		555	76.50		631.50	730
4260	Polished graphite		6.14	1.303		555	76.50		631.50	730
5000	Sillcock, compact, brass, IPS or copper to hose	↓	24	.333	↓	9.70	19.55		29.25	41
6000	Stop and waste valves, bronze									
6100	Angle, solder end 1/2"	1 Plum	24	.333	Ea.	12.05	19.55		31.60	44
6110	3/4"		20	.400		14.15	23.50		37.65	52
6300	Straightway, solder end 3/8"		24	.333		12.35	19.55		31.90	44
6310	1/2"		24	.333		12.35	19.55		31.90	44
6320	3/4"		20	.400		13.30	23.50		36.80	51
6410	Straightway, threaded 1/2"		24	.333		15.95	19.55		35.50	48
6420	3/4"		20	.400		18.30	23.50		41.80	56.50
6430	1"		19	.421		13.40	24.50		37.90	53.50
7800	Water closet, wax gasket		96	.083		1.38	4.89		6.27	9.15
7820	Gasket toilet tank to bowl		32	.250		2.38	14.70		17.08	25.50
7830	Replacement diaphragm washer assy for ballcock valve		12	.667		2.38	39		41.38	63.50
7850	Dual flush valve	↓	12	.667	↓	20	39		59	83
8000	Water supply stops, polished chrome plate									
8200	Angle, 3/8"	1 Plum	24	.333	Ea.	12.90	19.55		32.45	44.50
8300	1/2"		22	.364		12.90	21.50		34.40	47.50
8400	Straight, 3/8"		26	.308		13.55	18.05		31.60	43
8500	1/2"		24	.333		13.55	19.55		33.10	45.50
8600	Water closet, angle, w/flex riser, 3/8"		24	.333	↓	44	19.55		63.55	79
9000	Minimum labor/equipment charge	↓	4	2	Job		117		117	183

22 41 39.70 Washer/Dryer Accessories

	22 41 39.70 Washer/Dryer Accessories	Crew	Daily Output	Labor-Hours	Unit	Material	Labor	Equipment	Total	Total Incl O&P
0010	**WASHER/DRYER ACCESSORIES**									
1020	Valves ball type single lever									
1030	1/2" diam., IPS	1 Plum	21	.381	Ea.	54.50	22.50		77	95
1040	1/2" diam., solder	"	21	.381	"	54.50	22.50		77	95
1050	Recessed box, 16 ga., two hose valves and drain									
1060	1/2" size, 1-1/2" drain	1 Plum	18	.444	Ea.	116	26		142	169
1070	1/2" size, 2" drain	"	17	.471	"	108	27.50		135.50	162
1080	With grounding electric receptacle									
1090	1/2" size, 1-1/2" drain	1 Plum	18	.444	Ea.	127	26		153	181
1100	1/2" size, 2" drain	"	17	.471	"	138	27.50		165.50	194
1110	With grounding and dryer receptacle									
1120	1/2" size, 1-1/2" drain	1 Plum	18	.444	Ea.	157	26		183	214
1130	1/2" size, 2" drain	"	17	.471	"	159	27.50		186.50	218
1140	Recessed box 16 ga., ball valves with single lever and drain									
1150	1/2" size, 1-1/2" drain	1 Plum	19	.421	Ea.	219	24.50		243.50	280
1160	1/2" size, 2" drain	"	18	.444	"	197	26		223	258
1170	With grounding electric receptacle									
1180	1/2" size, 1-1/2" drain	1 Plum	19	.421	Ea.	190	24.50		214.50	248
1190	1/2" size, 2" drain	"	18	.444	"	211	26		237	273
1200	With grounding and dryer receptacles									
1210	1/2" size, 1-1/2" drain	1 Plum	19	.421	Ea.	208	24.50		232.50	267
1220	1/2" size, 2" drain	"	18	.444	"	230	26		256	295
1300	Recessed box, 20 ga., two hose valves and drain (economy type)									
1310	1/2" size, 1-1/2" drain	1 Plum	19	.421	Ea.	90.50	24.50		115	138
1320	1/2" size, 2" drain		18	.444		86	26		112	135
1330	Box with drain only	↓	24	.333	↓	52.50	19.55		72.05	88

22 41 Residential Plumbing Fixtures

22 41 39 – Residential Faucets, Supplies and Trim

22 41 39.70 Washer/Dryer Accessories		Crew	Daily Output	Labor-Hours	Unit	2015 Bare Costs Material	Labor	Equipment	Total	Total Incl O&P
1340	1/2" size, 1-1/2" ABS/PVC drain	1 Plum	19	.421	Ea.	100	24.50		124.50	149
1350	1/2" size, 2" ABS/PVC drain		18	.444		108	26		134	159
1352	Box with drain and 15 A receptacle		24	.333		90.50	19.55		110.05	130
1360	1/2" size, 2" drain ABS/PVC, 15 A receptacle		24	.333		127	19.55		146.55	171
1400	Wall mounted									
1410	1/2" size, 1-1/2" plastic drain	1 Plum	19	.421	Ea.	24.50	24.50		49	65.50
1420	1/2" size, 2" plastic drain	"	18	.444	"	21	26		47	63.50
1500	Dryer vent kit									
1510	8' flex duct, clamps and outside hood	1 Plum	20	.400	Ea.	15.30	23.50		38.80	53.50
1980	Rough-in, supply, waste, and vent for washer boxes		3.46	2.310		244	136		380	480
9605	Washing machine valve assembly, hot & cold water supply, recessed		8	1		70	58.50		128.50	169
9610	Washing machine valve assembly, hot & cold water supply, mounted		8	1		54.50	58.50		113	152

22 42 Commercial Plumbing Fixtures

22 42 13 – Commercial Water Closets, Urinals, and Bidets

22 42 13.13 Water Closets

Line	Description	Crew	Daily Output	Labor-Hours	Unit	Material	Labor	Equipment	Total	Total Incl O&P
0010	**WATER CLOSETS**									
3000	Bowl only, with flush valve, seat, 1.6 gpf unless noted									
3100	Wall hung	Q-1	5.80	2.759	Ea.	945	146		1,091	1,275
3200	For rough-in, supply, waste and vent, single WC		2.56	6.250		1,075	330		1,405	1,700
3300	Floor mounted		5.80	2.759		315	146		461	570
3350	With wall outlet		5.80	2.759		545	146		691	825
3360	With floor outlet, 1.28 gpf G		5.80	2.759		555	146		701	840
3362	With floor outlet, 1.28 gpf, ADA G		5.80	2.759		580	146		726	860
3370	For rough-in, supply, waste and vent, single WC		2.84	5.634		370	298		668	870
3390	Floor mounted children's size, 10-3/4" high									
3392	With automatic flush sensor, 1.6 gpf	Q-1	6.20	2.581	Ea.	615	136		751	895
3396	With automatic flush sensor, 1.28 gpf		6.20	2.581		620	136		756	895
3400	For rough-in, supply, waste and vent, single WC		2.84	5.634		370	298		668	870
3500	Gang side by side carrier system, rough-in, supply, waste & vent									
3510	For single hook-up	Q-1	1.97	8.122	Ea.	1,250	430		1,680	2,050
3520	For each additional hook-up, add	"	2.14	7.477	"	1,175	395		1,570	1,900
3550	Gang back to back carrier system, rough-in, supply, waste & vent									
3560	For pair hook-up	Q-1	1.76	9.091	Pr.	1,975	480		2,455	2,925
3570	For each additional pair hook-up, add		1.81	8.840	"	1,875	465		2,340	2,800
9000	Minimum labor/equipment charge		4	4	Job		211		211	330

22 42 13.16 Urinals

Line	Description	Crew	Daily Output	Labor-Hours	Unit	Material	Labor	Equipment	Total	Total Incl O&P
0010	**URINALS** R224000-30									
0102	For automatic flush see Line 22 42 39.10 0972									
3000	Wall hung, vitreous china, with self-closing valve									
3100	Siphon jet type	Q-1	3	5.333	Ea.	282	282		564	750
3120	Blowout type		3	5.333		465	282		747	950
3140	Water saving .5 gpf G		3	5.333		550	282		832	1,050
3300	Rough-in, supply, waste & vent		2.83	5.654		595	299		894	1,125
5000	Stall type, vitreous china, includes valve		2.50	6.400		740	340		1,080	1,325
6980	Rough-in, supply, waste and vent		1.99	8.040		365	425		790	1,075
8000	Waterless (no flush) urinal									
8010	Wall hung									
8014	Fiberglass reinforced polyester									
8020	Standard unit G	Q-1	21.30	.751	Ea.	385	39.50		424.50	480
8030	ADA compliant unit G	"	21.30	.751		400	39.50		439.50	500

22 42 Commercial Plumbing Fixtures

22 42 13 – Commercial Water Closets, Urinals, and Bidets

22 42 13.16 Urinals		Crew	Daily Output	Labor-Hours	Unit	2015 Bare Costs Material	Labor	Equipment	Total	Total Incl O&P
8070	For solid color, add G				Ea.	48			48	53
8080	For 2" brass flange, (new const.), add G	Q-1	96	.167		19.20	8.80		28	35
8200	Vitreous china									
8220	ADA compliant unit, 14" G	Q-1	21.30	.751	Ea.	198	39.50		237.50	280
8240	ADA compliant unit, 18" G		21.30	.751		320	39.50		359.50	410
8250	ADA compliant unit, 15.5"		21.30	.751		272	39.50		311.50	360
8270	For solid color, add G					48			48	53
8290	Rough-in, supply, waste & vent G	Q-1	2.92	5.479		560	289		849	1,075
8400	Trap liquid									
8410	1 quart G				Ea.	15.95			15.95	17.55
8420	1 gallon G				"	58			58	64
9000	Minimum labor/equipment charge	Q-1	4	4	Job		211		211	330

22 42 16 – Commercial Lavatories and Sinks

22 42 16.13 Lavatories

Line	Description	Crew	Daily Output	Labor-Hours	Unit	Material	Labor	Equipment	Total	Total Incl O&P
0010	**LAVATORIES**, With trim, white unless noted otherwise									
0020	Commercial lavatories same as residential. See Section 22 41 16									

22 42 16.34 Laboratory Countertops and Sinks

Line	Description	Crew	Daily Output	Labor-Hours	Unit	Material	Labor	Equipment	Total	Total Incl O&P
0010	**LABORATORY COUNTERTOPS AND SINKS**									
0050	Laboratory sinks, corrosion resistant									
1000	Stainless steel sink, bench mounted, with									
1020	plug & waste fitting with 1-1/2" straight threads									
1030	Single bowl, 2 drainboards, backnut & strainer									
1050	18-1/2" x 15-1/2" x 12-1/2" sink, 54" x 24" O.D.	Q-1	3	5.333	Ea.	1,700	282		1,982	2,325
1100	Single bowl, single drainboard, backnut & strainer									
1130	18-1/2" x 15-1/2" x 12-1/2" sink, 47" x 24" O.D.	Q-1	3	5.333	Ea.	1,200	282		1,482	1,750
1146	Double bowl, single drainboard, backnut & strainer									
1150	18-1/2" x 15-1/2" x 12-1/2" sink, 70" x 24" O.D.	Q-1	3	5.333	Ea.	1,875	282		2,157	2,525
1280	Polypropylene									
1290	Flanged 1-1/4" wide, rectangular with strainer									
1300	plug & waste fitting, 1-1/2" straight threads									
1320	12" x 12" x 8" sink, 14-1/2" x 14-1/2" O.D.	Q-1	4	4	Ea.	234	211		445	585
1340	16" x 16" x 8" sink, 18-1/2" x 18-1/2" O.D.		4	4		335	211		546	700
1360	21" x 18" x 10" sink, 23-1/2" x 20-1/2" O.D.		4	4		355	211		566	720
1490	For rough-in, supply, waste & vent, add		2.02	7.921		208	420		628	885
1600	Polypropylene									
1620	Cup sink, oval, integral strainers									
1640	6" x 3" I.D., 7" x 4" O.D.	Q-1	6	2.667	Ea.	117	141		258	350
1660	9" x 3" I.D., 10" x 4-1/2" O.D.	"	6	2.667		139	141		280	370
1740	1-1/2" diam. x 11" long					34.50			34.50	38
1980	For rough-in, supply, waste & vent, add	Q-1	1.70	9.412		211	495		706	1,000

22 42 16.40 Service Sinks

Line	Description	Crew	Daily Output	Labor-Hours	Unit	Material	Labor	Equipment	Total	Total Incl O&P
0010	**SERVICE SINKS**									
6650	Service, floor, corner, P.E. on C.I., 28" x 28"	Q-1	4.40	3.636	Ea.	1,000	192		1,192	1,400
6750	Vinyl coated rim guard, add					67.50			67.50	74
6760	Mop sink, molded stone, 24" x 36"	1 Plum	3.33	2.402		273	141		414	520
6770	Mop sink, molded stone, 24" x 36", w/rim 3 sides	"	3.33	2.402		278	141		419	525
6790	For rough-in, supply, waste & vent, floor service sinks	Q-1	1.64	9.756		705	515		1,220	1,575
7000	Service, wall, P.E. on C.I., roll rim, 22" x 18"		4	4		770	211		981	1,175
7100	24" x 20"		4	4		850	211		1,061	1,275
7600	For stainless steel rim guard, two sides only, add					88.50			88.50	97.50
7800	For stainless steel rim guard, front only, add					56			56	62
8600	Vitreous china, 22" x 20"	Q-1	4	4		610	211		821	1,000

22 42 Commercial Plumbing Fixtures

22 42 16 – Commercial Lavatories and Sinks

22 42 16.40 Service Sinks		Crew	Daily Output	Labor-Hours	Unit	2015 Bare Costs Material	Labor	Equipment	Total	Total Incl O&P
8960	For stainless steel rim guard, front or one side, add				Ea.	56			56	62
8980	For rough-in, supply, waste & vent, wall service sinks	Q-1	1.30	12.308		1,100	650		1,750	2,225
9000	Minimum labor/equipment charge	″	4	4	Job		211		211	330

22 42 23 – Commercial Showers

22 42 23.30 Group Showers		Crew	Daily Output	Labor-Hours	Unit	Material	Labor	Equipment	Total	Total Incl O&P
0010	GROUP SHOWERS									
6000	Group, w/pressure balancing valve, rough-in and rigging not included									
6800	Column, 6 heads, no receptors, less partitions	Q-1	3	5.333	Ea.	9,200	282		9,482	10,500
6900	With stainless steel partitions		1	16		11,900	845		12,745	14,400
7600	5 heads, no receptors, less partitions		3	5.333		6,350	282		6,632	7,425
7620	4 heads (1 handicap) no receptors, less partitions		3	5.333		5,650	282		5,932	6,650
7700	With stainless steel partitions		1	16		5,650	845		6,495	7,525
8000	Wall, 2 heads, no receptors, less partitions		4	4		2,725	211		2,936	3,300
8100	With stainless steel partitions		2	8		5,975	425		6,400	7,200
9000	Minimum labor/equipment charge		4	4	Job		211		211	330

22 42 33 – Wash Fountains

22 42 33.20 Commercial Wash Fountains		Crew	Daily Output	Labor-Hours	Unit	Material	Labor	Equipment	Total	Total Incl O&P
0010	COMMERCIAL WASH FOUNTAINS									
1900	Group, foot control									
2000	Precast terrazzo, circular, 36″ diam., 5 or 6 persons	Q-2	3	8	Ea.	7,275	440		7,715	8,675
2100	54″ diameter for 8 or 10 persons		2.50	9.600		9,050	525		9,575	10,800
2400	Semi-circular, 36″ diam. for 3 persons		3	8		6,375	440		6,815	7,700
2500	54″ diam. for 4 or 5 persons		2.50	9.600		8,575	525		9,100	10,200
2700	Quarter circle (corner), 54″ for 3 persons		3.50	6.857		7,850	375		8,225	9,225
3000	Stainless steel, circular, 36″ diameter		3.50	6.857		6,550	375		6,925	7,775
3100	54″ diameter		2.80	8.571		8,050	470		8,520	9,575
3400	Semi-circular, 36″ diameter		3.50	6.857		5,025	375		5,400	6,125
3500	54″ diameter		2.80	8.571		7,000	470		7,470	8,450
5000	Thermoplastic, pre-assembled, circular, 36″ diameter		6	4		4,450	219		4,669	5,250
5100	54″ diameter		4	6		5,175	330		5,505	6,225
5400	Semi-circular, 36″ diameter		6	4		4,100	219		4,319	4,850
5600	54″ diameter		4	6		5,000	330		5,330	6,025
5610	Group, infrared control, barrier free									
5614	Precast terrazzo									
5620	Semi-circular 36″ diam. for 3 persons	Q-2	3	8	Ea.	7,250	440		7,690	8,650
5630	46″ diam. for 4 persons		2.80	8.571		7,825	470		8,295	9,325
5640	Circular, 54″ diam. for 8 persons, button control		2.50	9.600		9,475	525		10,000	11,200
5700	Rough-in, supply, waste and vent for above wash fountains	Q-1	1.82	8.791		370	465		835	1,125
6200	Duo for small washrooms, stainless steel		2	8		2,725	425		3,150	3,625
6400	Bowl with backsplash		2	8		2,250	425		2,675	3,125
6500	Rough-in, supply, waste & vent for duo fountains		2.02	7.921		209	420		629	885
9000	Minimum labor/equipment charge	Q-2	3	8	Job		440		440	685

22 42 39 – Commercial Faucets, Supplies, and Trim

22 42 39.10 Faucets and Fittings		Crew	Daily Output	Labor-Hours	Unit	Material	Labor	Equipment	Total	Total Incl O&P
0010	FAUCETS AND FITTINGS									
0840	Flush valves, with vacuum breaker									
0850	Water closet									
0860	Exposed, rear spud	1 Plum	8	1	Ea.	148	58.50		206.50	255
0870	Top spud		8	1		149	58.50		207.50	255
0880	Concealed, rear spud		8	1		197	58.50		255.50	310
0890	Top spud		8	1		159	58.50		217.50	267

22 42 Commercial Plumbing Fixtures

22 42 39 – Commercial Faucets, Supplies, and Trim

22 42 39.10 Faucets and Fittings		Crew	Daily Output	Labor-Hours	Unit	Material	Labor	Equipment	Total	Total Incl O&P
						2015 Bare Costs				
0900	Wall hung	1 Plum	8	1	Ea.	177	58.50		235.50	287
0910	Dual flush flushometer		12	.667		197	39		236	277
0912	Flushometer retrofit kit		18	.444		16.30	26		42.30	58.50
0920	Urinal									
0930	Exposed, stall	1 Plum	8	1	Ea.	149	58.50		207.50	256
0940	Wall, (washout)		8	1		149	58.50		207.50	255
0950	Pedestal, top spud		8	1		143	58.50		201.50	249
0960	Concealed, stall		8	1		156	58.50		214.50	264
0970	Wall (washout)		8	1		168	58.50		226.50	277
0971	Automatic flush sensor and operator for									
0972	urinals or water closets, standard G	1 Plum	8	1	Ea.	450	58.50		508.50	585
0980	High efficiency water saving									
0984	Water closets, 1.28 gpf G	1 Plum	8	1	Ea.	415	58.50		473.50	550
0988	Urinals, .5 gpf G	"	8	1	"	415	58.50		473.50	550
2790	Faucets for lavatories									
2800	Self-closing, center set	1 Plum	10	.800	Ea.	131	47		178	217
2810	Automatic sensor and operator, with faucet head		6.15	1.301		450	76.50		526.50	615
3000	Service sink faucet, cast spout, pail hook, hose end		14	.571		80	33.50		113.50	141

22 42 39.30 Carriers and Supports		Crew	Daily Output	Labor-Hours	Unit	Material	Labor	Equipment	Total	Total Incl O&P
0010	**CARRIERS AND SUPPORTS**, For plumbing fixtures									
0500	Drinking fountain, wall mounted									
0600	Plate type with studs, top back plate	1 Plum	7	1.143	Ea.	95	67		162	210
0700	Top front and back plate		7	1.143		116	67		183	233
0800	Top & bottom, front & back plates, w/bearing jacks		7	1.143		193	67		260	320
3000	Lavatory, concealed arm									
3050	Floor mounted, single									
3100	High back fixture	1 Plum	6	1.333	Ea.	525	78.50		603.50	700
3200	Flat slab fixture		6	1.333		455	78.50		533.50	620
3220	Paraplegic		6	1.333		590	78.50		668.50	770
3250	Floor mounted, back to back									
3300	High back fixtures	1 Plum	5	1.600	Ea.	750	94		844	970
3400	Flat slab fixtures		5	1.600		925	94		1,019	1,175
3430	Paraplegic		5	1.600		860	94		954	1,100
3500	Wall mounted, in stud or masonry									
3600	High back fixture	1 Plum	6	1.333	Ea.	320	78.50		398.50	470
3700	Flat slab fixture	"	6	1.333	"	270	78.50		348.50	420
4000	Exposed arm type, floor mounted									
4100	Single high back or flat slab fixture	1 Plum	6	1.333	Ea.	750	78.50		828.50	945
4200	Back to back, high back or flat slab fixtures		5	1.600		1,175	94		1,269	1,450
4300	Wall mounted, High back or flat slab lavatory		6	1.333		520	78.50		598.50	695
4600	Sink, floor mounted									
4650	Exposed arm system									
4700	Single heavy fixture	1 Plum	5	1.600	Ea.	880	94		974	1,125
4750	Single heavy sink with slab		5	1.600		1,075	94		1,169	1,350
4800	Back to back, standard fixtures		5	1.600		650	94		744	855
4850	Back to back, heavy fixtures		5	1.600		920	94		1,014	1,150
4900	Back to back, heavy sink with slab		5	1.600		920	94		1,014	1,150
4950	Exposed offset arm system									
5000	Single heavy deep fixture	1 Plum	5	1.600	Ea.	840	94		934	1,075
5100	Plate type system									
5200	With bearing jacks, single fixture	1 Plum	5	1.600	Ea.	885	94		979	1,125
5300	With exposed arms, single heavy fixture		5	1.600		1,225	94		1,319	1,500

22 42 39.30 Carriers and Supports		Crew	Daily Output	Labor-Hours	Unit	Material	Labor	Equipment	Total	Total Incl O&P
						2015 Bare Costs				
5400	Wall mounted, exposed arms, single heavy fixture	1 Plum	5	1.600	Ea.	450	94		544	640
6000	Urinal, floor mounted, 2″ or 3″ coupling, blowout type		6	1.333		560	78.50		638.50	735
6100	With fixture or hanger bolts, blowout or washout		6	1.333		390	78.50		468.50	550
6200	With bearing plate		6	1.333		445	78.50		523.50	610
6300	Wall mounted, plate type system		6	1.333		345	78.50		423.50	500
6980	Water closet, siphon jet									
7000	Horizontal, adjustable, caulk									
7040	Single, 4″ pipe size	1 Plum	5.33	1.501	Ea.	890	88		978	1,100
7050	4″ pipe size, paraplegic		5.33	1.501		890	88		978	1,100
7060	5″ pipe size		5.33	1.501		965	88		1,053	1,175
7100	Double, 4″ pipe size		5	1.600		1,575	94		1,669	1,875
7110	4″ pipe size, paraplegic		5	1.600		1,575	94		1,669	1,875
7120	5″ pipe size		5	1.600		1,700	94		1,794	2,025
7160	Horizontal, adjustable, extended, caulk									
7180	Single, 4″ pipe size	1 Plum	5.33	1.501	Ea.	1,000	88		1,088	1,225
7200	5″ pipe size		5.33	1.501		1,275	88		1,363	1,525
7240	Double, 4″ pipe size		5	1.600		1,750	94		1,844	2,075
7260	5″ pipe size		5	1.600		2,125	94		2,219	2,500
7400	Vertical, adjustable, caulk or thread									
7440	Single, 4″ pipe size	1 Plum	5.33	1.501	Ea.	890	88		978	1,100
7460	5″ pipe size		5.33	1.501		1,125	88		1,213	1,350
7480	6″ pipe size		5	1.600		1,325	94		1,419	1,600
7520	Double, 4″ pipe size		5	1.600		1,525	94		1,619	1,850
7540	5″ pipe size		5	1.600		1,775	94		1,869	2,100
7560	6″ pipe size		4	2		1,950	117		2,067	2,325
7600	Vertical, adjustable, extended, caulk									
7620	Single, 4″ pipe size	1 Plum	5.33	1.501	Ea.	1,025	88		1,113	1,250
7640	5″ pipe size		5.33	1.501		1,275	88		1,363	1,525
7680	6″ pipe size		5	1.600		1,450	94		1,544	1,750
7720	Double, 4″ pipe size		5	1.600		1,675	94		1,769	2,000
7740	5″ pipe size		5	1.600		1,925	94		2,019	2,275
7760	6″ pipe size		4	2		2,100	117		2,217	2,500
7780	Water closet, blow out									
7800	Vertical offset, caulk or thread									
7820	Single, 4″ pipe size	1 Plum	5.33	1.501	Ea.	760	88		848	970
7840	Double, 4″ pipe size	″	5	1.600	″	1,300	94		1,394	1,575
7880	Vertical offset, extended, caulk									
7900	Single, 4″ pipe size	1 Plum	5.33	1.501	Ea.	950	88		1,038	1,175
7920	Double, 4″ pipe size	″	5	1.600	″	1,500	94		1,594	1,800
7960	Vertical, for floor mounted back-outlet									
7980	Single, 4″ thread, 2″ vent	1 Plum	5.33	1.501	Ea.	670	88		758	870
8000	Double, 4″ thread, 2″ vent	″	6	1.333	″	1,975	78.50		2,053.50	2,275
8040	Vertical, for floor mounted back-outlet, extended									
8060	Single, 4″ caulk, 2″ vent	1 Plum	6	1.333	Ea.	670	78.50		748.50	855
8080	Double, 4″ caulk, 2″ vent	″	6	1.333	″	1,975	78.50		2,053.50	2,275
8200	Water closet, residential									
8220	Vertical centerline, floor mount									
8240	Single, 3″ caulk, 2″ or 3″ vent	1 Plum	6	1.333	Ea.	610	78.50		688.50	790
8260	4″ caulk, 2″ or 4″ vent		6	1.333		785	78.50		863.50	980
8280	3″ copper sweat, 3″ vent		6	1.333		545	78.50		623.50	720
8300	4″ copper sweat, 4″ vent		6	1.333		660	78.50		738.50	845
8400	Vertical offset, floor mount									
8420	Single, 3″ or 4″ caulk, vent	1 Plum	4	2	Ea.	760	117		877	1,025

22 42 Commercial Plumbing Fixtures

22 42 39 – Commercial Faucets, Supplies, and Trim

22 42 39.30 Carriers and Supports		Crew	Daily Output	Labor-Hours	Unit	Material	2015 Bare Costs Labor	Equipment	Total	Total Incl O&P
8440	3″ or 4″ copper sweat, vent	1 Plum	5	1.600	Ea.	760	94		854	980
8460	Double, 3″ or 4″ caulk, vent		4	2		1,300	117		1,417	1,600
8480	3″ or 4″ copper sweat, vent		5	1.600		1,300	94		1,394	1,575
9000	Water cooler (electric), floor mounted									
9100	Plate type with bearing plate, single	1 Plum	6	1.333	Ea.	385	78.50		463.50	545
9140	Plate type with bearing plate, back to back		4	2	"	535	117		652	775
9990	Minimum labor/equipment charge		3.50	2.286	Job		134		134	209

22 43 Healthcare Plumbing Fixtures

22 43 13 – Healthcare Water Closets

22 43 13.40 Water Closets		Crew	Daily Output	Labor-Hours	Unit	Material	Labor	Equipment	Total	Total Incl O&P
0010	**WATER CLOSETS**									
1000	Bowl only, 1 piece, w/seat and flush valve, ADA compliant 18″ high									
1030	Floor mounted									
1150	With wall outlet	Q-1	5.30	3.019	Ea.	385	159		544	675
1180	For rough-in, supply, waste and vent		2.84	5.634		370	298		668	870
1200	With floor outlet		5.30	3.019		310	159		469	590
1800	For rough-in, supply, waste and vent		3.05	5.246		330	277		607	795
3100	Wall hung		5.80	2.759		945	146		1,091	1,275
3150	Hospital type, slotted rim for bed pan									
3156	Elongated bowl, top spud	Q-1	5.80	2.759	Ea.	600	146		746	885
3160	Elongated bowl, rear spud		5.80	2.759		360	146		506	620
3200	For rough-in, supply, waste and vent, single WC		2.56	6.250		1,075	330		1,405	1,700
3300	Floor mounted									
3320	Bariatric, (1,200 lb. capacity), elongated bowl, ADA	Q-1	4.60	3.478	Ea.	2,775	184		2,959	3,325
3360	Hospital type, slotted rim for bed pan									
3370	Elongated bowl, top spud	Q-1	5	3.200	Ea.	320	169		489	615
3380	Elongated bowl, rear spud		5	3.200		410	169		579	715
3500	For rough-in, supply, waste and vent		3.05	5.246		330	277		607	795

22 43 16 – Healthcare Sinks

22 43 16.10 Sinks		Crew	Daily Output	Labor-Hours	Unit	Material	Labor	Equipment	Total	Total Incl O&P
0010	**SINKS**									
0020	Vitreous china									
6702	Hospital type, without trim (see Section 22 41 39.10)									
6710	20″ x 18″, contoured splash shield	Q-1	8	2	Ea.	93.50	106		199.50	268
6730	28″ x 20″, surgeon, side decks		8	2		575	106		681	800
6740	28″ x 22″, surgeon scrub-up, deep bowl		8	2		825	106		931	1,075
6750	20″ x 27″, patient, wheelchair		7	2.286		605	121		726	855
6760	30″ x 22″, all purpose		7	2.286		685	121		806	945
6770	30″ x 22″, plaster work		7	2.286		685	121		806	945
6820	20″ x 24″ clinic service, liquid/solid waste		6	2.667		965	141		1,106	1,275

22 43 19 – Healthcare Bathtubs

22 43 19.10 Bathtubs		Crew	Daily Output	Labor-Hours	Unit	Material	Labor	Equipment	Total	Total Incl O&P
0010	**BATHTUBS**									
5002	Hospital type, with trim, see Section 22 41 39.10									
5050	Bathing pool, porcelain enamel on cast iron, grab bars									
5060	pop-up drain, 72″ x 36″	Q-1	3	5.333	Ea.	3,750	282		4,032	4,575
5100	Perineal (sitz), vitreous china		3	5.333		1,300	282		1,582	1,875
5120	For pedestal, vitreous china, add		8	2		251	106		357	440
5300	Whirlpool, porcelain enamel on cast iron, 72″ x 36″		1	16		3,575	845		4,420	5,250

22 43 Healthcare Plumbing Fixtures

22 43 23 – Healthcare Showers

22 43 23.10 Showers		Crew	Daily Output	Labor-Hours	Unit	Material	Labor	Equipment	Total	Total Incl O&P
						2015 Bare Costs				
0010	**SHOWERS**									
5950	Module, handicap, SS panel, fixed & hand held head, control									
5960	valves, grab bar, curtain & rod, folding seat	1 Plum	4	2	Ea.	2,875	117		2,992	3,325

22 43 39 – Healthcare Faucets

22 43 39.10 Faucets and Fittings

		Crew	Daily Output	Labor-Hours	Unit	Material	Labor	Equipment	Total	Total Incl O&P
0010	**FAUCETS AND FITTINGS**									
2850	Medical, bedpan cleanser, with pedal valve,	1 Plum	12	.667	Ea.	760	39		799	895
2860	With screwdriver stop valve		12	.667		395	39		434	490
2870	With self-closing spray valve		12	.667		242	39		281	325
2900	Faucet, gooseneck spout, wrist handles, grid drain		10	.800		194	47		241	286
2940	Mixing valve, knee action, screwdriver stops		4	2		425	117		542	650

22 45 Emergency Plumbing Fixtures

22 45 13 – Emergency Showers

22 45 13.10 Emergency Showers

		Crew	Daily Output	Labor-Hours	Unit	Material	Labor	Equipment	Total	Total Incl O&P
0010	**EMERGENCY SHOWERS**, Rough-in not included									
5000	Shower, single head, drench, ball valve, pull, freestanding	Q-1	4	4	Ea.	380	211		591	745
5200	Horizontal or vertical supply		4	4		555	211		766	940
6000	Multi-nozzle, eye/face wash combination		4	4		660	211		871	1,050
6400	Multi-nozzle, 12 spray, shower only		4	4		2,000	211		2,211	2,525
6600	For freeze-proof, add		6	2.667		465	141		606	735
8000	Walk-thru decontamination with eye-face wash		2	8		3,375	425		3,800	4,350
8200	For freeze proof, add		4	4		565	211		776	955
9000	Minimum labor/equipment charge		3	5.333	Job		282		282	440

22 45 16 – Eyewash Equipment

22 45 16.10 Eyewash Safety Equipment

		Crew	Daily Output	Labor-Hours	Unit	Material	Labor	Equipment	Total	Total Incl O&P
0010	**EYEWASH SAFETY EQUIPMENT**, Rough-in not included									
1000	Eye wash fountain									
1400	Plastic bowl, pedestal mounted	Q-1	4	4	Ea.	282	211		493	640
1600	Unmounted		4	4		247	211		458	600
1800	Wall mounted		4	4		455	211		666	830
2000	Stainless steel, pedestal mounted		4	4		350	211		561	715
2200	Unmounted		4	4		272	211		483	630
2400	Wall mounted		4	4		288	211		499	645

22 45 19 – Self-Contained Eyewash Equipment

22 45 19.10 Self-Contained Eyewash Safety Equipment

		Crew	Daily Output	Labor-Hours	Unit	Material	Labor	Equipment	Total	Total Incl O&P
0010	**SELF-CONTAINED EYEWASH SAFETY EQUIPMENT**									
3000	Eye wash, portable, self-contained				Ea.	1,025			1,025	1,125

22 45 26 – Eye/Face Wash Equipment

22 45 26.10 Eye/Face Wash Safety Equipment

		Crew	Daily Output	Labor-Hours	Unit	Material	Labor	Equipment	Total	Total Incl O&P
0010	**EYE/FACE WASH SAFETY EQUIPMENT**, Rough-in not included									
4000	Eye and face wash, combination fountain									
4200	Stainless steel, pedestal mounted	Q-1	4	4	Ea.	1,075	211		1,286	1,500
4400	Unmounted		4	4		272	211		483	630
4600	Wall mounted		4	4		246	211		457	600

22 46 Security Plumbing Fixtures

22 46 13 – Security Water Closets and Urinals

22 46 13.10 Security Water Closets and Urinals		Crew	Daily Output	Labor-Hours	Unit	Material	2015 Bare Costs Labor	Equipment	Total	Total Incl O&P
0010	**SECURITY WATER CLOSETS AND URINALS**, Stainless steel									
2000	Urinal, back supply and flush									
2200	Wall hung	Q-1	4	4	Ea.	1,225	211		1,436	1,675
2240	Stall		2.50	6.400		2,100	340		2,440	2,825
2300	For urinal rough-in, supply, waste and vent		1.49	10.738		320	565		885	1,225
3000	Water closet, integral seat, back supply and flush									
3300	Wall hung, wall outlet	Q-1	5.80	2.759	Ea.	985	146		1,131	1,300
3400	Floor mount, wall outlet		5.80	2.759		1,275	146		1,421	1,625
3440	Floor mount, floor outlet		5.80	2.759		1,300	146		1,446	1,650
3480	For recessed tissue holder, add					88			88	96.50
3500	For water closet rough-in, supply, waste and vent	Q-1	1.19	13.445		345	710		1,055	1,475
5000	Water closet and lavatory units, push button filler valves,									
5010	soap & paper holders, seat									
5300	Wall hung	Q-1	5	3.200	Ea.	2,100	169		2,269	2,600
5400	Floor mount		5	3.200		2,000	169		2,169	2,475
6300	For unit rough-in, supply, waste and vent		1	16		390	845		1,235	1,750

22 46 16 – Security Lavatories and Sinks

22 46 16.13 Security Lavatories		Crew	Daily Output	Labor-Hours	Unit	Material	Labor	Equipment	Total	Total Incl O&P
0010	**SECURITY LAVATORIES**, Stainless steel									
1000	Lavatory, wall hung, push button filler valve									
1100	Rectangular bowl	Q-1	8	2	Ea.	1,075	106		1,181	1,375
1200	Oval bowl		8	2		1,125	106		1,231	1,400
1240	Oval bowl, corner mount		8	2		1,225	106		1,331	1,525
1300	For lavatory rough-in, supply, waste and vent		1.50	10.667		282	565		847	1,200

22 46 63 – Security Service Sink

22 46 63.10 Security Service Sink		Crew	Daily Output	Labor-Hours	Unit	Material	Labor	Equipment	Total	Total Incl O&P
0010	**SECURITY SERVICE SINK**, Stainless steel									
1700	Service sink, with soap dish									
1740	24" x 19" size	Q-1	3	5.333	Ea.	2,175	282		2,457	2,850
1790	For sink rough-in, supply, waste and vent	"	.89	17.978	"	620	950		1,570	2,150

22 46 73 – Security Shower

22 46 73.10 Security Shower		Crew	Daily Output	Labor-Hours	Unit	Material	Labor	Equipment	Total	Total Incl O&P
0010	**SECURITY SHOWER**, Stainless steel									
1800	Shower cabinet, unitized									
1840	36" x 36" x 88"	Q-1	2.20	7.273	Ea.	4,950	385		5,335	6,050
1900	Shower package for built-in									
1940	Hot & cold valves, recessed soap dish	Q-1	6	2.667	Ea.	540	141		681	810

22 47 Drinking Fountains and Water Coolers

22 47 13 – Drinking Fountains

22 47 13.10 Drinking Water Fountains		Crew	Daily Output	Labor-Hours	Unit	Material	Labor	Equipment	Total	Total Incl O&P
0010	**DRINKING WATER FOUNTAINS**, For connection to cold water supply R224000-30									
0802	For remote water chiller, see Section 22 47 23.10									
1000	Wall mounted, non-recessed									
1200	Aluminum,									
1280	Dual bubbler type	1 Plum	3.20	2.500	Ea.	2,450	147		2,597	2,925
1400	Bronze, with no back		4	2		1,000	117		1,117	1,275
1600	Cast iron, enameled, low back, single bubbler		4	2		945	117		1,062	1,225
1640	Dual bubbler type		3.20	2.500		1,525	147		1,672	1,900

22 47 Drinking Fountains and Water Coolers

22 47 13 – Drinking Fountains

22 47 13.10 Drinking Water Fountains		Crew	Daily Output	Labor-Hours	Unit	Material	2015 Bare Costs Labor	Equipment	Total	Total Incl O&P
1680	Triple bubbler type	1 Plum	3.20	2.500	Ea.	1,975	147		2,122	2,400
1800	Cast aluminum, enameled, for correctional institutions		4	2		1,675	117		1,792	2,000
2000	Fiberglass, 12" back, single bubbler unit		4	2		1,925	117		2,042	2,300
2040	Dual bubbler		3.20	2.500		2,225	147		2,372	2,650
2080	Triple bubbler		3.20	2.500		2,475	147		2,622	2,950
2200	Polymarble, no back, single bubbler		4	2		705	117		822	960
2240	Dual bubbler		3.20	2.500		1,800	147		1,947	2,200
2280	Triple bubbler		3.20	2.500		2,125	147		2,272	2,575
2400	Precast stone, no back		4	2		915	117		1,032	1,175
2700	Stainless steel, single bubbler, no back		4	2		1,050	117		1,167	1,325
2740	With back		4	2		565	117		682	805
2780	Dual handle & wheelchair projection type		4	2		745	117		862	1,000
2820	Dual level for handicapped type		3.20	2.500		1,575	147		1,722	1,950
2840	Vandal resistant type		4	2		650	117		767	900
3300	Vitreous china									
3340	7" back	1 Plum	4	2	Ea.	620	117		737	865
3940	For vandal-resistant bottom plate, add					77			77	84.50
3960	For freeze-proof valve system, add	1 Plum	2	4		705	235		940	1,150
3980	For rough-in, supply and waste, add	"	2.21	3.620		174	212		386	520
4000	Wall mounted, semi-recessed									
4200	Poly-marble, single bubbler	1 Plum	4	2	Ea.	955	117		1,072	1,225
4600	Stainless steel, satin finish, single bubbler		4	2		1,150	117		1,267	1,450
4900	Vitreous china, single bubbler		4	2		895	117		1,012	1,175
5980	For rough-in, supply and waste, add		1.83	4.372		174	257		431	590
6000	Wall mounted, fully recessed									
6400	Poly-marble, single bubbler	1 Plum	4	2	Ea.	1,600	117		1,717	1,925
6440	For water glass filler, add					86.50			86.50	95
6800	Stainless steel, single bubbler	1 Plum	4	2		1,550	117		1,667	1,875
6900	Fountain and cuspidor combination		2	4		3,000	235		3,235	3,675
7560	For freeze-proof valve system, add		2	4		795	235		1,030	1,250
7580	For rough-in, supply and waste, add		1.83	4.372		174	257		431	590
7600	Floor mounted, pedestal type									
7700	Aluminum, architectural style, C.I. base	1 Plum	2	4	Ea.	2,250	235		2,485	2,850
7780	Wheelchair handicap unit		2	4		1,625	235		1,860	2,150
8000	Bronze, architectural style		2	4		1,925	235		2,160	2,500
8040	Enameled steel cylindrical column style		2	4		2,200	235		2,435	2,800
8200	Precast stone/concrete, cylindrical column		1	8		1,375	470		1,845	2,225
8240	Wheelchair handicap unit		1	8		2,725	470		3,195	3,725
8400	Stainless steel, architectural style		2	4		1,750	235		1,985	2,300
8600	Enameled iron, heavy duty service, 2 bubblers		2	4		2,575	235		2,810	3,225
8660	4 bubblers		2	4		3,950	235		4,185	4,700
8880	For freeze-proof valve system, add		2	4		705	235		940	1,150
8900	For rough-in, supply and waste, add		1.83	4.372		174	257		431	590
9000	Minimum labor/equipment charge		2	4	Job		235		235	365
9100	Deck mounted									
9500	Stainless steel, circular receptor	1 Plum	4	2	Ea.	435	117		552	665
9540	14" x 9" receptor		4	2		360	117		477	580
9580	25" x 17" deep receptor, with water glass filler		3	2.667		305	157		462	580
9760	White enameled steel, 14" x 9" receptor		4	2		375	117		492	600
9860	White enameled cast iron, 24" x 16" receptor		3	2.667		460	157		617	750
9980	For rough-in, supply and waste, add		1.83	4.372		174	257		431	590

22 47 16 – Pressure Water Coolers

22 47 16.10 Electric Water Coolers		Crew	Daily Output	Labor-Hours	Unit	2015 Bare Costs Material	Labor	Equipment	Total	Total Incl O&P
0010	**ELECTRIC WATER COOLERS** R224000-30									
0100	Wall mounted, non-recessed									
0140	4 GPH	Q-1	4	4	Ea.	680	211		891	1,075
0160	8 GPH, barrier free, sensor operated		4	4		1,025	211		1,236	1,475
0180	8.2 GPH		4	4		750	211		961	1,150
0220	14.3 GPH		4	4		875	211		1,086	1,300
0600	8 GPH hot and cold water		4	4		1,050	211		1,261	1,500
0640	For stainless steel cabinet, add					90			90	99.50
1000	Dual height, 8.2 GPH	Q-1	3.80	4.211		2,075	222		2,297	2,625
1040	14.3 GPH	"	3.80	4.211		975	222		1,197	1,425
1240	For stainless steel cabinet, add					171			171	188
2600	Wheelchair type, 8 GPH	Q-1	4	4		915	211		1,126	1,325
3000	Simulated recessed, 8 GPH		4	4		665	211		876	1,050
3040	11.5 GPH		4	4		1,025	211		1,236	1,450
3200	For glass filler, add					99			99	109
3240	For stainless steel cabinet, add					73.50			73.50	81
3300	Semi-recessed, 8.1 GPH	Q-1	4	4		750	211		961	1,150
3320	12 GPH	"	4	4		865	211		1,076	1,275
3340	For glass filler, add					129			129	142
3360	For stainless steel cabinet, add					145			145	160
3400	Full recessed, stainless steel, 8 GPH	Q-1	3.50	4.571		2,025	242		2,267	2,600
3420	11.5 GPH	"	3.50	4.571		1,650	242		1,892	2,175
3460	For glass filler, add					146			146	161
3600	For mounting can only					243			243	268
4600	Floor mounted, flush-to-wall									
4640	4 GPH	1 Plum	3	2.667	Ea.	740	157		897	1,050
4680	8.2 GPH		3	2.667		780	157		937	1,100
4720	14.3 GPH		3	2.667		890	157		1,047	1,225
4960	14 GPH hot and cold water		3	2.667		1,100	157		1,257	1,475
4980	For stainless steel cabinet, add					134			134	148
5000	Dual height, 8.2 GPH	1 Plum	2	4		1,175	235		1,410	1,650
5040	14.3 GPH	"	2	4		1,200	235		1,435	1,700
5120	For stainless steel cabinet, add					196			196	215
5600	Explosion Proof, 16 GPH	1 Plum	3	2.667		2,075	157		2,232	2,550
6000	Refrigerator Compartment Type, 4.5 GPH		3	2.667		1,500	157		1,657	1,900
6600	Bottle Supply Type, 1.0 GPH		4	2		395	117		512	620
6640	Hot and cold, 1.0 GPH		4	2		535	117		652	775
9000	Minimum labor/equipment charge		2	4	Job		235		235	365
9800	For supply, waste & vent, all coolers		2.21	3.620	Ea.	174	212		386	520

22 47 23 – Remote Water Coolers

22 47 23.10 Remote Water Coolers		Crew	Daily Output	Labor-Hours	Unit	Material	Labor	Equipment	Total	Total Incl O&P
0010	**REMOTE WATER COOLERS**, 80°F inlet									
0100	Air cooled, 50°F outlet, 115 V, 4.1 GPH	1 Plum	6	1.333	Ea.	525	78.50		603.50	700
0200	5.7 GPH		5.50	1.455		835	85.50		920.50	1,050
0300	8.0 GPH		5	1.600		625	94		719	835
0400	10.0 GPH		4.50	1.778		920	104		1,024	1,175
0500	13.4 GPH		4	2		1,425	117		1,542	1,750
0700	29 GPH	Q-1	5	3.200		1,675	169		1,844	2,125
1000	230V, 32 GPH	"	5	3.200		1,750	169		1,919	2,200

22 51 Swimming Pool Plumbing Systems

22 51 19 – Swimming Pool Water Treatment Equipment

22 51 19.50 Swimming Pool Filtration Equipment		Crew	Daily Output	Labor-Hours	Unit	Material	2015 Bare Costs Labor	Equipment	Total	Total Incl O&P
0010	**SWIMMING POOL FILTRATION EQUIPMENT**									
0900	Filter system, sand or diatomite type, incl. pump, 6,000 gal./hr.	2 Plum	1.80	8.889	Total	1,975	520		2,495	3,000
1020	Add for chlorination system, 800 S.F. pool		3	5.333	Ea.	181	315		496	690
1040	5,000 S.F. pool		3	5.333	″	1,850	315		2,165	2,525

22 52 Fountain Plumbing Systems

22 52 16 – Fountain Pumps

22 52 16.10 Fountain Water Pumps

		Crew	Daily Output	Labor-Hours	Unit	Material	Labor	Equipment	Total	Total Incl O&P
0010	**FOUNTAIN WATER PUMPS**									
0100	Pump w/controls									
0200	Single phase, 100′ cord, 1/2 H.P. pump	2 Skwk	4.40	3.636	Ea.	1,275	177		1,452	1,675
0300	3/4 H.P. pump		4.30	3.721		2,175	181		2,356	2,700
0400	1 H.P. pump		4.20	3.810		2,350	185		2,535	2,900
0500	1-1/2 H.P. pump		4.10	3.902		2,800	190		2,990	3,375
0600	2 H.P. pump		4	4		3,775	195		3,970	4,500
0700	Three phase, 200′ cord, 5 H.P. pump		3.90	4.103		5,175	200		5,375	6,025
0800	7-1/2 H.P. pump		3.80	4.211		9,000	205		9,205	10,200
0900	10 H.P. pump		3.70	4.324		13,300	210		13,510	15,000
1000	15 H.P. pump		3.60	4.444		16,800	216		17,016	18,900
2000	DESIGN NOTE: Use two horsepower per surface acre.									

22 52 33 – Fountain Ancillary

22 52 33.10 Fountain Miscellaneous

		Crew	Daily Output	Labor-Hours	Unit	Material	Labor	Equipment	Total	Total Incl O&P
0010	**FOUNTAIN MISCELLANEOUS**									
1300	Lights w/mounting kits, 200 watt	2 Skwk	18	.889	Ea.	1,050	43		1,093	1,225
1400	300 watt		18	.889		1,275	43		1,318	1,475
1500	500 watt		18	.889		1,425	43		1,468	1,650
1600	Color blender		12	1.333		555	65		620	715

22 62 Vacuum Systems for Laboratory and Healthcare Facilities

22 62 19 – Vacuum Equipment for Laboratory and Healthcare Facilities

22 62 19.70 Healthcare Vacuum Equipment

		Crew	Daily Output	Labor-Hours	Unit	Material	Labor	Equipment	Total	Total Incl O&P
0010	**HEALTHCARE VACUUM EQUIPMENT**									
0300	Dental oral									
0310	Duplex									
0330	165 SCFM with 77 gal. separator	Q-2	1.30	18.462	Ea.	52,000	1,000		53,000	58,500
1100	Vacuum system									
1110	Vacuum outlet alarm panel	1 Plum	3.20	2.500	Ea.	1,100	147		1,247	1,450
2000	Medical, with receiver									
2100	Rotary vane type, lubricated, with controls									
2110	Simplex									
2120	1.5 HP, 80 gal tank	Q-1	8	2	Ea.	6,525	106		6,631	7,350
2130	2 HP, 80 gal tank		7	2.286		6,750	121		6,871	7,625
2140	3 HP, 80 gal tank		6.60	2.424		7,125	128		7,253	8,025
2150	5 HP, 80 gal tank		6	2.667		7,775	141		7,916	8,775
2200	Duplex									
2210	1 HP, 80 gal tank	Q-1	8.40	1.905	Ea.	10,400	101		10,501	11,700
2220	1.5 HP, 80 gal tank		7.80	2.051		10,700	108		10,808	12,000
2230	2 HP, 80 gal tank		6.80	2.353		11,600	124		11,724	12,900
2240	3 HP, 120 gal tank		6	2.667		12,600	141		12,741	14,000

22 62 Vacuum Systems for Laboratory and Healthcare Facilities

22 62 19 – Vacuum Equipment for Laboratory and Healthcare Facilities

22 62 19.70 Healthcare Vacuum Equipment		Crew	Daily Output	Labor-Hours	Unit	2015 Bare Costs Material	Labor	Equipment	Total	Total Incl O&P
2250	5 HP, 120 gal tank	Q-1	5.40	2.963	Ea.	14,300	157		14,457	15,900
2260	7.5 HP, 200 gal tank	Q-2	7	3.429		20,400	188		20,588	22,800
2270	10 HP, 200 gal tank		6.40	3.750		23,600	205		23,805	26,300
2280	15 HP, 200 gal tank		5.80	4.138		42,200	227		42,427	46,900
2290	20 HP, 200 gal tank		5	4.800		48,100	263		48,363	53,500
2300	25 HP, 200 gal tank	↓	4	6	↓	56,000	330		56,330	62,000
2400	Triplex									
2410	7.5 HP, 200 gal tank	Q-2	6.40	3.750	Ea.	32,400	205		32,605	36,000
2420	10 HP, 200 gal tank		5.70	4.211		37,800	231		38,031	41,900
2430	15 HP, 200 gal tank		4.90	4.898		64,000	268		64,268	70,500
2440	20 HP, 200 gal tank		4	6		73,000	330		73,330	80,500
2450	25 HP, 200 gal tank	↓	3.70	6.486	↓	84,000	355		84,355	93,000
2500	Quadruplex									
2510	7.5 HP, 200 gal tank	Q-2	5.30	4.528	Ea.	40,400	248		40,648	44,800
2520	10 HP, 200 gal tank		4.70	5.106		46,600	280		46,880	51,500
2530	15 HP, 200 gal tank		4	6		84,000	330		84,330	93,000
2540	20 HP, 200 gal tank		3.60	6.667		96,000	365		96,365	106,000
2550	25 HP, 200 gal tank	↓	3.20	7.500	↓	111,500	410		111,910	123,000
4000	Liquid ring type , water sealed, with controls									
4200	Duplex									
4210	1.5 HP, 120 gal tank	Q-1	6	2.667	Ea.	20,600	141		20,741	22,900
4220	3 HP, 120 gal tank	"	5.50	2.909		21,500	154		21,654	23,900
4230	4 HP, 120 gal tank	Q-2	6.80	3.529		22,900	193		23,093	25,500
4240	5 HP, 120 gal tank		6.50	3.692		24,200	202		24,402	26,900
4250	7.5 HP, 200 gal tank		6.20	3.871		29,400	212		29,612	32,700
4260	10 HP, 200 gal tank		5.80	4.138		36,700	227		36,927	40,800
4270	15 HP, 200 gal tank		5.10	4.706		44,500	258		44,758	49,400
4280	20 HP, 200 gal tank		4.60	5.217		57,500	286		57,786	64,000
4290	30 HP, 200 gal tank	↓	4	6	↓	69,000	330		69,330	76,000

22 63 Gas Systems for Laboratory and Healthcare Facilities

22 63 13 – Gas Piping for Laboratory and Healthcare Facilities

22 63 13.70 Healthcare Gas Piping		Crew	Daily Output	Labor-Hours	Unit	Material	Labor	Equipment	Total	Total Incl O&P
0010	**HEALTHCARE GAS PIPING**									
0030	Air compressor intake filter									
0034	Rooftop									
0036	Filter/silencer									
0040	1"	1 Stpi	10	.800	Ea.	210	48		258	305
0044	1-1/4"		9.60	.833		216	50		266	315
0048	1-1/2"		9.20	.870		232	52		284	335
0052	2"		8.80	.909		232	54.50		286.50	340
0056	2-1/2"		8.40	.952		310	57		367	430
0060	3"		8	1		370	60		430	505
0064	4"		7.60	1.053		385	63		448	525
0068	5"	↓	7.40	1.081	↓	540	64.50		604.50	695
0076	Inline									
0080	Filter/silencer									
0084	2"	1 Stpi	8.20	.976	Ea.	700	58.50		758.50	860
0088	2-1/2"		8	1		720	60		780	890
0090	3"		7.60	1.053		795	63		858	975
0092	4"	↓	7.20	1.111	↓	930	66.50		996.50	1,125

22 63 13 – Gas Piping for Laboratory and Healthcare Facilities

22 63 13.70 Healthcare Gas Piping		Crew	Daily Output	Labor-Hours	Unit	2015 Bare Costs Material	Labor	Equipment	Total	Total Incl O&P
0094	5"	1 Stpi	6.80	1.176	Ea.	1,000	70.50		1,070.50	1,200
0096	6"		6.40	1.250		1,125	74.50		1,199.50	1,375
1000	Nitrogen or oxygen system									
1010	Cylinder manifold									
1020	5 cylinder	1 Plum	.80	10	Ea.	5,950	585		6,535	7,475
1026	10 cylinder	"	.40	20	"	7,175	1,175		8,350	9,700
1900	Vaporizers									
1910	LOX vaporizers									
1920	Nominal capacity									
1930	1410 SCFM	Q-1	2	8	Ea.	2,000	425		2,425	2,850
1940	5650 SCFM		1.40	11.429		4,000	605		4,605	5,350
1950	12703 SCFM		.80	20		6,000	1,050		7,050	8,250
1980	Removal of LOX vaporizers									
1982	Nominal capacity									
1986	1410 SCFM	Q-1	4	4	Ea.		211		211	330
1990	5650 SCFM		2.80	5.714			300		300	470
1994	12703 SCFM		1.60	10			530		530	825
3000	Outlets and valves									
3010	Recessed, wall mounted									
3012	Single outlet	1 Plum	3.20	2.500	Ea.	71.50	147		218.50	310
3100	Ceiling outlet									
3190	Zone valve with box									
3192	Cleaned for oxygen service, not including gauges									
3194	1/2" valve size	1 Plum	4.60	1.739	Ea.	214	102		316	395
3196	3/4" valve size		4.30	1.860		236	109		345	430
3198	1" valve size		4	2		261	117		378	470
3202	1-1/4" valve size		3.80	2.105		285	124		409	510
3206	1-1/2" valve size		3.60	2.222		320	130		450	555
3210	2" valve size		3.20	2.500		420	147		567	690
3214	2-1/2" valve size		3.10	2.581		970	151		1,121	1,300
3218	3" valve size		3	2.667		1,350	157		1,507	1,725
3224	Gauges for zone valve box									
3226	0-100 PSI (02, air, N2, CO_2)	1 Plum	16	.500	Ea.	18.75	29.50		48.25	66.50
3228	Vacuum, WAGD		16	.500		18.75	29.50		48.25	66.50
3230	0-300 PSI (Nitrogen)		16	.500		18.75	29.50		48.25	66.50
4000	Alarm panel, medical gases and vacuum									
4010	Alarm panel	1 Plum	3.20	2.500	Ea.	1,100	147		1,247	1,450
4030	Master alarm panel									
4034	can also monitor area alarms									
4038	and communicate with PC-based alarm monitor.									
4040	10 Signal	1 Elec	3.20	2.500	Ea.	980	137		1,117	1,275
4044	20 Signal		3	2.667		1,125	146		1,271	1,475
4048	30 Signal		2.80	2.857		1,500	156		1,656	1,900
4052	40 Signal		2.60	3.077		1,675	168		1,843	2,075
4056	50 Signal		2.40	3.333		1,825	182		2,007	2,300
4060	60 Signal		2.20	3.636		2,150	199		2,349	2,650
4100	Area alarm panel									
4104	does not include specific gas transducers.									
4108	3 Module alarm panel									
4112	P-P-P	1 Elec	2.80	2.857	Ea.	1,300	156		1,456	1,675
4116	P-P-V		2.80	2.857		875	156		1,031	1,200
4120	D-D-D		2.80	2.857		960	156		1,116	1,300
4130	6 Module alarm panel									

22 63 13 – Gas Piping for Laboratory and Healthcare Facilities

22 63 13.70 Healthcare Gas Piping		Crew	Daily Output	Labor-Hours	Unit	2015 Bare Costs Material	Labor	Equipment	Total	Total Incl O&P
4132	4-P, 3-V	1 Elec	2.20	3.636	Ea.	1,400	199		1,599	1,850
4136	3-P, 2-V, B		2.20	3.636		1,250	199		1,449	1,675
4140	5-D, B		2.20	3.636		1,350	199		1,549	1,775
4144	6-D		2.20	3.636		1,775	199		1,974	2,275
4148	3-P, 3-V		2.20	3.636		1,650	199		1,849	2,125
4170	Note: P = pressure, V = vacuum, B = blank, D = dual display									
4180	Alarm transducers, gas specific									
4182	Oxygen	1 Elec	24	.333	Ea.	148	18.25		166.25	191
4184	Vacuum		24	.333		148	18.25		166.25	191
4186	Nitrous oxide		24	.333		148	18.25		166.25	191
4188	Medical air		24	.333		148	18.25		166.25	191
4190	Carbon dioxide		24	.333		148	18.25		166.25	191
4192	Nitrogen		24	.333		148	18.25		166.25	191
4194	WAGD		24	.333		158	18.25		176.25	201
4300	Ball valves cleaned for oxygen service									
4310	with copper extensions and gauge port									
4320	1/4" diam.	1 Plum	24	.333	Ea.	52.50	19.55		72.05	88.50
4330	1/2" diam.		22	.364		58.50	21.50		80	98
4334	3/4" diam.		20	.400		82.50	23.50		106	128
4338	1" diam.		19	.421		105	24.50		129.50	155
4342	1-1/4" diam.		15	.533		127	31.50		158.50	188
4346	1-1/2" diam.		13	.615		176	36		212	251
4350	2" diam.		11	.727		254	42.50		296.50	345
4354	2-1/2" diam.	Q-1	15	1.067		700	56.50		756.50	855
4358	3" diam.		13	1.231		1,050	65		1,115	1,250
4362	4" diam.		10	1.600		1,975	84.50		2,059.50	2,300
5000	Manifold									
5010	Automatic switchover type									
5020	Note: Both a control panel and header assembly are required.									
5030	Control panel									
5040	Oxygen	Q-1	4	4	Ea.	4,000	211		4,211	4,725
5060	Header assembly									
5066	Oxygen									
5070	2 x 2	Q-1	6	2.667	Ea.	655	141		796	940
5074	3 x 3		5.50	2.909		770	154		924	1,075
5078	4 x 4		5	3.200		1,000	169		1,169	1,375
5082	5 x 5		4.50	3.556		1,125	188		1,313	1,525
5086	6 x 6		4	4		1,400	211		1,611	1,875
5090	7 x 7		3.50	4.571		1,525	242		1,767	2,050
7000	Medical air compressors									
7020	Oil-less with inlet filter, duplexed dryers and aftercoolers									
7030	Duplex systems, horizontal tank, 208/230/460/575 V, 3 Ph.									
7040	1 HP, 80 gal. tank, 3.8 ACFM @50PSIG, 2.7 ACFM @100PSIG	Q-5	4	4	Ea.	35,800	215		36,015	39,600
7050	1.5 HP, 80 gal. tank, 6.8 ACFM @50PSIG, 4.5 ACFM @100PSIG		3.80	4.211		35,800	226		36,026	39,700
7060	2 HP, 80 gal. tank, 8.2 ACFM @50PSIG, 6.3 ACFM @100PSIG		3.60	4.444		36,000	239		36,239	40,000
7070	3 HP, 120 gal. tank, 11.2 ACFM @50PSIG, 9.4 ACFM @100PSIG		3.20	5		39,600	269		39,869	43,900
7080	5 HP, 120 gal. tank, 17.4 ACFM @50PSIG, 15.6 ACFM @100PSIG		2.80	5.714		47,700	305		48,005	53,000
7090	7.5 HP, 250 gal. tank, 31.6 ACFM @50PSIG, 25.9 ACFM @100PSIG		2.40	6.667		62,000	360		62,360	68,500
7100	10 HP, 250 gal. tank, 43 ACFM @50PSIG, 35.2 ACFM @100PSIG		2	8		64,500	430		64,930	71,500
7110	15 HP, 250 gal. tank, 69 ACFM @50PSIG, 56.5 ACFM @100PSIG		1.80	8.889		78,000	480		78,480	86,000
7200	Aftercooler, air-cooled									
7210	Steel manifold, copper tube, aluminum fins									
7220	35 SCFM @ 100PSIG	Q-5	4	4	Ea.	895	215		1,110	1,325

22 63 Gas Systems for Laboratory and Healthcare Facilities

22 63 13 – Gas Piping for Laboratory and Healthcare Facilities

22 63 13.70 Healthcare Gas Piping		Crew	Daily Output	Labor-Hours	Unit	2015 Bare Costs Material	Labor	Equipment	Total	Total Incl O&P
7300	Air dryer system									
7310	Refrigerated type									
7320	Flow @ 125 PSI									
7330	20 SCFM	Q-5	6	2.667	Ea.	1,475	143		1,618	1,850
7332	25 SCFM		5.80	2.759		1,550	148		1,698	1,925
7334	35 SCFM		5.40	2.963		1,875	159		2,034	2,325
7336	50 SCFM		5	3.200		2,350	172		2,522	2,850
7338	75 SCFM		4.60	3.478		2,950	187		3,137	3,525
7340	100 SCFM		4.30	3.721		3,450	200		3,650	4,100
7342	125 SCFM		4	4		4,450	215		4,665	5,225
7360	Desiccant type									
7362	Flow with energy saving controls									
7366	40 SCFM	Q-5	3.60	4.444	Ea.	6,025	239		6,264	7,000
7368	60 SCFM		3.40	4.706		6,550	253		6,803	7,625
7370	90 SCFM		3.20	5		7,600	269		7,869	8,800
7372	115 SCFM		3	5.333		8,175	287		8,462	9,450
7374	165 SCFM		2.90	5.517		8,800	297		9,097	10,100
7378	260 SCFM	Q-6	3.60	6.667		10,400	370		10,770	12,000
7382	370 SCFM		3.40	7.059		12,100	395		12,495	13,900
7386	590 SCFM		3.20	7.500		14,900	420		15,320	17,100
7390	1130 SCFM		3	8		23,100	445		23,545	26,200
7460	Dew point monitor									
7466	LCD readout, high dew point alarm, probe included									
7470	Monitor	1 Stpi	2	4	Ea.	2,675	239		2,914	3,325

22 66 Chemical-Waste Systems for Lab. and Healthcare Facilities

22 66 53 – Laboratory Chemical-Waste and Vent Piping

22 66 53.30 Glass Pipe		Crew	Daily Output	Labor-Hours	Unit	Material	Labor	Equipment	Total	Total Incl O&P
0010	**GLASS PIPE**, Borosilicate, couplings & clevis hanger assemblies, 10′ O.C.									
0020	Drainage									
1100	1-1/2″ diameter	Q-1	52	.308	L.F.	11.40	16.25		27.65	38
1120	2″ diameter		44	.364		14.65	19.20		33.85	46
1140	3″ diameter		39	.410		19.50	21.50		41	55.50
1160	4″ diameter		30	.533		34	28		62	81.50
1180	6″ diameter		26	.615		58.50	32.50		91	115
1870	To delete coupling & hanger, subtract									
1880	1-1/2″ diam. to 2″ diam.					19%	22%			
1890	3″ diam. to 6″ diam.					20%	17%			
2000	Process supply (pressure), beaded joints									
2040	1/2″ diameter	1 Plum	36	.222	L.F.	6	13.05		19.05	27
2060	3/4″ diameter		31	.258		6.65	15.15		21.80	31
2080	1″ diameter		27	.296		16.55	17.40		33.95	45
2100	1-1/2″ diameter	Q-1	47	.340		14.70	18		32.70	44
2120	2″ diameter		39	.410		19.70	21.50		41.20	55.50
2140	3″ diameter		34	.471		26	25		51	67.50
2160	4″ diameter		25	.640		39	34		73	95.50
2180	6″ diameter		21	.762		107	40.50		147.50	181
2860	To delete coupling & hanger, subtract									
2870	1/2″ diam. to 1″ diam.					25%	33%			
2880	1-1/2″ diam. to 3″ diam.					22%	21%			
2890	4″ diam. to 6″ diam.					23%	15%			

22 66 53.30 Glass Pipe		Crew	Daily Output	Labor-Hours	Unit	2015 Bare Costs Material	Labor	Equipment	Total	Total Incl O&P
3800	Conical joint, transparent									
3980	6″ diameter	Q-1	21	.762	L.F.	144	40.50		184.50	221
4500	To delete couplings & hangers, subtract									
4530	6″ diam.					22%	26%			
9000	Minimum labor/equipment charge	1 Plum	4	2	Job		117		117	183

22 66 53.40 Pipe Fittings, Glass

		Crew	Daily Output	Labor-Hours	Unit	Material	Labor	Equipment	Total	Total Incl O&P
0010	**PIPE FITTINGS, GLASS**									
0020	Drainage, beaded ends									
0040	Coupling & labor required at joints not incl. in fitting									
0050	price. Add 1 per joint for installed price									
0070	90° Bend or sweep, 1-1/2″				Ea.	31.50			31.50	35
0090	2″					40			40	44
0100	3″					66			66	72.50
0110	4″					105			105	116
0120	6″ (sweep only)					330			330	360
0200	45° Bend or sweep same as 90°									
0350	Tee, single sanitary, 1-1/2″				Ea.	51			51	56.50
0370	2″					51			51	56.50
0380	3″					76.50			76.50	84
0390	4″					137			137	150
0400	6″					365			365	405
0410	Tee, straight, 1-1/2″					63.50			63.50	69.50
0430	2″					63.50			63.50	69.50
0440	3″					91.50			91.50	101
0450	4″					107			107	117
0460	6″					395			395	435
0500	Coupling, stainless steel, TFE seal ring									
0520	1-1/2″	Q-1	32	.500	Ea.	23.50	26.50		50	66.50
0530	2″		30	.533		29.50	28		57.50	76
0540	3″		25	.640		39.50	34		73.50	96
0550	4″		23	.696		67.50	37		104.50	132
0560	6″		20	.800		152	42.50		194.50	233
0600	Coupling, stainless steel, bead to plain end									
0610	1-1/2″	Q-1	36	.444	Ea.	30	23.50		53.50	69.50
0620	2″		34	.471		37	25		62	80
0630	3″		29	.552		63	29		92	115
0640	4″		27	.593		94	31.50		125.50	153
0650	6″		24	.667		320	35		355	405
2350	Coupling, Viton liner, for temperatures to 400°F									
2370	1/2″	Q-1	40	.400	Ea.	46.50	21		67.50	84.50
2380	3/4″		37	.432		53	23		76	94
2390	1″		35	.457		63	24		87	107
2400	1-1/2″		32	.500		23	26.50		49.50	66.50
2410	2″		30	.533		29.50	28		57.50	76.50
2420	3″		25	.640		39.50	34		73.50	96
2430	4″		23	.696		67.50	37		104.50	132
2440	6″		20	.800		152	42.50		194.50	233
2550	For beaded joint armored fittings, add					200%				
2600	Conical ends. Flange set, gasket & labor not incl. in fitting									
2620	price. Add 1 per joint for installed price.									
2650	90° Sweep elbow, 1″				Ea.	105			105	116
2670	1-1/2″					211			211	232

22 66 53 – Laboratory Chemical-Waste and Vent Piping

22 66 53.40 Pipe Fittings, Glass		Crew	Daily Output	Labor-Hours	Unit	Material	2015 Bare Costs Labor	Equipment	Total	Total Incl O&P
2680	2″				Ea.	219			219	241
2690	3″					340			340	375
2700	4″					605			605	665
2710	6″					1,050			1,050	1,175
2750	Cross (straight), add					55%				
2850	Tee, add					20%				
9000	Minimum labor/equipment charge	1 Plum	4	2	Job		117		117	183

22 66 53.60 Corrosion Resistant Pipe		Crew	Daily Output	Labor-Hours	Unit	Material	2015 Bare Costs Labor	Equipment	Total	Total Incl O&P
0010	**CORROSION RESISTANT PIPE**, No couplings or hangers									
0020	Iron alloy, drain, mechanical joint									
1000	1-1/2″ diameter	Q-1	70	.229	L.F.	46	12.10		58.10	69 50
1100	2″ diameter		66	.242		47	12.80		59.80	71.50
1120	3″ diameter		60	.267		60.50	14.10		74.60	88.50
1140	4″ diameter		52	.308		77.50	16.25		93.75	111
1980	Iron alloy, drain, B&S joint									
2000	2″ diameter	Q-1	54	.296	L.F.	55.50	15.65		71.15	86
2100	3″ diameter		52	.308		71	16.25		87.25	104
2120	4″ diameter		48	.333		98	17.60		115.60	136
2140	6″ diameter	Q-2	59	.407		158	22.50		180.50	208
2160	8″ diameter	″	54	.444		297	24.50		321.50	365
2980	Plastic, epoxy, fiberglass filament wound, B&S joint									
3000	2″ diameter	Q-1	62	.258	L.F.	12	13.65		25.65	34.50
3100	3″ diameter		51	.314		14	16.55		30.55	41.50
3120	4″ diameter		45	.356		20	18.80		38.80	51.50
3140	6″ diameter		32	.500		28	26.50		54.50	72
3160	8″ diameter	Q-2	38	.632		44	34.50		78.50	103
3180	10″ diameter		32	.750		60	41		101	130
3200	12″ diameter		28	.857		72	47		119	153
3980	Polyester, fiberglass filament wound, B&S joint									
4000	2″ diameter	Q-1	62	.258	L.F.	13.05	13.65		26.70	36
4100	3″ diameter		51	.314		17	16.55		33.55	44.50
4120	4″ diameter		45	.356		25	18.80		43.80	57
4140	6″ diameter		32	.500		36	26.50		62.50	80.50
4160	8″ diameter	Q-2	38	.632		85.50	34.50		120	148
4180	10″ diameter		32	.750		105	41		146	179
4200	12″ diameter		28	.857		125	47		172	211
4980	Polypropylene, acid resistant, fire retardant, schedule 40									
5000	1-1/2″ diameter	Q-1	68	.235	L.F.	7.90	12.45		20.35	28
5100	2″ diameter		62	.258		12.45	13.65		26.10	35
5120	3″ diameter		51	.314		22	16.55		38.55	50.50
5140	4″ diameter		45	.356		28	18.80		46.80	60.50
5160	6″ diameter		32	.500		56.50	26.50		83	103
5980	Proxylene, fire retardant, Schedule 40									
6000	1-1/2″ diameter	Q-1	68	.235	L.F.	12.40	12.45		24.85	33
6100	2″ diameter		62	.258		17	13.65		30.65	40
6120	3″ diameter		51	.314		30.50	16.55		47.05	60
6140	4″ diameter		45	.356		43.50	18.80		62.30	77
6160	6″ diameter		32	.500		73.50	26.50		100	122
6820	For Schedule 80, add					35%	2%			
9800	Minimum labor/equipment charge	1 Plum	4	2	Job		117		117	183

22 66 53.70 Pipe Fittings, Corrosion Resistant		Crew	Daily Output	Labor-Hours	Unit	Material	Labor	Equipment	Total	Total Incl O&P
						2015 Bare Costs				
0010	**PIPE FITTINGS, CORROSION RESISTANT**									
0030	Iron alloy									
0050	Mechanical joint									
0060	1/4 Bend, 1-1/2"	Q-1	12	1.333	Ea.	77	70.50		147.50	195
0080	2"		10	1.600		126	84.50		210.50	271
0090	3"		9	1.778		151	94		245	310
0100	4"		8	2		174	106		280	355
0110	1/8 Bend, 1-1/2"		12	1.333		49	70.50		119.50	164
0130	2"		10	1.600		84	84.50		168.50	225
0140	3"		9	1.778		112	94		206	269
0150	4"		8	2		150	106		256	330
0160	Tee and Y, sanitary, straight									
0170	1-1/2"	Q-1	8	2	Ea.	84	106		190	258
0180	2"		7	2.286		112	121		233	310
0190	3"		6	2.667		174	141		315	410
0200	4"		5	3.200		320	169		489	615
0360	Coupling, 1-1/2"		14	1.143		47.50	60.50		108	147
0380	2"		12	1.333		54	70.50		124.50	170
0390	3"		11	1.455		56.50	77		133.50	183
0400	4"		10	1.600		64.50	84.50		149	203
0500	Bell & Spigot									
0510	1/4 and 1/16 bend, 2"	Q-1	16	1	Ea.	99.50	53		152.50	192
0520	3"		14	1.143		232	60.50		292.50	350
0530	4"		13	1.231		238	65		303	365
0540	6"	Q-2	17	1.412		470	77.50		547.50	640
0550	8"	"	12	2		1,950	110		2,060	2,300
0620	1/8 bend, 2"	Q-1	16	1		112	53		165	206
0640	3"		14	1.143		207	60.50		267.50	320
0650	4"		13	1.231		238	65		303	365
0660	6"	Q-2	17	1.412		395	77.50		472.50	555
0680	8"	"	12	2		1,550	110		1,660	1,875
0700	Tee, sanitary, 2"	Q-1	10	1.600		213	84.50		297.50	365
0710	3"		9	1.778		690	94		784	905
0720	4"		8	2		575	106		681	795
0730	6"	Q-2	11	2.182		715	120		835	975
0740	8"	"	8	3		1,950	164		2,114	2,400
1800	Y, sanitary, 2"	Q-1	10	1.600		224	84.50		308.50	380
1820	3"		9	1.778		400	94		494	585
1830	4"		8	2		360	106		466	560
1840	6"	Q-2	11	2.182		1,200	120		1,320	1,500
1850	8"	"	8	3		3,300	164		3,464	3,875
3000	Epoxy, filament wound									
3030	Quick-lock joint									
3040	90° Elbow, 2"	Q-1	28	.571	Ea.	97.50	30		127.50	154
3060	3"		16	1		112	53		165	206
3070	4"		13	1.231		153	65		218	269
3080	6"		8	2		223	106		329	410
3090	8"	Q-2	9	2.667		410	146		556	680
3100	10"		7	3.429		520	188		708	865
3110	12"		6	4		740	219		959	1,150
3120	45° Elbow, 2"	Q-1	28	.571		75	30		105	130
3130	3"		16	1		106	53		159	200

22 66 53 – Laboratory Chemical-Waste and Vent Piping

22 66 53.70	Pipe Fittings, Corrosion Resistant	Crew	Daily Output	Labor-Hours	Unit	Material	Labor	Equipment	Total	Total Incl O&P
						2015 Bare Costs				
3140	4"	Q-1	13	1.231	Ea.	109	65		174	221
3150	6"		8	2		223	106		329	410
3160	8"	Q-2	9	2.667		410	146		556	680
3170	10"		7	3.429		515	188		703	865
3180	12"		6	4		740	219		959	1,150
3190	Tee, 2"	Q-1	19	.842		233	44.50		277.50	325
3200	3"		11	1.455		280	77		357	430
3210	4"		9	1.778		335	94		429	515
3220	6"		5	3.200		575	169		744	895
3230	8"	Q-2	6	4		645	219		864	1,050
3240	10"		5	4.800		900	263		1,163	1,400
3250	12"		4	6		1,275	330		1,605	1,925
4000	Polypropylene, acid resistant									
4020	Non-pressure, electrofusion joints									
4050	1/4 bend, 1-1/2"	1 Plum	16	.500	Ea.	13.95	29.50		43.45	61.50
4060	2"	Q-1	28	.571		27	30		57	77
4080	3"		17	.941		29.50	49.50		79	110
4090	4"		14	1.143		47.50	60.50		108	147
4110	6"		8	2		114	106		220	290
4150	1/4 Bend, long sweep									
4170	1-1/2"	1 Plum	16	.500	Ea.	15.10	29.50		44.60	62.50
4180	2"	Q-1	28	.571		27	30		57	77
4200	3"		17	.941		34	49.50		83.50	115
4210	4"		14	1.143		49	60.50		109.50	148
4250	1/8 bend, 1-1/2"	1 Plum	16	.500		15.90	29.50		45.40	63.50
4260	2"	Q-1	28	.571		16.90	30		46.90	65.50
4280	3"		17	.941		31	49.50		80.50	112
4290	4"		14	1.143		34.50	60.50		95	132
4310	6"		8	2		95.50	106		201.50	270
4400	Tee, sanitary									
4420	1-1/2"	1 Plum	10	.800	Ea.	17.60	47		64.60	92.50
4430	2"	Q-1	17	.941		20.50	49.50		70	100
4450	3"		11	1.455		41	77		118	165
4460	4"		9	1.778		61.50	94		155.50	214
4480	6"		5	3.200		415	169		584	720
4490	Tee, sanitary reducing, 2" x 2" x 1-1/2"		17	.941		20.50	49.50		70	100
4492	3" x 3" x 2"		11	1.455		41	77		118	165
4494	4" x 4" x 3"		9	1.778		64	94		158	216
4496	6" x 6" x 4"		5	3.200		202	169		371	485
4650	Wye 45°, 1-1/2"	1 Plum	10	.800		19.10	47		66.10	94
4652	2"	Q-1	17	.941		26.50	49.50		76	107
4653	3"		11	1.455		44	77		121	169
4654	4"		9	1.778		63.50	94		157.50	216
4656	6"		5	3.200		165	169		334	445
4678	Combination Y & 1/8 bend									
4681	1-1/2"	1 Plum	10	.800	Ea.	23	47		70	98.50
4683	2"	Q-1	17	.941		29.50	49.50		79	110
4684	3"		11	1.455		50	77		127	175
4685	4"		9	1.778		68	94		162	221
9000	Minimum labor/equipment charge	1 Plum	4	2	Job		117		117	183

22 66 83 – Chemical-Waste Tanks

22 66 83.13 Chemical-Waste Dilution Tanks		Crew	Daily Output	Labor-Hours	Unit	2015 Bare Costs Material	Labor	Equipment	Total	Total Incl O&P
0010	**CHEMICAL-WASTE DILUTION TANKS**									
7000	Tanks, covers included									
7800	Polypropylene									
7810	Continuous service to 200°F									
7830	2 gallon, 8" x 8" x 8"	Q-1	20	.800	Ea.	124	42.50		166.50	202
7850	7 gallon, 12" x 12" x 12"		20	.800		189	42.50		231.50	274
7870	16 gallon, 18" x 12" x 18"		17	.941		243	49.50		292.50	345
8010	33 gallon, 24" x 18" x 18"		12	1.333		315	70.50		385.50	455
8070	44 gallon, 24" x 18" x 24"		10	1.600		380	84.50		464.50	550
8080	89 gallon, 36" x 24" x 24"		8	2		525	106		631	740
8150	Polyethylene, heavy duty walls									
8160	Continuous service to 180°F									
8180	5 gallon, 12" X 6" X 18"	Q-1	20	.800	Ea.	48.50	42.50		91	120
8210	15 gallon, 14" I.D. x 27" deep		17	.941		73.50	49.50		123	159
8230	55 gallon, 22" I.D. x 36" deep		10	1.600		154	84.50		238.50	300
8250	100 gallon 28" I.D. x 42" deep		8	2		375	106		481	575
8270	200 gallon 36" I.D. x 48" deep		6	2.667		605	141		746	885
8290	360 gallon 48" I.D. x 48" deep		5	3.200		700	169		869	1,025

Division Notes

		CREW	DAILY OUTPUT	LABOR-HOURS	UNIT	BARE COSTS MAT.	BARE COSTS LABOR	BARE COSTS EQUIP.	BARE COSTS TOTAL	TOTAL INCL O&P

Estimating Tips

The labor adjustment factors listed in Subdivision 22 01 02.20 also apply to Division 23.

23 10 00 Facility Fuel Systems

- The prices in this subdivision for above- and below-ground storage tanks do not include foundations or hold-down slabs, unless noted. The estimator should refer to Divisions 3 and 31 for foundation system pricing. In addition to the foundations, required tank accessories, such as tank gauges, leak detection devices, and additional manholes and piping, must be added to the tank prices.

23 50 00 Central Heating Equipment

- When estimating the cost of an HVAC system, check to see who is responsible for providing and installing the temperature control system. It is possible to overlook controls, assuming that they would be included in the electrical estimate.
- When looking up a boiler, be careful on specified capacity. Some manufacturers rate their products on output while others use input.
- Include HVAC insulation for pipe, boiler, and duct (wrap and liner).
- Be careful when looking up mechanical items to get the correct pressure rating and connection type (thread, weld, flange).

23 70 00 Central HVAC Equipment

- Combination heating and cooling units are sized by the air conditioning requirements. (See Reference No. R236000-20 for preliminary sizing guide.)
- A ton of air conditioning is nominally 400 CFM.
- Rectangular duct is taken off by the linear foot for each size, but its cost is usually estimated by the pound. Remember that SMACNA standards now base duct on internal pressure.
- Prefabricated duct is estimated and purchased like pipe: straight sections and fittings.
- Note that cranes or other lifting equipment are not included on any lines in Division 23. For example, if a crane is required to lift a heavy piece of pipe into place high above a gym floor, or to put a rooftop unit on the roof of a four-story building, etc., it must be added. Due to the potential for extreme variation—from nothing additional required to a major crane or helicopter—we feel that including a nominal amount for "lifting contingency" would be useless and detract from the accuracy of the estimate. When using equipment rental cost data from RSMeans, do not forget to include the cost of the operator(s).

Reference Numbers

Reference numbers are shown in shaded boxes at the beginning of some major classifications. These numbers refer to related items in the Reference Section. The reference information may be an estimating procedure, an alternate pricing method, or technical information.

Note: Not all subdivisions listed here necessarily appear in this publication. ■

*Note: **Trade Service**, in part, has been used as a reference source for some of the material prices used in Division 23.*

23 05 02 – HVAC General

23 05 02.10 Air Conditioning, General		Crew	Daily Output	Labor-Hours	Unit	Material	2015 Bare Costs Labor	Equipment	Total	Total Incl O&P
0010	**AIR CONDITIONING, GENERAL** Prices are for standard efficiencies (SEER 13)									
0020	for upgrade to SEER 14 add					10%				

23 05 05 – Selective Demolition for HVAC

23 05 05.10 HVAC Demolition		Crew	Daily Output	Labor-Hours	Unit	Material	2015 Bare Costs Labor	Equipment	Total	Total Incl O&P
0010	**HVAC DEMOLITION** R220105-10									
0100	Air conditioner, split unit, 3 ton	Q-5	2	8	Ea.		430		430	670
0150	Package unit, 3 ton R024119-10	Q-6	3	8			445		445	695
0190	Rooftop, self contained, up to 5 ton	1 Plum	1.20	6.667			390		390	610
0250	Air curtain	Q-9	20	.800	L.F.		40.50		40.50	63.50
0254	Air filters, up thru 16,000 CFM		20	.800	Ea.		40.50		40.50	63.50
0256	20,000 thru 60,000 CFM		16	1			50.50		50.50	79.50
0297	Boiler blowdown	Q-5	8	2			108		108	168
0298	Boilers									
0300	Electric, up thru 148 kW	Q-19	2	12	Ea.		650		650	1,000
0310	150 thru 518 kW	"	1	24			1,300		1,300	2,025
0320	550 thru 2000 kW	Q-21	.40	80			4,450		4,450	6,925
0330	2070 kW and up	"	.30	106			5,925		5,925	9,225
0340	Gas and/or oil, up thru 150 MBH	Q-7	2.20	14.545			825		825	1,300
0350	160 thru 2000 MBH		.80	40			2,275		2,275	3,550
0360	2100 thru 4500 MBH		.50	64			3,650		3,650	5,675
0370	4600 thru 7000 MBH		.30	106			6,075		6,075	9,475
0380	7100 thru 12,000 MBH		.16	200			11,400		11,400	17,700
0390	12,200 thru 25,000 MBH		.12	266			15,200		15,200	23,700
0400	Central station air handler unit, up thru 15 ton	Q-5	1.60	10			540		540	840
0410	17.5 thru 30 ton	"	.80	20			1,075		1,075	1,675
0430	Computer room unit									
0434	Air cooled split, up thru 10 ton	Q-5	.67	23.881	Ea.		1,275		1,275	2,000
0436	12 thru 23 ton		.53	30.189			1,625		1,625	2,525
0440	Chilled water, up thru 10 ton		1.30	12.308			660		660	1,025
0444	12 thru 23 ton		1	16			860		860	1,350
0450	Glycol system, up thru 10 ton		.53	30.189			1,625		1,625	2,525
0454	12 thru 23 ton		.40	40			2,150		2,150	3,350
0460	Water cooled, not including condenser, up thru 10 ton		.80	20			1,075		1,075	1,675
0464	12 thru 23 ton		.60	26.667			1,425		1,425	2,225
0600	Condenser, up thru 50 ton		1	16			860		860	1,350
0610	51 thru 100 ton	Q-6	.90	26.667			1,475		1,475	2,325
0620	101 thru 1000 ton	"	.70	34.286			1,900		1,900	2,975
0660	Condensing unit, up thru 10 ton	Q-5	1.25	12.800			690		690	1,075
0670	11 thru 50 ton	"	.40	40			2,150		2,150	3,350
0680	60 thru 100 ton	Q-6	.30	80			4,450		4,450	6,950
0700	Cooling tower, up thru 400 ton		.80	30			1,675		1,675	2,600
0710	450 thru 600 ton		.53	45.283			2,525		2,525	3,950
0720	700 thru 1300 ton		.40	60			3,350		3,350	5,225
0780	Dehumidifier, up thru 155 lb./hr.	Q-1	8	2			106		106	165
0790	240 lb./hr. and up	"	2	8			425		425	660
1560	Ductwork									
1570	Metal, steel, sst, fabricated	Q-9	1000	.016	Lb.		.81		.81	1.27
1580	Aluminum, fabricated		485	.033	"		1.66		1.66	2.62
1590	Spiral, prefabricated		400	.040	L.F.		2.01		2.01	3.18
1600	Fiberglass, prefabricated		400	.040			2.01		2.01	3.18
1610	Flex, prefabricated		500	.032			1.61		1.61	2.54
1620	Glass fiber reinforced plastic, prefabricated		280	.057			2.88		2.88	4.54

23 05 05.10 HVAC Demolition		Crew	Daily Output	Labor-Hours	Unit	2015 Bare Costs Material	Labor	Equipment	Total	Total Incl O&P
1630	Diffusers, registers or grills, up thru 20" max dimension	1 Shee	50	.160	Ea.		8.95		8.95	14.15
1640	21 thru 36" max dimension		36	.222			12.45		12.45	19.60
1650	Above 36" max dimension		30	.267			14.90		14.90	23.50
1700	Evaporator, up thru 12,000 BTUH	Q-5	5.30	3.019			162		162	253
1710	12,500 thru 30,000 BTUH	"	2.70	5.926			320		320	495
1720	31,000 BTUH and up	Q-6	1.50	16			890		890	1,400
1730	Evaporative cooler, up thru 5 H.P.	Q-9	2.70	5.926			298		298	470
1740	10 thru 30 H.P.	"	.67	23.881			1,200		1,200	1,900
1750	Exhaust systems									
1760	Exhaust components	1 Shee	8	1	System		56		56	88.50
1770	Weld fume hoods	"	20	.400	Ea.		22.50		22.50	35.50
1850	Minimum labor/equipment charge	1 Clab	3	2.667	Job		100		100	164
2120	Fans, up thru 1 H.P. or 2000 CFM	Q-9	8	2	Ea.		101		101	159
2124	1-1/2 thru 10 H.P. or 20,000 CFM		5.30	3.019			152		152	240
2128	15 thru 30 H.P. or above 20,000 CFM		4	4			201		201	320
2150	Fan coil air conditioner, chilled water, up thru 7.5 ton	Q-5	14	1.143			61.50		61.50	96
2154	Direct expansion, up thru 10 ton		8	2			108		108	168
2158	11 thru 30 ton		2	8			430		430	670
2170	Flue shutter damper	Q-9	8	2			101		101	159
2200	Furnace, electric	Q-20	2	10			510		510	805
2300	Gas or oil, under 120 MBH	Q-9	4	4			201		201	320
2340	Over 120 MBH	"	3	5.333			269		269	425
2730	Heating and ventilating unit	Q-5	2.70	5.926			320		320	495
2740	Heater, electric, wall, baseboard and quartz	1 Elec	10	.800			44		44	68
2750	Heater, electric, unit, cabinet, fan and convector	"	8	1			54.50		54.50	84.50
2760	Heat exchanger, shell and tube type	Q-5	1.60	10			540		540	840
2770	Plate type	Q-6	.60	40			2,225		2,225	3,475
2810	Heat pump									
2820	Air source, split, up thru 10 ton	Q-5	.90	17.778	Ea.		955		955	1,500
2830	15 thru 25 ton	Q-6	.80	30			1,675		1,675	2,600
2850	Single package, up thru 12 ton	Q-5	1	16			860		860	1,350
2860	Water source, up thru 15 ton	"	.90	17.778			955		955	1,500
2870	20 thru 50 ton	Q-6	.80	30			1,675		1,675	2,600
2910	Heat recovery package, up thru 20,000 CFM	Q-5	2	8			430		430	670
2920	25,000 CFM and up		1.20	13.333			715		715	1,125
2930	Heat transfer package, up thru 130 GPM		.80	20			1,075		1,075	1,675
2934	255 thru 800 GPM		.42	38.095			2,050		2,050	3,200
2940	Humidifier		10.60	1.509			81		81	127
2961	Hydronic unit heaters, up thru 200 MBH		14	1.143			61.50		61.50	96
2962	Above 200 MBH		8	2			108		108	168
2964	Valance units		32	.500			27		27	42
2966	Radiant floor heating									
2967	System valves, controls, manifolds	Q-5	16	1	Ea.		54		54	84
2968	Per room distribution		8	2			108		108	168
2970	Hydronic heating, baseboard radiation		16	1			54		54	84
2976	Convectors and free standing radiators		18	.889			48		48	74.50
2980	Induced draft fan, up thru 1 H.P.	Q-9	4.60	3.478			175		175	276
2984	1-1/2 H.P. thru 7-1/2 H.P.	"	2.20	7.273			365		365	580
2988	Infrared unit	Q-5	16	1			54		54	84
2992	Louvers	1 Shee	46	.174	S.F.		9.75		9.75	15.35
3000	Mechanical equipment, light items. Unit is weight, not cooling.	Q-5	.90	17.778	Ton		955		955	1,500
3600	Heavy items		1.10	14.545	"		780		780	1,225
3720	Make-up air unit, up thru 6000 CFM		3	5.333	Ea.		287		287	445

23 05 Common Work Results for HVAC

23 05 05 – Selective Demolition for HVAC

23 05 05.10 HVAC Demolition		Crew	Daily Output	Labor-Hours	Unit	Material	2015 Bare Costs Labor	Equipment	Total	Total Incl O&P
3730	6500 thru 30,000 CFM	Q-5	1.60	10	Ea.		540		540	840
3740	35,000 thru 75,000 CFM	Q-6	1	24			1,350		1,350	2,100
3800	Mixing boxes, constant and VAV	Q-9	18	.889			45		45	70.50
4000	Packaged terminal air conditioner, up thru 18,000 BTUH	Q-5	8	2			108		108	168
4010	24,000 thru 48,000 BTUH	"	2.80	5.714			305		305	480
5000	Refrigerant compressor, reciprocating or scroll									
5010	Up thru 5 ton	1 Stpi	6	1.333	Ea.		79.50		79.50	124
5020	5.08 thru 10 ton	Q-5	6	2.667			143		143	224
5030	15 thru 50 ton	"	3	5.333			287		287	445
5040	60 thru 130 ton	Q-6	2.80	8.571			480		480	745
5090	Remove refrigerant from system	1 Stpi	40	.200	Lb.		11.95		11.95	18.65
5100	Rooftop air conditioner, up thru 10 ton	Q-5	1.40	11.429	Ea.		615		615	960
5110	12 thru 40 ton	Q-6	1	24			1,350		1,350	2,100
5120	50 thru 140 ton		.50	48			2,675		2,675	4,175
5130	150 thru 300 ton		.30	80			4,450		4,450	6,950
6000	Self contained single package air conditioner, up thru 10 ton	Q-5	1.60	10			540		540	840
6010	15 thru 60 ton	Q-6	1.20	20			1,125		1,125	1,750
6100	Space heaters, up thru 200 MBH	Q-5	10	1.600			86		86	134
6110	Over 200 MBH		5	3.200			172		172	268
6200	Split ductless, both sections		8	2			108		108	168
6300	Steam condensate meter	1 Stpi	11	.727			43.50		43.50	68
6600	Thru-the-wall air conditioner	L-2	8	2			82.50		82.50	135
7000	Vent chimney, prefabricated, up thru 12" diameter	Q-9	94	.170	V.L.F.		8.55		8.55	13.50
7010	14" thru 36" diameter		40	.400			20		20	32
7020	38" thru 48" diameter		32	.500			25		25	39.50
7030	54" thru 60" diameter	Q-10	14	1.714			89.50		89.50	141
7400	Ventilators, up thru 14" neck diameter	Q-9	58	.276	Ea.		13.90		13.90	22
7410	16" thru 50" neck diameter		40	.400			20		20	32
7450	Relief vent, up thru 24" x 96"		22	.727			36.50		36.50	58
7460	48" x 60" thru 96" x 144"		10	1.600			80.50		80.50	127
8000	Water chiller up thru 10 ton	Q-5	2.50	6.400			345		345	535
8010	15 thru 100 ton	Q-6	.48	50			2,800		2,800	4,350
8020	110 thru 500 ton	Q-7	.29	110			6,275		6,275	9,800
8030	600 thru 1000 ton		.23	139			7,925		7,925	12,300
8040	1100 ton and up		.20	160			9,100		9,100	14,200
8400	Window air conditioner	1 Carp	16	.500			23.50		23.50	38.50
9000	Minimum labor/equipment charge	Q-6	3	8	Job		445		445	695

23 05 23 – General-Duty Valves for HVAC Piping

23 05 23.20 Valves, Bronze/Brass		Crew	Daily Output	Labor-Hours	Unit	Material	Labor	Equipment	Total	Total Incl O&P
0010	**VALVES, BRONZE/BRASS**									
0020	Brass									
1300	Ball combination valves, shut-off and union									
1310	Solder, with strainer, drain and PT ports									
1320	1/2"	1 Stpi	17	.471	Ea.	75	28		103	127
1330	3/4"		16	.500		78	30		108	133
1340	1"		14	.571		123	34		157	189
1350	1-1/4"		12	.667		142	40		182	218
1360	1-1/2"		10	.800		221	48		269	320
1370	2"		8	1		260	60		320	380
1410	Threaded, with strainer, drain and PT ports									
1420	1/2"	1 Stpi	20	.400	Ea.	75	24		99	120
1430	3/4"		18	.444		78	26.50		104.50	128

23 05 23 – General-Duty Valves for HVAC Piping

23 05 23.20 Valves, Bronze/Brass		Crew	Daily Output	Labor-Hours	Unit	Material	2015 Bare Costs Labor	Equipment	Total	Total Incl O&P
1440	1"	1 Stpi	17	.471	Ea.	123	28		151	179
1450	1-1/4"		12	.667		142	40		182	218
1460	1-1/2"		11	.727		221	43.50		264.50	310
1470	2"		9	.889		260	53		313	370

23 05 23.30 Valves, Iron Body		Crew	Daily Output	Labor-Hours	Unit	Material	Labor	Equipment	Total	Total Incl O&P
0010	**VALVES, IRON BODY** R220523-90									
0022	For grooved joint, see Section 22 11 13.48									
0560	Butterfly, lug type, pneumatic operator, 2" size	1 Stpi	14	.571	Ea.	475	34		509	580
0570	3"	Q-1	8	2		510	106		616	730
0580	4"	"	5	3.200		610	169		779	935
0590	6"	Q-2	5	4.800		790	263		1,053	1,275
0600	8"		4.50	5.333		1,050	292		1,342	1,600
0610	10"		4	6		1,300	330		1,630	1,950
0620	12"		3	8		1,925	440		2,365	2,775
0630	14"		2.30	10.435		2,175	570		2,745	3,300
0640	18"		1.50	16		3,700	875		4,575	5,450
0650	20"		1	24		4,725	1,325		6,050	7,250
0790	Butterfly, lug type, gear operated, 2" size, 200 lb. except noted	1 Plum	14	.571		390	33.50		423.50	485
0800	2-1/2"	Q-1	9	1.778		395	94		489	580
0810	3"		8	2		415	106		521	620
0820	4"		5	3.200		470	169		639	780
0830	5"	Q-2	5	4.800		580	263		843	1,050
0840	6"		5	4.800		645	263		908	1,125
0850	8"		4.50	5.333		835	292		1,127	1,375
0860	10"		4	6		1,175	330		1,505	1,800
0870	12"		3	8		1,600	440		2,040	2,450
0880	14", 150 lb.		2.30	10.435		3,075	570		3,645	4,275
0890	16", 150 lb.		1.75	13.714		4,700	750		5,450	6,325
0900	18", 150 lb.		1.50	16		5,900	875		6,775	7,850
0910	20", 150 lb.		1	24		7,875	1,325		9,200	10,700
0930	24", 150 lb.		.75	32		13,400	1,750		15,150	17,400
1020	Butterfly, wafer type, gear actuator, 200 lb.									
1030	2"	1 Plum	14	.571	Ea.	96	33.50		129.50	158
1040	2-1/2"	Q-1	9	1.778		97.50	94		191.50	253
1050	3"		8	2		101	106		207	276
1060	4"		5	3.200		113	169		282	390
1070	5"	Q-2	5	4.800		126	263		389	550
1080	6"		5	4.800		143	263		406	570
1090	8"		4.50	5.333		178	292		470	650
1100	10"		4	6		239	330		569	780
1110	12"		3	8		360	440		800	1,075
1200	Wafer type, lever actuator, 200 lb.									
1220	2"	1 Plum	14	.571	Ea.	156	33.50		189.50	224
1230	2-1/2"	Q-1	9	1.778		161	94		255	325
1240	3"		8	2		190	106		296	375
1250	4"		5	3.200		207	169		376	490
1260	5"	Q-2	5	4.800		320	263		583	765
1270	6"		5	4.800		350	263		613	795
1280	8"		4.50	5.333		540	292		832	1,050
1290	10"		4	6		755	330		1,085	1,350
1300	12"		3	8		1,250	440		1,690	2,050
1650	Gate, 125 lb., N.R.S.									

23 05 23.30 Valves, Iron Body		Crew	Daily Output	Labor-Hours	Unit	Material	2015 Bare Costs Labor	Equipment	Total	Total Incl O&P
2150	Flanged									
2200	2"	1 Plum	5	1.600	Ea.	715	94		809	935
2240	2-1/2"	Q-1	5	3.200		735	169		904	1,075
2260	3"		4.50	3.556		825	188		1,013	1,200
2280	4"		3	5.333		1,175	282		1,457	1,750
2290	5"	Q-2	3.40	7.059		2,025	385		2,410	2,825
2300	6"		3	8		2,025	440		2,465	2,900
2320	8"		2.50	9.600		3,450	525		3,975	4,625
2340	10"		2.20	10.909		6,075	600		6,675	7,600
2360	12"		1.70	14.118		8,325	775		9,100	10,400
2370	14"		1.30	18.462		12,000	1,000		13,000	14,800
2380	16"		1	24		17,100	1,325		18,425	20,900
2420	For 250 lb. flanged, add					200%	10%			
3550	OS&Y, 125 lb., flanged									
3600	2"	1 Plum	5	1.600	Ea.	475	94		569	670
3640	2-1/2"	Q-1	5	3.200		480	169		649	795
3660	3"		4.50	3.556		530	188		718	875
3670	3-1/2"		3	5.333		680	282		962	1,200
3680	4"		3	5.333		770	282		1,052	1,300
3690	5"	Q-2	3.40	7.059		1,250	385		1,635	1,975
3700	6"		3	8		1,250	440		1,690	2,050
3720	8"		2.50	9.600		2,175	525		2,700	3,225
3740	10"		2.20	10.909		4,050	600		4,650	5,375
3760	12"		1.70	14.118		5,500	775		6,275	7,275
3770	14"		1.30	18.462		12,000	1,000		13,000	14,800
3780	16"		1	24		16,900	1,325		18,225	20,700
3790	18"		.80	30		20,700	1,650		22,350	25,400
3800	20"		.60	40		22,900	2,200		25,100	28,600
3830	24"		.50	48		24,900	2,625		27,525	31,500
3900	For 175 lb., flanged, add					200%	10%			
4350	Globe, OS&Y									
4540	Class 125, flanged									
4550	2"	1 Plum	5	1.600	Ea.	935	94		1,029	1,175
4560	2-1/2"	Q-1	5	3.200		940	169		1,109	1,300
4570	3"		4.50	3.556		1,150	188		1,338	1,550
4580	4"		3	5.333		1,650	282		1,932	2,250
4590	5"	Q-2	3.40	7.059		2,975	385		3,360	3,875
4600	6"		3	8		2,975	440		3,415	3,950
4610	8"		2.50	9.600		5,825	525		6,350	7,250
5040	Class 250, flanged									
5050	2"	1 Plum	4.50	1.778	Ea.	1,500	104		1,604	1,825
5060	2-1/2"	Q-1	4.50	3.556		1,950	188		2,138	2,450
5070	3"		4	4		2,000	211		2,211	2,550
5080	4"		2.70	5.926		2,950	315		3,265	3,750
5090	5"	Q-2	3	8		5,475	440		5,915	6,700
5100	6"		2.70	8.889		5,275	485		5,760	6,575
5110	8"		2.20	10.909		8,975	600		9,575	10,800
5120	10"		2	12		13,400	655		14,055	15,800
5130	12"		1.60	15		19,600	820		20,420	22,800
5450	Swing check, 125 lb., threaded									
5500	2"	1 Plum	11	.727	Ea.	430	42.50		472.50	540
5540	2-1/2"	Q-1	15	1.067		555	56.50		611.50	700
5550	3"		13	1.231		590	65		655	750

23 05 23 – General-Duty Valves for HVAC Piping

23 05 23.30 Valves, Iron Body		Crew	Daily Output	Labor-Hours	Unit	Material	2015 Bare Costs Labor	Equipment	Total	Total Incl O&P
5560	4″	Q-1	10	1.600	Ea.	955	84.50		1,039.50	1,175
5950	Flanged									
6000	2″	1 Plum	5	1.600	Ea.	395	94		489	580
6040	2-1/2″	Q-1	5	3.200		380	169		549	685
6050	3″		4.50	3.556		410	188		598	745
6060	4″		3	5.333		605	282		887	1,100
6070	6″	Q-2	3	8		1,025	440		1,465	1,825
6080	8″		2.50	9.600		1,950	525		2,475	2,975
6090	10″		2.20	10.909		3,325	600		3,925	4,575
6100	12″		1.70	14.118		5,175	775		5,950	6,900
6160	For 250 lb. flanged, add					200%	20%			
6600	Silent check, bronze trim									
6610	Compact wafer type, for 125 or 150 lb. flanges									
6630	1-1/2″	1 Plum	11	.727	Ea.	136	42.50		178.50	216
6640	2″	″	9	.889		165	52		217	264
6650	2-1/2″	Q-1	9	1.778		180	94		274	345
6660	3″		8	2		192	106		298	375
6670	4″		5	3.200		245	169		414	535
6680	5″	Q-2	6	4		325	219		544	695
6690	6″		6	4		590	219		809	985
6700	8″		4.50	5.333		730	292		1,022	1,250
6710	10″		4	6		1,300	330		1,630	1,950
6720	12″		3	8		2,425	440		2,865	3,325
9000	Minimum labor/equipment charge	1 Plum	3	2.667	Job		157		157	244

23 05 23.70 Valves, Semi-Steel		Crew	Daily Output	Labor-Hours	Unit	Material	Labor	Equipment	Total	Total Incl O&P
0010	**VALVES, SEMI-STEEL**									
1020	Lubricated plug valve, threaded, 200 psi									
1030	1/2″	1 Plum	18	.444	Ea.	560	26		586	655
1040	3/4″		16	.500		560	29.50		589.50	660
1050	1″		14	.571		375	33.50		408.50	465
1060	1-1/4″		12	.667		415	39		454	515
1070	1-1/2″		11	.727		440	42.50		482.50	550
1080	2″		8	1		535	58.50		593.50	680
1090	2-1/2″	Q-1	5	3.200		590	169		759	910
1100	3″	″	4.50	3.556		845	188		1,033	1,225
6990	Flanged, 200 psi									
7000	2″	1 Plum	8	1	Ea.	217	58.50		275.50	330
7010	2-1/2″	Q-1	5	3.200		310	169		479	605
7020	3″		4.50	3.556		375	188		563	710
7030	4″		3	5.333		470	282		752	960
7036	5″		2.50	6.400		840	340		1,180	1,450
7040	6″	Q-2	3	8		1,100	440		1,540	1,875
7050	8″		2.50	9.600		1,575	525		2,100	2,550
7060	10″		2.20	10.909		2,450	600		3,050	3,600
7070	12″		1.70	14.118		4,700	775		5,475	6,375

23 05 23.80 Valves, Steel		Crew	Daily Output	Labor-Hours	Unit	Material	Labor	Equipment	Total	Total Incl O&P
0010	**VALVES, STEEL** R220523-90									
0800	Cast									
1350	Check valve, swing type, 150 lb., flanged									
1370	1″	1 Plum	10	.800	Ea.	375	47		422	490
1400	2″	″	8	1		685	58.50		743.50	840
1440	2-1/2″	Q-1	5	3.200		890	169		1,059	1,250

23 05 23 – General-Duty Valves for HVAC Piping

23 05 23.80 Valves, Steel		Crew	Daily Output	Labor-Hours	Unit	Material	2015 Bare Costs Labor	Equipment	Total	Total Incl O&P
1450	3"	Q-1	4.50	3.556	Ea.	805	188		993	1,175
1460	4"	↓	3	5.333		1,225	282		1,507	1,775
1470	6"	Q-2	3	8		1,900	440		2,340	2,750
1480	8"		2.50	9.600		3,300	525		3,825	4,450
1490	10"		2.20	10.909		4,875	600		5,475	6,300
1500	12"		1.70	14.118		7,100	775		7,875	9,025
1510	14"		1.30	18.462		10,600	1,000		11,600	13,300
1520	16"	↓	1	24		11,900	1,325		13,225	15,200
1540	For 300 lb., flanged, add					50%	15%			
1548	For 600 lb., flanged, add					110%	20%			
1571	300 lb., 2"	1 Plum	7.40	1.081		775	63.50		838.50	950
1572	2-1/2"	Q-1	4.20	3.810		1,150	201		1,351	1,600
1573	3"		4	4		1,150	211		1,361	1,600
1574	4"	↓	2.80	5.714		1,575	300		1,875	2,225
1575	6"	Q-2	2.90	8.276		3,075	455		3,530	4,075
1576	8"		2.40	10		4,525	550		5,075	5,825
1577	10"		2.10	11.429		6,800	625		7,425	8,475
1578	12"		1.60	15		9,850	820		10,670	12,100
1579	14"		1.20	20		14,700	1,100		15,800	17,800
1581	16"	↓	.90	26.667	↓	18,700	1,450		20,150	22,900
1950	Gate valve, 150 lb., flanged									
2000	2"	1 Plum	8	1	Ea.	780	58.50		838.50	950
2040	2-1/2"	Q-1	5	3.200		1,100	169		1,269	1,500
2050	3"		4.50	3.556		1,100	188		1,288	1,525
2060	4"	↓	3	5.333		1,350	282		1,632	1,950
2070	6"	Q-2	3	8		2,250	440		2,690	3,150
2080	8"		2.50	9.600		3,425	525		3,950	4,575
2090	10"		2.20	10.909		5,050	600		5,650	6,475
2100	12"		1.70	14.118		6,900	775		7,675	8,775
2110	14"		1.30	18.462		11,300	1,000		12,300	14,000
2120	16"		1	24		14,600	1,325		15,925	18,100
2130	18"		.80	30		19,100	1,650		20,750	23,600
2140	20"	↓	.60	40	↓	22,000	2,200		24,200	27,600
2650	300 lb., flanged									
2700	2"	1 Plum	7.40	1.081	Ea.	1,075	63.50		1,138.50	1,275
2740	2-1/2"	Q-1	4.20	3.810		1,475	201		1,676	1,925
2750	3"		4	4		1,475	211		1,686	1,925
2760	4"	↓	2.80	5.714		2,050	300		2,350	2,725
2770	6"	Q-2	2.90	8.276		3,450	455		3,905	4,500
2780	8"		2.40	10		5,400	550		5,950	6,800
2790	10"		2.10	11.429		7,250	625		7,875	8,950
2800	12"		1.60	15		10,500	820		11,320	12,800
2810	14"		1.20	20		20,900	1,100		22,000	24,700
2820	16"		.90	26.667		26,200	1,450		27,650	31,100
2830	18"		.70	34.286		37,400	1,875		39,275	44,000
2840	20"	↓	.50	48	↓	41,600	2,625		44,225	49,900
3650	Globe valve, 150 lb., flanged									
3700	2"	1 Plum	8	1	Ea.	980	58.50		1,038.50	1,175
3740	2-1/2"	Q-1	5	3.200		1,250	169		1,419	1,650
3750	3"		4.50	3.556		1,250	188		1,438	1,675
3760	4"	↓	3	5.333		1,825	282		2,107	2,450
3770	6"	Q-2	3	8		2,875	440		3,315	3,825
3780	8"	↓	2.50	9.600	↓	5,350	525		5,875	6,700

23 05 23.80 Valves, Steel		Crew	Daily Output	Labor-Hours	Unit	2015 Bare Costs Material	Labor	Equipment	Total	Total Incl O&P
3790	10″	Q-2	2.20	10.909	Ea.	9,825	600		10,425	11,700
3800	12″		1.70	14.118		13,500	775		14,275	16,000
4080	300 lb., flanged									
4100	2″	1 Plum	7.40	1.081	Ea.	1,325	63.50		1,388.50	1,550
4140	2-1/2″	Q-1	4.20	3.810		1,775	201		1,976	2,300
4150	3″		4	4		1,775	211		1,986	2,300
4160	4″		2.80	5.714		2,475	300		2,775	3,200
4170	6″	Q-2	2.90	8.276		4,450	455		4,905	5,600
4180	8″		2.40	10		7,575	550		8,125	9,175
4190	10″		2.10	11.429		14,500	625		15,125	17,000
4200	12″		1.60	15		17,100	820		17,920	20,100
4680	600 lb., flanged									
4700	2″	1 Plum	7	1.143	Ea.	1,825	67		1,892	2,100
4740	2-1/2″	Q-1	4	4		2,925	211		3,136	3,550
4750	3″		3.60	4.444		2,925	235		3,160	3,600
4760	4″		2.50	6.400		4,500	340		4,840	5,475
4770	6″	Q-2	2.60	9.231		9,500	505		10,005	11,300
4780	8″	″	2.10	11.429		15,900	625		16,525	18,500
5150	Forged									
5340	Ball valve, 1500 psi, threaded, 1/4″ size	1 Plum	24	.333	Ea.	62	19.55		81.55	98.50
5350	3/8″		24	.333		76	19.55		95.55	114
5360	1/2″		24	.333		76.50	19.55		96.05	115
5370	3/4″		20	.400		102	23.50		125.50	149
5380	1″		19	.421		132	24.50		156.50	184
5390	1-1/4″		15	.533		208	31.50		239.50	277
5400	1-1/2″		13	.615		221	36		257	300
5410	2″		11	.727		305	42.50		347.50	400
5460	Ball valve, 800 lb., socket weld, 1/4″ size	Q-15	19	.842		76	44.50	3.04	123.54	156
5470	3/8″		19	.842		76	44.50	3.04	123.54	156
5480	1/2″		19	.842		94	44.50	3.04	141.54	176
5490	3/4″		19	.842		125	44.50	3.04	172.54	210
5500	1″		15	1.067		156	56.50	3.85	216.35	263
5510	1-1/4″		13	1.231		208	65	4.44	277.44	335
5520	1-1/2″		11	1.455		271	77	5.25	353.25	425
5530	2″		8.50	1.882		360	99.50	6.80	466.30	555
5650	Check valve, class 800, horizontal, socket									
5651	,Socket									
5698	Threaded									
5700	1/4″	1 Plum	24	.333	Ea.	94.50	19.55		114.05	135
5720	3/8″		24	.333		94.50	19.55		114.05	135
5730	1/2″		24	.333		94.50	19.55		114.05	135
5740	3/4″		20	.400		101	23.50		124.50	148
5750	1″		19	.421		119	24.50		143.50	170
5760	1-1/4″		15	.533		233	31.50		264.50	305
5770	1-1/2″		13	.615		233	36		269	315
5780	2″		11	.727		325	42.50		367.50	425
5840	For class 150, flanged, add					100%	15%			
5860	For class 300, flanged, add					120%	20%			
6100	Gate, class 800, OS&Y, socket									
6102	3/8″	Q-15	19	.842	Ea.	65.50	44.50	3.04	113.04	145
6103	1/2″		19	.842		65.50	44.50	3.04	113.04	145
6104	3/4″		19	.842		72	44.50	3.04	119.54	152
6105	1″		15	1.067		87	56.50	3.85	147.35	188

23 05 23.80 Valves, Steel		Crew	Daily Output	Labor-Hours	Unit	Material	Labor	Equipment	Total	Total Incl O&P
						2015 Bare Costs				
6106	1-1/4″	Q-15	13	1.231	Ea.	165	65	4.44	234.44	287
6107	1-1/2″		11	1.455		165	77	5.25	247.25	305
6108	2″		8.50	1.882		185	99.50	6.80	291.30	365
6118	Threaded									
6120	3/8″	1 Plum	24	.333	Ea.	65.50	19.55		85.05	103
6130	1/2″		24	.333		65.50	19.55		85.05	103
6140	3/4″		20	.400		72	23.50		95.50	116
6150	1″		19	.421		87	24.50		111.50	135
6160	1-1/4″		15	.533		165	31.50		196.50	230
6170	1-1/2″		13	.615		165	36		201	238
6180	2″		11	.727		185	42.50		227.50	271
6260	For OS&Y, flanged, add					100%	20%			
6700	Globe, OS&Y, class 800, socket									
6710	1/4″	Q-15	19	.842	Ea.	101	44.50	3.04	148.54	184
6720	3/8″		19	.842		101	44.50	3.04	148.54	184
6730	1/2″		19	.842		101	44.50	3.04	148.54	184
6740	3/4″		19	.842		116	44.50	3.04	163.54	201
6750	1″		15	1.067		151	56.50	3.85	211.35	258
6760	1-1/4″		13	1.231		295	65	4.44	364.44	430
6770	1-1/2″		11	1.455		295	77	5.25	377.25	450
6780	2″		8.50	1.882		380	99.50	6.80	486.30	575

23 05 23.90 Valves, Stainless Steel		Crew	Daily Output	Labor-Hours	Unit	Material	Labor	Equipment	Total	Total Incl O&P
0010	**VALVES, STAINLESS STEEL** R220523-90									
1610	Ball, threaded 1/4″	1 Stpi	24	.333	Ea.	50	19.90		69.90	86
1620	3/8″		24	.333		50	19.90		69.90	86
1630	1/2″		22	.364		50	21.50		71.50	89
1640	3/4″		20	.400		82.50	24		106.50	129
1650	1″		19	.421		101	25		126	150
1660	1-1/4		15	.533		192	32		224	261
1670	1-1/2″		13	.615		200	37		237	278
1680	2″		11	.727		260	43.50		303.50	355
1700	Check, 200 lb., threaded									
1710	1/4″	1 Plum	24	.333	Ea.	143	19.55		162.55	189
1720	1/2″		22	.364		143	21.50		164.50	192
1730	3/4″		20	.400		148	23.50		171.50	199
1750	1″		19	.421		192	24.50		216.50	250
1760	1-1/2″		13	.615		360	36		396	450
1770	2″		11	.727		610	42.50		652.50	735
1800	150 lb., flanged									
1810	2-1/2″	Q-1	5	3.200	Ea.	1,875	169		2,044	2,325
1820	3″		4.50	3.556		1,875	188		2,063	2,350
1830	4″		3	5.333		2,775	282		3,057	3,525
1840	6″	Q-2	3	8		4,925	440		5,365	6,100
1850	8″	″	2.50	9.600		10,100	525		10,625	11,900
2100	Gate, OS&Y, 150 lb., flanged									
2120	1/2″	1 Plum	18	.444	Ea.	595	26		621	695
2140	3/4″		16	.500		575	29.50		604.50	680
2150	1″		14	.571		725	33.50		758.50	855
2160	1-1/2″		11	.727		1,400	42.50		1,442.50	1,625
2170	2″		8	1		1,650	58.50		1,708.50	1,925
2180	2-1/2″	Q-1	5	3.200		2,025	169		2,194	2,500
2190	3″		4.50	3.556		2,025	188		2,213	2,525

23 05 23 – General-Duty Valves for HVAC Piping

23 05 23.90	Valves, Stainless Steel	Crew	Daily Output	Labor-Hours	Unit	2015 Bare Costs Material	Labor	Equipment	Total	Total Incl O&P
2200	4"	Q-1	3	5.333	Ea.	2,975	282		3,257	3,725
2205	5"		2.80	5.714		4,800	300		5,100	5,750
2210	6"	Q-2	3	8		5,575	440		6,015	6,800
2220	8"		2.50	9.600		9,675	525		10,200	11,400
2230	10"		2.30	10.435		16,800	570		17,370	19,400
2240	12"		1.90	12.632		22,500	690		23,190	25,900
2260	For 300 lb., flanged, add					120%	15%			
2600	600 lb., flanged									
2620	1/2"	1 Plum	16	.500	Ea.	213	29.50		242.50	280
2640	3/4"		14	.571		230	33.50		263.50	305
2650	1"		12	.667		277	39		316	365
2660	1-1/2"		10	.800		445	47		492	560
2670	2"		7	1.143		610	67		677	775
2680	2-1/2"	Q-1	4	4		7,375	211		7,586	8,450
2690	3"	"	3.60	4.444		7,375	235		7,610	8,500
3100	Globe, OS&Y, 150 lb., flanged									
3120	1/2"	1 Plum	18	.444	Ea.	570	26		596	665
3140	3/4"		16	.500		620	29.50		649.50	725
3150	1"		14	.571		810	33.50		843.50	945
3160	1-1/2"		11	.727		1,275	42.50		1,317.50	1,475
3170	2"		8	1		1,650	58.50		1,708.50	1,925
3180	2-1/2"	Q-1	5	3.200		3,425	169		3,594	4,025
3190	3"		4.50	3.556		3,425	188		3,613	4,050
3200	4"		3	5.333		5,450	282		5,732	6,450
3210	6"	Q-2	3	8		9,200	440		9,640	10,800

23 05 23.94 Hospital Type Valves

Line	Description	Crew	Daily Output	Labor-Hours	Unit	Material	Labor	Equipment	Total	Total Incl O&P
0010	**HOSPITAL TYPE VALVES**									
0300	Chiller valves									
0330	Manual operation balancing valve									
0340	2-1/2" line size	Q-1	9	1.778	Ea.	495	94		589	685
0350	3" line size		8	2		545	106		651	765
0360	4" line size		5	3.200		730	169		899	1,075
0370	5" line size	Q-2	5	4.800		920	263		1,183	1,425
0380	6" line size		5	4.800		1,150	263		1,413	1,650
0390	8" line size		4.50	5.333		2,325	292		2,617	3,000
0400	10" line size		4	6		3,725	330		4,055	4,600
0410	12" line size		3	8		5,175	440		5,615	6,350
0600	Automatic flow limiting valve									
0620	2-1/2" line size	Q-1	8	2	Ea.	500	106		606	715
0630	3" line size		7	2.286		790	121		911	1,050
0640	4" line size		6	2.667		1,125	141		1,266	1,475
0645	5" line size	Q-2	5	4.800		1,525	263		1,788	2,075
0650	6" line size		5	4.800		1,875	263		2,138	2,450
0660	8" line size		4.50	5.333		2,825	292		3,117	3,575
0670	10" line size		4	6		5,100	330		5,430	6,125
0680	12" line size		3	8		7,575	440		8,015	9,000

23 05 93 – Testing, Adjusting, and Balancing for HVAC

23 05 93.10 Balancing, Air

Line	Description	Crew	Daily Output	Labor-Hours	Unit	Material	Labor	Equipment	Total	Total Incl O&P
0010	**BALANCING, AIR** (Subcontractor's quote incl. material and labor)									
0900	Heating and ventilating equipment									
1000	Centrifugal fans, utility sets				Ea.				410	410
1100	Heating and ventilating unit								615	615

23 05 93 – Testing, Adjusting, and Balancing for HVAC

23 05 93.10 Balancing, Air		Crew	Daily Output	Labor-Hours	Unit	2015 Bare Costs Material	Labor	Equipment	Total	Total Incl O&P
1200	In-line fan				Ea.				615	615
1300	Propeller and wall fan								116	116
1400	Roof exhaust fan								274	274
2000	Air conditioning equipment, central station								890	890
2100	Built-up low pressure unit								820	820
2200	Built-up high pressure unit								960	960
2300	Built-up high pressure dual duct								1,500	1,500
2400	Built-up variable volume								1,775	1,775
2500	Multi-zone A.C. and heating unit								615	615
2600	For each zone over one, add								137	137
2700	Package A.C. unit								340	340
2800	Rooftop heating and cooling unit								480	480
3000	Supply, return, exhaust, registers & diffusers, avg. height ceiling								82	82
3100	High ceiling								123	123
3200	Floor height								68.50	68.50
3300	Off mixing box								54.50	54.50
3500	Induction unit								89	89
3600	Lab fume hood								410	410
3700	Linear supply								205	205
3800	Linear supply high								239	239
4000	Linear return								68.50	68.50
4100	Light troffers								82	82
4200	Moduline - master								82	82
4300	Moduline - slaves								41	41
4400	Regenerators								545	545
4500	Taps into ceiling plenums								103	103
4600	Variable volume boxes				↓				82	82

23 05 93.20 Balancing, Water

Line	Description	Crew	Daily Output	Labor-Hours	Unit	Material	Labor	Equipment	Total	Total Incl O&P
0010	**BALANCING, WATER** (Subcontractor's quote incl. material and labor)									
0050	Air cooled condenser				Ea.				253	253
0080	Boiler								510	510
0100	Cabinet unit heater								86.50	86.50
0200	Chiller								615	615
0300	Convector								72	72
0400	Converter								360	360
0500	Cooling tower								470	470
0600	Fan coil unit, unit ventilator								130	130
0700	Fin tube and radiant panels								144	144
0800	Main and duct re-heat coils								134	134
0810	Heat exchanger								134	134
0900	Main balancing cocks								108	108
1000	Pumps								320	320
1100	Unit heater				↓				101	101

23 05 93.50 Piping, Testing

Line	Description	Crew	Daily Output	Labor-Hours	Unit	Material	Labor	Equipment	Total	Total Incl O&P
0010	**PIPING, TESTING**									
0100	Nondestructive testing									
0110	Nondestructive hydraulic pressure test, isolate & 1 hr. hold									
0120	1" - 4" pipe									
0140	0 – 250 L.F.	1 Stpi	1.33	6.015	Ea.		360		360	560
0160	250 – 500 L.F.	"	.80	10			600		600	930
0180	500 – 1000 L.F.	Q-5	1.14	14.035			755		755	1,175
0200	1000 – 2000 L.F.	"	.80	20	↓		1,075		1,075	1,675

23 05 Common Work Results for HVAC

23 05 93 – Testing, Adjusting, and Balancing for HVAC

23 05 93.50 Piping, Testing		Crew	Daily Output	Labor-Hours	Unit	Material	2015 Bare Costs Labor	Equipment	Total	Total Incl O&P
0300	6" - 10" pipe									
0320	0 – 250 L.F.	Q-5	1	16	Ea.		860		860	1,350
0340	250 – 500 L.F.		.73	21.918			1,175		1,175	1,850
0360	500 – 1000 L.F.		.53	30.189			1,625		1,625	2,525
0380	1000 – 2000 L.F.		.38	42.105			2,275		2,275	3,525
2000	X-Ray of welds									
2110	2" diam.	1 Stpi	8	1	Ea.	13.50	60		73.50	108
2120	3" diam.		8	1		13.50	60		73.50	108
2130	4" diam.		8	1		20.50	60		80.50	116
2140	6" diam.		8	1		20.50	60		80.50	116
2150	8" diam.		6.60	1.212		20.50	72.50		93	136
2160	10" diam.		6	1.333		27	79.50		106.50	154

23 07 HVAC Insulation

23 07 13 – Duct Insulation

23 07 13.10 Duct Thermal Insulation		Crew	Daily Output	Labor-Hours	Unit	Material	Labor	Equipment	Total	Total Incl O&P
0010	**DUCT THERMAL INSULATION**									
0100	Rule of thumb, as a percentage of total mechanical costs				Job				10%	10%
0110	Insulation req'd. is based on the surface size/area to be covered									
3000	Ductwork									
3020	Blanket type, fiberglass, flexible									
3030	Fire rated for grease and hazardous exhaust ducts									
3060	1-1/2" thick	Q-14	300	.053	S.F.	4.54	2.51		7.05	9.05
3090	Fire rated for plenums									
3100	1/2" x24" x 25'	Q-14	7.20	2.222	Roll	167	105		272	350
3110	1/2" x24" x 25'		360	.044	S.F.	3.35	2.09		5.44	7.05
3120	1/2" x 48" x 25'		3.80	4.211	Roll	335	198		533	690
3126	1/2" x 48" x 25'		380	.042	S.F.	3.35	1.98		5.33	6.85
3140	FSK vapor barrier wrap, .75 lb. density									
3160	1" thick G	Q-14	350	.046	S.F.	.18	2.15		2.33	3.66
3170	1-1/2" thick G		320	.050		.22	2.36		2.58	4.03
3180	2" thick G		300	.053		.26	2.51		2.77	4.33
3190	3" thick G		260	.062		.36	2.90		3.26	5.05
3200	4" thick G		242	.066		.51	3.12		3.63	5.55
3210	Vinyl jacket, same as FSK									
3280	Unfaced, 1 lb. density									
3310	1" thick G	Q-14	360	.044	S.F.	.19	2.09		2.28	3.58
3320	1-1/2" thick G		330	.048		.29	2.28		2.57	3.99
3330	2" thick G		310	.052		.35	2.43		2.78	4.30
3400	FSK facing, 1 lb. density									
3420	1-1/2" thick G	Q-14	310	.052	S.F.	.23	2.43		2.66	4.16
3430	2" thick G	"	300	.053	"	.35	2.51		2.86	4.43
3450	FSK facing, 1.5 lb. density									
3470	1-1/2" thick G	Q-14	300	.053	S.F.	.36	2.51		2.87	4.44
3480	2" thick G	"	290	.055	"	.43	2.60		3.03	4.65
3730	Sheet insulation									
3760	Polyethylene foam, closed cell, UV resistant									
3770	Standard temperature (-90°F to +212°F)									
3771	1/4" thick G	Q-14	450	.036	S.F.	2.01	1.68		3.69	4.90
3772	3/8" thick G		440	.036		2.48	1.71		4.19	5.50
3773	1/2" thick G		420	.038		3.05	1.80		4.85	6.25

23 07 HVAC Insulation

23 07 13 – Duct Insulation

23 07 13.10 Duct Thermal Insulation		Crew	Daily Output	Labor-Hours	Unit	Material	2015 Bare Costs Labor	Equipment	Total	Total Incl O&P
3774	3/4" thick	G	Q-14	400	.040	S.F.	4.36	1.89		6.25	7.85
3775	1" thick	G		380	.042		5.90	1.98		7.88	9.65
3776	1-1/2" thick	G		360	.044		9.30	2.09		11.39	13.60
3777	2" thick	G		340	.047		12.30	2.22		14.52	17.10
3778	2-1/2" thick	G		320	.050		15.80	2.36		18.16	21
3779	Adhesive (see line 7878)										
3780	Foam, rubber										
3782	1" thick	G	1 Stpi	50	.160	S.F.	2.89	9.55		12.44	18.10
3795	Finishes										
3800	Stainless steel woven mesh		Q-14	100	.160	S.F.	.74	7.55		8.29	12.90
3810	For .010" stainless steel, add			160	.100		2.88	4.71		7.59	10.75
3820	18 oz. fiberglass cloth, pasted on			170	.094		.67	4.44		5.11	7.90
3900	8 oz. canvas, pasted on			180	.089		.21	4.19		4.40	7
3940	For .016" aluminum jacket, add			200	.080		.92	3.77		4.69	7.05
7878	Contact cement, quart can					Ea.	10.50			10.50	11.55
9600	Minimum labor/equipment charge		1 Stpi	4	2	Job		120		120	186

23 07 16 – HVAC Equipment Insulation

23 07 16.10 HVAC Equipment Thermal Insulation		Crew	Daily Output	Labor-Hours	Unit	Material	Labor	Equipment	Total	Total Incl O&P
0010	HVAC EQUIPMENT THERMAL INSULATION										
0100	Rule of thumb, as a percentage of total mechanical costs					Job				10%	10%
0110	Insulation req'd. is based on the surface size/area to be covered										
1000	Boiler, 1-1/2" calcium silicate only	G	Q-14	110	.145	S.F.	4.64	6.85		11.49	16.10
1020	Plus 2" fiberglass	G	"	80	.200	"	5.65	9.45		15.10	21.50
2000	Breeching, 2" calcium silicate										
2020	Rectangular	G	Q-14	42	.381	S.F.	9.10	17.95		27.05	39
2040	Round	G	"	38.70	.413	"	9.50	19.50		29	42
2300	Calcium silicate block, + 200°F to + 1200°F										
2310	On irregular surfaces, valves and fittings										
2340	1" thick	G	Q-14	30	.533	S.F.	4.12	25		29.12	45
2360	1-1/2" thick	G		25	.640		4.64	30		34.64	53.50
2380	2" thick	G		22	.727		6.10	34.50		40.60	61.50
2400	3" thick	G		18	.889		9.35	42		51.35	78
2410	On plane surfaces										
2420	1" thick	G	Q-14	126	.127	S.F.	4.12	6		10.12	14.15
2430	1-1/2" thick	G		120	.133		4.64	6.30		10.94	15.20
2440	2" thick	G		100	.160		6.10	7.55		13.65	18.80
2450	3" thick	G		70	.229		9.35	10.75		20.10	27.50
9610	Minimum labor/equipment charge		1 Stpi	4	2	Job		120		120	186

23 09 Instrumentation and Control for HVAC

23 09 13 – Instrumentation and Control Devices for HVAC

23 09 13.60 Water Level Controls	Crew	Daily Output	Labor-Hours	Unit	Material	Labor	Equipment	Total	Total Incl O&P
0010 WATER LEVEL CONTROLS									
1000 Electric water feeder	1 Stpi	12	.667	Ea.	293	40		333	380
2000 Feeder cut-off combination									
2100 Steam system up to 5000 sq. ft.	1 Stpi	12	.667	Ea.	625	40		665	750
2200 Steam system above 5000 sq. ft.		12	.667		775	40		815	910
2300 Steam and hot water, high pressure		10	.800		910	48		958	1,075
3000 Low water cut-off for hot water boiler, 50 psi maximum									
3100 1" top & bottom equalizing pipes, manual reset	1 Stpi	14	.571	Ea.	350	34		384	440

23 09 13 – Instrumentation and Control Devices for HVAC

23 09 13.60 Water Level Controls		Crew	Daily Output	Labor-Hours	Unit	Material	Labor	Equipment	Total	Total Incl O&P
						2015 Bare Costs				
3200	1" top & bottom equalizing pipes	1 Stpi	14	.571	Ea.	350	34		384	440
3300	2-1/2" side connection for nipple-to-boiler	↓	14	.571	↓	315	34		349	400
4000	Low water cut-off for low pressure steam with quick hook-up ftgs.									
4100	For installation in gauge glass tappings	1 Stpi	16	.500	Ea.	255	30		285	330
4200	Built-in type, 2-1/2" tap - 3-1/8" insertion		16	.500		210	30		240	278
4300	Built-in type, 2-1/2" tap - 1-3/4" insertion		16	.500		225	30		255	295
4400	Side connection to 2-1/2" tapping		16	.500		240	30		270	310
5000	Pump control, low water cut-off and alarm switch	↓	14	.571	↓	675	34		709	795

23 09 23 – Direct-Digital Control System for HVAC

23 09 23.10 Control Components/DDC Systems		Crew	Daily Output	Labor-Hours	Unit	Material	Labor	Equipment	Total	Total Incl O&P
0010	**CONTROL COMPONENTS/DDC SYSTEMS** (Sub's quote incl. M & L)									
0100	Analog inputs									
0110	Sensors (avg. 50' run in 1/2" EMT)									
0120	Duct temperature				Ea.				415	415
0130	Space temperature								665	665
0140	Duct humidity, +/- 3%								695	695
0150	Space humidity, +/- 2%								1,075	1,075
0160	Duct static pressure								565	565
0170	CFM/Transducer								765	765
0172	Water temperature								655	655
0174	Water flow								2,400	2,400
0176	Water pressure differential								975	975
0177	Steam flow								2,400	2,400
0178	Steam pressure								1,025	1,025
0180	KW/Transducer								1,350	1,350
0182	KWH totalization (not incl. elec. meter pulse xmtr.)								625	625
0190	Space static pressure				↓				1,075	1,075
1000	Analog outputs (avg. 50' run in 1/2" EMT)									
1010	P/I Transducer				Ea.				635	635
1020	Analog output, matl. in MUX								305	305
1030	Pneumatic (not incl. control device)								645	645
1040	Electric (not incl. control device)				↓				380	380
2000	Status (Alarms)									
2100	Digital inputs (avg. 50' run in 1/2" EMT)									
2110	Freeze				Ea.				435	435
2120	Fire								395	395
2130	Differential pressure, (air)								595	595
2140	Differential pressure, (water)								975	975
2150	Current sensor								435	435
2160	Duct high temperature thermostat								570	570
2170	Duct smoke detector				↓				705	705
2200	Digital output (avg. 50' run in 1/2" EMT)									
2210	Start/stop				Ea.				340	340
2220	On/off (maintained contact)				"				585	585
3000	Controller MUX panel, incl. function boards									
3100	48 point				Ea.				5,275	5,275
3110	128 point				"				7,225	7,225
3200	DDC controller (avg. 50' run in conduit)									
3210	Mechanical room									
3214	16 point controller (incl. 120 volt/1 phase power supply)				Ea.				3,275	3,275
3229	32 point controller (incl. 120 volt/1 phase power supply)				"				5,425	5,425
3230	Includes software programming and checkout									

23 09 23 – Direct-Digital Control System for HVAC

23 09 23.10 Control Components/DDC Systems		Crew	Daily Output	Labor-Hours	Unit	Material	2015 Bare Costs Labor	Equipment	Total	Total Incl O&P
3260	Space									
3266	VAV terminal box (incl. space temp. sensor)				Ea.				840	840
3280	Host computer (avg. 50′ run in conduit)									
3281	Package complete with PC, keyboard,									
3282	printer, monitor, basic software				Ea.				3,150	3,150
4000	Front end costs									
4100	Computer (P.C.) with software program				Ea.				6,350	6,350
4200	Color graphics software								3,925	3,925
4300	Color graphics slides								490	490
4350	Additional printer								980	980
4400	Communications trunk cable				L.F.				3.80	3.80
4500	Engineering labor, (not incl. dftg.)				Point				94	94
4600	Calibration labor								120	120
4700	Start-up, checkout labor								120	120
4800	Programming labor, as req'd									
5000	Communications bus (data transmission cable)									
5010	#18 twisted shielded pair in 1/2″ EMT conduit				C.L.F.				380	380
8000	Applications software									
8050	Basic maintenance manager software (not incl. data base entry)				Ea.				1,950	1,950
8100	Time program				Point				6.85	6.85
8120	Duty cycle								13.65	13.65
8140	Optimum start/stop								41.50	41.50
8160	Demand limiting								20.50	20.50
8180	Enthalpy program								41.50	41.50
8200	Boiler optimization				Ea.				1,225	1,225
8220	Chiller optimization				"				1,625	1,625
8240	Custom applications									
8260	Cost varies with complexity									

23 09 33 – Electric and Electronic Control System for HVAC

23 09 33.10 Electronic Control Systems		Crew	Daily Output	Labor-Hours	Unit	Material	Labor	Equipment	Total	Total Incl O&P
0010	**ELECTRONIC CONTROL SYSTEMS**									
0020	For electronic costs, add to Section 23 09 43.10				Ea.				15%	15%
9000	Minimum labor/equipment charge	1 Plum	8	1	Job		58.50		58.50	91.50

23 09 43 – Pneumatic Control System for HVAC

23 09 43.10 Pneumatic Control Systems		Crew	Daily Output	Labor-Hours	Unit	Material	Labor	Equipment	Total	Total Incl O&P
0010	**PNEUMATIC CONTROL SYSTEMS**									
0011	Including a nominal 50 ft. of tubing. Add control panelboard if req'd.									
0100	Heating and ventilating, split system									
0200	Mixed air control, economizer cycle, panel readout, tubing									
0220	Up to 10 tons G	Q-19	.68	35.294	Ea.	4,200	1,900		6,100	7,600
0240	For 10 to 20 tons G		.63	37.915		4,500	2,050		6,550	8,150
0260	For over 20 tons G		.58	41.096		4,875	2,225		7,100	8,800
0270	Enthalpy cycle, up to 10 tons		.50	48.387		4,650	2,625		7,275	9,175
0280	For 10 to 20 tons		.46	52.174		5,000	2,825		7,825	9,900
0290	For over 20 tons		.42	56.604		5,425	3,050		8,475	10,800
0300	Heating coil, hot water, 3 way valve,									
0320	Freezestat, limit control on discharge, readout	Q-5	.69	23.088	Ea.	3,125	1,250		4,375	5,350
0500	Cooling coil, chilled water, room									
0520	Thermostat, 3 way valve	Q-5	2	8	Ea.	1,400	430		1,830	2,200
0600	Cooling tower, fan cycle, damper control,									
0620	Control system including water readout in/out at panel	Q-19	.67	35.821	Ea.	5,525	1,925		7,450	9,100
1000	Unit ventilator, day/night operation,									

23 09 43 – Pneumatic Control System for HVAC

23 09 43.10 Pneumatic Control Systems		Crew	Daily Output	Labor-Hours	Unit	Material	Labor	Equipment	Total	Total Incl O&P
						2015 Bare Costs				
1100	freezestat, ASHRAE, cycle 2	Q-19	.91	26.374	Ea.	3,050	1,425		4,475	5,600
2000	Compensated hot water from boiler, valve control,									
2100	readout and reset at panel, up to 60 GPM	Q-19	.55	43.956	Ea.	5,725	2,375		8,100	10,000
2120	For 120 GPM		.51	47.059		6,125	2,550		8,675	10,700
2140	For 240 GPM		.49	49.180		6,400	2,650		9,050	11,200
3000	Boiler room combustion air, damper to 5 S.F., controls		1.37	17.582		2,750	950		3,700	4,500
3500	Fan coil, heating and cooling valves, 4 pipe control system		3	8		1,250	435		1,685	2,050
3600	Heat exchanger system controls		.86	27.907		2,675	1,500		4,175	5,300
3900	Multizone control (one per zone), includes thermostat, damper									
3910	motor and reset of discharge temperature	Q-5	.51	31.373	Ea.	2,750	1,675		4,425	5,675
4000	Pneumatic thermostat, including controlling room radiator valve	"	2.43	6.593		830	355		1,185	1,475
4040	Program energy saving optimizer G	Q-19	1.21	19.786		6,900	1,075		7,975	9,275
4060	Pump control system	"	3	8		1,275	435		1,710	2,075
4080	Reheat coil control system, not incl. coil	Q-5	2.43	6.593		1,075	355		1,430	1,750
4500	Air supply for pneumatic control system									
4600	Tank mounted duplex compressor, starter, alternator,									
4620	piping, dryer, PRV station and filter									
4630	1/2 HP	Q-19	.68	35.139	Ea.	10,300	1,900		12,200	14,300
4640	3/4 HP		.64	37.383		10,800	2,025		12,825	15,000
4650	1 HP		.61	39.539		11,800	2,150		13,950	16,200
4660	1-1/2 HP		.58	41.739		12,500	2,250		14,750	17,300
4680	3 HP		.55	43.956		17,100	2,375		19,475	22,500
4690	5 HP		.42	57.143		29,800	3,100		32,900	37,600
4800	Main air supply, includes 3/8" copper main and labor	Q-5	1.82	8.791	C.L.F.	345	475		820	1,125
4810	If poly tubing used, deduct								30%	30%
7000	Static pressure control for air handling unit, includes pressure									
7010	sensor, receiver controller, readout and damper motors	Q-19	.64	37.383	Ea.	8,175	2,025		10,200	12,100
7020	If return air fan requires control, add								70%	70%
8600	VAV boxes, incl. thermostat, damper motor, reheat coil & tubing	Q-5	1.46	10.989		1,275	590		1,865	2,325
8610	If no reheat coil, deduct								204	204

23 09 53 – Pneumatic and Electric Control System for HVAC

23 09 53.10 Control Components		Crew	Daily Output	Labor-Hours	Unit	Material	Labor	Equipment	Total	Total Incl O&P
0010	**CONTROL COMPONENTS**									
0600	Carbon monoxide detector system									
0606	Panel	1 Stpi	4	2	Ea.	1,075	120		1,195	1,350
0610	Sensor		7.30	1.096		720	65.50		785.50	895
0680	Controller for VAV box, includes actuator		7.30	1.096		278	65.50		343.50	405
0700	Controller, receiver									
0730	Pneumatic, panel mount, single input	1 Plum	8	1	Ea.	515	58.50		573.50	655
0740	With conversion mounting bracket		8	1		515	58.50		573.50	655
0750	Dual input, with control point adjustment		7	1.143		715	67		782	890
0850	Electric, single snap switch	1 Elec	4	2		465	109		574	680
0860	Dual snap switches	"	3	2.667		625	146		771	915
1000	Enthalpy control, boiler water temperature control									
1010	governed by outdoor temperature, with timer	1 Elec	3	2.667	Ea.	420	146		566	685
2000	Gauges, pressure or vacuum									
2100	2" diameter dial	1 Stpi	32	.250	Ea.	8.20	14.95		23.15	32.50
2200	2-1/2" diameter dial		32	.250		8.40	14.95		23.35	33
2300	3-1/2" diameter dial		32	.250		10.40	14.95		25.35	35
2400	4-1/2" diameter dial		32	.250		10.70	14.95		25.65	35.50
2700	Flanged iron case, black ring									
2800	3-1/2" diameter dial	1 Stpi	32	.250	Ea.	102	14.95		116.95	136

23 09 53 – Pneumatic and Electric Control System for HVAC

23 09 53.10 Control Components		Crew	Daily Output	Labor-Hours	Unit	Material	2015 Bare Costs Labor	Equipment	Total	Total Incl O&P
2900	4-1/2" diameter dial	1 Stpi	32	.250	Ea.	105	14.95		119.95	140
3000	6" diameter dial		32	.250		167	14.95		181.95	207
3300	For compound pressure-vacuum, add					18%				
3350	Humidistat									
3390	Electric operated	1 Shee	8	1	Ea.	70.50	56		126.50	166
3400	Relays									
3430	Pneumatic/electric	1 Plum	16	.500	Ea.	395	29.50		424.50	480
3440	Pneumatic proportioning		8	1		273	58.50		331.50	390
3450	Pneumatic switching		12	.667		179	39		218	258
3460	Selector, 3 point		6	1.333		128	78.50		206.50	262
3470	Pneumatic time delay		8	1		345	58.50		403.50	470
3500	Sensor, air operated									
3520	Humidity	1 Plum	16	.500	Ea.	440	29.50		469.50	530
3540	Pressure		16	.500		67.50	29.50		97	120
3560	Temperature		12	.667		192	39		231	272
3600	Electric operated									
3620	Humidity	1 Elec	8	1	Ea.	157	54.50		211.50	258
3650	Pressure		8	1		248	54.50		302.50	360
3680	Temperature		10	.800		122	44		166	203
4000	Thermometers									
4100	Dial type, 3-1/2" diameter, vapor type, union connection	1 Stpi	32	.250	Ea.	335	14.95		349.95	395
4120	Liquid type, union connection		32	.250		650	14.95		664.95	740
4500	Stem type, 6-1/2" case, 2" stem, 1/2" NPT		32	.250		68	14.95		82.95	98
4520	4" stem, 1/2" NPT		32	.250		83	14.95		97.95	115
4600	9" case, 3-1/2" stem, 3/4" NPT		28	.286		115	17.05		132.05	153
4620	6" stem, 3/4" NPT		28	.286		133	17.05		150.05	173
4640	8" stem, 3/4" NPT		28	.286		217	17.05		234.05	266
4660	12" stem, 1" NPT		26	.308		184	18.40		202.40	231
5000	Thermostats									
5030	Manual	1 Stpi	8	1	Ea.	51	60		111	149
5040	1 set back, electric, timed G		8	1		37	60		97	134
5050	2 set back, electric, timed G		8	1		186	60		246	298
5100	Locking cover		20	.400		18.80	24		42.80	58
5200	24 hour, automatic, clock G	1 Shee	8	1		165	56		221	271
5220	Electric, low voltage, 2 wire	1 Elec	13	.615		49.50	33.50		83	107
5230	3 wire	"	10	.800		36	44		80	108
5300	Transmitter, pneumatic									
5320	Temperature averaging element	Q-1	8	2	Ea.	151	106		257	330
5350	Pressure differential	1 Plum	7	1.143		1,275	67		1,342	1,500
5370	Humidity, duct		8	1		400	58.50		458.50	530
5380	Room		12	.667		405	39		444	505
5390	Temperature, with averaging element		6	1.333		207	78.50		285.50	350
6000	Valves, motorized zone									
6100	Sweat connections, 1/2" C x C	1 Stpi	20	.400	Ea.	168	24		192	222
6110	3/4" C x C		20	.400		168	24		192	223
6120	1" C x C		19	.421		219	25		244	280
6140	1/2" C x C, with end switch, 2 wire		20	.400		166	24		190	220
6150	3/4" C x C, with end switch, 2 wire		20	.400		177	24		201	233
6160	1" C x C, with end switch, 2 wire		19	.421		202	25		227	261
7090	Valves, motor controlled, including actuator									
7100	Electric motor actuated									
7200	Brass, two way, screwed									
7210	1/2" pipe size	L-6	36	.333	Ea.	296	19.10		315.10	355

23 09 53 – Pneumatic and Electric Control System for HVAC

23 09 53.10 Control Components		Crew	Daily Output	Labor-Hours	Unit	Material	2015 Bare Costs Labor	Equipment	Total	Total Incl O&P
7220	3/4" pipe size	L-6	30	.400	Ea.	525	23		548	610
7230	1" pipe size		28	.429		605	24.50		629.50	705
7240	1-1/2" pipe size		19	.632		665	36		701	785
7250	2" pipe size		16	.750		975	43		1,018	1,150
7350	Brass, three way, screwed									
7360	1/2" pipe size	L-6	33	.364	Ea.	320	21		341	390
7370	3/4" pipe size		27	.444		375	25.50		400.50	450
7380	1" pipe size		25.50	.471		485	27		512	575
7384	1-1/4" pipe size		21	.571		620	33		653	730
7390	1-1/2" pipe size		17	.706		705	40.50		745.50	840
7400	2" pipe size		14	.857		945	49		994	1,100
7550	Iron body, two way, flanged									
7560	2-1/2" pipe size	L-6	4	3	Ea.	1,400	172		1,572	1,825
7570	3" pipe size		3	4		1,475	229		1,704	1,975
7580	4" pipe size		2	6		2,550	345		2,895	3,350
7850	Iron body, three way, flanged									
7860	2-1/2" pipe size	L-6	3	4	Ea.	1,425	229		1,654	1,900
7870	3" pipe size		2.50	4.800		1,650	275		1,925	2,225
7880	4" pipe size		2	6		2,050	345		2,395	2,775
8000	Pneumatic, air operated									
8050	Brass, two way, screwed									
8060	1/2" pipe size, class 250	1 Plum	24	.333	Ea.	207	19.55		226.55	259
8070	3/4" pipe size, class 250		20	.400		248	23.50		271.50	310
8080	1" pipe size, class 250		19	.421		290	24.50		314.50	360
8090	1-1/4" pipe size, class 125		15	.533		360	31.50		391.50	445
8100	1-1/2" pipe size, class 125		13	.615		460	36		496	560
8110	2" pipe size, class 125		11	.727		535	42.50		577.50	650
8180	Brass, three way, screwed									
8190	1/2" pipe size, class 250	1 Plum	22	.364	Ea.	228	21.50		249.50	285
8200	3/4" pipe size, class 250		18	.444		282	26		308	350
8210	1" pipe size, class 250		17	.471		330	27.50		357.50	410
8214	1-1/4" pipe size, class 250		14	.571		465	33.50		498.50	565
8220	1-1/2" pipe size, class 125		11	.727		550	42.50		592.50	670
8230	2" pipe size, class 125		9	.889		665	52		717	810
8450	Iron body, two way, flanged									
8560	Iron body, three way, flanged									
8570	2-1/2" pipe size, class 125	Q-1	4.50	3.556	Ea.	1,225	188		1,413	1,650
8580	3" pipe size, class 125		4	4		1,400	211		1,611	1,875
8590	4" pipe size, class 125		2.50	6.400		2,825	340		3,165	3,625
8600	6" pipe size, class 125	Q-2	3	8		4,300	440		4,740	5,400
9000	Minimum labor/equipment charge	1 Plum	4	2	Job		117		117	183

23 11 13 – Facility Fuel-Oil Piping

23 11 13.10 Fuel Oil Specialties		Crew	Daily Output	Labor-Hours	Unit	Material	2015 Bare Costs Labor	Equipment	Total	Total Incl O&P
0010	**FUEL OIL SPECIALTIES**									
0020	Foot valve, single poppet, metal to metal construction									
0040	Bevel seat, 1/2" diameter	1 Stpi	20	.400	Ea.	61.50	24		85.50	106
0060	3/4" diameter		18	.444		71	26.50		97.50	120
0080	1" diameter		16	.500		91	30		121	147
0100	1-1/4" diameter		15	.533		118	32		150	180
0120	1-1/2" diameter		13	.615		163	37		200	238
0140	2" diameter		11	.727		169	43.50		212.50	254
0400	Fuel fill box, flush type									
0408	Nonlocking, watertight									
0410	1-1/2" diameter	1 Stpi	12	.667	Ea.	22	40		62	86
0440	2" diameter		10	.800		18.75	48		66.75	95
0450	2-1/2" diameter		9	.889		62.50	53		115.50	152
0460	3" diameter		7	1.143		55.50	68.50		124	168
0470	4" diameter		5	1.600		76.50	95.50		172	234
0500	Locking inner cover									
0510	2" diameter	1 Stpi	8	1	Ea.	84	60		144	186
0520	2-1/2" diameter		7	1.143		113	68.50		181.50	231
0530	3" diameter		5	1.600		121	95.50		216.50	282
0540	4" diameter		4	2		155	120		275	355
0600	Fuel system components									
0620	Spill container	1 Stpi	4	2	Ea.	790	120		910	1,050
0640	Fill adapter, 4", straight drop		8	1		69	60		129	169
0680	Fill cap, 4"		30	.267		37	15.95		52.95	66
0700	Extractor fitting, 4" x 1-1/2"		8	1		425	60		485	565
0740	Vapor hose adapter, 4"		8	1		111	60		171	215
0760	Wood gage stick, 10'					14.50			14.50	15.95
1000	Oil filters, 3/8" IPT, 20 gal. per hour	1 Stpi	20	.400		34.50	24		58.50	75.50
1020	32 gal. per hour		18	.444		46.50	26.50		73	93
1040	40 gal. per hour		16	.500		52	30		82	104
1060	50 gal. per hour		14	.571		60	34		94	120
2000	Remote tank gauging system, self contained									
2100	Single tank kit/8 sensors inputs w/printer	1 Stpi	2.50	3.200	Ea.	3,800	191		3,991	4,475
2120	Two tank kit/8 sensors inputs w/printer		2	4		5,125	239		5,364	6,000
3000	Valve, ball check, globe type, 3/8" diameter		24	.333		12.80	19.90		32.70	45
3500	Fusible, 3/8" diameter		24	.333		13.10	19.90		33	45.50
3600	1/2" diameter		24	.333		31.50	19.90		51.40	66
3610	3/4" diameter		20	.400		70	24		94	115
3620	1" diameter		19	.421		205	25		230	265
4000	Nonfusible, 3/8" diameter		24	.333		22.50	19.90		42.40	56
4500	Shutoff, gate type, lever handle, spring-fusible kit									
4520	1/4" diameter	1 Stpi	14	.571	Ea.	34.50	34		68.50	91.50
4540	3/8" diameter		12	.667		33.50	40		73.50	99
4560	1/2" diameter		10	.800		49.50	48		97.50	129
4570	3/4" diameter		8	1		86	60		146	188
4580	Lever handle, requires weight and fusible kit									
4600	1" diameter	1 Stpi	9	.889	Ea.	261	53		314	370
4620	1-1/4" diameter		8	1		320	60		380	445
4640	1-1/2" diameter		7	1.143		300	68.50		368.50	440
4660	2" diameter		6	1.333		545	79.50		624.50	725
4680	For fusible link, weight and braided wire, add					5%				
5000	Vent alarm, whistling signal					26			26	28.50
5500	Vent protector/breather, 1-1/4" diameter	1 Stpi	32	.250		13.60	14.95		28.55	38.50

23 11 Facility Fuel Piping

23 11 13 – Facility Fuel-Oil Piping

23 11 13.10 Fuel Oil Specialties		Crew	Daily Output	Labor-Hours	Unit	2015 Bare Costs Material	Labor	Equipment	Total	Total Incl O&P
5520	1-1/2" diameter	1 Stpi	32	.250	Ea.	15.20	14.95		30.15	40
5540	2" diameter		32	.250		30.50	14.95		45.45	57
5560	3" diameter		28	.286		46.50	17.05		63.55	77.50
5580	4" diameter		24	.333		56	19.90		75.90	92.50
5600	Dust cap, breather, 2"		40	.200		10.90	11.95		22.85	30.50
8000	Fuel oil and tank heaters									
8020	Electric, capacity rated at 230 volts									
8040	Immersion element in steel manifold									
8060	96 GPH at 50°F rise	Q-5	6.40	2.500	Ea.	1,175	134		1,309	1,500
8070	128 GPH at 50°F rise		6.20	2.581		1,225	139		1,364	1,575
8080	160 GPH at 50°F rise		5.90	2.712		1,350	146		1,496	1,725
8090	192 GPH at 50°F rise		5.50	2.909		1,475	156		1,631	1,875
8100	240 GPH at 50°F rise		5.10	3.137		1,700	169		1,869	2,150
8110	288 GPH at 50°F rise		4.60	3.478		1,925	187		2,112	2,425
8120	384 GPH at 50°F rise		3.10	5.161		2,250	278		2,528	2,900
8130	480 GPH at 50°F rise		2.30	6.957		2,825	375		3,200	3,675
8140	576 GPH at 50°F rise		2.10	7.619		2,825	410		3,235	3,750
8300	Suction stub, immersion type									
8320	75" long, 750 watts	1 Stpi	14	.571	Ea.	805	34		839	940
8330	99" long, 2000 watts		12	.667		1,125	40		1,165	1,300
8340	123" long, 3000 watts		10	.800		1,225	48		1,273	1,400
8660	Steam, cross flow, rated at 5 PSIG									
8680	42 GPH	Q-5	7	2.286	Ea.	1,850	123		1,973	2,250
8690	73 GPH		6.70	2.388		2,150	128		2,278	2,575
8700	112 GPH		6.20	2.581		2,250	139		2,389	2,700
8710	158 GPH		5.80	2.759		2,450	148		2,598	2,925
8720	187 GPH		4	4		3,350	215		3,565	4,025
8730	270 GPH		3.60	4.444		3,625	239		3,864	4,350
8740	365 GPH	Q-6	4.90	4.898		4,250	273		4,523	5,100
8750	635 GPH		3.70	6.486		5,375	360		5,735	6,475
8760	845 GPH		2.50	9.600		7,850	535		8,385	9,475
8770	1420 GPH		1.60	15		12,800	835		13,635	15,400
8780	2100 GPH		1.10	21.818		17,600	1,225		18,825	21,200

23 11 23 – Facility Natural-Gas Piping

23 11 23.10 Gas Meters

		Crew	Daily Output	Labor-Hours	Unit	Material	Labor	Equipment	Total	Total Incl O&P
0010	**GAS METERS**									
4000	Residential									
4010	Gas meter, residential, 3/4" pipe size	1 Plum	14	.571	Ea.	228	33.50		261.50	305
4020	Gas meter, residential, 1" pipe size		12	.667		183	39		222	262
4030	Gas meter, residential, 1-1/4" pipe size		10	.800		183	47		230	274

23 11 23.20 Gas Piping, Flexible (Csst)

		Crew	Daily Output	Labor-Hours	Unit	Material	Labor	Equipment	Total	Total Incl O&P
0010	**GAS PIPING, FLEXIBLE (CSST)**									
0100	Tubing with lightning protection									
0110	3/8"	1 Stpi	65	.123	L.F.	2.47	7.35		9.82	14.15
0120	1/2"		62	.129		2.77	7.70		10.47	15.10
0130	3/4"		60	.133		3.60	7.95		11.55	16.40
0140	1"		55	.145		5.15	8.70		13.85	19.20
0150	1-1/4"		50	.160		6.45	9.55		16	22
0160	1-1/2"		45	.178		11.30	10.60		21.90	29
0170	2"		40	.200		16.25	11.95		28.20	36.50
0200	Tubing for underground/underslab burial									
0210	3/8"	1 Stpi	65	.123	L.F.	3.43	7.35		10.78	15.20

23 11 23 – Facility Natural-Gas Piping

23 11 23.20 Gas Piping, Flexible (Csst)		Crew	Daily Output	Labor-Hours	Unit	2015 Bare Costs Material	Labor	Equipment	Total	Total Incl O&P
0220	1/2"	1 Stpi	62	.129	L.F.	3.91	7.70		11.61	16.35
0230	3/4"		60	.133		4.96	7.95		12.91	17.90
0240	1"		55	.145		6.70	8.70		15.40	21
0250	1-1/4"		50	.160		8.95	9.55		18.50	25
0260	1-1/2"		45	.178		16.80	10.60		27.40	35
0270	2"		40	.200		20	11.95		31.95	40.50
3000	Fittings									
3010	Straight									
3100	Tube to NPT									
3110	3/8"	1 Stpi	29	.276	Ea.	9.90	16.50		26.40	36.50
3120	1/2"		27	.296		10.65	17.70		28.35	39
3130	3/4"		25	.320		14.50	19.10		33.60	46
3140	1"		23	.348		22.50	21		43.50	57
3150	1-1/4"		20	.400		50	24		74	92.50
3160	1-1/2"		17	.471		102	28		130	157
3170	2"		15	.533		174	32		206	242
3200	Coupling									
3210	3/8"	1 Stpi	29	.276	Ea.	17.20	16.50		33.70	44.50
3220	1/2"		27	.296		19.60	17.70		37.30	49
3230	3/4"		25	.320		27	19.10		46.10	59.50
3240	1"		23	.348		41.50	21		62.50	78.50
3250	1-1/4"		20	.400		92.50	24		116.50	140
3260	1-1/2"		17	.471		194	28		222	257
3270	2"		15	.533		330	32		362	415
3300	Flange fitting									
3310	3/8"	1 Stpi	25	.320	Ea.	15.15	19.10		34.25	46.50
3320	1/2"		22	.364		14.45	21.50		35.95	50
3330	3/4"		19	.421		18.75	25		43.75	59.50
3340	1"		16	.500		26	30		56	75
3350	1-1/4"		12	.667		60	40		100	128
3400	90° Flange valve									
3410	3/8"	1 Stpi	25	.320	Ea.	29	19.10		48.10	62
3420	1/2"		22	.364		29.50	21.50		51	66.50
3430	3/4"		19	.421		36	25		61	78.50
4000	Tee									
4120	1/2"	1 Stpi	20.50	.390	Ea.	31	23.50		54.50	70.50
4130	3/4"		19	.421		41	25		66	84
4140	1"		17.50	.457		74	27.50		101.50	124
5000	Reducing									
5110	Tube to NPT									
5120	3/4" to 1/2" NPT	1 Stpi	26	.308	Ea.	15.60	18.40		34	45.50
5130	1" to 3/4" NPT	"	24	.333	"	24	19.90		43.90	57.50
5200	Reducing tee									
5210	1/2" x 3/8" x 3/8"	1 Stpi	21	.381	Ea.	40.50	23		63.50	80.50
5220	3/4" x 1/2" x 1/2"		20	.400		38.50	24		62.50	79.50
5230	1" x 3/4" x 1/2"		18	.444		63.50	26.50		90	111
5240	1-1/4" x 1-1/4" x 1"		15.60	.513		169	30.50		199.50	234
5250	1-1/2" x 1-1/2" x 1-1/4"		13.30	.602		258	36		294	340
5260	2" x 2" x 1-1/2"		11.80	.678		390	40.50		430.50	490
5300	Manifold with four ports and mounting bracket									
5302	Labor to mount manifold does not include making pipe									
5304	connections which are included in fitting labor.									
5310	3/4" x 1/2" x 1/2"(4)	1 Stpi	76	.105	Ea.	33.50	6.30		39.80	46.50

23 11 Facility Fuel Piping

23 11 23 – Facility Natural-Gas Piping

23 11 23.20 Gas Piping, Flexible (Csst)		Crew	Daily Output	Labor-Hours	Unit	Material	Labor	Equipment	Total	Total Incl O&P
						2015 Bare Costs				
5330	1-1/4" x 1" x 3/4"(4)	1 Stpi	72	.111	Ea.	48	6.65		54.65	63.50
5350	2" x 1-1/2" x 1"(4)	↓	68	.118	↓	60	7.05		67.05	77
5600	Protective striker plate									
5610	Quarter plate, 3" x 2"	1 Stpi	88	.091	Ea.	.77	5.45		6.22	9.30
5620	Half plate, 3" x 7"		82	.098		1.50	5.85		7.35	10.75
5630	Full plate, 3" x 12"	↓	78	.103	↓	3.11	6.15		9.26	12.95

23 12 Facility Fuel Pumps

23 12 13 – Facility Fuel-Oil Pumps

23 12 13.10 Pump and Motor Sets

		Crew	Daily Output	Labor-Hours	Unit	Material	Labor	Equipment	Total	Total Incl O&P
0010	**PUMP AND MOTOR SETS**									
1810	Light fuel and diesel oils									
1820	20 GPH 1/3 HP	Q-5	6	2.667	Ea.	1,050	143		1,193	1,375
1830	27 GPH, 1/3 HP		6	2.667		1,050	143		1,193	1,375
1840	80 GPH, 1/3 HP		5	3.200		1,050	172		1,222	1,425
1850	145 GPH, 1/2 HP		4	4		1,100	215		1,315	1,550
1860	277 GPH, 1 HP		4	4		1,225	215		1,440	1,675
1870	700 GPH, 1-1/2 HP		3	5.333		2,750	287		3,037	3,475
1880	1000 GPH, 2 HP		3	5.333		3,225	287		3,512	4,000
1890	1800 GPH, 5 HP	↓	1.80	8.889	↓	4,550	480		5,030	5,775

23 13 Facility Fuel-Storage Tanks

23 13 13 – Facility Underground Fuel-Oil, Storage Tanks

23 13 13.09 Single-Wall Steel Fuel-Oil Tanks

		Crew	Daily Output	Labor-Hours	Unit	Material	Labor	Equipment	Total	Total Incl O&P
0010	**SINGLE-WALL STEEL FUEL-OIL TANKS**									
5000	Tanks, steel ugnd., sti-p3, not incl. hold-down bars									
5500	Excavation, pad, pumps and piping not included									
5510	Single wall, 500 gallon capacity, 7 ga. shell	Q-5	2.70	5.926	Ea.	2,050	320		2,370	2,750
5520	1,000 gallon capacity, 7 ga. shell	"	2.50	6.400		3,825	345		4,170	4,725
5530	2,000 gallon capacity, 1/4" thick shell	Q-7	4.60	6.957		6,200	395		6,595	7,450
5535	2,500 gallon capacity, 7 ga. shell	Q-5	3	5.333		6,850	287		7,137	7,975
5540	5,000 gallon capacity, 1/4" thick shell	Q-7	3.20	10		11,500	570		12,070	13,600
5560	10,000 gallon capacity, 1/4" thick shell		2	16		12,300	910		13,210	15,000
5580	15,000 gallon capacity, 5/16" thick shell		1.70	18.824		19,000	1,075		20,075	22,600
5600	20,000 gallon capacity, 5/16" thick shell		1.50	21.333		26,700	1,225		27,925	31,200
5610	25,000 gallon capacity, 3/8" thick shell		1.30	24.615		37,900	1,400		39,300	43,900
5620	30,000 gallon capacity, 3/8" thick shell		1.10	29.091		38,400	1,650		40,050	44,800
5630	40,000 gallon capacity, 3/8" thick shell		.90	35.556		42,000	2,025		44,025	49,400
5640	50,000 gallon capacity, 3/8" thick shell	↓	.80	40	↓	46,600	2,275		48,875	55,000

23 13 13.13 Dbl-Wall Steel, Undrgrnd Fuel-Oil, Stor. Tanks

		Crew	Daily Output	Labor-Hours	Unit	Material	Labor	Equipment	Total	Total Incl O&P
0010	**DOUBLE-WALL STEEL, UNDERGROUND FUEL-OIL, STORAGE TANKS**									
6200	Steel, underground, 360°, double wall, U.L. listed,									
6210	with sti-P3 corrosion protection,									
6220	(dielectric coating, cathodic protection, electrical									
6230	isolation) 30 year warranty,									
6240	not incl. manholes or hold-downs.									
6250	500 gallon capacity	Q-5	2.40	6.667	Ea.	5,575	360		5,935	6,700
6260	1,000 gallon capactiy	"	2.25	7.111		6,900	380		7,280	8,175
6270	2,000 gallon capacity	Q-7	4.16	7.692	↓	9,050	440		9,490	10,600

23 13 13 – Facility Underground Fuel-Oil, Storage Tanks

23 13 13.13 Dbl-Wall Steel, Undrgrnd Fuel-Oil, Stor. Tanks		Crew	Daily Output	Labor-Hours	Unit	2015 Bare Costs Material	Labor	Equipment	Total	Total Incl O&P
6280	3,000 gallon capacity	Q-7	3.90	8.205	Ea.	10,900	465		11,365	12,700
6290	4,000 gallon capacity		3.64	8.791		13,800	500		14,300	15,900
6300	5,000 gallon capacity		2.91	10.997		19,900	625		20,525	22,900
6310	6,000 gallon capacity		2.42	13.223		25,600	750		26,350	29,300
6320	8,000 gallon capacity		2.08	15.385		24,600	875		25,475	28,400
6330	10,000 gallon capacity		1.82	17.582		27,700	1,000		28,700	32,100
6340	12,000 gallon capacity		1.70	18.824		30,600	1,075		31,675	35,400
6350	15,000 gallon capacity		1.33	24.060		31,900	1,375		33,275	37,200
6360	20,000 gallon capacity		1.33	24.060		32,700	1,375		34,075	38,000
6370	25,000 gallon capacity		1.16	27.586		72,500	1,575		74,075	82,500
6380	30,000 gallon capacity		1.03	31.068		87,500	1,775		89,275	99,000
6390	40,000 gallon capacity		.80	40		112,500	2,275		114,775	127,000
6395	50,000 gallon capacity		.73	43.836		136,500	2,500		139,000	154,500
6400	For hold-downs 500-2000 gal., add		16	2	Set	183	114		297	380
6410	For hold-downs 3000-6000 gal., add		12	2.667		370	152		522	645
6420	For hold-downs 8000-12,000 gal., add		11	2.909		445	166		611	750
6430	For hold-downs 15,000 gal., add		9	3.556		640	202		842	1,025
6440	For hold-downs 20,000 gal., add		8	4		735	228		963	1,175
6450	For hold-downs 20,000 gal. plus, add		6	5.333		960	305		1,265	1,525
6500	For manways, add				Ea.	1,800			1,800	2,000
6600	In place with hold-downs									
6652	550 gallon capacity	Q-5	1.84	8.696	Ea.	5,775	470		6,245	7,075

23 13 13.23 Glass-Fiber-Reinfcd-Plastic, Fuel-Oil, Storage		Crew	Daily Output	Labor-Hours	Unit	Material	Labor	Equipment	Total	Total Incl O&P
0010	GLASS-FIBER-REINFCD-PLASTIC, UNDERGRND. FUEL-OIL, STORAGE									
0210	Fiberglass, underground, single wall, U.L. listed, not including									
0220	manway or hold-down strap									
0225	550 gallon capacity	Q-5	2.67	5.993	Ea.	3,200	320		3,520	4,000
0230	1,000 gallon capacity	"	2.46	6.504		4,125	350		4,475	5,075
0240	2,000 gallon capacity	Q-7	4.57	7.002		5,950	400		6,350	7,175
0245	3,000 gallon capacity		3.90	8.205		7,125	465		7,590	8,575
0250	4,000 gallon capacity		3.55	9.014		8,275	515		8,790	9,900
0255	5,000 gallon capacity		3.20	10		9,200	570		9,770	11,000
0260	6,000 gallon capacity		2.67	11.985		9,400	680		10,080	11,400
0270	8,000 gallon capacity		2.29	13.974		11,200	795		11,995	13,700
0280	10,000 gallon capacity		2	16		12,800	910		13,710	15,500
0282	12,000 gallon capacity		1.88	17.021		14,500	970		15,470	17,500
0284	15,000 gallon capacity		1.68	19.048		20,500	1,075		21,575	24,200
0290	20,000 gallon capacity		1.45	22.069		26,300	1,250		27,550	31,000
0300	25,000 gallon capacity		1.28	25		54,000	1,425		55,425	61,500
0320	30,000 gallon capacity		1.14	28.070		77,500	1,600		79,100	87,500
0340	40,000 gallon capacity		.89	35.955		86,000	2,050		88,050	97,500
0360	48,000 gallon capacity		.81	39.506		128,000	2,250		130,250	144,000
0500	For manway, fittings and hold-downs, add					20%	15%			
0600	For manways, add					2,250			2,250	2,475
1000	For helical heating coil, add	Q-5	2.50	6.400		4,775	345		5,120	5,775
1020	Fiberglass, underground, double wall, U.L. listed									
1030	includes manways, not incl. hold-down straps									
1040	600 gallon capacity	Q-5	2.42	6.612	Ea.	7,200	355		7,555	8,475
1050	1,000 gallon capacity	"	2.25	7.111		9,825	380		10,205	11,400
1060	2,500 gallon capacity	Q-7	4.16	7.692		14,800	440		15,240	17,000
1070	3,000 gallon capacity		3.90	8.205		16,000	465		16,465	18,300
1080	4,000 gallon capacity		3.64	8.791		16,200	500		16,700	18,700

23 13 Facility Fuel-Storage Tanks

23 13 13 – Facility Underground Fuel-Oil, Storage Tanks

23 13 13.23 Glass-Fiber-Reinfcd-Plastic, Fuel-Oil, Storage		Crew	Daily Output	Labor-Hours	Unit	2015 Bare Costs Material	Labor	Equipment	Total	Total Incl O&P
1090	6,000 gallon capacity	Q-7	2.42	13.223	Ea.	21,100	750		21,850	24,500
1100	8,000 gallon capacity		2.08	15.385		23,500	875		24,375	27,300
1110	10,000 gallon capacity		1.82	17.582		27,300	1,000		28,300	31,600
1120	12,000 gallon capacity		1.70	18.824		34,100	1,075		35,175	39,200
1122	15,000 gallon capacity		1.52	21.053		45,300	1,200		46,500	51,500
1124	20,000 gallon capacity		1.33	24.060		56,000	1,375		57,375	63,500
1126	25,000 gallon capacity		1.16	27.586		76,500	1,575		78,075	87,000
1128	30,000 gallon capacity	↓	1.03	31.068		92,000	1,775		93,775	104,500
1140	For hold-down straps, add				↓	2%	10%			
1150	For hold-downs 500-4000 gal., add	Q-7	16	2	Set	445	114		559	665
1160	For hold-downs 5000-15000 gal., add		8	4		895	228		1,123	1,350
1170	For hold-downs 20,000 gal., add		5.33	6.004		1,350	340		1,690	2,000
1180	For hold-downs 25,000 gal., add		4	8		1,800	455		2,255	2,675
1190	For hold-downs 30,000 gal., add	↓	2.60	12.308	↓	2,675	700		3,375	4,050
2210	Fiberglass, underground, single wall, U.L. listed, including									
2220	hold-down straps, no manways									
2225	550 gallon capacity	Q-5	2	8	Ea.	3,625	430		4,055	4,675
2230	1,000 gallon capacity	"	1.88	8.511		4,575	460		5,035	5,750
2240	2,000 gallon capacity	Q-7	3.55	9.014		6,400	515		6,915	7,850
2250	4,000 gallon capacity		2.90	11.034		8,725	630		9,355	10,600
2260	6,000 gallon capacity		2	16		10,300	910		11,210	12,700
2270	8,000 gallon capacity		1.78	17.978		12,100	1,025		13,125	14,900
2280	10,000 gallon capacity		1.60	20		13,700	1,150		14,850	16,900
2282	12,000 gallon capacity		1.52	21.053		15,400	1,200		16,600	18,900
2284	15,000 gallon capacity		1.39	23.022		21,400	1,300		22,700	25,600
2290	20,000 gallon capacity		1.14	28.070		27,700	1,600		29,300	32,900
2300	25,000 gallon capacity		.96	33.333		56,000	1,900		57,900	64,500
2320	30,000 gallon capacity	↓	.80	40	↓	80,000	2,275		82,275	91,500
3020	Fiberglass, underground, double wall, U.L. listed									
3030	includes manways and hold-down straps									
3040	600 gallon capacity	Q-5	1.86	8.602	Ea.	7,650	465		8,115	9,125
3050	1,000 gallon capacity	"	1.70	9.412		10,300	505		10,805	12,100
3060	2,500 gallon capacity	Q-7	3.29	9.726		15,200	555		15,755	17,600
3070	3,000 gallon capacity		3.13	10.224		16,500	580		17,080	19,000
3080	4,000 gallon capacity		2.93	10.922		16,700	620		17,320	19,400
3090	6,000 gallon capacity		1.86	17.204		22,000	980		22,980	25,700
3100	8,000 gallon capacity		1.65	19.394		24,400	1,100		25,500	28,600
3110	10,000 gallon capacity		1.48	21.622		28,200	1,225		29,425	32,900
3120	12,000 gallon capacity		1.40	22.857		34,900	1,300		36,200	40,400
3122	15,000 gallon capacity		1.28	25		46,200	1,425		47,625	53,000
3124	20,000 gallon capacity		1.06	30.189		57,000	1,725		58,725	65,500
3126	25,000 gallon capacity		.90	35.556		78,500	2,025		80,525	89,500
3128	30,000 gallon capacity	↓	.74	43.243	↓	95,000	2,450		97,450	108,000

23 13 23 – Facility Aboveground Fuel-Oil, Storage Tanks

23 13 23.13 Vertical, Steel, Abvground Fuel-Oil, Stor. Tanks

		Crew	Daily Output	Labor-Hours	Unit	Material	Labor	Equipment	Total	Total Incl O&P
0010	**VERTICAL, STEEL, ABOVEGROUND FUEL-OIL, STORAGE TANKS**									
4000	Fixed roof oil storage tanks, steel, (1 BBL=42 gal. w/foundation 3'D x 1'W)									
4200	5,000 barrels				Ea.				194,000	213,500
4300	24,000 barrels								333,500	367,000
4500	56,000 barrels								729,000	802,000
4600	110,000 barrels								1,060,000	1,166,000
4800	143,000 barrels				↓				1,250,000	1,375,000

23 13 23 – Facility Aboveground Fuel-Oil, Storage Tanks

23 13 23.13 Vertical, Steel, Abvground Fuel-Oil, Stor. Tanks		Crew	Daily Output	Labor-Hours	Unit	2015 Bare Costs Material	Labor	Equipment	Total	Total Incl O&P
4900	225,000 barrels				Ea.				1,360,000	1,496,000
5100	Floating roof gasoline tanks, steel, 5,000 barrels (w/foundation 3'D x 1'W)								204,000	225,000
5200	25,000 barrels								381,000	419,000
5400	55,000 barrels								839,000	923,000
5500	100,000 barrels								1,253,000	1,379,000
5700	150,000 barrels								1,532,000	1,685,000
5800	225,000 barrels				↓				2,300,000	2,783,000

23 13 23.16 Horizontal, Stl, Abvgrd Fuel-Oil, Storage Tanks		Crew	Daily Output	Labor-Hours	Unit	Material	Labor	Equipment	Total	Total Incl O&P
0010	**HORIZONTAL, STEEL, ABOVEGROUND FUEL-OIL, STORAGE TANKS**									
3000	Steel, storage, above ground, including cradles, coating,									
3020	fittings, not including foundation, pumps or piping									
3040	Single wall, 275 gallon	Q-5	5	3.200	Ea.	490	172		662	810
3060	550 gallon	"	2.70	5.926		3,750	320		4,070	4,625
3080	1,000 gallon	Q-7	5	6.400		4,025	365		4,390	5,000
3100	1,500 gallon		4.75	6.737		8,775	385		9,160	10,300
3120	2,000 gallon		4.60	6.957		10,600	395		10,995	12,300
3140	5,000 gallon		3.20	10		19,100	570		19,670	21,900
3150	10,000 gallon		2	16		35,100	910		36,010	40,000
3160	15,000 gallon		1.70	18.824		45,100	1,075		46,175	51,500
3170	20,000 gallon		1.45	22.069		58,500	1,250		59,750	66,000
3180	25,000 gallon		1.30	24.615		68,000	1,400		69,400	77,000
3190	30,000 gallon	↓	1.10	29.091		81,500	1,650		83,150	92,500
3320	Double wall, 500 gallon capacity	Q-5	2.40	6.667		2,625	360		2,985	3,425
3330	2000 gallon capacity	Q-7	4.15	7.711		9,975	440		10,415	11,700
3340	4000 gallon capacity		3.60	8.889		17,800	505		18,305	20,400
3350	6000 gallon capacity		2.40	13.333		21,000	760		21,760	24,300
3360	8000 gallon capacity		2	16		27,000	910		27,910	31,100
3370	10000 gallon capacity		1.80	17.778		30,200	1,000		31,200	34,800
3380	15000 gallon capacity		1.50	21.333		45,900	1,225		47,125	52,500
3390	20000 gallon capacity		1.30	24.615		52,500	1,400		53,900	59,500
3400	25000 gallon capacity		1.15	27.826		63,500	1,575		65,075	72,500
3410	30000 gallon capacity	↓	1	32	↓	69,500	1,825		71,325	79,500

23 13 23.26 Horizontal, Conc., Abvgrd Fuel-Oil, Stor. Tanks		Crew	Daily Output	Labor-Hours	Unit	Material	Labor	Equipment	Total	Total Incl O&P
0010	**HORIZONTAL, CONCRETE, ABOVEGROUND FUEL-OIL, STORAGE TANKS**									
0050	Concrete, storage, above ground, including pad & pump									
0100	500 gallon	F-3	2	20	Ea.	10,000	960	325	11,285	12,900
0200	1,000 gallon	"	2	20		14,000	960	325	15,285	17,300
0300	2,000 gallon	F-4	2	24		18,000	1,150	565	19,715	22,300
0400	4,000 gallon		2	24		23,000	1,150	565	24,715	27,800
0500	8,000 gallon		2	24		36,000	1,150	565	37,715	42,100
0600	12,000 gallon	↓	2	24	↓	48,000	1,150	565	49,715	55,500

23 21 Hydronic Piping and Pumps

23 21 20 – Hydronic HVAC Piping Specialties

23 21 20.10 Air Control		Crew	Daily Output	Labor-Hours	Unit	2015 Bare Costs Material	Labor	Equipment	Total	Total Incl O&P
0010	**AIR CONTROL**									
0030	Air separator, with strainer									
0040	2″ diameter	Q-5	6	2.667	Ea.	1,175	143		1,318	1,525
0080	2-1/2″ diameter		5	3.200		1,325	172		1,497	1,725
0100	3″ diameter		4	4		2,050	215		2,265	2,575
0120	4″ diameter		3	5.333		2,950	287		3,237	3,675
0130	5″ diameter	Q-6	3.60	6.667		3,750	370		4,120	4,700
0140	6″ diameter		3.40	7.059		4,500	395		4,895	5,575
0160	8″ diameter		3	8		6,700	445		7,145	8,075
0180	10″ diameter		2.20	10.909		10,400	610		11,010	12,400
0200	12″ diameter		1.70	14.118		17,300	785		18,085	20,200

23 21 20.18 Automatic Air Vent

23 21 20.18 Automatic Air Vent		Crew	Daily Output	Labor-Hours	Unit	Material	Labor	Equipment	Total	Total Incl O&P
0010	**AUTOMATIC AIR VENT**									
0020	Cast iron body, stainless steel internals, float type									
0060	1/2″ NPT inlet, 300 psi	1 Stpi	12	.667	Ea.	109	40		149	182
0140	3/4″ NPT inlet, 300 psi		12	.667		109	40		149	182
0180	1/2″ NPT inlet, 250 psi		10	.800		350	48		398	460
0220	3/4″ NPT inlet, 250 psi		10	.800		350	48		398	460
0260	1″ NPT inlet, 250 psi		10	.800		510	48		558	635
0340	1-1/2″ NPT inlet, 250 psi	Q-5	12	1.333		1,075	71.50		1,146.50	1,300
0380	2″ NPT inlet, 250 psi	″	12	1.333		1,075	71.50		1,146.50	1,300
0600	Forged steel body, stainless steel internals, float type									
0640	1/2″ NPT inlet, 750 psi	1 Stpi	12	.667	Ea.	1,150	40		1,190	1,325
0680	3/4″ NPT inlet, 750 psi		12	.667		1,150	40		1,190	1,325
0760	3/4″ NPT inlet, 1000 psi		10	.800		1,725	48		1,773	1,975
0800	1″ NPT inlet, 1000 psi	Q-5	12	1.333		1,725	71.50		1,796.50	2,000
0880	1-1/2″ NPT inlet, 1000 psi		10	1.600		4,875	86		4,961	5,475
0920	2″ NPT inlet, 1000 psi		10	1.600		4,875	86		4,961	5,475
1100	Formed steel body, noncorrosive									
1110	1/8″ NPT inlet 150 psi	1 Stpi	32	.250	Ea.	12.45	14.95		27.40	37
1120	1/4″ NPT inlet 150 psi		32	.250		42	14.95		56.95	70
1130	3/4″ NPT inlet 150 psi		32	.250		42	14.95		56.95	70
1300	Chrome plated brass, automatic/manual, for radiators									
1310	1/8″ NPT inlet, nickel plated brass	1 Stpi	32	.250	Ea.	7.25	14.95		22.20	31.50

23 21 20.22 Circuit Sensor

23 21 20.22 Circuit Sensor		Crew	Daily Output	Labor-Hours	Unit	Material	Labor	Equipment	Total	Total Incl O&P
0010	**CIRCUIT SENSOR**, Flow meter									
0020	Metering stations									
0040	Wafer orifice insert type									
0060	2-1/2″ pipe size	Q-5	12	1.333	Ea.	210	71.50		281.50	345
0100	3″ pipe size		11	1.455		238	78		316	385
0140	4″ pipe size		8	2		273	108		381	470
0180	5″ pipe size		7.30	2.192		345	118		463	565
0220	6″ pipe size		6.40	2.500		415	134		549	665
0260	8″ pipe size	Q-6	5.30	4.528		575	253		828	1,025
0280	10″ pipe size		4.60	5.217		660	291		951	1,175
0360	12″ pipe size		4.20	5.714		1,100	320		1,420	1,700

23 21 20.26 Circuit Setter

23 21 20.26 Circuit Setter		Crew	Daily Output	Labor-Hours	Unit	Material	Labor	Equipment	Total	Total Incl O&P
0010	**CIRCUIT SETTER**, Balance valve									
0018	Threaded									
0020	3/4″ pipe size	1 Stpi	20	.400	Ea.	79	24		103	125
0040	1″ pipe size		18	.444		102	26.50		128.50	154
0060	1-1/2″ pipe size		12	.667		177	40		217	257

23 21 20 – Hydronic HVAC Piping Specialties

23 21 20.26 Circuit Setter		Crew	Daily Output	Labor-Hours	Unit	Material	2015 Bare Costs Labor	Equipment	Total	Total Incl O&P
0080	2" pipe size	1 Stpi	10	.800	Ea.	253	48		301	355
0100	2-1/2" pipe size	Q-5	15	1.067		575	57.50		632.50	720
0120	3" pipe size	"	10	1.600	↓	805	86		891	1,025
0130	Cast iron body, flanged									
0136	3" pipe size, flanged	Q-5	4	4	Ea.	805	215		1,020	1,225
0140	4" pipe size	"	3	5.333		1,200	287		1,487	1,775
0200	For differential meter, accurate to 1%, add				↓	930			930	1,025

23 21 20.34 Dielectric Unions		Crew	Daily Output	Labor-Hours	Unit	Material	Labor	Equipment	Total	Total Incl O&P
0010	**DIELECTRIC UNIONS**, Standard gaskets for water and air									
0020	250 psi maximum pressure									
0280	Female IPT to sweat, straight									
0300	1/2" pipe size	1 Plum	24	.333	Ea.	9	19.55		28.55	40.50
0340	3/4" pipe size		20	.400		6.20	23.50		29.70	43.50
0360	1" pipe size		19	.421		8.45	24.50		32.95	48
0380	1-1/4" pipe size		15	.533		13.10	31.50		44.60	63.50
0400	1-1/2" pipe size		13	.615		19.70	36		55.70	78
0420	2" pipe size	↓	11	.727	↓	26.50	42.50		69	95.50
0580	Female IPT to brass pipe thread, straight									
0600	1/2" pipe size	1 Plum	24	.333	Ea.	12.10	19.55		31.65	44
0640	3/4" pipe size		20	.400		13.30	23.50		36.80	51
0660	1" pipe size		19	.421		24.50	24.50		49	65.50
0680	1-1/4" pipe size		15	.533		30	31.50		61.50	82
0700	1-1/2" pipe size		13	.615		44	36		80	105
0720	2" pipe size	↓	11	.727	↓	86	42.50		128.50	161
0780	Female IPT to female IPT, straight									
0800	1/2" pipe size	1 Plum	24	.333	Ea.	13.85	19.55		33.40	45.50
0840	3/4" pipe size		20	.400		15.60	23.50		39.10	53.50
0860	1" pipe size		19	.421		21	24.50		45.50	61.50
0880	1-1/4" pipe size		15	.533		28	31.50		59.50	80
0900	1-1/2" pipe size		13	.615		43	36		79	104
0920	2" pipe size	↓	11	.727	↓	63	42.50		105.50	136
2000	175 psi maximum pressure									
2180	Female IPT to sweat									
2240	2" pipe size	1 Plum	9	.889	Ea.	142	52		194	238
2260	2-1/2" pipe size	Q-1	15	1.067		155	56.50		211.50	259
2280	3" pipe size		14	1.143		213	60.50		273.50	330
2300	4" pipe size	↓	11	1.455	↓	565	77		642	740
2480	Female IPT to brass pipe									
2500	1-1/2" pipe size	1 Plum	11	.727	Ea.	176	42.50		218.50	261
2540	2" pipe size	"	9	.889		207	52		259	310
2560	2-1/2" pipe size	Q-1	15	1.067		305	56.50		361.50	425
2580	3" pipe size		14	1.143		355	60.50		415.50	485
2600	4" pipe size	↓	11	1.455	↓	635	77		712	820
9000	Minimum labor/equipment charge	1 Plum	4	2	Job		117		117	183

23 21 20.38 Expansion Couplings		Crew	Daily Output	Labor-Hours	Unit	Material	Labor	Equipment	Total	Total Incl O&P
0010	**EXPANSION COUPLINGS**, Hydronic									
0100	Copper to copper, sweat									
1000	Baseboard riser fitting, 5" stub by coupling 12" long									
1020	1/2" diameter	1 Stpi	24	.333	Ea.	20.50	19.90		40.40	53.50
1040	3/4" diameter		20	.400		29	24		53	69.50
1060	1" diameter		19	.421		40.50	25		65.50	83.50
1080	1-1/4" diameter	↓	15	.533	↓	52.50	32		84.50	108

23 21 20 – Hydronic HVAC Piping Specialties

23 21 20.38 Expansion Couplings	Crew	Daily Output	Labor-Hours	Unit	Material	Labor	Equipment	Total	Total Incl O&P
					2015 Bare Costs				
1180 9" Stub by tubing 8" long									
1200 1/2" diameter	1 Stpi	24	.333	Ea.	18.30	19.90		38.20	51
1220 3/4" diameter		20	.400		24.50	24		48.50	64
1240 1" diameter		19	.421		33.50	25		58.50	76
1260 1-1/4" diameter		15	.533		47	32		79	102

23 21 20.42 Expansion Joints	Crew	Daily Output	Labor-Hours	Unit	Material	Labor	Equipment	Total	Total Incl O&P
0010 EXPANSION JOINTS									
0100 Bellows type, neoprene cover, flanged spool									
0140 6" face to face, 1-1/4" diameter	1 Stpi	11	.727	Ea.	255	43.50		298.50	350
0160 1-1/2" diameter	"	10.60	.755		255	45		300	350
0180 2" diameter	Q-5	13.30	1.203		258	64.50		322.50	385
0190 2-1/2" diameter		12.40	1.290		267	69.50		336.50	400
0200 3" diameter		11.40	1.404		299	75.50		374.50	450
0480 10" face to face, 2" diameter		13	1.231		370	66		436	515
0500 2-1/2" diameter		12	1.333		390	71.50		461.50	540
0520 3" diameter		11	1.455		400	78		478	560
0540 4" diameter		8	2		455	108		563	670
0560 5" diameter		7	2.286		540	123		663	785
0580 6" diameter		6	2.667		560	143		703	840
0600 8" diameter		5	3.200		670	172		842	1,000
0620 10" diameter		4.60	3.478		735	187		922	1,100
0640 12" diameter		4	4		910	215		1,125	1,325
0660 14" diameter		3.80	4.211		1,125	226		1,351	1,600

23 21 20.46 Expansion Tanks	Crew	Daily Output	Labor-Hours	Unit	Material	Labor	Equipment	Total	Total Incl O&P
0010 EXPANSION TANKS									
1400 Plastic, corrosion resistant, see Plumbing Cost Data									
1507 Underground fuel-oil storage tanks, see Section 23 13 13									
1512 Tank leak detection systems, see Section 28 33 33.50									
2000 Steel, liquid expansion, ASME, painted, 15 gallon capacity	Q-5	17	.941	Ea.	640	50.50		690.50	780
2020 24 gallon capacity		14	1.143		715	61.50		776.50	880
2040 30 gallon capacity		12	1.333		715	71.50		786.50	895
2060 40 gallon capacity		10	1.600		835	86		921	1,050
2080 60 gallon capacity		8	2		1,000	108		1,108	1,275
2100 80 gallon capacity		7	2.286		1,075	123		1,198	1,375
2120 100 gallon capacity		6	2.667		1,450	143		1,593	1,825
2130 120 gallon capacity		5	3.200		1,550	172		1,722	1,975
2140 135 gallon capacity		4.50	3.556		1,625	191		1,816	2,075
2150 175 gallon capacity		4	4		2,550	215		2,765	3,125
2160 220 gallon capacity		3.60	4.444		2,875	239		3,114	3,550
2170 240 gallon capacity		3.30	4.848		3,000	261		3,261	3,700
2180 305 gallon capacity		3	5.333		4,225	287		4,512	5,100
2190 400 gallon capacity		2.80	5.714		5,200	305		5,505	6,200
2360 Galvanized									
2370 15 gallon capacity	Q-5	17	.941	Ea.	1,125	50.50		1,175.50	1,300
2380 24 gallon capacity		14	1.143		1,150	61.50		1,211.50	1,375
2390 30 gallon capacity		12	1.333		1,350	71.50		1,421.50	1,575
2400 40 gallon capacity		10	1.600		1,575	86		1,661	1,850
2410 60 gallon capacity		8	2		1,825	108		1,933	2,175
2420 80 gallon capacity		7	2.286		2,075	123		2,198	2,475
2430 100 gallon capacity		6	2.667		2,700	143		2,843	3,200
2440 120 gallon capacity		5	3.200		2,900	172		3,072	3,475
2450 135 gallon capacity		4.50	3.556		3,050	191		3,241	3,675

23 21 Hydronic Piping and Pumps

23 21 20 – Hydronic HVAC Piping Specialties

23 21 20.46 Expansion Tanks		Crew	Daily Output	Labor-Hours	Unit	Material	Labor (2015 Bare Costs)	Equipment	Total	Total Incl O&P
2460	175 gallon capacity	Q-5	4	4	Ea.	4,950	215		5,165	5,775
2470	220 gallon capacity		3.60	4.444		5,675	239		5,914	6,600
2480	240 gallon capacity		3.30	4.848		5,900	261		6,161	6,875
2490	305 gallon capacity		3	5.333		8,600	287		8,887	9,925
2500	400 gallon capacity		2.80	5.714		10,600	305		10,905	12,200
3000	Steel ASME expansion, rubber diaphragm, 19 gal. cap. accept.		12	1.333		2,425	71.50		2,496.50	2,775
3020	31 gallon capacity		8	2		2,700	108		2,808	3,150
3040	61 gallon capacity		6	2.667		3,800	143		3,943	4,400
3060	79 gallon capacity		5	3.200		3,875	172		4,047	4,525
3080	119 gallon capacity		4	4		4,100	215		4,315	4,825
3100	158 gallon capacity		3.80	4.211		5,675	226		5,901	6,600
3120	211 gallon capacity		3.30	4.848		6,575	261		6,836	7,625
3140	317 gallon capacity		2.80	5.714		8,575	305		8,880	9,900
3160	422 gallon capacity		2.60	6.154		12,700	330		13,030	14,500
3180	528 gallon capacity		2.40	6.667		13,900	360		14,260	15,900
9000	Minimum labor/equipment charge		4	4	Job		215		215	335

23 21 20.58 Hydronic Heating Control Valves		Crew	Daily Output	Labor-Hours	Unit	Material	Labor	Equipment	Total	Total Incl O&P
0010	**HYDRONIC HEATING CONTROL VALVES**									
0050	Hot water, nonelectric, thermostatic									
0100	Radiator supply, 1/2" diameter	1 Stpi	24	.333	Ea.	68.50	19.90		88.40	106
0120	3/4" diameter		20	.400		71.50	24		95.50	116
0140	1" diameter		19	.421		88	25		113	136
0160	1-1/4" diameter		15	.533		120	32		152	182
0500	For low pressure steam, add					25%				
1000	Manual, radiator supply									
1010	1/2" pipe size, angle union	1 Stpi	24	.333	Ea.	52.50	19.90		72.40	89
1020	3/4" pipe size, angle union		20	.400		66	24		90	111
1030	1" pipe size, angle union		19	.421		85.50	25		110.50	133
1100	Radiator, balancing, straight, sweat connections									
1110	1/2" pipe size	1 Stpi	24	.333	Ea.	20	19.90		39.90	53
1120	3/4" pipe size		20	.400		28	24		52	68.50
1130	1" pipe size		19	.421		45.50	25		70.50	89
1140	Balance and stop valve 1/2" size		22	.364		51.50	21.50		73	90.50
1150	3/4" size		20	.400		55.50	24		79.50	99
1160	1" size		19	.421		64.50	25		89.50	110
1170	1-1/4" size		15	.533		82.50	32		114.50	141
1200	Steam, radiator, supply									
1210	1/2" pipe size, angle union	1 Stpi	24	.333	Ea.	49	19.90		68.90	85
1220	3/4" pipe size, angle union		20	.400		55.50	24		79.50	99
1230	1" pipe size, angle union		19	.421		64.50	25		89.50	110
1240	1-1/4" pipe size, angle union		15	.533		82.50	32		114.50	141
8000	System balancing and shut-off									
8020	Butterfly, quarter turn, calibrated, threaded or solder									
8040	Bronze, -30°F to +350°F, pressure to 175 psi									
8060	1/2" size	1 Stpi	22	.364	Ea.	18.65	21.50		40.15	54.50
8070	3/4" size		20	.400		29.50	24		53.50	70
8080	1" size		19	.421		36	25		61	78.50
8090	1-1/4" size		15	.533		58	32		90	114
8100	1-1/2" size		13	.615		75	37		112	140
8110	2" size		11	.727		94	43.50		137.50	171

23 21 20 – Hydronic HVAC Piping Specialties

23 21 20.70 Steam Traps

	23 21 20.70 Steam Traps	Crew	Daily Output	Labor-Hours	Unit	Material	Labor	Equipment	Total	Total Incl O&P
0010	**STEAM TRAPS**									
0030	Cast iron body, threaded									
0040	Inverted bucket									
0050	1/2" pipe size	1 Stpi	12	.667	Ea.	157	40		197	234
0070	3/4" pipe size		10	.800		278	48		326	380
0100	1" pipe size		9	.889		420	53		473	550
0120	1-1/4" pipe size		8	1		635	60		695	795
1000	Float & thermostatic, 15 psi									
1010	3/4" pipe size	1 Stpi	16	.500	Ea.	141	30		171	202
1020	1" pipe size		15	.533		169	32		201	236
1030	1-1/4" pipe size		13	.615		187	37		224	264
1040	1-1/2" pipe size		9	.889		298	53		351	415
1060	2" pipe size		6	1.333		590	79.50		669.50	775
1290	Brass body, threaded									
1300	Thermostatic, angle union, 25 psi									
1310	1/2" pipe size	1 Stpi	24	.333	Ea.	57	19.90		76.90	93.50
1320	3/4" pipe size		20	.400		91	24		115	138
1330	1" pipe size		19	.421		139	25		164	192
9000	Minimum labor/equipment charge		4	2	Job		120		120	186

23 21 20.74 Strainers, Basket Type

	23 21 20.74 Strainers, Basket Type	Crew	Daily Output	Labor-Hours	Unit	Material	Labor	Equipment	Total	Total Incl O&P
0010	**STRAINERS, BASKET TYPE**, Perforated stainless steel basket									
0100	Brass or monel available									
2000	Simplex style									
2300	Bronze body									
2320	Screwed, 3/8" pipe size	1 Stpi	22	.364	Ea.	172	21.50		193.50	223
2340	1/2" pipe size		20	.400		295	24		319	365
2360	3/4" pipe size		17	.471		400	28		428	485
2380	1" pipe size		15	.533		400	32		432	490
2400	1-1/4" pipe size		13	.615		575	37		612	695
2420	1-1/2" pipe size		12	.667		580	40		620	695
2440	2" pipe size		10	.800		855	48		903	1,025
2460	2-1/2" pipe size	Q-5	15	1.067		770	57.50		827.50	940
2480	3" pipe size	"	14	1.143		1,125	61.50		1,186.50	1,325
2600	Flanged, 2" pipe size	1 Stpi	6	1.333		820	79.50		899.50	1,025
2620	2-1/2" pipe size	Q-5	4.50	3.556		1,275	191		1,466	1,725
2640	3" pipe size		3.50	4.571		1,450	246		1,696	1,975
2660	4" pipe size		3	5.333		2,450	287		2,737	3,150
2680	5" pipe size	Q-6	3.40	7.059		3,750	395		4,145	4,750
2700	6" pipe size		3	8		4,775	445		5,220	5,950
2710	8" pipe size		2.50	9.600		7,750	535		8,285	9,350
3600	Iron body									
3700	Screwed, 3/8" pipe size	1 Stpi	22	.364	Ea.	98.50	21.50		120	143
3720	1/2" pipe size		20	.400		102	24		126	150
3740	3/4" pipe size		17	.471		130	28		158	187
3760	1" pipe size		15	.533		133	32		165	196
3780	1-1/4" pipe size		13	.615		173	37		210	248
3800	1-1/2" pipe size		12	.667		191	40		231	272
3820	2" pipe size		10	.800		228	48		276	325
3840	2-1/2" pipe size	Q-5	15	1.067		305	57.50		362.50	425
3860	3" pipe size	"	14	1.143		370	61.50		431.50	500
4000	Flanged, 2" pipe size	1 Stpi	6	1.333		355	79.50		434.50	515
4020	2-1/2" pipe size	Q-5	4.50	3.556		475	191		666	825

23 21 20 – Hydronic HVAC Piping Specialties

23 21 20.74 Strainers, Basket Type		Crew	Daily Output	Labor-Hours	Unit	Material	Labor	Equipment	Total	Total Incl O&P
						2015 Bare Costs				
4040	3" pipe size	Q-5	3.50	4.571	Ea.	500	246		746	935
4060	4" pipe size	↓	3	5.333		760	287		1,047	1,275
4080	5" pipe size	Q-6	3.40	7.059		1,150	395		1,545	1,900
4100	6" pipe size		3	8		1,475	445		1,920	2,325
4120	8" pipe size		2.50	9.600		2,675	535		3,210	3,775
4140	10" pipe size	↓	2.20	10.909	↓	5,875	610		6,485	7,425
6000	Cast steel body									
6400	Screwed, 1" pipe size	1 Stpi	15	.533	Ea.	195	32		227	265
6410	1-1/4" pipe size		13	.615		295	37		332	385
6420	1-1/2" pipe size		12	.667		320	40		360	410
6440	2" pipe size	↓	10	.800		405	48		453	520
6460	2-1/2" pipe size	Q-5	15	1.067		570	57.50		627.50	720
6480	3" pipe size	"	14	1.143		750	61.50		811.50	920
6560	Flanged, 2" pipe size	1 Stpi	6	1.333		685	79.50		764.50	880
6580	2-1/2" pipe size	Q-5	4.50	3.556		1,025	191		1,216	1,450
6600	3" pipe size		3.50	4.571		1,100	246		1,346	1,600
6620	4" pipe size	↓	3	5.333		1,575	287		1,862	2,175
6640	6" pipe size	Q-6	3	8		2,875	445		3,320	3,850
6660	8" pipe size	"	2.50	9.600	↓	4,625	535		5,160	5,925
7000	Stainless steel body									
7200	Screwed, 1" pipe size	1 Stpi	15	.533	Ea.	283	32		315	360
7210	1-1/4" pipe size		13	.615		440	37		477	545
7220	1-1/2" pipe size		12	.667		440	40		480	545
7240	2" pipe size	↓	10	.800		655	48		703	795
7260	2-1/2" pipe size	Q-5	15	1.067		925	57.50		982.50	1,125
7280	3" pipe size	"	14	1.143		1,275	61.50		1,336.50	1,500
7400	Flanged, 2" pipe size	1 Stpi	6	1.333		1,125	79.50		1,204.50	1,350
7420	2-1/2" pipe size	Q-5	4.50	3.556		2,025	191		2,216	2,550
7440	3" pipe size		3.50	4.571		2,075	246		2,321	2,675
7460	4" pipe size	↓	3	5.333		3,275	287		3,562	4,050
7480	6" pipe size	Q-6	3	8		5,700	445		6,145	6,975
7500	8" pipe size	"	2.50	9.600	↓	8,300	535		8,835	9,950
8100	Duplex style									
8200	Bronze body									
8240	Screwed, 3/4" pipe size	1 Stpi	16	.500	Ea.	1,175	30		1,205	1,325
8260	1" pipe size		14	.571		1,175	34		1,209	1,325
8280	1-1/4" pipe size		12	.667		2,825	40		2,865	3,150
8300	1-1/2" pipe size		11	.727		2,825	43.50		2,868.50	3,175
8320	2" pipe size	↓	9	.889		3,700	53		3,753	4,125
8340	2-1/2" pipe size	Q-5	14	1.143		4,775	61.50		4,836.50	5,350
8420	Flanged, 2" pipe size	1 Stpi	6	1.333		3,975	79.50		4,054.50	4,500
8440	2-1/2" pipe size	Q-5	4.50	3.556		5,475	191		5,666	6,325
8460	3" pipe size		3.50	4.571		5,975	246		6,221	6,950
8480	4" pipe size	↓	3	5.333		8,575	287		8,862	9,875
8500	5" pipe size	Q-6	3.40	7.059		18,600	395		18,995	21,000
8520	6" pipe size	"	3	8	↓	19,100	445		19,545	21,700
8700	Iron body									
8740	Screwed, 3/4" pipe size	1 Stpi	16	.500	Ea.	1,050	30		1,080	1,200
8760	1" pipe size		14	.571		1,050	34		1,084	1,200
8780	1-1/4" pipe size		12	.667		1,175	40		1,215	1,325
8800	1-1/2" pipe size		11	.727		1,175	43.50		1,218.50	1,350
8820	2" pipe size	↓	9	.889		1,975	53		2,028	2,250
8840	2-1/2" pipe size	Q-5	14	1.143	↓	2,175	61.50		2,236.50	2,500

23 21 20.74 Strainers, Basket Type		Crew	Daily Output	Labor-Hours	Unit	Material	2015 Bare Costs Labor	Equipment	Total	Total Incl O&P
9000	Flanged, 2" pipe size	1 Stpi	6	1.333	Ea.	2,100	79.50		2,179.50	2,450
9020	2-1/2" pipe size	Q-5	4.50	3.556		2,250	191		2,441	2,775
9040	3" pipe size		3.50	4.571		2,450	246		2,696	3,075
9060	4" pipe size		3	5.333		4,150	287		4,437	5,000
9080	5" pipe size	Q-6	3.40	7.059		9,025	395		9,420	10,500
9100	6" pipe size		3	8		9,025	445		9,470	10,600
9120	8" pipe size		2.50	9.600		16,700	535		17,235	19,200
9140	10" pipe size		2.20	10.909		23,900	610		24,510	27,200
9160	12" pipe size		1.70	14.118		26,200	785		26,985	30,100
9170	14" pipe size		1.40	17.143		31,000	955		31,955	35,600
9180	16" pipe size		1	24		38,200	1,350		39,550	44,100
9300	Cast steel body									
9340	Screwed, 1" pipe size	1 Stpi	14	.571	Ea.	1,350	34		1,384	1,525
9360	1-1/2" pipe size		11	.727		2,175	43.50		2,218.50	2,475
9380	2" pipe size		9	.889		2,875	53		2,928	3,250
9460	Flanged, 2" pipe size		6	1.333		3,225	79.50		3,304.50	3,675
9480	2-1/2" pipe size	Q-5	4.50	3.556		5,200	191		5,391	6,000
9500	3" pipe size		3.50	4.571		5,650	246		5,896	6,600
9520	4" pipe size		3	5.333		6,975	287		7,262	8,125
9540	6" pipe size	Q-6	3	8		13,100	445		13,545	15,100
9560	8" pipe size	"	2.50	9.600		31,500	535		32,035	35,400
9700	Stainless steel body									
9740	Screwed, 1" pipe size	1 Stpi	14	.571	Ea.	1,900	34		1,934	2,125
9760	1-1/2" pipe size		11	.727		2,850	43.50		2,893.50	3,225
9780	2" pipe size		9	.889		4,000	53		4,053	4,475
9860	Flanged, 2" pipe size		6	1.333		4,350	79.50		4,429.50	4,925
9880	2-1/2" pipe size	Q-5	4.50	3.556		7,300	191		7,491	8,325
9900	3" pipe size		3.50	4.571		7,950	246		8,196	9,125
9920	4" pipe size		3	5.333		10,400	287		10,687	11,800
9940	6" pipe size	Q-6	3	8		15,800	445		16,245	18,100
9960	8" pipe size	"	2.50	9.600		44,200	535		44,735	49,400

23 21 20.76 Strainers, Y Type, Bronze Body		Crew	Daily Output	Labor-Hours	Unit	Material	Labor	Equipment	Total	Total Incl O&P
0010	**STRAINERS, Y TYPE, BRONZE BODY**									
0050	Screwed, 125 lb., 1/4" pipe size	1 Stpi	24	.333	Ea.	23	19.90		42.90	56.50
0070	3/8" pipe size		24	.333		27.50	19.90		47.40	61.50
0100	1/2" pipe size		20	.400		27.50	24		51.50	68
0120	3/4" pipe size		19	.421		34	25		59	76.50
0140	1" pipe size		17	.471		50	28		78	99
0150	1-1/4" pipe size		15	.533		81	32		113	139
0160	1-1/2" pipe size		14	.571		108	34		142	173
0180	2" pipe size		13	.615		144	37		181	216
0182	3" pipe size		12	.667		845	40		885	990
0200	300 lb., 2-1/2" pipe size	Q-5	17	.941		510	50.50		560.50	640
0220	3" pipe size		16	1		1,000	54		1,054	1,175
0240	4" pipe size		15	1.067		2,300	57.50		2,357.50	2,625
0500	For 300 lb. rating 1/4" thru 2", add					15%				
1000	Flanged, 150 lb., 1-1/2" pipe size	1 Stpi	11	.727	Ea.	520	43.50		563.50	645
1020	2" pipe size	"	8	1		705	60		765	870
1030	2-1/2" pipe size	Q-5	5	3.200		950	172		1,122	1,325
1040	3" pipe size		4.50	3.556		1,175	191		1,366	1,575
1060	4" pipe size		3	5.333		1,775	287		2,062	2,400
1080	5" pipe size	Q-6	3.40	7.059		1,775	395		2,170	2,575

23 21 20 – Hydronic HVAC Piping Specialties

23 21 20.76 Strainers, Y Type, Bronze Body		Crew	Daily Output	Labor-Hours	Unit	Material	2015 Bare Costs Labor	Equipment	Total	Total Incl O&P
1100	6" pipe size	Q-6	3	8	Ea.	3,400	445		3,845	4,425
1106	8" pipe size	↓	2.60	9.231	↓	3,725	515		4,240	4,900
1500	For 300 lb. rating, add					40%				
9000	Minimum labor/equipment charge	1 Stpi	3.75	2.133	Job		127		127	199

23 21 20.78 Strainers, Y Type, Iron Body		Crew	Daily Output	Labor-Hours	Unit	Material	Labor	Equipment	Total	Total Incl O&P
0010	**STRAINERS, Y TYPE, IRON BODY**									
0050	Screwed, 250 lb., 1/4" pipe size	1 Stpi	20	.400	Ea.	11.05	24		35.05	49.50
0070	3/8" pipe size		20	.400		11.05	24		35.05	49.50
0100	1/2" pipe size		20	.400		11.05	24		35.05	49.50
0120	3/4" pipe size		18	.444		12.95	26.50		39.45	56
0140	1" pipe size		16	.500		18.25	30		48.25	66.50
0150	1-1/4" pipe size		15	.533		24	32		56	76
0160	1-1/2" pipe size		12	.667		30	40		70	94.50
0180	2" pipe size	↓	8	1		44.50	60		104.50	142
0200	2-1/2" pipe size	Q-5	12	1.333		268	71.50		339.50	405
0220	3" pipe size		11	1.455		289	78		367	440
0240	4" pipe size	↓	5	3.200	↓	490	172		662	810
0500	For galvanized body, add					50%				
1000	Flanged, 125 lb., 1-1/2" pipe size	1 Stpi	11	.727	Ea.	107	43.50		150.50	185
1020	2" pipe size	"	8	1		113	60		173	217
1030	2-1/2" pipe size	Q-5	5	3.200		115	172		287	395
1040	3" pipe size		4.50	3.556		186	191		377	500
1060	4" pipe size	↓	3	5.333		305	287		592	780
1080	5" pipe size	Q-6	3.40	7.059		385	395		780	1,050
1100	6" pipe size		3	8		615	445		1,060	1,375
1120	8" pipe size		2.50	9.600		830	535		1,365	1,750
1140	10" pipe size		2	12		1,875	670		2,545	3,100
1160	12" pipe size		1.70	14.118		2,275	785		3,060	3,725
1170	14" pipe size		1.30	18.462		4,175	1,025		5,200	6,200
1180	16" pipe size	↓	1	24	↓	5,925	1,350		7,275	8,600
1500	For 250 lb. rating, add					20%				
2000	For galvanized body, add					50%				
2500	For steel body, add					40%				

23 21 20.84 Thermoflo Indicator		Crew	Daily Output	Labor-Hours	Unit	Material	Labor	Equipment	Total	Total Incl O&P
0010	**THERMOFLO INDICATOR**, For balancing									
1000	Sweat connections, 1-1/4" pipe size	1 Stpi	12	.667	Ea.	650	40		690	775
1020	1-1/2" pipe size		10	.800		660	48		708	805
1040	2" pipe size		8	1		695	60		755	855
1060	2-1/2" pipe size	↓	7	1.143		1,050	68.50		1,118.50	1,275
2000	Flange connections, 3" pipe size	Q-5	5	3.200		1,275	172		1,447	1,675
2020	4" pipe size		4	4		1,525	215		1,740	2,000
2030	5" pipe size		3.50	4.571		1,900	246		2,146	2,475
2040	6" pipe size		3	5.333		2,025	287		2,312	2,700
2060	8" pipe size	↓	2	8	↓	2,475	430		2,905	3,400

23 21 20.88 Venturi Flow		Crew	Daily Output	Labor-Hours	Unit	Material	Labor	Equipment	Total	Total Incl O&P
0010	**VENTURI FLOW**, Measuring device									
0050	1/2" diameter	1 Stpi	24	.333	Ea.	281	19.90		300.90	340
0100	3/4" diameter		20	.400		257	24		281	320
0120	1" diameter		19	.421		276	25		301	345
0140	1-1/4" diameter		15	.533		340	32		372	425
0160	1-1/2" diameter		13	.615		355	37		392	450
0180	2" diameter	↓	11	.727	↓	365	43.50		408.50	475

23 21 20 – Hydronic HVAC Piping Specialties

23 21 20.88 Venturi Flow		Crew	Daily Output	Labor-Hours	Unit	Material	Labor (2015 Bare Costs)	Equipment	Total	Total Incl O&P
0200	2-1/2" diameter	Q-5	16	1	Ea.	500	54		554	635
0220	3" diameter		14	1.143		515	61.50		576.50	665
0240	4" diameter		11	1.455		775	78		853	970
0260	5" diameter	Q-6	4	6		1,025	335		1,360	1,650
0280	6" diameter		3.50	6.857		1,125	380		1,505	1,850
0300	8" diameter		3	8		1,450	445		1,895	2,300
0320	10" diameter		2	12		3,425	670		4,095	4,825
0500	For meter, add					2,125			2,125	2,350

23 21 20.94 Weld End Ball Joints		Crew	Daily Output	Labor-Hours	Unit	Material	Labor	Equipment	Total	Total Incl O&P
0010	**WELD END BALL JOINTS**, Steel									
0050	2-1/2" diameter	Q-17	13	1.231	Ea.	615	66	4.44	685.44	790
0100	3" diameter		12	1.333		720	71.50	4.81	796.31	905
0120	4" diameter		11	1.455		1,175	78	5.25	1,258.25	1,400
0140	5" diameter	Q-18	14	1.714		1,350	95.50	4.11	1,449.61	1,625
0160	6" diameter		12	2		2,000	112	4.80	2,116.80	2,375
0180	8" diameter		9	2.667		3,025	149	6.40	3,180.40	3,575
0200	10" diameter		8	3		4,075	167	7.20	4,249.20	4,750
0220	12" diameter		6	4		6,500	223	9.60	6,732.60	7,500

23 21 23 – Hydronic Pumps

23 21 23.13 In-Line Centrifugal Hydronic Pumps		Crew	Daily Output	Labor-Hours	Unit	Material	Labor	Equipment	Total	Total Incl O&P
0010	**IN-LINE CENTRIFUGAL HYDRONIC PUMPS**									
0600	Bronze, sweat connections, 1/40 HP, in line									
0640	3/4" size	Q-1	16	1	Ea.	218	53		271	325
1000	Flange connection, 3/4" to 1-1/2" size									
1040	1/12 HP	Q-1	6	2.667	Ea.	565	141		706	840
1060	1/8 HP		6	2.667		970	141		1,111	1,300
1100	1/3 HP		6	2.667		1,075	141		1,216	1,425
1140	2" size, 1/6 HP		5	3.200		1,400	169		1,569	1,800
1180	2-1/2" size, 1/4 HP		5	3.200		1,775	169		1,944	2,225
1220	3" size, 1/4 HP		4	4		1,850	211		2,061	2,375
1260	1/3 HP		4	4		2,275	211		2,486	2,825
1300	1/2 HP		4	4		2,325	211		2,536	2,875
1340	3/4 HP		4	4		2,550	211		2,761	3,125
1380	1 HP		4	4		4,100	211		4,311	4,825
2000	Cast iron, flange connection									
2040	3/4" to 1-1/2" size, in line, 1/12 HP	Q-1	6	2.667	Ea.	365	141		506	620
2060	1/8 HP		6	2.667		610	141		751	890
2100	1/3 HP		6	2.667		680	141		821	965
2140	2" size, 1/6 HP		5	3.200		745	169		914	1,075
2180	2-1/2" size, 1/4 HP		5	3.200		960	169		1,129	1,325
2220	3" size, 1/4 HP		4	4		975	211		1,186	1,400
2260	1/3 HP		4	4		1,325	211		1,536	1,775
2300	1/2 HP		4	4		1,375	211		1,586	1,825
2340	3/4 HP		4	4		1,575	211		1,786	2,075
2380	1 HP		4	4		2,275	211		2,486	2,825
2600	For nonferrous impeller, add					3%				
3000	High head, bronze impeller									
3030	1-1/2" size 1/2 HP	Q-1	5	3.200	Ea.	1,150	169		1,319	1,550
3040	1-1/2" size 3/4 HP		5	3.200		1,250	169		1,419	1,650
3050	2" size 1 HP		4	4		1,525	211		1,736	2,000
3090	2" size 1-1/2 HP		4	4		1,900	211		2,111	2,400
4000	Close coupled, end suction, bronze impeller									

23 21 Hydronic Piping and Pumps

23 21 23 – Hydronic Pumps

23 21 23.13 In-Line Centrifugal Hydronic Pumps		Crew	Daily Output	Labor-Hours	Unit	Material	2015 Bare Costs Labor	Equipment	Total	Total Incl O&P
4040	1-1/2" size, 1-1/2 HP, to 40 GPM	Q-1	3	5.333	Ea.	2,250	282		2,532	2,925
4090	2" size, 2 HP, to 50 GPM		3	5.333		2,675	282		2,957	3,375
4100	2" size, 3 HP, to 90 GPM		2.30	6.957		2,775	370		3,145	3,625
4190	2-1/2" size, 3 HP, to 150 GPM		2	8		3,000	425		3,425	3,950
4300	3" size, 5 HP, to 225 GPM		1.80	8.889		3,425	470		3,895	4,500
4410	3" size, 10 HP, to 350 GPM		1.60	10		5,025	530		5,555	6,350
4420	4" size, 7-1/2 HP, to 350 GPM		1.60	10		5,075	530		5,605	6,400
4520	4" size, 10 HP, to 600 GPM	Q-2	1.70	14.118		5,125	775		5,900	6,825
4530	5" size, 15 HP, to 1000 GPM		1.70	14.118		5,125	775		5,900	6,850
4610	5" size, 20 HP, to 1350 GPM		1.50	16		5,450	875		6,325	7,375
4620	5" size, 25 HP, to 1550 GPM		1.50	16		7,450	875		8,325	9,575
5000	Base mounted, bronze impeller, coupling guard									
5040	1-1/2" size, 1-1/2 HP, to 40 GPM	Q-1	2.30	6.957	Ea.	5,950	370		6,320	7,125
5090	2" size, 2 HP, to 50 GPM		2.30	6.957		6,675	370		7,045	7,925
5100	2" size, 3 HP, to 90 GPM		2	8		7,325	425		7,750	8,725
5190	2-1/2" size, 3 HP, to 150 GPM		1.80	8.889		7,900	470		8,370	9,400
5300	3" size, 5 HP, to 225 GPM		1.60	10		9,050	530		9,580	10,800
5410	4" size, 5 HP, to 350 GPM		1.50	10.667		9,850	565		10,415	11,700
5420	4" size, 7-1/2 HP, to 350 GPM		1.50	10.667		11,100	565		11,665	13,100
5520	5" size, 10 HP, to 600 GPM	Q-2	1.60	15		14,400	820		15,220	17,200
5530	5" size, 15 HP, to 1000 GPM		1.60	15		15,900	820		16,720	18,800
5610	6" size, 20 HP, to 1350 GPM		1.40	17.143		17,600	940		18,540	20,800
5620	6" size, 25 HP, to 1550 GPM		1.40	17.143		19,700	940		20,640	23,200
9000	Minimum labor/equipment charge	Q-1	3.25	4.923	Job		260		260	405

23 21 29 – Automatic Condensate Pump Units

23 21 29.10 Condensate Removal Pump System		Crew	Daily Output	Labor-Hours	Unit	Material	Labor	Equipment	Total	Total Incl O&P
0010	**CONDENSATE REMOVAL PUMP SYSTEM**									
0020	Pump with 1 gal. ABS tank									
0100	115 V									
0120	1/50 HP, 200 GPH G	1 Stpi	12	.667	Ea.	197	40		237	279
0140	1/18 HP, 270 GPH G		10	.800		210	48		258	305
0160	1/5 HP, 450 GPH G		8	1		470	60		530	610
0200	230 V									
0260	1/5 HP, 450 GPH G	1 Stpi	8	1	Ea.	520	60		580	665

23 22 Steam and Condensate Piping and Pumps

23 22 13 – Steam and Condensate Heating Piping

23 22 13.23 Aboveground Steam and Condensate Piping		Crew	Daily Output	Labor-Hours	Unit	Material	Labor	Equipment	Total	Total Incl O&P
0010	**ABOVEGROUND STEAM AND CONDENSATE HEATING PIPING**									
0020	Condensate meter									
0100	500 lb. per hour	1 Stpi	14	.571	Ea.	3,275	34		3,309	3,650
0140	1500 lb. per hour		7	1.143		4,100	68.50		4,168.50	4,600
0160	3000 lb. per hour		5	1.600		4,525	95.50		4,620.50	5,125
0200	12,000 lb. per hour	Q-5	3.50	4.571		5,400	246		5,646	6,300

23 22 Steam and Condensate Piping and Pumps

23 22 23 – Steam Condensate Pumps

23 22 23.10 Condensate Return System		Crew	Daily Output	Labor-Hours	Unit	2015 Bare Costs Material	Labor	Equipment	Total	Total Incl O&P
0010	**CONDENSATE RETURN SYSTEM**									
2000	Simplex									
2010	With pump, motor, CI receiver, float switch									
2020	3/4 HP, 15 GPM	Q-1	1.80	8.889	Ea.	6,425	470		6,895	7,775
2100	Duplex									
2110	With 2 pumps and motors, CI receiver, float switch, alternator									
2120	3/4 HP, 15 GPM, 15 gal. CI rcvr.	Q-1	1.40	11.429	Ea.	7,000	605		7,605	8,650
2130	1 HP, 25 GPM		1.20	13.333		8,400	705		9,105	10,400
2140	1-1/2 HP, 45 GPM		1	16		9,750	845		10,595	12,000
2150	1-1/2 HP, 60 GPM		1	16		10,900	845		11,745	13,300

23 23 Refrigerant Piping

23 23 13 – Refrigerant Piping Valves

23 23 13.10 Valves		Crew	Daily Output	Labor-Hours	Unit	Material	Labor	Equipment	Total	Total Incl O&P
0010	**VALVES**									
8100	Check valve, soldered									
8110	5/8"	1 Stpi	36	.222	Ea.	87	13.30		100.30	117
8114	7/8"		26	.308		93	18.40		111.40	131
8118	1-1/8"		18	.444		124	26.50		150.50	178
8122	1-3/8"		14	.571		199	34		233	273
8126	1-5/8"		13	.615		243	37		280	325
8130	2-1/8"		12	.667		284	40		324	370
8134	2-5/8"	Q-5	22	.727		405	39		444	505
8138	3-1/8"	"	20	.800		540	43		583	655
8500	Refrigeration valve, packless, soldered									
8510	1/2"	1 Stpi	38	.211	Ea.	46.50	12.60		59.10	70.50
8514	5/8"		36	.222		46.50	13.30		59.80	71.50
8518	7/8"		26	.308		113	18.40		131.40	153
8520	Packed, soldered									
8522	1-1/8"	1 Stpi	18	.444	Ea.	161	26.50		187.50	219
8526	1-3/8"		14	.571		284	34		318	365
8530	1-5/8"		13	.615		315	37		352	405
8534	2-1/8"		12	.667		500	40		540	610
8538	2-5/8"	Q-5	22	.727		700	39		739	825
8542	3-1/8"		20	.800		785	43		828	930
8546	4-1/8"		18	.889		1,375	48		1,423	1,600
8600	Solenoid valve, flange/solder									
8610	1/2"	1 Stpi	38	.211	Ea.	186	12.60		198.60	225
8614	5/8"		36	.222		253	13.30		266.30	299
8618	3/4"		30	.267		297	15.95		312.95	350
8622	7/8"		26	.308		365	18.40		383.40	430
8626	1-1/8"		18	.444		475	26.50		501.50	560
8630	1-3/8"		14	.571		590	34		624	705
8634	1-5/8"		13	.615		825	37		862	965
8638	2-1/8"		12	.667		900	40		940	1,050
8800	Thermostatic valve, flange/solder									
8810	1/2 – 3 ton, 3/8" x 5/8"	1 Stpi	9	.889	Ea.	182	53		235	283
8814	4 – 5 ton, 1/2" x 7/8"		7	1.143		213	68.50		281.50	340
8818	6 – 8 ton, 5/8" x 7/8"		5	1.600		213	95.50		308.50	385
8822	7 – 12 ton, 7/8" x 1-1/8"		4	2		249	120		369	460

23 23 Refrigerant Piping

23 23 13 – Refrigerant Piping Valves

23 23 13.10 Valves		Crew	Daily Output	Labor-Hours	Unit	Material	Labor	Equipment	Total	Total Incl O&P
						2015 Bare Costs				
8826	15 – 20 ton, 7/8" x 1-3/8"	1 Stpi	3.20	2.500	Ea.	249	149		398	505

23 23 16 – Refrigerant Piping Specialties

23 23 16.10 Refrigerant Piping Component Specialties

		Crew	Daily Output	Labor-Hours	Unit	Material	Labor	Equipment	Total	Total Incl O&P
0010	REFRIGERANT PIPING COMPONENT SPECIALTIES									
0600	Accumulator									
0610	3/4"	1 Stpi	8.80	.909	Ea.	70.50	54.50		125	162
0614	7/8"		6.40	1.250		80.50	74.50		155	206
0618	1-1/8"		4.80	1.667		111	99.50		210.50	277
0622	1-3/8"		4	2		145	120		265	345
0626	1-5/8"		3.20	2.500		157	149		306	405
0630	2-1/8"		2.40	3.333		375	199		574	725
0700	Condensate drip/drain pan, 10" x 9" x 2-3/4"		24	.333		143	19.90		162.90	189
1000	Filter dryer									
1010	Replaceable core type, solder									
1020	1/2"	1 Stpi	20	.400	Ea.	96.50	24		120.50	144
1030	5/8"		19	.421		96.50	25		121.50	145
1040	7/8"		18	.444		127	26.50		153.50	181
1050	1-1/8"		15	.533		134	32		166	197
1060	1-3/8"		14	.571		136	34		170	203
1070	1-5/8"		12	.667		133	40		173	208
1080	2-1/8"		10	.800		286	48		334	390
1090	2-5/8"		9	.889		575	53		628	715
1100	3-1/8"		8	1		620	60		680	775
1200	Sealed in-line, solder									
1210	1/4"	1 Stpi	22	.364	Ea.	30.50	21.50		52	67.50
1220	3/8"		21	.381		25.50	23		48.50	63.50
1230	1/2"		20	.400		30.50	24		54.50	71
1260	5/8"		19	.421		43.50	25		68.50	86.50
1270	7/8"		18	.444		45	26.50		71.50	91
1290	1-1/8"		15	.533		68.50	32		100.50	125
4000	P-Trap, suction line, solder									
4010	5/8"	1 Stpi	19	.421	Ea.	33.50	25		58.50	75.50
4020	3/4"		19	.421		49.50	25		74.50	93.50
4030	7/8"		18	.444		37	26.50		63.50	82.50
4040	1-1/8"		15	.533		56.50	32		88.50	112
4050	1-3/8"		14	.571		102	34		136	166
4060	1-5/8"		12	.667		214	40		254	298
4070	2-1/8"		10	.800		440	48		488	555
5000	Sightglass									
5010	Moisture and liquid indicator, solder									
5020	1/4"	1 Stpi	22	.364	Ea.	18	21.50		39.50	54
5030	3/8"		21	.381		18.15	23		41.15	55.50
5040	1/2"		20	.400		23	24		47	62.50
5050	5/8"		19	.421		23.50	25		48.50	65
5060	7/8"		18	.444		35	26.50		61.50	80
5070	1-1/8"		15	.533		38	32		70	91
5080	1-3/8"		14	.571		66.50	34		100.50	127
5090	1-5/8"		12	.667		75	40		115	145
5100	2-1/8"		10	.800		94.50	48		142.50	179
7400	Vacuum pump set									
7410	Two stage, high vacuum continuous duty	1 Stpi	8	1	Ea.	3,875	60		3,935	4,375

23 23 Refrigerant Piping

23 23 16 – Refrigerant Piping Specialties

23 23 16.16 Refrigerant Line Sets		Crew	Daily Output	Labor-Hours	Unit	2015 Bare Costs Material	Labor	Equipment	Total	Total Incl O&P
0010	**REFRIGERANT LINE SETS**									
0100	Copper tube									
0110	1/2" insulation, both tubes									
0120	Combination 1/4" and 1/2" tubes									
0130	10' set	Q-5	42	.381	Ea.	70.50	20.50		91	110
0140	20' set		40	.400		110	21.50		131.50	155
0150	30' set		37	.432		148	23.50		171.50	199
0160	40' set		35	.457		193	24.50		217.50	252
0170	50' set		32	.500		229	27		256	294
0180	100' set		22	.727		520	39		559	635
0300	Combination 1/4" and 3/4" tubes									
0310	10' set	Q-5	40	.400	Ea.	78.50	21.50		100	120
0320	20' set		38	.421		139	22.50		161.50	189
0330	30' set		35	.457		208	24.50		232.50	268
0340	40' set		33	.485		272	26		298	340
0350	50' set		30	.533		345	28.50		373.50	420
0380	100' set		20	.800		780	43		823	925
0500	Combination 3/8" & 3/4" tubes									
0510	10' set	Q-5	28	.571	Ea.	90.50	30.50		121	148
0520	20' set		36	.444		141	24		165	193
0530	30' set		34	.471		191	25.50		216.50	251
0540	40' set		31	.516		252	28		280	320
0550	50' set		28	.571		296	30.50		326.50	375
0580	100' set		18	.889		875	48		923	1,025
0700	Combination 3/8" & 1-1/8" tubes									
0710	10' set	Q-5	36	.444	Ea.	158	24		182	212
0720	20' set		33	.485		265	26		291	330
0730	30' set		31	.516		355	28		383	435
0740	40' set		28	.571		535	30.50		565.50	635
0750	50' set		26	.615		615	33		648	730
0900	Combination 1/2" & 3/4" tubes									
0910	10' set	Q-5	37	.432	Ea.	95.50	23.50		119	142
0920	20' set		35	.457		169	24.50		193.50	225
0930	30' set		33	.485		255	26		281	320
0940	40' set		30	.533		335	28.50		363.50	415
0950	50' set		27	.593		420	32		452	515
0980	100' set		17	.941		950	50.50		1,000.50	1,125
2100	Combination 1/2" & 1-1/8" tubes									
2110	10' set	Q-5	35	.457	Ea.	163	24.50		187.50	218
2120	20' set		31	.516		279	28		307	350
2130	30' set		29	.552		420	29.50		449.50	505
2140	40' set		25	.640		565	34.50		599.50	675
2150	50' set		14	1.143		625	61.50		686.50	780
2300	For 1" thick insulation add					30%	15%			

23 23 23 – Refrigerants

23 23 23.10 Anti-Freeze		Crew	Daily Output	Labor-Hours	Unit	Material	Labor	Equipment	Total	Total Incl O&P
0010	**ANTI-FREEZE**, Inhibited									
0900	Ethylene glycol concentrated									
1000	55 gallon drums, small quantities				Gal.	11.95			11.95	13.10
1200	Large quantities					9.40			9.40	10.35
2000	Propylene glycol, for solar heat, small quantities					15.55			15.55	17.10
2100	Large quantities					10.35			10.35	11.35

23 23 Refrigerant Piping

23 23 23 – Refrigerants

23 23 23.20 Refrigerant		Crew	Daily Output	Labor-Hours	Unit	2015 Bare Costs Material	Labor	Equipment	Total	Total Incl O&P
0010	**REFRIGERANT**									
4420	R-22, 30 lb. disposable cylinder				Lb.	35			35	38.50
4428	R-134A, 30 lb. disposable cylinder					13.45			13.45	14.80
4434	R-407C, 30 lb. disposable cylinder					28			28	30.50
4440	R-408A, 25 lb. disposable cylinder					31			31	34
4450	R-410A, 25 lb. disposable cylinder					22.50			22.50	24.50
4470	R-507, 25 lb. disposable cylinder				↓	17.75			17.75	19.55

23 31 HVAC Ducts and Casings

23 31 13 – Metal Ducts

23 31 13.13 Rectangular Metal Ducts		Crew	Daily Output	Labor-Hours	Unit	Material	Labor	Equipment	Total	Total Incl O&P
0010	**RECTANGULAR METAL DUCTS** R233100-40									
0020	Fabricated rectangular, includes fittings, joints, supports,									
0021	allowance for flexible connections and field sketches.									
0030	Does not include "as-built dwgs." or insulation.									
0031	NOTE: Fabrication and installation are combined									
0040	as LABOR cost. Approx. 25% fittings assumed.									
0042	Fabrication/Inst. is to commercial quality standards									
0043	(SMACNA or equiv.) for structure, sealing, leak testing, etc.									
0050	Add to labor for elevated installation									
0051	of fabricated ductwork									
0052	10′ to 15′ high						6%			
0053	15′ to 20′ high						12%			
0054	20′ to 25′ high						15%			
0055	25′ to 30′ high						21%			
0056	30′ to 35′ high						24%			
0057	35′ to 40′ high						30%			
0058	Over 40′ high						33%			
0072	For duct insulation see Line 23 07 13.10 3000									
0100	Aluminum, alloy 3003-H14, under 100 lb.	Q-10	75	.320	Lb.	3.20	16.70		19.90	30
0110	100 to 500 lb.		80	.300		1.88	15.65		17.53	26.50
0120	500 to 1,000 lb.		95	.253		1.82	13.20		15.02	23
0140	1,000 to 2,000 lb.		120	.200		1.77	10.45		12.22	18.45
0150	2,000 to 5,000 lb.		130	.185		1.77	9.65		11.42	17.15
0160	Over 5,000 lb.		145	.166		1.77	8.65		10.42	15.60
0500	Galvanized steel, under 200 lb.		235	.102		.65	5.35		6	9.10
0520	200 to 500 lb.		245	.098		.64	5.10		5.74	8.75
0540	500 to 1,000 lb.		255	.094		.62	4.91		5.53	8.45
0560	1,000 to 2,000 lb.		265	.091		.61	4.73		5.34	8.10
0570	2,000 to 5,000 lb.		275	.087		.61	4.56		5.17	7.85
0580	Over 5,000 lb.	↓	285	.084	↓	.61	4.40		5.01	7.60
0600	For large quantities special prices available from supplier									
1000	Stainless steel, type 304, under 100 lb.	Q-10	165	.145	Lb.	6.35	7.60		13.95	18.95
1020	100 to 500 lb.		175	.137		4.05	7.15		11.20	15.75
1030	500 to 1,000 lb.		190	.126		2.95	6.60		9.55	13.65
1040	1,000 to 2,000 lb.		200	.120		2.89	6.25		9.14	13.10
1050	2,000 to 5,000 lb.		225	.107		2.39	5.55		7.94	11.45
1060	Over 5,000 lb.	↓	235	.102	↓	1.96	5.35		7.31	10.55
1080	Note: Minimum order cost exceeds per lb. cost for min. wt.									
1100	For medium pressure ductwork, add				Lb.		15%			

23 31 13.13 Rectangular Metal Ducts		Crew	Daily Output	Labor-Hours	Unit	Material	2015 Bare Costs Labor	Equipment	Total	Total Incl O&P
1200	For high pressure ductwork, add				Lb.		40%			
1210	For welded ductwork, add						85%			
1220	For 30% fittings, add						11%			
1224	For 40% fittings, add						34%			
1228	For 50% fittings, add						56%			
1232	For 60% fittings, add						79%			
1236	For 70% fittings, add						101%			
1240	For 80% fittings, add						124%			
1244	For 90% fittings, add						147%			
1248	For 100% fittings, add						169%			
1252	Note: Fittings add includes time for detailing and installation.									

23 31 13.16 Round and Flat-Oval Spiral Ducts

		Crew	Daily Output	Labor-Hours	Unit	Material	Labor	Equipment	Total	Total Incl O&P
0010	**ROUND AND FLAT-OVAL SPIRAL DUCTS**									
1280	Add to labor for elevated installation									
1282	of prefabricated (purchased) ductwork									
1283	10′ to 15′ high						10%			
1284	15′ to 20′ high						20%			
1285	20′ to 25′ high						25%			
1286	25′ to 30′ high						35%			
1287	30′ to 35′ high						40%			
1288	35′ to 40′ high						50%			
1289	Over 40′ high						55%			
5400	Spiral preformed, steel, galv., straight lengths, Max 10″ spwg.									
5410	4″ diameter, 26 ga.	Q-9	360	.044	L.F.	1.70	2.24		3.94	5.40
5416	5″ diameter, 26 ga.		320	.050		1.82	2.52		4.34	5.95
5420	6″ diameter, 26 ga.		280	.057		1.82	2.88		4.70	6.55
5425	7″ diameter, 26 ga.		240	.067		2.12	3.36		5.48	7.65
5430	8″ diameter, 26 ga.		200	.080		2.42	4.03		6.45	9
5440	10″ diameter, 26 ga.		160	.100		3.02	5.05		8.07	11.25
5450	12″ diameter, 26 ga.		120	.133		3.63	6.70		10.33	14.60
5460	14″ diameter, 26 ga.		80	.200		4.23	10.05		14.28	20.50
5480	16″ diameter, 24 ga.		60	.267		5.65	13.45		19.10	27.50
5490	18″ diameter, 24 ga.		50	.320		6.35	16.10		22.45	32.50
5500	20″ diameter, 24 ga.	Q-10	65	.369		7.10	19.30		26.40	38.50
5510	22″ diameter, 24 ga.		60	.400		7.80	21		28.80	41.50
5520	24″ diameter, 24 ga.		55	.436		8.50	23		31.50	45.50
5540	30″ diameter, 22 ga.		45	.533		12.40	28		40.40	57.50
5600	36″ diameter, 22 ga.		40	.600		14.90	31.50		46.40	66
5800	Connector, 4″ diameter	Q-9	100	.160	Ea.	3.20	8.05		11.25	16.20
5810	5″ diameter		94	.170		3.20	8.55		11.75	17
5820	6″ diameter		88	.182		3.20	9.15		12.35	17.95
5840	8″ diameter		78	.205		3.60	10.35		13.95	20.50
5860	10″ diameter		70	.229		4.20	11.50		15.70	23
5880	12″ diameter		50	.320		4.70	16.10		20.80	30.50
5900	14″ diameter		44	.364		5.05	18.30		23.35	34.50
5920	16″ diameter		40	.400		5.50	20		25.50	38
5930	18″ diameter		37	.432		5.85	22		27.85	41
5940	20″ diameter		34	.471		6.65	23.50		30.15	45
5950	22″ diameter		31	.516		7.80	26		33.80	49.50
5960	24″ diameter		28	.571		8.25	29		37.25	54.50
5980	30″ diameter		22	.727		14.70	36.50		51.20	74
6000	36″ diameter		18	.889		17.90	45		62.90	90

23 31 13.16 Round and Flat-Oval Spiral Ducts		Crew	Daily Output	Labor-Hours	Unit	2015 Bare Costs Material	Labor	Equipment	Total	Total Incl O&P
6300	Elbow, 45°, 4" diameter	Q-9	60	.267	Ea.	6.05	13.45		19.50	27.50
6310	5" diameter		52	.308		6.70	15.50		22.20	32
6320	6" diameter		44	.364		6.70	18.30		25	36.50
6340	8" diameter		28	.571		8.10	29		37.10	54.50
6360	10" diameter		18	.889		8.50	45		53.50	80
6380	12" diameter		13	1.231		9.20	62		71.20	108
6400	14" diameter		11	1.455		12.75	73		85.75	130
6420	16" diameter		10	1.600		20	80.50		100.50	149
6430	18" diameter		9.60	1.667		50	84		134	187
6440	20" diameter	Q-10	14	1.714		55	89.50		144.50	202
6450	22" diameter		13	1.846		65	96.50		161.50	224
6460	24" diameter		12	2		75	104		179	248
6480	30" diameter		9	2.667		108	139		247	340
6500	36" diameter		7	3.429		132	179		311	430
6600	Elbow, 90°, 4" diameter	Q-9	60	.267		4.60	13.45		18.05	26
6610	5" diameter		52	.308		4.60	15.50		20.10	29.50
6620	6" diameter		44	.364		4.60	18.30		22.90	34
6625	7" diameter		36	.444		5.15	22.50		27.65	41
6630	8" diameter		28	.571		5.65	29		34.65	51.50
6640	10" diameter		18	.889		7.40	45		52.40	78.50
6650	12" diameter		13	1.231		9.25	62		71.25	108
6660	14" diameter		11	1.455		15.85	73		88.85	133
6670	16" diameter		10	1.600		18	80.50		98.50	147
6676	18" diameter		9.60	1.667		65	84		149	204
6680	20" diameter	Q-10	14	1.714		70	89.50		159.50	218
6684	22" diameter		13	1.846		75	96.50		171.50	235
6690	24" diameter		12	2		80	104		184	253
6700	30" diameter		9	2.667		144	139		283	380
6710	36" diameter		7	3.429		223	179		402	530
6800	Reducing coupling, 6" x 4"	Q-9	46	.348		18.80	17.50		36.30	48
6820	8" x 6"		40	.400		18.80	20		38.80	52.50
6840	10" x 8"		32	.500		20.50	25		45.50	62
6860	12" x 10"		24	.667		23.50	33.50		57	79
6880	14" x 12"		20	.800		25	40.50		65.50	91
6900	16" x 14"		18	.889		32	45		77	106
6920	18" x 16"		16	1		39	50.50		89.50	123
6940	20" x 18"	Q-10	24	1		45.50	52		97.50	133
6950	22" x 20"		23	1.043		45	54.50		99.50	136
6960	24" x 22"		22	1.091		58.50	57		115.50	155
6980	30" x 28"		18	1.333		112	69.50		181.50	234
7000	36" x 34"		16	1.500		143	78.50		221.50	281
7100	Tee, 90°, 4" diameter	Q-9	40	.400		12.65	20		32.65	46
7110	5" diameter		34.60	.462		19	23.50		42.50	57.50
7120	6" diameter		30	.533		19	27		46	63.50
7130	8" diameter		19	.842		24.50	42.50		67	94
7140	10" diameter		12	1.333		34	67		101	144
7150	12" diameter		9	1.778		41	89.50		130.50	186
7160	14" diameter		7	2.286		62.50	115		177.50	251
7170	16" diameter		6.70	2.388		72	120		192	270
7176	18" diameter		6	2.667		105	134		239	330
7180	20" diameter	Q-10	8.70	2.759		125	144		269	365
7190	24" diameter		7	3.429		196	179		375	500
7200	30" diameter		6	4		239	209		448	595

23 31 13.16 Round and Flat-Oval Spiral Ducts		Crew	Daily Output	Labor-Hours	Unit	Material	2015 Bare Costs Labor	Equipment	Total	Total Incl O&P
7210	36″ diameter	Q-10	5	4.800	Ea.	305	251		556	730
7250	Saddle Tap									
7254	4″ dia. on 6″ dia.	Q-9	36	.444	Ea.	25	22.50		47.50	62.50
7256	5″ dia on 7″ dia		35.80	.447		25	22.50		47.50	62.50
7258	6″ dia. on 8″ dia.		35.40	.452		22	23		45	60
7260	6″ dia. on 10″ dia.		35.20	.455		22	23		45	60
7262	6″ dia. on 12″ dia.		35	.457		22.50	23		45.50	61.50
7264	7″ dia. on 9″ dia.		35.20	.455		25.50	23		48.50	64
7266	8″ dia. on 10″ dia.		35	.457		25.50	23		48.50	64.50
7268	8″ dia. on 12″ dia.		34.80	.460		22.50	23		45.50	61.50
7270	8″ dia. on 14″ dia.		34.50	.464		28.50	23.50		52	68.50
7272	10″ dia. on 12″ dia.		34.20	.468		26	23.50		49.50	65.50
7274	10″ dia. on 14″ dia.		33.80	.473		28.50	24		52.50	69
7276	10″ dia. on 16″ dia.		33.40	.479		28.50	24		52.50	69.50
7278	10″ dia. on 18″ dia.		33	.485		28.50	24.50		53	70
7280	12″ dia. on 14″ dia.		32.60	.491		33	24.50		57.50	75.50
7282	12″ dia. on 18″ dia.		32.20	.497		36.50	25		61.50	79.50
7284	14″ dia. on 16″ dia.		31.60	.506		38	25.50		63.50	82
7286	14″ dia. on 20″ dia.		31	.516		38.50	26		64.50	83.50
7288	16″ dia. on 20″ dia.		30	.533		45	27		72	92
7290	18″ dia. on 20″ dia.		28	.571		52.50	29		81.50	103
7292	20″ dia. on 22″ dia.		26	.615		69.50	31		100.50	125
7294	22″ dia. on 24″ dia.		24	.667		74.50	33.50		108	135
7400	Steel, PVC coated both sides, straight lengths									
7410	4″ diameter, 26 ga.	Q-9	360	.044	L.F.	4.97	2.24		7.21	9
7416	5″ diameter, 26 ga.		240	.067		3.60	3.36		6.96	9.25
7420	6″ diameter, 26 ga.		280	.057		3.60	2.88		6.48	8.50
7440	8″ diameter, 26 ga.		200	.080		4.90	4.03		8.93	11.75
7460	10″ diameter, 26 ga.		160	.100		6.10	5.05		11.15	14.65
7480	12″ diameter, 26 ga.		120	.133		7.30	6.70		14	18.65
7500	14″ diameter, 26 ga.		80	.200		8.50	10.05		18.55	25.50
7520	16″ diameter, 24 ga.		60	.267		9.70	13.45		23.15	31.50
7540	18″ diameter, 24 ga.		45	.356		10.90	17.90		28.80	40.50
7560	20″ diameter, 24 ga.	Q-10	65	.369		12.10	19.30		31.40	44
7580	24″ diameter, 24 ga.		55	.436		14.50	23		37.50	52
7600	30″ diameter, 22 ga.		45	.533		22	28		50	68
7620	36″ diameter, 22 ga.		40	.600		26	31.50		57.50	78.50
7890	Connector, 4″ diameter	Q-9	100	.160	Ea.	2.60	8.05		10.65	15.55
7896	5″ diameter		83	.193		2.90	9.70		12.60	18.50
7900	6″ diameter		88	.182		3.30	9.15		12.45	18.10
7920	8″ diameter		78	.205		3.90	10.35		14.25	20.50
7940	10″ diameter		70	.229		4.50	11.50		16	23
7960	12″ diameter		50	.320		5.30	16.10		21.40	31.50
7980	14″ diameter		44	.364		5.60	18.30		23.90	35
8000	16″ diameter		40	.400		6.10	20		26.10	38.50
8020	18″ diameter		37	.432		6.80	22		28.80	42
8040	20″ diameter		34	.471		8.90	23.50		32.40	47.50
8060	24″ diameter		28	.571		11	29		40	57.50
8080	30″ diameter		22	.727		14.70	36.50		51.20	74
8100	36″ diameter		18	.889		17.90	45		62.90	90
8390	Elbow, 45°, 4″ diameter		60	.267		23	13.45		36.45	46.50
8396	5″ diameter		36	.444		23	22.50		45.50	61
8400	6″ diameter		44	.364		23	18.30		41.30	54.50

23 31 13 – Metal Ducts

23 31 13.16 Round and Flat-Oval Spiral Ducts		Crew	Daily Output	Labor-Hours	Unit	2015 Bare Costs Material	Labor	Equipment	Total	Total Incl O&P
8420	8″ diameter	Q-9	28	.571	Ea.	23	29		52	71
8440	10″ diameter		18	.889		23	45		68	96
8460	12″ diameter		13	1.231		23	62		85	124
8480	14″ diameter		11	1.455		23	73		96	142
8500	16″ diameter		10	1.600		23	80.50		103.50	153
8520	18″ diameter		9	1.778		25	89.50		114.50	169
8540	20″ diameter	Q-10	13	1.846		75.50	96.50		172	236
8560	24″ diameter		11	2.182		97.50	114		211.50	287
8580	30″ diameter		9	2.667		129	139		268	360
8600	36″ diameter		7	3.429		157	179		336	455
8800	Elbow, 90°, 4″ diameter	Q-9	60	.267		25	13.45		38.45	48.50
8804	5″ diameter		52	.308		25	15.50		40.50	52
8806	6″ diameter		44	.364		25	18.30		43.30	56.50
8808	8″ diameter		28	.571		25	29		54	73
8810	10″ diameter		18	.889		25	45		70	98
8812	12″ diameter		13	1.231		25	62		87	126
8814	14″ diameter		11	1.455		25	73		98	144
8816	16″ diameter		10	1.600		29	80.50		109.50	159
8818	18″ diameter		9.60	1.667		35.50	84		119.50	171
8820	20″ diameter	Q-10	14	1.714		45	89.50		134.50	191
8822	24″ diameter		12	2		104	104		208	279
8824	30″ diameter		9	2.667		175	139		314	410
8826	36″ diameter		7	3.429		285	179		464	600
9000	Reducing coupling									
9040	6″ x 4″	Q-9	46	.348	Ea.	19.50	17.50		37	49
9060	8″ x 6″		40	.400		19.50	20		39.50	53.50
9080	10″ x 8″		32	.500		20.50	25		45.50	62
9100	12″ x 10″		24	.667		23.50	33.50		57	79
9120	14″ x 12″		20	.800		25	40.50		65.50	91
9140	16″ x 14″		18	.889		32	45		77	106
9160	18″ x 16″		16	1		39	50.50		89.50	123
9180	20″ x 18″	Q-10	24	1		45.50	52		97.50	133
9200	24″ x 22″		22	1.091		58.50	57		115.50	155
9220	30″ x 28″		18	1.333		104	69.50		173.50	224
9240	36″ x 34″		16	1.500		143	78.50		221.50	281
9800	Steel,stainless, straight lengths									
9810	3″ diameter, 26 Ga.	Q-9	400	.040	L.F.	14.20	2.01		16.21	18.80
9820	4″ diameter, 26 Ga.		360	.044		16.45	2.24		18.69	21.50
9830	5″ diameter, 26 Ga.		320	.050		17.75	2.52		20.27	23.50
9840	6″ diameter, 26 Ga.		280	.057		17.75	2.88		20.63	24
9850	7″ diameter, 26 Ga.		240	.067		29.50	3.36		32.86	38
9860	8″ diameter, 26 Ga.		200	.080		23	4.03		27.03	32
9870	9″ diameter, 26 Ga.		180	.089		33.50	4.48		37.98	44
9880	10″ diameter, 26 Ga.		160	.100		43.50	5.05		48.55	56
9890	12″ diameter, 26 Ga.		120	.133		61.50	6.70		68.20	78.50
9900	14″ diameter, 26 Ga.		80	.200		65	10.05		75.05	87.50
9990	Minimum labor/equipment charge	1 Shee	3	2.667	Job		149		149	235

23 31 13.17 Round Grease Duct		Crew	Daily Output	Labor-Hours	Unit	Material	2015 Bare Costs Labor	Equipment	Total	Total Incl O&P
0010	**ROUND GREASE DUCT**									
1280	Add to labor for elevated installation									
1282	of prefabricated (purchased) ductwork									
1283	10′ to 15′ high						10%			
1284	15′ to 20′ high						20%			
1285	20′ to 25′ high						25%			
1286	25′ to 30′ high						35%			
1287	30′ to 35′ high						40%			
1288	35′ to 40′ high						50%			
1289	Over 40′ high						55%			
4020	Zero clearance, 2 hr. fire rated, UL listed, 3″ double wall									
4030	304 interior, aluminized exterior, all sizes are ID									
4032	For 316 interior and aluminized exterior, add					6%				
4034	For 304 interior and 304 exterior, add					21%				
4036	For 316 interior and 316 exterior, add					40%				
4040	Straight									
4052	6″	Q-9	34.20	.468	L.F.	74.50	23.50		98	119
4054	8″		29.64	.540		85.50	27		112.50	137
4056	10″		27.36	.585		95.50	29.50		125	152
4058	12″		25.10	.637		110	32		142	171
4060	14″		23.94	.668		123	33.50		156.50	188
4062	16″		22.80	.702		138	35.50		173.50	208
4064	18″		21.66	.739		157	37		194	231
4066	20″	Q-10	20.52	1.170		178	61		239	293
4068	22″		19.38	1.238		201	64.50		265.50	325
4070	24″		18.24	1.316		227	68.50		295.50	360
4072	26″		17.67	1.358		245	71		316	380
4074	28″		17.10	1.404		261	73.50		334.50	405
4076	30″		15.96	1.504		274	78.50		352.50	425
4078	32″		15.39	1.559		295	81.50		376.50	455
4080	36″		14.25	1.684		330	88		418	500
4100	Adjustable section, 30″ long									
4104	6″	Q-9	17.10	.936	Ea.	226	47		273	325
4106	8″		14.82	1.080		282	54.50		336.50	395
4108	10″		13.68	1.170		320	59		379	445
4110	12″		12.54	1.276		360	64		424	495
4112	14″		11.97	1.337		405	67.50		472.50	550
4114	16″		11.40	1.404		455	70.50		525.50	610
4116	18″		10.83	1.477		515	74.50		589.50	680
4118	20″	Q-10	10.26	2.339		585	122		707	840
4120	22″		9.69	2.477		660	129		789	930
4122	24″		9.12	2.632		750	137		887	1,050
4124	26″		8.83	2.718		805	142		947	1,100
4126	28″		8.55	2.807		860	147		1,007	1,175
4128	30″		8.26	2.906		915	152		1,067	1,250
4130	32″		7.98	3.008		970	157		1,127	1,325
4132	36″		6.84	3.509		1,075	183		1,258	1,500
4140	Tee, 90°, grease									
4144	6″	Q-9	13.68	1.170	Ea.	400	59		459	535
4146	8″		12.54	1.276		435	64		499	580
4148	10″		11.97	1.337		490	67.50		557.50	645
4150	12″		11.40	1.404		570	70.50		640.50	735

23 31 13.17 Round Grease Duct		Crew	Daily Output	Labor-Hours	Unit	Material	2015 Bare Costs Labor	Equipment	Total	Total Incl O&P
4152	14"	Q-9	10.26	1.559	Ea.	645	78.50		723.50	835
4154	16"		9.12	1.754		715	88.50		803.50	925
4156	18"		7.98	2.005		840	101		941	1,075
4158	20"		9.69	1.651		970	83		1,053	1,200
4160	22"		8.26	1.937		1,075	97.50		1,172.50	1,350
4162	24"		6.84	2.339		1,200	118		1,318	1,500
4164	26"		6.55	2.443		1,325	123		1,448	1,650
4166	28"	Q-10	6.27	3.828		1,450	200		1,650	1,925
4168	30"		5.98	4.013		1,725	210		1,935	2,225
4170	32"		5.70	4.211		2,025	220		2,245	2,575
4172	36"		5.13	4.678		2,250	244		2,494	2,850
4180	Cleanout Tee Cap									
4184	6"	Q-9	21.09	.759	Ea.	81.50	38		119.50	150
4186	8"		19.38	.826		86	41.50		127.50	160
4188	10"		18.24	.877		92.50	44		136.50	172
4190	12"		17.10	.936		99	47		146	184
4192	14"		15.96	1.003		105	50.50		155.50	195
4194	16"		14.25	1.123		116	56.50		172.50	216
4196	18"		13.68	1.170		122	59		181	227
4198	20"	Q-10	15.39	1.559		135	81.50		216.50	278
4200	22"		12.54	1.914		146	100		246	320
4202	24"		11.97	2.005		160	105		265	340
4204	26"		11.40	2.105		174	110		284	365
4206	28"		10.83	2.216		189	116		305	390
4208	30"		10.26	2.339		206	122		328	420
4210	32"		9.69	2.477		224	129		353	450
4212	36"		8.55	2.807		263	147		410	520
4220	Elbow, 45°									
4224	6"	Q-9	17.10	.936	Ea.	282	47		329	385
4226	8"		14.82	1.080		315	54.50		369.50	435
4228	10"		13.68	1.170		360	59		419	490
4230	12"		12.54	1.276		410	64		474	550
4232	14"		11.97	1.337		460	67.50		527.50	610
4234	16"		11.40	1.404		520	70.50		590.50	685
4236	18"		10.83	1.477		590	74.50		664.50	765
4238	20"	Q-10	10.26	2.339		670	122		792	930
4240	22"		9.69	2.477		745	129		874	1,025
4242	24"		9.12	2.632		855	137		992	1,150
4244	26"		8.83	2.718		905	142		1,047	1,225
4246	28"		8.55	2.807		1,075	147		1,222	1,400
4248	30"		8.26	2.906		1,075	152		1,227	1,425
4250	32"		7.98	3.008		1,300	157		1,457	1,675
4252	36"		6.84	3.509		1,725	183		1,908	2,200
4260	Elbow, 90°									
4264	6"	Q-9	17.10	.936	Ea.	565	47		612	695
4266	8"		14.82	1.080		635	54.50		689.50	785
4268	10"		13.68	1.170		720	59		779	890
4270	12"		12.54	1.276		815	64		879	1,000
4272	14"		11.97	1.337		925	67.50		992.50	1,125
4274	16"		11.40	1.404		1,050	70.50		1,120.50	1,250
4276	18"		10.83	1.477		1,275	74.50		1,349.50	1,525
4278	20"	Q-10	10.26	2.339		1,325	122		1,447	1,675
4280	22"		9.69	2.477		1,500	129		1,629	1,850

23 31 HVAC Ducts and Casings

23 31 13 – Metal Ducts

23 31 13.17	Round Grease Duct	Crew	Daily Output	Labor-Hours	Unit	Material	2015 Bare Costs Labor	Equipment	Total	Total Incl O&P
4282	24″	Q-10	9.12	2.632	Ea.	1,700	137		1,837	2,100
4284	26″		8.83	2.718		1,825	142		1,967	2,225
4286	28″		8.55	2.807		1,925	147		2,072	2,325
4288	30″		8.26	2.906		2,150	152		2,302	2,600
4290	32″		7.98	3.008		2,575	157		2,732	3,100
4292	36″		6.84	3.509		3,475	183		3,658	4,125
4300	Support strap									
4304	6″	Q-9	25.50	.627	Ea.	99	31.50		130.50	159
4306	8″		22	.727		116	36.50		152.50	185
4308	10″		20.50	.780		119	39.50		158.50	193
4310	12″		18.80	.851		128	43		171	208
4312	14″		17.90	.894		138	45		183	222
4314	16″		17	.941		139	47.50		186.50	227
4316	18″		16.20	.988		141	49.50		190.50	234
4318	20″	Q-10	15.40	1.558		156	81.50		237.50	300
4320	22″	″	14.50	1.655		173	86.50		259.50	325
4340	Plate Support Assembly									
4344	6″	Q-9	14.82	1.080	Ea.	139	54.50		193.50	238
4346	8″		12.54	1.276		162	64		226	279
4348	10″		11.40	1.404		176	70.50		246.50	305
4350	12″		10.26	1.559		185	78.50		263.50	325
4352	14″		9.69	1.651		219	83		302	370
4354	16″		9.12	1.754		231	88.50		319.50	395
4356	18″		8.55	1.871		243	94		337	415
4358	20″	Q-10	9.12	2.632		255	137		392	500
4360	22″		8.55	2.807		261	147		408	520
4362	24″		7.98	3.008		265	157		422	540
4364	26″		7.69	3.121		298	163		461	585
4366	28″		7.41	3.239		325	169		494	625
4368	30″		7.12	3.371		365	176		541	680
4370	32″		6.84	3.509		400	183		583	730
4372	36″		5.70	4.211		480	220		700	875
4380	Wall Guide Assembly									
4384	6″	Q-9	19.10	.838	Ea.	184	42		226	269
4386	8″		16.50	.970		208	49		257	305
4388	10″		15.30	1.046		220	52.50		272.50	325
4390	12″		14.10	1.135		230	57		287	345
4392	14″		13.40	1.194		249	60		309	370
4394	16″		12.75	1.255		252	63		315	375
4396	18″		12.10	1.322		255	66.50		321.50	385
4398	20″	Q-10	11.50	2.087		285	109		394	485
4400	22″		10.90	2.202		305	115		420	520
4402	24″		10.20	2.353		345	123		468	575
4404	26″		9.90	2.424		380	127		507	620
4406	28″		9.60	2.500		395	131		526	640
4408	30″		9.30	2.581		450	135		585	705
4410	32″		9	2.667		475	139		614	745
4412	36″		7.60	3.158		525	165		690	835
4420	Hood Transition, Flanged or Unflanged									
4424	6″	Q-9	15.60	1.026	Ea.	78	51.50		129.50	168
4426	8″		13.50	1.185		89	59.50		148.50	192
4428	10″		12.50	1.280		99	64.50		163.50	211
4430	12″		11.40	1.404		119	70.50		189.50	243

23 31 13 – Metal Ducts

23 31 13.17 Round Grease Duct		Crew	Daily Output	Labor-Hours	Unit	Material	2015 Bare Costs Labor	Equipment	Total	Total Incl O&P
4432	14"	Q-9	10.90	1.468	Ea.	128	74		202	257
4434	16"		10.40	1.538		138	77.50		215.50	273
4436	18"		9.80	1.633		157	82		239	305
4438	20"	Q-10	9.40	2.553		167	133		300	395
4440	22"		8.80	2.727		176	142		318	420
4442	24"		8.30	2.892		197	151		348	455
4444	26"		8.10	2.963		206	155		361	470
4446	28"		7.80	3.077		217	161		378	490
4448	30"		7.50	3.200		235	167		402	525
4450	32"		7.30	3.288		246	172		418	540
4452	36"		6.20	3.871		274	202		476	620
4460	Fan Adapter									
4464	6"	Q-9	15	1.067	Ea.	219	53.50		272.50	325
4466	8"		13	1.231		249	62		311	370
4468	10"		12	1.333		272	67		339	405
4470	12"		11	1.455		305	73		378	450
4472	14"		10.50	1.524		330	76.50		406.50	485
4474	16"		10	1.600		360	80.50		440.50	520
4476	18"		9.50	1.684		385	85		470	560
4478	20"	Q-10	9	2.667		415	139		554	675
4480	22"		8.50	2.824		445	147		592	725
4482	24"		8	3		470	157		627	760
4484	26"		7.70	3.117		495	163		658	800
4486	28"		7.50	3.200		525	167		692	845
4488	30"		7.20	3.333		555	174		729	885
4490	32"		7	3.429		585	179		764	925
4492	36"		6	4		640	209		849	1,025
4500	Tapered Increaser/Reducer									
4530	6" x 10" dia	Q-9	13.40	1.194	Ea.	282	60		342	405
4540	6" x 20" dia		10.10	1.584		282	80		362	435
4550	8" x 10" dia		13.20	1.212		340	61		401	470
4560	8" x 20" dia		9.90	1.616		340	81.50		421.50	505
4570	10" x 20" dia		10.60	1.509		395	76		471	555
4580	10" x 28" dia	Q-10	8.60	2.791		395	146		541	665
4590	12" x 20" dia		8.80	2.727		455	142		597	725
4600	14" x 28" dia		8.40	2.857		515	149		664	800
4610	16" x 20" dia		8.60	2.791		570	146		716	855
4620	16" x 30" dia		8.30	2.892		570	151		721	865
4630	18" x 24" dia		8.80	2.727		625	142		767	915
4640	18" x 30" dia		8.10	2.963		625	155		780	935
4650	20" x 24" dia		8.30	2.892		680	151		831	990
4660	20" x 32" dia		8	3		680	157		837	995
4670	24" x 26" dia		8	3		800	157		957	1,125
4680	24" x 36" dia		7.30	3.288		800	172		972	1,150
4690	28" x 30" dia		7.90	3.038		915	159		1,074	1,250
4700	28" x 36" dia		7.10	3.380		915	177		1,092	1,275
4710	30" x 36" dia		6.90	3.478		970	182		1,152	1,350
4730	Many intermediate standard sizes of tapered fittings are available									

23 31 13.19 Metal Duct Fittings		Crew	Daily Output	Labor-Hours	Unit	Material	Labor	Equipment	Total	Total Incl O&P
0010	**METAL DUCT FITTINGS**									
2000	Fabrics for flexible connections, with metal edge	1 Shee	100	.080	L.F.	3.44	4.48		7.92	10.85
2100	Without metal edge	"	160	.050	"	2.46	2.80		5.26	7.15

23 31 16 – Nonmetal Ducts

23 31 16.13 Fibrous-Glass Ducts

	23 31 16.13 Fibrous-Glass Ducts	Crew	Daily Output	Labor-Hours	Unit	2015 Bare Costs Material	Labor	Equipment	Total	Total Incl O&P
0010	**FIBROUS-GLASS DUCTS** R233100-40									
1280	Add to labor for elevated installation									
1282	of prefabricated (purchased) ductwork									
1283	10′ to 15′ high						10%			
1284	15′ to 20′ high						20%			
1285	20′ to 25′ high						25%			
1286	25′ to 30′ high						35%			
1287	30′ to 35′ high						40%			
1288	35′ to 40′ high						50%			
1289	Over 40′ high						55%			
3490	Rigid fiberglass duct board, foil reinf. kraft facing									
3500	Rectangular, 1″ thick, alum. faced, (FRK), std. weight	Q-10	350	.069	SF Surf	.79	3.58		4.37	6.50
9990	Minimum labor/equipment charge	1 Shee	3	2.667	Job		149		149	235

23 31 16.16 Thermoset Fiberglass-Reinforced Plastic Ducts

	23 31 16.16 Thermoset Fiberglass-Reinforced Plastic Ducts	Crew	Daily Output	Labor-Hours	Unit	Material	Labor	Equipment	Total	Total Incl O&P
0010	**THERMOSET FIBERGLASS-REINFORCED PLASTIC DUCTS**									
1280	Add to labor for elevated installation									
1282	of prefabricated (purchased) ductwork									
1283	10′ to 15′ high						10%			
1284	15′ to 20′ high						20%			
1285	20′ to 25′ high						25%			
1286	25′ to 30′ high						35%			
1287	30′ to 35′ high						40%			
1288	35′ to 40′ high						50%			
1289	Over 40′ high						55%			
3550	Rigid fiberglass reinforced plastic, FM approved									
3552	for acid fume and smoke exhaust system, nonflammable									
3554	Straight, 4″ diameter	Q-9	200	.080	L.F.	10.90	4.03		14.93	18.35
3555	6″ diameter		146	.110		14.25	5.50		19.75	24.50
3556	8″ diameter		106	.151		17.30	7.60		24.90	31
3557	10″ diameter		85	.188		21	9.50		30.50	38
3558	12″ diameter		68	.235		24.50	11.85		36.35	45
3561	18″ diameter		31	.516		44.50	26		70.50	90
3564	24″ diameter	Q-10	37	.649		57	34		91	117
3584	Note: joints are cemented with									
3586	fiberglass resin, included in material cost.									
3590	Elbow, 90°, 4″ diameter	Q-9	20.30	.788	Ea.	52.50	39.50		92	121
3591	6″ diameter		13.80	1.159		73.50	58.50		132	173
3592	8″ diameter		10	1.600		81	80.50		161.50	216
3593	10″ diameter		7.80	2.051		106	103		209	280
3594	12″ diameter		6.50	2.462		135	124		259	345
3597	18″ diameter		4.40	3.636		281	183		464	600
3600	24″ diameter	Q-10	4.90	4.898		420	256		676	865
3626	Elbow, 45°, 4″ diameter	Q-9	22.30	.717		50.50	36		86.50	113
3627	6″ diameter		15.20	1.053		70	53		123	161
3628	8″ diameter		11	1.455		74.50	73		147.50	198
3629	10″ diameter		8.60	1.860		78.50	93.50		172	235
3630	12″ diameter		7.20	2.222		86	112		198	272
3633	18″ diameter		4.84	3.306		147	166		313	425
3636	24″ diameter	Q-10	5.40	4.444		264	232		496	655
3660	Tee, 90°, 4″ diameter	Q-9	14.06	1.138		29	57.50		86.50	122
3661	6″ diameter		9.52	1.681		44	84.50		128.50	183
3662	8″ diameter		6.98	2.292		60	115		175	248

23 31 HVAC Ducts and Casings

23 31 16 – Nonmetal Ducts

	23 31 16.16 Thermoset Fiberglass-Reinforced Plastic Ducts	Crew	Daily Output	Labor-Hours	Unit	2015 Bare Costs Material	Labor	Equipment	Total	Total Incl O&P
3663	10″ diameter	Q-9	5.45	2.936	Ea.	78.50	148		226.50	320
3664	12″ diameter		4.60	3.478		98	175		273	385
3667	18″ diameter		3.05	5.246		201	264		465	635
3670	24″ diameter	Q-10	3.46	6.936		305	360		665	905
3690	For Y @ 45°, add					22%				

23 31 16.19 PVC Ducts

		Crew	Daily Output	Labor-Hours	Unit	Material	Labor	Equipment	Total	Total Incl O&P
0010	**PVC DUCTS**									
4000	Rigid plastic, corrosive fume resistant PVC									
4020	Straight, 6″ diameter	Q-9	220	.073	L.F.	11.55	3.66		15.21	18.50
4040	8″ diameter		160	.100		16.75	5.05		21.80	26.50
4060	10″ diameter		120	.133		21	6.70		27.70	33.50
4070	12″ diameter		100	.160		24.50	8.05		32.55	39.50
4080	14″ diameter		70	.229		30.50	11.50		42	51.50
4090	16″ diameter		62	.258		35.50	13		48.50	59.50
4100	18″ diameter		58	.276		48.50	13.90		62.40	75.50
4110	20″ diameter	Q-10	75	.320		59.50	16.70		76.20	92
4130	24″ diameter	″	55	.436		62	23		85	105
4250	Coupling, 6″ diameter	Q-9	88	.182	Ea.	23.50	9.15		32.65	40.50
4270	8″ diameter		78	.205		30	10.35		40.35	49.50
4290	10″ diameter		70	.229		35.50	11.50		47	57
4300	12″ diameter		55	.291		38.50	14.65		53.15	65.50
4310	14″ diameter		44	.364		50	18.30		68.30	84
4320	16″ diameter		40	.400		52.50	20		72.50	90
4330	18″ diameter		38	.421		68	21		89	109
4340	20″ diameter	Q-10	55	.436		82	23		105	126
4360	24″ diameter	″	45	.533		103	28		131	157
4470	Elbow, 90°, 6″ diameter	Q-9	44	.364		108	18.30		126.30	147
4490	8″ diameter		28	.571		121	29		150	179
4510	10″ diameter		18	.889		147	45		192	232
4520	12″ diameter		15	1.067		167	53.50		220.50	268
4530	14″ diameter		11	1.455		193	73		266	330
4540	16″ diameter		10	1.600		251	80.50		331.50	405
4550	18″ diameter	Q-10	15	1.600		350	83.50		433.50	510
4560	20″ diameter		14	1.714		520	89.50		609.50	710
4580	24″ diameter		12	2		640	104		744	865
4750	Elbow 45°, use 90° and deduct					25%				
9990	Minimum labor/equipment charge	1 Shee	3	2.667	Job		149		149	235

23 31 16.21 Polypropylene Ducts

		Crew	Daily Output	Labor-Hours	Unit	Material	Labor	Equipment	Total	Total Incl O&P
0010	**POLYPROPYLENE DUCTS**									
0020	Rigid plastic, corrosive fume resistant,flame retardant, PP									
0110	Straight, 4″ diameter	Q-9	200	.080	L.F.	7.55	4.03		11.58	14.65
0120	6″ diameter		146	.110		10.75	5.50		16.25	20.50
0140	8″ diameter		106	.151		13.75	7.60		21.35	27
0160	10″ diameter		85	.188		20.50	9.50		30	37.50
0170	12″ diameter		68	.235		39	11.85		50.85	61.50
0180	14″ diameter		50	.320		44	16.10		60.10	74
0190	16″ diameter		40	.400		59.50	20		79.50	97.50
0200	18″ diameter		31	.516		97	26		123	147
0210	20″ diameter	Q-10	44	.545		123	28.50		151.50	180
0220	22″ diameter		40	.600		137	31.50		168.50	201
0230	24″ diameter		37	.649		186	34		220	259
0250	28″ diameter		28	.857		221	45		266	315

23 31 16.21 Polypropylene Ducts		Crew	Daily Output	Labor-Hours	Unit	Material	2015 Bare Costs Labor	Equipment	Total	Total Incl O&P
0270	32" diameter	Q-10	24	1	L.F.	249	52		301	355
0400	Coupling									
0410	4" diameter	Q-9	23	.696	Ea.	21	35		56	78.50
0420	6" diameter		16	1		21.50	50.50		72	104
0430	8" diameter		11.40	1.404		31	70.50		101.50	147
0440	10" diameter		9	1.778		41	89.50		130.50	186
0450	12" diameter		8	2		48	101		149	212
0460	14" diameter		6.60	2.424		93.50	122		215.50	296
0470	16" diameter		5.70	2.807		103	141		244	335
0480	18" diameter		5.20	3.077		123	155		278	380
0490	20" diameter	Q-10	6.80	3.529		134	184		318	440
0500	22" diameter		6.20	3.871		340	202		542	695
0510	24" diameter		5.80	4.138		435	216		651	820
0520	28" diameter		5.50	4.364		440	228		668	845
0530	32" diameter		5.20	4.615		490	241		731	920
0800	Elbow, 90°									
0810	4" diameter	Q-9	22	.727	Ea.	35	36.50		71.50	96.50
0820	6" diameter		15.40	1.039		49.50	52.50		102	137
0830	8" diameter		11	1.455		60	73		133	182
0840	10" diameter		8.60	1.860		86.50	93.50		180	243
0850	12" diameter		7.70	2.078		225	105		330	410
0860	14" diameter		6.40	2.500		260	126		386	485
0870	16" diameter		5.40	2.963		273	149		422	535
0880	18" diameter	Q-10	7.50	3.200		705	167		872	1,050
0890	20" diameter		6.60	3.636		775	190		965	1,150
0896	22" diameter		6.10	3.934		1,225	205		1,430	1,675
0900	24" diameter		5.60	4.286		1,100	224		1,324	1,575
0910	28" diameter		5.40	4.444		2,200	232		2,432	2,775
0920	32" diameter		5.20	4.615		2,850	241		3,091	3,525
1000	Elbow, 45°									
1020	4" diameter	Q-9	22	.727	Ea.	23	36.50		59.50	83.50
1030	6" diameter		15.40	1.039		41.50	52.50		94	129
1040	8" diameter		11	1.455		48	73		121	169
1050	10" diameter		8.60	1.860		65	93.50		158.50	220
1060	12" diameter		7.70	2.078		174	105		279	355
1070	14" diameter		6.40	2.500		194	126		320	415
1080	16" diameter		5.40	2.963		226	149		375	485
1090	18" diameter	Q-10	7.50	3.200		445	167		612	755
1100	20" diameter		6.60	3.636		575	190		765	935
1106	22" diameter		6.10	3.934		800	205		1,005	1,200
1110	24" diameter		5.60	4.286		760	224		984	1,200
1120	28" diameter		5.40	4.444		1,475	232		1,707	1,975
1130	32" diameter		5.20	4.615		1,850	241		2,091	2,400
1200	Tee									
1210	4" diameter	Q-9	16.50	.970	Ea.	150	49		199	242
1220	6" diameter		11.60	1.379		183	69.50		252.50	310
1230	8" diameter		8.30	1.928		228	97		325	405
1240	10" diameter		6.50	2.462		315	124		439	540
1250	12" diameter		5.80	2.759		415	139		554	675
1260	14" diameter		4.80	3.333		540	168		708	860
1270	16" diameter		4.05	3.951		600	199		799	970
1280	18" diameter	Q-10	5.60	4.286		790	224		1,014	1,225
1290	20" diameter		4.90	4.898		855	256		1,111	1,350

23 31 HVAC Ducts and Casings

23 31 16 – Nonmetal Ducts

23 31 16.21 Polypropylene Ducts		Crew	Daily Output	Labor-Hours	Unit	2015 Bare Costs Material	Labor	Equipment	Total	Total Incl O&P
1300	22″ diameter	Q-10	4.40	5.455	Ea.	1,025	285		1,310	1,575
1310	24″ diameter		4.20	5.714		1,300	298		1,598	1,900
1320	28″ diameter		4.10	5.854		2,300	305		2,605	3,000
1330	32″ diameter		4	6		2,650	315		2,965	3,425

23 33 Air Duct Accessories

23 33 13 – Dampers

23 33 13.13 Volume-Control Dampers		Crew	Daily Output	Labor-Hours	Unit	Material	Labor	Equipment	Total	Total Incl O&P
0010	**VOLUME-CONTROL DAMPERS**									
5990	Multi-blade dampers, opposed blade, 8″ x 6″	1 Shee	24	.333	Ea.	21.50	18.65		40.15	53.50
5994	8″ x 8″		22	.364		22.50	20.50		43	56.50
5996	10″ x 10″		21	.381		26.50	21.50		48	63
6000	12″ x 12″		21	.381		29.50	21.50		51	66
6020	12″ x 18″		18	.444		39.50	25		64.50	82.50
6030	14″ x 10″		20	.400		29	22.50		51.50	67
6031	14″ x 14″		17	.471		35.50	26.50		62	80.50
6033	16″ x 12″		17	.471		35.50	26.50		62	80.50
6035	16″ x 16″		16	.500		44	28		72	92.50
6037	18″ x 16″		15	.533		48	30		78	100
6038	18″ x 18″		15	.533		52	30		82	104
6040	18″ x 24″		12	.667		67	37.50		104.50	133
6060	18″ x 28″		10	.800		78	45		123	156
6070	20″ x 16″		14	.571		52	32		84	108
6072	20″ x 20″		13	.615		62.50	34.50		97	124
6074	22″ x 18″		14	.571		62.50	32		94.50	120
6076	24″ x 16″		11	.727		61	40.50		101.50	132
6078	24″ x 20″		8	1		72.50	56		128.50	169
6080	24″ x 24″		8	1		85	56		141	182
6100	24″ x 28″		6	1.333		98	74.50		172.50	226
6110	26″ x 26″		6	1.333		95	74.50		169.50	223
6120	28″ x 28″	Q-9	11	1.455		107	73		180	233
6130	30″ x 18″		10	1.600		84	80.50		164.50	220
6132	30″ x 24″		7	2.286		107	115		222	299
6133	30″ x 30″		6.60	2.424		137	122		259	345
6135	32″ x 32″		6.40	2.500		154	126		280	370
6151	36″ x 12″		10	1.600		70	80.50		150.50	204
6152	36″ x 16″		8	2		96	101		197	264
6158	36″ x 36″		6	2.667		189	134		323	420
6160	44″ x 28″		5.80	2.759		194	139		333	435
6180	48″ x 36″		5.60	2.857		253	144		397	505
6200	56″ x 36″		5.40	2.963		310	149		459	575
6220	60″ x 36″		5.20	3.077		330	155		485	610
6240	60″ x 44″		5	3.200		390	161		551	685
7500	Variable volume modulating motorized damper, incl. elect. mtr.									
7504	8″ x 6″	1 Shee	15	.533	Ea.	118	30		148	177
7506	10″ x 6″		14	.571		118	32		150	181
7510	10″ x 10″		13	.615		122	34.50		156.50	189
7520	12″ x 12″		12	.667		127	37.50		164.50	198
7522	12″ x 16″		11	.727		127	40.50		167.50	204
7524	16″ x 10″		12	.667		125	37.50		162.50	196
7526	16″ x 14″		10	.800		130	45		175	214

23 33 13 – Dampers

23 33 13.13 Volume-Control Dampers		Crew	Daily Output	Labor-Hours	Unit	Material	Labor	Equipment	Total	Total Incl O&P
7528	16" x 18"	1 Shee	9	.889	Ea.	135	49.50		184.50	227
7540	18" x 12"		10	.800		130	45		175	214
7542	18" x 18"		8	1		138	56		194	240
7544	20" x 14"		8	1		141	56		197	244
7546	20" x 18"		7	1.143		147	64		211	263
7560	24" x 12"		8	1		143	56		199	246
7562	24" x 18"		7	1.143		154	64		218	270
7564	24" x 24"		6	1.333		160	74.50		234.50	294
7568	28" x 10"		7	1.143		143	64		207	258
7580	28" x 16"		6	1.333		163	74.50		237.50	297
7590	30" x 14"		5	1.600		151	89.50		240.50	305
7600	30" x 18"		4	2		198	112		310	395
7610	30" x 24"		3.80	2.105		264	118		382	475
7700	For thermostat, add		8	1		31	56		87	123
7800	For transformer 40 VA capacity, add		16	.500		19.45	28		47.45	65.50
8000	Multi-blade dampers, parallel blade									
8100	8" x 8"	1 Shee	24	.333	Ea.	82	18.65		100.65	120
8120	12" x 8"		22	.364		82	20.50		102.50	122
8140	16" x 10"		20	.400		103	22.50		125.50	150
8160	18" x 12"		18	.444		113	25		138	163
8180	22" x 12"		15	.533		122	30		152	181
8200	24" x 16"		11	.727		134	40.50		174.50	212
8220	28" x 16"		10	.800		150	45		195	236
8240	30" x 16"		8	1		152	56		208	257
8260	30" x 18"		7	1.143		186	64		250	305

Column group "2015 Bare Costs" covers Material, Labor, Equipment, Total.

23 33 13.16 Fire Dampers		Crew	Daily Output	Labor-Hours	Unit	Material	Labor	Equipment	Total	Total Incl O&P
0010	**FIRE DAMPERS**									
3000	Fire damper, curtain type, 1-1/2 hr. rated, vertical, 6" x 6"	1 Shee	24	.333	Ea.	24	18.65		42.65	56
3020	8" x 6"		22	.364		24	20.50		44.50	58.50
3040	12" x 6"		22	.364		24	20.50		44.50	58.50
3060	20" x 6"		18	.444		32	25		57	74
3080	12" x 8"		22	.364		24	20.50		44.50	58.50
3100	24" x 8"		16	.500		36.50	28		64.50	84
3120	12" x 10"		21	.381		24	21.50		45.50	60
3140	24" x 10"		15	.533		39	30		69	90
3160	36" x 10"		12	.667		50.50	37.50		88	115
3180	16" x 12"		20	.400		33	22.50		55.50	72
3200	24" x 12"		13	.615		41.50	34.50		76	100
3220	48" x 12"		10	.800		64	45		109	141
3240	16" x 14"		18	.444		44	25		69	87.50
3260	24" x 14"		12	.667		53	37.50		90.50	117
3280	30" x 14"		11	.727		50.50	40.50		91	120
3300	18" x 16"		17	.471		43	26.50		69.50	88.50
3320	24" x 16"		11	.727		46	40.50		86.50	115
3340	36" x 16"		10	.800		58.50	45		103.50	135
3360	24" x 18"		10	.800		54	45		99	130
3380	48" x 18"		8	1		77.50	56		133.50	174
3400	24" x 20"		8	1		54	56		110	148
3420	36" x 20"		7	1.143		65	64		129	173
3440	24" x 22"		7	1.143		55.50	64		119.50	162
3460	30" x 22"		6	1.333		60.50	74.50		135	185
3480	26" x 24"		7	1.143		56.50	64		120.50	163

23 33 13 – Dampers

23 33 13.16 Fire Dampers		Crew	Daily Output	Labor-Hours	Unit	Material	2015 Bare Costs Labor	Equipment	Total	Total Incl O&P
3500	48" x 24"	Q-9	12	1.333	Ea.	87	67		154	202
3520	28" x 26"		13	1.231		63.50	62		125.50	168
3540	30" x 28"		12	1.333		70	67		137	183
3560	48" x 30"		11	1.455		102	73		175	228
3580	48" x 36"		10	1.600		136	80.50		216.50	276
3600	44" x 44"		10	1.600		144	80.50		224.50	286
3620	48" x 48"		8	2		161	101		262	335
3700	U.L. label included in above									
3800	For horizontal operation, add				Ea.	20%				
3900	For cap for blades out of air stream, add					20%				
4000	For 10" 22 ga., U.L. approved sleeve, add					35%				
4100	For 10" 22 ga., U.L. sleeve, 100% free area, add					65%				
4200	For oversize openings group dampers									
4502	Power open-spring close,sleeve and actuator motor mounted, 1-1/2 hour									
4510	8" x 8"	1 Shee	22	.364	Ea.	214	20.50		234.50	267
4520	16" x 8"		20	.400		340	22.50		362.50	405
4540	18" x 8"		18	.444		345	25		370	420
4560	20" x 8"		16	.500		345	28		373	425
4580	10" x 10"		21	.381		310	21.50		331.50	375
4600	24" x 10"		15	.533		360	30		390	440
4620	30" x 10"		12	.667		370	37.50		407.50	470
4640	12" x 12"		20	.400		340	22.50		362.50	410
4660	18" x 12"		18	.444		355	25		380	430
4680	24" x 12"		13	.615		365	34.50		399.50	455
4700	30" x 12"		11	.727		375	40.50		415.50	480
4720	14" x 14"		17	.471		365	26.50		391.50	440
4740	16" x 14"		18	.444		365	25		390	445
4760	20" x 14"		14	.571		350	32		382	435
4780	24" x 14"		12	.667		375	37.50		412.50	475
4800	30" x 14"		10	.800		400	45		445	510
4820	16" x 16"		16	.500		375	28		403	460
4840	20" x 16"		14	.571		380	32		412	465
4860	24" x 16"		11	.727		385	40.50		425.50	490
4880	30" x 16"		8	1		415	56		471	545
4900	18" x 18"		15	.533		400	30		430	480
5000	24" x 18"		10	.800		405	45		450	515
5020	36" x 18"		7	1.143		525	64		589	675
5040	20" x 20"		13	.615		405	34.50		439.50	500
5060	24" x 20"		8	1		410	56		466	540
5080	30" x 20"		8	1		430	56		486	560
5100	36" x 20"		7	1.143		445	64		509	590
5120	24" x 24"		8	1		425	56		481	560
5130	30" x 24"		7.60	1.053		455	59		514	595
5140	36" x 24"	Q-9	12	1.333		475	67		542	625
5141	36" x 30"		10	1.600		540	80.50		620.50	720
5142	40" x 36"		8	2		890	101		991	1,150
5143	48" x 48"		4	4		1,700	201		1,901	2,200
5150	Damper operator motor, 24 or 120 volt	1 Shee	16	.500		540	28		568	635

23 33 13.28 Splitter Damper Assembly		Crew	Daily Output	Labor-Hours	Unit	Material	Labor	Equipment	Total	Total Incl O&P
0010	SPLITTER DAMPER ASSEMBLY									
7000	Self locking, 1' rod	1 Shee	24	.333	Ea.	23.50	18.65		42.15	55
7020	3' rod		22	.364		30	20.50		50.50	65

23 33 Air Duct Accessories

23 33 13 – Dampers

23 33 13.28 Splitter Damper Assembly		Crew	Daily Output	Labor-Hours	Unit	2015 Bare Costs Material	Labor	Equipment	Total	Total Incl O&P
7040	4′ rod	1 Shee	20	.400	Ea.	33.50	22.50		56	72.50
7060	6′ rod		18	.444		41	25		66	84

23 33 19 – Duct Silencers

23 33 19.10 Duct Silencers		Crew	Daily Output	Labor-Hours	Unit	Material	Labor	Equipment	Total	Total Incl O&P
0010	**DUCT SILENCERS**									
9000	Silencers, noise control for air flow, duct				MCFM	57			57	63

23 33 23 – Turning Vanes

23 33 23.13 Air Turning Vanes		Crew	Daily Output	Labor-Hours	Unit	Material	Labor	Equipment	Total	Total Incl O&P
0010	**AIR TURNING VANES**									
9400	Turning vane components									
9410	Turning vane rail	1 Shee	160	.050	L.F.	.75	2.80		3.55	5.25
9420	Double thick, factory fab. vane		300	.027		1.17	1.49		2.66	3.65
9428	12″ high set		170	.047		1.68	2.63		4.31	6
9432	14″ high set		160	.050		1.96	2.80		4.76	6.60
9434	16″ high set		150	.053		2.24	2.98		5.22	7.15
9436	18″ high set		144	.056		2.52	3.11		5.63	7.70
9438	20″ high set		138	.058		2.80	3.24		6.04	8.20
9440	22″ high set		130	.062		3.08	3.44		6.52	8.85
9442	24″ high set		124	.065		3.36	3.61		6.97	9.40
9444	26″ high set		116	.069		3.64	3.86		7.50	10.10
9446	30″ high set		112	.071		4.20	4		8.20	10.90
9900	Minimum labor/equipment charge		4	2	Job		112		112	177

23 33 33 – Duct-Mounting Access Doors

23 33 33.13 Duct Access Doors		Crew	Daily Output	Labor-Hours	Unit	Material	Labor	Equipment	Total	Total Incl O&P
0010	**DUCT ACCESS DOORS**									
1000	Duct access door, insulated, 6″ x 6″	1 Shee	14	.571	Ea.	16.05	32		48.05	68
1020	10″ x 10″		11	.727		18.45	40.50		58.95	84.50
1040	12″ x 12″		10	.800		20	45		65	92.50
1050	12″ x 18″		9	.889		37	49.50		86.50	119
1060	16″ x 12″		9	.889		29.50	49.50		79	111
1070	18″ x 18″		8	1		32.50	56		88.50	124
1074	24″ x 18″		8	1		44	56		100	137
1080	24″ x 24″		8	1		47	56		103	141

23 33 46 – Flexible Ducts

23 33 46.10 Flexible Air Ducts		Crew	Daily Output	Labor-Hours	Unit	Material	Labor	Equipment	Total	Total Incl O&P
0010	**FLEXIBLE AIR DUCTS** R233100-40									
1280	Add to labor for elevated installation									
1282	of prefabricated (purchased) ductwork									
1283	10′ to 15′ high						10%			
1284	15′ to 20′ high						20%			
1285	20′ to 25′ high						25%			
1286	25′ to 30′ high						35%			
1287	30′ to 35′ high						40%			
1288	35′ to 40′ high						50%			
1289	Over 40′ high						55%			
1300	Flexible, coated fiberglass fabric on corr. resist. metal helix									
1400	pressure to 12″ (WG) UL-181									
1500	Noninsulated, 3″ diameter	Q-9	400	.040	L.F.	1.12	2.01		3.13	4.41
1520	4″ diameter		360	.044		1.16	2.24		3.40	4.81
1540	5″ diameter		320	.050		1.30	2.52		3.82	5.40

23 33 46 – Flexible Ducts

23 33 46.10 Flexible Air Ducts			Crew	Daily Output	Labor-Hours	Unit	Material	Labor	Equipment	Total	Total Incl O&P
							2015 Bare Costs				
1560	6″ diameter		Q-9	280	.057	L.F.	1.50	2.88		4.38	6.20
1580	7″ diameter			240	.067		1.53	3.36		4.89	7
1600	8″ diameter			200	.080		1.91	4.03		5.94	8.45
1620	9″ diameter			180	.089		2.05	4.48		6.53	9.30
1640	10″ diameter			160	.100		2.46	5.05		7.51	10.65
1660	12″ diameter			120	.133		2.94	6.70		9.64	13.85
1680	14″ diameter			80	.200		3.56	10.05		13.61	19.80
1700	16″ diameter		↓	60	.267	↓	5.05	13.45		18.50	26.50
1800	For plastic cable tie, add					Ea.	.77			.77	.85
1900	Insulated, 1″ thick, PE jacket, 3″ diameter	G	Q-9	380	.042	L.F.	2.60	2.12		4.72	6.20
1910	4″ diameter	G		340	.047		2.60	2.37		4.97	6.60
1920	5″ diameter	G		300	.053		2.60	2.69		5.29	7.10
1940	6″ diameter	G		260	.062		2.94	3.10		6.04	8.10
1960	7″ diameter	G		220	.073		3.20	3.66		6.86	9.30
1980	8″ diameter	G		180	.089		3.49	4.48		7.97	10.90
2000	9″ diameter	G		160	.100		3.81	5.05		8.86	12.15
2020	10″ diameter	G		140	.114		4.25	5.75		10	13.80
2040	12″ diameter	G		100	.160		4.90	8.05		12.95	18.10
2060	14″ diameter	G		80	.200		6.05	10.05		16.10	22.50
2080	16″ diameter	G		60	.267		8.35	13.45		21.80	30
2100	18″ diameter	G	↓	45	.356		9.85	17.90		27.75	39.50
2120	20″ diameter	G	Q-10	65	.369	↓	10.85	19.30		30.15	42.50
2500	Insulated, heavy duty, coated fiberglass fabric										
2520	4″ diameter	G	Q-9	340	.047	L.F.	3.36	2.37		5.73	7.45
2540	5″ diameter	G		300	.053		3.54	2.69		6.23	8.15
2560	6″ diameter	G		260	.062		4.20	3.10		7.30	9.50
2580	7″ diameter	G		220	.073		4.58	3.66		8.24	10.85
2600	8″ diameter	G		180	.089		5.10	4.48		9.58	12.70
2620	9″ diameter	G		160	.100		5.55	5.05		10.60	14.10
2640	10″ diameter	G		140	.114		6.40	5.75		12.15	16.15
2660	12″ diameter	G		100	.160		7.50	8.05		15.55	21
2680	14″ diameter	G		80	.200		8.90	10.05		18.95	25.50
2700	16″ diameter	G		60	.267		12.20	13.45		25.65	34.50
2720	18″ diameter	G	↓	45	.356	↓	14.45	17.90		32.35	44.50
5000	Flexible, aluminum, acoustical, pressure to 2″ (wg), NFPA-90A										
5010	Fiberglass insulation 1-1/2″ thick, 1/2 lb. density										
5020	Polyethylene jacket, UL approved										
5026	5″ diameter	G	Q-9	300	.053	L.F.	4.29	2.69		6.98	8.95
5030	6″ diameter	G		260	.062		5.05	3.10		8.15	10.45
5034	7″ diameter	G		220	.073		6.05	3.66		9.71	12.45
5038	8″ diameter	G		180	.089		6.15	4.48		10.63	13.85
5042	9″ diameter	G		160	.100		7.25	5.05		12.30	15.95
5046	10″ diameter	G		140	.114		9.45	5.75		15.20	19.50
5050	12″ diameter	G		100	.160		10.45	8.05		18.50	24
5054	14″ diameter	G		80	.200		11.20	10.05		21.25	28.50
5058	16″ diameter	G	↓	60	.267	↓	15.85	13.45		29.30	38.50
5072	Hospital grade, PE jacket, UL approved										
5076	5″ diameter	G	Q-9	300	.053	L.F.	4.62	2.69		7.31	9.35
5080	6″ diameter	G		260	.062		5.50	3.10		8.60	10.95
5084	7″ diameter	G		220	.073		6.60	3.66		10.26	13.05
5088	8″ diameter	G		180	.089		6.80	4.48		11.28	14.55
5092	9″ diameter	G		160	.100		7.90	5.05		12.95	16.65
5096	10″ diameter	G	↓	140	.114	↓	10.35	5.75		16.10	20.50

23 33 Air Duct Accessories

23 33 46 – Flexible Ducts

23 33 46.10 Flexible Air Ducts			Crew	Daily Output	Labor-Hours	Unit	2015 Bare Costs Material	Labor	Equipment	Total	Total Incl O&P
5100	12″ diameter	G	Q-9	100	.160	L.F.	11.45	8.05		19.50	25.50
5104	14″ diameter	G		80	.200		12.30	10.05		22.35	29.50
5108	16″ diameter	G		60	.267		17.25	13.45		30.70	40
5140	Flexible, aluminum, silencer, pressure to 12″ (wg), NFPA-90A										
5144	Fiberglass insulation 1-1/2″ thick, 1/2 lb. density										
5148	Aluminum outer shell, UL approved										
5152	5″ diameter	G	Q-9	290	.055	L.F.	11.85	2.78		14.63	17.40
5156	6″ diameter	G		250	.064		13	3.22		16.22	19.40
5160	7″ diameter	G		210	.076		13.75	3.84		17.59	21
5164	8″ diameter	G		170	.094		14.70	4.74		19.44	23.50
5168	9″ diameter	G		150	.107		17.80	5.35		23.15	28
5172	10″ diameter	G		130	.123		20	6.20		26.20	32
5176	12″ diameter	G		90	.178		22.50	8.95		31.45	38.50
5180	14″ diameter	G		70	.229		26.50	11.50		38	47
5184	16″ diameter	G		50	.320		34	16.10		50.10	63
5200	Hospital grade, aluminum shell, UL approved										
5204	5″ diameter	G	Q-9	280	.057	L.F.	12.25	2.88		15.13	18.05
5208	6″ diameter	G		240	.067		13.50	3.36		16.86	20
5212	7″ diameter	G		200	.080		14.30	4.03		18.33	22
5216	8″ diameter	G		160	.100		15.30	5.05		20.35	25
5220	9″ diameter	G		140	.114		18.55	5.75		24.30	29.50
5224	10″ diameter	G		130	.123		21	6.20		27.20	33
5228	12″ diameter	G		80	.200		24.50	10.05		34.55	42.50
5232	14″ diameter	G		60	.267		27.50	13.45		40.95	51.50
5236	16″ diameter	G		40	.400		35.50	20		55.50	71
9990	Minimum labor/equipment charge		1 Shee	3	2.667	Job		149		149	235

23 33 53 – Duct Liners

23 33 53.10 Duct Liner Board			Crew	Daily Output	Labor-Hours	Unit	Material	Labor	Equipment	Total	Total Incl O&P
0010	**DUCT LINER BOARD**										
3340	Board type fiberglass liner, FSK, 1-1/2 lb. density										
3344	1″ thick	G	Q-14	150	.107	S.F.	.62	5.05		5.67	8.80
3345	1-1/2″ thick	G		130	.123		.68	5.80		6.48	10.05
3346	2″ thick	G		120	.133		.79	6.30		7.09	10.95
3348	3″ thick	G		110	.145		1.02	6.85		7.87	12.10
3350	4″ thick	G		100	.160		1.25	7.55		8.80	13.50
3356	3 lb. density, 1″ thick	G		150	.107		.79	5.05		5.84	8.95
3358	1-1/2″ thick	G		130	.123		1	5.80		6.80	10.40
3360	2″ thick	G		120	.133		1.22	6.30		7.52	11.45
3362	2-1/2″ thick	G		110	.145		1.43	6.85		8.28	12.55
3364	3″ thick	G		100	.160		1.64	7.55		9.19	13.90
3366	4″ thick	G		90	.178		2.06	8.40		10.46	15.70
3370	6 lb. density, 1″ thick	G		140	.114		1.12	5.40		6.52	9.90
3374	1-1/2″ thick	G		120	.133		1.50	6.30		7.80	11.75
3378	2″ thick	G		100	.160		1.88	7.55		9.43	14.15
3490	Board type, fiberglass liner, 3 lb. density										
3680	No finish										
3700	1″ thick	G	Q-14	170	.094	S.F.	.44	4.44		4.88	7.65
3710	1-1/2″ thick	G		140	.114		.66	5.40		6.06	9.40
3720	2″ thick	G		130	.123		.88	5.80		6.68	10.25
3940	Board type, non-fibrous foam										
3950	Temperature, bacteria and fungi resistant										
3960	1″ thick	G	Q-14	150	.107	S.F.	2.38	5.05		7.43	10.70

23 33 Air Duct Accessories

23 33 53 – Duct Liners

23 33 53.10 Duct Liner Board			Crew	Daily Output	Labor-Hours	Unit	2015 Bare Costs Material	Labor	Equipment	Total	Total Incl O&P
3970	1-1/2" thick	G	Q-14	130	.123	S.F.	3.28	5.80		9.08	12.90
3980	2" thick	G	↓	120	.133	↓	3.98	6.30		10.28	14.50

23 34 HVAC Fans

23 34 13 – Axial HVAC Fans

23 34 13.10 Axial Flow HVAC Fans

		Crew	Daily Output	Labor-Hours	Unit	Material	Labor	Equipment	Total	Total Incl O&P
0010	**AXIAL FLOW HVAC FANS**									
0020	Air conditioning and process air handling									
1500	Vaneaxial, low pressure, 2000 CFM, 1/2 HP	Q-20	3.60	5.556	Ea.	2,225	285		2,510	2,900
1520	4,000 CFM, 1 HP		3.20	6.250		2,600	320		2,920	3,350
1540	8,000 CFM, 2 HP		2.80	7.143		3,275	365		3,640	4,200
1560	16,000 CFM, 5 HP	↓	2.40	8.333	↓	4,600	425		5,025	5,750

23 34 14 – Blower HVAC Fans

23 34 14.10 Blower Type HVAC Fans

		Crew	Daily Output	Labor-Hours	Unit	Material	Labor	Equipment	Total	Total Incl O&P
0010	**BLOWER TYPE HVAC FANS**									
2000	Blowers, direct drive with motor, complete									
2020	1045 CFM @ .5" S.P., 1/5 HP	Q-20	18	1.111	Ea.	340	57		397	465
2040	1385 CFM @ .5" S.P., 1/4 HP		18	1.111		335	57		392	460
2060	1640 CFM @ .5" S.P., 1/3 HP		18	1.111		305	57		362	425
2080	1760 CFM @ .5" S.P., 1/2 HP	↓	18	1.111	↓	315	57		372	440
2090	4 speed									
2100	1164 to 1739 CFM @ .5" S.P., 1/3 HP	Q-20	16	1.250	Ea.	320	64		384	455
2120	1467 to 2218 CFM @ 1.0" S.P., 3/4 HP	"	14	1.429	"	370	73		443	520
2500	Ceiling fan, right angle, extra quiet, 0.10" S.P.									
2520	95 CFM	Q-20	20	1	Ea.	300	51		351	410
2540	210 CFM		19	1.053		355	54		409	475
2560	385 CFM		18	1.111		450	57		507	585
2580	885 CFM		16	1.250		890	64		954	1,075
2600	1,650 CFM		13	1.538		1,225	79		1,304	1,475
2620	2,960 CFM	↓	11	1.818		1,650	93		1,743	1,950
2640	For wall or roof cap, add	1 Shee	16	.500		300	28		328	375
2660	For straight thru fan, add					10%				
2680	For speed control switch, add	1 Elec	16	.500	↓	164	27.50		191.50	224
7500	Utility set, steel construction, pedestal, 1/4" S.P.									
7520	Direct drive, 150 CFM, 1/8 HP	Q-20	6.40	3.125	Ea.	870	160		1,030	1,200
7540	485 CFM, 1/6 HP		5.80	3.448		1,100	177		1,277	1,475
7560	1950 CFM, 1/2 HP		4.80	4.167		1,275	213		1,488	1,725
7580	2410 CFM, 3/4 HP		4.40	4.545		2,375	233		2,608	2,975
7600	3328 CFM, 1-1/2 HP	↓	3	6.667	↓	2,625	340		2,965	3,425
7680	V-belt drive, drive cover, 3 phase									
7700	800 CFM, 1/4 HP	Q-20	6	3.333	Ea.	980	171		1,151	1,350
7720	1,300 CFM, 1/3 HP		5	4		1,025	205		1,230	1,450
7740	2,000 CFM, 1 HP		4.60	4.348		1,225	223		1,448	1,700
7760	2,900 CFM, 3/4 HP		4.20	4.762		1,650	244		1,894	2,175
7780	3,600 CFM, 3/4 HP		4	5		2,025	256		2,281	2,625
7800	4,800 CFM, 1 HP		3.50	5.714		2,375	293		2,668	3,075
7820	6,700 CFM, 1-1/2 HP		3	6.667		2,950	340		3,290	3,750
7830	7,500 CFM, 2 HP		2.50	8		4,000	410		4,410	5,050
7840	11,000 CFM, 3 HP		2	10		5,325	510		5,835	6,675
7860	13,000 CFM, 3 HP	↓	1.60	12.500	↓	5,425	640		6,065	6,950

23 34 HVAC Fans

23 34 14 – Blower HVAC Fans

23 34 14.10 Blower Type HVAC Fans		Crew	Daily Output	Labor-Hours	Unit	2015 Bare Costs Material	Labor	Equipment	Total	Total Incl O&P
7880	15,000 CFM, 5 HP	Q-20	1	20	Ea.	5,600	1,025		6,625	7,775
7900	17,000 CFM, 7-1/2 HP		.80	25		6,000	1,275		7,275	8,625
7920	20,000 CFM, 7-1/2 HP		.80	25		7,150	1,275		8,425	9,900

23 34 16 – Centrifugal HVAC Fans

23 34 16.10 Centrifugal Type HVAC Fans		Crew	Daily Output	Labor-Hours	Unit	Material	Labor	Equipment	Total	Total Incl O&P
0010	**CENTRIFUGAL TYPE HVAC FANS**									
0200	In-line centrifugal, supply/exhaust booster									
0220	aluminum wheel/hub, disconnect switch, 1/4" S.P.									
0240	500 CFM, 10" diameter connection	Q-20	3	6.667	Ea.	1,300	340		1,640	1,950
0260	1,380 CFM, 12" diameter connection		2	10		1,375	510		1,885	2,325
0280	1,520 CFM, 16" diameter connection		2	10		1,500	510		2,010	2,450
0300	2,560 CFM, 18" diameter connection		1	20		1,625	1,025		2,650	3,400
0320	3,480 CFM, 20" diameter connection		.80	25		1,925	1,275		3,200	4,150
0326	5,080 CFM, 20" diameter connection		.75	26.667		2,100	1,375		3,475	4,475
3500	Centrifugal, airfoil, motor and drive, complete									
3520	1000 CFM, 1/2 HP	Q-20	2.50	8	Ea.	1,900	410		2,310	2,725
3540	2,000 CFM, 1 HP		2	10		2,150	510		2,660	3,150
3560	4,000 CFM, 3 HP		1.80	11.111		2,725	570		3,295	3,900
3580	8,000 CFM, 7-1/2 HP		1.40	14.286		4,100	730		4,830	5,650
3600	12,000 CFM, 10 HP		1	20		5,450	1,025		6,475	7,600
4000	Single width, belt drive, not incl. motor, capacities									
4020	at 2000 fpm, 2.5" S.P. for indicated motor									
4040	6900 CFM, 5 HP	Q-9	2.40	6.667	Ea.	3,775	335		4,110	4,700
4060	10,340 CFM, 7-1/2 HP		2.20	7.273		5,275	365		5,640	6,375
4080	15,320 CFM, 10 HP		2	8		6,200	405		6,605	7,450
4100	22,780 CFM, 15 HP		1.80	8.889		9,225	450		9,675	10,900
4120	33,840 CFM, 20 HP		1.60	10		12,300	505		12,805	14,300
4140	41,400 CFM, 25 HP		1.40	11.429		16,200	575		16,775	18,800
4160	50,100 CFM, 30 HP		.80	20		20,100	1,000		21,100	23,700
4200	Double width wheel, 12,420 CFM, 7.5 HP		2.20	7.273		5,400	365		5,765	6,525
4220	18,620 CFM, 15 HP		2	8		7,750	405		8,155	9,150
4240	27,580 CFM, 20 HP		1.80	8.889		9,475	450		9,925	11,100
4260	40,980 CFM, 25 HP		1.50	10.667		14,800	535		15,335	17,100
4280	60,920 CFM, 40 HP		1	16		20,000	805		20,805	23,300
4300	74,520 CFM, 50 HP		.80	20		23,700	1,000		24,700	27,600
4320	90,160 CFM, 50 HP		.70	22.857		32,400	1,150		33,550	37,400
4340	110,300 CFM, 60 HP		.50	32		43,800	1,600		45,400	51,000
4360	134,960 CFM, 75 HP		.40	40		58,000	2,025		60,025	67,000
5000	Utility set, centrifugal, V belt drive, motor									
5020	1/4" S.P., 1200 CFM, 1/4 HP	Q-20	6	3.333	Ea.	1,750	171		1,921	2,200
5040	1520 CFM, 1/3 HP		5	4		2,225	205		2,430	2,775
5060	1850 CFM, 1/2 HP		4	5		2,200	256		2,456	2,825
5080	2180 CFM, 3/4 HP		3	6.667		2,600	340		2,940	3,400
5100	1/2" S.P., 3600 CFM, 1 HP		2	10		2,700	510		3,210	3,775
5120	4250 CFM, 1-1/2 HP		1.60	12.500		3,300	640		3,940	4,625
5140	4800 CFM, 2 HP		1.40	14.286		4,000	730		4,730	5,525
5160	6920 CFM, 5 HP		1.30	15.385		4,875	790		5,665	6,625
5180	7700 CFM, 7-1/2 HP		1.20	16.667		5,875	855		6,730	7,825
5200	For explosion proof motor, add					15%				
5500	Fans, industrial exhauster, for air which may contain granular matl.									
5520	1000 CFM, 1-1/2 HP	Q-20	2.50	8	Ea.	2,875	410		3,285	3,825
5540	2000 CFM, 3 HP		2	10		3,500	510		4,010	4,650

23 34 HVAC Fans

23 34 16 – Centrifugal HVAC Fans

23 34 16.10 Centrifugal Type HVAC Fans		Crew	Daily Output	Labor-Hours	Unit	2015 Bare Costs Material	Labor	Equipment	Total	Total Incl O&P
5560	4000 CFM, 7-1/2 HP	Q-20	1.80	11.111	Ea.	4,875	570		5,445	6,250
5580	8000 CFM, 15 HP		1.40	14.286		6,275	730		7,005	8,050
5600	12,000 CFM, 30 HP		1	20		10,200	1,025		11,225	12,800
7000	Roof exhauster, centrifugal, aluminum housing, 12" galvanized									
7020	curb, bird screen, back draft damper, 1/4" S.P.									
7100	Direct drive, 320 CFM, 11" sq. damper	Q-20	7	2.857	Ea.	705	146		851	1,000
7120	600 CFM, 11" sq. damper		6	3.333		900	171		1,071	1,250
7140	815 CFM, 13" sq. damper		5	4		900	205		1,105	1,300
7160	1450 CFM, 13" sq. damper		4.20	4.762		1,450	244		1,694	1,975
7180	2050 CFM, 16" sq. damper		4	5		1,725	256		1,981	2,300
7200	V-belt drive, 1650 CFM, 12" sq. damper		6	3.333		1,300	171		1,471	1,700
7220	2750 CFM, 21" sq. damper		5	4		1,550	205		1,755	2,025
7230	3500 CFM, 21" sq. damper		4.50	4.444		1,725	228		1,953	2,250
7240	4910 CFM, 23" sq. damper		4	5		2,125	256		2,381	2,725
7260	8525 CFM, 28" sq. damper		3	6.667		2,800	340		3,140	3,600
7280	13,760 CFM, 35" sq. damper		2	10		3,925	510		4,435	5,125
7300	20,558 CFM, 43" sq. damper		1	20		7,875	1,025		8,900	10,300
7320	For 2 speed winding, add					15%				
7340	For explosionproof motor, add					600			600	660
7360	For belt driven, top discharge, add					15%				
8500	Wall exhausters, centrifugal, auto damper, 1/8" S.P.									
8520	Direct drive, 610 CFM, 1/20 HP	Q-20	14	1.429	Ea.	425	73		498	580
8540	796 CFM, 1/12 HP		13	1.538		880	79		959	1,100
8560	822 CFM, 1/6 HP		12	1.667		1,075	85.50		1,160.50	1,300
8580	1,320 CFM, 1/4 HP		12	1.667		1,250	85.50		1,335.50	1,500
8600	1756 CFM, 1/4 HP		11	1.818		1,250	93		1,343	1,525
8620	1983 CFM, 1/4 HP		10	2		1,300	102		1,402	1,575
8640	2900 CFM, 1/2 HP		9	2.222		1,375	114		1,489	1,675
8660	3307 CFM, 3/4 HP		8	2.500		1,475	128		1,603	1,825
9500	V-belt drive, 3 phase									
9520	2,800 CFM, 1/4 HP	Q-20	9	2.222	Ea.	1,925	114		2,039	2,300
9540	3,740 CFM, 1/2 HP		8	2.500		2,000	128		2,128	2,400
9560	4400 CFM, 3/4 HP		7	2.857		2,025	146		2,171	2,450
9580	5700 CFM, 1-1/2 HP		6	3.333		2,100	171		2,271	2,575
9900	Minimum labor/equipment charge	1 Elec	4	2	Job		109		109	169

23 34 23 – HVAC Power Ventilators

23 34 23.10 HVAC Power Circulators and Ventilators		Crew	Daily Output	Labor-Hours	Unit	Material	Labor	Equipment	Total	Total Incl O&P
0010	**HVAC POWER CIRCULATORS AND VENTILATORS**									
3000	Paddle blade air circulator, 3 speed switch									
3020	42", 5,000 CFM high, 3000 CFM low G	1 Elec	2.40	3.333	Ea.	163	182		345	460
3040	52", 6,500 CFM high, 4000 CFM low G	"	2.20	3.636	"	170	199		369	495
3100	For antique white motor, same cost									
3200	For brass plated motor, same cost									
3300	For light adaptor kit, add G				Ea.	41			41	45
4000	High volume, low speed (HVLS), paddle blade air circulator									
4010	Variable speed, reversible,1 HP motor, motor control panel,									
4020	motor drive cable, control cable with remote, and safety cable.									
4140	8' diameter	Q-5	1.50	10.667	Ea.	4,725	575		5,300	6,100
4150	10' diameter		1.45	11.034		4,750	595		5,345	6,150
4160	12' diameter		1.40	11.429		4,975	615		5,590	6,425
4170	14' diameter		1.35	11.852		5,050	635		5,685	6,575
4180	16' diameter		1.30	12.308		5,100	660		5,760	6,625

23 34 23.10 HVAC Power Circulators and Ventilators		Crew	Daily Output	Labor-Hours	Unit	Material	2015 Bare Costs Labor	Equipment	Total	Total Incl O&P
4190	18′ diameter	Q-5	1.25	12.800	Ea.	5,150	690		5,840	6,725
4200	20′ diameter		1.20	13.333		5,475	715		6,190	7,150
4210	24′ diameter	↓	1	16	↓	5,600	860		6,460	7,525
6000	Propeller exhaust, wall shutter									
6020	Direct drive, one speed, .075″ S.P.									
6100	653 CFM, 1/30 HP	Q-20	10	2	Ea.	192	102		294	370
6120	1033 CFM, 1/20 HP		9	2.222		287	114		401	495
6140	1323 CFM, 1/15 HP		8	2.500		320	128		448	550
6160	2444 CFM, 1/4 HP	↓	7	2.857	↓	400	146		546	670
6300	V-belt drive, 3 phase									
6320	6175 CFM, 3/4 HP	Q-20	5	4	Ea.	2,950	205		3,155	3,575
6340	7500 CFM, 3/4 HP		5	4		3,025	205		3,230	3,650
6360	10,100 CFM, 1 HP		4.50	4.444		3,175	228		3,403	3,850
6380	14,300 CFM, 1-1/2 HP		4	5		3,450	256		3,706	4,175
6400	19,800 CFM, 2 HP		3	6.667		1,950	340		2,290	2,650
6420	26,250 CFM, 3 HP		2.60	7.692		2,150	395		2,545	3,000
6440	38,500 CFM, 5 HP		2.20	9.091		2,550	465		3,015	3,525
6460	46,000 CFM, 7-1/2 HP		2	10		2,725	510		3,235	3,800
6480	51,500 CFM, 10 HP	↓	1.80	11.111	↓	2,825	570		3,395	4,025
6490	V-belt drive, 115 V, residential, whole house									
6500	Ceiling-wall, 5200 CFM, 1/4 HP, 30″ x 30″	1 Shee	6	1.333	Ea.	545	74.50		619.50	715
6530	13,200 CFM, 1/3 HP, 48″ x 48″		4	2		655	112		767	895
6540	15,445 CFM, 1/2 HP, 48″ x 48″		4	2		680	112		792	925
6550	17,025 CFM, 1/2 HP, 54″ x 54″	↓	4	2	↓	1,325	112		1,437	1,625
6560	For two speed motor, add					20%				
6570	Shutter, automatic, ceiling/wall									
6580	30″ x 30″	1 Shee	8	1	Ea.	177	56		233	283
6590	36″ x 36″		8	1		202	56		258	310
6600	42″ x 42″		8	1		252	56		308	365
6610	48″ x 48″		7	1.143		274	64		338	400
6620	54″ x 54″		6	1.333		360	74.50		434.50	515
6630	Timer, shut off, to 12 hour	↓	20	.400	↓	58	22.50		80.50	99.50
6650	Residential, bath exhaust, grille, back draft damper									
6660	50 CFM	Q-20	24	.833	Ea.	63.50	42.50		106	137
6670	110 CFM		22	.909		98	46.50		144.50	181
6680	Light combination, squirrel cage, 100 watt, 70 CFM	↓	24	.833	↓	112	42.50		154.50	190
6700	Light/heater combination, ceiling mounted									
6710	70 CFM, 1450 watt	Q-20	24	.833	Ea.	162	42.50		204.50	245
6800	Heater combination, recessed, 70 CFM		24	.833		67.50	42.50		110	141
6820	With 2 infrared bulbs		23	.870		105	44.50		149.50	186
6900	Kitchen exhaust, grille, complete, 160 CFM		22	.909		108	46.50		154.50	192
6910	180 CFM		20	1		90.50	51		141.50	180
6920	270 CFM		18	1.111		171	57		228	278
6930	350 CFM	↓	16	1.250	↓	129	64		193	243
6940	Residential roof jacks and wall caps									
6944	Wall cap with back draft damper									
6946	3″ & 4″ diam. round duct	1 Shee	11	.727	Ea.	26	40.50		66.50	92.50
6948	6″ diam. round duct	″	11	.727	″	64	40.50		104.50	135
6958	Roof jack with bird screen and back draft damper									
6960	3″ & 4″ diam. round duct	1 Shee	11	.727	Ea.	26	40.50		66.50	92.50
6962	3-1/4″ x 10″ rectangular duct	″	10	.800	″	48.50	45		93.50	124
6980	Transition									
6982	3-1/4″ x 10″ to 6″ diam. round	1 Shee	20	.400	Ea.	32	22.50		54.50	71

23 34 HVAC Fans

23 34 23 – HVAC Power Ventilators

23 34 23.10 HVAC Power Circulators and Ventilators		Crew	Daily Output	Labor-Hours	Unit	Material	Labor	Equipment	Total	Total Incl O&P
						2015 Bare Costs				
8020	Attic, roof type									
8030	Aluminum dome, damper & curb									
8080	12" diameter, 1000 CFM (gravity)	1 Elec	10	.800	Ea.	575	44		619	700
8090	16" diameter, 1500 CFM (gravity)		9	.889		690	48.50		738.50	835
8100	20" diameter, 2500 CFM (gravity)		8	1		850	54.50		904.50	1,025
8110	26" diameter, 4000 CFM (gravity)		7	1.143		1,025	62.50		1,087.50	1,225
8120	32" diameter, 6500 CFM (gravity)		6	1.333		1,400	73		1,473	1,675
8130	38" diameter, 8000 CFM (gravity)		5	1.600		2,100	87.50		2,187.50	2,425
8140	50" diameter, 13,000 CFM (gravity)		4	2		3,025	109		3,134	3,500
8160	Plastic, ABS dome									
8180	1050 CFM	1 Elec	14	.571	Ea.	168	31.50		199.50	234
8200	1600 CFM	"	12	.667	"	252	36.50		288.50	335
8240	Attic, wall type, with shutter, one speed									
8250	12" diameter, 1000 CFM	1 Elec	14	.571	Ea.	395	31.50		426.50	485
8260	14" diameter, 1500 CFM		12	.667		430	36.50		466.50	530
8270	16" diameter, 2000 CFM		9	.889		485	48.50		533.50	610
8290	Whole house, wall type, with shutter, one speed									
8300	30" diameter, 4800 CFM	1 Elec	7	1.143	Ea.	1,050	62.50		1,112.50	1,250
8310	36" diameter, 7000 CFM		6	1.333		1,125	73		1,198	1,375
8320	42" diameter, 10,000 CFM		5	1.600		1,275	87.50		1,362.50	1,525
8330	48" diameter, 16,000 CFM		4	2		1,575	109		1,684	1,925
8340	For two speed, add					95			95	105
8350	Whole house, lay-down type, with shutter, one speed									
8360	30" diameter, 4500 CFM	1 Elec	8	1	Ea.	1,100	54.50		1,154.50	1,300
8370	36" diameter, 6500 CFM		7	1.143		1,200	62.50		1,262.50	1,400
8380	42" diameter, 9000 CFM		6	1.333		1,300	73		1,373	1,575
8390	48" diameter, 12,000 CFM		5	1.600		1,500	87.50		1,587.50	1,750
8440	For two speed, add					71.50			71.50	78.50
8450	For 12 hour timer switch, add	1 Elec	32	.250		71.50	13.70		85.20	99.50

23 34 33 – Air Curtains

23 34 33.10 Air Barrier Curtains		Crew	Daily Output	Labor-Hours	Unit	Material	Labor	Equipment	Total	Total Incl O&P
0010	**AIR BARRIER CURTAINS**, Incl. motor starters, transformers,									
0050	and door switches									
2450	Conveyor openings or service windows									
3000	Service window, 5' high x 25" wide	2 Shee	5	3.200	Ea.	305	179		484	620
3100	Environmental separation									
3110	Door heights up to 8', low profile, super quiet									
3120	Unheated, variable speed									
3130	36" wide	2 Shee	4	4	Ea.	610	224		834	1,025
3134	42" wide		3.80	4.211		635	236		871	1,075
3138	48" wide		3.60	4.444		655	249		904	1,125
3142	60" wide		3.40	4.706		685	263		948	1,175
3146	72" wide	Q-3	4.60	6.957		850	390		1,240	1,550
3150	96" wide		4.40	7.273		1,300	405		1,705	2,050
3154	120" wide		4.20	7.619		1,400	425		1,825	2,200
3158	144" wide		4	8		1,700	445		2,145	2,575
3200	Door heights up to 10'									
3210	Unheated									
3230	36" wide	2 Shee	3.80	4.211	Ea.	645	236		881	1,075
3234	42" wide		3.60	4.444		665	249		914	1,125
3238	48" wide		3.40	4.706		685	263		948	1,175
3242	60" wide		3.20	5		1,025	280		1,305	1,575

23 34 HVAC Fans

23 34 33 – Air Curtains

23 34 33.10 Air Barrier Curtains		Crew	Daily Output	Labor-Hours	Unit	Material	2015 Bare Costs Labor	Equipment	Total	Total Incl O&P
3246	72″ wide	Q-3	4.40	7.273	Ea.	1,075	405		1,480	1,825
3250	96″ wide		4.20	7.619		1,275	425		1,700	2,075
3254	120″ wide		4	8		1,775	445		2,220	2,650
3258	144″ wide		3.80	8.421		1,925	470		2,395	2,825
3300	Door heights up to 12′									
3310	Unheated									
3334	42″ wide	2 Shee	3.40	4.706	Ea.	920	263		1,183	1,425
3338	48″ wide		3.20	5		925	280		1,205	1,475
3342	60″ wide		3	5.333		945	298		1,243	1,525
3346	72″ wide	Q-3	4.20	7.619		1,675	425		2,100	2,500
3350	96″ wide		4	8		1,825	445		2,270	2,700
3354	120″ wide		3.80	8.421		2,225	470		2,695	3,175
3358	144″ wide		3.60	8.889		2,375	495		2,870	3,400
3400	Door heights up to 16′									
3410	Unheated									
3438	48″ wide	2 Shee	3	5.333	Ea.	1,100	298		1,398	1,700
3442	60″ wide	″	2.80	5.714		1,175	320		1,495	1,775
3446	72″ wide	Q-3	3.80	8.421		2,000	470		2,470	2,925
3450	96″ wide		3.60	8.889		2,075	495		2,570	3,075
3454	120″ wide		3.40	9.412		2,825	525		3,350	3,925
3458	144″ wide		3.20	10		2,925	560		3,485	4,075
3470	Heated, electric									
3474	48″ wide	2 Shee	2.90	5.517	Ea.	1,850	310		2,160	2,500
3478	60″ wide	″	2.70	5.926		1,875	330		2,205	2,600
3482	72″ wide	Q-3	3.70	8.649		3,175	485		3,660	4,250
3486	96″ wide		3.50	9.143		3,275	510		3,785	4,400
3490	120″ wide		3.30	9.697		3,325	540		3,865	4,525
3494	144″ wide		3.10	10.323		4,475	575		5,050	5,825

23 35 Special Exhaust Systems

23 35 16 – Engine Exhaust Systems

23 35 16.10 Engine Exhaust Removal Systems

		Crew	Daily Output	Labor-Hours	Unit	Material	Labor	Equipment	Total	Total Incl O&P
0010	**ENGINE EXHAUST REMOVAL SYSTEMS**									
0500	Engine exhaust, garage, in-floor system									
0510	Single tube outlet assemblies									
0520	For transite pipe ducting, self-storing tube									
0530	3″ tubing adapter plate	1 Shee	16	.500	Ea.	240	28		268	310
0540	4″ tubing adapter plate		16	.500		245	28		273	315
0550	5″ tubing adapter plate		16	.500		245	28		273	315
0600	For vitrified tile ducting									
0610	3″ tubing adapter plate, self-storing tube	1 Shee	16	.500	Ea.	241	28		269	310
0620	4″ tubing adapter plate, self-storing tube		16	.500		245	28		273	315
0660	5″ tubing adapter plate, self-storing tube		16	.500		245	28		273	315
0800	Two tube outlet assemblies									
0810	For transite pipe ducting, self-storing tube									
0820	3″ tubing, dual exhaust adapter plate	1 Shee	16	.500	Ea.	240	28		268	310
0850	For vitrified tile ducting									
0860	3″ tubing, dual exhaust, self-storing tube	1 Shee	16	.500	Ea.	298	28		326	375
0870	3″ tubing, double outlet, non-storing tubes	″	16	.500	″	298	28		326	375
0900	Accessories for metal tubing, (overhead systems also)									
0910	Adapters, for metal tubing end									

23 35 16 – Engine Exhaust Systems

23 35 16.10 Engine Exhaust Removal Systems		Crew	Daily Output	Labor-Hours	Unit	Material	Labor	Equipment	Total	Total Incl O&P
						2015 Bare Costs				
0920	3″ tail pipe type				Ea.	49.50			49.50	54.50
0930	4″ tail pipe type					53			53	58.50
0940	5″ tail pipe type					54			54	59.50
0990	5″ diesel stack type					315			315	350
1000	6″ diesel stack type				↓	330			330	360
1100	Bullnose (guide) required for in-floor assemblies									
1110	3″ tubing size				Ea.	28			28	30.50
1120	4″ tubing size					29.50			29.50	32.50
1130	5″ tubing size				↓	31.50			31.50	34.50
1150	Plain rings, for tubing end									
1160	3″ tubing size				Ea.	24.50			24.50	26.50
1170	4″ tubing size				″	41.50			41.50	45.50
1200	Tubing, galvanized, flexible, (for overhead systems also)									
1210	3″ ID				L.F.	11.60			11.60	12.75
1220	4″ ID					14.20			14.20	15.65
1230	5″ ID					16.75			16.75	18.40
1240	6″ ID				↓	19.35			19.35	21.50
1250	Stainless steel, flexible, (for overhead system, also)									
1260	3″ ID				L.F.	25.50			25.50	28
1270	4″ ID					35			35	38.50
1280	5″ ID					39.50			39.50	43.50
1290	6″ ID				↓	46			46	50.50
1500	Engine exhaust, garage, overhead components, for neoprene tubing									
1510	Alternate metal tubing & accessories see above									
1550	Adapters, for neoprene tubing end									
1560	3″ tail pipe, adjustable, neoprene				Ea.	56			56	61.50
1570	3″ tail pipe, heavy wall neoprene					61			61	67.50
1580	4″ tail pipe, heavy wall neoprene					100			100	110
1590	5″ tail pipe, heavy wall neoprene				↓	105			105	116
1650	Connectors, tubing									
1660	3″ interior, aluminum				Ea.	24.50			24.50	26.50
1670	4″ interior, aluminum					41.50			41.50	45.50
1710	5″ interior, neoprene					61			61	67.50
1750	3″ spiralock, neoprene					24.50			24.50	26.50
1760	4″ spiralock, neoprene					41.50			41.50	45.50
1780	Y for 3″ ID tubing, neoprene, dual exhaust					195			195	215
1790	Y for 4″ ID tubing, aluminum, dual exhaust				↓	195			195	215
1850	Elbows, aluminum, splice into tubing for strap									
1860	3″ neoprene tubing size				Ea.	48.50			48.50	53.50
1870	4″ neoprene tubing size					40.50			40.50	44.50
1900	Flange assemblies, connect tubing to overhead duct				↓	73			73	80
2000	Hardware and accessories									
2020	Cable, galvanized, 1/8″ diameter				L.F.	.51			.51	.56
2040	Cleat, tie down cable or rope				Ea.	5.40			5.40	5.90
2060	Pulley					7.40			7.40	8.15
2080	Pulley hook, universal				↓	5.40			5.40	5.90
2100	Rope, nylon, 1/4″ diameter				L.F.	.41			.41	.45
2120	Winch, 1″ diameter				Ea.	114			114	126
2150	Lifting strap, mounts on neoprene									
2160	3″ tubing size				Ea.	27			27	29.50
2170	4″ tubing size					27			27	29.50
2180	5″ tubing size					27			27	29.50
2190	6″ tubing size				↓	27			27	29.50

23 35 Special Exhaust Systems

23 35 16 – Engine Exhaust Systems

23 35 16.10 Engine Exhaust Removal Systems		Crew	Daily Output	Labor-Hours	Unit	Material	2015 Bare Costs Labor	Equipment	Total	Total Incl O&P
2200	Tubing, neoprene, 11′ lengths									
2210	3″ ID				L.F.	10.90			10.90	11.95
2220	4″ ID					18.15			18.15	20
2230	5″ ID					27			27	29.50
2500	Engine exhaust, thru-door outlet									
2510	3″ tube size	1 Carp	16	.500	Ea.	59.50	23.50		83	104
2530	4″ tube size	"	16	.500	"	74	23.50		97.50	120

23 35 43 – Welding Fume Elimination Systems

23 35 43.10 Welding Fume Elimination System Components		Crew	Daily Output	Labor-Hours	Unit	Material	Labor	Equipment	Total	Total Incl O&P
0010	**WELDING FUME ELIMINATION SYSTEM COMPONENTS**									
7500	Welding fume elimination accessories for garage exhaust systems									
7600	Cut off (blast gate)									
7610	3″ tubing size, 3″ x 6″ opening	1 Shee	24	.333	Ea.	26	18.65		44.65	58
7620	4″ tubing size, 4″ x 8″ opening		24	.333		26	18.65		44.65	58
7630	5″ tubing size, 5″ x 10″ opening		24	.333		31.50	18.65		50.15	64
7640	6″ tubing size		24	.333		32.50	18.65		51.15	65
7650	8″ tubing size		24	.333		45	18.65		63.65	79
7700	Hoods, magnetic, with handle & screen									
7710	3″ tubing size, 3″ x 6″ opening	1 Shee	24	.333	Ea.	107	18.65		125.65	148
7720	4″ tubing size, 4″ x 8″ opening		24	.333		97	18.65		115.65	137
7730	5″ tubing size, 5″ x 10″ opening		24	.333		97	18.65		115.65	137

23 36 Air Terminal Units

23 36 13 – Constant-Air-Volume Units

23 36 13.10 Constant Volume Mixing Boxes		Crew	Daily Output	Labor-Hours	Unit	Material	Labor	Equipment	Total	Total Incl O&P
0010	**CONSTANT VOLUME MIXING BOXES**									
5180	Mixing box, includes electric or pneumatic motor									
5192	Recommend use with silencer, see Line 23 33 19.10 0010									
5200	Constant volume, 150 to 270 CFM	Q-9	12	1.333	Ea.	700	67		767	875
5210	270 to 600 CFM		11	1.455		720	73		793	910
5230	550 to 1000 CFM		9	1.778		720	89.50		809.50	935
5240	1000 to 1600 CFM		8	2		740	101		841	975
5250	1300 to 1900 CFM		6	2.667		755	134		889	1,050
5260	550 to 2640 CFM		5.60	2.857		835	144		979	1,150
5270	650 to 3120 CFM		5.20	3.077		885	155		1,040	1,225

23 36 16 – Variable-Air-Volume Units

23 36 16.10 Variable Volume Mixing Boxes		Crew	Daily Output	Labor-Hours	Unit	Material	Labor	Equipment	Total	Total Incl O&P
0010	**VARIABLE VOLUME MIXING BOXES**									
5180	Mixing box, includes electric or pneumatic motor									
5192	Recommend use with attenuator, see Line 23 33 19.10 0010									
5500	VAV Cool only, pneumatic, pressure independent 300 to 600 CFM	Q-9	11	1.455	Ea.	730	73		803	920
5510	500 to 1000 CFM		9	1.778		750	89.50		839.50	965
5520	800 to 1600 CFM		9	1.778		775	89.50		864.50	995
5530	1100 to 2000 CFM		8	2		795	101		896	1,025
5540	1500 to 3000 CFM		7	2.286		845	115		960	1,100
5550	2000 to 4000 CFM		6	2.667		855	134		989	1,150
5560	For electric, w/thermostat, pressure dependent, add					23			23	25.50
5600	VAV, HW coils, damper, actuator and t'stat									
5610	200 CFM	Q-9	11	1.455	Ea.	790	73		863	985
5620	400 CFM		10	1.600		800	80.50		880.50	1,000

23 36 Air Terminal Units

23 36 16 – Variable-Air-Volume Units

23 36 16.10 Variable Volume Mixing Boxes		Crew	Daily Output	Labor-Hours	Unit	Material (2015 Bare Costs)	Labor (2015 Bare Costs)	Equipment (2015 Bare Costs)	Total (2015 Bare Costs)	Total Incl O&P
5630	600 CFM	Q-9	10	1.600	Ea.	800	80.50		880.50	1,000
5640	800 CFM		8	2		820	101		921	1,075
5650	1000 CFM		8	2		820	101		921	1,075
5660	1250 CFM		6	2.667		900	134		1,034	1,200
5670	1500 CFM		6	2.667		900	134		1,034	1,200
5680	2000 CFM		4	4		975	201		1,176	1,400
5684	3000 CFM		3.80	4.211		1,075	212		1,287	1,500
5700	VAV Cool only, fan powered, damper, actuator, thermostat									
5710	200 CFM	Q-9	10	1.600	Ea.	1,075	80.50		1,155.50	1,325
5720	400 CFM		9	1.778		1,150	89.50		1,239.50	1,425
5730	600 CFM		9	1.778		1,150	89.50		1,239.50	1,425
5740	800 CFM		7	2.286		1,225	115		1,340	1,525
5750	1000 CFM		7	2.286		1,225	115		1,340	1,525
5760	1250 CFM		5	3.200		1,325	161		1,486	1,725
5770	1500 CFM		5	3.200		1,325	161		1,486	1,725
5780	2000 CFM		4	4		1,475	201		1,676	1,950
5800	VAV Fan powr'd, with HW coils, dampers, actuators, t'stat									
5810	200 CFM	Q-9	10	1.600	Ea.	1,350	80.50		1,430.50	1,600
5820	400 CFM		9	1.778		1,375	89.50		1,464.50	1,675
5830	600 CFM		9	1.778		1,375	89.50		1,464.50	1,675
5840	800 CFM		7	2.286		1,475	115		1,590	1,775
5850	1000 CFM		7	2.286		1,475	115		1,590	1,775
5860	1250 CFM		5	3.200		1,675	161		1,836	2,075
5870	1500 CFM		5	3.200		1,675	161		1,836	2,075
5880	2000 CFM		3.50	4.571		1,800	230		2,030	2,350

23 37 Air Outlets and Inlets

23 37 13 – Diffusers, Registers, and Grilles

23 37 13.10 Diffusers		Crew	Daily Output	Labor-Hours	Unit	Material	Labor	Equipment	Total	Total Incl O&P
0010	**DIFFUSERS**, Aluminum, opposed blade damper unless noted									
0100	Ceiling, linear, also for sidewall									
0120	2" wide	1 Shee	32	.250	L.F.	16.20	14		30.20	40
0140	3" wide		30	.267		18.45	14.90		33.35	44
0160	4" wide		26	.308		21.50	17.20		38.70	50.50
0180	6" wide		24	.333		26.50	18.65		45.15	59
0200	8" wide		22	.364		31	20.50		51.50	66.50
0220	10" wide		20	.400		35.50	22.50		58	74.50
0240	12" wide		18	.444		40	25		65	83
0260	For floor or sill application, add					15%				
0500	Perforated, 24" x 24" lay-in panel size, 6" x 6"	1 Shee	16	.500	Ea.	151	28		179	210
0520	8" x 8"		15	.533		159	30		189	222
0530	9" x 9"		14	.571		161	32		193	228
0540	10" x 10"		14	.571		162	32		194	229
0560	12" x 12"		12	.667		168	37.50		205.50	244
0580	15" x 15"		11	.727		171	40.50		211.50	252
0590	16" x 16"		11	.727		189	40.50		229.50	271
0600	18" x 18"		10	.800		202	45		247	293
0610	20" x 20"		10	.800		218	45		263	310
0620	24" x 24"		9	.889		239	49.50		288.50	340
1000	Rectangular, 1 to 4 way blow, 6" x 6"		16	.500		50	28		78	99
1010	8" x 8"		15	.533		58	30		88	111

23 37 13 – Diffusers, Registers, and Grilles

23 37 13.10 Diffusers		Crew	Daily Output	Labor-Hours	Unit	Material	2015 Bare Costs Labor	Equipment	Total	Total Incl O&P
1014	9″ x 9″	1 Shee	15	.533	Ea.	66	30		96	120
1016	10″ x 10″		15	.533		80	30		110	135
1020	12″ x 6″		15	.533		71.50	30		101.50	126
1040	12″ x 9″		14	.571		75.50	32		107.50	134
1060	12″ x 12″		12	.667		84.50	37.50		122	152
1070	14″ x 6″		13	.615		77.50	34.50		112	140
1074	14″ x 14″		12	.667		128	37.50		165.50	199
1080	18″ x 12″		11	.727		137	40.50		177.50	215
1120	15″ x 15″		10	.800		110	45		155	192
1140	18″ x 15″		9	.889		155	49.50		204.50	250
1150	18″ x 18″		9	.889		138	49.50		187.50	230
1160	21″ x 21″		8	1		215	56		271	325
1170	24″ x 12″		10	.800		163	45		208	251
1180	24″ x 24″		7	1.143		270	64		334	395
1500	Round, butterfly damper, steel, diffuser size, 6″ diameter		18	.444		9	25		34	49
1520	8″ diameter		16	.500		9.55	28		37.55	54.50
1540	10″ diameter		14	.571		11.85	32		43.85	63.50
1560	12″ diameter		12	.667		15.65	37.50		53.15	76
1580	14″ diameter		10	.800		19.55	45		64.55	92
1600	18″ diameter		9	.889		47.50	49.50		97	131
2000	T bar mounting, 24″ x 24″ lay-in frame, 6″ x 6″		16	.500		61	28		89	111
2020	8″ x 8″		14	.571		62	32		94	119
2040	12″ x 12″		12	.667		74	37.50		111.50	140
2060	16″ x 16″		11	.727		94.50	40.50		135	168
2080	18″ x 18″	↓	10	.800	↓	106	45		151	188
2500	Combination supply and return									
2520	21″ x 21″ supply, 15″ x 15″ return	Q-9	10	1.600	Ea.	111	80.50		191.50	249
2540	24″ x 24″ supply, 18″ x 18″ return		9.50	1.684		140	85		225	287
2560	27″ x 27″ supply, 18″ x 18″ return		9	1.778		170	89.50		259.50	330
2580	30″ x 30″ supply, 21″ x 21″ return		8.50	1.882		215	95		310	385
2600	33″ x 33″ supply, 21″ x 21″ return		8	2		252	101		353	435
2620	36″ x 36″ supply, 24″ x 24″ return	↓	7.50	2.133	↓	296	107		403	495
3000	Baseboard, white enameled steel									
3100	18″ long	1 Shee	20	.400	Ea.	4.43	22.50		26.93	40.50
3120	24″ long		18	.444		8.65	25		33.65	48.50
3140	48″ long	↓	16	.500		15.80	28		43.80	61.50
3400	For matching return, deduct				↓	10%				
4000	Floor, steel, adjustable pattern									
4100	2″ x 10″	1 Shee	34	.235	Ea.	5.95	13.15		19.10	27.50
4120	2″ x 12″		32	.250		6.55	14		20.55	29
4140	2″ x 14″		30	.267		7.40	14.90		22.30	31.50
4200	4″ x 10″		28	.286		6.90	16		22.90	32.50
4220	4″ x 12″		26	.308		7.60	17.20		24.80	35.50
4240	4″ x 14″		25	.320		8.60	17.90		26.50	38
4260	6″ x 10″		26	.308		10.35	17.20		27.55	38.50
4280	6″ x 12″		24	.333		10.80	18.65		29.45	41.50
4300	6″ x 14″	↓	22	.364	↓	11.30	20.50		31.80	44.50
5000	Sidewall, aluminum, 3 way dispersion									
5100	8″ x 4″	1 Shee	24	.333	Ea.	3.25	18.65		21.90	33
5110	8″ x 6″		24	.333		3.48	18.65		22.13	33.50
5120	10″ x 4″		22	.364		3.42	20.50		23.92	36
5130	10″ x 6″		22	.364		3.66	20.50		24.16	36
5140	10″ x 8″	↓	20	.400	↓	5	22.50		27.50	41

23 37 13 – Diffusers, Registers, and Grilles

23 37 13.10 Diffusers		Crew	Daily Output	Labor-Hours	Unit	2015 Bare Costs Material	Labor	Equipment	Total	Total Incl O&P
5160	12" x 4"	1 Shee	18	.444	Ea.	3.69	25		28.69	43
5170	12" x 6"		17	.471		4.07	26.50		30.57	46
5180	12" x 8"		16	.500		5.30	28		33.30	50
5200	14" x 4"		15	.533		4.11	30		34.11	51.50
5220	14" x 6"		14	.571		4.53	32		36.53	55.50
5240	14" x 8"		13	.615		5.90	34.50		40.40	61
5260	16" x 6"		12.50	.640		6.45	36		42.45	63.50
6000	For steel diffusers instead of aluminum, deduct					10%				
9000	Minimum labor/equipment charge	1 Shee	4	2	Job		112		112	177

23 37 13.30 Grilles		Crew	Daily Output	Labor-Hours	Unit	Material	Labor	Equipment	Total	Total Incl O&P
0010	**GRILLES**									
0020	Aluminum, unless noted otherwise									
0100	Air supply, single deflection, adjustable									
0120	8" x 4"	1 Shee	30	.267	Ea.	6.70	14.90		21.60	31
0140	8" x 8"		28	.286		8.15	16		24.15	34
0160	10" x 4"		24	.333		7.40	18.65		26.05	37.50
0180	10" x 10"		23	.348		9.40	19.45		28.85	41
0200	12" x 6"		23	.348		8.75	19.45		28.20	40
0220	12" x 12"		22	.364		11.50	20.50		32	44.50
0230	14" x 6"		23	.348		9.75	19.45		29.20	41.50
0240	14" x 8"		23	.348		10.45	19.45		29.90	42
0260	14" x 14"		22	.364		14.10	20.50		34.60	47.50
0270	18" x 8"		23	.348		12.05	19.45		31.50	44
0280	18" x 10"		23	.348		13.80	19.45		33.25	45.50
0300	18" x 18"		21	.381		21.50	21.50		43	57
0320	20" x 12"		22	.364		15.65	20.50		36.15	49.50
0340	20" x 20"		21	.381		26	21.50		47.50	62
0350	24" x 8"		20	.400		15.05	22.50		37.55	52
0360	24" x 14"		18	.444		21	25		46	62.50
0380	24" x 24"		15	.533		37	30		67	87.50
0400	30" x 8"		20	.400		19.35	22.50		41.85	57
0420	30" x 10"		19	.421		21.50	23.50		45	60.50
0440	30" x 12"		18	.444		23.50	25		48.50	65
0460	30" x 16"		17	.471		30.50	26.50		57	75
0480	30" x 18"		17	.471		34	26.50		60.50	79
0500	30" x 30"		14	.571		61	32		93	118
0520	36" x 12"		17	.471		29	26.50		55.50	73.50
0540	36" x 16"		16	.500		37	28		65	85
0560	36" x 18"		15	.533		41	30		71	92.50
0570	36" x 20"		15	.533		45	30		75	96.50
0580	36" x 24"		14	.571		58.50	32		90.50	115
0600	36" x 28"		13	.615		72	34.50		106.50	134
0620	36" x 30"		12	.667		79	37.50		116.50	146
0640	36" x 32"		12	.667		86.50	37.50		124	154
0660	36" x 34"		11	.727		94	40.50		134.50	168
0680	36" x 36"		11	.727		102	40.50		142.50	176
0700	For double deflecting, add					70%				
1000	Air return, steel, 6" x 6"	1 Shee	26	.308		18.70	17.20		35.90	47.50
1020	10" x 6"		24	.333		18.70	18.65		37.35	50
1040	14" x 6"		23	.348		22.50	19.45		41.95	55.50
1060	10" x 8"		23	.348		22.50	19.45		41.95	55.50
1080	16" x 8"		22	.364		26.50	20.50		47	61

23 37 Air Outlets and Inlets

23 37 13 – Diffusers, Registers, and Grilles

23 37 13.30 Grilles		Crew	Daily Output	Labor-Hours	Unit	Material	2015 Bare Costs Labor	Equipment	Total	Total Incl O&P
1100	12" x 12"	1 Shee	22	.364	Ea.	26.50	20.50		47	61
1120	24" x 12"		18	.444		36	25		61	78.50
1140	30" x 12"		16	.500		46	28		74	95
1160	14" x 14"		22	.364		32	20.50		52.50	67
1180	16" x 16"		22	.364		34	20.50		54.50	69.50
1200	18" x 18"		21	.381		38	21.50		59.50	75.50
1220	24" x 18"		16	.500		43.50	28		71.50	92
1240	36" x 18"		15	.533		68	30		98	122
1260	24" x 24"		15	.533		53	30		83	105
1280	36" x 24"		14	.571		75	32		107	133
1300	48" x 24"		12	.667		128	37.50		165.50	199
1320	48" x 30"		11	.727		164	40.50		204.50	245
1340	36" x 36"		13	.615		100	34.50		134.50	165
1360	48" x 36"		11	.727		205	40.50		245.50	290
1380	48" x 48"		8	1		250	56		306	365
2000	Door grilles, 12" x 12"		22	.364		68	20.50		88.50	107
2020	18" x 12"		22	.364		78.50	20.50		99	119
2040	24" x 12"		18	.444		88.50	25		113.50	137
2060	18" x 18"		18	.444		109	25		134	159
2080	24" x 18"		16	.500		124	28		152	180
2100	24" x 24"		15	.533		175	30		205	240
3000	Filter grille with filter, 12" x 12"		24	.333		53	18.65		71.65	87.50
3020	18" x 12"		20	.400		71	22.50		93.50	114
3040	24" x 18"		18	.444		83.50	25		108.50	131
3060	24" x 24"		16	.500		98	28		126	152
3080	30" x 24"		14	.571		119	32		151	182
3100	30" x 30"		13	.615		141	34.50		175.50	210
3950	Eggcrate, framed, 6" x 6" opening		26	.308		15.40	17.20		32.60	44
3954	8" x 8" opening		24	.333		18.15	18.65		36.80	49.50
3960	10" x 10" opening		23	.348		23	19.45		42.45	56
3970	12" x 12" opening		22	.364		26	20.50		46.50	60.50
3980	14" x 14" opening		22	.364		33.50	20.50		54	69
3984	16" x 16" opening		21	.381		37.50	21.50		59	74.50
3990	18" x 18" opening		21	.381		45	21.50		66.50	83
4020	22" x 22" opening		17	.471		53	26.50		79.50	99.50
4040	24" x 24" opening		15	.533		68	30		98	122
4044	28" x 28" opening		14	.571		102	32		134	164
4048	36" x 36" opening		12	.667		117	37.50		154.50	188
4050	48" x 24" opening		12	.667		149	37.50		186.50	222
4060	Eggcrate, lay-in, T-bar system									
4070	48" x 24" sheet	1 Shee	40	.200	Ea.	71	11.20		82.20	95.50
5000	Transfer grille, vision proof, 8" x 4"		30	.267		12.35	14.90		27.25	37
5020	8" x 6"		28	.286		12.35	16		28.35	38.50
5040	8" x 8"		26	.308		13.85	17.20		31.05	42
5060	10" x 6"		24	.333		12.85	18.65		31.50	43.50
5080	10" x 10"		23	.348		16.30	19.45		35.75	48.50
5090	12" x 6"		23	.348		13.85	19.45		33.30	45.50
5100	12" x 10"		22	.364		17.80	20.50		38.30	51.50
5110	12" x 12"		22	.364		19.25	20.50		39.75	53
5120	14" x 10"		22	.364		19.25	20.50		39.75	53
5140	16" x 8"		21	.381		19.25	21.50		40.75	54.50
5160	18" x 12"		20	.400		25	22.50		47.50	63
5170	18" x 18"		21	.381		35.50	21.50		57	72.50

23 37 13.30 Grilles		Crew	Daily Output	Labor-Hours	Unit	Material	Labor (2015 Bare Costs)	Equipment	Total	Total Incl O&P
5180	20" x 12"	1 Shee	20	.400	Ea.	27	22.50		49.50	65.50
5200	20" x 20"		19	.421		43	23.50		66.50	84
5220	24" x 12"		17	.471		30.50	26.50		57	75
5240	24" x 24"		16	.500		58.50	28		86.50	109
5260	30" x 6"		15	.533		27	30		57	77
5280	30" x 8"		15	.533		31.50	30		61.50	81.50
5300	30" x 12"		14	.571		39.50	32		71.50	94
5320	30" x 16"		13	.615		47	34.50		81.50	106
5340	30" x 20"		12	.667		59	37.50		96.50	124
5360	30" x 24"		11	.727		73	40.50		113.50	145
5380	30" x 30"		10	.800		96	45		141	177
6000	For steel grilles instead of aluminum in above, deduct					10%				
6200	Plastic, eggcrate, lay-in, T-bar system									
6210	48" x 24" sheet	1 Shee	50	.160	Ea.	19.25	8.95		28.20	35
6250	Steel door louver									
6270	With fire link, steel only									
6350	12" x 18"	1 Shee	22	.364	Ea.	208	20.50		228.50	261
6410	12" x 24"		21	.381		232	21.50		253.50	289
6470	18" x 18"		20	.400		231	22.50		253.50	290
6530	18" x 24"		19	.421		266	23.50		289.50	330
6660	24" x 24"		15	.533		269	30		299	345
9000	Minimum labor/equipment charge		4	2	Job		112		112	177

23 37 13.60 Registers		Crew	Daily Output	Labor-Hours	Unit	Material	Labor	Equipment	Total	Total Incl O&P
0010	**REGISTERS**									
0980	Air supply									
1000	Ceiling/wall, O.B. damper, anodized aluminum									
1010	One or two way deflection, adj. curved face bars									
1014	6" x 6"	1 Shee	24	.333	Ea.	11.55	18.65		30.20	42.50
1020	8" x 4"		26	.308		11.55	17.20		28.75	40
1040	8" x 8"		24	.333		14.45	18.65		33.10	45.50
1060	10" x 6"		20	.400		13.30	22.50		35.80	50
1080	10" x 10"		19	.421		17.90	23.50		41.40	56.50
1100	12" x 6"		19	.421		13.45	23.50		36.95	52
1120	12" x 12"		18	.444		21	25		46	62.50
1140	14" x 8"		17	.471		18.40	26.50		44.90	62
1160	14" x 14"		18	.444		24.50	25		49.50	66
1170	16" x 16"		17	.471		27.50	26.50		54	71.50
1180	18" x 8"		18	.444		20.50	25		45.50	61.50
1200	18" x 18"		17	.471		37.50	26.50		64	83
1220	20" x 4"		19	.421		18.90	23.50		42.40	58
1240	20" x 6"		18	.444		18.90	25		43.90	60
1260	20" x 8"		18	.444		20.50	25		45.50	62
1280	20" x 20"		17	.471		45.50	26.50		72	91.50
1290	22" x 22"		15	.533		52.50	30		82.50	105
1300	24" x 4"		17	.471		23.50	26.50		50	67
1320	24" x 6"		16	.500		23.50	28		51.50	69.50
1340	24" x 8"		13	.615		25.50	34.50		60	82.50
1350	24" x 18"		12	.667		47.50	37.50		85	111
1360	24" x 24"		11	.727		64.50	40.50		105	135
1380	30" x 4"		16	.500		29	28		57	76
1400	30" x 6"		15	.533		29	30		59	79
1420	30" x 8"		14	.571		33.50	32		65.50	87.50

23 37 13 – Diffusers, Registers, and Grilles

23 37 13.60 Registers		Crew	Daily Output	Labor-Hours	Unit	Material	Labor	Equipment	Total	Total Incl O&P
						2015 Bare Costs				
1440	30″ x 24″	1 Shee	12	.667	Ea.	83	37.50		120.50	151
1460	30″ x 30″		10	.800		107	45		152	188
1504	4 way deflection, adjustable curved face bars									
1510	6″ x 6″	1 Shee	26	.308	Ea.	13.90	17.20		31.10	42.50
1514	8″ x 8″		24	.333		17.35	18.65		36	48.50
1518	10″ x 10″		19	.421		21.50	23.50		45	60.50
1522	12″ x 6″		19	.421		16.10	23.50		39.60	55
1526	12″ x 12″		18	.444		25.50	25		50.50	67
1530	14″ x 14″		18	.444		29.50	25		54.50	71.50
1534	16″ x 16″		17	.471		33	26.50		59.50	77.50
1538	18″ x 18″		17	.471		45	26.50		71.50	91
1542	22″ x 22″		16	.500		63	28		91	114
1980	One way deflection, adj. vert. or horiz. face bars									
1990	6″ x 6″	1 Shee	26	.308	Ea.	12.65	17.20		29.85	41
2000	8″ x 4″		26	.308		11.85	17.20		29.05	40
2020	8″ x 8″		24	.333		14.10	18.65		32.75	45
2040	10″ x 6″		20	.400		14.20	22.50		36.70	51
2060	10″ x 10″		19	.421		16.60	23.50		40.10	55.50
2080	12″ x 6″		19	.421		15.35	23.50		38.85	54
2100	12″ x 12″		18	.444		20	25		45	61
2120	14″ x 8″		17	.471		18.35	26.50		44.85	61.50
2140	14″ x 12″		18	.444		22	25		47	63
2160	14″ x 14″		18	.444		25	25		50	66
2180	16″ x 6″		18	.444		18.35	25		43.35	59
2200	16″ x 12″		18	.444		24	25		49	65.50
2220	16″ x 16″		17	.471		32	26.50		58.50	76.50
2240	18″ x 8″		18	.444		21.50	25		46.50	62.50
2260	18″ x 12″		17	.471		26	26.50		52.50	70
2280	18″ x 18″		16	.500		37.50	28		65.50	85.50
2300	20″ x 10″		18	.444		25.50	25		50.50	67
2320	20″ x 16″		16	.500		35	28		63	82.50
2340	20″ x 20″		16	.500		46	28		74	94.50
2360	24″ x 12″		15	.533		32	30		62	82
2380	24″ x 16″		14	.571		42.50	32		74.50	97.50
2400	24″ x 20″		12	.667		53	37.50		90.50	118
2420	24″ x 24″		10	.800		65	45		110	142
2440	30″ x 12″		13	.615		41.50	34.50		76	100
2460	30″ x 16″		13	.615		53.50	34.50		88	114
2480	30″ x 24″		11	.727		84	40.50		124.50	156
2500	30″ x 30″		9	.889		107	49.50		156.50	197
2520	36″ x 12″		11	.727		51.50	40.50		92	121
2540	36″ x 24″		10	.800		103	45		148	185
2560	36″ x 36″		8	1		173	56		229	279
2600	For 2 way deflect., adj. vert. or horiz. face bars, add					40%				
2700	Above registers in steel instead of aluminum, deduct					10%				
3000	Baseboard, hand adj. damper, enameled steel									
3012	8″ x 6″	1 Shee	26	.308	Ea.	4.85	17.20		22.05	32.50
3020	10″ x 6″		24	.333		5.30	18.65		23.95	35.50
3040	12″ x 5″		23	.348		5.60	19.45		25.05	36.50
3060	12″ x 6″		23	.348		5.75	19.45		25.20	37
3080	12″ x 8″		22	.364		8.30	20.50		28.80	41
3100	14″ x 6″		20	.400		6.20	22.50		28.70	42.50
4000	Floor, toe operated damper, enameled steel									

23 37 Air Outlets and Inlets

23 37 13 – Diffusers, Registers, and Grilles

23 37 13.60 Registers		Crew	Daily Output	Labor-Hours	Unit	Material	Labor	Equipment	Total	Total Incl O&P
4020	4" x 8"	1 Shee	32	.250	Ea.	9	14		23	32
4040	4" x 12"		26	.308		10.55	17.20		27.75	38.50
4060	6" x 8"		28	.286		9.65	16		25.65	35.50
4080	6" x 14"		22	.364		12.20	20.50		32.70	45.50
4100	8" x 10"		22	.364		11	20.50		31.50	44
4120	8" x 16"		20	.400		14.75	22.50		37.25	51.50
4140	10" x 10"		20	.400		13.15	22.50		35.65	50
4160	10" x 16"		18	.444		19.90	25		44.90	61
4180	12" x 12"		18	.444		16.15	25		41.15	57
4200	12" x 24"		16	.500		27	28		55	73.50
4220	14" x 14"		16	.500		42	28		70	90.50
4240	14" x 20"		15	.533		49.50	30		79.50	101
4300	Spiral pipe supply register									
4310	Steel, with air scoop									
4320	4" x 12", for 8" thru 13" diameter duct	1 Shee	25	.320	Ea.	74.50	17.90		92.40	110
4330	4" x 18", for 8" thru 13" diameter duct		18	.444		87	25		112	135
4340	6" x 12", for 14" thru 21" diameter duct		19	.421		78.50	23.50		102	124
4350	6" x 16", for 14" thru 21" diameter duct		18	.444		89	25		114	137
4360	6" x 20", for 14" thru 21" diameter duct		17	.471		98.50	26.50		125	150
4370	6" x 24", for 14" thru 21" diameter duct		16	.500		116	28		144	171
4380	8" x 16", for 22" thru 31" diameter duct		19	.421		94	23.50		117.50	140
4390	8" x 18", for 22" thru 31" diameter duct		18	.444		98.50	25		123.50	147
4400	8" x 24", for 22" thru 31" diameter duct		15	.533		122	30		152	181
4980	Air return									
5000	Ceiling or wall, fixed 45° face blades									
5010	Adjustable O.B. damper, anodized aluminum									
5020	4" x 8"	1 Shee	26	.308	Ea.	11.85	17.20		29.05	40
5040	6" x 8"		24	.333		13	18.65		31.65	44
5060	6" x 10"		19	.421		14.20	23.50		37.70	52.50
5080	6" x 16"		18	.444		18.35	25		43.35	59
5100	8" x 10"		19	.421		15.45	23.50		38.95	54
5120	8" x 12"		16	.500		17.25	28		45.25	63
5140	10" x 10"		18	.444		16.60	25		41.60	57.50
5160	10" x 16"		17	.471		22	26.50		48.50	65.50
5180	12" x 18"		18	.444		26	25		51	67.50
5200	12" x 30"		12	.667		41.50	37.50		79	105
5220	16" x 16"		17	.471		32	26.50		58.50	76.50
5240	18" x 18"		16	.500		37.50	28		65.50	85.50
5260	18" x 36"		10	.800		72.50	45		117.50	151
5280	24" x 24"		11	.727		65	40.50		105.50	136
5300	24" x 36"		8	1		110	56		166	210
5320	24" x 48"		6	1.333		139	74.50		213.50	271
6000	For steel construction instead of aluminum, deduct					10%				
9000	Minimum labor/equipment charge	1 Shee	4	2	Job		112		112	177

23 37 15 – Louvers

23 37 15.40 HVAC Louvers		Crew	Daily Output	Labor-Hours	Unit	Material	Labor	Equipment	Total	Total Incl O&P
0010	**HVAC LOUVERS**									
0100	Aluminum, extruded, with screen, mill finish									
1002	Brick vent, see also Section 04 05 23.19									
1100	Standard, 4" deep, 8" wide, 5" high	1 Shee	24	.333	Ea.	34.50	18.65		53.15	67
1200	Modular, 4" deep, 7-3/4" wide, 5" high		24	.333		36	18.65		54.65	69
1300	Speed brick, 4" deep, 11-5/8" wide, 3-7/8" high		24	.333		36	18.65		54.65	69

23 37 15 – Louvers

23 37 15.40 HVAC Louvers		Crew	Daily Output	Labor-Hours	Unit	2015 Bare Costs Material	Labor	Equipment	Total	Total Incl O&P
1400	Fuel oil brick, 4″ deep, 8″ wide, 5″ high	1 Shee	24	.333	Ea.	61.50	18.65		80.15	97.50
2000	Cooling tower and mechanical equip., screens, light weight		40	.200	S.F.	15.40	11.20		26.60	34.50
2020	Standard weight		35	.229		41	12.80		53.80	65
2500	Dual combination, automatic, intake or exhaust		20	.400		56	22.50		78.50	97
2520	Manual operation		20	.400		41.50	22.50		64	81.50
2540	Electric or pneumatic operation		20	.400		41.50	22.50		64	81.50
2560	Motor, for electric or pneumatic		14	.571	Ea.	480	32		512	580
3000	Fixed blade, continuous line									
3100	Mullion type, stormproof	1 Shee	28	.286	S.F.	41.50	16		57.50	71
3200	Stormproof		28	.286		41.50	16		57.50	71
3300	Vertical line		28	.286		49.50	16		65.50	79
3500	For damper to use with above, add					50%	30%			
3520	Motor, for damper, electric or pneumatic	1 Shee	14	.571	Ea.	480	32		512	580
4000	Operating, 45°, manual, electric or pneumatic		24	.333	S.F.	50	18.65		68.65	84.50
4100	Motor, for electric or pneumatic		14	.571	Ea.	480	32		512	580
4200	Penthouse, roof		56	.143	S.F.	24.50	8		32.50	39.50
4300	Walls		40	.200		58	11.20		69.20	81.50
5000	Thinline, under 4″ thick, fixed blade		40	.200		24	11.20		35.20	44
5010	Finishes, applied by mfr. at additional cost, available in colors									
5020	Prime coat only, add				S.F.	3.30			3.30	3.63
5040	Baked enamel finish coating, add					6.10			6.10	6.70
5060	Anodized finish, add					6.60			6.60	7.25
5080	Duranodic finish, add					12			12	13.20
5100	Fluoropolymer finish coating, add					18.90			18.90	21
9000	Stainless steel, fixed blade, continuous line									
9010	Hospital grade									
9110	20 Ga.	1 Shee	22	.364	S.F.	40	20.50		60.50	76
9980	For small orders (under 10 pieces), add				"	25%				

23 37 23 – HVAC Gravity Ventilators

23 37 23.10 HVAC Gravity Air Ventilators		Crew	Daily Output	Labor-Hours	Unit	Material	Labor	Equipment	Total	Total Incl O&P
0010	**HVAC GRAVITY AIR VENTILATORS**, Includes base									
1280	Rotary ventilators, wind driven, galvanized									
1300	4″ neck diameter	Q-9	20	.800	Ea.	64.50	40.50		105	135
1320	5″ neck diameter		18	.889		64.50	45		109.50	142
1340	6″ neck diameter		16	1		64.50	50.50		115	151
1360	8″ neck diameter		14	1.143		74	57.50		131.50	172
1380	10″ neck diameter		12	1.333		83.50	67		150.50	198
1400	12″ neck diameter		10	1.600		91	80.50		171.50	227
1420	14″ neck diameter		10	1.600		153	80.50		233.50	295
1440	16″ neck diameter		9	1.778		179	89.50		268.50	335
1460	18″ neck diameter		9	1.778		215	89.50		304.50	380
1480	20″ neck diameter		8	2		256	101		357	440
1500	24″ neck diameter		8	2		390	101		491	585
1520	30″ neck diameter		7	2.286		445	115		560	670
1540	36″ neck diameter		6	2.667		655	134		789	930
1600	For aluminum, add					300%				
2000	Stationary, gravity, syphon, galvanized									
2100	3″ neck diameter, 40 CFM	Q-9	24	.667	Ea.	38	33.50		71.50	94.50
2120	4″ neck diameter, 50 CFM		20	.800		40.50	40.50		81	108
2140	5″ neck diameter, 58 CFM		18	.889		43.50	45		88.50	119
2160	6″ neck diameter, 66 CFM		16	1		45	50.50		95.50	129
2180	7″ neck diameter, 86 CFM		15	1.067		59.50	53.50		113	151

23 37 23 – HVAC Gravity Ventilators

23 37 23.10 HVAC Gravity Air Ventilators		Crew	Daily Output	Labor-Hours	Unit	Material	Labor	Equipment	Total	Total Incl O&P
						2015 Bare Costs				
2200	8″ neck diameter, 110 CFM	Q-9	14	1.143	Ea.	61	57.50		118.50	158
2220	10″ neck diameter, 140 CFM		12	1.333		85.50	67		152.50	200
2240	12″ neck diameter, 160 CFM		10	1.600		106	80.50		186.50	244
2260	14″ neck diameter, 250 CFM		10	1.600		168	80.50		248.50	310
2280	16″ neck diameter, 380 CFM		9	1.778		216	89.50		305.50	380
2300	18″ neck diameter, 500 CFM		9	1.778		244	89.50		333.50	410
2320	20″ neck diameter, 625 CFM		8	2		290	101		391	480
2340	24″ neck diameter, 900 CFM		8	2		360	101		461	555
2360	30″ neck diameter, 1375 CFM		7	2.286		400	115		515	620
2380	36″ neck diameter, 2,000 CFM		6	2.667		440	134		574	695
2400	42″ neck diameter, 3000 CFM		4	4		660	201		861	1,050
2500	For aluminum, add					300%				
2520	For stainless steel, add					600%				
3000	Rotating chimney cap, galvanized, 4″ neck diameter	Q-9	20	.800		24.50	40.50		65	90.50
3020	5″ neck diameter		18	.889		24.50	45		69.50	97.50
3040	6″ neck diameter		16	1		24.50	50.50		75	107
3060	7″ neck diameter		15	1.067		31.50	53.50		85	120
3080	8″ neck diameter		14	1.143		31.50	57.50		89	126
3100	10″ neck diameter		12	1.333		38	67		105	148
3600	Stationary chimney rain cap, galvanized, 3″ neck diameter		24	.667		4.32	33.50		37.82	58
3620	4″ neck diameter		20	.800		4.72	40.50		45.22	68.50
3640	6″ neck diameter		16	1		5.60	50.50		56.10	85.50
3680	8″ neck diameter		14	1.143		7.05	57.50		64.55	99
3700	10″ neck diameter		12	1.333		8.50	67		75.50	115
3720	12″ neck diameter		10	1.600		12.55	80.50		93.05	141
3740	14″ neck diameter		10	1.600		19.90	80.50		100.40	149
3760	16″ neck diameter		9	1.778		21	89.50		110.50	164
4200	Stationary mushroom, aluminum, 16″ orifice diameter		10	1.600		620	80.50		700.50	810
4220	26″ orifice diameter		6.15	2.602		915	131		1,046	1,200
4230	30″ orifice diameter		5.71	2.802		1,350	141		1,491	1,700
4240	38″ orifice diameter		5	3.200		1,925	161		2,086	2,350
4250	42″ orifice diameter		4.70	3.404		2,550	171		2,721	3,075
4260	50″ orifice diameter		4.44	3.604		3,025	181		3,206	3,600
5000	Relief vent									
5500	Rectangular, aluminum, galvanized curb									
5510	intake/exhaust, 0.033″ SP									
5580	500 CFM, 12″ x 12″	Q-9	8.60	1.860	Ea.	700	93.50		793.50	920
5600	600 CFM, 12″ x 16″		8	2		785	101		886	1,025
5620	750 CFM, 12″ x 20″		7.20	2.222		840	112		952	1,100
5640	1000 CFM, 12″ x 24″		6.60	2.424		880	122		1,002	1,150
5660	1500 CFM, 12″ x 36″		5.80	2.759		1,175	139		1,314	1,525
5680	3000 CFM, 20″ x 42″		4	4		1,550	201		1,751	2,025
5700	5000 CFM, 20″ x 72″		3	5.333		2,225	269		2,494	2,875
5720	6000 CFM, 24″ x 72″		2.30	6.957		2,400	350		2,750	3,200
5740	10,000 CFM, 48″ x 60″		1.80	8.889		3,050	450		3,500	4,050
5760	12,000 CFM, 48″ x 72″		1.60	10		3,600	505		4,105	4,750
5780	13,750 CFM, 60″ x 66″		1.40	11.429		3,975	575		4,550	5,275
5800	15,000 CFM, 60″ x 72″		1.30	12.308		4,250	620		4,870	5,650
5820	18,000 CFM, 72″ x 72″		1.20	13.333		4,825	670		5,495	6,350
5880	Size is throat area, volume is at 500 fpm									
7000	Note: sizes based on exhaust. Intake, with 0.125″ SP									
7100	loss, approximately twice listed capacity.									
9000	Minimum labor/equipment charge	1 Plum	2	4	Job		235		235	365

23 38 Ventilation Hoods

23 38 13 – Commercial-Kitchen Hoods

23 38 13.10 Hood and Ventilation Equipment		Crew	Daily Output	Labor-Hours	Unit	Material	Labor	Equipment	Total	Total Incl O&P
						2015 Bare Costs				
0010	HOOD AND VENTILATION EQUIPMENT									
2970	Exhaust hood, sst, gutter on all sides, 4′ x 4′ x 2′	1 Carp	1.80	4.444	Ea.	4,725	209		4,934	5,550
2980	4′ x 4′ x 7′	"	1.60	5		7,525	235		7,760	8,650
7800	Vent hood, wall canopy with fire protection, 30″	L-3A	9	1.333		425	68.50		493.50	575
7810	Without fire protection, 36″		10	1.200		460	61.50		521.50	605
7820	Island canopy with fire protection, 30″		7	1.714		790	88		878	1,000
7830	Without fire protection, 36″		8	1.500		800	77		877	1,000
7840	Back shelf with fire protection, 30″		11	1.091		400	56		456	530
7850	Without fire protection, black, 36″		12	1		400	51.50		451.50	525
7852	Without fire protection, stainless steel, 36″		12	1		500	51.50		551.50	635
7860	Range hood & CO_2 system, 30″	1 Carp	2.50	3.200		3,975	150		4,125	4,625
7950	Hood fire protection system, electric stove	Q-1	3	5.333		1,875	282		2,157	2,525
7952	Hood fire protection system, gas stove	"	3	5.333		2,200	282		2,482	2,875

23 41 Particulate Air Filtration

23 41 13 – Panel Air Filters

23 41 13.10 Panel Type Air Filters		Crew	Daily Output	Labor-Hours	Unit	Material	Labor	Equipment	Total	Total Incl O&P
0010	PANEL TYPE AIR FILTERS									
2950	Mechanical media filtration units									
3000	High efficiency type, with frame, non-supported G				MCFM	35			35	38.50
3100	Supported type G				"	45			45	49.50
5500	Throwaway glass or paper media type				Ea.	3.42			3.42	3.76

23 41 16 – Renewable-Media Air Filters

23 41 16.10 Disposable Media Air Filters		Crew	Daily Output	Labor-Hours	Unit	Material	Labor	Equipment	Total	Total Incl O&P
0010	DISPOSABLE MEDIA AIR FILTERS									
5000	Renewable disposable roll				C.S.F.	1.54			1.54	1.70

23 41 19 – Washable Air Filters

23 41 19.10 Permanent Air Filters		Crew	Daily Output	Labor-Hours	Unit	Material	Labor	Equipment	Total	Total Incl O&P
0010	PERMANENT AIR FILTERS									
4500	Permanent washable G				MCFM	20			20	22

23 41 23 – Extended Surface Filters

23 41 23.10 Expanded Surface Filters		Crew	Daily Output	Labor-Hours	Unit	Material	Labor	Equipment	Total	Total Incl O&P
0010	EXPANDED SURFACE FILTERS									
4000	Medium efficiency, extended surface G				MCFM	5.50			5.50	6.05

23 42 Gas-Phase Air Filtration

23 42 13 – Activated-Carbon Air Filtration

23 42 13.10 Charcoal Type Air Filtration		Crew	Daily Output	Labor-Hours	Unit	Material	Labor	Equipment	Total	Total Incl O&P
0010	CHARCOAL TYPE AIR FILTRATION									
0050	Activated charcoal type, full flow				MCFM	600			600	660
0060	Full flow, impregnated media 12″ deep					225			225	248
0070	HEPA filter & frame for field erection					350			350	385
0080	HEPA filter-diffuser, ceiling install.					300			300	330

23 43 Electronic Air Cleaners

23 43 13 – Washable Electronic Air Cleaners

	23 43 13.10 Electronic Air Cleaners	Crew	Daily Output	Labor-Hours	Unit	Material	2015 Bare Costs Labor	Equipment	Total	Total Incl O&P
0010	**ELECTRONIC AIR CLEANERS**									
2000	Electronic air cleaner, duct mounted									
2150	1000 CFM	1 Shee	4	2	Ea.	420	112		532	640
2200	1200 CFM		3.80	2.105		505	118		623	740
2250	1400 CFM		3.60	2.222		520	124		644	770

23 51 Breechings, Chimneys, and Stacks

23 51 13 – Draft Control Devices

23 51 13.13 Draft-Induction Fans

	Description	Crew	Daily Output	Labor-Hours	Unit	Material	Labor	Equipment	Total	Total Incl O&P
0010	**DRAFT-INDUCTION FANS**									
1000	Breeching installation									
1800	Hot gas, 600°F, variable pitch pulley and motor									
1840	6" diam. inlet, 1/4 H.P., 1 phase, 400 CFM	Q-9	6	2.667	Ea.	1,500	134		1,634	1,850
1860	8" diam. inlet, 1/4 H.P., 1phase, 1120 CFM		4	4		2,150	201		2,351	2,675
1870	9" diam. inlet, 3/4 H.P., 1 phase, 1440 CFM		3.60	4.444		2,650	224		2,874	3,250
1880	10" diam. inlet, 3/4 H.P., 1 phase, 2000 CFM		3.30	4.848		2,650	244		2,894	3,275
1900	12" diam. inlet, 3/4 H.P., 3 phase, 2960 CFM		3	5.333		2,925	269		3,194	3,625
1910	14" diam. inlet, 1 H.P., 3 phase, 4160 CFM		2.60	6.154		2,950	310		3,260	3,725
1920	16" diam. inlet, 2 H.P., 3 phase, 5500 CFM		2.30	6.957		3,250	350		3,600	4,125
1950	20" diam. inlet, 3 H.P., 3 phase, 9760 CFM		1.50	10.667		4,675	535		5,210	6,000
1960	22" diam. inlet, 5 H.P., 3 phase, 13,360 CFM		1	16		7,625	805		8,430	9,650
1980	24" diam. inlet, 7-1/2 H.P., 3 phase, 17,760 CFM		.80	20		8,375	1,000		9,375	10,800
2300	For multi-blade damper at fan inlet, add					20%				
3600	Chimney-top installation									
3700	6" size	1 Shee	8	1	Ea.	1,375	56		1,431	1,625
3740	8" size		7	1.143		1,400	64		1,464	1,625
3750	10" size		6.50	1.231		1,825	69		1,894	2,100
3780	13" size		6	1.333		1,875	74.50		1,949.50	2,175
3880	For speed control switch, add					106			106	116
3920	For thermal fan control, add					106			106	116
5500	Flue blade style damper for draft control,									
5510	locking quadrant blade									
5550	8" size	Q-9	8	2	Ea.	405	101		506	605
5560	9" size		7.50	2.133		405	107		512	615
5570	10" size		7	2.286		415	115		530	635
5580	12" size		6.50	2.462		420	124		544	660
5590	14" size		6	2.667		430	134		564	685
5600	16" size		5.50	2.909		445	146		591	720
5610	18" size		5	3.200		465	161		626	765
5620	20" size		4.50	3.556		490	179		669	820
5630	22" size		4	4		515	201		716	885
5640	24" size		3.50	4.571		545	230		775	960
5650	27" size		3	5.333		570	269		839	1,050
5660	30" size		2.50	6.400		625	320		945	1,200
5670	32" size		2	8		655	405		1,060	1,350
5680	36" size		1.50	10.667		725	535		1,260	1,650

23 51 13.16 Vent Dampers

	Description	Crew	Daily Output	Labor-Hours	Unit	Material	Labor	Equipment	Total	Total Incl O&P
0010	**VENT DAMPERS**									
5000	Vent damper, bi-metal, gas, 3" diameter	Q-9	24	.667	Ea.	44	33.50		77.50	102
5010	4" diameter		24	.667		44	33.50		77.50	102
5020	5" diameter		23	.696		44	35		79	104

23 51 Breechings, Chimneys, and Stacks

23 51 13 – Draft Control Devices

23 51 13.16 Vent Dampers		Crew	Daily Output	Labor-Hours	Unit	2015 Bare Costs Material	Labor	Equipment	Total	Total Incl O&P
5030	6″ diameter	Q-9	22	.727	Ea.	44	36.50		80.50	107
5040	7″ diameter		21	.762		45.50	38.50		84	111
5050	8″ diameter		20	.800		49	40.50		89.50	118
9000	Minimum labor/equipment charge	1 Shee	4	2	Job		112		112	177

23 51 13.19 Barometric Dampers		Crew	Daily Output	Labor-Hours	Unit	Material	Labor	Equipment	Total	Total Incl O&P
0010	**BAROMETRIC DAMPERS**									
1000	Barometric, gas fired system only, 6″ size for 5″ and 6″ pipes	1 Shee	20	.400	Ea.	105	22.50		127.50	151
1020	7″ size, for 6″ and 7″ pipes		19	.421		112	23.50		135.50	160
1040	8″ size, for 7″ and 8″ pipes		18	.444		144	25		169	198
1060	9″ size, for 8″ and 9″ pipes		16	.500		162	28		190	222
2000	All fuel, oil, oil/gas, coal									
2020	10″ for 9″ and 10″ pipes	1 Shee	15	.533	Ea.	242	30		272	315
2040	12″ for 11″ and 12″ pipes		15	.533		315	30		345	390
2060	14″ for 13″ and 14″ pipes		14	.571		410	32		442	500
2080	16″ for 15″ and 16″ pipes		13	.615		575	34.50		609.50	685
2100	18″ for 17″ and 18″ pipes		12	.667		765	37.50		802.50	900
2120	20″ for 19″ and 21″ pipes		10	.800		915	45		960	1,075
2140	24″ for 22″ and 25″ pipes	Q-9	12	1.333		1,125	67		1,192	1,325
2160	28″ for 26″ and 30″ pipes		10	1.600		1,375	80.50		1,455.50	1,650
2180	32″ for 31″ and 34″ pipes		8	2		1,775	101		1,876	2,100
3260	For thermal switch for above, add	1 Shee	24	.333		100	18.65		118.65	140

23 51 23 – Gas Vents

23 51 23.10 Gas Chimney Vents		Crew	Daily Output	Labor-Hours	Unit	Material	Labor	Equipment	Total	Total Incl O&P
0010	**GAS CHIMNEY VENTS**, Prefab metal, U.L. listed									
0020	Gas, double wall, galvanized steel									
0080	3″ diameter	Q-9	72	.222	V.L.F.	5.55	11.20		16.75	24
0100	4″ diameter		68	.235		7.20	11.85		19.05	26.50
0120	5″ diameter		64	.250		7.80	12.60		20.40	28.50
0140	6″ diameter		60	.267		9.45	13.45		22.90	31.50
0160	7″ diameter		56	.286		15.10	14.40		29.50	39
0180	8″ diameter		52	.308		17.40	15.50		32.90	43.50
0200	10″ diameter		48	.333		36	16.80		52.80	66.50
0220	12″ diameter		44	.364		43	18.30		61.30	76.50
0240	14″ diameter		42	.381		72	19.20		91.20	110
0260	16″ diameter		40	.400		103	20		123	146
0280	18″ diameter		38	.421		128	21		149	174
0300	20″ diameter	Q-10	36	.667		150	35		185	220
0320	22″ diameter		34	.706		187	37		224	263
0340	24″ diameter		32	.750		236	39		275	320
0600	For 4″, 5″ and 6″ oval, add					50%				
0650	Gas, double wall, galvanized steel, fittings									
0660	Elbow 45°, 3″ diameter	Q-9	36	.444	Ea.	13.50	22.50		36	50.50
0670	4″ diameter		34	.471		16.35	23.50		39.85	55.50
0680	5″ diameter		32	.500		19.10	25		44.10	60.50
0690	6″ diameter		30	.533		23.50	27		50.50	68.50
0700	7″ diameter		28	.571		37.50	29		66.50	87
0710	8″ diameter		26	.615		50	31		81	104
0720	10″ diameter		24	.667		107	33.50		140.50	171
0730	12″ diameter		22	.727		114	36.50		150.50	183
0740	14″ diameter		21	.762		179	38.50		217.50	258
0750	16″ diameter		20	.800		232	40.50		272.50	320
0760	18″ diameter		19	.842		305	42.50		347.50	400

23 51 23 – Gas Vents

23 51 23.10 Gas Chimney Vents		Crew	Daily Output	Labor-Hours	Unit	Material	2015 Bare Costs Labor	Equipment	Total	Total Incl O&P
0770	20″ diameter	Q-10	18	1.333	Ea.	340	69.50		409.50	485
0780	22″ diameter		17	1.412		545	73.50		618.50	715
0790	24″ diameter		16	1.500		695	78.50		773.50	890
0950	Elbow 90°, adjustable, 3″ diameter	Q-9	36	.444		23	22.50		45.50	61
0960	4″ diameter		34	.471		27	23.50		50.50	67
0970	5″ diameter		32	.500		33.50	25		58.50	76.50
0980	6″ diameter		30	.533		39	27		66	85.50
0990	7″ diameter		28	.571		67	29		96	120
1010	8″ diameter		26	.615		68	31		99	124
1020	Wall thimble, 4 to 7″ adjustable, 3″ diameter		36	.444		14.45	22.50		36.95	51.50
1022	4″ diameter		34	.471		16.25	23.50		39.75	55.50
1024	5″ diameter		32	.500		17.10	25		42.10	58.50
1026	6″ diameter		30	.533		17.55	27		44.55	62
1028	7″ diameter		28	.571		36.50	29		65.50	85.50
1030	8″ diameter		26	.615		51.50	31		82.50	106
1040	Roof flashing, 3″ diameter		36	.444		8.50	22.50		31	45
1050	4″ diameter		34	.471		9.90	23.50		33.40	48.50
1060	5″ diameter		32	.500		29.50	25		54.50	72
1070	6″ diameter		30	.533		23.50	27		50.50	68.50
1080	7″ diameter		28	.571		31	29		60	80
1090	8″ diameter		26	.615		33.50	31		64.50	85.50
1100	10″ diameter		24	.667		45	33.50		78.50	103
1110	12″ diameter		22	.727		63	36.50		99.50	127
1120	14″ diameter		20	.800		143	40.50		183.50	221
1130	16″ diameter		18	.889		201	45		246	292
1140	18″ diameter		16	1		229	50.50		279.50	330
1150	20″ diameter	Q-10	18	1.333		305	69.50		374.50	445
1160	22″ diameter		14	1.714		375	89.50		464.50	555
1170	24″ diameter		12	2		440	104		544	645
1200	Tee, 3″ diameter	Q-9	27	.593		35	30		65	86
1210	4″ diameter		26	.615		38.50	31		69.50	91.50
1220	5″ diameter		25	.640		40	32		72	95
1230	6″ diameter		24	.667		45.50	33.50		79	103
1240	7″ diameter		23	.696		65.50	35		100.50	128
1250	8″ diameter		22	.727		71.50	36.50		108	137
1260	10″ diameter		21	.762		191	38.50		229.50	271
1270	12″ diameter		20	.800		196	40.50		236.50	279
1280	14″ diameter		18	.889		350	45		395	455
1290	16″ diameter		16	1		515	50.50		565.50	650
1300	18″ diameter		14	1.143		625	57.50		682.50	775
1310	20″ diameter	Q-10	17	1.412		855	73.50		928.50	1,050
1320	22″ diameter		13	1.846		1,100	96.50		1,196.50	1,350
1330	24″ diameter		12	2		1,250	104		1,354	1,550
1460	Tee cap, 3″ diameter	Q-9	45	.356		2.36	17.90		20.26	31
1470	4″ diameter		42	.381		2.54	19.20		21.74	33.50
1490	6″ diameter		37	.432		4.73	22		26.73	39.50
1510	8″ diameter		34	.471		8.70	23.50		32.20	47
1530	12″ diameter		30	.533		70	27		97	120
1550	16″ diameter		25	.640		74.50	32		106.50	133
1570	20″ diameter	Q-10	27	.889		99.50	46.50		146	182
1590	24″ diameter	"	21	1.143		340	59.50		399.50	470
1750	Top, 3″ diameter	Q-9	46	.348		13.05	17.50		30.55	42
1760	4″ diameter		44	.364		13.75	18.30		32.05	44

23 51 Breechings, Chimneys, and Stacks

23 51 23 – Gas Vents

23 51 23.10 Gas Chimney Vents		Crew	Daily Output	Labor-Hours	Unit	2015 Bare Costs Material	Labor	Equipment	Total	Total Incl O&P
1780	6″ diameter	Q-9	40	.400	Ea.	23.50	20		43.50	58
1800	8″ diameter		36	.444		46.50	22.50		69	86.50
1820	12″ diameter		32	.500		154	25		179	209
1840	16″ diameter		28	.571		263	29		292	335
1860	20″ diameter	Q-10	28	.857		550	45		595	675
1880	24″ diameter	″	20	1.200		1,050	62.50		1,112.50	1,250

23 51 26 – All-Fuel Vent Chimneys

23 51 26.10 All-Fuel Vent Chimneys, Press. Tight, Dbl. Wall		Crew	Daily Output	Labor-Hours	Unit	Material	Labor	Equipment	Total	Total Incl O&P
0010	**ALL-FUEL VENT CHIMNEYS, PRESSURE TIGHT, DOUBLE WALL**									
3200	All fuel, pressure tight, double wall, 1″ insulation, U.L. listed, 1400°F.									
3210	304 stainless steel liner, aluminized steel outer jacket									
3220	6″ diameter	Q-9	60	.267	L.F.	54	13.45		67.45	80.50
3221	8″ diameter		52	.308		61	15.50		76.50	91.50
3222	10″ diameter		48	.333		68	16.80		84.80	102
3223	12″ diameter		44	.364		78	18.30		96.30	115
3224	14″ diameter		42	.381		87.50	19.20		106.70	127
3225	16″ diameter		40	.400		99	20		119	141
3226	18″ diameter		38	.421		112	21		133	158
3227	20″ diameter	Q-10	36	.667		127	35		162	195
3228	24″ diameter		32	.750		163	39		202	241
3229	28″ diameter		30	.800		187	42		229	272
3230	32″ diameter		27	.889		211	46.50		257.50	305
3231	36″ diameter		25	.960		236	50		286	340
3232	42″ diameter		22	1.091		276	57		333	395
3233	48″ diameter		19	1.263		315	66		381	450
3260	For 316 stainless steel liner add					30%				
3280	All fuel, pressure tight, double wall fittings									
3284	304 stainless steel inner, aluminized steel jacket									
3288	Adjustable 20″/29″ section									
3292	6″ diameter	Q-9	30	.533	Ea.	193	27		220	255
3293	8″ diameter		26	.615		201	31		232	270
3294	10″ diameter		24	.667		227	33.50		260.50	305
3295	12″ diameter		22	.727		257	36.50		293.50	340
3296	14″ diameter		21	.762		290	38.50		328.50	380
3297	16″ diameter		20	.800		325	40.50		365.50	425
3298	18″ diameter		19	.842		370	42.50		412.50	470
3299	20″ diameter	Q-10	18	1.333		415	69.50		484.50	570
3300	24″ diameter		16	1.500		535	78.50		613.50	715
3301	28″ diameter		15	1.600		575	83.50		658.50	765
3302	32″ diameter		14	1.714		695	89.50		784.50	905
3303	36″ diameter		12	2		775	104		879	1,025
3304	42″ diameter		11	2.182		910	114		1,024	1,175
3305	48″ diameter		10	2.400		1,025	125		1,150	1,325
3350	Elbow 90° fixed									
3354	6″ diameter	Q-9	30	.533	Ea.	405	27		432	490
3355	8″ diameter		26	.615		455	31		486	550
3356	10″ diameter		24	.667		515	33.50		548.50	620
3357	12″ diameter		22	.727		585	36.50		621.50	705
3358	14″ diameter		21	.762		660	38.50		698.50	785
3359	16″ diameter		20	.800		745	40.50		785.50	885
3360	18″ diameter		19	.842		840	42.50		882.50	990
3361	20″ diameter	Q-10	18	1.333		955	69.50		1,024.50	1,150

23 51 26 – All-Fuel Vent Chimneys

	23 51 26.10 All-Fuel Vent Chimneys, Press. Tight, Dbl. Wall	Crew	Daily Output	Labor-Hours	Unit	Material	2015 Bare Costs Labor	Equipment	Total	Total Incl O&P
3362	24" diameter	Q-10	16	1.500	Ea.	1,225	78.50		1,303.50	1,475
3363	28" diameter		15	1.600		1,375	83.50		1,458.50	1,625
3364	32" diameter		14	1.714		1,850	89.50		1,939.50	2,175
3365	36" diameter		12	2		2,475	104		2,579	2,900
3366	42" diameter		11	2.182		2,950	114		3,064	3,425
3367	48" diameter		10	2.400		3,300	125		3,425	3,825
3380	For 316 stainless steel liner, add					30%				
3400	Elbow 45°									
3404	6" diameter	Q-9	30	.533	Ea.	201	27		228	264
3405	8" diameter		26	.615		227	31		258	298
3406	10" diameter		24	.667		258	33.50		291.50	335
3407	12" diameter		22	.727		294	36.50		330.50	385
3408	14" diameter		21	.762		330	38.50		368.50	425
3409	16" diameter		20	.800		370	40.50		410.50	475
3410	18" diameter		19	.842		420	42.50		462.50	530
3411	20" diameter	Q-10	18	1.333		480	69.50		549.50	635
3412	24" diameter		16	1.500		610	78.50		688.50	800
3413	28" diameter		15	1.600		685	83.50		768.50	885
3414	32" diameter		14	1.714		925	89.50		1,014.50	1,175
3415	36" diameter		12	2		1,250	104		1,354	1,550
3416	42" diameter		11	2.182		1,450	114		1,564	1,775
3417	48" diameter		10	2.400		1,650	125		1,775	2,025
3430	For 316 stainless steel liner, add					30%				
3450	Tee 90°									
3454	6" diameter	Q-9	24	.667	Ea.	237	33.50		270.50	315
3455	8" diameter		22	.727		259	36.50		295.50	345
3456	10" diameter		21	.762		292	38.50		330.50	380
3457	12" diameter		20	.800		340	40.50		380.50	435
3458	14" diameter		18	.889		385	45		430	495
3459	16" diameter		16	1		425	50.50		475.50	545
3460	18" diameter		14	1.143		500	57.50		557.50	640
3461	20" diameter	Q-10	17	1.412		580	73.50		653.50	750
3462	24" diameter		12	2		720	104		824	960
3463	28" diameter		11	2.182		855	114		969	1,125
3464	32" diameter		10	2.400		1,200	125		1,325	1,500
3465	36" diameter		9	2.667		1,350	139		1,489	1,700
3466	42" diameter		6	4		1,575	209		1,784	2,050
3467	48" diameter		5	4.800		1,800	251		2,051	2,375
3480	For Tee Cap, add					35%	20%			
3500	For 316 stainless steel liner, add					30%				
3520	Plate support, galvanized									
3524	6" diameter	Q-9	26	.615	Ea.	118	31		149	179
3525	8" diameter		22	.727		138	36.50		174.50	210
3526	10" diameter		20	.800		150	40.50		190.50	229
3527	12" diameter		18	.889		158	45		203	245
3528	14" diameter		17	.941		187	47.50		234.50	281
3529	16" diameter		16	1		197	50.50		247.50	297
3530	18" diameter		15	1.067		208	53.50		261.50	315
3531	20" diameter	Q-10	16	1.500		218	78.50		296.50	365
3532	24" diameter		14	1.714		227	89.50		316.50	390
3533	28" diameter		13	1.846		278	96.50		374.50	455
3534	32" diameter		12	2		340	104		444	540
3535	36" diameter		10	2.400		410	125		535	650

23 51 26 – All-Fuel Vent Chimneys

23 51 26.10 All-Fuel Vent Chimneys, Press. Tight, Dbl. Wall		Crew	Daily Output	Labor-Hours	Unit	Material	Labor	Equipment	Total	Total Incl O&P
						2015 Bare Costs				
3536	42" diameter	Q-10	9	2.667	Ea.	480	139		619	750
3537	48" diameter	↓	8	3	↓	545	157		702	845
3570	Bellows, lined									
3574	6" diameter	Q-9	30	.533	Ea.	1,300	27		1,327	1,475
3575	8" diameter		26	.615		1,350	31		1,381	1,550
3576	10" diameter		24	.667		1,375	33.50		1,408.50	1,575
3577	12" diameter		22	.727		1,425	36.50		1,461.50	1,600
3578	14" diameter		21	.762		1,450	38.50		1,488.50	1,650
3579	16" diameter		20	.800		1,525	40.50		1,565.50	1,750
3580	18" diameter	↓	19	.842		1,575	42.50		1,617.50	1,800
3581	20" diameter	Q-10	18	1.333		1,625	69.50		1,694.50	1,875
3582	24" diameter		16	1.500		1,875	78.50		1,953.50	2,200
3583	28" diameter		15	1.600		2,000	83.50		2,083.50	2,325
3584	32" diameter		14	1.714		2,250	89.50		2,339.50	2,625
3585	36" diameter		12	2		2,450	104		2,554	2,875
3586	42" diameter		11	2.182		2,750	114		2,864	3,200
3587	48" diameter	↓	10	2.400		3,175	125		3,300	3,700
3590	For all 316 stainless steel construction, add				↓	55%				
3600	Ventilated roof thimble, 304 stainless steel									
3620	6" diameter	Q-9	26	.615	Ea.	247	31		278	320
3624	8" diameter		22	.727		257	36.50		293.50	340
3625	10" diameter		20	.800		264	40.50		304.50	355
3626	12" diameter		18	.889		272	45		317	370
3627	14" diameter		17	.941		282	47.50		329.50	385
3628	16" diameter		16	1		305	50.50		355.50	415
3629	18" diameter	↓	15	1.067		330	53.50		383.50	445
3630	20" diameter	Q-10	16	1.500		345	78.50		423.50	505
3631	24" diameter		14	1.714		380	89.50		469.50	555
3632	28" diameter		13	1.846		410	96.50		506.50	600
3633	32" diameter		12	2		445	104		549	650
3634	36" diameter		10	2.400		475	125		600	725
3635	42" diameter		9	2.667		525	139		664	795
3636	48" diameter	↓	8	3	↓	575	157		732	880
3650	For 316 stainless steel, add					30%				
3670	Exit cone, 316 stainless steel only									
3674	6" diameter	Q-9	46	.348	Ea.	187	17.50		204.50	234
3675	8" diameter		42	.381		193	19.20		212.20	243
3676	10" diameter		40	.400		204	20		224	256
3677	12" diameter		38	.421		217	21		238	273
3678	14" diameter		37	.432		232	22		254	290
3679	16" diameter		36	.444		291	22.50		313.50	355
3680	18" diameter	↓	35	.457		315	23		338	380
3681	20" diameter	Q-10	28	.857		375	45		420	485
3682	24" diameter		26	.923		495	48		543	620
3683	28" diameter		25	.960		560	50		610	695
3684	32" diameter		24	1		695	52		747	850
3685	36" diameter		22	1.091		845	57		902	1,025
3686	42" diameter		21	1.143		985	59.50		1,044.50	1,175
3687	48" diameter	↓	20	1.200	↓	1,125	62.50		1,187.50	1,325
3720	Roof guide, 304 stainless steel									
3724	6" diameter	Q-9	25	.640	Ea.	85.50	32		117.50	145
3725	8" diameter		21	.762		99.50	38.50		138	171
3726	10" diameter	↓	19	.842	↓	109	42.50		151.50	187

23 51 26 – All-Fuel Vent Chimneys

23 51 26.10 All-Fuel Vent Chimneys, Press. Tight, Dbl. Wall		Crew	Daily Output	Labor-Hours	Unit	Material	2015 Bare Costs Labor	Equipment	Total	Total Incl O&P
3727	12" diameter	Q-9	17	.941	Ea.	113	47.50		160.50	199
3728	14" diameter		16	1		132	50.50		182.50	225
3729	16" diameter		15	1.067		139	53.50		192.50	238
3730	18" diameter		14	1.143		148	57.50		205.50	253
3731	20" diameter	Q-10	15	1.600		155	83.50		238.50	305
3732	24" diameter		13	1.846		162	96.50		258.50	330
3733	28" diameter		12	2		197	104		301	380
3734	32" diameter		11	2.182		243	114		357	445
3735	36" diameter		9	2.667		290	139		429	540
3736	42" diameter		8	3		335	157		492	615
3737	48" diameter		7	3.429		385	179		564	710
3750	For 316 stainless steel, add					30%				
3770	Rain cap with bird screen									
3774	6" diameter	Q-9	46	.348	Ea.	290	17.50		307.50	350
3775	8" diameter		42	.381		335	19.20		354.20	400
3776	10" diameter		40	.400		390	20		410	460
3777	12" diameter		38	.421		450	21		471	530
3778	14" diameter		37	.432		510	22		532	600
3779	16" diameter		36	.444		580	22.50		602.50	675
3780	18" diameter		35	.457		675	23		698	775
3781	20" diameter	Q-10	28	.857		765	45		810	910
3782	24" diameter		26	.923		920	48		968	1,075
3783	28" diameter		25	.960		1,100	50		1,150	1,275
3784	32" diameter		24	1		1,250	52		1,302	1,450
3785	36" diameter		22	1.091		1,475	57		1,532	1,725
3786	42" diameter		21	1.143		1,725	59.50		1,784.50	2,000
3787	48" diameter		20	1.200		1,950	62.50		2,012.50	2,250

23 51 26.30 All-Fuel Vent Chimneys, Double Wall, St. Stl.

		Crew	Daily Output	Labor-Hours	Unit	Material	Labor	Equipment	Total	Total Incl O&P
0010	**ALL-FUEL VENT CHIMNEYS, DOUBLE WALL, STAINLESS STEEL**									
7780	All fuel, pressure tight, double wall, 4" insulation, U.L. listed, 1400°F.									
7790	304 stainless steel liner, aluminized steel outer jacket									
7800	6" diameter	Q-9	60	.267	V.L.F.	64	13.45		77.45	91
7804	8" diameter		52	.308		73	15.50		88.50	105
7806	10" diameter		48	.333		81.50	16.80		98.30	117
7808	12" diameter		44	.364		93.50	18.30		111.80	132
7810	14" diameter		42	.381		105	19.20		124.20	146
7880	For 316 stainless steel liner add				L.F.	30%				
8000	All fuel, double wall, stainless steel fittings									
8010	Roof support 6" diameter	Q-9	30	.533	Ea.	113	27		140	167
8030	8" diameter		26	.615		132	31		163	194
8040	10" diameter		24	.667		139	33.50		172.50	206
8050	12" diameter		22	.727		148	36.50		184.50	220
8060	14" diameter		21	.762		155	38.50		193.50	232
8100	Elbow 45°, 6" diameter		30	.533		241	27		268	310
8140	8" diameter		26	.615		271	31		302	345
8160	10" diameter		24	.667		310	33.50		343.50	395
8180	12" diameter		22	.727		350	36.50		386.50	445
8200	14" diameter		21	.762		395	38.50		433.50	495
8300	Insulated tee, 6" diameter		30	.533		283	27		310	355
8360	8" diameter		26	.615		310	31		341	390
8380	10" diameter		24	.667		350	33.50		383.50	440
8400	12" diameter		22	.727		395	36.50		431.50	495

23 51 Breechings, Chimneys, and Stacks

23 51 26 – All-Fuel Vent Chimneys

23 51 26.30 All-Fuel Vent Chimneys, Double Wall, St. Stl.		Crew	Daily Output	Labor-Hours	Unit	Material	2015 Bare Costs Labor	Equipment	Total	Total Incl O&P
8420	14" diameter	Q-9	20	.800	Ea.	460	40.50		500.50	575
8500	Boot tee, 6" diameter		28	.571		540	29		569	640
8520	8" diameter		24	.667		590	33.50		623.50	700
8530	10" diameter		22	.727		685	36.50		721.50	815
8540	12" diameter		20	.800		810	40.50		850.50	955
8550	14" diameter		18	.889		925	45		970	1,100
8600	Rain cap with bird screen, 6" diameter		30	.533		290	27		317	365
8640	8" diameter		26	.615		335	31		366	420
8660	10" diameter		24	.667		390	33.50		423.50	485
8680	12" diameter		22	.727		450	36.50		486.50	555
8700	14" diameter		21	.762		510	38.50		548.50	625
8800	Flat roof flashing, 6" diameter		30	.533		106	27		133	160
8840	8" diameter		26	.615		116	31		147	176
8860	10" diameter		24	.667		124	33.50		157.50	189
8880	12" diameter		22	.727		136	36.50		172.50	208
8900	14" diameter	▼	21	.762	▼	139	38.50		177.50	214

23 52 Heating Boilers

23 52 13 – Electric Boilers

23 52 13.10 Electric Boilers, ASME		Crew	Daily Output	Labor-Hours	Unit	Material	Labor	Equipment	Total	Total Incl O&P
0010	**ELECTRIC BOILERS, ASME**, Standard controls and trim									
1000	Steam, 6 KW, 20.5 MBH	Q-19	1.20	20	Ea.	3,950	1,075		5,025	6,025
1040	9 KW, 30.7 MBH		1.20	20		4,025	1,075		5,100	6,100
1080	24 KW, 81.8 MBH		1.10	21.818		4,800	1,175		5,975	7,125
1160	60 KW, 205 MBH		1	24		6,650	1,300		7,950	9,350
1240	148 KW, 505 MBH		.65	36.923		9,600	2,000		11,600	13,700
1280	222 KW, 758 MBH		.55	43.636		23,800	2,350		26,150	29,900
1320	300 KW, 1023 MBH		.40	60		26,400	3,250		29,650	34,100
1360	444 KW, 1515 MBH	▼	.30	80		31,400	4,325		35,725	41,200
1400	592 KW, 2020 MBH	Q-21	.34	94.118		35,400	5,225		40,625	47,000
1480	814 KW, 2778 MBH		.25	128		40,600	7,100		47,700	56,000
1520	1036 KW, 3536 MBH		.20	160		46,700	8,875		55,575	65,500
1560	2070 KW, 7063 MBH		.18	177		68,000	9,875		77,875	90,500
1600	2,340 KW, 7984 MBH	▼	.16	200		86,500	11,100		97,600	112,500
2000	Hot water, 7.5 KW, 25.6 MBH	Q-19	1.30	18.462		4,975	1,000		5,975	7,025
2040	30 KW, 102 MBH		1.20	20		5,325	1,075		6,400	7,525
2070	60 KW, 205 MBH		1.20	20		5,525	1,075		6,600	7,750
2100	90 KW, 307 MBH		1.10	21.818		6,000	1,175		7,175	8,425
2140	120 KW, 410 MBH		.90	26.667		6,375	1,450		7,825	9,275
2180	150 KW, 512 MBH		.65	36.923		7,500	2,000		9,500	11,300
2220	296 KW, 1010 MBH		.55	43.636		16,100	2,350		18,450	21,400
2300	444 KW, 1515 MBH	▼	.35	68.571		20,800	3,700		24,500	28,700
2340	518 KW, 1768 MBH	Q-21	.44	72.727		23,100	4,025		27,125	31,700
2420	740 KW, 2526 MBH		.39	82.051		28,500	4,550		33,050	38,400
2460	888 KW, 3031 MBH		.37	86.486		31,100	4,800		35,900	41,700
2500	1036 KW, 3536 MBH		.34	94.118		35,500	5,225		40,725	47,200
2540	1440 KW, 4915 MBH		.32	100		43,800	5,550		49,350	57,000
2580	1680 KW, 5733 MBH		.30	106		50,000	5,925		55,925	64,000
2620	1980 KW, 6757 MBH		.28	114		58,000	6,350		64,350	74,000
2660	2220 KW, 7576 MBH		.26	123		65,000	6,825		71,825	82,000
2700	2610 KW, 8905 MBH	▼	.24	133	▼	71,000	7,400		78,400	89,500

23 52 Heating Boilers

23 52 13 – Electric Boilers

23 52 13.10 Electric Boilers, ASME		Crew	Daily Output	Labor-Hours	Unit	2015 Bare Costs Material	Labor	Equipment	Total	Total Incl O&P
2740	2970 KW, 10133 MBH	Q-21	.21	152	Ea.	77,000	8,450		85,450	98,000
2780	3240 KW, 11055 MBH		.18	177		84,500	9,875		94,375	108,500
2820	3600 KW, 12,283 MBH		.16	200		94,000	11,100		105,100	121,000
9000	Minimum labor/equipment charge	Q-20	1	20	Job		1,025		1,025	1,600

23 52 16 – Condensing Boilers

23 52 16.24 Condensing Boilers		Crew	Daily Output	Labor-Hours	Unit	Material	Labor	Equipment	Total	Total Incl O&P
0010	**CONDENSING BOILERS**, Cast iron, high efficiency									
0020	Packaged with standard controls, circulator and trim									
0030	Intermittent (spark) pilot, natural or LP gas									
0040	Hot water, DOE MBH output, (AFUE)									
0100	42 MBH, (84.0%) G	Q-5	1.80	8.889	Ea.	1,575	480		2,055	2,500
0120	57 MBH, (84.3%) G		1.60	10		1,775	540		2,315	2,800
0140	85 MBH, (84.0%) G		1.40	11.429		1,950	615		2,565	3,075
0160	112 MBH, (83.7%) G		1.20	13.333		2,175	715		2,890	3,525
0180	140 MBH, (83.3%) G	Q-6	1.60	15		2,450	835		3,285	4,000
0200	167 MBH, (83.0%) G		1.40	17.143		2,775	955		3,730	4,550
0220	194 MBH, (82.7%) G		1.20	20		3,025	1,125		4,150	5,075

23 52 19 – Pulse Combustion Boilers

23 52 19.20 Pulse Type Combustion Boilers		Crew	Daily Output	Labor-Hours	Unit	Material	Labor	Equipment	Total	Total Incl O&P
0010	**PULSE TYPE COMBUSTION BOILERS**, High efficiency									
7990	Special feature gas fired boilers									
8000	Pulse combustion, standard controls/trim									
8010	Hot water, DOE MBH output, (AFUE %)									
8030	71 MBH, 95.2%	Q-5	1.60	10	Ea.	3,450	540		3,990	4,650
8050	94 MBH, 95.3% G		1.40	11.429		3,825	615		4,440	5,150
8080	139 MBH, 95.6% G		1.20	13.333		4,300	715		5,015	5,850
8090	207 MBH, 95.4%		1.16	13.793		4,850	740		5,590	6,500
8120	270 MBH, 96.4%		1.12	14.286		6,725	770		7,495	8,600
8130	365 MBH, 91.7%		1.07	14.953		7,600	805		8,405	9,600

23 52 23 – Cast-Iron Boilers

23 52 23.20 Gas-Fired Boilers		Crew	Daily Output	Labor-Hours	Unit	Material	Labor	Equipment	Total	Total Incl O&P
0010	**GAS-FIRED BOILERS**, Natural or propane, standard controls, packaged									
1000	Cast iron, with insulated jacket									
2000	Steam, gross output, 81 MBH	Q-7	1.40	22.857	Ea.	2,450	1,300		3,750	4,725
2020	102 MBH		1.30	24.615		2,675	1,400		4,075	5,125
2040	122 MBH		1	32		2,925	1,825		4,750	6,075
2060	163 MBH		.90	35.556		3,275	2,025		5,300	6,750
2080	203 MBH		.90	35.556		3,825	2,025		5,850	7,350
2100	240 MBH		.85	37.647		3,925	2,150		6,075	7,650
2120	280 MBH		.80	40		4,700	2,275		6,975	8,725
2140	320 MBH		.70	45.714		5,000	2,600		7,600	9,550
2160	360 MBH		.63	51.200		5,575	2,925		8,500	10,700
2180	400 MBH		.56	56.838		5,900	3,225		9,125	11,600
2200	440 MBH		.51	62.500		6,350	3,550		9,900	12,600
2220	544 MBH		.45	71.588		10,100	4,075		14,175	17,500
2240	765 MBH		.43	74.419		12,200	4,225		16,425	20,000
2260	892 MBH		.38	84.211		14,100	4,800		18,900	23,000
2280	1275 MBH		.34	94.118		17,300	5,350		22,650	27,400
2300	1530 MBH		.32	100		18,500	5,700		24,200	29,300
2320	1,875 MBH		.30	106		25,200	6,075		31,275	37,200
2340	2170 MBH		.26	122		27,000	6,950		33,950	40,400

23 52 Heating Boilers

23 52 23 – Cast-Iron Boilers

23 52 23.20 Gas-Fired Boilers		Crew	Daily Output	Labor-Hours	Unit	Material	2015 Bare Costs Labor	Equipment	Total	Total Incl O&P
2360	2675 MBH	Q-7	.20	163	Ea.	28,300	9,300		37,600	45,600
2380	3060 MBH		.19	172		29,300	9,800		39,100	47,600
2400	3570 MBH		.18	181		35,800	10,300		46,100	55,500
2420	4207 MBH		.16	205		39,200	11,700		50,900	61,500
2440	4,720 MBH		.15	207		68,000	11,800		79,800	93,500
2460	5660 MBH		.15	220		77,000	12,600		89,600	104,000
2480	6,100 MBH		.13	246		88,000	14,000		102,000	119,000
2500	6390 MBH		.12	266		93,000	15,200		108,200	125,500
2520	6680 MBH		.11	290		94,000	16,500		110,500	129,500
2540	6,970 MBH		.10	320		98,500	18,200		116,700	137,000
3000	Hot water, gross output, 80 MBH		1.46	21.918		1,975	1,250		3,225	4,125
3020	100 MBH		1.35	23.704		2,550	1,350		3,900	4,925
3040	122 MBH		1.10	29.091		2,675	1,650		4,325	5,500
3060	163 MBH		1	32		3,150	1,825		4,975	6,325
3080	203 MBH		1	32		3,625	1,825		5,450	6,850
3100	240 MBH		.95	33.684		3,700	1,925		5,625	7,050
3120	280 MBH		.90	35.556		4,325	2,025		6,350	7,900
3140	320 MBH		.80	40		4,675	2,275		6,950	8,700
3160	360 MBH		.71	45.070		5,400	2,575		7,975	9,950
3180	400 MBH		.64	50		5,575	2,850		8,425	10,600
3200	440 MBH		.58	54.983		5,950	3,125		9,075	11,400
3220	544 MBH		.51	62.992		9,800	3,575		13,375	16,400
3240	765 MBH		.46	70.022		11,900	3,975		15,875	19,200
3260	1,088 MBH		.40	80		13,500	4,550		18,050	22,000
3280	1,275 MBH		.36	89.888		17,300	5,125		22,425	27,000
3300	1,530 MBH		.31	104		18,200	5,975		24,175	29,300
3320	2,000 MBH		.26	125		21,400	7,100		28,500	34,600
3340	2,312 MBH		.22	148		24,400	8,425		32,825	40,000
3360	2,856 MBH		.20	160		30,800	9,100		39,900	48,100
3380	3,264 MBH		.18	179		32,700	10,200		42,900	52,000
3400	3,996 MBH		.16	195		35,800	11,100		46,900	56,500
3420	4,488 MBH		.15	210		39,400	12,000		51,400	62,000
3440	4,720 MBH		.15	220		74,000	12,600		86,600	101,000
3460	5,520 MBH		.14	228		95,000	13,000		108,000	125,000
3480	6,100 MBH		.13	250		115,500	14,200		129,700	149,000
3500	6,390 MBH		.11	285		121,500	16,300		137,800	159,500
3520	6,680 MBH		.10	310		126,500	17,700		144,200	167,000
3540	6,970 MBH		.09	359		118,500	20,500		139,000	162,500
7000	For tankless water heater, add					10%				
7050	For additional zone valves up to 312 MBH add					194			194	214
9900	Minimum labor/equipment charge	Q-6	1	24	Job		1,350		1,350	2,100

23 52 23.30 Gas/Oil Fired Boilers		Crew	Daily Output	Labor-Hours	Unit	Material	Labor	Equipment	Total	Total Incl O&P
0010	**GAS/OIL FIRED BOILERS**, Combination with burners and controls, packaged									
1000	Cast iron with insulated jacket									
2000	Steam, gross output, 720 MBH	Q-7	.43	74.074	Ea.	14,700	4,225		18,925	22,800
2020	810 MBH		.38	83.990		14,700	4,775		19,475	23,700
2040	1,084 MBH		.34	93.023		17,000	5,300		22,300	27,000
2060	1,360 MBH		.33	98.160		19,500	5,575		25,075	30,100
2080	1,600 MBH		.30	107		20,900	6,100		27,000	32,500
2100	2,040 MBH		.25	130		24,900	7,400		32,300	38,900
2120	2,450 MBH		.21	156		26,600	8,875		35,475	43,100
2140	2,700 MBH		.19	165		28,100	9,425		37,525	45,600

23 52 23 – Cast-Iron Boilers

23 52 23.30 Gas/Oil Fired Boilers		Crew	Daily Output	Labor-Hours	Unit	Material	Labor	Equipment	Total	Total Incl O&P
						2015 Bare Costs				
2160	3,000 MBH	Q-7	.18	175	Ea.	30,100	10,000		40,100	48,700
2180	3,270 MBH		.17	183		32,900	10,500		43,400	52,500
2200	3,770 MBH		.17	191		67,500	10,900		78,400	91,000
2220	4,070 MBH		.16	200		71,000	11,400		82,400	95,500
2240	4,650 MBH		.15	210		75,000	12,000		87,000	101,000
2260	5,230 MBH		.14	223		83,000	12,700		95,700	111,000
2280	5,520 MBH		.14	235		90,500	13,400		103,900	120,500
2300	5,810 MBH		.13	248		92,000	14,100		106,100	123,000
2320	6,100 MBH		.12	260		94,000	14,800		108,800	126,500
2340	6,390 MBH		.11	296		97,000	16,900		113,900	133,000
2360	6,680 MBH		.10	320		100,000	18,200		118,200	138,500
2380	6,970 MBH		.09	372		102,500	21,200		123,700	146,000
2900	Hot water, gross output									
2910	200 MBH	Q-6	.62	39.024	Ea.	10,500	2,175		12,675	15,000
2920	300 MBH		.49	49.080		10,500	2,725		13,225	15,900
2930	400 MBH		.41	57.971		12,300	3,225		15,525	18,600
2940	500 MBH		.36	67.039		13,300	3,750		17,050	20,400
3000	584 MBH	Q-7	.44	72.072		14,700	4,100		18,800	22,500
3020	876 MBH		.41	79.012		19,400	4,500		23,900	28,300
3040	1,168 MBH		.31	103		28,500	5,900		34,400	40,500
3060	1,460 MBH		.28	113		36,400	6,425		42,825	50,000
3080	2,044 MBH		.26	122		41,300	6,950		48,250	56,000
3100	2,628 MBH		.21	150		45,300	8,550		53,850	63,000
3120	3,210 MBH		.18	174		48,400	9,950		58,350	68,500
3140	3,796 MBH		.17	186		54,000	10,600		64,600	76,000
3160	4,088 MBH		.16	195		61,500	11,100		72,600	85,000
3180	4,672 MBH		.16	203		67,000	11,600		78,600	92,000
3200	5,256 MBH		.15	217		75,000	12,400		87,400	102,000
3220	6,000 MBH, 179 BHP		.13	256		108,000	14,600		122,600	141,000
3240	7,130 MBH, 213 BHP		.08	385		113,000	21,900		134,900	158,000
3260	9,800 MBH, 286 BHP		.06	533		130,000	30,300		160,300	190,500
3280	10,900 MBH, 325.6 BHP		.05	592		157,500	33,700		191,200	226,000
3290	12,200 MBH, 364.5 BHP		.05	666		172,000	37,900		209,900	248,000
3300	13,500 MBH, 403.3 BHP		.04	727		190,500	41,400		231,900	274,000

23 52 23.40 Oil-Fired Boilers

		Crew	Daily Output	Labor-Hours	Unit	Material	Labor	Equipment	Total	Total Incl O&P
0010	**OIL-FIRED BOILERS**, Standard controls, flame retention burner, packaged									
1000	Cast iron, with insulated flush jacket									
2000	Steam, gross output, 109 MBH	Q-7	1.20	26.667	Ea.	2,325	1,525		3,850	4,925
2020	144 MBH		1.10	29.091		2,625	1,650		4,275	5,450
2040	173 MBH		1	32		2,925	1,825		4,750	6,075
2060	207 MBH		.90	35.556		3,175	2,025		5,200	6,650
2080	236 MBH		.85	37.647		3,725	2,150		5,875	7,450
2100	300 MBH		.70	45.714		4,700	2,600		7,300	9,225
2120	480 MBH		.50	64		6,075	3,650		9,725	12,400
2140	665 MBH		.45	71.111		8,625	4,050		12,675	15,800
2160	794 MBH		.41	78.049		9,575	4,450		14,025	17,400
2180	1,084 MBH		.38	85.106		10,900	4,850		15,750	19,600
2200	1,360 MBH		.33	98.160		12,500	5,575		18,075	22,400
2220	1,600 MBH		.26	122		14,000	6,950		20,950	26,200
2240	2,175 MBH		.24	133		18,100	7,625		25,725	31,900
2260	2,480 MBH		.21	156		20,600	8,875		29,475	36,500
2280	3,000 MBH		.19	170		23,700	9,675		33,375	41,200

23 52 Heating Boilers

23 52 23 – Cast-Iron Boilers

23 52 23.40 Oil-Fired Boilers		Crew	Daily Output	Labor-Hours	Unit	Material	2015 Bare Costs Labor	Equipment	Total	Total Incl O&P
2300	3,550 MBH	Q-7	.17	187	Ea.	27,400	10,600		38,000	46,700
2320	3,820 MBH		.16	200		39,000	11,400		50,400	60,500
2340	4,360 MBH		.15	214		42,800	12,200		55,000	66,000
2360	4,940 MBH		.14	225		67,000	12,800		79,800	93,500
2380	5,520 MBH		.14	235		78,500	13,400		91,900	107,500
2400	6,100 MBH		.13	256		90,000	14,600		104,600	121,500
2420	6,390 MBH		.11	290		95,000	16,500		111,500	130,500
2440	6,680 MBH		.10	313		97,500	17,800		115,300	135,500
2460	6,970 MBH	↓	.09	363	↓	101,000	20,700		121,700	143,500
3000	Hot water, same price as steam									
4000	For tankless coil in smaller sizes, add				Ea.	15%				

23 52 23.60 Solid-Fuel Boilers		Crew	Daily Output	Labor-Hours	Unit	Material	Labor	Equipment	Total	Total Incl O&P
0010	**SOLID-FUEL BOILERS**									
3000	Stoker fired (coal) cast iron with flush jacket and									
3400	insulation, steam or water, gross output, 1280 MBH	Q-6	.36	66.667	Ea.	43,100	3,725		46,825	53,000
3420	1460 MBH		.30	80		47,300	4,450		51,750	59,000
3440	1640 MBH		.28	85.714		50,500	4,775		55,275	63,000
3460	1820 MBH		.26	92.308		54,000	5,150		59,150	67,500
3480	2000 MBH		.25	96		57,500	5,350		62,850	71,500
3500	2360 MBH		.23	104		65,000	5,825		70,825	80,500
3540	2725 MBH	↓	.20	120		72,500	6,700		79,200	90,000
3800	2950 MBH	Q-7	.16	200		121,000	11,400		132,400	150,500
3820	3210 MBH		.15	213		127,000	12,100		139,100	158,500
3840	3480 MBH		.14	228		132,500	13,000		145,500	166,500
3860	3745 MBH		.14	228		139,000	13,000		152,000	173,500
3880	4000 MBH		.13	246		144,500	14,000		158,500	181,000
3900	4200 MBH		.13	246		151,000	14,000		165,000	188,000
3920	4400 MBH		.12	266		157,000	15,200		172,200	196,000
3940	4600 MBH	↓	.12	266	↓	163,000	15,200		178,200	202,500

23 52 26 – Steel Boilers

23 52 26.40 Oil-Fired Boilers		Crew	Daily Output	Labor-Hours	Unit	Material	Labor	Equipment	Total	Total Incl O&P
0010	**OIL-FIRED BOILERS**, Standard controls, flame retention burner									
5000	Steel, with insulated flush jacket									
7000	Hot water, gross output, 103 MBH	Q-6	1.60	15	Ea.	1,775	835		2,610	3,250
7020	122 MBH		1.45	16.506		1,875	920		2,795	3,500
7040	137 MBH		1.36	17.595		2,000	980		2,980	3,725
7060	168 MBH		1.30	18.405		2,100	1,025		3,125	3,925
7080	225 MBH		1.22	19.704		2,575	1,100		3,675	4,550
7100	315 MBH		.96	25.105		6,550	1,400		7,950	9,400
7120	420 MBH		.70	34.483		6,900	1,925		8,825	10,600
7140	525 MBH		.57	42.403		8,150	2,375		10,525	12,700
7180	735 MBH		.48	50.104		9,575	2,800		12,375	14,900
7220	1,050 MBH		.37	65.753		19,000	3,675		22,675	26,600
7280	2,310 MBH		.21	114		25,800	6,400		32,200	38,400
7320	3,150 MBH	↓	.13	184		30,000	10,300		40,300	49,100
7340	For tankless coil in steam or hot water, add				↓	7%				
9000	Minimum labor/equipment charge	Q-6	1.75	13.714	Job		765		765	1,200

23 52 26.70 Packaged Water Tube Boilers		Crew	Daily Output	Labor-Hours	Unit	Material	Labor	Equipment	Total	Total Incl O&P
0010	**PACKAGED WATER TUBE BOILERS**									
2000	Packaged water tube, #2 oil, steam or hot water, gross output									
2040	1200 MBH	Q-7	.50	64	Ea.	18,500	3,650		22,150	26,000
2060	1600 MBH	↓	.40	80	↓	29,200	4,550		33,750	39,200

23 52 Heating Boilers

23 52 26 – Steel Boilers

23 52 26.70 Packaged Water Tube Boilers		Crew	Daily Output	Labor-Hours	Unit	2015 Bare Costs Material	Labor	Equipment	Total	Total Incl O&P
2080	2400 MBH	Q-7	.30	106	Ea.	39,300	6,075		45,375	53,000
2100	3200 MBH		.25	128		44,900	7,275		52,175	61,000
2120	4800 MBH		.20	160		58,500	9,100		67,600	78,500
2200	Gas fired									
2204	200 MBH	Q-6	1.50	16	Ea.	5,650	890		6,540	7,600
2208	275 MBH		1.40	17.143		5,975	955		6,930	8,050
2212	360 MBH		1.10	21.818		6,400	1,225		7,625	8,925
2216	520 MBH		.65	36.923		7,700	2,050		9,750	11,700
2220	600 MBH		.60	40		7,875	2,225		10,100	12,100
2224	720 MBH		.55	43.636		9,350	2,425		11,775	14,100
2228	960 MBH		.48	50		11,600	2,800		14,400	17,100
2232	1220 MBH	Q-7	.50	64		17,300	3,650		20,950	24,700
2236	1440 MBH		.45	71.111		17,800	4,050		21,850	25,900
2240	1680 MBH		.40	80		26,500	4,550		31,050	36,300
2244	1920 MBH		.35	91.429		26,600	5,200		31,800	37,400
2248	2160 MBH		.33	96.970		27,000	5,525		32,525	38,300
2252	2400 MBH		.30	106		30,400	6,075		36,475	42,900

23 52 28 – Swimming Pool Boilers

23 52 28.10 Swimming Pool Heaters		Crew	Daily Output	Labor-Hours	Unit	Material	Labor	Equipment	Total	Total Incl O&P
0010	**SWIMMING POOL HEATERS**, Not including wiring, external									
0020	piping, base or pad,									
0160	Gas fired, input, 155 MBH	Q-6	1.50	16	Ea.	2,000	890		2,890	3,600
0200	199 MBH		1	24		2,125	1,350		3,475	4,450
0220	250 MBH		.70	34.286		2,325	1,900		4,225	5,525
0240	300 MBH		.60	40		2,425	2,225		4,650	6,150
0260	399 MBH		.50	48		2,725	2,675		5,400	7,175
0280	500 MBH		.40	60		8,900	3,350		12,250	15,000
0300	650 MBH		.35	68.571		9,450	3,825		13,275	16,400
0320	750 MBH		.33	72.727		10,300	4,050		14,350	17,700
0360	990 MBH		.22	109		13,900	6,075		19,975	24,700
0370	1,260 MBH		.21	114		16,300	6,375		22,675	27,900
0380	1,440 MBH		.19	126		17,400	7,050		24,450	30,100
0400	1,800 MBH		.14	171		19,300	9,550		28,850	36,200
0410	2,070 MBH		.13	184		22,800	10,300		33,100	41,200
2000	Electric, 12 KW, 4,800 gallon pool	Q-19	3	8		2,075	435		2,510	2,950
2020	15 KW, 7,200 gallon pool		2.80	8.571		2,100	465		2,565	3,050
2040	24 KW, 9,600 gallon pool		2.40	10		2,425	540		2,965	3,525
2060	30 KW, 12,000 gallon pool		2	12		2,475	650		3,125	3,725
2080	36 KW, 14,400 gallon pool		1.60	15		2,850	810		3,660	4,375
2100	57 KW, 24,000 gallon pool		1.20	20		3,575	1,075		4,650	5,600
9000	To select pool heater: 12 BTUH x S.F. pool area									
9010	X temperature differential = required output									
9050	For electric, KW = gallons x 2.5 divided by 1000									
9100	For family home type pool, double the									
9110	Rated gallon capacity = 1/2°F rise per hour									

23 52 39 – Fire-Tube Boilers

23 52 39.13 Scotch Marine Boilers		Crew	Daily Output	Labor-Hours	Unit	Material	Labor	Equipment	Total	Total Incl O&P
0010	**SCOTCH MARINE BOILERS**									
1000	Packaged fire tube, #2 oil, gross output									
1006	15 PSI steam									
1020	3348 MBH, 100 HP	Q-7	.21	152	Ea.	69,500	8,675		78,175	89,500
1040	6696 MBH, 200 HP		.14	223		104,000	12,700		116,700	134,000

23 52 Heating Boilers

23 52 39 – Fire-Tube Boilers

23 52 39.13 Scotch Marine Boilers		Crew	Daily Output	Labor-Hours	Unit	2015 Bare Costs Material	Labor	Equipment	Total	Total Incl O&P
1060	10,044 MBH, 300 HP	Q-7	.13	251	Ea.	132,000	14,300		146,300	168,000
1080	16,740 MBH, 500 HP		.08	380		179,500	21,700		201,200	231,500
1100	23,435 MBH, 700 HP		.07	484		220,000	27,600		247,600	285,000
1102	26,780 MBH, 800 HP		.07	484		220,000	27,600		247,600	285,000
1104	30,125 MBH, 900 HP		.06	500		401,500	28,400		429,900	486,000
1106	33,475 MBH, 1,000 HP		.06	524		425,000	29,800		454,800	514,000
1107	36,825 MBH, 1,100 HP		.06	551		456,000	31,400		487,400	551,000
1108	40,170 MBH, 1,200 HP		.06	581		478,500	33,100		511,600	577,500
1110	46,865 MBH, 1,400 HP		.05	603		504,000	34,300		538,300	608,000
1112	53,560 MBH, 1,600 HP		.05	627		547,500	35,700		583,200	658,000
1114	60,250 MBH, 1,800 HP		.05	666		565,500	37,900		603,400	681,000
1116	66,950 MBH, 2,000 HP		.05	695		599,000	39,600		638,600	720,500
1118	73,650 MBH, 2,200 HP		.04	744		647,000	42,300		689,300	778,000
1120	To fire #6, add		.83	38.554		13,300	2,200		15,500	18,000
1140	To fire #6, and gas, add		.42	76.190		19,600	4,325		23,925	28,300
1160	For high pressure, add					21,700			21,700	23,900
1180	For duplex package feed system									
1200	To 3348 MBH boiler, add	Q-7	.54	59.259	Ea.	10,500	3,375		13,875	16,900
1220	To 6696 MBH boiler, add		.41	78.049		12,500	4,450		16,950	20,600
1240	To 10,044 MBH boiler, add		.38	84.211		14,300	4,800		19,100	23,300
1260	To 16,740 MBH boiler, add		.28	114		16,300	6,500		22,800	28,100
1280	To 23,435 MBH boiler, add		.25	128		19,300	7,275		26,575	32,600

23 52 84 – Boiler Blowdown

23 52 84.10 Boiler Blowdown Systems		Crew	Daily Output	Labor-Hours	Unit	Material	Labor	Equipment	Total	Total Incl O&P
0010	BOILER BLOWDOWN SYSTEMS									
1010	Boiler blowdown, auto/manual to 2000 MBH	Q-5	3.75	4.267	Ea.	4,850	229		5,079	5,675
1020	7300 MBH	"	3	5.333	"	5,550	287		5,837	6,550

23 52 88 – Burners

23 52 88.10 Replacement Type Burners		Crew	Daily Output	Labor-Hours	Unit	Material	Labor	Equipment	Total	Total Incl O&P
0010	REPLACEMENT TYPE BURNERS									
0990	Residential, conversion, gas fired, LP or natural									
1000	Gun type, atmospheric input 50 to 225 MBH	Q-1	2.50	6.400	Ea.	720	340		1,060	1,325
1020	100 to 400 MBH		2	8		1,200	425		1,625	1,975
1040	300 to 1000 MBH		1.70	9.412		3,625	495		4,120	4,750
2000	Commercial and industrial, gas/oil, input									
2050	400 MBH	Q-1	1.50	10.667	Ea.	4,175	565		4,740	5,475
2090	670 MBH		1.40	11.429		4,175	605		4,780	5,550
2140	1155 MBH		1.30	12.308		4,650	650		5,300	6,150
2200	1800 MBH		1.20	13.333		4,675	705		5,380	6,225
2260	3000 MBH		1.10	14.545		4,675	770		5,445	6,350
2320	4100 MBH		1	16		4,675	845		5,520	6,475
3000	Flame retention oil fired assembly, input									
3020	.50 to 2.25 GPH	Q-1	2.40	6.667	Ea.	284	350		634	860
3040	2.0 to 5.0 GPH		2	8		315	425		740	1,000
3060	3.0 to 7.0 GPH		1.80	8.889		570	470		1,040	1,350
3080	6.0 to 12.0 GPH		1.60	10		915	530		1,445	1,825
4600	Gas safety, shut off valve, 3/4" threaded	1 Stpi	20	.400		178	24		202	234
4610	1" threaded		19	.421		172	25		197	229
4620	1-1/4" threaded		15	.533		194	32		226	263
4630	1-1/2" threaded		13	.615		210	37		247	289
4640	2" threaded		11	.727		235	43.50		278.50	325
4650	2-1/2" threaded	Q-1	15	1.067		269	56.50		325.50	385

23 52 Heating Boilers

23 52 88 – Burners

23 52 88.10 Replacement Type Burners		Crew	Daily Output	Labor-Hours	Unit	Material	2015 Bare Costs Labor	Equipment	Total	Total Incl O&P
4660	3″ threaded	Q-1	13	1.231	Ea.	370	65		435	505
4670	4″ flanged	↓	3	5.333		2,625	282		2,907	3,350
4680	6″ flanged	Q-2	3	8	↓	5,875	440		6,315	7,150

23 54 Furnaces

23 54 13 – Electric-Resistance Furnaces

23 54 13.10 Electric Furnaces		Crew	Daily Output	Labor-Hours	Unit	Material	Labor	Equipment	Total	Total Incl O&P
0010	**ELECTRIC FURNACES**, Hot air, blowers, std. controls									
0011	not including gas, oil or flue piping									
1000	Electric, UL listed									
1100	34.1 MBH	Q-20	4.40	4.545	Ea.	455	233		688	865
1120	51.6 MBH		4.20	4.762		495	244		739	930
1140	68.3 MBH		4	5		515	256		771	970
1160	85.3 MBH	↓	3.80	5.263	↓	520	270		790	1,000

23 54 16 – Fuel-Fired Furnaces

23 54 16.13 Gas-Fired Furnaces		Crew	Daily Output	Labor-Hours	Unit	Material	Labor	Equipment	Total	Total Incl O&P
0010	**GAS-FIRED FURNACES**									
3000	Gas, AGA certified, upflow, direct drive models									
3020	45 MBH input	Q-9	4	4	Ea.	535	201		736	910
3040	60 MBH input		3.80	4.211		535	212		747	925
3060	75 MBH input		3.60	4.444		575	224		799	990
3100	100 MBH input		3.20	5		625	252		877	1,075
3120	125 MBH input		3	5.333		655	269		924	1,150
3130	150 MBH input		2.80	5.714		670	288		958	1,200
3140	200 MBH input		2.60	6.154		2,800	310		3,110	3,575
3160	300 MBH input		2.30	6.957		3,100	350		3,450	3,975
3180	400 MBH input	↓	2	8	↓	3,450	405		3,855	4,425

23 54 16.14 Condensing Furnaces		Crew	Daily Output	Labor-Hours	Unit	Material	Labor	Equipment	Total	Total Incl O&P
0010	**CONDENSING FURNACES**, High efficiency									
0020	Oil fired, packaged, complete									
0030	Upflow									
0040	Output @ 95% A.F.U.E.									
0100	49 MBH @ 1000 CFM	Q-9	3.70	4.324	Ea.	4,450	218		4,668	5,250
0110	73.5 MBH @ 2000 CFM		3.60	4.444		4,650	224		4,874	5,475
0120	96 MBH @ 2000 CFM		3.40	4.706		4,650	237		4,887	5,500
0130	115.6 MBH @ 2000 CFM		3.40	4.706		4,650	237		4,887	5,500
0140	147 MBH @ 2000 CFM		3.30	4.848		10,400	244		10,644	11,900
0150	192 MBH @ 4000 CFM		2.60	6.154		10,800	310		11,110	12,400
0170	231.5 MBH @ 4000 CFM	↓	2.30	6.957		10,800	350		11,150	12,500
0260	For variable speed motor, add				↓	565			565	625
0270	Note: Also available in horizontal, counterflow and lowboy configurations.									

23 54 16.16 Oil-Fired Furnaces		Crew	Daily Output	Labor-Hours	Unit	Material	Labor	Equipment	Total	Total Incl O&P
0010	**OIL-FIRED FURNACES**									
6000	Oil, UL listed, atomizing gun type burner									
6020	56 MBH output	Q-9	3.60	4.444	Ea.	1,725	224		1,949	2,250
6030	84 MBH output		3.50	4.571		1,850	230		2,080	2,425
6040	95 MBH output		3.40	4.706		1,875	237		2,112	2,450
6060	134 MBH output		3.20	5		2,175	252		2,427	2,800
6080	151 MBH output		3	5.333		2,250	269		2,519	2,900
6100	200 MBH input	↓	2.60	6.154	↓	2,450	310		2,760	3,200

23 54 Furnaces

23 54 16 – Fuel-Fired Furnaces

23 54 16.16 Oil-Fired Furnaces		Crew	Daily Output	Labor-Hours	Unit	2015 Bare Costs Material	Labor	Equipment	Total	Total Incl O&P
6120	300 MBH input	Q-9	2.30	6.957	Ea.	3,600	350		3,950	4,525
6140	400 MBH input		2	8		3,900	405		4,305	4,925
9000	Minimum labor/equipment charge		2.75	5.818	Job		293		293	460

23 54 24 – Furnace Components for Cooling

23 54 24.10 Furnace Components and Combinations		Crew	Daily Output	Labor-Hours	Unit	Material	Labor	Equipment	Total	Total Incl O&P
0010	FURNACE COMPONENTS AND COMBINATIONS									
0080	Coils, A.C. evaporator, for gas or oil furnaces									
0090	Add-on, with holding charge									
0100	Upflow									
0120	1-1/2 ton cooling	Q-5	4	4	Ea.	220	215		435	575
0130	2 ton cooling		3.70	4.324		241	233		474	630
0140	3 ton cooling		3.30	4.848		375	261		636	820
0150	4 ton cooling		3	5.333		465	287		752	960
0160	5 ton cooling		2.70	5.926		505	320		825	1,050
0300	Downflow									
0330	2-1/2 ton cooling	Q-5	3	5.333	Ea.	295	287		582	770
0340	3-1/2 ton cooling		2.60	6.154		435	330		765	990
0350	5 ton cooling		2.20	7.273		505	390		895	1,175
0600	Horizontal									
0630	2 ton cooling	Q-5	3.90	4.103	Ea.	320	221		541	700
0640	3 ton cooling		3.50	4.571		370	246		616	790
0650	4 ton cooling		3.20	5		460	269		729	925
0660	5 ton cooling		2.90	5.517		460	297		757	970
2000	Cased evaporator coils for air handlers									
2100	1-1/2 ton cooling	Q-5	4.40	3.636	Ea.	267	196		463	600
2110	2 ton cooling		4.10	3.902		300	210		510	655
2120	2-1/2 ton cooling		3.90	4.103		320	221		541	695
2130	3 ton cooling		3.70	4.324		360	233		593	760
2140	3-1/2 ton cooling		3.50	4.571		485	246		731	920
2150	4 ton cooling		3.20	5		515	269		784	990
2160	5 ton cooling		2.90	5.517		525	297		822	1,050
3010	Air handler, modular									
3100	With cased evaporator cooling coil									
3120	1-1/2 ton cooling	Q-5	3.80	4.211	Ea.	850	226		1,076	1,300
3130	2 ton cooling		3.50	4.571		940	246		1,186	1,400
3140	2-1/2 ton cooling		3.30	4.848		995	261		1,256	1,500
3150	3 ton cooling		3.10	5.161		1,150	278		1,428	1,675
3160	3-1/2 ton cooling		2.90	5.517		1,325	297		1,622	1,925
3170	4 ton cooling		2.50	6.400		1,425	345		1,770	2,075
3180	5 ton cooling		2.10	7.619		1,700	410		2,110	2,500
3500	With no cooling coil									
3520	1-1/2 ton coil size	Q-5	12	1.333	Ea.	620	71.50		691.50	790
3530	2 ton coil size		10	1.600		620	86		706	820
3540	2-1/2 ton coil size		10	1.600		725	86		811	930
3554	3 ton coil size		9	1.778		825	95.50		920.50	1,050
3560	3-1/2 ton coil size		9	1.778		845	95.50		940.50	1,075
3570	4 ton coil size		8.50	1.882		1,050	101		1,151	1,325
3580	5 ton coil size		8	2		1,075	108		1,183	1,350
4000	With heater									
4120	5 kW, 17.1 MBH	Q-5	16	1	Ea.	670	54		724	820
4130	7.5 kW, 25.6 MBH		15.60	1.026		680	55		735	830
4140	10 kW, 34.2 MBH		15.20	1.053		845	56.50		901.50	1,025

23 54 Furnaces

23 54 24 – Furnace Components for Cooling

23 54 24.10 Furnace Components and Combinations		Crew	Daily Output	Labor-Hours	Unit	2015 Bare Costs Material	Labor	Equipment	Total	Total Incl O&P
4150	12.5 KW, 42.7 MBH	Q-5	14.80	1.081	Ea.	950	58		1,008	1,150
4160	15 KW, 51.2 MBH		14.40	1.111		1,025	60		1,085	1,225
4170	25 KW, 85.4 MBH		14	1.143		1,375	61.50		1,436.50	1,600
4180	30 KW, 102 MBH		13	1.231		1,600	66		1,666	1,850

23 55 Fuel-Fired Heaters

23 55 13 – Fuel-Fired Duct Heaters

23 55 13.16 Gas-Fired Duct Heaters

		Crew	Daily Output	Labor-Hours	Unit	Material	Labor	Equipment	Total	Total Incl O&P
0010	**GAS-FIRED DUCT HEATERS**, Includes burner, controls, stainless steel									
0020	heat exchanger. Gas fired, electric ignition									
0030	Indoor installation									
0080	100 MBH output	Q-5	5	3.200	Ea.	2,900	172		3,072	3,450
0100	120 MBH output		4	4		3,175	215		3,390	3,825
0130	200 MBH output		2.70	5.926		4,025	320		4,345	4,925
0140	240 MBH output		2.30	6.957		4,225	375		4,600	5,225
0160	280 MBH output		2	8		4,600	430		5,030	5,725
0180	320 MBH output		1.60	10		5,150	540		5,690	6,525
0300	For powered venter and adapter, add					525			525	575
0502	For required flue pipe, see Section 23 51 23.10									
1000	Outdoor installation, with power venter									
1020	75 MBH output	Q-5	4	4	Ea.	3,525	215		3,740	4,200
1040	94 MBH output		4	4		3,625	215		3,840	4,300
1060	120 MBH output		4	4		3,875	215		4,090	4,575
1080	157 MBH output		3.50	4.571		4,200	246		4,446	4,975
1100	187 MBH output		3	5.333		4,800	287		5,087	5,725
1120	225 MBH output		2.50	6.400		5,000	345		5,345	6,025
1140	300 MBH output		1.80	8.889		7,750	480		8,230	9,275
1160	375 MBH output		1.60	10		8,150	540		8,690	9,825
1180	450 MBH output		1.40	11.429		9,350	615		9,965	11,300
1200	600 MBH output		1	16		10,600	860		11,460	13,000
1300	Aluminized exchanger, subtract					15%				
1500	For two stage gas valve, add					755			755	830

23 55 33 – Fuel-Fired Unit Heaters

23 55 33.13 Oil-Fired Unit Heaters

		Crew	Daily Output	Labor-Hours	Unit	Material	Labor	Equipment	Total	Total Incl O&P
0010	**OIL-FIRED UNIT HEATERS**, Cabinet, grilles, fan, ctrl., burner, no piping									
6000	Oil fired, suspension mounted, 94 MBH output	Q-5	4	4	Ea.	4,825	215		5,040	5,625
6040	140 MBH output		3	5.333		5,075	287		5,362	6,025
6060	184 MBH output		3	5.333		5,375	287		5,662	6,375

23 55 33.16 Gas-Fired Unit Heaters

		Crew	Daily Output	Labor-Hours	Unit	Material	Labor	Equipment	Total	Total Incl O&P
0010	**GAS-FIRED UNIT HEATERS**, Cabinet, grilles, fan, ctrls., burner, no piping									
0022	thermostat, no piping. For flue see Section 23 51 23.10									
1000	Gas fired, floor mounted									
1100	60 MBH output	Q-5	10	1.600	Ea.	870	86		956	1,100
1120	80 MBH output		9	1.778		880	95.50		975.50	1,125
1140	100 MBH output		8	2		960	108		1,068	1,225
1160	120 MBH output		7	2.286		1,150	123		1,273	1,450
1180	180 MBH output		6	2.667		1,375	143		1,518	1,725
1500	Rooftop mounted, power vent, stainless steel exchanger									
1520	75 MBH input	Q-6	4	6	Ea.	5,625	335		5,960	6,725
1540	100 MBH input		3.60	6.667		5,875	370		6,245	7,050

	23 55 33.16 Gas-Fired Unit Heaters	Crew	Daily Output	Labor-Hours	Unit	Material	Labor	Equipment	Total	Total Incl O&P
						2015 Bare Costs				
1560	125 MBH input	Q-6	3.30	7.273	Ea.	6,125	405		6,530	7,375
1580	150 MBH input		3	8		6,600	445		7,045	7,975
1600	175 MBH input		2.60	9.231		6,725	515		7,240	8,200
1620	225 MBH input		2.30	10.435		7,725	580		8,305	9,400
1640	300 MBH input		1.90	12.632		8,950	705		9,655	11,000
1660	350 MBH input		1.40	17.143		9,750	955		10,705	12,200
1680	450 MBH input		1.20	20		10,200	1,125		11,325	13,100
1720	700 MBH input		.80	30		10,600	1,675		12,275	14,300
1760	1200 MBH input		.30	80		11,000	4,450		15,450	19,100
1900	For aluminized steel exchanger, subtract					10%				
2000	Suspension mounted, propeller fan, 20 MBH output	Q-5	8.50	1.882		1,125	101		1,226	1,400
2020	40 MBH output		7.50	2.133		1,550	115		1,665	1,875
2040	60 MBH output		7	2.286		1,700	123		1,823	2,075
2060	80 MBH output		6	2.667		1,875	143		2,018	2,275
2080	100 MBH output		5.50	2.909		2,050	156		2,206	2,500
2100	130 MBH output		5	3.200		2,225	172		2,397	2,725
2120	140 MBH output		4.50	3.556		2,350	191		2,541	2,875
2140	160 MBH output		4	4		2,450	215		2,665	3,000
2160	180 MBH output		3.50	4.571		2,625	246		2,871	3,275
2180	200 MBH output		3	5.333		2,900	287		3,187	3,625
2200	240 MBH output		2.70	5.926		3,175	320		3,495	4,000
2220	280 MBH output		2.30	6.957		3,500	375		3,875	4,425
2240	320 MBH output		2	8		4,100	430		4,530	5,200
2500	For powered venter and adapter, add					490			490	540
3000	Suspension mounted, blower type, 40 MBH output	Q-5	6.80	2.353		935	127		1,062	1,225
3020	60 MBH output		6.60	2.424		960	130		1,090	1,250
3040	84 MBH output		5.80	2.759		995	148		1,143	1,325
3060	104 MBH output		5.20	3.077		1,025	165		1,190	1,375
3080	140 MBH output		4.30	3.721		1,100	200		1,300	1,500
3100	180 MBH output		3.30	4.848		1,300	261		1,561	1,825
3120	240 MBH output		2.50	6.400		1,500	345		1,845	2,175
3140	280 MBH output		2	8		1,675	430		2,105	2,525
4000	Suspension mounted, sealed combustion system,									
4020	Aluminized steel exchanger, powered vent									
4040	100 MBH output	Q-5	5	3.200	Ea.	3,025	172		3,197	3,600
4060	120 MBH output		4.70	3.404		3,600	183		3,783	4,225
4080	160 MBH output		3.70	4.324		4,150	233		4,383	4,950
4100	200 MBH output		2.90	5.517		5,100	297		5,397	6,075
4120	240 MBH output		2.50	6.400		5,450	345		5,795	6,525
4140	320 MBH output		1.70	9.412		6,525	505		7,030	7,975
5000	Wall furnace, 17.5 MBH output		6	2.667		745	143		888	1,050
5020	24 MBH output		5	3.200		745	172		917	1,100
5040	35 MBH output		4	4		790	215		1,005	1,200
9000	Minimum labor/equipment charge		3.50	4.571	Job		246		246	385

23 56 Solar Energy Heating Equipment

23 56 16 – Packaged Solar Heating Equipment

23 56 16.40 Solar Heating Systems			Crew	Daily Output	Labor-Hours	Unit	2015 Bare Costs Material	Labor	Equipment	Total	Total Incl O&P
0010	SOLAR HEATING SYSTEMS										
0020	System/Package prices, not including connecting										
0030	pipe, insulation, or special heating/plumbing fixtures										
0152	For solar ultraviolet pipe insulation see Section 22 07 19.10										
0500	Hot water, standard package, low temperature										
0540	1 collector, circulator, fittings, 65 gal. tank	G	Q-1	.50	32	Ea.	3,675	1,700		5,375	6,675
0580	2 collectors, circulator, fittings, 120 gal. tank	G		.40	40		5,050	2,125		7,175	8,850
0620	3 collectors, circulator, fittings, 120 gal. tank	G		.34	47.059		6,875	2,475		9,350	11,500
0700	Medium temperature package										
0720	1 collector, circulator, fittings, 80 gal. tank	G	Q-1	.50	32	Ea.	5,025	1,700		6,725	8,150
0740	2 collectors, circulator, fittings, 120 gal. tank	G		.40	40		6,450	2,125		8,575	10,400
0780	3 collectors, circulator, fittings, 120 gal. tank	G		.30	53.333		7,325	2,825		10,150	12,500
0980	For each additional 120 gal. tank, add	G					1,750			1,750	1,925

23 56 19 – Solar Heating Components

23 56 19.50 Solar Heating Ancillary			Crew	Daily Output	Labor-Hours	Unit	Material	Labor	Equipment	Total	Total Incl O&P
0010	SOLAR HEATING ANCILLARY										
2300	Circulators, air										
2310	Blowers										
2330	100-300 S.F. system, 1/10 HP	G	Q-9	16	1	Ea.	249	50.50		299.50	355
2340	300-500 S.F. system, 1/5 HP	G		15	1.067		330	53.50		383.50	445
2350	Two speed, 100-300 S.F., 1/10 HP	G		14	1.143		142	57.50		199.50	248
2400	Reversible fan, 20" diameter, 2 speed	G		18	.889		113	45		158	195
2550	Booster fan 6" diameter, 120 CFM	G		16	1		36.50	50.50		87	120
2570	6" diameter, 225 CFM	G		16	1		45	50.50		95.50	129
2580	8" diameter, 150 CFM	G		16	1		41	50.50		91.50	125
2590	8" diameter, 310 CFM	G		14	1.143		63.50	57.50		121	161
2600	8" diameter, 425 CFM	G		14	1.143		71	57.50		128.50	169
2650	Rheostat	G		32	.500		15.45	25		40.45	56.50
2660	Shutter/damper	G		12	1.333		56	67		123	168
2670	Shutter motor	G		16	1		139	50.50		189.50	233
2800	Circulators, liquid, 1/25 HP, 5.3 GPM	G	Q-1	14	1.143		114	60.50		174.50	220
2820	1/20 HP, 17 GPM	G		12	1.333		161	70.50		231.50	287
2850	1/20 HP, 17 GPM, stainless steel	G		12	1.333		254	70.50		324.50	390
2870	1/12 HP, 30 GPM	G		10	1.600		345	84.50		429.50	510
3000	Collector panels, air with aluminum absorber plate										
3010	Wall or roof mount										
3040	Flat black, plastic glazing										
3080	4' x 8'	G	Q-9	6	2.667	Ea.	660	134		794	940
3100	4' x 10'	G		5	3.200	"	815	161		976	1,150
3200	Flush roof mount, 10' to 16' x 22" wide	G		96	.167	L.F.	132	8.40		140.40	158
3210	Manifold, by L.F. width of collectors	G		160	.100	"	144	5.05		149.05	167
3300	Collector panels, liquid with copper absorber plate										
3320	Black chrome, tempered glass glazing										
3330	Alum. frame, 4' x 8', 5/32" single glazing	G	Q-1	9.50	1.684	Ea.	995	89		1,084	1,250
3390	Alum. frame, 4' x 10', 5/32" single glazing	G		6	2.667		1,150	141		1,291	1,475
3450	Flat black, alum. frame, 3.5' x 7.5'	G		9	1.778		880	94		974	1,100
3500	4' x 8'	G		5.50	2.909		1,050	154		1,204	1,400
3520	4' x 10'	G		10	1.600		1,250	84.50		1,334.50	1,500
3540	4' x 12.5'	G		5	3.200		1,250	169		1,419	1,650
3550	Liquid with fin tube absorber plate										
3560	Alum. frame 4' x 8' tempered glass	G	Q-1	10	1.600	Ea.	580	84.50		664.50	770
3580	Liquid with vacuum tubes, 4' x 6'-10"	G		9	1.778		895	94		989	1,125

23 56 Solar Energy Heating Equipment

23 56 19 – Solar Heating Components

23 56 19.50 Solar Heating Ancillary			Crew	Daily Output	Labor-Hours	Unit	Material	2015 Bare Costs Labor	Equipment	Total	Total Incl O&P
3600	Liquid, full wetted, plastic, alum. frame, 4′ x 10′	G	Q-1	5	3.200	Ea.	320	169		489	620
3650	Collector panel mounting, flat roof or ground rack	G		7	2.286		244	121		365	455
3670	Roof clamps	G		70	.229	Set	2.80	12.10		14.90	22
3700	Roof strap, teflon	G	1 Plum	205	.039	L.F.	23.50	2.29		25.79	29.50
3900	Differential controller with two sensors										
3930	Thermostat, hard wired	G	1 Plum	8	1	Ea.	101	58.50		159.50	203
3950	Line cord and receptacle	G		12	.667		130	39		169	204
4050	Pool valve system	G		2.50	3.200		161	188		349	470
4070	With 12 VAC actuator	G		2	4		325	235		560	720
4080	Pool pump system, 2″ pipe size	G		6	1.333		195	78.50		273.50	335
4100	Five station with digital read-out	G		3	2.667		263	157		420	535
4150	Sensors										
4200	Brass plug, 1/2″ MPT	G	1 Plum	32	.250	Ea.	16	14.70		30.70	40.50
4210	Brass plug, reversed	G		32	.250		26.50	14.70		41.20	52
4220	Freeze prevention	G		32	.250		22.50	14.70		37.20	48
4240	Screw attached	G		32	.250		9.85	14.70		24.55	34
4250	Brass, immersion	G		32	.250		13.55	14.70		28.25	38
4300	Heat exchanger										
4315	includes coil, blower, circulator										
4316	and controller for DHW and space hot air										
4330	Fluid to air coil, up flow, 45 MBH	G	Q-1	4	4	Ea.	315	211		526	680
4380	70 MBH	G		3.50	4.571		355	242		597	770
4400	80 MBH	G		3	5.333		475	282		757	965
4580	Fluid to fluid package includes two circulating pumps										
4590	expansion tank, check valve, relief valve										
4600	controller, high temperature cutoff and sensors	G	Q-1	2.50	6.400	Ea.	800	340		1,140	1,400
4650	Heat transfer fluid										
4700	Propylene glycol, inhibited anti-freeze	G	1 Plum	28	.286	Gal.	15.55	16.75		32.30	43
4800	Solar storage tanks, knocked down										
4810	Air, galvanized steel clad, double wall, 4″ fiberglass insulation										
5120	45 Mil reinforced polypropylene lining,										
5140	4′ high, 4′ x 4′ = 64 C.F./450 gallons	G	Q-9	2	8	Ea.	3,650	405		4,055	4,625
5150	4′ x 8′ = 128 C.F./900 gallons	G		1.50	10.667		5,450	535		5,985	6,850
5160	4′ x 12′ = 190 C.F./1300 gallons	G		1.30	12.308		7,275	620		7,895	8,975
5170	8′ x 8′ = 250 C.F./1700 gallons	G		1	16		7,275	805		8,080	9,275
5190	6′-3″ high, 7′ x 7′ = 306 C.F./2000 gallons	G	Q-10	1.20	20		14,200	1,050		15,250	17,400
5200	7′ x 10′-6″ = 459 C.F./3000 gallons	G		.80	30		17,800	1,575		19,375	22,100
5210	7′ x 14′ = 613 C.F./4000 gallons	G		.60	40		21,400	2,100		23,500	26,800
5220	10′-6″ x 10′-6″ = 689 C.F./4500 gallons	G		.50	48		21,400	2,500		23,900	27,500
5230	10′-6″ x 14′ = 919 C.F./6000 gallons	G		.40	60		24,900	3,125		28,025	32,400
5240	14′ x 14′ = 1225 C.F./8000 gallons	G	Q-11	.40	80		28,500	4,250		32,750	38,000
5250	14′ x 17′-6″ = 1531 C.F./10,000 gallons	G		.30	106		32,100	5,675		37,775	44,300
5260	17′-6″ x 17′-6″ = 1914 C.F./12,500 gallons	G		.25	128		35,600	6,825		42,425	50,000
5270	17′-6″ x 21′ = 2297 C.F./15,000 gallons	G		.20	160		39,100	8,525		47,625	56,500
5280	21′ x 21′ = 2756 C.F./18,000 gallons	G		.18	177		42,700	9,475		52,175	62,000
5290	30 Mil reinforced Hypalon lining, add						.02%				
7000	Solar control valves and vents										
7050	Air purger, 1″ pipe size	G	1 Plum	12	.667	Ea.	46.50	39		85.50	113
7070	Air eliminator, automatic 3/4″ size	G		32	.250		30	14.70		44.70	56
7090	Air vent, automatic, 1/8″ fitting	G		32	.250		17.25	14.70		31.95	42
7100	Manual, 1/8″ NPT	G		32	.250		2.44	14.70		17.14	25.50
7120	Backflow preventer, 1/2″ pipe size	G		16	.500		75.50	29.50		105	129
7130	3/4″ pipe size	G		16	.500		77	29.50		106.50	131

23 56 Solar Energy Heating Equipment

23 56 19 – Solar Heating Components

23 56 19.50 Solar Heating Ancillary		Crew	Daily Output	Labor-Hours	Unit	Material	Labor	Equipment	Total	Total Incl O&P
						2015 Bare Costs				
7150	Balancing valve, 3/4" pipe size G	1 Plum	20	.400	Ea.	63	23.50		86.50	106
7180	Draindown valve, 1/2" copper tube G		9	.889		212	52		264	315
7200	Flow control valve, 1/2" pipe size G		22	.364		120	21.50		141.50	166
7220	Expansion tank, up to 5 gal. G		32	.250		68	14.70		82.70	98
7250	Hydronic controller (aquastat) G		8	1		228	58.50		286.50	345
7400	Pressure gauge, 2" dial G		32	.250		24	14.70		38.70	49.50
7450	Relief valve, temp. and pressure 3/4" pipe size G		30	.267		21	15.65		36.65	47.50
7500	Solenoid valve, normally closed									
7520	Brass, 3/4" NPT, 24V G	1 Plum	9	.889	Ea.	123	52		175	217
7530	1" NPT, 24V G		9	.889		193	52		245	294
7750	Vacuum relief valve, 3/4" pipe size G		32	.250		29.50	14.70		44.20	55.50
7800	Thermometers									
7820	Digital temperature monitoring, 4 locations G	1 Plum	2.50	3.200	Ea.	137	188		325	445
7900	Upright, 1/2" NPT G		8	1		39.50	58.50		98	135
7970	Remote probe, 2" dial G		8	1		33	58.50		91.50	128
7990	Stem, 2" dial, 9" stem G		16	.500		21.50	29.50		51	70
8250	Water storage tank with heat exchanger and electric element									
8270	66 gal. with 2" x 2 lb. density insulation G	1 Plum	1.60	5	Ea.	1,600	294		1,894	2,225
8300	80 gal. with 2" x 2 lb. density insulation G		1.60	5		1,600	294		1,894	2,225
8380	120 gal. with 2" x 2 lb. density insulation G		1.40	5.714		1,825	335		2,160	2,550
8400	120 gal. with 2" x 2 lb. density insul., 40 S.F. heat coil G		1.40	5.714		2,325	335		2,660	3,075
8500	Water storage module, plastic									
8600	Tubular, 12" diameter, 4' high G	1 Carp	48	.167	Ea.	114	7.85		121.85	138
8610	12" diameter, 8' high G		40	.200		175	9.40		184.40	208
8620	18" diameter, 5' high G		38	.211		193	9.90		202.90	228
8630	18" diameter, 10' high G		32	.250		255	11.75		266.75	300
8640	58" diameter, 5' high G	2 Carp	32	.500		495	23.50		518.50	585
8650	Cap, 12" diameter G					22			22	24
8660	18" diameter G					28			28	31
9000	Minimum labor/equipment charge	1 Plum	2	4	Job		235		235	365

23 57 Heat Exchangers for HVAC

23 57 16 – Steam-to-Water Heat Exchangers

23 57 16.10 Shell/Tube Type Steam-to-Water Heat Exch.		Crew	Daily Output	Labor-Hours	Unit	Material	Labor	Equipment	Total	Total Incl O&P
0010	**SHELL AND TUBE TYPE STEAM-TO-WATER HEAT EXCHANGERS**									
0016	Shell & tube type, 2 or 4 pass, 3/4" O.D. copper tubes,									
0020	C.I. heads, C.I. tube sheet, steel shell									
0100	Hot water 40°F to 180°F, by steam at 10 PSI									
0120	8 GPM	Q-5	6	2.667	Ea.	2,150	143		2,293	2,600
0140	10 GPM		5	3.200		3,250	172		3,422	3,850
0160	40 GPM		4	4		5,025	215		5,240	5,850
0180	64 GPM		2	8		7,700	430		8,130	9,125
0200	96 GPM		1	16		10,300	860		11,160	12,700
0220	120 GPM	Q-6	1.50	16		13,500	890		14,390	16,300
0240	168 GPM		1	24		16,600	1,350		17,950	20,300
0260	240 GPM		.80	30		25,900	1,675		27,575	31,100
0300	600 GPM		.70	34.286		55,500	1,900		57,400	64,000
0500	For bronze head and tube sheet, add					50%				

23 57 Heat Exchangers for HVAC

23 57 19 – Liquid-to-Liquid Heat Exchangers

23 57 19.13 Plate-Type, Liquid-to-Liquid Heat Exchangers		Crew	Daily Output	Labor-Hours	Unit	2015 Bare Costs Material	Labor	Equipment	Total	Total Incl O&P
0010	PLATE-TYPE, LIQUID-TO-LIQUID HEAT EXCHANGERS									
3000	Plate type,									
3100	400 GPM	Q-6	.80	30	Ea.	36,700	1,675		38,375	43,000
3120	800 GPM	"	.50	48		63,500	2,675		66,175	73,500
3140	1200 GPM	Q-7	.34	94.118		94,000	5,350		99,350	112,000
3160	1800 GPM	"	.24	133		125,000	7,575		132,575	149,500

23 57 19.16 Shell-Type, Liquid-to-Liquid Heat Exchangers		Crew	Daily Output	Labor-Hours	Unit	Material	Labor	Equipment	Total	Total Incl O&P
0010	SHELL-TYPE, LIQUID-TO-LIQUID HEAT EXCHANGERS									
1000	Hot water 40°F to 140°F, by water at 200°F									
1020	7 GPM	Q-5	6	2.667	Ea.	2,650	143		2,793	3,125
1040	16 GPM		5	3.200		3,750	172		3,922	4,400
1060	34 GPM		4	4		5,675	215		5,890	6,550
1080	55 GPM		3	5.333		8,225	287		8,512	9,500
1100	74 GPM		1.50	10.667		10,300	575		10,875	12,200
1120	86 GPM		1.40	11.429		13,700	615		14,315	16,100
1140	112 GPM	Q-6	2	12		17,100	670		17,770	19,900
1160	126 GPM		1.80	13.333		21,400	745		22,145	24,700
1180	152 GPM		1	24		27,300	1,350		28,650	32,100

23 61 Refrigerant Compressors

23 61 15 – Rotary Refrigerant Compressors

23 61 15.10 Rotary Compressors		Crew	Daily Output	Labor-Hours	Unit	Material	Labor	Equipment	Total	Total Incl O&P
0010	ROTARY COMPRESSORS									
0100	Refrigeration, hermetic, switches and protective devices									
0210	1.25 ton	1 Stpi	3	2.667	Ea.	246	159		405	520
0220	1.42 ton		3	2.667		248	159		407	520
0230	1.68 ton		2.80	2.857		270	171		441	565
0240	2.00 ton		2.60	3.077		278	184		462	590
0250	2.37 ton		2.50	3.200		315	191		506	645
0260	2.67 ton		2.40	3.333		330	199		529	670
0270	3.53 ton		2.30	3.478		360	208		568	720
0280	4.43 ton		2.20	3.636		460	217		677	845
0290	5.08 ton		2.10	3.810		505	228		733	910
0300	5.22 ton		2	4		715	239		954	1,150
0310	7.31 ton	Q-5	3	5.333		1,150	287		1,437	1,700
0320	9.95 ton		2.60	6.154		1,350	330		1,680	2,000
0330	11.6 ton		2.50	6.400		1,725	345		2,070	2,425
0340	15.25 ton		2.30	6.957		2,350	375		2,725	3,175
0350	17.7 ton		2.10	7.619		2,525	410		2,935	3,425

23 61 16 – Reciprocating Refrigerant Compressors

23 61 16.10 Reciprocating Compressors		Crew	Daily Output	Labor-Hours	Unit	Material	Labor	Equipment	Total	Total Incl O&P
0010	RECIPROCATING COMPRESSORS									
0990	Refrigeration, recip. hermetic, switches & protective devices									
1000	10 ton	Q-5	1	16	Ea.	12,800	860		13,660	15,500
1100	20 ton	Q-6	.72	33.333		27,200	1,850		29,050	32,800
1200	30 ton		.64	37.500		28,300	2,100		30,400	34,500
1300	40 ton		.44	54.545		30,500	3,050		33,550	38,300
1400	50 ton		.20	120		32,100	6,700		38,800	45,700
1500	75 ton	Q-7	.27	118		35,500	6,750		42,250	49,500
1600	130 ton	"	.21	152		40,500	8,675		49,175	58,000

23 61 Refrigerant Compressors

23 61 19 – Scroll Refrigerant Compressors

23 61 19.10 Scroll Compressors		Crew	Daily Output	Labor-Hours	Unit	Material	2015 Bare Costs Labor	Equipment	Total	Total Incl O&P
0010	**SCROLL COMPRESSORS**									
1800	Refrigeration, scroll type									
1810	1.5 ton	1 Stpi	2.70	2.963	Ea.	885	177		1,062	1,250
1820	2.5 ton		2.50	3.200		885	191		1,076	1,275
1830	2.75 ton		2.35	3.404		935	203		1,138	1,350
1840	3 ton		2.30	3.478		940	208		1,148	1,350
1850	3.5 ton		2.25	3.556		965	212		1,177	1,375
1860	4 ton		2.20	3.636		1,075	217		1,292	1,525
1870	5 ton		2.10	3.810		1,175	228		1,403	1,625
1880	6 ton	Q-5	3.70	4.324		1,550	233		1,783	2,075
1900	7.5 ton		3.20	5		2,600	269		2,869	3,275
1910	10 ton		3.10	5.161		2,750	278		3,028	3,450
1920	13 ton		2.90	5.517		3,050	297		3,347	3,825
1960	25 ton		2.20	7.273		5,075	390		5,465	6,200

23 62 Packaged Compressor and Condenser Units

23 62 13 – Packaged Air-Cooled Refrigerant Compressor and Condenser Units

23 62 13.10 Packaged Air-Cooled Refrig. Condensing Units		Crew	Daily Output	Labor-Hours	Unit	Material	Labor	Equipment	Total	Total Incl O&P
0010	**PACKAGED AIR-COOLED REFRIGERANT CONDENSING UNITS**									
0020	Condensing unit									
0030	Air cooled, compressor, standard controls									
0050	1.5 ton	Q-5	2.50	6.400	Ea.	1,250	345		1,595	1,900
0100	2 ton		2.10	7.619		1,350	410		1,760	2,125
0200	2.5 ton		1.70	9.412		1,425	505		1,930	2,375
0300	3 ton		1.30	12.308		1,500	660		2,160	2,675
0350	3.5 ton		1.10	14.545		1,725	780		2,505	3,125
0400	4 ton		.90	17.778		1,975	955		2,930	3,675
0500	5 ton		.60	26.667		2,375	1,425		3,800	4,825
0550	7.5 ton		.55	29.091		3,775	1,575		5,350	6,600
0560	8.5 ton		.53	30.189		4,675	1,625		6,300	7,675
0600	10 ton		.50	32		5,000	1,725		6,725	8,175
0620	12.5 ton		.48	33.333		6,025	1,800		7,825	9,425
0650	15 ton		.40	40		7,950	2,150		10,100	12,100
0700	20 ton	Q-6	.40	60		11,600	3,350		14,950	17,900
0720	25 ton		.35	68.571		17,000	3,825		20,825	24,700
0750	30 ton		.30	80		19,600	4,450		24,050	28,600
0800	40 ton		.20	120		26,900	6,700		33,600	40,000
0840	50 ton		.18	133		29,200	7,425		36,625	43,700
0860	60 ton		.16	150		33,800	8,375		42,175	50,000
0900	70 ton		.14	171		38,700	9,550		48,250	57,500
1000	80 ton		.12	200		43,000	11,200		54,200	64,500
1010	90 ton		.09	266		47,700	14,900		62,600	75,500
1100	100 ton		.09	266		52,500	14,900		67,400	80,500

23 62 23 – Packaged Water-Cooled Refrigerant Compressor and Condenser Units

23 62 23.10 Packaged Water-Cooled Refrigerant Condensing Units		Crew	Daily Output	Labor-Hours	Unit	Material	Labor	Equipment	Total	Total Incl O&P
0010	**PACKAGED WATER-COOLED REFRIGERANT CONDENSING UNITS**									
2000	Water cooled, compressor, heat exchanger, controls									
2100	5 ton	Q-5	.70	22.857	Ea.	13,600	1,225		14,825	16,900
2200	15 ton	"	.50	32		20,300	1,725		22,025	25,100
2300	20 ton	Q-6	.40	60		25,000	3,350		28,350	32,700

23 62 Packaged Compressor and Condenser Units

23 62 23 – Packaged Water-Cooled Refrigerant Compressor and Condenser Units

23 62 23.10 Packaged Water-Cooled Refrigerant Condensing Units		Crew	Daily Output	Labor-Hours	Unit	Material	2015 Bare Costs Labor	Equipment	Total	Total Incl O&P
2400	40 ton	Q-6	.20	120	Ea.	31,200	6,700		37,900	44,700
9000	Minimum labor/equipment charge	Q-5	1.50	10.667	Job		575		575	895

23 63 Refrigerant Condensers

23 63 13 – Air-Cooled Refrigerant Condensers

23 63 13.10 Air-Cooled Refrig. Condensers		Crew	Daily Output	Labor-Hours	Unit	Material	2015 Bare Costs Labor	Equipment	Total	Total Incl O&P
0010	**AIR-COOLED REFRIG. CONDENSERS**									
0080	Air cooled, belt drive, propeller fan									
0220	45 ton	Q-6	.70	34.286	Ea.	9,800	1,900		11,700	13,800
0240	50 ton		.69	34.985		10,600	1,950		12,550	14,700
0260	54 ton		.64	37.795		11,500	2,100		13,600	16,000
0280	59 ton		.58	41.308		12,700	2,300		15,000	17,600
0300	65 ton		.53	45.541		15,200	2,550		17,750	20,700
0320	73 ton		.47	51.173		16,600	2,850		19,450	22,800
0340	81 ton		.42	56.738		18,300	3,175		21,475	25,000
0360	86 ton		.40	60.302		19,200	3,375		22,575	26,400
0380	88 ton		.39	61.697		20,600	3,450		24,050	28,100
0400	101 ton	Q-7	.45	70.640		22,700	4,025		26,725	31,200
0500	159 ton		.31	102		24,000	5,850		29,850	35,500
0600	228 ton		.22	148		51,000	8,425		59,425	69,000
1500	May be specified single or multi-circuit									
1550	Air cooled, direct drive, propeller fan									
1590	1 ton	Q-5	3.80	4.211	Ea.	1,650	226		1,876	2,175
1600	1-1/2 ton		3.60	4.444		1,975	239		2,214	2,550
1620	2 ton		3.20	5		2,175	269		2,444	2,825
1630	3 ton		2.40	6.667		2,450	360		2,810	3,225
1640	5 ton		2	8		5,400	430		5,830	6,600
1650	8 ton		1.80	8.889		5,625	480		6,105	6,925
1660	10 ton		1.40	11.429		6,325	615		6,940	7,900
1670	12 ton		1.30	12.308		7,700	660		8,360	9,500
1680	14 ton		1.20	13.333		8,800	715		9,515	10,800
1690	16 ton		1.10	14.545		10,200	780		10,980	12,400
1700	21 ton		1	16		11,100	860		11,960	13,600
1720	26 ton		.84	19.002		12,500	1,025		13,525	15,400
1740	30 ton		.70	22.792		16,000	1,225		17,225	19,500
1760	41 ton	Q-6	.77	31.008		17,600	1,725		19,325	22,100
1780	52 ton		.66	36.419		25,100	2,025		27,125	30,800
1800	63 ton		.55	44.037		27,700	2,450		30,150	34,300
1820	76 ton		.45	52.980		31,400	2,950		34,350	39,200
1840	86 ton		.40	60		38,900	3,350		42,250	47,900
1860	97 ton		.35	67.989		44,200	3,800		48,000	54,500
1880	105 ton	Q-7	.44	73.563		51,000	4,175		55,175	62,500
1890	118 ton		.39	82.687		55,000	4,700		59,700	68,000
1900	126 ton		.36	88.154		64,000	5,025		69,025	78,000
1910	136 ton		.34	95.238		71,500	5,425		76,925	87,000
1920	142 ton		.32	99.379		73,500	5,650		79,150	89,500

23 63 Refrigerant Condensers

23 63 30 – Lubricants

23 63 30.10 Lubricant Oils		Crew	Daily Output	Labor-Hours	Unit	2015 Bare Costs Material	Labor	Equipment	Total	Total Incl O&P
0010	**LUBRICANT OILS**									
8000	Oils									
8100	Lubricating									
8120	Oil, lubricating				Oz.	.39			.39	.43
8500	Refrigeration									
8520	Oil, refrigeration				Gal.	41			41	45
8525	Oil, refrigeration				Qt.	10.20			10.20	11.20

23 63 33 – Evaporative Refrigerant Condensers

23 63 33.10 Evaporative Condensers		Crew	Daily Output	Labor-Hours	Unit	Material	Labor	Equipment	Total	Total Incl O&P
0010	**EVAPORATIVE CONDENSERS**									
3400	Evaporative, copper coil, pump, fan motor									
3440	10 ton	Q-5	.54	29.630	Ea.	8,350	1,600		9,950	11,700
3460	15 ton		.50	32		8,550	1,725		10,275	12,100
3480	20 ton		.47	34.043		8,800	1,825		10,625	12,500
3500	25 ton		.45	35.556		9,000	1,900		10,900	12,900
3520	30 ton		.42	38.095		9,450	2,050		11,500	13,600
3540	40 ton	Q-6	.49	48.980		10,300	2,725		13,025	15,600
3560	50 ton		.39	61.538		11,700	3,425		15,125	18,300
3580	65 ton		.35	68.571		12,500	3,825		16,325	19,700
3600	80 ton		.33	72.727		13,800	4,050		17,850	21,500
3620	90 ton		.29	82.759		15,100	4,625		19,725	23,800
3640	100 ton	Q-7	.36	88.889		16,600	5,050		21,650	26,200
3660	110 ton		.33	96.970		17,500	5,525		23,025	27,800
3680	125 ton		.30	106		18,900	6,075		24,975	30,300
3700	135 ton		.28	114		19,900	6,500		26,400	32,000
3720	150 ton		.25	128		21,600	7,275		28,875	35,200
3740	165 ton		.23	139		22,900	7,925		30,825	37,500
3760	185 ton		.22	145		25,900	8,275		34,175	41,400
3860	For fan damper control, add	Q-5	2	8		785	430		1,215	1,525

23 64 Packaged Water Chillers

23 64 13 – Absorption Water Chillers

23 64 13.13 Direct-Fired Absorption Water Chillers		Crew	Daily Output	Labor-Hours	Unit	Material	Labor	Equipment	Total	Total Incl O&P
0010	**DIRECT-FIRED ABSORPTION WATER CHILLERS**									
3000	Gas fired, air cooled									
3220	5 ton	Q-5	.60	26.667	Ea.	8,850	1,425		10,275	12,000
3270	10 ton	"	.40	40	"	22,700	2,150		24,850	28,400
4000	Water cooled, duplex									
4130	100 ton	Q-7	.13	246	Ea.	136,000	14,000		150,000	171,500
4140	200 ton		.11	283		183,000	16,100		199,100	226,500
4150	300 ton		.11	299		219,500	17,000		236,500	268,000
4160	400 ton		.10	316		285,500	18,000		303,500	342,000
4170	500 ton		.10	329		359,500	18,800		378,300	424,500
4180	600 ton		.09	340		424,500	19,400		443,900	496,500
4190	700 ton		.09	359		470,500	20,500		491,000	549,500
4200	800 ton		.08	380		568,000	21,700		589,700	659,000
4210	900 ton		.08	400		658,000	22,800		680,800	759,500
4220	1000 ton		.08	421		731,500	24,000		755,500	842,000

23 64 Packaged Water Chillers

23 64 13 – Absorption Water Chillers

23 64 13.16 Indirect-Fired Absorption Water Chillers		Crew	Daily Output	Labor-Hours	Unit	Material	2015 Bare Costs Labor	Equipment	Total	Total Incl O&P
0010	**INDIRECT-FIRED ABSORPTION WATER CHILLERS**									
0020	Steam or hot water, water cooled									
0050	100 ton	Q-7	.13	240	Ea.	125,000	13,700		138,700	159,000
0100	148 ton		.12	258		219,000	14,700		233,700	264,000
0200	200 ton		.12	275		244,000	15,700		259,700	292,500
0300	354 ton		.11	304		335,000	17,300		352,300	395,500
0400	420 ton		.10	323		366,500	18,400		384,900	431,500
0500	665 ton		.09	340		514,500	19,400		533,900	595,500
0600	750 ton		.09	363		601,000	20,700		621,700	693,500
0700	850 ton		.08	385		641,500	21,900		663,400	739,500
0800	955 ton		.08	410		686,000	23,300		709,300	791,000
0900	1125 ton		.08	421		791,500	24,000		815,500	908,500
1000	1250 ton		.07	444		851,500	25,300		876,800	976,000
1100	1465 ton		.07	463		1,009,500	26,400		1,035,900	1,151,500
1200	1660 ton	↓	.07	477	↓	1,187,500	27,200		1,214,700	1,349,000
2000	For two stage unit, add					80%	25%			
9000	Minimum labor/equipment charge	Q-5	1	16	Job		860		860	1,350

23 64 16 – Centrifugal Water Chillers

23 64 16.10 Centrifugal Type Water Chillers		Crew	Daily Output	Labor-Hours	Unit	Material	Labor	Equipment	Total	Total Incl O&P
0010	**CENTRIFUGAL TYPE WATER CHILLERS**, With standard controls									
0020	Centrifugal liquid chiller, water cooled									
0030	not including water tower									
0100	2000 ton (twin 1000 ton units)	Q-7	.07	477	Ea.	699,000	27,200		726,200	811,500
0274	Centrifugal, packaged unit, water cooled, not incl. tower									
0278	200 ton	Q-7	.19	172	Ea.	90,000	9,800		99,800	114,500
0279	300 ton		.14	233		94,500	13,300		107,800	124,500
0280	400 ton		.11	283		123,000	16,100		139,100	160,500
0320	1000 ton		.09	372		349,500	21,200		370,700	417,500
0340	1300 ton		.08	410		413,000	23,300		436,300	491,000
0360	1500 ton	↓	.08	426	↓	469,000	24,300		493,300	554,000

23 64 19 – Reciprocating Water Chillers

23 64 19.10 Reciprocating Type Water Chillers		Crew	Daily Output	Labor-Hours	Unit	Material	Labor	Equipment	Total	Total Incl O&P
0010	**RECIPROCATING TYPE WATER CHILLERS**, With standard controls									
0494	Water chillers, integral air cooled condenser									
0546	70 ton cooling	Q-7	.27	119	Ea.	52,500	6,825		59,325	68,000
0554	90 ton cooling		.25	125		62,500	7,175		69,675	80,000
0600	100 ton cooling		.25	129		73,500	7,350		80,850	92,500
0620	110 ton cooling		.24	132		79,000	7,525		86,525	98,500
0630	130 ton cooling		.24	135		90,000	7,675		97,675	111,000
0640	150 ton cooling		.23	137		100,500	7,850		108,350	122,500
0650	175 ton cooling		.23	140		113,500	8,025		121,525	137,000
0654	190 ton cooling		.22	144		125,000	8,200		133,200	150,500
0660	210 ton cooling		.22	148		132,500	8,425		140,925	158,500
0662	250 ton cooling		.21	151		153,500	8,625		162,125	182,500
0664	275 ton cooling		.21	156		166,500	8,875		175,375	197,000
0666	300 ton cooling		.20	160		179,500	9,100		188,600	211,500
0668	330 ton cooling		.20	164		202,500	9,325		211,825	237,000
0670	360 ton cooling		.19	168		224,000	9,575		233,575	261,000
0672	390 ton cooling		.19	172		248,500	9,850		258,350	289,000
0674	420 ton cooling	↓	.18	177	↓	261,000	10,100		271,100	303,500
0980	Water cooled, multiple compressor, semi-hermetic, tower not incl.									

23 64 19 – Reciprocating Water Chillers

23 64 19.10 Reciprocating Type Water Chillers		Crew	Daily Output	Labor-Hours	Unit	Material	Labor	Equipment	Total	Total Incl O&P
						2015 Bare Costs				
1000	15 ton cooling	Q-6	.36	65.934	Ea.	18,900	3,675		22,575	26,500
1020	25 ton cooling	Q-7	.41	78.049		19,500	4,450		23,950	28,400
1040	30 ton cooling		.36	89.888		22,700	5,125		27,825	33,000
1060	35 ton cooling		.31	101		24,300	5,800		30,100	35,900
1080	40 ton cooling		.30	108		26,000	6,150		32,150	38,200
1090	45 ton cooling		.29	111		28,200	6,325		34,525	40,900
1100	50 ton cooling		.28	113		35,800	6,475		42,275	49,500
1120	60 ton cooling		.25	125		37,000	7,175		44,175	52,000
1130	75 ton cooling		.23	139		52,500	7,925		60,425	70,000
1140	85 ton cooling		.21	151		57,000	8,625		65,625	76,500
1150	95 ton cooling		.19	164		59,000	9,375		68,375	79,500
1160	100 ton cooling		.18	179		60,500	10,200		70,700	82,500
1170	115 ton cooling		.17	190		61,000	10,800		71,800	84,000
1180	125 ton cooling		.16	196		64,000	11,200		75,200	87,500
1200	145 ton cooling		.16	202		71,500	11,500		83,000	96,500
1210	155 ton cooling		.15	210		75,000	12,000		87,000	101,000
1300	Water cooled, single compressor, semi-hermetic, tower not incl.									
1320	40 ton cooling,	Q-7	.30	108	Ea.	17,300	6,150		23,450	28,600
1340	50 ton cooling		.28	113		21,500	6,475		27,975	33,700
1360	60 ton cooling		.25	125		23,800	7,175		30,975	37,400
1451	Water cooled, dual compressors, semi-hermetic, tower not incl.									
1500	80 ton cooling	Q-7	.14	222	Ea.	29,600	12,600		42,200	52,500
1520	100 ton cooling		.14	228		40,000	13,000		53,000	64,500
1540	120 ton cooling		.14	231		48,500	13,200		61,700	74,000
4000	Packaged chiller, remote air cooled condensers not incl.									
4020	15 ton cooling	Q-7	.30	108	Ea.	15,800	6,150		21,950	27,000
4030	20 ton cooling		.28	115		17,200	6,600		23,800	29,200
4040	25 ton cooling		.25	125		18,200	7,175		25,375	31,200
4050	35 ton cooling		.24	133		19,100	7,625		26,725	32,900
4060	40 ton cooling		.22	144		20,600	8,200		28,800	35,400
4066	45 ton cooling		.21	149		21,700	8,500		30,200	37,100
4070	50 ton cooling		.21	153		30,800	8,750		39,550	47,600
4080	60 ton cooling		.20	164		33,300	9,325		42,625	51,500
4090	75 ton cooling		.18	173		42,700	9,900		52,600	62,500
4100	85 ton cooling		.17	183		44,900	10,500		55,400	65,500
4110	95 ton cooling		.17	193		45,700	11,000		56,700	67,000
4120	105 ton cooling		.16	203		50,000	11,600		61,600	73,000
4130	115 ton cooling		.15	213		51,500	12,100		63,600	75,500
4140	125 ton cooling		.14	223		52,000	12,700		64,700	77,000
4150	145 ton cooling		.14	233		58,000	13,300		71,300	84,000

23 64 23 – Scroll Water Chillers

23 64 23.10 Scroll Water Chillers		Crew	Daily Output	Labor-Hours	Unit	Material	Labor	Equipment	Total	Total Incl O&P
0010	**SCROLL WATER CHILLERS**, With standard controls									
0480	Packaged w/integral air cooled condenser									
0482	10 ton cooling	Q-7	.34	94.118	Ea.	20,300	5,350		25,650	30,800
0490	15 ton cooling		.37	86.486		20,800	4,925		25,725	30,600
0500	20 ton cooling		.34	94.118		21,500	5,350		26,850	32,000
0510	25 ton cooling		.34	94.118		25,300	5,350		30,650	36,200
0515	30 ton cooling		.31	101		27,100	5,800		32,900	38,900
0517	35 ton cooling		.31	104		30,200	5,975		36,175	42,500
0520	40 ton cooling		.30	108		32,900	6,150		39,050	45,800
0528	45 ton cooling		.29	109		36,500	6,250		42,750	49,900

23 64 Packaged Water Chillers

23 64 23 – Scroll Water Chillers

23 64 23.10 Scroll Water Chillers		Crew	Daily Output	Labor-Hours	Unit	Material	Labor	Equipment	Total	Total Incl O&P
						2015 Bare Costs				
0536	50 ton cooling	Q-7	.28	113	Ea.	38,200	6,425		44,625	52,000
0680	Scroll water cooled, single compressor, hermetic, tower not incl.									
0700	2 ton cooling	Q-5	.57	28.070	Ea.	3,500	1,500		5,000	6,200
0710	5 ton cooling		.57	28.070		4,175	1,500		5,675	6,950
0720	6 ton cooling		.42	38.005		5,250	2,050		7,300	8,975
0740	8 ton cooling		.31	52.117		5,750	2,800		8,550	10,700
0760	10 ton cooling	Q-6	.36	67.039		6,625	3,750		10,375	13,100
0780	15 ton cooling	"	.33	72.727		10,400	4,050		14,450	17,800
0800	20 ton cooling	Q-7	.38	83.990		11,800	4,775		16,575	20,500
0820	30 ton cooling	"	.33	96.096		13,200	5,475		18,675	23,100

23 64 26 – Rotary-Screw Water Chillers

23 64 26.10 Rotary-Screw Type Water Chillers		Crew	Daily Output	Labor-Hours	Unit	Material	Labor	Equipment	Total	Total Incl O&P
0010	**ROTARY-SCREW TYPE WATER CHILLERS**, With standard controls									
0110	Screw, liquid chiller, air cooled, insulated evaporator									
0120	130 ton	Q-7	.14	228	Ea.	91,500	13,000		104,500	121,000
0124	160 ton		.13	246		112,500	14,000		126,500	145,500
0128	180 ton		.13	250		126,500	14,200		140,700	161,000
0132	210 ton		.12	258		139,000	14,700		153,700	176,000
0136	270 ton		.12	266		159,000	15,200		174,200	198,500
0140	320 ton		.12	275		199,500	15,700		215,200	244,000
0200	Packaged unit, water cooled, not incl. tower									
0240	200 ton	Q-7	.13	251	Ea.	82,000	14,300		96,300	112,500
1450	Water cooled, tower not included									
1560	135 ton cooling, screw compressors	Q-7	.14	235	Ea.	59,500	13,400		72,900	86,500
1580	150 ton cooling, screw compressors		.13	240		69,000	13,700		82,700	97,000
1620	200 ton cooling, screw compressors		.13	250		95,000	14,200		109,200	126,500
1660	291 ton cooling, screw compressors		.12	260		99,000	14,800		113,800	132,000

23 64 33 – Direct Expansion Water Chillers

23 64 33.10 Direct Expansion Type Water Chillers		Crew	Daily Output	Labor-Hours	Unit	Material	Labor	Equipment	Total	Total Incl O&P
0010	**DIRECT EXPANSION TYPE WATER CHILLERS**, With standard controls									
8000	Direct expansion, shell and tube type, for built up systems									
8020	1 ton	Q-5	2	8	Ea.	7,300	430		7,730	8,700
8030	5 ton		1.90	8.421		12,100	455		12,555	14,000
8040	10 ton		1.70	9.412		15,000	505		15,505	17,300
9000	Minimum labor/equipment charge	Q-6	1	24	Job		1,350		1,350	2,100

23 65 Cooling Towers

23 65 13 – Forced-Draft Cooling Towers

23 65 13.10 Forced-Draft Type Cooling Towers		Crew	Daily Output	Labor-Hours	Unit	Material	Labor	Equipment	Total	Total Incl O&P
0010	**FORCED-DRAFT TYPE COOLING TOWERS**, Packaged units									
0070	Galvanized steel									
0080	Induced draft, crossflow									
0100	Vertical, belt drive, 61 tons	Q-6	90	.267	TonAC	215	14.85		229.85	260
0150	100 ton		100	.240		208	13.40		221.40	249
0200	115 ton		109	.220		180	12.30		192.30	218
0250	131 ton		120	.200		217	11.15		228.15	255
0260	162 ton		132	.182		175	10.15		185.15	209
1000	For higher capacities, use multiples									
1500	Induced air, double flow									
1900	Vertical, gear drive, 167 ton	Q-6	126	.190	TonAC	170	10.60		180.60	204

23 65 Cooling Towers

23 65 13 – Forced-Draft Cooling Towers

23 65 13.10 Forced-Draft Type Cooling Towers		Crew	Daily Output	Labor-Hours	Unit	2015 Bare Costs Material	Labor	Equipment	Total	Total Incl O&P
2000	297 ton	Q-6	129	.186	TonAC	105	10.40		115.40	131
2100	582 ton		132	.182		58.50	10.15		68.65	80
2150	849 ton		142	.169		78.50	9.45		87.95	101
2200	1016 ton	↓	150	.160	↓	79	8.90		87.90	101
3000	For higher capacities, use multiples									
3500	For pumps and piping, add	Q-6	38	.632	TonAC	108	35		143	173
4000	For absorption systems, add				"	75%	75%			
4100	Cooling water chemical feeder	Q-5	3	5.333	Ea.	365	287		652	845
5000	Fiberglass tower on galvanized steel support structure									
5010	Draw thru									
5100	100 ton	Q-6	1.40	17.143	Ea.	13,900	955		14,855	16,800
5120	120 ton		1.20	20		16,200	1,125		17,325	19,600
5140	140 ton		1	24		17,500	1,350		18,850	21,300
5160	160 ton		.80	30		19,400	1,675		21,075	24,000
5180	180 ton		.65	36.923		22,100	2,050		24,150	27,500
5200	200 ton	↓	.48	50	↓	25,000	2,800		27,800	31,900
5300	For stainless steel support structure, add					30%				
5360	For higher capacities, use multiples of each size									
6000	Stainless steel									
6010	Induced draft, crossflow, horizontal, belt drive									
6100	57 ton	Q-6	1.50	16	Ea.	27,100	890		27,990	31,200
6120	91 ton		.99	24.242		32,700	1,350		34,050	38,100
6140	111 ton		.43	55.814		41,600	3,125		44,725	50,500
6160	126 ton	↓	.22	109	↓	41,600	6,075		47,675	55,500
6170	Induced draft, crossflow, vertical, gear drive									
6172	167 ton	Q-6	.75	32	Ea.	53,000	1,775		54,775	61,500
6174	297 ton		.43	55.814		59,000	3,125		62,125	70,000
6176	582 ton		.23	104		93,500	5,825		99,325	112,000
6178	849 ton		.17	141		132,000	7,875		139,875	158,000
6180	1016 ton		.15	160	↓	159,000	8,925		167,925	188,500
9000	Minimum labor/equipment charge	↓	1	24	Job		1,350		1,350	2,100

23 65 33 – Liquid Coolers

23 65 33.10 Liquid Cooler, Closed Circuit Type Cooling Towers		Crew	Daily Output	Labor-Hours	Unit	Material	Labor	Equipment	Total	Total Incl O&P
0010	**LIQUID COOLER, CLOSED CIRCUIT TYPE COOLING TOWERS**									
0070	Packaged options include: vibration switch, sump sweeper piping,									
0076	grooved coil connections, fan motor inverter capable,									
0080	304 St. St. cold water basin, external platform w/ladder.									
0100	50 ton	Q-6	2	12	Ea.	32,200	670		32,870	36,500
0150	100 ton	"	1.13	21.239		64,500	1,175		65,675	73,000
0200	150 ton	Q-7	.98	32.653		96,500	1,850		98,350	109,000
0250	200 ton		.78	41.026		128,500	2,325		130,825	145,000
0300	250 ton	↓	.63	50.794	↓	161,000	2,900		163,900	181,500

23 72 Air-to-Air Energy Recovery Equipment

23 72 13 – Heat-Wheel Air-to-Air Energy-Recovery Equipment

23 72 13.10 Heat-Wheel Air-to-Air Energy Recovery Equip.			Crew	Daily Output	Labor-Hours	Unit	Material	2015 Bare Costs Labor	Equipment	Total	Total Incl O&P
0010	HEAT-WHEEL AIR-TO-AIR ENERGY RECOVERY EQUIPMENT										
0100	Air to air										
4000	Enthalpy recovery wheel										
4010	1000 max CFM	G	Q-9	1.20	13.333	Ea.	6,700	670		7,370	8,425
4020	2000 max CFM	G		1	16		7,850	805		8,655	9,900
4030	4000 max CFM	G		.80	20		9,075	1,000		10,075	11,600
4040	6000 max CFM	G		.70	22.857		10,600	1,150		11,750	13,500
4050	8000 max CFM	G	Q-10	1	24		11,700	1,250		12,950	14,900
4060	10,000 max CFM	G		.90	26.667		14,000	1,400		15,400	17,600
4070	20,000 max CFM	G		.80	30		25,400	1,575		26,975	30,500
4080	25,000 max CFM	G		.70	34.286		31,100	1,800		32,900	37,000
4090	30,000 max CFM	G		.50	48		34,500	2,500		37,000	41,900
4100	40,000 max CFM	G		.45	53.333		47,600	2,775		50,375	57,000
4110	50,000 max CFM	G		.40	60		55,500	3,125		58,625	66,000
9000	Minimum labor/equipment charge		1 Shee	4	2	Job		112		112	177

23 73 Indoor Central-Station Air-Handling Units

23 73 13 – Modular Indoor Central-Station Air-Handling Units

23 73 13.20 Air Handling Units, Packaged Indoor Type		Crew	Daily Output	Labor-Hours	Unit	Material	Labor	Equipment	Total	Total Incl O&P
0010	AIR HANDLING UNITS, PACKAGED INDOOR TYPE									
0040	Cooling coils may be chilled water or DX									
0050	Heating coils may be hot water, steam or electric									
1090	Constant volume									
1200	2000 CFM	Q-5	1.10	14.545	Ea.	11,000	780		11,780	13,300
1400	5000 CFM	Q-6	.80	30		17,600	1,675		19,275	21,900
1550	10,000 CFM		.54	44.444		25,900	2,475		28,375	32,400
1670	15,000 CFM		.38	63.158		53,000	3,525		56,525	64,000
1700	20,000 CFM		.31	77.419		67,500	4,325		71,825	80,500
1710	30,000 CFM		.27	88.889		91,000	4,950		95,950	107,500
2300	Variable air volume									
2330	2000 CFM	Q-5	1	16	Ea.	15,400	860		16,260	18,400
2340	5000 CFM	Q-6	.70	34.286		21,500	1,900		23,400	26,700
2350	10,000 CFM		.50	48		42,000	2,675		44,675	50,500
2360	15,000 CFM		.40	60		61,000	3,350		64,350	72,000
2370	20,000 CFM		.29	82.759		78,000	4,625		82,625	92,500
2380	30,000 CFM		.20	120		94,000	6,700		100,700	114,000

23 73 39 – Indoor, Direct Gas-Fired Heating and Ventilating Units

23 73 39.10 Make-Up Air Unit		Crew	Daily Output	Labor-Hours	Unit	Material	Labor	Equipment	Total	Total Incl O&P
0010	MAKE-UP AIR UNIT									
0020	Indoor suspension, natural/LP gas, direct fired,									
0032	standard control. For flue see Section 23 51 23.10									
0040	70°F temperature rise, MBH is input									
0100	75 MBH input	Q-6	3.60	6.667	Ea.	4,925	370		5,295	6,000
0120	100 MBH input		3.40	7.059		5,125	395		5,520	6,250
0140	125 MBH input		3.20	7.500		5,425	420		5,845	6,625
0160	150 MBH input		3	8		5,975	445		6,420	7,275
0180	175 MBH input		2.80	8.571		6,125	480		6,605	7,500
0200	200 MBH input		2.60	9.231		6,475	515		6,990	7,925
0220	225 MBH input		2.40	10		6,750	560		7,310	8,275
0240	250 MBH input		2	12		7,725	670		8,395	9,550

23 73 Indoor Central-Station Air-Handling Units

23 73 39 – Indoor, Direct Gas-Fired Heating and Ventilating Units

23 73 39.10 Make-Up Air Unit		Crew	Daily Output	Labor-Hours	Unit	Material	2015 Bare Costs Labor	Equipment	Total	Total Incl O&P
0260	300 MBH input	Q-6	1.90	12.632	Ea.	7,925	705		8,630	9,825
0280	350 MBH input		1.75	13.714		8,875	765		9,640	11,000
0300	400 MBH input		1.60	15		15,400	835		16,235	18,200
0320	840 MBH input		1.20	20		29,100	1,125		30,225	33,800
0340	1200 MBH input		.88	27.273		31,700	1,525		33,225	37,300
0600	For discharge louver assembly, add					5%				
0700	For filters, add					10%				
0800	For air shut-off damper section, add					30%				

23 74 Packaged Outdoor HVAC Equipment

23 74 13 – Packaged, Outdoor, Central-Station Air-Handling Units

23 74 13.10 Packaged, Outdoor Type, Central-Station AHU		Crew	Daily Output	Labor-Hours	Unit	Material	Labor	Equipment	Total	Total Incl O&P
0010	PACKAGED, OUTDOOR TYPE, CENTRAL-STATION AIR-HANDLING UNITS									
2900	Cooling coils may be chilled water or DX									
2910	Heating coils may be hot water, steam or electric									
3000	Weathertight, ground or rooftop									
3010	Constant volume									
3100	2000 CFM	Q-6	1	24	Ea.	23,700	1,350		25,050	28,100
3140	5000 CFM		.79	30.380		28,100	1,700		29,800	33,600
3150	10,000 CFM		.52	46.154		57,500	2,575		60,075	67,500
3160	15,000 CFM		.41	58.537		84,500	3,275		87,775	97,500
3170	20,000 CFM		.30	80		115,500	4,450		119,950	134,000
3180	30,000 CFM		.24	100		147,000	5,575		152,575	170,500
3200	Variable air volume									
3210	2000 CFM	Q-6	.96	25	Ea.	15,200	1,400		16,600	18,900
3240	5000 CFM		.73	32.877		30,800	1,825		32,625	36,700
3250	10,000 CFM		.50	48		66,000	2,675		68,675	77,000
3260	15,000 CFM		.40	60		100,500	3,350		103,850	115,500
3270	20,000 CFM		.29	82.759		130,500	4,625		135,125	150,500
3280	30,000 CFM		.20	120		154,500	6,700		161,200	180,500

23 74 23 – Packaged, Outdoor, Heating-Only Makeup-Air Units

23 74 23.16 Packaged, Ind.-Fired, In/Outdoor		Crew	Daily Output	Labor-Hours	Unit	Material	Labor	Equipment	Total	Total Incl O&P
0010	PACKAGED, IND.-FIRED, IN/OUTDOOR, Heat-Only Makeup-Air Units									
1000	Rooftop unit, natural gas, gravity vent, S.S. exchanger									
1010	70°F temperature rise, MBH is input									
1020	250 MBH	Q-6	4	6	Ea.	15,400	335		15,735	17,400
1040	400 MBH		3.60	6.667		17,300	370		17,670	19,600
1060	550 MBH		3.30	7.273		17,800	405		18,205	20,200
1080	750 MBH		3	8		21,700	445		22,145	24,500
1100	1000 MBH		2.60	9.231		24,300	515		24,815	27,500
1120	1750 MBH		2.30	10.435		30,400	580		30,980	34,300
1140	2500 MBH		1.90	12.632		41,600	705		42,305	46,900
1160	3250 MBH		1.40	17.143		50,000	955		50,955	56,500
1180	4000 MBH		1.20	20		65,000	1,125		66,125	73,500
1200	6000 MBH		1	24		82,000	1,350		83,350	92,000
1600	For cleanable filters, add					5%				
1700	For electric modulating gas control, add					10%				
9000	Minimum labor/equipment charge	Q-6	2.75	8.727	Job		485		485	760

23 74 33 – Dedicated Outdoor-Air Units

23 74 33.10 Rooftop Air Conditioners		Crew	Daily Output	Labor-Hours	Unit	2015 Bare Costs Material	Labor	Equipment	Total	Total Incl O&P
0010	**ROOFTOP AIR CONDITIONERS**, Standard controls, curb, economizer									
1000	Single zone, electric cool, gas heat									
1100	3 ton cooling, 60 MBH heating	Q-5	.70	22.857	Ea.	3,925	1,225		5,150	6,250
1120	4 ton cooling, 95 MBH heating		.61	26.403		4,575	1,425		6,000	7,275
1140	5 ton cooling, 112 MBH heating		.56	28.521		5,125	1,525		6,650	8,025
1145	6 ton cooling, 140 MBH heating		.52	30.769		5,875	1,650		7,525	9,050
1150	7.5 ton cooling, 170 MBH heating		.50	32.258		6,825	1,725		8,550	10,200
1156	8.5 ton cooling, 170 MBH heating		.46	34.783		8,125	1,875		10,000	11,900
1160	10 ton cooling, 200 MBH heating	Q-6	.67	35.982		9,425	2,000		11,425	13,500
1170	12.5 ton cooling, 230 MBH heating		.63	37.975		11,900	2,125		14,025	16,400
1180	15 ton cooling, 270 MBH heating		.57	42.032		13,600	2,350		15,950	18,600
1190	17.5 ton cooling, 330 MBH heating		.52	45.889		16,100	2,550		18,650	21,800
1200	20 ton cooling, 360 MBH heating	Q-7	.67	47.976		23,800	2,725		26,525	30,500
1210	25 ton cooling, 450 MBH heating		.56	57.554		27,100	3,275		30,375	34,900
1220	30 ton cooling, 540 MBH heating		.47	68.376		30,800	3,900		34,700	40,000
1240	40 ton cooling, 675 MBH heating		.35	91.168		40,200	5,175		45,375	52,500
2000	Multizone, electric cool, gas heat, economizer									
2100	15 ton cooling, 360 MBH heating	Q-7	.61	52.545	Ea.	65,000	3,000		68,000	76,000
2120	20 ton cooling, 360 MBH heating		.53	60.038		70,000	3,425		73,425	82,500
2140	25 ton cooling, 450 MBH heating		.45	71.910		83,500	4,100		87,600	98,500
2160	28 ton cooling, 450 MBH heating		.41	79.012		95,500	4,500		100,000	112,000
2180	30 ton cooling, 540 MBH heating		.37	85.562		107,000	4,875		111,875	125,000
2200	40 ton cooling, 540 MBH heating		.28	113		122,500	6,475		128,975	144,500
2210	50 ton cooling, 540 MBH heating		.23	142		152,000	8,100		160,100	180,000
2220	70 ton cooling, 1500 MBH heating		.16	198		164,500	11,300		175,800	198,500
2240	80 ton cooling, 1500 MBH heating		.14	228		188,000	13,000		201,000	227,500
2260	90 ton cooling, 1500 MBH heating		.13	256		197,500	14,600		212,100	239,500
2280	105 ton cooling, 1500 MBH heating		.11	290		217,500	16,500		234,000	265,000
2400	For hot water heat coil, deduct					5%				
2500	For steam heat coil, deduct					2%				
2600	For electric heat, deduct					3%	5%			
5000	Single zone, electric cool only									
5050	3 ton cooling	Q-5	.88	18.203	Ea.	2,700	980		3,680	4,500
5060	4 ton cooling		.76	21.108		3,225	1,125		4,350	5,325
5070	5 ton cooling		.70	22.792		3,575	1,225		4,800	5,850
5080	6 ton cooling		.65	24.502		4,375	1,325		5,700	6,875
5090	7.5 ton cooling		.62	25.806		4,725	1,400		6,125	7,375
5100	8.5 ton cooling		.59	26.981		5,450	1,450		6,900	8,275
5110	10 ton cooling	Q-6	.83	28.812		6,375	1,600		7,975	9,525
5120	12.5 ton cooling		.80	30		8,775	1,675		10,450	12,300
5130	15 ton cooling		.71	33.613		11,800	1,875		13,675	15,900
5140	17.5 ton cooling		.65	36.697		12,600	2,050		14,650	17,000
5150	20 ton cooling	Q-7	.83	38.415		23,000	2,175		25,175	28,700
5160	25 ton cooling		.70	45.977		24,400	2,625		27,025	31,000
5170	30 ton cooling		.59	54.701		28,200	3,100		31,300	35,900
5180	40 ton cooling		.44	73.059		37,200	4,150		41,350	47,400
5190	50 ton cooling		.38	84.211		43,500	4,800		48,300	55,500
5200	60 ton cooling		.28	114		53,500	6,500		60,000	69,000
5400	For low heat, add					7%				
5410	For high heat, add					10%				
6000	Single zone electric cooling with variable volume distribution									
6020	20 ton cooling	Q-7	.72	44.199	Ea.	26,100	2,525		28,625	32,600

23 74 Packaged Outdoor HVAC Equipment

23 74 33 – Dedicated Outdoor-Air Units

23 74 33.10 Rooftop Air Conditioners		Crew	Daily Output	Labor-Hours	Unit	2015 Bare Costs Material	Labor	Equipment	Total	Total Incl O&P
6030	25 ton cooling	Q-7	.61	52.893	Ea.	29,000	3,000		32,000	36,600
6040	30 ton cooling		.51	62.868		33,100	3,575		36,675	42,100
6050	40 ton cooling		.38	83.990		38,100	4,775		42,875	49,400
6060	50 ton cooling		.31	104		44,300	5,975		50,275	58,000
6070	60 ton cooling		.25	125		53,500	7,175		60,675	69,500
6200	For low heat, add					7%				
6210	For high heat, add					10%				
7000	Multizone, cool/heat, variable volume distribution									
7100	50 ton cooling	Q-7	.20	164	Ea.	113,000	9,325		122,325	138,500
7110	70 ton cooling		.14	228		158,000	13,000		171,000	194,500
7120	90 ton cooling		.11	296		181,500	16,900		198,400	226,500
7130	105 ton cooling		.10	333		203,000	19,000		222,000	253,000
7140	120 ton cooling		.08	380		232,500	21,700		254,200	289,500
7150	140 ton cooling		.07	444		248,500	25,300		273,800	313,000
7300	Penthouse unit, cool/heat, variable volume distribution									
7400	150 ton cooling	Q-7	.10	310	Ea.	406,500	17,700		424,200	474,500
7410	170 ton cooling		.10	320		434,500	18,200		452,700	506,500
7420	200 ton cooling		.10	326		511,000	18,600		529,600	591,000
7430	225 ton cooling		.09	340		559,000	19,400		578,400	645,000
7440	250 ton cooling		.09	359		577,000	20,500		597,500	666,500
7450	270 ton cooling		.09	363		623,000	20,700		643,700	718,000
7460	300 ton cooling		.09	372		692,500	21,200		713,700	794,500
9000	Minimum labor/equipment charge	Q-5	1.50	10.667	Job		575		575	895
9400	Sub assemblies for assembly systems									
9410	Ductwork, per ton, rooftop 1 zone units	Q-9	.96	16.667	Ton	177	840		1,017	1,525
9420	Ductwork, per ton, rooftop multizone units		.50	32		268	1,600		1,868	2,850
9430	Ductwork, VAV cooling/ton, rooftop multizone, 1 zone/ton		.32	50		425	2,525		2,950	4,450
9440	Ductwork per ton, packaged water & air cooled units		1.26	12.698		69	640		709	1,075
9450	Ductwork per ton, split system remote condensing units		1.32	12.121		65	610		675	1,025
9500	Ductwork package for residential units, 1 ton		.80	20	Ea.	280	1,000		1,280	1,900
9520	2 ton		.40	40		560	2,025		2,585	3,800
9530	3 ton		.27	59.259		840	2,975		3,815	5,625
9540	4 ton	Q-10	.30	80		1,125	4,175		5,300	7,825
9550	5 ton		.23	104		1,400	5,450		6,850	10,200
9560	6 ton		.20	120		1,675	6,275		7,950	11,800
9570	7 ton		.17	141		1,950	7,375		9,325	13,800
9580	8 ton		.15	160		2,250	8,350		10,600	15,700

23 76 Evaporative Air-Cooling Equipment

23 76 13 – Direct Evaporative Air Coolers

23 76 13.10 Evaporative Coolers			Crew	Daily Output	Labor-Hours	Unit	Material	Labor	Equipment	Total	Total Incl O&P
0010	**EVAPORATIVE COOLERS,** (Swamp Coolers) ducted, not incl. duct.										
0100	Side discharge style, capacities at .25" S.P.										
0120	1785 CFM, 1/3 HP, 115 V	G	Q-9	5	3.200	Ea.	460	161		621	760
0140	2740 CFM, 1/3 HP, 115 V	G		4.50	3.556		545	179		724	880
0160	3235 CFM, 1/2 HP, 115 V	G		4	4		550	201		751	925
0180	3615 CFM, 1/2 HP, 230 V	G		3.60	4.444		680	224		904	1,100
0200	4215 CFM, 3/4 HP, 230 V	G		3.20	5		720	252		972	1,175
0220	5255 CFM. 1 HP, 115/230 V	G		3	5.333		1,275	269		1,544	1,825
0240	6090 CFM, 1 HP, 230/460 V	G		2.80	5.714		1,900	288		2,188	2,550
0260	8300 CFM, 1-1/2 HP, 230/460 V	G		2.60	6.154		1,900	310		2,210	2,600

23 76 13 – Direct Evaporative Air Coolers

23 76 13.10 Evaporative Coolers			Crew	Daily Output	Labor-Hours	Unit	2015 Bare Costs Material	Labor	Equipment	Total	Total Incl O&P
0280	8360 CFM, 1-1/2 HP, 230/460 V	G	Q-9	2.20	7.273	Ea.	1,900	365		2,265	2,675
0300	9725 CFM, 2 HP, 230/460 V	G		1.80	8.889		2,150	450		2,600	3,075
0320	11,715 CFM, 3 HP, 230/460 V	G		1.40	11.429		2,400	575		2,975	3,550
0340	14,410 CFM, 5 HP, 230/460 V	G		1	16		2,600	805		3,405	4,125
0400	For two-speed motor, add						5%				
0500	For down discharge style, add						10%				

23 76 26 – Fan Coil Evaporators for Coolers/Freezers

23 76 26.10 Evaporators		Crew	Daily Output	Labor-Hours	Unit	Material	Labor	Equipment	Total	Total Incl O&P
0010	**EVAPORATORS**, DX coils, remote compressors not included.									
1000	Coolers, reach-in type, above freezing temperatures									
1300	Shallow depth, wall mount, 7 fins per inch, air defrost									
1310	600 BTUH, 8″ fan, rust-proof core	Q-5	3.80	4.211	Ea.	680	226		906	1,100
1320	900 BTUH, 8″ fan, rust-proof core		3.30	4.848		790	261		1,051	1,275
1330	1200 BTUH, 8″ fan, rust-proof core		3	5.333		890	287		1,177	1,425
1340	1800 BTUH, 8″ fan, rust-proof core		2.40	6.667		995	360		1,355	1,650
1350	2500 BTUH, 10″ fan, rust-proof core		2.20	7.273		1,050	390		1,440	1,775
1360	3500 BTUH, 10″ fan		2	8		1,200	430		1,630	2,000
1370	4500 BTUH, 10″ fan		1.90	8.421		1,250	455		1,705	2,075
1390	For pan drain, add					59			59	64.50
1600	Undercounter refrigerators, ceiling or wall mount,									
1610	8 fins per inch, air defrost, rust-proof core									
1630	800 BTUH, one 6″ fan	Q-5	5	3.200	Ea.	305	172		477	605
1640	1300 BTUH, two 6″ fans		4	4		370	215		585	740
1650	1700 BTUH, two 6″ fans		3.60	4.444		415	239		654	830
2000	Coolers, reach-in and walk-in types, above freezing temperatures									
2600	Two-way discharge, ceiling mount, 150-4100 CFM,									
2610	above 34°F applications, air defrost									
2630	900 BTUH, 7 fins per inch, 8″ fan	Q-5	4	4	Ea.	550	215		765	940
2660	2500 BTUH, 7 fins per inch, 10″ fan		2.60	6.154		865	330		1,195	1,475
2690	5500 BTUH, 7 fins per inch, 12″ fan		1.70	9.412		1,100	505		1,605	2,000
2720	8500 BTUH, 8 fins per inch, 16″ fan		1.20	13.333		1,575	715		2,290	2,850
2750	15,000 BTUH, 7 fins per inch, 18″ fan		1.10	14.545		2,500	780		3,280	3,975
2770	24,000 BTUH, 7 fins per inch, two 16″ fans		1	16		3,750	860		4,610	5,450
2790	30,000 BTUH, 7 fins per inch, two 18″ fans		.90	17.778		4,650	955		5,605	6,600
2850	Two-way discharge, low profile, ceiling mount,									
2860	8 fins per inch, 200-570 CFM, air defrost									
2880	800 BTUH, one 6″ fan	Q-5	4	4	Ea.	370	215		585	740
2890	1300 BTUH, two 6″ fans		3.50	4.571		440	246		686	865
2900	1800 BTUH, two 6″ fans		3.30	4.848		460	261		721	910
2910	2700 BTUH, three 6″ fans		2.70	5.926		540	320		860	1,100
3000	Coolers, walk-in type, above freezing temperatures									
3300	General use, ceiling mount, 108-2080 CFM									
3320	600 BTUH, 7 fins per inch, 6″ fan	Q-5	5.20	3.077	Ea.	325	165		490	620
3340	1200 BTUH, 7 fins per inch, 8″ fan		3.70	4.324		450	233		683	860
3360	1800 BTUH, 7 fins per inch, 10″ fan		2.70	5.926		630	320		950	1,200
3380	3500 BTUH, 7 fins per inch, 12″ fan		2.40	6.667		930	360		1,290	1,575
3400	5500 BTUH, 8 fins per inch, 12″ fan		2	8		1,025	430		1,455	1,825
3430	8500 BTUH, 8 fins per inch, 16″ fan		1.40	11.429		1,550	615		2,165	2,650
3460	15,000 BTUH, 7 fins per inch, two 16″ fans		1.10	14.545		2,475	780		3,255	3,950
3640	Low velocity, high latent load, ceiling mount,									
3650	1050-2420 CFM, air defrost, 6 fins per inch									
3670	6700 BTUH, two 10″ fans	Q-5	1.30	12.308	Ea.	2,025	660		2,685	3,275

23 76 26 – Fan Coil Evaporators for Coolers/Freezers

23 76 26.10 Evaporators		Crew	Daily Output	Labor-Hours	Unit	Material	2015 Bare Costs Labor	Equipment	Total	Total Incl O&P
3680	10,000 BTUH, three 10" fans	Q-5	1	16	Ea.	2,575	860		3,435	4,200
3690	13,500 BTUH, three 10" fans		1	16		2,950	860		3,810	4,600
3700	18,000 BTUH, four 10" fans		1	16		3,975	860		4,835	5,725
3710	26,500 BTUH, four 10" fans		.80	20		4,650	1,075		5,725	6,800
3730	For electric defrost, add					17%				
5000	Freezers and coolers, reach-in type, above 34°F									
5030	to sub-freezing temperature range, low latent load,									
5050	air defrost, 7 fins per inch									
5070	1200 BTUH, 8" fan, rustproof core	Q-5	4.40	3.636	Ea.	605	196		801	970
5080	1500 BTUH, 8" fan, rustproof core		4.20	3.810		615	205		820	1,000
5090	1800 BTUH, 8" fan, rustproof core		3.70	4.324		635	233		868	1,075
5100	2500 BTUH, 10" fan, rustproof core		2.90	5.517		690	297		987	1,225
5110	3500 BTUH, 10" fan, rustproof core		2.50	6.400		865	345		1,210	1,475
5120	4500 BTUH, 12" fan, rustproof core		2.10	7.619		1,075	410		1,485	1,825
6000	Freezers and coolers, walk-in type									
6050	1960-18,000 CFM, medium profile									
6060	Standard motor, 6 fins per inch, to -30°F, all aluminum									
6080	10,500 BTUH, air defrost, one 18" fan	Q-5	1.40	11.429	Ea.	1,650	615		2,265	2,750
6090	12,500 BTUH, air defrost, one 18" fan		1.40	11.429		1,775	615		2,390	2,900
6100	16,400 BTUH, air defrost, one 18" fan		1.30	12.308		1,800	660		2,460	3,000
6110	20,900 BTUH, air defrost, one 18" fans		1	16		2,100	860		2,960	3,650
6120	27,000 BTUH, air defrost, two 18" fans		1	16		3,100	860		3,960	4,750
6130	32,900 BTUH, air defrost, two 20" fans	Q-6	1.30	18.462		3,450	1,025		4,475	5,400
6140	39,000 BTUH, air defrost, two 20" fans		1.25	19.200		4,025	1,075		5,100	6,100
6150	44,100 BTUH, air defrost, three 20" fans		1	24		4,550	1,350		5,900	7,125
6155	53,000 BTUH, air defrost, three 20" fans		1	24		4,550	1,350		5,900	7,125
6160	66,200 BTUH, air defrost, four 20" fans		1	24		5,175	1,350		6,525	7,800
6170	78,000 BTUH, air defrost, four 20" fans		.80	30		6,450	1,675		8,125	9,700
6180	88,200 BTUH, air defrost, four 24" fans		.60	40		6,875	2,225		9,100	11,000
6190	110,000 BTUH, air defrost, four 24" fans		.50	48		7,775	2,675		10,450	12,700
6330	Hot gas defrost, standard motor units, add					35%				
6370	Electric defrost, 230V, standard motor, add					16%				
6410	460V, standard or high capacity motor, add					25%				
6800	Eight fins per inch, increases BTUH 75%									
6810	Standard capacity, add				Ea.	5%				
6830	Hot gas & electric defrost not recommended									
7000	800-4150 CFM, 12" fans									
7010	Four fins per inch									
7030	3400 BTUH, 1 fan	Q-5	2.80	5.714	Ea.	555	305		860	1,100
7040	4200 BTUH, 1 fan		2.40	6.667		635	360		995	1,250
7050	5300 BTUH, 1 fan		1.90	8.421		825	455		1,280	1,625
7060	6800 BTUH, 1 fan		1.70	9.412		990	505		1,495	1,900
7070	8400 BTUH, 2 fans		1.60	10		1,125	540		1,665	2,100
7080	10,500 BTUH, 2 fans		1.40	11.429		1,200	615		1,815	2,275
7090	13,000 BTUH, 3 fans		1.30	12.308		1,600	660		2,260	2,775
7100	17,000 BTUH, 4 fans		1.20	13.333		2,025	715		2,740	3,350
7110	21,500 BTUH, 5 fans		1	16		2,525	860		3,385	4,125
7160	Six fins per inch increases BTUH 35%, add					5%				
7180	Eight fins per inch increases BTUH 53%,									
7190	not recommended for below 35°F, add				Ea.	8%				
7210	For 12°F temperature differential, add					20%				
7230	For adjustable thermostat control, add					141			141	155
7240	For hot gas defrost, except 8 fin models									

23 76 Evaporative Air-Cooling Equipment

23 76 26 – Fan Coil Evaporators for Coolers/Freezers

23 76 26.10 Evaporators		Crew	Daily Output	Labor-Hours	Unit	Material	2015 Bare Costs Labor	Equipment	Total	Total Incl O&P
7250	on applications below 35°F, add				Ea.	58%				
7260	For electric defrost, except 8 fin models									
7270	on applications below 35°F, add				Ea.	36%				
8000	Freezers, pass-thru door uprights, walk-in storage									
8300	Low temperature, thin profile, electric defrost									
8310	138-1800 CFM, 5 fins per inch									
8320	900 BTUH, one 6″ fan	Q-5	3.30	4.848	Ea.	430	261		691	875
8340	1500 BTUH, two 6″ fans		2.20	7.273		640	390		1,030	1,325
8360	2600 BTUH, three 6″ fans		1.90	8.421		920	455		1,375	1,700
8370	3300 BTUH, four 6″ fans		1.70	9.412		1,025	505		1,530	1,925
8380	4400 BTUH, five 6″ fans		1.40	11.429		1,425	615		2,040	2,525
8400	7000 BTUH, three 10″ fans		1.30	12.308		1,725	660		2,385	2,925
8410	8700 BTUH, four 10″ fans	↓	1.10	14.545		2,050	780		2,830	3,500
8450	For air defrost, deduct				↓	255			255	281

23 81 Decentralized Unitary HVAC Equipment

23 81 13 – Packaged Terminal Air-Conditioners

23 81 13.10 Packaged Cabinet Type Air-Conditioners

		Crew	Daily Output	Labor-Hours	Unit	Material	Labor	Equipment	Total	Total Incl O&P
0010	**PACKAGED CABINET TYPE AIR-CONDITIONERS**, Cabinet, wall sleeve,									
0100	louver, electric heat, thermostat, manual changeover, 208 V									
0200	6,000 BTUH cooling, 8800 BTU heat	Q-5	6	2.667	Ea.	775	143		918	1,075
0220	9,000 BTUH cooling, 13,900 BTU heat		5	3.200		1,200	172		1,372	1,600
0240	12,000 BTUH cooling, 13,900 BTU heat		4	4		1,350	215		1,565	1,825
0260	15,000 BTUH cooling, 13,900 BTU heat		3	5.333		1,450	287		1,737	2,050
0280	18,000 BTUH cooling, 10 KW heat		2.40	6.667		1,900	360		2,260	2,650
0300	24,000 BTUH cooling, 10 KW heat		1.90	8.421		1,925	455		2,380	2,825
0320	30,000 BTUH cooling, 10 KW heat		1.40	11.429		2,125	615		2,740	3,300
0340	36,000 BTUH cooling, 10 KW heat		1.25	12.800		2,250	690		2,940	3,550
0360	42,000 BTUH cooling, 10 KW heat		1	16		2,750	860		3,610	4,375
0380	48,000 BTUH cooling, 10 KW heat	↓	.90	17.778		2,900	955		3,855	4,700
0500	For hot water coil, increase heat by 10%, add					5%	10%			
1000	For steam, increase heat output by 30%, add				↓	8%	10%			

23 81 19 – Self-Contained Air-Conditioners

23 81 19.10 Window Unit Air Conditioners

		Crew	Daily Output	Labor-Hours	Unit	Material	Labor	Equipment	Total	Total Incl O&P
0010	**WINDOW UNIT AIR CONDITIONERS**									
4000	Portable/window, 15 amp 125 V grounded receptacle required									
4060	5000 BTUH	1 Carp	8	1	Ea.	295	47		342	400
4340	6000 BTUH		8	1		340	47		387	450
4480	8000 BTUH		6	1.333		460	62.50		522.50	615
4500	10,000 BTUH	↓	6	1.333		590	62.50		652.50	755
4520	12,000 BTUH	L-2	8	2	↓	725	82.50		807.50	930
4600	Window/thru-the-wall, 15 amp 230 V grounded receptacle required									
4780	18,000 BTUH	L-2	6	2.667	Ea.	970	110		1,080	1,250
4940	25,000 BTUH		4	4		1,175	165		1,340	1,575
4960	29,000 BTUH	↓	4	4	↓	1,325	165		1,490	1,750
9000	Minimum labor/equipment charge	1 Carp	2	4	Job		188		188	310

23 81 19 – Self-Contained Air-Conditioners

	23 81 19.20 Self-Contained Single Package	Crew	Daily Output	Labor-Hours	Unit	Material	2015 Bare Costs Labor	Equipment	Total	Total Incl O&P
0010	SELF-CONTAINED SINGLE PACKAGE									
0100	Air cooled, for free blow or duct, not incl. remote condenser									
0110	Constant volume									
0200	3 ton cooling	Q-5	1	16	Ea.	3,750	860		4,610	5,450
0210	4 ton cooling	″	.80	20		4,075	1,075		5,150	6,150
0220	5 ton cooling	Q-6	1.20	20		4,450	1,125		5,575	6,650
0230	7.5 ton cooling	″	.90	26.667		6,475	1,475		7,950	9,450
0240	10 ton cooling	Q-7	1	32		7,125	1,825		8,950	10,700
0250	15 ton cooling		.95	33.684		9,050	1,925		10,975	13,000
0260	20 ton cooling		.90	35.556		12,800	2,025		14,825	17,300
0270	25 ton cooling		.85	37.647		19,000	2,150		21,150	24,300
0280	30 ton cooling		.80	40		25,300	2,275		27,575	31,400
0300	40 ton cooling		.60	53.333		36,100	3,025		39,125	44,400
0320	50 ton cooling		.50	64		47,200	3,650		50,850	57,500
0340	60 ton cooling	Q-8	.40	80		56,000	4,550	144	60,694	69,000
0490	For duct mounting no price change									
0500	For steam heating coils, add				Ea.	10%	10%			
1000	Water cooled for free blow or duct, not including tower									
1010	Constant volume									
1100	3 ton cooling	Q-6	1	24	Ea.	3,725	1,350		5,075	6,200
1120	5 ton cooling		1	24		4,850	1,350		6,200	7,425
1130	7.5 ton cooling		.80	30		6,850	1,675		8,525	10,200
1140	10 ton cooling	Q-7	.90	35.556		9,450	2,025		11,475	13,600
1150	15 ton cooling		.85	37.647		13,700	2,150		15,850	18,500
1160	20 ton cooling		.80	40		28,600	2,275		30,875	35,000
1170	25 ton cooling		.75	42.667		33,700	2,425		36,125	40,900
1180	30 ton cooling		.70	45.714		37,900	2,600		40,500	45,800
1200	40 ton cooling		.40	80		48,000	4,550		52,550	60,000
1220	50 ton cooling		.30	106		59,500	6,075		65,575	75,000
1240	60 ton cooling	Q-8	.30	106		69,500	6,075	192	75,767	86,000
1300	For hot water or steam heat coils, add					12%	10%			
9000	Minimum labor/equipment charge	Q-5	1	16	Job		860		860	1,350

23 81 23 – Computer-Room Air-Conditioners

	23 81 23.10 Computer Room Units	Crew	Daily Output	Labor-Hours	Unit	Material	Labor	Equipment	Total	Total Incl O&P
0010	COMPUTER ROOM UNITS									
1000	Air cooled, includes remote condenser but not									
1020	interconnecting tubing or refrigerant R236000-30									
1080	3 ton	Q-5	.50	32	Ea.	18,800	1,725		20,525	23,400
1120	5 ton		.45	35.556		20,100	1,900		22,000	25,100
1160	6 ton		.30	53.333		37,000	2,875		39,875	45,200
1200	8 ton		.27	59.259		37,400	3,175		40,575	46,200
1240	10 ton		.25	64		39,100	3,450		42,550	48,500
1260	12 ton		.24	66.667		40,500	3,575		44,075	50,000
1280	15 ton		.22	72.727		43,100	3,900		47,000	53,500
1290	18 ton		.20	80		49,300	4,300		53,600	61,000
1300	20 ton	Q-6	.26	92.308		51,500	5,150		56,650	64,500
1320	22 ton		.24	100		52,000	5,575		57,575	66,000
1360	30 ton		.21	114		64,500	6,375		70,875	80,500
2200	Chilled water, for connection to									
2220	existing chiller system of adequate capacity									
2260	5 ton	Q-5	.74	21.622	Ea.	14,300	1,175		15,475	17,600
2280	6 ton		.52	30.769		14,400	1,650		16,050	18,500

23 81 Decentralized Unitary HVAC Equipment

23 81 23 – Computer-Room Air-Conditioners

23 81 23.10 Computer Room Units		Crew	Daily Output	Labor-Hours	Unit	2015 Bare Costs Material	Labor	Equipment	Total	Total Incl O&P
2300	8 ton	Q-5	.50	32	Ea.	14,600	1,725		16,325	18,800
2320	10 ton		.49	32.653		14,700	1,750		16,450	19,000
2330	12 ton		.49	32.990		15,100	1,775		16,875	19,400
2360	15 ton		.48	33.333		15,400	1,800		17,200	19,700
2400	20 ton		.46	34.783		16,300	1,875		18,175	20,900
2440	25 ton	Q-6	.63	38.095		17,400	2,125		19,525	22,400
2460	30 ton		.57	42.105		18,900	2,350		21,250	24,500
2480	35 ton		.52	46.154		19,100	2,575		21,675	25,000
2500	45 ton		.46	52.174		20,200	2,900		23,100	26,800
2520	50 ton		.42	57.143		23,100	3,175		26,275	30,400
2540	60 ton		.38	63.158		24,600	3,525		28,125	32,600
2560	77 ton		.35	68.571		41,600	3,825		45,425	51,500
2580	95 ton		.32	75		48,500	4,175		52,675	60,000
4000	Glycol system, complete except for interconnecting tubing									
4060	3 ton	Q-5	.40	40	Ea.	23,600	2,150		25,750	29,300
4100	5 ton		.38	42.105		25,800	2,275		28,075	31,900
4120	6 ton		.25	64		38,300	3,450		41,750	47,500
4140	8 ton		.23	69.565		41,400	3,750		45,150	51,500
4160	10 ton		.21	76.190		44,000	4,100		48,100	55,000
4180	12 ton		.19	84.211		47,700	4,525		52,225	59,500
4200	15 ton	Q-6	.26	92.308		54,500	5,150		59,650	68,000
4240	20 ton		.24	100		60,000	5,575		65,575	74,500
4280	22 ton		.23	104		63,000	5,825		68,825	78,500
4430	30 ton		.22	109		78,000	6,075		84,075	95,500
8000	Water cooled system, not including condenser,									
8020	water supply or cooling tower									
8060	3 ton	Q-5	.62	25.806	Ea.	19,100	1,400		20,500	23,200
8100	5 ton		.54	29.630		20,800	1,600		22,400	25,400
8120	6 ton		.35	45.714		29,700	2,450		32,150	36,500
8140	8 ton		.33	48.485		34,100	2,600		36,700	41,600
8160	10 ton		.31	51.613		35,200	2,775		37,975	43,000
8180	12 ton		.29	55.172		38,200	2,975		41,175	46,600
8200	15 ton		.27	59.259		41,300	3,175		44,475	50,500
8240	20 ton	Q-6	.38	63.158		44,300	3,525		47,825	54,500
8280	22 ton		.35	68.571		46,600	3,825		50,425	57,000
8300	30 ton		.30	80		58,000	4,450		62,450	70,500

23 81 26 – Split-System Air-Conditioners

23 81 26.10 Split Ductless Systems		Crew	Daily Output	Labor-Hours	Unit	Material	Labor	Equipment	Total	Total Incl O&P
0010	SPLIT DUCTLESS SYSTEMS									
0100	Cooling only, single zone									
0110	Wall mount									
0120	3/4 ton cooling	Q-5	2	8	Ea.	1,100	430		1,530	1,900
0130	1 ton cooling		1.80	8.889		1,225	480		1,705	2,100
0140	1-1/2 ton cooling		1.60	10		1,975	540		2,515	3,025
0150	2 ton cooling		1.40	11.429		2,250	615		2,865	3,425
1000	Ceiling mount									
1020	2 ton cooling	Q-5	1.40	11.429	Ea.	2,000	615		2,615	3,150
1030	3 ton cooling	"	1.20	13.333	"	2,550	715		3,265	3,925
3000	Multizone									
3010	Wall mount									
3020	2 @ 3/4 ton cooling	Q-5	1.80	8.889	Ea.	3,375	480		3,855	4,450
5000	Cooling/Heating									

23 81 Decentralized Unitary HVAC Equipment

23 81 26 – Split-System Air-Conditioners

23 81 26.10 Split Ductless Systems		Crew	Daily Output	Labor-Hours	Unit	Material	2015 Bare Costs Labor	Equipment	Total	Total Incl O&P
5010	Wall mount									
5110	1 ton cooling	Q-5	1.70	9.412	Ea.	1,300	505		1,805	2,225
5120	1-1/2 ton cooling	"	1.50	10.667	"	1,975	575		2,550	3,075
7000	Accessories for all split ductless systems									
7010	Add for ambient frost control	Q-5	8	2	Ea.	126	108		234	305
7020	Add for tube/wiring kit, (Line sets)									
7030	15′ kit	Q-5	32	.500	Ea.	103	27		130	155
7036	25′ kit		28	.571		166	30.50		196.50	230
7040	35′ kit		24	.667		174	36		210	247
7050	50′ kit	↓	20	.800	↓	261	43		304	355

23 81 29 – Variable Refrigerant Flow HVAC Systems

23 81 29.10 Heat Pump, Gas Driven		Crew	Daily Output	Labor-Hours	Unit	Material	Labor	Equipment	Total	Total Incl O&P
0010	**HEAT PUMP, GAS DRIVEN**, Variable refrigerant volume (VRV) type									
0020	Not including interconnecting tubing or multi-zone controls									
1000	For indoor fan VRV type AHU see 23 82 19.40									
1010	Multi-zone split									
1020	Outdoor unit									
1100	8 tons cooling, for up to 17 zones	Q-5	1.30	12.308	Ea.	29,100	660		29,760	33,000
1110	Isolation rails		2.60	6.154	Pair	1,025	330		1,355	1,650
1160	15 tons cooling, for up to 33 zones		1	16	Ea.	38,600	860		39,460	43,800
1170	Isolation rails	↓	2.60	6.154	Pair	1,200	330		1,530	1,850
2000	Packaged unit									
2020	Outdoor unit									
2200	11 tons cooling	Q-5	1.30	12.308	Ea.	37,200	660		37,860	41,900
2210	Roof curb adapter	"	2.60	6.154		645	330		975	1,225
2220	Thermostat	1 Stpi	1.30	6.154	↓	385	370		755	1,000

23 81 43 – Air-Source Unitary Heat Pumps

23 81 43.10 Air-Source Heat Pumps		Crew	Daily Output	Labor-Hours	Unit	Material	Labor	Equipment	Total	Total Incl O&P
0010	**AIR-SOURCE HEAT PUMPS**, Not including interconnecting tubing									
1000	Air to air, split system, not including curbs, pads, fan coil and ductwork									
1012	Outside condensing unit only, for fan coil see Section 23 82 19.10									
1015	1.5 ton cooling, 7 MBH heat @ 0°F	Q-5	2.40	6.667	Ea.	2,175	360		2,535	2,950
1020	2 ton cooling, 8.5 MBH heat @ 0°F		2	8		2,400	430		2,830	3,325
1030	2.5 ton cooling, 10 MBH heat @ 0°F		1.60	10		2,525	540		3,065	3,625
1040	3 ton cooling, 13 MBH heat @ 0°F		1.20	13.333		2,775	715		3,490	4,175
1050	3.5 ton cooling, 18 MBH heat @ 0°F		1	16		3,000	860		3,860	4,650
1054	4 ton cooling, 24 MBH heat @ 0°F		.80	20		3,250	1,075		4,325	5,250
1060	5 ton cooling, 27 MBH heat @ 0°F		.50	32		3,600	1,725		5,325	6,625
1080	7.5 ton cooling, 33 MBH heat @ 0°F	↓	.45	35.556		6,425	1,900		8,325	10,100
1100	10 ton cooling, 50 MBH heat @ 0°F	Q-6	.64	37.500		8,500	2,100		10,600	12,600
1120	15 ton cooling, 64 MBH heat @ 0°F		.50	48		11,800	2,675		14,475	17,200
1130	20 ton cooling, 85 MBH heat @ 0°F		.35	68.571		17,000	3,825		20,825	24,700
1140	25 ton cooling, 119 MBH heat @ 0°F	↓	.25	96	↓	20,400	5,350		25,750	30,900
1500	Single package, not including curbs, pads, or plenums									
1502	1/2 ton cooling, supplementary heat included	Q-5	8	2	Ea.	2,450	108		2,558	2,875
1504	3/4 ton cooling, supplementary heat supplementary heat included		6	2.667		2,575	143		2,718	3,050
1506	1 ton cooling, supplementary heat included		4	4		2,975	215		3,190	3,600
1510	1.5 ton cooling, 5 MBH heat @ 0°F		1.55	10.323		3,050	555		3,605	4,225
1520	2 ton cooling, 6.5 MBH heat @ 0°F		1.50	10.667		3,100	575		3,675	4,325
1540	2.5 ton cooling, 8 MBH heat @ 0°F		1.40	11.429		3,175	615		3,790	4,450
1560	3 ton cooling, 10 MBH heat @ 0°F		1.20	13.333		3,575	715		4,290	5,050
1570	3.5 ton cooling, 11 MBH heat @ 0°F	↓	1	16	↓	3,900	860		4,760	5,625

23 81 43 – Air-Source Unitary Heat Pumps

23 81 43.10 Air-Source Heat Pumps		Crew	Daily Output	Labor-Hours	Unit	2015 Bare Costs Material	Labor	Equipment	Total	Total Incl O&P
1580	4 ton cooling, 13 MBH heat @ 0°F	Q-5	.96	16.667	Ea.	4,200	895		5,095	6,025
1620	5 ton cooling, 27 MBH heat @ 0°F		.65	24.615		4,800	1,325		6,125	7,350
1640	7.5 ton cooling, 35 MBH heat @ 0°F		.40	40		7,325	2,150		9,475	11,400
1648	10 ton cooling, 45 MBH heat @ 0°F	Q-6	.40	60		9,925	3,350		13,275	16,100
1652	12 ton cooling, 50 MBH heat @ 0°F	"	.36	66.667		12,200	3,725		15,925	19,200
1696	Supplementary electric heat coil incl., unless noted otherwise									
6000	Air to water, single package, excluding. storage tank and ductwork									
6010	Includes circulating water pump, air duct connections, digital temperature									
6020	controller with remote tank temp. probe and sensor for storage tank.									
6040	Water heating - air cooling capacity									
6110	35.5 MBH heat water, 2.3 Ton cool air	Q-5	1.60	10	Ea.	11,400	540		11,940	13,400
6120	58 MBH heat water, 3.8 Ton cool air		1.10	14.545		13,000	780		13,780	15,500
6130	76 MBH heat water, 4.9 Ton cool air		.87	18.391		15,500	990		16,490	18,700
6140	98 MBH heat water, 6.5 Ton cool air		.62	25.806		20,100	1,400		21,500	24,300
6150	113 MBH heat water, 7.4 Ton cool air		.59	27.119		22,000	1,450		23,450	26,500
6160	142 MBH heat water, 9.2 Ton cool air		.52	30.769		26,700	1,650		28,350	32,000
6170	171 MBH heat water, 11.1 Ton cool air		.49	32.653		31,400	1,750		33,150	37,300

23 81 46 – Water-Source Unitary Heat Pumps

23 81 46.10 Water Source Heat Pumps		Crew	Daily Output	Labor-Hours	Unit	Material	Labor	Equipment	Total	Total Incl O&P
0010	**WATER SOURCE HEAT PUMPS**, Not incl. connecting tubing or water source									
2000	Water source to air, single package									
2100	1 ton cooling, 13 MBH heat @ 75°F	Q-5	2	8	Ea.	2,075	430		2,505	2,975
2120	1.5 ton cooling, 17 MBH heat @ 75°F		1.80	8.889		2,400	480		2,880	3,375
2140	2 ton cooling, 19 MBH heat @ 75°F		1.70	9.412		2,575	505		3,080	3,625
2160	2.5 ton cooling, 25 MBH heat @ 75°F		1.60	10		2,800	540		3,340	3,950
2180	3 ton cooling, 27 MBH heat @ 75°F		1.40	11.429		2,875	615		3,490	4,100
2190	3.5 ton cooling, 29 MBH heat @ 75°F		1.30	12.308		3,100	660		3,760	4,450
2200	4 ton cooling, 31 MBH heat @ 75°F		1.20	13.333		3,475	715		4,190	4,950
2220	5 ton cooling, 29 MBH heat @ 75°F		.90	17.778		3,825	955		4,780	5,700
2240	7.5 ton cooling, 35 MBH heat @ 75°F		.60	26.667		7,275	1,425		8,700	10,200
2250	8.5 ton cooling, 40 MBH heat @ 75°F		.58	27.586		8,800	1,475		10,275	12,000
2260	10 ton cooling, 50 MBH heat @ 75°F		.53	30.189		9,225	1,625		10,850	12,600
2280	15 ton cooling, 64 MBH heat @ 75°F	Q-6	.47	51.064		14,000	2,850		16,850	19,900
2300	20 ton cooling, 100 MBH heat @ 75°F		.41	58.537		15,300	3,275		18,575	21,900
2310	25 ton cooling, 100 MBH heat @ 75°F		.32	75		20,800	4,175		24,975	29,400
2320	30 ton cooling, 128 MBH heat @ 75°F		.24	102		22,800	5,700		28,500	34,000
2340	40 ton cooling, 200 MBH heat @ 75°F		.21	117		32,200	6,525		38,725	45,600
2360	50 ton cooling, 200 MBH heat @ 75°F		.15	160		36,400	8,925		45,325	54,000
3960	For supplementary heat coil, add					10%				
4000	For increase in capacity thru use									
4020	of solar collector, size boiler at 60%									
9000	Minimum labor/equipment charge	Q-5	1.75	9.143	Job		490		490	765

23 82 13 – Valance Heating and Cooling Units

23 82 13.16 Valance Units		Crew	Daily Output	Labor-Hours	Unit	2015 Bare Costs Material	Labor	Equipment	Total	Total Incl O&P
0010	**VALANCE UNITS**									
6000	Valance units, complete with 1/2" cooling coil, enclosure									
6020	2 tube	Q-5	18	.889	L.F.	33.50	48		81.50	112
6040	3 tube		16	1		40	54		94	128
6060	4 tube		16	1		48.50	54		102.50	138
6080	5 tube		15	1.067		52	57.50		109.50	147
6100	6 tube		15	1.067		57.50	57.50		115	153
6120	8 tube		14	1.143		83.50	61.50		145	188
6200	For 3/4" cooling coil, add					10%				

23 82 16 – Air Coils

23 82 16.10 Flanged Coils		Crew	Daily Output	Labor-Hours	Unit	Material	Labor	Equipment	Total	Total Incl O&P
0010	**FLANGED COILS**									
0500	Chilled water cooling, 6 rows, 24" x 48"	Q-5	3.20	5	Ea.	4,100	269		4,369	4,925
1000	Direct expansion cooling, 6 rows, 24" x 48"		2.80	5.714		4,450	305		4,755	5,375
1500	Hot water heating, 1 row, 24" x 48"		4	4		1,625	215		1,840	2,125
2000	Steam heating, 1 row, 24" x 48"		3.06	5.229		2,325	281		2,606	3,000
9000	Minimum labor/equipment charge	1 Plum	1	8	Job		470		470	730

23 82 16.20 Duct Heaters		Crew	Daily Output	Labor-Hours	Unit	Material	Labor	Equipment	Total	Total Incl O&P
0010	**DUCT HEATERS**, Electric, 480 V, 3 Ph.									
0020	Finned tubular insert, 500°F									
0100	8" wide x 6" high, 4.0 kW	Q-20	16	1.250	Ea.	765	64		829	945
0120	12" high, 8.0 kW		15	1.333		1,275	68.50		1,343.50	1,500
0140	18" high, 12.0 kW		14	1.429		1,775	73		1,848	2,075
0160	24" high, 16.0 kW		13	1.538		2,300	79		2,379	2,650
0180	30" high, 20.0 kW		12	1.667		2,800	85.50		2,885.50	3,200
0300	12" wide x 6" high, 6.7 kW		15	1.333		815	68.50		883.50	1,000
0320	12" high, 13.3 kW		14	1.429		1,325	73		1,398	1,575
0340	18" high, 20.0 kW		13	1.538		1,850	79		1,929	2,150
0360	24" high, 26.7 kW		12	1.667		2,375	85.50		2,460.50	2,725
0380	30" high, 33.3 kW		11	1.818		2,900	93		2,993	3,325
0500	18" wide x 6" high, 13.3 kW		14	1.429		880	73		953	1,075
0520	12" high, 26.7 kW		13	1.538		1,500	79		1,579	1,775
0540	18" high, 40.0 kW		12	1.667		2,000	85.50		2,085.50	2,325
0560	24" high, 53.3 kW		11	1.818		2,675	93		2,768	3,075
0580	30" high, 66.7 kW		10	2		3,325	102		3,427	3,825
0700	24" wide x 6" high, 17.8 kW		13	1.538		965	79		1,044	1,200
0720	12" high, 35.6 kW		12	1.667		1,625	85.50		1,710.50	1,925
0740	18" high, 53.3 kW		11	1.818		2,300	93		2,393	2,675
0760	24" high, 71.1 kW		10	2		2,950	102		3,052	3,400
0780	30" high, 88.9 kW		9	2.222		3,625	114		3,739	4,175
0900	30" wide x 6" high, 22.2 kW		12	1.667		1,025	85.50		1,110.50	1,250
0920	12" high, 44.4 kW		11	1.818		1,725	93		1,818	2,050
0940	18" high, 66.7 kW		10	2		2,425	102		2,527	2,825
0960	24" high, 88.9 kW		9	2.222		3,125	114		3,239	3,625
0980	30" high, 111.0 kW		8	2.500		3,825	128		3,953	4,425
1400	Note decreased kW available for									
1410	each duct size at same cost									
1420	See line 5000 for modifications and accessories									
2000	Finned tubular flange with insulated									
2020	terminal box, 500°F									
2100	12" wide x 36" high, 54 kW	Q-20	10	2	Ea.	3,775	102		3,877	4,300
2120	40" high, 60 kW		9	2.222		4,350	114		4,464	4,950

23 82 Convection Heating and Cooling Units

23 82 16 – Air Coils

23 82 16.20 Duct Heaters		Crew	Daily Output	Labor-Hours	Unit	Material	2015 Bare Costs Labor	Equipment	Total	Total Incl O&P
2200	24" wide x 36" high, 118.8 kW	Q-20	9	2.222	Ea.	4,825	114		4,939	5,500
2220	40" high, 132 kW		8	2.500		4,825	128		4,953	5,500
2400	36" wide x 8" high, 40 kW		11	1.818		1,925	93		2,018	2,275
2420	16" high, 80 kW		10	2		2,575	102		2,677	2,975
2440	24" high, 120 kW		9	2.222		3,350	114		3,464	3,875
2460	32" high, 160 kW		8	2.500		4,300	128		4,428	4,925
2480	36" high, 180 kW		7	2.857		5,300	146		5,446	6,050
2500	40" high, 200 kW		6	3.333		5,875	171		6,046	6,725
2600	40" wide x 8" high, 45 kW		11	1.818		2,000	93		2,093	2,350
2620	16" high, 90 kW		10	2		2,775	102		2,877	3,200
2640	24" high, 135 kW		9	2.222		3,500	114		3,614	4,025
2660	32" high, 180 kW		8	2.500		4,675	128		4,803	5,350
2680	36" high, 202.5 kW		7	2.857		5,375	146		5,521	6,150
2700	40" high, 225 kW		6	3.333		6,125	171		6,296	7,025
2800	48" wide x 8" high, 54.8 kW		10	2		2,100	102		2,202	2,475
2820	16" high, 109.8 kW		9	2.222		2,925	114		3,039	3,375
2840	24" high, 164.4 kW		8	2.500		3,725	128		3,853	4,275
2860	32" high, 219.2 kW		7	2.857		4,950	146		5,096	5,675
2880	36" high, 246.6 kW		6	3.333		5,700	171		5,871	6,550
2900	40" high, 274 kW		5	4		6,450	205		6,655	7,400
3000	56" wide x 8" high, 64 kW		9	2.222		2,400	114		2,514	2,800
3020	16" high, 128 kW		8	2.500		3,325	128		3,453	3,850
3040	24" high, 192 kW		7	2.857		4,025	146		4,171	4,650
3060	32" high, 256 kW		6	3.333		5,625	171		5,796	6,450
3080	36" high, 288kW		5	4		6,450	205		6,655	7,425
3100	40" high, 320kW		4	5		7,125	256		7,381	8,250
3200	64" wide x 8" high, 74kW		8	2.500		2,475	128		2,603	2,900
3220	16" high, 148kW		7	2.857		3,425	146		3,571	3,975
3240	24" high, 222kW		6	3.333		4,350	171		4,521	5,050
3260	32" high, 296kW		5	4		5,825	205		6,030	6,725
3280	36" high, 333kW		4	5		6,950	256		7,206	8,050
3300	40" high, 370kW	↓	3	6.667	↓	7,675	340		8,015	8,950
3800	Note decreased kW available for									
3820	each duct size at same cost									
5000	Duct heater modifications and accessories									
5120	T.C.O. limit auto or manual reset	Q-20	42	.476	Ea.	126	24.50		150.50	177
5140	Thermostat		28	.714		535	36.50		571.50	650
5160	Overheat thermocouple (removable)		7	2.857		760	146		906	1,075
5180	Fan interlock relay		18	1.111		183	57		240	292
5200	Air flow switch		20	1		158	51		209	255
5220	Split terminal box cover	↓	100	.200	↓	52.50	10.25		62.75	73.50
8000	To obtain BTU multiply kW by 3413									
9900	Minimum labor/equipment charge	1 Shee	4	2	Job		112		112	177

23 82 19 – Fan Coil Units

23 82 19.10 Fan Coil Air Conditioning		Crew	Daily Output	Labor-Hours	Unit	Material	Labor	Equipment	Total	Total Incl O&P
0010	**FAN COIL AIR CONDITIONING**									
0030	Fan coil AC, cabinet mounted, filters and controls									
0100	Chilled water, 1/2 ton cooling	Q-5	8	2	Ea.	815	108		923	1,075
0110	3/4 ton cooling		7	2.286		850	123		973	1,125
0120	1 ton cooling		6	2.667		950	143		1,093	1,275
0140	1.5 ton cooling		5.50	2.909		1,050	156		1,206	1,400
0150	2 ton cooling	↓	5.25	3.048	↓	1,350	164		1,514	1,725

23 82 19 – Fan Coil Units

23 82 19.10 Fan Coil Air Conditioning		Crew	Daily Output	Labor-Hours	Unit	Material	Labor	Equipment	Total	Total Incl O&P
						2015 Bare Costs				
0160	2.5 ton cooling	Q-5	5	3.200	Ea.	1,975	172		2,147	2,450
0180	3 ton cooling	↓	4	4	↓	2,175	215		2,390	2,725
0262	For hot water coil, add					40%	10%			
0940	Direct expansion, for use w/air cooled condensing unit, 1.5 ton cooling	Q-5	5	3.200	Ea.	680	172		852	1,025
0950	2 ton cooling		4.80	3.333		730	179		909	1,075
0960	2.5 ton cooling		4.40	3.636		775	196		971	1,150
0970	3 ton cooling		3.80	4.211		960	226		1,186	1,400
0980	3.5 ton cooling		3.60	4.444		970	239		1,209	1,450
0990	4 ton cooling		3.40	4.706		1,100	253		1,353	1,600
1000	5 ton cooling		3	5.333		1,300	287		1,587	1,900
1020	7.5 ton cooling	↓	3	5.333		2,450	287		2,737	3,150
1040	10 ton cooling	Q-6	2.60	9.231		2,800	515		3,315	3,875
1042	12.5 ton cooling		2.30	10.435		3,550	580		4,130	4,800
1050	15 ton cooling		1.60	15		3,950	835		4,785	5,625
1060	20 ton cooling		.70	34.286		5,400	1,900		7,300	8,925
1070	25 ton cooling		.65	36.923		6,475	2,050		8,525	10,300
1080	30 ton cooling	↓	.60	40		7,050	2,225		9,275	11,200
1500	For hot water coil, add				↓	40%	10%			
1512	For condensing unit add see Section 23 62									
9000	Minimum labor/equipment charge	Q-5	3.75	4.267	Job		229		229	360

23 82 19.20 Heating and Ventilating Units		Crew	Daily Output	Labor-Hours	Unit	Material	Labor	Equipment	Total	Total Incl O&P
0010	**HEATING AND VENTILATING UNITS**, Classroom units									
0020	Includes filter, heating/cooling coils, standard controls									
0080	750 CFM, 2 tons cooling	Q-6	2	12	Ea.	4,050	670		4,720	5,525
0100	1000 CFM, 2-1/2 tons cooling		1.60	15		4,800	835		5,635	6,575
0120	1250 CFM, 3 tons cooling		1.40	17.143		4,950	955		5,905	6,950
0140	1500 CFM, 4 tons cooling		.80	30		5,300	1,675		6,975	8,425
0160	2000 CFM 5 tons cooling	↓	.50	48		6,225	2,675		8,900	11,000
0500	For electric heat, add					35%				
1000	For no cooling, deduct				↓	25%	10%			

23 82 19.40 Fan Coil Air Conditioning			Crew	Daily Output	Labor-Hours	Unit	Material	Labor	Equipment	Total	Total Incl O&P
0010	**FAN COIL AIR CONDITIONING**, Variable refrigerant volume (VRV) type										
0020	Not including interconnecting tubing or multi-zone controls										
0030	For VRV condensing unit see section 23 81 29.10										
0050	Indoor type, ducted										
0100	Vertical concealed										
0130	1 ton cooling	G	Q-5	2.97	5.387	Ea.	2,475	290		2,765	3,175
0140	1.5 ton cooling	G		2.77	5.776		2,525	310		2,835	3,250
0150	2 ton cooling	G		2.70	5.926		2,725	320		3,045	3,500
0160	2.5 ton cooling	G		2.60	6.154		2,875	330		3,205	3,675
0170	3 ton cooling	G		2.50	6.400		2,925	345		3,270	3,725
0180	3.5 ton cooling	G		2.30	6.957		3,075	375		3,450	3,950
0190	4 ton cooling	G		2.19	7.306		3,075	395		3,470	4,025
0200	4.5 ton cooling	G	↓	2.08	7.692	↓	3,400	415		3,815	4,375
0230	Outside air connection possible										
1100	Ceiling concealed										
1130	.6 ton cooling	G	Q-5	2.60	6.154	Ea.	1,625	330		1,955	2,300
1140	.75 ton cooling	G		2.60	6.154		1,675	330		2,005	2,375
1150	1 ton cooling	G	↓	2.60	6.154		1,775	330		2,105	2,475
1152	Screening door	G	1 Stpi	5.20	1.538		98	92		190	251
1160	1.5 ton cooling	G	Q-5	2.60	6.154		1,825	330		2,155	2,550
1162	Screening door	G	1 Stpi	5.20	1.538	↓	113	92		205	268

23 82 19.40 Fan Coil Air Conditioning		Crew	Daily Output	Labor-Hours	Unit	Material	Labor	Equipment	Total	Total Incl O&P
1170	2 ton cooling	G	Q-5	2.60	6.154	Ea.	2,150	330		2,480	2,875
1172	Screening door	G	1 Stpi	5.20	1.538		98	92		190	251
1180	2.5 ton cooling	G	Q-5	2.60	6.154		2,500	330		2,830	3,275
1182	Screening door	G	1 Stpi	5.20	1.538		170	92		262	330
1190	3 ton cooling	G	Q-5	2.60	6.154		2,650	330		2,980	3,450
1192	Screening door	G	1 Stpi	5.20	1.538		170	92		262	330
1200	4 ton cooling	G	Q-5	2.30	6.957		2,725	375		3,100	3,550
1202	Screening door	G	1 Stpi	5.20	1.538		170	92		262	330
1220	6 ton cooling	G	Q-5	2.08	7.692		4,625	415		5,040	5,725
1230	8 ton cooling	G	"	1.89	8.466	↓	5,200	455		5,655	6,425
1260	Outside air connection possible										
4050	Indoor type, duct-free										
4100	Ceiling mounted cassette										
4130	.75 ton cooling	G	Q-5	3.46	4.624	Ea.	1,875	249		2,124	2,475
4140	1 ton cooling	G		2.97	5.387		2,000	290		2,290	2,650
4150	1.5 ton cooling	G		2.77	5.776		2,050	310		2,360	2,725
4160	2 ton cooling	G		2.60	6.154		2,200	330		2,530	2,950
4170	2.5 ton cooling	G		2.50	6.400		2,250	345		2,595	3,000
4180	3 ton cooling	G		2.30	6.957		2,425	375		2,800	3,225
4190	4 ton cooling	G	↓	2.08	7.692		2,700	415		3,115	3,625
4198	For ceiling cassette decoration panel, add	G	1 Stpi	5.20	1.538	↓	294	92		386	470
4230	Outside air connection possible										
4500	Wall mounted										
4530	.6 ton cooling	G	Q-5	5.20	3.077	Ea.	995	165		1,160	1,350
4540	.75 ton cooling	G		5.20	3.077		1,050	165		1,215	1,400
4550	1 ton cooling	G		4.16	3.846		1,200	207		1,407	1,625
4560	1.5 ton cooling	G		3.77	4.244		1,300	228		1,528	1,800
4570	2 ton cooling	G	↓	3.46	4.624		1,400	249		1,649	1,950
4590	Condensate pump for the above air handlers	G	1 Stpi	6.46	1.238	↓	211	74		285	345
4700	Floor standing unit										
4730	1 ton cooling	G	Q-5	2.60	6.154	Ea.	1,675	330		2,005	2,375
4740	1.5 ton cooling	G		2.60	6.154		1,900	330		2,230	2,625
4750	2 ton cooling	G	↓	2.60	6.154	↓	2,050	330		2,380	2,775
4800	Floor standing unit, concealed										
4830	1 ton cooling	G	Q-5	2.60	6.154	Ea.	1,600	330		1,930	2,300
4840	1.5 ton cooling	G		2.60	6.154		1,825	330		2,155	2,525
4850	2 ton cooling	G	↓	2.60	6.154	↓	1,950	330		2,280	2,675
4880	Outside air connection possible										
8000	Accessories										
8100	Branch divergence pipe fitting										
8110	Capacity under 76 MBH		1 Stpi	2.50	3.200	Ea.	151	191		342	465
8120	76 MBH to 112 MBH			2.50	3.200		178	191		369	495
8130	112 MBH to 234 MBH		↓	2.50	3.200	↓	505	191		696	855
8200	Header pipe fitting										
8210	Max 4 branches										
8220	Capacity under 76 MBH		1 Stpi	1.70	4.706	Ea.	221	281		502	685
8260	Max 8 branches										
8270	76 MBH to 112 MBH		1 Stpi	1.70	4.706	Ea.	450	281		731	935
8280	112 MBH to 234 MBH		"	1.30	6.154	"	720	370		1,090	1,375

23 82 Convection Heating and Cooling Units

23 82 27 – Infrared Units

23 82 27.10 Infrared Type Heating Units		Crew	Daily Output	Labor-Hours	Unit	Material	2015 Bare Costs Labor	Equipment	Total	Total Incl O&P
0010	**INFRARED TYPE HEATING UNITS**									
0020	Gas fired, unvented, electric ignition, 100% shutoff.									
0030	Piping and wiring not included									
0100	Input, 30 MBH	Q-5	6	2.667	Ea.	945	143		1,088	1,250
0120	45 MBH		5	3.200		950	172		1,122	1,325
0140	50 MBH		4.50	3.556		950	191		1,141	1,350
0160	60 MBH		4	4		950	215		1,165	1,375
0180	75 MBH		3	5.333		950	287		1,237	1,500
0200	90 MBH		2.50	6.400		950	345		1,295	1,575
0220	105 MBH		2	8		950	430		1,380	1,725
0240	120 MBH		2	8		1,575	430		2,005	2,400
1000	Gas fired, vented, electric ignition, tubular									
1020	Piping and wiring not included, 20' to 80' lengths									
1030	Single stage, input, 60 MBH	Q-6	4.50	5.333	Ea.	1,425	297		1,722	2,050
1040	80 MBH		3.90	6.154		1,425	345		1,770	2,100
1050	100 MBH		3.40	7.059		1,425	395		1,820	2,200
1060	125 MBH		2.90	8.276		1,425	460		1,885	2,300
1070	150 MBH		2.70	8.889		1,425	495		1,920	2,350
1080	170 MBH		2.50	9.600		1,425	535		1,960	2,400
1090	200 MBH		2.20	10.909		1,625	610		2,235	2,750
1100	Note: Final pricing may vary due to									
1110	tube length and configuration package selected									
1130	Two stage, input, 60 MBH high, 45 MBH low	Q-6	4.50	5.333	Ea.	1,750	297		2,047	2,400
1140	80 MBH high, 60 MBH low		3.90	6.154		1,750	345		2,095	2,450
1150	100 MBH high, 65 MBH low		3.40	7.059		1,750	395		2,145	2,550
1160	125 MBH high, 95 MBH low		2.90	8.276		1,750	460		2,210	2,650
1170	150 MBH high, 100 MBH low		2.70	8.889		1,750	495		2,245	2,700
1180	170 MBH high, 125 MBH low		2.50	9.600		1,950	535		2,485	2,975
1190	200 MBH high, 150 MBH low		2.20	10.909		1,950	610		2,560	3,100
1220	Note: Final pricing may vary due to									
1230	tube length and configuration package selected									
2000	Electric, single or three phase									
2050	6 kW, 20,478 BTU	1 Elec	2.30	3.478	Ea.	705	190		895	1,075
2100	13.5 KW, 40,956 BTU		2.20	3.636		870	199		1,069	1,275
2150	24 KW, 81,912 BTU		2	4		1,750	219		1,969	2,275
3000	Oil fired, pump, controls, fusible valve, oil supply tank									
3050	91,000 BTU	Q-5	2.50	6.400	Ea.	6,425	345		6,770	7,600
3080	105,000 BTU		2.25	7.111		6,800	380		7,180	8,100
3110	119,000 BTU		2	8		7,450	430		7,880	8,875

23 82 29 – Radiators

23 82 29.10 Hydronic Heating		Crew	Daily Output	Labor-Hours	Unit	Material	Labor	Equipment	Total	Total Incl O&P
0010	**HYDRONIC HEATING**, Terminal units, not incl. main supply pipe									
1000	Radiation									
1100	Panel, baseboard, C.I., including supports, no covers	Q-5	46	.348	L.F.	38	18.70		56.70	70.50
3000	Radiators, cast iron									
3100	Free standing or wall hung, 6 tube, 25" high	Q-5	96	.167	Section	44	8.95		52.95	62.50
3150	4 tube 25" high		96	.167		34	8.95		42.95	51.50
3200	4 tube, 19" high		96	.167		31.50	8.95		40.45	48.50
3250	Adj. brackets, 2 per wall radiator up to 30 sections	1 Stpi	32	.250	Ea.	53	14.95		67.95	82
3500	Recessed, 20" high x 5" deep, without grille	Q-5	60	.267	Section	37	14.35		51.35	63
3600	For inlet grille, add				"	2.58			2.58	2.84
9000	Minimum labor/equipment charge	Q-5	3.20	5	Job		269		269	420

23 82 Convection Heating and Cooling Units

23 82 29 – Radiators

23 82 29.10 Hydronic Heating		Crew	Daily Output	Labor-Hours	Unit	Material	Labor	Equipment	Total	Total Incl O&P
						2015 Bare Costs				
9500	To convert SFR to BTU rating: Hot water, 150 x SFR									
9510	Forced hot water, 180 x SFR; steam, 240 x SFR									

23 82 33 – Convectors

23 82 33.10 Convector Units		Crew	Daily Output	Labor-Hours	Unit	Material	Labor	Equipment	Total	Total Incl O&P
0010	**CONVECTOR UNITS**, Terminal units, not incl. main supply pipe									
2204	Convector, multifin, 2 pipe w/cabinet									
2210	17" H x 24" L	Q-5	10	1.600	Ea.	99	86		185	243
2214	17" H x 36" L		8.60	1.860		148	100		248	320
2218	17" H x 48" L		7.40	2.162		198	116		314	400
2222	21" H x 24" L		9	1.778		101	95.50		196.50	260
2226	21" H x 36" L		8.20	1.951		151	105		256	330
2228	21" H x 48" L	↓	6.80	2.353	↓	201	127		328	420
2240	For knob operated damper, add					140%				
2241	For metal trim strips, add	Q-5	64	.250	Ea.	12.90	13.45		26.35	35
2243	For snap-on inlet grille, add					10%	10%			
2245	For hinged access door, add	Q-5	64	.250	Ea.	35.50	13.45		48.95	60
2246	For air chamber, auto-venting, add	"	58	.276	"	7.75	14.85		22.60	31.50

23 82 36 – Finned-Tube Radiation Heaters

23 82 36.10 Finned Tube Radiation		Crew	Daily Output	Labor-Hours	Unit	Material	Labor	Equipment	Total	Total Incl O&P
0010	**FINNED TUBE RADIATION**, Terminal units, not incl. main supply pipe									
1150	Fin tube, wall hung, 14" slope top cover, with damper									
1200	1-1/4" copper tube, 4-1/4" alum. fin	Q-5	38	.421	L.F.	43	22.50		65.50	82.50
1250	1-1/4" steel tube, 4-1/4" steel fin		36	.444		39	24		63	80.50
1255	2" steel tube, 4-1/4" steel fin		32	.500		43	27		70	89
1310	Baseboard, pkgd, 1/2" copper tube, alum. fin, 7" high		60	.267		11.20	14.35		25.55	35
1320	3/4" copper tube, alum. fin, 7" high		58	.276		7.35	14.85		22.20	31
1340	1" copper tube, alum. fin, 8-7/8" high		56	.286		18.85	15.35		34.20	44.50
1360	1-1/4" copper tube, alum. fin, 8-7/8" high		54	.296		28	15.95		43.95	55.50
1380	1-1/4" IPS steel tube with steel fins	↓	52	.308	↓	27.50	16.55		44.05	56.50
1500	Note: fin tube may also require corners, caps, etc.									

23 82 39 – Unit Heaters

23 82 39.16 Propeller Unit Heaters		Crew	Daily Output	Labor-Hours	Unit	Material	Labor	Equipment	Total	Total Incl O&P
0010	**PROPELLER UNIT HEATERS**									
3950	Unit heaters, propeller, 115 V 2 psi steam, 60°F entering air									
4000	Horizontal, 12 MBH	Q-5	12	1.333	Ea.	360	71.50		431.50	505
4020	28.2 MBH		10	1.600		470	86		556	650
4040	36.5 MBH		8	2		545	108		653	770
4060	43.9 MBH		8	2		545	108		653	770
4080	56.5 MBH		7.50	2.133		585	115		700	820
4100	65.6 MBH		7	2.286		600	123		723	850
4120	87.6 MBH		6.50	2.462		650	132		782	920
4140	96.8 MBH		6	2.667		795	143		938	1,100
4160	133.3 MBH		5	3.200		865	172		1,037	1,225
4180	157.6 MBH		4	4		1,050	215		1,265	1,500
4200	197.7 MBH		3	5.333		1,200	287		1,487	1,775
4220	257.2 MBH		2.50	6.400		1,475	345		1,820	2,150
4240	286.9 MBH		2	8		1,650	430		2,080	2,500
4260	364 MBH		1.80	8.889		2,075	480		2,555	3,025
4270	404 MBH	↓	1.60	10	↓	2,150	540		2,690	3,225
4300	Vertical diffuser same price									
4310	Vertical flow, 40 MBH	Q-5	11	1.455	Ea.	540	78		618	715

23 82 Convection Heating and Cooling Units

23 82 39 – Unit Heaters

23 82 39.16 Propeller Unit Heaters		Crew	Daily Output	Labor-Hours	Unit	Material	Labor	Equipment	Total	Total Incl O&P
						2015 Bare Costs				
4314	58.5 MBH	Q-5	8	2	Ea.	560	108		668	785
4318	92 MBH		7	2.286		715	123		838	975
4322	109.7 MBH		6	2.667		845	143		988	1,150
4326	131.0 MBH		4	4		845	215		1,060	1,275
4330	160.0 MBH		3	5.333		895	287		1,182	1,425
4334	194.0 MBH		2.20	7.273		1,025	390		1,415	1,750
4338	212.0 MBH		2.10	7.619		1,275	410		1,685	2,050
4342	247.0 MBH		1.96	8.163		1,525	440		1,965	2,350
4346	297.0 MBH	↓	1.80	8.889		1,600	480		2,080	2,525
4350	333.0 MBH	Q-6	1.90	12.632		1,750	705		2,455	3,025
4354	420 MBH,(460 V)		1.80	13.333		2,150	745		2,895	3,525
4358	500 MBH, (460 V)		1.71	14.035		2,875	785		3,660	4,375
4362	570 MBH, (460 V)		1.40	17.143		3,925	955		4,880	5,825
4366	620 MBH, (460 V)		1.30	18.462		3,925	1,025		4,950	5,925
4370	960 MBH, (460 V)	↓	1.10	21.818	↓	7,450	1,225		8,675	10,100

23 83 Radiant Heating Units

23 83 16 – Radiant-Heating Hydronic Piping

23 83 16.10 Radiant Floor Heating		Crew	Daily Output	Labor-Hours	Unit	Material	Labor	Equipment	Total	Total Incl O&P
0010	**RADIANT FLOOR HEATING**									
0100	Tubing, PEX (cross-linked polyethylene)									
0110	Oxygen barrier type for systems with ferrous materials									
0120	1/2"	Q-5	800	.020	L.F.	.96	1.08		2.04	2.74
0130	3/4"		535	.030		1.35	1.61		2.96	4
0140	1"	↓	400	.040	↓	2.11	2.15		4.26	5.70
0200	Non barrier type for ferrous free systems									
0210	1/2"	Q-5	800	.020	L.F.	.61	1.08		1.69	2.35
0220	3/4"		535	.030		1.11	1.61		2.72	3.73
0230	1"	↓	400	.040	↓	1.90	2.15		4.05	5.45
1000	Manifolds									
1110	Brass									
1120	With supply and return valves, flow meter, thermometer,									
1122	auto air vent and drain/fill valve.									
1130	1", 2 circuit	Q-5	14	1.143	Ea.	257	61.50		318.50	380
1140	1", 3 circuit		13.50	1.185		294	63.50		357.50	425
1150	1", 4 circuit		13	1.231		320	66		386	455
1154	1", 5 circuit		12.50	1.280		365	69		434	510
1158	1", 6 circuit		12	1.333		415	71.50		486.50	565
1162	1", 7 circuit		11.50	1.391		455	75		530	615
1166	1", 8 circuit		11	1.455		500	78		578	670
1172	1", 9 circuit		10.50	1.524		540	82		622	725
1174	1", 10 circuit		10	1.600		580	86		666	775
1178	1", 11 circuit		9.50	1.684		605	90.50		695.50	805
1182	1", 12 circuit	↓	9	1.778	↓	670	95.50		765.50	885
1610	Copper manifold header, (cut to size)									
1620	1" header, 12 – 1/2" sweat outlets	Q-5	3.33	4.805	Ea.	91	258		349	505
1630	1-1/4" header, 12 – 1/2" sweat outlets		3.20	5		106	269		375	535
1640	1-1/4" header, 12 – 3/4" sweat outlets		3	5.333		114	287		401	570
1650	1-1/2" header, 12 – 3/4" sweat outlets		3.10	5.161		137	278		415	585
1660	2" header, 12 – 3/4" sweat outlets	↓	2.90	5.517	↓	201	297		498	685
3000	Valves									

23 83 Radiant Heating Units

23 83 16 – Radiant-Heating Hydronic Piping

23 83 16.10 Radiant Floor Heating		Crew	Daily Output	Labor-Hours	Unit	2015 Bare Costs Material	Labor	Equipment	Total	Total Incl O&P
3110	Thermostatic zone valve actuator with end switch	Q-5	40	.400	Ea.	38.50	21.50		60	76
3114	Thermostatic zone valve actuator	"	36	.444	"	81	24		105	127
3120	Motorized straight zone valve with operator complete									
3130	3/4"	Q-5	35	.457	Ea.	130	24.50		154.50	182
3140	1"		32	.500		141	27		168	197
3150	1-1/4"		29.60	.541		179	29		208	243
3500	4 Way mixing valve, manual, brass									
3530	1"	Q-5	13.30	1.203	Ea.	180	64.50		244.50	299
3540	1-1/4"		11.40	1.404		195	75.50		270.50	330
3550	1-1/2"		11	1.455		249	78		327	395
3560	2"		10.60	1.509		350	81		431	515
3800	Mixing valve motor, 4 way for valves, 1" and 1-1/4"		34	.471		310	25.50		335.50	380
3810	Mixing valve motor, 4 way for valves, 1-1/2" and 2"		30	.533		355	28.50		383.50	435
5000	Radiant floor heating, zone control panel									
5120	4 Zone actuator valve control, expandable	Q-5	20	.800	Ea.	163	43		206	246
5130	6 Zone actuator valve control, expandable		18	.889		226	48		274	325
6070	Thermal track, straight panel for long continuous runs, 5.333 S.F.		40	.400		26.50	21.50		48	63
6080	Thermal track, utility panel, for direction reverse at run end, 5.333 S.F.		40	.400		26.50	21.50		48	63
6090	Combination panel, for direction reverse plus straight run, 5.333 S.F.		40	.400		26.50	21.50		48	63
7000	PEX tubing fittings									
7100	Compression type									
7116	Coupling									
7120	1/2" x 1/2"	1 Stpi	27	.296	Ea.	6.70	17.70		24.40	35
7124	3/4" x 3/4"	"	23	.348	"	10.60	21		31.60	44
7130	Adapter									
7132	1/2" x female sweat 1/2"	1 Stpi	27	.296	Ea.	4.36	17.70		22.06	32.50
7134	1/2" x female sweat 3/4"		26	.308		4.88	18.40		23.28	34
7136	5/8" x female sweat 3/4"		24	.333		7	19.90		26.90	38.50
7140	Elbow									
7142	1/2" x female sweat 1/2"	1 Stpi	27	.296	Ea.	6.65	17.70		24.35	35
7144	1/2" x female sweat 3/4"		26	.308		7.80	18.40		26.20	37
7146	5/8" x female sweat 3/4"		24	.333		8.75	19.90		28.65	40.50
7200	Insert type									
7206	PEX x male NPT									
7210	1/2" x 1/2"	1 Stpi	29	.276	Ea.	2.59	16.50		19.09	28.50
7220	3/4" x 3/4"		27	.296		3.81	17.70		21.51	31.50
7230	1" x 1"		26	.308		6.45	18.40		24.85	35.50
7300	PEX coupling									
7310	1/2" x 1/2"	1 Stpi	30	.267	Ea.	1.76	15.95		17.71	27
7320	3/4" x 3/4"		29	.276		2.15	16.50		18.65	28
7330	1" x 1"		28	.286		5.85	17.05		22.90	33
7400	PEX stainless crimp ring									
7410	1/2" x 1/2"	1 Stpi	86	.093	Ea.	.37	5.55		5.92	9.05
7420	3/4" x 3/4"		84	.095		.51	5.70		6.21	9.45
7430	1" x 1"		82	.098		.73	5.85		6.58	9.90

23 83 33 – Electric Radiant Heaters

23 83 33.10 Electric Heating		Crew	Daily Output	Labor-Hours	Unit	Material	Labor	Equipment	Total	Total Incl O&P
0010	**ELECTRIC HEATING**, not incl. conduit or feed wiring									
1100	Rule of thumb: Baseboard units, including control	1 Elec	4.40	1.818	kW	104	99.50		203.50	268
1300	Baseboard heaters, 2' long, 350 watt		8	1	Ea.	28.50	54.50		83	116
1400	3' long, 750 watt		8	1		33	54.50		87.50	121
1600	4' long, 1000 watt		6.70	1.194		39	65.50		104.50	144

23 83 33 – Electric Radiant Heaters

23 83 33.10 Electric Heating		Crew	Daily Output	Labor-Hours	Unit	2015 Bare Costs Material	Labor	Equipment	Total	Total Incl O&P
1800	5′ long, 935 watt	1 Elec	5.70	1.404	Ea.	48	77		125	172
2000	6′ long, 1500 watt		5	1.600		54.50	87.50		142	196
2200	7′ long, 1310 watt		4.40	1.818		63	99.50		162.50	224
2400	8′ long, 2000 watt		4	2		66	109		175	242
2600	9′ long, 1680 watt		3.60	2.222		102	122		224	300
2800	10′ long, 1875 watt		3.30	2.424		190	133		323	415
2950	Wall heaters with fan, 120 to 277 volt									
3160	Recessed, residential, 750 watt	1 Elec	6	1.333	Ea.	134	73		207	260
3170	1000 watt		6	1.333		134	73		207	260
3180	1250 watt		5	1.600		134	87.50		221.50	283
3190	1500 watt		4	2		134	109		243	315
3210	2000 watt		4	2		134	109		243	315
3230	2500 watt		3.50	2.286		335	125		460	565
3240	3000 watt		3	2.667		495	146		641	765
3250	4000 watt		2.70	2.963		495	162		657	795
3260	Commercial, 750 watt		6	1.333		192	73		265	325
3270	1000 watt		6	1.333		192	73		265	325
3280	1250 watt		5	1.600		192	87.50		279.50	350
3290	1500 watt		4	2		192	109		301	380
3300	2000 watt		4	2		192	109		301	380
3310	2500 watt		3.50	2.286		192	125		317	405
3320	3000 watt		3	2.667		345	146		491	605
3330	4000 watt		2.70	2.963		345	162		507	630
3600	Thermostats, integral		16	.500		30	27.50		57.50	75.50
3800	Line voltage, 1 pole		8	1		19.20	54.50		73.70	106
3810	2 pole		8	1		29.50	54.50		84	117
4000	Heat trace system, 400 degree									
4020	115 V, 2.5 watts per L.F.	1 Elec	530	.015	L.F.	7.90	.83		8.73	9.95
4030	5 watts per L.F.		530	.015		7.90	.83		8.73	9.95
4050	10 watts per L.F.		530	.015		7.90	.83		8.73	9.95
4060	208 V, 5 watts per L.F.		530	.015		7.90	.83		8.73	9.95
4080	480 V, 8 watts per L.F.		530	.015		7.90	.83		8.73	9.95
4200	Heater raceway									
4260	Heat transfer cement									
4280	1 gallon				Ea.	60			60	66
4300	5 gallon				"	234			234	257
4320	Snap band, clamp									
4340	3/4" pipe size	1 Elec	470	.017	Ea.		.93		.93	1.44
4360	1" pipe size		444	.018			.99		.99	1.53
4380	1-1/4" pipe size		400	.020			1.09		1.09	1.69
4400	1-1/2" pipe size		355	.023			1.23		1.23	1.91
4420	2" pipe size		320	.025			1.37		1.37	2.12
4440	3" pipe size		160	.050			2.74		2.74	4.24
4460	4" pipe size		100	.080			4.38		4.38	6.80
4480	Thermostat NEMA 3R, 22 amp, 0-150 Deg, 10′ cap.		8	1		241	54.50		295.50	350
4500	Thermostat NEMA 4X, 25 amp, 40 Deg, 5-1/2′ cap.		7	1.143		241	62.50		303.50	360
4520	Thermostat NEMA 4X, 22 amp, 25-325 Deg, 10′ cap.		7	1.143		650	62.50		712.50	810
4540	Thermostat NEMA 4X, 22 amp, 15-140 Deg,		6	1.333		545	73		618	715
4580	Thermostat NEMA 4,7,9, 22 amp, 25-325 Deg, 10′ cap.		3.60	2.222		800	122		922	1,075
4600	Thermostat NEMA 4,7,9, 22 amp, 15-140 Deg,		3	2.667		795	146		941	1,100
4720	Fiberglass application tape, 36 yard roll		11	.727		77.50	40		117.50	147
5000	Radiant heating ceiling panels, 2′ x 4′, 500 watt		16	.500		305	27.50		332.50	380
5050	750 watt		16	.500		310	27.50		337.50	390

23 83 33 – Electric Radiant Heaters

23 83 33.10 Electric Heating		Crew	Daily Output	Labor-Hours	Unit	2015 Bare Costs Material	Labor	Equipment	Total	Total Incl O&P
5200	For recessed plaster frame, add	1 Elec	32	.250	Ea.	111	13.70		124.70	143
5300	Infrared quartz heaters, 120 volts, 1000 watts		6.70	1.194		284	65.50		349.50	410
5350	1500 watt		5	1.600		284	87.50		371.50	445
5400	240 volts, 1500 watt		5	1.600		284	87.50		371.50	445
5450	2000 watt		4	2		284	109		393	480
5500	3000 watt		3	2.667		325	146		471	580
5550	4000 watt		2.60	3.077		325	168		493	615
5570	Modulating control		.80	10		146	545		691	1,000
5600	Unit heaters, heavy duty, with fan & mounting bracket									
5650	Single phase, 208-240-277 volt, 3 kW	1 Elec	6	1.333	Ea.	445	73		518	605
5750	5 kW		5.50	1.455		460	79.50		539.50	630
5800	7 kW		5	1.600		710	87.50		797.50	915
5850	10 kW		4	2		810	109		919	1,050
5950	15 kW		3.80	2.105		1,300	115		1,415	1,600
6000	480 volt, 3 kW		6	1.333		425	73		498	580
6020	4 kW		5.80	1.379		445	75.50		520.50	605
6040	5 kW		5.50	1.455		460	79.50		539.50	630
6060	7 kW		5	1.600		710	87.50		797.50	915
6080	10 kW		4	2		780	109		889	1,025
6100	13 kW		3.80	2.105		1,300	115		1,415	1,600
6120	15 kW		3.70	2.162		1,300	118		1,418	1,600
6140	20 kW		3.50	2.286		1,700	125		1,825	2,050
6300	3 phase, 208-240 volt, 5 kW		5.50	1.455		435	79.50		514.50	605
6320	7 kW		5	1.600		670	87.50		757.50	870
6340	10 kW		4	2		720	109		829	965
6360	15 kW		3.70	2.162		1,200	118		1,318	1,500
6380	20 kW		3.50	2.286		1,725	125		1,850	2,100
6400	25 kW		3.30	2.424		2,025	133		2,158	2,425
6500	480 volt, 5 kW		5.50	1.455		605	79.50		684.50	795
6520	7 kW		5	1.600		765	87.50		852.50	975
6540	10 kW		4	2		810	109		919	1,050
6560	13 kW		3.80	2.105		1,300	115		1,415	1,600
6580	15 kW		3.70	2.162		1,700	118		1,818	2,025
6600	20 kW		3.50	2.286		1,700	125		1,825	2,050
6620	25 kW		3.30	2.424		2,025	133		2,158	2,425
6630	30 kW		3	2.667		2,350	146		2,496	2,825
6640	40 kW		2	4		3,000	219		3,219	3,650
6650	50 kW		1.60	5		3,625	274		3,899	4,400
6800	Vertical discharge heaters, with fan									
6820	Single phase, 208-240-277 volt, 10 kW	1 Elec	4	2	Ea.	750	109		859	995
6840	15 kW		3.70	2.162		1,250	118		1,368	1,550
6900	3 phase, 208-240 volt, 10 kW		4	2		720	109		829	965
6920	15 kW		3.70	2.162		1,200	118		1,318	1,500
6940	20 kW		3.50	2.286		1,725	125		1,850	2,100
6960	25 kW		3.30	2.424		2,025	133		2,158	2,425
6980	30 kW		3	2.667		2,350	146		2,496	2,825
7000	40 kW		2	4		3,000	219		3,219	3,650
7020	50 kW		1.60	5		3,625	274		3,899	4,400
7100	480 volt, 10 kW		4	2		810	109		919	1,050
7120	15 kW		3.70	2.162		1,300	118		1,418	1,600
7140	20 kW		3.50	2.286		1,700	125		1,825	2,050
7160	25 kW		3.30	2.424		2,025	133		2,158	2,425
7180	30 kW		3	2.667		2,350	146		2,496	2,825

23 83 33 – Electric Radiant Heaters

23 83 33.10 Electric Heating		Crew	Daily Output	Labor-Hours	Unit	Material	2015 Bare Costs Labor	Equipment	Total	Total Incl O&P
7200	40 kW	1 Elec	2	4	Ea.	3,000	219		3,219	3,650
7220	50 kW		1.60	5		3,625	274		3,899	4,400
7410	Sill height convector heaters, 5" high x 2' long, 500 watt		6.70	1.194		287	65.50		352.50	415
7420	3' long, 750 watt		6.50	1.231		340	67.50		407.50	475
7430	4' long, 1000 watt		6.20	1.290		390	70.50		460.50	540
7440	5' long, 1250 watt		5.50	1.455		445	79.50		524.50	615
7450	6' long, 1500 watt		4.80	1.667		505	91		596	695
7460	8' long, 2000 watt		3.60	2.222		685	122		807	945
7470	10' long, 2500 watt		3	2.667		850	146		996	1,150
7900	Cabinet convector heaters, 240 volt, three phase,									
7920	3' long, 2000 watt	1 Elec	5.30	1.509	Ea.	1,900	82.50		1,982.50	2,225
7940	3000 watt		5.30	1.509		2,000	82.50		2,082.50	2,300
7960	4000 watt		5.30	1.509		2,050	82.50		2,132.50	2,375
7980	6000 watt		4.60	1.739		2,125	95		2,220	2,475
8000	8000 watt		4.60	1.739		2,200	95		2,295	2,575
8020	4' long, 4000 watt		4.60	1.739		2,075	95		2,170	2,425
8040	6000 watt		4	2		2,150	109		2,259	2,550
8060	8000 watt		4	2		2,225	109		2,334	2,625
8080	10,000 watt		4	2		2,250	109		2,359	2,650
8100	Available also in 208 or 277 volt									
8200	Cabinet unit heaters, 120 to 277 volt, 1 pole,									
8220	wall mounted, 2 kW	1 Elec	4.60	1.739	Ea.	1,875	95		1,970	2,200
8230	3 kW		4.60	1.739		1,950	95		2,045	2,300
8240	4 kW		4.40	1.818		2,000	99.50		2,099.50	2,350
8250	5 kW		4.40	1.818		2,075	99.50		2,174.50	2,450
8260	6 kW		4.20	1.905		2,125	104		2,229	2,475
8270	8 kW		4	2		2,150	109		2,259	2,550
8280	10 kW		3.80	2.105		2,225	115		2,340	2,600
8290	12 kW		3.50	2.286		2,250	125		2,375	2,675
8300	13.5 kW		2.90	2.759		2,700	151		2,851	3,200
8310	16 kW		2.70	2.963		2,725	162		2,887	3,250
8320	20 kW		2.30	3.478		3,475	190		3,665	4,125
8330	24 kW		1.90	4.211		3,525	230		3,755	4,225
8350	Recessed, 2 kW		4.40	1.818		1,875	99.50		1,974.50	2,200
8370	3 kW		4.40	1.818		1,950	99.50		2,049.50	2,300
8380	4 kW		4.20	1.905		2,000	104		2,104	2,350
8390	5 kW		4.20	1.905		2,075	104		2,179	2,425
8400	6 kW		4	2		2,100	109		2,209	2,500
8410	8 kW		3.80	2.105		2,200	115		2,315	2,600
8420	10 kW		3.50	2.286		2,675	125		2,800	3,150
8430	12 kW		2.90	2.759		2,675	151		2,826	3,175
8440	13.5 kW		2.70	2.963		2,700	162		2,862	3,225
8450	16 kW		2.30	3.478		2,875	190		3,065	3,475
8460	20 kW		1.90	4.211		3,275	230		3,505	3,950
8470	24 kW		1.60	5		3,325	274		3,599	4,075
8490	Ceiling mounted, 2 kW		3.20	2.500		1,875	137		2,012	2,250
8510	3 kW		3.20	2.500		1,950	137		2,087	2,350
8520	4 kW		3	2.667		2,000	146		2,146	2,425
8530	5 kW		3	2.667		2,075	146		2,221	2,500
8540	6 kW		2.80	2.857		2,075	156		2,231	2,550
8550	8 kW		2.40	3.333		2,150	182		2,332	2,650
8560	10 kW		2.20	3.636		2,200	199		2,399	2,725
8570	12 kW		2	4		2,250	219		2,469	2,825

23 83 33 – Electric Radiant Heaters

23 83 33.10 Electric Heating		Crew	Daily Output	Labor-Hours	Unit	Material	2015 Bare Costs Labor	Equipment	Total	Total Incl O&P
8580	13.5 kW	1 Elec	1.50	5.333	Ea.	2,275	292		2,567	2,975
8590	16 kW		1.30	6.154		2,350	335		2,685	3,100
8600	20 kW		.90	8.889		3,275	485		3,760	4,350
8610	24 kW		.60	13.333		3,325	730		4,055	4,775
8630	208 to 480 V, 3 pole									
8650	Wall mounted, 2 kW	1 Elec	4.60	1.739	Ea.	1,975	95		2,070	2,325
8670	3 kW		4.60	1.739		2,075	95		2,170	2,425
8680	4 kW		4.40	1.818		2,125	99.50		2,224.50	2,475
8690	5 kW		4.40	1.818		2,175	99.50		2,274.50	2,550
8700	6 kW		4.20	1.905		2,200	104		2,304	2,575
8710	8 kW		4	2		2,275	109		2,384	2,675
8720	10 kW		3.80	2.105		2,325	115		2,440	2,725
8730	12 kW		3.50	2.286		2,375	125		2,500	2,800
8740	13.5 kW		2.90	2.759		2,400	151		2,551	2,875
8750	16 kW		2.70	2.963		2,450	162		2,612	2,950
8760	20 kW		2.30	3.478		3,400	190		3,590	4,025
8770	24 kW		1.90	4.211		3,600	230		3,830	4,300
8790	Recessed, 2 kW		4.40	1.818		1,975	99.50		2,074.50	2,325
8810	3 kW		4.40	1.818		2,000	99.50		2,099.50	2,350
8820	4 kW		4.20	1.905		2,075	104		2,179	2,450
8830	5 kW		4.20	1.905		2,100	104		2,204	2,475
8840	6 kW		4	2		2,125	109		2,234	2,500
8850	8 kW		3.80	2.105		2,200	115		2,315	2,600
8860	10 kW		3.50	2.286		2,350	125		2,475	2,775
8870	12 kW		2.90	2.759		2,450	151		2,601	2,925
8880	13.5 kW		2.70	2.963		2,500	162		2,662	3,000
8890	16 kW		2.30	3.478		3,625	190		3,815	4,275
8900	20 kW		1.90	4.211		3,675	230		3,905	4,400
8920	24 kW		1.60	5		3,725	274		3,999	4,525
8940	Ceiling mount, 2 kW		3.20	2.500		2,075	137		2,212	2,475
8950	3 kW		3.20	2.500		2,200	137		2,337	2,625
8960	4 kW		3	2.667		2,200	146		2,346	2,650
8970	5 kW		3	2.667		2,275	146		2,421	2,725
8980	6 kW		2.80	2.857		2,300	156		2,456	2,775
8990	8 kW		2.40	3.333		2,375	182		2,557	2,875
9000	10 kW		2.20	3.636		2,425	199		2,624	2,950
9020	13.5 kW		1.50	5.333		2,500	292		2,792	3,200
9030	16 kW		1.30	6.154		3,625	335		3,960	4,500
9040	20 kW		.90	8.889		3,675	485		4,160	4,800
9060	24 kW		.60	13.333		4,075	730		4,805	5,625
9990	Minimum labor/equipment charge		4	2	Job		109		109	169

23 84 Humidity Control Equipment

23 84 13 – Humidifiers

23 84 13.10 Humidifier Units		Crew	Daily Output	Labor-Hours	Unit	2015 Bare Costs Material	Labor	Equipment	Total	Total Incl O&P
0010	**HUMIDIFIER UNITS**									
0520	Steam, room or duct, filter, regulators, auto. controls, 220 V									
0540	11 lb. per hour R236000-30	Q-5	6	2.667	Ea.	2,700	143		2,843	3,200
0560	22 lb. per hour		5	3.200		2,975	172		3,147	3,550
0580	33 lb. per hour		4	4		3,050	215		3,265	3,700
0600	50 lb. per hour		4	4		3,550	215		3,765	4,225
0620	100 lb. per hour		3	5.333		4,475	287		4,762	5,375
0640	150 lb. per hour		2.50	6.400		6,400	345		6,745	7,575
0660	200 lb. per hour		2	8		8,025	430		8,455	9,500
0700	With blower									
0720	11 lb. per hour	Q-5	5.50	2.909	Ea.	4,300	156		4,456	4,975
0740	22 lb. per hour		4.75	3.368		4,525	181		4,706	5,250
0760	33 lb. per hour		3.75	4.267		4,650	229		4,879	5,475
0780	50 lb. per hour		3.50	4.571		5,325	246		5,571	6,225
0800	100 lb. per hour		2.75	5.818		5,950	315		6,265	7,050
0820	150 lb. per hour		2	8		8,675	430		9,105	10,200
0840	200 lb. per hour		1.50	10.667		9,700	575		10,275	11,600
5000	Furnace type, wheel bypass									
5020	10 GPD	1 Stpi	4	2	Ea.	260	120		380	470
5040	14 GPD		3.80	2.105		293	126		419	515
5060	19 GPD		3.60	2.222		405	133		538	650
9000	Minimum labor/equipment charge	Q-5	3.50	4.571	Job		246		246	385

23 84 16 – Mechanical Dehumidification Units

23 84 16.10 Dehumidifier Units		Crew	Daily Output	Labor-Hours	Unit	Material	Labor	Equipment	Total	Total Incl O&P
0010	**DEHUMIDIFIER UNITS** R236000-30									
6000	Self contained with filters and standard controls									
6040	1.5 lb./hr., 50 CFM	1 Plum	8	1	Ea.	7,950	58.50		8,008.50	8,825
6060	3 lb./hr., 150 CFM	Q-1	12	1.333		8,300	70.50		8,370.50	9,225
6065	6 lb./hr., 150 CFM		9	1.778		13,500	94		13,594	15,000
6070	16 to 20 lb./hr., 600 CFM		5	3.200		25,700	169		25,869	28,600
6080	30 to 40 lb./hr., 1125 CFM		4	4		40,000	211		40,211	44,300
6090	60 to 75 lb./hr., 2250 CFM		3	5.333		52,000	282		52,282	58,000
6100	120 to 155 lb./hr., 4500 CFM		2	8		92,500	425		92,925	102,000
6110	240 to 310 lb./hr., 9000 CFM		1.50	10.667		132,500	565		133,065	146,500
6120	400 to 515 lb./hr., 15,000 CFM	Q-2	1.60	15		166,500	820		167,320	184,500
6130	530 to 690 lb./hr., 20,000 CFM		1.40	17.143		180,500	940		181,440	200,000
6140	800 to 1030 lb./hr., 30,000 CFM		1.20	20		205,500	1,100		206,600	227,500
6150	1060 to 1375 lb./hr., 40,000 CFM		1	24		286,000	1,325		287,325	316,500

Estimating Tips

26 05 00 Common Work Results for Electrical

- Conduit should be taken off in three main categories—power distribution, branch power, and branch lighting—so the estimator can concentrate on systems and components, therefore making it easier to ensure all items have been accounted for.
- For cost modifications for elevated conduit installation, add the percentages to labor according to the height of installation, and only to the quantities exceeding the different height levels, not to the total conduit quantities.
- Remember that aluminum wiring of equal ampacity is larger in diameter than copper and may require larger conduit.
- If more than three wires at a time are being pulled, deduct percentages from the labor hours of that grouping of wires.
- When taking off grounding systems, identify separately the type and size of wire, and list each unique type of ground connection.
- The estimator should take the weights of materials into consideration when completing a takeoff. Topics to consider include: How will the materials be supported? What methods of support are available? How high will the support structure have to reach? Will the final support structure be able to withstand the total burden? Is the support material included or separate from the fixture, equipment, and material specified?
- Do not overlook the costs for equipment used in the installation. If scaffolding or highlifts are available in the field, contractors may use them in lieu of the proposed ladders and rolling staging.

26 20 00 Low-Voltage Electrical Transmission

- Supports and concrete pads may be shown on drawings for the larger equipment, or the support system may be only a piece of plywood for the back of a panelboard. In either case, it must be included in the costs.

26 40 00 Electrical and Cathodic Protection

- When taking off cathodic protections systems, identify the type and size of cable, and list each unique type of anode connection.

26 50 00 Lighting

- Fixtures should be taken off room by room, using the fixture schedule, specifications, and the ceiling plan. For large concentrations of lighting fixtures in the same area, deduct the percentages from labor hours.

Reference Numbers

Reference numbers are shown in shaded boxes at the beginning of some major classifications. These numbers refer to related items in the Reference Section. The reference information may be an estimating procedure, an alternate pricing method, or technical information.

Note: Not all subdivisions listed here necessarily appear in this publication. ■

*Note: **Trade Service**, in part, has been used as a reference source for some of the material prices used in Division 26.*

26 01 Operation and Maintenance of Electrical Systems

26 01 40 – Operation and Maintenance of Electrical Protection Systems

26 01 40.51 Electrical Systems Repair and Replacement		Crew	Daily Output	Labor-Hours	Unit	Material	2015 Bare Costs Labor	Equipment	Total	Total Incl O&P
0010	**ELECTRICAL SYSTEMS REPAIR AND REPLACEMENT**									
3000	Remove and replace (reinstall), switch cover	1 Elec	60	.133	Ea.		7.30		7.30	11.30
3020	Outlet cover	″	60	.133	″		7.30		7.30	11.30

26 01 50 – Operation and Maintenance of Lighting

26 01 50.81 Luminaire Replacement		Crew	Daily Output	Labor-Hours	Unit	Material	Labor	Equipment	Total	Total Incl O&P
0010	**LUMINAIRE REPLACEMENT**									
0962	Remove and install new replacement lens, 2′ x 2′ troffer	1 Elec	16	.500	Ea.	15.90	27.50		43.40	60
0964	2′ x 4′ troffer		16	.500		14.90	27.50		42.40	59
3200	Remove and replace (reinstall), lighting fixture		4	2			109		109	169

26 05 Common Work Results for Electrical

26 05 05 – Selective Demolition for Electrical

26 05 05.10 Electrical Demolition		Crew	Daily Output	Labor-Hours	Unit	Material	Labor	Equipment	Total	Total Incl O&P
0010	**ELECTRICAL DEMOLITION** R260105-30									
0020	Conduit to 15′ high, including fittings & hangers									
0100	Rigid galvanized steel, 1/2″ to 1″ diameter R024119-10	1 Elec	242	.033	L.F.		1.81		1.81	2.80
0120	1-1/4″ to 2″	″	200	.040			2.19		2.19	3.39
0140	2-1/2″ to 3-1/2″	2 Elec	302	.053			2.90		2.90	4.49
0160	4″ to 6″	″	160	.100			5.45		5.45	8.45
0170	PVC #40, 1/2″ to 1″	1 Elec	410	.020			1.07		1.07	1.65
0172	1-1/4″ to 2″		350	.023			1.25		1.25	1.94
0174	2-1/2″		250	.032			1.75		1.75	2.71
0176	3″ to 3-1/2″	2 Elec	340	.047			2.57		2.57	3.99
0178	4″ to 6″	″	230	.070			3.81		3.81	5.90
0200	Electric metallic tubing (EMT), 1/2″ to 1″	1 Elec	394	.020			1.11		1.11	1.72
0220	1-1/4″ to 1-1/2″		326	.025			1.34		1.34	2.08
0240	2″ to 3″		236	.034			1.85		1.85	2.87
0260	3-1/2″ to 4″	2 Elec	310	.052			2.82		2.82	4.37
0270	Armored cable, (BX) avg. 50′ runs									
0280	#14, 2 wire	1 Elec	690	.012	L.F.		.63		.63	.98
0290	#14, 3 wire		571	.014			.77		.77	1.19
0300	#12, 2 wire		605	.013			.72		.72	1.12
0310	#12, 3 wire		514	.016			.85		.85	1.32
0320	#10, 2 wire		514	.016			.85		.85	1.32
0330	#10, 3 wire		425	.019			1.03		1.03	1.59
0340	#8, 3 wire		342	.023			1.28		1.28	1.98
0350	Non metallic sheathed cable (Romex)									
0360	#14, 2 wire	1 Elec	720	.011	L.F.		.61		.61	.94
0370	#14, 3 wire		657	.012			.67		.67	1.03
0380	#12, 2 wire		629	.013			.70		.70	1.08
0390	#10, 3 wire		450	.018			.97		.97	1.51
0400	Wiremold raceway, including fittings & hangers									
0420	No. 3000	1 Elec	250	.032	L.F.		1.75		1.75	2.71
0440	No. 4000		217	.037			2.02		2.02	3.12
0460	No. 6000		166	.048			2.64		2.64	4.08
0462	Plugmold with receptacle		114	.070			3.84		3.84	5.95
0465	Telephone/power pole		12	.667	Ea.		36.50		36.50	56.50
0470	Non-metallic, straight section		480	.017	L.F.		.91		.91	1.41
0500	Channels, steel, including fittings & hangers									
0520	3/4″ x 1-1/2″	1 Elec	308	.026	L.F.		1.42		1.42	2.20

26 05 05 – Selective Demolition for Electrical

26 05 05.10 Electrical Demolition		Crew	Daily Output	Labor-Hours	Unit	Material	2015 Bare Costs Labor	Equipment	Total	Total Incl O&P
0540	1-1/2" x 1-1/2"	1 Elec	269	.030	L.F.		1.63		1.63	2.52
0560	1-1/2" x 1-7/8"	↓	229	.035	↓		1.91		1.91	2.96
0600	Copper bus duct, indoor, 3 phase									
0610	Including hangers & supports									
0620	225 amp	2 Elec	135	.119	L.F.		6.50		6.50	10.05
0640	400 amp		106	.151			8.25		8.25	12.80
0660	600 amp		86	.186			10.20		10.20	15.75
0680	1000 amp		60	.267			14.60		14.60	22.50
0700	1600 amp		40	.400			22		22	34
0720	3000 amp	↓	10	1.600	↓		87.50		87.50	136
0800	Plug-in switches, 600V 3 ph., incl. disconnecting									
0820	wire, conduit terminations, 30 amp	1 Elec	15.50	.516	Ea.		28		28	43.50
0840	60 amp		13.90	.576			31.50		31.50	49
0850	100 amp		10.40	.769			42		42	65
0860	200 amp	↓	6.20	1.290			70.50		70.50	109
0880	400 amp	2 Elec	5.40	2.963			162		162	251
0900	600 amp		3.40	4.706			257		257	400
0920	800 amp		2.60	6.154			335		335	520
0940	1200 amp		2	8			440		440	680
0960	1600 amp	↓	1.70	9.412	↓		515		515	795
1010	Safety switches, 250 or 600V, incl. disconnection									
1050	of wire & conduit terminations									
1100	30 amp	1 Elec	12.30	.650	Ea.		35.50		35.50	55
1120	60 amp		8.80	.909			49.50		49.50	77
1140	100 amp		7.30	1.096			60		60	93
1160	200 amp	↓	5	1.600			87.50		87.50	136
1180	400 amp	2 Elec	6.80	2.353			129		129	199
1200	600 amp	"	4.60	3.478			190		190	295
1202	Explosion proof, 60 amp	1 Elec	8.80	.909			49.50		49.50	77
1204	100 amp		5	1.600			87.50		87.50	136
1206	200 amp	↓	3.33	2.402	↓		131		131	203
1210	Panel boards, incl. removal of all breakers,									
1220	conduit terminations & wire connections									
1230	3 wire, 120/240 V, 100A, to 20 circuits	1 Elec	2.60	3.077	Ea.		168		168	261
1240	200 amps, to 42 circuits	2 Elec	2.60	6.154			335		335	520
1250	400 amps, to 42 circuits	"	2.20	7.273			400		400	615
1260	4 wire, 120/208 V, 125A, to 20 circuits	1 Elec	2.40	3.333			182		182	282
1270	200 amps, to 42 circuits	2 Elec	2.40	6.667			365		365	565
1280	400 amps, to 42 circuits		1.92	8.333			455		455	705
1285	600 amps, to 42 circuits	↓	1.60	10	↓		545		545	845
1300	Transformer, dry type, 1 phase, incl. removal of									
1320	supports, wire & conduit terminations									
1340	1 kVA	1 Elec	7.70	1.039	Ea.		57		57	88
1360	5 kVA		4.70	1.702			93		93	144
1380	10 kVA	↓	3.60	2.222			122		122	188
1400	37.5 kVA	2 Elec	3	5.333			292		292	450
1420	75 kVA	"	2.50	6.400	↓		350		350	540
1440	3 phase to 600V, primary									
1460	3 kVA	1 Elec	3.87	2.067	Ea.		113		113	175
1480	15 kVA	2 Elec	3.67	4.360			238		238	370
1490	25 kVA		3.51	4.558			249		249	385
1500	30 kVA		3.42	4.678			256		256	395
1510	45 kVA	↓	3.18	5.031	↓		275		275	425

26 05 Common Work Results for Electrical

26 05 05 – Selective Demolition for Electrical

26 05 05.10 Electrical Demolition		Crew	Daily Output	Labor-Hours	Unit	Material	2015 Bare Costs Labor	Equipment	Total	Total Incl O&P
1520	75 kVA	2 Elec	2.69	5.948	Ea.		325		325	505
1530	112.5 kVA	R-3	2.90	6.897			375	47.50	422.50	635
1540	150 kVA		2.70	7.407			400	51	451	680
1560	500 kVA		1.40	14.286			775	98.50	873.50	1,300
1570	750 kVA		1.10	18.182			985	125	1,110	1,675
1600	Pull boxes & cabinets, sheet metal, incl. removal									
1620	of supports and conduit terminations									
1640	6" x 6" x 4"	1 Elec	31.10	.257	Ea.		14.05		14.05	22
1660	12" x 12" x 4"		23.30	.343			18.80		18.80	29
1680	24" x 24" x 6"		12.30	.650			35.50		35.50	55
1700	36" x 36" x 8"		7.70	1.039			57		57	88
1720	Junction boxes, 4" sq. & oct.		80	.100			5.45		5.45	8.45
1740	Handy box		107	.075			4.09		4.09	6.35
1760	Switch box		107	.075			4.09		4.09	6.35
1780	Receptacle & switch plates		257	.031			1.70		1.70	2.64
1790	Receptacles & switches, 15 to 30 amp		135	.059			3.24		3.24	5
1800	Wire, THW-THWN-THHN, removed from									
1810	in place conduit, to 15' high									
1830	#14	1 Elec	65	.123	C.L.F.		6.75		6.75	10.40
1840	#12		55	.145			7.95		7.95	12.30
1850	#10		45.50	.176			9.60		9.60	14.90
1860	#8		40.40	.198			10.85		10.85	16.75
1870	#6		32.60	.245			13.40		13.40	21
1880	#4	2 Elec	53	.302			16.50		16.50	25.50
1890	#3		50	.320			17.50		17.50	27
1900	#2		44.60	.359			19.60		19.60	30.50
1910	1/0		33.20	.482			26.50		26.50	41
1920	2/0		29.20	.548			30		30	46.50
1930	3/0		25	.640			35		35	54
1940	4/0		22	.727			40		40	61.50
1950	250 kcmil		20	.800			44		44	68
1960	300 kcmil		19	.842			46		46	71.50
1970	350 kcmil		18	.889			48.50		48.50	75.50
1980	400 kcmil		17	.941			51.50		51.50	79.50
1990	500 kcmil		16.20	.988			54		54	83.50
2000	Interior fluorescent fixtures, incl. supports									
2010	& whips, to 15' high									
2100	Recessed drop-in 2' x 2', 2 lamp	2 Elec	35	.457	Ea.		25		25	38.50
2120	2' x 4', 2 lamp		33	.485			26.50		26.50	41
2140	2' x 4', 4 lamp		30	.533			29		29	45
2160	4' x 4', 4 lamp		20	.800			44		44	68
2180	Surface mount, acrylic lens & hinged frame									
2200	1' x 4', 2 lamp	2 Elec	44	.364	Ea.		19.90		19.90	31
2220	2' x 2', 2 lamp		44	.364			19.90		19.90	31
2260	2' x 4', 4 lamp		33	.485			26.50		26.50	41
2280	4' x 4', 4 lamp		23	.696			38		38	59
2300	Strip fixtures, surface mount									
2320	4' long, 1 lamp	2 Elec	53	.302	Ea.		16.50		16.50	25.50
2340	4' long, 2 lamp		50	.320			17.50		17.50	27
2360	8' long, 1 lamp		42	.381			21		21	32.50
2380	8' long, 2 lamp		40	.400			22		22	34
2400	Pendant mount, industrial, incl. removal									
2410	of chain or rod hangers, to 15' high									

26 05 05 – Selective Demolition for Electrical

26 05 05.10 Electrical Demolition

	26 05 05.10 Electrical Demolition	Crew	Daily Output	Labor-Hours	Unit	Material	Labor	Equipment	Total	Total Incl O&P
						2015 Bare Costs				
2420	4′ long, 2 lamp	2 Elec	35	.457	Ea.		25		25	38.50
2440	8′ long, 2 lamp	"	27	.593	"		32.50		32.50	50
2460	Interior incandescent, surface, ceiling									
2470	or wall mount, to 12′ high									
2480	Metal cylinder type, 75 Watt	2 Elec	62	.258	Ea.		14.10		14.10	22
2500	150 Watt		62	.258			14.10		14.10	22
2502	300 Watt		53	.302			16.50		16.50	25.50
2520	Metal halide, high bay									
2540	400 Watt	2 Elec	15	1.067	Ea.		58.50		58.50	90.50
2560	1000 Watt		12	1.333			73		73	113
2580	150 Watt, low bay		20	.800			44		44	68
2585	Globe type, ceiling mount		62	.258			14.10		14.10	22
2586	Can type, recessed mount		62	.258			14.10		14.10	22
2600	Exterior fixtures, incandescent, wall mount									
2620	100 Watt	2 Elec	50	.320	Ea.		17.50		17.50	27
2640	Quartz, 500 Watt		33	.485			26.50		26.50	41
2660	1500 Watt		27	.593			32.50		32.50	50
2680	Wall pack, mercury vapor									
2700	175 Watt	2 Elec	25	.640	Ea.		35		35	54
2720	250 Watt	"	25	.640			35		35	54
7000	Weatherhead/mast, 2″	1 Elec	16	.500			27.50		27.50	42.50
7002	3″		10	.800			44		44	68
7004	3-1/2″		8.50	.941			51.50		51.50	79.50
7006	4″		8	1			54.50		54.50	84.50
7100	Service entry cable, #6, +#6 neutral		420	.019	L.F.		1.04		1.04	1.61
7102	#4, +#4 neutral		360	.022			1.22		1.22	1.88
7104	#2, +#4 neutral		345	.023			1.27		1.27	1.96
7106	#2, +#2 neutral		330	.024			1.33		1.33	2.05
9000	Minimum labor/equipment charge		4	2	Job		109		109	169
9900	Add to labor for higher elevated installation									
9910	15′ to 20′ high, add						10%			
9920	20′ to 25′ high, add						20%			
9930	25′ to 30′ high, add						25%			
9940	30′ to 35′ high, add						30%			
9950	35′ to 40′ high, add						35%			
9960	Over 40′ high, add						40%			

26 05 05.15 Electrical Demolition, Grounding

	26 05 05.15 Electrical Demolition, Grounding	Crew	Daily Output	Labor-Hours	Unit	Material	Labor	Equipment	Total	Total Incl O&P
0010	**ELECTRICAL DEMOLITION, GROUNDING** Addition									
0100	Ground clamp, bronze	1 Elec	64	.125	Ea.		6.85		6.85	10.60
0140	Water pipe ground clamp, bronze, heavy duty		24	.333			18.25		18.25	28
0150	Ground rod, 8′ to 10′		13	.615			33.50		33.50	52
0200	Ground wire, bare armored	2 Elec	900	.018	L.F.		.97		.97	1.51
0240	Bare copper or aluminum	"	2800	.006	"		.31		.31	.48

26 05 05.20 Electrical Demolition, Wiring Methods

	26 05 05.20 Electrical Demolition, Wiring Methods	Crew	Daily Output	Labor-Hours	Unit	Material	Labor	Equipment	Total	Total Incl O&P
0010	**ELECTRICAL DEMOLITION, WIRING METHODS** Addition R024119-10									
0100	Armored cable, w/PVC jacket, in cable tray									
0110	#6	1 Elec	9.30	.860	C.L.F.		47		47	73
0120	#4	2 Elec	16.20	.988			54		54	83.50
0130	#2		13.80	1.159			63.50		63.50	98
0140	#1		12	1.333			73		73	113
0150	1/0		10.80	1.481			81		81	125
0160	2/0		10.20	1.569			86		86	133

26 05 05 – Selective Demolition for Electrical

26 05 05.20 Electrical Demolition, Wiring Methods		Crew	Daily Output	Labor-Hours	Unit	Material	2015 Bare Costs Labor	Equipment	Total	Total Incl O&P
0170	3/0	2 Elec	9.60	1.667	C.L.F.		91		91	141
0180	4/0		9	1.778			97		97	151
0190	250 kcmil	3 Elec	10.80	2.222			122		122	188
0210	350 kcmil		9.90	2.424			133		133	205
0230	500 kcmil		9	2.667			146		146	226
0240	750 kcmil		8.10	2.963			162		162	251
1100	Control cable, 600 V or less									
1110	3 wires	1 Elec	24	.333	C.L.F.		18.25		18.25	28
1120	5 wires		20	.400			22		22	34
1130	7 wires		16.50	.485			26.50		26.50	41
1140	9 wires		15	.533			29		29	45
1150	12 wires	2 Elec	26	.615			33.50		33.50	52
1160	15 wires		22	.727			40		40	61.50
1170	19 wires		19	.842			46		46	71.50
1180	25 wires		16	1			54.50		54.50	84.50
1500	Mineral insulated (MI) cable, 600 V									
1510	#10	1 Elec	4.80	1.667	C.L.F.		91		91	141
1520	#8		4.50	1.778			97		97	151
1530	#6		4.20	1.905			104		104	161
1540	#4	2 Elec	7.20	2.222			122		122	188
1550	#2		6.60	2.424			133		133	205
1560	#1		6.30	2.540			139		139	215
1570	1/0		6	2.667			146		146	226
1580	2/0		5.70	2.807			154		154	238
1590	3/0		5.40	2.963			162		162	251
1600	4/0		4.80	3.333			182		182	282
1610	250 kcmil	3 Elec	7.20	3.333			182		182	282
1620	500 kcmil	″	5.90	4.068			223		223	345
2000	Shielded cable, XLP shielding, to 35 kV									
2010	#4	2 Elec	13.20	1.212	C.L.F.		66.50		66.50	103
2020	#1		12	1.333			73		73	113
2030	1/0		11.40	1.404			77		77	119
2040	2/0		10.80	1.481			81		81	125
2050	4/0		9.60	1.667			91		91	141
2060	250 kcmil	3 Elec	13.50	1.778			97		97	151
2070	350 kcmil		11.70	2.051			112		112	174
2080	500 kcmil		11.20	2.143			117		117	182
2090	750 kcmil		10.80	2.222			122		122	188
3000	Modular flexible wiring									
3010	Cable set	1 Elec	120	.067	Ea.		3.65		3.65	5.65
3020	Conversion module		48	.167			9.10		9.10	14.10
3030	Switching assembly		96	.083			4.56		4.56	7.05
3200	Undercarpet									
3210	Power or telephone, flat cable	1 Elec	1320	.006	L.F.		.33		.33	.51
3220	Transition block assemblies w/fitting		75	.107	Ea.		5.85		5.85	9.05
3230	Floor box with fitting		60	.133			7.30		7.30	11.30
3300	Data system, cable with connection		50	.160			8.75		8.75	13.55
4000	Cable tray, including fitting & support									
4010	Galvanized steel, 6″ wide	2 Elec	310	.052	L.F.		2.82		2.82	4.37
4020	9″ wide		295	.054			2.97		2.97	4.59
4030	12″ wide		285	.056			3.07		3.07	4.76
4040	18″ wide		270	.059			3.24		3.24	5
4050	24″ wide		260	.062			3.37		3.37	5.20

26 05 05 – Selective Demolition for Electrical

26 05 05.20 Electrical Demolition, Wiring Methods		Crew	Daily Output	Labor-Hours	Unit	Material	2015 Bare Costs Labor	Equipment	Total	Total Incl O&P
4060	30″ wide	2 Elec	240	.067	L.F.		3.65		3.65	5.65
4070	36″ wide		220	.073			3.98		3.98	6.15
4110	Aluminum, 6″ wide		420	.038			2.08		2.08	3.23
4120	9″ wide		415	.039			2.11		2.11	3.27
4130	12″ wide		390	.041			2.24		2.24	3.48
4140	18″ wide		370	.043			2.37		2.37	3.66
4150	24″ wide		350	.046			2.50		2.50	3.87
4160	30″ wide		325	.049			2.69		2.69	4.17
4170	36″ wide		300	.053			2.92		2.92	4.52
4310	Cable channel, aluminum, 4″ wide straight	1 Elec	240	.033			1.82		1.82	2.82
4320	Cable tray fittings, 6″ to 9″ wide	2 Elec	27	.593	Ea.		32.50		32.50	50
4330	12″ to 18″ wide	″	19	.842	″		46		46	71.50
5000	Conduit nipples, with locknuts and bushings									
5020	1/2″	1 Elec	108	.074	Ea.		4.05		4.05	6.25
5040	3/4″		96	.083			4.56		4.56	7.05
5060	1″		81	.099			5.40		5.40	8.35
5080	1-1/4″		69	.116			6.35		6.35	9.80
5100	1-1/2″		60	.133			7.30		7.30	11.30
5120	2″		54	.148			8.10		8.10	12.55
5140	2-1/2″		45	.178			9.70		9.70	15.05
5160	3″		36	.222			12.15		12.15	18.80
5180	3-1/2″		33	.242			13.25		13.25	20.50
5200	4″		27	.296			16.20		16.20	25
5220	5″		21	.381			21		21	32.50
5240	6″		18	.444			24.50		24.50	37.50
5500	Electric nonmetallic tubing (ENT), flexible, 1/2″ to 1″ diameter		690	.012	L.F.		.63		.63	.98
5510	1-1/4″ to 2″ diameter		300	.027			1.46		1.46	2.26
5600	Flexible metallic tubing, steel, 3/8″ to 3/4″ diameter		600	.013			.73		.73	1.13
5610	1″ to 1-1/4″ diameter		260	.031			1.68		1.68	2.61
5620	1-1/2″ to 2″ diameter		140	.057			3.13		3.13	4.84
5630	2-1/2″ diameter		100	.080			4.38		4.38	6.80
5640	3″ to 3-1/2″ diameter	2 Elec	150	.107			5.85		5.85	9.05
5650	4″ diameter	″	100	.160			8.75		8.75	13.55
5700	Sealtite flexible conduit, 3/8″ to 3/4″ diameter	1 Elec	420	.019			1.04		1.04	1.61
5710	1″ to 1-1/4″ diameter		180	.044			2.43		2.43	3.76
5720	1-1/2″ to 2″ diameter		120	.067			3.65		3.65	5.65
5730	2-1/2″ diameter		80	.100			5.45		5.45	8.45
5740	3″ diameter	2 Elec	150	.107			5.85		5.85	9.05
5750	4″ diameter	″	100	.160			8.75		8.75	13.55
5800	Wiring duct, plastic, 1-1/2″ to 2-1/2″ wide	1 Elec	360	.022			1.22		1.22	1.88
5810	3″ wide		330	.024			1.33		1.33	2.05
5820	4″ wide		300	.027			1.46		1.46	2.26
6000	Floor box and carpet flange		9	.889	Ea.		48.50		48.50	75.50
6300	Wireway, with fittings and supports, to 15′ high									
6310	2-1/2″ x 2-1/2″	1 Elec	180	.044	L.F.		2.43		2.43	3.76
6320	4″ x 4″	″	160	.050			2.74		2.74	4.24
6330	6″ x 6″	2 Elec	240	.067			3.65		3.65	5.65
6340	8″ x 8″		160	.100			5.45		5.45	8.45
6350	10″ x 10″		120	.133			7.30		7.30	11.30
6360	12″ x 12″		80	.200			10.95		10.95	16.95
6400	Cable reel with receptacle, 120 V/208 V		13	1.231	Ea.		67.50		67.50	104
6500	Equipment connection, to 10 HP	1 Elec	23	.348			19.05		19.05	29.50
6510	To 15 to 30 HP		18	.444			24.50		24.50	37.50

26 05 05 – Selective Demolition for Electrical

26 05 05.20 Electrical Demolition, Wiring Methods		Crew	Daily Output	Labor-Hours	Unit	Material	2015 Bare Costs Labor	Equipment	Total	Total Incl O&P
6520	To 40 to 60 HP	1 Elec	15	.533	Ea.		29		29	45
6530	To 75 to 100 HP		10	.800			44		44	68
7008	Receptacle, explosionproof, 120 V, to 30 A		16	.500			27.50		27.50	42.50
7020	Dimmer switch, 2000 W or less		62	.129			7.05		7.05	10.95

26 05 05.25 Electrical Demolition, Electrical Power		Crew	Daily Output	Labor-Hours	Unit	Material	2015 Bare Costs Labor	Equipment	Total	Total Incl O&P
0010	**ELECTRICAL DEMOLITION, ELECTRICAL POWER** R024119-10									
0100	Meter centers and sockets									
0120	Meter socket, 4 terminal R260105-30	1 Elec	8.40	.952	Ea.		52		52	80.50
0140	Trans-socket, 13 terminal, 400 A	"	3.60	2.222			122		122	188
0160	800 A	2 Elec	4.40	3.636			199		199	310
0200	Meter center, 400 A		5.80	2.759			151		151	234
0210	600 A		4	4			219		219	340
0220	800 A		3.30	4.848			265		265	410
0230	1200 A		2.80	5.714			315		315	485
0240	1600 A		2.50	6.400			350		350	540
0300	Base meter devices, 3 meter		3.60	4.444			243		243	375
0310	4 meter		3.30	4.848			265		265	410
0320	5 meter		2.90	5.517			300		300	465
0330	6 meter		2.20	7.273			400		400	615
0340	7 meter		2	8			440		440	680
0350	8 meter		1.90	8.421			460		460	715
0400	Branch meter devices									
0410	Socket w/circuit breaker 200 A, 2 meter	2 Elec	3.30	4.848	Ea.		265		265	410
0420	3 meter		2.90	5.517			300		300	465
0430	4 meter		2.60	6.154			335		335	520
0450	Main circuit breaker, 400 A		5.80	2.759			151		151	234
0460	600 A		4	4			219		219	340
0470	800 A		3.30	4.848			265		265	410
0480	1200 A		2.80	5.714			315		315	485
0490	1600 A		2.50	6.400			350		350	540
0500	Main lug terminal box, 800 A		3.40	4.706			257		257	400
0510	1200 A		2.60	6.154			335		335	520
1000	Motors, 230/460 V, 60 Hz, 3/4 HP	1 Elec	10.70	.748			41		41	63.50
1010	5 HP		9	.889			48.50		48.50	75.50
1020	10 HP		8	1			54.50		54.50	84.50
1030	15 HP		6.40	1.250			68.50		68.50	106
1040	20 HP	2 Elec	10.40	1.538			84		84	130
1050	50 HP		9.60	1.667			91		91	141
1060	75 HP		5.60	2.857			156		156	242
1070	100 HP	3 Elec	5.40	4.444			243		243	375
1080	150 HP		3.60	6.667			365		365	565
1090	200 HP		3	8			440		440	680
1200	Variable frequency drive, 460 V, for 5 HP motor size	1 Elec	3.20	2.500			137		137	212
1210	10 HP motor size	"	2.70	2.963			162		162	251
1220	20 HP motor size	2 Elec	3.60	4.444			243		243	375
1230	50 HP motor size	"	2.10	7.619			415		415	645
1240	75 HP motor size	R-3	2.20	9.091			495	62.50	557.50	840
1250	100 HP motor size		2	10			545	69	614	920
1260	150 HP motor size		2	10			545	69	614	920
1270	200 HP motor size		1.70	11.765			640	81	721	1,075
2000	Generator set w/accessories, 3 phase 4 wire, 277/480 V									
2010	7.5 kW	3 Elec	2	12	Ea.		655		655	1,025

26 05 05 – Selective Demolition for Electrical

26 05 05.25 Electrical Demolition, Electrical Power		Crew	Daily Output	Labor-Hours	Unit	Material	2015 Bare Costs Labor	Equipment	Total	Total Incl O&P
2020	20 kW	3 Elec	1.70	14.118	Ea.		770		770	1,200
2040	30 kW		1.33	18.045			985		985	1,525
2060	50 kW		1	24			1,325		1,325	2,025
2080	100 kW		.75	32			1,750		1,750	2,700
2100	150 kW		.63	38.095			2,075		2,075	3,225
2120	250 kW		.59	40.678			2,225		2,225	3,450
2140	400 kW		.50	48			2,625		2,625	4,075
2160	500 kW		.44	54.545			2,975		2,975	4,625
2180	750 kW	4 Elec	.56	57.143			3,125		3,125	4,850
2200	1000 kW	"	.46	69.565			3,800		3,800	5,900
3000	Uninterruptible power supply system (UPS)									
3010	Single phase, 120 V, 1 kVA	1 Elec	3.20	2.500	Ea.		137		137	212
3020	2 kVA	2 Elec	3.60	4.444			243		243	375
3030	5 kVA	3 Elec	2.50	9.600			525		525	815
3040	10 kVA		2.30	10.435			570		570	885
3050	15 kVA		1.80	13.333			730		730	1,125
4000	Transformer, incl support, wire & conduit termination									
4010	Buck-boost, single phase, 120/240 V, 0.1 kVA	1 Elec	25	.320	Ea.		17.50		17.50	27
4020	0.5 kVA		12.50	.640			35		35	54
4030	1 kVA		6.30	1.270			69.50		69.50	108
4040	5 kVA		3.80	2.105			115		115	178
4800	5 kV or 15 kV primary, 277/480 V second, 112.5 kVA	R-3	3.50	5.714			310	39.50	349.50	530
4810	150 kVA		2.70	7.407			400	51	451	680
4820	225 kVA		2.30	8.696			470	60	530	800
4830	500 kVA		1.45	13.793			750	95	845	1,275
4840	750 kVA		1.33	15.038			815	104	919	1,400
4850	1000 kVA		1.25	16			870	110	980	1,475
4860	2000 kVA		1.05	19.048			1,025	131	1,156	1,750
4870	3000 kVA		.75	26.667			1,450	184	1,634	2,450
6010	Isolation panel, 3 kVA	1 Elec	2.30	3.478			190		190	295
6020	5 kVA		2.20	3.636			199		199	310
6030	7.5 kVA		2.10	3.810			208		208	325
6040	10 kVA		1.80	4.444			243		243	375
6050	15 kVA		1.40	5.714			315		315	485
7000	Power filters & conditioners									
7100	Automatic voltage regulator	2 Elec	6	2.667	Ea.		146		146	226
7200	Capacitor, 1 kVAR	1 Elec	8.60	.930			51		51	79
7210	5 kVAR		5.80	1.379			75.50		75.50	117
7220	10 kVAR		4.80	1.667			91		91	141
7230	15 kVAR		4.20	1.905			104		104	161
7240	20 kVAR		3.50	2.286			125		125	194
7250	30 kVAR		3.40	2.353			129		129	199
7260	50 kVAR		3.20	2.500			137		137	212
7400	Computer isolator transformer									
7410	Single phase, 120/240 V, 0.5 kVAR	1 Elec	12.80	.625	Ea.		34		34	53
7420	1 kVAR		8.50	.941			51.50		51.50	79.50
7430	2.5 kVAR		6.40	1.250			68.50		68.50	106
7440	5 kVAR		3.70	2.162			118		118	183
7500	Computer regulator transformer									
7510	Single phase, 240 V, 0.5 kVAR	1 Elec	8.50	.941	Ea.		51.50		51.50	79.50
7520	1 kVAR		6.40	1.250			68.50		68.50	106
7530	2 kVAR		3.20	2.500			137		137	212
7540	Single phase, plug-in unit 120 V, 0.5 kVAR		26	.308			16.85		16.85	26

26 05 05 – Selective Demolition for Electrical

26 05 05.25 Electrical Demolition, Electrical Power

	26 05 05.25 Electrical Demolition, Electrical Power	Crew	Daily Output	Labor-Hours	Unit	Material	2015 Bare Costs Labor	Equipment	Total	Total Incl O&P
7550	1 kVAR	1 Elec	17	.471	Ea.		25.50		25.50	40
7600	Power conditioner transformer									
7610	Single phase 115 V - 240 V, 3 kVA	2 Elec	5.10	3.137	Ea.		172		172	266
7620	5 kVA	"	3.70	4.324			237		237	365
7630	7.5 kVA	3 Elec	4.80	5			274		274	425
7640	10 kVA	"	4.30	5.581			305		305	475
7700	Transient suppressor/voltage regulator									
7710	Single phase 115 or 220 V, 1 kVA	1 Elec	8.50	.941	Ea.		51.50		51.50	79.50
7720	2 kVA		7.30	1.096			60		60	93
7730	4 kVA		6.80	1.176			64.50		64.50	99.50
7800	Transient voltage suppressor transformer									
7810	Single phase, 115 or 220 V, 3.6 kVA	1 Elec	12.80	.625	Ea.		34		34	53
7820	7.2 kVA		11.50	.696			38		38	59
7830	14.4 kVA		10.20	.784			43		43	66.50
7840	Single phase, plug-in, 120 V, 1.8 kVA		26	.308			16.85		16.85	26
8000	Power measurement & control									
8010	Switchboard instruments, 3 phase 4 wire, indicating unit	1 Elec	25	.320	Ea.		17.50		17.50	27
8020	Recording unit		12	.667			36.50		36.50	56.50
8100	3 current transformers, 3 phase 4 wire, 5 to 800 A		6.40	1.250			68.50		68.50	106
8110	1000 to 1500 A		4.20	1.905			104		104	161
8120	2000 to 4000 A		3.20	2.500			137		137	212

26 05 05.30 Electrical Demolition, Transmission and Distribution

	26 05 05.30 Electrical Demolition, Transmission and Distribution	Crew	Daily Output	Labor-Hours	Unit	Material	Labor	Equipment	Total	Total Incl O&P
0010	**ELECTRICAL DEMOLITION, TRANSMISSION & DISTRIBUTION**									
0100	Load interrupter switch, 600 A, NEMA 1, 4.8 kV	R-3	1.33	15.038	Ea.		815	104	919	1,400
0120	13.8 kV R024119-10	"	1.27	15.748			855	109	964	1,450
0200	Lightning arrester, 4.8 kV	1 Elec	9	.889			48.50		48.50	75.50
0220	13.8 kV R260105-30		6.67	1.199			65.50		65.50	102
0300	Alarm or option items		3.33	2.402			131		131	203

26 05 05.35 Electrical Demolition, L.V. Distribution

	26 05 05.35 Electrical Demolition, L.V. Distribution	Crew	Daily Output	Labor-Hours	Unit	Material	Labor	Equipment	Total	Total Incl O&P
0010	**ELECTRICAL DEMOLITION, L.V. DISTRIBUTION** Addition R024119-10									
0100	Circuit breakers in enclosure									
0120	Enclosed (NEMA 1), 600 V, 3 pole, 30 A R260105-30	1 Elec	12.30	.650	Ea.		35.50		35.50	55
0140	60 A		8.80	.909			49.50		49.50	77
0160	100 A		7.30	1.096			60		60	93
0180	225 A		5	1.600			87.50		87.50	136
0200	400 A	2 Elec	6.80	2.353			129		129	199
0220	600 A		4.60	3.478			190		190	295
0240	800 A		3.60	4.444			243		243	375
0260	1200 A		3.10	5.161			282		282	435
0280	1600 A		2.80	5.714			315		315	485
0300	2000 A		2.50	6.400			350		350	540
0400	Enclosed (NEMA 7), 600 V, 3 pole, 50 A	1 Elec	8.80	.909			49.50		49.50	77
0410	100 A		5.80	1.379			75.50		75.50	117
0420	150 A		3.80	2.105			115		115	178
0430	250 A	2 Elec	6.20	2.581			141		141	219
0440	400 A	"	4.60	3.478			190		190	295
0460	Manual motor starter, NEMA 1	1 Elec	24	.333			18.25		18.25	28
0480	NEMA 4 or NEMA 7		15	.533			29		29	45
0500	Time switches, single pole single throw		15.40	.519			28.50		28.50	44
0520	Photo cell		30	.267			14.60		14.60	22.50
0540	Load management device, 4 loads		7.70	1.039			57		57	88
0550	8 loads		3.80	2.105			115		115	178

26 05 05.35 Electrical Demolition, L.V. Distribution		Crew	Daily Output	Labor-Hours	Unit	Material	2015 Bare Costs Labor	Equipment	Total	Total Incl O&P
0560	Master light control panel	2 Elec	2	8	Ea.		440		440	680
0600	Transfer switches, enclosed, 30 A	1 Elec	12	.667			36.50		36.50	56.50
0610	60 A		9.50	.842			46		46	71.50
0620	100 A		6.50	1.231			67.50		67.50	104
0630	150 A	2 Elec	12	1.333			73		73	113
0640	260 A		10	1.600			87.50		87.50	136
0660	400 A		8	2			109		109	169
0670	600 A		5	3.200			175		175	271
0680	800 A		4	4			219		219	340
0690	1200 A		3.50	4.571			250		250	385
0700	1600 A		3	5.333			292		292	450
0710	2000 A		2.50	6.400			350		350	540
0800	Air terminal with base and connector	1 Elec	24	.333			18.25		18.25	28
1000	Enclosed controller, NEMA 1, 30 A		13.80	.580			31.50		31.50	49
1020	60 A		11.50	.696			38		38	59
1040	100 A		9.60	.833			45.50		45.50	70.50
1060	150 A		7.70	1.039			57		57	88
1080	200 A		5.40	1.481			81		81	125
1100	400 A	2 Elec	6.90	2.319			127		127	196
1120	600 A		4.60	3.478			190		190	295
1140	800 A		3.80	4.211			230		230	355
1160	1200 A		3.10	5.161			282		282	435
1200	Control station, NEMA 1	1 Elec	30	.267			14.60		14.60	22.50
1220	NEMA 7		23	.348			19.05		19.05	29.50
1230	Control switches, push button		69	.116			6.35		6.35	9.80
1240	Indicating light unit		120	.067			3.65		3.65	5.65
1250	Relay		15	.533			29		29	45
2000	Motor control center components									
2020	Starter, NEMA 1, size 1	1 Elec	9	.889	Ea.		48.50		48.50	75.50
2040	Size 2	2 Elec	13.30	1.203			66		66	102
2060	Size 3		6.70	2.388			131		131	202
2080	Size 4		5.30	3.019			165		165	256
2100	Size 5		3.30	4.848			265		265	410
2120	NEMA 7, size 1	1 Elec	8.70	.920			50.50		50.50	78
2140	Size 2	2 Elec	12.70	1.260			69		69	107
2160	Size 3		6.30	2.540			139		139	215
2180	Size 4		5	3.200			175		175	271
2200	Size 5		3.20	5			274		274	425
2300	Fuse, light contactor, NEMA 1, 30 A	1 Elec	9	.889			48.50		48.50	75.50
2310	60 A		6.70	1.194			65.50		65.50	101
2320	100 A		3.30	2.424			133		133	205
2330	200 A		2.70	2.963			162		162	251
2350	Motor control center, incoming section	2 Elec	4	4			219		219	340
2352	Structure per section	"	4.20	3.810			208		208	325
2400	Starter & structure, 10 HP	1 Elec	9	.889			48.50		48.50	75.50
2410	25 HP	2 Elec	13.30	1.203			66		66	102
2420	50 HP		6.70	2.388			131		131	202
2430	75 HP		5.30	3.019			165		165	256
2440	100 HP		4.70	3.404			186		186	288
2450	200 HP		3.30	4.848			265		265	410
2460	400 HP		2.70	5.926			325		325	500
2500	Motor starter & control									
2510	Motor starter, NEMA 1, 5 HP	1 Elec	8.90	.899	Ea.		49		49	76

26 05 Common Work Results for Electrical

26 05 05 – Selective Demolition for Electrical

26 05 05.35 Electrical Demolition, L.V. Distribution		Crew	Daily Output	Labor-Hours	Unit	Material	2015 Bare Costs Labor	Equipment	Total	Total Incl O&P
2520	10 HP	1 Elec	6.20	1.290	Ea.		70.50		70.50	109
2530	25 HP	2 Elec	8.40	1.905			104		104	161
2540	50 HP		6.90	2.319			127		127	196
2550	100 HP		4.60	3.478			190		190	295
2560	200 HP		3.50	4.571			250		250	385
2570	400 HP		3.10	5.161			282		282	435
2610	Motor starter, NEMA 7, 5 HP	1 Elec	6.20	1.290			70.50		70.50	109
2620	10 HP	″	4.20	1.905			104		104	161
2630	25 HP	2 Elec	6.90	2.319			127		127	196
2640	50 HP		4.60	3.478			190		190	295
2650	100 HP		3.50	4.571			250		250	385
2660	200 HP		1.90	8.421			460		460	715
2710	Combination control unit, NEMA 1, 5 HP	1 Elec	6.90	1.159			63.50		63.50	98
2720	10 HP	″	5	1.600			87.50		87.50	136
2730	25 HP	2 Elec	7.70	2.078			114		114	176
2740	50 HP		5.10	3.137			172		172	266
2750	100 HP		3.10	5.161			282		282	435
2810	NEMA 7, 5 HP	1 Elec	5	1.600			87.50		87.50	136
2820	10 HP	″	3.90	2.051			112		112	174
2830	25 HP	2 Elec	5.10	3.137			172		172	266
2840	50 HP		3.10	5.161			282		282	435
2850	100 HP		2.30	6.957			380		380	590
2860	200 HP		1.50	10.667			585		585	905
3000	Panelboard or load center circuit breaker									
3010	Bolt-on or plug in, 15 A to 50 A	1 Elec	20	.400	Ea.		22		22	34
3020	60 A to 70 A		16	.500			27.50		27.50	42.50
3030	Bolt-on, 80 A to 100 A		14	.571			31.50		31.50	48.50
3040	Up to 250 A		7.30	1.096			60		60	93
3050	Motor operated, 30 A		13	.615			33.50		33.50	52
3060	60 A		10	.800			44		44	68
3070	100 A		8	1			54.50		54.50	84.50
3200	Switchboard circuit breaker									
3210	15 A - 60 A	1 Elec	19	.421	Ea.		23		23	35.50
3220	70 A - 100 A		14	.571			31.50		31.50	48.50
3230	125 A - 400 A		10	.800			44		44	68
3240	450 A - 600 A		5.30	1.509			82.50		82.50	128
3250	700 A - 800 A		4.30	1.860			102		102	158
3260	1000 A		3.30	2.424			133		133	205
3270	1200 A		2.70	2.963			162		162	251
3500	Switchboard, incoming section, 400 A	2 Elec	3.70	4.324			237		237	365
3510	600 A		3.30	4.848			265		265	410
3520	800 A		2.90	5.517			300		300	465
3530	1200 A		2.40	6.667			365		365	565
3540	1600 A		2.20	7.273			400		400	615
3550	2000 A		2.10	7.619			415		415	645
3560	3000 A		1.90	8.421			460		460	715
3570	4000 A		1.70	9.412			515		515	795
3610	Distribution section, 600 A		4	4			219		219	340
3620	800 A		3.60	4.444			243		243	375
3630	1200 A		3.10	5.161			282		282	435
3640	1600 A		2.90	5.517			300		300	465
3650	2000 A		2.70	5.926			325		325	500
3710	Transition section, 600 A		3.80	4.211			230		230	355

26 05 05 – Selective Demolition for Electrical

26 05 05.35 Electrical Demolition, L.V. Distribution		Crew	Daily Output	Labor-Hours	Unit	Material	2015 Bare Costs Labor	Equipment	Total	Total Incl O&P
3720	800 A	2 Elec	3.30	4.848	Ea.		265		265	410
3730	1200 A		2.70	5.926			325		325	500
3740	1600 A		2.40	6.667			365		365	565
3750	2000 A		2.20	7.273			400		400	615
3760	2500 A		2.10	7.619			415		415	645
3770	3000 A		1.90	8.421			460		460	715
4000	Bus duct, aluminum or copper, 30 A	1 Elec	200	.040	L.F.		2.19		2.19	3.39
4020	60 A		160	.050			2.74		2.74	4.24
4040	100 A		140	.057			3.13		3.13	4.84
5000	Feedrail, trolley busway, up to 60 A		160	.050			2.74		2.74	4.24
5020	100 A		120	.067			3.65		3.65	5.65
5040	Busway, 50 A		200	.040			2.19		2.19	3.39
6000	Fuse, 30 A		133	.060	Ea.		3.29		3.29	5.10
6010	60 A		133	.060			3.29		3.29	5.10
6020	100 A		105	.076			4.17		4.17	6.45
6030	200 A		95	.084			4.61		4.61	7.15
6040	400 A		80	.100			5.45		5.45	8.45
6050	600 A		53	.151			8.25		8.25	12.80
6060	601 A - 1200 A		42	.190			10.40		10.40	16.15
6070	1500 A - 1600 A		34	.235			12.85		12.85	19.95
6080	1800 A - 2500 A		26	.308			16.85		16.85	26
6090	4000 A		21	.381			21		21	32.50
6100	4500 A - 5000 A		18	.444			24.50		24.50	37.50
6120	6000 A		15	.533			29		29	45
6200	Fuse plug or fustat		100	.080			4.38		4.38	6.80

26 05 05.50 Electrical Demolition, Lighting		Crew	Daily Output	Labor-Hours	Unit	Material	Labor	Equipment	Total	Total Incl O&P
0010	**ELECTRICAL DEMOLITION, LIGHTING** Addition R024119-10									
0100	Fixture hanger, flexible, 1/2" diameter	1 Elec	36	.222	Ea.		12.15		12.15	18.80
0120	3/4" diameter R260105-30	"	30	.267	"		14.60		14.60	22.50
3000	Light pole, anchor base, excl concrete bases									
3010	Metal light pole, 10'	2 Elec	24	.667	Ea.		36.50		36.50	56.50
3020	16'	"	18	.889			48.50		48.50	75.50
3030	20'	R-3	8.70	2.299			125	15.85	140.85	211
3040	40'	"	6	3.333			181	23	204	305
3100	Wood light pole, 10'	2 Elec	36	.444			24.50		24.50	37.50
3120	20'	"	24	.667			36.50		36.50	56.50
3140	Bollard light, 42"	1 Elec	9	.889			48.50		48.50	75.50
3160	Walkway luminaire	"	8.10	.988			54		54	83.50
4000	Explosionproof									
4010	Metal halide, 175 W	1 Elec	8.70	.920	Ea.		50.50		50.50	78
4020	250 W	"	8.10	.988			54		54	83.50
4030	400 W	2 Elec	14.40	1.111			61		61	94
4050	High pressure sodium, 70 W	1 Elec	9	.889			48.50		48.50	75.50
4060	100 W		9	.889			48.50		48.50	75.50
4070	150 W		8.10	.988			54		54	83.50
4100	Incandescent		8.70	.920			50.50		50.50	78
4200	Fluorescent		8.10	.988			54		54	83.50
5000	Ballast, fluorescent fixture		24	.333			18.25		18.25	28
5040	High intensity discharge fixture		24	.333			18.25		18.25	28
5300	Exit and emergency lighting									
5310	Exit light	1 Elec	24	.333	Ea.		18.25		18.25	28
5320	Emergency battery pack lighting unit		12	.667			36.50		36.50	56.50

26 05 Common Work Results for Electrical

26 05 05 – Selective Demolition for Electrical

26 05 05.50 Electrical Demolition, Lighting		Crew	Daily Output	Labor-Hours	Unit	Material	2015 Bare Costs Labor	Equipment	Total	Total Incl O&P
5330	Remote lamp only	1 Elec	80	.100	Ea.		5.45		5.45	8.45
5340	Self-contained fluorescent lamp pack		30	.267			14.60		14.60	22.50
5500	Track lighting, 8′ section	2 Elec	40	.400			22		22	34
5510	Track lighting fixture	1 Elec	64	.125			6.85		6.85	10.60
5800	Energy saving devices									
5810	Occupancy sensor	1 Elec	21	.381	Ea.		21		21	32.50
5820	Automatic wall switch		72	.111			6.10		6.10	9.40
5830	Remote power pack		30	.267			14.60		14.60	22.50
5840	Photoelectric control		24	.333			18.25		18.25	28
5850	Fixture whip		100	.080			4.38		4.38	6.80
6000	Lamps									
6010	Fluorescent	1 Elec	200	.040	Ea.		2.19		2.19	3.39
6030	High intensity discharge lamp, up to 400 W		68	.118			6.45		6.45	9.95
6040	Up to 1000 W		45	.178			9.70		9.70	15.05
6050	Quartz		90	.089			4.86		4.86	7.55
6070	Incandescent		360	.022			1.22		1.22	1.88
6080	Exterior, PAR		290	.028			1.51		1.51	2.34
6090	Guards for fluorescent lamp		24	.333			18.25		18.25	28

26 05 13 – Medium-Voltage Cables

26 05 13.10 Cable Terminations		Crew	Daily Output	Labor-Hours	Unit	Material	Labor	Equipment	Total	Total Incl O&P
0010	**CABLE TERMINATIONS**, 5 kV to 35 kV									
0100	Indoor, insulation diameter range .64″ to 1.08″									
0300	Padmount, 5 kV	1 Elec	8	1	Ea.	72.50	54.50		127	165
0400	10 kV		6.40	1.250		84.50	68.50		153	199
0500	15 kV		6	1.333		108	73		181	232
0600	25 kV		5.60	1.429		147	78		225	283
0700	insulation diameter range 1.05″ to 1.8″									
0800	Padmount, 5 kV	1 Elec	8	1	Ea.	102	54.50		156.50	197
0900	10 kV		6	1.333		117	73		190	242
1000	15 kV		5.60	1.429		128	78		206	262
1100	25 kV		5.30	1.509		168	82.50		250.50	315
1200	insulation diameter range 1.53″ to 2.32″									
1300	Padmount, 5 KV	1 Elec	7.40	1.081	Ea.	121	59		180	225
1400	10 kV		5.60	1.429		139	78		217	274
1500	15 kV		5.30	1.509		166	82.50		248.50	310
1600	25 kV		5	1.600		238	87.50		325.50	395
1700	Outdoor systems, #4 stranded to 1/0 stranded									
1800	5 kV	1 Elec	7.40	1.081	Ea.	91	59		150	192
1900	15 kV		5.30	1.509		134	82.50		216.50	276
2000	25 kV		5	1.600		180	87.50		267.50	335
2100	35 kV		4.80	1.667		248	91		339	415
2200	#1 solid to 4/0 stranded, 5 kV		6.90	1.159		123	63.50		186.50	234
2300	15 kV		5	1.600		172	87.50		259.50	325
2400	25 kV		4.80	1.667		232	91		323	395
2500	35 kV		4.60	1.739		264	95		359	435
2600	2/0 solid to 350 kcmil stranded, 5 kV		6.40	1.250		164	68.50		232.50	286
2700	15 kV		4.80	1.667		194	91		285	355
2800	25 kV		4.60	1.739		300	95		395	475
2900	35 kV		4.40	1.818		305	99.50		404.50	490
3000	400 kcmil compact to 750 kcmil stranded, 5 kV		6	1.333		196	73		269	330
3100	15 kV		4.60	1.739		234	95		329	405
3200	25 kV		4.40	1.818		340	99.50		439.50	530

26 05 13.10 Cable Terminations

	26 05 13.10 Cable Terminations	Crew	Daily Output	Labor-Hours	Unit	Material	2015 Bare Costs Labor	Equipment	Total	Total Incl O&P
3300	35 kV	1 Elec	4.20	1.905	Ea.	350	104		454	545
3400	1000 kcmil, 5 kV		5.60	1.429		266	78		344	415
3500	15 kV		4.40	1.818		315	99.50		414.50	505
3600	25 kV		4.20	1.905		350	104		454	545
3700	35 kV		4	2		350	109		459	555

26 05 13.16 Medium-Voltage, Single Cable

	26 05 13.16 Medium-Voltage, Single Cable	Crew	Daily Output	Labor-Hours	Unit	Material	Labor	Equipment	Total	Total Incl O&P
0010	**MEDIUM-VOLTAGE, SINGLE CABLE** Splicing & terminations not included									
0040	Copper, XLP shielding, 5 kV, #6	2 Elec	4.40	3.636	C.L.F.	160	199		359	485
0050	#4		4.40	3.636		207	199		406	540
0100	#2		4	4		228	219		447	590
0200	#1		4	4		281	219		500	650
0400	1/0		3.80	4.211		315	230		545	700
0600	2/0		3.60	4.444		385	243		628	795
0800	4/0		3.20	5		520	274		794	995
1000	250 kcmil	3 Elec	4.50	5.333		595	292		887	1,100
1200	350 kcmil		3.90	6.154		780	335		1,115	1,375
1400	500 kcmil		3.60	6.667		955	365		1,320	1,625
1600	15 kV, ungrounded neutral, #1	2 Elec	4	4		340	219		559	715
1800	1/0		3.80	4.211		410	230		640	805
2000	2/0		3.60	4.444		465	243		708	885
2200	4/0		3.20	5		620	274		894	1,100
2400	250 kcmil	3 Elec	4.50	5.333		685	292		977	1,200
2600	350 kcmil		3.90	6.154		865	335		1,200	1,475
2800	500 kcmil		3.60	6.667		1,075	365		1,440	1,750
3000	25 kV, grounded neutral, #1/0	2 Elec	3.60	4.444		565	243		808	995
3200	2/0		3.40	4.706		620	257		877	1,075
3400	4/0		3	5.333		780	292		1,072	1,300
3600	250 kcmil	3 Elec	4.20	5.714		970	315		1,285	1,550
3800	350 kcmil		3.60	6.667		1,125	365		1,490	1,825
3900	500 kcmil		3.30	7.273		1,325	400		1,725	2,100
4000	35 kV, grounded neutral, #1/0	2 Elec	3.40	4.706		600	257		857	1,050
4200	2/0		3.20	5		705	274		979	1,200
4400	4/0		2.80	5.714		885	315		1,200	1,450
4600	250 kcmil	3 Elec	3.90	6.154		1,025	335		1,360	1,675
4800	350 kcmil		3.30	7.273		1,250	400		1,650	2,000
5000	500 kcmil		3	8		1,475	440		1,915	2,275
5050	Aluminum, XLP shielding, 5 kV, #2	2 Elec	5	3.200		170	175		345	460
5070	#1		4.40	3.636		176	199		375	505
5090	1/0		4	4		204	219		423	565
5100	2/0		3.80	4.211		229	230		459	605
5150	4/0		3.60	4.444		271	243		514	675
5200	250 kcmil	3 Elec	4.80	5		330	274		604	785
5220	350 kcmil		4.50	5.333		385	292		677	875
5240	500 kcmil		3.90	6.154		495	335		830	1,075
5260	750 kcmil		3.60	6.667		655	365		1,020	1,275
5300	15 kV aluminum, XLP, #1	2 Elec	4.40	3.636		218	199		417	550
5320	1/0		4	4		226	219		445	590
5340	2/0		3.80	4.211		269	230		499	650
5360	4/0		3.60	4.444		299	243		542	705
5380	250 kcmil	3 Elec	4.80	5		355	274		629	815
5400	350 kcmil		4.50	5.333		400	292		692	890
5420	500 kcmil		3.90	6.154		555	335		890	1,125

26 05 13 – Medium-Voltage Cables

26 05 13.16 Medium-Voltage, Single Cable		Crew	Daily Output	Labor-Hours	Unit	Material	Labor	Equipment	Total	Total Incl O&P
						2015 Bare Costs				
5440	750 kcmil	3 Elec	3.60	6.667	C.L.F.	735	365		1,100	1,375

26 05 19 – Low-Voltage Electrical Power Conductors and Cables

26 05 19.13 Undercarpet Electrical Power Cables

	26 05 19.13 Undercarpet Electrical Power Cables	Crew	Daily Output	Labor-Hours	Unit	Material	Labor	Equipment	Total	Total Incl O&P
0010	**UNDERCARPET ELECTRICAL POWER CABLES** R260519-80									
0020	Power System									
0100	Cable flat, 3 conductor, #12, w/attached bottom shield	1 Elec	982	.008	L.F.	5.15	.45		5.60	6.35
0200	Shield, top, steel		1768	.005	"	5.50	.25		5.75	6.45
0250	Splice, 3 conductor		48	.167	Ea.	16.75	9.10		25.85	32.50
0300	Top shield		96	.083		1.49	4.56		6.05	8.70
0350	Tap		40	.200		21.50	10.95		32.45	40.50
0400	Insulating patch, splice, tap, & end		48	.167		53	9.10		62.10	72
0450	Fold		230	.035			1.90		1.90	2.95
0500	Top shield, tap & fold		96	.083		1.49	4.56		6.05	8.70
0700	Transition, block assembly		77	.104		77	5.70		82.70	94
0750	Receptacle frame & base		32	.250		42.50	13.70		56.20	67.50
0800	Cover receptacle		120	.067		3.61	3.65		7.26	9.60
0850	Cover blank		160	.050		4.24	2.74		6.98	8.90
0860	Receptacle, direct connected, single		25	.320		90.50	17.50		108	127
0870	Dual		16	.500		149	27.50		176.50	206
0880	Combination high & low, tension		21	.381		109	21		130	153
0900	Box, floor with cover		20	.400		92	22		114	135
0920	Floor service w/barrier		4	2		259	109		368	455
1000	Wall, surface, with cover		20	.400		60.50	22		82.50	101
1100	Wall, flush, with cover		20	.400		42.50	22		64.50	80.50
1450	Cable flat, 5 conductor #12, w/attached bottom shield		800	.010	L.F.	8.45	.55		9	10.10
1550	Shield, top, steel		1768	.005	"	8.35	.25		8.60	9.60
1600	Splice, 5 conductor		48	.167	Ea.	27	9.10		36.10	44
1650	Top shield		96	.083		1.49	4.56		6.05	8.70
1700	Tap		48	.167		35.50	9.10		44.60	53
1750	Insulating patch, splice tap, & end		83	.096		53	5.25		58.25	66
1800	Transition, block assembly		77	.104		55.50	5.70		61.20	70
1850	Box, wall, flush with cover		20	.400		56.50	22		78.50	96
1900	Cable flat, 4 conductor, #12		933	.009	L.F.	6.80	.47		7.27	8.25
1950	3 conductor #10		982	.008		5.90	.45		6.35	7.20
1960	4 conductor #10		933	.009		7.70	.47		8.17	9.25
1970	5 conductor #10		884	.009		9.45	.50		9.95	11.10
2500	Telephone System									
2510	Transition fitting wall box, surface	1 Elec	24	.333	Ea.	45.50	18.25		63.75	78
2520	Flush		24	.333		45.50	18.25		63.75	78
2530	Flush, for PC board		24	.333		45.50	18.25		63.75	78
2540	Floor service box		4	2		235	109		344	430
2550	Cover, surface					14.45			14.45	15.90
2560	Flush					14.45			14.45	15.90
2570	Flush for PC board					14.45			14.45	15.90
2700	Floor fitting w/duplex jack & cover	1 Elec	21	.381		44.50	21		65.50	81.50
2720	Low profile		53	.151		15.55	8.25		23.80	30
2740	Miniature w/duplex jack		53	.151		24	8.25		32.25	39.50
2760	25 pair kit		21	.381		47	21		68	84
2780	Low profile		53	.151		15.90	8.25		24.15	30.50
2800	Call director kit for 5 cable		19	.421		74.50	23		97.50	118
2820	4 pair kit		19	.421		89	23		112	134
2840	3 pair kit		19	.421		94	23		117	140

26 05 19.13 Undercarpet Electrical Power Cables		Crew	Daily Output	Labor-Hours	Unit	2015 Bare Costs Material	Labor	Equipment	Total	Total Incl O&P
2860	Comb. 25 pair & 3 conductor power	1 Elec	21	.381	Ea.	74.50	21		95.50	115
2880	5 conductor power		21	.381		85.50	21		106.50	127
2900	PC board, 8 per 3 pair		161	.050		64.50	2.72		67.22	75
2920	6 per 4 pair		161	.050		64.50	2.72		67.22	75
2940	3 pair adapter		161	.050		59	2.72		61.72	69
2950	Plug		77	.104		2.74	5.70		8.44	11.80
2960	Couplers		321	.025		7.75	1.36		9.11	10.60
3000	Bottom shield for 25 pair cable		4420	.002	L.F.	.78	.10		.88	1.01
3020	4 pair		4420	.002		.37	.10		.47	.56
3040	Top shield for 25 pair cable		4420	.002		.78	.10		.88	1.01
3100	Cable assembly, double-end, 50', 25 pair		11.80	.678	Ea.	227	37		264	310
3110	3 pair		23.60	.339		67	18.55		85.55	103
3120	4 pair		23.60	.339		74.50	18.55		93.05	111
3140	Bulk 3 pair		1473	.005	L.F.	1.15	.30		1.45	1.73
3160	4 pair		1473	.005	"	1.43	.30		1.73	2.03
3500	Data System									
3520	Cable 25 conductor w/connection 40', 75 ohm	1 Elec	14.50	.552	Ea.	67.50	30		97.50	121
3530	Single lead		22	.364		184	19.90		203.90	233
3540	Dual lead		22	.364		229	19.90		248.90	282
3560	Shields same for 25 conductor as 25 pair telephone									
3570	Single & dual, none required									
3590	BNC coax connectors, Plug	1 Elec	40	.200	Ea.	10.05	10.95		21	28
3600	TNC coax connectors, Plug	"	40	.200	"	12.90	10.95		23.85	31
3700	Cable-bulk									
3710	Single lead	1 Elec	1473	.005	L.F.	2.45	.30		2.75	3.16
3720	Dual lead	"	1473	.005	"	3.55	.30		3.85	4.37
3730	Hand tool crimp				Ea.	510			510	560
3740	Hand tool notch				"	17.55			17.55	19.30
3750	Boxes & floor fitting same as telephone									
3790	Data cable notching, 90°	1 Elec	97	.082	Ea.		4.51		4.51	7
3800	180°		60	.133			7.30		7.30	11.30
8100	Drill floor		160	.050		1.96	2.74		4.70	6.40
8200	Marking floor		1600	.005	L.F.		.27		.27	.42
8300	Tape, hold down		6400	.001	"	.16	.07		.23	.29
8350	Tape primer, 500 ft. per can		96	.083	Ea.	35.50	4.56		40.06	46.50
8400	Tool, splicing				"	201			201	221

26 05 19.20 Armored Cable		Crew	Daily Output	Labor-Hours	Unit	Material	Labor	Equipment	Total	Total Incl O&P
0010	**ARMORED CABLE**									
0050	600 volt, copper (BX), #14, 2 conductor, solid	1 Elec	2.40	3.333	C.L.F.	46	182		228	335
0100	3 conductor, solid		2.20	3.636		71	199		270	390
0120	4 conductor, solid		2	4		98.50	219		317.50	450
0150	#12, 2 conductor, solid		2.30	3.478		46	190		236	345
0200	3 conductor, solid		2	4		75.50	219		294.50	425
0220	4 conductor, solid		1.80	4.444		103	243		346	490
0250	#10, 2 conductor, solid		2	4		85	219		304	435
0300	3 conductor, solid		1.60	5		118	274		392	555
0320	4 conductor, solid		1.40	5.714		182	315		497	685
0340	#8, 2 conductor, stranded		1.50	5.333		222	292		514	695
0350	3 conductor, stranded		1.30	6.154		222	335		557	765
0370	4 conductor, stranded		1.10	7.273		315	400		715	965
0380	#6, 2 conductor, stranded		1.30	6.154		233	335		568	775
0390	#4, 3 conductor, stranded		1.40	5.714		555	315		870	1,100

26 05 19.20 Armored Cable		Crew	Daily Output	Labor-Hours	Unit	Material	2015 Bare Costs Labor	Equipment	Total	Total Incl O&P
0400	3 conductor with PVC jacket, in cable tray, #6	1 Elec	3.10	2.581	C.L.F.	540	141		681	810
0450	#4	2 Elec	5.40	2.963		655	162		817	970
0500	#2		4.60	3.478		795	190		985	1,175
0550	#1		4	4		1,025	219		1,244	1,475
0600	1/0		3.60	4.444		1,050	243		1,293	1,525
0650	2/0		3.40	4.706		1,250	257		1,507	1,775
0700	3/0		3.20	5		1,700	274		1,974	2,275
0750	4/0		3	5.333		2,000	292		2,292	2,650
0800	250 kcmil	3 Elec	3.60	6.667		2,325	365		2,690	3,150
0850	350 kcmil		3.30	7.273		3,150	400		3,550	4,075
0900	500 kcmil		3	8		4,350	440		4,790	5,450
0910	4 conductor with PVC jacket, in cable tray, #6	1 Elec	2.70	2.963		670	162		832	985
0920	#4	2 Elec	4.60	3.478		840	190		1,030	1,225
0930	#2		4	4		1,025	219		1,244	1,475
0940	#1		3.60	4.444		1,300	243		1,543	1,825
0950	1/0		3.40	4.706		1,350	257		1,607	1,875
0960	2/0		3.20	5		1,650	274		1,924	2,250
0970	3/0		3	5.333		2,200	292		2,492	2,850
0980	4/0		2.40	6.667		2,550	365		2,915	3,375
0990	250 kcmil	3 Elec	3.30	7.273		3,050	400		3,450	4,000
1000	350 kcmil		3	8		4,125	440		4,565	5,200
1010	500 kcmil		2.70	8.889		5,700	485		6,185	7,025
1050	5 kV, copper, 3 conductor with PVC jacket,									
1060	non-shielded, in cable tray, #4	2 Elec	380	.042	L.F.	7.40	2.30		9.70	11.70
1100	#2		360	.044		9.65	2.43		12.08	14.35
1200	#1		300	.053		12.30	2.92		15.22	18
1400	1/0		290	.055		14.20	3.02		17.22	20.50
1600	2/0		260	.062		16.40	3.37		19.77	23.50
2000	4/0		240	.067		22	3.65		25.65	29.50
2100	250 kcmil	3 Elec	330	.073		30	3.98		33.98	39
2150	350 kcmil		315	.076		37	4.17		41.17	47
2200	500 kcmil		270	.089		51.50	4.86		56.36	64.50
2400	15 kV, copper, 3 conductor with PVC jacket galv., steel armored									
2500	grounded neutral, in cable tray, #2	2 Elec	300	.053	L.F.	15.60	2.92		18.52	21.50
2600	#1		280	.057		16.60	3.13		19.73	23
2800	1/0		260	.062		19	3.37		22.37	26
2900	2/0		220	.073		25	3.98		28.98	34
3000	4/0		190	.084		28.50	4.61		33.11	38.50
3100	250 kcmil	3 Elec	270	.089		32	4.86		36.86	42.50
3150	350 kcmil		240	.100		37.50	5.45		42.95	50
3200	500 kcmil		210	.114		50.50	6.25		56.75	65
3400	15 kV, copper, 3 conductor with PVC jacket,									
3450	ungrounded neutral, in cable tray, #2	2 Elec	260	.062	L.F.	16.80	3.37		20.17	23.50
3500	#1		230	.070		18.60	3.81		22.41	26.50
3600	1/0		200	.080		21.50	4.38		25.88	30.50
3700	2/0		190	.084		26	4.61		30.61	36
3800	4/0		160	.100		31.50	5.45		36.95	43
4000	250 kcmil	3 Elec	210	.114		37	6.25		43.25	50
4050	350 kcmil		195	.123		48.50	6.75		55.25	64
4100	500 kcmil		180	.133		59	7.30		66.30	76.50
4200	600 volt, aluminum, 3 conductor in cable tray with PVC jacket									
4300	#2	2 Elec	540	.030	L.F.	3.96	1.62		5.58	6.85
4400	#1		460	.035		4.38	1.90		6.28	7.75

26 05 19.20 Armored Cable		Crew	Daily Output	Labor-Hours	Unit	2015 Bare Costs Material	Labor	Equipment	Total	Total Incl O&P
4500	#1/0	2 Elec	400	.040	L.F.	5.45	2.19		7.64	9.40
4600	#2/0		360	.044		5.55	2.43		7.98	9.85
4700	#3/0		340	.047		6.50	2.57		9.07	11.15
4800	#4/0		320	.050		7.85	2.74		10.59	12.90
4900	250 kcmil	3 Elec	450	.053		9.45	2.92		12.37	14.85
5000	350 kcmil		360	.067		11.25	3.65		14.90	18
5200	500 kcmil		330	.073		13.95	3.98		17.93	21.50
5300	750 kcmil		285	.084		18	4.61		22.61	27
5400	600 volt, aluminum, 4 conductor in cable tray with PVC jacket									
5410	#2	2 Elec	520	.031	L.F.	4.55	1.68		6.23	7.60
5430	#1		440	.036		5.55	1.99		7.54	9.20
5450	1/0		380	.042		6.50	2.30		8.80	10.70
5470	2/0		340	.047		6.60	2.57		9.17	11.30
5480	3/0		320	.050		7.80	2.74		10.54	12.80
5500	4/0		300	.053		9.15	2.92		12.07	14.55
5520	250 kcmil	3 Elec	420	.057		9.80	3.13		12.93	15.65
5540	350 kcmil		330	.073		12.65	3.98		16.63	20
5560	500 kcmil		300	.080		15.85	4.38		20.23	24.50
5580	750 kcmil		270	.089		22	4.86		26.86	31.50
5600	5 kV, aluminum, unshielded in cable tray, #2 with PVC jacket	2 Elec	380	.042		5.55	2.30		7.85	9.65
5700	#1 with PVC jacket		360	.044		6.20	2.43		8.63	10.55
5800	1/0 with PVC jacket		300	.053		6.35	2.92		9.27	11.45
6000	2/0 with PVC jacket		290	.055		6.50	3.02		9.52	11.80
6200	3/0 with PVC jacket		260	.062		7.80	3.37		11.17	13.75
6300	4/0 with PVC jacket		240	.067		9.15	3.65		12.80	15.70
6400	250 kcmil with PVC jacket	3 Elec	330	.073		10	3.98		13.98	17.15
6500	350 kcmil with PVC jacket		315	.076		11.75	4.17		15.92	19.35
6600	500 kcmil with PVC jacket		300	.080		13.90	4.38		18.28	22
6800	750 kcmil with PVC jacket		270	.089		17.15	4.86		22.01	26.50
6900	15 kV, aluminum, shielded-grounded, #2 with PVC jacket	2 Elec	320	.050		12.60	2.74		15.34	18.10
7000	#1 with PVC jacket		300	.053		12.95	2.92		15.87	18.75
7200	1/0 with PVC jacket		280	.057		13.95	3.13		17.08	20
7300	2/0 with PVC jacket		260	.062		14.20	3.37		17.57	21
7400	3/0 with PVC jacket		240	.067		15.90	3.65		19.55	23
7500	4/0 with PVC jacket		220	.073		16.35	3.98		20.33	24
7600	250 kcmil with PVC jacket	3 Elec	300	.080		17.95	4.38		22.33	26.50
7700	350 kcmil with PVC jacket		270	.089		21	4.86		25.86	30.50
7800	500 kcmil with PVC jacket		240	.100		25	5.45		30.45	36
8000	750 kcmil with PVC jacket		204	.118		30	6.45		36.45	43
8200	15 kV, aluminum, shielded-ungrounded, #1 with PVC jacket	2 Elec	250	.064		15.80	3.50		19.30	23
8300	1/0 with PVC jacket		230	.070		16.35	3.81		20.16	24
8400	2/0 with PVC jacket		210	.076		17.95	4.17		22.12	26
8500	3/0 with PVC jacket		200	.080		18.20	4.38		22.58	27
8600	4/0 with PVC jacket		190	.084		19.80	4.61		24.41	29
8700	250 kcmil with PVC jacket	3 Elec	270	.089		21.50	4.86		26.36	31
8800	350 kcmil with PVC jacket		240	.100		24.50	5.45		29.95	35
8900	500 kcmil with PVC jacket		210	.114		29.50	6.25		35.75	41.50
8950	750 kcmil with PVC jacket		174	.138		36.50	7.55		44.05	51.50
9010	600 volt, copper (MC) steel clad, #14, 2 wire	1 Elec	2.40	3.333	C.L.F.	46.50	182		228.50	335
9020	3 wire		2.20	3.636		72	199		271	390
9030	4 wire		2	4		100	219		319	450
9040	#12, 2 wire		2.30	3.478		47	190		237	345
9050	3 wire		2	4		79.50	219		298.50	430

26 05 19 – Low-Voltage Electrical Power Conductors and Cables

26 05 19.20 Armored Cable		Crew	Daily Output	Labor-Hours	Unit	2015 Bare Costs Material	Labor	Equipment	Total	Total Incl O&P
9060	4 wire	1 Elec	1.80	4.444	C.L.F.	107	243		350	495
9070	#10, 2 wire		2	4		98.50	219		317.50	450
9080	3 wire		1.60	5		138	274		412	575
9090	4 wire		1.40	5.714		216	315		531	720
9100	#8, 2 wire, stranded		1.80	4.444		190	243		433	585
9110	3 wire, stranded		1.30	6.154		267	335		602	815
9120	4 wire, stranded		1.10	7.273		360	400		760	1,000
9130	#6, 2 wire, stranded		1.30	6.154		281	335		616	830
9200	600 volt, copper (MC) aluminum clad, #14, 2 wire		2.65	3.019		45.50	165		210.50	305
9210	3 wire		2.45	3.265		71	179		250	355
9220	4 wire		2.20	3.636		99	199		298	420
9230	#12, 2 wire		2.55	3.137		47	172		219	320
9240	3 wire		2.20	3.636		78.50	199		277.50	395
9250	4 wire		2	4		106	219		325	455
9260	#10, 2 wire		2.20	3.636		97.50	199		296.50	415
9270	3 wire		1.80	4.444		137	243		380	525
9280	4 wire		1.55	5.161		215	282		497	670
9600	Alum (MC) aluminum clad, #6, 3 conductor w/#6 grnd		1.67	4.790		267	262		529	700
9610	4 conductor w/#6 grnd		1.64	4.878		290	267		557	735
9620	#4, 3 conductor w/#6 grnd	2 Elec	2.86	5.594		277	305		582	780
9630	4 conductor w/#6 grnd		2.82	5.674		300	310		610	810
9640	#2, 3 conductor w/#4 grnd		2.50	6.400		380	350		730	960
9650	4 conductor w/#4 grnd		2.47	6.478		430	355		785	1,025
9660	#1, 3 conductor w/#4 grnd		2	8		410	440		850	1,125
9670	4 conductor w/#4 grnd		1.98	8.081		580	440		1,020	1,325
9680	1/0, 3 conductor w/#4 grnd		1.82	8.791		470	480		950	1,250
9690	4 conductor w/#4 grnd		1.79	8.939		570	490		1,060	1,375
9700	2/0, 3 conductor w/#4 grnd		1.75	9.143		590	500		1,090	1,425
9710	4 conductor w/#4 grnd		1.71	9.357		640	510		1,150	1,500
9720	3/0, 3 conductor w/#4 grnd		1.71	9.357		755	510		1,265	1,625
9730	4 conductor w/#4 grnd		1.67	9.581		795	525		1,320	1,675
9740	4/0, 3 conductor w/#2 grnd		1.67	9.581		850	525		1,375	1,750
9750	4 conductor w/#2 grnd		1.61	9.938		975	545		1,520	1,925
9760	250 kcmil, 3 conductor w/#1 grnd	3 Elec	2.42	9.917		1,075	540		1,615	2,050
9770	4 conductor w/#1 grnd		2.36	10.169		1,475	555		2,030	2,475
9775	300 kcmil, 4 conductor w/#1 grnd		2.22	10.811		1,550	590		2,140	2,650
9780	350 kcmil, 3 conductor w/1/0 grnd		2.19	10.959		1,225	600		1,825	2,275
9790	4 conductor w/1/0 grnd		2.10	11.429		1,500	625		2,125	2,650
9800	500 kcmil, 3 conductor w/#1 grnd		2.10	11.429		1,800	625		2,425	2,950
9810	4 conductor w/2/0 grnd		2	12		1,850	655		2,505	3,050
9840	750 kcmil, 3 conductor w/1/0 grnd		1.95	12.308		2,575	675		3,250	3,900
9850	4 conductor w/3/0 grnd		1.89	12.698		2,800	695		3,495	4,175
9900	Minimum labor/equipment charge	1 Elec	4	2	Job		109		109	169

26 05 19.25 Cable Connectors		Crew	Daily Output	Labor-Hours	Unit	Material	Labor	Equipment	Total	Total Incl O&P
0010	**CABLE CONNECTORS**									
0100	600 volt, nonmetallic, #14-2 wire	1 Elec	160	.050	Ea.	1.44	2.74		4.18	5.80
0200	#14-3 wire to #12-2 wire		133	.060		1.44	3.29		4.73	6.70
0300	#12-3 wire to #10-2 wire		114	.070		1.44	3.84		5.28	7.55
0400	#10-3 wire to #14-4 and #12-4 wire		100	.080		1.44	4.38		5.82	8.40
0500	#8-3 wire to #10-4 wire		80	.100		2.71	5.45		8.16	11.45
0600	#6-3 wire		40	.200		3.65	10.95		14.60	21
0800	SER, 3 #8 insulated + 1 #8 ground		32	.250		2.32	13.70		16.02	23.50

26 05 19.25 Cable Connectors		Crew	Daily Output	Labor-Hours	Unit	Material	Labor	Equipment	Total	Total Incl O&P
						2015 Bare Costs				
0900	3 #6 + 1 #6 ground	1 Elec	24	.333	Ea.	2.90	18.25		21.15	31
1000	3 #4 + 1 #6 ground		22	.364		3.63	19.90		23.53	35
1100	3 #2 + 1 #4 ground		20	.400		6.60	22		28.60	41.50
1200	3 1/0 + 1 #2 ground		18	.444		17.10	24.50		41.60	56.50
1400	3 2/0 + 1 #1 ground		16	.500		20.50	27.50		48	65
1600	3 4/0 + 1 #2/0 ground		14	.571		27	31.50		58.50	78
1800	600 volt, armored, #14-2 wire		80	.100		1.01	5.45		6.46	9.55
2200	#14-4, #12-3 and #10-2 wire		40	.200		.91	10.95		11.86	17.95
2400	#12-4, #10-3 and #8-2 wire		32	.250		2.37	13.70		16.07	23.50
2600	#8-3 and #10-4 wire		26	.308		3.58	16.85		20.43	30
2650	#8-4 wire		22	.364		5.15	19.90		25.05	36.50
2652	Non-PVC jacket connector, #6-3 wire, #6-4 wire		22	.364		44	19.90		63.90	79
2660	1/0-3 wire		11	.727		60	40		100	128
2670	300 kcmil-3 wire		6	1.333		91	73		164	213
2680	700-kcmil-3 wire		4	2		135	109		244	315
2700	PVC jacket connector, #6-3 wire, #6-4 wire		16	.500		8	27.50		35.50	51.50
2800	#4-3 wire, #4-4 wire		16	.500		8	27.50		35.50	51.50
2900	#2-3 wire		12	.667		8	36.50		44.50	65.50
3000	#1-3 wire, #2-4 wire		12	.667		13.95	36.50		50.45	72
3200	1/0-3 wire		11	.727		13.95	40		53.95	77
3400	2/0-3 wire, 1/0-4 wire		10	.800		13.95	44		57.95	83.50
3500	3/0-3 wire, 2/0-4 wire		9	.889		20.50	48.50		69	98
3600	4/0-3 wire, 3/0-4 wire		7	1.143		20.50	62.50		83	120
3800	250 kcmil-3 wire, 4/0-4 wire		6	1.333		37.50	73		110.50	154
4000	350 kcmil-3 wire, 250 kcmil-4 wire		5	1.600		37.50	87.50		125	177
4100	350 kcmil-4 wire		4	2		190	109		299	380
4200	500 kcmil-3 wire		4	2		190	109		299	380
4250	500 kcmil-4 wire, 750 kcmil-3 wire		3.50	2.286		259	125		384	480
4300	750 kcmil-4 wire		3	2.667		259	146		405	510
4400	5 kV, armored, #4		8	1		67	54.50		121.50	159
4600	#2		8	1		67	54.50		121.50	159
4800	#1		8	1		86.50	54.50		141	180
5000	1/0		6.40	1.250		107	68.50		175.50	224
5200	2/0		5.30	1.509		107	82.50		189.50	246
5500	4/0		4	2		143	109		252	325
5600	250 kcmil		3.60	2.222		143	122		265	345
5650	350 kcmil		3.20	2.500		172	137		309	400
5700	500 kcmil		2.50	3.200		216	175		391	510
5720	750 kcmil		2.20	3.636		263	199		462	600
5750	1000 kcmil		2	4		325	219		544	695
5800	15 kV, armored, #1		4	2		129	109		238	310
5900	1/0		4	2		161	109		270	345
6000	3/0		3.60	2.222		214	122		336	425
6100	4/0		3.40	2.353		243	129		372	465
6200	250 kcmil		3.20	2.500		274	137		411	510
6300	350 kcmil		2.70	2.963		287	162		449	565
6400	500 kcmil	↓	2	4	↓	325	219		544	695

26 05 19.30 Cable Splicing		Crew	Daily Output	Labor-Hours	Unit	Material	Labor	Equipment	Total	Total Incl O&P
0010	**CABLE SPLICING** URD or similar, ideal conditions									
0100	#6 stranded to #1 stranded, 5 kV	1 Elec	4	2	Ea.	117	109		226	298
0120	15 kV		3.60	2.222		164	122		286	370
0140	25 kV	↓	3.20	2.500	↓	184	137		321	415

26 05 19 – Low-Voltage Electrical Power Conductors and Cables

26 05 19.30 Cable Splicing		Crew	Daily Output	Labor-Hours	Unit	Material	2015 Bare Costs Labor	Equipment	Total	Total Incl O&P
0200	#1 stranded to 4/0 stranded, 5 kV	1 Elec	3.60	2.222	Ea.	127	122		249	330
0210	15 kV		3.20	2.500		174	137		311	405
0220	25 kV		2.80	2.857		184	156		340	445
0300	4/0 stranded to 500 kcmil stranded, 5 kV		3.30	2.424		127	133		260	345
0310	15 kV		2.90	2.759		261	151		412	520
0320	25 kV		2.50	3.200		287	175		462	585
0400	500 kcmil, 5 kV		3.20	2.500		181	137		318	410
0410	15 kV		2.80	2.857		320	156		476	590
0420	25 kV		2.30	3.478		287	190		477	610
0500	600 kcmil, 5 kV		2.90	2.759		181	151		332	435
0510	15 kV		2.40	3.333		330	182		512	640
0520	25 kV		2	4		330	219		549	705
0600	750 kcmil, 5 kV		2.60	3.077		181	168		349	460
0610	15 kV		2.20	3.636		330	199		529	670
0620	25 kV		1.90	4.211		330	230		560	720
0700	1000 kcmil, 5 kV		2.30	3.478		181	190		371	495
0710	15 kV		1.90	4.211		355	230		585	750
0720	25 kV	↓	1.60	5	↓	380	274		654	840

26 05 19.35 Cable Terminations		Crew	Daily Output	Labor-Hours	Unit	Material	Labor	Equipment	Total	Total Incl O&P
0010	**CABLE TERMINATIONS**									
0015	Wire connectors, screw type, #22 to #14	1 Elec	260	.031	Ea.	.06	1.68		1.74	2.68
0020	#18 to #12		240	.033		.07	1.82		1.89	2.90
0025	#18 to #10		240	.033		.12	1.82		1.94	2.95
0030	Screw-on connectors, insulated, #18 to #12		240	.033		.25	1.82		2.07	3.10
0035	#16 to #10		230	.035		.27	1.90		2.17	3.25
0040	#14 to #8		210	.038		.30	2.08		2.38	3.56
0045	#12 to #6		180	.044		.62	2.43		3.05	4.44
0050	Terminal lugs, solderless, #16 to #10		50	.160		.37	8.75		9.12	13.95
0100	#8 to #4		30	.267		.73	14.60		15.33	23.50
0150	#2 to #1		22	.364		1.03	19.90		20.93	32
0200	1/0 to 2/0		16	.500		1.64	27.50		29.14	44.50
0250	3/0		12	.667		3.30	36.50		39.80	60
0300	4/0		11	.727		4.05	40		44.05	66
0350	250 kcmil		9	.889		3.41	48.50		51.91	79.50
0400	350 kcmil		7	1.143		4.44	62.50		66.94	102
0450	500 kcmil		6	1.333		8.20	73		81.20	122
0500	600 kcmil		5.80	1.379		9.30	75.50		84.80	127
0550	750 kcmil		5.20	1.538		10.95	84		94.95	142
0600	Split bolt connectors, tapped, #6		16	.500		3.78	27.50		31.28	46.50
0650	#4		14	.571		3.87	31.50		35.37	53
0700	#2		12	.667		5.70	36.50		42.20	63
0750	#1		11	.727		7.45	40		47.45	69.50
0800	1/0		10	.800		7.45	44		51.45	76
0850	2/0		9	.889		11.70	48.50		60.20	88.50
0900	3/0		7.20	1.111		16.25	61		77.25	112
1000	4/0		6.40	1.250		19.75	68.50		88.25	128
1100	250 kcmil		5.70	1.404		19.75	77		96.75	141
1200	300 kcmil		5.30	1.509		36	82.50		118.50	168
1400	350 kcmil		4.60	1.739		36	95		131	187
1500	500 kcmil	↓	4	2	↓	62.50	109		171.50	238
1600	Crimp 1 hole lugs, copper or aluminum, 600 volt									
1620	#14	1 Elec	60	.133	Ea.	.56	7.30		7.86	11.90

26 05 19.35 Cable Terminations		Crew	Daily Output	Labor-Hours	Unit	Material	Labor	Equipment	Total	Total Incl O&P
						2015 Bare Costs				
1630	#12	1 Elec	50	.160	Ea.	.90	8.75		9.65	14.55
1640	#10		45	.178		.90	9.70		10.60	16.05
1780	#8		36	.222		1.86	12.15		14.01	21
1800	#6		30	.267		2.11	14.60		16.71	25
2000	#4		27	.296		2.88	16.20		19.08	28
2200	#2		24	.333		4.64	18.25		22.89	33
2400	#1		20	.400		4.84	22		26.84	39.50
2500	1/0		17.50	.457		5.20	25		30.20	44
2600	2/0		15	.533		6.30	29		35.30	52
2800	3/0		12	.667		7	36.50		43.50	64
3000	4/0		11	.727		7.80	40		47.80	70
3200	250 kcmil		9	.889		9.10	48.50		57.60	85.50
3400	300 kcmil		8	1		11.05	54.50		65.55	96.50
3500	350 kcmil		7	1.143		11.50	62.50		74	110
3600	400 kcmil		6.50	1.231		13.75	67.50		81.25	119
3800	500 kcmil		6	1.333		15.70	73		88.70	130
4000	600 kcmil		5.80	1.379		27	75.50		102.50	147
4200	700 kcmil		5.50	1.455		32	79.50		111.50	158
4400	750 kcmil		5.20	1.538		32	84		116	165
4500	Crimp 2-way connectors, copper or alum., 600 volt,									
4510	#14	1 Elec	60	.133	Ea.	2.19	7.30		9.49	13.70
4520	#12		50	.160		2.33	8.75		11.08	16.10
4530	#10		45	.178		2.45	9.70		12.15	17.75
4540	#8		27	.296		2.48	16.20		18.68	27.50
4600	#6		25	.320		4.17	17.50		21.67	31.50
4800	#4		23	.348		4.76	19.05		23.81	35
5000	#2		20	.400		5.85	22		27.85	40.50
5200	#1		16	.500		5.90	27.50		33.40	49
5400	1/0		13	.615		6.50	33.50		40	59
5420	2/0		12	.667		8	36.50		44.50	65.50
5440	3/0		11	.727		9.45	40		49.45	72
5460	4/0		10	.800		10.55	44		54.55	79.50
5480	250 kcmil		9	.889		12.70	48.50		61.20	89.50
5500	300 kcmil		8.50	.941		14.30	51.50		65.80	95.50
5520	350 kcmil		8	1		14.75	54.50		69.25	101
5540	400 kcmil		7.30	1.096		22	60		82	117
5560	500 kcmil		6.20	1.290		26	70.50		96.50	138
5580	600 kcmil		5.50	1.455		30.50	79.50		110	157
5600	700 kcmil		4.50	1.778		45	97		142	201
5620	750 kcmil		4	2		41.50	109		150.50	215
7000	Compression equipment adapter, aluminum wire, #6		30	.267		10.30	14.60		24.90	34
7020	#4		27	.296		9.95	16.20		26.15	36
7040	#2		24	.333		10.40	18.25		28.65	39.50
7060	#1		20	.400		12.30	22		34.30	47.50
7080	1/0		18	.444		13.30	24.50		37.80	52
7100	2/0		15	.533		19.35	29		48.35	66.50
7140	4/0		11	.727		23	40		63	87
7160	250 kcmil		9	.889		26.50	48.50		75	105
7180	300 kcmil		8	1		26.50	54.50		81	114
7200	350 kcmil		7	1.143		30.50	62.50		93	131
7220	400 kcmil		6.50	1.231		39.50	67.50		107	148
7240	500 kcmil		6	1.333		41	73		114	159
7260	600 kcmil		5.80	1.379		46	75.50		121.50	168

26 05 19.35 Cable Terminations		Crew	Daily Output	Labor-Hours	Unit	2015 Bare Costs Material	Labor	Equipment	Total	Total Incl O&P
7280	750 kcmil	1 Elec	5.20	1.538	Ea.	50.50	84		134.50	186
8000	Compression tool, hand					1,200			1,200	1,325
8100	Hydraulic					1,700			1,700	1,875
8500	Hydraulic dies					325			325	360

26 05 19.50 Mineral Insulated Cable		Crew	Daily Output	Labor-Hours	Unit	2015 Bare Costs Material	Labor	Equipment	Total	Total Incl O&P
0010	**MINERAL INSULATED CABLE** 600 volt									
0100	1 conductor, #12	1 Elec	1.60	5	C.L.F.	365	274		639	825
0200	#10		1.60	5		470	274		744	940
0400	#8		1.50	5.333		520	292		812	1,025
0500	#6		1.40	5.714		620	315		935	1,175
0600	#4	2 Elec	2.40	6.667		835	365		1,200	1,475
0800	#2		2.20	7.273		1,175	400		1,575	1,925
0900	#1		2.10	7.619		1,350	415		1,765	2,150
1000	1/0		2	8		1,600	440		2,040	2,425
1100	2/0		1.90	8.421		1,900	460		2,360	2,825
1200	3/0		1.80	8.889		2,275	485		2,760	3,250
1400	4/0		1.60	10		2,625	545		3,170	3,750
1410	250 kcmil	3 Elec	2.40	10		2,975	545		3,520	4,125
1420	350 kcmil		1.95	12.308		3,400	675		4,075	4,775
1430	500 kcmil		1.95	12.308		4,375	675		5,050	5,850
1500	2 conductor, #12	1 Elec	1.40	5.714		760	315		1,075	1,325
1600	#10		1.20	6.667		935	365		1,300	1,600
1800	#8		1.10	7.273		1,075	400		1,475	1,800
2000	#6		1.05	7.619		1,475	415		1,890	2,275
2100	#4	2 Elec	2	8		2,000	440		2,440	2,875
2200	3 conductor, #12	1 Elec	1.20	6.667		950	365		1,315	1,625
2400	#10		1.10	7.273		1,075	400		1,475	1,800
2600	#8		1.05	7.619		1,350	415		1,765	2,125
2800	#6		1	8		1,775	440		2,215	2,625
3000	#4	2 Elec	1.80	8.889		2,225	485		2,710	3,200
3100	4 conductor, #12	1 Elec	1.20	6.667		1,000	365		1,365	1,675
3200	#10		1.10	7.273		1,200	400		1,600	1,950
3400	#8		1	8		1,575	440		2,015	2,400
3600	#6		.90	8.889		2,050	485		2,535	3,000
3620	7 conductor, #12		1.10	7.273		1,275	400		1,675	2,025
3640	#10		1	8		1,675	440		2,115	2,525
3800	Terminations, 600 volt, 1 conductor, #12		8	1	Ea.	16.05	54.50		70.55	102
4000	#10		7.60	1.053		16.05	57.50		73.55	107
4100	#8		7.30	1.096		16.05	60		76.05	111
4200	#6		6.70	1.194		16.05	65.50		81.55	119
4400	#4		6.20	1.290		16.05	70.50		86.55	127
4600	#2		5.70	1.404		24	77		101	146
4800	#1		5.30	1.509		24	82.50		106.50	155
5000	1/0		5	1.600		24	87.50		111.50	163
5100	2/0		4.70	1.702		24	93		117	171
5200	3/0		4.30	1.860		24	102		126	185
5400	4/0		4	2		53.50	109		162.50	228
5410	250 kcmil		4	2		53.50	109		162.50	228
5420	350 kcmil		4	2		84.50	109		193.50	262
5430	500 kcmil		4	2		84.50	109		193.50	262
5500	2 conductor, #12		6.70	1.194		16.05	65.50		81.55	119
5600	#10		6.40	1.250		24	68.50		92.50	133

26 05 19 – Low-Voltage Electrical Power Conductors and Cables

26 05 19.50 Mineral Insulated Cable		Crew	Daily Output	Labor-Hours	Unit	Material	2015 Bare Costs Labor	Equipment	Total	Total Incl O&P
5800	#8	1 Elec	6.20	1.290	Ea.	24	70.50		94.50	136
6000	#6		5.70	1.404		24	77		101	146
6200	#4		5.30	1.509		53.50	82.50		136	187
6400	3 conductor, #12		5.70	1.404		24	77		101	146
6500	#10		5.50	1.455		24	79.50		103.50	150
6600	#8		5.20	1.538		24	84		108	157
6800	#6		4.80	1.667		24	91		115	168
7200	#4		4.60	1.739		53.50	95		148.50	206
7400	4 conductor, #12		4.60	1.739		27	95		122	177
7500	#10		4.40	1.818		27	99.50		126.50	184
7600	#8		4.20	1.905		27	104		131	191
8400	#6		4	2		56.50	109		165.50	231
8500	7 conductor, #12		3.50	2.286		27	125		152	224
8600	#10		3	2.667		58	146		204	290
8800	Crimping tool, plier type					60			60	66
9000	Stripping tool					227			227	249
9200	Hand vise					67.50			67.50	74.50
9500	Minimum labor/equipment charge	1 Elec	4	2	Job		109		109	169

26 05 19.55 Non-Metallic Sheathed Cable		Crew	Daily Output	Labor-Hours	Unit	Material	Labor	Equipment	Total	Total Incl O&P
0010	**NON-METALLIC SHEATHED CABLE** 600 volt									
0100	Copper with ground wire, (Romex)									
0150	#14, 2 conductor	1 Elec	2.70	2.963	C.L.F.	24	162		186	278
0200	3 conductor		2.40	3.333		34	182		216	320
0220	4 conductor		2.20	3.636		49	199		248	365
0250	#12, 2 conductor		2.50	3.200		36.50	175		211.50	310
0300	3 conductor		2.20	3.636		52	199		251	370
0320	4 conductor		2	4		76.50	219		295.50	425
0350	#10, 2 conductor		2.20	3.636		56.50	199		255.50	370
0400	3 conductor		1.80	4.444		82.50	243		325.50	465
0420	4 conductor		1.60	5		119	274		393	555
0430	#8, 2 conductor		1.60	5		88.50	274		362.50	525
0450	3 conductor		1.50	5.333		133	292		425	595
0500	#6, 3 conductor		1.40	5.714		215	315		530	720
0520	#4, 3 conductor	2 Elec	2.40	6.667		550	365		915	1,175
0540	#2, 3 conductor	"	2.20	7.273		790	400		1,190	1,475
0550	SE type SER aluminum cable, 3 RHW and									
0600	1 bare neutral, 3 #8 & 1 #8	1 Elec	1.60	5	C.L.F.	158	274		432	600
0650	3 #6 & 1 #6	"	1.40	5.714		179	315		494	680
0700	3 #4 & 1 #6	2 Elec	2.40	6.667		165	365		530	745
0750	3 #2 & 1 #4		2.20	7.273		296	400		696	940
0800	3 #1/0 & 1 #2		2	8		450	440		890	1,175
0850	3 #2/0 & 1 #1		1.80	8.889		530	485		1,015	1,325
0900	3 #4/0 & 1 #2/0		1.60	10		755	545		1,300	1,675
1000	URD - triplex underground distribution cable, alum. 2 #4 + #4 neutral		2.80	5.714		101	315		416	595
1010	2 #2 + #4 neutral		2.65	6.038		126	330		456	650
1020	2 #2 + #2 neutral		2.55	6.275		126	345		471	670
1030	2 1/0 + #2 neutral		2.40	6.667		152	365		517	735
1040	2 1/0 + 1/0 neutral		2.30	6.957		166	380		546	775
1050	2 2/0 + #1 neutral		2.20	7.273		181	400		581	815
1060	2 2/0 + 2/0 neutral		2.10	7.619		197	415		612	860
1070	2 3/0 + 1/0 neutral		1.95	8.205		218	450		668	935
1080	2 3/0 + 3/0 neutral		1.95	8.205		250	450		700	970

26 05 19 – Low-Voltage Electrical Power Conductors and Cables

26 05 19.55 Non-Metallic Sheathed Cable		Crew	Daily Output	Labor-Hours	Unit	Material	Labor	Equipment	Total	Total Incl O&P
						2015 Bare Costs				
1090	2 4/0 + 2/0 neutral	2 Elec	1.85	8.649	C.L.F.	261	475		736	1,025
1100	2 4/0 + 4/0 neutral		1.85	8.649		276	475		751	1,050
1450	UF underground feeder cable, copper with ground, #14, 2 conductor	1 Elec	4	2		27	109		136	199
1500	#12, 2 conductor		3.50	2.286		41.50	125		166.50	240
1550	#10, 2 conductor		3	2.667		63.50	146		209.50	296
1600	#14, 3 conductor		3.50	2.286		39.50	125		164.50	238
1650	#12, 3 conductor		3	2.667		59.50	146		205.50	292
1700	#10, 3 conductor		2.50	3.200		93.50	175		268.50	375
1710	#8, 3 conductor		2	4		154	219		373	510
1720	#6, 3 conductor		1.80	4.444		248	243		491	650
2400	SEU service entrance cable, copper 2 conductors, #8 + #8 neutral		1.50	5.333		115	292		407	575
2600	#6 + #8 neutral		1.30	6.154		170	335		505	705
2800	#6 + #6 neutral		1.30	6.154		193	335		528	730
3000	#4 + #6 neutral	2 Elec	2.20	7.273		266	400		666	910
3200	#4 + #4 neutral		2.20	7.273		300	400		700	945
3400	#3 + #5 neutral		2.10	7.619		370	415		785	1,050
3600	#3 + #3 neutral		2.10	7.619		400	415		815	1,075
3800	#2 + #4 neutral		2	8		430	440		870	1,150
4000	#1 + #1 neutral		1.90	8.421		470	460		930	1,225
4200	1/0 + 1/0 neutral		1.80	8.889		750	485		1,235	1,575
4400	2/0 + 2/0 neutral		1.70	9.412		935	515		1,450	1,825
4600	3/0 + 3/0 neutral		1.60	10		1,175	545		1,720	2,150
4620	4/0 + 4/0 neutral		1.45	11.034		1,325	605		1,930	2,375
4800	Aluminum 2 conductors, #8 + #8 neutral	1 Elec	1.60	5		131	274		405	570
5000	#6 + #6 neutral	"	1.40	5.714		132	315		447	630
5100	#4 + #6 neutral	2 Elec	2.50	6.400		155	350		505	710
5200	#4 + #4 neutral		2.40	6.667		169	365		534	750
5300	#2 + #4 neutral		2.30	6.957		207	380		587	820
5400	#2 + #2 neutral		2.20	7.273		226	400		626	865
5450	1/0 + #2 neutral		2.10	7.619		330	415		745	1,000
5500	1/0 + 1/0 neutral		2	8		345	440		785	1,050
5550	2/0 + #1 neutral		1.90	8.421		370	460		830	1,125
5600	2/0 + 2/0 neutral		1.80	8.889		395	485		880	1,200
5800	3/0 + 1/0 neutral		1.70	9.412		430	515		945	1,275
6000	3/0 + 3/0 neutral		1.70	9.412		470	515		985	1,325
6200	4/0 + 2/0 neutral		1.60	10		510	545		1,055	1,400
6400	4/0 + 4/0 neutral		1.60	10		555	545		1,100	1,450
6500	Service entrance cap for copper SEU									
6600	100 amp	1 Elec	12	.667	Ea.	8.75	36.50		45.25	66
6700	150 amp		10	.800		16.10	44		60.10	85.50
6800	200 amp		8	1		24.50	54.50		79	112
9000	Minimum labor/equipment charge		4	2	Job		109		109	169

26 05 19.70 Portable Cord		Crew	Daily Output	Labor-Hours	Unit	Material	Labor	Equipment	Total	Total Incl O&P
0010	**PORTABLE CORD** 600 volt									
0100	Type SO, #18, 2 conductor	1 Elec	980	.008	L.F.	.55	.45		1	1.30
0110	3 conductor		980	.008		.67	.45		1.12	1.43
0120	#16, 2 conductor		840	.010		.65	.52		1.17	1.53
0130	3 conductor		840	.010		.93	.52		1.45	1.83
0140	4 conductor		840	.010		1.21	.52		1.73	2.14
0240	#14, 2 conductor		840	.010		1.07	.52		1.59	1.99
0250	3 conductor		840	.010		1.33	.52		1.85	2.27
0260	4 conductor		840	.010		1.55	.52		2.07	2.52

26 05 19 – Low-Voltage Electrical Power Conductors and Cables

26 05 19.70 Portable Cord		Crew	Daily Output	Labor-Hours	Unit	2015 Bare Costs Material	Labor	Equipment	Total	Total Incl O&P
0280	#12, 2 conductor	1 Elec	840	.010	L.F.	1.45	.52		1.97	2.41
0290	3 conductor		840	.010		1.95	.52		2.47	2.96
0300	4 conductor		840	.010		2.30	.52		2.82	3.34
0320	#10, 2 conductor		765	.010		2.21	.57		2.78	3.32
0330	3 conductor		765	.010		2.65	.57		3.22	3.81
0340	4 conductor		765	.010		3	.57		3.57	4.19
0360	#8, 2 conductor		555	.014		4.44	.79		5.23	6.10
0370	3 conductor		540	.015		3.89	.81		4.70	5.55
0380	4 conductor		525	.015		4.57	.83		5.40	6.35
0400	#6, 2 conductor		525	.015		3.64	.83		4.47	5.30
0410	3 conductor		490	.016		5.95	.89		6.84	7.95
0420	4 conductor		415	.019		6.50	1.05		7.55	8.80
0440	#4, 2 conductor	2 Elec	830	.019		6.15	1.05		7.20	8.40
0450	3 conductor		700	.023		8.20	1.25		9.45	11
0460	4 conductor		660	.024		11.85	1.33		13.18	15.05
0480	#2, 2 conductor		450	.036		7.30	1.95		9.25	11.05
0490	3 conductor		350	.046		7.30	2.50		9.80	11.90
0500	4 conductor		280	.057		16.15	3.13		19.28	22.50
2000	See 26 27 26.20 for Wiring Devices Elements									

26 05 19.75 Modular Flexible Wiring System		Crew	Daily Output	Labor-Hours	Unit	Material	Labor	Equipment	Total	Total Incl O&P
0010	**MODULAR FLEXIBLE WIRING SYSTEM**									
0020	Commercial system grid ceiling									
0100	Conversion Module	1 Elec	32	.250	Ea.	8.25	13.70		21.95	30
0120	Fixture cable, for fixture to fixture, 3 conductor, 15′ long		40	.200		33	10.95		43.95	53.50
0150	Extender cable, 3 conductor, 15′ long		48	.167		32	9.10		41.10	49
0200	Switch drop, 1 level, 9′ long		32	.250		33	13.70		46.70	57
0220	2 level, 9′ long		32	.250		30	13.70		43.70	54.50
0250	Power tee, 9′ long		32	.250		34	13.70		47.70	58
1020	Industrial system open ceiling									
1100	Converter, interface between hardwiring and modular wiring	1 Elec	32	.250	Ea.	15.65	13.70		29.35	38
1120	Fixture cable, for fixture to fixture, 3 conductor, 21′ long		32	.250		100	13.70		113.70	131
1125	3 conductor, 25′ long		24	.333		116	18.25		134.25	155
1130	3 conductor, 31′ long		19	.421		103	23		126	149
1150	Fixture cord drop, 10′ long		40	.200		27.50	10.95		38.45	47.50

26 05 19.90 Wire		Crew	Daily Output	Labor-Hours	Unit	Material	Labor	Equipment	Total	Total Incl O&P
0010	**WIRE** R260519-90									
0020	600 volt, copper type THW, solid, #14	1 Elec	13	.615	C.L.F.	7.80	33.50		41.30	60.50
0030	#12		11	.727		11.95	40		51.95	74.50
0040	#10		10	.800		18.65	44		62.65	88.50
0050	Stranded, #14 R260519-92		13	.615		9.35	33.50		42.85	62.50
0100	#12		11	.727		14.35	40		54.35	77.50
0120	#10 R260533-22		10	.800		22.50	44		66.50	93
0140	#8		8	1		37.50	54.50		92	126
0160	#6		6.50	1.231		63.50	67.50		131	174
0180	#4	2 Elec	10.60	1.509		100	82.50		182.50	238
0200	#3		10	1.600		126	87.50		213.50	274
0220	#2		9	1.778		158	97		255	325
0240	#1		8	2		200	109		309	390
0260	1/0		6.60	2.424		250	133		383	480
0280	2/0		5.80	2.759		315	151		466	580
0300	3/0		5	3.200		395	175		570	705
0350	4/0		4.40	3.636		500	199		699	860

26 05 19 – Low-Voltage Electrical Power Conductors and Cables

26 05 19.90 Wire		Crew	Daily Output	Labor-Hours	Unit	Material	2015 Bare Costs Labor	Equipment	Total	Total Incl O&P
0400	250 kcmil	3 Elec	6	4	C.L.F.	585	219		804	985
0420	300 kcmil		5.70	4.211		700	230		930	1,125
0450	350 kcmil		5.40	4.444		855	243		1,098	1,325
0480	400 kcmil		5.10	4.706		985	257		1,242	1,475
0490	500 kcmil		4.80	5		1,150	274		1,424	1,700
0500	600 kcmil		3.90	6.154		1,400	335		1,735	2,050
0510	750 kcmil		3.30	7.273		1,975	400		2,375	2,775
0520	1000 kcmil		2.70	8.889		2,825	485		3,310	3,850
0540	600 volt, aluminum type THHN, stranded, #6	1 Elec	8	1		43	54.50		97.50	132
0560	#4	2 Elec	13	1.231		53.50	67.50		121	163
0580	#2		10.60	1.509		72.50	82.50		155	208
0600	#1		9	1.778		106	97		203	267
0620	1/0		8	2		127	109		236	310
0640	2/0		7.20	2.222		150	122		272	355
0680	3/0		6.60	2.424		186	133		319	410
0700	4/0		6.20	2.581		207	141		348	445
0720	250 kcmil	3 Elec	8.70	2.759		253	151		404	510
0740	300 kcmil		8.10	2.963		350	162		512	635
0760	350 kcmil		7.50	3.200		355	175		530	660
0780	400 kcmil		6.90	3.478		415	190		605	750
0800	500 kcmil		6	4		460	219		679	845
0850	600 kcmil		5.70	4.211		580	230		810	995
0880	700 kcmil		5.10	4.706		670	257		927	1,125
0900	750 kcmil		4.80	5		695	274		969	1,175
0910	1000 kcmil		3.78	6.349		1,025	345		1,370	1,675
0920	600 volt, copper type THWN-THHN, solid, #14	1 Elec	13	.615		7.80	33.50		41.30	60.50
0940	#12		11	.727		11.95	40		51.95	74.50
0960	#10		10	.800		18.65	44		62.65	88.50
1000	Stranded, #14		13	.615		8.90	33.50		42.40	62
1200	#12		11	.727		13.30	40		53.30	76
1250	#10		10	.800		20.50	44		64.50	90.50
1300	#8		8	1		33.50	54.50		88	122
1350	#6		6.50	1.231		57.50	67.50		125	168
1400	#4	2 Elec	10.60	1.509		87.50	82.50		170	224
1450	#3		10	1.600		112	87.50		199.50	259
1500	#2		9	1.778		140	97		237	305
1550	#1		8	2		181	109		290	370
1600	1/0		6.60	2.424		217	133		350	445
1650	2/0		5.80	2.759		271	151		422	535
1700	3/0		5	3.200		340	175		515	645
2000	4/0		4.40	3.636		430	199		629	780
2200	250 kcmil	3 Elec	6	4		520	219		739	910
2400	300 kcmil		5.70	4.211		625	230		855	1,050
2600	350 kcmil		5.40	4.444		730	243		973	1,175
2700	400 kcmil		5.10	4.706		825	257		1,082	1,300
2800	500 kcmil		4.80	5		1,025	274		1,299	1,550
2802	600 kcmil		3.90	6.154		1,300	335		1,635	1,950
2804	750 kcmil		3.30	7.273		2,175	400		2,575	3,000
2805	1000 kcmil		1.94	12.371		2,875	675		3,550	4,225
2900	600 volt, copper type XHHW, solid, #14	1 Elec	13	.615		12.60	33.50		46.10	66
2920	#12		11	.727		19.90	40		59.90	83.50
2940	#10		10	.800		31	44		75	102
3000	Stranded, #14		13	.615		11.05	33.50		44.55	64

26 05 19 – Low-Voltage Electrical Power Conductors and Cables

26 05 19.90 Wire		Crew	Daily Output	Labor-Hours	Unit	Material	Labor	Equipment	Total	Total Incl O&P
						2015 Bare Costs				
3020	#12	1 Elec	11	.727	C.L.F.	16.60	40		56.60	80
3040	#10		10	.800		25	44		69	95.50
3060	#8		8	1		38.50	54.50		93	127
3080	#6		6.50	1.231		63.50	67.50		131	174
3100	#4	2 Elec	10.60	1.509		97.50	82.50		180	235
3120	#2		9	1.778		152	97		249	320
3140	#1		8	2		210	109		319	400
3160	1/0		6.60	2.424		260	133		393	490
3180	2/0		5.80	2.759		325	151		476	595
3200	3/0		5	3.200		405	175		580	720
3220	4/0		4.40	3.636		510	199		709	870
3240	250 kcmil	3 Elec	6	4		565	219		784	965
3260	300 kcmil		5.70	4.211		640	230		870	1,050
3280	350 kcmil		5.40	4.444		745	243		988	1,200
3300	400 kcmil		5.10	4.706		850	257		1,107	1,325
3320	500 kcmil		4.80	5		1,050	274		1,324	1,575
3340	600 kcmil		3.90	6.154		1,350	335		1,685	2,000
3360	750 kcmil		3.30	7.273		2,250	400		2,650	3,100
3380	1000 kcmil		2.40	10		2,925	545		3,470	4,075
5020	600 volt, aluminum type XHHW, stranded, #6	1 Elec	8	1		41	54.50		95.50	130
5040	#4	2 Elec	13	1.231		50.50	67.50		118	160
5060	#2		10.60	1.509		69	82.50		151.50	204
5080	#1		9	1.778		100	97		197	262
5100	1/0		8	2		121	109		230	300
5120	2/0		7.20	2.222		143	122		265	345
5140	3/0		6.60	2.424		177	133		310	400
5160	4/0		6.20	2.581		197	141		338	435
5180	250 kcmil	3 Elec	8.70	2.759		240	151		391	500
5200	300 kcmil		8.10	2.963		330	162		492	615
5220	350 kcmil		7.50	3.200		340	175		515	640
5240	400 kcmil		6.90	3.478		395	190		585	730
5260	500 kcmil		6	4		435	219		654	820
5280	600 kcmil		5.70	4.211		550	230		780	960
5300	700 kcmil		5.40	4.444		635	243		878	1,075
5320	750 kcmil		5.10	4.706		645	257		902	1,100
5340	1000 kcmil		3.60	6.667		955	365		1,320	1,625
5390	600 volt, copper type XLPE-USE (RHW), solid, #14	1 Elec	12	.667		14.40	36.50		50.90	72.50
5400	#12		11	.727		18.85	40		58.85	82.50
5420	#10		10	.800		27.50	44		71.50	98
5440	Stranded, #14		13	.615		18.30	33.50		51.80	72
5460	#12		11	.727		22	40		62	85.50
5480	#10		10	.800		32	44		76	103
5500	#8		8	1		43	54.50		97.50	132
5520	#6		6.50	1.231		72.50	67.50		140	184
5540	#4	2 Elec	10.60	1.509		110	82.50		192.50	249
5560	#2		9	1.778		178	97		275	345
5580	#1		8	2		241	109		350	435
5600	1/0		6.60	2.424		300	133		433	535
5620	2/0		5.80	2.759		390	151		541	665
5640	3/0		5	3.200		490	175		665	810
5660	4/0		4.40	3.636		540	199		739	900
5680	250 kcmil	3 Elec	6	4		590	219		809	990
5700	300 kcmil		5.70	4.211		765	230		995	1,200

26 05 19 – Low-Voltage Electrical Power Conductors and Cables

26 05 19.90 Wire		Crew	Daily Output	Labor-Hours	Unit	Material	2015 Bare Costs Labor	Equipment	Total	Total Incl O&P
5720	350 kcmil	3 Elec	5.40	4.444	C.L.F.	870	243		1,113	1,325
5740	400 kcmil		5.10	4.706		990	257		1,247	1,500
5760	500 kcmil		4.80	5		1,050	274		1,324	1,600
5780	600 kcmil		3.90	6.154		1,400	335		1,735	2,050
5800	750 kcmil		3.30	7.273		2,500	400		2,900	3,375
5820	1000 kcmil		2.70	8.889		3,300	485		3,785	4,375
5840	600 volt, aluminum type XLPE-USE (RHW), stranded, #6	1 Elec	8	1		49.50	54.50		104	139
5860	#4	2 Elec	13	1.231		57.50	67.50		125	167
5880	#2		10.60	1.509		79.50	82.50		162	216
5900	#1		9	1.778		110	97		207	272
5920	1/0		8	2		136	109		245	320
5940	2/0		7.20	2.222		159	122		281	365
5960	3/0		6.60	2.424		188	133		321	410
5980	4/0		6.20	2.581		206	141		347	445
6000	250 kcmil	3 Elec	8.70	2.759		283	151		434	545
6020	300 kcmil		8.10	2.963		370	162		532	655
6040	350 kcmil		7.50	3.200		375	175		550	685
6060	400 kcmil		6.90	3.478		460	190		650	800
6080	500 kcmil		6	4		505	219		724	895
6100	600 kcmil		5.70	4.211		650	230		880	1,075
6110	700 kcmil		5.40	4.444		745	243		988	1,200
6120	750 kcmil		5.10	4.706		755	257		1,012	1,225
9000	Minimum labor/equipment charge	1 Elec	4	2	Job		109		109	169

26 05 23 – Control-Voltage Electrical Power Cables

26 05 23.10 Control Cable

		Crew	Daily Output	Labor-Hours	Unit	Material	Labor	Equipment	Total	Total Incl O&P
0010	**CONTROL CABLE**									
0020	600 volt, copper, #14 THWN wire with PVC jacket, 2 wires	1 Elec	9	.889	C.L.F.	30.50	48.50		79	109
0030	3 wires		8	1		42.50	54.50		97	132
0100	4 wires		7	1.143		52	62.50		114.50	155
0150	5 wires		6.50	1.231		65	67.50		132.50	176
0200	6 wires		6	1.333		85.50	73		158.50	207
0300	8 wires		5.30	1.509		106	82.50		188.50	245
0400	10 wires		4.80	1.667		126	91		217	280
0500	12 wires		4.30	1.860		148	102		250	320
0600	14 wires		3.80	2.105		176	115		291	370
0700	16 wires		3.50	2.286		182	125		307	395
0800	18 wires		3.30	2.424		199	133		332	425
0810	19 wires		3.10	2.581		224	141		365	465
0900	20 wires		3	2.667		242	146		388	490
1000	22 wires		2.80	2.857		249	156		405	515

26 05 23.20 Special Wires and Fittings

		Crew	Daily Output	Labor-Hours	Unit	Material	Labor	Equipment	Total	Total Incl O&P
0010	**SPECIAL WIRES & FITTINGS**									
0100	Fixture TFFN 600 volt 90°C stranded, #18	1 Elec	13	.615	C.L.F.	9.50	33.50		43	62.50
0150	#16		13	.615		12.40	33.50		45.90	65.50
0500	Thermostat, jacket non-plenum, twisted, #18-2 conductor		8	1		16.45	54.50		70.95	103
0550	#18-3 conductor		7	1.143		21.50	62.50		84	121
0600	#18-4 conductor		6.50	1.231		32	67.50		99.50	139
0650	#18-5 conductor		6	1.333		35.50	73		108.50	152
0700	#18-6 conductor		5.50	1.455		44.50	79.50		124	172
0750	#18-7 conductor		5	1.600		51.50	87.50		139	193
0800	#18-8 conductor		4.80	1.667		58.50	91		149.50	206
2460	Tray cable, type TC, copper, #16-2 conductor		9.40	.851		24	46.50		70.50	98.50

26 05 23.20 Special Wires and Fittings		Crew	Daily Output	Labor-Hours	Unit	Material	2015 Bare Costs Labor	Equipment	Total	Total Incl O&P
2464	#16-3 conductor	1 Elec	8.40	.952	C.L.F.	31	52		83	115
2468	#16-4 conductor		7.30	1.096		39	60		99	136
2472	#16-5 conductor		6.70	1.194		42.50	65.50		108	148
2476	#16-7 conductor		5.50	1.455		58	79.50		137.50	187
2480	#16-9 conductor		5	1.600		72.50	87.50		160	216
2484	#16-12 conductor	2 Elec	8.80	1.818		103	99.50		202.50	267
2488	#16-15 conductor		7.20	2.222		126	122		248	325
2492	#16-19 conductor		6.60	2.424		156	133		289	375
2496	#16-25 conductor		6.40	2.500		220	137		357	455
2500	#14-2 conductor	1 Elec	9	.889		29.50	48.50		78	108
2520	#14-3 conductor		8	1		41	54.50		95.50	130
2540	#14-4 conductor		7	1.143		50.50	62.50		113	153
2560	#14-5 conductor		6.50	1.231		62.50	67.50		130	173
2564	#14-7 conductor		5.30	1.509		91.50	82.50		174	229
2568	#14-9 conductor	2 Elec	9.60	1.667		117	91		208	270
2572	#14-12 conductor		8.60	1.860		159	102		261	335
2576	#14-15 conductor		7.20	2.222		173	122		295	380
2578	#14-19 conductor		6.20	2.581		218	141		359	460
2582	#14-25 conductor		5.30	3.019		286	165		451	570
2590	#12-2 conductor	1 Elec	8.40	.952		42	52		94	127
2592	#12-3 conductor		7.60	1.053		49.50	57.50		107	144
2594	#12-4 conductor		6.60	1.212		77.50	66.50		144	189
2596	#12-5 conductor		6.20	1.290		94.50	70.50		165	213
2598	#12-7 conductor	2 Elec	10.40	1.538		139	84		223	283
2602	#12-9 conductor		8.80	1.818		181	99.50		280.50	355
2604	#12-12 conductor		7.80	2.051		236	112		348	435
2606	#12-15 conductor		6.80	2.353		261	129		390	485
2608	#12-19 conductor		6	2.667		320	146		466	575
2610	#12-25 conductor	3 Elec	7.70	3.117		430	170		600	740
2618	#10-2 conductor	1 Elec	8	1		63	54.50		117.50	154
2622	#10-3 conductor	"	7.30	1.096		84	60		144	185
2624	#10-4 conductor	2 Elec	12.80	1.250		120	68.50		188.50	238
2626	#10-5 conductor		11.80	1.356		135	74		209	264
2628	#10-7 conductor		9.40	1.702		195	93		288	360
2630	#10-9 conductor		8.40	1.905		227	104		331	410
2632	#10-12 conductor		7.20	2.222		242	122		364	455
2640	300 V, copper braided shield, PVC jacket									
2650	2 conductor #18 stranded	1 Elec	7	1.143	C.L.F.	47.50	62.50		110	150
2660	3 conductor #18 stranded	"	6	1.333	"	67	73		140	187
3000	Strain relief grip for cable									
3050	Cord, top, #12-3	1 Elec	40	.200	Ea.	10.55	10.95		21.50	28.50
3060	#12-4		40	.200		10.55	10.95		21.50	28.50
3070	#12-5		39	.205		11.25	11.20		22.45	30
3100	#10-3		39	.205		11.25	11.20		22.45	30
3110	#10-4		38	.211		11.25	11.50		22.75	30.50
3120	#10-5		38	.211		13.40	11.50		24.90	32.50
3200	Bottom, #12-3		40	.200		26.50	10.95		37.45	46
3210	#12-4		40	.200		26.50	10.95		37.45	46
3220	#12-5		39	.205		26.50	11.20		37.70	47
3230	#10-3		39	.205		26.50	11.20		37.70	47
3300	#10-4		38	.211		26.50	11.50		38	47.50
3310	#10-5		38	.211		33.50	11.50		45	54.50
3400	Cable ties, standard, 4" length		190	.042		.17	2.30		2.47	3.76

26 05 Common Work Results for Electrical

26 05 23 – Control-Voltage Electrical Power Cables

26 05 23.20 Special Wires and Fittings		Crew	Daily Output	Labor-Hours	Unit	Material	Labor (2015 Bare Costs)	Equipment	Total	Total Incl O&P
3410	7" length	1 Elec	160	.050	Ea.	.19	2.74		2.93	4.45
3420	14.5" length		90	.089		.46	4.86		5.32	8.05
3430	Heavy, 14.5" length		80	.100		.64	5.45		6.09	9.15
9000	Minimum labor/equipment charge		4	2	Job		109		109	169

26 05 26 – Grounding and Bonding for Electrical Systems

26 05 26.80 Grounding

		Crew	Daily Output	Labor-Hours	Unit	Material	Labor	Equipment	Total	Total Incl O&P
0010	**GROUNDING**									
0030	Rod, copper clad, 8' long, 1/2" diameter	1 Elec	5.50	1.455	Ea.	20	79.50		99.50	145
0040	5/8" diameter		5.50	1.455		19.20	79.50		98.70	144
0050	3/4" diameter		5.30	1.509		33.50	82.50		116	165
0080	10' long, 1/2" diameter		4.80	1.667		22	91		113	166
0090	5/8" diameter		4.60	1.739		24	95		119	174
0100	3/4" diameter		4.40	1.818		38.50	99.50		138	197
0130	15' long, 3/4" diameter		4	2		54	109		163	229
0150	Coupling, bronze, 1/2" diameter					4.07			4.07	4.48
0160	5/8" diameter					5.35			5.35	5.90
0170	3/4" diameter					13.20			13.20	14.50
0190	Drive studs, 1/2" diameter					10.70			10.70	11.75
0210	5/8" diameter					15.55			15.55	17.10
0220	3/4" diameter					18.15			18.15	20
0230	Clamp, bronze, 1/2" diameter	1 Elec	32	.250		4.65	13.70		18.35	26
0240	5/8" diameter		32	.250		5.25	13.70		18.95	27
0250	3/4" diameter		32	.250		5.90	13.70		19.60	27.50
0260	Wire ground bare armored, #8-1 conductor		2	4	C.L.F.	76	219		295	425
0270	#6-1 conductor		1.80	4.444		91.50	243		334.50	475
0280	#4-1 conductor		1.60	5		128	274		402	565
0320	Bare copper wire, #14 solid		14	.571		7.35	31.50		38.85	56.50
0330	#12		13	.615		12.60	33.50		46.10	66
0340	#10		12	.667		18	36.50		54.50	76.50
0350	#8		11	.727		29	40		69	93.50
0360	#6		10	.800		51	44		95	125
0370	#4		8	1		117	54.50		171.50	213
0380	#2		5	1.600		130	87.50		217.50	279
0390	Bare copper wire, stranded, #8		11	.727		29.50	40		69.50	94
0400	#6		10	.800		52	44		96	126
0450	#4	2 Elec	16	1		87.50	54.50		142	181
0600	#2		10	1.600		138	87.50		225.50	288
0650	#1		9	1.778		182	97		279	350
0700	1/0		8	2		197	109		306	385
0750	2/0		7.20	2.222		248	122		370	460
0800	3/0		6.60	2.424		310	133		443	545
1000	4/0		5.70	2.807		395	154		549	675
1200	250 kcmil	3 Elec	7.20	3.333		465	182		647	790
1210	300 kcmil		6.60	3.636		525	199		724	890
1220	350 kcmil		6	4		660	219		879	1,075
1230	400 kcmil		5.70	4.211		770	230		1,000	1,200
1240	500 kcmil		5.10	4.706		945	257		1,202	1,450
1260	750 kcmil		3.60	6.667		1,600	365		1,965	2,350
1270	1000 kcmil		3	8		2,125	440		2,565	3,025
1360	Bare aluminum, stranded, #6	1 Elec	9	.889		14.60	48.50		63.10	91.50
1370	#4	2 Elec	16	1		25.50	54.50		80	113
1380	#2		13	1.231		36.50	67.50		104	145

26 05 26 – Grounding and Bonding for Electrical Systems

26 05 26.80 Grounding		Crew	Daily Output	Labor-Hours	Unit	Material	2015 Bare Costs Labor	Equipment	Total	Total Incl O&P
1390	#1	2 Elec	10.60	1.509	C.L.F.	40.50	82.50		123	173
1400	1/0		9	1.778		46.50	97		143.50	202
1410	2/0		8	2		59	109		168	234
1420	3/0		7.20	2.222		79.50	122		201.50	275
1430	4/0		6.60	2.424		91	133		224	305
1440	250 kcmil	3 Elec	9.30	2.581		119	141		260	350
1450	300 kcmil		8.70	2.759		140	151		291	390
1460	400 kcmil		7.50	3.200		180	175		355	470
1470	500 kcmil		6.90	3.478		214	190		404	530
1480	600 kcmil		6	4		243	219		462	610
1490	700 kcmil		5.70	4.211		267	230		497	650
1500	750 kcmil		5.10	4.706		277	257		534	705
1510	1000 kcmil		4.80	5		350	274		624	810
1800	Water pipe ground clamps, heavy duty									
2000	Bronze, 1/2" to 1" diameter	1 Elec	8	1	Ea.	24	54.50		78.50	111
2100	1-1/4" to 2" diameter		8	1		33	54.50		87.50	121
2200	2-1/2" to 3" diameter		6	1.333		44.50	73		117.50	162
2730	Exothermic weld, 4/0 wire to 1" ground rod		7	1.143		10.70	62.50		73.20	109
2740	4/0 wire to building steel		7	1.143		10.70	62.50		73.20	109
2750	4/0 wire to motor frame		7	1.143		10.70	62.50		73.20	109
2760	4/0 wire to 4/0 wire		7	1.143		10.70	62.50		73.20	109
2770	4/0 wire to #4 wire		7	1.143		10.70	62.50		73.20	109
2780	4/0 wire to #8 wire		7	1.143		10.70	62.50		73.20	109
2790	Mold, reusable, for above					138			138	152
2800	Brazed connections, #6 wire	1 Elec	12	.667		16.15	36.50		52.65	74.50
3000	#2 wire		10	.800		21.50	44		65.50	92
3100	3/0 wire		8	1		32.50	54.50		87	120
3200	4/0 wire		7	1.143		37	62.50		99.50	138
3400	250 kcmil wire		5	1.600		43.50	87.50		131	184
3600	500 kcmil wire		4	2		53.50	109		162.50	228
3700	Insulated ground wire, copper #14		13	.615	C.L.F.	8.90	33.50		42.40	62
3710	#12		11	.727		13.30	40		53.30	76
3720	#10		10	.800		20.50	44		64.50	90.50
3730	#8		8	1		33.50	54.50		88	122
3740	#6		6.50	1.231		57.50	67.50		125	168
3750	#4	2 Elec	10.60	1.509		87.50	82.50		170	224
3770	#2		9	1.778		140	97		237	305
3780	#1		8	2		181	109		290	370
3790	1/0		6.60	2.424		217	133		350	445
3800	2/0		5.80	2.759		271	151		422	535
3810	3/0		5	3.200		340	175		515	645
3820	4/0		4.40	3.636		430	199		629	780
3830	250 kcmil	3 Elec	6	4		520	219		739	910
3840	300 kcmil		5.70	4.211		625	230		855	1,050
3850	350 kcmil		5.40	4.444		730	243		973	1,175
3860	400 kcmil		5.10	4.706		825	257		1,082	1,300
3870	500 kcmil		4.80	5		1,025	274		1,299	1,550
3880	600 kcmil		3.90	6.154		1,400	335		1,735	2,050
3890	750 kcmil		3.30	7.273		1,975	400		2,375	2,775
3900	1000 kcmil		2.70	8.889		2,825	485		3,310	3,850
3960	Insulated ground wire, aluminum, #6	1 Elec	8	1		43	54.50		97.50	132
3970	#4	2 Elec	13	1.231		53.50	67.50		121	163
3980	#2		10.60	1.509		72.50	82.50		155	208

26 05 26 – Grounding and Bonding for Electrical Systems

26 05 26.80 Grounding		Crew	Daily Output	Labor-Hours	Unit	2015 Bare Costs Material	Labor	Equipment	Total	Total Incl O&P
3990	#1	2 Elec	9	1.778	C.L.F.	106	97		203	267
4000	1/0		8	2		127	109		236	310
4010	2/0		7.20	2.222		150	122		272	355
4020	3/0		6.60	2.424		186	133		319	410
4030	4/0		6.20	2.581		207	141		348	445
4040	250 kcmil	3 Elec	8.70	2.759		253	151		404	510
4050	300 kcmil		8.10	2.963		350	162		512	635
4060	350 kcmil		7.50	3.200		355	175		530	660
4070	400 kcmil		6.90	3.478		415	190		605	750
4080	500 kcmil		6	4		460	219		679	845
4090	600 kcmil		5.70	4.211		580	230		810	995
4100	700 kcmil		5.10	4.706		670	257		927	1,125
4110	750 kcmil		4.80	5		695	274		969	1,175
5000	Copper Electrolytic ground rod system									
5010	Includes augering hole, mixing bentonite clay,									
5020	Installing rod, and terminating ground wire									
5100	Straight Vertical type, 2" diam.									
5120	8.5' long, clamp connection	1 Elec	2.67	2.996	Ea.	770	164		934	1,100
5130	With exothermic weld connection		1.95	4.103		770	224		994	1,200
5140	10' long		2.35	3.404		900	186		1,086	1,275
5150	With exothermic weld connection		1.78	4.494		900	246		1,146	1,375
5160	12' long		2.16	3.704		1,025	203		1,228	1,450
5170	With exothermic weld connection		1.67	4.790		1,025	262		1,287	1,525
5180	20' long		1.74	4.598		1,475	251		1,726	2,000
5190	With exothermic weld connection		1.40	5.714		1,775	315		2,090	2,425
5195	40' long with exothermic weld connection	2 Elec	2	8		3,075	440		3,515	4,050
5200	L-Shaped, 2" diam.									
5220	4' Vert. x 10' Horz., clamp connection	1 Elec	5.33	1.501	Ea.	1,275	82		1,357	1,550
5230	With exothermic weld connection	"	3.08	2.597	"	1,275	142		1,417	1,650
5300	Protective Box at grade level, with breather slots									
5320	Round 12" long, fiberlyte	1 Elec	32	.250	Ea.	65	13.70		78.70	92.50
5330	Concrete	"	16	.500		101	27.50		128.50	154
5400	Bentonite Clay, 50# bag, 1 per 10' of rod					41.50			41.50	45.50
5500	Equipotential earthing bar	1 Elec	2	4		176	219		395	535
9000	Minimum labor/equipment charge	"	4	2	Job		109		109	169

26 05 29 – Hangers and Supports for Electrical Systems

26 05 29.20 Hangers

26 05 29.20 Hangers		Crew	Daily Output	Labor-Hours	Unit	Material	Labor	Equipment	Total	Total Incl O&P
0010	**HANGERS** R260533-21									
0015	See section 22 05 29.10 for additional items									
0030	Conduit supports									
0050	Strap w/2 holes, rigid steel conduit									
0100	1/2" diameter	1 Elec	470	.017	Ea.	.13	.93		1.06	1.58
0150	3/4" diameter		440	.018		.14	.99		1.13	1.69
0200	1" diameter		400	.020		.22	1.09		1.31	1.93
0300	1-1/4" diameter		355	.023		.35	1.23		1.58	2.30
0350	1-1/2" diameter		320	.025		.40	1.37		1.77	2.56
0400	2" diameter		266	.030		.44	1.65		2.09	3.03
0500	2-1/2" diameter		160	.050		.84	2.74		3.58	5.15
0550	3" diameter		133	.060		1.08	3.29		4.37	6.30
0600	3-1/2" diameter		100	.080		2.81	4.38		7.19	9.90
0650	4" diameter		80	.100		1.46	5.45		6.91	10.05
0700	EMT, 1/2" diameter		470	.017		.14	.93		1.07	1.59

26 05 Common Work Results for Electrical

26 05 29 – Hangers and Supports for Electrical Systems

26 05 29.20 Hangers		Crew	Daily Output	Labor-Hours	Unit	Material	2015 Bare Costs Labor	Equipment	Total	Total Incl O&P
0800	3/4" diameter	1 Elec	440	.018	Ea.	.18	.99		1.17	1.74
0850	1" diameter		400	.020		.31	1.09		1.40	2.03
0900	1-1/4" diameter		355	.023		.51	1.23		1.74	2.47
0950	1-1/2" diameter		320	.025		.55	1.37		1.92	2.73
1000	2" diameter		266	.030		.80	1.65		2.45	3.43
1100	2-1/2" diameter		160	.050		1.34	2.74		4.08	5.70
1150	3" diameter		133	.060		1.88	3.29		5.17	7.15
1200	3-1/2" diameter		100	.080		2.10	4.38		6.48	9.10
1250	4" diameter		80	.100		2.44	5.45		7.89	11.15
1400	Hanger, with bolt, 1/2" diameter		200	.040		.53	2.19		2.72	3.97
1450	3/4" diameter		190	.042		.54	2.30		2.84	4.16
1500	1" diameter		176	.045		.94	2.49		3.43	4.88
1550	1-1/4" diameter		160	.050		1.26	2.74		4	5.65
1600	1-1/2" diameter		140	.057		1.57	3.13		4.70	6.55
1650	2" diameter		130	.062		2.04	3.37		5.41	7.45
1700	2-1/2" diameter		100	.080		2.11	4.38		6.49	9.10
1750	3" diameter		64	.125		2.91	6.85		9.76	13.80
1800	3-1/2" diameter		50	.160		3.68	8.75		12.43	17.60
1850	4" diameter		40	.200		8.35	10.95		19.30	26
1900	Riser clamps, conduit, 1/2" diameter		40	.200		8.40	10.95		19.35	26
1950	3/4" diameter		36	.222		9.10	12.15		21.25	29
2000	1" diameter		30	.267		10.15	14.60		24.75	33.50
2100	1-1/4" diameter		27	.296		11.20	16.20		27.40	37.50
2150	1-1/2" diameter		27	.296		11.75	16.20		27.95	38
2200	2" diameter		20	.400		12.30	22		34.30	47.50
2250	2-1/2" diameter		20	.400		13	22		35	48.50
2300	3" diameter		18	.444		14.15	24.50		38.65	53
2350	3-1/2" diameter		18	.444		16.05	24.50		40.55	55
2400	4" diameter		14	.571		19.20	31.50		50.70	69.50
2500	Threaded rod, painted, 1/4" diameter		260	.031	L.F.	1.79	1.68		3.47	4.58
2600	3/8" diameter		200	.040		2.89	2.19		5.08	6.55
2700	1/2" diameter		140	.057		3.77	3.13		6.90	9
2800	5/8" diameter		100	.080		5.05	4.38		9.43	12.40
2900	3/4" diameter		60	.133		9	7.30		16.30	21
2940	Couplings painted, 1/4" diameter				C	565			565	620
2960	3/8" diameter					655			655	720
2970	1/2" diameter					1,250			1,250	1,375
2980	5/8" diameter					1,950			1,950	2,125
2990	3/4" diameter					2,025			2,025	2,225
3000	Nuts, galvanized, 1/4" diameter					15.15			15.15	16.65
3050	3/8" diameter					22			22	24
3100	1/2" diameter					49			49	53.50
3150	5/8" diameter					125			125	138
3200	3/4" diameter					185			185	203
3250	Washers, galvanized, 1/4" diameter					15.15			15.15	16.65
3300	3/8" diameter					20.50			20.50	23
3350	1/2" diameter					33			33	36.50
3400	5/8" diameter					103			103	113
3450	3/4" diameter					150			150	165
3500	Lock washers, galvanized, 1/4" diameter					15.15			15.15	16.65
3550	3/8" diameter					17.75			17.75	19.50
3600	1/2" diameter					20.50			20.50	22.50
3650	5/8" diameter					42			42	46.50

26 05 29 – Hangers and Supports for Electrical Systems

26 05 29.20 Hangers		Crew	Daily Output	Labor-Hours	Unit	Material	2015 Bare Costs Labor	Equipment	Total	Total Incl O&P
3700	3/4" diameter				C	80.50			80.50	88.50
3800	Channels, steel, 3/4" x 1-1/2", 14 Ga	1 Elec	80	.100	L.F.	4.10	5.45		9.55	12.95
3900	1-1/2" x 1-1/2", 12 Ga		70	.114		5.40	6.25		11.65	15.65
4000	1-7/8" x 1-1/2"		60	.133		23	7.30		30.30	37
4100	3" x 1-1/2"		50	.160		39.50	8.75		48.25	57
4110	1-5/8" x 13/16", 16 Ga.		80	.100		5.20	5.45		10.65	14.15
4114	Trapeze channel support, 12" wide, steel, 12 Ga		8.80	.909	Ea.	29	49.50		78.50	109
4116	18" wide, steel		8.30	.964	"	31.50	52.50		84	116
4120	1-5/8" x 1-5/8", 14 Ga.		70	.114	L.F.	6.50	6.25		12.75	16.85
4130	1-5/8" x 7/8", 12 Ga.		70	.114		5.50	6.25		11.75	15.75
4140	1-5/8" x 1-3/8", 12 Ga.		60	.133		7.65	7.30		14.95	19.70
4150	1-5/8" x 1-5/8", 12 Ga.		50	.160		6.90	8.75		15.65	21
4160	Flat plate fitting, 2 hole, 3-1/2", material only				Ea.	3.40			3.40	3.74
4170	3 hole, 1-7/16" x 4-1/8"					4.17			4.17	4.59
4200	Spring nuts, long, 1/4"	1 Elec	120	.067		1.17	3.65		4.82	6.95
4250	3/8"		100	.080		1.23	4.38		5.61	8.15
4300	1/2"		80	.100		1.33	5.45		6.78	9.90
4350	Spring nuts, short, 1/4"		120	.067		1.16	3.65		4.81	6.95
4400	3/8"		100	.080		1.20	4.38		5.58	8.10
4450	1/2"		80	.100		1.67	5.45		7.12	10.30
4500	Closure strip		200	.040	L.F.	4.67	2.19		6.86	8.55
4550	End cap		60	.133	Ea.	1.63	7.30		8.93	13.10
4600	End connector 3/4" conduit		40	.200		6.60	10.95		17.55	24
4650	Junction box, 1 channel		16	.500		57.50	27.50		85	106
4700	2 channel		14	.571		71.50	31.50		103	127
4750	3 channel		12	.667		78.50	36.50		115	143
4800	4 channel		10	.800		82.50	44		126.50	159
4850	Splice plate		40	.200		9.45	10.95		20.40	27.50
4900	Continuous concrete insert, 1-1/2" deep, 1' long		16	.500		19.85	27.50		47.35	64.50
4950	2' long		14	.571		24.50	31.50		56	75.50
5000	3' long		12	.667		23	36.50		59.50	81.50
5050	4' long		10	.800		34	44		78	106
5100	6' long		8	1		51	54.50		105.50	141
5150	3/4" deep, 1' long		16	.500		14.30	27.50		41.80	58.50
5200	2' long		14	.571		17.80	31.50		49.30	68
5250	3' long		12	.667		18.50	36.50		55	77
5300	4' long		10	.800		24.50	44		68.50	95
5350	6' long		8	1		37	54.50		91.50	125
5400	90° angle fitting 2-1/8" x 2-1/8"		60	.133		3.80	7.30		11.10	15.50
5450	Supports, suspension rod type, small		60	.133		24	7.30		31.30	38
5500	Large		40	.200		26.50	10.95		37.45	46
5550	Beam clamp, small		60	.133		9.90	7.30		17.20	22
5600	Large		40	.200		11.40	10.95		22.35	29.50
5650	U-support, small		60	.133		6.05	7.30		13.35	17.95
5700	Large		40	.200		12	10.95		22.95	30
5750	Concrete insert, cast, for up to 1/2" threaded rod		16	.500		6	27.50		33.50	49
5800	Beam clamp, 1/4" clamp, for 1/4" threaded drop rod		32	.250		2.67	13.70		16.37	24
5900	3/8" clamp, for 3/8" threaded drop rod		32	.250		5.80	13.70		19.50	27.50
6000	Strap, rigid conduit, 1/2" diameter		540	.015		1.64	.81		2.45	3.05
6050	3/4" diameter		440	.018		1.26	.99		2.25	2.93
6100	1" diameter		420	.019		2.02	1.04		3.06	3.83
6150	1-1/4" diameter		400	.020		1.58	1.09		2.67	3.43
6200	1-1/2" diameter		400	.020		1.89	1.09		2.98	3.77

26 05 29 – Hangers and Supports for Electrical Systems

26 05 29.20 Hangers		Crew	Daily Output	Labor-Hours	Unit	Material	Labor	Equipment	Total	Total Incl O&P
						2015 Bare Costs				
6250	2" diameter	1 Elec	267	.030	Ea.	2.04	1.64		3.68	4.78
6300	2-1/2" diameter		267	.030		2.74	1.64		4.38	5.55
6350	3" diameter		160	.050		2.59	2.74		5.33	7.10
6400	3-1/2" diameter		133	.060		3.49	3.29		6.78	8.95
6450	4" diameter		100	.080		3.61	4.38		7.99	10.75
6500	5" diameter		80	.100		9.95	5.45		15.40	19.35
6550	6" diameter		60	.133		19.40	7.30		26.70	33
6600	EMT, 1/2" diameter		540	.015		1.13	.81		1.94	2.49
6650	3/4" diameter		440	.018		1.21	.99		2.20	2.87
6700	1" diameter		420	.019		1.34	1.04		2.38	3.08
6750	1-1/4" diameter		400	.020		1.51	1.09		2.60	3.35
6800	1-1/2" diameter		400	.020		1.86	1.09		2.95	3.74
6850	2" diameter		267	.030		2.11	1.64		3.75	4.86
6900	2-1/2" diameter		267	.030		2.25	1.64		3.89	5
6950	3" diameter		160	.050		2.43	2.74		5.17	6.90
6970	3-1/2" diameter		133	.060		2.72	3.29		6.01	8.10
6990	4" diameter		100	.080		3.14	4.38		7.52	10.25
7000	Clip, 1 hole for rigid conduit, 1/2" diameter		500	.016		.51	.88		1.39	1.92
7050	3/4" diameter		470	.017		.75	.93		1.68	2.27
7100	1" diameter		440	.018		.93	.99		1.92	2.56
7150	1-1/4" diameter		400	.020		1.97	1.09		3.06	3.86
7200	1-1/2" diameter		355	.023		2.15	1.23		3.38	4.28
7250	2" diameter		320	.025		4.19	1.37		5.56	6.75
7300	2-1/2" diameter		266	.030		9.35	1.65		11	12.80
7350	3" diameter		160	.050		13.35	2.74		16.09	18.90
7400	3-1/2" diameter		133	.060		18.95	3.29		22.24	26
7450	4" diameter		100	.080		42.50	4.38		46.88	53.50
7500	5" diameter		80	.100		136	5.45		141.45	158
7550	6" diameter		60	.133		143	7.30		150.30	169
7820	Conduit hangers, with bolt & 12" rod, 1/2" diameter		150	.053		3.42	2.92		6.34	8.30
7830	3/4" diameter		145	.055		3.43	3.02		6.45	8.45
7840	1" diameter		135	.059		3.83	3.24		7.07	9.20
7850	1-1/4" diameter		120	.067		4.15	3.65		7.80	10.20
7860	1-1/2" diameter		110	.073		4.46	3.98		8.44	11.05
7870	2" diameter		100	.080		5.80	4.38		10.18	13.20
7880	2-1/2" diameter		80	.100		5.90	5.45		11.35	14.90
7890	3" diameter		60	.133		6.70	7.30		14	18.65
7900	3-1/2" diameter		45	.178		7.45	9.70		17.15	23.50
7910	4" diameter		35	.229		12.15	12.50		24.65	32.50
7920	5" diameter		30	.267		13.45	14.60		28.05	37.50
7930	6" diameter		25	.320		27.50	17.50		45	57
7950	Jay clamp, 1/2" diameter		32	.250		3.75	13.70		17.45	25
7960	3/4" diameter		32	.250		4.85	13.70		18.55	26.50
7970	1" diameter		32	.250		6.85	13.70		20.55	28.50
7980	1-1/4" diameter		30	.267		5.65	14.60		20.25	29
7990	1-1/2" diameter		30	.267		9.25	14.60		23.85	32.50
8000	2" diameter		30	.267		13.70	14.60		28.30	37.50
8010	2-1/2" diameter		28	.286		19.70	15.65		35.35	45.50
8020	3" diameter		28	.286		19.70	15.65		35.35	45.50
8030	3-1/2" diameter		25	.320		23.50	17.50		41	53
8040	4" diameter		25	.320		23.50	17.50		41	53
8050	5" diameter		20	.400		99.50	22		121.50	143
8060	6" diameter		16	.500		187	27.50		214.50	249

26 05 29 – Hangers and Supports for Electrical Systems

26 05 29.20 Hangers		Crew	Daily Output	Labor-Hours	Unit	2015 Bare Costs Material	Labor	Equipment	Total	Total Incl O&P
8070	Channels, 3/4″ x 1-1/2″ w/12″ rods for 1/2″ to 1″ conduit	1 Elec	30	.267	Ea.	9.05	14.60		23.65	32.50
8080	1-1/2″ x 1-1/2″ w/12″ rods for 1-1/4″ to 2″ conduit		28	.286		9.65	15.65		25.30	34.50
8090	1-1/2″ x 1-1/2″ w/12″ rods for 2-1/2″ to 4″ conduit		26	.308		11.80	16.85		28.65	39
8100	1-1/2″ x 1-7/8″ w/12″ rods for 5″ to 6″ conduit		24	.333		55.50	18.25		73.75	89
8110	Beam clamp, conduit, plastic coated steel, 1/2″ diam.		30	.267		21	14.60		35.60	45.50
8120	3/4″ diameter		30	.267		22	14.60		36.60	46.50
8130	1″ diameter		30	.267		25	14.60		39.60	50
8140	1-1/4″ diameter		28	.286		30.50	15.65		46.15	57.50
8150	1-1/2″ diameter		28	.286		37.50	15.65		53.15	65
8160	2″ diameter		28	.286		52	15.65		67.65	81
8170	2-1/2″ diameter		26	.308		53	16.85		69.85	84.50
8180	3″ diameter		26	.308		57	16.85		73.85	89
8190	3-1/2″ diameter		23	.348		61.50	19.05		80.55	97
8200	4″ diameter		23	.348		67	19.05		86.05	103
8210	5″ diameter	↓	18	.444	↓	209	24.50		233.50	267
8220	Channels, plastic coated									
8250	3/4″ x 1-1/2″, w/12″ rods for 1/2″ to 1″ conduit	1 Elec	28	.286	Ea.	35	15.65		50.65	62
8260	1-1/2″ x 1-1/2″, w/12″ rods for 1-1/4″ to 2″ conduit		26	.308		39	16.85		55.85	69
8270	1-1/2″ x 1-1/2″, w/12″ rods for 2-1/2″ to 3-1/2″ conduit		24	.333		42.50	18.25		60.75	74.50
8280	1-1/2″ x 1-7/8″, w/12″ rods for 4″ to 5″ conduit		22	.364		70.50	19.90		90.40	109
8290	1-1/2″ x 1-7/8″, w/12″ rods for 6″ conduit		20	.400		77.50	22		99.50	119
8320	Conduit hangers, plastic coated steel, with bolt & 12″ rod, 1/2″ diam.		140	.057		20	3.13		23.13	27
8330	3/4″ diameter		135	.059		20.50	3.24		23.74	27.50
8340	1″ diameter		125	.064		21	3.50		24.50	28.50
8350	1-1/4″ diameter		110	.073		22	3.98		25.98	30.50
8360	1-1/2″ diameter		100	.080		25	4.38		29.38	34.50
8370	2″ diameter		90	.089		27.50	4.86		32.36	37.50
8380	2-1/2″ diameter		70	.114		32.50	6.25		38.75	45.50
8390	3″ diameter		50	.160		39.50	8.75		48.25	57
8400	3-1/2″ diameter		35	.229		40	12.50		52.50	63.50
8410	4″ diameter		25	.320		59	17.50		76.50	91.50
8420	5″ diameter		20	.400		64.50	22		86.50	105
9000	Parallel type, conduit beam clamp, 1/2″		32	.250		3.82	13.70		17.52	25
9010	3/4″		32	.250		3.97	13.70		17.67	25.50
9020	1″		32	.250		4.38	13.70		18.08	26
9030	1-1/4″		30	.267		5.70	14.60		20.30	29
9040	1-1/2″		30	.267		6.50	14.60		21.10	29.50
9050	2″		30	.267		9.10	14.60		23.70	32.50
9060	2-1/2″		28	.286		9.65	15.65		25.30	34.50
9070	3″		28	.286		13.10	15.65		28.75	38.50
9090	4″		25	.320		15.70	17.50		33.20	44.50
9110	Right angle, conduit beam clamp, 1/2″		32	.250		2.39	13.70		16.09	23.50
9120	3/4″		32	.250		2.51	13.70		16.21	24
9130	1″		32	.250		2.79	13.70		16.49	24
9140	1-1/4″		30	.267		3.20	14.60		17.80	26
9150	1-1/2″		30	.267		3.59	14.60		18.19	26.50
9160	2″		30	.267		5.25	14.60		19.85	28.50
9170	2-1/2″		28	.286		6.40	15.65		22.05	31
9180	3″		28	.286		6.90	15.65		22.55	31.50
9190	3-1/2″		25	.320		8.35	17.50		25.85	36
9200	4″		25	.320		9	17.50		26.50	37
9230	Adjustable, conduit hanger, 1/2″		32	.250		4.19	13.70		17.89	25.50
9240	3/4″	↓	32	.250	↓	5.55	13.70		19.25	27

26 05 Common Work Results for Electrical

26 05 29 – Hangers and Supports for Electrical Systems

26 05 29.20 Hangers		Crew	Daily Output	Labor-Hours	Unit	2015 Bare Costs Material	Labor	Equipment	Total	Total Incl O&P
9250	1"	1 Elec	32	.250	Ea.	7	13.70		20.70	28.50
9260	1-1/4"		30	.267		8.35	14.60		22.95	31.50
9270	1-1/2"		30	.267		9.70	14.60		24.30	33
9280	2"		30	.267		12.45	14.60		27.05	36
9290	2-1/2"		28	.286		15.30	15.65		30.95	41
9300	3"		28	.286		17.95	15.65		33.60	44
9310	3-1/2"		25	.320		20.50	17.50		38	50
9320	4"		25	.320		23.50	17.50		41	53
9330	5"		20	.400		29	22		51	66
9340	6"		16	.500		34.50	27.50		62	80.50
9350	Combination conduit hanger, 3/8"		32	.250		11.25	13.70		24.95	33.50
9360	Adjustable flange 3/8"		32	.250		13.20	13.70		26.90	35.50

26 05 33 – Raceway and Boxes for Electrical Systems

26 05 33.13 Conduit		Crew	Daily Output	Labor-Hours	Unit	Material	Labor	Equipment	Total	Total Incl O&P
0010	**CONDUIT** To 15' high, includes 2 terminations, 2 elbows, R260533-20									
0020	11 beam clamps, and 11 couplings per 100 L.F.									
0300	Aluminum, 1/2" diameter	1 Elec	100	.080	L.F.	1.75	4.38		6.13	8.70
0500	3/4" diameter		90	.089		2.41	4.86		7.27	10.20
0700	1" diameter		80	.100		3.46	5.45		8.91	12.25
1000	1-1/4" diameter		70	.114		4.27	6.25		10.52	14.40
1030	1-1/2" diameter		65	.123		5.45	6.75		12.20	16.40
1050	2" diameter		60	.133		7.70	7.30		15	19.75
1070	2-1/2" diameter		50	.160		11.50	8.75		20.25	26
1100	3" diameter	2 Elec	90	.178		15.70	9.70		25.40	32.50
1130	3-1/2" diameter		80	.200		21.50	10.95		32.45	40.50
1140	4" diameter		70	.229		25	12.50		37.50	47
1150	5" diameter		50	.320		55	17.50		72.50	87.50
1160	6" diameter		40	.400		92.50	22		114.50	136
1161	Field bends, 45° to 90°, 1/2" diameter	1 Elec	53	.151	Ea.		8.25		8.25	12.80
1162	3/4" diameter		47	.170			9.30		9.30	14.40
1163	1" diameter		44	.182			9.95		9.95	15.40
1164	1-1/4" diameter		23	.348			19.05		19.05	29.50
1165	1-1/2" diameter		21	.381			21		21	32.50
1166	2" diameter		16	.500			27.50		27.50	42.50
1170	Elbows, 1/2" diameter		40	.200		8.60	10.95		19.55	26.50
1200	3/4" diameter		32	.250		10.95	13.70		24.65	33
1230	1" diameter		28	.286		15.75	15.65		31.40	41.50
1250	1-1/4" diameter		24	.333		26.50	18.25		44.75	57
1270	1-1/2" diameter		20	.400		34.50	22		56.50	72
1300	2" diameter		16	.500		50	27.50		77.50	97.50
1330	2-1/2" diameter		12	.667		94	36.50		130.50	160
1350	3" diameter		8	1		139	54.50		193.50	237
1370	3-1/2" diameter		6	1.333		243	73		316	380
1400	4" diameter		5	1.600		280	87.50		367.50	445
1410	5" diameter		4	2		745	109		854	990
1420	6" diameter		2.50	3.200		1,300	175		1,475	1,700
1430	Couplings, 1/2" diameter		320	.025		2.69	1.37		4.06	5.10
1450	3/4" diameter		192	.042		4.01	2.28		6.29	7.95
1470	1" diameter		160	.050		5.30	2.74		8.04	10.05
1500	1-1/4" diameter		120	.067		6.80	3.65		10.45	13.15
1530	1-1/2" diameter		96	.083		7.60	4.56		12.16	15.40
1550	2" diameter		80	.100		11.15	5.45		16.60	20.50

26 05 33.13 Conduit		Crew	Daily Output	Labor-Hours	Unit	2015 Bare Costs Material	Labor	Equipment	Total	Total Incl O&P
1570	2-1/2" diameter	1 Elec	69	.116	Ea.	26.50	6.35		32.85	39
1600	3" diameter		64	.125		33.50	6.85		40.35	47
1630	3-1/2" diameter		56	.143		52.50	7.80		60.30	69.50
1650	4" diameter		53	.151		56	8.25		64.25	74.50
1670	5" diameter		48	.167		180	9.10		189.10	212
1690	6" diameter		46	.174		355	9.50		364.50	405
1691	See note on line 26 05 33.13 9995 R260533-30									
1750	Rigid galvanized steel, 1/2" diameter	1 Elec	90	.089	L.F.	2.49	4.86		7.35	10.30
1770	3/4" diameter		80	.100		2.74	5.45		8.19	11.45
1800	1" diameter		65	.123		3.95	6.75		10.70	14.75
1830	1-1/4" diameter		60	.133		5.15	7.30		12.45	16.95
1850	1-1/2" diameter		55	.145		6.10	7.95		14.05	19.05
1870	2" diameter		45	.178		7.90	9.70		17.60	24
1900	2-1/2" diameter		35	.229		13.75	12.50		26.25	34.50
1930	3" diameter	2 Elec	50	.320		15.95	17.50		33.45	44.50
1950	3-1/2" diameter		44	.364		20	19.90		39.90	53
1970	4" diameter		40	.400		23	22		45	59.50
1980	5" diameter		30	.533		46	29		75	95.50
1990	6" diameter		20	.800		66.50	44		110.50	141
1991	Field bends, 45° to 90°, 1/2" diameter	1 Elec	44	.182	Ea.		9.95		9.95	15.40
1992	3/4" diameter		40	.200			10.95		10.95	16.95
1993	1" diameter		36	.222			12.15		12.15	18.80
1994	1-1/4" diameter		19	.421			23		23	35.50
1995	1-1/2" diameter		18	.444			24.50		24.50	37.50
1996	2" diameter		13	.615			33.50		33.50	52
2000	Elbows, 1/2" diameter		32	.250		6.65	13.70		20.35	28.50
2030	3/4" diameter		28	.286		7	15.65		22.65	31.50
2050	1" diameter		24	.333		10.75	18.25		29	40
2070	1-1/4" diameter		18	.444		14.75	24.50		39.25	53.50
2100	1-1/2" diameter		16	.500		18.25	27.50		45.75	62.50
2130	2" diameter		12	.667		26.50	36.50		63	85.50
2150	2-1/2" diameter		8	1		49	54.50		103.50	139
2170	3" diameter		6	1.333		68	73		141	188
2200	3-1/2" diameter		4.20	1.905		108	104		212	280
2220	4" diameter		4	2		122	109		231	305
2230	5" diameter		3.50	2.286		335	125		460	565
2240	6" diameter		2	4		510	219		729	900
2250	Couplings, 1/2" diameter		267	.030		1.64	1.64		3.28	4.34
2270	3/4" diameter		160	.050		2.01	2.74		4.75	6.45
2300	1" diameter		133	.060		2.96	3.29		6.25	8.35
2330	1-1/4" diameter		100	.080		3.72	4.38		8.10	10.90
2350	1-1/2" diameter		80	.100		4.70	5.45		10.15	13.60
2370	2" diameter		67	.119		6.20	6.55		12.75	16.90
2400	2-1/2" diameter		57	.140		15.35	7.70		23.05	29
2430	3" diameter		53	.151		19.90	8.25		28.15	35
2450	3-1/2" diameter		47	.170		26.50	9.30		35.80	44
2470	4" diameter		44	.182		27	9.95		36.95	45
2480	5" diameter		40	.200		62	10.95		72.95	85.50
2490	6" diameter		38	.211		86.50	11.50		98	113
2491	See note on line 26 05 33.13 9995 R260533-30									
2500	Steel, intermediate conduit (IMC), 1/2" diameter	1 Elec	100	.080	L.F.	1.79	4.38		6.17	8.75
2530	3/4" diameter		90	.089		2.22	4.86		7.08	10
2550	1" diameter		70	.114		3.28	6.25		9.53	13.30

26 05 33 – Raceway and Boxes for Electrical Systems

26 05 33.13 Conduit		Crew	Daily Output	Labor-Hours	Unit	Material	Labor (2015 Bare Costs)	Equipment	Total	Total Incl O&P
2570	1-1/4" diameter	1 Elec	65	.123	L.F.	4.01	6.75		10.76	14.80
2600	1-1/2" diameter		60	.133		5.35	7.30		12.65	17.20
2630	2" diameter		50	.160		6.40	8.75		15.15	20.50
2650	2-1/2" diameter		40	.200		11.35	10.95		22.30	29.50
2670	3" diameter	2 Elec	60	.267		15.40	14.60		30	39.50
2700	3-1/2" diameter		54	.296		20.50	16.20		36.70	47.50
2730	4" diameter		50	.320		22	17.50		39.50	51.50
2731	Field bends, 45° to 90°, 1/2" diameter	1 Elec	44	.182	Ea.		9.95		9.95	15.40
2732	3/4" diameter		40	.200			10.95		10.95	16.95
2733	1" diameter		36	.222			12.15		12.15	18.80
2734	1-1/4" diameter		19	.421			23		23	35.50
2735	1-1/2" diameter		18	.444			24.50		24.50	37.50
2736	2" diameter		13	.615			33.50		33.50	52
2750	Elbows, 1/2" diameter		32	.250		12.10	13.70		25.80	34.50
2770	3/4" diameter		28	.286		12.90	15.65		28.55	38
2800	1" diameter		24	.333		19	18.25		37.25	49
2830	1-1/4" diameter		18	.444		29	24.50		53.50	69
2850	1-1/2" diameter		16	.500		35.50	27.50		63	81.50
2870	2" diameter		12	.667		38	36.50		74.50	98
2900	2-1/2" diameter		8	1		68.50	54.50		123	160
2930	3" diameter		6	1.333		99.50	73		172.50	223
2950	3-1/2" diameter		4.20	1.905		213	104		317	395
2970	4" diameter		4	2		203	109		312	395
3000	Couplings, 1/2" diameter		293	.027		1.64	1.49		3.13	4.11
3030	3/4" diameter		176	.045		2.01	2.49		4.50	6.05
3050	1" diameter		147	.054		2.96	2.98		5.94	7.85
3070	1-1/4" diameter		110	.073		3.72	3.98		7.70	10.25
3100	1-1/2" diameter		88	.091		4.70	4.97		9.67	12.85
3130	2" diameter		73	.110		6.20	6		12.20	16.10
3150	2-1/2" diameter		63	.127		15.35	6.95		22.30	27.50
3170	3" diameter		59	.136		19.90	7.40		27.30	33.50
3200	3-1/2" diameter		52	.154		26.50	8.40		34.90	42.50
3230	4" diameter		49	.163		27	8.95		35.95	43.50
3231	See note on line 26 05 33.13 9995 R260533-30									
4100	Rigid steel, plastic coated, 40 mil thick									
4130	1/2" diameter	1 Elec	80	.100	L.F.	6.95	5.45		12.40	16.10
4150	3/4" diameter		70	.114		7.50	6.25		13.75	17.90
4170	1" diameter		55	.145		9.25	7.95		17.20	22.50
4200	1-1/4" diameter		50	.160		13.25	8.75		22	28
4230	1-1/2" diameter		45	.178		14	9.70		23.70	30.50
4250	2" diameter		35	.229		18.85	12.50		31.35	40
4270	2-1/2" diameter		25	.320		31	17.50		48.50	61
4300	3" diameter	2 Elec	44	.364		35	19.90		54.90	69.50
4330	3-1/2" diameter		40	.400		46.50	22		68.50	85.50
4350	4" diameter		36	.444		48.50	24.50		73	91
4370	5" diameter		30	.533		121	29		150	178
4400	Elbows, 1/2" diameter	1 Elec	28	.286	Ea.	18.25	15.65		33.90	44
4430	3/4" diameter		24	.333		15.25	18.25		33.50	45
4450	1" diameter		18	.444		17.45	24.50		41.95	56.50
4470	1-1/4" diameter		16	.500		26.50	27.50		54	71.50
4500	1-1/2" diameter		12	.667		26.50	36.50		63	85.50
4530	2" diameter		8	1		37	54.50		91.50	125
4550	2-1/2" diameter		6	1.333		80	73		153	201

26 05 33.13 Conduit		Crew	Daily Output	Labor-Hours	Unit	2015 Bare Costs Material	Labor	Equipment	Total	Total Incl O&P
4570	3" diameter	1 Elec	4.20	1.905	Ea.	132	104		236	305
4600	3-1/2" diameter		4	2		163	109		272	350
4630	4" diameter		3.80	2.105		178	115		293	375
4650	5" diameter		3.50	2.286		440	125		565	680
4680	Couplings, 1/2" diameter		213	.038		5.35	2.05		7.40	9.05
4700	3/4" diameter		128	.063		4.27	3.42		7.69	10
4730	1" diameter		107	.075		5.55	4.09		9.64	12.45
4750	1-1/4" diameter		80	.100		8.30	5.45		13.75	17.60
4770	1-1/2" diameter		64	.125		7.70	6.85		14.55	19.05
4800	2" diameter		53	.151		11.25	8.25		19.50	25
4830	2-1/2" diameter		46	.174		33	9.50		42.50	51
4850	3" diameter		43	.186		40.50	10.20		50.70	60.50
4870	3-1/2" diameter		38	.211		50	11.50		61.50	73
4900	4" diameter		36	.222		59	12.15		71.15	84
4950	5" diameter		32	.250		188	13.70		201.70	227
4951	See note on line 26 05 33.13 9995 R260533-30									
5000	Electric metallic tubing (EMT), 1/2" diameter	1 Elec	170	.047	L.F.	.67	2.57		3.24	4.73
5020	3/4" diameter		130	.062		.94	3.37		4.31	6.25
5040	1" diameter		115	.070		1.60	3.81		5.41	7.65
5060	1-1/4" diameter		100	.080		2.61	4.38		6.99	9.65
5080	1-1/2" diameter		90	.089		3.35	4.86		8.21	11.25
5100	2" diameter		80	.100		4.20	5.45		9.65	13.05
5120	2-1/2" diameter		60	.133		9.05	7.30		16.35	21.50
5140	3" diameter	2 Elec	100	.160		10.60	8.75		19.35	25.50
5160	3-1/2" diameter		90	.178		13.25	9.70		22.95	29.50
5180	4" diameter		80	.200		14.50	10.95		25.45	33
5200	Field bends, 45° to 90°, 1/2" diameter	1 Elec	89	.090	Ea.		4.92		4.92	7.60
5220	3/4" diameter		80	.100			5.45		5.45	8.45
5240	1" diameter		73	.110			6		6	9.30
5260	1-1/4" diameter		38	.211			11.50		11.50	17.85
5280	1-1/2" diameter		36	.222			12.15		12.15	18.80
5300	2" diameter		26	.308			16.85		16.85	26
5320	Offsets, 1/2" diameter		65	.123			6.75		6.75	10.40
5340	3/4" diameter		62	.129			7.05		7.05	10.95
5360	1" diameter		53	.151			8.25		8.25	12.80
5380	1-1/4" diameter		30	.267			14.60		14.60	22.50
5400	1-1/2" diameter		28	.286			15.65		15.65	24
5420	2" diameter		20	.400			22		22	34
5700	Elbows, 1" diameter		40	.200		5.35	10.95		16.30	23
5720	1-1/4" diameter		32	.250		6.65	13.70		20.35	28.50
5740	1-1/2" diameter		24	.333		7.75	18.25		26	36.50
5760	2" diameter		20	.400		11.35	22		33.35	46.50
5780	2-1/2" diameter		12	.667		31	36.50		67.50	90.50
5800	3" diameter		9	.889		41.50	48.50		90	121
5820	3-1/2" diameter		7	1.143		55.50	62.50		118	158
5840	4" diameter		6	1.333		65.50	73		138.50	185
6200	Couplings, set screw, steel, 1/2" diameter		235	.034		1.29	1.86		3.15	4.30
6220	3/4" diameter		188	.043		1.97	2.33		4.30	5.75
6240	1" diameter		157	.051		3.30	2.79		6.09	7.95
6260	1-1/4" diameter		118	.068		6.85	3.71		10.56	13.30
6280	1-1/2" diameter		94	.085		9.60	4.66		14.26	17.75
6300	2" diameter		78	.103		12.85	5.60		18.45	23
6320	2-1/2" diameter		67	.119		37	6.55		43.55	50.50

26 05 33 – Raceway and Boxes for Electrical Systems

26 05 33.13 Conduit		Crew	Daily Output	Labor-Hours	Unit	Material	Labor	Equipment	Total	Total Incl O&P
						2015 Bare Costs				
6340	3" diameter	1 Elec	59	.136	Ea.	40.50	7.40		47.90	56
6360	3-1/2" diameter		47	.170		45.50	9.30		54.80	64.50
6380	4" diameter		43	.186		50	10.20		60.20	71
6381	See note on line 26 05 33.13 9995 R260533-30									
6500	Box connectors, set screw, steel, 1/2" diameter	1 Elec	120	.067	Ea.	.99	3.65		4.64	6.75
6520	3/4" diameter		110	.073		1.62	3.98		5.60	7.95
6540	1" diameter		90	.089		3.05	4.86		7.91	10.90
6560	1-1/4" diameter		70	.114		5.95	6.25		12.20	16.25
6580	1-1/2" diameter		60	.133		8.60	7.30		15.90	21
6600	2" diameter		50	.160		11.80	8.75		20.55	26.50
6620	2-1/2" diameter		36	.222		41.50	12.15		53.65	64.50
6640	3" diameter		27	.296		49.50	16.20		65.70	79.50
6680	3-1/2" diameter		21	.381		69	21		90	109
6700	4" diameter		16	.500		77	27.50		104.50	127
6740	Insulated box connectors, set screw, steel, 1/2" diameter		120	.067		1.36	3.65		5.01	7.15
6760	3/4" diameter		110	.073		2.14	3.98		6.12	8.50
6780	1" diameter		90	.089		3.94	4.86		8.80	11.90
6800	1-1/4" diameter		70	.114		7.15	6.25		13.40	17.55
6820	1-1/2" diameter		60	.133		10.15	7.30		17.45	22.50
6840	2" diameter		50	.160		14.75	8.75		23.50	30
6860	2-1/2" diameter		36	.222		69.50	12.15		81.65	95.50
6880	3" diameter		27	.296		82.50	16.20		98.70	116
6900	3-1/2" diameter		21	.381		115	21		136	159
6920	4" diameter		16	.500		125	27.50		152.50	181
7000	EMT to conduit adapters, 1/2" diameter (compression)		70	.114		3.85	6.25		10.10	13.95
7020	3/4" diameter		60	.133		5.40	7.30		12.70	17.20
7040	1" diameter		50	.160		8.40	8.75		17.15	23
7060	1-1/4" diameter		40	.200		15.65	10.95		26.60	34
7080	1-1/2" diameter		30	.267		19.20	14.60		33.80	43.50
7100	2" diameter		25	.320		28	17.50		45.50	57.50
7200	EMT to Greenfield adapters, 1/2" to 3/8" diameter (compression)		90	.089		2.58	4.86		7.44	10.40
7220	1/2" diameter		90	.089		4.64	4.86		9.50	12.65
7240	3/4" diameter		80	.100		6.05	5.45		11.50	15.10
7260	1" diameter		70	.114		17	6.25		23.25	28.50
7270	1-1/4" diameter		60	.133		19.50	7.30		26.80	33
7280	1-1/2" diameter		50	.160		23	8.75		31.75	39
7290	2" diameter		40	.200		33.50	10.95		44.45	54
7400	EMT,LB, LR or LL fittings with covers, 1/2" dia, set screw		24	.333		6.15	18.25		24.40	35
7420	3/4" diameter		20	.400		8.95	22		30.95	44
7440	1" diameter		16	.500		11.10	27.50		38.60	54.50
7450	1-1/4" diameter		13	.615		15.15	33.50		48.65	68.50
7460	1-1/2" diameter		11	.727		20.50	40		60.50	84
7470	2" diameter		9	.889		32.50	48.50		81	111
7600	EMT, "T" fittings with covers, 1/2" diameter, set screw		16	.500		7.90	27.50		35.40	51
7620	3/4" diameter		15	.533		10.15	29		39.15	56
7640	1" diameter		12	.667		13.90	36.50		50.40	72
7650	1-1/4" diameter		11	.727		20	40		60	83.50
7660	1-1/2" diameter		10	.800		24	44		68	94
7670	2" diameter		8	1		31.50	54.50		86	119
8000	EMT, expansion fittings, no jumper, 1/2" diameter		24	.333		70.50	18.25		88.75	106
8020	3/4" diameter		20	.400		82	22		104	125
8040	1" diameter		16	.500		97.50	27.50		125	150
8060	1-1/4" diameter		13	.615		132	33.50		165.50	197

26 05 33.13 Conduit		Crew	Daily Output	Labor-Hours	Unit	2015 Bare Costs Material	Labor	Equipment	Total	Total Incl O&P
8080	1-1/2" diameter	1 Elec	11	.727	Ea.	180	40		220	260
8100	2" diameter		9	.889		266	48.50		314.50	370
8110	2-1/2" diameter		7	1.143		425	62.50		487.50	560
8120	3" diameter		6	1.333		520	73		593	685
8140	4" diameter		5	1.600		905	87.50		992.50	1,125
8200	Split adapter, 1/2" diameter		110	.073		2.49	3.98		6.47	8.90
8210	3/4" diameter		90	.089		2.03	4.86		6.89	9.80
8220	1" diameter		70	.114		2.78	6.25		9.03	12.75
8230	1-1/4" diameter		60	.133		4.49	7.30		11.79	16.25
8240	1-1/2" diameter		50	.160		6.90	8.75		15.65	21
8250	2" diameter		36	.222		20.50	12.15		32.65	41.50
8300	1 hole clips, 1/2" diameter		500	.016		.39	.88		1.27	1.79
8320	3/4" diameter		470	.017		.50	.93		1.43	1.99
8340	1" diameter		444	.018		.92	.99		1.91	2.54
8360	1-1/4" diameter		400	.020		1.26	1.09		2.35	3.08
8380	1-1/2" diameter		355	.023		1.90	1.23		3.13	4
8400	2" diameter		320	.025		3.01	1.37		4.38	5.45
8420	2-1/2" diameter		266	.030		5.55	1.65		7.20	8.70
8440	3" diameter		160	.050		7	2.74		9.74	11.95
8460	3-1/2" diameter		133	.060		11.05	3.29		14.34	17.30
8480	4" diameter		100	.080		14.35	4.38		18.73	22.50
8500	Clamp back spacers, 1/2" diameter		500	.016		1.18	.88		2.06	2.66
8510	3/4" diameter		470	.017		1.40	.93		2.33	2.98
8520	1" diameter		444	.018		2.44	.99		3.43	4.21
8530	1-1/4" diameter		400	.020		4.36	1.09		5.45	6.50
8540	1-1/2" diameter		355	.023		4.87	1.23		6.10	7.25
8550	2" diameter		320	.025		8.20	1.37		9.57	11.10
8560	2-1/2" diameter		266	.030		15.85	1.65		17.50	20
8570	3" diameter		160	.050		22.50	2.74		25.24	29
8580	3-1/2" diameter		133	.060		30.50	3.29		33.79	38.50
8590	4" diameter		100	.080		66.50	4.38		70.88	80
8600	Offset connectors, 1/2" diameter		40	.200		4.28	10.95		15.23	21.50
8610	3/4" diameter		32	.250		6.20	13.70		19.90	28
8620	1" diameter		24	.333		6.90	18.25		25.15	35.50
8650	90° pulling elbows, female, 1/2" diameter, with gasket		24	.333		6.75	18.25		25	35.50
8660	3/4" diameter		20	.400		11.55	22		33.55	46.50
8700	Couplings, compression, 1/2" diameter, steel		78	.103		3.11	5.60		8.71	12.10
8710	3/4" diameter		67	.119		4.33	6.55		10.88	14.85
8720	1" diameter		59	.136		7.40	7.40		14.80	19.65
8730	1-1/4" diameter		47	.170		14.10	9.30		23.40	30
8740	1-1/2" diameter		38	.211		20.50	11.50		32	40.50
8750	2" diameter		31	.258		28	14.10		42.10	52.50
8760	2-1/2" diameter		24	.333		125	18.25		143.25	166
8770	3" diameter		21	.381		156	21		177	205
8780	3-1/2" diameter		19	.421		254	23		277	315
8790	4" diameter	↓	16	.500	↓	260	27.50		287.50	330
8791	See note on line 26 05 33.13 9995 R260533-30									
8800	Box connectors, compression, 1/2" diam., steel	1 Elec	120	.067	Ea.	2.59	3.65		6.24	8.50
8810	3/4" diameter		110	.073		3.60	3.98		7.58	10.10
8820	1" diameter		90	.089		5.35	4.86		10.21	13.45
8830	1-1/4" diameter		70	.114		13.60	6.25		19.85	24.50
8840	1-1/2" diameter		60	.133		16	7.30		23.30	29
8850	2" diameter	↓	50	.160	↓	23	8.75		31.75	39

26 05 33.13 Conduit		Crew	Daily Output	Labor-Hours	Unit	Material	2015 Bare Costs Labor	Equipment	Total	Total Incl O&P
8860	2-1/2" diameter	1 Elec	36	.222	Ea.	57	12.15		69.15	81.50
8870	3" diameter		27	.296		78.50	16.20		94.70	112
8880	3-1/2" diameter		21	.381		119	21		140	164
8890	4" diameter		16	.500		122	27.50		149.50	177
8900	Box connectors, insulated compression, 1/2" diam., steel		120	.067		2.51	3.65		6.16	8.40
8910	3/4" diameter		110	.073		3.43	3.98		7.41	9.90
8920	1" diameter		90	.089		5.80	4.86		10.66	13.90
8930	1-1/4" diameter		70	.114		11.65	6.25		17.90	22.50
8940	1-1/2" diameter		60	.133		17.55	7.30		24.85	30.50
8950	2" diameter		50	.160		25.50	8.75		34.25	41.50
8960	2-1/2" diameter		36	.222		66.50	12.15		78.65	92
8970	3" diameter		27	.296		87	16.20		103.20	121
8980	3-1/2" diameter		21	.381		127	21		148	172
8990	4" diameter		16	.500		129	27.50		156.50	185
9100	PVC, schedule 40, 1/2" diameter		190	.042	L.F.	.90	2.30		3.20	4.56
9110	3/4" diameter		145	.055		.98	3.02		4	5.75
9120	1" diameter		125	.064		2.11	3.50		5.61	7.70
9130	1-1/4" diameter		110	.073		2.63	3.98		6.61	9.05
9140	1-1/2" diameter		100	.080		2.99	4.38		7.37	10.10
9150	2" diameter		90	.089		3.90	4.86		8.76	11.85
9160	2-1/2" diameter		65	.123		5.45	6.75		12.20	16.40
9170	3" diameter	2 Elec	110	.145		6.30	7.95		14.25	19.20
9180	3-1/2" diameter		100	.160		8.25	8.75		17	22.50
9190	4" diameter		90	.178		9.65	9.70		19.35	25.50
9200	5" diameter		70	.229		12.45	12.50		24.95	33
9210	6" diameter		60	.267		15.95	14.60		30.55	40
9220	Elbows, 1/2" diameter	1 Elec	50	.160	Ea.	.90	8.75		9.65	14.55
9225	3/4" diameter		42	.190		.88	10.40		11.28	17.10
9230	1" diameter		35	.229		1.33	12.50		13.83	21
9235	1-1/4" diameter		28	.286		2.04	15.65		17.69	26
9240	1-1/2" diameter		20	.400		2.65	22		24.65	37
9245	2" diameter		16	.500		3.59	27.50		31.09	46.50
9250	2-1/2" diameter		11	.727		6.45	40		46.45	68.50
9255	3" diameter		9	.889		10.85	48.50		59.35	87.50
9260	3-1/2" diameter		7	1.143		14	62.50		76.50	112
9265	4" diameter		6	1.333		17.25	73		90.25	132
9270	5" diameter		4	2		28	109		137	200
9275	6" diameter		3	2.667		64.50	146		210.50	297
9312	Couplings, 1/2" diameter		50	.160		.29	8.75		9.04	13.85
9314	3/4" diameter		42	.190		.32	10.40		10.72	16.50
9316	1" diameter		35	.229		.36	12.50		12.86	19.75
9318	1-1/4" diameter		28	.286		.51	15.65		16.16	24.50
9320	1-1/2" diameter		20	.400		.54	22		22.54	34.50
9322	2" diameter		16	.500		1.07	27.50		28.57	43.50
9324	2-1/2" diameter		11	.727		1.80	40		41.80	63.50
9326	3" diameter		9	.889		1.97	48.50		50.47	77.50
9328	3-1/2" diameter		7	1.143		2.51	62.50		65.01	100
9330	4" diameter		6	1.333		3.07	73		76.07	116
9332	5" diameter		4	2		7.55	109		116.55	177
9334	6" diameter		3	2.667		10.60	146		156.60	238
9335	See note on line 26 05 33.13 9995 R260533-30									
9340	Field bends, 45° & 90°, 1/2" diameter	1 Elec	45	.178	Ea.		9.70		9.70	15.05
9350	3/4" diameter		40	.200			10.95		10.95	16.95

26 05 33 – Raceway and Boxes for Electrical Systems

26 05 33.13 Conduit		Crew	Daily Output	Labor-Hours	Unit	2015 Bare Costs Material	Labor	Equipment	Total	Total Incl O&P
9360	1" diameter	1 Elec	35	.229	Ea.		12.50		12.50	19.35
9370	1-1/4" diameter		32	.250			13.70		13.70	21
9380	1-1/2" diameter		27	.296			16.20		16.20	25
9390	2" diameter		20	.400			22		22	34
9400	2-1/2" diameter		16	.500			27.50		27.50	42.50
9410	3" diameter		13	.615			33.50		33.50	52
9420	3-1/2" diameter		12	.667			36.50		36.50	56.50
9430	4" diameter		10	.800			44		44	68
9440	5" diameter		9	.889			48.50		48.50	75.50
9450	6" diameter		8	1			54.50		54.50	84.50
9460	PVC adapters, 1/2" diameter		50	.160		.25	8.75		9	13.85
9470	3/4" diameter		42	.190		.41	10.40		10.81	16.60
9480	1" diameter		38	.211		.56	11.50		12.06	18.45
9490	1-1/4" diameter		35	.229		.74	12.50		13.24	20
9500	1-1/2" diameter		32	.250		.98	13.70		14.68	22
9510	2" diameter		27	.296		1.22	16.20		17.42	26.50
9520	2-1/2" diameter		23	.348		2.14	19.05		21.19	32
9530	3" diameter		18	.444		2.97	24.50		27.47	41
9540	3-1/2" diameter		13	.615		4.57	33.50		38.07	57
9550	4" diameter		11	.727		5.30	40		45.30	67.50
9560	5" diameter		8	1		11.20	54.50		65.70	97
9570	6" diameter	▼	6	1.333	▼	18.70	73		91.70	134
9580	PVC-LB, LR or LL fittings & covers									
9590	1/2" diameter	1 Elec	20	.400	Ea.	3.04	22		25.04	37.50
9600	3/4" diameter		16	.500		4.60	27.50		32.10	47.50
9610	1" diameter		12	.667		5.05	36.50		41.55	62
9620	1-1/4" diameter		9	.889		7.80	48.50		56.30	84
9630	1-1/2" diameter		7	1.143		9.30	62.50		71.80	107
9640	2" diameter		6	1.333		14.75	73		87.75	129
9650	2-1/2" diameter		6	1.333		52	73		125	171
9660	3" diameter		5	1.600		50.50	87.50		138	192
9670	3-1/2" diameter		4	2		53.50	109		162.50	228
9680	4" diameter	▼	3	2.667	▼	70.50	146		216.50	305
9690	PVC-tee fitting & cover									
9700	1/2"	1 Elec	14	.571	Ea.	5.85	31.50		37.35	55
9710	3/4"		13	.615		5.85	33.50		39.35	58.50
9720	1"		10	.800		6.55	44		50.55	75
9730	1-1/4"		9	.889		11.50	48.50		60	88
9740	1-1/2"		8	1		13.50	54.50		68	99.50
9750	2"	▼	7	1.143		20.50	62.50		83	120
9760	PVC-reducers, 3/4" x 1/2" diameter					1.41			1.41	1.55
9770	1" x 1/2" diameter					3.09			3.09	3.40
9780	1" x 3/4" diameter					3.34			3.34	3.67
9790	1-1/4" x 3/4" diameter					4.10			4.10	4.51
9800	1-1/4" x 1" diameter					4.34			4.34	4.77
9810	1-1/2" x 1-1/4" diameter					4.53			4.53	4.98
9820	2" x 1-1/4" diameter					5.45			5.45	5.95
9830	2-1/2" x 2" diameter					16.85			16.85	18.55
9840	3" x 2" diameter					17.40			17.40	19.15
9850	4" x 3" diameter					20			20	22
9860	Cement, quart					16.60			16.60	18.30
9870	Gallon					104			104	114
9880	Heat bender, to 6" diameter				▼	1,850			1,850	2,025

26 05 33 – Raceway and Boxes for Electrical Systems

26 05 33.13 Conduit		Crew	Daily Output	Labor-Hours	Unit	Material	2015 Bare Costs Labor	Equipment	Total	Total Incl O&P
9900	Add to labor for higher elevated installation									
9910	15′ to 20′ high, add						10%			
9920	20′ to 25′ high, add						20%			
9930	25′ to 30′ high, add						25%			
9940	30′ to 35′ high, add						30%			
9950	35′ to 40′ high, add						35%			
9960	Over 40′ high, add						40%			
9990	Minimum labor/equipment charge	1 Elec	4	2	Job		109		109	169
9995	Do not include labor when adding couplings to a fitting installation R260533-30									

26 05 33.14 Conduit		Crew	Daily Output	Labor-Hours	Unit	Material	Labor	Equipment	Total	Total Incl O&P
0010	**CONDUIT** To 15′ high, includes 11 couplings per 100′									
0200	Electric metallic tubing, 1/2″diameter	1 Elec	435	.018	L.F.	.48	1.01		1.49	2.09
0220	3/4″ diameter		253	.032		.76	1.73		2.49	3.51
0240	1″ diameter		207	.039		1.26	2.11		3.37	4.66
0260	1-1/4″ diamcter		173	.046		2.18	2.53		4.71	6.30
0280	1-1/2″ diameter		153	.052		2.83	2.86		5.69	7.55
0300	2″ diameter		130	.062		3.52	3.37		6.89	9.10
0320	2-1/2″ diameter		92	.087		7.20	4.76		11.96	15.25
0340	3″ diameter	2 Elec	148	.108		8.35	5.90		14.25	18.35
0360	3-1/2″ diameter		134	.119		10.25	6.55		16.80	21.50
0380	4″ diameter		114	.140		10.90	7.70		18.60	24
0500	Steel rigid galvanized, 1/2″ diameter	1 Elec	146	.055		1.93	3		4.93	6.75
0520	3/4″ diameter		125	.064		2.05	3.50		5.55	7.65
0540	1″ diameter		93	.086		2.95	4.71		7.66	10.55
0560	1-1/4″ diameter		88	.091		4.20	4.97		9.17	12.30
0580	1-1/2″ diameter		80	.100		4.67	5.45		10.12	13.60
0600	2″ diameter		65	.123		5.75	6.75		12.50	16.75
0620	2-1/2″ diameter		48	.167		11.20	9.10		20.30	26.50
0640	3″ diameter	2 Elec	64	.250		12.90	13.70		26.60	35
0660	3-1/2″ diameter		60	.267		16.55	14.60		31.15	40.50
0680	4″ diameter		52	.308		18.15	16.85		35	46
0700	5″ diameter		50	.320		36	17.50		53.50	66.50
0720	6″ diameter		48	.333		47.50	18.25		65.75	80.50
1000	Steel intermediate conduit (IMC), 1/2 diameter	1 Elec	155	.052		1.19	2.82		4.01	5.70
1010	3/4″ diameter		130	.062		1.44	3.37		4.81	6.80
1020	1″ diameter		100	.080		2.17	4.38		6.55	9.20
1030	1-1/4″ diameter		93	.086		2.81	4.71		7.52	10.40
1040	1-1/2″ diameter		85	.094		3.64	5.15		8.79	11.95
1050	2″ diameter		70	.114		4.51	6.25		10.76	14.65
1060	2-1/2″ diameter		53	.151		9.20	8.25		17.45	23
1070	3″ diameter	2 Elec	80	.200		12.15	10.95		23.10	30.50
1080	3-1/2″ diameter		70	.229		14.65	12.50		27.15	35.50
1090	4″ diameter		60	.267		16.10	14.60		30.70	40

26 05 33.15 Conduit Nipples		Crew	Daily Output	Labor-Hours	Unit	Material	Labor	Equipment	Total	Total Incl O&P
0010	**CONDUIT NIPPLES** With locknuts and bushings									
0100	Aluminum, 1/2″ diameter, close	1 Elec	36	.222	Ea.	8.45	12.15		20.60	28
0120	1-1/2″ long		36	.222		8.60	12.15		20.75	28.50
0140	2″ long		36	.222		8.95	12.15		21.10	28.50
0160	2-1/2″ long		36	.222		9.75	12.15		21.90	29.50
0180	3″ long		36	.222		9.90	12.15		22.05	29.50
0200	3-1/2″ long		36	.222		10.65	12.15		22.80	30.50
0220	4″ long		36	.222		10.85	12.15		23	31

26 05 33.15 Conduit Nipples		Crew	Daily Output	Labor-Hours	Unit	Material	2015 Bare Costs Labor	Equipment	Total	Total Incl O&P
0240	5" long	1 Elec	36	.222	Ea.	12.40	12.15		24.55	32.50
0260	6" long		36	.222		12.60	12.15		24.75	32.50
0280	8" long		36	.222		13.10	12.15		25.25	33
0300	10" long		36	.222		17.20	12.15		29.35	38
0320	12" long		36	.222		16.60	12.15		28.75	37
0340	3/4" diameter, close		32	.250		10.75	13.70		24.45	33
0360	1-1/2" long		32	.250		10.90	13.70		24.60	33
0380	2" long		32	.250		11.05	13.70		24.75	33
0400	2-1/2" long		32	.250		11.95	13.70		25.65	34
0420	3" long		32	.250		12.20	13.70		25.90	34.50
0440	3-1/2" long		32	.250		12.90	13.70		26.60	35
0460	4" long		32	.250		13.80	13.70		27.50	36
0480	5" long		32	.250		14.70	13.70		28.40	37
0500	6" long		32	.250		15.40	13.70		29.10	38
0520	8" long		32	.250		17.80	13.70		31.50	40.50
0540	10" long		32	.250		21	13.70		34.70	44
0560	12" long		32	.250		29.50	13.70		43.20	53.50
0580	1" diameter, close		27	.296		16.45	16.20		32.65	43
0600	2" long		27	.296		18.75	16.20		34.95	45.50
0620	2-1/2" long		27	.296		17.80	16.20		34	44.50
0640	3" long		27	.296		21	16.20		37.20	48
0660	3-1/2" long		27	.296		20.50	16.20		36.70	47.50
0680	4" long		27	.296		23.50	16.20		39.70	50.50
0700	5" long		27	.296		23	16.20		39.20	50
0720	6" long		27	.296		24.50	16.20		40.70	52
0740	8" long		27	.296		28.50	16.20		44.70	56.50
0760	10" long		27	.296		34	16.20		50.20	62
0780	12" long		27	.296		37	16.20		53.20	65.50
0800	1-1/4" diameter, close		23	.348		24	19.05		43.05	56
0820	2" long		23	.348		24.50	19.05		43.55	56.50
0840	2-1/2" long		23	.348		25.50	19.05		44.55	57.50
0860	3" long		23	.348		26.50	19.05		45.55	58.50
0880	3-1/2" long		23	.348		28.50	19.05		47.55	61
0900	4" long		23	.348		29.50	19.05		48.55	62
0920	5" long		23	.348		32.50	19.05		51.55	65.50
0940	6" long		23	.348		35	19.05		54.05	68
0960	8" long		23	.348		44.50	19.05		63.55	78.50
0980	10" long		23	.348		42.50	19.05		61.55	76
1000	12" long		23	.348		62	19.05		81.05	97.50
1020	1-1/2" diameter, close		20	.400		32	22		54	69
1040	2" long		20	.400		33	22		55	70
1060	2-1/2" long		20	.400		33.50	22		55.50	71
1080	3" long		20	.400		35	22		57	72.50
1100	3-1/2" long		20	.400		38	22		60	75.50
1120	4" long		20	.400		38.50	22		60.50	76
1140	5" long		20	.400		42	22		64	80.50
1160	6" long		20	.400		43.50	22		65.50	82
1180	8" long		20	.400		47.50	22		69.50	86.50
1200	10" long		20	.400		62.50	22		84.50	103
1220	12" long		20	.400		75.50	22		97.50	117
1240	2" diameter, close		18	.444		44	24.50		68.50	85.50
1260	2-1/2" long		18	.444		46.50	24.50		71	88.50
1280	3" long		18	.444		48	24.50		72.50	90.50

26 05 33 – Raceway and Boxes for Electrical Systems

26 05 33.15 Conduit Nipples		Crew	Daily Output	Labor-Hours	Unit	Material	Labor	Equipment	Total	Total Incl O&P
						2015 Bare Costs				
1300	3-1/2" long	1 Elec	18	.444	Ea.	50	24.50		74.50	92.50
1320	4" long		18	.444		51	24.50		75.50	94
1340	5" long		18	.444		56.50	24.50		81	99.50
1360	6" long		18	.444		59	24.50		83.50	103
1380	8" long		18	.444		76.50	24.50		101	122
1400	10" long		18	.444		64	24.50		88.50	108
1420	12" long		18	.444		86.50	24.50		111	133
1440	2-1/2" diameter, close		15	.533		107	29		136	162
1460	3" long		15	.533		108	29		137	164
1480	3-1/2" long		15	.533		110	29		139	166
1500	4" long		15	.533		113	29		142	170
1520	5" long		15	.533		119	29		148	176
1540	6" long		15	.533		118	29		147	175
1560	8" long		15	.533		129	29		158	187
1580	10" long		15	.533		140	29		169	198
1600	12" long		15	.533		149	29		178	208
1620	3" diameter, close		12	.667		117	36.50		153.50	186
1640	3" long		12	.667		127	36.50		163.50	196
1660	3-1/2" long		12	.667		122	36.50		158.50	191
1680	4" long		12	.667		135	36.50		171.50	206
1700	5" long		12	.667		130	36.50		166.50	200
1720	6" long		12	.667		141	36.50		177.50	212
1740	8" long		12	.667		175	36.50		211.50	250
1760	10" long		12	.667		173	36.50		209.50	248
1780	12" long		12	.667		198	36.50		234.50	275
1800	3-1/2" diameter, close		11	.727		242	40		282	330
1820	4" long		11	.727		254	40		294	340
1840	5" long		11	.727		257	40		297	345
1860	6" long		11	.727		262	40		302	350
1880	8" long		11	.727		278	40		318	365
1900	10" long		11	.727		295	40		335	385
1920	12" long		11	.727		350	40		390	450
1940	4" diameter, close		9	.889		264	48.50		312.50	365
1960	4" long		9	.889		291	48.50		339.50	395
1980	5" long		9	.889		300	48.50		348.50	405
2000	6" long		9	.889		289	48.50		337.50	395
2020	8" long		9	.889		345	48.50		393.50	450
2040	10" long		9	.889		355	48.50		403.50	465
2060	12" long		9	.889		345	48.50		393.50	455
2080	5" diameter, close		7	1.143		450	62.50		512.50	590
2100	5" long		7	1.143		485	62.50		547.50	630
2120	6" long		7	1.143		495	62.50		557.50	640
2140	8" long		7	1.143		525	62.50		587.50	670
2160	10" long		7	1.143		565	62.50		627.50	715
2180	12" long		7	1.143		585	62.50		647.50	740
2200	6" diameter, close		6	1.333		780	73		853	970
2220	5" long		6	1.333		800	73		873	995
2240	6" long		6	1.333		815	73		888	1,000
2260	8" long		6	1.333		855	73		928	1,050
2280	10" long		6	1.333		910	73		983	1,125
2300	12" long		6	1.333		925	73		998	1,150
2320	Rigid galvanized steel, 1/2" diameter, close		32	.250		2.98	13.70		16.68	24.50
2340	1-1/2" long		32	.250		3.27	13.70		16.97	24.50

26 05 33 – Raceway and Boxes for Electrical Systems

26 05 33.15 Conduit Nipples		Crew	Daily Output	Labor-Hours	Unit	Material	Labor	Equipment	Total	Total Incl O&P
						2015 Bare Costs				
2360	2" long	1 Elec	32	.250	Ea.	3.45	13.70		17.15	25
2380	2-1/2" long		32	.250		3.58	13.70		17.28	25
2400	3" long		32	.250		3.76	13.70		17.46	25
2420	3-1/2" long		32	.250		3.94	13.70		17.64	25.50
2440	4" long		32	.250		4.12	13.70		17.82	25.50
2460	5" long		32	.250		4.41	13.70		18.11	26
2480	6" long		32	.250		4.95	13.70		18.65	26.50
2500	8" long		32	.250		7.30	13.70		21	29
2520	10" long		32	.250		8.15	13.70		21.85	30
2540	12" long		32	.250		9.15	13.70		22.85	31
2560	3/4" diameter, close		27	.296		3.92	16.20		20.12	29.50
2580	2" long		27	.296		4.24	16.20		20.44	29.50
2600	2-1/2" long		27	.296		4.47	16.20		20.67	30
2620	3" long		27	.296		4.63	16.20		20.83	30
2640	3-1/2" long		27	.296		4.74	16.20		20.94	30
2660	4" long		27	.296		5.10	16.20		21.30	30.50
2680	5" long		27	.296		5.50	16.20		21.70	31
2700	6" long		27	.296		6	16.20		22.20	31.50
2720	8" long		27	.296		8.50	16.20		24.70	34.50
2740	10" long		27	.296		9.70	16.20		25.90	35.50
2760	12" long		27	.296		10.70	16.20		26.90	37
2780	1" diameter, close		23	.348		6.55	19.05		25.60	36.50
2800	2" long		23	.348		6.80	19.05		25.85	37
2820	2-1/2" long		23	.348		7	19.05		26.05	37
2840	3" long		23	.348		7.35	19.05		26.40	37.50
2860	3-1/2" long		23	.348		7.80	19.05		26.85	38
2880	4" long		23	.348		8.10	19.05		27.15	38.50
2900	5" long		23	.348		8.65	19.05		27.70	39
2920	6" long		23	.348		9.05	19.05		28.10	39.50
2940	8" long		23	.348		11.90	19.05		30.95	42.50
2960	10" long		23	.348		14.25	19.05		33.30	45
2980	12" long		23	.348		15.55	19.05		34.60	46.50
3000	1-1/4" diameter, close		20	.400		8.85	22		30.85	43.50
3020	2" long		20	.400		9.05	22		31.05	44
3040	3" long		20	.400		9.70	22		31.70	44.50
3060	3-1/2" long		20	.400		10.25	22		32.25	45.50
3080	4" long		20	.400		10.55	22		32.55	45.50
3100	5" long		20	.400		11.35	22		33.35	46.50
3120	6" long		20	.400		12.05	22		34.05	47.50
3140	8" long		20	.400		16.15	22		38.15	52
3160	10" long		20	.400		18.90	22		40.90	55
3180	12" long		20	.400		21	22		43	57
3200	1-1/2" diameter, close		18	.444		12.65	24.50		37.15	51.50
3220	2" long		18	.444		12.85	24.50		37.35	51.50
3240	2-1/2" long		18	.444		13.40	24.50		37.90	52.50
3260	3" long		18	.444		13.75	24.50		38.25	52.50
3280	3-1/2" nipple		18	.444		14.50	24.50		39	53.50
3300	4" long		18	.444		15	24.50		39.50	54
3320	5" long		18	.444		15.75	24.50		40.25	55
3340	6" long		18	.444		17.40	24.50		41.90	56.50
3360	8" long		18	.444		22	24.50		46.50	61.50
3380	10" long		18	.444		24.50	24.50		49	64.50
3400	12" long		18	.444		26	24.50		50.50	66

26 05 33 – Raceway and Boxes for Electrical Systems

26 05 33.15 Conduit Nipples		Crew	Daily Output	Labor-Hours	Unit	Material	2015 Bare Costs Labor	Equipment	Total	Total Incl O&P
3420	2" diameter, close	1 Elec	16	.500	Ea.	15	27.50		42.50	59
3440	2-1/2" long		16	.500		15.85	27.50		43.35	60
3460	3" long		16	.500		16.75	27.50		44.25	61
3480	3-1/2" long		16	.500		17.65	27.50		45.15	62
3500	4" long		16	.500		18.40	27.50		45.90	63
3520	5" long		16	.500		19.80	27.50		47.30	64.50
3540	6" long		16	.500		21	27.50		48.50	65.50
3560	8" long		16	.500		26	27.50		53.50	71
3580	10" long		16	.500		29	27.50		56.50	74.50
3600	12" long		16	.500		31.50	27.50		59	77
3620	2-1/2" diameter, close		13	.615		46	33.50		79.50	103
3640	3" long		13	.615		45.50	33.50		79	102
3660	3-1/2" long		13	.615		48.50	33.50		82	106
3680	4" long		13	.615		49	33.50		82.50	106
3700	5" long		13	.615		52.50	33.50		86	110
3720	6" long		13	.615		55	33.50		88.50	113
3740	8" long		13	.615		62	33.50		95.50	120
3760	10" long		13	.615		66.50	33.50		100	125
3780	12" long		13	.615		72.50	33.50		106	132
3800	3" diameter, close		12	.667		54.50	36.50		91	117
3820	3" long		12	.667		56.50	36.50		93	119
3900	3-1/2" long		12	.667		57.50	36.50		94	120
3920	4" long		12	.667		59	36.50		95.50	122
3940	5" long		12	.667		62.50	36.50		99	125
3960	6" long		12	.667		66	36.50		102.50	129
3980	8" long		12	.667		73.50	36.50		110	138
4000	10" long		12	.667		80	36.50		116.50	145
4020	12" long		12	.667		89	36.50		125.50	155
4040	3-1/2" diameter, close		10	.800		97	44		141	175
4060	4" long		10	.800		102	44		146	181
4080	5" long		10	.800		106	44		150	184
4100	6" long		10	.800		110	44		154	189
4120	8" long		10	.800		118	44		162	197
4140	10" long		10	.800		126	44		170	206
4160	12" long		10	.800		134	44		178	216
4180	4" diameter, close		8	1		109	54.50		163.50	205
4200	4" long		8	1		114	54.50		168.50	211
4220	5" long		8	1		119	54.50		173.50	216
4240	6" long		8	1		123	54.50		177.50	220
4260	8" long		8	1		131	54.50		185.50	229
4280	10" long		8	1		142	54.50		196.50	241
4300	12" long		8	1		153	54.50		207.50	253
4320	5" diameter, close		6	1.333		217	73		290	350
4340	5" long		6	1.333		237	73		310	375
4360	6" long		6	1.333		242	73		315	380
4380	8" long		6	1.333		255	73		328	395
4400	10" long		6	1.333		269	73		342	410
4420	12" long		6	1.333		290	73		363	435
4440	6" diameter, close		5	1.600		455	87.50		542.50	635
4460	5" long		5	1.600		480	87.50		567.50	660
4480	6" long		5	1.600		485	87.50		572.50	670
4500	8" long		5	1.600		500	87.50		587.50	685
4520	10" long		5	1.600		525	87.50		612.50	715

26 05 33.15 Conduit Nipples		Crew	Daily Output	Labor-Hours	Unit	2015 Bare Costs Material	Labor	Equipment	Total	Total Incl O&P
4540	12" long	1 Elec	5	1.600	Ea.	540	87.50		627.50	730
4560	Plastic coated, 40 mil thick, 1/2" diameter, 2" long		32	.250		17.60	13.70		31.30	40.50
4580	2-1/2" long		32	.250		20.50	13.70		34.20	43.50
4600	3" long		32	.250		20.50	13.70		34.20	43.50
4680	3-1/2" long		32	.250		23.50	13.70		37.20	47
4700	4" long		32	.250		23	13.70		36.70	46.50
4720	5" long		32	.250		23	13.70		36.70	46.50
4740	6" long		32	.250		24.50	13.70		38.20	48
4760	8" long		32	.250		23	13.70		36.70	46.50
4780	10" long		32	.250		24	13.70		37.70	47.50
4800	12" long		32	.250		25	13.70		38.70	48.50
4820	3/4" diameter, 2" long		26	.308		18.75	16.85		35.60	46.50
4840	2-1/2" long		26	.308		21.50	16.85		38.35	49.50
4860	3" long		26	.308		20.50	16.85		37.35	49
4880	3-1/2" long		26	.308		24.50	16.85		41.35	52.50
4900	4" long		26	.308		24	16.85		40.85	52.50
4920	5" long		26	.308		25	16.85		41.85	53.50
4940	6" long		26	.308		24	16.85		40.85	52.50
4960	8" long		26	.308		24	16.85		40.85	52.50
4980	10" long		26	.308		25	16.85		41.85	53.50
5000	12" long		26	.308		29	16.85		45.85	58
5020	1" diameter, 2" long		22	.364		20.50	19.90		40.40	54
5040	2-1/2" long		22	.364		25	19.90		44.90	58.50
5060	3" long		22	.364		24	19.90		43.90	57.50
5080	3-1/2" long		22	.364		25.50	19.90		45.40	59
5100	4" long		22	.364		28	19.90		47.90	61.50
5120	5" long		22	.364		28.50	19.90		48.40	62.50
5140	6" long		22	.364		26.50	19.90		46.40	60.50
5160	8" long		22	.364		27.50	19.90		47.40	61.50
5180	10" long		22	.364		29	19.90		48.90	63
5200	12" long		22	.364		35	19.90		54.90	69.50
5220	1-1/4" diameter, 2" long		18	.444		26	24.50		50.50	66
5240	2-1/2" long		18	.444		28.50	24.50		53	68.50
5260	3" long		18	.444		30	24.50		54.50	70.50
5280	3-1/2" long		18	.444		30	24.50		54.50	70.50
5300	4" long		18	.444		31	24.50		55.50	71.50
5320	5" long		18	.444		31.50	24.50		56	72.50
5340	6" long		18	.444		35.50	24.50		60	77
5360	8" long		18	.444		33	24.50		57.50	74
5380	10" long		18	.444		37	24.50		61.50	78
5400	12" long		18	.444		47.50	24.50		72	90
5420	1-1/2" diameter, 2" long		16	.500		30	27.50		57.50	75.50
5440	2-1/2" long		16	.500		32.50	27.50		60	78
5460	3" long		16	.500		32	27.50		59.50	77.50
5480	3-1/2" long		16	.500		34	27.50		61.50	80
5500	4" long		16	.500		38	27.50		65.50	84
5520	5" long		16	.500		39	27.50		66.50	85.50
5540	6" long		16	.500		40.50	27.50		68	87.50
5560	8" long		16	.500		41.50	27.50		69	88
5580	10" long		16	.500		49	27.50		76.50	96.50
5600	12" long		16	.500		62	27.50		89.50	111
5620	2" diameter, 2-1/2" long		14	.571		39	31.50		70.50	91.50
5640	3" long		14	.571		39	31.50		70.50	91

26 05 33 – Raceway and Boxes for Electrical Systems

26 05 33.15 Conduit Nipples		Crew	Daily Output	Labor-Hours	Unit	Material	Labor	Equipment	Total	Total Incl O&P
						2015 Bare Costs				
5660	3-1/2" long	1 Elec	14	.571	Ea.	40	31.50		71.50	92.50
5680	4" long		14	.571		43	31.50		74.50	96
5700	5" long		14	.571		47	31.50		78.50	100
5720	6" long		14	.571		48.50	31.50		80	102
5740	8" long		14	.571		52	31.50		83.50	106
5760	10" long		14	.571		62.50	31.50		94	118
5780	12" long		14	.571		75.50	31.50		107	132
5800	2-1/2" diameter, 3-1/2" long		12	.667		80	36.50		116.50	145
5820	4" long		12	.667		81.50	36.50		118	146
5840	5" long		12	.667		93.50	36.50		130	160
5860	6" long		12	.667		102	36.50		138.50	170
5880	8" long		12	.667		108	36.50		144.50	175
5900	10" long		12	.667		117	36.50		153.50	186
5920	12" long		12	.667		132	36.50		168.50	202
5940	3" diameter, 3-1/2" long		11	.727		97.50	40		137.50	169
5960	4" long		11	.727		97.50	40		137.50	169
5980	5" long		11	.727		104	40		144	176
6000	6" long		11	.727		113	40		153	186
6020	8" long		11	.727		125	40		165	200
6040	10" long		11	.727		143	40		183	219
6060	12" long		11	.727		163	40		203	241
6080	3-1/2" diameter, 4" long		9	.889		151	48.50		199.50	242
6100	5" long		9	.889		152	48.50		200.50	243
6120	6" long		9	.889		169	48.50		217.50	262
6140	8" long		9	.889		180	48.50		228.50	274
6160	10" long		9	.889		198	48.50		246.50	293
6180	12" long		9	.889		215	48.50		263.50	310
6200	4" diameter, 4" long		7.50	1.067		164	58.50		222.50	272
6220	5" long		7.50	1.067		168	58.50		226.50	276
6240	6" long		7.50	1.067		182	58.50		240.50	291
6260	8" long		7.50	1.067		202	58.50		260.50	315
6280	10" long		7.50	1.067		238	58.50		296.50	355
6300	12" long		7.50	1.067		247	58.50		305.50	365
6320	5" diameter, 5" long		5.50	1.455		262	79.50		341.50	410
6340	6" long		5.50	1.455		276	79.50		355.50	430
6360	8" long		5.50	1.455		286	79.50		365.50	440
6380	10" long		5.50	1.455		305	79.50		384.50	460
6400	12" long		5.50	1.455		310	79.50		389.50	465
6420	6" diameter, 5" long		4.50	1.778		515	97		612	715
6440	6" long		4.50	1.778		525	97		622	730
6460	8" long		4.50	1.778		540	97		637	745
6480	10" long		4.50	1.778		560	97		657	765
6500	12" long	↓	4.50	1.778	↓	580	97		677	790

26 05 33.16 Outlet Boxes		Crew	Daily Output	Labor-Hours	Unit	Material	Labor	Equipment	Total	Total Incl O&P
0010	**OUTLET BOXES**									
0020	Pressed steel, octagon, 4"	1 Elec	20	.400	Ea.	2.61	22		24.61	37
0040	For Romex or BX		20	.400		3.91	22		25.91	38.50
0050	For Romex or BX, with bracket		20	.400		6.15	22		28.15	41
0060	Covers, blank		64	.125		1.10	6.85		7.95	11.80
0100	Extension rings		40	.200		4.33	10.95		15.28	21.50
0150	Square, 4"		20	.400		2.42	22		24.42	36.50
0160	For Romex or BX	↓	20	.400	↓	7	22		29	41.50

26 05 33 – Raceway and Boxes for Electrical Systems

26 05 33.16 Outlet Boxes		Crew	Daily Output	Labor-Hours	Unit	Material	2015 Bare Costs Labor	Equipment	Total	Total Incl O&P
0170	For Romex or BX, with bracket	1 Elec	20	.400	Ea.	7.60	22		29.60	42.50
0200	Extension rings		40	.200		4.37	10.95		15.32	22
0220	2-1/8" deep, 1" KO		20	.400		5.80	22		27.80	40.50
0250	Covers, blank		64	.125		1.18	6.85		8.03	11.90
0260	Raised device		64	.125		2.21	6.85		9.06	13.05
0300	Plaster rings		64	.125		2.40	6.85		9.25	13.25
0350	Square, 4-11/16"		20	.400		6.10	22		28.10	41
0370	2-1/8" deep, 3/4" to 1-1/4" KO		20	.400		7.35	22		29.35	42
0400	Extension rings		40	.200		9.85	10.95		20.80	28
0450	Covers, blank		53	.151		2.20	8.25		10.45	15.20
0460	Raised device		53	.151		6.60	8.25		14.85	20
0500	Plaster rings		53	.151		6.55	8.25		14.80	20
0550	Handy box		27	.296		2.88	16.20		19.08	28
0560	Covers, device		64	.125		1.01	6.85		7.86	11.70
0600	Extension rings		54	.148		3.44	8.10		11.54	16.35
0652	Switchbox		24	.333		4.18	18.25		22.43	32.50
0660	Romex or BX		27	.296		6	16.20		22.20	31.50
0670	with bracket		27	.296		6.95	16.20		23.15	32.50
0680	Partition, metal		27	.296		2.62	16.20		18.82	28
0700	Masonry, 1 gang, 2-1/2" deep		27	.296		7.95	16.20		24.15	34
0710	3-1/2" deep		27	.296		8	16.20		24.20	34
0750	2 gang, 2-1/2" deep		20	.400		15	22		37	50.50
0760	3-1/2" deep		20	.400		12.05	22		34.05	47.50
0800	3 gang, 2-1/2" deep		13	.615		17.90	33.50		51.40	71.50
0850	4 gang, 2-1/2" deep		10	.800		20.50	44		64.50	91
0860	5 gang, 2-1/2" deep		9	.889		22	48.50		70.50	99.50
0870	6 gang, 2-1/2" deep		8	1		45	54.50		99.50	134
0880	Masonry thru-the-wall, 1 gang, 4" block		16	.500		29.50	27.50		57	75
0890	6" block		16	.500		45	27.50		72.50	92
0900	8" block		16	.500		50.50	27.50		78	98
0920	2 gang, 6" block		16	.500		67	27.50		94.50	117
0940	Bar hanger with 3/8" stud, for wood and masonry boxes		53	.151		5.60	8.25		13.85	18.95
0950	Concrete, set flush, 4" deep		20	.400		12.30	22		34.30	47.50
1000	Plate with 3/8" stud		80	.100		8.10	5.45		13.55	17.35
1100	Concrete, floor, 1 gang		5.30	1.509		89.50	82.50		172	227
1150	2 gang		4	2		137	109		246	320
1200	3 gang		2.70	2.963		203	162		365	475
1250	For duplex receptacle, pedestal mounted, add		24	.333		97.50	18.25		115.75	135
1270	Flush mounted, add		27	.296		30.50	16.20		46.70	58.50
1300	For telephone, pedestal mounted, add		30	.267		113	14.60		127.60	148
1350	Carpet flange, 1 gang		53	.151		50.50	8.25		58.75	68.50
1400	Cast, 1 gang, FS (2" deep), 1/2" hub		12	.667		18.25	36.50		54.75	76.50
1410	3/4" hub		12	.667		19.30	36.50		55.80	77.50
1420	FD (2-11/16" deep), 1/2" hub		12	.667		17.70	36.50		54.20	76
1430	3/4" hub		12	.667		19.25	36.50		55.75	77.50
1450	2 gang, FS, 1/2" hub		10	.800		33	44		77	104
1460	3/4" hub		10	.800		35.50	44		79.50	107
1470	FD, 1/2" hub		10	.800		39.50	44		83.50	112
1480	3/4" hub		10	.800		39.50	44		83.50	112
1500	3 gang, FS, 3/4" hub		9	.889		56	48.50		104.50	137
1510	Switch cover, 1 gang, FS		64	.125		4.83	6.85		11.68	15.90
1520	2 gang		53	.151		7.35	8.25		15.60	21
1530	Duplex receptacle cover, 1 gang, FS	↓	64	.125	↓	4.93	6.85		11.78	16

26 05 33 – Raceway and Boxes for Electrical Systems

26 05 33.16 Outlet Boxes		Crew	Daily Output	Labor-Hours	Unit	Material	2015 Bare Costs Labor	Equipment	Total	Total Incl O&P
1540	2 gang, FS	1 Elec	53	.151	Ea.	7.50	8.25		15.75	21
1542	Weatherproof blank cover, 1 gang		64	.125		1.39	6.85		8.24	12.15
1544	2 gang		53	.151		2.96	8.25		11.21	16.05
1550	Weatherproof switch cover, 1 gang		64	.125		6.40	6.85		13.25	17.65
1554	2 gang		53	.151		12.45	8.25		20.70	26.50
1600	Weatherproof receptacle cover, 1 gang		64	.125		4.82	6.85		11.67	15.90
1604	2 gang		53	.151		12.50	8.25		20.75	26.50
1620	Weatherproof receptacle cover, tamper resistant, 1 gang		58	.138		13.50	7.55		21.05	26.50
1624	2 gang		48	.167		27	9.10		36.10	43.50
1750	FSC, 1 gang, 1/2" hub		11	.727		20	40		60	83.50
1760	3/4" hub		11	.727		22.50	40		62.50	86
1770	2 gang, 1/2" hub		9	.889		36	48.50		84.50	115
1780	3/4" hub		9	.889		39	48.50		87.50	119
1790	FDC, 1 gang, 1/2" hub		11	.727		23	40		63	86.50
1800	3/4" hub		11	.727		25.50	40		65.50	89.50
1810	2 gang, 1/2" hub		9	.889		47.50	48.50		96	128
1820	3/4" hub		9	.889		42	48.50		90.50	122
1850	Weatherproof in-use cover, 1 gang		64	.125		24	6.85		30.85	37
1870	2 gang		53	.151		27	8.25		35.25	42.50
2000	Poke-thru fitting, fire rated, for 3-3/4" floor		6.80	1.176		131	64.50		195.50	244
2040	For 7" floor		6.80	1.176		165	64.50		229.50	281
2100	Pedestal, 15 amp, duplex receptacle & blank plate		5.25	1.524		138	83.50		221.50	280
2120	Duplex receptacle and telephone plate		5.25	1.524		138	83.50		221.50	280
2140	Pedestal, 20 amp, duplex recept. & phone plate		5	1.600		139	87.50		226.50	288
2160	Telephone plate, both sides		5.25	1.524		131	83.50		214.50	273
2200	Abandonment plate		32	.250		38.50	13.70		52.20	63.50
9000	Minimum labor/equipment charge		4	2	Job		109		109	169

26 05 33.17 Outlet Boxes, Plastic		Crew	Daily Output	Labor-Hours	Unit	Material	Labor	Equipment	Total	Total Incl O&P
0010	**OUTLET BOXES, PLASTIC**									
0050	4" diameter, round with 2 mounting nails	1 Elec	25	.320	Ea.	2.39	17.50		19.89	29.50
0100	Bar hanger mounted		25	.320		5.10	17.50		22.60	32.50
0200	4", square with 2 mounting nails		25	.320		4.63	17.50		22.13	32
0300	Plaster ring		64	.125		1.98	6.85		8.83	12.80
0400	Switch box with 2 mounting nails, 1 gang		30	.267		4.39	14.60		18.99	27.50
0500	2 gang		25	.320		3.33	17.50		20.83	30.50
0600	3 gang		20	.400		4.93	22		26.93	39.50
0700	Old work box		30	.267		5.70	14.60		20.30	29
1400	PVC, FSS, 1 gang, 1/2" hub		14	.571		16.35	31.50		47.85	66.50
1410	3/4" hub		14	.571		13.35	31.50		44.85	63
1420	FD, 1 gang for variable terminations		14	.571		11	31.50		42.50	60.50
1450	FS, 2 gang for variable terminations		12	.667		13	36.50		49.50	71
1480	Weatherproof blank cover, FS, 1 gang		64	.125		4.46	6.85		11.31	15.50
1500	2 gang		53	.151		4.88	8.25		13.13	18.15
1510	Weatherproof switch cover, FS, 1 gang		64	.125		9.85	6.85		16.70	21.50
1520	2 gang		53	.151		16.20	8.25		24.45	30.50
1530	Weatherproof duplex receptacle cover, FS, 1 gang		64	.125		13.55	6.85		20.40	25.50
1540	2 gang		53	.151		14.40	8.25		22.65	28.50
1750	FSC, 1 gang, 1/2" hub		13	.615		10.85	33.50		44.35	64
1760	3/4" hub		13	.615		11.65	33.50		45.15	65
1770	FSC, 2 gang, 1/2" hub		11	.727		16.60	40		56.60	80
1780	3/4" hub		11	.727		16.50	40		56.50	79.50
1790	FDC, 1 gang, 1/2" hub		13	.615		11.15	33.50		44.65	64.50

26 05 33 – Raceway and Boxes for Electrical Systems

26 05 33.17 Outlet Boxes, Plastic		Crew	Daily Output	Labor-Hours	Unit	Material	2015 Bare Costs Labor	Equipment	Total	Total Incl O&P
1800	3/4" hub	1 Elec	13	.615	Ea.	12.20	33.50		45.70	65.50
1810	Weatherproof, T box w/3 holes		14	.571		9.90	31.50		41.40	59.50
1820	4" diameter round w/5 holes		14	.571		9.45	31.50		40.95	59
1850	In-use cover, 1 gang		64	.125		8.75	6.85		15.60	20.50
1870	2 gang		53	.151		11	8.25		19.25	25
9000	Minimum labor/equipment charge		4	2	Job		109		109	169

26 05 33.18 Pull Boxes		Crew	Daily Output	Labor-Hours	Unit	Material	Labor	Equipment	Total	Total Incl O&P
0010	**PULL BOXES**									
0100	Steel, pull box, NEMA 1, type SC, 6" W x 6" H x 4" D	1 Elec	8	1	Ea.	9.65	54.50		64.15	95
0180	8" W x 6" H x 4" D		8	1		13.30	54.50		67.80	99
0200	8" W x 8" H x 4" D		8	1		14.50	54.50		69	100
0210	10" W x 10" H x 4" D		7	1.143		17.65	62.50		80.15	116
0220	12" W x 12" H x 4" D		6.50	1.231		22.50	67.50		90	129
0230	15" W x 15" H x 4" D		5.20	1.538		42	84		126	177
0240	18" W x 18" H x 4" D		4.40	1.818		42.50	99.50		142	201
0250	6" W x 6" H x 6" D		8	1		12.95	54.50		67.45	99
0260	8" W x 8" H x 6" D		7.50	1.067		16.15	58.50		74.65	108
0270	10" W x 10" H x 6" D		5.50	1.455		21	79.50		100.50	146
0300	10" W x 12" H x 6" D		5.30	1.509		24.50	82.50		107	155
0310	12" W x 12" H x 6" D		5.20	1.538		26.50	84		110.50	159
0320	15" W x 15" H x 6" D		4.60	1.739		37	95		132	188
0330	18" W x 18" H x 6" D		4.20	1.905		47.50	104		151.50	214
0340	24" W x 24" H x 6" D		3.20	2.500		95.50	137		232.50	315
0350	12" W x 12" H x 8" D		5	1.600		30.50	87.50		118	170
0360	15" W x 15" H x 8" D		4.50	1.778		45.50	97		142.50	201
0370	18" W x 18" H x 8" D		4	2		61	109		170	236
0380	24" W x 18" H x 6" D		3.70	2.162		87.50	118		205.50	280
0400	16" W x 20" H x 8" D		4	2		84.50	109		193.50	262
0500	20" W x 24" H x 8" D		3.20	2.500		98	137		235	320
0510	24" W x 24" H x 8" D		3	2.667		108	146		254	345
0600	24" W x 36" H x 8" D		2.70	2.963		151	162		313	415
0610	30" W x 30" H x 8" D		2.70	2.963		194	162		356	465
0620	36" W x 36" H x 8" D		2	4		223	219		442	585
0630	24" W x 24" H x 10" D		2.50	3.200		186	175		361	475
0650	Pull box, hinged , NEMA 1, 6" W x 6" H x 4" D		8	1		11.60	54.50		66.10	97.50
0660	8" W x 8" H x 4" D		8	1		16.45	54.50		70.95	103
0670	10" W x 10" H x 4" D		7	1.143		19.15	62.50		81.65	118
0680	12" W x 12" H x 4" D		6	1.333		27.50	73		100.50	143
0690	15" W x 15" H x 4" D		5.20	1.538		28.50	84		112.50	162
0700	18" W x 18" H x 4" D		4.40	1.818		35	99.50		134.50	193
0710	6" W x 6" H x 6" D		8	1		15.10	54.50		69.60	101
0720	8" W x 8" H x 6" D		7.50	1.067		20	58.50		78.50	113
0730	10" W x 10" H x 6" D		5.50	1.455		27	79.50		106.50	153
0740	12" W x 12" H x 6" D		5.20	1.538		32	84		116	166
0800	12" W x 16" H x 6" D		4.70	1.702		49.50	93		142.50	199
0810	15" W x 15" H x 6" D		4.60	1.739		32	95		127	183
0820	18" W x 18" H x 6" D		4.20	1.905		58.50	104		162.50	225
1000	20" W x 20" H x 6" D		3.60	2.222		86	122		208	283
1010	24" W x 24" H x 6" D		3.20	2.500		115	137		252	340
1020	12" W x 12" H x 8" D		5	1.600		69.50	87.50		157	213
1030	15" W x 15" H x 8" D		4.50	1.778		85.50	97		182.50	245
1040	18" W x 18" H x 8" D		4	2		120	109		229	300

26 05 33.18 Pull Boxes		Crew	Daily Output	Labor-Hours	Unit	2015 Bare Costs Material	Labor	Equipment	Total	Total Incl O&P
1200	20" W x 20" H x 8" D	1 Elec	3.20	2.500	Ea.	149	137		286	375
1210	24" W x 24" H x 8" D		3	2.667		169	146		315	410
1220	30" W x 30" H x 8" D		2.70	2.963		244	162		406	520
1400	24" W x 36" H x 8" D		2.70	2.963		237	162		399	510
1600	24" W x 42" H x 8" D		2	4		350	219		569	725
1610	36" W x 36" H x 8" D		2	4		350	219		569	725
2100	Pull box, NEMA 3R, type SC, raintight & weatherproof									
2150	6" L x 6" W x 6" D	1 Elec	10	.800	Ea.	14.75	44		58.75	84.50
2200	8" L x 6" W x 6" D		8	1		20.50	54.50		75	107
2250	10" L x 6" W x 6" D		7	1.143		32	62.50		94.50	132
2300	12" L x 12" W x 6" D		5	1.600		54.50	87.50		142	196
2350	16" L x 16" W x 6" D		4.50	1.778		77.50	97		174.50	237
2400	20" L x 20" W x 6" D		4	2		101	109		210	280
2450	24" L x 18" W x 8" D		3	2.667		123	146		269	360
2500	24" L x 24" W x 10" D		2.50	3.200		291	175		466	590
2550	30" L x 24" W x 12" D		2	4		385	219		604	760
2600	36" L x 36" W x 12" D		1.50	5.333		430	292		722	925
2800	Cast iron, pull boxes for surface mounting									
3000	NEMA 4, watertight & dust tight									
3050	6" L x 6" W x 6" D	1 Elec	4	2	Ea.	257	109		366	450
3100	8" L x 6" W x 6" D		3.20	2.500		360	137		497	605
3150	10" L x 6" W x 6" D		2.50	3.200		400	175		575	710
3200	12" L x 12" W x 6" D		2.30	3.478		730	190		920	1,100
3250	16" L x 16" W x 6" D		1.30	6.154		960	335		1,295	1,575
3300	20" L x 20" W x 6" D		.80	10		1,800	545		2,345	2,825
3350	24" L x 18" W x 8" D		.70	11.429		2,625	625		3,250	3,875
3400	24" L x 24" W x 10" D		.50	16		4,975	875		5,850	6,825
3450	30" L x 24" W x 12" D		.40	20		5,450	1,100		6,550	7,700
3500	36" L x 36" W x 12" D		.20	40		5,975	2,200		8,175	9,975
3510	NEMA 4 clamp cover, 6" L x 6" W x 4" D		4	2		168	109		277	355
3520	8" L x 6" W x 4" D		4	2		209	109		318	400
4000	NEMA 7, explosionproof									
4050	6" L x 6" W x 6" D	1 Elec	2	4	Ea.	690	219		909	1,100
4100	8" L x 6" W x 6" D		1.80	4.444		955	243		1,198	1,425
4150	10" L x 6" W x 6" D		1.60	5		1,275	274		1,549	1,825
4200	12" L x 12" W x 6" D		1	8		2,225	440		2,665	3,125
4250	16" L x 14" W x 6" D		.60	13.333		3,050	730		3,780	4,500
4300	18" L x 18" W x 8" D		.50	16		5,800	875		6,675	7,725
4350	24" L x 18" W x 8" D		.40	20		7,375	1,100		8,475	9,800
4400	24" L x 24" W x 10" D		.30	26.667		10,000	1,450		11,450	13,400
4450	30" L x 24" W x 12" D		.20	40		14,300	2,200		16,500	19,100
5000	NEMA 9, dust tight 6" L x 6" W x 6" D		3.20	2.500		380	137		517	625
5050	8" L x 6" W x 6" D		2.70	2.963		455	162		617	750
5100	10" L x 6" W x 6" D		2	4		600	219		819	1,000
5150	12" L x 12" W x 6" D		1.60	5		1,150	274		1,424	1,700
5200	16" L x 16" W x 6" D		1	8		2,000	440		2,440	2,875
5250	18" L x 18" W x 8" D		.70	11.429		3,125	625		3,750	4,400
5300	24" L x 18" W x 8" D		.60	13.333		4,425	730		5,155	5,975
5350	24" L x 24" W x 10" D		.40	20		5,850	1,100		6,950	8,125
5400	30" L x 24" W x 12" D		.30	26.667		8,950	1,450		10,400	12,100
6000	J.I.C. wiring boxes, NEMA 12, dust tight & drip tight									
6050	6" L x 8" W x 4" D	1 Elec	10	.800	Ea.	51.50	44		95.50	125
6100	8" L x 10" W x 4" D		8	1		64	54.50		118.50	155

26 05 33 – Raceway and Boxes for Electrical Systems

26 05 33.18 Pull Boxes		Crew	Daily Output	Labor-Hours	Unit	Material	2015 Bare Costs Labor	Equipment	Total	Total Incl O&P
6150	12" L x 14" W x 6" D	1 Elec	5.30	1.509	Ea.	124	82.50		206.50	265
6200	14" L x 16" W x 6" D		4.70	1.702		147	93		240	305
6250	16" L x 20" W x 6" D		4.40	1.818		226	99.50		325.50	400
6300	24" L x 30" W x 6" D		3.20	2.500		330	137		467	575
6350	24" L x 30" W x 8" D		2.90	2.759		340	151		491	610
6400	24" L x 36" W x 8" D		2.70	2.963		380	162		542	665
6450	24" L x 42" W x 8" D		2.30	3.478		425	190		615	765
6500	24" L x 48" W x 8" D	↓	2	4	↓	465	219		684	850

26 05 33.23 Wireway		Crew	Daily Output	Labor-Hours	Unit	Material	Labor	Equipment	Total	Total Incl O&P
0010	**WIREWAY** to 15' high									
0100	NEMA 1, Screw cover w/fittings and supports, 2-1/2" x 2-1/2"	1 Elec	45	.178	L.F.	10.85	9.70		20.55	27
0200	4" x 4"	"	40	.200		11.35	10.95		22.30	29.50
0400	6" x 6"	2 Elec	60	.267		19.60	14.60		34.20	44
0600	8" x 8"		40	.400		33.50	22		55.50	71
0620	10" x 10"		30	.533		48	29		77	97.50
0640	12" x 12"	↓	20	.800	↓	61	44		105	135
0800	Elbows, 90°, 2-1/2"	1 Elec	24	.333	Ea.	32.50	18.25		50.75	63.50
1000	4"		20	.400		36.50	22		58.50	74.50
1200	6"		18	.444		41.50	24.50		66	83
1400	8"		16	.500		66.50	27.50		94	116
1420	10"		12	.667		95	36.50		131.50	161
1440	12"		10	.800		121	44		165	201
1500	Elbows, 45°, 2-1/2"		24	.333		32.50	18.25		50.75	63.50
1510	4"		20	.400		39.50	22		61.50	77.50
1520	6"		18	.444		41.50	24.50		66	83
1530	8"		16	.500		66.50	27.50		94	116
1540	10"		12	.667		95	36.50		131.50	161
1550	12"		10	.800		169	44		213	253
1600	"T" box, 2-1/2"		18	.444		37	24.50		61.50	78.50
1800	4"		16	.500		45	27.50		72.50	92
2000	6"		14	.571		50.50	31.50		82	104
2200	8"		12	.667		93	36.50		129.50	159
2220	10"		10	.800		124	44		168	204
2240	12"		8	1		175	54.50		229.50	278
2300	Cross, 2-1/2"		16	.500		41.50	27.50		69	88
2310	4"		14	.571		50.50	31.50		82	104
2320	6"		12	.667		62.50	36.50		99	125
2400	Panel adapter, 2-1/2"		24	.333		11.15	18.25		29.40	40.50
2600	4"		20	.400		13.50	22		35.50	49
2800	6"		18	.444		14.45	24.50		38.95	53.50
3000	8"		16	.500		19.90	27.50		47.40	64.50
3020	10"		14	.571		37	31.50		68.50	89
3040	12"		12	.667		46.50	36.50		83	108
3200	Reducer, 4" to 2-1/2"		24	.333		15.05	18.25		33.30	44.50
3400	6" to 4"		20	.400		30	22		52	67
3600	8" to 6"		18	.444		32.50	24.50		57	73
3620	10" to 8"		16	.500		41	27.50		68.50	87.50
3640	12" to 10"		14	.571		49.50	31.50		81	103
3780	End cap, 2-1/2"		24	.333		4.62	18.25		22.87	33
3800	4"		20	.400		5.70	22		27.70	40.50
4000	6"		18	.444		6.85	24.50		31.35	45
4200	8"	↓	16	.500	↓	9.30	27.50		36.80	53

26 05 33 – Raceway and Boxes for Electrical Systems

26 05 33.23 Wireway		Crew	Daily Output	Labor-Hours	Unit	2015 Bare Costs Material	Labor	Equipment	Total	Total Incl O&P
4220	10"	1 Elec	14	.571	Ea.	15.10	31.50		46.60	65
4240	12"		12	.667		21	36.50		57.50	79.50
4300	U-connector, 2-1/2"		200	.040		4.62	2.19		6.81	8.50
4320	4"		200	.040		5.70	2.19		7.89	9.65
4340	6"		180	.044		6.85	2.43		9.28	11.25
4360	8"		170	.047		13.85	2.57		16.42	19.25
4380	10"		150	.053		20	2.92		22.92	26.50
4400	12"		130	.062		28	3.37		31.37	35.50
4420	Hanger, 2-1/2"		100	.080		12.60	4.38		16.98	20.50
4430	4"		100	.080		12.75	4.38		17.13	21
4440	6"		80	.100		16.30	5.45		21.75	26.50
4450	8"		65	.123		24.50	6.75		31.25	37.50
4460	10"		50	.160		44.50	8.75		53.25	62.50
4470	12"		40	.200		65	10.95		75.95	88.50
4475	NEMA 3R, Screw cover w/fittings and supports, 4" x 4"		36	.222	L.F.	16.50	12.15		28.65	37
4480	6" x 6"	2 Elec	55	.291		22	15.90		37.90	49
4485	8" x 8"		36	.444		34	24.50		58.50	75
4490	12" x 12"		18	.889		54.50	48.50		103	136
4500	Hinged cover, with fittings and supports, 2-1/2" x 2-1/2"	1 Elec	60	.133		15.05	7.30		22.35	28
4520	4" x 4"	"	45	.178		19.15	9.70		28.85	36
4540	6" x 6"	2 Elec	80	.200		30.50	10.95		41.45	51
4560	8" x 8"		60	.267		50	14.60		64.60	77.50
4580	10" x 10"		50	.320		63.50	17.50		81	96.50
4600	12" x 12"		24	.667		91.50	36.50		128	158
4700	Elbows 90°, 2-1/2" x 2-1/2"	1 Elec	32	.250	Ea.	46.50	13.70		60.20	72.50
4720	4"		27	.296		62.50	16.20		78.70	93.50
4730	6"		23	.348		71.50	19.05		90.55	109
4740	8"		18	.444		90.50	24.50		115	137
4750	10"		14	.571		109	31.50		140.50	169
4760	12"		12	.667		175	36.50		211.50	250
4800	Tee box, hinged cover, 2-1/2" x 2-1/2"		23	.348		54	19.05		73.05	89
4810	4"		20	.400		74	22		96	116
4820	6"		18	.444		82.50	24.50		107	128
4830	8"		16	.500		154	27.50		181.50	213
4840	10"		12	.667		157	36.50		193.50	229
4860	12"		10	.800		196	44		240	284
4880	Cross box, hinged cover, 2-1/2" x 2-1/2"		18	.444		60	24.50		84.50	104
4900	4"		16	.500		84	27.50		111.50	135
4920	6"		13	.615		106	33.50		139.50	169
4940	8"		11	.727		154	40		194	231
4960	10"		10	.800		231	44		275	325
4980	12"		9	.889		253	48.50		301.50	355
5000	NEMA 12, Hinged cover, 2-1/2" x 2-1/2"		40	.200	L.F.	37.50	10.95		48.45	58
5020	4" x 4"		35	.229		43.50	12.50		56	67.50
5040	6" x 6"	2 Elec	60	.267		62.50	14.60		77.10	91.50
5060	8" x 8"	"	50	.320		85.50	17.50		103	121
5120	Elbows 90°, flanged, 2-1/2" x 2-1/2"	1 Elec	23	.348	Ea.	83.50	19.05		102.55	121
5140	4"		20	.400		105	22		127	149
5160	6"		18	.444		133	24.50		157.50	185
5180	8"		15	.533		194	29		223	258
5240	Tee box, flanged, 2-1/2" x 2-1/2"		18	.444		114	24.50		138.50	163
5260	4"		16	.500		133	27.50		160.50	190
5280	6"		15	.533		180	29		209	243

26 05 33 – Raceway and Boxes for Electrical Systems

26 05 33.23 Wireway		Crew	Daily Output	Labor-Hours	Unit	Material	Labor	Equipment	Total	Total Incl O&P
						2015 Bare Costs				
5300	8"	1 Elec	13	.615	Ea.	262	33.50		295.50	340
5360	Cross box, flanged, 2-1/2" x 2-1/2"		15	.533		154	29		183	214
5380	4"		13	.615		195	33.50		228.50	267
5400	6"		12	.667		254	36.50		290.50	335
5420	8"		10	.800		310	44		354	410
5480	Flange gasket, 2-1/2"		160	.050		4.12	2.74		6.86	8.75
5500	4"		80	.100		5.70	5.45		11.15	14.75
5520	6"		53	.151		7.90	8.25		16.15	21.50
5530	8"		40	.200		10.25	10.95		21.20	28

26 05 33.25 Conduit Fittings for Rigid Galvanized Steel		Crew	Daily Output	Labor-Hours	Unit	Material	Labor	Equipment	Total	Total Incl O&P
0010	**CONDUIT FITTINGS FOR RIGID GALVANIZED STEEL**									
0050	Standard, locknuts, 1/2" diameter				Ea.	.21			.21	.23
0100	3/4" diameter					.37			.37	.41
0300	1" diameter					.63			.63	.69
0500	1-1/4" diameter					.72			.72	.79
0700	1-1/2" diameter					1.29			1.29	1.42
1000	2" diameter					1.80			1.80	1.98
1030	2-1/2" diameter					4.55			4.55	5
1050	3" diameter					5.80			5.80	6.40
1070	3-1/2" diameter					8.85			8.85	9.70
1100	4" diameter					11.70			11.70	12.90
1110	5" diameter					25			25	27.50
1120	6" diameter					67.50			67.50	74.50
1130	Bushings, plastic, 1/2" diameter	1 Elec	40	.200		.16	10.95		11.11	17.15
1150	3/4" diameter		32	.250		.27	13.70		13.97	21.50
1170	1" diameter		28	.286		.43	15.65		16.08	24.50
1200	1-1/4" diameter		24	.333		.65	18.25		18.90	28.50
1230	1-1/2" diameter		18	.444		.89	24.50		25.39	38.50
1250	2" diameter		15	.533		1.64	29		30.64	47
1270	2-1/2" diameter		13	.615		3.80	33.50		37.30	56
1300	3" diameter		12	.667		3.94	36.50		40.44	61
1330	3-1/2" diameter		11	.727		5.20	40		45.20	67
1350	4" diameter		9	.889		6.40	48.50		54.90	82.50
1360	5" diameter		7	1.143		14.10	62.50		76.60	113
1370	6" diameter		5	1.600		31	87.50		118.50	170
1390	Steel, 1/2" diameter		40	.200		.65	10.95		11.60	17.65
1400	3/4" diameter		32	.250		.77	13.70		14.47	22
1430	1" diameter		28	.286		1.40	15.65		17.05	25.50
1450	Steel insulated, 1-1/4" diameter		24	.333		4.66	18.25		22.91	33
1470	1-1/2" diameter		18	.444		5.80	24.50		30.30	44
1500	2" diameter		15	.533		8.30	29		37.30	54
1530	2-1/2" diameter		13	.615		16.15	33.50		49.65	70
1550	3" diameter		12	.667		22	36.50		58.50	81
1570	3-1/2" diameter		11	.727		29	40		69	93.50
1600	4" diameter		9	.889		35	48.50		83.50	114
1610	5" diameter		7	1.143		73	62.50		135.50	178
1620	6" diameter		5	1.600		167	87.50		254.50	320
1630	Sealing locknuts, 1/2" diameter		40	.200		1.31	10.95		12.26	18.40
1650	3/4" diameter		32	.250		1.41	13.70		15.11	22.50
1670	1" diameter		28	.286		2.08	15.65		17.73	26.50
1700	1-1/4" diameter		24	.333		3.13	18.25		21.38	31.50
1730	1-1/2" diameter		18	.444		3.98	24.50		28.48	42

26 05 33.25 Conduit Fittings for Rigid Galvanized Steel		Crew	Daily Output	Labor-Hours	Unit	2015 Bare Costs Material	Labor	Equipment	Total	Total Incl O&P
1750	2" diameter	1 Elec	15	.533	Ea.	5.10	29		34.10	50.50
1760	Grounding bushing, insulated, 1/2" diameter		32	.250		5.85	13.70		19.55	27.50
1770	3/4" diameter		28	.286		6.65	15.65		22.30	31.50
1780	1" diameter		20	.400		8.30	22		30.30	43
1800	1-1/4" diameter		18	.444		10.75	24.50		35.25	49.50
1830	1-1/2" diameter		16	.500		12.40	27.50		39.90	56
1850	2" diameter		13	.615		17.25	33.50		50.75	71
1870	2-1/2" diameter		12	.667		28	36.50		64.50	87.50
1900	3" diameter		11	.727		29	40		69	93.50
1930	3-1/2" diameter		9	.889		33.50	48.50		82	112
1950	4" diameter		8	1		48.50	54.50		103	138
1960	5" diameter		6	1.333		87.50	73		160.50	210
1970	6" diameter		4	2		123	109		232	305
1990	Coupling, with set screw, 1/2" diameter		50	.160		3.52	8.75		12.27	17.40
2000	3/4" diameter		40	.200		4.54	10.95		15.49	22
2030	1" diameter		35	.229		7.35	12.50		19.85	27.50
2050	1-1/4" diameter		28	.286		12.55	15.65		28.20	38
2070	1-1/2" diameter		23	.348		16.05	19.05		35.10	47
2090	2" diameter		20	.400		36	22		58	73.50
2100	2-1/2" diameter		18	.444		78.50	24.50		103	124
2110	3" diameter		15	.533		91.50	29		120.50	146
2120	3-1/2" diameter		12	.667		131	36.50		167.50	201
2130	4" diameter		10	.800		174	44		218	259
2140	5" diameter		9	.889		305	48.50		353.50	415
2150	6" diameter		8	1		540	54.50		594.50	680
2160	Box connector with set screw, plain, 1/2" diameter		70	.114		2.40	6.25		8.65	12.35
2170	3/4" diameter		60	.133		3.32	7.30		10.62	14.95
2180	1" diameter		50	.160		5.15	8.75		13.90	19.25
2190	Insulated, 1-1/4" diameter		40	.200		13.15	10.95		24.10	31.50
2200	1-1/2" diameter		30	.267		19.35	14.60		33.95	44
2210	2" diameter		20	.400		39	22		61	77
2220	2-1/2" diameter		18	.444		104	24.50		128.50	152
2230	3" diameter		15	.533		150	29		179	209
2240	3-1/2" diameter		12	.667		210	36.50		246.50	288
2250	4" diameter		10	.800		225	44		269	315
2260	5" diameter		9	.889		266	48.50		314.50	370
2270	6" diameter		8	1		280	54.50		334.50	395
2280	LB, LR or LL fittings & covers, 1/2" diameter		16	.500		8.60	27.50		36.10	52
2290	3/4" diameter		13	.615		10.65	33.50		44.15	64
2300	1" diameter		11	.727		15.55	40		55.55	78.50
2330	1-1/4" diameter		8	1		24.50	54.50		79	111
2350	1-1/2" diameter		6	1.333		32	73		105	149
2370	2" diameter		5	1.600		52.50	87.50		140	194
2380	2-1/2" diameter		4	2		111	109		220	291
2390	3" diameter		3.50	2.286		142	125		267	350
2400	3-1/2" diameter		3	2.667		247	146		393	500
2410	4" diameter		2.50	3.200		270	175		445	570
2420	T fittings, with cover, 1/2" diameter		12	.667		11	36.50		47.50	68.50
2430	3/4" diameter		11	.727		12.70	40		52.70	75.50
2440	1" diameter		9	.889		21.50	48.50		70	99.50
2450	1-1/4" diameter		6	1.333		28	73		101	144
2470	1-1/2" diameter		5	1.600		38.50	87.50		126	179
2500	2" diameter	↓	4	2	↓	56.50	109		165.50	232

26 05 33.25 Conduit Fittings for Rigid Galvanized Steel		Crew	Daily Output	Labor-Hours	Unit	2015 Bare Costs Material	Labor	Equipment	Total	Total Incl O&P
2510	2-1/2" diameter	1 Elec	3.50	2.286	Ea.	124	125		249	330
2520	3" diameter		3	2.667		173	146		319	415
2530	3-1/2" diameter		2.50	3.200		300	175		475	600
2540	4" diameter		2	4		360	219		579	735
2550	Nipples chase, plain, 1/2" diameter		40	.200		.79	10.95		11.74	17.80
2560	3/4" diameter		32	.250		.93	13.70		14.63	22
2570	1" diameter		28	.286		1.90	15.65		17.55	26
2600	Insulated, 1-1/4" diameter		24	.333		6.55	18.25		24.80	35
2630	1-1/2" diameter		18	.444		8.70	24.50		33.20	47
2650	2" diameter		15	.533		13.05	29		42.05	59.50
2660	2-1/2" diameter		12	.667		26.50	36.50		63	85.50
2670	3" diameter		10	.800		34	44		78	106
2680	3-1/2" diameter		9	.889		73	48.50		121.50	156
2690	4" diameter		8	1		142	54.50		196.50	242
2700	5" diameter		7	1.143		212	62.50		274.50	330
2710	6" diameter		6	1.333		330	73		403	475
2720	Nipples offset, plain, 1/2" diameter		40	.200		4.42	10.95		15.37	22
2730	3/4" diameter		32	.250		4.60	13.70		18.30	26
2740	1" diameter		24	.333		6.05	18.25		24.30	34.50
2750	Insulated, 1-1/4" diameter		20	.400		24.50	22		46.50	61
2760	1-1/2" diameter		18	.444		31	24.50		55.50	71.50
2770	2" diameter		16	.500		48	27.50		75.50	95
2780	3" diameter		14	.571		74	31.50		105.50	130
2850	Coupling, expansion, 1/2" diameter		12	.667		43.50	36.50		80	105
2880	3/4" diameter		10	.800		51	44		95	124
2900	1" diameter		8	1		59.50	54.50		114	150
2920	1-1/4" diameter		6.40	1.250		72.50	68.50		141	186
2940	1-1/2" diameter		5.30	1.509		112	82.50		194.50	252
2960	2" diameter		4.60	1.739		169	95		264	330
2980	2-1/2" diameter		3.60	2.222		236	122		358	445
3000	3" diameter		3	2.667		315	146		461	570
3020	3-1/2" diameter		2.80	2.857		385	156		541	665
3040	4" diameter		2.40	3.333		580	182		762	920
3060	5" diameter		2	4		905	219		1,124	1,325
3080	6" diameter		1.80	4.444		1,350	243		1,593	1,850
3100	Expansion deflection, 1/2" diameter		12	.667		249	36.50		285.50	330
3120	3/4" diameter		12	.667		280	36.50		316.50	365
3140	1" diameter		10	.800		298	44		342	400
3160	1-1/4" diameter		6.40	1.250		355	68.50		423.50	495
3180	1-1/2" diameter		5.30	1.509		415	82.50		497.50	585
3200	2" diameter		4.60	1.739		515	95		610	710
3220	2-1/2" diameter		3.60	2.222		700	122		822	960
3240	3" diameter		3	2.667		860	146		1,006	1,175
3260	3-1/2" diameter		2.80	2.857		1,100	156		1,256	1,475
3280	4" diameter		2.40	3.333		1,275	182		1,457	1,675
3300	5" diameter		2	4		1,925	219		2,144	2,475
3320	6" diameter		1.80	4.444		3,350	243		3,593	4,050
3340	Ericson, 1/2" diameter		16	.500		4.95	27.50		32.45	48
3360	3/4" diameter		14	.571		6.30	31.50		37.80	55.50
3380	1" diameter		11	.727		11.70	40		51.70	74.50
3400	1-1/4" diameter		8	1		25	54.50		79.50	112
3420	1-1/2" diameter		7	1.143		30	62.50		92.50	130
3440	2" diameter		5	1.600		60	87.50		147.50	202

26 05 33.25	Conduit Fittings for Rigid Galvanized Steel	Crew	Daily Output	Labor-Hours	Unit	2015 Bare Costs Material	Labor	Equipment	Total	Total Incl O&P
3460	2-1/2" diameter	1 Elec	4	2	Ea.	145	109		254	330
3480	3" diameter		3.50	2.286		221	125		346	435
3500	3-1/2" diameter		3	2.667		370	146		516	630
3520	4" diameter		2.70	2.963		430	162		592	725
3540	5" diameter		2.50	3.200		800	175		975	1,150
3560	6" diameter		2.30	3.478		1,075	190		1,265	1,475
3580	Split, 1/2" diameter		32	.250		3.60	13.70		17.30	25
3600	3/4" diameter		27	.296		4.50	16.20		20.70	30
3620	1" diameter		20	.400		8.10	22		30.10	43
3640	1-1/4" diameter		16	.500		12.45	27.50		39.95	56
3660	1-1/2" diameter		14	.571		14.15	31.50		45.65	64
3680	2" diameter		12	.667		28.50	36.50		65	87.50
3700	2-1/2" diameter		10	.800		64	44		108	138
3720	3" diameter		9	.889		96	48.50		144.50	182
3740	3-1/2" diameter		8	1		155	54.50		209.50	255
3760	4" diameter		7	1.143		181	62.50		243.50	296
3780	5" diameter		6	1.333		320	73		393	465
3800	6" diameter		5	1.600		425	87.50		512.50	600
4600	Reducing bushings, 3/4" to 1/2" diameter		54	.148		2.11	8.10		10.21	14.85
4620	1" to 3/4" diameter		46	.174		2.24	9.50		11.74	17.20
4640	1-1/4" to 1" diameter		40	.200		5.20	10.95		16.15	22.50
4660	1-1/2" to 1-1/4" diameter		36	.222		7.45	12.15		19.60	27
4680	2" to 1-1/2" diameter		32	.250		13.55	13.70		27.25	36
4740	2-1/2" to 2" diameter		30	.267		15.05	14.60		29.65	39
4760	3" to 2-1/2" diameter		28	.286		18.70	15.65		34.35	44.50
4800	Through-wall seal, 1/2" diameter		8	1		217	54.50		271.50	325
4820	3/4" diameter		7.50	1.067		239	58.50		297.50	355
4840	1" diameter		6.50	1.231		239	67.50		306.50	365
4860	1-1/4" diameter		5.50	1.455		284	79.50		363.50	435
4880	1-1/2" diameter		5	1.600		370	87.50		457.50	540
4900	2" diameter		4.20	1.905		335	104		439	525
4920	2-1/2" diameter		3.50	2.286		450	125		575	690
4940	3" diameter		3	2.667		480	146		626	750
4960	3-1/2" diameter		2.50	3.200		725	175		900	1,075
4980	4" diameter		2	4		735	219		954	1,150
5000	5" diameter		1.50	5.333		950	292		1,242	1,500
5020	6" diameter		1	8		950	440		1,390	1,725
5100	Cable supports, 2 or more wires									
5120	1-1/2" diameter	1 Elec	8	1	Ea.	139	54.50		193.50	238
5140	2" diameter		6	1.333		190	73		263	320
5160	2-1/2" diameter		4	2		204	109		313	395
5180	3" diameter		3.50	2.286		261	125		386	480
5200	3-1/2" diameter		2.60	3.077		340	168		508	635
5220	4" diameter		2	4		430	219		649	810
5240	5" diameter		1.50	5.333		585	292		877	1,100
5260	6" diameter		1	8		950	440		1,390	1,725
5280	Service entrance cap, 1/2" diameter		16	.500		5.65	27.50		33.15	48.50
5300	3/4" diameter		13	.615		6.50	33.50		40	59
5320	1" diameter		10	.800		7.85	44		51.85	76.50
5340	1-1/4" diameter		8	1		10.55	54.50		65.05	96
5360	1-1/2" diameter		6.50	1.231		21	67.50		88.50	127
5380	2" diameter		5.50	1.455		31.50	79.50		111	158
5400	2-1/2" diameter		4	2		84.50	109		193.50	262

26 05 33.25 Conduit Fittings for Rigid Galvanized Steel		Crew	Daily Output	Labor-Hours	Unit	Material	2015 Bare Costs Labor	Equipment	Total	Total Incl O&P
5420	3" diameter	1 Elec	3.40	2.353	Ea.	138	129		267	350
5440	3-1/2" diameter		3	2.667		197	146		343	445
5460	4" diameter		2.70	2.963		256	162		418	530
5750	90° pull elbows steel, female, 1/2" diameter		16	.500		6.80	27.50		34.30	50
5760	3/4" diameter		13	.615		7.65	33.50		41.15	60.50
5780	1" diameter		11	.727		14.05	40		54.05	77
5800	1-1/4" diameter		8	1		19.20	54.50		73.70	106
5820	1-1/2" diameter		6	1.333		27.50	73		100.50	143
5840	2" diameter		5	1.600		44.50	87.50		132	185
6000	Explosion proof, flexible coupling									
6010	1/2" diameter, 4" long	1 Elec	12	.667	Ea.	119	36.50		155.50	188
6020	6" long		12	.667		107	36.50		143.50	175
6050	12" long		12	.667		162	36.50		198.50	235
6070	18" long		12	.667		192	36.50		228.50	268
6090	24" long		12	.667		239	36.50		275.50	320
6110	30" long		12	.667		315	36.50		351.50	405
6130	36" long		12	.667		305	36.50		341.50	390
6140	3/4" diameter, 4" long		10	.800		146	44		190	229
6150	6" long		10	.800		130	44		174	211
6180	12" long		10	.800		194	44		238	281
6200	18" long		10	.800		243	44		287	335
6220	24" long		10	.800		305	44		349	405
6240	30" long		10	.800		415	44		459	525
6260	36" long		10	.800		360	44		404	465
6270	1" diameter, 6" long		8	1		340	54.50		394.50	455
6300	12" long		8	1		330	54.50		384.50	445
6320	18" long		8	1		405	54.50		459.50	530
6340	24" long		8	1		525	54.50		579.50	660
6360	30" long		8	1		700	54.50		754.50	855
6380	36" long		8	1		765	54.50		819.50	930
6390	1-1/4" diameter, 12" long		6.40	1.250		590	68.50		658.50	755
6410	18" long		6.40	1.250		705	68.50		773.50	880
6430	24" long		6.40	1.250		810	68.50		878.50	1,000
6450	30" long		6.40	1.250		855	68.50		923.50	1,050
6470	36" long		6.40	1.250		1,300	68.50		1,368.50	1,550
6480	1-1/2" diameter, 12" long		5.30	1.509		620	82.50		702.50	815
6500	18" long		5.30	1.509		755	82.50		837.50	960
6520	24" long		5.30	1.509		1,075	82.50		1,157.50	1,325
6540	30" long		5.30	1.509		1,225	82.50		1,307.50	1,475
6560	36" long		5.30	1.509		1,325	82.50		1,407.50	1,575
6570	2" diameter, 12" long		4.60	1.739		1,050	95		1,145	1,300
6590	18" long		4.60	1.739		1,350	95		1,445	1,650
6610	24" long		4.60	1.739		1,475	95		1,570	1,775
6630	30" long		4.60	1.739		1,400	95		1,495	1,700
6650	36" long		4.60	1.739		1,875	95		1,970	2,200
7000	Close up plug, 1/2" diameter, explosion proof		40	.200		2.24	10.95		13.19	19.40
7010	3/4" diameter		32	.250		2.63	13.70		16.33	24
7020	1" diameter		28	.286		3.07	15.65		18.72	27.50
7030	1-1/4" diameter		24	.333		3.39	18.25		21.64	31.50
7040	1-1/2" diameter		18	.444		4.68	24.50		29.18	42.50
7050	2" diameter		15	.533		8.05	29		37.05	54
7060	2-1/2" diameter		13	.615		13.30	33.50		46.80	66.50
7070	3" diameter		12	.667		19.75	36.50		56.25	78

26 05 33 – Raceway and Boxes for Electrical Systems

26 05 33.25 Conduit Fittings for Rigid Galvanized Steel		Crew	Daily Output	Labor-Hours	Unit	Material	2015 Bare Costs Labor	Equipment	Total	Total Incl O&P
7080	3-1/2" diameter	1 Elec	11	.727	Ea.	24.50	40		64.50	88.50
7090	4" diameter		9	.889		30.50	48.50		79	110
7091	Elbow female, 45°, 1/2"		16	.500		11.15	27.50		38.65	55
7092	3/4"		13	.615		13.85	33.50		47.35	67.50
7093	1"		11	.727		16.55	40		56.55	80
7094	1-1/4"		8	1		25	54.50		79.50	112
7095	1-1/2"		6	1.333		27	73		100	143
7096	2"		5	1.600		31.50	87.50		119	171
7097	2-1/2"		4.50	1.778		92	97		189	252
7098	3"		4.20	1.905		93	104		197	264
7099	3-1/2"		4	2		144	109		253	325
7100	4"		3.80	2.105		158	115		273	350
7101	90°, 1/2"		16	.500		11.25	27.50		38.75	55
7102	3/4"		13	.615		12.15	33.50		45.65	65.50
7103	1"		11	.727		16.60	40		56.60	80
7104	1-1/4"		8	1		26.50	54.50		81	114
7105	1-1/2"		6	1.333		37	73		110	154
7106	2"		5	1.600		58	87.50		145.50	200
7107	2-1/2"		4.50	1.778		136	97		233	300
7110	Elbows 90°, long male & female, 1/2" diameter, explosion proof		16	.500		16.45	27.50		43.95	60.50
7120	3/4" diameter		13	.615		19	33.50		52.50	73
7130	1" diameter		11	.727		24.50	40		64.50	88.50
7140	1-1/4" diameter		8	1		29	54.50		83.50	117
7150	1-1/2" diameter		6	1.333		36.50	73		109.50	153
7160	2" diameter		5	1.600		54	87.50		141.50	196
7170	Capped elbow, 1/2" diameter, explosion proof		11	.727		17.95	40		57.95	81.50
7180	3/4" diameter		8	1		22	54.50		76.50	109
7190	1" diameter		6	1.333		24	73		97	140
7200	1-1/4" diameter		5	1.600		48	87.50		135.50	189
7210	Pulling elbow, 1/2" diameter, explosion proof		11	.727		71	40		111	140
7220	3/4" diameter		8	1		126	54.50		180.50	223
7230	1" diameter		6	1.333		181	73		254	310
7240	1-1/4" diameter		5	1.600		209	87.50		296.50	365
7250	1-1/2" diameter		5	1.600		237	87.50		324.50	395
7260	2" diameter		4	2		293	109		402	490
7270	2-1/2" diameter		3.50	2.286		725	125		850	995
7280	3" diameter		3	2.667		995	146		1,141	1,325
7290	3-1/2" diameter		2.50	3.200		1,250	175		1,425	1,650
7300	4" diameter		2.20	3.636		1,300	199		1,499	1,725
7310	LB conduit body, 1/2" diameter		11	.727		41.50	40		81.50	108
7320	3/4" diameter		8	1		50.50	54.50		105	141
7330	T conduit body, 1/2" diameter		9	.889		45.50	48.50		94	126
7340	3/4" diameter		6	1.333		53.50	73		126.50	172
7350	Explosion proof, round box w/cover, 3 threaded hubs, 1/2" diameter		8	1		45.50	54.50		100	135
7351	3/4" diameter		8	1		65.50	54.50		120	157
7352	1" diameter		7.50	1.067		69.50	58.50		128	167
7353	1-1/4" diameter		7	1.143		126	62.50		188.50	236
7354	1-1/2" diameter		7	1.143		184	62.50		246.50	299
7355	2" diameter		6	1.333		190	73		263	320
7356	Round box w/cover & mtng flange, 3 threaded hubs, 1/2" diameter		8	1		51	54.50		105.50	141
7357	3/4" diameter		8	1		57.50	54.50		112	148
7358	4 threaded hubs, 1" diameter		7	1.143		63	62.50		125.50	166
7400	Unions, 1/2" diameter		20	.400		11.20	22		33.20	46.50

26 05 33.25 Conduit Fittings for Rigid Galvanized Steel		Crew	Daily Output	Labor-Hours	Unit	2015 Bare Costs Material	Labor	Equipment	Total	Total Incl O&P
7410	3/4" - 1/2" diameter	1 Elec	16	.500	Ea.	16.20	27.50		43.70	60.50
7420	3/4" diameter		16	.500		15.40	27.50		42.90	59.50
7430	1" diameter		14	.571		28	31.50		59.50	79.50
7440	1-1/4" diameter		12	.667		41	36.50		77.50	102
7450	1-1/2" diameter		10	.800		52.50	44		96.50	126
7460	2" diameter		8.50	.941		67.50	51.50		119	154
7480	2-1/2" diameter		8	1		105	54.50		159.50	201
7490	3" diameter		7	1.143		137	62.50		199.50	248
7500	3-1/2" diameter		6	1.333		254	73		327	390
7510	4" diameter		5	1.600		256	87.50		343.50	420
7680	Reducer, 3/4" to 1/2"		54	.148		2.36	8.10		10.46	15.15
7690	1" to 1/2"		46	.174		4.26	9.50		13.76	19.45
7700	1" to 3/4"		46	.174		4.36	9.50		13.86	19.55
7710	1-1/4" to 3/4"		40	.200		5.35	10.95		16.30	23
7720	1-1/4" to 1"		40	.200		6.55	10.95		17.50	24
7730	1-1/2" to 1"		36	.222		10.05	12.15		22.20	30
7740	1-1/2" to 1-1/4"		36	.222		10.65	12.15		22.80	30.50
7750	2" to 3/4"		32	.250		9.90	13.70		23.60	32
7760	2" to 1-1/4"		32	.250		11.80	13.70		25.50	34
7770	2" to 1-1/2"		32	.250		12.55	13.70		26.25	35
7780	2-1/2" to 1-1/2"		30	.267		21	14.60		35.60	45.50
7790	3" to 2"		30	.267		21.50	14.60		36.10	46
7800	3-1/2" to 2-1/2"		28	.286		43.50	15.65		59.15	72
7810	4" to 3"		28	.286		52.50	15.65		68.15	82
7820	Sealing fitting, vertical/horizontal, 1/2" diameter		14.50	.552		15.10	30		45.10	63
7830	3/4" diameter		13.30	.602		17.60	33		50.60	70.50
7840	1" diameter		11.40	.702		22	38.50		60.50	84
7850	1-1/4" diameter		10	.800		28	44		72	99
7860	1-1/2" diameter		8.80	.909		42	49.50		91.50	124
7870	2" diameter		8	1		55.50	54.50		110	146
7880	2-1/2" diameter		6.70	1.194		76	65.50		141.50	185
7890	3" diameter		5.70	1.404		98	77		175	227
7900	3-1/2" diameter		4.70	1.702		260	93		353	430
7910	4" diameter		4	2		430	109		539	640
7920	Sealing hubs, 1" by 1-1/2"		12	.667		31.50	36.50		68	91.50
7930	1-1/4" by 2"		10	.800		47	44		91	120
7940	1-1/2" by 2"		9	.889		62	48.50		110.50	144
7950	2" by 2-1/2"		8	1		81	54.50		135.50	174
7960	3" by 4"		7	1.143		137	62.50		199.50	248
7970	4" by 5"		6	1.333		285	73		358	430
7980	Drain, 1/2"		32	.250		83.50	13.70		97.20	113
7990	Breather, 1/2"	↓	32	.250	↓	83.50	13.70		97.20	113
8000	Plastic coated 40 mil thick									
8010	LB, LR or LL conduit body w/cover, 1/2" diameter	1 Elec	13	.615	Ea.	49.50	33.50		83	107
8020	3/4" diameter		11	.727		60.50	40		100.50	128
8030	1" diameter		8	1		72	54.50		126.50	164
8040	1-1/4" diameter		6	1.333		107	73		180	231
8050	1-1/2" diameter		5	1.600		130	87.50		217.50	279
8060	2" diameter		4.50	1.778		190	97		287	360
8070	2-1/2" diameter		4	2		300	109		409	500
8080	3" diameter		3.50	2.286		440	125		565	680
8090	3-1/2" diameter		3	2.667		550	146		696	825
8100	4" diameter	↓	2.50	3.200	↓	615	175		790	945

26 05 33.25 Conduit Fittings for Rigid Galvanized Steel		Crew	Daily Output	Labor-Hours	Unit	Material	Labor	Equipment	Total	Total Incl O&P
						2015 Bare Costs				
8150	T conduit body with cover, 1/2" diameter	1 Elec	11	.727	Ea.	61	40		101	129
8160	3/4" diameter		9	.889		57.50	48.50		106	139
8170	1" diameter		6	1.333		79.50	73		152.50	201
8180	1-1/4" diameter		5	1.600		122	87.50		209.50	271
8190	1-1/2" diameter		4.50	1.778		149	97		246	315
8200	2" diameter		4	2		203	109		312	395
8210	2-1/2" diameter		3.50	2.286		395	125		520	625
8220	3" diameter		3	2.667		510	146		656	790
8230	3-1/2" diameter		2.50	3.200		805	175		980	1,150
8240	4" diameter		2	4		875	219		1,094	1,300
8300	FS conduit body, 1 gang, 3/4" diameter		11	.727		59	40		99	127
8310	1" diameter		10	.800		61.50	44		105.50	136
8350	2 gang, 3/4" diameter		9	.889		107	48.50		155.50	194
8360	1" diameter		8	1		123	54.50		177.50	220
8400	Duplex receptacle cover		64	.125		43	6.85		49.85	58
8410	Switch cover		64	.125		62.50	6.85		69.35	79.50
8420	Switch, vaportight cover		53	.151		183	8.25		191.25	214
8430	Blank, cover		64	.125		41.50	6.85		48.35	56.50
8520	FSC conduit body, 1 gang, 3/4" diameter		10	.800		66.50	44		110.50	141
8530	1" diameter		9	.889		74	48.50		122.50	157
8550	2 gang, 3/4" diameter		8	1		121	54.50		175.50	218
8560	1" diameter		7	1.143		116	62.50		178.50	224
8590	Conduit hubs, 1/2" diameter		18	.444		37	24.50		61.50	78
8600	3/4" diameter		16	.500		37.50	27.50		65	84
8610	1" diameter		14	.571		47	31.50		78.50	101
8620	1-1/4" diameter		12	.667		63.50	36.50		100	126
8630	1-1/2" diameter		10	.800		76	44		120	152
8640	2" diameter		8.80	.909		89	49.50		138.50	175
8650	2-1/2" diameter		8.50	.941		149	51.50		200.50	244
8660	3" diameter		8	1		191	54.50		245.50	295
8670	3-1/2" diameter		7.50	1.067		288	58.50		346.50	405
8680	4" diameter		7	1.143		325	62.50		387.50	455
8690	5" diameter		6	1.333		425	73		498	585
8700	Plastic coated 40 mil thick									
8710	Pipe strap, stamped 1 hole, 1/2" diameter	1 Elec	470	.017	Ea.	9.80	.93		10.73	12.20
8720	3/4" diameter		440	.018		10.25	.99		11.24	12.80
8730	1" diameter		400	.020		10.55	1.09		11.64	13.30
8740	1-1/4" diameter		355	.023		18.10	1.23		19.33	22
8750	1-1/2" diameter		320	.025		19.05	1.37		20.42	23
8760	2" diameter		266	.030		27	1.65		28.65	32
8770	2-1/2" diameter		200	.040		36.50	2.19		38.69	43.50
8780	3" diameter		133	.060		48	3.29		51.29	58
8790	3-1/2" diameter		110	.073		73.50	3.98		77.48	86.50
8800	4" diameter		90	.089		82	4.86		86.86	97.50
8810	5" diameter		70	.114		129	6.25		135.25	152
8840	Clamp back spacers, 3/4" diameter		440	.018		17.45	.99		18.44	20.50
8850	1" diameter		400	.020		20	1.09		21.09	23.50
8860	1-1/4" diameter		355	.023		28.50	1.23		29.73	33.50
8870	1-1/2" diameter		320	.025		39	1.37		40.37	45
8880	2" diameter		266	.030		60	1.65		61.65	68
8900	3" diameter		133	.060		82	3.29		85.29	95
8920	4" diameter		90	.089		103	4.86		107.86	122
8950	Touch-up plastic coating, spray, 12 oz.					52			52	57

26 05 33.25 Conduit Fittings for Rigid Galvanized Steel

	26 05 33.25 Conduit Fittings for Rigid Galvanized Steel	Crew	Daily Output	Labor-Hours	Unit	2015 Bare Costs Material	Labor	Equipment	Total	Total Incl O&P
8960	Sealing fittings, 1/2" diameter	1 Elec	11	.727	Ea.	62	40		102	130
8970	3/4" diameter		9	.889		63.50	48.50		112	145
8980	1" diameter		7.50	1.067		74.50	58.50		133	173
8990	1-1/4" diameter		6.50	1.231		96.50	67.50		164	210
9000	1-1/2" diameter		5.50	1.455		126	79.50		205.50	262
9010	2" diameter		4.80	1.667		151	91		242	305
9020	2-1/2" diameter		4	2		261	109		370	455
9030	3" diameter		3.50	2.286		279	125		404	500
9040	3-1/2" diameter		3	2.667		755	146		901	1,050
9050	4" diameter		2.50	3.200		1,025	175		1,200	1,400
9060	5" diameter		1.70	4.706		2,100	257		2,357	2,725
9070	Unions, 1/2" diameter		18	.444		55	24.50		79.50	98
9080	3/4" diameter		15	.533		48.50	29		77.50	98.50
9090	1" diameter		13	.615		65	33.50		98.50	124
9100	1-1/4" diameter		11	.727		124	40		164	199
9110	1-1/2" diameter		9.50	.842		150	46		196	237
9120	2" diameter		8	1		169	54.50		223.50	271
9130	2-1/2" diameter		7.50	1.067		296	58.50		354.50	415
9140	3" diameter		6.80	1.176		385	64.50		449.50	525
9150	3-1/2" diameter		5.80	1.379		510	75.50		585.50	675
9160	4" diameter		4.80	1.667		615	91		706	815
9170	5" diameter		4	2		945	109		1,054	1,225

26 05 33.30 Electrical Nonmetallic Tubing (ENT)

	26 05 33.30 Electrical Nonmetallic Tubing (ENT)	Crew	Daily Output	Labor-Hours	Unit	Material	Labor	Equipment	Total	Total Incl O&P
0010	**ELECTRICAL NONMETALLIC TUBING (ENT)**									
0050	Flexible, 1/2" diameter	1 Elec	270	.030	L.F.	.74	1.62		2.36	3.32
0100	3/4" diameter		230	.035		1.21	1.90		3.11	4.28
0200	1" diameter		145	.055		2.03	3.02		5.05	6.90
0210	1-1/4" diameter		125	.064		1.98	3.50		5.48	7.60
0220	1-1/2" diameter		100	.080		2.93	4.38		7.31	10
0230	2" diameter		75	.107		3.89	5.85		9.74	13.35
0300	Connectors, to outlet box, 1/2" diameter		230	.035	Ea.	1.02	1.90		2.92	4.07
0310	3/4" diameter		210	.038		1.82	2.08		3.90	5.25
0320	1" diameter		200	.040		2.78	2.19		4.97	6.45
0400	Couplings, to conduit, 1/2" diameter		145	.055		1.08	3.02		4.10	5.85
0410	3/4" diameter		130	.062		1.29	3.37		4.66	6.60
0420	1" diameter		125	.064		2.68	3.50		6.18	8.35

26 05 33.35 Flexible Metallic Conduit

	26 05 33.35 Flexible Metallic Conduit	Crew	Daily Output	Labor-Hours	Unit	Material	Labor	Equipment	Total	Total Incl O&P
0010	**FLEXIBLE METALLIC CONDUIT**									
0050	Steel, 3/8" diameter	1 Elec	200	.040	L.F.	.43	2.19		2.62	3.86
0100	1/2" diameter		200	.040		.48	2.19		2.67	3.92
0200	3/4" diameter		160	.050		.67	2.74		3.41	4.98
0250	1" diameter		100	.080		1.21	4.38		5.59	8.15
0300	1-1/4" diameter		70	.114		1.57	6.25		7.82	11.45
0350	1-1/2" diameter		50	.160		2.55	8.75		11.30	16.35
0370	2" diameter		40	.200		3.11	10.95		14.06	20.50
0380	2-1/2" diameter		30	.267		3.77	14.60		18.37	26.50
0390	3" diameter	2 Elec	50	.320		6.60	17.50		24.10	34.50
0400	3-1/2" diameter		40	.400		7.45	22		29.45	42
0410	4" diameter		30	.533		8.50	29		37.50	54.50
0420	Connectors, plain, 3/8" diameter	1 Elec	100	.080	Ea.	1.97	4.38		6.35	8.95
0430	1/2" diameter		80	.100		2.28	5.45		7.73	10.95
0440	3/4" diameter		70	.114		2.59	6.25		8.84	12.55

26 05 33 – Raceway and Boxes for Electrical Systems

26 05 33.35 Flexible Metallic Conduit		Crew	Daily Output	Labor-Hours	Unit	Material	2015 Bare Costs Labor	Equipment	Total	Total Incl O&P
0450	1" diameter	1 Elec	50	.160	Ea.	5.50	8.75		14.25	19.60
0452	1-1/4" diameter		45	.178		6.75	9.70		16.45	22.50
0454	1-1/2" diameter		40	.200		10.05	10.95		21	28
0456	2" diameter		28	.286		15.05	15.65		30.70	40.50
0458	2-1/2" diameter		25	.320		25.50	17.50		43	55
0460	3" diameter		20	.400		36.50	22		58.50	74
0462	3-1/2" diameter		16	.500		107	27.50		134.50	161
0464	4" diameter		13	.615		136	33.50		169.50	202
0490	Insulated, 1" diameter		40	.200		6.05	10.95		17	23.50
0500	1-1/4" diameter		40	.200		13.50	10.95		24.45	32
0550	1-1/2" diameter		32	.250		20.50	13.70		34.20	43.50
0600	2" diameter		23	.348		31	19.05		50.05	63.50
0610	2-1/2" diameter		20	.400		72.50	22		94.50	114
0620	3" diameter		17	.471		96	25.50		121.50	146
0630	3-1/2" diameter		13	.615		195	33.50		228.50	266
0640	4" diameter		10	.800		256	44		300	350
0650	Connectors 90°, plain, 3/8" diameter		80	.100		2.84	5.45		8.29	11.55
0660	1/2" diameter		60	.133		4.77	7.30		12.07	16.55
0700	3/4" diameter		50	.160		7.85	8.75		16.60	22
0750	1" diameter		40	.200		13.40	10.95		24.35	31.50
0790	Insulated, 1" diameter		40	.200		12.40	10.95		23.35	30.50
0800	1-1/4" diameter		30	.267		29	14.60		43.60	54.50
0850	1-1/2" diameter		23	.348		54.50	19.05		73.55	89.50
0900	2" diameter		18	.444		66.50	24.50		91	111
0910	2-1/2" diameter		16	.500		128	27.50		155.50	183
0920	3" diameter		14	.571		163	31.50		194.50	228
0930	3-1/2" diameter		11	.727		495	40		535	605
0940	4" diameter		8	1		690	54.50		744.50	845
0960	Couplings, to flexible conduit, 1/2" diameter		50	.160		1.48	8.75		10.23	15.20
0970	3/4" diameter		40	.200		2.53	10.95		13.48	19.75
0980	1" diameter		35	.229		4.42	12.50		16.92	24
0990	1-1/4" diameter		28	.286		9.70	15.65		25.35	34.50
1000	1-1/2" diameter		23	.348		12.50	19.05		31.55	43.50
1010	2" diameter		20	.400		25.50	22		47.50	62
1020	2-1/2" diameter		18	.444		35	24.50		59.50	76
1030	3" diameter		15	.533		88	29		117	142
1032	Aluminum, 3/8" diameter		210	.038	L.F.	.44	2.08		2.52	3.71
1034	1/2" diameter		210	.038		.51	2.08		2.59	3.79
1036	3/4" diameter		165	.048		.71	2.65		3.36	4.89
1038	1" diameter		105	.076		1.33	4.17		5.50	7.90
1040	1-1/4" diameter		75	.107		1.85	5.85		7.70	11.10
1042	1-1/2" diameter		53	.151		2.65	8.25		10.90	15.70
1044	2" diameter		42	.190		3.38	10.40		13.78	19.85
1046	2-1/2" diameter		32	.250		4.35	13.70		18.05	26
1048	3" diameter	2 Elec	53	.302		8.20	16.50		24.70	34.50
1050	3-1/2" diameter		42	.381		8.70	21		29.70	42
1052	4" diameter		32	.500		9.15	27.50		36.65	52.50
1070	Sealtite, 3/8" diameter	1 Elec	140	.057		1.05	3.13		4.18	6
1080	1/2" diameter		140	.057		1.13	3.13		4.26	6.10
1090	3/4" diameter		100	.080		1.60	4.38		5.98	8.55
1100	1" diameter		70	.114		2.42	6.25		8.67	12.35
1200	1-1/4" diameter		50	.160		3.30	8.75		12.05	17.20
1300	1-1/2" diameter		40	.200		3.81	10.95		14.76	21

26 05 33.35 Flexible Metallic Conduit		Crew	Daily Output	Labor-Hours	Unit	2015 Bare Costs Material	Labor	Equipment	Total	Total Incl O&P
1400	2″ diameter	1 Elec	30	.267	L.F.	4.82	14.60		19.42	28
1410	2-1/2″ diameter		27	.296		8.20	16.20		24.40	34
1420	3″ diameter	2 Elec	50	.320		11.35	17.50		28.85	39.50
1440	4″ diameter	″	30	.533		17.15	29		46.15	64
1490	Connectors, plain, 3/8″ diameter	1 Elec	70	.114	Ea.	2.76	6.25		9.01	12.75
1500	1/2″ diameter		70	.114		3.77	6.25		10.02	13.85
1700	3/4″ diameter		50	.160		5.55	8.75		14.30	19.65
1900	1″ diameter		40	.200		9.75	10.95		20.70	27.50
1910	Insulated, 1″ diameter		40	.200		12.40	10.95		23.35	30.50
2000	1-1/4″ diameter		32	.250		19.35	13.70		33.05	42.50
2100	1-1/2″ diameter		27	.296		25.50	16.20		41.70	53
2200	2″ diameter		20	.400		45	22		67	84
2210	2-1/2″ diameter		15	.533		266	29		295	340
2220	3″ diameter		12	.667		297	36.50		333.50	380
2240	4″ diameter		8	1		370	54.50		424.50	490
2290	Connectors, 90°, 3/8″ diameter		70	.114		4.65	6.25		10.90	14.80
2300	1/2″ diameter		70	.114		5.70	6.25		11.95	15.95
2400	3/4″ diameter		50	.160		6.70	8.75		15.45	21
2600	1″ diameter		40	.200		13.65	10.95		24.60	32
2790	Insulated, 1″ diameter		40	.200		22	10.95		32.95	41
2800	1-1/4″ diameter		32	.250		35.50	13.70		49.20	60.50
3000	1-1/2″ diameter		27	.296		44	16.20		60.20	73.50
3100	2″ diameter		20	.400		61	22		83	102
3110	2-1/2″ diameter		14	.571		288	31.50		319.50	365
3120	3″ diameter		11	.727		335	40		375	430
3140	4″ diameter		7	1.143		430	62.50		492.50	570
4300	Coupling sealtite to rigid, 1/2″ diameter		20	.400		3.58	22		25.58	38
4500	3/4″ diameter		18	.444		5.25	24.50		29.75	43.50
4800	1″ diameter		14	.571		6.90	31.50		38.40	56
4900	1-1/4″ diameter		12	.667		11.50	36.50		48	69
5000	1-1/2″ diameter		11	.727		34.50	40		74.50	99.50
5100	2″ diameter		10	.800		35.50	44		79.50	108
5110	2-1/2″ diameter		9.50	.842		169	46		215	258
5120	3″ diameter		9	.889		187	48.50		235.50	282
5130	3-1/2″ diameter		9	.889		209	48.50		257.50	305
5140	4″ diameter		8.50	.941		232	51.50		283.50	335

26 05 33.95 Cutting and Drilling		Crew	Daily Output	Labor-Hours	Unit	Material	Labor	Equipment	Total	Total Incl O&P
0010	**CUTTING AND DRILLING**									
0100	Hole drilling to 10′ high, concrete wall									
0110	8″ thick, 1/2″ pipe size	R-31	12	.667	Ea.	.24	36.50	4.59	41.33	62
0120	3/4″ pipe size		12	.667		.24	36.50	4.59	41.33	62
0130	1″ pipe size		9.50	.842		.38	46	5.80	52.18	78.50
0140	1-1/4″ pipe size		9.50	.842		.38	46	5.80	52.18	78.50
0150	1-1/2″ pipe size		9.50	.842		.38	46	5.80	52.18	78.50
0160	2″ pipe size		4.40	1.818		.51	99.50	12.55	112.56	168
0170	2-1/2″ pipe size		4.40	1.818		.51	99.50	12.55	112.56	168
0180	3″ pipe size		4.40	1.818		.51	99.50	12.55	112.56	168
0190	3-1/2″ pipe size		3.30	2.424		.66	133	16.70	150.36	224
0200	4″ pipe size		3.30	2.424		.66	133	16.70	150.36	224
0500	12″ thick, 1/2″ pipe size		9.40	.851		.35	46.50	5.85	52.70	79
0520	3/4″ pipe size		9.40	.851		.35	46.50	5.85	52.70	79
0540	1″ pipe size		7.30	1.096		.56	60	7.55	68.11	102

26 05 33.95 Cutting and Drilling		Crew	Daily Output	Labor-Hours	Unit	2015 Bare Costs Material	Labor	Equipment	Total	Total Incl O&P
0560	1-1/4" pipe size	R-31	7.30	1.096	Ea.	.56	60	7.55	68.11	102
0570	1-1/2" pipe size		7.30	1.096		.56	60	7.55	68.11	102
0580	2" pipe size		3.60	2.222		.76	122	15.30	138.06	206
0590	2-1/2" pipe size		3.60	2.222		.76	122	15.30	138.06	206
0600	3" pipe size		3.60	2.222		.76	122	15.30	138.06	206
0610	3-1/2" pipe size		2.80	2.857		.99	156	19.70	176.69	265
0630	4" pipe size		2.50	3.200		.99	175	22	197.99	297
0650	16" thick, 1/2" pipe size		7.60	1.053		.47	57.50	7.25	65.22	97.50
0670	3/4" pipe size		7	1.143		.47	62.50	7.85	70.82	106
0690	1" pipe size		6	1.333		.75	73	9.20	82.95	124
0710	1-1/4" pipe size		5.50	1.455		.75	79.50	10	90.25	135
0730	1-1/2" pipe size		5.50	1.455		.75	79.50	10	90.25	135
0750	2" pipe size		3	2.667		1.02	146	18.35	165.37	247
0770	2-1/2" pipe size		2.70	2.963		1.02	162	20.50	183.52	275
0790	3" pipe size		2.50	3.200		1.02	175	22	198.02	297
0810	3-1/2" pipe size		2.30	3.478		1.32	190	24	215.32	325
0830	4" pipe size		2	4		1.32	219	27.50	247.82	370
0850	20" thick, 1/2" pipe size		6.40	1.250		.59	68.50	8.60	77.69	116
0870	3/4" pipe size		6	1.333		.59	73	9.20	82.79	124
0890	1" pipe size		5	1.600		.94	87.50	11	99.44	149
0910	1-1/4" pipe size		4.80	1.667		.94	91	11.50	103.44	155
0930	1-1/2" pipe size		4.60	1.739		.94	95	12	107.94	161
0950	2" pipe size		2.70	2.963		1.27	162	20.50	183.77	275
0970	2-1/2" pipe size		2.40	3.333		1.27	182	23	206.27	310
0990	3" pipe size		2.20	3.636		1.27	199	25	225.27	340
1010	3-1/2" pipe size		2	4		1.64	219	27.50	248.14	370
1030	4" pipe size		1.70	4.706		1.64	257	32.50	291.14	435
1050	24" thick, 1/2" pipe size		5.50	1.455		.71	79.50	10	90.21	135
1070	3/4" pipe size		5.10	1.569		.71	86	10.80	97.51	146
1090	1" pipe size		4.30	1.860		1.13	102	12.80	115.93	173
1110	1-1/4" pipe size		4	2		1.13	109	13.80	123.93	185
1130	1-1/2" pipe size		4	2		1.13	109	13.80	123.93	185
1150	2" pipe size		2.40	3.333		1.52	182	23	206.52	310
1170	2-1/2" pipe size		2.20	3.636		1.52	199	25	225.52	340
1190	3" pipe size		2	4		1.52	219	27.50	248.02	370
1210	3-1/2" pipe size		1.80	4.444		1.97	243	30.50	275.47	410
1230	4" pipe size		1.50	5.333		1.97	292	37	330.97	495
1500	Brick wall, 8" thick, 1/2" pipe size		18	.444		.24	24.50	3.06	27.80	41
1520	3/4" pipe size		18	.444		.24	24.50	3.06	27.80	41
1540	1" pipe size		13.30	.602		.38	33	4.14	37.52	56
1560	1-1/4" pipe size		13.30	.602		.38	33	4.14	37.52	56
1580	1-1/2" pipe size		13.30	.602		.38	33	4.14	37.52	56
1600	2" pipe size		5.70	1.404		.51	77	9.65	87.16	130
1620	2-1/2" pipe size		5.70	1.404		.51	77	9.65	87.16	130
1640	3" pipe size		5.70	1.404		.51	77	9.65	87.16	130
1660	3-1/2" pipe size		4.40	1.818		.66	99.50	12.55	112.71	169
1680	4" pipe size		4	2		.66	109	13.80	123.46	185
1700	12" thick, 1/2" pipe size		14.50	.552		.35	30	3.80	34.15	51
1720	3/4" pipe size		14.50	.552		.35	30	3.80	34.15	51
1740	1" pipe size		11	.727		.56	40	5	45.56	67.50
1760	1-1/4" pipe size		11	.727		.56	40	5	45.56	67.50
1780	1-1/2" pipe size		11	.727		.56	40	5	45.56	67.50
1800	2" pipe size		5	1.600		.76	87.50	11	99.26	149

26 05 33 – Raceway and Boxes for Electrical Systems

26 05 33.95 Cutting and Drilling		Crew	Daily Output	Labor-Hours	Unit	2015 Bare Costs Material	Labor	Equipment	Total	Total Incl O&P
1820	2-1/2″ pipe size	R-31	5	1.600	Ea.	.76	87.50	11	99.26	149
1840	3″ pipe size		5	1.600		.76	87.50	11	99.26	149
1860	3-1/2″ pipe size		3.80	2.105		.99	115	14.50	130.49	195
1880	4″ pipe size		3.30	2.424		.99	133	16.70	150.69	224
1900	16″ thick, 1/2″ pipe size		12.30	.650		.47	35.50	4.48	40.45	60.50
1920	3/4″ pipe size		12.30	.650		.47	35.50	4.48	40.45	60.50
1940	1″ pipe size		9.30	.860		.75	47	5.95	53.70	80.50
1960	1-1/4″ pipe size		9.30	.860		.75	47	5.95	53.70	80.50
1980	1-1/2″ pipe size		9.30	.860		.75	47	5.95	53.70	80.50
2000	2″ pipe size		4.40	1.818		1.02	99.50	12.55	113.07	169
2010	2-1/2″ pipe size		4.40	1.818		1.02	99.50	12.55	113.07	169
2030	3″ pipe size		4.40	1.818		1.02	99.50	12.55	113.07	169
2050	3-1/2″ pipe size		3.30	2.424		1.32	133	16.70	151.02	225
2070	4″ pipe size		3	2.667		1.32	146	18.35	165.67	247
2090	20″ thick, 1/2″ pipe size		10.70	.748		.59	41	5.15	46.74	70
2110	3/4″ pipe size		10.70	.748		.59	41	5.15	46.74	70
2130	1″ pipe size		8	1		.94	54.50	6.90	62.34	93
2150	1-1/4″ pipe size		8	1		.94	54.50	6.90	62.34	93
2170	1-1/2″ pipe size		8	1		.94	54.50	6.90	62.34	93
2190	2″ pipe size		4	2		1.27	109	13.80	124.07	186
2210	2-1/2″ pipe size		4	2		1.27	109	13.80	124.07	186
2230	3″ pipe size		4	2		1.27	109	13.80	124.07	186
2250	3-1/2″ pipe size		3	2.667		1.64	146	18.35	165.99	248
2270	4″ pipe size		2.70	2.963		1.64	162	20.50	184.14	275
2290	24″ thick, 1/2″ pipe size		9.40	.851		.71	46.50	5.85	53.06	79
2310	3/4″ pipe size		9.40	.851		.71	46.50	5.85	53.06	79
2330	1″ pipe size		7.10	1.127		1.13	61.50	7.75	70.38	105
2350	1-1/4″ pipe size		7.10	1.127		1.13	61.50	7.75	70.38	105
2370	1-1/2″ pipe size		7.10	1.127		1.13	61.50	7.75	70.38	105
2390	2″ pipe size		3.60	2.222		1.52	122	15.30	138.82	207
2410	2-1/2″ pipe size		3.60	2.222		1.52	122	15.30	138.82	207
2430	3″ pipe size		3.60	2.222		1.52	122	15.30	138.82	207
2450	3-1/2″ pipe size		2.80	2.857		1.97	156	19.70	177.67	266
2470	4″ pipe size		2.50	3.200		1.97	175	22	198.97	298
3000	Knockouts to 8′ high, metal boxes & enclosures									
3020	With hole saw, 1/2″ pipe size	1 Elec	53	.151	Ea.		8.25		8.25	12.80
3040	3/4″ pipe size		47	.170			9.30		9.30	14.40
3050	1″ pipe size		40	.200			10.95		10.95	16.95
3060	1-1/4″ pipe size		36	.222			12.15		12.15	18.80
3070	1-1/2″ pipe size		32	.250			13.70		13.70	21
3080	2″ pipe size		27	.296			16.20		16.20	25
3090	2-1/2″ pipe size		20	.400			22		22	34
4010	3″ pipe size		16	.500			27.50		27.50	42.50
4030	3-1/2″ pipe size		13	.615			33.50		33.50	52
4050	4″ pipe size		11	.727			40		40	61.50

26 05 36 – Cable Trays for Electrical Systems

26 05 36.10 Cable Tray Ladder Type

		Crew	Daily Output	Labor-Hours	Unit	Material	Labor	Equipment	Total	Total Incl O&P
0010	**CABLE TRAY LADDER TYPE** w/ftngs. & supports, 4″ dp., to 15′ elev.									
0100	For higher elevations, see Section 26 05 36.40									
0160	Galvanized steel tray									
0170	4″ rung spacing, 6″ wide	2 Elec	98	.163	L.F.	18.15	8.95		27.10	34
0180	9″ wide		92	.174		19.80	9.50		29.30	37

26 05 36 – Cable Trays for Electrical Systems

26 05 36.10 Cable Tray Ladder Type		Crew	Daily Output	Labor-Hours	Unit	Material	Labor	Equipment	Total	Total Incl O&P
						2015 Bare Costs				
0200	12″ wide	2 Elec	86	.186	L.F.	22	10.20		32.20	40
0400	18″ wide		82	.195		25.50	10.65		36.15	44.50
0600	24″ wide		78	.205		29	11.20		40.20	49.50
0650	30″ wide		68	.235		37	12.85		49.85	60.50
0700	36″ wide		60	.267		41	14.60		55.60	67.50
0800	6″ rung spacing, 6″ wide		100	.160		16.50	8.75		25.25	31.50
0850	9″ wide		94	.170		18.15	9.30		27.45	34.50
0860	12″ wide		88	.182		19.05	9.95		29	36.50
0870	18″ wide		84	.190		21.50	10.40		31.90	40
0880	24″ wide		80	.200		24.50	10.95		35.45	43.50
0890	30″ wide		70	.229		29	12.50		41.50	51.50
0900	36″ wide		64	.250		31.50	13.70		45.20	55.50
0910	9″ rung spacing, 6″ wide		102	.157		15.55	8.60		24.15	30.50
0920	9″ wide		98	.163		15.90	8.95		24.85	31.50
0930	12″ wide		94	.170		18.30	9.30		27.60	34.50
0940	18″ wide		90	.178		20.50	9.70		30.20	37.50
0950	24″ wide		86	.186		23.50	10.20		33.70	42
0960	30″ wide		80	.200		25.50	10.95		36.45	45.50
0970	36″ wide		74	.216		27.50	11.85		39.35	48.50
0980	12″ rung spacing, 6″ wide		106	.151		15.30	8.25		23.55	29.50
0990	9″ wide		104	.154		15.65	8.40		24.05	30.50
1000	12″ wide		100	.160		16.50	8.75		25.25	31.50
1010	18″ wide		96	.167		17.75	9.10		26.85	33.50
1020	24″ wide		94	.170		19.05	9.30		28.35	35.50
1030	30″ wide		88	.182		21.50	9.95		31.45	39.50
1040	36″ wide		84	.190		23	10.40		33.40	41
1041	18″ rung spacing, 6″ wide		108	.148		15.15	8.10		23.25	29.50
1042	9″ wide		106	.151		15.40	8.25		23.65	30
1043	12″ wide		102	.157		16.40	8.60		25	31.50
1044	18″ wide		98	.163		17.25	8.95		26.20	33
1045	24″ wide		96	.167		17.75	9.10		26.85	33.50
1046	30″ wide		90	.178		18.70	9.70		28.40	35.50
1047	36″ wide		86	.186		19.25	10.20		29.45	37
1050	Elbows horiz. 9″ rung spacing, 90°, 12″ radius, 6″ wide		9.60	1.667	Ea.	70.50	91		161.50	219
1060	9″ wide		8.40	1.905		75	104		179	244
1070	12″ wide		7.60	2.105		81.50	115		196.50	268
1080	18″ wide		6.20	2.581		107	141		248	335
1090	24″ wide		5.40	2.963		120	162		282	385
1100	30″ wide		4.80	3.333		158	182		340	455
1110	36″ wide		4.20	3.810		184	208		392	525
1120	90° 24″ radius, 6″ wide		9.20	1.739		146	95		241	305
1130	9″ wide		8	2		150	109		259	335
1140	12″ wide		7.20	2.222		154	122		276	360
1150	18″ wide		5.80	2.759		167	151		318	420
1160	24″ wide		5	3.200		180	175		355	470
1170	30″ wide		4.40	3.636		195	199		394	525
1180	36″ wide		3.80	4.211		231	230		461	610
1190	90° 36″ radius, 6″ wide		8.80	1.818		161	99.50		260.50	330
1200	9″ wide		7.60	2.105		169	115		284	365
1210	12″ wide		6.80	2.353		178	129		307	395
1220	18″ wide		5.40	2.963		203	162		365	475
1230	24″ wide		4.60	3.478		223	190		413	540
1240	30″ wide		4	4		268	219		487	635

26 05 36 – Cable Trays for Electrical Systems

26 05 36.10 Cable Tray Ladder Type		Crew	Daily Output	Labor-Hours	Unit	Material	2015 Bare Costs Labor	Equipment	Total	Total Incl O&P
1250	36" wide	2 Elec	3.40	4.706	Ea.	291	257		548	720
1260	45° 12" radius, 6" wide		13.20	1.212		54.50	66.50		121	163
1270	9" wide		11	1.455		56.50	79.50		136	186
1280	12" wide		9.60	1.667		59	91		150	206
1290	18" wide		7.60	2.105		62.50	115		177.50	247
1300	24" wide		6.20	2.581		80	141		221	305
1310	30" wide		5.40	2.963		102	162		264	365
1320	36" wide		4.60	3.478		107	190		297	410
1330	45° 24" radius, 6" wide		12.80	1.250		69	68.50		137.50	182
1340	9" wide		10.60	1.509		73.50	82.50		156	209
1350	12" wide		9.20	1.739		82.50	95		177.50	238
1360	18" wide		7.20	2.222		84.50	122		206.50	281
1370	24" wide		5.80	2.759		102	151		253	345
1380	30" wide		5	3.200		122	175		297	405
1390	36" wide		4.20	3.810		140	208		348	480
1400	45° 36" radius, 6" wide		12.40	1.290		99	70.50		169.50	218
1410	9" wide		10.20	1.569		105	86		191	248
1420	12" wide		8.80	1.818		107	99.50		206.50	271
1430	18" wide		6.80	2.353		116	129		245	325
1440	24" wide		5.40	2.963		122	162		284	385
1450	30" wide		4.60	3.478		158	190		348	470
1460	36" wide		3.80	4.211		176	230		406	550
1470	Elbows horizontal, 4" rung spacing, use 9" rung x 1.50									
1480	6" rung spacing use 9" rung x 1.20									
1490	12" rung spacing use 9" rung x .93									
1500	Elbows vert. 9" rung spacing, 90°, 12" radius, 6" wide	2 Elec	9.60	1.667	Ea.	97	91		188	248
1510	9" wide		8.40	1.905		99	104		203	270
1520	12" wide		7.60	2.105		102	115		217	291
1530	18" wide		6.20	2.581		107	141		248	335
1540	24" wide		5.40	2.963		111	162		273	375
1550	30" wide		4.80	3.333		120	182		302	415
1560	36" wide		4.20	3.810		124	208		332	460
1570	24" radius, 6" wide		9.20	1.739		137	95		232	298
1580	9" wide		8	2		141	109		250	325
1590	12" wide		7.20	2.222		146	122		268	350
1600	18" wide		5.80	2.759		154	151		305	405
1610	24" wide		5	3.200		163	175		338	450
1620	30" wide		4.40	3.636		169	199		368	495
1630	36" wide		3.80	4.211		191	230		421	565
1640	36" radius, 6" wide		8.80	1.818		184	99.50		283.50	355
1650	9" wide		7.60	2.105		191	115		306	390
1660	12" wide		6.80	2.353		193	129		322	410
1670	18" wide		5.40	2.963		210	162		372	480
1680	24" wide		4.60	3.478		212	190		402	530
1690	30" wide		4	4		240	219		459	605
1700	36" wide		3.40	4.706		257	257		514	685
1710	Elbows vertical, 4" rung spacing, use 9" rung x 1.25									
1720	6" rung spacing, use 9" rung x 1.15									
1730	12" rung spacing, use 9" rung x .90									
1740	Tee horizontal, 9" rung spacing, 12" radius, 6" wide	2 Elec	5	3.200	Ea.	150	175		325	435
1750	9" wide		4.60	3.478		161	190		351	470
1760	12" wide		4.40	3.636		169	199		368	495
1770	18" wide		4	4		193	219		412	550

26 05 36 – Cable Trays for Electrical Systems

26 05 36.10 Cable Tray Ladder Type		Crew	Daily Output	Labor-Hours	Unit	Material	2015 Bare Costs Labor	Equipment	Total	Total Incl O&P
1780	24" wide	2 Elec	3.60	4.444	Ea.	227	243		470	625
1790	30" wide		3.40	4.706		272	257		529	700
1800	36" wide		3	5.333		310	292		602	790
1810	24" radius, 6" wide		4.60	3.478		233	190		423	550
1820	9" wide		4.20	3.810		244	208		452	595
1830	12" wide		4	4		253	219		472	620
1840	18" wide		3.60	4.444		274	243		517	675
1850	24" wide		3.20	5		298	274		572	750
1860	30" wide		3	5.333		325	292		617	805
1870	36" wide		2.60	6.154		415	335		750	975
1880	36" radius, 6" wide		4.20	3.810		350	208		558	710
1890	9" wide		3.80	4.211		365	230		595	755
1900	12" wide		3.60	4.444		385	243		628	800
1910	18" wide		3.20	5		415	274		689	880
1920	24" wide		2.80	5.714		470	315		785	1,000
1930	30" wide		2.60	6.154		510	335		845	1,075
1940	36" wide		2.20	7.273		550	400		950	1,225
1980	Tee vertical, 9" rung spacing, 12" radius, 6" wide		5.40	2.963		255	162		417	530
1990	9" wide		5.20	3.077		257	168		425	545
2000	12" wide		5	3.200		259	175		434	555
2010	18" wide		4.60	3.478		261	190		451	580
2020	24" wide		4.40	3.636		268	199		467	605
2030	30" wide		4	4		285	219		504	655
2040	36" wide		3.60	4.444		295	243		538	700
2050	24" radius, 6" wide		5	3.200		460	175		635	775
2060	9" wide		4.80	3.333		465	182		647	790
2070	12" wide		4.60	3.478		470	190		660	815
2080	18" wide		4.20	3.810		485	208		693	860
2090	24" wide		4	4		500	219		719	890
2100	30" wide		3.60	4.444		510	243		753	940
2110	36" wide		3.20	5		535	274		809	1,025
2120	36" radius, 6" wide		4.60	3.478		855	190		1,045	1,225
2130	9" wide		4.40	3.636		870	199		1,069	1,275
2140	12" wide		4.20	3.810		880	208		1,088	1,300
2150	18" wide		3.80	4.211		890	230		1,120	1,325
2160	24" wide		3.60	4.444		900	243		1,143	1,375
2170	30" wide		3.20	5		935	274		1,209	1,450
2180	36" wide		2.80	5.714		945	315		1,260	1,525
2190	Tee, 4" rung spacing, use 9" rung x 1.30									
2200	6" rung spacing, use 9" rung x 1.20									
2210	12" rung spacing, use 9" rung x .90									
2220	Cross horizontal, 9" rung spacing, 12" radius, 6" wide	2 Elec	4	4	Ea.	185	219		404	545
2230	9" wide		3.80	4.211		194	230		424	570
2240	12" wide		3.60	4.444		212	243		455	610
2250	18" wide		3.40	4.706		231	257		488	655
2260	24" wide		3	5.333		257	292		549	735
2270	30" wide		2.80	5.714		310	315		625	825
2280	36" wide		2.60	6.154		345	335		680	900
2290	24" radius, 6" wide		3.60	4.444		360	243		603	770
2300	9" wide		3.40	4.706		370	257		627	805
2310	12" wide		3.20	5		380	274		654	845
2320	18" wide		3	5.333		405	292		697	895
2330	24" wide		2.60	6.154		430	335		765	995

26 05 36 – Cable Trays for Electrical Systems

26 05 36.10 Cable Tray Ladder Type		Crew	Daily Output	Labor-Hours	Unit	Material	2015 Bare Costs Labor	Equipment	Total	Total Incl O&P
2340	30″ wide	2 Elec	2.40	6.667	Ea.	460	365		825	1,075
2350	36″ wide		2.20	7.273		535	400		935	1,200
2360	36″ radius, 6″ wide		3.20	5		500	274		774	975
2370	9″ wide		3	5.333		520	292		812	1,025
2380	12″ wide		2.80	5.714		545	315		860	1,075
2390	18″ wide		2.60	6.154		600	335		935	1,175
2400	24″ wide		2.20	7.273		670	400		1,070	1,350
2410	30″ wide		2	8		725	440		1,165	1,475
2420	36″ wide		1.80	8.889		780	485		1,265	1,600
2430	Cross horizontal, 4″ rung spacing, use 9″ rung x 1.30									
2440	6″ rung spacing, use 9″ rung x 1.20									
2450	12″ rung spacing, use 9″ rung x .90									
2460	Reducer, 9″ to 6″ wide tray	2 Elec	13	1.231	Ea.	103	67.50		170.50	217
2470	12″ to 9″ wide tray		12	1.333		105	73		178	228
2480	18″ to 12″ wide tray		10.40	1.538		105	84		189	245
2490	24″ to 18″ wide tray		9	1.778		107	97		204	269
2500	30″ to 24″ wide tray		8	2		109	109		218	289
2510	36″ to 30″ wide tray		7	2.286		122	125		247	330
2511	Reducer, 18″ to 6″ wide tray		10.40	1.538		109	84		193	250
2512	24″ to 12″ wide tray		9	1.778		111	97		208	273
2513	30″ to 18″ wide tray		8	2		116	109		225	296
2514	30″ to 12″ wide tray		8	2		120	109		229	300
2515	36″ to 24″ wide tray		7	2.286		124	125		249	330
2516	36″ to 18″ wide tray		7	2.286		124	125		249	330
2517	36″ to 12″ wide tray		7	2.286		124	125		249	330
2520	Dropout or end plate, 6″ wide		32	.500		11.45	27.50		38.95	55
2530	9″ wide		28	.571		13.50	31.50		45	63.50
2540	12″ wide		26	.615		14.75	33.50		48.25	68.50
2550	18″ wide		22	.727		17.35	40		57.35	80.50
2560	24″ wide		20	.800		21	44		65	91
2570	30″ wide		18	.889		23	48.50		71.50	101
2580	36″ wide		16	1		26.50	54.50		81	114
2590	Tray connector		48	.333		20.50	18.25		38.75	50.50
3200	Aluminum tray, 4″ deep, 6″ rung spacing, 6″ wide		134	.119	L.F.	18	6.55		24.55	30
3210	9″ wide		128	.125		19.10	6.85		25.95	31.50
3220	12″ wide		124	.129		20	7.05		27.05	33
3230	18″ wide		114	.140		22.50	7.70		30.20	36.50
3240	24″ wide		106	.151		26	8.25		34.25	41.50
3250	30″ wide		100	.160		29	8.75		37.75	45
3260	36″ wide		94	.170		31	9.30		40.30	49
3270	9″ rung spacing, 6″ wide		140	.114		14.65	6.25		20.90	26
3280	9″ wide		134	.119		15.35	6.55		21.90	27
3290	12″ wide		130	.123		15.60	6.75		22.35	27.50
3300	18″ wide		122	.131		18.20	7.15		25.35	31
3310	24″ wide		116	.138		21	7.55		28.55	35
3320	30″ wide		108	.148		23.50	8.10		31.60	38.50
3330	36″ wide		100	.160		25	8.75		33.75	41
3340	12″ rung spacing, 6″ wide		146	.110		15.50	6		21.50	26.50
3350	9″ wide		140	.114		15.70	6.25		21.95	27
3360	12″ wide		134	.119		16.20	6.55		22.75	28
3370	18″ wide		128	.125		17.15	6.85		24	29.50
3380	24″ wide		124	.129		19.10	7.05		26.15	32
3390	30″ wide		114	.140		20.50	7.70		28.20	34.50

26 05 36 – Cable Trays for Electrical Systems

26 05 36.10 Cable Tray Ladder Type		Crew	Daily Output	Labor-Hours	Unit	Material	Labor	Equipment	Total	Total Incl O&P
						2015 Bare Costs				
3400	36" wide	2 Elec	106	.151	L.F.	21.50	8.25		29.75	36.50
3401	18" rung spacing, 6" wide		150	.107		15.40	5.85		21.25	26
3402	9" wide tray		144	.111		15.60	6.10		21.70	26.50
3403	12" wide tray		140	.114		16.20	6.25		22.45	27.50
3404	18" wide tray		134	.119		17.05	6.55		23.60	29
3405	24" wide tray		130	.123		18.35	6.75		25.10	30.50
3406	30" wide tray		120	.133		19.05	7.30		26.35	32.50
3407	36" wide tray		110	.145		19.90	7.95		27.85	34.50
3410	Elbows horiz. 9" rung spacing, 90° 12" radius, 6" wide		9.60	1.667	Ea.	69.50	91		160.50	217
3420	9" wide		8.40	1.905		71	104		175	240
3430	12" wide		7.60	2.105		84.50	115		199.50	271
3440	18" wide		6.20	2.581		95	141		236	325
3450	24" wide		5.40	2.963		118	162		280	380
3460	30" wide		4.80	3.333		125	182		307	420
3470	36" wide		4.20	3.810		150	208		358	490
3480	24" radius, 6" wide		9.20	1.739		123	95		218	283
3490	9" wide		8	2		128	109		237	310
3500	12" wide		7.20	2.222		134	122		256	335
3510	18" wide		5.80	2.759		146	151		297	395
3520	24" wide		5	3.200		161	175		336	450
3530	30" wide		4.40	3.636		174	199		373	500
3540	36" wide		3.80	4.211		198	230		428	575
3550	90°, 36" radius, 6" wide		8.80	1.818		135	99.50		234.50	305
3560	9" wide		7.60	2.105		139	115		254	330
3570	12" wide		6.80	2.353		155	129		284	370
3580	18" wide		5.40	2.963		174	162		336	440
3590	24" wide		4.60	3.478		196	190		386	510
3600	30" wide		4	4		216	219		435	580
3610	36" wide		3.40	4.706		248	257		505	675
3620	45° 12" radius, 6" wide		13.20	1.212		45.50	66.50		112	153
3630	9" wide		11	1.455		47.50	79.50		127	175
3640	12" wide		9.60	1.667		49	91		140	195
3650	18" wide		7.60	2.105		58	115		173	242
3660	24" wide		6.20	2.581		72	141		213	298
3670	30" wide		5.40	2.963		83	162		245	340
3680	36" wide		4.60	3.478		86	190		276	390
3690	45° 24" radius, 6" wide		12.80	1.250		61	68.50		129.50	173
3700	9" wide		10.60	1.509		69.50	82.50		152	204
3710	12" wide		9.20	1.739		71	95		166	225
3720	18" wide		7.20	2.222		76	122		198	272
3730	24" wide		5.80	2.759		88	151		239	330
3740	30" wide		5	3.200		105	175		280	385
3750	36" wide		4.20	3.810		112	208		320	450
3760	45° 36" radius, 6" wide		12.40	1.290		83	70.50		153.50	200
3770	9" wide		10.20	1.569		84.50	86		170.50	226
3780	12" wide		8.80	1.818		86	99.50		185.50	249
3790	18" wide		6.80	2.353		94.50	129		223.50	305
3800	24" wide		5.40	2.963		117	162		279	380
3810	30" wide		4.60	3.478		123	190		313	430
3820	36" wide		3.80	4.211		135	230		365	505
3830	Elbows horizontal, 4" rung spacing, use 9" rung x 1.50									
3840	6" rung spacing, use 9" rung x 1.20									
3850	12" rung spacing, use 9" rung x .93									

26 05 36.10 Cable Tray Ladder Type		Crew	Daily Output	Labor-Hours	Unit	Material	Labor	Equipment	Total	Total Incl O&P
						2015 Bare Costs				
3860	Elbows vertical 9" rung spacing, 90° 12" radius, 6" wide	2 Elec	9.60	1.667	Ea.	96.50	91		187.50	247
3870	9" wide		8.40	1.905		101	104		205	273
3880	12" wide		7.60	2.105		103	115		218	291
3890	18" wide		6.20	2.581		106	141		247	335
3900	24" wide		5.40	2.963		113	162		275	375
3910	30" wide		4.80	3.333		118	182		300	410
3920	36" wide		4.20	3.810		122	208		330	460
3930	24" radius, 6" wide		9.20	1.739		105	95		200	262
3940	9" wide		8	2		108	109		217	288
3950	12" wide		7.20	2.222		112	122		234	310
3960	18" wide		5.80	2.759		118	151		269	365
3970	24" wide		5	3.200		125	175		300	410
3980	30" wide		4.40	3.636		132	199		331	455
3990	36" wide		3.80	4.211		140	230		370	510
4000	36" radius, 6" wide		8.80	1.818		162	99.50		261.50	330
4010	9" wide		7.60	2.105		166	115		281	360
4020	12" wide		6.80	2.353		176	129		305	390
4030	18" wide		5.40	2.963		183	162		345	450
4040	24" wide		4.60	3.478		189	190		379	505
4050	30" wide		4	4		205	219		424	565
4060	36" wide	↓	3.40	4.706	↓	213	257		470	635
4070	Elbows vertical, 4" rung spacing, use 9" rung x 1.25									
4080	6" rung spacing, use 9" rung x 1.15									
4090	12" rung spacing, use 9" rung x .90									
4100	Tee horizontal 9" rung spacing, 12" radius, 6" wide	2 Elec	5	3.200	Ea.	107	175		282	390
4110	9" wide		4.60	3.478		115	190		305	420
4120	12" wide		4.40	3.636		127	199		326	450
4130	18" wide		4.20	3.810		139	208		347	475
4140	24" wide		4	4		152	219		371	505
4150	30" wide		3.60	4.444		177	243		420	570
4160	36" wide		3.40	4.706		213	257		470	635
4170	24" radius, 6" wide		4.60	3.478		183	190		373	495
4180	9" wide		4.20	3.810		193	208		401	535
4190	12" wide		4	4		198	219		417	560
4200	18" wide		3.80	4.211		216	230		446	595
4210	24" wide		3.60	4.444		237	243		480	635
4220	30" wide		3.20	5		257	274		531	710
4230	36" wide		3	5.333		395	292		687	885
4240	36" radius, 6" wide		4.20	3.810		259	208		467	610
4250	9" wide		3.80	4.211		265	230		495	645
4260	12" wide		3.60	4.444		291	243		534	695
4270	18" wide		3.40	4.706		305	257		562	735
4280	24" wide		3.20	5		340	274		614	795
4290	30" wide		2.80	5.714		430	315		745	955
4300	36" wide		2.60	6.154		475	335		810	1,050
4310	Tee vertical 9" rung spacing, 12" radius, 6" wide		5.40	2.963		240	162		402	515
4320	9" wide		5.20	3.077		248	168		416	535
4330	12" wide		5	3.200		252	175		427	550
4340	18" wide		4.60	3.478		255	190		445	575
4350	24" wide		4.40	3.636		275	199		474	615
4360	30" wide		4.20	3.810		281	208		489	635
4370	36" wide		4	4		291	219		510	660
4380	24" radius, 6" wide	↓	5	3.200	↓	405	175		580	720

26 05 36 – Cable Trays for Electrical Systems

26 05 36.10 Cable Tray Ladder Type		Crew	Daily Output	Labor-Hours	Unit	Material	Labor (2015 Bare Costs)	Equipment (2015 Bare Costs)	Total (2015 Bare Costs)	Total Incl O&P
4390	9" wide	2 Elec	4.80	3.333	Ea.	410	182		592	735
4400	12" wide		4.60	3.478		420	190		610	755
4410	18" wide		4.20	3.810		430	208		638	795
4420	24" wide		4	4		440	219		659	825
4430	30" wide		3.80	4.211		455	230		685	855
4440	36" wide		3.60	4.444		490	243		733	915
4450	36" radius, 6" wide		4.60	3.478		895	190		1,085	1,275
4460	9" wide		4.40	3.636		915	199		1,114	1,300
4470	12" wide		4.20	3.810		950	208		1,158	1,375
4480	18" wide		3.80	4.211		960	230		1,190	1,400
4490	24" wide		3.60	4.444		985	243		1,228	1,450
4500	30" wide		3.40	4.706		995	257		1,252	1,500
4510	36" wide	↓	3.20	5	↓	1,000	274		1,274	1,525
4520	Tees, 4" rung spacing, use 9" rung x 1.30									
4530	6" rung spacing, use 9" rung x 1.20									
4540	12" rung spacing, use 9" rung x .90									
4550	Cross horizontal 9" rung spacing, 12" radius, 6" wide	2 Elec	4.40	3.636	Ea.	149	199		348	475
4560	9" wide		4.20	3.810		166	208		374	505
4570	12" wide		4	4		184	219		403	545
4580	18" wide		3.60	4.444		203	243		446	600
4590	24" wide		3.40	4.706		210	257		467	630
4600	30" wide		3	5.333		247	292		539	720
4610	36" wide		2.80	5.714		310	315		625	825
4620	24" radius, 6" wide		4	4		286	219		505	655
4630	9" wide		3.80	4.211		296	230		526	680
4640	12" wide		3.60	4.444		305	243		548	710
4650	18" wide		3.20	5		320	274		594	775
4660	24" wide		3	5.333		350	292		642	835
4670	30" wide		2.60	6.154		370	335		705	930
4680	36" wide		2.40	6.667		420	365		785	1,025
4690	36" radius, 6" wide		3.60	4.444		340	243		583	745
4700	9" wide		3.40	4.706		350	257		607	785
4710	12" wide		3.20	5		370	274		644	835
4720	18" wide		2.80	5.714		395	315		710	920
4730	24" wide		2.60	6.154		490	335		825	1,050
4740	30" wide		2.20	7.273		545	400		945	1,225
4750	36" wide	↓	2	8	↓	605	440		1,045	1,350
4760	Cross horizontal, 4" rung spacing, use 9" rung x 1.30									
4770	6" rung spacing, use 9" rung x 1.20									
4780	12" rung spacing, use 9" rung x .90									
4790	Reducer, 9" to 6" wide tray	2 Elec	16	1	Ea.	79.50	54.50		134	172
4800	12" to 9" wide tray		14	1.143		83	62.50		145.50	188
4810	18" to 12" wide tray		12.40	1.290		83	70.50		153.50	200
4820	24" to 18" wide tray		10.60	1.509		84.50	82.50		167	221
4830	30" to 24" wide tray		9.20	1.739		86	95		181	242
4840	36" to 30" wide tray		8	2		91.50	109		200.50	269
4841	Reducer, 18" to 6" wide tray		12.40	1.290		83	70.50		153.50	200
4842	24" to 12" wide tray		10.60	1.509		84.50	82.50		167	221
4843	30" to 18" wide tray		9.20	1.739		86	95		181	242
4844	30" to 12" wide tray		9.20	1.739		86	95		181	242
4845	36" to 24" wide tray		8	2		91.50	109		200.50	269
4846	36" to 18" wide tray		8	2		91.50	109		200.50	269
4847	36" to 12" wide tray	↓	8	2	↓	91.50	109		200.50	269

26 05 36 – Cable Trays for Electrical Systems

26 05 36.10 Cable Tray Ladder Type		Crew	Daily Output	Labor-Hours	Unit	2015 Bare Costs Material	Labor	Equipment	Total	Total Incl O&P
4850	Dropout or end plate, 6" wide	2 Elec	32	.500	Ea.	10.45	27.50		37.95	54
4860	9" wide tray		28	.571		11.15	31.50		42.65	61
4870	12" wide tray		26	.615		12.15	33.50		45.65	65.50
4880	18" wide tray		22	.727		16.25	40		56.25	79.50
4890	24" wide tray		20	.800		19.10	44		63.10	89
4900	30" wide tray		18	.889		22.50	48.50		71	101
4910	36" wide tray		16	1		25	54.50		79.50	112
4920	Tray connector		48	.333		15.20	18.25		33.45	45
9200	Splice plate	1 Elec	48	.167	Pr.	15.85	9.10		24.95	31.50
9210	Expansion joint		48	.167		19.10	9.10		28.20	35
9220	Horizontal hinged		48	.167		15.85	9.10		24.95	31.50
9230	Vertical hinged		48	.167		19.10	9.10		28.20	35
9240	Ladder hanger, vertical		28	.286	Ea.	4.73	15.65		20.38	29
9250	Ladder to channel connector		24	.333		55.50	18.25		73.75	89
9260	Ladder to box connector 30" wide		19	.421		55.50	23		78.50	96.50
9270	24" wide		20	.400		55.50	22		77.50	95
9280	18" wide		21	.381		53.50	21		74.50	91.50
9290	12" wide		22	.364		50	19.90		69.90	86
9300	9" wide		23	.348		45	19.05		64.05	79
9310	6" wide		24	.333		40	18.25		58.25	72
9320	Ladder floor flange		24	.333		36	18.25		54.25	67.50
9330	Cable roller for tray 30" wide		10	.800		295	44		339	395
9340	24" wide		11	.727		251	40		291	340
9350	18" wide		12	.667		239	36.50		275.50	320
9360	12" wide		13	.615		188	33.50		221.50	259
9370	9" wide		14	.571		167	31.50		198.50	232
9380	6" wide		15	.533		135	29		164	194
9390	Pulley, single wheel		12	.667		310	36.50		346.50	395
9400	Triple wheel		10	.800		615	44		659	745
9440	Nylon cable tie, 14" long		80	.100		.46	5.45		5.91	8.95
9450	Ladder, hold down clamp		60	.133		10.15	7.30		17.45	22.50
9460	Cable clamp		60	.133		10.30	7.30		17.60	22.50
9470	Wall bracket, 30" wide tray		19	.421		54.50	23		77.50	95.50
9480	24" wide tray		20	.400		42.50	22		64.50	80.50
9490	18" wide tray		21	.381		38	21		59	74
9500	12" wide tray		22	.364		35.50	19.90		55.40	70
9510	9" wide tray		23	.348		31	19.05		50.05	64
9520	6" wide tray		24	.333		30.50	18.25		48.75	61.50

26 05 36.20 Cable Tray Solid Bottom

		Crew	Daily Output	Labor-Hours	Unit	Material	Labor	Equipment	Total	Total Incl O&P
0010	**CABLE TRAY SOLID BOTTOM** w/ftngs. & supports, 3" dp, to 15' high									
0200	For higher elevations, see Section 26 05 36.40									
0220	Galvanized steel, tray, 6" wide	2 Elec	120	.133	L.F.	14.65	7.30		21.95	27.50
0240	12" wide		100	.160		18.80	8.75		27.55	34
0260	18" wide		70	.229		23	12.50		35.50	44.50
0280	24" wide		60	.267		27	14.60		41.60	52
0300	30" wide		50	.320		32	17.50		49.50	62
0320	36" wide		44	.364		45.50	19.90		65.40	81
0340	Elbow horizontal 90°, 12" radius, 6" wide		9.60	1.667	Ea.	111	91		202	263
0360	12" wide		6.80	2.353		125	129		254	335
0370	18" wide		5.40	2.963		143	162		305	410
0380	24" wide		4.40	3.636		173	199		372	500
0390	30" wide		3.80	4.211		208	230		438	585

26 05 36 – Cable Trays for Electrical Systems

26 05 36.20 Cable Tray Solid Bottom		Crew	Daily Output	Labor-Hours	Unit	Material	Labor	Equipment	Total	Total Incl O&P
						2015 Bare Costs				
0400	36″ wide	2 Elec	3.40	4.706	Ea.	235	257		492	660
0420	24″ radius, 6″ wide		9.20	1.739		158	95		253	320
0440	12″ wide		6.40	2.500		184	137		321	415
0450	18″ wide		5	3.200		214	175		389	505
0460	24″ wide		4	4		248	219		467	615
0470	30″ wide		3.40	4.706		291	257		548	720
0480	36″ wide		3	5.333		325	292		617	810
0500	36″ radius, 6″ wide		8.80	1.818		231	99.50		330.50	410
0520	12″ wide		6	2.667		261	146		407	515
0530	18″ wide		4.60	3.478		320	190		510	645
0540	24″ wide		3.60	4.444		345	243		588	755
0550	30″ wide		3	5.333		415	292		707	910
0560	36″ wide		2.60	6.154		470	335		805	1,025
0580	Elbow vertical 90°, 12″ radius, 6″ wide		9.60	1.667		135	91		226	289
0600	12″ wide		6.80	2.353		146	129		275	360
0610	18″ wide		5.40	2.963		158	162		320	425
0620	24″ wide		4.40	3.636		169	199		368	495
0630	30″ wide		3.80	4.211		180	230		410	555
0640	36″ wide		3.40	4.706		186	257		443	605
0670	24″ radius, 6″ wide		9.20	1.739		191	95		286	355
0690	12″ wide		6.40	2.500		210	137		347	445
0700	18″ wide		5	3.200		227	175		402	520
0710	24″ wide		4	4		246	219		465	610
0720	30″ wide		3.40	4.706		261	257		518	685
0730	36″ wide		3	5.333		285	292		577	765
0750	36″ radius, 6″ wide		8.80	1.818		261	99.50		360.50	440
0770	12″ wide		6.60	2.424		291	133		424	525
0780	18″ wide		4.60	3.478		320	190		510	645
0790	24″ wide		3.60	4.444		350	243		593	760
0800	30″ wide		3	5.333		380	292		672	870
0810	36″ wide		2.60	6.154		415	335		750	975
0840	Tee horizontal, 12″ radius, 6″ wide		5	3.200		156	175		331	445
0860	12″ wide		4	4		169	219		388	525
0870	18″ wide		3.40	4.706		201	257		458	620
0880	24″ wide		2.80	5.714		225	315		540	730
0890	30″ wide		2.60	6.154		257	335		592	805
0900	36″ wide		2.20	7.273		293	400		693	940
0940	24″ radius, 6″ wide		4.60	3.478		242	190		432	560
0960	12″ wide		3.60	4.444		283	243		526	685
0970	18″ wide		3	5.333		315	292		607	795
0980	24″ wide		2.40	6.667		425	365		790	1,025
0990	30″ wide		2.20	7.273		460	400		860	1,125
1000	36″ wide		1.80	8.889		500	485		985	1,300
1020	36″ radius, 6″ wide		4.20	3.810		380	208		588	745
1040	12″ wide		3.20	5		425	274		699	895
1050	18″ wide		2.60	6.154		460	335		795	1,025
1060	24″ wide		2.20	7.273		595	400		995	1,275
1070	30″ wide		2	8		645	440		1,085	1,400
1080	36″ wide		1.60	10		670	545		1,215	1,575
1100	Tee vertical, 12″ radius, 6″ wide		5	3.200		257	175		432	555
1120	12″ wide		4	4		268	219		487	635
1130	18″ wide		3.60	4.444		272	243		515	675
1140	24″ wide		3.40	4.706		298	257		555	725

26 05 36 – Cable Trays for Electrical Systems

26 05 36.20 Cable Tray Solid Bottom		Crew	Daily Output	Labor-Hours	Unit	2015 Bare Costs Material	Labor	Equipment	Total	Total Incl O&P
1150	30″ wide	2 Elec	3	5.333	Ea.	320	292		612	800
1160	36″ wide		2.60	6.154		330	335		665	885
1180	24″ radius, 6″ wide		4.60	3.478		380	190		570	710
1200	12″ wide		3.60	4.444		395	243		638	810
1210	18″ wide		3.20	5		415	274		689	880
1220	24″ wide		3	5.333		435	292		727	930
1230	30″ wide		2.60	6.154		480	335		815	1,050
1240	36″ wide		2.20	7.273		500	400		900	1,175
1260	36″ radius, 6″ wide		4.20	3.810		590	208		798	975
1280	12″ wide		3.20	5		600	274		874	1,075
1290	18″ wide		2.80	5.714		645	315		960	1,200
1300	24″ wide		2.60	6.154		670	335		1,005	1,250
1310	30″ wide		2.20	7.273		780	400		1,180	1,475
1320	36″ wide		2	8		825	440		1,265	1,575
1340	Cross horizontal, 12″ radius, 6″ wide		4	4		186	219		405	545
1360	12″ wide		3.40	4.706		203	257		460	625
1370	18″ wide		2.80	5.714		238	315		553	745
1380	24″ wide		2.40	6.667		268	365		633	860
1390	30″ wide		2	8		298	440		738	1,000
1400	36″ wide		1.80	8.889		325	485		810	1,125
1420	24″ radius, 6″ wide		3.60	4.444		340	243		583	750
1440	12″ wide		3	5.333		380	292		672	870
1450	18″ wide		2.40	6.667		425	365		790	1,025
1460	24″ wide		2	8		545	440		985	1,275
1470	30″ wide		1.80	8.889		590	485		1,075	1,400
1480	36″ wide		1.60	10		625	545		1,170	1,525
1500	36″ radius, 6″ wide		3.20	5		555	274		829	1,025
1520	12″ wide		2.60	6.154		610	335		945	1,200
1530	18″ wide		2	8		670	440		1,110	1,425
1540	24″ wide		1.80	8.889		790	485		1,275	1,625
1550	30″ wide		1.60	10		855	545		1,400	1,775
1560	36″ wide		1.40	11.429		910	625		1,535	1,975
1580	Drop out or end plate, 6″ wide		32	.500		22.50	27.50		50	67.50
1600	12″ wide		26	.615		28.50	33.50		62	83
1610	18″ wide		22	.727		30.50	40		70.50	95
1620	24″ wide		20	.800		35.50	44		79.50	108
1630	30″ wide		18	.889		40	48.50		88.50	120
1640	36″ wide		16	1		43	54.50		97.50	132
1660	Reducer, 12″ to 6″ wide		12	1.333		109	73		182	233
1680	18″ to 12″ wide		10.60	1.509		111	82.50		193.50	250
1700	18″ to 6″ wide		10.60	1.509		111	82.50		193.50	250
1720	24″ to 18″ wide		9.20	1.739		116	95		211	274
1740	24″ to 12″ wide		9.20	1.739		116	95		211	274
1760	30″ to 24″ wide		8	2		124	109		233	305
1780	30″ to 18″ wide		8	2		124	109		233	305
1800	30″ to 12″ wide		8	2		124	109		233	305
1820	36″ to 30″ wide		7.20	2.222		128	122		250	330
1840	36″ to 24″ wide		7.20	2.222		128	122		250	330
1860	36″ to 18″ wide		7.20	2.222		128	122		250	330
1880	36″ to 12″ wide		7.20	2.222		128	122		250	330
2000	Aluminum tray, 6″ wide		150	.107	L.F.	14.05	5.85		19.90	24.50
2020	12″ wide		130	.123		18.35	6.75		25.10	30.50
2030	18″ wide		100	.160		23	8.75		31.75	39

26 05 36 – Cable Trays for Electrical Systems

26 05 36.20 Cable Tray Solid Bottom		Crew	Daily Output	Labor-Hours	Unit	Material	Labor	Equipment	Total	Total Incl O&P
						2015 Bare Costs				
2040	24" wide	2 Elec	90	.178	L.F.	29	9.70		38.70	47
2050	30" wide		70	.229		34	12.50		46.50	57
2060	36" wide		64	.250		43	13.70		56.70	68
2080	Elbow horizontal 90°, 12" radius, 6" wide		9.60	1.667	Ea.	113	91		204	265
2100	12" wide		7.60	2.105		127	115		242	315
2110	18" wide		6.80	2.353		155	129		284	370
2120	24" wide		5.80	2.759		177	151		328	430
2130	30" wide		5	3.200		216	175		391	510
2140	36" wide		4.40	3.636		243	199		442	580
2160	24" radius, 6" wide		9.20	1.739		161	95		256	325
2180	12" wide		7.20	2.222		189	122		311	395
2190	18" wide		6.40	2.500		216	137		353	450
2200	24" wide		5.40	2.963		267	162		429	545
2210	30" wide		4.60	3.478		305	190		495	630
2220	36" wide		4	4		350	219		569	725
2240	36" radius, 6" wide		8.80	1.818		237	99.50		336.50	415
2260	12" wide		6.80	2.353		284	129		413	510
2270	18" wide		6	2.667		340	146		486	595
2280	24" wide		5	3.200		360	175		535	670
2290	30" wide		4.20	3.810		410	208		618	775
2300	36" wide		3.60	4.444		480	243		723	905
2320	Elbow vertical 90°, 12" radius, 6" wide		9.60	1.667		134	91		225	288
2340	12" wide		7.60	2.105		139	115		254	330
2350	18" wide		6.80	2.353		157	129		286	370
2360	24" wide		5.80	2.759		166	151		317	415
2370	30" wide		5	3.200		176	175		351	465
2380	36" wide		4.40	3.636		181	199		380	510
2400	24" radius, 6" wide		9.20	1.739		189	95		284	355
2420	12" wide		7.20	2.222		206	122		328	415
2430	18" wide		6.40	2.500		221	137		358	455
2440	24" wide		5.40	2.963		240	162		402	515
2450	30" wide		4.60	3.478		255	190		445	575
2460	36" wide		4	4		281	219		500	650
2480	36" radius, 6" wide		8.80	1.818		250	99.50		349.50	430
2500	12" wide		6.80	2.353		281	129		410	510
2510	18" wide		6	2.667		305	146		451	560
2520	24" wide		5	3.200		320	175		495	620
2530	30" wide		4.20	3.810		350	208		558	710
2540	36" wide		3.60	4.444		365	243		608	775
2560	Tee horizontal, 12" radius, 6" wide		5	3.200		174	175		349	460
2580	12" wide		4.40	3.636		211	199		410	540
2590	18" wide		4	4		237	219		456	600
2600	24" wide		3.60	4.444		265	243		508	665
2610	30" wide		3	5.333		305	292		597	785
2620	36" wide		2.40	6.667		350	365		715	950
2640	24" radius, 6" wide		4.60	3.478		287	190		477	610
2660	12" wide		4	4		325	219		544	695
2670	18" wide		3.60	4.444		365	243		608	775
2680	24" wide		3	5.333		450	292		742	945
2690	30" wide		2.40	6.667		510	365		875	1,125
2700	36" wide		2.20	7.273		570	400		970	1,250
2720	36" radius, 6" wide		4.20	3.810		465	208		673	840
2740	12" wide		3.60	4.444		510	243		753	935

26 05 36.20 Cable Tray Solid Bottom		Crew	Daily Output	Labor-Hours	Unit	2015 Bare Costs Material	Labor	Equipment	Total	Total Incl O&P
2750	18" wide	2 Elec	3.20	5	Ea.	590	274		864	1,075
2760	24" wide		2.60	6.154		685	335		1,020	1,275
2770	30" wide		2	8		755	440		1,195	1,500
2780	36" wide		1.80	8.889		845	485		1,330	1,675
2800	Tee vertical, 12" radius, 6" wide		5	3.200		248	175		423	545
2820	12" wide		4.40	3.636		254	199		453	590
2830	18" wide		4.20	3.810		264	208		472	615
2840	24" wide		4	4		281	219		500	650
2850	30" wide		3.60	4.444		305	243		548	710
2860	36" wide		3	5.333		325	292		617	805
2880	24" radius, 6" wide		4.60	3.478		365	190		555	695
2900	12" wide		4	4		385	219		604	765
2910	18" wide		3.80	4.211		410	230		640	810
2920	24" wide		3.60	4.444		450	243		693	870
2930	30" wide		3.20	5		490	274		764	965
2940	36" wide		2.60	6.154		520	335		855	1,100
2960	36" radius, 6" wide		4.20	3.810		570	208		778	955
2980	12" wide		3.40	4.706		600	257		857	1,050
2990	18" wide		3.40	4.706		630	257		887	1,100
3000	24" wide		3.20	5		650	274		924	1,150
3010	30" wide		2.80	5.714		705	315		1,020	1,250
3020	36" wide		2.20	7.273		775	400		1,175	1,475
3040	Cross horizontal, 12" radius, 6" wide		4.40	3.636		221	199		420	555
3060	12" wide		4	4		250	219		469	615
3070	18" wide		3.40	4.706		284	257		541	710
3080	24" wide		2.80	5.714		325	315		640	840
3090	30" wide		2.60	6.154		365	335		700	920
3100	36" wide		2.20	7.273		405	400		805	1,050
3120	24" radius, 6" wide		4	4		385	219		604	765
3140	12" wide		3.60	4.444		450	243		693	870
3150	18" wide		3	5.333		485	292		777	985
3160	24" wide		2.40	6.667		590	365		955	1,200
3170	30" wide		2.20	7.273		650	400		1,050	1,325
3180	36" wide		1.80	8.889		675	485		1,160	1,500
3200	36" radius, 6" wide		3.60	4.444		675	243		918	1,125
3220	12" wide		3.20	5		710	274		984	1,200
3230	18" wide		2.60	6.154		790	335		1,125	1,400
3240	24" wide		2	8		860	440		1,300	1,625
3250	30" wide		1.80	8.889		930	485		1,415	1,775
3260	36" wide		1.60	10		1,100	545		1,645	2,050
3280	Dropout, or end plate, 6" wide		32	.500		22.50	27.50		50	67
3300	12" wide		26	.615		25	33.50		58.50	79.50
3310	18" wide		22	.727		30	40		70	94.50
3320	24" wide		20	.800		33	44		77	105
3330	30" wide		18	.889		38.50	48.50		87	118
3340	36" wide		16	1		45	54.50		99.50	134
3380	Reducer, 12" to 6" wide		14	1.143		109	62.50		171.50	217
3400	18" to 12" wide		12	1.333		113	73		186	237
3420	18" to 6" wide		12	1.333		113	73		186	237
3440	24" to 18" wide		10.60	1.509		119	82.50		201.50	259
3460	24" to 12" wide		10.60	1.509		119	82.50		201.50	259
3480	30" to 24" wide		9.20	1.739		130	95		225	290
3500	30" to 18" wide		9.20	1.739		130	95		225	290

26 05 36 – Cable Trays for Electrical Systems

26 05 36.20 Cable Tray Solid Bottom		Crew	Daily Output	Labor-Hours	Unit	Material	2015 Bare Costs Labor	Equipment	Total	Total Incl O&P
3520	30″ to 12″ wide	2 Elec	9.20	1.739	Ea.	130	95		225	290
3540	36″ to 30″ wide		8	2		137	109		246	320
3560	36″ to 24″ wide		8	2		139	109		248	320
3580	36″ to 18″ wide		8	2		139	109		248	320
3600	36″ to 12″ wide	↓	8	2	↓	142	109		251	325

26 05 36.30 Cable Tray Trough		Crew	Daily Output	Labor-Hours	Unit	Material	Labor	Equipment	Total	Total Incl O&P
0010	**CABLE TRAY TROUGH** vented, w/ftngs. & supports, 6″ dp, to 15′ high									
0020	For higher elevations, see Section 26 05 36.40									
0200	Galvanized steel, tray, 6″ wide	2 Elec	90	.178	L.F.	16.20	9.70		25.90	33
0240	12″ wide		80	.200		16.80	10.95		27.75	35.50
0260	18″ wide		70	.229		20.50	12.50		33	42
0280	24″ wide		60	.267		35.50	14.60		50.10	61.50
0300	30″ wide		50	.320		48	17.50		65.50	79.50
0320	36″ wide		40	.400	↓	59.50	22		81.50	99.50
0340	Elbow horizontal 90°, 12″ radius, 6″ wide		7.60	2.105	Ea.	119	115		234	310
0360	12″ wide		5.60	2.857		141	156		297	395
0370	18″ wide		4.40	3.636		163	199		362	490
0380	24″ wide		3.60	4.444		201	243		444	595
0390	30″ wide		3.20	5		227	274		501	675
0400	36″ wide		2.80	5.714		263	315		578	775
0420	24″ radius, 6″ wide		7.20	2.222		176	122		298	380
0440	12″ wide		5.20	3.077		203	168		371	485
0450	18″ wide		4	4		244	219		463	610
0460	24″ wide		3.20	5		276	274		550	730
0470	30″ wide		2.80	5.714		320	315		635	835
0480	36″ wide		2.40	6.667		370	365		735	970
0500	36″ radius, 6″ wide		6.80	2.353		263	129		392	490
0520	12″ wide		4.80	3.333		300	182		482	610
0530	18″ wide		3.60	4.444		360	243		603	770
0540	24″ wide		2.80	5.714		380	315		695	905
0550	30″ wide		2.40	6.667		430	365		795	1,025
0560	36″ wide		2	8		475	440		915	1,200
0580	Elbow vertical 90°, 12″ radius, 6″ wide		7.60	2.105		154	115		269	350
0600	12″ wide		5.60	2.857		171	156		327	430
0610	18″ wide		4.40	3.636		173	199		372	500
0620	24″ wide		3.60	4.444		199	243		442	595
0630	30″ wide		3.20	5		203	274		477	650
0640	36″ wide		2.80	5.714		214	315		529	720
0660	24″ radius, 6″ wide		7.20	2.222		220	122		342	430
0680	12″ wide		5.20	3.077		233	168		401	520
0690	18″ wide		4	4		259	219		478	625
0700	24″ wide		3.20	5		268	274		542	720
0710	30″ wide		2.80	5.714		300	315		615	815
0720	36″ wide		2.40	6.667		315	365		680	910
0740	36″ radius, 6″ wide		6.80	2.353		298	129		427	525
0760	12″ wide		4.80	3.333		315	182		497	630
0770	18″ wide		3.60	4.444		350	243		593	760
0780	24″ wide		2.80	5.714		375	315		690	900
0790	30″ wide		2.40	6.667		385	365		750	990
0800	36″ wide		2	8		430	440		870	1,150
0820	Tee horizontal, 12″ radius, 6″ wide		4	4		180	219		399	540
0840	12″ wide	↓	3.20	5	↓	201	274		475	645

26 05 36 – Cable Trays for Electrical Systems

26 05 36.30 Cable Tray Trough		Crew	Daily Output	Labor-Hours	Unit	Material	Labor	Equipment	Total	Total Incl O&P
						2015 Bare Costs				
0850	18" wide	2 Elec	2.80	5.714	Ea.	227	315		542	735
0860	24" wide		2.40	6.667		263	365		628	855
0870	30" wide		2.20	7.273		300	400		700	945
0880	36" wide		2	8		330	440		770	1,050
0900	24" radius, 6" wide		3.60	4.444		298	243		541	700
0920	12" wide		2.80	5.714		325	315		640	845
0930	18" wide		2.40	6.667		365	365		730	965
0940	24" wide		2	8		470	440		910	1,200
0950	30" wide		1.80	8.889		510	485		995	1,325
0960	36" wide		1.60	10		565	545		1,110	1,475
0980	36" radius, 6" wide		3.20	5		430	274		704	895
1000	12" wide		2.40	6.667		490	365		855	1,100
1010	18" wide		2	8		535	440		975	1,275
1020	24" wide		1.60	10		640	545		1,185	1,550
1030	30" wide		1.40	11.429		680	625		1,305	1,725
1040	36" wide		1.20	13.333		800	730		1,530	2,000
1060	Tee vertical, 12" radius, 6" wide		4	4		300	219		519	670
1080	12" wide		3.20	5		305	274		579	760
1090	18" wide		3	5.333		315	292		607	795
1100	24" wide		2.80	5.714		330	315		645	850
1110	30" wide		2.60	6.154		350	335		685	905
1120	36" wide		2.20	7.273		355	400		755	1,000
1140	24" radius, 6" wide		3.60	4.444		400	243		643	820
1160	12" wide		2.80	5.714		425	315		740	950
1170	18" wide		2.60	6.154		445	335		780	1,000
1180	24" wide		2.40	6.667		475	365		840	1,100
1190	30" wide		2.20	7.273		500	400		900	1,175
1200	36" wide		1.80	8.889		545	485		1,030	1,350
1220	36" radius, 6" wide		3.20	5		640	274		914	1,125
1240	12" wide		2.40	6.667		655	365		1,020	1,300
1250	18" wide		2.20	7.273		680	400		1,080	1,375
1260	24" wide		2	8		720	440		1,160	1,475
1270	30" wide		1.80	8.889		835	485		1,320	1,675
1280	36" wide		1.40	11.429		845	625		1,470	1,900
1300	Cross horizontal, 12" radius, 6" wide		3.20	5		231	274		505	680
1320	12" wide		2.80	5.714		233	315		548	740
1330	18" wide		2.40	6.667		261	365		626	850
1340	24" wide		2	8		274	440		714	980
1350	30" wide		1.80	8.889		310	485		795	1,100
1360	36" wide		1.60	10		345	545		890	1,225
1380	24" radius, 6" wide		2.80	5.714		365	315		680	885
1400	12" wide		2.40	6.667		380	365		745	985
1410	18" wide		2	8		420	440		860	1,150
1420	24" wide		1.60	10		535	545		1,080	1,425
1430	30" wide		1.40	11.429		590	625		1,215	1,625
1440	36" wide		1.20	13.333		625	730		1,355	1,800
1460	36" radius, 6" wide		2.40	6.667		625	365		990	1,250
1480	12" wide		2	8		645	440		1,085	1,400
1490	18" wide		1.60	10		655	545		1,200	1,575
1500	24" wide		1.20	13.333		790	730		1,520	2,000
1510	30" wide		1	16		870	875		1,745	2,300
1520	36" wide		.80	20		935	1,100		2,035	2,725
1540	Dropout or end plate, 6" wide		26	.615		27	33.50		60.50	81.50

26 05 36 – Cable Trays for Electrical Systems

26 05 36.30 Cable Tray Trough		Crew	Daily Output	Labor-Hours	Unit	Material	Labor	Equipment	Total	Total Incl O&P
						2015 Bare Costs				
1560	12″ wide	2 Elec	22	.727	Ea.	33	40		73	97.50
1580	18″ wide		20	.800		36	44		80	108
1600	24″ wide		18	.889		41.50	48.50		90	121
1620	30″ wide		16	1		44.50	54.50		99	133
1640	36″ wide		13.40	1.194		49	65.50		114.50	155
1660	Reducer, 12″ to 6″ wide		9.40	1.702		112	93		205	268
1680	18″ to 12″ wide		8.40	1.905		117	104		221	290
1700	18″ to 6″ wide		8.40	1.905		117	104		221	290
1720	24″ to 18″ wide		7.20	2.222		125	122		247	325
1740	24″ to 12″ wide		7.20	2.222		125	122		247	325
1760	30″ to 24″ wide		6.40	2.500		129	137		266	355
1780	30″ to 18″ wide		6.40	2.500		129	137		266	355
1800	30″ to 12″ wide		6.40	2.500		129	137		266	355
1820	36″ to 30″ wide		5.80	2.759		138	151		289	385
1840	36″ to 24″ wide		5.80	2.759		138	151		289	385
1860	36″ to 18″ wide		5.80	2.759		140	151		291	390
1880	36″ to 12″ wide		5.80	2.759		140	151		291	390
2000	Aluminum, tray, vented, 6″ wide		120	.133	L.F.	18.35	7.30		25.65	31.50
2010	9″ wide		110	.145		21	7.95		28.95	35.50
2020	12″ wide		100	.160		23	8.75		31.75	39
2030	18″ wide		90	.178		27.50	9.70		37.20	45.50
2040	24″ wide		80	.200		33	10.95		43.95	53
2050	30″ wide		70	.229		43.50	12.50		56	67.50
2060	36″ wide		60	.267		48	14.60		62.60	75
2080	Elbow horiz. 90°, 12″ radius, 6″ wide		7.60	2.105	Ea.	121	115		236	310
2090	9″ wide		7	2.286		132	125		257	340
2100	12″ wide		6.20	2.581		140	141		281	375
2110	18″ wide		5.60	2.857		164	156		320	420
2120	24″ wide		4.60	3.478		189	190		379	505
2130	30″ wide		4	4		233	219		452	595
2140	36″ wide		3.60	4.444		255	243		498	655
2160	24″ radius, 6″ wide		7.20	2.222		176	122		298	380
2180	12″ wide		5.80	2.759		206	151		357	460
2190	18″ wide		5.20	3.077		237	168		405	520
2200	24″ wide		4.20	3.810		270	208		478	620
2210	30″ wide		3.60	4.444		310	243		553	715
2220	36″ wide		3.20	5		335	274		609	795
2240	36″ radius, 6″ wide		6.80	2.353		248	129		377	470
2260	12″ wide		5.40	2.963		282	162		444	560
2270	18″ wide		4.80	3.333		320	182		502	635
2280	24″ wide		3.80	4.211		350	230		580	740
2290	30″ wide		3.40	4.706		410	257		667	855
2300	36″ wide		2.80	5.714		450	315		765	980
2320	Elbow vertical 90°, 12″ radius, 6″ wide		7.60	2.105		147	115		262	340
2330	9″ wide		7	2.286		157	125		282	365
2340	12″ wide		6.20	2.581		159	141		300	395
2350	18″ wide		5.60	2.857		164	156		320	420
2360	24″ wide		4.60	3.478		181	190		371	495
2370	30″ wide		4	4		189	219		408	550
2380	36″ wide		3.60	4.444		191	243		434	585
2400	24″ radius, 6″ wide		7.20	2.222		199	122		321	405
2420	12″ wide		5.80	2.759		216	151		367	470
2430	18″ wide		5.20	3.077		235	168		403	520

26 05 36.30 Cable Tray Trough		Crew	Daily Output	Labor-Hours	Unit	2015 Bare Costs Material	Labor	Equipment	Total	Total Incl O&P
2440	24" wide	2 Elec	4.20	3.810	Ea.	238	208		446	585
2450	30" wide		3.60	4.444		252	243		495	650
2460	36" wide		3.20	5		264	274		538	715
2480	36" radius, 6" wide		6.80	2.353		247	129		376	470
2500	12" wide		5.40	2.963		264	162		426	540
2510	18" wide		4.80	3.333		296	182		478	605
2520	24" wide		3.80	4.211		320	230		550	705
2530	30" wide		3.40	4.706		350	257		607	785
2540	36" wide		2.80	5.714		360	315		675	885
2560	Tee horizontal, 12" radius, 6" wide		4	4		184	219		403	545
2570	9" wide		3.80	4.211		189	230		419	565
2580	12" wide		3.60	4.444		206	243		449	600
2590	18" wide		3.20	5		238	274		512	685
2600	24" wide		2.80	5.714		274	315		589	785
2610	30" wide		2.40	6.667		294	365		659	890
2620	36" wide		2.20	7.273		335	400		735	985
2640	24" radius, 6" wide		3.60	4.444		281	243		524	685
2660	12" wide		3.20	5		320	274		594	775
2670	18" wide		2.80	5.714		350	315		665	870
2680	24" wide		2.40	6.667		440	365		805	1,050
2690	30" wide		2	8		475	440		915	1,200
2700	36" wide		1.80	8.889		555	485		1,040	1,375
2720	36" radius, 6" wide		3.20	5		450	274		724	920
2740	12" wide		2.80	5.714		520	315		835	1,050
2750	18" wide		2.40	6.667		590	365		955	1,200
2760	24" wide		2	8		685	440		1,125	1,425
2770	30" wide		1.60	10		730	545		1,275	1,650
2780	36" wide		1.40	11.429		845	625		1,470	1,900
2800	Tee vertical, 12" radius, 6" wide		4	4		254	219		473	620
2810	9" wide		3.80	4.211		255	230		485	635
2820	12" wide		3.60	4.444		274	243		517	675
2830	18" wide		3.40	4.706		275	257		532	705
2840	24" wide		3.20	5		282	274		556	735
2850	30" wide		3	5.333		296	292		588	775
2860	36" wide		2.60	6.154		310	335		645	860
2880	24" radius, 6" wide		3.60	4.444		360	243		603	770
2900	12" wide		3.20	5		385	274		659	850
2910	18" wide		3	5.333		410	292		702	905
2920	24" wide		2.80	5.714		440	315		755	970
2930	30" wide		2.60	6.154		475	335		810	1,050
2940	36" wide		2.20	7.273		485	400		885	1,150
2960	36" radius, 6" wide		3.20	5		555	274		829	1,025
2980	12" wide		2.80	5.714		590	315		905	1,125
2990	18" wide		2.60	6.154		605	335		940	1,175
3000	24" wide		2.40	6.667		650	365		1,015	1,275
3010	30" wide		2.20	7.273		705	400		1,105	1,400
3020	36" wide		1.80	8.889		740	485		1,225	1,575
3040	Cross horizontal, 12" radius, 6" wide		3.60	4.444		230	243		473	630
3050	9" wide		3.40	4.706		247	257		504	670
3060	12" wide		3.20	5		254	274		528	705
3070	18" wide		2.80	5.714		265	315		580	775
3080	24" wide		2.40	6.667		296	365		661	890
3090	30" wide		2.20	7.273		360	400		760	1,000

26 05 36 – Cable Trays for Electrical Systems

26 05 36.30 Cable Tray Trough		Crew	Daily Output	Labor-Hours	Unit	2015 Bare Costs Material	Labor	Equipment	Total	Total Incl O&P
3100	36" wide	2 Elec	1.80	8.889	Ea.	435	485		920	1,225
3120	24" radius, 6" wide		3.20	5		410	274		684	880
3140	12" wide		2.80	5.714		450	315		765	980
3150	18" wide		2.40	6.667		485	365		850	1,100
3160	24" wide		2	8		565	440		1,005	1,300
3170	30" wide		1.80	8.889		605	485		1,090	1,425
3180	36" wide		1.40	11.429		685	625		1,310	1,725
3200	36" radius, 6" wide		2.80	5.714		685	315		1,000	1,250
3220	12" wide		2.40	6.667		720	365		1,085	1,350
3230	18" wide		2	8		785	440		1,225	1,550
3240	24" wide		1.60	10		905	545		1,450	1,850
3250	30" wide		1.40	11.429		985	625		1,610	2,050
3260	36" wide		1.20	13.333		1,150	730		1,880	2,375
3280	Dropout, or end plate, 6" wide		26	.615		23	33.50		56.50	77.50
3300	12" wide		22	.727		27	40		67	91.50
3310	18" wide		20	.800		33	44		77	104
3320	24" wide		18	.889		39.50	48.50		88	119
3330	30" wide		16	1		42.50	54.50		97	131
3340	36" wide		14	1.143		48.50	62.50		111	150
3370	Reducer, 9" to 6" wide		12	1.333		113	73		186	237
3380	12" to 6" wide		11.40	1.404		118	77		195	249
3390	12" to 9" wide		11.40	1.404		118	77		195	249
3400	18" to 12" wide		9.60	1.667		127	91		218	280
3420	18" to 6" wide		9.60	1.667		127	91		218	280
3430	18" to 9" wide		9.60	1.667		127	91		218	280
3440	24" to 18" wide		8.40	1.905		135	104		239	310
3460	24" to 12" wide		8.40	1.905		135	104		239	310
3470	24" to 9" wide		8.40	1.905		137	104		241	310
3475	24" to 6" wide		8.40	1.905		137	104		241	310
3480	30" to 24" wide		7.20	2.222		139	122		261	340
3500	30" to 18" wide		7.20	2.222		139	122		261	340
3520	30" to 12" wide		7.20	2.222		142	122		264	345
3540	36" to 30" wide		6.40	2.500		144	137		281	370
3560	36" to 24" wide		6.40	2.500		144	137		281	370
3580	36" to 18" wide		6.40	2.500		144	137		281	370
3600	36" to 12" wide		6.40	2.500		144	137		281	370
3610	Elbow horizontal 60°, 12" radius, 6" wide		7.80	2.051		98	112		210	282
3620	9" wide		7.20	2.222		107	122		229	305
3630	12" wide		6.40	2.500		118	137		255	340
3640	18" wide		5.80	2.759		127	151		278	375
3650	24" wide		4.80	3.333		156	182		338	455
3680	Elbow horizontal 45°, 12" radius, 6" wide		8	2		84.50	109		193.50	262
3690	9" wide		7.40	2.162		89.50	118		207.50	282
3700	12" wide		6.60	2.424		94.50	133		227.50	310
3710	18" wide		6	2.667		106	146		252	345
3720	24" wide		5	3.200		123	175		298	405
3750	Elbow horizontal, 30° 12" radius, 6" wide		8.20	1.951		74	107		181	246
3760	9" wide		7.60	2.105		77.50	115		192.50	263
3770	12" wide		6.80	2.353		83	129		212	290
3780	18" wide		6.20	2.581		89.50	141		230.50	320
3790	24" wide		5.20	3.077		98	168		266	370
3820	Elbow vertical 60° in/outside, 12" radius, 6" wide		7.80	2.051		122	112		234	310
3830	9" wide		7.20	2.222		123	122		245	325

26 05 36 – Cable Trays for Electrical Systems

26 05 36.30 Cable Tray Trough		Crew	Daily Output	Labor-Hours	Unit	Material	Labor	Equipment	Total	Total Incl O&P
						2015 Bare Costs				
3840	12" wide	2 Elec	6.40	2.500	Ea.	125	137		262	350
3850	18" wide		5.80	2.759		130	151		281	375
3860	24" wide		4.80	3.333		135	182		317	430
3890	Elbow vertical 45° in/outside, 12" radius, 6" wide		8	2		98	109		207	277
3900	9" wide		7.40	2.162		104	118		222	297
3910	12" wide		6.60	2.424		107	133		240	325
3920	18" wide		6	2.667		111	146		257	350
3930	24" wide		5	3.200		121	175		296	405
3960	Elbow vertical 30° in/outside, 12" radius, 6" wide		8.20	1.951		84.50	107		191.50	258
3970	9" wide		7.60	2.105		89.50	115		204.50	277
3980	12" wide		6.80	2.353		91.50	129		220.50	299
3990	18" wide		6.20	2.581		95	141		236	325
4000	24" wide		5.20	3.077		96.50	168		264.50	365
4250	Reducer, left or right hand, 24" to 18" wide		8.40	1.905		128	104		232	300
4260	24" to 12" wide		8.40	1.905		128	104		232	300
4270	24" to 9" wide		8.40	1.905		128	104		232	300
4280	24" to 6" wide		8.40	1.905		130	104		234	305
4290	18" to 12" wide		9.60	1.667		120	91		211	273
4300	18" to 9" wide		9.60	1.667		120	91		211	273
4310	18" to 6" wide		9.60	1.667		120	91		211	273
4320	12" to 9" wide		11.40	1.404		115	77		192	245
4330	12" to 6" wide		11.40	1.404		115	77		192	245
4340	9" to 6" wide		12	1.333		112	73		185	236
4350	Splice plate	1 Elec	48	.167		10.65	9.10		19.75	26
4360	Splice plate, expansion joint		48	.167		10.90	9.10		20	26
4370	Splice plate, hinged, horizontal		48	.167		8.80	9.10		17.90	24
4380	Vertical		48	.167		12.50	9.10		21.60	28
4390	Trough, hanger, vertical		28	.286		35	15.65		50.65	62.50
4400	Box connector, 24" wide		20	.400		47.50	22		69.50	86
4410	18" wide		21	.381		39.50	21		60.50	76
4420	12" wide		22	.364		38	19.90		57.90	72.50
4430	9" wide		23	.348		36	19.05		55.05	69
4440	6" wide		24	.333		34.50	18.25		52.75	65.50
4450	Floor flange		24	.333		36	18.25		54.25	67.50
4460	Hold down clamp		60	.133		4.13	7.30		11.43	15.85
4520	Wall bracket, 24" wide tray		20	.400		36.50	22		58.50	74
4530	18" wide tray		21	.381		35.50	21		56.50	71.50
4540	12" wide tray		22	.364		18.90	19.90		38.80	52
4550	9" wide tray		23	.348		16.90	19.05		35.95	48
4560	6" wide tray		24	.333		15.35	18.25		33.60	45
5000	Cable channel aluminum, vented, 1-1/4" deep, 4" wide, straight		80	.100	L.F.	13.35	5.45		18.80	23
5010	Elbow horizontal, 36" radius, 90°		5	1.600	Ea.	241	87.50		328.50	400
5020	60°		5.50	1.455		184	79.50		263.50	325
5030	45°		6	1.333		150	73		223	278
5040	30°		6.50	1.231		129	67.50		196.50	246
5050	Adjustable		6	1.333		123	73		196	249
5060	Elbow vertical, 36" radius, 90°		5	1.600		259	87.50		346.50	420
5070	60°		5.50	1.455		202	79.50		281.50	345
5080	45°		6	1.333		167	73		240	296
5090	30°		6.50	1.231		147	67.50		214.50	266
5100	Adjustable		6	1.333		123	73		196	249
5110	Splice plate, hinged, horizontal		48	.167		9.20	9.10		18.30	24.50
5120	Splice plate, hinged, vertical		48	.167		13.10	9.10		22.20	28.50

26 05 36 – Cable Trays for Electrical Systems

26 05 36.30 Cable Tray Trough		Crew	Daily Output	Labor-Hours	Unit	Material	2015 Bare Costs Labor	Equipment	Total	Total Incl O&P
5130	Hanger, vertical	1 Elec	28	.286	Ea.	14.40	15.65		30.05	40
5140	Single		28	.286		22	15.65		37.65	48
5150	Double		20	.400		22.50	22		44.50	58.50
5160	Channel to box connector		24	.333		29.50	18.25		47.75	60.50
5170	Hold down clip		80	.100		4.16	5.45		9.61	13.05
5180	Wall bracket, single		28	.286		11.95	15.65		27.60	37
5190	Double		20	.400		15.35	22		37.35	51
5200	Cable roller		16	.500		188	27.50		215.50	250
5210	Splice plate		48	.167		5.80	9.10		14.90	20.50

26 05 36.40 Cable Tray, Covers and Dividers		Crew	Daily Output	Labor-Hours	Unit	Material	Labor	Equipment	Total	Total Incl O&P
0010	**CABLE TRAY, COVERS AND DIVIDERS** To 15′ high									
0011	For higher elevations, see lines 9900 – 9960									
0100	Covers, ventilated galv. steel, straight, 6″ wide tray size	2 Elec	520	.031	L.F.	5.95	1.68		7.63	9.10
0200	9″ wide tray size		460	.035		8.60	1.90		10.50	12.40
0300	12″ wide tray size		400	.040		8.75	2.19		10.94	13.05
0400	18″ wide tray size		300	.053		11.90	2.92		14.82	17.60
0500	24″ wide tray size		220	.073		14.60	3.98		18.58	22
0600	30″ wide tray size		180	.089		21	4.86		25.86	30.50
0700	36″ wide tray size		160	.100		24	5.45		29.45	35
1000	Elbow horizontal 90°, 12″ radius, 6″ wide tray size		150	.107	Ea.	52	5.85		57.85	66.50
1020	9″ wide tray size		128	.125		58	6.85		64.85	74
1040	12″ wide tray size		108	.148		61	8.10		69.10	80
1060	18″ wide tray size		84	.190		84.50	10.40		94.90	109
1080	24″ wide tray size		66	.242		102	13.25		115.25	134
1100	30″ wide tray size		60	.267		127	14.60		141.60	163
1120	36″ wide tray size		50	.320		154	17.50		171.50	196
1160	24″ radius, 6″ wide tray size		136	.118		87	6.45		93.45	105
1180	9″ wide tray size		116	.138		89	7.55		96.55	110
1200	12″ wide tray size		96	.167		99	9.10		108.10	123
1220	18″ wide tray size		76	.211		122	11.50		133.50	153
1240	24″ wide tray size		60	.267		149	14.60		163.60	187
1260	30″ wide tray size		52	.308		196	16.85		212.85	241
1280	36″ wide tray size		44	.364		227	19.90		246.90	281
1320	36″ radius, 6″ wide tray size		120	.133		127	7.30		134.30	151
1340	9″ wide tray size		104	.154		140	8.40		148.40	167
1360	12″ wide tray size		84	.190		150	10.40		160.40	181
1380	18″ wide tray size		72	.222		191	12.15		203.15	229
1400	24″ wide tray size		52	.308		227	16.85		243.85	276
1420	30″ wide tray size		46	.348		270	19.05		289.05	325
1440	36″ wide tray size		40	.400		315	22		337	385
1480	Elbow horizontal 45°, 12″ radius, 6″ wide tray size		150	.107		37	5.85		42.85	50
1500	9″ wide tray size		128	.125		44.50	6.85		51.35	59.50
1520	12″ wide tray size		108	.148		48.50	8.10		56.60	65.50
1540	18″ wide tray size		88	.182		59	9.95		68.95	80.50
1560	24″ wide tray size		76	.211		69	11.50		80.50	94
1580	30″ wide tray size		66	.242		81.50	13.25		94.75	110
1600	36″ wide tray size		60	.267		91	14.60		105.60	123
1640	24″ radius, 6″ wide tray size		136	.118		54.50	6.45		60.95	70
1660	9″ wide tray size		116	.138		61	7.55		68.55	79
1680	12″ wide tray size		96	.167		69	9.10		78.10	90
1700	18″ wide tray size		80	.200		79	10.95		89.95	104
1720	24″ wide tray size		70	.229		91	12.50		103.50	119

26 05 36.40 Cable Tray, Covers and Dividers		Crew	Daily Output	Labor-Hours	Unit	Material	2015 Bare Costs Labor	Equipment	Total	Total Incl O&P
1740	30" wide tray size	2 Elec	60	.267	Ea.	116	14.60		130.60	150
1760	36" wide tray size		52	.308		127	16.85		143.85	166
1800	36" radius, 6" wide tray size		120	.133		79	7.30		86.30	98.50
1820	9" wide tray size		104	.154		89	8.40		97.40	111
1840	12" wide tray size		84	.190		98	10.40		108.40	124
1860	18" wide tray size		76	.211		112	11.50		123.50	142
1880	24" wide tray size		62	.258		137	14.10		151.10	173
1900	30" wide tray size		52	.308		149	16.85		165.85	190
1920	36" wide tray size		48	.333		180	18.25		198.25	226
1960	Elbow vertical 90°, 12" radius, 6" wide tray size		150	.107		43.50	5.85		49.35	57
1980	9" wide tray size		128	.125		44.50	6.85		51.35	59.50
2000	12" wide tray size		108	.148		48	8.10		56.10	65.50
2020	18" wide tray size		88	.182		54.50	9.95		64.45	75.50
2040	24" wide tray size		68	.235		55.50	12.85		68.35	81
2060	30" wide tray size		60	.267		61	14.60		75.60	90
2080	36" wide tray size		50	.320		79	17.50		96.50	114
2120	24" radius, 6" wide tray size		136	.118		53.50	6.45		59.95	69
2140	9" wide tray size		116	.138		59	7.55		66.55	76.50
2160	12" wide tray size		96	.167		61	9.10		70.10	81.50
2180	18" wide tray size		80	.200		81.50	10.95		92.45	106
2200	24" wide tray size		62	.258		91	14.10		105.10	122
2220	30" wide tray size		52	.308		104	16.85		120.85	140
2240	36" wide tray size		44	.364		118	19.90		137.90	161
2280	36" radius, 6" wide tray size		120	.133		62.50	7.30		69.80	80
2300	9" wide tray size		104	.154		79	8.40		87.40	100
2320	12" wide tray size		84	.190		89	10.40		99.40	114
2340	18" wide tray size		76	.211		112	11.50		123.50	142
2350	24" wide tray size		54	.296		125	16.20		141.20	162
2360	30" wide tray size		46	.348		149	19.05		168.05	194
2370	36" wide tray size		40	.400		176	22		198	228
2400	Tee horizontal, 12" radius, 6" wide tray size		92	.174		79	9.50		88.50	102
2410	9" wide tray size		80	.200		81.50	10.95		92.45	106
2420	12" wide tray size		68	.235		92	12.85		104.85	121
2430	18" wide tray size		60	.267		111	14.60		125.60	145
2440	24" wide tray size		52	.308		140	16.85		156.85	180
2460	30" wide tray size		36	.444		165	24.50		189.50	219
2470	36" wide tray size		30	.533		199	29		228	264
2500	24" radius, 6" wide tray size		88	.182		124	9.95		133.95	152
2510	9" wide tray size		76	.211		143	11.50		154.50	175
2520	12" wide tray size		64	.250		147	13.70		160.70	183
2530	18" wide tray size		56	.286		188	15.65		203.65	231
2540	24" wide tray size		48	.333		289	18.25		307.25	350
2560	30" wide tray size		32	.500		330	27.50		357.50	410
2570	36" wide tray size		26	.615		365	33.50		398.50	450
2600	36" radius, 6" wide tray size		84	.190		225	10.40		235.40	263
2610	9" wide tray size		72	.222		229	12.15		241.15	271
2620	12" wide tray size		60	.267		255	14.60		269.60	305
2630	18" wide tray size		52	.308		298	16.85		314.85	350
2640	24" wide tray size		44	.364		390	19.90		409.90	460
2660	30" wide tray size		28	.571		425	31.50		456.50	520
2670	36" wide tray size		22	.727		485	40		525	590
2700	Cross horizontal, 12" radius, 6" wide tray size		68	.235		118	12.85		130.85	150
2710	9" wide tray size		64	.250		125	13.70		138.70	158

26 05 36 – Cable Trays for Electrical Systems

26 05 36.40 Cable Tray, Covers and Dividers		Crew	Daily Output	Labor-Hours	Unit	Material	Labor	Equipment	Total	Total Incl O&P
						2015 Bare Costs				
2720	12″ wide tray size	2 Elec	60	.267	Ea.	139	14.60		153.60	176
2730	18″ wide tray size		52	.308		165	16.85		181.85	207
2740	24″ wide tray size		36	.444		201	24.50		225.50	259
2760	30″ wide tray size		30	.533		229	29		258	297
2770	36″ wide tray size		28	.571		268	31.50		299.50	345
2800	24″ radius, 6″ wide tray size		64	.250		225	13.70		238.70	268
2810	9″ wide tray size		60	.267		244	14.60		258.60	291
2820	12″ wide tray size		56	.286		263	15.65		278.65	315
2830	18″ wide tray size		48	.333		315	18.25		333.25	375
2840	24″ wide tray size		32	.500		385	27.50		412.50	465
2860	30″ wide tray size		26	.615		430	33.50		463.50	525
2870	36″ wide tray size		24	.667		485	36.50		521.50	590
2900	36″ radius, 6″ wide tray size		60	.267		375	14.60		389.60	440
2910	9″ wide tray size		56	.286		390	15.65		405.65	455
2920	12″ wide tray size		52	.308		410	16.85		426.85	475
2930	18″ wide tray size		44	.364		470	19.90		489.90	550
2940	24″ wide tray size		28	.571		600	31.50		631.50	710
2960	30″ wide tray size		22	.727		645	40		685	770
2970	36″ wide tray size		20	.800		690	44		734	830
3000	Reducer, 9″ to 6″ wide tray size		128	.125		50	6.85		56.85	65.50
3010	12″ to 6″ wide tray size		108	.148		52	8.10		60.10	70
3020	12″ to 9″ wide tray size		108	.148		52	8.10		60.10	70
3030	18″ to 12″ wide tray size		88	.182		55.50	9.95		65.45	76.50
3050	18″ to 6″ wide tray size		88	.182		55.50	9.95		65.45	76.50
3060	24″ to 18″ wide tray size		80	.200		78	10.95		88.95	102
3070	24″ to 12″ wide tray size		80	.200		71.50	10.95		82.45	95.50
3090	30″ to 24″ wide tray size		70	.229		82.50	12.50		95	110
3100	30″ to 18″ wide tray size		70	.229		82.50	12.50		95	110
3110	30″ to 12″ wide tray size		70	.229		71.50	12.50		84	98
3140	36″ to 30″ wide tray size		64	.250		90	13.70		103.70	120
3150	36″ to 24″ wide tray size		64	.250		90	13.70		103.70	120
3160	36″ to 18″ wide tray size		64	.250		90	13.70		103.70	120
3170	36″ to 12″ wide tray size		64	.250		90	13.70		103.70	120
3250	Covers, aluminum, straight, 6″ wide tray size		520	.031	L.F.	5.40	1.68		7.08	8.55
3270	9″ wide tray size		460	.035		6.50	1.90		8.40	10.10
3290	12″ wide tray size		400	.040		7.80	2.19		9.99	11.95
3310	18″ wide tray size		320	.050		10.30	2.74		13.04	15.60
3330	24″ wide tray size		260	.062		12.85	3.37		16.22	19.35
3350	30″ wide tray size		200	.080		14.20	4.38		18.58	22.50
3370	36″ wide tray size		180	.089		15.05	4.86		19.91	24
3400	Elbow horizontal 90°, 12″ radius, 6″ wide tray size		150	.107	Ea.	40.50	5.85		46.35	53.50
3410	9″ wide tray size		128	.125		42.50	6.85		49.35	57
3420	12″ wide tray size		108	.148		45.50	8.10		53.60	62.50
3430	18″ wide tray size		88	.182		60	9.95		69.95	81.50
3440	24″ wide tray size		70	.229		75.50	12.50		88	102
3460	30″ wide tray size		64	.250		90.50	13.70		104.20	121
3470	36″ wide tray size		54	.296		111	16.20		127.20	147
3500	24″ radius, 6″ wide tray size		136	.118		55.50	6.45		61.95	71
3510	9″ wide tray size		116	.138		68.50	7.55		76.05	87
3520	12″ wide tray size		96	.167		75.50	9.10		84.60	97
3530	18″ wide tray size		80	.200		89.50	10.95		100.45	115
3540	24″ wide tray size		64	.250		111	13.70		124.70	143
3560	30″ wide tray size		56	.286		135	15.65		150.65	173

26 05 36 – Cable Trays for Electrical Systems

26 05 36.40 Cable Tray, Covers and Dividers		Crew	Daily Output	Labor-Hours	Unit	Material	Labor	Equipment	Total	Total Incl O&P
						2015 Bare Costs				
3570	36" wide tray size	2 Elec	48	.333	Ea.	167	18.25		185.25	212
3600	36" radius, 6" wide tray size		120	.133		95	7.30		102.30	116
3610	9" wide tray size		104	.154		104	8.40		112.40	127
3620	12" wide tray size		84	.190		118	10.40		128.40	146
3630	18" wide tray size		76	.211		140	11.50		151.50	172
3640	24" wide tray size		56	.286		171	15.65		186.65	212
3660	30" wide tray size		50	.320		196	17.50		213.50	243
3670	36" wide tray size		44	.364		230	19.90		249.90	284
3700	Elbow horizontal 45°, 12" radius, 6" wide tray size		150	.107		30	5.85		35.85	42
3710	9" wide tray size		128	.125		30.50	6.85		37.35	44.50
3720	12" wide tray size		108	.148		34.50	8.10		42.60	50.50
3730	18" wide tray size		88	.182		39.50	9.95		49.45	59
3740	24" wide tray size		80	.200		44	10.95		54.95	65.50
3760	30" wide tray size		70	.229		55.50	12.50		68	80.50
3770	36" wide tray size		64	.250		67	13.70		80.70	94.50
3800	24" radius, 6" wide tray size		136	.118		34.50	6.45		40.95	48
3810	9" wide tray size		116	.138		44	7.55		51.55	60
3820	12" wide tray size		96	.167		45.50	9.10		54.60	64.50
3830	18" wide tray size		80	.200		55.50	10.95		66.45	78
3840	24" wide tray size		72	.222		68.50	12.15		80.65	94.50
3860	30" wide tray size		64	.250		79	13.70		92.70	108
3870	36" wide tray size		56	.286		91.50	15.65		107.15	125
3900	36" radius, 6" wide tray size		120	.133		60	7.30		67.30	77.50
3910	9" wide tray size		104	.154		65	8.40		73.40	84.50
3920	12" wide tray size		84	.190		68.50	10.40		78.90	91.50
3930	18" wide tray size		76	.211		82.50	11.50		94	109
3940	24" wide tray size		64	.250		100	13.70		113.70	131
3960	30" wide tray size		56	.286		114	15.65		129.65	150
3970	36" wide tray size		50	.320		130	17.50		147.50	170
4000	Elbow vertical 90°, 12" radius, 6" wide tray size		150	.107		34.50	5.85		40.35	47
4010	9" wide tray size		128	.125		34.50	6.85		41.35	48.50
4020	12" wide tray size		108	.148		38	8.10		46.10	54
4030	18" wide tray size		88	.182		44	9.95		53.95	64
4040	24" wide tray size		70	.229		45.50	12.50		58	70
4060	30" wide tray size		64	.250		46.50	13.70		60.20	72.50
4070	36" wide tray size		54	.296		56.50	16.20		72.70	87
4100	24" radius, 6" wide tray size		136	.118		39.50	6.45		45.95	53.50
4110	9" wide tray size		116	.138		43	7.55		50.55	59
4120	12" wide tray size		96	.167		46.50	9.10		55.60	65.50
4130	18" wide tray size		80	.200		56.50	10.95		67.45	79
4140	24" wide tray size		64	.250		61.50	13.70		75.20	88.50
4160	30" wide tray size		56	.286		79	15.65		94.65	111
4170	36" wide tray size		48	.333		86	18.25		104.25	123
4200	36" radius, 6" wide tray size		120	.133		45.50	7.30		52.80	62
4210	9" wide tray size		104	.154		56.50	8.40		64.90	75
4220	12" wide tray size		84	.190		65	10.40		75.40	87.50
4230	18" wide tray size		76	.211		79	11.50		90.50	105
4240	24" wide tray size		56	.286		93	15.65		108.65	126
4260	30" wide tray size		50	.320		118	17.50		135.50	157
4270	36" wide tray size		44	.364		127	19.90		146.90	170
4300	Tee horizontal, 12" radius, 6" wide tray size		108	.148		56.50	8.10		64.60	74.50
4310	9" wide tray size		88	.182		60	9.95		69.95	81.50
4320	12" wide tray size		80	.200		67	10.95		77.95	90.50

26 05 36.40 Cable Tray, Covers and Dividers		Crew	Daily Output	Labor-Hours	Unit	Material	Labor	Equipment	Total	Total Incl O&P
						2015 Bare Costs				
4330	18" wide tray size	2 Elec	68	.235	Ea.	79	12.85		91.85	107
4340	24" wide tray size		56	.286		100	15.65		115.65	134
4360	30" wide tray size		44	.364		118	19.90		137.90	161
4370	36" wide tray size		36	.444		142	24.50		166.50	195
4400	24" radius, 6" wide tray size		96	.167		91.50	9.10		100.60	115
4410	9" wide tray size		80	.200		102	10.95		112.95	129
4420	12" wide tray size		72	.222		114	12.15		126.15	145
4430	18" wide tray size		60	.267		134	14.60		148.60	170
4440	24" wide tray size		48	.333		213	18.25		231.25	262
4460	30" wide tray size		40	.400		235	22		257	292
4470	36" wide tray size		32	.500		262	27.50		289.50	330
4500	36" radius, 6" wide tray size		88	.182		167	9.95		176.95	199
4510	9" wide tray size		72	.222		171	12.15		183.15	207
4520	12" wide tray size		64	.250		188	13.70		201.70	227
4530	18" wide tray size		56	.286		213	15.65		228.65	258
4540	24" wide tray size		44	.364		270	19.90		289.90	330
4560	30" wide tray size		36	.444		305	24.50		329.50	375
4570	36" wide tray size		28	.571		345	31.50		376.50	425
4600	Cross horizontal, 12" radius, 6" wide tray size		80	.200		86	10.95		96.95	112
4610	9" wide tray size		72	.222		91.50	12.15		103.65	119
4620	12" wide tray size		64	.250		101	13.70		114.70	133
4630	18" wide tray size		56	.286		123	15.65		138.65	160
4640	24" wide tray size		48	.333		142	18.25		160.25	184
4660	30" wide tray size		40	.400		169	22		191	220
4670	36" wide tray size		32	.500		194	27.50		221.50	257
4700	24" radius, 6" wide tray size		72	.222		169	12.15		181.15	205
4710	9" wide tray size		64	.250		181	13.70		194.70	220
4720	12" wide tray size		56	.286		194	15.65		209.65	238
4730	18" wide tray size		48	.333		226	18.25		244.25	277
4740	24" wide tray size		40	.400		270	22		292	330
4760	30" wide tray size		32	.500		315	27.50		342.50	395
4770	36" wide tray size		24	.667		345	36.50		381.50	430
4800	36" radius, 6" wide tray size		64	.250		270	13.70		283.70	320
4810	9" wide tray size		56	.286		282	15.65		297.65	335
4820	12" wide tray size		50	.320		305	17.50		322.50	360
4830	18" wide tray size		44	.364		335	19.90		354.90	400
4840	24" wide tray size		36	.444		410	24.50		434.50	495
4860	30" wide tray size		28	.571		475	31.50		506.50	570
4870	36" wide tray size		22	.727		520	40		560	630
4900	Reducer, 9" to 6" wide tray size		128	.125		43	6.85		49.85	58
4910	12" to 6" wide tray size		108	.148		45	8.10		53.10	62
4920	12" to 9" wide tray size		108	.148		45	8.10		53.10	62
4930	18" to 12" wide tray size		88	.182		49	9.95		58.95	69.50
4950	18" to 6" wide tray size		88	.182		49	9.95		58.95	69.50
4960	24" to 18" wide tray size		80	.200		63	10.95		73.95	86.50
4970	24" to 12" wide tray size		80	.200		54.50	10.95		65.45	77
4990	30" to 24" wide tray size		70	.229		66	12.50		78.50	92
5000	30" to 18" wide tray size		70	.229		66	12.50		78.50	92
5010	30" to 12" wide tray size		70	.229		66	12.50		78.50	92
5040	36" to 30" wide tray size		64	.250		73	13.70		86.70	102
5050	36" to 24" wide tray size		64	.250		73	13.70		86.70	102
5060	36" to 18" wide tray size		64	.250		73	13.70		86.70	102
5070	36" to 12" wide tray size		64	.250		73	13.70		86.70	102

26 05 36 – Cable Trays for Electrical Systems

26 05 36.40 Cable Tray, Covers and Dividers		Crew	Daily Output	Labor-Hours	Unit	2015 Bare Costs Material	Labor	Equipment	Total	Total Incl O&P
5710	Tray cover hold down clamp	1 Elec	60	.133	Ea.	11.70	7.30		19	24
8000	Divider strip, straight, galvanized, 3" deep		200	.040	L.F.	6.25	2.19		8.44	10.25
8020	4" deep		180	.044		7.60	2.43		10.03	12.15
8040	6" deep		160	.050		9.95	2.74		12.69	15.20
8060	Aluminum, straight, 3" deep		210	.038		6.25	2.08		8.33	10.10
8080	4" deep		190	.042		7.70	2.30		10	12.05
8100	6" deep	↓	170	.047	↓	9.85	2.57		12.42	14.80
8110	Divider strip vertical fitting 3" deep									
8120	12" radius, galvanized, 30°	1 Elec	28	.286	Ea.	30	15.65		45.65	57
8140	45°		27	.296		37.50	16.20		53.70	66.50
8160	60°		26	.308		40	16.85		56.85	70
8180	90°		25	.320		50	17.50		67.50	82
8200	Aluminum, 30°		29	.276		20	15.10		35.10	45.50
8220	45°		28	.286		23.50	15.65		39.15	50
8240	60°		27	.296		27.50	16.20		43.70	55.50
8260	90°		26	.308		35	16.85		51.85	64.50
8280	24" radius, galvanized, 30°		25	.320		45.50	17.50		63	77
8300	45°		24	.333		51	18.25		69.25	84.50
8320	60°		23	.348		64.50	19.05		83.55	101
8340	90°		22	.364		88	19.90		107.90	128
8360	Aluminum, 30°		26	.308		33.50	16.85		50.35	63
8380	45°		25	.320		38.50	17.50		56	69.50
8400	60°		24	.333		48.50	18.25		66.75	81
8420	90°		23	.348		66	19.05		85.05	102
8440	36" radius, galvanized, 30°		22	.364		61	19.90		80.90	98.50
8460	45°		21	.381		71.50	21		92.50	111
8480	60°		20	.400		83.50	22		105.50	126
8500	90°		19	.421		116	23		139	163
8520	Aluminum, 30°		23	.348		51	19.05		70.05	85.50
8540	45°		22	.364		66	19.90		85.90	104
8560	60°		21	.381		85.50	21		106.50	127
8570	90°	↓	20	.400	↓	112	22		134	157
8590	Divider strip vertical fitting 4" deep									
8600	12" radius, galvanized, 30°	1 Elec	27	.296	Ea.	38.50	16.20		54.70	67
8610	45°		26	.308		44.50	16.85		61.35	75
8620	60°		25	.320		50	17.50		67.50	82
8630	90°		24	.333		61	18.25		79.25	95.50
8640	Aluminum, 30°		28	.286		28.50	15.65		44.15	55
8650	45°		27	.296		33	16.20		49.20	61.50
8660	60°		26	.308		38	16.85		54.85	67.50
8670	90°		25	.320		45	17.50		62.50	76.50
8680	24" radius, galvanized, 30°		24	.333		61	18.25		79.25	95.50
8690	45°		23	.348		77	19.05		96.05	114
8700	60°		22	.364		87	19.90		106.90	127
8710	90°		21	.381		116	21		137	160
8720	Aluminum, 30°		25	.320		45	17.50		62.50	76.50
8730	45°		24	.333		54.50	18.25		72.75	88
8740	60°		23	.348		64	19.05		83.05	100
8750	90°		22	.364		88	19.90		107.90	128
8760	36" radius, galvanized, 30°		23	.348		71.50	19.05		90.55	108
8770	45°		22	.364		81.50	19.90		101.40	121
8780	60°		21	.381		102	21		123	146
8790	90°	↓	20	.400	↓	139	22		161	187

26 05 36 – Cable Trays for Electrical Systems

26 05 36.40 Cable Tray, Covers and Dividers		Crew	Daily Output	Labor-Hours	Unit	Material	2015 Bare Costs Labor	Equipment	Total	Total Incl O&P
8800	Aluminum, 30°	1 Elec	24	.333	Ea.	68.50	18.25		86.75	104
8810	45°		23	.348		88	19.05		107.05	126
8820	60°		22	.364		107	19.90		126.90	149
8830	90°		21	.381		135	21		156	182
8840	Divider strip vertical fitting 6" deep									
8850	12" radius, galvanized, 30°	1 Elec	24	.333	Ea.	42	18.25		60.25	74
8860	45°		23	.348		46.50	19.05		65.55	81
8870	60°		22	.364		53.50	19.90		73.40	90
8880	90°		21	.381		67	21		88	106
8890	Aluminum, 30°		25	.320		31	17.50		48.50	61
8900	45°		24	.333		37	18.25		55.25	68.50
8910	60°		23	.348		38.50	19.05		57.55	72
8920	90°		22	.364		45.50	19.90		65.40	81.50
8930	24" radius, galvanized, 30°		23	.348		61	19.05		80.05	97
8940	45°		22	.364		77	19.90		96.90	116
8950	60°		21	.381		88	21		109	130
8960	90°		20	.400		116	22		138	161
8970	Aluminum, 30°		24	.333		45.50	18.25		63.75	78.50
8980	45°		23	.348		62.50	19.05		81.55	98
8990	60°		22	.364		68.50	19.90		88.40	107
9000	90°		21	.381		93	21		114	135
9010	36" radius, galvanized, 30°		22	.364		71.50	19.90		91.40	110
9020	45°		21	.381		88	21		109	130
9030	60°		20	.400		116	22		138	161
9040	90°		19	.421		149	23		172	200
9050	Aluminum, 30°		23	.348		69.50	19.05		88.55	106
9060	45°		22	.364		93	19.90		112.90	133
9070	60°		21	.381		110	21		131	154
9080	90°		20	.400		132	22		154	179
9120	Divider strip, horizontal fitting, galvanized, 3" deep		33	.242		43	13.25		56.25	67.50
9130	4" deep		30	.267		48	14.60		62.60	75.50
9140	6" deep		27	.296		61	16.20		77.20	92.50
9150	Aluminum, 3" deep		35	.229		32	12.50		44.50	54.50
9160	4" deep		32	.250		35	13.70		48.70	59.50
9170	6" deep		29	.276		46.50	15.10		61.60	75
9300	Divider strip protector		300	.027	L.F.	3.82	1.46		5.28	6.45
9310	Fastener, ladder tray				Ea.	.68			.68	.75
9320	Trough or solid bottom tray				"	.46			.46	.51
9900	Add to labor for higher elevated installation									
9910	15' to 20' high add						10%			
9920	20' to 25' high add						20%			
9930	25' to 30' high add						25%			
9940	30' to 35' high add						30%			
9960	Over 40' high add						40%			

26 05 39 – Underfloor Raceways for Electrical Systems

26 05 39.30 Conduit In Concrete Slab		Crew	Daily Output	Labor-Hours	Unit	Material	Labor	Equipment	Total	Total Incl O&P
0010	**CONDUIT IN CONCRETE SLAB** Including terminations,									
0020	fittings and supports									
3230	PVC, schedule 40, 1/2" diameter	1 Elec	270	.030	L.F.	.57	1.62		2.19	3.14
3250	3/4" diameter		230	.035		.66	1.90		2.56	3.67
3270	1" diameter		200	.040		.87	2.19		3.06	4.34
3300	1-1/4" diameter		170	.047		1.17	2.57		3.74	5.30

26 05 39.30 Conduit In Concrete Slab		Crew	Daily Output	Labor-Hours	Unit	Material	2015 Bare Costs Labor	Equipment	Total	Total Incl O&P
3330	1-1/2" diameter	1 Elec	140	.057	L.F.	1.42	3.13		4.55	6.40
3350	2" diameter		120	.067		1.78	3.65		5.43	7.60
3370	2-1/2" diameter		90	.089		3.07	4.86		7.93	10.90
3400	3" diameter	2 Elec	160	.100		3.90	5.45		9.35	12.75
3430	3-1/2" diameter		120	.133		4.94	7.30		12.24	16.75
3440	4" diameter		100	.160		5.40	8.75		14.15	19.50
3450	5" diameter		80	.200		8.35	10.95		19.30	26
3460	6" diameter		60	.267		11.95	14.60		26.55	35.50
3530	Sweeps, 1" diameter, 30" radius	1 Elec	32	.250	Ea.	41	13.70		54.70	66
3550	1-1/4" diameter		24	.333		44.50	18.25		62.75	77
3570	1-1/2" diameter		21	.381		47	21		68	84
3600	2" diameter		18	.444		52	24.50		76.50	94.50
3630	2-1/2" diameter		14	.571		67.50	31.50		99	123
3650	3" diameter		10	.800		86	44		130	163
3670	3-1/2" diameter		8	1		105	54.50		159.50	201
3700	4" diameter		7	1.143		147	62.50		209.50	259
3710	5" diameter		6	1.333		168	73		241	298
3730	Couplings, 1/2" diameter					.21			.21	.23
3750	3/4" diameter					.24			.24	.26
3770	1" diameter					.36			.36	.40
3800	1-1/4" diameter					.55			.55	.61
3830	1-1/2" diameter					.68			.68	.75
3850	2" diameter					.88			.88	.97
3870	2-1/2" diameter					1.57			1.57	1.73
3900	3" diameter					2.54			2.54	2.79
3930	3-1/2" diameter					3.21			3.21	3.53
3950	4" diameter					3.59			3.59	3.95
3960	5" diameter					9.10			9.10	10
3970	6" diameter					12.35			12.35	13.60
4030	End bells 1" diameter, PVC	1 Elec	60	.133		3.90	7.30		11.20	15.60
4050	1-1/4" diameter		53	.151		4.80	8.25		13.05	18.10
4100	1-1/2" diameter		48	.167		5.95	9.10		15.05	20.50
4150	2" diameter		34	.235		7.15	12.85		20	28
4170	2-1/2" diameter		27	.296		7.90	16.20		24.10	33.50
4200	3" diameter		20	.400		8.30	22		30.30	43
4250	3-1/2" diameter		16	.500		9.90	27.50		37.40	53.50
4300	4" diameter		14	.571		10.75	31.50		42.25	60.50
4310	5" diameter		12	.667		11.35	36.50		47.85	69
4320	6" diameter		9	.889		13.55	48.50		62.05	90.50
4350	Rigid galvanized steel, 1/2" diameter		200	.040	L.F.	2.48	2.19		4.67	6.10
4400	3/4" diameter		170	.047		2.62	2.57		5.19	6.85
4450	1" diameter		130	.062		3.60	3.37		6.97	9.15
4500	1-1/4" diameter		110	.073		4.94	3.98		8.92	11.60
4600	1-1/2" diameter		100	.080		5.50	4.38		9.88	12.85
4800	2" diameter		90	.089		6.80	4.86		11.66	15.05
9000	Minimum labor/equipment charge		4	2	Job		109		109	169

26 05 39.40 Conduit In Trench		Crew	Daily Output	Labor-Hours	Unit	Material	Labor	Equipment	Total	Total Incl O&P
0010	**CONDUIT IN TRENCH** Includes terminations and fittings									
0020	Does not include excavation or backfill, see Section 31 23 16.00									
0200	Rigid galvanized steel, 2" diameter	1 Elec	150	.053	L.F.	6.40	2.92		9.32	11.55
0400	2-1/2" diameter	"	100	.080		12.45	4.38		16.83	20.50
0600	3" diameter	2 Elec	160	.100		14.55	5.45		20	24.50

26 05 39 – Underfloor Raceways for Electrical Systems

26 05 39.40 Conduit In Trench		Crew	Daily Output	Labor-Hours	Unit	2015 Bare Costs Material	Labor	Equipment	Total	Total Incl O&P
0800	3-1/2" diameter	2 Elec	140	.114	L.F.	19.20	6.25		25.45	30.50
1000	4" diameter		100	.160		21	8.75		29.75	37
1200	5" diameter		80	.200		44	10.95		54.95	65
1400	6" diameter	↓	60	.267	↓	61	14.60		75.60	90
9000	Minimum labor/equipment charge	1 Elec	4	2	Job		109		109	169

26 05 43 – Underground Ducts and Raceways for Electrical Systems

26 05 43.10 Trench Duct		Crew	Daily Output	Labor-Hours	Unit	Material	Labor	Equipment	Total	Total Incl O&P
0010	**TRENCH DUCT** Steel with cover									
0020	Standard adjustable, depths to 4"									
0100	Straight, single compartment, 9" wide	2 Elec	40	.400	L.F.	112	22		134	157
0200	12" wide		32	.500		136	27.50		163.50	192
0400	18" wide		26	.615		167	33.50		200.50	235
0600	24" wide		22	.727		199	40		239	281
0700	27" wide		21	.762		213	41.50		254.50	299
0800	30" wide		20	.800		234	44		278	325
1000	36" wide		16	1		265	54.50		319.50	375
1020	Two compartment, 9" wide		38	.421		118	23		141	166
1030	12" wide		30	.533		135	29		164	193
1040	18" wide		24	.667		165	36.50		201.50	239
1050	24" wide		20	.800		216	44		260	305
1060	30" wide		18	.889		263	48.50		311.50	365
1070	36" wide		14	1.143		305	62.50		367.50	430
1090	Three compartment, 9" wide		36	.444		135	24.50		159.50	186
1100	12" wide		28	.571		149	31.50		180.50	213
1110	18" wide		22	.727		187	40		227	268
1120	24" wide		18	.889		235	48.50		283.50	335
1130	30" wide		16	1		284	54.50		338.50	395
1140	36" wide		12	1.333	↓	330	73		403	480
1200	Horizontal elbow, 9" wide		5.40	2.963	Ea.	385	162		547	675
1400	12" wide		4.60	3.478		445	190		635	785
1600	18" wide		4	4		570	219		789	970
1800	24" wide		3.20	5		800	274		1,074	1,300
1900	27" wide		3	5.333		910	292		1,202	1,450
2000	30" wide		2.60	6.154		1,075	335		1,410	1,700
2200	36" wide		2.40	6.667		1,400	365		1,765	2,125
2220	Two compartment, 9" wide		3.80	4.211		625	230		855	1,050
2230	12" wide		3	5.333		690	292		982	1,200
2240	18" wide		2.40	6.667		830	365		1,195	1,475
2250	24" wide		2	8		1,050	440		1,490	1,825
2260	30" wide		1.80	8.889		1,400	485		1,885	2,300
2270	36" wide		1.60	10		1,700	545		2,245	2,725
2290	Three compartment, 9" wide		3.60	4.444		655	243		898	1,100
2300	12" wide		2.80	5.714		730	315		1,045	1,300
2310	18" wide		2.20	7.273		890	400		1,290	1,600
2320	24" wide		1.80	8.889		1,125	485		1,610	2,000
2330	30" wide		1.60	10		1,475	545		2,020	2,450
2350	36" wide		1.40	11.429		1,775	625		2,400	2,925
2400	Vertical elbow, 9" wide		5.40	2.963		135	162		297	400
2600	12" wide		4.60	3.478		146	190		336	455
2800	18" wide		4	4		168	219		387	525
3000	24" wide		3.20	5		209	274		483	655
3100	27" wide	↓	3	5.333	↓	218	292		510	690

26 05 43 – Underground Ducts and Raceways for Electrical Systems

26 05 43.10 Trench Duct		Crew	Daily Output	Labor-Hours	Unit	Material	Labor (2015 Bare Costs)	Equipment	Total	Total Incl O&P
3200	30″ wide	2 Elec	2.60	6.154	Ea.	230	335		565	775
3400	36″ wide		2.40	6.667		253	365		618	845
3600	Cross, 9″ wide		4	4		635	219		854	1,025
3800	12″ wide		3.20	5		670	274		944	1,150
4000	18″ wide		2.60	6.154		800	335		1,135	1,400
4200	24″ wide		2.20	7.273		1,025	400		1,425	1,750
4300	27″ wide		2.20	7.273		1,200	400		1,600	1,950
4400	30″ wide		2	8		1,325	440		1,765	2,125
4600	36″ wide		1.80	8.889		1,650	485		2,135	2,575
4620	Two compartment, 9″ wide		3.80	4.211		660	230		890	1,075
4630	12″ wide		3	5.333		695	292		987	1,225
4640	18″ wide		2.40	6.667		840	365		1,205	1,500
4650	24″ wide		2	8		1,075	440		1,515	1,850
4660	30″ wide		1.80	8.889		1,400	485		1,885	2,300
4670	36″ wide		1.60	10		1,700	545		2,245	2,725
4690	Three compartment, 9″ wide		3.60	4.444		670	243		913	1,100
4700	12″ wide		2.80	5.714		765	315		1,080	1,325
4710	18″ wide		2.20	7.273		910	400		1,310	1,625
4720	24″ wide		1.80	8.889		1,125	485		1,610	2,000
4730	30″ wide		1.60	10		1,475	545		2,020	2,475
4740	36″ wide		1.40	11.429		1,825	625		2,450	2,975
4800	End closure, 9″ wide		14.40	1.111		39.50	61		100.50	138
5000	12″ wide		12	1.333		45.50	73		118.50	163
5200	18″ wide		10	1.600		69.50	87.50		157	213
5400	24″ wide		8	2		91.50	109		200.50	270
5500	27″ wide		7	2.286		106	125		231	310
5600	30″ wide		6.60	2.424		115	133		248	330
5800	36″ wide		5.80	2.759		136	151		287	385
6000	Tees, 9″ wide		4	4		385	219		604	765
6200	12″ wide		3.60	4.444		445	243		688	865
6400	18″ wide		3.20	5		570	274		844	1,050
6600	24″ wide		3	5.333		820	292		1,112	1,350
6700	27″ wide		2.80	5.714		910	315		1,225	1,475
6800	30″ wide		2.60	6.154		1,075	335		1,410	1,700
7000	36″ wide		2	8		1,400	440		1,840	2,225
7020	Two compartment, 9″ wide		3.80	4.211		445	230		675	845
7030	12″ wide		3.40	4.706		480	257		737	930
7040	18″ wide		3	5.333		640	292		932	1,150
7050	24″ wide		2.80	5.714		865	315		1,180	1,450
7060	30″ wide		2.40	6.667		1,175	365		1,540	1,875
7070	36″ wide		1.90	8.421		1,475	460		1,935	2,350
7090	Three compartment, 9″ wide		3.60	4.444		510	243		753	935
7100	12″ wide		3.20	5		535	274		809	1,025
7110	18″ wide		2.80	5.714		680	315		995	1,225
7120	24″ wide		2.60	6.154		935	335		1,270	1,550
7130	30″ wide		2.20	7.273		1,225	400		1,625	1,975
7140	36″ wide		1.80	8.889		1,550	485		2,035	2,475
7200	Riser, and cabinet connector, 9″ wide		5.40	2.963		168	162		330	435
7400	12″ wide		4.60	3.478		196	190		386	510
7600	18″ wide		4	4		241	219		460	605
7800	24″ wide		3.20	5		291	274		565	745
7900	27″ wide		3	5.333		300	292		592	780
8000	30″ wide		2.60	6.154		335	335		670	890

26 05 43 – Underground Ducts and Raceways for Electrical Systems

26 05 43.10 Trench Duct		Crew	Daily Output	Labor-Hours	Unit	Material	Labor	Equipment	Total	Total Incl O&P
8200	36" wide	2 Elec	2	8	Ea.	390	440		830	1,100
8400	Insert assembly, cell to conduit adapter, 1-1/4"	1 Elec	16	.500	↓	66.50	27.50		94	116
8500	Adjustable partition	"	320	.025	L.F.	23.50	1.37		24.87	28
8600	Depth of duct over 4", per 1", add					10.30			10.30	11.35
8700	Support post	1 Elec	240	.033		23.50	1.82		25.32	29
8800	Cover double tile trim, 2 sides					36.50			36.50	40
8900	4 sides					108			108	119
9160	Trench duct 3-1/2" x 4-1/2", add					9.80			9.80	10.80
9170	Trench duct 4" x 5", add					9.80			9.80	10.80
9200	For carpet trim, add					33			33	36.50
9210	For double carpet trim, add				↓	101			101	111

26 05 43.20 Underfloor Duct		Crew	Daily Output	Labor-Hours	Unit	Material	Labor	Equipment	Total	Total Incl O&P
0010	**UNDERFLOOR DUCT**									
0100	Duct, 1-3/8" x 3-1/8" blank, standard	2 Elec	160	.100	L.F.	14.25	5.45		19.70	24
0200	1-3/8" x 7-1/4" blank, super duct		120	.133		28.50	7.30		35.80	43
0400	7/8" or 1-3/8" insert type, 24" O.C., 1-3/8" x 3-1/8", std.		140	.114		19.05	6.25		25.30	30.50
0600	1-3/8" x 7-1/4", super duct	↓	100	.160	↓	33.50	8.75		42.25	50
0800	Junction box, single duct, 1 level, 3-1/8"	1 Elec	4	2	Ea.	430	109		539	640
0820	3-1/8" x 7-1/4"		4	2		500	109		609	720
0840	2 level, 3-1/8" upper & lower		3.20	2.500		500	137		637	760
0860	3-1/8" upper, 7-1/4" lower		2.70	2.963		500	162		662	800
0880	Carpet pan for above		80	.100		355	5.45		360.45	400
0900	Terrazzo pan for above		67	.119		830	6.55		836.55	920
1000	Junction box, single duct, 1 level, 7-1/4"		2.70	2.963		500	162		662	800
1020	2 level, 7-1/4" upper & lower		2.70	2.963		570	162		732	880
1040	2 duct, two 3-1/8" upper & lower		3.20	2.500		755	137		892	1,050
1200	1 level, 2 duct, 3-1/8"		3.20	2.500		570	137		707	840
1220	Carpet pan for above boxes		80	.100		355	5.45		360.45	400
1240	Terrazzo pan for above boxes		67	.119		830	6.55		836.55	920
1260	Junction box, 1 level, two 3-1/8" x one 3-1/8" + one 7-1/4"		2.30	3.478		970	190		1,160	1,375
1280	2 level, two 3-1/8" upper, one 3-1/8" + one 7-1/4" lower		2	4		1,075	219		1,294	1,525
1300	Carpet pan for above boxes		80	.100		355	5.45		360.45	400
1320	Terrazzo pan for above boxes		67	.119		830	6.55		836.55	920
1400	Junction box, 1 level, 2 duct, 7-1/4"		2.30	3.478		1,475	190		1,665	1,925
1420	Two 3-1/8" + one 7-1/4"		2	4		1,475	219		1,694	1,975
1440	Carpet pan for above		80	.100		355	5.45		360.45	400
1460	Terrazzo pan for above		67	.119		830	6.55		836.55	920
1580	Junction box, 1 level, one 3-1/8" + one 7-1/4" x same		2.30	3.478		970	190		1,160	1,375
1600	Triple duct, 3-1/8"		2.30	3.478		970	190		1,160	1,375
1700	Junction box, 1 level, one 3-1/8" + two 7-1/4"		2	4		1,625	219		1,844	2,150
1720	Carpet pan for above		80	.100		355	5.45		360.45	400
1740	Terrazzo pan for above		67	.119		830	6.55		836.55	920
1800	Insert to conduit adapter, 3/4" & 1"		32	.250		35	13.70		48.70	59.50
2000	Support, single cell		27	.296		52.50	16.20		68.70	83
2200	Super duct		16	.500		52.50	27.50		80	101
2400	Double cell		16	.500		52.50	27.50		80	101
2600	Triple cell		11	.727		52.50	40		92.50	120
2800	Vertical elbow, standard duct		10	.800		94.50	44		138.50	172
3000	Super duct		8	1		94.50	54.50		149	189
3200	Cabinet connector, standard duct		32	.250		71	13.70		84.70	99
3400	Super duct		27	.296		71	16.20		87.20	103
3600	Conduit adapter, 1" to 1-1/4"	↓	32	.250	↓	71	13.70		84.70	99

Material, Labor and Equipment are 2015 Bare Costs.

26 05 43 – Underground Ducts and Raceways for Electrical Systems

26 05 43.20 Underfloor Duct		Crew	Daily Output	Labor-Hours	Unit	Material	Labor	Equipment	Total	Total Incl O&P
						2015 Bare Costs				
3800	2" to 1-1/4"	1 Elec	27	.296	Ea.	85	16.20		101.20	119
4000	Outlet, low tension (tele, computer, etc.)		8	1		99.50	54.50		154	195
4200	High tension, receptacle (120 volt)		8	1		99.50	54.50		154	195
4300	End closure, standard duct		160	.050		3.98	2.74		6.72	8.60
4310	Super duct		160	.050		7.55	2.74		10.29	12.55
4350	Elbow, horiz., standard duct		26	.308		238	16.85		254.85	287
4360	Super duct		26	.308		238	16.85		254.85	287
4380	Elbow, offset, standard duct		26	.308		94.50	16.85		111.35	130
4390	Super duct		26	.308		94.50	16.85		111.35	130
4400	Marker screw assembly for inserts		50	.160		16.80	8.75		25.55	32
4410	Y take off, standard duct		26	.308		142	16.85		158.85	182
4420	Super duct		26	.308		142	16.85		158.85	182
4430	Box opening plug, standard duct		160	.050		13.30	2.74		16.04	18.90
4440	Super duct		160	.050		13.30	2.74		16.04	18.90
4450	Sleeve coupling, standard duct		160	.050		35.50	2.74		38.24	43
4460	Super duct		160	.050		43.50	2.74		46.24	52
4470	Conduit adapter, standard duct, 3/4"		32	.250		71	13.70		84.70	99
4480	1" or 1-1/4"		32	.250		71	13.70		84.70	99
4500	1-1/2"	↓	32	.250	↓	71	13.70		84.70	99

26 05 80 – Wiring Connections

26 05 80.10 Motor Connections		Crew	Daily Output	Labor-Hours	Unit	Material	Labor	Equipment	Total	Total Incl O&P
0010	**MOTOR CONNECTIONS**									
0020	Flexible conduit and fittings, 115 volt, 1 phase, up to 1 HP motor	1 Elec	8	1	Ea.	6.35	54.50		60.85	91.50
0050	2 HP motor		6.50	1.231		12.55	67.50		80.05	118
0100	3 HP motor		5.50	1.455		11.05	79.50		90.55	135
0110	230 volt, 3 phase, 3 HP motor		6.78	1.180		7.65	64.50		72.15	108
0112	5 HP motor		5.47	1.463		6.50	80		86.50	131
0114	7-1/2 HP motor		4.61	1.735		9.55	95		104.55	158
0120	10 HP motor		4.20	1.905		21	104		125	185
0150	15 HP motor		3.30	2.424		21	133		154	229
0200	25 HP motor		2.70	2.963		29	162		191	283
0400	50 HP motor		2.20	3.636		59	199		258	375
0600	100 HP motor		1.50	5.333		140	292		432	605
1500	460 volt, 5 HP motor, 3 phase		8	1		6.85	54.50		61.35	92
1520	10 HP motor		8	1		6.85	54.50		61.35	92
1530	25 HP motor		6	1.333		13.45	73		86.45	128
1540	30 HP motor		6	1.333		13.45	73		86.45	128
1550	40 HP motor		5	1.600		20	87.50		107.50	158
1560	50 HP motor		5	1.600		23	87.50		110.50	161
1570	60 HP motor		3.80	2.105		25	115		140	206
1580	75 HP motor		3.50	2.286		34.50	125		159.50	232
1590	100 HP motor		2.50	3.200		53	175		228	330
1600	125 HP motor		2	4		68	219		287	415
1610	150 HP motor		1.80	4.444		72.50	243		315.50	455
1620	200 HP motor		1.50	5.333		116	292		408	580
2005	460 Volt, 5 HP motor, 3 Phase, w/sealtite		8	1		10.80	54.50		65.30	96.50
2010	10 HP motor		8	1		10.80	54.50		65.30	96.50
2015	25 HP motor		6	1.333		21	73		94	136
2020	30 HP motor		6	1.333		21	73		94	136
2025	40 HP motor		5	1.600		37.50	87.50		125	178
2030	50 HP motor		5	1.600		38.50	87.50		126	178
2035	60 HP motor	↓	3.80	2.105	↓	59.50	115		174.50	243

26 05 80 – Wiring Connections

26 05 80.10 Motor Connections		Crew	Daily Output	Labor-Hours	Unit	2015 Bare Costs Material	Labor	Equipment	Total	Total Incl O&P
2040	75 HP motor	1 Elec	3.50	2.286	Ea.	62.50	125		187.50	263
2045	100 HP motor		2.50	3.200		85.50	175		260.50	365
2055	150 HP motor		1.80	4.444		91	243		334	475
2060	200 HP motor		1.50	5.333		605	292		897	1,125
9000	Minimum labor/equipment charge		4	2	Job		109		109	169

26 05 90 – Residential Applications

26 05 90.10 Residential Wiring

		Crew	Daily Output	Labor-Hours	Unit	Material	Labor	Equipment	Total	Total Incl O&P
0010	**RESIDENTIAL WIRING**									
0020	20' avg. runs and #14/2 wiring incl. unless otherwise noted									
1000	Service & panel, includes 24' SE-AL cable, service eye, meter,									
1010	Socket, panel board, main bkr., ground rod, 15 or 20 amp									
1020	1-pole circuit breakers, and misc. hardware									
1100	100 amp, with 10 branch breakers	1 Elec	1.19	6.723	Ea.	560	370		930	1,175
1110	With PVC conduit and wire		.92	8.696		605	475		1,080	1,400
1120	With RGS conduit and wire		.73	10.959		765	600		1,365	1,775
1150	150 amp, with 14 branch breakers		1.03	7.767		860	425		1,285	1,600
1170	With PVC conduit and wire		.82	9.756		955	535		1,490	1,875
1180	With RGS conduit and wire		.67	11.940		1,275	655		1,930	2,400
1200	200 amp, with 18 branch breakers	2 Elec	1.80	8.889		1,150	485		1,635	2,025
1220	With PVC conduit and wire		1.46	10.959		1,250	600		1,850	2,300
1230	With RGS conduit and wire		1.24	12.903		1,650	705		2,355	2,900
1800	Lightning surge suppressor	1 Elec	32	.250		50.50	13.70		64.20	77
2000	Switch devices									
2100	Single pole, 15 amp, Ivory, with a 1-gang box, cover plate,									
2110	Type NM (Romex) cable	1 Elec	17.10	.468	Ea.	12.35	25.50		37.85	53
2120	Type MC (BX) cable		14.30	.559		21.50	30.50		52	71
2130	EMT & wire		5.71	1.401		30	76.50		106.50	152
2150	3-way, #14/3, type NM cable		14.55	.550		14.85	30		44.85	63
2170	Type MC cable		12.31	.650		27	35.50		62.50	84.50
2180	EMT & wire		5	1.600		32	87.50		119.50	171
2200	4-way, #14/3, type NM cable		14.55	.550		21	30		51	69.50
2220	Type MC cable		12.31	.650		33	35.50		68.50	91.50
2230	EMT & wire		5	1.600		38.50	87.50		126	178
2250	S.P., 20 amp, #12/2, type NM cable		13.33	.600		22	33		55	75
2270	Type MC cable		11.43	.700		28.50	38.50		67	91
2280	EMT & wire		4.85	1.649		41	90		131	185
2290	S.P. rotary dimmer, 600W, no wiring		17	.471		29.50	25.50		55	72.50
2300	S.P. rotary dimmer, 600W, type NM cable		14.55	.550		34.50	30		64.50	84.50
2320	Type MC cable		12.31	.650		43.50	35.50		79	103
2330	EMT & wire		5	1.600		53.50	87.50		141	195
2350	3-way rotary dimmer, type NM cable		13.33	.600		28.50	33		61.50	82.50
2370	Type MC cable		11.43	.700		37.50	38.50		76	101
2380	EMT & wire		4.85	1.649		47.50	90		137.50	193
2400	Interval timer wall switch, 20 amp, 1-30 min., #12/2									
2410	Type NM cable	1 Elec	14.55	.550	Ea.	56	30		86	108
2420	Type MC cable		12.31	.650		60	35.50		95.50	121
2430	EMT & wire		5	1.600		75	87.50		162.50	219
2500	Decorator style									
2510	S.P., 15 amp, type NM cable	1 Elec	17.10	.468	Ea.	16.25	25.50		41.75	57.50
2520	Type MC cable		14.30	.559		25	30.50		55.50	75
2530	EMT & wire		5.71	1.401		34	76.50		110.50	156
2550	3-way, #14/3, type NM cable		14.55	.550		18.75	30		48.75	67

26 05 90 – Residential Applications

26 05 90.10 Residential Wiring		Crew	Daily Output	Labor-Hours	Unit	Material	2015 Bare Costs Labor	Equipment	Total	Total Incl O&P
2570	Type MC cable	1 Elec	12.31	.650	Ea.	30.50	35.50		66	88.50
2580	EMT & wire		5	1.600		36	87.50		123.50	176
2600	4-way, #14/3, type NM cable		14.55	.550		25	30		55	74
2620	Type MC cable		12.31	.650		37	35.50		72.50	95.50
2630	EMT & wire		5	1.600		42	87.50		129.50	183
2650	S.P., 20 amp, #12/2, type NM cable		13.33	.600		26	33		59	79.50
2670	Type MC cable		11.43	.700		32.50	38.50		71	95
2680	EMT & wire		4.85	1.649		45	90		135	190
2700	S.P., slide dimmer, type NM cable		17.10	.468		28.50	25.50		54	71
2720	Type MC cable		14.30	.559		37.50	30.50		68	89
2730	EMT & wire		5.71	1.401		47.50	76.50		124	172
2750	S.P., touch dimmer, type NM cable		17.10	.468		32	25.50		57.50	75
2770	Type MC cable		14.30	.559		41	30.50		71.50	93
2780	EMT & wire		5.71	1.401		51.50	76.50		128	176
2800	3-way touch dimmer, type NM cable		13.33	.600		52	33		85	108
2820	Type MC cable		11.43	.700		60.50	38.50		99	127
2830	EMT & wire	↓	4.85	1.649	↓	71	90		161	218
3000	Combination devices									
3100	S.P. switch/15 amp recpt., Ivory, 1-gang box, plate									
3110	Type NM cable	1 Elec	11.43	.700	Ea.	24.50	38.50		63	86.50
3120	Type MC cable		10	.800		33.50	44		77.50	105
3130	EMT & wire		4.40	1.818		43.50	99.50		143	202
3150	S.P. switch/pilot light, type NM cable		11.43	.700		24.50	38.50		63	86.50
3170	Type MC cable		10	.800		33.50	44		77.50	105
3180	EMT & wire		4.43	1.806		43.50	99		142.50	201
3190	2-S.P. switches, 2-#14/2, no wiring		14	.571		6.90	31.50		38.40	56
3200	2-S.P. switches, 2-#14/2, type NM cables		10	.800		27.50	44		71.50	98
3220	Type MC cable		8.89	.900		41	49		90	121
3230	EMT & wire		4.10	1.951		46.50	107		153.50	216
3250	3-way switch/15 amp recpt., #14/3, type NM cable		10	.800		32	44		76	103
3270	Type MC cable		8.89	.900		44	49		93	124
3280	EMT & wire		4.10	1.951		49	107		156	219
3300	2-3 way switches, 2-#14/3, type NM cables		8.89	.900		41.50	49		90.50	122
3320	Type MC cable		8	1		60.50	54.50		115	151
3330	EMT & wire		4	2		56.50	109		165.50	231
3350	S.P. switch/20 amp recpt., #12/2, type NM cable		10	.800		32	44		76	103
3370	Type MC cable		8.89	.900		36	49		85	116
3380	EMT & wire	↓	4.10	1.951	↓	51	107		158	221
3400	Decorator style									
3410	S.P. switch/15 amp recpt., type NM cable	1 Elec	11.43	.700	Ea.	28.50	38.50		67	90.50
3420	Type MC cable		10	.800		37	44		81	109
3430	EMT & wire		4.40	1.818		47.50	99.50		147	206
3450	S.P. switch/pilot light, type NM cable		11.43	.700		28.50	38.50		67	90.50
3470	Type MC cable		10	.800		37.50	44		81.50	109
3480	EMT & wire		4.40	1.818		47.50	99.50		147	207
3500	2-S.P. switches, 2-#14/2, type NM cables		10	.800		31.50	44		75.50	103
3520	Type MC cable		8.89	.900		44.50	49		93.50	125
3530	EMT & wire		4.10	1.951		50.50	107		157.50	221
3550	3-way/15 amp recpt., #14/3, type NM cable		10	.800		36	44		80	108
3570	Type MC cable		8.89	.900		47.50	49		96.50	129
3580	EMT & wire		4.10	1.951		53	107		160	223
3650	2-3 way switches, 2-#14/3, type NM cables		8.89	.900		45.50	49		94.50	126
3670	Type MC cable	↓	8	1	↓	64.50	54.50		119	156

26 05 90.10 Residential Wiring		Crew	Daily Output	Labor-Hours	Unit	Material	Labor	Equipment	Total	Total Incl O&P
						2015 Bare Costs				
3680	EMT & wire	1 Elec	4	2	Ea.	60.50	109		169.50	236
3700	S.P. switch/20 amp recpt., #12/2, type NM cable		10	.800		35.50	44		79.50	108
3720	Type MC cable		8.89	.900		39.50	49		88.50	120
3730	EMT & wire	↓	4.10	1.951	↓	55	107		162	226
4000	Receptacle devices									
4010	Duplex outlet, 15 amp recpt., Ivory, 1-gang box, plate									
4015	Type NM cable	1 Elec	14.55	.550	Ea.	10.80	30		40.80	58.50
4020	Type MC cable		12.31	.650		19.75	35.50		55.25	76.50
4030	EMT & wire		5.33	1.501		28.50	82		110.50	159
4050	With #12/2, type NM cable		12.31	.650		13.30	35.50		48.80	69.50
4070	Type MC cable		10.67	.750		19.85	41		60.85	85.50
4080	EMT & wire		4.71	1.699		32.50	93		125.50	180
4100	20 amp recpt., #12/2, type NM cable		12.31	.650		19.95	35.50		55.45	77
4120	Type MC cable		10.67	.750		26.50	41		67.50	92.50
4130	EMT & wire	↓	4.71	1.699	↓	39	93		132	187
4140	For GFI see Section 26 05 90.10 line 4300 below									
4150	Decorator style, 15 amp recpt., type NM cable	1 Elec	14.55	.550	Ea.	14.70	30		44.70	62.50
4170	Type MC cable		12.31	.650		23.50	35.50		59	81
4180	EMT & wire		5.33	1.501		32.50	82		114.50	163
4200	With #12/2, type NM cable		12.31	.650		17.20	35.50		52.70	74
4220	Type MC cable		10.67	.750		23.50	41		64.50	89.50
4230	EMT & wire		4.71	1.699		36.50	93		129.50	184
4250	20 amp recpt. #12/2, type NM cable		12.31	.650		24	35.50		59.50	81
4270	Type MC cable		10.67	.750		30.50	41		71.50	97
4280	EMT & wire		4.71	1.699		43	93		136	192
4300	GFI, 15 amp recpt., type NM cable		12.31	.650		41.50	35.50		77	101
4320	Type MC cable		10.67	.750		50.50	41		91.50	119
4330	EMT & wire		4.71	1.699		59	93		152	209
4350	GFI with #12/2, type NM cable		10.67	.750		44	41		85	112
4370	Type MC cable		9.20	.870		50.50	47.50		98	129
4380	EMT & wire		4.21	1.900		63	104		167	231
4400	20 amp recpt., #12/2 type NM cable		10.67	.750		52	41		93	121
4420	Type MC cable		9.20	.870		58.50	47.50		106	138
4430	EMT & wire		4.21	1.900		71.50	104		175.50	240
4500	Weather-proof cover for above receptacles, add	↓	32	.250	↓	3.46	13.70		17.16	25
4550	Air conditioner outlet, 20 amp-240 volt recpt.									
4560	30′ of #12/2, 2 pole circuit breaker									
4570	Type NM cable	1 Elec	10	.800	Ea.	63	44		107	138
4580	Type MC cable		9	.889		70.50	48.50		119	153
4590	EMT & wire		4	2		82.50	109		191.50	260
4600	Decorator style, type NM cable		10	.800		68	44		112	143
4620	Type MC cable		9	.889		75.50	48.50		124	159
4630	EMT & wire	↓	4	2	↓	87	109		196	265
4650	Dryer outlet, 30 amp-240 volt recpt., 20′ of #10/3									
4660	2 pole circuit breaker									
4670	Type NM cable	1 Elec	6.41	1.248	Ea.	61.50	68.50		130	174
4680	Type MC cable		5.71	1.401		64.50	76.50		141	190
4690	EMT & wire	↓	3.48	2.299	↓	76	126		202	279
4700	Range outlet, 50 amp-240 volt recpt., 30′ of #8/3									
4710	Type NM cable	1 Elec	4.21	1.900	Ea.	90.50	104		194.50	261
4720	Type MC cable		4	2		124	109		233	305
4730	EMT & wire		2.96	2.703		108	148		256	350
4750	Central vacuum outlet, Type NM cable	↓	6.40	1.250	↓	58.50	68.50		127	171

26 05 90.10 Residential Wiring		Crew	Daily Output	Labor-Hours	Unit	Material	Labor	Equipment	Total	Total Incl O&P
						2015 Bare Costs				
4770	Type MC cable	1 Elec	5.71	1.401	Ea.	69.50	76.50		146	196
4780	EMT & wire	↓	3.48	2.299	↓	83.50	126		209.50	287
4800	30 amp-110 volt locking recpt., #10/2 circ. bkr.									
4810	Type NM cable	1 Elec	6.20	1.290	Ea.	67	70.50		137.50	183
4820	Type MC cable		5.40	1.481		80	81		161	213
4830	EMT & wire	↓	3.20	2.500	↓	94	137		231	315
4900	Low voltage outlets									
4910	Telephone recpt., 20′ of 4/C phone wire	1 Elec	26	.308	Ea.	9.30	16.85		26.15	36.50
4920	TV recpt., 20′ of RG59U coax wire, F type connector	"	16	.500	"	18.75	27.50		46.25	63
4950	Door bell chime, transformer, 2 buttons, 60′ of bellwire									
4970	Economy model	1 Elec	11.50	.696	Ea.	51.50	38		89.50	116
4980	Custom model		11.50	.696		99	38		137	168
4990	Luxury model, 3 buttons	↓	9.50	.842	↓	281	46		327	380
6000	Lighting outlets									
6050	Wire only (for fixture), type NM cable	1 Elec	32	.250	Ea.	6.90	13.70		20.60	28.50
6070	Type MC cable		24	.333		12.30	18.25		30.55	41.50
6080	EMT & wire		10	.800		20	44		64	90
6100	Box (4″), and wire (for fixture), type NM cable		25	.320		14.45	17.50		31.95	43
6120	Type MC cable		20	.400		19.90	22		41.90	56
6130	EMT & wire	↓	11	.727	↓	27.50	40		67.50	92
6200	Fixtures (use with lines 6050 or 6100 above)									
6210	Canopy style, economy grade	1 Elec	40	.200	Ea.	32	10.95		42.95	52
6220	Custom grade		40	.200		53.50	10.95		64.45	76
6250	Dining room chandelier, economy grade		19	.421		79.50	23		102.50	123
6260	Custom grade		19	.421		315	23		338	380
6270	Luxury grade		15	.533		715	29		744	830
6310	Kitchen fixture (fluorescent), economy grade		30	.267		71.50	14.60		86.10	102
6320	Custom grade		25	.320		219	17.50		236.50	268
6350	Outdoor, wall mounted, economy grade		30	.267		30	14.60		44.60	55.50
6360	Custom grade		30	.267		120	14.60		134.60	155
6370	Luxury grade		25	.320		248	17.50		265.50	300
6410	Outdoor PAR floodlights, 1 lamp, 150 watt		20	.400		32.50	22		54.50	69.50
6420	2 lamp, 150 watt each		20	.400		53.50	22		75.50	93
6425	Motion sensing, 2 lamp, 150 watt each		20	.400		87	22		109	130
6430	For infrared security sensor, add		32	.250		132	13.70		145.70	166
6450	Outdoor, quartz-halogen, 300 watt flood		20	.400		39	22		61	76.50
6600	Recessed downlight, round, pre-wired, 50 or 75 watt trim		30	.267		80	14.60		94.60	111
6610	With shower light trim		30	.267		89	14.60		103.60	121
6620	With wall washer trim		28	.286		99.50	15.65		115.15	133
6630	With eye-ball trim		28	.286		99.50	15.65		115.15	133
6700	Porcelain lamp holder		40	.200		2.96	10.95		13.91	20
6710	With pull switch		40	.200		6.55	10.95		17.50	24
6750	Fluorescent strip, 2-20 watt tube, wrap around diffuser, 24″		24	.333		49.50	18.25		67.75	82.50
6760	1-34 watt tube, 48″		24	.333		87	18.25		105.25	124
6770	2-34 watt tubes, 48″		20	.400		103	22		125	147
6800	Bathroom heat lamp, 1-250 watt		28	.286		44	15.65		59.65	72
6810	2-250 watt lamps	↓	28	.286	↓	70	15.65		85.65	101
6820	For timer switch, see Section 26 05 90.10 line 2400									
6900	Outdoor post lamp, incl. post, fixture, 35′ of #14/2									
6910	Type NMC cable	1 Elec	3.50	2.286	Ea.	258	125		383	480
6920	Photo-eye, add		27	.296		32	16.20		48.20	60.50
6950	Clock dial time switch, 24 hr., w/enclosure, type NM cable		11.43	.700		73.50	38.50		112	141
6970	Type MC cable	↓	11	.727	↓	82.50	40		122.50	152

26 05 90.10 Residential Wiring		Crew	Daily Output	Labor-Hours	Unit	Material	Labor	Equipment	Total	Total Incl O&P
						2015 Bare Costs				
6980	EMT & wire	1 Elec	4.85	1.649	Ea.	91	90		181	240
7000	Alarm systems									
7050	Smoke detectors, box, #14/3, type NM cable	1 Elec	14.55	.550	Ea.	34.50	30		64.50	84.50
7070	Type MC cable		12.31	.650		44	35.50		79.50	104
7080	EMT & wire		5	1.600		49.50	87.50		137	191
7090	For relay output to security system, add					12			12	13.20
8000	Residential equipment									
8050	Disposal hook-up, incl. switch, outlet box, 3′ of flex									
8060	20 amp-1 pole circ. bkr., and 25′ of #12/2									
8070	Type NM cable	1 Elec	10	.800	Ea.	32.50	44		76.50	104
8080	Type MC cable		8	1		39.50	54.50		94	128
8090	EMT & wire		5	1.600		54.50	87.50		142	196
8100	Trash compactor or dishwasher hook-up, incl. outlet box,									
8110	3′ of flex, 15 amp-1 pole circ. bkr., and 25′ of #14/2									
8120	Type NM cable	1 Elec	10	.800	Ea.	24	44		68	94
8130	Type MC cable		8	1		34	54.50		88.50	122
8140	EMT & wire		5	1.600		46	87.50		133.50	187
8150	Hot water sink dispensor hook-up, use line 8100									
8200	Vent/exhaust fan hook-up, type NM cable	1 Elec	32	.250	Ea.	6.90	13.70		20.60	28.50
8220	Type MC cable		24	.333		12.30	18.25		30.55	41.50
8230	EMT & wire		10	.800		20	44		64	90
8250	Bathroom vent fan, 50 CFM (use with above hook-up)									
8260	Economy model	1 Elec	15	.533	Ea.	23	29		52	70.50
8270	Low noise model		15	.533		41	29		70	90
8280	Custom model		12	.667		127	36.50		163.50	197
8300	Bathroom or kitchen vent fan, 110 CFM									
8310	Economy model	1 Elec	15	.533	Ea.	67.50	29		96.50	120
8320	Low noise model	"	15	.533	"	92	29		121	146
8350	Paddle fan, variable speed (w/o lights)									
8360	Economy model (AC motor)	1 Elec	10	.800	Ea.	109	44		153	188
8362	With light kit		10	.800		150	44		194	233
8370	Custom model (AC motor)		10	.800		227	44		271	320
8372	With light kit		10	.800		268	44		312	365
8380	Luxury model (DC motor)		8	1		330	54.50		384.50	450
8382	With light kit		8	1		375	54.50		429.50	495
8390	Remote speed switch for above, add		12	.667		37.50	36.50		74	97.50
8500	Whole house exhaust fan, ceiling mount, 36", variable speed									
8510	Remote switch, incl. shutters, 20 amp-1 pole circ. bkr.									
8520	30′ of #12/2, type NM cable	1 Elec	4	2	Ea.	1,300	109		1,409	1,600
8530	Type MC cable		3.50	2.286		1,300	125		1,425	1,625
8540	EMT & wire		3	2.667		1,325	146		1,471	1,675
8600	Whirlpool tub hook-up, incl. timer switch, outlet box									
8610	3′ of flex, 20 amp-1 pole GFI circ. bkr.									
8620	30′ of #12/2, type NM cable	1 Elec	5	1.600	Ea.	142	87.50		229.50	292
8630	Type MC cable		4.20	1.905		145	104		249	320
8640	EMT & wire		3.40	2.353		158	129		287	370
8650	Hot water heater hook-up, incl. 1-2 pole circ. bkr., box;									
8660	3′ of flex, 20′ of #10/2, type NM cable	1 Elec	5	1.600	Ea.	33	87.50		120.50	173
8670	Type MC cable		4.20	1.905		43	104		147	209
8680	EMT & wire		3.40	2.353		48.50	129		177.50	252
9000	Heating/air conditioning									
9050	Furnace/boiler hook-up, incl. firestat, local on-off switch									
9060	Emergency switch, and 40′ of type NM cable	1 Elec	4	2	Ea.	52.50	109		161.50	227

26 05 Common Work Results for Electrical

26 05 90 – Residential Applications

26 05 90.10 Residential Wiring		Crew	Daily Output	Labor-Hours	Unit	Material	2015 Bare Costs Labor	Equipment	Total	Total Incl O&P
9070	Type MC cable	1 Elec	3.50	2.286	Ea.	65.50	125		190.50	267
9080	EMT & wire	↓	1.50	5.333	↓	83	292		375	540
9100	Air conditioner hook-up, incl. local 60 amp disc. switch									
9110	3′ sealtite, 40 amp, 2 pole circuit breaker									
9130	40′ of #8/2, type NM cable	1 Elec	3.50	2.286	Ea.	163	125		288	375
9140	Type MC cable		3	2.667		216	146		362	465
9150	EMT & wire	↓	1.30	6.154	↓	204	335		539	745
9200	Heat pump hook-up, 1-40 & 1-100 amp 2 pole circ. bkr.									
9210	Local disconnect switch, 3′ sealtite									
9220	40′ of #8/2 & 30′ of #3/2									
9230	Type NM cable	1 Elec	1.30	6.154	Ea.	520	335		855	1,100
9240	Type MC cable		1.08	7.407		535	405		940	1,200
9250	EMT & wire	↓	.94	8.511	↓	585	465		1,050	1,375
9500	Thermostat hook-up, using low voltage wire									
9520	Heating only, 25′ of #18-3	1 Elec	24	.333	Ea.	8.80	18.25		27.05	37.50
9530	Heating/cooling, 25′ of #18-4	"	20	.400	"	11.45	22		33.45	46.50

26 09 Instrumentation and Control for Electrical Systems

26 09 13 – Electrical Power Monitoring

26 09 13.10 Switchboard Instruments

		Crew	Daily Output	Labor-Hours	Unit	Material	Labor	Equipment	Total	Total Incl O&P
0010	**SWITCHBOARD INSTRUMENTS** 3 phase, 4 wire									
0100	AC indicating, ammeter & switch	1 Elec	8	1	Ea.	2,500	54.50		2,554.50	2,825
0200	Voltmeter & switch		8	1		2,500	54.50		2,554.50	2,825
0300	Wattmeter		8	1		4,250	54.50		4,304.50	4,750
0400	AC recording, ammeter		4	2		7,550	109		7,659	8,500
0500	Voltmeter		4	2		7,550	109		7,659	8,500
0600	Ground fault protection, zero sequence		2.70	2.963		6,675	162		6,837	7,600
0700	Ground return path		2.70	2.963		6,675	162		6,837	7,600
0800	3 current transformers, 5 to 800 amp		2	4		3,100	219		3,319	3,775
0900	1000 to 1500 amp		1.30	6.154		4,475	335		4,810	5,450
1200	2000 to 4000 amp		1	8		5,275	440		5,715	6,475
1300	Fused potential transformer, maximum 600 volt	↓	8	1	↓	1,175	54.50		1,229.50	1,350

26 09 13.20 Voltage Monitor Systems

		Crew	Daily Output	Labor-Hours	Unit	Material	Labor	Equipment	Total	Total Incl O&P
0010	**VOLTAGE MONITOR SYSTEMS** (test equipment)									
0100	AC voltage monitor system, 120/240 V, one-channel				Ea.	2,750			2,750	3,025
0110	Modem adapter					345			345	380
0120	Add-on detector only					1,450			1,450	1,600
0150	AC voltage remote monitor sys., 3 channel, 120, 230, or 480 V					5,000			5,000	5,500
0160	With internal modem					5,275			5,275	5,825
0170	Combination temperature and humidity probe					775			775	855
0180	Add-on detector only					3,625			3,625	4,000
0190	With internal modem				↓	3,950			3,950	4,350

26 09 13.30 Smart Metering

		Crew	Daily Output	Labor-Hours	Unit	Material	Labor	Equipment	Total	Total Incl O&P
0010	**SMART METERING**, In panel									
0100	Single phase, 120/208 volt, 100 amp G	1 Elec	8.78	.911	Ea.	375	50		425	485
0120	200 amp G		8.78	.911		375	50		425	485
0200	277 volt, 100 amp G		8.78	.911		400	50		450	515
0220	200 amp G		8.78	.911		400	50		450	515
1100	Three phase, 120/208 volt, 100 amp G		4.69	1.706		690	93.50		783.50	905
1120	200 amp G	↓	4.69	1.706	↓	690	93.50		783.50	905

26 09 Instrumentation and Control for Electrical Systems

26 09 13 – Electrical Power Monitoring

26 09 13.30 Smart Metering			Crew	Daily Output	Labor-Hours	Unit	Material	2015 Bare Costs Labor	Equipment	Total	Total Incl O&P
1130	400 amp	G	1 Elec	4.69	1.706	Ea.	690	93.50		783.50	905
1140	800 amp	G		4.69	1.706		690	93.50		783.50	905
1150	1600 amp	G		4.69	1.706		690	93.50		783.50	905
1200	277/480 volt, 100 amp	G		4.69	1.706		775	93.50		868.50	995
1220	200 amp	G		4.69	1.706		775	93.50		868.50	995
1230	400 amp	G		4.69	1.706		775	93.50		868.50	995
1240	800 amp	G		4.69	1.706		775	93.50		868.50	995
1250	1600 amp	G		4.69	1.706		785	93.50		878.50	1,000
2000	Data recorder, 8 meters	G		10.97	.729		1,400	40		1,440	1,575
2100	16 meters	G	↓	8.53	.938		1,950	51.50		2,001.50	2,225
3000	Software package, per meter, basic	G					236			236	260
3100	Premium	G				↓	610			610	675

26 09 23 – Lighting Control Devices

26 09 23.10 Energy Saving Lighting Devices			Crew	Daily Output	Labor-Hours	Unit	Material	Labor	Equipment	Total	Total Incl O&P
0010	**ENERGY SAVING LIGHTING DEVICES**										
0100	Occupancy sensors, passive infrared ceiling mounted	G	1 Elec	7	1.143	Ea.	81	62.50		143.50	187
0110	Ultrasonic ceiling mounted	G		7	1.143		89.50	62.50		152	196
0120	Dual technology ceiling mounted	G		6.50	1.231		131	67.50		198.50	248
0150	Automatic wall switches	G		24	.333		64.50	18.25		82.75	99
0160	Daylighting sensor, manual control, ceiling mounted	G		7	1.143		111	62.50		173.50	219
0170	Remote and dimming control with remote controller	G		6.50	1.231		152	67.50		219.50	271
0200	Remote power pack	G		10	.800		31	44		75	102
0250	Photoelectric control, S.P.S.T. 120 V	G		8	1		19.35	54.50		73.85	106
0300	S.P.S.T. 208 V/277 V	G		8	1		28.50	54.50		83	116
0350	D.P.S.T. 120 V	G		6	1.333		217	73		290	350
0400	D.P.S.T. 208 V/277 V	G		6	1.333		176	73		249	305
0450	S.P.D.T. 208 V/277 V	G		6	1.333		211	73		284	345
0460	Daylight level sensor, wall mounted, on/off or dimming	G	↓	8	1	↓	135	54.50		189.50	234

26 09 36 – Modular Dimming Controls

26 09 36.13 Manual Modular Dimming Controls			Crew	Daily Output	Labor-Hours	Unit	Material	Labor	Equipment	Total	Total Incl O&P
0010	**MANUAL MODULAR DIMMING CONTROLS**										
2000	Lighting control module	G	1 Elec	2	4	Ea.	355	219		574	730

26 12 Medium-Voltage Transformers

26 12 19 – Pad-Mounted, Liquid-Filled, Medium-Voltage Transformers

26 12 19.10 Transformer, Oil-Filled		Crew	Daily Output	Labor-Hours	Unit	Material	Labor	Equipment	Total	Total Incl O&P
0010	**TRANSFORMER, OIL-FILLED** primary delta or Y,									
0050	Pad mounted 5 kV or 15 kV, with taps, 277/480 V secondary, 3 phase									
0100	150 kVA	R-3	.65	30.769	Ea.	9,250	1,675	212	11,137	13,000
0110	225 kVA		.55	36.364		10,500	1,975	251	12,726	15,000
0200	300 kVA		.45	44.444		13,200	2,425	305	15,930	18,600
0300	500 kVA		.40	50		18,700	2,725	345	21,770	25,200
0400	750 kVA		.38	52.632		23,700	2,850	365	26,915	31,000
0500	1000 kVA		.26	76.923		28,100	4,175	530	32,805	38,000
0600	1500 kVA		.23	86.957		33,400	4,725	600	38,725	44,700
0700	2000 kVA		.20	100		42,100	5,425	690	48,215	55,500
0710	2500 kVA		.19	105		51,000	5,725	725	57,450	65,500
0720	3000 kVA		.17	117		61,500	6,400	810	68,710	78,500
0800	3750 kVA	↓	.16	125	↓	79,000	6,800	860	86,660	98,500

26 12 Medium-Voltage Transformers

26 12 19 – Pad-Mounted, Liquid-Filled, Medium-Voltage Transformers

26 12 19.20 Transformer, Liquid-Filled		Crew	Daily Output	Labor-Hours	Unit	2015 Bare Costs Material	Labor	Equipment	Total	Total Incl O&P
0010	**TRANSFORMER, LIQUID-FILLED** Pad mounted									
0020	5 kV or 15 kV primary, 277/480 volt secondary, 3 phase									
0050	225 kVA	R-3	.55	36.364	Ea.	13,600	1,975	251	15,826	18,400
0100	300 kVA		.45	44.444		16,200	2,425	305	18,930	21,900
0200	500 kVA		.40	50		20,400	2,725	345	23,470	27,100
0250	750 kVA		.38	52.632		26,400	2,850	365	29,615	33,900
0300	1000 kVA		.26	76.923		30,600	4,175	530	35,305	40,800
0350	1500 kVA		.23	86.957		35,800	4,725	600	41,125	47,300
0400	2000 kVA		.20	100		44,300	5,425	690	50,415	58,000
0450	2500 kVA		.19	105		53,000	5,725	725	59,450	67,500

26 13 Medium-Voltage Switchgear

26 13 16 – Medium-Voltage Fusible Interrupter Switchgear

26 13 16.10 Switchgear		Crew	Daily Output	Labor-Hours	Unit	Material	Labor	Equipment	Total	Total Incl O&P
0010	**SWITCHGEAR**, Incorporate switch with cable connections, transformer,									
0100	& Low Voltage section									
0200	Load interrupter switch, 600 amp, 2 position									
0300	NEMA 1, 4.8 kV, 300 kVA & below w/CLF fuses	R-3	.40	50	Ea.	21,400	2,725	345	24,470	28,200
0400	400 kVA & above w/CLF fuses		.38	52.632		23,900	2,850	365	27,115	31,200
0500	Non fusible		.41	48.780		17,400	2,650	335	20,385	23,600
0600	13.8 kV, 300 kVA & below w/CLF fuses		.38	52.632		26,900	2,850	365	30,115	34,500
0700	400 kVA & above w/CLF fuses		.36	55.556		26,900	3,025	385	30,310	34,700
0800	Non fusible		.40	50		20,400	2,725	345	23,470	27,100
0900	Cable lugs for 2 feeders 4.8 kV or 13.8 kV	1 Elec	8	1		655	54.50		709.50	805
1000	Pothead, one 3 conductor or three 1 conductor		4	2		3,125	109		3,234	3,625
1100	Two 3 conductor or six 1 conductor		2	4		6,200	219		6,419	7,150
1200	Key interlocks		8	1		725	54.50		779.50	880
1300	Lightning arresters, Distribution class (no charge)									
1400	Intermediate class or line type 4.8 kV	1 Elec	2.70	2.963	Ea.	3,525	162		3,687	4,125
1500	13.8 kV		2	4		4,675	219		4,894	5,475
1600	Station class, 4.8 kV		2.70	2.963		6,025	162		6,187	6,875
1700	13.8 kV		2	4		10,400	219		10,619	11,700
1800	Transformers, 4800 volts to 480/277 volts, 75 kVA	R-3	.68	29.412		18,300	1,600	203	20,103	22,900
1900	112.5 kVA		.65	30.769		22,400	1,675	212	24,287	27,400
2000	150 kVA		.57	35.088		25,500	1,900	242	27,642	31,200
2100	225 kVA		.48	41.667		29,300	2,275	287	31,862	36,000
2200	300 kVA		.41	48.780		32,700	2,650	335	35,685	40,500
2300	500 kVA		.36	55.556		43,100	3,025	385	46,510	52,500
2400	750 kVA		.29	68.966		48,900	3,750	475	53,125	60,000
2500	13,800 volts to 480/277 volts, 75 kVA		.61	32.787		25,800	1,775	226	27,801	31,400
2600	112.5 kVA		.55	36.364		34,300	1,975	251	36,526	41,100
2700	150 kVA		.49	40.816		34,600	2,225	281	37,106	41,900
2800	225 kVA		.41	48.780		40,000	2,650	335	42,985	48,500
2900	300 kVA		.37	54.054		40,800	2,925	370	44,095	49,900
3000	500 kVA		.31	64.516		45,100	3,500	445	49,045	55,500
3100	750 kVA		.26	76.923		49,700	4,175	530	54,405	61,500
3200	Forced air cooling & temperature alarm	1 Elec	1	8		4,000	440		4,440	5,075
3300	Low voltage components									
3400	Maximum panel height 49-1/2", single or twin row									
3500	Breaker heights, type FA or FH, 6"									

26 13 Medium-Voltage Switchgear

26 13 16 – Medium-Voltage Fusible Interrupter Switchgear

26 13 16.10 Switchgear		Crew	Daily Output	Labor-Hours	Unit	Material	2015 Bare Costs Labor	Equipment	Total	Total Incl O&P
3600	type KA or KH, 8″									
3700	type LA, 11″									
3800	type MA, 14″									
3900	Breakers, 2 pole, 15 to 60 amp, type FA	1 Elec	5.60	1.429	Ea.	350	78		428	505
4000	70 to 100 amp, type FA		4.20	1.905		445	104		549	650
4100	15 to 60 amp, type FH		5.60	1.429		580	78		658	760
4200	70 to 100 amp, type FH		4.20	1.905		675	104		779	905
4300	125 to 225 amp, type KA		3.40	2.353		1,025	129		1,154	1,350
4400	125 to 225 amp, type KH		3.40	2.353		2,525	129		2,654	3,000
4500	125 to 400 amp, type LA		2.50	3.200		1,950	175		2,125	2,400
4600	125 to 600 amp, type MA		1.80	4.444		2,975	243		3,218	3,650
4700	700 & 800 amp, type MA		1.50	5.333		3,850	292		4,142	4,700
4800	3 pole, 15 to 60 amp, type FA		5.30	1.509		455	82.50		537.50	630
4900	70 to 100 amp, type FA		4	2		560	109		669	785
5000	15 to 60 amp, type FH		5.30	1.509		680	82.50		762.50	875
5100	70 to 100 amp, type FH		4	2		770	109		879	1,025
5200	125 to 225 amp, type KA		3.20	2.500		1,300	137		1,437	1,625
5300	125 to 225 amp, type KH		3.20	2.500		2,925	137		3,062	3,425
5400	125 to 400 amp, type LA		2.30	3.478		2,400	190		2,590	2,950
5500	125 to 600 amp, type MA		1.60	5		4,075	274		4,349	4,925
5600	700 & 800 amp, type MA		1.30	6.154		5,300	335		5,635	6,350

26 22 Low-Voltage Transformers

26 22 13 – Low-Voltage Distribution Transformers

26 22 13.10 Transformer, Dry-Type		Crew	Daily Output	Labor-Hours	Unit	Material	Labor	Equipment	Total	Total Incl O&P
0010	**TRANSFORMER, DRY-TYPE**									
0050	Single phase, 240/480 volt primary, 120/240 volt secondary									
0100	1 kVA	1 Elec	2	4	Ea.	330	219		549	700
0300	2 kVA		1.60	5		490	274		764	965
0500	3 kVA		1.40	5.714		610	315		925	1,150
0700	5 kVA		1.20	6.667		835	365		1,200	1,475
0900	7.5 kVA	2 Elec	2.20	7.273		1,175	400		1,575	1,900
1100	10 kVA		1.60	10		1,450	545		1,995	2,450
1300	15 kVA		1.20	13.333		1,700	730		2,430	3,000
1500	25 kVA		1	16		2,125	875		3,000	3,675
1700	37.5 kVA		.80	20		2,750	1,100		3,850	4,725
1900	50 kVA		.70	22.857		3,250	1,250		4,500	5,525
2100	75 kVA		.65	24.615		4,325	1,350		5,675	6,825
2110	100 kVA	R-3	.90	22.222		5,625	1,200	153	6,978	8,225
2120	167 kVA	″	.80	25		9,325	1,350	172	10,847	12,600
2190	480 V primary 120/240 V secondary, nonvent., 15 kVA	2 Elec	1.20	13.333		1,575	730		2,305	2,850
2200	25 kVA		.90	17.778		2,300	970		3,270	4,050
2210	37 kVA		.75	21.333		2,750	1,175		3,925	4,825
2220	50 kVA		.65	24.615		3,250	1,350		4,600	5,675
2230	75 kVA		.60	26.667		4,325	1,450		5,775	7,000
2240	100 kVA		.50	32		5,625	1,750		7,375	8,875
2250	Low operating temperature(80°C), 25 kVA		1	16		4,200	875		5,075	5,975
2260	37 kVA		.80	20		4,525	1,100		5,625	6,675
2270	50 kVA		.70	22.857		5,900	1,250		7,150	8,400
2280	75 kVA		.65	24.615		9,425	1,350		10,775	12,500
2290	100 kVA		.55	29.091		9,800	1,600		11,400	13,300

26 22 13 – Low-Voltage Distribution Transformers

26 22 13.10 Transformer, Dry-Type		Crew	Daily Output	Labor-Hours	Unit	Material	2015 Bare Costs Labor	Equipment	Total	Total Incl O&P
2300	3 phase, 480 volt primary 120/208 volt secondary									
2310	Ventilated, 3 kVA	1 Elec	1	8	Ea.	910	440		1,350	1,675
2700	6 kVA		.80	10		1,025	545		1,570	1,975
2900	9 kVA		.70	11.429		1,075	625		1,700	2,175
3100	15 kVA	2 Elec	1.10	14.545		1,300	795		2,095	2,650
3300	30 kVA		.90	17.778		1,425	970		2,395	3,075
3500	45 kVA		.80	20		1,700	1,100		2,800	3,575
3700	75 kVA		.70	22.857		2,375	1,250		3,625	4,525
3900	112.5 kVA	R-3	.90	22.222		3,400	1,200	153	4,753	5,800
4100	150 kVA		.85	23.529		4,450	1,275	162	5,887	7,075
4300	225 kVA		.65	30.769		6,075	1,675	212	7,962	9,500
4500	300 kVA		.55	36.364		7,600	1,975	251	9,826	11,700
4700	500 kVA		.45	44.444		12,600	2,425	305	15,330	18,000
4800	750 kVA		.35	57.143		21,100	3,100	395	24,595	28,600
4820	1000 kVA		.32	62.500		24,900	3,400	430	28,730	33,200
4850	K-4 rated, 15 kVA	2 Elec	1.10	14.545		3,325	795		4,120	4,875
4855	30 kVA		.90	17.778		5,000	970		5,970	6,975
4860	45 kVA		.80	20		6,000	1,100		7,100	8,300
4865	75 kVA		.70	22.857		9,050	1,250		10,300	11,900
4870	112.5 kVA	R-3	.90	22.222		12,000	1,200	153	13,353	15,300
4875	150 kVA		.85	23.529		15,700	1,275	162	17,137	19,500
4880	225 kVA		.65	30.769		21,900	1,675	212	23,787	26,900
4885	300 kVA		.55	36.364		28,700	1,975	251	30,926	34,900
4890	500 kVA		.45	44.444		40,100	2,425	305	42,830	48,200
4900	K-13 rated, 15 kVA	2 Elec	1.10	14.545		3,775	795		4,570	5,375
4905	30 kVA		.90	17.778		5,675	970		6,645	7,750
4910	45 kVA		.80	20		6,825	1,100		7,925	9,200
4915	75 kVA		.70	22.857		10,300	1,250		11,550	13,200
4920	112.5 kVA	R-3	.90	22.222		13,700	1,200	153	15,053	17,100
4925	150 kVA		.85	23.529		17,900	1,275	162	19,337	21,900
4930	225 kVA		.65	30.769		24,400	1,675	212	26,287	29,600
4935	300 kVA		.55	36.364		32,500	1,975	251	34,726	39,100
4940	500 kVA		.45	44.444		54,000	2,425	305	56,730	63,500
5020	480 volt primary 120/208 volt secondary									
5030	Nonventilated, 15 kVA	2 Elec	1.10	14.545	Ea.	3,225	795		4,020	4,775
5040	30 kVA		.80	20		4,600	1,100		5,700	6,775
5050	45 kVA		.70	22.857		6,650	1,250		7,900	9,225
5060	75 kVA		.65	24.615		8,875	1,350		10,225	11,800
5070	112.5 kVA	R-3	.85	23.529		13,300	1,275	162	14,737	16,800
5081	150 kVA		.85	23.529		16,000	1,275	162	17,437	19,800
5090	225 kVA		.60	33.333		17,900	1,800	230	19,930	22,800
5100	300 kVA		.50	40		20,300	2,175	276	22,751	26,000
5200	Low operating temperature (80°C), 30 kVA	2 Elec	.90	17.778		4,000	970		4,970	5,900
5210	45 kVA		.80	20		5,450	1,100		6,550	7,700
5220	75 kVA		.70	22.857		7,975	1,250		9,225	10,700
5230	112.5 kVA	R-3	.90	22.222		11,000	1,200	153	12,353	14,000
5240	150 kVA		.85	23.529		14,000	1,275	162	15,437	17,600
5250	225 kVA		.65	30.769		19,100	1,675	212	20,987	23,800
5260	300 kVA		.55	36.364		27,000	1,975	251	29,226	33,100
5270	500 kVA		.45	44.444		34,500	2,425	305	37,230	42,100
5380	3 phase, 5 kV primary 277/480 volt secondary									
5400	High voltage, 112.5 kVA	R-3	.85	23.529	Ea.	17,100	1,275	162	18,537	21,000
5410	150 kVA		.65	30.769		18,600	1,675	212	20,487	23,200

26 22 Low-Voltage Transformers

26 22 13 – Low-Voltage Distribution Transformers

26 22 13.10 Transformer, Dry-Type		Crew	Daily Output	Labor-Hours	Unit	2015 Bare Costs Material	Labor	Equipment	Total	Total Incl O&P
5420	225 kVA	R-3	.55	36.364	Ea.	21,900	1,975	251	24,126	27,500
5430	300 kVA		.45	44.444		28,000	2,425	305	30,730	34,900
5440	500 kVA		.35	57.143		36,100	3,100	395	39,595	45,000
5450	750 kVA		.32	62.500		56,500	3,400	430	60,330	68,000
5460	1000 kVA		.30	66.667		66,000	3,625	460	70,085	78,500
5470	1500 kVA		.27	74.074		77,000	4,025	510	81,535	91,500
5480	2000 kVA		.25	80		90,000	4,350	550	94,900	106,500
5490	2500 kVA		.20	100		101,500	5,425	690	107,615	120,500
5500	3000 kVA		.18	111		133,000	6,025	765	139,790	156,500
5590	15 kV primary 277/480 volt secondary									
5600	High voltage, 112.5 kVA	R-3	.85	23.529	Ea.	26,000	1,275	162	27,437	30,800
5610	150 kVA		.65	30.769		30,400	1,675	212	32,287	36,200
5620	225 kVA		.55	36.364		33,600	1,975	251	35,826	40,300
5630	300 kVA		.45	44.444		39,500	2,425	305	42,230	47,600
5640	500 kVA		.35	57.143		50,000	3,100	395	53,495	60,500
5650	750 kVA		.32	62.500		65,500	3,400	430	69,330	78,000
5660	1000 kVA		.30	66.667		74,500	3,625	460	78,585	88,000
5670	1500 kVA		.27	74.074		86,000	4,025	510	90,535	101,500
5680	2000 kVA		.25	80		95,500	4,350	550	100,400	112,500
5690	2500 kVA		.20	100		110,500	5,425	690	116,615	130,500
5700	3000 kVA		.18	111		131,500	6,025	765	138,290	154,500
6000	2400 volt primary, 480 volt secondary, 300 kVA		.45	44.444		28,000	2,425	305	30,730	34,900
6010	500 kVA		.35	57.143		36,100	3,100	395	39,595	45,000
6020	750 kVA		.32	62.500		52,000	3,400	430	55,830	63,000
9000	Minimum labor/equipment charge	1 Elec	1	8	Job		440		440	680

26 22 13.20 Isolating Panels

		Crew	Daily Output	Labor-Hours	Unit	Material	Labor	Equipment	Total	Total Incl O&P
0010	**ISOLATING PANELS** used with isolating transformers									
0020	For hospital applications									
0100	Critical care area, 8 circuit, 3 kVA	1 Elec	.58	13.793	Ea.	6,750	755		7,505	8,600
0200	5 kVA		.54	14.815		6,850	810		7,660	8,775
0400	7.5 kVA		.52	15.385		7,025	840		7,865	9,025
0600	10 kVA		.44	18.182		7,325	995		8,320	9,600
0800	Operating room power & lighting, 8 circuit, 3 kVA		.58	13.793		4,575	755		5,330	6,200
1000	5 kVA		.54	14.815		5,100	810		5,910	6,850
1200	7.5 kVA		.52	15.385		5,750	840		6,590	7,625
1400	10 kVA		.44	18.182		6,425	995		7,420	8,600
1600	X-ray systems, 15 kVA, 90 amp		.44	18.182		12,200	995		13,195	15,100
1800	25 kVA, 125 amp		.36	22.222		14,800	1,225		16,025	18,200

26 22 13.30 Isolating Transformer

		Crew	Daily Output	Labor-Hours	Unit	Material	Labor	Equipment	Total	Total Incl O&P
0010	**ISOLATING TRANSFORMER**									
0100	Single phase, 120/240 volt primary, 120/240 volt secondary									
0200	0.50 kVA	1 Elec	4	2	Ea.	380	109		489	585
0400	1 kVA		2	4		540	219		759	935
0600	2 kVA		1.60	5		805	274		1,079	1,300
0800	3 kVA		1.40	5.714		810	315		1,125	1,375
1000	5 kVA		1.20	6.667		1,075	365		1,440	1,750
1200	7.5 kVA		1.10	7.273		1,350	400		1,750	2,100
1400	10 kVA		.80	10		1,725	545		2,270	2,750
1600	15 kVA		.60	13.333		2,200	730		2,930	3,550
1800	25 kVA		.50	16		3,175	875		4,050	4,850
1810	37.5 kVA	2 Elec	.80	20		5,250	1,100		6,350	7,475
1820	75 kVA	"	.65	24.615		7,900	1,350		9,250	10,800

26 22 Low-Voltage Transformers

26 22 13 – Low-Voltage Distribution Transformers

26 22 13.30 Isolating Transformer		Crew	Daily Output	Labor-Hours	Unit	2015 Bare Costs Material	Labor	Equipment	Total	Total Incl O&P
1830	3 phase, 120/240 V primary, 120/240 V secondary, 112.5 kVA	R-3	.90	22.222	Ea.	9,950	1,200	153	11,303	13,000
1840	150 kVA		.85	23.529		12,700	1,275	162	14,137	16,100
1850	225 kVA		.65	30.769		17,600	1,675	212	19,487	22,200
1860	300 kVA		.55	36.364		23,400	1,975	251	25,626	29,200
1870	500 kVA		.45	44.444		39,000	2,425	305	41,730	47,100
1880	750 kVA		.35	57.143		39,400	3,100	395	42,895	48,700

26 22 13.90 Transformer Handling		Crew	Daily Output	Labor-Hours	Unit	Material	Labor	Equipment	Total	Total Incl O&P
0010	**TRANSFORMER HANDLING** Add to normal labor cost in restricted areas									
5000	Transformers									
5150	15 kVA, approximately 200 pounds	2 Elec	2.70	5.926	Ea.		325		325	500
5160	25 kVA, approximately 300 pounds		2.50	6.400			350		350	540
5170	37.5 kVA, approximately 400 pounds		2.30	6.957			380		380	590
5180	50 kVA, approximately 500 pounds		2	8			440		440	680
5190	75 kVA, approximately 600 pounds		1.80	8.889			485		485	755
5200	100 kVA, approximately 700 pounds		1.60	10			545		545	845
5210	112.5 kVA, approximately 800 pounds	3 Elec	2.20	10.909			595		595	925
5220	125 kVA, approximately 900 pounds		2	12			655		655	1,025
5230	150 kVA, approximately 1000 pounds		1.80	13.333			730		730	1,125
5240	167 kVA, approximately 1200 pounds		1.60	15			820		820	1,275
5250	200 kVA, approximately 1400 pounds		1.40	17.143			940		940	1,450
5260	225 kVA, approximately 1600 pounds		1.30	18.462			1,000		1,000	1,575
5270	250 kVA, approximately 1800 pounds		1.10	21.818			1,200		1,200	1,850
5280	300 kVA, approximately 2000 pounds		1	24			1,325		1,325	2,025
5290	500 kVA, approximately 3000 pounds		.75	32			1,750		1,750	2,700
5300	600 kVA, approximately 3500 pounds		.67	35.821			1,950		1,950	3,025
5310	750 kVA, approximately 4000 pounds		.60	40			2,200		2,200	3,400
5320	1000 kVA, approximately 5000 pounds		.50	48			2,625		2,625	4,075

26 22 16 – Low-Voltage Buck-Boost Transformers

26 22 16.10 Buck-Boost Transformer		Crew	Daily Output	Labor-Hours	Unit	Material	Labor	Equipment	Total	Total Incl O&P
0010	**BUCK-BOOST TRANSFORMER**									
0100	Single phase, 120/240 V primary, 12/24 V secondary									
0200	0.10 kVA	1 Elec	8	1	Ea.	110	54.50		164.50	206
0400	0.25 kVA		5.70	1.404		161	77		238	297
0600	0.50 kVA		4	2		220	109		329	410
0800	0.75 kVA		3.10	2.581		284	141		425	535
1000	1.0 kVA		2	4		355	219		574	730
1200	1.5 kVA		1.80	4.444		430	243		673	850
1400	2.0 kVA		1.60	5		525	274		799	1,000
1600	3.0 kVA		1.40	5.714		700	315		1,015	1,250
1800	5.0 kVA		1.20	6.667		925	365		1,290	1,600
2000	3 phase, 240 V primary, 208/120 V secondary, 15 kVA	2 Elec	2.40	6.667		2,125	365		2,490	2,925
2200	30 kVA		1.60	10		2,375	545		2,920	3,475
2400	45 kVA		1.40	11.429		2,850	625		3,475	4,125
2600	75 kVA		1.20	13.333		3,475	730		4,205	4,925
2800	112.5 kVA	R-3	1.40	14.286		4,300	775	98.50	5,173.50	6,025
3000	150 kVA		1.10	18.182		5,725	985	125	6,835	7,975
3200	225 kVA		1	20		7,450	1,075	138	8,663	10,100
3400	300 kVA		.90	22.222		10,100	1,200	153	11,453	13,100

26 24 13 – Switchboards

26 24 13.10 Incoming Switchboards		Crew	Daily Output	Labor-Hours	Unit	Material	2015 Bare Costs Labor	Equipment	Total	Total Incl O&P
0010	**INCOMING SWITCHBOARDS** main service section									
0100	Aluminum bus bars, not including CT's or PT's									
0200	No main disconnect, includes CT compartment									
0300	120/208 volt, 4 wire, 600 amp	2 Elec	1	16	Ea.	4,225	875		5,100	6,000
0400	800 amp		.88	18.182		4,225	995		5,220	6,200
0500	1000 amp		.80	20		5,075	1,100		6,175	7,275
0600	1200 amp		.72	22.222		5,075	1,225		6,300	7,450
0700	1600 amp		.66	24.242		5,075	1,325		6,400	7,625
0800	2000 amp		.62	25.806		5,450	1,400		6,850	8,175
1000	3000 amp		.56	28.571		7,200	1,575		8,775	10,400
1200	277/480 volt, 4 wire, 600 amp		1	16		4,225	875		5,100	6,000
1300	800 amp		.88	18.182		4,225	995		5,220	6,200
1400	1000 amp		.80	20		5,350	1,100		6,450	7,600
1500	1200 amp		.72	22.222		5,350	1,225		6,575	7,775
1600	1600 amp		.66	24.242		5,350	1,325		6,675	7,950
1700	2000 amp		.62	25.806		5,450	1,400		6,850	8,175
1800	3000 amp		.56	28.571		7,200	1,575		8,775	10,400
1900	4000 amp		.52	30.769		8,925	1,675		10,600	12,400
2000	Fused switch & CT compartment									
2100	120/208 volt, 4 wire, 400 amp	2 Elec	1.12	14.286	Ea.	2,825	780		3,605	4,300
2200	600 amp		.94	17.021		3,350	930		4,280	5,125
2300	800 amp		.84	19.048		11,400	1,050		12,450	14,200
2400	1200 amp		.68	23.529		14,800	1,275		16,075	18,300
2500	277/480 volt, 4 wire, 400 amp		1.14	14.035		3,200	770		3,970	4,725
2600	600 amp		.94	17.021		3,675	930		4,605	5,475
2700	800 amp		.84	19.048		11,400	1,050		12,450	14,200
2800	1200 amp		.68	23.529		14,800	1,275		16,075	18,300
2900	Pressure switch & CT compartment									
3000	120/208 volt, 4 wire, 800 amp	2 Elec	.80	20	Ea.	10,200	1,100		11,300	13,000
3100	1200 amp		.66	24.242		19,800	1,325		21,125	23,900
3200	1600 amp		.62	25.806		21,100	1,400		22,500	25,400
3300	2000 amp		.56	28.571		22,400	1,575		23,975	27,100
3310	2500 amp		.50	32		27,600	1,750		29,350	33,000
3320	3000 amp		.44	36.364		37,200	2,000		39,200	44,000
3330	4000 amp		.40	40		47,700	2,200		49,900	56,000
3340	120/208 volt, 4 wire, 800 amp, with ground fault		.80	20		16,900	1,100		18,000	20,300
3350	1200 amp, with ground fault		.66	24.242		21,900	1,325		23,225	26,200
3360	1600 amp, with ground fault		.62	25.806		23,800	1,400		25,200	28,300
3370	2000 amp, with ground fault		.56	28.571		25,700	1,575		27,275	30,600
3400	277/480 volt, 4 wire, 800 amp, with ground fault		.80	20		16,900	1,100		18,000	20,300
3600	1200 amp, with ground fault		.66	24.242		21,900	1,325		23,225	26,200
4000	1600 amp, with ground fault		.62	25.806		23,800	1,400		25,200	28,300
4200	2000 amp, with ground fault		.56	28.571		25,700	1,575		27,275	30,600
4400	Circuit breaker, molded case & CT compartment									
4600	3 pole, 4 wire, 600 amp	2 Elec	.94	17.021	Ea.	8,825	930		9,755	11,200
4800	800 amp		.84	19.048		10,600	1,050		11,650	13,200
5000	1200 amp		.68	23.529		14,400	1,275		15,675	17,800
5100	Copper bus bars, not incl. CT's or PT's, add, minimum					15%				

26 24 13.20 In Plant Distribution Switchboards		Crew	Daily Output	Labor-Hours	Unit	Material	Labor	Equipment	Total	Total Incl O&P
0010	**IN PLANT DISTRIBUTION SWITCHBOARDS**									
0100	Main lugs only, to 600 volt, 3 pole, 3 wire, 200 amp	2 Elec	1.20	13.333	Ea.	1,325	730		2,055	2,575
0110	400 amp		1.20	13.333		1,325	730		2,055	2,575

26 24 13 – Switchboards

26 24 13.20 In Plant Distribution Switchboards		Crew	Daily Output	Labor-Hours	Unit	Material	2015 Bare Costs Labor	Equipment	Total	Total Incl O&P
0120	600 amp	2 Elec	1.20	13.333	Ea.	1,350	730		2,080	2,625
0130	800 amp		1.08	14.815		1,475	810		2,285	2,875
0140	1200 amp		.92	17.391		1,800	950		2,750	3,450
0150	1600 amp		.86	18.605		2,350	1,025		3,375	4,150
0160	2000 amp		.82	19.512		2,600	1,075		3,675	4,525
0250	To 480 volt, 3 pole, 4 wire, 200 amp		1.20	13.333		1,150	730		1,880	2,375
0260	400 amp		1.20	13.333		1,325	730		2,055	2,575
0270	600 amp		1.20	13.333		1,450	730		2,180	2,700
0280	800 amp		1.08	14.815		1,575	810		2,385	2,975
0290	1200 amp		.92	17.391		1,950	950		2,900	3,625
0300	1600 amp		.86	18.605		2,225	1,025		3,250	4,000
0310	2000 amp		.82	19.512		2,575	1,075		3,650	4,500
0400	Main circuit breaker, to 600 volt, 3 pole, 3 wire, 200 amp		1.20	13.333		3,150	730		3,880	4,600
0410	400 amp		1.14	14.035		3,150	770		3,920	4,675
0420	600 amp		1.10	14.545		4,000	795		4,795	5,625
0430	800 amp		1.04	15.385		6,675	840		7,515	8,650
0440	1200 amp		.88	18.182		8,750	995		9,745	11,200
0450	1600 amp		.84	19.048		14,000	1,050		15,050	17,000
0460	2000 amp		.80	20		15,000	1,100		16,100	18,200
0550	277/480 volt, 3 pole, 4 wire, 200 amp		1.20	13.333		3,325	730		4,055	4,775
0560	400 amp		1.14	14.035		3,325	770		4,095	4,850
0570	600 amp		1.10	14.545		4,150	795		4,945	5,800
0580	800 amp		1.04	15.385		7,000	840		7,840	9,000
0590	1200 amp		.88	18.182		9,050	995		10,045	11,500
0600	1600 amp		.84	19.048		14,000	1,050		15,050	17,000
0610	2000 amp		.80	20		15,000	1,100		16,100	18,200
0700	Main fusible switch w/fuse, 208/240 volt, 3 pole, 3 wire, 200 amp		1.20	13.333		3,475	730		4,205	4,950
0710	400 amp		1.14	14.035		3,475	770		4,245	5,025
0720	600 amp		1.10	14.545		4,225	795		5,020	5,875
0730	800 amp		1.04	15.385		8,650	840		9,490	10,800
0740	1200 amp		.88	18.182		10,100	995		11,095	12,700
0800	120/208, 120/240 volt, 3 pole, 4 wire, 200 amp		1.20	13.333		3,075	730		3,805	4,525
0810	400 amp		1.14	14.035		3,075	770		3,845	4,600
0820	600 amp		1.10	14.545		4,025	795		4,820	5,650
0830	800 amp		1.04	15.385		6,425	840		7,265	8,350
0840	1200 amp		.88	18.182		7,450	995		8,445	9,750
0900	480 or 600 volt, 3 pole, 3 wire, 200 amp		1.20	13.333		3,350	730		4,080	4,800
0910	400 amp		1.14	14.035		3,350	770		4,120	4,875
0920	600 amp		1.10	14.545		4,000	795		4,795	5,625
0930	800 amp		1.04	15.385		6,175	840		7,015	8,100
0940	1200 amp		.88	18.182		7,175	995		8,170	9,425
1000	277 or 480 volt, 3 pole, 4 wire, 200 amp		1.20	13.333		3,475	730		4,205	4,950
1010	400 amp		1.14	14.035		3,475	770		4,245	5,025
1020	600 amp		1.10	14.545		4,200	795		4,995	5,825
1030	800 amp		1.04	15.385		6,425	840		7,265	8,375
1040	1200 amp		.88	18.182		7,450	995		8,445	9,750
1120	1600 amp		.76	21.053		13,600	1,150		14,750	16,800
1130	2000 amp		.68	23.529		17,900	1,275		19,175	21,700
1150	Pressure switch, bolted, 3 pole, 208/240 volt, 3 wire, 800 amp		.96	16.667		10,100	910		11,010	12,500
1160	1200 amp		.80	20		12,900	1,100		14,000	15,900
1170	1600 amp		.76	21.053		14,700	1,150		15,850	18,000
1180	2000 amp		.68	23.529		16,800	1,275		18,075	20,400
1200	120/208 or 120/240 volt, 3 pole, 4 wire, 800 amp		.96	16.667		7,975	910		8,885	10,200

26 24 13.20 In Plant Distribution Switchboards		Crew	Daily Output	Labor-Hours	Unit	Material	2015 Bare Costs Labor	Equipment	Total	Total Incl O&P
1210	1200 amp	2 Elec	.80	20	Ea.	9,300	1,100		10,400	11,900
1220	1600 amp		.76	21.053		14,700	1,150		15,850	18,000
1230	2000 amp		.68	23.529		16,800	1,275		18,075	20,400
1300	480 or 600 volt, 3 wire, 800 amp		.96	16.667		10,100	910		11,010	12,500
1310	1200 amp		.80	20		14,100	1,100		15,200	17,200
1320	1600 amp		.76	21.053		15,900	1,150		17,050	19,300
1330	2000 amp		.68	23.529		17,900	1,275		19,175	21,700
1400	277-480 volt, 4 wire, 800 amp		.96	16.667		10,100	910		11,010	12,500
1410	1200 amp		.80	20		14,100	1,100		15,200	17,200
1420	1600 amp		.76	21.053		15,900	1,150		17,050	19,300
1430	2000 amp		.68	23.529		17,900	1,275		19,175	21,700
1500	Main ground fault protector, 1200-2000 amp		5.40	2.963		3,350	162		3,512	3,950
1600	Busway connection, 200 amp		5.40	2.963		505	162		667	805
1610	400 amp		4.60	3.478		505	190		695	850
1620	600 amp		4	4		505	219		724	895
1630	800 amp		3.20	5		505	274		779	980
1640	1200 amp		2.60	6.154		505	335		840	1,075
1650	1600 amp		2.40	6.667		1,050	365		1,415	1,725
1660	2000 amp		2	8		1,050	440		1,490	1,825
1700	Shunt trip for remote operation 200 amp		8	2		670	109		779	905
1710	400 amp		8	2		1,075	109		1,184	1,350
1720	600 amp		8	2		1,250	109		1,359	1,550
1730	800 amp		8	2		1,625	109		1,734	1,950
1740	1200-2000 amp		8	2		3,450	109		3,559	3,950
1800	Motor operated main breaker 200 amp		8	2		3,325	109		3,434	3,825
1810	400 amp		8	2		3,325	109		3,434	3,825
1820	600 amp		8	2		3,400	109		3,509	3,925
1830	800 amp		8	2		3,400	109		3,509	3,925
1840	1200-2000 amp		8	2		3,500	109		3,609	4,025
1900	Current/potential transformer metering compartment 200-800 amp		5.40	2.963		2,900	162		3,062	3,425
1940	1200 amp		5.40	2.963		4,775	162		4,937	5,525
1950	1600-2000 amp		5.40	2.963		6,175	162		6,337	7,025
2000	With watt meter 200-800 amp		4	4		8,525	219		8,744	9,725
2040	1200 amp		4	4		10,500	219		10,719	11,900
2050	1600-2000 amp		4	4		10,500	219		10,719	11,900
2100	Split bus 60-200 amp	1 Elec	5.30	1.509		202	82.50		284.50	350
2130	400 amp	2 Elec	4.60	3.478		350	190		540	680
2140	600 amp		3.60	4.444		430	243		673	845
2150	800 amp		2.60	6.154		545	335		880	1,125
2170	1200 amp		2	8		620	440		1,060	1,375
2250	Contactor control 60 amp	1 Elec	2	4		1,325	219		1,544	1,800
2260	100 amp		1.50	5.333		1,500	292		1,792	2,100
2270	200 amp		1	8		2,250	440		2,690	3,150
2280	400 amp	2 Elec	1	16		6,825	875		7,700	8,850
2290	600 amp		.84	19.048		7,625	1,050		8,675	10,000
2300	800 amp		.72	22.222		9,050	1,225		10,275	11,800
2500	Modifier, two distribution sections, add		.80	20		2,825	1,100		3,925	4,825
2520	Three distribution sections, add		.40	40		6,350	2,200		8,550	10,400
2560	Auxiliary pull section, 20", add		2	8		1,125	440		1,565	1,925
2580	24", add		1.80	8.889		1,125	485		1,610	2,000
2600	30", add		1.60	10		1,125	545		1,670	2,100
2620	36", add		1.40	11.429		1,350	625		1,975	2,475
2640	Dog house, 12", add		2.40	6.667		233	365		598	820

26 24 13 – Switchboards

26 24 13.20 In Plant Distribution Switchboards		Crew	Daily Output	Labor-Hours	Unit	Material	2015 Bare Costs Labor	Equipment	Total	Total Incl O&P
2660	18", add	2 Elec	2	8	Ea.	465	440		905	1,200
3000	Transition section between switchboard and transformer									
3050	or motor control center, 4 wire alum. bus, 600 amp	2 Elec	1.14	14.035	Ea.	2,300	770		3,070	3,725
3100	800 amp		1	16		2,575	875		3,450	4,175
3150	1000 amp		.88	18.182		2,900	995		3,895	4,725
3200	1200 amp		.80	20		3,175	1,100		4,275	5,200
3250	1600 amp		.72	22.222		3,750	1,225		4,975	6,025
3300	2000 amp		.66	24.242		4,325	1,325		5,650	6,800
3350	2500 amp		.62	25.806		5,050	1,400		6,450	7,725
3400	3000 amp		.56	28.571		5,800	1,575		7,375	8,800
4000	Weatherproof construction, per vertical section		1.76	9.091		2,650	495		3,145	3,700

26 24 13.30 Distribution Switchboards Section		Crew	Daily Output	Labor-Hours	Unit	Material	Labor	Equipment	Total	Total Incl O&P
0010	**DISTRIBUTION SWITCHBOARDS SECTION**									
0100	Aluminum bus bars, not including breakers									
0160	Subfeed lug-rated at 60 amp	2 Elec	1.30	12.308	Ea.	1,050	675		1,725	2,200
0170	100 amp		1.26	12.698		1,200	695		1,895	2,400
0180	200 amp		1.20	13.333		1,225	730		1,955	2,475
0190	400 amp		1.10	14.545		1,225	795		2,020	2,575
0195	120/208 or 277/480 volt, 4 wire, 400 amp		1.10	14.545		1,300	795		2,095	2,675
0200	600 amp		1	16		1,625	875		2,500	3,125
0300	800 amp		.88	18.182		2,100	995		3,095	3,875
0400	1000 amp		.80	20		2,625	1,100		3,725	4,600
0500	1200 amp		.72	22.222		3,125	1,225		4,350	5,325
0600	1600 amp		.66	24.242		3,575	1,325		4,900	6,000
0700	2000 amp		.62	25.806		4,200	1,400		5,600	6,775
0800	2500 amp		.60	26.667		4,725	1,450		6,175	7,450
0900	3000 amp		.56	28.571		5,725	1,575		7,300	8,725
0950	4000 amp		.52	30.769		8,375	1,675		10,050	11,800

26 24 13.40 Switchboards Feeder Section		Crew	Daily Output	Labor-Hours	Unit	Material	Labor	Equipment	Total	Total Incl O&P
0010	**SWITCHBOARDS FEEDER SECTION** group mounted devices									
0030	Circuit breakers									
0160	FA frame, 15 to 60 amp, 240 volt, 1 pole	1 Elec	8	1	Ea.	120	54.50		174.50	217
0170	2 pole		7	1.143		205	62.50		267.50	320
0180	3 pole		5.30	1.509		305	82.50		387.50	465
0210	480 volt, 1 pole		8	1		151	54.50		205.50	251
0220	2 pole		7	1.143		375	62.50		437.50	505
0230	3 pole		5.30	1.509		485	82.50		567.50	660
0260	600 volt, 2 pole		7	1.143		435	62.50		497.50	570
0270	3 pole		5.30	1.509		560	82.50		642.50	745
0280	FA frame, 70 to 100 amp, 240 volt, 1 pole		7	1.143		191	62.50		253.50	305
0310	2 pole		5	1.600		335	87.50		422.50	500
0320	3 pole		4	2		435	109		544	645
0330	480 volt, 1 pole		7	1.143		228	62.50		290.50	350
0360	2 pole		5	1.600		485	87.50		572.50	670
0370	3 pole		4	2		570	109		679	795
0380	600 volt, 2 pole		5	1.600		550	87.50		637.50	740
0410	3 pole		4	2		685	109		794	925
0420	KA frame, 70 to 225 amp		3.20	2.500		1,250	137		1,387	1,575
0430	LA frame, 125 to 400 amp		2.30	3.478		2,475	190		2,665	3,025
0460	MA frame, 450 to 600 amp		1.60	5		5,000	274		5,274	5,925
0470	700 to 800 amp		1.30	6.154		6,500	335		6,835	7,675
0480	MAL frame, 1000 amp		1	8		6,725	440		7,165	8,075

26 24 13.40 Switchboards Feeder Section		Crew	Daily Output	Labor-Hours	Unit	Material	2015 Bare Costs Labor	Equipment	Total	Total Incl O&P
0490	PA frame, 1200 amp	1 Elec	.80	10	Ea.	13,700	545		14,245	15,900
0500	Branch circuit, fusible switch, 600 volt, double 30/30 amp		4	2		895	109		1,004	1,150
0550	60/60 amp		3.20	2.500		920	137		1,057	1,200
0600	100/100 amp		2.70	2.963		1,150	162		1,312	1,525
0650	Single, 30 amp		5.30	1.509		735	82.50		817.50	940
0700	60 amp		4.70	1.702		815	93		908	1,050
0750	100 amp		4	2		1,075	109		1,184	1,350
0800	200 amp		2.70	2.963		1,425	162		1,587	1,825
0850	400 amp		2.30	3.478		2,625	190		2,815	3,175
0900	600 amp		1.80	4.444		3,200	243		3,443	3,900
0950	800 amp		1.30	6.154		5,375	335		5,710	6,425
1000	1200 amp		.80	10		6,150	545		6,695	7,625
1080	Branch circuit, circuit breakers, high interrupting capacity									
1100	60 amp, 240, 480 or 600 volt, 1 pole	1 Elec	8	1	Ea.	259	54.50		313.50	370
1120	2 pole		7	1.143		665	62.50		727.50	830
1140	3 pole		5.30	1.509		675	82.50		757.50	870
1150	100 amp, 240, 480 or 600 volt, 1 pole		7	1.143		292	62.50		354.50	415
1160	2 pole		5	1.600		775	87.50		862.50	990
1180	3 pole		4	2		730	109		839	975
1200	225 amp, 240, 480 or 600 volt, 2 pole		3.50	2.286		2,150	125		2,275	2,550
1220	3 pole		3.20	2.500		2,275	137		2,412	2,700
1240	400 amp, 240, 480 or 600 volt, 2 pole		2.50	3.200		3,425	175		3,600	4,025
1260	3 pole		2.30	3.478		3,075	190		3,265	3,675
1280	600 amp, 240, 480 or 600 volt, 2 pole		1.80	4.444		3,975	243		4,218	4,750
1300	3 pole		1.60	5		3,750	274		4,024	4,550
1320	800 amp, 240, 480 or 600 volt, 2 pole		1.50	5.333		5,100	292		5,392	6,075
1340	3 pole		1.30	6.154		5,625	335		5,960	6,700
1360	1000 amp, 240, 480 or 600 volt, 2 pole		1.10	7.273		6,675	400		7,075	7,950
1380	3 pole		1	8		7,300	440		7,740	8,700
1400	1200 amp, 240, 480 or 600 volt, 2 pole		.90	8.889		6,925	485		7,410	8,375
1420	3 pole		.80	10		7,925	545		8,470	9,575
1700	Fusible switch, 240 V, 60 amp, 2 pole		3.20	2.500		430	137		567	680
1720	3 pole		3	2.667		590	146		736	875
1740	100 amp, 2 pole		2.70	2.963		435	162		597	730
1760	3 pole		2.50	3.200		660	175		835	995
1780	200 amp, 2 pole		2	4		910	219		1,129	1,350
1800	3 pole		1.90	4.211		1,075	230		1,305	1,525
1820	400 amp, 2 pole		1.50	5.333		1,700	292		1,992	2,325
1840	3 pole		1.30	6.154		2,125	335		2,460	2,875
1860	600 amp, 2 pole		1	8		2,425	440		2,865	3,350
1880	3 pole		.90	8.889		2,975	485		3,460	4,025
1900	240-600 V, 800 amp, 2 pole		.70	11.429		5,400	625		6,025	6,925
1920	3 pole		.60	13.333		6,625	730		7,355	8,400
2000	600 V, 60 amp, 2 pole		3.20	2.500		650	137		787	925
2040	100 amp, 2 pole		2.70	2.963		670	162		832	985
2080	200 amp, 2 pole		2	4		1,150	219		1,369	1,625
2120	400 amp, 2 pole		1.50	5.333		2,300	292		2,592	2,975
2160	600 amp, 2 pole		1	8		2,800	440		3,240	3,750
2500	Branch circuit, circuit breakers, 60 amp, 600 volt, 3 pole		5.30	1.509		525	82.50		607.50	705
2520	240, 480 or 600 volt, 1 pole		8	1		100	54.50		154.50	195
2540	240 volt, 2 pole		7	1.143		195	62.50		257.50	310
2560	480 or 600 volt, 2 pole		7	1.143		365	62.50		427.50	495
2580	240 volt, 3 pole		5.30	1.509		286	82.50		368.50	445

26 24 13 – Switchboards

26 24 13.40 Switchboards Feeder Section		Crew	Daily Output	Labor-Hours	Unit	Material	2015 Bare Costs Labor	Equipment	Total	Total Incl O&P
2600	480 volt, 3 pole	1 Elec	5.30	1.509	Ea.	485	82.50		567.50	665
2620	100 amp, 600 volt, 2 pole		5	1.600		465	87.50		552.50	650
2640	3 pole		4	2		590	109		699	820
2660	480 volt, 2 pole		5	1.600		400	87.50		487.50	575
2680	240 volt, 2 pole		5	1.600		229	87.50		316.50	390
2700	3 pole		4	2		370	109		479	575
2720	480 volt, 3 pole		4	2		540	109		649	765
2740	225 amp, 240, 480 or 600 volt, 2 pole		3.50	2.286		610	125		735	870
2760	3 pole		3.20	2.500		705	137		842	985
2780	400 amp, 240, 480 or 600 volt, 2 pole		2.50	3.200		1,400	175		1,575	1,825
2800	3 pole		2.30	3.478		1,625	190		1,815	2,100
2820	600 amp, 240 or 480 volt, 2 pole		1.80	4.444		2,300	243		2,543	2,900
2840	3 pole		1.60	5		2,825	274		3,099	3,525
2860	800 amp, 240, 480 volt or 600 volt, 2 pole		1.50	5.333		3,450	292		3,742	4,250
2880	3 pole		1.30	6.154		4,025	335		4,360	4,950
2900	1000 amp, 240, 480 or 600 volt, 2 pole		1.10	7.273		4,225	400		4,625	5,275
2920	480 volt, 600 volt, 3 pole		1	8		4,850	440		5,290	6,025
2940	1200 amp, 240, 480 or 600 volt, 2 pole		.90	8.889		5,975	485		6,460	7,325
2960	3 pole		.80	10		6,525	545		7,070	8,025
2980	600 volt, 3 pole		.80	10		6,525	545		7,070	8,025

26 24 16 – Panelboards

26 24 16.10 Load Centers

		Crew	Daily Output	Labor-Hours	Unit	Material	Labor	Equipment	Total	Total Incl O&P
0010	**LOAD CENTERS** (residential type) R262416-50									
0100	3 wire, 120/240 V, 1 phase, including 1 pole plug-in breakers									
0200	100 amp main lugs, indoor, 8 circuits	1 Elec	1.40	5.714	Ea.	178	315		493	680
0300	12 circuits		1.20	6.667		249	365		614	840
0400	Rainproof, 8 circuits		1.40	5.714		214	315		529	720
0500	12 circuits		1.20	6.667		300	365		665	895
0600	200 amp main lugs, indoor, 16 circuits	R-1A	1.80	8.889		330	400		730	1,000
0700	20 circuits		1.50	10.667		415	480		895	1,225
0800	24 circuits		1.30	12.308		545	555		1,100	1,475
0900	30 circuits		1.20	13.333		630	600		1,230	1,650
1000	40 circuits		.80	20		870	900		1,770	2,375
1200	Rainproof, 16 circuits		1.80	8.889		395	400		795	1,075
1300	20 circuits		1.50	10.667		475	480		955	1,275
1400	24 circuits		1.30	12.308		710	555		1,265	1,650
1500	30 circuits		1.20	13.333		790	600		1,390	1,825
1600	40 circuits		.80	20		1,025	900		1,925	2,575
1800	400 amp main lugs, indoor, 42 circuits		.72	22.222		1,225	1,000		2,225	2,950
1900	Rainproof, 42 circuit		.72	22.222		1,300	1,000		2,300	3,025
2200	Plug in breakers, 20 amp, 1 pole, 4 wire, 120/208 volts									
2210	125 amp main lugs, indoor, 12 circuits	1 Elec	1.20	6.667	Ea.	310	365		675	910
2300	18 circuits		.80	10		440	545		985	1,325
2400	Rainproof, 12 circuits		1.20	6.667		360	365		725	960
2500	18 circuits		.80	10		480	545		1,025	1,375
2600	200 amp main lugs, indoor, 24 circuits	R-1A	1.30	12.308		595	555		1,150	1,525
2700	30 circuits		1.20	13.333		680	600		1,280	1,700
2800	36 circuits		1	16		820	720		1,540	2,050
2900	42 circuits		.80	20		900	900		1,800	2,425
3000	Rainproof, 24 circuits		1.30	12.308		660	555		1,215	1,600
3100	30 circuits		1.20	13.333		745	600		1,345	1,775
3200	36 circuits		1	16		1,025	720		1,745	2,275

26 24 16.10 Load Centers

	26 24 16.10 Load Centers	Crew	Daily Output	Labor-Hours	Unit	Material	Labor	Equipment	Total	Total Incl O&P
						2015 Bare Costs				
3300	42 circuits	R-1A	.80	20	Ea.	1,100	900		2,000	2,650
3500	400 amp main lugs, indoor, 42 circuits		.72	22.222		1,350	1,000		2,350	3,100
3600	Rainproof, 42 circuits	↓	.72	22.222	↓	1,625	1,000		2,625	3,375
3700	Plug-in breakers, 20 amp, 1 pole, 3 wire, 120/240 volts									
3800	100 amp main breaker, indoor, 12 circuits	1 Elec	1.20	6.667	Ea.	335	365		700	930
3900	18 circuits	"	.80	10		450	545		995	1,350
4000	200 amp main breaker, indoor, 20 circuits	R-1A	1.50	10.667		605	480		1,085	1,425
4200	24 circuits		1.30	12.308		730	555		1,285	1,675
4300	30 circuits		1.20	13.333		815	600		1,415	1,850
4400	40 circuits		.90	17.778		1,025	800		1,825	2,400
4500	Rainproof, 20 circuits		1.50	10.667		660	480		1,140	1,500
4600	24 circuits		1.30	12.308		820	555		1,375	1,775
4700	30 circuits		1.20	13.333		900	600		1,500	1,950
4800	40 circuits		.90	17.778		1,100	800		1,900	2,500
5000	400 amp main breaker, indoor, 42 circuits		.72	22.222		2,575	1,000		3,575	4,425
5100	Rainproof, 42 circuits	↓	.72	22.222	↓	3,075	1,000		4,075	4,975
5300	Plug in breakers, 20 amp, 1 pole, 4 wire, 120/208 volts									
5400	200 amp main breaker, indoor, 30 circuits	R-1A	1.20	13.333	Ea.	1,225	600		1,825	2,300
5500	42 circuits		.80	20		1,500	900		2,400	3,075
5600	Rainproof, 30 circuits		1.20	13.333		1,325	600		1,925	2,400
5700	42 circuits	↓	.80	20	↓	1,600	900		2,500	3,175

26 24 16.20 Panelboard and Load Center Circuit Breakers

	26 24 16.20 Panelboard and Load Center Circuit Breakers	Crew	Daily Output	Labor-Hours	Unit	Material	Labor	Equipment	Total	Total Incl O&P
0010	**PANELBOARD AND LOAD CENTER CIRCUIT BREAKERS**									
0050	Bolt-on, 10,000 amp I.C., 120 volt, 1 pole									
0100	15 to 50 amp	1 Elec	10	.800	Ea.	16.65	44		60.65	86.50
0200	60 amp		8	1		19.20	54.50		73.70	106
0300	70 amp	↓	8	1	↓	28	54.50		82.50	116
0350	240 volt, 2 pole									
0400	15 to 50 amp	1 Elec	8	1	Ea.	36	54.50		90.50	124
0500	60 amp		7.50	1.067		49	58.50		107.50	144
0600	80 to 100 amp		5	1.600		93.50	87.50		181	239
0700	3 pole, 15 to 60 amp		6.20	1.290		115	70.50		185.50	236
0800	70 amp		5	1.600		146	87.50		233.50	297
0900	80 to 100 amp		3.60	2.222		166	122		288	370
1000	22,000 amp I.C., 240 volt, 2 pole, 70 – 225 amp		2.70	2.963		630	162		792	945
1100	3 pole, 70 – 225 amp		2.30	3.478		700	190		890	1,075
1200	14,000 amp I.C., 277 volts, 1 pole, 15 – 30 amp		8	1		44	54.50		98.50	133
1300	22,000 amp I.C., 480 volts, 2 pole, 70 – 225 amp		2.70	2.963		630	162		792	945
1400	3 pole, 70 – 225 amp		2.30	3.478		780	190		970	1,150
2000	Plug-in panel or load center, 120/240 volt, to 60 amp, 1 pole		12	.667		11.45	36.50		47.95	69
2004	Circuit breaker, 120/240 volt, 20 A, 1 pole with NM cable		6.50	1.231		18.75	67.50		86.25	125
2006	30 A, 1 pole with NM cable		6.50	1.231		18.75	67.50		86.25	125
2010	2 pole		9	.889		26	48.50		74.50	104
2014	50 A, 2 pole with NM cable		5.50	1.455		33	79.50		112.50	160
2020	3 pole		7.50	1.067		89.50	58.50		148	189
2030	100 amp, 2 pole		6	1.333		120	73		193	245
2040	3 pole		4.50	1.778		133	97		230	297
2050	150 to 200 amp, 2 pole		3	2.667		218	146		364	465
2060	Plug-in tandem, 120/240 V, 2-15 A, 1 pole		11	.727		24.50	40		64.50	88.50
2070	1-15 A & 1-20 A		11	.727		24.50	40		64.50	88.50
2080	2-20 A		11	.727		24.50	40		64.50	88.50
2082	Arc fault circuit interrupter, 120/240 V, 1-15 A & 1-20 A, 1 pole	↓	11	.727	↓	65	40		105	133

26 24 16.20 Panelboard and Load Center Circuit Breakers		Crew	Daily Output	Labor-Hours	Unit	Material	2015 Bare Costs Labor	Equipment	Total	Total Incl O&P
2100	High interrupting capacity, 120/240 volt, plug-in, 30 amp, 1 pole	1 Elec	12	.667	Ea.	26	36.50		62.50	85
2110	60 amp, 2 pole		9	.889		61.50	48.50		110	143
2120	3 pole		7.50	1.067		141	58.50		199.50	246
2130	100 amp, 2 pole		6	1.333		153	73		226	282
2140	3 pole		4.50	1.778		223	97		320	395
2150	125 amp, 2 pole		3	2.667		395	146		541	655
2200	Bolt-on, 30 amp, 1 pole		10	.800		33	44		77	105
2210	60 amp, 2 pole		7.50	1.067		70	58.50		128.50	168
2220	3 pole		6.20	1.290		180	70.50		250.50	305
2230	100 amp, 2 pole		5	1.600		196	87.50		283.50	350
2240	3 pole		3.60	2.222		260	122		382	475
2300	Ground fault, 240 volt, 30 amp, 1 pole		7	1.143		102	62.50		164.50	209
2310	2 pole		6	1.333		182	73		255	315
2350	Key operated, 240 volt, 1 pole, 30 amp		7	1.143		84	62.50		146.50	190
2360	Switched neutral, 240 volt, 30 amp, 2 pole		6	1.333		42	73		115	160
2370	3 pole		5.50	1.455		63.50	79.50		143	193
2400	Shunt trip, for 240 volt breaker, 60 amp, 1 pole		4	2		72.50	109		181.50	249
2410	2 pole		3.50	2.286		72.50	125		197.50	274
2420	3 pole		3	2.667		72.50	146		218.50	305
2430	100 amp, 2 pole		3	2.667		72.50	146		218.50	305
2440	3 pole		2.50	3.200		72.50	175		247.50	350
2450	150 amp, 2 pole		2	4		172	219		391	530
2500	Auxiliary switch, for 240 volt breaker, 60 amp, 1 pole		4	2		65.50	109		174.50	242
2510	2 pole		3.50	2.286		65.50	125		190.50	267
2520	3 pole		3	2.667		65.50	146		211.50	299
2530	100 amp, 2 pole		3	2.667		65.50	146		211.50	299
2540	3 pole		2.50	3.200		65.50	175		240.50	345
2550	150 amp, 2 pole		2	4		109	219		328	460
2600	Panel or load center, 277/480 volt, plug-in, 30 amp, 1 pole		12	.667		53	36.50		89.50	115
2610	60 amp, 2 pole		9	.889		159	48.50		207.50	251
2620	3 pole		7.50	1.067		233	58.50		291.50	350
2650	Bolt-on, 60 amp, 2 pole		7.50	1.067		159	58.50		217.50	266
2660	3 pole		6.20	1.290		233	70.50		303.50	365
2700	I-line, 277/480 volt, 30 amp, 1 pole		8	1		61	54.50		115.50	152
2710	60 amp, 2 pole		7.50	1.067		214	58.50		272.50	325
2720	3 pole		6.20	1.290		278	70.50		348.50	415
2730	100 amp, 1 pole		7.50	1.067		117	58.50		175.50	220
2740	2 pole		5	1.600		278	87.50		365.50	440
2750	3 pole		3.50	2.286		330	125		455	555
2800	High interrupting capacity, 277/480 volt, plug-in, 30 amp, 1 pole		12	.667		305	36.50		341.50	395
2810	60 amp, 2 pole		9	.889		475	48.50		523.50	595
2820	3 pole		7	1.143		550	62.50		612.50	700
2830	Bolt-on, 30 amp, 1 pole		8	1		305	54.50		359.50	425
2840	60 amp, 2 pole		7.50	1.067		525	58.50		583.50	670
2850	3 pole		6.20	1.290		610	70.50		680.50	785
2900	I-line, 30 amp, 1 pole		8	1		305	54.50		359.50	425
2910	60 amp, 2 pole		7.50	1.067		440	58.50		498.50	570
2920	3 pole		6.20	1.290		485	70.50		555.50	645
2930	100 amp, 1 pole		7.50	1.067		495	58.50		553.50	635
2940	2 pole		5	1.600		495	87.50		582.50	680
2950	3 pole		3.60	2.222		550	122		672	790
2960	Shunt trip, 277/480 volt breaker, remote oper., 30 amp, 1 pole		4	2		330	109		439	530
2970	60 amp, 2 pole		3.50	2.286		480	125		605	725

26 24 16.20 Panelboard and Load Center Circuit Breakers	Crew	Daily Output	Labor-Hours	Unit	Material	Labor	Equipment	Total	Total Incl O&P
2980 3 pole	1 Elec	3	2.667	Ea.	545	146		691	825
2990 100 amp, 1 pole		3.50	2.286		385	125		510	620
3000 2 pole		3	2.667		545	146		691	825
3010 3 pole		2.50	3.200		595	175		770	925
3050 Under voltage trip, 277/480 volt breaker, 30 amp, 1 pole		4	2		330	109		439	530
3060 60 amp, 2 pole		3.50	2.286		480	125		605	725
3070 3 pole		3	2.667		545	146		691	825
3080 100 amp, 1 pole		3.50	2.286		385	125		510	620
3090 2 pole		3	2.667		545	146		691	825
3100 3 pole		2.50	3.200		595	175		770	925
3150 Motor operated, 277/480 volt breaker, 30 amp, 1 pole		4	2		585	109		694	815
3160 60 amp, 2 pole		3.50	2.286		740	125		865	1,000
3170 3 pole		3	2.667		800	146		946	1,100
3180 100 amp, 1 pole		3.50	2.286		640	125		765	900
3190 2 pole		3	2.667		800	146		946	1,100
3200 3 pole		2.50	3.200		850	175		1,025	1,200
3250 Panelboard spacers, per pole		40	.200		4.43	10.95		15.38	22
9000 Minimum labor/equipment charge		3	2.667	Job		146		146	226

26 24 16.30 Panelboards Commercial Applications	Crew	Daily Output	Labor-Hours	Unit	Material	Labor	Equipment	Total	Total Incl O&P
0010 **PANELBOARDS COMMERCIAL APPLICATIONS** R262416-50									
0050 NQOD, w/20 amp 1 pole bolt-on circuit breakers									
0100 3 wire, 120/240 volts, 100 amp main lugs									
0150 10 circuits	1 Elec	1	8	Ea.	545	440		985	1,275
0200 14 circuits		.88	9.091		660	495		1,155	1,500
0250 18 circuits		.75	10.667		720	585		1,305	1,700
0300 20 circuits		.65	12.308		805	675		1,480	1,925
0350 225 amp main lugs, 24 circuits	2 Elec	1.20	13.333		910	730		1,640	2,125
0400 30 circuits		.90	17.778		1,050	970		2,020	2,650
0450 36 circuits		.80	20		1,200	1,100		2,300	3,025
0500 38 circuits		.72	22.222		1,275	1,225		2,500	3,275
0550 42 circuits		.66	24.242		1,350	1,325		2,675	3,525
0600 4 wire, 120/208 volts, 100 amp main lugs, 12 circuits	1 Elec	1	8		640	440		1,080	1,375
0650 16 circuits		.75	10.667		730	585		1,315	1,700
0700 20 circuits		.65	12.308		845	675		1,520	1,975
0750 24 circuits		.60	13.333		875	730		1,605	2,100
0800 30 circuits		.53	15.094		1,050	825		1,875	2,425
0850 225 amp main lugs, 32 circuits	2 Elec	.90	17.778		1,175	970		2,145	2,800
0900 34 circuits		.84	19.048		1,200	1,050		2,250	2,950
0950 36 circuits		.80	20		1,225	1,100		2,325	3,050
1000 42 circuits		.68	23.529		1,375	1,275		2,650	3,500
1040 225 amp main lugs, NEMA 7, 12 circuits		1	16		3,700	875		4,575	5,425
1100 24 circuits		.40	40		4,450	2,200		6,650	8,275
1200 NEHB,w/20 amp, 1 pole bolt-on circuit breakers									
1250 4 wire, 277/480 volts, 100 amp main lugs, 12 circuits	1 Elec	.88	9.091	Ea.	1,225	495		1,720	2,125
1300 20 circuits	"	.60	13.333		1,825	730		2,555	3,125
1350 225 amp main lugs, 24 circuits	2 Elec	.90	17.778		2,075	970		3,045	3,775
1400 30 circuits		.80	20		2,475	1,100		3,575	4,425
1450 36 circuits		.72	22.222		2,875	1,225		4,100	5,050
1500 42 circuits		.60	26.667		3,275	1,450		4,725	5,850
1510 225 amp main lugs, NEMA 7, 12 circuits		.90	17.778		5,500	970		6,470	7,550
1590 24 circuits		.30	53.333		7,425	2,925		10,350	12,700
1600 NQOD panel, w/20 amp, 1 pole, circuit breakers									

Material, Labor, Equipment and Total columns are 2015 Bare Costs.

26 24 16.30 Panelboards Commercial Applications		Crew	Daily Output	Labor-Hours	Unit	2015 Bare Costs Material	Labor	Equipment	Total	Total Incl O&P
1650	3 wire, 120/240 volt with main circuit breaker									
1700	100 amp main, 12 circuits	1 Elec	.80	10	Ea.	800	545		1,345	1,725
1750	20 circuits	"	.60	13.333		1,025	730		1,755	2,250
1800	225 amp main, 30 circuits	2 Elec	.68	23.529		1,900	1,275		3,175	4,100
1850	42 circuits		.52	30.769		2,200	1,675		3,875	5,025
1900	400 amp main, 30 circuits		.54	29.630		2,625	1,625		4,250	5,400
1950	42 circuits		.50	32		2,925	1,750		4,675	5,925
2000	4 wire, 120/208 volts with main circuit breaker									
2050	100 amp main, 24 circuits	1 Elec	.47	17.021	Ea.	1,175	930		2,105	2,750
2100	30 circuits	"	.40	20		1,325	1,100		2,425	3,150
2200	225 amp main, 32 circuits	2 Elec	.72	22.222		2,225	1,225		3,450	4,300
2250	42 circuits		.56	28.571		2,425	1,575		4,000	5,100
2300	400 amp main, 42 circuits		.48	33.333		3,250	1,825		5,075	6,400
2350	600 amp main, 42 circuits		.40	40		4,825	2,200		7,025	8,700
2400	NEHB, with 20 amp, 1 pole circuit breaker									
2450	4 wire, 277/480 volts with main circuit breaker									
2500	100 amp main, 24 circuits	1 Elec	.42	19.048	Ea.	2,375	1,050		3,425	4,250
2550	30 circuits	"	.38	21.053		2,775	1,150		3,925	4,850
2600	225 amp main, 30 circuits	2 Elec	.72	22.222		3,500	1,225		4,725	5,725
2650	42 circuits		.56	28.571		4,300	1,575		5,875	7,175
2700	400 amp main, 42 circuits		.46	34.783		5,175	1,900		7,075	8,650
2750	600 amp main, 42 circuits		.38	42.105		7,075	2,300		9,375	11,400
2900	Note: the following line items don't include branch circuit breakers									
2910	For branch circuit breakers information, see Section 26 24 16.20									
3010	Main lug, no main breaker, 240 volt, 1 pole, 3 wire, 100 amp	1 Elec	2.30	3.478	Ea.	535	190		725	885
3020	225 amp	2 Elec	2.40	6.667		640	365		1,005	1,275
3030	400 amp	"	1.80	8.889		940	485		1,425	1,775
3060	3 pole, 3 wire, 100 amp	1 Elec	2.30	3.478		575	190		765	930
3070	225 amp	2 Elec	2.40	6.667		680	365		1,045	1,300
3080	400 amp		1.80	8.889		1,025	485		1,510	1,875
3090	600 amp		1.60	10		1,175	545		1,720	2,150
3110	3 pole, 4 wire, 100 amp	1 Elec	2.30	3.478		635	190		825	995
3120	225 amp	2 Elec	2.40	6.667		795	365		1,160	1,450
3130	400 amp		1.80	8.889		1,025	485		1,510	1,875
3140	600 amp		1.60	10		1,175	545		1,720	2,150
3160	480 volt, 3 pole, 3 wire, 100 amp	1 Elec	2.30	3.478		760	190		950	1,125
3170	225 amp	2 Elec	2.40	6.667		920	365		1,285	1,575
3180	400 amp		1.80	8.889		1,275	485		1,760	2,150
3190	600 amp		1.60	10		1,425	545		1,970	2,425
3210	277/480 volt, 3 pole, 4 wire, 100 amp	1 Elec	2.30	3.478		740	190		930	1,100
3220	225 amp	2 Elec	2.40	6.667		900	365		1,265	1,550
3230	400 amp		1.80	8.889		1,250	485		1,735	2,125
3240	600 amp		1.60	10		1,400	545		1,945	2,375
3260	Main circuit breaker, 240 volt, 1 pole, 3 wire, 100 amp	1 Elec	2	4		715	219		934	1,125
3270	225 amp	2 Elec	2	8		1,475	440		1,915	2,275
3280	400 amp	"	1.60	10		2,300	545		2,845	3,375
3310	3 pole, 3 wire, 100 amp	1 Elec	2	4		820	219		1,039	1,250
3320	225 amp	2 Elec	2	8		1,675	440		2,115	2,525
3330	400 amp		1.60	10		2,625	545		3,170	3,750
3360	120/208 volt, 3 pole, 4 wire, 100 amp		4	4		820	219		1,039	1,250
3370	225 amp		2	8		1,675	440		2,115	2,525
3380	400 amp		1.60	10		2,625	545		3,170	3,750
3410	480 volt, 3 pole, 3 wire, 100 amp	1 Elec	2	4		1,150	219		1,369	1,625

26 24 16 – Panelboards

26 24 16.30 Panelboards Commercial Applications		Crew	Daily Output	Labor-Hours	Unit	Material	Labor	Equipment	Total	Total Incl O&P
						2015 Bare Costs				
3420	225 amp	2 Elec	2	8	Ea.	1,975	440		2,415	2,850
3430	400 amp		1.60	10		2,950	545		3,495	4,100
3460	277/480 volt, 3 pole, 4 wire, 100 amp		4	4		1,125	219		1,344	1,575
3470	225 amp		2	8		1,925	440		2,365	2,800
3480	400 amp		1.60	10		2,975	545		3,520	4,125
3510	Main circuit breaker, HIC, 240 volt, 1 pole, 3 wire, 100 amp	1 Elec	2	4		1,100	219		1,319	1,550
3520	225 amp	2 Elec	2	8		2,875	440		3,315	3,825
3530	400 amp	"	1.60	10		3,850	545		4,395	5,075
3560	3 pole, 3 wire, 100 amp	1 Elec	2	4		1,225	219		1,444	1,700
3570	225 amp	2 Elec	2	8		3,225	440		3,665	4,225
3580	400 amp	"	1.60	10		4,275	545		4,820	5,550
3610	120/208 volt, 3 pole, 4 wire, 100 amp	1 Elec	2	4		1,225	219		1,444	1,700
3620	225 amp	2 Elec	2	8		3,225	440		3,665	4,225
3630	400 amp	"	1.60	10		4,275	545		4,820	5,550
3660	480 volt, 3 pole, 3 wire, 100 amp	1 Elec	2	4		1,825	219		2,044	2,350
3670	225 amp	2 Elec	2	8		3,600	440		4,040	4,625
3680	400 amp	"	1.60	10		4,575	545		5,120	5,875
3710	277/480 volt, 3 pole, 4 wire, 100 amp	1 Elec	2	4		1,725	219		1,944	2,250
3720	225 amp	2 Elec	2	8		3,425	440		3,865	4,450
3730	400 amp	"	1.60	10		4,550	545		5,095	5,850
3760	Main circuit breaker, shunt trip, 100 amp	1 Elec	1.20	6.667		920	365		1,285	1,575
3770	225 amp	2 Elec	1.60	10		2,125	545		2,670	3,175
3780	400 amp	"	1.40	11.429		3,075	625		3,700	4,350
9000	Minimum labor/equipment charge	1 Elec	1	8	Job		440		440	680

26 24 19 – Motor-Control Centers

26 24 19.20 Motor Control Center Components		Crew	Daily Output	Labor-Hours	Unit	Material	Labor	Equipment	Total	Total Incl O&P
0010	**MOTOR CONTROL CENTER COMPONENTS**									
0100	Starter, size 1, FVNR, NEMA 1, type A, fusible	1 Elec	2.70	2.963	Ea.	1,575	162		1,737	2,000
0120	Circuit breaker		2.70	2.963		1,725	162		1,887	2,150
0140	Type B, fusible		2.70	2.963		1,750	162		1,912	2,150
0160	Circuit breaker		2.70	2.963		1,900	162		2,062	2,325
0180	NEMA 12, type A, fusible		2.60	3.077		1,600	168		1,768	2,025
0200	Circuit breaker		2.60	3.077		1,750	168		1,918	2,175
0220	Type B, fusible		2.60	3.077		1,775	168		1,943	2,200
0240	Circuit breaker		2.60	3.077		1,925	168		2,093	2,350
0300	Starter, size 1, FVR, NEMA 1, type A, fusible		2	4		2,275	219		2,494	2,850
0320	Circuit breaker		2	4		2,275	219		2,494	2,850
0340	Type B, fusible		2	4		2,500	219		2,719	3,100
0360	Circuit breaker		2	4		2,500	219		2,719	3,100
0380	NEMA 12, type A, fusible		1.90	4.211		2,300	230		2,530	2,875
0400	Circuit breaker		1.90	4.211		2,300	230		2,530	2,875
0420	Type B, fusible		1.90	4.211		2,525	230		2,755	3,125
0440	Circuit breaker		1.90	4.211		2,525	230		2,755	3,125
0490	Starter size 1, 2 speed, separate winding									
0500	NEMA 1, type A, fusible	1 Elec	2.60	3.077	Ea.	2,975	168		3,143	3,525
0520	Circuit breaker		2.60	3.077		2,975	168		3,143	3,525
0540	Type B, fusible		2.60	3.077		3,275	168		3,443	3,850
0560	Circuit breaker		2.60	3.077		3,275	168		3,443	3,850
0580	NEMA 12, type A, fusible		2.50	3.200		3,050	175		3,225	3,625
0600	Circuit breaker		2.50	3.200		3,050	175		3,225	3,625
0620	Type B, fusible		2.50	3.200		3,350	175		3,525	3,950
0640	Circuit breaker		2.50	3.200		3,350	175		3,525	3,950

26 24 19.20 Motor Control Center Components		Crew	Daily Output	Labor-Hours	Unit	Material	2015 Bare Costs Labor	Equipment	Total	Total Incl O&P
0650	Starter size 1, 2 speed, consequent pole									
0660	NEMA 1, type A, fusible	1 Elec	2.60	3.077	Ea.	2,975	168		3,143	3,525
0680	Circuit breaker		2.60	3.077		2,975	168		3,143	3,525
0700	Type B, fusible		2.60	3.077		3,275	168		3,443	3,850
0720	Circuit breaker		2.60	3.077		3,275	168		3,443	3,850
0740	NEMA 12, type A, fusible		2.50	3.200		3,050	175		3,225	3,625
0760	Circuit breaker		2.50	3.200		3,050	175		3,225	3,625
0780	Type B, fusible		2.50	3.200		3,325	175		3,500	3,925
0800	Circuit breaker		2.50	3.200		3,350	175		3,525	3,950
0810	Starter size 1, 2 speed, space only									
0820	NEMA 1, type A, fusible	1 Elec	16	.500	Ea.	670	27.50		697.50	780
0840	Circuit breaker		16	.500		670	27.50		697.50	780
0860	Type B, fusible		16	.500		670	27.50		697.50	780
0880	Circuit breaker		16	.500		670	27.50		697.50	780
0900	NEMA 12, type A, fusible		15	.533		695	29		724	810
0920	Circuit breaker		15	.533		695	29		724	810
0940	Type B, fusible		15	.533		695	29		724	810
0960	Circuit breaker		15	.533		695	29		724	810
1100	Starter size 2, FVNR, NEMA 1, type A, fusible	2 Elec	4	4		1,775	219		1,994	2,325
1120	Circuit breaker		4	4		1,950	219		2,169	2,500
1140	Type B, fusible		4	4		1,975	219		2,194	2,500
1160	Circuit breaker		4	4		2,150	219		2,369	2,700
1180	NEMA 12, type A, fusible		3.80	4.211		1,825	230		2,055	2,350
1200	Circuit breaker		3.80	4.211		1,975	230		2,205	2,525
1220	Type B, fusible		3.80	4.211		1,975	230		2,205	2,525
1240	Circuit breaker		3.80	4.211		2,175	230		2,405	2,725
1300	FVR, NEMA 1, type A, fusible		3.20	5		3,025	274		3,299	3,750
1320	Circuit breaker		3.20	5		3,025	274		3,299	3,750
1340	Type B, fusible		3.20	5		3,325	274		3,599	4,075
1360	Circuit breaker		3.20	5		3,325	274		3,599	4,075
1380	NEMA type 12, type A, fusible		3	5.333		3,075	292		3,367	3,825
1400	Circuit breaker		3	5.333		3,075	292		3,367	3,825
1420	Type B, fusible		3	5.333		3,375	292		3,667	4,175
1440	Circuit breaker		3	5.333		3,375	292		3,667	4,175
1490	Starter size 2, 2 speed, separate winding									
1500	NEMA 1, type A, fusible	2 Elec	3.80	4.211	Ea.	3,375	230		3,605	4,075
1520	Circuit breaker		3.80	4.211		3,375	230		3,605	4,075
1540	Type B, fusible		3.80	4.211		3,700	230		3,930	4,425
1560	Circuit breaker		3.80	4.211		3,700	230		3,930	4,425
1570	NEMA 12, type A, fusible		3.60	4.444		3,425	243		3,668	4,150
1580	Circuit breaker		3.60	4.444		3,425	243		3,668	4,150
1600	Type B, fusible		3.60	4.444		3,775	243		4,018	4,525
1620	Circuit breaker		3.60	4.444		3,775	243		4,018	4,525
1630	Starter size 2, 2 speed, consequent pole									
1640	NEMA 1, type A, fusible	2 Elec	3.80	4.211	Ea.	3,825	230		4,055	4,550
1660	Circuit breaker		3.80	4.211		3,825	230		4,055	4,550
1680	Type B, fusible		3.80	4.211		4,050	230		4,280	4,800
1700	Circuit breaker		3.80	4.211		4,050	230		4,280	4,800
1720	NEMA 12, type A, fusible		3.80	4.211		3,850	230		4,080	4,575
1740	Circuit breaker		3.60	4.444		3,875	243		4,118	4,650
1760	Type B, fusible		3.60	4.444		4,100	243		4,343	4,900
1780	Circuit breaker		3.60	4.444		4,100	243		4,343	4,900
1830	Starter size 2, autotransformer									

26 24 19.20 Motor Control Center Components		Crew	Daily Output	Labor-Hours	Unit	Material	2015 Bare Costs Labor	Equipment	Total	Total Incl O&P
1840	NEMA 1, type A, fusible	2 Elec	3.40	4.706	Ea.	5,975	257		6,232	6,975
1860	Circuit breaker		3.40	4.706		6,125	257		6,382	7,125
1880	Type B, fusible		3.40	4.706		6,525	257		6,782	7,575
1900	Circuit breaker		3.40	4.706		6,525	257		6,782	7,575
1920	NEMA 12, type A, fusible		3.20	5		6,075	274		6,349	7,125
1940	Circuit breaker		3.20	5		6,075	274		6,349	7,125
1960	Type B, fusible		3.20	5		6,625	274		6,899	7,725
1980	Circuit breaker		3.20	5		6,625	274		6,899	7,725
2030	Starter size 2, space only									
2040	NEMA 1, type A, fusible	1 Elec	16	.500	Ea.	670	27.50		697.50	780
2060	Circuit breaker		16	.500		670	27.50		697.50	780
2080	Type B, fusible		16	.500		670	27.50		697.50	780
2100	Circuit breaker		16	.500		670	27.50		697.50	780
2120	NEMA 12, type A, fusible		15	.533		695	29		724	810
2140	Circuit breaker		15	.533		695	29		724	810
2160	Type B, fusible		15	.533		695	29		724	810
2180	Circuit breaker		15	.533		695	29		724	810
2300	Starter size 3, FVNR, NEMA 1, type A, fusible	2 Elec	2	8		3,400	440		3,840	4,425
2320	Circuit breaker		2	8		3,025	440		3,465	4,000
2340	Type B, fusible		2	8		3,750	440		4,190	4,800
2360	Circuit breaker		2	8		3,325	440		3,765	4,350
2380	NEMA 12, type A, fusible		1.90	8.421		3,475	460		3,935	4,550
2400	Circuit breaker		1.90	8.421		3,075	460		3,535	4,100
2420	Type B, fusible		1.90	8.421		3,875	460		4,335	4,975
2440	Circuit breaker		1.90	8.421		3,375	460		3,835	4,450
2500	Starter size 3, FVR, NEMA 1, type A, fusible		1.60	10		4,625	545		5,170	5,950
2520	Circuit breaker		1.60	10		4,425	545		4,970	5,725
2540	Type B, fusible		1.60	10		5,050	545		5,595	6,400
2560	Circuit breaker		1.60	10		4,850	545		5,395	6,175
2580	NEMA 12, type A, fusible		1.50	10.667		4,725	585		5,310	6,100
2600	Circuit breaker		1.50	10.667		4,500	585		5,085	5,875
2620	Type B, fusible		1.50	10.667		5,150	585		5,735	6,550
2640	Circuit breaker		1.50	10.667		5,325	585		5,910	6,775
2690	Starter size 3, 2 speed, separate winding									
2700	NEMA 1, type A, fusible	2 Elec	2	8	Ea.	5,325	440		5,765	6,550
2720	Circuit breaker		2	8		4,700	440		5,140	5,850
2740	Type B, fusible		2	8		5,850	440		6,290	7,100
2760	Circuit breaker		2	8		5,150	440		5,590	6,350
2780	NEMA 12, type A, fusible		1.90	8.421		5,450	460		5,910	6,700
2800	Circuit breaker		1.90	8.421		4,800	460		5,260	6,000
2820	Type B, fusible		1.90	8.421		5,950	460		6,410	7,275
2840	Circuit breaker		1.90	8.421		5,250	460		5,710	6,500
2850	Starter size 3, 2 speed, consequent pole									
2860	NEMA 1, type A, fusible	2 Elec	2	8	Ea.	5,950	440		6,390	7,225
2880	Circuit breaker		2	8		5,325	440		5,765	6,525
2900	Type B, fusible		2	8		6,450	440		6,890	7,775
2920	Circuit breaker		2	8		5,775	440		6,215	7,025
2940	NEMA 12, type A, fusible		1.90	8.421		6,050	460		6,510	7,400
2960	Circuit breaker		1.90	8.421		5,425	460		5,885	6,700
2980	Type B, fusible		1.90	8.421		6,575	460		7,035	7,950
3000	Circuit breaker		1.90	8.421		5,875	460		6,335	7,175
3100	Starter size 3, autotransformer, NEMA 1, type A, fusible		1.60	10		7,475	545		8,020	9,075
3120	Circuit breaker		1.60	10		7,400	545		7,945	9,000

26 24 19 – Motor-Control Centers

26 24 19.20 Motor Control Center Components		Crew	Daily Output	Labor-Hours	Unit	Material	Labor	Equipment	Total	Total Incl O&P
						2015 Bare Costs				
3140	Type B, fusible	2 Elec	1.60	10	Ea.	8,100	545		8,645	9,750
3160	Circuit breaker		1.60	10		8,100	545		8,645	9,750
3180	NEMA 12, type A, fusible		1.50	10.667		7,550	585		8,135	9,200
3200	Circuit breaker		1.50	10.667		7,550	585		8,135	9,200
3220	Type B, fusible		1.50	10.667		8,225	585		8,810	9,950
3240	Circuit breaker		1.50	10.667		8,225	585		8,810	9,950
3260	Starter size 3, space only, NEMA 1, type A, fusible	1 Elec	15	.533		1,150	29		1,179	1,325
3280	Circuit breaker		15	.533		895	29		924	1,025
3300	Type B, fusible		15	.533		1,150	29		1,179	1,325
3320	Circuit breaker		15	.533		895	29		924	1,025
3340	NEMA 12, type A, fusible		14	.571		1,225	31.50		1,256.50	1,400
3360	Circuit breaker		14	.571		950	31.50		981.50	1,100
3380	Type B, fusible		14	.571		1,225	31.50		1,256.50	1,400
3400	Circuit breaker		14	.571		950	31.50		981.50	1,100
3500	Starter size 4, FVNR, NEMA 1, type A, fusible	2 Elec	1.60	10		4,425	545		4,970	5,725
3520	Circuit breaker		1.60	10		4,025	545		4,570	5,275
3540	Type B, fusible		1.60	10		4,850	545		5,395	6,175
3560	Circuit breaker		1.60	10		4,425	545		4,970	5,725
3580	NEMA 12, type A, fusible		1.50	10.667		4,550	585		5,135	5,900
3600	Circuit breaker		1.50	10.667		4,075	585		4,660	5,400
3620	Type B, fusible		1.50	10.667		4,950	585		5,535	6,350
3640	Circuit breaker		1.50	10.667		4,500	585		5,085	5,850
3700	Starter size 4, FVR, NEMA 1, type A, fusible		1.20	13.333		6,050	730		6,780	7,800
3720	Circuit breaker		1.20	13.333		5,450	730		6,180	7,125
3740	Type B, fusible		1.20	13.333		6,600	730		7,330	8,400
3760	Circuit breaker		1.20	13.333		6,000	730		6,730	7,725
3780	NEMA 12, type A, fusible		1.16	13.793		6,200	755		6,955	8,000
3800	Circuit breaker		1.16	13.793		5,550	755		6,305	7,275
3820	Type B, fusible		1.16	13.793		5,600	755		6,355	7,325
3840	Circuit breaker		1.16	13.793		6,100	755		6,855	7,875
3890	Starter size 4, 2 speed, separate windings									
3900	NEMA 1, type A, fusible	2 Elec	1.60	10	Ea.	7,650	545		8,195	9,250
3920	Circuit breaker		1.60	10		5,750	545		6,295	7,175
3940	Type B, fusible		1.60	10		8,400	545		8,945	10,100
3960	Circuit breaker		1.60	10		6,300	545		6,845	7,775
3980	NEMA 12, type A, fusible		1.50	10.667		7,775	585		8,360	9,450
4000	Circuit breaker		1.50	10.667		5,850	585		6,435	7,325
4020	Type B, fusible		1.50	10.667		8,525	585		9,110	10,300
4040	Circuit breaker		1.50	10.667		6,375	585		6,960	7,925
4050	Starter size 4, 2 speed, consequent pole									
4060	NEMA 1, type A, fusible	2 Elec	1.60	10	Ea.	8,925	545		9,470	10,700
4080	Circuit breaker		1.60	10		6,650	545		7,195	8,150
4100	Type B, fusible		1.60	10		9,825	545		10,370	11,600
4120	Circuit breaker		1.60	10		7,275	545		7,820	8,850
4140	NEMA 12, type A, fusible		1.50	10.667		9,075	585		9,660	10,900
4160	Circuit breaker		1.50	10.667		6,750	585		7,335	8,300
4180	Type B, fusible		1.50	10.667		9,950	585		10,535	11,900
4200	Circuit breaker		1.50	10.667		7,375	585		7,960	9,025
4300	Starter size 4, autotransformer, NEMA 1, type A, fusible		1.30	12.308		8,800	675		9,475	10,700
4320	Circuit breaker		1.30	12.308		8,875	675		9,550	10,800
4340	Type B, fusible		1.30	12.308		9,675	675		10,350	11,800
4360	Circuit breaker		1.30	12.308		9,675	675		10,350	11,800
4380	NEMA 12, type A, fusible		1.24	12.903		8,950	705		9,655	10,900

26 24 19.20 Motor Control Center Components		Crew	Daily Output	Labor-Hours	Unit	Material	2015 Bare Costs Labor	Equipment	Total	Total Incl O&P
4400	Circuit breaker	2 Elec	1.24	12.903	Ea.	8,975	705		9,680	11,000
4420	Type B, fusible		1.24	12.903		9,825	705		10,530	11,900
4440	Circuit breaker		1.24	12.903		9,800	705		10,505	11,900
4500	Starter size 4, space only, NEMA 1, type A, fusible	1 Elec	14	.571		1,600	31.50		1,631.50	1,800
4520	Circuit breaker		14	.571		1,150	31.50		1,181.50	1,325
4540	Type B, fusible		14	.571		1,600	31.50		1,631.50	1,800
4560	Circuit breaker		14	.571		1,150	31.50		1,181.50	1,325
4580	NEMA 12, type A, fusible		13	.615		1,700	33.50		1,733.50	1,900
4600	Circuit breaker		13	.615		1,225	33.50		1,258.50	1,400
4620	Type B, fusible		13	.615		1,700	33.50		1,733.50	1,900
4640	Circuit breaker		13	.615		1,225	33.50		1,258.50	1,400
4800	Starter size 5, FVNR, NEMA 1, type A, fusible	2 Elec	1	16		9,150	875		10,025	11,500
4820	Circuit breaker		1	16		6,500	875		7,375	8,500
4840	Type B, fusible		1	16		10,000	875		10,875	12,400
4860	Circuit breaker		1	16		7,150	875		8,025	9,200
4880	NEMA 12, type A, fusible		.96	16.667		9,275	910		10,185	11,600
4900	Circuit breaker		.96	16.667		6,625	910		7,535	8,675
4920	Type B, fusible		.96	16.667		10,200	910		11,110	12,600
4940	Circuit breaker		.96	16.667		7,275	910		8,185	9,400
5000	Starter size 5, FVR, NEMA 1, type A, fusible		.80	20		13,500	1,100		14,600	16,500
5020	Circuit breaker		.80	20		10,600	1,100		11,700	13,300
5040	Type B, fusible		.80	20		14,800	1,100		15,900	17,900
5060	Circuit breaker		.80	20		11,600	1,100		12,700	14,500
5080	NEMA 12, type A, fusible		.76	21.053		13,700	1,150		14,850	16,900
5100	Circuit breaker		.76	21.053		10,700	1,150		11,850	13,600
5120	Type B, fusible		.76	21.053		15,000	1,150		16,150	18,300
5140	Circuit breaker		.76	21.053		11,800	1,150		12,950	14,700
5190	Starter size 5, 2 speed, separate windings									
5200	NEMA 1, type A, fusible	2 Elec	1	16	Ea.	18,600	875		19,475	21,800
5220	Circuit breaker		1	16		13,900	875		14,775	16,600
5240	Type B, fusible		1	16		20,400	875		21,275	23,800
5260	Circuit breaker		1	16		15,200	875		16,075	18,200
5280	NEMA 12, type A, fusible		.96	16.667		18,800	910		19,710	22,100
5300	Circuit breaker		.96	16.667		14,000	910		14,910	16,800
5320	Type B, fusible		.96	16.667		20,700	910		21,610	24,100
5340	Circuit breaker		.96	16.667		15,400	910		16,310	18,300
5400	Starter size 5, autotransformer, NEMA 1, type A, fusible		.70	22.857		15,200	1,250		16,450	18,600
5420	Circuit breaker		.70	22.857		12,600	1,250		13,850	15,800
5440	Type B, fusible		.70	22.857		16,700	1,250		17,950	20,300
5460	Circuit breaker		.70	22.857		13,900	1,250		15,150	17,200
5480	NEMA 12, type A, fusible		.68	23.529		15,400	1,275		16,675	18,900
5500	Circuit breaker		.68	23.529		12,800	1,275		14,075	16,000
5520	Type B, fusible		.68	23.529		16,900	1,275		18,175	20,500
5540	Circuit breakers		.68	23.529		14,000	1,275		15,275	17,400
5600	Starter size 5, space only, NEMA 1, type A, fusible	1 Elec	12	.667		2,050	36.50		2,086.50	2,300
5620	Circuit breaker		12	.667		2,050	36.50		2,086.50	2,300
5640	Type B, fusible		12	.667		2,050	36.50		2,086.50	2,300
5660	Circuit breaker		12	.667		1,350	36.50		1,386.50	1,525
5680	NEMA 12, type A, fusible		11	.727		2,175	40		2,215	2,425
5700	Circuit breaker		11	.727		1,425	40		1,465	1,625
5720	Type B, fusible		11	.727		2,175	40		2,215	2,425
5740	Circuit breaker		11	.727		1,425	40		1,465	1,625
5800	Fuse, light contactor NEMA 1, type A, 30 amp		2.70	2.963		1,575	162		1,737	2,000

26 24 19.20 Motor Control Center Components		Crew	Daily Output	Labor-Hours	Unit	Material	2015 Bare Costs Labor	Equipment	Total	Total Incl O&P
5820	60 amp	1 Elec	2	4	Ea.	1,775	219		1,994	2,325
5840	100 amp		1	8		3,400	440		3,840	4,425
5860	200 amp		.80	10		7,900	545		8,445	9,550
5880	Type B, 30 amp		2.70	2.963		1,725	162		1,887	2,150
5900	60 amp		2	4		1,925	219		2,144	2,475
5920	100 amp		1	8		3,750	440		4,190	4,800
5940	200 amp	2 Elec	1.60	10		8,650	545		9,195	10,400
5960	NEMA 12, type A, 30 amp	1 Elec	2.60	3.077		1,600	168		1,768	2,025
5980	60 amp		1.90	4.211		1,825	230		2,055	2,350
6000	100 amp		.95	8.421		3,475	460		3,935	4,550
6020	200 amp	2 Elec	1.50	10.667		8,025	585		8,610	9,725
6040	Type B, 30 amp	1 Elec	2.60	3.077		1,750	168		1,918	2,175
6060	60 amp		1.90	4.211		1,950	230		2,180	2,500
6080	100 amp		.95	8.421		3,800	460		4,260	4,925
6100	200 amp	2 Elec	1.50	10.667		8,775	585		9,360	10,600
6200	Circuit breaker, light contactor NEMA 1, type A, 30 amp	1 Elec	2.70	2.963		1,725	162		1,887	2,150
6220	60 amp		2	4		1,950	219		2,169	2,500
6240	100 amp		1	8		3,025	440		3,465	4,000
6260	200 amp	2 Elec	1.60	10		6,550	545		7,095	8,050
6280	Type B, 30 amp	1 Elec	2.70	2.963		1,900	162		2,062	2,325
6300	60 amp		2	4		2,175	219		2,394	2,725
6320	100 amp		1	8		3,300	440		3,740	4,300
6340	200 amp	2 Elec	1.60	10		7,150	545		7,695	8,700
6360	NEMA 12, type A, 30 amp	1 Elec	2.60	3.077		1,750	168		1,918	2,175
6380	60 amp		1.90	4.211		1,975	230		2,205	2,525
6400	100 amp		.95	8.421		3,075	460		3,535	4,100
6420	200 amp	2 Elec	1.50	10.667		6,575	585		7,160	8,125
6440	Type B, 30 amp	1 Elec	2.60	3.077		2,050	168		2,218	2,500
6460	60 amp		1.90	4.211		2,200	230		2,430	2,750
6480	100 amp		.95	8.421		3,350	460		3,810	4,400
6500	200 amp	2 Elec	1.50	10.667		7,225	585		7,810	8,850
6600	Fusible switch, NEMA 1, type A, 30 amp	1 Elec	5.30	1.509		1,050	82.50		1,132.50	1,275
6620	60 amp		5	1.600		1,125	87.50		1,212.50	1,350
6640	100 amp		4	2		1,225	109		1,334	1,525
6660	200 amp		3.20	2.500		2,125	137		2,262	2,550
6680	400 amp	2 Elec	4.60	3.478		5,025	190		5,215	5,825
6700	600 amp		3.20	5		5,425	274		5,699	6,400
6720	800 amp		2.60	6.154		14,900	335		15,235	16,900
6740	NEMA 12, type A, 30 amp	1 Elec	5.20	1.538		1,075	84		1,159	1,300
6760	60 amp		4.90	1.633		1,150	89.50		1,239.50	1,400
6780	100 amp		3.90	2.051		1,275	112		1,387	1,575
6800	200 amp		3.10	2.581		2,175	141		2,316	2,625
6820	400 amp	2 Elec	4.40	3.636		5,125	199		5,324	5,925
6840	600 amp		3	5.333		5,575	292		5,867	6,600
6860	800 amp		2.40	6.667		15,000	365		15,365	17,100
6900	Circuit breaker, NEMA 1, type A, 30 amp	1 Elec	5.30	1.509		960	82.50		1,042.50	1,175
6920	60 amp		5	1.600		960	87.50		1,047.50	1,175
6940	100 amp		4	2		960	109		1,069	1,225
6960	225 amp		3.20	2.500		1,725	137		1,862	2,100
6980	400 amp	2 Elec	4.60	3.478		3,250	190		3,440	3,900
7000	600 amp		3.20	5		3,800	274		4,074	4,625
7020	800 amp		2.60	6.154		7,975	335		8,310	9,300
7040	NEMA 12, type A, 30 amp	1 Elec	5.20	1.538		990	84		1,074	1,225

26 24 19.20 Motor Control Center Components		Crew	Daily Output	Labor-Hours	Unit	Material	2015 Bare Costs Labor	Equipment	Total	Total Incl O&P
7060	60 amp	1 Elec	4.90	1.633	Ea.	990	89.50		1,079.50	1,250
7080	100 amp		3.90	2.051		990	112		1,102	1,275
7100	225 amp		3.10	2.581		1,750	141		1,891	2,150
7120	400 amp	2 Elec	4.40	3.636		3,325	199		3,524	3,950
7140	600 amp		3	5.333		3,875	292		4,167	4,700
7160	800 amp		2.40	6.667		8,125	365		8,490	9,500
7300	Incoming line, main lug only, 600 amp, alum., NEMA 1		1.60	10		1,225	545		1,770	2,175
7320	NEMA 12		1.50	10.667		1,250	585		1,835	2,275
7340	Copper, NEMA 1		1.60	10		1,275	545		1,820	2,250
7360	800 amp, alum., NEMA 1		1.50	10.667		3,050	585		3,635	4,250
7380	NEMA 12		1.40	11.429		3,100	625		3,725	4,375
7400	Copper, NEMA 1		1.50	10.667		3,175	585		3,760	4,400
7420	1200 amp, copper, NEMA 1		1.40	11.429		3,275	625		3,900	4,575
7440	Incoming line, fusible switch, 400 amp, alum., NEMA 1		1.20	13.333		4,200	730		4,930	5,725
7460	NEMA 12		1.10	14.545		4,275	795		5,070	5,925
7480	Copper, NEMA 1		1.20	13.333		4,250	730		4,980	5,800
7500	600 amp, alum., NEMA 1		1.10	14.545		5,125	795		5,920	6,850
7520	NEMA 12		1	16		5,200	875		6,075	7,075
7540	Copper, NEMA 1		1.10	14.545		5,175	795		5,970	6,925
7560	Incoming line, circuit breaker, 225 amp, alum., NEMA 1		1.20	13.333		2,175	730		2,905	3,525
7580	NEMA 12		1.10	14.545		2,225	795		3,020	3,675
7600	Copper, NEMA 1		1.20	13.333		2,250	730		2,980	3,600
7620	400 amp, alum., NEMA 1		1.20	13.333		3,250	730		3,980	4,725
7640	NEMA 12		1.10	14.545		3,325	795		4,120	4,875
7660	Copper, NEMA 1		1.20	13.333		3,325	730		4,055	4,775
7680	600 amp, alum., NEMA 1		1.10	14.545		3,800	795		4,595	5,425
7700	NEMA 12		1	16		3,875	875		4,750	5,600
7720	Copper, NEMA 1		1.10	14.545		3,875	795		4,670	5,500
7740	800 amp, copper, NEMA 1		.90	17.778		7,975	970		8,945	10,300
7760	Incoming line, for copper bus, add					120			120	132
7780	For 65000 amp bus bracing, add					179			179	196
7800	For NEMA 3R enclosure, add					5,600			5,600	6,175
7820	For NEMA 12 enclosure, add					155			155	170
7840	For 1/4" x 1" ground bus, add	1 Elec	16	.500		99.50	27.50		127	153
7860	For 1/4" x 2" ground bus, add	"	12	.667		99.50	36.50		136	167
7900	Main rating basic section, alum., NEMA 1, 800 amp	2 Elec	1.40	11.429		325	625		950	1,325
7920	1200 amp	"	1.20	13.333		640	730		1,370	1,825
7940	For copper bus, add					480			480	530
7960	For 65000 amp bus bracing, add					325			325	355
7980	For NEMA 3R enclosure, add					5,600			5,600	6,175
8000	For NEMA 12, enclosure, add					155			155	170
8020	For 1/4" x 1" ground bus, add	1 Elec	16	.500		99.50	27.50		127	153
8040	For 1/4" x 2" ground bus, add		12	.667		99.50	36.50		136	167
8060	Unit devices, pilot light, standard		16	.500		86	27.50		113.50	137
8080	Pilot light, push to test		16	.500		120	27.50		147.50	175
8100	Pilot light, standard, and push button		12	.667		206	36.50		242.50	284
8120	Pilot light, push to test, and push button		12	.667		240	36.50		276.50	320
8140	Pilot light, standard, and select switch		12	.667		206	36.50		242.50	284
8160	Pilot light, push to test, and select switch		12	.667		240	36.50		276.50	320

26 24 19.30 Motor Control Center		Crew	Daily Output	Labor-Hours	Unit	Material	Labor (2015 Bare Costs)	Equipment	Total	Total Incl O&P
0010	**MOTOR CONTROL CENTER** Consists of starters & structures									
0050	Starters, class 1, type B, comb. MCP, FVNR, with									
0100	control transformer, 10 HP, size 1, 12″ high	1 Elec	2.70	2.963	Ea.	1,725	162		1,887	2,150
0200	25 HP, size 2, 18″ high	2 Elec	4	4		1,950	219		2,169	2,500
0300	50 HP, size 3, 24″ high		2	8		3,025	440		3,465	4,000
0350	75 HP, size 4, 24″ high		1.60	10		4,025	545		4,570	5,275
0400	100 HP, size 4, 30″ high		1.40	11.429		5,300	625		5,925	6,800
0500	200 HP, size 5, 48″ high		1	16		7,900	875		8,775	10,100
0600	400 HP, size 6, 72″ high		.80	20		16,300	1,100		17,400	19,600
0800	Structures, 600 amp, 22,000 rms, takes any									
0900	combination of starters up to 72″ high	2 Elec	1.60	10	Ea.	1,925	545		2,470	2,975
1000	Back to back, 72″ front & 66″ back	″	1.20	13.333		2,600	730		3,330	3,975
1100	For copper bus add per structure					268			268	295
1200	For NEMA 12, add per structure					155			155	170
1300	For 42,000 rms, add per structure					201			201	221
1400	For 100,000 rms, size 1 & 2, add					685			685	755
1500	Size 3, add					1,100			1,100	1,225
1600	Size 4, add					895			895	980
1700	For pilot lights, add per starter	1 Elec	16	.500		120	27.50		147.50	175
1800	For push button, add per starter		16	.500		120	27.50		147.50	175
1900	For auxiliary contacts, add per starter		16	.500		179	27.50		206.50	239

26 24 19.40 Motor Starters and Controls		Crew	Daily Output	Labor-Hours	Unit	Material	Labor	Equipment	Total	Total Incl O&P
0010	**MOTOR STARTERS AND CONTROLS**									
0050	Magnetic, FVNR, with enclosure and heaters, 480 volt									
0080	2 HP, size 00	1 Elec	3.50	2.286	Ea.	203	125		328	415
0100	5 HP, size 0		2.30	3.478		272	190		462	595
0200	10 HP, size 1		1.60	5		275	274		549	730
0300	25 HP, size 2	2 Elec	2.20	7.273		520	400		920	1,175
0400	50 HP, size 3		1.80	8.889		845	485		1,330	1,675
0500	100 HP, size 4		1.20	13.333		1,875	730		2,605	3,175
0600	200 HP, size 5		.90	17.778		4,375	970		5,345	6,325
0610	400 HP, size 6		.80	20		19,000	1,100		20,100	22,600
0620	NEMA 7, 5 HP, size 0	1 Elec	1.60	5		1,400	274		1,674	1,975
0630	10 HP, size 1	″	1.10	7.273		1,475	400		1,875	2,250
0640	25 HP, size 2	2 Elec	1.80	8.889		2,375	485		2,860	3,375
0650	50 HP, size 3		1.20	13.333		3,575	730		4,305	5,050
0660	100 HP, size 4		.90	17.778		5,775	970		6,745	7,850
0670	200 HP, size 5		.50	32		13,800	1,750		15,550	17,900
0700	Combination, with motor circuit protectors, 5 HP, size 0	1 Elec	1.80	4.444		880	243		1,123	1,350
0800	10 HP, size 1	″	1.30	6.154		915	335		1,250	1,525
0900	25 HP, size 2	2 Elec	2	8		1,275	440		1,715	2,075
1000	50 HP, size 3		1.32	12.121		1,850	665		2,515	3,050
1200	100 HP, size 4		.80	20		4,000	1,100		5,100	6,100
1220	NEMA 7, 5 HP, size 0	1 Elec	1.30	6.154		2,825	335		3,160	3,625
1230	10 HP, size 1	″	1	8		2,900	440		3,340	3,850
1240	25 HP, size 2	2 Elec	1.32	12.121		3,850	665		4,515	5,275
1250	50 HP, size 3		.80	20		6,375	1,100		7,475	8,725
1260	100 HP, size 4		.60	26.667		9,925	1,450		11,375	13,200
1270	200 HP, size 5		.40	40		21,600	2,200		23,800	27,200
1400	Combination, with fused switch, 5 HP, size 0	1 Elec	1.80	4.444		610	243		853	1,050
1600	10 HP, size 1	″	1.30	6.154		650	335		985	1,225
1800	25 HP, size 2	2 Elec	2	8		1,050	440		1,490	1,825

26 24 19.40 Motor Starters and Controls		Crew	Daily Output	Labor-Hours	Unit	Material	Labor	Equipment	Total	Total Incl O&P
2000	50 HP, size 3	2 Elec	1.32	12.121	Ea.	1,775	665		2,440	3,000
2200	100 HP, size 4		.80	20		3,125	1,100		4,225	5,125
2610	NEMA 4, with start-stop pushbutton size 1	1 Elec	1.30	6.154		1,625	335		1,960	2,325
2620	Size 2	2 Elec	2	8		2,200	440		2,640	3,075
2630	Size 3		1.32	12.121		3,475	665		4,140	4,825
2640	Size 4		.80	20		5,325	1,100		6,425	7,550
2650	NEMA 4, FVNR, including control transformer									
2660	Size 1	2 Elec	2.60	6.154	Ea.	1,500	335		1,835	2,175
2670	Size 2		2	8		2,150	440		2,590	3,025
2680	Size 3		1.32	12.121		3,425	665		4,090	4,775
2690	Size 4		.80	20		5,275	1,100		6,375	7,500
2710	Magnetic, FVR, control circuit transformer, NEMA 1, size 1	1 Elec	1.30	6.154		885	335		1,220	1,500
2720	Size 2	2 Elec	2	8		1,425	440		1,865	2,225
2730	Size 3		1.32	12.121		2,150	665		2,815	3,400
2740	Size 4		.80	20		4,700	1,100		5,800	6,850
2760	NEMA 4, size 1	1 Elec	1.10	7.273		1,275	400		1,675	2,025
2770	Size 2	2 Elec	1.60	10		2,050	545		2,595	3,100
2780	Size 3		1.20	13.333		3,075	730		3,805	4,525
2790	Size 4		.70	22.857		6,325	1,250		7,575	8,900
2820	NEMA 12, size 1	1 Elec	1.10	7.273		1,050	400		1,450	1,775
2830	Size 2	2 Elec	1.60	10		1,650	545		2,195	2,675
2840	Size 3		1.20	13.333		2,600	730		3,330	3,975
2850	Size 4		.70	22.857		5,350	1,250		6,600	7,825
2870	Combination FVR, fused, w/control XFMR & PB, NEMA 1, size 1	1 Elec	1	8		1,525	440		1,965	2,350
2880	Size 2	2 Elec	1.50	10.667		2,175	585		2,760	3,300
2890	Size 3		1.10	14.545		3,225	795		4,020	4,775
2900	Size 4		.70	22.857		6,550	1,250		7,800	9,150
2910	NEMA 4, size 1	1 Elec	.90	8.889		2,275	485		2,760	3,250
2920	Size 2	2 Elec	1.40	11.429		3,300	625		3,925	4,600
2930	Size 3		1	16		5,200	875		6,075	7,075
2940	Size 4		.60	26.667		8,650	1,450		10,100	11,800
2950	NEMA 12, size 1	1 Elec	1	8		1,725	440		2,165	2,575
2960	Size 2	2 Elec	1.40	11.429		2,450	625		3,075	3,650
2970	Size 3		1	16		3,575	875		4,450	5,300
2980	Size 4		.60	26.667		7,125	1,450		8,575	10,100
3010	Manual, single phase, w/pilot, 1 pole 120 V NEMA 1	1 Elec	6.40	1.250		70	68.50		138.50	183
3020	NEMA 4		4	2		355	109		464	565
3030	2 pole, 120/240 V, NEMA 1		6.40	1.250		81.50	68.50		150	196
3040	NEMA 4		4	2		320	109		429	520
3041	3 phase, 3 pole 600 V, NEMA 1		5.50	1.455		248	79.50		327.50	395
3042	NEMA 4		3.50	2.286		485	125		610	730
3043	NEMA 12		3.50	2.286		273	125		398	495
3070	Auxiliary contact, normally open					84			84	92.50
3500	Magnetic FVNR with NEMA 12, enclosure & heaters, 480 volt									
3600	5 HP, size 0	1 Elec	2.20	3.636	Ea.	231	199		430	565
3700	10 HP, size 1	"	1.50	5.333		350	292		642	835
3800	25 HP, size 2	2 Elec	2	8		650	440		1,090	1,400
3900	50 HP, size 3		1.60	10		1,000	545		1,545	1,950
4000	100 HP, size 4		1	16		2,400	875		3,275	3,975
4100	200 HP, size 5		.80	20		5,750	1,100		6,850	8,025
4200	Combination, with motor circuit protectors, 5 HP, size 0	1 Elec	1.70	4.706		765	257		1,022	1,250
4300	10 HP, size 1	"	1.20	6.667		795	365		1,160	1,450
4400	25 HP, size 2	2 Elec	1.80	8.889		1,200	485		1,685	2,050

26 24 19.40 Motor Starters and Controls		Crew	Daily Output	Labor-Hours	Unit	2015 Bare Costs Material	Labor	Equipment	Total	Total Incl O&P
4500	50 HP, size 3	2 Elec	1.20	13.333	Ea.	1,950	730		2,680	3,250
4600	100 HP, size 4		.74	21.622		4,375	1,175		5,550	6,650
4700	Combination, with fused switch, 5 HP, size 0	1 Elec	1.70	4.706		740	257		997	1,225
4800	10 HP, size 1	"	1.20	6.667		770	365		1,135	1,425
4900	25 HP, size 2	2 Elec	1.80	8.889		1,175	485		1,660	2,050
5000	50 HP, size 3		1.20	13.333		1,875	730		2,605	3,200
5100	100 HP, size 4		.74	21.622		3,825	1,175		5,000	6,025
5200	Factory installed controls, adders to size 0 thru 5									
5300	Start-stop push button	1 Elec	32	.250	Ea.	48.50	13.70		62.20	74.50
5400	Hand-off-auto-selector switch		32	.250		48.50	13.70		62.20	74.50
5500	Pilot light		32	.250		91	13.70		104.70	121
5600	Start-stop-pilot		32	.250		139	13.70		152.70	174
5700	Auxiliary contact, NO or NC		32	.250		66.50	13.70		80.20	94
5800	NO-NC		32	.250		133	13.70		146.70	167
5810	Magnetic FVR, NEMA 7, w/heaters, size 1		.66	12.121		2,700	665		3,365	3,975
5830	Size 2	2 Elec	1.10	14.545		4,500	795		5,295	6,175
5840	Size 3		.70	22.857		7,225	1,250		8,475	9,875
5850	Size 4		.60	26.667		8,225	1,450		9,675	11,300
5860	Combination w/circuit breakers, heaters, control XFMR PB, size 1	1 Elec	.60	13.333		1,700	730		2,430	3,000
5870	Size 2	2 Elec	.80	20		2,175	1,100		3,275	4,100
5880	Size 3		.50	32		2,900	1,750		4,650	5,900
5890	Size 4		.40	40		5,750	2,200		7,950	9,725
5900	Manual, 240 volt, .75 HP motor	1 Elec	4	2		51.50	109		160.50	226
5910	2 HP motor		4	2		139	109		248	320
6000	Magnetic, 240 volt, 1 or 2 pole, .75 HP motor		4	2		203	109		312	390
6020	2 HP motor		4	2		224	109		333	415
6040	5 HP motor		3	2.667		320	146		466	580
6060	10 HP motor		2.30	3.478		795	190		985	1,175
6100	3 pole, .75 HP motor		3	2.667		203	146		349	450
6120	5 HP motor		2.30	3.478		275	190		465	600
6140	10 HP motor		1.60	5		520	274		794	995
6160	15 HP motor		1.60	5		520	274		794	995
6180	20 HP motor		1.10	7.273		845	400		1,245	1,550
6200	25 HP motor	2 Elec	2.20	7.273		845	400		1,245	1,550
6210	30 HP motor		1.80	8.889		845	485		1,330	1,675
6220	40 HP motor		1.80	8.889		1,875	485		2,360	2,800
6230	50 HP motor		1.80	8.889		1,875	485		2,360	2,800
6240	60 HP motor		1.20	13.333		4,375	730		5,105	5,950
6250	75 HP motor		1.20	13.333		4,375	730		5,105	5,950
6260	100 HP motor		1.20	13.333		4,375	730		5,105	5,950
6270	125 HP motor		.90	17.778		12,300	970		13,270	15,000
6280	150 HP motor		.90	17.778		12,300	970		13,270	15,000
6290	200 HP motor		.90	17.778		12,300	970		13,270	15,000
6400	Starter & nonfused disconnect, 240 volt, 1-2 pole, .75 HP motor	1 Elec	2	4		262	219		481	630
6410	2 HP motor		2	4		283	219		502	650
6420	5 HP motor		1.80	4.444		380	243		623	795
6430	10 HP motor		1.40	5.714		875	315		1,190	1,450
6440	3 pole, .75 HP motor		1.60	5		262	274		536	715
6450	5 HP motor		1.40	5.714		335	315		650	855
6460	10 HP motor		1.10	7.273		595	400		995	1,275
6470	15 HP motor		1	8		595	440		1,035	1,325
6480	20 HP motor	2 Elec	1.50	10.667		1,025	585		1,610	2,025
6490	25 HP motor		1.50	10.667		1,025	585		1,610	2,025

26 24 19.40 Motor Starters and Controls		Crew	Daily Output	Labor-Hours	Unit	Material	2015 Bare Costs Labor	Equipment	Total	Total Incl O&P
6500	30 HP motor	2 Elec	1.30	12.308	Ea.	1,025	675		1,700	2,175
6510	40 HP motor		1.24	12.903		2,200	705		2,905	3,525
6520	50 HP motor		1.12	14.286		2,200	780		2,980	3,625
6530	60 HP motor		.90	17.778		4,725	970		5,695	6,700
6540	75 HP motor		.76	21.053		4,725	1,150		5,875	6,975
6550	100 HP motor		.70	22.857		4,725	1,250		5,975	7,125
6560	125 HP motor		.60	26.667		13,100	1,450		14,550	16,700
6570	150 HP motor		.52	30.769		13,100	1,675		14,775	17,000
6580	200 HP motor		.50	32		13,100	1,750		14,850	17,100
6600	Starter & fused disconnect, 240 volt, 1-2 pole, .75 HP motor	1 Elec	2	4		276	219		495	645
6610	2 HP motor		2	4		297	219		516	665
6620	5 HP motor		1.80	4.444		395	243		638	810
6630	10 HP motor		1.40	5.714		920	315		1,235	1,475
6640	3 pole, .75 HP motor		1.60	5		276	274		550	730
6650	5 HP motor		1.40	5.714		350	315		665	870
6660	10 HP motor		1.10	7.273		640	400		1,040	1,325
6690	15 HP motor		1	8		640	440		1,080	1,375
6700	20 HP motor	2 Elec	1.60	10		1,050	545		1,595	2,025
6710	25 HP motor		1.60	10		1,050	545		1,595	2,025
6720	30 HP motor		1.40	11.429		1,050	625		1,675	2,150
6730	40 HP motor		1.20	13.333		2,325	730		3,055	3,675
6740	50 HP motor		1.20	13.333		2,325	730		3,055	3,675
6750	60 HP motor		.90	17.778		4,825	970		5,795	6,825
6760	75 HP motor		.90	17.778		4,825	970		5,795	6,825
6770	100 HP motor		.70	22.857		4,825	1,250		6,075	7,250
6780	125 HP motor		.54	29.630		13,400	1,625		15,025	17,300
6790	Combination starter & nonfusible disconnect									
6800	240 volt, 1-2 pole, .75 HP motor	1 Elec	2	4	Ea.	640	219		859	1,050
6810	2 HP motor		2	4		640	219		859	1,050
6820	5 HP motor		1.50	5.333		675	292		967	1,200
6830	10 HP motor		1.20	6.667		945	365		1,310	1,625
6840	3 pole, .75 HP motor		1.80	4.444		580	243		823	1,000
6850	5 HP motor		1.30	6.154		610	335		945	1,200
6860	10 HP motor		1	8		945	440		1,385	1,725
6870	15 HP motor		1	8		945	440		1,385	1,725
6880	20 HP motor	2 Elec	1.32	12.121		1,550	665		2,215	2,750
6890	25 HP motor		1.32	12.121		1,550	665		2,215	2,750
6900	30 HP motor		1.32	12.121		1,550	665		2,215	2,750
6910	40 HP motor		.80	20		2,975	1,100		4,075	4,975
6920	50 HP motor		.80	20		2,975	1,100		4,075	4,975
6930	60 HP motor		.70	22.857		6,650	1,250		7,900	9,250
6940	75 HP motor		.70	22.857		6,650	1,250		7,900	9,250
6950	100 HP motor		.70	22.857		6,650	1,250		7,900	9,250
6960	125 HP motor		.60	26.667		17,500	1,450		18,950	21,600
6970	150 HP motor		.60	26.667		17,500	1,450		18,950	21,600
6980	200 HP motor		.60	26.667		17,500	1,450		18,950	21,600
6990	Combination starter and fused disconnect									
7000	240 volt, 1-2 pole, .75 HP motor	1 Elec	2	4	Ea.	595	219		814	995
7010	2 HP motor		2	4		595	219		814	995
7020	5 HP motor		1.50	5.333		625	292		917	1,150
7030	10 HP motor		1.20	6.667		970	365		1,335	1,650
7040	3 pole, .75 HP motor		1.80	4.444		595	243		838	1,025
7050	5 HP motor		1.30	6.154		625	335		960	1,200

26 24 19.40 Motor Starters and Controls		Crew	Daily Output	Labor-Hours	Unit	2015 Bare Costs Material	Labor	Equipment	Total	Total Incl O&P
7060	10 HP motor	1 Elec	1	8	Ea.	970	440		1,410	1,750
7070	15 HP motor		1	8		970	440		1,410	1,750
7080	20 HP motor	2 Elec	1.32	12.121		1,625	665		2,290	2,800
7090	25 HP motor		1.32	12.121		1,625	665		2,290	2,800
7100	30 HP motor		1.32	12.121		1,625	665		2,290	2,800
7110	40 HP motor		.80	20		3,075	1,100		4,175	5,100
7120	50 HP motor		.80	20		3,075	1,100		4,175	5,100
7130	60 HP motor		.80	20		6,875	1,100		7,975	9,275
7140	75 HP motor		.70	22.857		6,875	1,250		8,125	9,500
7150	100 HP motor		.70	22.857		6,875	1,250		8,125	9,500
7160	125 HP motor		.70	22.857		18,200	1,250		19,450	21,900
7170	150 HP motor		.60	26.667		18,200	1,450		19,650	22,300
7180	200 HP motor		.60	26.667		18,200	1,450		19,650	22,300
7190	Combination starter & circuit breaker disconnect									
7200	240 volt, 1-2 pole, .75 HP motor	1 Elec	2	4	Ea.	600	219		819	1,000
7210	2 HP motor		2	4		600	219		819	1,000
7220	5 HP motor		1.50	5.333		635	292		927	1,150
7230	10 HP motor		1.20	6.667		965	365		1,330	1,625
7240	3 pole, .75 HP motor		1.80	4.444		620	243		863	1,050
7250	5 HP motor		1.30	6.154		650	335		985	1,225
7260	10 HP motor		1	8		985	440		1,425	1,750
7270	15 HP motor		1	8		985	440		1,425	1,750
7280	20 HP motor	2 Elec	1.32	12.121		1,675	665		2,340	2,875
7290	25 HP motor		1.32	12.121		1,675	665		2,340	2,875
7300	30 HP motor		1.32	12.121		1,675	665		2,340	2,875
7310	40 HP motor		.80	20		3,650	1,100		4,750	5,700
7320	50 HP motor		.80	20		3,650	1,100		4,750	5,700
7330	60 HP motor		.80	20		8,400	1,100		9,500	11,000
7340	75 HP motor		.70	22.857		8,400	1,250		9,650	11,200
7350	100 HP motor		.70	22.857		8,400	1,250		9,650	11,200
7360	125 HP motor		.70	22.857		18,200	1,250		19,450	21,900
7370	150 HP motor		.60	26.667		18,200	1,450		19,650	22,300
7380	200 HP motor		.60	26.667		18,200	1,450		19,650	22,300
7400	Magnetic FVNR with enclosure & heaters, 2 pole,									
7410	230 volt, 1 HP size 00	1 Elec	4	2	Ea.	185	109		294	370
7420	2 HP, size 0		4	2		206	109		315	395
7430	3 HP, size 1		3	2.667		236	146		382	485
7440	5 HP, size 1p		3	2.667		305	146		451	560
7450	115 volt, 1/3 HP, size 00		4	2		185	109		294	370
7460	1 HP, size 0		4	2		206	109		315	395
7470	2 HP, size 1		3	2.667		236	146		382	485
7480	3 HP, size 1P		3	2.667		294	146		440	550
7500	3 pole, 480 volt, 600 HP, size 7	2 Elec	.70	22.857		16,300	1,250		17,550	19,900
7590	Magnetic FVNR with heater, NEMA 1									
7600	600 volt, 3 pole, 5 HP motor	1 Elec	2.30	3.478	Ea.	245	190		435	565
7610	10 HP motor	"	1.60	5		275	274		549	730
7620	25 HP motor	2 Elec	2.20	7.273		520	400		920	1,175
7630	30 HP motor		1.80	8.889		845	485		1,330	1,675
7640	40 HP motor		1.80	8.889		845	485		1,330	1,675
7650	50 HP motor		1.80	8.889		845	485		1,330	1,675
7660	60 HP motor		1.20	13.333		1,875	730		2,605	3,175
7670	75 HP motor		1.20	13.333		2,125	730		2,855	3,475
7680	100 HP motor		1.20	13.333		2,475	730		3,205	3,850

26 24 19 – Motor-Control Centers

26 24 19.40 Motor Starters and Controls		Crew	Daily Output	Labor-Hours	Unit	2015 Bare Costs Material	Labor	Equipment	Total	Total Incl O&P
7690	125 HP motor	2 Elec	.90	17.778	Ea.	4,275	970		5,245	6,225
7700	150 HP motor		.90	17.778		4,500	970		5,470	6,450
7710	200 HP motor		.90	17.778		4,700	970		5,670	6,650
7750	Starter & nonfused disconnect, 600 volt, 3 pole, 5 HP motor	1 Elec	1.40	5.714		360	315		675	880
7760	10 HP motor	"	1.10	7.273		390	400		790	1,050
7770	25 HP motor	2 Elec	1.50	10.667		630	585		1,215	1,600
7780	30 HP motor		1.30	12.308		1,025	675		1,700	2,175
7790	40 HP motor		1.30	12.308		1,025	675		1,700	2,175
7800	50 HP motor		1.30	12.308		1,025	675		1,700	2,175
7810	60 HP motor		.92	17.391		2,175	950		3,125	3,850
7820	75 HP motor		.92	17.391		2,425	950		3,375	4,150
7830	100 HP motor		.84	19.048		2,775	1,050		3,825	4,675
7840	125 HP motor		.70	22.857		4,725	1,250		5,975	7,125
7850	150 HP motor		.70	22.857		4,950	1,250		6,200	7,350
7860	200 HP motor		.60	26.667		5,150	1,450		6,600	7,900
7870	Starter & fused disconnect, 600 volt, 3 pole, 5 HP motor	1 Elec	1.40	5.714		445	315		760	975
7880	10 HP motor	"	1.10	7.273		475	400		875	1,150
7890	25 HP motor	2 Elec	1.50	10.667		720	585		1,305	1,700
7900	30 HP motor		1.30	12.308		1,075	675		1,750	2,250
7910	40 HP motor		1.30	12.308		1,075	675		1,750	2,250
7920	50 HP motor		1.30	12.308		1,075	675		1,750	2,250
7930	60 HP motor		.92	17.391		2,325	950		3,275	4,025
7940	75 HP motor		.92	17.391		2,575	950		3,525	4,300
7950	100 HP motor		.84	19.048		2,925	1,050		3,975	4,825
7960	125 HP motor		.70	22.857		4,925	1,250		6,175	7,350
7970	150 HP motor		.70	22.857		5,125	1,250		6,375	7,575
7980	200 HP motor		.60	26.667		5,325	1,450		6,775	8,125
7990	Combination starter and nonfusible disconnect									
8000	600 volt, 3 pole, 5 HP motor	1 Elec	1.80	4.444	Ea.	645	243		888	1,075
8010	10 HP motor	"	1.30	6.154		720	335		1,055	1,300
8020	25 HP motor	2 Elec	2	8		1,050	440		1,490	1,825
8030	30 HP motor		1.32	12.121		1,625	665		2,290	2,800
8040	40 HP motor		1.32	12.121		1,625	665		2,290	2,800
8050	50 HP motor		1.32	12.121		1,925	665		2,590	3,125
8060	60 HP motor		.80	20		3,075	1,100		4,175	5,100
8070	75 HP motor		.80	20		3,075	1,100		4,175	5,100
8080	100 HP motor		.80	20		3,825	1,100		4,925	5,900
8090	125 HP motor		.70	22.857		6,675	1,250		7,925	9,275
8100	150 HP motor		.70	22.857		6,975	1,250		8,225	9,600
8110	200 HP motor		.70	22.857		7,275	1,250		8,525	9,925
8140	Combination starter and fused disconnect									
8150	600 volt, 3 pole, 5 HP motor	1 Elec	1.80	4.444	Ea.	595	243		838	1,025
8160	10 HP motor	"	1.30	6.154		650	335		985	1,225
8170	25 HP motor	2 Elec	2	8		1,050	440		1,490	1,825
8180	30 HP motor		1.32	12.121		1,650	665		2,315	2,850
8190	40 HP motor		1.32	12.121		1,775	665		2,440	2,975
8200	50 HP motor		1.32	12.121		1,775	665		2,440	2,975
8210	60 HP motor		.80	20		3,100	1,100		4,200	5,125
8220	75 HP motor		.80	20		3,100	1,100		4,200	5,125
8230	100 HP motor		.80	20		3,100	1,100		4,200	5,125
8240	125 HP motor		.70	22.857		6,875	1,250		8,125	9,500
8250	150 HP motor		.70	22.857		6,875	1,250		8,125	9,500
8260	200 HP motor		.70	22.857		6,875	1,250		8,125	9,500

26 24 19.40 Motor Starters and Controls		Crew	Daily Output	Labor-Hours	Unit	Material	2015 Bare Costs Labor	Equipment	Total	Total Incl O&P
8290	Combination starter & circuit breaker disconnect									
8300	600 volt, 3 pole, 5 HP motor	1 Elec	1.80	4.444	Ea.	880	243		1,123	1,350
8310	10 HP motor	"	1.30	6.154		1,075	335		1,410	1,725
8320	25 HP motor	2 Elec	2	8		1,275	440		1,715	2,100
8330	30 HP motor		1.32	12.121		1,675	665		2,340	2,875
8340	40 HP motor		1.32	12.121		1,675	665		2,340	2,875
8350	50 HP motor		1.32	12.121		1,675	665		2,340	2,875
8360	60 HP motor		.80	20		3,650	1,100		4,750	5,700
8370	75 HP motor		.80	20		3,650	1,100		4,750	5,700
8380	100 HP motor		.80	20		3,650	1,100		4,750	5,700
8390	125 HP motor		.70	22.857		8,400	1,250		9,650	11,200
8400	150 HP motor		.70	22.857		8,400	1,250		9,650	11,200
8410	200 HP motor		.70	22.857		8,400	1,250		9,650	11,200
8430	Starter & circuit breaker disconnect									
8440	600 volt, 3 pole, 5 HP motor	1 Elec	1.40	5.714	Ea.	745	315		1,060	1,300
8450	10 HP motor	"	1.10	7.273		780	400		1,180	1,475
8460	25 HP motor	2 Elec	1.50	10.667		1,025	585		1,610	2,025
8470	30 HP motor		1.30	12.308		1,450	675		2,125	2,650
8480	40 HP motor		1.30	12.308		1,450	675		2,125	2,650
8490	50 HP motor		1.30	12.308		1,450	675		2,125	2,650
8500	60 HP motor		.92	17.391		3,500	950		4,450	5,325
8510	75 HP motor		.92	17.391		3,775	950		4,725	5,625
8520	100 HP motor		.84	19.048		4,100	1,050		5,150	6,150
8530	125 HP motor		.70	22.857		5,925	1,250		7,175	8,425
8540	150 HP motor		.70	22.857		6,125	1,250		7,375	8,650
8550	200 HP motor		.60	26.667		7,475	1,450		8,925	10,500
8900	240 volt, 1-2 pole, .75 HP motor	1 Elec	2	4		705	219		924	1,125
8910	2 HP motor		2	4		725	219		944	1,150
8920	5 HP motor		1.80	4.444		825	243		1,068	1,275
8930	10 HP motor		1.40	5.714		1,425	315		1,740	2,025
8950	3 pole, .75 HP motor		1.60	5		705	274		979	1,200
8970	5 HP motor		1.40	5.714		780	315		1,095	1,350
8980	10 HP motor		1.10	7.273		1,125	400		1,525	1,875
8990	15 HP motor		1	8		1,125	440		1,565	1,925
9100	20 HP motor	2 Elec	1.50	10.667		1,550	585		2,135	2,600
9110	25 HP motor		1.50	10.667		1,550	585		2,135	2,600
9120	30 HP motor		1.30	12.308		1,550	675		2,225	2,750
9130	40 HP motor		1.24	12.903		3,500	705		4,205	4,950
9140	50 HP motor		1.12	14.286		3,500	780		4,280	5,050
9150	60 HP motor		.90	17.778		7,175	970		8,145	9,375
9160	75 HP motor		.76	21.053		7,175	1,150		8,325	9,650
9170	100 HP motor		.70	22.857		7,175	1,250		8,425	9,800
9180	125 HP motor		.60	26.667		16,300	1,450		17,750	20,300
9190	150 HP motor		.52	30.769		16,300	1,675		17,975	20,600
9200	200 HP motor		.50	32		16,300	1,750		18,050	20,700

26 25 13 – Bus Duct/Busway and Fittings

26 25 13.10 Aluminum Bus Duct		Crew	Daily Output	Labor-Hours	Unit	Material	Labor	Equipment	Total	Total Incl O&P
						2015 Bare Costs				
0010	**ALUMINUM BUS DUCT** 10 ft. long R262513-10									
0050	Indoor 3 pole 4 wire, plug-in, straight section, 225 amp	2 Elec	44	.364	L.F.	125	19.90		144.90	168
0100	400 amp		36	.444		146	24.50		170.50	199
0150	600 amp		32	.500		173	27.50		200.50	233
0200	800 amp		26	.615		198	33.50		231.50	269
0250	1000 amp		24	.667		222	36.50		258.50	300
0270	1200 amp		23	.696		245	38		283	330
0300	1350 amp		22	.727		264	40		304	350
0310	1600 amp		18	.889		297	48.50		345.50	400
0320	2000 amp		16	1		345	54.50		399.50	465
0330	2500 amp		14	1.143		410	62.50		472.50	545
0340	3000 amp		12	1.333		475	73		548	635
0350	Feeder, 600 amp		34	.471		110	25.50		135.50	161
0400	800 amp		28	.571		129	31.50		160.50	191
0450	1000 amp		26	.615		147	33.50		180.50	214
0455	1200 amp		25	.640		147	35		182	216
0500	1350 amp		24	.667		230	36.50		266.50	310
0550	1600 amp		20	.800		267	44		311	360
0600	2000 amp		18	.889		315	48.50		363.50	420
0620	2500 amp		14	1.143		405	62.50		467.50	540
0630	3000 amp		12	1.333		460	73		533	620
0640	4000 amp		10	1.600		665	87.50		752.50	865
0650	Elbow, 225 amp		4.40	3.636	Ea.	735	199		934	1,125
0700	400 amp		3.80	4.211		745	230		975	1,175
0750	600 amp		3.40	4.706		745	257		1,002	1,225
0800	800 amp		3	5.333		775	292		1,067	1,300
0850	1000 amp		2.80	5.714		995	315		1,310	1,575
0870	1200 amp		2.70	5.926		1,250	325		1,575	1,875
0900	1350 amp		2.60	6.154		1,300	335		1,635	1,975
0950	1600 amp		2.40	6.667		1,400	365		1,765	2,100
1000	2000 amp		2	8		1,525	440		1,965	2,350
1020	2500 amp		1.80	8.889		1,800	485		2,285	2,725
1030	3000 amp		1.60	10		2,075	545		2,620	3,150
1040	4000 amp		1.40	11.429		3,350	625		3,975	4,675
1100	Cable tap box end, 225 amp		3.60	4.444		1,450	243		1,693	1,950
1150	400 amp		3.20	5		1,475	274		1,749	2,050
1200	600 amp		2.60	6.154		1,500	335		1,835	2,175
1250	800 amp		2.20	7.273		1,525	400		1,925	2,325
1300	1000 amp		2	8		1,575	440		2,015	2,400
1320	1200 amp		2	8		1,600	440		2,040	2,450
1350	1350 amp		1.60	10		1,625	545		2,170	2,650
1400	1600 amp		1.40	11.429		1,675	625		2,300	2,825
1450	2000 amp		1.20	13.333		1,750	730		2,480	3,050
1460	2500 amp		1	16		1,825	875		2,700	3,350
1470	3000 amp		.80	20		1,925	1,100		3,025	3,800
1480	4000 amp		.60	26.667		2,075	1,450		3,525	4,550
1500	Switchboard stub, 225 amp		5.80	2.759		1,475	151		1,626	1,850
1550	400 amp		5.40	2.963		1,500	162		1,662	1,900
1600	600 amp		4.60	3.478		1,525	190		1,715	1,975
1650	800 amp		4	4		1,550	219		1,769	2,050
1700	1000 amp		3.20	5		1,550	274		1,824	2,150
1720	1200 amp		3.10	5.161		1,600	282		1,882	2,175
1750	1350 amp		3	5.333		1,600	292		1,892	2,225

26 25 13.10 Aluminum Bus Duct		Crew	Daily Output	Labor-Hours	Unit	2015 Bare Costs Material	Labor	Equipment	Total	Total Incl O&P
1800	1600 amp	2 Elec	2.60	6.154	Ea.	1,650	335		1,985	2,325
1850	2000 amp		2.40	6.667		1,700	365		2,065	2,425
1860	2500 amp		2.20	7.273		1,750	400		2,150	2,550
1870	3000 amp		2	8		1,825	440		2,265	2,675
1880	4000 amp		1.80	8.889		1,950	485		2,435	2,875
1890	Tee fittings, 225 amp		3.20	5		875	274		1,149	1,375
1900	400 amp		2.80	5.714		875	315		1,190	1,450
1950	600 amp		2.60	6.154		875	335		1,210	1,475
2000	800 amp		2.40	6.667		930	365		1,295	1,600
2050	1000 amp		2.20	7.273		985	400		1,385	1,700
2070	1200 amp		2.10	7.619		1,150	415		1,565	1,925
2100	1350 amp		2	8		1,600	440		2,040	2,425
2150	1600 amp		1.60	10		1,925	545		2,470	2,950
2200	2000 amp		1.20	13.333		2,125	730		2,855	3,450
2220	2500 amp		1	16		2,525	875		3,400	4,150
2230	3000 amp		.80	20		2,900	1,100		4,000	4,875
2240	4000 amp		.60	26.667		4,800	1,450		6,250	7,550
2300	Wall flange, 600 amp		20	.800		196	44		240	283
2310	800 amp		16	1		196	54.50		250.50	300
2320	1000 amp		13	1.231		196	67.50		263.50	320
2325	1200 amp		12	1.333		196	73		269	330
2330	1350 amp		10.80	1.481		196	81		277	340
2340	1600 amp		9	1.778		196	97		293	365
2350	2000 amp		8	2		196	109		305	385
2360	2500 amp		6.60	2.424		196	133		329	420
2370	3000 amp		5.40	2.963		196	162		358	465
2380	4000 amp		4	4		284	219		503	650
2390	5000 amp		3	5.333		284	292		576	760
2400	Vapor barrier		8	2		360	109		469	565
2420	Roof flange kit		4	4		640	219		859	1,050
2600	Expansion fitting, 225 amp		10	1.600		1,250	87.50		1,337.50	1,475
2610	400 amp		8	2		1,250	109		1,359	1,550
2620	600 amp		6	2.667		1,250	146		1,396	1,600
2630	800 amp		4.60	3.478		1,400	190		1,590	1,850
2640	1000 amp		4	4		1,600	219		1,819	2,100
2650	1350 amp		3.60	4.444		2,025	243		2,268	2,600
2660	1600 amp		3.20	5		2,425	274		2,699	3,100
2670	2000 amp		2.80	5.714		2,700	315		3,015	3,450
2680	2500 amp		2.40	6.667		3,275	365		3,640	4,175
2690	3000 amp		2	8		3,775	440		4,215	4,825
2700	4000 amp		1.60	10		4,950	545		5,495	6,300
2800	Reducer nonfused, 400 amp		8	2		875	109		984	1,125
2810	600 amp		6	2.667		875	146		1,021	1,200
2820	800 amp		4.60	3.478		1,050	190		1,240	1,450
2830	1000 amp		4	4		1,225	219		1,444	1,700
2840	1350 amp		3.60	4.444		1,600	243		1,843	2,150
2850	1600 amp		3.20	5		2,175	274		2,449	2,825
2860	2000 amp		2.80	5.714		2,500	315		2,815	3,225
2870	2500 amp		2.40	6.667		3,150	365		3,515	4,025
2880	3000 amp		2	8		3,650	440		4,090	4,675
2890	4000 amp		1.60	10		4,850	545		5,395	6,200
2950	Reducer fuse included, 225 amp		4.40	3.636		2,575	199		2,774	3,125
2960	400 amp		4.20	3.810		2,625	208		2,833	3,200

26 25 13 – Bus Duct/Busway and Fittings

26 25 13.10 Aluminum Bus Duct		Crew	Daily Output	Labor-Hours	Unit	Material	Labor	Equipment	Total	Total Incl O&P
						2015 Bare Costs				
2970	600 amp	2 Elec	3.60	4.444	Ea.	3,075	243		3,318	3,775
2980	800 amp		3.20	5		4,900	274		5,174	5,825
2990	1000 amp		3	5.333		5,600	292		5,892	6,625
3000	1200 amp		2.80	5.714		5,600	315		5,915	6,650
3010	1600 amp		2.20	7.273		12,800	400		13,200	14,700
3020	2000 amp		1.80	8.889		14,200	485		14,685	16,400
3100	Reducer circuit breaker, 225 amp		4.40	3.636		2,525	199		2,724	3,075
3110	400 amp		4.20	3.810		3,075	208		3,283	3,725
3120	600 amp		3.60	4.444		4,375	243		4,618	5,175
3130	800 amp		3.20	5		5,125	274		5,399	6,075
3140	1000 amp		3	5.333		5,825	292		6,117	6,850
3150	1200 amp		2.80	5.714		7,000	315		7,315	8,175
3160	1600 amp		2.20	7.273		10,300	400		10,700	12,000
3170	2000 amp		1.80	8.889		11,300	485		11,785	13,300
3250	Reducer circuit breaker, 75,000 AIC, 225 amp		4.40	3.636		3,950	199		4,149	4,650
3260	400 amp		4.20	3.810		3,950	208		4,158	4,675
3270	600 amp		3.60	4.444		5,300	243		5,543	6,200
3280	800 amp		3.20	5		5,800	274		6,074	6,825
3290	1000 amp		3	5.333		9,350	292		9,642	10,800
3300	1200 amp		2.80	5.714		9,350	315		9,665	10,800
3310	1600 amp		2.20	7.273		10,300	400		10,700	12,000
3320	2000 amp		1.80	8.889		11,300	485		11,785	13,300
3400	Reducer circuit breaker CLF 225 amp		4.40	3.636		4,075	199		4,274	4,775
3410	400 amp		4.20	3.810		4,825	208		5,033	5,650
3420	600 amp		3.60	4.444		7,225	243		7,468	8,325
3430	800 amp		3.20	5		7,550	274		7,824	8,725
3440	1000 amp		3	5.333		7,875	292		8,167	9,100
3450	1200 amp		2.80	5.714		10,300	315		10,615	11,800
3460	1600 amp		2.20	7.273		10,300	400		10,700	12,000
3470	2000 amp		1.80	8.889		11,300	485		11,785	13,300
3550	Ground bus added to bus duct, 225 amp		320	.050	L.F.	32.50	2.74		35.24	40
3560	400 amp		320	.050		32.50	2.74		35.24	40
3570	600 amp		280	.057		32.50	3.13		35.63	41
3580	800 amp		240	.067		32.50	3.65		36.15	41.50
3590	1000 amp		200	.080		32.50	4.38		36.88	43
3600	1350 amp		180	.089		32.50	4.86		37.36	43.50
3610	1600 amp		160	.100		32.50	5.45		37.95	44.50
3620	2000 amp		160	.100		32.50	5.45		37.95	44.50
3630	2500 amp		140	.114		32.50	6.25		38.75	45.50
3640	3000 amp		120	.133		32.50	7.30		39.80	47.50
3650	4000 amp		100	.160		32.50	8.75		41.25	49.50
3810	High short circuit, 400 amp		36	.444		99.50	24.50		124	148
3820	600 amp		32	.500		125	27.50		152.50	180
3830	800 amp		26	.615		144	33.50		177.50	211
3840	1000 amp		24	.667		156	36.50		192.50	229
3850	1350 amp		22	.727		184	40		224	264
3860	1600 amp		18	.889		212	48.50		260.50	310
3870	2000 amp		16	1		248	54.50		302.50	360
3880	2500 amp		14	1.143		415	62.50		477.50	550
3890	3000 amp		12	1.333		470	73		543	630
3920	Cross, 225 amp		5.60	2.857	Ea.	1,300	156		1,456	1,700
3930	400 amp		4.60	3.478		1,300	190		1,490	1,750
3940	600 amp		4	4		1,300	219		1,519	1,800

26 25 13.10 Aluminum Bus Duct		Crew	Daily Output	Labor-Hours	Unit	Material	2015 Bare Costs Labor	Equipment	Total	Total Incl O&P
3950	800 amp	2 Elec	3.40	4.706	Ea.	1,375	257		1,632	1,925
3960	1000 amp		3	5.333		1,450	292		1,742	2,050
3970	1350 amp		2.80	5.714		2,350	315		2,665	3,050
3980	1600 amp		2.20	7.273		2,775	400		3,175	3,675
3990	2000 amp		1.80	8.889		3,050	485		3,535	4,100
4000	2500 amp		1.60	10		3,625	545		4,170	4,825
4010	3000 amp		1.20	13.333		4,150	730		4,880	5,700
4020	4000 amp		1	16		6,350	875		7,225	8,350
4040	Cable tap box center, 225 amp		3.60	4.444		885	243		1,128	1,350
4050	400 amp		3.20	5		945	274		1,219	1,475
4060	600 amp		2.60	6.154		990	335		1,325	1,625
4070	800 amp		2.20	7.273		1,050	400		1,450	1,775
4080	1000 amp		2	8		1,150	440		1,590	1,925
4090	1350 amp		1.60	10		1,425	545		1,970	2,400
4100	1600 amp		1.40	11.429		1,600	625		2,225	2,725
4110	2000 amp		1.20	13.333		1,825	730		2,555	3,125
4120	2500 amp		1	16		2,225	875		3,100	3,775
4130	3000 amp		.80	20		2,450	1,100		3,550	4,400
4140	4000 amp		.60	26.667		3,250	1,450		4,700	5,825
4500	Weatherproof 3 pole 4 wire, feeder, 600 amp		30	.533	L.F.	132	29		161	190
4520	800 amp		24	.667		155	36.50		191.50	227
4540	1000 amp		22	.727		177	40		217	256
4550	1200 amp		21	.762		244	41.50		285.50	335
4560	1350 amp		20	.800		276	44		320	375
4580	1600 amp		17	.941		320	51.50		371.50	430
4600	2000 amp		16	1		375	54.50		429.50	500
4620	2500 amp		12	1.333		485	73		558	650
4640	3000 amp		10	1.600		550	87.50		637.50	740
4660	4000 amp		8	2		795	109		904	1,050
5000	Indoor 3 pole, 3 wire, feeder, 600 amp		40	.400		101	22		123	145
5010	800 amp		32	.500		120	27.50		147.50	175
5020	1000 amp		30	.533		129	29		158	187
5025	1200 amp		29	.552		129	30		159	189
5030	1350 amp		28	.571		175	31.50		206.50	241
5040	1600 amp		24	.667		202	36.50		238.50	280
5050	2000 amp		20	.800		239	44		283	330
5060	2500 amp		16	1		330	54.50		384.50	450
5070	3000 amp		14	1.143		395	62.50		457.50	530
5080	4000 amp		12	1.333		480	73		553	640
5200	Plug-in type, 225 amp		50	.320		109	17.50		126.50	147
5210	400 amp		42	.381		109	21		130	153
5220	600 amp		36	.444		109	24.50		133.50	158
5230	800 amp		30	.533		128	29		157	185
5240	1000 amp		28	.571		137	31.50		168.50	199
5245	1200 amp		27	.593		156	32.50		188.50	222
5250	1350 amp		26	.615		184	33.50		217.50	254
5260	1600 amp		20	.800		212	44		256	300
5270	2000 amp		18	.889		248	48.50		296.50	350
5280	2500 amp		16	1		340	54.50		394.50	460
5290	3000 amp		14	1.143		405	62.50		467.50	540
5300	4000 amp		12	1.333		490	73		563	650
5330	High short circuit, 400 amp		42	.381		109	21		130	153
5340	600 amp		36	.444		109	24.50		133.50	158

26 25 Enclosed Bus Assemblies

26 25 13 – Bus Duct/Busway and Fittings

26 25 13.10	Aluminum Bus Duct	Crew	Daily Output	Labor-Hours	Unit	2015 Bare Costs Material	Labor	Equipment	Total	Total Incl O&P
5350	800 amp	2 Elec	30	.533	L.F.	128	29		157	185
5360	1000 amp		28	.571		137	31.50		168.50	199
5370	1350 amp		26	.615		184	33.50		217.50	254
5380	1600 amp		20	.800		212	44		256	300
5390	2000 amp		18	.889		248	48.50		296.50	350
5400	2500 amp		16	1		390	54.50		444.50	515
5410	3000 amp		14	1.143		405	62.50		467.50	540
5440	Elbow, 225 amp		5	3.200	Ea.	555	175		730	880
5450	400 amp		4.40	3.636		555	199		754	920
5460	600 amp		4	4		555	219		774	950
5470	800 amp		3.40	4.706		595	257		852	1,050
5480	1000 amp		3.20	5		610	274		884	1,100
5485	1200 amp		3.10	5.161		650	282		932	1,150
5490	1350 amp		3	5.333		710	292		1,002	1,225
5500	1600 amp		2.80	5.714		1,100	315		1,415	1,675
5510	2000 amp		2.40	6.667		1,200	365		1,565	1,900
5520	2500 amp		2	8		1,475	440		1,915	2,300
5530	3000 amp		1.80	8.889		1,750	485		2,235	2,675
5540	4000 amp		1.60	10		2,325	545		2,870	3,425
5560	Tee fittings, 225 amp		3.60	4.444		780	243		1,023	1,225
5570	400 amp		3.20	5		780	274		1,054	1,275
5580	600 amp		3	5.333		780	292		1,072	1,300
5590	800 amp		2.80	5.714		835	315		1,150	1,400
5600	1000 amp		2.60	6.154		865	335		1,200	1,475
5605	1200 amp		2.50	6.400		920	350		1,270	1,575
5610	1350 amp		2.40	6.667		1,275	365		1,640	1,975
5620	1600 amp		1.80	8.889		1,500	485		1,985	2,400
5630	2000 amp		1.40	11.429		1,675	625		2,300	2,825
5640	2500 amp		1.20	13.333		2,100	730		2,830	3,425
5650	3000 amp		1	16		2,450	875		3,325	4,050
5660	4000 amp		.70	22.857		3,600	1,250		4,850	5,875
5680	Cross, 225 amp		6.40	2.500		1,275	137		1,412	1,600
5690	400 amp		5.40	2.963		1,275	162		1,437	1,650
5700	600 amp		4.60	3.478		1,275	190		1,465	1,700
5710	800 amp		4	4		1,350	219		1,569	1,825
5720	1000 amp		3.60	4.444		1,375	243		1,618	1,900
5730	1350 amp		3.20	5		2,050	274		2,324	2,675
5740	1600 amp		2.60	6.154		2,400	335		2,735	3,150
5750	2000 amp		2.20	7.273		2,625	400		3,025	3,500
5760	2500 amp		1.80	8.889		3,150	485		3,635	4,225
5770	3000 amp		1.40	11.429		3,775	625		4,400	5,125
5780	4000 amp		1.20	13.333		5,250	730		5,980	6,900
5800	Expansion fitting, 225 amp		11.60	1.379		910	75.50		985.50	1,125
5810	400 amp		9.20	1.739		910	95		1,005	1,150
5820	600 amp		7	2.286		910	125		1,035	1,200
5830	800 amp		5.20	3.077		1,075	168		1,243	1,425
5840	1000 amp		4.60	3.478		1,175	190		1,365	1,600
5850	1350 amp		4.20	3.810		1,450	208		1,658	1,900
5860	1600 amp		3.60	4.444		1,800	243		2,043	2,350
5870	2000 amp		3.20	5		2,125	274		2,399	2,750
5880	2500 amp		2.80	5.714		2,400	315		2,715	3,100
5890	3000 amp		2.40	6.667		2,800	365		3,165	3,650
5900	4000 amp		1.80	8.889		3,825	485		4,310	4,975

26 25 13 – Bus Duct/Busway and Fittings

26 25 13.10 Aluminum Bus Duct		Crew	Daily Output	Labor-Hours	Unit	Material	Labor	Equipment	Total	Total Incl O&P
						2015 Bare Costs				
5940	Reducer, nonfused, 400 amp	2 Elec	9.20	1.739	Ea.	750	95		845	970
5950	600 amp		7	2.286		750	125		875	1,025
5960	800 amp		5.20	3.077		790	168		958	1,125
5970	1000 amp		4.60	3.478		970	190		1,160	1,375
5980	1350 amp		4.20	3.810		1,450	208		1,658	1,900
5990	1600 amp		3.60	4.444		1,650	243		1,893	2,175
6000	2000 amp		3.20	5		1,925	274		2,199	2,525
6010	2500 amp		2.80	5.714		2,450	315		2,765	3,175
6020	3000 amp		2.20	7.273		2,875	400		3,275	3,775
6030	4000 amp		1.80	8.889		3,925	485		4,410	5,075
6050	Reducer, fuse included, 225 amp		5	3.200		1,975	175		2,150	2,450
6060	400 amp		4.80	3.333		2,625	182		2,807	3,150
6070	600 amp		4.20	3.810		3,350	208		3,558	4,000
6080	800 amp		3.60	4.444		5,175	243		5,418	6,075
6090	1000 amp		3.40	4.706		5,650	257		5,907	6,625
6100	1350 amp		3.20	5		10,300	274		10,574	11,800
6110	1600 amp		2.60	6.154		12,200	335		12,535	14,000
6120	2000 amp		2	8		14,200	440		14,640	16,300
6160	Reducer, circuit breaker, 225 amp		5	3.200		2,425	175		2,600	2,950
6170	400 amp		4.80	3.333		2,975	182		3,157	3,550
6180	600 amp		4.20	3.810		4,250	208		4,458	5,000
6190	800 amp		3.60	4.444		4,975	243		5,218	5,850
6200	1000 amp		3.40	4.706		5,650	257		5,907	6,625
6210	1350 amp		3.20	5		6,825	274		7,099	7,950
6220	1600 amp		2.60	6.154		10,200	335		10,535	11,700
6230	2000 amp		2	8		11,100	440		11,540	13,000
6270	Cable tap box center, 225 amp		4.20	3.810		875	208		1,083	1,275
6280	400 amp		3.60	4.444		875	243		1,118	1,325
6290	600 amp		3	5.333		875	292		1,167	1,400
6300	800 amp		2.60	6.154		955	335		1,290	1,575
6310	1000 amp		2.40	6.667		1,000	365		1,365	1,675
6320	1350 amp		1.80	8.889		1,225	485		1,710	2,100
6330	1600 amp		1.60	10		1,375	545		1,920	2,375
6340	2000 amp		1.40	11.429		1,575	625		2,200	2,725
6350	2500 amp		1.20	13.333		1,975	730		2,705	3,300
6360	3000 amp		1	16		2,225	875		3,100	3,800
6370	4000 amp		.70	22.857		2,650	1,250		3,900	4,825
6390	Cable tap box end, 225 amp		4.20	3.810		535	208		743	910
6400	400 amp		3.60	4.444		535	243		778	960
6410	600 amp		3	5.333		535	292		827	1,025
6420	800 amp		2.60	6.154		585	335		920	1,150
6430	1000 amp		2.40	6.667		630	365		995	1,250
6435	1200 amp		2.10	7.619		685	415		1,100	1,400
6440	1350 amp		1.80	8.889		765	485		1,250	1,600
6450	1600 amp		1.60	10		870	545		1,415	1,800
6460	2000 amp		1.40	11.429		980	625		1,605	2,050
6470	2500 amp		1.20	13.333		1,125	730		1,855	2,375
6480	3000 amp		1	16		1,350	875		2,225	2,825
6490	4000 amp		.70	22.857		1,625	1,250		2,875	3,700
7000	Weatherproof 3 pole 3 wire, feeder, 600 amp		34	.471	L.F.	131	25.50		156.50	184
7020	800 amp		28	.571		144	31.50		175.50	207
7040	1000 amp		26	.615		155	33.50		188.50	222
7050	1200 amp		25	.640		181	35		216	253

26 25 13.10 Aluminum Bus Duct		Crew	Daily Output	Labor-Hours	Unit	Material	2015 Bare Costs Labor	Equipment	Total	Total Incl O&P
7060	1350 amp	2 Elec	24	.667	L.F.	210	36.50		246.50	288
7080	1600 amp		20	.800		243	44		287	335
7100	2000 amp		18	.889		287	48.50		335.50	390
7120	2500 amp		14	1.143		395	62.50		457.50	530
7140	3000 amp		12	1.333		475	73		548	640
7160	4000 amp		10	1.600		575	87.50		662.50	765

26 25 13.20 Bus Duct		Crew	Daily Output	Labor-Hours	Unit	Material	Labor	Equipment	Total	Total Incl O&P
0010	**BUS DUCT** 100 amp and less, aluminum or copper, plug-in									
0080	Bus duct, 3 pole 3 wire, 100 amp	1 Elec	42	.190	L.F.	23.50	10.40		33.90	42
0110	Elbow		4	2	Ea.	57.50	109		166.50	233
0120	Tee		2	4		83.50	219		302.50	430
0130	Wall flange		8	1		8.35	54.50		62.85	93.50
0140	Ground kit		16	.500		17.75	27.50		45.25	62
0180	3 pole 4 wire, 100 amp		40	.200	L.F.	30.50	10.95		41.45	50.50
0200	Cable tap box		3.10	2.581	Ea.	78	141		219	305
0300	End closure		16	.500		10.40	27.50		37.90	54
0400	Elbow		4	2		65	109		174	241
0500	Tee		2	4		94.50	219		313.50	445
0600	Hangers		10	.800		6.25	44		50.25	75
0700	Circuit breakers, 15 to 50 amp, 1 pole		8	1		151	54.50		205.50	252
0800	15 to 60 amp, 2 pole		6.70	1.194		365	65.50		430.50	505
0900	3 pole		5.30	1.509		365	82.50		447.50	530
1000	60 to 100 amp, 1 pole		6.70	1.194		590	65.50		655.50	750
1100	70 to 100 amp, 2 pole		5.30	1.509		590	82.50		672.50	780
1200	3 pole		4.50	1.778		660	97		757	880
1220	Switch, nonfused, 3 pole, 4 wire		8	1		70.50	54.50		125	162
1240	Fused, 3 fuses, 4 wire, 30 amp		8	1		255	54.50		309.50	365
1260	60 amp		5.30	1.509		267	82.50		349.50	420
1280	100 amp		4.50	1.778		445	97		542	640
1300	Plug, fusible, 3 pole 250 volt, 30 amp		5.30	1.509		255	82.50		337.50	410
1310	60 amp		5.30	1.509		300	82.50		382.50	460
1320	100 amp		4.50	1.778		445	97		542	640
1330	3 pole 480 volt, 30 amp		5.30	1.509		291	82.50		373.50	450
1340	60 amp		5.30	1.509		315	82.50		397.50	475
1350	100 amp		4.50	1.778		455	97		552	650
1360	Circuit breaker, 3 pole 250 volt, 60 amp		5.30	1.509		515	82.50		597.50	700
1370	3 pole 480 volt, 100 amp		4.50	1.778		515	97		612	720
2000	Bus duct, 2 wire, 250 volt, 30 amp		60	.133	L.F.	2.82	7.30		10.12	14.40
2100	60 amp		50	.160		2.82	8.75		11.57	16.65
2200	300 volt, 30 amp		60	.133		2.82	7.30		10.12	14.40
2300	60 amp		50	.160		2.82	8.75		11.57	16.65
2400	3 wire, 250 volt, 30 amp		60	.133		2.82	7.30		10.12	14.40
2500	60 amp		50	.160		2.82	8.75		11.57	16.65
2600	480/277 volt, 30 amp		60	.133		2.82	7.30		10.12	14.40
2700	60 amp		50	.160		2.82	8.75		11.57	16.65
2750	End feed, 300 volt 2 wire max. 30 amp		6	1.333	Ea.	42	73		115	159
2800	60 amp		5.50	1.455		42	79.50		121.50	169
2850	30 amp miniature		6	1.333		42	73		115	159
2900	3 wire, 30 amp		6	1.333		53	73		126	171
2950	60 amp		5.50	1.455		53	79.50		132.50	181
3000	30 amp miniature		6	1.333		53	73		126	171
3050	Center feed, 300 volt 2 wire, 30 amp		6	1.333		58	73		131	177

26 25 13 – Bus Duct/Busway and Fittings

26 25 13.20 Bus Duct		Crew	Daily Output	Labor-Hours	Unit	Material	Labor	Equipment	Total	Total Incl O&P
						2015 Bare Costs				
3100	60 amp	1 Elec	5.50	1.455	Ea.	58	79.50		137.50	187
3150	3 wire, 30 amp		6	1.333		66	73		139	186
3200	60 amp		5.50	1.455		66	79.50		145.50	196
3220	Elbow, 30 amp		6	1.333		40.50	73		113.50	158
3240	60 amp		5.50	1.455		40.50	79.50		120	168
3260	End cap		40	.200		8.10	10.95		19.05	26
3280	Strength beam, 10 ft.		15	.533		22.50	29		51.50	70
3300	Hanger		24	.333		4.84	18.25		23.09	33.50
3320	Tap box, nonfusible		6.30	1.270		68.50	69.50		138	184
3340	Fusible switch 30 amp, 1 fuse		6	1.333		299	73		372	445
3360	2 fuse		6	1.333		305	73		378	450
3380	3 fuse		6	1.333		305	73		378	450
3400	Circuit breaker handle on cover, 1 pole		6	1.333		47.50	73		120.50	165
3420	2 pole		6	1.333		65.50	73		138.50	186
3440	3 pole		6	1.333		86	73		159	208
3460	Circuit breaker external operhandle, 1 pole		6	1.333		49	73		122	167
3480	2 pole		6	1.333		68	73		141	188
3500	3 pole		6	1.333		94.50	73		167.50	217
3520	Terminal plug only		16	.500		9.90	27.50		37.40	53.50
3540	Terminal with receptacle		16	.500		12.80	27.50		40.30	56.50
3560	Fixture plug		16	.500		8.80	27.50		36.30	52
4000	Copper bus duct, lighting, 2 wire 300 volt, 20 amp		70	.114	L.F.	5.20	6.25		11.45	15.40
4020	35 amp		60	.133		5.20	7.30		12.50	17
4040	50 amp		55	.145		5.20	7.95		13.15	18
4060	60 amp		50	.160		5.20	8.75		13.95	19.25
4080	3 wire 300 volt, 20 amp		70	.114		5.65	6.25		11.90	15.90
4100	35 amp		60	.133		5.65	7.30		12.95	17.50
4120	50 amp		55	.145		5.65	7.95		13.60	18.50
4140	60 amp		50	.160		5.65	8.75		14.40	19.75
4160	Feeder in box, end, 1 circuit		6	1.333	Ea.	78	73		151	199
4180	2 circuit		5.50	1.455		80.50	79.50		160	212
4200	Center, 1 circuit		6	1.333		106	73		179	230
4220	2 circuit		5.50	1.455		110	79.50		189.50	244
4240	End cap		40	.200		12.95	10.95		23.90	31
4260	Hanger, surface mount		24	.333		7.75	18.25		26	36.50
4280	Coupling		40	.200		9.85	10.95		20.80	28

26 25 13.30 Copper Bus Duct		Crew	Daily Output	Labor-Hours	Unit	Material	Labor	Equipment	Total	Total Incl O&P
0010	**COPPER BUS DUCT**									
0100	Weatherproof 3 pole 4 wire, feeder duct, 600 amp	2 Elec	24	.667	L.F.	256	36.50		292.50	340
0110	800 amp		18	.889		350	48.50		398.50	460
0120	1000 amp		17	.941		390	51.50		441.50	510
0125	1200 amp		16.50	.970		455	53		508	585
0130	1350 amp		16	1		480	54.50		534.50	610
0140	1600 amp		12	1.333		515	73		588	680
0150	2000 amp		10	1.600		570	87.50		657.50	765
0160	2500 amp		7	2.286		620	125		745	880
0170	3000 amp		5	3.200		860	175		1,035	1,225
0180	4000 amp		3.60	4.444		1,350	243		1,593	1,850
0200	Indoor 3 pole 4 wire, plug-in, bus duct high short circuit, 400 amp		32	.500		246	27.50		273.50	315
0210	600 amp		26	.615		246	33.50		279.50	325
0220	800 amp		20	.800		292	44		336	390
0230	1000 amp		18	.889		325	48.50		373.50	430

26 25 13.30 Copper Bus Duct		Crew	Daily Output	Labor-Hours	Unit	Material	2015 Bare Costs Labor	Equipment	Total	Total Incl O&P
0240	1350 amp	2 Elec	16	1	L.F.	440	54.50		494.50	565
0250	1600 amp		12	1.333		495	73		568	660
0260	2000 amp		10	1.600		630	87.50		717.50	825
0270	2500 amp		8	2		775	109		884	1,025
0280	3000 amp		6	2.667		885	146		1,031	1,200
0310	Cross, 225 amp		3	5.333	Ea.	2,925	292		3,217	3,650
0320	400 amp		2.80	5.714		2,925	315		3,240	3,675
0330	600 amp		2.60	6.154		2,925	335		3,260	3,725
0340	800 amp		2.20	7.273		3,150	400		3,550	4,100
0350	1000 amp		2	8		3,525	440		3,965	4,550
0360	1350 amp		1.80	8.889		3,800	485		4,285	4,925
0370	1600 amp		1.70	9.412		4,225	515		4,740	5,425
0380	2000 amp		1.60	10		6,925	545		7,470	8,475
0390	2500 amp		1.40	11.429		8,375	625		9,000	10,200
0400	3000 amp		1.20	13.333		9,150	730		9,880	11,200
0410	4000 amp		1	16		11,900	875		12,775	14,500
0430	Expansion fitting, 225 amp		5.40	2.963		1,525	162		1,687	1,925
0440	400 amp		4.60	3.478		1,725	190		1,915	2,200
0450	600 amp		4	4		2,150	219		2,369	2,700
0460	800 amp		3.40	4.706		2,525	257		2,782	3,175
0470	1000 amp		3	5.333		2,900	292		3,192	3,625
0480	1350 amp		2.80	5.714		3,450	315		3,765	4,275
0490	1600 amp		2.60	6.154		4,825	335		5,160	5,825
0500	2000 amp		2.20	7.273		5,550	400		5,950	6,725
0510	2500 amp		1.80	8.889		6,750	485		7,235	8,175
0520	3000 amp		1.60	10		7,550	545		8,095	9,150
0530	4000 amp		1.20	13.333		9,725	730		10,455	11,800
0550	Reducer nonfused, 225 amp		5.40	2.963		1,800	162		1,962	2,225
0560	400 amp		4.60	3.478		1,800	190		1,990	2,275
0570	600 amp		4	4		1,800	219		2,019	2,325
0580	800 amp		3.40	4.706		2,175	257		2,432	2,800
0590	1000 amp		3	5.333		2,600	292		2,892	3,325
0600	1350 amp		2.80	5.714		3,750	315		4,065	4,600
0610	1600 amp		2.60	6.154		4,300	335		4,635	5,275
0620	2000 amp		2.20	7.273		5,200	400		5,600	6,325
0630	2500 amp		1.80	8.889		6,550	485		7,035	7,950
0640	3000 amp		1.60	10		7,375	545		7,920	8,950
0650	4000 amp		1.20	13.333		9,600	730		10,330	11,700
0670	Reducer fuse included, 225 amp		4.40	3.636		3,825	199		4,024	4,525
0680	400 amp		4.20	3.810		4,800	208		5,008	5,600
0690	600 amp		3.60	4.444		5,900	243		6,143	6,875
0700	800 amp		3.20	5		8,350	274		8,624	9,600
0710	1000 amp		3	5.333		10,500	292		10,792	12,000
0720	1350 amp		2.80	5.714		16,300	315		16,615	18,400
0730	1600 amp		2.20	7.273		22,000	400		22,400	24,800
0740	2000 amp		1.80	8.889		24,800	485		25,285	28,100
0790	Reducer, circuit breaker, 225 amp		4.40	3.636		4,850	199		5,049	5,625
0800	400 amp		4.20	3.810		5,650	208		5,858	6,550
0810	600 amp		3.60	4.444		8,025	243		8,268	9,225
0820	800 amp		3.20	5		9,425	274		9,699	10,800
0830	1000 amp		3	5.333		10,700	292		10,992	12,300
0840	1350 amp		2.80	5.714		12,200	315		12,515	13,900
0850	1600 amp		2.20	7.273		18,000	400		18,400	20,400

26 25 13 – Bus Duct/Busway and Fittings

26 25 13.30 Copper Bus Duct		Crew	Daily Output	Labor-Hours	Unit	Material	Labor	Equipment	Total	Total Incl O&P
						2015 Bare Costs				
0860	2000 amp	2 Elec	1.80	8.889	Ea.	19,800	485		20,285	22,600
0910	Cable tap box, center, 225 amp		3.20	5		1,775	274		2,049	2,375
0920	400 amp		2.60	6.154		1,775	335		2,110	2,475
0930	600 amp		2.20	7.273		1,775	400		2,175	2,575
0940	800 amp		2	8		1,950	440		2,390	2,825
0950	1000 amp		1.60	10		2,100	545		2,645	3,175
0960	1350 amp		1.40	11.429		2,525	625		3,150	3,775
0970	1600 amp		1.20	13.333		2,825	730		3,555	4,225
0980	2000 amp		1	16		3,400	875		4,275	5,100
1040	2500 amp		.80	20		4,025	1,100		5,125	6,125
1060	3000 amp		.60	26.667		4,425	1,450		5,875	7,100
1080	4000 amp		.40	40		5,550	2,200		7,750	9,525
1800	Weatherproof 3 pole 3 wire, feeder duct, 600 amp		28	.571	L.F.	258	31.50		289.50	335
1820	800 amp		22	.727		315	40		355	405
1840	1000 amp		20	.800		350	44		394	455
1850	1200 amp		19	.842		365	46		411	470
1860	1350 amp		18	.889		490	48.50		538.50	615
1880	1600 amp		14	1.143		560	62.50		622.50	710
1900	2000 amp		12	1.333		720	73		793	905
1920	2500 amp		8	2		895	109		1,004	1,150
1940	3000 amp		6	2.667		1,025	146		1,171	1,350
1960	4000 amp		4	4		1,350	219		1,569	1,850
2000	Indoor 3 pole 3 wire, feeder duct, 600 amp		32	.500		215	27.50		242.50	280
2010	800 amp		26	.615		261	33.50		294.50	340
2020	1000 amp		24	.667		292	36.50		328.50	375
2025	1200 amp		22	.727		385	40		425	485
2030	1350 amp		20	.800		410	44		454	520
2040	1600 amp		16	1		465	54.50		519.50	600
2050	2000 amp		14	1.143		600	62.50		662.50	755
2060	2500 amp		10	1.600		745	87.50		832.50	955
2070	3000 amp		8	2		860	109		969	1,125
2080	4000 amp		6	2.667		1,125	146		1,271	1,475
2090	5000 amp		5	3.200		1,375	175		1,550	1,800
2200	Indoor 3 pole 3 wire, bus duct plug-in, 225 amp		46	.348		246	19.05		265.05	300
2210	400 amp		36	.444		246	24.50		270.50	310
2220	600 amp		30	.533		246	29		275	315
2230	800 amp		24	.667		292	36.50		328.50	375
2240	1000 amp		20	.800		325	44		369	425
2250	1350 amp		18	.889		440	48.50		488.50	555
2260	1600 amp		14	1.143		495	62.50		557.50	640
2270	2000 amp		12	1.333		630	73		703	805
2280	2500 amp		10	1.600		775	87.50		862.50	985
2290	3000 amp		8	2		885	109		994	1,150
2330	High short circuit, 400 amp		36	.444		246	24.50		270.50	310
2340	600 amp		30	.533		246	29		275	315
2350	800 amp		24	.667		292	36.50		328.50	375
2360	1000 amp		20	.800		325	44		369	425
2370	1350 amp		18	.889		440	48.50		488.50	555
2380	1600 amp		14	1.143		495	62.50		557.50	640
2390	2000 amp		12	1.333		630	73		703	805
2400	2500 amp		10	1.600		775	87.50		862.50	985
2410	3000 amp		8	2		885	109		994	1,150
2440	Elbows, 225 amp		4.60	3.478	Ea.	1,225	190		1,415	1,650

26 25 13.30 Copper Bus Duct		Crew	Daily Output	Labor-Hours	Unit	2015 Bare Costs Material	Labor	Equipment	Total	Total Incl O&P
2450	400 amp	2 Elec	4.20	3.810	Ea.	1,225	208		1,433	1,675
2460	600 amp		3.60	4.444		1,225	243		1,468	1,725
2470	800 amp		3.20	5		1,325	274		1,599	1,875
2480	1000 amp		3	5.333		1,375	292		1,667	1,975
2485	1200 amp		2.90	5.517		1,500	300		1,800	2,125
2490	1350 amp		2.80	5.714		1,575	315		1,890	2,200
2500	1600 amp		2.60	6.154		1,700	335		2,035	2,400
2510	2000 amp		2	8		2,050	440		2,490	2,925
2520	2500 amp		1.80	8.889		3,125	485		3,610	4,175
2530	3000 amp		1.60	10		3,425	545		3,970	4,600
2540	4000 amp		1.40	11.429		4,375	625		5,000	5,775
2560	Tee fittings, 225 amp		2.80	5.714		1,450	315		1,765	2,075
2570	400 amp		2.40	6.667		1,450	365		1,815	2,175
2580	600 amp		2	8		1,450	440		1,890	2,275
2590	800 amp		1.80	8.889		1,575	485		2,060	2,500
2600	1000 amp		1.60	10		1,775	545		2,320	2,800
2605	1200 amp		1.50	10.667		1,875	585		2,460	2,975
2610	1350 amp		1.40	11.429		1,975	625		2,600	3,150
2620	1600 amp		1.20	13.333		2,300	730		3,030	3,650
2630	2000 amp		1	16		3,800	875		4,675	5,525
2640	2500 amp		.70	22.857		4,450	1,250		5,700	6,825
2650	3000 amp		.60	26.667		4,900	1,450		6,350	7,650
2660	4000 amp		.50	32		6,350	1,750		8,100	9,675
2680	Cross, 225 amp		3.60	4.444		2,225	243		2,468	2,800
2690	400 amp		3.20	5		2,225	274		2,499	2,850
2700	600 amp		3	5.333		2,225	292		2,517	2,875
2710	800 amp		2.60	6.154		2,650	335		2,985	3,425
2720	1000 amp		2.40	6.667		2,775	365		3,140	3,625
2730	1350 amp		2.20	7.273		3,150	400		3,550	4,100
2740	1600 amp		2	8		3,400	440		3,840	4,400
2750	2000 amp		1.80	8.889		5,375	485		5,860	6,650
2760	2500 amp		1.60	10		6,250	545		6,795	7,725
2770	3000 amp		1.40	11.429		6,825	625		7,450	8,500
2780	4000 amp		1	16		8,725	875		9,600	11,000
2800	Expansion fitting, 225 amp		6.40	2.500		1,775	137		1,912	2,150
2810	400 amp		5.40	2.963		1,775	162		1,937	2,200
2820	600 amp		4.60	3.478		1,775	190		1,965	2,250
2830	800 amp		4	4		2,125	219		2,344	2,675
2840	1000 amp		3.60	4.444		2,325	243		2,568	2,925
2850	1350 amp		3.20	5		2,750	274		3,024	3,450
2860	1600 amp		3	5.333		3,025	292		3,317	3,775
2870	2000 amp		2.60	6.154		3,650	335		3,985	4,525
2880	2500 amp		2.20	7.273		5,075	400		5,475	6,225
2890	3000 amp		1.80	8.889		5,675	485		6,160	7,000
2900	4000 amp		1.40	11.429		7,250	625		7,875	8,950
2920	Reducer nonfused, 225 amp		6.40	2.500		1,475	137		1,612	1,825
2930	400 amp		5.40	2.963		1,475	162		1,637	1,875
2940	600 amp		4.60	3.478		1,475	190		1,665	1,925
2950	800 amp		4	4		1,725	219		1,944	2,225
2960	1000 amp		3.60	4.444		1,925	243		2,168	2,500
2970	1350 amp		3.20	5		2,425	274		2,699	3,100
2980	1600 amp		3	5.333		2,750	292		3,042	3,475
2990	2000 amp	↓	2.60	6.154	↓	3,300	335		3,635	4,150

26 25 Enclosed Bus Assemblies

26 25 13 – Bus Duct/Busway and Fittings

26 25 13.30 Copper Bus Duct		Crew	Daily Output	Labor-Hours	Unit	2015 Bare Costs Material	Labor	Equipment	Total	Total Incl O&P
3000	2500 amp	2 Elec	2.20	7.273	Ea.	4,825	400		5,225	5,925
3010	3000 amp		1.80	8.889		5,400	485		5,885	6,700
3020	4000 amp		1.40	11.429		7,025	625		7,650	8,700
3040	Reducer fuse included, 225 amp		5	3.200		3,475	175		3,650	4,100
3050	400 amp		4.80	3.333		4,625	182		4,807	5,375
3060	600 amp		4.20	3.810		5,525	208		5,733	6,400
3070	800 amp		3.60	4.444		7,875	243		8,118	9,025
3080	1000 amp		3.40	4.706		9,325	257		9,582	10,600
3090	1350 amp		3.20	5		9,025	274		9,299	10,400
3100	1600 amp		2.60	6.154		21,000	335		21,335	23,600
3110	2000 amp		2	8		23,800	440		24,240	26,900
3160	Reducer circuit breaker, 225 amp		5	3.200		4,450	175		4,625	5,175
3170	400 amp		4.80	3.333		5,450	182		5,632	6,275
3180	600 amp		4.20	3.810		7,825	208		8,033	8,925
3190	800 amp		3.60	4.444		9,150	243		9,393	10,500
3200	1000 amp		3.40	4.706		10,400	257		10,657	11,800
3210	1350 amp		3.20	5		11,900	274		12,174	13,500
3220	1600 amp		2.60	6.154		17,800	335		18,135	20,100
3230	2000 amp		2	8		19,400	440		19,840	22,100
3280	3 pole, 3 wire, cable tap box center, 225 amp		3.60	4.444		2,000	243		2,243	2,575
3290	400 amp		3	5.333		2,000	292		2,292	2,650
3300	600 amp		2.60	6.154		2,000	335		2,335	2,725
3310	800 amp		2.40	6.667		2,250	365		2,615	3,050
3320	1000 amp		1.80	8.889		2,425	485		2,910	3,425
3330	1350 amp		1.60	10		2,975	545		3,520	4,125
3340	1600 amp		1.40	11.429		3,325	625		3,950	4,625
3350	2000 amp		1.20	13.333		4,025	730		4,755	5,550
3360	2500 amp		1	16		4,800	875		5,675	6,625
3370	3000 amp		.70	22.857		5,300	1,250		6,550	7,750
3380	4000 amp		.50	32		6,725	1,750		8,475	10,100
3400	Cable tap box end, 225 amp		3.60	4.444		1,100	243		1,343	1,575
3410	400 amp		3	5.333		1,100	292		1,392	1,650
3420	600 amp		2.60	6.154		1,225	335		1,560	1,875
3430	800 amp		2.40	6.667		1,225	365		1,590	1,925
3440	1000 amp		1.80	8.889		1,350	485		1,835	2,225
3445	1200 amp		1.70	9.412		1,500	515		2,015	2,450
3450	1350 amp		1.60	10		1,600	545		2,145	2,600
3460	1600 amp		1.40	11.429		1,825	625		2,450	3,000
3470	2000 amp		1.20	13.333		2,100	730		2,830	3,425
3480	2500 amp		1	16		2,500	875		3,375	4,100
3490	3000 amp		.70	22.857		2,725	1,250		3,975	4,925
3500	4000 amp		.50	32		3,475	1,750		5,225	6,525
4600	Plug-in, fusible switch w/3 fuses, 3 pole, 250 volt, 30 amp	1 Elec	4	2		435	109		544	650
4610	60 amp		3.60	2.222		570	122		692	815
4620	100 amp		2.70	2.963		815	162		977	1,150
4630	200 amp	2 Elec	3.20	5		1,375	274		1,649	1,925
4640	400 amp		1.40	11.429		3,575	625		4,200	4,900
4650	600 amp		.90	17.778		4,925	970		5,895	6,925
4700	4 pole, 120/208 volt, 30 amp	1 Elec	3.90	2.051		595	112		707	830
4710	60 amp		3.50	2.286		640	125		765	900
4720	100 amp		2.60	3.077		895	168		1,063	1,250
4730	200 amp	2 Elec	3	5.333		1,500	292		1,792	2,100
4740	400 amp		1.30	12.308		3,525	675		4,200	4,925

26 25 13.30 Copper Bus Duct		Crew	Daily Output	Labor-Hours	Unit	Material	Labor	Equipment	Total	Total Incl O&P
						2015 Bare Costs				
4750	600 amp	2 Elec	.80	20	Ea.	4,950	1,100		6,050	7,125
4800	3 pole, 480 volt, 30 amp	1 Elec	4	2		445	109		554	660
4810	60 amp		3.60	2.222		470	122		592	710
4820	100 amp		2.70	2.963		795	162		957	1,125
4830	200 amp	2 Elec	3.20	5		1,375	274		1,649	1,925
4840	400 amp		1.40	11.429		3,200	625		3,825	4,475
4850	600 amp		.90	17.778		4,525	970		5,495	6,475
4860	800 amp		.66	24.242		15,600	1,325		16,925	19,300
4870	1000 amp		.60	26.667		16,600	1,450		18,050	20,500
4880	1200 amp		.50	32		17,400	1,750		19,150	21,900
4890	1600 amp		.44	36.364		19,200	2,000		21,200	24,200
4900	4 pole, 277/480 volt, 30 amp	1 Elec	3.90	2.051		645	112		757	885
4910	60 amp		3.50	2.286		690	125		815	950
4920	100 amp		2.60	3.077		1,000	168		1,168	1,350
4930	200 amp	2 Elec	3	5.333		2,000	292		2,292	2,650
4940	400 amp		1.30	12.308		3,775	675		4,450	5,200
4950	600 amp		.80	20		5,150	1,100		6,250	7,375
5050	800 amp		.60	26.667		17,000	1,450		18,450	21,000
5060	1000 amp		.56	28.571		19,500	1,575		21,075	23,900
5070	1200 amp		.48	33.333		19,700	1,825		21,525	24,500
5080	1600 amp		.42	38.095		21,700	2,075		23,775	27,100
5150	Fusible with starter, 3 pole 250 volt, 30 amp	1 Elec	3.50	2.286		2,400	125		2,525	2,850
5160	60 amp		3.20	2.500		2,550	137		2,687	3,000
5170	100 amp		2.50	3.200		2,875	175		3,050	3,450
5180	200 amp	2 Elec	2.80	5.714		4,750	315		5,065	5,700
5200	3 pole 480 volt, 30 amp	1 Elec	3.50	2.286		2,400	125		2,525	2,850
5210	60 amp		3.20	2.500		2,550	137		2,687	3,000
5220	100 amp		2.50	3.200		2,875	175		3,050	3,450
5230	200 amp	2 Elec	2.80	5.714		4,750	315		5,065	5,700
5300	Fusible with contactor, 3 pole 250 volt, 30 amp	1 Elec	3.50	2.286		2,350	125		2,475	2,775
5310	60 amp		3.20	2.500		2,975	137		3,112	3,475
5320	100 amp		2.50	3.200		4,175	175		4,350	4,875
5330	200 amp	2 Elec	2.80	5.714		4,775	315		5,090	5,725
5400	3 pole 480 volt, 30 amp	1 Elec	3.50	2.286		2,525	125		2,650	2,975
5410	60 amp		3.20	2.500		3,550	137		3,687	4,100
5420	100 amp		2.50	3.200		4,875	175		5,050	5,650
5430	200 amp	2 Elec	2.80	5.714		5,000	315		5,315	5,975
5450	Fusible with capacitor, 3 pole 250 volt, 30 amp	1 Elec	3	2.667		6,075	146		6,221	6,925
5460	60 amp		2	4		7,075	219		7,294	8,150
5500	3 pole 480 volt, 30 amp		3	2.667		5,100	146		5,246	5,825
5510	60 amp		2	4		6,375	219		6,594	7,350
5600	Circuit breaker, 3 pole, 250 volt, 60 amp		4.50	1.778		610	97		707	825
5610	100 amp		3.20	2.500		750	137		887	1,025
5650	4 pole, 120/208 volt, 60 amp		4.40	1.818		690	99.50		789.50	915
5660	100 amp		3.10	2.581		820	141		961	1,125
5700	3 pole, 4 wire 277/480 volt, 60 amp		4.30	1.860		930	102		1,032	1,175
5710	100 amp		3	2.667		1,025	146		1,171	1,375
5720	225 amp	2 Elec	3.20	5		2,300	274		2,574	2,950
5730	400 amp		1.20	13.333		4,775	730		5,505	6,375
5740	600 amp		.96	16.667		6,400	910		7,310	8,450
5750	700 amp		.60	26.667		8,125	1,450		9,575	11,200
5760	800 amp		.60	26.667		8,125	1,450		9,575	11,200
5770	900 amp		.54	29.630		10,700	1,625		12,325	14,300

26 25 13.30 Copper Bus Duct		Crew	Daily Output	Labor-Hours	Unit	Material	Labor	Equipment	Total	Total Incl O&P
						2015 Bare Costs				
5780	1000 amp	2 Elec	.54	29.630	Ea.	10,700	1,625		12,325	14,300
5790	1200 amp		.42	38.095		12,900	2,075		14,975	17,300
5810	Circuit breaker w/HIC fuses, 3 pole 480 volt, 60 amp	1 Elec	4.40	1.818		1,175	99.50		1,274.50	1,450
5820	100 amp	"	3.10	2.581		1,275	141		1,416	1,650
5830	225 amp	2 Elec	3.40	4.706		4,100	257		4,357	4,900
5840	400 amp		1.40	11.429		6,500	625		7,125	8,125
5850	600 amp		1	16		6,600	875		7,475	8,600
5860	700 amp		.64	25		8,750	1,375		10,125	11,800
5870	800 amp		.64	25		8,750	1,375		10,125	11,800
5880	900 amp		.56	28.571		18,900	1,575		20,475	23,200
5890	1000 amp		.56	28.571		18,900	1,575		20,475	23,200
5950	3 pole 4 wire 277/480 volt, 60 amp	1 Elec	4.30	1.860		1,175	102		1,277	1,450
5960	100 amp	"	3	2.667		1,275	146		1,421	1,650
5970	225 amp	2 Elec	3	5.333		4,100	292		4,392	4,950
5980	400 amp		1.10	14.545		6,500	795		7,295	8,375
5990	600 amp		.94	17.021		6,600	930		7,530	8,700
6000	700 amp		.58	27.586		8,750	1,500		10,250	12,000
6010	800 amp		.58	27.586		8,750	1,500		10,250	12,000
6020	900 amp		.52	30.769		18,900	1,675		20,575	23,400
6030	1000 amp		.52	30.769		18,900	1,675		20,575	23,400
6040	1200 amp		.40	40		18,900	2,200		21,100	24,200
6100	Circuit breaker with starter, 3 pole 250 volt, 60 amp	1 Elec	3.20	2.500		1,675	137		1,812	2,025
6110	100 amp	"	2.50	3.200		2,300	175		2,475	2,825
6120	225 amp	2 Elec	3	5.333		2,975	292		3,267	3,725
6130	3 pole 480 volt, 60 amp	1 Elec	3.20	2.500		1,675	137		1,812	2,025
6140	100 amp	"	2.50	3.200		2,300	175		2,475	2,825
6150	225 amp	2 Elec	3	5.333		2,825	292		3,117	3,575
6200	Circuit breaker with contactor, 3 pole 250 volt, 60 amp	1 Elec	3.20	2.500		1,575	137		1,712	1,925
6210	100 amp	"	2.50	3.200		2,150	175		2,325	2,625
6220	225 amp	2 Elec	3	5.333		2,750	292		3,042	3,475
6250	3 pole 480 volt, 60 amp	1 Elec	3.20	2.500		1,575	137		1,712	1,925
6260	100 amp	"	2.50	3.200		2,150	175		2,325	2,625
6270	225 amp	2 Elec	3	5.333		2,450	292		2,742	3,125
6300	Circuit breaker with capacitor, 3 pole 250 volt, 60 amp	1 Elec	2	4		7,900	219		8,119	9,050
6310	3 pole 480 volt, 60 amp		2	4		7,925	219		8,144	9,075
6400	Add control transformer with pilot light to above starter		16	.500		470	27.50		497.50	560
6410	Switch, fusible, mechanically held contactor optional		16	.500		1,250	27.50		1,277.50	1,400
6430	Circuit breaker, mechanically held contactor optional		16	.500		1,250	27.50		1,277.50	1,400
6450	Ground neutralizer, 3 pole		16	.500		57	27.50		84.50	106

26 25 13.40 Copper Bus Duct		Crew	Daily Output	Labor-Hours	Unit	Material	Labor	Equipment	Total	Total Incl O&P
0010	**COPPER BUS DUCT** 10 ft. long									
0050	Indoor 3 pole 4 wire, plug-in, straight section, 225 amp	2 Elec	40	.400	L.F.	230	22		252	287
1000	400 amp		32	.500		230	27.50		257.50	296
1500	600 amp		26	.615		230	33.50		263.50	305
2400	800 amp		20	.800		273	44		317	370
2450	1000 amp		18	.889		300	48.50		348.50	405
2470	1200 amp		17	.941		395	51.50		446.50	510
2500	1350 amp		16	1		415	54.50		469.50	540
2510	1600 amp		12	1.333		470	73		543	630
2520	2000 amp		10	1.600		595	87.50		682.50	790
2530	2500 amp		8	2		730	109		839	975
2540	3000 amp		6	2.667		840	146		986	1,150

26 25 Enclosed Bus Assemblies

26 25 13 – Bus Duct/Busway and Fittings

26 25 13.40 Copper Bus Duct		Crew	Daily Output	Labor-Hours	Unit	Material	Labor	Equipment	Total	Total Incl O&P
						2015 Bare Costs				
2550	Feeder, 600 amp	2 Elec	28	.571	L.F.	204	31.50		235.50	273
2600	800 amp		22	.727		247	40		287	335
2700	1000 amp		20	.800		276	44		320	375
2750	1200 amp		19	.842		365	46		411	470
2800	1350 amp		18	.889		385	48.50		433.50	500
2900	1600 amp		14	1.143		440	62.50		502.50	580
3000	2000 amp		12	1.333		565	73		638	740
3010	2500 amp		8	2		705	109		814	945
3020	3000 amp		6	2.667		810	146		956	1,125
3030	4000 amp		4	4		1,075	219		1,294	1,525
3040	5000 amp		2	8		1,300	440		1,740	2,125
3100	Elbows, 225 amp		4	4	Ea.	1,375	219		1,594	1,850
3200	400 amp		3.60	4.444		1,375	243		1,618	1,875
3300	600 amp		3.20	5		1,375	274		1,649	1,925
3400	800 amp		2.80	5.714		1,475	315		1,790	2,100
3500	1000 amp		2.60	6.154		1,650	335		1,985	2,350
3550	1200 amp		2.50	6.400		1,850	350		2,200	2,575
3600	1350 amp		2.40	6.667		1,850	365		2,215	2,625
3700	1600 amp		2.20	7.273		2,025	400		2,425	2,850
3800	2000 amp		1.80	8.889		2,500	485		2,985	3,500
3810	2500 amp		1.60	10		3,950	545		4,495	5,200
3820	3000 amp		1.40	11.429		4,325	625		4,950	5,750
3830	4000 amp		1.20	13.333		5,600	730		6,330	7,300
3840	5000 amp		1	16		9,050	875		9,925	11,300
4000	End box, 225 amp		34	.471		165	25.50		190.50	222
4100	400 amp		32	.500		186	27.50		213.50	248
4200	600 amp		28	.571		186	31.50		217.50	254
4300	800 amp		26	.615		186	33.50		219.50	257
4400	1000 amp		24	.667		186	36.50		222.50	262
4410	1200 amp		23	.696		187	38		225	264
4500	1350 amp		22	.727		177	40		217	256
4600	1600 amp		20	.800		177	44		221	262
4700	2000 amp		18	.889		217	48.50		265.50	315
4710	2500 amp		16	1		217	54.50		271.50	325
4720	3000 amp		14	1.143		205	62.50		267.50	325
4730	4000 amp		12	1.333		249	73		322	385
4740	5000 amp		10	1.600		249	87.50		336.50	410
4800	Cable tap box end, 225 amp		3.20	5		1,100	274		1,374	1,625
5000	400 amp		2.60	6.154		1,200	335		1,535	1,850
5100	600 amp		2.20	7.273		1,400	400		1,800	2,175
5200	800 amp		2	8		1,475	440		1,915	2,300
5300	1000 amp		1.60	10		1,500	545		2,045	2,500
5350	1200 amp		1.50	10.667		1,975	585		2,560	3,075
5400	1350 amp		1.40	11.429		2,200	625		2,825	3,400
5500	1600 amp		1.20	13.333		2,475	730		3,205	3,850
5600	2000 amp		1	16		2,750	875		3,625	4,375
5610	2500 amp		.80	20		3,050	1,100		4,150	5,050
5620	3000 amp		.60	26.667		3,550	1,450		5,000	6,175
5630	4000 amp		.40	40		4,100	2,200		6,300	7,925
5640	5000 amp		.20	80		4,850	4,375		9,225	12,100
5700	Switchboard stub, 225 amp		5.40	2.963		1,250	162		1,412	1,650
5800	400 amp		4.60	3.478		1,325	190		1,515	1,750
5900	600 amp		4	4		1,375	219		1,594	1,875

26 25 13 – Bus Duct/Busway and Fittings

26 25 13.40 Copper Bus Duct		Crew	Daily Output	Labor-Hours	Unit	Material	2015 Bare Costs Labor	Equipment	Total	Total Incl O&P
6000	800 amp	2 Elec	3.20	5	Ea.	1,675	274		1,949	2,250
6100	1000 amp		3	5.333		1,925	292		2,217	2,575
6150	1200 amp		2.80	5.714		2,275	315		2,590	2,975
6200	1350 amp		2.60	6.154		2,400	335		2,735	3,150
6300	1600 amp		2.40	6.667		2,700	365		3,065	3,550
6400	2000 amp		2	8		3,275	440		3,715	4,275
6410	2500 amp		1.80	8.889		3,975	485		4,460	5,125
6420	3000 amp		1.60	10		4,425	545		4,970	5,700
6430	4000 amp		1.40	11.429		5,750	625		6,375	7,300
6440	5000 amp		1.20	13.333		7,050	730		7,780	8,875
6490	Tee fittings, 225 amp		2.40	6.667		1,900	365		2,265	2,650
6500	400 amp		2	8		1,900	440		2,340	2,750
6600	600 amp		1.80	8.889		1,900	485		2,385	2,825
6700	800 amp		1.60	10		2,175	545		2,720	3,250
6750	1000 amp		1.40	11.429		2,525	625		3,150	3,750
6770	1200 amp		1.30	12.308		2,825	675		3,500	4,175
6800	1350 amp		1.20	13.333		3,000	730		3,730	4,425
7000	1600 amp		1	16		3,400	875		4,275	5,100
7100	2000 amp		.80	20		4,025	1,100		5,125	6,150
7110	2500 amp		.60	26.667		4,950	1,450		6,400	7,700
7120	3000 amp		.50	32		5,475	1,750		7,225	8,725
7130	4000 amp		.40	40		7,075	2,200		9,275	11,200
7140	5000 amp		.20	80		8,375	4,375		12,750	16,000
7200	Plug-in fusible switches w/3 fuses, 600 volt, 3 pole, 30 amp	1 Elec	4	2		850	109		959	1,100
7300	60 amp		3.60	2.222		955	122		1,077	1,250
7400	100 amp		2.70	2.963		1,450	162		1,612	1,850
7500	200 amp	2 Elec	3.20	5		2,600	274		2,874	3,300
7600	400 amp		1.40	11.429		7,625	625		8,250	9,350
7700	600 amp		.90	17.778		8,650	970		9,620	11,000
7800	800 amp		.66	24.242		12,100	1,325		13,425	15,400
7900	1200 amp		.50	32		22,800	1,750		24,550	27,700
7910	1600 amp		.44	36.364		21,600	2,000		23,600	26,900
8000	Plug-in circuit breakers, molded case, 15 to 50 amp	1 Elec	4.40	1.818		805	99.50		904.50	1,050
8100	70 to 100 amp	"	3.10	2.581		895	141		1,036	1,200
8200	150 to 225 amp	2 Elec	3.40	4.706		2,425	257		2,682	3,075
8300	250 to 400 amp		1.40	11.429		4,250	625		4,875	5,650
8400	500 to 600 amp		1	16		5,750	875		6,625	7,650
8500	700 to 800 amp		.64	25		7,075	1,375		8,450	9,900
8600	900 to 1000 amp		.56	28.571		10,100	1,575		11,675	13,500
8700	1200 amp		.44	36.364		12,200	2,000		14,200	16,500
8720	1400 amp		.40	40		17,100	2,200		19,300	22,200
8730	1600 amp		.40	40		18,700	2,200		20,900	24,000
8750	Circuit breakers, with current limiting fuse, 15 to 50 amp	1 Elec	4.40	1.818		1,600	99.50		1,699.50	1,925
8760	70 to 100 amp	"	3.10	2.581		1,900	141		2,041	2,325
8770	150 to 225 amp	2 Elec	3.40	4.706		4,100	257		4,357	4,900
8780	250 to 400 amp		1.40	11.429		6,325	625		6,950	7,950
8790	500 to 600 amp		1	16		7,300	875		8,175	9,375
8800	700 to 800 amp		.64	25		12,000	1,375		13,375	15,300
8810	900 to 1000 amp		.56	28.571		13,700	1,575		15,275	17,500
8850	Combination starter FVNR, fusible switch, NEMA size 0, 30 amp	1 Elec	2	4		2,175	219		2,394	2,750
8860	NEMA size 1, 60 amp		1.80	4.444		2,300	243		2,543	2,900
8870	NEMA size 2, 100 amp		1.30	6.154		2,900	335		3,235	3,725
8880	NEMA size 3, 200 amp	2 Elec	2	8		4,625	440		5,065	5,750

26 25 Enclosed Bus Assemblies

26 25 13 – Bus Duct/Busway and Fittings

26 25 13.40 Copper Bus Duct		Crew	Daily Output	Labor-Hours	Unit	Material	2015 Bare Costs Labor	Equipment	Total	Total Incl O&P
8900	Circuit breaker, NEMA size 0, 30 amp	1 Elec	2	4	Ea.	2,225	219		2,444	2,800
8910	NEMA size 1, 60 amp		1.80	4.444		2,325	243		2,568	2,925
8920	NEMA size 2, 100 amp		1.30	6.154		3,325	335		3,660	4,200
8930	NEMA size 3, 200 amp	2 Elec	2	8		4,225	440		4,665	5,325
8950	Combination contactor, fusible switch, NEMA size 0, 30 amp	1 Elec	2	4		1,275	219		1,494	1,750
8960	NEMA size 1, 60 amp		1.80	4.444		1,300	243		1,543	1,800
8970	NEMA size 2, 100 amp		1.30	6.154		1,925	335		2,260	2,650
8980	NEMA size 3, 200 amp	2 Elec	2	8		2,225	440		2,665	3,125
9000	Circuit breaker, NEMA size 0, 30 amp	1 Elec	2	4		1,450	219		1,669	1,950
9010	NEMA size 1, 60 amp		1.80	4.444		1,500	243		1,743	2,025
9020	NEMA size 2, 100 amp		1.30	6.154		2,300	335		2,635	3,050
9030	NEMA size 3, 200 amp	2 Elec	2	8		2,800	440		3,240	3,750
9050	Control transformer for above, NEMA size 0, 30 amp	1 Elec	8	1		246	54.50		300.50	355
9060	NEMA size 1, 60 amp		8	1		246	54.50		300.50	355
9070	NEMA size 2, 100 amp		7	1.143		345	62.50		407.50	470
9080	NEMA size 3, 200 amp	2 Elec	14	1.143		475	62.50		537.50	620
9100	Comb. fusible switch & lighting control, electrically held, 30 amp	1 Elec	2	4		1,000	219		1,219	1,450
9110	60 amp		1.80	4.444		1,450	243		1,693	1,975
9120	100 amp		1.30	6.154		1,875	335		2,210	2,575
9130	200 amp	2 Elec	2	8		4,575	440		5,015	5,725
9150	Mechanically held, 30 amp	1 Elec	2	4		1,250	219		1,469	1,725
9160	60 amp		1.80	4.444		1,875	243		2,118	2,450
9170	100 amp		1.30	6.154		2,425	335		2,760	3,175
9180	200 amp	2 Elec	2	8		4,925	440		5,365	6,075
9200	Ground bus added to bus duct, 225 amp		320	.050	L.F.	43.50	2.74		46.24	52
9210	400 amp		240	.067		43.50	3.65		47.15	53.50
9220	600 amp		240	.067		43.50	3.65		47.15	53.50
9230	800 amp		160	.100		51	5.45		56.45	64.50
9240	1000 amp		160	.100		58	5.45		63.45	72.50
9250	1350 amp		140	.114		83	6.25		89.25	101
9260	1600 amp		120	.133		90	7.30		97.30	110
9270	2000 amp		110	.145		117	7.95		124.95	141
9280	2500 amp		100	.160		145	8.75		153.75	174
9290	3000 amp		90	.178		164	9.70		173.70	195
9300	4000 amp		80	.200		216	10.95		226.95	255
9310	5000 amp		70	.229		262	12.50		274.50	305
9320	High short circuit bracing, add					18.25			18.25	20

26 25 13.60 Copper or Aluminum Bus Duct Fittings		Crew	Daily Output	Labor-Hours	Unit	Material	Labor	Equipment	Total	Total Incl O&P
0010	**COPPER OR ALUMINUM BUS DUCT FITTINGS**									
0100	Flange, wall, with vapor barrier, 225 amp	2 Elec	6.20	2.581	Ea.	795	141		936	1,100
0110	400 amp		6	2.667		795	146		941	1,100
0120	600 amp		5.80	2.759		795	151		946	1,100
0130	800 amp		5.40	2.963		795	162		957	1,125
0140	1000 amp		5	3.200		795	175		970	1,150
0145	1200 amp		4.80	3.333		795	182		977	1,150
0150	1350 amp		4.60	3.478		795	190		985	1,175
0160	1600 amp		4.20	3.810		795	208		1,003	1,200
0170	2000 amp		4	4		795	219		1,014	1,225
0180	2500 amp		3.60	4.444		795	243		1,038	1,250
0190	3000 amp		3.20	5		795	274		1,069	1,300
0200	4000 amp		2.60	6.154		795	335		1,130	1,400
0300	Roof, 225 amp		6.20	2.581		915	141		1,056	1,225

26 25 Enclosed Bus Assemblies

26 25 13 – Bus Duct/Busway and Fittings

26 25 13.60 Copper or Aluminum Bus Duct Fittings		Crew	Daily Output	Labor-Hours	Unit	2015 Bare Costs Material	Labor	Equipment	Total	Total Incl O&P
0310	400 amp	2 Elec	6	2.667	Ea.	915	146		1,061	1,225
0320	600 amp		5.80	2.759		915	151		1,066	1,225
0330	800 amp		5.40	2.963		915	162		1,077	1,250
0340	1000 amp		5	3.200		915	175		1,090	1,275
0345	1200 amp		4.80	3.333		915	182		1,097	1,275
0350	1350 amp		4.60	3.478		915	190		1,105	1,300
0360	1600 amp		4.20	3.810		915	208		1,123	1,325
0370	2000 amp		4	4		915	219		1,134	1,350
0380	2500 amp		3.60	4.444		915	243		1,158	1,375
0390	3000 amp		3.20	5		915	274		1,189	1,425
0400	4000 amp		2.60	6.154		915	335		1,250	1,525
0420	Support, floor mounted, 225 amp		20	.800		167	44		211	251
0430	400 amp		20	.800		167	44		211	251
0440	600 amp		18	.889		167	48.50		215.50	259
0450	800 amp		16	1		167	54.50		221.50	268
0460	1000 amp		13	1.231		167	67.50		234.50	287
0465	1200 amp		11.80	1.356		167	74		241	298
0470	1350 amp		10.60	1.509		167	82.50		249.50	310
0480	1600 amp		9.20	1.739		167	95		262	330
0490	2000 amp		8	2		167	109		276	350
0500	2500 amp		6.40	2.500		167	137		304	395
0510	3000 amp		5.40	2.963		167	162		329	435
0520	4000 amp		4	4		167	219		386	525
0540	Weather stop, 225 amp		12	1.333		515	73		588	680
0550	400 amp		10	1.600		515	87.50		602.50	700
0560	600 amp		9	1.778		515	97		612	715
0570	800 amp		8	2		515	109		624	735
0580	1000 amp		6.40	2.500		515	137		652	775
0585	1200 amp		5.90	2.712		515	148		663	795
0590	1350 amp		5.40	2.963		515	162		677	815
0600	1600 amp		4.60	3.478		515	190		705	860
0610	2000 amp		4	4		515	219		734	905
0620	2500 amp		3.20	5		515	274		789	990
0630	3000 amp		2.60	6.154		515	335		850	1,075
0640	4000 amp		2	8		515	440		955	1,250
0660	End closure, 225 amp		34	.471		169	25.50		194.50	225
0670	400 amp		32	.500		169	27.50		196.50	228
0680	600 amp		28	.571		169	31.50		200.50	234
0690	800 amp		26	.615		169	33.50		202.50	237
0700	1000 amp		24	.667		169	36.50		205.50	242
0705	1200 amp		23	.696		165	38		203	241
0710	1350 amp		22	.727		165	40		205	244
0720	1600 amp		20	.800		169	44		213	253
0730	2000 amp		18	.889		234	48.50		282.50	335
0740	2500 amp		16	1		245	54.50		299.50	355
0750	3000 amp		14	1.143		245	62.50		307.50	365
0760	4000 amp		12	1.333		245	73		318	385
0780	Switchboard stub, 3 pole 3 wire, 225 amp		6	2.667		995	146		1,141	1,325
0790	400 amp		5.20	3.077		995	168		1,163	1,350
0800	600 amp		4.60	3.478		995	190		1,185	1,400
0810	800 amp		3.60	4.444		1,200	243		1,443	1,700
0820	1000 amp		3.40	4.706		1,400	257		1,657	1,950
0825	1200 amp		3.20	5		1,400	274		1,674	1,975

26 25 13.60 Copper or Aluminum Bus Duct Fittings		Crew	Daily Output	Labor-Hours	Unit	Material	2015 Bare Costs Labor	Equipment	Total	Total Incl O&P
0830	1350 amp	2 Elec	3	5.333	Ea.	1,700	292		1,992	2,300
0840	1600 amp		2.80	5.714		1,975	315		2,290	2,650
0850	2000 amp		2.40	6.667		2,325	365		2,690	3,125
0860	2500 amp		2	8		2,825	440		3,265	3,775
0870	3000 amp		1.80	8.889		3,225	485		3,710	4,300
0880	4000 amp		1.60	10		4,175	545		4,720	5,425
0890	5000 amp		1.40	11.429		5,125	625		5,750	6,600
0900	3 pole 4 wire, 225 amp		5.40	2.963		1,250	162		1,412	1,625
0910	400 amp		4.60	3.478		1,250	190		1,440	1,675
0920	600 amp		4	4		1,250	219		1,469	1,725
0930	800 amp		3.20	5		1,500	274		1,774	2,075
0940	1000 amp		3	5.333		1,750	292		2,042	2,350
0950	1350 amp		2.60	6.154		2,225	335		2,560	2,975
0960	1600 amp		2.40	6.667		2,525	365		2,890	3,350
0970	2000 amp		2	8		3,050	440		3,490	4,050
0980	2500 amp		1.80	8.889		3,725	485		4,210	4,850
0990	3000 amp		1.60	10		4,350	545		4,895	5,650
1000	4000 amp		1.40	11.429		5,650	625		6,275	7,200
1050	Service head, weatherproof, 3 pole 3 wire, 225 amp		3	5.333		1,700	292		1,992	2,300
1060	400 amp		2.80	5.714		1,700	315		2,015	2,325
1070	600 amp		2.60	6.154		1,700	335		2,035	2,375
1080	800 amp		2.40	6.667		1,900	365		2,265	2,650
1090	1000 amp		2	8		2,050	440		2,490	2,925
1100	1350 amp		1.80	8.889		2,650	485		3,135	3,675
1110	1600 amp		1.60	10		2,975	545		3,520	4,125
1120	2000 amp		1.40	11.429		3,625	625		4,250	4,950
1130	2500 amp		1.20	13.333		4,325	730		5,055	5,875
1140	3000 amp		.90	17.778		5,050	970		6,020	7,050
1150	4000 amp		.70	22.857		6,400	1,250		7,650	8,975
1200	3 pole 4 wire, 225 amp		2.60	6.154		1,900	335		2,235	2,600
1210	400 amp		2.40	6.667		1,900	365		2,265	2,650
1220	600 amp		2.20	7.273		1,900	400		2,300	2,700
1230	800 amp		2	8		2,200	440		2,640	3,100
1240	1000 amp		1.70	9.412		2,525	515		3,040	3,575
1250	1350 amp		1.50	10.667		3,000	585		3,585	4,200
1260	1600 amp		1.40	11.429		3,275	625		3,900	4,575
1270	2000 amp		1.20	13.333		4,375	730		5,105	5,950
1280	2500 amp		1	16		5,400	875		6,275	7,275
1290	3000 amp		.80	20		6,325	1,100		7,425	8,650
1300	4000 amp		.60	26.667		8,200	1,450		9,650	11,300
1350	Flanged end, 3 pole 3 wire, 225 amp		6	2.667		920	146		1,066	1,250
1360	400 amp		5.20	3.077		920	168		1,088	1,275
1370	600 amp		4.60	3.478		920	190		1,110	1,325
1380	800 amp		3.60	4.444		1,025	243		1,268	1,500
1390	1000 amp		3.40	4.706		1,150	257		1,407	1,650
1395	1200 amp		3.20	5		1,275	274		1,549	1,825
1400	1350 amp		3	5.333		1,375	292		1,667	1,950
1410	1600 amp		2.80	5.714		1,550	315		1,865	2,200
1420	2000 amp		2.40	6.667		1,825	365		2,190	2,600
1430	2500 amp		2	8		2,150	440		2,590	3,025
1440	3000 amp		1.80	8.889		2,450	485		2,935	3,450
1450	4000 amp		1.60	10		3,050	545		3,595	4,200
1500	3 pole 4 wire, 225 amp	↓	5.40	2.963	↓	1,050	162		1,212	1,400

26 25 13.60 Copper or Aluminum Bus Duct Fittings		Crew	Daily Output	Labor-Hours	Unit	Material	Labor	Equipment	Total	Total Incl O&P
						2015 Bare Costs				
1510	400 amp	2 Elec	4.60	3.478	Ea.	1,050	190		1,240	1,450
1520	600 amp		4	4		1,050	219		1,269	1,500
1530	800 amp		3.20	5		1,250	274		1,524	1,800
1540	1000 amp		3	5.333		1,400	292		1,692	1,975
1545	1200 amp		2.80	5.714		1,575	315		1,890	2,200
1550	1350 amp		2.60	6.154		1,700	335		2,035	2,400
1560	1600 amp		2.40	6.667		1,950	365		2,315	2,725
1570	2000 amp		2	8		2,300	440		2,740	3,200
1580	2500 amp		1.80	8.889		2,725	485		3,210	3,750
1590	3000 amp		1.60	10		3,125	545		3,670	4,275
1600	4000 amp		1.40	11.429		4,025	625		4,650	5,425
1650	Hanger, standard, 225 amp		64	.250		22.50	13.70		36.20	45.50
1660	400 amp		48	.333		22.50	18.25		40.75	52.50
1670	600 amp		40	.400		22.50	22		44.50	58.50
1680	800 amp		32	.500		22.50	27.50		50	67
1690	1000 amp		24	.667		22.50	36.50		59	81
1695	1200 amp		22	.727		22.50	40		62.50	86
1700	1350 amp		20	.800		22.50	44		66.50	92.50
1710	1600 amp		20	.800		22.50	44		66.50	92.50
1720	2000 amp		18	.889		22.50	48.50		71	100
1730	2500 amp		16	1		22.50	54.50		77	109
1740	3000 amp		16	1		22.50	54.50		77	109
1750	4000 amp		16	1		22.50	54.50		77	109
1800	Spring type, 225 amp		16	1		86.50	54.50		141	180
1810	400 amp		14	1.143		86.50	62.50		149	192
1820	600 amp		14	1.143		86.50	62.50		149	192
1830	800 amp		14	1.143		86.50	62.50		149	192
1840	1000 amp		14	1.143		86.50	62.50		149	192
1845	1200 amp		14	1.143		86.50	62.50		149	192
1850	1350 amp		14	1.143		86.50	62.50		149	192
1860	1600 amp		12	1.333		86.50	73		159.50	208
1870	2000 amp		12	1.333		86.50	73		159.50	208
1880	2500 amp		12	1.333		86.50	73		159.50	208
1890	3000 amp		10	1.600		86.50	87.50		174	231
1900	4000 amp		10	1.600		86.50	87.50		174	231

26 25 13.70 Feedrail		Crew	Daily Output	Labor-Hours	Unit	Material	Labor	Equipment	Total	Total Incl O&P
0010	**FEEDRAIL**, 12 foot mounting									
0050	Trolley busway, 3 pole									
0100	300 volt 60 amp, plain, 10 ft. lengths	1 Elec	50	.160	L.F.	25.50	8.75		34.25	41.50
0300	Door track		50	.160		41	8.75		49.75	58.50
0500	Curved track		30	.267		445	14.60		459.60	515
0700	Coupling				Ea.	16.70			16.70	18.35
0900	Center feed	1 Elec	5.30	1.509		78	82.50		160.50	214
1100	End feed		5.30	1.509		68	82.50		150.50	203
1300	Hanger set		24	.333		4.48	18.25		22.73	33
3000	600 volt 100 amp, plain, 10 ft. lengths		35	.229	L.F.	58.50	12.50		71	84
3300	Door track		35	.229	"	86	12.50		98.50	114
3700	Coupling				Ea.	45.50			45.50	50
4000	End cap	1 Elec	40	.200		44.50	10.95		55.45	66
4200	End feed		4	2		199	109		308	390
4500	Trolley, 600 volt, 20 amp		5.30	1.509		425	82.50		507.50	600
4700	30 amp		5.30	1.509		425	82.50		507.50	600

26 25 Enclosed Bus Assemblies

26 25 13 – Bus Duct/Busway and Fittings

26 25 13.70 Feedrail		Crew	Daily Output	Labor-Hours	Unit	Material	2015 Bare Costs Labor	Equipment	Total	Total Incl O&P
4900	Duplex, 40 amp	1 Elec	4	2	Ea.	715	109		824	955
5000	60 amp		4	2		715	109		824	955
5300	Fusible, 20 amp		4	2		995	109		1,104	1,275
5500	30 amp		4	2		995	109		1,104	1,275
5900	300 volt, 20 amp		5.30	1.509		281	82.50		363.50	440
6000	30 amp		5.30	1.509		355	82.50		437.50	520
6300	Fusible, 20 amp		4.70	1.702		345	93		438	525
6500	30 amp		4.70	1.702		480	93		573	675
7300	Busway, 250 volt 50 amp, 2 wire		70	.114	L.F.	20.50	6.25		26.75	32
7330	Coupling				Ea.	42.50			42.50	47
7340	Center feed	1 Elec	6	1.333		495	73		568	660
7350	End feed		6	1.333		109	73		182	233
7360	End cap		40	.200		29	10.95		39.95	49
7370	Hanger set		24	.333		2.80	18.25		21.05	31
7400	125/250 volt 50 amp, 3 wire		60	.133	L.F.	20.50	7.30		27.80	34
7430	Coupling		6	1.333	Ea.	61.50	73		134.50	181
7440	Center feed		6	1.333		495	73		568	660
7450	End feed		6	1.333		109	73		182	233
7460	End cap		40	.200		29	10.95		39.95	49
7470	Hanger set		24	.333		3.75	18.25		22	32
7480	Trolley, 250 volt, 2 pole, 20 amp		6	1.333		37	73		110	154
7490	30 amp		6	1.333		37	73		110	154
7500	125/250 volt, 3 pole, 20 amp		6	1.333		37	73		110	154
7510	30 amp		6	1.333		37	73		110	154
8000	Cleaning tools, 300 volt, dust remover					107			107	118
8100	Bus bar cleaner					202			202	222
8300	600 volt, dust remover, 60 amp					297			297	325
8400	100 amp					650			650	715
8600	Bus bar cleaner, 60 amp					675			675	745
8700	100 amp					800			800	880

26 27 Low-Voltage Distribution Equipment

26 27 13 – Electricity Metering

26 27 13.10 Meter Centers and Sockets

		Crew	Daily Output	Labor-Hours	Unit	Material	Labor	Equipment	Total	Total Incl O&P
0010	**METER CENTERS AND SOCKETS**									
0100	Sockets, single position, 4 terminal, 100 amp	1 Elec	3.20	2.500	Ea.	43	137		180	260
0200	150 amp		2.30	3.478		48	190		238	350
0300	200 amp		1.90	4.211		92	230		322	455
0400	Transformer rated, 20 amp		3.20	2.500		152	137		289	380
0500	Double position, 4 terminal, 100 amp		2.80	2.857		204	156		360	465
0600	150 amp		2.10	3.810		240	208		448	590
0700	200 amp		1.70	4.706		450	257		707	895
0800	Trans-socket, 13 terminal, 3 CT mounts, 400 amp		1	8		1,325	440		1,765	2,125
0900	800 amp	2 Elec	1.20	13.333		1,550	730		2,280	2,825
1100	Meter centers and sockets, three phase, single pos, 7 terminal, 100 amp	1 Elec	2.80	2.857		164	156		320	425
1200	200 amp		2.10	3.810		295	208		503	650
1400	400 amp		1.70	4.706		720	257		977	1,200
2000	Meter center, main fusible switch, 1P 3W 120/240 volt									
2030	400 amp	2 Elec	1.60	10	Ea.	1,675	545		2,220	2,700
2040	600 amp		1.10	14.545		2,600	795		3,395	4,075
2050	800 amp		.90	17.778		4,325	970		5,295	6,275

26 27 13.10 Meter Centers and Sockets		Crew	Daily Output	Labor-Hours	Unit	Material	2015 Bare Costs Labor	Equipment	Total	Total Incl O&P
2060	Rainproof 1P 3W 120/240 volt, 400 amp	2 Elec	1.60	10	Ea.	1,625	545		2,170	2,625
2070	600 amp		1.10	14.545		2,825	795		3,620	4,325
2080	800 amp		.90	17.778		4,400	970		5,370	6,350
2100	3P 4W 120/208 V, 400 amp		1.60	10		2,125	545		2,670	3,200
2110	600 amp		1.10	14.545		3,350	795		4,145	4,925
2120	800 amp		.90	17.778		5,425	970		6,395	7,475
2130	Rainproof 3P 4W 120/208 V, 400 amp		1.60	10		1,850	545		2,395	2,875
2140	600 amp		1.10	14.545		3,450	795		4,245	5,025
2150	800 amp		.90	17.778		6,425	970		7,395	8,550
2170	Main circuit breaker, 1P 3W 120/240 V									
2180	400 amp	2 Elec	1.60	10	Ea.	2,475	545		3,020	3,575
2190	600 amp		1.10	14.545		3,300	795		4,095	4,875
2200	800 amp		.90	17.778		5,225	970		6,195	7,250
2210	1000 amp		.80	20		6,250	1,100		7,350	8,575
2220	1200 amp		.76	21.053		8,700	1,150		9,850	11,400
2230	1600 amp		.68	23.529		18,900	1,275		20,175	22,800
2240	Rainproof 1P 3W 120/240 V, 400 amp		1.60	10		2,900	545		3,445	4,025
2250	600 amp		1.10	14.545		3,900	795		4,695	5,500
2260	800 amp		.90	17.778		4,550	970		5,520	6,500
2270	1000 amp		.80	20		6,250	1,100		7,350	8,575
2280	1200 amp		.76	21.053		8,450	1,150		9,600	11,100
2300	3P 4W 120/208 V, 400 amp		1.60	10		3,300	545		3,845	4,475
2310	600 amp		1.10	14.545		4,625	795		5,420	6,325
2320	800 amp		.90	17.778		5,500	970		6,470	7,575
2330	1000 amp		.80	20		7,250	1,100		8,350	9,675
2340	1200 amp		.76	21.053		9,250	1,150		10,400	12,000
2350	1600 amp		.68	23.529		18,900	1,275		20,175	22,800
2360	Rainproof 3P 4W 120/208 V, 400 amp		1.60	10		3,300	545		3,845	4,475
2370	600 amp		1.10	14.545		4,625	795		5,420	6,325
2380	800 amp		.90	17.778		5,500	970		6,470	7,575
2390	1000 amp		.76	21.053		7,250	1,150		8,400	9,750
2400	1200 amp		.68	23.529		9,250	1,275		10,525	12,200
2420	Main lugs terminal box, 1P 3W 120/240 V									
2430	800 amp	2 Elec	.94	17.021	Ea.	460	930		1,390	1,950
2440	1200 amp		.72	22.222		925	1,225		2,150	2,900
2450	Rainproof 1P 3W 120/240 V, 225 amp		2.40	6.667		370	365		735	970
2460	800 amp		.94	17.021		545	930		1,475	2,050
2470	1200 amp		.72	22.222		1,100	1,225		2,325	3,075
2500	3P 4W 120/208 V, 800 amp		.94	17.021		605	930		1,535	2,125
2510	1200 amp		.72	22.222		1,225	1,225		2,450	3,200
2520	Rainproof 3P 4W 120/208 V, 225 amp		2.40	6.667		370	365		735	970
2530	800 amp		.94	17.021		605	930		1,535	2,125
2540	1200 amp		.72	22.222		1,225	1,225		2,450	3,200
2590	Basic meter device									
2600	1P 3W 120/240 V 4 jaw 125A sockets, 3 meter	2 Elec	1	16	Ea.	620	875		1,495	2,025
2610	4 meter		.90	17.778		745	970		1,715	2,325
2620	5 meter		.80	20		930	1,100		2,030	2,725
2630	6 meter		.60	26.667		1,075	1,450		2,525	3,425
2640	7 meter		.56	28.571		1,375	1,575		2,950	3,925
2650	8 meter		.52	30.769		1,500	1,675		3,175	4,250
2660	10 meter		.48	33.333		1,875	1,825		3,700	4,875
2680	Rainproof 1P 3W 120/240 V 4 jaw 125A sockets									
2690	3 meter	2 Elec	1	16	Ea.	620	875		1,495	2,025

26 27 Low-Voltage Distribution Equipment

26 27 13 – Electricity Metering

26 27 13.10	Meter Centers and Sockets	Crew	Daily Output	Labor-Hours	Unit	Material	2015 Bare Costs Labor	Equipment	Total	Total Incl O&P
2700	4 meter	2 Elec	.90	17.778	Ea.	745	970		1,715	2,325
2710	6 meter		.60	26.667		1,075	1,450		2,525	3,425
2720	7 meter		.56	28.571		1,375	1,575		2,950	3,925
2730	8 meter		.52	30.769		1,500	1,675		3,175	4,250
2750	1P 3W 120/240 V 4 jaw sockets									
2760	with 125A circuit breaker, 3 meter	2 Elec	1	16	Ea.	1,175	875		2,050	2,625
2770	4 meter		.90	17.778		1,475	970		2,445	3,125
2780	5 meter		.80	20		1,825	1,100		2,925	3,725
2790	6 meter		.60	26.667		2,150	1,450		3,600	4,625
2800	7 meter		.56	28.571		2,625	1,575		4,200	5,325
2810	8 meter		.52	30.769		2,925	1,675		4,600	5,825
2820	10 meter		.48	33.333		3,675	1,825		5,500	6,875
2830	Rainproof 1P 3W 120/240 V 4 jaw sockets									
2840	with 125A circuit breaker, 3 meter	2 Elec	1	16	Ea.	1,175	875		2,050	2,625
2850	4 meter		.90	17.778		1,475	970		2,445	3,125
2870	6 meter		.60	26.667		2,150	1,450		3,600	4,625
2880	7 meter		.56	28.571		2,625	1,575		4,200	5,325
2890	8 meter		.52	30.769		2,925	1,675		4,600	5,825
2920	1P 3W on 3P 4W 120/208 V system 5 jaw									
2930	125A sockets, 3 meter	2 Elec	1	16	Ea.	620	875		1,495	2,025
2940	4 meter		.90	17.778		745	970		1,715	2,325
2950	5 meter		.80	20		930	1,100		2,030	2,725
2960	6 meter		.60	26.667		1,075	1,450		2,525	3,425
2970	7 meter		.56	28.571		1,375	1,575		2,950	3,925
2980	8 meter		.52	30.769		1,500	1,675		3,175	4,250
2990	10 meter		.48	33.333		1,875	1,825		3,700	4,875
3000	Rainproof 1P 3W on 3P 4W 120/208 V system									
3020	5 jaw 125A sockets, 3 meter	2 Elec	1	16	Ea.	620	875		1,495	2,025
3030	4 meter		.90	17.778		745	970		1,715	2,325
3050	6 meter		.60	26.667		1,075	1,450		2,525	3,425
3060	7 meter		.56	28.571		1,375	1,575		2,950	3,925
3070	8 meter		.52	30.769		1,500	1,675		3,175	4,250
3090	1P 3W on 3P 4W 120/208 V system 5 jaw sockets									
3100	With 125A circuit breaker, 3 meter	2 Elec	1	16	Ea.	1,175	875		2,050	2,625
3110	4 meter		.90	17.778		1,475	970		2,445	3,125
3120	5 meter		.80	20		1,825	1,100		2,925	3,725
3130	6 meter		.60	26.667		2,150	1,450		3,600	4,625
3140	7 meter		.56	28.571		2,625	1,575		4,200	5,325
3150	8 meter		.52	30.769		2,925	1,675		4,600	5,825
3160	10 meter		.48	33.333		3,675	1,825		5,500	6,875
3170	Rainproof 1P 3W on 3P 4W 120/208 V system									
3180	5 jaw sockets w/125A circuit breaker, 3 meter	2 Elec	1	16	Ea.	1,175	875		2,050	2,625
3190	4 meter		.90	17.778		1,475	970		2,445	3,125
3210	6 meter		.60	26.667		2,150	1,450		3,600	4,625
3220	7 meter		.56	28.571		2,625	1,575		4,200	5,325
3230	8 meter		.52	30.769		2,925	1,675		4,600	5,825
3250	1P 3W 120/240 V 4 jaw sockets									
3260	with 200A circuit breaker, 3 meter	2 Elec	1	16	Ea.	1,750	875		2,625	3,275
3270	4 meter		.90	17.778		2,350	970		3,320	4,100
3290	6 meter		.60	26.667		3,500	1,450		4,950	6,100
3300	7 meter		.56	28.571		4,100	1,575		5,675	6,950
3310	8 meter		.56	28.571		4,725	1,575		6,300	7,625
3330	Rainproof 1P 3W 120/240 V 4 jaw sockets									

26 27 13 – Electricity Metering

26 27 13.10 Meter Centers and Sockets		Crew	Daily Output	Labor-Hours	Unit	Material	2015 Bare Costs Labor	Equipment	Total	Total Incl O&P
3350	with 200A circuit breaker, 3 meter	2 Elec	1	16	Ea.	1,750	875		2,625	3,275
3360	4 meter		.90	17.778		2,350	970		3,320	4,100
3380	6 meter		.60	26.667		3,500	1,450		4,950	6,100
3390	7 meter		.56	28.571		4,100	1,575		5,675	6,950
3400	8 meter		.52	30.769		4,725	1,675		6,400	7,800
3420	1P 3W on 3P 4W 120/208 V 5 jaw sockets									
3430	with 200A circuit breaker, 3 meter	2 Elec	1	16	Ea.	1,750	875		2,625	3,275
3440	4 meter		.90	17.778		2,350	970		3,320	4,100
3460	6 meter		.60	26.667		3,500	1,450		4,950	6,100
3470	7 meter		.56	28.571		4,100	1,575		5,675	6,950
3480	8 meter		.52	30.769		4,725	1,675		6,400	7,800
3500	Rainproof 1P 3W on 3P 4W 120/208 V 5 jaw socket									
3510	with 200A circuit breaker, 3 meter	2 Elec	1	16	Ea.	1,750	875		2,625	3,275
3520	4 meter		.90	17.778		2,350	970		3,320	4,100
3540	6 meter		.60	26.667		3,500	1,450		4,950	6,100
3550	7 meter		.56	28.571		4,100	1,575		5,675	6,950
3560	8 meter		.52	30.769		4,725	1,675		6,400	7,800
3600	Automatic circuit closing, add					67.50			67.50	74
3610	Manual circuit closing, add					77			77	84.50
3650	Branch meter device									
3660	3P 4W 208/120 or 240/120 V 7 jaw sockets									
3670	with 200A circuit breaker, 2 meter	2 Elec	.90	17.778	Ea.	2,925	970		3,895	4,700
3680	3 meter		.80	20		4,375	1,100		5,475	6,500
3690	4 meter		.70	22.857		5,825	1,250		7,075	8,325
3700	Main circuit breaker 42,000 rms, 400 amp		1.60	10		2,175	545		2,720	3,225
3710	600 amp		1.10	14.545		4,100	795		4,895	5,750
3720	800 amp		.90	17.778		5,500	970		6,470	7,575
3730	Rainproof main circ. breaker 42,000 rms, 400 amp		1.60	10		2,550	545		3,095	3,650
3740	600 amp		1.10	14.545		4,100	795		4,895	5,750
3750	800 amp		.90	17.778		5,500	970		6,470	7,575
3760	Main circuit breaker 65,000 rms, 400 amp		1.60	10		3,450	545		3,995	4,650
3770	600 amp		1.10	14.545		4,875	795		5,670	6,575
3780	800 amp		.90	17.778		5,500	970		6,470	7,575
3790	1000 amp		.80	20		7,250	1,100		8,350	9,675
3800	1200 amp		.76	21.053		9,250	1,150		10,400	12,000
3810	1600 amp		.68	23.529		18,900	1,275		20,175	22,800
3820	Rainproof main circ. breaker 65,000 rms, 400 amp		1.60	10		3,450	545		3,995	4,650
3830	600 amp		1.10	14.545		4,875	795		5,670	6,575
3840	800 amp		.90	17.778		5,500	970		6,470	7,550
3850	1000 amp		.80	20		7,250	1,100		8,350	9,675
3860	1200 amp		.76	21.053		9,250	1,150		10,400	12,000
3880	Main circuit breaker 100,000 rms, 400 amp		1.60	10		3,450	545		3,995	4,650
3890	600 amp		1.10	14.545		4,875	795		5,670	6,575
3900	800 amp		.90	17.778		5,725	970		6,695	7,775
3910	Rainproof main circ. breaker 100,000 rms, 400 amp		1.60	10		3,450	545		3,995	4,650
3920	600 amp		1.10	14.545		4,875	795		5,670	6,575
3930	800 amp		.90	17.778		5,725	970		6,695	7,775
3940	Main lugs terminal box, 800 amp		.94	17.021		605	930		1,535	2,125
3950	1600 amp		.72	22.222		2,025	1,225		3,250	4,125
3960	Rainproof, 800 amp		.94	17.021		605	930		1,535	2,125
3970	1600 amp		.72	22.222		2,025	1,225		3,250	4,125
9000	Minimum labor/equipment charge	1 Elec	3	2.667	Job		146		146	226

26 27 16.10 Cabinets		Crew	Daily Output	Labor-Hours	Unit	Material	2015 Bare Costs Labor	Equipment	Total	Total Incl O&P
0010	**CABINETS**									
7000	Cabinets, current transformer									
7050	Single door, 24" H x 24" W x 10" D	1 Elec	1.60	5	Ea.	152	274		426	590
7100	30" H x 24" W x 10" D		1.30	6.154		165	335		500	700
7150	36" H x 24" W x 10" D		1.10	7.273		177	400		577	810
7200	30" H x 30" W x 10" D		1	8		223	440		663	925
7250	36" H x 30" W x 10" D		.90	8.889		262	485		747	1,050
7300	36" H x 36" W x 10" D		.80	10		270	545		815	1,150
7500	Double door, 48" H x 36" W x 10" D		.60	13.333		590	730		1,320	1,775
7550	24" H x 24" W x 12" D		1	8		173	440		613	870
8000	NEMA 12, double door, floor mounted									
8020	54" H x 42" W x 8" D	2 Elec	6	2.667	Ea.	1,350	146		1,496	1,725
8040	60" H x 48" W x 8" D		5.40	2.963		1,800	162		1,962	2,250
8060	60" H x 48" W x 10" D		5.40	2.963		1,875	162		2,037	2,300
8080	60" H x 60" W x 10" D		5	3.200		2,075	175		2,250	2,575
8100	72" H x 60" W x 10" D		4	4		2,450	219		2,669	3,050
8120	72" H x 72" W x 10" D		3.40	4.706		2,700	257		2,957	3,375
8140	60" H x 48" W x 12" D		3.40	4.706		1,900	257		2,157	2,475
8160	60" H x 60" W x 12" D		3.20	5		2,150	274		2,424	2,800
8180	72" H x 60" W x 12" D		3	5.333		2,425	292		2,717	3,125
8200	72" H x 72" W x 12" D		3	5.333		2,725	292		3,017	3,450
8220	60" H x 48" W x 16" D		3.20	5		1,975	274		2,249	2,600
8240	72" H x 72" W x 16" D		2.60	6.154		2,850	335		3,185	3,675
8260	60" H x 48" W x 20" D		3	5.333		2,175	292		2,467	2,825
8280	72" H x 72" W x 20" D		2.20	7.273		3,075	400		3,475	4,000
8300	60" H x 48" W x 24" D		2.60	6.154		2,300	335		2,635	3,050
8320	72" H x 72" W x 24" D		2	8		3,250	440		3,690	4,250
8340	Pushbutton enclosure, oiltight									
8360	3-1/2" H x 3-1/4" W x 2-3/4" D, for 1 P.B.	1 Elec	12	.667	Ea.	56.50	36.50		93	119
8380	5-3/4" H x 3-1/4" W x 2-3/4" D, for 2 P.B.		11	.727		62	40		102	130
8400	8" H x 3-1/4" W x 2-3/4" D, for 3 P.B.		10.50	.762		67.50	41.50		109	139
8420	10-1/4" H x 3-1/4" W x 2-3/4" D, for 4 P.B.		10.50	.762		74	41.50		115.50	146
8460	12-1/2" H x 3-1/4" W x 3" D, for 5 P.B.		9	.889		88.50	48.50		137	173
8480	9-1/2" H x 6-1/4" W x 3" D, for 6 P.B.		8.50	.941		96.50	51.50		148	186
8500	9-1/2" H x 8-1/2" W x 3" D, for 9 P.B.		8	1		105	54.50		159.50	201
8510	11-3/4" H x 8-1/2" W x 3" D, for 12 P.B.		7	1.143		115	62.50		177.50	224
8520	11-3/4" H x 10-3/4" W x 3" D, for 16 P.B.		6.50	1.231		126	67.50		193.50	242
8540	14" H x 10-3/4" W x 3" D, for 20 P.B.		5	1.600		137	87.50		224.50	287
8560	14" H x 13" W x 3" D, for 25 P.B.		4.50	1.778		150	97		247	315
8580	Sloping front pushbutton enclosures									
8600	3-1/2" H x 7-3/4" W x 4-7/8" D, for 3 P.B.	1 Elec	10	.800	Ea.	88	44		132	165
8620	7-1/4" H x 8-1/2" W x 6-3/4" D, for 6 P.B.		8	1		129	54.50		183.50	227
8640	9-1/2" H x 8-1/2" W x 7-7/8" D, for 9 P.B.		7	1.143		153	62.50		215.50	266
8660	11-1/4" H x 8-1/2" W x 9" D, for 12 P.B.		5	1.600		177	87.50		264.50	330
8680	11-3/4" H x 10" W x 9" D, for 16 P.B.		5	1.600		191	87.50		278.50	345
8700	11-3/4" H x 13" W x 9" D, for 20 P.B.		5	1.600		220	87.50		307.50	380
8720	14" H x 13" W x 10-1/8" D, for 25 P.B.		4.50	1.778		249	97		346	425
8740	Pedestals, not including P.B. enclosure or base									
8760	Straight column 4" x 4"	1 Elec	4.50	1.778	Ea.	265	97		362	440
8780	6" x 6"		4	2		410	109		519	625
8800	Angled column 4" x 4"		4.50	1.778		305	97		402	485
8820	6" x 6"		4	2		455	109		564	675

26 27 16.10 Cabinets		Crew	Daily Output	Labor-Hours	Unit	Material	2015 Bare Costs Labor	Equipment	Total	Total Incl O&P
8840	Pedestal, base 18" x 18"	1 Elec	10	.800	Ea.	156	44		200	240
8860	24" x 24"	↓	9	.889	↓	360	48.50		408.50	470
8900	Electronic rack enclosures									
8920	72" H x 19" W x 24" D	1 Elec	1.50	5.333	Ea.	1,925	292		2,217	2,575
8940	72" H x 23" W x 24" D		1.50	5.333		2,175	292		2,467	2,850
8960	72" H x 19" W x 30" D		1.30	6.154		2,300	335		2,635	3,075
8980	72" H x 19" W x 36" D		1.20	6.667		2,650	365		3,015	3,475
9000	72" H x 23" W x 36" D	↓	1.20	6.667	↓	2,925	365		3,290	3,800
9020	NEMA 12 & 4 enclosure panels									
9040	12" x 24"	1 Elec	20	.400	Ea.	36	22		58	74
9060	16" x 12"		20	.400		26.50	22		48.50	63
9080	20" x 16"		20	.400		38.50	22		60.50	76
9100	20" x 20"		19	.421		46	23		69	86
9120	24" x 20"		18	.444		59.50	24.50		84	103
9140	24" x 24"		17	.471		68.50	25.50		94	115
9160	30" x 20"		16	.500		75.50	27.50		103	126
9180	30" x 24"		16	.500		83	27.50		110.50	134
9200	36" x 24"		15	.533		98.50	29		127.50	153
9220	36" x 30"		15	.533		132	29		161	190
9240	42" x 24"		15	.533		112	29		141	168
9260	42" x 30"		14	.571		149	31.50		180.50	213
9280	42" x 36"		14	.571		175	31.50		206.50	241
9300	48" x 24"		14	.571		134	31.50		165.50	196
9320	48" x 30"		14	.571		174	31.50		205.50	240
9340	48" x 36"		13	.615		201	33.50		234.50	273
9360	60" x 36"	↓	12	.667	↓	245	36.50		281.50	325
9400	Wiring trough steel JIC, clamp cover									
9490	4" x 4", 12" long	1 Elec	12	.667	Ea.	94	36.50		130.50	161
9510	24" long		10	.800		119	44		163	199
9530	36" long		8	1		149	54.50		203.50	248
9540	48" long		7	1.143		172	62.50		234.50	286
9550	60" long		6	1.333		202	73		275	335
9560	6" x 6", 12" long		11	.727		125	40		165	199
9580	24" long		9	.889		165	48.50		213.50	258
9600	36" long		7	1.143		202	62.50		264.50	320
9610	48" long		6	1.333		246	73		319	385
9620	60" long		5	1.600	↓	290	87.50		377.50	455
9990	Minimum labor/equipment charge	↓	2	4	Job		219		219	340

26 27 16.20 Cabinets and Enclosures		Crew	Daily Output	Labor-Hours	Unit	Material	Labor	Equipment	Total	Total Incl O&P
0010	**CABINETS AND ENCLOSURES** Nonmetallic									
0080	Enclosures fiberglass NEMA 4X									
0100	Wall mount, quick release latch door, 20"H x 16"W x 6"D	1 Elec	4.80	1.667	Ea.	615	91		706	815
0110	20"H x 20"W x 6"D		4.50	1.778		735	97		832	960
0120	24"H x 20"W x 6"D		4.20	1.905		785	104		889	1,025
0130	20"H x 16"W x 8"D		4.50	1.778		700	97		797	920
0140	20"H x 20"W x 8"D		4.20	1.905		785	104		889	1,025
0150	24"H x 24"W x 8"D		3.80	2.105		890	115		1,005	1,150
0160	30"H x 24"W x 8"D		3.20	2.500		960	137		1,097	1,250
0170	36"H x 30"W x 8"D		3	2.667		1,350	146		1,496	1,700
0180	20"H x 16"W x 10"D		3.50	2.286		815	125		940	1,100
0190	20"H x 20"W x 10"D		3.20	2.500		870	137		1,007	1,175
0200	24"H x 20"W x 10"D	↓	3	2.667	↓	925	146		1,071	1,250

26 27 16.20 Cabinets and Enclosures		Crew	Daily Output	Labor-Hours	Unit	2015 Bare Costs Material	Labor	Equipment	Total	Total Incl O&P
0210	30"H x 24"W x 10"D	1 Elec	2.80	2.857	Ea.	1,050	156		1,206	1,400
0220	20"H x 16"W x 12"D		3	2.667		865	146		1,011	1,175
0230	20"H x 20"W x 12"D		2.80	2.857		930	156		1,086	1,275
0240	24"H x 24"W x 12"D		2.60	3.077		1,025	168		1,193	1,375
0250	30"H x 24"W x 12"D		2.40	3.333		1,150	182		1,332	1,525
0260	36"H x 30"W x 12"D		2.20	3.636		1,575	199		1,774	2,050
0270	36"H x 36"W x 12"D		2.10	3.810		1,775	208		1,983	2,275
0280	48"H x 36"W x 12"D		2	4		2,000	219		2,219	2,550
0290	60"H x 36"W x 12"D		1.80	4.444		2,275	243		2,518	2,875
0300	30"H x 24"W x 16"D		1.40	5.714		1,325	315		1,640	1,950
0310	48"H x 36"W x 16"D		1.20	6.667		2,225	365		2,590	3,025
0320	60"H x 36"W x 16"D		1	8		2,525	440		2,965	3,450
0480	Freestanding, one door, 72"H x 25"W x 25"D		.80	10		4,275	545		4,820	5,550
0490	Two doors with two panels, 72"H x 49"W x 24"D		.50	16		10,700	875		11,575	13,100
0500	Floor stand kits, for NEMA 4 & 12, 20"W or more, 6"H x 8"D		24	.333		161	18.25		179.25	205
0510	6"H x 10"D		24	.333		176	18.25		194.25	221
0520	6"H x 12"D		24	.333		196	18.25		214.25	244
0530	6"H x 18"D		24	.333		228	18.25		246.25	279
0540	12"H x 8"D		22	.364		203	19.90		222.90	254
0550	12"H x 10"D		22	.364		215	19.90		234.90	267
0560	12"H x 12"D		22	.364		229	19.90		248.90	283
0570	12"H x 16"D		22	.364		267	19.90		286.90	325
0580	12"H x 18"D		22	.364		273	19.90		292.90	330
0590	12"H x 20"D		22	.364		291	19.90		310.90	350
0600	18"H x 8"D		20	.400		242	22		264	300
0610	18"H x 10"D		20	.400		259	22		281	320
0620	18"H x 12"D		20	.400		278	22		300	340
0630	18"H x 16"D		20	.400		310	22		332	375
0640	24"H x 8"D		16	.500		299	27.50		326.50	375
0650	24"H x 10"D		16	.500		310	27.50		337.50	385
0660	24"H x 12"D		16	.500		335	27.50		362.50	410
0670	24"H x 16"D		16	.500		350	27.50		377.50	430
0680	Small, screw cover, 5-1/2"H x 4"W x 4-15/16"D		12	.667		85.50	36.50		122	151
0690	7-1/2"H x 4"W x 4-15/16"D		12	.667		91	36.50		127.50	157
0700	7-1/2"H x 6"W x 5-3/16"D		10	.800		106	44		150	184
0710	9-1/2"H x 6"W x 5-11/16"D		10	.800		108	44		152	187
0720	11-1/2"H x 8"W x 6-11/16"D		8	1		161	54.50		215.50	262
0730	13-1/2"H x 10"W x 7-3/16"D		7	1.143		193	62.50		255.50	310
0740	15-1/2"H x 12"W x 8-3/16"D		6	1.333		253	73		326	390
0750	17-1/2"H x 14"W x 8-11/16"D		5	1.600		300	87.50		387.50	465
0760	Screw cover with window, 6"H x 4"W x 5"D		12	.667		159	36.50		195.50	232
0770	8"H x 4"W x 5"D		11	.727		167	40		207	246
0780	8"H x 6"W x 5"D		11	.727		218	40		258	300
0790	10"H x 6"W x 6"D		10	.800		235	44		279	325
0800	12"H x 8"W x 7"D		8	1		315	54.50		369.50	430
0810	14"H x 10"W x 7"D		7	1.143		345	62.50		407.50	475
0820	16"H x 12"W x 8"D		6	1.333		430	73		503	590
0830	18"H x 14"W x 9"D		5	1.600		520	87.50		607.50	705
0840	Quick-release latch cover, 5-1/2"H x 4"W x 5"D		12	.667		116	36.50		152.50	185
0850	7-1/2"H x 4"W x 5"D		12	.667		124	36.50		160.50	194
0860	7-1/2"H x 6"W x 5-1/4"D		10	.800		134	44		178	216
0870	9-1/2"H x 6"W x 5-3/4"D		10	.800		147	44		191	230
0880	11-1/2"H x 8"W x 6-3/4"D		8	1		210	54.50		264.50	315

26 27 16.20 Cabinets and Enclosures		Crew	Daily Output	Labor-Hours	Unit	Material	2015 Bare Costs Labor	Equipment	Total	Total Incl O&P
0890	13-1/2"H x 10"W x 7-1/4"D	1 Elec	7	1.143	Ea.	263	62.50		325.50	385
0900	15-1/2"H x 12"W x 8-1/4"D		6	1.333		325	73		398	470
0910	17-1/2"H x 14"W x 8-3/4"D		5	1.600		390	87.50		477.50	565
0920	Pushbutton, 1 hole 5-1/2"H x 4"W x 4-15/16"D		12	.667		81	36.50		117.50	146
0930	2 hole 7-1/2"H x 4"W x 4-15/16"D		11	.727		92.50	40		132.50	164
0940	4 hole 7-1/2"H x 6"W x 5-3/16"D		10.50	.762		109	41.50		150.50	185
0950	6 hole 9-1/2"H x 6"W x 5-11/16"D		9	.889		134	48.50		182.50	223
0960	8 hole 11-1/2"H x 8"W x 6-11/16"D		8.50	.941		170	51.50		221.50	267
0970	12 hole 13-1/2"H x 10"W x 7-3/16"D		8	1		219	54.50		273.50	325
0980	20 hole 15-1/2"H x 12"W x 8-3/16"D		5	1.600		299	87.50		386.50	465
0990	30 hole 17-1/2"H x 14"W x 8-11/16"D	↓	4.50	1.778	↓	330	97		427	515
1450	Enclosures polyester NEMA 4X									
1460	Small, screw cover,									
1500	3-15/16"H x 3-15/16"W x 3-1/16"D	1 Elec	12	.667	Ea.	68.50	36.50		105	132
1510	5-3/16"H x 3-5/16"W x 3-1/16"D		12	.667		67.50	36.50		104	131
1520	5-7/8"H x 3-7/8"W x 4-3/16"D		12	.667		71	36.50		107.50	135
1530	5-7/8"H x 5-7/8"W x 4-3/16"D		12	.667		80	36.50		116.50	145
1540	7-5/8"H x 3-5/16"W x 3-1/16"D		12	.667		73.50	36.50		110	138
1550	10-3/16"H x 3-5/16"W x 3-1/16"D		10	.800		87.50	44		131.50	164
1560	Clear cover, 3-15/16"H x 3-15/16"W x 2-7/8"D		12	.667		77.50	36.50		114	142
1570	5-3/16"H x 3-5/16"W x 2-7/8"D		12	.667		82.50	36.50		119	147
1580	5-7/8"H x 3-7/8"W x 4"D		12	.667		102	36.50		138.50	170
1590	5-7/8"H x 5-7/8"W x 4"D		12	.667		126	36.50		162.50	196
1600	7-5/8"H x 3-5/16"W x 2-7/8"D		12	.667		92.50	36.50		129	159
1610	10-3/16"H x 3-5/16"W x 2-7/8"D		10	.800		116	44		160	196
1620	Pushbutton, 1 hole, 5-5/16"H x 3-5/16"W x 3-1/16"D		12	.667		60	36.50		96.50	123
1630	2 hole, 7-5/8"H x 3-5/16"W x 3-1/8"D		11	.727		67.50	40		107.50	136
1640	3 hole, 10-3/16"H x 3-5/16"W x 3-1/16"D		10.50	.762		80	41.50		121.50	153
8000	Wireway fiberglass, straight sect. screwcover, 12" L, 4" W x 4" D		40	.200		194	10.95		204.95	231
8010	6" W x 6" D		30	.267		258	14.60		272.60	305
8020	24" L, 4" W x 4" D		20	.400		236	22		258	293
8030	6" W x 6" D		15	.533		390	29		419	475
8040	36" L, 4" W x 4" D		13.30	.602		295	33		328	375
8050	6" W x 6" D		10	.800		460	44		504	575
8060	48" L, 4" W x 4" D		10	.800		355	44		399	460
8070	6" W x 6" D		7.50	1.067		580	58.50		638.50	730
8080	60" L, 4" W x 4" D		8	1		610	54.50		664.50	755
8090	6" W x 6" D		6	1.333		675	73		748	855
8100	Elbow, 90°, 4" W x 4" D		20	.400		199	22		221	253
8110	6" W x 6" D		18	.444		400	24.50		424.50	480
8120	Elbow, 45°, 4" W x 4" D		20	.400		189	22		211	242
8130	6" W x 6" D		18	.444		375	24.50		399.50	455
8140	Tee, 4" W x 4" D		16	.500		252	27.50		279.50	320
8150	6" W x 6" D		14	.571		460	31.50		491.50	555
8160	Cross, 4" W x 4" D		14	.571		360	31.50		391.50	445
8170	6" W x 6" D		12	.667		775	36.50		811.50	910
8180	Cut-off fitting, w/flange & adhesive, 4" W x 4" D		18	.444		141	24.50		165.50	193
8190	6" W x 6" D		16	.500		305	27.50		332.50	385
8200	Flexible ftng., hvy. neoprene coated nylon, 4" W x 4" D		20	.400		274	22		296	335
8210	6" W x 6" D		18	.444		405	24.50		429.50	485
8220	Closure plate, fiberglass, 4" W x 4" D		20	.400		55	22		77	94.50
8230	6" W x 6" D		18	.444		59.50	24.50		84	103
8240	Box connector, stainless steel type 304, 4" W x 4" D	↓	20	.400	↓	88.50	22		110.50	132

26 27 Low-Voltage Distribution Equipment

26 27 16 – Electrical Cabinets and Enclosures

26 27 16.20 Cabinets and Enclosures		Crew	Daily Output	Labor-Hours	Unit	Material	2015 Bare Costs Labor	Equipment	Total	Total Incl O&P
8250	6" W x 6" D	1 Elec	18	.444	Ea.	104	24.50		128.50	152
8260	Hanger, 4" W x 4" D		100	.080		23	4.38		27.38	32.50
8270	6" W x 6" D		80	.100		29	5.45		34.45	40.50
8280	Straight tube section fiberglass, 4"W x 4"D, 12" long		40	.200		168	10.95		178.95	202
8290	24" long		20	.400		202	22		224	256
8300	36" long		13.30	.602		252	33		285	330
8310	48" long		10	.800		245	44		289	335
8320	60" long		8	1		300	54.50		354.50	415
8330	120" long	↓	4	2	↓	460	109		569	680

26 27 19 – Multi-Outlet Assemblies

26 27 19.10 Wiring Duct		Crew	Daily Output	Labor-Hours	Unit	Material	Labor	Equipment	Total	Total Incl O&P
0010	**WIRING DUCT** Plastic									
1250	PVC, snap-in slots, adhesive backed									
1270	1-1/2"W x 2"H	2 Elec	120	.133	L.F.	4.86	7.30		12.16	16.65
1280	1-1/2"W x 3"H		120	.133		5.85	7.30		13.15	17.70
1290	1-1/2"W x 4"H		120	.133		6.90	7.30		14.20	18.90
1300	2"W x 1"H		120	.133		4.36	7.30		11.66	16.10
1310	2"W x 1-1/2"H		120	.133		4.73	7.30		12.03	16.50
1320	2"W x 2"H		120	.133		4.75	7.30		12.05	16.55
1340	2"W x 3"H		120	.133		6.20	7.30		13.50	18.10
1350	2"W x 4"H		120	.133		7.50	7.30		14.80	19.55
1360	2-1/2"W x 3"H		120	.133		6.65	7.30		13.95	18.65
1370	3"W x 1"H		110	.145		4.95	7.95		12.90	17.75
1390	3"W x 2"H		110	.145		5.90	7.95		13.85	18.80
1400	3"W x 3"H		110	.145		7.15	7.95		15.10	20
1410	3"W x 4"H		110	.145		8.95	7.95		16.90	22
1420	3"W x 5"H		110	.145		11.75	7.95		19.70	25.50
1430	4"W x 1-1/2"H		100	.160		5.90	8.75		14.65	20
1440	4"W x 2"H		100	.160		6.90	8.75		15.65	21
1450	4"W x 3"H		100	.160		7.80	8.75		16.55	22
1460	4"W x 4"H		100	.160		9.45	8.75		18.20	24
1470	4"W x 5"H		100	.160		13.30	8.75		22.05	28
1550	Cover, 1-1/2"W		200	.080		.99	4.38		5.37	7.90
1560	2"W		200	.080		1.22	4.38		5.60	8.15
1570	2-1/2"W		200	.080		1.53	4.38		5.91	8.50
1580	3"W		200	.080		1.86	4.38		6.24	8.85
1590	4"W	↓	200	.080	↓	2.24	4.38		6.62	9.25

26 27 23 – Indoor Service Poles

26 27 23.40 Surface Raceway		Crew	Daily Output	Labor-Hours	Unit	Material	Labor	Equipment	Total	Total Incl O&P
0010	**SURFACE RACEWAY**									
0090	Metal, straight section									
0100	No. 500	1 Elec	100	.080	L.F.	1.04	4.38		5.42	7.95
0110	No. 700		100	.080		1.17	4.38		5.55	8.10
0400	No. 1500, small pancake		90	.089		2.15	4.86		7.01	9.90
0600	No. 2000, base & cover, blank		90	.089		2.19	4.86		7.05	9.95
0610	Receptacle, 6" O.C.		40	.200		17.55	10.95		28.50	36.50
0620	12" O.C.		44	.182		11.40	9.95		21.35	28
0630	18" O.C.		46	.174		9.45	9.50		18.95	25
0650	30" O.C.		50	.160		4.94	8.75		13.69	19
0660	60" O.C.		50	.160		4.56	8.75		13.31	18.55
0670	No. 2400, base & cover, blank		80	.100		1.89	5.45		7.34	10.55
0680	Receptacle, 6" O.C.	↓	42	.190	↓	34.50	10.40		44.90	54

26 27 23.40 Surface Raceway		Crew	Daily Output	Labor-Hours	Unit	Material	2015 Bare Costs Labor	Equipment	Total	Total Incl O&P
0690	12" O.C.	1 Elec	53	.151	L.F.	23.50	8.25		31.75	38.50
0700	18" O.C.		55	.145		16.95	7.95		24.90	31
0710	24" O.C.		57	.140		9.50	7.70		17.20	22.50
0720	30" O.C.		59	.136		6.95	7.40		14.35	19.15
0730	60" O.C.		61	.131		5.20	7.15		12.35	16.85
0800	No. 3000, base & cover, blank		75	.107		4.18	5.85		10.03	13.65
0810	Receptacle, 6" O.C.		45	.178		37.50	9.70		47.20	56.50
0820	12" O.C.		62	.129		21	7.05		28.05	34.50
0830	18" O.C.		64	.125		17.15	6.85		24	29.50
0840	24" O.C.		66	.121		12.85	6.65		19.50	24.50
0850	30" O.C.		68	.118		11.55	6.45		18	22.50
0860	60" O.C.		70	.114		8.45	6.25		14.70	19
1000	No. 4000, base & cover, blank		65	.123		6.80	6.75		13.55	17.90
1010	Receptacle, 6" O.C.		41	.195		50.50	10.65		61.15	72
1020	12" O.C.		52	.154		31.50	8.40		39.90	47.50
1030	18" O.C.		54	.148		25.50	8.10		33.60	40.50
1040	24" O.C.		56	.143		21	7.80		28.80	35
1050	30" O.C.		58	.138		19	7.55		26.55	32.50
1060	60" O.C.		60	.133		14.60	7.30		21.90	27.50
1200	No. 6000, base & cover, blank		50	.160		11.40	8.75		20.15	26
1210	Receptacle, 6" O.C.		30	.267		61.50	14.60		76.10	90.50
1220	12" O.C.		37	.216		41	11.85		52.85	63.50
1230	18" O.C.		39	.205		34.50	11.20		45.70	55.50
1240	24" O.C.		41	.195		28.50	10.65		39.15	47.50
1250	30" O.C.		43	.186		27	10.20		37.20	45.50
1260	60" O.C.		45	.178	↓	21	9.70		30.70	38
2400	Fittings, elbows, No. 500		40	.200	Ea.	1.90	10.95		12.85	19.05
2800	Elbow cover, No. 2000		40	.200		3.57	10.95		14.52	21
2880	Tee, No. 500		42	.190		3.66	10.40		14.06	20
2900	No. 2000		27	.296		11.85	16.20		28.05	38
3000	Switch box, No. 500		16	.500		11	27.50		38.50	54.50
3400	Telephone outlet, No. 1500		16	.500		14.10	27.50		41.60	58
3600	Junction box, No. 1500	↓	16	.500	↓	9.65	27.50		37.15	53
3800	Plugmold wired sections, No. 2000									
4000	1 circuit, 6 outlets, 3 ft. long	1 Elec	8	1	Ea.	36	54.50		90.50	124
4100	2 circuits, 8 outlets, 6 ft. long		5.30	1.509		53	82.50		135.50	187
4110	Tele-power pole, alum, w/2 recept, 10'		4	2		205	109		314	395
4120	12'		3.85	2.078		196	114		310	390
4130	15'		3.70	2.162		310	118		428	525
4140	Steel, w/2 recept, 10'		4	2		130	109		239	310
4150	One phone fitting, 10'		4	2		139	109		248	320
4160	Alum, 4 outlets, 10'	↓	3.70	2.162	↓	262	118		380	470
4300	Overhead distribution systems, 125 volt									
4800	No. 2000, entrance end fitting	1 Elec	20	.400	Ea.	5.50	22		27.50	40
5000	Blank end fitting		40	.200		2.25	10.95		13.20	19.45
5200	Supporting clip		40	.200		1.33	10.95		12.28	18.40
5800	No. 3000, entrance end fitting		20	.400		9.20	22		31.20	44
6000	Blank end fitting		40	.200		2.62	10.95		13.57	19.85
6020	Internal elbow		20	.400		12.10	22		34.10	47.50
6030	External elbow		20	.400		17.10	22		39.10	53
6040	Device bracket		53	.151		4.10	8.25		12.35	17.30
6400	Hanger clamp		32	.250	↓	6.25	13.70		19.95	28
7000	No. 4000 Base	↓	90	.089	L.F.	4.40	4.86		9.26	12.40

26 27 23 – Indoor Service Poles

26 27 23.40 Surface Raceway		Crew	Daily Output	Labor-Hours	Unit	Material	2015 Bare Costs Labor	Equipment	Total	Total Incl O&P
7200	Divider	1 Elec	100	.080	L.F.	.85	4.38		5.23	7.75
7400	Entrance end fitting		16	.500	Ea.	22.50	27.50		50	67
7600	Blank end fitting		40	.200		6.20	10.95		17.15	24
7610	Recpt. & tele. cover		53	.151		10.55	8.25		18.80	24.50
7620	External elbow		16	.500		34.50	27.50		62	80.50
7630	Coupling		53	.151		5.25	8.25		13.50	18.60
7640	Divider clip & coupling		80	.100		1.04	5.45		6.49	9.60
7650	Panel connector		16	.500		22	27.50		49.50	66.50
7800	Take off connector		16	.500		72.50	27.50		100	123
8000	No. 6000, take off connector		16	.500		86.50	27.50		114	138
8100	Take off fitting		16	.500		64	27.50		91.50	113
8200	Hanger clamp		32	.250		15.55	13.70		29.25	38
8230	Coupling					8.50			8.50	9.35
8240	One gang device plate	1 Elec	53	.151		8.30	8.25		16.55	22
8250	Two gang device plate		40	.200		10.15	10.95		21.10	28
8260	Blank end fitting		40	.200		8.80	10.95		19.75	26.50
8270	Combination elbow		14	.571		36	31.50		67.50	88
8300	Panel connector		16	.500		15.75	27.50		43.25	60
8500	Chan-L-Wire system installed in 1-5/8" x 1-5/8" strut. Strut									
8600	not incl., 30 amp, 4 wire, 3 phase	1 Elec	200	.040	L.F.	4.95	2.19		7.14	8.85
8700	Junction box		8	1	Ea.	33	54.50		87.50	121
8800	Insulating end cap		40	.200		8.80	10.95		19.75	26.50
8900	Strut splice plate		40	.200		11.65	10.95		22.60	30
9000	Tap		40	.200		23	10.95		33.95	42.50
9100	Fixture hanger		60	.133		10	7.30		17.30	22.50
9200	Pulling tool					89			89	97.50
9300	Non-metallic, straight section									
9310	7/16" x 7/8", base & cover, blank	1 Elec	160	.050	L.F.	1.72	2.74		4.46	6.15
9320	Base & cover w/adhesive		160	.050		1.42	2.74		4.16	5.80
9340	7/16" x 1-5/16", base & cover, blank		145	.055		1.88	3.02		4.90	6.75
9350	Base & cover w/adhesive		145	.055		2.17	3.02		5.19	7.05
9370	11/16" x 2-1/4", base & cover, blank		130	.062		2.66	3.37		6.03	8.15
9380	Base & cover w/adhesive		130	.062		3.01	3.37		6.38	8.50
9385	1-11/16" x 5-1/4", two compartment base & cover w/screws		80	.100		7.50	5.45		12.95	16.70
9400	Fittings, elbows, 7/16" x 7/8"		50	.160	Ea.	1.88	8.75		10.63	15.60
9410	7/16" x 1-5/16"		45	.178		1.95	9.70		11.65	17.20
9420	11/16" x 2-1/4"		40	.200		2.11	10.95		13.06	19.25
9425	1-11/16" x 5-1/4"		28	.286		11.05	15.65		26.70	36
9430	Tees, 7/16" x 7/8"		35	.229		2.42	12.50		14.92	22
9440	7/16" x 1-5/16"		32	.250		2.49	13.70		16.19	23.50
9450	11/16" x 2-1/4"		30	.267		2.56	14.60		17.16	25.50
9455	1-11/16" x 5-1/4"		24	.333		17.75	18.25		36	47.50
9460	Cover clip, 7/16" x 7/8"		80	.100		.49	5.45		5.94	9
9470	7/16" x 1-5/16"		72	.111		.44	6.10		6.54	9.90
9480	11/16" x 2-1/4"		64	.125		.75	6.85		7.60	11.45
9484	1-11/16" x 5-1/4"		42	.190		2.61	10.40		13.01	19
9486	Wire clip, 1-11/16" x 5-1/4"		68	.118		.45	6.45		6.90	10.45
9490	Blank end, 7/16" x 7/8"		50	.160		.70	8.75		9.45	14.30
9500	7/16" x 1-5/16"		45	.178		.78	9.70		10.48	15.90
9510	11/16" x 2-1/4"		40	.200		1.18	10.95		12.13	18.25
9515	1-11/16" x 5-1/4"		38	.211		5.45	11.50		16.95	24
9520	Round fixture box, 5.5" dia x 1"		25	.320		11.10	17.50		28.60	39
9530	Device box, 1 gang		30	.267		5	14.60		19.60	28

26 27 Low-Voltage Distribution Equipment

26 27 23 – Indoor Service Poles

26 27 23.40 Surface Raceway		Crew	Daily Output	Labor-Hours	Unit	Material	2015 Bare Costs Labor	Equipment	Total	Total Incl O&P
9540	2 gang	1 Elec	25	.320	Ea.	7.30	17.50		24.80	35
9990	Minimum labor/equipment charge		5	1.600	Job		87.50		87.50	136

26 27 26 – Wiring Devices

26 27 26.10 Low Voltage Switching

	26 27 26.10 Low Voltage Switching	Crew	Daily Output	Labor-Hours	Unit	Material	Labor	Equipment	Total	Total Incl O&P
0010	**LOW VOLTAGE SWITCHING**									
3600	Relays, 120 V or 277 V standard	1 Elec	12	.667	Ea.	41.50	36.50		78	103
3800	Flush switch, standard		40	.200		11.50	10.95		22.45	29.50
4000	Interchangeable		40	.200		15.05	10.95		26	33.50
4100	Surface switch, standard		40	.200		8.15	10.95		19.10	26
4200	Transformer 115 V to 25 V		12	.667		130	36.50		166.50	200
4400	Master control, 12 circuit, manual		4	2		126	109		235	310
4500	25 circuit, motorized		4	2		140	109		249	325
4600	Rectifier, silicon		12	.667		45.50	36.50		82	107
4800	Switchplates, 1 gang, 1, 2 or 3 switch, plastic		80	.100		5	5.45		10.45	13.95
5000	Stainless steel		80	.100		11.35	5.45		16.80	21
5400	2 gang, 3 switch, stainless steel		53	.151		23	8.25		31.25	38
5500	4 switch, plastic		53	.151		10.35	8.25		18.60	24
5600	2 gang, 4 switch, stainless steel		53	.151		21.50	8.25		29.75	36.50
5700	6 switch, stainless steel		53	.151		43	8.25		51.25	60.50
5800	3 gang, 9 switch, stainless steel		32	.250		64.50	13.70		78.20	92
5900	Receptacle, triple, 1 return, 1 feed		26	.308		42.50	16.85		59.35	73
6000	2 feed		20	.400		42.50	22		64.50	81
6100	Relay gang boxes, flush or surface, 6 gang		5.30	1.509		89.50	82.50		172	227
6200	12 gang		4.70	1.702		101	93		194	255
6400	18 gang		4	2		112	109		221	292
6500	Frame, to hold up to 6 relays		12	.667		83	36.50		119.50	148
7200	Control wire, 2 conductor		6.30	1.270	C.L.F.	29.50	69.50		99	141
7400	3 conductor		5	1.600		41	87.50		128.50	181
7600	19 conductor		2.50	3.200		320	175		495	620
7800	26 conductor		2	4		430	219		649	815
8000	Weatherproof, 3 conductor		5	1.600		86	87.50		173.50	231

26 27 26.20 Wiring Devices Elements

	26 27 26.20 Wiring Devices Elements	Crew	Daily Output	Labor-Hours	Unit	Material	Labor	Equipment	Total	Total Incl O&P
0010	**WIRING DEVICES ELEMENTS** R262726-90									
0200	Toggle switch, quiet type, single pole, 15 amp	1 Elec	40	.200	Ea.	6.55	10.95		17.50	24
0500	20 amp		27	.296		9.85	16.20		26.05	36
0510	30 amp		23	.348		19.50	19.05		38.55	51
0530	Lock handle, 20 amp		27	.296		23.50	16.20		39.70	50.50
0540	Security key, 20 amp		26	.308		116	16.85		132.85	153
0550	Rocker, 15 amp		40	.200		4.10	10.95		15.05	21.50
0560	20 amp		27	.296		9.50	16.20		25.70	35.50
0600	3 way, 15 amp		23	.348		5.05	19.05		24.10	35
0800	20 amp		18	.444		9.45	24.50		33.95	48
0810	30 amp		9	.889		24	48.50		72.50	102
0830	Lock handle, 20 amp		18	.444		26	24.50		50.50	66.50
0840	Security key, 20 amp		17	.471		110	25.50		135.50	161
0850	Rocker, 15 amp		23	.348		5.75	19.05		24.80	36
0860	20 amp		18	.444		13.65	24.50		38.15	52.50
0900	4 way, 15 amp		15	.533		9.10	29		38.10	55
1000	20 amp		11	.727		42.50	40		82.50	109
1020	Lock handle, 20 amp		11	.727		46.50	40		86.50	113
1030	Rocker, 15 amp		15	.533		19.15	29		48.15	66
1040	20 amp		11	.727		39	40		79	105

26 27 Low-Voltage Distribution Equipment

26 27 26 – Wiring Devices

26 27 26.20 Wiring Devices Elements			Crew	Daily Output	Labor-Hours	Unit	Material	Labor	Equipment	Total	Total Incl O&P
							2015 Bare Costs				
1100	Toggle switch, quiet type, double pole, 15 amp		1 Elec	15	.533	Ea.	12.30	29		41.30	58.50
1200	20 amp			11	.727		17.10	40		57.10	80.50
1210	30 amp			9	.889		24.50	48.50		73	103
1230	Lock handle, 20 amp			11	.727		18.45	40		58.45	82
1250	Security key, 20 amp			10	.800		113	44		157	192
1420	Toggle switch quiet type, 1 pole, 2 throw center off, 15 amp			23	.348		58	19.05		77.05	93
1440	20 amp			18	.444		51	24.50		75.50	93.50
1460	2 pole, 2 throw center off, lock handle, 20 amp			11	.727		60	40		100	128
1480	1 pole, momentary contact, 15 amp			23	.348		17.65	19.05		36.70	49
1500	20 amp			18	.444		29.50	24.50		54	70
1520	Momentary contact, lock handle, 20 amp			18	.444		31	24.50		55.50	71.50
1650	Dimmer switch, 120 volt, incandescent, 600 watt, 1 pole	G		16	.500		21	27.50		48.50	65.50
1700	600 watt, 3 way	G		12	.667		15.05	36.50		51.55	73
1750	1000 watt, 1 pole	G		16	.500		38	27.50		65.50	84
1800	1000 watt, 3 way	G		12	.667		73	36.50		109.50	137
2000	1500 watt, 1 pole	G		11	.727		84	40		124	154
2100	2000 watt, 1 pole	G		8	1		129	54.50		183.50	227
2110	Fluorescent, 600 watt	G		15	.533		109	29		138	165
2120	1000 watt	G		15	.533		133	29		162	191
2130	1500 watt	G		10	.800		246	44		290	340
2160	Explosionproof, toggle switch, wall, single pole 20 amp			5.30	1.509		187	82.50		269.50	335
2180	Receptacle, single outlet, 20 amp			5.30	1.509		330	82.50		412.50	490
2190	30 amp			4	2		685	109		794	920
2290	60 amp			2.50	3.200		775	175		950	1,125
2360	Plug, 20 amp			16	.500		173	27.50		200.50	233
2370	30 amp			12	.667		294	36.50		330.50	380
2380	60 amp			8	1		380	54.50		434.50	505
2410	Furnace, thermal cutoff switch with plate			26	.308		15.75	16.85		32.60	43.50
2460	Receptacle, duplex, 120 volt, grounded, 15 amp			40	.200		1.26	10.95		12.21	18.35
2470	20 amp			27	.296		7.90	16.20		24.10	33.50
2480	Ground fault interrupting, 15 amp			27	.296		32	16.20		48.20	60
2482	20 amp			27	.296		40	16.20		56.20	69
2490	Dryer, 30 amp			15	.533		4.39	29		33.39	50
2500	Range, 50 amp			11	.727		12.15	40		52.15	75
2540	Isolated ground receptacle, duplex, 20 amp			27	.296		21	16.20		37.20	48
2542	Quad, 20 amp			20	.400		32	22		54	69
2550	Simplex, 20 amp			27	.296		23	16.20		39.20	50
2560	Simplex, 30 amp			15	.533		27.50	29		56.50	75.50
2570	Cable reel w/receptacle 50′ w/3#12, 120 V, 20 A		2 Elec	2.67	5.999		920	330		1,250	1,500
2600	Wall plates, stainless steel, 1 gang		1 Elec	80	.100		2.56	5.45		8.01	11.25
2800	2 gang			53	.151		4.33	8.25		12.58	17.55
3000	3 gang			32	.250		8.45	13.70		22.15	30.50
3100	4 gang			27	.296		11	16.20		27.20	37
3110	Brown plastic, 1 gang			80	.100		.38	5.45		5.83	8.85
3120	2 gang			53	.151		.75	8.25		9	13.65
3130	3 gang			32	.250		1.11	13.70		14.81	22
3140	4 gang			27	.296		2.86	16.20		19.06	28
3150	Brushed brass, 1 gang			80	.100		4.26	5.45		9.71	13.15
3160	Anodized aluminum, 1 gang			80	.100		3.01	5.45		8.46	11.75
3170	Switch cover, weatherproof, 1 gang			60	.133		7	7.30		14.30	19
3180	Vandal proof lock, 1 gang			60	.133		14.05	7.30		21.35	27
3200	Lampholder, keyless			26	.308		12.65	16.85		29.50	40
3400	Pullchain with receptacle			22	.364		20.50	19.90		40.40	53.50

26 27 26.20 Wiring Devices Elements		Crew	Daily Output	Labor-Hours	Unit	Material	2015 Bare Costs Labor	Equipment	Total	Total Incl O&P
3500	Pilot light, neon with jewel	1 Elec	27	.296	Ea.	9.55	16.20		25.75	35.50
3600	Receptacle, 20 amp, 250 volt, NEMA 6		27	.296		21.50	16.20		37.70	48.50
3620	277 volt NEMA 7		27	.296		15.95	16.20		32.15	42.50
3640	125/250 volt NEMA 10		27	.296		17.95	16.20		34.15	45
3680	125/250 volt NEMA 14		25	.320		23	17.50		40.50	52
3700	3 pole, 250 volt NEMA 15		25	.320		23	17.50		40.50	52.50
3720	120/208 volt NEMA 18		25	.320		24	17.50		41.50	53
3740	30 amp, 125 volt NEMA 5		15	.533		17.80	29		46.80	64.50
3760	250 volt NEMA 6		15	.533		20.50	29		49.50	67.50
3780	277 volt NEMA 7		15	.533		26.50	29		55.50	74.50
3820	125/250 volt NEMA 14		14	.571		46.50	31.50		78	100
3840	3 pole, 250 volt NEMA 15		14	.571		46	31.50		77.50	99
3880	50 amp, 125 volt NEMA 5		11	.727		22.50	40		62.50	86.50
3900	250 volt NEMA 6		11	.727		23	40		63	87
3920	277 volt NEMA 7		11	.727		26.50	40		66.50	90.50
3960	125/250 volt NEMA 14		10	.800		61	44		105	135
3980	3 pole 250 volt NEMA 15		10	.800		62	44		106	136
4020	60 amp, 125/250 volt, NEMA 14		8	1		70.50	54.50		125	162
4040	3 pole, 250 volt NEMA 15		8	1		74.50	54.50		129	167
4060	120/208 volt NEMA 18		8	1		102	54.50		156.50	197
4100	Receptacle locking, 20 amp, 125 volt NEMA L5		27	.296		16.60	16.20		32.80	43.50
4120	250 volt NEMA L6		27	.296		16.50	16.20		32.70	43
4140	277 volt NEMA L7		27	.296		17.15	16.20		33.35	44
4150	3 pole, 250 volt, NEMA L11		27	.296		23	16.20		39.20	50.50
4160	20 amp, 480 volt NEMA L8		27	.296		19.35	16.20		35.55	46.50
4180	600 volt NEMA L9		27	.296		29	16.20		45.20	57
4200	125/250 volt NEMA L10		27	.296		23	16.20		39.20	50.50
4230	125/250 volt NEMA L14		25	.320		27	17.50		44.50	56.50
4280	250 volt NEMA L15		25	.320		29	17.50		46.50	59
4300	480 volt NEMA L16		25	.320		23.50	17.50		41	52.50
4320	3 phase, 120/208 volt NEMA L18		25	.320		27	17.50		44.50	57
4340	277/480 volt NEMA L19		25	.320		28	17.50		45.50	58
4360	347/600 volt NEMA L20		25	.320		27.50	17.50		45	57.50
4380	120/208 volt NEMA L21		23	.348		29.50	19.05		48.55	62
4400	277/480 volt NEMA L22		23	.348		30	19.05		49.05	62.50
4420	347/600 volt NEMA L23		23	.348		32	19.05		51.05	64.50
4440	30 amp, 125 volt NEMA L5		15	.533		23	29		52	70.50
4460	250 volt NEMA L6		15	.533		25.50	29		54.50	73
4480	277 volt NEMA L7		15	.533		25.50	29		54.50	73.50
4500	480 volt NEMA L8		15	.533		26.50	29		55.50	74.50
4520	600 volt NEMA L9		15	.533		27.50	29		56.50	75
4540	125/250 volt NEMA L10		15	.533		30	29		59	78
4560	3 phase, 250 volt NEMA L11		15	.533		27	29		56	75
4620	125/250 volt NEMA L14		14	.571		36.50	31.50		68	88.50
4640	250 volt NEMA L15		14	.571		37	31.50		68.50	89.50
4660	480 volt NEMA L16		14	.571		38.50	31.50		70	90.50
4680	600 volt NEMA L17		14	.571		38	31.50		69.50	90
4700	120/208 volt NEMA L18		14	.571		41.50	31.50		73	94
4720	277/480 volt NEMA L19		14	.571		42	31.50		73.50	94.50
4740	347/600 volt NEMA L20		14	.571		43.50	31.50		75	96
4760	120/208 volt NEMA L21		13	.615		40	33.50		73.50	96
4780	277/480 volt NEMA L22		13	.615		40	33.50		73.50	96
4800	347/600 volt NEMA L23		13	.615		53	33.50		86.50	110

26 27 26.20 Wiring Devices Elements		Crew	Daily Output	Labor-Hours	Unit	2015 Bare Costs Material	Labor	Equipment	Total	Total Incl O&P
4840	Receptacle, corrosion resistant, 15 or 20 amp, 125 volt NEMA L5	1 Elec	27	.296	Ea.	25.50	16.20		41.70	53
4860	250 volt NEMA L6		27	.296		21.50	16.20		37.70	48.50
4900	Receptacle, cover plate, phenolic plastic, NEMA 5 & 6		80	.100		.61	5.45		6.06	9.10
4910	NEMA 7-23		80	.100		.75	5.45		6.20	9.30
4920	Stainless steel, NEMA 5 & 6		80	.100		2.71	5.45		8.16	11.45
4930	NEMA 7-23		80	.100		2.59	5.45		8.04	11.30
4940	Brushed brass NEMA 5 & 6		80	.100		5.40	5.45		10.85	14.40
4950	NEMA 7-23		80	.100		5.85	5.45		11.30	14.90
4960	Anodized aluminum, NEMA 5 & 6		80	.100		3.51	5.45		8.96	12.30
4970	NEMA 7-23		80	.100		5.80	5.45		11.25	14.85
4980	Weatherproof NEMA 7-23		60	.133		33	7.30		40.30	47.50
5100	Plug, 20 amp, 250 volt, NEMA 6		30	.267		15.65	14.60		30.25	39.50
5110	277 volt NEMA 7		30	.267		19.20	14.60		33.80	43.50
5120	3 pole, 120/250 volt, NEMA 10		26	.308		18.15	16.85		35	46
5130	125/250 volt NEMA 14		26	.308		36.50	16.85		53.35	66.50
5140	250 volt NEMA 15		26	.308		41	16.85		57.85	71
5150	120/208 volt NEMA 8		26	.308		43.50	16.85		60.35	74
5160	30 amp, 125 volt NEMA 5		13	.615		55	33.50		88.50	113
5170	250 volt NEMA 6		13	.615		52	33.50		85.50	109
5180	277 volt NEMA 7		13	.615		60.50	33.50		94	119
5190	125/250 volt NEMA 14		13	.615		42.50	33.50		76	98.50
5200	3 pole, 250 volt NEMA 15		12	.667		44	36.50		80.50	105
5210	50 amp, 125 volt NEMA 5		9	.889		61.50	48.50		110	144
5220	250 volt NEMA 6		9	.889		64.50	48.50		113	147
5230	277 volt NEMA 7		9	.889		69	48.50		117.50	152
5240	125/250 volt NEMA 14		9	.889		54	48.50		102.50	135
5250	3 pole, 250 volt NEMA 15		8	1		55	54.50		109.50	145
5260	60 amp, 125/250 volt NEMA 14		7	1.143		61.50	62.50		124	165
5270	3 pole, 250 volt NEMA 15		7	1.143		65	62.50		127.50	169
5280	120/208 volt NEMA 18		7	1.143		82.50	62.50		145	188
5300	Plug angle, 20 amp, 250 volt NEMA 6		30	.267		24	14.60		38.60	49
5310	30 amp, 125 volt NEMA 5		13	.615		43	33.50		76.50	99.50
5320	250 volt NEMA 6		13	.615		44.50	33.50		78	101
5330	277 volt NEMA 7		13	.615		51	33.50		84.50	108
5340	125/250 volt NEMA 14		13	.615		47	33.50		80.50	104
5350	3 pole, 250 volt NEMA 15		12	.667		48.50	36.50		85	110
5360	50 amp, 125 volt NEMA 5		9	.889		45	48.50		93.50	125
5370	250 volt NEMA 6		9	.889		46	48.50		94.50	127
5380	277 volt NEMA 7		9	.889		52	48.50		100.50	133
5390	125/250 volt NEMA 14		9	.889		58.50	48.50		107	140
5400	3 pole, 250 volt NEMA 15		8	1		60.50	54.50		115	151
5410	60 amp, 125/250 volt NEMA 14		7	1.143		68	62.50		130.50	172
5420	3 pole, 250 volt NEMA 15		7	1.143		71	62.50		133.50	175
5430	120/208 volt NEMA 18		7	1.143		71.50	62.50		134	176
5500	Plug, locking, 20 amp, 125 volt NEMA L5		30	.267		13.20	14.60		27.80	37
5510	250 volt NEMA L6		30	.267		13.15	14.60		27.75	37
5520	277 volt NEMA L7		30	.267		13.45	14.60		28.05	37.50
5530	480 volt NEMA L8		30	.267		14.95	14.60		29.55	39
5540	600 volt NEMA L9		30	.267		15	14.60		29.60	39
5550	3 pole, 125/250 volt NEMA L10		26	.308		22	16.85		38.85	50
5560	250 volt NEMA L11		26	.308		22	16.85		38.85	50
5570	480 volt NEMA L12		26	.308		25.50	16.85		42.35	54
5580	125/250 volt NEMA L14		26	.308		19.90	16.85		36.75	48

26 27 26.20 Wiring Devices Elements		Crew	Daily Output	Labor-Hours	Unit	Material	2015 Bare Costs Labor	Equipment	Total	Total Incl O&P
5590	250 volt NEMA L15	1 Elec	26	.308	Ea.	21	16.85		37.85	49
5600	480 volt NEMA L16		26	.308		22.50	16.85		39.35	50.50
5610	4 pole, 120/208 volt NEMA L18		24	.333		25	18.25		43.25	55.50
5620	277/480 volt NEMA L19		24	.333		26	18.25		44.25	56.50
5630	347/600 volt NEMA L20		24	.333		29	18.25		47.25	59.50
5640	120/208 volt NEMA L21		24	.333		25.50	18.25		43.75	56
5650	277/480 volt NEMA L22		24	.333		27	18.25		45.25	57.50
5660	347/600 volt NEMA L23		24	.333		28	18.25		46.25	58.50
5670	30 amp, 125 volt NEMA L5		13	.615		20	33.50		53.50	74
5680	250 volt NEMA L6		13	.615		21	33.50		54.50	75
5690	277 volt NEMA L7		13	.615		20	33.50		53.50	74
5700	480 volt NEMA L8		13	.615		22	33.50		55.50	76
5710	600 volt NEMA L9		13	.615		21	33.50		54.50	75
5720	3 pole, 125/250 volt NEMA L10		11	.727		24	40		64	87.50
5730	250 volt NEMA L11		11	.727		24	40		64	87.50
5760	125/250 volt NEMA L14		11	.727		27	40		67	91.50
5770	250 volt NEMA L15		11	.727		27	40		67	91.50
5780	480 volt NEMA L16		11	.727		28.50	40		68.50	92.50
5790	600 volt NEMA L17		11	.727		29.50	40		69.50	94
5800	4 pole, 120/208 volt NEMA L18		10	.800		31	44		75	103
5810	120/208 volt NEMA L19		10	.800		31.50	44		75.50	103
5820	347/600 volt NEMA L20		10	.800		32.50	44		76.50	104
5830	120/208 volt NEMA L21		10	.800		30.50	44		74.50	102
5840	277/480 volt NEMA L22		10	.800		31	44		75	102
5850	347/600 volt NEMA L23		10	.800		35	44		79	107
6000	Connector, 20 amp, 250 volt NEMA 6		30	.267		24	14.60		38.60	49
6010	277 volt NEMA 7		30	.267		30.50	14.60		45.10	56.50
6020	3 pole, 120/250 volt NEMA 10		26	.308		34.50	16.85		51.35	64
6030	125/250 volt NEMA 14		26	.308		34.50	16.85		51.35	64
6040	250 volt NEMA 15		26	.308		35	16.85		51.85	64.50
6050	120/208 volt NEMA 18		26	.308		38	16.85		54.85	68
6060	30 amp, 125 volt NEMA 5		13	.615		60	33.50		93.50	118
6070	250 volt NEMA 6		13	.615		60	33.50		93.50	118
6080	277 volt NEMA 7		13	.615		60	33.50		93.50	118
6110	50 amp, 125 volt NEMA 5		9	.889		83	48.50		131.50	167
6120	250 volt NEMA 6		9	.889		83	48.50		131.50	167
6130	277 volt NEMA 7		9	.889		83	48.50		131.50	167
6200	Connector, locking, 20 amp, 125 volt NEMA L5		30	.267		20	14.60		34.60	44.50
6210	250 volt NEMA L6		30	.267		20.50	14.60		35.10	45
6220	277 volt NEMA L7		30	.267		20.50	14.60		35.10	45
6230	480 volt NEMA L8		30	.267		24	14.60		38.60	48.50
6240	600 volt NEMA L9		30	.267		24.50	14.60		39.10	49
6250	3 pole, 125/250 volt NEMA L10		26	.308		33.50	16.85		50.35	63
6260	250 volt NEMA L11		26	.308		33.50	16.85		50.35	63
6280	125/250 volt NEMA L14		26	.308		27.50	16.85		44.35	56.50
6290	250 volt NEMA L15		26	.308		28	16.85		44.85	56.50
6300	480 volt NEMA L16		26	.308		29.50	16.85		46.35	58.50
6310	4 pole, 120/208 volt NEMA L18		24	.333		35.50	18.25		53.75	67
6320	277/480 volt NEMA L19		24	.333		40.50	18.25		58.75	72.50
6330	347/600 volt NEMA L20		24	.333		40	18.25		58.25	72
6340	120/208 volt NEMA L21		24	.333		43	18.25		61.25	75.50
6350	277/480 volt NEMA L22		24	.333		45.50	18.25		63.75	78
6360	347/600 volt NEMA L23	▼	24	.333	▼	53	18.25		71.25	86.50

26 27 Low-Voltage Distribution Equipment

26 27 26 – Wiring Devices

26 27 26.20 Wiring Devices Elements		Crew	Daily Output	Labor-Hours	Unit	Material	2015 Bare Costs Labor	Equipment	Total	Total Incl O&P
6370	30 amp, 125 volt NEMA L5	1 Elec	13	.615	Ea.	41	33.50		74.50	97
6380	250 volt NEMA L6		13	.615		41.50	33.50		75	97.50
6390	277 volt NEMA L7		13	.615		41.50	33.50		75	98
6400	480 volt NEMA L8		13	.615		44	33.50		77.50	101
6410	600 volt NEMA L9		13	.615		43	33.50		76.50	99
6420	3 pole, 125/250 volt NEMA L10		11	.727		64.50	40		104.50	133
6430	250 volt NEMA L11		11	.727		64.50	40		104.50	133
6460	125/250 volt NEMA L14		11	.727		56	40		96	123
6470	250 volt NEMA L15		11	.727		56.50	40		96.50	124
6480	480 volt NEMA L16		11	.727		59.50	40		99.50	127
6490	600 volt NEMA L17		11	.727		61.50	40		101.50	129
6500	4 pole, 120/208 volt NEMA L18		10	.800		63	44		107	138
6510	120/208 volt NEMA L19		10	.800		65.50	44		109.50	140
6520	347/600 volt NEMA L20		10	.800		67.50	44		111.50	143
6530	120/208 volt NEMA L21		10	.800		57	44		101	131
6540	277/480 volt NEMA L22		10	.800		58	44		102	132
6550	347/600 volt NEMA L23		10	.800		66	44		110	141
7000	Receptacle computer, 250 volt, 15 amp, 3 pole 4 wire		8	1		82.50	54.50		137	176
7010	20 amp, 2 pole 3 wire		8	1		84	54.50		138.50	177
7020	30 amp, 2 pole 3 wire		6.50	1.231		137	67.50		204.50	254
7030	30 amp, 3 pole 4 wire		6.50	1.231		148	67.50		215.50	266
7040	60 amp, 3 pole 4 wire		4.50	1.778		249	97		346	425
7050	100 amp, 3 pole 4 wire		3	2.667		310	146		456	565
7100	Connector computer, 250 volt, 15 amp, 3 pole 4 wire		27	.296		130	16.20		146.20	168
7110	20 amp, 2 pole 3 wire		27	.296		118	16.20		134.20	154
7120	30 amp, 2 pole 3 wire		15	.533		181	29		210	244
7130	30 amp, 3 pole 4 wire		15	.533		185	29		214	249
7140	60 amp, 3 pole 4 wire		8	1		310	54.50		364.50	425
7150	100 amp, 3 pole 4 wire		4	2		425	109		534	640
7200	Plug, computer, 250 volt, 15 amp, 3 pole 4 wire		27	.296		116	16.20		132.20	153
7210	20 amp, 2 pole, 3 wire		27	.296		109	16.20		125.20	145
7220	30 amp, 2 pole, 3 wire		15	.533		190	29		219	254
7230	30 amp, 3 pole, 4 wire		15	.533		184	29		213	247
7240	60 amp, 3 pole, 4 wire		8	1		276	54.50		330.50	390
7250	100 amp, 3 pole, 4 wire		4	2		345	109		454	550
7300	Connector adapter to flexible conduit, 1/2"		60	.133		2.90	7.30		10.20	14.50
7310	3/4"		50	.160		4.24	8.75		12.99	18.20
7320	1-1/4"		30	.267		12.70	14.60		27.30	36.50
7330	1-1/2"		23	.348		18.15	19.05		37.20	49.50
9000	Minimum labor/equipment charge		4	2	Job		109		109	169

26 27 73 – Door Chimes

26 27 73.10 Doorbell System		Crew	Daily Output	Labor-Hours	Unit	Material	Labor	Equipment	Total	Total Incl O&P
0010	**DOORBELL SYSTEM**, incl. transformer, button & signal									
0100	6" bell	1 Elec	4	2	Ea.	129	109		238	310
0200	Buzzer		4	2		106	109		215	286
1000	Door chimes, 2 notes		16	.500		22.50	27.50		50	67.50
1020	with ambient light		12	.667		109	36.50		145.50	177
1100	Tube type, 3 tube system		12	.667		187	36.50		223.50	263
1180	4 tube system		10	.800		385	44		429	490
1900	For transformer & button, add		5	1.600		15.85	87.50		103.35	153
3000	For push button only		24	.333		2.58	18.25		20.83	31
3200	Bell transformer		16	.500		18.05	27.50		45.55	62.50

26 27 Low-Voltage Distribution Equipment

26 27 73 – Door Chimes

26 27 73.10 Doorbell System		Crew	Daily Output	Labor-Hours	Unit	Material	2015 Bare Costs Labor	Equipment	Total	Total Incl O&P
9000	Minimum labor/equipment charge	1 Elec	4	2	Job		109		109	169

26 28 Low-Voltage Circuit Protective Devices

26 28 13 – Fuses

26 28 13.10 Fuse Elements		Crew	Daily Output	Labor-Hours	Unit	Material	2015 Bare Costs Labor	Equipment	Total	Total Incl O&P
0010	**FUSE ELEMENTS**									
0020	Cartridge, nonrenewable									
0050	250 volt, 30 amp	1 Elec	50	.160	Ea.	2.40	8.75		11.15	16.20
0100	60 amp		50	.160		3.68	8.75		12.43	17.60
0150	100 amp		40	.200		16.10	10.95		27.05	34.50
0200	200 amp		36	.222		37	12.15		49.15	59.50
0250	400 amp		30	.267		74	14.60		88.60	104
0300	600 amp		24	.333		120	18.25		138.25	160
0400	600 volt, 30 amp		40	.200		8.65	10.95		19.60	26.50
0450	60 amp		40	.200		14.15	10.95		25.10	32.50
0500	100 amp		36	.222		28.50	12.15		40.65	50
0550	200 amp		30	.267		57.50	14.60		72.10	86
0600	400 amp		24	.333		140	18.25		158.25	182
0650	600 amp		20	.400		203	22		225	258
0800	Dual element, time delay, 250 volt, 30 amp		50	.160		9.45	8.75		18.20	24
0840	50 amp		50	.160		10.35	8.75		19.10	25
0850	60 amp		50	.160		11.60	8.75		20.35	26.50
0900	100 amp		40	.200		28	10.95		38.95	48
0950	200 amp		36	.222		61.50	12.15		73.65	86.50
1000	400 amp		30	.267		110	14.60		124.60	144
1050	600 amp		24	.333		181	18.25		199.25	227
1300	600 volt, 15 to 30 amp		40	.200		18.75	10.95		29.70	37.50
1350	35 to 60 amp		40	.200		31.50	10.95		42.45	51.50
1400	70 to 100 amp		36	.222		66.50	12.15		78.65	92
1450	110 to 200 amp		30	.267		129	14.60		143.60	165
1500	225 to 400 amp		24	.333		267	18.25		285.25	320
1550	600 amp		20	.400		355	22		377	430
1800	Class RK1, high capacity, 250 volt, 30 amp		50	.160		8	8.75		16.75	22.50
1850	60 amp		50	.160		14.75	8.75		23.50	30
1900	100 amp		40	.200		34.50	10.95		45.45	55
1950	200 amp		36	.222		78.50	12.15		90.65	105
2000	400 amp		30	.267		130	14.60		144.60	166
2050	600 amp		24	.333		182	18.25		200.25	228
2200	600 volt, 30 amp		40	.200		14.90	10.95		25.85	33.50
2250	60 amp		40	.200		28	10.95		38.95	48
2300	100 amp		36	.222		71	12.15		83.15	97.50
2350	200 amp		30	.267		137	14.60		151.60	174
2400	400 amp		24	.333		230	18.25		248.25	281
2450	600 amp		20	.400		263	22		285	325
2700	Class J, current limiting, 250 or 600 volt, 30 amp		40	.200		24.50	10.95		35.45	44
2750	60 amp		40	.200		40.50	10.95		51.45	61.50
2800	100 amp		36	.222		56	12.15		68.15	80.50
2850	200 amp		30	.267		111	14.60		125.60	145
2900	400 amp		24	.333		271	18.25		289.25	325
2950	600 amp		20	.400		405	22		427	480
3100	Class L, current limiting, 250 or 600 volt, 601 to 1200 amp		16	.500		630	27.50		657.50	735

26 28 Low-Voltage Circuit Protective Devices

26 28 13 – Fuses

26 28 13.10 Fuse Elements		Crew	Daily Output	Labor-Hours	Unit	Material	2015 Bare Costs Labor	Equipment	Total	Total Incl O&P
3150	1500-1600 amp	1 Elec	13	.615	Ea.	900	33.50		933.50	1,050
3200	1800-2000 amp		10	.800		1,025	44		1,069	1,200
3250	2500 amp		10	.800		1,225	44		1,269	1,425
3300	3000 amp		8	1		1,650	54.50		1,704.50	1,875
3350	3500-4000 amp		8	1		1,725	54.50		1,779.50	1,975
3400	4500-5000 amp		6.70	1.194		2,650	65.50		2,715.50	3,025
3450	6000 amp		5.70	1.404		3,550	77		3,627	4,050
3600	Plug, 120 volt, 1 to 10 amp		50	.160		3.13	8.75		11.88	17
3650	15 to 30 amp		50	.160		4.84	8.75		13.59	18.85
3700	Dual element 0.3 to 14 amp		50	.160		4.59	8.75		13.34	18.60
3750	15 to 30 amp		50	.160		6.65	8.75		15.40	21
3800	Fustat, 120 volt, 15 to 30 amp		50	.160		5.75	8.75		14.50	19.90
3850	0.3 to 14 amp		50	.160		7.10	8.75		15.85	21.50
3900	Adapters 0.3 to 10 amp		50	.160		5.65	8.75		14.40	19.75
3950	15 to 30 amp		50	.160		8.65	8.75		17.40	23

26 28 16 – Enclosed Switches and Circuit Breakers

26 28 16.10 Circuit Breakers		Crew	Daily Output	Labor-Hours	Unit	Material	Labor	Equipment	Total	Total Incl O&P
0010	**CIRCUIT BREAKERS** (in enclosure)									
0100	Enclosed (NEMA 1), 600 volt, 3 pole, 30 amp	1 Elec	3.20	2.500	Ea.	500	137		637	760
0200	60 amp		2.80	2.857		615	156		771	920
0400	100 amp		2.30	3.478		705	190		895	1,075
0500	200 amp		1.50	5.333		1,475	292		1,767	2,075
0600	225 amp		1.50	5.333		1,625	292		1,917	2,250
0700	400 amp	2 Elec	1.60	10		2,775	545		3,320	3,925
0800	600 amp		1.20	13.333		4,025	730		4,755	5,575
1000	800 amp		.94	17.021		5,250	930		6,180	7,225
1200	1000 amp		.84	19.048		6,625	1,050		7,675	8,925
1220	1200 amp		.80	20		8,500	1,100		9,600	11,100
1240	1600 amp		.72	22.222		15,600	1,225		16,825	19,100
1260	2000 amp		.64	25		16,900	1,375		18,275	20,700
1400	1200 amp with ground fault		.80	20		13,500	1,100		14,600	16,600
1600	1600 amp with ground fault		.72	22.222		17,700	1,225		18,925	21,400
1800	2000 amp with ground fault		.64	25		19,000	1,375		20,375	23,000
2000	Disconnect, 240 volt 3 pole, 5 HP motor	1 Elec	3.20	2.500		430	137		567	685
2020	10 HP motor		3.20	2.500		430	137		567	685
2040	15 HP motor		2.80	2.857		430	156		586	715
2060	20 HP motor		2.30	3.478		525	190		715	870
2080	25 HP motor		2.30	3.478		525	190		715	870
2100	30 HP motor		2.30	3.478		525	190		715	870
2120	40 HP motor		2	4		900	219		1,119	1,325
2140	50 HP motor		1.50	5.333		900	292		1,192	1,425
2160	60 HP motor		1.50	5.333		2,075	292		2,367	2,725
2180	75 HP motor	2 Elec	2	8		2,075	440		2,515	2,950
2200	100 HP motor		1.60	10		2,075	545		2,620	3,125
2220	125 HP motor		1.60	10		2,075	545		2,620	3,125
2240	150 HP motor		1.20	13.333		4,025	730		4,755	5,575
2260	200 HP motor		1.20	13.333		5,250	730		5,980	6,900
2300	Enclosed (NEMA 7), explosion proof, 600 volt 3 pole, 50 amp	1 Elec	2.30	3.478		1,525	190		1,715	1,975
2350	100 amp		1.50	5.333		1,575	292		1,867	2,200
2400	150 amp		1	8		3,825	440		4,265	4,875
2450	250 amp	2 Elec	1.60	10		4,775	545		5,320	6,100
2500	400 amp	″	1.20	13.333		5,300	730		6,030	6,950

26 28 16 – Enclosed Switches and Circuit Breakers

26 28 16.10 Circuit Breakers		Crew	Daily Output	Labor-Hours	Unit	Material	2015 Bare Costs Labor	Equipment	Total	Total Incl O&P
9000	Minimum labor/equipment charge	1 Elec	4	2	Job		109		109	169
26 28 16.20 Safety Switches										
0010	SAFETY SWITCHES									
0100	General duty 240 volt, 3 pole NEMA 1, fusible, 30 amp	1 Elec	3.20	2.500	Ea.	73.50	137		210.50	293
0200	60 amp		2.30	3.478		124	190		314	430
0300	100 amp		1.90	4.211		213	230		443	590
0400	200 amp		1.30	6.154		455	335		790	1,025
0500	400 amp	2 Elec	1.80	8.889		1,150	485		1,635	2,025
0600	600 amp	″	1.20	13.333		2,150	730		2,880	3,500
0610	Nonfusible, 30 amp	1 Elec	3.20	2.500		59	137		196	277
0650	60 amp		2.30	3.478		78.50	190		268.50	380
0700	100 amp		1.90	4.211		183	230		413	555
0750	200 amp		1.30	6.154		335	335		670	890
0800	400 amp	2 Elec	1.80	8.889		810	485		1,295	1,650
0850	600 amp	″	1.20	13.333		1,575	730		2,305	2,850
1100	Heavy duty, 600 volt, 3 pole NEMA 1 nonfused									
1110	30 amp	1 Elec	3.20	2.500	Ea.	113	137		250	335
1500	60 amp		2.30	3.478		187	190		377	500
1700	100 amp		1.90	4.211		297	230		527	680
1900	200 amp		1.30	6.154		450	335		785	1,025
2100	400 amp	2 Elec	1.80	8.889		1,000	485		1,485	1,850
2300	600 amp		1.20	13.333		1,800	730		2,530	3,100
2500	800 amp		.94	17.021		3,650	930		4,580	5,475
2700	1200 amp		.80	20		4,900	1,100		6,000	7,100
2900	Heavy duty, 240 volt, 3 pole NEMA 1 fusible									
2910	30 amp	1 Elec	3.20	2.500	Ea.	118	137		255	340
3000	60 amp		2.30	3.478		199	190		389	515
3300	100 amp		1.90	4.211		315	230		545	700
3500	200 amp		1.30	6.154		540	335		875	1,125
3700	400 amp	2 Elec	1.80	8.889		1,400	485		1,885	2,275
3900	600 amp		1.20	13.333		2,800	730		3,530	4,200
4100	800 amp		.94	17.021		5,600	930		6,530	7,600
4300	1200 amp		.80	20		7,325	1,100		8,425	9,750
4340	2 pole fusible, 30 amp	1 Elec	3.50	2.286		88.50	125		213.50	292
4350	600 volt, 3 pole, fusible, 30 amp		3.20	2.500		201	137		338	435
4380	60 amp		2.30	3.478		243	190		433	560
4400	100 amp		1.90	4.211		445	230		675	845
4420	200 amp		1.30	6.154		640	335		975	1,225
4440	400 amp	2 Elec	1.80	8.889		1,675	485		2,160	2,575
4450	600 amp		1.20	13.333		2,800	730		3,530	4,200
4460	800 amp		.94	17.021		5,600	930		6,530	7,600
4480	1200 amp		.80	20		7,325	1,100		8,425	9,750
4500	240 volt 3 pole NEMA 3R (no hubs), fusible									
4510	30 amp	1 Elec	3.10	2.581	Ea.	210	141		351	450
4700	60 amp		2.20	3.636		330	199		529	675
4900	100 amp		1.80	4.444		480	243		723	905
5100	200 amp		1.20	6.667		660	365		1,025	1,300
5300	400 amp	2 Elec	1.60	10		1,450	545		1,995	2,450
5500	600 amp	″	1	16		3,025	875		3,900	4,700
5510	Heavy duty, 600 volt, 3 pole 3ph. NEMA 3R fusible, 30 amp	1 Elec	3.10	2.581		335	141		476	590
5520	60 amp		2.20	3.636		390	199		589	740
5530	100 amp		1.80	4.444		615	243		858	1,050

26 28 16.20 Safety Switches		Crew	Daily Output	Labor-Hours	Unit	2015 Bare Costs Material	Labor	Equipment	Total	Total Incl O&P
5540	200 amp	1 Elec	1.20	6.667	Ea.	845	365		1,210	1,500
5550	400 amp	2 Elec	1.60	10		2,000	545		2,545	3,050
5700	600 volt, 3 pole NEMA 3R nonfused									
5710	30 amp	1 Elec	3.10	2.581	Ea.	187	141		328	425
5900	60 amp		2.20	3.636		330	199		529	670
6100	100 amp		1.80	4.444		455	243		698	875
6300	200 amp		1.20	6.667		555	365		920	1,175
6500	400 amp	2 Elec	1.60	10		1,375	545		1,920	2,375
6700	600 amp	″	1	16		2,775	875		3,650	4,400
6900	600 volt, 6 pole NEMA 3R nonfused, 30 amp	1 Elec	2.70	2.963		1,200	162		1,362	1,575
7100	60 amp		2	4		1,375	219		1,594	1,875
7300	100 amp		1.50	5.333		1,700	292		1,992	2,325
7500	200 amp		1.20	6.667		6,550	365		6,915	7,775
7600	600 volt, 3 pole NEMA 7 explosion proof nonfused									
7610	30 amp	1 Elec	2.20	3.636	Ea.	1,325	199		1,524	1,750
7620	60 amp		1.80	4.444		1,575	243		1,818	2,125
7630	100 amp		1.20	6.667		1,925	365		2,290	2,675
7640	200 amp		.80	10		3,975	545		4,520	5,225
7710	600 volt 6 pole, NEMA 3R fusible, 30 amp		2.70	2.963		1,525	162		1,687	1,925
7900	60 amp		2	4		1,600	219		1,819	2,100
8100	100 amp		1.50	5.333		2,250	292		2,542	2,925
8110	240 volt 3 pole, NEMA 12 fusible, 30 amp		3.10	2.581		260	141		401	505
8120	60 amp		2.20	3.636		415	199		614	770
8130	100 amp		1.80	4.444		510	243		753	935
8140	200 amp		1.20	6.667		735	365		1,100	1,375
8150	400 amp	2 Elec	1.60	10		1,675	545		2,220	2,700
8160	600 amp	″	1	16		3,000	875		3,875	4,650
8180	600 volt 3 pole, NEMA 12 fusible, 30 amp	1 Elec	3.10	2.581		340	141		481	595
8190	60 amp		2.20	3.636		405	199		604	755
8200	100 amp		1.80	4.444		655	243		898	1,100
8210	200 amp		1.20	6.667		995	365		1,360	1,675
8220	400 amp	2 Elec	1.60	10		2,025	545		2,570	3,075
8230	600 amp	″	1	16		4,525	875		5,400	6,350
8240	600 volt 3 pole, NEMA 12 nonfused, 30 amp	1 Elec	3.10	2.581		240	141		381	485
8250	60 amp		2.20	3.636		310	199		509	650
8260	100 amp		1.80	4.444		440	243		683	860
8270	200 amp		1.20	6.667		585	365		950	1,200
8280	400 amp	2 Elec	1.60	10		1,450	545		1,995	2,450
8290	600 amp	″	1	16		3,475	875		4,350	5,175
8310	600 volt, 3 pole NEMA 4 fusible, 30 amp	1 Elec	3	2.667		935	146		1,081	1,250
8320	60 amp		2.20	3.636		1,050	199		1,249	1,450
8330	100 amp		1.80	4.444		2,300	243		2,543	2,900
8340	200 amp		1.20	6.667		2,900	365		3,265	3,750
8350	400 amp	2 Elec	1.60	10		5,750	545		6,295	7,175
8360	600 volt 3 pole NEMA 4 nonfused, 30 amp	1 Elec	3	2.667		770	146		916	1,075
8370	60 amp		2.20	3.636		920	199		1,119	1,300
8380	100 amp		1.80	4.444		1,875	243		2,118	2,450
8390	200 amp		1.20	6.667		2,550	365		2,915	3,375
8400	400 amp	2 Elec	1.60	10		5,175	545		5,720	6,550
8490	Motor starters, manual, single phase, NEMA 1	1 Elec	6.40	1.250		65	68.50		133.50	178
8500	NEMA 4		4	2		182	109		291	370
8700	NEMA 7		4	2		200	109		309	390
8900	NEMA 1 with pilot		6.40	1.250		90	68.50		158.50	205

26 28 Low-Voltage Circuit Protective Devices

26 28 16 – Enclosed Switches and Circuit Breakers

26 28 16.20 Safety Switches		Crew	Daily Output	Labor-Hours	Unit	Material	2015 Bare Costs Labor	Equipment	Total	Total Incl O&P
8920	3 pole, NEMA 1, 230/460 volt, 5 HP, size 0	1 Elec	3.50	2.286	Ea.	207	125		332	420
8940	10 HP, size 1		2	4		245	219		464	610
9010	Disc. switch, 600 V 3 pole fusible, 30 amp, to 10 HP motor		3.20	2.500		355	137		492	600
9050	60 amp, to 30 HP motor		2.30	3.478		815	190		1,005	1,200
9070	100 amp, to 60 HP motor		1.90	4.211		815	230		1,045	1,250
9100	200 amp, to 125 HP motor		1.30	6.154		1,225	335		1,560	1,875
9110	400 amp, to 200 HP motor	2 Elec	1.80	8.889		3,075	485		3,560	4,150
9990	Minimum labor/equipment charge	1 Elec	3	2.667	Job		146		146	226

26 28 16.40 Time Switches		Crew	Daily Output	Labor-Hours	Unit	Material	Labor	Equipment	Total	Total Incl O&P
0010	**TIME SWITCHES**									
0100	Single pole, single throw, 24 hour dial	1 Elec	4	2	Ea.	145	109		254	330
0200	24 hour dial with reserve power		3.60	2.222		615	122		737	865
0300	Astronomic dial		3.60	2.222		209	122		331	420
0400	Astronomic dial with reserve power		3.30	2.424		665	133		798	935
0500	7 day calendar dial		3.30	2.424		150	133		283	370
0600	7 day calendar dial with reserve power		3.20	2.500		335	137		472	580
0700	Photo cell 2000 watt		8	1		28.50	54.50		83	116
1080	Load management device, 4 loads		2	4		995	219		1,214	1,450
1100	8 loads		1	8		2,325	440		2,765	3,225
9000	Minimum labor/equipment charge		3.50	2.286	Job		125		125	194

26 29 Low-Voltage Controllers

26 29 13 – Enclosed Controllers

26 29 13.10 Contactors, AC		Crew	Daily Output	Labor-Hours	Unit	Material	Labor	Equipment	Total	Total Incl O&P
0010	**CONTACTORS, AC** Enclosed (NEMA 1)									
0050	Lighting, 600 volt 3 pole, electrically held									
0100	20 amp	1 Elec	4	2	Ea.	315	109		424	515
0200	30 amp		3.60	2.222		360	122		482	585
0300	60 amp		3	2.667		675	146		821	965
0400	100 amp		2.50	3.200		1,000	175		1,175	1,400
0500	200 amp		1.40	5.714		2,525	315		2,840	3,250
0600	300 amp	2 Elec	1.60	10		6,775	545		7,320	8,300
0800	600 volt 3 pole, mechanically held, 30 amp	1 Elec	3.60	2.222		445	122		567	680
0900	60 amp		3	2.667		880	146		1,026	1,200
1000	75 amp		2.80	2.857		1,225	156		1,381	1,600
1100	100 amp		2.50	3.200		1,250	175		1,425	1,650
1200	150 amp		2	4		3,350	219		3,569	4,025
1300	200 amp		1.40	5.714		3,425	315		3,740	4,250
1500	Magnetic with auxiliary contact, size 00, 9 amp		4	2		187	109		296	375
1600	Size 0, 18 amp		4	2		223	109		332	415
1700	Size 1, 27 amp		3.60	2.222		253	122		375	465
1800	Size 2, 45 amp		3	2.667		470	146		616	740
1900	Size 3, 90 amp		2.50	3.200		760	175		935	1,100
2000	Size 4, 135 amp		2.30	3.478		1,725	190		1,915	2,200
2100	Size 5, 270 amp	2 Elec	1.80	8.889		3,650	485		4,135	4,775
2200	Size 6, 540 amp		1.20	13.333		10,700	730		11,430	12,800
2300	Size 7, 810 amp		1	16		14,400	875		15,275	17,200
2310	Size 8, 1215 amp		.80	20		22,300	1,100		23,400	26,300
2500	Magnetic, 240 volt, 1-2 pole, .75 HP motor	1 Elec	4	2		151	109		260	335
2520	2 HP motor		3.60	2.222		169	122		291	375
2540	5 HP motor		2.50	3.200		410	175		585	720

26 29 13 – Enclosed Controllers

26 29 13.10 Contactors, AC

	26 29 13.10 Contactors, AC	Crew	Daily Output	Labor-Hours	Unit	2015 Bare Costs Material	Labor	Equipment	Total	Total Incl O&P
2560	10 HP motor	1 Elec	1.40	5.714	Ea.	675	315		990	1,225
2600	240 volt or less, 3 pole, .75 HP motor		4	2		151	109		260	335
2620	5 HP motor		3.60	2.222		187	122		309	395
2640	10 HP motor		3.60	2.222		217	122		339	425
2660	15 HP motor		2.50	3.200		435	175		610	745
2700	25 HP motor		2.50	3.200		435	175		610	745
2720	30 HP motor	2 Elec	2.80	5.714		725	315		1,040	1,275
2740	40 HP motor		2.80	5.714		725	315		1,040	1,275
2760	50 HP motor		1.60	10		725	545		1,270	1,650
2800	75 HP motor		1.60	10		1,700	545		2,245	2,725
2820	100 HP motor		1	16		1,700	875		2,575	3,225
2860	150 HP motor		1	16		3,625	875		4,500	5,325
2880	200 HP motor		1	16		3,625	875		4,500	5,325
3000	600 volt, 3 pole, 5 HP motor	1 Elec	4	2		187	109		296	375
3020	10 HP motor		3.60	2.222		217	122		339	425
3040	25 HP motor		3	2.667		435	146		581	700
3100	50 HP motor		2.50	3.200		725	175		900	1,075
3160	100 HP motor	2 Elec	2.80	5.714		1,700	315		2,015	2,350
3220	200 HP motor	"	1.60	10		3,625	545		4,170	4,825

26 29 13.20 Control Stations

	26 29 13.20 Control Stations	Crew	Daily Output	Labor-Hours	Unit	Material	Labor	Equipment	Total	Total Incl O&P
0010	**CONTROL STATIONS**									
0050	NEMA 1, heavy duty, stop/start	1 Elec	8	1	Ea.	135	54.50		189.50	234
0100	Stop/start, pilot light		6.20	1.290		184	70.50		254.50	310
0200	Hand/off/automatic		6.20	1.290		100	70.50		170.50	219
0400	Stop/start/reverse		5.30	1.509		182	82.50		264.50	330
0500	NEMA 7, heavy duty, stop/start		6	1.333		425	73		498	580
0600	Stop/start, pilot light		4	2		515	109		624	740
0700	NEMA 7 or 9, 1 element		6	1.333		345	73		418	495
0800	2 element		6	1.333		445	73		518	605
0900	3 element		4	2		885	109		994	1,150
0910	Selector switch, 2 position		6	1.333		345	73		418	495
0920	3 position		4	2		345	109		454	550
0930	Oiltight, 1 element		8	1		93	54.50		147.50	187
0940	2 element		6.20	1.290		134	70.50		204.50	256
0950	3 element		5.30	1.509		180	82.50		262.50	325
0960	Selector switch, 2 position		6.20	1.290		99	70.50		169.50	218
0970	3 position		5.30	1.509		99	82.50		181.50	237

26 29 13.30 Control Switches

	26 29 13.30 Control Switches	Crew	Daily Output	Labor-Hours	Unit	Material	Labor	Equipment	Total	Total Incl O&P
0010	**CONTROL SWITCHES** Field installed									
6000	Push button 600 V 10A, momentary contact									
6150	Standard operator with colored button	1 Elec	34	.235	Ea.	15.70	12.85		28.55	37.50
6160	With single block 1NO 1NC		18	.444		33	24.50		57.50	74
6170	With double block 2NO 2NC		15	.533		50.50	29		79.50	101
6180	Std operator w/mushroom button 1-9/16" diam.		34	.235		33	12.85		45.85	56.50
6190	Std operator w/mushroom button 2-1/4" diam.									
6200	With single block 1NO 1NC	1 Elec	18	.444	Ea.	50.50	24.50		75	93
6210	With double block 2NO 2NC		15	.533		68	29		97	120
6500	Maintained contact, selector operator		34	.235		50.50	12.85		63.35	75.50
6510	With single block 1NO 1NC		18	.444		68	24.50		92.50	113
6520	With double block 2NO 2NC		15	.533		85.50	29		114.50	139
6560	Spring-return selector operator		34	.235		50.50	12.85		63.35	75.50
6570	With single block 1NO 1NC		18	.444		68	24.50		92.50	113

26 29 Low-Voltage Controllers

26 29 13 – Enclosed Controllers

26 29 13.30 Control Switches		Crew	Daily Output	Labor-Hours	Unit	Material	Labor (2015 Bare Costs)	Equipment	Total	Total Incl O&P
6580	With double block 2NO 2NC	1 Elec	15	.533	Ea.	85.50	29		114.50	139
6620	Transformer operator w/illuminated									
6630	button 6 V #12 lamp	1 Elec	32	.250	Ea.	85.50	13.70		99.20	115
6640	With single block 1NO 1NC w/guard		16	.500		103	27.50		130.50	156
6650	With double block 2NO 2NC w/guard		13	.615		120	33.50		153.50	185
6690	Combination operator		34	.235		50.50	12.85		63.35	75.50
6700	With single block 1NO 1NC		18	.444		68	24.50		92.50	113
6710	With double block 2NO 2NC		15	.533		85.50	29		114.50	139
9000	Indicating light unit, full voltage									
9010	110-125 V front mount	1 Elec	32	.250	Ea.	68	13.70		81.70	96
9020	130 V resistor type		32	.250		50.50	13.70		64.20	76.50
9030	6 V transformer type		32	.250		62.50	13.70		76.20	89.50

26 29 13.40 Relays		Crew	Daily Output	Labor-Hours	Unit	Material	Labor	Equipment	Total	Total Incl O&P
0010	**RELAYS** Enclosed (NEMA 1)									
0050	600 volt AC, 1 pole, 12 amp	1 Elec	5.30	1.509	Ea.	88.50	82.50		171	226
0100	2 pole, 12 amp		5	1.600		88.50	87.50		176	234
0200	4 pole, 10 amp		4.50	1.778		118	97		215	281
0500	250 volt DC, 1 pole, 15 amp		5.30	1.509		116	82.50		198.50	256
0600	2 pole, 10 amp		5	1.600		111	87.50		198.50	258
0700	4 pole, 4 amp		4.50	1.778		147	97		244	310

26 29 23 – Variable-Frequency Motor Controllers

26 29 23.10 Variable Frequency Drives/Adj. Frequency Drives			Crew	Daily Output	Labor-Hours	Unit	Material	Labor	Equipment	Total	Total Incl O&P
0010	**VARIABLE FREQUENCY DRIVES/ADJ. FREQUENCY DRIVES**										
0100	Enclosed (NEMA 1), 460 volt, for 3 HP motor size	G	1 Elec	.80	10	Ea.	1,700	545		2,245	2,725
0110	5 HP motor size	G		.80	10		1,925	545		2,470	2,950
0120	7.5 HP motor size	G		.67	11.940		2,300	655		2,955	3,525
0130	10 HP motor size	G		.67	11.940		2,675	655		3,330	3,925
0140	15 HP motor size	G	2 Elec	.89	17.978		3,250	985		4,235	5,100
0150	20 HP motor size	G		.89	17.978		3,800	985		4,785	5,725
0160	25 HP motor size	G		.67	23.881		4,700	1,300		6,000	7,175
0170	30 HP motor size	G		.67	23.881		5,900	1,300		7,200	8,500
0180	40 HP motor size	G		.67	23.881		7,025	1,300		8,325	9,775
0190	50 HP motor size	G		.53	30.189		8,675	1,650		10,325	12,100
0200	60 HP motor size	G	R-3	.56	35.714		10,500	1,950	246	12,696	14,800
0210	75 HP motor size	G		.56	35.714		12,100	1,950	246	14,296	16,600
0220	100 HP motor size	G		.50	40		14,100	2,175	276	16,551	19,200
0230	125 HP motor size	G		.50	40		15,700	2,175	276	18,151	21,000
0240	150 HP motor size	G		.50	40		17,800	2,175	276	20,251	23,300
0250	200 HP motor size	G		.42	47.619		23,400	2,575	330	26,305	30,200
1100	Custom-engineered, 460 volt, for 3 HP motor size	G	1 Elec	.56	14.286		2,725	780		3,505	4,200
1110	5 HP motor size	G		.56	14.286		2,725	780		3,505	4,200
1120	7.5 HP motor size	G		.47	17.021		2,875	930		3,805	4,600
1130	10 HP motor size	G		.47	17.021		3,000	930		3,930	4,750
1140	15 HP motor size	G	2 Elec	.62	25.806		3,750	1,400		5,150	6,300
1150	20 HP motor size	G		.62	25.806		4,225	1,400		5,625	6,825
1160	25 HP motor size	G		.47	34.043		4,950	1,850		6,800	8,300
1170	30 HP motor size	G		.47	34.043		6,175	1,850		8,025	9,675
1180	40 HP motor size	G		.47	34.043		7,300	1,850		9,150	10,900
1190	50 HP motor size	G		.37	43.243		8,375	2,375		10,750	12,900
1200	60 HP motor size	G	R-3	.39	51.282		12,600	2,775	355	15,730	18,600
1210	75 HP motor size	G		.39	51.282		13,400	2,775	355	16,530	19,500
1220	100 HP motor size	G		.35	57.143		14,700	3,100	395	18,195	21,500

26 29 Low-Voltage Controllers

26 29 23 – Variable-Frequency Motor Controllers

26 29 23.10 Variable Frequency Drives/Adj. Frequency Drives			Crew	Daily Output	Labor-Hours	Unit	2015 Bare Costs Material	Labor	Equipment	Total	Total Incl O&P
1230	125 HP motor size	G	R-3	.35	57.143	Ea.	15,700	3,100	395	19,195	22,600
1240	150 HP motor size	G		.35	57.143		18,000	3,100	395	21,495	25,100
1250	200 HP motor size	G		.29	68.966		23,500	3,750	475	27,725	32,200
2000	For complex & special design systems to meet specific										
2010	requirements, obtain quote from vendor.										

26 31 Photovoltaic Collectors

26 31 13 – Photovoltaics

26 31 13.50 Solar Energy - Photovoltaics

Line	Description		Crew	Daily Output	Labor-Hours	Unit	Material	Labor	Equipment	Total	Total Incl O&P
0010	SOLAR ENERGY - PHOTOVOLTAICS										
0220	Alt. energy source, photovoltaic module, 6 watt, 15 volts	G	1 Elec	8	1	Ea.	68	54.50		122.50	159
0230	10 watt, 16.3 volts	G		8	1		108	54.50		162.50	204
0240	20 watt, 14.5 volts	G		8	1		163	54.50		217.50	264
0250	36 watt, 17 volts	G		8	1		220	54.50		274.50	325
0260	55 watt, 17 volts	G		8	1		305	54.50		359.50	420
0270	75 watt, 17 volts	G		8	1		390	54.50		444.50	515
0280	130 watt, 33 volts	G		8	1		565	54.50		619.50	705
0290	140 watt, 33 volts	G		8	1		595	54.50		649.50	740
0300	150 watt, 33 volts	G		8	1		655	54.50		709.50	805
0310	DC to AC inverter for, 12 V 2,000 watt	G		4	2		1,175	109		1,284	1,475
0320	12 V, 2,500 watt	G		4	2		1,425	109		1,534	1,725
0330	24 V, 2,500 watt	G		4	2		2,000	109		2,109	2,375
0340	12 V, 3,000 watt	G		3	2.667		1,650	146		1,796	2,050
0350	24 V, 3,000 watt	G		3	2.667		2,425	146		2,571	2,900
0360	24 V, 4,000 watt	G		2	4		3,325	219		3,544	4,000
0370	48 V, 4,000 watt	G		2	4		3,325	219		3,544	4,000
0380	48 V, 5,500 watt	G		2	4		3,775	219		3,994	4,500
0390	PV components, combiner box, 10 lug, NEMA 3R enclosure	G		4	2		191	109		300	380
0400	Fuse, 15 A for combiner box	G		40	.200		18.75	10.95		29.70	37.50
0410	Battery charger controller w/temperature sensor	G		4	2		490	109		599	710
0420	Digital readout panel, displays hours, volts, amps, etc.	G		4	2		130	109		239	310
0430	Deep cycle solar battery, 6 V, 180 Ah (C/20)	G		8	1		235	54.50		289.50	345
0440	Battery interconn, 15" AWG #2/0, sealed w/copper ring lugs	G		16	.500		12.80	27.50		40.30	56.50
0442	Battery interconn, 24" AWG #2/0, sealed w/copper ring lugs	G		16	.500		18.35	27.50		45.85	62.50
0444	Battery interconn, 60" AWG #2/0, sealed w/copper ring lugs	G		16	.500		44	27.50		71.50	90.50
0446	Batt temp computer probe, RJ11 jack, 15' cord	G		16	.500		19.70	27.50		47.20	64
0450	System disconnect, DC 175 amp circuit breaker	G		8	1		140	54.50		194.50	239
0460	Conduit box for inverter	G		8	1		44.50	54.50		99	133
0470	Low voltage disconnect	G		8	1		40	54.50		94.50	129
0480	Vented battery enclosure, wood	G	1 Carp	2	4		174	188		362	500
0490	PV rack system, roof, non-penetrating ballast, 1 panel	G	R-1A	30.50	.525		895	23.50		918.50	1,025
0500	Penetrating surface mount, on steel framing, 1 panel			5.13	3.122		49	141		190	278
0510	On wood framing, 1 panel			11	1.455		47.50	65.50		113	156
0520	With standoff, 1 panel			11	1.455		55.50	65.50		121	165
0530	Ground, ballast, fixed, 3 panel			20.50	.780		1,150	35		1,185	1,325
0540	4 panel			20.50	.780		1,575	35		1,610	1,775
0550	5 panel			20.50	.780		2,075	35		2,110	2,325
0560	6 panel			20.50	.780		2,375	35		2,410	2,675
0570	Adjustable, 3 panel			20.50	.780		1,225	35		1,260	1,400
0580	4 panel			20.50	.780		1,675	35		1,710	1,875
0590	5 panel			20.50	.780		2,200	35		2,235	2,475

26 31 Photovoltaic Collectors

26 31 13 – Photovoltaics

26 31 13.50 Solar Energy - Photovoltaics		Crew	Daily Output	Labor-Hours	Unit	2015 Bare Costs Material	Labor	Equipment	Total	Total Incl O&P
0600	6 panel	R-1A	20.50	.780	Ea.	2,525	35		2,560	2,825
0605	Top of Pole, see 32 31 13.30 6710+ for poles									
0610	Passive tracking, 1 panel	R-1A	20.50	.780	Ea.	705	35		740	830
0620	2 panel		20.50	.780		1,425	35		1,460	1,625
0630	3 panel		20.50	.780		2,000	35		2,035	2,250
0640	4 panel		20.50	.780		2,150	35		2,185	2,400
0650	6 panel		20.50	.780		2,425	35		2,460	2,725
0660	8 panel	↓	20.50	.780	↓	3,600	35		3,635	4,000

26 32 Packaged Generator Assemblies

26 32 13 – Engine Generators

26 32 13.13 Diesel-Engine-Driven Generator Sets

		Crew	Daily Output	Labor-Hours	Unit	Material	Labor	Equipment	Total	Total Incl O&P
0010	**DIESEL-ENGINE-DRIVEN GENERATOR SETS**									
2000	Diesel engine, including battery, charger,									
2010	muffler, & day tank, 30 kW	R-3	.55	36.364	Ea.	14,000	1,975	251	16,226	18,800
2100	50 kW		.42	47.619		19,800	2,575	330	22,705	26,200
2110	60 kW		.39	51.282		21,100	2,775	355	24,230	27,900
2200	75 kW		.35	57.143		21,300	3,100	395	24,795	28,700
2300	100 kW		.31	64.516		29,900	3,500	445	33,845	38,800
2400	125 kW		.29	68.966		31,700	3,750	475	35,925	41,100
2500	150 kW		.26	76.923		33,800	4,175	530	38,505	44,300
2600	175 kW		.25	80		39,100	4,350	550	44,000	50,500
2700	200 kW		.24	83.333		42,200	4,525	575	47,300	54,000
2800	250 kW		.23	86.957		45,200	4,725	600	50,525	57,500
2850	275 kW		.22	90.909		47,300	4,925	625	52,850	60,500
2900	300 kW		.22	90.909		49,100	4,925	625	54,650	62,500
3000	350 kW		.20	100		55,500	5,425	690	61,615	70,000
3100	400 kW		.19	105		68,500	5,725	725	74,950	85,000
3200	500 kW		.18	111		86,500	6,025	765	93,290	105,000
3220	600 kW	↓	.17	117		112,500	6,400	810	119,710	135,000
3230	650 kW	R-13	.38	110		138,500	5,850	495	144,845	162,000
3240	750 kW		.38	110		140,000	5,850	495	146,345	163,500
3250	800 kW		.36	116		147,500	6,175	520	154,195	172,500
3260	900 kW		.31	135		170,500	7,150	605	178,255	199,500
3270	1000 kW	↓	.27	155	↓	174,500	8,225	695	183,420	205,500

26 32 13.16 Gas-Engine-Driven Generator Sets

		Crew	Daily Output	Labor-Hours	Unit	Material	Labor	Equipment	Total	Total Incl O&P
0010	**GAS-ENGINE-DRIVEN GENERATOR SETS**									
0020	Gas or gasoline operated, includes battery,									
0050	charger, & muffler									
0200	3 phase 4 wire, 277/480 volt, 7.5 kW	R-3	.83	24.096	Ea.	7,350	1,300	166	8,816	10,300
0300	11.5 kW		.71	28.169		10,400	1,525	194	12,119	14,100
0400	20 kW		.63	31.746		12,300	1,725	219	14,244	16,400
0500	35 kW		.55	36.364		14,600	1,975	251	16,826	19,500
0520	60 kW		.50	40		19,300	2,175	276	21,751	24,900
0600	80 kW		.40	50		24,000	2,725	345	27,070	31,000
0700	100 kW		.33	60.606		26,300	3,300	420	30,020	34,500
0800	125 kW		.28	71.429		54,000	3,875	490	58,365	65,500
0900	185 kW	↓	.25	80	↓	71,000	4,350	550	75,900	85,500

26 33 Battery Equipment

26 33 43 – Battery Chargers

26 33 43.55 Electric Vehicle Charging		Crew	Daily Output	Labor-Hours	Unit	Material	Labor	Equipment	Total	Total Incl O&P
						2015 Bare Costs				
0010	ELECTRIC VEHICLE CHARGING									
0020	Level 2, wall mounted									
2200	Heavy duty G	R-1A	15.36	1.042	Ea.	2,525	47		2,572	2,850
2210	with RFID G		12.29	1.302		2,700	58.50		2,758.50	3,075
2300	Free standing, single connector G		10.24	1.563		2,900	70.50		2,970.50	3,275
2310	with RFID G		8.78	1.822		3,625	82		3,707	4,100
2320	Double connector G		7.68	2.083		4,850	94		4,944	5,475
2330	with RFID G		6.83	2.343		6,300	106		6,406	7,100

26 33 53 – Static Uninterruptible Power Supply

26 33 53.10 Uninterruptible Power Supply/Conditioner Trans.		Crew	Daily Output	Labor-Hours	Unit	Material	Labor	Equipment	Total	Total Incl O&P
0010	UNINTERRUPTIBLE POWER SUPPLY/CONDITIONER TRANSFORMERS									
0100	Volt. regulating, isolating transf., w/invert. & 10 min. battery pack									
0110	Single-phase, 120 V, 0.35 kVA	1 Elec	2.29	3.493	Ea.	1,050	191		1,241	1,450
0120	0.5 kVA		2	4		1,100	219		1,319	1,550
0130	For additional 55 min. battery, add to 0.35 kVA		2.29	3.493		620	191		811	975
0140	Add to 0.5 kVA		1.14	7.018		650	385		1,035	1,300
0150	Single-phase, 120 V, 0.75 kVA		.80	10		1,400	545		1,945	2,375
0160	1.0 kVA		.80	10		2,000	545		2,545	3,050
0170	1.5 kVA	2 Elec	1.14	14.035		3,425	770		4,195	4,975
0180	2 kVA	"	.89	17.978		3,725	985		4,710	5,625
0190	3 kVA	R-3	.63	31.746		4,525	1,725	219	6,469	7,900
0200	5 kVA		.42	47.619		6,550	2,575	330	9,455	11,600
0210	7.5 kVA		.33	60.606		8,350	3,300	420	12,070	14,800
0220	10 kVA		.28	71.429		10,800	3,875	490	15,165	18,500
0230	15 kVA		.22	90.909		13,800	4,925	625	19,350	23,500
0240	3 phase, 120/208 V input 120/208 V output, 20 kVA, incl 17 min. battery		.21	95.238		25,300	5,175	655	31,130	36,700
0242	30 kVA, incl 11 min. battery		.20	100		28,400	5,425	690	34,515	40,400
0250	40 kVA, incl 15 min. battery		.20	100		37,800	5,425	690	43,915	51,000
0260	480 V input 277/480 V output, 60 kVA, incl 6 min. battery		.19	105		42,600	5,725	725	49,050	56,500
0262	80 kVA, incl 4 min. battery		.19	105		50,500	5,725	725	56,950	65,500
0400	For additional 34 min./15 min. battery, add to 40 kVA		.71	28.011		13,200	1,525	193	14,918	17,100
0600	For complex & special design systems to meet specific									
0610	requirements, obtain quote from vendor									

26 35 Power Filters and Conditioners

26 35 13 – Capacitors

26 35 13.10 Capacitors Indoor		Crew	Daily Output	Labor-Hours	Unit	Material	Labor	Equipment	Total	Total Incl O&P
0010	CAPACITORS INDOOR									
0020	240 volts, single & 3 phase, 0.5 kVAR	1 Elec	2.70	2.963	Ea.	430	162		592	720
0100	1.0 kVAR		2.70	2.963		515	162		677	820
0150	2.5 kVAR		2	4		580	219		799	980
0200	5.0 kVAR		1.80	4.444		790	243		1,033	1,250
0250	7.5 kVAR		1.60	5		885	274		1,159	1,400
0300	10 kVAR		1.50	5.333		1,025	292		1,317	1,575
0350	15 kVAR		1.30	6.154		1,325	335		1,660	1,975
0400	20 kVAR		1.10	7.273		1,600	400		2,000	2,375
0450	25 kVAR		1	8		1,825	440		2,265	2,700
1000	480 volts, single & 3 phase, 1 kVAR		2.70	2.963		390	162		552	680
1050	2 kVAR		2.70	2.963		450	162		612	745
1100	5 kVAR		2	4		565	219		784	960

26 35 Power Filters and Conditioners

26 35 13 – Capacitors

26 35 13.10 Capacitors Indoor		Crew	Daily Output	Labor-Hours	Unit	Material	2015 Bare Costs Labor	Equipment	Total	Total Incl O&P
1150	7.5 kVAR	1 Elec	2	4	Ea.	610	219		829	1,000
1200	10 kVAR		2	4		710	219		929	1,125
1250	15 kVAR		2	4		835	219		1,054	1,250
1300	20 kVAR		1.60	5		915	274		1,189	1,425
1350	30 kVAR		1.50	5.333		1,100	292		1,392	1,675
1400	40 kVAR		1.20	6.667		1,375	365		1,740	2,075
1450	50 kVAR		1.10	7.273		1,600	400		2,000	2,400
2000	600 volts, single & 3 phase, 1 kVAR		2.70	2.963		400	162		562	690
2050	2 kVAR		2.70	2.963		460	162		622	760
2100	5 kVAR		2	4		565	219		784	960
2150	7.5 kVAR		2	4		610	219		829	1,000
2200	10 kVAR		2	4		710	219		929	1,125
2250	15 kVAR		1.60	5		840	274		1,114	1,350
2300	20 kVAR		1.60	5		915	274		1,189	1,425
2350	25 kVAR		1.50	5.333		1,000	292		1,292	1,550
2400	35 kVAR		1.40	5.714		1,250	315		1,565	1,850
2450	50 kVAR		1.30	6.154		1,600	335		1,935	2,300

26 35 26 – Harmonic Filters

26 35 26.10 Computer Isolation Transformer		Crew	Daily Output	Labor-Hours	Unit	Material	Labor	Equipment	Total	Total Incl O&P
0010	**COMPUTER ISOLATION TRANSFORMER**									
0100	Computer grade									
0110	Single-phase, 120/240 V, 0.5 kVA	1 Elec	4	2	Ea.	380	109		489	585
0120	1.0 kVA		2.67	2.996		535	164		699	845
0130	2.5 kVA		2	4		815	219		1,034	1,225
0140	5 kVA		1.14	7.018		930	385		1,315	1,625

26 35 26.20 Computer Regulator Transformer		Crew	Daily Output	Labor-Hours	Unit	Material	Labor	Equipment	Total	Total Incl O&P
0010	**COMPUTER REGULATOR TRANSFORMER**									
0100	Ferro-resonant, constant voltage, variable transformer									
0110	Single-phase, 240 V, 0.5 kVA	1 Elec	2.67	2.996	Ea.	390	164		554	685
0120	1.0 kVA		2	4		535	219		754	930
0130	2.0 kVA		1	8		915	440		1,355	1,675
0210	Plug-in unit 120 V, 0.14 kVA		8	1		226	54.50		280.50	335
0220	0.25 kVA		8	1		264	54.50		318.50	375
0230	0.5 kVA		8	1		390	54.50		444.50	515
0240	1.0 kVA		5.33	1.501		535	82		617	715
0250	2.0 kVA		4	2		915	109		1,024	1,175

26 35 26.30 Power Conditioner Transformer		Crew	Daily Output	Labor-Hours	Unit	Material	Labor	Equipment	Total	Total Incl O&P
0010	**POWER CONDITIONER TRANSFORMER**									
0100	Electronic solid state, buck-boost, transformer, w/tap switch									
0110	Single-phase, 115 V, 3.0 kVA, + or - 3% accuracy	2 Elec	1.60	10	Ea.	2,100	545		2,645	3,175
0120	208, 220, 230, or 240 V, 5.0 kVA, + or - 1.5% accuracy	3 Elec	1.60	15		2,725	820		3,545	4,275
0130	5.0 kVA, + or - 6% accuracy	2 Elec	1.14	14.035		2,450	770		3,220	3,900
0140	7.5 kVA, + or - 1.5% accuracy	3 Elec	1.50	16		3,475	875		4,350	5,175
0150	7.5 kVA, + or - 6% accuracy		1.60	15		2,900	820		3,720	4,450
0160	10.0 kVA, + or - 1.5% accuracy		1.33	18.045		4,625	985		5,610	6,600
0170	10.0 kVA, + or - 6% accuracy		1.41	17.021		3,925	930		4,855	5,775

26 35 26.40 Transient Voltage Suppressor Transformer		Crew	Daily Output	Labor-Hours	Unit	Material	Labor	Equipment	Total	Total Incl O&P
0010	**TRANSIENT VOLTAGE SUPPRESSOR TRANSFORMER**									
0110	Single-phase, 120 V, 1.8 kVA	1 Elec	4	2	Ea.	855	109		964	1,100
0120	3.6 kVA		4	2		1,675	109		1,784	2,025
0130	7.2 kVA		3.20	2.500		2,300	137		2,437	2,725

26 35 Power Filters and Conditioners

26 35 26 – Harmonic Filters

26 35 26.40 Transient Voltage Suppressor Transformer		Crew	Daily Output	Labor-Hours	Unit	2015 Bare Costs Material	Labor	Equipment	Total	Total Incl O&P
0150	240 V, 3.6 kVA	1 Elec	4	2	Ea.	1,675	109		1,784	2,025
0160	7.2 kVA		4	2		2,175	109		2,284	2,575
0170	14.4 kVA		3.20	2.500		3,275	137		3,412	3,800
0210	Plug-in unit, 120 V, 1.8 kVA		8	1		650	54.50		704.50	800

26 35 53 – Voltage Regulators

26 35 53.10 Automatic Voltage Regulators

26 35 53.10 Automatic Voltage Regulators		Crew	Daily Output	Labor-Hours	Unit	Material	Labor	Equipment	Total	Total Incl O&P
0010	**AUTOMATIC VOLTAGE REGULATORS**									
0100	Computer grade, solid state, variable transf. volt. regulator									
0110	Single-phase, 120 V, 8.6 kVA	2 Elec	1.33	12.030	Ea.	4,900	660		5,560	6,400
0120	17.3 kVA		1.14	14.035		5,775	770		6,545	7,550
0130	208/240 V, 7.5/8.6 kVA		1.33	12.030		4,900	660		5,560	6,400
0140	13.5/15.6 kVA		1.33	12.030		5,775	660		6,435	7,375
0150	27.0/31.2 kVA		1.14	14.035		7,300	770		8,070	9,225
0210	Two-phase, single control, 208/240 V, 15.0/17.3 kVA		1.14	14.035		5,775	770		6,545	7,550
0220	Individual phase control, 15.0/17.3 kVA		1.14	14.035		5,775	770		6,545	7,550
0230	30.0/34.6 kVA	3 Elec	1.33	18.045		7,300	985		8,285	9,550
0310	Three-phase single control, 208/240 V, 26/30 kVA	2 Elec	1	16		5,775	875		6,650	7,700
0320	380/480 V, 24/30 kVA	"	1	16		5,775	875		6,650	7,700
0330	43/54 kVA	3 Elec	1.33	18.045		10,500	985		11,485	13,000
0340	Individual phase control, 208 V, 26 kVA	"	1.33	18.045		5,775	985		6,760	7,875
0350	52 kVA	R-3	.91	21.978		7,300	1,200	151	8,651	10,000
0360	340/480 V, 24/30 kVA	2 Elec	1	16		5,775	875		6,650	7,700
0370	43/54 kVA	"	1	16		7,300	875		8,175	9,375
0380	48/60 kVA	3 Elec	1.33	18.045		10,600	985		11,585	13,100
0390	86/108 kVA	R-3	.91	21.978		11,700	1,200	151	13,051	14,900
0500	Standard grade, solid state, variable transformer volt. regulator									
0510	Single-phase, 115 V, 2.3 kVA	1 Elec	2.29	3.493	Ea.	1,875	191		2,066	2,375
0520	4.2 kVA		2	4		3,025	219		3,244	3,675
0530	6.6 kVA		1.14	7.018		3,700	385		4,085	4,675
0540	13.0 kVA		1.14	7.018		6,375	385		6,760	7,625
0550	16.6 kVA	2 Elec	1.23	13.008		7,525	710		8,235	9,400
0610	230 V, 8.3 kVA		1.33	12.030		6,375	660		7,035	8,050
0620	21.4 kVA		1.23	13.008		7,525	710		8,235	9,400
0630	29.9 kVA		1.23	13.008		7,525	710		8,235	9,400
0710	460 V, 9.2 kVA		1.33	12.030		6,375	660		7,035	8,050
0720	20.7 kVA		1.23	13.008		7,525	710		8,235	9,400
0810	Three-phase, 230 V, 13.1 kVA	3 Elec	1.41	17.021		6,375	930		7,305	8,475
0820	19.1 kVA		1.41	17.021		7,525	930		8,455	9,750
0830	25.1 kVA		1.23	19.512		7,525	1,075		8,600	9,950
0840	57.8 kVA	R-3	.95	21.053		13,700	1,150	145	14,995	16,900
0850	74.9 kVA	"	.91	21.978		13,700	1,200	151	15,051	17,000
0910	460 V, 14.3 kVA	3 Elec	1.41	17.021		6,375	930		7,305	8,475
0920	19.1 kVA		1.41	17.021		7,525	930		8,455	9,750
0930	27.9 kVA		1.23	19.512		7,525	1,075		8,600	9,950
0940	59.8 kVA	R-3	1	20		13,700	1,075	138	14,913	16,900
0950	79.7 kVA		.95	21.053		15,300	1,150	145	16,595	18,700
0960	118 kVA		.95	21.053		16,100	1,150	145	17,395	19,600
1000	Laboratory grade, precision, electronic voltage regulator									
1110	Single-phase, 115 V, 0.5 kVA	1 Elec	2.29	3.493	Ea.	1,200	191		1,391	1,625
1120	1.0 kVA		2	4		1,275	219		1,494	1,750
1130	3.0 kVA		.80	10		1,775	545		2,320	2,825
1140	6.0 kVA	2 Elec	1.46	10.959		3,300	600		3,900	4,550

26 35 Power Filters and Conditioners

26 35 53 – Voltage Regulators

26 35 53.10 Automatic Voltage Regulators		Crew	Daily Output	Labor-Hours	Unit	Material	2015 Bare Costs Labor	Equipment	Total	Total Incl O&P
1150	10.0 kVA	3 Elec	1	24	Ea.	4,300	1,325		5,625	6,750
1160	15.0 kVA	"	1.50	16		4,900	875		5,775	6,750
1210	230 V, 3.0 kVA	1 Elec	.80	10		2,025	545		2,570	3,075
1220	6.0 kVA	2 Elec	1.46	10.959		3,375	600		3,975	4,650
1230	10.0 kVA	3 Elec	1.71	14.035		4,500	770		5,270	6,125
1240	15.0 kVA	"	1.60	15		5,075	820		5,895	6,850

26 35 53.30 Transient Suppressor/Voltage Regulator		Crew	Daily Output	Labor-Hours	Unit	Material	Labor	Equipment	Total	Total Incl O&P
0010	**TRANSIENT SUPPRESSOR/VOLTAGE REGULATOR** (without isolation)									
0110	Single-phase, 115 V, 1.0 kVA	1 Elec	2.67	2.996	Ea.	1,000	164		1,164	1,350
0120	2.0 kVA		2.29	3.493		1,375	191		1,566	1,825
0130	4.0 kVA		2.13	3.756		1,700	205		1,905	2,200
0140	220 V, 1.0 kVA		2.67	2.996		1,000	164		1,164	1,350
0150	2.0 kVA		2.29	3.493		1,375	191		1,566	1,825
0160	4.0 kVA		2.13	3.756		1,800	205		2,005	2,300
0210	Plug-in unit, 120 V, 1.0 kVA		8	1		965	54.50		1,019.50	1,125
0220	2.0 kVA		8	1		1,350	54.50		1,404.50	1,575

26 36 Transfer Switches

26 36 13 – Manual Transfer Switches

26 36 13.10 Non-Automatic Transfer Switches		Crew	Daily Output	Labor-Hours	Unit	Material	Labor	Equipment	Total	Total Incl O&P
0010	**NON-AUTOMATIC TRANSFER SWITCHES** enclosed									
0100	Manual operated, 480 volt 3 pole, 30 amp	1 Elec	2.30	3.478	Ea.	1,375	190		1,565	1,825
0150	60 amp		1.90	4.211		1,850	230		2,080	2,375
0200	100 amp		1.30	6.154		2,750	335		3,085	3,550
0250	200 amp	2 Elec	2	8		3,650	440		4,090	4,700
0300	400 amp		1.60	10		5,300	545		5,845	6,700
0350	600 amp		1	16		5,325	875		6,200	7,200
1000	250 volt 3 pole, 30 amp	1 Elec	2.30	3.478		1,000	190		1,190	1,400
1100	60 amp		1.90	4.211		1,450	230		1,680	1,950
1150	100 amp		1.30	6.154		1,825	335		2,160	2,525
1200	200 amp	2 Elec	2	8		3,600	440		4,040	4,650
1300	600 amp	"	1	16		8,100	875		8,975	10,300
1500	Electrically operated, 480 volt 3 pole, 60 amp	1 Elec	1.90	4.211		2,375	230		2,605	2,975
1600	100 amp	"	1.30	6.154		2,375	335		2,710	3,150
1650	200 amp	2 Elec	2	8		3,925	440		4,365	5,000
1700	400 amp		1.60	10		5,500	545		6,045	6,900
1750	600 amp		1	16		7,900	875		8,775	10,100
2000	250 volt 3 pole, 30 amp	1 Elec	2.30	3.478		2,350	190		2,540	2,900
2050	60 amp	"	1.90	4.211		2,350	230		2,580	2,950
2150	200 amp	2 Elec	2	8		3,900	440		4,340	4,950
2200	400 amp		1.60	10		5,425	545		5,970	6,825
2250	600 amp		1	16		7,825	875		8,700	9,975
2500	NEMA 3R, 480 volt 3 pole, 60 amp	1 Elec	1.80	4.444		2,775	243		3,018	3,425
2550	100 amp	"	1.20	6.667		3,575	365		3,940	4,500
2600	200 amp	2 Elec	1.80	8.889		4,375	485		4,860	5,550
2650	400 amp	"	1.40	11.429		6,000	625		6,625	7,575
2800	NEMA 3R, 250 volt 3 pole solid state, 100 amp	1 Elec	1.20	6.667		2,800	365		3,165	3,650
2850	150 amp	2 Elec	1.80	8.889		3,725	485		4,210	4,850
2900	250 volt 2 pole solid state, 100 amp	1 Elec	1.30	6.154		2,725	335		3,060	3,525
2950	150 amp	2 Elec	2	8		3,625	440		4,065	4,675

26 36 Transfer Switches

26 36 23 – Automatic Transfer Switches

26 36 23.10 Automatic Transfer Switch Devices		Crew	Daily Output	Labor-Hours	Unit	2015 Bare Costs Material	Labor	Equipment	Total	Total Incl O&P
0010	**AUTOMATIC TRANSFER SWITCH DEVICES**									
0015	Switches, enclosed 120/240 volt, 2 pole, 30 amp	1 Elec	2.40	3.333	Ea.	2,400	182		2,582	2,900
0020	70 amp		2	4		2,400	219		2,619	2,975
0030	100 amp		1.35	5.926		2,400	325		2,725	3,125
0040	225 amp	2 Elec	2.10	7.619		3,525	415		3,940	4,525
0050	400 amp		1.70	9.412		5,375	515		5,890	6,700
0060	600 amp		1.06	15.094		7,875	825		8,700	9,950
0070	800 amp		.84	19.048		9,250	1,050		10,300	11,800
0100	Switches, enclosed 480 volt, 3 pole, 30 amp	1 Elec	2.30	3.478		2,650	190		2,840	3,225
0200	60 amp		1.90	4.211		2,650	230		2,880	3,275
0300	100 amp		1.30	6.154		2,650	335		2,985	3,450
0400	150 amp	2 Elec	2.40	6.667		3,275	365		3,640	4,175
0500	225 amp		2	8		4,200	440		4,640	5,275
0600	260 amp		2	8		4,850	440		5,290	6,000
0700	400 amp		1.60	10		6,325	545		6,870	7,800
0800	600 amp		1	16		8,750	875		9,625	11,000
0900	800 amp		.80	20		10,300	1,100		11,400	13,000
1000	1000 amp		.76	21.053		13,600	1,150		14,750	16,700
1100	1200 amp		.70	22.857		18,600	1,250		19,850	22,400
1200	1600 amp		.60	26.667		21,200	1,450		22,650	25,600
1300	2000 amp		.50	32		23,700	1,750		25,450	28,700
1600	Accessories, time delay on engine starting					208			208	228
1700	Adjustable time delay on retransfer					208			208	228
1800	Shunt trips for customer connections					370			370	405
1900	Maintenance select switch					84.50			84.50	93
2000	Auxiliary contact when normal fails					97.50			97.50	107
2100	Pilot light-emergency					84.50			84.50	93
2200	Pilot light-normal					84.50			84.50	93
2300	Auxiliary contact-closed on normal					97.50			97.50	107
2400	Auxiliary contact-closed on emergency					97.50			97.50	107
2500	Emergency source sensing, frequency relay					430			430	470

26 41 Facility Lightning Protection

26 41 13 – Lightning Protection for Structures

26 41 13.13 Lightning Protection for Buildings

		Crew	Daily Output	Labor-Hours	Unit	Material	Labor	Equipment	Total	Total Incl O&P
0010	**LIGHTNING PROTECTION FOR BUILDINGS**									
0200	Air terminals & base, copper									
0400	3/8" diameter x 10" (to 75' high)	1 Elec	8	1	Ea.	22	54.50		76.50	109
0500	1/2" diameter x 12" (over 75' high)		8	1		25.50	54.50		80	113
0520	1/2" diameter x 24"		7.30	1.096		35	60		95	132
0540	1/2" diameter x 60"		6.70	1.194		64	65.50		129.50	171
1000	Aluminum, 1/2" diameter x 12" (to 75' high)		8	1		15.10	54.50		69.60	101
1020	1/2" diameter x 24"		7.30	1.096		17.25	60		77.25	112
1040	1/2" diameter x 60"		6.70	1.194		25.50	65.50		91	129
1100	5/8" diameter x 12" (over 75' high)		8	1		16.35	54.50		70.85	103
2000	Cable, copper, 220 lb. per thousand ft. (to 75' high)		320	.025	L.F.	3.03	1.37		4.40	5.45
2100	375 lb. per thousand ft. (over 75' high)		230	.035		5.50	1.90		7.40	9
2500	Aluminum, 101 lb. per thousand ft. (to 75' high)		280	.029		.95	1.56		2.51	3.47
2600	199 lb. per thousand ft. (over 75' high)		240	.033		1.40	1.82		3.22	4.36
3000	Arrester, 175 volt AC to ground		8	1	Ea.	103	54.50		157.50	198

26 41 Facility Lightning Protection

26 41 13 – Lightning Protection for Structures

26 41 13.13 Lightning Protection for Buildings		Crew	Daily Output	Labor-Hours	Unit	2015 Bare Costs Material	Labor	Equipment	Total	Total Incl O&P
3100	650 volt AC to ground	1 Elec	6.70	1.194	Ea.	89.50	65.50		155	200
4000	Air terminals, copper									
4010	3/8" x 10"	1 Elec	50	.160	Ea.	7.10	8.75		15.85	21.50
4020	1/2" x 12"		38	.211		10.65	11.50		22.15	29.50
4030	5/8" x 12"		33	.242		16	13.25		29.25	38
4040	1/2" x 24"		30	.267		20	14.60		34.60	44.50
4050	1/2" x 60"		19	.421		49	23		72	89.50
4060	Air terminals, aluminum									
4070	3/8" x 10"	1 Elec	50	.160	Ea.	2.87	8.75		11.62	16.70
4080	1/2" x 12"		38	.211		3.34	11.50		14.84	21.50
4090	5/8" x 12"		33	.242		4.62	13.25		17.87	25.50
4100	1/2" x 24"		30	.267		5.50	14.60		20.10	28.50
4110	1/2" x 60"		19	.421		13.70	23		36.70	50.50
4200	Air terminal bases, copper									
4210	Adhesive bases, 1/2"	1 Elec	15	.533	Ea.	16.75	29		45.75	63.50
4215	Adhesive base, lightning air terminal					16.75			16.75	18.45
4220	Bolted bases	1 Elec	9	.889		14.70	48.50		63.20	91.50
4230	Hinged bases		7	1.143		29.50	62.50		92	130
4240	Side mounted bases		9	.889		15.70	48.50		64.20	93
4250	Offset point support		9	.889		29.50	48.50		78	108
4260	Concealed base assembly		5	1.600		29	87.50		116.50	168
4270	Tee connector base		13	.615		37	33.50		70.50	93
4280	Tripod base, 36" for 60" air terminal		7	1.143		47.50	62.50		110	149
4290	Intermediate base, 1/2"		15	.533		24	29		53	71
4300	Air terminal bases, aluminum									
4310	Adhesive bases, 1/2"	1 Elec	15	.533	Ea.	7.65	29		36.65	53.50
4320	Bolted bases		9	.889		11.75	48.50		60.25	88.50
4330	Hinged bases		7	1.143		21.50	62.50		84	121
4340	Side mounted bases		9	.889		11.45	48.50		59.95	88
4350	Offset point support		9	.889		17.20	48.50		65.70	94.50
4360	Concealed base assembly		5	1.600		17.15	87.50		104.65	155
4370	Tee connector base		13	.615		11.60	33.50		45.10	65
4380	Tripod base, 36" for 60" air terminal		7	1.143		34	62.50		96.50	135
4390	Intermediate base, 1/2"		15	.533		12	29		41	58
4400	Connector cable, copper									
4410	Through wall connector	1 Elec	7	1.143	Ea.	75.50	62.50		138	180
4420	Through roof connector		5	1.600		80.50	87.50		168	225
4430	Beam connector		7	1.143		28.50	62.50		91	129
4440	Double bolt connector		38	.211		11.25	11.50		22.75	30.50
4450	Bar, copper		40	.200		12.40	10.95		23.35	30.50
4500	Connector cable, aluminum									
4510	Through wall connector	1 Elec	7	1.143	Ea.	30	62.50		92.50	130
4520	Through roof connector		5	1.600		24	87.50		111.50	163
4530	Beam connector		7	1.143		16.50	62.50		79	115
4540	Double bolt connector		38	.211		7.15	11.50		18.65	26
4600	Bonding plates, copper									
4610	I Beam, 8" square	1 Elec	7	1.143	Ea.	19.15	62.50		81.65	118
4620	Purlin, 8" square		7	1.143		28.50	62.50		91	129
4630	Large heavy duty, 16" square		5	1.600		28.50	87.50		116	167
4640	Aluminum, I Beam, 8" square		7	1.143		14.30	62.50		76.80	113
4650	Purlin, 8" square		7	1.143		13.15	62.50		75.65	111
4660	Large heavy duty, 16" square		5	1.600		17.05	87.50		104.55	155
4700	Cable support, copper, loop fastener		38	.211		.77	11.50		12.27	18.70

26 41 Facility Lightning Protection

26 41 13 – Lightning Protection for Structures

26 41 13.13 Lightning Protection for Buildings		Crew	Daily Output	Labor-Hours	Unit	2015 Bare Costs Material	Labor	Equipment	Total	Total Incl O&P
4710	Adhesive cable holder	1 Elec	38	.211	Ea.	1.38	11.50		12.88	19.35
4720	Aluminum, loop fastener		38	.211		.27	11.50		11.77	18.15
4730	Adhesive cable holder		38	.211		.49	11.50		11.99	18.40
4740	Bonding strap, copper, 3/4" x 9-1/2"		15	.533		8.55	29		37.55	54.50
4750	Aluminum, 3/4" x 9-1/2"		15	.533		4.62	29		33.62	50
4760	Swivel adapter, copper, 1/2"		73	.110		13.35	6		19.35	24
4770	Aluminum, 1/2"		73	.110		9.25	6		15.25	19.45

26 51 Interior Lighting

26 51 13 – Interior Lighting Fixtures, Lamps, and Ballasts

26 51 13.10 Fixture Hangers

		Crew	Daily Output	Labor-Hours	Unit	Material	Labor	Equipment	Total	Total Incl O&P
0010	**FIXTURE HANGERS**									
0220	Box hub cover	1 Elec	32	.250	Ea.	4.19	13.70		17.89	25.50
0240	Canopy		12	.667		8.95	36.50		45.45	66.50
0260	Connecting block		40	.200		2.53	10.95		13.48	19.75
0280	Cushion hanger		16	.500		22.50	27.50		50	67.50
0300	Box hanger, with mounting strap		8	1		9	54.50		63.50	94.50
0320	Connecting block		40	.200		2.17	10.95		13.12	19.35
0340	Flexible, 1/2" diameter, 4" long		12	.667		15.60	36.50		52.10	73.50
0360	6" long		12	.667		16.95	36.50		53.45	75
0380	8" long		12	.667		18.75	36.50		55.25	77
0400	10" long		12	.667		20	36.50		56.50	78.50
0420	12" long		12	.667		21	36.50		57.50	80
0440	15" long		12	.667		22	36.50		58.50	81
0460	18" long		12	.667		26	36.50		62.50	85
0480	3/4" diameter, 4" long		10	.800		19.25	44		63.25	89
0500	6" long		10	.800		21.50	44		65.50	91.50
0520	8" long		10	.800		23.50	44		67.50	93.50
0540	10" long		10	.800		23.50	44		67.50	94
0560	12" long		10	.800		25.50	44		69.50	96
0580	15" long		10	.800		28.50	44		72.50	99.50
0600	18" long		10	.800		32	44		76	103

26 51 13.40 Interior HID Fixtures

		Crew	Daily Output	Labor-Hours	Unit	Material	Labor	Equipment	Total	Total Incl O&P
0010	**INTERIOR HID FIXTURES** Incl. lamps, and mounting hardware									
0700	High pressure sodium, recessed, round, 70 watt	1 Elec	3.50	2.286	Ea.	430	125		555	670
0720	100 watt		3.50	2.286		460	125		585	700
0740	150 watt		3.20	2.500		515	137		652	775
0760	Square, 70 watt		3.60	2.222		430	122		552	665
0780	100 watt		3.60	2.222		460	122		582	695
0820	250 watt		3	2.667		680	146		826	970
0840	1000 watt	2 Elec	4.80	3.333		1,200	182		1,382	1,600
0860	Surface, round, 70 watt	1 Elec	3	2.667		615	146		761	900
0880	100 watt		3	2.667		630	146		776	920
0900	150 watt		2.70	2.963		660	162		822	975
0920	Square, 70 watt		3	2.667		585	146		731	865
0940	100 watt		3	2.667		615	146		761	900
0980	250 watt		2.50	3.200		665	175		840	1,000
1040	Pendent, round, 70 watt		3	2.667		640	146		786	925
1060	100 watt		3	2.667		610	146		756	900
1080	150 watt		2.70	2.963		625	162		787	935
1100	Square, 70 watt		3	2.667		665	146		811	955

26 51 13 – Interior Lighting Fixtures, Lamps, and Ballasts

26 51 13.40 Interior HID Fixtures		Crew	Daily Output	Labor-Hours	Unit	Material	Labor	Equipment	Total	Total Incl O&P
						2015 Bare Costs				
1120	100 watt	1 Elec	3	2.667	Ea.	675	146		821	965
1140	150 watt		2.70	2.963		690	162		852	1,000
1160	250 watt		2.50	3.200		935	175		1,110	1,300
1180	400 watt		2.40	3.333		985	182		1,167	1,350
1220	Wall, round, 70 watt		3	2.667		560	146		706	840
1240	100 watt		3	2.667		570	146		716	850
1260	150 watt		2.70	2.963		585	162		747	895
1300	Square, 70 watt		3	2.667		595	146		741	880
1320	100 watt		3	2.667		625	146		771	910
1340	150 watt		2.70	2.963		645	162		807	960
1360	250 watt		2.50	3.200		705	175		880	1,050
1380	400 watt	2 Elec	4.80	3.333		870	182		1,052	1,225
1400	1000 watt	"	3.60	4.444		1,250	243		1,493	1,750
1500	Metal halide, recessed, round, 175 watt	1 Elec	3.40	2.353		385	129		514	620
1520	250 watt	"	3.20	2.500		480	137		617	740
1540	400 watt	2 Elec	5.80	2.759		680	151		831	985
1580	Square, 175 watt	1 Elec	3.40	2.353		385	129		514	625
1640	Surface, round, 175 watt		2.90	2.759		535	151		686	825
1660	250 watt		2.70	2.963		850	162		1,012	1,175
1680	400 watt	2 Elec	4.80	3.333		940	182		1,122	1,300
1720	Square, 175 watt	1 Elec	2.90	2.759		580	151		731	875
1800	Pendent, round, 175 watt		2.90	2.759		705	151		856	1,000
1820	250 watt		2.70	2.963		920	162		1,082	1,250
1840	400 watt	2 Elec	4.80	3.333		1,000	182		1,182	1,375
1880	Square, 175 watt	1 Elec	2.90	2.759		505	151		656	790
1900	250 watt	"	2.70	2.963		605	162		767	915
1920	400 watt	2 Elec	4.80	3.333		975	182		1,157	1,350
1980	Wall, round, 175 watt	1 Elec	2.90	2.759		675	151		826	975
2000	250 watt	"	2.70	2.963		845	162		1,007	1,175
2020	400 watt	2 Elec	4.80	3.333		845	182		1,027	1,200
2060	Square, 175 watt	1 Elec	2.90	2.759		575	151		726	865
2080	250 watt	"	2.70	2.963		600	162		762	910
2100	400 watt	2 Elec	4.80	3.333		845	182		1,027	1,200
2800	High pressure sodium, vaporproof, recessed, 70 watt	1 Elec	3.50	2.286		565	125		690	820
2820	100 watt		3.50	2.286		580	125		705	830
2840	150 watt		3.20	2.500		595	137		732	865
2900	Surface, 70 watt		3	2.667		640	146		786	930
2920	100 watt		3	2.667		670	146		816	960
2940	150 watt		2.70	2.963		695	162		857	1,025
3000	Pendent, 70 watt		3	2.667		630	146		776	920
3020	100 watt		3	2.667		650	146		796	940
3040	150 watt		2.70	2.963		690	162		852	1,000
3100	Wall, 70 watt		3	2.667		680	146		826	975
3120	100 watt		3	2.667		710	146		856	1,000
3140	150 watt		2.70	2.963		740	162		902	1,050
3200	Metal halide, vaporproof, recessed, 175 watt		3.40	2.353		505	129		634	755
3220	250 watt		3.20	2.500		585	137		722	850
3240	400 watt	2 Elec	5.80	2.759		730	151		881	1,050
3260	1000 watt	"	4.80	3.333		1,325	182		1,507	1,750
3280	Surface, 175 watt	1 Elec	2.90	2.759		620	151		771	915
3300	250 watt	"	2.70	2.963		830	162		992	1,175
3320	400 watt	2 Elec	4.80	3.333		1,025	182		1,207	1,400
3340	1000 watt	"	3.60	4.444		1,500	243		1,743	2,025

26 51 13 – Interior Lighting Fixtures, Lamps, and Ballasts

26 51 13.40 Interior HID Fixtures		Crew	Daily Output	Labor-Hours	Unit	Material (2015 Bare Costs)	Labor (2015 Bare Costs)	Equipment (2015 Bare Costs)	Total (2015 Bare Costs)	Total Incl O&P
3360	Pendent, 175 watt	1 Elec	2.90	2.759	Ea.	645	151		796	945
3380	250 watt	"	2.70	2.963		855	162		1,017	1,200
3400	400 watt	2 Elec	4.80	3.333		1,025	182		1,207	1,400
3420	1000 watt	"	3.60	4.444		1,650	243		1,893	2,200
3440	Wall, 175 watt	1 Elec	2.90	2.759		700	151		851	1,000
3460	250 watt	"	2.70	2.963		905	162		1,067	1,250
3480	400 watt	2 Elec	4.80	3.333		1,075	182		1,257	1,450
3500	1000 watt	"	3.60	4.444	↓	1,725	243		1,968	2,250

26 51 13.50 Interior Lighting Fixtures

26 51 13.50 Interior Lighting Fixtures		Crew	Daily Output	Labor-Hours	Unit	Material	Labor	Equipment	Total	Total Incl O&P
0010	**INTERIOR LIGHTING FIXTURES** Including lamps, mounting R265113-40									
0030	hardware and connections									
0100	Fluorescent, C.W. lamps, troffer, recess mounted in grid, RS									
0130	Grid ceiling mount									
0200	Acrylic lens, 1'W x 4'L, two 40 watt	1 Elec	5.70	1.404	Ea.	47.50	77		124.50	171
0210	1'W x 4'L, three 40 watt		5.40	1.481		54	81		135	184
0300	2'W x 2'L, two U40 watt		5.70	1.404		51	77		128	176
0400	2'W x 4'L, two 40 watt		5.30	1.509		50	82.50		132.50	183
0500	2'W x 4'L, three 40 watt		5	1.600		55	87.50		142.50	197
0600	2'W x 4'L, four 40 watt	↓	4.70	1.702		57.50	93		150.50	208
0700	4'W x 4'L, four 40 watt	2 Elec	6.40	2.500		293	137		430	530
0800	4'W x 4'L, six 40 watt		6.20	2.581		305	141		446	555
0900	4'W x 4'L, eight 40 watt	↓	5.80	2.759		315	151		466	580
0910	Acrylic lens, 1'W x 4'L, two 32 watt T8 G	1 Elec	5.70	1.404		59.50	77		136.50	185
0930	2'W x 2'L, two U32 watt T8 G		5.70	1.404		81	77		158	208
0940	2'W x 4'L, two 32 watt T8 G		5.30	1.509		67	82.50		149.50	202
0950	2'W x 4'L, three 32 watt T8 G		5	1.600		68	87.50		155.50	211
0960	2'W x 4'L, four 32 watt T8 G	↓	4.70	1.702	↓	71	93		164	222
1000	Surface mounted, RS									
1030	Acrylic lens with hinged & latched door frame									
1100	1'W x 4'L, two 40 watt	1 Elec	7	1.143	Ea.	64	62.50		126.50	168
1110	1'W x 4'L, three 40 watt		6.70	1.194		66	65.50		131.50	174
1200	2'W x 2'L, two U40 watt		7	1.143		68.50	62.50		131	173
1300	2'W x 4'L, two 40 watt		6.20	1.290		78	70.50		148.50	195
1400	2'W x 4'L, three 40 watt		5.70	1.404		79	77		156	206
1500	2'W x 4'L, four 40 watt	↓	5.30	1.509		81	82.50		163.50	217
1600	4'W x 4'L, four 40 watt	2 Elec	7.20	2.222		400	122		522	630
1700	4'W x 4'L, six 40 watt		6.60	2.424		435	133		568	680
1800	4'W x 4'L, eight 40 watt		6.20	2.581		450	141		591	715
1900	2'W x 8'L, four 40 watt		6.40	2.500		159	137		296	385
2000	2'W x 8'L, eight 40 watt	↓	6.20	2.581	↓	171	141		312	405
2010	Acrylic wrap around lens									
2020	6"W x 4'L, one 40 watt	1 Elec	8	1	Ea.	65	54.50		119.50	156
2030	6"W x 8'L, two 40 watt	2 Elec	8	2		71.50	109		180.50	248
2040	11"W x 4'L, two 40 watt	1 Elec	7	1.143		43.50	62.50		106	145
2050	11"W x 8'L, four 40 watt	2 Elec	6.60	2.424		71.50	133		204.50	284
2060	16"W x 4'L, four 40 watt	1 Elec	5.30	1.509		71.50	82.50		154	207
2070	16"W x 8'L, eight 40 watt	2 Elec	6.40	2.500		156	137		293	385
2080	2'W x 2'L, two U40 watt	1 Elec	7	1.143	↓	87.50	62.50		150	194
2100	Strip fixture									
2130	Surface mounted									
2200	4' long, one 40 watt, RS	1 Elec	8.50	.941	Ea.	27.50	51.50		79	110
2300	4' long, two 40 watt, RS	↓	8	1	↓	38.50	54.50		93	127

26 51 13 – Interior Lighting Fixtures, Lamps, and Ballasts

26 51 13.50 Interior Lighting Fixtures			Crew	Daily Output	Labor-Hours	Unit	Material	2015 Bare Costs Labor	Equipment	Total	Total Incl O&P
2310	4′ long, two 32 watt T8, RS	G	1 Elec	8	1	Ea.	77.50	54.50		132	170
2400	4′ long, one 40 watt, SL			8	1		46.50	54.50		101	136
2500	4′ long, two 40 watt, SL			7	1.143		63	62.50		125.50	167
2580	8′ long, one 60 watt T8, SL	G	2 Elec	13.40	1.194		83.50	65.50		149	193
2590	8′ long, two 60 watt T8, SL	G		12.40	1.290		86.50	70.50		157	204
2600	8′ long, one 75 watt, SL			13.40	1.194		48	65.50		113.50	154
2700	8′ long, two 75 watt, SL			12.40	1.290		58	70.50		128.50	173
2800	4′ long, two 60 watt, HO		1 Elec	6.70	1.194		93.50	65.50		159	204
2810	4′ long, two 54 watt, T5HO	G	"	6.70	1.194		161	65.50		226.50	278
2900	8′ long, two 110 watt, HO		2 Elec	10.60	1.509		98.50	82.50		181	236
2910	4′ long, two 115 watt, VHO		1 Elec	6.50	1.231		130	67.50		197.50	247
2920	8′ long, two 215 watt, VHO		2 Elec	10.40	1.538		140	84		224	284
2950	High bay pendent mounted, 16″ W x 4′ L, four 54 watt, T5HO	G		8.90	1.798		229	98.50		327.50	405
2952	2′ W x 4′ L, six 54 watt, T5HO	G		8.50	1.882		305	103		408	495
2954	2′ W x 4′ L, six 32 watt, T8	G		8.50	1.882		179	103		282	355
3000	Strip, pendent mounted, industrial, white porcelain enamel										
3100	4′ long, two 40 watt, RS		1 Elec	5.70	1.404	Ea.	52.50	77		129.50	177
3110	4′ long, two 32 watt T8, RS	G		5.70	1.404		74	77		151	200
3200	4′ long, two 60 watt, HO			5	1.600		82.50	87.50		170	227
3290	8′ long, two 60 watt T8, SL	G	2 Elec	8.80	1.818		115	99.50		214.50	281
3300	8′ long, two 75 watt, SL			8.80	1.818		98	99.50		197.50	262
3400	8′ long, two 110 watt, HO			8	2		125	109		234	305
3410	Acrylic finish, 4′ long, two 40 watt, RS		1 Elec	5.70	1.404		86	77		163	214
3420	4′ long, two 60 watt, HO			5	1.600		160	87.50		247.50	310
3430	4′ long, two 115 watt, VHO			4.80	1.667		207	91		298	370
3440	8′ long, two 75 watt, SL		2 Elec	8.80	1.818		169	99.50		268.50	340
3450	8′ long, two 110 watt, HO			8	2		192	109		301	380
3460	8′ long, two 215 watt, VHO			7.60	2.105		271	115		386	475
3470	Troffer, air handling, 2′W x 4′L with four 32 watt T8	G	1 Elec	4	2		112	109		221	292
3480	2′W x 2′L with two U32 watt T8	G		5.50	1.455		108	79.50		187.50	242
3490	Air connector insulated, 5″ diameter			20	.400		67.50	22		89.50	108
3500	6″ diameter			20	.400		69	22		91	110
3502	Troffer, direct/indirect, 2′W x 4′L with two 32 W T8	G		5.30	1.509		273	82.50		355.50	430
3510	Troffer parabolic lay-in, 1′W x 4′L with one 32 W T8	G		5.70	1.404		114	77		191	245
3520	1′W x 4′L with two 32 W T8	G		5.30	1.509		137	82.50		219.50	278
3525	2′W x 2′L with two U32 W T8	G		5.70	1.404		115	77		192	246
3530	2′W x 4′L with three 32 W T8	G		5	1.600		128	87.50		215.50	276
3535	Downlight, recess mounted	G		8	1		148	54.50		202.50	247
3540	Wall wash reflector, recess mounted	G		8	1		105	54.50		159.50	201
3550	Direct/indirect, 4′ long, stl., pendent mtd.	G		5	1.600		162	87.50		249.50	315
3560	4′ long, alum., pendent mtd.	G		5	1.600		335	87.50		422.50	505
3565	Prefabricated cove, 4′ long, stl. continuous row	G		5	1.600		203	87.50		290.50	360
3570	4′ long, alum. continuous row	G		5	1.600		345	87.50		432.50	515
3580	Wet location, recess mounted, 2′W x 4′L with two 32 watt T8	G		5.30	1.509		237	82.50		319.50	390
3590	Pendent mounted, 2′W x 4′L with two 32 watt T8	G		5.70	1.404		365	77		442	525
4000	Induction lamp, integral ballast, ceiling mounted										
4110	High bay, aluminum reflector, 160 watt		1 Elec	3.20	2.500	Ea.	955	137		1,092	1,250
4120	320 watt		"	3	2.667		1,725	146		1,871	2,100
4130	480 watt		2 Elec	5.80	2.759		2,600	151		2,751	3,100
4150	Low bay, aluminum reflector, 250 watt		1 Elec	3.20	2.500		755	137		892	1,050
4170	Garage, aluminum reflector, 80 watt			3.60	2.222		755	122		877	1,025
4180	Vandalproof, aluminum reflector, 100 watt			3.20	2.500		535	137		672	795
4220	Metal halide, integral ballast, ceiling, recess mounted										

26 51 13 – Interior Lighting Fixtures, Lamps, and Ballasts

26 51 13.50 Interior Lighting Fixtures		Crew	Daily Output	Labor-Hours	Unit	2015 Bare Costs Material	Labor	Equipment	Total	Total Incl O&P
4230	prismatic glass lens, floating door									
4240	2'W x 2'L, 250 watt	1 Elec	3.20	2.500	Ea.	315	137		452	560
4250	2'W x 2'L, 400 watt	2 Elec	5.80	2.759		365	151		516	635
4260	Surface mounted, 2'W x 2'L, 250 watt	1 Elec	2.70	2.963		345	162		507	630
4270	400 watt	2 Elec	4.80	3.333		405	182		587	725
4280	High bay, aluminum reflector,									
4290	Single unit, 400 watt	2 Elec	4.60	3.478	Ea.	410	190		600	745
4300	Single unit, 1000 watt		4	4		590	219		809	990
4310	Twin unit, 400 watt		3.20	5		820	274		1,094	1,325
4320	Low bay, aluminum reflector, 250W DX lamp	1 Elec	3.20	2.500		360	137		497	610
4330	400 watt lamp	2 Elec	5	3.200		525	175		700	850
4340	High pressure sodium integral ballast ceiling, recess mounted									
4350	prismatic glass lens, floating door									
4360	2'W x 2'L, 150 watt lamp	1 Elec	3.20	2.500	Ea.	385	137		522	635
4370	2'W x 2'L, 400 watt lamp	2 Elec	5.80	2.759		460	151		611	740
4380	Surface mounted, 2'W x 2'L, 150 watt lamp	1 Elec	2.70	2.963		470	162		632	770
4390	400 watt lamp	2 Elec	4.80	3.333		525	182		707	860
4400	High bay, aluminum reflector,									
4410	Single unit, 400 watt lamp	2 Elec	4.60	3.478	Ea.	380	190		570	710
4430	Single unit, 1000 watt lamp	"	4	4		545	219		764	940
4440	Low bay, aluminum reflector, 150 watt lamp	1 Elec	3.20	2.500		325	137		462	570
4445	High bay H.I.D. quartz restrike	"	16	.500		163	27.50		190.50	223
4450	Incandescent, high hat can, round alzak reflector, prewired									
4470	100 watt	1 Elec	8	1	Ea.	71.50	54.50		126	163
4480	150 watt		8	1		104	54.50		158.50	199
4500	300 watt		6.70	1.194		241	65.50		306.50	365
4520	Round with reflector and baffles, 150 watt		8	1		51	54.50		105.50	141
4540	Round with concentric louver, 150 watt PAR		8	1		78.50	54.50		133	171
4600	Square glass lens with metal trim, prewired									
4630	100 watt	1 Elec	6.70	1.194	Ea.	55	65.50		120.50	162
4700	200 watt		6.70	1.194		97	65.50		162.50	208
4800	300 watt		5.70	1.404		145	77		222	278
4810	500 watt		5	1.600		285	87.50		372.50	450
4900	Ceiling/wall, surface mounted, metal cylinder, 75 watt		10	.800		56	44		100	130
4920	150 watt		10	.800		80.50	44		124.50	157
4930	300 watt		8	1		167	54.50		221.50	269
5000	500 watt		6.70	1.194		360	65.50		425.50	500
5010	Square, 100 watt		8	1		112	54.50		166.50	208
5020	150 watt		8	1		123	54.50		177.50	221
5030	300 watt		7	1.143		335	62.50		397.50	465
5040	500 watt		6	1.333		340	73		413	490
5200	Ceiling, surface mounted, opal glass drum									
5300	8", one 60 watt lamp	1 Elec	10	.800	Ea.	44.50	44		88.50	117
5400	10", two 60 watt lamps		8	1		50	54.50		104.50	140
5500	12", four 60 watt lamps		6.70	1.194		70.50	65.50		136	179
5510	Pendent, round, 100 watt		8	1		112	54.50		166.50	208
5520	150 watt		8	1		122	54.50		176.50	220
5530	300 watt		6.70	1.194		170	65.50		235.50	288
5540	500 watt		5.50	1.455		320	79.50		399.50	480
5550	Square, 100 watt		6.70	1.194		151	65.50		216.50	267
5560	150 watt		6.70	1.194		157	65.50		222.50	274
5570	300 watt		5.70	1.404		230	77		307	370
5580	500 watt		5	1.600		310	87.50		397.50	475

26 51 Interior Lighting

26 51 13 – Interior Lighting Fixtures, Lamps, and Ballasts

26 51 13.50 Interior Lighting Fixtures		Crew	Daily Output	Labor-Hours	Unit	Material	Labor	Equipment	Total	Total Incl O&P
						2015 Bare Costs				
5600	Wall, round, 100 watt	1 Elec	8	1	Ea.	65.50	54.50		120	157
5620	300 watt		8	1		131	54.50		185.50	229
5630	500 watt		6.70	1.194		375	65.50		440.50	515
5640	Square, 100 watt		8	1		103	54.50		157.50	198
5650	150 watt		8	1		105	54.50		159.50	201
5660	300 watt		7	1.143		165	62.50		227.50	279
5670	500 watt		6	1.333		291	73		364	435
6010	Vapor tight, incandescent, ceiling mounted, 200 watt		6.20	1.290		82.50	70.50		153	200
6020	Recessed, 200 watt		6.70	1.194		123	65.50		188.50	237
6030	Pendent, 200 watt		6.70	1.194		82.50	65.50		148	192
6040	Wall, 200 watt		8	1		79.50	54.50		134	172
6100	Fluorescent, surface mounted, 2 lamps, 4′L, RS, 40 watt		3.20	2.500		115	137		252	340
6110	Industrial, 2 lamps 4′ long in tandem, 430 MA		2.20	3.636		212	199		411	545
6130	2 lamps 4′ long, 800 MA		1.90	4.211		182	230		412	555
6160	Pendent, indust, 2 lamps 4′L in tandem, 430 MA		1.90	4.211		248	230		478	630
6170	2 lamps 4′ long, 430 MA		2.30	3.478		166	190		356	480
6180	2 lamps 4′ long, 800 MA		1.70	4.706		209	257		466	630
6850	Vandalproof, surface mounted, fluorescent, two 32 watt T8	G		3.20	2.500		252	137		389	490
6860	Incandescent, one 150 watt		8	1		98.50	54.50		153	193
6900	Mirror light, fluorescent, RS, acrylic enclosure, two 40 watt		8	1		128	54.50		182.50	225
6910	One 40 watt		8	1		106	54.50		160.50	201
6920	One 20 watt		12	.667		81.50	36.50		118	147
7000	Low bay, aluminum reflector. 70 watt, high pressure sodium		4	2		275	109		384	475
7010	250 watt		3.20	2.500		365	137		502	615
7020	400 watt	2 Elec	5	3.200		385	175		560	690
7500	Ballast replacement, by weight of ballast, to 15′ high									
7520	Indoor fluorescent, less than 2 lb.	1 Elec	10	.800	Ea.	26	44		70	96.50
7540	Two 40W, watt reducer, 2 to 5 lb.		9.40	.851		41	46.50		87.50	118
7560	Two F96 slimline, over 5 lb.		8	1		77	54.50		131.50	169
7580	Vaportite ballast, less than 2 lb.		9.40	.851		26	46.50		72.50	101
7600	2 lb. to 5 lb.		8.90	.899		41	49		90	122
7620	Over 5 lb.		7.60	1.053		77	57.50		134.50	174
7630	Electronic ballast for two tubes		8	1		37	54.50		91.50	126
7640	Dimmable ballast one lamp G		8	1		106	54.50		160.50	202
7650	Dimmable ballast two-lamp G		7.60	1.053		104	57.50		161.50	203
7690	Emergency ballast (factory installed in fixture)					157			157	173
7990	Decorator									
8000	Pendent RLM in colors, shallow dome, 12″ diam. 100 W	1 Elec	8	1	Ea.	80.50	54.50		135	173
8010	Regular dome, 12″ diam., 100 watt		8	1		83	54.50		137.50	176
8020	16″ diam., 200 watt		7	1.143		84.50	62.50		147	190
8030	18″ diam., 300 watt		6	1.333		93	73		166	215
8100	Picture framing light		16	.500		94	27.50		121.50	146
8150	Miniature low voltage, recessed, pinhole		8	1		138	54.50		192.50	237
8160	Star		8	1		134	54.50		188.50	232
8170	Adjustable cone		8	1		168	54.50		222.50	270
8180	Eyeball		8	1		126	54.50		180.50	223
8190	Cone		8	1		130	54.50		184.50	228
8200	Coilex baffle		8	1		127	54.50		181.50	225
8210	Surface mounted, adjustable cylinder		8	1		131	54.50		185.50	229
8250	Chandeliers, incandescent									
8260	24″ diam. x 42″ high, 6 light candle	1 Elec	6	1.333	Ea.	445	73		518	605
8270	24″ diam. x 42″ high, 6 light candle w/glass shade		6	1.333		455	73		528	615
8280	17″ diam. x 12″ high, 8 light w/glass panels		8	1		282	54.50		336.50	395

26 51 Interior Lighting

26 51 13 – Interior Lighting Fixtures, Lamps, and Ballasts

26 51 13.50 Interior Lighting Fixtures		Crew	Daily Output	Labor-Hours	Unit	Material	2015 Bare Costs Labor	Equipment	Total	Total Incl O&P
8300	27" diam. x 29"H, 10 light bohemian lead crystal	1 Elec	4	2	Ea.	590	109		699	820
8310	21" diam. x 9" high 6 light sculptured ice crystal		8	1		425	54.50		479.50	550
8500	Accent lights, on floor or edge, 0.5 W low volt incandescent									
8520	incl. transformer & fastenings, based on 100' lengths									
8550	Lights in clear tubing, 12" on center	1 Elec	230	.035	L.F.	8.80	1.90		10.70	12.65
8560	6" on center		160	.050		11.50	2.74		14.24	16.90
8570	4" on center		130	.062		17.55	3.37		20.92	24.50
8580	3" on center		125	.064		19.50	3.50		23	27
8590	2" on center		100	.080		28.50	4.38		32.88	38
8600	Carpet, lights both sides 6" OC, in alum. extrusion		270	.030		26.50	1.62		28.12	31.50
8610	In bronze extrusion		270	.030		30	1.62		31.62	35.50
8620	Carpet-bare floor, lights 18" OC, in alum. extrusion		270	.030		21	1.62		22.62	26
8630	In bronze extrusion		270	.030		25	1.62		26.62	30
8640	Carpet edge-wall, lights 6" OC in alum. extrusion		270	.030		26.50	1.62		28.12	31.50
8650	In bronze extrusion		270	.030		30	1.62		31.62	35.50
8660	Bare floor, lights 18" OC, in aluminum extrusion		300	.027		21	1.46		22.46	26
8670	In bronze extrusion		300	.027		25	1.46		26.46	30
8680	Bare floor conduit, aluminum extrusion		300	.027		7	1.46		8.46	9.95
8690	In bronze extrusion		300	.027		14.05	1.46		15.51	17.70
8700	Step edge to 36", lights 6" OC, in alum. extrusion		100	.080	Ea.	71	4.38		75.38	85
8710	In bronze extrusion		100	.080		73.50	4.38		77.88	88
8720	Step edge to 54", lights 6" OC, in alum. extrusion		100	.080		106	4.38		110.38	124
8730	In bronze extrusion		100	.080		112	4.38		116.38	130
8740	Step edge to 72", lights 6" OC, in alum. extrusion		100	.080		142	4.38		146.38	163
8750	In bronze extrusion		100	.080		155	4.38		159.38	177
8760	Connector, male		32	.250		2.62	13.70		16.32	24
8770	Female with pigtail		32	.250		5.50	13.70		19.20	27
8780	Clamps		400	.020		.51	1.09		1.60	2.25
8790	Transformers, 50 watt		8	1		79.50	54.50		134	172
8800	250 watt		4	2		262	109		371	455
8810	1000 watt		2.70	2.963		470	162		632	770
9000	Minimum labor/equipment charge		3	2.667	Job		146		146	226

26 51 13.55 Interior LED Fixtures

Line	Description	Crew	Daily Output	Labor-Hours	Unit	Material	Labor	Equipment	Total	Total Incl O&P
0010	**INTERIOR LED FIXTURES** Incl. lamps, and mounting hardware									
0100	Downlight, recess mounted, 7.5" diameter, 25 watt G	1 Elec	8	1	Ea.	335	54.50		389.50	450
0120	10" diameter, 36 watt G		8	1		360	54.50		414.50	480
0160	cylinder, 10 watts G		8	1		102	54.50		156.50	197
0180	20 watts G		8	1		585	54.50		639.50	730
1000	Troffer, recess mounted, 2' x 4', 3200 Lumens G		5.30	1.509		138	82.50		220.50	279
1010	4800 Lumens G		5	1.600		179	87.50		266.50	330
1020	6400 Lumens G		4.70	1.702		198	93		291	360
1100	Troffer retrofit lamp, 38 watt G		21	.381		238	21		259	294
1110	60 watt G		20	.400		340	22		362	410
1120	100 watt G		18	.444		510	24.50		534.50	600
1200	Troffer, volumetric recess mounted, 2' x 2' G		5.70	1.404		251	77		328	395
2000	Strip, surface mounted, one light bar 4' long, 3500K G		8.50	.941		299	51.50		350.50	410
2010	5000K G		8	1		299	54.50		353.50	415
2020	Two light bar 4' long, 5000K G		7	1.143		470	62.50		532.50	610
3000	Linear, suspended mounted, one light bar 4' long, 37 watt G		6.70	1.194		195	65.50		260.50	315
3010	One light bar 8' long, 74 watt G	2 Elec	12.20	1.311		360	71.50		431.50	510
3020	Two light bar 4' long, 74 watt G	1 Elec	5.70	1.404		390	77		467	545
3030	Two light bar 8' long, 148 watt G	2 Elec	8.80	1.818		450	99.50		549.50	650

26 51 Interior Lighting

26 51 13 – Interior Lighting Fixtures, Lamps, and Ballasts

26 51 13.55 Interior LED Fixtures			Crew	Daily Output	Labor-Hours	Unit	Material	Labor	Equipment	Total	Total Incl O&P
							2015 Bare Costs				
4000	High bay, surface mounted, round, 150 watts	G	2 Elec	5.41	2.959	Ea.	605	162		767	915
4010	2 bars,164 watts	G		5.41	2.959		570	162		732	875
4020	3 bars, 246 watts	G		5.01	3.197		730	175		905	1,075
4030	4 bars, 328 watts	G		4.60	3.478		895	190		1,085	1,275
4040	5 bars, 410 watts	G	3 Elec	4.20	5.716		1,050	315		1,365	1,625
4050	6 bars, 492 watts	G		3.80	6.324		1,200	345		1,545	1,825
4060	7 bars, 574 watts	G		3.39	7.075		1,350	385		1,735	2,100
4070	8 bars, 656 watts	G		2.99	8.029		1,500	440		1,940	2,325
5000	track, lighthead, 6 watt	G	1 Elec	32	.250		54.50	13.70		68.20	81
5010	9 watt	G	"	32	.250		61.50	13.70		75.20	88.50
6000	Garage, surface mount, 103 watts	G	2 Elec	6.50	2.462		970	135		1,105	1,275
6100	pendent mount, 80 watts	G		6.50	2.462		565	135		700	830
6200	95 watts	G		6.50	2.462		635	135		770	910
6300	125 watts	G		6.50	2.462		690	135		825	970

26 51 13.70 Residential Fixtures		Crew	Daily Output	Labor-Hours	Unit	Material	Labor	Equipment	Total	Total Incl O&P
0010	**RESIDENTIAL FIXTURES**									
0400	Fluorescent, interior, surface, circline, 32 watt & 40 watt	1 Elec	20	.400	Ea.	130	22		152	177
0500	2′ x 2′, two U-tube 32 watt T8		8	1		165	54.50		219.50	266
0700	Shallow under cabinet, two 20 watt		16	.500		62	27.50		89.50	111
0900	Wall mounted, 4′L, two 32 watt T8, with baffle		10	.800		133	44		177	214
2000	Incandescent, exterior lantern, wall mounted, 60 watt		16	.500		55.50	27.50		83	104
2100	Post light, 150W, with 7′ post		4	2		243	109		352	435
2500	Lamp holder, weatherproof with 150W PAR		16	.500		32	27.50		59.50	77.50
2550	With reflector and guard		12	.667		56	36.50		92.50	119
2600	Interior pendent, globe with shade, 150 watt		20	.400		201	22		223	255
9000	Minimum labor/equipment charge		4	2	Job		109		109	169

26 51 13.90 Ballast, Replacement HID		Crew	Daily Output	Labor-Hours	Unit	Material	Labor	Equipment	Total	Total Incl O&P
0010	**BALLAST, REPLACEMENT HID**									
7510	Multi-tap 120/208/240/277 volt									
7550	High pressure sodium, 70 watt	1 Elec	10	.800	Ea.	136	44		180	218
7560	100 watt		9.40	.851		142	46.50		188.50	228
7570	150 watt		9	.889		153	48.50		201.50	244
7580	250 watt		8.50	.941		227	51.50		278.50	330
7590	400 watt		7	1.143		258	62.50		320.50	380
7600	1000 watt		6	1.333		355	73		428	505
7610	Metal halide, 175 watt		8	1		83.50	54.50		138	176
7620	250 watt		8	1		108	54.50		162.50	204
7630	400 watt		7	1.143		135	62.50		197.50	245
7640	1000 watt		6	1.333		231	73		304	365
7650	1500 watt		5	1.600		286	87.50		373.50	450

26 52 Emergency Lighting

26 52 13 – Emergency Lighting Equipments

26 52 13.10 Emergency Lighting and Battery Units	Crew	Daily Output	Labor-Hours	Unit	2015 Bare Costs Material	Labor	Equipment	Total	Total Incl O&P
0010 EMERGENCY LIGHTING AND BATTERY UNITS									
0300 Emergency light units, battery operated									
0350 Twin sealed beam light, 25 watt, 6 volt each									
0500 Lead battery operated	1 Elec	4	2	Ea.	153	109		262	335
0700 Nickel cadmium battery operated		4	2		560	109		669	785
0780 Additional remote mount, sealed beam, 25 W 6 V		26.70	.300		29	16.40		45.40	57.50
0790 Twin sealed beam light, 25 W 6 V each		26.70	.300		54	16.40		70.40	85
0900 Self-contained fluorescent lamp pack		10	.800		156	44		200	239
9000 Minimum labor/equipment charge		4	2	Job		109		109	169

26 53 Exit Signs

26 53 13 – Exit Lighting

26 53 13.10 Exit Lighting Fixtures	Crew	Daily Output	Labor-Hours	Unit	Material	Labor	Equipment	Total	Total Incl O&P
0010 EXIT LIGHTING FIXTURES									
0080 Exit light ceiling or wall mount, incandescent, single face	1 Elec	8	1	Ea.	39	54.50		93.50	128
0100 Double face		6.70	1.194		39	65.50		104.50	144
0120 Explosion proof		3.80	2.105		595	115		710	835
0150 Fluorescent, single face		8	1		67	54.50		121.50	158
0160 Double face		6.70	1.194		75	65.50		140.50	184
0200 LED standard, single face G		8	1		77	54.50		131.50	169
0220 Double face G		6.70	1.194		77	65.50		142.50	186
0230 LED vandal-resistant, single face G		7.27	1.100		212	60		272	325
0240 LED w/battery unit, single face G		4.40	1.818		160	99.50		259.50	330
0260 Double face G		4	2		163	109		272	350
0262 LED w/battery unit, vandal-resistant, single face G		4.40	1.818		245	99.50		344.50	425
0270 Combination emergency light units and exit sign		4	2		179	109		288	365
0290 LED retrofit kits G		60	.133		51	7.30		58.30	67.50
1780 With emergency battery, explosion proof	R-19	7.70	2.597		3,625	142		3,767	4,200
9000 Minimum labor/equipment charge	1 Elec	4	2	Job		109		109	169

26 54 Classified Location Lighting

26 54 13 – Classified Lighting

26 54 13.20 Explosionproof	Crew	Daily Output	Labor-Hours	Unit	Material	Labor	Equipment	Total	Total Incl O&P
0010 **EXPLOSIONPROOF**, incl lamps, mounting hardware and connections									
6310 Metal halide with ballast, ceiling, surface mounted, 175 watt	1 Elec	2.90	2.759	Ea.	915	151		1,066	1,225
6320 250 watt	"	2.70	2.963		1,100	162		1,262	1,450
6330 400 watt	2 Elec	4.80	3.333		1,175	182		1,357	1,575
6340 Ceiling, pendent mounted, 175 watt	1 Elec	2.60	3.077		870	168		1,038	1,225
6350 250 watt	"	2.40	3.333		1,050	182		1,232	1,425
6360 400 watt	2 Elec	4.20	3.810		1,125	208		1,333	1,575
6370 Wall, surface mounted, 175 watt	1 Elec	2.90	2.759		980	151		1,131	1,300
6380 250 watt	"	2.70	2.963		1,150	162		1,312	1,525
6390 400 watt	2 Elec	4.80	3.333		1,250	182		1,432	1,650
6400 High pressure sodium, ceiling surface mounted, 70 watt	1 Elec	3	2.667		1,275	146		1,421	1,625
6410 100 watt		3	2.667		1,325	146		1,471	1,675
6420 150 watt		2.70	2.963		1,375	162		1,537	1,750
6430 Pendent mounted, 70 watt		2.70	2.963		1,225	162		1,387	1,600
6440 100 watt		2.70	2.963		1,275	162		1,437	1,650
6450 150 watt		2.40	3.333		1,300	182		1,482	1,725

26 54 Classified Location Lighting

26 54 13 – Classified Lighting

26 54 13.20 Explosionproof		Crew	Daily Output	Labor-Hours	Unit	2015 Bare Costs Material	Labor	Equipment	Total	Total Incl O&P
6460	Wall mounted, 70 watt	1 Elec	3	2.667	Ea.	1,375	146		1,521	1,725
6470	100 watt		3	2.667		1,425	146		1,571	1,775
6480	150 watt		2.70	2.963		1,450	162		1,612	1,850
6510	Incandescent, ceiling mounted, 200 watt		4	2		1,075	109		1,184	1,350
6520	Pendent mounted, 200 watt		3.50	2.286		925	125		1,050	1,225
6530	Wall mounted, 200 watt		4	2		1,075	109		1,184	1,350
6600	Fluorescent, RS, 4′ long, ceiling mounted, two 40 watt		2.70	2.963		2,850	162		3,012	3,375
6610	Three 40 watt		2.20	3.636		4,125	199		4,324	4,850
6620	Four 40 watt		1.90	4.211		5,300	230		5,530	6,175
6630	Pendent mounted, two 40 watt		2.30	3.478		3,325	190		3,515	3,950
6640	Three 40 watt		1.90	4.211		4,700	230		4,930	5,525
6650	Four 40 watt		1.70	4.706		6,200	257		6,457	7,225

26 55 Special Purpose Lighting

26 55 59 – Display Lighting

26 55 59.10 Track Lighting

		Crew	Daily Output	Labor-Hours	Unit	Material	Labor	Equipment	Total	Total Incl O&P
0010	**TRACK LIGHTING**									
0080	Track, 1 circuit, 4′ section	1 Elec	6.70	1.194	Ea.	44.50	65.50		110	150
0100	8′ section	2 Elec	10.60	1.509		69	82.50		151.50	204
0200	12′ section	"	8.80	1.818		104	99.50		203.50	268
0300	3 circuits, 4′ section	1 Elec	6.70	1.194		83	65.50		148.50	193
0400	8′ section	2 Elec	10.60	1.509		109	82.50		191.50	248
0500	12′ section	"	8.80	1.818		159	99.50		258.50	330
1000	Feed kit, surface mounting	1 Elec	16	.500		13	27.50		40.50	57
1100	End cover		24	.333		5.55	18.25		23.80	34
1200	Feed kit, stem mounting, 1 circuit		16	.500		39	27.50		66.50	85
1300	3 circuit		16	.500		39	27.50		66.50	85
2000	Electrical joiner, for continuous runs, 1 circuit		32	.250		21	13.70		34.70	44
2100	3 circuit		32	.250		59	13.70		72.70	86
2200	Fixtures, spotlight, 75W PAR halogen		16	.500		65.50	27.50		93	115
2210	50W MR16 halogen		16	.500		171	27.50		198.50	231
3000	Wall washer, 250 watt tungsten halogen		16	.500		132	27.50		159.50	188
3100	Low voltage, 25/50 watt, 1 circuit		16	.500		133	27.50		160.50	189
3120	3 circuit		16	.500		198	27.50		225.50	261
9000	Minimum labor/equipment charge		3	2.667	Job		146		146	226

26 55 61 – Theatrical Lighting

26 55 61.10 Lights

		Crew	Daily Output	Labor-Hours	Unit	Material	Labor	Equipment	Total	Total Incl O&P
0010	**LIGHTS**									
2000	Lights, border, quartz, reflector, vented,									
2100	colored or white	1 Elec	20	.400	L.F.	175	22		197	226
2500	Spotlight, follow spot, with transformer, 2,100 watt	"	4	2	Ea.	3,100	109		3,209	3,600
2600	For no transformer, deduct					920			920	1,025
3000	Stationary spot, fresnel quartz, 6″ lens	1 Elec	4	2		133	109		242	315
3100	8″ lens		4	2		234	109		343	425
3500	Ellipsoidal quartz, 1,000W, 6″ lens		4	2		340	109		449	545
3600	12″ lens		4	2		605	109		714	835
4000	Strobe light, 1 to 15 flashes per second, quartz		3	2.667		760	146		906	1,075
4500	Color wheel, portable, five hole, motorized		4	2		197	109		306	385

26 55 Special Purpose Lighting

26 55 63 – Detention Lighting

26 55 63.10 Detention Lighting Fixtures		Crew	Daily Output	Labor-Hours	Unit	2015 Bare Costs Material	Labor	Equipment	Total	Total Incl O&P
0010	**DETENTION LIGHTING FIXTURES**									
1000	Surface mounted, cold rolled steel, 14 ga., 1′ x 4′	1 Elec	5.40	1.481	Ea.	600	81		681	785
1010	2′ x 2′		5.40	1.481		710	81		791	905
1020	2′ x 4′		4.80	1.667		810	91		901	1,025
1100	12 ga., 1′ x 4′		5.40	1.481		845	81		926	1,050
1110	2′ x 2′		5.40	1.481		710	81		791	905
1120	2′ x 4′		4.80	1.667		885	91		976	1,100

26 56 Exterior Lighting

26 56 13 – Lighting Poles and Standards

26 56 13.10 Lighting Poles		Crew	Daily Output	Labor-Hours	Unit	2015 Bare Costs Material	Labor	Equipment	Total	Total Incl O&P
0010	**LIGHTING POLES**									
2800	Light poles, anchor base									
2820	not including concrete bases									
2840	Aluminum pole, 8′ high	1 Elec	4	2	Ea.	705	109		814	945
2850	10′ high		4	2		745	109		854	990
2860	12′ high		3.80	2.105		775	115		890	1,025
2870	14′ high		3.40	2.353		805	129		934	1,075
2880	16′ high		3	2.667		880	146		1,026	1,200
3000	20′ high	R-3	2.90	6.897		935	375	47.50	1,357.50	1,650
3200	30′ high		2.60	7.692		1,775	420	53	2,248	2,650
3400	35′ high		2.30	8.696		1,925	470	60	2,455	2,925
3600	40′ high		2	10		2,200	545	69	2,814	3,350
3800	Bracket arms, 1 arm	1 Elec	8	1		121	54.50		175.50	218
4000	2 arms		8	1		243	54.50		297.50	350
4200	3 arms		5.30	1.509		365	82.50		447.50	530
4400	4 arms		5.30	1.509		485	82.50		567.50	665
4500	Steel pole, galvanized, 8′ high		3.80	2.105		610	115		725	850
4510	10′ high		3.70	2.162		635	118		753	885
4520	12′ high		3.40	2.353		690	129		819	955
4530	14′ high		3.10	2.581		730	141		871	1,025
4540	16′ high		2.90	2.759		775	151		926	1,100
4550	18′ high		2.70	2.963		820	162		982	1,150
4600	20′ high	R-3	2.60	7.692		1,100	420	53	1,573	1,900
4800	30′ high		2.30	8.696		1,300	470	60	1,830	2,225
5000	35′ high		2.20	9.091		1,425	495	62.50	1,982.50	2,425
5200	40′ high		1.70	11.765		1,775	640	81	2,496	3,025
5400	Bracket arms, 1 arm	1 Elec	8	1		180	54.50		234.50	283
5600	2 arms		8	1		216	54.50		270.50	320
5800	3 arms		5.30	1.509		254	82.50		336.50	410
6000	4 arms		5.30	1.509		310	82.50		392.50	470
6100	Fiberglass pole, 1 or 2 fixtures, 20′ high	R-3	4	5		675	272	34.50	981.50	1,200
6200	30′ high		3.60	5.556		835	300	38.50	1,173.50	1,425
6300	35′ high		3.20	6.250		1,300	340	43	1,683	2,000
6400	40′ high		2.80	7.143		1,525	390	49	1,964	2,325
6420	Wood pole, 4-1/2″ x 5-1/8″, 8′ high	1 Elec	6	1.333		330	73		403	480
6430	10′ high		6	1.333		375	73		448	530
6440	12′ high		5.70	1.404		475	77		552	645
6450	15′ high		5	1.600		555	87.50		642.50	745
6460	20′ high		4	2		670	109		779	910

26 56 Exterior Lighting

26 56 13 – Lighting Poles and Standards

26 56 13.10 Lighting Poles		Crew	Daily Output	Labor-Hours	Unit	Material	2015 Bare Costs Labor	Equipment	Total	Total Incl O&P
7300	Transformer bases, not including concrete bases									
7320	Maximum pole size, steel, 40′ high	1 Elec	2	4	Ea.	1,425	219		1,644	1,900
7340	Cast aluminum, 30′ high		3	2.667		750	146		896	1,050
7350	40′ high		2.50	3.200		1,150	175		1,325	1,525

26 56 16 – Parking Lighting

26 56 16.55 Parking LED Lighting

			Crew	Daily Output	Labor-Hours	Unit	Material	Labor	Equipment	Total	Total Incl O&P
0010	**PARKING LED LIGHTING**										
0100	Round pole mounting, 88 lamp watts	G	1 Elec	2	4	Ea.	995	219		1,214	1,450
0110	Square pole mounting, 223 lamp watts	G	″	2	4	″	1,750	219		1,969	2,275

26 56 19 – Roadway Lighting

26 56 19.20 Roadway Luminaire

		Crew	Daily Output	Labor-Hours	Unit	Material	Labor	Equipment	Total	Total Incl O&P
0010	**ROADWAY LUMINAIRE**									
2650	Roadway area luminaire, low pressure sodium, 135 watt	1 Elec	2	4	Ea.	650	219		869	1,050
2700	180 watt	″	2	4		700	219		919	1,100
2750	Metal halide, 400 watt	2 Elec	4.40	3.636		555	199		754	920
2760	1000 watt		4	4		625	219		844	1,025
2780	High pressure sodium, 400 watt		4.40	3.636		580	199		779	950
2790	1000 watt		4	4		660	219		879	1,075

26 56 23 – Area Lighting

26 56 23.10 Exterior Fixtures

		Crew	Daily Output	Labor-Hours	Unit	Material	Labor	Equipment	Total	Total Incl O&P
0010	**EXTERIOR FIXTURES** With lamps									
0200	Wall mounted, incandescent, 100 watt	1 Elec	8	1	Ea.	35	54.50		89.50	123
0400	Quartz, 500 watt		5.30	1.509		50	82.50		132.50	183
0420	1500 watt		4.20	1.905		103	104		207	274
1100	Wall pack, low pressure sodium, 35 watt		4	2		227	109		336	420
1150	55 watt		4	2		270	109		379	465
1160	High pressure sodium, 70 watt		4	2		224	109		333	415
1170	150 watt		4	2		244	109		353	435
1180	Metal Halide, 175 watt		4	2		248	109		357	440
1190	250 watt		4	2		282	109		391	480
1195	400 watt		4	2		335	109		444	535
1250	Induction lamp, 40 watt		4	2		340	109		449	545
1260	80 watt		4	2		580	109		689	810
1278	LED, poly lens, 26 watt		4	2		330	109		439	530
1280	110 watt		4	2		1,300	109		1,409	1,625
1500	LED, glass lens, 13 watt		4	2		330	109		439	530

26 56 23.55 Exterior LED Fixtures

			Crew	Daily Output	Labor-Hours	Unit	Material	Labor	Equipment	Total	Total Incl O&P
0010	**EXTERIOR LED FIXTURES**										
0100	Wall mounted, indoor/outdoor, 12 watt	G	1 Elec	10	.800	Ea.	258	44		302	350
0110	32 watt	G		10	.800		350	44		394	455
0120	66 watt	G		10	.800		500	44		544	620
0200	outdoor, 110 watt	G		10	.800		765	44		809	910
0210	220 watt	G		10	.800		1,350	44		1,394	1,550
0300	modular, type IV, 120 V, 50 lamp watts	G		9	.889		930	48.50		978.50	1,100
0310	101 lamp watts	G		9	.889		1,050	48.50		1,098.50	1,225
0320	126 lamp watts	G		9	.889		1,325	48.50		1,373.50	1,525
0330	202 lamp watts	G		9	.889		1,500	48.50		1,548.50	1,725
0340	240 V, 50 lamp watts	G		8	1		970	54.50		1,024.50	1,150
0350	101 lamp watts	G		8	1		1,100	54.50		1,154.50	1,275
0360	126 lamp watts	G		8	1		1,350	54.50		1,404.50	1,575
0370	202 lamp watts	G		8	1		1,550	54.50		1,604.50	1,775

26 56 Exterior Lighting

26 56 23 – Area Lighting

26 56 23.55 Exterior LED Fixtures		Crew	Daily Output	Labor-Hours	Unit	Material (2015 Bare Costs)	Labor (2015 Bare Costs)	Equipment (2015 Bare Costs)	Total (2015 Bare Costs)	Total Incl O&P
0400	wall pack, glass, 13 lamp watts G	1 Elec	4	2	Ea.	360	109		469	565
0410	poly w/photocell, 26 lamp watts G		4	2		226	109		335	415
0420	50 lamp watts G		4	2		530	109		639	750
0430	replacement, 40 watts G		4	2		460	109		569	675
0440	60 watts G		4	2		580	109		689	810

26 56 26 – Landscape Lighting

26 56 26.20 Landscape Fixtures		Crew	Daily Output	Labor-Hours	Unit	Material	Labor	Equipment	Total	Total Incl O&P
0010	**LANDSCAPE FIXTURES**									
7380	Landscape recessed uplight, incl. housing, ballast, transformer									
7390	& reflector, not incl. conduit, wire, trench									
7420	Incandescent, 250 watt	1 Elec	5	1.600	Ea.	590	87.50		677.50	785
7440	Quartz, 250 watt		5	1.600		560	87.50		647.50	750
7460	500 watt		4	2		575	109		684	805

26 56 33 – Walkway Lighting

26 56 33.10 Walkway Luminaire		Crew	Daily Output	Labor-Hours	Unit	Material	Labor	Equipment	Total	Total Incl O&P
0010	**WALKWAY LUMINAIRE**									
6500	Bollard light, lamp & ballast, 42" high with polycarbonate lens									
6800	Metal halide, 175 watt	1 Elec	3	2.667	Ea.	815	146		961	1,125
6900	High pressure sodium, 70 watt		3	2.667		835	146		981	1,150
7000	100 watt		3	2.667		835	146		981	1,150
7100	150 watt		3	2.667		815	146		961	1,125
7200	Incandescent, 150 watt		3	2.667		595	146		741	880
7810	Walkway luminaire, square 16", metal halide 250 watt		2.70	2.963		635	162		797	950
7820	High pressure sodium, 70 watt		3	2.667		730	146		876	1,025
7830	100 watt		3	2.667		745	146		891	1,050
7840	150 watt		3	2.667		745	146		891	1,050
7850	200 watt		3	2.667		750	146		896	1,050
7910	Round 19", metal halide, 250 watt		2.70	2.963		930	162		1,092	1,275
7920	High pressure sodium, 70 watt		3	2.667		1,025	146		1,171	1,350
7930	100 watt		3	2.667		1,025	146		1,171	1,350
7940	150 watt		3	2.667		1,025	146		1,171	1,350
7950	250 watt		2.70	2.963		1,075	162		1,237	1,425
8000	Sphere 14" opal, incandescent, 200 watt		4	2		295	109		404	495
8020	Sphere 18" opal, incandescent, 300 watt		3.50	2.286		355	125		480	585
8040	Sphere 16" clear, high pressure sodium, 70 watt		3	2.667		615	146		761	905
8050	100 watt		3	2.667		660	146		806	950
8100	Cube 16" opal, incandescent, 300 watt		3.50	2.286		390	125		515	625
8120	High pressure sodium, 70 watt		3	2.667		570	146		716	855
8130	100 watt		3	2.667		585	146		731	870
8230	Lantern, high pressure sodium, 70 watt		3	2.667		510	146		656	790
8240	100 watt		3	2.667		550	146		696	830
8250	150 watt		3	2.667		515	146		661	790
8260	250 watt		2.70	2.963		720	162		882	1,050
8270	Incandescent, 300 watt		3.50	2.286		380	125		505	615
8330	Reflector 22" w/globe, high pressure sodium, 70 watt		3	2.667		490	146		636	765
8340	100 watt		3	2.667		500	146		646	775
8350	150 watt		3	2.667		505	146		651	780
8360	250 watt		2.70	2.963		645	162		807	960
9000	Minimum labor/equipment charge		3.75	2.133	Job		117		117	181

26 56 Exterior Lighting

26 56 36 – Flood Lighting

26 56 36.20 Floodlights		Crew	Daily Output	Labor-Hours	Unit	2015 Bare Costs Material	Labor	Equipment	Total	Total Incl O&P
0010	**FLOODLIGHTS** with ballast and lamp,									
1290	floor mtd, mount with swivel bracket									
1300	Induction lamp, 40 watt	1 Elec	3	2.667	Ea.	450	146		596	720
1310	80 watt		3	2.667		685	146		831	980
1320	150 watt		3	2.667		1,250	146		1,396	1,600
1400	Pole mounted, pole not included									
1950	Metal halide, 175 watt	1 Elec	2.70	2.963	Ea.	340	162		502	620
2000	400 watt	2 Elec	4.40	3.636		420	199		619	775
2200	1000 watt		4	4		580	219		799	975
2210	1500 watt		3.70	4.324		605	237		842	1,025
2250	Low pressure sodium, 55 watt	1 Elec	2.70	2.963		585	162		747	895
2270	90 watt		2	4		645	219		864	1,050
2290	180 watt		2	4		820	219		1,039	1,250
2340	High pressure sodium, 70 watt		2.70	2.963		246	162		408	520
2360	100 watt		2.70	2.963		252	162		414	530
2380	150 watt		2.70	2.963		290	162		452	570
2400	400 watt	2 Elec	4.40	3.636		380	199		579	730
2600	1000 watt	″	4	4		650	219		869	1,050
9005	Solar powered floodlight, w/motion det, incl batt pack for cloudy days G	1 Elec	8	1		119	54.50		173.50	216
9020	Battery pack for G	″	8	1		13.55	54.50		68.05	99.50

26 56 36.55 LED Floodlights

Line	Description	Crew	Daily Output	Labor-Hours	Unit	Material	Labor	Equipment	Total	Total Incl O&P
0010	**LED FLOODLIGHTS** with ballast and lamp,									
0020	Pole mounted, pole not included									
0100	11 watt G	1 Elec	4	2	Ea.	345	109		454	550
0110	46 watt G		4	2		1,050	109		1,159	1,325
0120	90 watt G		4	2		1,725	109		1,834	2,075
0130	288 watt G		4	2		2,125	109		2,234	2,500

26 61 Lighting Systems and Accessories

26 61 13 – Lighting Accessories

26 61 13.30 Fixture Whips

Line	Description	Crew	Daily Output	Labor-Hours	Unit	Material	Labor	Equipment	Total	Total Incl O&P
0010	**FIXTURE WHIPS**									
0080	3/8″ Greenfield, 2 connectors, 6′ long									
0100	TFFN wire, three #18	1 Elec	32	.250	Ea.	7.30	13.70		21	29
0150	Four #18		28	.286		7.75	15.65		23.40	32.50
0200	Three #16		32	.250		7.35	13.70		21.05	29
0250	Four #16		28	.286		7.85	15.65		23.50	32.50
0300	THHN wire, three #14		32	.250		8.55	13.70		22.25	30.50
0350	Four #14		28	.286		9.70	15.65		25.35	34.50
0360	Three #12		32	.250		11.10	13.70		24.80	33

26 61 23 – Lamps Applications

26 61 23.10 Lamps

Line	Description	Crew	Daily Output	Labor-Hours	Unit	Material	Labor	Equipment	Total	Total Incl O&P
0010	**LAMPS**									
0080	Fluorescent, rapid start, cool white, 2′ long, 20 watt	1 Elec	1	8	C	310	440		750	1,025
0100	4′ long, 40 watt		.90	8.889		236	485		721	1,025
0120	3′ long, 30 watt		.90	8.889		410	485		895	1,200
0125	3′ long, 25 watt energy saver G		.90	8.889		1,200	485		1,685	2,050
0150	U-40 watt		.80	10		1,175	545		1,720	2,150
0155	U-34 watt energy saver G		.80	10		1,200	545		1,745	2,175
0170	4′ long, 34 watt energy saver G		.90	8.889		430	485		915	1,225

26 61 23.10 Lamps			Crew	Daily Output	Labor-Hours	Unit	Material (2015 Bare Costs)	Labor (2015 Bare Costs)	Equipment (2015 Bare Costs)	Total (2015 Bare Costs)	Total Incl O&P
0176	2′ long, T8, 17 W energy saver	G	1 Elec	1	8	C	390	440		830	1,100
0178	3′ long, T8, 25 W energy saver	G		.90	8.889		415	485		900	1,225
0180	4′ long, T8, 32 watt energy saver	G		.90	8.889		305	485		790	1,100
0200	Slimline, 4′ long, 40 watt			.90	8.889		1,250	485		1,735	2,125
0210	4′ long, 30 watt energy saver	G		.90	8.889		1,250	485		1,735	2,125
0300	8′ long, 75 watt			.80	10		1,050	545		1,595	2,000
0350	8′ long, 60 watt energy saver	G		.80	10		450	545		995	1,350
0400	High output, 4′ long, 60 watt			.90	8.889		650	485		1,135	1,475
0410	8′ long, 95 watt energy saver	G		.80	10		640	545		1,185	1,550
0500	8′ long, 110 watt			.80	10		640	545		1,185	1,550
0512	2′ long, T5, 14 watt energy saver	G		1	8		860	440		1,300	1,625
0514	3′ long, T5, 21 watt energy saver	G		.90	8.889		1,000	485		1,485	1,875
0516	4′ long, T5, 28 watt energy saver	G		.90	8.889		1,300	485		1,785	2,175
0517	4′ long, T5, 54 watt energy saver	G		.90	8.889		1,575	485		2,060	2,475
0520	Very high output, 4′ long, 110 watt			.90	8.889		2,175	485		2,660	3,150
0525	8′ long, 195 watt energy saver	G		.70	11.429		1,775	625		2,400	2,925
0550	8′ long, 215 watt			.70	11.429		1,575	625		2,200	2,700
0554	Full spectrum, 4′ long, 60 watt			.90	8.889		1,250	485		1,735	2,125
0556	6′ long, 85 watt			.90	8.889		1,300	485		1,785	2,175
0558	8′ long, 110 watt			.80	10		1,825	545		2,370	2,850
0560	Twin tube compact lamp	G		.90	8.889		360	485		845	1,150
0570	Double twin tube compact lamp	G		.80	10		805	545		1,350	1,725
0600	Mercury vapor, mogul base, deluxe white, 100 watt			.30	26.667		3,650	1,450		5,100	6,275
0650	175 watt			.30	26.667		1,725	1,450		3,175	4,125
0700	250 watt			.30	26.667		3,000	1,450		4,450	5,550
0800	400 watt			.30	26.667		4,350	1,450		5,800	7,025
0900	1000 watt			.20	40		7,350	2,200		9,550	11,500
1000	Metal halide, mogul base, 175 watt			.30	26.667		2,550	1,450		4,000	5,075
1100	250 watt			.30	26.667		2,900	1,450		4,350	5,425
1200	400 watt			.30	26.667		4,750	1,450		6,200	7,475
1300	1000 watt			.20	40		6,625	2,200		8,825	10,700
1320	1000 watt, 125,000 initial lumens			.20	40		22,100	2,200		24,300	27,700
1330	1500 watt			.20	40		27,900	2,200		30,100	34,100
1350	High pressure sodium, 70 watt			.30	26.667		3,125	1,450		4,575	5,675
1360	100 watt			.30	26.667		3,250	1,450		4,700	5,825
1370	150 watt			.30	26.667		3,325	1,450		4,775	5,925
1380	250 watt			.30	26.667		4,225	1,450		5,675	6,900
1400	400 watt			.30	26.667		5,525	1,450		6,975	8,325
1450	1000 watt			.20	40		10,000	2,200		12,200	14,400
1500	Low pressure sodium, 35 watt			.30	26.667		12,000	1,450		13,450	15,500
1550	55 watt			.30	26.667		11,900	1,450		13,350	15,400
1600	90 watt			.30	26.667		13,400	1,450		14,850	17,100
1650	135 watt			.20	40		20,000	2,200		22,200	25,400
1700	180 watt			.20	40		23,100	2,200		25,300	28,800
1750	Quartz line, clear, 500 watt			1.10	7.273		1,375	400		1,775	2,150
1760	1500 watt			.20	40		3,050	2,200		5,250	6,750
1762	Spot, MR 16, 50 watt			1.30	6.154		1,075	335		1,410	1,700
1770	Tungsten halogen, T4, 400 watt			1.10	7.273		3,975	400		4,375	5,000
1775	T3, 1200 watt			.30	26.667		4,900	1,450		6,350	7,650
1778	PAR 30, 50 watt			1.30	6.154		1,150	335		1,485	1,775
1780	PAR 38, 90 watt			1.30	6.154		960	335		1,295	1,575
1800	Incandescent, interior, A21, 100 watt			1.60	5		233	274		507	680
1900	A21, 150 watt		↓	1.60	5	↓	166	274		440	610

26 61 Lighting Systems and Accessories

26 61 23 – Lamps Applications

26 61 23.10 Lamps		Crew	Daily Output	Labor-Hours	Unit	Material	2015 Bare Costs Labor	Equipment	Total	Total Incl O&P
2000	A23, 200 watt	1 Elec	1.60	5	C	330	274		604	790
2200	PS 35, 300 watt		1.60	5		860	274		1,134	1,375
2210	PS 35, 500 watt		1.60	5		1,425	274		1,699	1,975
2230	PS 52, 1000 watt		1.30	6.154		2,550	335		2,885	3,325
2240	PS 52, 1500 watt		1.30	6.154		6,825	335		7,160	8,050
2300	R30, 75 watt		1.30	6.154		620	335		955	1,200
2400	R40, 100 watt		1.30	6.154		620	335		955	1,200
2500	Exterior, PAR 38, 75 watt		1.30	6.154		1,850	335		2,185	2,575
2600	PAR 38, 150 watt		1.30	6.154		2,000	335		2,335	2,725
2700	PAR 46, 200 watt		1.10	7.273		3,875	400		4,275	4,875
2800	PAR 56, 300 watt		1.10	7.273		4,425	400		4,825	5,500
3000	Guards, fluorescent lamp, 4' long		1	8		1,400	440		1,840	2,225
3200	8' long		.90	8.889		2,800	485		3,285	3,850
9000	Minimum labor/equipment charge		4	2	Job		109		109	169

26 61 23.55 LED Lamps		Crew	Daily Output	Labor-Hours	Unit	Material	Labor	Equipment	Total	Total Incl O&P
0010	**LED LAMPS**									
0100	LED lamp, interior, shape A60, equal to 60 watt G	1 Elec	160	.050	Ea.	22.50	2.74		25.24	28.50
0200	Globe frosted A60, equal to 60 watt G		160	.050		22.50	2.74		25.24	28.50
0300	Globe earth, equal to 100 watt G		160	.050		74	2.74		76.74	85.50
1100	MR16, 3 watt, replacement of halogen lamp 25 watt G		130	.062		21.50	3.37		24.87	28.50
1200	6 watt replacement of halogen lamp 45 watt G		130	.062		47	3.37		50.37	56.50
2100	10 watt, PAR20, equal to 60 watt G		130	.062		47.50	3.37		50.87	57.50
2200	15 watt, PAR30, equal to 100 watt G		130	.062		79	3.37		82.37	92

26 71 Electrical Machines

26 71 13 – Motors Applications

26 71 13.10 Handling		Crew	Daily Output	Labor-Hours	Unit	Material	Labor	Equipment	Total	Total Incl O&P
0010	**HANDLING** Add to normal labor cost for restricted areas									
5000	Motors									
5100	1/2 HP, 23 pounds	1 Elec	4	2	Ea.		109		109	169
5110	3/4 HP, 28 pounds		4	2			109		109	169
5120	1 HP, 33 pounds		4	2			109		109	169
5130	1-1/2 HP, 44 pounds		3.20	2.500			137		137	212
5140	2 HP, 56 pounds		3	2.667			146		146	226
5150	3 HP, 71 pounds		2.30	3.478			190		190	295
5160	5 HP, 82 pounds		1.90	4.211			230		230	355
5170	7-1/2 HP, 124 pounds		1.50	5.333			292		292	450
5180	10 HP, 144 pounds		1.20	6.667			365		365	565
5190	15 HP, 185 pounds		1	8			440		440	680
5200	20 HP, 214 pounds	2 Elec	1.50	10.667			585		585	905
5210	25 HP, 266 pounds		1.40	11.429			625		625	970
5220	30 HP, 310 pounds		1.20	13.333			730		730	1,125
5230	40 HP, 400 pounds		1	16			875		875	1,350
5240	50 HP, 450 pounds		.90	17.778			970		970	1,500
5250	75 HP, 680 pounds		.80	20			1,100		1,100	1,700
5260	100 HP, 870 pounds	3 Elec	1	24			1,325		1,325	2,025
5270	125 HP, 940 pounds		.80	30			1,650		1,650	2,550
5280	150 HP, 1200 pounds		.70	34.286			1,875		1,875	2,900
5290	175 HP, 1300 pounds		.60	40			2,200		2,200	3,400
5300	200 HP, 1400 pounds		.50	48			2,625		2,625	4,075

26 71 13.20 Motors		Crew	Daily Output	Labor-Hours	Unit	2015 Bare Costs Material	Labor	Equipment	Total	Total Incl O&P
0010	**MOTORS** 230/460 volts, 60 HZ									
0050	Dripproof, premium efficiency, 1.15 service factor									
0060	1800 RPM, 1/4 HP	1 Elec	5.33	1.501	Ea.	190	82		272	335
0070	1/3 HP		5.33	1.501		195	82		277	340
0080	1/2 HP		5.33	1.501		218	82		300	365
0090	3/4 HP		5.33	1.501		248	82		330	400
0100	1 HP		4.50	1.778		279	97		376	455
0150	2 HP		4.50	1.778		335	97		432	520
0200	3 HP		4.50	1.778		560	97		657	765
0250	5 HP		4.50	1.778		610	97		707	820
0300	7.5 HP		4.20	1.905		810	104		914	1,050
0350	10 HP		4	2		990	109		1,099	1,275
0400	15 HP		3.20	2.500		1,375	137		1,512	1,725
0450	20 HP	2 Elec	5.20	3.077		1,750	168		1,918	2,175
0500	25 HP		5	3.200		2,050	175		2,225	2,550
0550	30 HP		4.80	3.333		2,250	182		2,432	2,750
0600	40 HP		4	4		2,925	219		3,144	3,575
0650	50 HP		3.20	5		3,200	274		3,474	3,950
0700	60 HP		2.80	5.714		4,075	315		4,390	4,950
0750	75 HP		2.40	6.667		4,150	365		4,515	5,150
0800	100 HP	3 Elec	2.70	8.889		5,350	485		5,835	6,625
0850	125 HP		2.10	11.429		6,100	625		6,725	7,700
0900	150 HP		1.80	13.333		7,925	730		8,655	9,850
0950	200 HP		1.50	16		9,650	875		10,525	12,000
1000	1200 RPM, 1 HP	1 Elec	4.50	1.778		475	97		572	670
1050	2 HP		4.50	1.778		560	97		657	765
1100	3 HP		4.50	1.778		720	97		817	940
1150	5 HP		4.50	1.778		945	97		1,042	1,175
1200	3600 RPM, 2 HP		4.50	1.778		465	97		562	665
1250	3 HP		4.50	1.778		475	97		572	670
1300	5 HP		4.50	1.778		515	97		612	720
1350	Totally enclosed, premium efficiency 1.15 service factor									
1360	1800 RPM, 1/4 HP	1 Elec	5.33	1.501	Ea.	263	82		345	415
1370	1/3 HP		5.33	1.501		291	82		373	445
1380	1/2 HP		5.33	1.501		340	82		422	500
1390	3/4 HP		5.33	1.501		375	82		457	535
1400	1 HP		4.50	1.778		485	97		582	680
1450	2 HP		4.50	1.778		575	97		672	780
1500	3 HP		4.50	1.778		615	97		712	825
1550	5 HP		4.50	1.778		730	97		827	955
1600	7.5 HP		4.20	1.905		960	104		1,064	1,200
1650	10 HP		4	2		1,100	109		1,209	1,375
1700	15 HP		3.20	2.500		1,625	137		1,762	2,000
1750	20 HP	2 Elec	5.20	3.077		2,225	168		2,393	2,675
1800	25 HP		5	3.200		2,575	175		2,750	3,100
1850	30 HP		4.80	3.333		2,725	182		2,907	3,250
1900	40 HP		4	4		3,325	219		3,544	4,000
1950	50 HP		3.20	5		3,575	274		3,849	4,375
2000	60 HP		2.80	5.714		5,275	315		5,590	6,275
2050	75 HP		2.40	6.667		6,425	365		6,790	7,650
2100	100 HP	3 Elec	2.70	8.889		8,200	485		8,685	9,750
2150	125 HP		2.10	11.429		11,500	625		12,125	13,600

26 71 Electrical Machines

26 71 13 – Motors Applications

26 71 13.20 Motors		Crew	Daily Output	Labor-Hours	Unit	2015 Bare Costs Material	Labor	Equipment	Total	Total Incl O&P
2200	150 HP	3 Elec	1.80	13.333	Ea.	12,600	730		13,330	14,900
2250	200 HP	↓	1.50	16		15,900	875		16,775	18,900
2300	1200 RPM, 1 HP	1 Elec	4.50	1.778		410	97		507	600
2350	2 HP		4.50	1.778		470	97		567	670
2400	3 HP		4.50	1.778		640	97		737	855
2450	5 HP		4.50	1.778		870	97		967	1,100
2500	3600 RPM, 2 HP		4.50	1.778		365	97		462	555
2550	3 HP		4.50	1.778		455	97		552	650
2600	5 HP	↓	4.50	1.778	↓	570	97		667	775

Estimating Tips

27 20 00 Data Communications
27 30 00 Voice Communications
27 40 00 Audio-Video Communications

When estimating material costs for special systems, it is always prudent to obtain manufacturers' quotations for equipment prices and special installation requirements which will affect the total costs.

Reference Numbers

Reference numbers are shown in shaded boxes at the beginning of some major classifications. These numbers refer to related items in the Reference Section. The reference information may be an estimating procedure, an alternate pricing method, or technical information.

Note: Not all subdivisions listed here necessarily appear in this publication. ■

*Note: **Trade Service**, in part, has been used as a reference source for some of the material prices used in Division 27.*

27 01 Operation and Maintenance of Communications Systems

27 01 30 – Operation and Maintenance of Voice Communications

27 01 30.51 Operation and Maintenance of Voice Equipment		Crew	Daily Output	Labor-Hours	Unit	Material	2015 Bare Costs Labor	Equipment	Total	Total Incl O&P
0010	OPERATION AND MAINTENANCE OF VOICE EQUIPMENT									
3400	Remove and replace (reinstall), speaker	1 Elec	6	1.333	Ea.		73		73	113

27 05 Common Work Results for Communications

27 05 05 – Selective Demolition for Communications

27 05 05.20 Electrical Demolition, Communications

27 05 05.20 Electrical Demolition, Communications		Crew	Daily Output	Labor-Hours	Unit	Material	2015 Bare Costs Labor	Equipment	Total	Total Incl O&P
0010	ELECTRICAL DEMOLITION, COMMUNICATIONS R024119-10									
0100	Fiber optics									
0120	Cable R260105-30	1 Elec	2400	.003	L.F.		.18		.18	.28
0160	Multi-channel rack enclosure		6	1.333	Ea.		73		73	113
0180	Patch panel		18	.444	″		24.50		24.50	37.50
0200	Communication cables & fittings									
0220	Voice/data outlet	1 Elec	140	.057	Ea.		3.13		3.13	4.84
0240	Telephone cable		2800	.003	L.F.		.16		.16	.24
0260	Phone jack		135	.059	Ea.		3.24		3.24	5
0300	High performance cable, 2 pair		3000	.003	L.F.		.15		.15	.23
0320	4 pair		2100	.004			.21		.21	.32
0340	25 pair		900	.009			.49		.49	.75
0400	Terminal cabinet		5	1.600	Ea.		87.50		87.50	136
1000	Nurse call system									
1020	Station	1 Elec	24	.333	Ea.		18.25		18.25	28
1040	Standard call button		24	.333			18.25		18.25	28
1060	Corridor dome light or zone indictor		24	.333			18.25		18.25	28
1080	Master control station	2 Elec	2	8			440		440	680

27 05 05.30 Electrical Demolition, Sound and Video

27 05 05.30 Electrical Demolition, Sound and Video		Crew	Daily Output	Labor-Hours	Unit	Material	2015 Bare Costs Labor	Equipment	Total	Total Incl O&P
0010	ELECTRICAL DEMOLITION, SOUND & VIDEO R024119-10									
0100	Cables									
0120	TV antenna lead-in cable R260105-30	1 Elec	2100	.004	L.F.		.21		.21	.32
0140	Sound cable		2400	.003			.18		.18	.28
0160	Microphone cable		2400	.003			.18		.18	.28
0180	Coaxial cable		2400	.003			.18		.18	.28
0200	Doorbell system, not including wires, cables, and conduit		16	.500	Ea.		27.50		27.50	42.50
0220	Door chime or devices		36	.222	″		12.15		12.15	18.80
0300	Public address system, not including wires, cables, and conduit									
0320	Conventional office	1 Elec	16	.500	Speaker		27.50		27.50	42.50
0340	Conventional industrial	″	8	1	″		54.50		54.50	84.50
0352	PA cabinet/panel	2 Elec	5	3.200	Ea.		175		175	271
0400	Sound system, not including wires, cables, and conduit									
0410	Components	1 Elec	24	.333	Ea.		18.25		18.25	28
0412	Speaker		24	.333			18.25		18.25	28
0416	Volume control		24	.333			18.25		18.25	28
0420	Intercom, master station		6	1.333			73		73	113
0440	Remote station		24	.333			18.25		18.25	28
0460	Intercom outlets		24	.333			18.25		18.25	28
0480	Handset		12	.667			36.50		36.50	56.50
0500	Emergency call system, not including wires, cables, and conduit									
0520	Annunciator	1 Elec	4	2	Ea.		109		109	169
0540	Devices	″	16	.500			27.50		27.50	42.50
0600	Master door, buzzer type unit	2 Elec	1.60	10			545		545	845
0800	TV System, not including wires, cables, and conduit									
0820	Master TV antenna system, per outlet	1 Elec	39	.205	Outlet		11.20		11.20	17.35

27 05 Common Work Results for Communications

27 05 05 – Selective Demolition for Communications

27 05 05.30 Electrical Demolition, Sound and Video		Crew	Daily Output	Labor-Hours	Unit	Material	2015 Bare Costs Labor	Equipment	Total	Total Incl O&P
0840	School application, per outlet	1 Elec	16	.500	Outlet		27.50		27.50	42.50
0860	Amplifier		12	.667	Ea.		36.50		36.50	56.50
0880	Antenna		6	1.333	"		73		73	113
0900	One camera & one monitor	2 Elec	7.80	2.051	Total		112		112	174
0920	One camera	1 Elec	8	1	Ea.		54.50		54.50	84.50

27 11 Communications Equipment Room Fittings

27 11 16 – Communications Cabinets, Racks, Frames and Enclosures

27 11 16.10 Public Phone

Code	Description	Crew	Daily Output	Labor-Hours	Unit	Material	Labor	Equipment	Total	Total Incl O&P
0010	**PUBLIC PHONE**									
7600	Telephone with wood backboard									
7620	Single door, 12" H x 12" W x 4" D	1 Elec	5.30	1.509	Ea.	88	82.50		170.50	225
7650	18" H x 12" W x 4" D		4.70	1.702		103	93		196	258
7700	24" H x 12" W x 4" D		4.20	1.905		113	104		217	285
7720	18" H x 18" W x 4" D		4.20	1.905		123	104		227	297
7750	24" H x 18" W x 4" D		4	2		175	109		284	360
7780	36" H x 36" W x 4" D		3.60	2.222		199	122		321	405
7800	24" H x 24" W x 6" D		3.60	2.222		218	122		340	425
7820	30" H x 24" W x 6" D		3.20	2.500		258	137		395	495
7850	30" H x 30" W x 6" D		2.70	2.963		300	162		462	585
7880	36" H x 30" W x 6" D		2.50	3.200		370	175		545	675
7900	48" H x 36" W x 6" D		2.20	3.636		610	199		809	980
7920	Double door, 48" H x 36" W x 6" D		2	4		850	219		1,069	1,275

27 11 19 – Communications Termination Blocks and Patch Panels

27 11 19.10 Termination Blocks and Patch Panels

Code	Description	Crew	Daily Output	Labor-Hours	Unit	Material	Labor	Equipment	Total	Total Incl O&P
0010	**TERMINATION BLOCKS AND PATCH PANELS**									
2960	Patch panel, RJ-45/110 type, 24 ports	2 Elec	6	2.667	Ea.	192	146		338	435
3000	48 ports	3 Elec	6	4		320	219		539	695
3040	96 ports	"	4	6		535	330		865	1,100
3100	Punch down termination per port	1 Elec	107	.075			4.09		4.09	6.35

27 13 Communications Backbone Cabling

27 13 23 – Communications Optical Fiber Backbone Cabling

27 13 23.13 Communications Optical Fiber

Code	Description	Crew	Daily Output	Labor-Hours	Unit	Material	Labor	Equipment	Total	Total Incl O&P
0010	**COMMUNICATIONS OPTICAL FIBER**									
0040	Specialized tools & techniques cause installation costs to vary.									
0070	Fiber optic, cable, bulk simplex, single mode R271323-40	1 Elec	8	1	C.L.F.	22.50	54.50		77	110
0080	Multi mode		8	1		42	54.50		96.50	131
0090	4 strand, single mode		7.34	1.090		38	59.50		97.50	134
0095	Multi mode		7.34	1.090		50.50	59.50		110	148
0100	12 strand, single mode		6.67	1.199		90	65.50		155.50	201
0105	Multi mode		6.67	1.199		96.50	65.50		162	208
0150	Jumper				Ea.	33			33	36.50
0200	Pigtail					33.50			33.50	37
0300	Connector	1 Elec	24	.333		23.50	18.25		41.75	53.50
0350	Finger splice		32	.250		32.50	13.70		46.20	57
0400	Transceiver (low cost bi-directional)		8	1		420	54.50		474.50	545
0450	Rack housing, 4 rack spaces, 12 panels (144 fibers)		2	4		500	219		719	890
0500	Patch panel, 12 ports		6	1.333		247	73		320	385

27 15 Communications Horizontal Cabling

27 15 10 – Special Communications Cabling

27 15 10.23 Sound and Video Cables and Fittings

	27 15 10.23 Sound and Video Cables and Fittings	Crew	Daily Output	Labor-Hours	Unit	Material	Labor (2015 Bare Costs)	Equipment	Total	Total Incl O&P
0010	**SOUND AND VIDEO CABLES & FITTINGS**									
0900	TV antenna lead-in, 300 ohm, #20-2 conductor	1 Elec	7	1.143	C.L.F.	19.50	62.50		82	119
0950	Coaxial, feeder outlet		7	1.143		17.35	62.50		79.85	116
1000	Coaxial, main riser		6	1.333		25	73		98	140
1100	Sound, shielded with drain, #22-2 conductor		8	1		13.95	54.50		68.45	100
1150	#22-3 conductor		7.50	1.067		17.55	58.50		76.05	110
1200	#22-4 conductor		6.50	1.231		22	67.50		89.50	129
1250	Nonshielded, #22-2 conductor		10	.800		14.25	44		58.25	83.50
1300	#22-3 conductor		9	.889		21.50	48.50		70	99
1350	#22-4 conductor		8	1		28.50	54.50		83	116
1400	Microphone cable		8	1		49.50	54.50		104	139

27 15 13 – Communications Copper Horizontal Cabling

27 15 13.13 Communication Cables

	27 15 13.13 Communication Cables		Crew	Daily Output	Labor-Hours	Unit	Material	Labor	Equipment	Total	Total Incl O&P
0010	**COMMUNICATION CABLES**										
2200	Telephone twisted, PVC insulation, #22-2 conductor		1 Elec	10	.800	C.L.F.	10.40	44		54.40	79.50
2250	#22-3 conductor			9	.889		12.45	48.50		60.95	89
2300	#22-4 conductor			8	1		15.90	54.50		70.40	102
2350	#18-2 conductor			9	.889		14.55	48.50		63.05	91.50
2370	Telephone jack, eight pins			32	.250	Ea.	4.12	13.70		17.82	25.50
5000	High performance unshielded twisted pair (UTP)										
5100	Cable, category 3, #24, 2 pair solid, PVC jacket	R271513-75	1 Elec	10	.800	C.L.F.	9.25	44		53.25	78
5200	4 pair solid, PVC jacket			7	1.143		13.90	62.50		76.40	112
5300	25 pair solid, PVC jacket			3	2.667		78.50	146		224.50	315
5400	2 pair solid, plenum			10	.800		13.05	44		57.05	82.50
5500	4 pair solid, plenum			7	1.143		14.15	62.50		76.65	113
5600	25 pair solid, plenum			3	2.667		111	146		257	350
5700	4 pair stranded, PVC jacket			7	1.143		30.50	62.50		93	131
7000	Category 5, #24, 4 pair solid, PVC jacket			7	1.143		14.95	62.50		77.45	113
7100	4 pair solid, plenum			7	1.143		45.50	62.50		108	147
7200	4 pair stranded, PVC jacket			7	1.143		22.50	62.50		85	122
7210	Category 5e, #24, 4 pair solid, PVC jacket			7	1.143		14.30	62.50		76.80	113
7212	4 pair solid, plenum			7	1.143		23.50	62.50		86	123
7214	4 pair stranded, PVC jacket			7	1.143		25.50	62.50		88	126
7240	Category 6, #24, 4 pair solid, PVC jacket			7	1.143		20.50	62.50		83	120
7242	4 pair solid, plenum			7	1.143		24.50	62.50		87	124
7244	4 pair stranded, PVC jacket			7	1.143		25.50	62.50		88	126
7300	Connector, RJ-45, category 5			80	.100	Ea.	1.33	5.45		6.78	9.90
7302	Shielded RJ-45, category 5			72	.111		3.39	6.10		9.49	13.15
7310	Jack, UTP RJ-45, category 3			72	.111		2.35	6.10		8.45	12
7312	Category 5			65	.123		5.05	6.75		11.80	15.95
7314	Category 5e			65	.123		3.29	6.75		10.04	14
7316	Category 6			65	.123		3.34	6.75		10.09	14.05
7322	Jack, shielded RJ-45, category 5			60	.133		6.10	7.30		13.40	18
7324	Category 5e			60	.133		6.10	7.30		13.40	18
7326	Category 6			60	.133		6.10	7.30		13.40	18
7400	Voice/data expansion module, category 5e			8	1		46.50	54.50		101	136
8000	Multipair unshielded non-plenum cable, 150 V PVC jacket										
8002	#22, 2 pair		1 Elec	8.40	.952	C.L.F.	16.75	52		68.75	99
8003	3 pair			8	1		22	54.50		76.50	109
8004	4 pair			7.30	1.096		42	60		102	139
8006	6 pair			6.20	1.290		42.50	70.50		113	156
8008	8 pair			5.70	1.404		55.50	77		132.50	180

27 15 Communications Horizontal Cabling

27 15 13 – Communications Copper Horizontal Cabling

27 15 13.13 Communication Cables		Crew	Daily Output	Labor-Hours	Unit	2015 Bare Costs Material	Labor	Equipment	Total	Total Incl O&P
8010	10 pair	1 Elec	5.30	1.509	C.L.F.	64.50	82.50		147	199
8012	12 pair		4.20	1.905		140	104		244	315
8015	15 pair		3.80	2.105		190	115		305	385
8020	20 pair	2 Elec	6.60	2.424		195	133		328	420
8025	25 pair		6	2.667		226	146		372	475
8030	30 pair		5.60	2.857		375	156		531	655
8040	40 pair		5.20	3.077		495	168		663	800
8050	50 pair		4.90	3.265		615	179		794	955
8100	Multipair unshielded non-plenum cable, 300 V PVC jacket									
8102	#20, 2 pair	1 Elec	7.30	1.096	C.L.F.	26	60		86	122
8103	3 pair		6.70	1.194		34	65.50		99.50	138
8104	4 pair		6	1.333		38	73		111	155
8106	6 pair		5.30	1.509		59	82.50		141.50	193
8108	8 pair		4.40	1.818		97	99.50		196.50	260
8110	10 pair	2 Elec	7.30	2.192		107	120		227	305
8112	12 pair		6.20	2.581		135	141		276	365
8115	15 pair		5.90	2.712		169	148		317	415
8201	#18, 1 pair	1 Elec	8	1		63	54.50		117.50	154
8202	2 pair		6.50	1.231		80	67.50		147.50	193
8203	3 pair		5.60	1.429		121	78		199	254
8204	4 pair		5	1.600		158	87.50		245.50	310
8206	6 pair		4.40	1.818		266	99.50		365.50	445
8208	8 pair		4	2		245	109		354	440
8215	15 pair	2 Elec	5	3.200		650	175		825	990
8300	Multipair shielded non-plenum cable, 300 V PVC jacket									
8303	#22, 3 pair	1 Elec	6.70	1.194	C.L.F.	62.50	65.50		128	170
8306	6 pair		5.70	1.404		125	77		202	256
8309	9 pair		5	1.600		195	87.50		282.50	350
8312	12 pair	2 Elec	8	2		490	109		599	710
8315	15 pair		7.30	2.192		505	120		625	745
8317	17 pair		6.80	2.353		690	129		819	955
8319	19 pair		6.40	2.500		690	137		827	970
8327	27 pair		5.90	2.712		1,250	148		1,398	1,600
8402	#20, 2 pair	1 Elec	6.70	1.194		77.50	65.50		143	187
8403	3 pair		6.20	1.290		97	70.50		167.50	215
8406	6 pair		4.40	1.818		244	99.50		343.50	420
8409	9 pair		3.60	2.222		280	122		402	500
8412	12 pair	2 Elec	5.90	2.712		565	148		713	850
8415	15 pair	"	5.70	2.807		475	154		629	760
8502	#18, 2 pair	1 Elec	6.20	1.290		133	70.50		203.50	255
8503	3 pair		5.30	1.509		208	82.50		290.50	355
8504	4 pair		4.70	1.702		247	93		340	415
8506	6 pair		4	2		385	109		494	595
8509	9 pair		3.40	2.353		590	129		719	850
8515	15 pair	2 Elec	4.80	3.333		985	182		1,167	1,350

27 15 33 – Communications Coaxial Horizontal Cabling

27 15 33.10 Coaxial Cable and Fittings		Crew	Daily Output	Labor-Hours	Unit	Material	Labor	Equipment	Total	Total Incl O&P
0010	**COAXIAL CABLE & FITTINGS**									
3500	Coaxial connectors, 50 ohm impedance quick disconnect									
3540	BNC plug, for RG A/U #58 cable	1 Elec	42	.190	Ea.	4.65	10.40		15.05	21.50
3550	RG A/U #59 cable		42	.190		4.65	10.40		15.05	21.50
3560	RG A/U #62 cable		42	.190		4.65	10.40		15.05	21.50

27 15 Communications Horizontal Cabling

27 15 33 – Communications Coaxial Horizontal Cabling

27 15 33.10 Coaxial Cable and Fittings		Crew	Daily Output	Labor-Hours	Unit	2015 Bare Costs Material	Labor	Equipment	Total	Total Incl O&P
3600	BNC jack, for RG A/U #58 cable	1 Elec	42	.190	Ea.	4.90	10.40		15.30	21.50
3610	RG A/U #59 cable		42	.190		4.90	10.40		15.30	21.50
3620	RG A/U #62 cable		42	.190		4.90	10.40		15.30	21.50
3660	BNC panel jack, for RG A/U #58 cable		40	.200		7.60	10.95		18.55	25.50
3670	RG A/U #59 cable		40	.200		7.60	10.95		18.55	25.50
3680	RG A/U #62 cable		40	.200		7.60	10.95		18.55	25.50
3720	BNC bulkhead jack, for RG A/U #58 cable		40	.200		7.85	10.95		18.80	25.50
3730	RG A/U #59 cable		40	.200		7.85	10.95		18.80	25.50
3740	RG A/U #62 cable		40	.200		7.85	10.95		18.80	25.50
3850	Coaxial cable, RG A/U 58, 50 ohm		8	1	C.L.F.	49.50	54.50		104	139
3860	RG A/U 59, 75 ohm		8	1		39	54.50		93.50	128
3870	RG A/U 62, 93 ohm		8	1		46.50	54.50		101	136
3875	RG 6/U, 75 ohm		8	1		28.50	54.50		83	116
3950	Fire rated, RG A/U 58, 50 ohm		8	1		94.50	54.50		149	189
3960	RG A/U 59, 75 ohm		8	1		125	54.50		179.50	222
3970	RG A/U 62, 93 ohm		8	1		110	54.50		164.50	206

27 15 43 – Communications Faceplates and Connectors

27 15 43.13 Communication Outlets		Crew	Daily Output	Labor-Hours	Unit	Material	Labor	Equipment	Total	Total Incl O&P
0010	COMMUNICATION OUTLETS									
0100	Voice/data devices not included									
0120	Voice/Data outlets, single opening	1 Elec	48	.167	Ea.	7.85	9.10		16.95	22.50
0140	Two jack openings		48	.167		2.56	9.10		11.66	16.90
0160	One jack & one 3/4" round opening		48	.167		6.95	9.10		16.05	21.50
0180	One jack & one twinaxial opening		48	.167		7.60	9.10		16.70	22.50
0200	One jack & one connector cabling opening		48	.167		6.95	9.10		16.05	21.50
0220	Two 3/8" coaxial openings		48	.167		6.95	9.10		16.05	21.50
0300	Data outlets, single opening		48	.167		6.95	9.10		16.05	21.50
0320	One 25-pin subminiature opening		48	.167		6.95	9.10		16.05	21.50
1000	Voice/Data wall plate plastic, 1 gang, 1-port		72	.111		2.08	6.10		8.18	11.70
1020	2-port		72	.111		2.08	6.10		8.18	11.70
1040	3-port		72	.111		2.08	6.10		8.18	11.70
1060	4-port		72	.111		2.08	6.10		8.18	11.70
1080	6-port		72	.111		2.08	6.10		8.18	11.70
1100	2 gang, 6-port		48	.167		6.25	9.10		15.35	21
1120	Voice/Data wall plate stainless steel, 1 gang, 1-port		72	.111		6.30	6.10		12.40	16.35
1140	2-port		72	.111		6.45	6.10		12.55	16.45
1160	3-port		72	.111		6.45	6.10		12.55	16.50
1180	4-port		72	.111		6.45	6.10		12.55	16.45
1200	2 gang, 6-port		48	.167		11.20	9.10		20.30	26.50

27 21 Data Communications Network Equipment

27 21 23 – Data Communications Switches and Hubs

27 21 23.10 Switching and Routing Equipment		Crew	Daily Output	Labor-Hours	Unit	Material	Labor	Equipment	Total	Total Incl O&P
0010	SWITCHING AND ROUTING EQUIPMENT									
1100	Network hub, dual speed, 24 ports, includes cabinet	3 Elec	.66	36.364	Ea.	1,725	2,000		3,725	4,950
2000	Network switch, 10/100/1000 Mbps, 24 ports		.75	32		1,875	1,750		3,625	4,750
2040	48 ports		.66	36.364		2,350	2,000		4,350	5,675

27 32 Voice Communications Terminal Equipment

27 32 36 – TTY Equipment

27 32 36.10 TTY Telephone Equipment		Crew	Daily Output	Labor-Hours	Unit	2015 Bare Costs Material	Labor	Equipment	Total	Total Incl O&P
0010	**TTY TELEPHONE EQUIPMENT**									
1620	Telephone, TTY, compact, pocket type				Ea.	253			253	278
1630	Advanced, desk type	2 Elec	20	.800		510	44		554	630
1640	Full-featured public, wall type	"	4	4		725	219		944	1,150

27 41 Audio-Video Systems

27 41 33 – Master Antenna Television Systems

27 41 33.10 T.V. Systems

Code	Description	Crew	Daily Output	Labor-Hours	Unit	Material	Labor	Equipment	Total	Total Incl O&P
0010	**T.V. SYSTEMS**, not including rough-in wires, cables & conduits									
0100	Master TV antenna system									
0200	VHF reception & distribution, 12 outlets	1 Elec	6	1.333	Outlet	133	73		206	259
0400	30 outlets		10	.800		141	44		185	223
0600	100 outlets		13	.615		144	33.50		177.50	210
0800	VHF & UHF reception & distribution, 12 outlets		6	1.333		213	73		286	345
1000	30 outlets		10	.800		141	44		185	223
1200	100 outlets		13	.615		144	33.50		177.50	210
1400	School and deluxe systems, 12 outlets		2.40	3.333		281	182		463	590
1600	30 outlets		4	2		246	109		355	440
1800	80 outlets		5.30	1.509		237	82.50		319.50	390
1900	Amplifier		4	2	Ea.	705	109		814	945

27 51 Distributed Audio-Video Communications Systems

27 51 16 – Public Address and Mass Notification Systems

27 51 16.10 Public Address System

Code	Description	Crew	Daily Output	Labor-Hours	Unit	Material	Labor	Equipment	Total	Total Incl O&P
0010	**PUBLIC ADDRESS SYSTEM**									
0100	Conventional, office	1 Elec	5.33	1.501	Speaker	135	82		217	276
0200	Industrial		2.70	2.963	"	261	162		423	540
9000	Minimum labor/equipment charge		3.50	2.286	Job		125		125	194

27 51 19 – Sound Masking Systems

27 51 19.10 Sound System

Code	Description	Crew	Daily Output	Labor-Hours	Unit	Material	Labor	Equipment	Total	Total Incl O&P
0010	**SOUND SYSTEM**, not including rough-in wires, cables & conduits									
0100	Components, projector outlet	1 Elec	8	1	Ea.	46	54.50		100.50	135
0200	Microphone		4	2		81.50	109		190.50	259
0400	Speakers, ceiling or wall		8	1		119	54.50		173.50	216
0600	Trumpets		4	2		222	109		331	415
0800	Privacy switch		8	1		88.50	54.50		143	182
1000	Monitor panel		4	2		395	109		504	605
1200	Antenna, AM/FM		4	2		138	109		247	320
1400	Volume control		8	1		90	54.50		144.50	184
1600	Amplifier, 250 watts		1	8		1,275	440		1,715	2,075
1800	Cabinets		1	8		860	440		1,300	1,625
2000	Intercom, 30 station capacity, master station	2 Elec	2	8		2,350	440		2,790	3,275
2020	10 station capacity	"	4	4		1,300	219		1,519	1,800
2200	Remote station	1 Elec	8	1		166	54.50		220.50	267
2400	Intercom outlets		8	1		97.50	54.50		152	192
2600	Handset		4	2		320	109		429	525
2800	Emergency call system, 12 zones, annunciator		1.30	6.154		970	335		1,305	1,600
3000	Bell		5.30	1.509		100	82.50		182.50	238
3200	Light or relay		8	1		50	54.50		104.50	140

27 51 Distributed Audio-Video Communications Systems

27 51 19 – Sound Masking Systems

27 51 19.10 Sound System		Crew	Daily Output	Labor-Hours	Unit	Material	Labor	Equipment	Total	Total Incl O&P
						2015 Bare Costs				
3400	Transformer	1 Elec	4	2	Ea.	220	109		329	410
3600	House telephone, talking station		1.60	5		475	274		749	945
3800	Press to talk, release to listen		5.30	1.509		110	82.50		192.50	249
4000	System-on button					66			66	72.50
4200	Door release	1 Elec	4	2		118	109		227	299
4400	Combination speaker and microphone		8	1		201	54.50		255.50	305
4600	Termination box		3.20	2.500		63	137		200	282
4800	Amplifier or power supply		5.30	1.509		725	82.50		807.50	930
5000	Vestibule door unit		16	.500	Name	133	27.50		160.50	190
5200	Strip cabinet		27	.296	Ea.	252	16.20		268.20	300
5400	Directory		16	.500		119	27.50		146.50	174
6000	Master door, button buzzer type, 100 unit	2 Elec	.54	29.630		1,200	1,625		2,825	3,825
6020	200 unit		.30	53.333		2,275	2,925		5,200	7,025
6040	300 unit		.20	80		3,475	4,375		7,850	10,600
6060	Transformer	1 Elec	8	1		31	54.50		85.50	119
6080	Door opener		5.30	1.509		44	82.50		126.50	177
6100	Buzzer with door release and plate		4	2		44	109		153	218
6200	Intercom type, 100 unit	2 Elec	.54	29.630		1,500	1,625		3,125	4,150
6220	200 unit		.30	53.333		2,925	2,925		5,850	7,750
6240	300 unit		.20	80		4,425	4,375		8,800	11,700
6260	Amplifier	1 Elec	2	4		220	219		439	585
6280	Speaker with door release		4	2		66	109		175	242
9000	Minimum labor/equipment charge		3.50	2.286	Job		125		125	194

27 52 Healthcare Communications and Monitoring Systems

27 52 23 – Nurse Call/Code Blue Systems

27 52 23.10 Nurse Call Systems		Crew	Daily Output	Labor-Hours	Unit	Material	Labor	Equipment	Total	Total Incl O&P
0010	**NURSE CALL SYSTEMS**									
0100	Single bedside call station	1 Elec	8	1	Ea.	231	54.50		285.50	340
0200	Ceiling speaker station		8	1		67.50	54.50		122	159
0400	Emergency call station		8	1		72	54.50		126.50	164
0600	Pillow speaker		8	1		177	54.50		231.50	280
0800	Double bedside call station		4	2		142	109		251	325
1000	Duty station		4	2		148	109		257	330
1200	Standard call button		8	1		87	54.50		141.50	180
1400	Lights, corridor, dome or zone indicator		8	1		49	54.50		103.50	139
1600	Master control station for 20 stations	2 Elec	.65	24.615	Total	3,975	1,350		5,325	6,450

27 53 Distributed Systems

27 53 13 – Clock Systems

27 53 13.50 Clock Equipments		Crew	Daily Output	Labor-Hours	Unit	Material	Labor	Equipment	Total	Total Incl O&P
0010	**CLOCK EQUIPMENTS**, not including wires & conduits									
0100	Time system components, master controller	1 Elec	.33	24.242	Ea.	1,850	1,325		3,175	4,100
0200	Program bell		8	1		86.50	54.50		141	180
0400	Combination clock & speaker		3.20	2.500		214	137		351	450
0600	Frequency generator		2	4		2,400	219		2,619	3,000
0800	Job time automatic stamp recorder		4	2		540	109		649	765
1600	Master time clock system, clocks & bells, 20 room	4 Elec	.20	160		6,200	8,750		14,950	20,400
1800	50 room	"	.08	400		12,700	21,900		34,600	47,900

27 53 Distributed Systems

27 53 13 – Clock Systems

27 53 13.50 Clock Equipments		Crew	Daily Output	Labor-Hours	Unit	2015 Bare Costs Material	Labor	Equipment	Total	Total Incl O&P
1900	Time clock	1 Elec	3.20	2.500	Ea.	445	137		582	700
2000	100 cards in & out, 1 color					9.15			9.15	10.10
2200	2 colors					9.15			9.15	10.10
2800	Metal rack for 25 cards	1 Elec	7	1.143		44	62.50		106.50	145
4000	Wireless time systems component, master controller	"	2	4	▼	605	219		824	1,000
4010	For transceiver and antenna, see Section 28 39 10.10									
4100	Wireless analog clock w/battery operated	1 Elec	8	1	Ea.	113	54.50		167.50	210
4200	Wireless digital clock w/battery operated	"	8	1	"	370	54.50		424.50	490

Division Notes

		CREW	DAILY OUTPUT	LABOR-HOURS	UNIT	BARE COSTS MAT.	BARE COSTS LABOR	BARE COSTS EQUIP.	BARE COSTS TOTAL	TOTAL INCL O&P

Division 28 — Electronic Safety & Security

Estimating Tips

- When estimating material costs for electronic safety and security systems, it is always prudent to obtain manufacturers' quotations for equipment prices and special installation requirements that affect the total cost.
- Fire alarm systems consist of control panels, annunciator panels, battery with rack, charger, and fire alarm actuating and indicating devices. Some fire alarm systems include speakers, telephone lines, door closer controls, and other components. Be careful not to overlook the costs related to installation for these items. Also be aware of costs for integrated automation instrumentation and terminal devices, control equipment, control wiring, and programming.
- Security equipment includes items such as CCTV, access control, and other detection and identification systems to perform alert and alarm functions. Be sure to consider the costs related to installation for this security equipment, such as for integrated automation instrumentation and terminal devices, control equipment, control wiring, and programming.

Reference Numbers

Reference numbers are shown in shaded boxes at the beginning of some major classifications. These numbers refer to related items in the Reference Section. The reference information may be an estimating procedure, an alternate pricing method, or technical information.

Note: Not all subdivisions listed here necessarily appear in this publication. ■

28 01 Operation and Maint. of Electronic Safety and Security

28 01 30 – Operation and Maint. of Electronic Detection and Alarm

28 01 30.51 Maint. and Admin. of Elec. Detection and Alarm		Crew	Daily Output	Labor-Hours	Unit	Material	2015 Bare Costs Labor	Equipment	Total	Total Incl O&P
0010	MAINT. AND ADMIN. OF ELEC. DETECTION AND ALARM									
3300	Remove and replace (reinstall), fire alarm device	1 Elec	5.33	1.501	Ea.		82		82	127

28 05 Common Work Results for Electronic Safety and Security

28 05 05 – Selective Demolition for Electronic Safety and Security

28 05 05.10 Safety and Security Demolition

Line	Description	Crew	Daily Output	Labor-Hours	Unit	Material	Labor	Equipment	Total	Total Incl O&P
0010	SAFETY AND SECURITY DEMOLITION									
1050	Finger print/card reader	1 Elec	8	1	Ea.		54.50		54.50	84.50
1060	Video camera		8	1			54.50		54.50	84.50
1070	Motion detector, multi-channel		7	1.143			62.50		62.50	97
1090	Infrared detector		7	1.143			62.50		62.50	97
1130	Video monitor, 19"		8	1			54.50		54.50	84.50
1210	Fire alarm horn and strobe light		16	.500			27.50		27.50	42.50
1220	Flame detector		23	.348			19.05		19.05	29.50
1230	Duct detector		10	.800			44		44	68
1240	Smoke detector		23	.348			19.05		19.05	29.50
1250	Fire alarm pull station		23	.348			19.05		19.05	29.50
1260	Fire alarm control panel, 4 to 8 zone	2 Elec	4	4			219		219	340
1270	12 to 16 zone	"	2.67	5.993			330		330	510
1280	Fire alarm annunciation panel, 4 to 8 zone	1 Elec	6	1.333			73		73	113
1290	12 to 16 zone	2 Elec	6	2.667			146		146	226

28 05 13 – Conductors and Cables for Electronic Safety and Security

28 05 13.23 Fire Alarm Communications Conductors and Cables

Line	Description	Crew	Daily Output	Labor-Hours	Unit	Material	Labor	Equipment	Total	Total Incl O&P
0010	FIRE ALARM COMMUNICATIONS CONDUCTORS AND CABLES									
1500	Fire alarm FEP teflon 150 volt to 200°C									
1550	#22, 1 pair	1 Elec	10	.800	C.L.F.	63	44		107	137
1600	2 pair		8	1		103	54.50		157.50	199
1650	4 pair		7	1.143		160	62.50		222.50	273
1700	6 pair		6	1.333		208	73		281	340
1750	8 pair		5.50	1.455		261	79.50		340.50	410
1800	10 pair		5	1.600		315	87.50		402.50	480
1850	#18, 1 pair		8	1		72	54.50		126.50	164
1900	2 pair		6.50	1.231		137	67.50		204.50	255
1950	4 pair		4.80	1.667		208	91		299	370
2000	6 pair		4	2		273	109		382	470
2050	8 pair		3.50	2.286		355	125		480	590
2100	10 pair		3	2.667		410	146		556	675

28 13 Access Control

28 13 53 – Security Access Detection

28 13 53.13 Security Access Metal Detectors

Line	Description	Crew	Daily Output	Labor-Hours	Unit	Material	Labor	Equipment	Total	Total Incl O&P
0010	SECURITY ACCESS METAL DETECTORS									
0240	Metal detector, hand-held, wand type, unit only				Ea.	89			89	98
0250	Metal detector, walk through portal type, single zone	1 Elec	2	4		3,700	219		3,919	4,425
0260	Multi-zone	"	2	4		4,700	219		4,919	5,525

28 13 Access Control

28 13 53 – Security Access Detection

28 13 53.16 Security Access X-Ray Equipment		Crew	Daily Output	Labor-Hours	Unit	Material	2015 Bare Costs Labor	Equipment	Total	Total Incl O&P
0010	**SECURITY ACCESS X-RAY EQUIPMENT**									
0290	X-ray machine, desk top, for mail/small packages/letters	1 Elec	4	2	Ea.	3,425	109		3,534	3,950
0300	Conveyor type, incl monitor		2	4		16,000	219		16,219	17,900
0310	Includes additional features		2	4		28,600	219		28,819	31,700
0320	X-ray machine, large unit, for airports, incl monitor	2 Elec	1	16		40,000	875		40,875	45,400
0330	Full console	"	.50	32		68,500	1,750		70,250	78,000

28 13 53.23 Security Access Explosive Detection Equipment		Crew	Daily Output	Labor-Hours	Unit	Material	Labor	Equipment	Total	Total Incl O&P
0010	**SECURITY ACCESS EXPLOSIVE DETECTION EQUIPMENT**									
0270	Explosives detector, walk through portal type	1 Elec	2	4	Ea.	44,200	219		44,419	48,900
0280	Hand-held, battery operated				"				25,500	28,100

28 16 Intrusion Detection

28 16 16 – Intrusion Detection Systems Infrastructure

28 16 16.50 Intrusion Detection		Crew	Daily Output	Labor-Hours	Unit	Material	Labor	Equipment	Total	Total Incl O&P
0010	**INTRUSION DETECTION**, not including wires & conduits									
0100	Burglar alarm, battery operated, mechanical trigger	1 Elec	4	2	Ea.	278	109		387	475
0200	Electrical trigger		4	2		330	109		439	535
0400	For outside key control, add		8	1		84	54.50		138.50	177
0800	Card reader, flush type, standard		2.70	2.963		930	162		1,092	1,275
1000	Multi-code		2.70	2.963		1,200	162		1,362	1,575
1010	Card reader, proximity type		2.70	2.963		315	162		477	600
1200	Door switches, hinge switch		5.30	1.509		58.50	82.50		141	193
1400	Magnetic switch		5.30	1.509		69	82.50		151.50	204
1600	Exit control locks, horn alarm		4	2		277	109		386	475
1800	Flashing light alarm		4	2		305	109		414	505
2000	Indicating panels, 1 channel		2.70	2.963		370	162		532	655
2200	10 channel	2 Elec	3.20	5		1,050	274		1,324	1,575
2400	20 channel		2	8		2,450	440		2,890	3,375
2600	40 channel		1.14	14.035		4,450	770		5,220	6,100
2800	Ultrasonic motion detector, 12 volt	1 Elec	2.30	3.478		230	190		420	550
3000	Infrared photoelectric detector		4	2		189	109		298	375
3200	Passive infrared detector		4	2		283	109		392	480
3400	Glass break alarm switch		8	1		93	54.50		147.50	187
3420	Switchmats, 30" x 5'		5.30	1.509		84.50	82.50		167	221
3440	30" x 25'		4	2		203	109		312	390
3460	Police connect panel		4	2		244	109		353	440
3480	Telephone dialer		5.30	1.509		385	82.50		467.50	555
3500	Alarm bell		4	2		104	109		213	284
3520	Siren		4	2		146	109		255	330
3540	Microwave detector, 10' to 200'		2	4		670	219		889	1,075
3560	10' to 350'		2	4		1,950	219		2,169	2,500

28 23 Video Surveillance

28 23 13 – Video Surveillance Control and Management Systems

28 23 13.10 Closed Circuit Television System		Crew	Daily Output	Labor-Hours	Unit	2015 Bare Costs Material	Labor	Equipment	Total	Total Incl O&P
0010	**CLOSED CIRCUIT TELEVISION SYSTEM**									
2000	Surveillance, one station (camera & monitor)	2 Elec	2.60	6.154	Total	1,350	335		1,685	2,000
2200	For additional camera stations, add	1 Elec	2.70	2.963	Ea.	755	162		917	1,075
2400	Industrial quality, one station (camera & monitor)	2 Elec	2.60	6.154	Total	2,800	335		3,135	3,600
2600	For additional camera stations, add	1 Elec	2.70	2.963	Ea.	1,725	162		1,887	2,125
2610	For low light, add		2.70	2.963		1,375	162		1,537	1,775
2620	For very low light, add		2.70	2.963		10,200	162		10,362	11,500
2800	For weatherproof camera station, add		1.30	6.154		1,050	335		1,385	1,700
3000	For pan and tilt, add		1.30	6.154		2,725	335		3,060	3,525
3200	For zoom lens - remote control, add		2	4		2,525	219		2,744	3,125
3400	Extended zoom lens		2	4		9,200	219		9,419	10,400
3410	For automatic iris for low light, add		2	4		2,200	219		2,419	2,775
3600	Educational T.V. studio, basic 3 camera system, black & white,									
3800	electrical & electronic equip. only	4 Elec	.80	40	Total	13,100	2,200		15,300	17,800
4000	Full console		.28	114		56,000	6,250		62,250	71,000
4100	As above, but color system		.28	114		74,000	6,250		80,250	90,500
4120	Full console		.12	266		321,000	14,600		335,600	375,500
4200	For film chain, black & white, add	1 Elec	1	8	Ea.	15,000	440		15,440	17,200
4250	Color, add		.25	32		18,200	1,750		19,950	22,700
4400	For video recorders, add		1	8		3,150	440		3,590	4,150
4600	Premium	4 Elec	.40	80		26,200	4,375		30,575	35,600

28 23 19 – Digital Video Recorders and Analog Recording Devices

28 23 19.10 Digital Video Recorder (DVR)		Crew	Daily Output	Labor-Hours	Unit	Material	Labor	Equipment	Total	Total Incl O&P
0010	**DIGITAL VIDEO RECORDER (DVR)**									
0100	Pentaplex hybrid, internet protocol, and hard drive									
0200	4 channel	1 Elec	1.33	6.015	Ea.	1,350	330		1,680	1,975
0300	8 channel		1	8		2,650	440		3,090	3,600
0400	16 channel		1	8		2,925	440		3,365	3,900

28 23 23 – Video Surveillance Systems Infrastructure

28 23 23.50 Video Surveillance Equipments		Crew	Daily Output	Labor-Hours	Unit	Material	Labor	Equipment	Total	Total Incl O&P
0010	**VIDEO SURVEILLANCE EQUIPMENTS**									
0200	Video cameras, wireless, hidden in exit signs, clocks, etc., incl. receiver	1 Elec	3	2.667	Ea.	108	146		254	345
0210	Accessories for video recorder, single camera		3	2.667		183	146		329	425
0220	For multiple cameras		3	2.667		1,750	146		1,896	2,150
0230	Video cameras, wireless, for under vehicle searching, complete		2	4		10,200	219		10,419	11,500
0400	Internet protocol network camera, day/night, color & power supply		2.60	3.077		1,025	168		1,193	1,375
0500	Monitor, color flat screen, liquid crystal display (LCD), 15"		2.70	2.963		695	162		857	1,025
0520	17"		2.70	2.963		820	162		982	1,150
0540	19"		2.70	2.963		920	162		1,082	1,275

28 31 Fire Detection and Alarm

28 31 23 – Fire Detection and Alarm Annunciation Panels and Fire Stations

28 31 23.50 Alarm Panels and Devices		Crew	Daily Output	Labor-Hours	Unit	2015 Bare Costs Material	Labor	Equipment	Total	Total Incl O&P
0010	**ALARM PANELS AND DEVICES**, not including wires & conduits									
3594	Fire, alarm control panel									
3600	4 zone	2 Elec	2	8	Ea.	400	440		840	1,125
3800	8 zone		1	16		780	875		1,655	2,200
4000	12 zone		.67	23.988		2,400	1,300		3,700	4,650
4020	Alarm device	1 Elec	8	1		238	54.50		292.50	345
4050	Actuating device	"	8	1		335	54.50		389.50	455
4160	Alarm control panel, addressable w/o voice, up to 200 points	2 Elec	1.14	13.998		4,475	765		5,240	6,100
4170	addressable w/voice, up to 400 points	"	.73	22.008		9,425	1,200		10,625	12,300
4175	Addressable interface device	1 Elec	7.25	1.103		135	60.50		195.50	242
4200	Battery and rack		4	2		410	109		519	625
4400	Automatic charger		8	1		585	54.50		639.50	725
4600	Signal bell		8	1		78.50	54.50		133	171
4800	Trouble buzzer or manual station		8	1		83.50	54.50		138	176
5600	Strobe and horn		5.30	1.509		152	82.50		234.50	295
5610	Strobe and horn (ADA type)		5.30	1.509		152	82.50		234.50	295
5620	Visual alarm (ADA type)		6.70	1.194		103	65.50		168.50	214
5800	Fire alarm horn		6.70	1.194		61	65.50		126.50	168
6000	Door holder, electro-magnetic		4	2		103	109		212	282
6200	Combination holder and closer		3.20	2.500		123	137		260	350
6600	Drill switch		8	1		370	54.50		424.50	495
6800	Master box		2.70	2.963		6,400	162		6,562	7,300
7000	Break glass station		8	1		55.50	54.50		110	146
7010	Break glass station, addressable		7.25	1.103		148	60.50		208.50	256
7800	Remote annunciator, 8 zone lamp		1.80	4.444		209	243		452	605
8000	12 zone lamp	2 Elec	2.60	6.154		335	335		670	890
8200	16 zone lamp	"	2.20	7.273		420	400		820	1,075

28 31 43 – Fire Detection Sensors

28 31 43.50 Fire and Heat Detectors		Crew	Daily Output	Labor-Hours	Unit	Material	Labor	Equipment	Total	Total Incl O&P
0010	**FIRE & HEAT DETECTORS**									
5000	Detector, rate of rise	1 Elec	8	1	Ea.	51	54.50		105.50	141
5010	Heat addressable type		7.25	1.103		257	60.50		317.50	375
5100	Fixed temperature		8	1		51	54.50		105.50	141

28 31 46 – Smoke Detection Sensors

28 31 46.50 Smoke Detectors		Crew	Daily Output	Labor-Hours	Unit	Material	Labor	Equipment	Total	Total Incl O&P
0010	**SMOKE DETECTORS**									
5200	Smoke detector, ceiling type	1 Elec	6.20	1.290	Ea.	110	70.50		180.50	230
5240	Smoke detector addressable type		6	1.333		221	73		294	355
5400	Duct type		3.20	2.500		325	137		462	570
5420	Duct addressable type		3.20	2.500		510	137		647	770

28 32 Radiation Detection and Alarm

28 32 33 – Radiation Detection Sensors

28 32 33.50 Radiation Detection Systems		Crew	Daily Output	Labor-Hours	Unit	Material	2015 Bare Costs Labor	Equipment	Total	Total Incl O&P
0010	**RADIATION DETECTION SYSTEMS**									
9410	Minimum labor/equipment charge	1 Elec	4	2	Job		109		109	169

28 33 Gas Detection and Alarm

28 33 33 – Gas Detection Sensors

28 33 33.50 Tank Leak Detection Systems		Crew	Daily Output	Labor-Hours	Unit	Material	Labor	Equipment	Total	Total Incl O&P
0010	**TANK LEAK DETECTION SYSTEMS** Liquid and vapor									
0100	For hydrocarbons and hazardous liquids/vapors									
0120	Controller, data acquisition, incl. printer, modem, RS232 port									
0140	24 channel, for use with all probes				Ea.	4,175			4,175	4,575
0160	9 channel, for external monitoring				"	915			915	1,000
0200	Probes									
0210	Well monitoring									
0220	Liquid phase detection				Ea.	475			475	525
0230	Hydrocarbon vapor, fixed position					480			480	530
0240	Hydrocarbon vapor, float mounted					480			480	530
0250	Both liquid and vapor hydrocarbon				↓	480			480	530
0300	Secondary containment, liquid phase									
0310	Pipe trench/manway sump				Ea.	820			820	905
0320	Double wall pipe and manual sump					825			825	905
0330	Double wall fiberglass annular space					325			325	355
0340	Double wall steel tank annular space				↓	335			335	370
0500	Accessories									
0510	Modem, non-dedicated phone line				Ea.	305			305	335
0600	Monitoring, internal									
0610	Automatic tank gauge, incl. overfill				Ea.	1,150			1,150	1,275
0620	Product line				"	1,150			1,150	1,275
0700	Monitoring, special									
0710	Cathodic protection				Ea.	725			725	795
0720	Annular space chemical monitor				"	985			985	1,075

28 39 Mass Notification Systems

28 39 10 – Notification Systems

28 39 10.10 Mass Notification System		Crew	Daily Output	Labor-Hours	Unit	Material	Labor	Equipment	Total	Total Incl O&P
0010	**MASS NOTIFICATION SYSTEM**									
0100	Wireless command center, 10,000 devices	2 Elec	1.33	12.030	Ea.	3,600	660		4,260	4,975
0200	Option, email notification					1,750			1,750	1,925
0210	Remote device supervision & monitor					2,450			2,450	2,675
0300	Antenna VHF or UHF, for medium range	1 Elec	4	2		129	109		238	310
0310	For high-power transmitter		2	4		670	219		889	1,075
0400	Transmitter, 25 watt		4	2		2,025	109		2,134	2,400
0410	40 watt		2.66	3.008		2,700	165		2,865	3,225
0420	100 watt		1.33	6.015		6,775	330		7,105	7,950
0500	Wireless receiver/control module for speaker		8	1		265	54.50		319.50	375
0600	Desktop paging controller, stand alone	↓	4	2	↓	370	109		479	580

Estimating Tips

31 05 00 Common Work Results for Earthwork

- Estimating the actual cost of performing earthwork requires careful consideration of the variables involved. This includes items such as type of soil, whether water will be encountered, dewatering, whether banks need bracing, disposal of excavated earth, and length of haul to fill or spoil sites, etc. If the project has large quantities of cut or fill, consider raising or lowering the site to reduce costs, while paying close attention to the effect on site drainage and utilities.
- If the project has large quantities of fill, creating a borrow pit on the site can significantly lower the costs.
- It is very important to consider what time of year the project is scheduled for completion. Bad weather can create large cost overruns from dewatering, site repair, and lost productivity from cold weather.

Reference Numbers

Reference numbers are shown in shaded boxes at the beginning of some major classifications. These numbers refer to related items in the Reference Section. The reference information may be an estimating procedure, an alternate pricing method, or technical information.

Note: Not all subdivisions listed here necessarily appear in this publication. ■

31 05 Common Work Results for Earthwork

31 05 23 – Cement and Concrete for Earthwork

31 05 23.30 Plant Mixed Bituminous Concrete		Crew	Daily Output	Labor-Hours	Unit	2015 Bare Costs Material	Labor	Equipment	Total	Total Incl O&P
0010	PLANT MIXED BITUMINOUS CONCRETE									
0020	Asphaltic concrete plant mix (145 lb. per C.F.)				Ton	70			70	77
0040	Asphaltic concrete less than 300 tons add trucking costs									
0050	See Section 31 23 23.20 for hauling costs									
0200	All weather patching mix, hot				Ton	69			69	76
0250	Cold patch					79.50			79.50	87.50
0300	Berm mix					69			69	76
0400	Base mix					70			70	77
0500	Binder mix					70			70	77
0600	Sand or sheet mix					67.50			67.50	74

31 05 23.40 Recycled Plant Mixed Bituminous Concrete

		Crew	Daily Output	Labor-Hours	Unit	Material	Labor	Equipment	Total	Total Incl O&P
0010	RECYCLED PLANT MIXED BITUMINOUS CONCRETE									
0200	Reclaimed pavement in stockpile G				Ton	53			53	58.50
0400	Recycled pavement, at plant, ratio old: new, 70:30 G					58			58	63.50
0600	Ratio old: new, 30:70 G					64			64	70.50

31 06 Schedules for Earthwork

31 06 60 – Schedules for Special Foundations and Load Bearing Elements

31 06 60.14 Piling Special Costs

		Crew	Daily Output	Labor-Hours	Unit	Material	Labor	Equipment	Total	Total Incl O&P
0010	PILING SPECIAL COSTS									
0011	Piling special costs, pile caps, see Section 03 30 53.40									
0500	Cutoffs, concrete piles, plain	1 Pile	5.50	1.455	Ea.		67		67	110
0600	With steel thin shell, add		38	.211			9.70		9.70	15.85
0700	Steel pile or "H" piles		19	.421			19.40		19.40	31.50
0800	Wood piles		38	.211			9.70		9.70	15.85
0900	Pre-augering up to 30′ deep, average soil, 24″ diameter	B-43	180	.267	L.F.		11.10	14.10	25.20	33.50
0920	36″ diameter		115	.417			17.35	22	39.35	52.50
0960	48″ diameter		70	.686			28.50	36.50	65	86
0980	60″ diameter		50	.960			40	51	91	121
1000	Testing, any type piles, test load is twice the design load									
1050	50 ton design load, 100 ton test				Ea.				14,000	15,500
1100	100 ton design load, 200 ton test								20,000	22,000
1150	150 ton design load, 300 ton test								26,000	28,500
1200	200 ton design load, 400 ton test								28,000	31,000
1250	400 ton design load, 800 ton test								32,000	35,000
1500	Wet conditions, soft damp ground									
1600	Requiring mats for crane, add								40%	40%
1700	Barge mounted driving rig, add								30%	30%

31 06 60.15 Mobilization

		Crew	Daily Output	Labor-Hours	Unit	Material	Labor	Equipment	Total	Total Incl O&P
0010	MOBILIZATION									
0020	Set up & remove, air compressor, 600 CFM	A-5	3.30	5.455	Ea.		206	18.60	224.60	355
0100	1200 CFM	"	2.20	8.182			310	28	338	535
0200	Crane, with pile leads and pile hammer, 75 ton	B-19	.60	106			5,075	2,900	7,975	11,400
0300	150 ton	"	.36	177			8,475	4,825	13,300	19,000
0500	Drill rig, for caissons, to 36″, minimum	B-43	2	24			995	1,275	2,270	3,025
0520	Maximum		.50	96			4,000	5,075	9,075	12,100
0600	Up to 84″		1	48			2,000	2,550	4,550	6,025
0800	Auxiliary boiler, for steam small	A-5	1.66	10.843			410	37	447	710
0900	Large	"	.83	21.687			820	74	894	1,425
1100	Rule of thumb: complete pile driving set up, small	B-19	.45	142			6,775	3,850	10,625	15,200

31 06 Schedules for Earthwork

31 06 60 – Schedules for Special Foundations and Load Bearing Elements

31 06 60.15 Mobilization		Crew	Daily Output	Labor-Hours	Unit	Material	2015 Bare Costs Labor	Equipment	Total	Total Incl O&P
1200	Large	B-19	.27	237	Ea.		11,300	6,425	17,725	25,300
1500	Mobilization, barge, by tug boat	B-83	25	.640	Mile		28	35	63	84
1600	Standby time for shore pile driving crew				Hr.				715	890
1700	Standby time for barge driving rig				"				1,000	1,250

31 11 Clearing and Grubbing

31 11 10 – Clearing and Grubbing Land

31 11 10.10 Clear and Grub Site

		Crew	Daily Output	Labor-Hours	Unit	Material	Labor	Equipment	Total	Total Incl O&P
0010	**CLEAR AND GRUB SITE**									
0020	Cut & chip light trees to 6" diam.	B-7	1	48	Acre		1,925	1,675	3,600	4,950
0150	Grub stumps and remove	B-30	2	12			525	1,200	1,725	2,175
0200	Cut & chip medium, trees to 12" diam.	B-7	.70	68.571			2,750	2,375	5,125	7,100
0250	Grub stumps and remove	B-30	1	24			1,050	2,400	3,450	4,325
0300	Cut & chip heavy, trees to 24" diam.	B-7	.30	160			6,425	5,575	12,000	16,500
0350	Grub stumps and remove	B-30	.50	48			2,100	4,825	6,925	8,650
0400	If burning is allowed, deduct cut & chip								40%	40%
3000	Chipping stumps, to 18" deep, 12" diam.	B-86	20	.400	Ea.		20	9.30	29.30	42
3040	18" diameter		16	.500			25.50	11.60	37.10	53
3080	24" diameter		14	.571			29	13.25	42.25	60.50
3100	30" diameter		12	.667			33.50	15.45	48.95	70.50
3120	36" diameter		10	.800			40.50	18.55	59.05	84.50
3160	48" diameter		8	1			50.50	23	73.50	106
5000	Tree thinning, feller buncher, conifer									
5080	Up to 8" diameter	B-93	240	.033	Ea.		1.69	3.49	5.18	6.50
5120	12" diameter		160	.050			2.53	5.25	7.78	9.75
5240	Hardwood, up to 4" diameter		240	.033			1.69	3.49	5.18	6.50
5280	8" diameter		180	.044			2.25	4.66	6.91	8.65
5320	12" diameter		120	.067			3.37	7	10.37	13.05
7000	Tree removal, congested area, aerial lift truck									
7040	8" diameter	B-85	7	5.714	Ea.		233	147	380	535
7080	12" diameter		6	6.667			271	172	443	630
7120	18" diameter		5	8			325	206	531	755
7160	24" diameter		4	10			405	258	663	945
7240	36" diameter		3	13.333			545	345	890	1,250
7280	48" diameter		2	20			815	515	1,330	1,900

31 13 Selective Tree and Shrub Removal and Trimming

31 13 13 – Selective Tree and Shrub Removal

31 13 13.20 Selective Tree Removal

		Crew	Daily Output	Labor-Hours	Unit	Material	Labor	Equipment	Total	Total Incl O&P
0010	**SELECTIVE TREE REMOVAL**									
0011	With tractor, large tract, firm									
0020	level terrain, no boulders, less than 12" diam. trees									
0300	300 HP dozer, up to 400 trees/acre, 0 to 25% hardwoods	B-10M	.75	16	Acre		740	2,525	3,265	3,950
0340	25% to 50% hardwoods		.60	20			925	3,150	4,075	4,950
0370	75% to 100% hardwoods		.45	26.667			1,225	4,225	5,450	6,600
0400	500 trees/acre, 0% to 25% hardwoods		.60	20			925	3,150	4,075	4,950
0440	25% to 50% hardwoods		.48	25			1,150	3,950	5,100	6,200
0470	75% to 100% hardwoods		.36	33.333			1,550	5,275	6,825	8,275
0500	More than 600 trees/acre, 0 to 25% hardwoods		.52	23.077			1,075	3,650	4,725	5,725

31 13 Selective Tree and Shrub Removal and Trimming

31 13 13 – Selective Tree and Shrub Removal

31 13 13.20 Selective Tree Removal		Crew	Daily Output	Labor-Hours	Unit	Material	2015 Bare Costs Labor	Equipment	Total	Total Incl O&P
0540	25% to 50% hardwoods	B-10M	.42	28.571	Acre		1,325	4,525	5,850	7,100
0570	75% to 100% hardwoods	↓	.31	38.710	↓		1,800	6,125	7,925	9,600
0900	Large tract clearing per tree									
1500	300 HP dozer, to 12" diameter, softwood	B-10M	320	.038	Ea.		1.74	5.95	7.69	9.30
1550	Hardwood		100	.120			5.55	18.95	24.50	30
1600	12" to 24" diameter, softwood		200	.060			2.78	9.50	12.28	14.90
1650	Hardwood		80	.150			6.95	23.50	30.45	37
1700	24" to 36" diameter, softwood		100	.120			5.55	18.95	24.50	30
1750	Hardwood		50	.240			11.10	38	49.10	59.50
1800	36" to 48" diameter, softwood		70	.171			7.95	27	34.95	42.50
1850	Hardwood	↓	35	.343	↓		15.85	54	69.85	85
2000	Stump removal on site by hydraulic backhoe, 1-1/2 C.Y.									
2040	4" to 6" diameter	B-17	60	.533	Ea.		22	12.90	34.90	49.50
2050	8" to 12" diameter	B-30	33	.727			31.50	73	104.50	132
2100	14" to 24" diameter		25	.960			42	96.50	138.50	173
2150	26" to 36" diameter	↓	16	1.500	↓		65.50	151	216.50	271
3000	Remove selective trees, on site using chain saws and chipper,									
3050	not incl. stumps, up to 6" diameter	B-7	18	2.667	Ea.		107	93	200	276
3100	8" to 12" diameter		12	4			160	139	299	415
3150	14" to 24" diameter		10	4.800			192	167	359	500
3200	26" to 36" diameter	↓	8	6			241	209	450	620
3300	Machine load, 2 mile haul to dump, 12" diam. tree	A-3B	8	2	↓		90.50	151	241.50	310

31 14 Earth Stripping and Stockpiling

31 14 13 – Soil Stripping and Stockpiling

31 14 13.23 Topsoil Stripping and Stockpiling

		Crew	Daily Output	Labor-Hours	Unit	Material	Labor	Equipment	Total	Total Incl O&P
0010	**TOPSOIL STRIPPING AND STOCKPILING**									
0020	200 H.P. dozer, ideal conditions	B-10B	2300	.005	C.Y.		.24	.60	.84	1.05
0100	Adverse conditions	"	1150	.010			.48	1.21	1.69	2.10
0200	300 H.P. dozer, ideal conditions	B-10M	3000	.004			.19	.63	.82	1
0300	Adverse conditions	"	1650	.007			.34	1.15	1.49	1.80
0400	400 H.P. dozer, ideal conditions	B-10X	3900	.003			.14	.62	.76	.91
0500	Adverse conditions	"	2000	.006			.28	1.20	1.48	1.76
0600	Clay, dry and soft, 200 H.P. dozer, ideal conditions	B-10B	1600	.008			.35	.87	1.22	1.51
0700	Adverse conditions	"	800	.015			.69	1.73	2.42	3.02
1000	Medium hard, 300 H.P. dozer, ideal conditions	B-10M	2000	.006			.28	.95	1.23	1.48
1100	Adverse conditions	"	1100	.011			.50	1.72	2.22	2.71
1200	Very hard, 400 H.P. dozer, ideal conditions	B-10X	2600	.005			.21	.93	1.14	1.36
1300	Adverse conditions	"	1340	.009	↓		.41	1.80	2.21	2.64
1500	Loam or topsoil, remove/stockpile on site									
1510	By hand, 6" deep, 50' haul, less than 100 S.Y.	B-1	100	.240	S.Y.		9.20		9.20	15.05
1520	By skid steer, 6" deep, 100' haul, 101-500 S.Y.	B-62	500	.048			1.98	.35	2.33	3.59
1530	100' haul, 501-900 S.Y.	"	900	.027			1.10	.19	1.29	1.99
1540	200' haul, 901-1100 S.Y.	B-63	1000	.040			1.59	.17	1.76	2.78
1550	By dozer, 200' haul, 1101-4000 S.Y.	B-10B	4000	.003	↓		.14	.35	.49	.60

31 22 Grading

31 22 13 – Rough Grading

31 22 13.20 Rough Grading Sites		Crew	Daily Output	Labor-Hours	Unit	Material	2015 Bare Costs Labor	Equipment	Total	Total Incl O&P
0010	**ROUGH GRADING SITES**									
0100	Rough grade sites 400 S.F. or less	B-1	2	12	Ea.		460		460	755
0120	410-1000 S.F.	"	1	24			920		920	1,500
0130	1100-3000 S.F.	B-62	1.50	16			660	116	776	1,200
0140	3100-5000 S.F.	"	1	24			990	174	1,164	1,800
0150	5100-8000 S.F.	B-63	1	40			1,600	174	1,774	2,800
0160	8100-10000 S.F.	"	.75	53.333			2,125	231	2,356	3,700
0170	8100-10000 S.F.	B-10L	1	12			555	470	1,025	1,400
0200	Rough grade open sites 10000-20000 S.F.	B-11L	1.80	8.889			390	410	800	1,075
0210	20100-25000 S.F.		1.40	11.429			505	525	1,030	1,400
0220	25100-30000 S.F.		1.20	13.333			590	615	1,205	1,625
0230	30100-35000 S.F.		1	16			705	735	1,440	1,925
0240	35100-40000 S.F.		.90	17.778			785	820	1,605	2,150
0250	40100-45000 S.F.		.80	20			880	920	1,800	2,425
0260	45100-50000 S.F.		.72	22.222			980	1,025	2,005	2,700
0270	50100-75000 S.F.		.50	32			1,400	1,475	2,875	3,900
0280	75100-100000 S.F.		.36	44.444			1,950	2,050	4,000	5,400

31 22 16 – Fine Grading

31 22 16.10 Finish Grading		Crew	Daily Output	Labor-Hours	Unit	Material	Labor	Equipment	Total	Total Incl O&P
0010	**FINISH GRADING**									
0012	Finish grading area to be paved with grader, small area	B-11L	400	.040	S.Y.		1.76	1.84	3.60	4.86
0100	Large area		2000	.008			.35	.37	.72	.97
0200	Grade subgrade for base course, roadways		3500	.005			.20	.21	.41	.55
1020	For large parking lots	B-32C	5000	.010			.43	.47	.90	1.21
1050	For small irregular areas	"	2000	.024			1.07	1.18	2.25	3.02
1100	Fine grade for slab on grade, machine	B-11L	1040	.015			.68	.71	1.39	1.87
1150	Hand grading	B-18	700	.034			1.31	.07	1.38	2.22
1200	Fine grade granular base for sidewalks and bikeways	B-62	1200	.020			.83	.14	.97	1.50
2550	Hand grade select gravel	2 Clab	60	.267	C.S.F.		10.05		10.05	16.45
3000	Hand grade select gravel, including compaction, 4" deep	B-18	555	.043	S.Y.		1.65	.08	1.73	2.80
3100	6" deep		400	.060			2.30	.12	2.42	3.89
3120	8" deep		300	.080			3.06	.16	3.22	5.15
3300	Finishing grading slopes, gentle	B-11L	8900	.002			.08	.08	.16	.22
3310	Steep slopes		7100	.002			.10	.10	.20	.27
3500	Finish grading lagoon bottoms		4	4	M.S.F.		176	184	360	485
9000	Minimum labor/equipment charge, hand grading	1 Clab	2	4	Job		150		150	247
9100	Minimum labor/equipment charge, machine grading	B-11L	2	8	"		355	370	725	975

31 23 Excavation and Fill

31 23 16 – Excavation

31 23 16.13 Excavating, Trench		Crew	Daily Output	Labor-Hours	Unit	Material	Labor	Equipment	Total	Total Incl O&P
0010	**EXCAVATING, TRENCH**									
0011	Or continuous footing									
0020	Common earth with no sheeting or dewatering included									
0050	1' to 4' deep, 3/8 C.Y. excavator	B-11C	150	.107	B.C.Y.		4.70	2.43	7.13	10.20
0060	1/2 C.Y. excavator	B-11M	200	.080			3.53	1.96	5.49	7.85
0090	4' to 6' deep, 1/2 C.Y. excavator	"	200	.080			3.53	1.96	5.49	7.85
0100	5/8 C.Y. excavator	B-12Q	250	.064			2.86	2.36	5.22	7.20
0300	1/2 C.Y. excavator, truck mounted	B-12J	200	.080			3.57	4.41	7.98	10.60
0500	6' to 10' deep, 3/4 C.Y. excavator	B-12F	225	.071			3.18	2.91	6.09	8.30

31 23 16.13 Excavating, Trench		Crew	Daily Output	Labor-Hours	Unit	Material	2015 Bare Costs Labor	Equipment	Total	Total Incl O&P
0600	1 C.Y. excavator, truck mounted	B-12K	400	.040	B.C.Y.		1.79	2.51	4.30	5.65
0900	10' to 14' deep, 3/4 C.Y. excavator	B-12F	200	.080			3.57	3.27	6.84	9.35
1000	1-1/2 C.Y. excavator	B-12B	540	.030			1.32	1.91	3.23	4.23
1300	14' to 20' deep, 1 C.Y. excavator	B-12A	320	.050			2.23	2.54	4.77	6.40
1340	20' to 24' deep, 1 C.Y. excavator	"	288	.056			2.48	2.82	5.30	7.10
1352	4' to 6' deep, 1/2 C.Y. excavator w/trench box	B-13H	188	.085			3.80	5.10	8.90	11.75
1354	5/8 C.Y. excavator	"	235	.068			3.04	4.10	7.14	9.40
1362	6' to 10' deep, 3/4 C.Y. excavator w/trench box	B-13G	212	.075			3.37	3.47	6.84	9.20
1374	10' to 14' deep, 3/4 C.Y. excavator w/trench box	"	188	.085			3.80	3.91	7.71	10.40
1376	1-1/2 C.Y. excavator	B-13E	508	.032			1.41	2.19	3.60	4.67
1381	14' to 20' deep, 1 C.Y. excavator w/trench box	B-13D	301	.053			2.37	2.97	5.34	7.10
1386	20' to 24' deep, 1 C.Y. excavator w/trench box	"	271	.059			2.64	3.30	5.94	7.85
1391	Shoring by S.F./day trench wall protected loose mat., 4' W	B-6	3200	.008	SF Wall	.53	.31	.11	.95	1.21
1392	Rent shoring per week per S.F. wall protected, loose mat., 4' W					1.46			1.46	1.60
1395	Hydraulic shoring, S.F. trench wall protected stable mat., 4' W	2 Clab	2700	.006		.19	.22		.41	.58
1397	semi-stable material, 4' W	"	2400	.007		.27	.25		.52	.71
1398	Rent hydraulic shoring per day/S.F. wall, stable mat., 4' W					.32			.32	.35
1399	semi-stable material					.40			.40	.44
1400	By hand with pick and shovel 2' to 6' deep, light soil	1 Clab	8	1	B.C.Y.		37.50		37.50	61.50
1500	Heavy soil	"	4	2	"		75		75	123
1700	For tamping backfilled trenches, air tamp, add	A-1G	100	.080	E.C.Y.		3.01	.56	3.57	5.55
1900	Vibrating plate, add	B-18	180	.133	"		5.10	.26	5.36	8.65
2100	Trim sides and bottom for concrete pours, common earth		1500	.016	S.F.		.61	.03	.64	1.03
2300	Hardpan		600	.040	"		1.53	.08	1.61	2.60
2400	Pier and spread footing excavation, add to above				B.C.Y.				30%	30%
3000	Backfill trench, F.E. loader, wheel mtd., 1 C.Y. bucket									
3020	Minimal haul	B-10R	400	.030	L.C.Y.		1.39	.75	2.14	3.04
3040	100' haul		200	.060			2.78	1.50	4.28	6.10
3060	200' haul		100	.120			5.55	2.99	8.54	12.20
5020	Loam & Sandy clay with no sheeting or dewatering included									
5050	1' to 4' deep, 3/8 C.Y. tractor loader/backhoe	B-11C	162	.099	B.C.Y.		4.36	2.25	6.61	9.45
5060	1/2 C.Y. excavator	B-11M	216	.074			3.27	1.81	5.08	7.25
5080	4' to 6' deep, 1/2 C.Y. excavator	"	216	.074			3.27	1.81	5.08	7.25
5090	5/8 C.Y. excavator	B-12Q	276	.058			2.59	2.14	4.73	6.50
5130	1/2 C.Y. excavator, truck mounted	B-12J	216	.074			3.31	4.08	7.39	9.80
5140	6' to 10' deep, 3/4 C.Y. excavator	B-12F	243	.066			2.94	2.69	5.63	7.70
5160	1 C.Y. excavator, truck mounted	B-12K	432	.037			1.65	2.32	3.97	5.20
5190	10' to 14' deep, 3/4 C.Y. excavator	B-12F	216	.074			3.31	3.03	6.34	8.65
5210	1-1/2 C.Y. excavator	B-12B	583	.027			1.23	1.77	3	3.91
5250	14' to 20' deep, 1 C.Y. excavator	B-12A	346	.046			2.06	2.35	4.41	5.90
5300	20' to 24' deep, 1 C.Y. excavator	"	311	.051			2.30	2.61	4.91	6.55
5352	4' to 6' deep, 1/2 C.Y. excavator w/trench box	B-13H	205	.078			3.48	4.70	8.18	10.75
5354	5/8 C.Y. excavator	"	257	.062			2.78	3.75	6.53	8.60
5362	6' to 10' deep, 3/4 C.Y. excavator w/trench box	B-13G	231	.069			3.09	3.19	6.28	8.50
5370	10' to 14' deep, 3/4 C.Y. excavator w/trench box	"	205	.078			3.48	3.59	7.07	9.55
5374	1-1/2 C.Y. excavator	B-13E	554	.029			1.29	2.01	3.30	4.29
5382	14' to 20' deep, 1 C.Y. excavator w/trench box	B-13D	329	.049			2.17	2.72	4.89	6.50
5392	20' to 24' deep, 1 C.Y. excavator w/trench box	"	295	.054			2.42	3.03	5.45	7.25
6020	Sand & gravel with no sheeting or dewatering included									
6050	1' to 4' deep, 3/8 C.Y. excavator	B-11C	165	.097	B.C.Y.		4.28	2.21	6.49	9.35
6060	1/2 C.Y. excavator	B-11M	220	.073			3.21	1.78	4.99	7.10
6080	4' to 6' deep, 1/2 C.Y. excavator	"	220	.073			3.21	1.78	4.99	7.10
6090	5/8 C.Y. excavator	B-12Q	275	.058			2.60	2.14	4.74	6.55

31 23 Excavation and Fill

31 23 16 – Excavation

31 23 16.13 Excavating, Trench		Crew	Daily Output	Labor-Hours	Unit	Material	2015 Bare Costs Labor	Equipment	Total	Total Incl O&P
6130	1/2 C.Y. excavator, truck mounted	B-12J	220	.073	B.C.Y.		3.25	4.01	7.26	9.65
6140	6' to 10' deep, 3/4 C.Y. excavator	B-12F	248	.065			2.88	2.64	5.52	7.55
6160	1 C.Y. excavator, truck mounted	B-12K	440	.036			1.62	2.28	3.90	5.10
6190	10' to 14' deep, 3/4 C.Y. excavator	B-12F	220	.073			3.25	2.98	6.23	8.50
6210	1-1/2 C.Y. excavator	B-12B	594	.027			1.20	1.73	2.93	3.85
6250	14' to 20' deep, 1 C.Y. excavator	B-12A	352	.045			2.03	2.31	4.34	5.80
6300	20' to 24' deep, 1 C.Y. excavator	"	317	.050			2.25	2.56	4.81	6.45
6352	4' to 6' deep, 1/2 C.Y. excavator w/trench box	B-13H	209	.077			3.42	4.61	8.03	10.55
6354	5/8 C.Y. excavator	"	261	.061			2.74	3.69	6.43	8.45
6362	6' to 10' deep, 3/4 C.Y. excavator w/trench box	B-13G	236	.068			3.03	3.12	6.15	8.30
6370	10' to 14' deep, 3/4 C.Y. excavator w/trench box	"	209	.077			3.42	3.52	6.94	9.35
6374	1-1/2 C.Y. excavator	B-13E	564	.028			1.27	1.97	3.24	4.21
6382	14' to 20' deep, 1 C.Y. excavator w/trench box	B-13D	334	.048			2.14	2.68	4.82	6.40
6392	20' to 24' deep, 1 C.Y. excavator w/trench box	"	301	.053			2.37	2.97	5.34	7.10
7020	Dense hard clay with no sheeting or dewatering included									
7050	1' to 4' deep, 3/8 C.Y. excavator	B-11C	132	.121	B.C.Y.		5.35	2.76	8.11	11.65
7060	1/2 C.Y. excavator	B-11M	176	.091			4.01	2.23	6.24	8.90
7080	4' to 6' deep, 1/2 C.Y. excavator	"	176	.091			4.01	2.23	6.24	8.90
7090	5/8 C.Y. excavator	B-12Q	220	.073			3.25	2.68	5.93	8.20
7130	1/2 C.Y. excavator, truck mounted	B-12J	176	.091			4.06	5	9.06	12.05
7140	6' to 10' deep, 3/4 C.Y. excavator	B-12F	198	.081			3.61	3.31	6.92	9.45
7160	1 C.Y. excavator, truck mounted	B-12K	352	.045			2.03	2.85	4.88	6.40
7190	10' to 14' deep, 3/4 C.Y. excavator	B-12F	176	.091			4.06	3.72	7.78	10.65
7210	1-1/2 C.Y. excavator	B-12B	475	.034			1.50	2.17	3.67	4.80
7250	14' to 20' deep, 1 C.Y. excavator	B-12A	282	.057			2.53	2.88	5.41	7.25
7300	20' to 24' deep, 1 C.Y. excavator	"	254	.063			2.81	3.20	6.01	8.05
9000	Minimum labor/equipment charge	1 Clab	4	2	Job		75		75	123

31 23 16.14 Excavating, Utility Trench		Crew	Daily Output	Labor-Hours	Unit	Material	Labor	Equipment	Total	Total Incl O&P
0010	**EXCAVATING, UTILITY TRENCH**									
0011	Common earth									
0050	Trenching with chain trencher, 12 H.P., operator walking									
0100	4" wide trench, 12" deep	B-53	800	.010	L.F.		.49	.09	.58	.86
0150	18" deep		750	.011			.52	.09	.61	.92
0200	24" deep		700	.011			.56	.10	.66	.99
0300	6" wide trench, 12" deep		650	.012			.60	.10	.70	1.07
0350	18" deep		600	.013			.65	.11	.76	1.15
0400	24" deep		550	.015			.71	.12	.83	1.26
0450	36" deep		450	.018			.86	.15	1.01	1.54
0600	8" wide trench, 12" deep		475	.017			.82	.14	.96	1.46
0650	18" deep		400	.020			.97	.17	1.14	1.73
0700	24" deep		350	.023			1.11	.19	1.30	1.97
0750	36" deep		300	.027			1.30	.23	1.53	2.31
0900	Minimum labor/equipment charge		2	4	Job		194	34	228	350
1000	Backfill by hand including compaction, add									
1050	4" wide trench, 12" deep	A-1G	800	.010	L.F.		.38	.07	.45	.70
1100	18" deep		530	.015			.57	.11	.68	1.05
1150	24" deep		400	.020			.75	.14	.89	1.39
1300	6" wide trench, 12" deep		540	.015			.56	.10	.66	1.02
1350	18" deep		405	.020			.74	.14	.88	1.37
1400	24" deep		270	.030			1.11	.21	1.32	2.06
1450	36" deep		180	.044			1.67	.31	1.98	3.08
1600	8" wide trench, 12" deep		400	.020			.75	.14	.89	1.39

31 23 16 – Excavation

31 23 16.14 Excavating, Utility Trench		Crew	Daily Output	Labor-Hours	Unit	Material	2015 Bare Costs Labor	Equipment	Total	Total Incl O&P
1650	18" deep	A-1G	265	.030	L.F.		1.14	.21	1.35	2.09
1700	24" deep		200	.040			1.50	.28	1.78	2.78
1750	36" deep		135	.059			2.23	.42	2.65	4.11
2000	Chain trencher, 40 H.P. operator riding									
2050	6" wide trench and backfill, 12" deep	B-54	1200	.007	L.F.		.32	.28	.60	.82
2100	18" deep		1000	.008			.39	.34	.73	.99
2150	24" deep		975	.008			.40	.34	.74	1.01
2200	36" deep		900	.009			.43	.37	.80	1.10
2250	48" deep		750	.011			.52	.45	.97	1.31
2300	60" deep		650	.012			.60	.52	1.12	1.52
2400	8" wide trench and backfill, 12" deep		1000	.008			.39	.34	.73	.99
2450	18" deep		950	.008			.41	.35	.76	1.04
2500	24" deep		900	.009			.43	.37	.80	1.10
2550	36" deep		800	.010			.49	.42	.91	1.23
2600	48" deep		650	.012			.60	.52	1.12	1.52
2700	12" wide trench and backfill, 12" deep		975	.008			.40	.34	.74	1.01
2750	18" deep		860	.009			.45	.39	.84	1.15
2800	24" deep		800	.010			.49	.42	.91	1.23
2850	36" deep		725	.011			.54	.46	1	1.36
3000	16" wide trench and backfill, 12" deep		835	.010			.47	.40	.87	1.18
3050	18" deep		750	.011			.52	.45	.97	1.31
3100	24" deep		700	.011			.56	.48	1.04	1.41
3200	Compaction with vibratory plate, add								35%	35%
5100	Hand excavate and trim for pipe bells after trench excavation									
5200	8" pipe	1 Clab	155	.052	L.F.		1.94		1.94	3.18
5300	18" pipe	"	130	.062	"		2.31		2.31	3.79
9000	Minimum labor/equipment charge	A-1G	4	2	Job		75	14.10	89.10	139
31 23 16.16 Structural Excavation for Minor Structures										
0010	**STRUCTURAL EXCAVATION FOR MINOR STRUCTURES**									
0015	Hand, pits to 6' deep, sandy soil	1 Clab	8	1	B.C.Y.		37.50		37.50	61.50
0100	Heavy soil or clay		4	2			75		75	123
0300	Pits 6' to 12' deep, sandy soil		5	1.600			60		60	98.50
0500	Heavy soil or clay		3	2.667			100		100	164
0700	Pits 12' to 18' deep, sandy soil		4	2			75		75	123
0900	Heavy soil or clay		2	4			150		150	247
1100	Hand loading trucks from stock pile, sandy soil		12	.667			25		25	41
1300	Heavy soil or clay		8	1			37.50		37.50	61.50
1500	For wet or muck hand excavation, add to above								50%	50%
1550	Excavation rock by hand/air tool	B-9	3.40	11.765	B.C.Y.		445	68.50	513.50	810
9000	Minimum labor/equipment charge	1 Clab	4	2	Job		75		75	123
31 23 16.26 Rock Removal										
0010	**ROCK REMOVAL**									
0015	Drilling only rock, 2" hole for rock bolts	B-47	316	.076	L.F.		3.18	5.10	8.28	10.75
0800	2-1/2" hole for pre-splitting		600	.040			1.68	2.68	4.36	5.65
4600	Quarry operations, 2-1/2" to 3-1/2" diameter		715	.034			1.41	2.25	3.66	4.75
31 23 16.30 Drilling and Blasting Rock										
0010	**DRILLING AND BLASTING ROCK**									
0020	Rock, open face, under 1500 C.Y.	B-47	225	.107	B.C.Y.	3.20	4.47	7.15	14.82	18.60
0100	Over 1500 C.Y.		300	.080		3.20	3.35	5.35	11.90	14.85
0200	Areas where blasting mats are required, under 1500 C.Y.		175	.137		3.20	5.75	9.20	18.15	23
0250	Over 1500 C.Y.		250	.096		3.20	4.03	6.45	13.68	17.10
0300	Bulk drilling and blasting, can vary greatly, average								9.65	12.20

31 23 Excavation and Fill

31 23 16 – Excavation

31 23 16.30 Drilling and Blasting Rock

	31 23 16.30 Drilling and Blasting Rock	Crew	Daily Output	Labor-Hours	Unit	Material	2015 Bare Costs Labor	Equipment	Total	Total Incl O&P
0500	Pits, average				B.C.Y.				25.50	31.50
1300	Deep hole method, up to 1500 C.Y.	B-47	50	.480		3.20	20	32	55.20	71.50
1400	Over 1500 C.Y.		66	.364		3.20	15.25	24.50	42.95	55
1900	Restricted areas, up to 1500 C.Y.		13	1.846		3.20	77.50	124	204.70	265
2000	Over 1500 C.Y.		20	1.200		3.20	50.50	80.50	134.20	174
2200	Trenches, up to 1500 C.Y.		22	1.091		9.30	45.50	73	127.80	165
2300	Over 1500 C.Y.		26	.923		9.30	38.50	62	109.80	141
2500	Pier holes, up to 1500 C.Y.		22	1.091		3.20	45.50	73	121.70	158
2600	Over 1500 C.Y.		31	.774		3.20	32.50	52	87.70	113
2800	Boulders under 1/2 C.Y., loaded on truck, no hauling	B-100	80	.150			6.95	11.95	18.90	24.50
2900	Boulders, drilled, blasted	B-47	100	.240		3.20	10.05	16.10	29.35	37.50
3100	Jackhammer operators with foreman compressor, air tools	B-9	1	40	Day		1,525	233	1,758	2,750
3300	Track drill, compressor, operator and foreman	B-47	1	24	"		1,000	1,600	2,600	3,400
3500	Blasting caps				Ea.	6.25			6.25	6.90
3700	Explosives					.48			.48	.52
3900	Blasting mats, rent, for first day					137			137	151
4000	Per added day					47			47	51.50
4200	Preblast survey for 6 room house, individual lot, minimum	A-6	2.40	6.667			315	23	338	535
4300	Maximum	"	1.35	11.852			555	40.50	595.50	945
4500	City block within zone of influence, minimum	A-8	25200	.001	S.F.		.06		.06	.10
4600	Maximum	"	15100	.002	"		.11		.11	.17
5000	Excavate and load boulders, less than 0.5 C.Y.	B-10T	80	.150	B.C.Y.		6.95	6.50	13.45	18.25
5020	0.5 C.Y. to 1 C.Y.	B-10U	100	.120			5.55	10.80	16.35	21
5200	Excavate and load blasted rock, 3 C.Y. power shovel	B-12T	1530	.010			.47	1.02	1.49	1.87
5400	Haul boulders, 25 Ton off-highway dump, 1 mile round trip	B-34E	330	.024			.97	4.09	5.06	6.05
5420	2 mile round trip		275	.029			1.17	4.91	6.08	7.30
5440	3 mile round trip		225	.036			1.42	6	7.42	8.90
5460	4 mile round trip		200	.040			1.60	6.75	8.35	10.05
5600	Bury boulders on site, less than 0.5 C.Y., 300 H.P. dozer									
5620	150′ haul	B-10M	310	.039	B.C.Y.		1.79	6.10	7.89	9.60
5640	300′ haul		210	.057			2.64	9.05	11.69	14.20
5800	0.5 to 1 C.Y., 300 H.P. dozer, 150′ haul		300	.040			1.85	6.30	8.15	9.90
5820	300′ haul		200	.060			2.78	9.50	12.28	14.90

31 23 16.32 Ripping

	31 23 16.32 Ripping	Crew	Daily Output	Labor-Hours	Unit	Material	Labor	Equipment	Total	Total Incl O&P
0010	**RIPPING**									
0020	Ripping, trap rock, soft, 300 HP dozer, ideal conditions	B-11S	700	.017	B.C.Y.		.79	2.83	3.62	4.38
1400	Ideal conditions		710	.017			.78	2.79	3.57	4.32
1500	Adverse conditions		660	.018			.84	3	3.84	4.65
1600	Medium hard, 300 HP dozer, ideal conditons		600	.020			.93	3.30	4.23	5.10
1700	Adverse conditions		540	.022			1.03	3.67	4.70	5.70
2000	Very hard, 410 HP dozer, ideal conditions	B-11T	350	.034			1.59	7.20	8.79	10.45
2100	Adverse conditions	"	310	.039			1.79	8.10	9.89	11.75
2200	Shale, soft, 300 HP dozer, ideal conditons	B-11S	1500	.008			.37	1.32	1.69	2.04
2300	Adverse conditions		1350	.009			.41	1.47	1.88	2.27
2400	Medium hard, 300 HP dozer, ideal conditons		1200	.010			.46	1.65	2.11	2.56
2500	Adverse conditions		1080	.011			.51	1.83	2.34	2.84
2600	Very hard, 410 HP dozer, ideal conditons	B-11T	800	.015			.69	3.14	3.83	4.57
2700	Adverse conditions	"	720	.017			.77	3.49	4.26	5.05
3000	Dozing ripped material, 200 HP, 100′ haul	B-10B	700	.017			.79	1.98	2.77	3.45
3050	300′ haul	"	250	.048			2.22	5.55	7.77	9.65
3200	300 HP, 100′ haul	B-10M	1150	.010			.48	1.65	2.13	2.58
3250	300′ haul	"	400	.030			1.39	4.74	6.13	7.40

31 23 16.32 Ripping		Crew	Daily Output	Labor-Hours	Unit	Material	2015 Bare Costs Labor	Equipment	Total	Total Incl O&P
3400	410 HP, 100' haul	B-10X	1680	.007	B.C.Y.		.33	1.43	1.76	2.10
3450	300' haul	"	600	.020			.93	4.01	4.94	5.90

31 23 16.42 Excavating, Bulk Bank Measure		Crew	Daily Output	Labor-Hours	Unit	Material	2015 Bare Costs Labor	Equipment	Total	Total Incl O&P
0010	**EXCAVATING, BULK BANK MEASURE**									
0011	Common earth piled									
0020	For loading onto trucks, add								15%	15%
0050	For mobilization and demobilization, see Section 01 54 36.50									
0100	For hauling, see Section 31 23 23.20									
0200	Excavator, hydraulic, crawler mtd., 1 C.Y. cap. = 100 C.Y./hr.	B-12A	800	.020	B.C.Y.		.89	1.02	1.91	2.56
0250	1-1/2 C.Y. cap. = 125 C.Y./hr.	B-12B	1000	.016			.71	1.03	1.74	2.28
0260	2 C.Y. cap. = 165 C.Y./hr.	B-12C	1320	.012			.54	.89	1.43	1.85
0300	3 C.Y. cap. = 260 C.Y./hr.	B-12D	2080	.008			.34	1.17	1.51	1.84
0305	3-1/2 C.Y. cap. = 300 C.Y./hr.	"	2400	.007			.30	1.02	1.32	1.60
0310	Wheel mounted, 1/2 C.Y. cap. = 40 C.Y./hr.	B-12E	320	.050			2.23	1.40	3.63	5.15
0360	3/4 C.Y. cap. = 60 C.Y./hr.	B-12F	480	.033			1.49	1.36	2.85	3.89
0500	Clamshell, 1/2 C.Y. cap. = 20 C.Y./hr.	B-12G	160	.100			4.47	4.43	8.90	12.05
0550	1 C.Y. cap. = 35 C.Y./hr.	B-12H	280	.057			2.55	4.28	6.83	8.80
0950	Dragline, 1/2 C.Y. cap. = 30 C.Y./hr.	B-12I	240	.067			2.98	3.68	6.66	8.85
1000	3/4 C.Y. cap. = 35 C.Y./hr.	"	280	.057			2.55	3.15	5.70	7.60
1050	1-1/2 C.Y. cap. = 65 C.Y./hr.	B-12P	520	.031			1.37	2.29	3.66	4.73
1100	3 C.Y. cap. = 112 C.Y./hr.	B-12V	900	.018			.79	1.68	2.47	3.13
1200	Front end loader, track mtd., 1-1/2 C.Y. cap. = 70 C.Y./hr.	B-10N	560	.021			.99	.93	1.92	2.62
1250	2-1/2 C.Y. cap. = 95 C.Y./hr.	B-10O	760	.016			.73	1.26	1.99	2.55
1300	3 C.Y. cap. = 130 C.Y./hr.	B-10P	1040	.012			.53	1.14	1.67	2.11
1350	5 C.Y. cap. = 160 C.Y./hr.	B-10Q	1280	.009			.43	1.22	1.65	2.03
1500	Wheel mounted, 3/4 C.Y. cap. = 45 C.Y./hr.	B-10R	360	.033			1.54	.83	2.37	3.38
1550	1-1/2 C.Y. cap. = 80 C.Y./hr.	B-10S	640	.019			.87	.59	1.46	2.04
1601	3 C.Y. cap. = 140 C.Y./hr.	B-10T	1120	.011			.50	.46	.96	1.30
1650	5 C.Y. cap. = 185 C.Y./hr.	B-10U	1480	.008			.38	.73	1.11	1.40
1800	Hydraulic excavator, truck mtd. 1/2 C.Y. = 30 C.Y./hr.	B-12J	240	.067			2.98	3.68	6.66	8.85
1850	48 inch bucket, 1 C.Y. = 45 C.Y./hr.	B-12K	360	.044			1.98	2.78	4.76	6.25
3700	Shovel, 1/2 C.Y. capacity = 55 C.Y./hr.	B-12L	440	.036			1.62	1.66	3.28	4.43
3750	3/4 C.Y. capacity = 85 C.Y./hr.	B-12M	680	.024			1.05	1.36	2.41	3.19
3800	1 C.Y. capacity = 120 C.Y./hr.	B-12N	960	.017			.74	1.27	2.01	2.60
3850	1-1/2 C.Y. capacity = 160 C.Y./hr.	B-12O	1280	.013			.56	.97	1.53	1.96
3900	3 C.Y. cap. = 250 C.Y./hr.	B-12T	2000	.008			.36	.78	1.14	1.43
4000	For soft soil or sand, deduct								15%	15%
4100	For heavy soil or stiff clay, add								60%	60%
4200	For wet excavation with clamshell or dragline, add								100%	100%
4250	All other equipment, add								50%	50%
4400	Clamshell in sheeting or cofferdam, minimum	B-12H	160	.100			4.47	7.50	11.97	15.45
4450	Maximum	"	60	.267			11.90	19.95	31.85	41
5000	Excavating, bulk bank measure, sandy clay & loam piled									
5020	For loading onto trucks, add								15%	15%
5100	Excavator, hydraulic, crawler mtd., 1 C.Y. cap. = 120 C.Y./hr.	B-12A	960	.017	B.C.Y.		.74	.85	1.59	2.13
5150	1-1/2 C.Y. cap. = 150 C.Y./hr.	B-12B	1200	.013			.60	.86	1.46	1.90
5300	2 C.Y. cap. = 195 C.Y./hr.	B-12C	1560	.010			.46	.75	1.21	1.57
5400	3 C.Y. cap. = 300 C.Y./hr.	B-12D	2400	.007			.30	1.02	1.32	1.60
5500	3.5 C.Y. cap. = 350 C.Y./hr.	"	2800	.006			.26	.87	1.13	1.37
5610	Wheel mounted, 1/2 C.Y. cap. = 44 C.Y./hr.	B-12E	352	.045			2.03	1.27	3.30	4.67
5660	3/4 C.Y. cap. = 66 C.Y./hr.	B-12F	528	.030			1.35	1.24	2.59	3.54
8000	For hauling excavated material, see Section 31 23 23.20									

31 23 Excavation and Fill

31 23 16 – Excavation

31 23 16.42 Excavating, Bulk Bank Measure		Crew	Daily Output	Labor-Hours	Unit	Material	2015 Bare Costs Labor	Equipment	Total	Total Incl O&P
9000	Minimum labor/equipment charge	B-10L	2	6	Job		278	236	514	705

31 23 16.46 Excavating, Bulk, Dozer

		Crew	Daily Output	Labor-Hours	Unit	Material	Labor	Equipment	Total	Total Incl O&P
0010	**EXCAVATING, BULK, DOZER**									
0011	Open site									
2000	80 H.P., 50′ haul, sand & gravel	B-10L	460	.026	B.C.Y.		1.21	1.03	2.24	3.06
2010	Sandy clay & loam		440	.027			1.26	1.07	2.33	3.20
2020	Common earth		400	.030			1.39	1.18	2.57	3.52
2040	Clay		250	.048			2.22	1.89	4.11	5.65
2200	150′ haul, sand & gravel		230	.052			2.41	2.05	4.46	6.10
2210	Sandy clay & loam		220	.055			2.52	2.15	4.67	6.40
2220	Common earth		200	.060			2.78	2.36	5.14	7.05
2240	Clay		125	.096			4.44	3.78	8.22	11.25
2400	300′ haul, sand & gravel		120	.100			4.63	3.94	8.57	11.75
2410	Sandy clay & loam		115	.104			4.83	4.11	8.94	12.25
2420	Common earth		100	.120			5.55	4.72	10.27	14.10
2440	Clay		65	.185			8.55	7.25	15.80	21.50
3000	105 H.P., 50′ haul, sand & gravel	B-10W	700	.017			.79	.86	1.65	2.22
3010	Sandy clay & loam		680	.018			.82	.89	1.71	2.29
3020	Common earth		610	.020			.91	.99	1.90	2.55
3040	Clay		385	.031			1.44	1.57	3.01	4.03
3200	150′ haul, sand & gravel		310	.039			1.79	1.94	3.73	5
3210	Sandy clay & loam		300	.040			1.85	2.01	3.86	5.15
3220	Common earth		270	.044			2.06	2.23	4.29	5.75
3240	Clay		170	.071			3.27	3.55	6.82	9.15
3300	300′ haul, sand & gravel		140	.086			3.97	4.31	8.28	11.10
3310	Sandy clay & loam		135	.089			4.11	4.46	8.57	11.50
3320	Common earth		120	.100			4.63	5	9.63	12.95
3340	Clay		100	.120			5.55	6.05	11.60	15.55
4000	200 H.P., 50′ haul, sand & gravel	B-10B	1400	.009			.40	.99	1.39	1.72
4010	Sandy clay & loam		1360	.009			.41	1.02	1.43	1.77
4020	Common earth		1230	.010			.45	1.13	1.58	1.96
4040	Clay		770	.016			.72	1.80	2.52	3.13
4200	150′ haul, sand & gravel		595	.020			.93	2.33	3.26	4.05
4210	Sandy clay & loam		580	.021			.96	2.39	3.35	4.16
4220	Common earth		516	.023			1.08	2.69	3.77	4.68
4240	Clay		325	.037			1.71	4.27	5.98	7.45
4400	300′ haul, sand & gravel		310	.039			1.79	4.47	6.26	7.80
4410	Sandy clay & loam		300	.040			1.85	4.62	6.47	8.05
4420	Common earth		270	.044			2.06	5.15	7.21	8.95
4440	Clay		170	.071			3.27	8.15	11.42	14.20
5000	300 H.P., 50′ haul, sand & gravel	B-10M	1900	.006			.29	1	1.29	1.57
5010	Sandy clay & loam		1850	.006			.30	1.03	1.33	1.61
5020	Common earth		1650	.007			.34	1.15	1.49	1.80
5040	Clay		1025	.012			.54	1.85	2.39	2.91
5200	150′ haul, sand & gravel		920	.013			.60	2.06	2.66	3.24
5210	Sandy clay & loam		895	.013			.62	2.12	2.74	3.32
5220	Common earth		800	.015			.69	2.37	3.06	3.72
5240	Clay		500	.024			1.11	3.79	4.90	5.95
5400	300′ haul, sand & gravel		470	.026			1.18	4.04	5.22	6.35
5410	Sandy clay & loam		455	.026			1.22	4.17	5.39	6.55
5420	Common earth		410	.029			1.35	4.63	5.98	7.25
5440	Clay		250	.048			2.22	7.60	9.82	11.90

31 23 16.50 Excavation, Bulk, Scrapers		Crew	Daily Output	Labor-Hours	Unit	Material	2015 Bare Costs Labor	Equipment	Total	Total Incl O&P
0010	EXCAVATION, BULK, SCRAPERS									
0100	Elev. scraper 11 C.Y., sand & gravel 1500′ haul, 1/4 dozer	B-33F	690	.020	B.C.Y.		.95	2.37	3.32	4.13
0150	3000′ haul		610	.023			1.08	2.68	3.76	4.67
0200	5000′ haul		505	.028			1.30	3.24	4.54	5.65
0300	Common earth, 1500′ haul		600	.023			1.09	2.73	3.82	4.75
0350	3000′ haul		530	.026			1.24	3.09	4.33	5.40
0400	5000′ haul		440	.032			1.49	3.72	5.21	6.50
0410	Sandy clay & loam, 1500′ haul		648	.022			1.01	2.53	3.54	4.40
0420	3000′ haul		572	.024			1.15	2.86	4.01	4.99
0430	5000′ haul		475	.029			1.38	3.45	4.83	6
0500	Clay, 1500′ haul		375	.037			1.75	4.37	6.12	7.60
0550	3000′ haul		330	.042			1.99	4.96	6.95	8.65
0600	5000′ haul		275	.051			2.39	5.95	8.34	10.35
1000	Self propelled scraper, 14 C.Y. 1/4 push dozer, sand									
1050	Sand and gravel, 1500′ haul	B-33D	920	.015	B.C.Y.		.71	2.56	3.27	3.96
1100	3000′ haul		805	.017			.82	2.93	3.75	4.52
1200	5000′ haul		645	.022			1.02	3.66	4.68	5.65
1300	Common earth, 1500′ haul		800	.018			.82	2.95	3.77	4.55
1350	3000′ haul		700	.020			.94	3.37	4.31	5.20
1400	5000′ haul		560	.025			1.17	4.21	5.38	6.50
1420	Sandy clay & loam, 1500′ haul		864	.016			.76	2.73	3.49	4.21
1430	3000′ haul		786	.018			.84	3	3.84	4.64
1440	5000′ haul		605	.023			1.09	3.90	4.99	6
1500	Clay, 1500′ haul		500	.028			1.31	4.72	6.03	7.30
1550	3000′ haul		440	.032			1.49	5.35	6.84	8.30
1600	5000′ haul		350	.040			1.88	6.75	8.63	10.40
2000	21 C.Y., 1/4 push dozer, sand & gravel, 1500′ haul	B-33E	1180	.012			.56	2.68	3.24	3.84
2100	3000′ haul		910	.015			.72	3.47	4.19	4.97
2200	5000′ haul		750	.019			.88	4.22	5.10	6.05
2300	Common earth, 1500′ haul		1030	.014			.64	3.07	3.71	4.40
2350	3000′ haul		790	.018			.83	4	4.83	5.75
2400	5000′ haul		650	.022			1.01	4.87	5.88	6.95
2420	Sandy clay & loam, 1500′ haul		1112	.013			.59	2.84	3.43	4.07
2430	3000′ haul		854	.016			.77	3.70	4.47	5.30
2440	5000′ haul		702	.020			.94	4.50	5.44	6.45
2500	Clay, 1500′ haul		645	.022			1.02	4.90	5.92	7.05
2550	3000′ haul		495	.028			1.33	6.40	7.73	9.15
2600	5000′ haul		405	.035			1.62	7.80	9.42	11.20
2700	Towed, 10 C.Y., 1/4 push dozer, sand & gravel, 1500′ haul	B-33B	560	.025			1.17	4.50	5.67	6.80
2720	3000′ haul		450	.031			1.46	5.60	7.06	8.50
2730	5000′ haul		365	.038			1.80	6.90	8.70	10.50
2750	Common earth, 1500′ haul		420	.033			1.56	6	7.56	9.10
2770	3000′ haul		400	.035			1.64	6.30	7.94	9.50
2780	5000′ haul		310	.045			2.12	8.10	10.22	12.35
2785	Sandy clay & Loam, 1500′ haul		454	.031			1.45	5.55	7	8.40
2790	3000′ haul		432	.032			1.52	5.85	7.37	8.85
2795	5000′ haul		340	.041			1.93	7.40	9.33	11.25
2800	Clay, 1500′ haul		315	.044			2.08	8	10.08	12.15
2820	3000′ haul		300	.047			2.19	8.40	10.59	12.75
2840	5000′ haul		225	.062			2.92	11.20	14.12	16.95
2900	15 C.Y., 1/4 push dozer, sand & gravel, 1500′ haul	B-33C	800	.018			.82	3.17	3.99	4.80
2920	3000′ haul		640	.022			1.03	3.96	4.99	6

31 23 16 – Excavation

31 23 16.50 Excavation, Bulk, Scrapers		Crew	Daily Output	Labor-Hours	Unit	Material	2015 Bare Costs Labor	Equipment	Total	Total Incl O&P
2940	5000′ haul	B-33C	520	.027	B.C.Y.		1.26	4.88	6.14	7.35
2960	Common earth, 1500′ haul		600	.023			1.09	4.23	5.32	6.40
2980	3000′ haul		560	.025			1.17	4.53	5.70	6.85
3000	5000′ haul		440	.032			1.49	5.75	7.24	8.75
3005	Sandy clay & Loam, 1500′ haul		648	.022			1.01	3.91	4.92	5.90
3010	3000′ haul		605	.023			1.09	4.19	5.28	6.35
3015	5000′ haul		475	.029			1.38	5.35	6.73	8.05
3020	Clay, 1500′ haul		450	.031			1.46	5.65	7.11	8.55
3040	3000′ haul		420	.033			1.56	6.05	7.61	9.15
3060	5000′ haul	↓	320	.044	↓		2.05	7.95	10	12

31 23 19 – Dewatering

31 23 19.20 Dewatering Systems		Crew	Daily Output	Labor-Hours	Unit	Material	Labor	Equipment	Total	Total Incl O&P
0010	**DEWATERING SYSTEMS**									
0020	Excavate drainage trench, 2′ wide, 2′ deep	B-11C	90	.178	C.Y.		7.85	4.05	11.90	17.05
0100	2′ wide, 3′ deep, with backhoe loader	"	135	.119			5.25	2.70	7.95	11.35
0200	Excavate sump pits by hand, light soil	1 Clab	7.10	1.127			42.50		42.50	69.50
0300	Heavy soil	"	3.50	2.286	↓		86		86	141
0500	Pumping 8 hr., attended 2 hrs. per day, including 20 L.F.									
0550	of suction hose & 100 L.F. discharge hose									
0600	2″ diaphragm pump used for 8 hours	B-10H	4	3	Day		139	18.70	157.70	243
0650	4″ diaphragm pump used for 8 hours	B-10I	4	3			139	30.50	169.50	256
0800	8 hrs. attended, 2″ diaphragm pump	B-10H	1	12			555	75	630	970
0900	3″ centrifugal pump	B-10J	1	12			555	84	639	985
1000	4″ diaphragm pump	B-10I	1	12			555	122	677	1,025
1100	6″ centrifugal pump	B-10K	1	12	↓		555	370	925	1,300
1300	CMP, incl. excavation 3′ deep, 12″ diameter	B-6	115	.209	L.F.	11.05	8.60	3.17	22.82	29.50
1400	18″ diameter		100	.240	"	16.70	9.90	3.64	30.24	38.50
1600	Sump hole construction, incl. excavation and gravel, pit		1250	.019	C.F.	1.09	.79	.29	2.17	2.80
1700	With 12″ gravel collar, 12″ pipe, corrugated, 16 ga.		70	.343	L.F.	21.50	14.15	5.20	40.85	52
1800	15″ pipe, corrugated, 16 ga.		55	.436		28	18	6.60	52.60	67
1900	18″ pipe, corrugated, 16 ga.		50	.480		32	19.80	7.30	59.10	75.50
2000	24″ pipe, corrugated, 14 ga.		40	.600	↓	38.50	25	9.10	72.60	92.50
2200	Wood lining, up to 4′ x 4′, add	↓	300	.080	SFCA	16.25	3.30	1.21	20.76	24.50
9950	See Section 31 23 19.40 for wellpoints									
9960	See Section 31 23 19.30 for deep well systems									

31 23 19.30 Wells		Crew	Daily Output	Labor-Hours	Unit	Material	Labor	Equipment	Total	Total Incl O&P
0010	**WELLS**									
0011	For dewatering 10′ to 20′ deep, 2′ diameter									
0020	with steel casing, minimum	B-6	165	.145	V.L.F.	38	6	2.21	46.21	53.50
0050	Average		98	.245		43	10.10	3.72	56.82	67.50
0100	Maximum	↓	49	.490	↓	47.50	20	7.45	74.95	93

31 23 19.40 Wellpoints		Crew	Daily Output	Labor-Hours	Unit	Material	Labor	Equipment	Total	Total Incl O&P
0010	**WELLPOINTS**									
0011	For equipment rental, see 01 54 33 in Reference Section									
0100	Installation and removal of single stage system									
0110	Labor only, .75 labor-hours per L.F.	1 Clab	10.70	.748	LF Hdr		28		28	46
0200	2.0 labor-hours per L.F.	"	4	2	"		75		75	123
0400	Pump operation, 4 @ 6 hr. shifts									
0410	Per 24 hour day	4 Eqlt	1.27	25.197	Day		1,225		1,225	1,950
0500	Per 168 hour week, 160 hr. straight, 8 hr. double time		.18	177	Week		8,650		8,650	13,700
0550	Per 4.3 week month	↓	.04	800	Month		38,900		38,900	61,500
0600	Complete installation, operation, equipment rental, fuel &									

31 23 Excavation and Fill

31 23 19 – Dewatering

31 23 19.40 Wellpoints		Crew	Daily Output	Labor-Hours	Unit	Material	2015 Bare Costs Labor	Equipment	Total	Total Incl O&P
0610	removal of system with 2" wellpoints 5' O.C.									
0700	100' long header, 6" diameter, first month	4 Eqlt	3.23	9.907	LF Hdr	159	480		639	940
0800	Thereafter, per month		4.13	7.748		127	375		502	740
1000	200' long header, 8" diameter, first month		6	5.333		145	259		404	570
1100	Thereafter, per month		8.39	3.814		71.50	185		256.50	375
1300	500' long header, 8" diameter, first month		10.63	3.010		55.50	146		201.50	293
1400	Thereafter, per month		20.91	1.530		39.50	74.50		114	162
1600	1,000' long header, 10" diameter, first month		11.62	2.754		47.50	134		181.50	265
1700	Thereafter, per month		41.81	.765		24	37		61	85
1900	Note: above figures include pumping 168 hrs. per week									
1910	and include the pump operator and one stand-by pump.									

31 23 23 – Fill

31 23 23.13 Backfill

		Crew	Daily Output	Labor-Hours	Unit	Material	Labor	Equipment	Total	Total Incl O&P
0010	**BACKFILL**									
0015	By hand, no compaction, light soil	1 Clab	14	.571	L.C.Y.		21.50		21.50	35
0100	Heavy soil		11	.727	"		27.50		27.50	45
0300	Compaction in 6" layers, hand tamp, add to above		20.60	.388	E.C.Y.		14.60		14.60	24
0400	Roller compaction operator walking, add	B-10A	100	.120			5.55	1.80	7.35	10.90
0500	Air tamp, add	B-9D	190	.211			8	1.40	9.40	14.65
0600	Vibrating plate, add	A-1D	60	.133			5	.60	5.60	8.85
0800	Compaction in 12" layers, hand tamp, add to above	1 Clab	34	.235			8.85		8.85	14.50
0900	Roller compaction operator walking, add	B-10A	150	.080			3.70	1.20	4.90	7.25
1000	Air tamp, add	B-9	285	.140			5.35	.82	6.17	9.65
1100	Vibrating plate, add	A-1E	90	.089			3.34	.52	3.86	6.05

31 23 23.14 Backfill, Structural

		Crew	Daily Output	Labor-Hours	Unit	Material	Labor	Equipment	Total	Total Incl O&P
0010	**BACKFILL, STRUCTURAL**									
0011	Dozer or F.E. loader									
0020	From existing stockpile, no compaction									
2000	80 H.P., 50' haul, sand & gravel	B-10L	1100	.011	L.C.Y.		.50	.43	.93	1.28
2010	Sandy clay & loam		1070	.011			.52	.44	.96	1.32
2020	Common earth		975	.012			.57	.48	1.05	1.44
2040	Clay		850	.014			.65	.56	1.21	1.66
2200	150' haul, sand & gravel		550	.022			1.01	.86	1.87	2.56
2210	Sandy clay & loam		535	.022			1.04	.88	1.92	2.63
2220	Common earth		490	.024			1.13	.96	2.09	2.87
2240	Clay		425	.028			1.31	1.11	2.42	3.31
2400	300' haul, sand & gravel		370	.032			1.50	1.28	2.78	3.80
2410	Sandy clay & loam		360	.033			1.54	1.31	2.85	3.91
2420	Common earth		330	.036			1.68	1.43	3.11	4.26
2440	Clay		290	.041			1.91	1.63	3.54	4.86
3000	105 H.P., 50' haul, sand & gravel	B-10W	1350	.009			.41	.45	.86	1.15
3010	Sandy clay & loam		1325	.009			.42	.46	.88	1.17
3020	Common earth		1225	.010			.45	.49	.94	1.27
3040	Clay		1100	.011			.50	.55	1.05	1.41
3200	150' haul, sand & gravel		670	.018			.83	.90	1.73	2.32
3210	Sandy clay & loam		655	.018			.85	.92	1.77	2.37
3220	Common earth		610	.020			.91	.99	1.90	2.55
3240	Clay		550	.022			1.01	1.10	2.11	2.83
3300	300' haul, sand & gravel		465	.026			1.19	1.30	2.49	3.34
3310	Sandy clay & loam		455	.026			1.22	1.32	2.54	3.41
3320	Common earth		415	.029			1.34	1.45	2.79	3.74
3340	Clay		370	.032			1.50	1.63	3.13	4.19

31 23 23.14 Backfill, Structural		Crew	Daily Output	Labor-Hours	Unit	Material	2015 Bare Costs Labor	Equipment	Total	Total Incl O&P
4000	200 H.P., 50′ haul, sand & gravel	B-10B	2500	.005	L.C.Y.		.22	.55	.77	.97
4010	Sandy clay & loam		2435	.005			.23	.57	.80	1
4020	Common earth		2200	.005			.25	.63	.88	1.09
4040	Clay		1950	.006			.28	.71	.99	1.24
4200	150′ haul, sand & gravel		1225	.010			.45	1.13	1.58	1.98
4210	Sandy clay & loam		1200	.010			.46	1.16	1.62	2.01
4220	Common earth		1100	.011			.50	1.26	1.76	2.20
4240	Clay		975	.012			.57	1.42	1.99	2.48
4400	300′ haul, sand & gravel		805	.015			.69	1.72	2.41	3
4410	Sandy clay & loam		790	.015			.70	1.76	2.46	3.06
4420	Common earth		735	.016			.76	1.89	2.65	3.29
4440	Clay		660	.018			.84	2.10	2.94	3.66
5000	300 H.P., 50′ haul, sand & gravel	B-10M	3170	.004			.18	.60	.78	.94
5010	Sandy clay & loam		3110	.004			.18	.61	.79	.96
5020	Common earth		2900	.004			.19	.65	.84	1.03
5040	Clay		2700	.004			.21	.70	.91	1.10
5200	150′ haul, sand & gravel		2200	.005			.25	.86	1.11	1.35
5210	Sandy clay & loam		2150	.006			.26	.88	1.14	1.38
5220	Common earth		1950	.006			.28	.97	1.25	1.53
5240	Clay		1700	.007			.33	1.12	1.45	1.75
5400	300′ haul, sand & gravel		1500	.008			.37	1.26	1.63	1.98
5410	Sandy clay & loam		1470	.008			.38	1.29	1.67	2.02
5420	Common earth		1350	.009			.41	1.41	1.82	2.21
5440	Clay		1225	.010			.45	1.55	2	2.43

31 23 23.15 Borrow, Loading And/Or Spreading		Crew	Daily Output	Labor-Hours	Unit	Material	Labor	Equipment	Total	Total Incl O&P
0010	**BORROW, LOADING AND/OR SPREADING**									
0020	Material only, bank run gravel				Ton	15.90			15.90	17.50
0100	Crushed stone, 1-1/2″ to 3/4″ size					14.85			14.85	16.35
0150	3/8″ size					19.65			19.65	21.50
0200	Dead or bank run sand					17.85			17.85	19.60
0500	Haul 2 mi. spread, 200 HP dozer, bank run gravel	B-15	1100	.025			1.09	2.52	3.61	4.52
0540	Crushed stone, 1-1/2″ to 3/4″ size		1150	.024			1.04	2.41	3.45	4.32
0560	3/8″ size		1150	.024			1.04	2.41	3.45	4.32
0600	Bank run or dead sand		1000	.028			1.20	2.77	3.97	4.97
1000	Hand spread, bank run gravel	A-5	33	.545			20.50	1.86	22.36	35.50
1040	Crushed stone, 1-1/2″ to 3/4″ size		38	.474			17.90	1.62	19.52	31.50
1060	3/8″ size		40	.450			17	1.53	18.53	29.50
1100	Bank run or dead sand		33	.545			20.50	1.86	22.36	35.50
1800	Delivery charge, minimum 20 tons, 1 hr. round trip, add	B-34B	130	.062			2.46	5.30	7.76	9.80
1820	1-1/2 hr. round trip, add		93	.086			3.45	7.45	10.90	13.70
1840	2 hr. round trip, add		65	.123			4.93	10.65	15.58	19.65
7000	Topsoil or loam from stockpile, shovel, 1 C.Y. bucket	B-12N	840	.019	B.C.Y.	24.50	.85	1.45	26.80	30
7010	1-1/2 C.Y. bucket	B-12O	1135	.014		24.50	.63	1.09	26.22	29
7020	3 C.Y. bucket	B-12T	1800	.009		24.50	.40	.87	25.77	28.50
7030	Front end loader, wheel mounted									
7050	3/4 C.Y. bucket	B-10R	550	.022	B.C.Y.	24.50	1.01	.54	26.05	29
7060	1-1/2 C.Y. bucket	B-10S	970	.012		24.50	.57	.39	25.46	28.50
7070	3 C.Y. bucket	B-10T	1575	.008		24.50	.35	.33	25.18	28
7080	5 C.Y. bucket	B-10U	2600	.005		24.50	.21	.42	25.13	28

31 23 Excavation and Fill

31 23 23 – Fill

31 23 23.16 Fill By Borrow and Utility Bedding		Crew	Daily Output	Labor-Hours	Unit	Material	2015 Bare Costs Labor	Equipment	Total	Total Incl O&P
0010	**FILL BY BORROW AND UTILITY BEDDING**									
0015	Fill by borrow, load, 1 mile haul, spread with dozer									
0020	for embankments	B-15	1200	.023	L.C.Y.	12.40	1	2.31	15.71	17.80
0035	Select fill for shoulders & embankments	"	1200	.023	"	21	1	2.31	24.31	27
0040	Fill,for hauling over 1 mile,add to above per C.Y., see Section 31 23 23.20				Mile				1.41	1.73
0049	Utility bedding, for pipe & conduit, not incl. compaction									
0050	Crushed or screened bank run gravel	B-6	150	.160	L.C.Y.	25.50	6.60	2.43	34.53	41.50
0100	Crushed stone 3/4" to 1/2"		150	.160		23.50	6.60	2.43	32.53	39.50
0200	Sand, dead or bank		150	.160		17.85	6.60	2.43	26.88	33
0500	Compacting bedding in trench	A-1D	90	.089	E.C.Y.		3.34	.40	3.74	5.95
0600	If material source exceeds 2 miles, add for extra mileage.									
0610	See Section 31 23 23.20 for hauling mileage add.									
31 23 23.17 General Fill										
0010	**GENERAL FILL**									
0011	Spread dumped material, no compaction									
0020	By dozer, no compaction	B-10B	1000	.012	L.C.Y.		.56	1.39	1.95	2.42
0100	By hand	1 Clab	12	.667	"		25		25	41
0150	Spread fill, from stockpile with 2-1/2 C.Y. F.E. loader									
0170	130 H.P., 300' haul	B-10P	600	.020	L.C.Y.		.93	1.98	2.91	3.66
0190	With dozer 300 H.P., 300' haul	B-10M	600	.020	"		.93	3.16	4.09	4.96
0500	Gravel fill, compacted, under floor slabs, 4" deep	B-37	10000	.005	S.F.	.42	.19	.02	.63	.79
0600	6" deep		8600	.006		.63	.22	.02	.87	1.07
0700	9" deep		7200	.007		1.05	.27	.02	1.34	1.60
0800	12" deep		6000	.008		1.47	.32	.03	1.82	2.17
1000	Alternate pricing method, 4" deep		120	.400	E.C.Y.	31.50	15.90	1.31	48.71	62
1100	6" deep		160	.300		31.50	11.95	.98	44.43	55
1200	9" deep		200	.240		31.50	9.55	.79	41.84	51
1300	12" deep		220	.218		31.50	8.70	.72	40.92	49.50
1500	For fill under exterior paving, see Section 32 11 23.23									
1600	For flowable fill, see Section 03 31 13.35									
9000	Minimum labor/equipment charge	1 Clab	4	2	Job		75		75	123
31 23 23.20 Hauling										
0010	**HAULING**									
0011	Excavated or borrow, loose cubic yards									
0012	no loading equipment, including hauling,waiting, loading/dumping									
0013	time per cycle (wait, load, travel, unload or dump & return)									
0014	8 C.Y. truck, 15 MPH ave, cycle 0.5 miles, 10 min. wait/Ld./Uld.	B-34A	320	.025	L.C.Y.		1	1.29	2.29	3.02
0016	cycle 1 mile		272	.029			1.18	1.51	2.69	3.56
0018	cycle 2 miles		208	.038			1.54	1.98	3.52	4.65
0020	cycle 4 miles		144	.056			2.23	2.86	5.09	6.75
0022	cycle 6 miles		112	.071			2.86	3.67	6.53	8.65
0024	cycle 8 miles		88	.091			3.64	4.67	8.31	11
0026	20 MPH ave,cycle 0.5 mile		336	.024			.95	1.22	2.17	2.89
0028	cycle 1 mile		296	.027			1.08	1.39	2.47	3.27
0030	cycle 2 miles		240	.033			1.33	1.71	3.04	4.03
0032	cycle 4 miles		176	.045			1.82	2.34	4.16	5.50
0034	cycle 6 miles		136	.059			2.36	3.02	5.38	7.15
0036	cycle 8 miles		112	.071			2.86	3.67	6.53	8.65
0044	25 MPH ave, cycle 4 miles		192	.042			1.67	2.14	3.81	5.05
0046	cycle 6 miles		160	.050			2	2.57	4.57	6.05
0048	cycle 8 miles		128	.063			2.50	3.21	5.71	7.55
0050	30 MPH ave, cycle 4 miles		216	.037			1.48	1.90	3.38	4.48

31 23 23.20 Hauling		Crew	Daily Output	Labor-Hours	Unit	Material	2015 Bare Costs Labor	Equipment	Total	Total Incl O&P
0052	cycle 6 miles	B-34A	176	.045	L.C.Y.		1.82	2.34	4.16	5.50
0054	cycle 8 miles		144	.056			2.23	2.86	5.09	6.75
0114	15 MPH ave, cycle 0.5 mile, 15 min. wait/Ld./Uld.		224	.036			1.43	1.84	3.27	4.33
0116	cycle 1 mile		200	.040			1.60	2.06	3.66	4.84
0118	cycle 2 miles		168	.048			1.91	2.45	4.36	5.75
0120	cycle 4 miles		120	.067			2.67	3.43	6.10	8.05
0122	cycle 6 miles		96	.083			3.34	4.28	7.62	10.10
0124	cycle 8 miles		80	.100			4.01	5.15	9.16	12.10
0126	20 MPH ave, cycle 0.5 mile		232	.034			1.38	1.77	3.15	4.18
0128	cycle 1 mile		208	.038			1.54	1.98	3.52	4.65
0130	cycle 2 miles		184	.043			1.74	2.23	3.97	5.25
0132	cycle 4 miles		144	.056			2.23	2.86	5.09	6.75
0134	cycle 6 miles		112	.071			2.86	3.67	6.53	8.65
0136	cycle 8 miles		96	.083			3.34	4.28	7.62	10.10
0144	25 MPH ave, cycle 4 miles		152	.053			2.11	2.71	4.82	6.40
0146	cycle 6 miles		128	.063			2.50	3.21	5.71	7.55
0148	cycle 8 miles		112	.071			2.86	3.67	6.53	8.65
0150	30 MPH ave, cycle 4 miles		168	.048			1.91	2.45	4.36	5.75
0152	cycle 6 miles		144	.056			2.23	2.86	5.09	6.75
0154	cycle 8 miles		120	.067			2.67	3.43	6.10	8.05
0214	15 MPH ave, cycle 0.5 mile, 20 min wait/Ld./Uld.		176	.045			1.82	2.34	4.16	5.50
0216	cycle 1 mile		160	.050			2	2.57	4.57	6.05
0218	cycle 2 miles		136	.059			2.36	3.02	5.38	7.15
0220	cycle 4 miles		104	.077			3.08	3.95	7.03	9.30
0222	cycle 6 miles		88	.091			3.64	4.67	8.31	11
0224	cycle 8 miles		72	.111			4.45	5.70	10.15	13.45
0226	20 MPH ave, cycle 0.5 mile		176	.045			1.82	2.34	4.16	5.50
0228	cycle 1 mile		168	.048			1.91	2.45	4.36	5.75
0230	cycle 2 miles		144	.056			2.23	2.86	5.09	6.75
0232	cycle 4 miles		120	.067			2.67	3.43	6.10	8.05
0234	cycle 6 miles		96	.083			3.34	4.28	7.62	10.10
0236	cycle 8 miles		88	.091			3.64	4.67	8.31	11
0244	25 MPH ave, cycle 4 miles		128	.063			2.50	3.21	5.71	7.55
0246	cycle 6 miles		112	.071			2.86	3.67	6.53	8.65
0248	cycle 8 miles		96	.083			3.34	4.28	7.62	10.10
0250	30 MPH ave, cycle 4 miles		136	.059			2.36	3.02	5.38	7.15
0252	cycle 6 miles		120	.067			2.67	3.43	6.10	8.05
0254	cycle 8 miles		104	.077			3.08	3.95	7.03	9.30
0314	15 MPH ave, cycle 0.5 mile, 25 min wait/Ld./Uld.		144	.056			2.23	2.86	5.09	6.75
0316	cycle 1 mile		128	.063			2.50	3.21	5.71	7.55
0318	cycle 2 miles		112	.071			2.86	3.67	6.53	8.65
0320	cycle 4 miles		96	.083			3.34	4.28	7.62	10.10
0322	cycle 6 miles		80	.100			4.01	5.15	9.16	12.10
0324	cycle 8 miles		64	.125			5	6.45	11.45	15.10
0326	20 MPH ave, cycle 0.5 mile		144	.056			2.23	2.86	5.09	6.75
0328	cycle 1 mile		136	.059			2.36	3.02	5.38	7.15
0330	cycle 2 miles		120	.067			2.67	3.43	6.10	8.05
0332	cycle 4 miles		104	.077			3.08	3.95	7.03	9.30
0334	cycle 6 miles		88	.091			3.64	4.67	8.31	11
0336	cycle 8 miles		80	.100			4.01	5.15	9.16	12.10
0344	25 MPH ave, cycle 4 miles		112	.071			2.86	3.67	6.53	8.65
0346	cycle 6 miles		96	.083			3.34	4.28	7.62	10.10
0348	cycle 8 miles		88	.091			3.64	4.67	8.31	11

31 23 23.20 Hauling		Crew	Daily Output	Labor-Hours	Unit	Material	2015 Bare Costs Labor	Equipment	Total	Total Incl O&P
0350	30 MPH ave, cycle 4 miles	B-34A	112	.071	L.C.Y.		2.86	3.67	6.53	8.65
0352	cycle 6 miles		104	.077			3.08	3.95	7.03	9.30
0354	cycle 8 miles		96	.083			3.34	4.28	7.62	10.10
0414	15 MPH ave, cycle 0.5 mile, 30 min wait/Ld./Uld.		120	.067			2.67	3.43	6.10	8.05
0416	cycle 1 mile		112	.071			2.86	3.67	6.53	8.65
0418	cycle 2 miles		96	.083			3.34	4.28	7.62	10.10
0420	cycle 4 miles		80	.100			4.01	5.15	9.16	12.10
0422	cycle 6 miles		72	.111			4.45	5.70	10.15	13.45
0424	cycle 8 miles		64	.125			5	6.45	11.45	15.10
0426	20 MPH ave, cycle 0.5 mile		120	.067			2.67	3.43	6.10	8.05
0428	cycle 1 mile		112	.071			2.86	3.67	6.53	8.65
0430	cycle 2 miles		104	.077			3.08	3.95	7.03	9.30
0432	cycle 4 miles		88	.091			3.64	4.67	8.31	11
0434	cycle 6 miles		80	.100			4.01	5.15	9.16	12.10
0436	cycle 8 miles		72	.111			4.45	5.70	10.15	13.45
0444	25 MPH ave, cycle 4 miles		96	.083			3.34	4.28	7.62	10.10
0446	cycle 6 miles		88	.091			3.64	4.67	8.31	11
0448	cycle 8 miles		80	.100			4.01	5.15	9.16	12.10
0450	30 MPH ave, cycle 4 miles		96	.083			3.34	4.28	7.62	10.10
0452	cycle 6 miles		88	.091			3.64	4.67	8.31	11
0454	cycle 8 miles		80	.100			4.01	5.15	9.16	12.10
0514	15 MPH ave, cycle 0.5 mile, 35 min wait/Ld./Uld.		104	.077			3.08	3.95	7.03	9.30
0516	cycle 1 mile		96	.083			3.34	4.28	7.62	10.10
0518	cycle 2 miles		88	.091			3.64	4.67	8.31	11
0520	cycle 4 miles		72	.111			4.45	5.70	10.15	13.45
0522	cycle 6 miles		64	.125			5	6.45	11.45	15.10
0524	cycle 8 miles		56	.143			5.70	7.35	13.05	17.30
0526	20 MPH ave, cycle 0.5 mile		104	.077			3.08	3.95	7.03	9.30
0528	cycle 1 mile		96	.083			3.34	4.28	7.62	10.10
0530	cycle 2 miles		96	.083			3.34	4.28	7.62	10.10
0532	cycle 4 miles		80	.100			4.01	5.15	9.16	12.10
0534	cycle 6 miles		72	.111			4.45	5.70	10.15	13.45
0536	cycle 8 miles		64	.125			5	6.45	11.45	15.10
0544	25 MPH ave, cycle 4 miles		88	.091			3.64	4.67	8.31	11
0546	cycle 6 miles		80	.100			4.01	5.15	9.16	12.10
0548	cycle 8 miles		72	.111			4.45	5.70	10.15	13.45
0550	30 MPH ave, cycle 4 miles		88	.091			3.64	4.67	8.31	11
0552	cycle 6 miles		80	.100			4.01	5.15	9.16	12.10
0554	cycle 8 miles		72	.111			4.45	5.70	10.15	13.45
1014	12 C.Y. truck, cycle 0.5 mile, 15 MPH ave, 15 min. wait/Ld./Uld.	B-34B	336	.024			.95	2.06	3.01	3.80
1016	cycle 1 mile		300	.027			1.07	2.30	3.37	4.25
1018	cycle 2 miles		252	.032			1.27	2.74	4.01	5.05
1020	cycle 4 miles		180	.044			1.78	3.84	5.62	7.10
1022	cycle 6 miles		144	.056			2.23	4.80	7.03	8.90
1024	cycle 8 miles		120	.067			2.67	5.75	8.42	10.65
1025	cycle 10 miles		96	.083			3.34	7.20	10.54	13.30
1026	20 MPH ave, cycle 0.5 mile		348	.023			.92	1.99	2.91	3.66
1028	cycle 1 mile		312	.026			1.03	2.21	3.24	4.10
1030	cycle 2 miles		276	.029			1.16	2.50	3.66	4.62
1032	cycle 4 miles		216	.037			1.48	3.20	4.68	5.90
1034	cycle 6 miles		168	.048			1.91	4.11	6.02	7.60
1036	cycle 8 miles		144	.056			2.23	4.80	7.03	8.90
1038	cycle 10 miles		120	.067			2.67	5.75	8.42	10.65

31 23 23 – Fill

31 23 23.20 Hauling		Crew	Daily Output	Labor-Hours	Unit	Material	2015 Bare Costs Labor	Equipment	Total	Total Incl O&P
1040	25 MPH ave, cycle 4 miles	B-34B	228	.035	L.C.Y.		1.41	3.03	4.44	5.60
1042	cycle 6 miles		192	.042			1.67	3.60	5.27	6.65
1044	cycle 8 miles		168	.048			1.91	4.11	6.02	7.60
1046	cycle 10 miles		144	.056			2.23	4.80	7.03	8.90
1050	30 MPH ave, cycle 4 miles		252	.032			1.27	2.74	4.01	5.05
1052	cycle 6 miles		216	.037			1.48	3.20	4.68	5.90
1054	cycle 8 miles		180	.044			1.78	3.84	5.62	7.10
1056	cycle 10 miles		156	.051			2.05	4.43	6.48	8.20
1060	35 MPH ave, cycle 4 miles		264	.030			1.21	2.62	3.83	4.84
1062	cycle 6 miles		228	.035			1.41	3.03	4.44	5.60
1064	cycle 8 miles		204	.039			1.57	3.39	4.96	6.25
1066	cycle 10 miles		180	.044			1.78	3.84	5.62	7.10
1068	cycle 20 miles		120	.067			2.67	5.75	8.42	10.65
1069	cycle 30 miles		84	.095			3.81	8.25	12.06	15.20
1070	cycle 40 miles		72	.111			4.45	9.60	14.05	17.70
1072	40 MPH ave, cycle 6 miles		240	.033			1.33	2.88	4.21	5.30
1074	cycle 8 miles		216	.037			1.48	3.20	4.68	5.90
1076	cycle 10 miles		192	.042			1.67	3.60	5.27	6.65
1078	cycle 20 miles		120	.067			2.67	5.75	8.42	10.65
1080	cycle 30 miles		96	.083			3.34	7.20	10.54	13.30
1082	cycle 40 miles		72	.111			4.45	9.60	14.05	17.70
1084	cycle 50 miles		60	.133			5.35	11.50	16.85	21.50
1094	45 MPH ave, cycle 8 miles		216	.037			1.48	3.20	4.68	5.90
1096	cycle 10 miles		204	.039			1.57	3.39	4.96	6.25
1098	cycle 20 miles		132	.061			2.43	5.25	7.68	9.65
1100	cycle 30 miles		108	.074			2.97	6.40	9.37	11.85
1102	cycle 40 miles		84	.095			3.81	8.25	12.06	15.20
1104	cycle 50 miles		72	.111			4.45	9.60	14.05	17.70
1106	50 MPH ave, cycle 10 miles		216	.037			1.48	3.20	4.68	5.90
1108	cycle 20 miles		144	.056			2.23	4.80	7.03	8.90
1110	cycle 30 miles		108	.074			2.97	6.40	9.37	11.85
1112	cycle 40 miles		84	.095			3.81	8.25	12.06	15.20
1114	cycle 50 miles		72	.111			4.45	9.60	14.05	17.70
1214	15 MPH ave, cycle 0.5 mile, 20 min. wait/Ld./Uld.		264	.030			1.21	2.62	3.83	4.84
1216	cycle 1 mile		240	.033			1.33	2.88	4.21	5.30
1218	cycle 2 miles		204	.039			1.57	3.39	4.96	6.25
1220	cycle 4 miles		156	.051			2.05	4.43	6.48	8.20
1222	cycle 6 miles		132	.061			2.43	5.25	7.68	9.65
1224	cycle 8 miles		108	.074			2.97	6.40	9.37	11.85
1225	cycle 10 miles		96	.083			3.34	7.20	10.54	13.30
1226	20 MPH ave, cycle 0.5 mile		264	.030			1.21	2.62	3.83	4.84
1228	cycle 1 mile		252	.032			1.27	2.74	4.01	5.05
1230	cycle 2 miles		216	.037			1.48	3.20	4.68	5.90
1232	cycle 4 miles		180	.044			1.78	3.84	5.62	7.10
1234	cycle 6 miles		144	.056			2.23	4.80	7.03	8.90
1236	cycle 8 miles		132	.061			2.43	5.25	7.68	9.65
1238	cycle 10 miles		108	.074			2.97	6.40	9.37	11.85
1240	25 MPH ave, cycle 4 miles		192	.042			1.67	3.60	5.27	6.65
1242	cycle 6 miles		168	.048			1.91	4.11	6.02	7.60
1244	cycle 8 miles		144	.056			2.23	4.80	7.03	8.90
1246	cycle 10 miles		132	.061			2.43	5.25	7.68	9.65
1250	30 MPH ave, cycle 4 miles		204	.039			1.57	3.39	4.96	6.25
1252	cycle 6 miles		180	.044			1.78	3.84	5.62	7.10

31 23 23.20 Hauling		Crew	Daily Output	Labor-Hours	Unit	Material	2015 Bare Costs Labor	Equipment	Total	Total Incl O&P
1254	cycle 8 miles	B-34B	156	.051	L.C.Y.		2.05	4.43	6.48	8.20
1256	cycle 10 miles		144	.056			2.23	4.80	7.03	8.90
1260	35 MPH ave, cycle 4 miles		216	.037			1.48	3.20	4.68	5.90
1262	cycle 6 miles		192	.042			1.67	3.60	5.27	6.65
1264	cycle 8 miles		168	.048			1.91	4.11	6.02	7.60
1266	cycle 10 miles		156	.051			2.05	4.43	6.48	8.20
1268	cycle 20 miles		108	.074			2.97	6.40	9.37	11.85
1269	cycle 30 miles		72	.111			4.45	9.60	14.05	17.70
1270	cycle 40 miles		60	.133			5.35	11.50	16.85	21.50
1272	40 MPH ave, cycle 6 miles		192	.042			1.67	3.60	5.27	6.65
1274	cycle 8 miles		180	.044			1.78	3.84	5.62	7.10
1276	cycle 10 miles		156	.051			2.05	4.43	6.48	8.20
1278	cycle 20 miles		108	.074			2.97	6.40	9.37	11.85
1280	cycle 30 miles		84	.095			3.81	8.25	12.06	15.20
1282	cycle 40 miles		72	.111			4.45	9.60	14.05	17.70
1284	cycle 50 miles		60	.133			5.35	11.50	16.85	21.50
1294	45 MPH ave, cycle 8 miles		180	.044			1.78	3.84	5.62	7.10
1296	cycle 10 miles		168	.048			1.91	4.11	6.02	7.60
1298	cycle 20 miles		120	.067			2.67	5.75	8.42	10.65
1300	cycle 30 miles		96	.083			3.34	7.20	10.54	13.30
1302	cycle 40 miles		72	.111			4.45	9.60	14.05	17.70
1304	cycle 50 miles		60	.133			5.35	11.50	16.85	21.50
1306	50 MPH ave, cycle 10 miles		180	.044			1.78	3.84	5.62	7.10
1308	cycle 20 miles		132	.061			2.43	5.25	7.68	9.65
1310	cycle 30 miles		96	.083			3.34	7.20	10.54	13.30
1312	cycle 40 miles		84	.095			3.81	8.25	12.06	15.20
1314	cycle 50 miles		72	.111			4.45	9.60	14.05	17.70
1414	15 MPH ave, cycle 0.5 mile, 25 min. wait/Ld./Uld.		204	.039			1.57	3.39	4.96	6.25
1416	cycle 1 mile		192	.042			1.67	3.60	5.27	6.65
1418	cycle 2 miles		168	.048			1.91	4.11	6.02	7.60
1420	cycle 4 miles		132	.061			2.43	5.25	7.68	9.65
1422	cycle 6 miles		120	.067			2.67	5.75	8.42	10.65
1424	cycle 8 miles		96	.083			3.34	7.20	10.54	13.30
1425	cycle 10 miles		84	.095			3.81	8.25	12.06	15.20
1426	20 MPH ave, cycle 0.5 mile		216	.037			1.48	3.20	4.68	5.90
1428	cycle 1 mile		204	.039			1.57	3.39	4.96	6.25
1430	cycle 2 miles		180	.044			1.78	3.84	5.62	7.10
1432	cycle 4 miles		156	.051			2.05	4.43	6.48	8.20
1434	cycle 6 miles		132	.061			2.43	5.25	7.68	9.65
1436	cycle 8 miles		120	.067			2.67	5.75	8.42	10.65
1438	cycle 10 miles		96	.083			3.34	7.20	10.54	13.30
1440	25 MPH ave, cycle 4 miles		168	.048			1.91	4.11	6.02	7.60
1442	cycle 6 miles		144	.056			2.23	4.80	7.03	8.90
1444	cycle 8 miles		132	.061			2.43	5.25	7.68	9.65
1446	cycle 10 miles		108	.074			2.97	6.40	9.37	11.85
1450	30 MPH ave, cycle 4 miles		168	.048			1.91	4.11	6.02	7.60
1452	cycle 6 miles		156	.051			2.05	4.43	6.48	8.20
1454	cycle 8 miles		132	.061			2.43	5.25	7.68	9.65
1456	cycle 10 miles		120	.067			2.67	5.75	8.42	10.65
1460	35 MPH ave, cycle 4 miles		180	.044			1.78	3.84	5.62	7.10
1462	cycle 6 miles		156	.051			2.05	4.43	6.48	8.20
1464	cycle 8 miles		144	.056			2.23	4.80	7.03	8.90
1466	cycle 10 miles	↓	132	.061	↓		2.43	5.25	7.68	9.65

31 23 23 – Fill

31 23 23.20 Hauling		Crew	Daily Output	Labor-Hours	Unit	Material	2015 Bare Costs Labor	Equipment	Total	Total Incl O&P
1468	cycle 20 miles	B-34B	96	.083	L.C.Y.		3.34	7.20	10.54	13.30
1469	cycle 30 miles		72	.111			4.45	9.60	14.05	17.70
1470	cycle 40 miles		60	.133			5.35	11.50	16.85	21.50
1472	40 MPH ave, cycle 6 miles		168	.048			1.91	4.11	6.02	7.60
1474	cycle 8 miles		156	.051			2.05	4.43	6.48	8.20
1476	cycle 10 miles		144	.056			2.23	4.80	7.03	8.90
1478	cycle 20 miles		96	.083			3.34	7.20	10.54	13.30
1480	cycle 30 miles		84	.095			3.81	8.25	12.06	15.20
1482	cycle 40 miles		60	.133			5.35	11.50	16.85	21.50
1484	cycle 50 miles		60	.133			5.35	11.50	16.85	21.50
1494	45 MPH ave, cycle 8 miles		156	.051			2.05	4.43	6.48	8.20
1496	cycle 10 miles		144	.056			2.23	4.80	7.03	8.90
1498	cycle 20 miles		108	.074			2.97	6.40	9.37	11.85
1500	cycle 30 miles		84	.095			3.81	8.25	12.06	15.20
1502	cycle 40 miles		72	.111			4.45	9.60	14.05	17.70
1504	cycle 50 miles		60	.133			5.35	11.50	16.85	21.50
1506	50 MPH ave, cycle 10 miles		156	.051			2.05	4.43	6.48	8.20
1508	cycle 20 miles		120	.067			2.67	5.75	8.42	10.65
1510	cycle 30 miles		96	.083			3.34	7.20	10.54	13.30
1512	cycle 40 miles		72	.111			4.45	9.60	14.05	17.70
1514	cycle 50 miles		60	.133			5.35	11.50	16.85	21.50
1614	15 MPH, cycle 0.5 mile, 30 min. wait/Ld./Uld.		180	.044			1.78	3.84	5.62	7.10
1616	cycle 1 mile		168	.048			1.91	4.11	6.02	7.60
1618	cycle 2 miles		144	.056			2.23	4.80	7.03	8.90
1620	cycle 4 miles		120	.067			2.67	5.75	8.42	10.65
1622	cycle 6 miles		108	.074			2.97	6.40	9.37	11.85
1624	cycle 8 miles		84	.095			3.81	8.25	12.06	15.20
1625	cycle 10 miles		84	.095			3.81	8.25	12.06	15.20
1626	20 MPH ave, cycle 0.5 mile		180	.044			1.78	3.84	5.62	7.10
1628	cycle 1 mile		168	.048			1.91	4.11	6.02	7.60
1630	cycle 2 miles		156	.051			2.05	4.43	6.48	8.20
1632	cycle 4 miles		132	.061			2.43	5.25	7.68	9.65
1634	cycle 6 miles		120	.067			2.67	5.75	8.42	10.65
1636	cycle 8 miles		108	.074			2.97	6.40	9.37	11.85
1638	cycle 10 miles		96	.083			3.34	7.20	10.54	13.30
1640	25 MPH ave, cycle 4 miles		144	.056			2.23	4.80	7.03	8.90
1642	cycle 6 miles		132	.061			2.43	5.25	7.68	9.65
1644	cycle 8 miles		108	.074			2.97	6.40	9.37	11.85
1646	cycle 10 miles		108	.074			2.97	6.40	9.37	11.85
1650	30 MPH ave, cycle 4 miles		144	.056			2.23	4.80	7.03	8.90
1652	cycle 6 miles		132	.061			2.43	5.25	7.68	9.65
1654	cycle 8 miles		120	.067			2.67	5.75	8.42	10.65
1656	cycle 10 miles		108	.074			2.97	6.40	9.37	11.85
1660	35 MPH ave, cycle 4 miles		156	.051			2.05	4.43	6.48	8.20
1662	cycle 6 miles		144	.056			2.23	4.80	7.03	8.90
1664	cycle 8 miles		132	.061			2.43	5.25	7.68	9.65
1666	cycle 10 miles		120	.067			2.67	5.75	8.42	10.65
1668	cycle 20 miles		84	.095			3.81	8.25	12.06	15.20
1669	cycle 30 miles		72	.111			4.45	9.60	14.05	17.70
1670	cycle 40 miles		60	.133			5.35	11.50	16.85	21.50
1672	40 MPH, cycle 6 miles		144	.056			2.23	4.80	7.03	8.90
1674	cycle 8 miles		132	.061			2.43	5.25	7.68	9.65
1676	cycle 10 miles		120	.067			2.67	5.75	8.42	10.65

31 23 23.20 Hauling		Crew	Daily Output	Labor-Hours	Unit	Material	2015 Bare Costs Labor	Equipment	Total	Total Incl O&P
1678	cycle 20 miles	B-34B	96	.083	L.C.Y.		3.34	7.20	10.54	13.30
1680	cycle 30 miles		72	.111			4.45	9.60	14.05	17.70
1682	cycle 40 miles		60	.133			5.35	11.50	16.85	21.50
1684	cycle 50 miles		48	.167			6.70	14.40	21.10	26.50
1694	45 MPH ave, cycle 8 miles		144	.056			2.23	4.80	7.03	8.90
1696	cycle 10 miles		132	.061			2.43	5.25	7.68	9.65
1698	cycle 20 miles		96	.083			3.34	7.20	10.54	13.30
1700	cycle 30 miles		84	.095			3.81	8.25	12.06	15.20
1702	cycle 40 miles		60	.133			5.35	11.50	16.85	21.50
1704	cycle 50 miles		60	.133			5.35	11.50	16.85	21.50
1706	50 MPH ave, cycle 10 miles		132	.061			2.43	5.25	7.68	9.65
1708	cycle 20 miles		108	.074			2.97	6.40	9.37	11.85
1710	cycle 30 miles		84	.095			3.81	8.25	12.06	15.20
1712	cycle 40 miles		72	.111			4.45	9.60	14.05	17.70
1714	cycle 50 miles		60	.133			5.35	11.50	16.85	21.50
2000	Hauling, 8 C.Y. truck, small project cost per hour	B-34A	8	1	Hr.		40	51.50	91.50	121
2100	12 C.Y. Truck	B-34B	8	1			40	86.50	126.50	160
2150	16.5 C.Y. Truck	B-34C	8	1			40	91	131	165
2175	18 C.Y. 8 wheel Truck	B-34I	8	1			40	109	149	184
2200	20 C.Y. Truck	B-34D	8	1			40	92.50	132.50	167
2300	Grading at dump, or embankment if required, by dozer	B-10B	1000	.012	L.C.Y.		.56	1.39	1.95	2.42
2310	Spotter at fill or cut, if required	1 Clab	8	1	Hr.		37.50		37.50	61.50
9014	18 C.Y. truck, 8 wheels,15 min. wait/Ld./Uld.,15 MPH, cycle 0.5 mi.	B-34I	504	.016	L.C.Y.		.64	1.72	2.36	2.92
9016	cycle 1 mile		450	.018			.71	1.93	2.64	3.27
9018	cycle 2 miles		378	.021			.85	2.30	3.15	3.90
9020	cycle 4 miles		270	.030			1.19	3.22	4.41	5.45
9022	cycle 6 miles		216	.037			1.48	4.02	5.50	6.80
9024	cycle 8 miles		180	.044			1.78	4.83	6.61	8.15
9025	cycle 10 miles		144	.056			2.23	6.05	8.28	10.25
9026	20 MPH ave, cycle 0.5 mile		522	.015			.61	1.67	2.28	2.82
9028	cycle 1 mile		468	.017			.68	1.86	2.54	3.14
9030	cycle 2 miles		414	.019			.77	2.10	2.87	3.56
9032	cycle 4 miles		324	.025			.99	2.68	3.67	4.54
9034	cycle 6 miles		252	.032			1.27	3.45	4.72	5.85
9036	cycle 8 miles		216	.037			1.48	4.02	5.50	6.80
9038	cycle 10 miles		180	.044			1.78	4.83	6.61	8.15
9040	25 MPH ave, cycle 4 miles		342	.023			.94	2.54	3.48	4.30
9042	cycle 6 miles		288	.028			1.11	3.02	4.13	5.10
9044	cycle 8 miles		252	.032			1.27	3.45	4.72	5.85
9046	cycle 10 miles		216	.037			1.48	4.02	5.50	6.80
9050	30 MPH ave, cycle 4 miles		378	.021			.85	2.30	3.15	3.90
9052	cycle 6 miles		324	.025			.99	2.68	3.67	4.54
9054	cycle 8 miles		270	.030			1.19	3.22	4.41	5.45
9056	cycle 10 miles		234	.034			1.37	3.71	5.08	6.30
9060	35 MPH ave, cycle 4 miles		396	.020			.81	2.19	3	3.71
9062	cycle 6 miles		342	.023			.94	2.54	3.48	4.30
9064	cycle 8 miles		288	.028			1.11	3.02	4.13	5.10
9066	cycle 10 miles		270	.030			1.19	3.22	4.41	5.45
9068	cycle 20 miles		162	.049			1.98	5.35	7.33	9.10
9070	cycle 30 miles		126	.063			2.54	6.90	9.44	11.70
9072	cycle 40 miles		90	.089			3.56	9.65	13.21	16.35
9074	40 MPH ave, cycle 6 miles		360	.022			.89	2.41	3.30	4.09
9076	cycle 8 miles		324	.025			.99	2.68	3.67	4.54

31 23 23 – Fill

31 23 23.20 Hauling		Crew	Daily Output	Labor-Hours	Unit	Material	2015 Bare Costs Labor	Equipment	Total	Total Incl O&P
9078	cycle 10 miles	B-34I	288	.028	L.C.Y.		1.11	3.02	4.13	5.10
9080	cycle 20 miles		180	.044			1.78	4.83	6.61	8.15
9082	cycle 30 miles		144	.056			2.23	6.05	8.28	10.25
9084	cycle 40 miles		108	.074			2.97	8.05	11.02	13.65
9086	cycle 50 miles		90	.089			3.56	9.65	13.21	16.35
9094	45 MPH ave, cycle 8 miles		324	.025			.99	2.68	3.67	4.54
9096	cycle 10 miles		306	.026			1.05	2.84	3.89	4.81
9098	cycle 20 miles		198	.040			1.62	4.39	6.01	7.45
9100	cycle 30 miles		144	.056			2.23	6.05	8.28	10.25
9102	cycle 40 miles		126	.063			2.54	6.90	9.44	11.70
9104	cycle 50 miles		108	.074			2.97	8.05	11.02	13.65
9106	50 MPH ave, cycle 10 miles		324	.025			.99	2.68	3.67	4.54
9108	cycle 20 miles		216	.037			1.48	4.02	5.50	6.80
9110	cycle 30 miles		162	.049			1.98	5.35	7.33	9.10
9112	cycle 40 miles		126	.063			2.54	6.90	9.44	11.70
9114	cycle 50 miles		108	.074			2.97	8.05	11.02	13.65
9214	20 min. wait/Ld./Uld.,15 MPH, cycle 0.5 mi.		396	.020			.81	2.19	3	3.71
9216	cycle 1 mile		360	.022			.89	2.41	3.30	4.09
9218	cycle 2 miles		306	.026			1.05	2.84	3.89	4.81
9220	cycle 4 miles		234	.034			1.37	3.71	5.08	6.30
9222	cycle 6 miles		198	.040			1.62	4.39	6.01	7.45
9224	cycle 8 miles		162	.049			1.98	5.35	7.33	9.10
9225	cycle 10 miles		144	.056			2.23	6.05	8.28	10.25
9226	20 MPH ave, cycle 0.5 mile		396	.020			.81	2.19	3	3.71
9228	cycle 1 mile		378	.021			.85	2.30	3.15	3.90
9230	cycle 2 miles		324	.025			.99	2.68	3.67	4.54
9232	cycle 4 miles		270	.030			1.19	3.22	4.41	5.45
9234	cycle 6 miles		216	.037			1.48	4.02	5.50	6.80
9236	cycle 8 miles		198	.040			1.62	4.39	6.01	7.45
9238	cycle 10 miles		162	.049			1.98	5.35	7.33	9.10
9240	25 MPH ave, cycle 4 miles		288	.028			1.11	3.02	4.13	5.10
9242	cycle 6 miles		252	.032			1.27	3.45	4.72	5.85
9244	cycle 8 miles		216	.037			1.48	4.02	5.50	6.80
9246	cycle 10 miles		198	.040			1.62	4.39	6.01	7.45
9250	30 MPH ave, cycle 4 miles		306	.026			1.05	2.84	3.89	4.81
9252	cycle 6 miles		270	.030			1.19	3.22	4.41	5.45
9254	cycle 8 miles		234	.034			1.37	3.71	5.08	6.30
9256	cycle 10 miles		216	.037			1.48	4.02	5.50	6.80
9260	35 MPH ave, cycle 4 miles		324	.025			.99	2.68	3.67	4.54
9262	cycle 6 miles		288	.028			1.11	3.02	4.13	5.10
9264	cycle 8 miles		252	.032			1.27	3.45	4.72	5.85
9266	cycle 10 miles		234	.034			1.37	3.71	5.08	6.30
9268	cycle 20 miles		162	.049			1.98	5.35	7.33	9.10
9270	cycle 30 miles		108	.074			2.97	8.05	11.02	13.65
9272	cycle 40 miles		90	.089			3.56	9.65	13.21	16.35
9274	40 MPH ave, cycle 6 miles		288	.028			1.11	3.02	4.13	5.10
9276	cycle 8 miles		270	.030			1.19	3.22	4.41	5.45
9278	cycle 10 miles		234	.034			1.37	3.71	5.08	6.30
9280	cycle 20 miles		162	.049			1.98	5.35	7.33	9.10
9282	cycle 30 miles		126	.063			2.54	6.90	9.44	11.70
9284	cycle 40 miles		108	.074			2.97	8.05	11.02	13.65
9286	cycle 50 miles		90	.089			3.56	9.65	13.21	16.35
9294	45 MPH ave, cycle 8 miles		270	.030			1.19	3.22	4.41	5.45

31 23 23.20 Hauling		Crew	Daily Output	Labor-Hours	Unit	Material	2015 Bare Costs Labor	Equipment	Total	Total Incl O&P
9296	cycle 10 miles	B-34I	252	.032	L.C.Y.		1.27	3.45	4.72	5.85
9298	cycle 20 miles		180	.044			1.78	4.83	6.61	8.15
9300	cycle 30 miles		144	.056			2.23	6.05	8.28	10.25
9302	cycle 40 miles		108	.074			2.97	8.05	11.02	13.65
9304	cycle 50 miles		90	.089			3.56	9.65	13.21	16.35
9306	50 MPH ave, cycle 10 miles		270	.030			1.19	3.22	4.41	5.45
9308	cycle 20 miles		198	.040			1.62	4.39	6.01	7.45
9310	cycle 30 miles		144	.056			2.23	6.05	8.28	10.25
9312	cycle 40 miles		126	.063			2.54	6.90	9.44	11.70
9314	cycle 50 miles		108	.074			2.97	8.05	11.02	13.65
9414	25 min. wait/Ld./Uld.,15 MPH, cycle 0.5 mi.		306	.026			1.05	2.84	3.89	4.81
9416	cycle 1 mile		288	.028			1.11	3.02	4.13	5.10
9418	cycle 2 miles		252	.032			1.27	3.45	4.72	5.85
9420	cycle 4 miles		198	.040			1.62	4.39	6.01	7.45
9422	cycle 6 miles		180	.044			1.78	4.83	6.61	8.15
9424	cycle 8 miles		144	.056			2.23	6.05	8.28	10.25
9425	cycle 10 miles		126	.063			2.54	6.90	9.44	11.70
9426	20 MPH ave, cycle 0.5 mile		324	.025			.99	2.68	3.67	4.54
9428	cycle 1 mile		306	.026			1.05	2.84	3.89	4.81
9430	cycle 2 miles		270	.030			1.19	3.22	4.41	5.45
9432	cycle 4 miles		234	.034			1.37	3.71	5.08	6.30
9434	cycle 6 miles		198	.040			1.62	4.39	6.01	7.45
9436	cycle 8 miles		180	.044			1.78	4.83	6.61	8.15
9438	cycle 10 miles		144	.056			2.23	6.05	8.28	10.25
9440	25 MPH ave, cycle 4 miles		252	.032			1.27	3.45	4.72	5.85
9442	cycle 6 miles		216	.037			1.48	4.02	5.50	6.80
9444	cycle 8 miles		198	.040			1.62	4.39	6.01	7.45
9446	cycle 10 miles		180	.044			1.78	4.83	6.61	8.15
9450	30 MPH ave, cycle 4 miles		252	.032			1.27	3.45	4.72	5.85
9452	cycle 6 miles		234	.034			1.37	3.71	5.08	6.30
9454	cycle 8 miles		198	.040			1.62	4.39	6.01	7.45
9456	cycle 10 miles		180	.044			1.78	4.83	6.61	8.15
9460	35 MPH ave, cycle 4 miles		270	.030			1.19	3.22	4.41	5.45
9462	cycle 6 miles		234	.034			1.37	3.71	5.08	6.30
9464	cycle 8 miles		216	.037			1.48	4.02	5.50	6.80
9466	cycle 10 miles		198	.040			1.62	4.39	6.01	7.45
9468	cycle 20 miles		144	.056			2.23	6.05	8.28	10.25
9470	cycle 30 miles		108	.074			2.97	8.05	11.02	13.65
9472	cycle 40 miles		90	.089			3.56	9.65	13.21	16.35
9474	40 MPH ave, cycle 6 miles		252	.032			1.27	3.45	4.72	5.85
9476	cycle 8 miles		234	.034			1.37	3.71	5.08	6.30
9478	cycle 10 miles		216	.037			1.48	4.02	5.50	6.80
9480	cycle 20 miles		144	.056			2.23	6.05	8.28	10.25
9482	cycle 30 miles		126	.063			2.54	6.90	9.44	11.70
9484	cycle 40 miles		90	.089			3.56	9.65	13.21	16.35
9486	cycle 50 miles		90	.089			3.56	9.65	13.21	16.35
9494	45 MPH ave, cycle 8 miles		234	.034			1.37	3.71	5.08	6.30
9496	cycle 10 miles		216	.037			1.48	4.02	5.50	6.80
9498	cycle 20 miles		162	.049			1.98	5.35	7.33	9.10
9500	cycle 30 miles		126	.063			2.54	6.90	9.44	11.70
9502	cycle 40 miles		108	.074			2.97	8.05	11.02	13.65
9504	cycle 50 miles		90	.089			3.56	9.65	13.21	16.35
9506	50 MPH ave, cycle 10 miles		234	.034			1.37	3.71	5.08	6.30

31 23 23.20 Hauling		Crew	Daily Output	Labor-Hours	Unit	Material	2015 Bare Costs Labor	Equipment	Total	Total Incl O&P
9508	cycle 20 miles	B-341	180	.044	L.C.Y.		1.78	4.83	6.61	8.15
9510	cycle 30 miles		144	.056			2.23	6.05	8.28	10.25
9512	cycle 40 miles		108	.074			2.97	8.05	11.02	13.65
9514	cycle 50 miles		90	.089			3.56	9.65	13.21	16.35
9614	30 min. wait/Ld./Uld.,15 MPH, cycle 0.5 mi.		270	.030			1.19	3.22	4.41	5.45
9616	cycle 1 mile		252	.032			1.27	3.45	4.72	5.85
9618	cycle 2 miles		216	.037			1.48	4.02	5.50	6.80
9620	cycle 4 miles		180	.044			1.78	4.83	6.61	8.15
9622	cycle 6 miles		162	.049			1.98	5.35	7.33	9.10
9624	cycle 8 miles		126	.063			2.54	6.90	9.44	11.70
9625	cycle 10 miles		126	.063			2.54	6.90	9.44	11.70
9626	20 MPH ave, cycle 0.5 mile		270	.030			1.19	3.22	4.41	5.45
9628	cycle 1 mile		252	.032			1.27	3.45	4.72	5.85
9630	cycle 2 miles		234	.034			1.37	3.71	5.08	6.30
9632	cycle 4 miles		198	.040			1.62	4.39	6.01	7.45
9634	cycle 6 miles		180	.044			1.78	4.83	6.61	8.15
9636	cycle 8 miles		162	.049			1.98	5.35	7.33	9.10
9638	cycle 10 miles		144	.056			2.23	6.05	8.28	10.25
9640	25 MPH ave, cycle 4 miles		216	.037			1.48	4.02	5.50	6.80
9642	cycle 6 miles		198	.040			1.62	4.39	6.01	7.45
9644	cycle 8 miles		180	.044			1.78	4.83	6.61	8.15
9646	cycle 10 miles		162	.049			1.98	5.35	7.33	9.10
9650	30 MPH ave, cycle 4 miles		216	.037			1.48	4.02	5.50	6.80
9652	cycle 6 miles		198	.040			1.62	4.39	6.01	7.45
9654	cycle 8 miles		180	.044			1.78	4.83	6.61	8.15
9656	cycle 10 miles		162	.049			1.98	5.35	7.33	9.10
9660	35 MPH ave, cycle 4 miles		234	.034			1.37	3.71	5.08	6.30
9662	cycle 6 miles		216	.037			1.48	4.02	5.50	6.80
9664	cycle 8 miles		198	.040			1.62	4.39	6.01	7.45
9666	cycle 10 miles		180	.044			1.78	4.83	6.61	8.15
9668	cycle 20 miles		126	.063			2.54	6.90	9.44	11.70
9670	cycle 30 miles		108	.074			2.97	8.05	11.02	13.65
9672	cycle 40 miles		90	.089			3.56	9.65	13.21	16.35
9674	40 MPH ave, cycle 6 miles		216	.037			1.48	4.02	5.50	6.80
9676	cycle 8 miles		198	.040			1.62	4.39	6.01	7.45
9678	cycle 10 miles		180	.044			1.78	4.83	6.61	8.15
9680	cycle 20 miles		144	.056			2.23	6.05	8.28	10.25
9682	cycle 30 miles		108	.074			2.97	8.05	11.02	13.65
9684	cycle 40 miles		90	.089			3.56	9.65	13.21	16.35
9686	cycle 50 miles		72	.111			4.45	12.05	16.50	20.50
9694	45 MPH ave, cycle 8 miles		216	.037			1.48	4.02	5.50	6.80
9696	cycle 10 miles		198	.040			1.62	4.39	6.01	7.45
9698	cycle 20 miles		144	.056			2.23	6.05	8.28	10.25
9700	cycle 30 miles		126	.063			2.54	6.90	9.44	11.70
9702	cycle 40 miles		108	.074			2.97	8.05	11.02	13.65
9704	cycle 50 miles		90	.089			3.56	9.65	13.21	16.35
9706	50 MPH ave, cycle 10 miles		198	.040			1.62	4.39	6.01	7.45
9708	cycle 20 miles		162	.049			1.98	5.35	7.33	9.10
9710	cycle 30 miles		126	.063			2.54	6.90	9.44	11.70
9712	cycle 40 miles		108	.074			2.97	8.05	11.02	13.65
9714	cycle 50 miles		90	.089			3.56	9.65	13.21	16.35

31 23 23.23 Compaction	Crew	Daily Output	Labor-Hours	Unit	Material	2015 Bare Costs Labor	Equipment	Total	Total Incl O&P
0010 **COMPACTION**									
5000 Riding, vibrating roller, 6" lifts, 2 passes	B-10Y	3000	.004	E.C.Y.		.19	.18	.37	.50
5020 3 passes		2300	.005			.24	.24	.48	.65
5040 4 passes		1900	.006			.29	.29	.58	.79
5050 8" lifts, 2 passes		4100	.003			.14	.13	.27	.37
5060 12" lifts, 2 passes		5200	.002			.11	.11	.22	.29
5080 3 passes		3500	.003			.16	.16	.32	.42
5100 4 passes		2600	.005			.21	.21	.42	.57
5600 Sheepsfoot or wobbly wheel roller, 6" lifts, 2 passes	B-10G	2400	.005			.23	.50	.73	.92
5620 3 passes		1735	.007			.32	.70	1.02	1.28
5640 4 passes		1300	.009			.43	.93	1.36	1.70
5680 12" lifts, 2 passes		5200	.002			.11	.23	.34	.43
5700 3 passes		3500	.003			.16	.34	.50	.63
5720 4 passes		2600	.005			.21	.46	.67	.85
7000 Walk behind, vibrating plate 18" wide, 6" lifts, 2 passes	A-1D	200	.040			1.50	.18	1.68	2.67
7020 3 passes		185	.043			1.63	.20	1.83	2.89
7040 4 passes		140	.057			2.15	.26	2.41	3.80
7200 12" lifts, 2 passes, 21" wide	A-1E	560	.014			.54	.08	.62	.97
7220 3 passes		375	.021			.80	.12	.92	1.46
7240 4 passes		280	.029			1.07	.17	1.24	1.94
7500 Vibrating roller 24" wide, 6" lifts, 2 passes	B-10A	420	.029			1.32	.43	1.75	2.59
7520 3 passes		280	.043			1.98	.64	2.62	3.89
7540 4 passes		210	.057			2.64	.86	3.50	5.15
7600 12" lifts, 2 passes		840	.014			.66	.21	.87	1.30
7620 3 passes		560	.021			.99	.32	1.31	1.94
7640 4 passes		420	.029			1.32	.43	1.75	2.59
8000 Rammer tamper, 6" to 11", 4" lifts, 2 passes	A-1F	130	.062			2.31	.38	2.69	4.21
8050 3 passes		97	.082			3.10	.52	3.62	5.65
8100 4 passes		65	.123			4.63	.77	5.40	8.45
8200 8" lifts, 2 passes		260	.031			1.16	.19	1.35	2.11
8250 3 passes		195	.041			1.54	.26	1.80	2.81
8300 4 passes		130	.062			2.31	.38	2.69	4.21
8400 13" to 18", 4" lifts, 2 passes	A-1G	390	.021			.77	.14	.91	1.42
8450 3 passes		290	.028			1.04	.19	1.23	1.91
8500 4 passes		195	.041			1.54	.29	1.83	2.85
8600 8" lifts, 2 passes		780	.010			.39	.07	.46	.71
8650 3 passes		585	.014			.51	.10	.61	.95
8700 4 passes		390	.021			.77	.14	.91	1.42
9000 Water, 3000 gal. truck, 3 mile haul	B-45	1888	.008		1.22	.38	.48	2.08	2.48
9010 6 mile haul		1444	.011		1.22	.50	.63	2.35	2.83
9020 12 mile haul		1000	.016		1.22	.73	.91	2.86	3.50
9030 6000 gal. wagon, 3 mile haul	B-59	2000	.004		1.22	.16	.25	1.63	1.88
9040 6 mile haul	"	1600	.005		1.22	.20	.31	1.73	2
9900 Minimum labor/equipment charge	1 Clab	4	2	Job		75		75	123

31 23 23.24 Compaction, Structural	Crew	Daily Output	Labor-Hours	Unit	Material	Labor	Equipment	Total	Total Incl O&P
0010 **COMPACTION, STRUCTURAL**									
0020 Steel wheel tandem roller, 5 tons	B-10E	8	1.500	Hr.		69.50	19.70	89.20	133

31 25 Erosion and Sedimentation Controls

31 25 14 – Stabilization Measures for Erosion and Sedimentation Control

31 25 14.16 Rolled Erosion Control Mats and Blankets			Crew	Daily Output	Labor-Hours	Unit	2015 Bare Costs Material	Labor	Equipment	Total	Total Incl O&P
0010	**ROLLED EROSION CONTROL MATS AND BLANKETS**										
0020	Jute mesh, 100 S.Y. per roll, 4′ wide, stapled	G	B-80A	2400	.010	S.Y.	1.03	.38	.13	1.54	1.89
0060	Polyethylene 3 dimensional geomatrix, 50 mil thick	G		700	.034		2.28	1.29	.43	4	5.10
0062	120 mil thick	G		515	.047		2.53	1.75	.59	4.87	6.30
0100	Plastic netting, stapled, 2″ x 1″ mesh, 20 mil	G	B-1	2500	.010		.24	.37		.61	.86
0120	Revegetation mat, webbed	G	2 Clab	1000	.016		4.48	.60		5.08	5.90
0200	Polypropylene mesh, stapled, 6.5 oz./S.Y.	G	B-1	2500	.010		2.30	.37		2.67	3.13
0300	Tobacco netting, or jute mesh #2, stapled	G	″	2500	.010		.19	.37		.56	.81
1000	Silt fence, install and maintain, remove	G	B-62	1300	.018	L.F.	.72	.76	.13	1.61	2.17
1100	Allow 25% per month for maintenance; 6-month max life										
1130	Cellular confinement, poly, 3-dimen, 8′ x 20′ panels, 4″ deep cell	G	B-6	1600	.015	S.F.	2	.62	.23	2.85	3.45
1140	8″ deep cells	G	″	1200	.020	″	2.72	.83	.30	3.85	4.66
1200	Place and remove hay bales	G	A-2	3	8	Ton	246	305	82	633	855
1250	Hay bales, staked	G	″	2500	.010	L.F.	9.85	.37	.10	10.32	11.50

31 31 Soil Treatment

31 31 16 – Termite Control

31 31 16.13 Chemical Termite Control		Crew	Daily Output	Labor-Hours	Unit	Material	Labor	Equipment	Total	Total Incl O&P
0010	**CHEMICAL TERMITE CONTROL**									
0020	Slab and walls, residential	1 Skwk	1200	.007	SF Flr.	.32	.32		.64	.88
0100	Commercial, minimum		2496	.003		.33	.16		.49	.61
0200	Maximum		1645	.005		.50	.24		.74	.92
0390	Minimum labor/equipment charge		4	2	Job		97.50		97.50	158
0400	Insecticides for termite control, minimum		14.20	.563	Gal.	68.50	27.50		96	120
0500	Maximum		11	.727	″	117	35.50		152.50	187
3000	Soil poisoning (sterilization)	1 Clab	4496	.002	S.F.	.44	.07		.51	.59
3100	Herbicide application from truck	B-59	19000	.001	S.Y.		.02	.03	.05	.06

31 32 Soil Stabilization

31 32 13 – Soil Mixing Stabilization

31 32 13.13 Asphalt Soil Stabilization		Crew	Daily Output	Labor-Hours	Unit	Material	Labor	Equipment	Total	Total Incl O&P
0010	**ASPHALT SOIL STABILIZATION**									
0011	Including scarifying and compaction									
0020	Asphalt, 1-1/2″ deep, 1/2 gal/S.Y.	B-75	4000	.014	S.Y.	.95	.64	1.50	3.09	3.72
0040	3/4 gal/S.Y.		4000	.014		1.43	.64	1.50	3.57	4.24
0100	3″ deep, 1 gal/S.Y.		3500	.016		1.90	.73	1.72	4.35	5.15
0140	1-1/2 gal/S.Y.		3500	.016		2.85	.73	1.72	5.30	6.20
0200	6″ deep, 2 gal/S.Y.		3000	.019		3.80	.85	2	6.65	7.75
0240	3 gal/S.Y.		3000	.019		5.70	.85	2	8.55	9.80
0300	8″ deep, 2-2/3 gal/S.Y.		2800	.020		5.05	.91	2.15	8.11	9.40
0340	4 gal/S.Y.		2800	.020		7.60	.91	2.15	10.66	12.15
0500	12″ deep, 4 gal/S.Y.		5000	.011		7.60	.51	1.20	9.31	10.50
0540	6 gal/S.Y.		2600	.022		11.40	.98	2.31	14.69	16.65

31 32 13.16 Cement Soil Stabilization		Crew	Daily Output	Labor-Hours	Unit	Material	Labor	Equipment	Total	Total Incl O&P
0010	**CEMENT SOIL STABILIZATION**									
0011	Including scarifying and compaction									
1020	Cement, 4% mix, by volume, 6″ deep	B-74	1100	.058	S.Y.	1.54	2.62	5.40	9.56	11.85
1030	8″ deep		1050	.061		2.01	2.74	5.70	10.45	12.85
1060	12″ deep		960	.067		3.01	3	6.20	12.21	14.95

31 32 Soil Stabilization

31 32 13 – Soil Mixing Stabilization

31 32 13.16 Cement Soil Stabilization

		Crew	Daily Output	Labor-Hours	Unit	2015 Bare Costs Material	Labor	Equipment	Total	Total Incl O&P
1100	6% mix, 6″ deep	B-74	1100	.058	S.Y.	2.21	2.62	5.40	10.23	12.60
1120	8″ deep		1050	.061		2.88	2.74	5.70	11.32	13.80
1160	12″ deep		960	.067		4.35	3	6.20	13.55	16.45
1200	9% mix, 6″ deep		1100	.058		3.35	2.62	5.40	11.37	13.85
1220	8″ deep		1050	.061		4.35	2.74	5.70	12.79	15.45
1260	12″ deep		960	.067		6.55	3	6.20	15.75	18.85
1300	12% mix, 6″ deep		1100	.058		4.35	2.62	5.40	12.37	14.95
1320	8″ deep		1050	.061		5.80	2.74	5.70	14.24	17.05
1360	12″ deep		960	.067		8.70	3	6.20	17.90	21

31 32 13.19 Lime Soil Stabilization

		Crew	Daily Output	Labor-Hours	Unit	Material	Labor	Equipment	Total	Total Incl O&P
0010	**LIME SOIL STABILIZATION**									
0011	Including scarifying and compaction									
2020	Hydrated lime, for base, 2% mix by weight, 6″ deep	B-74	1800	.036	S.Y.	2.34	1.60	3.31	7.25	8.75
2030	8″ deep		1700	.038		3.35	1.69	3.51	8.55	10.25
2060	12″ deep		1550	.041		4.73	1.86	3.85	10.44	12.40
2100	4% mix, 6″ deep		1800	.036		3.49	1.60	3.31	8.40	10.05
2120	8″ deep		1700	.038		4.72	1.69	3.51	9.92	11.75
2160	12″ deep		1550	.041		7.05	1.86	3.85	12.76	14.95
2200	6% mix, 6″ deep		1800	.036		5.30	1.60	3.31	10.21	12
2220	8″ deep		1700	.038		7.05	1.69	3.51	12.25	14.30
2260	12″ deep		1550	.041		10.50	1.86	3.85	16.21	18.75

31 32 13.30 Calcium Chloride

		Crew	Daily Output	Labor-Hours	Unit	Material	Labor	Equipment	Total	Total Incl O&P
0010	**CALCIUM CHLORIDE**									
0020	Calcium chloride delivered, 100 lb. bags, truckload lots				Ton	390			390	430
0030	Solution, 4 lb. flake per gallon, tank truck delivery				Gal.	1.48			1.48	1.63

31 32 19 – Geosynthetic Soil Stabilization and Layer Separation

31 32 19.16 Geotextile Soil Stabilization

		Crew	Daily Output	Labor-Hours	Unit	Material	Labor	Equipment	Total	Total Incl O&P
0010	**GEOTEXTILE SOIL STABILIZATION**									
1500	Geotextile fabric, woven, 200 lb. tensile strength	2 Clab	2500	.006	S.Y.	1.90	.24		2.14	2.48
1510	Heavy Duty, 600 lb. tensile strength		2400	.007		1.84	.25		2.09	2.43
1550	Non-woven, 120 lb. tensile strength		2500	.006		1.11	.24		1.35	1.61

31 33 Rock Stabilization

31 33 13 – Rock Bolting and Grouting

31 33 13.10 Rock Bolting

		Crew	Daily Output	Labor-Hours	Unit	Material	Labor	Equipment	Total	Total Incl O&P
0010	**ROCK BOLTING**									
2020	Hollow core, prestressable anchor, 1″ diameter, 5′ long	2 Skwk	32	.500	Ea.	182	24.50		206.50	240
2025	10′ long		24	.667		305	32.50		337.50	395
2060	2″ diameter, 5′ long		32	.500		590	24.50		614.50	690
2065	10′ long		24	.667		1,225	32.50		1,257.50	1,400
2100	Super high-tensile, 3/4″ diameter, 5′ long		32	.500		49.50	24.50		74	94
2105	10′ long		24	.667		136	32.50		168.50	202
2160	2″ diameter, 5′ long		32	.500		410	24.50		434.50	490
2165	10′ long		24	.667		690	32.50		722.50	810
4400	Drill hole for rock bolt, 1-3/4″ diam., 5′ long (for 3/4″ bolt)	B-56	17	.941			40.50	93.50	134	169
4405	10′ long		9	1.778			76.50	177	253.50	320
4420	2″ diameter, 5′ long (for 1″ bolt)		13	1.231			53	123	176	221
4425	10′ long		7	2.286			98.50	228	326.50	410
4460	3-1/2″ diameter, 5′ long (for 2″ bolt)		10	1.600			69	159	228	286
4465	10′ long		5	3.200			138	320	458	570

31 36 Gabions

31 36 13 – Gabion Boxes

31 36 13.10 Gabion Box Systems		Crew	Daily Output	Labor-Hours	Unit	Material	Labor	Equipment	Total	Total Incl O&P
						2015 Bare Costs				
0010	GABION BOX SYSTEMS									
0400	Gabions, galvanized steel mesh mats or boxes, stone filled, 6″ deep	B-13	200	.280	S.Y.	26.50	11.50	3.68	41.68	51.50
0500	9″ deep		163	.344		39	14.10	4.52	57.62	71
0600	12″ deep		153	.366		52.50	15	4.81	72.31	87.50
0700	18″ deep		102	.549		71.50	22.50	7.20	101.20	123
0800	36″ deep	↓	60	.933	↓	98	38.50	12.25	148.75	184

31 37 Riprap

31 37 13 – Machined Riprap

31 37 13.10 Riprap and Rock Lining		Crew	Daily Output	Labor-Hours	Unit	Material	Labor	Equipment	Total	Total Incl O&P
0010	RIPRAP AND ROCK LINING									
0011	Random, broken stone									
0100	Machine placed for slope protection	B-12G	62	.258	L.C.Y.	30	11.50	11.40	52.90	64
0110	3/8 to 1/4 C.Y. pieces, grouted	B-13	80	.700	S.Y.	59.50	28.50	9.20	97.20	122
0200	18″ minimum thickness, not grouted	″	53	1.057	″	18.90	43.50	13.90	76.30	106
0300	Dumped, 50 lb. average	B-11A	800	.020	Ton	26.50	.88	1.73	29.11	32.50
0350	100 lb. average		700	.023		26.50	1.01	1.98	29.49	33
0370	300 lb. average	↓	600	.027	↓	26.50	1.18	2.31	29.99	33.50

31 41 Shoring

31 41 13 – Timber Shoring

31 41 13.10 Building Shoring		Crew	Daily Output	Labor-Hours	Unit	Material	Labor	Equipment	Total	Total Incl O&P
0010	BUILDING SHORING									
0020	Shoring, existing building, with timber, no salvage allowance	B-51	2.20	21.818	M.B.F.	850	835	111	1,796	2,400
1000	On cribbing with 35 ton screw jacks, per box and jack		3.60	13.333	Jack	68.50	510	68	646.50	980
1090	Minimum labor/equipment charge	↓	2	24	Ea.		915	123	1,038	1,625

31 41 16 – Sheet Piling

31 41 16.10 Sheet Piling Systems		Crew	Daily Output	Labor-Hours	Unit	Material	Labor	Equipment	Total	Total Incl O&P
0010	SHEET PILING SYSTEMS									
0020	Sheet piling steel, not incl. wales, 22 psf, 15′ excav., left in place	B-40	10.81	5.920	Ton	1,600	282	345	2,227	2,575
0100	Drive, extract & salvage		6	10.667		510	510	625	1,645	2,075
0300	20′ deep excavation, 27 psf, left in place		12.95	4.942		1,600	235	289	2,124	2,450
0400	Drive, extract & salvage		6.55	9.771		510	465	570	1,545	1,950
0600	25′ deep excavation, 38 psf, left in place		19	3.368		1,600	160	197	1,957	2,225
0700	Drive, extract & salvage		10.50	6.095		510	290	355	1,155	1,425
0900	40′ deep excavation, 38 psf, left in place		21.20	3.019		1,600	144	177	1,921	2,175
1000	Drive, extract & salvage		12.25	5.224	↓	510	249	305	1,064	1,300
1200	15′ deep excavation, 22 psf, left in place		983	.065	S.F.	18.60	3.10	3.81	25.51	29.50
1300	Drive, extract & salvage		545	.117		5.70	5.60	6.85	18.15	23
1500	20′ deep excavation, 27 psf, left in place		960	.067		23.50	3.18	3.90	30.58	35
1600	Drive, extract & salvage		485	.132		7.45	6.30	7.70	21.45	27
1800	25′ deep excavation, 38 psf, left in place		1000	.064		34.50	3.05	3.74	41.29	47
1900	Drive, extract & salvage	↓	553	.116	↓	10.15	5.50	6.75	22.40	27.50
2100	Rent steel sheet piling and wales, first month				Ton	310			310	340
2200	Per added month					31			31	34
2300	Rental piling left in place, add to rental					1,150			1,150	1,275
2500	Wales, connections & struts, 2/3 salvage					480			480	530
2700	High strength piling, 50,000 psi, add					63.50			63.50	70
2800	55,000 psi, add				↓	82			82	90

31 41 Shoring

31 41 16 – Sheet Piling

31 41 16.10 Sheet Piling Systems		Crew	Daily Output	Labor-Hours	Unit	2015 Bare Costs Material	Labor	Equipment	Total	Total Incl O&P
3000	Tie rod, not upset, 1-1/2" to 4" diameter with turnbuckle				Ton	2,125			2,125	2,325
3100	No turnbuckle					1,650			1,650	1,825
3300	Upset, 1-3/4" to 4" diameter with turnbuckle					2,400			2,400	2,625
3400	No turnbuckle					2,075			2,075	2,275
3600	Lightweight, 18" to 28" wide, 7 ga., 9.22 psf, and									
3610	9 ga., 8.6 psf, minimum				Lb.	.77			.77	.85
3700	Average					.87			.87	.96
3750	Maximum					1.04			1.04	1.14
3900	Wood, solid sheeting, incl. wales, braces and spacers,									
3910	drive, extract & salvage, 8' deep excavation	B-31	330	.121	S.F.	1.89	4.83	.66	7.38	10.70
4000	10' deep, 50 S.F./hr. in & 150 S.F./hr. out		300	.133		1.95	5.30	.73	7.98	11.65
4100	12' deep, 45 S.F./hr. in & 135 S.F./hr. out		270	.148		2	5.90	.81	8.71	12.80
4200	14' deep, 42 S.F./hr. in & 126 S.F./hr. out		250	.160		2.06	6.40	.88	9.34	13.70
4300	16' deep, 40 S.F./hr. in & 120 S.F./hr. out		240	.167		2.13	6.65	.91	9.69	14.25
4400	18' deep, 38 S.F./hr. in & 114 S.F./hr. out		230	.174		2.20	6.95	.95	10.10	14.80
4500	20' deep, 35 S.F./hr. in & 105 S.F./hr. out		210	.190		2.27	7.60	1.04	10.91	16.10
4520	Left in place, 8' deep, 55 S.F./hr.		440	.091		3.41	3.62	.50	7.53	10.25
4540	10' deep, 50 S.F./hr.		400	.100		3.58	3.99	.55	8.12	11.10
4560	12' deep, 45 S.F./hr.		360	.111		3.78	4.43	.61	8.82	12.10
4565	14' deep, 42 S.F./hr.		335	.119		4.01	4.76	.65	9.42	12.95
4570	16' deep, 40 S.F./hr.		320	.125		4.26	4.98	.68	9.92	13.60
4580	18' deep, 38 S.F./hr.		305	.131		4.54	5.25	.72	10.51	14.35
4590	20' deep, 35 S.F./hr.		280	.143		4.86	5.70	.78	11.34	15.55
4700	Alternate pricing, left in place, 8' deep		1.76	22.727	M.B.F.	765	905	124	1,794	2,450
4800	Drive, extract and salvage, 8' deep		1.32	30.303	"	680	1,200	166	2,046	2,900
4990	Minimum labor/equipment charge		2	20	Job		795	109	904	1,425
5000	For treated lumber add cost of treatment to lumber									

31 43 Concrete Raising

31 43 13 – Pressure Grouting

31 43 13.13 Concrete Pressure Grouting		Crew	Daily Output	Labor-Hours	Unit	Material	Labor	Equipment	Total	Total Incl O&P
0010	**CONCRETE PRESSURE GROUTING**									
0020	Grouting, pressure, cement & sand, 1:1 mix, minimum	B-61	124	.323	Bag	12.10	12.95	2.83	27.88	37.50
0100	Maximum		51	.784	"	12.10	31.50	6.90	50.50	72.50
0200	Cement and sand, 1:1 mix, minimum		250	.160	C.F.	24	6.45	1.40	31.85	38.50
0300	Maximum		100	.400		36.50	16.10	3.51	56.11	70
0400	Epoxy cement grout, minimum		137	.292		700	11.75	2.56	714.31	790
0500	Maximum		57	.702		700	28	6.15	734.15	825
0700	Alternate pricing method: (Add for materials)									
0710	5 person crew and equipment	B-61	1	40	Day		1,600	350	1,950	3,000

31 45 Vibroflotation and Densification

31 45 13 – Vibroflotation

31 45 13.10 Vibroflotation Densification		Crew	Daily Output	Labor-Hours	Unit	Material	2015 Bare Costs Labor	Equipment	Total	Total Incl O&P
0010	VIBROFLOTATION DENSIFICATION									
0900	Vibroflotation compacted sand cylinder, minimum	B-60	750	.075	V.L.F.		3.30	2.91	6.21	8.50
0950	Maximum		325	.172			7.60	6.70	14.30	19.60
1100	Vibro replacement compacted stone cylinder, minimum		500	.112			4.94	4.36	9.30	12.75
1150	Maximum		250	.224			9.90	8.70	18.60	25.50
1300	Mobilization and demobilization, minimum		.47	119	Total		5,250	4,650	9,900	13,600
1400	Maximum		.14	400	"		17,700	15,600	33,300	45,500

31 46 Needle Beams

31 46 13 – Cantilever Needle Beams

31 46 13.10 Needle Beams

		Crew	Daily Output	Labor-Hours	Unit	Material	Labor	Equipment	Total	Total Incl O&P
0010	NEEDLE BEAMS									
0011	Incl. wood shoring 10′ x 10′ opening									
0400	Block, concrete, 8″ thick	B-9	7.10	5.634	Ea.	50	214	33	297	440
0420	12″ thick		6.70	5.970		60	227	35	322	475
0800	Brick, 4″ thick with 8″ backup block		5.70	7.018		60	267	41	368	545
1000	Brick, solid, 8″ thick		6.20	6.452		50	245	37.50	332.50	495
1040	12″ thick		4.90	8.163		60	310	47.50	417.50	630
1080	16″ thick		4.50	8.889		80.50	340	51.50	472	700
2000	Add for additional floors of shoring	B-1	6	4		50	153		203	305
9000	Minimum labor/equipment charge	"	2	12	Job		460		460	755

31 48 Underpinning

31 48 13 – Underpinning Piers

31 48 13.10 Underpinning Foundations

		Crew	Daily Output	Labor-Hours	Unit	Material	Labor	Equipment	Total	Total Incl O&P
0010	UNDERPINNING FOUNDATIONS									
0011	Including excavation,									
0020	forming, reinforcing, concrete and equipment									
0100	5′ to 16′ below grade, 100 to 500 C.Y.	B-52	2.30	24.348	C.Y.	283	1,050	258	1,591	2,325
0200	Over 500 C.Y.		2.50	22.400		255	975	238	1,468	2,150
0400	16′ to 25′ below grade, 100 to 500 C.Y.		2	28		310	1,225	297	1,832	2,650
0500	Over 500 C.Y.		2.10	26.667		295	1,175	283	1,753	2,525
0700	26′ to 40′ below grade, 100 to 500 C.Y.		1.60	35		340	1,525	370	2,235	3,250
0800	Over 500 C.Y.		1.80	31.111		310	1,350	330	1,990	2,900
0900	For under 50 C.Y., add					10%	40%			
1000	For 50 C.Y. to 100 C.Y., add					5%	20%			

31 52 Cofferdams

31 52 16 – Timber Cofferdams

31 52 16.10 Cofferdams

		Crew	Daily Output	Labor-Hours	Unit	Material	Labor	Equipment	Total	Total Incl O&P
0010	COFFERDAMS									
0011	Incl. mobilization and temporary sheeting									
0080	Soldier beams & lagging H piles with 3″ wood sheeting									
0090	horizontal between piles, including removal of wales & braces									
0100	No hydrostatic head, 15′ deep, 1 line of braces, minimum	B-50	545	.206	S.F.	8.50	9.30	4.48	22.28	29.50
0200	Maximum		495	.226		9.45	10.25	4.93	24.63	32.50
0400	15′ to 22′ deep with 2 lines of braces, 10″ H, minimum		360	.311		10	14.10	6.80	30.90	41.50

31 52 Cofferdams

31 52 16 – Timber Cofferdams

31 52 16.10 Cofferdams		Crew	Daily Output	Labor-Hours	Unit	Material	2015 Bare Costs Labor	Equipment	Total	Total Incl O&P
0500	Maximum	B-50	330	.339	S.F.	11.35	15.35	7.40	34.10	45.50
0700	23′ to 35′ deep with 3 lines of braces, 12″ H, minimum		325	.345		13.10	15.60	7.50	36.20	48
0800	Maximum		295	.380		14.20	17.20	8.30	39.70	52.50
1000	36′ to 45′ deep with 4 lines of braces, 14″ H, minimum		290	.386		14.70	17.50	8.40	40.60	54
1100	Maximum		265	.423		15.50	19.15	9.20	43.85	58
1300	No hydrostatic head, left in place, 15′ dp., 1 line of braces, min.		635	.176		11.35	8	3.85	23.20	29.50
1400	Maximum		575	.195		12.15	8.80	4.25	25.20	32.50
1600	15′ to 22′ deep with 2 lines of braces, minimum		455	.246		17.05	11.15	5.35	33.55	43
1700	Maximum		415	.270		18.95	12.25	5.90	37.10	47.50
1900	23′ to 35′ deep with 3 lines of braces, minimum		420	.267		20.50	12.10	5.80	38.40	48.50
2000	Maximum		380	.295		22.50	13.35	6.45	42.30	53
2200	36′ to 45′ deep with 4 lines of braces, minimum		385	.291		24.50	13.20	6.35	44.05	55.50
2300	Maximum	↓	350	.320		28.50	14.50	7	50	62
2350	Lagging only, 3″ thick wood between piles 8′ O.C., minimum	B-46	400	.120		1.89	5.05	.11	7.05	10.50
2370	Maximum		250	.192		2.84	8.10	.18	11.12	16.55
2400	Open sheeting no bracing, for trenches to 10′ deep, min.		1736	.028		.85	1.17	.03	2.05	2.88
2450	Maximum	↓	1510	.032		.95	1.34	.03	2.32	3.26
2500	Tie-back method, add to open sheeting, add, minimum								20%	20%
2550	Maximum				↓				60%	60%
2700	Tie-backs only, based on tie-backs total length, minimum	B-46	86.80	.553	L.F.	14.60	23.50	.52	38.62	54.50
2750	Maximum		38.50	1.247	"	25.50	52.50	1.17	79.17	116
3500	Tie-backs only, typical average, 25′ long		2	24	Ea.	645	1,000	22.50	1,667.50	2,375
3600	35′ long	↓	1.58	30.380	"	855	1,275	28.50	2,158.50	3,075
4500	Trench box, 7′ deep, 16′ x 8′, see 01 54 33 in Reference Section				Day				191	210
4600	20′ x 10′, see 01 54 33 in Reference Section				"				240	265
5200	Wood sheeting, in trench, jacks at 4′ O.C., 8′ deep	B-1	800	.030	S.F.	.66	1.15		1.81	2.60
5250	12′ deep		700	.034		.77	1.31		2.08	3
5300	15′ deep	↓	600	.040	↓	1.07	1.53		2.60	3.68

31 56 Slurry Walls

31 56 23 – Lean Concrete Slurry Walls

31 56 23.20 Slurry Trench		Crew	Daily Output	Labor-Hours	Unit	Material	Labor	Equipment	Total	Total Incl O&P
0010	**SLURRY TRENCH**									
0011	Excavated slurry trench in wet soils									
0020	backfilled with 3000 PSI concrete, no reinforcing steel									
0050	Minimum	C-7	333	.216	C.F.	7.55	8.85	3.65	20.05	26.50
0100	Maximum		200	.360	"	12.60	14.75	6.05	33.40	44.50
0200	Alternate pricing method, minimum		150	.480	S.F.	15.05	19.65	8.10	42.80	57.50
0300	Maximum	↓	120	.600		22.50	24.50	10.10	57.10	76
0500	Reinforced slurry trench, minimum	B-48	177	.316		11.30	13.45	16.40	41.15	52
0600	Maximum	"	69	.812	↓	37.50	34.50	42	114	144
0800	Haul for disposal, 2 mile haul, excavated material, add	B-34B	99	.081	C.Y.		3.24	7	10.24	12.90
0900	Haul bentonite castings for disposal, add	"	40	.200	"		8	17.30	25.30	32

31 62 13 – Concrete Piles

31 62 13.23 Prestressed Concrete Piles		Crew	Daily Output	Labor-Hours	Unit	2015 Bare Costs Material	Labor	Equipment	Total	Total Incl O&P
0010	**PRESTRESSED CONCRETE PILES**, 200 piles									
0020	Unless specified otherwise, not incl. pile caps or mobilization									
2200	Precast, prestressed, 50′ long, 12″ diam., 2-3/8″ wall	B-19	720	.089	V.L.F.	23.50	4.23	2.41	30.14	35.50
2300	14″ diameter, 2-1/2″ wall		680	.094		32	4.48	2.55	39.03	45
2500	16″ diameter, 3″ wall		640	.100		45.50	4.76	2.71	52.97	60.50
2600	18″ diameter, 3″ wall	B-19A	600	.107		57	5.10	3.61	65.71	74.50
2800	20″ diameter, 3-1/2″ wall		560	.114		67.50	5.45	3.87	76.82	87
2900	24″ diameter, 3-1/2″ wall		520	.123		76	5.85	4.16	86.01	97.50
2920	36″ diameter, 4-1/2″ wall	B-19	400	.160		102	7.60	4.34	113.94	129
2940	54″ diameter, 5″ wall		340	.188		133	8.95	5.10	147.05	166
2960	66″ diameter, 6″ wall		220	.291		207	13.85	7.90	228.75	259
3100	Precast, prestressed, 40′ long, 10″ thick, square		700	.091		18.35	4.36	2.48	25.19	30
3200	12″ thick, square		680	.094		22.50	4.48	2.55	29.53	34.50
3400	14″ thick, square		600	.107		25.50	5.10	2.90	33.50	39.50
3500	Octagonal		640	.100		34.50	4.76	2.71	41.97	48.50
3700	16″ thick, square		560	.114		41	5.45	3.10	49.55	57
3800	Octagonal		600	.107		41.50	5.10	2.90	49.50	57.50
4000	18″ thick, square	B-19A	520	.123		47.50	5.85	4.16	57.51	66
4100	Octagonal	B-19	560	.114		49.50	5.45	3.10	58.05	66
4300	20″ thick, square	B-19A	480	.133		60.50	6.35	4.51	71.36	81.50
4400	Octagonal	B-19	520	.123		54.50	5.85	3.34	63.69	73
4600	24″ thick, square	B-19A	440	.145		72	6.95	4.92	83.87	95.50
4700	Octagonal	B-19	480	.133		78.50	6.35	3.62	88.47	101
4730	Precast, prestressed, 60′ long, 10″ thick, square		700	.091		19.25	4.36	2.48	26.09	31
4740	12″ thick, square (60′ long)		680	.094		23	4.48	2.55	30.03	35.50
4750	Mobilization for 10,000 L.F. pile job, add		3300	.019			.92	.53	1.45	2.07
4800	25,000 L.F. pile job, add		8500	.008			.36	.20	.56	.80

31 62 16 – Steel Piles

31 62 16.13 Steel Piles		Crew	Daily Output	Labor-Hours	Unit	Material	Labor	Equipment	Total	Total Incl O&P
0010	**STEEL PILES**									
0100	Step tapered, round, concrete filled									
0110	8″ tip, 60 ton capacity, 30′ depth	B-19	760	.084	V.L.F.	11.45	4.01	2.29	17.75	21.50
0120	60′ depth		740	.086		12.95	4.12	2.35	19.42	23.50
0130	80′ depth		700	.091		13.40	4.36	2.48	20.24	24.50
0150	10″ tip, 90 ton capacity, 30′ depth		700	.091		13.75	4.36	2.48	20.59	25
0160	60′ depth		690	.093		14.20	4.42	2.52	21.14	25.50
0170	80′ depth		670	.096		15.30	4.55	2.59	22.44	27
0190	12″ tip, 120 ton capacity, 30′ depth		660	.097		18.35	4.62	2.63	25.60	30.50
0200	60′ depth, 12″ diameter		630	.102		17.65	4.84	2.76	25.25	30.50
0210	80′ depth		590	.108		17.15	5.15	2.94	25.24	30.50
0250	"H" Sections, 50′ long, HP8 x 36		640	.100		16.05	4.76	2.71	23.52	28.50
0400	HP10 X 42		610	.105		18.35	5	2.85	26.20	31
0500	HP10 X 57		610	.105		25.50	5	2.85	33.35	39
0700	HP12 X 53		590	.108		25	5.15	2.94	33.09	39
0800	HP12 X 74	B-19A	590	.108		34.50	5.15	3.67	43.32	50.50
1000	HP14 X 73		540	.119		35	5.65	4.01	44.66	52
1100	HP14 X 89		540	.119		41	5.65	4.01	50.66	58.50
1300	HP14 X 102		510	.125		47	6	4.24	57.24	66
1400	HP14 X 117		510	.125		54	6	4.24	64.24	73.50
1600	Splice on standard points, not in leads, 8″ or 10″	1 Sswl	5	1.600	Ea.	120	84		204	284
1700	12″ or 14″		4	2		177	105		282	385
1900	Heavy duty points, not in leads, 10″ wide		4	2		190	105		295	400

31 62 Driven Piles

31 62 16 – Steel Piles

31 62 16.13 Steel Piles		Crew	Daily Output	Labor-Hours	Unit	Material	Labor	Equipment	Total	Total Incl O&P
						2015 Bare Costs				
2100	14″ wide	1 Sswl	3.50	2.286	Ea.	238	120		358	480

31 62 19 – Timber Piles

31 62 19.10 Wood Piles

		Crew	Daily Output	Labor-Hours	Unit	Material	Labor	Equipment	Total	Total Incl O&P
0010	**WOOD PILES**									
0011	Friction or end bearing, not including									
0100	Untreated piles, up to 30′ long, 12″ butts, 8″ points	B-19	625	.102	V.L.F.	16.85	4.88	2.78	24.51	29.50
0200	30′ to 39′ long, 12″ butts, 8″ points		700	.091		16.85	4.36	2.48	23.69	28.50
0300	40′ to 49′ long, 12″ butts, 7″ points		720	.089		16.85	4.23	2.41	23.49	28
0400	50′ to 59′ long, 13″butts, 7″ points		800	.080		16.85	3.81	2.17	22.83	27
0500	60′ to 69′ long, 13″ butts, 7″ points		840	.076		18.95	3.63	2.07	24.65	29
0600	70′ to 80′ long, 13″ butts, 6″ points		840	.076		21	3.63	2.07	26.70	31
0800	Treated piles, 12 lb. per C.F.,									
0810	friction or end bearing, ASTM class B									
1000	Up to 30′ long, 12″ butts, 8″ points	B-19	625	.102	V.L.F.	15.85	4.88	2.78	23.51	28.50
1100	30′ to 39′ long, 12″ butts, 8″ points		700	.091		17.80	4.36	2.48	24.64	29.50
1200	40′ to 49′ long, 12″ butts, 7″ points		720	.089		19.35	4.23	2.41	25.99	31
1300	50′ to 59′ long, 13″ butts, 7″ points		800	.080		22	3.81	2.17	27.98	32.50
1400	60′ to 69′ long, 13″ butts, 6″ points	B-19A	840	.076		26	3.63	2.58	32.21	37
1500	70′ to 80′ long, 13″ butts, 6″ points	″	840	.076		28	3.63	2.58	34.21	39.50
1600	Treated piles, C.C.A., 2.5# per C.F.									
1610	8″ butts, 10′ long	B-19	400	.160	V.L.F.	18.40	7.60	4.34	30.34	37
1620	11′ to 16′ long		500	.128		18.40	6.10	3.47	27.97	33.50
1630	17′ to 20′ long		575	.111		18.40	5.30	3.02	26.72	32
1640	10″ butts, 10′ to 16′ long		500	.128		18.40	6.10	3.47	27.97	33.50
1650	17′ to 20′ long		575	.111		18.40	5.30	3.02	26.72	32
1660	21′ to 40′ long		700	.091		18.40	4.36	2.48	25.24	30
1670	12″ butts, 10′ to 20′ long		575	.111		18.40	5.30	3.02	26.72	32
1680	21′ to 35′ long		650	.098		18.40	4.69	2.67	25.76	30.50
1690	36′ to 40′ long		700	.091		18.40	4.36	2.48	25.24	30
1695	14″ butts, to 40′ long		700	.091		18.40	4.36	2.48	25.24	30
1700	Boot for pile tip, minimum	1 Pile	27	.296	Ea.	23	13.65		36.65	48
1800	Maximum		21	.381		69.50	17.55		87.05	105
2000	Point for pile tip, minimum		20	.400		23	18.45		41.45	55.50
2100	Maximum		15	.533		83.50	24.50		108	132
2300	Splice for piles over 50′ long, minimum	B-46	35	1.371		57	58	1.29	116.29	159
2400	Maximum		20	2.400		68.50	101	2.26	171.76	244
2600	Concrete encasement with wire mesh and tube		331	.145	V.L.F.	10.60	6.10	.14	16.84	22
2700	Mobilization for 10,000 L.F. pile job, add	B-19	3300	.019			.92	.53	1.45	2.07
2800	25,000 L.F. pile job, add	″	8500	.008			.36	.20	.56	.80

31 62 23 – Composite Piles

31 62 23.13 Concrete-Filled Steel Piles

		Crew	Daily Output	Labor-Hours	Unit	Material	Labor	Equipment	Total	Total Incl O&P
0010	**CONCRETE-FILLED STEEL PILES** no mobilization or demobilization									
2600	Pipe piles, 50′ lg. 8″ diam., 29 lb. per L.F., no concrete	B-19	500	.128	V.L.F.	20.50	6.10	3.47	30.07	36
2700	Concrete filled		460	.139		21.50	6.65	3.78	31.93	38.50
2900	10″ diameter, 34 lb. per L.F., no concrete		500	.128		27	6.10	3.47	36.57	43.50
3000	Concrete filled		450	.142		29	6.80	3.86	39.66	47
3200	12″ diameter, 44 lb. per L.F., no concrete		475	.135		33	6.40	3.66	43.06	51
3300	Concrete filled		415	.154		37	7.35	4.19	48.54	57.50
3500	14″ diameter, 46 lb. per L.F., no concrete		430	.149		35.50	7.10	4.04	46.64	55.50
3600	Concrete filled		355	.180		41.50	8.60	4.89	54.99	65
3800	16″ diameter, 52 lb. per L.F., no concrete		385	.166		39	7.90	4.51	51.41	61
3900	Concrete filled		335	.191		45.50	9.10	5.20	59.80	70.50

31 62 Driven Piles

31 62 23 – Composite Piles

31 62 23.13 Concrete-Filled Steel Piles		Crew	Daily Output	Labor-Hours	Unit	2015 Bare Costs Material	Labor	Equipment	Total	Total Incl O&P
4100	18″ diameter, 59 lb. per L.F., no concrete	B-19	355	.180	V.L.F.	48.50	8.60	4.89	61.99	73
4200	Concrete filled		310	.206		54.50	9.85	5.60	69.95	82
4400	Splices for pipe piles, not in leads, 8″ diameter	1 Sswl	4.67	1.713	Ea.	90.50	90		180.50	263
4500	14″ diameter		3.79	2.111		119	111		230	330
4600	16″ diameter		3.03	2.640		147	139		286	410
4650	18″ diameter		4.50	1.778		194	93.50		287.50	380
4800	Points, standard, 8″ diameter		4.61	1.735		148	91.50		239.50	330
4900	14″ diameter		4.05	1.975		206	104		310	415
5000	16″ diameter		3.37	2.374		251	125		376	500
5050	18″ diameter		5	1.600		390	84		474	580
5200	Points, heavy duty, 10″ diameter		2.89	2.768		296	146		442	590
5300	14″ or 16″ diameter		2.02	3.960		475	209		684	895
5500	For reinforcing steel, add		1150	.007	Lb.	.93	.37		1.30	1.68
5700	For thick wall sections, add				″	.96			.96	1.06
6020	Steel pipe pile end plates, 8″ diameter	1 Sswl	14	.571	Ea.	78	30		108	141
6050	10″ diameter		14	.571		82	30		112	145
6100	12″ diameter		12	.667		87	35		122	159
6150	14″ diameter		10	.800		90	42		132	175
6200	16″ diameter		9	.889		96	47		143	191
6250	18″ diameter		8	1		108	52.50		160.50	214
6300	Steel pipe pile shoes, 8″ diameter		12	.667		160	35		195	240
6350	10″ diameter		12	.667		166	35		201	246
6400	12″ diameter		10	.800		177	42		219	271
6450	14″ diameter		9	.889		204	47		251	310
6500	16″ diameter		8	1		227	52.50		279.50	345
6550	18″ diameter		6	1.333		254	70		324	405

31 62 33 – Drilled Micropiles

31 62 33.10 Drilled Micropiles Metal Pipe		Crew	Daily Output	Labor-Hours	Unit	Material	Labor	Equipment	Total	Total Incl O&P
0010	**DRILLED MICROPILES METAL PIPE**									
0011	No mobilization or demobilization									
5000	Pressure grouted pin pile, 5″ diam., cased, up to 50 ton,									
5040	End bearing, less than 20′	B-48	90	.622	V.L.F.	49.50	26.50	32.50	108.50	132
5080	More than 40′		135	.415		29.50	17.65	21.50	68.65	84
5120	Friction, loose sand and gravel		107	.523		49.50	22.50	27	99	120
5160	Dense sand and gravel		135	.415		29.50	17.65	21.50	68.65	84
5200	Uncased, up to 10 ton capacity, 20′		135	.415		29	17.65	21.50	68.15	84

31 63 Bored Piles

31 63 26 – Drilled Caissons

31 63 26.13 Fixed End Caisson Piles		Crew	Daily Output	Labor-Hours	Unit	Material	Labor	Equipment	Total	Total Incl O&P
0010	**FIXED END CAISSON PILES**									
0015	Including excavation, concrete, 50 lb. reinforcing									
0020	per C.Y., not incl. mobilization, boulder removal, disposal									
0100	Open style, machine drilled, to 50′ deep, in stable ground, no									
0110	casings or ground water, 18″ diam., 0.065 C.Y./L.F.	B-43	200	.240	V.L.F.	8.15	9.95	12.70	30.80	39
0200	24″ diameter, 0.116 C.Y./L.F.		190	.253		14.60	10.50	13.40	38.50	48
0300	30″ diameter, 0.182 C.Y./L.F.		150	.320		23	13.30	16.95	53.25	65
0400	36″ diameter, 0.262 C.Y./L.F.		125	.384		33	15.95	20.50	69.45	84.50
0500	48″ diameter, 0.465 C.Y./L.F.		100	.480		58.50	19.95	25.50	103.95	125
0600	60″ diameter, 0.727 C.Y./L.F.		90	.533		91.50	22	28.50	142	168

31 63 26.13 Fixed End Caisson Piles		Crew	Daily Output	Labor-Hours	Unit	2015 Bare Costs Material	Labor	Equipment	Total	Total Incl O&P
0700	72" diameter, 1.05 C.Y./L.F.	B-43	80	.600	V.L.F.	132	25	32	189	221
0800	84" diameter, 1.43 C.Y./L.F.		75	.640		180	26.50	34	240.50	279
1000	For bell excavation and concrete, add									
1020	4' bell diameter, 24" shaft, 0.444 C.Y.	B-43	20	2.400	Ea.	45	99.50	127	271.50	350
1040	6' bell diameter, 30" shaft, 1.57 C.Y.		5.70	8.421		160	350	445	955	1,225
1060	8' bell diameter, 36" shaft, 3.72 C.Y.		2.40	20		380	830	1,050	2,260	2,950
1080	9' bell diameter, 48" shaft, 4.48 C.Y.		2	24		455	995	1,275	2,725	3,525
1100	10' bell diameter, 60" shaft, 5.24 C.Y.		1.70	28.235		535	1,175	1,500	3,210	4,125
1120	12' bell diameter, 72" shaft, 8.74 C.Y.		1	48		890	2,000	2,550	5,440	7,000
1140	14' bell diameter, 84" shaft, 13.6 C.Y.		.70	68.571		1,375	2,850	3,625	7,850	10,200
1200	Open style, machine drilled, to 50' deep, in wet ground, pulled									
1300	casing and pumping, 18" diameter, 0.065 C.Y./L.F.	B-48	160	.350	V.L.F.	8.15	14.90	18.15	41.20	53
1400	24" diameter, 0.116 C.Y./L.F.		125	.448		14.60	19.05	23.50	57.15	72.50
1500	30" diameter, 0.182 C.Y./L.F.		85	.659		23	28	34	85	108
1600	36" diameter, 0.262 C.Y./L.F.		60	.933		33	39.50	48.50	121	154
1700	48" diameter, 0.465 C.Y./L.F.	B-49	55	1.600		58.50	71	66	195.50	252
1800	60" diameter, 0.727 C.Y./L.F.		35	2.514		91.50	111	104	306.50	395
1900	72" diameter, 1.05 C.Y./L.F.		30	2.933		132	130	121	383	490
2000	84" diameter, 1.43 C.Y./L.F.		25	3.520		180	156	146	482	610
2100	For bell excavation and concrete, add									
2120	4' bell diameter, 24" shaft, 0.444 C.Y.	B-48	19.80	2.828	Ea.	45	120	147	312	405
2140	6' bell diameter, 30" shaft, 1.57 C.Y.		5.70	9.825		160	420	510	1,090	1,400
2160	8' bell diameter, 36" shaft, 3.72 C.Y.		2.40	23.333		380	995	1,200	2,575	3,350
2180	9' bell diameter, 48" shaft, 4.48 C.Y.	B-49	3.30	26.667		455	1,175	1,100	2,730	3,625
2200	10' bell diameter, 60" shaft, 5.24 C.Y.		2.80	31.429		535	1,400	1,300	3,235	4,250
2220	12' bell diameter, 72" shaft, 8.74 C.Y.		1.60	55		890	2,425	2,275	5,590	7,400
2240	14' bell diameter, 84" shaft, 13.6 C.Y.		1	88		1,375	3,900	3,650	8,925	11,800
2300	Open style, machine drilled, to 50' deep, in soft rocks and									
2400	medium hard shales, 18" diameter, 0.065 C.Y./L.F.	B-49	50	1.760	V.L.F.	8.15	78	73	159.15	215
2500	24" diameter, 0.116 C.Y./L.F.		30	2.933		14.60	130	121	265.60	360
2600	30" diameter, 0.182 C.Y./L.F.		20	4.400		23	195	182	400	540
2700	36" diameter, 0.262 C.Y./L.F.		15	5.867		33	260	243	536	725
2800	48" diameter, 0.465 C.Y./L.F.		10	8.800		58.50	390	365	813.50	1,100
2900	60" diameter, 0.727 C.Y./L.F.		7	12.571		91.50	555	520	1,166.50	1,575
3000	72" diameter, 1.05 C.Y./L.F.		6	14.667		132	650	605	1,387	1,875
3100	84" diameter, 1.43 C.Y./L.F.		5	17.600		180	780	730	1,690	2,250
3200	For bell excavation and concrete, add									
3220	4' bell diameter, 24" shaft, 0.444 C.Y.	B-49	10.90	8.073	Ea.	45	355	335	735	995
3240	6' bell diameter, 30" shaft, 1.57 C.Y.		3.10	28.387		160	1,250	1,175	2,585	3,500
3260	8' bell diameter, 36" shaft, 3.72 C.Y.		1.30	67.692		380	3,000	2,800	6,180	8,325
3280	9' bell diameter, 48" shaft, 4.48 C.Y.		1.10	80		455	3,550	3,300	7,305	9,850
3300	10' bell diameter, 60" shaft, 5.24 C.Y.		.90	97.778		535	4,325	4,050	8,910	12,000
3320	12' bell diameter, 72" shaft, 8.74 C.Y.		.60	146		890	6,500	6,075	13,465	18,200
3340	14' bell diameter, 84" shaft, 13.6 C.Y.		.40	220		1,375	9,750	9,100	20,225	27,200
3600	For rock excavation, sockets, add, minimum		120	.733	C.F.		32.50	30.50	63	86
3650	Average		95	.926			41	38.50	79.50	108
3700	Maximum		48	1.833			81	76	157	215
3900	For 50' to 100' deep, add				V.L.F.				7%	7%
4000	For 100' to 150' deep, add								25%	25%
4100	For 150' to 200' deep, add								30%	30%
4200	For casings left in place, add				Lb.	1.19			1.19	1.31
4300	For other than 50 lb. reinf. per C.Y., add or deduct				"	1.16			1.16	1.28
4400	For steel "I" beam cores, add	B-49	8.30	10.602	Ton	2,125	470	440	3,035	3,575

31 63 Bored Piles

31 63 26 – Drilled Caissons

31 63 26.13 Fixed End Caisson Piles		Crew	Daily Output	Labor-Hours	Unit	2015 Bare Costs Material	Labor	Equipment	Total	Total Incl O&P
4500	Load and haul excess excavation, 2 miles	B-34B	178	.045	L.C.Y.		1.80	3.88	5.68	7.15

31 63 26.16 Concrete Caissons for Marine Construction

		Crew	Daily Output	Labor-Hours	Unit	Material	Labor	Equipment	Total	Total Incl O&P
0010	**CONCRETE CAISSONS FOR MARINE CONSTRUCTION**									
0100	Caissons, incl. mobilization and demobilization, up to 50 miles									
0200	Uncased shafts, 30 to 80 tons cap., 17" diam., 10' depth	B-44	88	.727	V.L.F.	20.50	34	20.50	75	100
0300	25' depth		165	.388		14.55	18.10	11	43.65	57.50
0400	80-150 ton capacity, 22" diameter, 10' depth		80	.800		25.50	37.50	22.50	85.50	114
0500	20' depth		130	.492		20.50	23	14	57.50	75.50
0700	Cased shafts, 10 to 30 ton capacity, 10-5/8" diam., 20' depth		175	.366		14.55	17.10	10.40	42.05	55
0800	30' depth		240	.267		13.55	12.45	7.55	33.55	43.50
0850	30 to 60 ton capacity, 12" diameter, 20' depth		160	.400		20.50	18.70	11.35	50.55	65.50
0900	40' depth		230	.278		15.65	13	7.90	36.55	47
1000	80 to 100 ton capacity, 16" diameter, 20' depth		160	.400		29	18.70	11.35	59.05	75
1100	40' depth		230	.278		27	13	7.90	47.90	59.50
1200	110 to 140 ton capacity, 17-5/8" diameter, 20' depth		160	.400		31.50	18.70	11.35	61.55	77.50
1300	40' depth		230	.278		29	13	7.90	49.90	61.50
1400	140 to 175 ton capacity, 19" diameter, 20' depth		130	.492		34	23	14	71	90.50
1500	40' depth		210	.305		31.50	14.25	8.65	54.40	67
1700	Over 30' long, L.F. cost tends to be lower									
1900	Maximum depth is about 90'									

31 63 29 – Drilled Concrete Piers and Shafts

31 63 29.13 Uncased Drilled Concrete Piers

		Crew	Daily Output	Labor-Hours	Unit	Material	Labor	Equipment	Total	Total Incl O&P
0010	**UNCASED DRILLED CONCRETE PIERS**									
0020	Unless specified otherwise, not incl. pile caps or mobilization									
0100	Cast in place, thin wall shell pile, straight sided,									
0110	not incl. reinforcing, 8" diam., 16 ga., 5.8 lb./L.F.	B-19	700	.091	V.L.F.	9.10	4.36	2.48	15.94	19.80
0200	10" diameter, 16 ga. corrugated, 7.3 lb./L.F.		650	.098		11.90	4.69	2.67	19.26	23.50
0300	12" diameter, 16 ga. corrugated, 8.7 lb./L.F.		600	.107		15.45	5.10	2.90	23.45	28.50
0400	14" diameter, 16 ga. corrugated, 10.0 lb./L.F.		550	.116		18.20	5.55	3.16	26.91	32.50
0500	16" diameter, 16 ga. corrugated, 11.6 lb./L.F.		500	.128		22.50	6.10	3.47	32.07	38
0800	Cast in place friction pile, 50' long, fluted,									
0810	tapered steel, 4000 psi concrete, no reinforcing									
0900	12" diameter, 7 ga.	B-19	600	.107	V.L.F.	29	5.10	2.90	37	43.50
1000	14" diameter, 7 ga.		560	.114		31.50	5.45	3.10	40.05	47
1100	16" diameter, 7 ga.		520	.123		37.50	5.85	3.34	46.69	54
1200	18" diameter, 7 ga.		480	.133		43.50	6.35	3.62	53.47	62
1300	End bearing, fluted, constant diameter,									
1320	4000 psi concrete, no reinforcing									
1340	12" diameter, 7 ga.	B-19	600	.107	V.L.F.	30.50	5.10	2.90	38.50	45
1360	14" diameter, 7 ga.		560	.114		38	5.45	3.10	46.55	54
1380	16" diameter, 7 ga.		520	.123		44	5.85	3.34	53.19	61.50
1400	18" diameter, 7 ga.		480	.133		48.50	6.35	3.62	58.47	67.50

31 63 29.20 Cast In Place Piles, Adds

		Crew	Daily Output	Labor-Hours	Unit	Material	Labor	Equipment	Total	Total Incl O&P
0010	**CAST IN PLACE PILES, ADDS**									
1500	For reinforcing steel, add				Lb.	.97			.97	1.06
1700	For ball or pedestal end, add	B-19	11	5.818	C.Y.	146	277	158	581	785
1900	For lengths above 60', concrete, add	"	11	5.818	"	152	277	158	587	790
2000	For steel thin shell, pipe only				Lb.	1.26			1.26	1.39

Division Notes

		CREW	DAILY OUTPUT	LABOR-HOURS	UNIT	BARE COSTS MAT.	LABOR	EQUIP.	TOTAL	TOTAL INCL O&P

Division 32 – Exterior Improvements

Estimating Tips

32 01 00 Operations and Maintenance of Exterior Improvements

- Recycling of asphalt pavement is becoming very popular and is an alternative to removal and replacement. It can be a good value engineering proposal if removed pavement can be recycled, either at the project site or at another site that is reasonably close to the project site. Sections on repair of flexible and rigid pavement are included.

32 10 00 Bases, Ballasts, and Paving

- When estimating paving, keep in mind the project schedule. Also note that prices for asphalt and concrete are generally higher in the cold seasons. Lines for pavement markings, including tactile warning systems and fence lines, are included.

32 90 00 Planting

- The timing of planting and guarantee specifications often dictate the costs for establishing tree and shrub growth and a stand of grass or ground cover. Establish the work performance schedule to coincide with the local planting season. Maintenance and growth guarantees can add from 20%–100% to the total landscaping cost and can be contractually cumbersome. The cost to replace trees and shrubs can be as high as 5% of the total cost, depending on the planting zone, soil conditions, and time of year.

Reference Numbers

Reference numbers are shown in shaded boxes at the beginning of some major classifications. These numbers refer to related items in the Reference Section. The reference information may be an estimating procedure, an alternate pricing method, or technical information.

Note: Not all subdivisions listed here necessarily appear in this publication. ■

32 01 13 – Flexible Paving Surface Treatment

32 01 13.61 Slurry Seal (Latex Modified)		Crew	Daily Output	Labor-Hours	Unit	2015 Bare Costs Material	Labor	Equipment	Total	Total Incl O&P
0010	**SLURRY SEAL (LATEX MODIFIED)**									
3600	Waterproofing, membrane, tar and fabric, small area	B-63	233	.172	S.Y.	14.05	6.85	.75	21.65	27.50
3640	Large area		1435	.028		12.95	1.11	.12	14.18	16.20
3680	Preformed rubberized asphalt, small area		100	.400		19.15	15.90	1.74	36.79	49
3720	Large area		367	.109		17.45	4.34	.47	22.26	26.50
3780	Rubberized asphalt (latex) seal	B-45	5000	.003		2.81	.15	.18	3.14	3.52
9000	Minimum labor/equipment charge	1 Clab	2	4	Job		150		150	247

32 01 13.62 Asphalt Surface Treatment		Crew	Daily Output	Labor-Hours	Unit	Material	Labor	Equipment	Total	Total Incl O&P
0010	**ASPHALT SURFACE TREATMENT**									
3000	Pavement overlay, polypropylene									
3040	6 oz. per S.Y., ideal conditions	B-63	10000	.004	S.Y.	.98	.16	.02	1.16	1.36
3080	Adverse conditions		1000	.040		1.31	1.59	.17	3.07	4.22
3120	4 oz. per S.Y., ideal conditions		10000	.004		.80	.16	.02	.98	1.16
3160	Adverse conditions		1000	.040		1.03	1.59	.17	2.79	3.92
3200	Tack coat, emulsion, .05 gal. per S.Y., 1000 S.Y.	B-45	2500	.006		.29	.29	.36	.94	1.18
3240	10,000 S.Y.		10000	.002		.23	.07	.09	.39	.48
3270	.10 gal. per S.Y., 1000 S.Y.		2500	.006		.54	.29	.36	1.19	1.46
3275	10,000 S.Y.		10000	.002		.44	.07	.09	.60	.70
3280	.15 gal. per S.Y., 1000 S.Y.		2500	.006		.80	.29	.36	1.45	1.74
3320	10,000 S.Y.		10000	.002		.64	.07	.09	.80	.92

32 01 13.64 Sand Seal		Crew	Daily Output	Labor-Hours	Unit	Material	Labor	Equipment	Total	Total Incl O&P
0010	**SAND SEAL**									
2080	Sand sealing, sharp sand, asphalt emulsion, small area	B-91	10000	.006	S.Y.	1.48	.29	.23	2	2.35
2120	Roadway or large area	"	18000	.004	"	1.27	.16	.13	1.56	1.80
3000	Sealing random cracks, min 1/2" wide, to 1-1/2", 1,000 L.F.	B-77	2800	.014	L.F.	1.53	.55	.20	2.28	2.79
3040	10,000 L.F.		4000	.010	"	1.06	.38	.14	1.58	1.96
3080	Alternate method, 1,000 L.F.		200	.200	Gal.	39.50	7.65	2.83	49.98	59
3120	10,000 L.F.		325	.123	"	32	4.71	1.74	38.45	44.50
3200	Multi-cracks (flooding), 1 coat, small area	B-92	460	.070	S.Y.	2.34	2.65	1.38	6.37	8.45
3240	Large area		2850	.011		2.08	.43	.22	2.73	3.23
3280	2 coat, small area		230	.139		14.50	5.30	2.76	22.56	27.50
3320	Large area		1425	.022		13.20	.86	.45	14.51	16.40
3360	Alternate method, small area		115	.278	Gal.	14.50	10.60	5.50	30.60	39.50
3400	Large area		715	.045	"	13.65	1.71	.89	16.25	18.80

32 01 13.66 Fog Seal		Crew	Daily Output	Labor-Hours	Unit	Material	Labor	Equipment	Total	Total Incl O&P
0010	**FOG SEAL**									
0012	Sealcoating, 2 coat coal tar pitch emulsion over 10,000 S.Y.	B-45	5000	.003	S.Y.	.90	.15	.18	1.23	1.42
0030	1000 to 10,000 S.Y.	"	3000	.005		.90	.24	.30	1.44	1.71
0100	Under 1000 S.Y.	B-1	1050	.023		.90	.87		1.77	2.42
0300	Petroleum resistant, over 10,000 S.Y.	B-45	5000	.003		1.30	.15	.18	1.63	1.86
0320	1000 to 10,000 S.Y.	"	3000	.005		1.30	.24	.30	1.84	2.15
0400	Under 1000 S.Y.	B-1	1050	.023		1.30	.87		2.17	2.86
0600	Non-skid pavement renewal, over 10,000 S.Y.	B-45	5000	.003		1.37	.15	.18	1.70	1.94
0620	1000 to 10,000 S.Y.	"	3000	.005		1.37	.24	.30	1.91	2.23
0700	Under 1000 S.Y.	B-1	1050	.023		1.37	.87		2.24	2.94
0800	Prepare and clean surface for above	A-2	8545	.003			.11	.03	.14	.20
1000	Hand seal asphalt curbing	B-1	4420	.005	L.F.	.62	.21		.83	1.02
1900	Asphalt surface treatment, single course, small area									
1901	0.30 gal/S.Y. asphalt material, 20#/S.Y. aggregate	B-91	5000	.013	S.Y.	1.30	.57	.46	2.33	2.86
1910	Roadway or large area		10000	.006		1.20	.29	.23	1.72	2.04
1950	Asphalt surface treatment, dbl. course for small area		3000	.021		2.94	.95	.77	4.66	5.60
1960	Roadway or large area		6000	.011		2.65	.48	.39	3.52	4.11

32 01 Operation and Maintenance of Exterior Improvements

32 01 13 – Flexible Paving Surface Treatment

32 01 13.66 Fog Seal		Crew	Daily Output	Labor-Hours	Unit	2015 Bare Costs Material	Labor	Equipment	Total	Total Incl O&P
1980	Asphalt surface treatment, single course, for shoulders	B-91	7500	.009	S.Y.	1.46	.38	.31	2.15	2.56

32 01 13.68 Slurry Seal

		Crew	Daily Output	Labor-Hours	Unit	Material	Labor	Equipment	Total	Total Incl O&P
0010	**SLURRY SEAL**									
0100	Slurry seal, type I, 8 lb. agg./S.Y., 1 coat, small or irregular area	B-90	2800	.023	S.Y.	1.86	.94	.79	3.59	4.44
0150	Roadway or large area		10000	.006		1.86	.26	.22	2.34	2.72
0200	Type II, 12 lb. aggregate/S.Y., 2 coats,small or irregular area		2000	.032		3.72	1.32	1.10	6.14	7.45
0250	Roadway or large area		8000	.008		3.72	.33	.28	4.33	4.92
0300	Type III, 20 lb. aggregate/S.Y., 2 coats, small or irregular area		1800	.036		4.39	1.47	1.23	7.09	8.55
0350	Roadway or large area		6000	.011		4.39	.44	.37	5.20	5.95
0400	Slurry seal, thermoplastic coal-tar, type I, small or irregular area		2400	.027		3.76	1.10	.92	5.78	6.95
0450	Roadway or large area		8000	.008		3.76	.33	.28	4.37	4.97
0500	Type II, small or irregular area		2400	.027		4.80	1.10	.92	6.82	8.10
0550	Roadway or large area	↓	7800	.008	↓	4.80	.34	.28	5.42	6.15

32 01 16 – Flexible Paving Rehabilitation

32 01 16.71 Cold Milling Asphalt Paving

		Crew	Daily Output	Labor-Hours	Unit	Material	Labor	Equipment	Total	Total Incl O&P
0010	**COLD MILLING ASPHALT PAVING**									
5200	Cold planing & cleaning, 1" to 3" asphalt pavmt., over 25,000 S.Y.	B-71	6000	.009	S.Y.		.41	1.14	1.55	1.90
5280	5,000 S.Y. to 10,000 S.Y.	"	4000	.014	"		.61	1.71	2.32	2.86
5300	Asphalt pavement removal from conc. base, no haul									
5320	Rip, load & sweep 1" to 3"	B-70	8000	.007	S.Y.		.30	.23	.53	.75
5330	3" to 6" deep	"	5000	.011			.49	.37	.86	1.20
5340	Profile grooving, asphalt pavement load & sweep, 1" deep	B-71	12500	.004			.19	.55	.74	.91
5350	3" deep		9000	.006			.27	.76	1.03	1.28
5360	6" deep	↓	5000	.011	↓		.49	1.37	1.86	2.30

32 01 16.73 In Place Cold Reused Asphalt Paving

		Crew	Daily Output	Labor-Hours	Unit	Material	Labor	Equipment	Total	Total Incl O&P
0010	**IN PLACE COLD REUSED ASPHALT PAVING**									
5000	Reclamation, pulverizing and blending with existing base									
5040	Aggregate base, 4" thick pavement, over 15,000 S.Y.	B-73	2400	.027	S.Y.		1.23	2.18	3.41	4.36
5080	5,000 S.Y. to 15,000 S.Y.		2200	.029			1.34	2.37	3.71	4.75
5120	8" thick pavement, over 15,000 S.Y.		2200	.029			1.34	2.37	3.71	4.75
5160	5,000 S.Y. to 15,000 S.Y.	↓	2000	.032	↓		1.47	2.61	4.08	5.25

32 01 16.74 In Place Hot Reused Asphalt Paving

		Crew	Daily Output	Labor-Hours	Unit	Material	Labor	Equipment	Total	Total Incl O&P
0010	**IN PLACE HOT REUSED ASPHALT PAVING**									
5500	Recycle asphalt pavement at site									
5520	Remove, rejuvenate and spread 4" deep G	B-72	2500	.026	S.Y.	4.73	1.14	4.56	10.43	12.05
5521	6" deep G	"	2000	.032	"	6.95	1.42	5.70	14.07	16.15

32 01 17 – Flexible Paving Repair

32 01 17.10 Repair of Asphalt Pavement Holes

		Crew	Daily Output	Labor-Hours	Unit	Material	Labor	Equipment	Total	Total Incl O&P
0010	**REPAIR OF ASPHALT PAVEMENT HOLES** (cold patch)									
0100	Flexible pavement repair holes, roadway,light traffic, 1 C.F. size	B-37A	24	1	Ea.	9	38	15.75	62.75	89
0150	Group of two, 1 C.F. size each		16	1.500	Set	18	57	23.50	98.50	139
0200	Group of three, 1 C.F. size each	↓	12	2	"	27	76	31.50	134.50	188
0300	Medium traffic, 1 C.F. size each	B-37B	24	1.333	Ea.	9	50.50	15.70	75.20	110
0350	Group of two, 1 C.F. size each		16	2	Set	18	76	23.50	117.50	170
0400	Group of three, 1 C.F. size each	↓	12	2.667	"	27	101	31.50	159.50	229
0500	Highway/heavy traffic, 1 C.F. size each	B-37C	24	1.333	Ea.	9	51	26	86	121
0550	Group of two, 1 C.F. size each		16	2	Set	18	76.50	39	133.50	188
0600	Group of three, 1 C.F. size each	↓	12	2.667	"	27	102	52	181	253
0700	Add police officer and car for traffic control				Hr.	45.50			45.50	50
1000	Flexible pavement repair holes,parking lot, bag material, 1 C.F. size each	B-37D	18	.889	Ea.	64.50	34	8	106.50	135
1010	Economy bag material, 1 C.F. each	↓	18	.889	"	42	34	8	84	111

32 01 17.10 Repair of Asphalt Pavement Holes

	32 01 17.10 Repair of Asphalt Pavement Holes	Crew	Daily Output	Labor-Hours	Unit	2015 Bare Costs Material	Labor	Equipment	Total	Total Incl O&P
1100	Group of two, 1 C.F. size each	B-37D	12	1.333	Set	129	51	12	192	238
1110	Group of two, economy bag, 1 C.F. size each		12	1.333		84	51	12	147	189
1200	Group of three 1 C.F. size each		10	1.600		193	61.50	14.40	268.90	330
1210	Economy bag, group of three, 1 C.F. size each		10	1.600		126	61.50	14.40	201.90	255
1300	Flexible pavement repair holes, parking lot, 1 C.F. single hole	A-3A	4	2	Ea.	64.50	97	39	200.50	268
1310	Economy material, 1 C.F. single hole		4	2	"	42	97	39	178	244
1400	Flexible pavement repair holes, parking lot, 1 C.F. four holes		2	4	Set	258	194	78	530	680
1410	Economy material, 1 C.F. four holes		2	4	"	168	194	78	440	580
1500	Flexible pavement repair holes,large parking lot, bulk matl , 1 C.F. size	B-37A	32	.750	Ea.	9	28.50	11.80	49.30	69.50

32 01 17.20 Repair of Asphalt Pavement Patches

	32 01 17.20 Repair of Asphalt Pavement Patches	Crew	Daily Output	Labor-Hours	Unit	Material	Labor	Equipment	Total	Total Incl O&P
0010	**REPAIR OF ASPHALT PAVEMENT PATCHES**									
0100	Flexible pavement patches, roadway,light traffic,sawcut 10-25 S.F.	B-89	12	1.333	Ea.		58.50	40.50	99	138
0150	Sawcut 26-60 S.F.		10	1.600			70	49	119	166
0200	Sawcut 61-100 S.F.		8	2			87.50	61	148.50	207
0210	Sawcut groups of small size patches		24	.667			29	20.50	49.50	69
0220	Large size patches		16	1			44	30.50	74.50	104
0300	Flexible pavement patches, roadway,light traffic,digout 10-25 S.F.	B-6	16	1.500			62	23	85	125
0350	Digout 26-60 S.F.		12	2			82.50	30.50	113	168
0400	Digout 61-100 S.F.		8	3			124	45.50	169.50	250
0450	Add 8 C.Y. truck, small project debris haulaway	B-34A	8	1	Hr.		40	51.50	91.50	121
0460	Add 12 C.Y. truck, small project debris haulaway	B-34B	8	1			40	86.50	126.50	160
0480	Add flagger for non-intersection medium traffic	1 Clab	8	1			37.50		37.50	61.50
0490	Add flasher truck for intersection medium traffic or heavy traffic	A-2B	8	1			39	30.50	69.50	96.50
0500	Flexible pavement patches, roadway, repave,cold, 15 S.F., 4" D	B-37	8	6	Ea.	43	239	19.70	301.70	460
0510	6" depth		8	6		65	239	19.70	323.70	485
0520	Repave,cold, 20 S.F., 4" depth		8	6		57	239	19.70	315.70	475
0530	6" depth		8	6		86.50	239	19.70	345.20	505
0540	Repave,cold, 25 S.F., 4" depth		8	6		71.50	239	19.70	330.20	490
0550	6" depth		8	6		108	239	19.70	366.70	530
0600	Repave,cold, 30 S.F., 4" depth		8	6		85.50	239	19.70	344.20	505
0610	6" depth		8	6		130	239	19.70	388.70	555
0640	Repave,cold, 40 S.F., 4" depth		8	6		114	239	19.70	372.70	535
0650	6" depth		8	6		173	239	19.70	431.70	600
0680	Repave,cold, 50 S.F., 4" depth		8	6		143	239	19.70	401.70	570
0690	6" depth		8	6		216	239	19.70	474.70	650
0720	Repave,cold, 60 S.F., 4" depth		8	6		171	239	19.70	429.70	600
0730	6" depth		8	6		259	239	19.70	517.70	695
0800	Repave,cold, 70 S.F., 4" depth		7	6.857		199	273	22.50	494.50	690
0810	6" depth		7	6.857		300	273	22.50	595.50	800
0820	Repave,cold, 75 S.F., 4" depth		7	6.857		213	273	22.50	508.50	705
0830	6" depth		7	6.857		325	273	22.50	620.50	825
0900	Repave,cold, 80 S.F., 4" depth		6	8		228	320	26	574	800
0910	6" depth		6	8		345	320	26	691	930
0940	Repave,cold, 90 S.F., 4" depth		6	8		256	320	26	602	830
0950	6" depth		6	8		390	320	26	736	980
0980	Repave,cold, 100 S.F., 4" depth		6	8		285	320	26	631	865
0990	6" depth		6	8		430	320	26	776	1,025
1000	Add flasher truck for paving operations in medium or heavy traffic	A-2B	8	1	Hr.		39	30.50	69.50	96.50
1100	Prime coat for repair 15-40 S.F., 25% overspray	B-37A	48	.500	Ea.	7.25	19.05	7.85	34.15	47.50
1150	41-60 S.F.		40	.600		14.50	23	9.45	46.95	64
1175	61-80 S.F.		32	.750		19.65	28.50	11.80	59.95	81
1200	81-100 S.F.		32	.750		24	28.50	11.80	64.30	86

32 01 Operation and Maintenance of Exterior Improvements

32 01 17 – Flexible Paving Repair

32 01 17.20 Repair of Asphalt Pavement Patches		Crew	Daily Output	Labor-Hours	Unit	Material	2015 Bare Costs Labor	Equipment	Total	Total Incl O&P
1210	Groups of patches w/25% overspray	B-37A	3600	.007	S.F.	.24	.25	.10	.59	.80
1300	Flexible pavement repair patches,street repave,hot, 60 S.F., 4″ D	B-37	8	6	Ea.	101	239	19.70	359.70	525
1310	6″ depth		8	6		153	239	19.70	411.70	580
1320	Repave,hot, 70 S.F., 4″ depth		7	6.857		118	273	22.50	413.50	600
1330	6″ depth		7	6.857		179	273	22.50	474.50	665
1340	Repave,hot, 75 S.F., 4″ depth		7	6.857		126	273	22.50	421.50	610
1350	6″ depth		7	6.857		192	273	22.50	487.50	680
1360	Repave,hot, 80 S.F., 4″ depth		6	8		135	320	26	481	695
1370	6″ depth		6	8		205	320	26	551	775
1380	Repave,hot, 90 S.F., 4″ depth		6	8		152	320	26	498	715
1390	6″ depth		6	8		230	320	26	576	805
1400	Repave,hot, 100 S.F., 4″ depth		6	8		169	320	26	515	735
1410	6″ depth		6	8		256	320	26	602	830
1420	Pave hot groups of patches, 4″ depth		900	.053	S.F.	1.68	2.12	.17	3.97	5.50
1430	6″ depth		900	.053	″	2.56	2.12	.17	4.85	6.45
1500	Add 8 C.Y. truck for hot asphalt paving operations	B-34A	1	8	Day		320	410	730	965
1550	Add flasher truck for hot paving operations in med/heavy traffic	A-2B	8	1	Hr.		39	30.50	69.50	96.50
2000	Add police officer and car for traffic control				″	45.50			45.50	50

32 01 17.61 Sealing Cracks In Asphalt Paving		Crew	Daily Output	Labor-Hours	Unit	Material	Labor	Equipment	Total	Total Incl O&P
0010	**SEALING CRACKS IN ASPHALT PAVING**									
0100	Sealing cracks in asphalt paving, 1/8″ wide x 1/2″ depth, slow set	B-37A	2000	.012	L.F.	.17	.46	.19	.82	1.15
0110	1/4″ wide x 1/2″ depth		2000	.012		.20	.46	.19	.85	1.18
0130	3/8″ wide x 1/2″ depth		2000	.012		.23	.46	.19	.88	1.21
0140	1/2″ wide x 1/2″ depth		1800	.013		.26	.51	.21	.98	1.34
0150	3/4″ wide x 1/2″ depth		1600	.015		.31	.57	.24	1.12	1.53
0160	1″ wide x 1/2″ depth		1600	.015		.37	.57	.24	1.18	1.59
0165	1/8″ wide x 1″ depth, rapid set	B-37F	2000	.016		.25	.61	.29	1.15	1.58
0170	1/4″ wide x 1″ depth		2000	.016		.48	.61	.29	1.38	1.84
0175	3/8″ wide x 1″ depth		1800	.018		.72	.68	.32	1.72	2.25
0180	1/2″ wide x 1″ depth		1800	.018		.97	.68	.32	1.97	2.51
0185	5/8″ wide x 1″ depth		1800	.018		1.21	.68	.32	2.21	2.78
0190	3/4″ wide x 1″ depth		1600	.020		1.45	.76	.36	2.57	3.23
0195	1″ wide x 1″ depth		1600	.020		1.93	.76	.36	3.05	3.76
0200	1/8″ wide x 2″ depth, rapid set		2000	.016		.48	.61	.29	1.38	1.84
0210	1/4″ wide x 2″ depth		2000	.016		.97	.61	.29	1.87	2.37
0220	3/8″ wide x 2″ depth		1800	.018		1.45	.68	.32	2.45	3.04
0230	1/2″ wide x 2″ depth		1800	.018		1.93	.68	.32	2.93	3.57
0240	5/8″ wide x 2″ depth		1800	.018		2.41	.68	.32	3.41	4.10
0250	3/4″ wide x 2″ depth		1600	.020		2.89	.76	.36	4.01	4.82
0260	1″ wide x 2″ depth		1600	.020		3.86	.76	.36	4.98	5.90
0300	Add flagger for non-intersection medium traffic	1 Clab	8	1	Hr.		37.50		37.50	61.50
0400	Add flasher truck for intersection medium traffic or heavy traffic	A-2B	8	1	″		39	30.50	69.50	96.50

32 01 29 – Rigid Paving Repair

32 01 29.61 Partial Depth Patching of Rigid Pavement		Crew	Daily Output	Labor-Hours	Unit	Material	Labor	Equipment	Total	Total Incl O&P
0010	**PARTIAL DEPTH PATCHING OF RIGID PAVEMENT**									
0100	Rigid pavement repair, roadway,light traffic, 25% pitting 15 S.F.	B-37F	16	2	Ea.	27	76	36	139	194
0110	Pitting 20 S.F.		16	2		36	76	36	148	204
0120	Pitting 25 S.F.		16	2		45	76	36	157	214
0130	Pitting 30 S.F.		16	2		54	76	36	166	224
0140	Pitting 40 S.F.		16	2		72	76	36	184	244
0150	Pitting 50 S.F.		12	2.667		90	101	48.50	239.50	320
0160	Pitting 60 S.F.		12	2.667		108	101	48.50	257.50	335

32 01 29 – Rigid Paving Repair

32 01 29.61 Partial Depth Patching of Rigid Pavement		Crew	Daily Output	Labor-Hours	Unit	Material	2015 Bare Costs Labor	Equipment	Total	Total Incl O&P
0170	Pitting 70 S.F.	B-37F	12	2.667	Ea.	126	101	48.50	275.50	355
0180	Pitting 75 S.F.		12	2.667		135	101	48.50	284.50	365
0190	Pitting 80 S.F.		8	4		144	152	72.50	368.50	485
0200	Pitting 90 S.F.		8	4		162	152	72.50	386.50	505
0210	Pitting 100 S.F.		8	4		180	152	72.50	404.50	525
0300	Roadway,light traffic, 50% pitting 15 S.F.		12	2.667		54	101	48.50	203.50	278
0310	Pitting 20 S.F.		12	2.667		72	101	48.50	221.50	298
0320	Pitting 25 S.F.		12	2.667		90	101	48.50	239.50	320
0330	Pitting 30 S.F.		12	2.667		108	101	48.50	257.50	335
0340	Pitting 40 S.F.		12	2.667		144	101	48.50	293.50	375
0350	Pitting 50 S.F.		8	4		180	152	72.50	404.50	525
0360	Pitting 60 S.F.		8	4		217	152	72.50	441.50	565
0370	Pitting 70 S.F.		8	4		253	152	72.50	477.50	605
0380	Pitting 75 S.F.		8	4		271	152	72.50	495.50	625
0390	Pitting 80 S.F.		6	5.333		289	203	96.50	588.50	755
0400	Pitting 90 S.F.		6	5.333		325	203	96.50	624.50	790
0410	Pitting 100 S.F.		6	5.333		360	203	96.50	659.50	830
1000	Rigid pavement repair,light traffic,surface patch, 2" deep, 15 S.F.		8	4		108	152	72.50	332.50	445
1010	Surface patch, 2" deep, 20 S.F.		8	4		144	152	72.50	368.50	485
1020	Surface patch, 2" deep, 25 S.F.		8	4		180	152	72.50	404.50	525
1030	Surface patch, 2" deep, 30 S.F.		8	4		217	152	72.50	441.50	565
1040	Surface patch, 2" deep, 40 S.F.		8	4		289	152	72.50	513.50	650
1050	Surface patch, 2" deep, 50 S.F.		6	5.333		360	203	96.50	659.50	830
1060	Surface patch, 2" deep, 60 S.F.		6	5.333		435	203	96.50	734.50	910
1070	Surface patch, 2" deep, 70 S.F.		6	5.333		505	203	96.50	804.50	990
1080	Surface patch, 2" deep, 75 S.F.		6	5.333		540	203	96.50	839.50	1,025
1090	Surface patch, 2" deep, 80 S.F.		5	6.400		580	243	116	939	1,150
1100	Surface patch, 2" deep, 90 S.F.		5	6.400		650	243	116	1,009	1,250
1200	Surface patch, 2" deep, 100 S.F.		5	6.400		720	243	116	1,079	1,325
2000	Add flagger for non-intersection medium traffic	1 Clab	8	1	Hr.		37.50		37.50	61.50
2100	Add flasher truck for intersection medium traffic or heavy traffic	A-2B	8	1			39	30.50	69.50	96.50
2200	Add police officer and car for traffic control					45.50			45.50	50

32 01 29.70 Full Depth Patching of Rigid Pavement		Crew	Daily Output	Labor-Hours	Unit	Material	Labor	Equipment	Total	Total Incl O&P
0010	**FULL DEPTH PATCHING OF RIGID PAVEMENT**									
0015	Pavement preparation includes sawcut, remove pavement and replace									
0020	6 inches of subbase and one layer of reinforcement									
0030	Trucking and haulaway of debris excluded									
0100	Rigid pavement replace, light traffic,prep, 15 S.F.,6" depth	B-37E	16	3.500	Ea.	19.65	145	64	228.65	325
0110	Replacement preparation 20 S.F., 6" depth		16	3.500		26	145	64	235	335
0115	25 S.F., 6" depth		16	3.500		32.50	145	64	241.50	340
0120	30 S.F., 6" depth		14	4		39.50	166	73	278.50	390
0125	35 S.F., 6" depth		14	4		46	166	73	285	400
0130	40 S.F., 6" depth		14	4		52.50	166	73	291.50	405
0135	45 S.F., 6" depth		14	4		59	166	73	298	415
0140	50 S.F., 6" depth		12	4.667		65.50	193	85	343.50	475
0145	55 S.F., 6" depth		12	4.667		72	193	85	350	485
0150	60 S.F., 6" depth		10	5.600		78.50	232	102	412.50	575
0155	65 S.F., 6" depth		10	5.600		85	232	102	419	580
0160	70 S.F., 6" depth		10	5.600		91.50	232	102	425.50	590
0165	75 S.F., 6" depth		10	5.600		98	232	102	432	595
0170	80 S.F., 6" depth		8	7		105	290	128	523	725
0175	85 S.F., 6" depth		8	7		111	290	128	529	735

32 01 29.70 Full Depth Patching of Rigid Pavement		Crew	Daily Output	Labor-Hours	Unit	Material	2015 Bare Costs Labor	Equipment	Total	Total Incl O&P
0180	90 S.F., 6" depth	B-37E	8	7	Ea.	118	290	128	536	740
0185	95 S.F., 6" depth		8	7		124	290	128	542	750
0190	100 S.F., 6" depth		8	7		131	290	128	549	755
0200	Pavement preparation for 8 inch rigid paving same as 6 inch									
0290	Pavement preparation for 9 inch rigid paving same as 10 inch									
0300	Rigid pavement replace, light traffic,prep, 15 S.F.,10" depth	B-37E	16	3.500	Ea.	32.50	145	64	241.50	340
0302	Pavement preparation 10 inch includes sawcut, remove pavement and									
0304	replace 6 inches of subbase and two layers of reinforcement									
0306	Trucking and haulaway of debris excluded									
0310	Replacement preparation 20 S.F., 10" depth	B-37E	16	3.500	Ea.	43.50	145	64	252.50	355
0315	25 S.F., 10" depth		16	3.500		54.50	145	64	263.50	365
0320	30 S.F., 10" depth		14	4		65.50	166	73	304.50	420
0325	35 S.F., 10" depth		14	4		50	166	73	289	405
0330	40 S.F., 10" depth		14	4		87	166	73	326	445
0335	45 S.F., 10" depth		12	4.667		98	193	85	376	510
0340	50 S.F., 10" depth		12	4.667		109	193	85	387	525
0345	55 S.F., 10" depth		12	4.667		120	193	85	398	535
0350	60 S.F., 10" depth		10	5.600		131	232	102	465	630
0355	65 S.F., 10" depth		10	5.600		142	232	102	476	645
0360	70 S.F., 10" depth		10	5.600		153	232	102	487	655
0365	75 S.F., 10" depth		8	7		163	290	128	581	790
0370	80 S.F., 10" depth		8	7		174	290	128	592	805
0375	85 S.F., 10" depth		8	7		185	290	128	603	815
0380	90 S.F., 10" depth		8	7		196	290	128	614	825
0385	95 S.F., 10" depth		8	7		207	290	128	625	840
0390	100 S.F., 10" depth		8	7		218	290	128	636	850
0500	Add 8 C.Y. truck, small project debris haulaway	B-34A	8	1	Hr.		40	51.50	91.50	121
0550	Add 12 C.Y. truck, small project debris haulaway	B-34B	8	1			40	86.50	126.50	160
0600	Add flagger for non-intersection medium traffic	1 Clab	8	1			37.50		37.50	61.50
0650	Add flasher truck for intersection medium traffic or heavy traffic	A-2B	8	1			39	30.50	69.50	96.50
0700	Add police officer and car for traffic control					45.50			45.50	50
0990	Concrete will replaced using quick set mix with aggregate									
1000	Rigid pavement replace, light traffic,repour, 15 S.F.,6" depth	B-37F	18	1.778	Ea.	161	67.50	32	260.50	325
1010	20 S.F., 6" depth		18	1.778		214	67.50	32	313.50	380
1015	25 S.F., 6" depth		18	1.778		268	67.50	32	367.50	440
1020	30 S.F., 6" depth		16	2		320	76	36	432	520
1025	35 S.F., 6" depth		16	2		375	76	36	487	580
1030	40 S.F., 6" depth		16	2		430	76	36	542	635
1035	45 S.F., 6" depth		16	2		480	76	36	592	695
1040	50 S.F., 6" depth		16	2		535	76	36	647	755
1045	55 S.F., 6" depth		12	2.667		590	101	48.50	739.50	870
1050	60 S.F., 6" depth		12	2.667		645	101	48.50	794.50	925
1055	65 S.F., 6" depth		12	2.667		695	101	48.50	844.50	985
1060	70 S.F., 6" depth		12	2.667		750	101	48.50	899.50	1,050
1065	75 S.F., 6" depth		10	3.200		805	122	58	985	1,150
1070	80 S.F., 6" depth		10	3.200		855	122	58	1,035	1,200
1075	85 S.F., 6" depth		8	4		910	152	72.50	1,134.50	1,325
1080	90 S.F., 6" depth		8	4		965	152	72.50	1,189.50	1,375
1085	95 S.F., 6" depth		8	4		1,025	152	72.50	1,249.50	1,450
1090	100 S.F., 6" depth		8	4		1,075	152	72.50	1,299.50	1,500
1099	Concrete will be replaced using 4500 PSI Concrete ready mix									
1100	Rigid pavement replace, light traffic,repour, 15 S.F.,6" depth	A-2	12	2	Ea.	278	76	20.50	374.50	450
1110	20-50 S.F., 6" depth		12	2		278	76	20.50	374.50	450

32 01 29 – Rigid Paving Repair

32 01 29.70 Full Depth Patching of Rigid Pavement		Crew	Daily Output	Labor-Hours	Unit	Material	2015 Bare Costs Labor	Equipment	Total	Total Incl O&P
1120	55-65 S.F., 6″ depth	A-2	10	2.400	Ea.	310	91.50	24.50	426	515
1130	70-80 S.F., 6″ depth		10	2.400		340	91.50	24.50	456	545
1140	85-100 S.F., 6″ depth		8	3		395	114	30.50	539.50	655
1200	Repour 15-40 S.F.,8″ depth		12	2		278	76	20.50	374.50	450
1210	45-50 S.F.,8″ depth		12	2		310	76	20.50	406.50	485
1220	55-60 S.F.,8″ depth		10	2.400		340	91.50	24.50	456	545
1230	65-80 S.F.,8″ depth		10	2.400		395	91.50	24.50	511	610
1240	85-90 S.F.,8″ depth		8	3		430	114	30.50	574.50	690
1250	95-100 S.F.,8″ depth		8	3		455	114	30.50	599.50	720
1300	Repour 15-30 S.F.,10″ depth		12	2		278	76	20.50	374.50	450
1310	35-40 S.F.,10″ depth		12	2		310	76	20.50	406.50	485
1320	45 S.F.,10″ depth		12	2		340	76	20.50	436.50	515
1330	50-65 S.F.,10″ depth		10	2.400		395	91.50	24.50	511	610
1340	70 S.F.,10″ depth		10	2.400		430	91.50	24.50	546	645
1350	75-80 S.F.,10″ depth		10	2.400		455	91.50	24.50	571	675
1360	85-95 S.F.,10″ depth		8	3		515	114	30.50	659.50	790
1370	100 S.F.,10″ depth		8	3		550	114	30.50	694.50	825

32 01 30 – Operation and Maintenance of Site Improvements

32 01 30.10 Site Maintenance		Crew	Daily Output	Labor-Hours	Unit	Material	Labor	Equipment	Total	Total Incl O&P
0010	**SITE MAINTENANCE** R019313-10									
0800	Flower bed maintenance									
0810	Cultivate bed, no mulch	1 Clab	14	.571	M.S.F.		21.50		21.50	35
0830	Fall clean-up of flower bed, including pick-up mulch for re-use		1	8			300		300	495
0840	Fertilize flower bed, dry granular 3 lb./C.S.F.		85	.094		8.75	3.54		12.29	15.45
1000	Mulch - see Section 32 91 13.16									
1100	Plant bed preparation see Section 32 91 13.16									
1110	Plant from flats	1 Clab	800	.010	S.F.		.38		.38	.62
1120	Planting, ground cover see Section 32 93 13.20									
1130	Police, hand pickup	1 Clab	30	.267	M.S.F.		10.05		10.05	16.45
1140	Vacuum (outside)		48	.167			6.25		6.25	10.30
1200	Spring prepare		2	4			150		150	247
1300	Weed mulched bed		20	.400			15.05		15.05	24.50
1310	Unmulched bed		8	1			37.50		37.50	61.50
1550	General site work maintenance									
1560	Clearing brush with brush saw & rake	1 Clab	565	.014	S.Y.		.53		.53	.87
1570	By hand	″	280	.029			1.07		1.07	1.76
1580	With dozer, ball and chain, light clearing	B-11A	3675	.004			.19	.38	.57	.72
1590	Medium clearing	″	3110	.005			.23	.45	.68	.85
1600	Ground cover planting see Section 32 93 13.20									
1610	Grounds, policing, incl. picking up trash & other debris	1 Clab	300	.027	M.S.F.		1		1	1.64
3000	Lawn maintenance									
3010	Aerate lawn, 18″ cultivating width, walk behind	A-1K	95	.084	M.S.F.		3.17	.76	3.93	6.05
3040	48″ cultivating width	B-66	750	.011			.52	.35	.87	1.21
3060	72″ cultivating width	″	1100	.007			.35	.24	.59	.82
3100	Edge lawn, by hand at walks	1 Clab	16	.500	C.L.F.		18.80		18.80	31
3150	At planting beds		7	1.143			43		43	70.50
3200	Using gas powered edger at walks		88	.091			3.42		3.42	5.60
3250	At planting beds		24	.333			12.55		12.55	20.50
3260	Vacuum, 30″ gas, outdoors with hose		96	.083	M.L.F.		3.13		3.13	5.15
3350	Edging material for lawns & planting beds see Section 32 94 13.20									
3400	Weed lawn, by hand	1 Clab	3	2.667	M.S.F.		100		100	164
3800	Lawn bed preparation see Section 32 91 13.23									

32 01 30.10 Site Maintenance		Crew	Daily Output	Labor-Hours	Unit	Material	Labor	Equipment	Total	Total Incl O&P
						2015 Bare Costs				
4500	Rake leaves or lawn, by hand	1 Clab	7.50	1.067	M.S.F.		40		40	66
4510	Power rake	″	45	.178	″		6.70		6.70	10.95
4700	Seeding lawn, see Section 32 92 19.14									
4750	Sodding, see Section 32 92 23.10									
5900	Road & walk maintenance									
5910	Asphaltic concrete paving, cold patch, 2″ thick	B-37	350	.137	S.Y.	11.35	5.45	.45	17.25	22
5913	3″ thick	″	260	.185	″	17.05	7.35	.61	25.01	31.50
5915	De-icing roads and walks									
5920	Calcium Chloride in truckload lots see Section 31 32 13.30									
6000	Ice melting comp., 90% Calc. Chlor., effec. to -30°F									
6010	50-80 lb. poly bags, med. applic. 19 lb./M.S.F., by hand	1 Clab	60	.133	M.S.F.	21.50	5		26.50	31.50
6050	With hand operated rotary spreader		110	.073		21	2.73		23.73	27.50
6100	Rock salt, med. applic. on road & walkway, by hand		60	.133		10	5		15	19.20
6110	With hand operated rotary spreader		110	.073		10	2.73		12.73	15.50
6130	Hosing, sidewalks & other paved areas		30	.267			10.05		10.05	16.45
6150	Painting lines on pavement see Section 32 17 23.13									
6250	Seal coating bituminous surfaces see Section 32 01 13.61									
6260	Sidewalk, brick pavers, steam cleaning	A-1H	950	.008	S.F.	.12	.32	.08	.52	.74
6400	Sweep walk by hand	1 Clab	15	.533	M.S.F.		20		20	33
6410	Power vacuum		100	.080			3.01		3.01	4.93
6420	Drives & parking areas with power vacuum		120	.067			2.51		2.51	4.11
6600	Shrub maintenance									
6640	Shrub bed fertilize dry granular 3 lb./M.S.F.	1 Clab	85	.094	M.S.F.	6.60	3.54		10.14	13.10
6800	Weed, by handhoe		8	1			37.50		37.50	61.50
6810	Spray out		32	.250			9.40		9.40	15.40
6820	Spray after mulch		48	.167			6.25		6.25	10.30
7100	Tree maintenance									
7140	Clear and grub trees, see Section 31 11 10.10									
7160	Cutting and piling trees, see Section 31 13 13.20									
7200	Fertilize, tablets, slow release, 30 gram/tree	1 Clab	100	.080	Ea.	.58	3.01		3.59	5.55
7280	Guying, including stakes, guy wire & wrap, see Section 32 94 50.10									
7300	Planting, trees, Deciduous, in prep. beds, see Section 32 93 43.20									
7400	Removal, trees see Section 32 96 43.20									
7420	Pest control, spray	1 Clab	24	.333	Ea.	26.50	12.55		39.05	49.50
7430	Systemic	″	48	.167	″	26.50	6.25		32.75	40

32 01 30.20 Snow Removal

		Crew	Daily Output	Labor-Hours	Unit	Material	Labor	Equipment	Total	Total Incl O&P
0010	**SNOW REMOVAL**									
0020	Plowing, 12 ton truck, 2″-4″ deep	B-34A	250	.032	M.S.F.		1.28	1.64	2.92	3.88
0040	4″-10″ deep		200	.040			1.60	2.06	3.66	4.84
0060	10″-15″ deep		150	.053			2.14	2.74	4.88	6.45
0080	Pickup truck, 2″-4″ deep	A-3A	175	.046			2.22	.89	3.11	4.51
0100	4″-10″ deep		130	.062			2.99	1.20	4.19	6.05
0120	10″-15″ deep		75	.107			5.20	2.07	7.27	10.55
0140	Load and haul snow, 1 mile round trip	A-3B	230	.070	C.Y.		3.15	5.25	8.40	10.85
0160	2 mile round trip		175	.091			4.14	6.90	11.04	14.20
0180	3 mile round trip		150	.107			4.84	8.05	12.89	16.65
0200	4 mile round trip		130	.123			5.60	9.30	14.90	19.15
0220	5 mile round trip		120	.133			6.05	10.10	16.15	21
0240	Clearing with wheeled skid steer loader, 1 C.Y.	A-3C	240	.033			1.62	1.33	2.95	4.03
0260	Spread sand and salt mix	B-34A	375	.021	M.S.F.	6.30	.85	1.10	8.25	9.55
0280	Sidewalks and drives, by hand	1 Clab	1200	.007	C.F.		.25		.25	.41
0300	Power, 24″ blower	A-1M	7000	.001	″		.04	.01	.05	.08

32 01 30 – Operation and Maintenance of Site Improvements

32 01 30.20 Snow Removal		Crew	Daily Output	Labor-Hours	Unit	Material	2015 Bare Costs Labor	Equipment	Total	Total Incl O&P
0320	2"-4" deep, single driveway (10' x 50')	A-1M	16	.500	Ea.		18.80	4.20	23	35.50
0340	Double driveway (20' x 50')		16	.500			18.80	4.20	23	35.50
0360	4"-10" deep, single driveway		16	.500			18.80	4.20	23	35.50
0380	Double driveway		16	.500			18.80	4.20	23	35.50
0400	10"-15" deep, single driveway		12	.667			25	5.60	30.60	47
0420	Double driveway		12	.667			25	5.60	30.60	47
0440	For heavy wet snow, add								20%	20%
9000	Minimum labor and equipment charge	A-3A	2	4	Job		194	78	272	395

32 01 90 – Operation and Maintenance of Planting

32 01 90.13 Fertilizing

		Crew	Daily Output	Labor-Hours	Unit	Material	Labor	Equipment	Total	Total Incl O&P
0010	**FERTILIZING**									
0100	Dry granular, 4#/M.S.F., hand spread	1 Clab	24	.333	M.S.F.	2.78	12.55		15.33	23.50
0110	Push rotary		140	.057	"	2.78	2.15		4.93	6.60
0112	Push rotary, per 1076 feet squared		130	.062	Ea.	2.78	2.31		5.09	6.85
0120	Tractor towed spreader, 8'	B-66	500	.016	M.S.F.	2.78	.78	.53	4.09	4.87
0130	12' spread		800	.010		2.78	.49	.33	3.60	4.19
0140	Truck whirlwind spreader		1200	.007		2.78	.32	.22	3.32	3.81
0180	Water soluable, hydro spread, 1.5#/M.S.F.	B-64	600	.027		2.85	1.02	.66	4.53	5.55
0190	Add for weed control					.48			.48	.53

32 01 90.19 Mowing

		Crew	Daily Output	Labor-Hours	Unit	Material	Labor	Equipment	Total	Total Incl O&P
0010	**MOWING**									
1650	Mowing brush, tractor with rotary mower									
1660	Light density	B-84	22	.364	M.S.F.		18.40	16.85	35.25	47.50
1670	Medium density		13	.615			31	28.50	59.50	81
1680	Heavy density		9	.889			45	41	86	117
2000	Mowing, brush/grass, tractor, rotary mower, highway/airport median		13	.615			31	28.50	59.50	81
2010	Traffic safety flashing truck for highway/airport median mowing	A-2B	1	8	Day		315	245	560	775
4050	Lawn mowing, power mower, 18" - 22"	1 Clab	65	.123	M.S.F.		4.63		4.63	7.60
4100	22" - 30"		110	.073			2.73		2.73	4.48
4150	30" - 32"		140	.057			2.15		2.15	3.52
4160	Riding mower, 36" - 44"	B-66	300	.027			1.30	.88	2.18	3.03
4170	48" - 58"	"	480	.017			.81	.55	1.36	1.89
4175	Mowing with tractor & attachments									
4180	3 gang reel, 7'	B-66	930	.009	M.S.F.		.42	.28	.70	.97
4190	5 gang reel, 12'		1200	.007			.32	.22	.54	.75
4200	Cutter or sickle-bar, 5', rough terrain		210	.038			1.85	1.25	3.10	4.32
4210	Cutter or sickle-bar, 5', smooth terrain		340	.024			1.14	.77	1.91	2.67
4220	Drainage channel, 5' sickle bar		5	1.600	Mile		78	52.50	130.50	181
4250	Lawnmower, rotary type, sharpen (all sizes)	1 Clab	10	.800	Ea.		30		30	49.50
4260	Repair or replace part		7	1.143	"		43		43	70.50
5000	Edge trimming with weed whacker		5760	.001	L.F.		.05		.05	.09

32 01 90.23 Pruning

		Crew	Daily Output	Labor-Hours	Unit	Material	Labor	Equipment	Total	Total Incl O&P
0010	**PRUNING**									
0020	1-1/2" caliper	1 Clab	84	.095	Ea.		3.58		3.58	5.85
0030	2" caliper		70	.114			4.30		4.30	7.05
0040	2-1/2" caliper		50	.160			6		6	9.85
0050	3" caliper		30	.267			10.05		10.05	16.45
0060	4" caliper, by hand	2 Clab	21	.762			28.50		28.50	47
0070	Aerial lift equipment	B-85	38	1.053			43	27	70	99.50
0100	6" caliper, by hand	2 Clab	12	1.333			50		50	82
0110	Aerial lift equipment	B-85	20	2			81.50	51.50	133	189
0200	9" caliper, by hand	2 Clab	7.50	2.133			80		80	132

32 01 Operation and Maintenance of Exterior Improvements

32 01 90 – Operation and Maintenance of Planting

32 01 90.23 Pruning		Crew	Daily Output	Labor-Hours	Unit	2015 Bare Costs Material	Labor	Equipment	Total	Total Incl O&P
0210	Aerial lift equipment	B-85	12.50	3.200	Ea.		130	82.50	212.50	300
0300	12" caliper, by hand	2 Clab	6.50	2.462			92.50		92.50	152
0310	Aerial lift equipment	B-85	10.80	3.704			151	95.50	246.50	350
0400	18" caliper by hand	2 Clab	5.60	2.857			107		107	176
0410	Aerial lift equipment	B-85	9.30	4.301			175	111	286	405
0500	24" caliper, by hand	2 Clab	4.60	3.478			131		131	214
0510	Aerial lift equipment	B-85	7.70	5.195			211	134	345	490
0600	30" caliper, by hand	2 Clab	3.70	4.324			163		163	267
0610	Aerial lift equipment	B-85	6.20	6.452			263	167	430	610
0700	36" caliper, by hand	2 Clab	2.70	5.926			223		223	365
0710	Aerial lift equipment	B-85	4.50	8.889			360	229	589	835
0800	48" caliper, by hand	2 Clab	1.70	9.412			355		355	580
0810	Aerial lift equipment	B-85	2.80	14.286			580	370	950	1,350

32 01 90.24 Shrub Pruning

		Crew	Daily Output	Labor-Hours	Unit	Material	Labor	Equipment	Total	Total Incl O&P
0010	**SHRUB PRUNING**									
6700	Prune, shrub bed	1 Clab	7	1.143	M.S.F.		43		43	70.50
6710	Shrub under 3' height		190	.042	Ea.		1.58		1.58	2.60
6720	4' height		90	.089			3.34		3.34	5.50
6730	Over 6'		50	.160			6		6	9.85
7350	Prune trees from ground		20	.400			15.05		15.05	24.50
7360	High work		8	1			37.50		37.50	61.50

32 01 90.26 Watering

		Crew	Daily Output	Labor-Hours	Unit	Material	Labor	Equipment	Total	Total Incl O&P
0010	**WATERING**									
4900	Water lawn or planting bed with hose, 1" of water	1 Clab	16	.500	M.S.F.		18.80		18.80	31
4910	50' soaker hoses, in place		82	.098			3.67		3.67	6
4920	60' soaker hoses, in place		89	.090			3.38		3.38	5.55
7500	Water trees or shrubs, under 1" caliper		32	.250	Ea.		9.40		9.40	15.40
7550	1" - 3" caliper		17	.471			17.70		17.70	29
7600	3" - 4" caliper		12	.667			25		25	41
7650	Over 4" caliper		10	.800			30		30	49.50
9000	For sprinkler irrigation systems, see Section 32 84 23.10									

32 01 90.29 Topsoil Preservation

		Crew	Daily Output	Labor-Hours	Unit	Material	Labor	Equipment	Total	Total Incl O&P
0010	**TOPSOIL PRESERVATION**									
0100	Weed planting bed	1 Clab	800	.010	S.Y.		.38		.38	.62

32 06 Schedules for Exterior Improvements

32 06 10 – Schedules for Bases, Ballasts, and Paving

32 06 10.10 Sidewalks, Driveways and Patios

		Crew	Daily Output	Labor-Hours	Unit	Material	Labor	Equipment	Total	Total Incl O&P
0010	**SIDEWALKS, DRIVEWAYS AND PATIOS** No base									
0020	Asphaltic concrete, 2" thick	B-37	720	.067	S.Y.	7.40	2.65	.22	10.27	12.70
0100	2-1/2" thick	"	660	.073	"	9.40	2.89	.24	12.53	15.30
0110	Bedding for brick or stone, mortar, 1" thick	D-1	300	.053	S.F.	.89	2.25		3.14	4.62
0120	2" thick	"	200	.080		2.21	3.37		5.58	7.95
0130	Sand, 2" thick	B-18	8000	.003		.28	.11	.01	.40	.51
0140	4" thick	"	4000	.006		.56	.23	.01	.80	1.01
0300	Concrete, 3000 psi, CIP, 6 x 6 - W1.4 x W1.4 mesh,									
0310	broomed finish, no base, 4" thick	B-24	600	.040	S.F.	1.69	1.73		3.42	4.66
0350	5" thick		545	.044		2.26	1.90		4.16	5.55
0400	6" thick		510	.047		2.64	2.03		4.67	6.20
0450	For bank run gravel base, 4" thick, add	B-18	2500	.010		.52	.37	.02	.91	1.19

32 06 10 – Schedules for Bases, Ballasts, and Paving

32 06 10.10 Sidewalks, Driveways and Patios		Crew	Daily Output	Labor-Hours	Unit	2015 Bare Costs Material	Labor	Equipment	Total	Total Incl O&P
0520	8" thick, add	B-18	1600	.015	S.F.	1.04	.57	.03	1.64	2.12
0550	Exposed aggregate finish, add to above, minimum	B-24	1875	.013		.11	.55		.66	1.02
0600	Maximum	"	455	.053		.36	2.28		2.64	4.09
0850	Splash block, precast concrete	1 Clab	150	.053	Ea.	11.65	2.01		13.66	16.10
0950	Concrete tree grate, 5' square	B-6	25	.960		420	39.50	14.55	474.05	545
0955	Tree well & cover, concrete, 3' square		25	.960		320	39.50	14.55	374.05	430
0960	Cast iron tree grate with frame, 2 piece, round, 5' diameter		25	.960		1,100	39.50	14.55	1,154.05	1,275
0980	Square, 5' side		25	.960		1,100	39.50	14.55	1,154.05	1,275
1000	Crushed stone, 1" thick, white marble	2 Clab	1700	.009	S.F.	.44	.35		.79	1.07
1050	Bluestone		1700	.009		.13	.35		.48	.72
1070	Granite chips		1700	.009		.22	.35		.57	.83
1660	Limestone pavers, 3" thick	D-1	72	.222		10.05	9.35		19.40	26.50
1670	4" thick		70	.229		13.30	9.60		22.90	30.50
1680	5" thick		68	.235		16.65	9.90		26.55	34.50
1700	Redwood, prefabricated, 4' x 4' sections	2 Carp	316	.051		4.77	2.38		7.15	9.15
1750	Redwood planks, 1" thick, on sleepers	"	240	.067		4.77	3.13		7.90	10.40
1830	1-1/2" thick	B-28	167	.144		7.65	6.30		13.95	18.75
1840	2" thick		167	.144		9.85	6.30		16.15	21
1850	3" thick		150	.160		14.50	7		21.50	27.50
1860	4" thick		150	.160		19.05	7		26.05	32.50
1870	5" thick		150	.160		24	7		31	38
2100	River or beach stone, stock	B-1	18	1.333	Ton	30	51		81	117
2150	Quarried	"	18	1.333	"	52	51		103	141
2160	Load, dump, and spread stone with skid steer, 100' haul	B-62	24	1	C.Y.		41.50	7.25	48.75	75
2165	200' haul		18	1.333			55	9.65	64.65	99.50
2168	300' haul		12	2			82.50	14.45	96.95	150
2170	Shale paver, 2-1/4" thick	D-1	200	.080	S.F.	3.13	3.37		6.50	8.95
2200	Coarse washed sand bed, 1"	B-62	1350	.018	S.Y.	1.77	.73	.13	2.63	3.28
2250	Stone dust, 4" thick	"	900	.027	"	3.33	1.10	.19	4.62	5.65
2300	Tile thinset pavers, 3/8" thick	D-1	300	.053	S.F.	3.53	2.25		5.78	7.55
2350	3/4" thick	"	280	.057	"	5.65	2.41		8.06	10.10
2400	Wood rounds, cypress	B-1	175	.137	Ea.	9.95	5.25		15.20	19.55
9000	Minimum labor/equipment charge	D-1	2	8	Job		335		335	550

32 06 10.20 Steps		Crew	Daily Output	Labor-Hours	Unit	Material	Labor	Equipment	Total	Total Incl O&P
0010	**STEPS**									
0011	Incl. excav., borrow & concrete base as required									
0100	Brick steps	B-24	35	.686	LF Riser	15.35	29.50		44.85	65
0200	Railroad ties	2 Clab	25	.640		3.59	24		27.59	43.50
0300	Bluestone treads, 12" x 2" or 12" x 1-1/2"	B-24	30	.800		33.50	34.50		68	92.50
0490	Minimum labor/equipment charge	D-1	2	8	Job		335		335	550
0600	Precast concrete, see Section 03 41 23.50									
4000	Edging, redwood, 2" x 4"	2 Carp	330	.048	L.F.	2.43	2.28		4.71	6.40
4025	Steel edge strips, incl. stakes, 1/4" x 5"	B-1	390	.062		3.92	2.36		6.28	8.15
4050	Edging, landscape timber or railroad ties, 6" x 8"	2 Carp	170	.094		2.79	4.42		7.21	10.30

32 11 Base Courses

32 11 23 – Aggregate Base Courses

32 11 23.23 Base Course Drainage Layers		Crew	Daily Output	Labor-Hours	Unit	2015 Bare Costs Material	Labor	Equipment	Total	Total Incl O&P
0010	**BASE COURSE DRAINAGE LAYERS**									
0011	For roadways and large areas									
0050	Crushed 3/4" stone base, compacted, 3" deep	B-36C	5200	.008	S.Y.	2.30	.36	.81	3.47	3.99
0100	6" deep		5000	.008		4.61	.37	.84	5.82	6.55
0200	9" deep		4600	.009		6.90	.40	.91	8.21	9.25
0300	12" deep		4200	.010		9.20	.44	1	10.64	11.95
0301	Crushed 1-1/2" stone base, compacted to 4" deep	B-36B	6000	.011		4.46	.48	.80	5.74	6.55
0302	6" deep		5400	.012		6.70	.53	.89	8.12	9.20
0303	8" deep		4500	.014		8.90	.64	1.07	10.61	12
0304	12" deep		3800	.017		13.35	.75	1.27	15.37	17.30
0310	Minimum labor/equipment charge	B-36	1	40	Job		1,725	1,675	3,400	4,625
0350	Bank run gravel, spread and compacted									
0370	6" deep	B-32	6000	.005	S.Y.	4.33	.25	.39	4.97	5.60
0390	9" deep		4900	.007		6.50	.31	.48	7.29	8.15
0400	12" deep		4200	.008		8.65	.36	.56	9.57	10.70
1500	Alternate method to figure base course									
1510	Crushed stone, 3/4", compacted, 3" deep	B-36C	435	.092	E.C.Y.	23.50	4.26	9.65	37.41	43.50
1511	6" deep	B-36B	835	.077		23.50	3.42	5.75	32.67	38
1512	9" deep		1150	.056		23.50	2.49	4.19	30.18	34.50
1513	12" deep		1400	.046		23.50	2.04	3.44	28.98	33
1520	Crushed stone, 1-1/2", compacted 4" deep		665	.096		23.50	4.30	7.25	35.05	41
1521	6" deep		900	.071		23.50	3.18	5.35	32.03	37
1522	8" deep		1000	.064		23.50	2.86	4.82	31.18	36
1523	12" deep		1265	.051		23.50	2.26	3.81	29.57	34
1530	Gravel, bank run, compacted, 6" deep	B-36C	835	.048		22	2.22	5	29.22	33.50
1531	9" deep		1150	.035		22	1.61	3.64	27.25	31
1532	12" deep		1400	.029		22	1.32	2.99	26.31	30
2010	Crushed stone, 3/4" maximum size, 3" deep	B-36	540	.074	Ton	14.30	3.20	3.08	20.58	24
2011	6" deep		1625	.025		14.30	1.06	1.02	16.38	18.55
2012	9" deep		1785	.022		14.30	.97	.93	16.20	18.30
2013	12" deep		1950	.021		14.30	.89	.85	16.04	18.05
2020	Crushed stone, 1-1/2" maximum size, 4" deep		720	.056		14.30	2.40	2.31	19.01	22
2021	6" deep		815	.049		14.30	2.12	2.04	18.46	21.50
2022	8" deep		835	.048		14.30	2.07	1.99	18.36	21
2023	12" deep		975	.041		14.30	1.77	1.71	17.78	20.50
2030	Bank run gravel, 6" deep	B-32A	875	.027		14.95	1.27	1.63	17.85	20.50
2031	9" deep		970	.025		14.95	1.14	1.47	17.56	19.90
2032	12" deep		1060	.023		14.95	1.05	1.34	17.34	19.60
6000	Stabilization fabric, polypropylene, 6 oz./S.Y.	B-6	10000	.002	S.Y.	1.28	.10	.04	1.42	1.61
6900	For small and irregular areas, add						50%	50%		
6990	Minimum labor/equipment charge	1 Clab	3	2.667	Job		100		100	164
7000	Prepare and roll sub-base, small areas to 2500 S.Y.	B-32A	1500	.016	S.Y.		.74	.95	1.69	2.23
8000	Large areas over 2500 S.Y.	"	3500	.007			.32	.41	.73	.96
8050	For roadways	B-32	4000	.008			.38	.59	.97	1.26
9000	Minimum labor/equipment charge	1 Clab	4	2	Job		75		75	123

32 11 26 – Asphaltic Base Courses

32 11 26.13 Plant Mix Asphaltic Base Courses		Crew	Daily Output	Labor-Hours	Unit	Material	Labor	Equipment	Total	Total Incl O&P
0010	**PLANT MIX ASPHALTIC BASE COURSES**									
0011	Roadways and large paved areas									
0500	Bituminous concrete, 4" thick	B-25	4545	.019	S.Y.	15.75	.80	.60	17.15	19.30
0550	6" thick		3700	.024		23	.98	.73	24.71	28
0560	8" thick		3000	.029		31	1.21	.90	33.11	37

32 11 Base Courses

32 11 26 – Asphaltic Base Courses

32 11 26.13 Plant Mix Asphaltic Base Courses		Crew	Daily Output	Labor-Hours	Unit	Material	2015 Bare Costs Labor	Equipment	Total	Total Incl O&P
0570	10″ thick	B-25	2545	.035	S.Y.	38	1.43	1.07	40.50	45.50
2000	Alternate method to figure base course									
2005	Bituminous concrete, 4″ thick	B-25	1000	.088	Ton	70	3.64	2.71	76.35	86
2006	6″ thick		1220	.072		70	2.98	2.22	75.20	84.50
2007	8″ thick		1320	.067		70	2.76	2.06	74.82	83.50
2008	10″ thick		1400	.063		70	2.60	1.94	74.54	83.50
8900	For small and irregular areas, add						50%	50%		

32 11 26.19 Bituminous-Stabilized Base Courses		Crew	Daily Output	Labor-Hours	Unit	Material	Labor	Equipment	Total	Total Incl O&P
0010	**BITUMINOUS-STABILIZED BASE COURSES**									
0020	And large paved areas									
0700	Liquid application to gravel base, asphalt emulsion	B-45	6000	.003	Gal.	4.29	.12	.15	4.56	5.10
0800	Prime and seal, cut back asphalt		6000	.003	″	5.05	.12	.15	5.32	5.90
1000	Macadam penetration crushed stone, 2 gal. per S.Y., 4″ thick		6000	.003	S.Y.	8.60	.12	.15	8.87	9.80
1100	6″ thick, 3 gal. per S.Y.		4000	.004		12.85	.18	.23	13.26	14.70
1200	8″ thick, 4 gal. per S.Y.		3000	.005		17.15	.24	.30	17.69	19.60
8900	For small and irregular areas, add						50%	50%		

32 12 Flexible Paving

32 12 16 – Asphalt Paving

32 12 16.13 Plant-Mix Asphalt Paving		Crew	Daily Output	Labor-Hours	Unit	Material	Labor	Equipment	Total	Total Incl O&P
0010	**PLANT-MIX ASPHALT PAVING**									
0020	And large paved areas with no hauling included									
0025	See Section 31 23 23.20 for hauling costs									
0080	Binder course, 1-1/2″ thick	B-25	7725	.011	S.Y.	5.70	.47	.35	6.52	7.45
0120	2″ thick		6345	.014		7.60	.57	.43	8.60	9.75
0130	2-1/2″ thick		5620	.016		9.50	.65	.48	10.63	12.05
0160	3″ thick		4905	.018		11.40	.74	.55	12.69	14.35
0170	3-1/2″ thick		4520	.019		13.30	.80	.60	14.70	16.60
0200	4″ thick		4140	.021		15.25	.88	.66	16.79	18.90
0300	Wearing course, 1″ thick	B-25B	10575	.009		3.78	.38	.28	4.44	5.10
0340	1-1/2″ thick		7725	.012		6.35	.52	.38	7.25	8.25
0380	2″ thick		6345	.015		8.50	.64	.46	9.60	10.95
0420	2-1/2″ thick		5480	.018		10.50	.74	.54	11.78	13.35
0460	3″ thick		4900	.020		12.55	.82	.60	13.97	15.80
0470	3-1/2″ thick		4520	.021		14.70	.89	.65	16.24	18.35
0480	4″ thick		4140	.023		16.80	.98	.71	18.49	21
0500	Open graded friction course	B-25C	5000	.010		2.13	.41	.47	3.01	3.52
0800	Alternate method of figuring paving costs									
0810	Binder course, 1-1/2″ thick	B-25	630	.140	Ton	70	5.75	4.31	80.06	91
0811	2″ thick		690	.128		70	5.25	3.93	79.18	90
0812	3″ thick		800	.110		70	4.55	3.39	77.94	88
0813	4″ thick		900	.098		70	4.04	3.01	77.05	87
0850	Wearing course, 1″ thick	B-25B	575	.167		69	7.05	5.15	81.20	93
0851	1-1/2″ thick		630	.152		69	6.40	4.68	80.08	91.50
0852	2″ thick		690	.139		69	5.85	4.28	79.13	90
0853	2-1/2″ thick		765	.125		69	5.30	3.86	78.16	89
0854	3″ thick		800	.120		69	5.05	3.69	77.74	88.50
1000	Pavement replacement over trench, 2″ thick	B-37	90	.533	S.Y.	7.85	21	1.75	30.60	45
1050	4″ thick		70	.686		15.55	27.50	2.25	45.30	64
1080	6″ thick		55	.873		25	34.50	2.86	62.36	86.50
3000	Prime coat, emulsion, .30 gal. per S.Y., 1000 S.Y.	B-45	2500	.006		1.74	.29	.36	2.39	2.77

32 12 Flexible Paving

32 12 16 – Asphalt Paving

32 12 16.13 Plant-Mix Asphalt Paving		Crew	Daily Output	Labor-Hours	Unit	Material	Labor	Equipment	Total	Total Incl O&P
						2015 Bare Costs				
3100	Tack coat, emulsion, .10 gal. per S.Y., 1000 S.Y.	B-45	2500	.006	S.Y.	.58	.29	.36	1.23	1.50

32 12 16.14 Asphaltic Concrete Paving

Code	Description	Crew	Daily Output	Labor-Hours	Unit	Material	Labor	Equipment	Total	Total Incl O&P
0011	**ASPHALTIC CONCRETE PAVING**, parking lots & driveways									
0015	No asphalt hauling included									
0018	Use 6.05 C.Y. per inch per M.S.F. for hauling									
0020	6″ stone base, 2″ binder course, 1″ topping	B-25C	9000	.005	S.F.	1.86	.23	.26	2.35	2.70
0025	2″ binder course, 2″ topping		9000	.005		2.28	.23	.26	2.77	3.16
0030	3″ binder course, 2″ topping		9000	.005		2.71	.23	.26	3.20	3.63
0035	4″ binder course, 2″ topping		9000	.005		3.13	.23	.26	3.62	4.09
0040	1.5″ binder course, 1″ topping		9000	.005		1.65	.23	.26	2.14	2.47
0042	3″ binder course, 1″ topping		9000	.005		2.29	.23	.26	2.78	3.17
0045	3″ binder course, 3″ topping		9000	.005		3.13	.23	.26	3.62	4.09
0050	4″ binder course, 3″ topping		9000	.005		3.55	.23	.26	4.04	4.55
0055	4″ binder course, 4″ topping		9000	.005		3.96	.23	.26	4.45	5
0300	Binder course, 1-1/2″ thick		35000	.001		.64	.06	.07	.77	.86
0400	2″ thick		25000	.002		.82	.08	.09	.99	1.14
0500	3″ thick		15000	.003		1.27	.14	.16	1.57	1.79
0600	4″ thick		10800	.004		1.67	.19	.22	2.08	2.37
0800	Sand finish course, 3/4″ thick		41000	.001		.32	.05	.06	.43	.49
0900	1″ thick		34000	.001		.40	.06	.07	.53	.62
1000	Fill pot holes, hot mix, 2″ thick	B-16	4200	.008		.85	.30	.16	1.31	1.60
1100	4″ thick		3500	.009		1.25	.35	.20	1.80	2.17
1120	6″ thick		3100	.010		1.67	.40	.22	2.29	2.74
1140	Cold patch, 2″ thick	B-51	3000	.016		.99	.61	.08	1.68	2.18
1160	4″ thick		2700	.018		1.89	.68	.09	2.66	3.29
1180	6″ thick		1900	.025		2.94	.96	.13	4.03	4.96
3000	Prime coat, emulsion, .30 gal. per S.Y., 1000 S.Y.	B-45	2500	.006	S.Y.	1.74	.29	.36	2.39	2.77
3100	Tack coat, emulsion, .10 gal. per S.Y., 1000 S.Y.	″	2500	.006	″	.58	.29	.36	1.23	1.50

32 13 Rigid Paving

32 13 13 – Concrete Paving

32 13 13.23 Concrete Paving Surface Treatment

Code	Description	Crew	Daily Output	Labor-Hours	Unit	Material	Labor	Equipment	Total	Total Incl O&P
0010	**CONCRETE PAVING SURFACE TREATMENT**									
0015	Including joints, finishing and curing									
0020	Fixed form, 12′ pass, unreinforced, 6″ thick	B-26	3000	.029	S.Y.	22	1.24	1.18	24.42	27.50
0030	7″ thick		2850	.031		27	1.30	1.24	29.54	33
0100	8″ thick		2750	.032		30	1.35	1.28	32.63	36.50
0110	8″ thick, small area		1375	.064		30	2.70	2.57	35.27	40
0200	9″ thick		2500	.035		34.50	1.48	1.41	37.39	41.50
0300	10″ thick		2100	.042		37.50	1.77	1.68	40.95	46
0310	10″ thick, small area		1050	.084		37.50	3.54	3.36	44.40	51
0400	12″ thick		1800	.049		43	2.06	1.96	47.02	53
0410	Conc. pavement, w/jt.,fnsh.&curing,fix form,24′ pass,unreinforced,6″T		6000	.015		21	.62	.59	22.21	24.50
0420	7″ thick		5700	.015		25	.65	.62	26.27	29.50
0430	8″ thick		5500	.016		28.50	.67	.64	29.81	33.50
0440	9″ thick		5000	.018		32.50	.74	.71	33.95	38
0450	10″ thick		4200	.021		36	.88	.84	37.72	42
0460	12″ thick		3600	.024		41.50	1.03	.98	43.51	48.50
0470	15″ thick		3000	.029		54.50	1.24	1.18	56.92	63
0500	Fixed form 12′ pass 15″ thick		1500	.059		55	2.47	2.35	59.82	67
0510	For small irregular areas, add				%	10%	100%	100%		

32 13 Rigid Paving

32 13 13 – Concrete Paving

32 13 13.23 Concrete Paving Surface Treatment		Crew	Daily Output	Labor-Hours	Unit	Material	2015 Bare Costs Labor	Equipment	Total	Total Incl O&P
0520	Welded wire fabric, sheets for rigid paving 2.33 lb./S.Y.	2 Rodm	389	.041	S.Y.	1.31	2.16		3.47	4.97
0530	Reinforcing steel for rigid paving 12 lb./S.Y.		666	.024		6.10	1.26		7.36	8.75
0540	Reinforcing steel for rigid paving 18 lb./S.Y.		444	.036		9.15	1.89		11.04	13.15
0610	For under 10′ pass, add				%	10%	100%	100%		
0620	Slip form, 12′ pass, unreinforced, 6″ thick	B-26A	5600	.016	S.Y.	21	.66	.66	22.32	25
0622	7″ thick		5600	.016		26	.66	.66	27.32	30.50
0624	8″ thick		5300	.017		29	.70	.70	30.40	34
0626	9″ thick		4820	.018		33	.77	.77	34.54	38.50
0628	10″ thick		4050	.022		36.50	.92	.91	38.33	42.50
0630	12″ thick		3470	.025		42	1.07	1.07	44.14	49
0632	15″ thick		2890	.030		53	1.28	1.28	55.56	62
0640	Slip form, 24′ pass, unreinforced, 6″ thick		11200	.008		20.50	.33	.33	21.16	24
0642	7″ thick		11200	.008		24.50	.33	.33	25.16	28
0644	8″ thick		10600	.008		28	.35	.35	28.70	31.50
0646	9″ thick		9640	.009		32	.39	.38	32.77	36
0648	10″ thick		8100	.011		35	.46	.46	35.92	40
0650	12″ thick		6940	.013		41	.53	.53	42.06	46.50
0652	15″ thick		5780	.015		51.50	.64	.64	52.78	58
0700	Finishing, broom finish small areas	2 Cefi	120	.133			6		6	9.50
0710	Transverse joint support dowels	C-1	350	.091	Ea.	5.50	4.08		9.58	12.75
0720	Transverse contraction joints, saw cut & grind	A-1B	120	.067	L.F.		2.51	1.40	3.91	5.65
0730	Transverse expansion joints, incl. premolded bit. jt. filler	C-1	150	.213		2.35	9.50		11.85	18.20
0740	Transverse construction joint using bulkhead	″	73	.438		3.41	19.55		22.96	36
0750	Longitudinal joint tie bars, grouted	B-23	70	.571	Ea.	5.30	21.50	40.50	67.30	86
1000	Curing, with sprayed membrane by hand	2 Clab	1500	.011	S.Y.	1	.40		1.40	1.75

32 14 Unit Paving

32 14 13 – Precast Concrete Unit Paving

32 14 13.13 Interlocking Precast Concrete Unit Paving

		Crew	Daily Output	Labor-Hours	Unit	Material	Labor	Equipment	Total	Total Incl O&P
0010	**INTERLOCKING PRECAST CONCRETE UNIT PAVING**									
0020	"V" blocks for retaining soil	D-1	205	.078	S.F.	10	3.29		13.29	16.35

32 14 13.16 Precast Concrete Unit Paving Slabs

		Crew	Daily Output	Labor-Hours	Unit	Material	Labor	Equipment	Total	Total Incl O&P
0010	**PRECAST CONCRETE UNIT PAVING SLABS**									
0710	Precast concrete patio blocks, 2-3/8″ thick, colors, 8″ x 16″	D-1	265	.060	S.F.	1.85	2.54		4.39	6.20
0715	12″ x 12″		300	.053		2.41	2.25		4.66	6.30
0720	16″ x 16″		335	.048		2.65	2.01		4.66	6.20
0730	24″ x 24″		510	.031		3.29	1.32		4.61	5.75
0740	Green, 8″ x 16″		265	.060		2.39	2.54		4.93	6.75
0750	Exposed local aggregate, natural	2 Bric	250	.064		7.30	2.95		10.25	12.85
0800	Colors		250	.064		7.30	2.95		10.25	12.85
0850	Exposed granite or limestone aggregate		250	.064		7.30	2.95		10.25	12.85
0900	Exposed white tumblestone aggregate		250	.064		5.70	2.95		8.65	11.05

32 14 13.18 Precast Concrete Plantable Pavers

		Crew	Daily Output	Labor-Hours	Unit	Material	Labor	Equipment	Total	Total Incl O&P
0010	**PRECAST CONCRETE PLANTABLE PAVERS** (50% grass)									
0015	Subgrade preparation and grass planting not included									
0100	Precast concrete plantable pavers with topsoil, 24″ x 16″	B-63	800	.050	S.F.	4.15	1.99	.22	6.36	8.05
0200	Less than 600 Square Feet or irregular area	″	500	.080	″	4.15	3.18	.35	7.68	10.15
0300	3/4″ crushed stone base for plantable pavers, 6 inch depth	B-62	1000	.024	S.Y.	4.61	.99	.17	5.77	6.85
0400	8 inch depth		900	.027		6.15	1.10	.19	7.44	8.80
0500	10 inch depth		800	.030		7.70	1.24	.22	9.16	10.70

32 14 Unit Paving

32 14 13 – Precast Concrete Unit Paving

32 14 13.18 Precast Concrete Plantable Pavers		Crew	Daily Output	Labor-Hours	Unit	Material	Labor	Equipment	Total	Total Incl O&P
						2015 Bare Costs				
0600	12 inch depth	B-62	700	.034	S.Y.	9.20	1.42	.25	10.87	12.70
0700	Hydro seeding plantable pavers	B-81A	20	.800	M.S.F.	12.45	30.50	21.50	64.45	87
0800	Apply fertilizer and seed to plantable pavers	1 Clab	8	1	"	50.50	37.50		88	117

32 14 16 – Brick Unit Paving

32 14 16.10 Brick Paving		Crew	Daily Output	Labor-Hours	Unit	Material	Labor	Equipment	Total	Total Incl O&P
0010	**BRICK PAVING**									
0012	4" x 8" x 1-1/2", without joints (4.5 brick/S.F.)	D-1	110	.145	S.F.	2.21	6.10		8.31	12.40
0100	Grouted, 3/8" joint (3.9 brick/S.F.)		90	.178		1.95	7.50		9.45	14.35
0200	4" x 8" x 2-1/4", without joints (4.5 bricks/S.F.)		110	.145		2.25	6.10		8.35	12.45
0300	Grouted, 3/8" joint (3.9 brick/S.F.)		90	.178		1.95	7.50		9.45	14.35
0455	Pervious brick paving, 4" x 8" x 3-1/4", without joints (4.5 bricks/S.F.)		110	.145		3.38	6.10		9.48	13.65
0500	Bedding, asphalt, 3/4" thick	B-25	5130	.017		.70	.71	.53	1.94	2.50
0540	Course washed sand bed, 1" thick	B-18	5000	.005		.32	.18	.01	.51	.66
0580	Mortar, 1" thick	D-1	300	.053		.74	2.25		2.99	4.46
0620	2" thick		200	.080		1.48	3.37		4.85	7.10
2000	Brick pavers, laid on edge, 7.2 per S.F.		70	.229		3.86	9.60		13.46	19.90
2500	For 4" thick concrete bed and joints, add		595	.027		1.19	1.13		2.32	3.15
2800	For steam cleaning, add	A-1H	950	.008		.09	.32	.08	.49	.71
9000	Minimum labor/equipment charge	1 Bric	2	4	Job		185		185	300

32 14 23 – Asphalt Unit Paving

32 14 23.10 Asphalt Blocks		Crew	Daily Output	Labor-Hours	Unit	Material	Labor	Equipment	Total	Total Incl O&P
0010	**ASPHALT BLOCKS**									
0020	Rectangular, 6" x 12" x 1-1/4", w/bed & neopr. adhesive	D-1	135	.119	S.F.	9.30	4.99		14.29	18.30
0100	3" thick		130	.123		13	5.20		18.20	23
0300	Hexagonal tile, 8" wide, 1-1/4" thick		135	.119		9.30	4.99		14.29	18.30
0400	2" thick		130	.123		13	5.20		18.20	23
0500	Square, 8" x 8", 1-1/4" thick		135	.119		9.30	4.99		14.29	18.30
0600	2" thick		130	.123		13	5.20		18.20	23
0900	For exposed aggregate (ground finish) add					.62			.62	.68
0910	For colors, add					.47			.47	.52
9000	Minimum labor/equipment charge	1 Bric	2	4	Job		185		185	300

32 14 40 – Stone Paving

32 14 40.10 Stone Pavers		Crew	Daily Output	Labor-Hours	Unit	Material	Labor	Equipment	Total	Total Incl O&P
0010	**STONE PAVERS**									
1100	Flagging, bluestone, irregular, 1" thick,	D-1	81	.198	S.F.	6.40	8.30		14.70	20.50
1110	1-1/2" thick		90	.178		7.55	7.50		15.05	20.50
1120	Pavers, 1/2" thick		110	.145		10.65	6.10		16.75	21.50
1130	3/4" thick		95	.168		13.60	7.10		20.70	26.50
1140	1" thick		81	.198		14.55	8.30		22.85	29.50
1150	Snapped random rectangular, 1" thick		92	.174		9.70	7.30		17	22.50
1200	1-1/2" thick		85	.188		11.65	7.90		19.55	26
1250	2" thick		83	.193		13.60	8.10		21.70	28
1300	Slate, natural cleft, irregular, 3/4" thick		92	.174		9.20	7.30		16.50	22
1310	1" thick		85	.188		10.70	7.90		18.60	24.50
1350	Random rectangular, gauged, 1/2" thick		105	.152		19.85	6.40		26.25	32.50
1400	Random rectangular, butt joint, gauged, 1/4" thick		150	.107		21.50	4.49		25.99	31
1450	For sand rubbed finish, add					9.30			9.30	10.20
1500	For interior setting, add								25%	25%
1550	Granite blocks, 3-1/2" x 3-1/2" x 3-1/2"	D-1	92	.174	S.F.	12	7.30		19.30	25
1560	4" x 4" x 4"		95	.168		12.70	7.10		19.80	25.50
1600	4" to 12" long, 3" to 5" wide, 3" to 5" thick		98	.163		10	6.85		16.85	22.50

32 14 Unit Paving

32 14 40 – Stone Paving

32 14 40.10 Stone Pavers		Crew	Daily Output	Labor-Hours	Unit	Material	2015 Bare Costs Labor	Equipment	Total	Total Incl O&P
1650	6″ to 15″ long, 3″ to 6″ wide, 3″ to 5″ thick	D-1	105	.152	S.F.	5.35	6.40		11.75	16.35

32 16 Curbs, Gutters, Sidewalks, and Driveways

32 16 13 – Curbs and Gutters

32 16 13.13 Cast-in-Place Concrete Curbs and Gutters		Crew	Daily Output	Labor-Hours	Unit	Material	Labor	Equipment	Total	Total Incl O&P
0010	**CAST-IN-PLACE CONCRETE CURBS AND GUTTERS**									
0290	Forms only, no concrete									
0300	Concrete, wood forms, 6″ x 18″, straight	C-2	500	.096	L.F.	2.99	4.39		7.38	10.50
0400	6″ x 18″, radius	″	200	.240	″	3.11	11		14.11	21.50
0402	Forms and concrete complete									
0404	Concrete, wood forms, 6″ x 18″, straight & concrete	C-2A	500	.096	L.F.	5.85	4.36		10.21	13.50
0406	6″ x 18″, radius	″	200	.240		5.95	10.90		16.85	24.50
0415	Machine formed, 6″ x 18″, straight	B-69A	2000	.024		3.52	.99	.50	5.01	6.05
0416	6″ x 18″, radius	″	900	.053	↓	3.55	2.20	1.11	6.86	8.70
0421	Curb and gutter, straight									
0422	with 6″ high curb and 6″ thick gutter, wood forms									
0430	24″ wide, .055 C.Y. per L.F.	C-2A	375	.128	L.F.	15.35	5.80		21.15	26.50
0435	30″ wide, .066 C.Y. per L.F.	″	340	.141	″	16.85	6.40		23.25	29

32 16 13.23 Precast Concrete Curbs and Gutters		Crew	Daily Output	Labor-Hours	Unit	Material	Labor	Equipment	Total	Total Incl O&P
0010	**PRECAST CONCRETE CURBS AND GUTTERS**									
0550	Precast, 6″ x 18″, straight	B-29	700	.080	L.F.	10.30	3.28	1.26	14.84	18.05
0600	6″ x 18″, radius	″	325	.172	″	10.80	7.05	2.71	20.56	26.50

32 16 13.33 Asphalt Curbs		Crew	Daily Output	Labor-Hours	Unit	Material	Labor	Equipment	Total	Total Incl O&P
0010	**ASPHALT CURBS**									
0012	Curbs, asphaltic, machine formed, 8″ wide, 6″ high, 40 L.F./ton	B-27	1000	.032	L.F.	1.82	1.22	.30	3.34	4.33
0100	8″ wide, 8″ high, 30 L.F. per ton		900	.036		2.43	1.35	.33	4.11	5.25
0150	Asphaltic berm, 12″ W, 3″-6″ H, 35 L.F./ton, before pavement	↓	700	.046		.04	1.74	.42	2.20	3.37
0200	12″ W, 1-1/2″ to 4″ H, 60 L.F. per ton, laid with pavement	B-2	1050	.038	↓	.02	1.45		1.47	2.40

32 16 13.43 Stone Curbs		Crew	Daily Output	Labor-Hours	Unit	Material	Labor	Equipment	Total	Total Incl O&P
0010	**STONE CURBS**									
1000	Granite, split face, straight, 5″ x 16″	D-13	275	.175	L.F.	11.65	7.85	1.72	21.22	27.50
1300	Radius curbing, 6″ x 18″, over 10′ radius	B-29	260	.215	″	18.75	8.85	3.39	30.99	38.50
1400	Corners, 2′ radius	″	80	.700	Ea.	63	28.50	11.05	102.55	128
1600	Edging, 4-1/2″ x 12″, straight	d-13	300	.160	L.F.	5.85	7.15	1.58	14.58	19.80
1800	Curb inlets, (guttermouth) straight	B-29	41	1.366	Ea.	140	56	21.50	217.50	269
2000	Indian granite (belgian block)									
2100	Jumbo, 10-1/2″ x 7-1/2″ x 4″, grey	D-1	150	.107	L.F.	6.15	4.49		10.64	14.05
2150	Pink		150	.107		7.75	4.49		12.24	15.80
2200	Regular, 9″ x 4-1/2″ x 4-1/2″, grey		160	.100		4.67	4.21		8.88	12
2250	Pink		160	.100		6.15	4.21		10.36	13.60
2300	Cubes, 4″ x 4″ x 4″, grey		175	.091		3.96	3.85		7.81	10.60
2350	Pink		175	.091		3.81	3.85		7.66	10.45
2400	6″ x 6″ x 6″, pink	↓	155	.103	↓	12.60	4.35		16.95	21
2500	Alternate pricing method for indian granite									
2550	Jumbo, 10-1/2″ x 7-1/2″ x 4″ (30 lb.), grey				Ton	350			350	385
2600	Pink					450			450	495
2650	Regular, 9″ x 4-1/2″ x 4-1/2″ (20 lb.), grey					330			330	365
2700	Pink					430			430	475
2750	Cubes, 4″ x 4″ x 4″ (5 lb.), grey					480			480	530
2800	Pink				↓	490			490	540

32 16 Curbs, Gutters, Sidewalks, and Driveways

32 16 13 – Curbs and Gutters

32 16 13.43 Stone Curbs		Crew	Daily Output	Labor-Hours	Unit	Material	2015 Bare Costs Labor	Equipment	Total	Total Incl O&P
2850	6" x 6" x 6" (25 lb.), pink				Ton	490			490	540
2900	For pallets, add					22			22	24

32 17 Paving Specialties

32 17 13 – Parking Bumpers

32 17 13.13 Metal Parking Bumpers

		Crew	Daily Output	Labor-Hours	Unit	Material	Labor	Equipment	Total	Total Incl O&P
0010	**METAL PARKING BUMPERS**									
0015	Bumper rails for garages, 12 Ga. rail, 6" wide, with steel									
0020	posts 12'-6" O.C., minimum	E-4	190	.168	L.F.	18.25	8.95	.77	27.97	37
0030	Average		165	.194		23	10.30	.88	34.18	44.50
0100	Maximum		140	.229		27.50	12.15	1.04	40.69	53
0300	12" channel rail, minimum		160	.200		23	10.65	.91	34.56	45
0400	Maximum		120	.267		34	14.15	1.22	49.37	64.50
1300	Pipe bollards, conc. filled/paint, 8' L x 4' D hole, 6" diam.	B-6	20	1.200	Ea.	600	49.50	18.20	667.70	760
1400	8" diam.		15	1.600		685	66	24.50	775.50	885
1500	12" diam.		12	2		965	82.50	30.50	1,078	1,225
2030	Folding with individual padlocks	B-2	50	.800		605	30.50		635.50	715
8000	Parking lot control, see Section 11 12 13.10									
8900	Security bollards, SS, lighted, hyd., incl. controls, group of 3	L-7	.06	509	Ea.	48,400	23,100		71,500	91,000
8910	Group of 5	"	.04	682	"	65,000	31,000		96,000	122,000
9000	Minimum labor/equipment charge	E-4	2	16	Job		850	73	923	1,600

32 17 13.16 Plastic Parking Bumpers

		Crew	Daily Output	Labor-Hours	Unit	Material	Labor	Equipment	Total	Total Incl O&P
0010	**PLASTIC PARKING BUMPERS**									
1200	Thermoplastic, 6" x 10" x 6'-0"	B-2	120	.333	Ea.	52.50	12.65		65.15	79

32 17 13.19 Precast Concrete Parking Bumpers

		Crew	Daily Output	Labor-Hours	Unit	Material	Labor	Equipment	Total	Total Incl O&P
0010	**PRECAST CONCRETE PARKING BUMPERS**									
1000	Wheel stops, precast concrete incl. dowels, 6" x 10" x 6'-0"	B-2	120	.333	Ea.	39.50	12.65		52.15	64.50
1100	8" x 13" x 6'-0"	"	120	.333	"	46	12.65		58.65	71.50

32 17 13.26 Wood Parking Bumpers

		Crew	Daily Output	Labor-Hours	Unit	Material	Labor	Equipment	Total	Total Incl O&P
0010	**WOOD PARKING BUMPERS**									
0020	Parking barriers, timber w/saddles, treated type									
0100	4" x 4" for cars	B-2	520	.077	L.F.	2.92	2.92		5.84	8
0200	6" x 6" for trucks		520	.077	"	6.10	2.92		9.02	11.50
0600	Flexible fixed stanchion, 2' high, 3" diameter		100	.400	Ea.	40.50	15.20		55.70	69.50

32 17 23 – Pavement Markings

32 17 23.13 Painted Pavement Markings

		Crew	Daily Output	Labor-Hours	Unit	Material	Labor	Equipment	Total	Total Incl O&P
0010	**PAINTED PAVEMENT MARKINGS**									
0020	Acrylic waterborne, white or yellow, 4" wide, less than 3000 L.F.	B-78	20000	.002	L.F.	.15	.09	.03	.27	.35
0200	6" wide, less than 3000 L.F.		11000	.004		.23	.17	.05	.45	.58
0500	8" wide, less than 3000 L.F.		10000	.005		.31	.18	.06	.55	.71
0600	12" wide, less than 3000 L.F.		4000	.012		.46	.46	.15	1.07	1.42
0620	Arrows or gore lines		2300	.021	S.F.	.22	.80	.26	1.28	1.83
0640	Temporary paint, white or yellow, less than 3000 L.F.		15000	.003	L.F.	.08	.12	.04	.24	.33
0660	Removal	1 Clab	300	.027			1		1	1.64
0680	Temporary tape	2 Clab	1500	.011		.48	.40		.88	1.19
0710	Thermoplastic, white or yellow, 4" wide, less than 6000 L.F.	B-79	15000	.003		.31	.10	.10	.51	.62
0730	6" wide, less than 6000 L.F.		14000	.003		.47	.11	.11	.69	.81
0740	8" wide, less than 6000 L.F.		12000	.003		.62	.13	.12	.87	1.03
0750	12" wide, less than 6000 L.F.		6000	.007		.91	.26	.25	1.42	1.70
0760	Arrows		660	.061	S.F.	.61	2.32	2.24	5.17	6.90

32 17 Paving Specialties

32 17 23 – Pavement Markings

32 17 23.13 Painted Pavement Markings

	32 17 23.13 Painted Pavement Markings	Crew	Daily Output	Labor-Hours	Unit	2015 Bare Costs Material	Labor	Equipment	Total	Total Incl O&P
0770	Gore lines	B-79	2500	.016	S.F.	.61	.61	.59	1.81	2.32
0780	Letters		660	.061		.61	2.32	2.24	5.17	6.90
0782	Thermoplastic material, small users				Ton	1,700			1,700	1,875
0784	Glass beads, highway use, add				Lb.	.59			.59	.65
0786	Thermoplastic material, highway departments				Ton	1,475			1,475	1,625
1000	Airport painted markings									
1050	Traffic safety flashing truck for airport painting	A-2B	1	8	Day		315	245	560	775
1100	Painting, white or yellow, taxiway markings	B-78	4000	.012	S.F.	.27	.46	.15	.88	1.21
1110	with 12 lb. beads per 100 S.F.		4000	.012		.53	.46	.15	1.14	1.49
1200	Runway markings		3500	.014		.27	.52	.17	.96	1.35
1210	with 12 lb. beads per 100 S.F.		3500	.014		.53	.52	.17	1.22	1.63
1300	Pavement location or direction signs		2500	.019		.27	.73	.24	1.24	1.76
1310	with 12 lb. beads per 100 S.F.		2500	.019		.53	.73	.24	1.50	2.04
1350	Mobilization airport pavement painting		4	12	Ea.		460	150	610	915
1400	Paint markings or pavement signs removal daytime	B-78B	400	.045	S.F.		1.75	.94	2.69	3.88
1500	Removal nighttime		335	.054	"		2.09	1.12	3.21	4.63
1600	Mobilization pavement paint removal		4	4.500	Ea.		175	94	269	390

32 17 23.14 Pavement Parking Markings

	32 17 23.14 Pavement Parking Markings	Crew	Daily Output	Labor-Hours	Unit	Material	Labor	Equipment	Total	Total Incl O&P
0010	**PAVEMENT PARKING MARKINGS**									
0790	Layout of pavement marking	A-2	25000	.001	L.F.		.04	.01	.05	.07
0800	Lines on pvmt., parking stall, paint, white, 4" wide	B-78B	400	.045	Stall	4.44	1.75	.94	7.13	8.75
0825	Parking stall, small quantities	2 Pord	80	.200		8.90	8.05		16.95	22.50
0830	Lines on pvmt., parking stall, thermoplastic, white, 4" wide	B-79	300	.133		13.50	5.10	4.92	23.52	28.50
1000	Street letters and numbers	B-78B	1600	.011	S.F.	.67	.44	.23	1.34	1.71
1100	Pavement marking letter, 6"	2 Pord	400	.040	Ea.	10.60	1.61		12.21	14.25
1110	12" letter		272	.059		13.35	2.37		15.72	18.45
1120	24" letter		160	.100		29.50	4.04		33.54	39
1130	36" letter		84	.190		38.50	7.70		46.20	55
1140	42" letter		84	.190		51	7.70		58.70	68.50
1150	72" letter		40	.400		42	16.15		58.15	72
1200	Handicap symbol		40	.400		31	16.15		47.15	60
1210	Handicap Parking sign 12" x 18 " and post	A-2	12	2		124	76	20.50	220.50	283
1300	Pavement marking, thermoplastic tape including layout, 4 inch width	B-79B	320	.025	L.F.	2.53	.94	.48	3.95	4.84
1310	12 inch width		192	.042	"	6.80	1.57	.79	9.16	10.95
1320	Letters including layout, 4 inch		240	.033	Ea.	12.35	1.25	.63	14.23	16.35
1330	6 inch		160	.050		15.15	1.88	.95	17.98	21
1340	12 inch		120	.067		19.20	2.51	1.27	22.98	26.50
1350	48 inch		64	.125		71.50	4.70	2.38	78.58	89.50
1360	96 inch		32	.250		91	9.40	4.75	105.15	121
1380	4 letter words, 8 feet tall		8	1		340	37.50	19	396.50	460

32 18 13 – Synthetic Grass Surfacing

32 18 13.10 Artificial Grass Surfacing		Crew	Daily Output	Labor-Hours	Unit	2015 Bare Costs Material	Labor	Equipment	Total	Total Incl O&P
0010	ARTIFICIAL GRASS SURFACING									
0015	Not including asphalt base or drainage,									
0020	but including cushion pad, over 50,000 S.F.									
0200	1/2" pile and 5/16" cushion pad, standard	C-17	3200	.025	S.F.	10.15	1.23		11.38	13.20
0300	Deluxe		2560	.031		15	1.53		16.53	19
0500	1/2" pile and 5/8" cushion pad, standard		2844	.028		14.60	1.38		15.98	18.30
0600	Deluxe		2327	.034		16.10	1.69		17.79	20.50
0800	For asphaltic concrete base, 2-1/2" thick,									
0900	with 6" crushed stone sub-base, add	B-25	12000	.007	S.F.	1.69	.30	.23	2.22	2.60

32 18 16 – Synthetic Resilient Surfacing

32 18 16.13 Playground Protective Surfacing		Crew	Daily Output	Labor-Hours	Unit	Material	Labor	Equipment	Total	Total Incl O&P
0010	PLAYGROUND PROTECTIVE SURFACING									
0100	Resilient rubber surface, poured in place, 4" thick, black	2 Skwk	300	.053	S.F.	13.35	2.59		15.94	18.90
0150	2" thick topping, colors	"	2800	.006		7.25	.28		7.53	8.40
0200	Wood chip mulch, 6" deep	1 Clab	300	.027		.76	1		1.76	2.48

32 18 23 – Athletic Surfacing

32 18 23.33 Running Track Surfacing		Crew	Daily Output	Labor-Hours	Unit	Material	Labor	Equipment	Total	Total Incl O&P
0010	RUNNING TRACK SURFACING									
0020	Running track, asphalt, incl base, 3" thick	B-37	300	.160	S.Y.	13	6.35	.52	19.87	25
0102	Surface, latex rubber system, 1/2" thick, black	B-20	115	.209		41	8.75		49.75	60
0152	Colors		115	.209		50.50	8.75		59.25	70
0302	Urethane rubber system, 1/2" thick, black		110	.218		30.50	9.15		39.65	48.50
0402	Color coating		110	.218		37.50	9.15		46.65	56

32 18 23.53 Tennis Court Surfacing		Crew	Daily Output	Labor-Hours	Unit	Material	Labor	Equipment	Total	Total Incl O&P
0010	TENNIS COURT SURFACING									
0020	Tennis court, asphalt, incl. base, 2-1/2" thick, one court	B-37	450	.107	S.Y.	37.50	4.24	.35	42.09	49
0200	Two courts		675	.071		15.85	2.83	.23	18.91	22.50
0300	Clay courts		360	.133		42	5.30	.44	47.74	55.50
0400	Pulverized natural greenstone with 4" base, fast dry		250	.192		39.50	7.65	.63	47.78	56.50
0800	Rubber-acrylic base resilient pavement		600	.080		56	3.18	.26	59.44	67
1000	Colored sealer, acrylic emulsion, 3 coats	2 Clab	800	.020		6.05	.75		6.80	7.90
1100	3 coat, 2 colors	"	900	.018		8.45	.67		9.12	10.35
1200	For preparing old courts, add	1 Clab	825	.010			.36		.36	.60
1400	Posts for nets, 3-1/2" diameter with eye bolts	B-1	3.40	7.059	Pr.	305	270		575	780
1500	With pulley & reel		3.40	7.059	"	780	270		1,050	1,300
1700	Net, 42' long, nylon thread with binder		50	.480	Ea.	254	18.35		272.35	310
1800	All metal		6.50	3.692	"	490	141		631	770
2000	Paint markings on asphalt, 2 coats	1 Pord	1.78	4.494	Court	187	181		368	495
2200	Complete court with fence, etc., asphaltic conc., minimum	B-37	.20	240		28,700	9,550	785	39,035	47,900
2300	Maximum		.16	300		56,500	11,900	985	69,385	83,000
2800	Clay courts, minimum		.20	240		31,500	9,550	785	41,835	51,000
2900	Maximum		.16	300		58,000	11,900	985	70,885	84,500

32 31 Fences and Gates

32 31 13 – Chain Link Fences and Gates

32 31 13.10 Chain Link Gates and Fences		Crew	Daily Output	Labor-Hours	Unit	2015 Bare Costs Material	Labor	Equipment	Total	Total Incl O&P
0010	**CHAIN LINK GATES AND FENCES**									
4750	Gate, transom for 10′ fence, galv. steel, single, 3′ x 7′	B-80A	52	.462	Ea.	390	17.35	5.85	413.20	465
4752	4′ x 7′		10	2.400		425	90	30.50	545.50	650
4754	3′ x 10′		8	3		390	113	38	541	655
4756	4′ x 10′	↓	10	2.400		420	90	30.50	540.50	645
4758	Double transom, 10′ x 7′	B-80B	10	3.200		745	129	24.50	898.50	1,050
4760	12′ x 7′		6	5.333		785	215	40.50	1,040.50	1,250
4762	14′ x 7′		5	6.400		855	258	49	1,162	1,425
4764	10′ x 10′		4	8		1,075	325	61	1,461	1,775
4766	12′ x 10′		7	4.571		1,200	184	35	1,419	1,650
4768	14′ x 10′		7	4.571		1,275	184	35	1,494	1,750
4780	Vinyl clad, single transom, 3′ x 7′		10	3.200		510	129	24.50	663.50	795
4782	4′ x 7′		10	3.200		555	129	24.50	708.50	845
4784	3′ x 10′		8	4		610	161	30.50	801.50	965
4786	4′ x 10′		8	4		690	161	30.50	881.50	1,050
4788	Double transom, 10′ x 7′		10	3.200		855	129	24.50	1,008.50	1,175
4790	12′ x 7′		6	5.333		875	215	40.50	1,130.50	1,350
4792	14′ x 7′		5	6.400		1,075	258	49	1,382	1,650
4794	10′ x 10′		4	8		1,050	325	61	1,436	1,750
4798	12′ x 12′		7	4.571		1,300	184	35	1,519	1,800
4799	14′ x 14′	↓	7	4.571	↓	1,925	184	35	2,144	2,450

32 31 13.20 Fence, Chain Link Industrial		Crew	Daily Output	Labor-Hours	Unit	Material	Labor	Equipment	Total	Total Incl O&P
0010	**FENCE, CHAIN LINK INDUSTRIAL**									
0011	Schedule 40, including concrete									
0020	3 strands barb wire, 2″ post @ 10′ O.C., set in concrete, 6′ H									
0200	9 ga. wire, galv. steel, in concrete	B-80C	240	.100	L.F.	19.10	3.81	1.06	23.97	28.50
0248	Fence, add for vinyl coated fabric				S.F.	.66			.66	.73
0300	Aluminized steel	B-80C	240	.100	L.F.	20.50	3.81	1.06	25.37	30
0500	6 ga. wire, galv. steel		240	.100		21	3.81	1.06	25.87	30.50
0600	Aluminized steel		240	.100		30	3.81	1.06	34.87	40.50
0800	6 ga. wire, 6′ high but omit barbed wire, galv. steel		250	.096		19.55	3.66	1.02	24.23	28.50
0900	Aluminized steel, in concrete		250	.096		23.50	3.66	1.02	28.18	33
0920	8′ H, 6 ga. wire, 2-1/2″ line post, galv. steel, in concrete		180	.133		31	5.10	1.41	37.51	44
0940	Aluminized steel, in concrete		180	.133	↓	38	5.10	1.41	44.51	51.50
1400	Gate for 6′ high fence, 1-5/8″ frame, 3′ wide, galv. steel		10	2.400	Ea.	203	91.50	25.50	320	400
1500	Aluminized steel, in concrete	↓	10	2.400	″	205	91.50	25.50	322	400
2000	5′-0″ high fence, 9 ga., no barbed wire, 2″ line post, in concrete									
2010	10′ O.C., 1-5/8″ top rail, in concrete									
2100	Galvanized steel, in concrete	B-80C	300	.080	L.F.	18.45	3.05	.85	22.35	26.50
2200	Aluminized steel, in concrete		300	.080	″	18.75	3.05	.85	22.65	26.50
2400	Gate, 4′ wide, 5′ high, 2″ frame, galv. steel, in concrete		10	2.400	Ea.	188	91.50	25.50	305	385
2500	Aluminized steel, in concrete		10	2.400	″	200	91.50	25.50	317	395
3100	Overhead slide gate, chain link, 6′ high, to 18′ wide, in concrete	↓	38	.632	L.F.	97	24	6.70	127.70	153
3110	Cantilever type, in concrete	B-80	48	.667		129	27.50	15	171.50	203
3120	8′ high, in concrete		24	1.333		155	55	30	240	293
3130	10′ high, in concrete	↓	18	1.778	↓	190	73.50	40	303.50	370
5000	Double swing gates, incl. posts & hardware, in concrete									
5010	5′ high, 12′ opening, in concrete	B-80C	3.40	7.059	Opng.	395	269	74.50	738.50	955
5020	20′ opening, in concrete		2.80	8.571		520	325	90.50	935.50	1,200
5060	6′ high, 12′ opening, in concrete		3.20	7.500		455	286	79.50	820.50	1,050
5070	20′ opening, in concrete	↓	2.60	9.231		655	350	97.50	1,102.50	1,400
5080	8′ high, 12′ opening, in concrete	B-80	2.13	15.002	↓	460	620	340	1,420	1,875

32 31 Fences and Gates

32 31 13 – Chain Link Fences and Gates

	32 31 13.20 Fence, Chain Link Industrial	Crew	Daily Output	Labor-Hours	Unit	2015 Bare Costs Material	Labor	Equipment	Total	Total Incl O&P
5090	20′ opening, in concrete	B-80	1.45	22.069	Opng.	685	910	495	2,090	2,775
5100	10′ high, 12′ opening, in concrete		1.31	24.427		825	1,000	550	2,375	3,125
5110	20′ opening, in concrete		1.03	31.068		865	1,275	700	2,840	3,800
5120	12′ high, 12′ opening, in concrete		1.05	30.476		1,175	1,250	685	3,110	4,075
5130	20′ opening, in concrete		.85	37.647		1,225	1,550	845	3,620	4,775
5190	For aluminized steel add					20%				
7055	Braces, galv. steel	B-80A	960	.025	L.F.	2.56	.94	.32	3.82	4.70
7056	Aluminized steel	"	960	.025	"	3.07	.94	.32	4.33	5.25
7071	Privacy slats, vertical, vinyl	1 Clab	500	.016	S.F.	1.44	.60		2.04	2.57
7072	Redwood		450	.018		1.41	.67		2.08	2.65
7073	Diagonal, aluminum		300	.027		4.55	1		5.55	6.65
9000	Minimum labor/equipment charge	B-80	2	16	Job		660	360	1,020	1,475

32 31 13.25 Fence, Chain Link Residential

	32 31 13.25 Fence, Chain Link Residential	Crew	Daily Output	Labor-Hours	Unit	Material	Labor	Equipment	Total	Total Incl O&P
0010	**FENCE, CHAIN LINK RESIDENTIAL**									
0011	Schedule 20, 11 ga. wire, 1-5/8″ post									
0020	10′ O.C., 1-3/8″ top rail, 2″ corner post, galv. stl. 3′ high	B-80C	500	.048	L.F.	2.12	1.83	.51	4.46	5.85
0050	4′ high		400	.060		7.10	2.29	.63	10.02	12.25
0100	6′ high		200	.120		9.45	4.57	1.27	15.29	19.25
0150	Add for gate 3′ wide, 1-3/8″ frame, 3′ high		12	2	Ea.	81.50	76	21	178.50	237
0170	4′ high		10	2.400		87.50	91.50	25.50	204.50	273
0190	6′ high		10	2.400		109	91.50	25.50	226	297
0200	Add for gate 4′ wide, 1-3/8″ frame, 3′ high		9	2.667		91.50	102	28	221.50	298
0220	4′ high		9	2.667		97.50	102	28	227.50	305
0240	6′ high		8	3		123	114	31.50	268.50	355
0350	Aluminized steel, 11 ga. wire, 3′ high		500	.048	L.F.	8.80	1.83	.51	11.14	13.25
0380	4′ high		400	.060		9.25	2.29	.63	12.17	14.60
0400	6′ high		200	.120		11.15	4.57	1.27	16.99	21
0450	Add for gate 3′ wide, 1-3/8″ frame, 3′ high		12	2	Ea.	95	76	21	192	252
0470	4′ high		10	2.400		101	91.50	25.50	218	288
0490	6′ high		10	2.400		125	91.50	25.50	242	315
0500	Add for gate 4′ wide, 1-3/8″ frame, 3′ high		10	2.400		105	91.50	25.50	222	293
0520	4′ high		9	2.667		121	102	28	251	330
0540	6′ high		8	3		130	114	31.50	275.50	365
0620	Vinyl covered, 9 ga. wire, 3′ high		500	.048	L.F.	7.80	1.83	.51	10.14	12.15
0640	4′ high		400	.060		8.15	2.29	.63	11.07	13.45
0660	6′ high		200	.120		10.15	4.57	1.27	15.99	20
0720	Add for gate 3′ wide, 1-3/8″ frame, 3′ high		12	2	Ea.	94.50	76	21	191.50	252
0740	4′ high		10	2.400		101	91.50	25.50	218	288
0760	6′ high		10	2.400		120	91.50	25.50	237	310
0780	Add for gate 4′ wide, 1-3/8″ frame, 3′ high		10	2.400		99.50	91.50	25.50	216.50	286
0800	4′ high		9	2.667		103	102	28	233	310
0820	6′ high		8	3		128	114	31.50	273.50	360
7076	Fence, for small jobs 100 L.F. fence or less w/or wo gate, add				S.F.	20%				
9000	Minimum labor/equipment charge	B-1	2	12	Job		460		460	755

32 31 13.26 Tennis Court Fences and Gates

	32 31 13.26 Tennis Court Fences and Gates	Crew	Daily Output	Labor-Hours	Unit	Material	Labor	Equipment	Total	Total Incl O&P
0010	**TENNIS COURT FENCES AND GATES**									
0860	Tennis courts, 11 ga. wire, 2-1/2″ post set									
0870	in concrete, 10′ O.C., 1-5/8″ top rail									
0900	10′ high	B-80	190	.168	L.F.	22.50	6.95	3.79	33.24	40
0920	12′ high		170	.188	"	23.50	7.75	4.24	35.49	43
1000	Add for gate 4′ wide, 1-5/8″ frame 7′ high		10	3.200	Ea.	242	132	72	446	560
1040	Aluminized steel, 11 ga. wire 10′ high		190	.168	L.F.	21	6.95	3.79	31.74	38.50

32 31 Fences and Gates

32 31 13 – Chain Link Fences and Gates

32 31 13.26 Tennis Court Fences and Gates		Crew	Daily Output	Labor-Hours	Unit	Material	Labor	Equipment	Total	Total Incl O&P
						2015 Bare Costs				
1100	12′ high	B-80	170	.188	L.F.	23	7.75	4.24	34.99	42
1140	Add for gate 4′ wide, 1-5/8″ frame, 7′ high		10	3.200	Ea.	262	132	72	466	580
1250	Vinyl covered, 9 ga. wire, 10′ high		190	.168	L.F.	21.50	6.95	3.79	32.24	39
1300	12′ high		170	.188	"	25.50	7.75	4.24	37.49	45
1310	Fence, CL, tennis court, transom gate, single, galv., 4′ x 7′	B-80A	8.72	2.752	Ea.	310	103	35	448	550
1400	Add for gate 4′ wide, 1-5/8″ frame, 7′ high	B-80	10	3.200	"	315	132	72	519	635

32 31 13.30 Fence, Chain Link, Gates and Posts

Code	Description	Crew	Daily Output	Labor-Hours	Unit	Material	Labor	Equipment	Total	Total Incl O&P
0010	**FENCE, CHAIN LINK, GATES & POSTS**									
0011	(1/3 post length in ground)									
6580	Line posts, galvanized, 2-1/2″ OD, set in conc., 4′	B-80	80	.400	Ea.	27.50	16.50	9	53	67
6585	5′		76	.421		34.50	17.35	9.50	61.35	76.50
6590	6′		74	.432		36.50	17.85	9.75	64.10	79.50
6595	7′		72	.444		44.50	18.30	10	72.80	89
6600	8′		69	.464		48.50	19.10	10.45	78.05	96
6610	H-beam, 1-7/8″, 4′		83	.386		34.50	15.90	8.70	59.10	73
6615	5′		81	.395		40	16.30	8.90	65.20	80.50
6620	6′		78	.410		45.50	16.90	9.25	71.65	88
6625	7′		75	.427		50	17.60	9.60	77.20	94
6630	8′		73	.438		56	18.05	9.85	83.90	102
6635	Vinyl coated, 2-1/2″ OD, set in conc., 4′		79	.405		39	16.70	9.10	64.80	80
6640	5′		77	.416		39.50	17.15	9.35	66	81.50
6645	6′		74	.432		53.50	17.85	9.75	81.10	98
6650	7′		72	.444		65	18.30	10	93.30	112
6655	8′		69	.464		73	19.10	10.45	102.55	123
6660	End gate post, steel, 3″ OD, set in conc.,4′		68	.471		45.50	19.40	10.60	75.50	93.50
6665	5′		65	.492		49	20.50	11.10	80.60	99
6670	6′		63	.508		51	21	11.45	83.45	103
6675	7′		61	.525		62	21.50	11.80	95.30	116
6680	8′		59	.542		70.50	22.50	12.20	105.20	127
6685	Vinyl, 4′		68	.471		50.50	19.40	10.60	80.50	98.50
6690	5′		65	.492		59	20.50	11.10	90.60	110
6695	6′		63	.508		92.50	21	11.45	124.95	149
6700	7′		61	.525		111	21.50	11.80	144.30	170
6705	8′		59	.542		124	22.50	12.20	158.70	186
6710	Corner post, galv. steel, 4″ OD, set in conc., 4′		65	.492		97	20.50	11.10	128.60	152
6715	6′		63	.508		106	21	11.45	138.45	164
6720	7′		61	.525		131	21.50	11.80	164.30	192
6725	8′		65	.492		141	20.50	11.10	172.60	200
6730	Vinyl, 5′		65	.492		91.50	20.50	11.10	123.10	145
6735	6′		63	.508		140	21	11.45	172.45	201
6740	7′		61	.525		163	21.50	11.80	196.30	227
6745	8′		59	.542		182	22.50	12.20	216.70	249
7031	For corner, end, & pull post bracing, add					20%	15%			
7770	Gates, sliding w/overhead support, 4′ high	B-80B	35	.914	L.F.	150	37	6.95	193.95	233
7775	5′ high		32	1		168	40.50	7.60	216.10	259
7780	6′ high		28	1.143		151	46	8.70	205.70	251
7785	7′ high		25	1.280		194	51.50	9.75	255.25	310
7790	8′ high		23	1.391		208	56	10.60	274.60	330
7795	Cantilever, manual, exp. roller, (pr), 40′ wide x 8′ high	B-22	1	30	Ea.	5,500	1,325	207	7,032	8,400
7800	30′ wide x 8′ high		1	30		4,125	1,325	207	5,657	6,875
7805	24′ wide x 8′ high		1	30		3,375	1,325	207	4,907	6,075
7810	Motor operators for gates, (no elec wiring), 3′ wide swing	2 Skwk	.50	32		1,150	1,550		2,700	3,800

32 31 Fences and Gates

32 31 13 – Chain Link Fences and Gates

32 31 13.30 Fence, Chain Link, Gates and Posts		Crew	Daily Output	Labor-Hours	Unit	Material	2015 Bare Costs Labor	Equipment	Total	Total Incl O&P
7815	Up to 20′ wide swing	2 Skwk	.50	32	Ea.	1,500	1,550		3,050	4,175
7820	Up to 45′ sliding		.50	32		2,675	1,550		4,225	5,475
7825	Overhead gate, 6′ to 18′ wide, sliding/cantilever		45	.356	L.F.	293	17.30		310.30	350
7830	Gate operators, digital receiver		7	2.286	Ea.	74	111		185	262
7835	Two button transmitter		24	.667		23	32.50		55.50	78
7840	3 button station		14	1.143		39	55.50		94.50	134
7845	Master slave system		4	4		168	195		363	500
7900	Auger fence post hole, 3′ deep, medium soil, by hand	1 Clab	30	.267			10.05		10.05	16.45
7925	By machine	B-80	175	.183			7.55	4.12	11.67	16.75
7950	Rock, with jackhammer	B-9	32	1.250			47.50	7.30	54.80	86
7975	With rock drill	B-47C	65	.246			10.60	25	35.60	44.50

32 31 13.33 Chain Link Backstops

		Crew	Daily Output	Labor-Hours	Unit	Material	Labor	Equipment	Total	Total Incl O&P
0010	**CHAIN LINK BACKSTOPS**									
0015	Backstops, baseball, prefabricated, 30′ wide, 12′ high & 1 overhang	B-1	1	24	Ea.	2,575	920		3,495	4,350
0100	40′ wide, 12′ high & 2 overhangs	"	.75	32		6,775	1,225		8,000	9,450
0110	Regulation, galvanized	B-13	2.40	23.333		14,900	955	305	16,160	18,300
0120	Vinyl coated	"	2.40	23.333		16,000	955	305	17,260	19,500
0180	Softball, prefabricated, no overhang	B-1	1.60	15		2,525	575		3,100	3,750
0200	Softball, regulation, galv., 3″ posts @8′, 6 & 9 ga. mesh, 14′ H	B-13	8.80	6.364		2,775	261	83.50	3,119.50	3,575
0205	18′ high		.75	74.667		6,875	3,050	980	10,905	13,600
0210	20′ high		.72	77.778		7,475	3,200	1,025	11,700	14,500
0215	22′ high		.70	80		7,925	3,275	1,050	12,250	15,200
0220	24′ high		.60	92.838		8,025	3,800	1,225	13,050	16,400
0250	Vinyl coated, 14′ high		1.20	46.667		8,675	1,925	615	11,215	13,300
0255	18′ high		.75	74.667		8,750	3,050	980	12,780	15,700
0260	20′ high		.72	77.778		12,900	3,200	1,025	17,125	20,500
0265	22′ high		.70	80		13,900	3,275	1,050	18,225	21,800
0270	24′ high		.60	93.333		15,000	3,825	1,225	20,050	24,100
0300	Basketball, steel, single goal		3.04	18.421		1,425	755	242	2,422	3,075
0400	Double goal		1.92	29.167		1,925	1,200	385	3,510	4,500
0600	Tennis, wire mesh with pair of ends	B-1	2.48	9.677	Set	2,650	370		3,020	3,500
0700	Enclosed court	"	1.30	18.462	Ea.	8,975	705		9,680	11,000

32 31 13.40 Fence, Fabric and Accessories

		Crew	Daily Output	Labor-Hours	Unit	Material	Labor	Equipment	Total	Total Incl O&P
0010	**FENCE, FABRIC & ACCESSORIES**									
1000	Fabric, 9 ga., galv., 1.2 oz. coat, 2″ chain link, 4′	B-80A	304	.079	L.F.	3.44	2.97	1	7.41	9.75
1150	5′		285	.084		4.15	3.17	1.06	8.38	10.95
1200	6′		266	.090		8.05	3.39	1.14	12.58	15.65
1250	7′		247	.097		9.70	3.65	1.23	14.58	18.05
1300	8′		228	.105		12.40	3.96	1.33	17.69	21.50
1400	9 ga., fused, 4′		304	.079		4.04	2.97	1	8.01	10.40
1450	5′		285	.084		4.60	3.17	1.06	8.83	11.40
1500	6′		266	.090		4.72	3.39	1.14	9.25	12
1550	7′		247	.097		5.95	3.65	1.23	10.83	13.90
1600	8′		228	.105		9.20	3.96	1.33	14.49	18.05
1650	Barbed wire, galv., cost per strand		2280	.011		.13	.40	.13	.66	.94
1700	Vinyl coated		2280	.011		.17	.40	.13	.70	.99
1750	Extension arms, 3 strands		143	.168	Ea.	4.13	6.30	2.12	12.55	17.20
1800	6 strands, 2-3/8″		119	.202		10.15	7.60	2.55	20.30	26.50
1850	Eye tops, 2-3/8″		143	.168		1.72	6.30	2.12	10.14	14.55
1900	Top rail, incl. tie wires, 1-5/8″, galv.		912	.026	L.F.	4.65	.99	.33	5.97	7.10
1950	Vinyl coated		912	.026		5.25	.99	.33	6.57	7.80
2100	Rail, middle/bottom, w/tie wire, 1-5/8″, galv.		912	.026		4.65	.99	.33	5.97	7.10

32 31 Fences and Gates

32 31 13 – Chain Link Fences and Gates

32 31 13.40 Fence, Fabric and Accessories		Crew	Daily Output	Labor-Hours	Unit	2015 Bare Costs Material	Labor	Equipment	Total	Total Incl O&P
2150	Vinyl coated	B-80A	912	.026	L.F.	5.25	.99	.33	6.57	7.80
2200	Reinforcing wire, coiled spring, 7 ga. galv.		2279	.011		.17	.40	.13	.70	.99
2250	9 ga., vinyl coated		2282	.011		.52	.40	.13	1.05	1.37
2300	Steel T-post, galvanized with clips, 5′, common earth, flat		200	.120	Ea.	9.60	4.51	1.52	15.63	19.60
2310	Clay		176	.136		9.60	5.15	1.72	16.47	21
2320	Soil & rock		144	.167		9.60	6.25	2.11	17.96	23
2330	5.5′, common earth, flat		200	.120		10.80	4.51	1.52	16.83	21
2340	Clay		176	.136		10.80	5.15	1.72	17.67	22
2350	Soil & rock		144	.167		10.80	6.25	2.11	19.16	24.50
2360	6′, common earth, flat		200	.120		11.25	4.51	1.52	17.28	21.50
2370	Clay		176	.136		11.25	5.15	1.72	18.12	22.50
2375	Soil & rock		144	.167		11.25	6.25	2.11	19.61	25
2600	Steel T-post, galvanized with clips, 5′, common earth, hills		180	.133		9.60	5	1.68	16.28	20.50
2610	Clay		160	.150		9.60	5.65	1.89	17.14	22
2620	Soil & rock		130	.185		9.60	6.95	2.33	18.88	24.50
2630	5.5′, common earth, hills		180	.133		10.80	5	1.68	17.48	22
2640	Clay		160	.150		10.80	5.65	1.89	18.34	23
2650	Soil & rock		130	.185		10.80	6.95	2.33	20.08	26
2660	6′, common earth, hills		180	.133		11.25	5	1.68	17.93	22.50
2670	Clay		160	.150		11.25	5.65	1.89	18.79	23.50
2675	Soil & rock		130	.185		11.25	6.95	2.33	20.53	26.50

32 31 13.53 High-Security Chain Link Fences, Gates and Sys.

		Crew	Daily Output	Labor-Hours	Unit	Material	Labor	Equipment	Total	Total Incl O&P
0010	**HIGH-SECURITY CHAIN LINK FENCES, GATES AND SYSTEMS**									
0100	Fence, chain link, security, 7′ H, standard FE-7, incl excavation & posts	B-80C	480	.050	L.F.	44	1.91	.53	46.44	52
0200	Fence, barbed wire, security, 7′ high, with 3 wire barbed wire arm	"	400	.060	"	7.75	2.29	.63	10.67	12.95
0300	Complete systems, including material and installation									
0310	Taunt wire fence detection system				M.L.F.				25,100	27,600
0410	Microwave fence detection system								41,300	45,400
0510	Passive magnetic fence detection system								19,500	21,400
0610	Infrared fence detection system								12,900	14,400
0710	Strain relief fence detection system								25,100	27,600
0810	Electro-shock fence detection system								35,900	39,500
0910	Photo-electric fence detection system								16,300	18,000

32 31 19 – Decorative Metal Fences and Gates

32 31 19.10 Decorative Fence

		Crew	Daily Output	Labor-Hours	Unit	Material	Labor	Equipment	Total	Total Incl O&P
0010	**DECORATIVE FENCE**									
5300	Tubular picket, steel, 6′ sections, 1-9/16″ posts, 4′ high	B-80C	300	.080	L.F.	31	3.05	.85	34.90	40
5400	2″ posts, 5′ high		240	.100		35	3.81	1.06	39.87	46
5600	2″ posts, 6′ high		200	.120		42	4.57	1.27	47.84	55.50
5700	Staggered picket 1-9/16″ posts, 4′ high		300	.080		31	3.05	.85	34.90	40
5800	2″ posts, 5′ high		240	.100		35	3.81	1.06	39.87	46
5900	2″ posts, 6′ high		200	.120		42	4.57	1.27	47.84	55.50
6200	Gates, 4′ high, 3′ wide	B-1	10	2.400	Ea.	282	92		374	460
6300	5′ high, 3′ wide		10	2.400		350	92		442	535
6400	6′ high, 3′ wide		10	2.400		415	92		507	605
6500	4′ wide		10	2.400		420	92		512	615

32 31 Fences and Gates

32 31 23 – Plastic Fences and Gates

32 31 23.10 Fence, Vinyl		Crew	Daily Output	Labor-Hours	Unit	Material	Labor	Equipment	Total	Total Incl O&P
						2015 Bare Costs				
0010	**FENCE, VINYL**									
0011	White, steel reinforced, stainless steel fasteners									
0020	Picket, 4" x 4" posts @ 6' - 0" OC, 3' high	B-1	140	.171	L.F.	22.50	6.55		29.05	35.50
0030	4' high		130	.185		24.50	7.05		31.55	38.50
0040	5' high		120	.200		27.50	7.65		35.15	42.50
0100	Board (semi-privacy), 5" x 5" posts @ 7' - 6" OC, 5' high		130	.185		25	7.05		32.05	39
0120	6' high		125	.192		28.50	7.35		35.85	43
0200	Basketweave, 5" x 5" posts @ 7' - 6" OC, 5' high		160	.150		25.50	5.75		31.25	37.50
0220	6' high		150	.160		28.50	6.10		34.60	41
0300	Privacy, 5" x 5" posts @ 7' - 6" OC, 5' high		130	.185		24.50	7.05		31.55	38.50
0320	6' high		150	.160		28	6.10		34.10	41
0350	Gate, 5' high		9	2.667	Ea.	310	102		412	505
0360	6' high		9	2.667		355	102		457	555
0400	For posts set in concrete, add		25	.960		8.45	36.50		44.95	69.50
0500	Post and rail fence, 2 rail		150	.160	L.F.	5.90	6.10		12	16.55
0510	3 rail		150	.160		7.60	6.10		13.70	18.40
0515	4 rail		150	.160		9.85	6.10		15.95	21

32 31 26 – Wire Fences and Gates

32 31 26.10 Fences, Misc. Metal

		Crew	Daily Output	Labor-Hours	Unit	Material	Labor	Equipment	Total	Total Incl O&P
0010	**FENCES, MISC. METAL**									
0012	Chicken wire, posts @ 4', 1" mesh, 4' high	B-80C	410	.059	L.F.	3.30	2.23	.62	6.15	7.95
0100	2" mesh, 6' high		350	.069		3.82	2.61	.73	7.16	9.25
0200	Galv. steel, 12 ga., 2" x 4" mesh, posts 5' O.C., 3' high		300	.080		2.78	3.05	.85	6.68	8.95
0300	5' high		300	.080		3.38	3.05	.85	7.28	9.60
0400	14 ga., 1" x 2" mesh, 3' high		300	.080		3.42	3.05	.85	7.32	9.65
0500	5' high		300	.080		4.57	3.05	.85	8.47	10.95
1000	Kennel fencing, 1-1/2" mesh, 6' long, 3'-6" wide, 6'-2" high	2 Clab	4	4	Ea.	510	150		660	805
1050	12' long		4	4		720	150		870	1,025
1200	Top covers, 1-1/2" mesh, 6' long		15	1.067		136	40		176	216
1250	12' long		12	1.333		191	50		241	292
1300	For kennel doors, see Section 08 31 13.40									
4500	Security fence, prison grade, set in concrete, 12' high	B-80	25	1.280	L.F.	61.50	53	29	143.50	185
4600	16' high	"	20	1.600	"	79	66	36	181	234

32 31 26.20 Wire Fencing, General

		Crew	Daily Output	Labor-Hours	Unit	Material	Labor	Equipment	Total	Total Incl O&P
0010	**WIRE FENCING, GENERAL**									
0015	Barbed wire, galvanized, domestic steel, hi-tensile 15-1/2 ga.				M.L.F.	98.50			98.50	108
0020	Standard, 12-3/4 ga.					111			111	122
0210	Barbless wire, 2-strand galvanized, 12-1/2 ga.					111			111	122
0500	Helical razor ribbon, stainless steel, 18" dia x 18" spacing				C.L.F.	164			164	180
0600	Hardware cloth galv., 1/4" mesh, 23 ga., 2' wide				C.S.F.	60			60	66
0700	3' wide					43.50			43.50	48
0900	1/2" mesh, 19 ga., 2' wide					35.50			35.50	39
1000	4' wide					23.50			23.50	26
1200	Chain link fabric, steel, 2" mesh, 6 ga., galvanized					152			152	167
1300	9 ga., galvanized					86			86	94.50
1350	Vinyl coated					82			82	90
1360	Aluminized					79.50			79.50	87.50
1400	2-1/4" mesh, 11.5 ga., galvanized					54.50			54.50	60
1600	1-3/4" mesh (tennis courts), 11.5 ga. (core), vinyl coated					61			61	67
1700	9 ga., galvanized					82.50			82.50	90.50
2100	Welded wire fabric, galvanized, 1" x 2", 14 ga.					59.50			59.50	65.50

32 31 26 – Wire Fences and Gates

32 31 26.20 Wire Fencing, General		Crew	Daily Output	Labor-Hours	Unit	2015 Bare Costs Material	Labor	Equipment	Total	Total Incl O&P
2200	2" x 4", 12-1/2 ga.				C.S.F.	57			57	62.50

32 31 29 – Wood Fences and Gates

32 31 29.10 Fence, Wood		Crew	Daily Output	Labor-Hours	Unit	2015 Bare Costs Material	Labor	Equipment	Total	Total Incl O&P
0010	**FENCE, WOOD**									
0011	Basket weave, 3/8" x 4" boards, 2" x 4"									
0020	stringers on spreaders, 4" x 4" posts									
0050	No. 1 cedar, 6' high	B-80C	160	.150	L.F.	26	5.70	1.59	33.29	39.50
0070	Treated pine, 6' high	"	150	.160	"	34.50	6.10	1.69	42.29	50
0200	Board fence, 1" x 4" boards, 2" x 4" rails, 4" x 4" post									
0220	Preservative treated, 2 rail, 3' high	B-80C	145	.166	L.F.	9.30	6.30	1.75	17.35	22.50
0240	4' high		135	.178		10.90	6.75	1.88	19.53	25
0260	3 rail, 5' high		130	.185		11.85	7.05	1.95	20.85	26.50
0300	6' high		125	.192		13.35	7.30	2.03	22.68	29
0320	No. 2 grade western cedar, 2 rail, 3' high		145	.166		10.20	6.30	1.75	18.25	23.50
0340	4' high		135	.178		11.25	6.75	1.88	19.88	25.50
0360	3 rail, 5' high		130	.185		12.60	7.05	1.95	21.60	27.50
0400	6' high		125	.192		13.45	7.30	2.03	22.78	29
0420	No. 1 grade cedar, 2 rail, 3' high		145	.166		12.30	6.30	1.75	20.35	25.50
0440	4' high		135	.178		13.95	6.75	1.88	22.58	28.50
0460	3 rail, 5' high		130	.185		17	7.05	1.95	26	32.50
0500	6' high		125	.192		18.75	7.30	2.03	28.08	34.50
0540	Shadow box, 1" x 6" board, 2" x 4" rail, 4" x 4"post									
0560	Pine, pressure treated, 3 rail, 6' high	B-80C	150	.160	L.F.	17.20	6.10	1.69	24.99	30.50
0600	Gate, 3'-6" wide		8	3	Ea.	107	114	31.50	252.50	340
0620	No. 1 cedar, 3 rail, 4' high		130	.185	L.F.	18	7.05	1.95	27	33.50
0640	6' high		125	.192		23	7.30	2.03	32.33	39
0860	Open rail fence, split rails, 2 rail 3' high, no. 1 cedar		160	.150		11.75	5.70	1.59	19.04	24
0870	No. 2 cedar		160	.150		10.75	5.70	1.59	18.04	23
0880	3 rail, 4' high, no. 1 cedar		150	.160		12.10	6.10	1.69	19.89	25
0890	No. 2 cedar		150	.160		8.55	6.10	1.69	16.34	21
0920	Rustic rails, 2 rail 3' high, no. 1 cedar		160	.150		9.35	5.70	1.59	16.64	21.50
0930	No. 2 cedar		160	.150		8.75	5.70	1.59	16.04	20.50
0940	3 rail, 4' high		150	.160		8.95	6.10	1.69	16.74	21.50
0950	No. 2 cedar		150	.160		7.30	6.10	1.69	15.09	19.85
0960	Picket fence, gothic, pressure treated pine									
1000	2 rail, 3' high	B-80C	140	.171	L.F.	7.15	6.55	1.81	15.51	20.50
1020	3 rail, 4' high		130	.185	"	8.15	7.05	1.95	17.15	22.50
1040	Gate, 3'-6" wide		9	2.667	Ea.	68	102	28	198	272
1060	No. 2 cedar, 2 rail, 3' high		140	.171	L.F.	8.25	6.55	1.81	16.61	21.50
1100	3 rail, 4' high		130	.185	"	8.35	7.05	1.95	17.35	23
1120	Gate, 3'-6" wide		9	2.667	Ea.	72.50	102	28	202.50	277
1140	No. 1 cedar, 2 rail 3' high		140	.171	L.F.	12.55	6.55	1.81	20.91	26.50
1160	3 rail, 4' high		130	.185	"	16.35	7.05	1.95	25.35	31.50
1170	Gate, 3'-6" wide		9	2.667	Ea.	232	102	28	362	450
1200	Rustic picket, molded pine, 2 rail, 3' high		140	.171	L.F.	7.75	6.55	1.81	16.11	21
1220	No. 1 cedar, 2 rail, 3' high		140	.171		8.95	6.55	1.81	17.31	22.50
1240	Stockade fence, no. 1 cedar, 3-1/4" rails, 6' high		160	.150		12.05	5.70	1.59	19.34	24.50
1260	8' high		155	.155		16.60	5.90	1.64	24.14	29.50
1300	No. 2 cedar, treated wood rails, 6' high		160	.150		12.35	5.70	1.59	19.64	24.50
1320	Gate, 3'-6" wide		8	3	Ea.	82	114	31.50	227.50	310
1360	Treated pine, treated rails, 6' high		160	.150	L.F.	12.80	5.70	1.59	20.09	25
1400	8' high		150	.160	"	19	6.10	1.69	26.79	33

32 31 Fences and Gates

32 31 29 – Wood Fences and Gates

32 31 29.20 Fence, Wood Rail		Crew	Daily Output	Labor-Hours	Unit	2015 Bare Costs Material	Labor	Equipment	Total	Total Incl O&P
0010	**FENCE, WOOD RAIL**									
0012	Picket, No. 2 cedar, Gothic, 2 rail, 3′ high	B-1	160	.150	L.F.	7.60	5.75		13.35	17.75
0050	Gate, 3′-6″ wide	B-80C	9	2.667	Ea.	77	102	28	207	282
0400	3 rail, 4′ high		150	.160	L.F.	8.50	6.10	1.69	16.29	21
0500	Gate, 3′-6″ wide		9	2.667	Ea.	95	102	28	225	300
5000	Fence rail, redwood, 2″ x 4″, merch. grade 8′	B-1	2400	.010	L.F.	2.42	.38		2.80	3.29
5050	Select grade, 8′		2400	.010	"	5.35	.38		5.73	6.50
6000	Fence post, select redwood, earthpacked & treated, 4″ x 4″ x 6′		96	.250	Ea.	13.60	9.55		23.15	30.50
6010	4″ x 4″ x 8′		96	.250		18.70	9.55		28.25	36
6020	Set in concrete, 4″ x 4″ x 6′		50	.480		21	18.35		39.35	53.50
6030	4″ x 4″ x 8′		50	.480		22.50	18.35		40.85	54.50
6040	Wood post, 4′ high, set in concrete, incl. concrete		50	.480		13.80	18.35		32.15	45
6050	Earth packed		96	.250		16.70	9.55		26.25	34
6060	6′ high, set in concrete, incl. concrete		50	.480		17.20	18.35		35.55	49
6070	Earth packed		96	.250		13.10	9.55		22.65	30
9000	Minimum labor/equipment charge	1 Clab	2	4	Job		150		150	247

32 32 Retaining Walls

32 32 13 – Cast-in-Place Concrete Retaining Walls

32 32 13.10 Retaining Walls, Cast Concrete		Crew	Daily Output	Labor-Hours	Unit	Material	Labor	Equipment	Total	Total Incl O&P
0010	**RETAINING WALLS, CAST CONCRETE**									
1800	Concrete gravity wall with vertical face including excavation & backfill									
1850	No reinforcing									
1900	6′ high, level embankment	C-17C	36	2.306	L.F.	74.50	113	16.90	204.40	285
2000	33° slope embankment		32	2.594		86.50	127	19.05	232.55	325
2200	8′ high, no surcharge		27	3.074		92.50	151	22.50	266	370
2300	33° slope embankment		24	3.458		112	170	25.50	307.50	425
2500	10′ high, level embankment		19	4.368		132	215	32	379	530
2600	33° slope embankment		18	4.611		183	227	34	444	610
2800	Reinforced concrete cantilever, incl. excavation, backfill & reinf.									
2900	6′ high, 33° slope embankment	C-17C	35	2.371	L.F.	68	117	17.40	202.40	283
3000	8′ high, 33° slope embankment		29	2.862		78	141	21	240	335
3100	10′ high, 33° slope embankment		20	4.150		102	204	30.50	336.50	475
3200	20′ high, 500 lb. per L.F. surcharge		7.50	11.067		305	545	81	931	1,300
3500	Concrete cribbing, incl. excavation and backfill									
3700	12′ high, open face	B-13	210	.267	S.F.	34	10.95	3.51	48.46	59
3900	Closed face	"	210	.267	"	32	10.95	3.51	46.46	56.50
4100	Concrete filled slurry trench, see Section 31 56 23.20									

32 32 23 – Segmental Retaining Walls

32 32 23.13 Segmental Conc. Unit Masonry Retaining Walls		Crew	Daily Output	Labor-Hours	Unit	Material	Labor	Equipment	Total	Total Incl O&P
0010	**SEGMENTAL CONC. UNIT MASONRY RETAINING WALLS**									
7100	Segmental Retaining Wall system, incl. pins, and void fill									
7120	base and backfill not included									
7140	Large unit, 8″ high x 18″ wide x 20″ deep, 3 plane split	B-62	300	.080	S.F.	13.50	3.30	.58	17.38	21
7150	Straight split		300	.080		13.60	3.30	.58	17.48	21
7160	Medium, lt. wt., 8″ high x 18″ wide x 12″ deep, 3 plane split		400	.060		10.50	2.48	.43	13.41	16.05
7170	Straight split		400	.060		10.40	2.48	.43	13.31	15.95
7180	Small unit, 4″ x 18″ x 10″ deep, 3 plane split		400	.060		13.40	2.48	.43	16.31	19.25
7190	Straight split		400	.060		13.20	2.48	.43	16.11	19
7200	Cap unit, 3 plane split		300	.080		13.80	3.30	.58	17.68	21

32 32 23 – Segmental Retaining Walls

32 32 23.13 Segmental Conc. Unit Masonry Retaining Walls		Crew	Daily Output	Labor-Hours	Unit	2015 Bare Costs Material	Labor	Equipment	Total	Total Incl O&P
7210	Cap unit, straight split	B-62	300	.080	S.F.	13.80	3.30	.58	17.68	21
7250	Geo-grid soil reinforcement 4′ x 50′	2 Clab	22500	.001		.76	.03		.79	.88
7255	Geo-grid soil reinforcement 6′ x 150′	"	22500	.001		.60	.03		.63	.70
8000	For higher walls, add components as necessary									

32 32 26 – Metal Crib Retaining Walls

32 32 26.10 Metal Bin Retaining Walls

Code	Description	Crew	Daily Output	Labor-Hours	Unit	Material	Labor	Equipment	Total	Total Incl O&P
0010	**METAL BIN RETAINING WALLS**									
0011	Aluminized steel bin, excavation									
0020	and backfill not included, 10′ wide									
0100	4′ high, 5.5′ deep	B-13	650	.086	S.F.	27	3.53	1.13	31.66	36.50
0200	8′ high, 5.5′ deep		615	.091		31	3.73	1.20	35.93	41.50
0300	10′ high, 7.7′ deep		580	.097		34.50	3.96	1.27	39.73	46
0400	12′ high, 7.7′ deep		530	.106		37	4.33	1.39	42.72	49.50
0500	16′ high, 7.7′ deep		515	.109		39.50	4.46	1.43	45.39	52
0600	16′ high, 9.9′ deep		500	.112		41.50	4.59	1.47	47.56	54.50
0700	20′ high, 9.9′ deep		470	.119		46.50	4.88	1.57	52.95	61
0800	20′ high, 12.1′ deep		460	.122		42	4.99	1.60	48.59	56
0900	24′ high, 12.1′ deep		455	.123		44.50	5.05	1.62	51.17	59
1000	24′ high, 14.3′ deep		450	.124		52.50	5.10	1.64	59.24	67.50
1100	28′ high, 14.3′ deep		440	.127		54.50	5.20	1.67	61.37	70.50
1300	For plain galvanized bin type walls, deduct					10%				

32 32 29 – Timber Retaining Walls

32 32 29.10 Landscape Timber Retaining Walls

Code	Description	Crew	Daily Output	Labor-Hours	Unit	Material	Labor	Equipment	Total	Total Incl O&P
0010	**LANDSCAPE TIMBER RETAINING WALLS**									
0100	Treated timbers, 6" x 6"	1 Clab	265	.030	L.F.	2.01	1.14		3.15	4.07
0110	6" x 8"	"	200	.040	"	2.65	1.50		4.15	5.40
0120	Drilling holes in timbers for fastening, 1/2"	1 Carp	450	.018	Inch		.83		.83	1.37
0130	5/8"	"	450	.018	"		.83		.83	1.37
0140	Reinforcing rods for fastening, 1/2"	1 Clab	312	.026	L.F.	.36	.96		1.32	1.97
0150	5/8"	"	312	.026	"	.56	.96		1.52	2.19
0160	Reinforcing fabric	2 Clab	2500	.006	S.Y.	1.90	.24		2.14	2.48
0170	Gravel backfill		28	.571	C.Y.	20	21.50		41.50	57
0180	Perforated pipe, 4" diameter with silt sock		1200	.013	L.F.	1.25	.50		1.75	2.20
0190	Galvanized 60d common nails	1 Clab	625	.013	Ea.	.16	.48		.64	.97
0200	20d common nails	"	3800	.002	"	.04	.08		.12	.17

32 32 36 – Gabion Retaining Walls

32 32 36.10 Stone Gabion Retaining Walls

Code	Description	Crew	Daily Output	Labor-Hours	Unit	Material	Labor	Equipment	Total	Total Incl O&P
0010	**STONE GABION RETAINING WALLS**									
4300	Stone filled gabions, not incl. excavation,									
4310	Stone, delivered, 3′ wide									
4340	Galvanized, 6′ long, 1′ high	B-13	113	.496	Ea.	105	20.50	6.50	132	156
4400	1′-6" high		50	1.120		164	46	14.75	224.75	271
4490	3′-0" high		13	4.308		244	177	56.50	477.50	615
4590	9′ long, 1′ high		50	1.120		198	46	14.75	258.75	310
4650	1′-6" high		22	2.545		212	104	33.50	349.50	440
4690	3′-0" high		6	9.333		375	385	123	883	1,175
4890	12′ long, 1′ high		28	2		238	82	26.50	346.50	425
4950	1′-6" high		13	4.308		315	177	56.50	548.50	695
4990	3′-0" high		3	18.667		450	765	245	1,460	2,025
5200	PVC coated, 6′ long, 1′ high		113	.496		110	20.50	6.50	137	161
5250	1′-6" high		50	1.120		168	46	14.75	228.75	276

32 32 36 – Gabion Retaining Walls

32 32 36.10 Stone Gabion Retaining Walls		Crew	Daily Output	Labor-Hours	Unit	2015 Bare Costs Material	Labor	Equipment	Total	Total Incl O&P
5300	3′ high	B-13	13	4.308	Ea.	261	177	56.50	494.50	635
5500	9′ long, 1′ high		50	1.120		209	46	14.75	269.75	320
5550	1′-6″ high		22	2.545		233	104	33.50	370.50	460
5600	3′ high		6	9.333		410	385	123	918	1,200
5800	12′ long, 1′ high		28	2		278	82	26.50	386.50	465
5850	1′-6″ high		13	4.308		335	177	56.50	568.50	720
5900	3′ high		3	18.667		480	765	245	1,490	2,050
6000	Galvanized, 6′ long, 1′ high		75	.747	C.Y.	95.50	30.50	9.80	135.80	165
6010	1′-6″ high		50	1.120		164	46	14.75	224.75	271
6020	3′-0″ high		25	2.240		122	92	29.50	243.50	315
6030	9′ long, 1′ high		50	1.120		198	46	14.75	258.75	310
6040	1′-6″ high		33.30	1.682		141	69	22	232	292
6050	3′-0″ high		16.70	3.353		125	137	44	306	410
6060	12′ long, 1′ high		37.50	1.493		178	61	19.65	258.65	315
6070	1′-6″ high		25	2.240		157	92	29.50	278.50	355
6080	3′-0″ high		12.50	4.480		112	184	59	355	485
6100	PVC coated, 6′ long, 1′ high		75	.747		165	30.50	9.80	205.30	242
6110	1′-6″ high		50	1.120		168	46	14.75	228.75	276
6120	3′ high		25	2.240		130	92	29.50	251.50	325
6130	9′ long, 1′ high		50	1.120		209	46	14.75	269.75	320
6140	1′-6″ high		33.30	1.682		155	69	22	246	310
6150	3′ high		16.67	3.359		136	138	44	318	420
6160	12′ long, 1′ high		37.50	1.493		209	61	19.65	289.65	350
6170	1′-6″ high		25	2.240		149	92	29.50	270.50	345
6180	3′ high		12.50	4.480		120	184	59	363	495

32 32 53 – Stone Retaining Walls

32 32 53.10 Retaining Walls, Stone		Crew	Daily Output	Labor-Hours	Unit	Material	Labor	Equipment	Total	Total Incl O&P
0010	**RETAINING WALLS, STONE**									
0015	Including excavation, concrete footing and									
0020	stone 3′ below grade. Price is exposed face area.									
0200	Decorative random stone, to 6′ high, 1′-6″ thick, dry set	D-1	35	.457	S.F.	58.50	19.25		77.75	96
0300	Mortar set		40	.400		60.50	16.85		77.35	94
0500	Cut stone, to 6′ high, 1′-6″ thick, dry set		35	.457		60.50	19.25		79.75	98.50
0600	Mortar set		40	.400		61.50	16.85		78.35	95
0800	Random stone, 6′ to 10′ high, 2′ thick, dry set		45	.356		66.50	14.95		81.45	97.50
0900	Mortar set		50	.320		69	13.45		82.45	97.50
1100	Cut stone, 6′ to 10′ high, 2′ thick, dry set		45	.356		67	14.95		81.95	98
1200	Mortar set		50	.320		69	13.45		82.45	98
5100	Setting stone, dry		100	.160	C.F.		6.75		6.75	10.95
5600	With mortar		120	.133	″		5.60		5.60	9.15
9000	Minimum labor/equipment charge		2	8	Job		335		335	550

32 33 Site Furnishings

32 33 33 – Site Manufactured Planters

32 33 33.10 Planters		Crew	Daily Output	Labor-Hours	Unit	Material	Labor	Equipment	Total	Total Incl O&P
						2015 Bare Costs				
0010	**PLANTERS**									
0012	Concrete, sandblasted, precast, 48" diameter, 24" high	2 Clab	15	1.067	Ea.	630	40		670	760
0100	Fluted, precast, 7' diameter, 36" high		10	1.600		1,575	60		1,635	1,825
0300	Fiberglass, circular, 36" diameter, 24" high		15	1.067		725	40		765	860
0320	36" diameter, 27" high		12	1.333		725	50		775	875
0330	33" high		15	1.067		770	40		810	910
0335	24" diameter, 36" high		15	1.067		425	40		465	535
0340	60" diameter, 39" high		8	2		1,275	75		1,350	1,525
0400	60" diameter, 24" high		10	1.600		1,125	60		1,185	1,325
0600	Square, 24" side, 36" high		15	1.067		620	40		660	745
0610	24" side, 27" high		12	1.333		675	50		725	820
0620	24" side, 16" high		20	.800		320	30		350	405
0700	48" side, 36" high		15	1.067		1,025	40		1,065	1,200
0900	Planter/bench, 72" square, 36" high		5	3.200		1,800	120		1,920	2,175
1000	96" square, 27" high		5	3.200		2,225	120		2,345	2,650
1200	Wood, square, 48" side, 24" high		15	1.067		1,400	40		1,440	1,625
1300	Circular, 48" diameter, 30" high		10	1.600		985	60		1,045	1,175
1500	72" diameter, 30" high		10	1.600		1,725	60		1,785	2,000
1600	Planter/bench, 72"	↓	5	3.200	↓	3,275	120		3,395	3,800
9000	Minimum labor/equipment charge	1 Clab	2	4	Job		150		150	247

32 33 43 – Site Seating and Tables

32 33 43.13 Site Seating		Crew	Daily Output	Labor-Hours	Unit	Material	Labor	Equipment	Total	Total Incl O&P
0010	**SITE SEATING**									
0012	Seating, benches, park, precast conc., w/backs, wood rails, 4' long	2 Clab	5	3.200	Ea.	595	120		715	850
0100	8' long		4	4		955	150		1,105	1,300
0300	Fiberglass, without back, one piece, 4' long		10	1.600		620	60		680	780
0400	8' long		7	2.286		815	86		901	1,050
0500	Steel barstock pedestals w/backs, 2" x 3" wood rails, 4' long		10	1.600		1,125	60		1,185	1,325
0510	8' long		7	2.286		1,425	86		1,511	1,725
0515	Powder coated steel, 4" x 4" plastic slats, 6' L		8	2		490	75		565	660
0520	3" x 8" wood plank, 4' long		10	1.600		1,200	60		1,260	1,425
0530	8' long		7	2.286		1,500	86		1,586	1,800
0540	Backless, 4" x 4" wood plank, 4' square		10	1.600		940	60		1,000	1,125
0550	8' long		7	2.286		1,000	86		1,086	1,250
0560	Powder coated steel, with back and 2 anti-vagrant dividers, 6' long		8	2		1,075	75		1,150	1,325
0600	Aluminum pedestals, with backs, aluminum slats, 8' long		8	2		480	75		555	650
0610	15' long		5	3.200		975	120		1,095	1,275
0620	Portable, aluminum slats, 8' long		8	2		465	75		540	635
0630	15' long		5	3.200		570	120		690	825
0800	Cast iron pedestals, back & arms, wood slats, 4' long		8	2		385	75		460	545
0820	8' long		5	3.200		1,075	120		1,195	1,375
0840	Backless, wood slats, 4' long		8	2		590	75		665	770
0860	8' long		5	3.200		1,150	120		1,270	1,475
1700	Steel frame, fir seat, 10' long		10	1.600		355	60		415	495
2000	Benches, park, with back, galv. stl. frame, 4" x 4" plastic slats, 6' L		7	2.286	↓	490	86		576	675
9000	Minimum labor/equipment charge	↓	2	8	Job		300		300	495

32 34 Fabricated Bridges

32 34 20 – Fabricated Pedestrian Bridges

32 34 20.10 Bridges, Pedestrian		Crew	Daily Output	Labor-Hours	Unit	2015 Bare Costs Material	Labor	Equipment	Total	Total Incl O&P
0010	**BRIDGES, PEDESTRIAN**									
0011	Spans over streams, roadways, etc.									
0020	including erection, not including foundations									
0050	Precast concrete, complete in place, 8′ wide, 60′ span	E-2	215	.260	S.F.	116	13.50	7.05	136.55	159
0100	100′ span		185	.303		127	15.65	8.15	150.80	177
0150	120′ span		160	.350		138	18.10	9.45	165.55	194
0200	150′ span		145	.386		144	20	10.40	174.40	204
0300	Steel, trussed or arch spans, compl. in place, 8′ wide, 40′ span		320	.175		117	9.05	4.72	130.77	150
0400	50′ span		395	.142		105	7.35	3.83	116.18	133
0500	60′ span		465	.120		105	6.25	3.25	114.50	130
0600	80′ span		570	.098		125	5.10	2.65	132.75	150
0700	100′ span		465	.120		176	6.25	3.25	185.50	208
0800	120′ span		365	.153		223	7.95	4.14	235.09	263
0900	150′ span		310	.181		237	9.35	4.87	251.22	282
1000	160′ span		255	.220		237	11.35	5.95	254.30	286
1100	10′ wide, 80′ span		640	.088		125	4.53	2.36	131.89	148
1200	120′ span		415	.135		162	7	3.64	172.64	194
1300	150′ span		445	.126		182	6.50	3.40	191.90	215
1400	200′ span		205	.273		194	14.15	7.35	215.50	246
1600	Wood, laminated type, complete in place, 80′ span	C-12	203	.236		86.50	11	3.22	100.72	117
1700	130′ span	"	153	.314		90.50	14.60	4.27	109.37	128

32 84 Planting Irrigation

32 84 23 – Underground Sprinklers

32 84 23.10 Sprinkler Irrigation System		Crew	Daily Output	Labor-Hours	Unit	Material	Labor	Equipment	Total	Total Incl O&P
0010	**SPRINKLER IRRIGATION SYSTEM**									
0011	For lawns									
0100	Golf course with fully automatic system	C-17	.05	1600	9 holes	100,000	78,500		178,500	237,500
0200	24′ diam. head at 15′ O.C. incl. piping, auto oper., minimum	B-20	70	.343	Head	27	14.40		41.40	53
0300	Maximum		40	.600		45	25		70	90.50
0600	Sprinkler irrigation sys, golf course, auto sys, 60′ dia HD		23	1.043		150	44		194	237
0800	Residential system, custom, 1″ supply		2000	.012	S.F.	.26	.50		.76	1.11
0900	1-1/2″ supply		1800	.013	"	.49	.56		1.05	1.45
1020	Pop up spray head w/risers, hi-pop, full circle pattern, 4″	2 Skwk	76	.211	Ea.	5.15	10.25		15.40	22.50
1030	1/2 circle pattern, 4″		76	.211		5.15	10.25		15.40	22.50
1040	6″, full circle pattern		76	.211		9.50	10.25		19.75	27
1050	1/2 circle pattern, 6″		76	.211		9.50	10.25		19.75	27
1060	12″, full circle pattern		76	.211		11.15	10.25		21.40	29
1070	1/2 circle pattern, 12″		76	.211		13.75	10.25		24	32
1080	Pop up bubbler head w/risers, hi-pop bubbler head, 4″		76	.211		4.40	10.25		14.65	21.50
1090	6″		76	.211		9.20	10.25		19.45	27
1100	12″		76	.211		11	10.25		21.25	29
1110	Impact full/part circle sprinklers, 28′-54′ 25-60 PSI		37	.432		17.65	21		38.65	53.50
1120	Spaced 37′-49′ @ 25-50 PSI		37	.432		22	21		43	58.50
1130	Spaced 43′-61′ @ 30-60 PSI		37	.432		61.50	21		82.50	102
1140	Spaced 54′-78′ @ 40-80 PSI		37	.432		106	21		127	151
1145	Impact rotor pop-up full/part commercial circle sprinklers									
1150	Spaced 42′-65′ 35-80 PSI	2 Skwk	25	.640	Ea.	15.30	31		46.30	67.50
1160	Spaced 48′-76′ 45-85 PSI	"	25	.640	"	16.85	31		47.85	69
1165	Impact rotor pop-up part. circle comm., 53′-75′, 55-100 PSI, w/accessories									
1170	Plastic case, metal cover	2 Skwk	25	.640	Ea.	74	31		105	132

32 84 23 – Underground Sprinklers

32 84 23.10 Sprinkler Irrigation System		Crew	Daily Output	Labor-Hours	Unit	2015 Bare Costs Material	Labor	Equipment	Total	Total Incl O&P
1180	Rubber cover	2 Skwk	25	.640	Ea.	56.50	31		87.50	113
1190	Iron case, metal cover		22	.727		122	35.50		157.50	192
1200	Rubber cover		22	.727		129	35.50		164.50	200
1250	Plastic case, 2 nozzle, metal cover		25	.640		89.50	31		120.50	149
1260	Rubber cover		25	.640		91.50	31		122.50	151
1270	Iron case, 2 nozzle, metal cover		22	.727		130	35.50		165.50	200
1280	Rubber cover		22	.727		130	35.50		165.50	200
1282	Impact rotor pop-up full circle commercial, 39'-99', 30-100 PSI									
1284	Plastic case, metal cover	2 Skwk	25	.640	Ea.	83	31		114	142
1286	Rubber cover		25	.640		94.50	31		125.50	155
1288	Iron case, metal cover		22	.727		122	35.50		157.50	192
1290	Rubber cover		22	.727		126	35.50		161.50	197
1292	Plastic case, 2 nozzle, metal cover		22	.727		90	35.50		125.50	157
1294	Rubber cover		22	.727		90	35.50		125.50	157
1296	Iron case, 2 nozzle, metal cover		20	.800		119	39		158	193
1298	Rubber cover		20	.800		123	39		162	198
1305	Electric remote control valve, plastic, 3/4"		18	.889		23.50	43		66.50	96.50
1310	1"		18	.889		23.50	43		66.50	96.50
1320	1-1/2"		18	.889		91	43		134	171
1330	2"		18	.889		109	43		152	191
1335	Quick coupling valves, brass, locking cover									
1340	Inlet coupling valve, 3/4"	2 Skwk	18.75	.853	Ea.	21.50	41.50		63	91
1350	1"		18.75	.853		30	41.50		71.50	101
1360	Controller valve boxes, 6" round boxes		18.75	.853		7.10	41.50		48.60	75.50
1370	10" round boxes		14.25	1.123		11.40	54.50		65.90	102
1380	12" square box		9.75	1.641		16.05	80		96.05	148
1388	Electromech. control, 14 day 3-60 min., auto start to 23/day									
1390	4 station	2 Skwk	1.04	15.385	Ea.	75	750		825	1,300
1400	7 station		.64	25		140	1,225		1,365	2,125
1410	12 station		.40	40		170	1,950		2,120	3,325
1420	Dual programs, 18 station		.24	66.667		200	3,250		3,450	5,500
1430	23 station		.16	100		220	4,875		5,095	8,150
1435	Backflow preventer, bronze, 0-175 PSI, w/valves, test cocks									
1440	3/4"	2 Skwk	6	2.667	Ea.	82	130		212	300
1450	1"		6	2.667		94	130		224	315
1460	1-1/2"		6	2.667		224	130		354	455
1470	2"		6	2.667		276	130		406	515
1475	Pressure vacuum breaker, brass, 15-150 PSI									
1480	3/4"	2 Skwk	6	2.667	Ea.	25	130		155	239
1490	1"		6	2.667		30	130		160	244
1500	1-1/2"		6	2.667		70	130		200	288
1510	2"		6	2.667		120	130		250	345

32 91 13 – Soil Preparation

32 91 13.16 Mulching		Crew	Daily Output	Labor-Hours	Unit	Material	2015 Bare Costs Labor	Equipment	Total	Total Incl O&P
0010	**MULCHING**									
0100	Aged barks, 3" deep, hand spread	1 Clab	100	.080	S.Y.	3.55	3.01		6.56	8.85
0150	Skid steer loader	B-63	13.50	2.963	M.S.F.	395	118	12.85	525.85	640
0200	Hay, 1" deep, hand spread	1 Clab	475	.017	S.Y.	.52	.63		1.15	1.61
0250	Power mulcher, small	B-64	180	.089	M.S.F.	58	3.41	2.21	63.62	71.50
0350	Large	B-65	530	.030	"	58	1.16	1.07	60.23	66.50
0400	Humus peat, 1" deep, hand spread	1 Clab	700	.011	S.Y.	2.42	.43		2.85	3.36
0450	Push spreader	"	2500	.003	"	2.42	.12		2.54	2.86
0550	Tractor spreader	B-66	700	.011	M.S.F.	269	.56	.38	269.94	297
0600	Oat straw, 1" deep, hand spread	1 Clab	475	.017	S.Y.	.60	.63		1.23	1.70
0650	Power mulcher, small	B-64	180	.089	M.S.F.	66.50	3.41	2.21	72.12	81.50
0700	Large	B-65	530	.030	"	66.50	1.16	1.07	68.73	76.50
0750	Add for asphaltic emulsion	B-45	1770	.009	Gal.	5.80	.41	.51	6.72	7.60
0800	Peat moss, 1" deep, hand spread	1 Clab	900	.009	S.Y.	2.80	.33		3.13	3.63
0850	Push spreader	"	2500	.003	"	2.80	.12		2.92	3.28
0950	Tractor spreader	B-66	700	.011	M.S.F.	310	.56	.38	310.94	340
1000	Polyethylene film, 6 mil	2 Clab	2000	.008	S.Y.	.46	.30		.76	1
1100	Redwood nuggets, 3" deep, hand spread	1 Clab	150	.053	"	2.77	2.01		4.78	6.35
1150	Skid steer loader	B-63	13.50	2.963	M.S.F.	310	118	12.85	440.85	545
1200	Stone mulch, hand spread, ceramic chips, economy	1 Clab	125	.064	S.Y.	6.85	2.41		9.26	11.50
1250	Deluxe	"	95	.084	"	10.60	3.17		13.77	16.85
1300	Granite chips	B-1	10	2.400	C.Y.	64	92		156	222
1400	Marble chips		10	2.400		152	92		244	320
1600	Pea gravel		28	.857		108	33		141	172
1700	Quartz		10	2.400		188	92		280	360
1800	Tar paper, 15 lb. felt	1 Clab	800	.010	S.Y.	.49	.38		.87	1.15
1900	Wood chips, 2" deep, hand spread	"	220	.036	"	1.60	1.37		2.97	4
1950	Skid steer loader	B-63	20.30	1.970	M.S.F.	178	78.50	8.55	265.05	335

32 91 13.23 Structural Soil Mixing		Crew	Daily Output	Labor-Hours	Unit	Material	Labor	Equipment	Total	Total Incl O&P
0010	**STRUCTURAL SOIL MIXING**									
0100	Rake topsoil, site material, harley rock rake, ideal	B-6	33	.727	M.S.F.		30	11.05	41.05	60.50
0200	Adverse	"	7	3.429			142	52	194	286
0300	Screened loam, york rake and finish, ideal	B-62	24	1			41.50	7.25	48.75	75
0400	Adverse	"	20	1.200			49.50	8.70	58.20	89.50
1000	Remove topsoil & stock pile on site, 75 HP dozer, 6" deep, 50' haul	B-10L	30	.400			18.50	15.75	34.25	47
1050	300' haul		6.10	1.967			91	77.50	168.50	231
1100	12" deep, 50' haul		15.50	.774			36	30.50	66.50	91
1150	300' haul		3.10	3.871			179	152	331	455
1200	200 HP dozer, 6" deep, 50' haul	B-10B	125	.096			4.44	11.10	15.54	19.30
1250	300' haul		30.70	.391			18.10	45	63.10	78.50
1300	12" deep, 50' haul		62	.194			8.95	22.50	31.45	39
1350	300' haul		15.40	.779			36	90	126	157
1400	Alternate method, 75 HP dozer, 50' haul	B-10L	860	.014	C.Y.		.65	.55	1.20	1.63
1450	300' haul	"	114	.105			4.87	4.14	9.01	12.35
1500	200 HP dozer, 50' haul	B-10B	2660	.005			.21	.52	.73	.90
1600	300' haul	"	570	.021			.97	2.43	3.40	4.24
1800	Rolling topsoil, hand push roller	1 Clab	3200	.003	S.F.		.09		.09	.15
1850	Tractor drawn roller	B-66	10666	.001	"		.04	.02	.06	.09
2000	Root raking and loading, residential, no boulders	B-6	53.30	.450	M.S.F.		18.60	6.85	25.45	37.50
2100	With boulders		32	.750			31	11.40	42.40	62.50
2200	Municipal, no boulders		200	.120			4.95	1.82	6.77	10
2300	With boulders		120	.200			8.25	3.04	11.29	16.70

32 91 13 – Soil Preparation

32 91 13.23 Structural Soil Mixing		Crew	Daily Output	Labor-Hours	Unit	Material	2015 Bare Costs Labor	Equipment	Total	Total Incl O&P
2400	Large commercial, no boulders	B-10B	400	.030	M.S.F.		1.39	3.47	4.86	6.05
2500	With boulders	″	240	.050			2.31	5.80	8.11	10.05
3000	Scarify subsoil, residential, skid steer loader w/scarifiers, 50 HP	B-66	32	.250			12.15	8.25	20.40	28.50
3050	Municipal, skid steer loader w/scarifiers, 50 HP	″	120	.067			3.24	2.19	5.43	7.55
3100	Large commercial, 75 HP, dozer w/scarifier	B-10L	240	.050			2.31	1.97	4.28	5.85
3500	Screen topsoil from stockpile, vibrating screen, wet material (organic)	B-10P	200	.060	C.Y.		2.78	5.95	8.73	11
3550	Dry material	″	300	.040			1.85	3.96	5.81	7.30
3600	Mixing with conditioners, manure and peat	B-10R	550	.022			1.01	.54	1.55	2.22
3650	Mobilization add for 2 days or less operation	B-34K	3	2.667	Job		107	315	422	520
3800	Spread conditioned topsoil, 6″ deep, by hand	B-1	360	.067	S.Y.	5.45	2.55		8	10.20
3850	300 HP dozer	B-10M	27	.444	M.S.F.	590	20.50	70.50	681	755
4000	Spread soil conditioners, alum. sulfate, 1#/S.Y., hand push spreader	1 Clab	17500	.001	S.Y.	20	.02		20.02	22
4050	Tractor spreader	B-66	700	.011	M.S.F.	2,225	.56	.38	2,225.94	2,450
4100	Fertilizer, 0.2#/S.Y., push spreader	1 Clab	17500	.001	S.Y.	.12	.02		.14	.16
4150	Tractor spreader	B-66	700	.011	M.S.F.	13.35	.56	.38	14.29	15.95
4200	Ground limestone, 1#/S.Y., push spreader	1 Clab	17500	.001	S.Y.	.11	.02		.13	.15
4250	Tractor spreader	B-66	700	.011	M.S.F.	12.20	.56	.38	13.14	14.75
4400	Manure, 18#/S.Y., push spreader	1 Clab	2500	.003	S.Y.	8.15	.12		8.27	9.15
4450	Tractor spreader	B-66	280	.029	M.S.F.	905	1.39	.94	907.33	1,000
4500	Perlite, 1″ deep, push spreader	1 Clab	17500	.001	S.Y.	10.35	.02		10.37	11.45
4550	Tractor spreader	B-66	700	.011	M.S.F.	1,150	.56	.38	1,150.94	1,275
4600	Vermiculite, push spreader	1 Clab	17500	.001	S.Y.	8.45	.02		8.47	9.35
4650	Tractor spreader	B-66	700	.011	M.S.F.	940	.56	.38	940.94	1,025
5000	Spread topsoil, skid steer loader and hand dress	B-62	270	.089	C.Y.	24.50	3.67	.64	28.81	33.50
5100	Articulated loader and hand dress	B-100	320	.038		24.50	1.74	2.99	29.23	33
5200	Articulated loader and 75HP dozer	B-10M	500	.024		24.50	1.11	3.79	29.40	33
5300	Road grader and hand dress	B-11L	1000	.016		24.50	.71	.74	25.95	29
6000	Tilling topsoil, 20 HP tractor, disk harrow, 2″ deep	B-66	450	.018	M.S.F.		.86	.59	1.45	2.01
6050	4″ deep		360	.022			1.08	.73	1.81	2.51
6100	6″ deep		270	.030			1.44	.97	2.41	3.36
6150	26″ rototiller, 2″ deep	A-1J	1250	.006	S.Y.		.24	.04	.28	.43
6200	4″ deep		1000	.008			.30	.05	.35	.54
6250	6″ deep		750	.011			.40	.06	.46	.73
7000	Lawn maintenance, see Section 32 01 30.10									

32 91 13.26 Planting Beds		Crew	Daily Output	Labor-Hours	Unit	Material	Labor	Equipment	Total	Total Incl O&P
0010	**PLANTING BEDS**									
0100	Backfill planting pit, by hand, on site topsoil	2 Clab	18	.889	C.Y.		33.50		33.50	55
0200	Prepared planting mix, by hand	″	24	.667			25		25	41
0300	Skid steer loader, on site topsoil	B-62	340	.071			2.91	.51	3.42	5.30
0400	Prepared planting mix	″	410	.059			2.42	.42	2.84	4.38
1000	Excavate planting pit, by hand, sandy soil	2 Clab	16	1			37.50		37.50	61.50
1100	Heavy soil or clay	″	8	2			75		75	123
1200	1/2 C.Y. backhoe, sandy soil	B-11C	150	.107			4.70	2.43	7.13	10.20
1300	Heavy soil or clay	″	115	.139			6.15	3.17	9.32	13.40
2000	Mix planting soil, incl. loam, manure, peat, by hand	2 Clab	60	.267		43	10.05		53.05	63.50
2100	Skid steer loader	B-62	150	.160		43	6.60	1.16	50.76	59
3000	Pile sod, skid steer loader	″	2800	.009	S.Y.		.35	.06	.41	.64
3100	By hand	2 Clab	400	.040			1.50		1.50	2.47
4000	Remove sod, F.E. loader	B-10S	2000	.006			.28	.19	.47	.65
4100	Sod cutter	B-12K	3200	.005			.22	.31	.53	.70
4200	By hand	2 Clab	240	.067			2.51		2.51	4.11

32 92 19.14 Seeding, Athletic Fields

	32 92 19.14 Seeding, Athletic Fields	Crew	Daily Output	Labor-Hours	Unit	Material	Labor	Equipment	Total	Total Incl O&P
						2015 Bare Costs				
0010	**SEEDING, ATHLETIC FIELDS** R329219-50									
0020	Seeding, athletic fields, athletic field mix, 8#/M.S.F. push spreader	1 Clab	8	1	M.S.F.	11.30	37.50		48.80	74
0100	Tractor spreader	B-66	52	.154		11.30	7.50	5.05	23.85	30
0200	Hydro or air seeding, with mulch and fertilizer	B-81	80	.300		12.45	12.85	8.65	33.95	43.50
0400	Birdsfoot trefoil, .45#/M.S.F., push spreader	1 Clab	8	1		4	37.50		41.50	66
0500	Tractor spreader	B-66	52	.154		4	7.50	5.05	16.55	22
0600	Hydro or air seeding, with mulch and fertilizer	B-81	80	.300		7.70	12.85	8.65	29.20	38.50
0800	Bluegrass, 4#/M.S.F., common, push spreader	1 Clab	8	1		12.80	37.50		50.30	75.50
0900	Tractor spreader	B-66	52	.154		12.80	7.50	5.05	25.35	31.50
1000	Hydro or air seeding, with mulch and fertilizer	B-81	80	.300		21	12.85	8.65	42.50	53
1100	Baron, push spreader	1 Clab	8	1		13.30	37.50		50.80	76
1200	Tractor spreader	B-66	52	.154		13.30	7.50	5.05	25.85	32
1300	Hydro or air seeding, with mulch and fertilizer	B-81	80	.300		18.30	12.85	8.65	39.80	50
1500	Clover, 0.67#/M.S.F., white, push spreader	1 Clab	8	1		3.08	37.50		40.58	65
1600	Tractor spreader	B-66	52	.154		3.08	7.50	5.05	15.63	21
1700	Hydro or air seeding, with mulch and fertilizer	B-81	80	.300		16.95	12.85	8.65	38.45	48.50
1800	Ladino, push spreader	1 Clab	8	1		3.08	37.50		40.58	65
1900	Tractor spreader	B-66	52	.154		3.08	7.50	5.05	15.63	21
2000	Hydro or air seeding, with mulch and fertilizer	B-81	80	.300		13.55	12.85	8.65	35.05	45
2200	Fescue 5.5#/M.S.F., tall, push spreader	1 Clab	8	1		7.75	37.50		45.25	70
2300	Tractor spreader	B-66	52	.154		7.75	7.50	5.05	20.30	26
2400	Hydro or air seeding, with mulch and fertilizer	B-81	80	.300		25.50	12.85	8.65	47	58
2500	Chewing, push spreader	1 Clab	8	1		8.25	37.50		45.75	70.50
2600	Tractor spreader	B-66	52	.154		8.25	7.50	5.05	20.80	26.50
2700	Hydro or air seeding, with mulch and fertilizer	B-81	80	.300		27	12.85	8.65	48.50	60
2900	Crown vetch, 4#/M.S.F., push spreader	1 Clab	8	1		105	37.50		142.50	178
3000	Tractor spreader	B-66	52	.154		105	7.50	5.05	117.55	133
3100	Hydro or air seeding, with mulch and fertilizer	B-81	80	.300		144	12.85	8.65	165.50	189
3300	Rye, 10#/M.S.F., annual, push spreader	1 Clab	8	1		15.15	37.50		52.65	78
3400	Tractor spreader	B-66	52	.154		15.15	7.50	5.05	27.70	34
3500	Hydro or air seeding, with mulch and fertilizer	B-81	80	.300		33.50	12.85	8.65	55	66.50
3600	Fine textured, push spreader	1 Clab	8	1		13.20	37.50		50.70	76
3700	Tractor spreader	B-66	52	.154		13.20	7.50	5.05	25.75	32
3800	Hydro or air seeding, with mulch and fertilizer	B-81	80	.300		29	12.85	8.65	50.50	62
4000	Shade mix, 6#/M.S.F., push spreader	1 Clab	8	1		10.45	37.50		47.95	73
4100	Tractor spreader	B-66	52	.154		10.45	7.50	5.05	23	29
4200	Hydro or air seeding, with mulch and fertilizer	B-81	80	.300		23	12.85	8.65	44.50	55.50
4400	Slope mix, 6#/M.S.F., push spreader	1 Clab	8	1		11.75	37.50		49.25	74.50
4500	Tractor spreader	B-66	52	.154		11.75	7.50	5.05	24.30	30.50
4600	Hydro or air seeding, with mulch and fertilizer	B-81	80	.300		29.50	12.85	8.65	51	62.50
4800	Turf mix, 4#/M.S.F., push spreader	1 Clab	8	1		11.25	37.50		48.75	74
4900	Tractor spreader	B-66	52	.154		11.25	7.50	5.05	23.80	30
5000	Hydro or air seeding, with mulch and fertilizer	B-81	80	.300		28	12.85	8.65	49.50	61
5200	Utility mix, 7#/M.S.F., push spreader	1 Clab	8	1		9.40	37.50		46.90	72
5300	Tractor spreader	B-66	52	.154		9.40	7.50	5.05	21.95	28
5400	Hydro or air seeding, with mulch and fertilizer	B-81	80	.300		35.50	12.85	8.65	57	69
5600	Wildflower, .10#/M.S.F., push spreader	1 Clab	8	1		1.86	37.50		39.36	63.50
5700	Tractor spreader	B-66	52	.154		1.86	7.50	5.05	14.41	19.45
5800	Hydro or air seeding, with mulch and fertilizer	B-81	80	.300		10.25	12.85	8.65	31.75	41.50
7025	Fertilizer, mechanical spread	1 Clab	1.75	4.571	Acre	5.40	172		177.40	288
7060	Limestone, mechanical spread		1.74	4.598	"	86	173		259	380
9000	Minimum labor/equipment charge		4	2	Job		75		75	123

32 92 Turf and Grasses

32 92 23 – Sodding

32 92 23.10 Sodding Systems		Crew	Daily Output	Labor-Hours	Unit	Material	Labor	Equipment	Total	Total Incl O&P
						2015 Bare Costs				
0010	**SODDING SYSTEMS**									
0020	Sodding, 1" deep, bluegrass sod, on level ground, over 8 M.S.F.	B-63	22	1.818	M.S.F.	243	72.50	7.90	323.40	395
0200	4 M.S.F.		17	2.353		254	93.50	10.20	357.70	440
0300	1000 S.F.		13.50	2.963		295	118	12.85	425.85	530
0500	Sloped ground, over 8 M.S.F.		6	6.667		243	265	29	537	730
0600	4 M.S.F.		5	8		254	320	34.50	608.50	835
0700	1000 S.F.		4	10		295	400	43.50	738.50	1,025
1000	Bent grass sod, on level ground, over 6 M.S.F.		20	2		252	79.50	8.70	340.20	420
1100	3 M.S.F.		18	2.222		261	88.50	9.65	359.15	440
1200	Sodding 1000 S.F. or less		14	2.857		286	114	12.40	412.40	515
1500	Sloped ground, over 6 M.S.F.		15	2.667		252	106	11.55	369.55	465
1600	3 M.S.F.		13.50	2.963		261	118	12.85	391.85	495
1700	1000 S.F.	▼	12	3.333	▼	286	133	14.45	433.45	545

32 93 Plants

32 93 10 – General Planting Costs

32 93 10.12 Travel		Crew	Daily Output	Labor-Hours	Unit	Material	Labor	Equipment	Total	Total Incl O&P
0010	**TRAVEL** add to all nursery items									
0015	10 to 20 miles one way, add				All				5%	5%
0100	30 to 50 miles one way, add				"				10%	10%

32 93 13 – Ground Covers

32 93 13.20 Ground Cover and Vines		Crew	Daily Output	Labor-Hours	Unit	Material	Labor	Equipment	Total	Total Incl O&P
0010	**GROUND COVER AND VINES** Planting only, no preparation									
0100	Ajuga, 1 year, bare root	B-1	9	2.667	C	325	102		427	520
0150	Potted, 2 year		6	4		810	153		963	1,150
0200	Berberis, potted, 2 year		6	4		795	153		948	1,125
0250	Cotoneaster, 15" to 18", shady areas, B & B		.60	40		2,575	1,525		4,100	5,350
0300	Boston ivy, on bank, 1 year, bare root		6	4		300	153		453	580
0350	Potted, 2 year		6	4		520	153		673	825
0400	English ivy, 1 year, bare root		9	2.667		82.50	102		184.50	258
0450	Potted, 2 year		6	4		269	153		422	545
0500	Halls honeysuckle, 1 year, bare root		5	4.800		167	184		351	485
0550	Potted, 2 year		4	6		430	230		660	845
0600	Memorial rose, 9" to 12", 1 gallon container		3	8		119	305		424	630
0650	Potted, 2 gallon container		2	12		202	460		662	975
0700	Pachysandra, 1 year, bare root		10	2.400		80	92		172	239
0750	Potted, 2 year		6	4		116	153		269	380
0800	Vinca minor, 1 year, bare root		10	2.400		65	92		157	223
0850	Potted, 2 year		6	4		199	153		352	470
0900	Woodbine, on bank, 1/2 year, bare root		6	4		54	153		207	310
0950	Potted, 2 year	▼	4	6	▼	136	230		366	525
2000	Alternate method of figuring									
2100	Ajuga, field division, 4000/M.S.F.	B-1	.23	104	M.S.F.	4,875	4,000		8,875	11,900
2300	Boston ivy, 1 year, 60/M.S.F.		10	2.400		182	92		274	350
2400	English ivy, 1 yr., 500/M.S.F.		1.80	13.333		465	510		975	1,350
2500	Halls honeysuckle, 1 yr., 333/M.S.F.		1.50	16		490	610		1,100	1,550
2600	Memorial rose, 9"-12", 1 gal., 333/M.S.F.		.90	26.667		920	1,025		1,945	2,675
2700	Pachysandra, 1 yr., 4000/M.S.F.		.25	96		1,525	3,675		5,200	7,700
2800	Vinca minor, rooted cutting, 2000/M.S.F.		1	24		810	920		1,730	2,400
2900	Woodbine, 1 yr., 60/M.S.F.	▼	10	2.400	▼	32.50	92		124.50	187

32 93 Plants

32 93 33 – Shrubs

32 93 33.10 Shrubs and Trees		Crew	Daily Output	Labor-Hours	Unit	Material	2015 Bare Costs Labor	Equipment	Total	Total Incl O&P
0010	**SHRUBS AND TREES**									
0011	Evergreen, in prepared beds, B & B									
0100	Arborvitae pyramidal, 4′-5′	B-17	30	1.067	Ea.	100	43.50	26	169.50	210
0150	Globe, 12″-15″	B-1	96	.250		21.50	9.55		31.05	39
0200	Balsam, fraser, 6′ - 7′	B-17	30	1.067		137	43.50	26	206.50	250
0300	Cedar, blue, 8′-10′		18	1.778		231	73	43	347	420
0350	Japanese, 4′ - 5′		55	.582		108	24	14.10	146.10	173
0400	Cypress, hinoki, 15″ - 18″	B-1	80	.300		76.50	11.50		88	103
0500	Hemlock, Canadian, 2-1/2′-3′		36	.667		31	25.50		56.50	76.50
0550	Holly, Savannah, 8′ - 10′ H		9.68	2.479		257	95		352	440
0600	Juniper, andorra, 18″-24″		80	.300		36.50	11.50		48	59
0620	Wiltoni, 15″-18″		80	.300		26	11.50		37.50	47.50
0640	Skyrocket, 4-1/2′-5′	B-17	55	.582		107	24	14.10	145.10	171
0660	Blue pfitzer, 2′-2-1/2′	B-1	44	.545		38.50	21		59.50	76.50
0680	Ketleerie, 2-1/2′-3′		50	.480		52	18.35		70.35	87
0700	Pine, black, 2-1/2′-3′		50	.480		59.50	18.35		77.85	95
0720	Mugo, 18″-24″		60	.400		59	15.30		74.30	90
0740	White, 4′-5′	B-17	75	.427		51	17.50	10.35	78.85	96
0800	Spruce, blue, 18″-24″	B-1	60	.400		66.50	15.30		81.80	98.50
0820	Dwarf alberta, 18″ - 24″	″	60	.400		57	15.30		72.30	87.50
0840	Norway, 4′-5′	B-17	75	.427		83	17.50	10.35	110.85	131
0900	Yew, denisforma, 12″-15″	B-1	60	.400		35	15.30		50.30	63.50
1000	Capitata, 18″-24″		30	.800		32.50	30.50		63	86
1100	Hicksi, 2′-2-1/2′		30	.800		78.50	30.50		109	137

32 93 33.20 Shrubs

32 93 33.20 Shrubs		Crew	Daily Output	Labor-Hours	Unit	Material	Labor	Equipment	Total	Total Incl O&P
0010	**SHRUBS**									
0011	Broadleaf Evergreen, planted in prepared beds									
0100	Andromeda, 15″-18″, container	B-1	96	.250	Ea.	32.50	9.55		42.05	51.50
0200	Azalea, 15″ - 18″, container		96	.250		29.50	9.55		39.05	48
0300	Barberry, 9″-12″, container		130	.185		17.60	7.05		24.65	31
0400	Boxwood, 15″-18″, B&B		96	.250		43	9.55		52.55	62.50
0500	Euonymus, emerald gaiety, 12″ to 15″, container		115	.209		24	8		32	39
0600	Holly, 15″-18″, B & B		96	.250		36	9.55		45.55	55.50
0700	Leucothoe, 15″ - 18″, container		96	.250		17.75	9.55		27.30	35.50
0800	Mahonia, 18″ - 24″, container		80	.300		32	11.50		43.50	54
0900	Mount laurel, 18″ - 24″, B & B		80	.300		70.50	11.50		82	96.50
1000	Paxistema, 9 – 12″ high		130	.185		21	7.05		28.05	35
1100	Rhododendron, 18″-24″, container		48	.500		37.50	19.15		56.65	73
1200	Rosemary, 1 gal. container		600	.040		17.70	1.53		19.23	22
2000	Deciduous, planted in prepared beds, amelanchier, 2′-3′, B & B		57	.421		119	16.10		135.10	158
2100	Azalea, 15″-18″, B & B		96	.250		29	9.55		38.55	47.50
2200	Barberry, 2′ - 3′, B & B		57	.421		23	16.10		39.10	51.50
2300	Bayberry, 2′-3′, B & B		57	.421		24	16.10		40.10	53
2400	Boston ivy, 2 year, container		600	.040		19.40	1.53		20.93	24
2500	Corylus, 3′ - 4′, B & B	B-17	75	.427		23	17.50	10.35	50.85	65
2600	Cotoneaster, 15″-18″, B & B	B-1	80	.300		26.50	11.50		38	48
2700	Deutzia, 12″ - 15″, B & B	″	96	.250		12.25	9.55		21.80	29
2800	Dogwood, 3′-4′, B & B	B-17	40	.800		32	33	19.40	84.40	110
2900	Euonymus, alatus compacta, 15″ to 18″, container	B-1	80	.300		26	11.50		37.50	48
3000	Flowering almond, 2′ - 3′, container	″	36	.667		19.50	25.50		45	63.50
3100	Flowering currant, 3′ - 4′, container	B-17	75	.427		26	17.50	10.35	53.85	68.50
3200	Forsythia, 2′-3′, container	B-1	60	.400		18.45	15.30		33.75	45.50

32 93 33 – Shrubs

32 93 33.20 Shrubs		Crew	Daily Output	Labor-Hours	Unit	Material	Labor	Equipment	Total	Total Incl O&P
						2015 Bare Costs				
3300	Hibiscus, 3′-4′, B & B	B-17	75	.427	Ea.	48	17.50	10.35	75.85	93
3400	Honeysuckle, 3′-4′, B & B	B-1	60	.400		26.50	15.30		41.80	54
3500	Hydrangea, 2′-3′, B & B	"	57	.421		30	16.10		46.10	59.50
3600	Lilac, 3′-4′, B & B	B-17	40	.800		28	33	19.40	80.40	105
3700	Mockorange, 3′ - 4′, B & B	B-1	36	.667		28	25.50		53.50	73
3800	Osier willow, 2′ - 3′, B & B		57	.421		34	16.10		50.10	64
3900	Privet, bare root, 18″-24″		80	.300		14.50	11.50		26	35
4000	Pyracantha, 2′ - 3′, container		80	.300		39	11.50		50.50	62
4100	Quince, 2′-3′, B & B		57	.421		28	16.10		44.10	57.50
4200	Russian olive, 3′-4′, B & B	B-17	75	.427		29	17.50	10.35	56.85	71.50
4300	Snowberry, 2′ - 3′, B & B	B-1	57	.421		19.35	16.10		35.45	48
4400	Spirea, 3′-4′, B & B	"	70	.343		20.50	13.10		33.60	44
4500	Viburnum, 3′-4′, B & B	B-17	40	.800		25	33	19.40	77.40	102
4600	Weigela, 3′ - 4′, B & B	B-1	70	.343		14	13.10		27.10	37

32 93 43 – Trees

32 93 43.10 Planting

		Crew	Daily Output	Labor-Hours	Unit	Material	Labor	Equipment	Total	Total Incl O&P
0010	**PLANTING**									
0011	Trees, shrubs and ground cover									
0100	Light soil									
0110	Bare root seedlings, 3″ to 5″ height	1 Clab	960	.008	Ea.		.31		.31	.51
0120	6″ to 10″		520	.015			.58		.58	.95
0130	11″ to 16″		370	.022			.81		.81	1.33
0140	17″ to 24″		210	.038			1.43		1.43	2.35
0200	Potted, 2-1/4″ diameter		840	.010			.36		.36	.59
0210	3″ diameter		700	.011			.43		.43	.70
0220	4″ diameter		620	.013			.49		.49	.80
0300	Container, 1 gallon	2 Clab	84	.190			7.15		7.15	11.75
0310	2 gallon		52	.308			11.55		11.55	18.95
0320	3 gallon		40	.400			15.05		15.05	24.50
0330	5 gallon		29	.552			20.50		20.50	34

32 93 43.20 Trees

			Crew	Daily Output	Labor-Hours	Unit	Material	Labor	Equipment	Total	Total Incl O&P
0010	**TREES**										
0011	Deciduous, in prep. beds, balled & burlapped (B&B)										
0100	Ash, 2″ caliper	G	B-17	8	4	Ea.	186	164	97	447	575
0200	Beech, 5′-6′	G		50	.640		187	26	15.50	228.50	266
0300	Birch, 6′-8′, 3 stems	G		20	1.600		168	65.50	39	272.50	335
0400	Cherry, 6′ - 8′, 1″ caliper	G		24	1.333		84	54.50	32.50	171	217
0500	Crabapple, 6′-8′	G		20	1.600		139	65.50	39	243.50	300
0600	Dogwood, 4′-5′	G		40	.800		131	33	19.40	183.40	219
0700	Eastern redbud 4′-5′	G		40	.800		149	33	19.40	201.40	239
0800	Elm, 8′-10′	G		20	1.600		258	65.50	39	362.50	435
0900	Ginkgo, 6′-7′	G		24	1.333		148	54.50	32.50	235	287
1000	Hawthorn, 8′-10′, 1″ caliper	G		20	1.600		162	65.50	39	266.50	325
1100	Honeylocust, 10′-12′, 1-1/2″ caliper	G		10	3.200		206	131	77.50	414.50	525
1200	Laburnum, 6′ - 8′, 1″ caliper	G		24	1.333		61	54.50	32.50	148	191
1300	Larch, 8′	G		32	1		127	41	24	192	233
1400	Linden, 8′-10′, 1″ caliper	G		20	1.600		144	65.50	39	248.50	305
1500	Magnolia, 4′-5′	G		20	1.600		101	65.50	39	205.50	260
1600	Maple, red, 8′-10′, 1-1/2″ caliper	G		10	3.200		202	131	77.50	410.50	520
1700	Mountain ash, 8′-10′, 1″ caliper	G		16	2		176	82	48.50	306.50	380
1800	Oak, 2-1/2″-3″ caliper	G		6	5.333		325	218	129	672	855
1900	Pagoda, 6′ - 8′	G		20	1.600		152	65.50	39	256.50	315

32 93 Plants

32 93 43 – Trees

32 93 43.20 Trees			Crew	Daily Output	Labor-Hours	Unit	Material	Labor	Equipment	Total	Total Incl O&P
							2015 Bare Costs				
2000	Pear, 6′ - 8′, 1″ caliper	G	B-17	20	1.600	Ea.	118	65.50	39	222.50	279
2100	Planetree, 9′-11′, 1-1/4″ caliper	G		10	3.200		236	131	77.50	444.50	560
2200	Plum, 6′-8′, 1″ caliper	G		20	1.600		82	65.50	39	186.50	239
2300	Poplar, 9′-11′, 1-1/4″ caliper	G		10	3.200		144	131	77.50	352.50	455
2400	Shadbush, 4′ - 5′	G		60	.533		66	22	12.90	100.90	122
2500	Sumac, 2′-3′	G		75	.427		44	17.50	10.35	71.85	88.50
2600	Tupelo, 5′ - 6′	G		40	.800		96	33	19.40	148.40	181
2700	Tulip, 5′-6′	G		40	.800		47	33	19.40	99.40	126
2800	Willow, 6′-8′, 1″ caliper	G		20	1.600		96	65.50	39	200.50	255
9000	Minimum labor/equipment charge		1 Clab	4	2	Job		75		75	123

32 94 Planting Accessories

32 94 13 – Landscape Edging

32 94 13.20 Edging		Crew	Daily Output	Labor-Hours	Unit	Material	Labor	Equipment	Total	Total Incl O&P
0010	**EDGING**									
0050	Aluminum alloy, including stakes, 1/8″ x 4″, mill finish	B-1	390	.062	L.F.	2.30	2.36		4.66	6.40
0051	Black paint		390	.062		2.67	2.36		5.03	6.80
0052	Black anodized		390	.062		3.08	2.36		5.44	7.25
0060	3/16″ x 4″, mill finish		380	.063		3.31	2.42		5.73	7.60
0061	Black paint		380	.063		3.82	2.42		6.24	8.15
0062	Black anodized		380	.063		4.46	2.42		6.88	8.85
0070	1/8″ x 5-1/2″ mill finish		370	.065		3.34	2.48		5.82	7.75
0071	Black paint		370	.065		3.96	2.48		6.44	8.40
0072	Black anodized		370	.065		4.53	2.48		7.01	9.05
0080	3/16″ x 5-1/2″ mill finish		360	.067		4.46	2.55		7.01	9.10
0081	Black paint		360	.067		5.05	2.55		7.60	9.75
0082	Black anodized		360	.067		5.80	2.55		8.35	10.60
0100	Brick, set horizontally, 1-1/2 bricks per L.F.	D-1	370	.043		1.22	1.82		3.04	4.31
0150	Set vertically, 3 bricks per L.F.	″	135	.119		3.68	4.99		8.67	12.15
0200	Corrugated aluminum, roll, 4″ wide	1 Carp	650	.012		1.93	.58		2.51	3.07
0250	6″ wide	″	550	.015		2.41	.68		3.09	3.77
0300	Concrete, cast in place, see Section 03 30 53.40									
0350	Granite, 5″ x 16″, straight	B-29	300	.187	L.F.	11.65	7.65	2.94	22.24	28.50
0400	Polyethylene grass barrier, 5″ x 1/8″	D-1	400	.040		.81	1.68		2.49	3.63
0500	Precast scallops, green, 2″ x 8″ x 16″		400	.040		2.08	1.68		3.76	5.05
0550	2″ x 8″ x 16″ other than green		400	.040		1.66	1.68		3.34	4.57
0600	Railroad ties, 6″ x 8″	2 Carp	170	.094		2.79	4.42		7.21	10.30
0650	7″ x 9″		136	.118		3.10	5.50		8.60	12.45
0750	Redwood 2″ x 4″		330	.048		2.25	2.28		4.53	6.20
0800	Steel edge strips, incl. stakes, 1/4″ x 5″	B-1	390	.062		3.92	2.36		6.28	8.15
0850	3/16″ x 4″	″	390	.062		3.10	2.36		5.46	7.25
0900	Hardwood, pressure treated, 4″ x 6″	2 Carp	250	.064		2.12	3		5.12	7.25
0940	6″ x 6″		200	.080		3.03	3.76		6.79	9.50
0980	6″ x 8″		170	.094		3.46	4.42		7.88	11.05
1000	Pine, pressure treated, 1″ x 4″		500	.032		.54	1.50		2.04	3.05
1040	2″ x 4″		330	.048		.91	2.28		3.19	4.73
1080	4″ x 6″		250	.064		3.02	3		6.02	8.25
1100	6″ x 6″		200	.080		4.54	3.76		8.30	11.15
1140	6″ x 8″		170	.094		5.95	4.42		10.37	13.80
9000	Minimum labor/equipment charge	1 Carp	4	2	Job		94		94	154

32 94 Planting Accessories

32 94 50 – Tree Guying

32 94 50.10 Tree Guying Systems		Crew	Daily Output	Labor-Hours	Unit	Material	2015 Bare Costs Labor	Equipment	Total	Total Incl O&P
0010	**TREE GUYING SYSTEMS**									
0015	Tree guying Including stakes, guy wire and wrap									
0100	Less than 3" caliper, 2 stakes	2 Clab	35	.457	Ea.	14.25	17.20		31.45	43.50
0200	3" to 4" caliper, 3 stakes	"	21	.762	"	19.25	28.50		47.75	68
1000	Including arrowhead anchor, cable, turnbuckles and wrap									
1100	Less than 3" caliper, 3" anchors	2 Clab	20	.800	Ea.	44.50	30		74.50	98.50
1200	3" to 6" caliper, 4" anchors		15	1.067		38.50	40		78.50	108
1300	6" caliper, 6" anchors		12	1.333		44.50	50		94.50	131
1400	8" caliper, 8" anchors		9	1.778		129	67		196	252

32 96 Transplanting

32 96 23 – Plant and Bulb Transplanting

32 96 23.23 Planting		Crew	Daily Output	Labor-Hours	Unit	Material	Labor	Equipment	Total	Total Incl O&P
0010	**PLANTING**									
0012	Moving shrubs on site, 12" ball	B-62	28	.857	Ea.		35.50	6.20	41.70	64.50
0100	24" ball	"	22	1.091	"		45	7.90	52.90	81.50
32 96 23.43 Moving Trees										
0010	**MOVING TREES**, On site									
0300	Moving trees on site, 36" ball	B-6	3.75	6.400	Ea.		264	97	361	535
0400	60" ball	"	1	24	"		990	365	1,355	2,000

32 96 43 – Tree Transplanting

32 96 43.20 Tree Removal		Crew	Daily Output	Labor-Hours	Unit	Material	Labor	Equipment	Total	Total Incl O&P
0010	**TREE REMOVAL**									
0100	Dig & lace, shrubs, broadleaf evergreen, 18"-24" high	B-1	55	.436	Ea.		16.70		16.70	27.50
0200	2'-3'	"	35	.686			26		26	43
0300	3'-4'	B-6	30	.800			33	12.15	45.15	67
0400	4'-5'	"	20	1.200			49.50	18.20	67.70	100
1000	Deciduous, 12"-15"	B-1	110	.218			8.35		8.35	13.70
1100	18"-24"		65	.369			14.15		14.15	23
1200	2'-3'		55	.436			16.70		16.70	27.50
1300	3'-4'	B-6	50	.480			19.80	7.30	27.10	40
2000	Evergreen, 18"-24"	B-1	55	.436			16.70		16.70	27.50
2100	2'-0" to 2'-6"		50	.480			18.35		18.35	30
2200	2'-6" to 3'-0"		35	.686			26		26	43
2300	3'-0" to 3'-6"		20	1.200			46		46	75.50
3000	Trees, deciduous, small, 2'-3'		55	.436			16.70		16.70	27.50
3100	3'-4'	B-6	50	.480			19.80	7.30	27.10	40
3200	4'-5'		35	.686			28.50	10.40	38.90	57.50
3300	5'-6'		30	.800			33	12.15	45.15	67
4000	Shade, 5'-6'		50	.480			19.80	7.30	27.10	40
4100	6'-8'		35	.686			28.50	10.40	38.90	57.50
4200	8'-10'		25	.960			39.50	14.55	54.05	80
4300	2" caliper		12	2			82.50	30.50	113	168
5000	Evergreen, 4'-5'		35	.686			28.50	10.40	38.90	57.50
5100	5'-6'		25	.960			39.50	14.55	54.05	80
5200	6'-7'		19	1.263			52	19.15	71.15	106
5300	7'-8'		15	1.600			66	24.50	90.50	134
5400	8'-10'		11	2.182			90	33	123	183

Estimating Tips

33 10 00 Water Utilities
33 30 00 Sanitary Sewerage Utilities
33 40 00 Storm Drainage Utilities

- Never assume that the water, sewer, and drainage lines will go in at the early stages of the project. Consider the site access needs before dividing the site in half with open trenches, loose pipe, and machinery obstructions. Always inspect the site to establish that the site drawings are complete. Check off all existing utilities on your drawings as you locate them. Be especially careful with underground utilities because appurtenances are sometimes buried during regrading or repaving operations. If you find any discrepancies, mark up the site plan for further research. Differing site conditions can be very costly if discovered later in the project.
- See also Section 33 01 00 for restoration of pipe where removal/replacement may be undesirable. Use of new types of piping materials can reduce the overall project cost. Owners/design engineers should consider the installing contractor as a valuable source of current information on utility products and local conditions that could lead to significant cost savings.

Reference Numbers

Reference numbers are shown in shaded boxes at the beginning of some major classifications. These numbers refer to related items in the Reference Section. The reference information may be an estimating procedure, an alternate pricing method, or technical information.

Note: Not all subdivisions listed here necessarily appear in this publication. ■

Note: ***Trade Service****, in part, has been used as a reference source for some of the material prices used in Division 33.*

33 01 10.10 Corrosion Resistance		Crew	Daily Output	Labor-Hours	Unit	2015 Bare Costs Material	Labor	Equipment	Total	Total Incl O&P
0010	**CORROSION RESISTANCE**									
0012	Wrap & coat, add to pipe, 4" diameter				L.F.	2.21			2.21	2.43
0020	5" diameter					2.66			2.66	2.93
0040	6" diameter					3.28			3.28	3.61
0060	8" diameter					4.04			4.04	4.44
0080	10" diameter					4.91			4.91	5.40
0100	12" diameter					6.20			6.20	6.80
0120	14" diameter					7.30			7.30	8
0140	16" diameter					9.75			9.75	10.70
0160	18" diameter					10			10	11
0180	20" diameter					10.50			10.50	11.55
0200	24" diameter					12.95			12.95	14.25
0220	Small diameter pipe, 1" diameter, add					1.02			1.02	1.12
0240	2" diameter					1.38			1.38	1.52
0260	2-1/2" diameter					1.62			1.62	1.78
0280	3" diameter				↓	2.01			2.01	2.21
0300	Fittings, field covered, add				S.F.	10.40			10.40	11.45
0500	Coating, bituminous, per diameter inch, 1 coat, add				L.F.	.63			.63	.69
0540	3 coat					1.92			1.92	2.11
0560	Coal tar epoxy, per diameter inch, 1 coat, add					.22			.22	.24
0600	3 coat					.68			.68	.75
1000	Polyethylene H.D. extruded, .025" thk., 1/2" diameter add					.10			.10	.11
1020	3/4" diameter					.15			.15	.17
1040	1" diameter					.20			.20	.22
1060	1-1/4" diameter					.25			.25	.28
1080	1-1/2" diameter					.30			.30	.33
1100	.030" thk., 2" diameter					.40			.40	.44
1120	2-1/2" diameter					.50			.50	.55
1140	.035" thk., 3" diameter					.60			.60	.66
1160	3-1/2" diameter					.70			.70	.77
1180	4" diameter					.80			.80	.88
1200	.040" thk, 5" diameter					1			1	1.10
1220	6" diameter					1.20			1.20	1.32
1240	8" diameter					1.60			1.60	1.76
1260	10" diameter					2			2	2.20
1280	12" diameter					2.40			2.40	2.64
1300	.060" thk., 14" diameter					2.80			2.80	3.08
1320	16" diameter					3.20			3.20	3.52
1340	18" diameter					3.60			3.60	3.96
1360	20" diameter				↓	4			4	4.40
1380	Fittings, field wrapped, add				S.F.	4.79			4.79	5.25

33 01 10.20 Pipe Repair		Crew	Daily Output	Labor-Hours	Unit	Material	Labor	Equipment	Total	Total Incl O&P
0010	**PIPE REPAIR**									
0020	Not including excavation or backfill									
0100	Clamp, stainless steel, lightweight, for steel pipe									
0110	3" long, 1/2" diameter pipe	1 Plum	34	.235	Ea.	12.80	13.80		26.60	35.50
0120	3/4" diameter pipe		32	.250		13.05	14.70		27.75	37.50
0130	1" diameter pipe		30	.267		14.25	15.65		29.90	40
0140	1-1/4" diameter pipe		28	.286		14.45	16.75		31.20	42
0150	1-1/2" diameter pipe		26	.308		15.35	18.05		33.40	45
0160	2" diameter pipe		24	.333		16.50	19.55		36.05	48.50
0170	2-1/2" diameter pipe	↓	23	.348	↓	17.50	20.50		38	51.50

33 01 Operation and Maintenance of Utilities

33 01 10 – Operation and Maintenance of Water Utilities

33 01 10.20 Pipe Repair		Crew	Daily Output	Labor-Hours	Unit	2015 Bare Costs Material	Labor	Equipment	Total	Total Incl O&P
0180	3″ diameter pipe	1 Plum	22	.364	Ea.	21	21.50		42.50	56.50
0190	3-1/2″ diameter pipe		21	.381		25	22.50		47.50	62.50
0200	4″ diameter pipe	B-20	44	.545		25	23		48	65
0210	5″ diameter pipe		42	.571		29.50	24		53.50	71.50
0220	6″ diameter pipe		38	.632		34	26.50		60.50	81
0230	8″ diameter pipe		30	.800		40.50	33.50		74	99.50
0240	10″ diameter pipe		28	.857		115	36		151	186
0250	12″ diameter pipe		24	1		120	42		162	201
0260	14″ diameter pipe		22	1.091		125	46		171	213
0270	16″ diameter pipe		20	1.200		135	50.50		185.50	231
0280	18″ diameter pipe		18	1.333		145	56		201	252
0290	20″ diameter pipe		16	1.500		150	63		213	268
0300	24″ diameter pipe		14	1.714		170	72		242	305
0360	For 6″ long, add					100%	40%			
0370	For 9″ long, add					200%	100%			
0380	For 12″ long, add					300%	150%			
0390	For 18″ long, add					500%	200%			
0400	Pipe Freezing for live repairs of systems 3/8 inch to 6 inch									
0410	Note: Pipe Freezing can also be used to install a valve into a live system									
0420	Pipe Freezing each side 3/8 inch	2 Skwk	8	2	Ea.	540	97.50		637.50	755
0425	Pipe Freezing each side 3/8 inch, second location same kit		8	2		22.50	97.50		120	183
0430	Pipe Freezing each side 3/4 inch		8	2		495	97.50		592.50	705
0435	Pipe Freezing each side 3/4 inch, second location same kit		8	2		22.50	97.50		120	183
0440	Pipe Freezing each side 1-1/2 inch		6	2.667		495	130		625	755
0445	Pipe Freezing each side 1-1/2 inch , second location same kit		6	2.667		22.50	130		152.50	236
0450	Pipe Freezing each side 2 inch		6	2.667		890	130		1,020	1,200
0455	Pipe Freezing each side 2 inch, second location same kit		6	2.667		22.50	130		152.50	236
0460	Pipe Freezing each side 2-1/2 inch-3 inch		6	2.667		925	130		1,055	1,225
0465	Pipe Freeze each side 2-1/2 -3 inch, second location same kit		6	2.667		45.50	130		175.50	261
0470	Pipe Freezing each side 4 inch		4	4		1,475	195		1,670	1,950
0475	Pipe Freezing each side 4 inch, second location same kit		4	4		72	195		267	395
0480	Pipe Freezing each side 5-6 inch		4	4		4,000	195		4,195	4,725
0485	Pipe Freezing each side 5-6 inch, second location same kit		4	4		216	195		411	555
0490	Pipe Freezing extra 20 lb. CO_2 cylinders (3/8″ to 2″ - 1 ea, 3″ -2 ea)					222			222	244
0500	Pipe Freezing extra 50 lb. CO_2 cylinders (4″ - 2 ea, 5″-6″ -6 ea)					440			440	485
1000	Clamp, stainless steel, with threaded service tap									
1040	Full seal for iron, steel, PVC pipe									
1100	6″ long, 2″ diameter pipe	1 Plum	17	.471	Ea.	87	27.50		114.50	139
1110	2-1/2″ diameter pipe		16	.500		90	29.50		119.50	145
1120	3″ diameter pipe		15.60	.513		104	30		134	162
1130	3-1/2″ diameter pipe		15	.533		110	31.50		141.50	170
1140	4″ diameter pipe	B-20	32	.750		117	31.50		148.50	180
1150	6″ diameter pipe		28	.857		145	36		181	219
1160	8″ diameter pipe		21	1.143		171	48		219	267
1170	10″ diameter pipe		20	1.200		224	50.50		274.50	330
1180	12″ diameter pipe		17	1.412		261	59		320	385
1200	8″ long, 2″ diameter pipe	1 Plum	11.72	.683		178	40		218	259
1210	2-1/2″ diameter pipe		11	.727		183	42.50		225.50	268
1220	3″ diameter pipe		10.75	.744		189	43.50		232.50	276
1230	3-1/2″ diameter pipe		10.34	.774		199	45.50		244.50	290
1240	4″ diameter pipe	B-20	22	1.091		206	46		252	300
1250	6″ diameter pipe		19.31	1.243		234	52		286	345
1260	8″ diameter pipe		14.48	1.657		264	69.50		333.50	405

33 01 10.20 Pipe Repair		Crew	Daily Output	Labor-Hours	Unit	2015 Bare Costs Material	Labor	Equipment	Total	Total Incl O&P
1270	10″ diameter pipe	B-20	13.80	1.739	Ea.	330	73		403	485
1280	12″ diameter pipe		11.72	2.048		370	86		456	545
1300	12″ long, 2″ diameter pipe	1 Plum	9.44	.847		257	50		307	360
1310	2-1/2″ diameter pipe		8.89	.900		262	53		315	370
1320	3″ diameter pipe		8.67	.923		269	54		323	380
1330	3-1/2″ diameter pipe		8.33	.960		280	56.50		336.50	400
1340	4″ diameter pipe	B-20	17.78	1.350		450	56.50		506.50	590
1350	6″ diameter pipe		15.56	1.542		340	64.50		404.50	475
1360	8″ diameter pipe		11.67	2.057		390	86.50		476.50	565
1370	10″ diameter pipe		11.11	2.160		485	90.50		575.50	685
1380	12″ diameter pipe		9.44	2.542		555	107		662	785
1400	20″ long, 2″ diameter pipe	1 Plum	8.10	.988		299	58		357	420
1410	2-1/2″ diameter pipe		7.62	1.050		325	61.50		386.50	450
1420	3″ diameter pipe		7.43	1.077		365	63		428	500
1430	3-1/2″ diameter pipe		7.14	1.120		380	66		446	525
1440	4″ diameter pipe	B-20	15.24	1.575		450	66		516	605
1450	6″ diameter pipe		13.33	1.800		525	75.50		600.50	700
1460	8″ diameter pipe		10	2.400		590	101		691	815
1470	10″ diameter pipe		9.52	2.521		715	106		821	960
1480	12″ diameter pipe		8.10	2.963		835	124		959	1,125
1600	Clamp, stainless steel, single section									
1640	Full seal for iron, steel, PVC pipe									
1700	6″ long, 2″ diameter pipe	1 Plum	17	.471	Ea.	87	27.50		114.50	139
1710	2-1/2″ diameter pipe		16	.500		90	29.50		119.50	146
1720	3″ diameter pipe		15.60	.513		104	30		134	162
1730	3-1/2″ diameter pipe		15	.533		110	31.50		141.50	170
1740	4″ diameter pipe	B-20	32	.750		117	31.50		148.50	180
1750	6″ diameter pipe		27	.889		145	37.50		182.50	221
1760	8″ diameter pipe		21	1.143		171	48		219	267
1770	10″ diameter pipe		20	1.200		224	50.50		274.50	330
1780	12″ diameter pipe		17	1.412		261	59		320	385
1800	8″ long, 2″ diameter pipe	1 Plum	11.72	.683		118	40		158	193
1805	2-1/2″ diameter pipe		11.03	.725		123	42.50		165.50	202
1810	3″ diameter pipe		10.76	.743		129	43.50		172.50	210
1815	3-1/2″ diameter pipe		10.34	.774		139	45.50		184.50	224
1820	4″ diameter pipe	B-20	22.07	1.087		146	45.50		191.50	235
1825	6″ diameter pipe		19.31	1.243		174	52		226	277
1830	8″ diameter pipe		14.48	1.657		204	69.50		273.50	340
1835	10″ diameter pipe		13.79	1.740		272	73		345	420
1840	12″ diameter pipe		11.72	2.048		310	86		396	480
1850	12″ long, 2″ diameter pipe	1 Plum	9.44	.847		197	50		247	294
1855	2-1/2″ diameter pipe		8.89	.900		201	53		254	305
1860	3″ diameter pipe		8.67	.923		209	54		263	315
1865	3-1/2″ diameter pipe		8.33	.960		220	56.50		276.50	330
1870	4″ diameter pipe	B-20	17.78	1.350		231	56.50		287.50	345
1875	6″ diameter pipe		15.56	1.542		278	64.50		342.50	410
1880	8″ diameter pipe		11.67	2.057		330	86.50		416.50	500
1885	10″ diameter pipe		11.11	2.160		425	90.50		515.50	620
1890	12″ diameter pipe		9.44	2.542		495	107		602	720
1900	20″ long, 2″ diameter pipe	1 Plum	8.10	.988		239	58		297	355
1905	2-1/2″ diameter pipe		7.62	1.050		264	61.50		325.50	385
1910	3″ diameter pipe		7.43	1.077		305	63		368	435
1915	3-1/2″ diameter pipe		7.14	1.120		320	66		386	455

33 01 10 – Operation and Maintenance of Water Utilities

33 01 10.20 Pipe Repair		Crew	Daily Output	Labor-Hours	Unit	Material	Labor	Equipment	Total	Total Incl O&P
						2015 Bare Costs				
1920	4" diameter pipe	B-20	15.24	1.575	Ea.	390	66		456	540
1925	6" diameter pipe		13.33	1.800		465	75.50		540.50	635
1930	8" diameter pipe		10	2.400		530	101		631	750
1935	10" diameter pipe		9.52	2.521		655	106		761	895
1940	12" diameter pipe		8.10	2.963		775	124		899	1,050
2000	Clamp, stainless steel, two section									
2040	Full seal, for iron, steel, PVC pipe									
2100	6" long, 4" diameter pipe	B-20	24	1	Ea.	230	42		272	320
2110	6" diameter pipe		20	1.200		264	50.50		314.50	370
2120	8" diameter pipe		13	1.846		300	77.50		377.50	455
2130	10" diameter pipe		12	2		305	84		389	470
2140	12" diameter pipe		10	2.400		390	101		491	595
2200	9" long, 4" diameter pipe		16	1.500		300	63		363	435
2210	6" diameter pipe		13	1.846		340	77.50		417.50	500
2220	8" diameter pipe		9	2.667		375	112		487	600
2230	10" diameter pipe		8	3		495	126		621	750
2240	12" diameter pipe		7	3.429		565	144		709	860
2250	14" diameter pipe		6.40	3.750		760	157		917	1,100
2260	16" diameter pipe		6	4		705	168		873	1,050
2270	18" diameter pipe		5	4.800		800	201		1,001	1,200
2280	20" diameter pipe		4.60	5.217		885	219		1,104	1,325
2290	24" diameter pipe		4	6		1,175	252		1,427	1,700
2320	For 12" long, add to 9"					15%	25%			
2330	For 18" long, add to 9"					70%	55%			
8000	For internal cleaning and inspection, see Section 33 01 30.16									
8100	For pipe testing, see Section 23 05 93.50									

33 01 30 – Operation and Maintenance of Sewer Utilities

33 01 30.16 TV Inspection of Sewer Pipelines

Line	Description	Crew	Daily Output	Labor-Hours	Unit	Material	Labor	Equipment	Total	Total Incl O&P
0010	**TV INSPECTION OF SEWER PIPELINES**									
0100	Pipe internal cleaning & inspection, cleaning, pressure pipe systems									
0120	Pig method, lengths 1000' to 10,000'									
0140	4" diameter thru 24" diameter, minimum				L.F.				3.60	4.14
0160	Maximum				"				18	21
6000	Sewage/sanitary systems									
6100	Power rodder with header & cutters									
6110	Mobilization charge, minimum				Total				695	800
6120	Mobilization charge, maximum				"				9,125	10,600
6140	Cleaning 4"-12" diameter				L.F.				3.39	3.90
6190	14"-24" diameter								3.98	4.59
6240	30" diameter								5.80	6.65
6250	36" diameter								6.75	7.75
6260	48" diameter								7.70	8.85
6270	60" diameter								8.70	9.95
6280	72" diameter								9.65	11.10
9000	Inspection, television camera with video									
9060	up to 500 linear feet				Total				715	820

33 01 30.71 Rehabilitation of Sewer Utilities

Line	Description	Crew	Daily Output	Labor-Hours	Unit	Material	Labor	Equipment	Total	Total Incl O&P
0010	**REHABILITATION OF SEWER UTILITIES**									
0011	300' runs, replace with HDPE pipe									
0020	Not including excavation, backfill, shoring, or dewatering									
0100	6" to 15" diameter, minimum				L.F.				104	114
0200	Maximum								214	236

33 01 30 – Operation and Maintenance of Sewer Utilities

33 01 30.71 Rehabilitation of Sewer Utilities		Crew	Daily Output	Labor-Hours	Unit	Material	2015 Bare Costs Labor	Equipment	Total	Total Incl O&P
0300	18″ to 36″ diameter, minimum				L.F.				208	229
0400	Maximum				↓				435	510
0500	Mobilize and demobilize, minimum				Job				3,000	3,300
0600	Maximum				"				32,100	35,400

33 01 30.72 Relining Sewers		Crew	Daily Output	Labor-Hours	Unit	Material	Labor	Equipment	Total	Total Incl O&P
0010	**RELINING SEWERS**									
0011	With cement incl. bypass & cleaning									
0020	Less than 10,000 L.F., urban, 6″ to 10″	C-17E	130	.615	L.F.	9.55	30	.74	40.29	60.50
0050	10″ to 12″		125	.640		11.70	31.50	.77	43.97	65
0070	12″ to 16″		115	.696		12.05	34	.84	46.89	69.50
0100	16″ to 20″		95	.842		14.15	41.50	1.02	56.67	83.50
0200	24″ to 36″		90	.889		15.25	43.50	1.08	59.83	89
0300	48″ to 72″		80	1		24.50	49	1.21	74.71	107
0500	Rural, 6″ to 10″		180	.444		9.55	22	.54	32.09	46.50
0550	10″ to 12″		175	.457		11.70	22.50	.55	34.75	50
0570	12″ to 16″		160	.500		12.05	24.50	.61	37.16	54
0600	16″ to 20″		135	.593		14.15	29	.72	43.87	63.50
0700	24″ to 36″		125	.640		15.25	31.50	.77	47.52	68.50
0800	48″ to 72″		100	.800		24.50	39	.97	64.47	91.50
1000	Greater than 10,000 L.F., urban, 6″ to 10″		160	.500		9.55	24.50	.61	34.66	51
1050	10″ to 12″		155	.516		11.70	25.50	.62	37.82	54.50
1070	12″ to 16″		140	.571		12.05	28	.69	40.74	59.50
1100	16″ to 20″		120	.667		14.15	32.50	.81	47.46	69.50
1200	24″ to 36″		115	.696		15.25	34	.84	50.09	73
1300	48″ to 72″		95	.842		24.50	41.50	1.02	67.02	94.50
1500	Rural, 6″ to 10″		215	.372		9.55	18.25	.45	28.25	40.50
1550	10″ to 12″		210	.381		11.70	18.70	.46	30.86	44
1570	12″ to 16″		185	.432		12.05	21	.52	33.57	48.50
1600	16″ to 20″		150	.533		14.15	26	.65	40.80	59
1700	24″ to 36″		140	.571		15.25	28	.69	43.94	63
1800	48″ to 72″	↓	120	.667	↓	24.50	32.50	.81	57.81	80.50
2000	Cured in place pipe, non-pressure, flexible felt resin, 400′ runs									
2100	6″ diameter				L.F.				25.50	28
2200	8″ diameter								26.50	29
2300	10″ diameter								29	32
2400	12″ diameter								34.50	38
2500	15″ diameter								59	65
2600	18″ diameter								82	90
2700	21″ diameter								105	115
2800	24″ diameter								177	195
2900	30″ diameter								195	215
3000	36″ diameter								205	225
3100	48″ diameter				↓				218	240

33 01 30.74 HDPE Pipe Lining		Crew	Daily Output	Labor-Hours	Unit	Material	Labor	Equipment	Total	Total Incl O&P
0010	**HDPE PIPE LINING**, excludes cleaning and video inspection									
0020	Pipe relined with one pipe size smaller than original (4″ for 6″)									
0100	6″ diameter, original size	B-6B	600	.080	L.F.	2.47	3.06	1.50	7.03	9.40
0150	8″ diameter, original size		600	.080		5.85	3.06	1.50	10.41	13.10
0200	10″ diameter, original size		600	.080		8.55	3.06	1.50	13.11	16.05
0250	12″ diameter, original size		400	.120		14.65	4.59	2.26	21.50	26
0300	14″ diameter, original size	↓	400	.120	↓	17.75	4.59	2.26	24.60	29.50

33 05 16 – Utility Structures

33 05 16.13 Precast Concrete Utility Boxes

	33 05 16.13 Precast Concrete Utility Boxes	Crew	Daily Output	Labor-Hours	Unit	Material (2015 Bare Costs)	Labor (2015 Bare Costs)	Equipment (2015 Bare Costs)	Total (2015 Bare Costs)	Total Incl O&P
0010	**PRECAST CONCRETE UTILITY BOXES**, 6" thick									
0040	4' x 6' x 6' high, I.D.	B-13	2	28	Ea.	3,025	1,150	370	4,545	5,575
0050	5' x 10' x 6' high, I.D.		2	28		3,750	1,150	370	5,270	6,375
0100	6' x 10' x 6' high, I.D.		2	28		3,900	1,150	370	5,420	6,525
0150	5' x 12' x 6' high, I.D.		2	28		4,125	1,150	370	5,645	6,775
0200	6' x 12' x 6' high, I.D.		1.80	31.111		4,600	1,275	410	6,285	7,600
0250	6' x 13' x 6' high, I.D.		1.50	37.333		6,050	1,525	490	8,065	9,675
0300	8' x 14' x 7' high, I.D.	↓	1	56	↓	6,525	2,300	735	9,560	11,700
0350	Hand hole, precast concrete, 1-1/2" thick									
0400	1'-0" x 2'-0" x 1'-9", I.D., light duty	B-1	4	6	Ea.	400	230		630	815
0450	4'-6" x 3'-2" x 2'-0", O.D., heavy duty	B-6	3	8		1,475	330	121	1,926	2,275
0460	Meter pit, 4' x 4', 4' deep		2	12		1,400	495	182	2,077	2,550
0470	6' deep		1.60	15		2,000	620	228	2,848	3,450
0480	8' deep		1.40	17.143		2,650	705	260	3,615	4,325
0490	10' deep		1.20	20		3,350	825	305	4,480	5,350
0500	15' deep		1	24		4,900	990	365	6,255	7,400
0510	6' x 6', 4' deep		1.40	17.143		2,275	705	260	3,240	3,925
0520	6' deep		1.20	20		3,425	825	305	4,555	5,425
0530	8' deep		1	24		4,575	990	365	5,930	7,025
0540	10' deep		.80	30		5,700	1,250	455	7,405	8,775
0550	15' deep	↓	.60	40	↓	8,675	1,650	605	10,930	12,900

33 05 23 – Trenchless Utility Installation

33 05 23.19 Microtunneling

	33 05 23.19 Microtunneling	Crew	Daily Output	Labor-Hours	Unit	Material	Labor	Equipment	Total	Total Incl O&P
0010	**MICROTUNNELING**									
0011	Not including excavation, backfill, shoring,									
0020	or dewatering, average 50'/day, slurry method									
0100	24" to 48" outside diameter, minimum				L.F.				875	965
0110	Adverse conditions, add				%				50%	50%
1000	Rent microtunneling machine, average monthly lease				Month				97,500	107,000
1010	Operating technician				Day				630	705
1100	Mobilization and demobilization, minimum				Job				41,200	45,900
1110	Maximum				"				445,500	490,500

33 05 23.20 Horizontal Boring

	33 05 23.20 Horizontal Boring	Crew	Daily Output	Labor-Hours	Unit	Material	Labor	Equipment	Total	Total Incl O&P
0010	**HORIZONTAL BORING**									
0011	Casing only, 100' minimum,									
0020	not incl. jacking pits or dewatering									
0100	Roadwork, 1/2" thick wall, 24" diameter casing	B-42	20	3.200	L.F.	121	136	68	325	430
0200	36" diameter		16	4		223	170	85	478	620
0300	48" diameter		15	4.267		310	181	90.50	581.50	740
0500	Railroad work, 24" diameter		15	4.267		121	181	90.50	392.50	530
0600	36" diameter		14	4.571		223	194	97	514	670
0700	48" diameter	↓	12	5.333		310	226	113	649	840
0900	For ledge, add								20%	20%
1000	Small diameter boring, 3", sandy soil	B-82	900	.018		22	.77	.09	22.86	25.50
1040	Rocky soil	"	500	.032	↓	22	1.38	.17	23.55	26.50

33 05 26 – Utility Identification

33 05 26.05 Utility Connection

	33 05 26.05 Utility Connection	Crew	Daily Output	Labor-Hours	Unit	Material	Labor	Equipment	Total	Total Incl O&P
0010	**UTILITY CONNECTION**									
0020	Water, sanitary, stormwater, gas, single connection	B-14	1	48	Ea.	3,225	1,900	365	5,490	7,050
0030	Telecommunication	"	1	48	"	395	1,900	365	2,660	3,925

33 05 Common Work Results for Utilities

33 05 26 – Utility Identification

33 05 26.10 Utility Accessories		Crew	Daily Output	Labor-Hours	Unit	Material	2015 Bare Costs Labor	Equipment	Total	Total Incl O&P
0010	**UTILITY ACCESSORIES**									
0400	Underground tape, detectable, reinforced, alum. foil core, 2"	1 Clab	150	.053	C.L.F.	6.50	2.01		8.51	10.45
0500	6"		140	.057	"	27.50	2.15		29.65	33.50
9000	Minimum labor/equipment charge	↓	4	2	Job		75		75	123

33 11 Water Utility Distribution Piping

33 11 13 – Public Water Utility Distribution Piping

33 11 13.10 Water Supply, Concrete Pipe

		Crew	Daily Output	Labor-Hours	Unit	Material	Labor	Equipment	Total	Total Incl O&P
0010	**WATER SUPPLY, CONCRETE PIPE**									
0020	Not including excavation or backfill									
3000	Prestressed Conc. Pipe (PCCP), 150 PSI, 12" diam.	B-13	192	.292	L.F.	45	11.95	3.84	60.79	73
3010	24" diameter	"	128	.438		45	17.95	5.75	68.70	85
3040	36" diameter	B-13B	96	.583		70	24	11.75	105.75	129
3050	48" diameter		64	.875		100	36	17.60	153.60	187
3070	72" diameter		60	.933		195	38.50	18.80	252.30	298
3080	84" diameter	↓	40	1.400		252	57.50	28	337.50	400
3090	96" diameter	B-13C	40	1.400		370	57.50	41.50	469	545
3100	108" diameter		32	1.750		525	71.50	52	648.50	755
3102	120" diameter		16	3.500		940	143	104	1,187	1,375
3104	144" diameter	↓	16	3.500	↓	920	143	104	1,167	1,350
3110	Prestressed Concrete Pipe (PCCP), 150 PSI, elbow, 90°, 12" diameter	B-13	24	2.333	Ea.	890	95.50	30.50	1,016	1,175
3140	24" diameter	"	6	9.333		1,750	385	123	2,258	2,675
3150	36" diameter	B-13B	4	14		3,400	575	282	4,257	5,000
3160	48" diameter		3	18.667		6,725	765	375	7,865	9,075
3180	72" diameter		1.60	35		19,200	1,425	705	21,330	24,200
3190	84" diameter		1.30	43.077		23,300	1,775	870	25,945	29,500
3200	96" diameter	↓	1	56		31,300	2,300	1,125	34,725	39,400
3210	108" diameter	B-13C	.66	84.848		46,700	3,475	2,525	52,700	60,000
3220	120" diameter		.40	140		47,200	5,750	4,175	57,125	66,000
3225	144" diameter	↓	.30	184		74,500	7,550	5,475	87,525	100,000
3230	Precast Concrete Pipe (PCCP), 150 PSI, elbow, 45°, 12" diameter	B-13	24	2.333		555	95.50	30.50	681	800
3250	24" diameter	"	6	9.333		1,125	385	123	1,633	2,000
3260	36" diameter	B-13B	4	14		1,900	575	282	2,757	3,350
3270	48" diameter		3	18.667		3,475	765	375	4,615	5,500
3290	72" diameter		1.60	35		11,900	1,425	705	14,030	16,200
3300	84" diameter		1.30	42.945		18,200	1,750	865	20,815	23,800
3310	96" diameter	↓	1	56		22,500	2,300	1,125	25,925	29,800
3320	108" diameter	B-13C	.66	84.337		28,400	3,450	2,500	34,350	39,600
3330	120" diameter		.40	140		28,700	5,750	4,175	38,625	45,500
3340	144" diameter	↓	.30	184	↓	45,900	7,550	5,475	58,925	68,500

33 11 13.15 Water Supply, Ductile Iron Pipe

		Crew	Daily Output	Labor-Hours	Unit	Material	Labor	Equipment	Total	Total Incl O&P
0010	**WATER SUPPLY, DUCTILE IRON PIPE** R331113-80									
0011	Cement lined									
0020	Not including excavation or backfill									
2000	Pipe, class 50 water piping, 18' lengths									
2020	Mechanical joint, 4" diameter	B-21A	200	.200	L.F.	30.50	9.40	2.37	42.27	51
2040	6" diameter		160	.250		32	11.75	2.96	46.71	57
2060	8" diameter		133.33	.300		44.50	14.05	3.55	62.10	75.50
2080	10" diameter		114.29	.350		58.50	16.40	4.14	79.04	94.50
2100	12" diameter	↓	105.26	.380	↓	79	17.85	4.50	101.35	120

33 11 13.15 Water Supply, Ductile Iron Pipe		Crew	Daily Output	Labor-Hours	Unit	2015 Bare Costs Material	Labor	Equipment	Total	Total Incl O&P
2120	14" diameter	B-21A	100	.400	L.F.	93	18.75	4.74	116.49	137
2140	16" diameter		72.73	.550		94.50	26	6.50	127	152
2160	18" diameter		68.97	.580		126	27	6.85	159.85	190
2170	20" diameter		57.14	.700		127	33	8.30	168.30	202
2180	24" diameter		47.06	.850		141	40	10.05	191.05	230
3000	Push-on joint, 4" diameter		400	.100		21	4.69	1.18	26.87	32
3020	6" diameter		333.33	.120		17	5.65	1.42	24.07	29
3040	8" diameter		200	.200		23	9.40	2.37	34.77	43
3060	10" diameter		181.82	.220		36.50	10.30	2.60	49.40	60
3080	12" diameter		160	.250		38.50	11.75	2.96	53.21	64.50
3100	14" diameter		133.33	.300		42.50	14.05	3.55	60.10	73.50
3120	16" diameter		114.29	.350		51.50	16.40	4.14	72.04	87
3140	18" diameter		100	.400		57	18.75	4.74	80.49	97.50
3160	20" diameter		88.89	.450		59.50	21	5.35	85.85	104
3180	24" diameter		76.92	.520		59	24.50	6.15	89.65	111
8000	Piping, fittings, mechanical joint, AWWA C110									
8006	90° bend, 4" diameter	B-20A	16	2	Ea.	155	91.50		246.50	315
8020	6" diameter		12.80	2.500		229	114		343	435
8040	8" diameter		10.67	2.999		450	137		587	715
8060	10" diameter	B-21A	11.43	3.500		620	164	41.50	825.50	985
8080	12" diameter		10.53	3.799		880	178	45	1,103	1,300
8100	14" diameter		10	4		1,200	188	47.50	1,435.50	1,675
8120	16" diameter		7.27	5.502		1,525	258	65	1,848	2,150
8140	18" diameter		6.90	5.797		2,125	272	68.50	2,465.50	2,850
8160	20" diameter		5.71	7.005		2,650	330	83	3,063	3,550
8180	24" diameter		4.70	8.511		4,200	400	101	4,701	5,350
8200	Wye or tee, 4" diameter	B-20A	10.67	2.999		355	137		492	610
8220	6" diameter		8.53	3.751		535	171		706	865
8240	8" diameter		7.11	4.501		850	206		1,056	1,275
8260	10" diameter	B-21A	7.62	5.249		1,225	246	62	1,533	1,800
8280	12" diameter		7.02	5.698		1,625	267	67.50	1,959.50	2,275
8300	14" diameter		6.67	5.997		2,600	281	71	2,952	3,400
8320	16" diameter		4.85	8.247		2,900	385	97.50	3,382.50	3,900
8340	18" diameter		4.60	8.696		3,900	410	103	4,413	5,075
8360	20" diameter		3.81	10.499		5,475	490	124	6,089	6,950
8380	24" diameter		3.14	12.739		9,275	600	151	10,026	11,300
8398	45° bends, 4" diameter	B-20A	16	2		255	91.50		346.50	425
8400	6" diameter		12.80	2.500		199	114		313	400
8405	8" diameter		10.67	2.999		291	137		428	540
8410	12" diameter	B-21A	10.53	3.799		730	178	45	953	1,150
8420	16" diameter		7.27	5.502		1,275	258	65	1,598	1,875
8430	20" diameter		5.71	7.005		2,075	330	83	2,488	2,900
8440	24" diameter		4.70	8.511		2,900	400	101	3,401	3,925
8450	Decreaser, 6" x 4" diameter	B-20A	14.22	2.250		207	103		310	390
8460	8" x 6" diameter	"	11.64	2.749		310	126		436	540
8470	10" x 6" diameter	B-21A	13.33	3.001		390	141	35.50	566.50	695
8480	12" x 6" diameter		12.70	3.150		550	148	37.50	735.50	880
8490	16" x 6" diameter		10	4		890	188	47.50	1,125.50	1,325
8500	20" x 6" diameter		8.42	4.751		1,625	223	56.50	1,904.50	2,200
8552	For water utility valves see Section 33 12 16									
8700	Joint restraint, ductile iron mechanical joints									
8710	4" diameter	B-20A	32	1	Ea.	29	45.50		74.50	105
8720	6" diameter		25.60	1.250		31.50	57		88.50	126

33 11 13 – Public Water Utility Distribution Piping

33 11 13.15 Water Supply, Ductile Iron Pipe		Crew	Daily Output	Labor-Hours	Unit	Material	Labor	Equipment	Total	Total Incl O&P
						2015 Bare Costs				
8730	8" diameter	B-20A	21.33	1.500	Ea.	30	68.50		98.50	142
8740	10" diameter		18.28	1.751		38.50	80		118.50	171
8750	12" diameter		16.84	1.900		50	87		137	193
8760	14" diameter		16	2		53	91.50		144.50	205
8770	16" diameter		11.64	2.749		58	126		184	264
8780	18" diameter		11.03	2.901		73	133		206	291
8785	20" diameter		9.14	3.501		78	160		238	340
8790	24" diameter	↓	7.53	4.250		104	194		298	425
9600	Steel sleeve with tap, 4" diameter	B-20	3	8		435	335		770	1,025
9620	6" diameter		2	12		475	505		980	1,350
9630	8" diameter	↓	2	12	↓	515	505		1,020	1,375

33 11 13.20 Water Supply, Polyethylene Pipe, C901		Crew	Daily Output	Labor-Hours	Unit	Material	Labor	Equipment	Total	Total Incl O&P
0010	**WATER SUPPLY, POLYETHYLENE PIPE, C901** R331113-80									
0020	Not including excavation or backfill									
1000	Piping, 160 PSI, 3/4" diameter	Q-1A	525	.019	L.F.	.46	1.13		1.59	2.27
1120	1" diameter		485	.021		.60	1.22		1.82	2.56
1140	1-1/2" diameter		450	.022		1.35	1.31		2.66	3.54
1160	2" diameter	↓	365	.027	↓	2.20	1.62		3.82	4.95

33 11 13.25 Water Supply, Polyvinyl Chloride Pipe		Crew	Daily Output	Labor-Hours	Unit	Material	Labor	Equipment	Total	Total Incl O&P
0010	**WATER SUPPLY, POLYVINYL CHLORIDE PIPE** R331113-80									
0020	Not including excavation or backfill, unless specified									
2100	PVC pipe, Class 150, 1-1/2" diameter	Q-1A	750	.013	L.F.	.45	.79		1.24	1.73
2120	2" diameter		686	.015		.68	.86		1.54	2.09
2140	2-1/2" diameter	↓	500	.020		1.30	1.18		2.48	3.27
2160	3" diameter	B-20	430	.056	↓	1.40	2.34		3.74	5.35
3010	AWWA C905, PR 100, DR 25									
3030	14" diameter	B-20A	213	.150	L.F.	13.65	6.85		20.50	26
3040	16" diameter		200	.160		19.25	7.30		26.55	32.50
3050	18" diameter		160	.200		24.50	9.15		33.65	41
3060	20" diameter		133	.241		30	11		41	50.50
3070	24" diameter		107	.299		43	13.65		56.65	69.50
3080	30" diameter		80	.400		81	18.30		99.30	118
3090	36" diameter		80	.400		126	18.30		144.30	167
3100	42" diameter		60	.533		168	24.50		192.50	224
3200	48" diameter		60	.533		220	24.50		244.50	281
4520	Pressure pipe Class 150, SDR 18, AWWA C900, 4" diameter		380	.084		2.73	3.85		6.58	9.15
4530	6" diameter		316	.101		5.35	4.63		9.98	13.30
4540	8" diameter		264	.121		8.80	5.55		14.35	18.50
4550	10" diameter		220	.145		14.05	6.65		20.70	26
4560	12" diameter	↓	186	.172	↓	19.90	7.85		27.75	34.50
8000	Fittings with rubber gasket									
8003	Class 150, DR 18									
8006	90° Bend , 4" diameter	B-20	100	.240	Ea.	42.50	10.05		52.55	63
8020	6" diameter		90	.267		75	11.20		86.20	101
8040	8" diameter		80	.300		145	12.60		157.60	180
8060	10" diameter		50	.480		330	20		350	400
8080	12" diameter		30	.800		425	33.50		458.50	520
8100	Tee, 4" diameter		90	.267		58.50	11.20		69.70	83
8120	6" diameter		80	.300		144	12.60		156.60	179
8140	8" diameter		70	.343		185	14.40		199.40	227
8160	10" diameter		40	.600		590	25		615	690
8180	12" diameter	↓	20	1.200	↓	765	50.50		815.50	920

33 11 13 – Public Water Utility Distribution Piping

33 11 13.25 Water Supply, Polyvinyl Chloride Pipe		Crew	Daily Output	Labor-Hours	Unit	2015 Bare Costs Material	Labor	Equipment	Total	Total Incl O&P
8200	45° Bend, 4" diameter	B-20	100	.240	Ea.	42	10.05		52.05	63
8220	6" diameter		90	.267		73	11.20		84.20	98.50
8240	8" diameter		50	.480		139	20		159	185
8260	10" diameter		50	.480		278	20		298	340
8280	12" diameter		30	.800		360	33.50		393.50	450
8300	Reducing tee 6" x 4"		100	.240		104	10.05		114.05	131
8320	8" x 6"		90	.267		166	11.20		177.20	200
8330	10" x 6"		90	.267		196	11.20		207.20	233
8340	10" x 8"		90	.267		216	11.20		227.20	256
8350	12" x 6"		90	.267		245	11.20		256.20	287
8360	12" x 8"		90	.267		265	11.20		276.20	310
8400	Tapped service tee (threaded type) 6" x 6" x 3/4"		100	.240		95.50	10.05		105.55	121
8430	6" x 6" x 1"		90	.267		95.50	11.20		106.70	123
8440	6" x 6" x 1-1/2"		90	.267		95.50	11.20		106.70	123
8450	6" x 6" x 2"		90	.267		95.50	11.20		106.70	123
8460	8" x 8" x 3/4"		90	.267		140	11.20		151.20	172
8470	8" x 8" x 1"		90	.267		140	11.20		151.20	172
8480	8" x 8" x 1-1/2"		90	.267		140	11.20		151.20	172
8490	8" x 8" x 2"		90	.267		140	11.20		151.20	172
8500	Repair coupling 4"		100	.240		26	10.05		36.05	45
8520	6" diameter		90	.267		40.50	11.20		51.70	63
8540	8" diameter		50	.480		96.50	20		116.50	139
8560	10" diameter		50	.480		203	20		223	256
8580	12" diameter		50	.480		296	20		316	360
8600	Plug end 4"		100	.240		23	10.05		33.05	42
8620	6" diameter		90	.267		41	11.20		52.20	63.50
8640	8" diameter		50	.480		69	20		89	109
8660	10" diameter		50	.480		97	20		117	139
8680	12" diameter		50	.480		119	20		139	164
8700	PVC pipe, joint restraint									
8710	4" diameter	B-20A	32	1	Ea.	45.50	45.50		91	123
8720	6" diameter		25.60	1.250		56	57		113	153
8730	8" diameter		21.33	1.500		74.50	68.50		143	191
8740	10" diameter		18.28	1.751		130	80		210	271
8750	12" diameter		16.84	1.900		145	87		232	297
8760	14" diameter		16	2		204	91.50		295.50	370
8770	16" diameter		11.64	2.749		265	126		391	490
8780	18" diameter		11.03	2.901		340	133		473	585
8785	20" diameter		9.14	3.501		420	160		580	715
8790	24" diameter		7.53	4.250		490	194		684	850

33 11 13.35 Water Supply, HDPE

33 11 13.35 Water Supply, HDPE		Crew	Daily Output	Labor-Hours	Unit	Material	Labor	Equipment	Total	Total Incl O&P
0010	**WATER SUPPLY, HDPE** R331113-80									
0011	Butt fusion joints, SDR 21 40' lengths not including excavation or backfill									
0100	4" diameter	B-22A	400	.100	L.F.	2.47	4.30	1.64	8.41	11.55
0200	6" diameter		380	.105		5.85	4.53	1.73	12.11	15.70
0300	8" diameter		320	.125		8.55	5.40	2.06	16.01	20.50
0400	10" diameter		300	.133		14.65	5.75	2.19	22.59	28
0500	12" diameter		260	.154		17.75	6.60	2.53	26.88	33
0600	14" diameter	B-22B	220	.182		27.50	7.80	5	40.30	48.50
0700	16" diameter		180	.222		32.50	9.55	6.15	48.20	58
0800	18" diameter		140	.286		38	12.30	7.90	58.20	70
0900	24" diameter		100	.400		59	17.20	11.05	87.25	105

33 11 Water Utility Distribution Piping

33 11 13 – Public Water Utility Distribution Piping

33 11 13.35 Water Supply, HDPE		Crew	Daily Output	Labor-Hours	Unit	2015 Bare Costs Material	Labor	Equipment	Total	Total Incl O&P
1000	Fittings									
1100	Elbows, 90 degrees									
1200	4″ diameter	B-22A	32	1.250	Ea.	18.45	54	20.50	92.95	131
1300	6″ diameter		28	1.429		59	61.50	23.50	144	191
1400	8″ diameter		24	1.667		121	71.50	27.50	220	280
1500	10″ diameter		18	2.222		385	95.50	36.50	517	620
1600	12″ diameter		12	3.333		400	143	55	598	735
1700	14″ diameter	B-22B	9	4.444		490	191	123	804	980
1800	16″ diameter		6	6.667		570	287	184	1,041	1,300
1900	18″ diameter		4	10		660	430	276	1,366	1,725
2000	24″ diameter		3	13.333		985	575	370	1,930	2,400
2100	Tees									
2200	4″ diameter	B-22A	30	1.333	Ea.	23.50	57.50	22	103	143
2300	6″ diameter		26	1.538		50	66	25.50	149.50	199
2400	8″ diameter		22	1.818		146	78	30	254	320
2500	10″ diameter		15	2.667		193	115	44	352	445
2600	12″ diameter		10	4		545	172	66	783	950
2700	14″ diameter	B-22B	8	5		640	215	138	993	1,200
2800	16″ diameter		6	6.667		755	287	184	1,226	1,500
2900	18″ diameter		4	10		985	430	276	1,691	2,075
3000	24″ diameter		2	20		1,500	860	555	2,915	3,650
4100	Caps									
4110	4″ diameter	B-22A	34	1.176	Ea.	13.60	50.50	19.35	83.45	118
4120	6″ diameter		30	1.333		35.50	57.50	22	115	156
4130	8″ diameter		26	1.538		52	66	25.50	143.50	192
4150	10″ diameter		20	2		111	86	33	230	298
4160	12″ diameter		14	2.857		124	123	47	294	390

33 11 13.45 Water Supply, Copper Pipe		Crew	Daily Output	Labor-Hours	Unit	Material	Labor	Equipment	Total	Total Incl O&P
0010	**WATER SUPPLY, COPPER PIPE**									
0020	Not including excavation or backfill									
2000	Tubing, type K, 20′ joints, 3/4″ diameter	Q-1	400	.040	L.F.	6.30	2.11		8.41	10.25
2200	1″ diameter		320	.050		8.40	2.64		11.04	13.35
3000	1-1/2″ diameter		265	.060		13.40	3.19		16.59	19.70
3020	2″ diameter		230	.070		20.50	3.68		24.18	28.50
3040	2-1/2″ diameter		146	.110		30.50	5.80		36.30	42.50
3060	3″ diameter		134	.119		42	6.30		48.30	56.50
4012	4″ diameter		95	.168		69.50	8.90		78.40	90.50
4016	6″ diameter	Q-2	80	.300		150	16.45		166.45	191
5000	Tubing, type L									
5108	2″ diameter	Q-1	230	.070	L.F.	16.05	3.68		19.73	23.50
6010	3″ diameter		134	.119		31.50	6.30		37.80	44.50
6012	4″ diameter		95	.168		55.50	8.90		64.40	75
6016	6″ diameter	Q-2	80	.300		132	16.45		148.45	171
7165	Fittings, brass, corporation stops, no lead, 3/4″ diameter	1 Plum	19	.421	Ea.	82	24.50		106.50	129
7166	1″ diameter		16	.500		108	29.50		137.50	165
7167	1-1/2″ diameter		13	.615		240	36		276	320
7168	2″ diameter		11	.727		350	42.50		392.50	450
7170	Curb stops, no lead, 3/4″ diameter		19	.421		112	24.50		136.50	162
7171	1″ diameter		16	.500		156	29.50		185.50	218
7172	1-1/2″ diameter		13	.615		296	36		332	380
7173	2″ diameter		11	.727		315	42.50		357.50	410
7180	Curb box, cast iron, 1/2″ to 1″ curb stops		12	.667		51.50	39		90.50	118

33 11 Water Utility Distribution Piping

33 11 13 – Public Water Utility Distribution Piping

33 11 13.45 Water Supply, Copper Pipe		Crew	Daily Output	Labor-Hours	Unit	Material	Labor	Equipment	Total	Total Incl O&P
						2015 Bare Costs				
7200	1-1/4" to 2" curb stops	1 Plum	8	1	Ea.	84.50	58.50		143	185
7220	Saddles, 3/4" & 1" diameter, add					76.50			76.50	84.50
7240	1-1/2" to 2" diameter, add					117			117	129

33 11 13.90 Water Supply, Thrust Blocks

		Crew	Daily Output	Labor-Hours	Unit	Material	Labor	Equipment	Total	Total Incl O&P
0010	**WATER SUPPLY, THRUST BLOCKS**									
0110	Thrust block for 90 elbow, 4" Diameter	C-30	41	.195	Ea.	17.15	7.35	4.20	28.70	35.50
0115	6" Diameter		23	.348		29.50	13.10	7.50	50.10	62
0120	8" Diameter		14	.571		45	21.50	12.30	78.80	98
0125	10" Diameter		9	.889		64.50	33.50	19.15	117.15	147
0130	12" Diameter		7	1.143		87.50	43	24.50	155	194
0135	14" Diameter		5	1.600		117	60	34.50	211.50	266
0140	16" Diameter		4	2		148	75	43	266	335
0145	18" Diameter		3	2.667		182	100	57.50	339.50	425
0150	20" Diameter		2.50	3.200		219	120	69	408	515
0155	24" Diameter		2	4		315	150	86	551	685
0210	Thrust block for tee or deadend, 4" Diameter		65	.123		11.85	4.63	2.65	19.13	23.50
0215	6" Diameter		35	.229		21	8.60	4.93	34.53	42.50
0220	8" Diameter		21	.381		32.50	14.30	8.20	55	68
0225	10" Diameter		14	.571		46.50	21.50	12.30	80.30	99.50
0230	12" Diameter		10	.800		63	30	17.25	110.25	138
0235	14" Diameter		7	1.143		84.50	43	24.50	152	191
0240	16" Diameter		5	1.600		106	60	34.50	200.50	254
0245	18" Diameter		4	2		131	75	43	249	315
0250	20" Diameter		3.50	2.286		158	86	49.50	293.50	370
0255	24" Diameter		2.50	3.200		223	120	69	412	520

33 12 Water Utility Distribution Equipment

33 12 13 – Water Service Connections

33 12 13.15 Tapping, Crosses and Sleeves

		Crew	Daily Output	Labor-Hours	Unit	Material	Labor	Equipment	Total	Total Incl O&P
0010	**TAPPING, CROSSES AND SLEEVES**									
4000	Drill and tap pressurized main (labor only)									
4100	6" main, 1" to 2" service	Q-1	3	5.333	Ea.		282		282	440
4150	8" main, 1" to 2" service	"	2.75	5.818	"		305		305	480
4500	Tap and insert gate valve									
4600	8" main, 4" branch	B-21	3.20	8.750	Ea.		380	43	423	665
4650	6" branch		2.70	10.370			450	51	501	785
4700	10" Main, 4" branch		2.70	10.370			450	51	501	785
4750	6" branch		2.35	11.915			515	58.50	573.50	905
4800	12" main, 6" branch		2.35	11.915			515	58.50	573.50	905
4850	8" branch		2.35	11.915			515	58.50	573.50	905
7000	Tapping crosses, sleeves, valves; with rubber gaskets									
7020	Crosses, 4" x 4"	B-21	37	.757	Ea.	1,375	33	3.72	1,411.72	1,550
7060	8" x 6"		21	1.333		1,500	58	6.55	1,564.55	1,750
7080	8" x 8"		21	1.333		1,800	58	6.55	1,864.55	2,075
7100	10" x 6"		21	1.333		2,600	58	6.55	2,664.55	2,950
7160	12" x 12"		18	1.556		3,575	67.50	7.65	3,650.15	4,075
7180	14" x 6"		16	1.750		6,475	76	8.60	6,559.60	7,250
7240	16" x 10"		14	2		7,000	86.50	9.85	7,096.35	7,850
7280	18" x 6"		10	2.800		9,875	121	13.80	10,009.80	11,100
7320	18" x 18"		10	2.800		9,650	121	13.80	9,784.80	10,800
7340	20" x 6"		8	3.500		8,475	152	17.20	8,644.20	9,600

33 12 13 – Water Service Connections

33 12 13.15 Tapping, Crosses and Sleeves		Crew	Daily Output	Labor-Hours	Unit	2015 Bare Costs Material	Labor	Equipment	Total	Total Incl O&P
7360	20″ x 12″	B-21	8	3.500	Ea.	8,800	152	17.20	8,969.20	9,950
7420	24″ x 12″		6	4.667		11,600	202	23	11,825	13,200
7440	24″ x 18″		6	4.667		17,400	202	23	17,625	19,600
7600	Cut-in sleeves with rubber gaskets, 4″		18	1.556		430	67.50	7.65	505.15	595
7640	8″		10	2.800		650	121	13.80	784.80	925
7680	12″		9	3.111		1,225	135	15.30	1,375.30	1,575
7800	Cut-in valves with rubber gaskets, 4″		18	1.556		505	67.50	7.65	580.15	675
7840	8″		10	2.800		1,050	121	13.80	1,184.80	1,375
7880	12″		9	3.111		1,300	135	15.30	1,450.30	1,650
7900	Tapping Valve 4 inch, MJ, ductile iron		18	1.556		555	67.50	7.65	630.15	730
7920	6 inch, MJ, ductile iron		12	2.333		780	101	11.50	892.50	1,025
8000	Sleeves with rubber gaskets, 4″ x 4″		37	.757		845	33	3.72	881.72	990
8030	8″ x 4″		21	1.333		960	58	6.55	1,024.55	1,150
8040	8″ x 6″		21	1.333		1,075	58	6.55	1,139.55	1,275
8060	8″ x 8″		21	1.333		1,250	58	6.55	1,314.55	1,475
8070	10″ x 4″		21	1.333		1,075	58	6.55	1,139.55	1,300
8080	10″ x 6″		21	1.333		1,200	58	6.55	1,264.55	1,425
8090	10″ x 8″		21	1.333		1,325	58	6.55	1,389.55	1,550
8140	12″ x 12″		18	1.556		2,350	67.50	7.65	2,425.15	2,725
8160	14″ x 6″		16	1.750		1,900	76	8.60	1,984.60	2,225
8220	16″ x 10″		14	2		3,525	86.50	9.85	3,621.35	4,025
8260	18″ x 6″		10	2.800		2,250	121	13.80	2,384.80	2,675
8300	18″ x 18″		10	2.800		11,600	121	13.80	11,734.80	13,000
8320	20″ x 6″		8	3.500		2,550	152	17.20	2,719.20	3,100
8340	20″ x 12″		8	3.500		4,400	152	17.20	4,569.20	5,125
8400	24″ x 12″		6	4.667		4,400	202	23	4,625	5,200
8420	24″ x 18″		6	4.667		11,700	202	23	11,925	13,300
8800	Hydrant valve box, 6′ long	B-20	20	1.200		280	50.50		330.50	390
8820	8′ long		18	1.333		355	56		411	480
8830	Valve box w/lid 4′ deep		14	1.714		107	72		179	234
8840	Valve box and large base w/lid		14	1.714		325	72		397	475

33 12 16 – Water Utility Distribution Valves

33 12 16.10 Valves		Crew	Daily Output	Labor-Hours	Unit	Material	Labor	Equipment	Total	Total Incl O&P
0010	**VALVES**, water distribution									
0011	See Sections 22 05 23.20 and 22 05 23.60									
3000	Butterfly valves with boxes, cast iron, mech. jt.									
3100	4″ diameter	B-6	6	4	Ea.	575	165	60.50	800.50	970
3180	8″ diameter		6	4		940	165	60.50	1,165.50	1,350
3340	12″ diameter		6	4		1,700	165	60.50	1,925.50	2,200
3400	14″ diameter		4	6		3,400	248	91	3,739	4,225
3460	18″ diameter		4	6		6,225	248	91	6,564	7,350
3480	20″ diameter		4	6		8,200	248	91	8,539	9,500
3500	24″ diameter		4	6		13,700	248	91	14,039	15,600
3600	With lever operator									
3610	4″ diameter	B-6	6	4	Ea.	470	165	60.50	695.50	850
3616	8″ diameter		6	4		835	165	60.50	1,060.50	1,250
3620	12″ diameter		6	4		1,600	165	60.50	1,825.50	2,100
3624	16″ diameter		4	6		4,700	248	91	5,039	5,650
3630	24″ diameter		4	6		13,400	248	91	13,739	15,200
3700	Check valves, flanged									
3710	4″ diameter	B-6	6	4	Ea.	605	165	60.50	830.50	1,000
3714	6″ diameter		6	4		1,025	165	60.50	1,250.50	1,475

33 12 16 – Water Utility Distribution Valves

33 12 16.10 Valves		Crew	Daily Output	Labor-Hours	Unit	2015 Bare Costs Material	Labor	Equipment	Total	Total Incl O&P
3716	8″ diameter	B-6	6	4	Ea.	1,950	165	60.50	2,175.50	2,475
3720	12″ diameter		6	4		5,175	165	60.50	5,400.50	6,025
3726	18″ diameter		4	6		23,700	248	91	24,039	26,600
3730	24″ diameter	↓	4	6	↓	52,000	248	91	52,339	58,000
3800	Gate valves, C.I., 250 PSI, mechanical joint, w/boxes									
3810	4″ diameter	B-6	6	4	Ea.	880	165	60.50	1,105.50	1,300
3814	6″ diameter		6	4		1,350	165	60.50	1,575.50	1,800
3816	8″ diameter		6	4		2,300	165	60.50	2,525.50	2,850
3820	12″ diameter		6	4		5,625	165	60.50	5,850.50	6,500
3824	16″ diameter		4	6		17,300	248	91	17,639	19,500
3828	20″ diameter		4	6		23,200	248	91	23,539	26,000
3830	24″ diameter	↓	4	6	↓	25,200	248	91	25,539	28,200

33 12 16.20 Valves		Crew	Daily Output	Labor-Hours	Unit	Material	Labor	Equipment	Total	Total Incl O&P
0010	**VALVES**									
0011	Special trim or use									
0150	Altitude valve, single acting, modulating, 2.5″ diameter	B-6	6	4	Ea.	2,925	165	60.50	3,150.50	3,550
0155	3″ diameter		6	4		2,950	165	60.50	3,175.50	3,575
0160	4″ diameter		6	4		3,450	165	60.50	3,675.50	4,125
0165	6″ diameter		6	4		4,275	165	60.50	4,500.50	5,050
0170	8″ diameter		6	4		6,150	165	60.50	6,375.50	7,100
0175	10″ diameter		6	4		6,800	165	60.50	7,025.50	7,800
0180	12″ diameter		6	4		10,300	165	60.50	10,525.50	11,700
0250	Altitude valve, single acting, non-modulating, 2.5″ diameter		6	4		2,925	165	60.50	3,150.50	3,550
0255	3″ diameter		6	4		2,950	165	60.50	3,175.50	3,575
0260	4″ diameter		6	4		3,200	165	60.50	3,425.50	3,825
0265	6″ diameter		6	4		4,275	165	60.50	4,500.50	5,050
0270	8″ diameter		6	4		6,150	165	60.50	6,375.50	7,100
0275	10″ diameter		6	4		6,800	165	60.50	7,025.50	7,800
0280	12″ diameter		6	4		10,300	165	60.50	10,525.50	11,700
0350	Altitude valve, double acting, non-modulating, 2.5″ diameter		6	4		3,800	165	60.50	4,025.50	4,500
0355	3″ diameter		6	4		39,800	165	60.50	40,025.50	44,100
0360	4″ diameter		6	4		4,500	165	60.50	4,725.50	5,275
0365	6″ diameter		6	4		5,550	165	60.50	5,775.50	6,425
0370	8″ diameter		6	4		8,000	165	60.50	8,225.50	9,125
0375	10″ diameter		6	4		8,825	165	60.50	9,050.50	10,000
0380	12″ diameter	↓	6	4		13,400	165	60.50	13,625.50	15,100
1000	Air release valve for water, 1/2″ inlet	1 Plum	16	.500		60.50	29.50		90	113
1005	3/4″ inlet		16	.500		60.50	29.50		90	113
1010	1″ inlet		14	.571		60.50	33.50		94	120
1020	2″ inlet		9	.889		182	52		234	282
1100	Air release & vacuum valve for water, 1/2″ inlet		16	.500		266	29.50		295.50	340
1105	3/4″ inlet		16	.500		266	29.50		295.50	340
1110	1″ inlet		14	.571		266	33.50		299.50	345
1120	2″ inlet	↓	9	.889		395	52		447	510
1130	3″ inlet	Q-1	8	2		1,500	106		1,606	1,825
1140	4″ inlet	″	5	3.200		1,875	169		2,044	2,350
9000	Valves, gate valve, N.R.S. PIV with post, 4″ diameter	B-6	6	4		1,875	165	60.50	2,100.50	2,375
9040	8″ diameter		6	4		3,275	165	60.50	3,500.50	3,950
9080	12″ diameter		6	4		6,600	165	60.50	6,825.50	7,600
9120	OS&Y, 4″ diameter		6	4		730	165	60.50	955.50	1,150
9160	8″ diameter		6	4		1,550	165	60.50	1,775.50	2,025
9200	12″ diameter	↓	6	4	↓	3,725	165	60.50	3,950.50	4,425

33 12 Water Utility Distribution Equipment

33 12 16 – Water Utility Distribution Valves

33 12 16.20 Valves		Crew	Daily Output	Labor-Hours	Unit	2015 Bare Costs Material	Labor	Equipment	Total	Total Incl O&P
9220	14″ diameter	B-6	4	6	Ea.	5,075	248	91	5,414	6,100
9400	Check valves, rubber disc, 2-1/2″ diameter		6	4		710	165	60.50	935.50	1,125
9440	4″ diameter		6	4		605	165	60.50	830.50	1,000
9500	8″ diameter		6	4		1,950	165	60.50	2,175.50	2,475
9540	12″ diameter		6	4		5,175	165	60.50	5,400.50	6,025
9700	Detector check valves, reducing, 4″ diameter		6	4		1,350	165	60.50	1,575.50	1,800
9740	8″ diameter		6	4		3,050	165	60.50	3,275.50	3,675
9800	Galvanized, 4″ diameter		6	4		1,575	165	60.50	1,800.50	2,075
9840	8″ diameter		6	4		3,350	165	60.50	3,575.50	4,000

33 12 19 – Water Utility Distribution Fire Hydrants

33 12 19.10 Fire Hydrants

33 12 19.10 Fire Hydrants		Crew	Daily Output	Labor-Hours	Unit	Material	Labor	Equipment	Total	Total Incl O&P
0010	**FIRE HYDRANTS**									
0020	Mechanical joints unless otherwise noted									
1000	Fire hydrants, two way; excavation and backfill not incl.									
1100	4-1/2″ valve size, depth 2′-0″	B-21	10	2.800	Ea.	1,650	121	13.80	1,784.80	2,025
1200	4′-6″		9	3.111		1,975	135	15.30	2,125.30	2,400
1260	6′-0″		7	4		2,100	173	19.70	2,292.70	2,600
1340	8′-0″		6	4.667		2,225	202	23	2,450	2,800
1420	10′-0″		5	5.600		2,375	243	27.50	2,645.50	3,050
2000	5-1/4″ valve size, depth 2′-0″		10	2.800		1,775	121	13.80	1,909.80	2,150
2080	4′-0″		9	3.111		1,900	135	15.30	2,050.30	2,325
2160	6′-0″		7	4		2,050	173	19.70	2,242.70	2,550
2240	8′-0″		6	4.667		2,175	202	23	2,400	2,750
2320	10′-0″		5	5.600		2,500	243	27.50	2,770.50	3,175
2350	For threeway valves, add					7%				
2400	Lower barrel extensions with stems, 1′-0″	B-20	14	1.714		320	72		392	465
2440	2′-0″		13	1.846		390	77.50		467.50	555
2480	3′-0″		12	2		770	84		854	980
2520	4′-0″		10	2.400		740	101		841	980
5000	Indicator post									
5020	Adjustable, valve size 4″ to 14″, 4′ bury	B-21	10	2.800	Ea.	815	121	13.80	949.80	1,100
5060	8′ bury		7	4		1,100	173	19.70	1,292.70	1,500
5080	10′ bury		6	4.667		1,150	202	23	1,375	1,625
5100	12′ bury		5	5.600		1,400	243	27.50	1,670.50	1,975
5120	14′ bury		4	7		1,525	305	34.50	1,864.50	2,200
5500	Non-adjustable, valve size 4″ to 14″, 3′ bury		10	2.800		815	121	13.80	949.80	1,100
5520	3′-6″ bury		10	2.800		815	121	13.80	949.80	1,100
5540	4′ bury		9	3.111		815	135	15.30	965.30	1,125

33 16 Water Utility Storage Tanks

33 16 13 – Aboveground Water Utility Storage Tanks

33 16 13.13 Steel Water Storage Tanks

33 16 13.13 Steel Water Storage Tanks		Crew	Daily Output	Labor-Hours	Unit	Material	Labor	Equipment	Total	Total Incl O&P
0010	**STEEL WATER STORAGE TANKS**									
0910	Steel, ground level, ht./diam. less than 1, not incl. fdn., 100,000 gallons				Ea.				202,000	244,500
1000	250,000 gallons								295,500	324,000
1200	500,000 gallons								417,000	458,500
1250	750,000 gallons								538,000	591,500
1300	1,000,000 gallons								558,000	725,500
1500	2,000,000 gallons								1,043,000	1,148,000
1600	4,000,000 gallons								2,121,000	2,333,000

33 16 13 – Aboveground Water Utility Storage Tanks

33 16 13.13 Steel Water Storage Tanks		Crew	Daily Output	Labor-Hours	Unit	Material (2015 Bare Costs)	Labor (2015 Bare Costs)	Equipment (2015 Bare Costs)	Total (2015 Bare Costs)	Total Incl O&P
1800	6,000,000 gallons				Ea.				3,095,000	3,405,000
1850	8,000,000 gallons								4,068,000	4,475,000
1910	10,000,000 gallons				↓				5,050,000	5,554,500
2100	Steel standpipes, ht./diam. more than 1, 100' to overflow, no fdn.									
2200	500,000 gallons				Ea.				546,500	600,500
2400	750,000 gallons								722,500	794,500
2500	1,000,000 gallons								1,060,500	1,167,000
2700	1,500,000 gallons								1,749,000	1,923,000
2800	2,000,000 gallons				↓				2,327,000	2,559,000

33 16 13.16 Prestressed Conc. Water Storage Tanks		Crew	Daily Output	Labor-Hours	Unit	Material	Labor	Equipment	Total	Total Incl O&P
0010	**PRESTRESSED CONC. WATER STORAGE TANKS**									
0020	Not including fdn., pipe or pumps, 250,000 gallons				Ea.				299,000	329,500
0100	500,000 gallons								487,000	536,000
0300	1,000,000 gallons								707,000	807,500
0400	2,000,000 gallons								1,072,000	1,179,000
0600	4,000,000 gallons								1,706,000	1,877,000
0700	6,000,000 gallons								2,266,000	2,493,000
0750	8,000,000 gallons								2,924,000	3,216,000
0800	10,000,000 gallons				↓				3,533,000	3,886,000

33 16 13.23 Plastic-Coated Fabric Pillow Water Tanks		Crew	Daily Output	Labor-Hours	Unit	Material	Labor	Equipment	Total	Total Incl O&P
0010	**PLASTIC-COATED FABRIC PILLOW WATER TANKS**									
7000	Water tanks, vinyl coated fabric pillow tanks, freestanding, 5,000 gallons	4 Clab	4	8	Ea.	3,675	300		3,975	4,550
7100	Supporting embankment not included, 25,000 gallons	6 Clab	2	24		13,200	900		14,100	16,000
7200	50,000 gallons	8 Clab	1.50	42.667		18,500	1,600		20,100	22,900
7300	100,000 gallons	9 Clab	.90	80		42,300	3,000		45,300	51,500
7400	150,000 gallons		.50	144		60,500	5,425		65,925	76,000
7500	200,000 gallons		.40	180		75,000	6,775		81,775	93,500
7600	250,000 gallons	↓	.30	240	↓	105,500	9,025		114,525	131,000

33 16 19 – Elevated Water Utility Storage Tanks

33 16 19.50 Elevated Water Storage Tanks		Crew	Daily Output	Labor-Hours	Unit	Material	Labor	Equipment	Total	Total Incl O&P
0010	**ELEVATED WATER STORAGE TANKS**									
0011	Not incl. pipe, pumps or foundation									
3000	Elevated water tanks, 100' to bottom capacity line, incl. painting									
3010	50,000 gallons				Ea.				185,000	204,000
3300	100,000 gallons								280,000	307,500
3400	250,000 gallons								751,500	826,500
3600	500,000 gallons								1,336,000	1,470,000
3700	750,000 gallons								1,622,000	1,783,500
3900	1,000,000 gallons				↓				2,322,000	2,556,000

33 21 13.10 Wells and Accessories		Crew	Daily Output	Labor-Hours	Unit	Material	2015 Bare Costs Labor	Equipment	Total	Total Incl O&P
0010	**WELLS & ACCESSORIES**									
0011	Domestic									
0100	Drilled, 4" to 6" diameter	B-23	120	.333	L.F.		12.65	23.50	36.15	47
0200	8" diameter	"	95.20	.420	"		15.95	30	45.95	59
0400	Gravel pack well, 40' deep, incl. gravel & casing, complete									
0500	24" diameter casing x 18" diameter screen	B-23	.13	307	Total	38,800	11,700	21,900	72,400	86,000
0600	36" diameter casing x 18" diameter screen		.12	333	"	40,000	12,700	23,700	76,400	91,000
0800	Observation wells, 1-1/4" riser pipe		163	.245	V.L.F.	20.50	9.35	17.45	47.30	57
0900	For flush Buffalo roadway box, add	1 Skwk	16.60	.482	Ea.	51	23.50		74.50	94
1200	Test well, 2-1/2" diameter, up to 50' deep (15 to 50 GPM)	B-23	1.51	26.490	"	805	1,000	1,875	3,680	4,625
1300	Over 50' deep, add	"	121.80	.328	L.F.	21.50	12.50	23.50	57.50	69.50
1500	Pumps, installed in wells to 100' deep, 4" submersible									
1510	1/2 H.P.	Q-1	3.22	4.969	Ea.	350	263		613	795
1520	3/4 H.P.		2.66	6.015		400	320		720	935
1600	1 H.P.		2.29	6.987		425	370		795	1,050
1700	1-1/2 H.P.	Q-22	1.60	10		800	530	410	1,740	2,150
1800	2 H.P.		1.33	12.030		830	635	490	1,955	2,450
1900	3 H.P.		1.14	14.035		1,150	740	575	2,465	3,050
2000	5 H.P.		1.14	14.035		1,600	740	575	2,915	3,525
2050	Remove and install motor only, 4 H.P.		1.14	14.035		1,100	740	575	2,415	2,975
3000	Pump, 6" submersible, 25' to 150' deep, 25 H.P., 249 to 297 GPM		.89	17.978		3,350	950	735	5,035	5,950
3100	25' to 500' deep, 30 H.P., 100 to 300 GPM		.73	21.918		4,100	1,150	895	6,145	7,275
8000	Steel well casing	B-23A	3020	.008	Lb.	1.21	.34	.89	2.44	2.86
8110	Well screen assembly, stainless steel, 2" diameter		273	.088	L.F.	81	3.75	9.85	94.60	106
8120	3" diameter		253	.095		142	4.04	10.60	156.64	174
8130	4" diameter		200	.120		177	5.10	13.45	195.55	217
8140	5" diameter		168	.143		187	6.10	16	209.10	232
8150	6" diameter		126	.190		212	8.10	21.50	241.60	270
8160	8" diameter		98.50	.244		288	10.40	27.50	325.90	360
8170	10" diameter		73	.329		360	14	37	411	460
8180	12" diameter		62.50	.384		410	16.35	43	469.35	525
8190	14" diameter		54.30	.442		460	18.85	49.50	528.35	590
8200	16" diameter		48.30	.497		510	21	55.50	586.50	655
8210	18" diameter		39.20	.612		635	26	68.50	729.50	820
8220	20" diameter		31.20	.769		710	33	86	829	935
8230	24" diameter		23.80	1.008		855	43	113	1,011	1,125
8240	26" diameter		21	1.143		930	48.50	128	1,106.50	1,250
8244	Well casing or drop pipe, PVC, 0.5" dia		550	.044		1.55	1.86	4.88	8.29	10.05
8245	0.75" diameter		550	.044		1.13	1.86	4.88	7.87	9.60
8246	1" diameter		550	.044		1.17	1.86	4.88	7.91	9.65
8247	1.25" diameter		520	.046		1.35	1.97	5.15	8.47	10.35
8248	1.5" diameter		490	.049		1.58	2.09	5.50	9.17	11.15
8249	1.75" diameter		380	.063		1.48	2.69	7.05	11.22	13.80
8250	2" diameter		280	.086		1.65	3.65	9.60	14.90	18.25
8252	3" diameter		260	.092		3.39	3.93	10.35	17.67	21.50
8254	4" diameter		205	.117		4.90	4.99	13.10	22.99	28
8255	5" diameter		170	.141		3.20	6	15.80	25	30.50
8256	6" diameter		130	.185		8.50	7.85	20.50	36.85	44.50
8258	8" diameter		100	.240		12.95	10.20	27	50.15	60.50
8260	10" diameter		73	.329		9	14	37	60	73
8262	12" diameter		62	.387		12	16.50	43.50	72	87
8300	Slotted PVC, 1-1/4" diameter		521	.046		2.40	1.96	5.15	9.51	11.45
8310	1-1/2" diameter		488	.049		2.85	2.10	5.50	10.45	12.60

33 21 Water Supply Wells

33 21 13 – Public Water Supply Wells

33 21 13.10 Wells and Accessories		Crew	Daily Output	Labor-Hours	Unit	Material	2015 Bare Costs Labor	Equipment	Total	Total Incl O&P
8320	2″ diameter	B-23A	273	.088	L.F.	3.67	3.75	9.85	17.27	21
8330	3″ diameter		253	.095		5.85	4.04	10.60	20.49	24.50
8340	4″ diameter		200	.120		6.75	5.10	13.45	25.30	30.50
8350	5″ diameter		168	.143		14.05	6.10	16	36.15	43
8360	6″ diameter		126	.190		16.65	8.10	21.50	46.25	55
8370	8″ diameter		98.50	.244		26	10.40	27.50	63.90	75.50
8400	Artificial gravel pack, 2″ screen, 6″ casing	B-23B	174	.138		3.60	5.90	17.45	26.95	32.50
8405	8″ casing		111	.216		4.63	9.20	27.50	41.33	50
8410	10″ casing		74.50	.322		6.60	13.70	40.50	60.80	74.50
8415	12″ casing		60	.400		9.25	17.05	50.50	76.80	93
8420	14″ casing		50.20	.478		10.75	20.50	60.50	91.75	111
8425	16″ casing		40.70	.590		13.65	25	74.50	113.15	138
8430	18″ casing		36	.667		16.40	28.50	84.50	129.40	157
8435	20″ casing		29.50	.814		19.25	34.50	103	156.75	190
8440	24″ casing		25.70	.934		22	40	118	180	219
8445	26″ casing		24.60	.976		24.50	41.50	123	189	231
8450	30″ casing		20	1.200		28	51	152	231	280
8455	36″ casing		16.40	1.463		31.50	62.50	185	279	340
8500	Develop well		8	3	Hr.	262	128	380	770	910
8550	Pump test well		8	3		82.50	128	380	590.50	715
8560	Standby well	B-23A	8	3		78	128	335	541	665
8570	Standby, drill rig		8	3			128	335	463	575
8580	Surface seal well, concrete filled		1	24	Ea.	850	1,025	2,675	4,550	5,525
8590	Well test pump, install & remove	B-23	1	40			1,525	2,850	4,375	5,625
8600	Well sterilization, chlorine	2 Clab	1	16		122	600		722	1,125
8610	Well water pressure switch	1 Clab	12	.667		50.50	25		75.50	96.50
8630	Well, water pressure switch with manual reset	"	12	.667		35.50	25		60.50	80
9000	Minimum labor/equipment charge	B-21	1.80	15.556	Job		675	76.50	751.50	1,175
9950	See Section 31 23 19.40 for wellpoints									
9960	See Section 31 23 19.30 for drainage wells									

33 21 13.20 Water Supply Wells, Pumps

		Crew	Daily Output	Labor-Hours	Unit	Material	Labor	Equipment	Total	Total Incl O&P
0010	**WATER SUPPLY WELLS, PUMPS**									
0011	With pressure control									
1000	Deep well, jet, 42 gal. galvanized tank									
1040	3/4 HP	1 Plum	.80	10	Ea.	1,100	585		1,685	2,150
3000	Shallow well, jet, 30 gal. galvanized tank									
3040	1/2 HP	1 Plum	2	4	Ea.	895	235		1,130	1,350

33 31 Sanitary Utility Sewerage Piping

33 31 13 – Public Sanitary Utility Sewerage Piping

33 31 13.13 Sewage Collection, Vent Cast Iron Pipe

		Crew	Daily Output	Labor-Hours	Unit	Material	Labor	Equipment	Total	Total Incl O&P
0010	**SEWAGE COLLECTION, VENT CAST IRON PIPE**									
0020	Not including excavation or backfill									
2022	Sewage vent cast iron, B&S, 4″ diameter	Q-1	66	.242	L.F.	15.95	12.80		28.75	37.50
2024	5″ diameter	Q-2	88	.273		23.50	14.95		38.45	49.50
2026	6″ diameter	"	84	.286		28	15.65		43.65	55.50
2028	8″ diameter	Q-3	70	.457		46	25.50		71.50	90.50
2030	10″ diameter		66	.485		76	27		103	126
2032	12″ diameter		57	.561		109	31.50		140.50	169
2034	15″ diameter		49	.653		156	36.50		192.50	228
8001	Fittings, bends and elbows									

33 31 Sanitary Utility Sewerage Piping

33 31 13 – Public Sanitary Utility Sewerage Piping

33 31 13.13 Sewage Collection, Vent Cast Iron Pipe		Crew	Daily Output	Labor-Hours	Unit	Material	Labor	Equipment	Total	Total Incl O&P
						2015 Bare Costs				
8110	4" diameter	Q-1	13	1.231	Ea.	56.50	65		121.50	164
8112	5" diameter	Q-2	18	1.333		82	73		155	204
8114	6" diameter	"	17	1.412		97	77.50		174.50	228
8116	8" diameter	Q-3	11	2.909		272	163		435	555
8118	10" diameter		10	3.200		415	179		594	735
8120	12" diameter		9	3.556		550	199		749	920
8122	15" diameter		7	4.571		1,625	256		1,881	2,200
8500	Wyes and tees									
8510	4" diameter	Q-1	8	2	Ea.	93.50	106		199.50	268
8512	5" diameter	Q-2	12	2		159	110		269	345
8514	6" diameter	"	11	2.182		194	120		314	400
8516	8" diameter	Q-3	7	4.571		460	256		716	910
8518	10" diameter		6	5.333		760	298		1,058	1,300
8520	12" diameter		4	8		1,550	445		1,995	2,400
8522	15" diameter		3	10.667		3,150	595		3,745	4,400

33 31 13.15 Sewage Collection, Concrete Pipe

0010	**SEWAGE COLLECTION, CONCRETE PIPE**									
0020	See Section 33 41 13.60 for sewage/drainage collection, concrete pipe									

33 31 13.20 Sewage Collection, Plastic Pipe

		Crew	Daily Output	Labor-Hours	Unit	Material	Labor	Equipment	Total	Total Incl O&P
0010	**SEWAGE COLLECTION, PLASTIC PIPE**									
0020	Not including excavation & backfill									
1100	Piping, DWV Sch 40 ABS, 4" diameter	B-20	375	.064	L.F.	3.94	2.68		6.62	8.70
1110	6" diameter		350	.069	"	6.85	2.88		9.73	12.25
1120	Fitting, 1/4 bend, 4"		19	1.263	Ea.	22	53		75	111
1130	6"		15	1.600		15.95	67		82.95	128
1140	Tee, 4"		12	2		22	84		106	161
3000	Piping, HDPE Corrugated Type S with watertight gaskets, 4" diameter		425	.056	L.F.	.92	2.37		3.29	4.88
3020	6" diameter		400	.060		2.19	2.52		4.71	6.50
3040	8" diameter		380	.063		4.58	2.65		7.23	9.40
3060	10" diameter		370	.065		6.55	2.72		9.27	11.65
3080	12" diameter		340	.071		7.35	2.96		10.31	12.95
3100	15" diameter		300	.080		8.75	3.36		12.11	15.10
3120	18" diameter	B-21	275	.102		13.90	4.41	.50	18.81	23
3140	24" diameter		250	.112		18.45	4.85	.55	23.85	29
3160	30" diameter		200	.140		24	6.05	.69	30.74	37
3180	36" diameter		180	.156		32.50	6.75	.77	40.02	47.50
3200	42" diameter		175	.160		43.50	6.95	.79	51.24	60
3220	48" diameter		170	.165		53.50	7.15	.81	61.46	71.50
3240	54" diameter		160	.175		97	7.60	.86	105.46	120
3260	60" diameter		150	.187		142	8.10	.92	151.02	170
3300	Watertight elbows 12" diameter	B-20	11	2.182	Ea.	72	91.50		163.50	230
3320	15" diameter	"	9	2.667		110	112		222	305
3340	18" diameter	B-21	9	3.111		181	135	15.30	331.30	435
3360	24" diameter		9	3.111		380	135	15.30	530.30	655
3380	30" diameter		8	3.500		610	152	17.20	779.20	935
3400	36" diameter		8	3.500		785	152	17.20	954.20	1,125
3420	42" diameter		6	4.667		990	202	23	1,215	1,450
3440	48" diameter		6	4.667		1,775	202	23	2,000	2,325
3460	Watertight tee 12" diameter	B-20	7	3.429		123	144		267	370
3480	15" diameter	"	6	4		184	168		352	475
3500	18" diameter	B-21	6	4.667		258	202	23	483	640
3520	24" diameter		5	5.600		355	243	27.50	625.50	815

33 31 Sanitary Utility Sewerage Piping

33 31 13 – Public Sanitary Utility Sewerage Piping

33 31 13.20 Sewage Collection, Plastic Pipe		Crew	Daily Output	Labor-Hours	Unit	2015 Bare Costs Material	Labor	Equipment	Total	Total Incl O&P
3540	30″ diameter	B-21	5	5.600	Ea.	705	243	27.50	975.50	1,200
3560	36″ diameter		4	7		795	305	34.50	1,134.50	1,400
3580	42″ diameter		4	7		870	305	34.50	1,209.50	1,500
3600	48″ diameter		4	7		1,500	305	34.50	1,839.50	2,175

33 31 13.25 Sewage Collection, Polyvinyl Chloride Pipe		Crew	Daily Output	Labor-Hours	Unit	Material	Labor	Equipment	Total	Total Incl O&P
0010	SEWAGE COLLECTION, POLYVINYL CHLORIDE PIPE									
0020	Not including excavation or backfill									
2000	20′ lengths, SDR 35, B&S, 4″ diameter	B-20	375	.064	L.F.	1.45	2.68		4.13	6
2040	6″ diameter		350	.069		3.28	2.88		6.16	8.30
2080	13′ lengths , SDR 35, B&S, 8″ diameter		335	.072		6.90	3.01		9.91	12.50
2120	10″ diameter	B-21	330	.085		11.40	3.68	.42	15.50	18.95
2160	12″ diameter		320	.088		12.75	3.79	.43	16.97	20.50
2200	15″ diameter		240	.117		12.95	5.05	.57	18.57	23
3040	Fittings, bends or elbows, 4″ diameter	2 Skwk	24	.667	Ea.	9.80	32.50		42.30	63.50
3080	6″ diameter		24	.667		32	32.50		64.50	87.50
3120	Tees, 4″ diameter		16	1		13.45	48.50		61.95	94
3160	6″ diameter		16	1		43	48.50		91.50	127
3200	Wyes, 4″ diameter		16	1		16.65	48.50		65.15	97.50
3240	6″ diameter		16	1		46	48.50		94.50	130
4000	Piping, DWV PVC, no exc./bkfill., 10′ L, Sch 40, 4″ diameter	B-20	375	.064	L.F.	3.84	2.68		6.52	8.60
4010	6″ diameter		350	.069		7.65	2.88		10.53	13.10
4020	8″ diameter		335	.072		12	3.01		15.01	18.10

33 36 Utility Septic Tanks

33 36 13 – Utility Septic Tank and Effluent Wet Wells

33 36 13.13 Concrete Utility Septic Tank		Crew	Daily Output	Labor-Hours	Unit	Material	Labor	Equipment	Total	Total Incl O&P
0010	CONCRETE UTILITY SEPTIC TANK									
0011	Not including excavation or piping									
0015	Septic tanks, precast, 1,000 gallon	B-21	8	3.500	Ea.	1,000	152	17.20	1,169.20	1,375
0020	1,250 gallon		8	3.500		1,050	152	17.20	1,219.20	1,425
0060	1,500 gallon		7	4		1,325	173	19.70	1,517.70	1,775
0100	2,000 gallon		5	5.600		1,925	243	27.50	2,195.50	2,525
0140	2,500 gallon		5	5.600		2,025	243	27.50	2,295.50	2,650
0180	4,000 gallon		4	7		5,550	305	34.50	5,889.50	6,625
0220	5,000 gal., 4 piece	B-13	3	18.667		9,575	765	245	10,585	12,000
0300	15,000 gallon, 4 piece	B-13B	1.70	32.941		20,000	1,350	665	22,015	24,900
0400	25,000 gallon, 4 piece		1.10	50.909		38,800	2,075	1,025	41,900	47,200
0500	40,000 gallon, 4 piece		.80	70		49,700	2,875	1,400	53,975	60,500
0520	50,000 gallon, 5 piece	B-13C	.60	93.333		57,000	3,825	2,775	63,600	72,500
0640	75,000 gallon, cast in place	C-14C	.25	448		69,500	20,200	125	89,825	109,500
0660	100,000 gallon	"	.15	746		86,000	33,700	209	119,909	149,500
1150	Leaching field chambers, 13′ x 3′-7″ x 1′-4″, standard	B-13	16	3.500		485	143	46	674	815
1200	Heavy duty, 8′ x 4′ x 1′-6″		14	4		284	164	52.50	500.50	635
1300	13′ x 3′-9″ x 1′-6″		12	4.667		1,125	191	61.50	1,377.50	1,600
1350	20′ x 4′ x 1′-6″		5	11.200		1,175	460	147	1,782	2,175
1600	Leaching pit, 6′-6″ diameter, 6′ deep	B-21	5	5.600		1,025	243	27.50	1,295.50	1,550
1620	8′ deep		4	7		1,200	305	34.50	1,539.50	1,850
1700	8′ diameter, H-20 load, 6′ deep		4	7		1,500	305	34.50	1,839.50	2,175
1720	8′ deep		3	9.333		2,500	405	46	2,951	3,450
2000	Velocity reducing pit, precast conc., 6′ diameter, 3′ deep		4.70	5.957		1,600	258	29.50	1,887.50	2,200

33 36 Utility Septic Tanks

33 36 13 – Utility Septic Tank and Effluent Wet Wells

	33 36 13.19 Polyethylene Utility Septic Tank	Crew	Daily Output	Labor-Hours	Unit	2015 Bare Costs Material	Labor	Equipment	Total	Total Incl O&P
0010	**POLYETHYLENE UTILITY SEPTIC TANK**									
0015	High density polyethylene, 1,000 gallon	B-21	8	3.500	Ea.	1,550	152	17.20	1,719.20	1,975
0020	1,250 gallon		8	3.500		1,200	152	17.20	1,369.20	1,600
0025	1,500 gallon		7	4		1,475	173	19.70	1,667.70	1,925

33 36 19 – Utility Septic Tank Effluent Filter

	33 36 19.13 Utility Septic Tank Effluent Tube Filter	Crew	Daily Output	Labor-Hours	Unit	Material	Labor	Equipment	Total	Total Incl O&P
0010	**UTILITY SEPTIC TANK EFFLUENT TUBE FILTER**									
3000	Effluent filter, 4″ diameter	1 Skwk	8	1	Ea.	42	48.50		90.50	125
3020	6″ diameter		7	1.143		249	55.50		304.50	365
3030	8″ diameter		7	1.143		225	55.50		280.50	340
3040	8″ diameter, very fine		7	1.143		460	55.50		515.50	595
3050	10″ diameter, very fine		6	1.333		235	65		300	365
3060	10″ diameter		6	1.333		280	65		345	415
3080	12″ diameter		6	1.333		670	65		735	845
3090	15″ diameter		5	1.600		1,150	78		1,228	1,375

33 36 33 – Utility Septic Tank Drainage Field

	33 36 33.13 Utility Septic Tank Tile Drainage Field	Crew	Daily Output	Labor-Hours	Unit	Material	Labor	Equipment	Total	Total Incl O&P
0010	**UTILITY SEPTIC TANK TILE DRAINAGE FIELD**									
0015	Distribution box, concrete, 5 outlets	2 Clab	20	.800	Ea.	80.50	30		110.50	138
0020	7 outlets		16	1		80.50	37.50		118	150
0025	9 outlets		8	2		490	75		565	665
0115	Distribution boxes, HDPE, 5 outlets		20	.800		64.50	30		94.50	121
0117	6 outlets		15	1.067		65	40		105	138
0118	7 outlets		15	1.067		65	40		105	138
0120	8 outlets		10	1.600		68	60		128	174
0240	Distribution boxes, Outlet Flow Leveler	1 Clab	50	.160		2.25	6		8.25	12.35
0300	Precast concrete, galley, 4′ x 4′ x 4′	B-21	16	1.750		250	76	8.60	334.60	405
0350	HDPE infiltration chamber 12″ H X 15″ W	2 Clab	300	.053	L.F.	7.30	2.01		9.31	11.35
0351	12″ H X 15″ W End Cap	1 Clab	32	.250	Ea.	19.75	9.40		29.15	37
0355	chamber 12″ H X 22″ W	2 Clab	300	.053	L.F.	6.70	2.01		8.71	10.65
0356	12″ H X 22″ W End Cap	1 Clab	32	.250	Ea.	16.85	9.40		26.25	34
0360	chamber 13″ H X 34″ W	2 Clab	300	.053	L.F.	15	2.01		17.01	19.80
0361	13″ H X 34″ W End Cap	1 Clab	32	.250	Ea.	53.50	9.40		62.90	74
0365	chamber 16″ H X 34″ W	2 Clab	300	.053	L.F.	10.70	2.01		12.71	15.10
0366	16″ H X 34″ W End Cap	1 Clab	32	.250	Ea.	8	9.40		17.40	24
0370	chamber 8″ H X 16″ W	2 Clab	300	.053	L.F.	9.65	2.01		11.66	13.90
0371	8″ H X 16″ W End Cap	1 Clab	32	.250	Ea.	10.80	9.40		20.20	27.50

33 36 50 – Drainage Field Systems

	33 36 50.10 Drainage Field Excavation and Fill	Crew	Daily Output	Labor-Hours	Unit	Material	Labor	Equipment	Total	Total Incl O&P
0010	**DRAINAGE FIELD EXCAVATION AND FILL**									
2200	Septic tank & drainage field excavation with 3/4 c.y backhoe	B-12F	145	.110	C.Y.		4.93	4.52	9.45	12.90
2400	4′ trench for disposal field, 3/4 C.Y. backhoe	″	335	.048	L.F.		2.13	1.95	4.08	5.60
2600	Gravel fill, run of bank	B-6	150	.160	C.Y.	20	6.60	2.43	29.03	35.50
2800	Crushed stone, 3/4″	″	150	.160	″	36.50	6.60	2.43	45.53	53.50

33 41 13.40 Piping, Storm Drainage, Corrugated Metal		Crew	Daily Output	Labor-Hours	Unit	2015 Bare Costs Material	Labor	Equipment	Total	Total Incl O&P
0010	PIPING, STORM DRAINAGE, CORRUGATED METAL									
0020	Not including excavation or backfill									
2000	Corrugated metal pipe, galvanized									
2020	Bituminous coated with paved invert, 20′ lengths									
2040	8″ diameter, 16 ga.	B-14	330	.145	L.F.	8.65	5.80	1.10	15.55	20
2060	10″ diameter, 16 ga.		260	.185		9	7.35	1.40	17.75	23.50
2080	12″ diameter, 16 ga.		210	.229		11.05	9.10	1.73	21.88	29
2100	15″ diameter, 16 ga.		200	.240		15.15	9.55	1.82	26.52	34
2120	18″ diameter, 16 ga.		190	.253		16.70	10.05	1.92	28.67	37
2140	24″ diameter, 14 ga.		160	.300		21.50	11.95	2.28	35.73	45.50
2160	30″ diameter, 14 ga.	B-13	120	.467		27.50	19.15	6.15	52.80	68.50
2180	36″ diameter, 12 ga.		120	.467		35.50	19.15	6.15	60.80	77
2200	48″ diameter, 12 ga.		100	.560		52.50	23	7.35	82.85	103
2220	60″ diameter, 10 ga.	B-13B	75	.747		80	30.50	15.05	125.55	154
2240	72″ diameter, 8 ga.	″	45	1.244		95.50	51	25	171.50	215
2250	End sections, 8″ diameter, 16 ga.	B-14	20	2.400	Ea.	43.50	95.50	18.20	157.20	223
2255	10″ diameter, 16 ga.		20	2.400		60	95.50	18.20	173.70	241
2260	12″ diameter, 16 ga.		18	2.667		120	106	20	246	330
2265	15″ diameter, 16 ga.		18	2.667		205	106	20	331	420
2270	18″ diameter, 16 ga.		16	3		240	119	23	382	485
2275	24″ diameter, 16 ga.	B-13	16	3.500		280	143	46	469	595
2280	30″ diameter, 16 ga.		14	4		490	164	52.50	706.50	860
2285	36″ diameter, 14 ga.		14	4		730	164	52.50	946.50	1,125
2290	48″ diameter, 14 ga.		10	5.600		1,850	230	73.50	2,153.50	2,475
2292	60″ diameter, 14 ga.		6	9.333		1,950	385	123	2,458	2,900
2294	72″ diameter, 14 ga.	B-13B	5	11.200		3,125	460	226	3,811	4,450
2300	Bends or elbows, 8″ diameter	B-14	28	1.714		117	68	13	198	254
2320	10″ diameter		25	1.920		146	76.50	14.55	237.05	300
2340	12″ diameter, 16 ga.		23	2.087		169	83	15.85	267.85	340
2342	18″ diameter, 16 ga.		20	2.400		244	95.50	18.20	357.70	445
2344	24″ diameter, 14 ga.		16	3		335	119	23	477	590
2346	30″ diameter, 14 ga.		15	3.200		430	127	24.50	581.50	705
2348	36″ diameter, 14 ga.	B-13	15	3.733		580	153	49	782	935
2350	48″ diameter, 12 ga.	″	12	4.667		805	191	61.50	1,057.50	1,275
2352	60″ diameter, 10 ga.	B-13B	10	5.600		1,075	230	113	1,418	1,675
2354	72″ diameter, 10 ga.	″	6	9.333		1,325	385	188	1,898	2,300
2360	Wyes or tees, 8″ diameter	B-14	25	1.920		164	76.50	14.55	255.05	320
2380	10″ diameter		21	2.286		205	91	17.35	313.35	390
2400	12″ diameter, 16 ga.		19	2.526		244	100	19.15	363.15	455
2410	18″ diameter, 16 ga.		16	3		325	119	23	467	575
2412	24″ diameter, 14 ga.		16	3		485	119	23	627	755
2414	30″ diameter, 14 ga.	B-13	12	4.667		645	191	61.50	897.50	1,100
2416	36″ diameter, 14 ga.		11	5.091		795	209	67	1,071	1,300
2418	48″ diameter, 12 ga.		10	5.600		1,175	230	73.50	1,478.50	1,725
2420	60″ diameter, 10 ga.	B-13B	8	7		1,700	287	141	2,128	2,500
2422	72″ diameter, 10 ga.	″	5	11.200		2,050	460	226	2,736	3,250
2500	Galvanized, uncoated, 20′ lengths									
2520	8″ diameter, 16 ga.	B-14	355	.135	L.F.	7.80	5.40	1.03	14.23	18.50
2540	10″ diameter, 16 ga.		280	.171		8.95	6.80	1.30	17.05	22.50
2560	12″ diameter, 16 ga.		220	.218		9.95	8.70	1.66	20.31	27
2580	15″ diameter, 16 ga.		220	.218		12.45	8.70	1.66	22.81	29.50
2600	18″ diameter, 16 ga.		205	.234		15.05	9.30	1.78	26.13	33.50
2620	24″ diameter, 14 ga.		175	.274		18.95	10.90	2.08	31.93	41

33 41 13.40 Piping, Storm Drainage, Corrugated Metal		Crew	Daily Output	Labor-Hours	Unit	2015 Bare Costs Material	Labor	Equipment	Total	Total Incl O&P
2640	30" diameter, 14 ga.	B-13	130	.431	L.F.	25	17.65	5.65	48.30	62.50
2660	36" diameter, 12 ga.		130	.431		32	17.65	5.65	55.30	70
2680	48" diameter, 12 ga.		110	.509		47.50	21	6.70	75.20	93.50
2690	60" diameter, 10 ga.	B-13B	78	.718		72	29.50	14.45	115.95	142
2695	72" diameter, 10 ga.	"	60	.933		86	38.50	18.80	143.30	177
2711	Bends or elbows, 12" diameter, 16 ga.	B-14	30	1.600	Ea.	145	63.50	12.15	220.65	277
2712	15" diameter, 16 ga.		25.04	1.917		180	76	14.55	270.55	340
2714	18" diameter, 16 ga.		20	2.400		202	95.50	18.20	315.70	400
2716	24" diameter, 14 ga.		16	3		293	119	23	435	540
2718	30" diameter, 14 ga.		15	3.200		340	127	24.50	491.50	610
2720	36" diameter, 14 ga.	B-13	15	3.733		515	153	49	717	865
2722	48" diameter, 12 ga.		12	4.667		680	191	61.50	932.50	1,125
2724	60" diameter, 10 ga.		10	5.600		1,050	230	73.50	1,353.50	1,600
2726	72" diameter, 10 ga.		6	9.333		1,350	385	123	1,858	2,225
2728	Wyes or tees, 12" diameter, 16 ga.	B-14	22.48	2.135		193	85	16.20	294.20	370
2730	18" diameter, 16 ga.		15	3.200		282	127	24.50	433.50	545
2732	24" diameter, 14 ga.		15	3.200		450	127	24.50	601.50	730
2734	30" diameter, 14 ga.		14	3.429		580	136	26	742	885
2736	36" diameter, 14 ga.	B-13	14	4		755	164	52.50	971.50	1,150
2738	48" diameter, 12 ga.		12	4.667		1,075	191	61.50	1,327.50	1,575
2740	60" diameter, 10 ga.		10	5.600		1,575	230	73.50	1,878.50	2,175
2742	72" diameter, 10 ga.		6	9.333		1,850	385	123	2,358	2,800
2780	End sections, 8" diameter	B-14	35	1.371		73	54.50	10.40	137.90	180
2785	10" diameter		35	1.371		77	54.50	10.40	141.90	185
2790	12" diameter		35	1.371		114	54.50	10.40	178.90	225
2800	18" diameter		30	1.600		115	63.50	12.15	190.65	243
2810	24" diameter	B-13	25	2.240		215	92	29.50	336.50	420
2820	30" diameter		25	2.240		330	92	29.50	451.50	540
2825	36" diameter		20	2.800		480	115	37	632	755
2830	48" diameter		10	5.600		950	230	73.50	1,253.50	1,500
2835	60" diameter	B-13B	5	11.200		1,650	460	226	2,336	2,800
2840	72" diameter	"	4	14		1,975	575	282	2,832	3,400
2850	Couplings, 12" diameter					10.55			10.55	11.60
2855	18" diameter					15.10			15.10	16.60
2860	24" diameter					19.95			19.95	22
2865	30" diameter					25			25	27.50
2870	36" diameter					32			32	35
2875	48" diameter					47.50			47.50	52
2880	60" diameter					60			60	66
2885	72" diameter					72.50			72.50	79.50

33 41 13.50 Piping, Drainage & Sewage, Corrug. HDPE Type S		Crew	Daily Output	Labor-Hours	Unit	Material	Labor	Equipment	Total	Total Incl O&P
0010	**PIPING, DRAINAGE & SEWAGE, CORRUGATED HDPE TYPE S**									
0020	Not including excavation & backfill, bell & spigot									
1000	With gaskets, 4" diameter	B-20	425	.056	L.F.	.80	2.37		3.17	4.75
1010	6" diameter		400	.060		1.90	2.52		4.42	6.20
1020	8" diameter		380	.063		3.98	2.65		6.63	8.70
1030	10" diameter		370	.065		5.70	2.72		8.42	10.70
1040	12" diameter		340	.071		6.40	2.96		9.36	11.90
1050	15" diameter		300	.080		7.60	3.36		10.96	13.85
1060	18" diameter	B-21	275	.102		12.10	4.41	.50	17.01	21
1070	24" diameter		250	.112		16.05	4.85	.55	21.45	26
1080	30" diameter		200	.140		21	6.05	.69	27.74	33.50

33 41 13 – Public Storm Utility Drainage Piping

33 41 13.50 Piping, Drainage & Sewage, Corrug. HDPE Type S		Crew	Daily Output	Labor-Hours	Unit	Material	Labor	Equipment	Total	Total Incl O&P
						2015 Bare Costs				
1090	36″ diameter	B-21	180	.156	L.F.	28	6.75	.77	35.52	43
1100	42″ diameter		175	.160		38	6.95	.79	45.74	54
1110	48″ diameter		170	.165		46.50	7.15	.81	54.46	63.50
1120	54″ diameter		160	.175		84.50	7.60	.86	92.96	106
1130	60″ diameter		150	.187		124	8.10	.92	133.02	150
1135	Add 15% to material pipe cost for water tight connection bell & spigot									
1140	HDPE type S, elbows 12″ diameter	B-20	11	2.182	Ea.	63	91.50		154.50	219
1150	15″ diameter	"	9	2.667		96	112		208	288
1160	18″ diameter	B-21	9	3.111		158	135	15.30	308.30	410
1170	24″ diameter		9	3.111		335	135	15.30	485.30	600
1180	30″ diameter		8	3.500		530	152	17.20	699.20	845
1190	36″ diameter		8	3.500		680	152	17.20	849.20	1,025
1200	42″ diameter		6	4.667		860	202	23	1,085	1,300
1220	48″ diameter		6	4.667		1,550	202	23	1,775	2,050
1240	HDPE type S, Tee 12″ diameter	B-20	7	3.429		107	144		251	355
1260	15″ diameter	"	6	4		160	168		328	450
1280	18″ diameter	B-21	6	4.667		224	202	23	449	605
1300	24″ diameter		5	5.600		310	243	27.50	580.50	765
1320	30″ diameter		5	5.600		615	243	27.50	885.50	1,100
1340	36″ diameter		4	7		690	305	34.50	1,029.50	1,300
1360	42″ diameter		4	7		755	305	34.50	1,094.50	1,375
1380	48″ diameter		4	7		1,300	305	34.50	1,639.50	1,950
1400	Add to basic installation cost for each split coupling joint									
1402	HDPE type S, split coupling, 12″ diameter	B-20	17	1.412	Ea.	7.45	59		66.45	105
1420	15″ diameter		15	1.600		12.45	67		79.45	124
1440	18″ diameter		13	1.846		22	77.50		99.50	151
1460	24″ diameter		12	2		31.50	84		115.50	172
1480	30″ diameter		10	2.400		70	101		171	241
1500	36″ diameter		9	2.667		133	112		245	330
1520	42″ diameter		8	3		157	126		283	380
1540	48″ diameter		8	3		168	126		294	390

33 41 13.60 Sewage/Drainage Collection, Concrete Pipe

Line	Description	Crew	Daily Output	Labor-Hours	Unit	Material	Labor	Equipment	Total	Total Incl O&P
0010	**SEWAGE/DRAINAGE COLLECTION, CONCRETE PIPE**									
0020	Not including excavation or backfill									
0100	Box culvert, precast, base price, 8′ long, 6′ x 3′	B-69	140	.343	L.F.	254	14.25	11.60	279.85	315
0150	6′ x 7′		125	.384		320	15.95	13	348.95	395
0200	8′ x 3′		113	.425		315	17.65	14.40	347.05	395
0250	8′ x 8′		100	.480		385	19.95	16.25	421.20	475
0300	10′ x 3′		110	.436		440	18.15	14.75	472.90	530
0350	10′ x 8′		80	.600		510	25	20.50	555.50	625
0400	12′ x 3′		100	.480		660	19.95	16.25	696.20	775
0450	12′ x 8′		67	.716		845	30	24.50	899.50	1,000
0500	Set up charge at plant, add to base price				Job	5,875			5,875	6,475
0510	Inserts and keyway, add				Ea.	545			545	600
0520	Sloped or skewed end, add				"	830			830	915
1000	Non-reinforced pipe, extra strength, B&S or T&G joints									
1010	6″ diameter	B-14	265.04	.181	L.F.	6	7.20	1.37	14.57	19.85
1020	8″ diameter		224	.214		6.60	8.50	1.63	16.73	23
1030	10″ diameter		216	.222		7.30	8.85	1.69	17.84	24.50
1040	12″ diameter		200	.240		8.20	9.55	1.82	19.57	26.50
1050	15″ diameter		180	.267		12	10.60	2.02	24.62	32.50
1060	18″ diameter		144	.333		15	13.25	2.53	30.78	41

	33 41 13.60 Sewage/Drainage Collection, Concrete Pipe	Crew	Daily Output	Labor-Hours	Unit	2015 Bare Costs Material	Labor	Equipment	Total	Total Incl O&P
1070	21″ diameter	B-14	112	.429	L.F.	17	17.05	3.25	37.30	50.50
1080	24″ diameter		100	.480		22	19.10	3.64	44.74	59
1560	Reinforced culvert, class 2, no gaskets									
1590	27″ diameter	B-21	88	.318	L.F.	37	13.80	1.57	52.37	64.50
1592	30″ diameter	B-13	80	.700		40	28.50	9.20	77.70	101
1594	36″ diameter	″	72	.778		56	32	10.25	98.25	124
2000	Reinforced culvert, class 3, no gaskets									
2010	12″ diameter	B-14	150	.320	L.F.	11	12.75	2.43	26.18	35.50
2020	15″ diameter		150	.320		14	12.75	2.43	29.18	38.50
2030	18″ diameter		132	.364		18	14.45	2.76	35.21	46.50
2035	21″ diameter		120	.400		22	15.90	3.04	40.94	53.50
2040	24″ diameter		100	.480		26	19.10	3.64	48.74	63.50
2045	27″ diameter	B-13	92	.609		37	25	8	70	90
2050	30″ diameter		88	.636		42	26	8.35	76.35	97.50
2060	36″ diameter		72	.778		56	32	10.25	98.25	124
2070	42″ diameter	B-13B	72	.778		75	32	15.65	122.65	151
2080	48″ diameter		64	.875		89	36	17.60	142.60	175
2090	60″ diameter		48	1.167		136	48	23.50	207.50	254
2100	72″ diameter		40	1.400		206	57.50	28	291.50	350
2120	84″ diameter		32	1.750		275	71.50	35.50	382	460
2140	96″ diameter		24	2.333		330	95.50	47	472.50	570
2200	With gaskets, class 3, 12″ diameter	B-21	168	.167		12.10	7.20	.82	20.12	26
2220	15″ diameter		160	.175		15.40	7.60	.86	23.86	30.50
2230	18″ diameter		152	.184		19.80	8	.91	28.71	36
2240	24″ diameter		136	.206		33	8.90	1.01	42.91	52
2260	30″ diameter	B-13	88	.636		50	26	8.35	84.35	107
2270	36″ diameter	″	72	.778		65.50	32	10.25	107.75	135
2290	48″ diameter	B-13B	64	.875		102	36	17.60	155.60	189
2310	72″ diameter	″	40	1.400		225	57.50	28	310.50	370
2330	Flared ends, 12″ diameter	B-21	31	.903	Ea.	226	39	4.44	269.44	315
2340	15″ diameter		25	1.120		267	48.50	5.50	321	380
2400	18″ diameter		20	1.400		305	60.50	6.90	372.40	445
2420	24″ diameter		14	2		370	86.50	9.85	466.35	560
2440	36″ diameter	B-13	10	5.600		790	230	73.50	1,093.50	1,325
2500	Class 4									
2510	12″ diameter	B-21	168	.167	L.F.	13	7.20	.82	21.02	27
2512	15″ diameter		160	.175		17	7.60	.86	25.46	32
2514	18″ diameter		152	.184		21	8	.91	29.91	37
2516	21″ diameter		144	.194		27	8.45	.96	36.41	44.50
2518	24″ diameter		136	.206		32	8.90	1.01	41.91	50.50
2520	27″ diameter		120	.233		43	10.10	1.15	54.25	65
2522	30″ diameter	B-13	88	.636		49	26	8.35	83.35	106
2524	36″ diameter	″	72	.778		66	32	10.25	108.25	135
2600	Class 5									
2610	12″ diameter	B-21	168	.167	L.F.	15	7.20	.82	23.02	29
2612	15″ diameter		160	.175		19	7.60	.86	27.46	34.50
2614	18″ diameter		152	.184		23.50	8	.91	32.41	40
2616	21″ diameter		144	.194		29.50	8.45	.96	38.91	47.50
2618	24″ diameter		136	.206		36.50	8.90	1.01	46.41	55.50
2620	27″ diameter		120	.233		47.50	10.10	1.15	58.75	70
2622	30″ diameter	B-13	88	.636		55	26	8.35	89.35	112
2624	36″ diameter	″	72	.778		74	32	10.25	116.25	144
2800	Add for rubber joints 12″- 36″ diameter					12%				

33 41 Storm Utility Drainage Piping

33 41 13 – Public Storm Utility Drainage Piping

33 41 13.60 Sewage/Drainage Collection, Concrete Pipe		Crew	Daily Output	Labor-Hours	Unit	2015 Bare Costs Material	Labor	Equipment	Total	Total Incl O&P
3080	Radius pipe, add to pipe prices, 12″ to 60″ diameter				L.F.	50%				
3090	Over 60″ diameter, add					20%				
3500	Reinforced elliptical, 8′ lengths, C507 class 3									
3520	14″ x 23″ inside, round equivalent 18″ diameter	B-21	82	.341	L.F.	41	14.80	1.68	57.48	71
3530	24″ x 38″ inside, round equivalent 30″ diameter	B-13	58	.966		62	39.50	12.70	114.20	146
3540	29″ x 45″ inside, round equivalent 36″ diameter		52	1.077		78	44	14.15	136.15	173
3550	38″ x 60″ inside, round equivalent 48″ diameter		38	1.474		134	60.50	19.40	213.90	267
3560	48″ x 76″ inside, round equivalent 60″ diameter		26	2.154		186	88.50	28.50	303	380
3570	58″ x 91″ inside, round equivalent 72″ diameter		22	2.545		272	104	33.50	409.50	505
3780	Concrete slotted pipe, class 4 mortar joint									
3800	12″ diameter	B-21	168	.167	L.F.	28	7.20	.82	36.02	43.50
3840	18″ diameter	″	152	.184	″	32	8	.91	40.91	49
3900	Concrete slotted pipe, Class 4 O-ring joint									
3940	12″ diameter	B-21	168	.167	L.F.	28	7.20	.82	36.02	43.50
3960	18″ diameter	″	152	.184	″	32	8	.91	40.91	49
6200	Gasket, conc. pipe joint, 12″				Ea.	4			4	4.40
6220	24″					7			7	7.70
6240	36″					9.50			9.50	10.45
6260	48″					13			13	14.30
6265	54″					14			14	15.40
6270	60″					15.50			15.50	17.05
6280	72″					19			19	21

33 42 Culverts

33 42 16 – Concrete Culverts

33 42 16.15 Oval Arch Culverts		Crew	Daily Output	Labor-Hours	Unit	Material	Labor	Equipment	Total	Total Incl O&P
0010	**OVAL ARCH CULVERTS**									
3000	Corrugated galvanized or aluminum, coated & paved									
3020	17″ x 13″, 16 ga., 15″ equivalent	B-14	200	.240	L.F.	13.20	9.55	1.82	24.57	32
3040	21″ x 15″, 16 ga., 18″ equivalent		150	.320		16.05	12.75	2.43	31.23	41
3060	28″ x 20″, 14 ga., 24″ equivalent		125	.384		24	15.25	2.91	42.16	54.50
3080	35″ x 24″, 14 ga., 30″ equivalent		100	.480		29.50	19.10	3.64	52.24	67.50
3100	42″ x 29″, 12 ga., 36″ equivalent	B-13	100	.560		35.50	23	7.35	65.85	84
3120	49″ x 33″, 12 ga., 42″ equivalent		90	.622		41	25.50	8.20	74.70	96
3140	57″ x 38″, 12 ga., 48″ equivalent		75	.747		57	30.50	9.80	97.30	123
3160	Steel, plain oval arch culverts, plain									
3180	17″ x 13″, 16 ga., 15″ equivalent	B-14	225	.213	L.F.	11.95	8.50	1.62	22.07	28.50
3200	21″ x 15″, 16 ga., 18″ equivalent		175	.274		14.45	10.90	2.08	27.43	36
3220	28″ x 20″, 14 ga., 24″ equivalent		150	.320		21.50	12.75	2.43	36.68	46.50
3240	35″ x 24″, 14 ga., 30″ equivalent	B-13	108	.519		26.50	21.50	6.80	54.80	71
3260	42″ x 29″, 12 ga., 36″ equivalent		108	.519		32	21.50	6.80	60.30	77.50
3280	49″ x 33″, 12 ga., 42″ equivalent		92	.609		37	25	8	70	90.50
3300	57″ x 38″, 12 ga., 48″ equivalent		75	.747		51.50	30.50	9.80	91.80	117
3320	End sections, 17″ x 13″		22	2.545	Ea.	144	104	33.50	281.50	365
3340	42″ x 29″		17	3.294	″	395	135	43.50	573.50	700
3360	Multi-plate arch, steel	B-20	1690	.014	Lb.	1.25	.60		1.85	2.35

33 44 Storm Utility Water Drains

33 44 13 – Utility Area Drains

33 44 13.13 Catchbasins		Crew	Daily Output	Labor-Hours	Unit	2015 Bare Costs Material	Labor	Equipment	Total	Total Incl O&P
0010	**CATCHBASINS**									
0011	Not including footing & excavation									
1600	Frames & grates, C.I., 24" square, 500 lb.	B-6	7.80	3.077	Ea.	340	127	46.50	513.50	635
1700	26" D shape, 600 lb.		7	3.429		505	142	52	699	840
1800	Light traffic, 18" diameter, 100 lb.		10	2.400		123	99	36.50	258.50	335
1900	24" diameter, 300 lb.		8.70	2.759		196	114	42	352	445
2000	36" diameter, 900 lb.		5.80	4.138		570	171	63	804	970
2100	Heavy traffic, 24" diameter, 400 lb.		7.80	3.077		244	127	46.50	417.50	525
2200	36" diameter, 1150 lb.		3	8		795	330	121	1,246	1,550
2300	Mass. State standard, 26" diameter, 475 lb.		7	3.429		266	142	52	460	580
2400	30" diameter, 620 lb.		7	3.429		345	142	52	539	665
2500	Watertight, 24" diameter, 350 lb.		7.80	3.077		320	127	46.50	493.50	610
2600	26" diameter, 500 lb.		7	3.429		425	142	52	619	755
2700	32" diameter, 575 lb.		6	4		850	165	60.50	1,075.50	1,275
2800	3 piece cover & frame, 10" deep,									
2900	1200 lb., for heavy equipment	B-6	3	8	Ea.	1,050	330	121	1,501	1,850
3000	Raised for paving 1-1/4" to 2" high									
3100	4 piece expansion ring									
3200	20" to 26" diameter	1 Clab	3	2.667	Ea.	158	100		258	340
3300	30" to 36" diameter	"	3	2.667	"	217	100		317	405
3320	Frames and covers, existing, raised for paving, 2", including									
3340	row of brick, concrete collar, up to 12" wide frame	B-6	18	1.333	Ea.	45	55	20	120	161
3360	20" to 26" wide frame		11	2.182		67.50	90	33	190.50	257
3380	30" to 36" wide frame		9	2.667		83.50	110	40.50	234	315
3400	Inverts, single channel brick	D-1	3	5.333		97	225		322	470
3500	Concrete		5	3.200		104	135		239	335
3600	Triple channel, brick		2	8		148	335		483	715
3700	Concrete		3	5.333		139	225		364	520

33 44 13.50 Stormwater Management		Crew	Daily Output	Labor-Hours	Unit	Material	Labor	Equipment	Total	Total Incl O&P
0010	**STORMWATER MANAGEMENT**									
0020	Allowance, add per SF of impervious surface				S.F.				3	3

33 46 Subdrainage

33 46 16 – Subdrainage Piping

33 46 16.25 Piping, Subdrainage, Corrugated Metal		Crew	Daily Output	Labor-Hours	Unit	Material	Labor	Equipment	Total	Total Incl O&P
0010	**PIPING, SUBDRAINAGE, CORRUGATED METAL**									
0021	Not including excavation and backfill									
2010	Aluminum, perforated									
2020	6" diameter, 18 ga.	B-20	380	.063	L.F.	6.50	2.65		9.15	11.50
2200	8" diameter, 16 ga.	"	370	.065		8.55	2.72		11.27	13.85
2220	10" diameter, 16 ga.	B-21	360	.078		10.70	3.37	.38	14.45	17.65
2240	12" diameter, 16 ga.		285	.098		11.95	4.26	.48	16.69	20.50
2260	18" diameter, 16 ga.		205	.137		17.95	5.90	.67	24.52	30
3000	Uncoated galvanized, perforated									
3020	6" diameter, 18 ga.	B-20	380	.063	L.F.	6.05	2.65		8.70	11
3200	8" diameter, 16 ga.	"	370	.065		8.30	2.72		11.02	13.60
3220	10" diameter, 16 ga.	B-21	360	.078		8.80	3.37	.38	12.55	15.60
3240	12" diameter, 16 ga.		285	.098		9.80	4.26	.48	14.54	18.20
3260	18" diameter, 16 ga.		205	.137		15	5.90	.67	21.57	27
4000	Steel, perforated, asphalt coated									
4020	6" diameter 18 ga.	B-20	380	.063	L.F.	6.50	2.65		9.15	11.50

33 46 Subdrainage

33 46 16 – Subdrainage Piping

33 46 16.25 Piping, Subdrainage, Corrugated Metal		Crew	Daily Output	Labor-Hours	Unit	Material	Labor	Equipment	Total	Total Incl O&P
						2015 Bare Costs				
4030	8″ diameter 18 ga.	B-20	370	.065	L.F.	8.55	2.72		11.27	13.85
4040	10″ diameter 16 ga.	B-21	360	.078		10.05	3.37	.38	13.80	16.95
4050	12″ diameter 16 ga.		285	.098		11.05	4.26	.48	15.79	19.60
4060	18″ diameter 16 ga.		205	.137		17.10	5.90	.67	23.67	29

33 46 16.30 Piping, Subdrainage, Plastic		Crew	Daily Output	Labor-Hours	Unit	Material	Labor	Equipment	Total	Total Incl O&P
0010	**PIPING, SUBDRAINAGE, PLASTIC**									
0020	Not including excavation and backfill									
2100	Perforated PVC, 4″ diameter	B-14	314	.153	L.F.	1.45	6.10	1.16	8.71	12.80
2110	6″ diameter		300	.160		3.28	6.35	1.21	10.84	15.30
2120	8″ diameter		290	.166		6.30	6.60	1.26	14.16	19
2130	10″ diameter		280	.171		9.10	6.80	1.30	17.20	22.50
2140	12″ diameter		270	.178		12.75	7.05	1.35	21.15	27

33 46 26 – Geotextile Subsurface Drainage Filtration

33 46 26.10 Geotextiles for Subsurface Drainage		Crew	Daily Output	Labor-Hours	Unit	Material	Labor	Equipment	Total	Total Incl O&P
0010	**GEOTEXTILES FOR SUBSURFACE DRAINAGE**									
0100	Fabric, laid in trench, polypropylene, ideal conditions	2 Clab	2400	.007	S.Y.	1.50	.25		1.75	2.06
0110	Adverse conditions		1600	.010	"	1.50	.38		1.88	2.27
0170	Fabric ply bonded to 3 dimen. nylon mat, .4″ thick, ideal conditions		2000	.008	S.F.	.23	.30		.53	.74
0180	Adverse conditions		1200	.013	"	.29	.50		.79	1.13
0185	Soil drainage mat on vertical wall, 0.44″ thick		265	.060	S.Y.	1.84	2.27		4.11	5.75
0188	0.25″ thick		300	.053	"	.81	2.01		2.82	4.18
0190	0.8″ thick, ideal conditions		2400	.007	S.F.	.22	.25		.47	.65
0200	Adverse conditions		1600	.010	"	.37	.38		.75	1.02
0300	Drainage material, 3/4″ gravel fill in trench	B-6	260	.092	C.Y.	24	3.81	1.40	29.21	34
0400	Pea stone	"	260	.092	"	23.50	3.81	1.40	28.71	33.50

33 49 Storm Drainage Structures

33 49 13 – Storm Drainage Manholes, Frames, and Covers

33 49 13.10 Storm Drainage Manholes, Frames and Covers		Crew	Daily Output	Labor-Hours	Unit	Material	Labor	Equipment	Total	Total Incl O&P
0010	**STORM DRAINAGE MANHOLES, FRAMES & COVERS**									
0020	Excludes footing, excavation, backfill (See line items for frame & cover)									
0050	Brick, 4′ inside diameter, 4′ deep	D-1	1	16	Ea.	520	675		1,195	1,675
0100	6′ deep		.70	22.857		740	960		1,700	2,400
0150	8′ deep		.50	32		955	1,350		2,305	3,250
0200	For depths over 8′, add		4	4	V.L.F.	83.50	168		251.50	365
0400	Concrete blocks (radial), 4′ I.D., 4′ deep		1.50	10.667	Ea.	415	450		865	1,200
0500	6′ deep		1	16		565	675		1,240	1,725
0600	8′ deep		.70	22.857		710	960		1,670	2,350
0700	For depths over 8′, add		5.50	2.909	V.L.F.	77	122		199	284
0800	Concrete, cast in place, 4′ x 4′, 8″ thick, 4′ deep	C-14H	2	24	Ea.	505	1,100	15.60	1,620.60	2,375
0900	6′ deep		1.50	32		730	1,475	21	2,226	3,250
1000	8′ deep		1	48		1,050	2,225	31	3,306	4,800
1100	For depths over 8′, add		8	6	V.L.F.	119	278	3.90	400.90	590
1110	Precast, 4′ I.D., 4′ deep	B-22	4.10	7.317	Ea.	725	320	50.50	1,095.50	1,375
1120	6′ deep		3	10		925	440	69	1,434	1,800
1130	8′ deep		2	15		1,075	660	103	1,838	2,375
1140	For depths over 8′, add		16	1.875	V.L.F.	127	82.50	12.90	222.40	288
1150	5′ I.D., 4′ deep	B-6	3	8	Ea.	1,650	330	121	2,101	2,475
1160	6′ deep		2	12		1,925	495	182	2,602	3,125
1170	8′ deep		1.50	16		2,400	660	243	3,303	4,000

33 49 Storm Drainage Structures

33 49 13 – Storm Drainage Manholes, Frames, and Covers

33 49 13.10 Storm Drainage Manholes, Frames and Covers		Crew	Daily Output	Labor-Hours	Unit	2015 Bare Costs Material	Labor	Equipment	Total	Total Incl O&P
1180	For depths over 8', add	B-6	12	2	V.L.F.	280	82.50	30.50	393	480
1190	6' I.D., 4' deep		2	12	Ea.	2,150	495	182	2,827	3,375
1200	6' deep		1.50	16		2,600	660	243	3,503	4,200
1210	8' deep		1	24		3,200	990	365	4,555	5,500
1220	For depths over 8', add		8	3	V.L.F.	380	124	45.50	549.50	665
1250	Slab tops, precast, 8" thick									
1300	4' diameter manhole	B-6	8	3	Ea.	252	124	45.50	421.50	525
1400	5' diameter manhole		7.50	3.200		410	132	48.50	590.50	725
1500	6' diameter manhole		7	3.429		635	142	52	829	985
3800	Steps, heavyweight cast iron, 7" x 9"	1 Bric	40	.200		19.25	9.25		28.50	36
3900	8" x 9"		40	.200		23	9.25		32.25	40.50
3928	12" x 10-1/2"		40	.200		27	9.25		36.25	44.50
4000	Standard sizes, galvanized steel		40	.200		22	9.25		31.25	39
4100	Aluminum		40	.200		24	9.25		33.25	41.50
4150	Polyethylene		40	.200		26	9.25		35.25	43.50
4210	Rubber boot 6" diam. or smaller	1 Clab	32	.250		82	9.40		91.40	105
4215	8" diam.		24	.333		93.50	12.55		106.05	124
4220	10" diam.		19	.421		107	15.85		122.85	144
4225	12" diam.		16	.500		136	18.80		154.80	181
4230	16" diam.		15	.533		174	20		194	224
4235	18" diam.		15	.533		201	20		221	254
4240	24" diam.		14	.571		230	21.50		251.50	288
4245	30" diam.		12	.667		300	25		325	370

33 51 Natural-Gas Distribution

33 51 13 – Natural-Gas Piping

33 51 13.10 Piping, Gas Service and Distribution, P.E.		Crew	Daily Output	Labor-Hours	Unit	Material	Labor	Equipment	Total	Total Incl O&P
0010	**PIPING, GAS SERVICE AND DISTRIBUTION, POLYETHYLENE**									
0020	Not including excavation or backfill									
1000	60 psi coils, compression coupling @ 100', 1/2" diameter, SDR 11	B-20A	608	.053	L.F.	.45	2.41		2.86	4.33
1010	1" diameter, SDR 11		544	.059		1.05	2.69		3.74	5.45
1040	1-1/4" diameter, SDR 11		544	.059		1.62	2.69		4.31	6.05
1100	2" diameter, SDR 11		488	.066		2.82	3		5.82	7.90
1160	3" diameter, SDR 11		408	.078		6.45	3.59		10.04	12.80
1500	60 PSI 40' joints with coupling, 3" diameter, SDR 11	B-21A	408	.098		10.05	4.60	1.16	15.81	19.65
1540	4" diameter, SDR 11		352	.114		14.25	5.35	1.35	20.95	25.50
1600	6" diameter, SDR 11		328	.122		33.50	5.70	1.44	40.64	47.50
1640	8" diameter, SDR 11		272	.147		51.50	6.90	1.74	60.14	69.50
9000	Minimum labor/equipment charge	B-20	2	12	Job		505		505	820

33 51 13.20 Piping, Gas Service and Distribution, Steel		Crew	Daily Output	Labor-Hours	Unit	Material	Labor	Equipment	Total	Total Incl O&P
0010	**PIPING, GAS SERVICE & DISTRIBUTION, STEEL**									
0020	Not including excavation or backfill, tar coated and wrapped									
4000	Pipe schedule 40, plain end									
4040	1" diameter	Q-4	300	.107	L.F.	4.75	5.95	.19	10.89	14.75
4080	2" diameter		280	.114		7.45	6.40	.21	14.06	18.40
4120	3" diameter		260	.123		12.35	6.90	.22	19.47	24.50
4160	4" diameter	B-35	255	.188		16.05	8.85	2.79	27.69	35
4200	5" diameter		220	.218		23.50	10.25	3.24	36.99	45.50
4240	6" diameter		180	.267		28.50	12.50	3.96	44.96	56
4280	8" diameter		140	.343		45	16.10	5.10	66.20	81
4320	10" diameter		100	.480		117	22.50	7.10	146.60	173

33 51 13.20 Piping, Gas Service and Distribution, Steel		Crew	Daily Output	Labor-Hours	Unit	Material	Labor	Equipment	Total	Total Incl O&P
						2015 Bare Costs				
4360	12″ diameter	B-35	80	.600	L.F.	130	28	8.90	166.90	198
4400	14″ diameter		75	.640		139	30	9.50	178.50	210
4440	16″ diameter		70	.686		152	32	10.20	194.20	230
4480	18″ diameter		65	.738		195	34.50	10.95	240.45	282
4520	20″ diameter		60	.800		305	37.50	11.85	354.35	410
4560	24″ diameter		50	.960		345	45	14.25	404.25	470
6000	Schedule 80, plain end									
6002	4″ diameter	B-35	144	.333	L.F.	46.50	15.65	4.95	67.10	82
6006	6″ diameter		126	.381		93.50	17.85	5.65	117	138
6008	8″ diameter		108	.444		125	21	6.60	152.60	178
6012	12″ diameter		72	.667		250	31.50	9.90	291.40	335
8008	Elbow, weld joint, standard weight									
8020	4″ diameter	Q-16	6.80	3.529	Ea.	96	193	8.45	297.45	415
8026	8″ diameter		3.40	7.059		400	385	16.95	801.95	1,075
8030	12″ diameter		2.30	10.435		885	570	25	1,480	1,900
8034	16″ diameter		1.50	16		2,000	875	38.50	2,913.50	3,625
8038	20″ diameter		1.20	20		3,675	1,100	48	4,823	5,800
8040	24″ diameter		1.02	23.529		5,200	1,300	56.50	6,556.50	7,775
8100	Extra heavy									
8102	4″ diameter	Q-16	5.30	4.528	Ea.	192	248	10.85	450.85	610
8108	8″ diameter		2.60	9.231		600	505	22	1,127	1,475
8112	12″ diameter		1.80	13.333		1,175	730	32	1,937	2,475
8116	16″ diameter		1.20	20		2,675	1,100	48	3,823	4,700
8120	20″ diameter		.94	25.532		4,900	1,400	61.50	6,361.50	7,625
8122	24″ diameter		.80	30		6,925	1,650	72	8,647	10,300
8200	Malleable, standard weight									
8202	4″ diameter	B-20	12	2	Ea.	545	84		629	735
8208	8″ diameter		6	4		1,475	168		1,643	1,900
8212	12″ diameter		4	6		1,575	252		1,827	2,125
8300	Extra heavy									
8302	4″ diameter	B-20	12	2	Ea.	1,100	84		1,184	1,325
8308	8″ diameter	B-21	6	4.667		1,875	202	23	2,100	2,400
8312	12″ diameter	″	4	7		2,525	305	34.50	2,864.50	3,300
8500	Tee weld, standard weight									
8510	4″ diameter	Q-16	4.50	5.333	Ea.	177	292	12.80	481.80	665
8514	6″ diameter		3	8		305	440	19.20	764.20	1,050
8516	8″ diameter		2.30	10.435		530	570	25	1,125	1,500
8520	12″ diameter		1.50	16		1,450	875	38.50	2,363.50	3,025
8524	16″ diameter		1	24		2,875	1,325	57.50	4,257.50	5,275
8528	20″ diameter		.80	30		7,150	1,650	72	8,872	10,500
8530	24″ diameter		.70	34.286		9,225	1,875	82.50	11,182.50	13,100
8810	Malleable, standard weight									
8812	4″ diameter	B-20	8	3	Ea.	925	126		1,051	1,225
8818	8″ diameter	B-21	4	7		1,600	305	34.50	1,939.50	2,275
8822	12″ diameter	″	2.70	10.370		2,925	450	51	3,426	3,975
8900	Extra heavy									
8902	4″ diameter	B-20	8	3	Ea.	1,400	126		1,526	1,725
8908	8″ diameter	B-21	4	7		2,000	305	34.50	2,339.50	2,700
8912	12″ diameter	″	2.70	10.370		2,925	450	51	3,426	3,975

33 51 Natural-Gas Distribution

33 51 33 – Natural-Gas Metering

33 51 33.10 Piping, Valves and Meters, Gas Distribution		Crew	Daily Output	Labor-Hours	Unit	Material	2015 Bare Costs Labor	Equipment	Total	Total Incl O&P
0010	PIPING, VALVES & METERS, GAS DISTRIBUTION									
0020	Not including excavation or backfill									
0100	Gas stops, with or without checks									
0140	1-1/4" size	1 Plum	12	.667	Ea.	64	39		103	132
0180	1-1/2" size		10	.800		80.50	47		127.50	162
0200	2" size		8	1		129	58.50		187.50	234
0600	Pressure regulator valves, iron and bronze									
0640	1-1/2" diameter	1 Plum	13	.615	Ea.	150	36		186	222
0680	2" diameter	"	11	.727		284	42.50		326.50	375
0700	3" diameter	Q-1	13	1.231		570	65		635	730
0740	4" diameter	"	8	2		1,925	106		2,031	2,300
2000	Lubricated semi-steel plug valve									
2040	3/4" diameter	1 Plum	16	.500	Ea.	560	29.50		589.50	660
2080	1" diameter		14	.571		375	33.50		408.50	465
2100	1-1/4" diameter		12	.667		415	39		454	515
2140	1-1/2" diameter		11	.727		440	42.50		482.50	550
2180	2" diameter		8	1		535	58.50		593.50	680
2300	2-1/2" diameter	Q-1	5	3.200		590	169		759	910
2340	3" diameter	"	4.50	3.556		845	188		1,033	1,225

33 52 Liquid Fuel Distribution

33 52 16 – Gasoline Distribution

33 52 16.13 Gasoline Piping		Crew	Daily Output	Labor-Hours	Unit	Material	Labor	Equipment	Total	Total Incl O&P
0010	GASOLINE PIPING									
0020	Primary containment pipe, fiberglass-reinforced									
0030	Plastic pipe 15' & 30' lengths									
0040	2" diameter	Q-6	425	.056	L.F.	5.95	3.15		9.10	11.40
0050	3" diameter		400	.060		10.35	3.35		13.70	16.60
0060	4" diameter		375	.064		13.65	3.57		17.22	20.50
0100	Fittings									
0110	Elbows, 90° & 45°, bell-ends, 2"	Q-6	24	1	Ea.	43.50	56		99.50	135
0120	3" diameter		22	1.091		55	61		116	156
0130	4" diameter		20	1.200		70	67		137	181
0200	Tees, bell ends, 2"		21	1.143		60.50	63.50		124	166
0210	3" diameter		18	1.333		64	74.50		138.50	187
0220	4" diameter		15	1.600		84	89		173	232
0230	Flanges bell ends, 2"		24	1		33.50	56		89.50	124
0240	3" diameter		22	1.091		39	61		100	138
0250	4" diameter		20	1.200		45	67		112	154
0260	Sleeve couplings, 2"		21	1.143		12.40	63.50		75.90	113
0270	3" diameter		18	1.333		17.80	74.50		92.30	136
0280	4" diameter		15	1.600		23	89		112	164
0290	Threaded adapters 2"		21	1.143		17.95	63.50		81.45	119
0300	3" diameter		18	1.333		34	74.50		108.50	154
0310	4" diameter		15	1.600		38	89		127	181
0320	Reducers, 2"		27	.889		27.50	49.50		77	108
0330	3" diameter		22	1.091		27.50	61		88.50	125
0340	4" diameter		20	1.200		37	67		104	145
1010	Gas station product line for secondary containment (double wall)									
1100	Fiberglass reinforced plastic pipe 25' lengths									

33 52 Liquid Fuel Distribution

33 52 16 – Gasoline Distribution

33 52 16.13 Gasoline Piping		Crew	Daily Output	Labor-Hours	Unit	Material	Labor	Equipment	Total	Total Incl O&P
						2015 Bare Costs				
1120	Pipe, plain end, 3" diameter	Q-6	375	.064	L.F.	26.50	3.57		30.07	34.50
1130	4" diameter		350	.069		32	3.82		35.82	41.50
1140	5" diameter		325	.074		35.50	4.12		39.62	45.50
1150	6" diameter	↓	300	.080	↓	39	4.46		43.46	50
1200	Fittings									
1230	Elbows, 90° & 45°, 3" diameter	Q-6	18	1.333	Ea.	136	74.50		210.50	266
1240	4" diameter		16	1.500		167	83.50		250.50	315
1250	5" diameter		14	1.714		184	95.50		279.50	350
1260	6" diameter		12	2		204	112		316	400
1270	Tees, 3" diameter		15	1.600		166	89		255	320
1280	4" diameter		12	2		202	112		314	395
1290	5" diameter		9	2.667		315	149		464	580
1300	6" diameter		6	4		375	223		598	765
1310	Couplings, 3" diameter		18	1.333		55	74.50		129.50	177
1320	4" diameter		16	1.500		119	83.50		202.50	261
1330	5" diameter		14	1.714		214	95.50		309.50	385
1340	6" diameter		12	2		315	112		427	525
1350	Cross-over nipples, 3" diameter		18	1.333		10.60	74.50		85.10	128
1360	4" diameter		16	1.500		12.70	83.50		96.20	144
1370	5" diameter		14	1.714		15.90	95.50		111.40	167
1380	6" diameter		12	2		19.05	112		131.05	195
1400	Telescoping, reducers, concentric 4" x 3"		18	1.333		48	74.50		122.50	169
1410	5" x 4"		17	1.412		95.50	78.50		174	228
1420	6" x 5"	↓	16	1.500	↓	234	83.50		317.50	385

33 61 Hydronic Energy Distribution

33 61 13 – Underground Hydronic Energy Distribution

33 61 13.20 Pipe Conduit, Prefabricated/Preinsulated		Crew	Daily Output	Labor-Hours	Unit	Material	Labor	Equipment	Total	Total Incl O&P
0010	**PIPE CONDUIT, PREFABRICATED/PREINSULATED**									
0020	Does not include trenching, fittings or crane.									
0300	For cathodic protection, add 12 to 14%									
0310	of total built-up price (casing plus service pipe)									
0580	Polyurethane insulated system, 250°F. max. temp.									
0620	Black steel service pipe, standard wt., 1/2" insulation									
0660	3/4" diam. pipe size	Q-17	54	.296	L.F.	57	15.95	1.07	74.02	89
0670	1" diam. pipe size		50	.320		63	17.20	1.16	81.36	97.50
0680	1-1/4" diam. pipe size		47	.340		70	18.30	1.23	89.53	107
0690	1-1/2" diam. pipe size		45	.356		76	19.10	1.28	96.38	115
0700	2" diam. pipe size		42	.381		79	20.50	1.38	100.88	120
0710	2-1/2" diam. pipe size		34	.471		80	25.50	1.70	107.20	129
0720	3" diam. pipe size		28	.571		93	30.50	2.06	125.56	152
0730	4" diam. pipe size		22	.727		117	39	2.63	158.63	193
0740	5" diam. pipe size	↓	18	.889		149	48	3.21	200.21	242
0750	6" diam. pipe size	Q-18	23	1.043		173	58	2.50	233.50	285
0760	8" diam. pipe size		19	1.263		254	70.50	3.03	327.53	395
0770	10" diam. pipe size		16	1.500		320	83.50	3.60	407.10	490
0780	12" diam. pipe size		13	1.846		400	103	4.43	507.43	605
0790	14" diam. pipe size		11	2.182		445	122	5.25	572.25	685
0800	16" diam. pipe size		10	2.400		510	134	5.75	649.75	780
0810	18" diam. pipe size		8	3		590	167	7.20	764.20	920
0820	20" diam. pipe size	↓	7	3.429	↓	655	191	8.25	854.25	1,025

33 61 13 – Underground Hydronic Energy Distribution

33 61 13.20 Pipe Conduit, Prefabricated/Preinsulated		Crew	Daily Output	Labor-Hours	Unit	Material	2015 Bare Costs Labor	Equipment	Total	Total Incl O&P
0830	24" diam. pipe size	Q-18	6	4	L.F.	805	223	9.60	1,037.60	1,250
0900	For 1" thick insulation, add					10%				
0940	For 1-1/2" thick insulation, add					13%				
0980	For 2" thick insulation, add					20%				
1500	Gland seal for system, 3/4" diam. pipe size	Q-17	32	.500	Ea.	775	27	1.81	803.81	895
1510	1" diam. pipe size		32	.500		775	27	1.81	803.81	895
1540	1-1/4" diam. pipe size		30	.533		825	28.50	1.93	855.43	955
1550	1-1/2" diam. pipe size		30	.533		825	28.50	1.93	855.43	955
1560	2" diam. pipe size		28	.571		980	30.50	2.06	1,012.56	1,125
1570	2-1/2" diam. pipe size		26	.615		1,050	33	2.22	1,085.22	1,225
1580	3" diam. pipe size		24	.667		1,125	36	2.41	1,163.41	1,300
1590	4" diam. pipe size		22	.727		1,350	39	2.63	1,391.63	1,550
1600	5" diam. pipe size		19	.842		1,650	45.50	3.04	1,698.54	1,900
1610	6" diam. pipe size	Q-18	26	.923		1,750	51.50	2.22	1,803.72	2,000
1620	8" diam. pipe size		25	.960		2,050	53.50	2.30	2,105.80	2,325
1630	10" diam. pipe size		23	1.043		2,375	58	2.50	2,435.50	2,725
1640	12" diam. pipe size		21	1.143		2,625	63.50	2.74	2,691.24	3,000
1650	14" diam. pipe size		19	1.263		2,950	70.50	3.03	3,023.53	3,375
1660	16" diam. pipe size		18	1.333		3,500	74.50	3.20	3,577.70	3,975
1670	18" diam. pipe size		16	1.500		3,700	83.50	3.60	3,787.10	4,200
1680	20" diam. pipe size		14	1.714		4,200	95.50	4.11	4,299.61	4,775
1690	24" diam. pipe size		12	2		4,675	112	4.80	4,791.80	5,325
2000	Elbow, 45° for system									
2020	3/4" diam. pipe size	Q-17	14	1.143	Ea.	485	61.50	4.13	550.63	635
2040	1" diam. pipe size		13	1.231		500	66	4.44	570.44	660
2050	1-1/4" diam. pipe size		11	1.455		570	78	5.25	653.25	755
2060	1-1/2" diam. pipe size		9	1.778		595	95.50	6.40	696.90	810
2070	2" diam. pipe size		6	2.667		620	143	9.65	772.65	920
2080	2-1/2" diam. pipe size		4	4		670	215	14.45	899.45	1,100
2090	3" diam. pipe size		3.50	4.571		775	246	16.50	1,037.50	1,250
2100	4" diam. pipe size		3	5.333		905	287	19.25	1,211.25	1,450
2110	5" diam. pipe size		2.80	5.714		1,175	305	20.50	1,500.50	1,775
2120	6" diam. pipe size	Q-18	4	6		1,325	335	14.40	1,674.40	1,975
2130	8" diam. pipe size		3	8		1,925	445	19.20	2,389.20	2,825
2140	10" diam. pipe size		2.40	10		2,350	560	24	2,934	3,475
2150	12" diam. pipe size		2	12		3,075	670	29	3,774	4,475
2160	14" diam. pipe size		1.80	13.333		3,825	745	32	4,602	5,400
2170	16" diam. pipe size		1.60	15		4,550	835	36	5,421	6,375
2180	18" diam. pipe size		1.30	18.462		5,725	1,025	44.50	6,794.50	7,950
2190	20" diam. pipe size		1	24		7,200	1,350	57.50	8,607.50	10,100
2200	24" diam. pipe size		.70	34.286		9,050	1,900	82.50	11,032.50	13,000
2260	For elbow, 90°, add					25%				
2300	For tee, straight, add					85%	30%			
2340	For tee, reducing, add					170%	30%			
2380	For weldolet, straight, add					50%				

33 63 13 – Underground Steam and Condensate Distribution Piping

33 63 13.10 Calcium Silicate Insulated System		Crew	Daily Output	Labor-Hours	Unit	Material (2015 Bare Costs)	Labor (2015 Bare Costs)	Equipment (2015 Bare Costs)	Total (2015 Bare Costs)	Total Incl O&P
0010	**CALCIUM SILICATE INSULATED SYSTEM**									
0011	High temp. (1200 degrees F)									
2840	Steel casing with protective exterior coating									
2850	6-5/8" diameter	Q-18	52	.462	L.F.	106	25.50	1.11	132.61	158
2860	8-5/8" diameter		50	.480		115	27	1.15	143.15	170
2870	10-3/4" diameter		47	.511		135	28.50	1.23	164.73	195
2880	12-3/4" diameter		44	.545		147	30.50	1.31	178.81	211
2890	14" diameter		41	.585		166	32.50	1.40	199.90	236
2900	16" diameter		39	.615		178	34.50	1.48	213.98	251
2910	18" diameter		36	.667		196	37	1.60	234.60	276
2920	20" diameter		34	.706		221	39.50	1.69	262.19	305
2930	22" diameter		32	.750		305	42	1.80	348.80	400
2940	24" diameter		29	.828		350	46	1.99	397.99	460
2950	26" diameter		26	.923		400	51.50	2.22	453.72	525
2960	28" diameter		23	1.043		495	58	2.50	555.50	640
2970	30" diameter		21	1.143		530	63.50	2.74	596.24	690
2980	32" diameter		19	1.263		595	70.50	3.03	668.53	770
2990	34" diameter		18	1.333		600	74.50	3.20	677.70	780
3000	36" diameter		16	1.500		645	83.50	3.60	732.10	845
3040	For multi-pipe casings, add					10%				
3060	For oversize casings, add					2%				
3400	Steel casing gland seal, single pipe									
3420	6-5/8" diameter	Q-18	25	.960	Ea.	1,475	53.50	2.30	1,530.80	1,700
3440	8-5/8" diameter		23	1.043		1,725	58	2.50	1,785.50	2,000
3450	10-3/4" diameter		21	1.143		1,925	63.50	2.74	1,991.24	2,225
3460	12-3/4" diameter		19	1.263		2,275	70.50	3.03	2,348.53	2,625
3470	14" diameter		17	1.412		2,525	78.50	3.39	2,606.89	2,900
3480	16" diameter		16	1.500		2,950	83.50	3.60	3,037.10	3,375
3490	18" diameter		15	1.600		3,275	89	3.84	3,367.84	3,750
3500	20" diameter		13	1.846		3,625	103	4.43	3,732.43	4,175
3510	22" diameter		12	2		4,075	112	4.80	4,191.80	4,650
3520	24" diameter		11	2.182		4,575	122	5.25	4,702.25	5,225
3530	26" diameter		10	2.400		5,225	134	5.75	5,364.75	5,975
3540	28" diameter		9.50	2.526		5,900	141	6.05	6,047.05	6,725
3550	30" diameter		9	2.667		6,000	149	6.40	6,155.40	6,850
3560	32" diameter		8.50	2.824		6,750	157	6.80	6,913.80	7,675
3570	34" diameter		8	3		7,350	167	7.20	7,524.20	8,350
3580	36" diameter		7	3.429		7,825	191	8.25	8,024.25	8,900
3620	For multi-pipe casings, add					5%				
4000	Steel casing anchors, single pipe									
4020	6-5/8" diameter	Q-18	8	3	Ea.	1,325	167	7.20	1,499.20	1,725
4040	8-5/8" diameter		7.50	3.200		1,375	178	7.70	1,560.70	1,800
4050	10-3/4" diameter		7	3.429		1,800	191	8.25	1,999.25	2,275
4060	12-3/4" diameter		6.50	3.692		1,925	206	8.85	2,139.85	2,450
4070	14" diameter		6	4		2,275	223	9.60	2,507.60	2,850
4080	16" diameter		5.50	4.364		2,625	243	10.45	2,878.45	3,300
4090	18" diameter		5	4.800		2,975	268	11.50	3,254.50	3,700
4100	20" diameter		4.50	5.333		3,275	297	12.80	3,584.80	4,075
4110	22" diameter		4	6		3,650	335	14.40	3,999.40	4,550
4120	24" diameter		3.50	6.857		3,975	380	16.45	4,371.45	5,000
4130	26" diameter		3	8		4,500	445	19.20	4,964.20	5,675
4140	28" diameter		2.50	9.600		4,950	535	23	5,508	6,300
4150	30" diameter		2	12		5,250	670	29	5,949	6,850

33 63 Steam Energy Distribution

33 63 13 – Underground Steam and Condensate Distribution Piping

33 63 13.10	Calcium Silicate Insulated System	Crew	Daily Output	Labor-Hours	Unit	2015 Bare Costs Material	Labor	Equipment	Total	Total Incl O&P
4160	32″ diameter	Q-18	1.50	16	Ea.	6,275	890	38.50	7,203.50	8,350
4170	34″ diameter		1	24		7,050	1,350	57.50	8,457.50	9,925
4180	36″ diameter		1	24		7,675	1,350	57.50	9,082.50	10,600
4220	For multi-pipe, add					5%	20%			
4800	Steel casing elbow									
4820	6-5/8″ diameter	Q-18	15	1.600	Ea.	1,800	89	3.84	1,892.84	2,125
4830	8-5/8″ diameter		15	1.600		1,925	89	3.84	2,017.84	2,275
4850	10-3/4″ diameter		14	1.714		2,325	95.50	4.11	2,424.61	2,700
4860	12-3/4″ diameter		13	1.846		2,750	103	4.43	2,857.43	3,200
4870	14″ diameter		12	2		2,950	112	4.80	3,066.80	3,425
4880	16″ diameter		11	2.182		3,225	122	5.25	3,352.25	3,750
4890	18″ diameter		10	2.400		3,675	134	5.75	3,814.75	4,275
4900	20″ diameter		9	2.667		3,925	149	6.40	4,080.40	4,575
4910	22″ diameter		8	3		4,225	167	7.20	4,399.20	4,925
4920	24″ diameter		7	3.429		4,650	191	8.25	4,849.25	5,425
4930	26″ diameter		6	4		5,100	223	9.60	5,332.60	5,950
4940	28″ diameter		5	4.800		5,500	268	11.50	5,779.50	6,475
4950	30″ diameter		4	6		5,525	335	14.40	5,874.40	6,600
4960	32″ diameter		3	8		6,275	445	19.20	6,739.20	7,625
4970	34″ diameter		2	12		6,900	670	29	7,599	8,675
4980	36″ diameter		2	12		7,350	670	29	8,049	9,150
5500	Black steel service pipe, std. wt., 1″ thick insulation									
5510	3/4″ diameter pipe size	Q-17	54	.296	L.F.	62	15.95	1.07	79.02	94
5540	1″ diameter pipe size		50	.320		64	17.20	1.16	82.36	99
5550	1-1/4″ diameter pipe size		47	.340		70	18.30	1.23	89.53	107
5560	1-1/2″ diameter pipe size		45	.356		74	19.10	1.28	94.38	113
5570	2″ diameter pipe size		42	.381		81	20.50	1.38	102.88	123
5580	2-1/2″ diameter pipe size		34	.471		85	25.50	1.70	112.20	135
5590	3″ diameter pipe size		28	.571		93	30.50	2.06	125.56	152
5600	4″ diameter pipe size		22	.727		98	39	2.63	139.63	172
5610	5″ diameter pipe size		18	.889		107	48	3.21	158.21	196
5620	6″ diameter pipe size	Q-18	23	1.043		115	58	2.50	175.50	221
6000	Black steel service pipe, std. wt., 1-1/2″ thick insul.									
6010	3/4″ diameter pipe size	Q-17	54	.296	L.F.	62	15.95	1.07	79.02	94
6040	1″ diameter pipe size		50	.320		67	17.20	1.16	85.36	102
6050	1-1/4″ diameter pipe size		47	.340		73	18.30	1.23	92.53	110
6060	1-1/2″ diameter pipe size		45	.356		79	19.10	1.28	99.38	118
6070	2″ diameter pipe size		42	.381		85	20.50	1.38	106.88	127
6080	2-1/2″ diameter pipe size		34	.471		89	25.50	1.70	116.20	139
6090	3″ diameter pipe size		28	.571		99	30.50	2.06	131.56	159
6100	4″ diameter pipe size		22	.727		107	39	2.63	148.63	182
6110	5″ diameter pipe size		18	.889		115	48	3.21	166.21	205
6120	6″ diameter pipe size	Q-18	23	1.043		130	58	2.50	190.50	237
6130	8″ diameter pipe size		19	1.263		180	70.50	3.03	253.53	310
6140	10″ diameter pipe size		16	1.500		240	83.50	3.60	327.10	400
6150	12″ diameter pipe size		13	1.846		300	103	4.43	407.43	495
6190	For 2″ thick insulation, add					15%				
6220	For 2-1/2″ thick insulation, add					25%				
6260	For 3″ thick insulation, add					30%				
6800	Black steel service pipe, ex. hvy. wt., 1″ thick insul.									
6820	3/4″ diameter pipe size	Q-17	50	.320	L.F.	64	17.20	1.16	82.36	99
6840	1″ diameter pipe size		47	.340		68	18.30	1.23	87.53	105
6850	1-1/4″ diameter pipe size		44	.364		76	19.55	1.31	96.86	115

33 63 Steam Energy Distribution

33 63 13 – Underground Steam and Condensate Distribution Piping

33 63 13.10 Calcium Silicate Insulated System		Crew	Daily Output	Labor-Hours	Unit	Material	2015 Bare Costs Labor	Equipment	Total	Total Incl O&P
6860	1-1/2" diameter pipe size	Q-17	42	.381	L.F.	78	20.50	1.38	99.88	120
6870	2" diameter pipe size		40	.400		85	21.50	1.44	107.94	129
6880	2-1/2" diameter pipe size		31	.516		89	28	1.86	118.86	144
6890	3" diameter pipe size		27	.593		97	32	2.14	131.14	159
6900	4" diameter pipe size		21	.762		102	41	2.75	145.75	179
6910	5" diameter pipe size		17	.941		110	50.50	3.40	163.90	204
6920	6" diameter pipe size	Q-18	22	1.091		120	61	2.62	183.62	230
7400	Black steel service pipe, ex. hvy. wt., 1-1/2" thick insul.									
7420	3/4" diameter pipe size	Q-17	50	.320	L.F.	55	17.20	1.16	73.36	89
7440	1" diameter pipe size		47	.340		60	18.30	1.23	79.53	96
7450	1-1/4" diameter pipe size		44	.364		67	19.55	1.31	87.86	105
7460	1-1/2" diameter pipe size		42	.381		72	20.50	1.38	93.88	113
7470	2" diameter pipe size		40	.400		77	21.50	1.44	99.94	120
7480	2-1/2" diameter pipe size		31	.516		82	28	1.86	111.86	136
7490	3" diameter pipe size		27	.593		91	32	2.14	125.14	152
7500	4" diameter pipe size		21	.762		110	41	2.75	153.75	188
7510	5" diameter pipe size		17	.941		152	50.50	3.40	205.90	250
7520	6" diameter pipe size	Q-18	22	1.091		160	61	2.62	223.62	274
7530	8" diameter pipe size		18	1.333		215	74.50	3.20	292.70	355
7540	10" diameter pipe size		15	1.600		275	89	3.84	367.84	450
7550	12" diameter pipe size		13	1.846		330	103	4.43	437.43	530
7590	For 2" thick insulation, add					13%				
7640	For 2-1/2" thick insulation, add					18%				
7680	For 3" thick insulation, add					24%				

33 71 Electrical Utility Transmission and Distribution

33 71 16 – Electrical Utility Poles

33 71 16.33 Wood Electrical Utility Poles		Crew	Daily Output	Labor-Hours	Unit	Material	Labor	Equipment	Total	Total Incl O&P
0010	**WOOD ELECTRICAL UTILITY POLES**									
0011	Excludes excavation, backfill and cast-in-place concrete									
6200	Electric & tel sitework, 20' high, treated wd., see Section 26 56 13.10	R-3	3.10	6.452	Ea.	224	350	44.50	618.50	840
6400	25' high		2.90	6.897		265	375	47.50	687.50	930
6600	30' high		2.60	7.692		435	420	53	908	1,175
6800	35' high		2.40	8.333		495	455	57.50	1,007.50	1,325
7000	40' high		2.30	8.696		710	470	60	1,240	1,575
7200	45' high		1.70	11.765		865	640	81	1,586	2,025
7400	Cross arms with hardware & insulators									
7600	4' long	1 Elec	2.50	3.200	Ea.	150	175		325	435
7800	5' long		2.40	3.333		172	182		354	470
8000	6' long		2.20	3.636		166	199		365	495

33 71 19 – Electrical Underground Ducts and Manholes

33 71 19.15 Underground Ducts and Manholes		Crew	Daily Output	Labor-Hours	Unit	Material	Labor	Equipment	Total	Total Incl O&P
0010	**UNDERGROUND DUCTS AND MANHOLES**									
0011	Not incl. excavation, backfill and concrete, in slab or duct bank									
1000	Direct burial									
1010	PVC, schedule 40, w/coupling, 1/2" diameter	1 Elec	340	.024	L.F.	.35	1.29		1.64	2.38
1020	3/4" diameter		290	.028		.47	1.51		1.98	2.86
1030	1" diameter		260	.031		.75	1.68		2.43	3.44
1040	1-1/2" diameter		210	.038		1.14	2.08		3.22	4.48
1050	2" diameter		180	.044		1.48	2.43		3.91	5.40

33 71 19.15 Underground Ducts and Manholes		Crew	Daily Output	Labor-Hours	Unit	Material	Labor (2015 Bare Costs)	Equipment	Total	Total Incl O&P
1060	3" diameter	2 Elec	240	.067	L.F.	2.82	3.65		6.47	8.75
1070	4" diameter		160	.100		3.95	5.45		9.40	12.80
1080	5" diameter		120	.133		5.75	7.30		13.05	17.60
1090	6" diameter		90	.178		7.60	9.70		17.30	23.50
1110	Elbows, 1/2" diameter	1 Elec	48	.167	Ea.	.90	9.10		10	15.10
1120	3/4" diameter		38	.211		.88	11.50		12.38	18.80
1130	1" diameter		32	.250		1.33	13.70		15.03	22.50
1140	1-1/2" diameter		21	.381		2.65	21		23.65	35.50
1150	2" diameter		16	.500		3.59	27.50		31.09	46.50
1160	3" diameter		12	.667		10.85	36.50		47.35	68.50
1170	4" diameter		9	.889		17.25	48.50		65.75	94.50
1180	5" diameter		8	1		28	54.50		82.50	116
1190	6" diameter		5	1.600		64.50	87.50		152	207
1210	Adapters, 1/2" diameter		52	.154		.25	8.40		8.65	13.35
1220	3/4" diameter		43	.186		.41	10.20		10.61	16.20
1230	1" diameter		39	.205		.56	11.20		11.76	17.95
1240	1-1/2" diameter		35	.229		.98	12.50		13.48	20.50
1250	2" diameter		26	.308		1.22	16.85		18.07	27.50
1260	3" diameter		20	.400		2.97	22		24.97	37.50
1270	4" diameter		14	.571		5.30	31.50		36.80	54.50
1280	5" diameter		12	.667		11.20	36.50		47.70	69
1290	6" diameter		9	.889		18.70	48.50		67.20	96
1340	Bell end & cap, 1-1/2" diameter		35	.229		8.55	12.50		21.05	29
1350	Bell end & plug, 2" diameter		26	.308		9.65	16.85		26.50	36.50
1360	3" diameter		20	.400		11.70	22		33.70	47
1370	4" diameter		14	.571		14.75	31.50		46.25	65
1380	5" diameter		12	.667		16.75	36.50		53.25	75
1390	6" diameter		9	.889		20	48.50		68.50	97.50
1450	Base spacer, 2" diameter		56	.143		1.55	7.80		9.35	13.80
1460	3" diameter		46	.174		1.58	9.50		11.08	16.50
1470	4" diameter		41	.195		1.64	10.65		12.29	18.35
1480	5" diameter		37	.216		2.02	11.85		13.87	20.50
1490	6" diameter		34	.235		2.75	12.85		15.60	23
1550	Intermediate spacer, 2" diameter		60	.133		1.50	7.30		8.80	12.95
1560	3" diameter		46	.174		1.69	9.50		11.19	16.60
1570	4" diameter		41	.195		1.55	10.65		12.20	18.25
1580	5" diameter		37	.216		1.85	11.85		13.70	20.50
1590	6" diameter		34	.235		2.70	12.85		15.55	23
4010	PVC, schedule 80, w/coupling, 1/2" diameter		215	.037	L.F.	.92	2.04		2.96	4.16
4020	3/4" diameter		180	.044		1.24	2.43		3.67	5.10
4030	1" diameter		145	.055		1.76	3.02		4.78	6.60
4040	1-1/2" diameter		120	.067		2.93	3.65		6.58	8.85
4050	2" diameter		100	.080		4.05	4.38		8.43	11.25
4060	3" diameter	2 Elec	130	.123		8.25	6.75		15	19.50
4070	4" diameter		90	.178		12	9.70		21.70	28.50
4080	5" diameter		70	.229		17.25	12.50		29.75	38.50
4090	6" diameter		50	.320		21.50	17.50		39	51
4110	Elbows, 1/2" diameter	1 Elec	29	.276	Ea.	1.33	15.10		16.43	25
4120	3/4" diameter		23	.348		2.57	19.05		21.62	32.50
4130	1" diameter		20	.400		3.18	22		25.18	37.50
4140	1-1/2" diameter		16	.500		6.95	27.50		34.45	50
4150	2" diameter		12	.667		8.25	36.50		44.75	65.50
4160	3" diameter		9	.889		23	48.50		71.50	101

33 71 19.15 Underground Ducts and Manholes		Crew	Daily Output	Labor-Hours	Unit	2015 Bare Costs Material	Labor	Equipment	Total	Total Incl O&P
4170	4" diameter	1 Elec	7	1.143	Ea.	46.50	62.50		109	149
4180	5" diameter		6	1.333		195	73		268	330
4190	6" diameter		4	2		231	109		340	425
4210	Adapter, 1/2" diameter		39	.205		.25	11.20		11.45	17.65
4220	3/4" diameter		33	.242		.41	13.25		13.66	21
4230	1" diameter		29	.276		.56	15.10		15.66	24
4240	1-1/2" diameter		26	.308		.98	16.85		17.83	27
4250	2" diameter		23	.348		1.22	19.05		20.27	31
4260	3" diameter		18	.444		2.97	24.50		27.47	41
4270	4" diameter		13	.615		5.30	33.50		38.80	58
4280	5" diameter		11	.727		11.20	40		51.20	74
4290	6" diameter		8	1		18.70	54.50		73.20	105
4310	Bell end & cap, 1-1/2" diameter		26	.308		8.55	16.85		25.40	35.50
4320	Bell end & plug, 2" diameter		23	.348		9.65	19.05		28.70	40
4330	3" diameter		18	.444		11.70	24.50		36.20	50.50
4340	4" diameter		13	.615		14.75	33.50		48.25	68.50
4350	5" diameter		11	.727		16.75	40		56.75	80
4360	6" diameter		8	1		20	54.50		74.50	107
4370	Base spacer, 2" diameter		42	.190		1.55	10.40		11.95	17.85
4380	3" diameter		33	.242		1.58	13.25		14.83	22
4390	4" diameter		29	.276		1.64	15.10		16.74	25.50
4400	5" diameter		26	.308		2.02	16.85		18.87	28
4410	6" diameter		25	.320		2.75	17.50		20.25	30
4420	Intermediate spacer, 2" diameter		45	.178		1.50	9.70		11.20	16.70
4430	3" diameter		34	.235		1.69	12.85		14.54	22
4440	4" diameter		31	.258		1.55	14.10		15.65	23.50
4450	5" diameter		28	.286		1.85	15.65		17.50	26
4460	6" diameter		25	.320		2.70	17.50		20.20	30

33 71 19.17 Electric and Telephone Underground		Crew	Daily Output	Labor-Hours	Unit	Material	Labor	Equipment	Total	Total Incl O&P
0010	**ELECTRIC AND TELEPHONE UNDERGROUND**									
0011	Not including excavation R337119-30									
0200	backfill and cast in place concrete									
0400	Hand holes, precast concrete, with concrete cover									
0600	2' x 2' x 3' deep	R-3	2.40	8.333	Ea.	405	455	57.50	917.50	1,225
0800	3' x 3' x 3' deep		1.90	10.526		525	570	72.50	1,167.50	1,550
1000	4' x 4' x 4' deep		1.40	14.286		1,400	775	98.50	2,273.50	2,850
1200	Manholes, precast with iron racks & pulling irons, C.I. frame									
1400	and cover, 4' x 6' x 7' deep	B-13	2	28	Ea.	6,050	1,150	370	7,570	8,900
1600	6' x 8' x 7' deep		1.90	29.474		6,800	1,200	390	8,390	9,850
1800	6' x 10' x 7' deep		1.80	31.111		7,625	1,275	410	9,310	10,900
4200	Underground duct, banks ready for concrete fill, min. of 7.5"									
4400	between conduits, center to center									
4580	PVC, type EB, 1 @ 2" diameter	2 Elec	480	.033	L.F.	.71	1.82		2.53	3.60
4600	2 @ 2" diameter		240	.067		1.41	3.65		5.06	7.20
4800	4 @ 2" diameter		120	.133		2.82	7.30		10.12	14.40
4900	1 @ 3" diameter		400	.040		.98	2.19		3.17	4.47
5000	2 @ 3" diameter		200	.080		1.96	4.38		6.34	8.95
5200	4 @ 3" diameter		100	.160		3.91	8.75		12.66	17.85
5300	1 @ 4" diameter		320	.050		1.52	2.74		4.26	5.90
5400	2 @ 4" diameter		160	.100		3.05	5.45		8.50	11.80
5600	4 @ 4" diameter		80	.200		6.10	10.95		17.05	23.50
5800	6 @ 4" diameter		54	.296		9.15	16.20		25.35	35

33 71 19.17 Electric and Telephone Underground		Crew	Daily Output	Labor-Hours	Unit	2015 Bare Costs Material	Labor	Equipment	Total	Total Incl O&P
5810	1 @ 5″ diameter	2 Elec	260	.062	L.F.	2.24	3.37		5.61	7.65
5820	2 @ 5″ diameter		130	.123		4.48	6.75		11.23	15.35
5840	4 @ 5″ diameter		70	.229		8.95	12.50		21.45	29
5860	6 @ 5″ diameter		50	.320		13.45	17.50		30.95	42
5870	1 @ 6″ diameter		200	.080		3.21	4.38		7.59	10.35
5880	2 @ 6″ diameter		100	.160		6.40	8.75		15.15	20.50
5900	4 @ 6″ diameter		50	.320		12.80	17.50		30.30	41
5920	6 @ 6″ diameter		30	.533		19.25	29		48.25	66
6200	Rigid galvanized steel, 2 @ 2″ diameter		180	.089		12.90	4.86		17.76	22
6400	4 @ 2″ diameter		90	.178		26	9.70		35.70	43.50
6800	2 @ 3″ diameter		100	.160		29	8.75		37.75	45
7000	4 @ 3″ diameter		50	.320		57.50	17.50		75	90.50
7200	2 @ 4″ diameter		70	.229		41.50	12.50		54	65
7400	4 @ 4″ diameter		34	.471		83	25.50		108.50	131
7600	6 @ 4″ diameter		22	.727		124	40		164	199
7620	2 @ 5″ diameter		60	.267		85.50	14.60		100.10	117
7640	4 @ 5″ diameter		30	.533		171	29		200	233
7660	6 @ 5″ diameter		18	.889		257	48.50		305.50	360
7680	2 @ 6″ diameter		40	.400		116	22		138	162
7700	4 @ 6″ diameter		20	.800		232	44		276	325
7720	6 @ 6″ diameter		14	1.143		350	62.50		412.50	480
7800	For cast-in-place concrete - Add									
7810	Under 1 C.Y.	C-6	16	3	C.Y.	163	118	3.90	284.90	375
7820	1 C.Y. - 5 C.Y.		19.20	2.500		147	98	3.25	248.25	325
7830	Over 5 C.Y.		24	2		122	78.50	2.60	203.10	265
7850	For reinforcing rods - Add									
7860	#4 to #7	2 Rodm	1.10	14.545	Ton	970	765		1,735	2,325
7870	#8 to #14	″	1.50	10.667	″	970	560		1,530	2,000
8000	Fittings, PVC type EB, elbow, 2″ diameter	1 Elec	16	.500	Ea.	15.90	27.50		43.40	60
8200	3″ diameter		14	.571		20.50	31.50		52	71
8400	4″ diameter		12	.667		28.50	36.50		65	88
8420	5″ diameter		10	.800		152	44		196	235
8440	6″ diameter		9	.889		173	48.50		221.50	266
8500	Coupling, 2″ diameter					.72			.72	.79
8600	3″ diameter					3.23			3.23	3.55
8700	4″ diameter					3.53			3.53	3.88
8720	5″ diameter					6.65			6.65	7.30
8740	6″ diameter					22.50			22.50	24.50
8800	Adapter, 2″ diameter	1 Elec	26	.308		1.02	16.85		17.87	27
9000	3″ diameter		20	.400		2.90	22		24.90	37
9200	4″ diameter		16	.500		4.23	27.50		31.73	47
9220	5″ diameter		13	.615		10.25	33.50		43.75	63.50
9240	6″ diameter		10	.800		13.30	44		57.30	82.50
9400	End bell, 2″ diameter		16	.500		1.30	27.50		28.80	44
9600	3″ diameter		14	.571		6.65	31.50		38.15	56
9800	4″ diameter		12	.667		4.26	36.50		40.76	61
9810	5″ diameter		10	.800		13	44		57	82.50
9820	6″ diameter		8	1		66	54.50		120.50	158
9830	5° angle coupling, 2″ diameter		26	.308		19.60	16.85		36.45	47.50
9840	3″ diameter		20	.400		29.50	22		51.50	66.50
9850	4″ diameter		16	.500		15.25	27.50		42.75	59.50
9860	5″ diameter		13	.615		14.75	33.50		48.25	68
9870	6″ diameter		10	.800		23	44		67	93

33 71 Electrical Utility Transmission and Distribution

33 71 19 – Electrical Underground Ducts and Manholes

33 71 19.17 Electric and Telephone Underground		Crew	Daily Output	Labor-Hours	Unit	2015 Bare Costs Material	Labor	Equipment	Total	Total Incl O&P
9880	Expansion joint, 2" diameter	1 Elec	16	.500	Ea.	33.50	27.50		61	79.50
9890	3" diameter		18	.444		64	24.50		88.50	108
9900	4" diameter		12	.667		85	36.50		121.50	150
9910	5" diameter		10	.800		114	44		158	194
9920	6" diameter		8	1		143	54.50		197.50	242
9930	Heat bender, 2" diameter					545			545	600
9940	6" diameter					1,850			1,850	2,025
9950	Cement, quart					16.60			16.60	18.30
9960	Nylon polyethylene pull rope, 1/4"	2 Elec	2000	.008	L.F.	.16	.44		.60	.86
9990	Minimum labor/equipment charge	1 Elec	3.50	2.286	Job		125		125	194

33 81 Communications Structures

33 81 13 – Communications Transmission Towers

33 81 13.10 Radio Towers		Crew	Daily Output	Labor-Hours	Unit	Material	Labor	Equipment	Total	Total Incl O&P
0010	**RADIO TOWERS**									
0020	Guyed, 50' H, 40 lb. sect., 70 MPH basic wind spd.	2 Sswk	1	16	Ea.	2,750	840		3,590	4,550
0100	Wind load 90 MPH basic wind speed	"	1	16		3,700	840		4,540	5,600
0300	190' high, 40 lb. section, wind load 70 MPH basic wind speed	K-2	.33	72.727		9,550	3,550	920	14,020	17,700
0400	200' high, 70 lb. section, wind load 90 MPH basic wind speed		.33	72.727		15,900	3,550	920	20,370	24,700
0600	300' high, 70 lb. section, wind load 70 MPH basic wind speed		.20	120		25,500	5,850	1,525	32,875	40,000
0700	270' high, 90 lb. section, wind load 90 MPH basic wind speed		.20	120		27,600	5,850	1,525	34,975	42,300
0800	400' high, 100 lb. section, wind load 70 MPH basic wind speed		.14	171		38,100	8,375	2,175	48,650	59,000
0900	Self-supporting, 60' high, wind load 70 MPH basic wind speed		.80	30		4,500	1,475	380	6,355	7,950
0910	60' high, wind load 90 MPH basic wind speed		.45	53.333		5,275	2,600	675	8,550	11,100
1000	120' high, wind load 70 MPH basic wind speed		.40	60		10,000	2,925	760	13,685	17,000
1200	190' high, wind load 90 MPH basic wind speed		.20	120		28,300	5,850	1,525	35,675	43,100
2000	For states west of Rocky Mountains, add for shipping					10%				

Division Notes

		CREW	DAILY OUTPUT	LABOR-HOURS	UNIT	BARE COSTS MAT.	BARE COSTS LABOR	BARE COSTS EQUIP.	BARE COSTS TOTAL	TOTAL INCL O&P

Division 34 – Transportation

Estimating Tips

34 11 00 Rail Tracks

This subdivision includes items that may involve either repair of existing, or construction of new, railroad tracks. Additional preparation work, such as the roadbed earthwork, would be found in Division 31. Additional new construction siding and turnouts are found in Subdivision 34 72. Maintenance of railroads is found under 34 01 23 Operation and Maintenance of Railways.

34 40 00 Traffic Signals

This subdivision includes traffic signal systems. Other traffic control devices such as traffic signs are found in Subdivision 10 14 53 Traffic Signage.

34 70 00 Vehicle Barriers

This subdivision includes security vehicle barriers, guide and guard rails, crash barriers, and delineators. The actual maintenance and construction of concrete and asphalt pavement is found in Division 32.

Reference Numbers

Reference numbers are shown in shaded boxes at the beginning of some major classifications. These numbers refer to related items in the Reference Section. The reference information may be an estimating procedure, an alternate pricing method, or technical information.

Note: Not all subdivisions listed here necessarily appear in this publication. ■

34 01 Operation and Maintenance of Transportation

34 01 23 – Operation and Maintenance of Railways

34 01 23.51 Maintenance of Railroads		Crew	Daily Output	Labor-Hours	Unit	2015 Bare Costs Material	Labor	Equipment	Total	Total Incl O&P
0010	MAINTENANCE OF RAILROADS									
0400	Resurface and realign existing track	B-14	200	.240	L.F.		9.55	1.82	11.37	17.55
0600	For crushed stone ballast, add	"	500	.096	"	13.40	3.82	.73	17.95	21.50

34 11 Rail Tracks

34 11 13 – Track Rails

34 11 13.23 Heavy Rail Track

Line	Description	Crew	Daily Output	Labor-Hours	Unit	Material	Labor	Equipment	Total	Total Incl O&P
0010	HEAVY RAIL TRACK R347216-10									
1000	Rail, 100 lb. prime grade				L.F.	35			35	38.50
1500	Relay rail				"	17.55			17.55	19.30

34 11 33 – Track Cross Ties

34 11 33.13 Concrete Track Cross Ties

Line	Description	Crew	Daily Output	Labor-Hours	Unit	Material	Labor	Equipment	Total	Total Incl O&P
0010	CONCRETE TRACK CROSS TIES									
1400	Ties, concrete, 8'-6" long, 30" O.C.	B-14	80	.600	Ea.	173	24	4.55	201.55	234

34 11 33.16 Timber Track Cross Ties

Line	Description	Crew	Daily Output	Labor-Hours	Unit	Material	Labor	Equipment	Total	Total Incl O&P
0010	TIMBER TRACK CROSS TIES									
1600	Wood, pressure treated, 6" x 8" x 8'-6", C.L. lots	B-14	90	.533	Ea.	52	21	4.05	77.05	96
1700	L.C.L. lots		90	.533		54.50	21	4.05	79.55	99
1900	Heavy duty, 7" x 9" x 8'-6", C.L. lots		70	.686		57	27.50	5.20	89.70	113
2000	L.C.L. lots	↓	70	.686	↓	57	27.50	5.20	89.70	113

34 11 33.17 Timber Switch Ties

Line	Description	Crew	Daily Output	Labor-Hours	Unit	Material	Labor	Equipment	Total	Total Incl O&P
0010	TIMBER SWITCH TIES									
1200	Switch timber, for a #8 switch, pressure treated	B-14	3.70	12.973	M.B.F.	3,100	515	98.50	3,713.50	4,375
1300	Complete set of timbers, 3.7 MBF for #8 switch	"	1	48	Total	12,000	1,900	365	14,265	16,700

34 11 93 – Track Appurtenances and Accessories

34 11 93.50 Track Accessories

Line	Description	Crew	Daily Output	Labor-Hours	Unit	Material	Labor	Equipment	Total	Total Incl O&P
0010	TRACK ACCESSORIES									
0020	Car bumpers, test	B-14	2	24	Ea.	3,750	955	182	4,887	5,875
0100	Heavy duty R347216-20		2	24		7,100	955	182	8,237	9,575
0200	Derails hand throw (sliding)		10	4.800		1,250	191	36.50	1,477.50	1,725
0300	Hand throw with standard timbers, open stand & target		8	6	↓	1,350	239	45.50	1,634.50	1,925
2400	Wheel stops, fixed		18	2.667	Pr.	920	106	20	1,046	1,225
2450	Hinged	↓	14	3.429	"	1,250	136	26	1,412	1,625

34 11 93.60 Track Material

Line	Description	Crew	Daily Output	Labor-Hours	Unit	Material	Labor	Equipment	Total	Total Incl O&P
0010	TRACK MATERIAL									
0020	Track bolts				Ea.	4.17			4.17	4.59
0100	Joint bars				Pr.	93			93	103
0200	Spikes				Ea.	2			2	2.20
0300	Tie plates				"	15.60			15.60	17.15

34 41 Roadway Signaling and Control Equipment

34 41 13 – Traffic Signals

34 41 13.10 Traffic Signals Systems		Crew	Daily Output	Labor-Hours	Unit	Material	2015 Bare Costs Labor	Equipment	Total	Total Incl O&P
0010	**TRAFFIC SIGNALS SYSTEMS**									
0020	Component costs									
0600	Crew employs crane / directional driller as required									
1000	Vertical mast with foundation									
1010	Mast sized for single arm to 40′; no lighting or power function	R-11	.50	112	Signal	10,000	5,825	1,725	17,550	22,000
1100	Horizontal arm									
1110	Per linear foot of arm	R-11	50	1.120	Signal	200	58	17.20	275.20	330
1200	Traffic signal									
1210	Includes signal, bracket, sensor, and wiring	R-11	2.50	22.400	Signal	1,000	1,175	345	2,520	3,300
1300	Pedestrian signals and callers									
1310	Includes four signals with brackets and two call buttons	R-11	2.50	22.400	Signal	3,000	1,175	345	4,520	5,500
1400	Controller, design, and underground conduit									
1410	Includes miscellaneous signage and adjacent surface work	R-11	.25	224	Signal	20,000	11,600	3,450	35,050	44,000

34 43 Airfield Signaling and Control Equipment

34 43 13 – Airfield Signals and Lighting

34 43 13.16 Airport Lighting		Crew	Daily Output	Labor-Hours	Unit	Material	Labor	Equipment	Total	Total Incl O&P
0010	**AIRPORT LIGHTING**									
0100	Runway centerline, bidir., semi-flush, 200 W, w/shallow insert base	R-22	12.40	3.006	Ea.	1,725	140		1,865	2,125
0110	for mounting in base housing		18.60	2.004		865	93		958	1,100
0120	Flush, 200 W, w/shallow insert base		12.40	3.006		1,700	140		1,840	2,100
0130	for mounting in base housing		18.64	2		885	93		978	1,125
0150	Touchdown zone light, unidirectional, 200 W, w/shallow insert base		12.40	3.006		1,575	140		1,715	1,950
0160	115 W		12.40	3.006		1,525	140		1,665	1,900
0180	Unidirectional, 200 W, for mounting in base housing		18.60	2.004		700	93		793	920
0190	115 W		18.60	2.004		755	93		848	975
0210	Runway edge & threshold light, bidir., 200 W, for base housing		9.36	3.983		940	185		1,125	1,325
0240	Threshold & approach light, unidir., 200 W, for base housing		9.36	3.983		545	185		730	895
0260	Runway edge, bi-directional, 2-115 W, for base housing		12.40	3.006		1,450	140		1,590	1,800
0280	Runway threshold & end, bidir., 2-115 W, for base housing		12.40	3.006		1,600	140		1,740	1,975
0370	45 W, flush, for mounting in base housing		18.64	2		780	93		873	1,000
0380	115 W		18.64	2		795	93		888	1,025

34 43 23 – Weather Observation Equipment

34 43 23.16 Airfield Wind Cones		Crew	Daily Output	Labor-Hours	Unit	Material	Labor	Equipment	Total	Total Incl O&P
0010	**AIRFIELD WIND CONES**									
1200	Wind cone, 12′ lighted assembly, rigid, w/obstruction light	R-21	1.36	24.118	Ea.	9,200	1,325	44	10,569	12,200
1210	Without obstruction light		1.52	21.579		7,725	1,175	39.50	8,939.50	10,400
1220	Unlighted assembly, w/obstruction light		1.68	19.524		8,425	1,075	35.50	9,535.50	10,900
1230	Without obstruction light		1.84	17.826		7,725	975	32.50	8,732.50	10,000
1240	Wind cone slip fitter, 2-1/2″ pipe		21.84	1.502		102	82	2.75	186.75	242
1250	Wind cone sock, 12′ x 3′, cotton		6.56	5		560	274	9.15	843.15	1,050
1260	Nylon		6.56	5		565	274	9.15	848.15	1,050

34 71 13.17 Security Vehicle Barriers		Crew	Daily Output	Labor-Hours	Unit	2015 Bare Costs Material	Labor	Equipment	Total	Total Incl O&P
0010	**SECURITY VEHICLE BARRIERS**									
0020	Security planters excludes filling material									
0100	Concrete security planter, exposed aggregate 36″ diam. x 30″ high	B-11M	8	2	Ea.	565	88	49	702	815
0200	48″ diam. x 36″ high		8	2		845	88	49	982	1,125
0300	53″ diam. x 18″ high		8	2		945	88	49	1,082	1,250
0400	72″ diam. x 18″ high with seats		8	2		2,275	88	49	2,412	2,700
0450	84″ diam. x 36″ high		8	2		2,250	88	49	2,387	2,675
0500	36″ x 36″ x 24″ high square		8	2		510	88	49	647	760
0600	36″ x 36″ x 30″ high square		8	2		820	88	49	957	1,100
0700	48″ L x 24″ W x 30″ H rectangle		8	2		525	88	49	662	770
0800	72″ L x 24″ W x 30″ H rectangle		8	2		610	88	49	747	870
0900	96″ L x 24″ W x 30″ H rectangle		8	2		780	88	49	917	1,050
0950	Decorative geometric concrete barrier, 96″ L x 24″ W x 36″ H		8	2		830	88	49	967	1,100
1000	Concrete security planter, filling with washed sand or gravel <1 C.Y.		8	2		39	88	49	176	239
1050	2 C.Y. or less		6	2.667		78	118	65.50	261.50	345
1200	Jersey concrete barrier, 10′ L x 2′ by 0.5′ W x 30″ H	B-21B	16	2.500		385	102	41	528	630
1300	10 or more same site		24	1.667		385	68	27.50	480.50	560
1400	10′ L x 2′ by 0.5′ W x 32″ H		16	2.500		405	102	41	548	655
1500	10 or more same site		24	1.667		405	68	27.50	500.50	585
1600	20′ L x 2′ by 0.5′ W x 30″ H		12	3.333		565	136	54.50	755.50	900
1700	10 or more same site		18	2.222		565	90.50	36.50	692	810
1800	20′ L x 2′ by 0.5′ W x 32″ H		12	3.333		680	136	54.50	870.50	1,025
1900	10 or more same site		18	2.222		680	90.50	36.50	807	940
2000	GFRC decorative security barrier per 10 feet section including concrete		4	10		3,200	410	164	3,774	4,350
2100	Per 12 feet section including concrete		4	10		3,825	410	164	4,399	5,075
2210	GFRC decorative security barrier will stop 30 MPH, 4000 lb. vehicle									
2300	High security barrier base prep per 12 feet section on bare ground	B-11C	4	4	Ea.	21.50	176	91	288.50	410
2310	GFRC barrier base prep does not include haul away of excavated matl.									
2400	GFRC decorative high security barrier per 12 feet section w/concrete	B-21B	3	13.333	Ea.	6,025	545	218	6,788	7,750
2410	GFRC decorative high security barrier will stop 50 MPH, 15000 lb. vehicle									
2500	GFRC decorative impaler security barrier per 10 feet section w/prep	B-6	4	6	Ea.	2,300	248	91	2,639	3,025
2600	Per 12 feet section w/prep	″	4	6	″	2,750	248	91	3,089	3,525
2610	Impaler barrier should stop 50 MPH, 15000 lb. vehicle w/some penetr.									
2700	Pipe bollards, steel, concrete filled/painted, 8′ L x 4′ D hole, 8″ diam.	B-6	10	2.400	Ea.	875	99	36.50	1,010.50	1,175
2710	Schedule 80 concrete bollards will stop 4000 lb. vehicle @ 30 MPH									
2800	GFRC decorative jersey barrier cover per 10 feet section excludes soil	B-6	8	3	Ea.	990	124	45.50	1,159.50	1,350
2900	Per 12 feet section excludes soil		8	3		1,175	124	45.50	1,344.50	1,550
3000	GFRC decorative 8″ diameter bollard cover		12	2		515	82.50	30.50	628	735
3100	GFRC decorative barrier face 10 foot section excludes earth backing		10	2.400		1,075	99	36.50	1,210.50	1,400
3200	Drop arm crash barrier, 15000 lb. vehicle @ 50 MPH									
3205	Includes all material, labor for complete installation									
3210	12′ width				Ea.				64,500	71,000
3310	24′ width								86,000	95,000
3410	Wedge crash barrier, 10′ width								98,000	108,000
3510	12.5′ width								108,000	119,000
3520	15′ width								118,500	130,000
3610	Sliding crash barrier, 12′ width								178,500	196,000
3710	Sliding roller crash barrier, 20′ width								198,000	217,500
3810	Sliding cantilever crash barrier, 20′ width								198,000	217,500
3890	Note: Raised bollard crash barriers should be used w/tire shredders									
3910	Raised bollard crash barrier, 10′ width				Ea.				38,800	42,900
4010	12′ width								44,900	49,000
4110	Raised bollard crash barrier, 10′ width, solar powered								46,900	52,000

34 71 Roadway Construction

34 71 13 – Vehicle Barriers

34 71 13.17 Security Vehicle Barriers		Crew	Daily Output	Labor-Hours	Unit	Material	2015 Bare Costs Labor	Equipment	Total	Total Incl O&P
4210	12′ width				Ea.				53,000	58,000
4310	In ground tire shredder, 16′ width								43,900	47,900

34 71 13.26 Vehicle Guide Rails

Line	Description	Crew	Daily Output	Labor-Hours	Unit	Material	Labor	Equipment	Total	Total Incl O&P
0010	**VEHICLE GUIDE RAILS**									
0012	Corrugated stl., galv. stl. posts, 6′-3″ O.C.	B-80	850	.038	L.F.	23	1.55	.85	25.40	28.50
0200	End sections, galvanized, flared		50	.640	Ea.	98.50	26.50	14.40	139.40	166
0300	Wrap around end		50	.640	″	141	26.50	14.40	181.90	213
0400	Timber guide rail, 4″ x 8″ with 6″ x 8″ wood posts, treated		960	.033	L.F.	12.05	1.37	.75	14.17	16.30
0600	Cable guide rail, 3 at 3/4″ cables, steel posts, single face		900	.036		10.75	1.47	.80	13.02	15.10
0700	Wood posts		950	.034		12.95	1.39	.76	15.10	17.35
0900	Guide rail, steel box beam, 6″ x 6″		120	.267		34	11	6	51	62
1100	Median barrier, steel box beam, 6″ x 8″		215	.149		45	6.15	3.35	54.50	63
1400	Resilient guide fence and light shield, 6′ high	B-2	130	.308		38	11.70		49.70	61
1500	Concrete posts, individual, 6′-5″, triangular	B-80	110	.291	Ea.	72.50	12	6.55	91.05	107
1550	Square	″	110	.291	″	78	12	6.55	96.55	113
2000	Median, precast concrete, 3′-6″ high, 2′ wide, single face	B-29	380	.147	L.F.	54.50	6.05	2.32	62.87	72
2200	Double face	″	340	.165	″	62.50	6.75	2.59	71.84	82.50
2400	Speed bumps, thermoplastic, 10-1/2″ x 2-1/4″ x 48″ long	B-2	120	.333	Ea.	138	12.65		150.65	172
3030	Impact barrier, UTMCD, barrel type	B-16	30	1.067	″	510	41.50	23	574.50	660

34 71 19 – Vehicle Delineators

34 71 19.13 Fixed Vehicle Delineators

Line	Description	Crew	Daily Output	Labor-Hours	Unit	Material	Labor	Equipment	Total	Total Incl O&P
0010	**FIXED VEHICLE DELINEATORS**									
0020	Crash barriers									
0100	Traffic channelizing pavement markers, layout only	A-7	2000	.012	Ea.		.62	.03	.65	1.02
0110	13″ x 7-1/2″ x 2-1/2″ high, non-plowable install	2 Clab	96	.167		24.50	6.25		30.75	37
0200	8″ x 8″ x 3-1/4″ high, non-plowable, install		96	.167		22.50	6.25		28.75	35.50
0230	4″ x 4″ x 3/4″ high, non-plowable, install		120	.133		1.79	5		6.79	10.15
0240	9-1/4″ x 5-7/8″ x 1/4″ high, plowable, concrete pavmt.	A-2A	70	.343		18.20	13.05	5.90	37.15	48
0250	9-1/4″ x 5-7/8″ x 1/4″ high, plowable, asphalt pav′t	″	120	.200		3.83	7.60	3.44	14.87	20.50
0300	Barrier and curb delineators, reflectorized, 2″ x 4″	2 Clab	150	.107		2.04	4.01		6.05	8.85
0310	3″ x 5″	″	150	.107		3.98	4.01		7.99	11
0500	Rumble strip, polycarbonate									
0510	24″ x 3-1/2″ x 1/2″ high	2 Clab	50	.320	Ea.	8.40	12.05		20.45	29

34 72 Railway Construction

34 72 16 – Railway Siding

34 72 16.50 Railroad Sidings

Line	Description	Crew	Daily Output	Labor-Hours	Unit	Material	Labor	Equipment	Total	Total Incl O&P
0010	**RAILROAD SIDINGS** R347216-10									
0800	Siding, yard spur, level grade									
0820	100 lb. new rail	B-14	57	.842	L.F.	133	33.50	6.40	172.90	208
1020	100 lb. new rail	″	22	2.182	″	181	87	16.55	284.55	360

34 72 16.60 Railroad Turnouts

Line	Description	Crew	Daily Output	Labor-Hours	Unit	Material	Labor	Equipment	Total	Total Incl O&P
0010	**RAILROAD TURNOUTS**									
2200	Turnout, #8 complete, w/rails, plates, bars, frog, switch point,									
2250	timbers, and ballast to 6″ below bottom of ties									
2280	90 lb. rails	B-13	.25	224	Ea.	40,300	9,175	2,950	52,425	62,500
2290	90 lb. relay rails		.25	224		27,100	9,175	2,950	39,225	48,000
2300	100 lb. rails		.25	224		44,900	9,175	2,950	57,025	67,500
2310	100 lb. relay rails		.25	224		30,200	9,175	2,950	42,325	51,500
2320	110 lb. rails		.25	224		49,500	9,175	2,950	61,625	72,500

34 72 Railway Construction

34 72 16 – Railway Siding

34 72 16.60 Railroad Turnouts		Crew	Daily Output	Labor-Hours	Unit	2015 Bare Costs Material	Labor	Equipment	Total	Total Incl O&P
2330	110 lb. relay rails	B-13	.25	224	Ea.	33,300	9,175	2,950	45,425	55,000
2340	115 lb. rails		.25	224		54,500	9,175	2,950	66,625	77,500
2350	115 lb. relay rails		.25	224		36,100	9,175	2,950	48,225	58,000
2360	132 lb. rails		.25	224		62,000	9,175	2,950	74,125	86,000
2370	132 lb. relay rails	▼	.25	224	▼	41,100	9,175	2,950	53,225	63,500

Division 35 – Waterway & Marine

Estimating Tips

35 01 50 Operation and Maintenance of Marine Construction

Includes unit price lines for pile cleaning and pile wrapping for protection.

35 20 16 Hydraulic Gates

This subdivision includes various types of gates that are commonly used in waterway and canal construction. Various earthwork items and structural support is found in Division 31, and concrete work in Division 3.

35 20 23 Dredging

This subdivision includes barge and shore dredging systems for rivers, canals, and channels.

35 31 00 Shoreline Protection

This subdivision includes breakwaters, bulkheads, and revetments for ocean and river inlets. Additional earthwork may be required from Division 31, and concrete work from Division 3.

35 41 00 Levees

Information on levee construction, including estimated cost of clay cone material.

35 49 00 Waterway Structures

This subdivision includes breakwaters and bulkheads for canals.

35 51 00 Floating Construction

This section includes floating piers, docks, and dock accessories. Fixed Pier Timber Construction is found in 06 13 33. Driven piles are found in Division 31, as well as sheet piling, cofferdams, and riprap.

Reference Numbers

Reference numbers are shown in shaded boxes at the beginning of some major classifications. These numbers refer to related items in the Reference Section. The reference information may be an estimating procedure, an alternate pricing method, or technical information.

Note: Not all subdivisions listed here necessarily appear in this publication. ■

35 20 16.26 Hydraulic Sluice Gates

	35 20 16.26 Hydraulic Sluice Gates	Crew	Daily Output	Labor-Hours	Unit	2015 Bare Costs Material	Labor	Equipment	Total	Total Incl O&P
0010	**HYDRAULIC SLUICE GATES**									
0100	Heavy duty, self contained w/crank oper. gate, 18" x 18"	L-5A	1.70	18.824	Ea.	8,750	995	355	10,100	11,800
0110	24" x 24"		1.20	26.667		12,800	1,400	500	14,700	17,100
0120	30" x 30"		1	32		13,600	1,700	600	15,900	18,500
0130	36" x 36"		.90	35.556		16,400	1,875	665	18,940	22,100
0140	42" x 42"		.80	40		18,900	2,125	750	21,775	25,200
0150	48" x 48"		.50	64		20,800	3,375	1,200	25,375	30,200
0160	54" x 54"		.40	80		28,200	4,225	1,500	33,925	40,200
0170	60" x 60"		.30	106		30,900	5,650	2,000	38,550	46,000
0180	66" x 66"		.30	106		36,200	5,650	2,000	43,850	52,000
0190	72" x 72"		.20	160		40,200	8,475	3,000	51,675	62,500
0200	78" x 78"		.20	160		51,500	8,475	3,000	62,975	74,500
0210	84" x 84"		.10	320		56,000	16,900	6,000	78,900	98,500
0220	90" x 90"	E-20	.30	213		67,000	11,100	3,875	81,975	97,000
0230	96" x 96"		.30	213		71,000	11,100	3,875	85,975	102,000
0240	108" x 108"		.20	320		81,000	16,600	5,825	103,425	124,500
0250	120" x 120"		.10	640		89,500	33,200	11,600	134,300	170,000
0260	132" x 132"		.10	640		110,000	33,200	11,600	154,800	192,500

35 20 16.63 Canal Gates

	35 20 16.63 Canal Gates	Crew	Daily Output	Labor-Hours	Unit	Material	Labor	Equipment	Total	Total Incl O&P
0010	**CANAL GATES**									
0011	Cast iron body, fabricated frame									
0100	12" diameter	L-5A	4.60	6.957	Ea.	1,250	370	130	1,750	2,175
0110	18" diameter		4	8		1,450	425	150	2,025	2,500
0120	24" diameter		3.50	9.143		2,700	485	171	3,356	4,000
0130	30" diameter		2.80	11.429		3,075	605	214	3,894	4,675
0140	36" diameter		2.30	13.913		5,750	735	261	6,746	7,900
0150	42" diameter		1.70	18.824		11,000	995	355	12,350	14,200
0160	48" diameter		1.20	26.667		12,000	1,400	500	13,900	16,200
0170	54" diameter		.90	35.556		12,600	1,875	665	15,140	17,900
0180	60" diameter		.50	64		14,400	3,375	1,200	18,975	23,100
0190	66" diameter		.50	64		18,000	3,375	1,200	22,575	27,100
0200	72" diameter		.40	80		24,000	4,225	1,500	29,725	35,500

35 20 16.66 Flap Gates

	35 20 16.66 Flap Gates	Crew	Daily Output	Labor-Hours	Unit	Material	Labor	Equipment	Total	Total Incl O&P
0010	**FLAP GATES**									
0100	Aluminum, 18" diameter	L-5A	5	6.400	Ea.	2,250	340	120	2,710	3,200
0110	24" diameter		4	8		2,850	425	150	3,425	4,025
0120	30" diameter		3.50	9.143		3,900	485	171	4,556	5,350
0130	36" diameter		2.80	11.429		4,850	605	214	5,669	6,600
0140	42" diameter		2.30	13.913		5,575	735	261	6,571	7,700
0150	48" diameter		1.70	18.824		6,375	995	355	7,725	9,175
0160	54" diameter		1.20	26.667		7,050	1,400	500	8,950	10,800
0170	60" diameter		.80	40		7,850	2,125	750	10,725	13,200
0180	66" diameter		.50	64		8,425	3,375	1,200	13,000	16,500
0190	72" diameter		.40	80		9,925	4,225	1,500	15,650	20,000

35 20 16.69 Knife Gates

	35 20 16.69 Knife Gates	Crew	Daily Output	Labor-Hours	Unit	Material	Labor	Equipment	Total	Total Incl O&P
0010	**KNIFE GATES**									
0100	Incl. handwheel operator for hub, 6" diameter	Q-23	7.70	3.117	Ea.	1,400	177	123	1,700	1,950
0110	8" diameter		7.20	3.333		1,725	189	132	2,046	2,350
0120	10" diameter		4.80	5		2,525	283	198	3,006	3,425
0130	12" diameter		3.60	6.667		3,525	380	264	4,169	4,750
0140	14" diameter		3.40	7.059		4,850	400	279	5,529	6,275
0150	16" diameter		3.20	7.500		6,500	425	297	7,222	8,150

35 20 Waterway and Marine Construction and Equipment

35 20 16 – Hydraulic Gates

35 20 16.69 Knife Gates		Crew	Daily Output	Labor-Hours	Unit	Material	2015 Bare Costs Labor	Equipment	Total	Total Incl O&P
0160	18″ diameter	Q-23	3	8	Ea.	8,850	455	315	9,620	10,800
0170	20″ diameter		2.70	8.889		12,200	505	350	13,055	14,700
0180	24″ diameter		2.40	10		15,800	565	395	16,760	18,700
0190	30″ diameter		1.80	13.333		32,800	755	525	34,080	37,900
0200	36″ diameter	↓	1.20	20	↓	43,200	1,125	790	45,115	50,000

35 20 16.73 Slide Gates		Crew	Daily Output	Labor-Hours	Unit	Material	Labor	Equipment	Total	Total Incl O&P
0010	**SLIDE GATES**									
0100	Steel, self contained incl. anchor bolts and grout, 12″ x 12″	L-5A	4.60	6.957	Ea.	4,950	370	130	5,450	6,250
0110	18″ x 18″		4	8		5,125	425	150	5,700	6,525
0120	24″ x 24″		3.50	9.143		5,825	485	171	6,481	7,450
0130	30″ x 30″		2.80	11.429		6,250	605	214	7,069	8,150
0140	36″ x 36″		2.30	13.913		6,575	735	261	7,571	8,825
0150	42″ x 42″		1.70	18.824		7,225	995	355	8,575	10,100
0160	48″ x 48″		1.20	26.667		7,725	1,400	500	9,625	11,500
0170	54″ x 54″		.90	35.556		8,400	1,875	665	10,940	13,300
0180	60″ x 60″		.55	58.182		8,600	3,075	1,100	12,775	16,100
0190	72″ x 72″	↓	.36	88.889	↓	10,900	4,700	1,675	17,275	22,100

35 20 23 – Dredging

35 20 23.13 Mechanical Dredging		Crew	Daily Output	Labor-Hours	Unit	Material	Labor	Equipment	Total	Total Incl O&P
0010	**MECHANICAL DREDGING**									
0020	Dredging mobilization and demobilization, add to below, minimum	B-8	.53	120	Total		5,150	6,250	11,400	15,200
0100	Maximum	″	.10	640	″		27,300	33,100	60,400	80,500
0300	Barge mounted clamshell excavation into scows									
0310	Dumped 20 miles at sea, minimum	B-57	310	.155	B.C.Y.		6.70	5.40	12.10	16.75
0400	Maximum	″	213	.225	″		9.80	7.90	17.70	24.50
0500	Barge mounted dragline or clamshell, hopper dumped,									
0510	pumped 1000′ to shore dump, minimum	B-57	340	.141	B.C.Y.		6.10	4.94	11.04	15.30
0525	All pumping uses 2000 gallons of water per cubic yard									
0600	Maximum	B-57	243	.198	B.C.Y.		8.55	6.90	15.45	21.50

35 20 23.23 Hydraulic Dredging		Crew	Daily Output	Labor-Hours	Unit	Material	Labor	Equipment	Total	Total Incl O&P
0010	**HYDRAULIC DREDGING**									
1000	Hydraulic method, pumped 1000′ to shore dump, minimum	B-57	460	.104	B.C.Y.		4.53	3.65	8.18	11.30
1100	Maximum		310	.155			6.70	5.40	12.10	16.75
1400	Into scows dumped 20 miles, minimum		425	.113			4.90	3.95	8.85	12.25
1500	Maximum	↓	243	.198			8.55	6.90	15.45	21.50
1600	For inland rivers and canals in South, deduct				↓				30%	30%

35 31 Shoreline Protection

35 31 16 – Seawalls

35 31 16.13 Concrete Seawalls		Crew	Daily Output	Labor-Hours	Unit	Material	Labor	Equipment	Total	Total Incl O&P
0010	**CONCRETE SEAWALLS**									
0011	Reinforced concrete									
0015	include footing and tie-backs									
0020	Up to 6′ high, minimum	C-17C	28	2.964	L.F.	51	146	22	219	315
0060	Maximum		24.25	3.423		81.50	168	25	274.50	390
0100	12′ high, minimum		20	4.150		132	204	30.50	366.50	510
0160	Maximum	↓	18.50	4.486	↓	153	221	33	407	565

35 31 Shoreline Protection

35 31 16 – Seawalls

35 31 16.19 Steel Sheet Piling Seawalls		Crew	Daily Output	Labor-Hours	Unit	2015 Bare Costs Material	Labor	Equipment	Total	Total Incl O&P
0010	**STEEL SHEET PILING SEAWALLS**									
0200	Steel sheeting, with 4′ x 4′ x 8″ concrete deadmen, @ 10′ O.C.									
0210	12′ high, shore driven	B-40	27	2.370	L.F.	107	113	139	359	450
0260	Barge driven	B-76	15	4.800	"	160	228	221	609	790

35 31 19 – Revetments

35 31 19.18 Revetments, Concrete		Crew	Daily Output	Labor-Hours	Unit	Material	Labor	Equipment	Total	Total Incl O&P
0010	**REVETMENTS, CONCRETE**									
0100	Concrete revetment matt 8′ x 20′ x 4 1/2″ excluding site preparation									
0110	Includes all labor, material and equip. for complete installation				Ea.	3,350			3,350	3,700

35 51 Floating Construction

35 51 13 – Floating Piers

35 51 13.23 Floating Wood Piers		Crew	Daily Output	Labor-Hours	Unit	Material	Labor	Equipment	Total	Total Incl O&P
0010	**FLOATING WOOD PIERS**									
0020	Polyethylene encased polystyrene, no pilings included	F-3	330	.121	S.F.	29	5.80	1.98	36.78	43.50
0200	Pile supported, shore constructed, bare, 3″ decking		130	.308		28	14.75	5.05	47.80	60
0250	4″ decking		120	.333		29	15.95	5.45	50.40	64
0400	Floating, small boat, prefab, no shore facilities, minimum		250	.160		24.50	7.65	2.62	34.77	42.50
0500	Maximum		150	.267	↓	52	12.75	4.36	69.11	83
0700	Per slip, minimum (180 S.F. each)		1.59	25.157	Ea.	5,000	1,200	410	6,610	7,900
0800	Maximum	↓	1.40	28.571	"	8,400	1,375	465	10,240	12,000

Estimating Tips

Products such as conveyors, material handling cranes and hoists, as well as other items specified in this division, require trained installers. The general contractor may not have any choice as to who will perform the installation or when it will be performed. Long lead times are often required for these products, making early decisions in purchasing and scheduling necessary. The installation of this type of equipment may require the embedment of mounting hardware during construction of floors, structural walls, or interior walls/partitions. Electrical connections will require coordination with the electrical contractor.

Reference Numbers

Reference numbers are shown in shaded boxes at the beginning of some major classifications. These numbers refer to related items in the Reference Section. The reference information may be an estimating procedure, an alternate pricing method, or technical information.

Note: Not all subdivisions listed here necessarily appear in this publication. ■

41 21 Conveyors

41 21 23 – Piece Material Conveyors

41 21 23.16 Container Piece Material Conveyors		Crew	Daily Output	Labor-Hours	Unit	2015 Bare Costs Material	Labor	Equipment	Total	Total Incl O&P
0010	CONTAINER PIECE MATERIAL CONVEYORS									
0020	Gravity fed, 2" rollers, 3" O.C.									
0050	10' sections with 2 supports, 600 lb. capacity, 18" wide				Ea.	480			480	530
0100	24" wide					545			545	595
0150	1400 lb. capacity, 18" wide					640			640	705
0200	24" wide					545			545	595
0350	Horizontal belt, center drive and takeup, 60 fpm									
0400	16" belt, 26.5' length	2 Mill	.50	32	Ea.	3,350	1,575		4,925	6,175
0450	24" belt, 41.5' length		.40	40		5,100	1,975		7,075	8,725
0500	61.5' length		.30	53.333		7,200	2,625		9,825	12,100
0600	Inclined belt, 10' rise with horizontal loader and									
0620	End idler assembly, 27.5' length, 18" belt	2 Mill	.30	53.333	Ea.	7,300	2,625		9,925	12,200
0700	24" belt	"	.15	106	"	8,925	5,250		14,175	18,100
3600	Monorail, overhead, manual, channel type									
3700	125 lb. per L.F.	1 Mill	26	.308	L.F.	19.40	15.10		34.50	45.50
3900	500 lb. per L.F.	"	21	.381	"	20.50	18.70		39.20	52
4000	Trolleys for above, 2 wheel, 125 lb. capacity				Ea.	81.50			81.50	89.50
4200	4 wheel, 250 lb. capacity					330			330	365
4300	8 wheel, 500 lb. capacity					770			770	845

41 22 Cranes and Hoists

41 22 13 – Cranes

41 22 13.10 Crane Rail

		Crew	Daily Output	Labor-Hours	Unit	Material	Labor	Equipment	Total	Total Incl O&P
0010	CRANE RAIL									
0020	Box beam bridge, no equipment included	E-4	3400	.009	Lb.	1.33	.50	.04	1.87	2.41
0210	Running track only, 104 lb. per yard, 20' piece	"	160	.200	L.F.	23	10.65	.91	34.56	45.50

41 22 13.13 Bridge Cranes

		Crew	Daily Output	Labor-Hours	Unit	Material	Labor	Equipment	Total	Total Incl O&P
0010	BRIDGE CRANES									
0100	1 girder, 20' span, 3 ton	M-3	1	34	Ea.	25,400	1,950	139	27,489	31,200
0125	5 ton		1	34		27,900	1,950	139	29,989	33,900
0150	7.5 ton		1	34		33,100	1,950	139	35,189	39,600
0175	10 ton		.80	42.500		43,900	2,450	173	46,523	52,500
0200	15 ton		.80	42.500		56,500	2,450	173	59,123	66,000
0225	30' span, 3 ton		1	34		26,500	1,950	139	28,589	32,300
0250	5 ton		1	34		29,100	1,950	139	31,189	35,200
0275	7.5 ton		1	34		34,800	1,950	139	36,889	41,500
0300	10 ton		.80	42.500		45,400	2,450	173	48,023	54,000
0325	15 ton		.80	42.500		59,000	2,450	173	61,623	69,000
0350	2 girder, 40' span, 3 ton	M-4	.50	72		43,600	4,100	350	48,050	54,500
0375	5 ton		.50	72		45,600	4,100	350	50,050	57,000
0400	7.5 ton		.50	72		50,000	4,100	350	54,450	62,000
0425	10 ton		.40	90		58,500	5,125	440	64,065	73,000
0450	15 ton		.40	90		80,000	5,125	440	85,565	96,000
0475	25 ton		.30	120		94,000	6,825	585	101,410	115,000
0500	50' span, 3 ton		.50	72		49,700	4,100	350	54,150	61,500
0525	5 ton		.50	72		51,500	4,100	350	55,950	64,000
0550	7.5 ton		.50	72		55,500	4,100	350	59,950	68,000
0575	10 ton		.40	90		64,000	5,125	440	69,565	78,500
0600	15 ton		.40	90		83,500	5,125	440	89,065	100,500
0625	25 ton		.30	120		98,500	6,825	585	105,910	120,000

41 22 Cranes and Hoists

41 22 13 – Cranes

41 22 13.19 Jib Cranes		Crew	Daily Output	Labor-Hours	Unit	Material	2015 Bare Costs Labor	Equipment	Total	Total Incl O&P
0010	**JIB CRANES**									
0020	Jib crane, wall cantilever, 500 lb. capacity, 8′ span	2 Mill	1	16	Ea.	1,400	785		2,185	2,750
0040	12′ span		1	16		1,525	785		2,310	2,900
0060	16′ span		1	16		1,700	785		2,485	3,100
0080	20′ span		1	16		2,150	785		2,935	3,600
0100	1000 lb. capacity, 8′ span		1	16		1,500	785		2,285	2,875
0120	12′ span		1	16		1,625	785		2,410	3,025
0130	16′ span		1	16		2,150	785		2,935	3,600
0150	20′ span		1	16		2,525	785		3,310	4,000

41 22 23 – Hoists

41 22 23.10 Material Handling		Crew	Daily Output	Labor-Hours	Unit	Material	2015 Bare Costs Labor	Equipment	Total	Total Incl O&P
0010	**MATERIAL HANDLING**, cranes, hoists and lifts									
1500	Cranes, portable hydraulic, floor type, 2,000 lb. capacity				Ea.	3,500			3,500	3,875
1600	4,000 lb. capacity					4,025			4,025	4,425
1800	Movable gantry type, 12′ to 15′ range, 2,000 lb. capacity					3,950			3,950	4,325
1900	6,000 lb. capacity					5,975			5,975	6,575
2100	Hoists, electric overhead, chain, hook hung, 15′ lift, 1 ton cap.					2,500			2,500	2,750
2200	3 ton capacity					3,425			3,425	3,775
2500	5 ton capacity					6,925			6,925	7,625
2600	For hand-pushed trolley, add					15%				
2700	For geared trolley, add					30%				
2800	For motor trolley, add					75%				
3000	For lifts over 15′, 1 ton, add				L.F.	24.50			24.50	27
3100	5 ton, add				"	56.50			56.50	62
3110	Air powered hoist, 500 lb. capacity	2 Mill	2.67	5.993	Ea.	4,950	295		5,245	5,925
3120	1000 lb. capacity		2.46	6.504		5,225	320		5,545	6,250
3130	2000 lb. capacity		2.29	6.987		7,100	345		7,445	8,350
3140	4000 lb. capacity		2	8		9,300	395		9,695	10,800
3160	6000 lb. capacity		1.88	8.511		11,500	420		11,920	13,300
3180	8000 lb. capacity		1.68	9.524		11,600	470		12,070	13,500
3190	10000 lb. capacity		1.60	10		12,200	490		12,690	14,200
3300	Lifts, scissor type, portable, electric, 36″ high, 2,000 lb.					3,475			3,475	3,825
3400	48″ high, 4,000 lb.					4,175			4,175	4,600

Division Notes

	CREW	DAILY OUTPUT	LABOR-HOURS	UNIT	BARE COSTS MAT.	LABOR	EQUIP.	TOTAL	TOTAL INCL O&P

Estimating Tips

This section involves equipment and construction costs for air noise and odor pollution control systems. These systems may be interrelated and care must be taken that the complete systems are estimated. For example, air pollution equipment may include dust and air-entrained particles that have to be collected. The vacuum systems could be noisy, requiring silencers to reduce noise pollution, and the collected solids have to be disposed of to prevent solid pollution.

Reference Numbers

Reference numbers are shown in shaded boxes at the beginning of some major classifications. These numbers refer to related items in the Reference Section. The reference information may be an estimating procedure, an alternate pricing method, or technical information.

Note: Not all subdivisions listed here necessarily appear in this publication. ■

44 11 Particulate Control Equipment

44 11 16 – Fugitive Dust Barrier Systems

44 11 16.10 Dust Collection Systems		Crew	Daily Output	Labor-Hours	Unit	2015 Bare Costs Material	Labor	Equipment	Total	Total Incl O&P
0010	**DUST COLLECTION SYSTEMS** Commercial/Industrial									
0120	Central vacuum units									
0130	Includes stand, filters and motorized shaker									
0200	500 CFM, 10" inlet, 2 HP	Q-20	2.40	8.333	Ea.	4,175	425		4,600	5,275
0220	1000 CFM, 10" inlet, 3 HP		2.20	9.091		4,400	465		4,865	5,550
0240	1500 CFM, 10" inlet, 5 HP		2	10		4,625	510		5,135	5,900
0260	3000 CFM, 13" inlet, 10 HP		1.50	13.333		12,900	685		13,585	15,300
0280	5000 CFM, 16" inlet, 2 @ 10 HP	↓	1	20	↓	13,700	1,025		14,725	16,700
1000	Vacuum tubing, galvanized									
1100	2-1/8" OD, 16 ga.	Q-9	440	.036	L.F.	2.95	1.83		4.78	6.15
1110	2-1/2" OD, 16 ga.		420	.038		3.24	1.92		5.16	6.60
1120	3" OD, 16 ga.		400	.040		4.13	2.01		6.14	7.70
1130	3-1/2" OD, 16 ga.		380	.042		5.85	2.12		7.97	9.80
1140	4" OD, 16 ga.		360	.044		6	2.24		8.24	10.15
1150	5" OD, 14 ga.		320	.050		10.80	2.52		13.32	15.80
1160	6" OD, 14 ga.		280	.057		12.10	2.88		14.98	17.85
1170	8" OD, 14 ga.		200	.080		18.30	4.03		22.33	26.50
1180	10" OD, 12 ga.		160	.100		35.50	5.05		40.55	47
1190	12" OD, 12 ga.		120	.133		47	6.70		53.70	62
1200	14" OD, 12 ga.	↓	80	.200	↓	53.50	10.05		63.55	74.50
1940	Hose, flexible wire reinforced rubber									
1956	3" diam.	Q-9	400	.040	L.F.	9.15	2.01		11.16	13.25
1960	4" diam.		360	.044		10.60	2.24		12.84	15.20
1970	5" diam.		320	.050		13.75	2.52		16.27	19.10
1980	6" diam.	↓	280	.057	↓	14.40	2.88		17.28	20.50
2000	90° Elbow, slip fit									
2110	2-1/8" diam.	Q-9	70	.229	Ea.	10.85	11.50		22.35	30
2120	2-1/2" diam.		65	.246		15.20	12.40		27.60	36.50
2130	3" diam.		60	.267		20.50	13.45		33.95	43.50
2140	3-1/2" diam.		55	.291		25.50	14.65		40.15	51
2150	4" diam.		50	.320		31.50	16.10		47.60	60
2160	5" diam.		45	.356		60	17.90		77.90	94.50
2170	6" diam.		40	.400		82.50	20		102.50	123
2180	8" diam.	↓	30	.533	↓	157	27		184	215
2400	45° Elbow, slip fit									
2410	2-1/8" diam.	Q-9	70	.229	Ea.	9.50	11.50		21	28.50
2420	2-1/2" diam.		65	.246		14	12.40		26.40	35
2430	3" diam.		60	.267		16.90	13.45		30.35	39.50
2440	3-1/2" diam.		55	.291		21.50	14.65		36.15	46.50
2450	4" diam.		50	.320		27.50	16.10		43.60	56
2460	5" diam.		45	.356		46.50	17.90		64.40	79.50
2470	6" diam.		40	.400		63	20		83	101
2480	8" diam.	↓	35	.457	↓	123	23		146	173
2800	90° TY, slip fit thru 6" diam.									
2810	2-1/8" diam.	Q-9	42	.381	Ea.	17.75	19.20		36.95	50
2820	2-1/2" diam.		39	.410		26.50	20.50		47	61.50
2830	3" diam.		36	.444		37	22.50		59.50	76
2840	3-1/2" diam.		33	.485		39	24.50		63.50	81
2850	4" diam.		30	.533		41.50	27		68.50	88.50
2860	5" diam.		27	.593		70.50	30		100.50	125
2870	6" diam.	↓	24	.667		101	33.50		134.50	164
2880	8" diam., butt end					155			155	170
2890	10" diam., butt end				↓	475			475	525

44 11 16.10 Dust Collection Systems		Crew	Daily Output	Labor-Hours	Unit	Material	Labor	Equipment	Total	Total Incl O&P
						2015 Bare Costs				
2900	12″ diam., butt end				Ea.	475			475	525
2910	14″ diam., butt end					855			855	940
2920	6″ x 4″ diam., butt end					106			106	116
2930	8″ x 4″ diam., butt end					200			200	220
2940	10″ x 4″ diam., butt end					263			263	289
2950	12″ x 4″ diam., butt end					300			300	330
3100	90° Elbow, butt end segmented									
3110	8″ diam., butt end, segmented				Ea.	157			157	172
3120	10″ diam., butt end, segmented					390			390	430
3130	12″ diam., butt end, segmented					495			495	545
3140	14″ diam., butt end, segmented					620			620	680
3200	45° Elbow, butt end segmented									
3210	8″ diam., butt end, segmented				Ea.	123			123	136
3220	10″ diam., butt end, segmented					282			282	310
3230	12″ diam., butt end, segmented					288			288	315
3240	14″ diam., butt end, segmented					585			585	645
3400	All butt end fittings require one coupling per joint.									
3410	Labor for fitting included with couplings.									
3460	Compression coupling, galvanized, neoprene gasket									
3470	2-1/8″ diam.	Q-9	44	.364	Ea.	10.85	18.30		29.15	41
3480	2-1/2″ diam.		44	.364		10.85	18.30		29.15	41
3490	3″ diam.		38	.421		15.60	21		36.60	50.50
3500	3-1/2″ diam.		35	.457		17.45	23		40.45	55.50
3510	4″ diam.		33	.485		18.95	24.50		43.45	59.50
3520	5″ diam.		29	.552		21.50	28		49.50	68
3530	6″ diam.		26	.615		25.50	31		56.50	77
3540	8″ diam.		22	.727		46	36.50		82.50	109
3550	10″ diam.		20	.800		67.50	40.50		108	138
3560	12″ diam.		18	.889		84	45		129	163
3570	14″ diam.		16	1		119	50.50		169.50	211
3800	Air gate valves, galvanized									
3810	2-1/8″ diam.	Q-9	30	.533	Ea.	111	27		138	165
3820	2-1/2″ diam.		28	.571		119	29		148	177
3830	3″ diam.		26	.615		127	31		158	189
3840	4″ diam.		23	.696		147	35		182	218
3850	6″ diam.		18	.889		205	45		250	296

Division Notes

		CREW	DAILY OUTPUT	LABOR-HOURS	UNIT	BARE COSTS MAT.	LABOR	EQUIP.	TOTAL	TOTAL INCL O&P

Division 46 – Water & Wastewater Equipment

Estimating Tips

This division contains information about water and wastewater equipment and systems, which was formerly located in Division 44. The main areas of focus are total wastewater treatment plants and components of wastewater treatment plants. In addition, there are assemblies such as sewage treatment lagoons that can be found in some publications under G30 Site Mechanical Utilities. Also included in this section are oil/water separators for wastewater treatment.

Reference Numbers

Reference numbers are shown in shaded boxes at the beginning of some major classifications. These numbers refer to related items in the Reference Section. The reference information may be an estimating procedure, an alternate pricing method, or technical information.

Note: Not all subdivisions listed here necessarily appear in this publication. ■

46 07 Packaged Water and Wastewater Treatment Equipment

46 07 53 – Packaged Wastewater Treatment Equipment

46 07 53.10 Biological Pkg. Wastewater Treatment Plants		Crew	Daily Output	Labor-Hours	Unit	2015 Bare Costs Material	Labor	Equipment	Total	Total Incl O&P
0010	**BIOLOGICAL PACKAGED WASTEWATER TREATMENT PLANTS**									
0011	Not including fencing or external piping									
0020	Steel packaged, blown air aeration plants									
0100	1,000 GPD				Gal.				55	60.50
0200	5,000 GPD								22	24
0300	15,000 GPD								22	24
0400	30,000 GPD								15.40	16.95
0500	50,000 GPD								11	12.10
0600	100,000 GPD								9.90	10.90
0700	200,000 GPD								8.80	9.70
0800	500,000 GPD								7.70	8.45
1000	Concrete, extended aeration, primary and secondary treatment									
1010	10,000 GPD				Gal.				22	24
1100	30,000 GPD								15.40	16.95
1200	50,000 GPD								11	12.10
1400	100,000 GPD								9.90	10.90
1500	500,000 GPD								7.70	8.45
1700	Municipal wastewater treatment facility									
1720	1.0 MGD				Gal.				11	12.10
1740	1.5 MGD								10.60	11.65
1760	2.0 MGD								10	11
1780	3.0 MGD								7.80	8.60
1800	5.0 MGD								5.80	6.70
2000	Holding tank system, not incl. excavation or backfill									
2010	Recirculating chemical water closet	2 Plum	4	4	Ea.	530	235		765	950
2100	For voltage converter, add	"	16	1		300	58.50		358.50	420
2200	For high level alarm, add	1 Plum	7.80	1.026		117	60		177	222

46 07 53.20 Wastewater Treatment System		Crew	Daily Output	Labor-Hours	Unit	Material	Labor	Equipment	Total	Total Incl O&P
0010	**WASTEWATER TREATMENT SYSTEM**									
0020	Fiberglass, 1,000 gallon	B-21	1.29	21.705	Ea.	4,250	940	107	5,297	6,325
0100	1,500 gallon	"	1.03	27.184	"	8,475	1,175	134	9,784	11,400

46 23 Grit Removal And Handling Equipment

46 23 23 – Vortex Grit Removal Equipment

46 23 23.10 Rainwater Filters		Crew	Daily Output	Labor-Hours	Unit	Material	Labor	Equipment	Total	Total Incl O&P
0010	**RAINWATER FILTERS**									
0100	42 gal./min	B-21	3.50	8	Ea.	45,100	345	39.50	45,484.50	50,000
0200	65 gal./min		3.50	8		45,200	345	39.50	45,584.50	50,500
0300	208 gal./min		3.50	8		45,200	345	39.50	45,584.50	50,500

46 25 Oil and Grease Separation and Removal Equipment

46 25 13 – Coalescing Oil-Water Separators

46 25 13.20 Oil/Water Separators		Crew	Daily Output	Labor-Hours	Unit	2015 Bare Costs Material	Labor	Equipment	Total	Total Incl O&P
0010	**OIL/WATER SEPARATORS**									
0020	Underground, tank only									
0030	Excludes excavation, backfill, & piping									
0100	200 GPM	B-21	3.50	8	Ea.	44,900	345	39.50	45,284.50	50,000
0110	400 GPM		3.25	8.615		62,500	375	42.50	62,917.50	69,500
0120	600 GPM		2.75	10.182		70,000	440	50	70,490	78,000
0130	800 GPM		2.50	11.200		79,000	485	55	79,540	88,000
0140	1000 GPM		2	14		88,500	605	69	89,174	98,500
0150	1200 GPM		1.50	18.667		101,000	810	92	101,902	112,500
0160	1500 GPM	↓	1	28	↓	119,500	1,225	138	120,863	133,000

Division Notes

		CREW	DAILY OUTPUT	LABOR-HOURS	UNIT	BARE COSTS MAT.	LABOR	EQUIP.	TOTAL	TOTAL INCL O&P

Division 48 – Electrical Power Generation

Estimating Tips

- When estimating costs for the installation of electrical power generation equipment, factors to review include access to the job site, access and setting at the installation site, required connections, uncrating pads, anchors, leveling, final assembly of the components, and temporary protection from physical damage, including from exposure to the environment.
- Be aware of the cost of equipment supports, concrete pads, and vibration isolators, and cross-reference to other trades' specifications. Also, review site and structural drawings for items that must be included in the estimates.
- It is important to include items that are not documented in the plans and specifications but must be priced. These items include, but are not limited to, testing, dust protection, roof penetration, core drilling concrete floors and walls, patching, cleanup, and final adjustments. Add a contingency or allowance for utility company fees for power hookups, if needed.
- The project size and scope of electrical power generation equipment will have a significant impact on cost. The intent of RSMeans cost data is to provide a benchmark cost so that owners, engineers, and electrical contractors will have a comfortable number with which to start a project. Additionally, there are many websites available to use for research and to obtain a vendor's quote to finalize costs.

Reference Numbers

Reference numbers are shown in shaded boxes at the beginning of some major classifications. These numbers refer to related items in the Reference Section. The reference information may be an estimating procedure, an alternate pricing method, or technical information.

Note: Not all subdivisions listed here necessarily appear in this publication. ■

48 15 Wind Energy Electrical Power Generation Equipment

48 15 13 – Wind Turbines

48 15 13.50 Wind Turbines and Components		Crew	Daily Output	Labor-Hours	Unit	2015 Bare Costs Material	Labor	Equipment	Total	Total Incl O&P
0010	**WIND TURBINES & COMPONENTS**									
0500	Complete system, grid connected									
1000	20 kW, 31' dia, incl. labor & material G				System				49,900	49,900
1010	Enhanced G								92,000	92,000
1500	10 kW, 23' dia, incl. labor & material G								74,000	74,000
2000	2.4 kW, 12' dia, incl. labor & material G				↓				18,000	18,000
2900	Component system									
3000	Turbine, 400 watt, 3' dia G	1 Elec	3.41	2.346	Ea.	650	128		778	915
3100	600 watt, 3' dia G		2.56	3.125		880	171		1,051	1,225
3200	1000 watt, 9' dia G	↓	2.05	3.902	↓	3,650	213		3,863	4,350
3400	Mounting hardware									
3500	30' guyed tower kit G	2 Clab	5.12	3.125	Ea.	415	118		533	655
3505	3' galvanized helical earth screw G	1 Clab	8	1		49.50	37.50		87	116
3510	Attic mount kit G	1 Rofc	2.56	3.125		154	125		279	395
3520	Roof mount kit G	1 Clab	3.41	2.346	↓	136	88		224	295
8900	Equipment									
9100	DC to AC inverter for, 48 V, 4,000 watt G	1 Elec	2	4	Ea.	2,325	219		2,544	2,925

Assemblies Section

Table of Contents

Table of Contents

How RSMeans Assemblies Data Works

Assemblies estimating provides a fast and reasonably accurate way to develop construction costs. An assembly is the grouping of individual work items, with appropriate quantities, to provide a cost for a major construction component in a convenient unit of measure.

An assemblies estimate is often used during early stages of design development to compare the cost impact of various design alternatives on total building cost.

Assemblies estimates are also used as an efficient tool to verify construction estimates.

Assemblies estimates do not require a completed design or detailed drawings. Instead, they are based on the general size of the structure and other known parameters of the project. The degree of accuracy of an assemblies estimate is generally within +/- 15%.

NARRATIVE FORMAT

C20 Stairs

C2010 Stair Construction

General Design: See reference section for code requirements. Maximum height between landings is 12'; usual stair angle is 20° to 50° with 30° to 35° best. Usual relation of riser to treads is:
Riser + tread = 17.5.
2x (Riser) + tread = 25.
Riser x tread = 70 or 75.
Maximum riser height is 7" for commercial, 8-1/4" for residential.
Usual riser height is 6-1/2" to 7-1/4".

Minimum tread width is 11" for commercial and 9" for residential.

For additional information please see reference section.

Cost Per Flight: Table below lists the cost per flight for 4'-0" wide stairs.
Side walls are not included.
Railings are included.

System Components	QUANTITY	UNIT	COST PER FLIGHT MAT.	INST.	TOTAL
SYSTEM C2010 110 0560					
STAIRS, C.I.P. CONCRETE WITH LANDING, 12 RISERS					
Concrete in place, free standing stairs not incl. safety treads	48.000	L.F.	297.60	1,988.16	2,285.76
Concrete in place, free standing stair landing	32.000	S.F.	159.04	551.04	710.08
Cast alum nosing insert, abr surface, pre-drilled, 3" wide x 4' long	12.000	Ea.	924	206.40	1,130.40
Industrial railing, welded, 2 rail 3'-6" high 1-1/2" pipe	18.000	L.F.	540	220.14	760.14
Wall railing with returns, steel pipe	17.000	L.F.	289.85	207.91	497.76
TOTAL			2,210.49	3,173.65	5,384.14

C2010 110	Stairs	COST PER FLIGHT MAT.	INST.	TOTAL
0470	Stairs, C.I.P. concrete, w/o landing, 12 risers, w/o nosing (RC2010-100)	1,125	2,425	3,550
0480	With nosing	2,050	2,625	4,675
0550	W/landing, 12 risers, w/o nosing	1,275	2,975	4,250
0560	With nosing	2,200	3,175	5,375
0570	16 risers, w/o nosing	1,575	3,725	5,300
0580	With nosing	2,800	4,000	6,800
0590	20 risers, w/o nosing	1,850	4,500	6,350
0600	With nosing	3,400	4,825	8,225
0610	24 risers, w/o nosing	2,150	5,250	7,400
0620	With nosing	4,000	5,650	9,650
0630	Steel, grate type w/nosing & rails, 12 risers, w/o landing	5,700	1,250	6,950
0640	With landing	7,975	1,725	9,700
0660	16 risers, with landing	9,875	2,125	12,000
0680	20 risers, with landing	11,800	2,550	14,350
0700	24 risers, with landing	13,700	2,950	16,650
0710	Concrete fill metal pan & picket rail, 12 risers, w/o landing	7,375	1,250	8,625
0720	With landing	9,725	1,875	11,600
0740	16 risers, with landing	12,200	2,300	14,500
0760	20 risers, with landing	14,600	2,700	17,300
0780	24 risers, with landing	17,100	3,100	20,200
0790	Cast iron tread & pipe rail, 12 risers, w/o landing	7,375	1,250	8,625
0800	With landing	9,725	1,875	11,600
0820	16 risers, with landing	12,200	2,300	14,500
0840	20 risers, with landing	14,600	2,700	17,300

RSMeans assemblies are identified by a **unique 12-character identifier**. The assemblies are numbered using UNIFORMAT II, ASTM Standard E1557. The first 6 characters represent this system to Level 4. The last 6 characters represent further breakdown by RSMeans in order to arrange items in understandable groups of similar tasks. Line numbers are consistent across all RSMeans publications, so a line number in any RSMeans assemblies data set will always refer to the same work.

Information is available in the Reference Section to assist the estimator with estimating procedures, alternate pricing methods, and additional technical information.

The **Reference Box** indicates the exact location of this information in the Reference Section at the back of the book. The "R" stands for "reference," and the remaining characters are the RSMeans line numbers.

RSMeans assemblies descriptions appear in two formats: narrative and table. **Narrative descriptions** are shown in a hierarchical structure to make them readable. In order to read a complete description, read up through the indents to the top of the section. Include everything that is above and to the left that is not contradicted by information below.

For supplemental customizable square foot estimating forms, visit: **http://www.reedconstructiondata.com/rsmeans/assemblies**

Most assemblies consist of three major elements: a graphic, the system components, and the cost data itself. The **Graphic** is a visual representation showing the typical appearance of the assembly in question, frequently accompanied by additional explanatory technical information describing the class of items. The **System Components** is a listing of the individual tasks that make up the assembly, including the quantity and unit of measure for each item, along with the cost of material and installation. The **Assemblies Data** below lists prices for other similar systems with dimensional and/or size variations.

All RSMeans assemblies costs represent the cost for the installing contractor. An allowance for profit has been added to all material, labor, and equipment rental costs. A markup for labor burdens, including workers' compensation, fixed overhead, and business overhead, is included with installation costs.

The data published in RSMeans print books represents a "national average" cost. This data should be modified to the project location using the **City Cost Indexes** or **Location Factors** tables found in the Reference Section in the back of the book.

TABLE FORMAT

A20 Basement Construction

A2020 Basement Walls

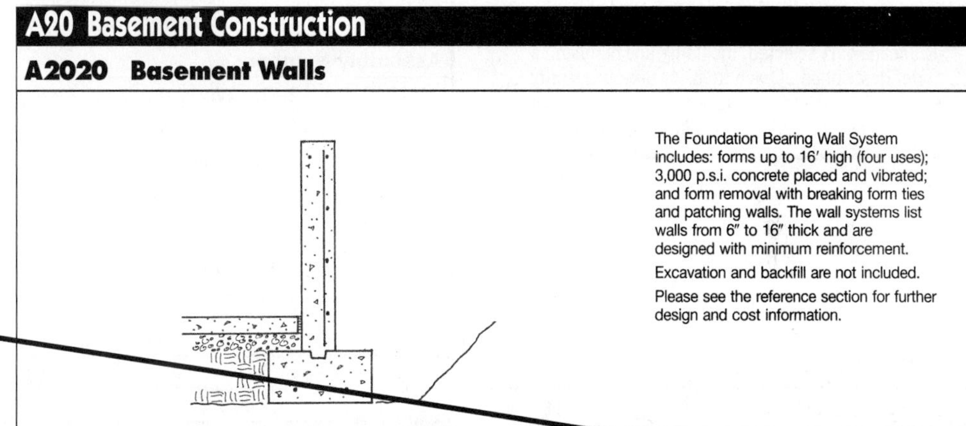

The Foundation Bearing Wall System includes: forms up to 16′ high (four uses); 3,000 p.s.i. concrete placed and vibrated; and form removal with breaking form ties and patching walls. The wall systems list walls from 6″ to 16″ thick and are designed with minimum reinforcement.

Excavation and backfill are not included.

Please see the reference section for further design and cost information.

All RSMeans assemblies data includes a typical **Unit of Measure** used for estimating that item. For instance, for continuous footings or foundation walls the unit is linear feet (L.F.). For spread footings the unit is each (Ea.). The estimator needs to take special care that the unit in the data matches the unit in the takeoff. Abbreviations and unit conversions can be found in the Reference Section.

System Components	QUANTITY	UNIT	COST PER L.F. MAT.	INST.	TOTAL
SYSTEM A2020 110 1500					
FOUNDATION WALL, CAST IN PLACE, DIRECT CHUTE, 4′ HIGH, 6″ THICK					
Formwork	8.000	SFCA	6.56	45.20	51.76
Reinforcing	3.300	Lb.	1.77	1.44	3.21
Unloading & sorting reinforcing	3.300	Lb.		.08	.08
Concrete, 3,000 psi	.074	C.Y.	8.29		8.29
Place concrete, direct chute	.074	C.Y.		2.43	2.43
Finish walls, break ties and patch voids, one side	4.000	S.F.	.16	3.96	4.12
TOTAL			16.78	53.11	69.89

System components are listed separately to detail what is included in the development of the total system price.

Table descriptions work in a similar fashion, except that if there is a blank in the column at a particular line number, read up to the description above in the same column.

A2020 110	Walls, Cast in Place							
	WALL HEIGHT (FT.)	PLACING METHOD	CONCRETE (C.Y. per L.F.)	REINFORCING (LBS. per L.F.)	WALL THICKNESS (IN.)	COST PER L.F. MAT.	INST.	TOTAL
1500	4′	direct chute	.074	3.3	6	16.80	53	69.80
1520			.099	4.8	8	20.50	54.50	75
1540			.123	6.0	10	23.50	55.50	79
1560	RA2020 -210		.148	7.2	12	27	56.50	83.50
1580			.173	8.1	14	30.50	57.50	88
1600			.197	9.44	16	34	59	93
1700	4′	pumped	.074	3.3	6	16.80	54	70.80
1720			.099	4.8	8	20.50	56.50	77
1740			.123	6.0	10	23.50	57.50	81
1760			.148	7.2	12	27	59	86
1780			.173	8.1	14	30.50	60	90.50
1800			.197	9.44	16	34	61.50	95.50
3000	6′	direct chute	.111	4.95	6	25	79.50	104.50
3020			.149	7.20	8	30.50	82	112.50
3040			.184	9.00	10	35.50	83	118.50
3060			.222	10.8	12	41	85	126
3080			.260	12.15	14	45.50	86.50	132
3100			.300	14.39	16	51.50	89	140.50

How RSMeans Assemblies Data Works (Continued)

Sample Estimate

This sample demonstrates the elements of an estimate, including a tally of the RSMeans data lines. Published assemblies costs include all markups for labor burden and profit for the installing contractor. This estimate adds a summary of the markups applied by a general contractor on the installing contractors' work. These figures represent the total cost to the owner. The RSMeans location factor is added at the bottom of the estimate to adjust the cost of the work to a specific location

Work performed: The body of the estimate shows the RSMeans data selected, including line numbers, a brief description of each item, its takeoff quantity and unit, and the total installed cost, including the installing contractor's overhead and profit.

Location Factor: RSMeans published data is based on national average costs. If necessary, adjust the total cost of the project using a location factor from the "Location Factor" table or the "City Cost Indexes" table, found in the Reference Section. Use location factors if the work is general, covering the work of multiple trades. If the work is by a single trade (e.g., masonry) use the more specific data found in the City Cost Indexes.

To adjust costs by location factors, multiply the base cost by the factor and divide by 100.

Contingency: A factor for contingency may be added to any estimate to represent the cost of unknowns that may occur between the time that the estimate is performed and the time the project is constructed. The amount of the allowance will depend on the stage of design at which the estimate is done, and the contractor's assessment of the risk invloved.

Project Name:	**Interior Fit-out, ABC Office**
Location:	**Anywhere, USA**
Assembly Number	**Description**
C1010 124 1200	Wood partition, 2 x 4 @ 16" OC w/5/8" FR gypsum board
C1020 114 1800	Metal door & frame, flush hollow core, 3'-0" x 7'-0"
C3010 230 0080	Painting, brushwork, primer & 2 coats
C3020 410 0140	Carpet, tufted, nylon, roll goods, 12' wide, 26 oz
C3030 210 6000	Acoustic ceilings, 24" x 48" tile, tee grid suspension
D5020 125 0560	Receptacles incl plate, box, conduit, wire, 20 A duplex
D5020 125 0720	Light switch incl plate, box, conduit, wire, 20 A single pole
D5020 210 0560	Fluorescent fixtures, recess mounted, 20 per 1000 SF
	Assembly Subtotal
	Sales Tax @
	General Requirements @
	Subtotal A
	GC Overhead @
	Subtotal B
	GC Profit @
	Subtotal C
	Adjusted by Location Factor
	Architects Fee @
	Contingency @
	Project Total Cost

This estimate is based on an interactive spreadsheet. A copy of this spreadsheet is located on the RSMeans website at **http:/www.reedconstructiondata.com/rsmeans/extras/546011**. You are free to download it and adjust it to your methodology.

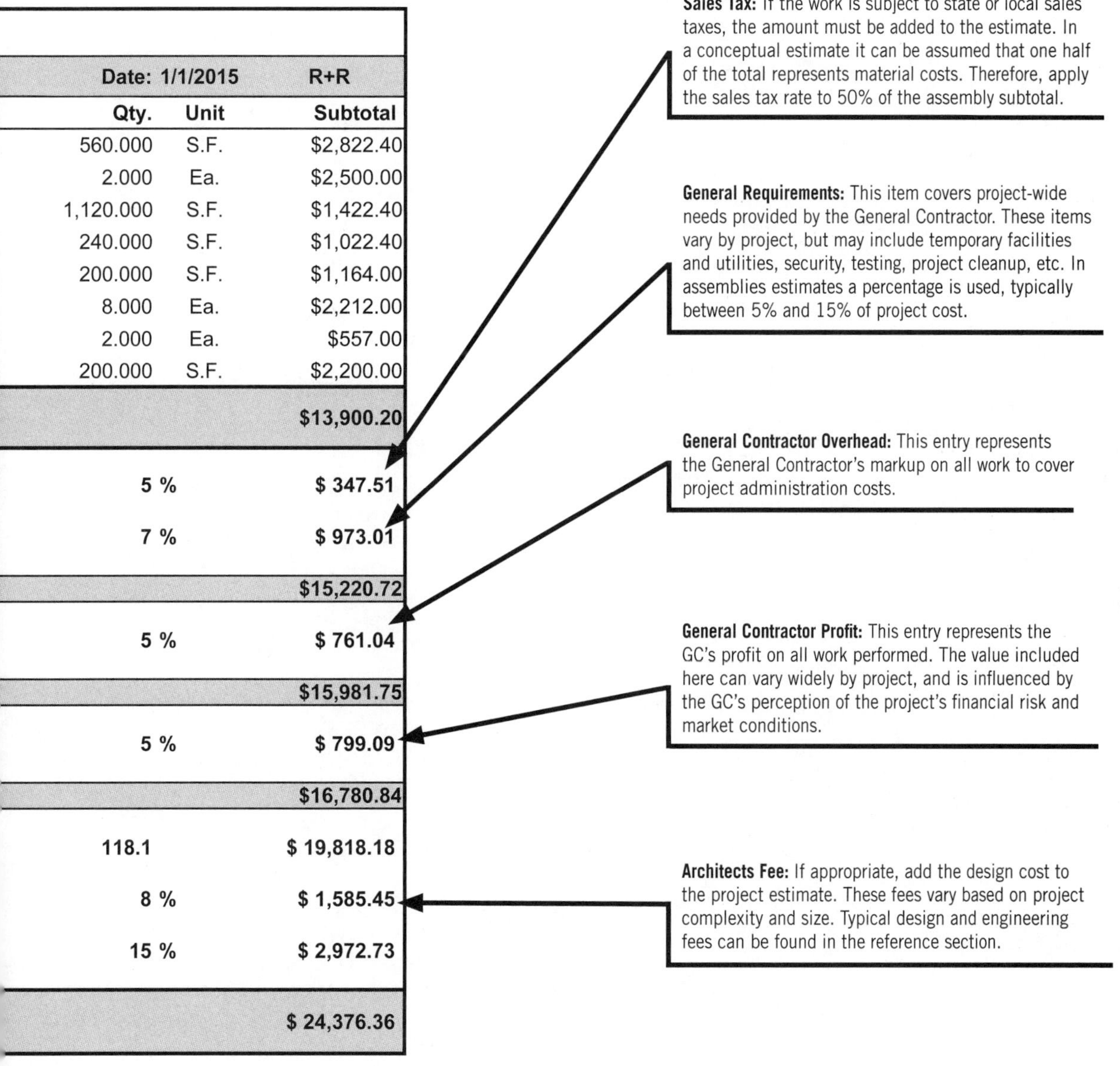

Date: 1/1/2015		R+R
Qty.	Unit	Subtotal
560.000	S.F.	$2,822.40
2.000	Ea.	$2,500.00
1,120.000	S.F.	$1,422.40
240.000	S.F.	$1,022.40
200.000	S.F.	$1,164.00
8.000	Ea.	$2,212.00
2.000	Ea.	$557.00
200.000	S.F.	$2,200.00
		$13,900.20
5 %		$ 347.51
7 %		$ 973.01
		$15,220.72
5 %		$ 761.04
		$15,981.75
5 %		$ 799.09
		$16,780.84
118.1		$ 19,818.18
8 %		$ 1,585.45
15 %		$ 2,972.73
		$ 24,376.36

Sales Tax: If the work is subject to state or local sales taxes, the amount must be added to the estimate. In a conceptual estimate it can be assumed that one half of the total represents material costs. Therefore, apply the sales tax rate to 50% of the assembly subtotal.

General Requirements: This item covers project-wide needs provided by the General Contractor. These items vary by project, but may include temporary facilities and utilities, security, testing, project cleanup, etc. In assemblies estimates a percentage is used, typically between 5% and 15% of project cost.

General Contractor Overhead: This entry represents the General Contractor's markup on all work to cover project administration costs.

General Contractor Profit: This entry represents the GC's profit on all work performed. The value included here can vary widely by project, and is influenced by the GC's perception of the project's financial risk and market conditions.

Architects Fee: If appropriate, add the design cost to the project estimate. These fees vary based on project complexity and size. Typical design and engineering fees can be found in the reference section.

A1010 Standard Foundations

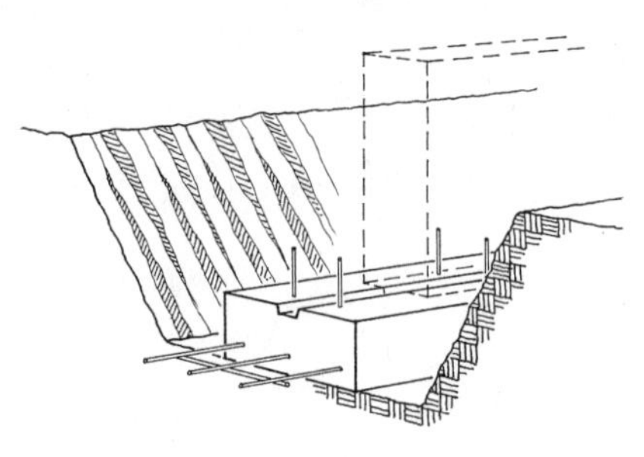

The Strip Footing System includes: excavation; hand trim; all forms needed for footing placement; forms for 2″ x 6″ keyway (four uses); dowels; and 3,000 p.s.i. concrete.

The footing size required varies for different soils. Soil bearing capacities are listed for 3 KSF and 6 KSF. Depths of the system range from 8″ and deeper. Widths range from 16″ and wider. Smaller strip footings may not require reinforcement.

Please see the reference section for further design and cost information.

System Components	QUANTITY	UNIT	COST PER L.F.		
			MAT.	INST.	TOTAL
SYSTEM A1010 110 2500					
STRIP FOOTING, LOAD 5.1 KLF, SOIL CAP. 3 KSF, 24″ WIDE X 12″ DEEP, REINF.					
Trench excavation	.148	C.Y.		1.52	1.52
Hand trim	2.000	S.F.		2.06	2.06
Compacted backfill	.074	C.Y.		.30	.30
Formwork, 4 uses	2.000	S.F.	4.88	9.66	14.54
Keyway form, 4 uses	1.000	L.F.	.36	1.23	1.59
Reinforcing, fy = 60000 psi	3.000	Lb.	1.65	1.95	3.60
Dowels	2.000	Ea.	1.56	5.72	7.28
Concrete, f'c = 3000 psi	.074	C.Y.	8.29		8.29
Place concrete, direct chute	.074	C.Y.		1.93	1.93
Screed finish	2.000	S.F.		.82	.82
TOTAL			16.74	25.19	41.93

A1010 110	Strip Footings	COST PER L.F.		
		MAT.	INST.	TOTAL
2100	Strip footing, load 2.6 KLF, soil capacity 3 KSF, 16″ wide x 8″ deep, plain	8.65	15.10	23.75
2300	Load 3.9 KLF, soil capacity, 3 KSF, 24″ wide x 8″ deep, plain	10.55	16.60	27.15
2500	Load 5.1 KLF, soil capacity 3 KSF, 24″ wide x 12″ deep, reinf.	16.75	25	41.75
2700	Load 11.1 KLF, soil capacity 6 KSF, 24″ wide x 12″ deep, reinf.	16.75	25	41.75
2900	Load 6.8 KLF, soil capacity 3 KSF, 32″ wide x 12″ deep, reinf.	20	27.50	47.50
3100	Load 14.8 KLF, soil capacity 6 KSF, 32″ wide x 12″ deep, reinf.	20	27.50	47.50
3300	Load 9.3 KLF, soil capacity 3 KSF, 40″ wide x 12″ deep, reinf.	23.50	29.50	53
3500	Load 18.4 KLF, soil capacity 6 KSF, 40″ wide x 12″ deep, reinf.	23.50	29.50	53
3700	Load 10.1 KLF, soil capacity 3 KSF, 48″ wide x 12″ deep, reinf.	26	32	58
3900	Load 22.1 KLF, soil capacity 6 KSF, 48″ wide x 12″ deep, reinf.	27.50	34	61.50
4100	Load 11.8 KLF, soil capacity 3 KSF, 56″ wide x 12″ deep, reinf.	30.50	35.50	66
4300	Load 25.8 KLF, soil capacity 6 KSF, 56″ wide x 12″ deep, reinf.	33	38	71
4500	Load 10 KLF, soil capacity 3 KSF, 48″ wide x 16″ deep, reinf.	33	37.50	70.50
4700	Load 22 KLF, soil capacity 6 KSF, 48″ wide, 16″ deep, reinf.	34	38.50	72.50
4900	Load 11.6 KLF, soil capacity 3 KSF, 56″ wide x 16″ deep, reinf.	37.50	53.50	91
5100	Load 25.6 KLF, soil capacity 6 KSF, 56″ wide x 16″ deep, reinf.	39.50	56	95.50
5300	Load 13.3 KLF, soil capacity 3 KSF, 64″ wide x 16″ deep, reinf.	43	44.50	87.50
5500	Load 29.3 KLF, soil capacity 6 KSF, 64″ wide x 16″ deep, reinf.	45.50	47.50	93
5700	Load 15 KLF, soil capacity 3 KSF, 72″ wide x 20″ deep, reinf.	57	53	110
5900	Load 33 KLF, soil capacity 6 KSF, 72″ wide x 20″ deep, reinf.	60	57	117
6100	Load 18.3 KLF, soil capacity 3 KSF, 88″ wide x 24″ deep, reinf.	80	68	148
6300	Load 40.3 KLF, soil capacity 6 KSF, 88″ wide x 24″ deep, reinf.	86.50	75	161.50
6500	Load 20 KLF, soil capacity 3 KSF, 96″ wide x 24″ deep, reinf.	87	72	159
6700	Load 44 KLF, soil capacity 6 KSF, 96″ wide x 24″ deep, reinf.	91.50	77.50	169

A1010 Standard Foundations

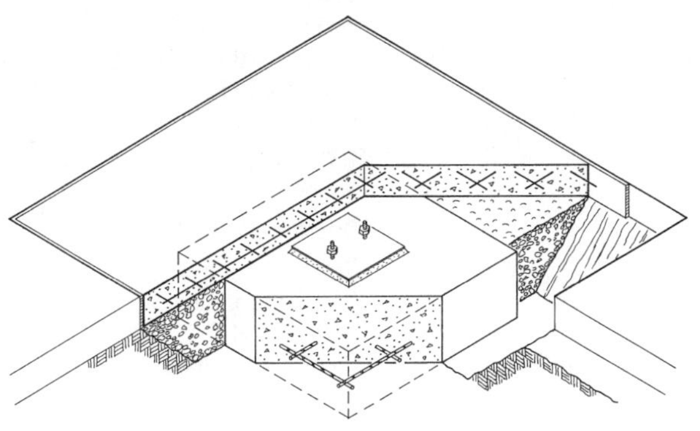

Column footing includes: Remove existing slab; excavation; forms for concrete; reinforcing steel; expansion joint; anchor bolts; grouting column base plate; and concrete placed and finished.

The footing size required varies for different soils.

Design assumptions:

Conc. slab f'c = 3.5 ksi
Conc. foundation f'c = 3.0 ksi
WWF 6 x 6 #10/#10
Reinforcement, fy = 60 ksi

System Components	QUANTITY	UNIT	COST EACH MAT.	COST EACH INST.	COST EACH TOTAL
SYSTEM A1010 260 0200					
CONC. COLUMN FOOTING IN EXISTING BUILDING, LOAD 10K, SOIL CAPACITY 3 KSF, 3′ SQ X 12″ DEEP, 3,000 PSI					
Saw cut slab	16.000	S.F.	16.32	162.24	178.56
Break up slab, 6″	16.000	S.F.		251.36	251.36
Remove concrete; hand	.755	C.Y.		30.96	30.96
Excavate; hand	.592	C.Y.		72.82	72.82
Forms, 1 use	12.000	SFCA	45.48	270	315.48
Reinforcing steel, #4 bar	15.000	Lb.	7.95	9.75	17.70
Concrete, 3,000 psi, for footing, incl. premium delv. chg.	.333	C.Y.	52.19		52.19
Place conc. footing	.333	C.Y.		26.44	26.44
Backfill by hand	.311	C.Y.		10.89	10.89
Compaction; hand tamp	.311	C.Y.		7.46	7.46
Welded wire fabric	.160	C.S.F.	2.55	6.24	8.79
Premolded, bituminous fiber, 1/2″ x 6″	16.000	L.F.	7.68	26.24	33.92
Concrete, 3500 psi, for slab, incl. premium delv. chg.	.166	C.Y.	26.68		26.68
Place conc. slab, 6″	.166	C.Y.		4.40	4.40
Finish slab; steel trowel	16.000	S.F.		15.68	15.68
Anchor bolts, 3/4″ x 12″	2.000	Ea.	10.10	9.84	19.94
Grout level plate	1.000	S.F.	7.15	16.25	23.40
TOTAL			176.10	920.57	1,096.67
COST per C.Y. of FOOTING			528.83	2,764.47	3,293.30

A1020 Special Foundations

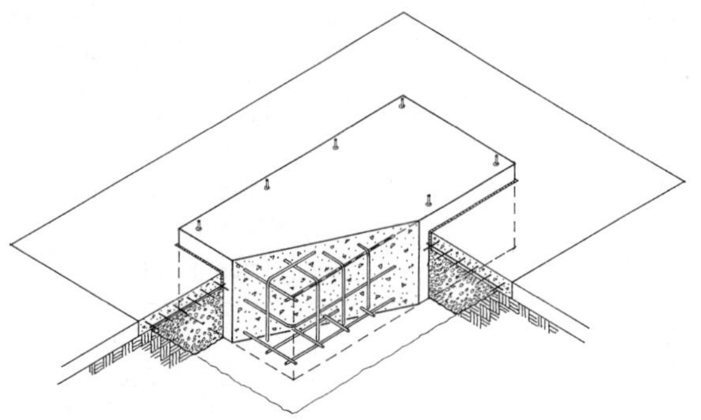

Equipment foundation includes: Remove existing slab; excavation; forms for concrete; reinforcing steel; expansion joint; anchor bolts; and concrete placed and screeded.

The foundation size required varies for different soils.

Design assumptions:

Conc. slab f'c = 3.5 ksi
Conc. foundation f'c = 3.0 ksi
WWF 6 x 6 #10/#10
Reinforcement, fy = 60 ksi

System Components	QUANTITY	UNIT	COST EACH MAT.	COST EACH INST.	COST EACH TOTAL
SYSTEM A1020 810 0170					
CONC. EQUIPMENT FOUNDATION, 4′ X 8′ X 30″ DEEP, 3,000 PSI.					
Saw cut slab	45.000	S.F.	28.56	283.92	312.48
Break up slab; 6″	45.000	S.F.		706.95	706.95
Remove concrete; hand	3.133	C.Y.		128.45	128.45
Excavate; hand	2.500	C.Y.		307.50	307.50
Forms, 1 use	60.000	SFCA	227.40	1,350	1,577.40
Reinforcing steel, #6 bar	318.000	Lb.	168.54	206.70	375.24
Concrete, 3000 psi, for foundation, incl. premium delv. chg.	3.000	C.Y.	464.58		464.58
Anchor bolts, 3/4″ x 12″	8.000	Ea.	40.40	39.36	79.76
Place concrete foundation	3.000	C.Y.		108.13	108.13
Backfill by hand	.866	C.Y.		30.31	30.31
Compaction in 6″ layers, hand tamp	.866	C.Y.		20.78	20.78
Premolded, bituminous fiber, 1/2″x6″	24.000	L.F.	11.52	39.36	50.88
Welded wire fabric; 6x6-#10/10	.130	C.S.F.	2.07	5.07	7.14
Concrete, 3500 psi, for slab, incl. premium delv. chg.	.241	C.Y.	38.76		38.76
Place conc. slab	.241	C.Y.		6.39	6.39
Finish slab; steel trowel	13.000	S.F.		12.74	12.74
Finish equip. foundation; float finish	32.000	S.F.		13.12	13.12
TOTAL			981.83	3,258.78	4,240.61
COST per C.Y. of FOUNDATION			327.28	1,086.26	1,413.54

B1010 Floor Construction

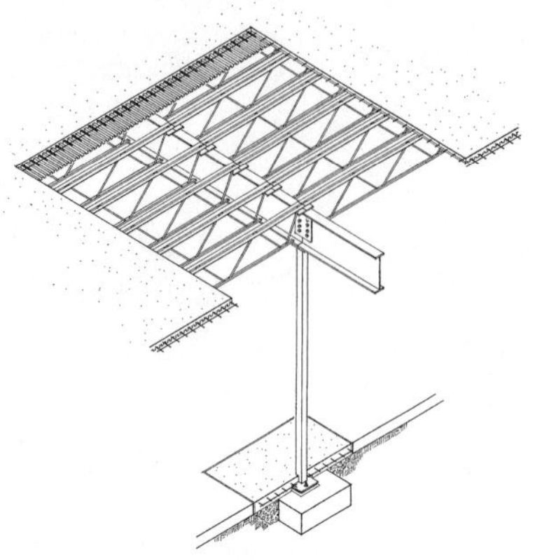

Mezzanine addition to existing building includes: Column footings; steel columns; structural steel; open web steel joists; uncoated 28 ga. steel slab forms; 2-1/2″ concrete slab reinforced with welded wire fabric; steel trowel finish.

Design assumptions:

Structural steel is A36, high strength bolted. Slab form is 28 gauge, galvanized.
WWF 6 x 6 #10/#10
Conc. slab f'c = 3 ksi

System Components	QUANTITY	UNIT	COST EACH MAT.	COST EACH INST.	COST EACH TOTAL
SYSTEM B1010 310 0142					
MEZZANINE ADDITION TO EXISTING BUILDING; 100 PSF SUPERIMPOSED LOAD, SLAB FORM DECK, MTL. JOISTS, 2.5″ CONC. SLAB, 3,000 PSI, 3,000 S.F.					
Saw cut existing slab, 6″ thick	192.000	L.F.	195.84	1,946.88	2,142.72
Cut out existing slab, 4′ x 4′ openings	192.000	S.F.		3,016.32	3,016.32
Remove concrete debris	9.060	C.Y.		371.46	371.46
Excavate by hand	7.104	C.Y.		873.79	873.79
Backfill by hand	3.732	C.Y.		130.62	130.62
Compaction of backfill	3.732	C.Y.		89.57	89.57
Forms for footing, 3′ x 3′ x 1′	144.000	SFCA	545.76	3,240	3,785.76
Reinforcing bar for footing, 3#5 E.W. x 30″ long	180.000	Lb.	95.40	117	212.40
Anchor bolts, 12″, 2 per column	24.000	Ea.	121.20	118.08	239.28
Concrete, 3000 psi, footing	5.592	C.Y.	626.30		626.30
Place concrete in footing, direct chute	5.592	C.Y.		317.35	317.35
Premolded, bituminous fiber, 1/2″ x 6″	192.000	L.F.	92.16	314.88	407.04
Reinforcing mesh for slab cutouts	1.920	C.S.F.	30.62	74.88	105.50
Concrete, 3500 psi, slab cutouts	2.784	C.Y.	320.16		320.16
Place concrete slab cutouts, direct chute	2.784	C.Y.		52.81	52.81
Machine trowel slab cutouts	192.000	S.F.		188.16	188.16
Grout for leveling plates	12.000	S.F.	85.80	195	280.80
Column, 4″ x 4″ x 1/4″ x12′	12.000	Ea.	2,880	1,392	4,272
Structural steel W 21x50	80.000	L.F.	9,600	1,039.20	10,639.20
Structural steel W 16 x 31	80.000	L.F.	5,940	900	6,840
Open web joists, H or K series	6.800	Ton	18,615	7,956	26,571
Metal decking, slab form, steel, 28 gauge, 9/16″ deep, type UFS, uncoated	30.000	C.S.F.	5,100	2,430	7,530
Concrete 3,000 psi, 2 1/2″ slab, incl. premium delv. chg.	23.500	C.Y.	3,948		3,948
Welded wire fabric 6x6 #10/#10 (w1.4/w1.4)	30.000	C.S.F.	478.50	1,170	1,648.50
Place concrete	23.500	C.Y.		1,274.29	1,274.29
Monolithic steel trowel finish	30.000	C.S.F.		2,940	2,940
Curing with sprayed membrane curing compound	30.000	C.S.F.	364.50	312	676.50
TOTAL			49,039.24	30,460.29	79,499.53
COST PER S.F.			16.35	10.15	26.50

B1010 Floor Construction

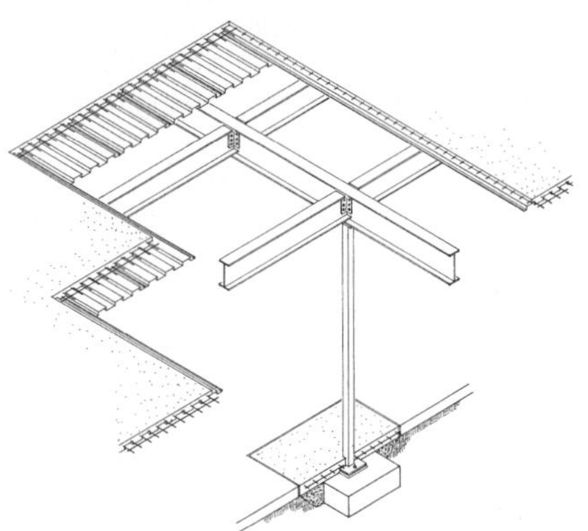

Mezzanine addition to existing building includes: Column footings; steel columns; structural steel; galvanized composite steel deck; 4″ concrete slab reinforced with welded wire fabric; steel trowel finish.

Design assumptions:

Structural steel is A36, high strength bolted.
Composite deck is 22 gauge, galvanized.
WWF 6 x 6 #10/#10
Conc. slab f'c = 3 ksi

System Components	QUANTITY	UNIT	COST EACH MAT.	COST EACH INST.	COST EACH TOTAL
SYSTEM B1010 320 0152					
MEZZANINE ADDITION TO EXISTING BUILDING; 100 PSF SUPERIMPOSED LOAD					
COMPOSITE DECK, BEAMS, 4″ CONC. SLAB, 3,000 PSI, 3,000 S.F.					
Saw cut existing slab, 6″ thick	192.000	L.F.	195.84	1,946.88	2,142.72
Remove existing slab, 4′ x 4′ openings	192.000	S.F.		3,016.32	3,016.32
Remove concrete debris	9.060	C.Y.		371.46	371.46
Excavate by hand	7.104	C.Y.		873.79	873.79
Backfill by hand, no compaction, light soil	3.732	C.Y.		130.62	130.62
Compaction of backfill	3.732	C.Y.		89.57	89.57
Forms in place, footing, 3′ x 3′ x 1′	144.000	SFCA	545.76	3,240	3,785.76
Footings, #4 to #7 bars, 3#5 E.W. x 30″ long	180.000	Lb.	95.40	117	212.40
Anchor bolts, 12″ long, 2 per column	24.000	Ea.	121.20	118.08	239.28
Concrete, 3000 psi, footing	5.592	C.Y.	626.30		626.30
Place concrete in footings, direct chute	5.592	C.Y.		317.35	317.35
Premolded, bituminous fiber, 1/2″ x 6″	192.000	L.F.	92.16	314.88	407.04
Structural steel W 12x19	225.000	L.F.	7,987.50	1,719	9,706.50
Reinforcing mesh for slab cutouts	1.920	C.S.F.	30.62	74.88	105.50
Concrete, 3500 psi, for slab cutouts	2.784	C.Y.	320.16		320.16
Structural steel W 14 x 30	375.000	L.F.	27,000	4,218.76	31,218.76
Place concrete, slab cutouts, direct chute	2.784	C.Y.		52.81	52.81
Structural steel W 16 x 31	80.000	L.F.	5,940	900	6,840
Machine trowel slab cutouts	192.000	S.F.		188.16	188.16
Grout for leveling plates	12.000	S.F.	85.80	195	280.80
Column, 4″ x 4″ x 1/4″ x 12′	12.000	Ea.	2,880	1,392	4,272
Structural steel W 18 x 50	80.000	L.F.	9,600	1,212	10,812
Metal deck, galvanized, 2″ deep, 22 gauge.	30.000	C.S.F.	6,360	2,520	8,880
Concrete, 3,000 p.s.i., 4″ slab, incl. premium delv. chg.	37.000	C.Y.	6,216		6,216
Welded wire fabric 6x6 #10/#10 (w1.4/w1.4)	30.000	C.S.F.	478.50	1,170	1,648.50
Place concrete	37.000	C.Y.		2,006.33	2,006.33
Monolithic steel trowel finish	30.000	C.S.F.		2,940	2,940
Curing with sprayed membrane curing compound	30.000	C.S.F.	364.50	312	676.50
TOTAL			68,939.74	29,436.89	98,376.63
COST PER S.F.			22.98	9.81	32.79

B1010 Floor Construction

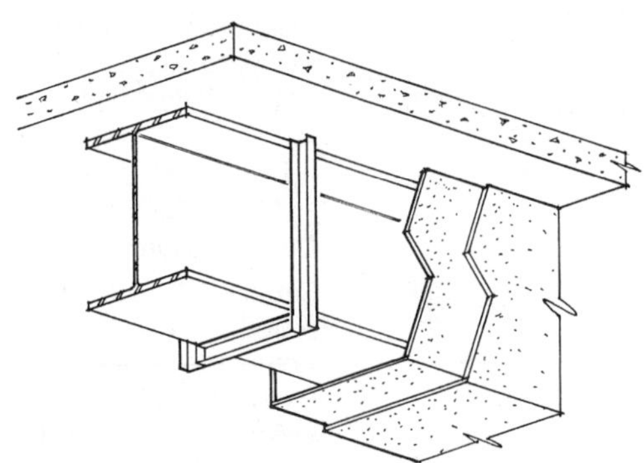

The table below lists fireproofing costs for steel beams by type, beam size, thickness and fire rating. Weights listed are for the fireproofing material only.

System Components	QUANTITY	UNIT	COST PER L.F. MAT.	COST PER L.F. INST.	COST PER L.F. TOTAL
SYSTEM B1010 710 1300					
FIREPROOFING, 5/8" F.R. GYP. BOARD, 12"X 4" BEAM, 2" THICK, 2 HR. F.R.					
1-1/4" x 1-1/4", galvanized	.020	C.L.F.	.36	3.52	3.88
L bead for drywall, galvanized	.020	C.L.F.	.42	4.10	4.52
Furring, beams & columns, 3/4" galv. channels, 24" O.C.	2.330	S.F.	.56	6.83	7.39
Drywall on beam, no finish, 2 layers at 5/8" thick	3.000	S.F.	2.55	12.30	14.85
Drywall, taping and finishing joints, add	3.000	S.F.	.15	1.86	2.01
TOTAL			4.04	28.61	32.65

B1010 710 Steel Beam Fireproofing

	ENCASEMENT SYSTEM	BEAM SIZE (IN.)	THICKNESS (IN.)	FIRE RATING (HRS.)	WEIGHT (P.L.F.)	COST PER L.F. MAT.	COST PER L.F. INST.	COST PER L.F. TOTAL
0400	Concrete	12x4	1	1	77	6.40	30	36.40
0450	3000 PSI		1-1/2	2	93	7.45	32.50	39.95
0500			2	3	121	8.35	35	43.35
0550		14x5	1	1	100	8.35	36.50	44.85
0600			1-1/2	2	122	9.65	40.50	50.15
0650			2	3	142	10.90	44.50	55.40
0700		16x7	1	1	147	9.75	39	48.75
0750			1-1/2	2	169	10.45	40	50.45
0800			2	3	195	11.85	44	55.85
0850		18x7-1/2	1	1	172	11.30	45	56.30
0900			1-1/2	2	196	12.70	49.50	62.20
0950			2	3	225	14.15	54	68.15
1000		24x9	1	1	264	15.10	54.50	69.60
1050			1-1/2	2	295	16.40	57.50	73.90
1100			2	3	328	17.70	61	78.70
1150		30x10-1/2	1	1	366	19.75	69	88.75
1200			1-1/2	2	404	21.50	73	94.50
1250			2	3	449	22.50	77	99.50
1300	5/8" fire rated	12x4	2	2	15	4.04	28.50	32.54
1350	Gypsum board		2-5/8	3	24	5.65	35.50	41.15
1400		14x5	2	2	17	4.35	29.50	33.85
1450			2-5/8	3	27	6.25	37.50	43.75
1500		16x7	2	2	20	4.89	33	37.89
1550			2-5/8	3	31	6.20	32.50	38.70
1600		18x7-1/2	2	2	22	5.25	35.50	40.75
1650			2-5/8	3	34	7.50	45	52.50

B1010 Floor Construction

B1010 710 Steel Beam Fireproofing

	ENCASEMENT SYSTEM	BEAM SIZE (IN.)	THICKNESS (IN.)	FIRE RATING (HRS.)	WEIGHT (P.L.F.)	COST PER L.F.		
						MAT.	INST.	TOTAL
1700	5/8" fire rated	24x9	2	2	27	6.55	44	50.55
1750	Gypsum board		2-5/8	3	42	9.30	55	64.30
1800		30x10-1/2	2	2	33	7.70	51.50	59.20
1850			2-5/8	3	51	10.90	64.50	75.40
1900	Gypsum	12x4	1-1/8	3	18	5.25	22	27.25
1950	Plaster on	14x5	1-1/8	3	21	5.95	24.50	30.45
2000	Metal lath	16x7	1-1/8	3	25	6.95	28	34.95
2050		18x7-1/2	1-1/8	3	32	7.95	31	38.95
2100		24x9	1-1/8	3	35	9.80	38	47.80
2150		30x10-1/2	1-1/8	3	44	11.80	46	57.80
2200	Perlite plaster	12x4	1-1/8	2	16	4.91	24.50	29.41
2250	On metal lath		1-1/4	3	20	5.25	25.50	30.75
2300			1-1/2	4	22	5.60	27	32.60
2350		14x5	1-1/8	2	18	6.40	30	36.40
2400			1-1/4	3	23	6.80	31	37.80
2450			1-1/2	4	25	7.50	33.50	41
2500		16x7	1-1/8	2	21	6	30.50	36.50
2550			1-1/4	3	26	6.30	31.50	37.80
2600			1-1/2	4	29	6.65	33	39.65
2650		18x7-1/2	1-1/8	2	26	7.40	35.50	42.90
2700			1-1/4	3	33	7.75	36.50	44.25
2750			1-1/2	4	36	8.50	39	47.50
2800		24x9	1-1/8	2	30	7.75	39.50	47.25
2850			1-1/4	3	38	8.05	41	49.05
2900			1-1/2	4	41	8.45	42	50.45
2950		30x10-1/2	1-1/8	2	37	10.05	49	59.05
3000			1-1/4	3	46	10.40	50.50	60.90
3050			1-1/2	4	51	11.15	52.50	63.65
3100	Sprayed Fiber	12x4	5/8	1	12	1.26	2.43	3.69
3150	(non asbestos)		1-1/8	2	21	2.26	4.37	6.63
3200			1-1/4	3	22	2.52	4.85	7.37
3250		14x5	5/8	1	14	1.26	2.43	3.69
3300			1-1/8	2	26	2.26	4.37	6.63
3350			1-1/4	3	29	2.52	4.85	7.37
3400		16x7	5/8	1	18	1.59	3.07	4.66
3450			1-1/8	2	32	2.86	5.55	8.41
3500			1-1/4	3	36	3.18	6.15	9.33
3550		18x7-1/2	5/8	1	20	1.76	3.39	5.15
3600			1-1/8	2	35	3.16	6.10	9.26
3650			1-1/4	3	39	3.49	6.75	10.24
3700		24x9	5/8	1	25	2.22	4.28	6.50
3750			1-1/8	2	45	4.02	7.75	11.77
3800			1-1/4	3	50	4.47	8.60	13.07
3850		30x10-1/2	5/8	1	31	2.69	5.20	7.89
3900			1-1/8	2	55	4.84	9.35	14.19
3950			1-1/4	3	61	5.40	10.40	15.80
4000	On Decking	Flat	1		6	.58	.70	1.28
4050	Per S.F.	Corrugated	1		7	.87	1.34	2.21
4100		Fluted	1		7	.87	1.34	2.21

B1010 Floor Construction

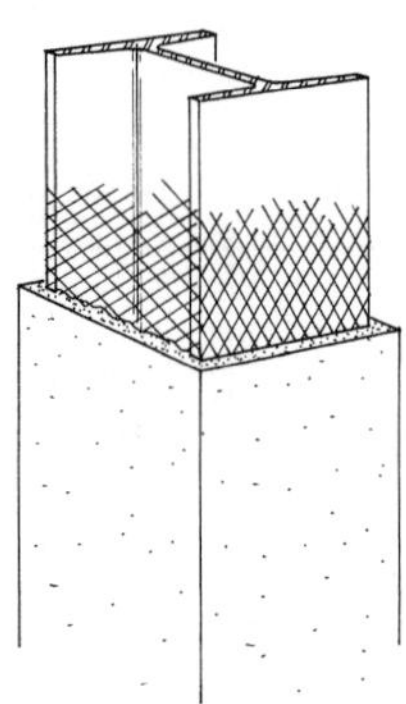

Listed below are costs per V.L.F. for fireproofing by material, column size, thickness and fire rating. Weights listed are for the fireproofing material only.

System Components	QUANTITY	UNIT	COST PER V.L.F. MAT.	COST PER V.L.F. INST.	COST PER V.L.F. TOTAL
SYSTEM B1010 720 3000					
CONCRETE FIREPROOFING, 8" STEEL COLUMN, 1" THICK, 1 HR. FIRE RATING					
Forms in place, columns, plywood, 4 uses	3.330	SFCA	2.76	31.14	33.90
Welded wire fabric, 2 x 2 #14 galv. 21 lb./C.S.F., column wrap	2.700	S.F.	1.24	5.70	6.94
Concrete ready mix, regular weight, 3000 psi	.621	C.F.	2.58		2.58
Place and vibrate concrete, 12" sq./round columns, pumped	.621	C.F.		1.94	1.94
TOTAL			6.58	38.78	45.36

B1010 720 Steel Column Fireproofing

	ENCASEMENT SYSTEM	COLUMN SIZE (IN.)	THICKNESS (IN.)	FIRE RATING (HRS.)	WEIGHT (P.L.F.)	COST PER V.L.F. MAT.	COST PER V.L.F. INST.	COST PER V.L.F. TOTAL
3000	Concrete	8	1	1	110	6.60	39	45.60
3050			1-1/2	2	133	7.55	43	50.55
3100			2	3	145	8.50	46.50	55
3150		10	1	1	145	8.60	48	56.60
3200			1-1/2	2	168	9.55	51.50	61.05
3250			2	3	196	10.60	55	65.60
3600	Gypsum board	8	1	3	14	5	30.50	35.50
3650	1/2" fire rated	10	1	3	17	5.35	32.50	37.85
3700	2 layers	14	1	3	22	5.55	33.50	39.05
3750	Gypsum board	8	1-1/2	3	23	6.65	38.50	45.15
3800	1/2" fire rated	10	1-1/2	3	27	7.50	42.50	50
3850	3 layers	14	1-1/2	3	35	8.30	46.50	54.80
3900	Sprayed fiber	8	1-1/2	2	6.3	3.93	7.60	11.53
3950	Direct application		2	3	8.3	5.40	10.50	15.90
4000			2-1/2	4	10.4	7	13.55	20.55
4050		10	1-1/2	2	7.9	4.75	9.15	13.90
4100			2	3	10.5	6.50	12.55	19.05
4150			2-1/2	4	13.1	8.40	16.20	24.60
4500	Perlite plaster	8	1	2	18	7.75	34	41.75
4550	On metal lath		1-3/8	3	23	8.45	38	46.45
4600			1-3/4	4	35	10	45.50	55.50
4650	Perlite plaster	10	1	2	21	8.90	39.50	48.40
4700			1-3/8	3	27	10.05	45	55.05
4750			1-3/4	4	41	11.30	52	63.30

B1010 Floor Construction

B1010 720 Steel Column Fireproofing

	ENCASEMENT SYSTEM	COLUMN SIZE (IN.)	THICKNESS (IN.)	FIRE RATING (HRS.)	WEIGHT (P.L.F.)	COST PER V.L.F.		
						MAT.	INST.	TOTAL
5100	5/8" gypsum plaster	8	1	1-1/2	20	6.10	28	34.10
5150	On 3/8" gypsum lath	10	1	1-1/2	24	7.20	33	40.20
5200		14	1	1-1/2	33	8.75	40.50	49.25
5250	1"perlite plaster	8	1-3/8	2	23	7.75	29.50	37.25
5300	On 3/8" gypsum lath	10	1-3/8	2	28	8.90	33.50	42.40
5350		14	1-3/8	2	37	11.25	42.50	53.75
5400	1-3/8" perlite plaster	8	1-3/4	3	27	7.30	33	40.30
5450	On 3/8" gypsum lath	10	1-3/4	3	33	8.30	38.50	46.80
5500		14	1-3/4	3	43	10.40	48	58.40
5550	Concrete masonry	8	4-3/4	4	126	11.60	36	47.60
5600	Units 4" thick	10	4-3/4	4	166	14.50	45	59.50
5650	75% solid	14	4-3/4	4	262	17.40	54	71.40

B2010 Exterior Walls

This page illustrates masonry cleaning and restoration systems, including staging, cleaning and repointing.

System Components	QUANTITY	UNIT	COST PER S.F. MAT.	COST PER S.F. INST.	COST PER S.F. TOTAL
SYSTEM B2010 143 1300					
Repoint existing building, brick, common bond, high pressure cleaning,					
Water only, soft old mortar.					
Scaffolding,steel tubular,bldg ext wall face, labor only, 1 to 5 stories	.010	C.S.F.		2.31	2.31
Cleaning masonry, high pressure wash, water only, avg soil, bio staining	1.100	S.F.		1.68	1.68
Repoint, brick common bond	1.000	S.F.	.61	6.25	6.86
TOTAL			.61	10.24	10.85

B2010 143	Masonry Restoration - Cleaning	COST PER S.F. MAT.	COST PER S.F. INST.	COST PER S.F. TOTAL
1000	Repoint existing building, high pressure cleaning, water only,			
1100	soft old mortar			
1200	For alternate masonry surfaces:			
1300	Brick, common bond	.61	10.25	10.86
1400	Flemish bond	.65	10.70	11.35
1500	English bond	.65	11.35	12
1700	Stone, 2′ x 2′ blocks	.82	7.75	8.57
1800	2′ x 4′ blocks	.62	6.85	7.47
2000	Add to above prices for alternate cleaning systems:			
2100	Chemical brush and wash	.05	.69	.74
2200	High pressure chemical and water	.08	.57	.65
2300	Sandblasting, wet system	.38	1.64	2.02
2400	Dry system	.36	.97	1.33
2500	Steam cleaning		.47	.47

B2020 Exterior Windows

B2020 108 Vinyl Clad Windows

	MATERIAL	TYPE	GLAZING	SIZE	DETAIL	COST PER UNIT		
						MAT.	INST.	TOTAL
1000	Vinyl clad	casement	insul. glass	1'-4" x 4'-0"		420	142	562
1010				2'-0" x 3'-0"		360	142	502
1020				2'-0" x 4'-0"		410	149	559
1030				2'-0" x 5'-0"		460	157	617
1040				2'-0" x 6'-0"		480	157	637
1050				2'-4" x 4'-0"		460	157	617
1060				2'-6" x 5'-0"		565	153	718
1070				3'-0" x 5'-0"		790	157	947
1080				4'-0" x 3'-0"		865	157	1,022
1090				4'-0" x 4'-0"		745	157	902
1100				4'-8" x 4'-0"		820	157	977
1110				4'-8" x 5'-0"		925	183	1,108
1120				4'-8" x 6'-0"		1,025	183	1,208
1130				6'-0" x 4'-0"		945	183	1,128
1140				6'-0" x 5'-0"		1,050	183	1,233
3000		double-hung		2'-0" x 4'-0"		440	136	576
3050				2'-0" x 5'-0"		500	136	636
3100				2'-4" x 4'-0"		455	142	597
3150				2'-4" x 4'-8"		455	142	597
3200				2'-4" x 6'-0"		515	142	657
3250				2'-6" x 4'-0"		455	142	597
3300				2'-8" x 4'-0"		680	183	863
3350				2'-8" x 5'-0"		535	157	692
3375				2'-8" x 6'-0"		545	157	702
3400				3'-0 x 3'-6"		430	142	572
3450				3'-0 x 4'-0"		495	149	644
3500				3'-0 x 4'-8"		530	149	679
3550				3'-0 x 5'-0"		545	157	702
3600				3'-0 x 6'-0"		490	157	647
3700				4'-0 x 5'-0"		680	168	848
3800				4'-0 x 6'-0"		830	168	998

B3010 Roof Coverings

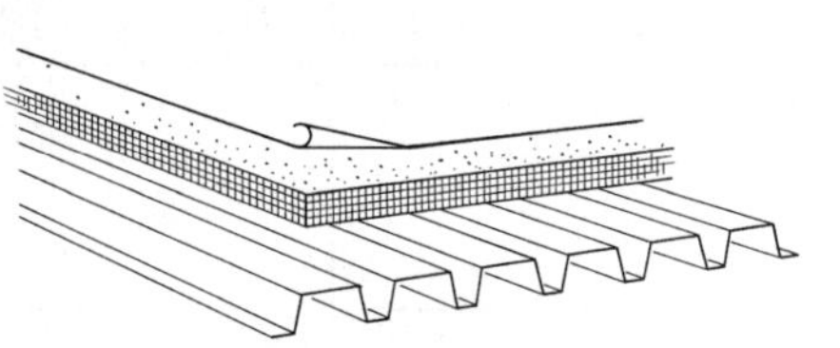

Fully Adhered

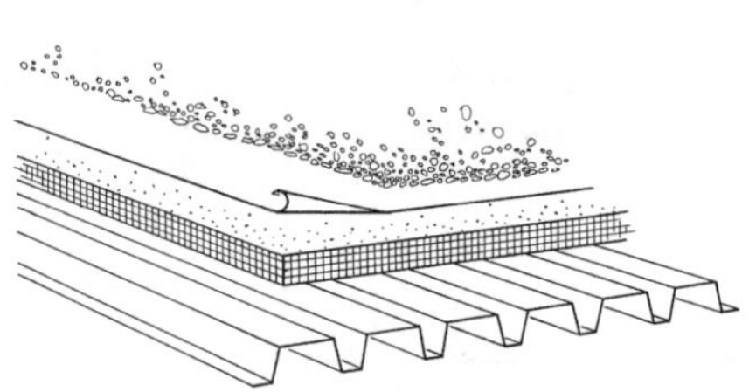

Ballasted

The systems listed below reflect only the cost for the single ply membrane.

B3010 120	Single Ply Membrane	COST PER S.F.		
		MAT.	INST.	TOTAL
1000	CSPE (Chlorosulfonated polyethylene), 45 mils, plate attached	3.40	.82	4.22
1100	Loosely laid with stone ballast	3.40	.82	4.22
2000	EPDM (Ethylene propylene diene monomer), 45 mils, fully adhered	1.22	1.09	2.31
2100	Loosely laid with stone ballast	.94	.56	1.50
2200	Mechanically fastened with batten strips	.87	.82	1.69
3300	60 mils, fully adhered	1.38	1.09	2.47
3400	Loosely laid with stone ballast	1.10	.56	1.66
3500	Mechanically fastened with batten strips	1.02	.82	1.84
4000	Modified bit., SBS modified, granule surface cap sheet, mopped, 150 mils	1.38	2.21	3.59
4100	Smooth surface cap sheet, mopped, 145 mils	.86	2.11	2.97
4500	APP modified, granule surface cap sheet, torched, 180 mils	.98	1.43	2.41
4600	Smooth surface cap sheet, torched, 170 mils	.77	1.36	2.13
5000	PIB (Polyisobutylene), 100 mils, fully adhered with contact cement	2.75	1.09	3.84
5100	Loosely laid with stone ballast	2.20	.56	2.76
5200	Partially adhered with adhesive	2.64	.82	3.46
5300	Hot asphalt attachment	2.53	.82	3.35
6000	Reinforced PVC, 48 mils, loose laid and ballasted with stone	1.30	.56	1.86
6100	Partially adhered with mechanical fasteners	1.22	.82	2.04
6200	Fully adhered with adhesive	1.68	1.09	2.77
6300	Reinforced PVC, 60 mils, loose laid and ballasted with stone	1.32	.56	1.88
6400	Partially adhered with mechanical fasteners	1.24	.82	2.06
6500	Fully adhered with adhesive	1.70	1.09	2.79

C1010 Partitions

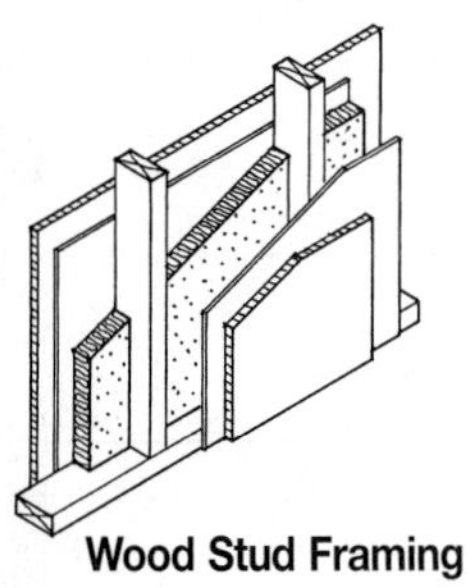

Wood Stud Framing

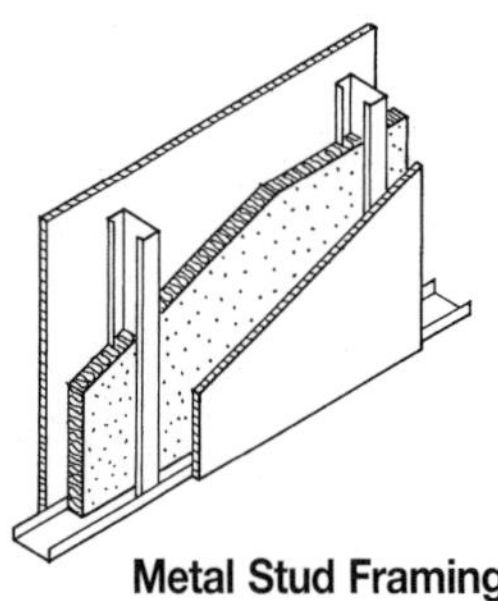

Metal Stud Framing

The Drywall Partitions/Stud Framing Systems are defined by type of drywall and number of layers, type and spacing of stud framing, and treatment on the opposite face. Components include taping and finishing.

Cost differences between regular and fire resistant drywall are negligible, and terminology is interchangeable. In some cases fiberglass insulation is included for additional sound deadening.

System Components	QUANTITY	UNIT	COST PER S.F. MAT.	INST.	TOTAL
SYSTEM C1010 124 1250					
DRYWALL PARTITION,5/8″ F.R.1 SIDE,5/8″ REG.1 SIDE,2″X4″STUDS,16″ O.C.					
Gypsum plasterboard, nailed/screwed to studs, 5/8″ fire resistant	1.000	S.F.	.37	.62	.99
Gypsum plasterboard, nailed/screwed to studs, 5/8″ regular	1.000	S.F.	.36	.62	.98
Taping and finishing joints	2.000	S.F.	.10	1.24	1.34
Framing, 2 x 4 studs @ 16″ O.C., 10′ high	1.000	S.F.	.49	1.23	1.72
TOTAL			1.32	3.71	5.03

C1010 124 Drywall Partitions/Wood Stud Framing

	FACE LAYER	BASE LAYER	FRAMING	OPPOSITE FACE	INSULATION	COST PER S.F. MAT.	INST.	TOTAL
1200	5/8″ FR drywall	none	2 x 4, @ 16″ O.C.	same	0	1.33	3.71	5.04
1250				5/8″ reg. drywall	0	1.32	3.71	5.03
1300				nothing	0	.91	2.47	3.38
1400		1/4″ SD gypsum	2 x 4 @ 16″ O.C.	same	1-1/2″ fiberglass	2.63	5.70	8.33
1450				5/8″ FR drywall	1-1/2″ fiberglass	2.20	5	7.20
1500				nothing	1-1/2″ fiberglass	1.78	3.77	5.55
1600		resil. channels	2 x 4 @ 16″, O.C.	same	1-1/2″ fiberglass	2.11	7.20	9.31
1650				5/8″ FR drywall	1-1/2″ fiberglass	1.94	5.80	7.74
1700				nothing	1-1/2″ fiberglass	1.52	4.54	6.06
1800		5/8″ FR drywall	2 x 4 @ 24″ O.C.	same	0	1.95	4.71	6.66
1850				5/8″ FR drywall	0	1.58	4.09	5.67
1900				nothing	0	1.16	2.85	4.01
1950		5/8″ FR drywall	2 x 4, 16″ O.C.	same	0	2.07	4.95	7.02
1955				5/8″ FR drywall	0	1.70	4.33	6.03
2000				nothing	0	1.28	3.09	4.37
2010		5/8″ FR drywall	staggered, 6″ plate	same	0	2.57	6.20	8.77
2015				5/8″ FR drywall	0	2.20	5.60	7.80
2020				nothing	0	1.78	4.34	6.12
2200		5/8″ FR drywall	2 rows-2 x 4	same	2″ fiberglass	3.06	6.80	9.86
2250			16″O.C.	5/8″ FR drywall	2″ fiberglass	2.69	6.20	8.89
2300				nothing	2″ fiberglass	2.27	4.94	7.21
2400	5/8″ WR drywall	none	2 x 4, @ 16″ O.C.	same	0	1.51	3.71	5.22
2450				5/8″ FR drywall	0	1.42	3.71	5.13
2500				nothing	0	1	2.47	3.47
2600		5/8″ FR drywall	2 x 4, @ 24″ O.C.	same	0	2.13	4.71	6.84
2650				5/8″ FR drywall	0	1.67	4.09	5.76
2700				nothing	0	1.25	2.85	4.10

C1010 Partitions

C1010 128	Drywall Components	COST PER S.F.		
		MAT.	INST.	TOTAL
0140	Metal studs, 24" O.C. including track, load bearing, 18 gage, 2-1/2"	.66	1.15	1.81
0160	3-5/8"	.78	1.18	1.96
0180	4"	.70	1.20	1.90
0200	6"	1.04	1.22	2.26
0220	16 gage, 2-1/2"	.77	1.31	2.08
0240	3-5/8"	.91	1.34	2.25
0260	4"	.96	1.37	2.33
0280	6"	1.20	1.40	2.60
0300	Non load bearing, 25 gage, 1-5/8"	.20	.81	1.01
0340	3-5/8"	.30	.83	1.13
0360	4"	.33	.83	1.16
0380	6"	.40	.85	1.25
0400	20 gage, 2-1/2"	.32	1.03	1.35
0420	3-5/8"	.37	1.04	1.41
0440	4"	.44	1.04	1.48
0460	6"	.51	1.06	1.57
0542	Wood studs including blocking, shoe and double top plate, 2"x4", 12"O.C.	.61	1.54	2.15
0562	16" O.C.	.49	1.23	1.72
0582	24" O.C.	.37	.99	1.36
0602	2"x6", 12" O.C.	.96	1.76	2.72
0622	16" O.C.	.77	1.37	2.14
0641	24" O.C.	.59	1.07	1.66
0642	Furring one side only, steel channels, 3/4", 12" O.C.	.40	2.30	2.70
0644	16" O.C.	.36	2.04	2.40
0646	24" O.C.	.24	1.55	1.79
0647	1-1/2" , 12" O.C.	.54	2.58	3.12
0648	16" O.C.	.48	2.26	2.74
0649	24" O.C.	.32	1.78	2.10
0656	Wood strips, 1" x 3", on wood, 12" O.C.	.42	1.12	1.54
0657	16" O.C.	.32	.84	1.16
0658	On masonry, 12" O.C.	.46	1.24	1.70
0659	16" O.C.	.35	.93	1.28
0660	On concrete, 12" O.C.	.46	2.37	2.83
0661	16" O.C.	.35	1.78	2.13
0662	Gypsum board, one face only, exterior sheathing, 1/2"	.50	1.09	1.59
0680	Interior, fire resistant, 1/2"	.39	.62	1.01
0700	5/8"	.37	.62	.99
0720	Sound deadening board 1/4"	.43	.68	1.11
0740	Standard drywall 3/8"	.37	.62	.99
0760	1/2"	.33	.62	.95
0780	5/8"	.36	.62	.98
0800	Tongue & groove coreboard 1"	.86	2.56	3.42
0820	Water resistant, 1/2"	.43	.62	1.05
0840	5/8"	.46	.62	1.08
0860	Add for the following:, foil backing	.17		.17
0880	Fiberglass insulation, 3-1/2"	.50	.46	.96
0900	6"	.66	.46	1.12
0920	Rigid insulation 1"	.57	.62	1.19
0940	Resilient furring @ 16" O.C.	.22	1.93	2.15
0960	Taping and finishing	.05	.62	.67
0980	Texture spray	.04	.68	.72
1000	Thin coat plaster	.12	.77	.89
1050	Sound wall framing, 2x6 plates, 2x4 staggered studs, 12" O.C.	.75	1.52	2.27

C1020 Interior Doors

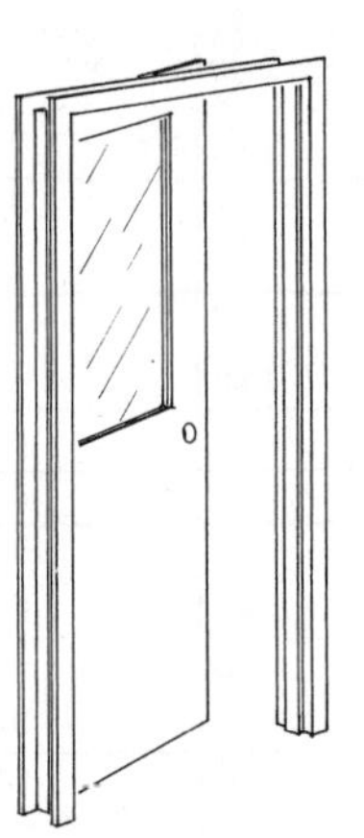

Steel Door, Half Glass
Steel Frame

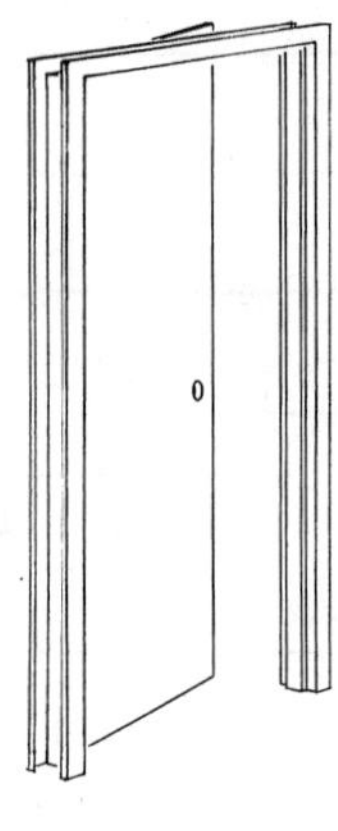

Steel Door, Flush
Steel Frame

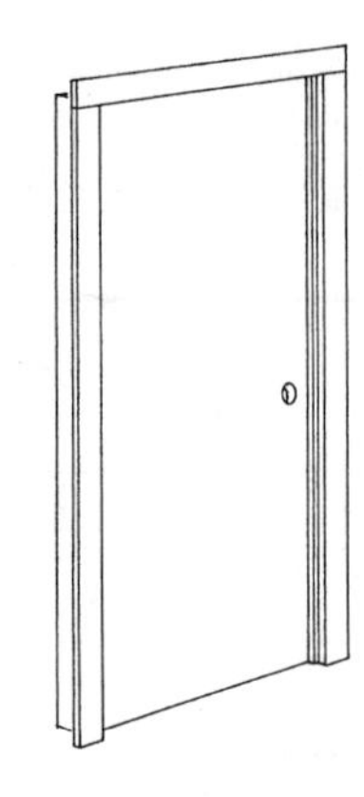

Wood Door, Flush
Wood Frame

The Metal Door/Metal Frame Systems are defined as follows: door type, design and size, frame type and depth. Included in the components for each system is painting the door and frame. No hardware has been included in the systems.

System Components	QUANTITY	UNIT	COST EACH MAT.	COST EACH INST.	COST EACH TOTAL
SYSTEM C1020 114 1200					
STEEL DOOR, HOLLOW, 20 GA., HALF GLASS, 3′-0″X7′-0″, D.W.FRAME, 4-7/8″ DP					
Steel door, flush, hollow core, 1-3/4″ thk, half gl, 20 Ga., 3′-0″ x 7′-0″	1.000	Ea.	650	68.50	718.50
Steel frame, 16 ga., up to 4-7/8″ deep, 3′-0″ x 7′-0″ single	1.000	Ea.	183	77	260
Hinges full mortise, avg. freq., steel base, USP, 4-1/2″ x 4-1/2″	1.500	Pr.	60		60
Float glass, 3/16″ thick, clear, tempered	5.000	S.F.	38.25	45	83.25
Paint door and frame each side, primer	1.000	Ea.	7.58	104	111.58
Paint door and frame each side, 2 coats	1.000	Ea.	11	173	184
TOTAL			949.83	467.50	1,417.33

C1020 114 Metal Door/Metal Frame

	TYPE	DESIGN	SIZE	FRAME	DEPTH	COST EACH MAT.	COST EACH INST.	COST EACH TOTAL
1000	Flush-hollow	20 ga. full panel	3′-0″ x 7′-0″	drywall K.D.	4-7/8″	740	425	1,165
1020				butt welded	8-3/4″	850	480	1,330
1160			6′-0″ x 7′-0″	drywall K.D.	4-7/8″	1,325	675	2,000
1180				butt welded	8-3/4″	1,450	730	2,180
1200		20 ga. half glass	3′-0″ x 7′-0″	drywall K.D.	4-7/8″	950	470	1,420
1220				butt welded	8-3/4″	1,050	525	1,575
1360			6′-0″ x 7′-0″	drywall K.D.	4-7/8″	1,725	760	2,485
1380				butt welded	8-3/4″	1,850	815	2,665
1800		18 ga. full panel	3′-0″ x 7′-0″	drywall K.D.	4-7/8″	825	425	1,250
1820				butt welded	8-3/4″	935	480	1,415
1960			6′-0″ x 7′-0″	drywall K.D.	4-7/8″	1,475	675	2,150
1980				butt welded	8-3/4″	1,600	730	2,330
2000		18 ga. half glass	3′-0″ x 7′-0″	drywall K.D.	4-7/8″	1,000	470	1,470
2020				butt welded	8-3/4″	1,125	525	1,650
2160			6′-0″ x 7′-0″	drywall K.D.	4-7/8″	1,850	765	2,615
2180				butt welded	8-3/4″	1,975	820	2,795

C1020 Interior Doors

C1020 114 Metal Door/Metal Frame

	TYPE	DESIGN	SIZE	FRAME	DEPTH	COST EACH		
						MAT.	INST.	TOTAL
5000	Flush-hollow	16 ga. full panel	3'-0" x 7'-0"	drywall K.D.	4-7/8"	875	425	1,300
5020				butt welded	8-3/4"	985	480	1,465
5160			6'-0" x 7'-0"	drywall K.D.	4-7/8"	1,575	670	2,245
5180				butt welded	8-3/4"	1,700	725	2,425

C1020 Interior Doors

C1020 310	Hardware	COST EACH		
		MAT.	INST.	TOTAL
0060	**HINGES**			
0080				
0100	Full mortise, low frequency, steel base, 4-1/2" x 4-1/2", USP	12.25		12.25
0120	5" x 5" USP	26		26
0140	6" x 6" USP	49		49
0160	Average frequency, steel base, 4-1/2" x 4-1/2" USP	20		20
0180	5" x 5" USP	31.50		31.50
0200	6" x 6" USP	65.50		65.50
0220	High frequency, steel base, 4-1/2" x 4-1/2", USP	34.50		34.50
0240	5" x 5" USP	27.50		27.50
0260	6" x 6" USP	74		74
0280				
0300	**LOCKSETS**			
0320				
0340	Heavy duty cylindrical, passage door			
0360	Non-keyed, passage	77	51.50	128.50
0380	Privacy	82.50	51.50	134
0400	Keyed, single cylinder function	118	61.50	179.50
0420	Hotel	220	77	297
0440				
0460	For re-core cylinder, add	55		55
0480				
0500	**CLOSERS**			
0520				
0540	Rack & pinion			
0560	Adjustable backcheck, 3 way mount, all sizes, regular arm	211	103	314
0580	Hold open arm	211	103	314
0600	Fusible link	182	94.50	276.50
0620	Non-sized, regular arm	211	103	314
0640	4 way mount, non-sized, regular arm	211	103	314
0660	Hold open arm	220	103	323
0680				
0700	**PUSH, PULL**			
0720				
0740	Push plate, aluminum	14.05	51.50	65.55
0760	Bronze	27.50	51.50	79
0780	Pull handle, push bar, aluminum	130	56	186
0800	Bronze	175	61.50	236.50
0810				
0820	**PANIC DEVICES**			
0840				
0860	Narrow stile, rim mounted, bar, exit only	670	103	773
0880	Outside key & pull	980	123	1,103
0900	Bar and vertical rod, exit only	1,000	123	1,123
0920	Outside key & pull	1,000	154	1,154
0940	Bar and concealed rod, exit only	1,175	205	1,380
0960	Mortise, bar, exit only	660	154	814
0980	Touch bar, exit only	845	154	999
1000				
1020	**WEATHERSTRIPPING**			
1040				
1060	Interlocking, 3' x 7', zinc	48.50	205	253.50
1080	Spring type, 3' x 7', bronze	51	205	256

C3010 Wall Finishes

C3010 230	Paint & Covering	COST PER S.F.		
		MAT.	INST.	TOTAL
0060	Painting, interior on plaster and drywall, brushwork, primer & 1 coat	.14	.80	.94
0080	Primer & 2 coats	.21	1.06	1.27
0100	Primer & 3 coats	.29	1.30	1.59
0120	Walls & ceilings, roller work, primer & 1 coat	.14	.53	.67
0140	Primer & 2 coats	.21	.68	.89
0160	Woodwork incl. puttying, brushwork, primer & 1 coat	.14	1.15	1.29
0180	Primer & 2 coats	.21	1.53	1.74
0200	Primer & 3 coats	.29	2.08	2.37
0260	Cabinets and casework, enamel, primer & 1 coat	.14	1.30	1.44
0280	Primer & 2 coats	.22	1.60	1.82
0300	Masonry or concrete, latex, brushwork, primer & 1 coat	.31	1.08	1.39
0320	Primer & 2 coats	.41	1.55	1.96
0340	Addition for block filler	.24	1.33	1.57
0380	Fireproof paints, intumescent, 1/8" thick 3/4 hour	2.43	1.06	3.49
0420	7/16" thick 2 hour	7	3.71	10.71
0440	1-1/16" thick 3 hour	11.40	7.40	18.80
0500	Gratings, primer & 1 coat	.36	1.62	1.98
0600	Pipes over 12" diameter	.83	5.20	6.03
0720	Heavy framing 50-100 S.F./Ton	.11	.96	1.07
0740	Spraywork, light framing 300-500 S.F./Ton	.12	.43	.55
0800	Varnish, interior wood trim, no sanding sealer & 1 coat	.08	1.30	1.38
0820	Hardwood floor, no sanding 2 coats	.17	.27	.44
0840	Wall coatings, acrylic glazed coatings, minimum	.34	.99	1.33
0860	Maximum	.72	1.70	2.42
0880	Epoxy coatings, solvent based	.44	.99	1.43
1080	Water based	1.16	3.58	4.74
1100	High build epoxy 50 mil, solvent based	.75	1.33	2.08
1120	Water based	1.27	5.45	6.70
1140	Laminated epoxy with fiberglass solvent based	.80	1.76	2.56
1160	Water based	1.45	3.58	5.03
1180	Sprayed perlite or vermiculite 1/16" thick, solvent based	.30	.18	.48
1200	Water based	.81	.81	1.62
1260	Wall coatings, vinyl plastic, solvent based	.36	.71	1.07
1280	Water based	.90	2.16	3.06
1300	Urethane on smooth surface, 2 coats, solvent based	.30	.46	.76
1320	Water based	.65	.78	1.43
1340	3 coats, solvent based	.39	.62	1.01
1360	Water based	.87	1.10	1.97
1380	Ceramic-like glazed coating, cementitious, solvent based	.53	1.18	1.71
1400	Water based	.89	1.50	2.39
1420	Resin base, solvent based	.36	.81	1.17
1440	Water based	.59	1.57	2.16
1460	Wall coverings, aluminum foil	1.14	1.91	3.05
1500	Vinyl backing	6.10	2.19	8.29
1940	Ceramic tile, thin set, 4-1/4" x 4-1/4"	2.48	5.10	7.58
1960	12" x 12"	4.78	6.05	10.83

C3020 Floor Finishes

C3020 410	Tile & Covering	COST PER S.F.		
		MAT.	INST.	TOTAL
0060	Carpet tile, nylon, fusion bonded, 18″ x 18″ or 24″ x 24″, 24 oz.	3.66	.75	4.41
0080	35 oz.	4.27	.75	5.02
0100	42 oz.	5.65	.75	6.40
0140	Carpet, tufted, nylon, roll goods, 12′ wide, 26 oz.	3.04	1.22	4.26
0160	36 oz.	4.22	1.22	5.44
0180	Woven, wool, 36 oz.	11.10	.86	11.96
0200	42 oz.	12.20	.86	13.06
0220	Padding, add to above, 2.7 density	.73	.40	1.13
0240	13.0 density	.98	.40	1.38
0260	Composition flooring, acrylic, 1/4″ thick	1.83	6.05	7.88
0280	3/8″ thick	2.31	6.95	9.26
0300	Epoxy, 3/8″ thick	3.09	4.64	7.73
0320	1/2″ thick	4.46	6.40	10.86
0340	Epoxy terrazzo, granite chips	6.55	6.60	13.15
0360	Recycled porcelain	9.95	8.85	18.80
0380	Mastic, hot laid, 1-1/2″ thick, minimum	4.57	4.54	9.11
0400	Maximum	5.85	6.05	11.90
0420	Neoprene 1/4″ thick, minimum	4.50	5.75	10.25
0440	Maximum	6.15	7.25	13.40
0460	Polyacrylate with ground granite 1/4″, granite chips	3.85	4.26	8.11
0480	Recycled porcelain	7.05	6.55	13.60
0500	Polyester with colored quartz chips 1/16″, minimum	3.48	2.93	6.40
0520	Maximum	5.20	4.64	9.84
0540	Polyurethane with vinyl chips, clear	8.35	2.93	11.28
0560	Pigmented	12.15	3.64	15.79
0600	Concrete topping, granolithic concrete, 1/2″ thick	.34	4.95	5.29
0620	1″ thick	.68	5.05	5.73
0640	2″ thick	1.35	5.85	7.20
0660	Heavy duty 3/4″ thick, minimum	.46	7.70	8.16
0680	Maximum	.88	9.15	10.03
0700	For colors, add to above, minimum	.48	1.77	2.25
0720	Maximum	.81	1.95	2.76
0740	Exposed aggregate finish, minimum	.51	.91	1.42
0760	Maximum	.96	1.22	2.18
0780	Abrasives, .25 P.S.F. add to above, minimum	.58	.67	1.25
0800	Maximum	.87	.67	1.54
0820	Dust on coloring, add, minimum	.48	.44	.92
0840	Maximum	.81	.91	1.72
0860	Floor coloring using 0.6 psf powdered color, 1/2″ integral, minimum	5.45	4.95	10.40
0880	Maximum	5.80	4.95	10.75
1020	Integral topping and finish, 1:1:2 mix, 3/16″ thick	.11	2.92	3.03
1040	1/2″ thick	.31	3.08	3.39
1060	3/4″ thick	.46	3.43	3.89
1080	1″ thick	.61	3.89	4.50
1100	Terrazzo, minimum	3.69	17.65	21.34
1120	Maximum	6.80	22	28.80

C3030 Ceiling Finishes

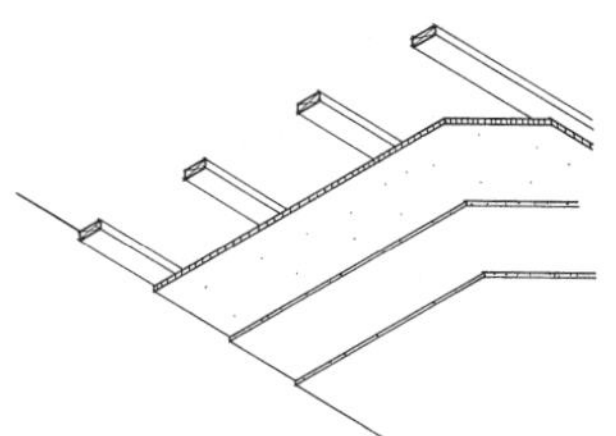

2 Coats of Plaster on Gypsum Lath on Wood Furring

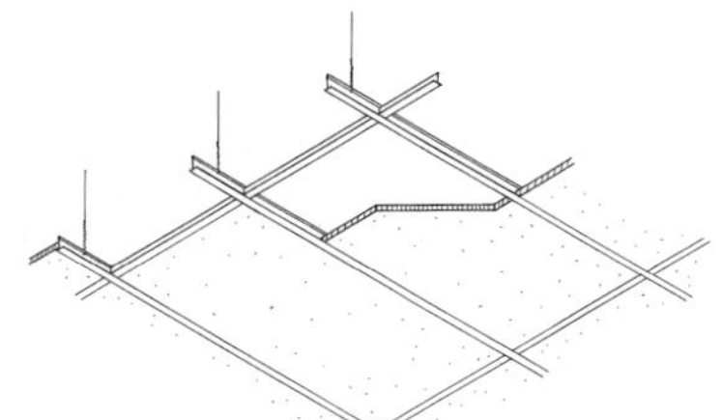

Fiberglass Board on Exposed Suspended Grid System

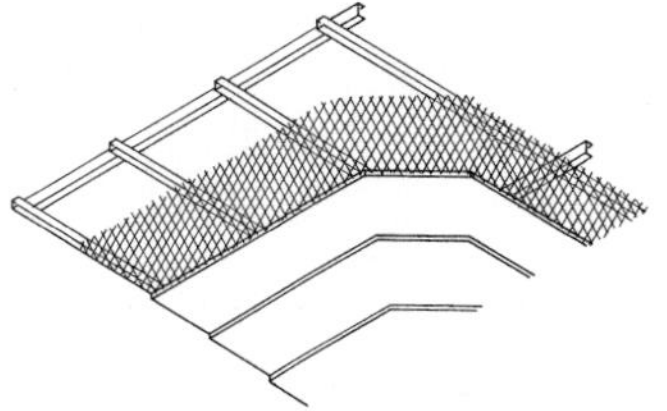

Plaster and Metal Lath on Metal Furring

System Components	QUANTITY	UNIT	COST PER S.F. MAT.	COST PER S.F. INST.	COST PER S.F. TOTAL
SYSTEM C3030 105 2400					
GYPSUM PLASTER, 2 COATS, 3/8″GYP. LATH, WOOD FURRING, FRAMING					
Gypsum plaster 2 coats no lath, on ceilings	.110	S.Y.	.44	3.35	3.79
Gypsum lath plain/perforated nailed, 3/8″ thick	.110	S.Y.	.37	.70	1.07
Add for ceiling installation	.110	S.Y.		.28	.28
Furring, 1″ x 3″ wood strips on wood joists	.750	L.F.	.32	1.32	1.64
Paint, primer and one coat	1.000	S.F.	.14	.53	.67
TOTAL			1.27	6.18	7.45

C3030 105 Plaster Ceilings

	TYPE	LATH	FURRING	SUPPORT		COST PER S.F. MAT.	COST PER S.F. INST.	COST PER S.F. TOTAL
2400	2 coat gypsum	3/8″ gypsum	1″x3″ wood, 16″ O.C.	wood		1.27	6.20	7.47
2500	Painted			masonry		1.30	6.30	7.60
2600				concrete		1.30	7.05	8.35
2700	3 coat gypsum	3.4# metal	1″x3″ wood, 16″ O.C.	wood		1.61	6.65	8.26
2800	Painted			masonry		1.64	6.75	8.39
2900				concrete		1.64	7.50	9.14
3000	2 coat perlite	3/8″ gypsum	1″x3″ wood, 16″ O.C.	wood		1.55	6.40	7.95
3100	Painted			masonry		1.58	6.55	8.13
3200				concrete		1.58	7.30	8.88
3300	3 coat perlite	3.4# metal	1″x3″ wood, 16″ O.C.	wood		1.70	6.55	8.25
3400	Painted			masonry		1.73	6.70	8.43
3500				concrete		1.73	7.45	9.18
3600	2 coat gypsum	3/8″ gypsum	3/4″ CRC, 12″ O.C.	1-1/2″ CRC, 48″O.C.		1.35	7.45	8.80
3700	Painted		3/4″ CRC, 16″ O.C.	1-1/2″ CRC, 48″O.C.		1.31	6.75	8.06
3800			3/4″ CRC, 24″ O.C.	1-1/2″ CRC, 48″O.C.		1.19	6.15	7.34
3900	2 coat perlite	3/8″ gypsum	3/4″ CRC, 12″ O.C.	1-1/2″ CRC, 48″O.C		1.63	7.95	9.58
4000	Painted		3/4″ CRC, 16″ O.C.	1-1/2″ CRC, 48″O.C.		1.59	7.25	8.84
4100			3/4″ CRC, 24″ O.C.	1-1/2″ CRC, 48″O.C.		1.47	6.65	8.12
4200	3 coat gypsum	3.4# metal	3/4″ CRC, 12″ O.C.	1-1/2″ CRC, 36″ O.C.		1.85	9.70	11.55
4300	Painted		3/4″ CRC, 16″ O.C.	1-1/2″ CRC, 36″ O.C.		1.81	9	10.81
4400			3/4″ CRC, 24″ O.C.	1-1/2″ CRC, 36″ O.C.		1.69	8.45	10.14
4500	3 coat perlite	3.4# metal	3/4″ CRC, 12″ O.C.	1-1/2″ CRC,36″ O.C.		2	10.65	12.65
4600	Painted		3/4″ CRC, 16″ O.C.	1-1/2″ CRC, 36″ O.C.		1.96	9.95	11.91
4700			3/4″ CRC, 24″ O.C.	1-1/2″ CRC, 36″ O.C.		1.84	9.35	11.19

D Services

D1010 Elevators and Lifts

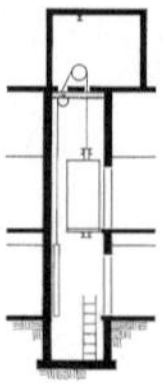

Geared traction elevators are the intermediate group both in cost and in operating areas. These are in buildings of four to fifteen floors and speed ranges from 200′ to 500′ per minute.

System Components	QUANTITY	UNIT	COST EACH MAT.	COST EACH INST.	COST EACH TOTAL
SYSTEM D1010 140 1600					
PASSENGER, 2500 LB., 5 FLOORS, 200 FPM					
Passenger elevator, geared, 2000 lb. capacity, 4 stop, 200 FPM	1.000	Ea.	101,000	37,800	138,800
Over 40′ travel height, passenger elevator electric, add	10.000	V.L.F.	7,650	2,610	10,260
Passenger elevator, electric, 2500 lb. cap., add	1.000	Ea.	4,175		4,175
Over 4 stops, passenger elevator, electric, add	1.000	Stop	3,400	7,000	10,400
Hall lantern	5.000	Ea.	2,850	1,180	4,030
Maintenance agreement for passenger elevator, 9 months	1.000	Ea.	4,400		4,400
Position indicator at lobby	1.000	Ea.	460	59	519
TOTAL			123,935	48,649	172,584

D1010 140	Traction Geared Elevators	COST EACH MAT.	COST EACH INST.	COST EACH TOTAL
1300	Passenger, 2000 Lb., 5 floors, 200 FPM	120,000	48,600	168,600
1500	15 floors, 350 FPM	245,500	147,000	392,500
1600	2500 Lb., 5 floors, 200 FPM	124,000	48,600	172,600
1800	15 floors, 350 FPM	250,000	147,000	397,000
2200	3500 Lb., 5 floors, 200 FPM	126,000	48,600	174,600
2400	15 floors, 350 FPM	252,000	147,000	399,000
2500	4000 Lb., 5 floors, 200 FPM	127,500	48,600	176,100
2700	15 floors, 350 FPM	253,000	147,000	400,000
2800	4500 Lb., 5 floors, 200 FPM	130,000	48,600	178,600
3000	15 floors, 350 FPM	255,500	147,000	402,500
3100	5000 Lb., 5 floors, 200 FPM	132,500	48,600	181,100
3300	15 floors, 350 FPM	258,000	147,000	405,000
4000	Hospital, 3500 Lb., 5 floors, 200 FPM	125,000	48,600	173,600
4200	15 floors, 350 FPM	296,500	147,000	443,500
4300	4000 Lb., 5 floors, 200 FPM	125,000	48,600	173,600
4500	15 floors, 350 FPM	296,500	147,000	443,500
4600	4500 Lb., 5 floors, 200 FPM	131,000	48,600	179,600
4800	15 floors, 350 FPM	302,500	147,000	449,500
4900	5000 Lb., 5 floors, 200 FPM	133,000	48,600	181,600
5100	15 floors, 350 FPM	304,500	147,000	451,500
6000	Freight, 4000 Lb., 5 floors, 50 FPM class 'B'	150,500	50,500	201,000
6200	15 floors, 200 FPM class 'B'	351,000	175,000	526,000
6300	8000 Lb., 5 floors, 50 FPM class 'B'	175,000	50,500	225,500
6500	15 floors, 200 FPM class 'B'	375,500	175,000	550,500
7000	10,000 Lb., 5 floors, 50 FPM class 'B'	202,500	50,500	253,000
7200	15 floors, 200 FPM class 'B'	587,000	175,000	762,000
8000	20,000 Lb., 5 floors, 50 FPM class 'B'	224,000	50,500	274,500
8200	15 floors, 200 FPM class 'B'	608,500	175,000	783,500

D1020 Escalators and Moving Walks

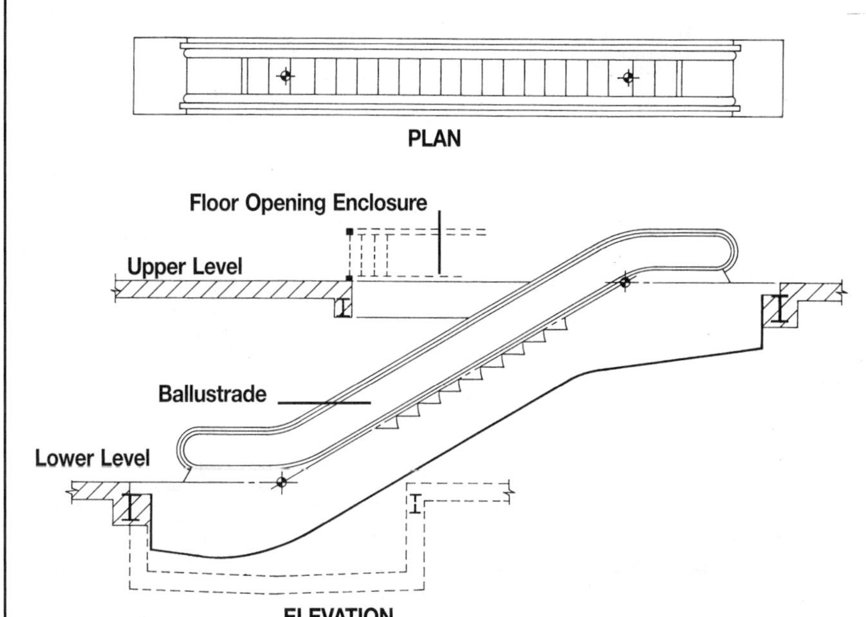

Moving stairs can be used for buildings where 600 or more people are to be carried to the second floor or beyond. Freight cannot be carried on escalators and at least one elevator must be available for this function.

Carrying capacity is 5000 to 8000 people per hour. Power requirement is 2 KW to 3 KW per hour and incline angle is 30°.

D1020 110 Moving Stairs

	TYPE	HEIGHT	WIDTH	BALUSTRADE MATERIAL		COST EACH		
						MAT.	INST.	TOTAL
0100	Escalator	10ft	32″	glass		95,000	52,000	147,000
0150				metal		103,000	52,000	155,000
0200			48″	glass		102,500	52,000	154,500
0250				metal		111,500	52,000	163,500
0300		15ft	32″	glass		100,000	61,000	161,000
0350				metal		108,500	61,000	169,500
0400			48″	glass		106,000	61,000	167,000
0450				metal		115,000	61,000	176,000
0500		20ft	32″	glass		106,500	74,000	180,500
0550				metal		115,500	74,000	189,500
0600			48″	glass		115,500	74,000	189,500
0650				metal		125,000	74,000	199,000
0700		25ft	32″	glass		116,000	91,000	207,000
0750				metal		125,500	91,000	216,500
0800			48″	glass		134,000	91,000	225,000
0850				metal		144,500	91,000	235,500

D1020 210 Moving Walks

	TYPE	STORY HEIGHT	DEGREE SLOPE	WIDTH	COST RANGE	COST PER L.F.		
						MAT.	INST.	TOTAL
1500	Ramp	10′-23′	12	3′-4″	minimum	2,575	1,025	3,600
1600					maximum	3,250	1,250	4,500
2000	Walk	0′	0	2′-0″	minimum	955	560	1,515
2500					maximum	1,325	820	2,145
3000				3′-4″	minimum	2,150	820	2,970
3500					maximum	2,525	955	3,480

D2010 Plumbing Fixtures

Systems are complete with trim seat and rough-in (supply, waste and vent) for connection to supply branches and waste mains.

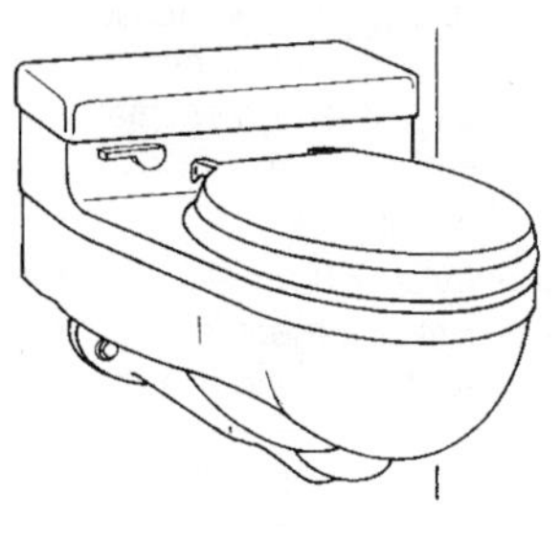

One Piece Wall Hung

Supply

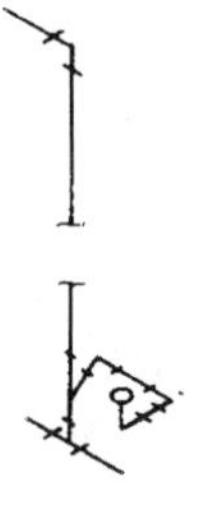

Waste/Vent

Floor Mount

System Components	QUANTITY	UNIT	COST EACH MAT.	COST EACH INST.	COST EACH TOTAL
SYSTEM D2010 110 1880					
WATER CLOSET, VITREOUS CHINA					
TANK TYPE, WALL HUNG, TWO PIECE					
Water closet, tank type vit china wall hung 2 pc. w/seat supply & stop	1.000	Ea.	695	249	944
Pipe Steel galvanized, schedule 40, threaded, 2″ diam.	4.000	L.F.	45.40	82	127.40
Pipe, CI soil, no hub, cplg 10′ OC, hanger 5′ OC, 4″ diam.	2.000	L.F.	42	45	87
Pipe, coupling, standard coupling, CI soil, no hub, 4″ diam.	2.000	Ea.	44	80	124
Copper tubing type L solder joint, hangar 10′ O.C., 1/2″ diam.	6.000	L.F.	24.42	54.30	78.72
Wrought copper 90° elbow for solder joints 1/2″ diam.	2.000	Ea.	2.42	73	75.42
Wrought copper Tee for solder joints 1/2″ diam.	1.000	Ea.	2.06	56.50	58.56
Supports/carrier, water closet, siphon jet, horiz, single, 4″ waste	1.000	Ea.	975	137	1,112
TOTAL			1,830.30	776.80	2,607.10

D2010 110	Water Closet Systems		COST EACH MAT.	COST EACH INST.	COST EACH TOTAL
1800	Water closet, vitreous china				
1840	Tank type, wall hung				
1880	Close coupled two piece	R224000 -30	1,825	775	2,600
1920	Floor mount, one piece		1,600	825	2,425
1960	One piece low profile		1,625	825	2,450
2000	Two piece close coupled		670	825	1,495
2040	Bowl only with flush valve				
2080	Wall hung		2,300	880	3,180
2120	Floor mount		795	840	1,635
2160	Floor mount, ADA compliant with 18″ high bowl		790	860	1,650

D2010 Plumbing Fixtures

Systems are complete with trim, flush valve and rough-in (supply, waste and vent) for connection to supply branches and waste mains.

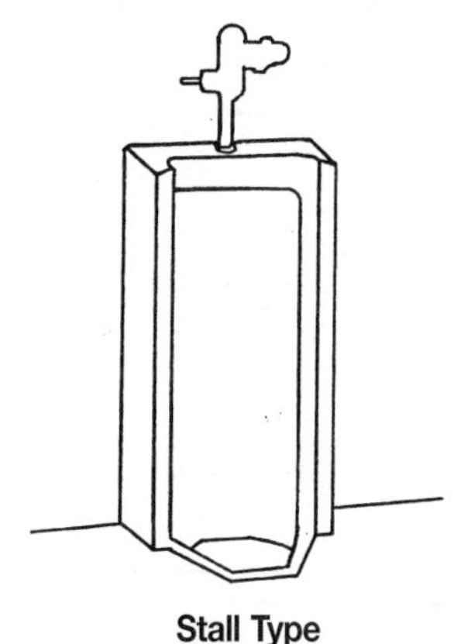

Stall Type

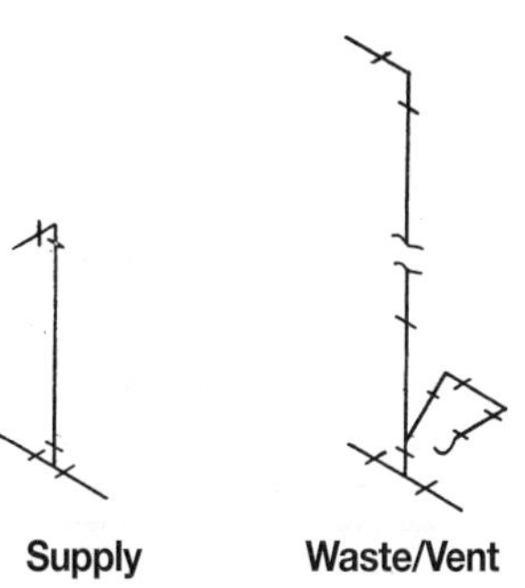

Supply   Waste/Vent

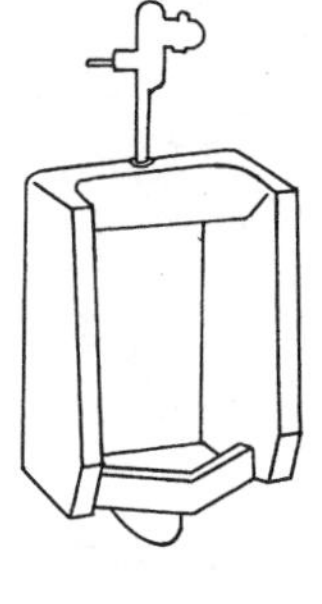

Wall Hung

System Components	QUANTITY	UNIT	COST EACH MAT.	COST EACH INST.	COST EACH TOTAL
SYSTEM D2010 210 2000					
URINAL, VITREOUS CHINA, WALL HUNG					
Urinal, wall hung, vitreous china, incl. hanger	1.000	Ea.	310	440	750
Pipe, steel, galvanized, schedule 40, threaded, 1-1/2" diam.	5.000	L.F.	44.25	82.50	126.75
Copper tubing type DWV, solder joint, hangers 10′ OC, 2" diam.	3.000	L.F.	48.30	49.95	98.25
Combination Y & 1/8 bend for CI soil pipe, no hub, 3" diam.	1.000	Ea.	21		21
Pipe, CI, no hub, cplg. 10′ OC, hanger 5′ OC, 3" diam.	4.000	L.F.	66.80	82	148.80
Pipe coupling standard, CI soil, no hub, 3" diam.	3.000	Ea.	37.60	69	106.60
Copper tubing type L, solder joint, hanger 10′ OC 3/4" diam.	5.000	L.F.	28.50	48.25	76.75
Wrought copper 90° elbow for solder joints 3/4" diam.	1.000	Ea.	2.71	38.50	41.21
Wrought copper Tee for solder joints, 3/4" diam.	1.000	Ea.	4.96	61	65.96
TOTAL			564.12	871.20	1,435.32

D2010 210	Urinal Systems		COST EACH MAT.	COST EACH INST.	COST EACH TOTAL
2000	Urinal, vitreous china, wall hung	R224000 -30	565	870	1,435
2040	Stall type		1,200	1,050	2,250

D2010 Plumbing Fixtures

Systems are complete with trim and rough-in (supply, waste and vent) to connect to supply branches and waste mains.

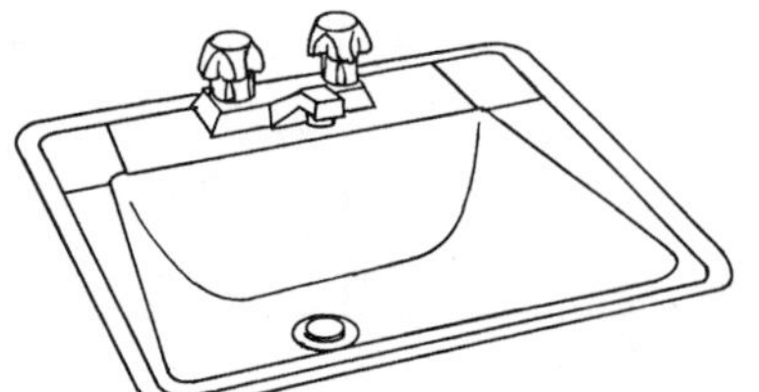

Vanity Top

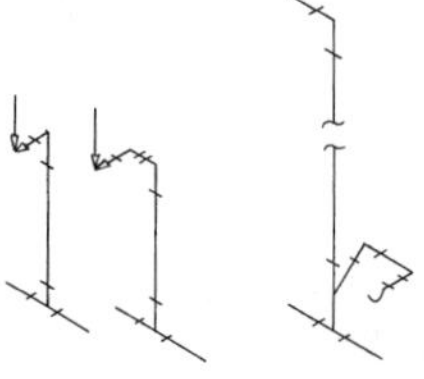

Supply Waste/Vent

Wall Hung

System Components	QUANTITY	UNIT	COST EACH MAT.	COST EACH INST.	COST EACH TOTAL
SYSTEM D2010 310 1560					
LAVATORY W/TRIM, VANITY TOP, P.E. ON C.I., 20″ X 18″					
Lavatory w/trim, PE on CI, white, vanity top, 20″ x 18″ oval	1.000	Ea.	370	206	576
Pipe, steel, galvanized, schedule 40, threaded, 1-1/4″ diam.	4.000	L.F.	30.80	59.20	90
Copper tubing type DWV, solder joint, hanger 10′ OC 1-1/4″ diam.	4.000	L.F.	40.20	48.80	89
Wrought copper DWV, Tee, sanitary, 1-1/4″ diam.	1.000	Ea.	23	81.50	104.50
P trap w/cleanout, 20 ga., 1-1/4″ diam.	1.000	Ea.	86	40.50	126.50
Copper tubing type L, solder joint, hanger 10′ OC 1/2″ diam.	10.000	L.F.	40.70	90.50	131.20
Wrought copper 90° elbow for solder joints 1/2″ diam.	2.000	Ea.	2.42	73	75.42
Wrought copper Tee for solder joints, 1/2″ diam.	2.000	Ea.	4.12	113	117.12
Stop, chrome, angle supply, 1/2″ diam.	2.000	Ea.	28.40	67	95.40
TOTAL			625.64	779.50	1,405.14

D2010 310	Lavatory Systems	COST EACH MAT.	COST EACH INST.	COST EACH TOTAL
1560	Lavatory w/trim, vanity top, PE on CI, 20″ x 18″, Vanity top by others.	625	780	1,405
1600	19″ x 16″ oval	445	780	1,225
1640	18″ round	735	780	1,515
1680	Cultured marble, 19″ x 17″	450	780	1,230
1720	25″ x 19″	485	780	1,265
1760	Stainless, self-rimming, 25″ x 22″	660	780	1,440
1800	17″ x 22″ R224000-30	645	780	1,425
1840	Steel enameled, 20″ x 17″	435	800	1,235
1880	19″ round	445	800	1,245
1920	Vitreous china, 20″ x 16″	540	820	1,360
1960	19″ x 16″	550	820	1,370
2000	22″ x 13″	550	820	1,370
2040	Wall hung, PE on CI, 18″ x 15″	945	860	1,805
2080	19″ x 17″	970	860	1,830
2120	20″ x 18″	805	860	1,665
2160	Vitreous china, 18″ x 15″	755	885	1,640
2200	19″ x 17″	730	885	1,615
2240	24″ x 20″	810	885	1,695
2300	20″ x 27″, handicap	1,475	955	2,430

D2010 Plumbing Fixtures

Systems are complete with trim and rough-in (supply, waste and vent) to connect to supply branches and waste mains.

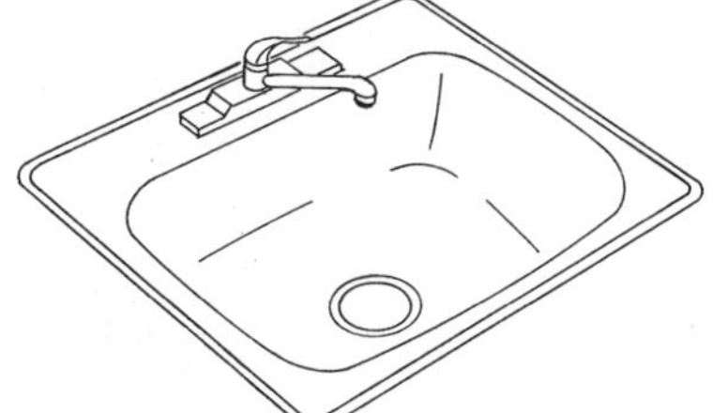

Countertop Single Bowl

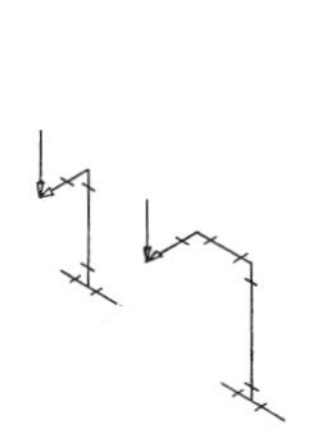

Supply

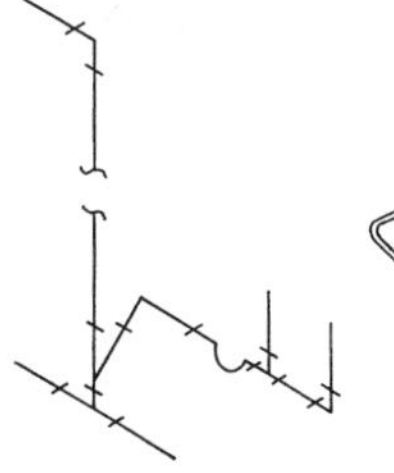

Waste/Vent

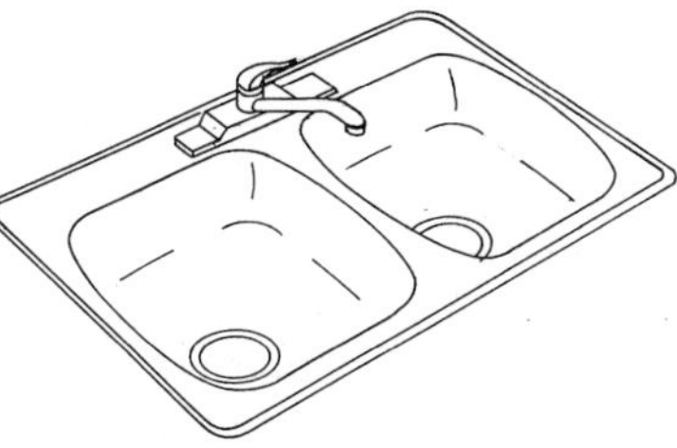

Countertop Double Bowl

System Components	QUANTITY	UNIT	COST EACH MAT.	COST EACH INST.	COST EACH TOTAL
SYSTEM D2010 410 1720					
KITCHEN SINK W/TRIM, COUNTERTOP, P.E. ON C.I., 24″ X 21″, SINGLE BOWL					
Kitchen sink, counter top, PE on CI, 1 bowl, 24″ x 21″ OD	1.000	Ea.	315	235	550
Pipe, steel, galvanized, schedule 40, threaded, 1-1/4″ diam.	4.000	L.F.	30.80	59.20	90
Copper tubing, type DWV, solder, hangers 10′ OC 1-1/2″ diam.	6.000	L.F.	73.80	81.30	155.10
Wrought copper, DWV, Tee, sanitary, 1-1/2″ diam.	1.000	Ea.	28.50	91.50	120
P trap, standard, copper, 1-1/2″ diam.	1.000	Ea.	79	43	122
Copper tubing, type L, solder joints, hangers 10′ OC 1/2″ diam.	10.000	L.F.	40.70	90.50	131.20
Wrought copper 90° elbow for solder joints 1/2″ diam.	2.000	Ea.	2.42	73	75.42
Wrought copper Tee for solder joints, 1/2″ diam.	2.000	Ea.	4.12	113	117.12
Stop, angle supply, chrome, 1/2″ CTS	2.000	Ea.	28.40	67	95.40
TOTAL			602.74	853.50	1,456.24

D2010 410	Kitchen Sink Systems	COST EACH MAT.	COST EACH INST.	COST EACH TOTAL
1720	Kitchen sink w/trim, countertop, PE on CI, 24″x21″, single bowl	605	855	1,460
1760	30″ x 21″ single bowl	970	855	1,825
1800	32″ x 21″ double bowl	700	920	1,620
1880	Stainless steel, 19″ x 18″ single bowl	940	855	1,795
1920	25″ x 22″ single bowl	1,025	855	1,880
1960	33″ x 22″ double bowl	1,350	920	2,270
2000	43″ x 22″ double bowl	1,550	935	2,485
2040	44″ x 22″ triple bowl	1,800	975	2,775
2080	44″ x 24″ corner double bowl	1,250	935	2,185
2120	Steel, enameled, 24″ x 21″ single bowl	850	855	1,705
2160	32″ x 21″ double bowl	855	920	1,775
2240	Raised deck, PE on CI, 32″ x 21″, dual level, double bowl	785	1,175	1,960
2280	42″ x 21″ dual level, triple bowl	1,625	1,275	2,900

D2010 Plumbing Fixtures

Systems are complete with trim and rough-in (supply, waste and vent) to connect to supply branches and waste mains.

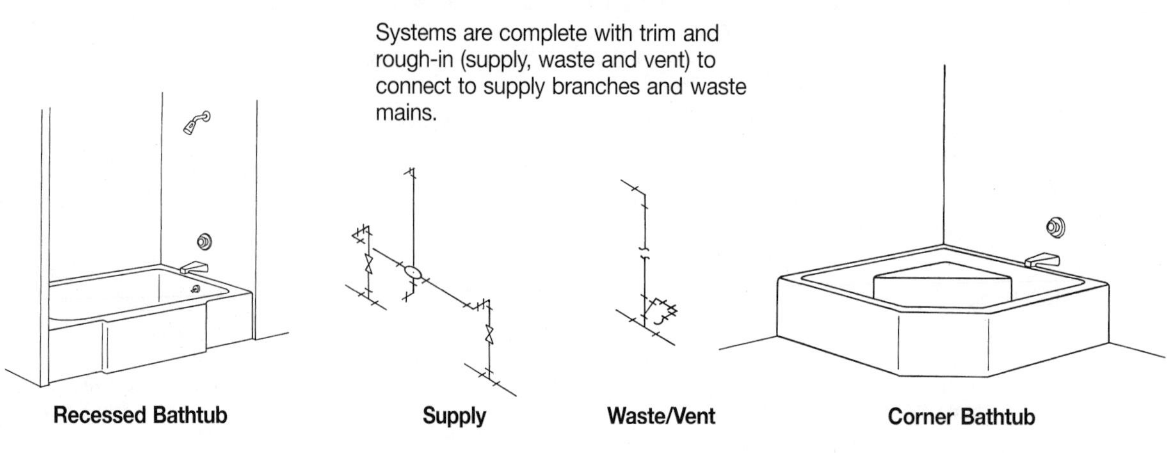

System Components	QUANTITY	UNIT	COST EACH MAT.	COST EACH INST.	COST EACH TOTAL
SYSTEM D2010 510 2000					
BATHTUB, RECESSED, PORCELAIN ENAMEL ON CAST IRON,, 48″ x 42″					
Bath tub, porcelain enamel on cast iron, w/fittings, 48″ x 42″	1.000	Ea.	2,900	330	3,230
Pipe, steel, galvanized, schedule 40, threaded, 1-1/4″ diam.	4.000	L.F.	30.80	59.20	90
Pipe, CI no hub soil w/couplings 10′ OC, hangers 5′ OC, 4″ diam.	3.000	L.F.	63	67.50	130.50
Combination Y and 1/8 bend for C.I. soil pipe, no hub, 4″ pipe size	1.000	Ea.	54		54
Drum trap, 3″ x 5″, copper, 1-1/2″ diam.	1.000	Ea.	117	46	163
Copper tubing type L, solder joints, hangers 10′ OC 1/2″ diam.	10.000	L.F.	40.70	90.50	131.20
Wrought copper 90° elbow, solder joints, 1/2″ diam.	2.000	Ea.	2.42	73	75.42
Wrought copper Tee, solder joints, 1/2″ diam.	2.000	Ea.	4.12	113	117.12
Stop, angle supply, 1/2″ diameter	2.000	Ea.	28.40	67	95.40
Copper tubing type DWV, solder joints, hanger 10′ OC 1-1/2″ diam.	3.000	L.F.	36.90	40.65	77.55
Pipe coupling, standard, C.I. soil no hub, 4″ pipe size	2.000	Ea.	44	80	124
TOTAL			3,321.34	966.85	4,288.19

D2010 510	Bathtub Systems		COST EACH MAT.	COST EACH INST.	COST EACH TOTAL
2000	Bathtub, recessed, P.E. on CI., 48″ x 42″	R224000 -30	3,325	965	4,290
2040	72″ x 36″		3,425	1,075	4,500
2080	Mat bottom, 5′ long		1,675	935	2,610
2120	5′-6″ long		2,375	965	3,340
2160	Corner, 48″ x 42″		3,325	935	4,260
2200	Formed steel, enameled, 4′-6″ long		965	865	1,830

D2010 Plumbing Fixtures

Systems are complete with trim, flush valve and rough-in (supply, waste and vent) for connection to supply branches and waste mains.

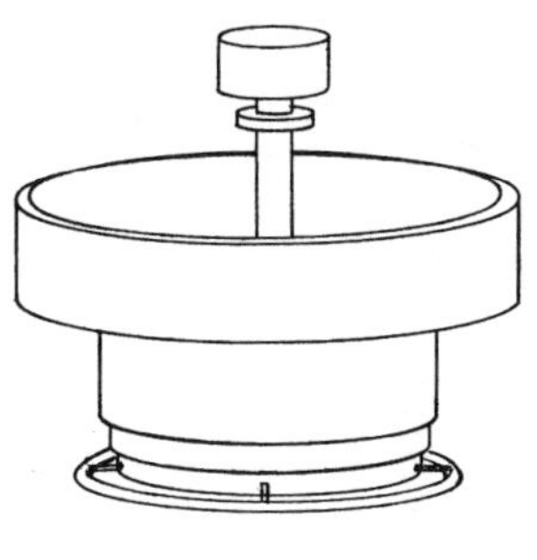

Circular Fountain

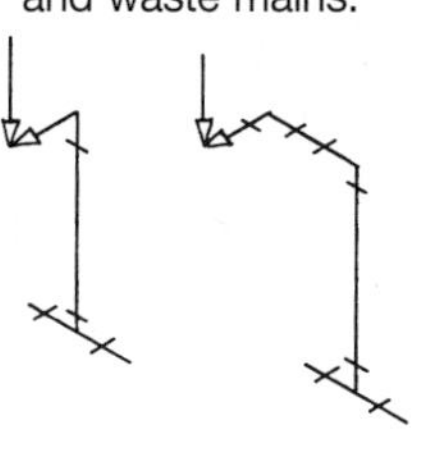

Supply

Waste/Vent

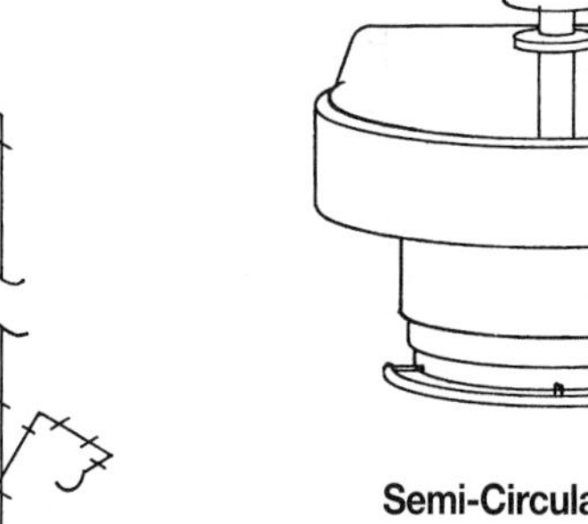

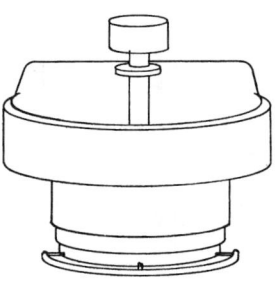

Semi-Circular Fountain

System Components	QUANTITY	UNIT	COST EACH MAT.	COST EACH INST.	COST EACH TOTAL
SYSTEM D2010 610 1760					
GROUP WASH FOUNTAIN, PRECAST TERRAZZO					
CIRCULAR, 36″ DIAMETER					
Wash fountain, group, precast terrazzo, foot control 36″ diam.	1.000	Ea.	8,000	685	8,685
Copper tubing type DWV, solder joint, hanger 10′ OC, 2″ diam.	10.000	L.F.	161	166.50	327.50
P trap, standard, copper, 2″ diam.	1.000	Ea.	122	49	171
Wrought copper, Tee, sanitary, 2″ diam.	1.000	Ea.	33	105	138
Copper tubing type L, solder joint, hanger 10′ OC 1/2″ diam.	20.000	L.F.	81.40	181	262.40
Wrought copper 90° elbow for solder joints 1/2″ diam.	3.000	Ea.	3.63	109.50	113.13
Wrought copper Tee for solder joints, 1/2″ diam.	2.000	Ea.	4.12	113	117.12
TOTAL			8,405.15	1,409	9,814.15

D2010 610	Group Wash Fountain Systems		COST EACH MAT.	COST EACH INST.	COST EACH TOTAL
1740	Group wash fountain, precast terrazzo				
1760	Circular, 36″ diameter		8,400	1,400	9,800
1800	54″ diameter	R224000 -30	10,400	1,550	11,950
1840	Semi-circular, 36″ diameter		7,425	1,400	8,825
1880	54″ diameter		9,825	1,550	11,375
1960	Stainless steel, circular, 36″ diameter		7,600	1,300	8,900
2000	54″ diameter		9,250	1,450	10,700
2040	Semi-circular, 36″ diameter		5,950	1,300	7,250
2080	54″ diameter		8,125	1,450	9,575
2160	Thermoplastic, circular, 36″ diameter		5,300	1,075	6,375
2200	54″ diameter		6,100	1,250	7,350
2240	Semi-circular, 36″ diameter		4,900	1,075	5,975
2280	54″ diameter		5,900	1,250	7,150

D2010 Plumbing Fixtures

Systems are complete with trim and rough-in (supply, waste and vent) for connection to supply branches and waste mains.

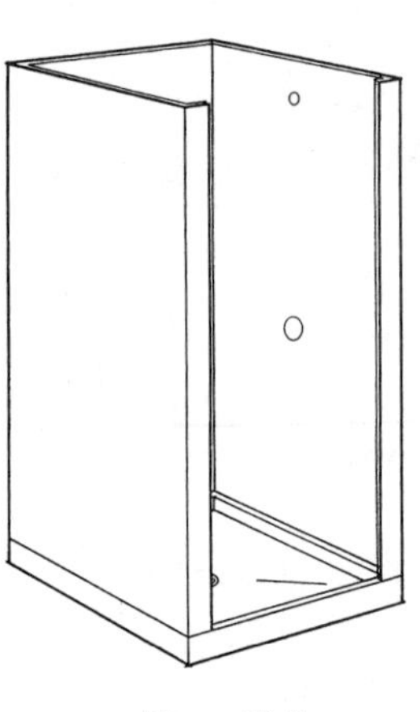

Three Wall

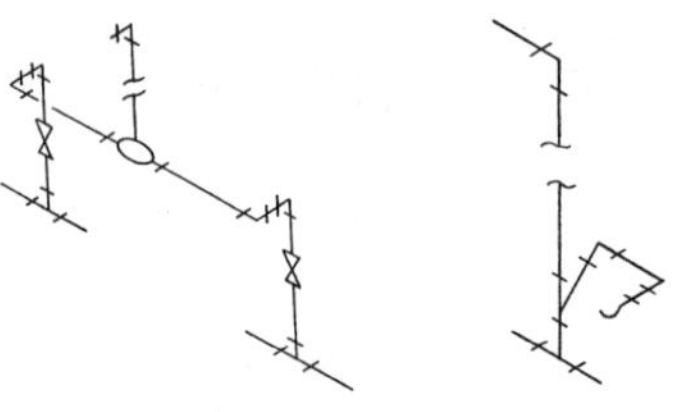

Supply

Waste/Vent

Corner Angle

System Components	QUANTITY	UNIT	COST EACH MAT.	COST EACH INST.	COST EACH TOTAL
SYSTEM D2010 710 1560					
SHOWER, STALL, BAKED ENAMEL, MOLDED STONE RECEPTOR, 30″ SQUARE					
Shower stall, enameled steel, molded stone receptor, 30″ square	1.000	Ea.	1,275	254	1,529
Copper tubing type DWV, solder joints, hangers 10′ OC, 2″ diam.	6.000	L.F.	73.80	81.30	155.10
Wrought copper DWV, Tee, sanitary, 2″ diam.	1.000	Ea.	28.50	91.50	120
Trap, standard, copper, 2″ diam.	1.000	Ea.	79	43	122
Copper tubing type L, solder joint, hanger 10′ OC 1/2″ diam.	16.000	L.F.	65.12	144.80	209.92
Wrought copper 90° elbow for solder joints 1/2″ diam.	3.000	Ea.	3.63	109.50	113.13
Wrought copper Tee for solder joints, 1/2″ diam.	2.000	Ea.	4.12	113	117.12
Stop and waste, straightway, bronze, solder joint 1/2″ diam.	2.000	Ea.	27.10	61	88.10
TOTAL			1,556.27	898.10	2,454.37

D2010 710	Shower Systems	COST EACH MAT.	COST EACH INST.	COST EACH TOTAL
1560	Shower, stall, baked enamel, molded stone receptor, 30″ square	1,550	900	2,450
1600	32″ square	1,550	910	2,460
1640	Terrazzo receptor, 32″ square (R224000-30)	1,750	910	2,660
1680	36″ square	1,900	920	2,820
1720	36″ corner angle	2,175	920	3,095
1800	Fiberglass one piece, three walls, 32″ square	820	885	1,705
1840	36″ square	835	885	1,720
1880	Polypropylene, molded stone receptor, 30″ square	980	1,300	2,280
1920	32″ square	995	1,300	2,295
1960	Built-in head, arm, bypass, stops and handles	112	340	452

D2010 Plumbing Fixtures

Systems are complete with trim and rough-in (supply, waste and vent) for connection to supply branches and waste mains.

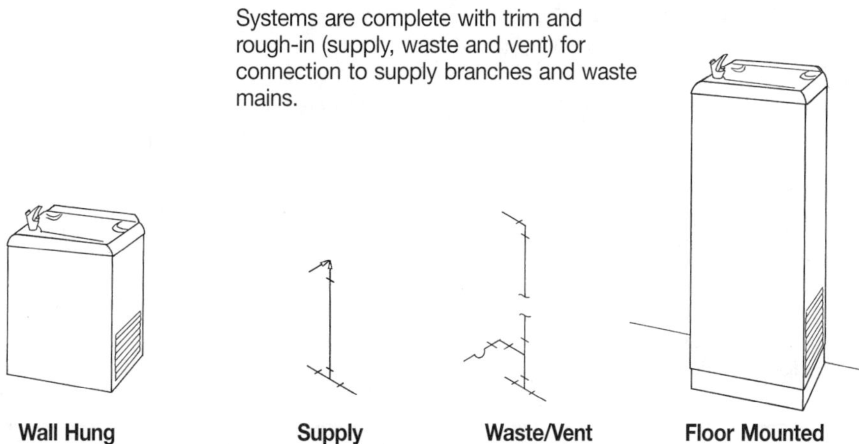

System Components	QUANTITY	UNIT	COST EACH MAT.	COST EACH INST.	COST EACH TOTAL
SYSTEM D2010 820 1840					
WATER COOLER, ELECTRIC, SELF CONTAINED, WALL HUNG, 8.2 G.P.H.					
Water cooler, wall mounted, 8.2 GPH	1.000	Ea.	825	330	1,155
Copper tubing type DWV, solder joint, hanger 10′ OC 1-1/4″ diam.	4.000	L.F.	40.20	48.80	89
Wrought copper DWV, Tee, sanitary 1-1/4″ diam.	1.000	Ea.	23	81.50	104.50
P trap, copper drainage, 1-1/4″ diam.	1.000	Ea.	86	40.50	126.50
Copper tubing type L, solder joint, hanger 10′ OC 3/8″ diam.	5.000	L.F.	19.25	43.50	62.75
Wrought copper 90° elbow for solder joints 3/8″ diam.	1.000	Ea.	3.42	33.50	36.92
Wrought copper Tee for solder joints, 3/8″ diam.	1.000	Ea.	5.80	52.50	58.30
Stop and waste, straightway, bronze, solder, 3/8″ diam.	1.000	Ea.	13.55	30.50	44.05
TOTAL			1,016.22	660.80	1,677.02

D2010 820	Water Cooler Systems		COST EACH MAT.	COST EACH INST.	COST EACH TOTAL
1840	Water cooler, electric, wall hung, 8.2 G.P.H.	R224000 -30	1,025	660	1,685
1880	Dual height, 14.3 G.P.H.		1,275	675	1,950
1920	Wheelchair type, 7.5 G.P.H.		1,200	660	1,860
1960	Semi recessed, 8.1 G.P.H.		1,025	660	1,685
2000	Full recessed, 8 G.P.H.		2,425	705	3,130
2040	Floor mounted, 14.3 G.P.H.		1,175	575	1,750
2080	Dual height, 14.3 G.P.H.		1,525	695	2,220
2120	Refrigerated compartment type, 1.5 G.P.H.		1,850	575	2,425

D2010 Plumbing Fixtures

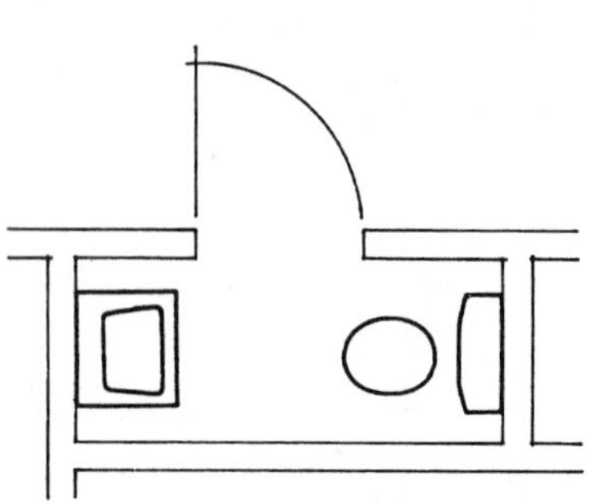

*Common wall is with an adjacent bathroom.

Two Fixture Bathroom Systems consisting of a lavatory, water closet, and rough-in service piping.

- Prices for plumbing and fixtures only.

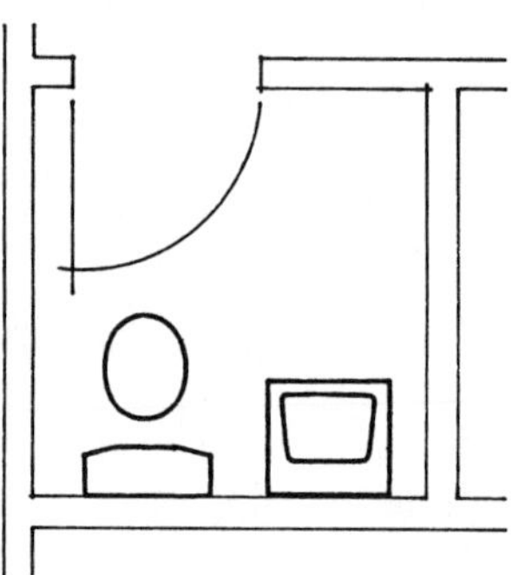

System Components	QUANTITY	UNIT	COST EACH		
			MAT.	INST.	TOTAL
SYSTEM D2010 920 1180					
BATHROOM, LAVATORY & WATER CLOSET, 2 WALL PLUMBING, STAND ALONE					
Water closet, two piece, close coupled	1.000	Ea.	261	249	510
Water closet, rough-in waste & vent	1.000	Set	365	430	795
Lavatory w/ftngs., wall hung, white, PE on CI, 20″ x 18″	1.000	Ea.	305	165	470
Lavatory, rough-in waste & vent	1.000	Set	500	795	1,295
Copper tubing type L, solder joint, hanger 10′ OC 1/2″ diam.	10.000	L.F.	40.70	90.50	131.20
Pipe, steel, galvanized, schedule 40, threaded, 2″ diam.	12.000	L.F.	136.20	246	382.20
Pipe, CI soil, no hub, coupling 10′ OC, hanger 5′ OC, 4″ diam.	7.000	L.F.	143.50	168	311.50
TOTAL			1,751.40	2,143.50	3,894.90

D2010 920	Two Fixture Bathroom, Two Wall Plumbing	COST EACH		
		MAT.	INST.	TOTAL
1180	Bathroom, lavatory & water closet, 2 wall plumbing, stand alone	1,750	2,150	3,900
1200	Share common plumbing wall*	1,600	1,850	3,450

D2010 922	Two Fixture Bathroom, One Wall Plumbing	COST EACH		
		MAT.	INST.	TOTAL
2220	Bathroom, lavatory & water closet, one wall plumbing, stand alone	1,650	1,925	3,575
2240	Share common plumbing wall*	1,425	1,650	3,075

D3030 Cooling Generating Systems

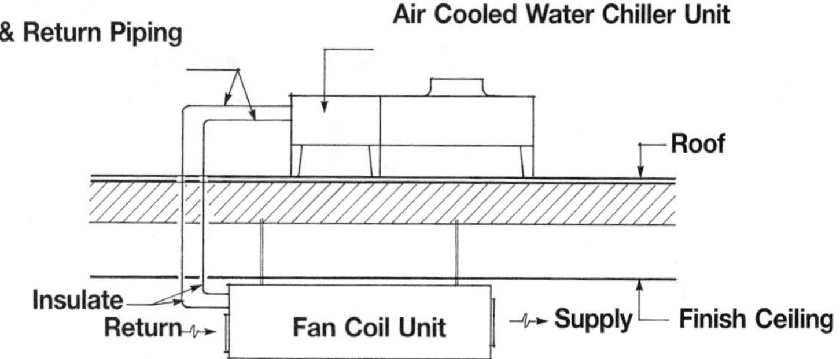

Design Assumptions: The chilled water, air cooled systems priced, utilize reciprocating hermetic compressors and propeller-type condenser fans. Piping with pumps and expansion tanks is included based on a two pipe system. No ducting is included and the fan-coil units are cooling only. Water treatment and balancing are not included. Chilled water piping is insulated. Area distribution is through the use of multiple fan coil units. Fewer but larger fan coil units with duct distribution would be approximately the same S.F. cost.

System Components	QUANTITY	UNIT	COST EACH		
			MAT.	INST.	TOTAL
SYSTEM D3030 110 1200					
PACKAGED CHILLER, AIR COOLED, WITH FAN COIL UNIT					
APARTMENT CORRIDORS, 3,000 S.F., 5.50 TON					
Fan coil air conditioning unit, cabinet mounted & filters chilled water	1.000	Ea.	4,399.20	614.06	5,013.26
Water chiller, air conditioning unit, air cooled	1.000	Ea.	6,490	2,296.25	8,786.25
Chilled water unit coil connections	1.000	Ea.	1,300	1,675	2,975
Chilled water distribution piping	440.000	L.F.	10,120	22,660	32,780
TOTAL			22,309.20	27,245.31	49,554.51
COST PER S.F.			7.44	9.08	16.52

D3030 110	Chilled Water, Air Cooled Condenser Systems	COST PER S.F.		
		MAT.	INST.	TOTAL
1180	Packaged chiller, air cooled, with fan coil unit			
1200	Apartment corridors, 3,000 S.F., 5.50 ton	7.43	9.07	16.50
1360	40,000 S.F., 73.33 ton	5.05	3.98	9.03
1440	Banks and libraries, 3,000 S.F., 12.50 ton	10.75	10.55	21.30
1560	20,000 S.F., 83.33 ton	8.55	5.35	13.90
1680	Bars and taverns, 3,000 S.F., 33.25 ton	18.80	12.35	31.15
1760	10,000 S.F., 110.83 ton	13.80	2.81	16.61
1920	Bowling alleys, 3,000 S.F., 17.00 ton	13.45	11.85	25.30
1960	6,000 S.F., 34.00 ton	10.50	8.25	18.75
2000	10,000 S.F., 56.66 ton	9.10	5.90	15
2040	20,000 S.F., 113.33 ton	9.80	5.40	15.20
2160	Department stores, 3,000 S.F., 8.75 ton	10.30	10.10	20.40
2320	40,000 S.F., 116.66 ton	6.40	4.09	10.49
2400	Drug stores, 3,000 S.F., 20.00 ton	15.20	12.20	27.40
2520	20,000 S.F., 133.33 ton	11.65	5.65	17.30
2640	Factories, 2,000 S.F., 10.00 ton	9.50	10.15	19.65
2800	40,000 S.F., 133.33 ton	6.90	4.19	11.09
2880	Food supermarkets, 3,000 S.F., 8.50 ton	10.10	10.05	20.15
3040	40,000 S.F., 113.33 ton	6.20	4.09	10.29
3120	Medical centers, 3,000 S.F., 7.00 ton	9.10	9.80	18.90
3280	40,000 S.F., 93.33 ton	5.65	4.03	9.68
3360	Offices, 3,000 S.F., 9.50 ton	9	10	19
3520	40,000 S.F., 126.66 ton	6.80	4.23	11.03
3600	Restaurants, 3,000 S.F., 15.00 ton	12	10.95	22.95
3720	20,000 S.F., 100.00 ton	10	5.70	15.70
3840	Schools and colleges, 3,000 S.F., 11.50 ton	10.25	10.40	20.65
3960	20,000 S.F., 76.66 ton	8.40	5.40	13.80

D3030 Cooling Generating Systems

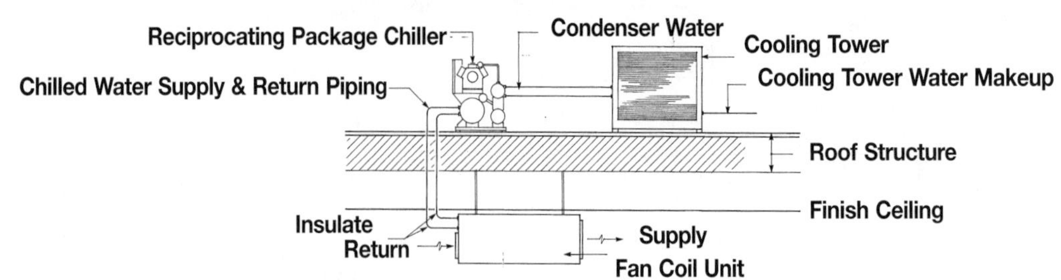

General: Water cooled chillers are available in the same sizes as air cooled units. They are also available in larger capacities.

Design Assumptions: The chilled water systems with water cooled condenser include reciprocating hermetic compressors, water cooling tower, pumps, piping and expansion tanks and are based on a two pipe system. Chilled water piping is insulated. No ducts are included and fan coil units are cooling only. Area distribution is through use of multiple fan coil units. Fewer but larger fan coil units with duct distribution would be approximately the same S.F. cost. Water treatment and balancing are not included.

System Components	QUANTITY	UNIT	COST EACH MAT.	COST EACH INST.	COST EACH TOTAL
SYSTEM D3030 115 1320					
PACKAGED CHILLER, WATER COOLED, WITH FAN COIL UNIT					
APARTMENT CORRIDORS, 4,000 S.F., 7.33 TON					
Fan coil air conditioner unit, cabinet mounted & filters, chilled water	2.000	Ea.	5,863.20	818.41	6,681.61
Water chiller, water cooled, 1 compressor, hermetic scroll,	1.000	Ea.	5,793.70	4,007.50	9,801.20
Cooling tower, draw thru single flow, belt drive	1.000	Ea.	1,737.21	168.59	1,905.80
Cooling tower pumps & piping	1.000	System	864.94	403.15	1,268.09
Chilled water unit coil connections	2.000	Ea.	2,600	3,350	5,950
Chilled water distribution piping	520.000	L.F.	11,960	26,780	38,740
TOTAL			28,819.05	35,527.65	64,346.70
COST PER S.F.			7.20	8.88	16.08

D3030 115	Chilled Water, Cooling Tower Systems	COST PER S.F. MAT.	COST PER S.F. INST.	COST PER S.F. TOTAL
1300	Packaged chiller, water cooled, with fan coil unit			
1320	Apartment corridors, 4,000 S.F., 7.33 ton	7.20	8.88	16.08
1600	Banks and libraries, 4,000 S.F., 16.66 ton	12.45	9.65	22.10
1800	60,000 S.F., 250.00 ton	9.25	7.70	16.95
1880	Bars and taverns, 4,000 S.F., 44.33 ton	19.80	12	31.80
2000	20,000 S.F., 221.66 ton	21.50	9.85	31.35
2160	Bowling alleys, 4,000 S.F., 22.66 ton	14.30	10.55	24.85
2320	40,000 S.F., 226.66 ton	12.20	7.20	19.40
2440	Department stores, 4,000 S.F., 11.66 ton	7.95	9.55	17.50
2640	60,000 S.F., 175.00 ton	8.20	7.10	15.30
2720	Drug stores, 4,000 S.F., 26.66 ton	15.55	10.85	26.40
2880	40,000 S.F., 266.67 ton	12.35	8.15	20.50
3000	Factories, 4,000 S.F., 13.33 ton	10.70	9.20	19.90
3200	60,000 S.F., 200.00 ton	8.35	7.40	15.75
3280	Food supermarkets, 4,000 S.F., 11.33 ton	7.80	9.50	17.30
3480	60,000 S.F., 170.00 ton	8.10	7.05	15.15
3560	Medical centers, 4.000 S.F., 9.33 ton	6.80	8.70	15.50
3760	60,000 S.F., 140.00 ton	6.90	7.15	14.05
3840	Offices, 4,000 S.F., 12.66 ton	10.40	9.10	19.50
4040	60,000 S.F., 190.00 ton	8.10	7.30	15.40
4120	Restaurants, 4,000 S.F., 20.00 ton	12.80	9.80	22.60
4320	60,000 S.F., 300.00 ton	10.30	7.95	18.25
4400	Schools and colleges, 4,000 S.F., 15.33 ton	11.75	9.45	21.20
4600	60,000 S.F., 230.00 ton	8.55	7.40	15.95

D3050 Terminal & Package Units

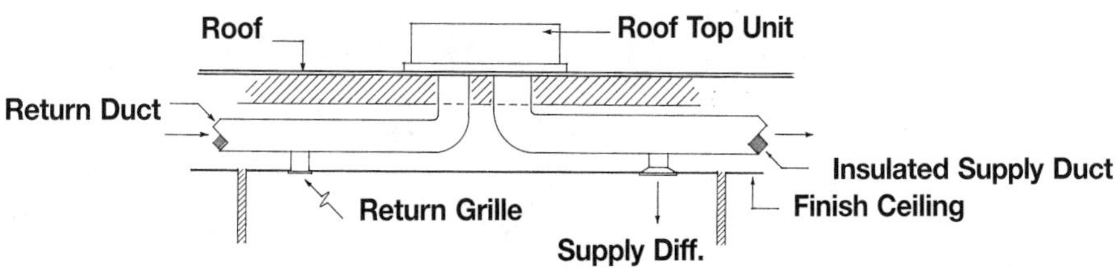

System Description: Rooftop single zone units are electric cooling and gas heat. Duct systems are low velocity, galvanized steel supply and return. Price variations between sizes are due to several factors. Jumps in the cost of the rooftop unit occur when the manufacturer shifts from the largest capacity unit on a small frame to the smallest capacity on the next larger frame, or changes from one compressor to two. As the unit capacity increases for larger areas the duct distribution grows in proportion. For most applications there is a tradeoff point where it is less expensive and more efficient to utilize smaller units with short simple distribution systems. Larger units also require larger initial supply and return ducts which can create a space problem. Supplemental heat may be desired in colder locations. The table below is based on one unit supplying the area listed. The 10,000 S.F. unit for bars and taverns is not listed because a nominal 110 ton unit would be required and this is above the normal single zone rooftop capacity.

System Components	QUANTITY	UNIT	COST EACH MAT.	COST EACH INST.	COST EACH TOTAL
SYSTEM D3050 150 1280					
ROOFTOP, SINGLE ZONE, AIR CONDITIONER					
APARTMENT CORRIDORS, 500 S.F., .92 TON					
Rooftop air conditioner, 1 zone, electric cool, standard controls, curb	1.000	Ea.	1,989.50	885.50	2,875
Ductwork package for rooftop single zone units	1.000	System	179.40	1,219	1,398.40
TOTAL			2,168.90	2,104.50	4,273.40
COST PER S.F.			4.34	4.21	8.55

D3050 150	Rooftop Single Zone Unit Systems	COST PER S.F. MAT.	COST PER S.F. INST.	COST PER S.F. TOTAL
1260	Rooftop, single zone, air conditioner			
1280	Apartment corridors, 500 S.F., .92 ton	4.35	4.20	8.55
1480	10,000 S.F., 18.33 ton	2.76	2.82	5.58
1560	Banks or libraries, 500 S.F., 2.08 ton	9.80	9.50	19.30
1760	10,000 S.F., 41.67 ton	5.40	6.35	11.75
1840	Bars and taverns, 500 S.F. 5.54 ton	13.55	12.65	26.20
2000	5,000 S.F., 55.42 ton	12.60	9.60	22.20
2080	Bowling alleys, 500 S.F., 2.83 ton	8.25	10.65	18.90
2280	10,000 S.F., 56.67 ton	7	8.65	15.65
2360	Department stores, 500 S.F., 1.46 ton	6.90	6.70	13.60
2560	10,000 S.F., 29.17 ton	3.87	4.46	8.33
2640	Drug stores, 500 S.F., 3.33 ton	9.70	12.55	22.25
2840	10,000 S.F., 66.67 ton	8.25	10.20	18.45
2920	Factories, 500 S.F., 1.67 ton	7.85	7.65	15.50
3120	10,000 S.F., 33.33 ton	4.42	5.10	9.52
3200	Food supermarkets, 500 S.F., 1.42 ton	6.65	6.50	13.15
3400	10,000 S.F., 28.33 ton	3.75	4.32	8.07
3480	Medical centers, 500 S.F., 1.17 ton	5.50	5.35	10.85
3680	10,000 S.F., 23.33 ton	3.51	3.59	7.10
3760	Offices, 500 S.F., 1.58 ton	7.45	7.25	14.70
3960	10,000 S.F., 31.67 ton	4.20	4.84	9.04
4000	Restaurants, 500 S.F., 2.50 ton	11.80	11.45	23.25
4200	10,000 S.F., 50.00 ton	6.20	7.65	13.85
4240	Schools and colleges, 500 S.F., 1.92 ton	9.05	8.80	17.85
4440	10,000 S.F., 38.33 ton	4.98	5.85	10.83

D3050 Terminal & Package Units

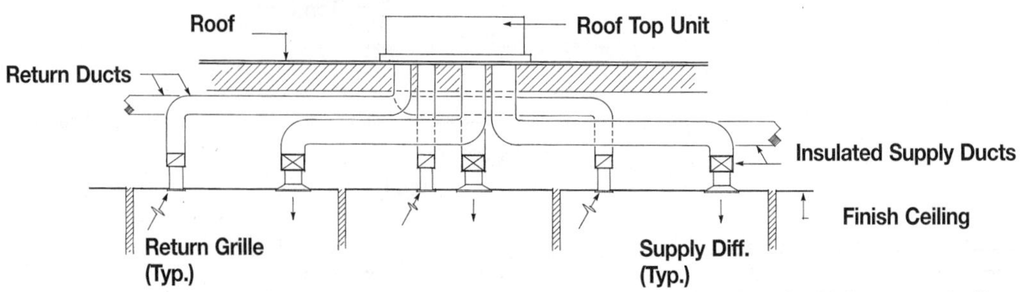

System Description: Rooftop units are multizone with up to 12 zones, and include electric cooling, gas heat, thermostats, filters, supply and return fans complete. Duct systems are low velocity, galvanized steel supply and return with insulated supplies.

Multizone units cost more per ton of cooling than single zone. However, they offer flexibility where load conditions are varied due to heat generating areas or exposure to radiational heating. For example, perimeter offices on the "sunny side" may require cooling at the same time "shady side" or central offices may require heating. It is possible to accomplish similar results using duct heaters in branches of the single zone unit. However, heater location could be a problem and total system operating energy efficiency could be lower.

System Components	QUANTITY	UNIT	COST EACH MAT.	COST EACH INST.	COST EACH TOTAL
SYSTEM D3050 155 1280					
ROOFTOP, MULTIZONE, AIR CONDITIONER					
APARTMENT CORRIDORS, 3,000 S.F., 5.50 TON					
Rooftop multizone unit, standard controls, curb	1.000	Ea.	31,460	2,057	33,517
Ductwork package for rooftop multizone units	1.000	System	1,617	14,025	15,642
TOTAL			33,077	16,082	49,159
COST PER S.F.			11.03	5.36	16.39
Note A: Small single zone unit recommended.					

D3050 155	Rooftop Multizone Unit Systems	COST PER S.F. MAT.	COST PER S.F. INST.	COST PER S.F. TOTAL
1240	Rooftop, multizone, air conditioner			
1260	Apartment corridors, 1,500 S.F., 2.75 ton. See Note A.			
1280	3,000 S.F., 5.50 ton	11.03	5.37	16.40
1440	25,000 S.F., 45.80 ton	7.20	5.15	12.35
1520	Banks or libraries, 1,500 S.F., 6.25 ton	25	12.20	37.20
1640	15,000 S.F., 62.50 ton	12	11.70	23.70
1720	25,000 S.F., 104.00 ton	10.70	11.65	22.35
1800	Bars and taverns, 1,500 S.F., 16.62 ton	54.50	17.60	72.10
1840	3,000 S.F., 33.24 ton	45	16.95	61.95
1880	10,000 S.F., 110.83 ton	27	16.85	43.85
2080	Bowling alleys, 1,500 S.F., 8.50 ton	34	16.55	50.55
2160	10,000 S.F., 56.70 ton	22.50	16	38.50
2240	20,000 S.F., 113.00 ton	14.50	15.80	30.30
2640	Drug stores, 1,500 S.F., 10.00 ton	40	19.50	59.50
2680	3,000 S.F., 20.00 ton	27.50	18.80	46.30
2760	15,000 S.F., 100.00 ton	17.15	18.65	35.80
3760	Offices, 1,500 S.F., 4.75 ton, See Note A.			
3880	15,000 S.F., 47.50 ton	12.45	8.95	21.40
3960	25,000 S.F., 79.16 ton	9.10	8.85	17.95
4000	Restaurants, 1,500 S.F., 7.50 ton	30	14.60	44.60
4080	10,000 S.F., 50.00 ton	19.65	14.10	33.75
4160	20,000 S.F., 100.00 ton	12.85	14	26.85
4240	Schools and colleges, 1,500 S.F., 5.75 ton	23	11.20	34.20
4360	15,000 S.F., 57.50 ton	11.05	10.75	21.80
4440	25,000 S.F., 95.83 ton	10.35	10.75	21.10

D3050 Terminal & Package Units

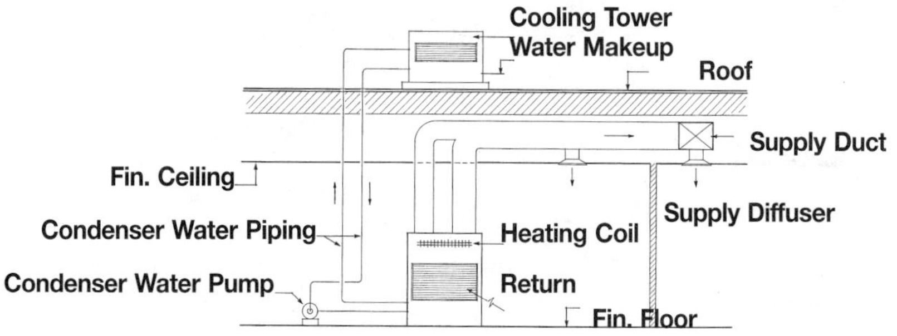

System Description: Self-contained, single package water cooled units include cooling tower, pump, piping allowance. Systems for 1000 S.F. and up include duct and diffusers to provide for even distribution of air. Smaller units distribute air through a supply air plenum, which is integral with the unit.

Returns are not ducted and supplies are not insulated.

Hot water or steam heating coils are included but piping to boiler and the boiler itself is not included.

Where local codes or conditions permit single pass cooling for the smaller units, deduct 10%.

System Components	QUANTITY	UNIT	COST EACH MAT.	COST EACH INST.	COST EACH TOTAL
SYSTEM D3050 160 1300					
SELF-CONTAINED, WATER COOLED UNIT					
APARTMENT CORRIDORS, 500 S.F., .92 TON					
Self contained, water cooled, single package air conditioner unit	1.000	Ea.	1,377.60	705.60	2,083.20
Ductwork package for water or air cooled packaged units	1.000	System	69.92	920	989.92
Cooling tower, draw thru single flow, belt drive	1.000	Ea.	218.04	21.16	239.20
Cooling tower pumps & piping	1.000	System	108.56	50.60	159.16
TOTAL			1,774.12	1,697.36	3,471.48
COST PER S.F.			3.55	3.39	6.94

D3050 160	Self-contained, Water Cooled Unit Systems	COST PER S.F. MAT.	COST PER S.F. INST.	COST PER S.F. TOTAL
1280	Self-contained, water cooled unit	3.56	3.39	6.95
1300	Apartment corridors, 500 S.F., .92 ton	3.55	3.40	6.95
1440	10,000 S.F., 18.33 ton	3.96	2.33	6.29
1520	Banks or libraries, 500 S.F., 2.08 ton	7.75	3.53	11.28
1680	10,000 S.F., 41.66 ton	7.85	5.30	13.15
1760	Bars and taverns, 500 S.F., 5.54 ton	16.85	5.95	22.80
1920	10,000 S.F., 110.00 ton	19.35	8.30	27.65
2000	Bowling alleys, 500 S.F., 2.83 ton	10.50	4.80	15.30
2160	10,000 S.F., 56.66 ton	10.40	7.10	17.50
2200	Department stores, 500 S.F., 1.46 ton	5.40	2.48	7.88
2360	10,000 S.F., 29.17 ton	5.70	3.58	9.28
2440	Drug stores, 500 S.F., 3.33 ton	12.40	5.65	18.05
2600	10,000 S.F., 66.66 ton	12.25	8.40	20.65
2680	Factories, 500 S.F., 1.66 ton	6.20	2.82	9.02
2840	10,000 S.F., 33.33 ton	6.55	4.08	10.63
2920	Food supermarkets, 500 S.F., 1.42 ton	5.25	2.41	7.66
3080	10,000 S.F., 28.33 ton	5.55	3.48	9.03
3160	Medical centers, 500 S.F., 1.17 ton	4.33	1.97	6.30
3320	10,000 S.F., 23.33 ton	5.05	2.97	8.02
3400	Offices, 500 S.F., 1.58 ton	5.90	2.68	8.58
3560	10,000 S.F., 31.67 ton	6.20	3.88	10.08
3640	Restaurants, 500 S.F., 2.50 ton	9.30	4.25	13.55
3800	10,000 S.F., 50.00 ton	6.75	5.85	12.60
3880	Schools and colleges, 500 S.F., 1.92 ton	7.10	3.25	10.35
4040	10,000 S.F., 38.33 ton	7.25	4.88	12.13

D3050 Terminal & Package Units

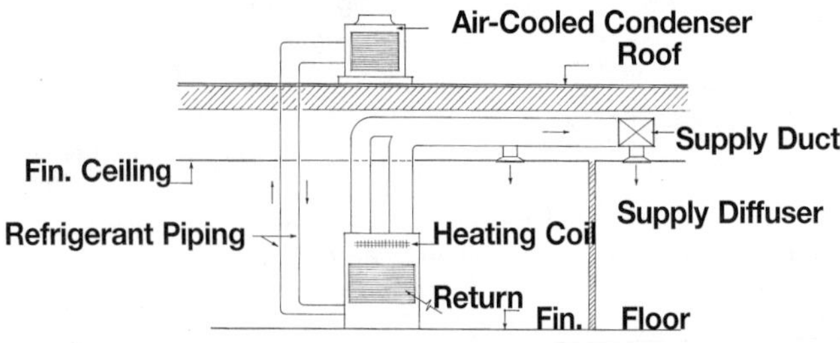

System Description: Self-contained air cooled units with remote air cooled condenser and interconnecting tubing. Systems for 1000 S.F. and up include duct and diffusers. Smaller units distribute air directly.

Returns are not ducted and supplies are not insulated.

Potential savings may be realized by using a single zone rooftop system or through-the-wall unit, especially in the smaller capacities, if the application permits.

Hot water or steam heating coils are included but piping to boiler and the boiler itself is not included.

Condenserless models are available for 15% less where remote refrigerant source is available.

System Components	QUANTITY	UNIT	COST EACH MAT.	COST EACH INST.	COST EACH TOTAL
SYSTEM D3050 165 1320					
SELF-CONTAINED, AIR COOLED UNIT					
APARTMENT CORRIDORS, 500 S.F., .92 TON					
Air cooled, package unit	1.000	Ea.	1,394	459	1,853
Ductwork package for water or air cooled packaged units	1.000	System	69.92	920	989.92
Refrigerant piping	1.000	System	381.80	616.40	998.20
Air cooled condenser, direct drive, propeller fan	1.000	Ea.	829.25	173.60	1,002.85
TOTAL			2,674.97	2,169	4,843.97
COST PER S.F.			5.35	4.34	9.69

D3050 165	Self-contained, Air Cooled Unit Systems	COST PER S.F. MAT.	COST PER S.F. INST.	COST PER S.F. TOTAL
1300	Self-contained, air cooled unit			
1320	Apartment corridors, 500 S.F., .92 ton	5.35	4.35	9.70
1480	10,000 S.F., 18.33 ton	3.38	3.50	6.88
1560	Banks or libraries, 500 S.F., 2.08 ton	11.60	5.60	17.20
1720	10,000 S.F., 41.66 ton	8.55	7.75	16.30
1800	Bars and taverns, 500 S.F., 5.54 ton	27.50	12.20	39.70
1960	10,000 S.F., 110.00 ton	23.50	15.05	38.55
2040	Bowling alleys, 500 S.F., 2.83 ton	15.90	7.65	23.55
2200	10,000 S.F., 56.66 ton	11.90	10.55	22.45
2240	Department stores, 500 S.F., 1.46 ton	8.20	3.93	12.13
2400	10,000 S.F., 29.17 ton	6.10	5.45	11.55
2480	Drug stores, 500 S.F., 3.33 ton	18.70	9	27.70
2640	10,000 S.F., 66.66 ton	14.15	12.45	26.60
2720	Factories, 500 S.F., 1.66 ton	9.50	4.53	14.03
2880	10,000 S.F., 33.33 ton	6.95	6.20	13.15
3200	Medical centers, 500 S.F., 1.17 ton	6.60	3.16	9.76
3360	10,000 S.F., 23.33 ton	4.32	4.45	8.77
3440	Offices, 500 S.F., 1.58 ton	8.90	4.28	13.18
3600	10,000 S.F., 31.66 ton	6.65	5.90	12.55
3680	Restaurants, 500 S.F., 2.50 ton	14	6.75	20.75
3840	10,000 S.F., 50.00 ton	10.85	9.25	20.10
3920	Schools and colleges, 500 S.F., 1.92 ton	10.75	5.20	15.95
4080	10,000 S.F., 38.33 ton	7.90	7.15	15.05

D3050 Terminal & Package Units

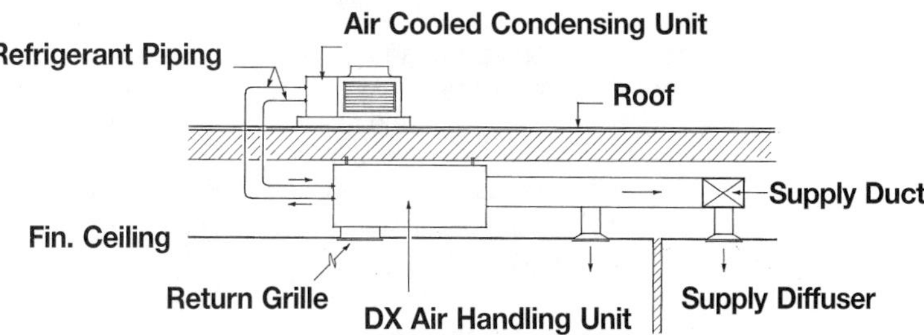

General: Split systems offer several important advantages which should be evaluated when a selection is to be made. They provide a greater degree of flexibility in component selection which permits an accurate match-up of the proper equipment size and type with the particular needs of the building. This allows for maximum use of modern energy saving concepts in heating and cooling. Outdoor installation of the air cooled condensing unit allows space savings in the building and also isolates the equipment operating sounds from building occupants.

Design Assumptions: The systems below are comprised of a direct expansion air handling unit and air cooled condensing unit with interconnecting copper tubing. Ducts and diffusers are also included for distribution of air. Systems are priced for cooling only. Heat can be added as desired either by putting hot water/steam coils into the air unit or into the duct supplying the particular area of need. Gas fired duct furnaces are also available. Refrigerant liquid line is insulated.

System Components	QUANTITY	UNIT	COST EACH MAT.	COST EACH INST.	COST EACH TOTAL
SYSTEM D3050 170 1280					
SPLIT SYSTEM, AIR COOLED CONDENSING UNIT					
APARTMENT CORRIDORS, 1,000 S.F., 1.80 TON					
Fan coil AC unit, cabinet mntd & filters direct expansion air cool	1.000	Ea.	530.70	162.87	693.57
Ductwork package, for split system, remote condensing unit	1.000	System	65.42	882.98	948.40
Refrigeration piping	1.000	System	353.19	1,024.80	1,377.99
Condensing unit, air cooled, incls compressor & standard controls	1.000	Ea.	1,677.50	652.70	2,330.20
TOTAL			2,626.81	2,723.35	5,350.16
COST PER S.F.			2.63	2.72	5.35
*Cooling requirements would lead to choosing a water cooled unit.					

D3050 170	Split Systems With Air Cooled Condensing Units	COST PER S.F. MAT.	COST PER S.F. INST.	COST PER S.F. TOTAL
1260	Split system, air cooled condensing unit			
1280	Apartment corridors, 1,000 S.F., 1.83 ton	2.63	2.73	5.36
1440	20,000 S.F., 36.66 ton	2.70	3.81	6.51
1520	Banks and libraries, 1,000 S.F., 4.17 ton	4.43	6.25	10.68
1680	20,000 S.F., 83.32 ton	6.75	9	15.75
1760	Bars and taverns, 1,000 S.F., 11.08 ton	10.95	12.30	23.25
1880	10,000 S.F., 110.84 ton	14.85	14.40	29.25
2000	Bowling alleys, 1,000 S.F., 5.66 ton	6.10	11.65	17.75
2160	20,000 S.F., 113.32 ton	9.70	12.80	22.50
2320	Department stores, 1,000 S.F., 2.92 ton	3.11	4.30	7.41
2480	20,000 S.F., 58.33 ton	4.29	6.05	10.34
2560	Drug stores, 1,000 S.F., 6.66 ton	7.15	13.70	20.85
2720	20,000 S.F., 133.32 ton*			
2800	Factories, 1,000 S.F., 3.33 ton	3.56	4.90	8.46
2960	20,000 S.F., 66.66 ton	4.90	7.25	12.15
3040	Food supermarkets, 1,000 S.F., 2.83 ton	3.03	4.18	7.21
3200	20,000 S.F., 56.66 ton	4.16	5.90	10.06
3280	Medical centers, 1,000 S.F., 2.33 ton	2.68	3.39	6.07
3440	20,000 S.F., 46.66 ton	3.43	4.85	8.28
3520	Offices, 1,000 S.F., 3.17 ton	3.38	4.66	8.04
3680	20,000 S.F., 63.32 ton	4.66	6.90	11.56
3760	Restaurants, 1,000 S.F., 5.00 ton	5.40	10.30	15.70
3920	20,000 S.F., 100.00 ton	8.55	11.30	19.85
4000	Schools and colleges, 1,000 S.F., 3.83 ton	4.07	5.80	9.87
4160	20,000 S.F., 76.66 ton	5.65	8.30	13.95

D3050 Terminal & Package Units

Computer rooms impose special requirements on air conditioning systems. A prime requirement is reliability, due to the potential monetary loss that could be incurred by a system failure. A second basic requirement is the tolerance of control with which temperature and humidity are regulated, and dust eliminated. As the air conditioning system reliability is so vital, the additional cost of reserve capacity and redundant components is often justified.

System Descriptions: Computer areas may be environmentally controlled by one of three methods as follows:

1. **Self-contained Units**
   These are units built to higher standards of performance and reliability. They usually contain alarms and controls to indicate component operation failure, filter change, etc. It should be remembered that these units in the room will occupy space that is relatively expensive to build and that all alterations and service of the equipment will also have to be accomplished within the computer area.
2. **Decentralized Air Handling Units**
   In operation these are similar to the self-contained units except that their cooling capability comes from remotely located refrigeration equipment as refrigerant or chilled water. As no compressors or refrigerating equipment are required in the air units, they are smaller and require less service than self-contained units. An added plus for this type of system occurs if some of the computer components themselves also require chilled water for cooling.
3. **Central System Supply**
   Cooling is obtained from a central source which, since it is not located within the computer room, may have excess capacity and permit greater flexibility without interfering with the computer components. System performance criteria must still be met.

Note: The costs shown below do not include an allowance for ductwork or piping.

D3050 190	Computer Room Cooling Units	COST EACH		
		MAT.	INST.	TOTAL
0560	Computer room unit, air cooled, includes remote condenser			
0580	3 ton	20,700	2,675	23,375
0600	5 ton	22,100	2,975	25,075
0620	8 ton	41,200	4,975	46,175
0640	10 ton	43,100	5,375	48,475
0660	15 ton	47,400	6,100	53,500
0680	20 ton	57,500	8,700	66,200
0700	23 ton	70,500	9,950	80,450
0800	Chilled water, for connection to existing chiller system			
0820	5 ton	15,800	1,825	17,625
0840	8 ton	16,100	2,675	18,775
0860	10 ton	16,200	2,750	18,950
0880	15 ton	16,900	2,800	19,700
0900	20 ton	18,000	2,925	20,925
0920	23 ton	19,100	3,325	22,425
1000	Glycol system, complete except for interconnecting tubing			
1020	3 ton	25,900	3,350	29,250
1040	5 ton	28,400	3,525	31,925
1060	8 ton	45,500	5,825	51,325
1080	10 ton	48,400	6,400	54,800
1100	15 ton	60,000	8,025	68,025
1120	20 ton	66,000	8,700	74,700
1140	23 ton	69,500	9,075	78,575
1240	Water cooled, not including condenser water supply or cooling tower			
1260	3 ton	21,000	2,175	23,175
1280	5 ton	22,900	2,475	25,375
1300	8 ton	37,500	4,075	41,575
1320	15 ton	45,400	4,975	50,375
1340	20 ton	48,800	5,500	54,300
1360	23 ton	51,000	5,975	56,975

D4010 Sprinklers

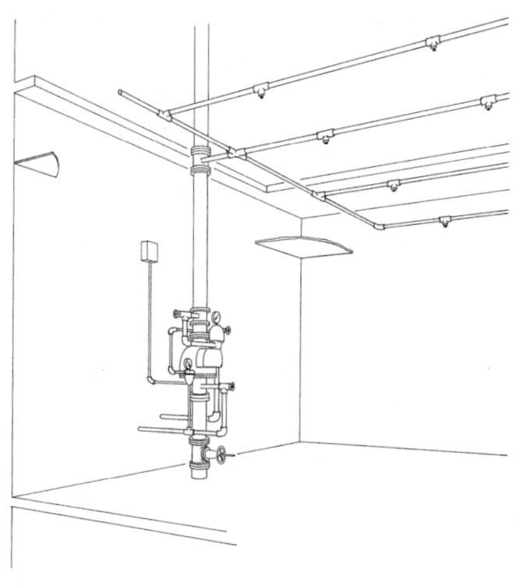

Dry Pipe System: A system employing automatic sprinklers attached to a piping system containing air under pressure, the release of which from the opening of sprinklers permits the water pressure to open a valve known as a "dry pipe valve". The water then flows into the piping system and out the opened sprinklers.

All areas are assumed to be open.

System Components	QUANTITY	UNIT	COST EACH MAT.	COST EACH INST.	COST EACH TOTAL
SYSTEM D4010 310 0580					
DRY PIPE SPRINKLER, STEEL, BLACK, SCH. 40 PIPE					
LIGHT HAZARD, ONE FLOOR, 2000 S.F.					
Valve, gate, iron body 125 lb., OS&Y, flanged, 4″ pipe size	1.000	Ea.	637.50	330	967.50
Valve, swing check, bronze, 125 lb, regrinding disc, 2-1/2″ pipe size	1.000	Ea.	660	66	726
Valve, angle, bronze, 150 lb., rising stem, threaded, 2″ pipe size	1.000	Ea.	622.50	49.88	672.38
*Alarm valve, 2-1/2″ pipe size	1.000	Ea.	1,350	315	1,665
Alarm, water motor, complete with gong	1.000	Ea.	341.25	132	473.25
Fire alarm horn, electric	1.000	Ea.	50.25	75.75	126
Valve swing check w/balldrip CI with brass trim, 4″ pipe size	1.000	Ea.	288.75	315	603.75
Pipe, steel, black, schedule 40, 4″ diam.	10.000	L.F.	180	279.15	459.15
Dry pipe valve, trim & gauges, 4″ pipe size	1.000	Ea.	2,362.50	956.25	3,318.75
Pipe, steel, black, schedule 40, threaded, cplg & hngr 10′OC 2-1/2″ diam.	20.000	L.F.	231	397.50	628.50
Pipe, steel, black, schedule 40, threaded, cplg & hngr 10′OC 2″ diam.	12.500	L.F.	92.81	192.19	285
Pipe, steel, black, schedule 40, threaded, cplg & hngr 10′OC 1-1/4″ diam.	37.500	L.F.	191.25	416.25	607.50
Pipe, steel, black, schedule 40, threaded, cplg & hngr 10′OC 1″ diam.	112.000	L.F.	466.20	1,159.20	1,625.40
Pipe Tee, malleable iron black, 150 lb. threaded, 4″ pipe size	2.000	Ea.	472.50	495	967.50
Pipe Tee, malleable iron black, 150 lb. threaded, 2-1/2″ pipe size	2.000	Ea.	132.75	219	351.75
Pipe Tee, malleable iron black, 150 lb. threaded, 2″ pipe size	1.000	Ea.	30.75	90	120.75
Pipe Tee, malleable iron black, 150 lb. threaded, 1-1/4″ pipe size	5.000	Ea.	72.56	352.50	425.06
Pipe Tee, malleable iron black, 150 lb. threaded, 1″ pipe size	4.000	Ea.	35.85	274.50	310.35
Pipe 90° elbow malleable iron black, 150 lb. threaded, 1″ pipe size	6.000	Ea.	34.20	254.25	288.45
Sprinkler head dry 1/2″ orifice 1″ NPT, 3″ to 4-3/4″ length	12.000	Ea.	1,632	600	2,232
Air compressor, 200 Gal sprinkler system capacity, 1/3 HP	1.000	Ea.	630	405	1,035
*Standpipe connection, wall, flush, brs. w/plug & chain 2-1/2″x2-1/2″	1.000	Ea.	131.25	189.75	321
Valve gate bronze, 300 psi, NRS, class 150, threaded, 1″ pipe size	1.000	Ea.	84	28.88	112.88
TOTAL			10,729.87	7,593.05	18,322.92
COST PER S.F.			5.36	3.80	9.16
*Not included in systems under 2000 S.F.					

D4010 310	Dry Pipe Sprinkler Systems		COST PER S.F. MAT.	COST PER S.F. INST.	COST PER S.F. TOTAL
0520	Dry pipe sprinkler systems, steel, black, sch. 40 pipe				
0530	Light hazard, one floor, 500 S.F.		9.75	6.40	16.15
0560	1000 S.F.	R211313 -20	5.55	3.77	9.32
0580	2000 S.F.		5.35	3.80	9.15
0600	5000 S.F.		2.85	2.59	5.44
0620	10,000 S.F.		2.02	2.13	4.15

D4010 310	Dry Pipe Sprinkler Systems	COST PER S.F.		
		MAT.	INST.	TOTAL
0640	50,000 S.F.	1.55	1.90	3.45
0660	Each additional floor, 500 S.F.	2.13	3.14	5.27
0680	1000 S.F.	1.86	2.60	4.46
0700	2000 S.F.	1.85	2.41	4.26
0720	5000 S.F.	1.51	2.06	3.57
0740	10,000 S.F.	1.39	1.89	3.28
0760	50,000 S.F.	1.26	1.69	2.95
1000	Ordinary hazard, one floor, 500 S.F.	9.95	6.50	16.45
1020	1000 S.F.	5.70	3.80	9.50
1040	2000 S.F.	5.50	3.95	9.45
1060	5000 S.F.	3.28	2.79	6.07
1080	10,000 S.F.	2.63	2.80	5.43
1100	50,000 S.F.	2.42	2.65	5.07
1140	Each additional floor, 500 S.F.	2.34	3.22	5.56
1160	1000 S.F.	2.17	2.91	5.08
1180	2000 S.F.	2.19	2.64	4.83
1200	5000 S.F.	2.02	2.30	4.32
1220	10,000 S.F.	1.84	2.21	4.05
1240	50,000 S.F.	1.79	1.94	3.73
1500	Extra hazard, one floor, 500 S.F.	13.85	8.10	21.95
1520	1000 S.F.	8.50	5.85	14.35
1540	2000 S.F.	6.15	5.10	11.25
1560	5000 S.F.	3.79	3.88	7.67
1580	10,000 S.F.	4	3.65	7.65
1600	50,000 S.F.	4.34	3.55	7.89
1660	Each additional floor, 500 S.F.	3.31	3.99	7.30
1680	1000 S.F.	3.24	3.80	7.04
1700	2000 S.F.	3	3.81	6.81
1720	5000 S.F.	2.55	3.35	5.90
1740	10,000 S.F.	3.04	3.02	6.06
1760	50,000 S.F.	3.10	2.91	6.01
2020	Grooved steel, black, sch. 40 pipe, light hazard, one floor, 2000 S.F.	5.15	3.26	8.41
2060	10,000 S.F.	2.01	1.84	3.85
2100	Each additional floor, 2000 S.F.	1.83	1.95	3.78
2150	10,000 S.F.	1.38	1.60	2.98
2200	Ordinary hazard, one floor, 2000 S.F.	5.45	3.47	8.92
2250	10,000 S.F.	2.48	2.38	4.86
2300	Each additional floor, 2000 S.F.	2.10	2.16	4.26
2350	10,000 S.F.	1.86	2.14	4
2400	Extra hazard, one floor, 2000 S.F.	6.15	4.33	10.48
2450	10,000 S.F.	3.50	3.06	6.56
2500	Each additional floor, 2000 S.F.	3.01	3.12	6.13
2550	10,000 S.F.	2.71	2.68	5.39
3050	Grooved steel, black, sch. 10 pipe, light hazard, one floor, 2000 S.F.	5.10	3.23	8.33
3100	10,000 S.F.	1.96	1.81	3.77
3150	Each additional floor, 2000 S.F.	1.75	1.92	3.67
3200	10,000 S.F.	1.33	1.57	2.90
3250	Ordinary hazard, one floor, 2000 S.F.	5.35	3.44	8.79
3300	10,000 S.F.	2.38	2.33	4.71
3350	Each additional floor, 2000 S.F.	2.02	2.13	4.15
3400	10,000 S.F.	1.76	2.09	3.85
3450	Extra hazard, one floor, 2000 S.F.	6.10	4.30	10.40
3500	10,000 S.F.	3.31	3.01	6.32
3550	Each additional floor, 2000 S.F.	2.95	3.09	6.04
3600	10,000 S.F.	2.63	2.64	5.27
4050	Copper tubing, type M, light hazard, one floor, 2000 S.F.	5.45	3.21	8.66
4100	10,000 S.F.	2.38	1.84	4.22
4150	Each additional floor, 2000 S.F.	2.14	1.94	4.08

D4010 Sprinklers

D4010 310	Dry Pipe Sprinkler Systems	COST PER S.F.		
		MAT.	INST.	TOTAL
4200	10,000 S.F.	1.76	1.61	3.37
4250	Ordinary hazard, one floor, 2000 S.F.	5.85	3.58	9.43
4300	10,000 S.F.	2.94	2.17	5.11
4350	Each additional floor, 2000 S.F.	2.79	2.24	5.03
4400	10,000 S.F.	2.26	1.89	4.15
4450	Extra hazard, one floor, 2000 S.F.	6.70	4.37	11.07
4500	10,000 S.F.	5.10	3.29	8.39
4550	Each additional floor, 2000 S.F.	3.57	3.16	6.73
4600	10,000 S.F.	3.76	2.89	6.65
5050	Copper tubing, type M, T-drill system, light hazard, one floor			
5060	2000 S.F.	5.45	2.99	8.44
5100	10,000 S.F.	2.26	1.53	3.79
5150	Each additional floor, 2000 S.F.	2.13	1.72	3.85
5200	10,000 S.F.	1.64	1.30	2.94
5250	Ordinary hazard, one floor, 2000 S.F.	5.65	3.06	8.71
5300	10,000 S.F.	2.84	1.94	4.78
5350	Each additional floor, 2000 S.F.	2.30	1.75	4.05
5400	10,000 S.F.	2.12	1.61	3.73
5450	Extra hazard, one floor, 2000 S.F.	6.20	3.62	9.82
5500	10,000 S.F.	4.32	2.42	6.74
5550	Each additional floor, 2000 S.F.	3.06	2.41	5.47
5600	10,000 S.F.	2.99	2.02	5.01

D4010 Sprinklers

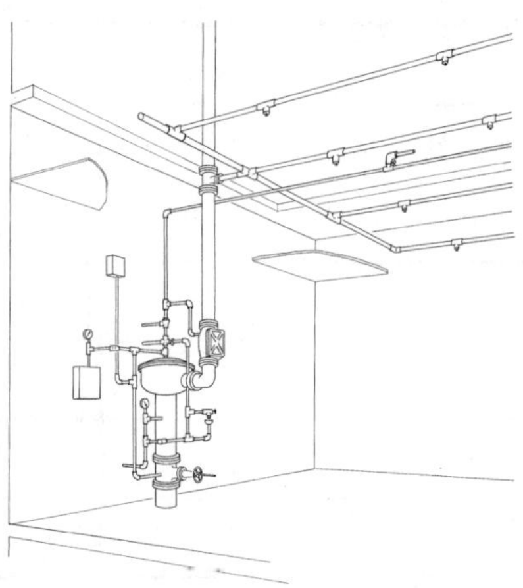

Pre-Action System: A system employing automatic sprinklers attached to a piping system containing air that may or may not be under pressure, with a supplemental heat responsive system of generally more sensitive characteristics than the automatic sprinklers themselves, installed in the same areas as the sprinklers. Actuation of the heat responsive system, as from a fire, opens a valve which permits water to flow into the sprinkler piping system and to be discharged from those sprinklers which were opened by heat from the fire.

All areas are assumed to be open.

System Components	QUANTITY	UNIT	COST EACH MAT.	COST EACH INST.	COST EACH TOTAL
SYSTEM D4010 350 0580					
PREACTION SPRINKLER SYSTEM, STEEL BLACK SCH. 40 PIPE					
LIGHT HAZARD, 1 FLOOR, 2000 S.F.					
Valve, gate, iron body 125 lb., OS&Y, flanged, 4″ pipe size	1.000	Ea.	637.50	330	967.50
*Valve, swing check w/ball drip CI with brass trim 4″ pipe size	1.000	Ea.	288.75	315	603.75
Valve, swing check, bronze, 125 lb, regrinding disc, 2-1/2″ pipe size	1.000	Ea.	660	66	726
Valve, angle, bronze, 150 lb., rising stem, threaded, 2″ pipe size	1.000	Ea.	622.50	49.88	672.38
*Alarm valve, 2-1/2″ pipe size	1.000	Ea.	1,350	315	1,665
Alarm, water motor, complete with gong	1.000	Ea.	341.25	132	473.25
Fire alarm horn, electric	1.000	Ea.	50.25	75.75	126
Thermostatic release for release line	2.000	Ea.	1,110	52.50	1,162.50
Pipe, steel, black, schedule 40, 4″ diam.	10.000	L.F.	180	279.15	459.15
Dry pipe valve, trim & gauges, 4″ pipe size	1.000	Ea.	2,362.50	956.25	3,318.75
Pipe, steel, black, schedule 40, threaded, cplg. & hngr. 10′OC 2-1/2″ diam.	20.000	L.F.	231	397.50	628.50
Pipe steel black, schedule 40, threaded, cplg. & hngr. 10′OC 2″ diam.	12.500	L.F.	92.81	192.19	285
Pipe, steel, black, schedule 40, threaded, cplg. & hngr. 10′OC 1-1/4″ diam.	37.500	L.F.	191.25	416.25	607.50
Pipe, steel, black, schedule 40, threaded, cplg. & hngr. 10′OC 1″ diam.	112.000	L.F.	466.20	1,159.20	1,625.40
Pipe, Tee, malleable iron, black, 150 lb. threaded, 4″ diam.	2.000	Ea.	472.50	495	967.50
Pipe, Tee, malleable iron, black, 150 lb. threaded, 2-1/2″ pipe size	2.000	Ea.	132.75	219	351.75
Pipe, Tee, malleable iron, black, 150 lb. threaded, 2″ pipe size	1.000	Ea.	30.75	90	120.75
Pipe, Tee, malleable iron, black, 150 lb. threaded, 1-1/4″ pipe size	5.000	Ea.	72.56	352.50	425.06
Pipe, Tee, malleable iron, black, 150 lb. threaded, 1″ pipe size	4.000	Ea.	35.85	274.50	310.35
Pipe, 90° elbow, malleable iron, blk., 150 lb. threaded, 1″ pipe size	6.000	Ea.	34.20	254.25	288.45
Sprinkler head, std. spray, brass 135°-286°F 1/2″ NPT, 3/8″ orifice	12.000	Ea.	203.40	528	731.40
Air compressor auto complete 200 Gal sprinkler sys. cap., 1/3 HP	1.000	Ea.	630	405	1,035
*Standpipe conn.,wall, flush, brass w/plug & chain 2-1/2″ x 2-1/2″	1.000	Ea.	131.25	189.75	321
Valve, gate, bronze, 300 psi, NRS, class 150, threaded, 1″ pipe size	1.000	Ea.	84	28.88	112.88
TOTAL			10,411.27	7,573.55	17,984.82
COST PER S.F.			5.21	3.79	9
*Not included in systems under 2000 S.F.					

D4010 350	Preaction Sprinkler Systems		COST PER S.F. MAT.	COST PER S.F. INST.	COST PER S.F. TOTAL
0520	Preaction sprinkler systems, steel, black, sch. 40 pipe	R211313-20			
0530	Light hazard, one floor, 500 S.F.		9.60	5.10	14.70
0560	1000 S.F.		5.60	3.86	9.46
0580	2000 S.F.		5.20	3.79	8.99

D4010 Sprinklers

D4010 350	Preaction Sprinkler Systems	COST PER S.F.		
		MAT.	INST.	TOTAL
0600	5000 S.F.	2.72	2.58	5.30
0620	10,000 S.F.	1.91	2.12	4.03
0640	50,000 S.F.	1.45	1.89	3.34
0660	Each additional floor, 500 S.F.	2.42	2.80	5.22
0680	1000 S.F.	1.89	2.60	4.49
0700	2000 S.F.	1.87	2.42	4.29
0720	5000 S.F.	1.38	2.05	3.43
0740	10,000 S.F.	1.28	1.88	3.16
0760	50,000 S.F.	1.26	1.74	3
1000	Ordinary hazard, one floor, 500 S.F.	9.90	5.55	15.45
1020	1000 S.F.	5.55	3.79	9.34
1040	2000 S.F.	5.65	3.95	9.60
1060	5000 S.F.	3	2.77	5.77
1080	10,000 S.F.	2.30	2.78	5.08
1100	50,000 S.F.	2.10	2.65	4.75
1140	Each additional floor, 500 S.F.	2.73	3.23	5.96
1160	1000 S.F.	1.88	2.62	4.50
1180	2000 S.F.	1.75	2.61	4.36
1200	5000 S.F.	1.86	2.47	4.33
1220	10,000 S.F.	1.68	2.54	4.22
1240	50,000 S.F.	1.63	2.30	3.93
1500	Extra hazard, one floor, 500 S.F.	13.40	7.10	20.50
1520	1000 S.F.	7.75	5.35	13.10
1540	2000 S.F.	5.60	5.05	10.65
1560	5000 S.F.	3.52	4.19	7.71
1580	10,000 S.F.	3.51	4.06	7.57
1600	50,000 S.F.	3.76	3.96	7.72
1660	Each additional floor, 500 S.F.	3.35	3.99	7.34
1680	1000 S.F.	2.73	3.78	6.51
1700	2000 S.F.	2.48	3.79	6.27
1720	5000 S.F.	2.05	3.36	5.41
1740	10,000 S.F.	2.39	3.06	5.45
1760	50,000 S.F.	2.33	2.88	5.21
2020	Grooved steel, black, sch. 40 pipe, light hazard, one floor, 2000 S.F.	5.20	3.27	8.47
2060	10,000 S.F.	1.90	1.83	3.73
2100	Each additional floor of 2000 S.F.	1.85	1.96	3.81
2150	10,000 S.F.	1.27	1.59	2.86
2200	Ordinary hazard, one floor, 2000 S.F.	5.25	3.46	8.71
2250	10,000 S.F.	2.15	2.36	4.51
2300	Each additional floor, 2000 S.F.	1.93	2.15	4.08
2350	10,000 S.F.	1.53	2.12	3.65
2400	Extra hazard, one floor, 2000 S.F.	5.60	4.31	9.91
2450	10,000 S.F.	2.85	3.03	5.88
2500	Each additional floor, 2000 S.F.	2.49	3.10	5.59
2550	10,000 S.F.	2.02	2.65	4.67
3050	Grooved steel, black, sch. 10 pipe light hazard, one floor, 2000 S.F.	5.10	3.24	8.34
3100	10,000 S.F.	1.85	1.80	3.65
3150	Each additional floor, 2000 S.F.	1.77	1.93	3.70
3200	10,000 S.F.	1.22	1.56	2.78
3250	Ordinary hazard, one floor, 2000 S.F.	5.10	3.22	8.32
3300	10,000 S.F.	1.77	2.30	4.07
3350	Each additional floor, 2000 S.F.	1.85	2.12	3.97
3400	10,000 S.F.	1.43	2.07	3.50
3450	Extra hazard, one floor, 2000 S.F.	5.55	4.28	9.83
3500	10,000 S.F.	2.62	2.98	5.60
3550	Each additional floor, 2000 S.F.	2.43	3.07	5.50
3600	10,000 S.F.	1.94	2.61	4.55
4050	Copper tubing, type M, light hazard, one floor, 2000 S.F.	5.45	3.22	8.67

D4010 Sprinklers

D4010 350	Preaction Sprinkler Systems	COST PER S.F.		
		MAT.	INST.	TOTAL
4100	10,000 S.F.	2.27	1.83	4.10
4150	Each additional floor, 2000 S.F.	2.17	1.95	4.12
4200	10,000 S.F.	1.37	1.59	2.96
4250	Ordinary hazard, one floor, 2000 S.F.	5.65	3.57	9.22
4300	10,000 S.F.	2.61	2.15	4.76
4350	Each additional floor, 2000 S.F.	2.13	1.98	4.11
4400	10,000 S.F.	1.76	1.72	3.48
4450	Extra hazard, one floor, 2000 S.F.	6.20	4.35	10.55
4500	10,000 S.F.	4.36	3.26	7.62
4550	Each additional floor, 2000 S.F.	3.05	3.14	6.19
4600	10,000 S.F.	3.07	2.86	5.93
5050	Copper tubing, type M, T-drill system, light hazard, one floor			
5060	2000 S.F.	5.45	3	8.45
5100	10,000 S.F.	2.15	1.52	3.67
5150	Each additional floor, 2000 S.F.	2.15	1.73	3.88
5200	10,000 S.F.	1.53	1.29	2.82
5250	Ordinary hazard, one floor, 2000 S.F.	5.45	3.05	8.50
5300	10,000 S.F.	2.51	1.92	4.43
5350	Each additional floor, 2000 S.F.	2.14	1.76	3.90
5400	10,000 S.F.	1.89	1.68	3.57
5450	Extra hazard, one floor, 2000 S.F.	5.65	3.60	9.25
5500	10,000 S.F.	3.59	2.39	5.98
5550	Each additional floor, 2000 S.F.	2.54	2.39	4.93
5600	10,000 S.F.	2.30	1.99	4.29

D4010 Sprinklers

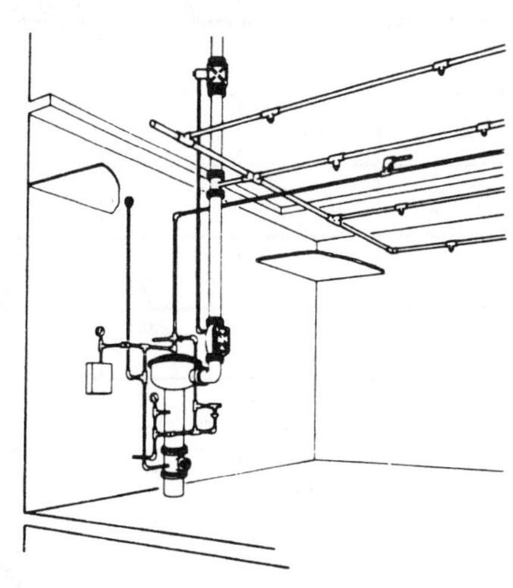

Deluge System: A system employing open sprinklers attached to a piping system connected to a water supply through a valve which is opened by the operation of a heat responsive system installed in the same areas as the sprinklers. When this valve opens, water flows into the piping system and discharges from all sprinklers attached thereto.

All areas are assumed to be open.

System Components	QUANTITY	UNIT	COST EACH MAT.	COST EACH INST.	COST EACH TOTAL
SYSTEM D4010 370 0580					
DELUGE SPRINKLER SYSTEM, STEEL BLACK SCH. 40 PIPE					
LIGHT HAZARD, 1 FLOOR, 2000 S.F.					
Valve, gate, iron body 125 lb., OS&Y, flanged, 4″ pipe size	1.000	Ea.	637.50	330	967.50
Valve, swing check w/ball drip, CI w/brass ftngs., 4″ pipe size	1.000	Ea.	288.75	315	603.75
Valve, swing check, bronze, 125 lb, regrinding disc, 2-1/2″ pipe size	1.000	Ea.	660	66	726
Valve, angle, bronze, 150 lb., rising stem, threaded, 2″ pipe size	1.000	Ea.	622.50	49.88	672.38
*Alarm valve, 2-1/2″ pipe size	1.000	Ea.	1,350	315	1,665
Alarm, water motor, complete with gong	1.000	Ea.	341.25	132	473.25
Fire alarm horn, electric	1.000	Ea.	50.25	75.75	126
Thermostatic release for release line	2.000	Ea.	1,110	52.50	1,162.50
Pipe, steel, black, schedule 40, 4″ diam.	10.000	L.F.	180	279.15	459.15
Deluge valve trim, pressure relief, emergency release, gauge, 4″ pipe size	1.000	Ea.	3,862.50	956.25	4,818.75
Deluge system, monitoring panel w/deluge valve & trim	1.000	Ea.	8,925	29.25	8,954.25
Pipe, steel, black, schedule 40, threaded, cplg & hngr 10′ OC 2-1/2″ diam.	20.000	L.F.	231	397.50	628.50
Pipe, steel, black, schedule 40, threaded, cplg & hngr 10′ OC 2″ diam.	12.500	L.F.	92.81	192.19	285
Pipe, steel, black, schedule 40, threaded, cplg & hngr 10′ OC 1-1/4″ diam.	37.500	L.F.	191.25	416.25	607.50
Pipe, steel, black, schedule 40, threaded, cplg & hngr 10′ OC 1″ diam.	112.000	L.F.	466.20	1,159.20	1,625.40
Pipe, Tee, malleable iron, black, 150 lb. threaded, 4″ pipe size	2.000	Ea.	472.50	495	967.50
Pipe, Tee, malleable iron, black, 150 lb. threaded, 2-1/2″ pipe size	2.000	Ea.	132.75	219	351.75
Pipe, Tee, malleable iron, black, 150 lb. threaded, 2″ pipe size	1.000	Ea.	30.75	90	120.75
Pipe, Tee, malleable iron, black, 150 lb. threaded, 1-1/4″ pipe size	5.000	Ea.	72.56	352.50	425.06
Pipe, Tee, malleable iron, black, 150 lb. threaded, 1″ pipe size	4.000	Ea.	35.85	274.50	310.35
Pipe, 90° elbow, malleable iron, black, 150 lb. threaded 1″ pipe size	6.000	Ea.	34.20	254.25	288.45
Sprinkler head, std spray, brass 135°-286°F 1/2″ NPT, 3/8″ orifice	9.720	Ea.	203.40	528	731.40
Air compressor, auto, complete, 200 Gal sprinkler sys. cap., 1/3 HP	1.000	Ea.	630	405	1,035
*Standpipe connection, wall, flush w/plug & chain 2-1/2″ x 2-1/2″	1.000	Ea.	131.25	189.75	321
Valve, gate, bronze, 300 psi, NRS, class 150, threaded, 1″ pipe size	1.000	Ea.	84	28.88	112.88
TOTAL			20,836.27	7,602.80	28,439.07
COST PER S.F.			10.42	3.80	14.22
*Not included in systems under 2000 S.F.					

D4010 370	Deluge Sprinkler Systems		COST PER S.F. MAT.	COST PER S.F. INST.	COST PER S.F. TOTAL
0520	Deluge sprinkler systems, steel, black, sch. 40 pipe	R211313 -20			
0530	Light hazard, one floor, 500 S.F.		29	5.15	34.15

D4010 Sprinklers

D4010 370	Deluge Sprinkler Systems	COST PER S.F.		
		MAT.	INST.	TOTAL
0560	1000 S.F.	15.35	3.71	19.06
0580	2000 S.F.	10.40	3.81	14.21
0600	5000 S.F.	4.80	2.59	7.39
0620	10,000 S.F.	2.95	2.12	5.07
0640	50,000 S.F.	1.66	1.89	3.55
0660	Each additional floor, 500 S.F.	2.42	2.80	5.22
0680	1000 S.F.	1.89	2.60	4.49
0700	2000 S.F.	1.87	2.42	4.29
0720	5000 S.F.	1.38	2.05	3.43
0740	10,000 S.F.	1.28	1.88	3.16
0760	50,000 S.F.	1.26	1.74	3
1000	Ordinary hazard, one floor, 500 S.F.	30.50	5.90	36.40
1020	1000 S.F.	15.35	3.82	19.17
1040	2000 S.F.	10.85	3.96	14.81
1060	5000 S.F.	5.10	2.78	7.88
1080	10,000 S.F.	3.34	2.78	6.12
1100	50,000 S.F.	2.36	2.69	5.05
1140	Each additional floor, 500 S.F.	2.73	3.23	5.96
1160	1000 S.F.	1.88	2.62	4.50
1180	2000 S.F.	1.75	2.61	4.36
1200	5000 S.F.	1.74	2.28	4.02
1220	10,000 S.F.	1.64	2.24	3.88
1240	50,000 S.F.	1.54	2.11	3.65
1500	Extra hazard, one floor, 500 S.F.	33	7.15	40.15
1520	1000 S.F.	18	5.55	23.55
1540	2000 S.F.	10.85	5.05	15.90
1560	5000 S.F.	5.35	3.86	9.21
1580	10,000 S.F.	4.41	3.71	8.12
1600	50,000 S.F.	4.46	3.64	8.10
1660	Each additional floor, 500 S.F.	3.35	3.99	7.34
1680	1000 S.F.	2.73	3.78	6.51
1700	2000 S.F.	2.48	3.79	6.27
1720	5000 S.F.	2.05	3.36	5.41
1740	10,000 S.F.	2.44	3.18	5.62
1760	50,000 S.F.	2.45	3.07	5.52
2000	Grooved steel, black, sch. 40 pipe, light hazard, one floor			
2020	2000 S.F.	10.40	3.28	13.68
2060	10,000 S.F.	2.96	1.85	4.81
2100	Each additional floor, 2,000 S.F.	1.85	1.96	3.81
2150	10,000 S.F.	1.27	1.59	2.86
2200	Ordinary hazard, one floor, 2000 S.F.	5.25	3.46	8.71
2250	10,000 S.F.	3.19	2.36	5.55
2300	Each additional floor, 2000 S.F.	1.93	2.15	4.08
2350	10,000 S.F.	1.53	2.12	3.65
2400	Extra hazard, one floor, 2000 S.F.	10.85	4.32	15.17
2450	10,000 S.F.	3.92	3.04	6.96
2500	Each additional floor, 2000 S.F.	2.49	3.10	5.59
2550	10,000 S.F.	2.02	2.65	4.67
3000	Grooved steel, black, sch. 10 pipe, light hazard, one floor			
3050	2000 S.F.	9.70	3.11	12.81
3100	10,000 S.F.	2.89	1.80	4.69
3150	Each additional floor, 2000 S.F.	1.77	1.93	3.70
3200	10,000 S.F.	1.22	1.56	2.78
3250	Ordinary hazard, one floor, 2000 S.F.	10.40	3.44	13.84
3300	10,000 S.F.	2.81	2.30	5.11
3350	Each additional floor, 2000 S.F.	1.85	2.12	3.97
3400	10,000 S.F.	1.43	2.07	3.50
3450	Extra hazard, one floor, 2000 S.F.	10.75	4.29	15.04

D4010 Sprinklers

D4010 370	Deluge Sprinkler Systems	COST PER S.F.		
		MAT.	INST.	TOTAL
3500	10,000 S.F.	3.66	2.98	6.64
3550	Each additional floor, 2000 S.F.	2.43	3.07	5.50
3600	10,000 S.F.	1.94	2.61	4.55
4000	Copper tubing, type M, light hazard, one floor			
4050	2000 S.F.	10.65	3.23	13.88
4100	10,000 S.F.	3.31	1.83	5.14
4150	Each additional floor, 2000 S.F.	2.15	1.94	4.09
4200	10,000 S.F.	1.37	1.59	2.96
4250	Ordinary hazard, one floor, 2000 S.F.	10.85	3.58	14.43
4300	10,000 S.F.	3.65	2.15	5.80
4350	Each additional floor, 2000 S.F.	2.13	1.98	4.11
4400	10,000 S.F.	1.76	1.72	3.48
4450	Extra hazard, one floor, 2000 S.F.	11.40	4.36	15.76
4500	10,000 S.F.	5.45	3.27	8.72
4550	Each additional floor, 2000 S.F.	3.05	3.14	6.19
4600	10,000 S.F.	3.07	2.86	5.93
5000	Copper tubing, type M, T-drill system, light hazard, one floor			
5050	2000 S.F.	10.65	3.01	13.66
5100	10,000 S.F.	3.19	1.52	4.71
5150	Each additional floor, 2000 S.F.	2.17	1.75	3.92
5200	10,000 S.F.	1.53	1.29	2.82
5250	Ordinary hazard, one floor, 2000 S.F.	10.65	3.06	13.71
5300	10,000 S.F.	3.55	1.92	5.47
5350	Each additional floor, 2000 S.F.	2.13	1.74	3.87
5400	10,000 S.F.	1.89	1.68	3.57
5450	Extra hazard, one floor, 2000 S.F.	10.90	3.61	14.51
5500	10,000 S.F.	4.63	2.39	7.02
5550	Each additional floor, 2000 S.F.	2.54	2.39	4.93
5600	10,000 S.F.	2.30	1.99	4.29

D4010 Sprinklers

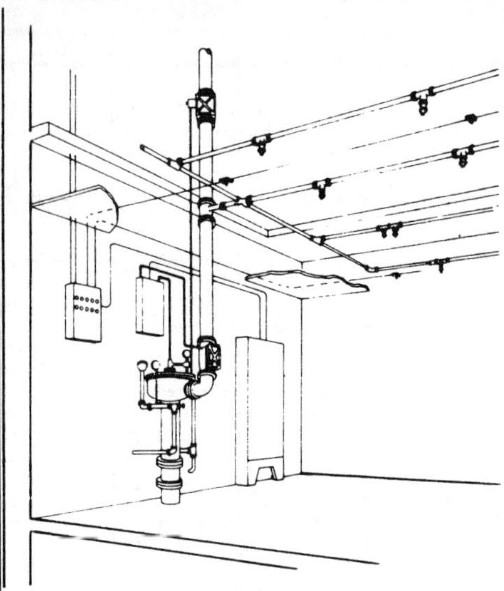

Firecycle is a fixed fire protection sprinkler system utilizing water as its extinguishing agent. It is a time delayed, recycling, preaction type which automatically shuts the water off when heat is reduced below the detector operating temperature and turns the water back on when that temperature is exceeded.

The system senses a fire condition through a closed circuit electrical detector system which controls water flow to the fire automatically. Batteries supply up to 90 hour emergency power supply for system operation. The piping system is dry (until water is required) and is monitored with pressurized air. Should any leak in the system piping occur, an alarm will sound, but water will not enter the system until heat is sensed by a Firecycle detector.

All areas are assumed to be open.

System Components	QUANTITY	UNIT	COST EACH		
			MAT.	INST.	TOTAL
SYSTEM D4010 390 0580					
FIRECYCLE SPRINKLER SYSTEM, STEEL BLACK SCH. 40 PIPE					
LIGHT HAZARD, ONE FLOOR, 2000 S.F.					
Valve, gate, iron body 125 lb., OS&Y, flanged, 4" pipe size	1.000	Ea.	637.50	330	967.50
Valve, angle, bronze, 150 lb., rising stem, threaded, 2" pipe size	1.000	Ea.	622.50	49.88	672.38
Valve, swing check, bronze, 125 lb, regrinding disc, 2-1/2" pipe size	1.000	Ea.	660	66	726
*Alarm valve, 2-1/2" pipe size	1.000	Ea.	1,350	315	1,665
Alarm, water motor, complete with gong	1.000	Ea.	341.25	132	473.25
Pipe, steel, black, schedule 40, 4" diam.	10.000	L.F.	180	279.15	459.15
Fire alarm, horn, electric	1.000	Ea.	50.25	75.75	126
Pipe, steel, black, schedule 40, threaded, cplg & hngr 10' OC 2-1/2" diam.	20.000	L.F.	231	397.50	628.50
Pipe, steel, black, schedule 40, threaded, cplg & hngr 10' OC 2" diam.	12.500	L.F.	92.81	192.19	285
Pipe, steel, black, schedule 40, threaded, cplg & hngr 10' OC 1-1/4" diam.	37.500	L.F.	191.25	416.25	607.50
Pipe, steel, black, schedule 40, threaded, cplg & hngr 10' OC 1" diam.	112.000	L.F.	466.20	1,159.20	1,625.40
Pipe, Tee, malleable iron, black, 150 lb. threaded, 4" pipe size	2.000	Ea.	472.50	495	967.50
Pipe, Tee, malleable iron, black, 150 lb. threaded, 2-1/2" pipe size	2.000	Ea.	132.75	219	351.75
Pipe, Tee, malleable iron, black, 150 lb. threaded, 2" pipe size	1.000	Ea.	30.75	90	120.75
Pipe, Tee, malleable iron, black, 150 lb. threaded, 1-1/4" pipe size	5.000	Ea.	72.56	352.50	425.06
Pipe, Tee, malleable iron, black, 150 lb. threaded, 1" pipe size	4.000	Ea.	35.85	274.50	310.35
Pipe, 90° elbow, malleable iron, black, 150 lb. threaded, 1" pipe size	6.000	Ea.	34.20	254.25	288.45
Sprinkler head std spray, brass 135°-286°F 1/2" NPT, 3/8" orifice	12.000	Ea.	203.40	528	731.40
Firecycle controls, incls panel, battery, solenoid valves, press switches	1.000	Ea.	17,250	2,006.25	19,256.25
Detector, firecycle system	2.000	Ea.	1,177.50	66	1,243.50
Firecycle pkg, swing check & flow control valves w/trim 4" pipe size	1.000	Ea.	4,668.75	956.25	5,625
Air compressor, auto, complete, 200 Gal sprinkler sys. cap., 1/3 HP	1.000	Ea.	630	405	1,035
*Standpipe connection, wall, flush, brass w/plug & chain 2-1/2"x2-1/2"	1.000	Ea.	131.25	189.75	321
Valve, gate, bronze 300 psi, NRS, class 150, threaded, 1" diam.	1.000	Ea.	84	28.88	112.88
TOTAL			29,746.27	9,278.30	39,024.57
COST PER S.F.			14.87	4.64	19.51

*Not included in systems under 2000 S.F.

D4010 390	Firecycle Sprinkler Systems		COST PER S.F.		
			MAT.	INST.	TOTAL
0520	Firecycle sprinkler systems, steel black sch. 40 pipe	R211313-20			
0530	Light hazard, one floor, 500 S.F.		49	10	59

D4010 Sprinklers

D4010 390	Firecycle Sprinkler Systems	COST PER S.F.		
		MAT.	INST.	TOTAL
0560	1000 S.F.	25	6.30	31.30
0580	2000 S.F.	14.85	4.65	19.50
0600	5000 S.F.	6.60	2.93	9.53
0620	10,000 S.F.	3.92	2.30	6.22
0640	50,000 S.F.	1.88	1.92	3.80
0660	Each additional floor of 500 S.F.	2.49	2.82	5.31
0680	1000 S.F.	1.92	2.60	4.52
0700	2000 S.F.	1.61	2.41	4.02
0720	5000 S.F.	1.41	2.06	3.47
0740	10,000 S.F.	1.36	1.89	3.25
0760	50,000 S.F.	1.30	1.73	3.03
1000	Ordinary hazard, one floor, 500 S.F.	49	10.45	59.45
1020	1000 S.F.	25	6.25	31.25
1040	2000 S.F.	15	4.78	19.78
1060	5000 S.F.	6.90	3.12	10.02
1080	10,000 S.F.	4.31	2.96	7.27
1100	50,000 S.F.	2.79	3	5.79
1140	Each additional floor, 500 S.F.	2.80	3.25	6.05
1160	1000 S.F.	1.91	2.62	4.53
1180	2000 S.F.	1.96	2.40	4.36
1200	5000 S.F.	1.77	2.29	4.06
1220	10,000 S.F.	1.59	2.20	3.79
1240	50,000 S.F.	1.55	1.99	3.54
1500	Extra hazard, one floor, 500 S.F.	52.50	12	64.50
1520	1000 S.F.	27	7.80	34.80
1540	2000 S.F.	15.30	5.90	21.20
1560	5000 S.F.	7.15	4.20	11.35
1580	10,000 S.F.	5.45	4.22	9.67
1600	50,000 S.F.	4.59	4.68	9.27
1660	Each additional floor, 500 S.F.	3.42	4.01	7.43
1680	1000 S.F.	2.76	3.78	6.54
1700	2000 S.F.	2.52	3.79	6.31
1720	5000 S.F.	2.08	3.37	5.45
1740	10,000 S.F.	2.47	3.07	5.54
1760	50,000 S.F.	2.46	2.96	5.42
2020	Grooved steel, black, sch. 40 pipe, light hazard, one floor			
2030	2000 S.F.	14.85	4.11	18.96
2060	10,000 S.F.	4.23	2.78	7.01
2100	Each additional floor, 2000 S.F.	1.89	1.96	3.85
2150	10,000 S.F.	1.35	1.60	2.95
2200	Ordinary hazard, one floor, 2000 S.F.	14.95	4.30	19.25
2250	10,000 S.F.	4.46	2.68	7.14
2300	Each additional floor, 2000 S.F.	1.97	2.15	4.12
2350	10,000 S.F.	1.61	2.13	3.74
2400	Extra hazard, one floor, 2000 S.F.	15.30	5.15	20.45
2450	10,000 S.F.	4.80	3.19	7.99
2500	Each additional floor, 2000 S.F.	2.53	3.10	5.63
2550	10,000 S.F.	2.10	2.66	4.76
3050	Grooved steel, black, sch. 10 pipe light hazard, one floor,			
3060	2000 S.F.	14.75	4.08	18.83
3100	10,000 S.F.	3.86	1.98	5.84
3150	Each additional floor, 2000 S.F.	1.81	1.93	3.74
3200	10,000 S.F.	1.30	1.57	2.87
3250	Ordinary hazard, one floor, 2000 S.F.	14.85	4.27	19.12
3300	10,000 S.F.	4.06	2.49	6.55
3350	Each additional floor, 2000 S.F.	1.89	2.12	4.01
3400	10,000 S.F.	1.51	2.08	3.59
3450	Extra hazard, one floor, 2000 S.F.	15.25	5.10	20.35

D4010 Sprinklers

D4010 390	Firecycle Sprinkler Systems	COST PER S.F.		
		MAT.	INST.	TOTAL
3500	10,000 S.F.	4.61	3.14	7.75
3550	Each additional floor, 2000 S.F.	2.47	3.07	5.54
3600	10,000 S.F.	2.02	2.62	4.64
4060	Copper tubing, type M, light hazard, one floor, 2000 S.F.	15.15	4.06	19.21
4100	10,000 S.F.	4.28	2.01	6.29
4150	Each additional floor, 2000 S.F.	2.20	1.95	4.15
4200	10,000 S.F.	1.73	1.61	3.34
4250	Ordinary hazard, one floor, 2000 S.F.	15.35	4.41	19.76
4300	10,000 S.F.	4.62	2.33	6.95
4350	Each additional floor, 2000 S.F.	2.17	1.98	4.15
4400	10,000 S.F.	1.81	1.70	3.51
4450	Extra hazard, one floor, 2000 S.F.	15.85	5.20	21.05
4500	10,000 S.F.	6.40	3.46	9.86
4550	Each additional floor, 2000 S.F.	3.09	3.14	6.23
4600	10,000 S.F.	3.15	2.87	6.02
5060	Copper tubing, type M, T-drill system, light hazard, one floor 2000 S.F.	15.10	3.84	18.94
5100	10,000 S.F.	4.16	1.70	5.86
5150	Each additional floor, 2000 S.F.	2.39	1.83	4.22
5200	10,000 S.F.	1.61	1.30	2.91
5250	Ordinary hazard, one floor, 2000 S.F.	15.15	3.89	19.04
5300	10,000 S.F.	4.52	2.10	6.62
5350	Each additional floor, 2000 S.F.	2.17	1.74	3.91
5400	10,000 S.F.	1.97	1.69	3.66
5450	Extra hazard, one floor, 2000 S.F.	15.35	4.44	19.79
5500	10,000 S.F.	5.60	2.55	8.15
5550	Each additional floor, 2000 S.F.	2.58	2.39	4.97
5600	10,000 S.F.	2.38	2	4.38

D4010 Sprinklers

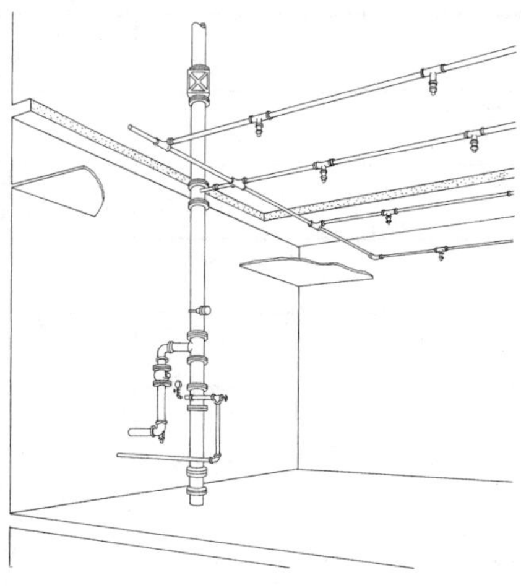

Wet Pipe System. A system employing automatic sprinklers attached to a piping system containing water and connected to a water supply so that water discharges immediately from sprinklers opened by heat from a fire.

All areas are assumed to be open.

System Components	QUANTITY	UNIT	COST EACH MAT.	COST EACH INST.	COST EACH TOTAL
SYSTEM D4010 410 0580					
WET PIPE SPRINKLER, STEEL, BLACK, SCH. 40 PIPE					
LIGHT HAZARD, ONE FLOOR, 2000 S.F.					
Valve, gate, iron body, 125 lb., OS&Y, flanged, 4″ diam.	1.000	Ea.	637.50	330	967.50
Valve, swing check, bronze, 125 lb, regrinding disc, 2-1/2″ pipe size	1.000	Ea.	660	66	726
Valve, angle, bronze, 150 lb., rising stem, threaded, 2″ diam.	1.000	Ea.	622.50	49.88	672.38
*Alarm valve, 2-1/2″ pipe size	1.000	Ea.	1,350	315	1,665
Alarm, water motor, complete with gong	1.000	Ea.	341.25	132	473.25
Valve, swing check, w/balldrip CI with brass trim 4″ pipe size	1.000	Ea.	288.75	315	603.75
Pipe, steel, black, schedule 40, 4″ diam.	10.000	L.F.	180	279.15	459.15
*Flow control valve, trim & gauges, 4″ pipe size	1.000	Set	4,762.50	716.25	5,478.75
Fire alarm horn, electric	1.000	Ea.	50.25	75.75	126
Pipe, steel, black, schedule 40, threaded, cplg & hngr 10′ OC, 2-1/2″ diam.	20.000	L.F.	231	397.50	628.50
Pipe, steel, black, schedule 40, threaded, cplg & hngr 10′ OC, 2″ diam.	12.500	L.F.	92.81	192.19	285
Pipe, steel, black, schedule 40, threaded, cplg & hngr 10′ OC, 1-1/4″ diam.	37.500	L.F.	191.25	416.25	607.50
Pipe, steel, black, schedule 40, threaded cplg & hngr 10′ OC, 1″ diam.	112.000	L.F.	466.20	1,159.20	1,625.40
Pipe Tee, malleable iron black, 150 lb. threaded, 4″ pipe size	2.000	Ea.	472.50	495	967.50
Pipe Tee, malleable iron black, 150 lb. threaded, 2-1/2″ pipe size	2.000	Ea.	132.75	219	351.75
Pipe Tee, malleable iron black, 150 lb. threaded, 2″ pipe size	1.000	Ea.	30.75	90	120.75
Pipe Tee, malleable iron black, 150 lb. threaded, 1-1/4″ pipe size	5.000	Ea.	72.56	352.50	425.06
Pipe Tee, malleable iron black, 150 lb. threaded, 1″ pipe size	4.000	Ea.	35.85	274.50	310.35
Pipe 90° elbow, malleable iron black, 150 lb. threaded, 1″ pipe size	6.000	Ea.	34.20	254.25	288.45
Sprinkler head, standard spray, brass 135°-286°F 1/2″ NPT, 3/8″ orifice	12.000	Ea.	203.40	528	731.40
Valve, gate, bronze, NRS, class 150, threaded, 1″ pipe size	1.000	Ea.	84	28.88	112.88
*Standpipe connection, wall, single, flush w/plug & chain 2-1/2″x2-1/2″	1.000	Ea.	131.25	189.75	321
TOTAL			11,071.27	6,876.05	17,947.32
COST PER S.F.			5.54	3.44	8.98

*Not included in systems under 2000 S.F.

D4010 410	Wet Pipe Sprinkler Systems		COST PER S.F. MAT.	COST PER S.F. INST.	COST PER S.F. TOTAL
0520	Wet pipe sprinkler systems, steel, black, sch. 40 pipe				
0530	Light hazard, one floor, 500 S.F.		2.95	3.30	6.25
0560	1000 S.F.	R211313 -20	6.15	3.42	9.57
0580	2000 S.F.		5.55	3.45	9
0600	5000 S.F.	R211313 -40	2.63	2.43	5.06
0620	10,000 S.F.		1.70	2.03	3.73

D4010 Sprinklers

D4010 410	Wet Pipe Sprinkler Systems	COST PER S.F.		
		MAT.	INST.	TOTAL
0640	50,000 S.F.	1.13	1.85	2.98
0660	Each additional floor, 500 S.F.	1.33	2.81	4.14
0680	1000 S.F.	1.35	2.63	3.98
0700	2000 S.F.	1.32	2.39	3.71
0720	5000 S.F.	.94	2.03	2.97
0740	10,000 S.F.	.89	1.86	2.75
0760	50,000 S.F.	.74	1.46	2.20
1000	Ordinary hazard, one floor, 500 S.F.	3.23	3.53	6.76
1020	1000 S.F.	6.10	3.35	9.45
1040	2000 S.F.	5.70	3.59	9.29
1060	5000 S.F.	2.91	2.62	5.53
1080	10,000 S.F.	2.09	2.69	4.78
1100	50,000 S.F.	1.75	2.58	4.33
1140	Each additional floor, 500 S.F.	1.62	3.18	4.80
1160	1000 S.F.	1.32	2.59	3.91
1180	2000 S.F.	1.47	2.60	4.07
1200	5000 S.F.	1.45	2.49	3.94
1220	10,000 S.F.	1.29	2.52	3.81
1240	50,000 S.F.	1.27	2.28	3.55
1500	Extra hazard, one floor, 500 S.F.	11.80	5.45	17.25
1520	1000 S.F.	7.35	4.76	12.11
1540	2000 S.F.	6	4.86	10.86
1560	5000 S.F.	3.63	4.27	7.90
1580	10,000 S.F.	3.23	4.02	7.25
1600	50,000 S.F.	3.57	3.90	7.47
1660	Each additional floor, 500 S.F.	2.24	3.94	6.18
1680	1000 S.F.	2.17	3.75	5.92
1700	2000 S.F.	1.93	3.76	5.69
1720	5000 S.F.	1.61	3.34	4.95
1740	10,000 S.F.	2	3.04	5.04
1760	50,000 S.F.	1.99	2.90	4.89
2020	Grooved steel, black sch. 40 pipe, light hazard, one floor, 2000 S.F.	5.50	2.92	8.42
2060	10,000 S.F.	2.16	1.81	3.97
2100	Each additional floor, 2000 S.F.	1.30	1.93	3.23
2150	10,000 S.F.	.88	1.57	2.45
2200	Ordinary hazard, one floor, 2000 S.F.	5.60	3.11	8.71
2250	10,000 S.F.	1.94	2.27	4.21
2300	Each additional floor, 2000 S.F.	1.38	2.12	3.50
2350	10,000 S.F.	1.14	2.10	3.24
2400	Extra hazard, one floor, 2000 S.F.	5.95	3.96	9.91
2450	10,000 S.F.	2.60	2.94	5.54
2500	Each additional floor, 2000 S.F.	1.94	3.07	5.01
2550	10,000 S.F.	1.63	2.63	4.26
3050	Grooved steel, black sch. 10 pipe, light hazard, one floor, 2000 S.F.	5.45	2.89	8.34
3100	10,000 S.F.	1.64	1.71	3.35
3150	Each additional floor, 2000 S.F.	1.22	1.90	3.12
3200	10,000 S.F.	.83	1.54	2.37
3250	Ordinary hazard, one floor, 2000 S.F.	5.50	3.08	8.58
3300	10,000 S.F.	1.84	2.22	4.06
3350	Each additional floor, 2000 S.F.	1.30	2.09	3.39
3400	10,000 S.F.	1.04	2.05	3.09
3450	Extra hazard, one floor, 2000 S.F.	5.90	3.93	9.83
3500	10,000 S.F.	2.41	2.89	5.30
3550	Each additional floor, 2000 S.F.	1.88	3.04	4.92
3600	10,000 S.F.	1.55	2.59	4.14
4050	Copper tubing, type M, light hazard, one floor, 2000 S.F.	5.80	2.87	8.67
4100	10,000 S.F.	2.06	1.74	3.80
4150	Each additional floor, 2000 S.F.	1.61	1.92	3.53

D4010 Sprinklers

D4010 410	Wet Pipe Sprinkler Systems	COST PER S.F.		
		MAT.	INST.	TOTAL
4200	10,000 S.F.	1.26	1.58	2.84
4250	Ordinary hazard, one floor, 2000 S.F.	6	3.22	9.22
4300	10,000 S.F.	2.40	2.06	4.46
4350	Each additional floor, 2000 S.F.	1.81	2.12	3.93
4400	10,000 S.F.	1.54	1.85	3.39
4450	Extra hazard, one floor, 2000 S.F.	6.50	4	10.50
4500	10,000 S.F.	4.15	3.17	7.32
4550	Each additional floor, 2000 S.F.	2.50	3.11	5.61
4600	10,000 S.F.	2.68	2.84	5.52
5050	Copper tubing, type M, T-drill system, light hazard, one floor			
5060	2000 S.F.	5.80	2.65	8.45
5100	10,000 S.F.	1.94	1.43	3.37
5150	Each additional floor, 2000 S.F.	1.60	1.70	3.30
5200	10,000 S.F.	1.14	1.27	2.41
5250	Ordinary hazard, one floor, 2000 S.F.	5.80	2.70	8.50
5300	10,000 S.F.	2.30	1.83	4.13
5350	Each additional floor, 2000 S.F.	1.58	1.71	3.29
5400	10,000 S.F.	1.50	1.66	3.16
5450	Extra hazard, one floor, 2000 S.F.	6	3.25	9.25
5500	10,000 S.F.	3.38	2.30	5.68
5550	Each additional floor, 2000 S.F.	2.09	2.41	4.50
5600	10,000 S.F.	1.91	1.97	3.88

D4020 Standpipes

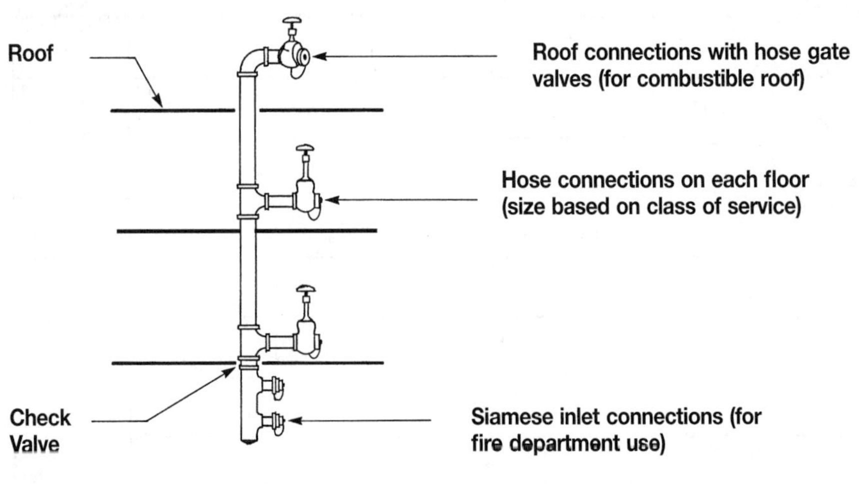

System Components	QUANTITY	UNIT	COST PER FLOOR MAT.	INST.	TOTAL
SYSTEM D4020 310 0560					
WET STANDPIPE RISER, CLASS I, STEEL, BLACK, SCH. 40 PIPE, 10′ HEIGHT					
4″ DIAMETER PIPE, ONE FLOOR					
Pipe, steel, black, schedule 40, threaded, 4″ diam.	20.000	L.F.	590	730	1,320
Pipe, Tee, malleable iron, black, 150 lb. threaded, 4″ pipe size	2.000	Ea.	630	660	1,290
Pipe, 90° elbow, malleable iron, black, 150 lb threaded 4″ pipe size	1.000	Ea.	200	220	420
Pipe, nipple, steel, black, schedule 40, 2-1/2″ pipe size x 3″ long	2.000	Ea.	24.50	165	189.50
Fire valve, gate, 300 lb., brass w/handwheel, 2-1/2″ pipe size	1.000	Ea.	208	100	308
Fire valve, pressure restricting, adj, rgh brs, 2-1/2″ pipe size	1.000	Ea.	352	200	552
Valve, swing check, w/ball drip, CI w/brs. ftngs., 4″ pipe size	1.000	Ea.	385	420	805
Standpipe conn wall dble. flush brs. w/plugs & chains 2-1/2″x2-1/2″x4″	1.000	Ea.	570	253	823
Valve, swing check, bronze, 125 lb, regrinding disc, 2-1/2″ pipe size	1.000	Ea.	880	88	968
Roof manifold, fire, w/valves & caps, horiz/vert brs 2-1/2″x2-1/2″x4″	1.000	Ea.	185	263	448
Fire, hydrolator, vent & drain, 2-1/2″ pipe size	1.000	Ea.	92.50	58.50	151
Valve, gate, iron body 125 lb., OS&Y, threaded, 4″ pipe size	1.000	Ea.	850	440	1,290
TOTAL			4,967	3,597.50	8,564.50

D4020 310	Wet Standpipe Risers, Class I	COST PER FLOOR MAT.	INST.	TOTAL
0550	Wet standpipe risers, Class I, steel, black sch. 40, 10′ height			
0560	4″ diameter pipe, one floor	4,975	3,600	8,575
0580	Additional floors	1,100	1,125	2,225
0600	6″ diameter pipe, one floor	8,550	6,325	14,875
0620	Additional floors R211226-20	2,075	1,775	3,850
0640	8″ diameter pipe, one floor	13,200	7,625	20,825
0660	Additional floors	3,100	2,150	5,250
0680				

D4020 310	Wet Standpipe Risers, Class II	COST PER FLOOR MAT.	INST.	TOTAL
1030	Wet standpipe risers, Class II, steel, black sch. 40, 10′ height			
1040	2″ diameter pipe, one floor	1,950	1,300	3,250
1060	Additional floors	570	500	1,070
1080	2-1/2″ diameter pipe, one floor	2,875	1,900	4,775
1100	Additional floors	660	585	1,245
1120				

D4020 Standpipes

D4020 310	Wet Standpipe Risers, Class III	COST PER FLOOR		
		MAT.	INST.	TOTAL
1530	Wet standpipe risers, Class III, steel, black sch. 40, 10′ height			
1540	4″ diameter pipe, one floor	5,100	3,600	8,700
1560	Additional floors	950	935	1,885
1580	6″ diameter pipe, one floor	8,675	6,325	15,000
1600	Additional floors	2,125	1,775	3,900
1620	8″ diameter pipe, one floor	13,300	7,625	20,925
1640	Additional floors	3,150	2,150	5,300

D4020 Standpipes

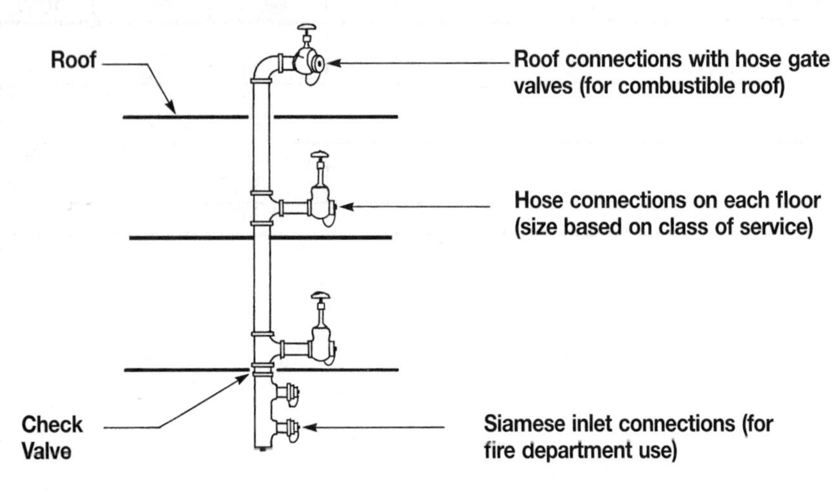

System Components	QUANTITY	UNIT	COST PER FLOOR MAT.	INST.	TOTAL
SYSTEM D4020 330 0540					
DRY STANDPIPE RISER, CLASS I, PIPE, STEEL, BLACK, SCH 40, 10′ HEIGHT					
4″ DIAMETER PIPE, ONE FLOOR					
Pipe, steel, black, schedule 40, threaded, 4″ diam.	20.000	L.F.	590	730	1,320
Pipe, Tee, malleable iron, black, 150 lb. threaded, 4″ pipe size	2.000	Ea.	630	660	1,290
Pipe, 90° elbow, malleable iron, black, 150 lb threaded 4″ pipe size	1.000	Ea.	200	220	420
Pipe, nipple, steel, black, schedule 40, 2-1/2″ pipe size x 3″ long	2.000	Ea.	24.50	165	189.50
Fire valve gate NRS 300 lb., brass w/handwheel, 2-1/2″ pipe size	1.000	Ea.	208	100	308
Fire valve, pressure restricting, adj, rgh brs, 2-1/2″ pipe size	1.000	Ea.	176	100	276
Standpipe conn wall dble. flush brs. w/plugs & chains 2-1/2″x2-1/2″x4″	1.000	Ea.	570	253	823
Valve swing check w/ball drip CI w/brs. ftngs., 4″pipe size	1.000	Ea.	385	420	805
Roof manifold, fire, w/valves & caps, horiz/vert brs 2-1/2″x2-1/2″x4″	1.000	Ea.	185	263	448
TOTAL			2,968.50	2,911	5,879.50

D4020 330	Dry Standpipe Risers, Class I	COST PER FLOOR MAT.	INST.	TOTAL
0530	Dry standpipe riser, Class I, steel, black sch. 40, 10′ height			
0540	4″ diameter pipe, one floor	2,975	2,900	5,875
0560	Additional floors R211226 -20	1,025	1,050	2,075
0580	6″ diameter pipe, one floor	6,100	5,000	11,100
0600	Additional floors	1,975	1,700	3,675
0620	8″ diameter pipe, one floor	9,700	6,075	15,775
0640	Additional floors	3,000	2,100	5,100
0660				

D4020 330	Dry Standpipe Risers, Class II	COST PER FLOOR MAT.	INST.	TOTAL
1030	Dry standpipe risers, Class II, steel, black sch. 40, 10′ height			
1040	2″ diameter pipe, one floor	1,575	1,350	2,925
1060	Additional floor	480	440	920
1080	2-1/2″ diameter pipe, one floor	2,250	1,575	3,825
1100	Additional floors	570	530	1,100
1120				

D4020 Standpipes

D4020 330	Dry Standpipe Risers, Class III	COST PER FLOOR		
		MAT.	INST.	TOTAL
1530	Dry standpipe risers, Class III, steel, black sch. 40, 10′ height			
1540	4″ diameter pipe, one floor	3,025	2,850	5,875
1560	Additional floors	870	960	1,830
1580	6″ diameter pipe, one floor	6,150	5,000	11,150
1600	Additional floors	2,025	1,700	3,725
1620	8″ diameter pipe, one floor	9,775	6,075	15,850
1640	Additional floor	3,075	2,100	5,175

D4020 Standpipes

D4020 410	Fire Hose Equipment	COST EACH		
		MAT.	INST.	TOTAL
0100	Adapters, reducing, 1 piece, FxM, hexagon, cast brass, 2-1/2" x 1-1/2"	61		61
0200	Pin lug, 1-1/2" x 1"	52.50		52.50
0250	3" x 2-1/2"	137		137
0300	For polished chrome, add 75% mat.			
0400	Cabinets, D.S. glass in door, recessed, steel box, not equipped			
0500	Single extinguisher, steel door & frame	128	158	286
0550	Stainless steel door & frame	226	158	384
0600	Valve, 2-1/2" angle, steel door & frame	149	105	254
0650	Aluminum door & frame	180	105	285
0700	Stainless steel door & frame	243	105	348
0750	Hose rack assy, 2-1/2" x 1-1/2" valve & 100' hose, steel door & frame	278	211	489
0800	Aluminum door & frame	410	211	621
0850	Stainless steel door & frame	545	211	756
0900	Hose rack assy & extinguisher,2-1/2"x1-1/2" valve & hose,steel door & frame	291	253	544
0950	Aluminum	525	253	778
1000	Stainless steel	590	253	843
1550	Compressor, air, dry pipe system, automatic, 200 gal., 3/4 H.P.	840	540	1,380
1600	520 gal., 1 H.P.	870	540	1,410
1650	Alarm, electric pressure switch (circuit closer)	97.50	27	124.50
2500	Couplings, hose, rocker lug, cast brass, 1-1/2"	58.50		58.50
2550	2-1/2"	77		77
3000	Escutcheon plate, for angle valves, polished brass, 1-1/2"	16.70		16.70
3050	2-1/2"	27		27
3500	Fire pump, electric, w/controller, fittings, relief valve			
3550	4" pump, 30 H.P., 500 G.P.M.	16,400	3,925	20,325
3600	5" pump, 40 H.P., 1000 G.P.M.	18,200	4,450	22,650
3650	5" pump, 100 H.P., 1000 G.P.M.	24,800	4,950	29,750
3700	For jockey pump system, add	2,850	630	3,480
3800	Fire pump, diesel, w/controller, 6" pump, 140 HP, 1500 GPM, 100 psi	61,000	6,625	67,625
3900	Fire pump, diesel, w/controller, 10" pump, 300 HP, 3500 GPM, 100 psi	85,000	11,900	96,900
4100	Fire pump, electric, w/controller, 6" pump, 139 HP, 1500 GPM, 1770 RPM	34,600	6,625	41,225
4200	Fire pump, electric, w/controller, 10" pump, 300 HP, 3500 GPM, 1770 RPM	66,500	11,100	77,600
5000	Hose, per linear foot, synthetic jacket, lined,			
5100	300 lb. test, 1-1/2" diameter	3.56	.49	4.05
5150	2-1/2" diameter	6.15	.57	6.72
5200	500 lb. test, 1-1/2" diameter	3.69	.49	4.18
5250	2-1/2" diameter	6.50	.57	7.07
5500	Nozzle, plain stream, polished brass, 1-1/2" x 10"	52.50		52.50
5550	2-1/2" x 15" x 13/16" or 1-1/2"	110		110
5600	Heavy duty combination adjustable fog and straight stream w/handle 1-1/2"	495		495
5650	2-1/2" direct connection	615		615
6000	Rack, for 1-1/2" diameter hose 100 ft. long, steel	67	63	130
6050	Brass	115	63	178
6500	Reel, steel, for 50 ft. long 1-1/2" diameter hose	144	90.50	234.50
6550	For 75 ft. long 2-1/2" diameter hose	234	90.50	324.50
7050	Siamese, w/plugs & chains, polished brass, sidewalk, 4" x 2-1/2" x 2-1/2"	695	505	1,200
7100	6" x 2-1/2" x 2-1/2"	860	630	1,490
7200	Wall type, flush, 4" x 2-1/2" x 2-1/2"	570	253	823
7250	6" x 2-1/2" x 2-1/2"	795	275	1,070
7300	Projecting, 4" x 2-1/2" x 2-1/2"	520	253	773
7350	6" x 2-1/2" x 2-1/2"	880	275	1,155
7400	For chrome plate, add 15% mat.			
8000	Valves, angle, wheel handle, 300 Lb., rough brass, 1-1/2"	104	58.50	162.50
8050	2-1/2"	192	100	292
8100	Combination pressure restricting, 1-1/2"	94	58.50	152.50
8150	2-1/2"	192	100	292
8200	Pressure restricting, adjustable, satin brass, 1-1/2"	117	58.50	175.50
8250	2-1/2"	176	100	276

D4020 Standpipes

D4020 410	Fire Hose Equipment	COST EACH		
		MAT.	INST.	TOTAL
8300	Hydrolator, vent and drain, rough brass, 1-1/2"	92.50	58.50	151
8350	2-1/2"	92.50	58.50	151
8400	Cabinet assy, incls. adapter, rack, hose, and nozzle	875	380	1,255

D4090 Other Fire Protection Systems

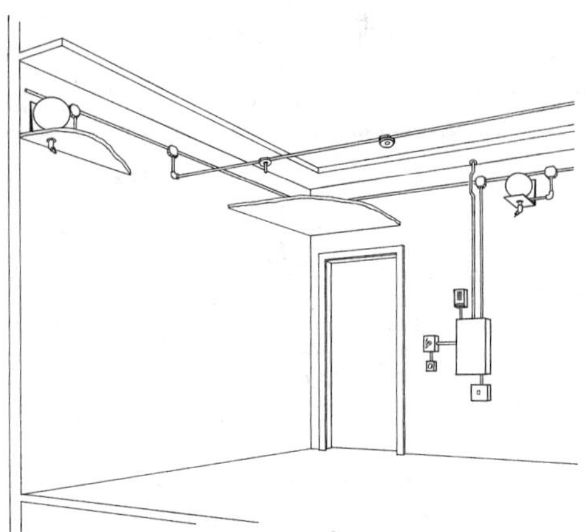

General: Automatic fire protection (suppression) systems other than water sprinklers may be desired for special environments, high risk areas, isolated locations or unusual hazards. Some typical applications would include:

Paint dip tanks
Securities vaults
Electronic data processing
Tape and data storage
Transformer rooms
Spray booths
Petroleum storage
High rack storage

Piping and wiring costs are dependent on the individual application and must be added to the component costs shown below.

All areas are assumed to be open.

D4090 910	Fire Suppression Unit Components	COST EACH		
		MAT.	INST.	TOTAL
0020	Detectors with brackets			
0040	Fixed temperature heat detector	56	84.50	140.50
0060	Rate of temperature rise detector	56	84.50	140.50
0080	Ion detector (smoke) detector	121	109	230
0200	Extinguisher agent			
0240	200 lb FM200, container	8,550	330	8,880
0280	75 lb carbon dioxide cylinder	1,400	220	1,620
0320	Dispersion nozzle			
0340	FM200 1-1/2" dispersion nozzle	73.50	52.50	126
0380	Carbon dioxide 3" x 5" dispersion nozzle	73.50	40.50	114
0420	Control station			
0440	Single zone control station with batteries	1,900	680	2,580
0470	Multizone (4) control station with batteries	3,600	1,350	4,950
0490				
0500	Electric mechanical release	184	355	539
0520				
0550	Manual pull station	66.50	122	188.50
0570				
0640	Battery standby power 10" x 10" x 17"	455	169	624
0700				
0740	Bell signalling device	86.50	84.50	171

D4090 920	FM200 Systems	COST PER C.F.		
		MAT.	INST.	TOTAL
0820	Average FM200 system, minimum			1.94
0840	Maximum			3.85

D5010 Electrical Service/Distribution

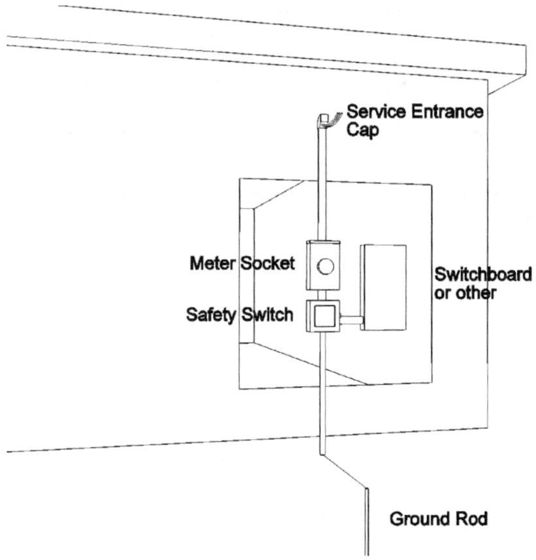

System Components	QUANTITY	UNIT	COST EACH MAT.	COST EACH INST.	COST EACH TOTAL
SYSTEM D5010 120 0220					
SERVICE INSTALLATION, INCLUDES BREAKERS, METERING, 20′ CONDUIT & WIRE					
3 PHASE, 4 WIRE, 60 A					
Wire, 600 volt, copper type XHHW, stranded #6	.600	C.L.F.	41.70	62.40	104.10
Rigid galvanized steel conduit, 3/4″ including fittings	20.000	L.F.	60.40	169	229.40
Service entrance cap, 3/4″ diameter	1.000	Ea.	7.15	52	59.15
Conduit LB fitting with cover, 3/4″ diameter	1.000	Ea.	11.75	52	63.75
Meter socket, three phase, 100 A	1.000	Ea.	181	242	423
Safety switches, heavy duty, 240 volt, 3 pole NEMA 1 fusible, 60 amp	1.000	Ea.	219	295	514
Grounding, wire ground bare armored, #8-1 conductor	.200	C.L.F.	16.80	68	84.80
Grounding, clamp, bronze, 3/4″ diameter	1.000	Ea.	6.50	21	27.50
Grounding, rod, copper clad, 8′ long, 3/4″ diameter	1.000	Ea.	37	128	165
Wireway w/fittings, 2-1/2″ x 2-1/2″	1.000	L.F.	11.95	15.05	27
TOTAL			593.25	1,104.45	1,697.70

D5010 120	Overhead Electric Service, 3 Phase - 4 Wire	COST EACH MAT.	COST EACH INST.	COST EACH TOTAL
0200	Service installation, includes breakers, metering, 20′ conduit & wire			
0220	3 phase, 4 wire, 120/208 volts, 60 A	595	1,100	1,695
0240	3 phase, 4 wire, 120/208 volts, 100 A	790	1,250	2,040
0245	100 A w/circuit breaker	1,575	1,550	3,125
0280	200 A	1,475	1,725	3,200
0285	200 A w/circuit breaker	3,100	2,175	5,275
0320	400 A	2,925	3,450	6,375
0325	400 A w/circuit breaker	6,000	4,275	10,275
0360	600 A	5,025	5,075	10,100
0365	600 A, w/switchboard	9,675	6,425	16,100
0400	800 A	6,950	6,075	13,025
0405	800 A, w/switchboard	11,600	7,625	19,225
0440	1000 A	8,625	7,500	16,125
0445	1000 A, w/switchboard	14,200	9,200	23,400
0480	1200 A	11,500	8,550	20,050
0485	1200 A, w/groundfault switchboard	35,600	10,600	46,200
0520	1600 A	14,200	11,200	25,400
0525	1600 A, w/groundfault switchboard	40,300	13,400	53,700

D5010 Electrical Service/Distribution

D5010 120	Overhead Electric Service, 3 Phase - 4 Wire	COST EACH		
		MAT.	INST.	TOTAL
0560	2000 A	19,000	13,600	32,600
0565	2000 A, w/groundfault switchboard	48,700	16,700	65,400
0610	1 phase, 3 wire, 120/240 volts, 100 A (no safety switch)	190	610	800
0615	100 A w/load center	555	1,175	1,730
0620	200 A	445	840	1,285
0625	200 A w/load center	1,250	1,725	2,975

D5010 Electrical Service/Distribution

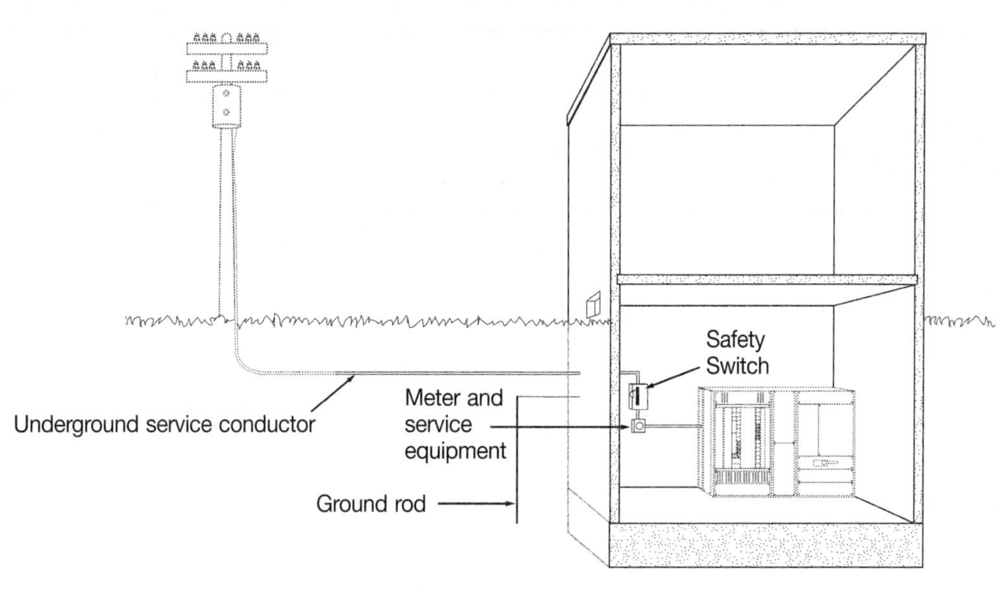

System Components	QUANTITY	UNIT	COST EACH MAT.	COST EACH INST.	COST EACH TOTAL
SYSTEM D5010 130 1000					
2000 AMP UNDERGROUND ELECTRIC SERVICE WITH GROUNDFAULT SWITCHBOARD INCLUDING EXCAVATION, BACKFILL, AND COMPACTION					
Excavate Trench	44.440	B.C.Y.		307.96	307.96
4 inch conduit bank	108.000	L.F.	1,085.40	2,700	3,785.40
4 inch fitting	8.000	Ea.	252	452	704
4 inch bells	4.000	Ea.	18.76	226	244.76
Concrete material	16.580	C.Y.	1,856.96		1,856.96
Concrete placement	16.580	C.Y.		432.24	432.24
Backfill trench	33.710	L.C.Y.		102.48	102.48
Compact fill material in trench	25.930	E.C.Y.		134.05	134.05
Dispose of excess fill material on-site	24.070	L.C.Y.		293.41	293.41
500 kcmil power cable	18.000	C.L.F.	22,950	7,650	30,600
Wire, 600 volt, type THW, copper, stranded, 1/0	6.000	C.L.F.	1,650	1,230	2,880
Saw cutting, concrete walls, plain, per inch of depth	64.000	L.F.	3.20	529.28	532.48
Meter centers and sockets, single pos, 4 terminal, 400 amp	5.000	Ea.	3,975	2,000	5,975
Safety switch, 400 amp	5.000	Ea.	7,625	3,775	11,400
Ground rod clamp	1.000	Ea.	6.50	21	27.50
Wireway	1.000	L.F.	11.95	15.05	27
600 volt stranded copper wire, 1/0	1.200	C.L.F.	330	246	576
Flexible metallic conduit	120.000	L.F.	88.80	508.80	597.60
Ground rod	6.000	Ea.	357	1,014	1,371
TOTAL			40,210.57	21,637.27	61,847.84

D5010 130	Underground Electric Service	COST EACH MAT.	COST EACH INST.	COST EACH TOTAL
0950	Underground electric service including excavation,backfill, and compaction			
1000	3 phase, 4 wire, 277/480 volts, 2000 A	40,200	21,600	61,800
1050	2000 A w/groundfault switch	68,500	24,200	92,700
1100	1600 A	31,900	17,500	49,400
1150	1600 A w/groundfault switchboard	58,000	19,700	77,700
1200	1200 A	24,400	13,900	38,300
1250	1200 A w/groundfault switchboard	48,500	16,000	64,500
1400	800 A	17,700	11,500	29,200

D5010 Electrical Service/Distribution

D5010 130	Underground Electric Service	COST EACH		
		MAT.	INST.	TOTAL
1450	800 A w/ switchboard	22,300	13,100	35,400
1500	600 A	11,400	10,600	22,000
1550	600 A w/ switchboard	16,000	11,900	27,900
1600	1 phase, 3 wire, 120/240 volts, 200 A	3,225	3,675	6,900
1650	200 A w/load center	4,050	4,550	8,600
1700	100 A	2,200	3,000	5,200
1750	100 A w/load center	3,000	3,900	6,900

D5010 Electrical Service/Distribution

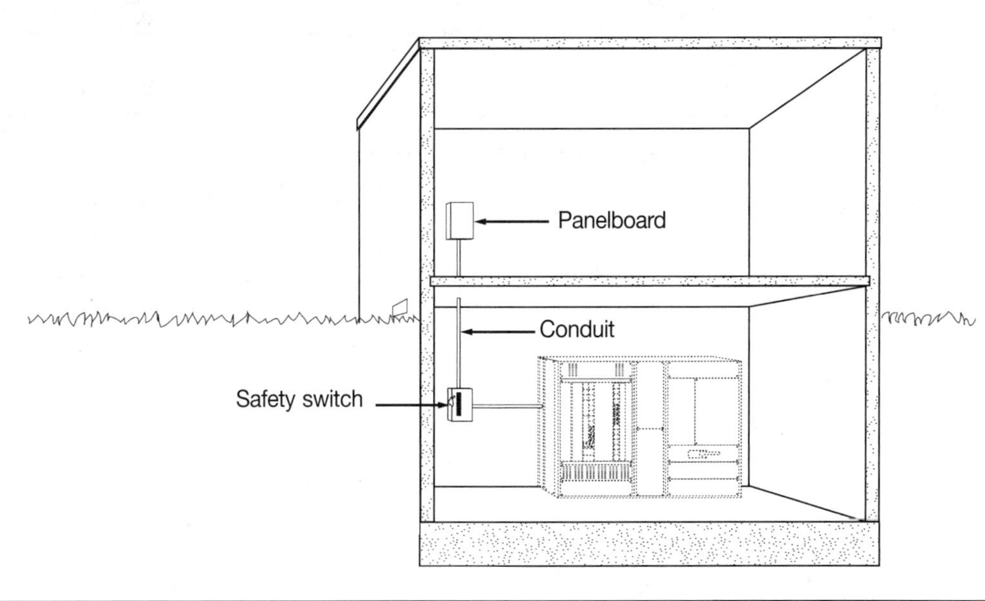

System Components	QUANTITY	UNIT	COST EACH MAT.	COST EACH INST.	COST EACH TOTAL
SYSTEM D5010 250 1020					
PANELBOARD INSTALLATION, INCLUDES PANELBOARD, CONDUCTOR, & CONDUIT					
Conduit, galvanized steel, 1-1/4"dia.	37.000	L.F.	209.05	418.10	627.15
Panelboards, NQOD, 4 wire, 120/208 volts, 100 amp main, 24 circuits	1.000	Ea.	1,300	1,450	2,750
Wire, 600 volt, type THW, copper, stranded, #3	1.480	C.L.F.	204.24	201.28	405.52
TOTAL			1,713.29	2,069.38	3,782.67

D5010 250	Panelboard	COST EACH MAT.	COST EACH INST.	COST EACH TOTAL
0900	Panelboards, NQOD, 4 wire, 120/208 volts w/conductor & conduit			
1000	100 A, 0 stories, 0′ horizontal	1,300	1,450	2,750
1020	1 stories, 25′ horizontal	1,725	2,075	3,800
1040	5 stories, 50′ horizontal	2,525	3,300	5,825
1060	10 stories, 75′ horizontal	3,475	4,725	8,200
1080	225A, 0 stories, 0′ horizontal	2,425	1,875	4,300
2000	1 stories, 25′ horizontal	3,650	3,000	6,650
2020	5 stories, 50′ horizontal	6,100	5,200	11,300
2040	10 stories, 75′ horizontal	8,925	7,750	16,675
2060	400A, 0 stories, 0′ horizontal	3,575	2,825	6,400
2080	1 stories, 25′ horizontal	5,275	4,625	9,900
3000	5 stories, 50′ horizontal	8,600	8,200	16,800
3020	10 stories, 75′ horizontal	18,000	16,000	34,000
3040	600 A, 0 stories, 0′ horizontal	5,300	3,400	8,700
3060	1 stories, 25′ horizontal	8,800	6,350	15,150
3080	5 stories, 50′ horizontal	15,700	12,100	27,800
4000	10 stories, 75′ horizontal	35,500	23,900	59,400
4010	Panelboards, NEHB, 4 wire, 277/480 volts w/conductor, conduit, & safety switch			
4020	100 A, 0 stories, 0′ horizontal, includes safety switch	2,975	1,975	4,950
4040	1 stories, 25′ horizontal	3,375	2,600	5,975
4060	5 stories, 50′ horizontal	4,200	3,825	8,025
4080	10 stories, 75′ horizontal	5,150	5,250	10,400
5000	225 A, 0 stories, 0′ horizontal	4,450	2,400	6,850
5020	1 stories, 25′ horizontal	5,675	3,500	9,175
5040	5 stories, 50′ horizontal	8,100	5,725	13,825
5060	10 stories, 75′ horizontal	10,900	8,275	19,175

D5010 Electrical Service/Distribution

D5010 250	Panelboard	COST EACH		
		MAT.	INST.	TOTAL
5080	400 A, 0 stories, 0' horizontal	7,225	3,700	10,925
6000	1 stories, 25' horizontal	8,925	5,525	14,450
6020	5 stories, 50' horizontal	12,300	9,075	21,375
6040	10 stories, 75' horizontal	21,700	16,900	38,600
6060	600 A, 0 stories, 0' horizontal	10,900	4,700	15,600
6080	1 stories, 25' horizontal	14,400	7,650	22,050
7000	5 stories, 50' horizontal	21,300	13,400	34,700
7020	10 stories, 75' horizontal	41,100	25,200	66,300

D5020 Lighting and Branch Wiring

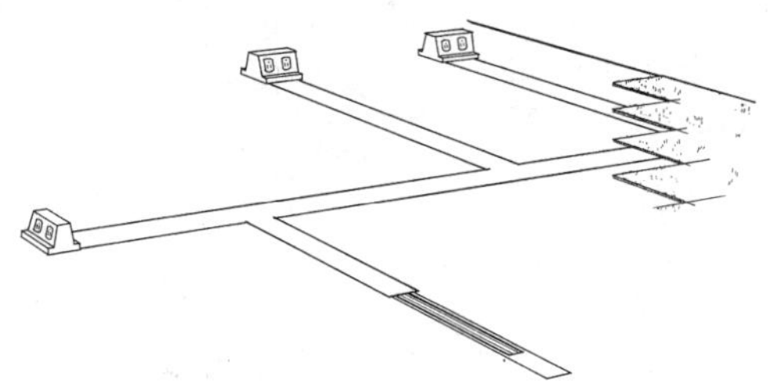

Underfloor Receptacle System

Description: Table D5020 115 includes installed costs of raceways and copper wire from panel to and including receptacle.

National Electrical Code prohibits use of undercarpet system in residential, school or hospital buildings. Can only be used with carpet squares.

Low density = (1) Outlet per 259 S.F. of floor area.

High density = (1) Outlet per 127 S.F. of floor area.

System Components	QUANTITY	UNIT	COST PER S.F. MAT.	COST PER S.F. INST.	COST PER S.F. TOTAL
SYSTEM D5020 115 0200					
RECEPTACLE SYSTEMS, UNDERFLOOR DUCT, 5′ ON CENTER, LOW DENSITY					
Underfloor duct 3-1/8″ x 7/8″ w/insert 24″ on center	.190	L.F.	3.99	1.84	5.83
Vertical elbow for underfloor duct, 3-1/8″, included					
Underfloor duct conduit adapter, 2″ x 1-1/4″, included					
Underfloor duct junction box, single duct, 3-1/8″	.003	Ea.	1.41	.51	1.92
Underfloor junction box carpet pan	.003	Ea.	1.17	.03	1.20
Underfloor duct outlet, high tension receptacle	.004	Ea.	.44	.34	.78
Wire 600V type THWN-THHN copper solid #12	.010	C.L.F.	.13	.62	.75
TOTAL			7.14	3.34	10.48

D5020 115	Receptacles, Floor	COST PER S.F. MAT.	COST PER S.F. INST.	COST PER S.F. TOTAL
0200	Receptacle systems, underfloor duct, 5′ on center, low density	7.15	3.34	10.49
0240	High density	7.70	4.29	11.99
0280	7′ on center, low density	5.65	2.87	8.52
0320	High density	6.20	3.82	10.02
0400	Poke thru fittings, low density	1.26	1.63	2.89
0440	High density	2.50	3.26	5.76
0520	Telepoles, using Romex, low density	1.19	1	2.19
0560	High density	2.38	2	4.38
0600	Using EMT, low density	1.26	1.32	2.58
0640	High density	2.50	2.64	5.14
0720	Conduit system with floor boxes, low density	1.17	1.13	2.30
0760	High density	2.34	2.27	4.61
0840	Undercarpet power system, 3 conductor with 5 conductor feeder, low density	1.55	.40	1.95
0880	High density	3.04	.84	3.88

D5020 Lighting and Branch Wiring

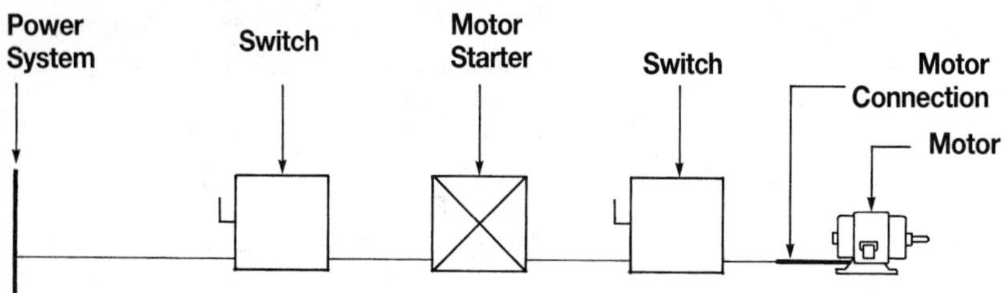

System D5020 145 installed cost of motor wiring using 50′ of rigid conduit and copper wire. **Cost and setting of motor not included.**

System Components	QUANTITY	UNIT	COST EACH MAT.	COST EACH INST.	COST EACH TOTAL
SYSTEM D5020 145 0200					
MOTOR INST., SINGLE PHASE, 115V, TO AND INCLUDING 1/3 HP MOTOR SIZE					
Wire 600V type THWN-THHN, copper solid #12	1.250	C.L.F.	16.38	76.88	93.26
Steel intermediate conduit, (IMC) 1/2″ diam.	50.000	L.F.	98.50	340	438.50
Magnetic FVNR, 115V, 1/3 HP, size 00 starter	1.000	Ea.	203	169	372
Safety switch, fused, heavy duty, 240V 2P 30 amp	1.000	Ea.	97.50	194	291.50
Safety switch, non fused, heavy duty, 600V, 3 phase, 30 A	1.000	Ea.	125	212	337
Flexible metallic conduit, Greenfield 1/2″ diam.	1.500	L.F.	.80	5.09	5.89
Connectors for flexible metallic conduit Greenfield 1/2″ diam.	1.000	Ea.	2.51	8.45	10.96
Coupling for Greenfield to conduit 1/2″ diam. flexible metalic conduit	1.000	Ea.	1.63	13.55	15.18
Fuse cartridge nonrenewable, 250V 30 amp	1.000	Ea.	2.64	13.55	16.19
TOTAL			547.96	1,032.52	1,580.48

D5020 145	Motor Installation	COST EACH MAT.	COST EACH INST.	COST EACH TOTAL
0200	Motor installation, single phase, 115V, 1/3 HP motor size	550	1,025	1,575
0240	1 HP motor size	570	1,025	1,595
0280	2 HP motor size	620	1,100	1,720
0320	3 HP motor size	710	1,125	1,835
0360	230V, 1 HP motor size	550	1,050	1,600
0400	2 HP motor size	590	1,050	1,640
0440	3 HP motor size	660	1,125	1,785
0520	Three phase, 200V, 1-1/2 HP motor size	640	1,150	1,790
0560	3 HP motor size	720	1,250	1,970
0600	5 HP motor size	760	1,400	2,160
0640	7-1/2 HP motor size	790	1,425	2,215
0680	10 HP motor size	1,200	1,775	2,975
0720	15 HP motor size	1,625	1,975	3,600
0760	20 HP motor size	2,075	2,275	4,350
0800	25 HP motor size	2,125	2,300	4,425
0840	30 HP motor size	3,300	2,700	6,000
0880	40 HP motor size	4,125	3,175	7,300
0920	50 HP motor size	7,250	3,700	10,950
0960	60 HP motor size	7,550	3,925	11,475
1000	75 HP motor size	9,500	4,475	13,975
1040	100 HP motor size	27,200	5,300	32,500
1080	125 HP motor size	27,500	5,800	33,300
1120	150 HP motor size	31,800	6,825	38,625
1160	200 HP motor size	32,800	8,075	40,875
1240	230V, 1-1/2 HP motor size	615	1,125	1,740
1280	3 HP motor size	690	1,225	1,915
1320	5 HP motor size	730	1,375	2,105
1360	7-1/2 HP motor size	730	1,375	2,105
1400	10 HP motor size	1,125	1,675	2,800
1440	15 HP motor size	1,275	1,850	3,125
1480	20 HP motor size	1,975	2,225	4,200
1520	25 HP motor size	2,075	2,275	4,350

D5020 Lighting and Branch Wiring

D5020 145	Motor Installation	COST EACH		
		MAT.	INST.	TOTAL
1560	30 HP motor size	2,100	2,300	4,400
1600	40 HP motor size	3,975	3,100	7,075
1640	50 HP motor size	4,300	3,275	7,575
1680	60 HP motor size	7,275	3,725	11,000
1720	75 HP motor size	8,750	4,225	12,975
1760	100 HP motor size	10,300	4,700	15,000
1800	125 HP motor size	27,700	5,475	33,175
1840	150 HP motor size	29,300	6,200	35,500
1880	200 HP motor size	31,500	6,875	38,375
1960	460V, 2 HP motor size	760	1,150	1,910
2000	5 HP motor size	835	1,250	2,085
2040	10 HP motor size	840	1,375	2,215
2080	15 HP motor size	1,125	1,575	2,700
2120	20 HP motor size	1,200	1,700	2,900
2160	25 HP motor size	1,275	1,775	3,050
2200	30 HP motor size	1,625	1,925	3,550
2240	40 HP motor size	2,100	2,050	4,150
2280	50 HP motor size	2,350	2,275	4,625
2320	60 HP motor size	3,500	2,675	6,175
2360	75 HP motor size	4,150	2,950	7,100
2400	100 HP motor size	4,625	3,300	7,925
2440	125 HP motor size	7,600	3,725	11,325
2480	150 HP motor size	9,425	4,175	13,600
2520	200 HP motor size	11,100	4,725	15,825
2600	575V, 2 HP motor size	760	1,150	1,910
2640	5 HP motor size	835	1,250	2,085
2680	10 HP motor size	840	1,375	2,215
2720	20 HP motor size	1,125	1,575	2,700
2760	25 HP motor size	1,200	1,700	2,900
2800	30 HP motor size	1,625	1,925	3,550
2840	50 HP motor size	1,750	2,000	3,750
2880	60 HP motor size	3,475	2,650	6,125
2920	75 HP motor size	3,500	2,675	6,175
2960	100 HP motor size	4,150	2,950	7,100
3000	125 HP motor size	7,400	3,675	11,075
3040	150 HP motor size	7,600	3,725	11,325
3080	200 HP motor size	9,575	4,225	13,800

D5020 Lighting and Branch Wiring

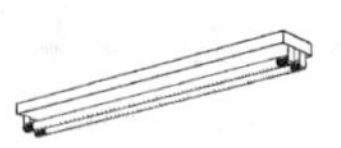

A. Strip Fixture

B. Surface Mounted

C. Recessed

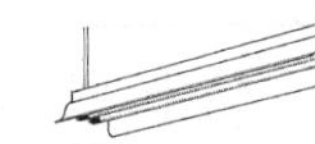

D. Pendent Mounted

Design Assumptions:

1. A 100 footcandle average maintained level of illumination.
2. Ceiling heights range from 9′ to 11′.
3. Average reflectance values are assumed for ceilings, walls and floors.
4. Cool white (CW) fluorescent lamps with 3150 lumens for 40 watt lamps and 6300 lumens for 8′ slimline lamps.
5. Four 40 watt lamps per 4′ fixture and two 8′ lamps per 8′ fixture.
6. Average fixture efficiency values and spacing to mounting height ratios.
7. Installation labor is average U.S. rate as of January 1.

System Components	QUANTITY	UNIT	COST PER S.F. MAT.	INST.	TOTAL
SYSTEM D5020 208 0520					
FLUORESCENT FIXTURES MOUNTED 9′-11″ ABOVE FLOOR, 100 FC					
TYPE A, 8 FIXTURES PER 400 S.F.					
Conduit, steel intermediate, 1/2″ diam.	.185	L.F.	.36	1.26	1.62
Wire, 600V, type THWN-THHN, copper, solid, #12	.004	C.L.F.	.05	.23	.28
Fluorescent strip fixture 8′ long, surface mounted, two 75W SL	.020	Ea.	1.27	2.18	3.45
Steel outlet box 4″ concrete	.020	Ea.	.27	.68	.95
Steel outlet box plate with stud, 4″ concrete	.020	Ea.	.18	.17	.35
Fixture hangers, flexible, 1/2″ diameter, 4″ long	.040	Ea.	.69	2.26	2.95
Fixture whip, THHN wire, three #12, 3/8″ Greenfield	.040	Ea.	.49	.84	1.33
TOTAL			3.31	7.62	10.93

D5020 208	Fluorescent Fixtures (by Type)	COST PER S.F. MAT.	INST.	TOTAL
0520	Fluorescent fixtures, type A, 8 fixtures per 400 S.F.	3.31	7.60	10.91
0560	11 fixtures per 600 S.F.	2.99	6.95	9.94
0600	17 fixtures per 1000 S.F.	2.80	6.45	9.25
0640	23 fixtures per 1600 S.F.	2.35	5.45	7.80
0680	28 fixtures per 2000 S.F.	2.31	5.35	7.66
0720	41 fixtures per 3000 S.F.	2.27	5.25	7.52
0800	53 fixtures per 4000 S.F.	2.19	5.05	7.24
0840	64 fixtures per 5000 S.F.	2.16	4.94	7.10
0880	Type B, 11 fixtures per 400 S.F.	5.25	10.85	16.10
0920	15 fixtures per 600 S.F.	4.71	9.80	14.51
0960	24 fixtures per 1000 S.F.	4.53	9.45	13.98
1000	35 fixtures per 1600 S.F.	4.15	8.60	12.75
1040	42 fixtures per 2000 S.F.	3.95	8.20	12.15
1080	61 fixtures per 3000 S.F.	3.80	7.90	11.70
1160	80 fixtures per 4000 S.F.	2.71	5.05	7.76
1200	98 fixtures per 5000 S.F.	3.74	7.70	11.44
1240	Type C, 11 fixtures per 400 S.F.	4.52	11.30	15.82
1280	14 fixtures per 600 S.F.	3.78	9.45	13.23
1320	23 fixtures per 1000 S.F.	3.74	9.35	13.09
1360	34 fixtures per 1600 S.F.	3.45	8.65	12.10
1400	43 fixtures per 2000 S.F.	3.56	8.85	12.41
1440	63 fixtures per 3000 S.F.	3.41	8.55	11.96
1520	81 fixtures per 4000 S.F.	3.29	8.25	11.54
1560	101 fixtures per 5000 S.F.	3.27	8.15	11.42
1600	Type D, 8 fixtures per 400 S.F.	4.20	8.50	12.70
1640	12 fixtures per 600 S.F.	4.20	8.50	12.70
1680	19 fixtures per 1000 S.F.	3.99	8.10	12.09
1720	27 fixtures per 1600 S.F.	3.56	7.20	10.76
1760	34 fixtures per 2000 S.F.	3.56	7.25	10.81
1800	48 fixtures per 3000 S.F.	3.36	6.80	10.16
1880	64 fixtures per 4000 S.F.	3.36	6.80	10.16
1920	79 fixtures per 5000 S.F.	3.36	6.80	10.16

D5020 Lighting and Branch Wiring

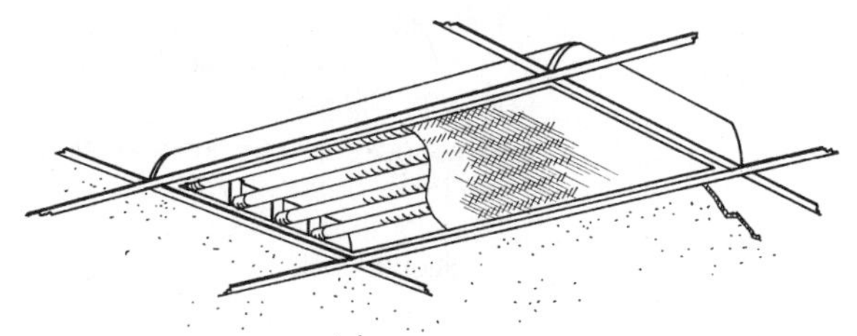

Type C. Recessed, mounted on grid ceiling suspension system, 2′ x 4′, four 40 watt lamps, acrylic prismatic diffusers.

5.3 watts per S.F. for 100 footcandles.

3 watts per S.F. for 57 footcandles.

System Components	QUANTITY	UNIT	COST PER S.F. MAT.	COST PER S.F. INST.	COST PER S.F. TOTAL
SYSTEM D5020 210 0200					
FLUORESCENT FIXTURES RECESS MOUNTED IN CEILING					
1 WATT PER S.F., 20 FC, 5 FIXTURES PER 1000 S.F.					
Steel intermediate conduit, (IMC) 1/2″ diam.	.128	L.F.	.25	.87	1.12
Wire, 600 volt, type THW, copper, solid, #12	.003	C.L.F.	.04	.18	.22
Fluorescent fixture, recessed, 2′x 4′, four 40W, w/lens, for grid ceiling	.005	Ea.	.32	.72	1.04
Steel outlet box 4″ square	.005	Ea.	.07	.17	.24
Fixture whip, Greenfield w/#12 THHN wire	.005	Ea.	.04	.04	.08
TOTAL			.72	1.98	2.70

D5020 210	Fluorescent Fixtures (by Wattage)	COST PER S.F. MAT.	COST PER S.F. INST.	COST PER S.F. TOTAL
0190	Fluorescent fixtures recess mounted in ceiling			
0195	T12, standard 40 watt lamps			
0200	1 watt per S.F., 20 FC, 5 fixtures @40 watts per 1000 S.F.	.72	1.98	2.70
0240	2 watt per S.F., 40 FC, 10 fixtures @40 watt per 1000 S.F.	1.44	3.91	5.35
0280	3 watt per S.F., 60 FC, 15 fixtures @40 watt per 1000 S.F	2.14	5.90	8.04
0320	4 watt per S.F., 80 FC, 20 fixtures @40 watt per 1000 S.F.	2.86	7.85	10.71
0400	5 watt per S.F., 100 FC, 25 fixtures @40 watt per 1000 S.F.	3.58	9.80	13.38
0450	T8, energy saver 32 watt lamps			
0500	0.8 watt per S.F., 20 FC, 5 fixtures @32 watt per 1000 S.F.	.79	1.98	2.77
0520	1.6 watt per S.F., 40 FC, 10 fixtures @32 watt per 1000 S.F.	1.58	3.91	5.49
0540	2.4 watt per S.F., 60 FC, 15 fixtures @ 32 watt per 1000 S.F	2.36	5.90	8.26
0560	3.2 watt per S.F., 80 FC, 20 fixtures @32 watt per 1000 S.F.	3.15	7.85	11
0580	4 watt per S.F., 100 FC, 25 fixtures @32 watt per 1000 S.F.	3.94	9.80	13.74

D5030 Communications and Security

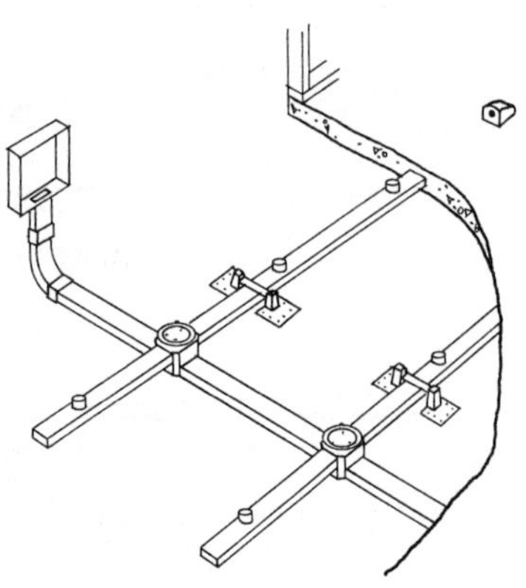

Description: System below includes telephone fitting installed. Does not include cable.

When poke thru fittings and telepoles are used for power, they can also be used for telephones at a negligible additional cost.

System Components	QUANTITY	UNIT	COST PER S.F. MAT.	COST PER S.F. INST.	COST PER S.F. TOTAL
SYSTEM D5030 310 0200					
TELEPHONE SYSTEMS, UNDERFLOOR DUCT, 5′ ON CENTER, LOW DENSITY					
Underfloor duct 7-1/4″ w/insert 2′ O.C. 1-3/8″ x 7-1/4″ super duct	.190	L.F.	6.94	2.57	9.51
Vertical elbow for underfloor superduct, 7-1/4″, included					
Underfloor duct conduit adapter, 2″ x 1-1/4″, included					
Underfloor duct junction box, single duct, 7-1/4″ x 3 1/8″	.003	Ea.	1.65	.51	2.16
Underfloor junction box carpet pan	.003	Ea.	1.17	.03	1.20
Underfloor duct outlet, low tension	.004	Ea.	.44	.34	.78
TOTAL			10.20	3.45	13.65

D5030 310	Telephone Systems	COST PER S.F. MAT.	COST PER S.F. INST.	COST PER S.F. TOTAL
0200	Telephone systems, underfloor duct, 5′ on center, low density	10.20	3.45	13.65
0240	5′ on center, high density	10.65	3.79	14.44
0280	7′ on center, low density	8.15	2.87	11.02
0320	7′ on center, high density	8.60	3.21	11.81
0400	Poke thru fittings, low density	1.18	1.23	2.41
0440	High density	2.36	2.45	4.81
0520	Telepoles, low density	1.15	.73	1.88
0560	High density	2.30	1.46	3.76
0640	Conduit system with floor boxes, low density	1.34	1.20	2.54
0680	High density	2.68	2.37	5.05
1020	Telephone wiring for offices & laboratories, 8 jacks/M.S.F.	.39	1.85	2.24

D5030 Communications and Security

D5030 810	Security & Detection Systems	COST EACH		
		MAT.	INST.	TOTAL
0200	Security system, head end equipment	29,500	4,525	34,025
0240	Security system, door /window contact biased, box, conduit & cable	265	515	780
0280	Security system, door /window contact balanced, box, conduit & cable	265	515	780
0440	Security system, proximity card reader, box, conduit & cable	470	680	1,150
1600	Card control entrance system, to 6 zone, including hardware for 100 doors	42,900	31,400	74,300

D5030 Communications and Security

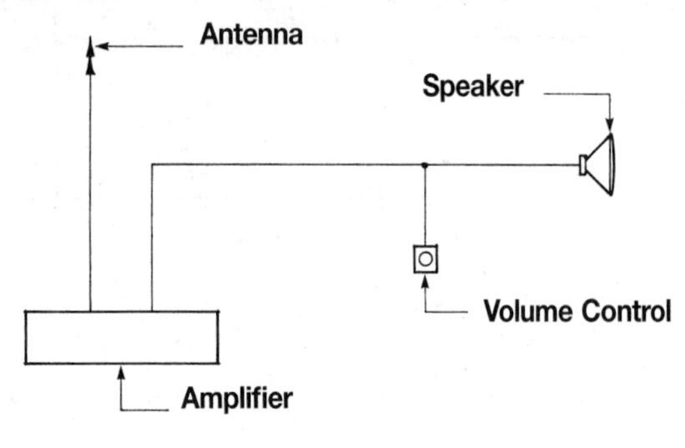

Sound System Includes AM–FM antenna, outlets, rigid conduit, and copper wire.
Fire Detection System Includes pull stations, signals, smoke and heat detectors, rigid conduit, and copper wire.
Intercom System Includes master and remote stations, rigid conduit, and copper wire.
Master Clock System Includes clocks, bells, rigid conduit, and copper wire.
Master TV Antenna Includes antenna, VHF–UHF reception and distribution, rigid conduit, and copper wire.

System Components	QUANTITY	UNIT	COST EACH MAT.	INST.	TOTAL
SYSTEM D5030 910 0220					
SOUND SYSTEM, INCLUDES OUTLETS, BOXES, CONDUIT & WIRE					
Steel intermediate conduit, (IMC) 1/2″ diam	1200.000	L.F.	2,364	8,160	10,524
Wire sound shielded w/drain, #22-2 conductor	15.500	C.L.F.	237.93	1,309.75	1,547.68
Sound system speakers ceiling or wall	12.000	Ea.	1,572	1,014	2,586
Sound system volume control	12.000	Ea.	1,188	1,014	2,202
Sound system amplifier, 250 Watts	1.000	Ea.	1,400	680	2,080
Sound system antenna, AM FM	1.000	Ea.	152	169	321
Sound system monitor panel	1.000	Ea.	435	169	604
Sound system cabinet	1.000	Ea.	945	680	1,625
Steel outlet box 4″ square	12.000	Ea.	31.92	408	439.92
Steel outlet box 4″ plaster rings	12.000	Ea.	31.68	127.20	158.88
TOTAL			8,357.53	13,730.95	22,088.48

D5030 910	Communication & Alarm Systems	COST EACH MAT.	INST.	TOTAL
0200	Communication & alarm systems, includes outlets, boxes, conduit & wire			
0210	Sound system, 6 outlets	5,800	8,575	14,375
0220	12 outlets	8,350	13,700	22,050
0240	30 outlets	14,900	25,900	40,800
0280	100 outlets	45,300	87,000	132,300
0320	Fire detection systems, non-addressable, 12 detectors	3,300	7,125	10,425
0360	25 detectors	5,800	12,000	17,800
0400	50 detectors	11,400	23,800	35,200
0440	100 detectors	21,200	43,400	64,600
0450	Addressable type, 12 detectors	4,975	7,175	12,150
0452	25 detectors	9,075	12,100	21,175
0454	50 detectors	17,500	23,900	41,400
0456	100 detectors	33,900	43,800	77,700
0458	Fire alarm control panel, 8 zone, excluding wire and conduit	860	1,350	2,210
0459	12 zone	2,625	2,025	4,650
0460	Fire alarm command center, addressable without voice, excl. wire & conduit	4,925	1,175	6,100
0462	Addressable with voice	10,400	1,875	12,275
0480	Intercom systems, 6 stations	4,000	5,875	9,875
0520	12 stations	7,700	11,700	19,400
0560	25 stations	12,900	22,500	35,400
0600	50 stations	24,900	42,300	67,200
0640	100 stations	49,200	82,500	131,700
0680	Master clock systems, 6 rooms	5,175	9,575	14,750
0720	12 rooms	8,100	16,300	24,400
0760	20 rooms	11,200	23,100	34,300
0800	30 rooms	18,300	42,600	60,900

D5030 Communications and Security

D5030 910	Communication & Alarm Systems	COST EACH		
		MAT.	INST.	TOTAL
0840	50 rooms	29,700	72,000	101,700
0880	100 rooms	57,500	142,500	200,000
0920	Master TV antenna systems, 6 outlets	2,425	6,075	8,500
0960	12 outlets	4,500	11,300	15,800
1000	30 outlets	11,200	26,200	37,400
1040	100 outlets	37,300	85,500	122,800

D5090 Other Electrical Systems

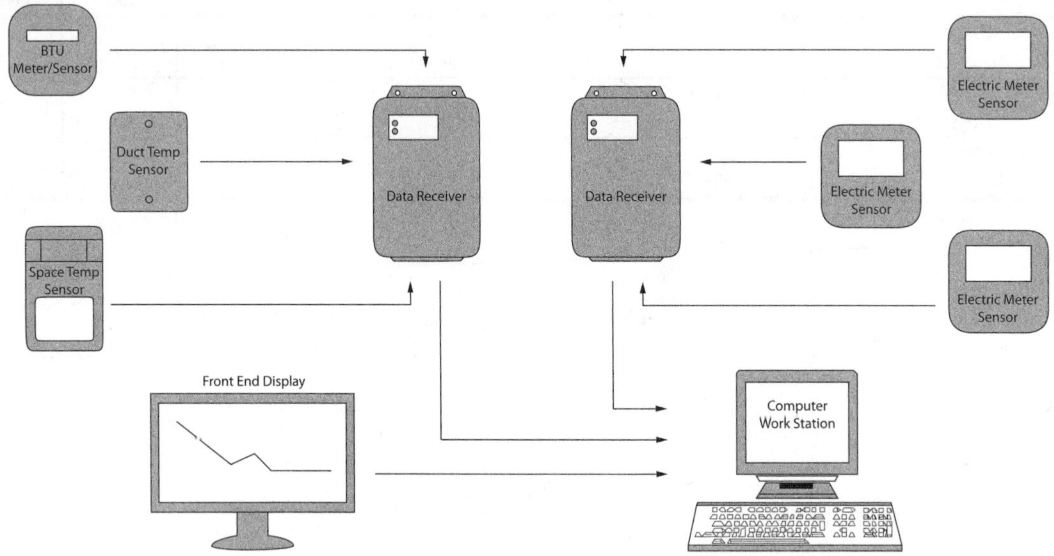

System Components	QUANTITY	UNIT	COST EACH MAT.	COST EACH INST.	COST EACH TOTAL
SYSTEM D5090 480 1100					
ELECTRICAL, SINGLE PHASE, 1 METER					
#18 twisted shielded pair in 1/2″ EMT conduit	.050	C.L.F.	8.10	11.25	19.35
Wire, 600 volt, type THW, copper, solid, #12	.150	C.L.F.	1.97	9.23	11.20
Conduit (EMT) , to 15′ H, incl 2 termn,2 elb&11 bm clp per 100′, 3/4″	5.000	L.F.	5.20	26	31.20
Outlet boxes, pressed steel, handy box	1.000	Ea.	3.17	25	28.17
Outlet boxes, pressed steel, handy box, covers, device	1.000	Ea.	1.11	10.60	11.71
Wiring devices, receptacle, duplex, 120 volt, ground, 20 amp	1.000	Ea.	8.70	25	33.70
Single phase, 277 volt, 200 amp	1.000	Ea.	440	77	517
Data recorder, 8 meters	1.000	Ea.	1,525	62	1,587
Software package, per meter, premium	1.000	Ea.	675		675
TOTAL			2,668.25	246.08	2,914.33

D5090 480	Energy Monitoring Systems	COST EACH MAT.	COST EACH INST.	COST EACH TOTAL
1000	Electrical			
1100	Single phase, 1 meter	2,675	246	2,921
1110	4 meters	6,800	1,850	8,650
1120	8 meters	13,700	3,825	17,525
1200	Three phase, 1 meter	3,000	315	3,315
1210	5 meters	10,000	2,925	12,925
1220	10 meters	20,500	5,950	26,450
1230	25 meters	46,700	11,800	58,500
2000	Mechanical			
2100	BTU, 1 meter	3,725	905	4,630
2110	w/1 duct sensor	3,925	1,125	5,050
2120	& 1 space sensor	4,450	1,275	5,725
2130	& 5 space sensors	6,525	1,875	8,400
2140	& 10 space sensors	9,750	2,650	12,400
2200	BTU, 3 meters	8,100	2,575	10,675
2210	w/3 duct sensors	8,725	3,225	11,950
2220	& 3 space sensors	10,300	3,675	13,975
2230	& 15 space sensors	16,500	5,500	22,000
2240	& 30 space sensors	25,000	7,775	32,775
9000	Front end display	700	132	832
9100	Computer workstation	1,450	1,750	3,200

E Equipment & Furnishings

E2020 Moveable Furnishings

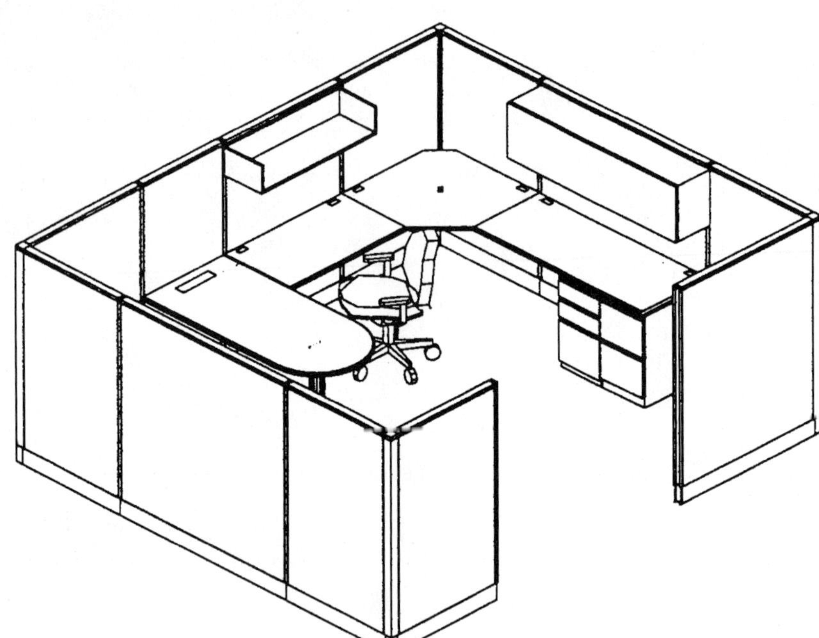

Note: 4′ raceway included in panels (not pre-wired)

The work station shown here represents one of many types and sizes on the market. Costs can vary considerably due to the various parameters involved. For instance, wood systems can cost up to 40% more than metal or fabric systems. Generally, the cost per work station will decrease for "quantity" purchases.

System Components	QUANTITY	UNIT	COST EACH MAT.	COST EACH INST.	COST EACH TOTAL
SYSTEM E2020 410 2600					
WORK STATION - MANAGER					
Acoustical panel, aluminum frame, fabric covered, 30″ wide, 64″ high	1.000	Ea.	465		465
36″ wide, 64″ high	6.000	Ea.	3,000		3,000
42″ wide, 64″ high	2.000	Ea.	1,160		1,160
48″ wide, 64″ high	1.000	Ea.	615		615
60″ wide, 64″ high	2.000	Ea.	1,350		1,350
Straight connector kit, 64″ high	7.000	Ea.	283.50		283.50
Ell connector kit, 64″ high, 90″ deep	4.000	Ea.	232		232
Panel end cover, 64″ high	2.000	Ea.	81		81
Worksurface, 24″ deep, 42″ wide	1.000	Ea.	237		237
Worksurface, 24″ deep, 60″ wide	1.000	Ea.	315		315
Corner worksurface, 24″ deep, 36″ wide	1.000	Ea.	465		465
Peninsula worksurface, 36″ wide, 66″ long	1.000	Ea.	605		605
Peninsula support column, 29-1/2″ high	1.000	Ea.	141		141
Cantilever bracket, 20″ deep, 24″ deep	3.000	Ea.	165		165
Pedestal spacer, 22″ deep, 15″ wide	2.000	Ea.	132		132
Pedestal, floorstanding, 2 box, 1 file	1.000	Ea.	460		460
Pedestal, floorstanding, 2 file, 22″ deep	1.000	Ea.	430		430
Overhead storage cabinet with door, 60″ wide	1.000	Ea.	580		580
Open bookshelf, 42″ wide	1.000	Ea.	171		171
Task light, recessed, 42″ - 48″ wide	1.000	Ea.	194		194
Task light, recessed, 60″ wide	1.000	Ea.	210		210
Electrical power harness, 36″ wide	2.000	Ea.	264		264
Electrical power harness, 42″ wide	1.000	Ea.	277		277
Electrical power harness, 60″ wide	1.000	Ea.	139		139
Base power in-feed cable	1.000	Ea.	147		147
Duplex receptacle, circuit 1	2.000	Ea.	40		40
Duplex receptacle, circuit 2	2.000	Ea.	40		40
Installation	1.000	Ea.		615	615
TOTAL			12,198.50	615	12,813.50

E2020 Moveable Furnishings

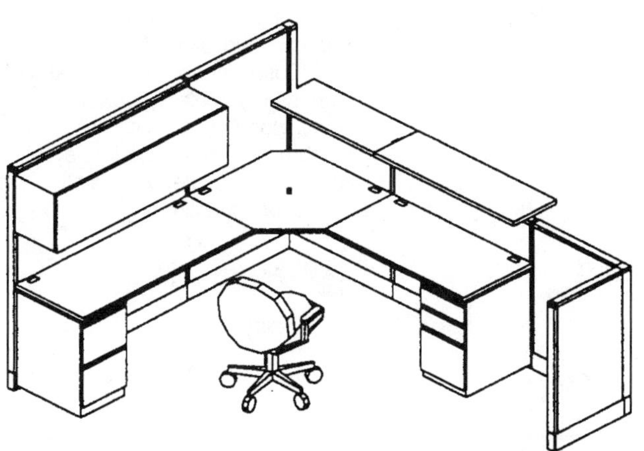

The work station shown here represents one of many types and sizes on the market. Costs can vary considerably due to the various parameters involved. For instance, wood systems can cost up to 40% more than metal or fabric systems. Generally, the cost per work station will decrease for "quantity" purchases.

System Components	QUANTITY	UNIT	COST EACH MAT.	COST EACH INST.	COST EACH TOTAL
SYSTEM E2020 420 2300					
WORK STATION - PROFESSIONAL/EXECUTIVE SECRETARY					
Acoustical panel, aluminum frame, fabric covered, 24" wide x 43" high	1.000	Ea.	350		350
30" wide x 43" high	1.000	Ea.	395		395
36" wide x 43" high	1.000	Ea.	435		435
48" wide x 43" high	1.000	Ea.	520		520
36" wide x 64" high	1.000	Ea.	500		500
60" wide x 64" high	1.000	Ea.	675		675
Straight connector kit, 43" high	2.000	Ea.	81		81
Straight connector kit, 64" high	1.000	Ea.	40.50		40.50
Ell connector kit, 43" high, 90" deep	1.000	Ea.	98		98
Ell connector kit, 64" high	1.000	Ea.	58		58
Panel end cover, 43" high	1.000	Ea.	40.50		40.50
Panel end cover, 64" high	1.000	Ea.	40.50		40.50
Finish end cover, variable height, 2-way	1.000	Ea.	60.50		60.50
Corner worksurface, 24" deep, 36" wide	1.000	Ea.	465		465
Worksurface, 24" deep, 48" wide	1.000	Ea.	253		253
Worksurface, 24" deep, 60" wide	1.000	Ea.	315		315
Cantilever bracket, 20" deep, 24" deep	2.000	Ea.	110		110
Countertop brackets, 1 pair	2.000	Ea.	46		46
Countertop, 15" deep, 36" wide	1.000	Ea.	186		186
Countertop, 15" deep, 48" wide	1.000	Ea.	207		207
Pedestal spacer, 22" deep, 15" wide	2.000	Ea.	132		132
Pedestal, floorstanding, 2 box, 1 file, 22" deep	1.000	Ea.	460		460
Pedestal, floorstanding, 2 file, 22" deep	1.000	Ea.	430		430
Overhead storage cabinet with door, 60" wide	1.000	Ea.	580		580
Task light, recessed, 60" wide	1.000	Ea.	210		210
Electrical power harness, 36" wide	2.000	Ea.	264		264
Electrical power harness, 60" wide	1.000	Ea.	139		139
Base power in-feed cable	1.000	Ea.	147		147
Duplex receptacle, circuit 1	2.000	Ea.	40		40
Duplex receptacle, circuit 2	1.000	Ea.	20		20
Installation	1.000	Ea.		375	375
TOTAL			7,298	375	7,673

E2020 Moveable Furnishings

The work station shown here represents one of many types and sizes on the market. Costs can vary considerably due to the various parameters involved. For instance, wood systems can cost up to 40% more than metal or fabric systems. Generally, the cost per work station will decrease for "quantity" purchases.

System Components	QUANTITY	UNIT	COST EACH MAT.	COST EACH INST.	COST EACH TOTAL
SYSTEM E2020 430 1500					
WORK STATION - SECRETARY/CLERK					
Acoustical panel, aluminum frame, fabric covered, 30" wide x 64" high	2.000	Ea.	930		930
36" wide x 64" high	2.000	Ea.	1,000		1,000
42" wide x 64" high	1.000	Ea.	580		580
Straight connector kit, 64" high	2.000	Ea.	81		81
Ell connector kit, 64" high, 90" deep	2.000	Ea.	116		116
Panel end cover, 64" high	2.000	Ea.	81		81
Worksurface, 24" deep, 42" wide	1.000	Ea.	237		237
Worksurface, 72" wide, 30" deep	1.000	Ea.	405		405
Worksurface bracket kit, pair	2.000	Ea.	46		46
Flat bracket, 20" deep, 24" deep	1.000	Ea.	28.50		28.50
Pedestal spacer 22" deep, 15" wide	1.000	Ea.	66		66
Pedestal spacer, 28" deep, 15" wide	1.000	Ea.	79.50		79.50
Pedestal, 2 box, 1 file, 28" deep	1.000	Ea.	480		480
Pedestal, 2 file, 22" deep	1.000	Ea.	430		430
Overhead storage cabinet with door, 42" wide	1.000	Ea.	380		380
Task light, recessed, 42" - 48" wide	1.000	Ea.	194		194
Installation	1.000	Ea.		246	246
TOTAL			5,134	246	5,380

E2020 Moveable Furnishings

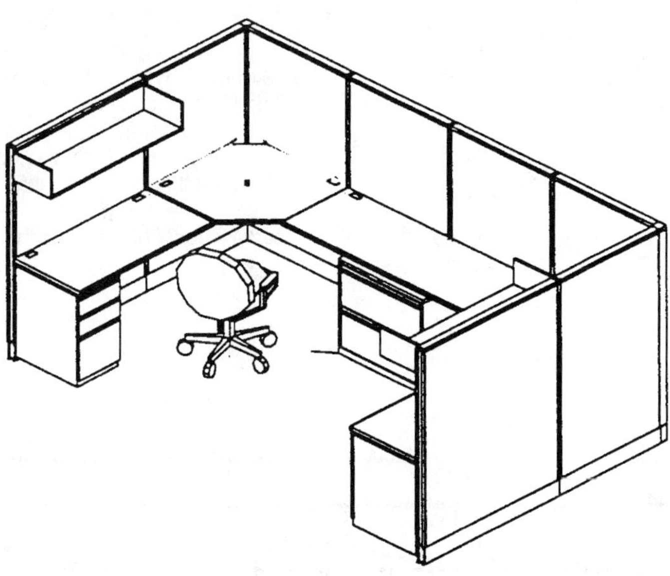

The work station shown here represents one of many types and sizes on the market. Costs can vary considerably due to the various parameters involved. For instance, wood systems can cost up to 40% more than metal or fabric systems. Generally, the cost per work station will decrease for "quantity" purchases.

System Components	QUANTITY	UNIT	COST EACH MAT.	COST EACH INST.	COST EACH TOTAL
SYSTEM E2020 440 1600					
WORK STATION - CLERICAL, TWO PERSON, UNIT					
Acoustical panel, aluminum frame, fabric covered, 36" wide, 64" high	6.000	Ea.	3,000		3,000
48" wide x 64" high	2.000	Ea.	1,230		1,230
Lateral file, 2 drawer, 30" wide	1.000	Ea.	610		610
Ell connector kit, 64" high, 90" deep	2.000	Ea.	116		116
Panel end cover, 64" high	2.000	Ea.	81		81
Corner worksurface, 36" wide x 24" deep	2.000	Ea.	930		930
Worksurface, 48" wide x 24" deep	2.000	Ea.	506		506
Worksurface, 72" wide x 24" deep	1.000	Ea.	360		360
Cantilever bracket, 20" deep, 24" deep	4.000	Ea.	220		220
Open bookshelf, 48" wide	2.000	Ea.	352		352
Pedestal spacer, 22" deep, 15" wide	2.000	Ea.	132		132
Pedestal, floorstanding, 2 box, 1 file, 22" deep	2.000	Ea.	920		920
Straight connector kit, 64" high	5.000	Ea.	202.50		202.50
Task light, recessed, 42" - 48" wide	2.000	Ea.	388		388
Electrical power harness, 36" wide	4.000	Ea.	528		528
Base power in-feed cable	1.000	Ea.	147		147
Duplex receptacle, circuit 1	2.000	Ea.	40		40
Duplex receptacle, circuit 2	2.000	Ea.	40		40
Installation	1.000	Ea.		490	490
TOTAL			9,802.50	490	10,292.50

E2020 Moveable Furnishings

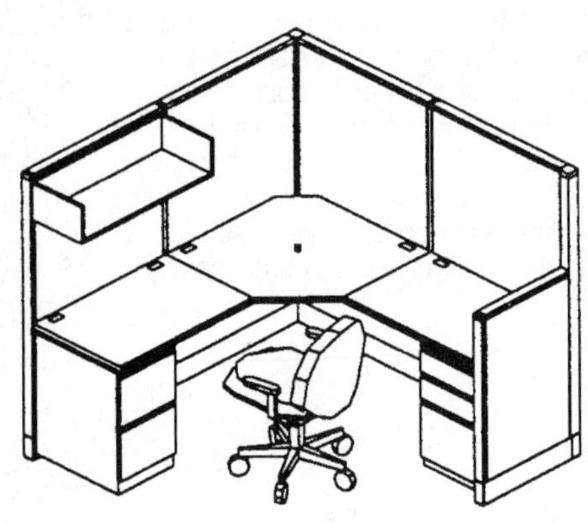

The work station shown here represents one of many types and sizes on the market. Costs can vary considerably due to the various parameters involved. For instance, wood systems can cost up to 40% more than metal or fabric systems. Generally, the cost per work station will decrease for "quantity" purchases.

System Components	QUANTITY	UNIT	COST EACH MAT.	COST EACH INST.	COST EACH TOTAL
SYSTEM E2020 450 1600					
WORK STATION - SECRETARY/WORD PROCESSING					
Acoustical panel, aluminum frame, fabric covered, 43" high, 24" wide	1.000	Ea.	350		350
Acoustical panel, 64" high, 36" wide	4.000	Ea.	2,000		2,000
Straight connector kit, 64" high	2.000	Ea.	81		81
Ell connector kit, 64" high, 90" deep	2.000	Ea.	116		116
Panel end cover, 43" high	1.000	Ea.	40.50		40.50
Panel end cover, 64" high	1.000	Ea.	40.50		40.50
Variable height finish end cover, 2-way	1.000	Ea.	60.50		60.50
Corner worksurface, 24" deep, 36" wide	1.000	Ea.	465		465
Worksurface, 24" deep, 36" wide	2.000	Ea.	384		384
Cantilever bracket, 20" deep, 24" deep	2.000	Ea.	110		110
Pedestal spacer, 22" deep, 15" wide	2.000	Ea.	132		132
Pedestal, floorstanding, 2 box, 1 file, 22" deep	1.000	Ea.	460		460
Pedestal, floorstanding, 2 file, 22" deep	1.000	Ea.	430		430
Open bookshelf, 36" wide	1.000	Ea.	163		163
Task light, recessed, 30" - 36" wide	1.000	Ea.	179		179
Electrical power harness, 36" wide	3.000	Ea.	396		396
Base power in-feed cable	1.000	Ea.	147		147
Duplex receptacle, circuit 1	2.000	Ea.	40		40
Duplex receptacle, circuit 2	1.000	Ea.	20		20
Installation	1.000	Ea.		310	310
TOTAL			5,614.50	310	5,924.50

F Special Construction

F1020 Integrated Construction

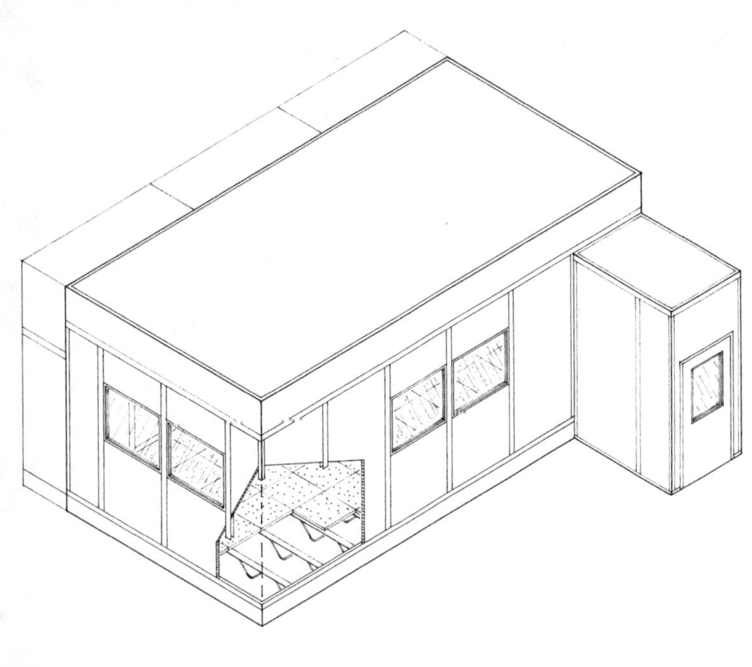

A clean room is a free standing enclosure designed to control particulate levels through air filtration and pressurization control, as well as temperature and humidity levels. These controlled environments are necessary for semi-conductor and electronics-related processes, precision machinery and metrology applications, laser and optic devices, pharmaceutical, biotechnology and medical device manufacture.

The class 10 and 100 prefab clean rooms include: 100% filtered ceiling coverage using air plenum modules, complete packaged HVAC with integrated electronic controls, fully insulated steel demountable modular wall partitions with integral utility chases, ante-room gown-up/airlock with interlocked doors and passthru.

All costs shown include factory pre-assembled and pre-tested components, field installed and tested, with only electrical power, water supply and drain connections required.

Costs are based on 288 S.F. net usable floor area.

System Components	QUANTITY	UNIT	COST EACH MAT.	COST EACH INST.	COST EACH TOTAL
CLASS 10 CLEAN ROOM SYSTEM					
Flooring, raised pedestal type w/steel perforated tiles, incl. steps	1.000	Ea.	6,700	890	7,590
Walls, modular baked enamel panels, insulated, all steel, incl. windows, doors, power outlets & light switches	1.000	Ea.	39,100	18,000	57,100
Ceiling, sealed plenum type, 99.9995% on .12 micron, HEPA, 100% coverage, 2" gel grid tee bar incl. lights	1.000	Ea.	23,200	8,500	31,700
HVAC sys. incl. pressure, temp. and humidity control system and recirc. fan	1.000	Ea.	132,500	48,100	180,600
Gown-up, incl. floor, walls, ceiling, filters, lights, doors	1.000	Ea.	4,500	8,275	12,775
Test, balance and certification	1.000	Ea.	895	3,350	4,245
TOTAL			206,895	87,115	294,010

System Components	QUANTITY	UNIT	COST EACH MAT.	COST EACH INST.	COST EACH TOTAL
CLASS 100 CLEAN ROOM SYSTEM					
Flooring, seamless vinyl with coving	1.000	Ea.	4,500	1,825	6,325
Walls, modular baked enamel insul. panels, all steel incl. windows, doors, power outlets and light switches	1.000	Ea.	39,100	18,000	57,100
Ceiling, sealed plenum type, 99.99% on .3 micron HEPA, 100% coverage, 2" gel grid tee bar incl. lights	1.000	Ea.	21,500	8,500	30,000
HVAC sys. incl. pressure, temp., and humidity control system, & recirc. fan	1.000	Ea.	116,500	36,700	153,200
Gown-up, incl. floor, walls, ceiling, filters, lights, doors with interlock	1.000	Ea.	4,500	8,275	12,775
Test, balance and certification	1.000	Ea.	895	3,350	4,245
TOTAL			186,995	76,650	263,645

F1020 Integrated Construction

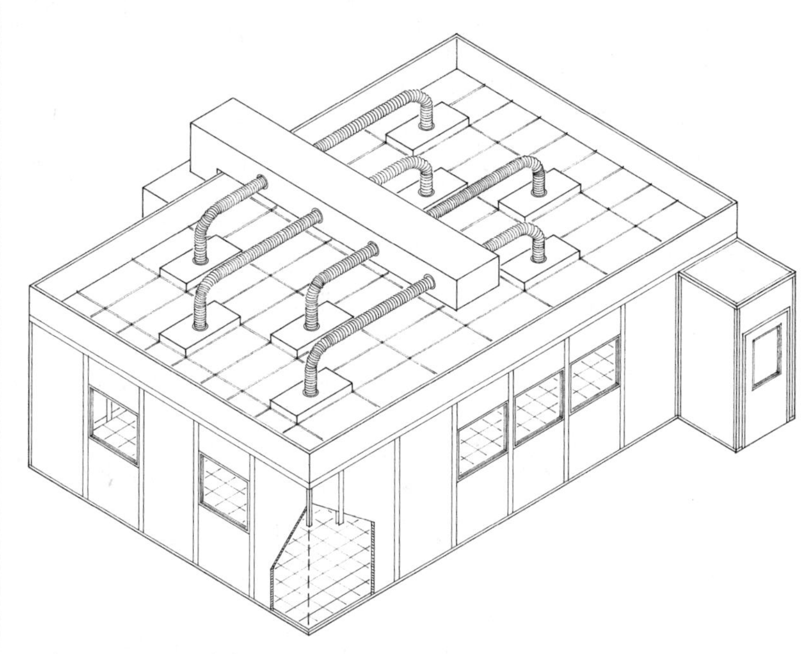

The Class 1000 through Class 100,000 prefab clean rooms use gasketed 2″ heavy duty tee-bar ceilings, ducted supply to and from the HVAC unit, sidewall returns, complete packaged HVAC and integrated electronic controls, fully insulated steel demountable modular wall partitions with integral utility chases, ante-room gown-up/airlock with interlocked doors and passthru.

All costs include factory pre-assembled and pre-tested components, field installation and certification, with the only connections required being electrical power, water supply, and drain.

Costs are based on 600 S.F. net usable floor area.

System Components	QUANTITY	UNIT	COST EACH MAT.	COST EACH INST.	COST EACH TOTAL
SYSTEM F1020 220 0310					
CLASS 1000 CLEAN ROOM SYSTEM					
Flooring, seamless vinyl with coving	1.000	Ea.	9,300	4,325	13,625
Walls, modular baked enamel insul panels, all steel incl. windows, doors, power outlets and light switch	1.000	Ea.	16,400	16,300	32,700
Ceiling,thermal ducted type 99.99% on .3 micron HEPA, steel blank tiles, 2″ gasketed grid tee bars including lights	1.000	Ea.	52,000	18,600	70,600
HVAC system incl. pressure, temp., and humidity control sys. and recirc fan	1.000	Ea.	104,000	37,600	141,600
Gown-up/air lock, incl. floor, walls, ceiling, filter, light, door	1.000	Ea.	3,225	4,525	7,750
Test, balance and certification	1.000	Ea.	455	2,250	2,705
TOTAL			185,380	83,600	268,980

F1020 220	Prefab Clean Room	COST EACH MAT.	COST EACH INST.	COST EACH TOTAL
0310	Prefab clean room, Class 1000	185,500	83,500	269,000
0320	Class 10,000	166,500	69,000	235,500
0330	Class 100,000	161,500	66,000	227,500

G Building Sitework

G1030 Site Earthwork

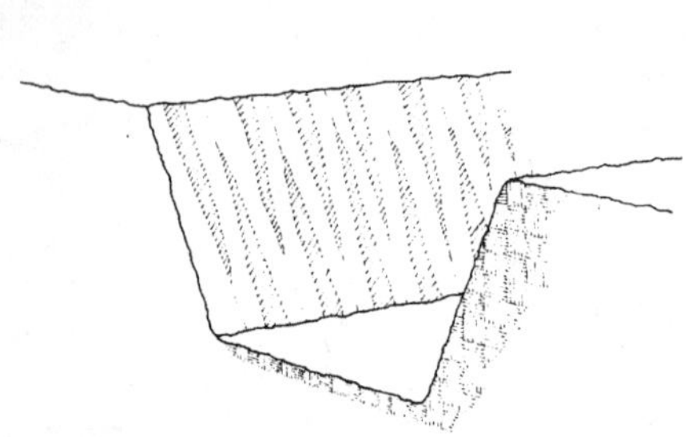

Trenching Systems are shown on a cost per linear foot basis. The systems include: excavation; backfill and removal of spoil; and compaction for various depths and trench bottom widths. The backfill has been reduced to accommodate a pipe of suitable diameter and bedding.

The slope for trench sides varies from none to 1:1.

The Expanded System Listing shows Trenching Systems that range from 2′ to 12′ in width. Depths range from 2′ to 25′.

System Components	QUANTITY	UNIT	COST PER L.F. EQUIP.	LABOR	TOTAL
SYSTEM G1030 805 1310					
TRENCHING COMMON EARTH, NO SLOPE, 2′ WIDE, 2′ DP, 3/8 C.Y. BUCKET					
Excavation, trench, hyd. backhoe, track mtd., 3/8 C.Y. bucket	.148	B.C.Y.	.40	1.12	1.52
Backfill and load spoil, from stockpile	.153	L.C.Y.	.13	.34	.47
Compaction by vibrating plate, 6″ lifts, 4 passes	.118	E.C.Y.	.03	.42	.45
Remove excess spoil, 8 C.Y. dump truck, 2 mile roundtrip	.040	L.C.Y.	.15	.17	.32
TOTAL			.71	2.05	2.76

G1030 805	Trenching Common Earth	COST PER L.F. EQUIP.	LABOR	TOTAL
1310	Trenching, common earth, no slope, 2′ wide, 2′ deep, 3/8 C.Y. bucket	.71	2.05	2.76
1320	3′ deep, 3/8 C.Y. bucket	1	3.09	4.09
1330	4′ deep, 3/8 C.Y. bucket	1.29	4.11	5.40
1340	6′ deep, 3/8 C.Y. bucket	1.67	5.35	7.02
1350	8′ deep, 1/2 C.Y. bucket	2.21	7.10	9.31
1360	10′ deep, 1 C.Y. bucket	3.46	8.40	11.86
1400	4′ wide, 2′ deep, 3/8 C.Y. bucket	1.60	4.03	5.63
1410	3′ deep, 3/8 C.Y. bucket	2.20	6.10	8.30
1420	4′ deep, 1/2 C.Y. bucket	2.56	6.85	9.41
1430	6′ deep, 1/2 C.Y. bucket	4.12	11	15.12
1440	8′ deep, 1/2 C.Y. bucket	6.60	14.15	20.75
1450	10′ deep, 1 C.Y. bucket	7.95	17.55	25.50
1460	12′ deep, 1 C.Y. bucket	10.25	22.50	32.75
1470	15′ deep, 1-1/2 C.Y. bucket	9.40	20	29.40
1480	18′ deep, 2-1/2 C.Y. bucket	13	28	41
1520	6′ wide, 6′ deep, 5/8 C.Y. bucket w/trench box	8.50	16.25	24.75
1530	8′ deep, 3/4 C.Y. bucket	11.30	21.50	32.80
1540	10′ deep, 1 C.Y. bucket	11.50	22	33.50
1550	12′ deep, 1-1/2 C.Y. bucket	12.35	24	36.35
1560	16′ deep, 2-1/2 C.Y. bucket	16.90	30	46.90
1570	20′ deep, 3-1/2 C.Y. bucket	22	35.50	57.50
1580	24′ deep, 3-1/2 C.Y. bucket	26	43	69
1640	8′ wide, 12′ deep, 1-1/2 C.Y. bucket w/trench box	17.30	30	47.30
1650	15′ deep, 1-1/2 C.Y. bucket	22.50	39.50	62
1660	18′ deep, 2-1/2 C.Y. bucket	24.50	40	64.50
1680	24′ deep, 3-1/2 C.Y. bucket	35.50	55.50	91
1730	10′ wide, 20′ deep, 3-1/2 C.Y. bucket w/trench box	28.50	52.50	81
1740	24′ deep, 3-1/2 C.Y. bucket	42.50	63	105.50
1780	12′ wide, 20′ deep, 3-1/2 C.Y. bucket w/trench box	45	66.50	111.50
1790	25′ deep, bucket	55.50	85	140.50
1800	1/2 to 1 slope, 2′ wide, 2′ deep, 3/8 C.Y. bucket	1	3.09	4.09
1810	3′ deep, 3/8 C.Y. bucket	1.67	5.40	7.07
1820	4′ deep, 3/8 C.Y. bucket	2.49	8.25	10.74
1840	6′ deep, 3/8 C.Y. bucket	4	13.45	17.45

G1030 Site Earthwork

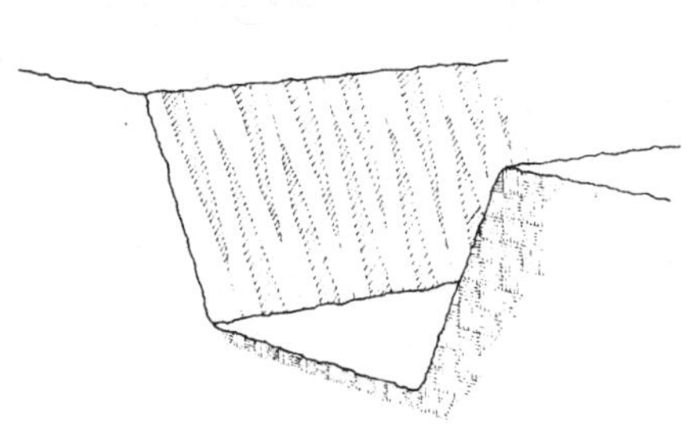

Trenching Systems are shown on a cost per linear foot basis. The systems include: excavation; backfill and removal of spoil; and compaction for various depths and trench bottom widths. The backfill has been reduced to accommodate a pipe of suitable diameter and bedding.

The slope for trench sides varies from none to 1:1.

The Expanded System Listing shows Trenching Systems that range from 2′ to 12′ in width. Depths range from 2′ to 25′.

System Components	QUANTITY	UNIT	COST PER L.F. EQUIP.	COST PER L.F. LABOR	COST PER L.F. TOTAL
SYSTEM G1030 806 1310					
TRENCHING LOAM & SANDY CLAY, NO SLOPE, 2′ WIDE, 2′ DP, 3/8 C.Y. BUCKET					
Excavation, trench, hyd. backhoe, track mtd., 3/8 C.Y. bucket	.148	B.C.Y.	.37	1.04	1.41
Backfill and load spoil, from stockpile	.165	L.C.Y.	.14	.37	.51
Compaction by vibrating plate 18″ wide, 6″ lifts, 4 passes	.118	E.C.Y.	.03	.42	.45
Remove excess spoil, 8 C.Y. dump truck, 2 mile roundtrip	.042	L.C.Y.	.16	.18	.34
TOTAL			.70	2.01	2.71

G1030 806	Trenching Loam & Sandy Clay	COST PER L.F. EQUIP.	COST PER L.F. LABOR	COST PER L.F. TOTAL
1310	Trenching, loam & sandy clay, no slope, 2′ wide, 2′ deep, 3/8 C.Y. bucket	.70	2.01	2.71
1320	3′ deep, 3/8 C.Y. bucket	1.08	3.30	4.38
1330	4′ deep, 3/8 C.Y. bucket	1.27	4.02	5.30
1340	6′ deep, 3/8 C.Y. bucket	1.80	4.78	6.60
1350	8′ deep, 1/2 C.Y. bucket	2.38	6.30	8.70
1360	10′ deep, 1 C.Y. bucket	2.69	6.75	9.45
1400	4′ wide, 2′ deep, 3/8 C.Y. bucket	1.60	3.96	5.55
1410	3′ deep, 3/8 C.Y. bucket	2.18	6	8.20
1420	4′ deep, 1/2 C.Y. bucket	2.55	6.75	9.30
1430	6′ deep, 1/2 C.Y. bucket	4.39	9.85	14.25
1440	8′ deep, 1/2 C.Y. bucket	6.45	13.95	20.50
1450	10′ deep, 1 C.Y. bucket	6.45	14.30	21
1460	12′ deep, 1 C.Y. bucket	8.10	17.80	26
1470	15′ deep, 1-1/2 C.Y. bucket	9.80	20.50	30.50
1480	18′ deep, 2-1/2 C.Y. bucket	11.50	22.50	34
1520	6′ wide, 6′ deep, 5/8 C.Y. bucket w/trench box	8.25	16	24.50
1530	8′ deep, 3/4 C.Y. bucket	10.90	21	32
1540	10′ deep, 1 C.Y. bucket	10.55	21.50	32
1550	12′ deep, 1-1/2 C.Y. bucket	12.05	24	36
1560	16′ deep, 2-1/2 C.Y. bucket	16.50	30	46.50
1570	20′ deep, 3-1/2 C.Y. bucket	20	35.50	55.50
1580	24′ deep, 3-1/2 C.Y. bucket	25.50	43.50	69
1640	8′ wide, 12′ deep, 1-1/4 C.Y. bucket w/trench box	17.05	30.50	47.50
1650	15′ deep, 1-1/2 C.Y. bucket	21	38.50	59.50
1660	18′ deep, 2-1/2 C.Y. bucket	25	43.50	68.50
1680	24′ deep, 3-1/2 C.Y. bucket	34.50	56	90.50
1730	10′ wide, 20′ deep, 3-1/2 C.Y. bucket w/trench box	34.50	56.50	91
1740	24′ deep, 3-1/2 C.Y. bucket	43.50	69.50	113
1780	12′ wide, 20′ deep, 3-1/2 C.Y. bucket w/trench box	42	67	109
1790	25′ deep, 3-1/2 C.Y. bucket	54.50	86.50	141
1800	1/2:1 slope, 2′ wide, 2′ deep, 3/8 C.Y. bucket	.98	3.01	3.99
1810	3′ deep, 3/8 C.Y. bucket	1.63	5.30	6.95
1820	4′ deep, 3/8 C.Y. bucket	2.43	8.05	10.50
1840	6′ deep, 3/8 C.Y. bucket	4.31	12	16.30

G1030 Site Earthwork

G1030 806	Trenching Loam & Sandy Clay	COST PER L.F.		
		EQUIP.	LABOR	TOTAL
1860	8' deep, 1/2 C.Y. bucket	6.85	19.10	26
1880	10' deep, 1 C.Y. bucket	9.15	24	33
2300	4' wide, 2' deep, 3/8 C.Y. bucket	1.88	4.98	6.85
2310	3' deep, 3/8 C.Y. bucket	2.83	8.25	11.10
2320	4' deep, 1/2 C.Y. bucket	3.57	10.25	13.80
2340	6' deep, 1/2 C.Y. bucket	7.35	17.50	25
2360	8' deep, 1/2 C.Y. bucket	12.40	28	40.50
2380	10' deep, 1 C.Y. bucket	14.10	33	47
2400	12' deep, 1 C.Y. bucket	23.50	54.50	78
2430	15' deep, 1-1/2 C.Y. bucket	27.50	60.50	88
2460	18' deep, 2-1/2 C.Y. bucket	45.50	92.50	138
2840	6' wide, 6' deep, 5/8 C.Y. bucket w/trench box	11.95	24	36
2860	8' deep, 3/4 C.Y. bucket	17.50	35.50	53
2880	10' deep, 1 C.Y. bucket	18.90	40	59
2900	12' deep, 1-1/2 C.Y. bucket	23.50	49	72.50
2940	16' deep, 2-1/2 C.Y. bucket	37.50	72	110
2980	20' deep, 3-1/2 C.Y. bucket	53	97	150
3020	24' deep, 3-1/2 C.Y. bucket	75	133	208
3100	8' wide, 12' deep, 1-1/2 C.Y. bucket w/trench box	28.50	55.50	84
3120	15' deep, 1-1/2 C.Y. bucket	39	78	117
3140	18' deep, 2-1/2 C.Y. bucket	52	96.50	149
3180	24' deep, 3-1/2 C.Y. bucket	84	146	230
3270	10' wide, 20' deep, 3-1/2 C.Y. bucket w/trench box	67.50	118	186
3280	24' deep, 3-1/2 C.Y. bucket	93	159	252
3320	12' wide, 20' deep, 3-1/2 C.Y. bucket w/trench box	75	128	203
3380	25' deep, 3-1/2 C.Y. bucket w/trench box	102	172	274
3500	1:1 slope, 2' wide, 2' deep, 3/8 C.Y. bucket	1.27	4.02	5.30
3520	3' deep, 3/8 C.Y. bucket	2.28	7.55	9.85
3540	4' deep, 3/8 C.Y. bucket	3.59	12.10	15.70
3560	6' deep, 1/2 C.Y. bucket	4.31	12	16.30
3580	8' deep, 1/2 C.Y. bucket	11.35	32	43.50
3600	10' deep, 1 C.Y. bucket	15.65	41	56.50
3800	4' wide, 2' deep, 3/8 C.Y. bucket	2.18	6	8.20
3820	3' deep, 1/2 C.Y. bucket	3.48	10.55	14.05
3840	4' deep, 1/2 C.Y. bucket	4.58	13.75	18.35
3860	6' deep, 1/2 C.Y. bucket	10.30	25	35.50
3880	8' deep, 1/2 C.Y. bucket	18.40	42.50	61
3900	10' deep, 1 C.Y. bucket	22	51.50	73.50
3920	12' deep, 1 C.Y. bucket	31.50	72.50	104
3940	15' deep, 1-1/2 C.Y. bucket	45.50	100	146
3960	18' deep, 2-1/2 C.Y. bucket	62	127	189
4030	6' wide, 6' deep, 5/8 C.Y. bucket w/trench box	15.70	32.50	48
4040	8' deep, 3/4 C.Y. bucket	24	50	74
4050	10' deep, 1 C.Y. bucket	27	59	86
4060	12' deep, 1-1/2 C.Y. bucket	35	74	109
4070	16' deep, 2-1/2 C.Y. bucket	59	114	173
4080	20' deep, 3-1/2 C.Y. bucket	86	159	245
4090	24' deep, 3-1/2 C.Y. bucket	125	223	350
4500	8' wide, 12' deep, 1-1/4 C.Y. bucket w/trench box	40	80.50	121
4550	15' deep, 1-1/2 C.Y. bucket	56.50	117	174
4600	18' deep, 2-1/2 C.Y. bucket	78.50	149	228
4650	24' deep, 3-1/2 C.Y. bucket	134	236	370
4800	10' wide, 20' deep, 3-1/2 C.Y. bucket w/trench box	100	179	279
4850	24' deep, 3-1/2 C.Y. bucket	143	249	390
4950	12' wide, 20' deep, 3-1/2 C.Y. bucket w/trench box	108	190	298
4980	25' deep, 3-1/2 C.Y. bucket	162	281	445

G1030 Site Earthwork

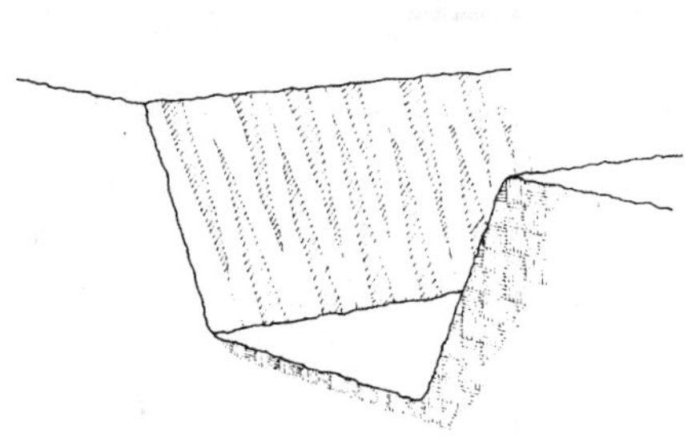

Trenching Systems are shown on a cost per linear foot basis. The systems include: excavation; backfill and removal of spoil; and compaction for various depths and trench bottom widths. The backfill has been reduced to accommodate a pipe of suitable diameter and bedding.

The slope for trench sides varies from none to 1:1.

The Expanded System Listing shows Trenching Systems that range from 2′ to 12′ in width. Depths range from 2′ to 25′.

System Components	QUANTITY	UNIT	COST PER L.F. EQUIP.	LABOR	TOTAL
SYSTEM G1030 807 1310					
TRENCHING SAND & GRAVEL, NO SLOPE, 2′ WIDE, 2′ DEEP, 3/8 C.Y. BUCKET					
Excavation, trench, hyd. backhoe, track mtd., 3/8 C.Y. bucket	.148	B.C.Y.	.36	1.02	1.38
Backfill and load spoil, from stockpile	.140	L.C.Y.	.11	.31	.42
Compaction by vibrating plate 18″ wide, 6″ lifts, 4 passes	.118	E.C.Y.	.03	.42	.45
Remove excess spoil, 8 C.Y. dump truck, 2 mile roundtrip	.035	L.C.Y.	.13	.15	.28
TOTAL			.63	1.90	2.53

G1030 807	Trenching Sand & Gravel	COST PER L.F. EQUIP.	LABOR	TOTAL
1310	Trenching, sand & gravel, no slope, 2′ wide, 2′ deep, 3/8 C.Y. bucket	.63	1.90	2.53
1320	3′ deep, 3/8 C.Y. bucket	1.01	3.18	4.19
1330	4′ deep, 3/8 C.Y. bucket	1.18	3.83	5
1340	6′ deep, 3/8 C.Y. bucket	1.70	4.56	6.25
1350	8′ deep, 1/2 C.Y. bucket	2.25	6.05	8.30
1360	10′ deep, 1 C.Y. bucket	2.50	6.35	8.85
1400	4′ wide, 2′ deep, 3/8 C.Y. bucket	1.46	3.73	5.20
1410	3′ deep, 3/8 C.Y. bucket	2.01	5.65	7.65
1420	4′ deep, 1/2 C.Y. bucket	2.33	6.35	8.70
1430	6′ deep, 1/2 C.Y. bucket	4.16	9.40	13.55
1440	8′ deep, 1/2 C.Y. bucket	6.05	13.25	19.30
1450	10′ deep, 1 C.Y. bucket	6.05	13.50	19.55
1460	12′ deep, 1 C.Y. bucket	7.65	16.85	24.50
1470	15′ deep, 1-1/2 C.Y. bucket	9.25	19.55	29
1480	18′ deep, 2-1/2 C.Y. bucket	10.85	21.50	32.50
1520	6′ wide, 6′ deep, 5/8 C.Y. bucket w/trench box	7.75	15.15	23
1530	8′ deep, 3/4 C.Y. bucket	10.30	19.95	30.50
1540	10′ deep, 1 C.Y. bucket	9.95	20	30
1550	12′ deep, 1-1/2 C.Y. bucket	11.40	22.50	34
1560	16′ deep, 2 C.Y. bucket	15.75	28.50	44.50
1570	20′ deep, 3-1/2 C.Y. bucket	19.10	33.50	52.50
1580	24′ deep, 3-1/2 C.Y. bucket	24	41	65
1640	8′ wide, 12′ deep, 1-1/2 C.Y. bucket w/trench box	16.05	28.50	44.50
1650	15′ deep, 1-1/2 C.Y. bucket	19.50	36.50	56
1660	18′ deep, 2-1/2 C.Y. bucket	23.50	41	64.50
1680	24′ deep, 3-1/2 C.Y. bucket	32.50	52.50	85
1730	10′ wide, 20′ deep, 3-1/2 C.Y. bucket w/trench box	32.50	53	85.50
1740	24′ deep, 3-1/2 C.Y. bucket	41	65	106
1780	12′ wide, 20′ deep, 3-1/2 C.Y. bucket w/trench box	39.50	62.50	102
1790	25′ deep, 3-1/2 C.Y. bucket	51	81	132
1800	1/2:1 slope, 2′ wide, 2′ deep, 3/8 C.Y. bucket	.91	2.86	3.77
1810	3′ deep, 3/8 C.Y. bucket	1.53	5.05	6.60
1820	4′ deep, 3/8 C.Y. bucket	2.27	7.70	9.95
1840	6′ deep, 3/8 C.Y. bucket	4.10	11.45	15.55

G1030 Site Earthwork

G1030 807	Trenching Sand & Gravel	COST PER L.F.		
		EQUIP.	LABOR	TOTAL
1860	8' deep, 1/2 C.Y. bucket	6.55	18.25	25
1880	10' deep, 1 C.Y. bucket	8.55	22.50	31
2300	4' wide, 2' deep, 3/8 C.Y. bucket	1.74	4.72	6.45
2310	3' deep, 3/8 C.Y. bucket	2.62	7.85	10.45
2320	4' deep, 1/2 C.Y. bucket	3.29	9.70	13
2340	6' deep, 1/2 C.Y. bucket	7	16.75	24
2360	8' deep, 1/2 C.Y. bucket	11.75	27	39
2380	10' deep, 1 C.Y. bucket	13.35	31	44.50
2400	12' deep, 1 C.Y. bucket	22.50	51.50	74
2430	15' deep, 1-1/2 C.Y. bucket	26	57	83
2460	18' deep, 2-1/2 C.Y. bucket	43	87	130
2840	6' wide, 6' deep, 5/8 C.Y. bucket w/trench box	11.35	23	34.50
2860	8' deep, 3/4 C.Y. bucket	16.60	34	50.50
2880	10' deep, 1 C.Y. bucket	17.85	38	56
2900	12' deep, 1-1/2 C.Y. bucket	22.50	46.50	69
2940	16' deep, 2 C.Y. bucket	36	68	104
2980	20' deep, 3-1/2 C.Y. bucket	50.50	91.50	142
3020	24' deep, 3-1/2 C.Y. bucket	71.50	125	197
3100	8' wide, 12' deep, 1-1/4 C.Y. bucket w/trench box	27	52.50	79.50
3120	15' deep, 1-1/2 C.Y. bucket	36.50	73.50	110
3140	18' deep, 2-1/2 C.Y. bucket	49	90.50	140
3180	24' deep, 3-1/2 C.Y. bucket	80	137	217
3270	10' wide, 20' deep, 3-1/2 C.Y. bucket w/trench box	64	111	175
3280	24' deep, 3-1/2 C.Y. bucket	88	149	237
3370	12' wide, 20' deep, 3-1/2 C.Y. bucket w/trench box	71.50	123	195
3380	25' deep, 3-1/2 C.Y. bucket	103	172	275
3500	1:1 slope, 2' wide, 2' deep, 3/8 C.Y. bucket	2.05	4.72	6.75
3520	3' deep, 3/8 C.Y. bucket	2.14	7.20	9.35
3540	4' deep, 3/8 C.Y. bucket	3.36	11.55	14.90
3560	6' deep, 3/8 C.Y. bucket	4.10	11.45	15.55
3580	8' deep, 1/2 C.Y. bucket	10.80	30.50	41.50
3600	10' deep, 1 C.Y. bucket	14.65	39	53.50
3800	4' wide, 2' deep, 3/8 C.Y. bucket	2.01	5.70	7.70
3820	3' deep, 3/8 C.Y. bucket	3.24	10	13.25
3840	4' deep, 1/2 C.Y. bucket	4.23	13.05	17.30
3860	6' deep, 1/2 C.Y. bucket	9.85	24	34
3880	8' deep, 1/2 C.Y. bucket	17.45	40.50	58
3900	10' deep, 1 C.Y. bucket	20.50	48.50	69
3920	12' deep, 1 C.Y. bucket	30	69	99
3940	15' deep, 1-1/2 C.Y. bucket	43	95	138
3960	18' deep, 2-1/2 C.Y. bucket	59	120	179
4030	6' wide, 6' deep, 5/8 C.Y. bucket w/trench box	14.90	31	46
4040	8' deep, 3/4 C.Y. bucket	23	47.50	70.50
4050	10' deep, 1 C.Y. bucket	25.50	55.50	81
4060	12' deep, 1-1/2 C.Y. bucket	33	70	103
4070	16' deep, 2 C.Y. bucket	56.50	107	164
4080	20' deep, 3-1/2 C.Y. bucket	81.50	149	231
4090	24' deep, 3-1/2 C.Y. bucket	119	209	330
4500	8' wide, 12' deep, 1-1/2 C.Y. bucket w/trench box	38	76	114
4550	15' deep, 1-1/2 C.Y. bucket	53.50	110	164
4600	18' deep, 2-1/2 C.Y. bucket	74.50	141	216
4650	24' deep, 3-1/2 C.Y. bucket	127	221	350
4800	10' wide, 20' deep, 3-1/2 C.Y. bucket w/trench box	95	168	263
4850	24' deep, 3-1/2 C.Y. bucket	135	234	370
4950	12' wide, 20' deep, 3-1/2 C.Y. bucket w/trench box	102	178	280
4980	25' deep, 3-1/2 C.Y. bucket	154	264	420

G2020 Parking Lots

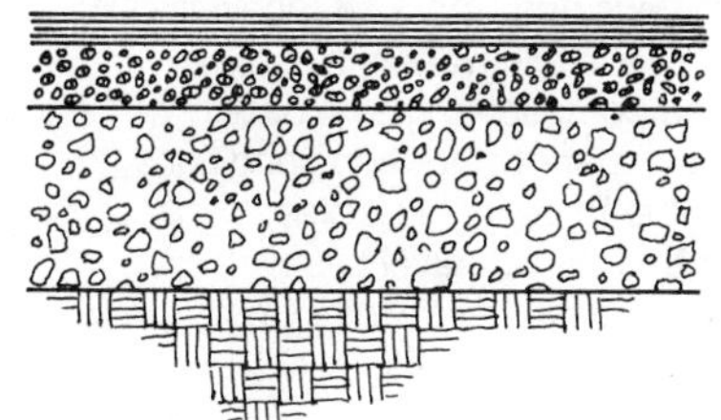

The Parking Lot System includes: compacted bank-run gravel; fine grading with a grader and roller; and bituminous concrete wearing course. All Parking Lot systems are on a cost per car basis. There are three basic types of systems: 90° angle, 60° angle, and 45° angle. The gravel base is compacted to 98%. Final stall design and lay-out of the parking lot with precast bumpers, sealcoating and white paint is also included.

The Expanded System Listing shows the three basic parking lot types with various depths of both gravel base and wearing course. The gravel base depths range from 6″ to 10″. The bituminous paving wearing course varies from a depth of 3″ to 6″.

System Components	QUANTITY	UNIT	COST PER CAR MAT.	INST.	TOTAL
SYSTEM G2020 220 1500					
PARKING LOT, 90° ANGLE PARKING, 3″ BITUMINOUS PAVING, 6″ GRAVEL BASE					
Surveying crew for layout, 4 man crew	.020	Day		52.70	52.70
Borrow, bank run gravel, haul 2 mi., spread w/dozer, no compaction	7.223	C.Y.	176.96	59.95	236.91
Grading, fine grade 3 passes with motor grader	43.333	S.Y.		106.92	106.92
Compact w/vibrating plate, 8″ lifts, granular mat'l. to 98%	7.223	C.Y.		51.50	51.50
Bituminous paving, 3″ thick	43.333	S.Y.	598	86.23	684.23
Seal coating, petroleum resistant under 1,000 S.Y.	43.333	S.Y.	61.97	61.97	123.94
Lines on pvmt., parking stall, paint, white, 4″ wide	1.000	Ea.	4.88	3.88	8.76
Precast concrete parking bar, 6″ x 10″ x 6′-0″	1.000	Ea.	43.50	21	64.50
TOTAL			885.31	444.15	1,329.46

G2020 220	Parking Lots	COST PER CAR MAT.	INST.	TOTAL
1500	Parking lot, 90° angle parking, 3″ bituminous paving, 6″ gravel base	885	445	1,330
1520	8″ gravel base	945	480	1,425
1540	10″ gravel base	1,000	520	1,520
1560	4″ bituminous paving, 6″ gravel base	1,050	440	1,490
1580	8″ gravel base	1,100	480	1,580
1600	10″ gravel base	1,150	520	1,670
1620	6″ bituminous paving, 6″ gravel base	1,400	460	1,860
1640	8″ gravel base	1,450	500	1,950
1661	10″ gravel base	1,500	535	2,035
1800	60° angle parking, 3″ bituminous paving, 6″ gravel base	885	445	1,330
1820	8″ gravel base	945	480	1,425
1840	10″ gravel base	1,000	520	1,520
1860	4″ bituminous paving, 6″ gravel base	1,050	440	1,490
1880	8″ gravel base	1,100	480	1,580
1900	10″ gravel base	1,150	520	1,670
1920	6″ bituminous paving, 6″ gravel base	1,400	460	1,860
1940	8″gravel base	1,450	500	1,950
1961	10″ gravel base	1,500	535	2,035
2200	45° angle parking, 3″ bituminous paving, 6″ gravel base	905	455	1,360
2220	8″ gravel base	965	495	1,460
2240	10″ gravel base	1,025	535	1,560
2260	4″ bituminous paving, 6″ gravel base	1,075	455	1,530
2280	8″ gravel base	1,125	495	1,620
2300	10″ gravel base	1,175	535	1,710
2320	6″ bituminous paving, 6″ gravel base	1,425	470	1,895
2340	8″ gravel base	1,475	510	1,985
2361	10″ gravel base	1,550	550	2,100

G2030 Pedestrian Paving

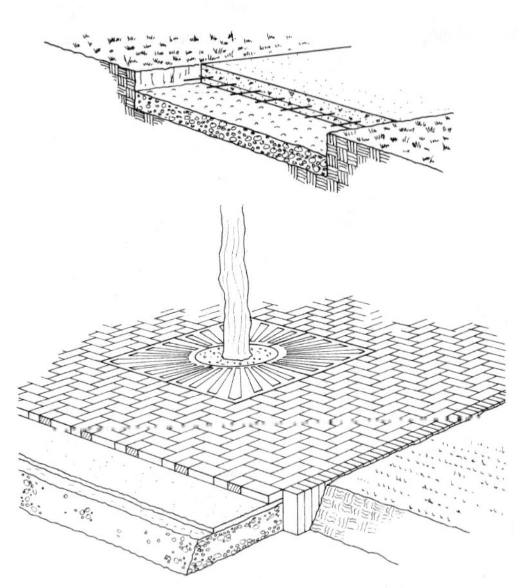

The Bituminous and Concrete Sidewalk Systems include excavation, hand grading and compacted gravel base. Pavements are shown for two conditions of each of the following variables; pavement thickness, gravel base thickness and pavement width. Costs are given on a linear foot basis.

The Plaza Systems listed include several brick and tile paving surfaces on two different bases: gravel and slab on grade. The type of bedding for the pavers depends on the base being used, and alternate bedding may be desirable. Also included in the paving costs are edging and precast grating costs and where concrete bases are involved, expansion joints. Costs are given on a square foot basis.

G2030 110	Bituminous Sidewalks	COST PER L.F.		
		MAT.	INST.	TOTAL
1580	Bituminous sidewalk, 1" thick paving, 4" gravel base, 3' width	2.27	4.83	7.10
1600	4' width	3.01	5.25	8.26
1640	6" gravel base, 3' width	2.73	5.10	7.83
1660	4' width	3.62	5.55	9.17
2120	2" thick paving, 4" gravel base, 3' width	3.62	5.80	9.42
2140	4' width	4.82	6.50	11.32
2180	6" gravel base, 3' width	4.08	6.05	10.13
2200	4' width	5.45	6.85	12.30

G2030 120	Concrete Sidewalks	COST PER L.F.		
		MAT.	INST.	TOTAL
1580	Concrete sidewalk, 4" thick, 4" gravel base, 3' wide	6.50	12.80	19.30
1600	4' wide	8.65	15.80	24.45
1640	6" gravel base, 3' wide	6.95	13	19.95
1660	4' wide	9.25	16.10	25.35
2120	6" thick concrete, 4" gravel base, 3' wide	9.60	14.60	24.20
2140	4' wide	12.80	18.05	30.85
2180	6" gravel base, 3' wide	10.05	14.85	24.90
2200	4' wide	13.40	18.45	31.85

G2030 150	Brick & Tile Plazas	COST PER S.F.		
		MAT.	INST.	TOTAL
2050	Brick pavers, 4" x 8" x 1-3/4", gravel base, stone dust bedding	5.75	6.35	12.10
2100	Slab on grade, asphalt bedding	8	9.30	17.30
3550	Concrete paving stone, 4" x 8" x 2-1/2", gravel base, sand bedding	4.48	4.49	8.97
3600	Slab on grade, asphalt bedding	6.30	6.85	13.15
4050	Concrete patio blocks, 8" x 16" x 2", gravel base, sand bedding	4.66	5.80	10.46
4100	Slab on grade, asphalt bedding	6.70	8.60	15.30
6050	Granite pavers, 3-1/2" x 3-1/2" x 3-1/2", gravel base, sand bedding	16	13.95	29.95
6100	Slab on grade, mortar bedding	19.05	22	41.05
7050	Limestone, 3" thick, gravel base, sand bedding	13.85	17.25	31.10
7100	Slab on grade, mortar bedding	15.90	22	37.90
8050	Slate flagging, 3/4" thick, gravel base, sand bedding	12.90	13.95	26.85
8100	Slab on grade, mastic bedding	25.50	20	45.50

G4020 Site Lighting

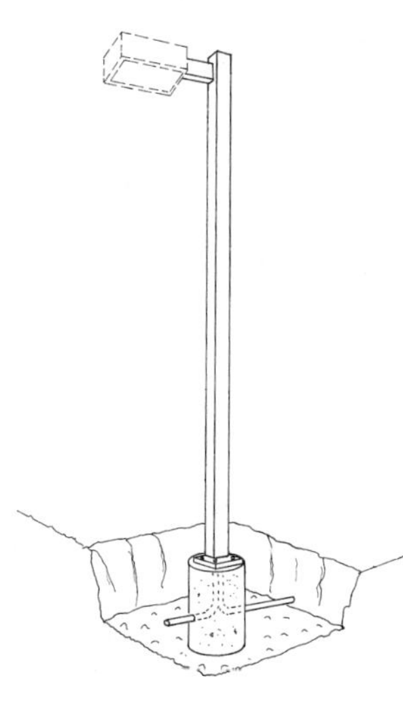

Table G4020 210 Procedure for Calculating Floodlights Required for Various Footcandles
Poles should not be spaced more than 4 times the fixture mounting height for good light distribution.

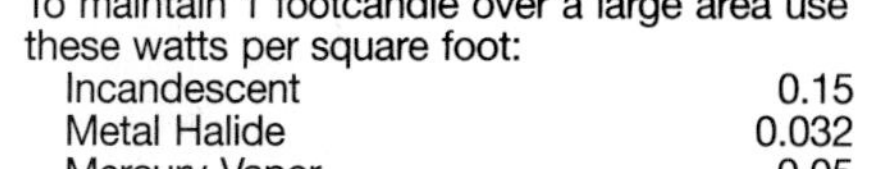

To maintain 1 footcandle over a large area use these watts per square foot:

Incandescent	0.15
Metal Halide	0.032
Mercury Vapor	0.05
High Pressure Sodium	0.024

Estimating Chart
Select Lamp type.

Determine total square feet.

Chart will show quantity of fixtures to provide 1 footcandle initial, at intersection of lines. Multiply fixture quantity by desired footcandle level.

Chart based on use of wide beam luminaires in an area whose dimensions are large compared to mounting height and is approximate only.

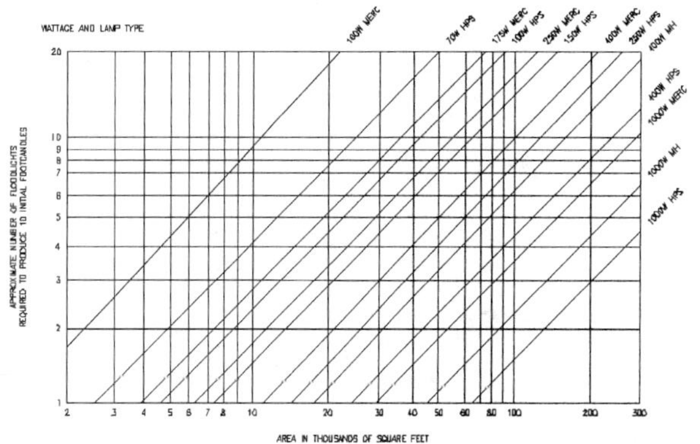

System Components	QUANTITY	UNIT	COST EACH MAT.	COST EACH INST.	COST EACH TOTAL
SYSTEM G4020 210 0200					
LIGHT POLES, ALUMINUM, 20′ HIGH, 1 ARM BRACKET					
Aluminum light pole, 20′, no concrete base	1.000	Ea.	1,025	637	1,662
Bracket arm for Aluminum light pole	1.000	Ea.	133	84.50	217.50
Excavation by hand, pits to 6′ deep, heavy soil or clay	2.368	C.Y.		291.26	291.26
Footing, concrete incl forms, reinforcing, spread, under 1 C.Y.	.465	C.Y.	85.10	137.76	222.86
Backfill by hand	1.903	C.Y.		85.64	85.64
Compaction vibrating plate	1.903	C.Y.		11.55	11.55
TOTAL			1,243.10	1,247.71	2,490.81

G4020 210	Light Pole (Installed)	COST EACH MAT.	COST EACH INST.	COST EACH TOTAL
0200	Light pole, aluminum, 20′ high, 1 arm bracket	1,250	1,250	2,500
0240	2 arm brackets	1,375	1,250	2,625
0280	3 arm brackets	1,500	1,300	2,800
0320	4 arm brackets	1,650	1,300	2,950
0360	30′ high, 1 arm bracket	2,200	1,575	3,775
0400	2 arm brackets	2,325	1,575	3,900
0440	3 arm brackets	2,450	1,600	4,050
0480	4 arm brackets	2,600	1,600	4,200
0680	40′ high, 1 arm bracket	2,675	2,100	4,775
0720	2 arm brackets	2,825	2,100	4,925
0760	3 arm brackets	2,950	2,150	5,100
0800	4 arm brackets	3,100	2,150	5,250
0840	Steel, 20′ high, 1 arm bracket	1,475	1,300	2,775
0880	2 arm brackets	1,525	1,300	2,825
0920	3 arm brackets	1,575	1,350	2,925
0960	4 arm brackets	1,625	1,350	2,975
1000	30′ high, 1 arm bracket	1,725	1,675	3,400
1040	2 arm brackets	1,775	1,675	3,450
1080	3 arm brackets	1,800	1,725	3,525
1120	4 arm brackets	1,875	1,725	3,600
1320	40′ high, 1 arm bracket	2,275	2,275	4,550
1360	2 arm brackets	2,325	2,275	4,600
1400	3 arm brackets	2,350	2,325	4,675
1440	4 arm brackets	2,425	2,325	4,750

Reference Section

All the reference information is in one section, making it easy to find what you need to know . . . and easy to use the book on a daily basis. This section is visually identified by a vertical black bar on the page edges.

In this Reference Section, we've included Equipment Rental Costs, a listing of rental and operating costs; Crew Listings, a full listing of all crews and equipment, and their costs; Historical Cost Indexes for cost comparisons over time; City Cost Indexes and Location Factors for adjusting costs to the region you are in; Reference Tables, where you will find explanations, estimating information and procedures, or technical data; Change Orders, information on pricing changes to contract documents; Square Foot Costs that allow you to make a rough estimate for the overall cost of a project; and an explanation of all the Abbreviations in the book.

Table of Contents

Estimating Tips

- This section contains the average costs to rent and operate hundreds of pieces of construction equipment. This is useful information when estimating the time and material requirements of any particular operation in order to establish a unit or total cost. Equipment costs include not only rental, but also operating costs for equipment under normal use.

Rental Costs

- Equipment rental rates are obtained from the following industry sources throughout North America: contractors, suppliers, dealers, manufacturers, and distributors.
- Rental rates vary throughout the country, with larger cities generally having lower rates. Lease plans for new equipment are available for periods in excess of six months, with a percentage of payments applying toward purchase.
- Monthly rental rates vary from 2% to 5% of the purchase price of the equipment depending on the anticipated life of the equipment and its wearing parts.
- Weekly rental rates are about 1/3 the monthly rates, and daily rental rates are about 1/3 the weekly rate.
- Rental rates can also be treated as reimbursement costs for contractor-owned equipment. Owned equipment costs include depreciation, loan payments, interest, taxes, insurance, storage, and major repairs.

Operating Costs

- The operating costs include parts and labor for routine servicing, such as repair and replacement of pumps, filters and worn lines. Normal operating expendables, such as fuel, lubricants, tires and electricity (where applicable), are also included.
- Extraordinary operating expendables with highly variable wear patterns, such as diamond bits and blades, are excluded. These costs can be found as material costs in the Unit Price section.
- The hourly operating costs listed do not include the operator's wages.

Equipment Cost/Day

- Any power equipment required by a crew is shown in the Crew Listings with a daily cost.
- The daily cost of equipment needed by a crew is based on dividing the weekly rental rate by 5 (number of working days in the week), and then adding the hourly operating cost times 8 (the number of hours in a day). This "Equipment Cost/Day" is shown in the far right column of the Equipment Rental pages.
- If equipment is needed for only one or two days, it is best to develop your own cost by including components for daily rent and hourly operating cost. This is important when the listed Crew for a task does not contain the equipment needed, such as a crane for lifting mechanical heating/cooling equipment up onto a roof.
- If the quantity of work is less than the crew's Daily Output shown for a Unit Price line item that includes a bare unit equipment cost, it is recommended to estimate one day's rental cost and operating cost for equipment shown in the Crew Listing for that line item.

Mobilization/ Demobilization

- The cost to move construction equipment from an equipment yard or rental company to the job site and back again is not included in equipment rental costs listed in the Reference Section, nor in the bare equipment cost of any unit price line item, nor in any equipment costs shown in the Crew listings.
- Mobilization (to the site) and demobilization (from the site) costs can be found in the Unit Price Section.
- If a piece of equipment is already at the job site, it is not appropriate to utilize mobilization/demobilization. costs again in an estimate. ■

01 54 33 | Equipment Rental

		01 54 33 Equipment Rental	UNIT	HOURLY OPER. COST	RENT PER DAY	RENT PER WEEK	RENT PER MONTH	EQUIPMENT COST/DAY	
10	0010	**CONCRETE EQUIPMENT RENTAL** without operators R015433-10							10
	0200	Bucket, concrete lightweight, 1/2 C.Y.	Ea.	.80	23.50	70	210	20.40	
	0300	1 C.Y.		.90	27.50	83	249	23.80	
	0400	1-1/2 C.Y.		1.10	36.50	110	330	30.80	
	0500	2 C.Y.		1.25	45	135	405	37	
	0580	8 C.Y.		6.10	257	770	2,300	202.80	
	0600	Cart, concrete, self-propelled, operator walking, 10 C.F.		3.25	56.50	170	510	60	
	0700	Operator riding, 18 C.F.		5.35	95	285	855	99.80	
	0800	Conveyer for concrete, portable, gas, 16″ wide, 26′ long		12.60	123	370	1,100	174.80	
	0900	46′ long		13.00	148	445	1,325	193	
	1000	56′ long		13.10	157	470	1,400	198.80	
	1100	Core drill, electric, 2-1/2 H.P., 1″ to 8″ bit diameter		1.77	68.50	205	615	55.15	
	1150	11 H.P., 8″ to 18″ cores		5.95	113	340	1,025	115.60	
	1200	Finisher, concrete floor, gas, riding trowel, 96″ wide		12.75	145	435	1,300	189	
	1300	Gas, walk-behind, 3 blade, 36″ trowel		2.15	20	60	180	29.20	
	1400	4 blade, 48″ trowel		4.20	27.50	83	249	50.20	
	1500	Float, hand-operated (Bull float) 48″ wide		.08	13.35	40	120	8.65	
	1570	Curb builder, 14 H.P., gas, single screw		14.65	248	745	2,225	266.20	
	1590	Double screw		15.35	290	870	2,600	296.80	
	1600	Floor grinder, concrete and terrazzo, electric, 22″ path		2.57	158	475	1,425	115.55	
	1700	Edger, concrete, electric, 7″ path		1.04	51.50	155	465	39.30	
	1750	Vacuum pick-up system for floor grinders, wet/dry		1.49	81.50	245	735	60.90	
	1800	Mixer, powered, mortar and concrete, gas, 6 C.F., 18 H.P.		8.65	118	355	1,075	140.20	
	1900	10 C.F., 25 H.P.		10.80	143	430	1,300	172.40	
	2000	16 C.F.		11.15	165	495	1,475	188.20	
	2100	Concrete, stationary, tilt drum, 2 C.Y.		6.90	232	695	2,075	194.20	
	2120	Pump, concrete, truck mounted 4″ line 80′ boom		25.00	865	2,600	7,800	720	
	2140	5″ line, 110′ boom		32.40	1,150	3,435	10,300	946.20	
	2160	Mud jack, 50 C.F. per hr.		7.30	125	375	1,125	133.40	
	2180	225 C.F. per hr.		9.50	145	435	1,300	163	
	2190	Shotcrete pump rig, 12 C.Y./hr.		14.85	220	660	1,975	250.80	
	2200	35 C.Y./hr.		17.50	235	705	2,125	281	
	2600	Saw, concrete, manual, gas, 18 H.P.		6.75	45	135	405	81	
	2650	Self-propelled, gas, 30 H.P.		13.30	102	305	915	167.40	
	2675	V-groove crack chaser, manual, gas, 6 H.P.		2.35	17.35	52	156	29.20	
	2700	Vibrators, concrete, electric, 60 cycle, 2 H.P.		.46	8.65	26	78	8.90	
	2800	3 H.P.		.60	11.65	35	105	11.80	
	2900	Gas engine, 5 H.P.		2.00	16	48	144	25.60	
	3000	8 H.P.		2.75	15.35	46	138	31.20	
	3050	Vibrating screed, gas engine, 8 H.P.		2.91	71.50	215	645	66.30	
	3120	Concrete transit mixer, 6 x 4, 250 H.P., 8 C.Y., rear discharge		61.55	570	1,715	5,150	835.40	
	3200	Front discharge		72.45	700	2,095	6,275	998.60	
	3300	6 x 6, 285 H.P., 12 C.Y., rear discharge		71.50	660	1,980	5,950	968	
	3400	Front discharge		74.00	705	2,120	6,350	1,016	
20	0010	**EARTHWORK EQUIPMENT RENTAL** without operators R015433-10							20
	0040	Aggregate spreader, push type 8′ to 12′ wide	Ea.	3.05	25.50	76	228	39.60	
	0045	Tailgate type, 8′ wide		2.90	32.50	98	294	42.80	
	0055	Earth auger, truck-mounted, for fence & sign posts, utility poles		19.15	440	1,320	3,950	417.20	
	0060	For borings and monitoring wells		46.65	675	2,030	6,100	779.20	
	0070	Portable, trailer mounted		3.30	32.50	98	294	46	
	0075	Truck-mounted, for caissons, water wells		100.15	2,900	8,705	26,100	2,542	
	0080	Horizontal boring machine, 12″ to 36″ diameter, 45 H.P.		24.20	192	575	1,725	308.60	
	0090	12″ to 48″ diameter, 65 H.P.		34.35	335	1,000	3,000	474.80	
	0095	Auger, for fence posts, gas engine, hand held		.60	6	18	54	8.40	
	0100	Excavator, diesel hydraulic, crawler mounted, 1/2 C.Y. cap.		24.65	420	1,255	3,775	448.20	
	0120	5/8 C.Y. capacity		32.55	550	1,645	4,925	589.40	
	0140	3/4 C.Y. capacity		35.60	615	1,850	5,550	654.80	
	0150	1 C.Y. capacity		49.70	690	2,075	6,225	812.60	

01 54 33 | Equipment Rental

			UNIT	HOURLY OPER. COST	RENT PER DAY	RENT PER WEEK	RENT PER MONTH	EQUIPMENT COST/DAY	
20	0200	1-1/2 C.Y. capacity	Ea.	59.80	920	2,760	8,275	1,030	20
	0300	2 C.Y. capacity		67.35	1,050	3,180	9,550	1,175	
	0320	2-1/2 C.Y. capacity		102.55	1,300	3,925	11,800	1,605	
	0325	3-1/2 C.Y. capacity		144.45	2,150	6,420	19,300	2,440	
	0330	4-1/2 C.Y. capacity		173.40	2,625	7,880	23,600	2,963	
	0335	6 C.Y. capacity		219.80	2,900	8,680	26,000	3,494	
	0340	7 C.Y. capacity		222.05	3,025	9,070	27,200	3,590	
	0342	Excavator attachments, bucket thumbs		3.20	245	735	2,200	172.60	
	0345	Grapples		2.75	193	580	1,750	138	
	0346	Hydraulic hammer for boom mounting, 4000 ft lb.		12.40	350	1,045	3,125	308.20	
	0347	5000 ft lb.		14.40	425	1,280	3,850	371.20	
	0348	8000 ft lb.		21.30	625	1,875	5,625	545.40	
	0349	12,000 ft lb.		23.35	750	2,245	6,725	635.80	
	0350	Gradall type, truck mounted, 3 ton @ 15′ radius, 5/8 C.Y.		43.65	890	2,665	8,000	882.20	
	0370	1 C.Y. capacity		47.85	1,025	3,095	9,275	1,002	
	0400	Backhoe-loader, 40 to 45 H.P., 5/8 C.Y. capacity		14.90	240	720	2,150	263.20	
	0450	45 H.P. to 60 H.P., 3/4 C.Y. capacity		23.40	295	885	2,650	364.20	
	0460	80 H.P., 1-1/4 C.Y. capacity		25.60	310	935	2,800	391.80	
	0470	112 H.P., 1-1/2 C.Y. capacity		40.95	645	1,930	5,800	713.60	
	0482	Backhoe-loader attachment, compactor, 20,000 lb.		5.80	142	425	1,275	131.40	
	0485	Hydraulic hammer, 750 ft lb.		3.30	96.50	290	870	84.40	
	0486	Hydraulic hammer, 1200 ft lb.		6.30	217	650	1,950	180.40	
	0500	Brush chipper, gas engine, 6″ cutter head, 35 H.P.		11.10	105	315	945	151.80	
	0550	Diesel engine, 12″ cutter head, 130 H.P.		27.55	285	855	2,575	391.40	
	0600	15″ cutter head, 165 H.P.		33.10	335	1,005	3,025	465.80	
	0750	Bucket, clamshell, general purpose, 3/8 C.Y.		1.30	38.50	115	345	33.40	
	0800	1/2 C.Y.		1.40	45	135	405	38.20	
	0850	3/4 C.Y.		1.55	55	165	495	45.40	
	0900	1 C.Y.		1.60	58.50	175	525	47.80	
	0950	1-1/2 C.Y.		2.55	81.50	245	735	69.40	
	1000	2 C.Y.		2.70	90	270	810	75.60	
	1010	Bucket, dragline, medium duty, 1/2 C.Y.		.75	23.50	70	210	20	
	1020	3/4 C.Y.		.75	24.50	73	219	20.60	
	1030	1 C.Y.		.80	26	78	234	22	
	1040	1-1/2 C.Y.		1.20	40	120	360	33.60	
	1050	2 C.Y.		1.30	45	135	405	37.40	
	1070	3 C.Y.		1.95	61.50	185	555	52.60	
	1200	Compactor, manually guided 2-drum vibratory smooth roller, 7.5 H.P.		7.25	203	610	1,825	180	
	1250	Rammer/tamper, gas, 8″		2.75	46.50	140	420	50	
	1260	15″		3.05	53.50	160	480	56.40	
	1300	Vibratory plate, gas, 18″ plate, 3000 lb. blow		2.70	24.50	73	219	36.20	
	1350	21″ plate, 5000 lb. blow		3.40	32.50	98	294	46.80	
	1370	Curb builder/extruder, 14 H.P., gas, single screw		14.65	248	745	2,225	266.20	
	1390	Double screw		15.35	290	870	2,600	296.80	
	1500	Disc harrow attachment, for tractor		.44	73.50	221	665	47.70	
	1810	Feller buncher, shearing & accumulating trees, 100 H.P.		48.30	755	2,260	6,775	838.40	
	1860	Grader, self-propelled, 25,000 lb.		39.50	640	1,925	5,775	701	
	1910	30,000 lb.		43.15	650	1,955	5,875	736.20	
	1920	40,000 lb.		64.90	1,100	3,275	9,825	1,174	
	1930	55,000 lb.		84.15	1,700	5,130	15,400	1,699	
	1950	Hammer, pavement breaker, self-propelled, diesel, 1000 to 1250 lb.		29.90	350	1,055	3,175	450.20	
	2000	1300 to 1500 lb.		44.85	705	2,110	6,325	780.80	
	2050	Pile driving hammer, steam or air, 4150 ft lb. @ 225 bpm		10.40	475	1,420	4,250	367.20	
	2100	8750 ft lb. @ 145 bpm		12.40	660	1,975	5,925	494.20	
	2150	15,000 ft lb. @ 60 bpm		14.00	790	2,370	7,100	586	
	2200	24,450 ft lb. @ 111 bpm		15.00	875	2,630	7,900	646	
	2250	Leads, 60′ high for pile driving hammers up to 20,000 ft lb.		3.30	80.50	242	725	74.80	
	2300	90′ high for hammers over 20,000 ft lb.		4.95	141	424	1,275	124.40	

01 54 33 | Equipment Rental

			UNIT	HOURLY OPER. COST	RENT PER DAY	RENT PER WEEK	RENT PER MONTH	EQUIPMENT COST/DAY	
20	2350	Diesel type hammer, 22,400 ft lb.	Ea.	22.55	460	1,380	4,150	456.40	20
	2400	41,300 ft lb.		32.00	540	1,625	4,875	581	
	2450	141,000 ft lb.		51.65	960	2,875	8,625	988.20	
	2500	Vib. elec. hammer/extractor, 200 kW diesel generator, 34 H.P.		54.90	635	1,900	5,700	819.20	
	2550	80 H.P.		100.65	925	2,775	8,325	1,360	
	2600	150 H.P.		190.90	1,775	5,300	15,900	2,587	
	2800	Log chipper, up to 22" diameter, 600 H.P.		63.30	640	1,915	5,750	889.40	
	2850	Logger, for skidding & stacking logs, 150 H.P.		52.75	815	2,445	7,325	911	
	2860	Mulcher, diesel powered, trailer mounted		24.05	212	635	1,900	319.40	
	2900	Rake, spring tooth, with tractor		18.23	345	1,042	3,125	354.25	
	3000	Roller, vibratory, tandem, smooth drum, 20 H.P.		8.80	145	435	1,300	157.40	
	3050	35 H.P.		11.10	247	740	2,225	236.80	
	3100	Towed type vibratory compactor, smooth drum, 50 H.P.		26.00	315	940	2,825	396	
	3150	Sheepsfoot, 50 H.P.		27.35	345	1,035	3,100	425.80	
	3170	Landfill compactor, 220 H.P.		91.00	1,475	4,440	13,300	1,616	
	3200	Pneumatic tire roller, 80 H.P.		16.80	350	1,050	3,150	344.40	
	3250	120 H.P.		25.20	600	1,800	5,400	561.60	
	3300	Sheepsfoot vibratory roller, 240 H.P.		68.95	1,100	3,270	9,800	1,206	
	3320	340 H.P.		100.85	1,575	4,730	14,200	1,753	
	3350	Smooth drum vibratory roller, 75 H.P.		24.85	585	1,755	5,275	549.80	
	3400	125 H.P.		32.55	710	2,135	6,400	687.40	
	3410	Rotary mower, brush, 60", with tractor		22.60	315	950	2,850	370.80	
	3420	Rototiller, walk-behind, gas, 5 H.P.		1.96	51.50	155	465	46.70	
	3422	8 H.P.		3.08	80	240	720	72.65	
	3440	Scrapers, towed type, 7 C.Y. capacity		5.80	113	340	1,025	114.40	
	3450	10 C.Y. capacity		6.60	157	470	1,400	146.80	
	3500	15 C.Y. capacity		7.10	180	540	1,625	164.80	
	3525	Self-propelled, single engine, 14 C.Y. capacity		114.76	1,600	4,830	14,500	1,884	
	3550	Dual engine, 21 C.Y. capacity		173.80	2,175	6,490	19,500	2,688	
	3600	31 C.Y. capacity		231.70	3,075	9,195	27,600	3,693	
	3640	44 C.Y. capacity		287.80	3,950	11,885	35,700	4,679	
	3650	Elevating type, single engine, 11 C.Y. capacity		71.45	985	2,955	8,875	1,163	
	3700	22 C.Y. capacity		139.55	2,350	7,025	21,100	2,521	
	3710	Screening plant 110 H.P. w/5' x 10' screen		25.35	400	1,200	3,600	442.80	
	3720	5' x 16' screen		29.65	505	1,515	4,550	540.20	
	3850	Shovel, crawler-mounted, front-loading, 7 C.Y. capacity		258.85	3,200	9,610	28,800	3,993	
	3855	12 C.Y. capacity		386.45	4,400	13,220	39,700	5,736	
	3860	Shovel/backhoe bucket, 1/2 C.Y.		2.50	66.50	200	600	60	
	3870	3/4 C.Y.		2.55	75	225	675	65.40	
	3880	1 C.Y.		2.65	83.50	250	750	71.20	
	3890	1-1/2 C.Y.		2.75	96.50	290	870	80	
	3910	3 C.Y.		3.15	130	390	1,175	103.20	
	3950	Stump chipper, 18" deep, 30 H.P.		7.82	205	615	1,850	185.55	
	4110	Dozer, crawler, torque converter, diesel 80 H.P.		29.05	400	1,200	3,600	472.40	
	4150	105 H.P.		35.60	530	1,590	4,775	602.80	
	4200	140 H.P.		51.25	795	2,380	7,150	886	
	4260	200 H.P.		77.25	1,275	3,845	11,500	1,387	
	4310	300 H.P.		100.65	1,825	5,460	16,400	1,897	
	4360	410 H.P.		132.60	2,250	6,720	20,200	2,405	
	4370	500 H.P.		171.15	3,225	9,650	29,000	3,299	
	4380	700 H.P.		280.90	4,550	13,660	41,000	4,979	
	4400	Loader, crawler, torque conv., diesel, 1-1/2 C.Y., 80 H.P.		31.25	455	1,360	4,075	522	
	4450	1-1/2 to 1-3/4 C.Y., 95 H.P.		34.60	590	1,765	5,300	629.80	
	4510	1-3/4 to 2-1/4 C.Y., 130 H.P.		53.80	875	2,630	7,900	956.40	
	4530	2-1/2 to 3-1/4 C.Y., 190 H.P.		65.95	1,100	3,300	9,900	1,188	
	4560	3-1/2 to 5 C.Y., 275 H.P.		87.80	1,425	4,295	12,900	1,561	
	4610	Front end loader, 4WD, articulated frame, diesel, 1 to 1-1/4 C.Y., 70 H.P.		19.75	235	705	2,125	299	
	4620	1-1/2 to 1-3/4 C.Y., 95 H.P.	↓	24.80	300	900	2,700	378.40	

01 54 33 | Equipment Rental

			UNIT	HOURLY OPER. COST	RENT PER DAY	RENT PER WEEK	RENT PER MONTH	EQUIPMENT COST/DAY	
20	4650	1-3/4 to 2 C.Y., 130 H.P.	Ea.	29.15	380	1,140	3,425	461.20	20
	4710	2-1/2 to 3-1/2 C.Y., 145 H.P.		33.10	425	1,275	3,825	519.80	
	4730	3 to 4-1/2 C.Y., 185 H.P.		42.40	550	1,650	4,950	669.20	
	4760	5-1/4 to 5-3/4 C.Y., 270 H.P.		67.50	905	2,710	8,125	1,082	
	4810	7 to 9 C.Y., 475 H.P.		114.05	1,700	5,125	15,400	1,937	
	4870	9 - 11 C.Y., 620 H.P.		160.55	2,775	8,290	24,900	2,942	
	4880	Skid steer loader, wheeled, 10 C.F., 30 H.P. gas		10.45	150	450	1,350	173.60	
	4890	1 C.Y., 78 H.P., diesel		20.20	262	785	2,350	318.60	
	4892	Skid-steer attachment, auger		.48	80.50	242	725	52.25	
	4893	Backhoe		.66	111	332	995	71.70	
	4894	Broom		.74	124	372	1,125	80.30	
	4895	Forks		.22	36	108	325	23.35	
	4896	Grapple		.56	92.50	278	835	60.10	
	4897	Concrete hammer		1.06	177	531	1,600	114.70	
	4898	Tree spade		1.03	172	515	1,550	111.25	
	4899	Trencher		.54	90	270	810	58.30	
	4900	Trencher, chain, boom type, gas, operator walking, 12 H.P.		5.00	46.50	140	420	68	
	4910	Operator riding, 40 H.P.		20.15	290	870	2,600	335.20	
	5000	Wheel type, diesel, 4′ deep, 12″ wide		85.95	845	2,535	7,600	1,195	
	5100	6′ deep, 20″ wide		92.75	1,925	5,740	17,200	1,890	
	5150	Chain type, diesel, 5′ deep, 8″ wide		38.35	560	1,675	5,025	641.80	
	5200	Diesel, 8′ deep, 16″ wide		155.95	3,600	10,770	32,300	3,402	
	5202	Rock trencher, wheel type, 6″ wide x 18″ deep		21.50	335	1,000	3,000	372	
	5206	Chain type, 18″ wide x 7′ deep		112.00	2,800	8,395	25,200	2,575	
	5210	Tree spade, self-propelled		14.38	267	800	2,400	275.05	
	5250	Truck, dump, 2-axle, 12 ton, 8 C.Y. payload, 220 H.P.		34.40	227	680	2,050	411.20	
	5300	Three axle dump, 16 ton, 12 C.Y. payload, 400 H.P.		61.00	340	1,015	3,050	691	
	5310	Four axle dump, 25 ton, 18 C.Y. payload, 450 H.P.		71.75	490	1,475	4,425	869	
	5350	Dump trailer only, rear dump, 16-1/2 C.Y.		5.45	138	415	1,250	126.60	
	5400	20 C.Y.		5.90	157	470	1,400	141.20	
	5450	Flatbed, single axle, 1-1/2 ton rating		25.55	68.50	205	615	245.40	
	5500	3 ton rating		30.65	96.50	290	870	303.20	
	5550	Off highway rear dump, 25 ton capacity		73.90	1,275	3,800	11,400	1,351	
	5600	35 ton capacity		82.50	1,425	4,260	12,800	1,512	
	5610	50 ton capacity		102.65	1,700	5,080	15,200	1,837	
	5620	65 ton capacity		105.65	1,700	5,090	15,300	1,863	
	5630	100 ton capacity		151.40	2,850	8,570	25,700	2,925	
	6000	Vibratory plow, 25 H.P., walking	↓	8.45	60	180	540	103.60	
40	0010	**GENERAL EQUIPMENT RENTAL** without operators R015433-10							40
	0150	Aerial lift, scissor type, to 15′ high, 1000 lb. cap., electric	Ea.	3.05	51.50	155	465	55.40	
	0160	To 25′ high, 2000 lb. capacity		3.45	66.50	200	600	67.60	
	0170	Telescoping boom to 40′ high, 500 lb. capacity, diesel		13.55	320	965	2,900	301.40	
	0180	To 45′ high, 500 lb. capacity		14.75	350	1,055	3,175	329	
	0190	To 60′ high, 600 lb. capacity		17.20	495	1,490	4,475	435.60	
	0195	Air compressor, portable, 6.5 CFM, electric		.67	12.65	38	114	12.95	
	0196	Gasoline		.81	19	57	171	17.90	
	0200	Towed type, gas engine, 60 CFM		13.55	48.50	145	435	137.40	
	0300	160 CFM		15.80	50	150	450	156.40	
	0400	Diesel engine, rotary screw, 250 CFM		17.05	108	325	975	201.40	
	0500	365 CFM		23.10	132	395	1,175	263.80	
	0550	450 CFM		29.35	165	495	1,475	333.80	
	0600	600 CFM		51.70	228	685	2,050	550.60	
	0700	750 CFM	↓	51.90	237	710	2,125	557.20	
	0800	For silenced models, small sizes, add to rent		3%	5%	5%	5%		
	0900	Large sizes, add to rent		5%	7%	7%	7%		
	0930	Air tools, breaker, pavement, 60 lb.	Ea.	.50	9.65	29	87	9.80	
	0940	80 lb.		.50	10.35	31	93	10.20	
	0950	Drills, hand (jackhammer) 65 lb.	↓	.60	17.35	52	156	15.20	

01 54 33 | Equipment Rental

			UNIT	HOURLY OPER. COST	RENT PER DAY	RENT PER WEEK	RENT PER MONTH	EQUIPMENT COST/DAY	
40	0960	Track or wagon, swing boom, 4" drifter	Ea.	62.50	880	2,640	7,925	1,028	40
	0970	5" drifter		74.45	1,075	3,235	9,700	1,243	
	0975	Track mounted quarry drill, 6" diameter drill		128.10	1,600	4,765	14,300	1,978	
	0980	Dust control per drill		1.02	23.50	71	213	22.35	
	0990	Hammer, chipping, 12 lb.		.55	26	78	234	20	
	1000	Hose, air with couplings, 50' long, 3/4" diameter		.03	5	15	45	3.25	
	1100	1" diameter		.04	6.35	19	57	4.10	
	1200	1-1/2" diameter		.05	9	27	81	5.80	
	1300	2" diameter		.07	12	36	108	7.75	
	1400	2-1/2" diameter		.11	19	57	171	12.30	
	1410	3" diameter		.14	23	69	207	14.90	
	1450	Drill, steel, 7/8" x 2'		.05	8.65	26	78	5.60	
	1460	7/8" x 6'		.05	9	27	81	5.80	
	1520	Moil points		.02	3.33	10	30	2.15	
	1525	Pneumatic nailer w/accessories		.58	38.50	115	345	27.65	
	1530	Sheeting driver for 60 lb. breaker		.04	6	18	54	3.90	
	1540	For 90 lb. breaker		.12	8	24	72	5.75	
	1550	Spade, 25 lb.		.45	6.65	20	60	7.60	
	1560	Tamper, single, 35 lb.		.55	36.50	109	325	26.20	
	1570	Triple, 140 lb.		.82	54.50	164	490	39.35	
	1580	Wrenches, impact, air powered, up to 3/4" bolt		.40	12.65	38	114	10.80	
	1590	Up to 1-1/4" bolt		.50	23.50	70	210	18	
	1600	Barricades, barrels, reflectorized, 1 to 99 barrels		.03	4.60	13.80	41.50	3	
	1610	100 to 200 barrels		.02	3.53	10.60	32	2.30	
	1620	Barrels with flashers, 1 to 99 barrels		.03	5.25	15.80	47.50	3.40	
	1630	100 to 200 barrels		.03	4.20	12.60	38	2.75	
	1640	Barrels with steady burn type C lights		.04	7	21	63	4.50	
	1650	Illuminated board, trailer mounted, with generator		3.50	130	390	1,175	106	
	1670	Portable barricade, stock, with flashers, 1 to 6 units		.03	5.25	15.80	47.50	3.40	
	1680	25 to 50 units		.03	4.90	14.70	44	3.20	
	1685	Butt fusion machine, wheeled, 1.5 HP electric, 2" - 8" diameter pipe		2.63	167	500	1,500	121.05	
	1690	Tracked, 20 HP diesel, 4"-12" diameter pipe		11.14	490	1,465	4,400	382.10	
	1695	83 HP diesel, 8" - 24" diameter pipe		30.46	975	2,930	8,800	829.70	
	1700	Carts, brick, hand powered, 1000 lb. capacity		.50	83.50	251	755	54.20	
	1800	Gas engine, 1500 lb., 7-1/2' lift		4.17	115	345	1,025	102.35	
	1822	Dehumidifier, medium, 6 lb./hr., 150 CFM		1.00	62	186	560	45.20	
	1824	Large, 18 lb./hr., 600 CFM		2.02	126	378	1,125	91.75	
	1830	Distributor, asphalt, trailer mounted, 2000 gal., 38 H.P. diesel		9.95	325	980	2,950	275.60	
	1840	3000 gal., 38 H.P. diesel		11.45	355	1,070	3,200	305.60	
	1850	Drill, rotary hammer, electric		1.02	26	78	234	23.75	
	1860	Carbide bit, 1-1/2" diameter, add to electric rotary hammer		.02	3.61	10.84	32.50	2.35	
	1865	Rotary, crawler, 250 H.P.		148.50	2,050	6,185	18,600	2,425	
	1870	Emulsion sprayer, 65 gal., 5 H.P. gas engine		2.91	97	291	875	81.50	
	1880	200 gal., 5 H.P. engine		7.85	162	485	1,450	159.80	
	1900	Floor auto-scrubbing machine, walk-behind, 28" path		4.98	325	970	2,900	233.85	
	1930	Floodlight, mercury vapor, or quartz, on tripod, 1000 watt		.44	20.50	62	186	15.90	
	1940	2000 watt		.83	41	123	370	31.25	
	1950	Floodlights, trailer mounted with generator, 1 - 300 watt light		3.65	71.50	215	645	72.20	
	1960	2 - 1000 watt lights		4.85	96.50	290	870	96.80	
	2000	4 - 300 watt lights		4.55	93.50	280	840	92.40	
	2005	Foam spray rig, incl. box trailer, compressor, generator, proportioner		35.03	490	1,465	4,400	573.25	
	2020	Forklift, straight mast, 12' lift, 5000 lb., 2 wheel drive, gas		26.35	202	605	1,825	331.80	
	2040	21' lift, 5000 lb., 4 wheel drive, diesel		20.90	240	720	2,150	311.20	
	2050	For rough terrain, 42' lift, 35' reach, 9000 lb., 110 H.P.		29.55	485	1,450	4,350	526.40	
	2060	For plant, 4 ton capacity, 80 H.P., 2 wheel drive, gas		16.15	93.50	280	840	185.20	
	2080	10 ton capacity, 120 H.P., 2 wheel drive, diesel		24.10	162	485	1,450	289.80	
	2100	Generator, electric, gas engine, 1.5 kW to 3 kW		3.75	11.35	34	102	36.80	
	2200	5 kW	↓	4.80	15.35	46	138	47.60	

01 54 33 | Equipment Rental

			UNIT	HOURLY OPER. COST	RENT PER DAY	RENT PER WEEK	RENT PER MONTH	EQUIPMENT COST/DAY	
40	2300	10 kW	Ea.	9.10	36.50	110	330	94.80	40
	2400	25 kW		10.60	85	255	765	135.80	
	2500	Diesel engine, 20 kW		12.10	66.50	200	600	136.80	
	2600	50 kW		23.35	103	310	930	248.80	
	2700	100 kW		43.30	128	385	1,150	423.40	
	2800	250 kW		85.15	235	705	2,125	822.20	
	2850	Hammer, hydraulic, for mounting on boom, to 500 ft lb.		2.55	75	225	675	65.40	
	2860	1000 ft lb.		4.35	127	380	1,150	110.80	
	2900	Heaters, space, oil or electric, 50 MBH		2.01	7.65	23	69	20.70	
	3000	100 MBH		3.76	10.65	32	96	36.50	
	3100	300 MBH		11.03	38.50	115	345	111.25	
	3150	500 MBH		18.15	45	135	405	172.20	
	3200	Hose, water, suction with coupling, 20′ long, 2″ diameter		.02	3	9	27	1.95	
	3210	3″ diameter		.03	4.33	13	39	2.85	
	3220	4″ diameter		.03	5	15	45	3.25	
	3230	6″ diameter		.11	17.65	53	159	11.50	
	3240	8″ diameter		.27	44.50	133	400	28.75	
	3250	Discharge hose with coupling, 50′ long, 2″ diameter		.01	1.33	4	12	.90	
	3260	3″ diameter		.01	2.33	7	21	1.50	
	3270	4″ diameter		.02	3.67	11	33	2.35	
	3280	6″ diameter		.06	9.35	28	84	6.10	
	3290	8″ diameter		.18	30	90	270	19.45	
	3295	Insulation blower		.78	6	18	54	9.85	
	3300	Ladders, extension type, 16′ to 36′ long		.14	22.50	68	204	14.70	
	3400	40′ to 60′ long		.19	31	93	279	20.10	
	3405	Lance for cutting concrete		2.35	71.50	215	645	61.80	
	3407	Lawn mower, rotary, 22″, 5 H.P.		1.80	49	147	440	43.80	
	3408	48″ self propelled		2.98	75	225	675	68.85	
	3410	Level, electronic, automatic, with tripod and leveling rod		1.20	75.50	226	680	54.80	
	3430	Laser type, for pipe and sewer line and grade		.73	48.50	145	435	34.85	
	3440	Rotating beam for interior control		.78	52	156	470	37.45	
	3460	Builder's optical transit, with tripod and rod		.09	14.65	44	132	9.50	
	3500	Light towers, towable, with diesel generator, 2000 watt		4.55	93.50	280	840	92.40	
	3600	4000 watt		4.85	96.50	290	870	96.80	
	3700	Mixer, powered, plaster and mortar, 6 C.F., 7 H.P.		2.70	20	60	180	33.60	
	3800	10 C.F., 9 H.P.		2.80	31.50	94	282	41.20	
	3850	Nailer, pneumatic		.58	38.50	115	345	27.65	
	3900	Paint sprayers complete, 8 CFM		1.04	69.50	208	625	49.90	
	4000	17 CFM		1.90	127	380	1,150	91.20	
	4020	Pavers, bituminous, rubber tires, 8′ wide, 50 H.P., diesel		31.15	490	1,465	4,400	542.20	
	4030	10′ wide, 150 H.P.		106.10	1,800	5,370	16,100	1,923	
	4050	Crawler, 8′ wide, 100 H.P., diesel		88.65	1,800	5,385	16,200	1,786	
	4060	10′ wide, 150 H.P.		113.15	2,200	6,610	19,800	2,227	
	4070	Concrete paver, 12′ to 24′ wide, 250 H.P.		101.40	1,550	4,675	14,000	1,746	
	4080	Placer-spreader-trimmer, 24′ wide, 300 H.P.		150.25	2,650	7,960	23,900	2,794	
	4100	Pump, centrifugal gas pump, 1-1/2″ diam., 65 GPM		3.90	48.50	145	435	60.20	
	4200	2″ diameter, 130 GPM		5.35	53.50	160	480	74.80	
	4300	3″ diameter, 250 GPM		5.65	55	165	495	78.20	
	4400	6″ diameter, 1500 GPM		30.45	172	515	1,550	346.60	
	4500	Submersible electric pump, 1-1/4″ diameter, 55 GPM		.39	16.35	49	147	12.90	
	4600	1-1/2″ diameter, 83 GPM		.43	18.65	56	168	14.65	
	4700	2″ diameter, 120 GPM		1.50	23.50	70	210	26	
	4800	3″ diameter, 300 GPM		2.69	41.50	125	375	46.50	
	4900	4″ diameter, 560 GPM		12.34	158	475	1,425	193.70	
	5000	6″ diameter, 1590 GPM		18.40	212	635	1,900	274.20	
	5100	Diaphragm pump, gas, single, 1-1/2″ diameter		1.18	49.50	148	445	39.05	
	5200	2″ diameter		4.25	61.50	185	555	71	
	5300	3″ diameter		4.25	61.50	185	555	71	

01 54 33 | Equipment Rental

	Line	Description	Unit	Hourly Oper. Cost	Rent Per Day	Rent Per Week	Rent Per Month	Equipment Cost/Day	
40	5400	Double, 4″ diameter	Ea.	6.40	105	315	945	114.20	40
	5450	Pressure washer 5 GPM, 3000 psi		4.90	51.50	155	465	70.20	
	5460	7 GPM, 3000 psi		6.45	60	180	540	87.60	
	5500	Trash pump, self-priming, gas, 2″ diameter		4.60	21	63	189	49.40	
	5600	Diesel, 4″ diameter		9.15	88.50	265	795	126.20	
	5650	Diesel, 6″ diameter		25.25	147	440	1,325	290	
	5655	Grout Pump		26.35	268	805	2,425	371.80	
	5700	Salamanders, L.P. gas fired, 100,000 Btu		3.96	13.65	41	123	39.90	
	5705	50,000 Btu		2.21	10.35	31	93	23.90	
	5720	Sandblaster, portable, open top, 3 C.F. capacity		.55	26.50	80	240	20.40	
	5730	6 C.F. capacity		.95	40	120	360	31.60	
	5740	Accessories for above		.13	21.50	65	195	14.05	
	5750	Sander, floor		.70	14.35	43	129	14.20	
	5760	Edger		.50	14.35	43	129	12.60	
	5800	Saw, chain, gas engine, 18″ long		2.30	21.50	64	192	31.20	
	5900	Hydraulic powered, 36″ long		.75	65	195	585	45	
	5950	60″ long		.75	66.50	200	600	46	
	6000	Masonry, table mounted, 14″ diameter, 5 H.P.		1.32	56.50	170	510	44.55	
	6050	Portable cut-off, 8 H.P.		2.50	33.50	100	300	40	
	6100	Circular, hand held, electric, 7-1/4″ diameter		.23	4.67	14	42	4.65	
	6200	12″ diameter		.23	8	24	72	6.65	
	6250	Wall saw, w/hydraulic power, 10 H.P.		9.70	61.50	185	555	114.60	
	6275	Shot blaster, walk-behind, 20″ wide		4.75	293	880	2,650	214	
	6280	Sidewalk broom, walk-behind		2.52	78.50	235	705	67.15	
	6300	Steam cleaner, 100 gallons per hour		3.70	76.50	230	690	75.60	
	6310	200 gallons per hour		5.35	95	285	855	99.80	
	6340	Tar Kettle/Pot, 400 gallons		15.60	75	225	675	169.80	
	6350	Torch, cutting, acetylene-oxygen, 150′ hose, excludes gases		.30	15	45	135	11.40	
	6360	Hourly operating cost includes tips and gas		19.00				152	
	6410	Toilet, portable chemical		.13	21	63	189	13.65	
	6420	Recycle flush type		.15	25	75	225	16.20	
	6430	Toilet, fresh water flush, garden hose,		.18	30.50	91	273	19.65	
	6440	Hoisted, non-flush, for high rise		.15	24.50	74	222	16	
	6465	Tractor, farm with attachment		21.50	297	890	2,675	350	
	6480	Trailers, platform, flush deck, 2 axle, 3 ton capacity		1.45	20	60	180	23.60	
	6500	25 ton capacity		5.45	117	350	1,050	113.60	
	6600	40 ton capacity		7.00	163	490	1,475	154	
	6700	3 axle, 50 ton capacity		7.55	180	540	1,625	168.40	
	6800	75 ton capacity		9.40	235	705	2,125	216.20	
	6810	Trailer mounted cable reel for high voltage line work		5.45	260	779	2,325	199.40	
	6820	Trailer mounted cable tensioning rig		10.85	515	1,550	4,650	396.80	
	6830	Cable pulling rig		73.05	2,900	8,680	26,000	2,320	
	6900	Water tank trailer, engine driven discharge, 5000 gallons		7.00	142	425	1,275	141	
	6925	10,000 gallons		9.50	197	590	1,775	194	
	6950	Water truck, off highway, 6000 gallons		88.15	770	2,315	6,950	1,168	
	7010	Tram car for high voltage line work, powered, 2 conductor		6.60	141	423	1,275	137.40	
	7020	Transit (builder's level) with tripod		.09	14.65	44	132	9.50	
	7030	Trench box, 3000 lb., 6′ x 8′		.56	93.50	280	840	60.50	
	7040	7200 lb., 6′ x 20′		.75	125	375	1,125	81	
	7050	8000 lb., 8′ x 16′		1.08	180	540	1,625	116.65	
	7060	9500 lb., 8′ x 20′		1.21	201	603	1,800	130.30	
	7065	11,000 lb., 8′ x 24′		1.27	211	633	1,900	136.75	
	7070	12,000 lb., 10′ x 20′		1.50	249	748	2,250	161.60	
	7100	Truck, pickup, 3/4 ton, 2 wheel drive		13.65	58.50	175	525	144.20	
	7200	4 wheel drive		13.95	73.50	220	660	155.60	
	7250	Crew carrier, 9 passenger		19.30	86.50	260	780	206.40	
	7290	Flat bed truck, 20,000 lb. GVW		21.70	125	375	1,125	248.60	
	7300	Tractor, 4 x 2, 220 H.P.	↓	30.20	197	590	1,775	359.60	

01 54 33 | Equipment Rental

			UNIT	HOURLY OPER. COST	RENT PER DAY	RENT PER WEEK	RENT PER MONTH	EQUIPMENT COST/DAY
40	7410	330 H.P.	Ea.	44.75	270	810	2,425	520
	7500	6 x 4, 380 H.P.		51.35	315	945	2,825	599.80
	7600	450 H.P.		62.30	380	1,145	3,425	727.40
	7610	Tractor, with A frame, boom and winch, 225 H.P.		33.00	272	815	2,450	427
	7620	Vacuum truck, hazardous material, 2500 gallons		12.05	290	870	2,600	270.40
	7625	5,000 gallons		14.26	405	1,220	3,650	358.10
	7650	Vacuum, HEPA, 16 gallon, wet/dry		.85	18	54	162	17.60
	7655	55 gallon, wet/dry		.80	27	81	243	22.60
	7660	Water tank, portable		.16	26.50	80	240	17.30
	7690	Sewer/catch basin vacuum, 14 C.Y., 1500 gallons		18.53	610	1,830	5,500	514.25
	7700	Welder, electric, 200 amp		3.88	16.35	49	147	40.85
	7800	300 amp		5.71	20	60	180	57.70
	7900	Gas engine, 200 amp		14.65	23.50	70	210	131.20
	8000	300 amp		16.38	24.50	74	222	145.85
	8100	Wheelbarrow, any size		.08	13	39	117	8.45
	8200	Wrecking ball, 4000 lb.	↓	2.35	68.50	205	615	59.80
50	0010	**HIGHWAY EQUIPMENT RENTAL** without operators (R015433 -10)						
	0050	Asphalt batch plant, portable drum mixer, 100 ton/hr.	Ea.	78.06	1,425	4,270	12,800	1,478
	0060	200 ton/hr.		89.18	1,500	4,530	13,600	1,619
	0070	300 ton/hr.		106.34	1,775	5,325	16,000	1,916
	0100	Backhoe attachment, long stick, up to 185 H.P., 10.5′ long		.35	23.50	70	210	16.80
	0140	Up to 250 H.P., 12′ long		.38	25	75	225	18.05
	0180	Over 250 H.P., 15′ long		.53	35	105	315	25.25
	0200	Special dipper arm, up to 100 H.P., 32′ long		1.08	71.50	215	645	51.65
	0240	Over 100 H.P., 33′ long		1.34	89.50	268	805	64.30
	0280	Catch basin/sewer cleaning truck, 3 ton, 9 C.Y., 1000 gal.		42.70	385	1,160	3,475	573.60
	0300	Concrete batch plant, portable, electric, 200 C.Y./hr.		22.87	515	1,550	4,650	492.95
	0520	Grader/dozer attachment, ripper/scarifier, rear mounted, up to 135 H.P.		2.90	56.50	170	510	57.20
	0540	Up to 180 H.P.		3.90	88.50	265	795	84.20
	0580	Up to 250 H.P.		5.00	117	350	1,050	110
	0700	Pvmt. removal bucket, for hyd. excavator, up to 90 H.P.		1.90	53.50	160	480	47.20
	0740	Up to 200 H.P.		2.15	73.50	220	660	61.20
	0780	Over 200 H.P.		2.25	85	255	765	69
	0900	Aggregate spreader, self-propelled, 187 H.P.		53.00	685	2,050	6,150	834
	1000	Chemical spreader, 3 C.Y.		3.30	43.50	130	390	52.40
	1900	Hammermill, traveling, 250 H.P.		78.83	2,075	6,240	18,700	1,879
	2000	Horizontal borer, 3″ diameter, 13 H.P. gas driven		6.35	55	165	495	83.80
	2150	Horizontal directional drill, 20,000 lb. thrust, 78 H.P. diesel		31.15	675	2,025	6,075	654.20
	2160	30,000 lb. thrust, 115 H.P.		38.90	1,025	3,100	9,300	931.20
	2170	50,000 lb. thrust, 170 H.P.		55.65	1,325	3,960	11,900	1,237
	2190	Mud trailer for HDD, 1500 gallons, 175 H.P., gas		31.90	152	455	1,375	346.20
	2200	Hydromulcher, diesel, 3000 gallon, for truck mounting		23.40	245	735	2,200	334.20
	2300	Gas, 600 gallon		8.60	98.50	295	885	127.80
	2400	Joint & crack cleaner, walk behind, 25 H.P.		3.95	50	150	450	61.60
	2500	Filler, trailer mounted, 400 gallons, 20 H.P.		9.50	210	630	1,900	202
	3000	Paint striper, self-propelled, 40 gallon, 22 H.P.		7.30	155	465	1,400	151.40
	3100	120 gallon, 120 H.P.		23.00	395	1,185	3,550	421
	3200	Post drivers, 6″ I-Beam frame, for truck mounting		19.05	380	1,135	3,400	379.40
	3400	Road sweeper, self-propelled, 8′ wide, 90 H.P.		39.80	575	1,720	5,150	662.40
	3450	Road sweeper, vacuum assisted, 4 C.Y., 220 gallons		75.75	635	1,900	5,700	986
	4000	Road mixer, self-propelled, 130 H.P.		46.85	765	2,295	6,875	833.80
	4100	310 H.P.		83.25	2,100	6,315	18,900	1,929
	4220	Cold mix paver, incl. pug mill and bitumen tank, 165 H.P.		97.00	2,275	6,855	20,600	2,147
	4240	Pavement brush, towed		3.20	93.50	280	840	81.60
	4250	Paver, asphalt, wheel or crawler, 130 H.P., diesel		96.65	2,275	6,795	20,400	2,132
	4300	Paver, road widener, gas 1′ to 6′, 67 H.P.		48.50	895	2,680	8,050	924
	4400	Diesel, 2′ to 14′, 88 H.P.		62.70	1,050	3,135	9,400	1,129
	4600	Slipform pavers, curb and gutter, 2 track, 75 H.P.	↓	57.20	900	2,700	8,100	997.60

01 54 33 | Equipment Rental

		01 54 33 | Equipment Rental	UNIT	HOURLY OPER. COST	RENT PER DAY	RENT PER WEEK	RENT PER MONTH	EQUIPMENT COST/DAY	
50	4700	4 track, 165 H.P.	Ea.	41.50	740	2,220	6,650	776	50
	4800	Median barrier, 215 H.P.		60.50	1,050	3,150	9,450	1,114	
	4901	Trailer, low bed, 75 ton capacity		10.10	235	705	2,125	221.80	
	5000	Road planer, walk behind, 10" cutting width, 10 H.P.		3.70	33.50	100	300	49.60	
	5100	Self-propelled, 12" cutting width, 64 H.P.		10.40	113	340	1,025	151.20	
	5120	Traffic line remover, metal ball blaster, truck mounted, 115 H.P.		47.95	730	2,190	6,575	821.60	
	5140	Grinder, truck mounted, 115 H.P.		54.60	790	2,375	7,125	911.80	
	5160	Walk-behind, 11 H.P.		4.20	53.50	160	480	65.60	
	5200	Pavement profiler, 4' to 6' wide, 450 H.P.		254.50	3,325	9,960	29,900	4,028	
	5300	8' to 10' wide, 750 H.P.		399.20	4,350	13,055	39,200	5,805	
	5400	Roadway plate, steel, 1" x 8' x 20'		.08	13.35	40	120	8.65	
	5600	Stabilizer, self-propelled, 150 H.P.		49.30	600	1,795	5,375	753.40	
	5700	310 H.P.		99.05	1,675	5,030	15,100	1,798	
	5800	Striper, truck mounted, 120 gallon paint, 460 H.P.		64.20	485	1,455	4,375	804.60	
	5900	Thermal paint heating kettle, 115 gallons		8.20	25.50	77	231	81	
	6000	Tar kettle, 330 gallon, trailer mounted		12.25	56.50	170	510	132	
	7000	Tunnel locomotive, diesel, 8 to 12 ton		31.85	585	1,750	5,250	604.80	
	7005	Electric, 10 ton		26.00	665	1,995	5,975	607	
	7010	Muck cars, 1/2 C.Y. capacity		2.00	24.50	73	219	30.60	
	7020	1 C.Y. capacity		2.25	32.50	97	291	37.40	
	7030	2 C.Y. capacity		2.35	36.50	110	330	40.80	
	7040	Side dump, 2 C.Y. capacity		2.60	45	135	405	47.80	
	7050	3 C.Y. capacity		3.45	50	150	450	57.60	
	7060	5 C.Y. capacity		4.95	63.50	190	570	77.60	
	7100	Ventilating blower for tunnel, 7-1/2 H.P.		2.05	51.50	155	465	47.40	
	7110	10 H.P.		2.28	51.50	155	465	49.25	
	7120	20 H.P.		3.48	67.50	202	605	68.25	
	7140	40 H.P.		5.76	96.50	290	870	104.10	
	7160	60 H.P.		8.77	152	455	1,375	161.15	
	7175	75 H.P.		11.66	207	620	1,850	217.30	
	7180	200 H.P.		23.65	305	910	2,725	371.20	
	7800	Windrow loader, elevating	↓	55.00	1,300	3,935	11,800	1,227	
60	0010	**LIFTING AND HOISTING EQUIPMENT RENTAL** without operators R015433-10							60
	0120	Aerial lift truck, 2 person, to 80'	Ea.	26.30	705	2,120	6,350	634.40	
	0140	Boom work platform, 40' snorkel R015433-15		12.00	275	825	2,475	261	
	0150	Crane, flatbed mounted, 3 ton capacity		16.35	188	565	1,700	243.80	
	0200	Crane, climbing, 106' jib, 6000 lb. capacity, 410 fpm R312316-45		39.35	1,650	4,960	14,900	1,307	
	0300	101' jib, 10,250 lb. capacity, 270 fpm		45.95	2,100	6,280	18,800	1,624	
	0500	Tower, static, 130' high, 106' jib, 6200 lb. capacity at 400 fpm		43.20	1,900	5,730	17,200	1,492	
	0600	Crawler mounted, lattice boom, 1/2 C.Y., 15 tons at 12' radius		36.98	625	1,870	5,600	669.85	
	0700	3/4 C.Y., 20 tons at 12' radius		49.31	780	2,340	7,025	862.50	
	0800	1 C.Y., 25 tons at 12' radius		65.75	1,050	3,120	9,350	1,150	
	0900	1-1/2 C.Y., 40 tons at 12' radius		65.75	1,050	3,150	9,450	1,156	
	1000	2 C.Y., 50 tons at 12' radius		69.70	1,225	3,680	11,000	1,294	
	1100	3 C.Y., 75 tons at 12' radius		74.70	1,425	4,305	12,900	1,459	
	1200	100 ton capacity, 60' boom		84.60	1,650	4,950	14,900	1,667	
	1300	165 ton capacity, 60' boom		107.75	1,925	5,785	17,400	2,019	
	1400	200 ton capacity, 70' boom		130.55	2,400	7,210	21,600	2,486	
	1500	350 ton capacity, 80' boom		182.90	3,625	10,845	32,500	3,632	
	1600	Truck mounted, lattice boom, 6 x 4, 20 tons at 10' radius		37.11	1,075	3,260	9,775	948.90	
	1700	25 tons at 10' radius		40.09	1,175	3,550	10,700	1,031	
	1800	8 x 4, 30 tons at 10' radius		43.61	1,250	3,780	11,300	1,105	
	1900	40 tons at 12' radius		46.70	1,325	3,950	11,900	1,164	
	2000	60 tons at 15' radius		53.07	1,400	4,180	12,500	1,261	
	2050	82 tons at 15' radius		59.80	1,475	4,460	13,400	1,370	
	2100	90 tons at 15' radius		67.36	1,625	4,860	14,600	1,511	
	2200	115 tons at 15' radius		76.11	1,800	5,430	16,300	1,695	
	2300	150 tons at 18' radius	↓	83.65	1,900	5,720	17,200	1,813	

01 54 33 | Equipment Rental

			Unit	Hourly Oper. Cost	Rent per Day	Rent per Week	Rent per Month	Equipment Cost/Day
60	2350	165 tons at 18′ radius	Ea.	90.05	2,025	6,060	18,200	1,932
	2400	Truck mounted, hydraulic, 12 ton capacity		42.60	520	1,565	4,700	653.80
	2500	25 ton capacity		44.95	630	1,885	5,650	736.60
	2550	33 ton capacity		45.50	645	1,930	5,800	750
	2560	40 ton capacity		58.85	750	2,250	6,750	920.80
	2600	55 ton capacity		76.80	855	2,570	7,700	1,128
	2700	80 ton capacity		100.45	1,375	4,105	12,300	1,625
	2720	100 ton capacity		94.20	1,425	4,250	12,800	1,604
	2740	120 ton capacity		100.05	1,525	4,575	13,700	1,715
	2760	150 ton capacity		127.60	2,000	5,995	18,000	2,220
	2800	Self-propelled, 4 x 4, with telescoping boom, 5 ton		17.55	225	675	2,025	275.40
	2900	12-1/2 ton capacity		32.30	360	1,075	3,225	473.40
	3000	15 ton capacity		33.00	375	1,130	3,400	490
	3050	20 ton capacity		35.95	445	1,335	4,000	554.60
	3100	25 ton capacity		37.55	500	1,495	4,475	599.40
	3150	40 ton capacity		46.00	560	1,675	5,025	703
	3200	Derricks, guy, 20 ton capacity, 60′ boom, 75′ mast		27.55	405	1,214	3,650	463.20
	3300	100′ boom, 115′ mast		43.21	695	2,090	6,275	763.70
	3400	Stiffleg, 20 ton capacity, 70′ boom, 37′ mast		30.11	525	1,580	4,750	556.90
	3500	100′ boom, 47′ mast		46.29	845	2,530	7,600	876.30
	3550	Helicopter, small, lift to 1250 lb. maximum, w/pilot		103.70	3,275	9,800	29,400	2,790
	3600	Hoists, chain type, overhead, manual, 3/4 ton		.10	.33	1	3	1
	3900	10 ton		.70	6	18	54	9.20
	4000	Hoist and tower, 5000 lb. cap., portable electric, 40′ high		4.95	233	699	2,100	179.40
	4100	For each added 10′ section, add		.11	18.35	55	165	11.90
	4200	Hoist and single tubular tower, 5000 lb. electric, 100′ high		6.70	325	976	2,925	248.80
	4300	For each added 6′-6″ section, add		.19	31.50	95	285	20.50
	4400	Hoist and double tubular tower, 5000 lb., 100′ high		7.19	360	1,075	3,225	272.50
	4500	For each added 6′-6″ section, add		.21	35	105	315	22.70
	4550	Hoist and tower, mast type, 6000 lb., 100′ high		7.76	370	1,115	3,350	285.10
	4570	For each added 10′ section, add		.13	21.50	65	195	14.05
	4600	Hoist and tower, personnel, electric, 2000 lb., 100′ @ 125 fpm		16.31	990	2,970	8,900	724.50
	4700	3000 lb., 100′ @ 200 fpm		18.62	1,125	3,360	10,100	820.95
	4800	3000 lb., 150′ @ 300 fpm		20.62	1,250	3,760	11,300	916.95
	4900	4000 lb., 100′ @ 300 fpm		21.38	1,275	3,840	11,500	939.05
	5000	6000 lb., 100′ @ 275 fpm		23.06	1,350	4,030	12,100	990.50
	5100	For added heights up to 500′, add	L.F.	.01	1.67	5	15	1.10
	5200	Jacks, hydraulic, 20 ton	Ea.	.05	2	6	18	1.60
	5500	100 ton		.40	11.65	35	105	10.20
	6100	Jacks, hydraulic, climbing w/50′ jackrods, control console, 30 ton cap.		2.01	134	402	1,200	96.50
	6150	For each added 10′ jackrod section, add		.05	3.33	10	30	2.40
	6300	50 ton capacity		3.23	215	646	1,950	155.05
	6350	For each added 10′ jackrod section, add		.06	4	12	36	2.90
	6500	125 ton capacity		8.45	565	1,690	5,075	405.60
	6550	For each added 10′ jackrod section, add		.58	38.50	115	345	27.65
	6600	Cable jack, 10 ton capacity with 200′ cable		1.68	112	336	1,000	80.65
	6650	For each added 50′ of cable, add		.20	13.35	40	120	9.60
70	0010	**WELLPOINT EQUIPMENT RENTAL** without operators R015433-10						
	0020	Based on 2 months rental						
	0100	Combination jetting & wellpoint pump, 60 H.P. diesel	Ea.	18.40	330	996	3,000	346.40
	0200	High pressure gas jet pump, 200 H.P., 300 psi	"	44.06	284	851	2,550	522.70
	0300	Discharge pipe, 8″ diameter	L.F.	.01	.54	1.62	4.86	.40
	0350	12″ diameter		.01	.79	2.38	7.15	.55
	0400	Header pipe, flows up to 150 GPM, 4″ diameter		.01	.49	1.47	4.41	.35
	0500	400 GPM, 6″ diameter		.01	.57	1.72	5.15	.40
	0600	800 GPM, 8″ diameter		.01	.79	2.38	7.15	.55
	0700	1500 GPM, 10″ diameter		.01	.84	2.51	7.55	.60

01 54 33 | Equipment Rental

		01 54 33 Equipment Rental	UNIT	HOURLY OPER. COST	RENT PER DAY	RENT PER WEEK	RENT PER MONTH	EQUIPMENT COST/DAY	
70	0800	2500 GPM, 12″ diameter	L.F.	.02	1.58	4.74	14.20	1.10	70
	0900	4500 GPM, 16″ diameter		.03	2.02	6.07	18.20	1.45	
	0950	For quick coupling aluminum and plastic pipe, add	↓	.03	2.09	6.28	18.85	1.50	
	1100	Wellpoint, 25′ long, with fittings & riser pipe, 1-1/2″ or 2″ diameter	Ea.	.06	4.18	12.54	37.50	3	
	1200	Wellpoint pump, diesel powered, 4″ suction, 20 H.P.		7.83	191	574	1,725	177.45	
	1300	6″ suction, 30 H.P.		10.70	237	712	2,125	228	
	1400	8″ suction, 40 H.P.		14.45	325	976	2,925	310.80	
	1500	10″ suction, 75 H.P.		22.27	380	1,141	3,425	406.35	
	1600	12″ suction, 100 H.P.		31.72	605	1,810	5,425	615.75	
	1700	12″ suction, 175 H.P.	↓	47.41	670	2,010	6,025	781.30	
80	0010	**MARINE EQUIPMENT RENTAL** without operators R015433-10							80
	0200	Barge, 400 Ton, 30′ wide x 90′ long	Ea.	17.50	1,050	3,180	9,550	776	
	0240	800 Ton, 45′ wide x 90′ long		21.25	1,300	3,870	11,600	944	
	2000	Tugboat, diesel, 100 H.P.		38.20	217	650	1,950	435.60	
	2040	250 H.P.		79.25	395	1,180	3,550	870	
	2080	380 H.P.		150.80	1,175	3,535	10,600	1,961	
	3000	Small work boat, gas, 16-foot, 50 H.P.		17.15	60	180	540	173.20	
	4000	Large, diesel, 48-foot, 200 H.P.	↓	89.40	1,250	3,740	11,200	1,463	

Crews

Crew No.	Bare Costs		Incl. Subs O&P		Cost Per Labor-Hour	
Crew A-1	**Hr.**	**Daily**	**Hr.**	**Daily**	**Bare Costs**	**Incl. O&P**
1 Building Laborer	$37.60	$300.80	$61.65	$493.20	$37.60	$61.65
1 Concrete Saw, Gas Manual		81.00		89.10	10.13	11.14
8 L.H., Daily Totals		$381.80		$582.30	$47.73	$72.79
Crew A-1A	**Hr.**	**Daily**	**Hr.**	**Daily**	**Bare Costs**	**Incl. O&P**
1 Skilled Worker	$48.65	$389.20	$79.05	$632.40	$48.65	$79.05
1 Shot Blaster, 20"		214.00		235.40	26.75	29.43
8 L.H., Daily Totals		$603.20		$867.80	$75.40	$108.47
Crew A-1B	**Hr.**	**Daily**	**Hr.**	**Daily**	**Bare Costs**	**Incl. O&P**
1 Building Laborer	$37.60	$300.80	$61.65	$493.20	$37.60	$61.65
1 Concrete Saw		167.40		184.14	20.93	23.02
8 L.H., Daily Totals		$468.20		$677.34	$58.52	$84.67
Crew A-1C	**Hr.**	**Daily**	**Hr.**	**Daily**	**Bare Costs**	**Incl. O&P**
1 Building Laborer	$37.60	$300.80	$61.65	$493.20	$37.60	$61.65
1 Chain Saw, Gas, 18"		31.20		34.32	3.90	4.29
8 L.H., Daily Totals		$332.00		$527.52	$41.50	$65.94
Crew A-1D	**Hr.**	**Daily**	**Hr.**	**Daily**	**Bare Costs**	**Incl. O&P**
1 Building Laborer	$37.60	$300.80	$61.65	$493.20	$37.60	$61.65
1 Vibrating Plate, Gas, 18"		36.20		39.82	4.53	4.98
8 L.H., Daily Totals		$337.00		$533.02	$42.13	$66.63
Crew A-1E	**Hr.**	**Daily**	**Hr.**	**Daily**	**Bare Costs**	**Incl. O&P**
1 Building Laborer	$37.60	$300.80	$61.65	$493.20	$37.60	$61.65
1 Vibrating Plate, Gas, 21"		46.80		51.48	5.85	6.43
8 L.H., Daily Totals		$347.60		$544.68	$43.45	$68.08
Crew A-1F	**Hr.**	**Daily**	**Hr.**	**Daily**	**Bare Costs**	**Incl. O&P**
1 Building Laborer	$37.60	$300.80	$61.65	$493.20	$37.60	$61.65
1 Rammer/Tamper, Gas, 8"		50.00		55.00	6.25	6.88
8 L.H., Daily Totals		$350.80		$548.20	$43.85	$68.53
Crew A-1G	**Hr.**	**Daily**	**Hr.**	**Daily**	**Bare Costs**	**Incl. O&P**
1 Building Laborer	$37.60	$300.80	$61.65	$493.20	$37.60	$61.65
1 Rammer/Tamper, Gas, 15"		56.40		62.04	7.05	7.75
8 L.H., Daily Totals		$357.20		$555.24	$44.65	$69.41
Crew A-1H	**Hr.**	**Daily**	**Hr.**	**Daily**	**Bare Costs**	**Incl. O&P**
1 Building Laborer	$37.60	$300.80	$61.65	$493.20	$37.60	$61.65
1 Exterior Steam Cleaner		75.60		83.16	9.45	10.40
8 L.H., Daily Totals		$376.40		$576.36	$47.05	$72.05
Crew A-1J	**Hr.**	**Daily**	**Hr.**	**Daily**	**Bare Costs**	**Incl. O&P**
1 Building Laborer	$37.60	$300.80	$61.65	$493.20	$37.60	$61.65
1 Cultivator, Walk-Behind, 5 H.P.		46.70		51.37	5.84	6.42
8 L.H., Daily Totals		$347.50		$544.57	$43.44	$68.07
Crew A-1K	**Hr.**	**Daily**	**Hr.**	**Daily**	**Bare Costs**	**Incl. O&P**
1 Building Laborer	$37.60	$300.80	$61.65	$493.20	$37.60	$61.65
1 Cultivator, Walk-Behind, 8 H.P.		72.65		79.92	9.08	9.99
8 L.H., Daily Totals		$373.45		$573.12	$46.68	$71.64
Crew A-1M	**Hr.**	**Daily**	**Hr.**	**Daily**	**Bare Costs**	**Incl. O&P**
1 Building Laborer	$37.60	$300.80	$61.65	$493.20	$37.60	$61.65
1 Snow Blower, Walk-Behind		67.15		73.86	8.39	9.23
8 L.H., Daily Totals		$367.95		$567.07	$45.99	$70.88

Crew No.	Bare Costs		Incl. Subs O&P		Cost Per Labor-Hour	
Crew A-2	**Hr.**	**Daily**	**Hr.**	**Daily**	**Bare Costs**	**Incl. O&P**
2 Laborers	$37.60	$601.60	$61.65	$986.40	$38.10	$62.12
1 Truck Driver (light)	39.10	312.80	63.05	504.40		
1 Flatbed Truck, Gas, 1.5 Ton		245.40		269.94	10.23	11.25
24 L.H., Daily Totals		$1159.80		$1760.74	$48.33	$73.36
Crew A-2A	**Hr.**	**Daily**	**Hr.**	**Daily**	**Bare Costs**	**Incl. O&P**
2 Laborers	$37.60	$601.60	$61.65	$986.40	$38.10	$62.12
1 Truck Driver (light)	39.10	312.80	63.05	504.40		
1 Flatbed Truck, Gas, 1.5 Ton		245.40		269.94		
1 Concrete Saw		167.40		184.14	17.20	18.92
24 L.H., Daily Totals		$1327.20		$1944.88	$55.30	$81.04
Crew A-2B	**Hr.**	**Daily**	**Hr.**	**Daily**	**Bare Costs**	**Incl. O&P**
1 Truck Driver (light)	$39.10	$312.80	$63.05	$504.40	$39.10	$63.05
1 Flatbed Truck, Gas, 1.5 Ton		245.40		269.94	30.68	33.74
8 L.H., Daily Totals		$558.20		$774.34	$69.78	$96.79
Crew A-3A	**Hr.**	**Daily**	**Hr.**	**Daily**	**Bare Costs**	**Incl. O&P**
1 Equip. Oper. (light)	$48.60	$388.80	$77.15	$617.20	$48.60	$77.15
1 Pickup Truck, 4 x 4, 3/4 Ton		155.60		171.16	19.45	21.40
8 L.H., Daily Totals		$544.40		$788.36	$68.05	$98.55
Crew A-3B	**Hr.**	**Daily**	**Hr.**	**Daily**	**Bare Costs**	**Incl. O&P**
1 Equip. Oper. (medium)	$50.60	$404.80	$80.30	$642.40	$45.33	$72.42
1 Truck Driver (heavy)	40.05	320.40	64.55	516.40		
1 Dump Truck, 12 C.Y., 400 H.P.		691.00		760.10		
1 F.E. Loader, W.M.,2.5 C.Y.		519.80		571.78	75.67	83.24
16 L.H., Daily Totals		$1936.00		$2490.68	$121.00	$155.67
Crew A-3C	**Hr.**	**Daily**	**Hr.**	**Daily**	**Bare Costs**	**Incl. O&P**
1 Equip. Oper. (light)	$48.60	$388.80	$77.15	$617.20	$48.60	$77.15
1 Loader, Skid Steer, 78 H.P.		318.60		350.46	39.83	43.81
8 L.H., Daily Totals		$707.40		$967.66	$88.42	$120.96
Crew A-3D	**Hr.**	**Daily**	**Hr.**	**Daily**	**Bare Costs**	**Incl. O&P**
1 Truck Driver (light)	$39.10	$312.80	$63.05	$504.40	$39.10	$63.05
1 Pickup Truck, 4 x 4, 3/4 Ton		155.60		171.16		
1 Flatbed Trailer, 25 Ton		113.60		124.96	33.65	37.02
8 L.H., Daily Totals		$582.00		$800.52	$72.75	$100.07
Crew A-3E	**Hr.**	**Daily**	**Hr.**	**Daily**	**Bare Costs**	**Incl. O&P**
1 Equip. Oper. (crane)	$51.70	$413.60	$82.05	$656.40	$45.88	$73.30
1 Truck Driver (heavy)	40.05	320.40	64.55	516.40		
1 Pickup Truck, 4 x 4, 3/4 Ton		155.60		171.16	9.72	10.70
16 L.H., Daily Totals		$889.60		$1343.96	$55.60	$84.00
Crew A-3F	**Hr.**	**Daily**	**Hr.**	**Daily**	**Bare Costs**	**Incl. O&P**
1 Equip. Oper. (crane)	$51.70	$413.60	$82.05	$656.40	$45.88	$73.30
1 Truck Driver (heavy)	40.05	320.40	64.55	516.40		
1 Pickup Truck, 4 x 4, 3/4 Ton		155.60		171.16		
1 Truck Tractor, 6x4, 380 H.P.		599.80		659.78		
1 Lowbed Trailer, 75 Ton		221.80		243.98	61.08	67.18
16 L.H., Daily Totals		$1711.20		$2247.72	$106.95	$140.48

Crew No. | Bare Costs | Incl. Subs O&P | Cost Per Labor-Hour

Crew A-3G	Hr.	Daily	Hr.	Daily	Bare Costs	Incl. O&P
1 Equip. Oper. (crane)	$51.70	$413.60	$82.05	$656.40	$45.88	$73.30
1 Truck Driver (heavy)	40.05	320.40	64.55	516.40		
1 Pickup Truck, 4 x 4, 3/4 Ton		155.60		171.16		
1 Truck Tractor, 6x4, 450 H.P.		727.40		800.14		
1 Lowbed Trailer, 75 Ton		221.80		243.98	69.05	75.95
16 L.H., Daily Totals		$1838.80		$2388.08	$114.93	$149.26

Crew A-3H	Hr.	Daily	Hr.	Daily	Bare Costs	Incl. O&P
1 Equip. Oper. (crane)	$51.70	$413.60	$82.05	$656.40	$51.70	$82.05
1 Hyd. Crane, 12 Ton (Daily)		860.80		946.88	107.60	118.36
8 L.H., Daily Totals		$1274.40		$1603.28	$159.30	$200.41

Crew A-3I	Hr.	Daily	Hr.	Daily	Bare Costs	Incl. O&P
1 Equip. Oper. (crane)	$51.70	$413.60	$82.05	$656.40	$51.70	$82.05
1 Hyd. Crane, 25 Ton (Daily)		989.60		1088.56	123.70	136.07
8 L.H., Daily Totals		$1403.20		$1744.96	$175.40	$218.12

Crew A-3J	Hr.	Daily	Hr.	Daily	Bare Costs	Incl. O&P
1 Equip. Oper. (crane)	$51.70	$413.60	$82.05	$656.40	$51.70	$82.05
1 Hyd. Crane, 40 Ton (Daily)		1221.00		1343.10	152.63	167.89
8 L.H., Daily Totals		$1634.60		$1999.50	$204.32	$249.94

Crew A-3K	Hr.	Daily	Hr.	Daily	Bare Costs	Incl. O&P
1 Equip. Oper. (crane)	$51.70	$413.60	$82.05	$656.40	$48.45	$76.90
1 Equip. Oper. (oiler)	45.20	361.60	71.75	574.00		
1 Hyd. Crane, 55 Ton (Daily)		1469.00		1615.90		
1 P/U Truck, 3/4 Ton (Daily)		167.20		183.92	102.26	112.49
16 L.H., Daily Totals		$2411.40		$3030.22	$150.71	$189.39

Crew A-3L	Hr.	Daily	Hr.	Daily	Bare Costs	Incl. O&P
1 Equip. Oper. (crane)	$51.70	$413.60	$82.05	$656.40	$48.45	$76.90
1 Equip. Oper. (oiler)	45.20	361.60	71.75	574.00		
1 Hyd. Crane, 80 Ton (Daily)		2174.00		2391.40		
1 P/U Truck, 3/4 Ton (Daily)		167.20		183.92	146.32	160.96
16 L.H., Daily Totals		$3116.40		$3805.72	$194.78	$237.86

Crew A-3M	Hr.	Daily	Hr.	Daily	Bare Costs	Incl. O&P
1 Equip. Oper. (crane)	$51.70	$413.60	$82.05	$656.40	$48.45	$76.90
1 Equip. Oper. (oiler)	45.20	361.60	71.75	574.00		
1 Hyd. Crane, 100 Ton (Daily)		2169.00		2385.90		
1 P/U Truck, 3/4 Ton (Daily)		167.20		183.92	146.01	160.61
16 L.H., Daily Totals		$3111.40		$3800.22	$194.46	$237.51

Crew A-3N	Hr.	Daily	Hr.	Daily	Bare Costs	Incl. O&P
1 Equip. Oper. (crane)	$51.70	$413.60	$82.05	$656.40	$51.70	$82.05
1 Tower Cane (monthly)		1128.00		1240.80	141.00	155.10
8 L.H., Daily Totals		$1541.60		$1897.20	$192.70	$237.15

Crew A-3P	Hr.	Daily	Hr.	Daily	Bare Costs	Incl. O&P
1 Equip. Oper. (light)	$48.60	$388.80	$77.15	$617.20	$48.60	$77.15
1 A.T. Forklift, 42' lift		526.40		579.04	65.80	72.38
8 L.H., Daily Totals		$915.20		$1196.24	$114.40	$149.53

Crew A-3Q	Hr.	Daily	Hr.	Daily	Bare Costs	Incl. O&P
1 Equip. Oper. (light)	$48.60	$388.80	$77.15	$617.20	$48.60	$77.15
1 Pickup Truck, 4 x 4, 3/4 Ton		155.60		171.16		
1 Flatbed trailer, 3 Ton		23.60		25.96	22.40	24.64
8 L.H., Daily Totals		$568.00		$814.32	$71.00	$101.79

Crew A-4	Hr.	Daily	Hr.	Daily	Bare Costs	Incl. O&P
2 Carpenters	$46.95	$751.20	$76.95	$1231.20	$44.75	$72.92
1 Painter, Ordinary	40.35	322.80	64.85	518.80		
24 L.H., Daily Totals		$1074.00		$1750.00	$44.75	$72.92

Crew A-5	Hr.	Daily	Hr.	Daily	Bare Costs	Incl. O&P
2 Laborers	$37.60	$601.60	$61.65	$986.40	$37.77	$61.81
.25 Truck Driver (light)	39.10	78.20	63.05	126.10		
.25 Flatbed Truck, Gas, 1.5 Ton		61.35		67.48	3.41	3.75
18 L.H., Daily Totals		$741.15		$1179.98	$41.17	$65.55

Crew A-6	Hr.	Daily	Hr.	Daily	Bare Costs	Incl. O&P
1 Instrument Man	$48.65	$389.20	$79.05	$632.40	$46.88	$76.13
1 Rodman/Chainman	45.10	360.80	73.20	585.60		
1 Level, Electronic		54.80		60.28	3.42	3.77
16 L.H., Daily Totals		$804.80		$1278.28	$50.30	$79.89

Crew A-7	Hr.	Daily	Hr.	Daily	Bare Costs	Incl. O&P
1 Chief of Party	$60.90	$487.20	$96.20	$769.60	$51.55	$82.82
1 Instrument Man	48.65	389.20	79.05	632.40		
1 Rodman/Chainman	45.10	360.80	73.20	585.60		
1 Level, Electronic		54.80		60.28	2.28	2.51
24 L.H., Daily Totals		$1292.00		$2047.88	$53.83	$85.33

Crew A-8	Hr.	Daily	Hr.	Daily	Bare Costs	Incl. O&P
1 Chief of Party	$60.90	$487.20	$96.20	$769.60	$49.94	$80.41
1 Instrument Man	48.65	389.20	79.05	632.40		
2 Rodmen/Chainmen	45.10	721.60	73.20	1171.20		
1 Level, Electronic		54.80		60.28	1.71	1.88
32 L.H., Daily Totals		$1652.80		$2633.48	$51.65	$82.30

Crew A-9	Hr.	Daily	Hr.	Daily	Bare Costs	Incl. O&P
1 Asbestos Foreman	$52.85	$422.80	$84.95	$679.60	$52.41	$84.25
7 Asbestos Workers	52.35	2931.60	84.15	4712.40		
64 L.H., Daily Totals		$3354.40		$5392.00	$52.41	$84.25

Crew A-10A	Hr.	Daily	Hr.	Daily	Bare Costs	Incl. O&P
1 Asbestos Foreman	$52.85	$422.80	$84.95	$679.60	$52.52	$84.42
2 Asbestos Workers	52.35	837.60	84.15	1346.40		
24 L.H., Daily Totals		$1260.40		$2026.00	$52.52	$84.42

Crew A-10B	Hr.	Daily	Hr.	Daily	Bare Costs	Incl. O&P
1 Asbestos Foreman	$52.85	$422.80	$84.95	$679.60	$52.48	$84.35
3 Asbestos Workers	52.35	1256.40	84.15	2019.60		
32 L.H., Daily Totals		$1679.20		$2699.20	$52.48	$84.35

Crew A-10C	Hr.	Daily	Hr.	Daily	Bare Costs	Incl. O&P
3 Asbestos Workers	$52.35	$1256.40	$84.15	$2019.60	$52.35	$84.15
1 Flatbed Truck, Gas, 1.5 Ton		245.40		269.94	10.23	11.25
24 L.H., Daily Totals		$1501.80		$2289.54	$62.58	$95.40

Crew A-10D	Hr.	Daily	Hr.	Daily	Bare Costs	Incl. O&P
2 Asbestos Workers	$52.35	$837.60	$84.15	$1346.40	$50.40	$80.53
1 Equip. Oper. (crane)	51.70	413.60	82.05	656.40		
1 Equip. Oper. (oiler)	45.20	361.60	71.75	574.00		
1 Hydraulic Crane, 33 Ton		750.00		825.00	23.44	25.78
32 L.H., Daily Totals		$2362.80		$3401.80	$73.84	$106.31

Crew A-11	Bare Costs Hr.	Bare Costs Daily	Incl. Subs O&P Hr.	Incl. Subs O&P Daily	Cost Per Labor-Hour Bare Costs	Cost Per Labor-Hour Incl. O&P
1 Asbestos Foreman	$52.85	$422.80	$84.95	$679.60	$52.41	$84.25
7 Asbestos Workers	52.35	2931.60	84.15	4712.40		
2 Chip. Hammers, 12 Lb., Elec.		40.00		44.00	.63	.69
64 L.H., Daily Totals		$3394.40		$5436.00	$53.04	$84.94

Crew A-12	Hr.	Daily	Hr.	Daily	Bare Costs	Incl. O&P
1 Asbestos Foreman	$52.85	$422.80	$84.95	$679.60	$52.41	$84.25
7 Asbestos Workers	52.35	2931.60	84.15	4712.40		
1 Trk-Mtd Vac, 14 CY, 1500 Gal.		514.25		565.67		
1 Flatbed Truck, 20,000 GVW		248.60		273.46	11.92	13.11
64 L.H., Daily Totals		$4117.25		$6231.14	$64.33	$97.36

Crew A-13	Hr.	Daily	Hr.	Daily	Bare Costs	Incl. O&P
1 Equip. Oper. (light)	$48.60	$388.80	$77.15	$617.20	$48.60	$77.15
1 Trk-Mtd Vac, 14 CY, 1500 Gal.		514.25		565.67		
1 Flatbed Truck, 20,000 GVW		248.60		273.46	95.36	104.89
8 L.H., Daily Totals		$1151.65		$1456.34	$143.96	$182.04

Crew B-1	Hr.	Daily	Hr.	Daily	Bare Costs	Incl. O&P
1 Labor Foreman (outside)	$39.60	$316.80	$64.90	$519.20	$38.27	$62.73
2 Laborers	37.60	601.60	61.65	986.40		
24 L.H., Daily Totals		$918.40		$1505.60	$38.27	$62.73

Crew B-1A	Hr.	Daily	Hr.	Daily	Bare Costs	Incl. O&P
1 Labor Foreman (outside)	$39.60	$316.80	$64.90	$519.20	$38.27	$62.73
2 Laborers	37.60	601.60	61.65	986.40		
2 Cutting Torches		22.80		25.08		
2 Sets of Gases		304.00		334.40	13.62	14.98
24 L.H., Daily Totals		$1245.20		$1865.08	$51.88	$77.71

Crew B-1B	Hr.	Daily	Hr.	Daily	Bare Costs	Incl. O&P
1 Labor Foreman (outside)	$39.60	$316.80	$64.90	$519.20	$41.63	$67.56
2 Laborers	37.60	601.60	61.65	986.40		
1 Equip. Oper. (crane)	51.70	413.60	82.05	656.40		
2 Cutting Torches		22.80		25.08		
2 Sets of Gases		304.00		334.40		
1 Hyd. Crane, 12 Ton		653.80		719.18	30.64	33.71
32 L.H., Daily Totals		$2312.60		$3240.66	$72.27	$101.27

Crew B-1C	Hr.	Daily	Hr.	Daily	Bare Costs	Incl. O&P
1 Labor Foreman (outside)	$39.60	$316.80	$64.90	$519.20	$38.27	$62.73
2 Laborers	37.60	601.60	61.65	986.40		
1 Aerial Lift Truck, 60' Boom		435.60		479.16	18.15	19.97
24 L.H., Daily Totals		$1354.00		$1984.76	$56.42	$82.70

Crew B-1D	Hr.	Daily	Hr.	Daily	Bare Costs	Incl. O&P
2 Laborers	$37.60	$601.60	$61.65	$986.40	$37.60	$61.65
1 Small Work Boat, Gas, 50 H.P.		173.20		190.52		
1 Pressure Washer, 7 GPM		87.60		96.36	16.30	17.93
16 L.H., Daily Totals		$862.40		$1273.28	$53.90	$79.58

Crew B-1E	Hr.	Daily	Hr.	Daily	Bare Costs	Incl. O&P
1 Labor Foreman (outside)	$39.60	$316.80	$64.90	$519.20	$38.10	$62.46
3 Laborers	37.60	902.40	61.65	1479.60		
1 Work Boat, Diesel, 200 H.P.		1463.00		1609.30		
2 Pressure Washer, 7 GPM		175.20		192.72	51.19	56.31
32 L.H., Daily Totals		$2857.40		$3800.82	$89.29	$118.78

Crew B-1F	Hr.	Daily	Hr.	Daily	Bare Costs	Incl. O&P
2 Skilled Workers	$48.65	$778.40	$79.05	$1264.80	$44.97	$73.25
1 Laborer	37.60	300.80	61.65	493.20		
1 Small Work Boat, Gas, 50 H.P.		173.20		190.52		
1 Pressure Washer, 7 GPM		87.60		96.36	10.87	11.95
24 L.H., Daily Totals		$1340.00		$2044.88	$55.83	$85.20

Crew B-1G	Hr.	Daily	Hr.	Daily	Bare Costs	Incl. O&P
2 Laborers	$37.60	$601.60	$61.65	$986.40	$37.60	$61.65
1 Small Work Boat, Gas, 50 H.P.		173.20		190.52	10.82	11.91
16 L.H., Daily Totals		$774.80		$1176.92	$48.42	$73.56

Crew B-1H	Hr.	Daily	Hr.	Daily	Bare Costs	Incl. O&P
2 Skilled Workers	$48.65	$778.40	$79.05	$1264.80	$44.97	$73.25
1 Laborer	37.60	300.80	61.65	493.20		
1 Small Work Boat, Gas, 50 H.P.		173.20		190.52	7.22	7.94
24 L.H., Daily Totals		$1252.40		$1948.52	$52.18	$81.19

Crew B-1J	Hr.	Daily	Hr.	Daily	Bare Costs	Incl. O&P
1 Labor Foreman (inside)	$38.10	$304.80	$62.45	$499.60	$37.85	$62.05
1 Laborer	37.60	300.80	61.65	493.20		
16 L.H., Daily Totals		$605.60		$992.80	$37.85	$62.05

Crew B-1K	Hr.	Daily	Hr.	Daily	Bare Costs	Incl. O&P
1 Carpenter Foreman (inside)	$47.45	$379.60	$77.75	$622.00	$47.20	$77.35
1 Carpenter	46.95	375.60	76.95	615.60		
16 L.H., Daily Totals		$755.20		$1237.60	$47.20	$77.35

Crew B-2	Hr.	Daily	Hr.	Daily	Bare Costs	Incl. O&P
1 Labor Foreman (outside)	$39.60	$316.80	$64.90	$519.20	$38.00	$62.30
4 Laborers	37.60	1203.20	61.65	1972.80		
40 L.H., Daily Totals		$1520.00		$2492.00	$38.00	$62.30

Crew B-2A	Hr.	Daily	Hr.	Daily	Bare Costs	Incl. O&P
1 Labor Foreman (outside)	$39.60	$316.80	$64.90	$519.20	$38.27	$62.73
2 Laborers	37.60	601.60	61.65	986.40		
1 Aerial Lift Truck, 60' Boom		435.60		479.16	18.15	19.97
24 L.H., Daily Totals		$1354.00		$1984.76	$56.42	$82.70

Crew B-3	Hr.	Daily	Hr.	Daily	Bare Costs	Incl. O&P
1 Labor Foreman (outside)	$39.60	$316.80	$64.90	$519.20	$40.92	$66.27
2 Laborers	37.60	601.60	61.65	986.40		
1 Equip. Oper. (medium)	50.60	404.80	80.30	642.40		
2 Truck Drivers (heavy)	40.05	640.80	64.55	1032.80		
1 Crawler Loader, 3 C.Y.		1188.00		1306.80		
2 Dump Trucks, 12 C.Y., 400 H.P.		1382.00		1520.20	53.54	58.90
48 L.H., Daily Totals		$4534.00		$6007.80	$94.46	$125.16

Crew B-3A	Hr.	Daily	Hr.	Daily	Bare Costs	Incl. O&P
4 Laborers	$37.60	$1203.20	$61.65	$1972.80	$40.20	$65.38
1 Equip. Oper. (medium)	50.60	404.80	80.30	642.40		
1 Hyd. Excavator, 1.5 C.Y.		1030.00		1133.00	25.75	28.32
40 L.H., Daily Totals		$2638.00		$3748.20	$65.95	$93.70

Crew B-3B	Hr.	Daily	Hr.	Daily	Bare Costs	Incl. O&P
2 Laborers	$37.60	$601.60	$61.65	$986.40	$41.46	$67.04
1 Equip. Oper. (medium)	50.60	404.80	80.30	642.40		
1 Truck Driver (heavy)	40.05	320.40	64.55	516.40		
1 Backhoe Loader, 80 H.P.		391.80		430.98		
1 Dump Truck, 12 C.Y., 400 H.P.		691.00		760.10	33.84	37.22
32 L.H., Daily Totals		$2409.60		$3336.28	$75.30	$104.26

Crew No.	Bare Costs		Incl. Subs O&P		Cost Per Labor-Hour	
Crew B-3C	**Hr.**	**Daily**	**Hr.**	**Daily**	**Bare Costs**	**Incl. O&P**
3 Laborers	$37.60	$902.40	$61.65	$1479.60	$40.85	$66.31
1 Equip. Oper. (medium)	50.60	404.80	80.30	642.40		
1 Crawler Loader, 4 C.Y.		1561.00		1717.10	48.78	53.66
32 L.H., Daily Totals		$2868.20		$3839.10	$89.63	$119.97

Crew No.	Bare Costs		Incl. Subs O&P		Cost Per Labor-Hour	
Crew B-4	**Hr.**	**Daily**	**Hr.**	**Daily**	**Bare Costs**	**Incl. O&P**
1 Labor Foreman (outside)	$39.60	$316.80	$64.90	$519.20	$38.34	$62.67
4 Laborers	37.60	1203.20	61.65	1972.80		
1 Truck Driver (heavy)	40.05	320.40	64.55	516.40		
1 Truck Tractor, 220 H.P.		359.60		395.56		
1 Flatbed Trailer, 40 Ton		154.00		169.40	10.70	11.77
48 L.H., Daily Totals		$2354.00		$3573.36	$49.04	$74.44

Crew No.	Bare Costs		Incl. Subs O&P		Cost Per Labor-Hour	
Crew B-5	**Hr.**	**Daily**	**Hr.**	**Daily**	**Bare Costs**	**Incl. O&P**
1 Labor Foreman (outside)	$39.60	$316.80	$64.90	$519.20	$41.60	$67.44
4 Laborers	37.60	1203.20	61.65	1972.80		
2 Equip. Oper. (medium)	50.60	809.60	80.30	1284.80		
1 Air Compressor, 250 cfm		201.40		221.54		
2 Breakers, Pavement, 60 lb.		19.60		21.56		
2 -50' Air Hoses, 1.5"		11.60		12.76		
1 Crawler Loader, 3 C.Y.		1188.00		1306.80	25.37	27.90
56 L.H., Daily Totals		$3750.20		$5339.46	$66.97	$95.35

Crew No.	Bare Costs		Incl. Subs O&P		Cost Per Labor-Hour	
Crew B-5D	**Hr.**	**Daily**	**Hr.**	**Daily**	**Bare Costs**	**Incl. O&P**
1 Labor Foreman (outside)	$39.60	$316.80	$64.90	$519.20	$41.41	$67.08
4 Laborers	37.60	1203.20	61.65	1972.80		
2 Equip. Oper. (medium)	50.60	809.60	80.30	1284.80		
1 Truck Driver (heavy)	40.05	320.40	64.55	516.40		
1 Air Compressor, 250 cfm		201.40		221.54		
2 Breakers, Pavement, 60 lb.		19.60		21.56		
2 -50' Air Hoses, 1.5"		11.60		12.76		
1 Crawler Loader, 3 C.Y.		1188.00		1306.80		
1 Dump Truck, 12 C.Y., 400 H.P.		691.00		760.10	32.99	36.29
64 L.H., Daily Totals		$4761.60		$6615.96	$74.40	$103.37

Crew No.	Bare Costs		Incl. Subs O&P		Cost Per Labor-Hour	
Crew B-6	**Hr.**	**Daily**	**Hr.**	**Daily**	**Bare Costs**	**Incl. O&P**
2 Laborers	$37.60	$601.60	$61.65	$986.40	$41.27	$66.82
1 Equip. Oper. (light)	48.60	388.80	77.15	617.20		
1 Backhoe Loader, 48 H.P.		364.20		400.62	15.18	16.69
24 L.H., Daily Totals		$1354.60		$2004.22	$56.44	$83.51

Crew No.	Bare Costs		Incl. Subs O&P		Cost Per Labor-Hour	
Crew B-6B	**Hr.**	**Daily**	**Hr.**	**Daily**	**Bare Costs**	**Incl. O&P**
2 Labor Foremen (outside)	$39.60	$633.60	$64.90	$1038.40	$38.27	$62.73
4 Laborers	37.60	1203.20	61.65	1972.80		
1 S.P. Crane, 4x4, 5 Ton		275.40		302.94		
1 Flatbed Truck, Gas, 1.5 Ton		245.40		269.94		
1 Butt Fusion Mach., 4"-12" diam.		382.10		420.31	18.81	20.69
48 L.H., Daily Totals		$2739.70		$4004.39	$57.08	$83.42

Crew No.	Bare Costs		Incl. Subs O&P		Cost Per Labor-Hour	
Crew B-6C	**Hr.**	**Daily**	**Hr.**	**Daily**	**Bare Costs**	**Incl. O&P**
2 Labor Foremen (outside)	$39.60	$633.60	$64.90	$1038.40	$38.27	$62.73
4 Laborers	37.60	1203.20	61.65	1972.80		
1 S.P. Crane, 4x4, 12 Ton		473.40		520.74		
1 Flatbed Truck, Gas, 3 Ton		303.20		333.52		
1 Butt Fusion Mach., 8"-24" diam.		829.70		912.67	33.46	36.81
48 L.H., Daily Totals		$3443.10		$4778.13	$71.73	$99.54

Crew No.	Bare Costs		Incl. Subs O&P		Cost Per Labor-Hour	
Crew B-7	**Hr.**	**Daily**	**Hr.**	**Daily**	**Bare Costs**	**Incl. O&P**
1 Labor Foreman (outside)	$39.60	$316.80	$64.90	$519.20	$40.10	$65.30
4 Laborers	37.60	1203.20	61.65	1972.80		
1 Equip. Oper. (medium)	50.60	404.80	80.30	642.40		
1 Brush Chipper, 12", 130 H.P.		391.40		430.54		
1 Crawler Loader, 3 C.Y.		1188.00		1306.80		
2 Chain Saws, Gas, 36" Long		90.00		99.00	34.78	38.26
48 L.H., Daily Totals		$3594.20		$4970.74	$74.88	$103.56

Crew No.	Bare Costs		Incl. Subs O&P		Cost Per Labor-Hour	
Crew B-7A	**Hr.**	**Daily**	**Hr.**	**Daily**	**Bare Costs**	**Incl. O&P**
2 Laborers	$37.60	$601.60	$61.65	$986.40	$41.27	$66.82
1 Equip. Oper. (light)	48.60	388.80	77.15	617.20		
1 Rake w/Tractor		354.25		389.68		
2 Chain Saw, Gas, 18"		62.40		68.64	17.36	19.10
24 L.H., Daily Totals		$1407.05		$2061.92	$58.63	$85.91

Crew No.	Bare Costs		Incl. Subs O&P		Cost Per Labor-Hour	
Crew B-7B	**Hr.**	**Daily**	**Hr.**	**Daily**	**Bare Costs**	**Incl. O&P**
1 Labor Foreman (outside)	$39.60	$316.80	$64.90	$519.20	$40.09	$65.19
4 Laborers	37.60	1203.20	61.65	1972.80		
1 Equip. Oper. (medium)	50.60	404.80	80.30	642.40		
1 Truck Driver (heavy)	40.05	320.40	64.55	516.40		
1 Brush Chipper, 12", 130 H.P.		391.40		430.54		
1 Crawler Loader, 3 C.Y.		1188.00		1306.80		
2 Chain Saws, Gas, 36" Long		90.00		99.00		
1 Dump Truck, 8 C.Y., 220 H.P.		411.20		452.32	37.15	40.87
56 L.H., Daily Totals		$4325.80		$5939.46	$77.25	$106.06

Crew No.	Bare Costs		Incl. Subs O&P		Cost Per Labor-Hour	
Crew B-7C	**Hr.**	**Daily**	**Hr.**	**Daily**	**Bare Costs**	**Incl. O&P**
1 Labor Foreman (outside)	$39.60	$316.80	$64.90	$519.20	$40.09	$65.19
4 Laborers	37.60	1203.20	61.65	1972.80		
1 Equip. Oper. (medium)	50.60	404.80	80.30	642.40		
1 Truck Driver (heavy)	40.05	320.40	64.55	516.40		
1 Brush Chipper, 12", 130 H.P.		391.40		430.54		
1 Crawler Loader, 3 C.Y.		1188.00		1306.80		
2 Chain Saws, Gas, 36" Long		90.00		99.00		
1 Dump Truck, 12 C.Y., 400 H.P.		691.00		760.10	42.15	46.37
56 L.H., Daily Totals		$4605.60		$6247.24	$82.24	$111.56

Crew No.	Bare Costs		Incl. Subs O&P		Cost Per Labor-Hour	
Crew B-8	**Hr.**	**Daily**	**Hr.**	**Daily**	**Bare Costs**	**Incl. O&P**
1 Labor Foreman (outside)	$39.60	$316.80	$64.90	$519.20	$42.66	$68.71
2 Laborers	37.60	601.60	61.65	986.40		
2 Equip. Oper. (medium)	50.60	809.60	80.30	1284.80		
1 Equip. Oper. (oiler)	45.20	361.60	71.75	574.00		
2 Truck Drivers (heavy)	40.05	640.80	64.55	1032.80		
1 Hyd. Crane, 25 Ton		736.60		810.26		
1 Crawler Loader, 3 C.Y.		1188.00		1306.80		
2 Dump Trucks, 12 C.Y., 400 H.P.		1382.00		1520.20	51.67	56.83
64 L.H., Daily Totals		$6037.00		$8034.46	$94.33	$125.54

Crew No.	Bare Costs		Incl. Subs O&P		Cost Per Labor-Hour	
Crew B-9	**Hr.**	**Daily**	**Hr.**	**Daily**	**Bare Costs**	**Incl. O&P**
1 Labor Foreman (outside)	$39.60	$316.80	$64.90	$519.20	$38.00	$62.30
4 Laborers	37.60	1203.20	61.65	1972.80		
1 Air Compressor, 250 cfm		201.40		221.54		
2 Breakers, Pavement, 60 lb.		19.60		21.56		
2 -50' Air Hoses, 1.5"		11.60		12.76	5.82	6.40
40 L.H., Daily Totals		$1752.60		$2747.86	$43.81	$68.70

Crew B-9A	Hr.	Daily	Hr.	Daily	Bare Costs	Incl. O&P
2 Laborers	$37.60	$601.60	$61.65	$986.40	$38.42	$62.62
1 Truck Driver (heavy)	40.05	320.40	64.55	516.40		
1 Water Tank Trailer, 5000 Gal.		141.00		155.10		
1 Truck Tractor, 220 H.P.		359.60		395.56		
2 -50' Discharge Hoses, 3"		3.00		3.30	20.98	23.08
24 L.H., Daily Totals		$1425.60		$2056.76	$59.40	$85.70

Crew B-9B	Hr.	Daily	Hr.	Daily	Bare Costs	Incl. O&P
2 Laborers	$37.60	$601.60	$61.65	$986.40	$38.42	$62.62
1 Truck Driver (heavy)	40.05	320.40	64.55	516.40		
2 -50' Discharge Hoses, 3"		3.00		3.30		
1 Water Tank Trailer, 5000 Gal.		141.00		155.10		
1 Truck Tractor, 220 H.P.		359.60		395.56		
1 Pressure Washer		70.20		77.22	23.91	26.30
24 L.H., Daily Totals		$1495.80		$2133.98	$62.33	$88.92

Crew B-9D	Hr.	Daily	Hr.	Daily	Bare Costs	Incl. O&P
1 Labor Foreman (outside)	$39.60	$316.80	$64.90	$519.20	$38.00	$62.30
4 Common Laborers	37.60	1203.20	61.65	1972.80		
1 Air Compressor, 250 cfm		201.40		221.54		
2 -50' Air Hoses, 1.5"		11.60		12.76		
2 Air Powered Tampers		52.40		57.64	6.63	7.30
40 L.H., Daily Totals		$1785.40		$2783.94	$44.63	$69.60

Crew B-10	Hr.	Daily	Hr.	Daily	Bare Costs	Incl. O&P
1 Equip. Oper. (medium)	$50.60	$404.80	$80.30	$642.40	$46.27	$74.08
.5 Laborer	37.60	150.40	61.65	246.60		
12 L.H., Daily Totals		$555.20		$889.00	$46.27	$74.08

Crew B-10A	Hr.	Daily	Hr.	Daily	Bare Costs	Incl. O&P
1 Equip. Oper. (medium)	$50.60	$404.80	$80.30	$642.40	$46.27	$74.08
.5 Laborer	37.60	150.40	61.65	246.60		
1 Roller, 2-Drum, W.B., 7.5 H.P.		180.00		198.00	15.00	16.50
12 L.H., Daily Totals		$735.20		$1087.00	$61.27	$90.58

Crew B-10B	Hr.	Daily	Hr.	Daily	Bare Costs	Incl. O&P
1 Equip. Oper. (medium)	$50.60	$404.80	$80.30	$642.40	$46.27	$74.08
.5 Laborer	37.60	150.40	61.65	246.60		
1 Dozer, 200 H.P.		1387.00		1525.70	115.58	127.14
12 L.H., Daily Totals		$1942.20		$2414.70	$161.85	$201.22

Crew B-10C	Hr.	Daily	Hr.	Daily	Bare Costs	Incl. O&P
1 Equip. Oper. (medium)	$50.60	$404.80	$80.30	$642.40	$46.27	$74.08
.5 Laborer	37.60	150.40	61.65	246.60		
1 Dozer, 200 H.P.		1387.00		1525.70		
1 Vibratory Roller, Towed, 23 Ton		396.00		435.60	148.58	163.44
12 L.H., Daily Totals		$2338.20		$2850.30	$194.85	$237.53

Crew B-10D	Hr.	Daily	Hr.	Daily	Bare Costs	Incl. O&P
1 Equip. Oper. (medium)	$50.60	$404.80	$80.30	$642.40	$46.27	$74.08
.5 Laborer	37.60	150.40	61.65	246.60		
1 Dozer, 200 H.P.		1387.00		1525.70		
1 Sheepsft. Roller, Towed		425.80		468.38	151.07	166.17
12 L.H., Daily Totals		$2368.00		$2883.08	$197.33	$240.26

Crew B-10E	Hr.	Daily	Hr.	Daily	Bare Costs	Incl. O&P
1 Equip. Oper. (medium)	$50.60	$404.80	$80.30	$642.40	$46.27	$74.08
.5 Laborer	37.60	150.40	61.65	246.60		
1 Tandem Roller, 5 Ton		157.40		173.14	13.12	14.43
12 L.H., Daily Totals		$712.60		$1062.14	$59.38	$88.51

Crew B-10F	Hr.	Daily	Hr.	Daily	Bare Costs	Incl. O&P
1 Equip. Oper. (medium)	$50.60	$404.80	$80.30	$642.40	$46.27	$74.08
.5 Laborer	37.60	150.40	61.65	246.60		
1 Tandem Roller, 10 Ton		236.80		260.48	19.73	21.71
12 L.H., Daily Totals		$792.00		$1149.48	$66.00	$95.79

Crew B-10G	Hr.	Daily	Hr.	Daily	Bare Costs	Incl. O&P
1 Equip. Oper. (medium)	$50.60	$404.80	$80.30	$642.40	$46.27	$74.08
.5 Laborer	37.60	150.40	61.65	246.60		
1 Sheepsfoot Roller, 240 H.P.		1206.00		1326.60	100.50	110.55
12 L.H., Daily Totals		$1761.20		$2215.60	$146.77	$184.63

Crew B-10H	Hr.	Daily	Hr.	Daily	Bare Costs	Incl. O&P
1 Equip. Oper. (medium)	$50.60	$404.80	$80.30	$642.40	$46.27	$74.08
.5 Laborer	37.60	150.40	61.65	246.60		
1 Diaphragm Water Pump, 2"		71.00		78.10		
1 -20' Suction Hose, 2"		1.95		2.15		
2 -50' Discharge Hoses, 2"		1.80		1.98	6.23	6.85
12 L.H., Daily Totals		$629.95		$971.23	$52.50	$80.94

Crew B-10I	Hr.	Daily	Hr.	Daily	Bare Costs	Incl. O&P
1 Equip. Oper. (medium)	$50.60	$404.80	$80.30	$642.40	$46.27	$74.08
.5 Laborer	37.60	150.40	61.65	246.60		
1 Diaphragm Water Pump, 4"		114.20		125.62		
1 -20' Suction Hose, 4"		3.25		3.58		
2 -50' Discharge Hoses, 4"		4.70		5.17	10.18	11.20
12 L.H., Daily Totals		$677.35		$1023.37	$56.45	$85.28

Crew B-10J	Hr.	Daily	Hr.	Daily	Bare Costs	Incl. O&P
1 Equip. Oper. (medium)	$50.60	$404.80	$80.30	$642.40	$46.27	$74.08
.5 Laborer	37.60	150.40	61.65	246.60		
1 Centrifugal Water Pump, 3"		78.20		86.02		
1 -20' Suction Hose, 3"		2.85		3.13		
2 -50' Discharge Hoses, 3"		3.00		3.30	7.00	7.70
12 L.H., Daily Totals		$639.25		$981.46	$53.27	$81.79

Crew B-10K	Hr.	Daily	Hr.	Daily	Bare Costs	Incl. O&P
1 Equip. Oper. (medium)	$50.60	$404.80	$80.30	$642.40	$46.27	$74.08
.5 Laborer	37.60	150.40	61.65	246.60		
1 Centr. Water Pump, 6"		346.60		381.26		
1 -20' Suction Hose, 6"		11.50		12.65		
2 -50' Discharge Hoses, 6"		12.20		13.42	30.86	33.94
12 L.H., Daily Totals		$925.50		$1296.33	$77.13	$108.03

Crew B-10L	Hr.	Daily	Hr.	Daily	Bare Costs	Incl. O&P
1 Equip. Oper. (medium)	$50.60	$404.80	$80.30	$642.40	$46.27	$74.08
.5 Laborer	37.60	150.40	61.65	246.60		
1 Dozer, 80 H.P.		472.40		519.64	39.37	43.30
12 L.H., Daily Totals		$1027.60		$1408.64	$85.63	$117.39

Crew B-10M	Hr.	Daily	Hr.	Daily	Bare Costs	Incl. O&P
1 Equip. Oper. (medium)	$50.60	$404.80	$80.30	$642.40	$46.27	$74.08
.5 Laborer	37.60	150.40	61.65	246.60		
1 Dozer, 300 H.P.		1897.00		2086.70	158.08	173.89
12 L.H., Daily Totals		$2452.20		$2975.70	$204.35	$247.97

Crew B-10N	Hr.	Daily	Hr.	Daily	Bare Costs	Incl. O&P
1 Equip. Oper. (medium)	$50.60	$404.80	$80.30	$642.40	$46.27	$74.08
.5 Laborer	37.60	150.40	61.65	246.60		
1 F.E. Loader, T.M., 1.5 C.Y		522.00		574.20	43.50	47.85
12 L.H., Daily Totals		$1077.20		$1463.20	$89.77	$121.93

Crews

Crew No.	Bare Costs		Incl. Subs O&P		Cost Per Labor-Hour	
Crew B-10O	**Hr.**	**Daily**	**Hr.**	**Daily**	**Bare Costs**	**Incl. O&P**
1 Equip. Oper. (medium)	$50.60	$404.80	$80.30	$642.40	$46.27	$74.08
.5 Laborer	37.60	150.40	61.65	246.60		
1 F.E. Loader, T.M., 2.25 C.Y.		956.40		1052.04	79.70	87.67
12 L.H., Daily Totals		$1511.60		$1941.04	$125.97	$161.75
Crew B-10P	**Hr.**	**Daily**	**Hr.**	**Daily**	**Bare Costs**	**Incl. O&P**
1 Equip. Oper. (medium)	$50.60	$404.80	$80.30	$642.40	$46.27	$74.08
.5 Laborer	37.60	150.40	61.65	246.60		
1 Crawler Loader, 3 C.Y.		1188.00		1306.80	99.00	108.90
12 L.H., Daily Totals		$1743.20		$2195.80	$145.27	$182.98
Crew B-10Q	**Hr.**	**Daily**	**Hr.**	**Daily**	**Bare Costs**	**Incl. O&P**
1 Equip. Oper. (medium)	$50.60	$404.80	$80.30	$642.40	$46.27	$74.08
.5 Laborer	37.60	150.40	61.65	246.60		
1 Crawler Loader, 4 C.Y.		1561.00		1717.10	130.08	143.09
12 L.H., Daily Totals		$2116.20		$2606.10	$176.35	$217.18
Crew B-10R	**Hr.**	**Daily**	**Hr.**	**Daily**	**Bare Costs**	**Incl. O&P**
1 Equip. Oper. (medium)	$50.60	$404.80	$80.30	$642.40	$46.27	$74.08
.5 Laborer	37.60	150.40	61.65	246.60		
1 F.E. Loader, W.M., 1 C.Y.		299.00		328.90	24.92	27.41
12 L.H., Daily Totals		$854.20		$1217.90	$71.18	$101.49
Crew B-10S	**Hr.**	**Daily**	**Hr.**	**Daily**	**Bare Costs**	**Incl. O&P**
1 Equip. Oper. (medium)	$50.60	$404.80	$80.30	$642.40	$46.27	$74.08
.5 Laborer	37.60	150.40	61.65	246.60		
1 F.E. Loader, W.M., 1.5 C.Y.		378.40		416.24	31.53	34.69
12 L.H., Daily Totals		$933.60		$1305.24	$77.80	$108.77
Crew B-10T	**Hr.**	**Daily**	**Hr.**	**Daily**	**Bare Costs**	**Incl. O&P**
1 Equip. Oper. (medium)	$50.60	$404.80	$80.30	$642.40	$46.27	$74.08
.5 Laborer	37.60	150.40	61.65	246.60		
1 F.E. Loader, W.M.,2.5 C.Y.		519.80		571.78	43.32	47.65
12 L.H., Daily Totals		$1075.00		$1460.78	$89.58	$121.73
Crew B-10U	**Hr.**	**Daily**	**Hr.**	**Daily**	**Bare Costs**	**Incl. O&P**
1 Equip. Oper. (medium)	$50.60	$404.80	$80.30	$642.40	$46.27	$74.08
.5 Laborer	37.60	150.40	61.65	246.60		
1 F.E. Loader, W.M., 5.5 C.Y.		1082.00		1190.20	90.17	99.18
12 L.H., Daily Totals		$1637.20		$2079.20	$136.43	$173.27
Crew B-10V	**Hr.**	**Daily**	**Hr.**	**Daily**	**Bare Costs**	**Incl. O&P**
1 Equip. Oper. (medium)	$50.60	$404.80	$80.30	$642.40	$46.27	$74.08
.5 Laborer	37.60	150.40	61.65	246.60		
1 Dozer, 700 H.P.		4979.00		5476.90	414.92	456.41
12 L.H., Daily Totals		$5534.20		$6365.90	$461.18	$530.49
Crew B-10W	**Hr.**	**Daily**	**Hr.**	**Daily**	**Bare Costs**	**Incl. O&P**
1 Equip. Oper. (medium)	$50.60	$404.80	$80.30	$642.40	$46.27	$74.08
.5 Laborer	37.60	150.40	61.65	246.60		
1 Dozer, 105 H.P.		602.80		663.08	50.23	55.26
12 L.H., Daily Totals		$1158.00		$1552.08	$96.50	$129.34
Crew B-10X	**Hr.**	**Daily**	**Hr.**	**Daily**	**Bare Costs**	**Incl. O&P**
1 Equip. Oper. (medium)	$50.60	$404.80	$80.30	$642.40	$46.27	$74.08
.5 Laborer	37.60	150.40	61.65	246.60		
1 Dozer, 410 H.P.		2405.00		2645.50	200.42	220.46
12 L.H., Daily Totals		$2960.20		$3534.50	$246.68	$294.54

Crew No.	Bare Costs		Incl. Subs O&P		Cost Per Labor-Hour	
Crew B-10Y	**Hr.**	**Daily**	**Hr.**	**Daily**	**Bare Costs**	**Incl. O&P**
1 Equip. Oper. (medium)	$50.60	$404.80	$80.30	$642.40	$46.27	$74.08
.5 Laborer	37.60	150.40	61.65	246.60		
1 Vibr. Roller, Towed, 12 Ton		549.80		604.78	45.82	50.40
12 L.H., Daily Totals		$1105.00		$1493.78	$92.08	$124.48
Crew B-11A	**Hr.**	**Daily**	**Hr.**	**Daily**	**Bare Costs**	**Incl. O&P**
1 Equipment Oper. (med.)	$50.60	$404.80	$80.30	$642.40	$44.10	$70.97
1 Laborer	37.60	300.80	61.65	493.20		
1 Dozer, 200 H.P.		1387.00		1525.70	86.69	95.36
16 L.H., Daily Totals		$2092.60		$2661.30	$130.79	$166.33
Crew B-11B	**Hr.**	**Daily**	**Hr.**	**Daily**	**Bare Costs**	**Incl. O&P**
1 Equipment Oper. (light)	$48.60	$388.80	$77.15	$617.20	$43.10	$69.40
1 Laborer	37.60	300.80	61.65	493.20		
1 Air Powered Tamper		26.20		28.82		
1 Air Compressor, 365 cfm		263.80		290.18		
2 -50' Air Hoses, 1.5"		11.60		12.76	18.85	20.73
16 L.H., Daily Totals		$991.20		$1442.16	$61.95	$90.14
Crew B-11C	**Hr.**	**Daily**	**Hr.**	**Daily**	**Bare Costs**	**Incl. O&P**
1 Equipment Oper. (med.)	$50.60	$404.80	$80.30	$642.40	$44.10	$70.97
1 Laborer	37.60	300.80	61.65	493.20		
1 Backhoe Loader, 48 H.P.		364.20		400.62	22.76	25.04
16 L.H., Daily Totals		$1069.80		$1536.22	$66.86	$96.01
Crew B-11K	**Hr.**	**Daily**	**Hr.**	**Daily**	**Bare Costs**	**Incl. O&P**
1 Equipment Oper. (med.)	$50.60	$404.80	$80.30	$642.40	$44.10	$70.97
1 Laborer	37.60	300.80	61.65	493.20		
1 Trencher, Chain Type, 8' D		3402.00		3742.20	212.63	233.89
16 L.H., Daily Totals		$4107.60		$4877.80	$256.73	$304.86
Crew B-11L	**Hr.**	**Daily**	**Hr.**	**Daily**	**Bare Costs**	**Incl. O&P**
1 Equipment Oper. (med.)	$50.60	$404.80	$80.30	$642.40	$44.10	$70.97
1 Laborer	37.60	300.80	61.65	493.20		
1 Grader, 30,000 Lbs.		736.20		809.82	46.01	50.61
16 L.H., Daily Totals		$1441.80		$1945.42	$90.11	$121.59
Crew B-11M	**Hr.**	**Daily**	**Hr.**	**Daily**	**Bare Costs**	**Incl. O&P**
1 Equipment Oper. (med.)	$50.60	$404.80	$80.30	$642.40	$44.10	$70.97
1 Laborer	37.60	300.80	61.65	493.20		
1 Backhoe Loader, 80 H.P.		391.80		430.98	24.49	26.94
16 L.H., Daily Totals		$1097.40		$1566.58	$68.59	$97.91
Crew B-11S	**Hr.**	**Daily**	**Hr.**	**Daily**	**Bare Costs**	**Incl. O&P**
1 Equipment Operator (med.)	$50.60	$404.80	$80.30	$642.40	$46.27	$74.08
.5 Laborer	37.60	150.40	61.65	246.60		
1 Dozer, 300 H.P.		1897.00		2086.70		
1 Ripper, Beam & 1 Shank		84.20		92.62	165.10	181.61
12 L.H., Daily Totals		$2536.40		$3068.32	$211.37	$255.69
Crew B-11T	**Hr.**	**Daily**	**Hr.**	**Daily**	**Bare Costs**	**Incl. O&P**
1 Equipment Operator (med.)	$50.60	$404.80	$80.30	$642.40	$46.27	$74.08
.5 Laborer	37.60	150.40	61.65	246.60		
1 Dozer, 410 H.P.		2405.00		2645.50		
1 Ripper, Beam & 2 Shanks		110.00		121.00	209.58	230.54
12 L.H., Daily Totals		$3070.20		$3655.50	$255.85	$304.63

Crew No.	Bare Costs		Incl. Subs O&P		Cost Per Labor-Hour	
Crew B-11W	**Hr.**	**Daily**	**Hr.**	**Daily**	**Bare Costs**	**Incl. O&P**
1 Equipment Operator (med.)	$50.60	$404.80	$80.30	$642.40	$40.73	$65.62
1 Common Laborer	37.60	300.80	61.65	493.20		
10 Truck Drivers (heavy)	40.05	3204.00	64.55	5164.00		
1 Dozer, 200 H.P.		1387.00		1525.70		
1 Vibratory Roller, Towed, 23 Ton		396.00		435.60		
10 Dump Trucks, 8 C.Y., 220 H.P.		4112.00		4523.20	61.41	67.55
96 L.H., Daily Totals		$9804.60		$12784.10	$102.13	$133.17

Crew B-11Y	Hr.	Daily	Hr.	Daily	Bare Costs	Incl. O&P
1 Labor Foreman (outside)	$39.60	$316.80	$64.90	$519.20	$42.16	$68.23
5 Common Laborers	37.60	1504.00	61.65	2466.00		
3 Equipment Operators (med.)	50.60	1214.40	80.30	1927.20		
1 Dozer, 80 H.P.		472.40		519.64		
2 Roller, 2-Drum, W.B., 7.5 H.P.		360.00		396.00		
4 Vibrating Plate, Gas, 21"		187.20		205.92	14.16	15.58
72 L.H., Daily Totals		$4054.80		$6033.96	$56.32	$83.81

Crew B-12A	Hr.	Daily	Hr.	Daily	Bare Costs	Incl. O&P
1 Equip. Oper. (crane)	$51.70	$413.60	$82.05	$656.40	$44.65	$71.85
1 Laborer	37.60	300.80	61.65	493.20		
1 Hyd. Excavator, 1 C.Y.		812.60		893.86	50.79	55.87
16 L.H., Daily Totals		$1527.00		$2043.46	$95.44	$127.72

Crew B-12B	Hr.	Daily	Hr.	Daily	Bare Costs	Incl. O&P
1 Equip. Oper. (crane)	$51.70	$413.60	$82.05	$656.40	$44.65	$71.85
1 Laborer	37.60	300.80	61.65	493.20		
1 Hyd. Excavator, 1.5 C.Y.		1030.00		1133.00	64.38	70.81
16 L.H., Daily Totals		$1744.40		$2282.60	$109.03	$142.66

Crew B-12C	Hr.	Daily	Hr.	Daily	Bare Costs	Incl. O&P
1 Equip. Oper. (crane)	$51.70	$413.60	$82.05	$656.40	$44.65	$71.85
1 Laborer	37.60	300.80	61.65	493.20		
1 Hyd. Excavator, 2 C.Y.		1175.00		1292.50	73.44	80.78
16 L.H., Daily Totals		$1889.40		$2442.10	$118.09	$152.63

Crew B-12D	Hr.	Daily	Hr.	Daily	Bare Costs	Incl. O&P
1 Equip. Oper. (crane)	$51.70	$413.60	$82.05	$656.40	$44.65	$71.85
1 Laborer	37.60	300.80	61.65	493.20		
1 Hyd. Excavator, 3.5 C.Y.		2440.00		2684.00	152.50	167.75
16 L.H., Daily Totals		$3154.40		$3833.60	$197.15	$239.60

Crew B-12E	Hr.	Daily	Hr.	Daily	Bare Costs	Incl. O&P
1 Equip. Oper. (crane)	$51.70	$413.60	$82.05	$656.40	$44.65	$71.85
1 Laborer	37.60	300.80	61.65	493.20		
1 Hyd. Excavator, .5 C.Y.		448.20		493.02	28.01	30.81
16 L.H., Daily Totals		$1162.60		$1642.62	$72.66	$102.66

Crew B-12F	Hr.	Daily	Hr.	Daily	Bare Costs	Incl. O&P
1 Equip. Oper. (crane)	$51.70	$413.60	$82.05	$656.40	$44.65	$71.85
1 Laborer	37.60	300.80	61.65	493.20		
1 Hyd. Excavator, .75 C.Y.		654.80		720.28	40.92	45.02
16 L.H., Daily Totals		$1369.20		$1869.88	$85.58	$116.87

Crew B-12G	Hr.	Daily	Hr.	Daily	Bare Costs	Incl. O&P
1 Equip. Oper. (crane)	$51.70	$413.60	$82.05	$656.40	$44.65	$71.85
1 Laborer	37.60	300.80	61.65	493.20		
1 Crawler Crane, 15 Ton		669.85		736.84		
1 Clamshell Bucket, .5 C.Y.		38.20		42.02	44.25	48.68
16 L.H., Daily Totals		$1422.45		$1928.45	$88.90	$120.53

Crew B-12H	Hr.	Daily	Hr.	Daily	Bare Costs	Incl. O&P
1 Equip. Oper. (crane)	$51.70	$413.60	$82.05	$656.40	$44.65	$71.85
1 Laborer	37.60	300.80	61.65	493.20		
1 Crawler Crane, 25 Ton		1150.00		1265.00		
1 Clamshell Bucket, 1 C.Y.		47.80		52.58	74.86	82.35
16 L.H., Daily Totals		$1912.20		$2467.18	$119.51	$154.20

Crew B-12I	Hr.	Daily	Hr.	Daily	Bare Costs	Incl. O&P
1 Equip. Oper. (crane)	$51.70	$413.60	$82.05	$656.40	$44.65	$71.85
1 Laborer	37.60	300.80	61.65	493.20		
1 Crawler Crane, 20 Ton		862.50		948.75		
1 Dragline Bucket, .75 C.Y.		20.60		22.66	55.19	60.71
16 L.H., Daily Totals		$1597.50		$2121.01	$99.84	$132.56

Crew B-12J	Hr.	Daily	Hr.	Daily	Bare Costs	Incl. O&P
1 Equip. Oper. (crane)	$51.70	$413.60	$82.05	$656.40	$44.65	$71.85
1 Laborer	37.60	300.80	61.65	493.20		
1 Gradall, 5/8 C.Y.		882.20		970.42	55.14	60.65
16 L.H., Daily Totals		$1596.60		$2120.02	$99.79	$132.50

Crew B-12K	Hr.	Daily	Hr.	Daily	Bare Costs	Incl. O&P
1 Equip. Oper. (crane)	$51.70	$413.60	$82.05	$656.40	$44.65	$71.85
1 Laborer	37.60	300.80	61.65	493.20		
1 Gradall, 3 Ton, 1 C.Y.		1002.00		1102.20	62.63	68.89
16 L.H., Daily Totals		$1716.40		$2251.80	$107.28	$140.74

Crew B-12L	Hr.	Daily	Hr.	Daily	Bare Costs	Incl. O&P
1 Equip. Oper. (crane)	$51.70	$413.60	$82.05	$656.40	$44.65	$71.85
1 Laborer	37.60	300.80	61.65	493.20		
1 Crawler Crane, 15 Ton		669.85		736.84		
1 F.E. Attachment, .5 C.Y.		60.00		66.00	45.62	50.18
16 L.H., Daily Totals		$1444.25		$1952.43	$90.27	$122.03

Crew B-12M	Hr.	Daily	Hr.	Daily	Bare Costs	Incl. O&P
1 Equip. Oper. (crane)	$51.70	$413.60	$82.05	$656.40	$44.65	$71.85
1 Laborer	37.60	300.80	61.65	493.20		
1 Crawler Crane, 20 Ton		862.50		948.75		
1 F.E. Attachment, .75 C.Y.		65.40		71.94	57.99	63.79
16 L.H., Daily Totals		$1642.30		$2170.29	$102.64	$135.64

Crew B-12N	Hr.	Daily	Hr.	Daily	Bare Costs	Incl. O&P
1 Equip. Oper. (crane)	$51.70	$413.60	$82.05	$656.40	$44.65	$71.85
1 Laborer	37.60	300.80	61.65	493.20		
1 Crawler Crane, 25 Ton		1150.00		1265.00		
1 F.E. Attachment, 1 C.Y.		71.20		78.32	76.33	83.96
16 L.H., Daily Totals		$1935.60		$2492.92	$120.97	$155.81

Crew B-12O	Hr.	Daily	Hr.	Daily	Bare Costs	Incl. O&P
1 Equip. Oper. (crane)	$51.70	$413.60	$82.05	$656.40	$44.65	$71.85
1 Laborer	37.60	300.80	61.65	493.20		
1 Crawler Crane, 40 Ton		1156.00		1271.60		
1 F.E. Attachment, 1.5 C.Y.		80.00		88.00	77.25	84.97
16 L.H., Daily Totals		$1950.40		$2509.20	$121.90	$156.82

Crew B-12P	Hr.	Daily	Hr.	Daily	Bare Costs	Incl. O&P
1 Equip. Oper. (crane)	$51.70	$413.60	$82.05	$656.40	$44.65	$71.85
1 Laborer	37.60	300.80	61.65	493.20		
1 Crawler Crane, 40 Ton		1156.00		1271.60		
1 Dragline Bucket, 1.5 C.Y.		33.60		36.96	74.35	81.78
16 L.H., Daily Totals		$1904.00		$2458.16	$119.00	$153.63

Crews

Crew B-12Q	Bare Costs Hr.	Bare Costs Daily	Incl. Subs O&P Hr.	Incl. Subs O&P Daily	Cost Per Labor-Hour Bare Costs	Cost Per Labor-Hour Incl. O&P
1 Equip. Oper. (crane)	$51.70	$413.60	$82.05	$656.40	$44.65	$71.85
1 Laborer	37.60	300.80	61.65	493.20		
1 Hyd. Excavator, 5/8 C.Y.		589.40		648.34	36.84	40.52
16 L.H., Daily Totals		$1303.80		$1797.94	$81.49	$112.37

Crew B-12S	Hr.	Daily	Hr.	Daily	Bare Costs	Incl. O&P
1 Equip. Oper. (crane)	$51.70	$413.60	$82.05	$656.40	$44.65	$71.85
1 Laborer	37.60	300.80	61.65	493.20		
1 Hyd. Excavator, 2.5 C.Y.		1605.00		1765.50	100.31	110.34
16 L.H., Daily Totals		$2319.40		$2915.10	$144.96	$182.19

Crew B-12T	Hr.	Daily	Hr.	Daily	Bare Costs	Incl. O&P
1 Equip. Oper. (crane)	$51.70	$413.60	$82.05	$656.40	$44.65	$71.85
1 Laborer	37.60	300.80	61.65	493.20		
1 Crawler Crane, 75 Ton		1459.00		1604.90		
1 F.E. Attachment, 3 C.Y.		103.20		113.52	97.64	107.40
16 L.H., Daily Totals		$2276.60		$2868.02	$142.29	$179.25

Crew B-12V	Hr.	Daily	Hr.	Daily	Bare Costs	Incl. O&P
1 Equip. Oper. (crane)	$51.70	$413.60	$82.05	$656.40	$44.65	$71.85
1 Laborer	37.60	300.80	61.65	493.20		
1 Crawler Crane, 75 Ton		1459.00		1604.90		
1 Dragline Bucket, 3 C.Y.		52.60		57.86	94.47	103.92
16 L.H., Daily Totals		$2226.00		$2812.36	$139.13	$175.77

Crew B-12Y	Hr.	Daily	Hr.	Daily	Bare Costs	Incl. O&P
1 Equip. Oper. (crane)	$51.70	$413.60	$82.05	$656.40	$42.30	$68.45
2 Laborers	37.60	601.60	61.65	986.40		
1 Hyd. Excavator, 3.5 C.Y.		2440.00		2684.00	101.67	111.83
24 L.H., Daily Totals		$3455.20		$4326.80	$143.97	$180.28

Crew B-12Z	Hr.	Daily	Hr.	Daily	Bare Costs	Incl. O&P
1 Equip. Oper. (crane)	$51.70	$413.60	$82.05	$656.40	$42.30	$68.45
2 Laborers	37.60	601.60	61.65	986.40		
1 Hyd. Excavator, 2.5 C.Y.		1605.00		1765.50	66.88	73.56
24 L.H., Daily Totals		$2620.20		$3408.30	$109.18	$142.01

Crew B-13	Hr.	Daily	Hr.	Daily	Bare Costs	Incl. O&P
1 Labor Foreman (outside)	$39.60	$316.80	$64.90	$519.20	$40.99	$66.47
4 Laborers	37.60	1203.20	61.65	1972.80		
1 Equip. Oper. (crane)	51.70	413.60	82.05	656.40		
1 Equip. Oper. (oiler)	45.20	361.60	71.75	574.00		
1 Hyd. Crane, 25 Ton		736.60		810.26	13.15	14.47
56 L.H., Daily Totals		$3031.80		$4532.66	$54.14	$80.94

Crew B-13A	Hr.	Daily	Hr.	Daily	Bare Costs	Incl. O&P
1 Labor Foreman (outside)	$39.60	$316.80	$64.90	$519.20	$42.30	$68.27
2 Laborers	37.60	601.60	61.65	986.40		
2 Equipment Operators (med.)	50.60	809.60	80.30	1284.80		
2 Truck Drivers (heavy)	40.05	640.80	64.55	1032.80		
1 Crawler Crane, 75 Ton		1459.00		1604.90		
1 Crawler Loader, 4 C.Y.		1561.00		1717.10		
2 Dump Trucks, 8 C.Y., 220 H.P.		822.40		904.64	68.61	75.48
56 L.H., Daily Totals		$6211.20		$8049.84	$110.91	$143.75

Crew B-13B	Hr.	Daily	Hr.	Daily	Bare Costs	Incl. O&P
1 Labor Foreman (outside)	$39.60	$316.80	$64.90	$519.20	$40.99	$66.47
4 Laborers	37.60	1203.20	61.65	1972.80		
1 Equip. Oper. (crane)	51.70	413.60	82.05	656.40		
1 Equip. Oper. (oiler)	45.20	361.60	71.75	574.00		
1 Hyd. Crane, 55 Ton		1128.00		1240.80	20.14	22.16
56 L.H., Daily Totals		$3423.20		$4963.20	$61.13	$88.63

Crew B-13C	Hr.	Daily	Hr.	Daily	Bare Costs	Incl. O&P
1 Labor Foreman (outside)	$39.60	$316.80	$64.90	$519.20	$40.99	$66.47
4 Laborers	37.60	1203.20	61.65	1972.80		
1 Equip. Oper. (crane)	51.70	413.60	82.05	656.40		
1 Equip. Oper. (oiler)	45.20	361.60	71.75	574.00		
1 Crawler Crane, 100 Ton		1667.00		1833.70	29.77	32.74
56 L.H., Daily Totals		$3962.20		$5556.10	$70.75	$99.22

Crew B-13D	Hr.	Daily	Hr.	Daily	Bare Costs	Incl. O&P
1 Laborer	$37.60	$300.80	$61.65	$493.20	$44.65	$71.85
1 Equip. Oper. (crane)	51.70	413.60	82.05	656.40		
1 Hyd. Excavator, 1 C.Y.		812.60		893.86		
1 Trench Box		81.00		89.10	55.85	61.44
16 L.H., Daily Totals		$1608.00		$2132.56	$100.50	$133.29

Crew B-13E	Hr.	Daily	Hr.	Daily	Bare Costs	Incl. O&P
1 Laborer	$37.60	$300.80	$61.65	$493.20	$44.65	$71.85
1 Equip. Oper. (crane)	51.70	413.60	82.05	656.40		
1 Hyd. Excavator, 1.5 C.Y.		1030.00		1133.00		
1 Trench Box		81.00		89.10	69.44	76.38
16 L.H., Daily Totals		$1825.40		$2371.70	$114.09	$148.23

Crew B-13F	Hr.	Daily	Hr.	Daily	Bare Costs	Incl. O&P
1 Laborer	$37.60	$300.80	$61.65	$493.20	$44.65	$71.85
1 Equip. Oper. (crane)	51.70	413.60	82.05	656.40		
1 Hyd. Excavator, 3.5 C.Y.		2440.00		2684.00		
1 Trench Box		81.00		89.10	157.56	173.32
16 L.H., Daily Totals		$3235.40		$3922.70	$202.21	$245.17

Crew B-13G	Hr.	Daily	Hr.	Daily	Bare Costs	Incl. O&P
1 Laborer	$37.60	$300.80	$61.65	$493.20	$44.65	$71.85
1 Equip. Oper. (crane)	51.70	413.60	82.05	656.40		
1 Hyd. Excavator, .75 C.Y.		654.80		720.28		
1 Trench Box		81.00		89.10	45.99	50.59
16 L.H., Daily Totals		$1450.20		$1958.98	$90.64	$122.44

Crew B-13H	Hr.	Daily	Hr.	Daily	Bare Costs	Incl. O&P
1 Laborer	$37.60	$300.80	$61.65	$493.20	$44.65	$71.85
1 Equip. Oper. (crane)	51.70	413.60	82.05	656.40		
1 Gradall, 5/8 C.Y.		882.20		970.42		
1 Trench Box		81.00		89.10	60.20	66.22
16 L.H., Daily Totals		$1677.60		$2209.12	$104.85	$138.07

Crew B-13I	Hr.	Daily	Hr.	Daily	Bare Costs	Incl. O&P
1 Laborer	$37.60	$300.80	$61.65	$493.20	$44.65	$71.85
1 Equip. Oper. (crane)	51.70	413.60	82.05	656.40		
1 Gradall, 3 Ton, 1 C.Y.		1002.00		1102.20		
1 Trench Box		81.00		89.10	67.69	74.46
16 L.H., Daily Totals		$1797.40		$2340.90	$112.34	$146.31

Crew B-13J	Bare Costs Hr.	Bare Costs Daily	Incl. Subs O&P Hr.	Incl. Subs O&P Daily	Cost Per Labor-Hour Bare Costs	Cost Per Labor-Hour Incl. O&P
1 Laborer	$37.60	$300.80	$61.65	$493.20	$44.65	$71.85
1 Equip. Oper. (crane)	51.70	413.60	82.05	656.40		
1 Hyd. Excavator, 2.5 C.Y.		1605.00		1765.50		
1 Trench Box		81.00		89.10	105.38	115.91
16 L.H., Daily Totals		$2400.40		$3004.20	$150.03	$187.76

Crew B-13K	Hr.	Daily	Hr.	Daily	Bare Costs	Incl. O&P
2 Equip. Opers. (crane)	$51.70	$827.20	$82.05	$1312.80	$51.70	$82.05
1 Hyd. Excavator, .75 C.Y.		654.80		720.28		
1 Hyd. Hammer, 4000 ft-lb		308.20		339.02		
1 Hyd. Excavator, .75 C.Y.		654.80		720.28	101.11	111.22
16 L.H., Daily Totals		$2445.00		$3092.38	$152.81	$193.27

Crew B-13L	Hr.	Daily	Hr.	Daily	Bare Costs	Incl. O&P
2 Equip. Opers. (crane)	$51.70	$827.20	$82.05	$1312.80	$51.70	$82.05
1 Hyd. Excavator, 1.5 C.Y.		1030.00		1133.00		
1 Hyd. Hammer, 5000 ft-lb		371.20		408.32		
1 Hyd. Excavator, .75 C.Y.		654.80		720.28	128.50	141.35
16 L.H., Daily Totals		$2883.20		$3574.40	$180.20	$223.40

Crew B-13M	Hr.	Daily	Hr.	Daily	Bare Costs	Incl. O&P
2 Equip. Opers. (crane)	$51.70	$827.20	$82.05	$1312.80	$51.70	$82.05
1 Hyd. Excavator, 2.5 C.Y.		1605.00		1765.50		
1 Hyd. Hammer, 8000 ft-lb		545.40		599.94		
1 Hyd. Excavator, 1.5 C.Y.		1030.00		1133.00	198.78	218.65
16 L.H., Daily Totals		$4007.60		$4811.24	$250.47	$300.70

Crew B-13N	Hr.	Daily	Hr.	Daily	Bare Costs	Incl. O&P
2 Equip. Opers. (crane)	$51.70	$827.20	$82.05	$1312.80	$51.70	$82.05
1 Hyd. Excavator, 3.5 C.Y.		2440.00		2684.00		
1 Hyd. Hammer, 12,000 ft-lb		635.80		699.38		
1 Hyd. Excavator, 1.5 C.Y.		1030.00		1133.00	256.61	282.27
16 L.H., Daily Totals		$4933.00		$5829.18	$308.31	$364.32

Crew B-14	Hr.	Daily	Hr.	Daily	Bare Costs	Incl. O&P
1 Labor Foreman (outside)	$39.60	$316.80	$64.90	$519.20	$39.77	$64.78
4 Laborers	37.60	1203.20	61.65	1972.80		
1 Equip. Oper. (light)	48.60	388.80	77.15	617.20		
1 Backhoe Loader, 48 H.P.		364.20		400.62	7.59	8.35
48 L.H., Daily Totals		$2273.00		$3509.82	$47.35	$73.12

Crew B-14A	Hr.	Daily	Hr.	Daily	Bare Costs	Incl. O&P
1 Equip. Oper. (crane)	$51.70	$413.60	$82.05	$656.40	$47.00	$75.25
.5 Laborer	37.60	150.40	61.65	246.60		
1 Hyd. Excavator, 4.5 C.Y.		2963.00		3259.30	246.92	271.61
12 L.H., Daily Totals		$3527.00		$4162.30	$293.92	$346.86

Crew B-14B	Hr.	Daily	Hr.	Daily	Bare Costs	Incl. O&P
1 Equip. Oper. (crane)	$51.70	$413.60	$82.05	$656.40	$47.00	$75.25
.5 Laborer	37.60	150.40	61.65	246.60		
1 Hyd. Excavator, 6 C.Y.		3494.00		3843.40	291.17	320.28
12 L.H., Daily Totals		$4058.00		$4746.40	$338.17	$395.53

Crew B-14C	Hr.	Daily	Hr.	Daily	Bare Costs	Incl. O&P
1 Equip. Oper. (crane)	$51.70	$413.60	$82.05	$656.40	$47.00	$75.25
.5 Laborer	37.60	150.40	61.65	246.60		
1 Hyd. Excavator, 7 C.Y.		3590.00		3949.00	299.17	329.08
12 L.H., Daily Totals		$4154.00		$4852.00	$346.17	$404.33

Crew B-14F	Hr.	Daily	Hr.	Daily	Bare Costs	Incl. O&P
1 Equip. Oper. (crane)	$51.70	$413.60	$82.05	$656.40	$47.00	$75.25
.5 Laborer	37.60	150.40	61.65	246.60		
1 Hyd. Shovel, 7 C.Y.		3993.00		4392.30	332.75	366.02
12 L.H., Daily Totals		$4557.00		$5295.30	$379.75	$441.27

Crew B-14G	Hr.	Daily	Hr.	Daily	Bare Costs	Incl. O&P
1 Equip. Oper. (crane)	$51.70	$413.60	$82.05	$656.40	$47.00	$75.25
.5 Laborer	37.60	150.40	61.65	246.60		
1 Hyd. Shovel, 12 C.Y.		5736.00		6309.60	478.00	525.80
12 L.H., Daily Totals		$6300.00		$7212.60	$525.00	$601.05

Crew B-14J	Hr.	Daily	Hr.	Daily	Bare Costs	Incl. O&P
1 Equip. Oper. (medium)	$50.60	$404.80	$80.30	$642.40	$46.27	$74.08
.5 Laborer	37.60	150.40	61.65	246.60		
1 F.E. Loader, 8 C.Y.		1937.00		2130.70	161.42	177.56
12 L.H., Daily Totals		$2492.20		$3019.70	$207.68	$251.64

Crew B-14K	Hr.	Daily	Hr.	Daily	Bare Costs	Incl. O&P
1 Equip. Oper. (medium)	$50.60	$404.80	$80.30	$642.40	$46.27	$74.08
.5 Laborer	37.60	150.40	61.65	246.60		
1 F.E. Loader, 10 C.Y.		2942.00		3236.20	245.17	269.68
12 L.H., Daily Totals		$3497.20		$4125.20	$291.43	$343.77

Crew B-15	Hr.	Daily	Hr.	Daily	Bare Costs	Incl. O&P
1 Equipment Oper. (med.)	$50.60	$404.80	$80.30	$642.40	$42.71	$68.64
.5 Laborer	37.60	150.40	61.65	246.60		
2 Truck Drivers (heavy)	40.05	640.80	64.55	1032.80		
2 Dump Trucks, 12 C.Y., 400 H.P.		1382.00		1520.20		
1 Dozer, 200 H.P.		1387.00		1525.70	98.89	108.78
28 L.H., Daily Totals		$3965.00		$4967.70	$141.61	$177.42

Crew B-16	Hr.	Daily	Hr.	Daily	Bare Costs	Incl. O&P
1 Labor Foreman (outside)	$39.60	$316.80	$64.90	$519.20	$38.71	$63.19
2 Laborers	37.60	601.60	61.65	986.40		
1 Truck Driver (heavy)	40.05	320.40	64.55	516.40		
1 Dump Truck, 12 C.Y., 400 H.P.		691.00		760.10	21.59	23.75
32 L.H., Daily Totals		$1929.80		$2782.10	$60.31	$86.94

Crew B-17	Hr.	Daily	Hr.	Daily	Bare Costs	Incl. O&P
2 Laborers	$37.60	$601.60	$61.65	$986.40	$40.96	$66.25
1 Equip. Oper. (light)	48.60	388.80	77.15	617.20		
1 Truck Driver (heavy)	40.05	320.40	64.55	516.40		
1 Backhoe Loader, 48 H.P.		364.20		400.62		
1 Dump Truck, 8 C.Y., 220 H.P.		411.20		452.32	24.23	26.65
32 L.H., Daily Totals		$2086.20		$2972.94	$65.19	$92.90

Crew B-17A	Hr.	Daily	Hr.	Daily	Bare Costs	Incl. O&P
2 Labor Foremen (outside)	$39.60	$633.60	$64.90	$1038.40	$40.41	$66.11
6 Laborers	37.60	1804.80	61.65	2959.20		
1 Skilled Worker Foreman (out)	50.65	405.20	82.30	658.40		
1 Skilled Worker	48.65	389.20	79.05	632.40		
80 L.H., Daily Totals		$3232.80		$5288.40	$40.41	$66.11

Crew B-17B	Hr.	Daily	Hr.	Daily	Bare Costs	Incl. O&P
2 Laborers	$37.60	$601.60	$61.65	$986.40	$40.96	$66.25
1 Equip. Oper. (light)	48.60	388.80	77.15	617.20		
1 Truck Driver (heavy)	40.05	320.40	64.55	516.40		
1 Backhoe Loader, 48 H.P.		364.20		400.62		
1 Dump Truck, 12 C.Y., 400 H.P.		691.00		760.10	32.98	36.27
32 L.H., Daily Totals		$2366.00		$3280.72	$73.94	$102.52

Crew No.	Bare Costs		Incl. Subs O&P		Cost Per Labor-Hour	
Crew B-18	**Hr.**	**Daily**	**Hr.**	**Daily**	**Bare Costs**	**Incl. O&P**
1 Labor Foreman (outside)	$39.60	$316.80	$64.90	$519.20	$38.27	$62.73
2 Laborers	37.60	601.60	61.65	986.40		
1 Vibrating Plate, Gas, 21"		46.80		51.48	1.95	2.15
24 L.H., Daily Totals		$965.20		$1557.08	$40.22	$64.88

Crew B-19	**Hr.**	**Daily**	**Hr.**	**Daily**	**Bare Costs**	**Incl. O&P**
1 Pile Driver Foreman (outside)	$48.10	$384.80	$78.55	$628.40	$47.64	$76.95
4 Pile Drivers	46.10	1475.20	75.30	2409.60		
2 Equip. Oper. (crane)	51.70	827.20	82.05	1312.80		
1 Equip. Oper. (oiler)	45.20	361.60	71.75	574.00		
1 Crawler Crane, 40 Ton		1156.00		1271.60		
1 Lead, 90' High		124.40		136.84		
1 Hammer, Diesel, 22k ft-lb		456.40		502.04	27.14	29.85
64 L.H., Daily Totals		$4785.60		$6835.28	$74.78	$106.80

Crew B-19A	**Hr.**	**Daily**	**Hr.**	**Daily**	**Bare Costs**	**Incl. O&P**
1 Pile Driver Foreman (outside)	$48.10	$384.80	$78.55	$628.40	$47.64	$76.95
4 Pile Drivers	46.10	1475.20	75.30	2409.60		
2 Equip. Oper. (crane)	51.70	827.20	82.05	1312.80		
1 Equip. Oper. (oiler)	45.20	361.60	71.75	574.00		
1 Crawler Crane, 75 Ton		1459.00		1604.90		
1 Lead, 90' high		124.40		136.84		
1 Hammer, Diesel, 41k ft-lb		581.00		639.10	33.82	37.20
64 L.H., Daily Totals		$5213.20		$7305.64	$81.46	$114.15

Crew B-19B	**Hr.**	**Daily**	**Hr.**	**Daily**	**Bare Costs**	**Incl. O&P**
1 Pile Driver Foreman (outside)	$48.10	$384.80	$78.55	$628.40	$47.64	$76.95
4 Pile Drivers	46.10	1475.20	75.30	2409.60		
2 Equip. Oper. (crane)	51.70	827.20	82.05	1312.80		
1 Equip. Oper. (oiler)	45.20	361.60	71.75	574.00		
1 Crawler Crane, 40 Ton		1156.00		1271.60		
1 Lead, 90' High		124.40		136.84		
1 Hammer, Diesel, 22k ft-lb		456.40		502.04		
1 Barge, 400 Ton		776.00		853.60	39.26	43.19
64 L.H., Daily Totals		$5561.60		$7688.88	$86.90	$120.14

Crew B-19C	**Hr.**	**Daily**	**Hr.**	**Daily**	**Bare Costs**	**Incl. O&P**
1 Pile Driver Foreman (outside)	$48.10	$384.80	$78.55	$628.40	$47.64	$76.95
4 Pile Drivers	46.10	1475.20	75.30	2409.60		
2 Equip. Oper. (crane)	51.70	827.20	82.05	1312.80		
1 Equip. Oper. (oiler)	45.20	361.60	71.75	574.00		
1 Crawler Crane, 75 Ton		1459.00		1604.90		
1 Lead, 90' High		124.40		136.84		
1 Hammer, Diesel, 41k ft-lb		581.00		639.10		
1 Barge, 400 Ton		776.00		853.60	45.94	50.54
64 L.H., Daily Totals		$5989.20		$8159.24	$93.58	$127.49

Crew B-20	**Hr.**	**Daily**	**Hr.**	**Daily**	**Bare Costs**	**Incl. O&P**
1 Labor Foreman (outside)	$39.60	$316.80	$64.90	$519.20	$41.95	$68.53
1 Skilled Worker	48.65	389.20	79.05	632.40		
1 Laborer	37.60	300.80	61.65	493.20		
24 L.H., Daily Totals		$1006.80		$1644.80	$41.95	$68.53

Crew B-20A	**Hr.**	**Daily**	**Hr.**	**Daily**	**Bare Costs**	**Incl. O&P**
1 Labor Foreman (outside)	$39.60	$316.80	$64.90	$519.20	$45.71	$72.84
1 Laborer	37.60	300.80	61.65	493.20		
1 Plumber	58.70	469.60	91.55	732.40		
1 Plumber Apprentice	46.95	375.60	73.25	586.00		
32 L.H., Daily Totals		$1462.80		$2330.80	$45.71	$72.84

Crew No.	Bare Costs		Incl. Subs O&P		Cost Per Labor-Hour	
Crew B-21	**Hr.**	**Daily**	**Hr.**	**Daily**	**Bare Costs**	**Incl. O&P**
1 Labor Foreman (outside)	$39.60	$316.80	$64.90	$519.20	$43.34	$70.46
1 Skilled Worker	48.65	389.20	79.05	632.40		
1 Laborer	37.60	300.80	61.65	493.20		
.5 Equip. Oper. (crane)	51.70	206.80	82.05	328.20		
.5 S.P. Crane, 4x4, 5 Ton		137.70		151.47	4.92	5.41
28 L.H., Daily Totals		$1351.30		$2124.47	$48.26	$75.87

Crew B-21A	**Hr.**	**Daily**	**Hr.**	**Daily**	**Bare Costs**	**Incl. O&P**
1 Labor Foreman (outside)	$39.60	$316.80	$64.90	$519.20	$46.91	$74.68
1 Laborer	37.60	300.80	61.65	493.20		
1 Plumber	58.70	469.60	91.55	732.40		
1 Plumber Apprentice	46.95	375.60	73.25	586.00		
1 Equip. Oper. (crane)	51.70	413.60	82.05	656.40		
1 S.P. Crane, 4x4, 12 Ton		473.40		520.74	11.84	13.02
40 L.H., Daily Totals		$2349.80		$3607.94	$58.74	$87.70

Crew B-21B	**Hr.**	**Daily**	**Hr.**	**Daily**	**Bare Costs**	**Incl. O&P**
1 Labor Foreman (outside)	$39.60	$316.80	$64.90	$519.20	$40.82	$66.38
3 Laborers	37.60	902.40	61.65	1479.60		
1 Equip. Oper. (crane)	51.70	413.60	82.05	656.40		
1 Hyd. Crane, 12 Ton		653.80		719.18	16.34	17.98
40 L.H., Daily Totals		$2286.60		$3374.38	$57.16	$84.36

Crew B-21C	**Hr.**	**Daily**	**Hr.**	**Daily**	**Bare Costs**	**Incl. O&P**
1 Labor Foreman (outside)	$39.60	$316.80	$64.90	$519.20	$40.99	$66.47
4 Laborers	37.60	1203.20	61.65	1972.80		
1 Equip. Oper. (crane)	51.70	413.60	82.05	656.40		
1 Equip. Oper. (oiler)	45.20	361.60	71.75	574.00		
2 Cutting Torches		22.80		25.08		
2 Sets of Gases		304.00		334.40		
1 Lattice Boom Crane, 90 Ton		1511.00		1662.10	32.82	36.10
56 L.H., Daily Totals		$4133.00		$5743.98	$73.80	$102.57

Crew B-22	**Hr.**	**Daily**	**Hr.**	**Daily**	**Bare Costs**	**Incl. O&P**
1 Labor Foreman (outside)	$39.60	$316.80	$64.90	$519.20	$43.90	$71.24
1 Skilled Worker	48.65	389.20	79.05	632.40		
1 Laborer	37.60	300.80	61.65	493.20		
.75 Equip. Oper. (crane)	51.70	310.20	82.05	492.30		
.75 S.P. Crane, 4x4, 5 Ton		206.55		227.21	6.88	7.57
30 L.H., Daily Totals		$1523.55		$2364.30	$50.78	$78.81

Crew B-22A	**Hr.**	**Daily**	**Hr.**	**Daily**	**Bare Costs**	**Incl. O&P**
1 Labor Foreman (outside)	$39.60	$316.80	$64.90	$519.20	$43.03	$69.86
1 Skilled Worker	48.65	389.20	79.05	632.40		
2 Laborers	37.60	601.60	61.65	986.40		
1 Equipment Operator, Crane	51.70	413.60	82.05	656.40		
1 S.P. Crane, 4x4, 5 Ton		275.40		302.94		
1 Butt Fusion Mach., 4"-12" diam.		382.10		420.31	16.44	18.08
40 L.H., Daily Totals		$2378.70		$3517.65	$59.47	$87.94

Crew B-22B	**Hr.**	**Daily**	**Hr.**	**Daily**	**Bare Costs**	**Incl. O&P**
1 Labor Foreman (outside)	$39.60	$316.80	$64.90	$519.20	$43.03	$69.86
1 Skilled Worker	48.65	389.20	79.05	632.40		
2 Laborers	37.60	601.60	61.65	986.40		
1 Equip. Oper. (crane)	51.70	413.60	82.05	656.40		
1 S.P. Crane, 4x4, 5 Ton		275.40		302.94		
1 Butt Fusion Mach., 8"-24" diam.		829.70		912.67	27.63	30.39
40 L.H., Daily Totals		$2826.30		$4010.01	$70.66	$100.25

Crew No.	Bare Costs		Incl. Subs O&P		Cost Per Labor-Hour	
Crew B-22C	**Hr.**	**Daily**	**Hr.**	**Daily**	**Bare Costs**	**Incl. O&P**
1 Skilled Worker	$48.65	$389.20	$79.05	$632.40	$43.13	$70.35
1 Laborer	37.60	300.80	61.65	493.20		
1 Butt Fusion Mach., 2"-8" diam.		121.05		133.16	7.57	8.32
16 L.H., Daily Totals		$811.05		$1258.76	$50.69	$78.67

Crew B-23	Hr.	Daily	Hr.	Daily	Bare Costs	Incl. O&P
1 Labor Foreman (outside)	$39.60	$316.80	$64.90	$519.20	$38.00	$62.30
4 Laborers	37.60	1203.20	61.65	1972.80		
1 Drill Rig, Truck-Mounted		2542.00		2796.20		
1 Flatbed Truck, Gas, 3 Ton		303.20		333.52	71.13	78.24
40 L.H., Daily Totals		$4365.20		$5621.72	$109.13	$140.54

Crew B-23A	Hr.	Daily	Hr.	Daily	Bare Costs	Incl. O&P
1 Labor Foreman (outside)	$39.60	$316.80	$64.90	$519.20	$42.60	$68.95
1 Laborer	37.60	300.80	61.65	493.20		
1 Equip. Oper. (medium)	50.60	404.80	80.30	642.40		
1 Drill Rig, Truck-Mounted		2542.00		2796.20		
1 Pickup Truck, 3/4 Ton		144.20		158.62	111.93	123.12
24 L.H., Daily Totals		$3708.60		$4609.62	$154.53	$192.07

Crew B-23B	Hr.	Daily	Hr.	Daily	Bare Costs	Incl. O&P
1 Labor Foreman (outside)	$39.60	$316.80	$64.90	$519.20	$42.60	$68.95
1 Laborer	37.60	300.80	61.65	493.20		
1 Equip. Oper. (medium)	50.60	404.80	80.30	642.40		
1 Drill Rig, Truck-Mounted		2542.00		2796.20		
1 Pickup Truck, 3/4 Ton		144.20		158.62		
1 Centr. Water Pump, 6"		346.60		381.26	126.37	139.00
24 L.H., Daily Totals		$4055.20		$4990.88	$168.97	$207.95

Crew B-24	Hr.	Daily	Hr.	Daily	Bare Costs	Incl. O&P
1 Cement Finisher	$45.00	$360.00	$71.15	$569.20	$43.18	$69.92
1 Laborer	37.60	300.80	61.65	493.20		
1 Carpenter	46.95	375.60	76.95	615.60		
24 L.H., Daily Totals		$1036.40		$1678.00	$43.18	$69.92

Crew B-25	Hr.	Daily	Hr.	Daily	Bare Costs	Incl. O&P
1 Labor Foreman (outside)	$39.60	$316.80	$64.90	$519.20	$41.33	$67.03
7 Laborers	37.60	2105.60	61.65	3452.40		
3 Equip. Oper. (medium)	50.60	1214.40	80.30	1927.20		
1 Asphalt Paver, 130 H.P.		2132.00		2345.20		
1 Tandem Roller, 10 Ton		236.80		260.48		
1 Roller, Pneum. Whl., 12 Ton		344.40		378.84	30.83	33.91
88 L.H., Daily Totals		$6350.00		$8883.32	$72.16	$100.95

Crew B-25B	Hr.	Daily	Hr.	Daily	Bare Costs	Incl. O&P
1 Labor Foreman (outside)	$39.60	$316.80	$64.90	$519.20	$42.10	$68.14
7 Laborers	37.60	2105.60	61.65	3452.40		
4 Equip. Oper. (medium)	50.60	1619.20	80.30	2569.60		
1 Asphalt Paver, 130 H.P.		2132.00		2345.20		
2 Tandem Rollers, 10 Ton		473.60		520.96		
1 Roller, Pneum. Whl., 12 Ton		344.40		378.84	30.73	33.80
96 L.H., Daily Totals		$6991.60		$9786.20	$72.83	$101.94

Crew B-25C	Hr.	Daily	Hr.	Daily	Bare Costs	Incl. O&P
1 Labor Foreman (outside)	$39.60	$316.80	$64.90	$519.20	$42.27	$68.41
3 Laborers	37.60	902.40	61.65	1479.60		
2 Equip. Oper. (medium)	50.60	809.60	80.30	1284.80		
1 Asphalt Paver, 130 H.P.		2132.00		2345.20		
1 Tandem Roller, 10 Ton		236.80		260.48	49.35	54.28
48 L.H., Daily Totals		$4397.60		$5889.28	$91.62	$122.69

Crew B-25D	Hr.	Daily	Hr.	Daily	Bare Costs	Incl. O&P
1 Labor Foreman (outside)	$39.60	$316.80	$64.90	$519.20	$42.39	$68.57
3 Laborers	37.60	902.40	61.65	1479.60		
2.125 Equip. Oper. (medium)	50.60	860.20	80.30	1365.10		
.125 Truck Driver (heavy)	40.05	40.05	64.55	64.55		
.125 Truck Tractor, 6x4, 380 H.P.		74.97		82.47		
.125 Dist. Tanker, 3000 Gallon		38.20		42.02		
1 Asphalt Paver, 130 H.P.		2132.00		2345.20		
1 Tandem Roller, 10 Ton		236.80		260.48	49.64	54.60
50 L.H., Daily Totals		$4601.43		$6158.62	$92.03	$123.17

Crew B-25E	Hr.	Daily	Hr.	Daily	Bare Costs	Incl. O&P
1 Labor Foreman (outside)	$39.60	$316.80	$64.90	$519.20	$42.50	$68.72
3 Laborers	37.60	902.40	61.65	1479.60		
2.250 Equip. Oper. (medium)	50.60	910.80	80.30	1445.40		
.25 Truck Driver (heavy)	40.05	80.10	64.55	129.10		
.25 Truck Tractor, 6x4, 380 H.P.		149.95		164.94		
.25 Dist. Tanker, 3000 Gallon		76.40		84.04		
1 Asphalt Paver, 130 H.P.		2132.00		2345.20		
1 Tandem Roller, 10 Ton		236.80		260.48	49.91	54.90
52 L.H., Daily Totals		$4805.25		$6427.97	$92.41	$123.61

Crew B-26	Hr.	Daily	Hr.	Daily	Bare Costs	Incl. O&P
1 Labor Foreman (outside)	$39.60	$316.80	$64.90	$519.20	$42.18	$68.39
6 Laborers	37.60	1804.80	61.65	2959.20		
2 Equip. Oper. (medium)	50.60	809.60	80.30	1284.80		
1 Rodman (reinf.)	52.55	420.40	85.75	686.00		
1 Cement Finisher	45.00	360.00	71.15	569.20		
1 Grader, 30,000 Lbs.		736.20		809.82		
1 Paving Mach. & Equip.		2794.00		3073.40	40.12	44.13
88 L.H., Daily Totals		$7241.80		$9901.62	$82.29	$112.52

Crew B-26A	Hr.	Daily	Hr.	Daily	Bare Costs	Incl. O&P
1 Labor Foreman (outside)	$39.60	$316.80	$64.90	$519.20	$42.18	$68.39
6 Laborers	37.60	1804.80	61.65	2959.20		
2 Equip. Oper. (medium)	50.60	809.60	80.30	1284.80		
1 Rodman (reinf.)	52.55	420.40	85.75	686.00		
1 Cement Finisher	45.00	360.00	71.15	569.20		
1 Grader, 30,000 Lbs.		736.20		809.82		
1 Paving Mach. & Equip.		2794.00		3073.40		
1 Concrete Saw		167.40		184.14	42.02	46.22
88 L.H., Daily Totals		$7409.20		$10085.76	$84.20	$114.61

Crew B-26B	Hr.	Daily	Hr.	Daily	Bare Costs	Incl. O&P
1 Labor Foreman (outside)	$39.60	$316.80	$64.90	$519.20	$42.88	$69.38
6 Laborers	37.60	1804.80	61.65	2959.20		
3 Equip. Oper. (medium)	50.60	1214.40	80.30	1927.20		
1 Rodman (reinf.)	52.55	420.40	85.75	686.00		
1 Cement Finisher	45.00	360.00	71.15	569.20		
1 Grader, 30,000 Lbs.		736.20		809.82		
1 Paving Mach. & Equip.		2794.00		3073.40		
1 Concrete Pump, 110' Boom		946.20		1040.82	46.63	51.29
96 L.H., Daily Totals		$8592.80		$11584.84	$89.51	$120.68

Crew B-26C	Hr.	Daily	Hr.	Daily	Bare Costs	Incl. O&P
1 Labor Foreman (outside)	$39.60	$316.80	$64.90	$519.20	$41.34	$67.20
6 Laborers	37.60	1804.80	61.65	2959.20		
1 Equip. Oper. (medium)	50.60	404.80	80.30	642.40		
1 Rodman (reinf.)	52.55	420.40	85.75	686.00		
1 Cement Finisher	45.00	360.00	71.15	569.20		
1 Paving Mach. & Equip.		2794.00		3073.40		
1 Concrete Saw		167.40		184.14	37.02	40.72
80 L.H., Daily Totals		$6268.20		$8633.54	$78.35	$107.92

Crew No.	Bare Costs		Incl. Subs O&P		Cost Per Labor-Hour	
Crew B-27	**Hr.**	**Daily**	**Hr.**	**Daily**	**Bare Costs**	**Incl. O&P**
1 Labor Foreman (outside)	$39.60	$316.80	$64.90	$519.20	$38.10	$62.46
3 Laborers	37.60	902.40	61.65	1479.60		
1 Berm Machine		296.80		326.48	9.28	10.20
32 L.H., Daily Totals		$1516.00		$2325.28	$47.38	$72.67

Crew B-28	**Hr.**	**Daily**	**Hr.**	**Daily**	**Bare Costs**	**Incl. O&P**
2 Carpenters	$46.95	$751.20	$76.95	$1231.20	$43.83	$71.85
1 Laborer	37.60	300.80	61.65	493.20		
24 L.H., Daily Totals		$1052.00		$1724.40	$43.83	$71.85

Crew B-29	**Hr.**	**Daily**	**Hr.**	**Daily**	**Bare Costs**	**Incl. O&P**
1 Labor Foreman (outside)	$39.60	$316.80	$64.90	$519.20	$40.99	$66.47
4 Laborers	37.60	1203.20	61.65	1972.80		
1 Equip. Oper. (crane)	51.70	413.60	82.05	656.40		
1 Equip. Oper. (oiler)	45.20	361.60	71.75	574.00		
1 Gradall, 5/8 C.Y.		882.20		970.42	15.75	17.33
56 L.H., Daily Totals		$3177.40		$4692.82	$56.74	$83.80

Crew B-30	**Hr.**	**Daily**	**Hr.**	**Daily**	**Bare Costs**	**Incl. O&P**
1 Equip. Oper. (medium)	$50.60	$404.80	$80.30	$642.40	$43.57	$69.80
2 Truck Drivers (heavy)	40.05	640.80	64.55	1032.80		
1 Hyd. Excavator, 1.5 C.Y.		1030.00		1133.00		
2 Dump Trucks, 12 C.Y., 400 H.P.		1382.00		1520.20	100.50	110.55
24 L.H., Daily Totals		$3457.60		$4328.40	$144.07	$180.35

Crew B-31	**Hr.**	**Daily**	**Hr.**	**Daily**	**Bare Costs**	**Incl. O&P**
1 Labor Foreman (outside)	$39.60	$316.80	$64.90	$519.20	$39.87	$65.36
3 Laborers	37.60	902.40	61.65	1479.60		
1 Carpenter	46.95	375.60	76.95	615.60		
1 Air Compressor, 250 cfm		201.40		221.54		
1 Sheeting Driver		5.75		6.33		
2 -50' Air Hoses, 1.5"		11.60		12.76	5.47	6.02
40 L.H., Daily Totals		$1813.55		$2855.03	$45.34	$71.38

Crew B-32	**Hr.**	**Daily**	**Hr.**	**Daily**	**Bare Costs**	**Incl. O&P**
1 Laborer	$37.60	$300.80	$61.65	$493.20	$47.35	$75.64
3 Equip. Oper. (medium)	50.60	1214.40	80.30	1927.20		
1 Grader, 30,000 Lbs.		736.20		809.82		
1 Tandem Roller, 10 Ton		236.80		260.48		
1 Dozer, 200 H.P.		1387.00		1525.70	73.75	81.13
32 L.H., Daily Totals		$3875.20		$5016.40	$121.10	$156.76

Crew B-32A	**Hr.**	**Daily**	**Hr.**	**Daily**	**Bare Costs**	**Incl. O&P**
1 Laborer	$37.60	$300.80	$61.65	$493.20	$46.27	$74.08
2 Equip. Oper. (medium)	50.60	809.60	80.30	1284.80		
1 Grader, 30,000 Lbs.		736.20		809.82		
1 Roller, Vibratory, 25 Ton		687.40		756.14	59.32	65.25
24 L.H., Daily Totals		$2534.00		$3343.96	$105.58	$139.33

Crew B-32B	**Hr.**	**Daily**	**Hr.**	**Daily**	**Bare Costs**	**Incl. O&P**
1 Laborer	$37.60	$300.80	$61.65	$493.20	$46.27	$74.08
2 Equip. Oper. (medium)	50.60	809.60	80.30	1284.80		
1 Dozer, 200 H.P.		1387.00		1525.70		
1 Roller, Vibratory, 25 Ton		687.40		756.14	86.43	95.08
24 L.H., Daily Totals		$3184.80		$4059.84	$132.70	$169.16

Crew No.	Bare Costs		Incl. Subs O&P		Cost Per Labor-Hour	
Crew B-32C	**Hr.**	**Daily**	**Hr.**	**Daily**	**Bare Costs**	**Incl. O&P**
1 Labor Foreman (outside)	$39.60	$316.80	$64.90	$519.20	$44.43	$71.52
2 Laborers	37.60	601.60	61.65	986.40		
3 Equip. Oper. (medium)	50.60	1214.40	80.30	1927.20		
1 Grader, 30,000 Lbs.		736.20		809.82		
1 Tandem Roller, 10 Ton		236.80		260.48		
1 Dozer, 200 H.P.		1387.00		1525.70	49.17	54.08
48 L.H., Daily Totals		$4492.80		$6028.80	$93.60	$125.60

Crew B-33A	**Hr.**	**Daily**	**Hr.**	**Daily**	**Bare Costs**	**Incl. O&P**
1 Equip. Oper. (medium)	$50.60	$404.80	$80.30	$642.40	$46.89	$74.97
.5 Laborer	37.60	150.40	61.65	246.60		
.25 Equip. Oper. (medium)	50.60	101.20	80.30	160.60		
1 Scraper, Towed, 7 C.Y.		114.40		125.84		
1.25 Dozers, 300 H.P.		2371.25		2608.38	177.55	195.30
14 L.H., Daily Totals		$3142.05		$3783.82	$224.43	$270.27

Crew B-33B	**Hr.**	**Daily**	**Hr.**	**Daily**	**Bare Costs**	**Incl. O&P**
1 Equip. Oper. (medium)	$50.60	$404.80	$80.30	$642.40	$46.89	$74.97
.5 Laborer	37.60	150.40	61.65	246.60		
.25 Equip. Oper. (medium)	50.60	101.20	80.30	160.60		
1 Scraper, Towed, 10 C.Y.		146.80		161.48		
1.25 Dozers, 300 H.P.		2371.25		2608.38	179.86	197.85
14 L.H., Daily Totals		$3174.45		$3819.45	$226.75	$272.82

Crew B-33C	**Hr.**	**Daily**	**Hr.**	**Daily**	**Bare Costs**	**Incl. O&P**
1 Equip. Oper. (medium)	$50.60	$404.80	$80.30	$642.40	$46.89	$74.97
.5 Laborer	37.60	150.40	61.65	246.60		
.25 Equip. Oper. (medium)	50.60	101.20	80.30	160.60		
1 Scraper, Towed, 15 C.Y.		164.80		181.28		
1.25 Dozers, 300 H.P.		2371.25		2608.38	181.15	199.26
14 L.H., Daily Totals		$3192.45		$3839.26	$228.03	$274.23

Crew B-33D	**Hr.**	**Daily**	**Hr.**	**Daily**	**Bare Costs**	**Incl. O&P**
1 Equip. Oper. (medium)	$50.60	$404.80	$80.30	$642.40	$46.89	$74.97
.5 Laborer	37.60	150.40	61.65	246.60		
.25 Equip. Oper. (medium)	50.60	101.20	80.30	160.60		
1 S.P. Scraper, 14 C.Y.		1884.00		2072.40		
.25 Dozer, 300 H.P.		474.25		521.67	168.45	185.29
14 L.H., Daily Totals		$3014.65		$3643.68	$215.33	$260.26

Crew B-33E	**Hr.**	**Daily**	**Hr.**	**Daily**	**Bare Costs**	**Incl. O&P**
1 Equip. Oper. (medium)	$50.60	$404.80	$80.30	$642.40	$46.89	$74.97
.5 Laborer	37.60	150.40	61.65	246.60		
.25 Equip. Oper. (medium)	50.60	101.20	80.30	160.60		
1 S.P. Scraper, 21 C.Y.		2688.00		2956.80		
.25 Dozer, 300 H.P.		474.25		521.67	225.88	248.46
14 L.H., Daily Totals		$3818.65		$4528.07	$272.76	$323.43

Crew B-33F	**Hr.**	**Daily**	**Hr.**	**Daily**	**Bare Costs**	**Incl. O&P**
1 Equip. Oper. (medium)	$50.60	$404.80	$80.30	$642.40	$46.89	$74.97
.5 Laborer	37.60	150.40	61.65	246.60		
.25 Equip. Oper. (medium)	50.60	101.20	80.30	160.60		
1 Elev. Scraper, 11 C.Y.		1163.00		1279.30		
.25 Dozer, 300 H.P.		474.25		521.67	116.95	128.64
14 L.H., Daily Totals		$2293.65		$2850.57	$163.83	$203.61

Crew No.	Bare Costs		Incl. Subs O&P		Cost Per Labor-Hour	

Crew B-33G	Hr.	Daily	Hr.	Daily	Bare Costs	Incl. O&P
1 Equip. Oper. (medium)	$50.60	$404.80	$80.30	$642.40	$46.89	$74.97
.5 Laborer	37.60	150.40	61.65	246.60		
.25 Equip. Oper. (medium)	50.60	101.20	80.30	160.60		
1 Elev. Scraper, 22 C.Y.		2521.00		2773.10		
.25 Dozer, 300 H.P.		474.25		521.67	213.95	235.34
14 L.H., Daily Totals		$3651.65		$4344.38	$260.83	$310.31

Crew B-33K	Hr.	Daily	Hr.	Daily	Bare Costs	Incl. O&P
1 Equipment Operator (med.)	$50.60	$404.80	$80.30	$642.40	$46.89	$74.97
.25 Equipment Operator (med.)	50.60	101.20	80.30	160.60		
.5 Laborer	37.60	150.40	61.65	246.60		
1 S.P. Scraper, 31 C.Y.		3693.00		4062.30		
.25 Dozer, 410 H.P.		601.25		661.38	306.73	337.41
14 L.H., Daily Totals		$4950.65		$5773.27	$353.62	$412.38

Crew B-34A	Hr.	Daily	Hr.	Daily	Bare Costs	Incl. O&P
1 Truck Driver (heavy)	$40.05	$320.40	$64.55	$516.40	$40.05	$64.55
1 Dump Truck, 8 C.Y., 220 H.P.		411.20		452.32	51.40	56.54
8 L.H., Daily Totals		$731.60		$968.72	$91.45	$121.09

Crew B-34B	Hr.	Daily	Hr.	Daily	Bare Costs	Incl. O&P
1 Truck Driver (heavy)	$40.05	$320.40	$64.55	$516.40	$40.05	$64.55
1 Dump Truck, 12 C.Y., 400 H.P.		691.00		760.10	86.38	95.01
8 L.H., Daily Totals		$1011.40		$1276.50	$126.43	$159.56

Crew B-34C	Hr.	Daily	Hr.	Daily	Bare Costs	Incl. O&P
1 Truck Driver (heavy)	$40.05	$320.40	$64.55	$516.40	$40.05	$64.55
1 Truck Tractor, 6x4, 380 H.P.		599.80		659.78		
1 Dump Trailer, 16.5 C.Y.		126.60		139.26	90.80	99.88
8 L.H., Daily Totals		$1046.80		$1315.44	$130.85	$164.43

Crew B-34D	Hr.	Daily	Hr.	Daily	Bare Costs	Incl. O&P
1 Truck Driver (heavy)	$40.05	$320.40	$64.55	$516.40	$40.05	$64.55
1 Truck Tractor, 6x4, 380 H.P.		599.80		659.78		
1 Dump Trailer, 20 C.Y.		141.20		155.32	92.63	101.89
8 L.H., Daily Totals		$1061.40		$1331.50	$132.68	$166.44

Crew B-34E	Hr.	Daily	Hr.	Daily	Bare Costs	Incl. O&P
1 Truck Driver (heavy)	$40.05	$320.40	$64.55	$516.40	$40.05	$64.55
1 Dump Truck, Off Hwy., 25 Ton		1351.00		1486.10	168.88	185.76
8 L.H., Daily Totals		$1671.40		$2002.50	$208.93	$250.31

Crew B-34F	Hr.	Daily	Hr.	Daily	Bare Costs	Incl. O&P
1 Truck Driver (heavy)	$40.05	$320.40	$64.55	$516.40	$40.05	$64.55
1 Dump Truck, Off Hwy., 35 Ton		1512.00		1663.20	189.00	207.90
8 L.H., Daily Totals		$1832.40		$2179.60	$229.05	$272.45

Crew B-34G	Hr.	Daily	Hr.	Daily	Bare Costs	Incl. O&P
1 Truck Driver (heavy)	$40.05	$320.40	$64.55	$516.40	$40.05	$64.55
1 Dump Truck, Off Hwy., 50 Ton		1837.00		2020.70	229.63	252.59
8 L.H., Daily Totals		$2157.40		$2537.10	$269.68	$317.14

Crew B-34H	Hr.	Daily	Hr.	Daily	Bare Costs	Incl. O&P
1 Truck Driver (heavy)	$40.05	$320.40	$64.55	$516.40	$40.05	$64.55
1 Dump Truck, Off Hwy., 65 Ton		1863.00		2049.30	232.88	256.16
8 L.H., Daily Totals		$2183.40		$2565.70	$272.93	$320.71

Crew B-34I	Hr.	Daily	Hr.	Daily	Bare Costs	Incl. O&P
1 Truck Driver (heavy)	$40.05	$320.40	$64.55	$516.40	$40.05	$64.55
1 Dump Truck, 18 C.Y., 450 H.P.		869.00		955.90	108.63	119.49
8 L.H., Daily Totals		$1189.40		$1472.30	$148.68	$184.04

Crew B-34J	Hr.	Daily	Hr.	Daily	Bare Costs	Incl. O&P
1 Truck Driver (heavy)	$40.05	$320.40	$64.55	$516.40	$40.05	$64.55
1 Dump Truck, Off Hwy., 100 Ton		2925.00		3217.50	365.63	402.19
8 L.H., Daily Totals		$3245.40		$3733.90	$405.68	$466.74

Crew B-34K	Hr.	Daily	Hr.	Daily	Bare Costs	Incl. O&P
1 Truck Driver (heavy)	$40.05	$320.40	$64.55	$516.40	$40.05	$64.55
1 Truck Tractor, 6x4, 450 H.P.		727.40		800.14		
1 Lowbed Trailer, 75 Ton		221.80		243.98	118.65	130.51
8 L.H., Daily Totals		$1269.60		$1560.52	$158.70	$195.07

Crew B-34L	Hr.	Daily	Hr.	Daily	Bare Costs	Incl. O&P
1 Equip. Oper. (light)	$48.60	$388.80	$77.15	$617.20	$48.60	$77.15
1 Flatbed Truck, Gas, 1.5 Ton		245.40		269.94	30.68	33.74
8 L.H., Daily Totals		$634.20		$887.14	$79.28	$110.89

Crew B-34M	Hr.	Daily	Hr.	Daily	Bare Costs	Incl. O&P
1 Equip. Oper. (light)	$48.60	$388.80	$77.15	$617.20	$48.60	$77.15
1 Flatbed Truck, Gas, 3 Ton		303.20		333.52	37.90	41.69
8 L.H., Daily Totals		$692.00		$950.72	$86.50	$118.84

Crew B-34N	Hr.	Daily	Hr.	Daily	Bare Costs	Incl. O&P
1 Truck Driver (heavy)	$40.05	$320.40	$64.55	$516.40	$45.33	$72.42
1 Equip. Oper. (medium)	50.60	404.80	80.30	642.40		
1 Truck Tractor, 6x4, 380 H.P.		599.80		659.78		
1 Flatbed Trailer, 40 Ton		154.00		169.40	47.11	51.82
16 L.H., Daily Totals		$1479.00		$1987.98	$92.44	$124.25

Crew B-34P	Hr.	Daily	Hr.	Daily	Bare Costs	Incl. O&P
1 Pipe Fitter	$59.75	$478.00	$93.20	$745.60	$49.82	$78.85
1 Truck Driver (light)	39.10	312.80	63.05	504.40		
1 Equip. Oper. (medium)	50.60	404.80	80.30	642.40		
1 Flatbed Truck, Gas, 3 Ton		303.20		333.52		
1 Backhoe Loader, 48 H.P.		364.20		400.62	27.81	30.59
24 L.H., Daily Totals		$1863.00		$2626.54	$77.63	$109.44

Crew B-34Q	Hr.	Daily	Hr.	Daily	Bare Costs	Incl. O&P
1 Pipe Fitter	$59.75	$478.00	$93.20	$745.60	$50.18	$79.43
1 Truck Driver (light)	39.10	312.80	63.05	504.40		
1 Equip. Oper. (crane)	51.70	413.60	82.05	656.40		
1 Flatbed Trailer, 25 Ton		113.60		124.96		
1 Dump Truck, 8 C.Y., 220 H.P.		411.20		452.32		
1 Hyd. Crane, 25 Ton		736.60		810.26	52.56	57.81
24 L.H., Daily Totals		$2465.80		$3293.94	$102.74	$137.25

Crew B-34R	Hr.	Daily	Hr.	Daily	Bare Costs	Incl. O&P
1 Pipe Fitter	$59.75	$478.00	$93.20	$745.60	$50.18	$79.43
1 Truck Driver (light)	39.10	312.80	63.05	504.40		
1 Equip. Oper. (crane)	51.70	413.60	82.05	656.40		
1 Flatbed Trailer, 25 Ton		113.60		124.96		
1 Dump Truck, 8 C.Y., 220 H.P.		411.20		452.32		
1 Hyd. Crane, 25 Ton		736.60		810.26		
1 Hyd. Excavator, 1 C.Y.		812.60		893.86	86.42	95.06
24 L.H., Daily Totals		$3278.40		$4187.80	$136.60	$174.49

Crew B-34S	Bare Costs Hr.	Bare Costs Daily	Incl. Subs O&P Hr.	Incl. Subs O&P Daily	Cost Per Labor-Hour Bare Costs	Cost Per Labor-Hour Incl. O&P
2 Pipe Fitters	$59.75	$956.00	$93.20	$1491.20	$52.81	$83.25
1 Truck Driver (heavy)	40.05	320.40	64.55	516.40		
1 Equip. Oper. (crane)	51.70	413.60	82.05	656.40		
1 Flatbed Trailer, 40 Ton		154.00		169.40		
1 Truck Tractor, 6x4, 380 H.P.		599.80		659.78		
1 Hyd. Crane, 80 Ton		1625.00		1787.50		
1 Hyd. Excavator, 2 C.Y.		1175.00		1292.50	111.06	122.16
32 L.H., Daily Totals		$5243.80		$6573.18	$163.87	$205.41

Crew B-34T	Hr.	Daily	Hr.	Daily	Bare Costs	Incl. O&P
2 Pipe Fitters	$59.75	$956.00	$93.20	$1491.20	$52.81	$83.25
1 Truck Driver (heavy)	40.05	320.40	64.55	516.40		
1 Equip. Oper. (crane)	51.70	413.60	82.05	656.40		
1 Flatbed Trailer, 40 Ton		154.00		169.40		
1 Truck Tractor, 6x4, 380 H.P.		599.80		659.78		
1 Hyd. Crane, 80 Ton		1625.00		1787.50	74.34	81.77
32 L.H., Daily Totals		$4068.80		$5280.68	$127.15	$165.02

Crew B-34U	Hr.	Daily	Hr.	Daily	Bare Costs	Incl. O&P
1 Truck Driver (heavy)	$40.05	$320.40	$64.55	$516.40	$44.33	$70.85
1 Equip. Oper. (light)	48.60	388.80	77.15	617.20		
1 Truck Tractor, 220 H.P.		359.60		395.56		
1 Flatbed Trailer, 25 Ton		113.60		124.96	29.57	32.53
16 L.H., Daily Totals		$1182.40		$1654.12	$73.90	$103.38

Crew B-34V	Hr.	Daily	Hr.	Daily	Bare Costs	Incl. O&P
1 Truck Driver (heavy)	$40.05	$320.40	$64.55	$516.40	$46.78	$74.58
1 Equip. Oper. (crane)	51.70	413.60	82.05	656.40		
1 Equip. Oper. (light)	48.60	388.80	77.15	617.20		
1 Truck Tractor, 6x4, 450 H.P.		727.40		800.14		
1 Equipment Trailer, 50 Ton		168.40		185.24		
1 Pickup Truck, 4 x 4, 3/4 Ton		155.60		171.16	43.81	48.19
24 L.H., Daily Totals		$2174.20		$2946.54	$90.59	$122.77

Crew B-34W	Hr.	Daily	Hr.	Daily	Bare Costs	Incl. O&P
5 Truck Drivers (heavy)	$40.05	$1602.00	$64.55	$2582.00	$43.66	$70.05
2 Equip. Opers. (crane)	51.70	827.20	82.05	1312.80		
1 Equip. Oper. (mechanic)	51.65	413.20	81.95	655.60		
1 Laborer	37.60	300.80	61.65	493.20		
4 Truck Tractor, 6x4, 380 H.P.		2399.20		2639.12		
2 Equipment Trailer, 50 Ton		336.80		370.48		
2 Flatbed Trailer, 40 Ton		308.00		338.80		
1 Pickup Truck, 4 x 4, 3/4 Ton		155.60		171.16		
1 S.P. Crane, 4x4, 20 Ton		554.60		610.06	52.14	57.36
72 L.H., Daily Totals		$6897.40		$9173.22	$95.80	$127.41

Crew B-35	Hr.	Daily	Hr.	Daily	Bare Costs	Incl. O&P
1 Labor Foreman (outside)	$39.60	$316.80	$64.90	$519.20	$46.91	$75.16
1 Skilled Worker	48.65	389.20	79.05	632.40		
1 Welder (plumber)	58.70	469.60	91.55	732.40		
1 Laborer	37.60	300.80	61.65	493.20		
1 Equip. Oper. (crane)	51.70	413.60	82.05	656.40		
1 Equip. Oper. (oiler)	45.20	361.60	71.75	574.00		
1 Welder, Electric, 300 amp		57.70		63.47		
1 Hyd. Excavator, .75 C.Y.		654.80		720.28	14.84	16.33
48 L.H., Daily Totals		$2964.10		$4391.35	$61.75	$91.49

Crew B-35A	Hr.	Daily	Hr.	Daily	Bare Costs	Incl. O&P
1 Labor Foreman (outside)	$39.60	$316.80	$64.90	$519.20	$45.58	$73.23
2 Laborers	37.60	601.60	61.65	986.40		
1 Skilled Worker	48.65	389.20	79.05	632.40		
1 Welder (plumber)	58.70	469.60	91.55	732.40		
1 Equip. Oper. (crane)	51.70	413.60	82.05	656.40		
1 Equip. Oper. (oiler)	45.20	361.60	71.75	574.00		
1 Welder, Gas Engine, 300 amp		145.85		160.44		
1 Crawler Crane, 75 Ton		1459.00		1604.90	28.66	31.52
56 L.H., Daily Totals		$4157.25		$5866.14	$74.24	$104.75

Crew B-36	Hr.	Daily	Hr.	Daily	Bare Costs	Incl. O&P
1 Labor Foreman (outside)	$39.60	$316.80	$64.90	$519.20	$43.20	$69.76
2 Laborers	37.60	601.60	61.65	986.40		
2 Equip. Oper. (medium)	50.60	809.60	80.30	1284.80		
1 Dozer, 200 H.P.		1387.00		1525.70		
1 Aggregate Spreader		39.60		43.56		
1 Tandem Roller, 10 Ton		236.80		260.48	41.59	45.74
40 L.H., Daily Totals		$3391.40		$4620.14	$84.78	$115.50

Crew B-36A	Hr.	Daily	Hr.	Daily	Bare Costs	Incl. O&P
1 Labor Foreman (outside)	$39.60	$316.80	$64.90	$519.20	$45.31	$72.77
2 Laborers	37.60	601.60	61.65	986.40		
4 Equip. Oper. (medium)	50.60	1619.20	80.30	2569.60		
1 Dozer, 200 H.P.		1387.00		1525.70		
1 Aggregate Spreader		39.60		43.56		
1 Tandem Roller, 10 Ton		236.80		260.48		
1 Roller, Pneum. Whl., 12 Ton		344.40		378.84	35.85	39.44
56 L.H., Daily Totals		$4545.40		$6283.78	$81.17	$112.21

Crew B-36B	Hr.	Daily	Hr.	Daily	Bare Costs	Incl. O&P
1 Labor Foreman (outside)	$39.60	$316.80	$64.90	$519.20	$44.66	$71.74
2 Laborers	37.60	601.60	61.65	986.40		
4 Equip. Oper. (medium)	50.60	1619.20	80.30	2569.60		
1 Truck Driver (heavy)	40.05	320.40	64.55	516.40		
1 Grader, 30,000 Lbs.		736.20		809.82		
1 F.E. Loader, Crl, 1.5 C.Y.		629.80		692.78		
1 Dozer, 300 H.P.		1897.00		2086.70		
1 Roller, Vibratory, 25 Ton		687.40		756.14		
1 Truck Tractor, 6x4, 450 H.P.		727.40		800.14		
1 Water Tank Trailer, 5000 Gal.		141.00		155.10	75.29	82.82
64 L.H., Daily Totals		$7676.80		$9892.28	$119.95	$154.57

Crew B-36C	Hr.	Daily	Hr.	Daily	Bare Costs	Incl. O&P
1 Labor Foreman (outside)	$39.60	$316.80	$64.90	$519.20	$46.29	$74.07
3 Equip. Oper. (medium)	50.60	1214.40	80.30	1927.20		
1 Truck Driver (heavy)	40.05	320.40	64.55	516.40		
1 Grader, 30,000 Lbs.		736.20		809.82		
1 Dozer, 300 H.P.		1897.00		2086.70		
1 Roller, Vibratory, 25 Ton		687.40		756.14		
1 Truck Tractor, 6x4, 450 H.P.		727.40		800.14		
1 Water Tank Trailer, 5000 Gal.		141.00		155.10	104.72	115.20
40 L.H., Daily Totals		$6040.60		$7570.70	$151.01	$189.27

Crew B-37	Hr.	Daily	Hr.	Daily	Bare Costs	Incl. O&P
1 Labor Foreman (outside)	$39.60	$316.80	$64.90	$519.20	$39.77	$64.78
4 Laborers	37.60	1203.20	61.65	1972.80		
1 Equip. Oper. (light)	48.60	388.80	77.15	617.20		
1 Tandem Roller, 5 Ton		157.40		173.14	3.28	3.61
48 L.H., Daily Totals		$2066.20		$3282.34	$43.05	$68.38

Crew No.	Bare Costs		Incl. Subs O&P		Cost Per Labor-Hour	
Crew B-37A	**Hr.**	**Daily**	**Hr.**	**Daily**	**Bare Costs**	**Incl. O&P**
2 Laborers	$37.60	$601.60	$61.65	$986.40	$38.10	$62.12
1 Truck Driver (light)	39.10	312.80	63.05	504.40		
1 Flatbed Truck, Gas, 1.5 Ton		245.40		269.94		
1 Tar Kettle, T.M.		132.00		145.20	15.73	17.30
24 L.H., Daily Totals		$1291.80		$1905.94	$53.83	$79.41
Crew B-37B	**Hr.**	**Daily**	**Hr.**	**Daily**	**Bare Costs**	**Incl. O&P**
3 Laborers	$37.60	$902.40	$61.65	$1479.60	$37.98	$62.00
1 Truck Driver (light)	39.10	312.80	63.05	504.40		
1 Flatbed Truck, Gas, 1.5 Ton		245.40		269.94		
1 Tar Kettle, T.M.		132.00		145.20	11.79	12.97
32 L.H., Daily Totals		$1592.60		$2399.14	$49.77	$74.97
Crew B-37C	**Hr.**	**Daily**	**Hr.**	**Daily**	**Bare Costs**	**Incl. O&P**
2 Laborers	$37.60	$601.60	$61.65	$986.40	$38.35	$62.35
2 Truck Drivers (light)	39.10	625.60	63.05	1008.80		
2 Flatbed Trucks, Gas, 1.5 Ton		490.80		539.88		
1 Tar Kettle, T.M.		132.00		145.20	19.46	21.41
32 L.H., Daily Totals		$1850.00		$2680.28	$57.81	$83.76
Crew B-37D	**Hr.**	**Daily**	**Hr.**	**Daily**	**Bare Costs**	**Incl. O&P**
1 Laborer	$37.60	$300.80	$61.65	$493.20	$38.35	$62.35
1 Truck Driver (light)	39.10	312.80	63.05	504.40		
1 Pickup Truck, 3/4 Ton		144.20		158.62	9.01	9.91
16 L.H., Daily Totals		$757.80		$1156.22	$47.36	$72.26
Crew B-37E	**Hr.**	**Daily**	**Hr.**	**Daily**	**Bare Costs**	**Incl. O&P**
3 Laborers	$37.60	$902.40	$61.65	$1479.60	$41.46	$66.93
1 Equip. Oper. (light)	48.60	388.80	77.15	617.20		
1 Equip. Oper. (medium)	50.60	404.80	80.30	642.40		
2 Truck Drivers (light)	39.10	625.60	63.05	1008.80		
4 Barrels w/ Flasher		13.60		14.96		
1 Concrete Saw		167.40		184.14		
1 Rotary Hammer Drill		23.75		26.13		
1 Hammer Drill Bit		2.35		2.59		
1 Loader, Skid Steer, 30 H.P.		173.60		190.96		
1 Conc. Hammer Attach.		114.70		126.17		
1 Vibrating Plate, Gas, 18"		36.20		39.82		
2 Flatbed Trucks, Gas, 1.5 Ton		490.80		539.88	18.26	20.08
56 L.H., Daily Totals		$3344.00		$4872.64	$59.71	$87.01
Crew B-37F	**Hr.**	**Daily**	**Hr.**	**Daily**	**Bare Costs**	**Incl. O&P**
3 Laborers	$37.60	$902.40	$61.65	$1479.60	$37.98	$62.00
1 Truck Driver (light)	39.10	312.80	63.05	504.40		
4 Barrels w/ Flasher		13.60		14.96		
1 Concrete Mixer, 10 C.F.		172.40		189.64		
1 Air Compressor, 60 cfm		137.40		151.14		
1 -50' Air Hose, 3/4"		3.25		3.58		
1 Spade (Chipper)		7.60		8.36		
1 Flatbed Truck, Gas, 1.5 Ton		245.40		269.94	18.11	19.93
32 L.H., Daily Totals		$1794.85		$2621.61	$56.09	$81.93
Crew B-37G	**Hr.**	**Daily**	**Hr.**	**Daily**	**Bare Costs**	**Incl. O&P**
1 Labor Foreman (outside)	$39.60	$316.80	$64.90	$519.20	$39.77	$64.78
4 Laborers	37.60	1203.20	61.65	1972.80		
1 Equip. Oper. (light)	48.60	388.80	77.15	617.20		
1 Berm Machine		296.80		326.48		
1 Tandem Roller, 5 Ton		157.40		173.14	9.46	10.41
48 L.H., Daily Totals		$2363.00		$3608.82	$49.23	$75.18

Crew No.	Bare Costs		Incl. Subs O&P		Cost Per Labor-Hour	
Crew B-37H	**Hr.**	**Daily**	**Hr.**	**Daily**	**Bare Costs**	**Incl. O&P**
1 Labor Foreman (outside)	$39.60	$316.80	$64.90	$519.20	$39.77	$64.78
4 Laborers	37.60	1203.20	61.65	1972.80		
1 Equip. Oper. (light)	48.60	388.80	77.15	617.20		
1 Tandem Roller, 5 Ton		157.40		173.14		
1 Flatbed Trucks, Gas, 1.5 Ton		245.40		269.94		
1 Tar Kettle, T.M.		132.00		145.20	11.14	12.26
48 L.H., Daily Totals		$2443.60		$3697.48	$50.91	$77.03
Crew B-37I	**Hr.**	**Daily**	**Hr.**	**Daily**	**Bare Costs**	**Incl. O&P**
3 Laborers	$37.60	$902.40	$61.65	$1479.60	$41.46	$66.93
1 Equip. Oper. (light)	48.60	388.80	77.15	617.20		
1 Equip. Oper. (medium)	50.60	404.80	80.30	642.40		
2 Truck Drivers (light)	39.10	625.60	63.05	1008.80		
4 Barrels w/ Flasher		13.60		14.96		
1 Concrete Saw		167.40		184.14		
1 Rotary Hammer Drill		23.75		26.13		
1 Hammer Drill Bit		2.35		2.59		
1 Air Compressor, 60 cfm		137.40		151.14		
1 -50' Air Hose, 3/4"		3.25		3.58		
1 Spade (Chipper)		7.60		8.36		
1 Loader, Skid Steer, 30 H.P.		173.60		190.96		
1 Conc. Hammer Attach.		114.70		126.17		
1 Concrete Mixer, 10 C.F.		172.40		189.64		
1 Vibrating Plate, Gas, 18"		36.20		39.82		
2 Flatbed Trucks, Gas, 1.5 Ton		490.80		539.88	23.98	26.38
56 L.H., Daily Totals		$3664.65		$5225.35	$65.44	$93.31
Crew B-37J	**Hr.**	**Daily**	**Hr.**	**Daily**	**Bare Costs**	**Incl. O&P**
1 Labor Foreman (outside)	$39.60	$316.80	$64.90	$519.20	$39.77	$64.78
4 Laborers	37.60	1203.20	61.65	1972.80		
1 Equip. Oper. (light)	48.60	388.80	77.15	617.20		
1 Air Compressor, 60 cfm		137.40		151.14		
1 -50' Air Hose, 3/4"		3.25		3.58		
2 Concrete Mixer, 10 C.F.		344.80		379.28		
2 Flatbed Trucks, Gas, 1.5 Ton		490.80		539.88		
1 Shot Blaster, 20"		214.00		235.40	24.80	27.28
48 L.H., Daily Totals		$3099.05		$4418.48	$64.56	$92.05
Crew B-37K	**Hr.**	**Daily**	**Hr.**	**Daily**	**Bare Costs**	**Incl. O&P**
1 Labor Foreman (outside)	$39.60	$316.80	$64.90	$519.20	$39.77	$64.78
4 Laborers	37.60	1203.20	61.65	1972.80		
1 Equip. Oper. (light)	48.60	388.80	77.15	617.20		
1 Air Compressor, 60 cfm		137.40		151.14		
1 -50' Air Hose, 3/4"		3.25		3.58		
2 Flatbed Trucks, Gas, 1.5 Ton		490.80		539.88		
1 Shot Blaster, 20"		214.00		235.40	17.61	19.37
48 L.H., Daily Totals		$2754.25		$4039.20	$57.38	$84.15
Crew B-38	**Hr.**	**Daily**	**Hr.**	**Daily**	**Bare Costs**	**Incl. O&P**
1 Labor Foreman (outside)	$39.60	$316.80	$64.90	$519.20	$42.80	$69.13
2 Laborers	37.60	601.60	61.65	986.40		
1 Equip. Oper. (light)	48.60	388.80	77.15	617.20		
1 Equip. Oper. (medium)	50.60	404.80	80.30	642.40		
1 Backhoe Loader, 48 H.P.		364.20		400.62		
1 Hyd. Hammer, (1200 lb.)		180.40		198.44		
1 F.E. Loader, W.M., 4 C.Y.		669.20		736.12		
1 Pvmt. Rem. Bucket		61.20		67.32	31.88	35.06
40 L.H., Daily Totals		$2987.00		$4167.70	$74.67	$104.19

Crew No.	Bare Costs		Incl. Subs O&P		Cost Per Labor-Hour	
Crew B-39	**Hr.**	**Daily**	**Hr.**	**Daily**	**Bare Costs**	**Incl. O&P**
1 Labor Foreman (outside)	$39.60	$316.80	$64.90	$519.20	$39.77	$64.78
4 Laborers	37.60	1203.20	61.65	1972.80		
1 Equip. Oper. (light)	48.60	388.80	77.15	617.20		
1 Air Compressor, 250 cfm		201.40		221.54		
2 Breakers, Pavement, 60 lb.		19.60		21.56		
2 -50' Air Hoses, 1.5"		11.60		12.76	4.85	5.33
48 L.H., Daily Totals		$2141.40		$3365.06	$44.61	$70.11

Crew B-40	Hr.	Daily	Hr.	Daily	Bare Costs	Incl. O&P
1 Pile Driver Foreman (outside)	$48.10	$384.80	$78.55	$628.40	$47.64	$76.95
4 Pile Drivers	46.10	1475.20	75.30	2409.60		
2 Equip. Oper. (crane)	51.70	827.20	82.05	1312.80		
1 Equip. Oper. (oiler)	45.20	361.60	71.75	574.00		
1 Crawler Crane, 40 Ton		1156.00		1271.60		
1 Vibratory Hammer & Gen.		2587.00		2845.70	58.48	64.33
64 L.H., Daily Totals		$6791.80		$9042.10	$106.12	$141.28

Crew B-40B	Hr.	Daily	Hr.	Daily	Bare Costs	Incl. O&P
1 Labor Foreman (outside)	$39.60	$316.80	$64.90	$519.20	$41.55	$67.28
3 Laborers	37.60	902.40	61.65	1479.60		
1 Equip. Oper. (crane)	51.70	413.60	82.05	656.40		
1 Equip. Oper. (oiler)	45.20	361.60	71.75	574.00		
1 Lattice Boom Crane, 40 Ton		1164.00		1280.40	24.25	26.68
48 L.H., Daily Totals		$3158.40		$4509.60	$65.80	$93.95

Crew B-41	Hr.	Daily	Hr.	Daily	Bare Costs	Incl. O&P
1 Labor Foreman (outside)	$39.60	$316.80	$64.90	$519.20	$38.95	$63.63
4 Laborers	37.60	1203.20	61.65	1972.80		
.25 Equip. Oper. (crane)	51.70	103.40	82.05	164.10		
.25 Equip. Oper. (oiler)	45.20	90.40	71.75	143.50		
.25 Crawler Crane, 40 Ton		289.00		317.90	6.57	7.22
44 L.H., Daily Totals		$2002.80		$3117.50	$45.52	$70.85

Crew B-42	Hr.	Daily	Hr.	Daily	Bare Costs	Incl. O&P
1 Labor Foreman (outside)	$39.60	$316.80	$64.90	$519.20	$42.44	$70.06
4 Laborers	37.60	1203.20	61.65	1972.80		
1 Equip. Oper. (crane)	51.70	413.60	82.05	656.40		
1 Equip. Oper. (oiler)	45.20	361.60	71.75	574.00		
1 Welder	52.65	421.20	95.15	761.20		
1 Hyd. Crane, 25 Ton		736.60		810.26		
1 Welder, Gas Engine, 300 amp		145.85		160.44		
1 Horz. Boring Csg. Mch.		474.80		522.28	21.21	23.33
64 L.H., Daily Totals		$4073.65		$5976.57	$63.65	$93.38

Crew B-43	Hr.	Daily	Hr.	Daily	Bare Costs	Incl. O&P
1 Labor Foreman (outside)	$39.60	$316.80	$64.90	$519.20	$41.55	$67.28
3 Laborers	37.60	902.40	61.65	1479.60		
1 Equip. Oper. (crane)	51.70	413.60	82.05	656.40		
1 Equip. Oper. (oiler)	45.20	361.60	71.75	574.00		
1 Drill Rig, Truck-Mounted		2542.00		2796.20	52.96	58.25
48 L.H., Daily Totals		$4536.40		$6025.40	$94.51	$125.53

Crew B-44	Hr.	Daily	Hr.	Daily	Bare Costs	Incl. O&P
1 Pile Driver Foreman (outside)	$48.10	$384.80	$78.55	$628.40	$46.69	$75.69
4 Pile Drivers	46.10	1475.20	75.30	2409.60		
2 Equip. Oper. (crane)	51.70	827.20	82.05	1312.80		
1 Laborer	37.60	300.80	61.65	493.20		
1 Crawler Crane, 40 Ton		1156.00		1271.60		
1 Lead, 60' High		74.80		82.28		
1 Hammer, Diesel, 15K ft.-lbs.		586.00		644.60	28.39	31.23
64 L.H., Daily Totals		$4804.80		$6842.48	$75.08	$106.91

Crew B-45	Hr.	Daily	Hr.	Daily	Bare Costs	Incl. O&P
1 Equip. Oper. (medium)	$50.60	$404.80	$80.30	$642.40	$45.33	$72.42
1 Truck Driver (heavy)	40.05	320.40	64.55	516.40		
1 Dist. Tanker, 3000 Gallon		305.60		336.16		
1 Truck Tractor, 6x4, 380 H.P.		599.80		659.78	56.59	62.25
16 L.H., Daily Totals		$1630.60		$2154.74	$101.91	$134.67

Crew B-46	Hr.	Daily	Hr.	Daily	Bare Costs	Incl. O&P
1 Pile Driver Foreman (outside)	$48.10	$384.80	$78.55	$628.40	$42.18	$69.02
2 Pile Drivers	46.10	737.60	75.30	1204.80		
3 Laborers	37.60	902.40	61.65	1479.60		
1 Chain Saw, Gas, 36" Long		45.00		49.50	.94	1.03
48 L.H., Daily Totals		$2069.80		$3362.30	$43.12	$70.05

Crew B-47	Hr.	Daily	Hr.	Daily	Bare Costs	Incl. O&P
1 Blast Foreman (outside)	$39.60	$316.80	$64.90	$519.20	$41.93	$67.90
1 Driller	37.60	300.80	61.65	493.20		
1 Equip. Oper. (light)	48.60	388.80	77.15	617.20		
1 Air Track Drill, 4"		1028.00		1130.80		
1 Air Compressor, 600 cfm		550.60		605.66		
2 -50' Air Hoses, 3"		29.80		32.78	67.02	73.72
24 L.H., Daily Totals		$2614.80		$3398.84	$108.95	$141.62

Crew B-47A	Hr.	Daily	Hr.	Daily	Bare Costs	Incl. O&P
1 Drilling Foreman (outside)	$39.60	$316.80	$64.90	$519.20	$45.50	$72.90
1 Equip. Oper. (heavy)	51.70	413.60	82.05	656.40		
1 Equip. Oper. (oiler)	45.20	361.60	71.75	574.00		
1 Air Track Drill, 5"		1243.00		1367.30	51.79	56.97
24 L.H., Daily Totals		$2335.00		$3116.90	$97.29	$129.87

Crew B-47C	Hr.	Daily	Hr.	Daily	Bare Costs	Incl. O&P
1 Laborer	$37.60	$300.80	$61.65	$493.20	$43.10	$69.40
1 Equip. Oper. (light)	48.60	388.80	77.15	617.20		
1 Air Compressor, 750 cfm		557.20		612.92		
2 -50' Air Hoses, 3"		29.80		32.78		
1 Air Track Drill, 4"		1028.00		1130.80	100.94	111.03
16 L.H., Daily Totals		$2304.60		$2886.90	$144.04	$180.43

Crew B-47E	Hr.	Daily	Hr.	Daily	Bare Costs	Incl. O&P
1 Labor Foreman (outside)	$39.60	$316.80	$64.90	$519.20	$38.10	$62.46
3 Laborers	37.60	902.40	61.65	1479.60		
1 Flatbed Truck, Gas, 3 Ton		303.20		333.52	9.47	10.42
32 L.H., Daily Totals		$1522.40		$2332.32	$47.58	$72.89

Crew B-47G	Hr.	Daily	Hr.	Daily	Bare Costs	Incl. O&P
1 Labor Foreman (outside)	$39.60	$316.80	$64.90	$519.20	$40.85	$66.34
2 Laborers	37.60	601.60	61.65	986.40		
1 Equip. Oper. (light)	48.60	388.80	77.15	617.20		
1 Air Track Drill, 4"		1028.00		1130.80		
1 Air Compressor, 600 cfm		550.60		605.66		
2 -50' Air Hoses, 3"		29.80		32.78		
1 Gunite Pump Rig		371.80		408.98	61.88	68.07
32 L.H., Daily Totals		$3287.40		$4301.02	$102.73	$134.41

Crew B-47H	Hr.	Daily	Hr.	Daily	Bare Costs	Incl. O&P
1 Skilled Worker Foreman (out)	$50.65	$405.20	$82.30	$658.40	$49.15	$79.86
3 Skilled Workers	48.65	1167.60	79.05	1897.20		
1 Flatbed Truck, Gas, 3 Ton		303.20		333.52	9.47	10.42
32 L.H., Daily Totals		$1876.00		$2889.12	$58.63	$90.28

Crews

Crew B-48	Hr.	Daily	Hr.	Daily	Bare Costs	Incl. O&P
1 Labor Foreman (outside)	$39.60	$316.80	$64.90	$519.20	$42.56	$68.69
3 Laborers	37.60	902.40	61.65	1479.60		
1 Equip. Oper. (crane)	51.70	413.60	82.05	656.40		
1 Equip. Oper. (oiler)	45.20	361.60	71.75	574.00		
1 Equip. Oper. (light)	48.60	388.80	77.15	617.20		
1 Centr. Water Pump, 6"		346.60		381.26		
1 -20' Suction Hose, 6"		11.50		12.65		
1 -50' Discharge Hose, 6"		6.10		6.71		
1 Drill Rig, Truck-Mounted		2542.00		2796.20	51.90	57.09
56 L.H., Daily Totals		$5289.40		$7043.22	$94.45	$125.77

Crew B-49	Hr.	Daily	Hr.	Daily	Bare Costs	Incl. O&P
1 Labor Foreman (outside)	$39.60	$316.80	$64.90	$519.20	$44.27	$71.38
3 Laborers	37.60	902.40	61.65	1479.60		
2 Equip. Oper. (crane)	51.70	827.20	82.05	1312.80		
2 Equip. Oper. (oilers)	45.20	723.20	71.75	1148.00		
1 Equip. Oper. (light)	48.60	388.80	77.15	617.20		
2 Pile Drivers	46.10	737.60	75.30	1204.80		
1 Hyd. Crane, 25 Ton		736.60		810.26		
1 Centr. Water Pump, 6"		346.60		381.26		
1 -20' Suction Hose, 6"		11.50		12.65		
1 -50' Discharge Hose, 6"		6.10		6.71		
1 Drill Rig, Truck-Mounted		2542.00		2796.20	41.40	45.53
88 L.H., Daily Totals		$7538.80		$10288.68	$85.67	$116.92

Crew B-50	Hr.	Daily	Hr.	Daily	Bare Costs	Incl. O&P
2 Pile Driver Foremen (outside)	$48.10	$769.60	$78.55	$1256.80	$45.30	$73.55
6 Pile Drivers	46.10	2212.80	75.30	3614.40		
2 Equip. Oper. (crane)	51.70	827.20	82.05	1312.80		
1 Equip. Oper. (oiler)	45.20	361.60	71.75	574.00		
3 Laborers	37.60	902.40	61.65	1479.60		
1 Crawler Crane, 40 Ton		1156.00		1271.60		
1 Lead, 60' High		74.80		82.28		
1 Hammer, Diesel, 15K ft.-lbs.		586.00		644.60		
1 Air Compressor, 600 cfm		550.60		605.66		
2 -50' Air Hoses, 3"		29.80		32.78		
1 Chain Saw, Gas, 36" Long		45.00		49.50	21.81	23.99
112 L.H., Daily Totals		$7515.80		$10924.02	$67.11	$97.54

Crew B-51	Hr.	Daily	Hr.	Daily	Bare Costs	Incl. O&P
1 Labor Foreman (outside)	$39.60	$316.80	$64.90	$519.20	$38.18	$62.42
4 Laborers	37.60	1203.20	61.65	1972.80		
1 Truck Driver (light)	39.10	312.80	63.05	504.40		
1 Flatbed Truck, Gas, 1.5 Ton		245.40		269.94	5.11	5.62
48 L.H., Daily Totals		$2078.20		$3266.34	$43.30	$68.05

Crew B-52	Hr.	Daily	Hr.	Daily	Bare Costs	Incl. O&P
1 Carpenter Foreman (outside)	$48.95	$391.60	$80.25	$642.00	$43.61	$70.90
1 Carpenter	46.95	375.60	76.95	615.60		
3 Laborers	37.60	902.40	61.65	1479.60		
1 Cement Finisher	45.00	360.00	71.15	569.20		
.5 Rodman (reinf.)	52.55	210.20	85.75	343.00		
.5 Equip. Oper. (medium)	50.60	202.40	80.30	321.20		
.5 Crawler Loader, 3 C.Y.		594.00		653.40	10.61	11.67
56 L.H., Daily Totals		$3036.20		$4624.00	$54.22	$82.57

Crew B-53	Hr.	Daily	Hr.	Daily	Bare Costs	Incl. O&P
1 Equip. Oper. (light)	$48.60	$388.80	$77.15	$617.20	$48.60	$77.15
1 Trencher, Chain, 12 H.P.		68.00		74.80	8.50	9.35
8 L.H., Daily Totals		$456.80		$692.00	$57.10	$86.50

Crew B-54	Hr.	Daily	Hr.	Daily	Bare Costs	Incl. O&P
1 Equip. Oper. (light)	$48.60	$388.80	$77.15	$617.20	$48.60	$77.15
1 Trencher, Chain, 40 H.P.		335.20		368.72	41.90	46.09
8 L.H., Daily Totals		$724.00		$985.92	$90.50	$123.24

Crew B-54A	Hr.	Daily	Hr.	Daily	Bare Costs	Incl. O&P
.17 Labor Foreman (outside)	$39.60	$53.86	$64.90	$88.26	$49.00	$78.06
1 Equipment Operator (med.)	50.60	404.80	80.30	642.40		
1 Wheel Trencher, 67 H.P.		1195.00		1314.50	127.67	140.44
9.36 L.H., Daily Totals		$1653.66		$2045.16	$176.67	$218.50

Crew B-54B	Hr.	Daily	Hr.	Daily	Bare Costs	Incl. O&P
.25 Labor Foreman (outside)	$39.60	$79.20	$64.90	$129.80	$48.40	$77.22
1 Equipment Operator (med.)	50.60	404.80	80.30	642.40		
1 Wheel Trencher, 150 H.P.		1890.00		2079.00	189.00	207.90
10 L.H., Daily Totals		$2374.00		$2851.20	$237.40	$285.12

Crew B-54D	Hr.	Daily	Hr.	Daily	Bare Costs	Incl. O&P
1 Laborer	$37.60	$300.80	$61.65	$493.20	$44.10	$70.97
1 Equipment Operator (med.)	50.60	404.80	80.30	642.40		
1 Rock Trencher, 6" Width		372.00		409.20	23.25	25.57
16 L.H., Daily Totals		$1077.60		$1544.80	$67.35	$96.55

Crew B-54E	Hr.	Daily	Hr.	Daily	Bare Costs	Incl. O&P
1 Laborer	$37.60	$300.80	$61.65	$493.20	$44.10	$70.97
1 Equipment Operator (med.)	50.60	404.80	80.30	642.40		
1 Rock Trencher, 18" Width		2575.00		2832.50	160.94	177.03
16 L.H., Daily Totals		$3280.60		$3968.10	$205.04	$248.01

Crew B-55	Hr.	Daily	Hr.	Daily	Bare Costs	Incl. O&P
2 Laborers	$37.60	$601.60	$61.65	$986.40	$38.10	$62.12
1 Truck Driver (light)	39.10	312.80	63.05	504.40		
1 Truck-Mounted Earth Auger		779.20		857.12		
1 Flatbed Truck, Gas, 3 Ton		303.20		333.52	45.10	49.61
24 L.H., Daily Totals		$1996.80		$2681.44	$83.20	$111.73

Crew B-56	Hr.	Daily	Hr.	Daily	Bare Costs	Incl. O&P
1 Laborer	$37.60	$300.80	$61.65	$493.20	$43.10	$69.40
1 Equip. Oper. (light)	48.60	388.80	77.15	617.20		
1 Air Track Drill, 4"		1028.00		1130.80		
1 Air Compressor, 600 cfm		550.60		605.66		
1 -50' Air Hose, 3"		14.90		16.39	99.59	109.55
16 L.H., Daily Totals		$2283.10		$2863.25	$142.69	$178.95

Crew B-57	Hr.	Daily	Hr.	Daily	Bare Costs	Incl. O&P
1 Labor Foreman (outside)	$39.60	$316.80	$64.90	$519.20	$43.38	$69.86
2 Laborers	37.60	601.60	61.65	986.40		
1 Equip. Oper. (crane)	51.70	413.60	82.05	656.40		
1 Equip. Oper. (light)	48.60	388.80	77.15	617.20		
1 Equip. Oper. (oiler)	45.20	361.60	71.75	574.00		
1 Crawler Crane, 25 Ton		1150.00		1265.00		
1 Clamshell Bucket, 1 C.Y.		47.80		52.58		
1 Centr. Water Pump, 6"		346.60		381.26		
1 -20' Suction Hose, 6"		11.50		12.65		
20 -50' Discharge Hoses, 6"		122.00		134.20	34.96	38.45
48 L.H., Daily Totals		$3760.30		$5198.89	$78.34	$108.31

Crew B-58	Hr.	Daily	Hr.	Daily	Bare Costs	Incl. O&P
2 Laborers	$37.60	$601.60	$61.65	$986.40	$41.27	$66.82
1 Equip. Oper. (light)	48.60	388.80	77.15	617.20		
1 Backhoe Loader, 48 H.P.		364.20		400.62		
1 Small Helicopter, w/ Pilot		2790.00		3069.00	131.43	144.57
24 L.H., Daily Totals		$4144.60		$5073.22	$172.69	$211.38

Crew B-59	Hr.	Daily	Hr.	Daily	Bare Costs	Incl. O&P
1 Truck Driver (heavy)	$40.05	$320.40	$64.55	$516.40	$40.05	$64.55
1 Truck Tractor, 220 H.P.		359.60		395.56		
1 Water Tank Trailer, 5000 Gal.		141.00		155.10	62.58	68.83
8 L.H., Daily Totals		$821.00		$1067.06	$102.63	$133.38

Crew B-59A	Hr.	Daily	Hr.	Daily	Bare Costs	Incl. O&P
2 Laborers	$37.60	$601.60	$61.65	$986.40	$38.42	$62.62
1 Truck Driver (heavy)	40.05	320.40	64.55	516.40		
1 Water Tank Trailer, 5000 Gal.		141.00		155.10		
1 Truck Tractor, 220 H.P.		359.60		395.56	20.86	22.94
24 L.H., Daily Totals		$1422.60		$2053.46	$59.27	$85.56

Crew B-60	Hr.	Daily	Hr.	Daily	Bare Costs	Incl. O&P
1 Labor Foreman (outside)	$39.60	$316.80	$64.90	$519.20	$44.13	$70.90
2 Laborers	37.60	601.60	61.65	986.40		
1 Equip. Oper. (crane)	51.70	413.60	82.05	656.40		
2 Equip. Oper. (light)	48.60	777.60	77.15	1234.40		
1 Equip. Oper. (oiler)	45.20	361.60	71.75	574.00		
1 Crawler Crane, 40 Ton		1156.00		1271.60		
1 Lead, 60' High		74.80		82.28		
1 Hammer, Diesel, 15K ft.-lbs.		586.00		644.60		
1 Backhoe Loader, 48 H.P.		364.20		400.62	38.95	42.84
56 L.H., Daily Totals		$4652.20		$6369.50	$83.08	$113.74

Crew B-61	Hr.	Daily	Hr.	Daily	Bare Costs	Incl. O&P
1 Labor Foreman (outside)	$39.60	$316.80	$64.90	$519.20	$40.20	$65.40
3 Laborers	37.60	902.40	61.65	1479.60		
1 Equip. Oper. (light)	48.60	388.80	77.15	617.20		
1 Cement Mixer, 2 C.Y.		194.20		213.62		
1 Air Compressor, 160 cfm		156.40		172.04	8.77	9.64
40 L.H., Daily Totals		$1958.60		$3001.66	$48.97	$75.04

Crew B-62	Hr.	Daily	Hr.	Daily	Bare Costs	Incl. O&P
2 Laborers	$37.60	$601.60	$61.65	$986.40	$41.27	$66.82
1 Equip. Oper. (light)	48.60	388.80	77.15	617.20		
1 Loader, Skid Steer, 30 H.P.		173.60		190.96	7.23	7.96
24 L.H., Daily Totals		$1164.00		$1794.56	$48.50	$74.77

Crew B-62A	Hr.	Daily	Hr.	Daily	Bare Costs	Incl. O&P
2 Laborers	$37.60	$601.60	$61.65	$986.40	$41.27	$66.82
1 Equip. Oper. (light)	48.60	388.80	77.15	617.20		
1 Loader, Skid Steer, 30 H.P.		173.60		190.96		
1 Trencher Attachment		58.30		64.13	9.66	10.63
24 L.H., Daily Totals		$1222.30		$1858.69	$50.93	$77.45

Crew B-63	Hr.	Daily	Hr.	Daily	Bare Costs	Incl. O&P
4 Laborers	$37.60	$1203.20	$61.65	$1972.80	$39.80	$64.75
1 Equip. Oper. (light)	48.60	388.80	77.15	617.20		
1 Loader, Skid Steer, 30 H.P.		173.60		190.96	4.34	4.77
40 L.H., Daily Totals		$1765.60		$2780.96	$44.14	$69.52

Crew B-63B	Hr.	Daily	Hr.	Daily	Bare Costs	Incl. O&P
1 Labor Foreman (inside)	$38.10	$304.80	$62.45	$499.60	$40.48	$65.72
2 Laborers	37.60	601.60	61.65	986.40		
1 Equip. Oper. (light)	48.60	388.80	77.15	617.20		
1 Loader, Skid Steer, 78 H.P.		318.60		350.46	9.96	10.95
32 L.H., Daily Totals		$1613.80		$2453.66	$50.43	$76.68

Crew B-64	Hr.	Daily	Hr.	Daily	Bare Costs	Incl. O&P
1 Laborer	$37.60	$300.80	$61.65	$493.20	$38.35	$62.35
1 Truck Driver (light)	39.10	312.80	63.05	504.40		
1 Power Mulcher (small)		151.80		166.98		
1 Flatbed Truck, Gas, 1.5 Ton		245.40		269.94	24.82	27.31
16 L.H., Daily Totals		$1010.80		$1434.52	$63.17	$89.66

Crew B-65	Hr.	Daily	Hr.	Daily	Bare Costs	Incl. O&P
1 Laborer	$37.60	$300.80	$61.65	$493.20	$38.35	$62.35
1 Truck Driver (light)	39.10	312.80	63.05	504.40		
1 Power Mulcher (Large)		319.40		351.34		
1 Flatbed Truck, Gas, 1.5 Ton		245.40		269.94	35.30	38.83
16 L.H., Daily Totals		$1178.40		$1618.88	$73.65	$101.18

Crew B-66	Hr.	Daily	Hr.	Daily	Bare Costs	Incl. O&P
1 Equip. Oper. (light)	$48.60	$388.80	$77.15	$617.20	$48.60	$77.15
1 Loader-Backhoe, 40 H.P.		263.20		289.52	32.90	36.19
8 L.H., Daily Totals		$652.00		$906.72	$81.50	$113.34

Crew B-67	Hr.	Daily	Hr.	Daily	Bare Costs	Incl. O&P
1 Millwright	$49.15	$393.20	$77.25	$618.00	$48.88	$77.20
1 Equip. Oper. (light)	48.60	388.80	77.15	617.20		
1 Forklift, R/T, 4,000 Lb.		311.20		342.32	19.45	21.40
16 L.H., Daily Totals		$1093.20		$1577.52	$68.33	$98.59

Crew B-67B	Hr.	Daily	Hr.	Daily	Bare Costs	Incl. O&P
1 Millwright Foreman (inside)	$49.65	$397.20	$78.05	$624.40	$49.40	$77.65
1 Millwright	49.15	393.20	77.25	618.00		
16 L.H., Daily Totals		$790.40		$1242.40	$49.40	$77.65

Crew B-68	Hr.	Daily	Hr.	Daily	Bare Costs	Incl. O&P
2 Millwrights	$49.15	$786.40	$77.25	$1236.00	$48.97	$77.22
1 Equip. Oper. (light)	48.60	388.80	77.15	617.20		
1 Forklift, R/T, 4,000 Lb.		311.20		342.32	12.97	14.26
24 L.H., Daily Totals		$1486.40		$2195.52	$61.93	$91.48

Crew B-68A	Hr.	Daily	Hr.	Daily	Bare Costs	Incl. O&P
1 Millwright Foreman (inside)	$49.65	$397.20	$78.05	$624.40	$49.32	$77.52
2 Millwrights	49.15	786.40	77.25	1236.00		
1 Forklift, 8,000 Lb.		185.20		203.72	7.72	8.49
24 L.H., Daily Totals		$1368.80		$2064.12	$57.03	$86.00

Crew B-68B	Hr.	Daily	Hr.	Daily	Bare Costs	Incl. O&P
1 Millwright Foreman (inside)	$49.65	$397.20	$78.05	$624.40	$53.54	$83.58
2 Millwrights	49.15	786.40	77.25	1236.00		
2 Electricians	54.70	875.20	84.70	1355.20		
2 Plumbers	58.70	939.20	91.55	1464.80		
1 Forklift, 5,000 Lb.		331.80		364.98	5.92	6.52
56 L.H., Daily Totals		$3329.80		$5045.38	$59.46	$90.10

Crew No.	Bare Costs		Incl. Subs O&P		Cost Per Labor-Hour	
Crew B-68C	**Hr.**	**Daily**	**Hr.**	**Daily**	**Bare Costs**	**Incl. O&P**
1 Millwright Foreman (inside)	$49.65	$397.20	$78.05	$624.40	$53.05	$82.89
1 Millwright	49.15	393.20	77.25	618.00		
1 Electrician	54.70	437.60	84.70	677.60		
1 Plumber	58.70	469.60	91.55	732.40		
1 Forklift, 5,000 Lb.		331.80		364.98	10.37	11.41
32 L.H., Daily Totals		$2029.40		$3017.38	$63.42	$94.29
Crew B-68D	**Hr.**	**Daily**	**Hr.**	**Daily**	**Bare Costs**	**Incl. O&P**
1 Labor Foreman (inside)	$38.10	$304.80	$62.45	$499.60	$41.43	$67.08
1 Laborer	37.60	300.80	61.65	493.20		
1 Equip. Oper. (light)	48.60	388.80	77.15	617.20		
1 Forklift, 5,000 Lb.		331.80		364.98	13.82	15.21
24 L.H., Daily Totals		$1326.20		$1974.98	$55.26	$82.29
Crew B-68E	**Hr.**	**Daily**	**Hr.**	**Daily**	**Bare Costs**	**Incl. O&P**
1 Struc. Steel Foreman (inside)	$53.15	$425.20	$96.05	$768.40	$52.75	$95.33
3 Struc. Steel Workers	52.65	1263.60	95.15	2283.60		
1 Welder	52.65	421.20	95.15	761.20		
1 Forklift, 8,000 Lb.		185.20		203.72	4.63	5.09
40 L.H., Daily Totals		$2295.20		$4016.92	$57.38	$100.42
Crew B-68F	**Hr.**	**Daily**	**Hr.**	**Daily**	**Bare Costs**	**Incl. O&P**
1 Skilled Worker Foreman (out)	$50.65	$405.20	$82.30	$658.40	$49.32	$80.13
2 Skilled Workers	48.65	778.40	79.05	1264.80		
1 Forklift, 5,000 Lb.		331.80		364.98	13.82	15.21
24 L.H., Daily Totals		$1515.40		$2288.18	$63.14	$95.34
Crew B-68G	**Hr.**	**Daily**	**Hr.**	**Daily**	**Bare Costs**	**Incl. O&P**
2 Structural Steel Workers	$52.65	$842.40	$95.15	$1522.40	$52.65	$95.15
1 Forklift, 5,000 Lb.		331.80		364.98	20.74	22.81
16 L.H., Daily Totals		$1174.20		$1887.38	$73.39	$117.96
Crew B-69	**Hr.**	**Daily**	**Hr.**	**Daily**	**Bare Costs**	**Incl. O&P**
1 Labor Foreman (outside)	$39.60	$316.80	$64.90	$519.20	$41.55	$67.28
3 Laborers	37.60	902.40	61.65	1479.60		
1 Equip. Oper. (crane)	51.70	413.60	82.05	656.40		
1 Equip. Oper. (oiler)	45.20	361.60	71.75	574.00		
1 Hyd. Crane, 80 Ton		1625.00		1787.50	33.85	37.24
48 L.H., Daily Totals		$3619.40		$5016.70	$75.40	$104.51
Crew B-69A	**Hr.**	**Daily**	**Hr.**	**Daily**	**Bare Costs**	**Incl. O&P**
1 Labor Foreman (outside)	$39.60	$316.80	$64.90	$519.20	$41.33	$66.88
3 Laborers	37.60	902.40	61.65	1479.60		
1 Equip. Oper. (medium)	50.60	404.80	80.30	642.40		
1 Concrete Finisher	45.00	360.00	71.15	569.20		
1 Curb/Gutter Paver, 2-Track		997.60		1097.36	20.78	22.86
48 L.H., Daily Totals		$2981.60		$4307.76	$62.12	$89.75
Crew B-69B	**Hr.**	**Daily**	**Hr.**	**Daily**	**Bare Costs**	**Incl. O&P**
1 Labor Foreman (outside)	$39.60	$316.80	$64.90	$519.20	$41.33	$66.88
3 Laborers	37.60	902.40	61.65	1479.60		
1 Equip. Oper. (medium)	50.60	404.80	80.30	642.40		
1 Cement Finisher	45.00	360.00	71.15	569.20		
1 Curb/Gutter Paver, 4-Track		776.00		853.60	16.17	17.78
48 L.H., Daily Totals		$2760.00		$4064.00	$57.50	$84.67

Crew No.	Bare Costs		Incl. Subs O&P		Cost Per Labor-Hour	
Crew B-70	**Hr.**	**Daily**	**Hr.**	**Daily**	**Bare Costs**	**Incl. O&P**
1 Labor Foreman (outside)	$39.60	$316.80	$64.90	$519.20	$43.46	$70.11
3 Laborers	37.60	902.40	61.65	1479.60		
3 Equip. Oper. (medium)	50.60	1214.40	80.30	1927.20		
1 Grader, 30,000 Lbs.		736.20		809.82		
1 Ripper, Beam & 1 Shank		84.20		92.62		
1 Road Sweeper, S.P., 8' wide		662.40		728.64		
1 F.E. Loader, W.M., 1.5 C.Y.		378.40		416.24	33.24	36.56
56 L.H., Daily Totals		$4294.80		$5973.32	$76.69	$106.67
Crew B-71	**Hr.**	**Daily**	**Hr.**	**Daily**	**Bare Costs**	**Incl. O&P**
1 Labor Foreman (outside)	$39.60	$316.80	$64.90	$519.20	$43.46	$70.11
3 Laborers	37.60	902.40	61.65	1479.60		
3 Equip. Oper. (medium)	50.60	1214.40	80.30	1927.20		
1 Pvmt. Profiler, 750 H.P.		5805.00		6385.50		
1 Road Sweeper, S.P., 8' wide		662.40		728.64		
1 F.E. Loader, W.M., 1.5 C.Y.		378.40		416.24	122.25	134.47
56 L.H., Daily Totals		$9279.40		$11456.38	$165.70	$204.58
Crew B-72	**Hr.**	**Daily**	**Hr.**	**Daily**	**Bare Costs**	**Incl. O&P**
1 Labor Foreman (outside)	$39.60	$316.80	$64.90	$519.20	$44.35	$71.38
3 Laborers	37.60	902.40	61.65	1479.60		
4 Equip. Oper. (medium)	50.60	1619.20	80.30	2569.60		
1 Pvmt. Profiler, 750 H.P.		5805.00		6385.50		
1 Hammermill, 250 H.P.		1879.00		2066.90		
1 Windrow Loader		1227.00		1349.70		
1 Mix Paver 165 H.P.		2147.00		2361.70		
1 Roller, Pneum. Whl., 12 Ton		344.40		378.84	178.16	195.98
64 L.H., Daily Totals		$14240.80		$17111.04	$222.51	$267.36
Crew B-73	**Hr.**	**Daily**	**Hr.**	**Daily**	**Bare Costs**	**Incl. O&P**
1 Labor Foreman (outside)	$39.60	$316.80	$64.90	$519.20	$45.98	$73.71
2 Laborers	37.60	601.60	61.65	986.40		
5 Equip. Oper. (medium)	50.60	2024.00	80.30	3212.00		
1 Road Mixer, 310 H.P.		1929.00		2121.90		
1 Tandem Roller, 10 Ton		236.80		260.48		
1 Hammermill, 250 H.P.		1879.00		2066.90		
1 Grader, 30,000 Lbs.		736.20		809.82		
.5 F.E. Loader, W.M., 1.5 C.Y.		189.20		208.12		
.5 Truck Tractor, 220 H.P.		179.80		197.78		
.5 Water Tank Trailer, 5000 Gal.		70.50		77.55	81.57	89.73
64 L.H., Daily Totals		$8162.90		$10460.15	$127.55	$163.44
Crew B-74	**Hr.**	**Daily**	**Hr.**	**Daily**	**Bare Costs**	**Incl. O&P**
1 Labor Foreman (outside)	$39.60	$316.80	$64.90	$519.20	$44.96	$72.11
1 Laborer	37.60	300.80	61.65	493.20		
4 Equip. Oper. (medium)	50.60	1619.20	80.30	2569.60		
2 Truck Drivers (heavy)	40.05	640.80	64.55	1032.80		
1 Grader, 30,000 Lbs.		736.20		809.82		
1 Ripper, Beam & 1 Shank		84.20		92.62		
2 Stabilizers, 310 H.P.		3596.00		3955.60		
1 Flatbed Truck, Gas, 3 Ton		303.20		333.52		
1 Chem. Spreader, Towed		52.40		57.64		
1 Roller, Vibratory, 25 Ton		687.40		756.14		
1 Water Tank Trailer, 5000 Gal.		141.00		155.10		
1 Truck Tractor, 220 H.P.		359.60		395.56	93.13	102.44
64 L.H., Daily Totals		$8837.60		$11170.80	$138.09	$174.54

Crew No.	Bare Costs		Incl. Subs O&P		Cost Per Labor-Hour	
Crew B-75	**Hr.**	**Daily**	**Hr.**	**Daily**	**Bare Costs**	**Incl. O&P**
1 Labor Foreman (outside)	$39.60	$316.80	$64.90	$519.20	$45.66	$73.19
1 Laborer	37.60	300.80	61.65	493.20		
4 Equip. Oper. (medium)	50.60	1619.20	80.30	2569.60		
1 Truck Driver (heavy)	40.05	320.40	64.55	516.40		
1 Grader, 30,000 Lbs.		736.20		809.82		
1 Ripper, Beam & 1 Shank		84.20		92.62		
2 Stabilizers, 310 H.P.		3596.00		3955.60		
1 Dist. Tanker, 3000 Gallon		305.60		336.16		
1 Truck Tractor, 6x4, 380 H.P.		599.80		659.78		
1 Roller, Vibratory, 25 Ton		687.40		756.14	107.31	118.04
56 L.H., Daily Totals		$8566.40		$10708.52	$152.97	$191.22

Crew B-76	**Hr.**	**Daily**	**Hr.**	**Daily**	**Bare Costs**	**Incl. O&P**
1 Dock Builder Foreman (outside)	$48.10	$384.80	$78.55	$628.40	$47.47	$76.77
5 Dock Builders	46.10	1844.00	75.30	3012.00		
2 Equip. Oper. (crane)	51.70	827.20	82.05	1312.80		
1 Equip. Oper. (oiler)	45.20	361.60	71.75	574.00		
1 Crawler Crane, 50 Ton		1294.00		1423.40		
1 Barge, 400 Ton		776.00		853.60		
1 Hammer, Diesel, 15K ft.-lbs.		586.00		644.60		
1 Lead, 60' High		74.80		82.28		
1 Air Compressor, 600 cfm		550.60		605.66		
2 -50' Air Hoses, 3"		29.80		32.78	45.99	50.59
72 L.H., Daily Totals		$6728.80		$9169.52	$93.46	$127.35

Crew B-76A	**Hr.**	**Daily**	**Hr.**	**Daily**	**Bare Costs**	**Incl. O&P**
1 Labor Foreman (outside)	$39.60	$316.80	$64.90	$519.20	$40.56	$65.87
5 Laborers	37.60	1504.00	61.65	2466.00		
1 Equip. Oper. (crane)	51.70	413.60	82.05	656.40		
1 Equip. Oper. (oiler)	45.20	361.60	71.75	574.00		
1 Crawler Crane, 50 Ton		1294.00		1423.40		
1 Barge, 400 Ton		776.00		853.60	32.34	35.58
64 L.H., Daily Totals		$4666.00		$6492.60	$72.91	$101.45

Crew B-77	**Hr.**	**Daily**	**Hr.**	**Daily**	**Bare Costs**	**Incl. O&P**
1 Labor Foreman (outside)	$39.60	$316.80	$64.90	$519.20	$38.30	$62.58
3 Laborers	37.60	902.40	61.65	1479.60		
1 Truck Driver (light)	39.10	312.80	63.05	504.40		
1 Crack Cleaner, 25 H.P.		61.60		67.76		
1 Crack Filler, Trailer Mtd.		202.00		222.20		
1 Flatbed Truck, Gas, 3 Ton		303.20		333.52	14.17	15.59
40 L.H., Daily Totals		$2098.80		$3126.68	$52.47	$78.17

Crew B-78	**Hr.**	**Daily**	**Hr.**	**Daily**	**Bare Costs**	**Incl. O&P**
1 Labor Foreman (outside)	$39.60	$316.80	$64.90	$519.20	$38.18	$62.42
4 Laborers	37.60	1203.20	61.65	1972.80		
1 Truck Driver (light)	39.10	312.80	63.05	504.40		
1 Paint Striper, S.P., 40 Gallon		151.40		166.54		
1 Flatbed Truck, Gas, 3 Ton		303.20		333.52		
1 Pickup Truck, 3/4 Ton		144.20		158.62	12.48	13.72
48 L.H., Daily Totals		$2431.60		$3655.08	$50.66	$76.15

Crew B-78A	**Hr.**	**Daily**	**Hr.**	**Daily**	**Bare Costs**	**Incl. O&P**
1 Equip. Oper. (light)	$48.60	$388.80	$77.15	$617.20	$48.60	$77.15
1 Line Rem. (Metal Balls) 115 H.P.		821.60		903.76	102.70	112.97
8 L.H., Daily Totals		$1210.40		$1520.96	$151.30	$190.12

Crew No.	Bare Costs		Incl. Subs O&P		Cost Per Labor-Hour	
Crew B-78B	**Hr.**	**Daily**	**Hr.**	**Daily**	**Bare Costs**	**Incl. O&P**
2 Laborers	$37.60	$601.60	$61.65	$986.40	$38.82	$63.37
.25 Equip. Oper. (light)	48.60	97.20	77.15	154.30		
1 Pickup Truck, 3/4 Ton		144.20		158.62		
1 Line Rem.,11 H.P.,Walk Behind		65.60		72.16		
.25 Road Sweeper, S.P., 8' wide		165.60		182.16	20.86	22.94
18 L.H., Daily Totals		$1074.20		$1553.64	$59.68	$86.31

Crew B-78C	**Hr.**	**Daily**	**Hr.**	**Daily**	**Bare Costs**	**Incl. O&P**
1 Labor Foreman (outside)	$39.60	$316.80	$64.90	$519.20	$38.18	$62.42
4 Laborers	37.60	1203.20	61.65	1972.80		
1 Truck Driver (light)	39.10	312.80	63.05	504.40		
1 Paint Striper, T.M., 120 Gal.		804.60		885.06		
1 Flatbed Truck, Gas, 3 Ton		303.20		333.52		
1 Pickup Truck, 3/4 Ton		144.20		158.62	26.08	28.69
48 L.H., Daily Totals		$3084.80		$4373.60	$64.27	$91.12

Crew B-78D	**Hr.**	**Daily**	**Hr.**	**Daily**	**Bare Costs**	**Incl. O&P**
2 Labor Foremen (outside)	$39.60	$633.60	$64.90	$1038.40	$38.15	$62.44
7 Laborers	37.60	2105.60	61.65	3452.40		
1 Truck Driver (light)	39.10	312.80	63.05	504.40		
1 Paint Striper, T.M., 120 Gal.		804.60		885.06		
1 Flatbed Truck, Gas, 3 Ton		303.20		333.52		
3 Pickup Trucks, 3/4 Ton		432.60		475.86		
1 Air Compressor, 60 cfm		137.40		151.14		
1 -50' Air Hose, 3/4"		3.25		3.58		
1 Breakers, Pavement, 60 lb.		9.80		10.78	21.14	23.25
80 L.H., Daily Totals		$4742.85		$6855.14	$59.29	$85.69

Crew B-78E	**Hr.**	**Daily**	**Hr.**	**Daily**	**Bare Costs**	**Incl. O&P**
2 Labor Foremen (outside)	$39.60	$633.60	$64.90	$1038.40	$38.06	$62.31
9 Laborers	37.60	2707.20	61.65	4438.80		
1 Truck Driver (light)	39.10	312.80	63.05	504.40		
1 Paint Striper, T.M., 120 Gal.		804.60		885.06		
1 Flatbed Truck, Gas, 3 Ton		303.20		333.52		
4 Pickup Trucks, 3/4 Ton		576.80		634.48		
2 Air Compressor, 60 cfm		274.80		302.28		
2 -50' Air Hose, 3/4"		6.50		7.15		
2 Breakers, Pavement, 60 lb.		19.60		21.56	20.68	22.75
96 L.H., Daily Totals		$5639.10		$8165.65	$58.74	$85.06

Crew B-78F	**Hr.**	**Daily**	**Hr.**	**Daily**	**Bare Costs**	**Incl. O&P**
2 Labor Foremen (outside)	$39.60	$633.60	$64.90	$1038.40	$37.99	$62.21
11 Laborers	37.60	3308.80	61.65	5425.20		
1 Truck Driver (light)	39.10	312.80	63.05	504.40		
1 Paint Striper, T.M., 120 Gal.		804.60		885.06		
1 Flatbed Truck, Gas, 3 Ton		303.20		333.52		
7 Pickup Trucks, 3/4 Ton		1009.40		1110.34		
3 Air Compressor, 60 cfm		412.20		453.42		
3 -50' Air Hose, 3/4"		9.75		10.73		
3 Breakers, Pavement, 60 lb.		29.40		32.34	22.93	25.23
112 L.H., Daily Totals		$6823.75		$9793.41	$60.93	$87.44

Crew B-79	**Hr.**	**Daily**	**Hr.**	**Daily**	**Bare Costs**	**Incl. O&P**
1 Labor Foreman (outside)	$39.60	$316.80	$64.90	$519.20	$38.30	$62.58
3 Laborers	37.60	902.40	61.65	1479.60		
1 Truck Driver (light)	39.10	312.80	63.05	504.40		
1 Paint Striper, T.M., 120 Gal.		804.60		885.06		
1 Heating Kettle, 115 Gallon		81.00		89.10		
1 Flatbed Truck, Gas, 3 Ton		303.20		333.52		
2 Pickup Trucks, 3/4 Ton		288.40		317.24	36.93	40.62
40 L.H., Daily Totals		$3009.20		$4128.12	$75.23	$103.20

Crew No.	Bare Costs		Incl. Subs O&P		Cost Per Labor-Hour	
Crew B-79A	**Hr.**	**Daily**	**Hr.**	**Daily**	**Bare Costs**	**Incl. O&P**
1.5 Equip. Oper. (light)	$48.60	$583.20	$77.15	$925.80	$48.60	$77.15
.5 Line Remov. (Grinder) 115 H.P.		455.90		501.49		
1 Line Rem. (Metal Balls) 115 H.P.		821.60		903.76	106.46	117.10
12 L.H., Daily Totals		$1860.70		$2331.05	$155.06	$194.25

Crew B-79B	Hr.	Daily	Hr.	Daily	Bare Costs	Incl. O&P
1 Laborer	$37.60	$300.80	$61.65	$493.20	$37.60	$61.65
1 Set of Gases		152.00		167.20	19.00	20.90
8 L.H., Daily Totals		$452.80		$660.40	$56.60	$82.55

Crew B-79C	Hr.	Daily	Hr.	Daily	Bare Costs	Incl. O&P
1 Labor Foreman (outside)	$39.60	$316.80	$64.90	$519.20	$38.10	$62.31
5 Laborers	37.60	1504.00	61.65	2466.00		
1 Truck Driver (light)	39.10	312.80	63.05	504.40		
1 Paint Striper, T.M., 120 Gal.		804.60		885.06		
1 Heating Kettle, 115 Gallon		81.00		89.10		
1 Flatbed Truck, Gas, 3 Ton		303.20		333.52		
3 Pickup Trucks, 3/4 Ton		432.60		475.86		
1 Air Compressor, 60 cfm		137.40		151.14		
1 -50' Air Hose, 3/4"		3.25		3.58		
1 Breakers, Pavement, 60 lb.		9.80		10.78	31.64	34.80
56 L.H., Daily Totals		$3905.45		$5438.64	$69.74	$97.12

Crew B-79D	Hr.	Daily	Hr.	Daily	Bare Costs	Incl. O&P
2 Labor Foremen (outside)	$39.60	$633.60	$64.90	$1038.40	$38.29	$62.64
5 Laborers	37.60	1504.00	61.65	2466.00		
1 Truck Driver (light)	39.10	312.80	63.05	504.40		
1 Paint Striper, T.M., 120 Gal.		804.60		885.06		
1 Heating Kettle, 115 Gallon		81.00		89.10		
1 Flatbed Truck, Gas, 3 Ton		303.20		333.52		
4 Pickup Trucks, 3/4 Ton		576.80		634.48		
1 Air Compressor, 60 cfm		137.40		151.14		
1 -50' Air Hose, 3/4"		3.25		3.58		
1 Breakers, Pavement, 60 lb.		9.80		10.78	29.94	32.93
64 L.H., Daily Totals		$4366.45		$6116.45	$68.23	$95.57

Crew B-79E	Hr.	Daily	Hr.	Daily	Bare Costs	Incl. O&P
2 Labor Foremen (outside)	$39.60	$633.60	$64.90	$1038.40	$38.15	$62.44
7 Laborers	37.60	2105.60	61.65	3452.40		
1 Truck Driver (light)	39.10	312.80	63.05	504.40		
1 Paint Striper, T.M., 120 Gal.		804.60		885.06		
1 Heating Kettle, 115 Gallon		81.00		89.10		
1 Flatbed Truck, Gas, 3 Ton		303.20		333.52		
5 Pickup Trucks, 3/4 Ton		721.00		793.10		
2 Air Compressors, 60 cfm		274.80		302.28		
2 -50' Air Hoses, 3/4"		6.50		7.15		
2 Breakers, Pavement, 60 lb.		19.60		21.56	27.63	30.40
80 L.H., Daily Totals		$5262.70		$7426.97	$65.78	$92.84

Crew B-80	Hr.	Daily	Hr.	Daily	Bare Costs	Incl. O&P
1 Labor Foreman (outside)	$39.60	$316.80	$64.90	$519.20	$41.23	$66.69
1 Laborer	37.60	300.80	61.65	493.20		
1 Truck Driver (light)	39.10	312.80	63.05	504.40		
1 Equip. Oper. (light)	48.60	388.80	77.15	617.20		
1 Flatbed Truck, Gas, 3 Ton		303.20		333.52		
1 Earth Auger, Truck-Mtd.		417.20		458.92	22.51	24.76
32 L.H., Daily Totals		$2039.60		$2926.44	$63.74	$91.45

Crew B-80A	Hr.	Daily	Hr.	Daily	Bare Costs	Incl. O&P
3 Laborers	$37.60	$902.40	$61.65	$1479.60	$37.60	$61.65
1 Flatbed Truck, Gas, 3 Ton		303.20		333.52	12.63	13.90
24 L.H., Daily Totals		$1205.60		$1813.12	$50.23	$75.55

Crew B-80B	Hr.	Daily	Hr.	Daily	Bare Costs	Incl. O&P
3 Laborers	$37.60	$902.40	$61.65	$1479.60	$40.35	$65.53
1 Equip. Oper. (light)	48.60	388.80	77.15	617.20		
1 Crane, Flatbed Mounted, 3 Ton		243.80		268.18	7.62	8.38
32 L.H., Daily Totals		$1535.00		$2364.98	$47.97	$73.91

Crew B-80C	Hr.	Daily	Hr.	Daily	Bare Costs	Incl. O&P
2 Laborers	$37.60	$601.60	$61.65	$986.40	$38.10	$62.12
1 Truck Driver (light)	39.10	312.80	63.05	504.40		
1 Flatbed Truck, Gas, 1.5 Ton		245.40		269.94		
1 Manual Fence Post Auger, Gas		8.40		9.24	10.57	11.63
24 L.H., Daily Totals		$1168.20		$1769.98	$48.67	$73.75

Crew B-81	Hr.	Daily	Hr.	Daily	Bare Costs	Incl. O&P
1 Laborer	$37.60	$300.80	$61.65	$493.20	$42.75	$68.83
1 Equip. Oper. (medium)	50.60	404.80	80.30	642.40		
1 Truck Driver (heavy)	40.05	320.40	64.55	516.40		
1 Hydromulcher, T.M., 3000 Gal.		334.20		367.62		
1 Truck Tractor, 220 H.P.		359.60		395.56	28.91	31.80
24 L.H., Daily Totals		$1719.80		$2415.18	$71.66	$100.63

Crew B-81A	Hr.	Daily	Hr.	Daily	Bare Costs	Incl. O&P
1 Laborer	$37.60	$300.80	$61.65	$493.20	$38.35	$62.35
1 Truck Driver (light)	39.10	312.80	63.05	504.40		
1 Hydromulcher, T.M., 600 Gal.		127.80		140.58		
1 Flatbed Truck, Gas, 3 Ton		303.20		333.52	26.94	29.63
16 L.H., Daily Totals		$1044.60		$1471.70	$65.29	$91.98

Crew B-82	Hr.	Daily	Hr.	Daily	Bare Costs	Incl. O&P
1 Laborer	$37.60	$300.80	$61.65	$493.20	$43.10	$69.40
1 Equip. Oper. (light)	48.60	388.80	77.15	617.20		
1 Horiz. Borer, 6 H.P.		83.80		92.18	5.24	5.76
16 L.H., Daily Totals		$773.40		$1202.58	$48.34	$75.16

Crew B-82A	Hr.	Daily	Hr.	Daily	Bare Costs	Incl. O&P
2 Laborers	$37.60	$601.60	$61.65	$986.40	$43.10	$69.40
2 Equip. Opers. (light)	48.60	777.60	77.15	1234.40		
2 Dump Truck, 8 C.Y., 220 H.P.		822.40		904.64		
1 Flatbed Trailer, 25 Ton		113.60		124.96		
1 Horiz. Dir. Drill, 20k lb. Thrust		654.20		719.62		
1 Mud Trailer for HDD, 1500 Gal.		346.20		380.82		
1 Pickup Truck, 4 x 4, 3/4 Ton		155.60		171.16		
1 Flatbed trailer, 3 Ton		23.60		25.96		
1 Loader, Skid Steer, 78 H.P.		318.60		350.46	76.07	83.68
32 L.H., Daily Totals		$3813.40		$4898.42	$119.17	$153.08

Crew B-82B	Hr.	Daily	Hr.	Daily	Bare Costs	Incl. O&P
2 Laborers	$37.60	$601.60	$61.65	$986.40	$43.10	$69.40
2 Equip. Opers. (light)	48.60	777.60	77.15	1234.40		
2 Dump Truck, 8 C.Y., 220 H.P.		822.40		904.64		
1 Flatbed Trailer, 25 Ton		113.60		124.96		
1 Horiz. Dir. Drill, 30k lb. Thrust		931.20		1024.32		
1 Mud Trailer for HDD, 1500 Gal.		346.20		380.82		
1 Pickup Truck, 4 x 4, 3/4 Ton		155.60		171.16		
1 Flatbed trailer, 3 Ton		23.60		25.96		
1 Loader, Skid Steer, 78 H.P.		318.60		350.46	84.72	93.20
32 L.H., Daily Totals		$4090.40		$5203.12	$127.83	$162.60

Crew B-82C	Bare Costs Hr.	Bare Costs Daily	Incl. Subs O&P Hr.	Incl. Subs O&P Daily	Bare Costs	Incl. O&P
2 Laborers	$37.60	$601.60	$61.65	$986.40	$43.10	$69.40
2 Equip. Opers. (light)	48.60	777.60	77.15	1234.40		
2 Dump Truck, 8 C.Y., 220 H.P.		822.40		904.64		
1 Flatbed Trailer, 25 Ton		113.60		124.96		
1 Horiz. Dir. Drill, 50k lb. Thrust		1237.00		1360.70		
1 Mud Trailer for HDD, 1500 Gal.		346.20		380.82		
1 Pickup Truck, 4 x 4, 3/4 Ton		155.60		171.16		
1 Flatbed trailer, 3 Ton		23.60		25.96		
1 Loader, Skid Steer, 78 H.P.		318.60		350.46	94.28	103.71
32 L.H., Daily Totals		$4396.20		$5539.50	$137.38	$173.11

Crew B-82D	Hr.	Daily	Hr.	Daily	Bare Costs	Incl. O&P
1 Equip. Oper. (light)	$48.60	$388.80	$77.15	$617.20	$48.60	$77.15
1 Mud Trailer for HDD, 1500 Gal.		346.20		380.82	43.27	47.60
8 L.H., Daily Totals		$735.00		$998.02	$91.88	$124.75

Crew B-83	Hr.	Daily	Hr.	Daily	Bare Costs	Incl. O&P
1 Tugboat Captain	$50.60	$404.80	$80.30	$642.40	$44.10	$70.97
1 Tugboat Hand	37.60	300.80	61.65	493.20		
1 Tugboat, 250 H.P.		870.00		957.00	54.38	59.81
16 L.H., Daily Totals		$1575.60		$2092.60	$98.47	$130.79

Crew B-84	Hr.	Daily	Hr.	Daily	Bare Costs	Incl. O&P
1 Equip. Oper. (medium)	$50.60	$404.80	$80.30	$642.40	$50.60	$80.30
1 Rotary Mower/Tractor		370.80		407.88	46.35	50.98
8 L.H., Daily Totals		$775.60		$1050.28	$96.95	$131.29

Crew B-85	Hr.	Daily	Hr.	Daily	Bare Costs	Incl. O&P
3 Laborers	$37.60	$902.40	$61.65	$1479.60	$40.69	$65.96
1 Equip. Oper. (medium)	50.60	404.80	80.30	642.40		
1 Truck Driver (heavy)	40.05	320.40	64.55	516.40		
1 Aerial Lift Truck, 80'		634.40		697.84		
1 Brush Chipper, 12", 130 H.P.		391.40		430.54		
1 Pruning Saw, Rotary		6.65		7.32	25.81	28.39
40 L.H., Daily Totals		$2660.05		$3774.09	$66.50	$94.35

Crew B-86	Hr.	Daily	Hr.	Daily	Bare Costs	Incl. O&P
1 Equip. Oper. (medium)	$50.60	$404.80	$80.30	$642.40	$50.60	$80.30
1 Stump Chipper, S.P.		185.55		204.10	23.19	25.51
8 L.H., Daily Totals		$590.35		$846.51	$73.79	$105.81

Crew B-86A	Hr.	Daily	Hr.	Daily	Bare Costs	Incl. O&P
1 Equip. Oper. (medium)	$50.60	$404.80	$80.30	$642.40	$50.60	$80.30
1 Grader, 30,000 Lbs.		736.20		809.82	92.03	101.23
8 L.H., Daily Totals		$1141.00		$1452.22	$142.63	$181.53

Crew B-86B	Hr.	Daily	Hr.	Daily	Bare Costs	Incl. O&P
1 Equip. Oper. (medium)	$50.60	$404.80	$80.30	$642.40	$50.60	$80.30
1 Dozer, 200 H.P.		1387.00		1525.70	173.38	190.71
8 L.H., Daily Totals		$1791.80		$2168.10	$223.97	$271.01

Crew B-87	Hr.	Daily	Hr.	Daily	Bare Costs	Incl. O&P
1 Laborer	$37.60	$300.80	$61.65	$493.20	$48.00	$76.57
4 Equip. Oper. (medium)	50.60	1619.20	80.30	2569.60		
2 Feller Bunchers, 100 H.P.		1676.80		1844.48		
1 Log Chipper, 22" Tree		889.40		978.34		
1 Dozer, 105 H.P.		602.80		663.08		
1 Chain Saw, Gas, 36" Long		45.00		49.50	80.35	88.39
40 L.H., Daily Totals		$5134.00		$6598.20	$128.35	$164.96

Crew B-88	Hr.	Daily	Hr.	Daily	Bare Costs	Incl. O&P
1 Laborer	$37.60	$300.80	$61.65	$493.20	$48.74	$77.64
6 Equip. Oper. (medium)	50.60	2428.80	80.30	3854.40		
2 Feller Bunchers, 100 H.P.		1676.80		1844.48		
1 Log Chipper, 22" Tree		889.40		978.34		
2 Log Skidders, 50 H.P.		1822.00		2004.20		
1 Dozer, 105 H.P.		602.80		663.08		
1 Chain Saw, Gas, 36" Long		45.00		49.50	89.93	98.92
56 L.H., Daily Totals		$7765.60		$9887.20	$138.67	$176.56

Crew B-89	Hr.	Daily	Hr.	Daily	Bare Costs	Incl. O&P
1 Equip. Oper. (light)	$48.60	$388.80	$77.15	$617.20	$43.85	$70.10
1 Truck Driver (light)	39.10	312.80	63.05	504.40		
1 Flatbed Truck, Gas, 3 Ton		303.20		333.52		
1 Concrete Saw		167.40		184.14		
1 Water Tank, 65 Gal.		17.30		19.03	30.49	33.54
16 L.H., Daily Totals		$1189.50		$1658.29	$74.34	$103.64

Crew B-89A	Hr.	Daily	Hr.	Daily	Bare Costs	Incl. O&P
1 Skilled Worker	$48.65	$389.20	$79.05	$632.40	$43.13	$70.35
1 Laborer	37.60	300.80	61.65	493.20		
1 Core Drill (Large)		115.60		127.16	7.22	7.95
16 L.H., Daily Totals		$805.60		$1252.76	$50.35	$78.30

Crew B-89B	Hr.	Daily	Hr.	Daily	Bare Costs	Incl. O&P
1 Equip. Oper. (light)	$48.60	$388.80	$77.15	$617.20	$43.85	$70.10
1 Truck Driver (light)	39.10	312.80	63.05	504.40		
1 Wall Saw, Hydraulic, 10 H.P.		114.60		126.06		
1 Generator, Diesel, 100 kW		423.40		465.74		
1 Water Tank, 65 Gal.		17.30		19.03		
1 Flatbed Truck, Gas, 3 Ton		303.20		333.52	53.66	59.02
16 L.H., Daily Totals		$1560.10		$2065.95	$97.51	$129.12

Crew B-90	Hr.	Daily	Hr.	Daily	Bare Costs	Incl. O&P
1 Labor Foreman (outside)	$39.60	$316.80	$64.90	$519.20	$41.21	$66.66
3 Laborers	37.60	902.40	61.65	1479.60		
2 Equip. Oper. (light)	48.60	777.60	77.15	1234.40		
2 Truck Drivers (heavy)	40.05	640.80	64.55	1032.80		
1 Road Mixer, 310 H.P.		1929.00		2121.90		
1 Dist. Truck, 2000 Gal.		275.60		303.16	34.45	37.89
64 L.H., Daily Totals		$4842.20		$6691.06	$75.66	$104.55

Crew B-90A	Hr.	Daily	Hr.	Daily	Bare Costs	Incl. O&P
1 Labor Foreman (outside)	$39.60	$316.80	$64.90	$519.20	$45.31	$72.77
2 Laborers	37.60	601.60	61.65	986.40		
4 Equip. Oper. (medium)	50.60	1619.20	80.30	2569.60		
2 Graders, 30,000 Lbs.		1472.40		1619.64		
1 Tandem Roller, 10 Ton		236.80		260.48		
1 Roller, Pneum. Whl., 12 Ton		344.40		378.84	36.67	40.34
56 L.H., Daily Totals		$4591.20		$6334.16	$81.99	$113.11

Crew B-90B	Hr.	Daily	Hr.	Daily	Bare Costs	Incl. O&P
1 Labor Foreman (outside)	$39.60	$316.80	$64.90	$519.20	$44.43	$71.52
2 Laborers	37.60	601.60	61.65	986.40		
3 Equip. Oper. (medium)	50.60	1214.40	80.30	1927.20		
1 Roller, Pneum. Whl., 12 Ton		344.40		378.84		
1 Road Mixer, 310 H.P.		1929.00		2121.90	47.36	52.10
48 L.H., Daily Totals		$4406.20		$5933.54	$91.80	$123.62

Crew No.	Bare Costs		Incl. Subs O&P		Cost Per Labor-Hour	
Crew B-90C	**Hr.**	**Daily**	**Hr.**	**Daily**	**Bare Costs**	**Incl. O&P**
1 Labor Foreman (outside)	$39.60	$316.80	$64.90	$519.20	$42.00	$67.82
4 Laborers	37.60	1203.20	61.65	1972.80		
3 Equip. Oper. (medium)	50.60	1214.40	80.30	1927.20		
3 Truck Drivers (heavy)	40.05	961.20	64.55	1549.20		
3 Road Mixers, 310 H.P.		5787.00		6365.70	65.76	72.34
88 L.H., Daily Totals		$9482.60		$12334.10	$107.76	$140.16
Crew B-90D	**Hr.**	**Daily**	**Hr.**	**Daily**	**Bare Costs**	**Incl. O&P**
1 Labor Foreman (outside)	$39.60	$316.80	$64.90	$519.20	$41.32	$66.87
6 Laborers	37.60	1804.80	61.65	2959.20		
3 Equip. Oper. (medium)	50.60	1214.40	80.30	1927.20		
3 Truck Drivers (heavy)	40.05	961.20	64.55	1549.20		
3 Road Mixers, 310 H.P.		5787.00		6365.70	55.64	61.21
104 L.H., Daily Totals		$10084.20		$13320.50	$96.96	$128.08
Crew B-90E	**Hr.**	**Daily**	**Hr.**	**Daily**	**Bare Costs**	**Incl. O&P**
1 Labor Foreman (outside)	$39.60	$316.80	$64.90	$519.20	$42.43	$68.55
4 Laborers	37.60	1203.20	61.65	1972.80		
3 Equip. Oper. (medium)	50.60	1214.40	80.30	1927.20		
1 Truck Driver (heavy)	40.05	320.40	64.55	516.40		
1 Road Mixers, 310 H.P.		1929.00		2121.90	26.79	29.47
72 L.H., Daily Totals		$4983.80		$7057.50	$69.22	$98.02
Crew B-91	**Hr.**	**Daily**	**Hr.**	**Daily**	**Bare Costs**	**Incl. O&P**
1 Labor Foreman (outside)	$39.60	$316.80	$64.90	$519.20	$44.66	$71.74
2 Laborers	37.60	601.60	61.65	986.40		
4 Equip. Oper. (medium)	50.60	1619.20	80.30	2569.60		
1 Truck Driver (heavy)	40.05	320.40	64.55	516.40		
1 Dist. Tanker, 3000 Gallon		305.60		336.16		
1 Truck Tractor, 6x4, 380 H.P.		599.80		659.78		
1 Aggreg. Spreader, S.P.		834.00		917.40		
1 Roller, Pneum. Whl., 12 Ton		344.40		378.84		
1 Tandem Roller, 10 Ton		236.80		260.48	36.26	39.89
64 L.H., Daily Totals		$5178.60		$7144.26	$80.92	$111.63
Crew B-91B	**Hr.**	**Daily**	**Hr.**	**Daily**	**Bare Costs**	**Incl. O&P**
1 Laborer	$37.60	$300.80	$61.65	$493.20	$44.10	$70.97
1 Equipment Oper. (med.)	50.60	404.80	80.30	642.40		
1 Road Sweeper, Vac. Assist.		986.00		1084.60	61.63	67.79
16 L.H., Daily Totals		$1691.60		$2220.20	$105.72	$138.76
Crew B-91C	**Hr.**	**Daily**	**Hr.**	**Daily**	**Bare Costs**	**Incl. O&P**
1 Laborer	$37.60	$300.80	$61.65	$493.20	$38.35	$62.35
1 Truck Driver (light)	39.10	312.80	63.05	504.40		
1 Catch Basin Cleaning Truck		573.60		630.96	35.85	39.44
16 L.H., Daily Totals		$1187.20		$1628.56	$74.20	$101.79
Crew B-91D	**Hr.**	**Daily**	**Hr.**	**Daily**	**Bare Costs**	**Incl. O&P**
1 Labor Foreman (outside)	$39.60	$316.80	$64.90	$519.20	$43.13	$69.52
5 Laborers	37.60	1504.00	61.65	2466.00		
5 Equip. Oper. (medium)	50.60	2024.00	80.30	3212.00		
2 Truck Drivers (heavy)	40.05	640.80	64.55	1032.80		
1 Aggreg. Spreader, S.P.		834.00		917.40		
2 Truck Tractor, 6x4, 380 H.P.		1199.60		1319.56		
2 Dist. Tanker, 3000 Gallon		611.20		672.32		
2 Pavement Brush, Towed		163.20		179.52		
2 Roller, Pneum. Whl., 12 Ton		688.80		757.68	33.62	36.99
104 L.H., Daily Totals		$7982.40		$11076.48	$76.75	$106.50

Crew No.	Bare Costs		Incl. Subs O&P		Cost Per Labor-Hour	
Crew B-92	**Hr.**	**Daily**	**Hr.**	**Daily**	**Bare Costs**	**Incl. O&P**
1 Labor Foreman (outside)	$39.60	$316.80	$64.90	$519.20	$38.10	$62.46
3 Laborers	37.60	902.40	61.65	1479.60		
1 Crack Cleaner, 25 H.P.		61.60		67.76		
1 Air Compressor, 60 cfm		137.40		151.14		
1 Tar Kettle, T.M.		132.00		145.20		
1 Flatbed Truck, Gas, 3 Ton		303.20		333.52	19.82	21.80
32 L.H., Daily Totals		$1853.40		$2696.42	$57.92	$84.26
Crew B-93	**Hr.**	**Daily**	**Hr.**	**Daily**	**Bare Costs**	**Incl. O&P**
1 Equip. Oper. (medium)	$50.60	$404.80	$80.30	$642.40	$50.60	$80.30
1 Feller Buncher, 100 H.P.		838.40		922.24	104.80	115.28
8 L.H., Daily Totals		$1243.20		$1564.64	$155.40	$195.58
Crew B-94A	**Hr.**	**Daily**	**Hr.**	**Daily**	**Bare Costs**	**Incl. O&P**
1 Laborer	$37.60	$300.80	$61.65	$493.20	$37.60	$61.65
1 Diaphragm Water Pump, 2"		71.00		78.10		
1 -20' Suction Hose, 2"		1.95		2.15		
2 -50' Discharge Hoses, 2"		1.80		1.98	9.34	10.28
8 L.H., Daily Totals		$375.55		$575.42	$46.94	$71.93
Crew B-94B	**Hr.**	**Daily**	**Hr.**	**Daily**	**Bare Costs**	**Incl. O&P**
1 Laborer	$37.60	$300.80	$61.65	$493.20	$37.60	$61.65
1 Diaphragm Water Pump, 4"		114.20		125.62		
1 -20' Suction Hose, 4"		3.25		3.58		
2 -50' Discharge Hoses, 4"		4.70		5.17	15.27	16.80
8 L.H., Daily Totals		$422.95		$627.57	$52.87	$78.45
Crew B-94C	**Hr.**	**Daily**	**Hr.**	**Daily**	**Bare Costs**	**Incl. O&P**
1 Laborer	$37.60	$300.80	$61.65	$493.20	$37.60	$61.65
1 Centrifugal Water Pump, 3"		78.20		86.02		
1 -20' Suction Hose, 3"		2.85		3.13		
2 -50' Discharge Hoses, 3"		3.00		3.30	10.51	11.56
8 L.H., Daily Totals		$384.85		$585.65	$48.11	$73.21
Crew B-94D	**Hr.**	**Daily**	**Hr.**	**Daily**	**Bare Costs**	**Incl. O&P**
1 Laborer	$37.60	$300.80	$61.65	$493.20	$37.60	$61.65
1 Centr. Water Pump, 6"		346.60		381.26		
1 -20' Suction Hose, 6"		11.50		12.65		
2 -50' Discharge Hoses, 6"		12.20		13.42	46.29	50.92
8 L.H., Daily Totals		$671.10		$900.53	$83.89	$112.57
Crew C-1	**Hr.**	**Daily**	**Hr.**	**Daily**	**Bare Costs**	**Incl. O&P**
3 Carpenters	$46.95	$1126.80	$76.95	$1846.80	$44.61	$73.13
1 Laborer	37.60	300.80	61.65	493.20		
32 L.H., Daily Totals		$1427.60		$2340.00	$44.61	$73.13
Crew C-2	**Hr.**	**Daily**	**Hr.**	**Daily**	**Bare Costs**	**Incl. O&P**
1 Carpenter Foreman (outside)	$48.95	$391.60	$80.25	$642.00	$45.73	$74.95
4 Carpenters	46.95	1502.40	76.95	2462.40		
1 Laborer	37.60	300.80	61.65	493.20		
48 L.H., Daily Totals		$2194.80		$3597.60	$45.73	$74.95
Crew C-2A	**Hr.**	**Daily**	**Hr.**	**Daily**	**Bare Costs**	**Incl. O&P**
1 Carpenter Foreman (outside)	$48.95	$391.60	$80.25	$642.00	$45.40	$73.98
3 Carpenters	46.95	1126.80	76.95	1846.80		
1 Cement Finisher	45.00	360.00	71.15	569.20		
1 Laborer	37.60	300.80	61.65	493.20		
48 L.H., Daily Totals		$2179.20		$3551.20	$45.40	$73.98

Crew No.	Bare Costs		Incl. Subs O&P		Cost Per Labor-Hour	
Crew C-3	**Hr.**	**Daily**	**Hr.**	**Daily**	**Bare Costs**	**Incl. O&P**
1 Rodman Foreman (outside)	$54.55	$436.40	$89.05	$712.40	$48.57	$79.06
4 Rodmen (reinf.)	52.55	1681.60	85.75	2744.00		
1 Equip. Oper. (light)	48.60	388.80	77.15	617.20		
2 Laborers	37.60	601.60	61.65	986.40		
3 Stressing Equipment		30.60		33.66		
.5 Grouting Equipment		81.50		89.65	1.75	1.93
64 L.H., Daily Totals		$3220.50		$5183.31	$50.32	$80.99

Crew No.	Bare Costs		Incl. Subs O&P		Cost Per Labor-Hour	
Crew C-4	**Hr.**	**Daily**	**Hr.**	**Daily**	**Bare Costs**	**Incl. O&P**
1 Rodman Foreman (outside)	$54.55	$436.40	$89.05	$712.40	$53.05	$86.58
3 Rodmen (reinf.)	52.55	1261.20	85.75	2058.00		
3 Stressing Equipment		30.60		33.66	.96	1.05
32 L.H., Daily Totals		$1728.20		$2804.06	$54.01	$87.63

Crew No.	Bare Costs		Incl. Subs O&P		Cost Per Labor-Hour	
Crew C-4A	**Hr.**	**Daily**	**Hr.**	**Daily**	**Bare Costs**	**Incl. O&P**
2 Rodmen (reinf.)	$52.55	$840.80	$85.75	$1372.00	$52.55	$85.75
4 Stressing Equipment		40.80		44.88	2.55	2.81
16 L.H., Daily Totals		$881.60		$1416.88	$55.10	$88.56

Crew No.	Bare Costs		Incl. Subs O&P		Cost Per Labor-Hour	
Crew C-5	**Hr.**	**Daily**	**Hr.**	**Daily**	**Bare Costs**	**Incl. O&P**
1 Rodman Foreman (outside)	$54.55	$436.40	$89.05	$712.40	$51.66	$83.69
4 Rodmen (reinf.)	52.55	1681.60	85.75	2744.00		
1 Equip. Oper. (crane)	51.70	413.60	82.05	656.40		
1 Equip. Oper. (oiler)	45.20	361.60	71.75	574.00		
1 Hyd. Crane, 25 Ton		736.60		810.26	13.15	14.47
56 L.H., Daily Totals		$3629.80		$5497.06	$64.82	$98.16

Crew No.	Bare Costs		Incl. Subs O&P		Cost Per Labor-Hour	
Crew C-6	**Hr.**	**Daily**	**Hr.**	**Daily**	**Bare Costs**	**Incl. O&P**
1 Labor Foreman (outside)	$39.60	$316.80	$64.90	$519.20	$39.17	$63.77
4 Laborers	37.60	1203.20	61.65	1972.80		
1 Cement Finisher	45.00	360.00	71.15	569.20		
2 Gas Engine Vibrators		62.40		68.64	1.30	1.43
48 L.H., Daily Totals		$1942.40		$3129.84	$40.47	$65.20

Crew No.	Bare Costs		Incl. Subs O&P		Cost Per Labor-Hour	
Crew C-7	**Hr.**	**Daily**	**Hr.**	**Daily**	**Bare Costs**	**Incl. O&P**
1 Labor Foreman (outside)	$39.60	$316.80	$64.90	$519.20	$40.93	$66.26
5 Laborers	37.60	1504.00	61.65	2466.00		
1 Cement Finisher	45.00	360.00	71.15	569.20		
1 Equip. Oper. (medium)	50.60	404.80	80.30	642.40		
1 Equip. Oper. (oiler)	45.20	361.60	71.75	574.00		
2 Gas Engine Vibrators		62.40		68.64		
1 Concrete Bucket, 1 C.Y.		23.80		26.18		
1 Hyd. Crane, 55 Ton		1128.00		1240.80	16.86	18.55
72 L.H., Daily Totals		$4161.40		$6106.42	$57.80	$84.81

Crew No.	Bare Costs		Incl. Subs O&P		Cost Per Labor-Hour	
Crew C-8	**Hr.**	**Daily**	**Hr.**	**Daily**	**Bare Costs**	**Incl. O&P**
1 Labor Foreman (outside)	$39.60	$316.80	$64.90	$519.20	$41.86	$67.49
3 Laborers	37.60	902.40	61.65	1479.60		
2 Cement Finishers	45.00	720.00	71.15	1138.40		
1 Equip. Oper. (medium)	50.60	404.80	80.30	642.40		
1 Concrete Pump (Small)		720.00		792.00	12.86	14.14
56 L.H., Daily Totals		$3064.00		$4571.60	$54.71	$81.64

Crew No.	Bare Costs		Incl. Subs O&P		Cost Per Labor-Hour	
Crew C-8A	**Hr.**	**Daily**	**Hr.**	**Daily**	**Bare Costs**	**Incl. O&P**
1 Labor Foreman (outside)	$39.60	$316.80	$64.90	$519.20	$40.40	$65.36
3 Laborers	37.60	902.40	61.65	1479.60		
2 Cement Finishers	45.00	720.00	71.15	1138.40		
48 L.H., Daily Totals		$1939.20		$3137.20	$40.40	$65.36

Crew No.	Bare Costs		Incl. Subs O&P		Cost Per Labor-Hour	
Crew C-8B	**Hr.**	**Daily**	**Hr.**	**Daily**	**Bare Costs**	**Incl. O&P**
1 Labor Foreman (outside)	$39.60	$316.80	$64.90	$519.20	$40.60	$66.03
3 Laborers	37.60	902.40	61.65	1479.60		
1 Equip. Oper. (medium)	50.60	404.80	80.30	642.40		
1 Vibrating Power Screed		66.30		72.93		
1 Roller, Vibratory, 25 Ton		687.40		756.14		
1 Dozer, 200 H.P.		1387.00		1525.70	53.52	58.87
40 L.H., Daily Totals		$3764.70		$4995.97	$94.12	$124.90

Crew No.	Bare Costs		Incl. Subs O&P		Cost Per Labor-Hour	
Crew C-8C	**Hr.**	**Daily**	**Hr.**	**Daily**	**Bare Costs**	**Incl. O&P**
1 Labor Foreman (outside)	$39.60	$316.80	$64.90	$519.20	$41.33	$66.88
3 Laborers	37.60	902.40	61.65	1479.60		
1 Cement Finisher	45.00	360.00	71.15	569.20		
1 Equip. Oper. (medium)	50.60	404.80	80.30	642.40		
1 Shotcrete Rig, 12 C.Y./hr		250.80		275.88		
1 Air Compressor, 160 cfm		156.40		172.04		
4 -50' Air Hoses, 1"		16.40		18.04		
4 -50' Air Hoses, 2"		31.00		34.10	9.47	10.42
48 L.H., Daily Totals		$2438.60		$3710.46	$50.80	$77.30

Crew No.	Bare Costs		Incl. Subs O&P		Cost Per Labor-Hour	
Crew C-8D	**Hr.**	**Daily**	**Hr.**	**Daily**	**Bare Costs**	**Incl. O&P**
1 Labor Foreman (outside)	$39.60	$316.80	$64.90	$519.20	$42.70	$68.71
1 Laborer	37.60	300.80	61.65	493.20		
1 Cement Finisher	45.00	360.00	71.15	569.20		
1 Equipment Oper. (light)	48.60	388.80	77.15	617.20		
1 Air Compressor, 250 cfm		201.40		221.54		
2 -50' Air Hoses, 1"		8.20		9.02	6.55	7.21
32 L.H., Daily Totals		$1576.00		$2429.36	$49.25	$75.92

Crew No.	Bare Costs		Incl. Subs O&P		Cost Per Labor-Hour	
Crew C-8E	**Hr.**	**Daily**	**Hr.**	**Daily**	**Bare Costs**	**Incl. O&P**
1 Labor Foreman (outside)	$39.60	$316.80	$64.90	$519.20	$41.00	$66.36
3 Laborers	37.60	902.40	61.65	1479.60		
1 Cement Finisher	45.00	360.00	71.15	569.20		
1 Equipment Oper. (light)	48.60	388.80	77.15	617.20		
1 Shotcrete Rig, 35 C.Y./hr		281.00		309.10		
1 Air Compressor, 250 cfm		201.40		221.54		
4 -50' Air Hoses, 1"		16.40		18.04		
4 -50' Air Hoses, 2"		31.00		34.10	11.04	12.14
48 L.H., Daily Totals		$2497.80		$3767.98	$52.04	$78.50

Crew No.	Bare Costs		Incl. Subs O&P		Cost Per Labor-Hour	
Crew C-10	**Hr.**	**Daily**	**Hr.**	**Daily**	**Bare Costs**	**Incl. O&P**
1 Laborer	$37.60	$300.80	$61.65	$493.20	$42.53	$67.98
2 Cement Finishers	45.00	720.00	71.15	1138.40		
24 L.H., Daily Totals		$1020.80		$1631.60	$42.53	$67.98

Crew No.	Bare Costs		Incl. Subs O&P		Cost Per Labor-Hour	
Crew C-10B	**Hr.**	**Daily**	**Hr.**	**Daily**	**Bare Costs**	**Incl. O&P**
3 Laborers	$37.60	$902.40	$61.65	$1479.60	$40.56	$65.45
2 Cement Finishers	45.00	720.00	71.15	1138.40		
1 Concrete Mixer, 10 C.F.		172.40		189.64		
2 Trowels, 48" Walk-Behind		100.40		110.44	6.82	7.50
40 L.H., Daily Totals		$1895.20		$2918.08	$47.38	$72.95

Crew No.	Bare Costs		Incl. Subs O&P		Cost Per Labor-Hour	
Crew C-10C	**Hr.**	**Daily**	**Hr.**	**Daily**	**Bare Costs**	**Incl. O&P**
1 Laborer	$37.60	$300.80	$61.65	$493.20	$42.53	$67.98
2 Cement Finishers	45.00	720.00	71.15	1138.40		
1 Trowel, 48" Walk-Behind		50.20		55.22	2.09	2.30
24 L.H., Daily Totals		$1071.00		$1686.82	$44.63	$70.28

Crew C-10D	Hr.	Daily	Hr.	Daily	Bare Costs	Incl. O&P
1 Laborer	$37.60	$300.80	$61.65	$493.20	$42.53	$67.98
2 Cement Finishers	45.00	720.00	71.15	1138.40		
1 Vibrating Power Screed		66.30		72.93		
1 Trowel, 48" Walk-Behind		50.20		55.22	4.85	5.34
24 L.H., Daily Totals		$1137.30		$1759.75	$47.39	$73.32

Crew C-10E	Hr.	Daily	Hr.	Daily	Bare Costs	Incl. O&P
1 Laborer	$37.60	$300.80	$61.65	$493.20	$42.53	$67.98
2 Cement Finishers	45.00	720.00	71.15	1138.40		
1 Vibrating Power Screed		66.30		72.93		
1 Cement Trowel, 96" Ride-On		189.00		207.90	10.64	11.70
24 L.H., Daily Totals		$1276.10		$1912.43	$53.17	$79.68

Crew C-10F	Hr.	Daily	Hr.	Daily	Bare Costs	Incl. O&P
1 Laborer	$37.60	$300.80	$61.65	$493.20	$42.53	$67.98
2 Cement Finishers	45.00	720.00	71.15	1138.40		
1 Aerial Lift Truck, 60' Boom		435.60		479.16	18.15	19.97
24 L.H., Daily Totals		$1456.40		$2110.76	$60.68	$87.95

Crew C-11	Hr.	Daily	Hr.	Daily	Bare Costs	Incl. O&P
1 Struc. Steel Foreman (outside)	$54.65	$437.20	$98.75	$790.00	$51.94	$91.49
6 Struc. Steel Workers	52.65	2527.20	95.15	4567.20		
1 Equip. Oper. (crane)	51.70	413.60	82.05	656.40		
1 Equip. Oper. (oiler)	45.20	361.60	71.75	574.00		
1 Lattice Boom Crane, 150 Ton		1813.00		1994.30	25.18	27.70
72 L.H., Daily Totals		$5552.60		$8581.90	$77.12	$119.19

Crew C-12	Hr.	Daily	Hr.	Daily	Bare Costs	Incl. O&P
1 Carpenter Foreman (outside)	$48.95	$391.60	$80.25	$642.00	$46.52	$75.80
3 Carpenters	46.95	1126.80	76.95	1846.80		
1 Laborer	37.60	300.80	61.65	493.20		
1 Equip. Oper. (crane)	51.70	413.60	82.05	656.40		
1 Hyd. Crane, 12 Ton		653.80		719.18	13.62	14.98
48 L.H., Daily Totals		$2886.60		$4357.58	$60.14	$90.78

Crew C-13	Hr.	Daily	Hr.	Daily	Bare Costs	Incl. O&P
1 Struc. Steel Worker	$52.65	$421.20	$95.15	$761.20	$50.75	$89.08
1 Welder	52.65	421.20	95.15	761.20		
1 Carpenter	46.95	375.60	76.95	615.60		
1 Welder, Gas Engine, 300 amp		145.85		160.44	6.08	6.68
24 L.H., Daily Totals		$1363.85		$2298.43	$56.83	$95.77

Crew C-14	Hr.	Daily	Hr.	Daily	Bare Costs	Incl. O&P
1 Carpenter Foreman (outside)	$48.95	$391.60	$80.25	$642.00	$46.18	$75.04
5 Carpenters	46.95	1878.00	76.95	3078.00		
4 Laborers	37.60	1203.20	61.65	1972.80		
4 Rodmen (reinf.)	52.55	1681.60	85.75	2744.00		
2 Cement Finishers	45.00	720.00	71.15	1138.40		
1 Equip. Oper. (crane)	51.70	413.60	82.05	656.40		
1 Equip. Oper. (oiler)	45.20	361.60	71.75	574.00		
1 Hyd. Crane, 80 Ton		1625.00		1787.50	11.28	12.41
144 L.H., Daily Totals		$8274.60		$12593.10	$57.46	$87.45

Crew C-14A	Hr.	Daily	Hr.	Daily	Bare Costs	Incl. O&P
1 Carpenter Foreman (outside)	$48.95	$391.60	$80.25	$642.00	$47.25	$77.17
16 Carpenters	46.95	6009.60	76.95	9849.60		
4 Rodmen (reinf.)	52.55	1681.60	85.75	2744.00		
2 Laborers	37.60	601.60	61.65	986.40		
1 Cement Finisher	45.00	360.00	71.15	569.20		
1 Equip. Oper. (medium)	50.60	404.80	80.30	642.40		
1 Gas Engine Vibrator		31.20		34.32		
1 Concrete Pump (Small)		720.00		792.00	3.76	4.13
200 L.H., Daily Totals		$10200.40		$16259.92	$51.00	$81.30

Crew C-14B	Hr.	Daily	Hr.	Daily	Bare Costs	Incl. O&P
1 Carpenter Foreman (outside)	$48.95	$391.60	$80.25	$642.00	$47.16	$76.94
16 Carpenters	46.95	6009.60	76.95	9849.60		
4 Rodmen (reinf.)	52.55	1681.60	85.75	2744.00		
2 Laborers	37.60	601.60	61.65	986.40		
2 Cement Finishers	45.00	720.00	71.15	1138.40		
1 Equip. Oper. (medium)	50.60	404.80	80.30	642.40		
1 Gas Engine Vibrator		31.20		34.32		
1 Concrete Pump (Small)		720.00		792.00	3.61	3.97
208 L.H., Daily Totals		$10560.40		$16829.12	$50.77	$80.91

Crew C-14C	Hr.	Daily	Hr.	Daily	Bare Costs	Incl. O&P
1 Carpenter Foreman (outside)	$48.95	$391.60	$80.25	$642.00	$45.08	$73.66
6 Carpenters	46.95	2253.60	76.95	3693.60		
2 Rodmen (reinf.)	52.55	840.80	85.75	1372.00		
4 Laborers	37.60	1203.20	61.65	1972.80		
1 Cement Finisher	45.00	360.00	71.15	569.20		
1 Gas Engine Vibrator		31.20		34.32	.28	.31
112 L.H., Daily Totals		$5080.40		$8283.92	$45.36	$73.96

Crew C-14D	Hr.	Daily	Hr.	Daily	Bare Costs	Incl. O&P
1 Carpenter Foreman (outside)	$48.95	$391.60	$80.25	$642.00	$46.80	$76.46
18 Carpenters	46.95	6760.80	76.95	11080.80		
2 Rodmen (reinf.)	52.55	840.80	85.75	1372.00		
2 Laborers	37.60	601.60	61.65	986.40		
1 Cement Finisher	45.00	360.00	71.15	569.20		
1 Equip. Oper. (medium)	50.60	404.80	80.30	642.40		
1 Gas Engine Vibrator		31.20		34.32		
1 Concrete Pump (Small)		720.00		792.00	3.76	4.13
200 L.H., Daily Totals		$10110.80		$16119.12	$50.55	$80.60

Crew C-14E	Hr.	Daily	Hr.	Daily	Bare Costs	Incl. O&P
1 Carpenter Foreman (outside)	$48.95	$391.60	$80.25	$642.00	$46.44	$75.75
2 Carpenters	46.95	751.20	76.95	1231.20		
4 Rodmen (reinf.)	52.55	1681.60	85.75	2744.00		
3 Laborers	37.60	902.40	61.65	1479.60		
1 Cement Finisher	45.00	360.00	71.15	569.20		
1 Gas Engine Vibrator		31.20		34.32	.35	.39
88 L.H., Daily Totals		$4118.00		$6700.32	$46.80	$76.14

Crew C-14F	Hr.	Daily	Hr.	Daily	Bare Costs	Incl. O&P
1 Labor Foreman (outside)	$39.60	$316.80	$64.90	$519.20	$42.76	$68.34
2 Laborers	37.60	601.60	61.65	986.40		
6 Cement Finishers	45.00	2160.00	71.15	3415.20		
1 Gas Engine Vibrator		31.20		34.32	.43	.48
72 L.H., Daily Totals		$3109.60		$4955.12	$43.19	$68.82

Crew No.	Bare Costs		Incl. Subs O&P		Cost Per Labor-Hour	
Crew C-14G	**Hr.**	**Daily**	**Hr.**	**Daily**	**Bare Costs**	**Incl. O&P**
1 Labor Foreman (outside)	$39.60	$316.80	$64.90	$519.20	$42.11	$67.54
2 Laborers	37.60	601.60	61.65	986.40		
4 Cement Finishers	45.00	1440.00	71.15	2276.80		
1 Gas Engine Vibrator		31.20		34.32	.56	.61
56 L.H., Daily Totals		$2389.60		$3816.72	$42.67	$68.16
Crew C-14H	**Hr.**	**Daily**	**Hr.**	**Daily**	**Bare Costs**	**Incl. O&P**
1 Carpenter Foreman (outside)	$48.95	$391.60	$80.25	$642.00	$46.33	$75.45
2 Carpenters	46.95	751.20	76.95	1231.20		
1 Rodman (reinf.)	52.55	420.40	85.75	686.00		
1 Laborer	37.60	300.80	61.65	493.20		
1 Cement Finisher	45.00	360.00	71.15	569.20		
1 Gas Engine Vibrator		31.20		34.32	.65	.71
48 L.H., Daily Totals		$2255.20		$3655.92	$46.98	$76.17
Crew C-14L	**Hr.**	**Daily**	**Hr.**	**Daily**	**Bare Costs**	**Incl. O&P**
1 Carpenter Foreman (outside)	$48.95	$391.60	$80.25	$642.00	$43.84	$71.64
6 Carpenters	46.95	2253.60	76.95	3693.60		
4 Laborers	37.60	1203.20	61.65	1972.80		
1 Cement Finisher	45.00	360.00	71.15	569.20		
1 Gas Engine Vibrator		31.20		34.32	.33	.36
96 L.H., Daily Totals		$4239.60		$6911.92	$44.16	$72.00
Crew C-14M	**Hr.**	**Daily**	**Hr.**	**Daily**	**Bare Costs**	**Incl. O&P**
1 Carpenter Foreman (outside)	$48.95	$391.60	$80.25	$642.00	$45.77	$74.33
2 Carpenters	46.95	751.20	76.95	1231.20		
1 Rodman (reinf.)	52.55	420.40	85.75	686.00		
2 Laborers	37.60	601.60	61.65	986.40		
1 Cement Finisher	45.00	360.00	71.15	569.20		
1 Equip. Oper. (medium)	50.60	404.80	80.30	642.40		
1 Gas Engine Vibrator		31.20		34.32		
1 Concrete Pump (Small)		720.00		792.00	11.74	12.91
64 L.H., Daily Totals		$3680.80		$5583.52	$57.51	$87.24
Crew C-15	**Hr.**	**Daily**	**Hr.**	**Daily**	**Bare Costs**	**Incl. O&P**
1 Carpenter Foreman (outside)	$48.95	$391.60	$80.25	$642.00	$44.24	$71.91
2 Carpenters	46.95	751.20	76.95	1231.20		
3 Laborers	37.60	902.40	61.65	1479.60		
2 Cement Finishers	45.00	720.00	71.15	1138.40		
1 Rodman (reinf.)	52.55	420.40	85.75	686.00		
72 L.H., Daily Totals		$3185.60		$5177.20	$44.24	$71.91
Crew C-16	**Hr.**	**Daily**	**Hr.**	**Daily**	**Bare Costs**	**Incl. O&P**
1 Labor Foreman (outside)	$39.60	$316.80	$64.90	$519.20	$41.86	$67.49
3 Laborers	37.60	902.40	61.65	1479.60		
2 Cement Finishers	45.00	720.00	71.15	1138.40		
1 Equip. Oper. (medium)	50.60	404.80	80.30	642.40		
1 Gunite Pump Rig		371.80		408.98		
2 -50' Air Hoses, 3/4"		6.50		7.15		
2 -50' Air Hoses, 2"		15.50		17.05	7.03	7.74
56 L.H., Daily Totals		$2737.80		$4212.78	$48.89	$75.23
Crew C-16A	**Hr.**	**Daily**	**Hr.**	**Daily**	**Bare Costs**	**Incl. O&P**
1 Laborer	$37.60	$300.80	$61.65	$493.20	$44.55	$71.06
2 Cement Finishers	45.00	720.00	71.15	1138.40		
1 Equip. Oper. (medium)	50.60	404.80	80.30	642.40		
1 Gunite Pump Rig		371.80		408.98		
2 -50' Air Hoses, 3/4"		6.50		7.15		
2 -50' Air Hoses, 2"		15.50		17.05		
1 Aerial Lift Truck, 60' Boom		435.60		479.16	25.92	28.51
32 L.H., Daily Totals		$2255.00		$3186.34	$70.47	$99.57

Crew No.	Bare Costs		Incl. Subs O&P		Cost Per Labor-Hour	
Crew C-17	**Hr.**	**Daily**	**Hr.**	**Daily**	**Bare Costs**	**Incl. O&P**
2 Skilled Worker Foremen (out)	$50.65	$810.40	$82.30	$1316.80	$49.05	$79.70
8 Skilled Workers	48.65	3113.60	79.05	5059.20		
80 L.H., Daily Totals		$3924.00		$6376.00	$49.05	$79.70
Crew C-17A	**Hr.**	**Daily**	**Hr.**	**Daily**	**Bare Costs**	**Incl. O&P**
2 Skilled Worker Foremen (out)	$50.65	$810.40	$82.30	$1316.80	$49.08	$79.73
8 Skilled Workers	48.65	3113.60	79.05	5059.20		
.125 Equip. Oper. (crane)	51.70	51.70	82.05	82.05		
.125 Hyd. Crane, 80 Ton		203.13		223.44	2.51	2.76
81 L.H., Daily Totals		$4178.82		$6681.49	$51.59	$82.49
Crew C-17B	**Hr.**	**Daily**	**Hr.**	**Daily**	**Bare Costs**	**Incl. O&P**
2 Skilled Worker Foremen (out)	$50.65	$810.40	$82.30	$1316.80	$49.11	$79.76
8 Skilled Workers	48.65	3113.60	79.05	5059.20		
.25 Equip. Oper. (crane)	51.70	103.40	82.05	164.10		
.25 Hyd. Crane, 80 Ton		406.25		446.88		
.25 Trowel, 48" Walk-Behind		12.55		13.81	5.11	5.62
82 L.H., Daily Totals		$4446.20		$7000.78	$54.22	$85.38
Crew C-17C	**Hr.**	**Daily**	**Hr.**	**Daily**	**Bare Costs**	**Incl. O&P**
2 Skilled Worker Foremen (out)	$50.65	$810.40	$82.30	$1316.80	$49.15	$79.78
8 Skilled Workers	48.65	3113.60	79.05	5059.20		
.375 Equip. Oper. (crane)	51.70	155.10	82.05	246.15		
.375 Hyd. Crane, 80 Ton		609.38		670.31	7.34	8.08
83 L.H., Daily Totals		$4688.48		$7292.46	$56.49	$87.86
Crew C-17D	**Hr.**	**Daily**	**Hr.**	**Daily**	**Bare Costs**	**Incl. O&P**
2 Skilled Worker Foremen (out)	$50.65	$810.40	$82.30	$1316.80	$49.18	$79.81
8 Skilled Workers	48.65	3113.60	79.05	5059.20		
.5 Equip. Oper. (crane)	51.70	206.80	82.05	328.20		
.5 Hyd. Crane, 80 Ton		812.50		893.75	9.67	10.64
84 L.H., Daily Totals		$4943.30		$7597.95	$58.85	$90.45
Crew C-17E	**Hr.**	**Daily**	**Hr.**	**Daily**	**Bare Costs**	**Incl. O&P**
2 Skilled Worker Foremen (out)	$50.65	$810.40	$82.30	$1316.80	$49.05	$79.70
8 Skilled Workers	48.65	3113.60	79.05	5059.20		
1 Hyd. Jack with Rods		96.50		106.15	1.21	1.33
80 L.H., Daily Totals		$4020.50		$6482.15	$50.26	$81.03
Crew C-18	**Hr.**	**Daily**	**Hr.**	**Daily**	**Bare Costs**	**Incl. O&P**
.125 Labor Foreman (outside)	$39.60	$39.60	$64.90	$64.90	$37.82	$62.01
1 Laborer	37.60	300.80	61.65	493.20		
1 Concrete Cart, 10 C.F.		60.00		66.00	6.67	7.33
9 L.H., Daily Totals		$400.40		$624.10	$44.49	$69.34
Crew C-19	**Hr.**	**Daily**	**Hr.**	**Daily**	**Bare Costs**	**Incl. O&P**
.125 Labor Foreman (outside)	$39.60	$39.60	$64.90	$64.90	$37.82	$62.01
1 Laborer	37.60	300.80	61.65	493.20		
1 Concrete Cart, 18 C.F.		99.80		109.78	11.09	12.20
9 L.H., Daily Totals		$440.20		$667.88	$48.91	$74.21
Crew C-20	**Hr.**	**Daily**	**Hr.**	**Daily**	**Bare Costs**	**Incl. O&P**
1 Labor Foreman (outside)	$39.60	$316.80	$64.90	$519.20	$40.40	$65.58
5 Laborers	37.60	1504.00	61.65	2466.00		
1 Cement Finisher	45.00	360.00	71.15	569.20		
1 Equip. Oper. (medium)	50.60	404.80	80.30	642.40		
2 Gas Engine Vibrators		62.40		68.64		
1 Concrete Pump (Small)		720.00		792.00	12.23	13.45
64 L.H., Daily Totals		$3368.00		$5057.44	$52.63	$79.02

Crew C-21	Bare Costs Hr.	Bare Costs Daily	Incl. Subs O&P Hr.	Incl. Subs O&P Daily	Cost Per Labor-Hour Bare Costs	Cost Per Labor-Hour Incl. O&P
1 Labor Foreman (outside)	$39.60	$316.80	$64.90	$519.20	$40.40	$65.58
5 Laborers	37.60	1504.00	61.65	2466.00		
1 Cement Finisher	45.00	360.00	71.15	569.20		
1 Equip. Oper. (medium)	50.60	404.80	80.30	642.40		
2 Gas Engine Vibrators		62.40		68.64		
1 Concrete Conveyer		198.80		218.68	4.08	4.49
64 L.H., Daily Totals		$2846.80		$4484.12	$44.48	$70.06

Crew C-22	Hr.	Daily	Hr.	Daily	Bare Costs	Incl. O&P
1 Rodman Foreman (outside)	$54.55	$436.40	$89.05	$712.40	$52.74	$85.96
4 Rodmen (reinf.)	52.55	1681.60	85.75	2744.00		
.125 Equip. Oper. (crane)	51.70	51.70	82.05	82.05		
.125 Equip. Oper. (oiler)	45.20	45.20	71.75	71.75		
.125 Hyd. Crane, 25 Ton		92.08		101.28	2.19	2.41
42 L.H., Daily Totals		$2306.97		$3711.48	$54.93	$88.37

Crew C-23	Hr.	Daily	Hr.	Daily	Bare Costs	Incl. O&P
2 Skilled Worker Foremen (out)	$50.65	$810.40	$82.30	$1316.80	$49.01	$79.27
6 Skilled Workers	48.65	2335.20	79.05	3794.40		
1 Equip. Oper. (crane)	51.70	413.60	82.05	656.40		
1 Equip. Oper. (oiler)	45.20	361.60	71.75	574.00		
1 Lattice Boom Crane, 90 Ton		1511.00		1662.10	18.89	20.78
80 L.H., Daily Totals		$5431.80		$8003.70	$67.90	$100.05

Crew C-24	Hr.	Daily	Hr.	Daily	Bare Costs	Incl. O&P
2 Skilled Worker Foremen (out)	$50.65	$810.40	$82.30	$1316.80	$49.01	$79.27
6 Skilled Workers	48.65	2335.20	79.05	3794.40		
1 Equip. Oper. (crane)	51.70	413.60	82.05	656.40		
1 Equip. Oper. (oiler)	45.20	361.60	71.75	574.00		
1 Lattice Boom Crane, 150 Ton		1813.00		1994.30	22.66	24.93
80 L.H., Daily Totals		$5733.80		$8335.90	$71.67	$104.20

Crew C-25	Hr.	Daily	Hr.	Daily	Bare Costs	Incl. O&P
2 Rodmen (reinf.)	$52.55	$840.80	$85.75	$1372.00	$41.30	$70.03
2 Rodmen Helpers	30.05	480.80	54.30	868.80		
32 L.H., Daily Totals		$1321.60		$2240.80	$41.30	$70.03

Crew C-27	Hr.	Daily	Hr.	Daily	Bare Costs	Incl. O&P
2 Cement Finishers	$45.00	$720.00	$71.15	$1138.40	$45.00	$71.15
1 Concrete Saw		167.40		184.14	10.46	11.51
16 L.H., Daily Totals		$887.40		$1322.54	$55.46	$82.66

Crew C-28	Hr.	Daily	Hr.	Daily	Bare Costs	Incl. O&P
1 Cement Finisher	$45.00	$360.00	$71.15	$569.20	$45.00	$71.15
1 Portable Air Compressor, Gas		17.90		19.69	2.24	2.46
8 L.H., Daily Totals		$377.90		$588.89	$47.24	$73.61

Crew C-29	Hr.	Daily	Hr.	Daily	Bare Costs	Incl. O&P
1 Laborer	$37.60	$300.80	$61.65	$493.20	$37.60	$61.65
1 Pressure Washer		70.20		77.22	8.78	9.65
8 L.H., Daily Totals		$371.00		$570.42	$46.38	$71.30

Crew C-30	Hr.	Daily	Hr.	Daily	Bare Costs	Incl. O&P
1 Laborer	$37.60	$300.80	$61.65	$493.20	$37.60	$61.65
1 Concrete Mixer, 10 C.F.		172.40		189.64	21.55	23.70
8 L.H., Daily Totals		$473.20		$682.84	$59.15	$85.36

Crew C-31	Hr.	Daily	Hr.	Daily	Bare Costs	Incl. O&P
1 Cement Finisher	$45.00	$360.00	$71.15	$569.20	$45.00	$71.15
1 Grout Pump		371.80		408.98	46.48	51.12
8 L.H., Daily Totals		$731.80		$978.18	$91.47	$122.27

Crew C-32	Hr.	Daily	Hr.	Daily	Bare Costs	Incl. O&P
1 Cement Finisher	$45.00	$360.00	$71.15	$569.20	$41.30	$66.40
1 Laborer	37.60	300.80	61.65	493.20		
1 Crack Chaser Saw, Gas, 6 H.P.		29.20		32.12		
1 Vacuum Pick-Up System		60.90		66.99	5.63	6.19
16 L.H., Daily Totals		$750.90		$1161.51	$46.93	$72.59

Crew D-1	Hr.	Daily	Hr.	Daily	Bare Costs	Incl. O&P
1 Bricklayer	$46.15	$369.20	$75.10	$600.80	$42.10	$68.50
1 Bricklayer Helper	38.05	304.40	61.90	495.20		
16 L.H., Daily Totals		$673.60		$1096.00	$42.10	$68.50

Crew D-2	Hr.	Daily	Hr.	Daily	Bare Costs	Incl. O&P
3 Bricklayers	$46.15	$1107.60	$75.10	$1802.40	$43.28	$70.47
2 Bricklayer Helpers	38.05	608.80	61.90	990.40		
.5 Carpenter	46.95	187.80	76.95	307.80		
44 L.H., Daily Totals		$1904.20		$3100.60	$43.28	$70.47

Crew D-3	Hr.	Daily	Hr.	Daily	Bare Costs	Incl. O&P
3 Bricklayers	$46.15	$1107.60	$75.10	$1802.40	$43.10	$70.16
2 Bricklayer Helpers	38.05	608.80	61.90	990.40		
.25 Carpenter	46.95	93.90	76.95	153.90		
42 L.H., Daily Totals		$1810.30		$2946.70	$43.10	$70.16

Crew D-4	Hr.	Daily	Hr.	Daily	Bare Costs	Incl. O&P
1 Bricklayer	$46.15	$369.20	$75.10	$600.80	$42.71	$69.01
2 Bricklayer Helpers	38.05	608.80	61.90	990.40		
1 Equip. Oper. (light)	48.60	388.80	77.15	617.20		
1 Grout Pump, 50 C.F./hr.		133.40		146.74	4.17	4.59
32 L.H., Daily Totals		$1500.20		$2355.14	$46.88	$73.60

Crew D-5	Hr.	Daily	Hr.	Daily	Bare Costs	Incl. O&P
1 Bricklayer	46.15	369.20	75.10	600.80	46.15	75.10
8 L.H., Daily Totals		$369.20		$600.80	$46.15	$75.10

Crew D-6	Hr.	Daily	Hr.	Daily	Bare Costs	Incl. O&P
3 Bricklayers	$46.15	$1107.60	$75.10	$1802.40	$42.29	$68.84
3 Bricklayer Helpers	38.05	913.20	61.90	1485.60		
.25 Carpenter	46.95	93.90	76.95	153.90		
50 L.H., Daily Totals		$2114.70		$3441.90	$42.29	$68.84

Crew D-7	Hr.	Daily	Hr.	Daily	Bare Costs	Incl. O&P
1 Tile Layer	$42.80	$342.40	$67.60	$540.80	$38.17	$60.30
1 Tile Layer Helper	33.55	268.40	53.00	424.00		
16 L.H., Daily Totals		$610.80		$964.80	$38.17	$60.30

Crew D-8	Hr.	Daily	Hr.	Daily	Bare Costs	Incl. O&P
3 Bricklayers	$46.15	$1107.60	$75.10	$1802.40	$42.91	$69.82
2 Bricklayer Helpers	38.05	608.80	61.90	990.40		
40 L.H., Daily Totals		$1716.40		$2792.80	$42.91	$69.82

Crew D-9	Hr.	Daily	Hr.	Daily	Bare Costs	Incl. O&P
3 Bricklayers	$46.15	$1107.60	$75.10	$1802.40	$42.10	$68.50
3 Bricklayer Helpers	38.05	913.20	61.90	1485.60		
48 L.H., Daily Totals		$2020.80		$3288.00	$42.10	$68.50

Crew No.	Bare Costs		Incl. Subs O&P		Cost Per Labor-Hour	
Crew D-10	**Hr.**	**Daily**	**Hr.**	**Daily**	**Bare Costs**	**Incl. O&P**
1 Bricklayer Foreman (outside)	$48.15	$385.20	$78.35	$626.80	$46.01	$74.35
1 Bricklayer	46.15	369.20	75.10	600.80		
1 Bricklayer Helper	38.05	304.40	61.90	495.20		
1 Equip. Oper. (crane)	51.70	413.60	82.05	656.40		
1 S.P. Crane, 4x4, 12 Ton		473.40		520.74	14.79	16.27
32 L.H., Daily Totals		$1945.80		$2899.94	$60.81	$90.62
Crew D-11	**Hr.**	**Daily**	**Hr.**	**Daily**	**Bare Costs**	**Incl. O&P**
1 Bricklayer Foreman (outside)	$48.15	$385.20	$78.35	$626.80	$44.12	$71.78
1 Bricklayer	46.15	369.20	75.10	600.80		
1 Bricklayer Helper	38.05	304.40	61.90	495.20		
24 L.H., Daily Totals		$1058.80		$1722.80	$44.12	$71.78
Crew D-12	**Hr.**	**Daily**	**Hr.**	**Daily**	**Bare Costs**	**Incl. O&P**
1 Bricklayer Foreman (outside)	$48.15	$385.20	$78.35	$626.80	$42.60	$69.31
1 Bricklayer	46.15	369.20	75.10	600.80		
2 Bricklayer Helpers	38.05	608.80	61.90	990.40		
32 L.H., Daily Totals		$1363.20		$2218.00	$42.60	$69.31
Crew D-13	**Hr.**	**Daily**	**Hr.**	**Daily**	**Bare Costs**	**Incl. O&P**
1 Bricklayer Foreman (outside)	$48.15	$385.20	$78.35	$626.80	$44.84	$72.71
1 Bricklayer	46.15	369.20	75.10	600.80		
2 Bricklayer Helpers	38.05	608.80	61.90	990.40		
1 Carpenter	46.95	375.60	76.95	615.60		
1 Equip. Oper. (crane)	51.70	413.60	82.05	656.40		
1 S.P. Crane, 4x4, 12 Ton		473.40		520.74	9.86	10.85
48 L.H., Daily Totals		$2625.80		$4010.74	$54.70	$83.56
Crew E-1	**Hr.**	**Daily**	**Hr.**	**Daily**	**Bare Costs**	**Incl. O&P**
1 Welder Foreman (outside)	$54.65	$437.20	$98.75	$790.00	$51.97	$90.35
1 Welder	52.65	421.20	95.15	761.20		
1 Equip. Oper. (light)	48.60	388.80	77.15	617.20		
1 Welder, Gas Engine, 300 amp		145.85		160.44	6.08	6.68
24 L.H., Daily Totals		$1393.05		$2328.84	$58.04	$97.03
Crew E-2	**Hr.**	**Daily**	**Hr.**	**Daily**	**Bare Costs**	**Incl. O&P**
1 Struc. Steel Foreman (outside)	$54.65	$437.20	$98.75	$790.00	$51.74	$90.45
4 Struc. Steel Workers	52.65	1684.80	95.15	3044.80		
1 Equip. Oper. (crane)	51.70	413.60	82.05	656.40		
1 Equip. Oper. (oiler)	45.20	361.60	71.75	574.00		
1 Lattice Boom Crane, 90 Ton		1511.00		1662.10	26.98	29.68
56 L.H., Daily Totals		$4408.20		$6727.30	$78.72	$120.13
Crew E-3	**Hr.**	**Daily**	**Hr.**	**Daily**	**Bare Costs**	**Incl. O&P**
1 Struc. Steel Foreman (outside)	$54.65	$437.20	$98.75	$790.00	$53.32	$96.35
1 Struc. Steel Worker	52.65	421.20	95.15	761.20		
1 Welder	52.65	421.20	95.15	761.20		
1 Welder, Gas Engine, 300 amp		145.85		160.44	6.08	6.68
24 L.H., Daily Totals		$1425.45		$2472.84	$59.39	$103.03
Crew E-3A	**Hr.**	**Daily**	**Hr.**	**Daily**	**Bare Costs**	**Incl. O&P**
1 Struc. Steel Foreman (outside)	$54.65	$437.20	$98.75	$790.00	$53.32	$96.35
1 Struc. Steel Worker	52.65	421.20	95.15	761.20		
1 Welder	52.65	421.20	95.15	761.20		
1 Welder, Gas Engine, 300 amp		145.85		160.44		
1 Aerial Lift Truck, 40' Boom		301.40		331.54	18.64	20.50
24 L.H., Daily Totals		$1726.85		$2804.38	$71.95	$116.85

Crew No.	Bare Costs		Incl. Subs O&P		Cost Per Labor-Hour	
Crew E-4	**Hr.**	**Daily**	**Hr.**	**Daily**	**Bare Costs**	**Incl. O&P**
1 Struc. Steel Foreman (outside)	$54.65	$437.20	$98.75	$790.00	$53.15	$96.05
3 Struc. Steel Workers	52.65	1263.60	95.15	2283.60		
1 Welder, Gas Engine, 300 amp		145.85		160.44	4.56	5.01
32 L.H., Daily Totals		$1846.65		$3234.03	$57.71	$101.06
Crew E-5	**Hr.**	**Daily**	**Hr.**	**Daily**	**Bare Costs**	**Incl. O&P**
2 Struc. Steel Foremen (outside)	$54.65	$874.40	$98.75	$1580.00	$52.21	$92.22
5 Struc. Steel Workers	52.65	2106.00	95.15	3806.00		
1 Equip. Oper. (crane)	51.70	413.60	82.05	656.40		
1 Welder	52.65	421.20	95.15	761.20		
1 Equip. Oper. (oiler)	45.20	361.60	71.75	574.00		
1 Lattice Boom Crane, 90 Ton		1511.00		1662.10		
1 Welder, Gas Engine, 300 amp		145.85		160.44	20.71	22.78
80 L.H., Daily Totals		$5833.65		$9200.14	$72.92	$115.00
Crew E-6	**Hr.**	**Daily**	**Hr.**	**Daily**	**Bare Costs**	**Incl. O&P**
3 Struc. Steel Foremen (outside)	$54.65	$1311.60	$98.75	$2370.00	$52.25	$92.42
9 Struc. Steel Workers	52.65	3790.80	95.15	6850.80		
1 Equip. Oper. (crane)	51.70	413.60	82.05	656.40		
1 Welder	52.65	421.20	95.15	761.20		
1 Equip. Oper. (oiler)	45.20	361.60	71.75	574.00		
1 Equip. Oper. (light)	48.60	388.80	77.15	617.20		
1 Lattice Boom Crane, 90 Ton		1511.00		1662.10		
1 Welder, Gas Engine, 300 amp		145.85		160.44		
1 Air Compressor, 160 cfm		156.40		172.04		
2 Impact Wrenches		36.00		39.60	14.45	15.89
128 L.H., Daily Totals		$8536.85		$13863.78	$66.69	$108.31
Crew E-7	**Hr.**	**Daily**	**Hr.**	**Daily**	**Bare Costs**	**Incl. O&P**
1 Struc. Steel Foreman (outside)	$54.65	$437.20	$98.75	$790.00	$52.21	$92.22
4 Struc. Steel Workers	52.65	1684.80	95.15	3044.80		
1 Equip. Oper. (crane)	51.70	413.60	82.05	656.40		
1 Equip. Oper. (oiler)	45.20	361.60	71.75	574.00		
1 Welder Foreman (outside)	54.65	437.20	98.75	790.00		
2 Welders	52.65	842.40	95.15	1522.40		
1 Lattice Boom Crane, 90 Ton		1511.00		1662.10		
2 Welder, Gas Engine, 300 amp		291.70		320.87	22.53	24.79
80 L.H., Daily Totals		$5979.50		$9360.57	$74.74	$117.01
Crew E-8	**Hr.**	**Daily**	**Hr.**	**Daily**	**Bare Costs**	**Incl. O&P**
1 Struc. Steel Foreman (outside)	$54.65	$437.20	$98.75	$790.00	$52.00	$91.51
4 Struc. Steel Workers	52.65	1684.80	95.15	3044.80		
1 Welder Foreman (outside)	54.65	437.20	98.75	790.00		
4 Welders	52.65	1684.80	95.15	3044.80		
1 Equip. Oper. (crane)	51.70	413.60	82.05	656.40		
1 Equip. Oper. (oiler)	45.20	361.60	71.75	574.00		
1 Equip. Oper. (light)	48.60	388.80	77.15	617.20		
1 Lattice Boom Crane, 90 Ton		1511.00		1662.10		
4 Welder, Gas Engine, 300 amp		583.40		641.74	20.14	22.15
104 L.H., Daily Totals		$7502.40		$11821.04	$72.14	$113.66
Crew E-9	**Hr.**	**Daily**	**Hr.**	**Daily**	**Bare Costs**	**Incl. O&P**
2 Struc. Steel Foremen (outside)	$54.65	$874.40	$98.75	$1580.00	$52.25	$92.42
5 Struc. Steel Workers	52.65	2106.00	95.15	3806.00		
1 Welder Foreman (outside)	54.65	437.20	98.75	790.00		
5 Welders	52.65	2106.00	95.15	3806.00		
1 Equip. Oper. (crane)	51.70	413.60	82.05	656.40		
1 Equip. Oper. (oiler)	45.20	361.60	71.75	574.00		
1 Equip. Oper. (light)	48.60	388.80	77.15	617.20		
1 Lattice Boom Crane, 90 Ton		1511.00		1662.10		
5 Welder, Gas Engine, 300 amp		729.25		802.17	17.50	19.25
128 L.H., Daily Totals		$8927.85		$14293.88	$69.75	$111.67

Crew E-10	Bare Costs Hr.	Bare Costs Daily	Incl. Subs O&P Hr.	Incl. Subs O&P Daily	Cost Per Labor-Hour Bare Costs	Cost Per Labor-Hour Incl. O&P
1 Welder Foreman (outside)	$54.65	$437.20	$98.75	$790.00	$53.65	$96.95
1 Welder	52.65	421.20	95.15	761.20		
1 Welder, Gas Engine, 300 amp		145.85		160.44		
1 Flatbed Truck, Gas, 3 Ton		303.20		333.52	28.07	30.87
16 L.H., Daily Totals		$1307.45		$2045.16	$81.72	$127.82

Crew E-11	Hr.	Daily	Hr.	Daily	Bare Costs	Incl. O&P
2 Painters, Struc. Steel	$41.45	$663.20	$77.10	$1233.60	$42.27	$73.25
1 Building Laborer	37.60	300.80	61.65	493.20		
1 Equip. Oper. (light)	48.60	388.80	77.15	617.20		
1 Air Compressor, 250 cfm		201.40		221.54		
1 Sandblaster, Portable, 3 C.F.		20.40		22.44		
1 Set Sand Blasting Accessories		14.05		15.46	7.37	8.11
32 L.H., Daily Totals		$1588.65		$2603.43	$49.65	$81.36

Crew E-11A	Hr.	Daily	Hr.	Daily	Bare Costs	Incl. O&P
2 Painters, Struc. Steel	$41.45	$663.20	$77.10	$1233.60	$42.27	$73.25
1 Building Laborer	37.60	300.80	61.65	493.20		
1 Equip. Oper. (light)	48.60	388.80	77.15	617.20		
1 Air Compressor, 250 cfm		201.40		221.54		
1 Sandblaster, Portable, 3 C.F.		20.40		22.44		
1 Set Sand Blasting Accessories		14.05		15.46		
1 Aerial Lift Truck, 60' Boom		435.60		479.16	20.98	23.08
32 L.H., Daily Totals		$2024.25		$3082.59	$63.26	$96.33

Crew E-11B	Hr.	Daily	Hr.	Daily	Bare Costs	Incl. O&P
2 Painters, Struc. Steel	$41.45	$663.20	$77.10	$1233.60	$40.17	$71.95
1 Building Laborer	37.60	300.80	61.65	493.20		
2 Paint Sprayer, 8 C.F.M.		99.80		109.78		
1 Aerial Lift Truck, 60' Boom		435.60		479.16	22.31	24.54
24 L.H., Daily Totals		$1499.40		$2315.74	$62.48	$96.49

Crew E-12	Hr.	Daily	Hr.	Daily	Bare Costs	Incl. O&P
1 Welder Foreman (outside)	$54.65	$437.20	$98.75	$790.00	$51.63	$87.95
1 Equip. Oper. (light)	48.60	388.80	77.15	617.20		
1 Welder, Gas Engine, 300 amp		145.85		160.44	9.12	10.03
16 L.H., Daily Totals		$971.85		$1567.64	$60.74	$97.98

Crew E-13	Hr.	Daily	Hr.	Daily	Bare Costs	Incl. O&P
1 Welder Foreman (outside)	$54.65	$437.20	$98.75	$790.00	$52.63	$91.55
.5 Equip. Oper. (light)	48.60	194.40	77.15	308.60		
1 Welder, Gas Engine, 300 amp		145.85		160.44	12.15	13.37
12 L.H., Daily Totals		$777.45		$1259.04	$64.79	$104.92

Crew E-14	Hr.	Daily	Hr.	Daily	Bare Costs	Incl. O&P
1 Welder Foreman (outside)	$54.65	$437.20	$98.75	$790.00	$54.65	$98.75
1 Welder, Gas Engine, 300 amp		145.85		160.44	18.23	20.05
8 L.H., Daily Totals		$583.05		$950.43	$72.88	$118.80

Crew E-16	Hr.	Daily	Hr.	Daily	Bare Costs	Incl. O&P
1 Welder Foreman (outside)	$54.65	$437.20	$98.75	$790.00	$53.65	$96.95
1 Welder	52.65	421.20	95.15	761.20		
1 Welder, Gas Engine, 300 amp		145.85		160.44	9.12	10.03
16 L.H., Daily Totals		$1004.25		$1711.64	$62.77	$106.98

Crew E-17	Hr.	Daily	Hr.	Daily	Bare Costs	Incl. O&P
1 Struc. Steel Foreman (outside)	$54.65	$437.20	$98.75	$790.00	$53.65	$96.95
1 Structural Steel Worker	52.65	421.20	95.15	761.20		
16 L.H., Daily Totals		$858.40		$1551.20	$53.65	$96.95

Crew E-18	Hr.	Daily	Hr.	Daily	Bare Costs	Incl. O&P
1 Struc. Steel Foreman (outside)	$54.65	$437.20	$98.75	$790.00	$52.64	$92.90
3 Structural Steel Workers	52.65	1263.60	95.15	2283.60		
1 Equipment Operator (med.)	50.60	404.80	80.30	642.40		
1 Lattice Boom Crane, 20 Ton		948.90		1043.79	23.72	26.09
40 L.H., Daily Totals		$3054.50		$4759.79	$76.36	$118.99

Crew E-19	Hr.	Daily	Hr.	Daily	Bare Costs	Incl. O&P
1 Struc. Steel Foreman (outside)	$54.65	$437.20	$98.75	$790.00	$51.97	$90.35
1 Structural Steel Worker	52.65	421.20	95.15	761.20		
1 Equip. Oper. (light)	48.60	388.80	77.15	617.20		
1 Lattice Boom Crane, 20 Ton		948.90		1043.79	39.54	43.49
24 L.H., Daily Totals		$2196.10		$3212.19	$91.50	$133.84

Crew E-20	Hr.	Daily	Hr.	Daily	Bare Costs	Incl. O&P
1 Struc. Steel Foreman (outside)	$54.65	$437.20	$98.75	$790.00	$51.85	$91.04
5 Structural Steel Workers	52.65	2106.00	95.15	3806.00		
1 Equip. Oper. (crane)	51.70	413.60	82.05	656.40		
1 Equip. Oper. (oiler)	45.20	361.60	71.75	574.00		
1 Lattice Boom Crane, 40 Ton		1164.00		1280.40	18.19	20.01
64 L.H., Daily Totals		$4482.40		$7106.80	$70.04	$111.04

Crew E-22	Hr.	Daily	Hr.	Daily	Bare Costs	Incl. O&P
1 Skilled Worker Foreman (out)	$50.65	$405.20	$82.30	$658.40	$49.32	$80.13
2 Skilled Workers	48.65	778.40	79.05	1264.80		
24 L.H., Daily Totals		$1183.60		$1923.20	$49.32	$80.13

Crew E-24	Hr.	Daily	Hr.	Daily	Bare Costs	Incl. O&P
3 Structural Steel Workers	$52.65	$1263.60	$95.15	$2283.60	$52.14	$91.44
1 Equipment Operator (med.)	50.60	404.80	80.30	642.40		
1 Hyd. Crane, 25 Ton		736.60		810.26	23.02	25.32
32 L.H., Daily Totals		$2405.00		$3736.26	$75.16	$116.76

Crew E-25	Hr.	Daily	Hr.	Daily	Bare Costs	Incl. O&P
1 Welder Foreman (outside)	$54.65	$437.20	$98.75	$790.00	$54.65	$98.75
1 Cutting Torch		11.40		12.54	1.43	1.57
8 L.H., Daily Totals		$448.60		$802.54	$56.08	$100.32

Crew F-3	Hr.	Daily	Hr.	Daily	Bare Costs	Incl. O&P
4 Carpenters	$46.95	$1502.40	$76.95	$2462.40	$47.90	$77.97
1 Equip. Oper. (crane)	51.70	413.60	82.05	656.40		
1 Hyd. Crane, 12 Ton		653.80		719.18	16.34	17.98
40 L.H., Daily Totals		$2569.80		$3837.98	$64.25	$95.95

Crew F-4	Hr.	Daily	Hr.	Daily	Bare Costs	Incl. O&P
4 Carpenters	$46.95	$1502.40	$76.95	$2462.40	$47.45	$76.93
1 Equip. Oper. (crane)	51.70	413.60	82.05	656.40		
1 Equip. Oper. (oiler)	45.20	361.60	71.75	574.00		
1 Hyd. Crane, 55 Ton		1128.00		1240.80	23.50	25.85
48 L.H., Daily Totals		$3405.60		$4933.60	$70.95	$102.78

Crew F-5	Hr.	Daily	Hr.	Daily	Bare Costs	Incl. O&P
1 Carpenter Foreman (outside)	$48.95	$391.60	$80.25	$642.00	$47.45	$77.78
3 Carpenters	46.95	1126.80	76.95	1846.80		
32 L.H., Daily Totals		$1518.40		$2488.80	$47.45	$77.78

Crew No.	Bare Costs		Incl. Subs O&P		Cost Per Labor-Hour	
Crew F-6	**Hr.**	**Daily**	**Hr.**	**Daily**	**Bare Costs**	**Incl. O&P**
2 Carpenters	$46.95	$751.20	$76.95	$1231.20	$44.16	$71.85
2 Building Laborers	37.60	601.60	61.65	986.40		
1 Equip. Oper. (crane)	51.70	413.60	82.05	656.40		
1 Hyd. Crane, 12 Ton		653.80		719.18	16.34	17.98
40 L.H., Daily Totals		$2420.20		$3593.18	$60.51	$89.83

Crew F-7	Hr.	Daily	Hr.	Daily	Bare Costs	Incl. O&P
2 Carpenters	$46.95	$751.20	$76.95	$1231.20	$42.27	$69.30
2 Building Laborers	37.60	601.60	61.65	986.40		
32 L.H., Daily Totals		$1352.80		$2217.60	$42.27	$69.30

Crew G-1	Hr.	Daily	Hr.	Daily	Bare Costs	Incl. O&P
1 Roofer Foreman (outside)	$42.10	$336.80	$76.05	$608.40	$37.51	$67.78
4 Roofers Composition	40.10	1283.20	72.45	2318.40		
2 Roofer Helpers	30.05	480.80	54.30	868.80		
1 Application Equipment		193.00		212.30		
1 Tar Kettle/Pot		169.80		186.78		
1 Crew Truck		206.40		227.04	10.16	11.18
56 L.H., Daily Totals		$2670.00		$4421.72	$47.68	$78.96

Crew G-2	Hr.	Daily	Hr.	Daily	Bare Costs	Incl. O&P
1 Plasterer	$42.95	$343.60	$68.70	$549.60	$39.55	$63.75
1 Plasterer Helper	38.10	304.80	60.90	487.20		
1 Building Laborer	37.60	300.80	61.65	493.20		
1 Grout Pump, 50 C.F./hr.		133.40		146.74	5.56	6.11
24 L.H., Daily Totals		$1082.60		$1676.74	$45.11	$69.86

Crew G-2A	Hr.	Daily	Hr.	Daily	Bare Costs	Incl. O&P
1 Roofer Composition	$40.10	$320.80	$72.45	$579.60	$35.92	$62.80
1 Roofer Helper	30.05	240.40	54.30	434.40		
1 Building Laborer	37.60	300.80	61.65	493.20		
1 Foam Spray Rig, Trailer-Mtd.		573.25		630.58		
1 Pickup Truck, 3/4 Ton		144.20		158.62	29.89	32.88
24 L.H., Daily Totals		$1579.45		$2296.40	$65.81	$95.68

Crew G-3	Hr.	Daily	Hr.	Daily	Bare Costs	Incl. O&P
2 Sheet Metal Workers	$55.95	$895.20	$88.30	$1412.80	$46.77	$74.97
2 Building Laborers	37.60	601.60	61.65	986.40		
32 L.H., Daily Totals		$1496.80		$2399.20	$46.77	$74.97

Crew G-4	Hr.	Daily	Hr.	Daily	Bare Costs	Incl. O&P
1 Labor Foreman (outside)	$39.60	$316.80	$64.90	$519.20	$38.27	$62.73
2 Building Laborers	37.60	601.60	61.65	986.40		
1 Flatbed Truck, Gas, 1.5 Ton		245.40		269.94		
1 Air Compressor, 160 cfm		156.40		172.04	16.74	18.42
24 L.H., Daily Totals		$1320.20		$1947.58	$55.01	$81.15

Crew G-5	Hr.	Daily	Hr.	Daily	Bare Costs	Incl. O&P
1 Roofer Foreman (outside)	$42.10	$336.80	$76.05	$608.40	$36.48	$65.91
2 Roofers Composition	40.10	641.60	72.45	1159.20		
2 Roofer Helpers	30.05	480.80	54.30	868.80		
1 Application Equipment		193.00		212.30	4.83	5.31
40 L.H., Daily Totals		$1652.20		$2848.70	$41.31	$71.22

Crew G-6A	Hr.	Daily	Hr.	Daily	Bare Costs	Incl. O&P
2 Roofers Composition	$40.10	$641.60	$72.45	$1159.20	$40.10	$72.45
1 Small Compressor, Electric		12.95		14.24		
2 Pneumatic Nailers		55.30		60.83	4.27	4.69
16 L.H., Daily Totals		$709.85		$1234.28	$44.37	$77.14

Crew No.	Bare Costs		Incl. Subs O&P		Cost Per Labor-Hour	
Crew G-7	**Hr.**	**Daily**	**Hr.**	**Daily**	**Bare Costs**	**Incl. O&P**
1 Carpenter	$46.95	$375.60	$76.95	$615.60	$46.95	$76.95
1 Small Compressor, Electric		12.95		14.24		
1 Pneumatic Nailer		27.65		30.41	5.08	5.58
8 L.H., Daily Totals		$416.20		$660.26	$52.02	$82.53

Crew H-1	Hr.	Daily	Hr.	Daily	Bare Costs	Incl. O&P
2 Glaziers	$45.10	$721.60	$73.20	$1171.20	$48.88	$84.17
2 Struc. Steel Workers	52.65	842.40	95.15	1522.40		
32 L.H., Daily Totals		$1564.00		$2693.60	$48.88	$84.17

Crew H-2	Hr.	Daily	Hr.	Daily	Bare Costs	Incl. O&P
2 Glaziers	$45.10	$721.60	$73.20	$1171.20	$42.60	$69.35
1 Building Laborer	37.60	300.80	61.65	493.20		
24 L.H., Daily Totals		$1022.40		$1664.40	$42.60	$69.35

Crew H-3	Hr.	Daily	Hr.	Daily	Bare Costs	Incl. O&P
1 Glazier	$45.10	$360.80	$73.20	$585.60	$40.27	$65.85
1 Helper	35.45	283.60	58.50	468.00		
16 L.H., Daily Totals		$644.40		$1053.60	$40.27	$65.85

Crew H-4	Hr.	Daily	Hr.	Daily	Bare Costs	Incl. O&P
1 Carpenter	$46.95	$375.60	$76.95	$615.60	$43.90	$71.12
1 Carpenter Helper	35.45	283.60	58.50	468.00		
.5 Electrician	54.70	218.80	84.70	338.80		
20 L.H., Daily Totals		$878.00		$1422.40	$43.90	$71.12

Crew J-1	Hr.	Daily	Hr.	Daily	Bare Costs	Incl. O&P
3 Plasterers	$42.95	$1030.80	$68.70	$1648.80	$41.01	$65.58
2 Plasterer Helpers	38.10	609.60	60.90	974.40		
1 Mixing Machine, 6 C.F.		140.20		154.22	3.50	3.86
40 L.H., Daily Totals		$1780.60		$2777.42	$44.52	$69.44

Crew J-2	Hr.	Daily	Hr.	Daily	Bare Costs	Incl. O&P
3 Plasterers	$42.95	$1030.80	$68.70	$1648.80	$41.34	$65.93
2 Plasterer Helpers	38.10	609.60	60.90	974.40		
1 Lather	43.00	344.00	67.70	541.60		
1 Mixing Machine, 6 C.F.		140.20		154.22	2.92	3.21
48 L.H., Daily Totals		$2124.60		$3319.02	$44.26	$69.15

Crew J-3	Hr.	Daily	Hr.	Daily	Bare Costs	Incl. O&P
1 Terrazzo Worker	$42.95	$343.60	$67.80	$542.40	$39.23	$61.92
1 Terrazzo Helper	35.50	284.00	56.05	448.40		
1 Floor Grinder, 22" Path		115.55		127.11		
1 Terrazzo Mixer		188.20		207.02	18.98	20.88
16 L.H., Daily Totals		$931.35		$1324.93	$58.21	$82.81

Crew J-4	Hr.	Daily	Hr.	Daily	Bare Costs	Incl. O&P
2 Cement Finishers	$45.00	$720.00	$71.15	$1138.40	$42.53	$67.98
1 Laborer	37.60	300.80	61.65	493.20		
1 Floor Grinder, 22" Path		115.55		127.11		
1 Floor Edger, 7" Path		39.30		43.23		
1 Vacuum Pick-Up System		60.90		66.99	8.99	9.89
24 L.H., Daily Totals		$1236.55		$1868.93	$51.52	$77.87

Crew No.	Bare Costs		Incl. Subs O&P		Cost Per Labor-Hour	
Crew J-4A	**Hr.**	**Daily**	**Hr.**	**Daily**	**Bare Costs**	**Incl. O&P**
2 Cement Finishers	$45.00	$720.00	$71.15	$1138.40	$41.30	$66.40
2 Laborers	37.60	601.60	61.65	986.40		
1 Floor Grinder, 22" Path		115.55		127.11		
1 Floor Edger, 7" Path		39.30		43.23		
1 Vacuum Pick-Up System		60.90		66.99		
1 Floor Auto Scrubber		233.85		257.24	14.05	15.46
32 L.H., Daily Totals		$1771.20		$2619.36	$55.35	$81.86

Crew J-4B	**Hr.**	**Daily**	**Hr.**	**Daily**	**Bare Costs**	**Incl. O&P**
1 Laborer	$37.60	$300.80	$61.65	$493.20	$37.60	$61.65
1 Floor Auto Scrubber		233.85		257.24	29.23	32.15
8 L.H., Daily Totals		$534.65		$750.43	$66.83	$93.80

Crew J-6	**Hr.**	**Daily**	**Hr.**	**Daily**	**Bare Costs**	**Incl. O&P**
2 Painters	$40.35	$645.60	$64.85	$1037.60	$41.73	$67.13
1 Building Laborer	37.60	300.80	61.65	493.20		
1 Equip. Oper. (light)	48.60	388.80	77.15	617.20		
1 Air Compressor, 250 cfm		201.40		221.54		
1 Sandblaster, Portable, 3 C.F.		20.40		22.44		
1 Set Sand Blasting Accessories		14.05		15.46	7.37	8.11
32 L.H., Daily Totals		$1571.05		$2407.43	$49.10	$75.23

Crew J-7	**Hr.**	**Daily**	**Hr.**	**Daily**	**Bare Costs**	**Incl. O&P**
2 Painters	$40.35	$645.60	$64.85	$1037.60	$40.35	$64.85
1 Floor Belt Sander		14.20		15.62		
1 Floor Sanding Edger		12.60		13.86	1.68	1.84
16 L.H., Daily Totals		$672.40		$1067.08	$42.02	$66.69

Crew K-1	**Hr.**	**Daily**	**Hr.**	**Daily**	**Bare Costs**	**Incl. O&P**
1 Carpenter	$46.95	$375.60	$76.95	$615.60	$43.02	$70.00
1 Truck Driver (light)	39.10	312.80	63.05	504.40		
1 Flatbed Truck, Gas, 3 Ton		303.20		333.52	18.95	20.84
16 L.H., Daily Totals		$991.60		$1453.52	$61.98	$90.84

Crew K-2	**Hr.**	**Daily**	**Hr.**	**Daily**	**Bare Costs**	**Incl. O&P**
1 Struc. Steel Foreman (outside)	$54.65	$437.20	$98.75	$790.00	$48.80	$85.65
1 Struc. Steel Worker	52.65	421.20	95.15	761.20		
1 Truck Driver (light)	39.10	312.80	63.05	504.40		
1 Flatbed Truck, Gas, 3 Ton		303.20		333.52	12.63	13.90
24 L.H., Daily Totals		$1474.40		$2389.12	$61.43	$99.55

Crew L-1	**Hr.**	**Daily**	**Hr.**	**Daily**	**Bare Costs**	**Incl. O&P**
1 Electrician	$54.70	$437.60	$84.70	$677.60	$56.70	$88.13
1 Plumber	58.70	469.60	91.55	732.40		
16 L.H., Daily Totals		$907.20		$1410.00	$56.70	$88.13

Crew L-2	**Hr.**	**Daily**	**Hr.**	**Daily**	**Bare Costs**	**Incl. O&P**
1 Carpenter	$46.95	$375.60	$76.95	$615.60	$41.20	$67.72
1 Carpenter Helper	35.45	283.60	58.50	468.00		
16 L.H., Daily Totals		$659.20		$1083.60	$41.20	$67.72

Crew L-3	**Hr.**	**Daily**	**Hr.**	**Daily**	**Bare Costs**	**Incl. O&P**
1 Carpenter	$46.95	$375.60	$76.95	$615.60	$51.14	$81.72
.5 Electrician	54.70	218.80	84.70	338.80		
.5 Sheet Metal Worker	55.95	223.80	88.30	353.20		
16 L.H., Daily Totals		$818.20		$1307.60	$51.14	$81.72

Crew No.	Bare Costs		Incl. Subs O&P		Cost Per Labor-Hour	
Crew L-3A	**Hr.**	**Daily**	**Hr.**	**Daily**	**Bare Costs**	**Incl. O&P**
1 Carpenter Foreman (outside)	$48.95	$391.60	$80.25	$642.00	$51.28	$82.93
.5 Sheet Metal Worker	55.95	223.80	88.30	353.20		
12 L.H., Daily Totals		$615.40		$995.20	$51.28	$82.93

Crew L-4	**Hr.**	**Daily**	**Hr.**	**Daily**	**Bare Costs**	**Incl. O&P**
2 Skilled Workers	$48.65	$778.40	$79.05	$1264.80	$44.25	$72.20
1 Helper	35.45	283.60	58.50	468.00		
24 L.H., Daily Totals		$1062.00		$1732.80	$44.25	$72.20

Crew L-5	**Hr.**	**Daily**	**Hr.**	**Daily**	**Bare Costs**	**Incl. O&P**
1 Struc. Steel Foreman (outside)	$54.65	$437.20	$98.75	$790.00	$52.80	$93.79
5 Struc. Steel Workers	52.65	2106.00	95.15	3806.00		
1 Equip. Oper. (crane)	51.70	413.60	82.05	656.40		
1 Hyd. Crane, 25 Ton		736.60		810.26	13.15	14.47
56 L.H., Daily Totals		$3693.40		$6062.66	$65.95	$108.26

Crew L-5A	**Hr.**	**Daily**	**Hr.**	**Daily**	**Bare Costs**	**Incl. O&P**
1 Struc. Steel Foreman (outside)	$54.65	$437.20	$98.75	$790.00	$52.91	$92.78
2 Structural Steel Workers	52.65	842.40	95.15	1522.40		
1 Equip. Oper. (crane)	51.70	413.60	82.05	656.40		
1 S.P. Crane, 4x4, 25 Ton		599.40		659.34	18.73	20.60
32 L.H., Daily Totals		$2292.60		$3628.14	$71.64	$113.38

Crew L-5B	**Hr.**	**Daily**	**Hr.**	**Daily**	**Bare Costs**	**Incl. O&P**
1 Struc. Steel Foreman (outside)	$54.65	$437.20	$98.75	$790.00	$53.97	$88.74
2 Structural Steel Workers	52.65	842.40	95.15	1522.40		
2 Electricians	54.70	875.20	84.70	1355.20		
2 Steamfitters/Pipefitters	59.75	956.00	93.20	1491.20		
1 Equip. Oper. (crane)	51.70	413.60	82.05	656.40		
1 Equip. Oper. (oiler)	45.20	361.60	71.75	574.00		
1 Hyd. Crane, 80 Ton		1625.00		1787.50	22.57	24.83
72 L.H., Daily Totals		$5511.00		$8176.70	$76.54	$113.57

Crew L-6	**Hr.**	**Daily**	**Hr.**	**Daily**	**Bare Costs**	**Incl. O&P**
1 Plumber	$58.70	$469.60	$91.55	$732.40	$57.37	$89.27
.5 Electrician	54.70	218.80	84.70	338.80		
12 L.H., Daily Totals		$688.40		$1071.20	$57.37	$89.27

Crew L-7	**Hr.**	**Daily**	**Hr.**	**Daily**	**Bare Costs**	**Incl. O&P**
2 Carpenters	$46.95	$751.20	$76.95	$1231.20	$45.39	$73.69
1 Building Laborer	37.60	300.80	61.65	493.20		
.5 Electrician	54.70	218.80	84.70	338.80		
28 L.H., Daily Totals		$1270.80		$2063.20	$45.39	$73.69

Crew L-8	**Hr.**	**Daily**	**Hr.**	**Daily**	**Bare Costs**	**Incl. O&P**
2 Carpenters	$46.95	$751.20	$76.95	$1231.20	$49.30	$79.87
.5 Plumber	58.70	234.80	91.55	366.20		
20 L.H., Daily Totals		$986.00		$1597.40	$49.30	$79.87

Crew L-9	**Hr.**	**Daily**	**Hr.**	**Daily**	**Bare Costs**	**Incl. O&P**
1 Labor Foreman (inside)	$38.10	$304.80	$62.45	$499.60	$42.96	$71.83
2 Building Laborers	37.60	601.60	61.65	986.40		
1 Struc. Steel Worker	52.65	421.20	95.15	761.20		
.5 Electrician	54.70	218.80	84.70	338.80		
36 L.H., Daily Totals		$1546.40		$2586.00	$42.96	$71.83

Crew No.	Bare Costs		Incl. Subs O&P		Cost Per Labor-Hour	
Crew L-10	**Hr.**	**Daily**	**Hr.**	**Daily**	**Bare Costs**	**Incl. O&P**
1 Struc. Steel Foreman (outside)	$54.65	$437.20	$98.75	$790.00	$53.00	$91.98
1 Structural Steel Worker	52.65	421.20	95.15	761.20		
1 Equip. Oper. (crane)	51.70	413.60	82.05	656.40		
1 Hyd. Crane, 12 Ton		653.80		719.18	27.24	29.97
24 L.H., Daily Totals		$1925.80		$2926.78	$80.24	$121.95
Crew L-11	**Hr.**	**Daily**	**Hr.**	**Daily**	**Bare Costs**	**Incl. O&P**
2 Wreckers	$37.60	$601.60	$65.90	$1054.40	$43.88	$72.75
1 Equip. Oper. (crane)	51.70	413.60	82.05	656.40		
1 Equip. Oper. (light)	48.60	388.80	77.15	617.20		
1 Hyd. Excavator, 2.5 C.Y.		1605.00		1765.50		
1 Loader, Skid Steer, 78 H.P.		318.60		350.46	60.11	66.12
32 L.H., Daily Totals		$3327.60		$4443.96	$103.99	$138.87
Crew M-1	**Hr.**	**Daily**	**Hr.**	**Daily**	**Bare Costs**	**Incl. O&P**
3 Elevator Constructors	$76.50	$1836.00	$118.05	$2833.20	$72.67	$112.15
1 Elevator Apprentice	61.20	489.60	94.45	755.60		
5 Hand Tools		46.00		50.60	1.44	1.58
32 L.H., Daily Totals		$2371.60		$3639.40	$74.11	$113.73
Crew M-3	**Hr.**	**Daily**	**Hr.**	**Daily**	**Bare Costs**	**Incl. O&P**
1 Electrician Foreman (outside)	$56.70	$453.60	$87.75	$702.00	$57.56	$89.88
1 Common Laborer	37.60	300.80	61.65	493.20		
.25 Equipment Operator (med.)	50.60	101.20	80.30	160.60		
1 Elevator Constructor	76.50	612.00	118.05	944.40		
1 Elevator Apprentice	61.20	489.60	94.45	755.60		
.25 S.P. Crane, 4x4, 20 Ton		138.65		152.51	4.08	4.49
34 L.H., Daily Totals		$2095.85		$3208.32	$61.64	$94.36
Crew M-4	**Hr.**	**Daily**	**Hr.**	**Daily**	**Bare Costs**	**Incl. O&P**
1 Electrician Foreman (outside)	$56.70	$453.60	$87.75	$702.00	$56.94	$88.97
1 Common Laborer	37.60	300.80	61.65	493.20		
.25 Equipment Operator, Crane	51.70	103.40	82.05	164.10		
.25 Equip. Oper. (oiler)	45.20	90.40	71.75	143.50		
1 Elevator Constructor	76.50	612.00	118.05	944.40		
1 Elevator Apprentice	61.20	489.60	94.45	755.60		
.25 S.P. Crane, 4x4, 40 Ton		175.75		193.32	4.88	5.37
36 L.H., Daily Totals		$2225.55		$3396.13	$61.82	$94.34
Crew Q-1	**Hr.**	**Daily**	**Hr.**	**Daily**	**Bare Costs**	**Incl. O&P**
1 Plumber	$58.70	$469.60	$91.55	$732.40	$52.83	$82.40
1 Plumber Apprentice	46.95	375.60	73.25	586.00		
16 L.H., Daily Totals		$845.20		$1318.40	$52.83	$82.40
Crew Q-1A	**Hr.**	**Daily**	**Hr.**	**Daily**	**Bare Costs**	**Incl. O&P**
.25 Plumber Foreman (outside)	$60.70	$121.40	$94.70	$189.40	$59.10	$92.18
1 Plumber	58.70	469.60	91.55	732.40		
10 L.H., Daily Totals		$591.00		$921.80	$59.10	$92.18
Crew Q-1C	**Hr.**	**Daily**	**Hr.**	**Daily**	**Bare Costs**	**Incl. O&P**
1 Plumber	$58.70	$469.60	$91.55	$732.40	$52.08	$81.70
1 Plumber Apprentice	46.95	375.60	73.25	586.00		
1 Equip. Oper. (medium)	50.60	404.80	80.30	642.40		
1 Trencher, Chain Type, 8' D		3402.00		3742.20	141.75	155.93
24 L.H., Daily Totals		$4652.00		$5703.00	$193.83	$237.63
Crew Q-2	**Hr.**	**Daily**	**Hr.**	**Daily**	**Bare Costs**	**Incl. O&P**
2 Plumbers	$58.70	$939.20	$91.55	$1464.80	$54.78	$85.45
1 Plumber Apprentice	46.95	375.60	73.25	586.00		
24 L.H., Daily Totals		$1314.80		$2050.80	$54.78	$85.45

Crew No.	Bare Costs		Incl. Subs O&P		Cost Per Labor-Hour	
Crew Q-3	**Hr.**	**Daily**	**Hr.**	**Daily**	**Bare Costs**	**Incl. O&P**
1 Plumber Foreman (inside)	$59.20	$473.60	$92.35	$738.80	$55.89	$87.17
2 Plumbers	58.70	939.20	91.55	1464.80		
1 Plumber Apprentice	46.95	375.60	73.25	586.00		
32 L.H., Daily Totals		$1788.40		$2789.60	$55.89	$87.17
Crew Q-4	**Hr.**	**Daily**	**Hr.**	**Daily**	**Bare Costs**	**Incl. O&P**
1 Plumber Foreman (inside)	$59.20	$473.60	$92.35	$738.80	$55.89	$87.17
1 Plumber	58.70	469.60	91.55	732.40		
1 Welder (plumber)	58.70	469.60	91.55	732.40		
1 Plumber Apprentice	46.95	375.60	73.25	586.00		
1 Welder, Electric, 300 amp		57.70		63.47	1.80	1.98
32 L.H., Daily Totals		$1846.10		$2853.07	$57.69	$89.16
Crew Q-5	**Hr.**	**Daily**	**Hr.**	**Daily**	**Bare Costs**	**Incl. O&P**
1 Steamfitter	$59.75	$478.00	$93.20	$745.60	$53.77	$83.88
1 Steamfitter Apprentice	47.80	382.40	74.55	596.40		
16 L.H., Daily Totals		$860.40		$1342.00	$53.77	$83.88
Crew Q-6	**Hr.**	**Daily**	**Hr.**	**Daily**	**Bare Costs**	**Incl. O&P**
2 Steamfitters	$59.75	$956.00	$93.20	$1491.20	$55.77	$86.98
1 Steamfitter Apprentice	47.80	382.40	74.55	596.40		
24 L.H., Daily Totals		$1338.40		$2087.60	$55.77	$86.98
Crew Q-7	**Hr.**	**Daily**	**Hr.**	**Daily**	**Bare Costs**	**Incl. O&P**
1 Steamfitter Foreman (inside)	$60.25	$482.00	$94.00	$752.00	$56.89	$88.74
2 Steamfitters	59.75	956.00	93.20	1491.20		
1 Steamfitter Apprentice	47.80	382.40	74.55	596.40		
32 L.H., Daily Totals		$1820.40		$2839.60	$56.89	$88.74
Crew Q-8	**Hr.**	**Daily**	**Hr.**	**Daily**	**Bare Costs**	**Incl. O&P**
1 Steamfitter Foreman (inside)	$60.25	$482.00	$94.00	$752.00	$56.89	$88.74
1 Steamfitter	59.75	478.00	93.20	745.60		
1 Welder (steamfitter)	59.75	478.00	93.20	745.60		
1 Steamfitter Apprentice	47.80	382.40	74.55	596.40		
1 Welder, Electric, 300 amp		57.70		63.47	1.80	1.98
32 L.H., Daily Totals		$1878.10		$2903.07	$58.69	$90.72
Crew Q-9	**Hr.**	**Daily**	**Hr.**	**Daily**	**Bare Costs**	**Incl. O&P**
1 Sheet Metal Worker	$55.95	$447.60	$88.30	$706.40	$50.35	$79.45
1 Sheet Metal Apprentice	44.75	358.00	70.60	564.80		
16 L.H., Daily Totals		$805.60		$1271.20	$50.35	$79.45
Crew Q-10	**Hr.**	**Daily**	**Hr.**	**Daily**	**Bare Costs**	**Incl. O&P**
2 Sheet Metal Workers	$55.95	$895.20	$88.30	$1412.80	$52.22	$82.40
1 Sheet Metal Apprentice	44.75	358.00	70.60	564.80		
24 L.H., Daily Totals		$1253.20		$1977.60	$52.22	$82.40
Crew Q-11	**Hr.**	**Daily**	**Hr.**	**Daily**	**Bare Costs**	**Incl. O&P**
1 Sheet Metal Foreman (inside)	$56.45	$451.60	$89.10	$712.80	$53.27	$84.08
2 Sheet Metal Workers	55.95	895.20	88.30	1412.80		
1 Sheet Metal Apprentice	44.75	358.00	70.60	564.80		
32 L.H., Daily Totals		$1704.80		$2690.40	$53.27	$84.08
Crew Q-12	**Hr.**	**Daily**	**Hr.**	**Daily**	**Bare Costs**	**Incl. O&P**
1 Sprinkler Installer	$56.15	$449.20	$87.75	$702.00	$50.52	$78.97
1 Sprinkler Apprentice	44.90	359.20	70.20	561.60		
16 L.H., Daily Totals		$808.40		$1263.60	$50.52	$78.97

Crew Q-13	Bare Costs Hr.	Bare Costs Daily	Incl. Subs O&P Hr.	Incl. Subs O&P Daily	Cost Per Labor-Hour Bare Costs	Cost Per Labor-Hour Incl. O&P
1 Sprinkler Foreman (inside)	$56.65	$453.20	$88.55	$708.40	$53.46	$83.56
2 Sprinkler Installers	56.15	898.40	87.75	1404.00		
1 Sprinkler Apprentice	44.90	359.20	70.20	561.60		
32 L.H., Daily Totals		$1710.80		$2674.00	$53.46	$83.56

Crew Q-14	Hr.	Daily	Hr.	Daily	Bare Costs	Incl. O&P
1 Asbestos Worker	$52.35	$418.80	$84.15	$673.20	$47.13	$75.75
1 Asbestos Apprentice	41.90	335.20	67.35	538.80		
16 L.H., Daily Totals		$754.00		$1212.00	$47.13	$75.75

Crew Q-15	Hr.	Daily	Hr.	Daily	Bare Costs	Incl. O&P
1 Plumber	$58.70	$469.60	$91.55	$732.40	$52.83	$82.40
1 Plumber Apprentice	46.95	375.60	73.25	586.00		
1 Welder, Electric, 300 amp		57.70		63.47	3.61	3.97
16 L.H., Daily Totals		$902.90		$1381.87	$56.43	$86.37

Crew Q-16	Hr.	Daily	Hr.	Daily	Bare Costs	Incl. O&P
2 Plumbers	$58.70	$939.20	$91.55	$1464.80	$54.78	$85.45
1 Plumber Apprentice	46.95	375.60	73.25	586.00		
1 Welder, Electric, 300 amp		57.70		63.47	2.40	2.64
24 L.H., Daily Totals		$1372.50		$2114.27	$57.19	$88.09

Crew Q-17	Hr.	Daily	Hr.	Daily	Bare Costs	Incl. O&P
1 Steamfitter	$59.75	$478.00	$93.20	$745.60	$53.77	$83.88
1 Steamfitter Apprentice	47.80	382.40	74.55	596.40		
1 Welder, Electric, 300 amp		57.70		63.47	3.61	3.97
16 L.H., Daily Totals		$918.10		$1405.47	$57.38	$87.84

Crew Q-17A	Hr.	Daily	Hr.	Daily	Bare Costs	Incl. O&P
1 Steamfitter	$59.75	$478.00	$93.20	$745.60	$53.08	$83.27
1 Steamfitter Apprentice	47.80	382.40	74.55	596.40		
1 Equip. Oper. (crane)	51.70	413.60	82.05	656.40		
1 Hyd. Crane, 12 Ton		653.80		719.18		
1 Welder, Electric, 300 amp		57.70		63.47	29.65	32.61
24 L.H., Daily Totals		$1985.50		$2781.05	$82.73	$115.88

Crew Q-18	Hr.	Daily	Hr.	Daily	Bare Costs	Incl. O&P
2 Steamfitters	$59.75	$956.00	$93.20	$1491.20	$55.77	$86.98
1 Steamfitter Apprentice	47.80	382.40	74.55	596.40		
1 Welder, Electric, 300 amp		57.70		63.47	2.40	2.64
24 L.H., Daily Totals		$1396.10		$2151.07	$58.17	$89.63

Crew Q-19	Hr.	Daily	Hr.	Daily	Bare Costs	Incl. O&P
1 Steamfitter	$59.75	$478.00	$93.20	$745.60	$54.08	$84.15
1 Steamfitter Apprentice	47.80	382.40	74.55	596.40		
1 Electrician	54.70	437.60	84.70	677.60		
24 L.H., Daily Totals		$1298.00		$2019.60	$54.08	$84.15

Crew Q-20	Hr.	Daily	Hr.	Daily	Bare Costs	Incl. O&P
1 Sheet Metal Worker	$55.95	$447.60	$88.30	$706.40	$51.22	$80.50
1 Sheet Metal Apprentice	44.75	358.00	70.60	564.80		
.5 Electrician	54.70	218.80	84.70	338.80		
20 L.H., Daily Totals		$1024.40		$1610.00	$51.22	$80.50

Crew Q-21	Hr.	Daily	Hr.	Daily	Bare Costs	Incl. O&P
2 Steamfitters	$59.75	$956.00	$93.20	$1491.20	$55.50	$86.41
1 Steamfitter Apprentice	47.80	382.40	74.55	596.40		
1 Electrician	54.70	437.60	84.70	677.60		
32 L.H., Daily Totals		$1776.00		$2765.20	$55.50	$86.41

Crew Q-22	Hr.	Daily	Hr.	Daily	Bare Costs	Incl. O&P
1 Plumber	$58.70	$469.60	$91.55	$732.40	$52.83	$82.40
1 Plumber Apprentice	46.95	375.60	73.25	586.00		
1 Hyd. Crane, 12 Ton		653.80		719.18	40.86	44.95
16 L.H., Daily Totals		$1499.00		$2037.58	$93.69	$127.35

Crew Q-22A	Hr.	Daily	Hr.	Daily	Bare Costs	Incl. O&P
1 Plumber	$58.70	$469.60	$91.55	$732.40	$48.74	$77.13
1 Plumber Apprentice	46.95	375.60	73.25	586.00		
1 Laborer	37.60	300.80	61.65	493.20		
1 Equip. Oper. (crane)	51.70	413.60	82.05	656.40		
1 Hyd. Crane, 12 Ton		653.80		719.18	20.43	22.47
32 L.H., Daily Totals		$2213.40		$3187.18	$69.17	$99.60

Crew Q-23	Hr.	Daily	Hr.	Daily	Bare Costs	Incl. O&P
1 Plumber Foreman (outside)	$60.70	$485.60	$94.70	$757.60	$56.67	$88.85
1 Plumber	58.70	469.60	91.55	732.40		
1 Equip. Oper. (medium)	50.60	404.80	80.30	642.40		
1 Lattice Boom Crane, 20 Ton		948.90		1043.79	39.54	43.49
24 L.H., Daily Totals		$2308.90		$3176.19	$96.20	$132.34

Crew R-1	Hr.	Daily	Hr.	Daily	Bare Costs	Incl. O&P
1 Electrician Foreman	$55.20	$441.60	$85.45	$683.60	$48.37	$76.09
3 Electricians	54.70	1312.80	84.70	2032.80		
2 Helpers	35.45	567.20	58.50	936.00		
48 L.H., Daily Totals		$2321.60		$3652.40	$48.37	$76.09

Crew R-1A	Hr.	Daily	Hr.	Daily	Bare Costs	Incl. O&P
1 Electrician	$54.70	$437.60	$84.70	$677.60	$45.08	$71.60
1 Helper	35.45	283.60	58.50	468.00		
16 L.H., Daily Totals		$721.20		$1145.60	$45.08	$71.60

Crew R-2	Hr.	Daily	Hr.	Daily	Bare Costs	Incl. O&P
1 Electrician Foreman	$55.20	$441.60	$85.45	$683.60	$48.84	$76.94
3 Electricians	54.70	1312.80	84.70	2032.80		
2 Helpers	35.45	567.20	58.50	936.00		
1 Equip. Oper. (crane)	51.70	413.60	82.05	656.40		
1 S.P. Crane, 4x4, 5 Ton		275.40		302.94	4.92	5.41
56 L.H., Daily Totals		$3010.60		$4611.74	$53.76	$82.35

Crew R-3	Hr.	Daily	Hr.	Daily	Bare Costs	Incl. O&P
1 Electrician Foreman	$55.20	$441.60	$85.45	$683.60	$54.30	$84.47
1 Electrician	54.70	437.60	84.70	677.60		
.5 Equip. Oper. (crane)	51.70	206.80	82.05	328.20		
.5 S.P. Crane, 4x4, 5 Ton		137.70		151.47	6.88	7.57
20 L.H., Daily Totals		$1223.70		$1840.87	$61.19	$92.04

Crew R-4	Hr.	Daily	Hr.	Daily	Bare Costs	Incl. O&P
1 Struc. Steel Foreman (outside)	$54.65	$437.20	$98.75	$790.00	$53.46	$93.78
3 Struc. Steel Workers	52.65	1263.60	95.15	2283.60		
1 Electrician	54.70	437.60	84.70	677.60		
1 Welder, Gas Engine, 300 amp		145.85		160.44	3.65	4.01
40 L.H., Daily Totals		$2284.25		$3911.64	$57.11	$97.79

Crew R-5	Hr.	Daily	Hr.	Daily	Bare Costs	Incl. O&P
1 Electrician Foreman	$55.20	$441.60	$85.45	$683.60	$47.75	$75.24
4 Electrician Linemen	54.70	1750.40	84.70	2710.40		
2 Electrician Operators	54.70	875.20	84.70	1355.20		
4 Electrician Groundmen	35.45	1134.40	58.50	1872.00		
1 Crew Truck		206.40		227.04		
1 Flatbed Truck, 20,000 GVW		248.60		273.46		
1 Pickup Truck, 3/4 Ton		144.20		158.62		
.2 Hyd. Crane, 55 Ton		225.60		248.16		
.2 Hyd. Crane, 12 Ton		130.76		143.84		
.2 Earth Auger, Truck-Mtd.		83.44		91.78		
1 Tractor w/Winch		427.00		469.70	16.66	18.32
88 L.H., Daily Totals		$5667.60		$8233.80	$64.40	$93.57

Crew R-6	Hr.	Daily	Hr.	Daily	Bare Costs	Incl. O&P
1 Electrician Foreman	$55.20	$441.60	$85.45	$683.60	$47.75	$75.24
4 Electrician Linemen	54.70	1750.40	84.70	2710.40		
2 Electrician Operators	54.70	875.20	84.70	1355.20		
4 Electrician Groundmen	35.45	1134.40	58.50	1872.00		
1 Crew Truck		206.40		227.04		
1 Flatbed Truck, 20,000 GVW		248.60		273.46		
1 Pickup Truck, 3/4 Ton		144.20		158.62		
.2 Hyd. Crane, 55 Ton		225.60		248.16		
.2 Hyd. Crane, 12 Ton		130.76		143.84		
.2 Earth Auger, Truck-Mtd.		83.44		91.78		
1 Tractor w/Winch		427.00		469.70		
3 Cable Trailers		598.20		658.02		
.5 Tensioning Rig		198.40		218.24		
.5 Cable Pulling Rig		1160.00		1276.00	38.89	42.78
88 L.H., Daily Totals		$7624.20		$10386.06	$86.64	$118.02

Crew R-7	Hr.	Daily	Hr.	Daily	Bare Costs	Incl. O&P
1 Electrician Foreman	$55.20	$441.60	$85.45	$683.60	$38.74	$62.99
5 Electrician Groundmen	35.45	1418.00	58.50	2340.00		
1 Crew Truck		206.40		227.04	4.30	4.73
48 L.H., Daily Totals		$2066.00		$3250.64	$43.04	$67.72

Crew R-8	Hr.	Daily	Hr.	Daily	Bare Costs	Incl. O&P
1 Electrician Foreman	$55.20	$441.60	$85.45	$683.60	$48.37	$76.09
3 Electrician Linemen	54.70	1312.80	84.70	2032.80		
2 Electrician Groundmen	35.45	567.20	58.50	936.00		
1 Pickup Truck, 3/4 Ton		144.20		158.62		
1 Crew Truck		206.40		227.04	7.30	8.03
48 L.H., Daily Totals		$2672.20		$4038.06	$55.67	$84.13

Crew R-9	Hr.	Daily	Hr.	Daily	Bare Costs	Incl. O&P
1 Electrician Foreman	$55.20	$441.60	$85.45	$683.60	$45.14	$71.69
1 Electrician Lineman	54.70	437.60	84.70	677.60		
2 Electrician Operators	54.70	875.20	84.70	1355.20		
4 Electrician Groundmen	35.45	1134.40	58.50	1872.00		
1 Pickup Truck, 3/4 Ton		144.20		158.62		
1 Crew Truck		206.40		227.04	5.48	6.03
64 L.H., Daily Totals		$3239.40		$4974.06	$50.62	$77.72

Crew R-10	Hr.	Daily	Hr.	Daily	Bare Costs	Incl. O&P
1 Electrician Foreman	$55.20	$441.60	$85.45	$683.60	$51.58	$80.46
4 Electrician Linemen	54.70	1750.40	84.70	2710.40		
1 Electrician Groundman	35.45	283.60	58.50	468.00		
1 Crew Truck		206.40		227.04		
3 Tram Cars		412.20		453.42	12.89	14.18
48 L.H., Daily Totals		$3094.20		$4542.46	$64.46	$94.63

Crew R-11	Hr.	Daily	Hr.	Daily	Bare Costs	Incl. O&P
1 Electrician Foreman	$55.20	$441.60	$85.45	$683.60	$51.90	$81.14
4 Electricians	54.70	1750.40	84.70	2710.40		
1 Equip. Oper. (crane)	51.70	413.60	82.05	656.40		
1 Common Laborer	37.60	300.80	61.65	493.20		
1 Crew Truck		206.40		227.04		
1 Hyd. Crane, 12 Ton		653.80		719.18	15.36	16.90
56 L.H., Daily Totals		$3766.60		$5489.82	$67.26	$98.03

Crew R-12	Hr.	Daily	Hr.	Daily	Bare Costs	Incl. O&P
1 Carpenter Foreman (inside)	$47.45	$379.60	$77.75	$622.00	$44.45	$73.42
4 Carpenters	46.95	1502.40	76.95	2462.40		
4 Common Laborers	37.60	1203.20	61.65	1972.80		
1 Equip. Oper. (medium)	50.60	404.80	80.30	642.40		
1 Steel Worker	52.65	421.20	95.15	761.20		
1 Dozer, 200 H.P.		1387.00		1525.70		
1 Pickup Truck, 3/4 Ton		144.20		158.62	17.40	19.14
88 L.H., Daily Totals		$5442.40		$8145.12	$61.85	$92.56

Crew R-13	Hr.	Daily	Hr.	Daily	Bare Costs	Incl. O&P
1 Electrician Foreman	$55.20	$441.60	$85.45	$683.60	$52.84	$82.25
3 Electricians	54.70	1312.80	84.70	2032.80		
.25 Equip. Oper. (crane)	51.70	103.40	82.05	164.10		
1 Equipment Oiler	45.20	361.60	71.75	574.00		
.25 Hydraulic Crane, 33 Ton		187.50		206.25	4.46	4.91
42 L.H., Daily Totals		$2406.90		$3660.75	$57.31	$87.16

Crew R-15	Hr.	Daily	Hr.	Daily	Bare Costs	Incl. O&P
1 Electrician Foreman	$55.20	$441.60	$85.45	$683.60	$53.77	$83.57
4 Electricians	54.70	1750.40	84.70	2710.40		
1 Equipment Oper. (light)	48.60	388.80	77.15	617.20		
1 Aerial Lift Truck, 40' Boom		301.40		331.54	6.28	6.91
48 L.H., Daily Totals		$2882.20		$4342.74	$60.05	$90.47

Crew R-18	Hr.	Daily	Hr.	Daily	Bare Costs	Incl. O&P
.25 Electrician Foreman	$55.20	$110.40	$85.45	$170.90	$42.89	$68.63
1 Electrician	54.70	437.60	84.70	677.60		
2 Helpers	35.45	567.20	58.50	936.00		
26 L.H., Daily Totals		$1115.20		$1784.50	$42.89	$68.63

Crew R-19	Hr.	Daily	Hr.	Daily	Bare Costs	Incl. O&P
.5 Electrician Foreman	$55.20	$220.80	$85.45	$341.80	$54.80	$84.85
2 Electricians	54.70	875.20	84.70	1355.20		
20 L.H., Daily Totals		$1096.00		$1697.00	$54.80	$84.85

Crew R-21	Hr.	Daily	Hr.	Daily	Bare Costs	Incl. O&P
1 Electrician Foreman	$55.20	$441.60	$85.45	$683.60	$54.72	$84.78
3 Electricians	54.70	1312.80	84.70	2032.80		
.1 Equip. Oper. (medium)	50.60	40.48	80.30	64.24		
.1 S.P. Crane, 4x4, 25 Ton		59.94		65.93	1.83	2.01
32.8 L.H., Daily Totals		$1854.82		$2846.57	$56.55	$86.79

Crew R-22	Hr.	Daily	Hr.	Daily	Bare Costs	Incl. O&P
.66 Electrician Foreman	$55.20	$291.46	$85.45	$451.18	$46.51	$73.56
2 Electricians	54.70	875.20	84.70	1355.20		
2 Helpers	35.45	567.20	58.50	936.00		
37.28 L.H., Daily Totals		$1733.86		$2742.38	$46.51	$73.56

Crew No.	Bare Costs		Incl. Subs O & P		Cost Per Labor-Hour	
Crew R-30	**Hr.**	**Daily**	**Hr.**	**Daily**	**Bare Costs**	**Incl. O&P**
.25 Electrician Foreman (outside)	$56.70	$113.40	$87.75	$175.50	$44.33	$70.75
1 Electrician	54.70	437.60	84.70	677.60		
2 Laborers, (Semi-Skilled)	37.60	601.60	61.65	986.40		
26 L.H., Daily Totals		$1152.60		$1839.50	$44.33	$70.75
Crew R-31	**Hr.**	**Daily**	**Hr.**	**Daily**	**Bare Costs**	**Incl. O&P**
1 Electrician	$54.70	$437.60	$84.70	$677.60	$54.70	$84.70
1 Core Drill, Electric, 2.5 H.P.		55.15		60.66	6.89	7.58
8 L.H., Daily Totals		$492.75		$738.26	$61.59	$92.28

Notes

Historical Cost Indexes

The table below lists both the RSMeans® historical cost index based on Jan. 1, 1993 = 100 as well as the computed value of an index based on Jan. 1, 2015 costs. Since the Jan. 1, 2015 figure is estimated, space is left to write in the actual index figures as they become available through either the quarterly *RSMeans Construction Cost Indexes* or as printed in the *Engineering News-Record.* To compute the actual index based on Jan. 1, 2015 = 100, divide the historical cost index for a particular year by the actual Jan. 1, 2015 construction cost index. Space has been left to advance the index figures as the year progresses.

Year	Historical Cost Index Jan. 1, 1993 = 100		Current Index Based on Jan. 1, 2015 = 100		Year	Historical Cost Index Jan. 1, 1993 = 100	Current Index Based on Jan. 1, 2015 = 100		Year	Historical Cost Index Jan. 1, 1993 = 100	Current Index Based on Jan. 1, 2015 = 100	
	Est.	Actual	Est.	Actual		Actual	Est.	Actual		Actual	Est.	Actual
Oct 2015*					July 2000	120.9	58.5		July 1982	76.1	36.8	
July 2015*					1999	117.6	56.9		1981	70.0	33.9	
April 2015*					1998	115.1	55.7		1980	62.9	30.4	
Jan 2015*	206.7		100.0	100.0	1997	112.8	54.6		1979	57.8	28.0	
July 2014		204.9	99.1		1996	110.2	53.3		1978	53.5	25.9	
2013		201.2	97.3		1995	107.6	52.1		1977	49.5	23.9	
2012		194.6	94.1		1994	104.4	50.5		1976	46.9	22.7	
2011		191.2	92.5		1993	101.7	49.2		1975	44.8	21.7	
2010		183.5	88.8		1992	99.4	48.1		1974	41.4	20.0	
2009		180.1	87.1		1991	96.8	46.8		1973	37.7	18.2	
2008		180.4	87.3		1990	94.3	45.6		1972	34.8	16.8	
2007		169.4	82.0		1989	92.1	44.6		1971	32.1	15.5	
2006		162.0	78.4		1988	89.9	43.5		1970	28.7	13.9	
2005		151.6	73.3		1987	87.7	42.4		1969	26.9	13.0	
2004		143.7	69.5		1986	84.2	40.7		1968	24.9	12.0	
2003		132.0	63.9		1985	82.6	40.0		1967	23.5	11.4	
2002		128.7	62.3		1984	82.0	39.7		1966	22.7	11.0	
2001		125.1	60.5		1983	80.2	38.8		1965	21.7	10.5	

Adjustments to Costs

The "Historical Cost Index" can be used to convert national average building costs at a particular time to the approximate building costs for some other time.

Time Adjustment Using the Historical Cost Indexes:

$$\frac{\text{Index for Year A}}{\text{Index for Year B}} \times \text{Cost in Year B} = \text{Cost in Year A}$$

Example:

Estimate and compare construction costs for different years in the same city.

To estimate the national average construction cost of a building in 1970, knowing that it cost $900,000 in 2015:

INDEX in 1970 = 28.7

INDEX in 2015 = 206.7

$$\frac{\text{INDEX 1970}}{\text{INDEX 2015}} \times \text{Cost 2015} = \text{Cost 1970}$$

$$\frac{28.7}{206.7} \times \$900{,}000 = .139 \times \$900{,}000 = \$125{,}100$$

The construction cost of the building in 1970 is $125,100.

Note: The city cost indexes for Canada can be used to convert U.S. national averages to local costs in Canadian dollars.

Example:

To estimate and compare the cost of a building in Toronto, ON in 2015 with the known cost of $600,000 (US$) in New York, NY in 2015:

INDEX Toronto = 110.9

INDEX New York = 131.8

$$\frac{\text{INDEX Toronto}}{\text{INDEX New York}} \times \text{Cost New York} = \text{Cost Toronto}$$

$$\frac{110.9}{131.8} \times \$600{,}000 = .841 \times \$600{,}000 = \$504{,}600$$

The construction cost of the building in Toronto is $504,600 (CN$).

*Historical Cost Index updates and other resources are provided on the following website.
http://info.thegordiangroup.com/RSMeans.html

How to Use the City Cost Indexes

What you should know before you begin

RSMeans City Cost Indexes (CCI) are an extremely useful tool to use when you want to compare costs from city to city and region to region.

This publication contains average construction cost indexes for 731 U.S. and Canadian cities covering over 930 three-digit zip code locations, as listed directly under each city.

Keep in mind that a City Cost Index number is a percentage ratio of a specific city's cost to the national average cost of the same item at a stated time period.

In other words, these index figures represent relative construction factors (or, if you prefer, multipliers) for Material and Installation costs, as well as the weighted average for Total In Place costs for each CSI MasterFormat division. Installation costs include both labor and equipment rental costs. When estimating equipment rental rates only, for a specific location, use 01 54 33 EQUIPMENT RENTAL COSTS in the Reference Section at the back of the book.

The 30 City Average Index is the average of 30 major U.S. cities and serves as a National Average.

Index figures for both material and installation are based on the 30 major city average of 100 and represent the cost relationship as of July 1, 2014. The index for each division is computed from representative material and labor quantities for that division. The weighted average for each city is a weighted total of the components listed above it, but does not include relative productivity between trades or cities.

As changes occur in local material prices, labor rates, and equipment rental rates (including fuel costs), the impact of these changes should be accurately measured by the change in the City Cost Index for each particular city (as compared to the 30 City Average).

Therefore, if you know (or have estimated) building costs in one city today, you can easily convert those costs to expected building costs in another city.

In addition, by using the Historical Cost Index, you can easily convert National Average building costs at a particular time to the approximate building costs for some other time. The City Cost Indexes can then be applied to calculate the costs for a particular city.

Quick Calculations

Location Adjustment Using the City Cost Indexes:

$$\frac{\text{Index for City A}}{\text{Index for City B}} \times \text{Cost in City B} = \text{Cost in City A}$$

Time Adjustment for the National Average Using the Historical Cost Index:

$$\frac{\text{Index for Year A}}{\text{Index for Year B}} \times \text{Cost in Year B} = \text{Cost in Year A}$$

Adjustment from the National Average:

$$\frac{\text{Index for City A}}{100} \times \text{National Average Cost} = \text{Cost in City A}$$

Since each of the other RSMeans publications contains many different items, any *one* item multiplied by the particular city index may give incorrect results. However, the larger the number of items compiled, the closer the results should be to actual costs for that particular city.

The City Cost Indexes for Canadian cities are calculated using Canadian material and equipment prices and labor rates, in Canadian dollars. Therefore, indexes for Canadian cities can be used to convert U.S. National Average prices to local costs in Canadian dollars.

How to use this section

1. Compare costs from city to city.

In using the RSMeans Indexes, remember that an index number is not a fixed number but a ratio: It's a percentage ratio of a building component's cost at any stated time to the National Average cost of that same component at the same time period. Put in the form of an equation:

$$\frac{\text{Specific City Cost}}{\text{National Average Cost}} \times 100 = \text{City Index Number}$$

Therefore, when making cost comparisons between cities, do not subtract one city's index number from the index number of another city and read the result as a percentage difference. Instead, divide one city's index number by that of the other city. The resulting number may then be used as a multiplier to calculate cost differences from city to city.

The formula used to find cost differences between cities for the purpose of comparison is as follows:

$$\frac{\text{City A Index}}{\text{City B Index}} \times \text{City B Cost (Known)} = \text{City A Cost (Unknown)}$$

In addition, you can use RSMeans CCI to calculate and compare costs division by division between cities using the same basic formula. (Just be sure that you're comparing similar divisions.)

2. Compare a specific city's construction costs with the National Average.

When you're studying construction location feasibility, it's advisable to compare a prospective project's cost index with an index of the National Average cost.

For example, divide the weighted average index of construction costs of a specific city by that of the 30 City Average, which = 100.

$$\frac{\text{City Index}}{100} = \text{\% of National Average}$$

As a result, you get a ratio that indicates the relative cost of construction in that city in comparison with the National Average.

3. Convert U.S. National Average to actual costs in Canadian City.

$$\frac{\text{Index for Canadian City}}{100} \times \text{National Average Cost} = \text{Cost in Canadian City in \$ CAN}$$

4. **Adjust construction cost data based on a National Average.**

When you use a source of construction cost data which is based on a National Average (such as RSMeans cost data publications), it is necessary to adjust those costs to a specific location.

$$\frac{\text{City Index}}{100} \times \frac{\text{"Book" Cost Based on}}{\text{National Average Costs}} = \frac{\text{City Cost}}{\text{(Unknown)}}$$

5. **When applying the City Cost Indexes to demolition projects, use the appropriate division installation index.** For example, for removal of existing doors and windows, use Division 8 (Openings) index.

What you might like to know about how we developed the Indexes

The information presented in the CCI is organized according to the Construction Specifications Institute (CSI) MasterFormat 2014 classification system.

To create a reliable index, RSMeans researched the building type most often constructed in the United States and Canada. Because it was concluded that no one type of building completely represented the building construction industry, nine different types of buildings were combined to create a composite model.

The exact material, labor, and equipment quantities are based on detailed analyses of these nine building types, and then each quantity is weighted in proportion to expected usage. These various material items, labor hours, and equipment rental rates are thus combined to form a composite building representing as closely as possible the actual usage of materials, labor, and equipment used in the North American building construction industry.

The following structures were chosen to make up that composite model:

1. Factory, 1 story
2. Office, 2–4 story
3. Store, Retail
4. Town Hall, 2–3 story
5. High School, 2–3 story
6. Hospital, 4–8 story
7. Garage, Parking
8. Apartment, 1–3 story
9. Hotel/Motel, 2–3 story

For the purposes of ensuring the timeliness of the data, the components of the index for the composite model have been streamlined. They currently consist of:

- specific quantities of 66 commonly used construction materials;
- specific labor-hours for 21 building construction trades; and
- specific days of equipment rental for 6 types of construction equipment (normally used to install the 66 material items by the 21 trades.) Fuel costs and routine maintenance costs are included in the equipment cost.

A sophisticated computer program handles the updating of all costs for each city on a quarterly basis. Material and equipment price quotations are gathered quarterly from cities in the United States and Canada. These prices and the latest negotiated labor wage rates for 21 different building trades are used to compile the quarterly update of the City Cost Index.

The 30 major U.S. cities used to calculate the National Average are:

Atlanta, GA	Memphis, TN
Baltimore, MD	Milwaukee, WI
Boston, MA	Minneapolis, MN
Buffalo, NY	Nashville, TN
Chicago, IL	New Orleans, LA
Cincinnati, OH	New York, NY
Cleveland, OH	Philadelphia, PA
Columbus, OH	Phoenix, AZ
Dallas, TX	Pittsburgh, PA
Denver, CO	St. Louis, MO
Detroit, MI	San Antonio, TX
Houston, TX	San Diego, CA
Indianapolis, IN	San Francisco, CA
Kansas City, MO	Seattle, WA
Los Angeles, CA	Washington, DC

What the CCI does not indicate

The weighted average for each city is a total of the divisional components weighted to reflect typical usage, but it does not include the productivity variations between trades or cities.

In addition, the CCI does not take into consideration factors such as the following:

- managerial efficiency
- competitive conditions
- automation
- restrictive union practices
- unique local requirements
- regional variations due to specific building codes

City Cost Indexes

	DIVISION	UNITED STATES			ALABAMA														
		30 CITY			ANNISTON			BIRMINGHAM			BUTLER			DECATUR			DOTHAN		
		AVERAGE			362			350 - 352			369			356			363		
		MAT.	INST.	TOTAL	MAT.	INST.	TOTAL	MAT.	INST.	TOTAL	MAT.	INST.	TOTAL	MAT.	INST.	TOTAL	MAT.	INST.	TOTAL
015433	**CONTRACTOR EQUIPMENT**		100.0	100.0		101.0	101.0		101.1	101.1		98.2	98.2		101.0	101.0		98.2	98.2
0241, 31 - 34	**SITE & INFRASTRUCTURE, DEMOLITION**	100.0	100.0	100.0	90.2	93.1	92.3	97.4	93.8	94.8	103.5	87.1	91.8	88.8	91.5	90.7	101.2	87.1	91.1
0310	Concrete Forming & Accessories	100.0	100.0	100.0	90.1	47.1	53.0	93.1	75.9	78.2	86.4	42.6	48.6	94.6	45.4	52.1	95.5	42.0	49.4
0320	Concrete Reinforcing	100.0	100.0	100.0	87.8	86.3	87.1	94.7	86.9	90.7	92.8	45.9	68.9	88.6	77.0	82.7	92.8	45.4	68.7
0330	Cast-in-Place Concrete	100.0	100.0	100.0	101.8	49.7	80.4	110.0	74.6	95.4	99.3	54.8	81.0	103.4	64.1	87.2	99.3	46.3	77.5
03	**CONCRETE**	100.0	100.0	100.0	102.1	57.0	79.9	102.2	78.4	90.5	103.0	49.3	76.6	98.3	59.3	79.1	102.3	46.0	74.7
04	**MASONRY**	100.0	100.0	100.0	100.1	66.8	79.3	98.8	76.0	84.6	105.9	48.6	70.2	97.3	51.3	68.6	107.1	39.7	65.1
05	**METALS**	100.0	100.0	100.0	104.0	91.5	100.1	104.0	94.1	101.0	102.8	76.6	94.8	106.2	86.1	100.0	102.9	75.3	94.4
06	**WOOD, PLASTICS & COMPOSITES**	100.0	100.0	100.0	90.5	42.7	63.7	97.4	76.1	85.4	85.0	42.4	61.1	98.9	42.9	67.6	97.4	42.4	66.6
07	**THERMAL & MOISTURE PROTECTION**	100.0	100.0	100.0	97.9	52.9	79.5	99.7	81.5	92.2	98.0	56.7	81.0	96.7	58.3	81.0	98.0	49.1	77.9
08	**OPENINGS**	100.0	100.0	100.0	98.6	50.1	87.3	103.2	76.0	96.8	98.6	44.2	85.9	106.5	50.4	93.5	98.6	44.2	86.0
0920	Plaster & Gypsum Board	100.0	100.0	100.0	90.8	41.4	57.4	94.8	75.8	81.9	87.9	41.0	56.2	95.8	41.6	59.2	98.6	41.0	59.7
0950, 0980	Ceilings & Acoustic Treatment	100.0	100.0	100.0	80.4	41.4	54.7	89.8	75.8	80.6	80.4	41.0	54.5	87.2	41.6	57.2	80.4	41.0	54.5
0960	Flooring	100.0	100.0	100.0	91.5	40.2	76.8	102.5	76.7	95.1	95.8	56.0	84.4	99.7	50.6	85.6	100.6	27.5	79.7
0970, 0990	Wall Finishes & Painting/Coating	100.0	100.0	100.0	103.7	31.5	60.1	103.9	66.5	81.3	103.7	53.3	73.3	99.6	66.8	79.8	103.7	53.3	73.3
09	**FINISHES**	100.0	100.0	100.0	87.0	43.0	62.8	95.4	74.8	84.1	89.9	45.7	65.5	91.9	47.7	67.6	92.5	40.1	63.6
COVERS	**DIVS. 10 - 14, 25, 28, 41, 43, 44, 46**	100.0	100.0	100.0	100.0	68.3	93.6	100.0	80.4	97.7	100.0	41.4	88.2	100.0	43.2	88.5	100.0	41.5	88.2
21, 22, 23	**FIRE SUPPRESSION, PLUMBING & HVAC**	100.0	100.0	100.0	100.0	55.4	82.0	100.0	70.4	88.1	97.3	33.3	71.5	100.0	40.8	76.1	97.3	33.2	71.5
26, 27, 3370	**ELECTRICAL, COMMUNICATIONS & UTIL.**	100.0	100.0	100.0	93.1	57.3	74.2	99.0	61.6	79.3	95.0	40.2	66.0	94.2	64.9	78.7	93.8	57.4	74.6
MF2014	**WEIGHTED AVERAGE**	100.0	100.0	100.0	98.6	61.8	82.6	100.6	76.7	90.2	98.9	49.4	77.3	99.8	57.9	81.5	99.0	49.4	77.4

	DIVISION	ALABAMA																	
		EVERGREEN			GADSDEN			HUNTSVILLE			JASPER			MOBILE			MONTGOMERY		
		364			359			357 - 358			355			365 - 366			360 - 361		
		MAT.	INST.	TOTAL	MAT.	INST.	TOTAL	MAT.	INST.	TOTAL	MAT.	INST.	TOTAL	MAT.	INST.	TOTAL	MAT.	INST.	TOTAL
015433	**CONTRACTOR EQUIPMENT**		98.2	98.2		101.0	101.0		101.0	101.0		101.0	101.0		98.2	98.2		98.2	98.2
0241, 31 - 34	**SITE & INFRASTRUCTURE, DEMOLITION**	104.0	87.4	92.2	94.9	92.7	93.3	88.6	92.7	91.5	94.6	92.4	93.1	96.4	88.1	90.5	96.5	88.1	90.5
0310	Concrete Forming & Accessories	83.0	44.2	49.5	86.9	42.5	48.6	94.6	69.9	73.3	92.1	34.6	42.5	94.5	54.0	59.6	94.1	44.1	51.0
0320	Concrete Reinforcing	92.9	45.9	69.0	94.0	86.8	90.3	88.6	81.6	85.1	88.6	85.6	87.1	90.6	83.6	87.0	95.8	86.1	90.9
0330	Cast-in-Place Concrete	99.3	57.3	82.1	103.4	66.7	88.3	100.7	69.5	87.9	114.5	44.1	85.6	104.1	66.4	88.6	105.2	56.5	85.2
03	**CONCRETE**	103.3	50.9	77.5	102.9	60.8	82.2	97.0	72.9	85.2	106.6	49.4	78.5	98.9	65.3	82.4	100.2	57.9	79.4
04	**MASONRY**	105.9	53.2	73.0	95.6	62.4	74.9	98.8	67.7	79.4	92.9	41.2	60.7	103.7	54.2	72.9	100.0	49.7	68.6
05	**METALS**	102.9	76.5	94.7	104.0	93.2	100.7	106.2	90.6	101.4	104.0	90.2	99.7	105.0	91.6	100.9	104.0	91.9	100.3
06	**WOOD, PLASTICS & COMPOSITES**	81.3	42.4	59.5	89.3	38.7	61.0	98.9	72.6	84.2	95.9	31.6	59.9	95.9	52.8	71.8	95.3	42.4	65.6
07	**THERMAL & MOISTURE PROTECTION**	97.9	56.9	81.1	96.8	71.7	86.5	96.6	77.6	88.8	96.8	45.4	75.7	97.6	70.1	86.3	97.9	66.6	85.0
08	**OPENINGS**	98.6	44.2	85.9	102.8	49.0	90.3	106.5	67.4	97.4	102.8	48.0	90.0	102.2	58.8	92.1	103.3	53.1	91.7
0920	Plaster & Gypsum Board	87.2	41.0	56.0	88.2	37.3	53.8	95.8	72.2	79.9	92.1	30.0	50.1	95.2	51.8	65.9	94.6	41.0	58.4
0950, 0980	Ceilings & Acoustic Treatment	80.4	41.0	54.5	83.8	37.3	53.2	89.0	72.2	78.0	83.8	30.0	48.4	85.6	51.8	63.4	88.0	41.0	57.1
0960	Flooring	93.9	56.0	83.1	95.5	76.7	90.1	99.7	76.7	93.1	97.8	40.2	81.3	100.5	58.5	88.5	97.3	58.5	86.2
0970, 0990	Wall Finishes & Painting/Coating	103.7	53.3	73.3	99.6	60.7	76.1	99.6	65.9	79.3	99.6	38.6	62.8	107.3	54.3	75.3	103.3	53.3	73.1
09	**FINISHES**	89.2	46.8	65.9	89.4	48.7	67.0	92.3	70.7	80.4	90.5	34.4	59.6	92.7	53.5	71.1	92.8	46.1	67.0
COVERS	**DIVS. 10 - 14, 25, 28, 41, 43, 44, 46**	100.0	42.9	88.5	100.0	79.8	95.9	100.0	85.3	97.0	100.0	39.8	87.8	100.0	83.1	96.6	100.0	79.6	95.9
21, 22, 23	**FIRE SUPPRESSION, PLUMBING & HVAC**	97.3	35.6	72.4	102.0	36.3	75.5	100.0	61.6	84.5	102.0	61.4	85.6	99.9	60.7	84.1	100.0	34.2	73.4
26, 27, 3370	**ELECTRICAL, COMMUNICATIONS & UTIL.**	92.5	40.2	64.8	94.3	61.6	77.0	95.3	64.9	79.3	93.8	61.3	76.6	95.7	59.1	76.3	96.3	60.9	77.6
MF2014	**WEIGHTED AVERAGE**	98.6	50.7	77.7	99.9	60.0	82.5	99.8	72.1	87.8	100.2	57.5	81.6	99.9	65.4	84.9	99.9	57.1	81.2

	DIVISION	ALABAMA									ALASKA								
		PHENIX CITY			SELMA			TUSCALOOSA			ANCHORAGE			FAIRBANKS			JUNEAU		
		368			367			354			995 - 996			997			998		
		MAT.	INST.	TOTAL	MAT.	INST.	TOTAL	MAT.	INST.	TOTAL	MAT.	INST.	TOTAL	MAT.	INST.	TOTAL	MAT.	INST.	TOTAL
015433	**CONTRACTOR EQUIPMENT**		98.2	98.2		98.2	98.2		101.0	101.0		114.7	114.7		114.7	114.7		114.7	114.7
0241, 31 - 34	**SITE & INFRASTRUCTURE, DEMOLITION**	107.8	88.2	93.9	101.0	88.1	91.8	89.1	92.9	91.8	127.2	129.6	128.9	119.2	129.6	126.6	135.6	129.6	131.3
0310	Concrete Forming & Accessories	90.1	39.1	46.1	87.6	42.6	48.8	94.5	47.9	54.3	123.7	119.8	120.3	131.6	119.9	121.5	130.1	119.8	121.2
0320	Concrete Reinforcing	92.8	61.4	76.8	92.8	85.2	89.0	88.6	86.8	87.7	149.7	110.7	129.9	150.9	110.7	130.5	135.7	110.7	123.0
0330	Cast-in-Place Concrete	99.3	57.0	81.9	99.3	46.5	77.6	104.9	68.4	89.9	132.0	117.8	126.2	128.3	118.2	124.1	133.2	117.8	126.9
03	**CONCRETE**	106.2	51.3	79.2	101.7	53.7	78.1	99.0	63.8	81.7	138.6	116.7	127.9	123.8	116.9	120.4	136.0	116.7	126.5
04	**MASONRY**	105.9	41.1	65.5	109.6	38.7	65.4	97.6	65.5	77.6	184.6	125.3	147.6	189.5	125.3	149.5	176.6	125.3	144.6
05	**METALS**	102.8	81.9	96.4	102.8	89.3	98.7	105.3	93.2	101.6	113.6	103.4	110.5	117.6	103.5	113.3	116.3	103.4	112.3
06	**WOOD, PLASTICS & COMPOSITES**	90.2	35.4	59.5	87.0	42.4	62.0	98.9	44.4	68.4	129.2	118.7	123.4	136.1	118.7	126.4	129.4	118.7	123.4
07	**THERMAL & MOISTURE PROTECTION**	98.3	63.8	84.1	97.8	52.1	79.1	96.7	73.7	87.3	159.5	118.0	142.5	166.5	119.1	147.1	168.0	118.0	147.5
08	**OPENINGS**	98.6	43.5	85.8	98.6	53.1	88.0	106.5	58.8	95.4	132.7	115.6	128.7	128.9	115.7	125.8	129.8	115.6	126.5
0920	Plaster & Gypsum Board	92.2	33.9	52.8	90.1	41.0	56.9	95.8	43.1	60.2	140.9	119.1	126.2	164.8	119.1	133.9	148.1	119.1	128.5
0950, 0980	Ceilings & Acoustic Treatment	80.4	33.9	49.8	80.4	41.0	54.5	89.0	43.1	58.9	119.6	119.1	119.3	119.2	119.1	119.2	125.9	119.1	121.4
0960	Flooring	97.6	58.5	86.4	96.2	28.5	76.8	99.7	76.7	93.1	134.6	133.0	134.1	127.2	133.0	128.9	131.0	133.0	131.6
0970, 0990	Wall Finishes & Painting/Coating	103.7	53.3	73.3	103.7	53.3	73.3	99.6	45.9	67.2	136.8	116.3	124.4	133.5	121.2	126.1	132.1	116.3	122.6
09	**FINISHES**	91.4	42.2	64.3	90.0	40.2	62.6	92.3	51.2	69.7	134.3	122.2	127.7	132.4	122.8	127.1	134.4	122.2	127.7
COVERS	**DIVS. 10 - 14, 25, 28, 41, 43, 44, 46**	100.0	79.3	95.8	100.0	41.4	88.2	100.0	81.4	96.2	100.0	112.8	102.6	100.0	112.8	102.6	100.0	112.8	102.6
21, 22, 23	**FIRE SUPPRESSION, PLUMBING & HVAC**	97.3	34.5	72.0	97.3	33.6	71.6	100.0	34.0	73.4	100.3	105.0	102.2	100.2	108.0	103.4	100.3	105.0	102.2
26, 27, 3370	**ELECTRICAL, COMMUNICATIONS & UTIL.**	94.4	69.5	81.3	93.6	40.2	65.4	94.8	61.6	77.3	117.7	117.8	117.7	130.0	117.8	123.5	119.9	117.8	118.8
MF2014	**WEIGHTED AVERAGE**	99.5	54.6	79.9	98.7	49.9	77.4	99.8	61.2	83.0	121.1	115.6	118.7	121.0	116.4	119.0	121.3	115.6	118.8

City Cost Indexes

DIVISION		ALASKA			ARIZONA															
		KETCHIKAN			CHAMBERS			FLAGSTAFF			GLOBE			KINGMAN			MESA/TEMPE		
		999			865			860			855			864			852		
		MAT.	INST.	TOTAL	MAT.	INST.	TOTAL	MAT.	INST.	TOTAL	MAT.	INST.	TOTAL	MAT.	INST.	TOTAL	MAT.	INST.	TOTAL
015433	**CONTRACTOR EQUIPMENT**		114.7	114.7		93.0	93.0		93.0	93.0		92.2	92.2		93.0	93.0		92.2	92.2
0241, 31 - 34	**SITE & INFRASTRUCTURE, DEMOLITION**	173.8	129.6	142.4	69.6	96.1	88.4	87.3	96.3	93.7	97.3	95.6	96.1	69.6	96.3	88.6	89.0	95.8	93.8
0310	Concrete Forming & Accessories	122.9	119.8	120.2	98.8	58.1	63.7	104.4	65.1	70.5	99.0	58.1	63.7	97.0	65.1	69.5	102.2	69.3	73.9
0320	Concrete Reinforcing	117.4	110.7	114.0	97.1	85.2	91.1	97.0	85.3	91.0	108.0	85.2	96.4	97.2	85.3	91.1	108.7	85.3	96.8
0330	Cast-in-Place Concrete	259.2	117.8	201.2	90.7	73.2	83.5	90.8	73.4	83.7	94.3	72.0	85.1	90.4	73.4	83.4	95.0	72.3	85.7
03	**CONCRETE**	208.7	116.7	163.5	95.0	68.6	82.0	114.1	71.8	93.3	110.0	68.2	89.5	94.6	71.8	83.4	101.8	73.3	87.8
04	**MASONRY**	198.4	125.3	152.8	92.5	62.3	73.7	92.6	62.4	73.8	109.9	62.2	80.1	92.5	62.4	73.7	110.1	62.2	80.3
05	**METALS**	117.8	103.4	113.4	96.2	75.6	89.8	96.7	76.3	90.4	93.3	76.0	88.0	96.8	76.3	90.5	93.6	76.8	88.4
06	**WOOD, PLASTICS & COMPOSITES**	126.3	118.7	122.1	101.0	55.0	75.2	107.3	64.2	83.1	97.6	55.0	73.8	96.1	64.2	78.2	101.1	69.8	83.6
07	**THERMAL & MOISTURE PROTECTION**	169.3	118.0	148.3	94.6	65.0	82.5	96.2	68.1	84.7	101.9	63.0	85.9	94.6	65.6	82.7	101.2	65.1	86.4
08	**OPENINGS**	130.8	115.6	127.3	108.1	65.3	98.2	108.3	70.3	99.4	100.0	65.3	91.9	108.3	70.3	99.5	100.1	73.3	93.8
0920	Plaster & Gypsum Board	152.7	119.1	130.0	90.2	53.7	65.5	93.6	63.2	73.0	96.6	53.7	67.6	82.9	63.2	69.6	98.7	68.9	78.6
0950, 0980	Ceilings & Acoustic Treatment	112.7	119.1	116.9	99.7	53.7	69.5	100.5	63.2	76.0	88.9	53.7	65.8	100.5	63.2	76.0	88.9	68.9	75.8
0960	Flooring	127.2	133.0	128.9	93.3	39.5	77.9	95.4	39.7	79.5	103.5	39.5	85.2	92.1	53.7	81.1	104.7	49.5	88.9
0970, 0990	Wall Finishes & Painting/Coating	133.5	116.3	123.1	98.3	55.1	72.2	98.3	55.1	72.2	103.4	55.1	74.3	98.3	55.1	72.2	103.4	55.1	74.3
09	**FINISHES**	133.4	122.2	127.2	93.7	52.9	71.2	96.6	58.4	75.6	97.1	53.0	72.8	92.5	60.7	75.0	96.8	63.6	78.5
COVERS	**DIVS. 10 - 14, 25, 28, 41, 43, 44, 46**	100.0	112.8	102.6	100.0	82.3	96.4	100.0	83.3	96.6	100.0	82.4	96.4	100.0	83.3	96.6	100.0	84.0	96.8
21, 22, 23	**FIRE SUPPRESSION, PLUMBING & HVAC**	98.4	105.0	101.1	97.0	78.9	89.7	100.2	79.0	91.6	95.2	78.9	88.7	97.0	79.0	89.7	100.0	79.0	91.5
26, 27, 3370	**ELECTRICAL, COMMUNICATIONS & UTIL.**	129.9	117.8	123.5	104.5	70.9	86.7	103.4	61.3	81.1	97.5	61.2	78.3	104.5	61.3	81.7	94.2	61.3	76.8
MF2014	**WEIGHTED AVERAGE**	132.0	115.6	124.8	97.6	71.5	86.2	101.3	71.8	88.4	98.8	70.0	86.3	97.6	72.0	86.4	98.6	72.8	87.4

DIVISION		ARIZONA												ARKANSAS					
		PHOENIX			PRESCOTT			SHOW LOW			TUCSON			BATESVILLE			CAMDEN		
		850,853			863			859			856 - 857			725			717		
		MAT.	INST.	TOTAL	MAT.	INST.	TOTAL	MAT.	INST.	TOTAL	MAT.	INST.	TOTAL	MAT.	INST.	TOTAL	MAT.	INST.	TOTAL
015433	**CONTRACTOR EQUIPMENT**		92.7	92.7		93.0	93.0		92.2	92.2		92.2	92.2		89.1	89.1		89.1	89.1
0241, 31 - 34	**SITE & INFRASTRUCTURE, DEMOLITION**	89.4	96.0	94.1	76.0	96.0	90.2	99.2	95.6	96.6	85.2	95.8	92.7	77.1	84.9	82.6	78.3	84.4	82.7
0310	Concrete Forming & Accessories	103.1	68.3	73.1	100.4	52.5	59.1	106.3	69.1	74.2	102.5	68.1	72.9	85.9	44.9	50.5	83.7	30.7	38.0
0320	Concrete Reinforcing	107.0	85.4	96.0	97.0	85.2	91.0	108.7	85.2	96.8	89.9	85.3	87.5	87.4	67.5	77.3	93.6	67.4	80.3
0330	Cast-in-Place Concrete	95.1	72.4	85.8	90.7	73.1	83.5	94.3	72.1	85.2	97.8	72.3	87.3	77.4	45.3	64.2	82.6	38.3	64.4
03	**CONCRETE**	101.4	73.0	87.4	100.1	66.1	83.4	112.4	73.1	93.1	100.0	72.8	86.6	83.5	50.1	67.1	87.5	41.3	64.8
04	**MASONRY**	97.5	64.1	76.7	92.6	62.3	73.7	109.9	62.2	80.1	95.5	62.2	74.8	99.8	40.7	63.0	116.6	31.4	63.5
05	**METALS**	95.1	77.6	89.7	96.7	75.3	90.1	93.1	76.1	87.8	94.3	76.7	88.9	99.5	66.9	89.4	103.9	66.5	92.4
06	**WOOD, PLASTICS & COMPOSITES**	102.1	68.3	83.2	102.4	47.4	71.6	105.8	69.8	85.6	101.4	68.3	82.8	89.9	45.5	65.0	90.3	29.7	56.4
07	**THERMAL & MOISTURE PROTECTION**	100.9	67.0	87.0	95.1	64.2	82.4	102.1	65.0	86.9	102.4	64.2	86.7	98.2	43.9	76.0	94.2	35.7	70.2
08	**OPENINGS**	102.0	72.5	95.2	108.3	61.1	97.3	99.2	73.3	93.2	96.3	72.5	90.7	97.7	45.8	85.6	102.1	40.6	87.8
0920	Plaster & Gypsum Board	101.0	67.4	78.2	90.3	45.9	60.3	100.9	68.9	79.3	104.0	67.4	79.2	83.3	44.3	56.9	85.0	28.0	46.5
0950, 0980	Ceilings & Acoustic Treatment	96.5	67.4	77.3	98.8	45.9	64.0	88.9	68.9	75.8	89.8	67.4	75.0	83.2	44.3	57.6	83.3	28.0	47.0
0960	Flooring	104.9	51.7	89.7	94.1	39.5	78.5	106.2	45.0	88.7	95.2	44.2	80.6	96.7	59.1	85.9	99.6	39.6	82.4
0970, 0990	Wall Finishes & Painting/Coating	103.4	61.7	78.3	98.3	55.1	72.2	103.4	55.1	74.3	104.5	55.1	74.7	107.4	41.2	67.4	104.4	50.4	71.8
09	**FINISHES**	98.9	63.8	79.6	94.2	48.5	69.0	98.7	62.7	78.8	94.9	61.6	76.5	86.6	46.7	64.6	88.5	33.5	58.2
COVERS	**DIVS. 10 - 14, 25, 28, 41, 43, 44, 46**	100.0	83.9	96.7	100.0	81.4	96.3	100.0	84.0	96.8	100.0	83.9	96.7	100.0	39.5	87.8	100.0	35.6	87.0
21, 22, 23	**FIRE SUPPRESSION, PLUMBING & HVAC**	99.9	79.0	91.5	100.2	78.9	91.6	95.2	79.0	88.7	100.0	79.0	91.5	95.3	51.6	77.7	95.3	53.1	78.3
26, 27, 3370	**ELECTRICAL, COMMUNICATIONS & UTIL.**	101.0	66.7	82.9	103.0	61.2	80.9	94.6	61.2	77.0	96.5	61.3	77.9	96.5	63.5	79.0	96.8	62.1	78.5
MF2014	**WEIGHTED AVERAGE**	99.2	73.8	88.1	99.2	68.9	86.0	98.9	72.5	87.4	97.4	72.4	86.5	94.6	54.6	77.1	97.1	50.1	76.6

DIVISION		ARKANSAS																	
		FAYETTEVILLE			FORT SMITH			HARRISON			HOT SPRINGS			JONESBORO			LITTLE ROCK		
		727			729			726			719			724			720 - 722		
		MAT.	INST.	TOTAL	MAT.	INST.	TOTAL	MAT.	INST.	TOTAL	MAT.	INST.	TOTAL	MAT.	INST.	TOTAL	MAT.	INST.	TOTAL
015433	**CONTRACTOR EQUIPMENT**		89.1	89.1		89.1	89.1		89.1	89.1		89.1	89.1		108.3	108.3		89.1	89.1
0241, 31 - 34	**SITE & INFRASTRUCTURE, DEMOLITION**	76.4	86.3	83.5	81.8	86.7	85.3	82.1	84.9	84.1	81.4	85.8	84.5	103.2	101.1	101.7	93.1	86.8	88.6
0310	Concrete Forming & Accessories	81.2	38.5	44.4	101.3	57.3	63.4	90.6	44.8	51.1	81.1	35.7	41.9	89.6	48.2	53.9	96.4	60.0	65.0
0320	Concrete Reinforcing	87.4	71.3	79.2	88.4	72.0	80.1	87.0	63.7	75.1	91.8	67.5	79.4	84.5	70.1	77.2	93.4	68.3	80.6
0330	Cast-in-Place Concrete	77.4	49.4	65.9	88.5	70.3	81.0	85.8	43.4	68.4	84.5	38.8	65.7	84.3	55.0	72.2	87.3	70.3	80.3
03	**CONCRETE**	83.2	49.3	66.5	91.3	65.0	78.4	90.8	48.7	70.1	91.2	43.7	67.9	88.1	56.0	72.3	92.8	65.5	79.4
04	**MASONRY**	90.5	41.4	59.9	96.9	50.9	68.3	100.1	39.0	62.0	87.6	33.5	53.9	91.9	41.9	60.7	96.8	50.9	68.2
05	**METALS**	99.5	67.8	89.7	101.8	70.6	92.2	100.6	66.5	90.1	103.8	66.9	92.5	95.9	80.0	91.0	102.0	69.4	92.0
06	**WOOD, PLASTICS & COMPOSITES**	85.8	35.5	57.6	107.6	59.0	80.4	96.0	45.5	67.7	87.5	35.7	58.5	93.9	48.3	68.4	99.9	62.5	78.9
07	**THERMAL & MOISTURE PROTECTION**	99.0	45.3	77.0	99.4	57.5	82.2	98.5	42.6	75.6	94.4	37.9	71.2	103.3	49.1	81.1	95.7	57.8	80.2
08	**OPENINGS**	97.7	44.7	85.4	98.5	56.2	88.7	98.5	46.5	86.4	102.1	42.3	88.2	100.6	54.0	89.7	99.5	58.4	89.9
0920	Plaster & Gypsum Board	81.8	34.0	49.5	89.4	58.3	68.4	88.3	44.3	58.6	84.0	34.2	50.4	97.7	46.9	63.4	94.6	61.8	72.5
0950, 0980	Ceilings & Acoustic Treatment	83.2	34.0	50.9	84.8	58.3	67.4	84.8	44.3	58.2	83.3	34.2	51.0	86.2	46.9	60.3	88.0	61.8	70.8
0960	Flooring	94.0	59.1	84.0	102.9	60.8	90.9	98.8	59.1	87.5	98.5	59.1	87.2	68.9	53.4	64.5	101.0	60.8	89.5
0970, 0990	Wall Finishes & Painting/Coating	107.4	29.4	60.3	107.4	53.8	75.1	107.4	41.2	67.4	104.4	55.2	74.7	94.9	48.4	66.8	108.8	55.2	76.5
09	**FINISHES**	85.7	40.0	60.5	89.9	57.6	72.1	88.6	46.8	65.6	88.4	41.0	62.3	83.7	48.5	64.3	92.7	59.8	74.6
COVERS	**DIVS. 10 - 14, 25, 28, 41, 43, 44, 46**	100.0	48.7	89.6	100.0	78.8	95.7	100.0	49.1	89.7	100.0	36.5	87.2	100.0	45.2	88.9	100.0	79.2	95.8
21, 22, 23	**FIRE SUPPRESSION, PLUMBING & HVAC**	95.3	50.2	77.1	100.1	50.6	80.1	95.3	49.4	76.8	95.3	48.7	76.5	100.3	52.5	81.0	99.9	56.2	82.3
26, 27, 3370	**ELECTRICAL, COMMUNICATIONS & UTIL.**	90.1	51.2	69.5	93.7	64.5	78.3	95.0	37.1	64.4	99.0	67.8	82.5	100.2	63.5	80.8	100.7	70.2	84.6
MF2014	**WEIGHTED AVERAGE**	93.4	52.0	75.4	97.2	61.6	81.7	95.9	50.4	76.0	96.4	51.8	76.9	96.7	59.1	80.3	98.5	64.0	83.4

	DIVISION	ARKANSAS												CALIFORNIA					
		PINE BLUFF			RUSSELLVILLE			TEXARKANA			WEST MEMPHIS			ALHAMBRA			ANAHEIM		
		716			728			718			723			917 - 918			928		
		MAT.	INST.	TOTAL	MAT.	INST.	TOTAL	MAT.	INST.	TOTAL	MAT.	INST.	TOTAL	MAT.	INST.	TOTAL	MAT.	INST.	TOTAL
015433	CONTRACTOR EQUIPMENT		89.1	89.1		89.1	89.1		89.9	89.9		108.3	108.3		100.1	100.1		100.8	100.8
0241, 31 - 34	SITE & INFRASTRUCTURE, DEMOLITION	83.7	86.8	85.9	78.3	84.9	83.0	94.9	87.1	89.3	110.5	101.1	103.8	97.9	110.9	107.1	99.4	108.3	105.7
0310	Concrete Forming & Accessories	80.7	60.0	62.8	86.6	54.7	59.1	87.3	40.1	46.6	95.5	48.5	54.9	117.1	116.4	116.5	105.4	124.2	121.6
0320	Concrete Reinforcing	93.5	68.2	80.7	88.0	67.3	77.5	93.1	67.5	80.1	84.5	70.1	77.2	106.6	113.3	110.0	93.8	113.2	103.7
0330	Cast-in-Place Concrete	84.5	70.3	78.7	81.1	45.3	66.4	92.1	43.3	72.0	88.4	55.1	74.7	95.1	120.8	105.6	92.9	123.3	105.4
03	CONCRETE	92.1	65.5	79.0	86.7	54.4	70.8	89.8	47.2	68.9	95.5	56.2	76.2	101.1	116.5	108.7	101.8	120.8	111.2
04	MASONRY	125.0	50.9	78.8	96.1	37.3	59.5	102.4	33.1	59.2	79.4	41.9	56.0	122.7	121.1	121.7	80.0	118.8	104.2
05	METALS	104.6	69.4	93.8	99.5	66.5	89.3	96.8	66.9	87.6	95.0	80.3	90.5	85.3	100.9	90.1	102.7	101.1	102.2
06	WOOD, PLASTICS & COMPOSITES	87.0	62.5	73.3	91.3	59.0	73.2	95.5	42.0	65.5	100.1	48.3	71.1	100.0	112.1	106.7	101.9	122.6	113.5
07	THERMAL & MOISTURE PROTECTION	94.5	57.8	79.5	99.2	43.9	76.5	95.2	44.5	74.4	103.7	49.1	81.3	95.1	118.1	104.5	99.4	122.0	108.7
08	OPENINGS	102.1	58.4	91.9	97.7	55.7	87.9	107.3	46.5	93.2	100.6	54.0	89.7	91.8	114.8	97.2	104.2	120.6	108.0
0920	Plaster & Gypsum Board	83.6	61.8	68.9	83.3	58.3	66.4	86.6	40.7	55.6	100.2	46.9	64.2	97.4	112.5	107.6	107.1	123.2	118.0
0950, 0980	Ceilings & Acoustic Treatment	83.3	61.8	69.2	83.2	58.3	66.8	86.6	40.7	56.4	84.3	46.9	59.7	100.2	112.5	108.3	105.0	123.2	117.0
0960	Flooring	98.2	60.8	87.5	96.2	59.1	85.6	100.4	51.0	86.3	71.0	53.4	66.0	95.4	110.2	99.6	101.9	112.0	104.8
0970, 0990	Wall Finishes & Painting/Coating	104.4	55.2	74.7	107.4	35.0	63.7	104.4	30.3	59.7	94.9	50.4	68.0	100.5	111.5	107.1	98.9	106.8	103.7
09	FINISHES	88.3	59.8	72.6	86.7	54.1	68.7	90.7	40.4	62.9	85.0	48.8	65.0	99.8	114.0	107.6	100.5	120.1	111.3
COVERS	DIVS. 10 - 14, 25, 28, 41, 43, 44, 46	100.0	79.2	95.8	100.0	41.0	88.1	100.0	32.6	86.4	100.0	45.2	88.9	100.0	110.5	102.1	100.0	112.0	102.4
21, 22, 23	FIRE SUPPRESSION, PLUMBING & HVAC	100.1	52.9	81.1	95.3	51.5	77.6	100.1	53.5	81.3	95.6	65.8	83.6	95.2	115.6	103.4	100.0	118.6	107.5
26, 27, 3370	ELECTRICAL, COMMUNICATIONS & UTIL.	97.0	70.2	82.8	93.7	44.7	67.8	98.9	36.0	65.7	101.9	65.7	82.8	118.4	120.5	119.5	92.2	105.7	99.3
MF2014	WEIGHTED AVERAGE	99.4	63.3	83.6	94.6	53.7	76.8	98.2	49.2	76.8	96.1	62.4	81.4	98.3	114.8	105.5	99.4	114.9	106.2

	DIVISION	CALIFORNIA																	
		BAKERSFIELD			BERKELEY			EUREKA			FRESNO			INGLEWOOD			LONG BEACH		
		932 - 933			947			955			936 - 938			903 - 905			906 - 908		
		MAT.	INST.	TOTAL	MAT.	INST.	TOTAL	MAT.	INST.	TOTAL	MAT.	INST.	TOTAL	MAT.	INST.	TOTAL	MAT.	INST.	TOTAL
015433	CONTRACTOR EQUIPMENT		98.8	98.8		100.3	100.3		98.5	98.5		98.8	98.8		96.5	96.5		96.5	96.5
0241, 31 - 34	SITE & INFRASTRUCTURE, DEMOLITION	100.4	106.0	104.4	118.4	107.6	110.7	110.8	103.9	105.9	103.1	105.4	104.7	89.3	103.5	99.4	95.9	103.5	101.3
0310	Concrete Forming & Accessories	104.0	124.3	121.5	115.9	149.9	145.2	114.8	133.6	131.0	104.3	134.2	130.1	111.8	115.4	114.9	106.5	115.4	114.1
0320	Concrete Reinforcing	103.4	113.2	108.4	94.5	114.4	104.6	102.5	114.1	108.4	80.6	113.7	97.5	102.7	113.3	108.1	101.9	113.3	107.7
0330	Cast-in-Place Concrete	94.0	122.6	105.7	123.4	124.8	124.0	100.8	118.0	107.9	99.4	118.9	107.4	84.5	121.4	99.7	96.2	121.4	106.5
03	CONCRETE	97.9	120.6	109.1	107.3	132.7	119.8	112.8	123.2	117.9	98.6	123.8	111.0	90.9	116.3	103.4	99.8	116.3	107.9
04	MASONRY	100.2	118.3	111.5	113.4	134.3	126.5	104.0	132.6	121.8	103.6	122.0	115.1	76.3	121.2	104.3	85.3	121.2	107.7
05	METALS	105.3	100.9	104.0	106.9	100.5	104.9	102.6	98.9	101.4	105.6	100.5	104.1	94.4	101.5	96.6	94.3	101.5	96.5
06	WOOD, PLASTICS & COMPOSITES	96.0	122.7	111.0	111.4	154.9	135.8	116.6	138.5	128.9	108.1	138.5	125.1	103.8	110.7	107.6	96.9	110.7	104.6
07	THERMAL & MOISTURE PROTECTION	98.1	116.1	105.5	105.2	137.2	118.3	103.4	119.8	110.1	88.7	116.0	99.9	98.3	118.0	106.4	98.5	118.0	106.5
08	OPENINGS	97.8	116.1	102.0	93.0	140.6	104.1	103.5	113.5	105.8	98.1	124.7	104.3	88.8	114.1	94.7	88.8	114.1	94.7
0920	Plaster & Gypsum Board	101.9	123.2	116.3	111.6	156.0	141.6	113.4	139.5	131.0	100.6	139.5	126.9	103.8	111.0	108.7	99.9	111.0	107.4
0950, 0980	Ceilings & Acoustic Treatment	96.3	123.2	114.0	106.2	156.0	139.0	110.1	139.5	129.4	92.8	139.5	123.5	100.3	111.0	107.4	100.3	111.0	107.4
0960	Flooring	106.8	110.2	107.7	108.2	127.6	113.8	106.0	116.5	109.0	113.3	133.5	119.1	104.4	110.2	106.0	101.8	110.2	104.2
0970, 0990	Wall Finishes & Painting/Coating	111.4	102.0	105.7	102.5	143.9	127.5	100.5	46.2	67.7	129.0	107.2	115.8	100.9	111.5	107.3	100.9	111.5	107.3
09	FINISHES	100.0	120.5	111.3	105.4	147.2	128.4	106.1	123.6	115.8	102.1	133.2	119.2	102.7	113.2	108.5	101.8	113.2	108.1
COVERS	DIVS. 10 - 14, 25, 28, 41, 43, 44, 46	100.0	111.9	102.4	100.0	126.4	105.3	100.0	122.2	104.5	100.0	122.2	104.5	100.0	110.4	102.1	100.0	110.4	102.1
21, 22, 23	FIRE SUPPRESSION, PLUMBING & HVAC	100.1	117.1	107.0	95.3	147.8	116.5	95.2	110.6	101.4	100.2	111.3	104.7	94.8	115.6	103.2	94.8	115.6	103.2
26, 27, 3370	ELECTRICAL, COMMUNICATIONS & UTIL.	103.6	102.5	103.1	109.4	145.0	128.2	99.1	116.7	108.4	93.8	100.8	97.5	103.8	120.5	112.6	103.5	120.5	112.5
MF2014	WEIGHTED AVERAGE	100.7	113.6	106.3	102.7	135.1	116.8	102.1	116.4	108.4	100.1	115.3	106.7	94.8	114.1	103.2	96.2	114.1	104.0

	DIVISION	CALIFORNIA																	
		LOS ANGELES			MARYSVILLE			MODESTO			MOJAVE			OAKLAND			OXNARD		
		900 - 902			959			953			935			946			930		
		MAT.	INST.	TOTAL	MAT.	INST.	TOTAL	MAT.	INST.	TOTAL	MAT.	INST.	TOTAL	MAT.	INST.	TOTAL	MAT.	INST.	TOTAL
015433	CONTRACTOR EQUIPMENT		100.1	100.1		98.5	98.5		98.5	98.5		98.8	98.8		100.3	100.3		97.7	97.7
0241, 31 - 34	SITE & INFRASTRUCTURE, DEMOLITION	96.6	107.1	104.1	107.2	105.0	105.7	102.2	105.1	104.3	96.4	106.0	103.2	124.7	107.6	112.5	103.3	104.0	103.8
0310	Concrete Forming & Accessories	108.6	124.3	122.2	104.6	134.4	130.3	100.7	134.3	129.7	116.1	112.1	112.6	104.9	149.9	143.7	107.5	124.3	122.0
0320	Concrete Reinforcing	103.6	113.4	108.6	102.5	113.8	108.2	106.2	113.8	110.0	97.3	113.1	105.3	96.7	114.4	105.7	95.5	113.1	104.4
0330	Cast-in-Place Concrete	91.2	122.0	103.9	112.6	119.2	115.3	100.8	119.2	108.4	86.0	122.4	101.0	117.0	124.8	120.2	100.1	122.7	109.4
03	CONCRETE	95.9	120.6	108.0	113.9	124.0	118.9	104.7	124.0	114.2	91.0	115.1	102.8	106.8	132.7	119.5	99.2	120.6	109.7
04	MASONRY	90.5	122.6	110.5	104.9	123.1	116.2	102.9	123.1	115.5	101.9	118.3	112.1	120.6	134.3	129.2	104.8	116.5	112.1
05	METALS	101.1	102.9	101.7	102.0	101.8	102.0	99.1	101.7	99.9	102.8	100.3	102.0	101.4	100.4	101.1	100.7	100.7	100.7
06	WOOD, PLASTICS & COMPOSITES	103.0	122.6	113.9	103.0	138.5	122.9	98.5	138.5	120.9	107.4	106.8	107.1	98.8	154.9	130.2	102.1	122.7	113.6
07	THERMAL & MOISTURE PROTECTION	97.6	121.2	107.3	102.9	121.5	110.5	102.5	118.7	109.1	94.4	110.9	101.2	103.1	137.2	117.1	97.4	119.8	106.6
08	OPENINGS	95.4	120.7	101.3	102.8	125.3	108.0	101.6	125.4	107.1	93.2	107.5	96.5	93.1	140.6	104.1	95.7	120.6	101.5
0920	Plaster & Gypsum Board	103.6	123.2	116.9	105.0	139.5	128.3	107.5	139.5	129.1	110.8	106.8	108.1	105.7	156.0	139.7	104.7	123.2	117.2
0950, 0980	Ceilings & Acoustic Treatment	110.5	123.2	118.8	109.2	139.5	129.1	105.0	139.5	127.7	94.2	106.8	102.5	108.9	156.0	139.9	96.6	123.2	114.1
0960	Flooring	102.0	112.0	104.9	102.0	117.6	106.4	102.4	112.8	105.3	112.6	110.2	111.9	103.4	127.6	110.3	104.8	112.0	106.9
0970, 0990	Wall Finishes & Painting/Coating	100.0	111.5	106.9	100.5	115.8	109.7	100.5	117.0	110.5	111.1	102.0	105.6	102.5	143.9	127.5	111.1	101.1	105.1
09	FINISHES	104.2	120.6	113.3	103.1	131.5	118.8	102.2	130.9	118.0	101.8	109.9	106.2	104.3	147.2	127.9	99.4	119.6	110.5
COVERS	DIVS. 10 - 14, 25, 28, 41, 43, 44, 46	100.0	111.9	102.4	100.0	122.2	104.5	100.0	122.2	104.5	100.0	107.8	101.6	100.0	126.3	105.3	100.0	112.2	102.5
21, 22, 23	FIRE SUPPRESSION, PLUMBING & HVAC	99.9	118.6	107.5	95.2	109.0	100.8	100.0	111.3	104.6	95.3	115.1	103.3	100.1	147.8	119.4	100.1	118.6	107.6
26, 27, 3370	ELECTRICAL, COMMUNICATIONS & UTIL.	102.0	121.3	112.2	95.7	107.4	101.9	98.2	104.0	101.3	92.0	99.9	96.2	108.5	139.3	124.8	97.7	107.3	102.8
MF2014	WEIGHTED AVERAGE	99.2	117.5	107.2	101.3	115.9	107.7	100.8	115.7	107.3	96.8	109.8	102.5	103.1	134.3	116.7	99.6	114.4	106.0

	DIVISION	CALIFORNIA																	
		PALM SPRINGS			PALO ALTO			PASADENA			REDDING			RICHMOND			RIVERSIDE		
		922			943			910 - 912			960			948			925		
		MAT.	INST.	TOTAL	MAT.	INST.	TOTAL	MAT.	INST.	TOTAL	MAT.	INST.	TOTAL	MAT.	INST.	TOTAL	MAT.	INST.	TOTAL
015433	CONTRACTOR EQUIPMENT		99.6	99.6		100.3	100.3		100.1	100.1		98.5	98.5		100.3	100.3		99.6	99.6
0241, 31 - 34	SITE & INFRASTRUCTURE, DEMOLITION	90.9	106.3	101.8	114.4	107.6	109.6	94.9	110.9	106.3	125.1	105.0	110.8	123.8	107.6	112.3	97.9	106.3	103.9
0310	Concrete Forming & Accessories	101.9	115.3	113.4	103.0	146.3	140.3	106.2	115.3	114.1	107.8	142.7	137.9	118.7	149.7	145.5	105.8	124.2	121.7
0320	Concrete Reinforcing	107.8	113.1	110.5	94.5	114.4	104.6	107.5	113.3	110.5	118.7	113.8	116.2	94.5	114.3	104.6	104.7	113.1	109.0
0330	Cast-in-Place Concrete	88.6	123.2	102.8	104.4	124.8	112.8	90.2	120.8	102.8	118.0	119.2	118.5	120.0	124.8	122.0	96.3	123.3	107.4
03	CONCRETE	96.6	116.8	106.5	96.5	131.1	113.5	96.6	116.0	106.2	122.7	127.7	125.2	108.8	132.6	120.5	102.6	120.8	111.6
04	MASONRY	77.6	118.5	103.1	97.4	134.3	120.4	106.5	121.1	115.6	130.0	123.1	125.7	113.2	132.8	125.4	78.6	118.1	103.2
05	METALS	103.3	100.8	102.5	98.9	100.3	99.3	85.3	100.8	90.1	101.6	101.8	101.7	98.9	100.2	99.3	102.8	101.0	102.3
06	WOOD, PLASTICS & COMPOSITES	96.6	110.8	104.5	96.1	150.2	126.4	86.5	110.6	100.0	110.6	149.8	132.6	115.3	154.9	137.5	101.9	122.6	113.5
07	THERMAL & MOISTURE PROTECTION	99.1	118.7	107.2	102.7	137.5	117.0	94.9	117.4	104.1	121.6	122.1	121.8	103.3	135.9	116.7	99.6	121.1	108.4
08	OPENINGS	100.1	114.1	103.4	93.1	136.9	103.3	91.8	114.0	97.0	114.1	131.5	118.1	93.1	139.4	103.9	102.8	120.6	107.0
0920	Plaster & Gypsum Board	102.0	111.0	108.1	103.9	151.1	135.9	90.2	111.0	104.3	106.2	151.1	136.6	113.6	156.0	142.3	106.3	123.2	117.7
0950, 0980	Ceilings & Acoustic Treatment	101.7	111.0	107.8	107.1	151.1	136.0	100.2	111.0	107.3	131.5	151.1	144.4	107.1	156.0	139.3	109.2	123.2	118.4
0960	Flooring	104.6	107.2	105.3	102.4	127.6	109.6	90.8	110.2	96.3	98.0	117.6	103.6	109.9	127.6	115.0	106.0	112.0	107.7
0970, 0990	Wall Finishes & Painting/Coating	97.2	111.3	105.7	102.5	143.9	127.5	100.5	111.5	107.1	115.9	115.8	115.8	102.5	143.9	127.5	97.2	106.8	103.0
09	FINISHES	99.1	112.8	106.6	102.7	144.4	125.6	97.3	113.2	106.0	109.3	138.2	125.3	106.9	147.2	129.1	102.1	120.1	112.0
COVERS	DIVS. 10 - 14, 25, 28, 41, 43, 44, 46	100.0	110.7	102.2	100.0	125.9	105.2	100.0	110.3	102.1	100.0	123.4	104.7	100.0	126.3	105.3	100.0	112.0	102.4
21, 22, 23	FIRE SUPPRESSION, PLUMBING & HVAC	95.2	115.5	103.4	95.3	145.5	115.6	95.2	115.6	103.4	100.2	109.0	103.7	95.3	147.8	116.5	100.0	118.6	107.5
26, 27, 3370	ELECTRICAL, COMMUNICATIONS & UTIL.	95.3	105.4	100.7	108.4	150.2	130.5	115.1	120.5	118.0	99.2	107.4	103.5	109.0	132.7	121.5	91.9	105.4	99.1
MF2014	WEIGHTED AVERAGE	97.2	112.0	103.6	98.9	134.5	114.4	96.4	114.6	104.3	107.8	117.7	112.1	101.8	133.1	115.4	99.4	114.6	106.0

	DIVISION	CALIFORNIA																	
		SACRAMENTO			SALINAS			SAN BERNARDINO			SAN DIEGO			SAN FRANCISCO			SAN JOSE		
		942,956 - 958			939			923 - 924			919 - 921			940 - 941			951		
		MAT.	INST.	TOTAL	MAT.	INST.	TOTAL	MAT.	INST.	TOTAL	MAT.	INST.	TOTAL	MAT.	INST.	TOTAL	MAT.	INST.	TOTAL
015433	CONTRACTOR EQUIPMENT		99.9	99.9		98.8	98.8		99.6	99.6		100.1	100.1		110.7	110.7		99.3	99.3
0241, 31 - 34	SITE & INFRASTRUCTURE, DEMOLITION	100.6	113.3	109.6	116.3	105.7	108.8	77.7	106.3	98.0	101.7	104.0	103.4	127.0	113.6	117.5	133.5	100.4	109.9
0310	Concrete Forming & Accessories	103.3	137.0	132.4	111.0	137.4	133.7	109.6	115.2	114.4	105.6	113.3	112.2	104.6	150.9	144.6	107.0	149.8	143.9
0320	Concrete Reinforcing	89.4	113.8	101.8	96.0	114.2	105.2	104.7	113.0	108.9	104.2	113.1	108.7	110.1	115.0	112.6	93.5	114.5	104.1
0330	Cast-in-Place Concrete	99.0	120.3	107.7	98.9	119.6	107.4	66.6	123.2	89.8	98.9	107.8	102.6	120.0	126.4	122.6	117.1	124.2	120.0
03	CONCRETE	97.0	125.2	110.9	108.5	125.6	116.9	76.8	116.8	96.4	101.8	110.7	106.2	110.3	134.4	122.1	110.8	132.9	121.7
04	MASONRY	98.4	123.1	113.8	101.6	128.1	118.1	84.9	115.9	104.2	97.4	115.6	108.7	121.1	141.5	133.8	131.4	134.4	133.3
05	METALS	97.0	96.1	96.7	105.5	102.9	104.7	102.8	100.5	102.1	101.1	101.4	101.2	107.6	110.8	108.6	97.5	107.5	100.6
06	WOOD, PLASTICS & COMPOSITES	92.9	141.7	120.2	107.3	141.5	126.4	105.8	110.8	108.6	100.6	110.2	106.0	98.8	155.1	130.3	111.3	154.7	135.6
07	THERMAL & MOISTURE PROTECTION	110.8	121.3	115.1	95.0	126.1	107.7	98.3	117.7	106.3	99.3	107.1	102.5	104.9	141.9	120.1	98.9	139.3	115.4
08	OPENINGS	106.3	127.2	111.1	96.9	133.3	105.4	100.2	114.1	103.4	99.1	111.8	102.1	97.3	140.7	107.4	92.8	140.5	103.9
0920	Plaster & Gypsum Board	101.0	142.6	129.1	105.8	142.6	130.6	108.2	111.0	110.1	97.8	110.3	106.3	108.3	156.0	140.6	102.7	156.0	138.7
0950, 0980	Ceilings & Acoustic Treatment	107.1	142.6	130.4	94.2	142.6	126.0	105.0	111.0	109.0	107.7	110.3	109.4	117.1	156.0	142.7	103.3	156.0	138.0
0960	Flooring	102.6	117.6	106.9	107.1	121.1	111.1	107.8	110.2	108.5	98.5	112.0	102.4	103.4	127.6	110.3	95.4	127.6	104.6
0970, 0990	Wall Finishes & Painting/Coating	100.2	116.7	110.1	112.3	143.9	131.4	97.2	104.0	101.3	97.1	111.5	105.8	102.5	153.2	133.1	100.8	143.9	126.8
09	FINISHES	101.4	133.6	119.1	101.4	137.0	121.0	100.5	112.5	107.1	102.9	112.9	108.4	106.5	148.3	129.6	100.7	147.0	126.2
COVERS	DIVS. 10 - 14, 25, 28, 41, 43, 44, 46	100.0	123.0	104.6	100.0	122.6	104.6	100.0	108.1	101.6	100.0	109.7	102.0	100.0	126.8	105.4	100.0	125.9	105.2
21, 22, 23	FIRE SUPPRESSION, PLUMBING & HVAC	100.0	120.8	108.4	95.3	117.7	104.4	95.2	115.6	103.4	99.9	116.9	106.8	100.1	175.0	130.3	100.0	147.3	119.1
26, 27, 3370	ELECTRICAL, COMMUNICATIONS & UTIL.	103.5	108.7	106.2	93.2	120.9	107.8	95.3	103.4	99.6	101.2	98.5	99.8	108.6	157.8	134.6	101.1	155.9	130.1
MF2014	WEIGHTED AVERAGE	100.5	119.3	108.7	100.2	121.6	109.5	95.1	111.3	102.1	100.5	109.6	104.5	105.2	145.4	122.7	102.5	136.6	117.4

	DIVISION	CALIFORNIA																	
		SAN LUIS OBISPO			SAN MATEO			SAN RAFAEL			SANTA ANA			SANTA BARBARA			SANTA CRUZ		
		934			944			949			926 - 927			931			950		
		MAT.	INST.	TOTAL	MAT.	INST.	TOTAL	MAT.	INST.	TOTAL	MAT.	INST.	TOTAL	MAT.	INST.	TOTAL	MAT.	INST.	TOTAL
015433	CONTRACTOR EQUIPMENT		98.8	98.8		100.3	100.3		100.3	100.3		99.6	99.6		98.8	98.8		99.3	99.3
0241, 31 - 34	SITE & INFRASTRUCTURE, DEMOLITION	108.4	106.0	106.7	121.3	107.6	111.6	113.2	113.4	113.4	89.4	106.3	101.4	103.3	106.0	105.2	133.1	100.0	109.6
0310	Concrete Forming & Accessories	117.9	115.4	115.8	109.0	149.8	144.2	112.9	149.9	144.9	109.9	115.3	114.5	108.3	124.2	122.1	107.0	137.6	133.4
0320	Concrete Reinforcing	97.3	113.1	105.4	94.5	114.6	104.7	95.1	114.6	105.1	108.3	113.1	110.8	95.5	113.2	104.5	115.7	114.2	114.9
0330	Cast-in-Place Concrete	106.3	122.5	112.9	116.2	124.8	119.7	135.2	124.1	130.7	85.0	123.2	100.7	99.8	122.6	109.1	116.3	121.4	118.4
03	CONCRETE	106.9	116.6	111.7	105.4	132.7	118.8	126.0	132.4	129.1	94.2	116.8	105.3	99.0	120.6	109.6	113.9	126.5	120.1
04	MASONRY	103.5	118.3	112.7	112.9	137.3	128.1	93.2	137.4	120.7	74.6	118.8	102.1	102.2	119.1	112.7	135.3	128.2	130.9
05	METALS	103.5	100.6	102.6	98.7	100.6	99.3	100.1	98.8	99.7	102.9	100.8	102.2	101.2	100.9	101.1	104.5	106.3	105.0
06	WOOD, PLASTICS & COMPOSITES	109.9	110.9	110.5	104.0	154.9	132.5	101.8	154.7	131.4	107.8	110.8	109.5	102.1	122.7	113.6	111.3	141.6	128.3
07	THERMAL & MOISTURE PROTECTION	95.1	116.9	104.0	103.1	138.8	117.7	107.2	138.9	120.2	99.4	118.3	107.2	94.6	118.9	104.5	98.8	129.2	111.3
08	OPENINGS	95.1	109.7	98.5	93.1	139.4	103.8	103.8	139.3	112.1	99.5	114.1	102.9	96.6	120.6	102.2	94.0	133.4	103.2
0920	Plaster & Gypsum Board	112.2	111.0	111.4	108.9	156.0	140.8	111.5	156.0	141.6	109.6	111.0	110.6	104.7	123.2	117.2	110.1	142.6	132.0
0950, 0980	Ceilings & Acoustic Treatment	94.2	111.0	105.3	107.1	156.0	139.3	115.4	156.0	142.1	105.0	111.0	109.0	96.6	123.2	114.1	106.7	142.6	130.3
0960	Flooring	113.5	110.2	112.6	105.1	127.6	111.5	115.0	121.1	116.8	108.3	110.2	108.8	106.2	108.6	106.9	99.6	121.1	105.7
0970, 0990	Wall Finishes & Painting/Coating	111.1	101.7	105.5	102.5	143.9	127.5	98.9	142.6	125.3	97.2	106.8	103.0	111.1	101.1	105.1	101.0	143.9	126.9
09	FINISHES	103.2	112.3	108.2	104.7	147.2	128.1	107.6	145.8	128.6	102.0	112.8	107.9	100.0	119.0	110.5	103.7	137.1	122.1
COVERS	DIVS. 10 - 14, 25, 28, 41, 43, 44, 46	100.0	120.6	104.2	100.0	126.4	105.3	100.0	125.7	105.2	100.0	110.7	102.2	100.0	112.2	102.5	100.0	122.9	104.6
21, 22, 23	FIRE SUPPRESSION, PLUMBING & HVAC	95.3	115.6	103.5	95.3	142.2	114.2	95.3	169.5	125.2	95.2	114.3	102.9	100.1	118.6	107.6	100.0	117.8	107.2
26, 27, 3370	ELECTRICAL, COMMUNICATIONS & UTIL.	92.0	106.4	99.6	108.4	148.0	129.3	105.4	120.0	113.1	95.4	105.7	100.8	91.0	110.5	101.3	100.3	120.9	111.2
MF2014	WEIGHTED AVERAGE	99.4	112.0	104.9	100.9	134.6	115.6	103.5	136.7	118.0	96.9	111.8	103.4	98.9	115.1	106.0	104.5	121.8	112.0

DIVISION		CALIFORNIA															COLORADO		
		SANTA ROSA			STOCKTON			SUSANVILLE			VALLEJO			VAN NUYS			ALAMOSA		
		954			952			961			945			913 - 916			811		
		MAT.	INST.	TOTAL	MAT.	INST.	TOTAL	MAT.	INST.	TOTAL	MAT.	INST.	TOTAL	MAT.	INST.	TOTAL	MAT.	INST.	TOTAL
015433	CONTRACTOR EQUIPMENT		99.0	99.0		98.5	98.5		98.5	98.5		100.3	100.3		100.1	100.1		93.8	93.8
0241, 31 - 34	SITE & INFRASTRUCTURE, DEMOLITION	103.0	105.1	104.5	102.0	105.1	104.2	131.8	105.0	112.8	100.5	113.2	109.5	111.4	110.9	111.0	134.7	88.0	101.5
0310	Concrete Forming & Accessories	102.9	148.1	141.9	104.8	136.6	132.2	109.1	142.8	138.1	103.4	147.7	141.6	112.7	115.3	115.0	105.0	67.5	72.7
0320	Concrete Reinforcing	103.4	114.8	109.2	106.2	113.8	110.1	118.7	113.2	115.9	96.3	114.5	105.6	107.5	113.3	110.5	105.1	76.4	90.5
0330	Cast-in-Place Concrete	110.6	120.8	114.8	98.2	119.2	106.9	107.4	119.3	112.3	107.7	121.5	113.4	95.1	120.8	105.7	101.1	78.4	91.8
03	CONCRETE	113.7	130.9	122.1	103.7	125.0	114.2	125.2	127.7	126.4	102.2	130.5	116.1	110.2	116.0	113.0	113.5	73.3	93.8
04	MASONRY	103.1	137.2	124.3	102.8	123.1	115.5	127.9	112.7	118.4	71.6	133.1	109.9	122.7	121.1	121.7	126.0	71.9	92.3
05	METALS	103.2	105.0	103.8	99.3	101.8	100.0	100.6	101.1	100.7	100.1	98.1	99.5	84.4	100.8	89.5	98.2	80.2	92.7
06	WOOD, PLASTICS & COMPOSITES	98.0	154.5	129.6	104.3	141.5	125.1	112.5	149.8	133.4	90.6	154.7	126.5	95.0	110.6	103.7	98.0	67.5	80.9
07	THERMAL & MOISTURE PROTECTION	99.8	136.2	114.7	102.6	119.9	109.7	122.2	120.4	121.4	105.0	135.3	117.4	95.7	117.4	104.6	104.9	78.1	93.9
08	OPENINGS	101.0	140.4	110.2	101.6	127.1	107.5	113.9	131.5	118.0	105.6	140.5	113.7	91.6	114.0	96.8	98.3	74.1	92.7
0920	Plaster & Gypsum Board	103.9	156.0	139.2	107.5	142.6	131.2	107.1	151.1	136.9	105.8	156.0	139.8	95.3	111.0	105.9	79.0	66.3	70.4
0950, 0980	Ceilings & Acoustic Treatment	105.0	156.0	138.6	112.5	142.6	132.3	124.1	151.1	141.9	117.3	156.0	142.8	97.7	111.0	106.5	96.3	66.3	76.6
0960	Flooring	105.0	115.2	107.9	102.4	112.8	105.3	98.4	122.1	105.2	110.6	127.6	115.4	93.2	110.2	98.0	113.5	54.8	96.7
0970, 0990	Wall Finishes & Painting/Coating	97.2	142.6	124.6	100.5	117.0	110.5	115.9	115.8	115.8	99.8	142.6	125.6	100.5	111.5	107.1	114.1	39.8	69.3
09	FINISHES	101.1	143.6	124.5	103.8	132.7	119.7	108.9	139.0	125.5	104.6	145.8	127.3	99.3	113.2	107.0	102.5	61.7	80.0
COVERS	DIVS. 10 - 14, 25, 28, 41, 43, 44, 46	100.0	123.8	104.8	100.0	122.5	104.6	100.0	123.5	104.7	100.0	124.3	104.9	100.0	110.3	102.1	100.0	88.5	97.7
21, 22, 23	FIRE SUPPRESSION, PLUMBING & HVAC	95.2	167.3	124.3	100.0	111.3	104.6	95.4	109.0	100.9	100.1	127.5	111.1	95.2	115.6	103.4	95.2	72.0	85.9
26, 27, 3370	ELECTRICAL, COMMUNICATIONS & UTIL.	95.7	114.3	105.5	98.2	110.6	104.8	99.6	118.4	109.5	100.9	122.4	112.3	115.1	120.5	118.0	99.1	72.5	85.0
MF2014	WEIGHTED AVERAGE	100.8	134.7	115.6	100.9	117.2	108.0	106.8	118.1	111.7	100.1	127.2	111.9	99.1	114.6	105.9	102.2	73.7	89.8

DIVISION		COLORADO																	
		BOULDER			COLORADO SPRINGS			DENVER			DURANGO			FORT COLLINS			FORT MORGAN		
		803			808 - 809			800 - 802			813			805			807		
		MAT.	INST.	TOTAL	MAT.	INST.	TOTAL	MAT.	INST.	TOTAL	MAT.	INST.	TOTAL	MAT.	INST.	TOTAL	MAT.	INST.	TOTAL
015433	CONTRACTOR EQUIPMENT		97.5	97.5		95.8	95.8		100.4	100.4		93.8	93.8		97.5	97.5		97.5	97.5
0241, 31 - 34	SITE & INFRASTRUCTURE, DEMOLITION	94.8	96.4	95.9	96.9	94.7	95.3	96.1	102.6	100.7	128.5	88.0	99.7	107.1	96.0	99.2	97.7	95.9	96.4
0310	Concrete Forming & Accessories	103.1	79.9	83.1	93.6	79.4	81.3	100.0	76.0	79.3	111.4	67.7	73.7	100.7	74.7	78.3	103.6	74.9	78.9
0320	Concrete Reinforcing	98.9	76.6	87.5	98.1	80.2	89.0	98.1	80.2	89.0	105.1	76.5	90.5	99.0	76.6	87.6	99.1	76.5	87.6
0330	Cast-in-Place Concrete	101.8	80.0	92.9	104.6	88.3	97.9	99.0	82.8	92.4	116.3	78.5	100.8	114.9	78.9	100.1	99.9	78.9	91.3
03	CONCRETE	103.1	79.6	91.5	106.3	82.9	94.8	100.7	79.5	90.3	116.1	73.4	95.1	113.5	76.7	95.4	101.5	76.8	89.4
04	MASONRY	91.0	72.1	79.2	91.3	81.9	85.4	93.3	72.2	80.1	113.6	71.9	87.6	108.5	75.7	88.1	104.9	72.2	84.5
05	METALS	96.0	83.5	92.2	99.1	85.7	95.0	101.6	85.6	96.7	98.2	80.4	92.7	97.2	80.4	92.0	95.7	80.3	91.0
06	WOOD, PLASTICS & COMPOSITES	102.2	83.0	91.4	91.7	77.3	83.6	99.8	77.2	87.2	107.6	67.5	85.1	99.6	77.2	87.0	102.2	77.2	88.2
07	THERMAL & MOISTURE PROTECTION	103.1	81.7	94.3	103.9	85.6	96.4	102.4	75.8	91.5	104.9	78.1	93.9	103.5	73.5	91.2	103.0	81.2	94.1
08	OPENINGS	98.2	82.5	94.5	102.4	80.5	97.3	103.0	80.4	97.8	105.3	74.1	98.0	98.2	79.4	93.8	98.1	79.4	93.8
0920	Plaster & Gypsum Board	105.1	82.7	89.9	87.9	76.7	80.3	100.7	76.8	84.5	92.5	66.3	74.8	99.1	76.7	84.0	105.1	76.7	85.9
0950, 0980	Ceilings & Acoustic Treatment	96.9	82.7	87.5	104.4	76.7	86.2	107.3	76.8	87.2	96.3	66.3	76.6	96.9	76.7	83.6	96.9	76.7	83.6
0960	Flooring	99.7	83.8	95.2	92.0	68.2	85.2	96.7	84.8	93.3	118.2	54.8	100.0	96.6	54.8	84.6	100.1	54.8	87.1
0970, 0990	Wall Finishes & Painting/Coating	101.7	66.6	80.5	101.4	40.5	64.7	101.7	75.7	86.0	114.1	39.8	69.3	101.7	40.2	64.6	101.7	53.6	72.6
09	FINISHES	101.7	79.4	89.4	98.8	72.8	84.5	102.4	77.3	88.6	105.0	61.7	81.1	100.5	67.4	82.3	101.8	68.9	83.7
COVERS	DIVS. 10 - 14, 25, 28, 41, 43, 44, 46	100.0	89.4	97.9	100.0	92.2	98.4	100.0	88.8	97.7	100.0	88.4	97.7	100.0	88.8	97.7	100.0	88.8	97.7
21, 22, 23	FIRE SUPPRESSION, PLUMBING & HVAC	95.3	77.3	88.1	100.2	88.2	95.4	100.0	79.9	91.9	95.2	83.1	90.3	100.1	77.2	90.9	95.3	77.2	88.0
26, 27, 3370	ELECTRICAL, COMMUNICATIONS & UTIL.	97.4	84.5	90.6	100.8	82.1	90.9	102.6	83.2	92.4	98.5	69.4	83.1	97.4	84.5	90.6	97.8	84.5	90.7
MF2014	WEIGHTED AVERAGE	97.8	81.3	90.6	100.4	84.0	93.2	100.8	81.7	92.5	102.7	75.7	90.9	101.3	79.0	91.6	98.3	79.0	89.9

DIVISION		COLORADO																	
		GLENWOOD SPRINGS			GOLDEN			GRAND JUNCTION			GREELEY			MONTROSE			PUEBLO		
		816			804			815			806			814			810		
		MAT.	INST.	TOTAL	MAT.	INST.	TOTAL	MAT.	INST.	TOTAL	MAT.	INST.	TOTAL	MAT.	INST.	TOTAL	MAT.	INST.	TOTAL
015433	CONTRACTOR EQUIPMENT		96.7	96.7		97.5	97.5		96.7	96.7		97.5	97.5		95.2	95.2		93.8	93.8
0241, 31 - 34	SITE & INFRASTRUCTURE, DEMOLITION	143.2	95.1	109.0	107.3	96.2	99.4	128.4	94.7	104.5	94.0	95.4	95.0	137.1	91.2	104.5	120.6	91.0	99.6
0310	Concrete Forming & Accessories	102.0	75.0	78.7	96.1	74.8	77.7	110.2	74.4	79.3	98.6	78.4	81.2	101.4	74.8	78.4	107.3	79.4	83.3
0320	Concrete Reinforcing	104.0	76.5	90.0	99.1	76.4	87.6	104.3	76.3	90.1	98.9	75.1	86.8	103.9	76.4	89.9	100.6	80.2	90.2
0330	Cast-in-Place Concrete	101.1	77.9	91.6	100.0	78.9	91.3	111.9	77.1	97.6	96.1	59.1	80.9	101.1	77.8	91.5	100.4	87.8	95.2
03	CONCRETE	118.5	76.5	97.9	111.9	76.7	94.6	112.5	75.9	94.5	98.3	71.3	85.0	109.7	76.3	93.3	102.7	82.7	92.9
04	MASONRY	98.9	72.1	82.2	107.7	71.8	85.3	132.6	71.5	94.5	102.6	49.0	69.2	106.2	71.9	84.9	95.4	81.7	86.9
05	METALS	97.9	80.9	92.6	95.9	80.1	91.0	99.5	79.2	93.2	97.2	77.5	91.1	97.1	80.0	91.8	101.1	86.2	96.5
06	WOOD, PLASTICS & COMPOSITES	93.2	77.3	84.3	94.3	77.2	84.7	105.3	77.3	89.6	96.8	83.0	89.1	94.3	77.4	84.8	100.9	77.6	87.8
07	THERMAL & MOISTURE PROTECTION	104.8	79.2	94.3	103.9	75.4	92.2	104.0	68.5	89.5	102.9	66.4	87.9	105.0	79.2	94.4	103.5	83.6	95.3
08	OPENINGS	104.2	79.4	98.5	98.2	78.8	93.7	105.0	78.8	98.9	98.1	82.4	94.5	105.5	79.5	99.4	100.2	80.6	95.6
0920	Plaster & Gypsum Board	117.8	76.7	90.0	96.9	76.7	83.3	131.0	76.7	94.3	97.6	82.7	87.5	78.3	76.7	77.2	83.9	76.7	79.1
0950, 0980	Ceilings & Acoustic Treatment	95.5	76.7	83.2	96.9	76.7	83.6	95.5	76.7	83.2	96.9	82.7	87.5	96.3	76.7	83.4	104.7	76.7	86.3
0960	Flooring	112.8	50.4	95.0	94.7	54.8	83.3	117.5	54.8	99.6	95.7	54.8	84.0	115.8	45.4	95.7	114.6	84.8	106.1
0970, 0990	Wall Finishes & Painting/Coating	114.1	66.5	85.4	101.7	66.6	80.5	114.1	66.6	85.4	101.7	25.0	55.4	114.1	39.8	69.3	114.1	37.5	67.9
09	FINISHES	107.9	69.6	86.8	100.4	71.0	84.2	109.3	70.4	87.9	99.3	69.2	82.7	103.2	65.7	82.5	103.7	75.6	88.2
COVERS	DIVS. 10 - 14, 25, 28, 41, 43, 44, 46	100.0	88.8	97.7	100.0	88.8	97.7	100.0	88.8	97.7	100.0	89.4	97.9	100.0	89.1	97.8	100.0	92.7	98.5
21, 22, 23	FIRE SUPPRESSION, PLUMBING & HVAC	95.2	83.0	90.3	95.3	76.7	87.8	99.9	82.5	92.9	100.1	77.2	90.8	95.2	83.0	90.3	99.9	76.8	90.6
26, 27, 3370	ELECTRICAL, COMMUNICATIONS & UTIL.	95.7	69.4	81.8	97.8	84.5	90.7	98.2	53.8	74.7	97.4	84.5	90.6	98.2	56.4	76.1	99.1	73.0	85.3
MF2014	WEIGHTED AVERAGE	102.3	78.1	91.8	99.8	78.9	90.7	104.8	75.3	91.9	98.9	75.5	88.7	101.4	75.4	90.1	101.1	80.3	92.0

	DIVISION	COLORADO			CONNECTICUT														
		SALIDA			BRIDGEPORT			BRISTOL			HARTFORD			MERIDEN			NEW BRITAIN		
		812			066			060			061			064			060		
		MAT.	INST.	TOTAL	MAT.	INST.	TOTAL	MAT.	INST.	TOTAL	MAT.	INST.	TOTAL	MAT.	INST.	TOTAL	MAT.	INST.	TOTAL
015433	CONTRACTOR EQUIPMENT		95.2	95.2		100.9	100.9		100.9	100.9		100.9	100.9		101.3	101.3		100.9	100.9
0241, 31 - 34	SITE & INFRASTRUCTURE, DEMOLITION	128.2	91.5	102.1	109.8	105.1	106.4	108.9	105.0	106.2	104.4	105.0	104.9	106.7	105.8	106.0	109.1	105.0	106.2
0310	Concrete Forming & Accessories	110.1	74.7	79.6	98.9	124.5	121.0	98.9	124.3	120.8	95.7	124.3	120.4	98.6	124.3	120.8	99.2	124.3	120.9
0320	Concrete Reinforcing	103.7	76.4	89.8	105.1	127.3	116.4	105.1	127.3	116.4	104.6	127.3	116.2	105.1	127.3	116.4	105.1	127.3	116.4
0330	Cast-in-Place Concrete	115.8	77.8	100.2	107.9	127.1	115.8	101.1	127.1	111.8	103.0	127.1	112.9	97.3	127.1	109.5	102.8	127.1	112.7
03	CONCRETE	111.2	76.3	94.1	106.6	125.7	116.0	103.4	125.6	114.3	104.0	125.6	114.6	101.6	125.6	113.4	104.2	125.6	114.7
04	MASONRY	133.9	71.9	95.3	111.8	134.0	125.6	102.9	134.0	122.3	107.7	134.0	124.1	102.5	134.0	122.1	105.2	134.0	123.1
05	METALS	96.7	79.9	91.6	100.5	125.5	108.2	100.5	125.4	108.1	105.6	125.4	111.7	97.4	125.4	106.0	96.5	125.4	105.4
06	WOOD, PLASTICS & COMPOSITES	101.9	77.4	88.2	98.8	123.4	112.6	98.8	123.4	112.6	92.3	123.4	109.7	98.8	123.4	112.6	98.8	123.4	112.6
07	THERMAL & MOISTURE PROTECTION	104.0	79.2	93.8	98.6	129.1	111.1	98.7	126.0	109.9	102.8	126.0	112.3	98.7	126.0	109.9	98.7	126.0	109.9
08	OPENINGS	98.4	79.5	94.0	101.2	132.1	108.4	101.2	132.1	108.4	100.3	132.1	107.7	103.6	132.1	110.2	101.2	132.1	108.4
0920	Plaster & Gypsum Board	78.6	76.7	77.3	101.7	123.5	116.4	101.7	123.5	116.4	97.4	123.5	115.0	103.2	123.5	116.9	101.7	123.5	116.4
0950, 0980	Ceilings & Acoustic Treatment	96.3	76.7	83.4	86.4	123.5	110.8	86.4	123.5	110.8	89.7	123.5	111.9	90.4	123.5	112.2	86.4	123.5	110.8
0960	Flooring	120.6	45.4	99.1	101.9	131.2	110.3	101.9	131.2	110.3	100.5	131.2	109.2	101.9	131.2	110.3	101.9	131.2	110.3
0970, 0990	Wall Finishes & Painting/Coating	114.1	41.7	70.4	104.9	122.9	115.8	104.9	122.9	115.8	105.9	122.9	116.2	104.9	122.9	115.8	104.9	122.9	115.8
09	FINISHES	103.6	65.9	82.8	97.2	125.0	112.6	97.3	125.0	112.6	96.3	125.0	112.1	98.4	125.0	113.1	97.3	125.0	112.6
COVERS	DIVS. 10 - 14, 25, 28, 41, 43, 44, 46	100.0	89.2	97.8	100.0	112.4	102.5	100.0	112.4	102.5	100.0	112.4	102.5	100.0	112.4	102.5	100.0	112.4	102.5
21, 22, 23	FIRE SUPPRESSION, PLUMBING & HVAC	95.2	72.0	85.8	100.0	116.4	106.6	100.0	116.4	106.6	100.0	116.4	106.6	95.2	116.4	103.8	100.0	116.4	106.6
26, 27, 3370	ELECTRICAL, COMMUNICATIONS & UTIL.	98.4	72.5	84.7	98.5	112.2	105.7	98.5	111.8	105.6	97.9	112.8	105.8	98.4	111.8	105.5	98.6	111.8	105.6
MF2014	WEIGHTED AVERAGE	101.9	75.3	90.3	101.3	120.8	109.8	100.5	120.6	109.3	101.4	120.8	109.8	98.9	120.7	108.4	100.1	120.6	109.1

	DIVISION	CONNECTICUT																	
		NEW HAVEN			NEW LONDON			NORWALK			STAMFORD			WATERBURY			WILLIMANTIC		
		065			063			068			069			067			062		
		MAT.	INST.	TOTAL	MAT.	INST.	TOTAL	MAT.	INST.	TOTAL	MAT.	INST.	TOTAL	MAT.	INST.	TOTAL	MAT.	INST.	TOTAL
015433	CONTRACTOR EQUIPMENT		101.3	101.3		101.3	101.3		100.9	100.9		100.9	100.9		100.9	100.9		100.9	100.9
0241, 31 - 34	SITE & INFRASTRUCTURE, DEMOLITION	108.9	105.8	106.7	101.1	105.8	104.4	109.5	105.1	106.4	110.2	105.1	106.6	109.5	105.0	106.3	109.5	105.0	106.3
0310	Concrete Forming & Accessories	98.6	124.3	120.8	98.6	124.3	120.8	98.9	124.9	121.3	98.9	124.9	121.3	98.9	124.3	120.8	98.9	124.1	120.6
0320	Concrete Reinforcing	105.1	127.3	116.4	82.4	127.3	105.2	105.1	127.4	116.5	105.1	127.4	116.5	105.1	127.3	116.4	105.1	127.2	116.4
0330	Cast-in-Place Concrete	104.5	127.1	113.8	89.1	127.1	104.7	106.2	128.6	115.4	107.9	128.6	116.4	107.9	127.1	115.8	100.8	125.8	111.0
03	CONCRETE	118.7	125.6	122.1	91.4	125.6	108.2	105.8	126.4	115.9	106.6	126.4	116.3	106.6	125.6	115.9	103.3	125.0	114.0
04	MASONRY	103.2	134.0	122.4	101.5	134.0	121.7	102.7	135.4	123.1	103.4	135.4	123.3	103.4	134.0	122.5	102.7	134.0	122.2
05	METALS	96.8	125.4	105.6	96.5	125.4	105.4	100.5	126.1	108.3	100.5	126.1	108.3	100.5	125.4	108.1	100.2	125.2	107.9
06	WOOD, PLASTICS & COMPOSITES	98.8	123.4	112.6	98.8	123.4	112.6	98.8	123.4	112.6	98.8	123.4	112.6	98.8	123.4	112.6	98.8	123.4	112.6
07	THERMAL & MOISTURE PROTECTION	98.8	126.1	110.0	98.7	126.0	109.9	98.8	129.6	111.4	98.7	129.6	111.4	98.7	126.1	110.0	98.9	124.6	109.5
08	OPENINGS	101.2	132.1	108.4	104.0	132.1	110.6	101.2	132.1	108.4	101.2	132.1	108.4	101.2	132.1	108.4	104.0	132.1	110.6
0920	Plaster & Gypsum Board	101.7	123.5	116.4	101.7	123.5	116.4	101.7	123.5	116.4	101.7	123.5	116.4	101.7	123.5	116.4	101.7	123.5	116.4
0950, 0980	Ceilings & Acoustic Treatment	86.4	123.5	110.8	84.5	123.5	110.2	86.4	123.5	110.8	86.4	123.5	110.8	86.4	123.5	110.8	84.5	123.5	110.2
0960	Flooring	101.9	131.2	110.3	101.9	131.2	110.3	101.9	131.2	110.3	101.9	131.2	110.3	101.9	131.2	110.3	101.9	133.2	110.9
0970, 0990	Wall Finishes & Painting/Coating	104.9	122.9	115.8	104.9	122.9	115.8	104.9	122.9	115.8	104.9	122.9	115.8	104.9	122.9	115.8	104.9	122.9	115.8
09	FINISHES	97.3	125.0	112.6	96.4	125.0	112.2	97.3	125.0	112.6	97.4	125.0	112.6	97.2	125.0	112.5	97.0	125.4	112.6
COVERS	DIVS. 10 - 14, 25, 28, 41, 43, 44, 46	100.0	112.4	102.5	100.0	112.4	102.5	100.0	112.6	102.6	100.0	112.6	102.6	100.0	112.4	102.5	100.0	112.4	102.5
21, 22, 23	FIRE SUPPRESSION, PLUMBING & HVAC	100.0	116.4	106.6	95.2	116.4	103.8	100.0	116.4	106.6	100.0	116.4	106.6	100.0	116.4	106.6	100.0	116.1	106.5
26, 27, 3370	ELECTRICAL, COMMUNICATIONS & UTIL.	98.4	111.8	105.5	95.3	111.8	104.0	98.5	167.2	134.8	98.5	167.2	134.8	98.0	112.2	105.5	98.5	109.6	104.4
MF2014	WEIGHTED AVERAGE	101.6	120.7	109.9	97.1	120.7	107.4	100.8	128.7	113.0	100.9	128.7	113.0	100.9	120.7	109.5	100.8	120.2	109.2

	DIVISION	D.C.			DELAWARE									FLORIDA					
		WASHINGTON			DOVER			NEWARK			WILMINGTON			DAYTONA BEACH			FORT LAUDERDALE		
		200 - 205			199			197			198			321			333		
		MAT.	INST.	TOTAL	MAT.	INST.	TOTAL	MAT.	INST.	TOTAL	MAT.	INST.	TOTAL	MAT.	INST.	TOTAL	MAT.	INST.	TOTAL
015433	CONTRACTOR EQUIPMENT		105.5	105.5		118.0	118.0		118.0	118.0		118.2	118.2		98.2	98.2		91.0	91.0
0241, 31 - 34	SITE & INFRASTRUCTURE, DEMOLITION	103.8	94.5	97.2	102.3	113.3	110.1	102.6	113.3	110.2	99.5	113.6	109.5	105.9	89.2	94.0	96.4	77.1	82.7
0310	Concrete Forming & Accessories	97.4	78.3	80.9	96.3	102.2	101.4	97.0	102.2	101.5	98.1	102.2	101.6	97.6	68.1	72.1	95.7	68.4	72.1
0320	Concrete Reinforcing	105.4	93.8	99.5	94.0	102.9	98.5	91.5	102.9	97.3	93.4	102.9	98.2	91.2	76.6	83.8	88.2	72.1	80.0
0330	Cast-in-Place Concrete	115.4	88.0	104.2	104.6	104.6	104.6	90.0	104.6	96.0	99.2	104.6	101.4	90.2	70.8	82.2	94.6	77.1	87.4
03	CONCRETE	107.7	85.9	97.0	102.2	104.0	103.1	94.6	104.0	99.2	99.6	104.0	101.8	91.6	71.8	81.9	94.5	73.3	84.0
04	MASONRY	98.2	80.0	86.9	107.5	98.1	101.6	103.2	98.1	100.0	108.5	98.1	102.0	99.1	65.7	78.3	102.8	68.3	81.3
05	METALS	99.9	108.6	102.6	102.0	116.7	106.5	103.5	116.7	107.5	102.0	116.7	106.5	104.7	91.1	100.5	102.0	89.6	98.2
06	WOOD, PLASTICS & COMPOSITES	95.3	76.2	84.6	94.0	101.9	98.4	95.8	101.9	99.2	92.0	101.9	97.6	96.3	69.2	81.1	84.3	67.2	74.7
07	THERMAL & MOISTURE PROTECTION	101.7	85.8	95.2	98.8	110.0	103.4	101.7	110.0	105.1	97.9	110.0	102.9	96.2	73.6	86.9	100.3	77.2	90.8
08	OPENINGS	99.9	87.9	97.1	91.3	110.9	95.9	91.6	110.9	96.1	89.9	110.9	94.8	98.6	67.4	91.3	97.5	65.5	90.1
0920	Plaster & Gypsum Board	108.0	75.5	86.0	97.1	101.8	100.3	99.5	101.8	101.1	101.7	101.8	101.8	95.6	68.7	77.4	105.4	66.6	79.2
0950, 0980	Ceilings & Acoustic Treatment	112.2	75.5	88.1	91.3	101.8	98.2	88.7	101.8	97.3	92.4	101.8	98.6	84.9	68.7	74.2	87.9	66.6	73.9
0960	Flooring	104.7	93.6	101.5	99.8	109.7	102.7	96.8	109.7	100.5	104.3	109.7	105.9	110.5	73.6	100.0	105.1	69.3	94.9
0970, 0990	Wall Finishes & Painting/Coating	108.0	82.2	92.4	97.2	106.4	102.8	97.3	106.4	102.8	95.0	106.4	101.9	104.5	74.1	86.2	98.9	70.2	81.6
09	FINISHES	100.8	80.5	89.6	96.2	103.6	100.3	95.0	103.6	99.7	98.7	103.6	101.4	96.9	69.3	81.7	94.9	68.1	80.1
COVERS	DIVS. 10 - 14, 25, 28, 41, 43, 44, 46	100.0	98.6	99.7	100.0	88.0	97.6	100.0	88.0	97.6	100.0	88.0	97.6	100.0	84.4	96.9	100.0	87.4	97.5
21, 22, 23	FIRE SUPPRESSION, PLUMBING & HVAC	100.1	92.1	96.9	100.0	117.4	107.0	100.1	117.4	107.1	100.1	117.4	107.1	99.9	76.1	90.3	100.0	66.5	86.5
26, 27, 3370	ELECTRICAL, COMMUNICATIONS & UTIL.	101.5	105.8	103.8	96.6	111.9	104.7	98.7	111.9	105.7	99.8	111.9	106.2	95.8	54.9	74.2	96.0	72.7	83.7
MF2014	WEIGHTED AVERAGE	101.1	91.9	97.1	99.3	109.2	103.6	98.8	109.2	103.3	99.3	109.2	103.6	99.0	72.9	87.6	98.6	72.6	87.2

DIVISION		FLORIDA																	
		FORT MYERS			GAINESVILLE			JACKSONVILLE			LAKELAND			MELBOURNE			MIAMI		
		339,341			326,344			320,322			338			329			330 - 332,340		
		MAT.	INST.	TOTAL	MAT.	INST.	TOTAL	MAT.	INST.	TOTAL	MAT.	INST.	TOTAL	MAT.	INST.	TOTAL	MAT.	INST.	TOTAL
015433	**CONTRACTOR EQUIPMENT**		98.2	98.2		98.2	98.2		98.2	98.2		98.2	98.2		98.2	98.2		91.0	91.0
0241, 31 - 34	**SITE & INFRASTRUCTURE, DEMOLITION**	107.3	88.3	93.8	113.9	88.3	95.7	106.0	88.6	93.6	109.2	88.7	94.6	112.7	88.5	95.5	99.5	76.9	83.4
0310	Concrete Forming & Accessories	91.5	74.8	77.1	92.7	54.3	59.6	97.4	54.7	60.5	88.0	75.3	77.1	93.8	69.8	73.1	100.7	68.7	73.1
0320	Concrete Reinforcing	89.3	92.7	91.0	96.8	64.8	80.5	91.2	64.9	77.8	91.5	93.6	92.5	92.3	76.6	84.3	94.3	72.2	83.0
0330	Cast-in-Place Concrete	98.7	68.2	86.2	103.6	62.6	86.7	91.1	68.3	81.8	100.9	69.6	88.1	108.7	73.1	94.1	95.5	78.1	88.3
03	**CONCRETE**	95.1	76.9	86.2	102.6	60.7	82.0	92.1	62.8	77.7	96.8	77.8	87.5	103.0	73.3	88.4	96.1	73.7	85.1
04	**MASONRY**	95.8	61.9	74.7	114.0	61.2	81.1	99.1	61.2	75.5	114.1	75.8	90.2	97.5	69.8	80.2	103.1	71.9	83.7
05	**METALS**	104.2	96.8	102.0	103.6	85.0	97.9	103.2	85.4	97.7	104.1	97.8	102.2	113.4	91.3	106.6	102.4	88.5	98.1
06	**WOOD, PLASTICS & COMPOSITES**	81.1	77.4	79.0	89.9	51.8	68.5	96.3	51.8	71.4	76.5	77.4	77.0	91.5	69.2	79.0	90.6	67.2	77.5
07	**THERMAL & MOISTURE PROTECTION**	100.1	80.0	91.9	96.5	61.7	82.2	96.4	62.3	82.4	100.1	84.6	93.7	96.6	76.0	88.2	101.6	71.6	89.3
08	**OPENINGS**	98.4	74.4	92.8	96.9	52.7	86.6	98.6	55.8	88.6	98.4	75.1	93.0	97.8	70.7	91.5	99.7	65.5	91.7
0920	Plaster & Gypsum Board	101.0	77.1	84.9	91.3	50.7	63.9	95.6	50.7	65.3	97.4	77.1	83.7	91.3	68.7	76.0	103.2	66.6	78.4
0950, 0980	Ceilings & Acoustic Treatment	82.7	77.1	79.0	79.5	50.7	60.6	84.9	50.7	62.5	82.7	77.1	79.0	83.3	68.7	73.7	92.2	66.6	75.4
0960	Flooring	102.3	55.1	88.8	108.1	43.1	89.5	110.5	64.5	97.4	100.3	56.5	87.8	108.3	73.6	98.4	107.3	73.0	97.5
0970, 0990	Wall Finishes & Painting/Coating	103.7	67.6	81.9	104.5	67.6	82.3	104.5	67.6	82.3	103.7	67.6	81.9	104.5	91.9	96.9	96.0	70.2	80.5
09	**FINISHES**	94.4	70.7	81.3	95.3	52.5	71.7	97.0	56.7	74.8	93.5	71.0	81.1	95.9	72.3	82.9	97.1	69.2	81.7
COVERS	**DIVS. 10 - 14, 25, 28, 41, 43, 44, 46**	100.0	72.6	94.5	100.0	82.8	96.5	100.0	81.0	96.2	100.0	72.6	94.5	100.0	85.8	97.1	100.0	88.0	97.6
21, 22, 23	**FIRE SUPPRESSION, PLUMBING & HVAC**	97.4	64.3	84.1	98.8	64.1	84.8	99.9	64.2	85.5	97.4	80.8	90.7	99.9	78.2	91.2	100.0	66.4	86.4
26, 27, 3370	**ELECTRICAL, COMMUNICATIONS & UTIL.**	98.1	62.8	79.5	96.1	71.8	83.3	95.5	62.2	77.9	96.3	61.6	77.9	97.0	67.8	81.6	100.1	74.9	86.8
MF2014	**WEIGHTED AVERAGE**	98.6	72.6	87.3	100.3	66.7	85.7	98.8	66.4	84.7	99.4	77.8	90.0	101.7	76.3	90.6	99.8	73.1	88.2

DIVISION		FLORIDA																	
		ORLANDO			PANAMA CITY			PENSACOLA			SARASOTA			ST. PETERSBURG			TALLAHASSEE		
		327 - 328,347			324			325			342			337			323		
		MAT.	INST.	TOTAL	MAT.	INST.	TOTAL	MAT.	INST.	TOTAL	MAT.	INST.	TOTAL	MAT.	INST.	TOTAL	MAT.	INST.	TOTAL
015433	**CONTRACTOR EQUIPMENT**		98.2	98.2		98.2	98.2		98.2	98.2		98.2	98.2		98.2	98.2		98.2	98.2
0241, 31 - 34	**SITE & INFRASTRUCTURE, DEMOLITION**	107.6	88.4	93.9	117.6	87.4	96.1	117.6	87.9	96.5	114.0	88.4	95.8	110.9	88.1	94.7	105.1	87.7	92.7
0310	Concrete Forming & Accessories	101.4	72.0	76.1	96.7	43.9	51.1	94.6	51.8	57.7	96.0	75.0	77.9	94.9	51.0	57.1	99.9	44.1	51.8
0320	Concrete Reinforcing	96.6	73.9	85.0	95.3	72.5	83.7	97.7	73.0	85.1	92.5	93.5	93.0	91.5	86.1	88.7	98.3	64.7	81.2
0330	Cast-in-Place Concrete	112.1	70.6	95.1	95.7	56.7	79.7	118.2	65.9	96.8	106.6	69.5	91.4	102.0	64.5	86.6	97.2	56.4	80.5
03	**CONCRETE**	103.4	73.0	88.4	100.8	55.4	78.5	110.7	62.2	86.9	100.2	77.5	89.1	98.5	63.8	81.4	97.3	54.0	76.0
04	**MASONRY**	100.6	65.7	78.8	103.9	47.3	68.6	124.7	54.5	80.9	99.8	75.8	84.8	157.9	49.0	90.0	103.6	53.8	72.6
05	**METALS**	102.4	89.6	98.5	104.4	86.5	98.9	105.6	87.9	100.2	105.2	97.5	102.8	105.0	92.9	101.3	102.2	84.6	96.8
06	**WOOD, PLASTICS & COMPOSITES**	95.8	75.2	84.2	95.1	41.8	65.2	92.7	51.5	69.7	95.7	77.4	85.4	85.5	50.1	65.7	95.0	41.3	64.9
07	**THERMAL & MOISTURE PROTECTION**	94.9	74.3	86.4	96.7	56.6	80.2	96.6	62.3	82.5	98.1	84.6	92.5	100.3	57.8	82.8	102.6	71.5	89.9
08	**OPENINGS**	101.4	69.2	93.9	96.5	46.5	84.9	96.5	56.9	87.3	99.7	74.2	93.8	98.4	60.9	89.6	100.2	47.3	87.9
0920	Plaster & Gypsum Board	99.7	74.9	82.9	94.5	40.4	57.9	97.4	50.5	65.7	97.8	77.1	83.8	103.5	49.0	66.7	108.1	40.0	62.0
0950, 0980	Ceilings & Acoustic Treatment	91.4	74.9	80.5	83.3	40.4	55.1	83.3	50.5	61.7	86.2	77.1	80.3	84.5	49.0	61.2	94.1	40.0	58.5
0960	Flooring	104.1	73.6	95.4	110.1	43.0	90.9	106.1	63.0	93.7	111.1	57.9	95.9	104.1	55.2	90.1	111.1	62.0	97.0
0970, 0990	Wall Finishes & Painting/Coating	103.4	70.2	83.4	104.5	64.9	80.6	104.5	67.6	82.3	109.5	67.6	84.2	103.7	60.8	77.8	99.5	67.6	80.3
09	**FINISHES**	98.2	72.4	84.0	97.5	44.8	68.5	96.4	54.8	73.5	99.9	71.2	84.1	95.9	51.7	71.6	100.7	48.5	72.0
COVERS	**DIVS. 10 - 14, 25, 28, 41, 43, 44, 46**	100.0	85.1	97.0	100.0	46.2	89.1	100.0	46.3	89.1	100.0	72.6	94.5	100.0	55.8	91.1	100.0	65.8	93.1
21, 22, 23	**FIRE SUPPRESSION, PLUMBING & HVAC**	99.9	56.6	82.4	99.9	52.3	80.7	99.9	52.5	80.8	99.9	64.8	85.7	100.0	58.4	83.2	100.0	38.9	75.4
26, 27, 3370	**ELECTRICAL, COMMUNICATIONS & UTIL.**	97.6	60.0	77.7	94.5	59.3	75.9	98.5	55.8	75.9	97.2	61.6	78.4	96.3	61.6	77.9	103.3	60.0	80.4
MF2014	**WEIGHTED AVERAGE**	100.5	70.0	87.2	100.2	57.8	81.7	102.7	61.2	84.6	100.8	74.3	89.2	102.6	63.3	85.5	100.8	56.9	81.7

DIVISION		FLORIDA						GEORGIA											
		TAMPA			WEST PALM BEACH			ALBANY			ATHENS			ATLANTA			AUGUSTA		
		335 - 336,346			334,349			317,398			306			300 - 303,399			308 - 309		
		MAT.	INST.	TOTAL	MAT.	INST.	TOTAL	MAT.	INST.	TOTAL	MAT.	INST.	TOTAL	MAT.	INST.	TOTAL	MAT.	INST.	TOTAL
015433	**CONTRACTOR EQUIPMENT**		98.2	98.2		91.0	91.0		91.9	91.9		94.2	94.2		94.7	94.7		94.2	94.2
0241, 31 - 34	**SITE & INFRASTRUCTURE, DEMOLITION**	111.4	88.6	95.2	93.2	77.1	81.8	98.8	78.7	84.5	100.9	93.4	95.6	97.6	94.8	95.7	94.5	93.2	93.6
0310	Concrete Forming & Accessories	97.7	75.6	78.6	99.2	68.1	72.4	90.4	43.6	50.0	92.9	45.9	52.4	96.7	73.5	76.7	94.2	65.5	69.4
0320	Concrete Reinforcing	88.2	93.6	91.0	90.7	71.8	81.1	90.6	80.2	85.3	95.3	77.9	86.5	94.6	81.2	87.8	95.7	71.8	83.6
0330	Cast-in-Place Concrete	99.7	69.7	87.4	90.0	74.3	83.5	92.4	54.5	76.8	107.8	55.6	86.4	107.8	70.8	92.6	101.9	49.9	80.5
03	**CONCRETE**	97.0	77.9	87.6	91.3	72.1	81.9	93.1	55.8	74.8	104.8	56.0	80.8	102.2	74.1	88.4	97.5	61.7	79.9
04	**MASONRY**	103.9	75.8	86.4	102.3	66.6	80.0	102.5	49.1	69.2	81.5	53.8	64.2	95.2	66.2	77.1	95.5	43.3	63.0
05	**METALS**	104.0	98.1	102.2	101.1	89.3	97.4	105.0	85.5	99.0	91.9	73.3	86.2	92.8	77.0	88.0	91.6	71.6	85.5
06	**WOOD, PLASTICS & COMPOSITES**	89.3	77.4	82.6	89.2	67.2	76.9	85.0	38.0	58.7	91.8	40.9	63.3	96.1	75.8	84.7	93.3	70.9	80.8
07	**THERMAL & MOISTURE PROTECTION**	100.5	84.6	94.0	100.1	70.2	87.8	95.9	60.3	81.3	94.1	52.8	77.2	94.0	72.0	85.0	93.7	57.0	78.7
08	**OPENINGS**	99.7	79.2	94.9	96.8	65.5	89.5	91.1	44.2	80.2	89.5	45.7	79.3	94.7	71.9	89.4	89.5	62.2	83.1
0920	Plaster & Gypsum Board	106.1	77.1	86.5	110.2	66.6	80.7	97.5	36.5	56.3	98.7	39.4	58.7	100.9	75.4	83.6	99.8	70.4	79.9
0950, 0980	Ceilings & Acoustic Treatment	87.9	77.1	80.8	82.7	66.6	72.1	84.0	36.5	52.8	97.2	39.4	59.2	97.2	75.4	82.9	98.1	70.4	79.9
0960	Flooring	105.1	56.5	91.2	106.8	66.4	95.3	110.9	47.8	92.9	96.4	53.9	84.2	97.7	65.2	88.4	96.6	46.6	82.3
0970, 0990	Wall Finishes & Painting/Coating	103.7	67.6	81.9	98.9	70.2	81.6	105.3	52.8	73.6	106.0	46.0	69.8	106.0	85.1	93.4	106.0	46.0	69.8
09	**FINISHES**	97.4	71.0	82.8	94.7	67.5	79.7	98.1	43.2	67.9	95.4	45.7	68.0	95.7	73.4	83.4	95.2	60.5	76.1
COVERS	**DIVS. 10 - 14, 25, 28, 41, 43, 44, 46**	100.0	85.1	97.0	100.0	87.4	97.5	100.0	80.2	96.0	100.0	77.3	95.4	100.0	85.6	97.1	100.0	78.5	95.7
21, 22, 23	**FIRE SUPPRESSION, PLUMBING & HVAC**	100.0	80.8	92.2	97.4	62.7	83.4	99.9	68.0	87.0	95.2	69.6	84.9	99.9	70.6	88.1	100.0	61.2	84.4
26, 27, 3370	**ELECTRICAL, COMMUNICATIONS & UTIL.**	96.0	61.6	77.8	97.2	72.7	84.3	96.9	58.7	76.7	99.7	69.5	83.8	99.0	71.9	84.7	100.4	61.8	80.0
MF2014	**WEIGHTED AVERAGE**	100.1	78.4	90.6	97.4	71.1	85.9	98.5	61.3	82.3	95.5	63.9	81.7	97.6	74.4	87.5	96.3	63.6	82.1

DIVISION		GEORGIA																
		COLUMBUS			DALTON			GAINESVILLE			MACON			SAVANNAH			STATESBORO		
		318 - 319			307			305			310 - 312			313 - 314			304		
		MAT.	INST.	TOTAL	MAT.	INST.	TOTAL	MAT.	INST.	TOTAL	MAT.	INST.	TOTAL	MAT.	INST.	TOTAL	MAT.	INST.	TOTAL
015433	CONTRACTOR EQUIPMENT		91.9	91.9		106.1	106.1		94.2	94.2		102.3	102.3		92.8	92.8		93.4	93.4
0241, 31 - 34	SITE & INFRASTRUCTURE, DEMOLITION	98.7	78.7	84.5	101.1	97.8	98.8	100.8	93.3	95.5	99.8	93.5	95.3	100.6	80.1	86.1	102.0	77.5	84.6
0310	Concrete Forming & Accessories	90.3	54.3	59.3	85.7	46.7	52.1	96.4	42.9	50.2	89.9	51.9	57.1	92.1	50.0	55.8	80.3	51.4	55.4
0320	Concrete Reinforcing	90.9	80.6	85.7	94.8	73.9	84.2	95.1	77.8	86.3	92.1	80.3	86.1	98.1	71.9	84.7	94.4	41.8	67.6
0330	Cast-in-Place Concrete	92.1	53.9	76.4	104.7	50.2	82.3	113.3	52.9	88.5	90.9	65.6	80.5	100.2	55.2	81.7	107.6	59.7	87.9
03	CONCRETE	93.0	60.4	77.0	103.9	54.6	79.7	106.7	53.7	80.6	92.6	63.3	78.2	98.0	57.3	78.0	104.2	54.0	79.6
04	MASONRY	102.6	55.2	73.1	83.8	36.7	54.4	89.5	54.2	67.5	115.9	44.5	71.4	98.8	50.9	69.0	84.7	40.6	57.2
05	METALS	104.5	86.3	98.9	92.8	82.8	89.7	91.2	72.5	85.4	100.1	85.9	95.7	101.3	82.4	95.5	96.3	72.6	89.0
06	WOOD, PLASTICS & COMPOSITES	85.0	52.1	66.6	75.2	47.5	59.7	95.9	37.9	63.4	91.8	51.2	69.0	96.0	45.9	68.0	68.5	53.9	60.3
07	THERMAL & MOISTURE PROTECTION	95.9	63.5	82.6	96.1	51.5	77.8	94.1	54.8	78.0	94.4	62.2	81.2	95.0	56.6	79.2	94.7	51.1	76.8
08	OPENINGS	91.1	59.0	83.7	90.1	49.5	80.6	89.5	40.0	77.9	90.0	52.8	81.3	94.9	48.4	84.1	91.1	40.6	79.4
0920	Plaster & Gypsum Board	97.5	51.1	66.1	86.2	46.2	59.2	100.9	36.3	57.3	103.0	50.1	67.2	95.6	44.7	61.2	86.9	52.8	63.9
0950, 0980	Ceilings & Acoustic Treatment	84.0	51.1	62.4	109.3	46.2	67.8	97.2	36.3	57.2	79.2	50.1	60.1	91.5	44.7	60.8	105.7	52.8	71.0
0960	Flooring	110.9	50.4	93.6	96.8	46.9	82.5	97.6	46.6	83.0	87.9	47.8	76.5	108.6	46.4	90.8	115.6	44.9	95.4
0970, 0990	Wall Finishes & Painting/Coating	105.3	66.5	81.9	96.3	59.8	74.2	106.0	46.0	69.8	107.7	52.8	74.6	103.8	58.3	76.3	103.5	39.2	64.7
09	FINISHES	98.0	53.5	73.5	103.0	47.1	72.2	95.9	41.6	66.0	86.7	49.9	66.4	98.2	48.9	71.0	106.9	49.0	75.0
COVERS	DIVS. 10 - 14, 25, 28, 41, 43, 44, 46	100.0	81.8	96.3	100.0	22.8	84.4	100.0	39.4	87.8	100.0	80.2	96.0	100.0	78.0	95.6	100.0	43.5	88.6
21, 22, 23	FIRE SUPPRESSION, PLUMBING & HVAC	99.9	63.4	85.2	95.2	57.0	79.8	95.2	69.5	84.8	99.9	66.7	86.6	100.0	61.8	84.6	95.7	56.7	80.0
26, 27, 3370	ELECTRICAL, COMMUNICATIONS & UTIL.	97.1	69.6	82.5	109.7	67.7	87.5	99.7	69.5	83.8	96.1	61.2	77.7	101.7	57.4	78.3	99.9	57.4	77.4
MF2014	WEIGHTED AVERAGE	98.4	65.4	84.0	97.2	59.0	80.5	96.0	61.6	81.0	97.2	64.7	83.0	99.2	60.9	82.5	97.3	55.8	79.2

DIVISION		GEORGIA						HAWAII									IDAHO		
		VALDOSTA			WAYCROSS			HILO			HONOLULU			STATES & POSS., GUAM			BOISE		
		316			315			967			968			969			836 - 837		
		MAT.	INST.	TOTAL	MAT.	INST.	TOTAL	MAT.	INST.	TOTAL	MAT.	INST.	TOTAL	MAT.	INST.	TOTAL	MAT.	INST.	TOTAL
015433	CONTRACTOR EQUIPMENT		91.9	91.9		91.9	91.9		99.5	99.5		99.5	99.5		164.5	164.5		98.2	98.2
0241, 31 - 34	SITE & INFRASTRUCTURE, DEMOLITION	107.9	78.8	87.2	104.7	77.4	85.3	144.8	106.5	117.6	155.4	106.5	120.6	184.9	103.7	127.2	85.5	96.3	93.2
0310	Concrete Forming & Accessories	81.0	44.1	49.2	82.8	64.7	67.2	111.2	135.6	132.2	123.6	135.6	134.0	113.7	63.6	70.5	100.8	77.4	80.7
0320	Concrete Reinforcing	92.8	76.4	84.4	92.8	72.7	82.5	123.5	117.9	120.7	132.5	117.9	125.1	214.4	30.2	120.7	99.2	80.4	89.6
0330	Cast-in-Place Concrete	90.5	55.9	76.3	102.2	48.2	80.0	198.3	125.7	168.5	163.7	125.7	148.1	171.9	105.5	144.7	91.6	88.4	90.3
03	CONCRETE	98.1	55.8	77.3	101.2	61.4	81.6	158.1	127.7	143.2	150.8	127.7	139.5	159.9	72.3	116.9	99.1	81.9	90.7
04	MASONRY	108.9	50.5	72.5	109.7	39.3	65.8	150.5	128.3	136.7	150.6	128.3	136.7	213.9	43.5	107.7	121.4	84.2	98.2
05	METALS	104.1	84.0	97.9	103.1	77.7	95.3	108.1	107.3	107.8	120.7	107.3	116.5	139.2	76.2	119.8	102.2	80.9	95.6
06	WOOD, PLASTICS & COMPOSITES	73.4	37.6	53.4	75.0	72.4	73.5	114.5	139.2	128.3	134.4	139.2	137.1	127.1	67.4	93.7	93.1	75.9	83.4
07	THERMAL & MOISTURE PROTECTION	96.1	62.4	82.3	95.9	50.3	77.2	120.8	124.6	122.4	138.4	124.6	132.7	140.4	65.2	109.6	94.6	81.4	89.2
08	OPENINGS	87.3	42.5	76.9	87.4	57.8	80.5	111.3	132.8	116.3	121.8	132.8	124.4	116.6	54.1	102.1	99.1	71.2	92.6
0920	Plaster & Gypsum Board	90.4	36.2	53.8	90.4	72.0	78.0	109.7	140.1	130.3	151.3	140.1	143.7	219.9	55.5	108.8	92.4	75.1	80.7
0950, 0980	Ceilings & Acoustic Treatment	81.5	36.2	51.7	79.6	72.0	74.6	121.5	140.1	133.8	132.0	140.1	137.3	237.3	55.5	117.8	99.1	75.1	83.3
0960	Flooring	104.7	47.8	88.4	105.9	29.9	84.1	116.0	139.9	122.8	132.2	139.9	134.4	135.5	46.4	110.0	96.3	83.7	92.7
0970, 0990	Wall Finishes & Painting/Coating	105.3	52.2	73.2	105.3	46.0	69.5	110.4	143.8	130.6	119.4	143.8	134.2	115.9	35.9	67.6	103.2	39.5	64.7
09	FINISHES	95.6	43.6	67.0	95.2	57.4	74.4	113.7	138.6	127.4	128.4	138.6	134.0	188.8	58.9	117.2	96.0	74.8	84.3
COVERS	DIVS. 10 - 14, 25, 28, 41, 43, 44, 46	100.0	77.1	95.4	100.0	51.5	90.2	100.0	116.0	103.2	100.0	116.0	103.2	100.0	75.2	95.0	100.0	87.4	97.5
21, 22, 23	FIRE SUPPRESSION, PLUMBING & HVAC	99.9	70.0	87.9	96.9	57.1	80.8	100.2	107.7	103.2	100.3	107.7	103.3	102.6	37.4	76.3	100.0	71.8	88.6
26, 27, 3370	ELECTRICAL, COMMUNICATIONS & UTIL.	95.0	56.2	74.5	99.6	57.4	77.3	106.5	121.1	114.2	107.9	121.1	114.9	153.5	41.2	94.1	98.5	72.8	84.9
MF2014	WEIGHTED AVERAGE	98.5	61.4	82.3	98.3	59.4	81.4	114.8	120.2	117.2	119.5	120.2	119.8	136.2	58.0	102.1	100.1	78.6	90.7

DIVISION		IDAHO															ILLINOIS		
		COEUR D'ALENE			IDAHO FALLS			LEWISTON			POCATELLO			TWIN FALLS			BLOOMINGTON		
		838			834			835			832			833			617		
		MAT.	INST.	TOTAL	MAT.	INST.	TOTAL	MAT.	INST.	TOTAL	MAT.	INST.	TOTAL	MAT.	INST.	TOTAL	MAT.	INST.	TOTAL
015433	CONTRACTOR EQUIPMENT		92.8	92.8		98.2	98.2		92.8	92.8		98.2	98.2		98.2	98.2		101.6	101.6
0241, 31 - 34	SITE & INFRASTRUCTURE, DEMOLITION	83.1	91.7	89.2	83.1	96.1	92.4	89.9	92.5	91.7	85.8	96.3	93.3	92.1	97.2	95.8	97.4	98.4	98.1
0310	Concrete Forming & Accessories	112.6	81.1	85.5	94.7	78.1	80.4	117.9	82.2	87.1	101.0	77.1	80.4	102.0	54.7	61.2	84.3	117.2	112.7
0320	Concrete Reinforcing	106.2	96.5	101.3	101.0	80.0	90.3	106.2	96.8	101.4	99.6	80.2	89.7	101.3	79.9	90.4	94.6	110.7	102.8
0330	Cast-in-Place Concrete	99.0	87.2	94.1	87.2	74.3	81.9	102.9	86.0	95.9	94.1	88.3	91.7	96.6	63.9	83.1	102.8	114.8	107.7
03	CONCRETE	105.7	86.1	96.1	91.3	77.2	84.4	109.3	86.2	98.0	98.3	81.7	90.1	105.8	63.3	84.9	100.1	115.2	107.5
04	MASONRY	122.9	83.9	98.6	116.5	81.0	94.4	123.4	85.9	100.0	118.8	81.1	95.3	121.6	81.1	96.4	120.3	118.3	119.1
05	METALS	96.3	86.9	93.4	110.0	79.1	100.5	95.7	88.0	93.4	110.1	80.3	100.9	110.1	78.9	100.5	97.6	113.6	102.5
06	WOOD, PLASTICS & COMPOSITES	96.8	81.0	88.0	86.8	78.8	82.3	102.4	81.0	90.4	93.1	75.9	83.4	94.2	47.2	67.9	86.4	115.4	102.6
07	THERMAL & MOISTURE PROTECTION	148.3	80.8	120.6	94.3	71.7	85.0	148.5	81.4	121.0	94.7	73.7	86.1	95.4	74.7	86.9	98.4	112.8	104.3
08	OPENINGS	117.8	73.0	107.4	102.8	68.0	94.7	117.8	75.6	108.0	99.8	66.3	92.0	102.8	48.8	90.2	94.9	104.9	97.2
0920	Plaster & Gypsum Board	161.4	80.5	106.7	78.5	78.1	78.2	162.5	80.5	107.0	80.7	75.1	76.9	82.3	45.6	57.5	91.3	115.8	107.9
0950, 0980	Ceilings & Acoustic Treatment	129.0	80.5	97.1	97.2	78.1	84.6	129.0	80.5	97.1	104.7	75.1	85.2	99.7	45.6	64.1	88.0	115.8	106.3
0960	Flooring	138.3	45.9	111.9	96.4	43.1	81.2	141.2	93.8	127.7	99.6	83.7	95.1	100.7	43.1	84.3	93.8	118.7	100.9
0970, 0990	Wall Finishes & Painting/Coating	123.8	67.8	90.0	103.3	40.3	65.3	123.8	67.8	90.0	103.1	41.1	65.7	103.3	37.3	63.4	93.8	126.4	113.5
09	FINISHES	161.8	73.3	113.0	93.3	68.7	79.7	162.9	82.8	118.8	96.6	75.0	84.7	96.7	49.7	70.8	93.3	118.9	107.4
COVERS	DIVS. 10 - 14, 25, 28, 41, 43, 44, 46	100.0	87.4	97.5	100.0	47.7	89.4	100.0	87.6	97.5	100.0	87.4	97.5	100.0	44.2	88.7	100.0	105.2	101.0
21, 22, 23	FIRE SUPPRESSION, PLUMBING & HVAC	99.5	82.6	92.7	100.9	71.7	89.1	100.7	85.6	94.6	99.9	71.8	88.6	99.9	69.3	87.6	95.1	108.8	100.7
26, 27, 3370	ELECTRICAL, COMMUNICATIONS & UTIL.	91.0	77.6	83.9	90.8	70.4	80.0	88.9	80.7	84.5	96.2	70.4	82.6	92.3	60.1	75.3	94.2	94.6	94.4
MF2014	WEIGHTED AVERAGE	107.9	82.2	96.7	99.8	74.7	88.9	108.7	84.9	98.3	101.1	77.4	90.8	102.2	67.3	87.0	97.5	109.4	102.7

DIVISION		ILLINOIS																	
		CARBONDALE			CENTRALIA			CHAMPAIGN			CHICAGO			DECATUR			EAST ST. LOUIS		
		629			628			618 - 619			606 - 608			625			620 - 622		
		MAT.	INST.	TOTAL	MAT.	INST.	TOTAL	MAT.	INST.	TOTAL	MAT.	INST.	TOTAL	MAT.	INST.	TOTAL	MAT.	INST.	TOTAL
015433	CONTRACTOR EQUIPMENT		108.0	108.0		108.0	108.0		102.4	102.4		94.3	94.3		102.4	102.4		108.0	108.0
0241, 31 - 34	SITE & INFRASTRUCTURE, DEMOLITION	98.7	99.5	99.2	99.0	100.1	99.8	106.4	99.0	101.2	107.0	95.7	99.0	93.2	99.1	97.4	101.2	99.8	100.2
0310	Concrete Forming & Accessories	90.0	109.0	106.4	91.6	112.8	109.9	90.7	115.2	111.8	96.6	156.7	148.4	91.8	114.9	111.8	87.6	114.8	111.1
0320	Concrete Reinforcing	91.2	111.6	101.6	91.2	111.8	101.7	94.6	105.1	99.9	99.0	158.1	129.1	89.3	102.7	96.1	91.1	109.9	100.6
0330	Cast-in-Place Concrete	97.2	102.5	99.4	97.7	117.3	105.7	119.2	109.5	115.2	107.5	149.4	124.7	106.1	110.5	107.9	99.3	117.0	106.6
03	CONCRETE	90.1	107.9	98.9	90.6	114.8	102.5	113.1	111.3	112.2	102.6	153.2	127.5	101.3	111.1	106.1	91.7	115.2	103.2
04	MASONRY	83.6	108.6	99.2	83.6	116.7	104.2	145.3	117.7	128.1	101.1	156.6	135.7	79.6	114.7	101.5	83.9	116.6	104.3
05	METALS	96.2	119.6	103.4	96.2	120.8	103.8	97.6	108.7	101.0	94.3	134.3	106.6	99.9	108.1	102.4	97.3	119.4	104.1
06	WOOD, PLASTICS & COMPOSITES	91.7	106.1	99.8	94.1	109.7	102.8	93.5	113.9	104.9	104.0	155.9	133.1	91.9	113.9	104.2	89.2	112.1	102.0
07	THERMAL & MOISTURE PROTECTION	97.1	101.5	98.9	97.2	111.7	103.1	99.0	113.2	104.9	98.0	145.1	117.3	103.2	109.3	105.7	97.2	110.2	102.5
08	OPENINGS	89.4	114.2	95.2	89.4	116.2	95.7	95.5	111.2	99.2	105.3	158.7	117.8	100.9	110.6	103.2	89.5	116.9	95.9
0920	Plaster & Gypsum Board	97.2	106.2	103.3	98.3	110.0	106.2	93.7	114.2	107.6	92.7	157.6	136.6	100.0	114.2	109.6	96.1	112.4	107.2
0950, 0980	Ceilings & Acoustic Treatment	91.4	106.2	101.2	91.4	110.0	103.6	88.0	114.2	105.2	99.2	157.6	137.6	98.1	114.2	108.7	91.4	112.4	105.2
0960	Flooring	122.2	120.2	121.6	123.2	115.7	121.0	96.7	120.2	103.4	96.9	148.0	111.5	109.1	116.7	111.3	121.2	115.7	119.6
0970, 0990	Wall Finishes & Painting/Coating	112.6	97.4	103.5	112.6	106.6	109.0	93.8	110.0	103.6	91.9	153.0	128.8	102.5	106.5	104.9	112.6	103.5	107.1
09	FINISHES	101.8	108.6	105.5	102.2	112.6	107.9	95.2	115.9	106.6	98.0	155.6	129.7	101.8	115.0	109.1	101.4	113.8	108.2
COVERS	DIVS. 10 - 14, 25, 28, 41, 43, 44, 46	100.0	102.6	100.5	100.0	104.1	100.8	100.0	104.4	100.9	100.0	124.5	105.0	100.0	104.3	100.9	100.0	104.4	100.9
21, 22, 23	FIRE SUPPRESSION, PLUMBING & HVAC	95.1	106.1	99.5	95.1	95.3	95.2	95.1	105.5	99.3	99.8	133.9	113.6	99.9	98.5	99.4	99.9	98.8	99.4
26, 27, 3370	ELECTRICAL, COMMUNICATIONS & UTIL.	95.4	107.7	101.9	96.8	107.7	102.6	97.4	94.9	96.1	98.3	133.9	117.1	99.5	90.4	94.7	96.4	103.5	100.1
MF2014	WEIGHTED AVERAGE	94.7	107.9	100.5	95.0	108.5	100.9	101.0	107.6	103.9	99.8	139.7	117.2	99.2	104.9	101.7	96.4	108.7	101.8

DIVISION		ILLINOIS																	
		EFFINGHAM			GALESBURG			JOLIET			KANKAKEE			LA SALLE			NORTH SUBURBAN		
		624			614			604			609			613			600 - 603		
		MAT.	INST.	TOTAL	MAT.	INST.	TOTAL	MAT.	INST.	TOTAL	MAT.	INST.	TOTAL	MAT.	INST.	TOTAL	MAT.	INST.	TOTAL
015433	CONTRACTOR EQUIPMENT		102.4	102.4		101.6	101.6		92.5	92.5		92.5	92.5		101.6	101.6		92.5	92.5
0241, 31 - 34	SITE & INFRASTRUCTURE, DEMOLITION	97.6	98.9	98.5	99.9	98.3	98.8	106.9	94.9	98.4	100.5	94.6	96.3	99.2	99.2	99.2	106.1	94.9	98.1
0310	Concrete Forming & Accessories	96.3	114.0	111.6	90.6	116.6	113.1	98.4	159.6	151.2	91.8	143.2	136.2	104.6	125.3	122.4	97.8	154.0	146.3
0320	Concrete Reinforcing	92.2	99.9	96.1	94.1	110.6	102.5	99.0	150.8	125.4	99.8	147.4	124.0	94.3	144.7	119.9	99.0	156.5	128.3
0330	Cast-in-Place Concrete	105.7	108.4	106.8	105.9	106.2	106.0	107.4	146.5	123.5	100.1	132.5	113.4	105.7	122.5	112.6	107.5	141.2	121.3
03	CONCRETE	102.1	109.4	105.7	103.2	112.0	107.5	102.7	152.1	126.9	96.7	139.3	117.7	104.1	127.9	115.8	102.7	148.8	125.4
04	MASONRY	88.2	108.5	100.8	120.5	118.0	118.9	104.3	148.5	131.8	100.6	141.2	125.9	120.5	125.1	123.4	101.1	144.1	127.9
05	METALS	97.0	105.6	99.7	97.6	113.0	102.4	92.2	129.1	103.6	92.2	126.6	102.8	97.7	131.6	108.1	93.4	131.4	105.1
06	WOOD, PLASTICS & COMPOSITES	94.2	113.9	105.2	93.3	115.5	105.7	105.7	161.0	136.7	97.8	142.2	122.7	108.8	123.7	117.1	104.0	155.7	132.9
07	THERMAL & MOISTURE PROTECTION	102.7	106.6	104.3	98.5	107.6	102.2	97.8	141.6	115.8	97.0	135.7	112.9	98.7	117.8	106.5	98.2	139.3	115.1
08	OPENINGS	94.9	109.9	98.4	94.9	111.0	98.6	102.8	159.6	116.0	95.4	148.4	107.7	94.9	129.5	102.9	102.9	158.2	115.8
0920	Plaster & Gypsum Board	99.7	114.2	109.5	93.7	115.9	108.7	89.9	162.9	139.2	86.7	143.4	125.0	100.1	124.3	116.5	92.7	157.3	136.4
0950, 0980	Ceilings & Acoustic Treatment	91.4	114.2	106.4	88.0	115.9	106.3	99.2	162.9	141.0	99.2	143.4	128.3	88.0	124.3	111.9	99.2	157.3	137.4
0960	Flooring	110.1	120.2	113.0	96.6	118.7	102.9	96.5	140.3	109.1	93.7	131.9	104.6	102.8	122.6	108.5	96.9	140.3	109.3
0970, 0990	Wall Finishes & Painting/Coating	102.5	104.2	103.5	93.8	95.1	94.5	90.1	152.5	127.8	90.1	126.4	112.0	93.8	126.4	113.5	91.9	152.5	128.4
09	FINISHES	100.6	115.2	108.7	94.6	115.4	106.1	97.4	156.9	130.2	95.7	137.9	119.0	97.3	124.3	112.2	97.9	152.7	128.1
COVERS	DIVS. 10 - 14, 25, 28, 41, 43, 44, 46	100.0	70.8	94.1	100.0	105.1	101.0	100.0	124.5	105.0	100.0	120.8	104.2	100.0	102.5	100.5	100.0	122.3	104.5
21, 22, 23	FIRE SUPPRESSION, PLUMBING & HVAC	95.2	102.8	98.2	95.1	105.3	99.2	99.9	131.2	112.5	95.1	129.0	108.7	95.1	123.4	106.6	99.8	128.6	111.4
26, 27, 3370	ELECTRICAL, COMMUNICATIONS & UTIL.	97.3	107.6	102.8	95.1	86.6	90.6	97.5	137.9	118.9	92.3	137.1	116.0	92.2	137.1	116.0	97.3	128.5	113.8
MF2014	WEIGHTED AVERAGE	97.3	106.0	101.1	98.2	106.7	101.9	99.2	138.4	116.3	95.6	131.7	111.3	98.4	124.4	109.7	99.3	135.1	114.9

DIVISION		ILLINOIS																	
		PEORIA			QUINCY			ROCK ISLAND			ROCKFORD			SOUTH SUBURBAN			SPRINGFIELD		
		615 - 616			623			612			610 - 611			605			626 - 627		
		MAT.	INST.	TOTAL	MAT.	INST.	TOTAL	MAT.	INST.	TOTAL	MAT.	INST.	TOTAL	MAT.	INST.	TOTAL	MAT.	INST.	TOTAL
015433	CONTRACTOR EQUIPMENT		101.6	101.6		102.4	102.4		101.6	101.6		101.6	101.6		92.5	92.5		102.4	102.4
0241, 31 - 34	SITE & INFRASTRUCTURE, DEMOLITION	100.4	98.3	98.9	96.5	98.7	98.0	98.0	97.4	97.5	99.8	99.7	99.7	106.1	94.9	98.1	99.0	99.1	99.1
0310	Concrete Forming & Accessories	93.7	117.4	114.2	94.2	111.8	109.4	92.3	104.2	102.5	98.1	131.3	126.7	97.8	154.0	146.3	92.6	115.3	112.1
0320	Concrete Reinforcing	91.7	110.8	101.4	91.8	105.3	98.7	94.1	103.7	99.0	86.7	136.9	112.2	99.0	156.5	128.3	94.2	105.1	99.7
0330	Cast-in-Place Concrete	102.7	114.1	107.3	105.9	102.9	104.7	103.6	97.3	101.0	105.1	126.4	113.8	107.5	141.2	121.3	100.8	109.2	104.2
03	CONCRETE	100.2	115.0	107.5	101.7	107.6	104.6	101.0	102.0	101.5	100.9	130.4	115.4	102.7	148.8	125.4	99.1	111.3	105.1
04	MASONRY	120.1	118.1	118.8	111.7	104.1	106.9	120.3	97.0	105.8	94.1	135.8	120.0	101.1	144.1	127.9	91.0	117.8	107.7
05	METALS	100.4	113.8	104.5	97.1	108.0	100.5	97.7	108.4	101.0	100.4	128.2	108.9	93.4	131.4	105.1	97.5	109.3	101.1
06	WOOD, PLASTICS & COMPOSITES	101.2	115.5	109.2	91.8	113.9	104.2	95.0	104.4	100.3	101.1	128.3	116.4	104.0	155.7	132.9	89.2	113.9	103.0
07	THERMAL & MOISTURE PROTECTION	99.3	112.4	104.7	102.7	103.5	103.0	98.5	98.3	98.4	101.8	129.0	113.0	98.2	139.3	115.1	104.4	112.2	107.6
08	OPENINGS	101.3	116.0	104.7	95.8	111.4	99.4	94.9	103.2	96.8	101.3	132.1	108.5	102.9	158.2	115.8	103.0	111.2	104.9
0920	Plaster & Gypsum Board	97.1	115.9	109.8	98.3	114.2	109.0	93.7	104.5	101.0	97.1	129.1	118.8	92.7	157.3	136.4	98.7	114.2	109.2
0950, 0980	Ceilings & Acoustic Treatment	93.0	115.9	108.1	91.4	114.2	106.4	88.0	104.5	98.8	93.0	129.1	116.7	99.2	157.3	137.4	102.1	114.2	110.1
0960	Flooring	99.9	118.7	105.3	109.1	108.1	108.8	97.7	105.6	100.0	99.9	122.6	106.4	96.9	140.3	109.3	113.5	107.8	111.9
0970, 0990	Wall Finishes & Painting/Coating	93.8	126.4	113.5	102.5	106.5	104.9	93.8	95.1	94.5	93.8	133.1	117.5	91.9	152.5	128.4	101.1	106.5	104.3
09	FINISHES	97.2	118.9	109.2	100.1	111.8	106.6	94.9	103.7	99.7	97.2	130.0	115.3	97.9	152.7	128.1	104.9	113.5	109.6
COVERS	DIVS. 10 - 14, 25, 28, 41, 43, 44, 46	100.0	105.2	101.0	100.0	71.2	94.2	100.0	99.0	99.8	100.0	114.3	102.9	100.0	122.3	104.5	100.0	104.5	100.9
21, 22, 23	FIRE SUPPRESSION, PLUMBING & HVAC	99.9	104.9	101.9	95.2	101.0	97.5	95.1	99.9	97.1	100.0	116.5	106.7	99.8	128.6	111.4	99.9	103.3	101.3
26, 27, 3370	ELECTRICAL, COMMUNICATIONS & UTIL.	96.1	96.7	96.4	94.7	80.8	87.3	87.3	95.5	91.6	96.4	131.1	114.7	97.3	128.5	113.8	102.4	92.0	96.9
MF2014	WEIGHTED AVERAGE	100.5	109.3	104.4	98.1	101.0	99.3	97.2	100.5	98.6	99.5	124.8	110.5	99.3	135.1	114.9	100.0	106.5	102.8

	DIVISION	INDIANA																	
		ANDERSON			BLOOMINGTON			COLUMBUS			EVANSVILLE			FORT WAYNE			GARY		
		460			474			472			476 - 477			467 - 468			463 - 464		
		MAT.	INST.	TOTAL	MAT.	INST.	TOTAL	MAT.	INST.	TOTAL	MAT.	INST.	TOTAL	MAT.	INST.	TOTAL	MAT.	INST.	TOTAL
015433	CONTRACTOR EQUIPMENT		97.0	97.0		86.5	86.5		86.5	86.5		116.0	116.0		97.0	97.0		97.0	97.0
0241, 31 - 34	SITE & INFRASTRUCTURE, DEMOLITION	93.9	96.1	95.5	85.8	94.4	91.9	82.2	94.3	90.8	91.0	123.9	114.4	94.8	96.0	95.7	94.5	99.6	98.1
0310	Concrete Forming & Accessories	97.8	81.5	83.7	101.0	80.7	83.5	95.0	78.6	80.8	94.4	82.7	84.3	96.1	75.2	78.1	97.9	116.3	113.8
0320	Concrete Reinforcing	95.8	82.8	89.2	86.8	80.9	83.8	87.2	80.9	84.0	95.1	78.1	86.4	95.8	75.5	85.5	95.8	110.8	103.5
0330	Cast-in-Place Concrete	109.4	80.2	97.4	103.6	78.2	93.1	103.1	71.2	90.0	99.0	87.9	94.4	116.3	82.8	102.5	114.3	113.9	114.1
03	CONCRETE	100.8	81.7	91.5	105.0	79.7	92.6	104.2	76.3	90.5	105.3	83.8	94.7	104.0	78.4	91.4	103.2	114.2	108.6
04	MASONRY	94.1	80.1	85.4	95.6	76.5	83.7	95.4	76.5	83.6	91.2	82.5	85.8	98.5	77.9	85.6	95.6	114.6	107.5
05	METALS	92.8	89.3	91.7	95.8	77.6	90.2	95.9	77.0	90.0	89.3	85.0	87.9	92.8	85.9	90.7	92.8	108.7	97.7
06	WOOD, PLASTICS & COMPOSITES	100.2	81.9	89.9	112.9	80.9	95.0	107.7	78.2	91.2	93.5	82.0	87.0	99.9	74.7	85.8	97.7	115.7	107.8
07	THERMAL & MOISTURE PROTECTION	107.7	75.9	94.7	95.1	78.3	88.2	94.7	78.1	87.9	99.4	83.7	93.0	107.5	77.7	95.3	106.3	108.6	107.2
08	OPENINGS	98.5	82.1	94.7	105.8	81.1	100.1	101.5	79.6	96.4	99.1	80.8	94.9	98.5	74.2	92.9	98.5	120.3	103.6
0920	Plaster & Gypsum Board	102.5	81.6	88.4	97.5	81.0	86.4	95.0	78.2	83.6	93.4	80.8	84.9	101.8	74.2	83.1	96.1	116.4	109.8
0950, 0980	Ceilings & Acoustic Treatment	87.9	81.6	83.7	80.9	81.0	81.0	80.9	78.2	79.1	85.5	80.8	82.4	87.9	74.2	78.9	87.9	116.4	106.6
0960	Flooring	100.6	85.0	96.2	105.5	75.0	96.8	100.8	75.0	93.4	100.3	80.7	94.7	100.6	78.8	94.4	100.6	123.4	107.1
0970, 0990	Wall Finishes & Painting/Coating	105.6	69.4	83.7	95.3	82.4	87.5	95.3	82.4	87.5	101.2	84.9	91.4	105.6	73.9	86.5	105.6	123.9	116.7
09	FINISHES	95.6	81.0	87.6	94.8	79.8	86.5	93.0	78.2	84.9	93.8	82.6	87.6	95.4	75.7	84.5	94.7	118.7	107.9
COVERS	DIVS. 10 - 14, 25, 28, 41, 43, 44, 46	100.0	91.8	98.3	100.0	90.8	98.1	100.0	90.5	98.1	100.0	95.9	99.2	100.0	91.7	98.3	100.0	106.1	101.2
21, 22, 23	FIRE SUPPRESSION, PLUMBING & HVAC	100.0	78.8	91.5	99.7	78.9	91.3	94.9	78.8	88.4	99.9	79.6	91.7	100.0	72.1	88.7	100.0	106.6	102.7
26, 27, 3370	ELECTRICAL, COMMUNICATIONS & UTIL.	87.0	88.4	87.7	98.9	88.0	93.2	98.1	87.9	92.7	95.1	88.3	91.5	87.7	79.2	83.2	98.7	107.6	103.4
MF2014	WEIGHTED AVERAGE	97.0	83.8	91.3	99.3	81.8	91.6	97.2	80.9	90.1	97.1	86.9	92.6	97.6	79.1	89.5	98.3	110.4	103.6

	DIVISION	INDIANA																	
		INDIANAPOLIS			KOKOMO			LAFAYETTE			LAWRENCEBURG			MUNCIE			NEW ALBANY		
		461 - 462			469			479			470			473			471		
		MAT.	INST.	TOTAL	MAT.	INST.	TOTAL	MAT.	INST.	TOTAL	MAT.	INST.	TOTAL	MAT.	INST.	TOTAL	MAT.	INST.	TOTAL
015433	CONTRACTOR EQUIPMENT		93.7	93.7		97.0	97.0		86.5	86.5		104.4	104.4		95.0	95.0		94.5	94.5
0241, 31 - 34	SITE & INFRASTRUCTURE, DEMOLITION	93.5	99.3	97.7	90.3	96.0	94.4	83.1	94.3	91.0	81.2	110.4	102.0	85.6	94.6	92.0	78.2	96.8	91.4
0310	Concrete Forming & Accessories	98.6	85.6	87.4	101.2	76.6	80.0	92.6	82.6	84.0	91.5	75.4	77.6	92.6	81.0	82.6	89.5	73.3	75.5
0320	Concrete Reinforcing	99.4	83.0	91.1	86.7	81.1	83.9	86.8	82.7	84.7	86.1	74.2	80.1	96.0	82.7	89.3	87.4	77.1	82.2
0330	Cast-in-Place Concrete	101.9	86.4	95.5	108.3	82.7	97.8	103.7	81.8	94.7	97.0	74.8	87.9	108.8	78.6	96.4	100.1	74.0	89.4
03	CONCRETE	100.9	85.1	93.1	97.5	80.1	89.0	104.5	82.1	93.5	97.4	75.5	86.6	103.5	81.0	92.4	102.8	74.6	89.0
04	MASONRY	95.8	80.3	86.2	93.7	79.1	84.6	101.2	80.3	88.2	79.9	74.7	76.7	97.7	80.1	86.7	86.9	68.3	75.3
05	METALS	93.5	80.5	89.5	89.3	88.3	89.0	94.3	78.4	89.4	91.0	83.5	88.7	97.6	89.2	95.0	92.9	81.0	89.3
06	WOOD, PLASTICS & COMPOSITES	99.7	86.2	92.2	103.4	75.0	87.5	104.7	83.1	92.6	91.9	75.2	82.6	106.3	81.5	92.4	94.1	74.3	83.0
07	THERMAL & MOISTURE PROTECTION	100.7	80.8	92.5	107.3	75.7	94.3	94.7	80.6	88.9	100.1	76.7	90.5	97.7	76.8	89.2	86.8	69.6	79.7
08	OPENINGS	105.7	84.5	100.8	93.3	77.9	89.8	99.9	82.8	95.9	101.4	75.3	95.3	99.0	81.9	95.0	98.7	76.3	93.5
0920	Plaster & Gypsum Board	97.0	85.8	89.5	107.5	74.5	85.2	92.2	83.3	86.1	71.5	75.0	73.9	93.4	81.6	85.4	91.6	73.8	79.6
0950, 0980	Ceilings & Acoustic Treatment	93.8	85.8	88.6	87.9	74.5	79.1	76.7	83.3	81.0	89.7	75.0	80.0	81.7	81.6	81.6	85.5	73.8	77.8
0960	Flooring	101.4	85.0	96.7	104.6	93.3	101.3	99.7	88.8	96.6	74.1	85.0	77.2	99.9	85.0	95.6	97.9	62.8	87.9
0970, 0990	Wall Finishes & Painting/Coating	103.6	82.4	90.8	105.6	71.7	85.2	95.3	93.1	94.0	96.2	71.6	81.4	95.3	69.4	79.7	101.2	82.7	90.0
09	FINISHES	96.1	85.4	90.2	97.3	78.8	87.1	91.4	85.0	87.9	83.7	77.2	80.1	92.4	80.6	85.9	93.1	72.7	81.8
COVERS	DIVS. 10 - 14, 25, 28, 41, 43, 44, 46	100.0	93.3	98.7	100.0	91.1	98.2	100.0	91.1	98.2	100.0	43.6	88.6	100.0	90.8	98.1	100.0	43.1	88.5
21, 22, 23	FIRE SUPPRESSION, PLUMBING & HVAC	99.9	79.4	91.6	95.2	79.0	88.7	94.9	79.4	88.6	95.8	74.1	87.1	99.7	78.7	91.2	95.2	75.2	87.1
26, 27, 3370	ELECTRICAL, COMMUNICATIONS & UTIL.	101.0	88.4	94.4	91.4	78.7	84.7	97.6	82.5	89.6	93.0	73.5	82.7	90.9	79.6	84.9	93.6	75.8	84.2
MF2014	WEIGHTED AVERAGE	99.1	84.8	92.9	94.8	81.5	89.0	96.9	82.7	90.7	94.0	77.7	86.9	97.9	82.3	91.1	95.1	75.4	86.5

	DIVISION	INDIANA									IOWA								
		SOUTH BEND			TERRE HAUTE			WASHINGTON			BURLINGTON			CARROLL			CEDAR RAPIDS		
		465 - 466			478			475			526			514			522 - 524		
		MAT.	INST.	TOTAL	MAT.	INST.	TOTAL	MAT.	INST.	TOTAL	MAT.	INST.	TOTAL	MAT.	INST.	TOTAL	MAT.	INST.	TOTAL
015433	CONTRACTOR EQUIPMENT		105.3	105.3		116.0	116.0		116.0	116.0		99.4	99.4		99.4	99.4		96.1	96.1
0241, 31 - 34	SITE & INFRASTRUCTURE, DEMOLITION	96.7	96.2	96.3	92.9	124.1	115.1	92.4	121.3	113.0	98.3	96.5	97.0	87.5	96.5	93.9	100.0	95.1	96.5
0310	Concrete Forming & Accessories	99.5	81.4	83.9	95.4	82.2	84.0	96.2	78.8	81.2	94.8	75.5	78.2	82.6	50.4	54.8	100.5	82.3	84.8
0320	Concrete Reinforcing	97.3	79.9	88.4	95.1	82.9	88.9	87.9	48.8	68.0	91.0	82.7	86.8	91.8	78.7	85.1	91.7	84.2	87.9
0330	Cast-in-Place Concrete	106.6	79.8	95.6	95.9	85.6	91.6	104.3	84.8	96.3	111.7	55.1	88.4	111.7	59.7	90.3	111.9	79.7	98.7
03	CONCRETE	98.3	81.8	90.2	108.4	83.7	96.2	114.3	75.3	95.1	102.6	70.8	86.9	101.3	60.2	81.1	102.7	82.3	92.7
04	MASONRY	105.7	77.8	88.3	98.9	79.6	86.9	91.3	80.8	84.8	102.3	64.9	79.0	104.2	74.1	85.4	108.2	78.0	89.4
05	METALS	92.8	99.9	95.0	90.0	87.4	89.2	84.6	68.2	79.5	88.8	93.0	90.1	88.9	88.7	88.8	91.3	92.2	91.5
06	WOOD, PLASTICS & COMPOSITES	94.8	81.6	87.4	95.7	82.1	88.1	96.1	78.9	86.5	95.1	76.6	84.8	81.6	44.9	61.1	101.9	82.6	91.1
07	THERMAL & MOISTURE PROTECTION	101.9	80.6	93.2	99.5	80.8	91.8	99.6	82.2	92.5	102.9	73.9	91.0	103.2	69.4	89.3	103.9	80.9	94.5
08	OPENINGS	96.0	80.5	92.4	99.7	82.2	95.6	96.4	67.6	89.7	94.8	70.4	89.1	99.3	53.4	88.6	100.3	81.9	96.0
0920	Plaster & Gypsum Board	90.8	81.3	84.4	93.4	81.0	85.0	93.4	77.7	82.8	102.3	75.9	84.5	97.7	43.2	60.9	106.9	82.3	90.3
0950, 0980	Ceilings & Acoustic Treatment	88.9	81.3	83.9	85.5	81.0	82.6	79.6	77.7	78.3	98.7	75.9	83.7	98.7	43.2	62.3	101.2	82.3	88.8
0960	Flooring	99.3	91.1	96.9	100.3	85.0	95.9	101.2	76.4	94.1	107.9	37.8	87.8	102.2	33.2	82.5	124.0	78.3	110.9
0970, 0990	Wall Finishes & Painting/Coating	99.2	87.1	91.9	101.2	83.1	90.3	101.2	83.4	90.5	106.3	78.5	89.5	106.3	78.7	89.6	107.6	72.2	86.3
09	FINISHES	95.4	83.9	89.1	93.8	82.8	87.7	93.1	79.6	85.6	102.5	68.9	84.0	98.7	48.3	71.0	108.4	80.6	93.1
COVERS	DIVS. 10 - 14, 25, 28, 41, 43, 44, 46	100.0	92.8	98.5	100.0	93.7	98.7	100.0	95.3	99.0	100.0	86.8	97.3	100.0	65.1	92.9	100.0	91.8	98.3
21, 22, 23	FIRE SUPPRESSION, PLUMBING & HVAC	99.9	77.1	90.7	99.9	79.6	91.7	95.2	77.6	88.1	95.4	75.2	87.3	95.4	72.6	86.2	100.2	80.5	92.2
26, 27, 3370	ELECTRICAL, COMMUNICATIONS & UTIL.	99.6	86.6	92.7	93.3	86.7	89.8	93.8	84.2	88.8	100.6	67.7	83.2	101.3	81.3	90.7	98.1	80.6	88.8
MF2014	WEIGHTED AVERAGE	98.0	84.4	92.1	97.9	86.5	92.9	95.8	81.8	89.7	97.2	75.3	87.6	97.0	71.2	85.8	100.0	83.2	92.7

DIVISION		IOWA																	
		COUNCIL BLUFFS			CRESTON			DAVENPORT			DECORAH			DES MOINES			DUBUQUE		
		515			508			527 - 528			521			500 - 503,509			520		
		MAT.	INST.	TOTAL	MAT.	INST.	TOTAL	MAT.	INST.	TOTAL	MAT.	INST.	TOTAL	MAT.	INST.	TOTAL	MAT.	INST.	TOTAL
015433	CONTRACTOR EQUIPMENT		95.5	95.5		99.4	99.4		99.4	99.4		99.4	99.4		101.1	101.1		94.9	94.9
0241, 31 - 34	SITE & INFRASTRUCTURE, DEMOLITION	103.8	91.5	95.0	93.9	95.5	95.1	98.7	98.6	98.6	96.9	95.5	95.9	102.9	99.8	100.7	97.9	92.4	94.0
0310	Concrete Forming & Accessories	82.0	71.2	72.7	78.4	65.8	67.5	100.0	91.3	92.5	92.3	45.4	51.8	95.4	81.8	83.6	83.3	74.9	76.1
0320	Concrete Reinforcing	93.6	79.0	86.1	89.3	82.4	85.8	91.7	97.0	94.4	91.0	76.5	83.6	98.2	86.7	92.3	90.4	84.1	87.2
0330	Cast-in-Place Concrete	116.4	73.6	98.8	115.2	64.1	94.2	107.8	90.1	100.5	108.6	58.0	87.8	101.4	87.5	95.7	109.6	98.6	105.1
03	CONCRETE	105.0	74.3	89.9	102.9	69.3	86.4	100.7	92.5	96.7	100.5	57.0	79.1	100.0	85.4	92.8	99.4	85.5	92.6
04	MASONRY	109.8	75.0	88.1	108.1	81.9	91.7	105.3	86.8	93.7	124.0	71.4	91.2	101.0	81.2	88.7	109.2	73.9	87.2
05	METALS	96.3	89.3	94.1	93.8	90.5	92.8	91.3	102.7	94.8	89.0	85.8	88.0	99.9	96.5	98.8	89.8	92.0	90.5
06	WOOD, PLASTICS & COMPOSITES	80.4	70.9	75.1	74.1	61.6	67.1	101.9	91.0	95.8	92.2	37.1	61.3	91.8	81.1	85.8	82.1	73.5	77.3
07	THERMAL & MOISTURE PROTECTION	103.3	67.1	88.5	104.2	77.6	93.3	103.4	87.5	96.8	103.1	53.4	82.7	98.5	79.4	90.6	103.6	73.9	91.4
08	OPENINGS	99.3	75.9	93.9	108.4	63.9	98.1	100.3	90.8	98.1	97.8	48.6	86.4	102.7	86.5	98.9	99.3	79.3	94.7
0920	Plaster & Gypsum Board	97.7	70.3	79.2	93.6	60.5	71.2	106.9	90.8	96.0	101.2	35.2	56.6	90.4	80.6	83.8	97.7	72.9	81.0
0950, 0980	Ceilings & Acoustic Treatment	98.7	70.3	80.0	90.8	60.5	70.9	101.2	90.8	94.4	98.7	35.2	57.0	94.2	80.6	85.2	98.7	72.9	81.8
0960	Flooring	100.8	80.7	95.0	97.4	33.2	79.1	110.3	97.0	106.5	107.6	47.2	90.3	102.5	89.6	98.8	114.9	77.0	104.1
0970, 0990	Wall Finishes & Painting/Coating	102.2	65.8	80.2	101.0	78.7	87.5	106.3	95.5	99.8	106.3	32.9	62.0	96.2	88.6	91.6	106.7	63.8	80.8
09	FINISHES	99.4	72.1	84.4	94.2	60.2	75.4	104.3	92.7	97.9	102.2	42.6	69.4	97.1	83.7	89.7	103.7	73.5	87.1
COVERS	DIVS. 10 - 14, 25, 28, 41, 43, 44, 46	100.0	88.6	97.7	100.0	69.5	93.8	100.0	94.5	98.9	100.0	82.9	96.5	100.0	92.1	98.4	100.0	90.2	98.0
21, 22, 23	FIRE SUPPRESSION, PLUMBING & HVAC	100.2	74.2	89.7	95.2	79.7	89.0	100.2	92.8	97.2	95.4	72.9	86.3	99.9	79.7	91.7	100.2	75.3	90.1
26, 27, 3370	ELECTRICAL, COMMUNICATIONS & UTIL.	103.6	81.5	91.9	93.8	81.3	87.2	95.6	91.0	93.2	98.1	44.3	69.7	105.4	81.3	92.7	102.1	78.2	89.4
MF2014	WEIGHTED AVERAGE	100.7	78.1	90.9	98.2	77.3	89.1	99.0	93.1	96.4	98.0	64.2	83.3	100.5	85.2	93.8	99.1	80.2	90.9

DIVISION		IOWA																	
		FORT DODGE			MASON CITY			OTTUMWA			SHENANDOAH			SIBLEY			SIOUX CITY		
		505			504			525			516			512			510 - 511		
		MAT.	INST.	TOTAL	MAT.	INST.	TOTAL	MAT.	INST.	TOTAL	MAT.	INST.	TOTAL	MAT.	INST.	TOTAL	MAT.	INST.	TOTAL
015433	CONTRACTOR EQUIPMENT		99.4	99.4		99.4	99.4		94.9	94.9		95.5	95.5		99.4	99.4		99.4	99.4
0241, 31 - 34	SITE & INFRASTRUCTURE, DEMOLITION	102.2	94.4	96.6	102.3	95.4	97.4	98.1	90.5	92.7	102.2	90.5	93.9	108.1	94.2	98.2	109.9	95.0	99.3
0310	Concrete Forming & Accessories	78.9	45.1	49.7	83.1	45.5	50.6	90.3	73.7	75.9	83.6	56.5	60.2	84.0	37.9	44.2	100.5	66.4	71.1
0320	Concrete Reinforcing	89.3	67.2	78.1	89.2	81.8	85.4	91.0	86.1	88.5	93.6	68.4	80.7	93.6	65.3	79.2	91.7	77.9	84.7
0330	Cast-in-Place Concrete	108.2	44.2	81.9	108.2	57.5	87.4	112.4	53.7	88.3	112.5	59.8	90.9	110.3	44.7	83.4	111.0	55.5	88.2
03	CONCRETE	98.2	50.4	74.7	98.5	57.9	78.5	102.1	69.9	86.3	102.4	61.0	82.1	101.4	47.0	74.6	102.0	65.8	84.2
04	MASONRY	106.9	37.9	63.9	120.5	69.8	88.9	105.8	57.8	75.9	109.4	75.0	88.0	128.6	38.1	72.2	102.2	54.6	72.6
05	METALS	93.9	81.3	90.1	94.0	88.7	92.4	88.8	91.7	89.7	95.3	83.3	91.6	89.1	79.7	86.2	91.3	87.4	90.1
06	WOOD, PLASTICS & COMPOSITES	74.5	45.2	58.1	78.4	37.1	55.3	89.3	80.9	84.6	82.1	52.8	65.7	82.9	35.9	56.6	101.9	66.6	82.1
07	THERMAL & MOISTURE PROTECTION	103.5	57.1	84.5	103.0	63.5	86.8	103.7	64.6	87.7	102.6	65.2	87.2	102.9	48.3	80.5	103.4	64.3	87.4
08	OPENINGS	101.9	49.5	89.7	93.5	50.1	83.4	99.3	77.7	94.3	90.3	54.7	82.0	95.9	43.8	83.8	100.3	65.9	92.3
0920	Plaster & Gypsum Board	93.6	43.5	59.8	93.6	35.2	54.1	98.8	80.6	86.5	97.7	51.6	66.5	97.7	33.9	54.6	106.9	65.7	79.0
0950, 0980	Ceilings & Acoustic Treatment	90.8	43.5	59.7	90.8	35.2	54.2	98.7	80.6	86.8	98.7	51.6	67.7	98.7	33.9	56.1	101.2	65.7	77.9
0960	Flooring	98.9	47.2	84.1	100.9	47.2	85.6	117.9	50.2	98.5	101.5	34.3	82.3	103.1	34.0	83.3	110.3	53.7	94.1
0970, 0990	Wall Finishes & Painting/Coating	101.0	61.7	77.3	101.0	30.1	58.2	106.7	78.7	89.8	102.2	65.8	80.2	106.3	61.7	79.4	106.3	63.5	80.5
09	FINISHES	96.1	46.1	68.5	96.7	42.0	66.5	104.9	70.4	85.9	99.5	52.3	73.4	101.9	38.0	66.7	105.8	63.6	82.5
COVERS	DIVS. 10 - 14, 25, 28, 41, 43, 44, 46	100.0	81.7	96.3	100.0	86.5	97.3	100.0	82.4	96.4	100.0	66.1	93.2	100.0	80.8	96.1	100.0	88.2	97.6
21, 22, 23	FIRE SUPPRESSION, PLUMBING & HVAC	95.2	65.5	83.3	95.2	70.5	85.2	95.4	70.9	85.5	95.4	73.5	86.6	95.4	68.0	84.3	100.2	75.8	90.4
26, 27, 3370	ELECTRICAL, COMMUNICATIONS & UTIL.	100.3	41.9	69.4	99.4	55.6	76.3	100.4	72.0	85.4	98.1	81.5	89.3	98.1	41.4	68.1	98.1	73.0	84.8
MF2014	WEIGHTED AVERAGE	97.9	58.2	80.6	97.6	65.8	83.8	97.9	73.7	87.3	97.6	71.2	86.1	98.3	56.4	80.0	99.7	72.6	87.9

DIVISION		IOWA						KANSAS											
		SPENCER			WATERLOO			BELLEVILLE			COLBY			DODGE CITY			EMPORIA		
		513			506 - 507			669			677			678			668		
		MAT.	INST.	TOTAL	MAT.	INST.	TOTAL	MAT.	INST.	TOTAL	MAT.	INST.	TOTAL	MAT.	INST.	TOTAL	MAT.	INST.	TOTAL
015433	CONTRACTOR EQUIPMENT		99.4	99.4		99.4	99.4		103.4	103.4		103.4	103.4		103.4	103.4		101.6	101.6
0241, 31 - 34	SITE & INFRASTRUCTURE, DEMOLITION	108.1	94.2	98.2	107.6	95.3	98.8	111.0	95.0	99.7	112.3	95.2	100.1	114.9	94.9	100.7	103.0	92.5	95.5
0310	Concrete Forming & Accessories	90.2	37.9	45.1	94.1	55.1	60.5	96.2	54.1	59.8	99.5	58.3	64.0	93.1	58.2	63.0	87.2	65.3	68.3
0320	Concrete Reinforcing	93.6	67.1	80.1	89.9	83.9	86.8	96.6	55.1	75.5	98.9	55.2	76.7	96.5	55.1	75.4	95.3	55.7	75.2
0330	Cast-in-Place Concrete	110.3	44.7	83.4	115.8	59.2	92.6	124.2	55.7	96.1	127.1	55.8	97.8	129.3	55.5	99.0	120.1	53.6	92.8
03	CONCRETE	101.8	47.3	75.0	104.4	63.1	84.1	119.3	56.2	88.3	119.9	58.1	89.5	121.3	57.9	90.2	111.5	60.5	86.5
04	MASONRY	128.6	38.1	72.2	107.7	73.1	86.1	99.2	58.7	73.9	108.1	59.9	78.1	118.9	58.9	81.5	105.5	68.5	82.4
05	METALS	89.0	80.5	86.4	96.4	91.1	94.7	95.6	77.3	90.0	96.0	77.7	90.4	97.4	76.7	91.1	95.3	78.9	90.3
06	WOOD, PLASTICS & COMPOSITES	89.2	35.9	59.3	91.6	48.0	67.2	97.9	52.1	72.3	103.6	57.9	78.0	95.8	57.9	74.5	88.7	66.6	76.3
07	THERMAL & MOISTURE PROTECTION	103.8	48.2	81.0	103.3	71.3	90.2	94.8	60.7	80.8	99.1	61.6	83.7	99.1	60.7	83.4	93.1	75.9	86.0
08	OPENINGS	107.5	44.4	92.9	94.4	58.2	86.0	98.6	48.4	86.9	103.2	51.5	91.2	103.1	51.5	91.1	96.3	56.3	87.0
0920	Plaster & Gypsum Board	98.8	33.9	54.9	101.9	46.5	64.4	97.2	50.6	65.7	99.7	56.5	70.5	94.0	56.5	68.7	94.4	65.5	74.9
0950, 0980	Ceilings & Acoustic Treatment	98.7	33.9	56.1	94.2	46.5	62.8	86.3	50.6	62.8	84.0	56.5	65.9	84.0	56.5	65.9	86.3	65.5	72.7
0960	Flooring	105.7	34.0	85.2	106.2	64.9	94.4	103.0	39.2	84.8	103.1	39.2	84.9	99.6	39.2	82.3	98.0	36.4	80.4
0970, 0990	Wall Finishes & Painting/Coating	106.3	61.7	79.4	101.0	77.8	87.0	100.0	38.9	63.1	104.7	38.9	65.0	104.7	38.9	65.0	100.0	38.9	63.1
09	FINISHES	102.8	38.0	67.1	100.3	57.6	76.8	96.9	49.3	70.6	96.6	52.7	72.4	94.9	52.7	71.6	94.0	57.3	73.8
COVERS	DIVS. 10 - 14, 25, 28, 41, 43, 44, 46	100.0	80.8	96.1	100.0	87.8	97.5	100.0	40.8	88.0	100.0	41.4	88.2	100.0	41.4	88.2	100.0	40.5	88.0
21, 22, 23	FIRE SUPPRESSION, PLUMBING & HVAC	95.4	68.0	84.3	100.0	79.8	91.9	95.2	71.6	85.7	95.2	69.0	84.6	100.0	69.0	87.5	95.2	73.4	86.4
26, 27, 3370	ELECTRICAL, COMMUNICATIONS & UTIL.	99.9	41.4	69.0	95.8	55.7	74.6	96.5	67.7	81.3	100.0	64.5	81.2	96.8	73.8	84.6	93.9	71.4	82.0
MF2014	WEIGHTED AVERAGE	99.9	56.6	81.0	99.5	71.8	87.4	99.6	64.9	84.4	101.1	65.0	85.4	102.7	66.0	86.7	97.9	69.2	85.4

	DIVISION	KANSAS																	
		FORT SCOTT			HAYS			HUTCHINSON			INDEPENDENCE			KANSAS CITY			LIBERAL		
		667			676			675			673			660 - 662			679		
		MAT.	INST.	TOTAL	MAT.	INST.	TOTAL	MAT.	INST.	TOTAL	MAT.	INST.	TOTAL	MAT.	INST.	TOTAL	MAT.	INST.	TOTAL
015433	CONTRACTOR EQUIPMENT		102.5	102.5		103.4	103.4		103.4	103.4		103.4	103.4		99.9	99.9		103.4	103.4
0241, 31 - 34	SITE & INFRASTRUCTURE, DEMOLITION	99.8	93.4	95.2	117.7	95.2	101.7	96.2	95.0	95.4	116.0	95.1	101.1	94.4	93.1	93.5	117.3	95.0	101.5
0310	Concrete Forming & Accessories	104.2	76.8	80.6	97.2	58.3	63.7	87.9	58.2	62.2	107.9	66.3	72.0	100.9	94.2	95.1	93.5	58.2	63.0
0320	Concrete Reinforcing	94.6	93.3	94.0	96.5	55.2	75.5	96.5	55.1	75.4	95.9	62.7	79.0	91.8	97.2	94.6	97.8	55.1	76.1
0330	Cast-in-Place Concrete	111.4	54.5	88.0	100.3	55.8	82.0	92.9	53.1	76.6	129.9	53.4	98.5	95.4	97.4	96.3	100.3	53.1	80.9
03	CONCRETE	106.6	72.9	90.1	110.4	58.1	84.7	92.4	57.1	75.0	122.6	62.2	92.9	98.4	96.2	97.3	112.5	57.1	85.2
04	MASONRY	106.7	60.3	77.8	118.0	59.9	81.8	107.9	58.6	77.2	105.3	64.5	79.9	107.8	99.4	102.6	116.6	58.6	80.4
05	METALS	95.3	92.8	94.5	95.6	77.7	90.1	95.4	76.7	89.6	95.3	80.5	90.8	102.9	100.1	102.1	95.9	76.7	90.0
06	WOOD, PLASTICS & COMPOSITES	108.7	83.5	94.6	100.6	57.9	76.7	90.9	57.9	72.4	113.7	68.0	88.1	104.4	94.5	98.8	96.4	57.9	74.8
07	THERMAL & MOISTURE PROTECTION	94.0	80.8	88.6	99.4	61.6	83.9	98.0	60.6	82.7	99.1	80.5	91.5	93.7	98.0	95.5	99.5	60.6	83.6
08	OPENINGS	96.3	78.2	92.1	103.1	51.5	91.1	103.0	51.5	91.0	100.7	58.8	90.9	97.7	89.0	95.7	103.1	51.5	91.1
0920	Plaster & Gypsum Board	99.7	82.9	88.4	97.3	56.5	69.7	93.0	56.5	68.3	106.9	67.0	79.9	92.9	94.2	93.8	94.8	56.5	68.9
0950, 0980	Ceilings & Acoustic Treatment	86.3	82.9	84.1	84.0	56.5	65.9	84.0	56.5	65.9	84.0	67.0	72.8	86.3	94.2	91.5	84.0	56.5	65.9
0960	Flooring	113.7	39.4	92.5	102.0	39.2	84.0	96.8	39.2	80.3	107.3	39.2	87.8	91.4	98.7	93.5	99.8	39.2	82.5
0970, 0990	Wall Finishes & Painting/Coating	101.8	38.9	63.8	104.7	38.9	65.0	104.7	38.9	65.0	104.7	38.9	65.0	108.5	67.7	83.9	104.7	38.9	65.0
09	FINISHES	99.6	66.9	81.6	96.4	52.7	72.3	92.4	52.7	70.5	98.9	58.7	76.8	94.3	91.0	92.5	95.7	52.7	72.0
COVERS	DIVS. 10 - 14, 25, 28, 41, 43, 44, 46	100.0	46.2	89.1	100.0	41.4	88.2	100.0	41.4	88.2	100.0	42.5	88.4	100.0	61.9	92.3	100.0	41.4	88.2
21, 22, 23	FIRE SUPPRESSION, PLUMBING & HVAC	95.2	68.3	84.3	95.2	69.0	84.6	95.2	69.0	84.6	95.2	70.1	85.1	99.9	96.4	98.5	95.2	67.2	83.9
26, 27, 3370	ELECTRICAL, COMMUNICATIONS & UTIL.	93.2	71.4	81.6	98.9	67.7	82.4	93.9	67.7	80.0	96.1	71.4	83.0	98.7	97.2	97.9	96.8	73.8	84.6
MF2014	WEIGHTED AVERAGE	97.9	73.0	87.0	100.5	65.4	85.2	96.5	65.0	82.8	100.9	69.2	87.1	99.5	94.8	97.5	100.4	65.5	85.2

	DIVISION	KANSAS									KENTUCKY								
		SALINA			TOPEKA			WICHITA			ASHLAND			BOWLING GREEN			CAMPTON		
		674			664 - 666			670 - 672			411 - 412			421 - 422			413 - 414		
		MAT.	INST.	TOTAL	MAT.	INST.	TOTAL	MAT.	INST.	TOTAL	MAT.	INST.	TOTAL	MAT.	INST.	TOTAL	MAT.	INST.	TOTAL
015433	CONTRACTOR EQUIPMENT		103.4	103.4		101.6	101.6		103.4	103.4		97.3	97.3		94.5	94.5		101.1	101.1
0241, 31 - 34	SITE & INFRASTRUCTURE, DEMOLITION	105.2	95.2	98.1	98.2	91.5	93.4	101.8	93.8	96.1	112.3	86.3	93.8	78.6	97.1	91.8	87.0	98.3	95.0
0310	Concrete Forming & Accessories	89.7	54.9	59.7	98.4	40.9	48.8	95.3	51.0	57.1	87.9	105.2	102.8	85.7	83.0	83.4	89.4	82.6	83.6
0320	Concrete Reinforcing	95.9	59.4	77.3	94.9	99.5	97.2	95.3	77.3	86.1	88.8	108.7	98.9	86.2	91.3	88.8	87.0	101.8	94.5
0330	Cast-in-Place Concrete	112.3	53.7	88.2	100.4	45.5	77.9	105.3	51.7	83.3	91.1	101.1	95.2	90.4	95.5	92.5	100.8	73.6	89.6
03	CONCRETE	107.3	56.7	82.5	99.4	55.1	77.7	101.6	57.4	79.9	98.6	104.8	101.6	96.8	88.9	92.9	100.7	83.2	92.1
04	MASONRY	134.8	59.9	88.2	102.5	55.9	73.4	106.3	50.2	71.4	97.8	103.7	101.5	99.7	80.2	87.5	96.9	69.0	79.5
05	METALS	97.3	80.6	92.1	99.4	96.9	98.6	99.4	84.2	94.7	92.1	110.3	97.7	93.6	87.3	91.7	92.9	92.0	92.6
06	WOOD, PLASTICS & COMPOSITES	92.4	52.1	69.8	97.1	37.3	63.6	95.1	51.3	70.6	76.0	105.2	92.3	88.6	83.1	85.5	87.1	89.0	88.2
07	THERMAL & MOISTURE PROTECTION	98.6	61.4	83.3	96.7	67.4	84.7	97.2	56.1	80.4	91.0	97.5	93.7	86.6	80.2	84.0	99.7	67.5	86.5
08	OPENINGS	103.1	48.5	90.4	100.8	55.4	90.2	105.3	56.2	93.9	97.9	99.6	98.3	98.7	79.3	94.2	100.2	87.6	97.3
0920	Plaster & Gypsum Board	93.0	50.6	64.3	96.8	35.3	55.2	91.8	49.8	63.4	60.9	105.4	91.0	87.4	82.9	84.3	87.4	88.1	87.9
0950, 0980	Ceilings & Acoustic Treatment	84.0	50.6	62.0	92.0	35.3	54.7	88.8	49.8	63.2	80.8	105.4	97.0	85.5	82.9	83.8	85.5	88.1	87.2
0960	Flooring	98.2	61.9	87.8	103.4	73.7	94.9	105.4	63.0	93.3	79.4	102.4	86.0	95.8	86.0	93.0	98.0	41.2	81.7
0970, 0990	Wall Finishes & Painting/Coating	104.7	38.9	65.0	101.8	52.5	72.1	103.6	48.0	70.0	103.1	101.2	102.0	101.2	72.7	84.0	101.2	65.4	79.6
09	FINISHES	93.6	53.6	71.6	98.5	45.9	69.5	97.7	52.3	72.7	80.9	104.8	94.1	91.8	82.8	86.8	92.6	74.0	82.4
COVERS	DIVS. 10 - 14, 25, 28, 41, 43, 44, 46	100.0	86.0	97.2	100.0	47.2	89.3	100.0	83.0	96.6	100.0	94.1	98.8	100.0	62.3	92.4	100.0	55.8	91.1
21, 22, 23	FIRE SUPPRESSION, PLUMBING & HVAC	100.0	69.1	87.5	100.0	69.1	87.5	99.8	64.8	85.7	95.0	92.5	94.0	99.9	82.7	93.0	95.2	79.8	89.0
26, 27, 3370	ELECTRICAL, COMMUNICATIONS & UTIL.	96.6	73.8	84.5	97.5	71.8	83.9	100.9	73.8	86.5	91.4	97.3	94.5	94.0	81.5	87.4	91.4	66.1	78.0
MF2014	WEIGHTED AVERAGE	101.5	67.6	86.7	99.5	66.0	84.9	100.7	66.0	85.6	94.6	99.3	96.6	96.2	83.9	90.8	95.8	78.6	88.3

	DIVISION	KENTUCKY																	
		CORBIN			COVINGTON			ELIZABETHTOWN			FRANKFORT			HAZARD			HENDERSON		
		407 - 409			410			427			406			417 - 418			424		
		MAT.	INST.	TOTAL	MAT.	INST.	TOTAL	MAT.	INST.	TOTAL	MAT.	INST.	TOTAL	MAT.	INST.	TOTAL	MAT.	INST.	TOTAL
015433	CONTRACTOR EQUIPMENT		101.1	101.1		104.4	104.4		94.5	94.5		101.1	101.1		101.1	101.1		116.0	116.0
0241, 31 - 34	SITE & INFRASTRUCTURE, DEMOLITION	87.9	98.1	95.1	82.6	112.3	103.7	73.1	97.2	90.2	88.9	99.2	96.2	84.8	99.6	95.3	81.1	124.0	111.6
0310	Concrete Forming & Accessories	84.5	70.6	72.5	85.0	82.0	82.4	80.7	80.5	80.5	94.2	71.8	74.8	86.1	83.5	83.8	92.5	81.5	83.0
0320	Concrete Reinforcing	86.0	69.8	77.7	85.7	91.1	88.5	86.6	90.8	88.8	97.7	90.8	94.2	87.4	107.6	97.7	86.3	84.8	85.6
0330	Cast-in-Place Concrete	93.6	58.6	79.2	96.5	97.6	96.9	81.8	73.0	78.2	89.4	69.1	81.1	97.0	82.5	91.0	79.9	90.6	84.3
03	CONCRETE	93.3	66.8	80.3	98.9	89.7	94.4	88.6	80.0	84.4	93.8	74.8	84.5	97.3	87.7	92.6	93.9	85.4	89.7
04	MASONRY	94.4	61.9	74.1	111.8	101.5	105.4	83.5	75.4	78.5	94.6	79.4	85.1	94.8	71.5	80.3	103.6	92.9	96.9
05	METALS	90.4	78.2	86.6	91.0	96.9	92.8	92.8	87.2	91.1	93.0	87.8	91.4	92.9	94.2	93.3	84.4	86.9	85.1
06	WOOD, PLASTICS & COMPOSITES	76.9	75.9	76.4	85.0	71.7	77.6	83.7	82.0	82.7	91.4	68.7	78.7	84.1	89.0	86.8	91.3	79.0	84.4
07	THERMAL & MOISTURE PROTECTION	99.3	63.8	84.7	100.3	92.3	97.0	86.3	76.9	82.5	101.0	75.1	90.4	99.7	69.5	87.3	99.0	91.3	95.9
08	OPENINGS	97.9	65.6	90.4	102.4	81.7	97.6	98.7	83.2	95.1	104.5	76.7	98.0	100.6	79.9	95.8	96.8	79.9	92.9
0920	Plaster & Gypsum Board	92.3	74.6	80.3	68.8	71.4	70.6	86.7	81.7	83.3	96.5	67.2	76.7	86.7	88.1	87.6	90.2	77.8	81.8
0950, 0980	Ceilings & Acoustic Treatment	80.4	74.6	76.6	88.8	71.4	77.4	85.5	81.7	83.0	89.6	67.2	74.8	85.5	88.1	87.2	79.6	77.8	78.4
0960	Flooring	97.2	41.2	81.2	71.8	88.6	76.6	93.4	86.0	91.3	106.0	54.4	91.2	96.4	42.0	80.9	99.5	86.0	95.6
0970, 0990	Wall Finishes & Painting/Coating	107.2	49.4	72.3	96.2	83.2	88.3	101.2	76.2	86.1	109.1	79.8	91.4	101.2	65.4	79.6	101.2	96.0	98.1
09	FINISHES	90.7	63.6	75.8	82.7	82.0	82.4	90.6	80.7	85.1	97.6	68.8	81.7	91.9	74.8	82.5	91.4	83.8	87.2
COVERS	DIVS. 10 - 14, 25, 28, 41, 43, 44, 46	100.0	48.4	89.6	100.0	98.7	99.7	100.0	84.2	96.8	100.0	65.5	93.0	100.0	56.6	91.2	100.0	65.4	93.0
21, 22, 23	FIRE SUPPRESSION, PLUMBING & HVAC	95.2	71.4	85.6	95.9	92.4	94.5	95.5	80.1	89.3	100.1	82.3	92.9	95.2	80.7	89.4	95.5	77.6	88.3
26, 27, 3370	ELECTRICAL, COMMUNICATIONS & UTIL.	91.4	80.1	85.4	95.2	79.3	86.8	91.2	80.2	85.4	101.3	87.1	93.8	91.4	58.4	73.9	93.4	80.3	86.5
MF2014	WEIGHTED AVERAGE	94.0	71.8	84.3	95.9	91.4	93.9	92.8	82.0	88.1	98.1	80.7	90.5	95.2	78.7	88.0	93.7	86.1	90.4

DIVISION		KENTUCKY																	
		LEXINGTON			LOUISVILLE			OWENSBORO			PADUCAH			PIKEVILLE			SOMERSET		
		403 - 405			400 - 402			423			420			415 - 416			425 - 426		
		MAT.	INST.	TOTAL	MAT.	INST.	TOTAL	MAT.	INST.	TOTAL	MAT.	INST.	TOTAL	MAT.	INST.	TOTAL	MAT.	INST.	TOTAL
015433	CONTRACTOR EQUIPMENT		101.1	101.1		94.5	94.5		116.0	116.0		116.0	116.0		97.3	97.3		101.1	101.1
0241, 31 - 34	SITE & INFRASTRUCTURE, DEMOLITION	90.1	100.6	97.5	84.8	97.3	93.7	91.0	124.5	114.8	83.7	123.4	111.9	123.2	85.3	96.3	77.9	98.8	92.8
0310	Concrete Forming & Accessories	96.4	73.0	76.2	95.8	82.0	83.9	90.9	85.1	85.9	88.9	79.9	81.1	96.8	88.9	90.0	87.0	75.9	77.4
0320	Concrete Reinforcing	94.4	92.2	93.3	96.8	92.6	94.7	86.3	91.8	89.1	86.8	86.2	86.5	89.3	108.4	99.0	86.6	90.8	88.7
0330	Cast-in-Place Concrete	95.8	92.7	94.5	96.4	73.9	87.2	93.2	91.1	92.4	85.2	86.5	85.7	100.1	96.7	98.7	79.9	98.3	87.5
03	CONCRETE	96.5	83.7	90.2	97.1	81.3	89.4	106.3	88.5	97.5	98.6	83.5	91.2	112.7	96.0	104.5	83.6	86.6	85.1
04	MASONRY	93.7	73.3	81.0	91.6	78.8	83.6	95.9	91.5	93.2	98.8	88.4	92.3	95.3	96.8	96.2	89.8	76.7	81.6
05	METALS	92.8	89.0	91.6	93.9	88.2	92.1	85.8	90.6	87.3	82.9	86.7	84.1	92.0	108.8	97.2	92.8	87.6	91.2
06	WOOD, PLASTICS & COMPOSITES	92.0	68.7	79.0	90.9	83.1	86.5	89.2	83.3	85.9	86.8	79.1	82.5	85.5	88.3	87.1	84.6	75.9	79.7
07	THERMAL & MOISTURE PROTECTION	99.5	84.1	93.2	96.3	78.2	88.9	99.4	92.8	96.7	99.1	78.8	90.8	91.7	79.5	86.7	99.0	71.2	87.6
08	OPENINGS	98.1	73.9	92.5	93.5	84.3	91.4	96.8	85.7	94.2	96.0	77.5	91.7	98.6	87.8	96.1	99.5	77.7	94.4
0920	Plaster & Gypsum Board	101.9	67.2	78.4	98.0	82.9	87.8	88.7	82.2	84.3	87.7	77.9	81.1	63.8	88.1	80.2	86.7	74.6	78.5
0950, 0980	Ceilings & Acoustic Treatment	84.5	67.2	73.1	88.9	82.9	84.9	79.6	82.2	81.3	79.6	77.9	78.5	80.8	88.1	85.6	85.5	74.6	78.3
0960	Flooring	101.9	70.3	92.9	98.6	86.0	95.0	98.9	86.0	95.2	97.8	59.4	86.8	83.4	102.4	88.8	96.7	41.2	80.8
0970, 0990	Wall Finishes & Painting/Coating	107.2	81.0	91.4	104.7	76.2	87.5	101.2	96.0	98.1	101.2	79.7	88.2	103.1	81.2	89.9	101.2	76.2	86.1
09	FINISHES	94.4	72.3	82.2	94.9	82.3	88.0	91.5	86.0	88.5	90.7	76.1	82.7	83.3	90.9	87.5	91.2	69.4	79.2
COVERS	DIVS. 10 - 14, 25, 28, 41, 43, 44, 46	100.0	86.5	97.3	100.0	84.9	96.9	100.0	102.9	100.6	100.0	61.8	92.3	100.0	56.3	91.2	100.0	58.5	91.6
21, 22, 23	FIRE SUPPRESSION, PLUMBING & HVAC	100.0	79.5	91.7	100.0	80.9	92.3	99.9	78.6	91.4	95.5	81.1	89.7	95.0	87.2	91.8	95.5	77.4	88.2
26, 27, 3370	ELECTRICAL, COMMUNICATIONS & UTIL.	94.2	81.5	87.5	99.4	81.5	89.9	93.5	81.4	87.1	95.8	79.9	87.4	94.4	73.4	83.3	91.8	80.2	85.6
MF2014	WEIGHTED AVERAGE	96.6	81.4	90.0	96.6	83.3	90.8	96.3	88.9	93.1	93.9	84.5	89.8	96.9	88.6	93.3	93.2	79.9	87.4

DIVISION		LOUISIANA																	
		ALEXANDRIA			BATON ROUGE			HAMMOND			LAFAYETTE			LAKE CHARLES			MONROE		
		713 - 714			707 - 708			704			705			706			712		
		MAT.	INST.	TOTAL	MAT.	INST.	TOTAL	MAT.	INST.	TOTAL	MAT.	INST.	TOTAL	MAT.	INST.	TOTAL	MAT.	INST.	TOTAL
015433	CONTRACTOR EQUIPMENT		89.9	89.9		89.5	89.5		90.1	90.1		90.1	90.1		89.5	89.5		89.9	89.9
0241, 31 - 34	SITE & INFRASTRUCTURE, DEMOLITION	101.2	87.1	91.1	104.0	87.2	92.0	102.1	87.8	91.9	103.2	88.1	92.5	103.9	87.0	91.9	101.2	87.0	91.1
0310	Concrete Forming & Accessories	82.9	43.4	48.9	97.6	60.8	65.8	78.4	46.2	50.7	96.0	54.1	59.9	96.8	56.7	62.2	82.4	43.3	48.7
0320	Concrete Reinforcing	94.9	54.3	74.2	101.1	58.9	79.6	97.9	59.3	78.3	99.3	58.9	78.7	99.3	59.1	78.9	93.8	54.0	73.6
0330	Cast-in-Place Concrete	96.2	49.9	77.2	97.2	58.8	81.4	95.8	44.0	74.5	95.3	47.9	75.8	100.3	67.3	86.8	96.2	56.8	80.0
03	CONCRETE	95.5	48.8	72.6	98.9	60.4	80.0	95.2	48.9	72.5	96.4	53.7	75.4	98.8	61.5	80.5	95.3	51.1	73.6
04	MASONRY	121.3	53.4	79.0	94.9	54.1	69.5	95.7	51.2	68.0	95.7	52.4	68.7	95.0	58.8	72.5	115.6	47.9	73.4
05	METALS	95.4	70.4	87.7	100.6	72.1	91.8	95.6	71.2	88.1	94.8	72.1	87.8	94.8	72.6	87.9	95.4	69.9	87.6
06	WOOD, PLASTICS & COMPOSITES	90.5	41.7	63.2	99.0	63.8	79.3	84.2	47.0	63.3	105.7	55.7	77.7	103.8	57.7	78.0	89.8	42.2	63.1
07	THERMAL & MOISTURE PROTECTION	95.6	62.9	82.2	96.2	65.4	83.5	96.5	61.4	82.1	97.1	63.8	83.4	96.1	66.8	84.1	95.6	60.8	81.3
08	OPENINGS	109.1	44.7	94.1	97.9	59.1	88.9	93.0	51.7	83.4	96.7	53.2	86.6	96.7	54.9	87.0	109.1	47.6	94.8
0920	Plaster & Gypsum Board	84.2	40.4	54.6	96.7	63.0	74.0	97.5	45.7	62.5	105.7	54.7	71.2	105.7	56.8	72.6	83.8	40.9	54.8
0950, 0980	Ceilings & Acoustic Treatment	84.1	40.4	55.4	91.3	63.0	72.7	94.1	45.7	62.3	92.4	54.7	67.6	93.3	56.8	69.3	84.1	40.9	55.7
0960	Flooring	98.5	65.7	89.1	101.8	65.7	91.5	96.3	65.7	87.5	104.9	65.7	93.7	104.9	72.9	95.7	98.1	57.1	86.4
0970, 0990	Wall Finishes & Painting/Coating	104.4	64.4	80.3	94.7	46.3	65.5	99.4	48.3	68.6	99.4	57.9	74.4	99.4	50.0	69.6	104.4	48.1	70.4
09	FINISHES	89.8	48.9	67.2	94.7	60.2	75.7	94.8	49.7	70.0	98.1	56.3	75.0	98.3	58.6	76.4	89.6	45.5	65.3
COVERS	DIVS. 10 - 14, 25, 28, 41, 43, 44, 46	100.0	50.0	89.9	100.0	80.8	96.1	100.0	45.4	89.0	100.0	79.4	95.8	100.0	80.3	96.0	100.0	45.7	89.0
21, 22, 23	FIRE SUPPRESSION, PLUMBING & HVAC	100.1	55.3	82.1	100.0	59.5	83.7	95.3	42.8	74.1	100.1	61.1	84.4	100.1	62.0	84.7	100.1	54.3	81.6
26, 27, 3370	ELECTRICAL, COMMUNICATIONS & UTIL.	95.7	57.9	75.7	100.8	61.6	80.1	96.7	56.0	75.2	97.9	67.4	81.8	97.4	65.0	80.3	97.7	59.6	77.6
MF2014	WEIGHTED AVERAGE	99.4	57.2	81.0	99.1	63.7	83.7	95.8	54.5	77.8	97.9	63.0	82.7	98.1	65.0	83.7	99.3	56.5	80.6

DIVISION		LOUISIANA									MAINE								
		NEW ORLEANS			SHREVEPORT			THIBODAUX			AUGUSTA			BANGOR			BATH		
		700 - 701			710 - 711			703			043			044			045		
		MAT.	INST.	TOTAL	MAT.	INST.	TOTAL	MAT.	INST.	TOTAL	MAT.	INST.	TOTAL	MAT.	INST.	TOTAL	MAT.	INST.	TOTAL
015433	CONTRACTOR EQUIPMENT		90.4	90.4		89.9	89.9		90.1	90.1		100.9	100.9		100.9	100.9		100.9	100.9
0241, 31 - 34	SITE & INFRASTRUCTURE, DEMOLITION	103.4	91.1	94.7	105.5	87.1	92.4	104.4	88.1	92.8	91.6	101.1	98.4	93.3	101.3	99.0	91.0	101.1	98.2
0310	Concrete Forming & Accessories	95.1	65.5	69.5	98.5	44.2	51.7	90.4	64.6	68.1	95.9	94.8	94.9	93.1	96.6	96.1	88.9	94.9	94.1
0320	Concrete Reinforcing	100.2	60.1	79.8	98.3	54.4	76.0	97.9	59.1	78.2	96.1	106.7	101.5	87.7	108.1	98.1	86.8	107.9	97.5
0330	Cast-in-Place Concrete	98.0	69.4	86.3	101.1	53.6	81.6	102.9	51.6	81.8	101.5	58.8	84.0	82.2	110.6	93.9	82.2	58.9	72.7
03	CONCRETE	99.0	66.3	82.9	100.4	50.5	75.9	100.5	59.7	80.4	103.1	84.1	93.8	94.1	102.9	98.4	94.1	84.4	89.4
04	MASONRY	97.7	59.8	74.1	107.1	50.5	71.8	120.5	48.8	75.8	107.4	63.6	80.1	118.0	103.9	109.2	125.6	86.1	101.0
05	METALS	110.3	73.3	98.9	99.5	70.6	90.6	95.6	71.6	88.2	106.0	86.6	100.1	96.1	89.3	94.0	94.6	88.4	92.7
06	WOOD, PLASTICS & COMPOSITES	97.9	66.3	80.2	101.7	42.7	68.7	92.3	69.7	79.6	91.1	105.3	99.0	90.1	96.4	93.6	84.6	105.3	96.2
07	THERMAL & MOISTURE PROTECTION	94.9	68.8	84.2	95.3	59.8	80.8	96.4	61.8	82.2	100.1	65.0	85.7	97.8	84.4	92.3	97.8	71.8	87.1
08	OPENINGS	97.8	65.4	90.2	108.0	45.3	93.4	97.6	64.1	89.8	106.3	87.9	102.0	102.8	83.0	98.2	102.8	87.8	99.3
0920	Plaster & Gypsum Board	101.4	65.7	77.2	94.0	41.5	58.5	99.7	69.2	79.0	100.6	104.8	103.5	99.4	95.7	96.9	95.5	104.8	101.8
0950, 0980	Ceilings & Acoustic Treatment	98.2	65.7	76.8	88.1	41.5	57.4	94.1	69.2	77.7	97.4	104.8	102.3	86.2	95.7	92.5	84.3	104.8	97.8
0960	Flooring	105.7	65.7	94.2	101.9	61.3	90.3	102.3	44.1	85.7	103.9	55.3	90.0	97.1	114.3	102.0	95.3	55.3	83.9
0970, 0990	Wall Finishes & Painting/Coating	101.9	64.1	79.1	100.4	44.6	66.7	100.7	49.6	69.9	112.7	63.3	82.9	105.0	44.1	68.3	105.0	44.1	68.3
09	FINISHES	101.2	65.1	81.3	94.5	46.3	67.9	97.2	59.8	76.6	100.2	85.8	92.3	95.7	94.8	95.2	94.2	83.7	88.4
COVERS	DIVS. 10 - 14, 25, 28, 41, 43, 44, 46	100.0	83.1	96.6	100.0	78.1	95.6	100.0	81.4	96.2	100.0	100.5	100.1	100.0	109.8	102.0	100.0	100.5	100.1
21, 22, 23	FIRE SUPPRESSION, PLUMBING & HVAC	100.0	65.1	85.9	99.9	58.6	83.3	95.3	61.7	81.8	99.9	64.2	85.5	100.1	74.3	89.7	95.4	64.3	82.8
26, 27, 3370	ELECTRICAL, COMMUNICATIONS & UTIL.	102.1	73.3	86.9	103.5	66.6	84.0	95.3	73.3	83.7	100.9	80.1	89.9	97.2	75.8	85.9	95.2	80.1	87.2
MF2014	WEIGHTED AVERAGE	101.4	69.5	87.5	101.0	59.6	83.0	98.2	65.4	83.9	102.1	79.5	92.3	99.0	89.5	94.9	97.5	81.9	90.7

	DIVISION	MAINE																	
		HOULTON			KITTERY			LEWISTON			MACHIAS			PORTLAND			ROCKLAND		
		047			039			042			046			040 - 041			048		
		MAT.	INST.	TOTAL	MAT.	INST.	TOTAL	MAT.	INST.	TOTAL	MAT.	INST.	TOTAL	MAT.	INST.	TOTAL	MAT.	INST.	TOTAL
015433	CONTRACTOR EQUIPMENT		100.9	100.9		100.9	100.9		100.9	100.9		100.9	100.9		100.9	100.9		100.9	100.9
0241, 31 - 34	SITE & INFRASTRUCTURE, DEMOLITION	93.0	102.2	99.5	84.8	101.2	96.5	90.9	101.3	98.3	92.3	102.2	99.4	90.1	101.3	98.1	88.7	102.2	98.3
0310	Concrete Forming & Accessories	96.7	101.6	101.0	89.2	95.5	94.6	98.5	96.7	96.9	93.9	101.6	100.5	98.5	96.7	96.9	94.9	101.6	100.7
0320	Concrete Reinforcing	87.7	106.5	97.2	84.1	107.9	96.2	107.8	108.1	107.9	87.7	106.5	97.2	102.2	108.1	105.2	87.7	106.5	97.2
0330	Cast-in-Place Concrete	82.3	71.7	77.9	81.2	59.8	72.4	83.9	110.7	94.9	82.2	71.1	77.7	96.7	110.7	102.4	83.9	71.1	78.7
03	CONCRETE	95.1	91.5	93.4	89.7	85.0	87.4	94.7	102.9	98.7	94.6	91.3	93.0	101.9	102.9	102.4	92.0	91.3	91.6
04	MASONRY	101.3	76.3	85.7	113.3	87.1	97.0	101.9	103.9	103.2	101.3	76.3	85.7	112.8	103.9	107.2	95.2	76.3	83.4
05	METALS	94.8	86.6	92.3	85.5	88.6	86.4	99.5	89.3	96.4	94.8	86.5	92.3	107.9	89.3	102.2	94.7	86.5	92.2
06	WOOD, PLASTICS & COMPOSITES	94.0	105.3	100.3	89.2	105.3	98.2	96.0	96.4	96.2	91.0	105.3	99.0	93.7	96.4	95.2	91.9	105.3	99.4
07	THERMAL & MOISTURE PROTECTION	97.9	72.6	87.5	99.2	71.1	87.7	97.6	84.4	92.2	97.8	71.6	87.1	100.1	84.4	93.7	97.5	71.7	86.9
08	OPENINGS	102.9	86.1	99.0	102.0	87.8	98.7	106.2	83.0	100.8	102.9	86.1	99.0	104.5	83.0	99.5	102.8	86.1	98.9
0920	Plaster & Gypsum Board	101.6	104.8	103.8	100.0	104.8	103.3	104.9	95.7	98.7	100.1	104.8	103.3	102.5	95.7	97.9	100.1	104.8	103.3
0950, 0980	Ceilings & Acoustic Treatment	84.3	104.8	97.8	94.2	104.8	101.2	95.4	95.7	95.6	84.3	104.8	97.8	99.1	95.7	96.9	84.3	104.8	97.8
0960	Flooring	98.3	51.7	85.0	99.8	57.7	87.7	100.0	114.3	104.1	97.5	51.7	84.4	98.6	114.3	103.1	97.9	51.7	84.7
0970, 0990	Wall Finishes & Painting/Coating	105.0	137.1	124.4	96.4	38.4	61.4	105.0	44.1	68.3	105.0	137.1	124.4	102.0	44.1	67.1	105.0	137.1	124.4
09	FINISHES	95.9	97.8	97.0	96.5	83.7	89.5	98.9	94.8	96.6	95.4	97.8	96.7	98.8	94.8	96.6	95.2	97.8	96.6
COVERS	DIVS. 10 - 14, 25, 28, 41, 43, 44, 46	100.0	106.4	101.3	100.0	100.8	100.2	100.0	109.8	102.0	100.0	106.4	101.3	100.0	109.8	102.0	100.0	106.4	101.3
21, 22, 23	FIRE SUPPRESSION, PLUMBING & HVAC	95.4	73.2	86.4	95.3	76.6	87.8	100.1	74.3	89.7	95.4	73.2	86.4	100.0	74.3	89.6	95.4	73.2	86.4
26, 27, 3370	ELECTRICAL, COMMUNICATIONS & UTIL.	99.2	80.1	89.1	97.0	80.1	88.1	99.3	80.1	89.1	99.2	80.1	89.1	98.9	80.1	89.0	99.1	80.1	89.1
MF2014	WEIGHTED AVERAGE	97.2	85.6	92.1	95.2	84.7	90.6	99.6	90.1	95.5	97.1	85.5	92.0	102.0	90.1	96.8	96.3	85.5	91.6

	DIVISION	MAINE			MARYLAND														
		WATERVILLE			ANNAPOLIS			BALTIMORE			COLLEGE PARK			CUMBERLAND			EASTON		
		049			214			210 - 212			207 - 208			215			216		
		MAT.	INST.	TOTAL	MAT.	INST.	TOTAL	MAT.	INST.	TOTAL	MAT.	INST.	TOTAL	MAT.	INST.	TOTAL	MAT.	INST.	TOTAL
015433	CONTRACTOR EQUIPMENT		100.9	100.9		99.8	99.8		103.4	103.4		105.5	105.5		99.8	99.8		99.8	99.8
0241, 31 - 34	SITE & INFRASTRUCTURE, DEMOLITION	92.8	101.1	98.7	103.0	91.5	94.9	101.5	95.6	97.3	100.5	94.5	96.2	94.3	91.8	92.5	101.4	88.6	92.3
0310	Concrete Forming & Accessories	88.4	94.8	93.9	97.2	73.6	76.8	101.4	71.1	75.3	84.3	70.2	72.2	91.4	81.0	82.4	89.2	70.2	72.8
0320	Concrete Reinforcing	87.7	106.7	97.3	100.3	80.7	90.3	105.6	80.7	92.9	105.4	78.1	91.5	86.6	72.3	79.3	86.0	79.0	82.4
0330	Cast-in-Place Concrete	82.3	58.8	72.6	110.0	75.7	95.9	107.5	76.4	94.7	116.5	77.5	100.4	91.7	85.6	89.2	101.8	48.1	79.8
03	CONCRETE	95.6	84.1	90.0	104.8	76.6	90.9	104.8	75.8	90.6	107.5	75.5	91.8	90.2	81.8	86.1	98.0	65.2	81.9
04	MASONRY	112.2	63.6	81.9	103.5	71.6	83.6	100.1	71.7	82.4	111.8	69.0	85.1	99.1	83.8	89.5	113.4	42.9	69.4
05	METALS	94.8	86.6	92.3	104.0	93.4	100.8	100.6	93.9	98.5	86.8	97.4	90.0	98.5	90.5	96.1	98.8	87.2	95.2
06	WOOD, PLASTICS & COMPOSITES	84.0	105.3	95.9	92.9	75.4	83.1	99.6	72.1	84.2	80.0	70.0	74.4	84.9	80.1	82.2	82.7	77.4	79.7
07	THERMAL & MOISTURE PROTECTION	97.9	65.0	84.4	100.2	78.8	91.4	101.9	78.6	92.3	101.9	78.7	92.4	100.2	78.9	91.5	100.3	59.0	83.4
08	OPENINGS	102.9	87.9	99.4	105.8	80.3	99.9	99.0	78.5	94.2	94.3	74.8	89.7	99.7	78.0	94.7	98.0	71.4	91.8
0920	Plaster & Gypsum Board	95.5	104.8	101.8	100.5	74.9	83.2	102.3	71.3	81.4	99.0	69.0	78.8	102.3	79.8	87.1	102.3	77.0	85.2
0950, 0980	Ceilings & Acoustic Treatment	84.3	104.8	97.8	91.3	74.9	80.6	94.0	71.3	79.1	102.2	69.0	80.4	93.3	79.8	84.4	93.3	77.0	82.6
0960	Flooring	95.0	55.3	83.6	99.8	77.2	93.3	100.7	77.2	94.0	97.8	79.7	92.6	95.5	90.8	94.2	94.7	51.7	82.4
0970, 0990	Wall Finishes & Painting/Coating	105.0	63.3	79.8	92.1	79.2	84.3	97.9	79.2	86.6	108.0	77.6	89.7	96.5	73.4	82.5	96.5	77.6	85.1
09	FINISHES	94.3	85.8	89.6	94.8	74.3	83.5	100.1	72.3	84.8	95.3	71.9	82.4	96.5	81.9	88.5	96.7	69.1	81.5
COVERS	DIVS. 10 - 14, 25, 28, 41, 43, 44, 46	100.0	100.5	100.1	100.0	86.2	97.2	100.0	86.2	97.2	100.0	85.5	97.1	100.0	90.5	98.1	100.0	75.3	95.0
21, 22, 23	FIRE SUPPRESSION, PLUMBING & HVAC	95.4	64.2	82.8	100.1	80.5	92.2	100.0	80.6	92.2	95.4	83.6	90.7	95.2	73.0	86.3	95.2	68.3	84.4
26, 27, 3370	ELECTRICAL, COMMUNICATIONS & UTIL.	99.2	80.1	89.1	99.5	91.9	95.5	102.0	91.9	96.6	101.1	99.7	100.4	98.3	81.8	89.6	97.8	65.2	80.5
MF2014	WEIGHTED AVERAGE	97.5	79.5	89.7	101.5	82.0	93.0	100.8	81.9	92.6	97.2	83.4	91.2	96.8	81.8	90.2	98.3	68.6	85.3

	DIVISION	MARYLAND															MASSACHUSETTS		
		ELKTON			HAGERSTOWN			SALISBURY			SILVER SPRING			WALDORF			BOSTON		
		219			217			218			209			206			020 - 022, 024		
		MAT.	INST.	TOTAL	MAT.	INST.	TOTAL	MAT.	INST.	TOTAL	MAT.	INST.	TOTAL	MAT.	INST.	TOTAL	MAT.	INST.	TOTAL
015433	CONTRACTOR EQUIPMENT		99.8	99.8		99.8	99.8		99.8	99.8		98.1	98.1		98.1	98.1		106.5	106.5
0241, 31 - 34	SITE & INFRASTRUCTURE, DEMOLITION	88.4	89.5	89.2	92.6	92.3	92.4	101.3	88.6	92.3	89.0	87.6	88.0	95.4	87.3	89.7	98.9	109.1	106.2
0310	Concrete Forming & Accessories	95.4	81.0	82.9	90.3	75.1	77.2	103.9	51.7	58.9	92.8	70.4	73.5	100.2	68.7	73.0	102.7	143.8	138.1
0320	Concrete Reinforcing	86.0	105.3	95.8	86.6	72.3	79.3	86.0	63.5	74.6	104.1	77.9	90.8	104.8	78.0	91.1	108.9	155.8	132.8
0330	Cast-in-Place Concrete	82.5	75.2	79.5	87.4	85.7	86.7	101.8	46.7	79.2	119.3	79.3	102.8	133.6	76.5	110.2	104.0	151.2	123.4
03	CONCRETE	83.3	84.3	83.8	86.8	79.2	83.0	99.0	53.7	76.8	105.9	76.1	91.2	116.5	74.4	95.8	105.7	147.5	126.2
04	MASONRY	98.3	58.9	73.8	105.3	83.8	91.9	113.1	47.3	72.1	110.9	72.3	86.8	95.4	67.2	77.9	106.6	164.2	142.5
05	METALS	98.8	101.3	99.6	98.7	90.7	96.2	98.8	81.2	93.4	91.0	93.8	91.9	91.0	93.9	91.9	100.2	132.4	110.1
06	WOOD, PLASTICS & COMPOSITES	90.0	86.8	88.3	84.0	71.9	77.2	100.6	55.1	75.1	86.5	69.2	76.8	94.0	69.2	80.1	98.6	143.3	123.6
07	THERMAL & MOISTURE PROTECTION	99.9	73.5	89.1	100.0	76.8	90.5	100.6	63.7	85.4	105.4	84.9	97.0	105.8	82.7	96.4	103.4	151.9	123.3
08	OPENINGS	98.0	81.3	94.1	98.0	73.0	92.2	98.2	61.7	89.7	85.7	74.3	83.0	86.2	74.3	83.5	101.4	146.2	111.8
0920	Plaster & Gypsum Board	105.2	86.7	92.7	102.3	71.3	81.4	111.6	54.0	72.6	105.1	69.0	80.7	108.3	69.0	81.8	107.5	144.1	132.2
0950, 0980	Ceilings & Acoustic Treatment	93.3	86.7	89.0	94.3	71.3	79.2	93.3	54.0	67.5	112.2	69.0	83.8	112.2	69.0	83.8	106.1	144.1	131.1
0960	Flooring	97.0	58.5	86.0	95.2	90.8	93.9	100.7	64.7	90.4	104.1	79.7	97.2	107.7	79.0	99.5	97.9	182.7	122.2
0970, 0990	Wall Finishes & Painting/Coating	96.5	77.6	85.1	96.5	77.6	85.1	96.5	77.6	85.1	115.8	77.6	92.8	115.8	77.6	92.8	104.4	155.9	135.4
09	FINISHES	96.9	76.9	85.9	96.4	77.5	86.0	99.8	56.3	75.8	96.9	71.4	82.8	98.6	70.7	83.2	104.2	152.7	130.9
COVERS	DIVS. 10 - 14, 25, 28, 41, 43, 44, 46	100.0	59.9	91.9	100.0	89.6	97.9	100.0	38.7	87.6	100.0	84.7	96.9	100.0	83.0	96.6	100.0	119.4	103.9
21, 22, 23	FIRE SUPPRESSION, PLUMBING & HVAC	95.2	80.5	89.3	100.0	86.2	94.4	95.2	59.9	81.0	95.4	84.8	91.2	95.4	82.3	90.1	100.1	132.4	113.1
26, 27, 3370	ELECTRICAL, COMMUNICATIONS & UTIL.	99.6	91.9	95.5	98.1	81.8	89.5	96.5	67.1	81.0	98.4	99.7	99.1	95.7	99.7	97.8	100.8	135.4	119.1
MF2014	WEIGHTED AVERAGE	95.9	81.9	89.8	97.6	83.3	91.4	98.7	62.1	82.7	96.5	83.2	90.7	97.1	81.7	90.4	101.6	139.5	118.1

	DIVISION	MASSACHUSETTS																	
		BROCKTON			BUZZARDS BAY			FALL RIVER			FITCHBURG			FRAMINGHAM			GREENFIELD		
		023			025			027			014			017			013		
		MAT.	INST.	TOTAL	MAT.	INST.	TOTAL	MAT.	INST.	TOTAL	MAT.	INST.	TOTAL	MAT.	INST.	TOTAL	MAT.	INST.	TOTAL
015433	CONTRACTOR EQUIPMENT		102.7	102.7		102.7	102.7		103.7	103.7		100.9	100.9		101.9	101.9		100.9	100.9
0241, 31 - 34	SITE & INFRASTRUCTURE, DEMOLITION	94.5	105.7	102.4	84.8	105.6	99.6	93.5	105.8	102.2	86.3	105.6	100.0	83.0	105.3	98.9	90.0	104.3	100.1
0310	Concrete Forming & Accessories	101.6	138.7	133.6	99.2	138.5	133.1	101.6	138.8	133.6	93.8	130.8	125.7	101.2	138.8	133.6	92.1	113.2	110.3
0320	Concrete Reinforcing	106.4	155.6	131.4	85.3	160.1	123.4	106.4	160.2	133.7	83.9	151.4	118.3	83.9	155.6	120.4	87.3	124.4	106.2
0330	Cast-in-Place Concrete	97.7	151.5	119.8	81.2	151.5	110.1	94.5	152.0	118.1	85.9	150.5	112.4	85.9	148.3	111.5	88.3	131.2	105.9
03	CONCRETE	101.8	145.2	123.1	86.2	145.9	115.5	100.3	146.2	122.9	85.9	140.4	112.7	88.7	144.1	116.0	89.6	120.9	104.9
04	MASONRY	101.7	160.0	138.0	94.4	160.0	135.3	102.6	159.9	138.3	101.0	159.0	137.2	107.2	160.1	140.1	105.4	136.0	124.4
05	METALS	97.4	130.0	107.5	92.3	131.7	104.4	97.4	132.1	108.1	95.5	125.5	104.7	95.5	130.0	106.2	97.8	110.4	101.7
06	WOOD, PLASTICS & COMPOSITES	99.1	138.4	121.1	96.1	138.4	119.8	99.1	138.7	121.3	93.7	128.3	113.1	100.1	138.2	121.4	91.5	110.9	102.4
07	THERMAL & MOISTURE PROTECTION	100.9	148.7	120.5	100.1	147.0	119.4	100.8	146.3	119.5	99.2	141.5	116.5	99.3	149.1	119.7	99.2	121.9	108.5
08	OPENINGS	102.2	143.5	111.9	97.9	139.2	107.5	102.2	139.3	110.9	104.6	137.0	112.2	94.9	143.4	106.2	104.7	115.0	107.1
0920	Plaster & Gypsum Board	93.7	139.0	124.3	89.5	139.0	123.0	93.7	139.0	124.3	100.3	128.6	119.4	102.8	139.0	127.3	101.0	110.7	107.5
0950, 0980	Ceilings & Acoustic Treatment	102.2	139.0	126.4	88.2	139.0	121.6	102.2	139.0	126.4	89.0	128.6	115.0	89.0	139.0	121.9	97.4	110.7	106.1
0960	Flooring	98.6	182.7	122.7	96.7	182.7	121.3	97.7	182.7	122.0	98.1	182.7	122.3	99.7	182.7	123.5	97.3	154.0	113.5
0970, 0990	Wall Finishes & Painting/Coating	99.9	155.1	133.2	99.9	155.1	133.2	99.9	155.1	133.2	99.5	155.1	133.1	100.5	155.1	133.5	99.5	115.8	109.4
09	FINISHES	98.9	149.1	126.6	94.2	149.1	124.4	98.7	149.3	126.6	94.6	143.1	121.3	95.3	148.9	124.8	96.7	121.0	110.0
COVERS	DIVS. 10 - 14, 25, 28, 41, 43, 44, 46	100.0	117.9	103.6	100.0	117.9	103.6	100.0	118.5	103.7	100.0	109.0	101.8	100.0	117.5	103.5	100.0	104.6	100.9
21, 22, 23	FIRE SUPPRESSION, PLUMBING & HVAC	100.1	111.1	104.6	95.4	111.1	101.7	100.1	111.2	104.6	96.0	113.6	103.1	96.0	126.1	108.1	96.0	102.8	98.7
26, 27, 3370	ELECTRICAL, COMMUNICATIONS & UTIL.	99.3	100.1	99.7	96.3	100.1	98.3	99.2	100.1	99.7	100.1	105.1	102.8	96.6	128.9	113.7	100.1	98.9	99.5
MF2014	WEIGHTED AVERAGE	99.9	128.1	112.2	94.4	128.1	109.1	99.7	128.3	112.2	96.5	126.5	109.6	95.7	135.1	112.9	97.7	112.3	104.1

	DIVISION	MASSACHUSETTS																	
		HYANNIS			LAWRENCE			LOWELL			NEW BEDFORD			PITTSFIELD			SPRINGFIELD		
		026			019			018			027			012			010 - 011		
		MAT.	INST.	TOTAL	MAT.	INST.	TOTAL	MAT.	INST.	TOTAL	MAT.	INST.	TOTAL	MAT.	INST.	TOTAL	MAT.	INST.	TOTAL
015433	CONTRACTOR EQUIPMENT		102.7	102.7		102.7	102.7		100.9	100.9		103.7	103.7		100.9	100.9		100.9	100.9
0241, 31 - 34	SITE & INFRASTRUCTURE, DEMOLITION	90.9	105.6	101.3	95.2	105.7	102.7	94.2	105.6	102.3	92.1	105.8	101.8	95.2	104.2	101.6	94.6	104.4	101.6
0310	Concrete Forming & Accessories	93.5	138.5	132.3	102.4	139.0	134.0	99.0	139.0	133.5	101.6	138.8	133.6	99.0	112.2	110.4	99.3	113.6	111.6
0320	Concrete Reinforcing	85.3	160.1	123.4	103.8	150.4	127.5	104.6	150.4	127.9	106.4	160.2	133.7	86.6	122.3	104.8	104.6	124.4	114.7
0330	Cast-in-Place Concrete	89.1	151.5	114.7	99.4	148.9	119.7	90.3	148.9	114.4	83.3	152.0	111.5	98.5	129.7	111.3	94.0	131.7	109.5
03	CONCRETE	92.4	145.9	118.7	103.7	143.5	123.2	95.1	143.3	118.8	95.0	146.2	120.2	96.2	119.5	107.6	96.9	121.2	108.8
04	MASONRY	100.6	160.0	137.6	112.3	160.7	142.5	100.2	160.0	137.5	100.6	159.9	137.6	100.9	133.2	121.0	100.5	136.9	123.2
05	METALS	93.7	131.7	105.4	98.2	128.3	107.5	98.2	126.0	106.7	97.4	132.1	108.1	98.0	109.5	101.5	100.9	110.4	103.8
06	WOOD, PLASTICS & COMPOSITES	89.3	138.4	116.8	100.6	138.4	121.8	99.8	138.4	121.4	99.1	138.7	121.3	99.8	110.9	106.0	99.8	110.9	106.0
07	THERMAL & MOISTURE PROTECTION	100.4	147.0	119.5	99.7	148.5	119.7	99.5	148.6	119.7	100.7	146.3	119.4	99.6	120.7	108.2	99.5	122.2	108.8
08	OPENINGS	98.5	139.2	108.0	99.0	142.1	109.0	105.9	142.1	114.4	102.2	139.3	110.9	105.9	114.4	107.9	105.9	115.0	108.0
0920	Plaster & Gypsum Board	85.2	139.0	121.6	105.6	139.0	128.2	105.6	139.0	128.2	93.7	139.0	124.3	105.6	110.7	109.0	105.6	110.7	109.0
0950, 0980	Ceilings & Acoustic Treatment	93.8	139.0	123.5	99.3	139.0	125.4	99.3	139.0	125.4	102.2	139.0	126.4	99.3	110.7	106.8	99.3	110.7	106.8
0960	Flooring	94.4	182.7	119.6	100.2	182.7	123.8	100.2	182.7	123.8	97.7	182.7	122.0	100.6	154.0	115.9	99.7	154.0	115.3
0970, 0990	Wall Finishes & Painting/Coating	99.9	155.1	133.2	99.6	155.1	133.1	99.5	155.1	133.1	99.9	155.1	133.2	99.5	115.8	109.4	100.9	115.8	109.9
09	FINISHES	94.4	149.1	124.6	98.8	149.1	126.5	98.7	149.1	126.5	98.6	149.3	126.5	98.8	120.3	110.6	98.7	121.2	111.1
COVERS	DIVS. 10 - 14, 25, 28, 41, 43, 44, 46	100.0	117.9	103.6	100.0	118.1	103.6	100.0	118.1	103.6	100.0	118.5	103.7	100.0	103.7	100.7	100.0	104.9	101.0
21, 22, 23	FIRE SUPPRESSION, PLUMBING & HVAC	100.1	111.1	104.6	100.1	124.8	110.0	100.1	125.7	110.4	100.1	111.2	104.6	100.1	101.4	100.6	100.1	103.3	101.4
26, 27, 3370	ELECTRICAL, COMMUNICATIONS & UTIL.	96.8	100.1	98.5	99.1	128.9	114.9	99.6	128.9	115.1	100.0	100.1	100.1	99.6	98.9	99.2	99.7	98.9	99.2
MF2014	WEIGHTED AVERAGE	97.0	128.1	110.6	100.3	134.6	115.3	99.5	134.5	114.8	99.1	128.3	111.8	99.7	111.3	104.8	100.2	112.6	105.6

	DIVISION	MASSACHUSETTS			MICHIGAN														
		WORCESTER			ANN ARBOR			BATTLE CREEK			BAY CITY			DEARBORN			DETROIT		
		015 - 016			481			490			487			481			482		
		MAT.	INST.	TOTAL	MAT.	INST.	TOTAL	MAT.	INST.	TOTAL	MAT.	INST.	TOTAL	MAT.	INST.	TOTAL	MAT.	INST.	TOTAL
015433	CONTRACTOR EQUIPMENT		100.9	100.9		110.7	110.7		102.8	102.8		110.7	110.7		110.7	110.7		98.8	98.8
0241, 31 - 34	SITE & INFRASTRUCTURE, DEMOLITION	94.6	105.6	102.4	82.3	98.8	94.0	93.3	87.5	89.2	73.9	97.8	90.9	82.1	98.9	94.1	94.7	100.6	98.9
0310	Concrete Forming & Accessories	99.6	130.7	126.5	98.3	111.6	109.8	97.6	87.6	89.0	98.4	87.1	88.6	98.2	115.8	113.3	100.8	115.8	113.7
0320	Concrete Reinforcing	104.6	150.6	128.0	92.3	120.1	106.5	91.4	93.3	92.4	92.3	119.2	106.0	92.3	120.2	106.5	94.9	120.1	107.8
0330	Cast-in-Place Concrete	93.5	150.5	116.9	88.8	107.4	96.4	99.0	100.4	99.6	85.0	90.6	87.3	86.8	109.9	96.3	94.1	109.9	100.6
03	CONCRETE	96.6	140.2	118.0	93.5	112.0	102.6	98.1	92.5	95.4	91.7	95.2	93.4	92.6	114.7	103.5	96.6	113.6	105.0
04	MASONRY	100.0	159.0	136.8	104.4	108.4	106.9	105.2	87.2	94.0	104.0	87.0	93.4	104.3	111.6	108.9	97.9	111.6	106.5
05	METALS	101.0	125.2	108.4	94.2	118.8	101.8	98.3	87.6	95.0	94.8	115.4	101.2	94.3	119.1	101.9	94.9	102.1	97.1
06	WOOD, PLASTICS & COMPOSITES	100.3	128.3	116.0	97.5	112.8	106.0	96.5	87.2	91.3	97.5	86.5	91.3	97.5	116.8	108.3	103.0	116.8	110.7
07	THERMAL & MOISTURE PROTECTION	99.5	141.5	116.7	100.4	108.2	103.6	94.6	84.8	90.6	98.2	92.4	95.8	98.9	115.7	105.8	97.7	115.7	105.1
08	OPENINGS	105.9	136.7	113.1	99.4	110.4	101.9	96.6	82.3	93.3	99.4	93.9	98.1	99.4	112.6	102.5	99.2	112.9	102.4
0920	Plaster & Gypsum Board	105.6	128.6	121.2	103.8	112.4	109.6	93.7	82.9	86.4	103.8	85.3	91.3	103.8	116.6	112.4	100.5	116.6	111.3
0950, 0980	Ceilings & Acoustic Treatment	99.3	128.6	118.5	88.7	112.4	104.3	96.3	82.9	87.5	89.7	85.3	86.8	88.7	116.6	107.0	91.2	116.6	107.9
0960	Flooring	100.2	179.8	123.0	97.5	117.6	103.2	105.7	91.9	101.7	97.5	81.5	92.9	96.9	113.9	101.8	95.9	113.9	101.1
0970, 0990	Wall Finishes & Painting/Coating	99.5	155.1	133.1	96.2	98.7	97.7	107.9	81.9	92.2	96.2	82.0	87.6	96.2	100.8	99.0	97.3	100.8	99.4
09	FINISHES	98.7	142.5	122.9	92.8	111.6	103.1	98.0	87.6	92.3	92.6	84.8	88.3	92.6	114.3	104.5	93.9	114.3	105.1
COVERS	DIVS. 10 - 14, 25, 28, 41, 43, 44, 46	100.0	109.0	101.8	100.0	105.8	101.2	100.0	96.4	99.3	100.0	94.1	98.8	100.0	107.1	101.4	100.0	107.1	101.4
21, 22, 23	FIRE SUPPRESSION, PLUMBING & HVAC	100.1	113.6	105.5	100.0	98.8	99.5	100.0	87.4	94.9	100.0	83.6	93.4	100.0	109.2	103.7	100.0	110.2	104.1
26, 27, 3370	ELECTRICAL, COMMUNICATIONS & UTIL.	99.7	105.1	102.5	96.3	108.8	102.9	94.4	83.7	88.8	95.2	90.3	92.6	96.3	107.0	101.9	98.2	107.0	102.9
MF2014	WEIGHTED AVERAGE	100.2	126.4	111.6	97.1	107.5	101.7	98.3	87.6	93.7	96.6	91.7	94.5	97.0	111.0	103.1	97.8	109.6	102.9

DIVISION		MICHIGAN																	
		FLINT			GAYLORD			GRAND RAPIDS			IRON MOUNTAIN			JACKSON			KALAMAZOO		
		484 - 485			497			493,495			498 - 499			492			491		
		MAT.	INST.	TOTAL	MAT.	INST.	TOTAL	MAT.	INST.	TOTAL	MAT.	INST.	TOTAL	MAT.	INST.	TOTAL	MAT.	INST.	TOTAL
015433	CONTRACTOR EQUIPMENT		110.7	110.7		104.7	104.7		102.8	102.8		93.7	93.7		104.7	104.7		102.8	102.8
0241, 31 - 34	SITE & INFRASTRUCTURE, DEMOLITION	71.7	98.0	90.4	88.5	85.0	86.0	93.2	87.6	89.2	96.1	93.7	94.4	109.4	87.3	93.7	93.6	87.5	89.3
0310	Concrete Forming & Accessories	101.3	89.9	91.5	95.5	77.3	79.8	97.1	84.6	86.3	87.8	85.0	85.4	92.5	86.7	87.5	97.6	87.3	88.7
0320	Concrete Reinforcing	92.3	119.6	106.2	85.2	118.6	102.2	97.8	93.2	95.5	85.0	99.8	92.5	82.8	119.4	101.4	91.4	93.4	92.4
0330	Cast-in-Place Concrete	89.4	92.4	90.6	98.7	86.1	93.5	104.1	97.6	101.4	116.8	85.4	103.9	98.6	92.2	96.0	101.0	98.8	100.1
03	CONCRETE	94.0	97.1	95.5	94.6	88.9	91.8	101.6	90.2	96.0	104.3	87.9	96.2	89.3	95.4	92.3	101.6	91.8	96.8
04	MASONRY	104.5	96.5	99.5	116.2	79.1	93.1	102.0	87.4	92.9	101.2	86.1	91.8	95.1	91.6	92.9	103.8	89.1	94.6
05	METALS	94.3	116.3	101.0	99.7	109.9	102.8	95.3	87.1	92.8	99.0	89.6	96.1	99.9	114.1	104.2	98.3	86.5	94.7
06	WOOD, PLASTICS & COMPOSITES	100.9	88.2	93.8	89.4	76.8	82.4	97.7	83.2	89.6	85.0	85.1	85.0	88.1	84.7	86.2	96.5	87.2	91.3
07	THERMAL & MOISTURE PROTECTION	98.2	96.3	97.4	93.2	74.5	85.5	97.4	76.6	88.9	96.7	81.4	90.4	92.5	92.6	92.5	94.6	85.1	90.7
08	OPENINGS	99.4	95.6	98.5	97.5	75.1	92.3	103.4	84.7	99.0	104.6	76.0	98.0	96.5	93.1	95.7	96.6	81.7	93.1
0920	Plaster & Gypsum Board	105.2	87.1	93.0	93.4	74.9	80.9	95.4	78.8	84.2	51.7	85.2	74.3	91.7	82.9	85.8	93.7	82.9	86.4
0950, 0980	Ceilings & Acoustic Treatment	88.7	87.1	87.6	95.4	74.9	81.9	104.3	78.8	87.5	94.4	85.2	88.3	95.4	82.9	87.2	96.3	82.9	87.5
0960	Flooring	97.5	95.8	97.0	97.8	92.4	96.2	105.5	85.7	99.8	122.4	93.8	114.2	96.5	82.7	92.6	105.7	82.7	99.1
0970, 0990	Wall Finishes & Painting/Coating	96.2	85.4	89.7	103.8	82.0	90.6	109.7	80.8	92.3	126.1	72.1	93.5	103.8	96.8	99.6	107.9	81.9	92.2
09	FINISHES	92.1	89.8	90.9	97.4	79.1	87.3	101.6	84.2	92.0	100.0	85.3	91.9	98.3	86.2	91.6	98.0	85.6	91.2
COVERS	DIVS. 10 - 14, 25, 28, 41, 43, 44, 46	100.0	95.5	99.1	100.0	88.6	97.7	100.0	96.4	99.3	100.0	87.2	97.4	100.0	100.9	100.2	100.0	96.4	99.3
21, 22, 23	FIRE SUPPRESSION, PLUMBING & HVAC	100.0	89.6	95.8	95.6	78.4	88.6	100.0	83.0	93.1	95.5	85.3	91.4	95.6	88.4	92.7	100.0	80.8	92.3
26, 27, 3370	ELECTRICAL, COMMUNICATIONS & UTIL.	96.3	96.0	96.1	92.4	79.1	85.4	99.7	79.1	88.8	98.7	84.3	91.1	96.3	108.8	102.9	94.3	81.8	87.7
MF2014	WEIGHTED AVERAGE	96.8	96.0	96.4	97.2	83.6	91.3	99.7	85.1	93.3	99.3	86.2	93.6	96.5	95.1	95.9	98.7	85.7	93.0

DIVISION		MICHIGAN															MINNESOTA		
		LANSING			MUSKEGON			ROYAL OAK			SAGINAW			TRAVERSE CITY			BEMIDJI		
		488 - 489			494			480,483			486			496			566		
		MAT.	INST.	TOTAL	MAT.	INST.	TOTAL	MAT.	INST.	TOTAL	MAT.	INST.	TOTAL	MAT.	INST.	TOTAL	MAT.	INST.	TOTAL
015433	CONTRACTOR EQUIPMENT		110.7	110.7		102.8	102.8		96.2	96.2		110.7	110.7		93.7	93.7		98.2	98.2
0241, 31 - 34	SITE & INFRASTRUCTURE, DEMOLITION	95.6	98.0	97.3	91.2	87.5	88.5	86.5	98.4	95.0	74.9	97.8	91.2	82.4	93.2	90.0	95.4	97.5	96.9
0310	Concrete Forming & Accessories	98.2	89.7	90.9	98.0	84.1	86.0	94.1	112.6	110.1	98.3	87.6	89.1	87.8	75.2	76.9	86.1	89.6	89.1
0320	Concrete Reinforcing	94.3	119.4	107.1	92.1	93.3	92.7	83.7	119.9	102.1	92.3	119.2	106.0	86.3	92.5	89.5	93.7	106.6	100.2
0330	Cast-in-Place Concrete	103.5	92.1	98.8	98.6	94.8	97.1	77.8	106.3	89.5	87.7	90.5	88.9	91.2	78.2	85.9	106.0	103.0	104.8
03	CONCRETE	100.8	96.9	98.9	96.4	89.1	92.8	80.5	110.8	95.4	93.0	95.4	94.2	85.8	79.8	82.9	97.6	98.5	98.1
04	MASONRY	106.6	94.2	98.9	102.2	88.5	93.7	98.2	112.0	106.8	106.0	87.0	94.1	99.2	82.4	88.7	102.5	103.7	103.3
05	METALS	93.9	115.6	100.6	96.0	87.2	93.3	97.3	98.0	97.5	94.3	115.2	100.7	99.0	88.2	95.7	92.1	120.0	100.7
06	WOOD, PLASTICS & COMPOSITES	96.3	88.8	92.1	93.2	83.4	87.7	93.0	113.1	104.3	93.8	87.6	90.3	85.0	75.1	79.4	71.3	85.5	79.3
07	THERMAL & MOISTURE PROTECTION	98.6	89.6	94.9	93.6	78.9	87.6	96.8	111.2	102.7	99.0	92.6	96.4	95.8	71.7	85.9	104.7	95.4	100.9
08	OPENINGS	103.2	95.1	101.3	95.8	84.5	93.2	99.3	109.4	101.7	97.4	94.5	96.7	104.6	70.6	96.7	100.2	107.2	101.8
0920	Plaster & Gypsum Board	96.9	87.7	90.7	75.7	79.0	77.9	101.3	112.7	109.0	103.8	86.4	92.1	51.7	74.9	67.3	102.2	85.3	90.8
0950, 0980	Ceilings & Acoustic Treatment	94.0	87.7	89.8	97.1	79.0	85.2	88.0	112.7	104.3	88.7	86.4	87.2	94.4	74.9	81.6	122.8	85.3	98.2
0960	Flooring	106.6	85.7	100.6	104.5	85.7	99.1	94.8	113.9	100.3	97.5	81.5	92.9	122.4	94.9	114.5	104.4	120.8	109.1
0970, 0990	Wall Finishes & Painting/Coating	108.0	80.8	91.6	106.3	80.8	90.9	97.6	96.8	97.1	96.2	82.0	87.6	126.1	44.7	76.9	102.2	95.9	98.4
09	FINISHES	98.3	87.6	92.4	94.7	83.2	88.4	91.7	111.5	102.6	92.4	85.5	88.6	99.0	74.1	85.3	106.2	95.2	100.1
COVERS	DIVS. 10 - 14, 25, 28, 41, 43, 44, 46	100.0	100.6	100.1	100.0	96.0	99.2	100.0	103.0	100.6	100.0	94.3	98.8	100.0	84.6	96.9	100.0	96.9	99.4
21, 22, 23	FIRE SUPPRESSION, PLUMBING & HVAC	99.9	88.6	95.3	99.9	82.3	92.8	95.6	107.2	100.3	100.0	83.1	93.2	95.5	77.7	88.3	95.6	83.4	90.6
26, 27, 3370	ELECTRICAL, COMMUNICATIONS & UTIL.	98.6	93.3	95.8	94.8	79.1	86.5	98.4	104.4	101.6	93.9	90.5	92.1	94.0	79.1	86.1	104.3	104.6	104.5
MF2014	WEIGHTED AVERAGE	99.3	94.7	97.3	97.2	84.8	91.8	94.9	106.9	100.1	96.4	91.8	94.4	96.3	80.1	89.3	98.2	98.2	98.2

DIVISION		MINNESOTA																	
		BRAINERD			DETROIT LAKES			DULUTH			MANKATO			MINNEAPOLIS			ROCHESTER		
		564			565			556 - 558			560			553 - 555			559		
		MAT.	INST.	TOTAL	MAT.	INST.	TOTAL	MAT.	INST.	TOTAL	MAT.	INST.	TOTAL	MAT.	INST.	TOTAL	MAT.	INST.	TOTAL
015433	CONTRACTOR EQUIPMENT		100.8	100.8		98.2	98.2		99.1	99.1		100.8	100.8		102.6	102.6		99.1	99.1
0241, 31 - 34	SITE & INFRASTRUCTURE, DEMOLITION	96.9	102.2	100.7	93.6	97.9	96.6	100.6	99.4	99.7	93.8	102.9	100.3	97.9	104.7	102.8	97.4	99.0	98.5
0310	Concrete Forming & Accessories	87.7	91.6	91.1	83.1	91.6	90.4	98.7	109.9	108.4	95.6	99.1	98.6	100.7	128.6	124.8	99.3	106.5	105.5
0320	Concrete Reinforcing	92.5	106.7	99.7	93.7	106.6	100.2	97.4	107.1	102.3	92.4	116.2	104.5	92.2	116.6	104.6	91.7	116.4	104.3
0330	Cast-in-Place Concrete	115.2	106.7	111.7	102.9	106.1	104.2	114.8	106.0	111.2	106.2	107.3	106.6	109.6	117.4	112.8	112.3	99.4	107.0
03	CONCRETE	101.7	100.7	101.2	95.2	100.4	97.8	106.5	108.7	107.6	97.2	106.1	101.5	103.4	122.8	112.9	102.5	106.8	104.6
04	MASONRY	126.4	112.0	117.4	126.3	110.1	116.2	107.6	117.6	113.9	114.5	110.4	112.0	109.0	126.7	120.1	104.5	114.1	110.5
05	METALS	93.2	120.0	101.5	92.1	120.0	100.6	101.7	122.4	108.0	93.1	125.5	103.0	99.1	128.9	108.3	100.4	127.3	108.7
06	WOOD, PLASTICS & COMPOSITES	89.4	85.3	87.1	68.4	85.5	78.0	98.2	109.3	104.4	98.8	96.4	97.5	106.7	127.4	118.3	103.2	105.3	104.4
07	THERMAL & MOISTURE PROTECTION	103.4	107.7	105.2	104.6	108.3	106.1	101.6	114.9	107.1	103.8	99.0	101.8	102.8	127.4	112.9	104.9	100.9	103.2
08	OPENINGS	86.9	107.1	91.6	100.2	107.2	101.8	102.5	116.5	105.8	91.5	115.5	97.1	97.1	133.6	105.6	96.4	121.5	102.2
0920	Plaster & Gypsum Board	86.9	85.3	85.8	101.1	85.3	90.5	90.2	110.0	103.6	91.5	96.8	95.1	95.8	128.6	117.9	97.6	105.8	103.2
0950, 0980	Ceilings & Acoustic Treatment	59.3	85.3	76.4	122.8	85.3	98.2	96.1	110.0	105.3	59.3	96.8	84.0	98.2	128.6	118.2	93.2	105.8	101.5
0960	Flooring	103.3	120.8	108.3	103.2	120.8	108.2	101.9	122.1	107.6	105.0	120.8	109.5	97.9	120.8	104.4	102.9	120.8	108.0
0970, 0990	Wall Finishes & Painting/Coating	96.0	95.9	95.9	102.2	95.9	98.4	101.9	110.3	107.0	108.3	104.9	106.3	98.2	126.1	115.1	97.2	104.9	101.9
09	FINISHES	89.4	96.4	93.2	105.5	96.5	100.6	98.3	112.5	106.1	91.1	103.4	97.9	98.6	127.6	114.6	97.4	109.1	103.8
COVERS	DIVS. 10 - 14, 25, 28, 41, 43, 44, 46	100.0	98.3	99.7	100.0	98.6	99.7	100.0	97.6	99.5	100.0	98.6	99.7	100.0	109.7	102.0	100.0	102.8	100.6
21, 22, 23	FIRE SUPPRESSION, PLUMBING & HVAC	94.8	86.1	91.3	95.6	86.0	91.7	99.8	96.5	98.5	94.8	87.7	91.9	99.9	115.6	106.2	99.9	96.2	98.4
26, 27, 3370	ELECTRICAL, COMMUNICATIONS & UTIL.	101.8	102.4	102.1	104.0	67.5	84.7	98.4	102.4	100.6	108.6	89.2	98.3	101.2	111.4	106.6	98.6	89.2	93.6
MF2014	WEIGHTED AVERAGE	96.9	100.5	98.5	98.9	95.1	97.3	101.3	107.3	103.9	97.2	101.3	99.0	100.4	120.0	108.9	99.9	104.7	102.0

DIVISION		MINNESOTA															MISSISSIPPI		
		SAINT PAUL			ST. CLOUD			THIEF RIVER FALLS			WILLMAR			WINDOM			BILOXI		
		550 - 551			563			567			562			561			395		
		MAT.	INST.	TOTAL	MAT.	INST.	TOTAL	MAT.	INST.	TOTAL	MAT.	INST.	TOTAL	MAT.	INST.	TOTAL	MAT.	INST.	TOTAL
015433	CONTRACTOR EQUIPMENT		99.1	99.1		100.8	100.8		98.2	98.2		100.8	100.8		100.8	100.8		98.8	98.8
0241, 31 - 34	SITE & INFRASTRUCTURE, DEMOLITION	95.2	100.1	98.7	92.5	104.3	100.9	94.4	97.5	96.6	91.8	102.0	99.1	86.0	101.0	96.6	100.6	88.4	91.9
0310	Concrete Forming & Accessories	93.9	123.0	119.0	84.9	117.7	113.2	86.8	89.0	88.7	84.7	91.1	90.2	89.1	84.8	85.4	94.7	52.9	58.7
0320	Concrete Reinforcing	98.9	116.6	107.9	92.6	116.4	104.7	94.0	106.5	100.4	92.2	116.0	104.3	92.2	115.3	104.0	95.0	58.3	76.4
0330	Cast-in-Place Concrete	124.0	116.7	121.0	101.7	114.8	107.1	105.1	88.1	98.1	103.3	85.3	95.9	89.4	66.3	79.9	108.4	54.0	86.0
03	CONCRETE	111.2	120.1	115.5	92.9	116.9	104.7	96.3	93.1	94.7	92.9	94.9	93.9	83.6	85.5	84.5	101.6	55.9	79.2
04	MASONRY	111.7	126.7	121.1	111.1	119.9	116.6	102.4	104.5	103.7	115.4	112.0	113.3	125.8	92.1	104.8	102.7	44.3	66.3
05	METALS	101.1	128.6	109.6	93.9	126.8	104.0	92.2	119.4	100.6	93.0	124.5	102.7	92.9	122.4	102.0	94.9	81.7	90.8
06	WOOD, PLASTICS & COMPOSITES	93.4	119.9	108.2	86.7	114.9	102.5	72.3	85.5	79.7	86.4	85.5	85.9	90.7	82.8	86.3	99.1	56.1	75.0
07	THERMAL & MOISTURE PROTECTION	104.1	126.4	113.3	103.6	117.5	109.3	105.6	98.4	102.7	103.4	108.6	105.5	103.4	89.6	97.7	97.4	56.1	80.4
08	OPENINGS	99.8	129.5	106.7	91.5	126.7	99.7	100.2	107.2	101.8	89.0	91.9	89.7	92.7	90.4	92.2	99.6	57.7	89.8
0920	Plaster & Gypsum Board	88.9	120.9	110.6	86.9	115.8	106.4	101.8	85.3	90.7	86.9	85.6	86.0	86.9	82.8	84.1	101.7	55.2	70.3
0950, 0980	Ceilings & Acoustic Treatment	95.6	120.9	112.3	59.3	115.8	96.4	122.8	85.3	98.2	59.3	85.6	76.6	59.3	82.8	74.7	87.3	55.2	66.2
0960	Flooring	103.7	120.8	108.6	100.0	120.8	105.9	104.1	120.8	108.8	101.6	120.8	107.1	103.9	120.8	108.8	101.4	55.1	88.1
0970, 0990	Wall Finishes & Painting/Coating	105.3	126.1	117.8	108.3	126.1	119.1	102.2	95.9	98.4	102.2	95.9	98.4	102.2	104.9	103.8	99.5	43.4	65.7
09	FINISHES	97.9	123.1	111.8	88.9	119.2	105.6	106.0	95.2	100.0	88.9	94.2	91.8	89.1	91.3	90.3	93.3	52.4	70.8
COVERS	DIVS. 10 - 14, 25, 28, 41, 43, 44, 46	100.0	108.6	101.7	100.0	104.3	100.9	100.0	96.8	99.3	100.0	98.2	99.6	100.0	93.7	98.7	100.0	55.4	91.0
21, 22, 23	FIRE SUPPRESSION, PLUMBING & HVAC	99.9	113.8	105.5	99.6	113.8	105.3	95.6	83.1	90.5	94.8	106.8	99.6	94.8	84.1	90.5	100.0	49.1	79.5
26, 27, 3370	ELECTRICAL, COMMUNICATIONS & UTIL.	97.1	111.4	104.7	101.8	111.4	106.9	101.3	67.5	83.5	101.8	85.2	93.0	108.6	89.2	98.3	103.8	56.9	79.0
MF2014	WEIGHTED AVERAGE	101.4	117.9	108.6	96.9	116.0	105.2	97.8	92.4	95.4	95.5	101.3	98.0	95.9	92.3	94.3	99.2	58.1	81.3

DIVISION		MISSISSIPPI																	
		CLARKSDALE			COLUMBUS			GREENVILLE			GREENWOOD			JACKSON			LAUREL		
		386			397			387			389			390 - 392			394		
		MAT.	INST.	TOTAL	MAT.	INST.	TOTAL	MAT.	INST.	TOTAL	MAT.	INST.	TOTAL	MAT.	INST.	TOTAL	MAT.	INST.	TOTAL
015433	CONTRACTOR EQUIPMENT		98.8	98.8		98.8	98.8		98.8	98.8		98.8	98.8		98.8	98.8		98.8	98.8
0241, 31 - 34	SITE & INFRASTRUCTURE, DEMOLITION	98.3	88.2	91.1	98.9	88.4	91.5	103.9	88.4	92.9	101.0	87.9	91.7	97.6	88.4	91.1	104.4	88.3	92.9
0310	Concrete Forming & Accessories	85.0	49.2	54.2	83.3	51.2	55.6	81.7	63.7	66.1	94.2	49.2	55.4	93.6	57.5	62.5	83.3	52.9	57.1
0320	Concrete Reinforcing	99.8	41.6	70.1	101.8	42.5	71.6	100.3	60.3	79.9	99.8	52.5	75.7	100.2	56.1	77.7	102.4	39.1	70.2
0330	Cast-in-Place Concrete	105.6	53.5	84.2	110.5	58.6	89.2	108.7	55.6	86.9	113.5	53.2	88.7	98.7	56.8	81.5	108.1	55.2	86.4
03	CONCRETE	98.7	50.9	75.2	103.2	53.7	78.9	104.2	61.6	83.3	105.2	52.7	79.4	97.7	58.5	78.5	106.0	52.7	79.8
04	MASONRY	100.6	45.0	66.0	131.6	51.6	81.7	149.8	49.5	87.3	101.3	44.9	66.1	109.2	49.5	72.0	127.1	48.1	77.9
05	METALS	95.4	73.4	88.6	91.9	74.2	86.5	96.4	82.5	92.1	95.4	74.8	89.0	101.8	80.8	95.4	92.0	73.3	86.3
06	WOOD, PLASTICS & COMPOSITES	82.1	51.2	64.8	83.9	52.2	66.1	78.7	68.1	72.8	95.0	51.2	70.5	96.3	59.6	75.8	85.1	54.6	68.0
07	THERMAL & MOISTURE PROTECTION	95.7	47.2	75.8	97.3	51.5	78.5	96.0	60.0	81.2	96.1	47.1	76.0	97.5	59.2	81.8	97.4	53.1	79.3
08	OPENINGS	97.0	49.4	85.9	98.9	49.5	87.4	97.0	61.4	88.7	97.0	50.7	86.2	103.7	55.6	92.5	95.6	50.3	85.0
0920	Plaster & Gypsum Board	90.2	50.2	63.1	91.3	51.2	64.2	89.8	67.6	74.8	100.9	50.2	66.6	95.0	58.8	70.5	91.3	53.7	65.8
0950, 0980	Ceilings & Acoustic Treatment	82.9	50.2	61.4	82.0	51.2	61.8	85.7	67.6	73.8	82.9	50.2	61.4	89.7	58.8	69.4	82.0	53.7	63.4
0960	Flooring	105.7	52.8	90.6	95.5	59.3	85.1	104.1	52.8	89.4	111.1	52.8	94.4	99.3	55.1	86.7	94.2	52.8	82.4
0970, 0990	Wall Finishes & Painting/Coating	104.4	46.1	69.2	99.5	53.0	71.4	104.4	65.9	81.2	104.4	46.1	69.2	96.5	65.9	78.0	99.5	60.9	76.2
09	FINISHES	94.0	49.9	69.7	89.1	53.2	69.3	94.6	62.9	77.1	97.4	49.9	71.2	93.2	58.3	74.0	89.2	54.3	70.0
COVERS	DIVS. 10 - 14, 25, 28, 41, 43, 44, 46	100.0	55.7	91.1	100.0	56.8	91.3	100.0	58.5	91.6	100.0	55.7	91.1	100.0	57.5	91.4	100.0	37.5	87.4
21, 22, 23	FIRE SUPPRESSION, PLUMBING & HVAC	97.9	41.3	75.1	97.4	37.5	73.3	100.0	47.6	78.8	97.9	39.3	74.3	100.1	57.0	82.7	97.5	45.8	76.6
26, 27, 3370	ELECTRICAL, COMMUNICATIONS & UTIL.	97.1	46.4	70.3	100.8	49.1	73.5	97.1	62.7	78.9	97.1	43.4	68.8	105.4	62.7	82.8	102.6	56.9	78.5
MF2014	WEIGHTED AVERAGE	97.3	52.6	77.8	98.8	53.9	79.2	101.0	61.7	83.9	98.4	52.2	78.3	100.7	62.2	83.9	98.9	55.8	80.1

DIVISION		MISSISSIPPI									MISSOURI								
		MCCOMB			MERIDIAN			TUPELO			BOWLING GREEN			CAPE GIRARDEAU			CHILLICOTHE		
		396			393			388			633			637			646		
		MAT.	INST.	TOTAL	MAT.	INST.	TOTAL	MAT.	INST.	TOTAL	MAT.	INST.	TOTAL	MAT.	INST.	TOTAL	MAT.	INST.	TOTAL
015433	CONTRACTOR EQUIPMENT		98.8	98.8		98.8	98.8		98.8	98.8		106.9	106.9		106.9	106.9		101.9	101.9
0241, 31 - 34	SITE & INFRASTRUCTURE, DEMOLITION	92.2	88.0	89.2	96.2	88.7	90.8	95.6	88.1	90.3	89.6	95.6	93.9	91.3	95.4	94.2	105.2	94.6	97.7
0310	Concrete Forming & Accessories	83.3	50.9	55.4	80.4	63.3	65.7	82.2	51.2	55.5	92.3	87.3	88.0	85.2	85.1	85.1	89.0	93.0	92.5
0320	Concrete Reinforcing	103.0	41.0	71.4	101.8	56.1	78.5	97.6	52.5	74.7	101.4	102.8	102.1	102.7	88.7	95.5	96.8	106.5	101.8
0330	Cast-in-Place Concrete	96.0	53.0	78.3	102.7	59.2	84.8	105.6	54.9	84.8	91.2	84.7	88.5	90.2	86.0	88.5	103.0	85.8	95.9
03	CONCRETE	92.9	51.3	72.5	97.2	61.9	79.8	98.2	54.2	76.6	95.8	90.5	93.2	94.9	87.3	91.2	105.6	93.6	99.7
04	MASONRY	133.0	44.4	77.7	102.4	53.5	71.9	138.4	47.9	82.0	117.6	98.7	105.8	113.9	80.0	92.8	102.9	98.8	100.3
05	METALS	92.2	71.6	85.8	93.0	81.0	89.3	95.3	75.1	89.1	94.6	114.4	100.7	95.6	107.0	99.1	95.1	108.5	99.2
06	WOOD, PLASTICS & COMPOSITES	83.9	54.0	67.2	81.2	65.1	72.2	79.3	52.2	64.1	90.8	86.0	88.1	83.5	85.5	84.6	93.6	93.3	93.4
07	THERMAL & MOISTURE PROTECTION	96.9	46.4	76.2	97.0	61.6	82.5	95.7	52.9	78.2	100.3	98.9	99.7	100.2	84.2	93.6	98.3	94.5	96.8
08	OPENINGS	99.0	51.0	87.8	98.9	64.1	90.8	97.0	50.6	86.2	97.7	97.5	97.6	97.7	82.9	94.2	95.0	95.7	95.2
0920	Plaster & Gypsum Board	91.3	53.0	65.4	91.3	64.5	73.2	89.8	51.2	63.7	95.2	85.6	88.7	94.2	85.1	88.0	102.0	92.9	95.8
0950, 0980	Ceilings & Acoustic Treatment	82.0	53.0	63.0	83.9	64.5	71.2	82.9	51.2	62.0	92.6	85.6	88.0	92.6	85.1	87.6	89.1	92.9	91.6
0960	Flooring	95.5	52.8	83.3	94.2	55.1	83.0	104.4	52.8	89.6	96.4	96.8	96.5	93.3	84.0	90.6	97.3	102.5	98.8
0970, 0990	Wall Finishes & Painting/Coating	99.5	42.3	65.0	99.5	76.1	85.4	104.4	58.7	76.8	97.1	102.8	100.5	97.1	73.2	82.7	87.9	106.3	99.0
09	FINISHES	88.5	51.0	67.8	88.8	63.7	75.0	93.5	52.6	71.0	97.8	89.7	93.4	96.7	83.0	89.2	100.2	96.1	97.9
COVERS	DIVS. 10 - 14, 25, 28, 41, 43, 44, 46	100.0	59.1	91.7	100.0	59.5	91.8	100.0	56.8	91.3	100.0	88.2	97.6	100.0	93.5	98.7	100.0	88.5	97.7
21, 22, 23	FIRE SUPPRESSION, PLUMBING & HVAC	97.4	34.6	72.1	100.0	59.1	83.5	98.1	43.2	76.0	95.2	96.2	95.6	99.9	96.8	98.7	95.3	97.0	96.0
26, 27, 3370	ELECTRICAL, COMMUNICATIONS & UTIL.	99.2	49.4	72.9	102.6	57.8	78.9	96.9	49.3	71.7	96.9	82.6	89.3	96.9	101.5	99.3	92.4	78.6	85.1
MF2014	WEIGHTED AVERAGE	97.4	51.7	77.5	97.6	64.1	83.0	98.8	54.9	79.7	97.3	94.4	96.0	98.2	92.4	95.7	97.6	94.5	96.3

DIVISION		MISSOURI																	
		COLUMBIA			FLAT RIVER			HANNIBAL			HARRISONVILLE			JEFFERSON CITY			JOPLIN		
		652			636			634			647			650 - 651			648		
		MAT.	INST.	TOTAL	MAT.	INST.	TOTAL	MAT.	INST.	TOTAL	MAT.	INST.	TOTAL	MAT.	INST.	TOTAL	MAT.	INST.	TOTAL
015433	CONTRACTOR EQUIPMENT		108.0	108.0		106.9	106.9		106.9	106.9		101.9	101.9		108.0	108.0		105.7	105.7
0241, 31 - 34	SITE & INFRASTRUCTURE, DEMOLITION	102.8	97.2	98.8	92.2	95.6	94.6	87.6	95.7	93.4	96.6	95.7	95.9	102.3	97.2	98.6	105.6	99.7	101.4
0310	Concrete Forming & Accessories	86.9	81.8	82.5	98.3	84.3	86.2	90.5	79.0	80.6	86.2	98.7	97.0	98.3	81.8	84.1	101.3	75.3	78.9
0320	Concrete Reinforcing	96.3	119.0	107.9	102.7	107.7	105.3	100.9	97.7	99.3	96.4	115.3	106.0	98.5	111.9	105.3	99.9	98.5	99.2
0330	Cast-in-Place Concrete	96.1	87.8	92.7	94.2	93.8	94.0	86.3	86.0	86.2	105.5	103.9	104.8	102.0	87.0	95.9	111.5	77.9	97.7
03	CONCRETE	93.2	92.2	92.7	98.8	93.2	96.0	92.1	86.3	89.2	101.8	104.0	102.9	98.9	90.6	94.8	105.2	81.4	93.5
04	MASONRY	154.1	90.0	114.2	114.5	81.4	93.9	109.0	101.3	104.2	97.4	103.4	101.2	112.0	90.0	98.3	96.5	85.9	89.9
05	METALS	98.4	121.2	105.4	94.5	115.6	101.0	94.6	111.9	99.9	95.6	113.2	101.0	98.0	118.3	104.2	98.2	100.5	98.9
06	WOOD, PLASTICS & COMPOSITES	91.7	78.7	84.4	99.1	82.6	89.9	88.9	73.8	80.4	89.9	97.5	94.2	100.0	78.7	88.1	106.0	74.2	88.2
07	THERMAL & MOISTURE PROTECTION	96.2	87.5	92.6	100.5	93.8	97.7	100.1	97.0	98.8	97.6	103.2	99.9	101.8	87.6	96.0	97.7	78.9	90.0
08	OPENINGS	99.5	95.0	98.4	97.7	97.2	97.6	97.7	82.3	94.1	95.1	102.8	96.9	99.1	93.1	97.7	96.2	78.5	92.1
0920	Plaster & Gypsum Board	87.2	78.0	81.0	101.3	82.1	88.3	94.5	73.0	79.9	97.3	97.3	97.3	91.3	78.0	82.3	108.2	73.2	84.5
0950, 0980	Ceilings & Acoustic Treatment	93.2	78.0	83.2	92.6	82.1	85.7	92.6	73.0	79.7	89.1	97.3	94.5	97.2	78.0	84.6	90.0	73.2	78.9
0960	Flooring	97.6	98.6	97.9	99.4	84.0	95.0	95.8	96.8	96.1	92.6	102.5	95.4	104.4	98.6	102.7	123.8	74.9	109.8
0970, 0990	Wall Finishes & Painting/Coating	105.2	77.1	88.2	97.1	78.4	85.8	97.1	102.8	100.5	92.5	103.8	99.3	102.3	77.1	87.1	87.5	74.9	79.9
09	FINISHES	92.8	83.0	87.4	99.7	82.2	90.1	97.4	83.1	89.5	97.8	99.5	98.7	97.2	83.0	89.4	107.0	75.3	89.6
COVERS	DIVS. 10 - 14, 25, 28, 41, 43, 44, 46	100.0	96.2	99.2	100.0	91.8	98.3	100.0	87.8	97.5	100.0	90.6	98.1	100.0	96.2	99.2	100.0	87.9	97.6
21, 22, 23	FIRE SUPPRESSION, PLUMBING & HVAC	99.8	96.4	98.4	95.2	97.8	96.2	95.2	97.5	96.1	95.2	100.0	97.2	99.8	98.1	99.1	100.1	72.1	88.8
26, 27, 3370	ELECTRICAL, COMMUNICATIONS & UTIL.	94.8	86.5	90.4	101.1	101.5	101.3	95.7	82.6	88.8	98.8	99.6	99.2	100.4	86.5	93.0	90.4	77.5	83.6
MF2014	WEIGHTED AVERAGE	100.2	93.9	97.5	98.1	95.0	96.8	96.2	92.4	94.6	97.1	101.5	99.1	99.9	93.7	97.2	99.5	81.7	91.8

DIVISION		MISSOURI																	
		KANSAS CITY			KIRKSVILLE			POPLAR BLUFF			ROLLA			SEDALIA			SIKESTON		
		640 - 641			635			639			654 - 655			653			638		
		MAT.	INST.	TOTAL	MAT.	INST.	TOTAL	MAT.	INST.	TOTAL	MAT.	INST.	TOTAL	MAT.	INST.	TOTAL	MAT.	INST.	TOTAL
015433	CONTRACTOR EQUIPMENT		103.3	103.3		99.1	99.1		101.5	101.5		108.0	108.0		99.9	99.9		101.5	101.5
0241, 31 - 34	SITE & INFRASTRUCTURE, DEMOLITION	98.9	98.2	98.4	91.8	91.4	91.5	79.1	95.4	90.7	101.5	97.1	98.3	99.0	92.8	94.6	82.4	95.4	91.7
0310	Concrete Forming & Accessories	100.1	105.4	104.6	83.2	80.6	81.0	83.3	85.0	84.7	95.0	91.9	92.3	92.4	81.4	82.9	84.3	84.9	84.9
0320	Concrete Reinforcing	94.9	119.8	107.6	101.6	98.8	100.2	104.7	94.7	99.6	96.7	94.7	95.7	95.3	106.2	100.8	104.0	98.4	101.2
0330	Cast-in-Place Concrete	103.7	106.6	104.9	94.1	83.8	89.9	72.4	86.5	78.2	98.3	96.2	97.5	102.7	83.8	94.9	77.4	86.5	81.1
03	CONCRETE	101.1	108.7	104.8	110.8	86.0	98.6	85.9	88.1	86.9	95.2	95.0	95.1	109.8	87.7	99.0	89.5	88.7	89.1
04	MASONRY	99.5	107.5	104.5	121.0	88.9	101.0	112.5	80.1	92.3	126.3	87.9	102.4	132.9	88.3	105.1	112.2	80.1	92.2
05	METALS	105.9	115.8	108.9	94.2	103.8	97.2	94.8	101.7	96.9	97.8	110.5	101.7	96.5	108.0	100.1	95.1	103.2	97.6
06	WOOD, PLASTICS & COMPOSITES	105.3	105.0	105.1	77.0	78.7	77.9	76.1	85.5	81.4	100.3	92.8	96.1	93.1	78.7	85.1	77.7	85.5	82.1
07	THERMAL & MOISTURE PROTECTION	97.8	106.8	101.5	106.8	88.2	99.2	105.2	85.9	97.3	96.4	93.4	95.1	102.5	93.1	98.7	105.3	84.5	96.8
08	OPENINGS	102.5	109.2	104.0	102.7	85.8	98.8	103.7	84.5	99.2	99.5	88.3	96.9	104.5	87.4	100.5	103.7	85.5	99.5
0920	Plaster & Gypsum Board	105.2	105.0	105.0	89.8	78.0	81.8	90.2	85.1	86.7	89.3	92.5	91.5	82.5	78.0	79.5	92.0	85.1	87.3
0950, 0980	Ceilings & Acoustic Treatment	96.7	105.0	102.1	90.9	78.0	82.4	92.6	85.1	87.6	93.2	92.5	92.7	93.2	78.0	83.2	92.6	85.1	87.6
0960	Flooring	98.0	109.0	101.2	74.4	95.8	80.5	88.3	82.0	86.5	101.2	95.8	99.6	78.8	95.8	83.7	88.7	82.0	86.8
0970, 0990	Wall Finishes & Painting/Coating	92.5	112.2	104.4	92.5	82.1	86.2	92.0	73.2	80.7	105.2	100.6	102.5	105.2	106.3	105.9	92.0	73.2	80.7
09	FINISHES	102.0	106.6	104.5	96.5	82.7	88.9	96.5	82.7	88.9	94.3	93.4	93.8	91.9	85.4	88.3	97.1	82.7	89.2
COVERS	DIVS. 10 - 14, 25, 28, 41, 43, 44, 46	100.0	100.6	100.1	100.0	87.1	97.4	100.0	93.6	98.7	100.0	89.0	97.8	100.0	90.2	98.0	100.0	94.2	98.8
21, 22, 23	FIRE SUPPRESSION, PLUMBING & HVAC	100.0	104.3	101.7	95.2	97.3	96.1	95.2	96.3	95.6	95.0	98.2	96.3	95.0	95.9	95.4	95.2	96.3	95.6
26, 27, 3370	ELECTRICAL, COMMUNICATIONS & UTIL.	100.3	100.1	100.2	95.8	82.5	88.8	96.1	101.5	99.0	93.2	86.5	89.6	94.7	128.8	112.7	95.3	101.5	98.6
MF2014	WEIGHTED AVERAGE	101.4	105.7	103.3	99.5	89.9	95.3	96.2	92.0	94.4	97.9	94.6	96.5	100.2	97.3	98.9	96.7	92.2	94.8

DIVISION		MISSOURI									MONTANA								
		SPRINGFIELD			ST. JOSEPH			ST. LOUIS			BILLINGS			BUTTE			GREAT FALLS		
		656 - 658			644 - 645			630 - 631			590 - 591			597			594		
		MAT.	INST.	TOTAL	MAT.	INST.	TOTAL	MAT.	INST.	TOTAL	MAT.	INST.	TOTAL	MAT.	INST.	TOTAL	MAT.	INST.	TOTAL
015433	CONTRACTOR EQUIPMENT		102.5	102.5		101.9	101.9		107.9	107.9		98.5	98.5		98.2	98.2		98.2	98.2
0241, 31 - 34	SITE & INFRASTRUCTURE, DEMOLITION	101.6	95.1	97.0	100.5	93.9	95.8	91.7	98.9	96.8	95.0	96.3	95.9	102.0	96.1	97.8	105.7	96.2	98.9
0310	Concrete Forming & Accessories	101.4	79.1	82.2	100.0	89.6	91.0	98.4	104.8	103.9	98.4	71.5	75.2	85.5	71.8	73.7	98.4	71.5	75.2
0320	Concrete Reinforcing	92.7	118.7	105.9	93.8	115.0	104.6	93.9	112.7	103.4	90.2	81.7	85.9	97.8	81.9	89.7	90.2	81.8	85.9
0330	Cast-in-Place Concrete	104.3	79.6	94.2	103.6	103.3	103.5	90.2	105.0	96.3	129.3	67.6	103.9	142.3	69.0	112.2	150.4	63.7	114.8
03	CONCRETE	105.8	87.4	96.8	100.9	99.7	100.3	94.4	107.1	100.7	109.8	72.9	91.7	113.8	73.5	94.0	119.5	71.5	96.0
04	MASONRY	102.9	89.8	94.8	99.3	96.1	97.3	95.4	111.9	105.7	128.0	73.7	94.2	123.5	76.6	94.3	128.0	75.4	95.2
05	METALS	102.8	108.8	104.7	101.9	112.4	105.2	100.3	120.2	106.4	105.8	90.0	100.9	100.1	90.1	97.1	103.2	90.0	99.2
06	WOOD, PLASTICS & COMPOSITES	101.8	77.1	88.0	106.0	87.6	95.7	99.0	103.0	101.2	98.6	70.5	82.8	85.7	70.8	77.4	99.9	70.5	83.4
07	THERMAL & MOISTURE PROTECTION	100.7	82.2	93.2	98.2	95.6	97.1	100.2	106.9	103.0	106.5	69.9	91.5	106.3	71.0	91.8	106.9	68.5	91.2
08	OPENINGS	107.2	84.3	101.9	100.6	98.6	100.1	97.6	110.9	100.7	97.3	68.7	90.6	95.6	70.0	89.6	98.5	68.0	91.4
0920	Plaster & Gypsum Board	90.4	76.3	80.9	109.7	87.0	94.4	102.2	103.1	102.8	98.0	69.8	78.9	98.7	70.2	79.4	107.3	69.8	82.0
0950, 0980	Ceilings & Acoustic Treatment	93.2	76.3	82.1	95.8	87.0	90.0	97.6	103.1	101.2	81.5	69.8	73.8	88.3	70.2	76.4	90.0	69.8	76.7
0960	Flooring	99.9	74.9	92.8	101.5	107.3	103.2	99.2	97.6	98.7	106.6	70.2	96.2	104.4	74.2	95.7	111.5	74.2	100.8
0970, 0990	Wall Finishes & Painting/Coating	99.0	74.9	84.4	87.9	86.7	87.2	97.1	106.0	102.5	105.2	86.6	94.0	103.4	51.2	71.9	103.4	86.6	93.3
09	FINISHES	96.7	78.0	86.4	103.2	91.7	96.9	100.8	103.1	102.1	96.6	72.6	83.4	97.2	69.7	82.0	101.1	73.3	85.8
COVERS	DIVS. 10 - 14, 25, 28, 41, 43, 44, 46	100.0	94.8	99.0	100.0	96.9	99.4	100.0	102.4	100.5	100.0	94.7	98.9	100.0	94.8	98.9	100.0	94.7	98.9
21, 22, 23	FIRE SUPPRESSION, PLUMBING & HVAC	99.8	74.6	89.6	100.1	89.0	95.6	100.0	107.4	103.0	100.1	74.7	89.9	100.2	75.6	90.3	100.2	70.7	88.3
26, 27, 3370	ELECTRICAL, COMMUNICATIONS & UTIL.	99.2	69.7	83.6	98.8	78.6	88.2	99.2	108.3	104.0	98.5	72.2	84.6	105.5	71.7	87.7	97.7	71.9	84.1
MF2014	WEIGHTED AVERAGE	101.7	83.7	93.8	100.6	93.5	97.5	98.7	107.8	102.7	102.7	77.1	91.5	102.6	77.3	91.6	104.1	76.2	91.9

DIVISION		MONTANA																	
		HAVRE			HELENA			KALISPELL			MILES CITY			MISSOULA			WOLF POINT		
		595			596			599			593			598			592		
		MAT.	INST.	TOTAL	MAT.	INST.	TOTAL	MAT.	INST.	TOTAL	MAT.	INST.	TOTAL	MAT.	INST.	TOTAL	MAT.	INST.	TOTAL
015433	**CONTRACTOR EQUIPMENT**		98.2	98.2		98.2	98.2		98.2	98.2		98.2	98.2		98.2	98.2		98.2	98.2
0241, 31 - 34	**SITE & INFRASTRUCTURE, DEMOLITION**	109.2	95.8	99.7	98.5	95.8	96.6	92.2	96.0	94.9	98.5	95.4	96.3	85.1	95.8	92.7	115.5	95.6	101.3
0310	Concrete Forming & Accessories	78.4	69.0	70.3	101.5	65.2	70.2	88.7	71.4	73.8	96.8	67.6	71.6	88.7	71.4	73.8	89.2	67.9	70.8
0320	Concrete Reinforcing	98.6	81.7	90.0	102.5	81.7	91.9	100.4	83.4	91.8	98.2	81.8	89.8	99.5	83.4	91.3	99.6	81.8	90.5
0330	Cast-in-Place Concrete	153.3	60.4	115.1	115.0	62.9	93.6	123.6	62.7	98.6	135.3	60.3	104.5	104.9	64.9	88.4	151.5	60.6	114.2
03	**CONCRETE**	122.7	69.3	96.5	106.7	68.4	87.9	102.6	71.5	87.3	110.4	68.6	89.9	90.0	72.2	81.3	126.5	68.9	98.2
04	**MASONRY**	124.5	69.8	90.4	120.1	75.7	92.4	122.2	77.0	94.0	129.7	65.3	89.6	147.9	76.6	103.4	131.0	66.0	90.5
05	**METALS**	96.3	89.6	94.2	102.1	88.8	98.0	96.1	90.6	94.4	95.4	89.9	93.7	96.6	90.7	94.8	95.5	89.2	93.6
06	**WOOD, PLASTICS & COMPOSITES**	77.4	70.5	73.5	102.9	62.3	80.2	89.1	70.5	78.7	96.9	70.5	82.1	89.1	70.5	78.7	88.5	70.5	78.4
07	**THERMAL & MOISTURE PROTECTION**	106.6	68.0	90.8	103.9	70.7	90.3	105.9	72.8	92.3	106.2	65.7	89.6	105.5	79.1	94.7	107.1	66.0	90.3
08	**OPENINGS**	95.6	68.0	89.2	99.8	65.2	91.7	95.6	68.5	89.3	95.1	68.0	88.8	95.6	68.3	89.2	95.1	67.4	88.7
0920	Plaster & Gypsum Board	94.8	69.8	77.9	96.0	61.4	72.6	98.7	69.8	79.2	106.7	69.8	81.8	98.7	69.8	79.2	102.1	69.8	80.3
0950, 0980	Ceilings & Acoustic Treatment	88.3	69.8	76.2	90.8	61.4	71.5	88.3	69.8	76.2	87.5	69.8	75.9	88.3	69.8	76.2	87.5	69.8	75.9
0960	Flooring	101.8	74.2	93.9	106.7	59.1	93.1	106.2	74.2	97.1	111.3	70.2	99.5	106.2	74.2	97.1	107.6	70.2	96.9
0970, 0990	Wall Finishes & Painting/Coating	103.4	48.6	70.4	99.7	47.2	68.0	103.4	47.2	69.5	103.4	47.2	69.5	103.4	62.0	78.4	103.4	48.6	70.4
09	**FINISHES**	96.5	67.7	80.6	98.6	61.6	78.2	97.2	69.1	81.7	100.0	65.7	81.1	96.7	70.7	82.3	99.6	66.0	81.1
COVERS	**DIVS. 10 - 14, 25, 28, 41, 43, 44, 46**	100.0	70.8	94.1	100.0	94.0	98.8	100.0	93.9	98.8	100.0	90.4	98.1	100.0	93.7	98.7	100.0	90.6	98.1
21, 22, 23	**FIRE SUPPRESSION, PLUMBING & HVAC**	95.4	67.8	84.3	100.1	70.1	88.0	95.4	69.2	84.8	95.4	69.6	85.0	100.2	69.2	87.7	95.4	70.0	85.1
26, 27, 3370	**ELECTRICAL, COMMUNICATIONS & UTIL.**	97.7	71.2	83.7	104.6	71.3	87.0	102.2	69.3	84.8	97.7	76.9	86.7	103.2	67.7	84.5	97.7	76.9	86.7
MF2014	**WEIGHTED AVERAGE**	101.2	73.1	89.0	102.4	73.8	89.9	99.0	75.3	88.7	100.1	74.0	88.7	99.9	75.5	89.3	102.3	74.2	90.0

DIVISION		NEBRASKA																	
		ALLIANCE			COLUMBUS			GRAND ISLAND			HASTINGS			LINCOLN			MCCOOK		
		693			686			688			689			683 - 685			690		
		MAT.	INST.	TOTAL	MAT.	INST.	TOTAL	MAT.	INST.	TOTAL	MAT.	INST.	TOTAL	MAT.	INST.	TOTAL	MAT.	INST.	TOTAL
015433	**CONTRACTOR EQUIPMENT**		97.8	97.8		101.6	101.6		101.6	101.6		101.6	101.6		101.6	101.6		101.6	101.6
0241, 31 - 34	**SITE & INFRASTRUCTURE, DEMOLITION**	100.5	98.8	99.3	100.2	92.3	94.6	105.0	93.1	96.5	103.7	92.3	95.6	92.7	93.0	92.9	103.8	92.2	95.6
0310	Concrete Forming & Accessories	87.3	56.4	60.7	95.5	67.4	71.2	95.1	72.4	75.5	98.2	71.9	75.5	95.7	68.2	71.9	93.0	56.1	61.2
0320	Concrete Reinforcing	108.8	85.1	96.7	98.3	76.5	87.2	97.7	76.4	86.9	97.7	77.1	87.2	96.7	76.0	86.1	101.8	76.2	88.8
0330	Cast-in-Place Concrete	113.7	60.0	91.6	115.7	60.0	92.8	122.5	67.2	99.8	122.5	63.2	98.2	92.4	69.6	83.1	122.6	59.3	96.6
03	**CONCRETE**	123.9	63.5	94.2	108.8	67.3	88.4	113.7	72.1	93.3	113.9	70.6	92.6	95.6	71.0	83.5	112.9	62.0	87.9
04	**MASONRY**	115.6	74.1	89.7	122.7	77.8	94.7	115.4	76.1	90.9	124.9	88.0	101.9	104.8	67.2	81.4	110.7	73.9	87.8
05	**METALS**	101.5	78.7	94.5	96.1	84.6	92.6	97.8	86.3	94.3	98.7	85.8	94.7	99.9	85.5	95.5	96.3	83.6	92.4
06	**WOOD, PLASTICS & COMPOSITES**	87.0	52.2	67.5	101.2	66.7	81.9	100.4	72.2	84.6	104.1	72.2	86.3	98.4	66.7	80.7	95.7	52.2	71.4
07	**THERMAL & MOISTURE PROTECTION**	104.3	67.8	89.4	100.8	69.8	88.1	100.9	72.8	89.4	100.9	83.4	93.8	98.1	69.9	86.5	98.9	66.6	85.6
08	**OPENINGS**	94.5	58.6	86.1	95.4	66.1	88.6	95.5	69.7	89.5	95.5	69.1	89.3	107.8	63.0	97.4	95.0	56.6	86.1
0920	Plaster & Gypsum Board	82.7	50.7	61.1	90.9	65.6	73.8	90.1	71.3	77.4	91.9	71.3	78.0	96.2	65.6	75.5	93.7	50.7	64.7
0950, 0980	Ceilings & Acoustic Treatment	91.9	50.7	64.8	80.2	65.6	70.6	80.2	71.3	74.3	80.2	71.3	74.3	90.4	65.6	74.1	88.0	50.7	63.5
0960	Flooring	99.9	82.1	94.8	92.9	100.4	95.0	92.7	107.2	96.8	93.9	100.4	95.7	101.0	83.6	96.0	97.6	82.1	93.1
0970, 0990	Wall Finishes & Painting/Coating	168.5	54.8	99.9	85.5	63.7	72.4	85.5	67.6	74.7	85.5	63.7	72.4	103.8	77.9	88.1	93.8	47.7	66.0
09	**FINISHES**	98.4	59.5	77.0	89.9	71.9	80.0	90.0	77.1	82.9	90.6	75.2	82.1	96.9	71.7	83.0	95.6	58.6	75.2
COVERS	**DIVS. 10 - 14, 25, 28, 41, 43, 44, 46**	100.0	65.3	93.0	100.0	86.1	97.2	100.0	88.7	97.7	100.0	86.7	97.3	100.0	88.1	97.6	100.0	64.5	92.8
21, 22, 23	**FIRE SUPPRESSION, PLUMBING & HVAC**	95.3	70.5	85.3	95.2	76.0	87.4	100.0	78.0	91.1	95.2	76.0	87.5	99.9	78.0	91.0	95.1	75.9	87.4
26, 27, 3370	**ELECTRICAL, COMMUNICATIONS & UTIL.**	92.3	69.4	80.2	93.6	82.3	87.6	92.2	67.0	78.9	91.6	83.6	87.4	105.4	67.0	85.1	94.4	69.4	81.2
MF2014	**WEIGHTED AVERAGE**	101.0	70.5	87.7	98.4	77.1	89.1	100.0	77.1	90.0	99.4	79.8	90.9	100.5	74.9	89.4	98.7	71.1	86.7

DIVISION		NEBRASKA												NEVADA					
		NORFOLK			NORTH PLATTE			OMAHA			VALENTINE			CARSON CITY			ELKO		
		687			691			680 - 681			692			897			898		
		MAT.	INST.	TOTAL	MAT.	INST.	TOTAL	MAT.	INST.	TOTAL	MAT.	INST.	TOTAL	MAT.	INST.	TOTAL	MAT.	INST.	TOTAL
015433	**CONTRACTOR EQUIPMENT**		92.8	92.8		101.6	101.6		92.8	92.8		95.9	95.9		98.2	98.2		98.2	98.2
0241, 31 - 34	**SITE & INFRASTRUCTURE, DEMOLITION**	83.6	91.5	89.2	105.2	92.3	96.0	90.0	92.1	91.5	88.6	96.5	94.2	84.5	99.0	94.8	66.9	97.6	88.8
0310	Concrete Forming & Accessories	81.7	71.1	72.5	95.4	71.5	74.8	94.3	72.4	75.4	83.7	55.8	59.6	103.3	92.8	94.2	109.2	81.3	85.2
0320	Concrete Reinforcing	98.4	66.5	82.2	101.2	77.1	88.9	99.7	76.6	87.9	101.8	66.1	83.7	103.2	118.4	111.0	102.7	113.2	108.1
0330	Cast-in-Place Concrete	116.7	62.4	94.4	122.6	64.5	98.7	101.1	75.5	90.6	108.4	57.2	87.4	100.4	85.7	94.4	95.1	80.2	89.0
03	**CONCRETE**	107.3	67.5	87.8	113.0	70.8	92.3	100.1	74.5	87.5	110.2	58.9	85.0	100.9	95.2	98.1	95.4	87.2	91.4
04	**MASONRY**	129.4	79.1	98.0	98.6	87.9	92.0	106.8	79.0	89.4	110.9	73.9	87.8	117.3	94.6	103.2	115.0	71.9	88.1
05	**METALS**	99.6	73.5	91.6	95.5	85.4	92.4	99.9	78.9	93.4	107.6	73.4	97.1	94.0	104.7	97.3	96.9	99.4	97.7
06	**WOOD, PLASTICS & COMPOSITES**	84.0	71.7	77.1	97.8	72.2	83.5	94.1	72.0	81.7	81.7	51.7	64.9	92.1	94.4	93.4	100.6	81.8	90.1
07	**THERMAL & MOISTURE PROTECTION**	100.8	73.6	89.6	98.9	82.8	92.3	95.5	78.7	88.6	99.6	67.6	86.5	103.5	92.3	99.0	99.9	76.1	90.1
08	**OPENINGS**	97.2	66.5	90.0	94.3	69.1	88.4	100.0	71.5	93.3	96.9	55.6	87.3	99.5	107.9	101.4	100.9	85.9	97.4
0920	Plaster & Gypsum Board	90.8	71.3	77.6	93.7	71.3	78.6	100.5	71.6	81.0	95.6	50.7	65.3	100.9	94.1	96.3	105.3	81.2	89.0
0950, 0980	Ceilings & Acoustic Treatment	92.0	71.3	78.4	88.0	71.3	77.0	93.0	71.6	78.9	104.9	50.7	69.3	94.0	94.1	94.1	94.4	81.2	85.7
0960	Flooring	117.4	100.4	112.5	98.5	100.4	99.1	113.4	81.9	104.4	127.4	79.8	113.8	101.9	107.9	103.6	105.4	51.1	89.9
0970, 0990	Wall Finishes & Painting/Coating	151.7	63.7	98.6	93.8	63.7	75.6	126.2	67.2	90.6	167.2	66.1	106.1	101.2	80.8	88.9	102.6	95.9	98.5
09	**FINISHES**	108.8	75.2	90.3	95.9	75.2	84.5	106.0	73.3	88.0	118.2	59.7	86.0	97.9	94.3	95.9	97.6	77.2	86.3
COVERS	**DIVS. 10 - 14, 25, 28, 41, 43, 44, 46**	100.0	85.5	97.1	100.0	66.8	93.3	100.0	87.6	97.5	100.0	62.5	92.4	100.0	89.8	97.9	100.0	62.8	92.5
21, 22, 23	**FIRE SUPPRESSION, PLUMBING & HVAC**	95.0	75.8	87.2	99.9	76.0	90.2	99.9	76.8	90.6	94.8	75.7	87.1	100.0	80.1	92.0	97.7	79.6	90.4
26, 27, 3370	**ELECTRICAL, COMMUNICATIONS & UTIL.**	92.4	82.3	87.1	92.6	77.1	84.4	99.1	82.3	90.2	89.7	86.6	88.0	103.2	96.7	99.7	100.7	96.6	98.6
MF2014	**WEIGHTED AVERAGE**	100.0	76.7	89.8	99.0	78.3	90.0	100.2	78.6	90.8	101.2	72.5	88.7	99.7	93.5	97.0	98.3	84.9	92.4

DIVISION		NEVADA									NEW HAMPSHIRE								
		ELY			LAS VEGAS			RENO			CHARLESTON			CLAREMONT			CONCORD		
		893			889 - 891			894 - 895			036			037			032 - 033		
		MAT.	INST.	TOTAL	MAT.	INST.	TOTAL	MAT.	INST.	TOTAL	MAT.	INST.	TOTAL	MAT.	INST.	TOTAL	MAT.	INST.	TOTAL
015433	CONTRACTOR EQUIPMENT		98.2	98.2		98.2	98.2		98.2	98.2		100.9	100.9		100.9	100.9		100.9	100.9
0241, 31 - 34	SITE & INFRASTRUCTURE, DEMOLITION	72.2	99.1	91.3	75.4	101.0	93.6	72.2	99.0	91.3	86.9	99.6	96.0	80.9	99.6	94.2	93.7	103.9	100.9
0310	Concrete Forming & Accessories	102.3	105.7	105.2	103.5	109.6	108.8	98.7	92.7	93.5	86.8	81.1	81.9	92.6	81.1	82.7	95.5	93.6	93.8
0320	Concrete Reinforcing	101.5	111.6	106.7	93.7	119.9	107.1	96.0	119.9	108.1	84.1	93.5	88.9	84.1	93.5	88.9	95.3	94.1	94.7
0330	Cast-in-Place Concrete	102.1	105.4	103.4	98.9	107.8	102.5	108.0	85.7	98.8	97.1	70.5	86.2	89.2	70.5	81.6	112.2	90.5	103.3
03	CONCRETE	103.4	106.4	104.9	98.7	110.6	104.6	103.1	95.4	99.3	99.1	80.0	89.7	91.1	80.0	85.6	106.6	92.5	99.6
04	MASONRY	119.7	102.5	109.0	108.3	104.0	105.6	114.3	94.6	102.0	97.5	84.7	89.5	96.9	84.7	89.3	110.9	104.2	106.7
05	METALS	96.9	103.3	98.9	104.5	108.6	105.8	98.4	105.1	100.5	92.8	90.1	92.0	92.8	90.1	92.0	98.8	93.8	97.3
06	WOOD, PLASTICS & COMPOSITES	91.3	106.8	100.0	89.8	107.9	99.9	85.9	94.4	90.6	87.4	90.2	89.0	93.6	90.2	91.7	95.4	92.4	93.7
07	THERMAL & MOISTURE PROTECTION	100.3	95.7	98.4	113.8	102.6	109.2	99.9	92.3	96.8	98.8	72.3	88.0	98.7	72.3	87.9	101.9	91.5	97.6
08	OPENINGS	100.8	98.8	100.3	99.9	115.6	103.5	98.7	102.3	99.6	102.0	76.4	96.0	103.3	76.4	97.0	104.9	87.2	100.8
0920	Plaster & Gypsum Board	101.0	107.0	105.1	97.5	108.1	104.7	92.0	94.1	93.4	100.0	89.3	92.8	101.0	89.3	93.1	103.9	91.6	95.6
0950, 0980	Ceilings & Acoustic Treatment	94.4	107.0	102.7	102.7	108.1	106.3	99.4	94.1	95.9	94.2	89.3	91.0	94.2	89.3	91.0	99.9	91.6	94.4
0960	Flooring	103.0	57.2	89.9	94.6	107.9	98.4	99.9	107.9	102.2	98.4	32.1	79.4	100.6	32.1	81.0	102.7	116.9	106.8
0970, 0990	Wall Finishes & Painting/Coating	102.6	121.0	113.7	105.4	121.0	114.8	102.6	80.8	89.4	96.4	45.7	65.8	96.4	46.0	65.9	95.8	95.4	95.6
09	FINISHES	96.8	98.9	98.0	95.8	110.4	103.9	95.6	94.3	94.9	95.3	69.7	81.2	95.6	69.8	81.4	97.4	97.9	97.7
COVERS	DIVS. 10 - 14, 25, 28, 41, 43, 44, 46	100.0	63.6	92.6	100.0	104.4	100.9	100.0	89.8	97.9	100.0	92.2	98.4	100.0	92.2	98.4	100.0	104.3	100.9
21, 22, 23	FIRE SUPPRESSION, PLUMBING & HVAC	97.7	104.5	100.5	100.1	104.4	101.8	100.0	80.1	92.0	95.3	39.6	72.8	95.3	39.7	72.9	99.9	84.4	93.7
26, 27, 3370	ELECTRICAL, COMMUNICATIONS & UTIL.	101.0	109.1	105.3	105.4	119.9	113.1	101.4	96.7	98.9	98.5	52.9	74.4	98.5	52.9	74.4	99.5	83.6	91.1
MF2014	WEIGHTED AVERAGE	99.4	102.2	100.6	100.9	108.7	104.3	99.6	93.3	96.8	96.7	69.5	84.8	95.9	69.5	84.4	101.1	92.5	97.4

DIVISION		NEW HAMPSHIRE															NEW JERSEY		
		KEENE			LITTLETON			MANCHESTER			NASHUA			PORTSMOUTH			ATLANTIC CITY		
		034			035			031			030			038			082,084		
		MAT.	INST.	TOTAL	MAT.	INST.	TOTAL	MAT.	INST.	TOTAL	MAT.	INST.	TOTAL	MAT.	INST.	TOTAL	MAT.	INST.	TOTAL
015433	CONTRACTOR EQUIPMENT		100.9	100.9		100.9	100.9		100.9	100.9		100.9	100.9		100.9	100.9		99.1	99.1
0241, 31 - 34	SITE & INFRASTRUCTURE, DEMOLITION	94.0	99.9	98.2	81.0	99.9	94.4	93.6	103.9	100.9	95.7	103.9	101.5	89.2	104.0	99.7	96.4	105.2	102.7
0310	Concrete Forming & Accessories	91.3	83.0	84.2	102.9	83.0	85.8	96.9	94.0	94.4	99.3	94.0	94.8	88.2	94.3	93.4	109.0	127.5	125.0
0320	Concrete Reinforcing	84.1	93.6	88.9	84.8	93.6	89.3	103.7	94.1	98.8	105.1	94.1	99.5	84.1	94.2	89.2	76.3	118.1	97.6
0330	Cast-in-Place Concrete	97.6	72.8	87.4	87.6	72.7	81.5	110.6	113.1	111.7	92.4	113.1	100.9	87.6	113.2	98.1	86.1	133.0	105.3
03	CONCRETE	98.7	81.6	90.3	90.5	81.6	86.1	107.2	100.4	103.9	98.9	100.4	99.7	90.6	100.6	95.5	93.1	126.5	109.5
04	MASONRY	101.3	88.3	93.2	107.9	88.3	95.7	102.9	104.2	103.7	101.4	104.2	103.1	97.2	104.2	101.5	111.8	132.0	124.4
05	METALS	93.5	90.7	92.6	93.5	90.7	92.6	101.3	94.3	99.2	98.8	94.3	97.5	95.0	94.9	94.9	95.6	106.1	98.8
06	WOOD, PLASTICS & COMPOSITES	92.0	90.2	91.0	103.5	90.2	96.0	95.3	92.4	93.7	101.9	92.4	96.6	88.7	92.4	90.8	109.6	127.2	119.4
07	THERMAL & MOISTURE PROTECTION	99.2	75.3	89.4	98.8	73.8	88.5	101.5	95.0	98.8	99.6	95.0	97.7	99.3	115.0	105.7	105.9	124.2	113.4
08	OPENINGS	100.4	81.2	96.0	104.4	77.1	98.0	105.2	87.2	101.0	105.4	87.2	101.2	106.2	84.0	101.0	101.7	124.4	107.0
0920	Plaster & Gypsum Board	100.3	89.3	92.9	114.6	89.3	97.5	104.7	91.6	95.8	110.0	91.6	97.5	100.0	91.6	94.3	105.4	127.5	120.3
0950, 0980	Ceilings & Acoustic Treatment	94.2	89.3	91.0	94.2	89.3	91.0	101.7	91.6	95.0	105.1	91.6	96.2	95.1	91.6	92.8	81.3	127.5	111.6
0960	Flooring	100.2	52.6	86.6	109.9	32.1	87.6	98.6	116.9	103.8	103.7	116.9	107.4	98.5	116.9	103.8	105.4	159.9	121.0
0970, 0990	Wall Finishes & Painting/Coating	96.4	45.7	65.8	96.4	59.9	74.4	100.7	95.4	97.5	96.4	95.4	95.8	96.4	95.4	95.8	93.9	130.2	115.8
09	FINISHES	97.0	74.5	84.6	100.2	72.2	84.8	98.6	97.9	98.2	101.7	97.9	99.6	96.2	97.9	97.1	96.0	134.5	117.2
COVERS	DIVS. 10 - 14, 25, 28, 41, 43, 44, 46	100.0	93.4	98.7	100.0	98.6	99.7	100.0	104.3	100.9	100.0	104.3	100.9	100.0	104.3	100.9	100.0	111.5	102.3
21, 22, 23	FIRE SUPPRESSION, PLUMBING & HVAC	95.3	43.3	74.3	95.3	63.8	82.6	99.9	84.4	93.7	100.1	84.4	93.8	100.1	84.4	93.8	99.7	123.6	109.3
26, 27, 3370	ELECTRICAL, COMMUNICATIONS & UTIL.	98.5	63.2	79.9	99.8	55.9	76.6	99.8	83.6	91.2	101.2	83.6	91.9	99.1	83.6	90.9	91.1	138.1	116.0
MF2014	WEIGHTED AVERAGE	97.2	73.3	86.7	97.1	76.3	88.0	101.4	93.8	98.1	100.5	93.8	97.6	97.9	94.3	96.4	98.2	124.8	109.8

DIVISION		NEW JERSEY																	
		CAMDEN			DOVER			ELIZABETH			HACKENSACK			JERSEY CITY			LONG BRANCH		
		081			078			072			076			073			077		
		MAT.	INST.	TOTAL	MAT.	INST.	TOTAL	MAT.	INST.	TOTAL	MAT.	INST.	TOTAL	MAT.	INST.	TOTAL	MAT.	INST.	TOTAL
015433	CONTRACTOR EQUIPMENT		99.1	99.1		100.9	100.9		100.9	100.9		100.9	100.9		99.1	99.1		98.7	98.7
0241, 31 - 34	SITE & INFRASTRUCTURE, DEMOLITION	97.8	105.5	103.3	102.0	106.3	105.1	106.1	106.3	106.3	103.0	106.3	105.3	93.4	106.3	102.5	97.5	106.0	103.6
0310	Concrete Forming & Accessories	100.2	127.5	123.8	97.8	128.3	124.1	110.2	128.4	125.9	97.8	128.3	124.1	101.8	128.3	124.7	102.4	128.0	124.5
0320	Concrete Reinforcing	100.1	118.3	109.4	77.2	137.2	107.7	77.2	137.2	107.7	77.2	137.2	107.7	100.1	137.2	119.0	77.2	137.1	107.7
0330	Cast-in-Place Concrete	83.5	132.9	103.8	101.5	127.7	112.2	87.1	131.4	105.3	99.2	131.4	112.4	79.2	127.7	99.1	88.0	132.9	106.4
03	CONCRETE	94.0	126.5	110.0	98.8	128.7	113.5	94.5	130.1	112.0	97.0	130.0	113.2	92.1	128.6	110.0	96.2	130.2	112.9
04	MASONRY	101.3	132.0	120.4	93.0	132.5	117.6	108.7	132.5	123.5	96.9	132.5	119.1	87.0	132.5	115.4	101.3	132.0	120.5
05	METALS	101.1	106.1	102.6	93.3	116.2	100.4	94.8	116.2	101.4	93.4	116.1	100.4	98.9	113.9	103.5	93.4	113.7	99.7
06	WOOD, PLASTICS & COMPOSITES	98.4	127.2	114.5	96.2	127.2	113.5	111.7	127.2	120.4	96.2	127.2	113.5	97.1	127.2	114.0	98.2	127.1	114.4
07	THERMAL & MOISTURE PROTECTION	105.8	123.3	113.0	100.6	132.8	113.8	100.8	133.3	114.1	100.4	125.5	110.7	100.2	132.8	113.5	100.3	124.4	110.2
08	OPENINGS	104.1	124.5	108.8	105.5	127.7	110.6	103.6	127.7	109.2	102.9	127.7	108.7	101.6	127.7	107.7	97.8	127.6	104.7
0920	Plaster & Gypsum Board	101.5	127.5	119.0	100.9	127.5	118.9	107.7	127.5	121.0	100.9	127.5	118.9	104.3	127.5	120.0	102.7	127.5	119.4
0950, 0980	Ceilings & Acoustic Treatment	91.5	127.5	115.1	83.8	127.5	112.5	85.7	127.5	113.1	83.8	127.5	112.5	94.9	127.5	116.3	83.8	127.5	112.5
0960	Flooring	101.5	159.9	118.2	93.6	178.1	117.8	98.5	178.1	121.3	93.6	178.1	117.8	94.6	178.1	118.5	94.8	178.1	118.6
0970, 0990	Wall Finishes & Painting/Coating	93.9	130.2	115.8	98.4	132.5	119.0	98.4	132.5	119.0	98.4	132.5	119.0	98.5	132.5	119.0	98.5	130.2	117.6
09	FINISHES	96.4	134.5	117.4	93.7	136.9	117.5	96.9	136.9	118.9	93.5	136.9	117.4	96.6	136.9	118.8	94.5	137.5	118.2
COVERS	DIVS. 10 - 14, 25, 28, 41, 43, 44, 46	100.0	111.5	102.3	100.0	118.1	103.7	100.0	118.1	103.7	100.0	118.1	103.7	100.0	118.1	103.7	100.0	111.3	102.3
21, 22, 23	FIRE SUPPRESSION, PLUMBING & HVAC	100.0	123.6	109.5	99.7	127.4	110.9	100.1	125.5	110.3	99.7	127.4	110.9	100.1	127.4	111.1	99.7	127.1	110.8
26, 27, 3370	ELECTRICAL, COMMUNICATIONS & UTIL.	96.0	138.1	118.2	92.8	137.7	116.5	93.4	137.7	116.8	92.8	140.1	117.8	97.6	140.1	120.1	92.5	130.8	112.7
MF2014	WEIGHTED AVERAGE	99.5	124.8	110.5	97.9	127.8	110.9	98.8	127.6	111.3	97.6	128.1	110.9	97.9	127.9	111.0	97.1	126.3	109.8

DIVISION		NEW JERSEY																	
		NEW BRUNSWICK			NEWARK			PATERSON			POINT PLEASANT			SUMMIT			TRENTON		
		088 - 089			070 - 071			074 - 075			087			079			085 - 086		
		MAT.	INST.	TOTAL	MAT.	INST.	TOTAL	MAT.	INST.	TOTAL	MAT.	INST.	TOTAL	MAT.	INST.	TOTAL	MAT.	INST.	TOTAL
015433	CONTRACTOR EQUIPMENT		98.7	98.7		100.9	100.9		100.9	100.9		98.7	98.7		100.9	100.9		98.7	98.7
0241, 31 - 34	SITE & INFRASTRUCTURE, DEMOLITION	109.2	106.0	107.0	107.8	106.3	106.8	105.0	106.3	105.9	110.9	106.0	107.4	103.7	106.3	105.6	95.8	106.0	103.0
0310	Concrete Forming & Accessories	103.3	128.3	124.8	97.6	128.4	124.2	99.9	128.2	124.3	97.7	127.9	123.7	100.6	128.4	124.5	98.6	127.7	123.7
0320	Concrete Reinforcing	77.2	137.2	107.7	99.8	137.2	118.8	100.1	137.2	119.0	77.2	137.1	107.7	77.2	137.2	107.7	99.8	112.4	106.2
0330	Cast-in-Place Concrete	106.3	133.2	117.4	108.8	131.5	118.1	100.8	131.4	113.4	106.3	132.8	117.2	84.3	131.4	103.6	101.8	132.8	114.5
03	CONCRETE	110.4	130.5	120.3	105.8	130.1	117.7	102.2	130.0	115.9	110.1	130.1	119.9	91.5	130.1	110.5	102.5	125.5	113.8
04	MASONRY	109.3	132.5	123.7	99.6	132.5	120.1	93.6	132.5	117.9	97.0	132.0	118.8	95.5	132.5	118.6	103.9	132.0	121.4
05	METALS	95.6	113.8	101.2	100.8	116.3	105.5	94.2	116.1	100.9	95.6	113.5	101.1	93.3	116.2	100.4	100.8	105.2	102.1
06	WOOD, PLASTICS & COMPOSITES	103.0	127.1	116.5	93.9	127.2	112.5	98.8	127.2	114.7	95.9	127.1	113.4	100.1	127.2	115.3	95.6	127.1	113.3
07	THERMAL & MOISTURE PROTECTION	106.1	132.0	116.7	101.1	133.3	114.3	100.7	125.5	110.9	106.2	124.4	113.7	101.0	133.3	114.2	105.5	124.3	113.2
08	OPENINGS	96.1	127.6	103.5	104.4	127.7	109.8	108.3	127.7	112.8	98.2	129.2	105.4	110.2	127.7	114.2	106.1	122.2	109.8
0920	Plaster & Gypsum Board	102.6	127.5	119.4	101.1	127.5	118.9	104.3	127.5	120.0	98.3	127.5	118.0	102.7	127.5	119.4	98.7	127.5	118.1
0950, 0980	Ceilings & Acoustic Treatment	81.3	127.5	111.6	97.6	127.5	117.2	94.9	127.5	116.3	81.3	127.5	111.6	83.8	127.5	112.5	93.1	127.5	115.7
0960	Flooring	103.0	178.1	124.5	95.7	178.1	119.2	94.6	178.1	118.5	100.5	159.9	117.5	94.9	178.1	118.7	102.1	172.2	122.1
0970, 0990	Wall Finishes & Painting/Coating	93.9	132.5	117.2	99.8	132.5	119.5	98.4	132.5	119.0	93.9	130.2	115.8	98.4	132.5	119.0	99.1	130.2	117.8
09	FINISHES	96.0	136.8	118.5	95.6	136.9	118.3	96.7	136.9	118.9	94.7	134.5	116.6	94.6	136.9	117.9	96.6	136.5	118.6
COVERS	DIVS. 10 - 14, 25, 28, 41, 43, 44, 46	100.0	118.0	103.6	100.0	118.1	103.7	100.0	118.1	103.7	100.0	108.8	101.8	100.0	118.1	103.7	100.0	111.3	102.3
21, 22, 23	FIRE SUPPRESSION, PLUMBING & HVAC	99.7	127.4	110.8	100.1	127.4	111.1	100.1	127.4	111.1	99.7	127.0	110.7	99.7	125.5	110.1	100.1	126.9	110.9
26, 27, 3370	ELECTRICAL, COMMUNICATIONS & UTIL.	91.8	136.8	115.6	101.2	140.1	121.8	97.6	137.7	118.8	91.1	130.8	112.1	93.4	137.7	116.8	99.3	135.4	118.4
MF2014	WEIGHTED AVERAGE	99.7	127.6	111.9	101.2	128.4	113.0	99.6	127.8	111.9	99.2	125.9	110.8	97.9	127.6	110.8	100.9	125.1	111.5

DIVISION		NEW JERSEY			NEW MEXICO														
		VINELAND			ALBUQUERQUE			CARRIZOZO			CLOVIS			FARMINGTON			GALLUP		
		080,083			870 - 872			883			881			874			873		
		MAT.	INST.	TOTAL	MAT.	INST.	TOTAL	MAT.	INST.	TOTAL	MAT.	INST.	TOTAL	MAT.	INST.	TOTAL	MAT.	INST.	TOTAL
015433	CONTRACTOR EQUIPMENT		99.1	99.1		110.1	110.1		110.1	110.1		110.1	110.1		110.1	110.1		110.1	110.1
0241, 31 - 34	SITE & INFRASTRUCTURE, DEMOLITION	100.7	105.5	104.1	87.6	103.3	98.7	105.8	103.3	104.0	94.1	103.3	100.6	94.1	103.3	100.6	102.1	103.3	102.9
0310	Concrete Forming & Accessories	95.2	127.6	123.1	101.7	64.6	69.7	99.2	64.6	69.3	99.2	64.5	69.2	101.7	64.6	69.7	101.7	64.6	69.7
0320	Concrete Reinforcing	76.3	116.1	96.6	98.4	68.9	83.4	107.0	68.9	87.6	108.2	68.9	88.2	107.6	68.9	87.9	103.0	68.9	85.6
0330	Cast-in-Place Concrete	92.8	133.0	109.3	98.9	70.4	87.2	95.8	70.4	85.4	95.7	70.3	85.3	99.8	70.4	87.7	93.9	70.4	84.3
03	CONCRETE	97.9	126.2	111.8	100.2	68.5	84.6	117.0	68.5	93.2	105.8	68.4	87.4	103.9	68.5	86.5	110.0	68.5	89.6
04	MASONRY	99.4	132.0	119.7	105.0	58.7	76.1	103.4	58.7	75.5	103.4	58.7	75.5	113.7	58.7	79.4	99.4	58.7	74.0
05	METALS	95.5	105.5	98.6	105.2	86.8	99.5	99.5	86.8	95.6	99.2	86.7	95.3	102.9	86.8	97.9	102.0	86.8	97.3
06	WOOD, PLASTICS & COMPOSITES	92.8	127.2	112.1	97.5	65.5	79.6	93.1	65.5	77.6	93.1	65.5	77.6	97.7	65.5	79.7	97.7	65.5	79.7
07	THERMAL & MOISTURE PROTECTION	105.7	124.2	113.3	94.8	73.5	86.0	100.3	73.5	89.3	99.2	73.5	88.7	94.9	73.5	86.2	95.9	73.5	86.7
08	OPENINGS	97.7	124.1	103.8	101.2	67.7	93.4	98.6	67.7	91.4	98.7	67.7	91.5	103.6	67.7	95.3	103.6	67.7	95.3
0920	Plaster & Gypsum Board	96.9	127.5	117.5	95.8	64.2	74.4	78.6	64.2	68.8	78.6	64.2	68.8	89.4	64.2	72.3	89.4	64.2	72.3
0950, 0980	Ceilings & Acoustic Treatment	81.3	127.5	111.6	95.5	64.2	74.9	96.3	64.2	75.2	96.3	64.2	75.2	92.6	64.2	73.9	92.6	64.2	73.9
0960	Flooring	99.6	159.9	116.8	101.9	66.0	91.6	101.2	66.0	91.2	101.2	66.0	91.2	103.7	66.0	92.9	103.7	66.0	92.9
0970, 0990	Wall Finishes & Painting/Coating	93.9	130.2	115.8	113.2	66.6	85.1	103.3	66.6	81.1	103.3	66.6	81.1	107.1	66.6	82.7	107.1	66.6	82.7
09	FINISHES	93.5	134.5	116.1	95.9	64.8	78.8	96.6	64.8	79.1	95.4	64.8	78.5	94.8	64.8	78.3	96.0	64.8	78.8
COVERS	DIVS. 10 - 14, 25, 28, 41, 43, 44, 46	100.0	111.5	102.3	100.0	82.9	96.5	100.0	82.9	96.5	100.0	82.9	96.5	100.0	82.9	96.5	100.0	82.9	96.5
21, 22, 23	FIRE SUPPRESSION, PLUMBING & HVAC	99.7	123.6	109.3	100.1	69.0	87.6	97.2	69.0	85.8	97.2	68.7	85.7	100.0	69.0	87.5	97.1	69.0	85.8
26, 27, 3370	ELECTRICAL, COMMUNICATIONS & UTIL.	91.1	138.1	116.0	91.2	71.2	80.6	92.8	71.2	81.4	90.2	71.2	80.2	88.8	71.2	79.5	88.0	71.2	79.1
MF2014	WEIGHTED AVERAGE	97.5	124.7	109.4	99.6	72.6	87.8	100.3	72.6	88.2	98.3	72.5	87.1	100.1	72.6	88.1	99.5	72.6	87.8

DIVISION		NEW MEXICO																	
		LAS CRUCES			LAS VEGAS			ROSWELL			SANTA FE			SOCORRO			TRUTH/CONSEQUENCES		
		880			877			882			875			878			879		
		MAT.	INST.	TOTAL	MAT.	INST.	TOTAL	MAT.	INST.	TOTAL	MAT.	INST.	TOTAL	MAT.	INST.	TOTAL	MAT.	INST.	TOTAL
015433	CONTRACTOR EQUIPMENT		86.0	86.0		110.1	110.1		110.1	110.1		110.1	110.1		110.1	110.1		86.0	86.0
0241, 31 - 34	SITE & INFRASTRUCTURE, DEMOLITION	94.0	82.9	86.1	93.6	103.3	100.5	96.3	103.3	101.3	98.8	103.3	102.0	89.8	103.3	99.4	108.3	82.9	90.3
0310	Concrete Forming & Accessories	95.7	63.4	67.8	101.7	64.6	69.7	99.2	64.6	69.3	100.4	64.6	69.5	101.7	64.6	69.7	99.3	63.4	68.3
0320	Concrete Reinforcing	104.6	68.7	86.4	104.7	68.9	86.5	108.2	68.9	88.2	103.8	68.9	86.1	106.8	68.9	87.5	100.8	68.7	84.5
0330	Cast-in-Place Concrete	90.4	62.7	79.0	97.1	70.4	86.1	95.8	70.4	85.3	105.3	70.4	91.0	95.1	70.4	85.0	104.2	62.7	87.1
03	CONCRETE	84.9	65.0	75.1	101.2	68.5	85.1	106.5	68.5	87.8	104.0	68.5	86.5	100.2	68.5	84.6	94.1	65.0	79.8
04	MASONRY	99.2	58.3	73.7	99.7	58.7	74.1	114.3	58.7	79.6	104.1	58.7	75.8	99.6	58.7	74.1	97.2	58.3	73.0
05	METALS	98.1	80.3	92.6	101.7	86.8	97.1	100.4	86.8	96.2	99.0	86.8	95.2	102.0	86.8	97.3	101.6	80.3	95.0
06	WOOD, PLASTICS & COMPOSITES	82.5	64.4	72.4	97.7	65.5	79.7	93.1	65.5	77.6	99.3	65.5	80.4	97.7	65.5	79.7	89.1	64.4	75.3
07	THERMAL & MOISTURE PROTECTION	86.0	68.6	78.8	94.5	73.5	85.9	99.3	73.5	88.7	96.9	73.5	87.3	94.5	73.5	85.9	83.3	68.6	77.2
08	OPENINGS	91.5	67.1	85.8	99.9	67.7	92.4	98.5	67.7	91.4	102.0	67.7	94.0	99.8	67.7	92.3	93.1	67.1	87.0
0920	Plaster & Gypsum Board	77.5	64.2	68.5	89.4	64.2	72.3	78.6	64.2	68.8	98.9	64.2	75.4	89.4	64.2	72.3	91.1	64.2	72.9
0950, 0980	Ceilings & Acoustic Treatment	84.2	64.2	71.0	92.6	64.2	73.9	96.3	64.2	75.2	93.0	64.2	74.0	92.6	64.2	73.9	82.4	64.2	70.4
0960	Flooring	131.2	66.0	112.5	103.7	66.0	92.9	101.2	66.0	91.2	110.8	66.0	98.0	103.7	66.0	92.9	135.3	66.0	115.4
0970, 0990	Wall Finishes & Painting/Coating	90.7	66.6	76.2	107.1	66.6	82.7	103.3	66.6	81.1	111.8	66.6	84.5	107.1	66.6	82.7	97.7	66.6	79.0
09	FINISHES	105.2	63.9	82.5	94.7	64.8	78.2	95.5	64.8	78.6	99.7	64.8	80.5	94.6	64.8	78.2	107.5	63.9	83.5
COVERS	DIVS. 10 - 14, 25, 28, 41, 43, 44, 46	100.0	80.2	96.0	100.0	82.9	96.5	100.0	82.9	96.5	100.0	82.9	96.5	100.0	82.9	96.5	100.0	80.2	96.0
21, 22, 23	FIRE SUPPRESSION, PLUMBING & HVAC	100.3	68.7	87.6	97.1	69.0	85.8	99.9	69.0	87.5	100.0	69.0	87.5	97.1	69.0	85.8	97.1	68.7	85.6
26, 27, 3370	ELECTRICAL, COMMUNICATIONS & UTIL.	92.1	71.2	81.0	90.6	71.2	80.4	91.7	71.2	80.9	102.9	71.2	86.2	88.5	71.2	79.4	92.3	71.2	81.2
MF2014	WEIGHTED AVERAGE	96.1	69.4	84.4	98.0	72.6	86.9	100.0	72.6	88.0	100.8	72.6	88.5	97.6	72.6	86.7	97.5	69.4	85.2

DIVISION		NEW MEXICO			NEW YORK															
		TUCUMCARI			ALBANY			BINGHAMTON			BRONX			BROOKLYN			BUFFALO			
		884			120 - 122			137 - 139			104			112			140 - 142			
		MAT.	INST.	TOTAL	MAT.	INST.	TOTAL	MAT.	INST.	TOTAL	MAT.	INST.	TOTAL	MAT.	INST.	TOTAL	MAT.	INST.	TOTAL	
015433	CONTRACTOR EQUIPMENT		110.1	110.1		112.9	112.9		114.1	114.1		110.6	110.6		113.3	113.3		97.1	97.1	
0241, 31 - 34	SITE & INFRASTRUCTURE, DEMOLITION	93.8	103.3	100.5	83.8	106.3	99.8	95.9	94.1	94.6	108.8	120.8	117.3	120.3	127.6	125.5	98.3	98.2	98.2	
0310	Concrete Forming & Accessories	99.2	64.5	69.2	100.1	100.6	100.5	100.8	88.5	90.2	98.4	175.3	164.8	107.2	182.9	172.5	97.3	116.5	113.9	
0320	Concrete Reinforcing	106.0	68.9	87.1	104.0	103.8	103.9	93.7	97.8	95.8	103.9	184.4	144.9	95.1	205.9	151.5	97.8	102.1	100.0	
0330	Cast-in-Place Concrete	95.7	70.3	85.3	91.9	111.3	99.9	102.7	127.9	113.1	95.9	173.4	127.7	104.8	172.1	132.4	106.7	119.7	112.0	
03	CONCRETE	105.0	68.4	87.0	99.0	105.4	102.1	95.8	105.2	100.4	96.1	174.8	134.8	107.5	181.3	143.7	103.0	114.1	108.4	
04	MASONRY	114.6	58.7	79.8	101.0	112.6	108.2	109.7	127.8	121.0	92.7	177.5	145.5	120.2	177.4	155.9	105.1	121.0	115.0	
05	METALS	99.2	86.7	95.3	104.1	110.3	106.0	96.3	118.9	103.3	99.9	151.6	115.8	104.4	150.1	118.4	99.6	95.4	98.3	
06	WOOD, PLASTICS & COMPOSITES	93.1	65.5	77.6	98.7	97.9	98.3	104.9	84.7	93.6	94.4	174.9	139.5	106.9	185.1	150.6	99.4	116.4	108.9	
07	THERMAL & MOISTURE PROTECTION	99.1	73.5	88.6	105.9	105.8	105.9	107.3	103.6	105.8	108.5	163.2	131.0	108.4	164.1	131.2	101.8	110.6	105.4	
08	OPENINGS	98.5	67.7	91.3	102.6	94.0	100.6	93.0	85.0	91.1	87.9	175.5	108.3	90.4	180.0	111.2	97.7	103.9	99.1	
0920	Plaster & Gypsum Board	78.6	64.2	68.8	97.2	97.6	97.5	107.4	83.8	91.4	99.2	176.9	151.7	102.8	187.6	160.2	98.3	116.7	110.7	
0950, 0980	Ceilings & Acoustic Treatment	96.3	64.2	75.2	92.4	97.6	95.8	91.1	83.8	86.3	83.8	176.9	145.0	87.1	187.6	153.2	102.0	116.7	111.6	
0960	Flooring	101.2	66.0	91.2	97.1	113.7	101.9	106.8	103.6	105.9	98.3	186.8	123.6	113.9	186.8	134.8	101.3	121.2	107.0	
0970, 0990	Wall Finishes & Painting/Coating	103.3	66.6	81.1	103.8	94.3	98.0	93.3	98.8	96.6	102.9	157.4	135.8	123.4	157.4	143.9	98.8	112.4	107.0	
09	FINISHES	95.4	64.8	78.5	94.7	101.9	98.7	94.9	91.1	92.8	94.0	175.9	139.1	108.4	181.9	148.9	100.6	117.9	110.2	
COVERS	DIVS. 10 - 14, 25, 28, 41, 43, 44, 46	100.0	82.9	96.5	100.0	99.0	99.8	100.0	96.2	99.2	100.0	135.0	107.1	100.0	135.4	107.1	100.0	105.6	101.1	
21, 22, 23	FIRE SUPPRESSION, PLUMBING & HVAC	97.2	68.7	85.7	100.0	102.8	101.1	100.5	90.2	96.3	100.2	165.5	126.5	99.7	165.4	126.2	100.0	96.7	98.7	
26, 27, 3370	ELECTRICAL, COMMUNICATIONS & UTIL.	92.8	71.2	81.4	98.7	104.1	101.6	99.9	105.4	102.8	97.0	181.9	141.9	99.7	181.9	143.1	100.2	102.7	101.5	
MF2014	WEIGHTED AVERAGE	99.0	72.5	87.4	100.1	104.7	102.1	98.5	101.4	99.8	97.7	166.1	127.5	102.8	168.5	131.4	100.3	106.2	102.9	

DIVISION		NEW YORK																	
		ELMIRA			FAR ROCKAWAY			FLUSHING			GLENS FALLS			HICKSVILLE			JAMAICA		
		148 - 149			116			113			128			115,117,118			114		
		MAT.	INST.	TOTAL	MAT.	INST.	TOTAL	MAT.	INST.	TOTAL	MAT.	INST.	TOTAL	MAT.	INST.	TOTAL	MAT.	INST.	TOTAL
015433	CONTRACTOR EQUIPMENT		116.0	116.0		113.3	113.3		113.3	113.3		112.9	112.9		113.3	113.3		113.3	113.3
0241, 31 - 34	SITE & INFRASTRUCTURE, DEMOLITION	97.1	94.1	95.0	123.4	127.6	126.4	123.4	127.6	126.4	73.6	105.8	96.5	113.3	126.3	122.6	117.7	127.6	124.7
0310	Concrete Forming & Accessories	81.3	93.2	91.6	93.5	175.1	163.9	97.3	175.1	164.4	85.7	90.3	89.7	90.0	154.5	145.6	97.3	175.1	164.4
0320	Concrete Reinforcing	97.4	95.7	96.5	95.1	205.9	151.5	96.7	205.9	152.3	95.4	94.4	94.9	95.1	210.2	153.7	95.1	205.9	151.5
0330	Cast-in-Place Concrete	93.7	103.2	97.6	113.3	172.1	137.5	113.3	172.1	137.5	85.3	106.7	94.1	96.3	165.5	124.7	104.8	172.1	132.4
03	CONCRETE	90.8	98.6	94.6	113.6	177.8	145.1	114.1	177.8	145.4	88.3	97.5	92.9	99.4	167.1	132.6	106.8	177.8	141.7
04	MASONRY	103.5	102.1	102.6	124.1	177.4	157.4	118.1	177.4	155.1	100.7	105.6	103.7	114.6	166.8	147.1	122.2	177.4	156.6
05	METALS	96.7	117.3	103.0	104.4	150.1	118.4	104.4	150.1	118.4	97.7	106.1	100.3	105.8	149.1	119.2	104.4	150.1	118.4
06	WOOD, PLASTICS & COMPOSITES	85.0	92.2	89.0	90.1	174.6	137.4	94.8	174.6	139.5	87.1	87.6	87.3	86.6	152.6	123.6	94.8	174.6	139.5
07	THERMAL & MOISTURE PROTECTION	103.9	94.1	99.9	108.3	163.0	130.7	108.3	163.0	130.7	98.8	97.2	98.2	107.9	156.3	127.8	108.1	163.0	130.6
08	OPENINGS	99.4	88.9	97.0	89.0	174.4	108.8	89.0	174.4	108.8	93.3	86.2	91.7	89.0	163.4	106.3	89.0	174.4	108.8
0920	Plaster & Gypsum Board	97.8	91.7	93.7	91.6	176.9	149.2	94.1	176.9	150.0	89.1	87.0	87.7	91.2	154.2	133.8	94.1	176.9	150.0
0950, 0980	Ceilings & Acoustic Treatment	95.4	91.7	93.0	76.2	176.9	142.4	76.2	176.9	142.4	82.1	87.0	85.3	75.2	154.2	127.2	76.2	176.9	142.4
0960	Flooring	94.6	103.6	97.1	109.2	186.8	131.4	110.6	186.8	132.4	86.2	111.3	93.3	108.3	185.1	130.2	110.6	186.8	132.4
0970, 0990	Wall Finishes & Painting/Coating	98.9	90.0	93.5	123.4	157.4	143.9	123.4	157.4	143.9	101.1	87.4	92.8	123.4	157.4	143.9	123.4	157.4	143.9
09	FINISHES	95.5	94.7	95.0	103.5	175.7	143.3	104.3	175.7	143.6	86.1	93.1	90.0	102.1	159.7	133.8	103.8	175.7	143.4
COVERS	DIVS. 10 - 14, 25, 28, 41, 43, 44, 46	100.0	99.4	99.9	100.0	134.2	106.9	100.0	134.2	106.9	100.0	96.2	99.2	100.0	128.2	105.7	100.0	134.2	106.9
21, 22, 23	FIRE SUPPRESSION, PLUMBING & HVAC	95.5	91.5	93.9	95.0	165.3	123.4	95.0	165.3	123.4	95.5	95.7	95.6	99.7	154.4	121.8	95.0	165.3	123.4
26, 27, 3370	ELECTRICAL, COMMUNICATIONS & UTIL.	96.5	96.2	96.3	107.0	181.9	146.6	107.0	181.9	146.6	93.6	100.5	97.3	99.0	143.4	122.5	97.9	181.9	142.3
MF2014	WEIGHTED AVERAGE	96.7	97.4	97.0	102.6	166.8	130.6	102.5	166.8	130.5	94.0	98.6	96.0	100.9	153.3	123.7	100.8	166.8	129.6

DIVISION		NEW YORK																	
		JAMESTOWN			KINGSTON			LONG ISLAND CITY			MONTICELLO			MOUNT VERNON			NEW ROCHELLE		
		147			124			111			127			105			108		
		MAT.	INST.	TOTAL	MAT.	INST.	TOTAL	MAT.	INST.	TOTAL	MAT.	INST.	TOTAL	MAT.	INST.	TOTAL	MAT.	INST.	TOTAL
015433	CONTRACTOR EQUIPMENT		93.9	93.9		113.3	113.3		113.3	113.3		113.3	113.3		110.6	110.6		110.6	110.6
0241, 31 - 34	SITE & INFRASTRUCTURE, DEMOLITION	98.4	94.2	95.4	140.6	122.9	128.0	121.3	127.6	125.8	135.5	122.7	126.4	115.2	118.5	117.6	114.6	118.5	117.4
0310	Concrete Forming & Accessories	81.3	86.9	86.1	87.1	104.4	102.0	101.6	175.1	165.0	94.7	102.9	101.8	88.8	139.1	132.1	103.6	139.0	134.2
0320	Concrete Reinforcing	97.6	98.9	98.2	95.8	140.6	118.6	95.1	205.9	151.5	95.1	140.1	118.0	102.9	183.3	143.8	103.0	183.3	143.9
0330	Cast-in-Place Concrete	97.2	102.1	99.2	115.2	135.1	123.4	108.1	172.1	134.4	107.8	124.8	114.8	107.1	141.3	121.1	107.0	141.3	121.1
03	CONCRETE	93.8	94.3	94.0	110.8	121.5	116.0	109.9	177.8	143.3	105.5	117.2	111.2	105.4	146.9	125.8	105.0	146.9	125.6
04	MASONRY	111.8	99.9	104.4	116.0	141.3	131.8	116.9	177.4	154.6	108.6	128.7	121.1	98.1	147.3	128.8	98.1	147.3	128.8
05	METALS	94.0	92.1	93.5	106.1	117.2	109.5	104.4	150.1	118.4	106.0	116.2	109.2	99.6	135.6	110.7	99.9	135.6	110.9
06	WOOD, PLASTICS & COMPOSITES	83.5	84.0	83.8	88.4	96.4	92.9	100.9	174.6	142.2	96.1	96.4	96.3	84.6	136.2	113.5	101.4	136.2	120.9
07	THERMAL & MOISTURE PROTECTION	103.4	93.5	99.3	122.3	137.2	128.4	108.3	163.0	130.7	122.0	132.1	126.1	109.4	144.1	123.6	109.5	144.1	123.7
08	OPENINGS	99.2	85.6	96.1	95.9	117.3	100.9	89.0	174.4	108.8	91.1	117.3	97.2	87.9	148.6	102.0	88.0	148.6	102.1
0920	Plaster & Gypsum Board	88.5	83.2	84.9	91.4	96.3	94.7	98.7	176.9	151.5	92.1	96.3	94.9	94.9	137.0	123.3	106.3	137.0	127.0
0950, 0980	Ceilings & Acoustic Treatment	92.0	83.2	86.2	72.9	96.3	88.3	76.2	176.9	142.4	72.9	96.3	88.3	82.1	137.0	118.2	82.1	137.0	118.2
0960	Flooring	97.7	103.6	99.4	103.8	72.5	94.9	112.1	186.8	133.5	106.1	72.5	96.5	90.2	186.8	117.8	97.1	186.8	122.8
0970, 0990	Wall Finishes & Painting/Coating	100.2	94.2	96.6	134.4	116.8	123.8	123.4	157.4	143.9	134.4	116.8	123.8	101.1	157.4	135.1	101.1	157.4	135.1
09	FINISHES	94.6	89.9	92.0	101.0	97.1	98.8	105.1	175.7	144.0	101.4	96.3	98.6	91.2	149.2	123.2	94.6	149.2	124.7
COVERS	DIVS. 10 - 14, 25, 28, 41, 43, 44, 46	100.0	98.6	99.7	100.0	111.9	102.4	100.0	134.2	106.9	100.0	110.8	102.2	100.0	125.2	105.1	100.0	121.9	104.4
21, 22, 23	FIRE SUPPRESSION, PLUMBING & HVAC	95.4	87.2	92.1	95.4	122.3	106.3	99.7	165.3	126.2	95.4	116.7	104.0	95.5	134.1	111.1	95.5	134.1	111.1
26, 27, 3370	ELECTRICAL, COMMUNICATIONS & UTIL.	95.4	96.3	95.9	95.1	111.0	103.5	98.5	181.9	142.6	95.1	111.0	103.5	95.2	155.2	126.9	95.2	155.2	126.9
MF2014	WEIGHTED AVERAGE	96.8	92.5	94.9	102.6	118.5	109.6	102.4	166.8	130.5	101.1	115.1	107.2	97.5	141.4	116.6	97.9	141.3	116.8

	DIVISION	NEW YORK																	
		NEW YORK			NIAGARA FALLS			PLATTSBURGH			POUGHKEEPSIE			QUEENS			RIVERHEAD		
		100 - 102			143			129			125 - 126			110			119		
		MAT.	INST.	TOTAL	MAT.	INST.	TOTAL	MAT.	INST.	TOTAL	MAT.	INST.	TOTAL	MAT.	INST.	TOTAL	MAT.	INST.	TOTAL
015433	CONTRACTOR EQUIPMENT		111.1	111.1		93.9	93.9		98.9	98.9		113.3	113.3		113.3	113.3		113.3	113.3
0241, 31 - 34	SITE & INFRASTRUCTURE, DEMOLITION	117.3	121.5	120.2	100.5	95.5	97.0	106.1	103.0	103.9	136.5	123.5	127.2	116.6	127.6	124.4	114.5	125.9	122.6
0310	Concrete Forming & Accessories	102.6	183.3	172.2	81.3	114.4	109.8	91.1	95.4	94.9	87.1	165.0	154.3	90.1	175.1	163.4	94.6	153.4	145.3
0320	Concrete Reinforcing	109.9	210.4	161.0	96.3	102.5	99.5	99.9	102.7	101.3	95.8	141.0	118.8	96.7	205.9	152.3	96.9	184.2	141.3
0330	Cast-in-Place Concrete	107.8	177.7	136.5	100.6	120.3	108.7	104.4	104.4	104.4	111.5	137.9	122.4	99.6	172.1	129.4	97.9	164.5	125.3
03	CONCRETE	106.2	184.2	144.5	95.9	113.3	104.4	102.5	99.3	101.0	107.9	149.5	128.3	102.6	177.8	139.6	100.1	161.4	130.2
04	MASONRY	102.5	177.5	149.2	119.9	127.5	124.6	95.0	101.6	99.1	108.6	144.5	131.0	111.1	177.4	152.4	119.9	166.3	148.8
05	METALS	113.1	151.9	125.0	96.7	93.8	95.8	102.0	91.2	98.7	106.1	119.8	110.3	104.4	150.1	118.4	106.3	138.4	116.2
06	WOOD, PLASTICS & COMPOSITES	98.5	185.4	147.2	83.4	110.3	98.5	93.8	93.5	93.6	88.4	174.6	136.7	86.7	174.6	135.9	91.8	152.6	125.9
07	THERMAL & MOISTURE PROTECTION	108.6	164.5	131.5	103.5	112.9	107.4	116.3	100.5	109.8	122.3	146.8	132.3	107.9	163.0	130.5	108.8	156.1	128.2
08	OPENINGS	93.7	180.9	114.0	99.2	100.7	99.6	101.6	90.8	99.1	95.9	159.5	110.7	89.0	174.4	108.8	89.0	157.5	104.9
0920	Plaster & Gypsum Board	106.0	187.6	161.2	88.5	110.3	103.3	109.4	92.6	98.0	91.4	176.9	149.2	91.2	176.9	149.1	92.5	154.2	134.2
0950, 0980	Ceilings & Acoustic Treatment	102.2	187.6	158.4	92.0	110.3	104.1	98.1	92.6	94.5	72.9	176.9	141.2	76.2	176.9	142.4	76.1	154.2	127.5
0960	Flooring	99.5	186.8	124.5	97.7	121.2	104.4	108.9	113.7	110.3	103.8	167.2	122.0	108.3	186.8	130.7	109.2	142.7	118.8
0970, 0990	Wall Finishes & Painting/Coating	102.9	157.4	135.8	100.2	112.4	107.6	130.0	91.4	106.7	134.4	117.2	124.0	123.4	157.4	143.9	123.4	157.4	143.9
09	FINISHES	99.9	182.1	145.2	94.7	115.9	106.4	97.9	98.0	98.0	100.8	163.0	135.1	102.5	175.7	142.8	102.7	151.3	129.5
COVERS	DIVS. 10 - 14, 25, 28, 41, 43, 44, 46	100.0	130.1	107.3	100.0	107.1	101.4	100.0	98.0	99.6	100.0	121.6	104.4	100.0	134.2	106.9	100.0	127.6	105.6
21, 22, 23	FIRE SUPPRESSION, PLUMBING & HVAC	100.1	165.5	126.5	95.4	100.0	97.2	95.4	96.6	95.9	95.4	127.0	108.1	99.7	165.3	126.2	99.9	151.4	120.7
26, 27, 3370	ELECTRICAL, COMMUNICATIONS & UTIL.	104.7	181.9	145.5	94.0	100.3	97.3	91.6	90.3	90.9	95.1	119.9	108.2	99.0	181.9	142.8	100.6	133.6	118.1
MF2014	WEIGHTED AVERAGE	103.4	168.7	131.8	97.8	106.4	101.5	99.1	96.7	98.0	101.8	136.7	117.0	100.9	166.8	129.6	101.6	148.2	121.9

	DIVISION	NEW YORK																	
		ROCHESTER			SCHENECTADY			STATEN ISLAND			SUFFERN			SYRACUSE			UTICA		
		144 - 146			123			103			109			130 - 132			133 - 135		
		MAT.	INST.	TOTAL	MAT.	INST.	TOTAL	MAT.	INST.	TOTAL	MAT.	INST.	TOTAL	MAT.	INST.	TOTAL	MAT.	INST.	TOTAL
015433	CONTRACTOR EQUIPMENT		116.7	116.7		112.9	112.9		110.6	110.6		110.6	110.6		112.9	112.9		112.9	112.9
0241, 31 - 34	SITE & INFRASTRUCTURE, DEMOLITION	85.9	109.9	102.9	84.2	106.3	99.9	119.9	120.8	120.6	111.4	116.3	114.9	94.8	105.6	102.4	73.5	104.3	95.4
0310	Concrete Forming & Accessories	98.7	98.2	98.3	103.0	100.6	100.9	88.3	183.3	170.2	96.9	135.2	129.9	99.8	91.0	92.2	101.0	87.9	89.7
0320	Concrete Reinforcing	100.1	95.7	97.9	94.4	103.8	99.2	103.9	210.4	158.1	103.0	140.9	122.3	94.7	96.4	95.6	94.7	95.7	95.2
0330	Cast-in-Place Concrete	94.2	104.4	98.4	102.9	111.3	106.4	107.1	173.5	134.4	103.8	138.2	117.9	95.4	105.9	99.7	87.3	104.7	94.4
03	CONCRETE	99.3	100.6	99.9	102.9	105.4	104.1	107.1	182.7	144.2	102.0	136.3	118.9	98.9	97.8	98.4	96.9	95.9	96.4
04	MASONRY	107.0	104.9	105.7	97.7	112.6	107.0	104.7	177.5	150.1	97.6	142.6	125.7	101.5	105.2	103.8	93.1	103.9	99.8
05	METALS	104.1	107.3	105.1	101.9	110.3	104.5	97.8	151.8	114.4	97.8	119.4	104.5	99.8	105.6	101.6	97.7	105.1	100.0
06	WOOD, PLASTICS & COMPOSITES	97.6	97.6	97.6	107.4	97.9	102.1	83.3	185.4	140.4	94.0	136.2	117.6	101.3	88.5	94.2	101.3	84.7	92.0
07	THERMAL & MOISTURE PROTECTION	103.5	101.7	102.7	100.3	105.8	102.6	108.9	164.3	131.6	109.3	142.1	122.8	102.3	97.6	100.4	90.6	97.6	93.5
08	OPENINGS	105.3	92.2	102.2	99.4	94.0	98.1	87.9	181.2	109.6	88.0	139.0	99.8	95.0	85.6	92.8	98.0	83.4	94.6
0920	Plaster & Gypsum Board	107.2	97.4	100.6	98.8	97.6	98.0	95.0	187.6	157.6	98.5	137.0	124.5	98.0	88.0	91.2	98.0	84.0	88.5
0950, 0980	Ceilings & Acoustic Treatment	100.4	97.4	98.5	88.7	97.6	94.6	83.8	187.6	152.0	82.1	137.0	118.2	91.1	88.0	89.0	91.1	84.0	86.4
0960	Flooring	93.8	113.3	99.4	92.9	113.7	98.9	94.2	186.8	120.7	93.5	185.1	119.7	94.6	102.1	96.8	92.1	102.2	95.0
0970, 0990	Wall Finishes & Painting/Coating	100.3	99.1	99.6	101.1	94.3	97.0	102.9	157.4	135.8	101.1	124.9	115.5	98.5	99.8	99.3	91.3	99.8	96.4
09	FINISHES	100.4	101.3	100.9	91.6	101.9	97.3	93.2	182.1	142.2	92.3	140.0	118.6	94.0	93.5	93.7	92.1	91.1	91.6
COVERS	DIVS. 10 - 14, 25, 28, 41, 43, 44, 46	100.0	99.8	100.0	100.0	99.0	99.8	100.0	136.1	107.3	100.0	123.6	104.8	100.0	96.5	99.3	100.0	96.2	99.2
21, 22, 23	FIRE SUPPRESSION, PLUMBING & HVAC	100.1	90.3	96.1	100.2	102.8	101.2	100.2	165.5	126.5	95.5	123.3	106.7	100.2	91.9	96.9	100.2	92.2	97.0
26, 27, 3370	ELECTRICAL, COMMUNICATIONS & UTIL.	98.7	94.6	96.6	98.1	104.1	101.3	97.0	181.9	141.9	102.8	119.9	111.8	100.0	102.2	101.1	98.1	102.2	100.3
MF2014	WEIGHTED AVERAGE	101.1	99.1	100.2	99.3	104.7	101.7	99.2	168.4	129.4	97.6	129.0	111.3	98.9	98.0	98.5	97.1	97.1	97.1

	DIVISION	NEW YORK									NORTH CAROLINA								
		WATERTOWN			WHITE PLAINS			YONKERS			ASHEVILLE			CHARLOTTE			DURHAM		
		136			106			107			287 - 288			281 - 282			277		
		MAT.	INST.	TOTAL	MAT.	INST.	TOTAL	MAT.	INST.	TOTAL	MAT.	INST.	TOTAL	MAT.	INST.	TOTAL	MAT.	INST.	TOTAL
015433	CONTRACTOR EQUIPMENT		112.9	112.9		110.6	110.6		110.6	110.6		96.3	96.3		96.3	96.3		101.7	101.7
0241, 31 - 34	SITE & INFRASTRUCTURE, DEMOLITION	80.9	105.7	98.5	108.5	118.5	115.6	116.7	118.5	118.0	102.4	76.9	84.3	105.6	76.9	85.2	100.4	85.7	89.9
0310	Concrete Forming & Accessories	85.9	94.3	93.1	101.9	139.1	134.0	102.2	143.9	138.2	96.0	41.3	48.8	99.7	42.8	50.6	100.2	44.3	52.0
0320	Concrete Reinforcing	95.3	96.5	95.9	103.0	183.3	143.9	107.1	183.3	145.9	93.0	63.0	77.7	98.8	58.5	78.3	91.2	57.9	74.2
0330	Cast-in-Place Concrete	101.5	107.6	104.1	95.1	141.3	114.1	106.3	141.4	120.7	115.4	51.3	89.1	120.2	49.8	91.3	101.6	47.1	79.2
03	CONCRETE	109.3	99.9	104.7	95.7	146.9	120.9	105.2	149.1	126.7	106.0	50.6	78.8	108.6	49.9	79.8	98.7	49.5	74.5
04	MASONRY	94.2	107.9	102.8	97.1	147.3	128.4	102.1	147.3	130.3	93.6	43.7	62.5	101.9	51.2	70.3	86.4	37.9	56.2
05	METALS	97.8	105.6	100.2	99.3	135.6	110.5	108.9	135.7	117.1	103.1	82.3	96.7	104.0	80.9	96.9	121.4	79.9	108.6
06	WOOD, PLASTICS & COMPOSITES	83.0	92.2	88.1	99.5	136.2	120.0	99.3	142.7	123.6	97.9	40.3	65.6	103.0	41.9	68.8	96.7	44.7	67.6
07	THERMAL & MOISTURE PROTECTION	90.9	100.1	94.7	109.2	144.1	123.5	109.5	144.8	124.0	106.7	43.7	80.8	100.8	46.6	78.6	106.9	45.7	81.8
08	OPENINGS	98.0	89.4	96.0	88.0	148.6	102.1	91.1	151.7	105.2	97.1	44.1	84.8	102.5	45.0	89.1	105.8	47.1	92.2
0920	Plaster & Gypsum Board	88.7	91.7	90.8	101.3	137.0	125.4	105.6	143.6	131.3	100.1	38.3	58.3	100.0	39.9	59.4	105.9	42.8	63.3
0950, 0980	Ceilings & Acoustic Treatment	91.1	91.7	91.5	82.1	137.0	118.2	100.5	143.6	128.9	85.7	38.3	54.5	88.8	39.9	56.7	88.1	42.8	58.3
0960	Flooring	86.0	102.2	90.6	95.6	186.8	121.7	95.2	186.8	121.4	102.2	42.7	85.2	101.6	43.4	84.9	103.9	42.7	86.4
0970, 0990	Wall Finishes & Painting/Coating	91.3	96.2	94.3	101.1	157.4	135.1	101.1	157.4	135.1	113.6	40.3	69.3	113.8	49.4	74.9	105.4	37.4	64.3
09	FINISHES	89.7	95.7	93.0	92.9	149.2	123.9	98.0	153.0	128.3	95.6	40.7	65.3	95.5	42.9	66.5	96.0	42.9	66.7
COVERS	DIVS. 10 - 14, 25, 28, 41, 43, 44, 46	100.0	97.5	99.5	100.0	125.2	105.1	100.0	126.3	105.3	100.0	77.7	95.5	100.0	78.1	95.6	100.0	72.1	94.4
21, 22, 23	FIRE SUPPRESSION, PLUMBING & HVAC	100.2	85.9	94.4	100.3	134.1	114.0	100.3	134.1	114.0	100.4	53.4	81.4	100.0	54.1	81.5	100.5	52.9	81.3
26, 27, 3370	ELECTRICAL, COMMUNICATIONS & UTIL.	100.0	90.4	94.9	95.2	155.2	126.9	102.9	163.8	135.1	102.2	55.7	77.6	101.2	58.5	78.6	96.1	55.6	74.7
MF2014	WEIGHTED AVERAGE	98.5	96.3	97.5	97.6	141.4	116.7	102.1	143.7	120.2	100.8	55.3	80.9	101.9	56.8	82.2	102.9	55.3	82.1

DIVISION		NORTH CAROLINA																	
		ELIZABETH CITY			FAYETTEVILLE			GASTONIA			GREENSBORO			HICKORY			KINSTON		
		279			283			280			270,272 - 274			286			285		
		MAT.	INST.	TOTAL	MAT.	INST.	TOTAL	MAT.	INST.	TOTAL	MAT.	INST.	TOTAL	MAT.	INST.	TOTAL	MAT.	INST.	TOTAL
015433	CONTRACTOR EQUIPMENT		106.0	106.0		101.7	101.7		96.3	96.3		101.7	101.7		101.7	101.7		101.7	101.7
0241, 31 - 34	SITE & INFRASTRUCTURE, DEMOLITION	104.6	87.4	92.4	101.7	85.6	90.3	102.3	76.9	84.2	100.3	85.7	89.9	101.2	85.5	90.1	100.3	85.6	89.8
0310	Concrete Forming & Accessories	85.3	42.5	48.4	95.6	60.3	65.2	103.1	38.7	47.5	99.9	44.4	52.1	92.0	37.1	44.6	88.2	42.1	48.4
0320	Concrete Reinforcing	89.2	45.9	67.2	96.8	58.0	77.0	93.5	56.5	74.7	90.1	58.0	73.8	93.0	56.1	74.2	92.5	45.8	68.7
0330	Cast-in-Place Concrete	101.8	47.0	79.3	121.0	48.3	91.2	112.8	52.7	88.1	100.8	47.6	79.0	115.4	48.1	87.7	111.4	44.8	84.0
03	CONCRETE	98.7	46.4	73.0	108.1	57.1	83.1	104.5	48.7	77.1	98.1	49.8	74.4	105.7	46.3	76.5	102.4	45.5	74.4
04	MASONRY	98.6	48.0	67.1	97.3	38.9	60.9	98.3	50.9	68.7	82.5	41.4	56.9	82.3	43.7	58.2	89.1	48.1	63.6
05	METALS	107.1	75.8	97.5	124.2	80.0	110.6	103.9	80.0	96.5	113.7	80.0	103.4	103.2	78.7	95.7	102.0	74.8	93.6
06	WOOD, PLASTICS & COMPOSITES	80.1	43.6	59.7	97.1	66.8	80.1	107.0	36.6	67.6	96.4	44.9	67.6	92.1	35.2	60.2	88.6	42.9	63.0
07	THERMAL & MOISTURE PROTECTION	106.2	43.3	80.4	106.3	45.9	81.5	106.9	45.2	81.6	106.7	42.8	80.5	107.0	41.7	80.2	106.8	42.8	80.6
08	OPENINGS	102.6	38.4	87.7	97.2	59.1	88.4	100.9	41.0	87.0	105.8	47.2	92.2	97.2	36.8	83.1	97.3	43.1	84.7
0920	Plaster & Gypsum Board	98.6	41.0	59.7	104.7	65.6	78.3	107.0	34.5	58.0	107.5	43.0	63.9	100.1	33.1	54.8	99.8	41.0	60.1
0950, 0980	Ceilings & Acoustic Treatment	88.1	41.0	57.1	86.5	65.6	72.7	89.0	34.5	53.2	88.1	43.0	58.4	85.7	33.1	51.1	89.0	41.0	57.4
0960	Flooring	95.7	23.2	75.0	102.4	42.7	85.3	105.4	42.7	87.4	103.9	39.7	85.6	102.1	32.8	82.3	99.6	22.6	77.6
0970, 0990	Wall Finishes & Painting/Coating	105.4	42.0	67.1	113.6	33.2	65.0	113.6	40.3	69.3	105.4	31.6	60.9	113.6	40.3	69.3	113.6	39.2	68.7
09	FINISHES	93.0	38.7	63.0	96.5	55.3	73.8	98.1	38.6	65.3	96.3	41.9	66.3	95.8	35.4	62.5	95.5	38.0	63.8
COVERS	DIVS. 10 - 14, 25, 28, 41, 43, 44, 46	100.0	79.4	95.8	100.0	74.3	94.8	100.0	77.3	95.4	100.0	75.9	95.1	100.0	77.1	95.4	100.0	71.6	94.3
21, 22, 23	FIRE SUPPRESSION, PLUMBING & HVAC	95.6	51.1	77.7	100.2	52.7	81.0	100.4	52.4	81.0	100.4	53.1	81.3	95.6	52.2	78.1	95.6	51.2	77.7
26, 27, 3370	ELECTRICAL, COMMUNICATIONS & UTIL.	95.9	34.5	63.4	101.4	50.7	74.6	101.6	57.4	78.2	95.2	55.7	74.3	99.6	57.4	77.3	99.4	46.1	71.2
MF2014	WEIGHTED AVERAGE	99.3	51.6	78.5	104.5	58.1	84.2	101.6	55.1	81.3	101.3	55.7	81.4	98.8	53.9	79.2	98.4	52.7	78.5

DIVISION		NORTH CAROLINA															NORTH DAKOTA		
		MURPHY			RALEIGH			ROCKY MOUNT			WILMINGTON			WINSTON-SALEM			BISMARCK		
		289			275 - 276			278			284			271			585		
		MAT.	INST.	TOTAL	MAT.	INST.	TOTAL	MAT.	INST.	TOTAL	MAT.	INST.	TOTAL	MAT.	INST.	TOTAL	MAT.	INST.	TOTAL
015433	CONTRACTOR EQUIPMENT		96.3	96.3		101.7	101.7		101.7	101.7		96.3	96.3		101.7	101.7		98.2	98.2
0241, 31 - 34	SITE & INFRASTRUCTURE, DEMOLITION	103.5	76.7	84.5	101.3	85.7	90.2	102.6	85.7	90.6	103.7	77.1	84.8	100.6	85.7	90.0	101.4	96.9	98.2
0310	Concrete Forming & Accessories	103.7	39.3	48.2	98.9	46.6	53.8	91.9	44.2	50.8	97.5	48.8	55.5	101.9	48.0	55.4	104.9	40.8	49.6
0320	Concrete Reinforcing	92.6	45.4	68.6	96.2	56.0	75.7	89.2	55.8	72.2	93.7	57.9	75.5	90.1	57.9	73.7	99.9	92.7	96.2
0330	Cast-in-Place Concrete	119.4	44.4	88.6	106.7	51.4	84.0	99.6	47.7	78.3	115.0	49.5	88.1	103.4	50.0	81.4	102.6	47.5	79.9
03	CONCRETE	109.2	44.1	77.2	101.7	51.7	77.1	99.5	49.3	74.9	105.9	52.4	79.6	99.4	52.2	76.2	101.5	54.2	78.3
04	MASONRY	85.3	41.2	57.8	86.6	42.2	58.9	76.7	39.5	53.5	82.8	41.7	57.2	82.7	38.6	55.2	114.6	56.4	78.3
05	METALS	100.9	74.2	92.7	105.7	79.2	97.5	106.3	78.4	97.7	102.7	79.9	95.7	110.8	79.9	101.3	103.4	88.4	98.8
06	WOOD, PLASTICS & COMPOSITES	107.8	39.7	69.7	95.1	47.5	68.5	87.4	44.7	63.5	100.1	49.0	71.5	96.4	49.4	70.1	95.3	34.7	61.4
07	THERMAL & MOISTURE PROTECTION	106.9	40.8	79.8	100.5	47.0	78.5	106.6	40.8	79.7	106.7	46.9	82.2	106.7	43.5	80.8	112.4	51.3	87.4
08	OPENINGS	97.1	38.9	83.5	104.8	47.4	91.4	101.8	41.9	87.9	97.3	51.0	86.5	105.8	49.7	92.8	108.0	50.1	94.5
0920	Plaster & Gypsum Board	106.1	37.6	59.8	99.8	45.7	63.3	100.5	42.8	61.5	102.4	47.2	65.1	107.5	47.6	67.0	98.8	33.0	54.3
0950, 0980	Ceilings & Acoustic Treatment	85.7	37.6	54.1	88.8	45.7	60.5	85.6	42.8	57.5	86.5	47.2	60.7	88.1	47.6	61.5	112.9	33.0	60.4
0960	Flooring	105.7	24.1	82.4	101.2	42.7	84.5	99.4	21.8	77.2	102.9	44.4	86.2	103.9	42.7	86.4	99.4	72.1	91.6
0970, 0990	Wall Finishes & Painting/Coating	113.6	39.4	68.8	104.2	36.9	63.5	105.4	38.5	65.0	113.6	38.7	68.4	105.4	36.7	64.0	101.9	29.4	58.1
09	FINISHES	97.7	36.1	63.7	95.5	44.7	67.5	93.9	39.3	63.8	96.3	46.8	69.0	96.3	45.7	68.4	102.7	42.8	69.7
COVERS	DIVS. 10 - 14, 25, 28, 41, 43, 44, 46	100.0	76.9	95.3	100.0	72.6	94.5	100.0	72.5	94.5	100.0	73.8	94.7	100.0	78.7	95.7	100.0	82.5	96.5
21, 22, 23	FIRE SUPPRESSION, PLUMBING & HVAC	95.6	50.9	77.6	100.0	52.5	80.8	95.6	52.7	78.3	100.4	54.9	82.0	100.4	53.3	81.4	100.1	71.9	88.7
26, 27, 3370	ELECTRICAL, COMMUNICATIONS & UTIL.	103.2	29.3	64.1	98.3	40.4	67.7	97.9	39.7	67.1	102.4	50.7	75.1	95.2	55.7	74.3	102.5	71.6	86.2
MF2014	WEIGHTED AVERAGE	99.6	48.3	77.2	100.5	54.1	80.3	98.5	52.3	78.3	100.3	55.9	81.0	101.0	56.5	81.6	103.1	66.2	87.0

DIVISION		NORTH DAKOTA																	
		DEVILS LAKE			DICKINSON			FARGO			GRAND FORKS			JAMESTOWN			MINOT		
		583			586			580 - 581			582			584			587		
		MAT.	INST.	TOTAL	MAT.	INST.	TOTAL	MAT.	INST.	TOTAL	MAT.	INST.	TOTAL	MAT.	INST.	TOTAL	MAT.	INST.	TOTAL
015433	CONTRACTOR EQUIPMENT		98.2	98.2		98.2	98.2		98.2	98.2		98.2	98.2		98.2	98.2		98.2	98.2
0241, 31 - 34	SITE & INFRASTRUCTURE, DEMOLITION	105.7	94.4	97.7	113.6	92.9	98.9	102.6	96.9	98.5	109.6	92.9	97.7	104.7	92.9	96.3	107.1	96.9	99.8
0310	Concrete Forming & Accessories	101.2	35.4	44.4	90.8	34.9	42.5	99.3	41.5	49.4	94.5	34.2	42.5	92.3	34.1	42.1	90.5	67.0	70.2
0320	Concrete Reinforcing	100.2	93.1	96.6	101.1	92.8	96.9	96.9	92.8	94.8	98.7	92.9	95.7	100.8	91.3	95.9	102.1	93.2	97.6
0330	Cast-in-Place Concrete	126.3	46.0	93.3	114.3	44.4	85.6	113.6	49.2	87.2	114.3	44.1	85.5	124.7	44.1	91.6	114.3	46.5	86.4
03	CONCRETE	110.6	51.3	81.5	109.6	50.2	80.4	105.9	55.1	80.9	106.8	49.8	78.8	109.1	48.7	79.5	105.5	65.6	85.9
04	MASONRY	121.7	64.9	86.3	124.3	60.4	84.5	114.2	56.4	78.2	115.9	64.4	83.8	135.1	32.8	71.3	114.2	65.4	83.8
05	METALS	99.9	87.4	96.0	99.8	82.5	94.5	103.0	88.7	98.6	99.8	82.1	94.4	99.8	63.7	88.7	100.1	89.3	96.8
06	WOOD, PLASTICS & COMPOSITES	95.2	32.0	59.8	83.1	32.0	54.5	92.9	35.2	60.6	87.4	32.0	56.4	85.0	32.0	55.3	82.8	70.1	75.7
07	THERMAL & MOISTURE PROTECTION	107.6	49.8	83.9	108.1	48.5	83.7	107.4	51.6	84.5	107.8	49.8	84.0	107.4	41.5	80.4	107.5	57.4	87.0
08	OPENINGS	100.7	42.7	87.2	100.7	42.7	87.2	100.5	50.3	88.9	100.7	42.7	87.2	100.7	32.2	84.8	100.9	69.3	93.5
0920	Plaster & Gypsum Board	118.4	30.2	58.8	109.1	30.2	55.8	96.6	33.5	54.0	110.5	30.2	56.2	110.2	30.2	56.1	109.1	69.4	82.3
0950, 0980	Ceilings & Acoustic Treatment	114.3	30.2	59.0	114.3	30.2	59.0	112.1	33.5	60.5	114.3	30.2	59.0	114.3	30.2	59.0	114.3	69.4	84.8
0960	Flooring	107.9	35.9	87.3	101.4	35.9	82.7	102.8	72.1	94.0	103.3	35.9	84.0	102.1	35.9	83.2	101.1	89.3	97.8
0970, 0990	Wall Finishes & Painting/Coating	104.4	22.5	54.9	104.4	31.5	60.4	101.2	70.7	82.8	104.4	28.2	58.4	104.4	22.5	54.9	104.4	27.9	58.2
09	FINISHES	108.2	32.2	66.3	105.9	33.2	65.8	103.8	47.7	72.9	106.1	32.8	65.7	105.3	32.2	65.0	105.0	67.0	84.1
COVERS	DIVS. 10 - 14, 25, 28, 41, 43, 44, 46	100.0	32.4	86.3	100.0	32.6	86.4	100.0	82.6	96.5	100.0	32.5	86.4	100.0	80.6	96.1	100.0	86.4	97.3
21, 22, 23	FIRE SUPPRESSION, PLUMBING & HVAC	95.6	74.0	86.9	95.6	66.2	83.7	100.1	77.0	90.8	100.4	33.3	73.3	95.6	35.0	71.2	100.4	60.2	84.2
26, 27, 3370	ELECTRICAL, COMMUNICATIONS & UTIL.	98.3	35.8	65.3	107.7	72.5	89.1	102.4	68.4	84.4	102.1	53.5	76.4	98.3	35.7	65.2	105.5	75.7	89.8
MF2014	WEIGHTED AVERAGE	102.0	58.6	83.1	102.8	60.9	84.6	102.7	67.6	87.4	102.7	51.6	80.4	102.1	45.2	77.3	102.7	71.3	89.0

DIVISION		NORTH DAKOTA			OHIO														
		WILLISTON			AKRON			ATHENS			CANTON			CHILLICOTHE			CINCINNATI		
		588			442 - 443			457			446 - 447			456			451 - 452		
		MAT.	INST.	TOTAL	MAT.	INST.	TOTAL	MAT.	INST.	TOTAL	MAT.	INST.	TOTAL	MAT.	INST.	TOTAL	MAT.	INST.	TOTAL
015433	**CONTRACTOR EQUIPMENT**		98.2	98.2		94.4	94.4		90.4	90.4		94.4	94.4		99.7	99.7		99.5	99.5
0241, 31 - 34	**SITE & INFRASTRUCTURE, DEMOLITION**	107.5	92.9	97.1	98.1	101.6	100.6	111.0	91.6	97.2	98.2	101.0	100.2	97.0	102.6	101.0	94.1	102.4	100.0
0310	Concrete Forming & Accessories	96.3	34.9	43.3	99.0	94.3	95.0	95.0	84.3	85.8	99.0	83.7	85.8	97.4	92.3	93.0	99.3	79.2	82.0
0320	Concrete Reinforcing	103.1	92.8	97.9	99.5	92.1	95.7	93.1	87.0	90.0	99.5	75.8	87.5	90.1	80.4	85.1	95.5	79.2	87.2
0330	Cast-in-Place Concrete	114.3	44.4	85.6	94.0	98.0	95.7	111.5	96.5	105.4	94.9	95.0	95.0	101.2	99.3	100.4	93.2	92.4	92.9
03	**CONCRETE**	106.9	50.2	79.1	97.3	94.4	95.9	107.6	88.6	98.3	97.7	85.7	91.8	101.9	92.4	97.2	96.3	84.0	90.3
04	**MASONRY**	108.8	60.4	78.7	93.3	96.2	95.1	86.4	90.4	88.9	94.0	84.3	87.9	94.0	100.5	98.1	93.7	85.0	88.2
05	**METALS**	100.0	82.5	94.6	95.4	81.6	91.1	103.0	80.0	95.9	95.4	74.5	88.9	94.9	85.7	92.1	97.2	84.6	93.3
06	**WOOD, PLASTICS & COMPOSITES**	88.8	32.0	57.0	99.1	93.6	96.0	85.7	84.9	85.3	99.5	82.8	90.2	98.4	89.6	93.5	100.9	76.8	87.4
07	**THERMAL & MOISTURE PROTECTION**	107.7	48.5	83.5	110.8	96.5	104.9	99.4	94.8	97.5	112.0	91.4	103.5	101.2	97.4	99.7	99.2	88.7	94.9
08	**OPENINGS**	100.8	42.7	87.3	110.7	93.3	106.7	101.7	82.3	97.2	104.3	77.9	98.2	93.8	83.5	91.4	102.0	77.4	96.3
0920	Plaster & Gypsum Board	110.5	30.2	56.2	96.9	93.2	94.4	91.4	84.1	86.5	98.0	82.1	87.2	93.3	89.5	90.7	94.9	76.3	82.3
0950, 0980	Ceilings & Acoustic Treatment	114.3	30.2	59.0	93.0	93.2	93.1	103.3	84.1	90.7	93.0	82.1	85.8	97.9	89.5	92.4	98.7	76.3	84.0
0960	Flooring	104.1	35.9	84.6	96.9	93.4	95.9	121.3	99.0	114.9	97.1	81.7	92.7	98.3	97.7	98.1	99.3	90.7	96.8
0970, 0990	Wall Finishes & Painting/Coating	104.4	31.5	60.4	96.3	106.3	102.3	100.5	99.1	99.6	96.3	83.2	88.4	97.8	92.9	94.9	97.8	83.9	89.4
09	**FINISHES**	106.2	33.2	66.0	98.0	95.3	96.5	101.8	89.0	94.8	98.2	82.9	89.8	98.8	93.3	95.8	99.2	81.1	89.3
COVERS	**DIVS. 10 - 14, 25, 28, 41, 43, 44, 46**	100.0	32.6	80.4	100.0	98.2	99.6	100.0	52.0	90.3	100.0	95.4	99.1	100.0	92.6	98.5	100.0	89.3	97.8
21, 22, 23	**FIRE SUPPRESSION, PLUMBING & HVAC**	95.6	66.2	83.7	100.1	94.6	97.9	95.3	51.3	77.6	100.1	81.4	92.5	95.8	95.2	95.6	100.0	83.3	93.3
26, 27, 3370	**ELECTRICAL, COMMUNICATIONS & UTIL.**	102.5	72.5	86.6	99.3	93.9	96.5	95.5	98.1	96.9	98.5	92.7	95.5	95.0	85.6	90.0	93.9	79.4	86.3
MF2014	**WEIGHTED AVERAGE**	101.2	60.9	83.7	99.9	94.2	97.4	99.6	80.4	91.2	99.3	85.6	93.3	96.8	93.0	95.1	98.3	84.5	92.2

DIVISION		OHIO																	
		CLEVELAND			COLUMBUS			DAYTON			HAMILTON			LIMA			LORAIN		
		441			430 - 432			453 - 454			450			458			440		
		MAT.	INST.	TOTAL	MAT.	INST.	TOTAL	MAT.	INST.	TOTAL	MAT.	INST.	TOTAL	MAT.	INST.	TOTAL	MAT.	INST.	TOTAL
015433	**CONTRACTOR EQUIPMENT**		94.7	94.7		93.6	93.6		94.7	94.7		99.7	99.7		92.9	92.9		94.4	94.4
0241, 31 - 34	**SITE & INFRASTRUCTURE, DEMOLITION**	98.0	102.2	101.0	95.5	98.3	97.5	92.9	101.8	99.2	92.8	102.1	99.4	104.6	91.6	95.4	97.5	102.9	101.4
0310	Concrete Forming & Accessories	99.1	99.6	99.5	100.3	84.5	86.7	99.3	79.0	81.8	99.4	79.4	82.1	95.0	86.9	88.1	99.1	86.8	88.5
0320	Concrete Reinforcing	100.0	92.5	96.2	102.7	82.6	92.5	95.5	81.3	88.3	95.5	79.2	87.2	93.1	81.5	87.2	99.5	92.4	95.9
0330	Cast-in-Place Concrete	92.2	106.5	98.1	94.1	93.7	93.9	86.7	86.4	86.6	92.9	92.7	92.8	102.4	97.5	100.4	89.5	104.1	95.5
03	**CONCRETE**	96.5	99.8	98.2	97.9	87.1	92.6	93.3	81.8	87.6	96.2	84.2	90.3	100.4	89.3	94.9	95.2	93.2	94.2
04	**MASONRY**	97.8	105.6	102.7	96.6	94.0	95.0	93.2	83.5	87.1	93.5	85.5	88.5	117.7	86.4	98.2	89.9	104.6	99.1
05	**METALS**	96.9	84.4	93.1	96.9	80.7	91.9	96.4	78.6	91.0	96.5	84.5	92.8	103.0	81.5	96.4	96.0	82.8	91.9
06	**WOOD, PLASTICS & COMPOSITES**	98.2	97.1	97.6	97.1	82.3	88.8	102.2	77.1	88.1	100.9	76.8	87.4	85.6	86.3	86.0	99.1	80.9	88.9
07	**THERMAL & MOISTURE PROTECTION**	109.5	108.6	109.1	100.9	94.9	98.4	104.7	87.6	97.7	101.4	88.8	96.2	99.0	95.0	97.3	111.9	103.0	108.2
08	**OPENINGS**	100.6	95.2	99.3	102.4	80.0	97.2	101.3	78.1	95.9	98.9	77.4	93.9	101.7	79.8	96.6	104.3	86.4	100.2
0920	Plaster & Gypsum Board	96.2	96.8	96.6	93.5	81.8	85.6	94.9	76.6	82.6	94.9	76.3	82.3	91.4	85.5	87.4	96.9	80.1	85.5
0950, 0980	Ceilings & Acoustic Treatment	91.3	96.8	94.9	97.7	81.8	87.3	99.7	76.6	84.5	98.7	76.3	84.0	102.4	85.5	91.3	93.0	80.1	84.5
0960	Flooring	96.7	105.0	99.1	92.0	90.3	91.5	102.0	80.7	95.9	99.3	90.7	96.8	120.3	92.2	112.3	97.1	105.0	99.3
0970, 0990	Wall Finishes & Painting/Coating	96.3	105.1	101.6	95.5	92.9	94.0	97.8	85.0	90.1	97.8	84.7	89.9	100.5	82.7	89.8	96.3	105.1	101.6
09	**FINISHES**	97.5	100.9	99.4	94.1	85.9	89.6	100.2	79.3	88.7	99.1	81.4	89.3	100.8	87.2	93.3	98.0	91.1	94.2
COVERS	**DIVS. 10 - 14, 25, 28, 41, 43, 44, 46**	100.0	102.5	100.5	100.0	93.3	98.6	100.0	89.1	97.8	100.0	89.5	97.9	100.0	95.2	99.0	100.0	100.3	100.1
21, 22, 23	**FIRE SUPPRESSION, PLUMBING & HVAC**	100.0	100.5	100.2	100.1	92.0	96.8	100.9	86.6	95.2	100.6	83.6	93.7	95.3	89.0	92.8	100.1	92.1	96.9
26, 27, 3370	**ELECTRICAL, COMMUNICATIONS & UTIL.**	98.8	105.0	102.1	97.6	87.8	92.4	92.6	83.0	87.5	93.0	83.3	87.9	95.8	80.3	87.6	98.7	88.0	93.0
MF2014	**WEIGHTED AVERAGE**	99.0	100.3	99.6	98.6	89.2	94.5	98.0	84.4	92.1	97.9	85.1	92.3	100.0	86.8	94.3	98.9	93.0	96.3

DIVISION		OHIO																	
		MANSFIELD			MARION			SPRINGFIELD			STEUBENVILLE			TOLEDO			YOUNGSTOWN		
		448 - 449			433			455			439			434 - 436			444 - 445		
		MAT.	INST.	TOTAL	MAT.	INST.	TOTAL	MAT.	INST.	TOTAL	MAT.	INST.	TOTAL	MAT.	INST.	TOTAL	MAT.	INST.	TOTAL
015433	**CONTRACTOR EQUIPMENT**		94.4	94.4		93.2	93.2		94.7	94.7		98.1	98.1		96.0	96.0		94.4	94.4
0241, 31 - 34	**SITE & INFRASTRUCTURE, DEMOLITION**	93.9	101.5	99.3	91.8	97.1	95.6	93.2	100.5	98.4	133.1	106.9	114.5	94.8	99.0	97.8	98.0	102.0	100.8
0310	Concrete Forming & Accessories	89.1	83.7	84.5	96.7	81.2	83.3	99.3	83.5	85.7	98.3	89.4	90.7	100.3	97.1	97.6	99.0	87.5	89.1
0320	Concrete Reinforcing	90.6	76.4	83.4	94.7	82.5	88.5	95.5	81.3	88.3	92.3	85.5	88.8	102.7	85.4	93.9	99.5	85.6	92.4
0330	Cast-in-Place Concrete	87.1	94.7	90.2	85.9	92.7	88.7	89.1	86.3	87.9	93.3	94.1	93.6	94.1	100.0	96.5	93.1	96.9	94.7
03	**CONCRETE**	89.7	85.7	87.7	89.9	85.1	87.6	94.4	83.8	89.2	93.8	89.7	91.8	97.9	95.5	96.7	96.9	89.8	93.4
04	**MASONRY**	92.4	97.0	95.3	98.8	95.5	96.7	93.4	83.5	87.2	86.2	94.2	91.2	104.8	99.9	101.7	93.6	93.6	93.6
05	**METALS**	96.2	75.5	89.8	96.0	78.7	90.6	96.4	78.4	90.9	92.5	79.5	88.5	96.7	85.5	93.3	95.4	78.7	90.3
06	**WOOD, PLASTICS & COMPOSITES**	86.7	80.9	83.4	92.8	79.8	85.5	103.6	83.6	92.4	88.4	88.3	88.3	97.1	96.9	97.0	99.1	86.1	91.8
07	**THERMAL & MOISTURE PROTECTION**	111.3	94.4	104.4	100.5	85.2	94.2	104.6	88.2	97.9	112.8	96.4	106.1	102.5	102.7	102.6	112.1	94.3	104.8
08	**OPENINGS**	104.9	76.0	98.2	96.6	76.5	91.9	99.3	79.0	94.6	97.0	83.7	94.0	99.7	91.3	97.7	104.3	85.1	99.8
0920	Plaster & Gypsum Board	90.7	80.1	83.5	91.4	79.3	83.2	94.9	83.3	87.1	90.7	87.5	88.6	93.5	96.9	95.8	96.9	85.5	89.2
0950, 0980	Ceilings & Acoustic Treatment	93.9	80.1	84.8	97.7	79.3	85.6	99.7	83.3	88.9	94.8	87.5	90.0	97.7	96.9	97.2	93.0	85.5	88.1
0960	Flooring	92.5	107.5	96.8	90.9	107.5	95.6	102.0	80.7	95.9	118.6	100.4	113.4	91.2	99.7	93.6	97.1	93.9	96.2
0970, 0990	Wall Finishes & Painting/Coating	96.3	88.5	91.6	95.5	47.5	66.5	97.8	85.0	90.1	107.8	103.1	104.9	95.6	102.0	99.4	96.3	92.6	94.1
09	**FINISHES**	95.8	88.2	91.6	93.1	82.6	87.3	100.2	83.2	90.8	110.0	92.6	100.4	93.8	98.1	96.2	98.1	88.8	93.0
COVERS	**DIVS. 10 - 14, 25, 28, 41, 43, 44, 46**	100.0	96.2	99.2	100.0	54.4	90.8	100.0	89.8	97.9	100.0	95.0	99.0	100.0	98.7	99.7	100.0	96.4	99.3
21, 22, 23	**FIRE SUPPRESSION, PLUMBING & HVAC**	95.3	88.9	92.7	95.3	91.0	93.6	100.9	81.3	93.0	95.7	92.0	94.2	100.1	100.4	100.2	100.1	87.1	94.8
26, 27, 3370	**ELECTRICAL, COMMUNICATIONS & UTIL.**	96.2	79.0	87.1	91.9	79.0	85.1	92.6	87.8	90.1	86.9	114.1	101.3	97.6	104.7	101.4	98.7	86.0	92.0
MF2014	**WEIGHTED AVERAGE**	96.7	87.3	92.6	95.1	85.3	90.8	98.0	84.7	92.2	96.8	94.9	96.0	98.7	98.1	98.4	99.2	89.0	94.7

DIVISION		OHIO			OKLAHOMA															
		ZANESVILLE			ARDMORE			CLINTON			DURANT			ENID			GUYMON			
		437 - 438			734			736			747			737			739			
		MAT.	INST.	TOTAL	MAT.	INST.	TOTAL	MAT.	INST.	TOTAL	MAT.	INST.	TOTAL	MAT.	INST.	TOTAL	MAT.	INST.	TOTAL	
015433	CONTRACTOR EQUIPMENT		93.2	93.2		82.7	82.7		81.8	81.8		81.8	81.8		81.8	81.8		81.8	81.8	
0241, 31 - 34	SITE & INFRASTRUCTURE, DEMOLITION	94.5	98.7	97.5	96.2	91.6	92.9	97.6	90.2	92.3	95.5	89.9	91.5	99.3	90.2	92.8	101.6	90.0	93.3	
0310	Concrete Forming & Accessories	93.7	81.3	83.0	94.0	40.4	47.8	92.5	46.6	53.0	85.5	44.1	49.8	96.2	34.1	42.7	99.8	44.9	52.4	
0320	Concrete Reinforcing	94.1	85.1	89.5	90.8	79.8	85.2	91.3	79.8	85.5	95.8	80.2	87.9	90.7	79.8	85.2	91.3	79.8	85.5	
0330	Cast-in-Place Concrete	90.5	91.7	91.0	97.5	43.4	75.3	94.3	45.1	74.1	91.7	42.8	71.6	94.3	46.4	74.6	94.3	43.1	73.3	
03	CONCRETE	93.5	85.3	89.5	95.2	49.3	72.7	94.7	52.7	74.1	92.3	50.8	71.9	95.3	47.6	71.8	98.2	51.2	75.1	
04	MASONRY	95.5	85.9	89.5	103.4	57.1	74.5	129.7	57.1	84.4	95.6	62.3	74.9	110.1	57.1	77.0	105.9	53.7	73.3	
05	METALS	97.4	80.4	92.1	102.5	70.1	92.5	102.6	70.1	92.6	93.5	70.7	86.5	104.0	70.0	93.6	103.2	69.8	92.9	
06	WOOD, PLASTICS & COMPOSITES	88.5	79.8	83.6	97.4	37.9	64.0	96.4	46.2	68.3	88.1	43.0	62.9	100.1	29.3	60.5	104.0	46.0	71.5	
07	THERMAL & MOISTURE PROTECTION	100.6	92.6	97.3	108.9	60.6	89.1	109.1	61.5	89.6	99.7	61.8	84.2	109.2	59.9	89.0	109.5	57.3	88.1	
08	OPENINGS	96.6	79.3	92.5	102.9	49.0	90.3	102.9	53.3	91.4	96.3	52.0	86.0	102.9	43.9	89.1	103.0	49.2	90.5	
0920	Plaster & Gypsum Board	88.2	79.3	82.2	89.8	36.6	53.8	89.4	45.2	59.6	79.4	41.9	54.1	90.5	27.8	48.1	90.7	44.9	59.8	
0950, 0980	Ceilings & Acoustic Treatment	97.7	79.3	85.6	85.7	36.6	53.4	85.7	45.2	59.1	82.4	41.9	55.8	85.7	27.8	47.6	86.5	44.9	59.2	
0960	Flooring	89.3	90.3	89.6	107.0	43.2	88.7	105.8	41.1	87.3	102.8	61.9	91.1	107.7	41.1	88.6	109.3	24.4	85.0	
0970, 0990	Wall Finishes & Painting/Coating	95.5	92.9	94.0	104.7	51.3	72.5	104.7	51.3	72.5	103.9	51.3	72.1	104.7	51.3	72.5	104.7	33.0	61.5	
09	FINISHES	92.4	83.6	87.5	92.9	40.1	63.8	92.7	44.7	66.3	90.2	46.6	66.2	93.5	34.6	61.1	94.5	38.9	63.9	
COVERS	DIVS. 10 - 14, 25, 28, 41, 43, 44, 46	100.0	88.5	97.7	100.0	77.6	95.5	100.0	78.6	95.7	100.0	77.9	95.5	100.0	76.7	95.3	100.0	77.4	95.4	
21, 22, 23	FIRE SUPPRESSION, PLUMBING & HVAC	95.3	89.9	93.1	95.6	65.8	83.6	95.6	65.9	83.6	95.5	65.4	83.4	100.3	65.8	86.4	95.6	63.8	82.7	
26, 27, 3370	ELECTRICAL, COMMUNICATIONS & UTIL.	92.3	83.4	87.5	93.6	70.9	81.6	94.6	70.9	82.1	96.5	70.9	83.0	94.6	70.9	82.1	96.2	61.5	77.9	
MF2014	WEIGHTED AVERAGE	95.6	86.5	91.6	98.3	61.9	82.4	99.6	63.1	83.7	95.1	63.5	81.3	100.2	60.5	82.9	99.4	59.7	82.1	

DIVISION		OKLAHOMA																	
		LAWTON			MCALESTER			MIAMI			MUSKOGEE			OKLAHOMA CITY			PONCA CITY		
		735			745			743			744			730 - 731			746		
		MAT.	INST.	TOTAL	MAT.	INST.	TOTAL	MAT.	INST.	TOTAL	MAT.	INST.	TOTAL	MAT.	INST.	TOTAL	MAT.	INST.	TOTAL
015433	CONTRACTOR EQUIPMENT		82.7	82.7		81.8	81.8		89.9	89.9		89.9	89.9		83.0	83.0		81.8	81.8
0241, 31 - 34	SITE & INFRASTRUCTURE, DEMOLITION	95.7	91.6	92.8	89.1	90.1	89.9	90.5	87.3	88.2	90.8	87.1	88.2	95.1	92.0	92.9	95.9	90.1	91.8
0310	Concrete Forming & Accessories	99.8	44.3	52.0	83.5	43.6	49.1	96.9	66.2	70.4	101.4	33.4	42.7	97.7	57.9	63.4	92.2	46.3	52.6
0320	Concrete Reinforcing	91.0	79.8	85.3	95.5	79.8	87.5	94.0	79.9	86.8	94.9	79.3	87.0	96.2	79.9	87.9	94.9	79.9	87.2
0330	Cast-in-Place Concrete	91.2	46.4	72.8	80.4	45.2	65.9	84.3	47.5	69.2	85.3	45.8	69.1	92.9	46.9	74.0	94.2	44.7	73.8
03	CONCRETE	91.5	52.1	72.1	82.9	51.3	67.4	87.6	63.0	75.5	89.4	47.7	68.9	94.9	58.3	76.9	94.4	52.4	73.8
04	MASONRY	105.8	57.1	75.4	114.3	56.1	78.1	98.4	56.3	72.1	116.9	45.2	72.2	109.8	56.4	76.5	90.9	56.1	69.2
05	METALS	108.0	70.1	96.4	93.4	70.0	86.2	93.4	82.1	89.9	94.8	80.4	90.4	100.7	70.2	91.3	93.4	70.3	86.3
06	WOOD, PLASTICS & COMPOSITES	103.0	43.1	69.5	85.6	43.0	61.7	101.0	72.9	85.3	105.6	30.7	63.6	97.0	61.8	77.3	96.3	46.0	68.1
07	THERMAL & MOISTURE PROTECTION	108.9	61.3	89.4	99.4	60.6	83.5	99.8	64.3	85.2	99.9	48.6	78.8	101.1	62.9	85.4	99.9	66.2	86.1
08	OPENINGS	104.6	51.4	92.2	96.2	51.8	85.9	96.2	68.3	89.7	96.2	42.4	83.7	104.0	61.6	94.2	96.2	53.1	86.2
0920	Plaster & Gypsum Board	92.4	42.0	58.3	78.4	41.9	53.7	84.8	72.6	76.5	86.9	29.0	47.8	98.0	61.2	73.1	83.7	44.9	57.5
0950, 0980	Ceilings & Acoustic Treatment	93.2	42.0	59.6	82.4	41.9	55.8	82.4	72.6	75.9	90.8	29.0	50.2	96.4	61.2	73.3	82.4	44.9	57.8
0960	Flooring	109.7	41.1	90.1	101.8	41.1	84.4	108.8	61.9	95.4	111.3	39.8	90.8	106.2	41.1	87.6	105.9	41.1	87.4
0970, 0990	Wall Finishes & Painting/Coating	104.7	51.3	72.5	103.9	37.8	64.0	103.9	77.5	87.9	103.9	34.2	61.8	106.4	51.3	73.2	103.9	51.3	72.1
09	FINISHES	95.5	42.8	66.5	89.2	41.0	62.6	92.0	67.3	78.4	94.9	33.0	60.8	97.6	53.7	73.4	91.8	44.3	65.6
COVERS	DIVS. 10 - 14, 25, 28, 41, 43, 44, 46	100.0	78.2	95.6	100.0	77.9	95.5	100.0	81.6	96.3	100.0	76.1	95.2	100.0	80.0	96.0	100.0	78.2	95.6
21, 22, 23	FIRE SUPPRESSION, PLUMBING & HVAC	100.3	65.9	86.4	95.5	61.6	81.8	95.5	61.9	81.9	100.3	60.6	84.3	100.1	65.5	86.2	95.5	61.8	81.9
26, 27, 3370	ELECTRICAL, COMMUNICATIONS & UTIL.	96.2	70.9	82.8	94.9	64.5	78.8	96.3	64.6	79.5	94.5	52.4	72.2	101.7	70.9	85.4	94.5	70.4	81.7
MF2014	WEIGHTED AVERAGE	100.7	62.8	84.2	94.5	60.4	79.7	94.8	67.5	82.9	97.4	55.7	79.2	100.3	65.7	85.2	95.1	62.1	80.7

DIVISION		OKLAHOMA												OREGON					
		POTEAU			SHAWNEE			TULSA			WOODWARD			BEND			EUGENE		
		749			748			740 - 741			738			977			974		
		MAT.	INST.	TOTAL	MAT.	INST.	TOTAL	MAT.	INST.	TOTAL	MAT.	INST.	TOTAL	MAT.	INST.	TOTAL	MAT.	INST.	TOTAL
015433	CONTRACTOR EQUIPMENT		89.1	89.1		81.8	81.8		89.9	89.9		81.8	81.8		98.8	98.8		98.8	98.8
0241, 31 - 34	SITE & INFRASTRUCTURE, DEMOLITION	77.2	85.8	83.3	99.0	90.1	92.7	97.1	87.5	90.3	97.9	90.2	92.4	105.4	102.9	103.6	96.1	102.9	100.9
0310	Concrete Forming & Accessories	90.1	41.0	47.8	85.4	43.7	49.5	101.5	40.3	48.7	92.6	46.8	53.1	110.1	98.8	100.4	106.5	98.7	99.8
0320	Concrete Reinforcing	95.9	79.9	87.8	94.9	79.8	87.2	95.1	79.8	87.3	90.7	79.9	85.2	93.3	99.6	96.5	97.4	99.6	98.5
0330	Cast-in-Place Concrete	84.3	44.9	68.1	97.2	42.9	74.9	92.9	46.2	73.7	94.3	45.3	74.2	104.4	102.7	103.7	101.1	102.6	101.7
03	CONCRETE	89.9	50.9	70.7	96.0	50.6	73.7	94.7	51.0	73.2	95.0	52.8	74.3	107.1	100.0	103.6	98.8	100.0	99.4
04	MASONRY	98.7	56.2	72.2	115.6	56.1	78.5	99.3	60.0	74.8	98.6	57.1	72.7	104.1	103.3	103.6	101.1	103.3	102.5
05	METALS	93.4	81.7	89.8	93.3	69.9	86.1	98.1	81.4	93.0	102.7	70.5	92.8	91.9	96.4	93.3	92.6	96.2	93.7
06	WOOD, PLASTICS & COMPOSITES	93.0	39.1	62.8	88.0	43.0	62.8	104.8	36.6	66.6	96.5	46.0	68.2	101.7	98.4	99.9	97.6	98.4	98.1
07	THERMAL & MOISTURE PROTECTION	99.9	60.5	83.7	99.9	59.3	83.2	99.8	62.2	84.4	109.1	66.6	91.7	107.9	94.9	102.6	107.2	91.6	100.8
08	OPENINGS	96.2	49.9	85.5	96.2	51.8	85.9	97.8	47.5	86.1	102.9	53.1	91.3	96.9	102.3	98.1	97.1	102.3	98.4
0920	Plaster & Gypsum Board	82.3	37.8	52.2	79.4	41.9	54.1	86.9	35.2	51.9	89.6	44.9	59.4	107.4	98.2	101.2	106.0	98.2	100.7
0950, 0980	Ceilings & Acoustic Treatment	82.4	37.8	53.1	82.4	41.9	55.8	90.8	35.2	54.3	86.5	44.9	59.2	91.3	98.2	95.8	92.3	98.2	96.1
0960	Flooring	105.2	61.9	92.8	102.8	32.7	82.8	110.0	42.8	90.8	105.8	43.2	87.9	110.5	103.4	108.5	108.9	103.4	107.3
0970, 0990	Wall Finishes & Painting/Coating	103.9	51.3	72.1	103.9	34.1	61.8	103.9	40.5	65.6	104.7	51.3	72.5	106.4	74.8	87.3	106.4	74.8	87.3
09	FINISHES	89.9	44.3	64.8	90.4	39.2	62.2	94.7	38.9	63.9	93.0	44.9	66.5	102.7	96.8	99.5	101.2	96.8	98.8
COVERS	DIVS. 10 - 14, 25, 28, 41, 43, 44, 46	100.0	77.7	95.5	100.0	77.9	95.5	100.0	78.8	95.7	100.0	78.5	95.7	100.0	99.6	99.9	100.0	99.6	99.9
21, 22, 23	FIRE SUPPRESSION, PLUMBING & HVAC	95.5	61.8	81.9	95.5	65.4	83.4	100.3	63.5	85.4	95.6	65.9	83.6	95.1	100.8	97.4	99.9	100.8	100.2
26, 27, 3370	ELECTRICAL, COMMUNICATIONS & UTIL.	94.6	64.6	78.7	96.6	70.9	83.0	96.5	64.6	79.6	96.1	70.9	82.8	101.6	96.1	98.7	100.1	96.1	98.0
MF2014	WEIGHTED AVERAGE	94.3	61.4	80.0	96.5	61.8	81.4	98.1	61.6	82.2	98.3	63.4	83.1	98.9	99.4	99.1	98.6	99.3	98.9

DIVISION		OREGON																	
		KLAMATH FALLS			MEDFORD			PENDLETON			PORTLAND			SALEM			VALE		
		976			975			978			970 - 972			973			979		
		MAT.	INST.	TOTAL	MAT.	INST.	TOTAL	MAT.	INST.	TOTAL	MAT.	INST.	TOTAL	MAT.	INST.	TOTAL	MAT.	INST.	TOTAL
015433	CONTRACTOR EQUIPMENT		98.8	98.8		98.8	98.8		96.2	96.2		98.8	98.8		98.8	98.8		96.2	96.2
0241, 31 - 34	SITE & INFRASTRUCTURE, DEMOLITION	109.2	102.9	104.7	103.4	102.9	103.0	102.5	96.4	98.2	98.5	102.9	101.6	91.9	102.9	99.7	90.5	96.4	94.7
0310	Concrete Forming & Accessories	102.9	98.6	99.2	102.0	98.6	99.1	103.3	99.0	99.6	107.7	98.9	100.1	106.0	98.8	99.8	109.9	98.7	100.2
0320	Concrete Reinforcing	93.3	99.6	96.5	94.9	99.6	97.3	92.6	99.7	96.2	98.1	99.7	98.9	103.6	99.6	101.6	90.4	99.6	95.1
0330	Cast-in-Place Concrete	104.4	102.6	103.7	104.4	102.6	103.7	105.2	103.9	104.7	103.9	102.7	103.4	97.9	102.7	99.9	82.9	103.8	91.5
03	CONCRETE	109.7	99.9	104.9	104.6	99.9	102.3	91.4	100.6	95.9	100.3	100.1	100.2	98.2	100.0	99.1	76.9	100.4	88.5
04	MASONRY	117.5	103.3	108.6	98.2	103.3	101.4	107.8	103.4	105.0	102.5	103.3	103.0	109.3	103.3	105.6	106.1	103.4	104.4
05	METALS	91.9	96.2	93.2	92.2	96.2	93.4	98.1	96.9	97.7	93.5	96.5	94.4	98.7	96.4	98.0	98.0	96.6	97.5
06	WOOD, PLASTICS & COMPOSITES	92.6	98.4	95.9	91.5	98.4	95.3	94.7	98.5	96.8	98.6	98.4	98.5	95.5	98.4	97.1	103.2	98.5	100.6
07	THERMAL & MOISTURE PROTECTION	108.2	92.9	101.9	107.8	92.9	101.7	101.1	93.5	98.0	107.1	96.7	102.9	104.9	94.9	100.8	100.5	94.4	98.0
08	OPENINGS	96.9	102.3	98.1	99.7	102.3	100.3	93.2	102.4	95.4	95.0	102.3	96.7	98.7	102.3	99.5	93.2	94.1	93.4
0920	Plaster & Gypsum Board	102.9	98.2	99.7	102.2	98.2	99.5	90.2	98.2	95.6	105.5	98.2	100.5	102.1	98.2	99.4	95.9	98.2	97.4
0950, 0980	Ceilings & Acoustic Treatment	98.9	98.2	98.4	105.4	98.2	100.7	64.4	98.2	86.6	94.2	98.2	96.8	99.3	98.2	98.5	64.4	98.2	86.6
0960	Flooring	107.6	103.4	106.4	107.1	103.4	106.1	74.7	103.4	82.9	106.4	103.4	105.5	107.2	103.4	106.1	76.7	103.4	84.4
0970, 0990	Wall Finishes & Painting/Coating	106.4	70.4	84.6	106.4	70.4	84.6	96.6	72.8	82.2	106.2	72.8	86.0	104.6	74.8	86.6	96.6	74.8	83.4
09	FINISHES	103.4	96.3	99.5	103.8	96.3	99.6	72.3	96.7	85.7	100.8	96.6	98.5	100.0	96.8	98.2	72.7	96.9	86.0
COVERS	DIVS. 10 - 14, 25, 28, 41, 43, 44, 46	100.0	99.5	99.9	100.0	99.5	99.9	100.0	90.1	98.0	100.0	99.6	99.9	100.0	99.6	99.9	100.0	99.9	100.0
21, 22, 23	FIRE SUPPRESSION, PLUMBING & HVAC	95.1	100.7	97.4	99.9	100.7	100.2	97.0	113.2	103.5	99.9	100.8	100.2	99.9	100.8	100.2	97.0	98.0	97.4
26, 27, 3370	ELECTRICAL, COMMUNICATIONS & UTIL.	100.2	80.8	89.9	103.9	80.8	91.6	92.3	97.1	94.8	100.4	103.2	101.9	108.0	96.1	101.7	92.3	97.0	94.8
MF2014	WEIGHTED AVERAGE	99.8	97.1	98.6	100.0	97.1	98.8	94.8	101.4	97.7	98.8	100.4	99.5	100.5	99.4	100.0	92.9	98.1	95.2

DIVISION		PENNSYLVANIA																	
		ALLENTOWN			ALTOONA			BEDFORD			BRADFORD			BUTLER			CHAMBERSBURG		
		181			166			155			167			160			172		
		MAT.	INST.	TOTAL	MAT.	INST.	TOTAL	MAT.	INST.	TOTAL	MAT.	INST.	TOTAL	MAT.	INST.	TOTAL	MAT.	INST.	TOTAL
015433	CONTRACTOR EQUIPMENT		112.9	112.9		112.9	112.9		109.6	109.6		112.9	112.9		112.9	112.9		112.1	112.1
0241, 31 - 34	SITE & INFRASTRUCTURE, DEMOLITION	93.6	104.5	101.3	96.8	104.4	102.2	101.4	100.1	100.5	92.4	103.5	100.3	88.1	105.8	100.7	89.2	101.2	97.8
0310	Concrete Forming & Accessories	99.2	113.0	111.1	84.2	80.4	80.9	84.1	80.8	81.3	86.4	81.9	82.5	85.7	96.0	94.6	90.2	80.0	81.4
0320	Concrete Reinforcing	94.7	108.1	101.5	91.8	102.7	97.3	93.1	80.9	86.9	93.7	103.2	98.5	92.4	109.1	100.9	91.7	103.0	97.4
0330	Cast-in-Place Concrete	86.5	104.9	94.0	96.3	86.1	92.1	106.7	70.1	91.6	92.0	92.6	92.3	85.1	96.7	89.9	91.1	69.8	82.3
03	CONCRETE	93.6	110.1	101.7	89.1	88.1	88.6	102.1	78.5	90.5	95.4	91.0	93.2	81.3	99.9	90.4	101.9	82.2	92.3
04	MASONRY	97.4	101.0	99.7	100.5	64.5	78.0	114.1	87.0	97.2	97.6	88.4	91.9	102.6	98.9	100.3	102.6	87.4	93.1
05	METALS	100.0	123.0	107.1	93.9	116.8	101.0	97.6	104.0	99.6	97.9	116.5	103.6	93.6	121.9	102.3	97.5	114.6	102.7
06	WOOD, PLASTICS & COMPOSITES	100.7	115.8	109.1	79.2	83.2	81.4	84.0	80.7	82.2	85.5	80.1	82.5	80.6	95.5	89.0	88.0	80.8	84.0
07	THERMAL & MOISTURE PROTECTION	102.3	119.1	109.2	101.4	91.0	97.1	103.0	87.7	96.7	102.3	91.7	97.9	101.1	100.7	100.9	99.3	72.5	88.3
08	OPENINGS	95.0	115.0	99.6	88.7	89.3	88.8	98.0	81.6	94.2	95.2	91.0	94.2	88.7	104.8	92.4	94.1	82.8	91.4
0920	Plaster & Gypsum Board	95.9	116.0	109.5	87.7	82.5	84.2	92.3	80.0	84.0	88.5	79.2	82.2	87.7	95.2	92.7	96.7	80.0	85.4
0950, 0980	Ceilings & Acoustic Treatment	82.8	116.0	104.6	87.0	82.5	84.0	95.0	80.0	85.1	85.4	79.2	81.3	87.9	95.2	92.7	84.0	80.0	81.4
0960	Flooring	94.6	95.9	95.0	88.2	97.5	90.9	93.1	88.2	91.7	89.4	105.3	94.0	89.0	81.0	86.7	97.9	46.2	83.1
0970, 0990	Wall Finishes & Painting/Coating	98.5	70.5	81.6	94.0	109.2	103.2	100.2	81.6	89.0	98.5	94.6	96.1	94.0	109.2	103.2	105.9	81.6	91.2
09	FINISHES	91.8	105.5	99.4	90.0	85.8	87.7	95.0	80.8	87.2	89.9	85.5	87.5	89.9	93.6	91.9	92.1	73.7	82.0
COVERS	DIVS. 10 - 14, 25, 28, 41, 43, 44, 46	100.0	105.6	101.1	100.0	95.4	99.1	100.0	98.3	99.7	100.0	99.2	99.8	100.0	102.4	100.5	100.0	96.1	99.2
21, 22, 23	FIRE SUPPRESSION, PLUMBING & HVAC	100.2	112.1	105.0	99.7	82.5	92.8	95.2	87.3	92.0	95.5	90.7	93.6	95.0	93.9	94.6	95.5	87.1	92.1
26, 27, 3370	ELECTRICAL, COMMUNICATIONS & UTIL.	99.1	98.4	98.8	89.3	112.2	101.4	94.3	112.2	103.8	92.7	112.2	103.0	89.9	111.2	101.1	89.6	86.3	87.8
MF2014	WEIGHTED AVERAGE	97.9	108.6	102.5	94.6	91.9	93.4	98.2	91.3	95.2	95.7	96.4	96.0	92.5	102.0	96.6	96.2	88.0	92.6

DIVISION		PENNSYLVANIA																	
		DOYLESTOWN			DUBOIS			ERIE			GREENSBURG			HARRISBURG			HAZLETON		
		189			158			164 - 165			156			170 - 171			182		
		MAT.	INST.	TOTAL	MAT.	INST.	TOTAL	MAT.	INST.	TOTAL	MAT.	INST.	TOTAL	MAT.	INST.	TOTAL	MAT.	INST.	TOTAL
015433	CONTRACTOR EQUIPMENT		93.9	93.9		109.6	109.6		112.9	112.9		109.6	109.6		112.1	112.1		112.9	112.9
0241, 31 - 34	SITE & INFRASTRUCTURE, DEMOLITION	106.9	90.2	95.0	106.1	100.5	102.1	93.7	105.2	101.9	97.8	102.9	101.4	89.6	103.5	99.5	86.9	104.7	99.6
0310	Concrete Forming & Accessories	83.2	128.2	122.0	83.5	84.6	84.4	98.5	88.1	89.5	90.7	95.9	95.2	100.4	87.9	89.6	80.8	88.7	87.6
0320	Concrete Reinforcing	91.6	130.9	111.6	92.6	103.5	98.1	93.7	103.3	98.6	92.6	109.1	101.0	101.4	106.2	103.9	91.9	107.1	99.6
0330	Cast-in-Place Concrete	81.7	86.7	83.8	102.9	92.9	98.8	94.7	81.8	89.4	99.0	96.3	97.9	93.2	95.8	94.3	81.7	93.1	86.4
03	CONCRETE	89.2	114.2	101.5	103.6	92.2	98.0	88.2	90.0	89.1	97.2	99.6	98.4	97.7	95.4	96.6	86.3	95.0	90.6
04	MASONRY	100.7	128.1	117.7	114.4	91.1	99.9	89.6	92.5	91.4	124.8	98.9	108.7	102.5	89.6	94.5	110.2	95.7	101.2
05	METALS	97.5	123.8	105.6	97.6	116.1	103.3	94.1	117.0	101.2	97.5	120.6	104.6	104.0	121.4	109.4	99.7	120.6	106.1
06	WOOD, PLASTICS & COMPOSITES	81.0	130.7	108.8	82.9	83.1	83.0	96.9	86.6	91.1	90.9	95.4	93.4	95.8	86.7	90.7	79.6	86.6	83.5
07	THERMAL & MOISTURE PROTECTION	100.1	131.4	112.9	103.2	96.5	100.5	101.9	92.6	98.1	102.8	100.7	102.0	102.5	110.2	105.7	101.8	106.6	103.8
08	OPENINGS	97.2	137.9	106.7	98.0	92.7	96.7	88.8	92.2	89.6	97.9	104.7	99.5	100.6	94.0	99.0	95.6	92.3	94.8
0920	Plaster & Gypsum Board	86.2	131.4	116.7	91.2	82.5	85.3	95.9	86.0	89.2	93.5	95.2	94.6	100.6	86.0	90.7	86.6	86.0	86.2
0950, 0980	Ceilings & Acoustic Treatment	82.0	131.4	114.5	95.0	82.5	86.7	82.8	86.0	84.9	94.2	95.2	94.8	92.6	86.0	88.3	83.7	86.0	85.2
0960	Flooring	79.4	134.3	95.1	92.9	105.3	96.4	93.2	91.8	92.8	96.2	68.0	88.2	103.2	91.7	99.9	87.0	94.6	89.2
0970, 0990	Wall Finishes & Painting/Coating	98.0	69.0	80.5	100.2	106.6	104.1	103.5	94.6	98.1	100.2	106.6	104.1	106.4	89.6	96.3	98.5	106.4	103.2
09	FINISHES	83.2	122.4	104.8	95.2	89.5	92.1	92.2	88.8	90.4	95.6	91.7	93.4	96.9	88.0	92.0	88.1	89.5	88.9
COVERS	DIVS. 10 - 14, 25, 28, 41, 43, 44, 46	100.0	70.6	94.1	100.0	99.1	99.8	100.0	100.8	100.2	100.0	102.2	100.4	100.0	97.4	99.5	100.0	101.0	100.2
21, 22, 23	FIRE SUPPRESSION, PLUMBING & HVAC	95.0	127.4	108.1	95.2	88.4	92.4	99.7	92.9	97.0	95.2	91.1	93.5	100.1	92.1	96.9	95.5	99.2	97.0
26, 27, 3370	ELECTRICAL, COMMUNICATIONS & UTIL.	92.1	128.1	111.1	94.9	112.2	104.0	91.1	96.7	94.0	94.9	112.2	104.1	96.8	88.3	92.3	93.6	91.7	92.6
MF2014	WEIGHTED AVERAGE	94.9	120.6	106.1	98.5	96.8	97.8	94.5	95.8	95.1	98.2	100.9	99.4	99.8	95.6	98.0	95.4	98.3	96.6

	DIVISION	PENNSYLVANIA																	
		INDIANA			JOHNSTOWN			KITTANNING			LANCASTER			LEHIGH VALLEY			MONTROSE		
		157			159			162			175 - 176			180			188		
		MAT.	INST.	TOTAL	MAT.	INST.	TOTAL	MAT.	INST.	TOTAL	MAT.	INST.	TOTAL	MAT.	INST.	TOTAL	MAT.	INST.	TOTAL
015433	CONTRACTOR EQUIPMENT		109.6	109.6		109.6	109.6		112.9	112.9		112.1	112.1		112.9	112.9		112.9	112.9
0241, 31 - 34	SITE & INFRASTRUCTURE, DEMOLITION	95.9	101.2	99.7	101.9	102.2	102.1	90.7	105.7	101.4	81.6	103.5	97.2	90.7	104.3	100.4	89.4	102.1	98.4
0310	Concrete Forming & Accessories	84.7	86.3	86.1	83.5	84.8	84.6	85.7	95.9	94.5	92.3	87.5	88.2	92.8	112.7	110.0	81.8	88.5	87.6
0320	Concrete Reinforcing	91.8	109.2	100.6	93.1	108.9	101.2	92.4	109.2	101.0	91.4	106.1	98.9	91.9	108.1	100.1	96.3	106.3	101.4
0330	Cast-in-Place Concrete	97.1	95.9	96.6	107.6	92.4	101.4	88.4	96.5	91.7	77.5	97.9	85.9	88.4	104.7	95.1	86.7	90.9	88.4
03	CONCRETE	94.7	95.2	94.9	102.9	93.2	98.2	83.8	99.7	91.6	89.8	96.0	92.8	92.6	109.8	101.1	91.3	93.7	92.5
04	MASONRY	110.1	100.1	103.9	110.9	90.6	98.3	105.4	100.1	102.1	108.2	90.1	97.0	97.3	101.0	99.6	97.3	95.1	95.9
05	METALS	97.7	120.1	104.6	97.6	118.9	104.2	93.7	121.9	102.4	97.5	120.8	104.7	99.7	122.0	106.5	98.0	114.1	102.9
06	WOOD, PLASTICS & COMPOSITES	84.8	83.1	83.8	82.9	83.1	83.0	80.6	95.5	89.0	90.8	86.7	88.5	92.2	115.8	105.4	80.4	88.4	84.9
07	THERMAL & MOISTURE PROTECTION	102.7	98.2	100.9	103.0	95.1	99.7	101.1	101.0	101.1	98.8	97.1	98.1	102.2	102.8	102.4	101.8	91.7	97.7
08	OPENINGS	98.0	94.4	97.1	98.0	90.9	96.3	88.7	101.2	91.6	94.1	98.6	95.1	95.6	115.0	100.1	92.2	93.1	92.4
0920	Plaster & Gypsum Board	92.7	82.5	85.8	91.0	82.5	85.2	87.7	95.2	92.7	98.5	86.0	90.1	89.5	116.0	107.4	87.1	87.8	87.6
0950, 0980	Ceilings & Acoustic Treatment	95.0	82.5	86.7	94.2	82.5	86.5	87.9	95.2	92.7	84.0	86.0	85.3	83.7	116.0	105.0	85.4	87.8	87.0
0960	Flooring	93.7	105.3	97.0	92.9	98.2	94.4	89.0	105.3	93.6	98.8	91.7	96.8	91.9	95.9	93.0	87.6	57.9	79.1
0970, 0990	Wall Finishes & Painting/Coating	100.2	106.6	104.1	100.2	109.2	105.6	94.0	106.6	101.6	105.9	57.1	76.5	98.5	66.4	79.1	98.5	106.4	103.2
09	FINISHES	94.8	90.0	92.1	94.6	88.8	91.4	90.1	97.4	94.1	92.0	84.5	87.9	90.2	104.1	97.9	88.9	84.7	86.6
COVERS	DIVS. 10 - 14, 25, 28, 41, 43, 44, 46	100.0	100.6	100.1	100.0	99.3	99.9	100.0	102.2	100.5	100.0	97.4	99.5	100.0	106.0	101.2	100.0	101.2	100.2
21, 22, 23	FIRE SUPPRESSION, PLUMBING & HVAC	95.2	90.8	93.4	95.2	88.7	92.5	95.0	96.9	95.8	95.5	92.3	94.2	95.5	112.0	102.2	95.5	98.4	96.7
26, 27, 3370	ELECTRICAL, COMMUNICATIONS & UTIL.	94.9	112.2	104.1	94.9	112.2	104.0	89.3	112.2	101.4	91.0	44.0	66.2	93.6	143.4	119.9	92.7	97.6	95.3
MF2014	WEIGHTED AVERAGE	97.1	99.3	98.0	98.1	97.2	97.7	92.9	103.3	97.4	95.1	89.0	92.4	95.8	114.0	103.7	94.7	96.9	95.7

	DIVISION	PENNSYLVANIA																	
		NEW CASTLE			NORRISTOWN			OIL CITY			PHILADELPHIA			PITTSBURGH			POTTSVILLE		
		161			194			163			190 - 191			150 - 152			179		
		MAT.	INST.	TOTAL	MAT.	INST.	TOTAL	MAT.	INST.	TOTAL	MAT.	INST.	TOTAL	MAT.	INST.	TOTAL	MAT.	INST.	TOTAL
015433	CONTRACTOR EQUIPMENT		112.9	112.9		98.8	98.8		112.9	112.9		98.2	98.2		110.8	110.8		112.1	112.1
0241, 31 - 34	SITE & INFRASTRUCTURE, DEMOLITION	88.5	105.8	100.8	95.8	100.6	99.2	87.1	103.7	98.9	101.6	100.0	100.5	101.3	104.9	103.8	84.4	103.4	97.9
0310	Concrete Forming & Accessories	85.7	95.6	94.3	83.2	129.7	123.3	85.7	83.4	83.7	98.9	140.1	134.4	98.5	96.7	97.0	83.3	89.7	88.8
0320	Concrete Reinforcing	91.3	92.5	91.9	90.2	137.9	114.5	92.4	92.3	92.4	100.6	137.9	119.6	93.6	109.4	101.6	90.7	102.7	96.8
0330	Cast-in-Place Concrete	85.9	96.5	90.2	85.0	127.1	102.3	83.4	95.5	88.4	98.0	132.2	112.1	102.9	96.5	100.3	82.6	100.9	90.1
03	CONCRETE	81.6	96.4	88.9	89.2	130.1	109.3	80.1	90.5	85.2	99.4	136.4	117.6	100.5	100.1	100.3	93.3	97.3	95.3
04	MASONRY	101.9	97.7	99.3	110.4	124.9	119.4	101.8	95.8	98.0	96.7	130.7	117.9	103.0	102.1	102.4	102.1	92.2	95.9
05	METALS	93.7	113.6	99.8	99.2	130.0	108.7	93.7	111.5	99.2	102.0	130.2	110.7	99.0	121.3	105.9	97.7	118.8	104.2
06	WOOD, PLASTICS & COMPOSITES	80.6	95.9	89.2	79.9	130.6	108.3	80.6	80.1	80.3	98.6	141.9	122.9	100.6	95.8	97.9	80.2	88.1	84.6
07	THERMAL & MOISTURE PROTECTION	101.1	98.0	99.8	101.4	131.1	113.5	101.0	94.4	98.3	102.8	135.2	116.1	103.1	101.6	102.4	98.9	106.7	102.1
08	OPENINGS	88.7	96.6	90.5	86.9	140.1	99.2	88.7	80.3	86.7	98.6	146.2	109.7	101.8	105.0	102.5	94.1	91.8	93.6
0920	Plaster & Gypsum Board	87.7	95.6	93.0	85.6	131.4	116.5	87.7	79.2	82.0	96.6	143.1	128.0	99.8	95.6	96.9	93.5	87.5	89.5
0950, 0980	Ceilings & Acoustic Treatment	87.9	95.6	92.9	84.5	131.4	115.3	87.9	79.2	82.2	94.0	143.1	126.2	95.0	95.6	95.4	84.0	87.5	86.3
0960	Flooring	89.0	57.5	80.0	92.9	139.2	106.1	89.0	105.3	93.6	98.4	139.2	110.1	99.5	106.6	101.5	95.2	91.7	94.2
0970, 0990	Wall Finishes & Painting/Coating	94.0	109.2	103.2	98.6	147.8	128.3	94.0	106.6	101.6	99.8	154.7	133.0	100.2	119.2	111.7	105.9	106.4	106.2
09	FINISHES	90.0	89.6	89.8	90.7	133.0	114.0	89.8	87.8	88.7	98.9	141.8	122.5	97.7	100.2	99.1	90.5	91.3	90.9
COVERS	DIVS. 10 - 14, 25, 28, 41, 43, 44, 46	100.0	102.4	100.5	100.0	117.5	103.5	100.0	100.6	100.1	100.0	120.8	104.2	100.0	102.2	100.4	100.0	98.7	99.7
21, 22, 23	FIRE SUPPRESSION, PLUMBING & HVAC	95.0	93.6	94.4	95.2	127.3	108.2	95.0	92.9	94.2	100.0	131.8	112.9	99.9	100.4	100.1	95.5	98.6	96.8
26, 27, 3370	ELECTRICAL, COMMUNICATIONS & UTIL.	89.9	98.1	94.2	93.0	148.4	122.3	91.8	111.1	102.0	97.1	148.4	124.2	97.2	112.2	105.1	89.2	95.5	92.5
MF2014	WEIGHTED AVERAGE	92.5	97.9	94.8	95.0	129.5	110.0	92.4	96.9	94.4	99.7	133.6	114.5	99.9	104.6	102.0	94.9	98.5	96.5

	DIVISION	PENNSYLVANIA																	
		READING			SCRANTON			STATE COLLEGE			STROUDSBURG			SUNBURY			UNIONTOWN		
		195 - 196			184 - 185			168			183			178			154		
		MAT.	INST.	TOTAL	MAT.	INST.	TOTAL	MAT.	INST.	TOTAL	MAT.	INST.	TOTAL	MAT.	INST.	TOTAL	MAT.	INST.	TOTAL
015433	CONTRACTOR EQUIPMENT		118.0	118.0		112.9	112.9		112.1	112.1		112.9	112.9		112.9	112.9		109.6	109.6
0241, 31 - 34	SITE & INFRASTRUCTURE, DEMOLITION	100.3	112.7	109.1	94.1	104.8	101.7	84.4	103.1	97.7	88.5	102.2	98.2	96.6	104.5	102.2	96.5	102.9	101.0
0310	Concrete Forming & Accessories	98.8	90.2	91.4	99.3	88.6	90.0	84.4	80.3	80.8	87.2	89.4	89.1	96.1	88.5	89.6	77.6	96.0	93.5
0320	Concrete Reinforcing	91.5	104.7	98.2	94.7	106.6	100.8	93.0	103.4	98.3	95.0	111.6	103.4	93.2	106.2	99.8	92.6	109.1	101.0
0330	Cast-in-Place Concrete	76.2	97.3	84.9	90.3	93.1	91.5	87.1	65.6	78.2	85.1	72.6	80.0	90.2	95.5	92.4	97.1	96.4	96.8
03	CONCRETE	88.2	96.7	92.4	95.4	94.9	95.1	95.5	81.0	88.4	90.1	88.9	89.5	96.7	95.5	96.1	94.4	99.7	97.0
04	MASONRY	99.5	93.0	95.4	97.7	95.7	96.4	103.1	77.9	87.4	95.1	100.4	98.4	102.3	89.7	94.5	127.0	98.9	109.5
05	METALS	99.5	121.2	106.2	102.1	120.5	107.8	97.7	117.6	103.8	99.7	115.9	104.7	97.4	120.0	104.4	97.4	120.6	104.5
06	WOOD, PLASTICS & COMPOSITES	98.5	88.1	92.6	100.7	86.2	92.6	87.6	83.2	85.1	86.4	88.4	87.5	88.9	88.1	88.5	76.7	95.4	87.2
07	THERMAL & MOISTURE PROTECTION	101.8	112.1	106.0	102.2	95.6	99.5	101.5	90.6	97.0	102.0	85.3	95.2	100.0	104.9	102.0	102.6	100.7	101.8
08	OPENINGS	91.3	98.9	93.1	95.0	91.9	94.3	92.0	89.3	91.4	95.6	88.2	93.9	94.2	92.8	93.9	97.9	104.7	99.5
0920	Plaster & Gypsum Board	96.8	87.5	90.5	98.0	85.6	89.6	89.6	82.5	84.8	88.0	87.8	87.9	92.7	87.5	89.2	89.2	95.2	93.2
0950, 0980	Ceilings & Acoustic Treatment	76.9	87.5	83.9	91.1	85.6	87.5	82.9	82.5	82.6	82.0	87.8	85.8	80.7	87.5	85.2	94.2	95.2	94.8
0960	Flooring	96.8	91.7	95.4	94.6	107.0	98.2	92.3	97.5	93.8	89.9	51.3	78.9	95.9	87.7	93.5	90.4	105.3	94.7
0970, 0990	Wall Finishes & Painting/Coating	97.3	106.4	102.8	98.5	106.4	103.2	98.5	109.2	105.0	98.5	64.6	78.0	105.9	94.3	98.9	100.2	109.2	105.6
09	FINISHES	92.0	90.8	91.3	93.9	93.1	93.4	89.4	85.8	87.4	88.9	80.0	84.0	91.3	88.7	89.9	93.2	97.5	95.6
COVERS	DIVS. 10 - 14, 25, 28, 41, 43, 44, 46	100.0	100.3	100.1	100.0	101.0	100.2	100.0	93.1	98.6	100.0	62.4	92.4	100.0	97.9	99.6	100.0	102.2	100.4
21, 22, 23	FIRE SUPPRESSION, PLUMBING & HVAC	100.1	108.8	103.7	100.2	99.2	99.8	95.5	85.9	91.6	95.5	100.0	97.3	95.5	92.0	94.1	95.2	91.1	93.5
26, 27, 3370	ELECTRICAL, COMMUNICATIONS & UTIL.	99.6	95.5	97.5	99.2	97.7	98.4	91.9	112.2	102.6	93.6	143.3	119.9	89.5	90.8	90.2	92.0	112.2	102.7
MF2014	WEIGHTED AVERAGE	97.1	102.1	99.3	98.6	99.2	98.9	95.3	92.8	94.2	95.3	101.4	97.9	95.8	95.8	95.8	97.3	101.7	99.2

DIVISION		PENNSYLVANIA																	
		WASHINGTON			WELLSBORO			WESTCHESTER			WILKES-BARRE			WILLIAMSPORT			YORK		
		153			169			193			186 - 187			177			173 - 174		
		MAT.	INST.	TOTAL	MAT.	INST.	TOTAL	MAT.	INST.	TOTAL	MAT.	INST.	TOTAL	MAT.	INST.	TOTAL	MAT.	INST.	TOTAL
015433	CONTRACTOR EQUIPMENT		109.6	109.6		112.9	112.9		98.8	98.8		112.9	112.9		112.9	112.9		112.1	112.1
0241, 31 - 34	SITE & INFRASTRUCTURE, DEMOLITION	96.6	102.9	101.1	95.9	102.0	100.3	101.6	97.8	98.9	86.5	104.7	99.5	87.9	103.4	98.9	85.2	103.5	98.2
0310	Concrete Forming & Accessories	84.8	96.2	94.6	85.8	86.7	86.6	89.8	128.2	122.9	89.9	88.9	89.1	92.5	61.0	65.3	86.9	88.1	87.9
0320	Concrete Reinforcing	92.6	109.3	101.1	93.0	106.2	99.7	89.3	113.6	101.7	93.7	107.1	100.6	92.5	64.6	78.3	93.2	106.2	99.8
0330	Cast-in-Place Concrete	97.1	96.4	96.8	91.2	88.9	90.3	94.1	126.4	107.4	81.7	93.0	86.4	76.4	74.2	75.5	83.0	98.1	89.2
03	CONCRETE	94.9	99.8	97.3	97.8	92.2	95.1	96.9	124.5	110.5	87.2	95.1	91.1	84.7	68.2	76.6	94.5	96.3	95.4
04	MASONRY	109.4	100.5	103.9	103.6	89.7	94.9	104.7	124.9	117.3	110.6	95.3	101.0	93.9	93.7	93.8	103.6	90.1	95.2
05	METALS	97.3	120.9	104.6	97.8	114.2	102.9	99.2	117.2	104.8	97.9	121.0	105.0	97.5	100.4	98.4	99.1	121.4	105.9
06	WOOD, PLASTICS & COMPOSITES	84.9	95.4	90.8	84.9	88.1	86.7	87.1	130.6	111.5	88.9	86.6	87.6	85.2	51.0	66.1	84.1	86.7	85.5
07	THERMAL & MOISTURE PROTECTION	102.7	101.1	102.1	102.5	89.6	97.2	101.7	129.9	113.3	101.8	106.4	103.7	99.4	102.4	100.6	99.0	110.4	103.7
08	OPENINGS	97.9	104.7	99.5	95.1	93.0	94.6	86.9	126.1	96.0	92.2	98.8	93.8	94.2	53.5	84.7	94.1	94.0	94.1
0920	Plaster & Gypsum Board	92.4	95.2	94.3	87.9	87.5	87.6	87.0	131.4	117.0	88.8	86.0	86.9	93.5	49.3	63.6	94.3	86.0	88.7
0950, 0980	Ceilings & Acoustic Treatment	94.2	95.2	94.8	82.9	87.5	85.9	84.5	131.4	115.3	85.4	86.0	85.8	84.0	49.3	61.2	83.1	86.0	85.0
0960	Flooring	93.8	105.3	97.1	89.1	50.8	78.2	95.7	139.2	108.1	90.7	94.6	91.8	94.8	49.7	81.9	96.4	91.7	95.1
0970, 0990	Wall Finishes & Painting/Coating	100.2	109.2	105.6	98.5	106.4	103.2	98.6	147.8	128.3	98.5	106.4	103.2	105.9	106.4	106.2	105.9	89.6	96.1
09	FINISHES	94.6	97.8	96.4	89.5	82.2	85.5	92.1	131.0	113.6	89.9	90.8	90.4	91.1	62.3	75.2	90.8	88.1	89.3
COVERS	DIVS. 10 - 14, 25, 28, 41, 43, 44, 46	100.0	102.2	100.4	100.0	97.9	99.6	100.0	117.3	103.5	100.0	100.8	100.2	100.0	94.6	98.9	100.0	97.6	99.5
21, 22, 23	FIRE SUPPRESSION, PLUMBING & HVAC	95.2	97.2	96.0	95.5	91.1	93.7	95.2	125.8	107.6	95.5	99.0	96.9	95.5	93.1	94.5	100.2	92.4	97.1
26, 27, 3370	ELECTRICAL, COMMUNICATIONS & UTIL.	94.3	112.2	103.8	92.7	86.5	89.5	92.9	114.9	104.5	93.6	91.7	92.6	90.0	74.6	81.8	91.0	88.3	89.5
MF2014	WEIGHTED AVERAGE	97.0	103.2	99.7	96.3	92.6	94.7	95.9	121.5	107.1	95.1	98.7	96.6	93.8	83.0	89.1	96.7	95.9	96.3

DIVISION		PUERTO RICO			RHODE ISLAND						SOUTH CAROLINA								
		SAN JUAN			NEWPORT			PROVIDENCE			AIKEN			BEAUFORT			CHARLESTON		
		009			028			029			298			299			294		
		MAT.	INST.	TOTAL	MAT.	INST.	TOTAL	MAT.	INST.	TOTAL	MAT.	INST.	TOTAL	MAT.	INST.	TOTAL	MAT.	INST.	TOTAL
015433	CONTRACTOR EQUIPMENT		90.5	90.5		102.4	102.4		102.4	102.4		101.3	101.3		101.3	101.3		101.3	101.3
0241, 31 - 34	SITE & INFRASTRUCTURE, DEMOLITION	133.8	91.2	103.5	89.2	105.0	100.4	91.4	105.0	101.1	118.9	87.4	96.5	114.3	85.6	93.9	99.7	86.3	90.2
0310	Concrete Forming & Accessories	92.4	17.8	28.1	101.5	122.4	119.5	99.8	122.4	119.3	97.5	67.4	71.5	96.4	37.9	46.0	95.4	63.0	67.5
0320	Concrete Reinforcing	188.3	12.7	98.9	106.4	149.2	128.2	102.3	149.2	126.2	93.7	65.2	79.2	92.8	25.8	58.7	92.7	59.3	75.7
0330	Cast-in-Place Concrete	103.8	30.4	73.7	79.5	124.1	97.8	95.5	124.1	107.2	79.2	69.8	75.4	79.2	47.2	66.1	92.9	49.5	75.1
03	CONCRETE	108.7	22.2	66.2	93.2	127.5	110.0	100.4	127.5	113.7	102.7	68.9	86.1	99.9	40.9	70.9	94.4	58.9	77.0
04	MASONRY	90.1	16.9	44.5	95.4	132.5	118.5	101.3	132.5	120.7	79.6	61.4	68.3	93.9	33.3	56.1	95.1	40.9	61.4
05	METALS	118.4	34.6	92.6	97.4	125.6	106.1	103.1	125.6	110.0	101.8	83.8	96.3	101.8	68.2	91.5	103.8	80.4	96.6
06	WOOD, PLASTICS & COMPOSITES	94.4	17.5	51.3	99.0	120.9	111.3	100.1	120.9	111.7	97.0	68.6	81.1	95.3	37.8	63.1	94.0	68.4	79.7
07	THERMAL & MOISTURE PROTECTION	129.1	21.6	85.0	100.6	121.5	109.2	100.5	121.5	109.1	102.7	66.4	87.8	102.4	38.8	76.3	101.6	47.3	79.3
08	OPENINGS	152.0	15.4	120.2	102.2	128.2	108.3	107.3	128.2	112.2	99.0	64.8	91.1	99.0	36.7	84.5	102.9	63.3	93.7
0920	Plaster & Gypsum Board	159.8	14.9	61.9	92.9	120.9	111.9	94.5	120.9	112.4	104.0	67.4	79.3	107.4	35.7	59.0	109.0	67.3	80.8
0950, 0980	Ceilings & Acoustic Treatment	225.7	14.9	87.2	93.2	120.9	111.4	90.1	120.9	110.4	86.4	67.4	73.9	89.8	35.7	54.3	89.8	67.3	75.0
0960	Flooring	224.3	16.6	164.9	97.7	141.5	110.2	98.0	141.5	110.4	105.4	68.4	94.8	106.8	51.2	90.9	106.5	58.2	92.7
0970, 0990	Wall Finishes & Painting/Coating	213.5	16.8	94.8	99.9	129.2	117.6	97.7	129.2	116.7	109.5	70.3	85.8	109.5	34.7	64.3	109.5	66.9	83.8
09	FINISHES	210.6	17.6	104.2	96.6	126.9	113.3	93.7	126.9	112.0	98.3	68.0	81.6	99.5	39.8	66.6	97.8	63.5	78.9
COVERS	DIVS. 10 - 14, 25, 28, 41, 43, 44, 46	100.0	17.5	83.3	100.0	108.6	101.7	100.0	108.6	101.7	100.0	71.8	94.3	100.0	70.4	94.0	100.0	69.0	93.7
21, 22, 23	FIRE SUPPRESSION, PLUMBING & HVAC	103.3	14.0	67.3	100.1	110.6	104.3	99.9	110.6	104.2	95.7	63.1	82.5	95.7	36.5	71.8	100.5	53.7	81.6
26, 27, 3370	ELECTRICAL, COMMUNICATIONS & UTIL.	125.9	13.2	66.3	100.0	99.0	99.5	99.6	99.0	99.2	96.9	65.9	80.5	100.7	33.6	65.2	99.0	88.8	93.6
MF2014	WEIGHTED AVERAGE	122.7	24.4	79.8	98.4	117.5	106.7	100.6	117.5	108.0	98.6	69.1	85.7	99.3	44.9	75.6	99.9	65.2	84.8

DIVISION		SOUTH CAROLINA															SOUTH DAKOTA		
		COLUMBIA			FLORENCE			GREENVILLE			ROCK HILL			SPARTANBURG			ABERDEEN		
		290 - 292			295			296			297			293			574		
		MAT.	INST.	TOTAL	MAT.	INST.	TOTAL	MAT.	INST.	TOTAL	MAT.	INST.	TOTAL	MAT.	INST.	TOTAL	MAT.	INST.	TOTAL
015433	CONTRACTOR EQUIPMENT		101.3	101.3		101.3	101.3		101.3	101.3		101.3	101.3		101.3	101.3		98.2	98.2
0241, 31 - 34	SITE & INFRASTRUCTURE, DEMOLITION	99.8	86.3	90.2	108.8	86.3	92.8	104.1	85.9	91.2	101.7	85.0	89.8	103.9	86.0	91.1	99.2	93.7	95.3
0310	Concrete Forming & Accessories	94.4	45.1	51.9	83.2	45.3	50.5	95.0	45.1	52.0	93.2	38.7	46.2	98.0	45.3	52.5	94.8	36.6	44.6
0320	Concrete Reinforcing	95.8	59.1	77.1	92.3	59.3	75.5	92.2	44.5	67.9	93.0	44.1	68.1	92.2	57.0	74.3	96.1	38.1	66.6
0330	Cast-in-Place Concrete	95.8	50.1	77.0	79.2	49.4	66.9	79.2	49.3	66.9	79.2	43.9	64.7	79.2	49.4	66.9	105.6	43.5	80.1
03	CONCRETE	96.2	51.1	74.1	94.2	51.0	73.0	93.0	48.2	71.0	90.9	43.3	67.5	93.2	50.6	72.3	102.8	40.7	72.3
04	MASONRY	91.5	37.6	57.9	79.8	40.9	55.6	77.4	40.9	54.7	101.6	34.6	59.8	79.8	40.9	55.6	115.9	55.4	78.2
05	METALS	100.9	79.6	94.3	102.6	80.0	95.6	102.6	74.5	93.9	101.8	71.2	92.4	102.6	79.1	95.4	95.9	63.6	86.0
06	WOOD, PLASTICS & COMPOSITES	98.0	44.8	68.2	79.7	44.8	60.2	93.7	44.8	66.4	92.0	38.9	62.3	98.0	44.8	68.2	99.6	36.0	64.0
07	THERMAL & MOISTURE PROTECTION	97.2	43.3	75.1	101.9	44.8	78.5	101.8	44.8	78.4	101.6	39.7	76.3	101.8	44.8	78.5	99.2	47.7	78.1
08	OPENINGS	104.6	50.5	92.0	99.1	50.5	87.8	99.0	47.0	86.9	99.0	39.4	85.2	99.0	50.0	87.6	99.1	36.7	84.6
0920	Plaster & Gypsum Board	102.3	42.9	62.2	97.1	42.9	60.5	102.6	42.9	62.3	101.9	36.8	57.9	105.5	42.9	63.2	103.7	34.3	56.8
0950, 0980	Ceilings & Acoustic Treatment	91.3	42.9	59.5	87.3	42.9	58.1	86.4	42.9	57.8	86.4	36.8	53.8	86.4	42.9	57.8	93.2	34.3	54.5
0960	Flooring	100.5	43.6	84.2	98.4	43.6	82.7	104.3	57.2	90.8	103.3	44.4	86.5	105.6	57.2	91.8	108.4	50.3	91.8
0970, 0990	Wall Finishes & Painting/Coating	106.2	66.9	82.5	109.5	66.9	83.8	109.5	66.9	83.8	109.5	39.2	67.0	109.5	66.9	83.8	102.6	37.1	63.1
09	FINISHES	95.9	46.6	68.7	94.4	47.1	68.3	96.3	49.3	70.4	95.7	39.4	64.6	97.1	49.3	70.8	100.3	38.8	66.4
COVERS	DIVS. 10 - 14, 25, 28, 41, 43, 44, 46	100.0	66.3	93.2	100.0	66.3	93.2	100.0	66.3	93.2	100.0	64.6	92.9	100.0	66.3	93.2	100.0	40.5	88.0
21, 22, 23	FIRE SUPPRESSION, PLUMBING & HVAC	100.0	53.0	81.1	100.5	53.1	81.4	100.5	53.0	81.3	95.7	44.2	74.9	100.5	53.1	81.3	100.1	39.8	75.8
26, 27, 3370	ELECTRICAL, COMMUNICATIONS & UTIL.	99.0	59.7	78.3	96.8	59.7	77.2	99.1	57.4	77.0	99.1	56.6	76.6	99.1	57.4	77.1	101.7	51.1	75.0
MF2014	WEIGHTED AVERAGE	99.3	56.5	80.6	98.2	56.9	80.2	98.3	55.8	79.8	97.8	50.3	77.1	98.6	56.7	80.3	100.5	49.6	78.3

	DIVISION	SOUTH DAKOTA																	
		MITCHELL			MOBRIDGE			PIERRE			RAPID CITY			SIOUX FALLS			WATERTOWN		
		573			576			575			577			570 - 571			572		
		MAT.	INST.	TOTAL	MAT.	INST.	TOTAL	MAT.	INST.	TOTAL	MAT.	INST.	TOTAL	MAT.	INST.	TOTAL	MAT.	INST.	TOTAL
015433	**CONTRACTOR EQUIPMENT**		98.2	98.2		98.2	98.2		98.2	98.2		98.2	98.2		99.2	99.2		98.2	98.2
0241, 31 - 34	**SITE & INFRASTRUCTURE, DEMOLITION**	96.0	93.7	94.3	95.9	93.7	94.3	100.5	93.7	95.7	97.7	93.8	94.9	94.3	95.4	95.0	95.8	93.7	94.3
0310	Concrete Forming & Accessories	94.0	37.0	44.8	84.9	36.8	43.4	97.7	38.4	46.6	102.2	36.7	45.7	98.8	40.5	48.5	81.6	36.6	42.8
0320	Concrete Reinforcing	95.5	46.3	70.5	98.0	38.1	67.5	99.0	71.1	84.8	89.9	71.3	80.4	98.0	71.2	84.3	92.9	38.2	65.1
0330	Cast-in-Place Concrete	102.6	42.3	77.8	102.6	43.5	78.3	98.5	42.2	75.4	101.8	42.8	77.6	92.3	42.9	72.0	102.6	45.9	79.3
03	**CONCRETE**	100.5	41.8	71.7	100.3	40.7	71.0	99.0	47.3	73.6	99.8	46.7	73.7	95.9	48.4	72.6	99.2	41.5	70.9
04	**MASONRY**	103.9	53.3	72.4	113.4	55.4	77.2	117.4	53.5	77.6	113.5	56.4	77.9	105.1	53.6	73.0	141.1	57.4	88.9
05	**METALS**	95.0	63.8	85.4	95.0	63.6	85.3	98.4	79.1	92.4	97.8	79.7	92.2	98.2	79.3	92.4	95.0	64.1	85.5
06	**WOOD, PLASTICS & COMPOSITES**	98.5	36.5	63.8	87.6	36.2	58.8	104.5	36.8	66.6	103.5	33.3	64.2	96.8	39.3	64.6	83.8	36.0	57.0
07	**THERMAL & MOISTURE PROTECTION**	98.9	45.6	77.1	99.0	47.8	78.0	101.8	45.9	78.9	99.6	47.8	78.3	101.9	48.2	79.9	98.8	48.3	78.1
08	**OPENINGS**	97.7	37.3	83.6	100.4	36.3	85.5	105.7	46.7	92.0	103.1	44.8	89.6	106.5	48.1	92.9	97.7	36.5	83.4
0920	Plaster & Gypsum Board	102.1	34.8	56.7	96.2	34.6	54.5	100.1	35.1	56.2	103.2	31.6	54.8	90.7	37.7	54.9	93.9	34.3	53.6
0950, 0980	Ceilings & Acoustic Treatment	89.9	34.8	53.7	93.2	34.6	54.7	93.3	35.1	55.0	95.7	31.6	53.6	90.6	37.7	55.8	89.9	34.3	53.3
0960	Flooring	107.9	50.3	91.4	103.5	50.3	88.3	107.4	35.8	86.9	107.6	78.6	99.3	103.8	74.9	95.5	102.1	50.3	87.3
0970, 0990	Wall Finishes & Painting/Coating	102.6	40.6	65.2	102.6	41.7	65.8	105.7	44.6	68.8	102.6	44.6	67.6	101.7	44.6	67.2	102.6	37.1	63.1
09	**FINISHES**	99.1	39.5	66.2	97.7	39.4	65.6	100.8	37.3	65.8	100.5	43.9	69.3	97.1	46.5	69.2	96.2	38.8	64.6
COVERS	**DIVS. 10 - 14, 25, 28, 41, 43, 44, 46**	100.0	37.6	87.4	100.0	40.5	88.0	100.0	75.8	95.1	100.0	75.7	95.1	100.0	76.2	95.2	100.0	40.5	88.0
21, 22, 23	**FIRE SUPPRESSION, PLUMBING & HVAC**	95.3	39.4	72.8	95.3	39.9	73.0	100.0	64.5	85.7	100.1	65.0	85.9	100.0	38.6	75.2	95.3	39.9	73.0
26, 27, 3370	**ELECTRICAL, COMMUNICATIONS & UTIL.**	99.9	44.0	70.3	101.7	44.8	71.7	104.8	51.8	76.8	98.0	51.8	73.6	101.3	75.1	87.5	99.0	44.8	70.4
MF2014	**WEIGHTED AVERAGE**	97.8	48.4	76.3	98.5	48.8	76.8	101.7	58.4	82.8	100.4	59.5	82.6	100.0	57.7	81.5	99.0	49.1	77.2

	DIVISION	TENNESSEE																	
		CHATTANOOGA			COLUMBIA			COOKEVILLE			JACKSON			JOHNSON CITY			KNOXVILLE		
		373 - 374			384			385			383			376			377 - 379		
		MAT.	INST.	TOTAL	MAT.	INST.	TOTAL	MAT.	INST.	TOTAL	MAT.	INST.	TOTAL	MAT.	INST.	TOTAL	MAT.	INST.	TOTAL
015433	**CONTRACTOR EQUIPMENT**		103.8	103.8		98.2	98.2		98.2	98.2		104.5	104.5		97.7	97.7		97.7	97.7
0241, 31 - 34	**SITE & INFRASTRUCTURE, DEMOLITION**	104.5	97.9	99.8	89.1	86.9	87.5	94.5	86.4	88.8	97.7	96.8	97.1	111.1	86.6	93.7	90.7	87.3	88.3
0310	Concrete Forming & Accessories	97.1	56.4	62.0	82.1	62.0	64.8	82.2	35.4	41.8	89.0	44.9	51.0	83.7	38.9	45.1	95.6	61.0	65.7
0320	Concrete Reinforcing	91.2	67.4	79.1	87.4	61.2	74.1	87.4	61.2	74.1	87.4	62.0	74.4	91.7	60.3	75.7	91.2	62.3	76.5
0330	Cast-in-Place Concrete	101.3	62.7	85.5	94.2	51.6	76.7	106.7	43.5	80.7	104.2	49.3	81.6	81.6	60.2	72.8	95.2	65.7	83.1
03	**CONCRETE**	96.0	62.3	79.4	96.7	59.8	78.6	106.9	45.2	76.6	97.7	51.6	75.1	103.3	52.3	78.3	93.4	64.4	79.1
04	**MASONRY**	109.7	51.7	73.5	127.7	54.6	82.1	121.9	43.8	73.2	127.9	47.2	77.6	125.7	44.8	75.3	86.6	55.3	67.1
05	**METALS**	98.7	89.2	95.8	96.4	86.0	93.2	96.5	85.9	93.2	98.8	85.9	94.8	95.9	85.2	92.6	99.3	87.0	95.5
06	**WOOD, PLASTICS & COMPOSITES**	104.0	57.3	77.8	70.1	64.4	66.9	70.3	33.7	49.8	85.4	45.4	63.0	76.9	36.9	54.5	91.0	60.6	73.9
07	**THERMAL & MOISTURE PROTECTION**	99.4	59.9	83.2	93.6	59.2	79.5	94.0	45.7	74.2	96.0	54.1	78.8	94.5	52.6	77.3	92.5	61.6	79.8
08	**OPENINGS**	99.7	57.6	89.9	92.6	57.9	84.6	92.7	42.2	80.9	100.3	51.6	89.0	96.2	45.1	84.3	93.3	58.6	85.2
0920	Plaster & Gypsum Board	82.8	56.5	65.0	87.7	63.7	71.5	87.7	32.1	50.1	89.3	44.1	58.8	102.4	35.5	57.2	109.0	59.8	75.8
0950, 0980	Ceilings & Acoustic Treatment	96.6	56.5	70.2	77.0	63.7	68.3	77.0	32.1	47.5	86.2	44.1	58.5	93.0	35.5	55.2	93.8	59.8	71.5
0960	Flooring	99.1	58.9	87.7	89.3	20.5	69.7	89.4	58.6	80.6	87.1	40.8	73.9	96.7	40.6	80.7	101.3	55.8	88.3
0970, 0990	Wall Finishes & Painting/Coating	102.5	70.7	83.3	94.0	33.4	57.4	94.0	36.5	59.3	95.8	49.6	67.9	99.6	42.1	64.9	99.6	81.2	88.5
09	**FINISHES**	97.2	58.2	75.7	89.8	52.0	69.0	90.3	38.4	61.7	89.7	43.7	64.4	101.9	38.1	66.7	94.5	62.0	76.6
COVERS	**DIVS. 10 - 14, 25, 28, 41, 43, 44, 46**	100.0	41.8	88.3	100.0	47.9	89.5	100.0	41.0	88.1	100.0	43.0	88.5	100.0	75.5	95.1	100.0	81.7	96.3
21, 22, 23	**FIRE SUPPRESSION, PLUMBING & HVAC**	100.2	62.6	85.0	97.4	78.1	89.6	97.4	72.6	87.4	100.1	68.0	87.1	99.9	59.1	83.4	99.9	67.9	87.0
26, 27, 3370	**ELECTRICAL, COMMUNICATIONS & UTIL.**	102.3	69.2	84.8	94.2	57.4	74.7	95.9	61.6	77.8	101.0	58.5	78.5	92.5	46.3	68.1	98.1	56.0	75.8
MF2014	**WEIGHTED AVERAGE**	100.0	66.2	85.2	96.9	66.1	83.5	98.1	59.1	81.1	99.9	61.2	83.0	99.8	56.5	80.9	96.6	66.9	83.7

	DIVISION	TENNESSEE									TEXAS								
		MCKENZIE			MEMPHIS			NASHVILLE			ABILENE			AMARILLO			AUSTIN		
		382			375,380 - 381			370 - 372			795 - 796			790 - 791			786 - 787		
		MAT.	INST.	TOTAL	MAT.	INST.	TOTAL	MAT.	INST.	TOTAL	MAT.	INST.	TOTAL	MAT.	INST.	TOTAL	MAT.	INST.	TOTAL
015433	**CONTRACTOR EQUIPMENT**		98.2	98.2		102.8	102.8		103.7	103.7		89.9	89.9		89.9	89.9		89.4	89.4
0241, 31 - 34	**SITE & INFRASTRUCTURE, DEMOLITION**	94.3	86.5	88.7	94.1	93.7	93.8	98.9	97.5	97.9	99.4	88.7	91.8	97.3	88.6	91.1	102.4	88.4	92.4
0310	Concrete Forming & Accessories	90.0	38.1	45.2	94.3	64.9	69.0	98.9	65.8	70.3	99.0	65.5	70.1	97.9	57.1	62.7	98.6	59.1	64.5
0320	Concrete Reinforcing	87.5	61.9	74.5	99.7	68.7	83.9	96.0	67.7	81.6	92.9	52.5	72.3	99.0	51.3	74.7	96.5	48.5	72.1
0330	Cast-in-Place Concrete	104.4	57.3	85.1	94.1	61.3	80.6	90.3	65.9	80.3	96.6	63.7	83.1	94.1	63.7	81.6	93.4	66.3	82.3
03	**CONCRETE**	105.5	51.3	78.9	94.1	65.8	80.2	92.4	67.6	80.2	95.0	63.0	79.3	96.7	59.0	78.2	95.7	60.2	78.3
04	**MASONRY**	126.2	46.2	76.3	99.8	62.7	76.7	93.8	59.0	72.1	104.2	63.6	78.9	109.7	63.6	81.0	107.6	56.5	75.7
05	**METALS**	96.5	86.1	93.3	100.1	90.6	97.2	100.4	89.9	97.2	105.3	69.8	94.4	101.6	69.2	91.7	100.5	66.3	90.0
06	**WOOD, PLASTICS & COMPOSITES**	79.1	36.3	55.1	95.4	67.0	79.5	101.6	66.6	82.0	101.1	68.5	82.8	100.8	57.1	76.4	94.4	59.9	75.0
07	**THERMAL & MOISTURE PROTECTION**	94.0	48.7	75.4	95.5	63.4	82.4	95.7	61.5	81.7	100.0	67.7	86.7	97.9	65.3	84.5	104.2	64.8	88.1
08	**OPENINGS**	92.7	44.1	81.4	100.1	66.0	92.2	98.5	65.9	90.9	97.1	63.3	89.2	101.1	56.8	90.8	104.3	55.6	92.9
0920	Plaster & Gypsum Board	90.5	34.8	52.8	95.6	66.3	75.8	95.8	66.0	75.6	84.0	68.0	73.2	92.1	56.3	67.9	86.7	59.2	68.1
0950, 0980	Ceilings & Acoustic Treatment	77.0	34.8	49.2	93.9	66.3	75.8	95.5	66.0	76.1	88.9	68.0	75.2	98.0	56.3	70.6	87.9	59.2	69.0
0960	Flooring	92.1	39.3	77.0	99.4	36.1	81.3	97.7	64.2	88.2	110.9	76.5	101.1	106.5	76.5	97.9	105.7	64.1	93.8
0970, 0990	Wall Finishes & Painting/Coating	94.0	49.6	67.2	97.0	55.7	72.1	100.5	72.0	83.3	106.6	56.1	76.1	100.8	56.1	73.8	104.3	48.8	70.8
09	**FINISHES**	91.3	38.2	62.1	97.3	58.5	75.9	100.2	66.0	81.4	93.7	67.1	79.0	96.8	60.3	76.7	95.6	58.7	75.2
COVERS	**DIVS. 10 - 14, 25, 28, 41, 43, 44, 46**	100.0	27.4	85.3	100.0	82.4	96.4	100.0	82.9	96.5	100.0	82.2	96.4	100.0	68.6	93.7	100.0	80.5	96.1
21, 22, 23	**FIRE SUPPRESSION, PLUMBING & HVAC**	97.4	68.3	85.6	100.0	73.9	89.5	100.0	84.0	93.5	100.3	49.0	79.6	100.0	54.9	81.8	100.1	60.1	84.0
26, 27, 3370	**ELECTRICAL, COMMUNICATIONS & UTIL.**	95.6	62.9	78.3	101.7	65.3	82.4	97.6	63.1	79.4	99.3	48.7	72.5	101.1	64.1	81.6	96.8	63.9	79.4
MF2014	**WEIGHTED AVERAGE**	98.2	59.2	81.2	99.0	71.3	86.9	98.4	74.1	87.8	99.7	62.2	83.3	100.2	63.2	84.1	100.0	63.6	84.1

	DIVISION	TEXAS																	
		BEAUMONT			BROWNWOOD			BRYAN			CHILDRESS			CORPUS CHRISTI			DALLAS		
		776 - 777			768			778			792			783 - 784			752 - 753		
		MAT.	INST.	TOTAL	MAT.	INST.	TOTAL	MAT.	INST.	TOTAL	MAT.	INST.	TOTAL	MAT.	INST.	TOTAL	MAT.	INST.	TOTAL
015433	**CONTRACTOR EQUIPMENT**		91.5	91.5		89.9	89.9		91.5	91.5		89.9	89.9		96.6	96.6		99.1	99.1
0241, 31 - 34	**SITE & INFRASTRUCTURE, DEMOLITION**	90.0	89.9	89.9	105.2	89.6	94.1	81.8	90.5	88.0	109.7	88.0	94.3	140.1	84.4	100.5	104.3	89.8	94.0
0310	Concrete Forming & Accessories	101.8	61.9	67.4	99.0	60.5	65.8	81.0	67.1	69.0	97.3	65.5	69.8	100.2	58.1	63.9	101.2	65.8	70.7
0320	Concrete Reinforcing	95.9	65.2	80.3	92.9	52.1	72.1	98.3	48.6	73.0	93.0	52.1	72.2	88.0	48.5	67.9	100.2	52.5	75.9
0330	Cast-in-Place Concrete	89.5	64.7	79.3	101.4	63.6	85.9	71.7	65.4	69.1	99.1	63.7	84.5	109.8	64.4	91.2	98.0	64.6	84.3
03	**CONCRETE**	96.3	64.2	80.5	101.2	60.7	81.3	80.9	63.7	72.5	104.0	62.9	83.8	100.1	59.9	80.4	99.0	64.1	81.9
04	**MASONRY**	106.8	63.9	80.0	141.8	56.6	88.7	146.3	60.4	92.7	108.5	56.6	76.2	91.0	56.6	69.6	102.4	56.7	73.9
05	**METALS**	95.2	75.5	89.1	99.9	69.1	90.5	94.8	70.1	87.2	102.7	69.2	92.4	94.7	77.9	89.5	100.4	80.0	94.1
06	**WOOD, PLASTICS & COMPOSITES**	106.1	61.8	81.3	99.5	61.9	78.4	73.5	68.5	70.7	100.4	68.5	82.6	116.0	58.4	83.7	104.1	68.6	84.2
07	**THERMAL & MOISTURE PROTECTION**	101.9	66.7	87.5	98.2	65.2	84.7	93.9	66.3	82.5	100.5	64.8	85.9	104.5	66.2	88.8	93.0	66.4	82.1
08	**OPENINGS**	94.0	62.3	86.7	94.7	59.7	86.6	97.2	61.6	88.9	94.2	63.3	87.1	109.1	54.7	96.5	99.5	63.5	91.1
0920	Plaster & Gypsum Board	102.3	61.2	74.5	87.8	61.2	69.8	90.3	68.0	75.2	83.6	68.0	73.1	98.4	57.5	70.7	96.7	68.0	77.3
0950, 0980	Ceilings & Acoustic Treatment	98.2	61.2	73.9	80.6	61.2	67.8	91.2	68.0	76.0	87.3	68.0	74.6	92.7	57.5	69.6	97.8	68.0	78.2
0960	Flooring	112.4	73.0	101.1	95.3	64.1	86.3	85.8	64.1	79.6	109.1	64.1	96.2	118.7	64.1	103.1	102.8	64.1	91.7
0970, 0990	Wall Finishes & Painting/Coating	95.1	59.4	73.6	104.2	56.1	75.1	92.5	62.2	74.2	106.6	56.1	76.1	120.1	56.5	81.7	107.0	56.1	76.3
09	**FINISHES**	93.7	63.4	77.0	87.7	60.7	72.8	82.3	66.2	73.4	94.0	64.6	77.8	105.9	58.7	79.9	100.0	64.7	80.5
COVERS	**DIVS. 10 - 14, 25, 28, 41, 43, 44, 46**	100.0	83.4	96.7	100.0	81.4	96.3	100.0	82.7	96.5	100.0	82.2	96.4	100.0	82.6	96.5	100.0	82.5	96.5
21, 22, 23	**FIRE SUPPRESSION, PLUMBING & HVAC**	100.2	64.1	85.6	95.4	50.0	77.1	95.4	66.5	83.8	95.5	54.9	79.1	100.2	58.9	83.5	100.0	62.4	84.9
26, 27, 3370	**ELECTRICAL, COMMUNICATIONS & UTIL.**	95.2	70.6	82.2	93.4	42.8	66.7	93.4	67.4	79.7	99.2	64.1	80.7	92.8	60.0	75.5	95.2	64.6	79.0
MF2014	**WEIGHTED AVERAGE**	97.4	68.6	84.9	98.9	59.4	81.7	94.9	68.2	83.3	99.3	64.3	84.0	100.9	63.5	84.6	99.5	67.4	85.5

	DIVISION	TEXAS																	
		DEL RIO			DENTON			EASTLAND			EL PASO			FORT WORTH			GALVESTON		
		788			762			764			798 - 799,885			760 - 761			775		
		MAT.	INST.	TOTAL	MAT.	INST.	TOTAL	MAT.	INST.	TOTAL	MAT.	INST.	TOTAL	MAT.	INST.	TOTAL	MAT.	INST.	TOTAL
015433	**CONTRACTOR EQUIPMENT**		89.4	89.4		95.8	95.8		89.9	89.9		89.9	89.9		89.9	89.9		100.5	100.5
0241, 31 - 34	**SITE & INFRASTRUCTURE, DEMOLITION**	121.8	88.4	98.0	105.5	81.3	88.3	108.0	87.6	93.5	101.7	87.6	91.7	102.1	88.7	92.6	107.2	88.4	93.8
0310	Concrete Forming & Accessories	96.5	57.8	63.1	106.2	65.3	70.9	99.8	65.3	70.0	97.8	63.4	68.1	98.0	65.6	70.1	89.8	64.6	68.1
0320	Concrete Reinforcing	88.7	48.4	68.2	94.3	52.1	72.8	93.1	50.9	71.6	99.2	52.0	75.2	98.4	52.4	75.0	97.8	62.4	79.8
0330	Cast-in-Place Concrete	118.2	63.2	95.6	78.5	64.6	72.8	107.2	63.6	89.3	88.4	63.7	78.3	95.0	63.7	82.1	95.2	66.4	83.4
03	**CONCRETE**	121.3	58.6	90.5	80.0	63.9	72.1	105.9	62.6	84.6	94.0	61.9	78.3	97.0	63.1	80.4	97.5	66.4	82.2
04	**MASONRY**	106.9	56.5	75.5	152.1	56.7	92.6	106.8	56.6	75.5	99.0	57.7	73.3	104.7	56.6	74.7	102.5	60.4	76.3
05	**METALS**	94.4	66.2	85.7	99.5	80.6	93.7	99.7	68.5	90.1	98.5	67.4	89.0	101.5	69.5	91.7	96.3	88.3	93.9
06	**WOOD, PLASTICS & COMPOSITES**	96.5	58.3	75.1	111.3	68.6	87.4	106.1	68.5	85.1	91.3	65.7	76.9	97.3	68.5	81.2	89.4	64.7	75.6
07	**THERMAL & MOISTURE PROTECTION**	101.3	65.3	86.5	96.0	66.7	84.0	98.6	65.9	85.1	99.8	64.4	85.3	96.0	65.9	83.6	93.0	67.1	82.4
08	**OPENINGS**	101.8	54.7	90.8	112.2	63.4	100.8	70.4	63.0	68.7	93.3	58.1	85.1	99.4	63.4	91.0	101.6	63.2	92.7
0920	Plaster & Gypsum Board	94.7	57.5	69.6	92.1	68.0	75.8	87.8	68.0	74.4	95.9	65.1	75.1	91.8	68.0	75.7	96.8	64.0	74.7
0950, 0980	Ceilings & Acoustic Treatment	89.1	57.5	68.3	84.7	68.0	73.8	80.6	68.0	72.3	91.4	65.1	74.1	90.5	68.0	75.7	94.6	64.0	74.5
0960	Flooring	100.7	64.1	90.2	90.4	64.1	82.9	121.4	64.1	105.0	110.4	64.1	97.1	117.9	64.1	102.5	100.1	64.1	89.8
0970, 0990	Wall Finishes & Painting/Coating	106.4	48.8	71.6	115.6	56.1	79.7	105.8	56.1	75.8	106.2	53.7	74.5	103.2	56.1	74.7	103.4	60.0	77.2
09	**FINISHES**	98.2	57.7	75.9	85.8	64.7	74.2	95.8	64.6	78.6	97.3	63.0	78.4	98.4	64.6	79.8	91.4	63.7	76.2
COVERS	**DIVS. 10 - 14, 25, 28, 41, 43, 44, 46**	100.0	80.5	96.1	100.0	82.5	96.5	100.0	82.2	96.4	100.0	81.0	96.2	100.0	82.3	96.4	100.0	84.5	96.9
21, 22, 23	**FIRE SUPPRESSION, PLUMBING & HVAC**	95.4	57.3	80.0	95.4	42.6	74.1	95.4	49.9	77.1	100.0	50.3	80.0	100.0	58.4	83.2	95.4	66.7	83.8
26, 27, 3370	**ELECTRICAL, COMMUNICATIONS & UTIL.**	94.9	68.8	81.1	95.9	64.6	79.3	93.3	64.5	78.1	99.1	57.4	77.0	94.7	64.5	78.8	95.0	67.9	80.7
MF2014	**WEIGHTED AVERAGE**	100.8	63.2	84.4	98.9	62.5	83.0	95.9	63.2	81.6	98.0	61.7	82.2	99.3	65.3	84.5	97.0	70.0	85.2

	DIVISION	TEXAS																	
		GIDDINGS			GREENVILLE			HOUSTON			HUNTSVILLE			LAREDO			LONGVIEW		
		789			754			770 - 772			773			780			756		
		MAT.	INST.	TOTAL	MAT.	INST.	TOTAL	MAT.	INST.	TOTAL	MAT.	INST.	TOTAL	MAT.	INST.	TOTAL	MAT.	INST.	TOTAL
015433	**CONTRACTOR EQUIPMENT**		89.4	89.4		96.7	96.7		100.4	100.4		91.5	91.5		89.4	89.4		91.4	91.4
0241, 31 - 34	**SITE & INFRASTRUCTURE, DEMOLITION**	107.8	88.4	94.0	97.9	85.4	89.0	106.0	88.2	93.4	96.6	89.9	91.8	102.1	88.4	92.3	95.8	92.3	93.3
0310	Concrete Forming & Accessories	94.1	57.9	62.9	92.3	65.3	69.0	91.8	64.6	68.4	87.8	61.6	65.2	96.6	58.0	63.3	88.4	64.9	68.1
0320	Concrete Reinforcing	89.2	45.9	67.2	100.7	52.1	76.0	97.5	61.4	79.1	98.5	50.6	74.1	88.7	48.8	68.4	99.6	52.0	75.4
0330	Cast-in-Place Concrete	100.2	63.2	85.0	89.3	63.9	78.9	92.3	66.4	81.7	98.6	64.6	84.6	84.6	63.3	75.8	103.8	62.8	86.9
03	**CONCRETE**	97.6	58.2	78.2	91.6	63.6	77.8	95.2	66.2	80.9	103.9	61.3	83.0	92.5	58.8	75.9	107.5	62.3	85.3
04	**MASONRY**	115.7	56.5	78.8	160.0	56.6	95.6	102.3	66.5	80.0	145.1	58.9	91.4	100.2	63.5	77.3	155.7	56.6	93.9
05	**METALS**	93.9	65.3	85.1	97.7	79.3	92.1	99.2	88.2	95.8	94.7	69.2	86.8	97.0	66.9	87.8	91.1	67.8	83.9
06	**WOOD, PLASTICS & COMPOSITES**	95.6	58.3	74.7	93.5	68.6	79.5	91.7	64.7	76.6	82.0	61.8	70.7	96.5	58.3	75.1	87.2	68.4	76.7
07	**THERMAL & MOISTURE PROTECTION**	102.0	64.5	86.6	93.0	66.0	81.9	92.6	67.8	82.4	94.9	65.9	83.0	100.2	66.5	86.4	94.5	65.2	82.5
08	**OPENINGS**	100.8	54.0	89.9	97.7	63.4	89.7	104.3	62.8	94.6	97.2	58.5	88.2	101.6	54.7	90.7	87.8	63.3	82.1
0920	Plaster & Gypsum Board	93.7	57.5	69.2	90.3	68.0	75.3	99.1	64.0	75.4	94.2	61.2	71.9	95.8	57.5	69.9	89.1	68.0	74.9
0950, 0980	Ceilings & Acoustic Treatment	89.1	57.5	68.3	93.6	68.0	76.8	99.6	64.0	76.2	91.2	61.2	71.5	93.3	57.5	69.8	90.3	68.0	75.7
0960	Flooring	101.1	64.1	90.5	98.6	64.1	88.7	101.2	64.1	90.6	89.4	64.1	82.1	100.5	64.1	90.1	103.5	64.1	92.2
0970, 0990	Wall Finishes & Painting/Coating	106.4	48.8	71.6	107.0	56.1	76.3	103.4	62.2	78.5	92.5	59.4	72.5	106.4	53.2	74.3	96.8	56.1	72.2
09	**FINISHES**	97.1	57.7	75.4	96.6	64.6	79.0	98.9	64.0	79.7	84.7	61.6	72.0	97.7	58.2	75.9	101.7	64.5	81.2
COVERS	**DIVS. 10 - 14, 25, 28, 41, 43, 44, 46**	100.0	80.8	96.1	100.0	82.3	96.4	100.0	84.5	96.9	100.0	81.0	96.2	100.0	80.5	96.1	100.0	82.0	96.4
21, 22, 23	**FIRE SUPPRESSION, PLUMBING & HVAC**	95.4	64.6	83.0	95.3	49.0	76.6	100.1	66.7	86.6	95.4	65.9	83.5	100.2	64.1	85.6	95.3	33.6	70.4
26, 27, 3370	**ELECTRICAL, COMMUNICATIONS & UTIL.**	91.8	63.9	77.1	92.0	64.6	77.5	96.9	67.5	81.4	93.4	68.0	80.0	95.1	60.9	77.0	92.2	55.6	72.9
MF2014	**WEIGHTED AVERAGE**	97.7	64.0	83.0	98.8	64.0	83.6	99.4	70.5	86.8	98.0	66.7	84.4	98.3	64.5	83.6	98.5	58.7	81.2

DIVISION		TEXAS																	
		LUBBOCK			LUFKIN			MCALLEN			MCKINNEY			MIDLAND			ODESSA		
		793 - 794			759			785			750			797			797		
		MAT.	INST.	TOTAL	MAT.	INST.	TOTAL	MAT.	INST.	TOTAL	MAT.	INST.	TOTAL	MAT.	INST.	TOTAL	MAT.	INST.	TOTAL
015433	CONTRACTOR EQUIPMENT		98.5	98.5		91.4	91.4		96.8	96.8		96.7	96.7		98.5	98.5		89.9	89.9
0241, 31 - 34	SITE & INFRASTRUCTURE, DEMOLITION	123.8	86.1	97.0	91.2	94.0	93.1	144.1	84.4	101.7	94.6	85.3	88.0	126.7	86.1	97.8	99.7	88.0	91.4
0310	Concrete Forming & Accessories	97.9	57.2	62.8	91.7	61.7	65.9	101.2	57.5	63.5	91.3	65.3	68.8	101.9	65.4	70.5	98.9	65.4	70.0
0320	Concrete Reinforcing	94.1	52.5	72.9	101.3	64.9	82.8	88.2	48.4	67.9	100.7	50.9	75.3	95.1	52.1	73.2	92.9	52.1	72.1
0330	Cast-in-Place Concrete	96.8	64.7	83.6	92.9	64.0	81.0	119.3	64.1	96.6	83.8	63.8	75.6	102.9	64.6	87.2	96.6	63.6	83.1
03	CONCRETE	93.7	60.4	77.4	99.8	63.6	82.0	107.9	59.5	84.1	87.0	63.3	75.3	98.2	64.0	81.4	95.0	62.8	79.2
04	MASONRY	103.6	64.0	78.9	119.4	58.8	81.6	106.8	56.4	75.4	172.2	56.6	100.2	121.5	56.7	81.1	104.2	56.6	74.6
05	METALS	109.0	81.3	100.5	97.9	72.4	90.0	94.3	77.5	89.2	97.6	78.8	91.8	107.2	80.7	99.0	104.6	69.1	93.7
06	WOOD, PLASTICS & COMPOSITES	101.1	57.3	76.5	95.3	61.8	76.5	114.6	58.4	83.1	92.3	68.6	79.0	106.0	68.6	85.1	101.1	68.5	82.8
07	THERMAL & MOISTURE PROTECTION	89.3	66.2	79.8	94.3	65.6	82.5	104.8	59.8	86.3	92.8	66.0	81.8	89.5	65.4	79.7	100.0	65.9	86.0
08	OPENINGS	108.5	57.2	96.6	66.7	63.0	65.9	105.7	54.7	93.9	97.7	63.0	89.6	107.5	63.4	97.2	97.1	63.3	89.2
0920	Plaster & Gypsum Board	84.4	56.3	65.4	88.0	61.2	69.9	99.4	57.5	71.1	90.0	68.0	75.1	85.9	68.0	73.8	84.0	68.0	73.2
0950, 0980	Ceilings & Acoustic Treatment	90.6	56.3	68.1	84.4	61.2	69.1	93.3	57.5	69.8	93.6	68.0	76.8	88.1	68.0	74.9	88.9	68.0	75.2
0960	Flooring	104.4	64.1	92.8	138.2	64.1	117.0	118.1	63.8	102.6	98.2	64.1	88.4	105.7	64.1	93.8	110.9	64.1	97.5
0970, 0990	Wall Finishes & Painting/Coating	119.0	56.1	81.0	96.8	56.1	72.2	120.1	48.8	77.1	107.0	56.1	76.3	119.0	56.1	81.0	106.6	56.1	76.1
09	FINISHES	96.4	58.0	75.2	110.2	61.2	83.2	106.4	57.8	79.6	96.2	64.6	78.8	96.8	64.7	79.1	93.7	64.6	77.7
COVERS	DIVS. 10 - 14, 25, 28, 41, 43, 44, 46	100.0	81.3	96.2	100.0	80.9	96.1	100.0	80.7	96.1	100.0	82.3	96.4	100.0	81.2	96.2	100.0	80.9	96.1
21, 22, 23	FIRE SUPPRESSION, PLUMBING & HVAC	99.8	50.7	80.0	95.3	64.4	82.8	95.4	58.7	80.6	95.3	45.4	75.2	95.0	49.1	76.5	100.3	49.1	79.6
26, 27, 3370	ELECTRICAL, COMMUNICATIONS & UTIL.	97.9	64.2	80.1	93.5	70.6	81.4	92.6	36.3	62.8	92.1	64.6	77.5	97.9	64.2	80.1	99.4	64.1	80.7
MF2014	WEIGHTED AVERAGE	101.6	63.6	85.0	95.6	67.8	83.5	101.0	59.8	83.0	98.7	63.1	83.2	101.5	64.2	85.2	99.6	63.1	83.7

DIVISION		TEXAS																	
		PALESTINE			SAN ANGELO			SAN ANTONIO			TEMPLE			TEXARKANA			TYLER		
		758			769			781 - 782			765			755			757		
		MAT.	INST.	TOTAL	MAT.	INST.	TOTAL	MAT.	INST.	TOTAL	MAT.	INST.	TOTAL	MAT.	INST.	TOTAL	MAT.	INST.	TOTAL
015433	CONTRACTOR EQUIPMENT		91.4	91.4		89.9	89.9		91.9	91.9		89.9	89.9		91.4	91.4		91.4	91.4
0241, 31 - 34	SITE & INFRASTRUCTURE, DEMOLITION	96.6	92.6	93.8	101.5	89.7	93.1	100.9	92.0	94.5	89.8	89.1	89.3	85.9	92.0	90.2	95.1	92.4	93.2
0310	Concrete Forming & Accessories	82.7	65.4	67.8	99.3	58.7	64.3	95.0	58.1	63.2	102.8	57.8	64.0	99.0	57.4	63.1	93.6	65.4	69.3
0320	Concrete Reinforcing	98.8	52.1	75.1	92.7	52.1	72.1	93.6	49.8	71.3	92.9	49.1	70.6	98.7	52.3	75.1	99.6	52.4	75.6
0330	Cast-in-Place Concrete	84.9	63.0	75.9	95.7	64.8	83.0	85.6	65.2	77.2	78.5	63.1	72.2	85.5	62.7	76.2	101.9	63.0	85.9
03	CONCRETE	101.6	62.6	82.4	96.7	60.3	78.8	91.7	59.7	76.0	82.9	58.7	71.0	92.8	59.0	76.2	107.1	62.7	85.3
04	MASONRY	113.8	56.6	78.2	138.0	58.7	88.5	96.8	65.3	77.2	151.0	56.4	92.1	176.5	56.6	101.8	166.0	56.6	97.8
05	METALS	97.6	68.3	88.6	100.1	69.1	90.6	96.9	68.6	88.2	99.8	66.5	89.6	91.0	67.8	83.9	97.5	68.4	88.5
06	WOOD, PLASTICS & COMPOSITES	85.3	68.4	75.8	99.8	58.3	76.5	94.7	57.4	73.8	108.9	58.3	80.6	99.7	58.2	76.5	97.0	68.4	81.0
07	THERMAL & MOISTURE PROTECTION	94.8	65.3	82.7	98.0	66.0	84.9	96.4	68.2	84.8	97.7	64.8	84.2	94.1	64.2	81.9	94.6	65.3	82.6
08	OPENINGS	66.7	63.3	65.9	94.7	57.8	86.1	102.7	54.4	91.5	67.1	57.0	64.7	87.7	57.8	80.8	66.6	63.4	65.9
0920	Plaster & Gypsum Board	85.2	68.0	73.6	87.8	57.5	67.3	98.0	56.5	70.0	87.8	57.5	67.3	93.4	57.5	69.1	88.0	68.0	74.5
0950, 0980	Ceilings & Acoustic Treatment	84.4	68.0	73.6	80.6	57.5	65.4	102.7	56.5	72.3	80.6	57.5	65.4	90.3	57.5	68.7	84.4	68.0	73.6
0960	Flooring	129.9	64.1	111.1	95.3	64.1	86.4	104.1	64.1	92.6	123.2	64.1	106.3	111.1	64.1	97.6	140.1	64.1	118.4
0970, 0990	Wall Finishes & Painting/Coating	96.8	56.1	72.2	104.2	56.1	75.1	106.3	53.2	74.2	105.8	48.8	71.4	96.8	56.1	72.2	96.8	56.1	72.2
09	FINISHES	108.0	64.5	84.0	87.5	59.1	71.8	103.4	58.1	78.4	95.1	57.7	74.5	104.0	58.5	78.9	111.1	64.5	85.4
COVERS	DIVS. 10 - 14, 25, 28, 41, 43, 44, 46	100.0	82.0	96.4	100.0	81.0	96.2	100.0	81.2	96.2	100.0	81.1	96.2	100.0	80.9	96.1	100.0	82.0	96.4
21, 22, 23	FIRE SUPPRESSION, PLUMBING & HVAC	95.3	59.6	80.9	95.4	50.1	77.1	100.0	63.2	85.1	95.4	54.7	79.0	95.3	37.0	71.8	95.3	62.4	82.0
26, 27, 3370	ELECTRICAL, COMMUNICATIONS & UTIL.	89.8	55.6	71.7	97.4	44.8	69.6	96.4	60.9	77.6	94.5	59.8	76.2	93.4	60.1	75.8	92.2	55.5	72.8
MF2014	WEIGHTED AVERAGE	95.0	64.4	81.7	98.5	59.6	81.5	98.5	65.1	84.0	94.7	61.7	80.3	98.0	58.4	80.7	98.5	65.0	83.9

DIVISION		TEXAS															UTAH		
		VICTORIA			WACO			WAXAHACKIE			WHARTON			WICHITA FALLS			LOGAN		
		779			766 - 767			751			774			763			843		
		MAT.	INST.	TOTAL	MAT.	INST.	TOTAL	MAT.	INST.	TOTAL	MAT.	INST.	TOTAL	MAT.	INST.	TOTAL	MAT.	INST.	TOTAL
015433	CONTRACTOR EQUIPMENT		99.3	99.3		89.9	89.9		96.7	96.7		100.5	100.5		89.9	89.9		97.5	97.5
0241, 31 - 34	SITE & INFRASTRUCTURE, DEMOLITION	111.8	85.7	93.2	98.6	88.7	91.6	96.0	85.4	88.4	117.1	87.8	96.2	99.3	88.7	91.8	96.0	95.7	95.8
0310	Concrete Forming & Accessories	90.0	59.0	63.3	101.2	65.5	70.4	91.3	65.5	69.0	84.9	61.7	64.9	101.2	65.5	70.4	104.7	58.1	64.5
0320	Concrete Reinforcing	93.9	48.5	70.8	92.5	48.5	70.1	100.7	50.9	75.4	97.7	50.6	73.7	92.5	51.3	71.5	99.6	81.3	90.3
0330	Cast-in-Place Concrete	106.9	65.6	89.9	84.9	66.8	77.4	88.4	63.9	78.3	109.8	65.5	91.6	90.7	63.7	79.6	87.3	73.4	81.6
03	CONCRETE	104.9	60.9	83.3	89.7	63.3	76.7	90.6	63.4	77.2	109.1	62.5	86.2	92.4	62.8	77.9	106.0	68.3	87.5
04	MASONRY	120.0	58.9	81.9	105.2	56.6	74.9	160.7	56.6	95.8	103.6	58.9	75.8	105.7	63.6	79.5	105.8	63.2	79.3
05	METALS	94.9	80.7	90.5	102.2	67.8	91.6	97.7	79.0	92.0	96.3	81.6	91.8	102.2	69.8	92.2	102.2	79.0	95.0
06	WOOD, PLASTICS & COMPOSITES	92.5	58.4	73.4	107.1	68.5	85.5	92.3	68.6	79.0	83.1	62.0	71.3	107.1	68.5	85.5	85.3	54.9	68.3
07	THERMAL & MOISTURE PROTECTION	96.8	66.9	84.6	98.5	65.8	85.1	92.9	66.0	81.9	93.4	66.6	82.4	98.5	67.7	85.9	97.4	67.9	85.3
08	OPENINGS	101.4	56.1	90.8	77.7	62.4	74.1	97.7	63.0	89.6	101.6	58.6	91.6	77.7	63.3	74.4	94.3	56.5	85.5
0920	Plaster & Gypsum Board	93.8	57.5	69.3	88.2	68.0	74.6	90.4	68.0	75.3	92.5	61.2	71.3	88.2	68.0	74.6	78.9	53.5	61.7
0950, 0980	Ceilings & Acoustic Treatment	95.4	57.5	70.5	82.2	68.0	72.9	95.3	68.0	77.4	94.6	61.2	72.6	82.2	68.0	72.9	97.2	53.5	68.5
0960	Flooring	99.5	64.1	89.4	122.4	64.1	105.8	98.2	64.1	88.4	97.7	64.1	88.1	123.4	76.5	110.0	101.2	47.7	85.9
0970, 0990	Wall Finishes & Painting/Coating	103.6	59.4	76.9	105.8	48.8	71.4	107.0	56.1	76.3	103.4	59.4	76.8	108.5	56.1	76.8	103.3	60.7	77.5
09	FINISHES	89.4	59.6	73.0	95.8	63.8	78.1	96.8	64.6	79.1	90.8	61.7	74.8	96.3	67.1	80.2	96.2	55.3	73.6
COVERS	DIVS. 10 - 14, 25, 28, 41, 43, 44, 46	100.0	80.8	96.1	100.0	82.3	96.4	100.0	82.3	96.4	100.0	81.3	96.2	100.0	69.8	93.9	100.0	85.1	97.0
21, 22, 23	FIRE SUPPRESSION, PLUMBING & HVAC	95.4	65.9	83.5	100.2	54.7	81.8	95.3	58.5	80.5	95.4	65.0	83.1	100.2	54.2	81.6	99.9	69.5	87.6
26, 27, 3370	ELECTRICAL, COMMUNICATIONS & UTIL.	99.8	57.6	77.5	97.6	63.6	79.6	92.1	64.6	77.5	98.6	68.0	82.4	99.3	60.1	78.6	97.5	73.0	84.6
MF2014	WEIGHTED AVERAGE	99.0	65.6	84.4	96.5	64.1	82.4	98.6	65.9	84.4	98.9	67.7	85.3	97.1	64.4	82.8	99.8	70.1	86.9

DIVISION		UTAH												VERMONT					
		OGDEN			PRICE			PROVO			SALT LAKE CITY			BELLOWS FALLS			BENNINGTON		
		842,844			845			846 - 847			840 - 841			051			052		
		MAT.	INST.	TOTAL	MAT.	INST.	TOTAL	MAT.	INST.	TOTAL	MAT.	INST.	TOTAL	MAT.	INST.	TOTAL	MAT.	INST.	TOTAL
015433	CONTRACTOR EQUIPMENT		97.5	97.5		96.6	96.6		96.6	96.6		97.5	97.5		100.9	100.9		100.9	100.9
0241, 31 - 34	SITE & INFRASTRUCTURE, DEMOLITION	84.9	95.7	92.6	93.3	93.8	93.7	92.3	94.1	93.6	84.5	95.6	92.4	88.2	100.5	96.9	87.5	100.5	96.7
0310	Concrete Forming & Accessories	104.7	58.1	64.5	107.1	48.3	56.4	106.3	58.0	64.7	107.1	58.0	64.8	98.3	85.5	87.3	95.9	106.6	105.2
0320	Concrete Reinforcing	99.2	81.3	90.1	106.8	81.1	93.7	107.7	81.3	94.3	101.5	81.3	91.2	83.2	85.7	84.5	83.2	85.7	84.5
0330	Cast-in-Place Concrete	88.7	73.4	82.4	87.4	59.5	75.9	87.4	73.3	81.6	96.9	73.3	87.2	87.0	110.7	96.7	87.0	110.7	96.7
03	CONCRETE	95.9	68.3	82.3	107.3	59.0	83.6	105.9	68.2	87.4	114.8	68.2	91.9	91.0	94.2	92.6	90.8	103.6	97.1
04	MASONRY	99.7	63.2	77.0	110.9	61.3	80.0	111.1	63.2	81.2	113.1	63.2	82.0	103.6	95.2	98.4	112.4	95.2	101.6
05	METALS	102.7	79.0	95.4	99.4	76.6	92.4	100.3	78.9	93.7	106.2	78.9	97.8	94.8	88.3	92.8	94.7	88.2	92.7
06	WOOD, PLASTICS & COMPOSITES	85.3	54.9	68.3	88.6	44.3	63.8	86.8	54.9	68.9	87.2	54.9	69.1	103.7	83.5	92.4	100.6	112.4	107.2
07	THERMAL & MOISTURE PROTECTION	96.4	67.9	84.7	99.0	59.9	82.9	99.0	67.9	86.2	103.4	67.9	88.9	97.0	82.7	91.1	97.0	81.1	90.4
08	OPENINGS	94.3	56.5	85.5	98.2	48.9	86.7	98.2	56.5	88.5	96.2	56.5	86.9	105.0	86.5	100.7	105.0	102.3	104.3
0920	Plaster & Gypsum Board	78.9	53.5	61.7	81.7	42.5	55.2	79.2	53.5	61.8	90.5	53.5	65.5	106.0	82.4	90.1	104.2	112.2	109.6
0950, 0980	Ceilings & Acoustic Treatment	97.2	53.5	68.5	97.2	42.5	61.3	97.2	53.5	68.5	91.5	53.5	66.5	91.5	82.4	85.6	91.5	112.2	105.1
0960	Flooring	99.0	47.7	84.3	102.3	35.5	83.2	102.0	47.7	86.5	103.2	47.7	87.3	100.6	104.6	101.8	99.7	104.6	101.1
0970, 0990	Wall Finishes & Painting/Coating	103.3	60.7	77.5	103.3	37.5	63.6	103.3	62.6	78.7	106.7	62.6	80.1	97.4	105.8	102.5	97.4	105.8	102.5
09	FINISHES	94.3	55.3	72.8	97.3	43.5	67.7	96.8	55.5	74.0	96.6	55.5	74.0	95.9	90.7	93.0	95.4	107.8	102.2
COVERS	DIVS. 10 - 14, 25, 28, 41, 43, 44, 46	100.0	85.1	97.0	100.0	45.8	89.0	100.0	85.0	97.0	100.0	85.0	97.0	100.0	97.6	99.5	100.0	100.7	100.1
21, 22, 23	FIRE SUPPRESSION, PLUMBING & HVAC	99.9	69.5	87.6	97.5	67.7	85.5	99.9	69.4	87.6	100.1	69.4	87.7	95.3	96.4	95.7	95.3	96.4	95.7
26, 27, 3370	ELECTRICAL, COMMUNICATIONS & UTIL.	97.9	73.0	84.7	103.5	73.0	87.4	98.4	59.8	78.0	100.5	59.8	79.0	101.5	89.3	95.1	101.5	61.2	80.2
MF2014	WEIGHTED AVERAGE	98.1	70.1	85.9	100.2	64.6	84.7	100.3	68.2	86.3	102.2	68.3	87.4	97.2	93.0	95.3	97.5	93.5	95.8

DIVISION		VERMONT																	
		BRATTLEBORO			BURLINGTON			GUILDHALL			MONTPELIER			RUTLAND			ST. JOHNSBURY		
		053			054			059			056			057			058		
		MAT.	INST.	TOTAL	MAT.	INST.	TOTAL	MAT.	INST.	TOTAL	MAT.	INST.	TOTAL	MAT.	INST.	TOTAL	MAT.	INST.	TOTAL
015433	CONTRACTOR EQUIPMENT		100.9	100.9		100.9	100.9		100.9	100.9		100.9	100.9		100.9	100.9		100.9	100.9
0241, 31 - 34	SITE & INFRASTRUCTURE, DEMOLITION	88.9	100.5	97.1	93.0	100.4	98.2	87.2	99.0	95.6	91.5	100.1	97.6	91.7	100.1	97.7	87.3	99.0	95.6
0310	Concrete Forming & Accessories	98.6	85.4	87.2	98.0	87.6	89.0	96.1	78.3	80.7	97.9	84.1	86.0	98.9	87.6	89.1	94.4	78.3	80.5
0320	Concrete Reinforcing	82.3	85.7	84.0	101.1	85.6	93.2	83.9	85.6	84.8	94.1	85.6	89.8	103.9	85.6	94.6	82.3	85.6	84.0
0330	Cast-in-Place Concrete	89.7	110.7	98.3	100.9	109.6	104.5	84.3	101.0	91.2	100.9	109.6	104.5	85.2	109.6	95.2	84.3	101.0	91.2
03	CONCRETE	93.0	94.1	93.6	101.2	94.7	98.0	88.6	87.6	88.1	100.1	93.1	96.7	94.3	94.7	94.5	88.2	87.6	87.9
04	MASONRY	111.5	95.2	101.3	115.7	93.7	102.0	111.8	78.2	90.8	109.6	93.7	99.7	93.9	93.7	93.8	138.8	78.2	101.0
05	METALS	94.7	88.2	92.7	103.5	87.5	98.6	94.8	87.3	92.5	99.6	87.4	95.9	100.6	87.4	96.5	94.8	87.3	92.5
06	WOOD, PLASTICS & COMPOSITES	104.0	83.5	92.5	99.9	88.2	93.3	100.0	83.5	90.8	97.7	83.5	89.7	104.2	88.2	95.2	94.8	83.5	88.5
07	THERMAL & MOISTURE PROTECTION	97.1	78.0	89.3	103.8	80.4	94.2	96.8	70.7	86.1	103.3	79.9	93.7	97.2	80.4	90.3	96.7	70.7	86.1
08	OPENINGS	105.0	86.5	100.7	108.4	85.0	103.0	105.0	82.4	99.7	105.8	82.4	100.3	108.4	85.0	102.9	105.0	82.4	99.7
0920	Plaster & Gypsum Board	106.0	82.4	90.1	109.7	87.2	94.5	112.4	82.4	92.2	108.9	82.4	91.0	106.6	87.2	93.5	113.8	82.4	92.6
0950, 0980	Ceilings & Acoustic Treatment	91.5	82.4	85.6	101.0	87.2	91.9	91.5	82.4	85.6	95.1	82.4	86.8	95.9	87.2	90.2	91.5	82.4	85.6
0960	Flooring	100.7	104.6	101.8	102.4	104.6	103.1	103.6	104.6	103.9	103.2	104.6	103.6	100.6	104.6	101.8	106.9	104.6	106.3
0970, 0990	Wall Finishes & Painting/Coating	97.4	105.8	102.5	102.9	85.2	92.2	97.4	85.2	90.0	101.7	85.2	91.7	97.4	85.2	90.0	97.4	85.2	90.0
09	FINISHES	96.0	90.7	93.1	99.7	90.8	94.8	97.5	84.2	90.2	97.9	88.0	92.5	97.1	90.8	93.6	98.7	84.2	90.7
COVERS	DIVS. 10 - 14, 25, 28, 41, 43, 44, 46	100.0	97.6	99.5	100.0	97.5	99.5	100.0	91.8	98.3	100.0	97.0	99.4	100.0	97.5	99.5	100.0	91.8	98.3
21, 22, 23	FIRE SUPPRESSION, PLUMBING & HVAC	95.3	96.4	95.7	99.9	71.8	88.6	95.3	64.0	82.7	95.1	71.8	85.7	100.1	71.8	88.7	95.3	64.0	82.7
26, 27, 3370	ELECTRICAL, COMMUNICATIONS & UTIL.	101.5	89.3	95.1	102.2	61.2	80.5	101.5	61.2	80.2	99.7	61.2	79.3	101.6	61.2	80.2	101.5	61.2	80.2
MF2014	WEIGHTED AVERAGE	97.8	92.8	95.6	102.4	83.6	94.2	97.4	77.8	88.8	99.5	82.8	92.2	99.7	83.6	92.7	98.6	77.8	89.6

DIVISION		VERMONT			VIRGINIA														
		WHITE RIVER JCT.			ALEXANDRIA			ARLINGTON			BRISTOL			CHARLOTTESVILLE			CULPEPER		
		050			223			222			242			229			227		
		MAT.	INST.	TOTAL	MAT.	INST.	TOTAL	MAT.	INST.	TOTAL	MAT.	INST.	TOTAL	MAT.	INST.	TOTAL	MAT.	INST.	TOTAL
015433	CONTRACTOR EQUIPMENT		100.9	100.9		103.0	103.0		101.7	101.7		101.7	101.7		106.1	106.1		101.7	101.7
0241, 31 - 34	SITE & INFRASTRUCTURE, DEMOLITION	91.5	99.2	97.0	114.3	89.8	96.9	124.5	87.7	98.3	108.8	86.1	92.6	113.5	88.1	95.4	111.8	87.6	94.6
0310	Concrete Forming & Accessories	93.1	78.9	80.8	92.2	72.8	75.5	91.2	72.0	74.7	87.1	42.5	48.6	85.4	48.7	53.7	82.5	71.4	72.9
0320	Concrete Reinforcing	83.2	85.6	84.4	83.0	87.0	85.1	93.7	84.4	88.9	93.7	68.4	80.8	93.1	72.8	82.7	93.7	84.2	88.9
0330	Cast-in-Place Concrete	89.7	101.9	94.7	105.1	81.8	95.6	102.3	80.3	93.3	101.8	48.5	79.9	105.9	55.5	85.2	104.8	69.2	90.2
03	CONCRETE	94.7	88.2	91.5	102.0	79.6	91.0	106.7	78.3	92.8	102.9	51.2	77.5	103.5	57.2	80.8	100.6	74.1	87.6
04	MASONRY	124.5	79.7	96.6	91.3	72.8	79.8	106.3	70.7	84.1	95.4	51.4	68.0	120.3	53.2	78.5	107.5	70.7	84.6
05	METALS	94.8	87.3	92.5	105.7	97.0	103.0	104.3	97.2	102.1	103.1	85.7	97.8	103.4	90.2	99.3	103.5	95.3	101.0
06	WOOD, PLASTICS & COMPOSITES	97.4	83.5	89.6	95.0	71.0	81.6	91.5	71.0	80.0	83.7	40.0	59.2	82.2	45.6	61.7	80.8	71.0	75.3
07	THERMAL & MOISTURE PROTECTION	97.1	71.4	86.6	102.7	81.5	94.0	104.7	80.6	94.8	104.2	60.1	86.1	103.8	67.7	89.0	104.0	78.1	93.4
08	OPENINGS	105.0	82.4	99.7	100.9	75.2	94.9	99.0	75.2	93.5	102.0	46.5	89.1	100.2	52.0	89.0	100.5	75.2	94.6
0920	Plaster & Gypsum Board	102.8	82.4	89.0	107.4	69.9	82.1	104.2	69.9	81.0	99.7	38.0	58.0	99.7	43.0	61.4	99.9	69.9	79.6
0950, 0980	Ceilings & Acoustic Treatment	91.5	82.4	85.6	91.4	69.9	77.3	89.8	69.9	76.7	88.9	38.0	55.4	88.9	43.0	58.8	89.8	69.9	76.7
0960	Flooring	98.6	104.6	100.4	106.0	82.9	99.4	104.5	81.9	98.0	101.0	66.5	91.1	99.7	66.5	90.2	99.7	81.9	94.6
0970, 0990	Wall Finishes & Painting/Coating	97.4	85.2	90.0	124.0	80.0	97.5	124.0	80.0	97.5	109.6	54.3	76.2	109.6	76.1	89.3	124.0	76.1	95.1
09	FINISHES	95.2	84.6	89.3	100.1	74.6	86.0	100.0	73.8	85.6	96.5	47.2	69.3	96.2	53.5	72.6	96.9	73.2	83.9
COVERS	DIVS. 10 - 14, 25, 28, 41, 43, 44, 46	100.0	92.3	98.4	100.0	88.6	97.7	100.0	87.6	97.5	100.0	67.8	93.5	100.0	79.3	95.8	100.0	87.6	97.5
21, 22, 23	FIRE SUPPRESSION, PLUMBING & HVAC	95.3	64.8	83.0	100.3	86.9	94.9	100.3	85.4	94.3	95.6	50.5	77.4	95.6	69.3	85.0	95.6	72.9	86.4
26, 27, 3370	ELECTRICAL, COMMUNICATIONS & UTIL.	101.5	61.2	80.2	97.1	102.4	99.9	94.7	99.2	97.1	96.8	35.4	64.3	96.7	71.9	83.6	99.3	99.2	99.2
MF2014	WEIGHTED AVERAGE	98.5	78.3	89.7	101.1	85.5	94.3	101.9	84.0	94.1	99.4	54.9	80.0	100.5	67.1	86.0	99.9	80.5	91.4

DIVISION		VIRGINIA																	
		FAIRFAX			FARMVILLE			FREDERICKSBURG			GRUNDY			HARRISONBURG			LYNCHBURG		
		220 - 221			239			224 - 225			246			228			245		
		MAT.	INST.	TOTAL	MAT.	INST.	TOTAL	MAT.	INST.	TOTAL	MAT.	INST.	TOTAL	MAT.	INST.	TOTAL	MAT.	INST.	TOTAL
015433	CONTRACTOR EQUIPMENT		101.7	101.7		106.1	106.1		101.7	101.7		101.7	101.7		101.7	101.7		101.7	101.7
0241, 31 - 34	SITE & INFRASTRUCTURE, DEMOLITION	123.1	87.8	98.0	109.2	87.6	93.8	111.4	87.7	94.5	106.6	84.8	91.1	120.1	86.3	96.1	107.5	86.3	92.4
0310	Concrete Forming & Accessories	85.4	72.0	73.9	98.0	43.7	51.2	85.4	69.4	71.6	90.3	36.4	43.8	81.4	42.6	47.9	87.1	60.3	64.0
0320	Concrete Reinforcing	93.7	87.0	90.3	90.9	49.9	70.0	94.4	87.0	90.6	92.4	50.4	71.0	93.7	64.9	79.0	93.1	68.5	80.5
0330	Cast-in-Place Concrete	102.3	80.6	93.4	103.1	53.5	82.7	103.9	69.3	89.7	101.8	47.5	79.5	102.3	58.0	84.1	101.8	57.7	83.7
03	CONCRETE	106.3	78.8	92.8	101.2	49.9	76.0	100.3	73.8	87.3	101.6	44.4	73.5	104.1	53.8	79.4	101.5	62.3	82.3
04	MASONRY	106.2	71.4	84.5	103.5	44.9	66.9	106.6	70.7	84.2	96.5	47.3	65.8	104.2	50.3	70.6	112.0	53.2	75.4
05	METALS	103.6	97.0	101.5	101.0	77.1	93.6	103.5	96.2	101.3	103.1	69.9	92.9	103.4	83.9	97.4	103.3	86.3	98.1
06	WOOD, PLASTICS & COMPOSITES	83.7	71.0	76.6	96.4	44.1	67.1	83.7	68.1	75.0	86.7	35.1	57.8	79.8	39.0	56.9	83.7	62.6	71.9
07	THERMAL & MOISTURE PROTECTION	104.6	73.7	91.9	104.3	49.3	81.8	104.0	78.5	93.6	104.2	45.3	80.0	104.4	63.9	87.8	104.0	63.9	87.6
08	OPENINGS	99.0	75.2	93.5	100.7	43.0	87.3	100.2	73.6	94.0	102.0	34.4	86.3	100.5	49.4	88.6	100.5	58.8	90.8
0920	Plaster & Gypsum Board	99.9	69.9	79.6	108.6	41.5	63.2	99.9	66.9	77.6	99.7	32.9	54.5	99.7	36.9	57.3	99.7	61.3	73.7
0950, 0980	Ceilings & Acoustic Treatment	89.8	69.9	76.7	84.8	41.5	56.3	89.8	66.9	74.8	88.9	32.9	52.1	88.9	36.9	54.8	88.9	61.3	70.8
0960	Flooring	101.4	81.9	95.8	106.5	60.3	93.3	101.4	81.9	95.8	102.3	33.7	82.6	99.4	74.0	92.1	101.0	66.5	91.1
0970, 0990	Wall Finishes & Painting/Coating	124.0	80.0	97.5	112.8	38.7	68.1	124.0	76.1	95.1	109.6	35.1	64.6	124.0	54.3	81.9	109.6	54.3	76.2
09	FINISHES	98.5	74.0	85.0	98.0	46.0	69.4	97.4	71.5	83.1	96.7	35.1	62.8	97.3	47.8	70.1	96.3	60.2	76.4
COVERS	DIVS. 10 - 14, 25, 28, 41, 43, 44, 46	100.0	87.8	97.5	100.0	45.7	89.0	100.0	83.4	96.6	100.0	43.7	88.6	100.0	67.6	93.5	100.0	70.9	94.1
21, 22, 23	FIRE SUPPRESSION, PLUMBING & HVAC	95.6	85.7	91.6	95.5	43.4	74.5	95.6	85.1	91.3	95.6	62.1	82.1	95.6	66.9	84.0	95.6	69.0	84.8
26, 27, 3370	ELECTRICAL, COMMUNICATIONS & UTIL.	97.9	99.2	98.6	90.1	48.2	68.0	94.9	99.2	97.2	96.8	44.6	69.2	96.9	96.4	96.7	97.9	52.7	74.0
MF2014	WEIGHTED AVERAGE	100.7	84.0	93.4	98.7	52.4	78.5	99.5	82.7	92.1	99.3	52.5	78.9	100.2	67.2	85.8	99.9	65.6	85.0

DIVISION		VIRGINIA																	
		NEWPORT NEWS			NORFOLK			PETERSBURG			PORTSMOUTH			PULASKI			RICHMOND		
		236			233 - 235			238			237			243			230 - 232		
		MAT.	INST.	TOTAL	MAT.	INST.	TOTAL	MAT.	INST.	TOTAL	MAT.	INST.	TOTAL	MAT.	INST.	TOTAL	MAT.	INST.	TOTAL
015433	CONTRACTOR EQUIPMENT		106.1	106.1		106.7	106.7		106.1	106.1		106.0	106.0		101.7	101.7		106.1	106.1
0241, 31 - 34	SITE & INFRASTRUCTURE, DEMOLITION	108.2	88.8	94.4	108.9	89.8	95.3	111.7	89.0	95.6	106.7	88.6	93.8	105.9	85.5	91.4	103.6	89.0	93.2
0310	Concrete Forming & Accessories	97.3	64.1	68.7	97.9	64.3	68.9	90.6	57.2	61.8	86.4	53.1	57.7	90.3	38.7	45.8	97.6	57.2	62.8
0320	Concrete Reinforcing	90.6	72.0	81.2	94.7	72.1	83.2	90.3	73.0	81.5	90.3	72.0	81.0	92.4	64.6	78.3	97.5	73.0	85.0
0330	Cast-in-Place Concrete	100.2	66.1	86.2	105.2	66.1	89.1	106.3	53.7	84.7	99.2	66.0	85.6	101.8	48.2	79.8	98.1	53.7	79.8
03	CONCRETE	98.4	67.6	83.3	101.5	67.7	84.9	103.6	60.4	82.4	97.1	62.6	80.2	101.6	48.7	75.6	98.5	60.4	79.8
04	MASONRY	97.8	53.9	70.5	103.8	53.9	72.7	111.9	52.6	75.0	103.8	53.9	72.7	91.5	48.2	64.5	102.5	52.6	71.4
05	METALS	103.3	89.8	99.2	103.1	89.9	99.1	101.1	90.7	97.9	102.2	89.1	98.2	103.2	82.5	96.8	107.1	90.7	102.0
06	WOOD, PLASTICS & COMPOSITES	95.3	66.2	79.0	93.2	66.2	78.1	86.4	58.6	70.8	82.9	51.5	65.3	86.7	36.8	58.7	95.0	58.6	74.6
07	THERMAL & MOISTURE PROTECTION	104.3	68.4	89.6	100.9	68.4	87.5	104.3	68.3	89.5	104.3	66.8	88.9	104.2	51.1	82.4	102.2	68.3	88.3
08	OPENINGS	101.1	62.2	92.1	100.4	62.2	91.5	100.4	59.0	90.8	101.2	54.1	90.2	102.0	42.2	88.1	104.8	59.0	94.1
0920	Plaster & Gypsum Board	109.4	64.3	78.9	105.5	64.3	77.6	101.3	56.4	71.0	101.6	49.1	66.1	99.7	34.7	55.7	104.2	56.4	71.9
0950, 0980	Ceilings & Acoustic Treatment	88.2	64.3	72.5	88.9	64.3	72.7	85.7	56.4	66.4	88.2	49.1	62.5	88.9	34.7	53.3	90.6	56.4	68.1
0960	Flooring	106.5	66.5	95.1	105.6	66.5	94.4	102.5	71.2	93.6	99.5	66.5	90.0	102.3	66.5	92.0	103.4	71.2	94.2
0970, 0990	Wall Finishes & Painting/Coating	112.8	42.0	70.1	113.0	76.1	90.7	112.8	76.1	90.6	112.8	76.1	90.6	109.6	54.3	76.2	110.0	76.1	89.5
09	FINISHES	98.7	62.2	78.6	98.2	66.0	80.5	96.3	61.2	76.9	95.6	56.7	74.2	96.7	43.7	67.5	98.5	61.2	77.9
COVERS	DIVS. 10 - 14, 25, 28, 41, 43, 44, 46	100.0	82.2	96.4	100.0	82.2	96.4	100.0	79.6	95.9	100.0	72.0	94.3	100.0	44.0	88.7	100.0	79.6	95.9
21, 22, 23	FIRE SUPPRESSION, PLUMBING & HVAC	100.3	63.6	85.5	100.0	64.3	85.6	95.5	67.5	84.2	100.3	64.3	85.8	95.6	62.2	82.1	99.9	67.5	86.8
26, 27, 3370	ELECTRICAL, COMMUNICATIONS & UTIL.	92.8	63.5	77.3	95.5	58.3	75.8	93.0	71.9	81.8	91.2	58.3	73.8	96.8	54.4	74.4	96.8	71.9	83.6
MF2014	WEIGHTED AVERAGE	100.0	68.2	86.1	100.5	68.2	86.4	99.4	68.7	86.0	99.4	65.3	84.5	99.0	57.3	80.9	101.3	68.7	87.1

DIVISION		VIRGINIA									WASHINGTON								
		ROANOKE			STAUNTON			WINCHESTER			CLARKSTON			EVERETT			OLYMPIA		
		240 - 241			244			226			994			982			985		
		MAT.	INST.	TOTAL	MAT.	INST.	TOTAL	MAT.	INST.	TOTAL	MAT.	INST.	TOTAL	MAT.	INST.	TOTAL	MAT.	INST.	TOTAL
015433	CONTRACTOR EQUIPMENT		101.7	101.7		106.1	106.1		101.7	101.7		91.2	91.2		101.6	101.6		101.6	101.6
0241, 31 - 34	SITE & INFRASTRUCTURE, DEMOLITION	106.4	86.3	92.1	109.9	87.7	94.1	118.7	87.6	96.6	99.2	90.1	92.7	91.5	108.4	103.5	93.4	108.4	104.1
0310	Concrete Forming & Accessories	96.8	60.5	65.5	89.9	50.0	55.5	83.8	69.5	71.4	112.7	67.4	73.6	113.4	100.4	102.2	99.0	100.3	100.1
0320	Concrete Reinforcing	93.4	68.5	80.7	93.1	48.9	70.6	93.1	86.4	89.7	112.0	87.2	99.4	104.2	98.8	101.5	110.8	98.7	104.6
0330	Cast-in-Place Concrete	115.7	57.7	91.9	105.9	51.5	83.6	102.3	69.3	88.7	96.3	81.9	90.4	103.5	106.2	104.6	98.9	106.2	101.9
03	CONCRETE	105.9	62.4	84.5	102.9	52.0	77.9	103.5	73.7	88.9	107.2	76.3	92.0	98.7	101.6	100.1	98.9	101.6	100.2
04	MASONRY	97.7	53.2	70.0	106.9	51.4	72.3	101.7	59.7	75.5	102.3	81.5	89.3	110.6	100.8	104.5	103.0	99.1	100.6
05	METALS	105.5	86.5	99.7	103.4	80.5	96.3	103.5	94.7	100.8	88.3	81.6	86.2	101.8	92.4	98.9	101.2	92.1	98.4
06	WOOD, PLASTICS & COMPOSITES	95.9	62.6	77.3	86.7	49.9	66.1	82.2	68.1	74.3	113.2	63.8	85.5	111.7	100.0	105.2	91.1	100.0	96.1
07	THERMAL & MOISTURE PROTECTION	103.8	66.8	88.7	103.8	52.0	82.6	104.5	75.6	92.7	143.8	77.2	116.5	103.7	101.7	102.9	103.5	96.8	100.7
08	OPENINGS	100.9	58.8	91.1	100.5	48.1	88.3	102.1	73.7	95.5	119.0	68.7	107.3	101.3	99.8	101.0	106.9	99.9	105.3
0920	Plaster & Gypsum Board	107.4	61.3	76.2	99.7	47.5	64.4	99.9	66.9	77.6	145.7	62.5	89.5	109.4	99.9	103.0	101.3	99.9	100.4
0950, 0980	Ceilings & Acoustic Treatment	91.4	61.3	71.6	88.9	47.5	61.7	89.8	66.9	74.8	97.6	62.5	74.5	99.8	99.9	99.9	98.1	99.9	99.3
0960	Flooring	106.0	66.5	94.7	101.9	39.5	84.0	100.7	81.9	95.4	91.6	47.1	78.9	113.8	91.6	107.4	103.8	91.7	100.3
0970, 0990	Wall Finishes & Painting/Coating	109.6	54.3	76.2	109.6	34.5	64.3	124.0	87.2	101.8	100.0	58.5	75.0	95.4	91.5	93.0	99.0	91.5	94.5
09	FINISHES	99.0	61.1	78.1	96.6	46.6	69.0	97.9	72.7	84.0	110.3	61.3	83.3	104.7	97.6	100.8	98.9	97.6	98.2
COVERS	DIVS. 10 - 14, 25, 28, 41, 43, 44, 46	100.0	70.9	94.1	100.0	70.9	94.1	100.0	87.3	97.4	100.0	75.4	95.0	100.0	99.0	99.8	100.0	100.5	100.1
21, 22, 23	FIRE SUPPRESSION, PLUMBING & HVAC	100.3	64.8	86.0	95.6	58.9	80.8	95.6	87.7	92.4	95.6	83.4	90.7	100.1	99.1	99.7	100.0	99.1	99.7
26, 27, 3370	ELECTRICAL, COMMUNICATIONS & UTIL.	96.8	54.4	74.4	95.6	74.9	84.7	95.3	99.2	97.4	92.4	94.8	93.7	104.9	98.5	101.5	104.4	99.1	101.6
MF2014	WEIGHTED AVERAGE	101.5	65.1	85.6	99.7	61.8	83.2	100.0	82.2	92.2	101.5	80.2	92.2	101.6	99.6	100.7	101.2	99.4	100.4

	DIVISION	WASHINGTON																	
		RICHLAND			SEATTLE			SPOKANE			TACOMA			VANCOUVER			WENATCHEE		
		993			980 - 981,987			990 - 992			983 - 984			986			988		
		MAT.	INST.	TOTAL	MAT.	INST.	TOTAL	MAT.	INST.	TOTAL	MAT.	INST.	TOTAL	MAT.	INST.	TOTAL	MAT.	INST.	TOTAL
015433	CONTRACTOR EQUIPMENT		91.2	91.2		101.6	101.6		91.2	91.2		101.6	101.6		96.7	96.7		101.6	101.6
0241, 31 - 34	SITE & INFRASTRUCTURE, DEMOLITION	101.8	91.0	94.1	95.1	106.7	103.4	101.1	91.0	93.9	94.5	108.4	104.4	104.2	96.1	98.4	103.2	107.1	106.0
0310	Concrete Forming & Accessories	112.8	78.5	83.2	108.3	100.9	102.0	118.2	78.1	83.6	104.2	100.4	100.9	104.9	90.5	92.5	105.8	77.6	81.5
0320	Concrete Reinforcing	107.4	87.4	97.2	105.1	98.9	101.9	108.2	87.3	97.6	103.1	98.8	100.9	103.9	98.4	101.1	103.9	87.7	95.6
0330	Cast-in-Place Concrete	96.5	84.4	91.5	100.9	106.4	103.2	100.3	84.2	93.7	106.4	106.2	106.3	118.8	98.4	110.4	108.6	77.3	95.7
03	CONCRETE	106.8	82.2	94.7	99.6	102.1	100.8	109.1	82.0	95.8	100.4	101.6	101.0	110.7	94.6	102.8	108.4	79.6	94.2
04	MASONRY	104.3	84.3	91.8	108.2	100.9	103.6	104.9	84.3	92.0	106.2	100.9	102.9	106.7	95.6	99.8	108.6	91.4	97.9
05	METALS	88.7	84.9	87.5	102.5	94.2	99.9	90.9	84.5	89.0	103.6	92.3	100.1	101.0	91.8	98.2	101.0	86.2	96.4
06	WOOD, PLASTICS & COMPOSITES	113.4	76.1	92.5	104.9	100.0	102.2	122.4	76.1	96.5	101.0	100.0	100.4	92.9	90.2	91.4	102.5	75.8	87.5
07	THERMAL & MOISTURE PROTECTION	145.0	78.8	117.8	102.1	101.8	102.0	141.4	79.5	116.0	103.5	98.5	101.4	104.0	91.6	98.9	104.1	81.5	94.8
08	OPENINGS	121.4	71.3	109.7	100.1	99.9	100.0	121.9	71.8	110.3	102.0	99.9	101.5	98.3	92.8	97.0	101.5	71.0	94.4
0920	Plaster & Gypsum Board	145.7	75.2	98.1	103.6	99.9	101.1	137.7	75.2	95.5	107.8	99.9	102.5	105.7	90.1	95.1	110.2	74.9	86.4
0950, 0980	Ceilings & Acoustic Treatment	104.2	75.2	85.1	110.2	99.9	103.5	99.5	75.2	83.5	103.1	99.9	101.0	100.2	90.1	93.5	95.4	74.9	82.0
0960	Flooring	92.0	80.1	88.6	108.3	106.8	107.8	91.2	85.3	89.5	107.2	91.7	102.8	113.5	102.2	110.2	109.9	62.6	96.4
0970, 0990	Wall Finishes & Painting/Coating	100.0	67.6	80.4	96.3	91.5	93.4	100.2	70.9	82.5	95.4	91.5	93.0	97.6	70.8	81.4	95.4	72.6	81.6
09	FINISHES	112.0	76.7	92.6	105.0	100.6	102.6	109.8	78.1	92.3	103.4	97.6	100.2	101.5	90.6	95.5	103.8	73.5	87.1
COVERS	DIVS. 10 - 14, 25, 28, 41, 43, 44, 46	100.0	85.4	97.0	100.0	100.5	100.1	100.0	85.3	97.0	100.0	100.5	100.1	100.0	60.7	92.1	100.0	77.2	95.4
21, 22, 23	FIRE SUPPRESSION, PLUMBING & HVAC	100.5	107.6	103.3	100.0	115.1	106.1	100.4	84.3	93.9	100.2	99.2	99.8	100.3	87.0	94.9	95.4	86.7	91.9
26, 27, 3370	ELECTRICAL, COMMUNICATIONS & UTIL.	89.7	94.8	92.4	102.6	107.0	104.9	88.0	79.0	83.3	104.7	99.1	101.7	109.7	101.0	105.1	105.5	98.4	101.8
MF2014	WEIGHTED AVERAGE	103.1	89.3	97.1	101.4	104.7	102.8	103.4	82.3	94.2	101.8	99.7	100.9	102.7	92.1	98.1	101.6	86.6	95.0

	DIVISION	WASHINGTON			WEST VIRGINIA														
		YAKIMA			BECKLEY			BLUEFIELD			BUCKHANNON			CHARLESTON			CLARKSBURG		
		989			258 - 259			247 - 248			262			250 - 253			263 - 264		
		MAT.	INST.	TOTAL	MAT.	INST.	TOTAL	MAT.	INST.	TOTAL	MAT.	INST.	TOTAL	MAT.	INST.	TOTAL	MAT.	INST.	TOTAL
015433	CONTRACTOR EQUIPMENT		101.6	101.6		101.7	101.7		101.7	101.7		101.7	101.7		101.7	101.7		101.7	101.7
0241, 31 - 34	SITE & INFRASTRUCTURE, DEMOLITION	96.8	107.8	104.7	100.5	89.9	92.9	100.6	89.9	92.9	106.8	89.9	94.8	101.5	90.7	93.8	107.4	89.9	95.0
0310	Concrete Forming & Accessories	104.6	95.8	97.0	83.5	93.6	92.2	87.0	93.6	92.7	86.4	94.3	93.2	96.7	94.1	94.4	83.8	94.2	92.8
0320	Concrete Reinforcing	103.5	87.4	95.3	92.7	89.0	90.8	92.0	84.1	88.0	92.6	84.2	88.3	101.0	89.1	94.9	92.6	84.1	88.3
0330	Cast-in-Place Concrete	113.6	83.9	101.4	98.1	103.0	100.1	99.6	102.9	101.0	99.3	101.8	100.3	96.8	103.0	99.3	108.8	100.0	105.2
03	CONCRETE	105.3	89.9	97.7	97.6	96.4	97.0	97.4	95.5	96.5	100.4	95.4	98.0	99.1	96.7	97.9	104.4	94.8	99.7
04	MASONRY	99.8	83.1	89.4	98.9	98.9	98.9	93.3	98.9	96.8	104.5	98.9	101.0	95.8	97.9	97.1	108.4	98.9	102.4
05	METALS	101.6	86.3	96.9	100.9	98.1	100.0	103.4	96.3	101.2	103.6	97.1	101.6	99.6	98.9	99.4	103.6	96.9	101.5
06	WOOD, PLASTICS & COMPOSITES	101.4	100.0	100.6	83.5	93.4	89.1	85.6	93.4	90.0	84.9	93.4	89.7	98.4	93.4	95.6	81.5	93.4	88.2
07	THERMAL & MOISTURE PROTECTION	103.6	82.7	95.0	102.9	90.6	97.9	104.0	90.6	98.5	104.2	91.7	99.1	99.4	90.4	95.7	104.1	90.6	98.6
08	OPENINGS	101.5	81.0	96.7	101.3	85.8	97.7	102.5	84.6	98.3	102.5	84.6	98.4	103.6	85.8	99.4	102.5	85.1	98.5
0920	Plaster & Gypsum Board	107.4	99.9	102.4	97.0	93.0	94.3	98.9	93.0	94.9	99.3	93.0	95.1	100.3	93.0	95.4	97.2	93.0	94.4
0950, 0980	Ceilings & Acoustic Treatment	97.3	99.9	99.0	83.9	93.0	89.9	87.3	93.0	91.1	88.9	93.0	91.6	92.2	93.0	92.7	88.9	93.0	91.6
0960	Flooring	108.1	58.0	93.8	102.4	118.0	106.8	98.9	118.0	104.3	98.6	118.0	104.1	106.4	118.0	109.7	97.6	118.0	103.4
0970, 0990	Wall Finishes & Painting/Coating	95.4	70.9	80.6	107.4	91.5	97.8	109.6	94.6	100.6	109.6	94.3	100.3	106.9	95.9	100.2	109.6	94.3	100.3
09	FINISHES	102.5	86.6	93.7	94.7	98.6	96.9	94.8	98.9	97.1	95.7	98.9	97.5	99.1	98.8	99.0	95.0	98.9	97.2
COVERS	DIVS. 10 - 14, 25, 28, 41, 43, 44, 46	100.0	97.6	99.5	100.0	56.9	91.3	100.0	56.9	91.3	100.0	101.5	100.3	100.0	101.2	100.2	100.0	101.5	100.3
21, 22, 23	FIRE SUPPRESSION, PLUMBING & HVAC	100.2	107.2	103.0	95.5	93.5	94.7	95.6	92.8	94.4	95.6	95.1	95.4	100.0	93.4	97.4	95.6	95.0	95.3
26, 27, 3370	ELECTRICAL, COMMUNICATIONS & UTIL.	107.8	94.9	101.0	93.9	91.1	92.4	95.7	91.1	93.3	97.1	96.1	96.6	99.1	91.1	94.8	97.1	96.1	96.6
MF2014	WEIGHTED AVERAGE	102.0	94.1	98.6	97.9	93.3	95.9	98.3	92.9	96.0	99.6	95.5	97.8	99.9	94.8	97.6	100.1	95.4	98.1

	DIVISION	WEST VIRGINIA																	
		GASSAWAY			HUNTINGTON			LEWISBURG			MARTINSBURG			MORGANTOWN			PARKERSBURG		
		266			255 - 257			249			254			265			261		
		MAT.	INST.	TOTAL	MAT.	INST.	TOTAL	MAT.	INST.	TOTAL	MAT.	INST.	TOTAL	MAT.	INST.	TOTAL	MAT.	INST.	TOTAL
015433	CONTRACTOR EQUIPMENT		101.7	101.7		101.7	101.7		101.7	101.7		101.7	101.7		101.7	101.7		101.7	101.7
0241, 31 - 34	SITE & INFRASTRUCTURE, DEMOLITION	104.2	89.9	94.0	105.1	90.9	95.0	116.3	89.8	97.5	104.2	90.2	94.3	101.6	90.5	93.7	109.8	90.6	96.2
0310	Concrete Forming & Accessories	85.8	92.2	91.4	95.3	95.9	95.8	84.2	93.5	92.2	83.6	82.8	82.9	84.2	93.9	92.6	88.6	90.3	90.1
0320	Concrete Reinforcing	92.6	84.0	88.3	94.1	92.9	93.5	92.6	83.9	88.2	92.7	80.6	86.6	92.6	84.1	88.3	92.0	83.9	87.9
0330	Cast-in-Place Concrete	104.0	103.5	103.7	106.9	102.5	105.1	99.7	102.9	101.0	102.8	96.7	100.3	99.3	101.6	100.2	101.5	96.4	99.4
03	CONCRETE	100.8	95.1	98.0	102.8	98.0	100.4	107.1	95.4	101.4	101.2	87.9	94.7	97.2	95.2	96.2	102.5	91.8	97.2
04	MASONRY	109.2	95.9	100.9	96.4	100.8	99.1	95.9	95.9	95.9	99.7	89.5	93.3	126.5	98.9	109.3	82.9	91.7	88.4
05	METALS	103.5	96.8	101.4	103.3	100.4	102.4	103.5	96.0	101.2	101.3	94.2	99.1	103.6	96.6	101.5	104.2	96.6	101.9
06	WOOD, PLASTICS & COMPOSITES	84.0	90.7	87.8	94.7	95.3	95.1	81.8	93.4	88.3	83.5	81.6	82.5	81.8	93.4	88.3	85.0	89.1	87.3
07	THERMAL & MOISTURE PROTECTION	104.0	90.6	98.5	103.0	90.8	98.0	104.8	89.8	98.7	103.1	79.3	93.4	104.0	90.6	98.5	103.9	89.0	97.8
08	OPENINGS	100.7	83.2	96.6	100.5	87.6	97.5	102.5	84.6	98.4	103.3	73.4	96.3	103.8	85.1	99.4	101.4	82.3	96.9
0920	Plaster & Gypsum Board	98.6	90.2	92.9	104.8	95.0	98.1	97.2	93.0	94.4	97.6	80.9	86.3	97.2	93.0	94.4	99.7	88.5	92.1
0950, 0980	Ceilings & Acoustic Treatment	88.9	90.2	89.8	86.4	95.0	92.0	88.9	93.0	91.6	86.4	80.9	82.8	88.9	93.0	91.6	88.9	88.5	88.7
0960	Flooring	98.4	118.0	104.0	109.8	121.7	113.2	97.7	118.0	103.5	102.4	118.0	106.8	97.7	118.0	103.5	101.5	118.0	106.2
0970, 0990	Wall Finishes & Painting/Coating	109.6	95.9	101.3	107.4	94.6	99.7	109.6	75.4	88.9	107.4	41.4	67.5	109.6	94.3	100.3	109.6	94.6	100.6
09	FINISHES	95.2	97.5	96.5	98.5	100.8	99.8	96.1	96.8	96.5	95.6	83.1	88.7	94.6	98.9	97.0	96.6	95.9	96.2
COVERS	DIVS. 10 - 14, 25, 28, 41, 43, 44, 46	100.0	101.2	100.2	100.0	101.8	100.4	100.0	56.9	91.3	100.0	97.7	99.5	100.0	65.8	93.1	100.0	100.5	100.1
21, 22, 23	FIRE SUPPRESSION, PLUMBING & HVAC	95.6	93.9	94.9	100.3	90.5	96.4	95.6	93.5	94.7	95.5	86.0	91.7	95.6	95.0	95.3	100.3	88.3	95.4
26, 27, 3370	ELECTRICAL, COMMUNICATIONS & UTIL.	97.1	91.1	93.9	97.8	95.3	96.5	93.0	91.1	92.0	100.1	81.8	90.4	97.3	96.1	96.7	97.2	92.7	94.8
MF2014	WEIGHTED AVERAGE	99.5	93.9	97.1	100.7	95.7	98.5	99.8	92.4	96.6	99.3	86.4	93.7	100.2	94.4	97.6	100.1	91.8	96.5

	DIVISION	WEST VIRGINIA									WISCONSIN								
		PETERSBURG			ROMNEY			WHEELING			BELOIT			EAU CLAIRE			GREEN BAY		
		268			267			260			535			547			541 - 543		
		MAT.	INST.	TOTAL	MAT.	INST.	TOTAL	MAT.	INST.	TOTAL	MAT.	INST.	TOTAL	MAT.	INST.	TOTAL	MAT.	INST.	TOTAL
015433	**CONTRACTOR EQUIPMENT**		101.7	101.7		101.7	101.7		101.7	101.7		102.3	102.3		100.4	100.4		98.2	98.2
0241, 31 - 34	**SITE & INFRASTRUCTURE, DEMOLITION**	100.9	90.5	93.5	103.6	90.5	94.3	110.6	90.4	96.2	94.7	108.3	104.3	96.1	102.9	101.0	99.7	98.9	99.2
0310	Concrete Forming & Accessories	87.4	91.6	91.0	83.4	91.8	90.6	90.3	93.2	92.8	100.2	100.3	100.3	97.1	99.4	99.1	105.5	98.7	99.7
0320	Concrete Reinforcing	92.0	82.5	87.2	92.6	81.0	86.7	91.4	84.0	87.7	92.3	116.0	104.3	90.4	99.2	94.9	88.7	92.4	90.6
0330	Cast-in-Place Concrete	99.3	99.1	99.2	104.0	99.1	102.0	101.5	100.1	100.9	100.6	100.9	100.7	104.0	99.2	102.0	107.6	99.7	104.3
03	**CONCRETE**	97.3	93.0	95.2	100.6	92.8	96.8	102.5	94.4	98.5	98.8	103.5	101.1	99.7	99.4	99.6	102.8	98.0	100.5
04	**MASONRY**	99.5	98.9	99.1	96.2	98.9	97.9	107.6	93.9	99.1	100.6	104.3	102.9	94.7	102.1	99.3	126.7	101.4	110.9
05	**METALS**	103.7	95.8	101.2	103.7	95.5	101.2	104.4	96.8	102.1	93.1	107.7	97.6	95.0	101.4	97.0	97.4	98.2	97.6
06	**WOOD, PLASTICS & COMPOSITES**	85.9	90.7	88.6	80.9	90.7	86.4	86.7	93.4	90.5	99.9	99.2	99.5	105.1	99.3	101.9	110.0	99.3	104.0
07	**THERMAL & MOISTURE PROTECTION**	104.0	86.0	96.6	104.1	83.4	95.6	104.3	90.0	98.4	102.1	96.8	99.9	102.4	86.7	95.9	104.3	86.0	96.8
08	**OPENINGS**	103.8	82.8	98.9	103.7	82.4	98.8	102.2	85.1	98.2	102.5	108.8	103.9	103.7	96.2	102.0	100.1	96.0	99.2
0920	Plaster & Gypsum Board	99.3	90.2	93.2	96.8	90.2	92.4	99.7	93.0	95.2	94.5	99.6	97.9	102.9	99.6	100.6	98.7	99.6	99.3
0950, 0980	Ceilings & Acoustic Treatment	88.9	90.2	89.8	88.9	90.2	89.8	88.9	93.0	91.6	88.1	99.6	95.6	90.4	99.6	96.4	81.9	99.6	93.5
0960	Flooring	99.4	118.0	104.7	97.5	118.0	103.4	102.3	118.0	106.8	99.8	118.1	105.1	91.4	114.1	97.9	110.2	114.1	111.3
0970, 0990	Wall Finishes & Painting/Coating	109.6	94.3	100.3	109.6	94.3	100.3	109.6	94.3	100.3	98.3	97.2	97.7	94.5	86.4	89.6	105.4	83.2	92.0
09	**FINISHES**	95.4	97.3	96.4	94.7	97.3	96.1	96.9	98.2	97.6	97.4	103.1	100.5	94.6	101.2	98.3	99.0	100.7	99.9
COVERS	**DIVS. 10 - 14, 25, 28, 41, 43, 44, 46**	100.0	65.5	93.0	100.0	101.2	100.2	100.0	95.4	99.1	100.0	96.6	99.3	100.0	96.8	99.3	100.0	96.5	99.3
21, 22, 23	**FIRE SUPPRESSION, PLUMBING & HVAC**	95.6	93.7	94.8	95.6	93.4	94.7	100.3	93.6	97.6	100.1	98.6	99.5	100.1	87.6	95.0	100.3	85.1	94.2
26, 27, 3370	**ELECTRICAL, COMMUNICATIONS & UTIL.**	100.5	81.8	90.6	99.8	81.8	90.3	94.4	96.1	95.3	101.0	86.7	93.4	103.8	84.9	93.8	98.2	82.0	89.6
MF2014	**WEIGHTED AVERAGE**	99.3	91.3	95.8	99.4	92.2	96.3	101.2	94.3	98.2	98.9	100.7	99.7	99.3	95.3	97.5	101.2	93.4	97.8

	DIVISION	WISCONSIN																	
		KENOSHA			LA CROSSE			LANCASTER			MADISON			MILWAUKEE			NEW RICHMOND		
		531			546			538			537			530,532			540		
		MAT.	INST.	TOTAL	MAT.	INST.	TOTAL	MAT.	INST.	TOTAL	MAT.	INST.	TOTAL	MAT.	INST.	TOTAL	MAT.	INST.	TOTAL
015433	**CONTRACTOR EQUIPMENT**		100.2	100.2		100.4	100.4		102.3	102.3		102.3	102.3		90.5	90.5		100.8	100.8
0241, 31 - 34	**SITE & INFRASTRUCTURE, DEMOLITION**	100.3	105.2	103.8	90.1	102.8	99.1	93.9	108.2	104.1	92.4	108.3	103.7	94.6	98.5	97.4	94.4	103.3	100.8
0310	Concrete Forming & Accessories	108.0	100.5	101.6	84.1	99.1	97.1	99.5	97.9	98.2	102.0	100.1	100.4	104.1	118.0	116.1	92.4	100.9	99.7
0320	Concrete Reinforcing	92.1	97.9	95.0	90.1	91.5	90.9	93.5	91.3	92.3	97.9	91.6	94.7	95.7	98.2	97.0	87.8	99.0	93.5
0330	Cast-in-Place Concrete	109.4	107.3	108.5	93.5	95.9	94.5	100.0	100.9	100.4	91.1	100.8	95.1	99.8	109.6	103.8	108.3	82.5	97.7
03	**CONCRETE**	103.5	102.4	103.0	90.8	96.8	93.7	98.5	97.8	98.1	95.7	98.8	97.2	99.7	110.5	105.0	97.5	94.3	95.9
04	**MASONRY**	97.9	114.1	108.0	93.8	105.9	101.4	100.7	104.3	102.9	101.5	104.3	103.2	104.9	120.6	114.7	121.6	102.3	109.6
05	**METALS**	93.9	101.2	96.2	94.9	98.2	95.9	90.6	95.0	92.0	91.9	97.2	93.6	97.5	94.8	96.7	95.1	99.7	96.5
06	**WOOD, PLASTICS & COMPOSITES**	105.0	96.4	100.2	90.1	99.3	95.3	99.1	99.2	99.2	96.7	99.2	98.1	104.7	118.5	112.5	94.0	102.5	98.7
07	**THERMAL & MOISTURE PROTECTION**	102.2	108.7	104.9	101.8	90.2	97.1	101.9	82.5	93.9	102.8	105.1	103.8	100.8	114.0	106.2	103.6	92.2	98.9
08	**OPENINGS**	96.9	101.1	97.9	103.7	85.0	99.4	97.9	85.8	95.1	107.9	100.5	106.2	101.0	113.2	103.8	89.1	97.5	91.1
0920	Plaster & Gypsum Board	85.3	96.7	93.0	97.7	99.6	99.0	93.3	99.6	97.5	99.6	99.6	99.6	101.0	119.3	113.4	87.7	103.0	98.1
0950, 0980	Ceilings & Acoustic Treatment	88.1	96.7	93.7	89.5	99.6	96.1	83.1	99.6	93.9	94.0	99.6	97.6	95.9	119.3	111.3	58.5	103.0	87.8
0960	Flooring	118.8	110.4	116.4	85.6	114.1	93.7	99.5	110.5	102.7	103.2	110.5	105.3	98.2	114.0	102.7	103.2	114.1	106.3
0970, 0990	Wall Finishes & Painting/Coating	108.3	115.5	112.6	94.5	77.9	84.5	98.3	66.4	79.1	102.4	97.2	99.3	94.4	122.9	111.6	108.3	86.4	95.1
09	**FINISHES**	102.5	103.6	103.1	91.6	100.3	96.4	96.0	96.5	96.3	98.6	101.9	100.4	100.9	118.6	110.7	90.0	101.3	96.2
COVERS	**DIVS. 10 - 14, 25, 28, 41, 43, 44, 46**	100.0	101.1	100.2	100.0	96.7	99.3	100.0	49.7	89.8	100.0	99.7	99.9	100.0	104.6	100.9	100.0	96.1	99.2
21, 22, 23	**FIRE SUPPRESSION, PLUMBING & HVAC**	100.3	97.0	98.9	100.1	87.5	95.0	95.3	87.3	92.1	100.0	95.6	98.2	100.0	103.5	101.4	94.8	86.9	91.6
26, 27, 3370	**ELECTRICAL, COMMUNICATIONS & UTIL.**	101.5	98.7	100.1	104.2	85.1	94.1	100.8	85.1	92.5	102.1	92.1	96.8	100.8	99.5	100.1	101.9	85.1	93.0
MF2014	**WEIGHTED AVERAGE**	99.5	102.2	100.7	97.8	94.5	96.3	96.7	92.5	94.8	99.1	99.0	99.1	100.0	107.2	103.1	96.8	94.6	95.9

	DIVISION	WISCONSIN																	
		OSHKOSH			PORTAGE			RACINE			RHINELANDER			SUPERIOR			WAUSAU		
		549			539			534			545			548			544		
		MAT.	INST.	TOTAL	MAT.	INST.	TOTAL	MAT.	INST.	TOTAL	MAT.	INST.	TOTAL	MAT.	INST.	TOTAL	MAT.	INST.	TOTAL
015433	**CONTRACTOR EQUIPMENT**		98.2	98.2		102.3	102.3		102.3	102.3		98.2	98.2		100.8	100.8		98.2	98.2
0241, 31 - 34	**SITE & INFRASTRUCTURE, DEMOLITION**	91.5	98.9	96.8	85.0	108.0	101.4	94.5	109.0	104.8	103.3	98.8	100.1	91.3	103.3	99.9	87.4	98.9	95.6
0310	Concrete Forming & Accessories	88.6	98.1	96.8	91.7	98.9	97.9	100.5	100.5	100.5	86.2	97.8	96.2	90.5	91.9	91.7	88.1	98.7	97.2
0320	Concrete Reinforcing	88.8	92.3	90.6	93.5	91.4	92.5	92.3	97.9	95.1	89.0	90.9	90.0	87.8	91.4	89.6	89.0	91.3	90.2
0330	Cast-in-Place Concrete	99.7	99.4	99.6	85.9	101.8	92.4	98.7	107.0	102.1	113.1	99.3	107.4	102.0	97.9	100.3	93.0	99.6	95.7
03	**CONCRETE**	92.4	97.6	94.9	86.4	98.5	92.4	97.9	102.3	100.1	104.7	97.2	101.0	91.8	94.2	93.0	87.3	97.8	92.5
04	**MASONRY**	107.9	101.4	103.8	99.6	104.3	102.5	100.6	114.0	109.0	125.3	101.4	110.4	120.8	104.5	110.6	107.4	101.4	103.6
05	**METALS**	95.4	96.7	95.8	91.3	96.0	92.7	94.8	101.2	96.8	95.3	96.7	95.7	96.1	98.2	96.8	95.1	97.7	95.9
06	**WOOD, PLASTICS & COMPOSITES**	89.8	99.3	95.1	89.2	99.2	94.8	100.2	96.4	98.1	87.3	99.3	94.0	92.3	90.1	91.0	89.2	99.3	94.9
07	**THERMAL & MOISTURE PROTECTION**	103.4	85.7	96.1	101.4	92.6	97.7	102.1	108.3	104.7	104.2	83.5	95.7	103.3	92.2	98.7	103.3	84.4	95.5
08	**OPENINGS**	95.9	95.5	95.8	98.1	92.3	96.8	102.5	101.1	102.2	96.0	87.1	93.9	88.6	91.1	89.2	96.1	95.7	96.0
0920	Plaster & Gypsum Board	86.2	99.6	95.2	86.4	99.6	95.3	94.5	96.7	96.0	86.2	99.6	95.2	87.6	90.2	89.4	86.2	99.6	95.2
0950, 0980	Ceilings & Acoustic Treatment	81.9	99.6	93.5	85.6	99.6	94.8	88.1	96.7	93.7	81.9	99.6	93.5	59.3	90.2	79.6	81.9	99.6	93.5
0960	Flooring	101.6	114.1	105.2	96.2	118.1	102.5	99.8	110.4	102.8	100.9	114.1	104.7	104.4	122.4	109.6	101.5	114.1	105.1
0970, 0990	Wall Finishes & Painting/Coating	102.2	101.4	101.7	98.3	66.4	79.1	98.3	115.5	108.7	102.2	60.4	76.9	96.0	102.8	100.1	102.2	83.2	90.7
09	**FINISHES**	94.0	100.9	97.8	94.1	98.6	96.6	97.4	103.6	100.8	94.8	97.0	96.0	89.3	98.6	94.4	93.7	100.7	97.5
COVERS	**DIVS. 10 - 14, 25, 28, 41, 43, 44, 46**	100.0	64.6	92.8	100.0	63.4	92.6	100.0	101.1	100.2	100.0	55.0	90.9	100.0	94.1	98.8	100.0	96.5	99.3
21, 22, 23	**FIRE SUPPRESSION, PLUMBING & HVAC**	95.6	85.0	91.3	95.3	95.3	95.3	100.1	97.0	98.8	95.6	86.7	92.0	94.8	89.4	92.6	95.6	86.4	91.9
26, 27, 3370	**ELECTRICAL, COMMUNICATIONS & UTIL.**	102.5	80.4	90.9	104.8	92.1	98.1	100.8	98.6	99.6	101.8	81.8	91.3	107.1	99.0	102.8	103.7	81.8	92.1
MF2014	**WEIGHTED AVERAGE**	96.9	91.9	94.7	95.3	96.6	95.9	99.1	102.4	100.5	99.3	91.2	95.8	96.6	96.3	96.5	96.2	93.5	95.0

	DIVISION	WYOMING																	
		CASPER			CHEYENNE			NEWCASTLE			RAWLINS			RIVERTON			ROCK SPRINGS		
		826			820			827			823			825			829 - 831		
		MAT.	INST.	TOTAL	MAT.	INST.	TOTAL	MAT.	INST.	TOTAL	MAT.	INST.	TOTAL	MAT.	INST.	TOTAL	MAT.	INST.	TOTAL
015433	CONTRACTOR EQUIPMENT		98.2	98.2		98.2	98.2		98.2	98.2		98.2	98.2		98.2	98.2		98.2	98.2
0241, 31 - 34	SITE & INFRASTRUCTURE, DEMOLITION	98.6	94.4	95.6	92.5	94.4	93.8	85.1	94.0	91.5	98.4	94.0	95.3	92.1	94.0	93.5	89.2	94.0	92.6
0310	Concrete Forming & Accessories	102.5	50.1	57.3	104.4	53.5	60.5	94.8	62.2	66.7	99.2	62.2	67.3	93.7	62.1	66.5	101.5	62.1	67.5
0320	Concrete Reinforcing	105.7	70.6	87.8	98.0	70.6	84.1	105.5	70.8	87.9	105.2	70.8	87.7	106.1	70.6	88.0	106.1	70.6	88.0
0330	Cast-in-Place Concrete	102.8	72.0	90.2	95.0	72.1	85.6	95.9	61.6	81.8	96.0	61.6	81.9	96.0	60.0	81.2	95.9	59.9	81.1
03	CONCRETE	101.7	62.1	82.3	98.6	63.7	81.4	99.0	63.9	81.8	111.8	63.9	88.3	106.7	63.3	85.4	99.5	63.2	81.7
04	MASONRY	108.2	51.9	73.1	103.7	51.9	71.4	100.3	54.8	71.9	100.3	54.8	71.9	100.3	50.5	69.3	160.4	50.4	91.8
05	METALS	99.3	72.1	90.9	101.7	72.1	92.6	97.8	72.0	89.9	97.9	72.0	89.9	98.0	71.6	89.8	98.7	71.1	90.2
06	WOOD, PLASTICS & COMPOSITES	97.2	49.7	70.6	96.5	54.3	72.9	86.4	66.9	75.5	90.4	66.9	77.2	85.3	66.9	75.0	95.4	66.9	79.4
07	THERMAL & MOISTURE PROTECTION	105.9	55.1	85.1	100.1	55.6	81.8	101.8	55.8	82.9	103.2	55.8	83.7	102.7	59.9	85.1	101.9	59.9	84.7
08	OPENINGS	108.0	53.3	95.3	108.3	55.8	96.1	112.6	61.7	100.8	112.3	61.7	100.5	112.5	61.7	100.7	113.0	61.3	101.0
0920	Plaster & Gypsum Board	97.4	48.1	64.1	89.3	52.9	64.7	85.2	65.8	72.1	85.5	65.8	72.2	85.2	65.8	72.1	97.3	65.8	76.0
0950, 0980	Ceilings & Acoustic Treatment	99.2	48.1	65.6	91.0	52.9	66.0	92.9	65.8	75.1	92.9	65.8	75.1	92.9	65.8	75.1	92.9	65.8	75.1
0960	Flooring	112.0	56.4	96.1	109.9	56.4	94.6	103.5	56.4	90.0	106.2	56.4	92.0	103.0	56.4	89.7	108.4	56.4	93.5
0970, 0990	Wall Finishes & Painting/Coating	104.7	43.1	67.5	111.2	43.1	70.0	107.1	55.4	75.9	107.1	55.4	75.9	107.1	55.4	75.9	107.1	55.4	75.9
09	FINISHES	103.2	49.6	73.6	100.0	52.3	73.7	95.0	60.8	76.1	97.2	60.8	77.1	95.7	60.8	76.5	98.1	60.8	77.5
COVERS	DIVS. 10 - 14, 25, 28, 41, 43, 44, 46	100.0	87.6	97.5	100.0	88.2	97.6	100.0	89.1	97.8	100.0	89.1	97.8	100.0	81.4	96.2	100.0	80.8	96.1
21, 22, 23	FIRE SUPPRESSION, PLUMBING & HVAC	99.9	64.1	85.4	99.9	64.1	85.4	97.4	63.4	83.7	97.4	63.4	83.7	97.4	63.4	83.7	99.8	63.3	85.1
26, 27, 3370	ELECTRICAL, COMMUNICATIONS & UTIL.	103.3	64.1	82.6	105.8	68.6	86.1	104.5	69.0	85.7	104.5	69.0	85.7	104.5	62.9	82.5	102.4	62.9	81.5
MF2014	WEIGHTED AVERAGE	102.0	63.9	85.4	101.5	65.2	85.7	99.9	66.9	85.5	101.8	66.9	86.6	101.0	65.4	85.5	103.7	65.3	87.0

	DIVISION	WYOMING												CANADA					
		SHERIDAN			WHEATLAND			WORLAND			YELLOWSTONE NAT'L PA			BARRIE, ONTARIO			BATHURST, NEW BRUNSWICK		
		828			822			824			821								
		MAT.	INST.	TOTAL	MAT.	INST.	TOTAL	MAT.	INST.	TOTAL	MAT.	INST.	TOTAL	MAT.	INST.	TOTAL	MAT.	INST.	TOTAL
015433	CONTRACTOR EQUIPMENT		98.2	98.2		98.2	98.2		98.2	98.2		98.2	98.2		102.1	102.1		102.2	102.2
0241, 31 - 34	SITE & INFRASTRUCTURE, DEMOLITION	92.4	94.4	93.8	89.6	94.0	92.8	87.1	94.0	92.0	87.2	94.2	92.2	118.5	100.4	105.6	103.4	96.6	98.6
0310	Concrete Forming & Accessories	102.2	50.6	57.7	96.8	52.7	58.8	96.9	62.0	66.8	96.9	63.1	67.8	124.4	89.4	94.2	104.4	63.7	69.3
0320	Concrete Reinforcing	106.1	70.9	88.2	105.5	70.8	87.8	106.1	70.6	88.0	107.9	70.6	88.9	173.2	83.8	127.7	137.0	56.5	96.0
0330	Cast-in-Place Concrete	99.2	72.0	88.1	100.2	61.5	84.3	95.9	59.9	81.1	95.9	61.6	81.8	165.6	89.0	134.2	131.6	61.9	103.0
03	CONCRETE	106.9	62.4	85.0	103.7	59.7	82.1	99.2	63.2	81.6	99.5	64.3	82.2	150.3	88.4	119.9	124.6	62.4	94.1
04	MASONRY	100.6	56.3	73.0	100.7	54.7	72.0	100.3	50.4	69.2	100.4	53.4	71.1	171.5	98.0	125.7	160.8	64.9	101.1
05	METALS	101.6	72.4	92.6	97.8	71.7	89.8	98.0	71.5	89.8	98.6	71.5	90.3	110.2	90.2	104.0	109.0	71.5	97.5
06	WOOD, PLASTICS & COMPOSITES	97.9	50.4	71.3	88.3	54.3	69.2	88.3	66.9	76.3	88.3	66.9	76.3	119.8	88.2	102.1	101.8	64.0	80.7
07	THERMAL & MOISTURE PROTECTION	102.9	56.3	83.8	102.1	55.8	83.1	101.9	57.5	83.7	101.4	58.8	83.9	111.8	89.1	102.5	106.0	62.2	88.1
08	OPENINGS	113.2	52.7	99.1	111.1	55.8	98.3	112.8	61.7	100.9	105.6	61.7	95.4	93.9	86.5	92.2	87.9	55.9	80.5
0920	Plaster & Gypsum Board	109.4	48.8	68.5	85.2	52.8	63.3	85.2	65.8	72.1	85.4	65.8	72.1	151.1	87.7	108.2	137.6	62.7	87.0
0950, 0980	Ceilings & Acoustic Treatment	95.7	48.8	64.9	92.9	52.8	66.6	92.9	65.8	75.1	93.7	65.8	75.4	91.5	87.7	89.0	106.4	62.7	77.7
0960	Flooring	107.5	56.4	92.9	105.0	56.4	91.1	105.0	56.4	91.1	105.0	56.4	91.1	127.2	98.9	119.1	107.9	47.2	90.6
0970, 0990	Wall Finishes & Painting/Coating	109.4	43.1	69.4	107.1	34.4	63.3	107.1	55.4	75.9	107.1	55.4	75.9	116.2	89.9	100.3	117.6	51.6	77.8
09	FINISHES	102.8	50.0	73.7	95.8	51.0	71.1	95.5	60.8	76.3	95.7	61.5	76.8	115.0	91.5	102.1	111.3	59.7	82.8
COVERS	DIVS. 10 - 14, 25, 28, 41, 43, 44, 46	100.0	87.7	97.5	100.0	80.3	96.0	100.0	81.4	96.2	100.0	82.4	96.4	139.2	74.4	126.1	131.1	65.3	117.8
21, 22, 23	FIRE SUPPRESSION, PLUMBING & HVAC	97.4	64.1	83.9	97.4	63.3	83.6	97.4	63.3	83.6	97.4	64.9	84.3	101.8	99.3	100.8	102.0	69.8	89.0
26, 27, 3370	ELECTRICAL, COMMUNICATIONS & UTIL.	107.4	63.0	83.9	104.5	68.6	85.5	104.5	62.9	82.5	103.2	62.9	81.9	116.8	90.5	102.9	112.7	61.8	85.8
MF2014	WEIGHTED AVERAGE	102.6	64.3	85.9	100.4	64.3	84.7	100.0	65.3	84.9	99.3	66.3	84.9	117.7	93.0	106.9	111.5	67.3	92.2

	DIVISION	CANADA																	
		BRANDON, MANITOBA			BRANTFORD, ONTARIO			BRIDGEWATER, NOVA SCOTIA			CALGARY, ALBERTA			CAP-DE-LA-MADELEINE, QUEBEC			CHARLESBOURG, QUEBEC		
		MAT.	INST.	TOTAL	MAT.	INST.	TOTAL	MAT.	INST.	TOTAL	MAT.	INST.	TOTAL	MAT.	INST.	TOTAL	MAT.	INST.	TOTAL
015433	CONTRACTOR EQUIPMENT		104.5	104.5		102.1	102.1		101.9	101.9		107.1	107.1		102.8	102.8		102.8	102.8
0241, 31 - 34	SITE & INFRASTRUCTURE, DEMOLITION	133.0	99.3	109.1	118.0	100.7	105.7	102.6	98.2	99.5	124.0	105.4	110.7	98.4	99.9	99.4	98.4	99.9	99.4
0310	Concrete Forming & Accessories	146.5	72.5	82.7	124.9	96.9	100.7	97.1	71.6	75.1	124.4	97.4	101.1	130.7	88.0	93.8	130.7	88.0	93.8
0320	Concrete Reinforcing	185.9	53.2	118.4	165.4	82.5	123.2	140.6	46.7	92.8	135.6	71.9	103.2	140.6	78.8	109.2	140.6	78.8	109.2
0330	Cast-in-Place Concrete	133.4	76.7	110.1	153.6	110.3	135.8	157.8	72.2	122.7	189.0	107.3	155.5	124.0	97.8	113.2	124.0	97.8	113.2
03	CONCRETE	145.6	71.0	109.0	141.8	98.8	120.7	139.4	67.9	104.3	159.1	96.1	128.2	124.3	89.8	107.4	124.3	89.8	107.4
04	MASONRY	233.5	67.6	130.1	168.0	102.6	127.2	163.4	71.9	106.4	196.3	91.4	130.9	163.9	87.1	116.0	163.9	87.1	116.0
05	METALS	125.6	76.8	110.6	109.1	90.9	103.5	108.2	74.2	97.7	137.0	89.3	122.3	107.4	87.6	101.3	107.4	87.6	101.3
06	WOOD, PLASTICS & COMPOSITES	154.2	73.5	109.0	123.7	95.7	108.1	92.9	71.1	80.7	102.3	97.2	99.4	135.3	88.0	108.8	135.3	88.0	108.8
07	THERMAL & MOISTURE PROTECTION	135.0	71.8	109.1	113.8	94.7	106.0	108.6	69.9	92.8	117.8	94.6	108.3	107.4	89.9	100.3	107.4	89.9	100.3
08	OPENINGS	105.5	64.8	96.1	92.4	92.7	92.5	86.1	64.9	81.1	89.9	86.2	89.1	93.6	81.1	90.7	93.6	81.1	90.7
0920	Plaster & Gypsum Board	119.3	72.2	87.4	130.1	95.5	106.7	130.7	70.0	89.7	139.4	96.6	110.5	157.0	87.4	109.9	157.0	87.4	109.9
0950, 0980	Ceilings & Acoustic Treatment	109.6	72.2	85.0	96.5	95.5	95.8	96.5	70.0	79.1	138.8	96.6	111.0	96.5	87.4	90.5	96.5	87.4	90.5
0960	Flooring	138.6	70.2	119.1	120.5	98.9	114.3	102.7	67.2	92.5	122.3	92.6	113.8	120.5	99.5	114.5	120.5	99.5	114.5
0970, 0990	Wall Finishes & Painting/Coating	126.5	58.7	85.6	118.3	98.6	106.4	118.3	64.1	85.6	127.4	112.4	118.3	118.3	91.7	102.2	118.3	91.7	102.2
09	FINISHES	122.5	71.0	94.1	112.4	97.9	104.4	106.6	70.3	86.6	125.5	98.7	110.7	115.3	90.8	101.8	115.3	90.8	101.8
COVERS	DIVS. 10 - 14, 25, 28, 41, 43, 44, 46	131.1	67.6	118.3	131.1	76.6	120.1	131.1	66.6	118.1	131.1	96.9	124.2	131.1	85.2	121.9	131.1	85.2	121.9
21, 22, 23	FIRE SUPPRESSION, PLUMBING & HVAC	102.1	84.1	94.9	102.0	102.3	102.1	102.0	84.4	94.9	101.2	92.9	97.8	102.3	91.0	97.8	102.3	91.0	97.8
26, 27, 3370	ELECTRICAL, COMMUNICATIONS & UTIL.	118.6	69.6	92.7	111.5	89.9	100.1	117.1	64.4	89.2	113.5	101.3	107.1	111.2	72.6	90.8	111.2	72.6	90.8
MF2014	WEIGHTED AVERAGE	125.2	75.8	103.7	114.9	96.9	107.1	112.9	74.6	96.2	123.5	95.7	111.4	112.4	87.6	101.6	112.4	87.6	101.6

	DIVISION	CHARLOTTETOWN, PRINCE EDWARD ISLAND			CHICOUTIMI, QUEBEC			CORNER BROOK, NEWFOUNDLAND			CORNWALL, ONTARIO			DALHOUSIE, NEW BRUNSWICK			DARTMOUTH, NOVA SCOTIA		
		CANADA																	
		MAT.	INST.	TOTAL	MAT.	INST.	TOTAL	MAT.	INST.	TOTAL	MAT.	INST.	TOTAL	MAT.	INST.	TOTAL	MAT.	INST.	TOTAL
015433	CONTRACTOR EQUIPMENT		101.8	101.8		102.8	102.8		103.3	103.3		102.1	102.1		102.2	102.2		101.9	101.9
0241, 31 - 34	SITE & INFRASTRUCTURE, DEMOLITION	121.3	95.7	103.1	100.7	99.1	99.6	137.7	97.1	108.9	116.1	100.2	104.8	100.9	96.6	97.8	124.8	98.2	105.9
0310	Concrete Forming & Accessories	110.4	58.2	65.4	133.2	93.1	98.6	121.8	61.7	69.9	122.6	89.6	94.1	103.5	63.9	69.3	111.8	71.6	77.2
0320	Concrete Reinforcing	141.5	46.8	93.3	103.3	95.9	99.5	170.3	48.6	108.4	165.4	82.2	123.1	142.7	56.6	98.9	178.2	46.7	111.3
0330	Cast-in-Place Concrete	166.8	61.2	123.4	124.0	99.8	114.0	155.3	70.6	120.5	138.2	100.6	122.7	131.8	62.0	103.2	149.8	72.2	118.0
03	CONCRETE	147.1	57.9	103.3	114.6	95.9	105.4	182.2	63.2	123.7	134.3	92.2	113.6	132.5	62.5	98.1	162.1	67.9	115.8
04	MASONRY	178.4	62.6	106.3	162.7	95.0	120.5	229.3	64.1	126.4	166.8	94.0	121.4	160.2	64.9	100.8	244.8	71.9	137.1
05	METALS	133.9	67.6	113.5	108.8	92.2	103.7	126.0	72.9	109.7	109.1	89.6	103.1	104.4	71.8	94.3	125.9	74.2	110.0
06	WOOD, PLASTICS & COMPOSITES	97.8	57.8	75.4	137.4	93.6	112.9	128.3	60.8	90.5	122.1	89.0	103.6	100.2	64.0	80.0	116.1	71.1	90.9
07	THERMAL & MOISTURE PROTECTION	121.3	60.4	96.3	106.5	96.4	102.3	139.8	61.8	107.8	113.6	89.3	103.6	110.8	62.2	90.9	136.8	69.9	109.4
08	OPENINGS	88.2	50.0	79.3	92.2	80.6	89.5	112.4	57.1	99.5	93.6	86.2	91.9	89.1	55.9	81.4	95.3	64.9	88.2
0920	Plaster & Gypsum Board	131.6	56.3	80.7	159.4	93.2	114.6	149.0	59.3	88.4	189.4	88.6	121.3	141.2	62.7	88.2	144.6	70.0	94.2
0950, 0980	Ceilings & Acoustic Treatment	120.2	56.3	78.2	105.6	93.2	97.4	110.5	59.3	76.9	99.0	88.6	92.2	98.9	62.7	75.1	117.2	70.0	86.2
0960	Flooring	111.7	62.9	97.8	123.6	99.5	116.8	121.0	57.2	102.8	120.5	97.5	113.9	106.5	72.1	96.7	116.1	67.2	102.1
0970, 0990	Wall Finishes & Painting/Coating	125.6	42.9	75.7	117.6	106.4	110.8	126.4	61.8	87.4	118.3	91.9	102.4	120.7	51.6	79.0	126.4	64.1	88.8
09	FINISHES	117.1	57.8	84.4	118.4	96.5	106.4	122.0	60.9	88.4	121.0	91.6	104.8	110.6	64.5	85.2	120.1	70.3	92.6
COVERS	DIVS. 10 - 14, 25, 28, 41, 43, 44, 46	131.1	64.8	117.7	131.1	86.5	122.1	131.1	65.6	117.9	131.1	74.0	119.6	131.1	65.3	117.8	131.1	66.6	118.1
21, 22, 23	FIRE SUPPRESSION, PLUMBING & HVAC	102.3	63.8	86.8	102.0	92.3	98.1	102.1	71.8	89.9	102.3	100.1	101.4	102.0	69.8	89.0	102.1	84.4	95.0
26, 27, 3370	ELECTRICAL, COMMUNICATIONS & UTIL.	110.6	52.3	79.8	109.7	86.8	97.6	116.0	58.3	85.5	112.4	90.9	101.0	114.5	58.4	84.8	120.7	64.4	90.9
MF2014	WEIGHTED AVERAGE	120.0	62.8	95.1	111.4	92.9	103.3	129.6	67.6	102.6	115.0	93.3	105.5	111.9	67.5	92.5	126.1	74.6	103.7

	DIVISION	EDMONTON, ALBERTA			FORT MCMURRAY, ALBERTA			FREDERICTON, NEW BRUNSWICK			GATINEAU, QUEBEC			GRANBY, QUEBEC			HALIFAX, NOVA SCOTIA		
		CANADA																	
		MAT.	INST.	TOTAL	MAT.	INST.	TOTAL	MAT.	INST.	TOTAL	MAT.	INST.	TOTAL	MAT.	INST.	TOTAL	MAT.	INST.	TOTAL
015433	CONTRACTOR EQUIPMENT		107.1	107.1		104.4	104.4		102.2	102.2		102.8	102.8		102.8	102.8		101.9	101.9
0241, 31 - 34	SITE & INFRASTRUCTURE, DEMOLITION	144.1	105.4	116.6	123.1	101.3	107.6	105.0	96.6	99.1	98.2	99.8	99.4	98.7	99.8	99.5	109.2	98.3	101.5
0310	Concrete Forming & Accessories	123.2	97.4	100.9	122.8	92.3	96.5	120.3	64.2	71.9	130.7	87.8	93.7	130.7	87.8	93.7	108.4	78.8	82.9
0320	Concrete Reinforcing	134.2	71.9	102.5	152.9	71.8	111.7	132.8	56.7	94.1	148.7	78.8	113.2	148.7	78.8	113.2	144.3	62.7	102.7
0330	Cast-in-Place Concrete	183.9	107.3	152.4	203.9	105.1	163.3	128.3	62.0	101.1	122.3	97.8	112.2	126.4	97.7	114.6	149.4	81.9	121.7
03	CONCRETE	156.4	96.1	126.8	163.6	93.1	129.0	126.6	62.7	95.2	124.8	89.8	107.6	126.7	89.7	108.5	136.4	77.3	107.4
04	MASONRY	187.2	91.4	127.5	207.9	89.1	133.8	182.6	66.5	110.2	163.7	87.1	116.0	164.1	87.1	116.1	185.9	84.7	122.8
05	METALS	136.9	89.3	122.3	135.4	88.9	121.1	129.1	72.3	111.6	107.4	87.4	101.2	107.4	87.4	101.2	134.9	81.5	118.4
06	WOOD, PLASTICS & COMPOSITES	105.2	97.2	100.7	117.6	91.8	103.2	116.7	64.0	87.2	135.3	88.0	108.8	135.3	88.0	108.8	97.1	77.8	86.3
07	THERMAL & MOISTURE PROTECTION	122.7	94.6	111.2	121.9	91.9	109.6	116.0	63.2	94.4	107.4	89.9	100.3	107.4	88.3	99.6	119.0	79.4	102.8
08	OPENINGS	89.2	86.2	88.5	93.6	83.3	91.2	89.3	54.8	81.2	93.6	76.3	89.6	93.6	76.3	89.6	93.5	72.0	88.5
0920	Plaster & Gypsum Board	134.3	96.6	108.8	128.2	91.1	103.1	137.4	62.7	86.9	126.6	87.4	100.1	129.0	87.4	100.8	127.0	77.0	93.2
0950, 0980	Ceilings & Acoustic Treatment	139.5	96.6	111.3	104.9	91.1	95.8	116.1	62.7	81.0	96.5	87.4	90.5	96.5	87.4	90.5	118.7	77.0	91.3
0960	Flooring	122.5	92.6	114.0	120.5	92.6	112.5	116.8	75.3	104.9	120.5	99.5	114.5	120.5	99.5	114.5	109.2	88.3	103.3
0970, 0990	Wall Finishes & Painting/Coating	122.4	112.4	116.3	118.4	95.5	104.5	123.1	65.8	88.5	118.3	91.7	102.2	118.3	91.7	102.2	124.9	87.4	102.2
09	FINISHES	126.6	98.7	111.2	115.3	93.0	103.0	117.7	66.7	89.6	111.2	90.8	100.0	111.5	90.8	100.1	115.6	81.8	97.0
COVERS	DIVS. 10 - 14, 25, 28, 41, 43, 44, 46	131.1	96.9	124.2	131.1	95.2	123.9	131.1	65.3	117.8	131.1	85.2	121.9	131.1	85.2	121.9	131.1	68.5	118.5
21, 22, 23	FIRE SUPPRESSION, PLUMBING & HVAC	101.1	92.9	97.8	102.4	99.0	101.0	102.3	79.3	93.0	102.3	91.0	97.8	102.0	91.0	97.6	100.6	77.6	91.3
26, 27, 3370	ELECTRICAL, COMMUNICATIONS & UTIL.	110.6	101.3	105.7	105.8	85.4	95.1	116.1	77.1	95.5	111.2	72.6	90.8	111.8	72.6	91.1	117.0	81.2	98.1
MF2014	WEIGHTED AVERAGE	123.1	95.7	111.2	123.6	92.8	110.2	117.4	72.5	97.9	112.1	87.4	101.3	112.3	87.4	101.4	119.6	80.9	102.7

	DIVISION	HAMILTON, ONTARIO			HULL, QUEBEC			JOLIETTE, QUEBEC			KAMLOOPS, BRITISH COLUMBIA			KINGSTON, ONTARIO			KITCHENER, ONTARIO		
		CANADA																	
		MAT.	INST.	TOTAL	MAT.	INST.	TOTAL	MAT.	INST.	TOTAL	MAT.	INST.	TOTAL	MAT.	INST.	TOTAL	MAT.	INST.	TOTAL
015433	CONTRACTOR EQUIPMENT		108.9	108.9		102.8	102.8		102.8	102.8		106.1	106.1		104.4	104.4		104.3	104.3
0241, 31 - 34	SITE & INFRASTRUCTURE, DEMOLITION	115.5	112.4	113.3	98.2	99.8	99.4	98.8	99.9	99.6	120.5	103.3	108.3	116.1	104.0	107.5	99.6	104.8	103.3
0310	Concrete Forming & Accessories	120.9	94.7	98.3	130.7	87.8	93.7	130.7	88.0	93.8	123.4	90.8	95.3	122.8	89.6	94.2	113.0	87.2	90.8
0320	Concrete Reinforcing	146.4	92.0	118.7	148.7	78.8	113.2	140.6	78.8	109.2	110.3	76.8	93.2	165.4	82.2	123.1	100.8	91.9	96.3
0330	Cast-in-Place Concrete	143.5	101.0	126.1	122.3	97.8	112.2	127.4	97.8	115.2	110.3	101.6	106.7	138.2	100.6	122.7	129.0	93.6	114.4
03	CONCRETE	132.8	96.3	114.8	124.8	89.8	107.6	125.9	89.8	108.2	132.9	92.0	112.8	136.2	92.2	114.6	110.8	90.4	100.8
04	MASONRY	183.0	101.2	132.0	163.7	87.1	116.0	164.1	87.1	116.1	170.7	95.0	123.5	173.5	94.1	124.0	150.8	97.6	117.6
05	METALS	123.4	92.5	113.9	107.4	87.4	101.2	107.4	87.6	101.3	109.8	87.5	102.9	110.8	89.5	104.3	117.4	92.2	109.6
06	WOOD, PLASTICS & COMPOSITES	100.4	93.9	96.8	135.3	88.0	108.8	135.3	88.0	108.8	105.2	89.4	96.4	122.1	89.2	103.7	109.7	85.5	96.1
07	THERMAL & MOISTURE PROTECTION	118.2	95.7	109.0	107.4	89.9	100.2	107.4	89.9	100.2	123.3	86.8	108.3	113.6	90.4	104.1	108.6	92.8	102.1
08	OPENINGS	91.3	91.6	91.4	93.6	76.3	89.6	93.6	81.1	90.7	90.2	85.7	89.2	93.6	85.9	91.8	84.7	85.3	84.8
0920	Plaster & Gypsum Board	149.5	93.6	111.7	126.6	87.4	100.1	157.0	87.4	109.9	113.6	88.5	96.6	192.8	88.7	122.4	120.2	84.9	96.3
0950, 0980	Ceilings & Acoustic Treatment	122.1	93.6	103.4	96.5	87.4	90.5	96.5	87.4	90.5	96.5	88.5	91.2	112.4	88.7	96.8	98.6	84.9	89.6
0960	Flooring	123.1	102.3	117.2	120.5	99.5	114.5	120.5	99.5	114.5	119.6	55.6	101.3	120.5	97.5	113.9	110.5	102.3	108.1
0970, 0990	Wall Finishes & Painting/Coating	121.6	102.6	110.2	118.3	91.7	102.2	118.3	91.7	102.2	118.3	84.0	97.6	118.3	84.9	98.1	115.3	92.6	101.6
09	FINISHES	122.0	97.0	108.2	111.2	90.8	100.0	115.3	90.8	101.8	111.7	84.2	96.6	124.4	90.9	105.9	106.4	90.2	97.5
COVERS	DIVS. 10 - 14, 25, 28, 41, 43, 44, 46	131.1	98.8	124.6	131.1	85.2	121.9	131.1	85.2	121.9	131.1	95.6	124.0	131.1	74.0	119.6	131.1	97.0	124.2
21, 22, 23	FIRE SUPPRESSION, PLUMBING & HVAC	101.3	89.2	96.4	102.0	91.0	97.6	102.0	91.0	97.6	102.0	94.4	98.9	102.3	100.3	101.5	100.4	87.7	95.3
26, 27, 3370	ELECTRICAL, COMMUNICATIONS & UTIL.	108.3	100.5	104.2	112.9	72.6	91.6	111.8	72.6	91.1	115.1	81.9	97.6	112.4	89.6	100.3	111.5	98.0	104.4
MF2014	WEIGHTED AVERAGE	117.0	96.7	108.2	112.2	87.4	101.4	112.5	87.6	101.7	114.5	90.6	104.1	116.0	93.4	106.2	109.6	92.9	102.3

		CANADA																	
	DIVISION	LAVAL, QUEBEC			LETHBRIDGE, ALBERTA			LLOYDMINSTER, ALBERTA			LONDON, ONTARIO			MEDICINE HAT, ALBERTA			MONCTON, NEW BRUNSWICK		
		MAT.	INST.	TOTAL	MAT.	INST.	TOTAL	MAT.	INST.	TOTAL	MAT.	INST.	TOTAL	MAT.	INST.	TOTAL	MAT.	INST.	TOTAL
015433	CONTRACTOR EQUIPMENT		102.8	102.8		104.4	104.4		104.4	104.4		104.4	104.4		104.4	104.4		102.2	102.2
0241, 31 - 34	SITE & INFRASTRUCTURE, DEMOLITION	98.7	99.8	99.5	115.6	101.9	105.9	115.4	101.3	105.4	113.1	104.9	107.3	114.2	101.4	105.1	102.8	96.8	98.5
0310	Concrete Forming & Accessories	130.8	87.8	93.8	124.1	92.4	96.8	122.3	82.6	88.1	119.1	88.5	92.7	124.1	82.6	88.3	104.4	64.5	69.9
0320	Concrete Reinforcing	148.7	78.8	113.2	152.9	71.8	111.7	152.9	71.8	111.7	132.8	91.9	112.0	152.9	71.8	111.7	137.0	61.5	98.6
0330	Cast-in-Place Concrete	126.4	97.8	114.6	153.0	105.1	133.3	141.9	101.4	125.3	147.4	98.6	127.4	141.9	101.4	125.3	126.8	78.7	107.0
03	CONCRETE	126.7	89.8	108.6	139.5	93.2	116.7	134.1	87.5	111.3	132.4	92.7	112.9	134.3	87.5	111.3	122.3	69.7	96.5
04	MASONRY	164.0	87.1	116.1	181.7	89.1	124.0	163.4	82.5	113.0	182.6	98.7	130.3	163.4	82.5	113.0	160.5	65.0	101.0
05	METALS	107.4	87.4	101.3	128.8	88.9	116.5	110.0	88.8	103.4	124.9	91.8	114.7	110.0	88.7	103.4	109.0	80.4	100.2
06	WOOD, PLASTICS & COMPOSITES	135.5	88.0	108.9	121.3	91.8	104.8	117.6	82.0	97.7	108.2	86.6	96.1	121.3	82.0	99.3	101.8	64.0	80.7
07	THERMAL & MOISTURE PROTECTION	107.9	89.9	100.5	119.1	91.9	108.0	116.0	87.5	104.3	119.1	93.5	108.6	122.3	87.5	108.0	110.2	65.3	91.8
08	OPENINGS	93.6	76.3	89.6	93.6	83.3	91.2	93.6	77.9	89.9	86.4	86.6	86.4	93.6	77.9	89.9	87.9	61.0	81.6
0920	Plaster & Gypsum Board	129.3	87.4	100.9	119.1	91.1	100.2	114.4	81.0	91.8	152.4	86.1	107.6	117.2	81.0	92.8	137.6	62.7	87.0
0950, 0980	Ceilings & Acoustic Treatment	96.5	87.4	90.5	104.1	91.1	95.5	96.5	81.0	86.3	121.1	86.1	98.1	96.5	81.0	86.3	106.4	62.7	77.7
0960	Flooring	120.5	99.5	114.5	120.5	92.6	112.5	120.5	92.6	112.5	117.2	102.3	113.0	120.5	92.6	112.5	107.9	72.1	97.7
0970, 0990	Wall Finishes & Painting/Coating	118.3	91.7	102.2	118.2	104.2	109.8	118.4	81.2	96.0	122.0	99.4	108.3	118.2	81.2	95.9	117.6	51.6	77.8
09	FINISHES	111.5	90.8	100.1	112.9	94.0	102.5	110.7	84.0	96.0	120.5	91.9	104.7	110.9	84.0	96.1	111.3	64.5	85.5
COVERS	DIVS. 10 - 14, 25, 28, 41, 43, 44, 46	131.1	85.2	121.9	131.1	95.2	123.9	131.1	91.9	123.2	131.1	97.5	124.3	131.1	91.9	123.2	131.1	65.3	117.8
21, 22, 23	FIRE SUPPRESSION, PLUMBING & HVAC	100.4	91.0	96.6	102.2	95.6	99.6	102.3	95.6	99.6	101.4	86.6	95.4	102.0	92.3	98.1	102.0	70.1	89.1
26, 27, 3370	ELECTRICAL, COMMUNICATIONS & UTIL.	112.9	72.6	91.6	107.1	85.4	95.7	104.9	85.4	94.6	107.0	97.5	102.0	104.9	85.4	94.6	117.2	61.8	88.0
MF2014	WEIGHTED AVERAGE	112.1	87.4	101.3	118.4	92.2	107.0	113.4	88.9	102.8	116.4	93.3	106.3	113.6	88.2	102.5	111.8	70.1	93.6

		CANADA																	
	DIVISION	MONTREAL, QUEBEC			MOOSE JAW, SASKATCHEWAN			NEW GLASGOW, NOVA SCOTIA			NEWCASTLE, NEW BRUNSWICK			NORTH BAY, ONTARIO			OSHAWA, ONTARIO		
		MAT.	INST.	TOTAL	MAT.	INST.	TOTAL	MAT.	INST.	TOTAL	MAT.	INST.	TOTAL	MAT.	INST.	TOTAL	MAT.	INST.	TOTAL
015433	CONTRACTOR EQUIPMENT		104.5	104.5		100.6	100.6		101.9	101.9		102.2	102.2		102.1	102.1		104.3	104.3
0241, 31 - 34	SITE & INFRASTRUCTURE, DEMOLITION	112.1	99.4	103.1	115.5	95.9	101.6	118.5	98.2	104.1	103.4	96.6	98.6	133.9	99.8	109.7	110.7	103.9	105.9
0310	Concrete Forming & Accessories	124.3	93.4	97.7	107.1	60.5	66.9	111.7	71.6	77.1	104.4	63.9	69.5	147.4	87.0	95.3	118.4	89.1	93.1
0320	Concrete Reinforcing	134.0	96.0	114.7	107.9	61.9	84.5	170.3	46.7	107.4	137.0	56.6	96.1	201.7	81.8	140.7	159.6	84.4	121.3
0330	Cast-in-Place Concrete	161.7	101.2	136.8	137.7	71.1	110.4	149.8	72.2	118.0	131.6	62.0	103.0	144.3	87.1	120.8	149.1	87.9	124.0
03	CONCRETE	139.7	96.5	118.4	117.9	65.1	92.0	160.9	67.9	115.2	124.6	62.5	94.1	163.4	86.3	125.5	135.6	88.0	112.2
04	MASONRY	167.9	95.1	122.5	162.0	64.4	101.1	228.9	71.9	131.1	160.8	64.9	101.1	235.3	89.9	144.7	153.8	94.4	116.8
05	METALS	128.0	92.7	117.1	106.3	73.9	96.3	123.6	74.2	108.4	109.0	71.8	97.5	124.6	89.2	113.7	108.6	91.4	103.3
06	WOOD, PLASTICS & COMPOSITES	114.4	93.9	102.9	102.7	59.1	78.3	116.1	71.1	90.9	101.8	64.0	80.7	157.4	87.6	118.3	116.7	88.2	100.7
07	THERMAL & MOISTURE PROTECTION	116.5	96.9	108.5	106.8	64.1	89.3	136.8	69.9	109.4	110.2	62.2	90.5	143.5	85.6	119.8	109.4	87.4	100.4
08	OPENINGS	90.1	82.4	88.3	89.3	56.8	81.7	95.3	64.9	88.2	87.9	55.9	80.5	104.2	83.7	99.4	90.4	87.9	89.8
0920	Plaster & Gypsum Board	135.1	93.2	106.8	110.7	57.8	74.9	142.7	70.0	93.6	137.6	62.7	87.0	140.3	87.1	104.4	124.7	87.7	99.7
0950, 0980	Ceilings & Acoustic Treatment	126.1	93.2	104.4	96.5	57.8	71.0	109.6	70.0	83.6	106.4	62.7	77.7	109.6	87.1	94.8	95.3	87.7	90.3
0960	Flooring	122.0	99.5	115.6	110.6	63.3	97.1	116.1	67.2	102.1	107.9	72.1	97.7	138.6	97.5	126.9	113.7	104.9	111.2
0970, 0990	Wall Finishes & Painting/Coating	120.8	106.4	112.1	118.3	66.8	87.2	126.4	64.1	88.8	117.6	51.6	77.8	126.4	91.2	105.1	115.3	106.7	110.1
09	FINISHES	119.9	96.7	107.1	107.1	61.4	81.9	118.2	70.3	91.8	111.3	64.5	85.5	125.1	89.8	105.7	107.9	93.9	100.2
COVERS	DIVS. 10 - 14, 25, 28, 41, 43, 44, 46	131.1	87.1	122.3	131.1	64.6	117.7	131.1	66.6	118.1	131.1	65.3	117.8	131.1	72.7	119.3	131.1	97.2	124.3
21, 22, 23	FIRE SUPPRESSION, PLUMBING & HVAC	101.2	92.4	97.7	102.3	76.1	91.8	102.1	84.4	95.0	102.0	69.8	89.0	102.1	98.1	100.5	100.4	100.7	100.5
26, 27, 3370	ELECTRICAL, COMMUNICATIONS & UTIL.	112.6	86.8	99.0	113.9	62.8	86.9	116.3	64.4	88.9	112.0	61.8	85.5	116.4	90.8	102.9	112.6	90.5	100.9
MF2014	WEIGHTED AVERAGE	117.8	93.2	107.1	110.7	69.5	92.7	124.1	74.6	102.5	111.6	67.9	92.5	127.3	91.1	111.5	112.3	94.3	104.4

		CANADA																	
	DIVISION	OTTAWA, ONTARIO			OWEN SOUND, ONTARIO			PETERBOROUGH, ONTARIO			PORTAGE LA PRAIRIE, MANITOBA			PRINCE ALBERT, SASKATCHEWAN			PRINCE GEORGE, BRITISH COLUMBIA		
		MAT.	INST.	TOTAL	MAT.	INST.	TOTAL	MAT.	INST.	TOTAL	MAT.	INST.	TOTAL	MAT.	INST.	TOTAL	MAT.	INST.	TOTAL
015433	CONTRACTOR EQUIPMENT		104.3	104.3		102.1	102.1		102.1	102.1		104.5	104.5		100.6	100.6		106.1	106.1
0241, 31 - 34	SITE & INFRASTRUCTURE, DEMOLITION	108.2	104.6	105.7	118.5	100.3	105.5	118.0	100.2	105.3	116.6	99.3	104.3	110.9	96.1	100.3	124.0	103.3	109.3
0310	Concrete Forming & Accessories	121.2	91.4	95.5	124.4	85.5	90.9	124.9	88.1	93.1	124.3	72.0	79.2	107.1	60.3	66.8	112.7	85.5	89.3
0320	Concrete Reinforcing	138.9	91.9	115.0	173.2	83.8	127.7	165.4	82.3	123.1	152.9	53.2	102.2	112.7	61.8	86.8	110.3	76.8	93.2
0330	Cast-in-Place Concrete	145.6	99.9	126.8	165.6	83.1	131.7	153.6	88.7	127.0	141.9	76.1	114.9	124.8	71.0	102.7	138.2	101.6	123.2
03	CONCRETE	132.6	94.4	113.8	150.3	84.6	118.1	141.8	87.4	115.1	127.8	70.6	99.7	112.6	65.0	89.2	145.4	89.7	118.0
04	MASONRY	168.0	98.8	124.9	171.5	95.6	124.2	168.0	96.7	123.5	166.5	66.6	104.2	161.1	64.4	100.8	172.9	95.0	124.3
05	METALS	127.9	91.7	116.8	110.2	90.0	104.0	109.1	89.7	103.1	110.0	76.7	99.7	106.3	73.7	96.3	109.8	87.6	102.9
06	WOOD, PLASTICS & COMPOSITES	109.4	90.6	98.9	119.8	84.3	99.9	123.7	86.3	102.8	121.1	73.5	94.4	102.7	59.1	78.3	105.2	82.1	92.3
07	THERMAL & MOISTURE PROTECTION	124.4	93.8	111.8	111.8	86.4	101.4	113.8	91.2	104.5	107.3	71.3	92.5	106.7	63.0	88.8	117.2	86.1	104.4
08	OPENINGS	93.9	88.7	92.7	93.9	83.1	91.4	92.4	85.5	90.8	93.6	64.8	86.9	88.2	56.8	80.9	90.2	81.8	88.2
0920	Plaster & Gypsum Board	168.3	90.2	115.5	151.1	83.7	105.6	130.1	85.8	100.1	114.2	72.2	85.8	110.7	57.8	74.9	113.6	81.0	91.6
0950, 0980	Ceilings & Acoustic Treatment	125.6	90.2	102.4	91.5	83.7	86.4	96.5	85.8	89.4	96.5	72.2	80.5	96.5	57.8	71.0	96.5	81.0	86.3
0960	Flooring	112.3	97.5	108.1	127.2	98.9	119.1	120.5	97.5	113.9	120.5	70.2	106.1	110.6	63.3	97.1	116.0	76.2	104.6
0970, 0990	Wall Finishes & Painting/Coating	123.2	94.2	105.7	116.2	89.9	100.3	118.3	93.5	103.3	118.4	58.7	82.4	118.3	57.0	81.3	118.3	84.0	97.6
09	FINISHES	122.6	92.9	106.2	115.0	88.6	100.5	112.4	90.5	100.3	110.5	70.8	88.6	107.1	60.3	81.3	110.6	83.4	95.6
COVERS	DIVS. 10 - 14, 25, 28, 41, 43, 44, 46	131.1	95.7	124.0	139.2	73.2	125.9	131.1	74.2	119.6	131.1	67.2	118.2	131.1	64.6	117.7	131.1	94.8	123.8
21, 22, 23	FIRE SUPPRESSION, PLUMBING & HVAC	101.4	87.2	95.7	101.8	98.1	100.3	102.0	101.7	101.9	102.0	83.6	94.5	102.3	68.6	88.7	102.0	94.4	98.9
26, 27, 3370	ELECTRICAL, COMMUNICATIONS & UTIL.	107.3	98.7	102.8	118.2	89.6	103.1	111.5	90.4	100.4	113.6	60.1	85.3	113.9	62.8	86.9	111.7	81.9	96.0
MF2014	WEIGHTED AVERAGE	117.3	94.0	107.2	117.8	91.2	106.2	114.9	93.1	105.4	113.4	74.2	96.3	109.9	67.7	91.5	115.4	89.9	104.3

City Cost Indexes

	DIVISION	CANADA																	
		QUEBEC, QUEBEC			RED DEER, ALBERTA			REGINA, SASKATCHEWAN			RIMOUSKI, QUEBEC			ROUYN-NORANDA, QUEBEC			SAINT HYACINTHE, QUEBEC		
		MAT.	INST.	TOTAL	MAT.	INST.	TOTAL	MAT.	INST.	TOTAL	MAT.	INST.	TOTAL	MAT.	INST.	TOTAL	MAT.	INST.	TOTAL
015433	CONTRACTOR EQUIPMENT		104.9	104.9		104.4	104.4		100.6	100.6		102.8	102.8		102.8	102.8		102.8	102.8
0241, 31 - 34	SITE & INFRASTRUCTURE, DEMOLITION	110.8	99.5	102.8	114.2	101.4	105.1	124.4	98.0	105.6	98.7	99.1	99.0	98.2	99.8	99.4	98.7	99.8	99.5
0310	Concrete Forming & Accessories	125.6	93.7	98.1	139.6	82.6	90.4	117.6	90.9	94.6	130.7	93.1	98.3	130.7	87.8	93.7	130.7	87.8	93.7
0320	Concrete Reinforcing	138.3	96.0	116.8	152.9	71.8	111.7	138.0	84.8	111.0	106.0	95.9	100.9	148.7	78.8	113.2	148.7	78.8	113.2
0330	Cast-in-Place Concrete	141.8	101.6	125.2	141.9	101.4	125.3	171.5	95.7	140.4	128.6	99.8	116.8	122.3	97.8	112.2	126.4	97.8	114.6
03	CONCRETE	131.0	96.7	114.1	135.3	87.5	111.8	148.9	91.4	120.6	121.1	95.9	108.7	124.8	89.8	107.6	126.7	89.8	108.5
04	MASONRY	163.0	95.1	120.7	163.4	82.5	113.0	183.5	96.7	129.4	163.6	95.0	120.9	163.7	87.1	116.0	164.0	87.1	116.1
05	METALS	128.0	93.0	117.2	110.0	88.7	103.4	129.3	86.4	116.1	106.9	92.2	102.4	107.4	87.4	101.2	107.4	87.4	101.2
06	WOOD, PLASTICS & COMPOSITES	115.8	93.9	103.6	121.3	82.0	99.3	101.9	91.9	96.3	135.3	93.6	112.0	135.3	88.0	108.8	135.3	88.0	108.8
07	THERMAL & MOISTURE PROTECTION	118.4	97.0	109.6	132.2	87.5	113.9	123.2	86.0	107.9	107.4	96.4	102.9	107.4	89.9	100.2	107.7	89.9	100.4
08	OPENINGS	93.0	90.1	92.3	93.6	77.9	89.9	91.9	80.7	89.3	93.2	80.6	90.2	93.6	76.3	89.6	93.6	76.3	89.6
0920	Plaster & Gypsum Board	142.1	93.2	109.0	117.2	81.0	92.8	130.5	91.6	104.2	156.8	93.2	113.8	126.4	87.4	100.0	128.7	87.4	100.8
0950, 0980	Ceilings & Acoustic Treatment	126.1	93.2	104.4	96.5	81.0	86.3	125.4	91.6	103.2	95.7	93.2	94.0	95.7	87.4	90.2	95.7	87.4	90.2
0960	Flooring	125.8	99.5	118.3	123.0	92.6	114.3	120.0	65.5	104.4	120.5	99.5	114.5	120.5	99.5	114.5	120.5	99.5	114.5
0970, 0990	Wall Finishes & Painting/Coating	127.8	106.4	114.9	118.2	81.2	95.9	123.9	92.9	105.2	118.3	106.4	111.1	118.3	91.7	102.2	118.3	91.7	102.2
09	FINISHES	122.2	96.8	108.2	111.7	84.0	96.4	122.7	87.2	103.1	115.1	96.5	104.9	111.0	90.8	99.9	111.3	90.8	100.0
COVERS	DIVS. 10 - 14, 25, 28, 41, 43, 44, 46	131.1	87.3	122.3	131.1	91.9	123.2	131.1	72.4	119.3	131.1	86.5	122.1	131.1	85.2	121.9	131.1	85.2	121.9
21, 22, 23	FIRE SUPPRESSION, PLUMBING & HVAC	101.2	92.4	97.7	102.0	92.3	98.1	100.7	84.8	94.3	102.0	92.3	98.1	102.0	91.0	97.6	98.8	91.0	95.7
26, 27, 3370	ELECTRICAL, COMMUNICATIONS & UTIL.	111.4	86.8	98.4	104.9	85.4	94.6	114.4	92.4	102.8	111.8	86.8	98.6	111.8	72.6	91.1	112.5	72.6	91.4
MF2014	WEIGHTED AVERAGE	117.1	93.6	106.8	114.1	88.2	102.8	120.7	89.0	106.9	111.8	92.9	103.6	112.0	87.4	101.3	111.6	87.4	101.1

	DIVISION	CANADA																	
		SAINT JOHN, NEW BRUNSWICK			SARNIA, ONTARIO			SASKATOON, SASKATCHEWAN			SAULT STE MARIE, ONTARIO			SHERBROOKE, QUEBEC			SOREL, QUEBEC		
		MAT.	INST.	TOTAL	MAT.	INST.	TOTAL	MAT.	INST.	TOTAL	MAT.	INST.	TOTAL	MAT.	INST.	TOTAL	MAT.	INST.	TOTAL
015433	CONTRACTOR EQUIPMENT		102.2	102.2		102.1	102.1		100.6	100.6		102.1	102.1		102.8	102.8		102.8	102.8
0241, 31 - 34	SITE & INFRASTRUCTURE, DEMOLITION	103.5	98.0	99.6	116.5	100.3	105.0	111.7	98.0	102.0	106.7	99.8	101.8	98.7	99.8	99.5	98.8	99.9	99.6
0310	Concrete Forming & Accessories	124.6	67.0	74.9	123.6	95.3	99.1	107.4	90.9	93.2	112.8	87.9	91.3	130.7	87.8	93.7	130.7	88.0	93.8
0320	Concrete Reinforcing	137.0	62.0	98.8	117.4	83.6	100.2	114.5	84.8	99.4	106.1	82.1	93.9	148.7	78.8	113.2	140.6	78.8	109.2
0330	Cast-in-Place Concrete	129.7	80.1	109.3	141.7	102.1	125.5	135.4	95.7	119.1	127.4	87.2	110.9	126.4	97.8	114.6	127.4	97.8	115.2
03	CONCRETE	125.2	71.4	98.8	128.6	95.4	112.3	119.6	91.4	105.7	113.1	86.8	100.2	126.7	89.8	108.5	125.9	89.8	108.2
04	MASONRY	180.8	73.7	114.0	179.1	99.5	129.5	168.8	96.7	123.8	164.5	94.4	120.8	164.1	87.1	116.1	164.1	87.1	116.1
05	METALS	108.9	81.7	100.5	109.1	90.1	103.2	104.5	86.4	99.0	108.3	90.2	102.7	107.4	87.4	101.2	107.4	87.6	101.3
06	WOOD, PLASTICS & COMPOSITES	125.7	65.4	91.9	122.6	94.7	107.0	100.5	91.9	95.7	109.6	88.3	97.6	135.3	88.0	108.8	135.3	88.0	108.8
07	THERMAL & MOISTURE PROTECTION	110.5	70.9	94.3	113.9	94.7	106.0	107.1	86.0	98.4	112.7	88.4	102.7	107.4	89.9	100.3	107.4	89.9	100.2
08	OPENINGS	87.8	60.6	81.5	94.7	89.4	93.4	88.9	80.7	87.0	86.5	85.6	86.3	93.6	76.3	89.6	93.6	81.1	90.7
0920	Plaster & Gypsum Board	152.7	64.2	92.9	150.6	94.4	112.6	123.5	91.6	101.9	119.6	87.8	98.1	128.7	87.4	100.8	156.8	87.4	109.9
0950, 0980	Ceilings & Acoustic Treatment	111.5	64.2	80.4	100.7	94.4	96.6	117.4	91.6	100.4	96.5	87.8	90.8	95.7	87.4	90.2	95.7	87.4	90.2
0960	Flooring	119.5	72.1	105.9	120.5	106.6	116.5	111.0	65.5	98.0	113.4	100.8	109.8	120.5	99.5	114.5	120.5	99.5	114.5
0970, 0990	Wall Finishes & Painting/Coating	117.6	80.2	95.0	118.3	105.7	110.7	120.7	92.9	103.9	118.3	97.9	106.0	118.3	91.7	102.2	118.3	91.7	102.2
09	FINISHES	117.9	69.0	91.0	116.1	98.9	106.6	114.3	87.2	99.3	108.1	91.5	99.0	111.3	90.8	100.0	115.1	90.8	101.7
COVERS	DIVS. 10 - 14, 25, 28, 41, 43, 44, 46	131.1	66.3	118.0	131.1	75.7	119.9	131.1	74.1	119.6	131.1	96.4	124.1	131.1	85.2	121.9	131.1	85.2	121.9
21, 22, 23	FIRE SUPPRESSION, PLUMBING & HVAC	102.0	79.8	93.0	102.0	107.8	104.3	100.5	84.8	94.2	102.0	94.8	99.1	102.3	91.0	97.8	102.0	91.0	97.6
26, 27, 3370	ELECTRICAL, COMMUNICATIONS & UTIL.	120.3	88.2	103.3	114.2	92.7	102.8	116.2	92.4	103.6	113.0	90.8	101.3	111.8	72.6	91.1	111.8	72.6	91.1
MF2014	WEIGHTED AVERAGE	114.0	77.9	98.3	114.8	97.5	107.3	111.1	89.1	101.5	110.3	92.1	102.3	112.4	87.4	101.5	112.5	87.6	101.7

	DIVISION	CANADA																	
		ST CATHARINES, ONTARIO			ST JEROME, QUEBEC			ST JOHNS, NEWFOUNDLAND			SUDBURY, ONTARIO			SUMMERSIDE, PRINCE EDWARD ISLAND			SYDNEY, NOVA SCOTIA		
		MAT.	INST.	TOTAL	MAT.	INST.	TOTAL	MAT.	INST.	TOTAL	MAT.	INST.	TOTAL	MAT.	INST.	TOTAL	MAT.	INST.	TOTAL
015433	CONTRACTOR EQUIPMENT		102.1	102.1		102.8	102.8		103.3	103.3		102.1	102.1		101.8	101.8		101.9	101.9
0241, 31 - 34	SITE & INFRASTRUCTURE, DEMOLITION	100.0	101.4	101.0	98.2	99.8	99.4	123.4	99.9	106.7	100.1	100.9	100.7	128.4	95.7	105.2	114.3	98.2	102.8
0310	Concrete Forming & Accessories	110.9	94.2	96.5	130.7	87.8	93.7	123.9	84.8	90.1	107.0	89.0	91.5	112.0	58.3	65.7	111.7	71.6	77.1
0320	Concrete Reinforcing	101.7	92.0	96.7	148.7	78.8	113.2	158.9	75.5	116.4	102.5	91.3	96.8	168.1	46.8	106.4	170.3	46.7	107.4
0330	Cast-in-Place Concrete	123.2	100.6	113.9	122.3	97.8	112.2	167.5	96.4	138.3	124.2	97.0	113.0	142.3	61.2	109.0	115.6	72.2	97.8
03	CONCRETE	108.0	95.9	102.1	124.8	89.8	107.6	152.6	87.2	120.5	108.4	92.2	100.4	170.1	57.9	115.0	144.7	67.9	106.9
04	MASONRY	150.3	100.7	119.4	163.7	87.1	116.0	187.7	92.4	128.3	150.4	96.6	116.8	228.1	62.6	125.0	226.2	71.9	130.1
05	METALS	107.7	92.2	102.9	107.4	87.4	101.2	136.3	84.3	120.3	107.7	91.5	102.7	123.6	67.6	106.4	123.6	74.2	108.4
06	WOOD, PLASTICS & COMPOSITES	107.4	93.9	99.8	135.3	88.0	108.8	114.7	82.8	96.8	103.6	88.3	95.0	116.4	57.8	83.6	116.1	71.1	90.9
07	THERMAL & MOISTURE PROTECTION	108.6	96.5	103.6	107.4	89.9	100.2	127.1	91.4	112.5	108.0	91.7	101.3	136.1	61.1	105.4	136.8	69.9	109.4
08	OPENINGS	84.1	90.2	85.5	93.6	76.3	89.6	93.7	74.5	89.2	84.8	85.6	85.0	108.0	50.0	94.5	95.3	64.9	88.2
0920	Plaster & Gypsum Board	108.5	93.6	98.4	126.4	87.4	100.0	143.9	82.1	102.1	112.9	87.8	95.9	143.1	56.3	84.4	142.7	70.0	93.6
0950, 0980	Ceilings & Acoustic Treatment	95.3	93.6	94.1	95.7	87.4	90.2	120.6	82.1	95.3	90.3	87.8	88.6	109.6	56.3	74.6	109.6	70.0	83.6
0960	Flooring	109.3	98.9	106.3	120.5	99.5	114.5	116.0	59.3	99.8	107.3	100.8	105.5	116.2	62.9	100.9	116.1	67.2	102.1
0970, 0990	Wall Finishes & Painting/Coating	115.3	116.0	115.7	118.3	91.7	102.2	120.4	95.9	105.6	115.3	93.6	102.2	126.4	42.9	76.0	126.4	64.1	88.8
09	FINISHES	103.8	97.8	100.4	111.0	90.8	99.9	122.0	81.4	99.6	102.7	91.6	96.5	119.2	57.8	85.4	118.2	70.3	91.8
COVERS	DIVS. 10 - 14, 25, 28, 41, 43, 44, 46	131.1	75.2	119.8	131.1	85.2	121.9	131.1	72.6	119.3	131.1	97.1	124.3	131.1	64.8	117.7	131.1	66.6	118.1
21, 22, 23	FIRE SUPPRESSION, PLUMBING & HVAC	100.4	87.8	95.3	102.0	91.0	97.6	100.8	82.1	93.3	101.9	87.8	96.2	102.1	63.8	86.7	102.1	84.4	95.0
26, 27, 3370	ELECTRICAL, COMMUNICATIONS & UTIL.	112.9	98.8	105.5	112.4	72.6	91.3	114.4	83.2	97.9	110.3	99.2	104.4	115.4	52.3	82.0	116.3	64.4	88.9
MF2014	WEIGHTED AVERAGE	107.6	94.5	101.9	112.1	87.4	101.3	122.7	85.2	106.4	107.7	93.0	101.3	126.7	62.8	98.8	122.1	74.6	101.4

DIVISION		CANADA																	
		THUNDER BAY, ONTARIO			TIMMINS, ONTARIO			TORONTO, ONTARIO			TROIS RIVIERES, QUEBEC			TRURO, NOVA SCOTIA			VANCOUVER, BRITISH COLUMBIA		
		MAT.	INST.	TOTAL	MAT.	INST.	TOTAL	MAT.	INST.	TOTAL	MAT.	INST.	TOTAL	MAT.	INST.	TOTAL	MAT.	INST.	TOTAL
015433	CONTRACTOR EQUIPMENT		102.1	102.1		102.1	102.1		104.4	104.4		102.8	102.8		101.9	101.9		112.7	112.7
0241, 31 - 34	SITE & INFRASTRUCTURE, DEMOLITION	104.8	101.3	102.3	118.0	99.8	105.1	133.8	105.6	113.7	114.8	99.9	104.2	102.8	98.2	99.5	118.2	107.6	110.6
0310	Concrete Forming & Accessories	118.3	92.2	95.8	124.9	87.0	92.2	124.3	100.8	104.0	155.5	88.0	97.3	97.1	71.6	75.1	121.1	91.0	95.1
0320	Concrete Reinforcing	90.9	90.8	90.9	165.4	81.8	122.9	146.0	94.4	119.7	170.3	78.8	123.8	140.6	46.7	92.8	134.1	79.2	106.2
0330	Cast-in-Place Concrete	135.6	99.7	120.8	153.6	87.1	126.3	136.8	111.8	126.5	119.7	97.8	110.7	159.5	72.2	123.7	143.5	97.9	124.8
03	CONCRETE	115.8	94.5	105.3	141.8	86.3	114.5	129.7	103.1	116.7	147.4	89.8	119.1	140.2	67.9	104.7	137.1	91.4	114.6
04	MASONRY	151.0	100.6	119.6	168.0	89.9	119.3	182.0	110.9	137.7	230.7	87.1	141.2	163.6	71.9	106.5	166.7	91.4	119.7
05	METALS	107.5	91.1	102.5	109.1	89.2	103.0	132.5	93.9	120.7	122.7	87.6	111.9	108.2	74.2	97.7	139.3	90.3	124.2
06	WOOD, PLASTICS & COMPOSITES	116.7	90.9	102.2	123.7	87.6	103.5	115.1	98.9	106.0	173.0	88.0	125.4	92.9	71.1	80.7	104.8	91.5	97.4
07	THERMAL & MOISTURE PROTECTION	108.8	93.8	102.7	113.8	85.6	102.2	123.9	102.8	115.2	135.2	89.9	116.6	108.6	69.9	92.8	125.2	86.9	109.5
08	OPENINGS	83.3	88.0	84.4	92.4	83.7	90.4	90.5	96.8	92.0	105.5	81.1	99.8	86.1	64.9	81.1	88.9	86.9	88.4
0920	Plaster & Gypsum Board	139.8	90.5	106.5	130.1	87.1	101.1	133.0	98.8	109.9	173.4	87.4	115.3	130.7	70.0	89.7	129.7	90.6	103.3
0950, 0980	Ceilings & Acoustic Treatment	90.3	90.5	90.4	96.5	87.1	90.3	114.9	98.8	104.3	108.8	87.4	94.7	96.5	70.0	79.1	132.2	90.6	104.8
0960	Flooring	113.7	105.9	111.4	120.5	97.5	113.9	117.0	108.4	114.6	138.6	99.5	127.5	102.7	67.2	92.5	124.6	96.0	116.4
0970, 0990	Wall Finishes & Painting/Coating	115.3	94.6	102.8	118.3	91.2	101.9	120.4	106.7	112.1	126.4	91.7	105.4	118.3	64.1	85.6	121.7	95.6	106.0
09	FINISHES	108.5	95.1	101.1	112.4	89.8	99.9	116.9	103.0	109.2	128.7	90.8	107.8	106.6	70.3	86.6	123.8	92.7	106.6
COVERS	DIVS. 10 - 14, 25, 28, 41, 43, 44, 46	131.1	75.2	119.8	131.1	72.7	119.3	131.1	101.2	125.1	131.1	85.2	121.9	131.1	66.6	118.1	131.1	94.9	123.8
21, 22, 23	FIRE SUPPRESSION, PLUMBING & HVAC	100.4	88.0	95.4	102.0	98.1	100.4	101.0	97.3	99.5	102.1	91.0	97.7	102.0	84.4	94.9	101.2	80.5	92.8
26, 27, 3370	ELECTRICAL, COMMUNICATIONS & UTIL.	111.5	97.0	103.8	113.0	90.8	101.3	108.8	101.0	104.7	116.6	72.6	93.4	111.5	64.4	86.6	109.0	83.3	95.4
MF2014	WEIGHTED AVERAGE	108.8	93.4	102.1	115.1	91.1	104.6	118.3	101.3	110.9	124.8	87.6	108.6	112.5	74.6	96.0	119.5	89.1	106.2

DIVISION		CANADA																	
		VICTORIA, BRITISH COLUMBIA			WHITEHORSE, YUKON			WINDSOR, ONTARIO			WINNIPEG, MANITOBA			YARMOUTH, NOVA SCOTIA			YELLOWKNIFE, NWT		
		MAT.	INST.	TOTAL	MAT.	INST.	TOTAL	MAT.	INST.	TOTAL	MAT.	INST.	TOTAL	MAT.	INST.	TOTAL	MAT.	INST.	TOTAL
015433	CONTRACTOR EQUIPMENT		109.3	109.3		102.1	102.1		102.1	102.1		106.5	106.5		101.9	101.9		101.9	101.9
0241, 31 - 34	SITE & INFRASTRUCTURE, DEMOLITION	123.0	107.1	111.7	142.3	96.7	109.9	95.9	101.3	99.7	118.2	100.9	105.9	118.3	98.2	104.0	154.1	101.0	116.4
0310	Concrete Forming & Accessories	112.7	90.4	93.5	129.4	59.6	69.2	118.3	90.5	94.3	128.5	69.3	77.4	111.7	71.6	77.1	132.7	80.2	87.4
0320	Concrete Reinforcing	112.0	79.1	95.2	160.8	60.0	109.5	99.6	91.9	95.7	132.9	57.9	94.7	170.3	46.7	107.4	151.4	62.7	106.2
0330	Cast-in-Place Concrete	140.3	97.2	122.6	209.3	70.3	152.2	126.2	101.9	116.2	163.9	75.0	127.4	148.3	72.2	117.0	211.2	88.1	160.6
03	CONCRETE	150.1	90.7	120.9	182.4	64.1	124.3	109.6	94.7	102.3	144.4	69.7	107.7	160.1	67.9	114.8	182.0	80.0	131.9
04	MASONRY	171.7	91.3	121.6	260.3	62.8	137.2	150.5	99.9	119.0	176.6	70.8	110.7	228.8	71.9	131.0	260.3	74.8	144.7
05	METALS	107.0	86.3	100.6	135.3	73.6	116.3	107.6	91.7	102.7	141.7	75.0	121.2	123.6	74.2	108.4	137.9	78.7	119.7
06	WOOD, PLASTICS & COMPOSITES	104.4	91.4	97.1	117.6	58.3	84.4	116.7	89.0	101.2	113.6	70.2	89.3	116.1	71.1	90.9	124.0	81.9	100.4
07	THERMAL & MOISTURE PROTECTION	116.9	86.7	104.5	139.1	62.0	107.5	108.6	93.9	102.6	120.8	71.8	100.7	136.8	69.9	109.4	151.5	78.4	121.5
08	OPENINGS	90.5	83.2	88.8	107.6	55.5	95.4	83.1	87.5	84.1	85.5	62.6	80.2	95.3	64.9	88.2	102.2	68.9	94.4
0920	Plaster & Gypsum Board	115.8	90.6	98.8	150.5	56.8	87.2	125.7	88.6	100.6	129.5	68.6	88.3	142.7	70.0	93.6	163.7	81.1	107.9
0950, 0980	Ceilings & Acoustic Treatment	98.1	90.6	93.2	144.6	56.8	86.9	90.3	88.6	89.2	136.2	68.6	91.8	109.6	70.0	83.6	138.9	81.1	100.9
0960	Flooring	116.4	76.2	104.9	141.0	61.3	118.2	113.7	103.1	110.6	117.0	75.6	105.2	116.1	67.2	102.1	135.5	92.5	123.2
0970, 0990	Wall Finishes & Painting/Coating	120.7	95.6	105.5	143.1	56.0	90.5	115.3	95.5	103.3	124.5	56.5	83.4	126.4	64.1	88.8	135.3	80.3	102.1
09	FINISHES	112.1	89.3	99.5	140.9	59.3	95.9	106.3	93.4	99.2	122.3	69.5	93.2	118.2	70.3	91.8	137.5	82.0	106.9
COVERS	DIVS. 10 - 14, 25, 28, 41, 43, 44, 46	131.1	71.5	119.1	131.1	63.9	117.6	131.1	74.7	119.7	131.1	68.7	118.5	131.1	66.6	118.1	131.1	67.1	118.2
21, 22, 23	FIRE SUPPRESSION, PLUMBING & HVAC	102.0	80.5	93.3	102.5	74.8	91.3	100.4	88.0	95.4	101.1	66.9	87.3	102.1	84.4	95.0	102.7	94.0	99.2
26, 27, 3370	ELECTRICAL, COMMUNICATIONS & UTIL.	113.1	82.7	97.0	133.5	62.2	95.8	116.0	99.0	107.0	113.3	68.0	89.4	116.3	64.4	88.9	130.5	85.2	106.5
MF2014	WEIGHTED AVERAGE	115.7	87.2	103.3	135.3	68.5	106.2	108.2	93.4	101.7	121.0	71.8	99.5	124.0	74.6	102.5	135.3	84.2	113.0

Location Factors

Costs shown in RSMeans cost data publications are based on national averages for materials and installation. To adjust these costs to a specific location, simply multiply the base cost by the factor and divide by 100 for that city. The data is arranged alphabetically by state and postal zip code numbers. For a city not listed, use the factor for a nearby city with similar economic characteristics.

STATE/ZIP	CITY	MAT.	INST.	TOTAL
ALABAMA				
350-352	Birmingham	100.6	76.7	90.2
354	Tuscaloosa	99.8	61.2	83.0
355	Jasper	100.2	57.5	81.6
356	Decatur	99.8	57.9	81.5
357-358	Huntsville	99.8	72.1	87.8
359	Gadsden	99.9	60.0	82.5
360-361	Montgomery	99.9	57.1	81.2
362	Anniston	98.6	61.8	82.6
363	Dothan	99.0	49.4	77.4
364	Evergreen	98.6	50.7	77.7
365-366	Mobile	99.9	65.4	84.9
367	Selma	98.7	49.9	77.4
368	Phenix City	99.5	54.6	79.9
369	Butler	98.9	49.4	77.3
ALASKA				
995-996	Anchorage	121.1	115.6	118.7
997	Fairbanks	121.0	116.4	119.0
998	Juneau	121.3	115.6	118.8
999	Ketchikan	132.0	115.6	124.8
ARIZONA				
850,853	Phoenix	99.2	73.8	88.1
851,852	Mesa/Tempe	98.6	72.8	87.4
855	Globe	98.8	70.0	86.3
856-857	Tucson	97.4	72.4	86.5
859	Show Low	98.9	72.5	87.4
860	Flagstaff	101.3	71.8	88.4
863	Prescott	99.2	68.9	86.0
864	Kingman	97.6	72.0	86.4
865	Chambers	97.6	71.5	86.2
ARKANSAS				
716	Pine Bluff	99.4	63.3	83.6
717	Camden	97.1	50.1	76.6
718	Texarkana	98.2	49.2	76.8
719	Hot Springs	96.4	51.8	76.9
720-722	Little Rock	98.5	64.0	83.4
723	West Memphis	96.1	62.4	81.4
724	Jonesboro	96.7	59.1	80.3
725	Batesville	94.6	54.6	77.1
726	Harrison	95.9	50.4	76.0
727	Fayetteville	93.4	52.0	75.4
728	Russellville	94.6	53.7	76.8
729	Fort Smith	97.2	61.6	81.7
CALIFORNIA				
900-902	Los Angeles	99.2	117.5	107.2
903-905	Inglewood	94.8	114.1	103.2
906-908	Long Beach	96.2	114.1	104.0
910-912	Pasadena	96.4	114.6	104.3
913-916	Van Nuys	99.1	114.6	105.9
917-918	Alhambra	98.3	114.8	105.5
919-921	San Diego	100.5	109.6	104.5
922	Palm Springs	97.2	112.0	103.6
923-924	San Bernardino	95.1	111.3	102.1
925	Riverside	99.4	114.6	106.0
926-927	Santa Ana	96.9	111.8	103.4
928	Anaheim	99.4	114.9	106.2
930	Oxnard	99.6	114.4	106.0
931	Santa Barbara	98.9	115.1	106.0
932-933	Bakersfield	100.7	113.6	106.3
934	San Luis Obispo	99.4	112.0	104.9
935	Mojave	96.8	109.8	102.5
936-938	Fresno	100.1	115.3	106.7
939	Salinas	100.2	121.6	109.5
940-941	San Francisco	105.2	145.4	122.7
942,956-958	Sacramento	100.5	119.3	108.7
943	Palo Alto	98.9	134.5	114.4
944	San Mateo	100.9	134.6	115.6
945	Vallejo	100.1	127.2	111.9
946	Oakland	103.1	134.3	116.7
947	Berkeley	102.7	135.1	116.8
948	Richmond	101.8	133.1	115.4
949	San Rafael	103.5	136.7	118.0
950	Santa Cruz	104.5	121.8	112.0

STATE/ZIP	CITY	MAT.	INST.	TOTAL
CALIFORNIA (CONT'D)				
951	San Jose	102.5	136.6	117.4
952	Stockton	100.9	117.2	108.0
953	Modesto	100.8	115.7	107.3
954	Santa Rosa	100.8	134.7	115.6
955	Eureka	102.1	116.4	108.4
959	Marysville	101.3	115.9	107.7
960	Redding	107.8	117.7	112.1
961	Susanville	106.8	118.1	111.7
COLORADO				
800-802	Denver	100.8	81.7	92.5
803	Boulder	97.8	81.3	90.6
804	Golden	99.8	78.9	90.7
805	Fort Collins	101.3	79.0	91.6
806	Greeley	98.9	75.5	88.7
807	Fort Morgan	98.3	79.0	89.9
808-809	Colorado Springs	100.4	84.0	93.2
810	Pueblo	101.1	80.3	92.0
811	Alamosa	102.2	73.7	89.8
812	Salida	101.9	75.3	90.3
813	Durango	102.7	75.7	90.9
814	Montrose	101.4	75.4	90.1
815	Grand Junction	104.8	75.3	91.9
816	Glenwood Springs	102.3	78.1	91.8
CONNECTICUT				
060	New Britain	100.1	120.6	109.1
061	Hartford	101.4	120.8	109.8
062	Willimantic	100.8	120.2	109.2
063	New London	97.1	120.7	107.4
064	Meriden	98.9	120.7	108.4
065	New Haven	101.6	120.7	109.9
066	Bridgeport	101.3	120.8	109.8
067	Waterbury	100.9	120.7	109.5
068	Norwalk	100.8	128.7	113.0
069	Stamford	100.9	128.7	113.0
D.C.				
200-205	Washington	101.1	91.9	97.1
DELAWARE				
197	Newark	98.8	109.2	103.3
198	Wilmington	99.3	109.2	103.6
199	Dover	99.3	109.2	103.6
FLORIDA				
320,322	Jacksonville	98.8	66.4	84.7
321	Daytona Beach	99.0	72.9	87.6
323	Tallahassee	100.8	56.9	81.7
324	Panama City	100.2	57.8	81.7
325	Pensacola	102.7	61.2	84.6
326,344	Gainesville	100.3	66.7	85.7
327-328,347	Orlando	100.5	70.0	87.2
329	Melbourne	101.7	76.3	90.6
330-332,340	Miami	99.8	73.1	88.2
333	Fort Lauderdale	98.6	72.6	87.2
334,349	West Palm Beach	97.4	71.1	85.9
335-336,346	Tampa	100.1	78.4	90.6
337	St. Petersburg	102.6	63.3	85.5
338	Lakeland	99.4	77.8	90.0
339,341	Fort Myers	98.6	72.6	87.3
342	Sarasota	100.8	74.3	89.2
GEORGIA				
300-303,399	Atlanta	97.6	74.4	87.5
304	Statesboro	97.3	55.8	79.2
305	Gainesville	96.0	61.6	81.0
306	Athens	95.5	63.9	81.7
307	Dalton	97.2	59.0	80.5
308-309	Augusta	96.3	63.6	82.1
310-312	Macon	97.2	64.7	83.0
313-314	Savannah	99.2	60.9	82.5
315	Waycross	98.3	59.4	81.4
316	Valdosta	98.5	61.4	82.3
317,398	Albany	98.5	61.3	82.3
318-319	Columbus	98.4	65.4	84.0

STATE/ZIP	CITY	MAT.	INST.	TOTAL
HAWAII				
967	Hilo	114.8	120.2	117.2
968	Honolulu	119.5	120.2	119.8
STATES & POSS.				
969	Guam	136.2	58.0	102.1
IDAHO				
832	Pocatello	101.1	77.4	90.8
833	Twin Falls	102.2	67.3	87.0
834	Idaho Falls	99.8	74.7	88.9
835	Lewiston	108.7	84.9	98.3
836-837	Boise	100.1	78.6	90.7
838	Coeur d'Alene	107.9	82.2	96.7
ILLINOIS				
600-603	North Suburban	99.3	135.1	114.9
604	Joliet	99.2	138.4	116.3
605	South Suburban	99.3	135.1	114.9
606-608	Chicago	99.8	139.7	117.2
609	Kankakee	95.6	131.7	111.3
610-611	Rockford	99.5	124.8	110.5
612	Rock Island	97.2	100.5	98.6
613	La Salle	98.4	124.4	109.7
614	Galesburg	98.2	106.7	101.9
615-616	Peoria	100.5	109.3	104.4
617	Bloomington	97.5	109.4	102.7
618-619	Champaign	101.0	107.6	103.9
620-622	East St. Louis	96.4	108.7	101.8
623	Quincy	98.1	101.0	99.3
624	Effingham	97.3	106.0	101.1
625	Decatur	99.2	104.9	101.7
626-627	Springfield	100.0	106.5	102.8
628	Centralia	95.0	108.5	100.9
629	Carbondale	94.7	107.9	100.5
INDIANA				
460	Anderson	97.0	83.8	91.3
461-462	Indianapolis	99.1	84.8	92.9
463-464	Gary	98.3	110.4	103.6
465-466	South Bend	98.0	84.4	92.1
467-468	Fort Wayne	97.6	79.1	89.5
469	Kokomo	94.8	81.5	89.0
470	Lawrenceburg	94.0	77.7	86.9
471	New Albany	95.1	75.4	86.5
472	Columbus	97.2	80.9	90.1
473	Muncie	97.9	82.3	91.1
474	Bloomington	99.3	81.8	91.6
475	Washington	95.8	81.8	89.7
476-477	Evansville	97.1	86.9	92.6
478	Terre Haute	97.9	86.5	92.9
479	Lafayette	96.9	82.7	90.7
IOWA				
500-503,509	Des Moines	100.5	85.2	93.8
504	Mason City	97.6	65.8	83.8
505	Fort Dodge	97.9	58.2	80.6
506-507	Waterloo	99.5	71.8	87.4
508	Creston	98.2	77.3	89.1
510-511	Sioux City	99.7	72.6	87.9
512	Sibley	98.3	56.4	80.0
513	Spencer	99.9	56.6	81.0
514	Carroll	97.0	71.2	85.8
515	Council Bluffs	100.7	78.1	90.9
516	Shenandoah	97.6	71.2	86.1
520	Dubuque	99.1	80.2	90.9
521	Decorah	98.0	64.2	83.3
522-524	Cedar Rapids	100.0	83.2	92.7
525	Ottumwa	97.9	73.7	87.3
526	Burlington	97.2	75.3	87.6
527-528	Davenport	99.0	93.1	96.4
KANSAS				
660-662	Kansas City	99.5	94.8	97.5
664-666	Topeka	99.5	66.0	84.9
667	Fort Scott	97.9	73.0	87.0
668	Emporia	97.9	69.2	85.4
669	Belleville	99.6	64.9	84.4
670-672	Wichita	100.7	66.0	85.6
673	Independence	100.9	69.2	87.1
674	Salina	101.5	67.6	86.7
675	Hutchinson	96.5	65.0	82.8
676	Hays	100.5	65.4	85.2
677	Colby	101.1	65.0	85.4

STATE/ZIP	CITY	MAT.	INST.	TOTAL
KANSAS (CONT'D)				
678	Dodge City	102.7	66.0	86.7
679	Liberal	100.4	65.5	85.2
KENTUCKY				
400-402	Louisville	96.6	83.3	90.8
403-405	Lexington	96.6	81.4	90.0
406	Frankfort	98.1	80.7	90.5
407-409	Corbin	94.0	71.8	84.3
410	Covington	95.9	91.4	93.9
411-412	Ashland	94.6	99.3	96.6
413-414	Campton	95.8	78.6	88.3
415-416	Pikeville	96.9	88.6	93.3
417-418	Hazard	95.2	78.7	88.0
420	Paducah	93.9	84.5	89.8
421-422	Bowling Green	96.2	83.9	90.8
423	Owensboro	96.3	88.9	93.1
424	Henderson	93.7	86.1	90.4
425-426	Somerset	93.2	79.9	87.4
427	Elizabethtown	92.8	82.0	88.1
LOUISIANA				
700-701	New Orleans	101.4	69.5	87.5
703	Thibodaux	98.2	65.4	83.9
704	Hammond	95.8	54.5	77.8
705	Lafayette	97.9	63.0	82.7
706	Lake Charles	98.1	65.0	83.7
707-708	Baton Rouge	99.1	63.7	83.7
710-711	Shreveport	101.0	59.6	83.0
712	Monroe	99.3	56.5	80.6
713-714	Alexandria	99.4	57.2	81.0
MAINE				
039	Kittery	95.2	84.7	90.6
040-041	Portland	102.0	90.1	96.8
042	Lewiston	99.6	90.1	95.5
043	Augusta	102.1	79.5	92.3
044	Bangor	99.0	89.5	94.9
045	Bath	97.5	81.9	90.7
046	Machias	97.1	85.5	92.0
047	Houlton	97.2	85.6	92.1
048	Rockland	96.3	85.5	91.6
049	Waterville	97.5	79.5	89.7
MARYLAND				
206	Waldorf	97.1	81.7	90.4
207-208	College Park	97.2	83.4	91.2
209	Silver Spring	96.5	83.2	90.7
210-212	Baltimore	100.8	81.9	92.6
214	Annapolis	101.5	82.0	93.0
215	Cumberland	96.8	81.8	90.2
216	Easton	98.3	68.6	85.3
217	Hagerstown	97.6	83.3	91.4
218	Salisbury	98.7	62.1	82.7
219	Elkton	95.9	81.9	89.8
MASSACHUSETTS				
010-011	Springfield	100.2	112.6	105.6
012	Pittsfield	99.7	111.3	104.8
013	Greenfield	97.7	112.3	104.1
014	Fitchburg	96.5	126.5	109.6
015-016	Worcester	100.2	126.4	111.6
017	Framingham	95.7	135.1	112.9
018	Lowell	99.5	134.5	114.8
019	Lawrence	100.3	134.6	115.3
020-022, 024	Boston	101.6	139.5	118.1
023	Brockton	99.9	128.1	112.2
025	Buzzards Bay	94.4	128.1	109.1
026	Hyannis	97.0	128.1	110.6
027	New Bedford	99.1	128.3	111.8
MICHIGAN				
480,483	Royal Oak	94.9	106.9	100.1
481	Ann Arbor	97.1	107.5	101.7
482	Detroit	97.8	109.6	102.9
484-485	Flint	96.8	96.0	96.4
486	Saginaw	96.4	91.8	94.4
487	Bay City	96.6	91.7	94.5
488-489	Lansing	99.3	94.7	97.3
490	Battle Creek	98.3	87.6	93.7
491	Kalamazoo	98.7	85.7	93.0
492	Jackson	96.5	95.1	95.9
493,495	Grand Rapids	99.7	85.1	93.3
494	Muskegon	97.2	84.8	91.8

Location Factors

STATE/ZIP	CITY	MAT.	INST.	TOTAL
MICHIGAN (CONT'D)				
496	Traverse City	96.3	80.1	89.3
497	Gaylord	97.2	83.6	91.3
498-499	Iron mountain	99.3	86.2	93.6
MINNESOTA				
550-551	Saint Paul	101.4	117.9	108.6
553-555	Minneapolis	100.4	120.0	108.9
556-558	Duluth	101.3	107.3	103.9
559	Rochester	99.9	104.7	102.0
560	Mankato	97.2	101.3	99.0
561	Windom	95.9	92.3	94.3
562	Willmar	95.5	101.3	98.0
563	St. Cloud	96.9	116.0	105.2
564	Brainerd	96.9	100.5	98.5
565	Detroit Lakes	98.9	95.1	97.3
566	Bemidji	98.2	98.2	98.2
567	Thief River Falls	97.8	92.4	95.4
MISSISSIPPI				
386	Clarksdale	97.3	52.6	77.8
387	Greenville	101.0	61.7	83.9
388	Tupelo	98.8	54.9	79.7
389	Greenwood	98.4	52.2	78.3
390-392	Jackson	100.7	62.2	83.9
393	Meridian	97.6	64.1	83.0
394	Laurel	98.9	55.8	80.1
395	Biloxi	99.2	58.1	81.3
396	Mccomb	97.4	51.7	77.5
397	Columbus	98.8	53.9	79.2
MISSOURI				
630-631	St. Louis	98.7	107.8	102.7
633	Bowling Green	97.3	94.4	96.0
634	Hannibal	96.2	92.4	94.6
635	Kirksville	99.5	89.9	95.3
636	Flat River	98.1	95.0	96.8
637	Cape Girardeau	98.2	92.4	95.7
638	Sikeston	96.7	92.2	94.8
639	Poplar Bluff	96.2	92.0	94.4
640-641	Kansas City	101.4	105.7	103.3
644-645	St. Joseph	100.6	93.5	97.5
646	Chillicothe	97.6	94.5	96.3
647	Harrisonville	97.1	101.5	99.1
648	Joplin	99.5	81.7	91.8
650-651	Jefferson City	99.9	93.7	97.2
652	Columbia	100.2	93.9	97.5
653	Sedalia	100.2	97.3	98.9
654-655	Rolla	97.9	94.6	96.5
656-658	Springfield	101.7	83.7	93.8
MONTANA				
590-591	Billings	102.7	77.1	91.5
592	Wolf Point	102.3	74.2	90.0
593	Miles City	100.1	74.0	88.7
594	Great Falls	104.1	76.2	91.9
595	Havre	101.2	73.1	89.0
596	Helena	102.4	73.8	89.9
597	Butte	102.6	77.3	91.6
598	Missoula	99.9	75.5	89.3
599	Kalispell	99.0	75.3	88.7
NEBRASKA				
680-681	Omaha	100.2	78.6	90.8
683-685	Lincoln	100.5	74.9	89.4
686	Columbus	98.4	77.1	89.1
687	Norfolk	100.0	76.7	89.8
688	Grand Island	100.0	77.1	90.0
689	Hastings	99.4	79.8	90.9
690	Mccook	98.7	71.1	86.7
691	North Platte	99.0	78.3	90.0
692	Valentine	101.2	72.5	88.7
693	Alliance	101.0	70.5	87.7
NEVADA				
889-891	Las Vegas	100.9	108.7	104.3
893	Ely	99.4	102.2	100.6
894-895	Reno	99.6	93.3	96.8
897	Carson City	99.7	93.5	97.0
898	Elko	98.3	84.9	92.4
NEW HAMPSHIRE				
030	Nashua	100.5	93.8	97.6
031	Manchester	101.4	93.8	98.1

STATE/ZIP	CITY	MAT.	INST.	TOTAL
NEW HAMPSHIRE (CONT'D)				
032-033	Concord	101.1	92.5	97.4
034	Keene	97.2	73.3	86.7
035	Littleton	97.1	76.3	88.0
036	Charleston	96.7	69.5	84.8
037	Claremont	95.9	69.5	84.4
038	Portsmouth	97.9	94.3	96.4
NEW JERSEY				
070-071	Newark	101.2	128.4	113.0
072	Elizabeth	98.8	127.6	111.3
073	Jersey City	97.9	127.9	111.0
074-075	Paterson	99.6	127.8	111.9
076	Hackensack	97.6	128.1	110.9
077	Long Branch	97.1	126.3	109.8
078	Dover	97.9	127.8	110.9
079	Summit	97.9	127.6	110.8
080,083	Vineland	97.5	124.7	109.4
081	Camden	99.5	124.8	110.5
082,084	Atlantic City	98.2	124.8	109.8
085-086	Trenton	100.9	125.1	111.5
087	Point Pleasant	99.2	125.9	110.8
088-089	New Brunswick	99.7	127.6	111.9
NEW MEXICO				
870-872	Albuquerque	99.6	72.6	87.8
873	Gallup	99.5	72.6	87.8
874	Farmington	100.1	72.6	88.1
875	Santa Fe	100.8	72.6	88.5
877	Las Vegas	98.0	72.6	86.9
878	Socorro	97.6	72.6	86.7
879	Truth/Consequences	97.5	69.4	85.2
880	Las Cruces	96.1	69.4	84.4
881	Clovis	98.3	72.5	87.1
882	Roswell	100.0	72.6	88.0
883	Carrizozo	100.3	72.6	88.2
884	Tucumcari	99.0	72.5	87.4
NEW YORK				
100-102	New York	103.4	168.7	131.8
103	Staten Island	99.2	168.4	129.4
104	Bronx	97.7	166.1	127.5
105	Mount Vernon	97.5	141.4	116.6
106	White Plains	97.6	141.4	116.7
107	Yonkers	102.1	143.7	120.2
108	New Rochelle	97.9	141.3	116.8
109	Suffern	97.6	129.0	111.3
110	Queens	100.9	166.8	129.6
111	Long Island City	102.4	166.8	130.5
112	Brooklyn	102.8	168.5	131.4
113	Flushing	102.5	166.8	130.5
114	Jamaica	100.8	166.8	129.6
115,117,118	Hicksville	100.9	153.3	123.7
116	Far Rockaway	102.6	166.8	130.6
119	Riverhead	101.6	148.2	121.9
120-122	Albany	100.1	104.7	102.1
123	Schenectady	99.3	104.7	101.7
124	Kingston	102.6	118.5	109.6
125-126	Poughkeepsie	101.8	136.7	117.0
127	Monticello	101.1	115.1	107.2
128	Glens Falls	94.0	98.6	96.0
129	Plattsburgh	99.1	96.7	98.0
130-132	Syracuse	98.9	98.0	98.5
133-135	Utica	97.1	97.1	97.1
136	Watertown	98.5	96.3	97.5
137-139	Binghamton	98.5	101.4	99.8
140-142	Buffalo	100.3	106.2	102.9
143	Niagara Falls	97.8	106.4	101.5
144-146	Rochester	101.1	99.1	100.2
147	Jamestown	96.8	92.5	94.9
148-149	Elmira	96.7	97.4	97.0
NORTH CAROLINA				
270,272-274	Greensboro	101.3	55.7	81.4
271	Winston-Salem	101.0	56.5	81.6
275-276	Raleigh	100.5	54.1	80.3
277	Durham	102.9	55.3	82.1
278	Rocky Mount	98.5	52.3	78.3
279	Elizabeth City	99.3	51.6	78.5
280	Gastonia	101.6	55.1	81.3
281-282	Charlotte	101.9	56.8	82.2
283	Fayetteville	104.5	58.1	84.2
284	Wilmington	100.3	55.9	81.0
285	Kinston	98.4	52.7	78.5

STATE/ZIP	CITY	MAT.	INST.	TOTAL
NORTH CAROLINA (CONT'D)				
286	Hickory	98.8	53.9	79.2
287-288	Asheville	100.8	55.3	80.9
289	Murphy	99.6	48.3	77.2
NORTH DAKOTA				
580-581	Fargo	102.7	67.6	87.4
582	Grand Forks	102.7	51.6	80.4
583	Devils Lake	102.0	58.6	83.1
584	Jamestown	102.1	45.2	77.3
585	Bismarck	103.1	66.2	87.0
586	Dickinson	102.8	60.9	84.6
587	Minot	102.7	71.3	89.0
588	Williston	101.2	60.9	83.7
OHIO				
430-432	Columbus	98.6	89.2	94.5
433	Marion	95.1	85.3	90.8
434-436	Toledo	98.7	98.1	98.4
437-438	Zanesville	95.6	86.5	91.6
439	Steubenville	96.8	94.9	96.0
440	Lorain	98.9	93.0	96.3
441	Cleveland	99.0	100.3	99.6
442-443	Akron	99.9	94.2	97.4
444-445	Youngstown	99.2	89.0	94.7
446-447	Canton	99.3	85.6	93.3
448-449	Mansfield	96.7	87.3	92.6
450	Hamilton	97.9	85.1	92.3
451-452	Cincinnati	98.3	84.5	92.2
453-454	Dayton	98.0	84.4	92.1
455	Springfield	98.0	84.7	92.2
456	Chillicothe	96.8	93.0	95.1
457	Athens	99.6	80.4	91.2
458	Lima	100.0	86.8	94.3
OKLAHOMA				
730-731	Oklahoma City	100.3	65.7	85.2
734	Ardmore	98.3	61.9	82.4
735	Lawton	100.7	62.8	84.2
736	Clinton	99.6	63.1	83.7
737	Enid	100.2	60.5	82.9
738	Woodward	98.3	63.4	83.1
739	Guymon	99.4	59.7	82.1
740-741	Tulsa	98.1	61.6	82.2
743	Miami	94.8	67.5	82.9
744	Muskogee	97.4	55.7	79.2
745	Mcalester	94.5	60.4	79.7
746	Ponca City	95.1	62.1	80.7
747	Durant	95.1	63.5	81.3
748	Shawnee	96.5	61.8	81.4
749	Poteau	94.3	61.4	80.0
OREGON				
970-972	Portland	98.8	100.4	99.5
973	Salem	100.5	99.4	100.0
974	Eugene	98.6	99.3	98.9
975	Medford	100.0	97.1	98.8
976	Klamath Falls	99.8	97.1	98.6
977	Bend	98.9	99.4	99.1
978	Pendleton	94.8	101.4	97.7
979	Vale	92.9	98.1	95.2
PENNSYLVANIA				
150-152	Pittsburgh	99.9	104.6	102.0
153	Washington	97.0	103.2	99.7
154	Uniontown	97.3	101.7	99.2
155	Bedford	98.2	91.3	95.2
156	Greensburg	98.2	100.9	99.4
157	Indiana	97.1	99.3	98.0
158	Dubois	98.5	96.8	97.8
159	Johnstown	98.1	97.2	97.7
160	Butler	92.5	102.0	96.6
161	New Castle	92.5	97.9	94.8
162	Kittanning	92.9	103.3	97.4
163	Oil City	92.4	96.9	94.4
164-165	Erie	94.5	95.8	95.1
166	Altoona	94.6	91.9	93.4
167	Bradford	95.7	96.4	96.0
168	State College	95.3	92.8	94.2
169	Wellsboro	96.3	92.6	94.7
170-171	Harrisburg	99.8	95.6	98.0
172	Chambersburg	96.2	88.0	92.6
173-174	York	96.7	95.9	96.3
175-176	Lancaster	95.1	89.0	92.4
PENNSYLVANIA (CONT'D)				
177	Williamsport	93.8	83.0	89.1
178	Sunbury	95.8	95.8	95.8
179	Pottsville	94.9	98.5	96.5
180	Lehigh Valley	95.8	114.0	103.7
181	Allentown	97.9	108.6	102.5
182	Hazleton	95.4	98.3	96.6
183	Stroudsburg	95.3	101.4	97.9
184-185	Scranton	98.6	99.2	98.9
186-187	Wilkes-Barre	95.1	98.7	96.6
188	Montrose	94.7	96.9	95.7
189	Doylestown	94.9	120.6	106.1
190-191	Philadelphia	99.7	133.6	114.5
193	Westchester	95.9	121.5	107.1
194	Norristown	95.0	129.5	110.0
195-196	Reading	97.1	102.1	99.3
PUERTO RICO				
009	San Juan	122.7	24.4	79.8
RHODE ISLAND				
028	Newport	98.4	117.5	106.7
029	Providence	100.6	117.5	108.0
SOUTH CAROLINA				
290-292	Columbia	99.3	56.5	80.6
293	Spartanburg	98.6	56.7	80.3
294	Charleston	99.9	65.2	84.8
295	Florence	98.2	56.9	80.2
296	Greenville	98.3	55.8	79.8
297	Rock Hill	97.8	50.3	77.1
298	Aiken	98.6	69.1	85.7
299	Beaufort	99.3	44.9	75.6
SOUTH DAKOTA				
570-571	Sioux Falls	100.0	57.7	81.5
572	Watertown	99.0	49.1	77.2
573	Mitchell	97.8	48.4	76.3
574	Aberdeen	100.5	49.6	78.3
575	Pierre	101.7	58.4	82.8
576	Mobridge	98.5	48.8	76.8
577	Rapid City	100.4	59.5	82.6
TENNESSEE				
370-372	Nashville	98.4	74.1	87.8
373-374	Chattanooga	100.0	66.2	85.2
375,380-381	Memphis	99.0	71.3	86.9
376	Johnson City	99.8	56.5	80.9
377-379	Knoxville	96.6	66.9	83.7
382	Mckenzie	98.2	59.2	81.2
383	Jackson	99.9	61.2	83.0
384	Columbia	96.9	66.1	83.5
385	Cookeville	98.1	59.1	81.1
TEXAS				
750	Mckinney	98.7	63.1	83.2
751	Waxahackie	98.6	65.9	84.4
752-753	Dallas	99.5	67.4	85.5
754	Greenville	98.8	64.0	83.6
755	Texarkana	98.0	58.4	80.7
756	Longview	98.5	58.7	81.2
757	Tyler	98.5	65.0	83.9
758	Palestine	95.0	64.4	81.7
759	Lufkin	95.6	67.8	83.5
760-761	Fort Worth	99.3	65.3	84.5
762	Denton	98.9	62.5	83.0
763	Wichita Falls	97.1	64.4	82.8
764	Eastland	95.9	63.2	81.6
765	Temple	94.7	61.7	80.3
766-767	Waco	96.5	64.1	82.4
768	Brownwood	98.9	59.4	81.7
769	San Angelo	98.5	59.6	81.5
770-772	Houston	99.4	70.5	86.8
773	Huntsville	98.0	66.7	84.4
774	Wharton	98.9	67.7	85.3
775	Galveston	97.0	70.0	85.2
776-777	Beaumont	97.4	68.6	84.9
778	Bryan	94.9	68.2	83.3
779	Victoria	99.0	65.6	84.4
780	Laredo	98.3	64.5	83.6
781-782	San Antonio	98.5	65.1	84.0
783-784	Corpus Christi	100.9	63.5	84.6
785	Mc Allen	101.0	59.8	83.0
786-787	Austin	100.0	63.6	84.1

Location Factors

STATE/ZIP	CITY	MAT.	INST.	TOTAL
TEXAS (CONT'D)				
788	Del Rio	100.8	63.2	84.4
789	Giddings	97.7	64.0	83.0
790-791	Amarillo	100.2	63.2	84.1
792	Childress	99.3	64.3	84.0
793-794	Lubbock	101.6	63.6	85.0
795-796	Abilene	99.7	62.2	83.3
797	Midland	101.5	64.2	85.2
798-799,885	El Paso	98.0	61.7	82.2
UTAH				
840-841	Salt Lake City	102.2	68.3	87.4
842,844	Ogden	98.1	70.1	85.9
843	Logan	99.8	70.1	86.9
845	Price	100.2	64.6	84.7
846-847	Provo	100.3	68.2	86.3
VERMONT				
050	White River Jct.	98.5	78.3	89.7
051	Bellows Falls	97.2	93.0	95.3
052	Bennington	97.5	93.5	95.8
053	Brattleboro	97.8	92.8	95.6
054	Burlington	102.4	83.6	94.2
056	Montpelier	99.5	82.8	92.2
057	Rutland	99.7	83.6	92.7
058	St. Johnsbury	98.6	77.8	89.6
059	Guildhall	97.4	77.8	88.8
VIRGINIA				
220-221	Fairfax	100.7	84.0	93.4
222	Arlington	101.9	84.0	94.1
223	Alexandria	101.1	85.5	94.3
224-225	Fredericksburg	99.5	82.7	92.1
226	Winchester	100.0	82.2	92.2
227	Culpeper	99.9	80.5	91.4
228	Harrisonburg	100.2	67.2	85.8
229	Charlottesville	100.5	67.1	86.0
230-232	Richmond	101.3	68.7	87.1
233-235	Norfolk	100.5	68.2	86.4
236	Newport News	100.0	68.2	86.1
237	Portsmouth	99.4	65.3	84.5
238	Petersburg	99.4	68.7	86.0
239	Farmville	98.7	52.4	78.5
240-241	Roanoke	101.5	65.1	85.6
242	Bristol	99.4	54.9	80.0
243	Pulaski	99.0	57.3	80.9
244	Staunton	99.7	61.8	83.2
245	Lynchburg	99.9	65.6	85.0
246	Grundy	99.3	52.5	78.9
WASHINGTON				
980-981,987	Seattle	101.4	104.7	102.8
982	Everett	101.6	99.6	100.7
983-984	Tacoma	101.8	99.7	100.9
985	Olympia	101.2	99.4	100.4
986	Vancouver	102.7	92.1	98.1
988	Wenatchee	101.6	86.6	95.0
989	Yakima	102.0	94.1	98.6
990-992	Spokane	103.4	82.3	94.2
993	Richland	103.1	89.3	97.1
994	Clarkston	101.5	80.2	92.2
WEST VIRGINIA				
247-248	Bluefield	98.3	92.9	96.0
249	Lewisburg	99.8	92.4	96.6
250-253	Charleston	99.9	94.8	97.6
254	Martinsburg	99.3	86.4	93.7
255-257	Huntington	100.7	95.7	98.5
258-259	Beckley	97.9	93.3	95.9
260	Wheeling	101.2	94.3	98.2
261	Parkersburg	100.1	91.8	96.5
262	Buckhannon	99.6	95.5	97.8
263-264	Clarksburg	100.1	95.4	98.1
265	Morgantown	100.2	94.4	97.6
266	Gassaway	99.5	93.9	97.1
267	Romney	99.4	92.2	96.3
268	Petersburg	99.3	91.3	95.8
WISCONSIN				
530,532	Milwaukee	100.0	107.2	103.1
531	Kenosha	99.5	102.2	100.7
534	Racine	99.1	102.4	100.5
535	Beloit	98.9	100.7	99.7
537	Madison	99.1	99.0	99.1

STATE/ZIP	CITY	MAT.	INST.	TOTAL
WISCONSIN (CONT'D)				
538	Lancaster	96.7	92.5	94.8
539	Portage	95.3	96.6	95.9
540	New Richmond	96.8	94.6	95.9
541-543	Green Bay	101.2	93.4	97.8
544	Wausau	96.2	93.5	95.0
545	Rhinelander	99.3	91.2	95.8
546	La Crosse	97.8	94.5	96.3
547	Eau Claire	99.3	95.3	97.5
548	Superior	96.6	96.3	96.5
549	Oshkosh	96.9	91.9	94.7
WYOMING				
820	Cheyenne	101.5	65.2	85.7
821	Yellowstone Nat'l Park	99.3	66.3	84.9
822	Wheatland	100.4	64.3	84.7
823	Rawlins	101.8	66.9	86.6
824	Worland	100.0	65.3	84.9
825	Riverton	101.0	65.4	85.5
826	Casper	102.0	63.9	85.4
827	Newcastle	99.9	66.9	85.5
828	Sheridan	102.6	64.3	85.9
829-831	Rock Springs	103.7	65.3	87.0
CANADIAN FACTORS (reflect Canadian currency)				
ALBERTA				
	Calgary	123.5	95.7	111.4
	Edmonton	123.1	95.7	111.2
	Fort McMurray	123.6	92.8	110.2
	Lethbridge	118.4	92.2	107.0
	Lloydminster	113.4	88.9	102.8
	Medicine Hat	113.6	88.2	102.5
	Red Deer	114.1	88.2	102.8
BRITISH COLUMBIA				
	Kamloops	114.5	90.6	104.1
	Prince George	115.4	89.9	104.3
	Vancouver	119.5	89.1	106.2
	Victoria	115.7	87.2	103.3
MANITOBA				
	Brandon	125.2	75.8	103.7
	Portage la Prairie	113.4	74.2	96.3
	Winnipeg	121.0	71.8	99.5
NEW BRUNSWICK				
	Bathurst	111.5	67.3	92.2
	Dalhousie	111.9	67.5	92.5
	Fredericton	117.4	72.5	97.9
	Moncton	111.8	70.1	93.6
	Newcastle	111.6	67.9	92.5
	St. John	114.0	77.9	98.3
NEWFOUNDLAND				
	Corner Brook	129.6	67.6	102.6
	St Johns	122.7	85.2	106.4
NORTHWEST TERRITORIES				
	Yellowknife	135.3	84.2	113.0
NOVA SCOTIA				
	Bridgewater	112.9	74.6	96.2
	Dartmouth	126.1	74.6	103.7
	Halifax	119.6	80.9	102.7
	New Glasgow	124.1	74.6	102.5
	Sydney	122.1	74.6	101.4
	Truro	112.5	74.6	96.0
	Yarmouth	124.0	74.6	102.5
ONTARIO				
	Barrie	117.7	93.0	106.9
	Brantford	114.9	96.9	107.1
	Cornwall	115.0	93.3	105.5
	Hamilton	117.0	96.7	108.2
	Kingston	116.0	93.4	106.2
	Kitchener	109.6	92.9	102.3
	London	116.4	93.3	106.3
	North Bay	127.3	91.1	111.5
	Oshawa	112.3	94.3	104.4
	Ottawa	117.3	94.0	107.2
	Owen Sound	117.8	91.2	106.2
	Peterborough	114.9	93.1	105.4
	Sarnia	114.8	97.5	107.3
	Sault Ste Marie	110.3	92.1	102.3

STATE/ZIP	CITY	MAT.	INST.	TOTAL
ONTARIO (CONT'D)				
	St. Catharines	107.6	94.5	101.9
	Sudbury	107.7	93.0	101.3
	Thunder Bay	108.8	93.4	102.1
	Timmins	115.1	91.1	104.6
	Toronto	118.3	101.3	110.9
	Windsor	108.2	93.4	101.7
PRINCE EDWARD ISLAND				
	Charlottetown	120.0	62.8	95.1
	Summerside	126.7	62.8	98.8
QUEBEC				
	Cap-de-la-Madeleine	112.4	87.6	101.6
	Charlesbourg	112.4	87.6	101.6
	Chicoutimi	111.4	92.9	103.3
	Gatineau	112.1	87.4	101.3
	Granby	112.3	87.4	101.4
	Hull	112.2	87.4	101.4
	Joliette	112.5	87.6	101.7
	Laval	112.1	87.4	101.3
	Montreal	117.8	93.2	107.1
	Quebec	117.1	93.6	106.8
	Rimouski	111.8	92.9	103.6
	Rouyn-Noranda	112.0	87.4	101.3
	Saint Hyacinthe	111.6	87.4	101.1
	Sherbrooke	112.4	87.4	101.5
	Sorel	112.5	87.6	101.7
	St Jerome	112.1	87.4	101.3
	Trois Rivieres	124.8	87.6	108.6
SASKATCHEWAN				
	Moose Jaw	110.7	69.5	92.7
	Prince Albert	109.9	67.7	91.5
	Regina	120.7	89.0	106.9
	Saskatoon	111.1	89.1	101.5
YUKON				
	Whitehorse	135.3	68.5	106.2

R011105-05 Tips for Accurate Estimating

1. Use pre-printed or columnar forms for orderly sequence of dimensions and locations and for recording telephone quotations.
2. Use only the front side of each paper or form except for certain pre-printed summary forms.
3. Be consistent in listing dimensions: For example, length x width x height. This helps in rechecking to ensure that, the total length of partitions is appropriate for the building area.
4. Use printed (rather than measured) dimensions where given.
5. Add up multiple printed dimensions for a single entry where possible.
6. Measure all other dimensions carefully.
7. Use each set of dimensions to calculate multiple related quantities.
8. Convert foot and inch measurements to decimal feet when listing. Memorize decimal equivalents to .01 parts of a foot (1/8″ equals approximately .01′).
9. Do not "round off" quantities until the final summary.
10. Mark drawings with different colors as items are taken off.
11. Keep similar items together, different items separate.
12. Identify location and drawing numbers to aid in future checking for completeness.
13. Measure or list everything on the drawings or mentioned in the specifications.
14. It may be necessary to list items not called for to make the job complete.
15. Be alert for: Notes on plans such as N.T.S. (not to scale); changes in scale throughout the drawings; reduced size drawings; discrepancies between the specifications and the drawings.
16. Develop a consistent pattern of performing an estimate. For example:
    a. Start the quantity takeoff at the lower floor and move to the next higher floor.
    b. Proceed from the main section of the building to the wings.
    c. Proceed from south to north or vice versa, clockwise or counterclockwise.
    d. Take off floor plan quantities first, elevations next, then detail drawings.
17. List all gross dimensions that can be either used again for different quantities, or used as a rough check of other quantities for verification (exterior perimeter, gross floor area, individual floor areas, etc.).
18. Utilize design symmetry or repetition (repetitive floors, repetitive wings, symmetrical design around a center line, similar room layouts, etc.). Note: Extreme caution is needed here so as not to omit or duplicate an area.
19. Do not convert units until the final total is obtained. For instance, when estimating concrete work, keep all units to the nearest cubic foot, then summarize and convert to cubic yards.
20. When figuring alternatives, it is best to total all items involved in the basic system, then total all items involved in the alternates. Therefore you work with positive numbers in all cases. When adds and deducts are used, it is often confusing whether to add or subtract a portion of an item; especially on a complicated or involved alternate.

R011105-50 Metric Conversion Factors

Description: This table is primarily for converting customary U.S. units in the left hand column to SI metric units in the right hand column. In addition, conversion factors for some commonly encountered Canadian and non-SI metric units are included.

	If You Know		Multiply By		To Find
Length	Inches	x	25.4[a]	=	Millimeters
	Feet	x	0.3048[a]	=	Meters
	Yards	x	0.9144[a]	=	Meters
	Miles (statute)	x	1.609	=	Kilometers
Area	Square inches	x	645.2	=	Square millimeters
	Square feet	x	0.0929	=	Square meters
	Square yards	x	0.8361	=	Square meters
Volume (Capacity)	Cubic inches	x	16,387	=	Cubic millimeters
	Cubic feet	x	0.02832	=	Cubic meters
	Cubic yards	x	0.7646	=	Cubic meters
	Gallons (U.S. liquids)[b]	x	0.003785	=	Cubic meters[c]
	Gallons (Canadian liquid)[b]	x	0.004546	=	Cubic meters[c]
	Ounces (U.S. liquid)[b]	x	29.57	=	Milliliters[c, d]
	Quarts (U.S. liquid)[b]	x	0.9464	=	Liters[c, d]
	Gallons (U.S. liquid)[b]	x	3.785	=	Liters[c, d]
Force	Kilograms force[d]	x	9.807	=	Newtons
	Pounds force	x	4.448	=	Newtons
	Pounds force	x	0.4536	=	Kilograms force[d]
	Kips	x	4448	=	Newtons
	Kips	x	453.6	=	Kilograms force[d]
Pressure, Stress, Strength (Force per unit area)	Kilograms force per square centimeter[d]	x	0.09807	=	Megapascals
	Pounds force per square inch (psi)	x	0.006895	=	Megapascals
	Kips per square inch	x	6.895	=	Megapascals
	Pounds force per square inch (psi)	x	0.07031	=	Kilograms force per square centimeter[d]
	Pounds force per square foot	x	47.88	=	Pascals
	Pounds force per square foot	x	4.882	=	Kilograms force per square meter[d]
Bending Moment Or Torque	Inch-pounds force	x	0.01152	=	Meter-kilograms force[d]
	Inch-pounds force	x	0.1130	=	Newton-meters
	Foot-pounds force	x	0.1383	=	Meter-kilograms force[d]
	Foot-pounds force	x	1.356	=	Newton-meters
	Meter-kilograms force[d]	x	9.807	=	Newton-meters
Mass	Ounces (avoirdupois)	x	28.35	=	Grams
	Pounds (avoirdupois)	x	0.4536	=	Kilograms
	Tons (metric)	x	1000	=	Kilograms
	Tons, short (2000 pounds)	x	907.2	=	Kilograms
	Tons, short (2000 pounds)	x	0.9072	=	Megagrams[e]
Mass per Unit Volume	Pounds mass per cubic foot	x	16.02	=	Kilograms per cubic meter
	Pounds mass per cubic yard	x	0.5933	=	Kilograms per cubic meter
	Pounds mass per gallon (U.S. liquid)[b]	x	119.8	=	Kilograms per cubic meter
	Pounds mass per gallon (Canadian liquid)[b]	x	99.78	=	Kilograms per cubic meter
Temperature	Degrees Fahrenheit	(F-32)/1.8		=	Degrees Celsius
	Degrees Fahrenheit	(F+459.67)/1.8		=	Degrees Kelvin
	Degrees Celsius	C+273.15		=	Degrees Kelvin

[a]The factor given is exact
[b]One U.S. gallon = 0.8327 Canadian gallon
[c]1 liter = 1000 milliliters = 1000 cubic centimeters
1 cubic decimeter = 0.001 cubic meter
[d]Metric but not SI unit
[e]Called "tonne" in England and "metric ton" in other metric countries

R011105-60 Weights and Measures

Measures of Length
1 Mile = 1760 Yards = 5280 Feet
1 Yard = 3 Feet = 36 inches
1 Foot = 12 Inches
1 Mil = 0.001 Inch
1 Fathom = 2 Yards = 6 Feet
1 Rod = 5.5 Yards = 16.5 Feet
1 Hand = 4 Inches
1 Span = 9 Inches
1 Micro-inch = One Millionth Inch or 0.000001 Inch
1 Micron = One Millionth Meter + 0.00003937 Inch

Surveyor's Measure
1 Mile = 8 Furlongs = 80 Chains
1 Furlong = 10 Chains = 220 Yards
1 Chain = 4 Rods = 22 Yards = 66 Feet = 100 Links
1 Link = 7.92 Inches

Square Measure
1 Square Mile = 640 Acres = 6400 Square Chains
1 Acre = 10 Square Chains = 4840 Square Yards = 43,560 Sq. Ft.
1 Square Chain = 16 Square Rods = 484 Square Yards = 4356 Sq. Ft.
1 Square Rod = 30.25 Square Yards = 272.25 Square Feet = 625 Square Lines
1 Square Yard = 9 Square Feet
1 Square Foot = 144 Square Inches
An Acre equals a Square 208.7 Feet per Side

Cubic Measure
1 Cubic Yard = 27 Cubic Feet
1 Cubic Foot = 1728 Cubic Inches
1 Cord of Wood = 4 x 4 x 8 Feet = 128 Cubic Feet
1 Perch of Masonry = 16½ x 1½ x 1 Foot = 24.75 Cubic Feet

Avoirdupois or Commercial Weight
1 Gross or Long Ton = 2240 Pounds
1 Net or Short ton = 2000 Pounds
1 Pound = 16 Ounces = 7000 Grains
1 Ounce = 16 Drachms = 437.5 Grains
1 Stone = 14 Pounds

Shipping Measure
For Measuring Internal Capacity of a Vessel:
1 Register Ton = 100 Cubic Feet

For Measurement of Cargo:
Approximately 40 Cubic Feet of Merchandise is considered a Shipping Ton, unless that bulk would weigh more than 2000 Pounds, in which case Freight Charge may be based upon weight.

40 Cubic Feet = 32.143 U.S. Bushels = 31.16 Imp. Bushels

Liquid Measure
1 Imperial Gallon = 1.2009 U.S. Gallon = 277.42 Cu. In.
1 Cubic Foot = 7.48 U.S. Gallons

R011110-10 Architectural Fees

Tabulated below are typical percentage fees by project size, for good professional architectural service. Fees may vary from those listed depending upon degree of design difficulty and economic conditions in any particular area.

Rates can be interpolated horizontally and vertically. Various portions of the same project requiring different rates should be adjusted proportionately. For alterations, add 50% to the fee for the first $500,000 of project cost and add 25% to the fee for project cost over $500,000.

Architectural fees tabulated below include Structural, Mechanical, and Electrical Engineering Fees. They do not include the fees for special consultants such as kitchen planning, security, acoustical, interior design, etc.

Civil Engineering fees are included in the Architectural fee for project sites requiring minimal design such as city sites. However, separate Civil Engineering fees must be added when utility connections require design, drainage calculations are needed, stepped foundations are required, or provisions are required to protect adjacent wetlands.

Building Types	Total Project Size in Thousands of Dollars						
	100	250	500	1,000	5,000	10,000	50,000
Factories, garages, warehouses, repetitive housing	9.0%	8.0%	7.0%	6.2%	5.3%	4.9%	4.5%
Apartments, banks, schools, libraries, offices, municipal buildings	12.2	12.3	9.2	8.0	7.0	6.6	6.2
Churches, hospitals, homes, laboratories, museums, research	15.0	13.6	12.7	11.9	9.5	8.8	8.0
Memorials, monumental work, decorative furnishings	—	16.0	14.5	13.1	10.0	9.0	8.3

General Requirements | R0111 Summary of Work

R011110-30 Engineering Fees

Typical **Structural Engineering Fees** based on type of construction and total project size. These fees are included in Architectural Fees.

Type of Construction	Total Project Size (in thousands of dollars)			
	$500	$500-$1,000	$1,000-$5,000	Over $5000
Industrial buildings, factories & warehouses	Technical payroll times 2.0 to 2.5	1.60%	1.25%	1.00%
Hotels, apartments, offices, dormitories, hospitals, public buildings, food stores	↓	2.00%	1.70%	1.20%
Museums, banks, churches and cathedrals	↓	2.00%	1.75%	1.25%
Thin shells, prestressed concrete, earthquake resistive	↓	2.00%	1.75%	1.50%
Parking ramps, auditoriums, stadiums, convention halls, hangars & boiler houses	↓	2.50%	2.00%	1.75%
Special buildings, major alterations, underpinning & future expansion	↓	Add to above 0.5%	Add to above 0.5%	Add to above 0.5%

For complex reinforced concrete or unusually complicated structures, add 20% to 50%.

Typical **Mechanical and Electrical Engineering Fees** are based on the size of the subcontract. The fee structure for both are shown below. These fees are included in Architectural Fees.

Type of Construction	Subcontract Size							
	$25,000	$50,000	$100,000	$225,000	$350,000	$500,000	$750,000	$1,000,000
Simple structures	6.4%	5.7%	4.8%	4.5%	4.4%	4.3%	4.2%	4.1%
Intermediate structures	8.0	7.3	6.5	5.6	5.1	5.0	4.9	4.8
Complex structures	10.1	9.0	9.0	8.0	7.5	7.5	7.0	7.0

For renovations, add 15% to 25% to applicable fee.

General Requirements | R0121 Allowances

R012153-10 Repair and Remodeling

Cost figures are based on new construction utilizing the most cost-effective combination of labor, equipment, and material with the work scheduled in proper sequence to allow the various trades to accomplish their work in an efficient manner.

The costs for repair and remodeling work must be modified due to the following factors that may be present in any given repair and remodeling project:

1. Equipment usage curtailment due to the physical limitations of the project, with only hand-operated equipment being used.
2. Increased requirement for shoring and bracing to hold up the building while structural changes are being made and to allow for temporary storage of construction materials on above-grade floors.
3. Material handling becomes more costly due to having to move within the confines of an enclosed building. For multi-story construction, low capacity elevators and stairwells may be the only access to the upper floors.
4. Large amount of cutting and patching and attempting to match the existing construction is required. It is often more economical to remove entire walls rather than create many new door and window openings. This sort of trade-off has to be carefully analyzed.
5. Cost of protection of completed work is increased since the usual sequence of construction usually cannot be accomplished.
6. Economies of scale usually associated with new construction may not be present. If small quantities of components must be custom fabricated due to job requirements, unit costs will naturally increase. Also, if only small work areas are available at a given time, job scheduling between trades becomes difficult and subcontractor quotations may reflect the excessive start-up and shut-down phases of the job.
7. Work may have to be done on other than normal shifts and may have to be done around an existing production facility which has to stay in production during the course of the repair and remodeling.
8. Dust and noise protection of adjoining non-construction areas can involve substantial special protection and alter usual construction methods.
9. Job may be delayed due to unexpected conditions discovered during demolition or removal. These delays ultimately increase construction costs.
10. Piping and ductwork runs may not be as simple as for new construction. Wiring may have to be snaked through walls and floors.
11. Matching "existing construction" may be impossible because materials may no longer be manufactured. Substitutions may be expensive.
12. Weather protection of existing structure requires additional temporary structures to protect building at openings.
13. On small projects, because of local conditions, it may be necessary to pay a tradesman for a minimum of four hours for a task that is completed in one hour.

All of the above areas can contribute to increased costs for a repair and remodeling project. Each of the above factors should be considered in the planning, bidding and construction stage in order to minimize the increased costs associated with repair and remodeling jobs.

R012157-20 Construction Time Requirements

Table at left is average construction time in months for different types of building projects. Table at right is the construction time in months for different size projects. Design time runs 25% to 40% of construction time.

Building Type	Size S.F.
Industrial/Warehouse	100,000 500,000 1,000,000
Offices/Retail	50,000 250,000 500,000
Institutional/Hospitals/Laboratory	200,000 500,000 750,000 1,000,000

Project Value	Construction Duration
$8,000,000	14 months
$32,000,000	19 months
$75,000,000	21 months
$7,000,000	15 months
$28,000,000	23 months
$58,000,000	34 months
$45,000,000	31 months
$110,000,000	52 months
$160,000,000	55 months
$210,000,000	60 months

R012909-80 Sales Tax by State

State sales tax on materials is tabulated below (5 states have no sales tax). Many states allow local jurisdictions, such as a county or city, to levy additional sales tax. Some projects may be sales tax exempt, particularly those constructed with public funds.

State	Tax (%)	State	Tax (%)	State	Tax (%)	State	Tax (%)
Alabama	4	Illinois	6.25	Montana	0	Rhode Island	7
Alaska	0	Indiana	7	Nebraska	5.5	South Carolina	6
Arizona	5.6	Iowa	6	Nevada	6.85	South Dakota	4
Arkansas	6	Kansas	6.3	New Hampshire	0	Tennessee	7
California	7.5	Kentucky	6	New Jersey	7	Texas	6.25
Colorado	2.9	Louisiana	4	New Mexico	5.13	Utah	4.7
Connecticut	6.35	Maine	5	New York	4	Vermont	6
Delaware	0	Maryland	6	North Carolina	4.75	Virginia	5
District of Columbia	6	Massachusetts	6.25	North Dakota	5	Washington	6.5
Florida	6	Michigan	6	Ohio	5.5	West Virginia	6
Georgia	4	Minnesota	6.88	Oklahoma	4.5	Wisconsin	5
Hawaii	4	Mississippi	7	Oregon	0	Wyoming	4
Idaho	6	Missouri	4.23	Pennsylvania	6	Average	5.04 %

Sales Tax by Province (Canada)

GST - a value-added tax, which the government imposes on most goods and services provided in or imported into Canada. PST - a retail sales tax, which three of the provinces impose on the price of most goods and some services. QST - a value-added tax, similar to the federal GST, which Quebec imposes. HST - Five provinces have combined their retail sales tax with the federal GST into one harmonized tax.

Province	PST (%)	QST (%)	GST(%)	HST(%)
Alberta	0	0	5	0
British Columbia	7	0	5	0
Manitoba	8	0	5	0
New Brunswick	0	0	0	13
Newfoundland	0	0	0	13
Northwest Territories	0	0	5	0
Nova Scotia	0	0	0	15
Ontario	0	0	0	13
Prince Edward Island	0	0	0	14
Quebec	0	9.975	5	0
Saskatchewan	5	0	5	0
Yukon	0	0	5	0

General Requirements — R0129 Payment Procedures

R012909-85 Unemployment Taxes and Social Security Taxes

State unemployment tax rates vary not only from state to state, but also with the experience rating of the contractor. The federal unemployment tax rate is 6.0% of the first $7,000 of wages. This is reduced by a credit of up to 5.4% for timely payment to the state. The minimum federal unemployment tax is 0.6% after all credits.

Social security (FICA) for 2015 is estimated at time of publication to be 7.65% of wages up to $117,000.

R012909-90 Overtime

One way to improve the completion date of a project or eliminate negative float from a schedule is to compress activity duration times. This can be achieved by increasing the crew size or working overtime with the proposed crew.

To determine the costs of working overtime to compress activity duration times, consider the following examples. Below is an overtime efficiency and cost chart based on a five, six, or seven day week with an eight through twelve hour day. Payroll percentage increases for time and one half and double time are shown for the various working days.

Days per Week	Hours per Day	Production Efficiency					Payroll Cost Factors	
		1st Week	2nd Week	3rd Week	4th Week	Average 4 Weeks	@ 1-1/2 Times	@ 2 Times
5	8	100%	100%	100%	100%	100 %	1.000	1.000
	9	100	100	95	90	96	1.056	1.111
	10	100	95	90	85	93	1.100	1.200
	11	95	90	75	65	81	1.136	1.273
	12	90	85	70	60	76	1.167	1.333
6	8	100	100	95	90	96	1.083	1.167
	9	100	95	90	85	93	1.130	1.259
	10	95	90	85	80	88	1.167	1.333
	11	95	85	70	65	79	1.197	1.394
	12	90	80	65	60	74	1.222	1.444
7	8	100	95	85	75	89	1.143	1.286
	9	95	90	80	70	84	1.183	1.365
	10	90	85	75	65	79	1.214	1.429
	11	85	80	65	60	73	1.240	1.481
	12	85	75	60	55	69	1.262	1.524

General Requirements — R0131 Project Management & Coordination

R013113-40 Builder's Risk Insurance

Builder's Risk Insurance is insurance on a building during construction. Premiums are paid by the owner or the contractor. Blasting, collapse, and underground insurance would raise total insurance costs above those listed. Floater policy for materials delivered to the job runs $.75 to $1.25 per $100 value. Contractor equipment insurance runs $.50 to $1.50 per $100 value. Insurance for miscellaneous tools to $1,500 value runs from $3.00 to $7.50 per $100 value.

Tabulated below are New England Builder's Risk insurance rates in dollars per $100 value for $1,000 deductible. For $25,000 deductible, rates can be reduced 13% to 34%. On contracts over $1,000,000, rates may be lower than those tabulated. Policies are written annually for the total completed value in place. For "all risk" insurance (excluding flood, earthquake, and certain other perils) add $.025 to total rates below.

Coverage	Frame Construction (Class 1)		Brick Construction (Class 4)		Fire Resistive (Class 6)	
	Range	Average	Range	Average	Range	Average
Fire Insurance	$.350 to $ .850	$.600	$.158 to $.189	$.174	$.052 to $.080	$.070
Extended Coverage	.115 to .200	.158	.080 to .105	.101	.081 to .105	.100
Vandalism	.012 to .016	.014	.008 to .011	.011	.008 to .011	.010
Total Annual Rate	$.477 to $1.066	$.772	$.246 to $.305	$.286	$.141 to $.196	$.180

R013113-50 General Contractor's Overhead

There are two distinct types of overhead on a construction project: Project Overhead and Main Office Overhead. Project Overhead includes those costs at a construction site not directly associated with the installation of construction materials. Examples of Project Overhead costs include the following:

1. Superintendent
2. Construction office and storage trailers
3. Temporary sanitary facilities
4. Temporary utilities
5. Security fencing
6. Photographs
7. Clean up
8. Performance and payment bonds

The above Project Overhead items are also referred to as General Requirements and therefore are estimated in Division 1. Division 1 is the first division listed in the CSI MasterFormat but it is usually the last division estimated. The sum of the costs in Divisions 1 through 49 is referred to as the sum of the direct costs.

All construction projects also include indirect costs. The primary components of indirect costs are the contractor's Main Office Overhead and Profit. The amount of the Main Office Overhead expense varies depending on the the following:

1. Owner's compensation
2. Project managers and estimator's wages
3. Clerical support wages
4. Office rent and utilities
5. Corporate legal and accounting costs
6. Advertising
7. Automobile expenses
8. Association dues
9. Travel and entertainment expenses

These costs are usually calculated as a percentage of annual sales volume. This percentage can range from 35% for a small contractor doing less than $500,000 to 5% for a large contractor with sales in excess of $100 million.

R013113-60 Workers' Compensation Insurance Rates by Trade

The table below tabulates the national averages for workers' compensation insurance rates by trade and type of building. The average "Insurance Rate" is multiplied by the "% of Building Cost" for each trade. This produces the "Workers' Compensation" cost by % of total labor cost, to be added for each trade by building type to determine the weighted average workers' compensation rate for the building types analyzed.

Trade	Insurance Rate (% Labor Cost)			% of Building Cost			Workers' Compensation		
	Range		Average	Office Bldgs.	Schools & Apts.	Mfg.	Office Bldgs.	Schools & Apts.	Mfg.
Excavation, Grading, etc.	3.4 %	to 21.2%	9.7%	4.8%	4.9%	4.5%	0.47%	0.48%	0.44%
Piles & Foundations	6.1	to 28.3	14.3	7.1	5.2	8.7	1.02	0.74	1.24
Concrete	4.2	to 35.8	12.9	5.0	14.8	3.7	0.65	1.91	0.48
Masonry	4.9	to 36.4	13.7	6.9	7.5	1.9	0.95	1.03	0.26
Structural Steel	6.1	to 101.1	31.7	10.7	3.9	17.6	3.39	1.24	5.58
Miscellaneous & Ornamental Metals	4.3	to 31.5	12.2	2.8	4.0	3.6	0.34	0.49	0.44
Carpentry & Millwork	5.0	to 36.8	14.9	3.7	4.0	0.5	0.55	0.60	0.07
Metal or Composition Siding	6.1	to 69.2	18.4	2.3	0.3	4.3	0.42	0.06	0.79
Roofing	6.1	to 100.3	31.7	2.3	2.6	3.1	0.73	0.82	0.98
Doors & Hardware	4.1	to 36.8	11.3	0.9	1.4	0.4	0.10	0.16	0.05
Sash & Glazing	4.8	to 30.7	13.3	3.5	4.0	1.0	0.47	0.53	0.13
Lath & Plaster	3.1	to 36.2	10.9	3.3	6.9	0.8	0.36	0.75	0.09
Tile, Marble & Floors	3	to 24.8	8.9	2.6	3.0	0.5	0.23	0.27	0.04
Acoustical Ceilings	3.8	to 33.8	8.4	2.4	0.2	0.3	0.20	0.02	0.03
Painting	4.5	to 36.1	11.7	1.5	1.6	1.6	0.18	0.19	0.19
Interior Partitions	5.0	to 36.8	14.9	3.9	4.3	4.4	0.58	0.64	0.66
Miscellaneous Items	2.5	to 132.3	13.4	5.2	3.7	9.7	0.70	0.50	1.30
Elevators	1.3	to 12.8	5.3	2.1	1.1	2.2	0.11	0.06	0.12
Sprinklers	2.9	to 18.5	7.3	0.5	—	2.0	0.04	—	0.15
Plumbing	2.2	to 19.3	7.0	4.9	7.2	5.2	0.34	0.50	0.36
Heat., Vent., Air Conditioning	3.8	to 18.1	8.8	13.5	11.0	12.9	1.19	0.97	1.14
Electrical	2.4	to 13.5	5.8	10.1	8.4	11.1	0.59	0.49	0.64
Total	1.3 %	to 132.3%	—	100.0%	100.0%	100.0%	13.61%	12.45%	15.18%
Overall Weighted Average 13.75%									

Workers' Compensation Insurance Rates by States

The table below lists the weighted average workers' compensation base rate for each state with a factor comparing this with the national average of 13.5.0%.

State	Weighted Average	Factor	State	Weighted Average	Factor	State	Weighted Average	Factor
Alabama	20.6%	153	Kentucky	12.5%	93	North Dakota	8.1%	60
Alaska	12.5	93	Louisiana	19.6	145	Ohio	8.1	60
Arizona	13.8	102	Maine	11.4	84	Oklahoma	10.3	76
Arkansas	7.7	57	Maryland	13.5	100	Oregon	11.3	84
California	27.1	201	Massachusetts	11.5	85	Pennsylvania	18.3	136
Colorado	7.3	54	Michigan	13.5	100	Rhode Island	14.7	109
Connecticut	25.9	192	Minnesota	22.8	169	South Carolina	16.9	125
Delaware	14.0	104	Mississippi	14.0	104	South Dakota	14.1	104
District of Columbia	10.4	77	Missouri	13.9	103	Tennessee	12.2	90
Florida	10.8	80	Montana	8.3	61	Texas	9.3	69
Georgia	29.5	219	Nebraska	17.0	126	Utah	8.8	65
Hawaii	8.7	64	Nevada	9.2	68	Vermont	13.3	99
Idaho	10.5	78	New Hampshire	18.8	139	Virginia	8.7	64
Illinois	26.9	199	New Jersey	15.7	116	Washington	9.5	70
Indiana	5.0	37	New Mexico	15.3	113	West Virginia	7.4	55
Iowa	15.5	115	New York	19.0	141	Wisconsin	13.2	98
Kansas	9.2	68	North Carolina	17.3	128	Wyoming	6.5	48
Weighted Average for U.S. is 13.7% of payroll = 100%								

The weighted average skilled worker rate for 35 trades is 13.5%. For bidding purposes, apply the full value of workers' compensation directly to total labor costs, or if labor is 38%, materials 42% and overhead and profit 20% of total cost, carry 38/80 × 13.5% =6.4% of cost (before overhead and profit) into overhead. Rates vary not only from state to state but also with the experience rating of the contractor.

Rates are the most current available at the time of publication.

R013113-80 Performance Bond

This table shows the cost of a Performance Bond for a construction job scheduled to be completed in 12 months. Add 1% of the premium cost per month for jobs requiring more than 12 months to complete. The rates are "standard" rates offered to contractors that the bonding company considers financially sound and capable of doing the work. Preferred rates are offered by some bonding companies based upon financial strength of the contractor. Actual rates vary from contractor to contractor and from bonding company to bonding company. Contractors should prequalify through a bonding agency before submitting a bid on a contract that requires a bond.

Contract Amount	Building Construction Class B Projects	Highways & Bridges	
		Class A New Construction	Class A-1 Highway Resurfacing
First $ 100,000 bid	$25.00 per M	$15.00 per M	$9.40 per M
Next 400,000 bid	$ 2,500 plus $15.00 per M	$ 1,500 plus $10.00 per M	$ 940 plus $7.20 per M
Next 2,000,000 bid	8,500 plus 10.00 per M	5,500 plus 7.00 per M	3,820 plus 5.00 per M
Next 2,500,000 bid	28,500 plus 7.50 per M	19,500 plus 5.50 per M	15,820 plus 4.50 per M
Next 2,500,000 bid	47,250 plus 7.00 per M	33,250 plus 5.00 per M	28,320 plus 4.50 per M
Over 7,500,000 bid	64,750 plus 6.00 per M	45,750 plus 4.50 per M	39,570 plus 4.00 per M

R015423-10 Steel Tubular Scaffolding

On new construction, tubular scaffolding is efficient up to 60′ high or five stories. Above this it is usually better to use a hung scaffolding if construction permits. Swing scaffolding operations may interfere with tenants. In this case, the tubular is more practical at all heights.

In repairing or cleaning the front of an existing building the cost of tubular scaffolding per S.F. of building front increases as the height increases above the first tier. The first tier cost is relatively high due to leveling and alignment.

The minimum efficient crew for erecting and dismantling is three workers. They can set up and remove 18 frame sections per day up to 5 stories high. For 6 to 12 stories high, a crew of four is most efficient. Use two or more on top and two on the bottom for handing up or hoisting. They can also set up and remove 18 frame sections per day. At 7' horizontal spacing, this will run about 800 S.F. per day of erecting and dismantling. Time for placing and removing planks must be added to the above. A crew of three can place and remove 72 planks per day up to 5 stories. For over 5 stories, a crew of four can place and remove 80 planks per day.

The table below shows the number of pieces required to erect tubular steel scaffolding for 1000 S.F. of building frontage. This area is made up of a scaffolding system that is 12 frames (11 bays) long by 2 frames high.

For jobs under twenty-five frames, add 50% to rental cost. Rental rates will be lower for jobs over three months duration. Large quantities for long periods can reduce rental rates by 20%.

Description of Component	Number of Pieces for 1000 S.F. of Building Front	Unit
5′ Wide Standard Frame, 6′-4″ High	24	Ea.
Leveling Jack & Plate	24	↓
Cross Brace	44	↓
Side Arm Bracket, 21″	12	↓
Guardrail Post	12	↓
Guardrail, 7′ section	22	↓
Stairway Section	2	↓
Stairway Starter Bar	1	↓
Stairway Inside Handrail	2	↓
Stairway Outside Handrail	2	↓
Walk-Thru Frame Guardrail	2	↓

Scaffolding is often used as falsework over 15′ high during construction of cast-in-place concrete beams and slabs. Two foot wide scaffolding is generally used for heavy beam construction. The span between frames depends upon the load to be carried with a maximum span of 5′.

Heavy duty shoring frames with a capacity of 10,000#/leg can be spaced up to 10′ O.C. depending upon form support design and loading.

Scaffolding used as horizontal shoring requires less than half the material required with conventional shoring.

On new construction, erection is done by carpenters.

Rolling towers supporting horizontal shores can reduce labor and speed the job. For maintenance work, catwalks with spans up to 70′ can be supported by the rolling towers.

R015423-20 Pump Staging

Pump staging is generally not available for rent. The table below shows the number of pieces required to erect pump staging for 2400 S.F. of building frontage. This area is made up of a pump jack system that is 3 poles (2 bays) wide by 2 poles high.

Item	Number of Pieces for 2400 S.F. of Building Front	Unit
Aluminum pole section, 24′ long	6	Ea.
Aluminum splice joint, 6′ long	3	↓
Aluminum foldable brace	3	↓
Aluminum pump jack	3	↓
Aluminum support for workbench/back safety rail	3	↓
Aluminum scaffold plank/workbench, 14″ wide x 24′ long	4	↓
Safety net, 22′ long	2	↓
Aluminum plank end safety rail	2	↓

The cost in place for this 2400 S.F. will depend on how many uses are realized during the life of the equipment.

R015433-10 Contractor Equipment

Rental Rates shown elsewhere in the book pertain to late model high quality machines in excellent working condition, rented from equipment dealers. Rental rates from contractors may be substantially lower than the rental rates from equipment dealers depending upon economic conditions; for older, less productive machines, reduce rates by a maximum of 15%. Any overtime must be added to the base rates. For shift work, rates are lower. Usual rule of thumb is 150% of one shift rate for two shifts; 200% for three shifts.

For periods of less than one week, operated equipment is usually more economical to rent than renting bare equipment and hiring an operator.

Costs to move equipment to a job site (mobilization) or from a job site (demobilization) are not included in rental rates, nor in any Equipment costs on any Unit Price line items or crew listings. These costs can be found elsewhere. If a piece of equipment is already at a job site, it is not appropriate to utilize mob/demob costs in an estimate again.

Rental rates vary throughout the country with larger cities generally having lower rates. Lease plans for new equipment are available for periods in excess of six months with a percentage of payments applying toward purchase.

Rental rates can also be treated as reimbursement costs for contractor-owned equipment. Owned equipment costs include depreciation, loan payments, interest, taxes, insurance, storage, and major repairs.

Monthly rental rates vary from 2% to 5% of the cost of the equipment depending on the anticipated life of the equipment and its wearing parts. Weekly rates are about 1/3 the monthly rates and daily rental rates about 1/3 the weekly rate.

The hourly operating costs for each piece of equipment include costs to the user such as fuel, oil, lubrication, normal expendables for the equipment, and a percentage of mechanic's wages chargeable to maintenance. The hourly operating costs listed do not include the operator's wages.

The daily cost for equipment used in the standard crews is figured by dividing the weekly rate by five, then adding eight times the hourly operating cost to give the total daily equipment cost, not including the operator. This figure is in the right hand column of the Equipment listings under Equipment Cost/Day.

Pile Driving rates shown for pile hammer and extractor do not include leads, crane, boiler or compressor. Vibratory pile driving requires an added field specialist during set-up and pile driving operation for the electric model. The hydraulic model requires a field specialist for set-up only. Up to 125 reuses of sheet piling are possible using vibratory drivers. For normal conditions, crane capacity for hammer type and size are as follows.

Crane Capacity	Hammer Type and Size		
	Air or Steam	Diesel	Vibratory
25 ton	to 8,750 ft.-lb.		70 H.P.
40 ton	15,000 ft.-lb.	to 32,000 ft.-lb.	170 H.P.
60 ton	25,000 ft.-lb.		300 H.P.
100 ton		112,000 ft.-lb.	

Cranes should be specified for the job by size, building and site characteristics, availability, performance characteristics, and duration of time required.

Backhoes & Shovels rent for about the same as equivalent size cranes but maintenance and operating expense is higher. Crane operators rate must be adjusted for high boom heights. Average adjustments: for 150′ boom add 2% per hour; over 185′, add 4% per hour; over 210′, add 6% per hour; over 250′, add 8% per hour and over 295′, add 12% per hour.

Tower Cranes of the climbing or static type have jibs from 50′ to 200′ and capacities at maximum reach range from 4,000 to 14,000 pounds. Lifting capacities increase up to maximum load as the hook radius decreases.

Typical rental rates, based on purchase price are about 2% to 3% per month.

Erection and dismantling runs between 500 and 2000 labor hours. Climbing operation takes 10 labor hours per 20′ climb. Crane dead time is about 5 hours per 40′ climb. If crane is bolted to side of the building add cost of ties and extra mast sections. Climbing cranes have from 80′ to 180′ of mast while static cranes have 80′ to 800′ of mast.

Truck Cranes can be converted to tower cranes by using tower attachments. Mast heights over 400′ have been used.

A single 100′ high material **Hoist and Tower** can be erected and dismantled in about 400 labor hours; a double 100′ high hoist and tower in about 600 labor hours. Erection times for additional heights are 3 and 4 labor hours per vertical foot respectively up to 150′, and 4 to 5 labor hours per vertical foot over 150′ high. A 40′ high portable Buck hoist takes about 160 labor hours to erect and dismantle. Additional heights take 2 labor hours per vertical foot to 80′ and 3 labor hours per vertical foot for the next 100′. Most material hoists do not meet local code requirements for carrying personnel.

A 150′ high **Personnel Hoist** requires about 500 to 800 labor hours to erect and dismantle. Budget erection time at 5 labor hours per vertical foot for all trades. Local code requirements or labor scarcity requiring overtime can add up to 50% to any of the above erection costs.

Earthmoving Equipment: The selection of earthmoving equipment depends upon the type and quantity of material, moisture content, haul distance, haul road, time available, and equipment available. Short haul cut and fill operations may require dozers only, while another operation may require excavators, a fleet of trucks, and spreading and compaction equipment. Stockpiled material and granular material are easily excavated with front end loaders. Scrapers are most economically used with hauls between 300′ and 1-1/2 miles if adequate haul roads can be maintained. Shovels are often used for blasted rock and any material where a vertical face of 8′ or more can be excavated. Special conditions may dictate the use of draglines, clamshells, or backhoes. Spreading and compaction equipment must be matched to the soil characteristics, the compaction required and the rate the fill is being supplied.

R015433-15 Heavy Lifting

Hydraulic Climbing Jacks

The use of hydraulic heavy lift systems is an alternative to conventional type crane equipment. The lifting, lowering, pushing, or pulling mechanism is a hydraulic climbing jack moving on a square steel jackrod from 1-5/8″ to 4″ square, or a steel cable. The jackrod or cable can be vertical or horizontal, stationary or movable, depending on the individual application. When the jackrod is stationary, the climbing jack will climb the rod and push or pull the load along with itself. When the climbing jack is stationary, the jackrod is movable with the load attached to the end and the climbing jack will lift or lower the jackrod with the attached load. The heavy lift system is normally operated by a single control lever located at the hydraulic pump.

The system is flexible in that one or more climbing jacks can be applied wherever a load support point is required, and the rate of lift synchronized.

Economic benefits have been demonstrated on projects such as: erection of ground assembled roofs and floors, complete bridge spans, girders and trusses, towers, chimney liners and steel vessels, storage tanks, and heavy machinery. Other uses are raising and lowering offshore work platforms, caissons, tunnel sections and pipelines.

R015436-50 Mobilization

Costs to move rented construction equipment to a job site from an equipment dealer's or contractor's yard (mobilization), or to move the equipment off the job site (demobilization), are not included in the rental or operating rates, nor in the equipment cost on a unit price line or in a crew listing. These costs can be found consolidated in the Mobilization section of the data and elsewhere in particular site work sections. If a piece of equipment is already on the job site, it is not appropriate to include mob/demob costs in a new estimate that requires use of that equipment. The following table identifies approximate sizes of rented construction equipment that would be hauled on a towed trailer. Because this listing is not all-encompassing, the user can infer as to what size trailer might be required for a piece of equipment not listed.

3-ton Trailer	20-ton Trailer	40-ton Trailer	50-ton Trailer
20 H.P. Excavator	110 H.P. Excavator	200 H.P. Excavator	270 H.P. Excavator
50 H.P. Skid Steer	165 H.P. Dozer	300 H.P. Dozer	Small Crawler Crane
35 H.P. Roller	150 H.P. Roller	400 H.P. Scraper	500 H.P. Scraper
40 H.P. Trencher	Backhoe	450 H.P. Art. Dump Truck	500 H.P. Art. Dump Truck

R019313-10 Facility Maintenance - Frequency Table

The following table lists "average" frequency data for selected facility maintenance activities. The frequencies given are for a normal standard of maintenance under average conditions.

Activity	Average Frequency	Notes
Acoustical tile, cleaning		
Heavy smoking	2-3 years	
Non-smoking	10 years	
Carpet, cleaning		Frequency depends on the type of carpet and the occupancy in the building.
Heavy traffic	Weekly	
Light traffic	Every 6 weeks	
Carpet, vacuuming		
Heavy traffic	Daily	6-8 passes by machine in key areas.
Light traffic	Twice weekly	
Offices	Weekly	
Corridor and lobby, policing		Includes picking up loose trash, removing cigarette butts from in and around jardinieres and/or sand urns. Frequency depends on occupancy in the building.
Main	4 times daily	
Secondary	Daily	
Elevator, cleaning		The frequency of elevator cleaning is dependent upon the amount of traffic, size of the elevator car and the occupancy in the building.
Passenger	Daily	
Freight	Weekly	
Elevator lobby, cleaning	Daily	Includes sweeping, mopping and rinsing.
Escalator, cleaning	Daily	Frequency given is for a normal standard of cleaning under average conditions. Approximately 40% of escalator treads are exposed when escalator is stopped.
Floors, mopping		Frequency given is for a normal standard of cleaning under average conditions.
Main corridors	Daily	
Secondary	Weekly	
Floors, sweeping	Daily	Frequency given is for a normal standard of cleaning under average conditions.
Floors, waxing and polishing		Frequency given is for a normal standard of cleaning under average conditions.
Office area	Every 9 weeks	
Open area	Every 9 weeks	
Flower beds		Frequency depends on geographic location.
Fall clean-up	Every year	
Fertilize	2 times per year	
Mulch	Every year	
Police-up	30 times per year	
Weed		
With mulch	15 times per year	
No mulch	25 times per year	
Lawn, mowing	30 times per year	Frequency depends on geographic location.
Fertilize	2 times per year	
Sweep	3 times per year	
Weed control	2 times per year	
Edge-trim		
Walks	30 times per year	
Shrub	10 times per year	
Light fixtures		Frequency depends on occupancy in the building.
Dusting	Every month	
Washing	Every 3 months	
Shrub Areas		Frequency depends on geographic location.
Fertilize	Every year	
Mulch	2 times per year	
Police-up	30 times per year	
Prune	5 times per year	
Weed	10 times per year	
Stairway		Frequency is dependent upon weather conditions, type of building, type of occupancy, and amount of traffic.
Sweeping and dusting	Daily	
Mopping or scrubbing	Weekly	
Toilet, cleaning	Daily	Includes toilet cleaning, collect waste, cleaning wash basins, urinals, water closets, partitions, walls and floors.
Trash collection	Daily	
Trees		Frequency depends on geographic location.
Fertilize	Every year	
Prune	2 times per year	
Pest control		
Spray	3 times per year	
Systemic	Every year	
Urn and Jardiniere cleaning	Daily	Includes the removal of refuse and debris, cleaning and polishing.
Walks, sweeping	30 times per year	
Walls, periodic cleaning	Every 6 months	Damp wipe, spot removal.
Windows, washing	Every month	Frequency given is for a normal standard of cleaning under average conditions.

R019313-20 Facility Maintenance Labor-Hours

This section lists minimum and maximum cleaning times per unit.
For more information, see *Means Facilities Maintenance Standards* book.

	Unit	
	Minimum	Maximum
Blast Cleaning		
White-metal	100 S.F./Hr	
Near-white	175 S.F./Hr.	
Commercial	370 S.F./Hr.	
Brush-off	870 S.F./Hr.	
Paint Application		
Brushing	125 S.F./Hr.	
Rolling	125 S.F./Hr	
Spraying	500 S.F./Hr.	
Plaster Cleaning		
Wall dusting	2 sec./S.F.	3 sec./S.F.
Vacuuming	4 sec./S.F.	5 sec./S.F.
Spot washing	125 S.F./Hr.	175 S.F./Hr.
Thorough cleaning	275 S.F./Hr.	
Plaster Repair		
Gypsum and lime repair	5 S.Y./Hr.	10 S.Y./Hr.
Ceramic Tile Repair		
General	7 S.F./Hr.	10 S.F./Hr.
Adhesive tile setting	9S.F./Hr	12 S.F./Hr.
Pointing tile joints	10 S.F./Hr.	15 S.F./Hr.
Floor Cleaning		
Manual sweeping	10 min./1000 S.F.	25 min./1000 S.F.
Dust mopping	5 min./1000 S.F.	20 min./1000 S.F.
Buffing	15 min./1000 S.F.	40 min./1000 S.F.
Spray buffing	20 min./1000 S.F.	50 min./1000 S.F.
Damp mopping	15 min./1000 S.F.	30 min./1000 S.F.
Wet mopping	30 min./1000 S.F.	50 min./1000 S.F.
Scrubbing	50 min./1000 S.F.	140 min./1000 S.F.
Scrubbing using electric floor machine		
General	15 min./1000 S.F.	30 min./1000 S.F.
Stripping	100 min./1000 S.F.	200 min./1000 S.F.
Waxing and Buffing using power maching		
Rewaxing	15 min./1000 S.F.	30 min./1000 S.F.
Stripping and rewaxing (two coats)	100 min./1000 S.F.	300 min./1000 S.F.
Waxing and buffing (one coat)	30 min./1000 S.F.	70 min./1000 S.F.
Carpets		
Dry vacuuming	15 min./1000 S.F.	40 min./1000 S.F.
Wet vacuuming	30 min./1000 S.F.	50 min./1000 S.F.
Carpet mopping	20 min./1000 S.F.	40 min./1000 S.F.
Shampooing	175 min./1000 S.F.	250 min./1000 S.F.
Resilient Floor Repair		
Grinding	50 S.F./Hr.	80 S.F./Hr.
Floor Replacement		
Removal (by hand)		
Tiles	100 S.F./Hr.	130 S.F./Hr.
Sheet	120 S.F./Hr	160 S.F./Hr.
Hardwood	40 S.F./Hr.	60 S.F./Hr.
Replacement		
Ceramic	10 S.F./Hr.	20 S.F./Hr.
Resilient	40 S.F./Hr.	70 S.F./Hr.
Hardwood	25 S.F./Hr.	35 S.F./Hr.
Add for related items:		
Replace wood subfloor	80 S.F./Hr.	100 S.F./Hr.
Replace underlayment	75 S.F./Hr.	90 S.F./Hr.
Replace floor moulding	10 S.F./Hr.	30 S.F./Hr.

Unit		
	Minimum	Maximum
Seamless Floor Repair	5 S.F./Hr.	15 S.F./Hr.
Wood Floors		
Sanding	40 S.F./Hr.	60 S.F./Hr.
Sealing	200 S.F./Hr.	300 S.F./Hr.
Waxing*		
Wood Floor Repair		
Loose boards or tiles	50 S.F./Hr.	250S.F./Hr.
Wood strip floor replacement	30 S.F./Hr.	60 S.F./Hr.
Window Washing	300 S.F./Hr.	450 S.F./Hr.
Venetian Blinds, cleaning	15 min./set of blinds	

This section lists average cleaning times per each, or square foot	
	Average
Floor Operations	
Sweeping	
Halls and Corridors	15min./1000 S.F.
General Rooms	30 min./100 S.F.
Dust Mop (unobstructed)	10 min./1000 S.F.
Dust Mop (obstructed)	15 min./1000 S.F.
Damp Mop (unobstructed)	20 min./1000 S.F.
Damp Mop (obstructed)	40 min./1000 S.F.
Wet Mop and Rinse	100 min./1000 S.F.
Hand Scrubbing 12″ brush	300 min./1000 S.F.
Deck Scrubbing	100 min./1000 S.F.
Machine Scrubbing	
12″ diameter	50 min./1000 S.F.
14″ diameter	40 min./1000 S.F.
16″ diameter	35 min./1000 S.F.
18″ diameter	31 min./1000 S.F.
19″ diameter	28 min./1000 S.F.
21″ diameter	25 min./1000 S.F.
23″ diameter	23 min./1000 S.F.
24″ diameter	20 min./1000 S.F.
32″ diameter	18 min./1000 S.F.
36″ diameter	15 min./1000 S.F.
Automatic Scrub Machine (24″)	5 min./1000 S.F.
Vacuum (unobstructed)	20 min./1000 S.F.
Vacuum (obstructed)	30 min./1000 S.F.
Waxing	30 min./1000 S.F.
Machine Polish (19″ machine)	15 min./1000 S.F.
Rectangular Machine (48″ plate)	5 min./1000 S.F.
Buff with steel wool	20 min./1000 S.F.
Strip and Rewax	150 min./1000 S.F.
Dry Strip and Rewax	120 min./1000 S.F.
Spray Buffing (unobstructed)	30 min./1000 S.F.
Spray Buffing (obstructed)	45 min./1000 S.F.
Carpeting	
Vacuuming (unobstructed)	20 min./1000 S.F.
Vacuuming (obstructed)	30 min./1000 S.F.
Spot Vacuuming	15 min./1000 S.F.
Shampoo (dry foam)	60 min. 1000 S.F.
Pile Lift	30 min 1000 S.F.
Lockers	.20 min
Radiators	.30 min.
Tables (medium)	.50 min.
Telephones	.15 min.
Towel dispensers	.12 min.

*See waxing and buffing using electric power machine.

R019313-20 Facility Maintenance Labor-Hours (cont.)

	Average
Towel Disposal Cans	.40 min.
Typewriter and Stand	.50 min
Wash Basin (office)	.60 min
Waste Basin (office)	.50 min.
Window Sill	.20 min.
Venetian Blinds, std. size	3.50 min.
Washrooms	
Cleaning commode	4 min.
Door (spot wash both sides)	1 min.
Mirrors	1 min.
Sanitary napkin dispenser	.50 min.
Urinals	3 min.
Wash basin-soap dispenser	3 min.
General cleaning	120 min./1000 S.F.
Wall Washing	
Painted walls (manual)	240 min./1000 S.F.
Painted walls (machine)	150 min./1000 S.F.
Marble walls (manual)	90 min./1000 S.F.
Ceiling Washing	
Ceiling washing (manual)	300 min./1000 S.F.
Ceiling washing (machine)	180 min./1000 S.F.
Window Washing	
Single pane	125 min./1000 S.F.
Multi-pane	170 min./1000 S.F.
Frosted single pane	190 min./1000 S.F.
Opaque glass	50 min./1000 S.F.
Plate glass	35 min./1000 S.F.
Office partitions (glass)	110 min./1000 S.F.
Dusting Lamps and Lighting Fixtures	
Wall fluorescent fixtures	.15 min.
Desk fluorescent lamp	.30 min.
Table lamp and shade	.60 min.
Floor lamp and shade	.60 min.
Washing Fluorescent Light Fixtures	
Ceiling fixtures (egg crate) 4′ ea.	9 min.
Ceiling fixtures (egg crate) 8′ ea.	12 min.
Dusting	
Air conditioners	.30 min.
Ash trays (desk)	.25 min.
Book cases (3-tier set)	.30 min.
Chairs	.30 min.
Cigarette stands	.40 min.

	Average
Couch	.25 min.
Desks	.80 min.
Desk Trays	.15 min.
File cabinets (4 drawer)	.40 min.
Fabric Upholstery Cleaning	
Whisk or vacuum armless chair	.50 min.
Armchair	1 min.
Couch	2 min.
Shampooing armless chair	4 min.
Armchair	7 min.
Couch	20 min.
Stairway Cleaning	
Sweep and dust, 1 flight, 15 steps	6 min.
Damp mop, 1 flight, 15 steps	5 min.
Scrubbing (hand)	20 min.
Carpentry	
Repair door surface closer	1-1/2 Hrs.
Repair concealed door closer	4 Hrs.
Repair door damage at shop	4 Hrs.
Repair and replace screens	1 Hr.
Repair broken glass	1-1/2 Hrs.
Repair and replace ceiling tile	2 Hrs.
Repair small drywall damage	4 Hrs.
Repair and replace sash balance	4 Hrs.
Change lock on door	1-1/2 Hrs.
Painting	
Repaint 20′ x 15′ room, 1 coat	34 Hrs.
Repaint small bathroom, 1 coat	11-1/2 Hrs.
Repaint exterior window	2-1/2 Hrs.
Plumbing and Steamfitting	
Clear stopped water closet	3 Hrs.
Clear stopped basin	1-1/2 Hrs.
Replace leaking radiator valve	6 Hrs.
Clear external sewer stoppage	9 Hrs.
Install replacement valve, faucet, or trap	10 Hrs.
Electrical	
Replace fluorescent lamp ballast	2 Hrs.
Replace fractional HP motor	10 Hrs.
Replace blown fuse or reset	
Circuit breaker	1-1/2 Hrs.
Repair exterior light damage	4 Hrs.
Control circuit problems	5 Hrs.

R019313-30 A Review of Major Building Materials–Advantages and Disadvantages

Material	Advantages	Disadvantages
Aluminum	Lightweight High resistance to corrosion Low electrical resistance Good conductor	Softness Limited strength for structural uses Low stiffness High rate of thermal expansion Low rate of fire resistance Relatively high cost
Concrete	High compressive resistance Durable Resistance to moisture, rot, insects, fire, and wear Watertight (depending on water-cement ratio)	Workability Lack of tensile strength Lack of resistance to many types of chemical exposure, such as salt Hard to remove stains
Copper	High resistance to corrosion Good electrical conductivity Workable Forms its own surface protection Thermal contraction and expansion not high	Costly Strength varies with treatment and mechanical working
Epoxy	Can have varied properties depending on composition Liquid use very applicable Controllable Excellent strength properties Small creep during curing Hard Tough Resistant to abrasion Resistant to corrosion, salts, acids, petroleum products, solvents, and other chemicals Adhesion to surfaces good	Color Form
Glass	Considerably strong Non-corrosive	Brittle, subject to shattering under shock Transmits energy at a rapid rate Large sizes expensive
Lead	Good resistance to corrosion	Heavy High coefficient of thermal expansion Difficult to hold in place
Masonry	Available in small units Appearance (available in many textures, sizes, and colors) Good insulator	Stains hard to remove Porous (absorption rate high) Shrinkage of mortar Thermal and expansion cracking Others — see concrete
Paper	Readily available Low in cost Used in conjunction with other materials	Susceptible to water damage Susceptible to rot Relatively weak Highly combustible
Plastics — general	Applicable to many uses Relatively low in cost Workable into many shapes High strength Lightweight Non-corrosive Others — see specific plastic entry	Lack of resistance to fire Low stiffness High rate of thermal expansion Low thermal conductivity Some cases of chemical or physical instability with time Non-salvageable Others — see specific plastic entry

R019313-30 A Review of Major Building Materials—Advantages and Disadvantages (cont.)

Material	Advantages	Disadvantages
Plastics — specific		
Acrylics	Transparent Hard Weather resistant Shatter resistant	Easily scratched
Alkyds	Water resistant Touch Good adhesive properties	
Melamines	Hard Durable Abrasive resistant Chemical and heat resistant	
Polyamides (nylon)	Hard Tough Wear resistant	Costly
Polyesters	Weather and chemical resistant Stiff Hard	
Polyethylene	Flexible Tough Translucent Low cost	Easily scratched
Polystyrene	Hard Clear Water and chemical resistant Low cost	Brittle
Vinyls	Tough Wear resistant Stain resistant	
Plywood	Stronger than standard lumber Durable High resistance to impact and load Not as affected by moisture changes as standard wood Workability See Wood for others	Dimensional stability not as good as that of standard structural lumber Poor quality glues are possible Thermal expansion and contraction more of a problem
Steel	Strong Most resistant to aging Most reliable in quality, non-combustible, non-rotting, dimensionally stable with time and moisture change Resistant to staining	Costly Resultant loss of strength when exposed to intense heat, heat, rapid heat gain and loss Corrosive when exposed to moisture and air or other corrosive conditions
Tin	Extremely workable Very resistant to corrosion	Heavy Loss of strength when exposed to heat
Wood	Readily available Relatively low in cost Simple to work with Available in many shapes and forms Good insulating properties	Paintability changes with moisture content Combustible Susceptible to rot and insect infestation Soft and easily damaged (porous) Dimensional changes due to changes in temperature and moisture Strength changes with changing moisture content
Zinc	Fairly high resistance to corrosion Forms its own protective surface Workable	Brittle

R024119-10 Demolition Defined

Whole Building Demolition - Demolition of the whole building with no concern for any particular building element, component, or material type being demolished. This type of demolition is accomplished with large pieces of construction equipment that break up the structure, load it into trucks, and haul it to a disposal site, but disposal or dump fees are not included. Demolition of below-grade foundation elements, such as footings, foundation walls, grade beams, slabs on grade, etc., is not included. Certain mechanical equipment containing flammable liquids or ozone-depleting refrigerants, electric lighting elements, communication equipment components, and other building elements may contain hazardous waste, and must be removed, either selectively or carefully, as hazardous waste before the building can be demolished.

Foundation Demolition - Demolition of below-grade foundation footings, foundation walls, grade beams, and slabs on grade. This type of demolition is accomplished by hand or pneumatic hand tools, and does not include saw cutting, or handling, loading, hauling, or disposal of the debris.

Gutting - Removal of building interior finishes and electrical/mechanical systems down to the load-bearing and sub-floor elements of the rough building frame, with no concern for any particular building element, component, or material type being demolished. This type of demolition is accomplished by hand or pneumatic hand tools, and includes loading into trucks, but not hauling, disposal or dump fees, scaffolding, or shoring. Certain mechanical equipment containing flammable liquids or ozone-depleting refrigerants, electric lighting elements, communication equipment components, and other building elements may contain hazardous waste, and must be removed, either selectively or carefully, as hazardous waste, before the building is gutted.

Selective Demolition - Demolition of a selected building element, component, or finish, with some concern for surrounding or adjacent elements, components, or finishes (see the first Subdivision(s) at the beginning of appropriate Divisions). This type of demolition is accomplished by hand or pneumatic hand tools, and does not include handling, loading, storing, hauling, or disposal of the debris, scaffolding, or shoring. "Gutting" methods may be used in order to save time, but damage that is caused to surrounding or adjacent elements, components, or finishes may have to be repaired at a later time.

Careful Removal - Removal of a piece of service equipment, building element or component, or material type, with great concern for both the removed item and surrounding or adjacent elements, components or finishes. The purpose of careful removal may be to protect the removed item for later re-use, preserve a higher salvage value of the removed item, or replace an item while taking care to protect surrounding or adjacent elements, components, connections, or finishes from cosmetic and/or structural damage. An approximation of the time required to perform this type of removal is 1/3 to 1/2 the time it would take to install a new item of like kind (see Reference Number R220105-10). This type of removal is accomplished by hand or pneumatic hand tools, and does not include loading, hauling, or storing the removed item, scaffolding, shoring, or lifting equipment.

Cutout Demolition - Demolition of a small quantity of floor, wall, roof, or other assembly, with concern for the appearance and structural integrity of the surrounding materials. This type of demolition is accomplished by hand or pneumatic hand tools, and does not include saw cutting, handling, loading, hauling, or disposal of debris, scaffolding, or shoring.

Rubbish Handling - Work activities that involve handling, loading or hauling of debris. Generally, the cost of rubbish handling must be added to the cost of all types of demolition, with the exception of whole building demolition.

Minor Site Demolition - Demolition of site elements outside the footprint of a building. This type of demolition is accomplished by hand or pneumatic hand tools, or with larger pieces of construction equipment, and may include loading a removed item onto a truck (check the Crew for equipment used). It does not include saw cutting, hauling, or disposal of debris, and, sometimes, handling or loading.

R024119-20 Dumpsters

Dumpster rental costs on construction sites are presented in two ways.

The cost per week rental includes the delivery of the dumpster, its pulling or emptying once per week, and its final removal. The assumption is made that the dumpster contractor could choose to empty a dumpster by simply bringing in an empty unit and removing the full one. These costs also include the disposal of the materials in the dumpster.

The Alternate Pricing can be used when actual planned conditions are not approximated by the weekly numbers. For example, these lines can be used when a dumpster is needed for 4 weeks and will need to be emptied 2 or 3 times per week. Conversely the Alternate Pricing lines can be used when a dumpster will be rented for several weeks or months but needs to be emptied only a few times over this period.

R024119-30 Rubbish Handling Chutes

To correctly estimate the cost of rubbish handling chute systems, the individual components must be priced separately. First choose the size of the system; a 30-inch diameter chute is quite common, but the sizes range from 18 to 36 inches in diameter. The 30-inch chute comes in a standard weight and two thinner weights. The thinner weight chutes are sometimes chosen for cost savings, but they are more easily damaged.

There are several types of major chute pieces that make up the chute system. The first component to consider is the top chute section (top intake hopper) where the material is dropped into the chute at the highest point. After determining the top chute, the intermediate chute pieces called the regular chute sections are priced. Next, the number of chute control door sections (intermediate intake hoppers) must be determined. In the more complex systems, a chute control door section is provided at each floor level. The last major component to consider is bolt down frames; these are usually provided at every other floor level.

There are a number of accessories to consider for safe operation and control. There are covers for the top chute and the chute door sections. The top chute can have a trough that allows for better loading of the chute. For the safest operation, a chute warning light system can be added that will warn the other chute intake locations not to load while another is being used. There are dust control devices that spray a water mist to keep down the dust as the debris is loaded into a Dumpster. There are special breakaway cords that are used to prevent damage to the chute if the dumpster is removed without disconnecting from the chute. There are chute liners that can be installed to protect the chute structure from physical damage from rough abrasive materials. Warning signs can be posted at each floor level that is provided with a chute control door section.

In summary, a complete rubbish handling chute system will include one top section, several intermediate regular sections, several intermediate control door (intake hopper) sections and bolt down frames at every other floor level starting with the top floor. If so desired, the system can also include covers and a light warning system for a safer operation. The bottom of the chute should always be above the Dumpster and should be tied off with a breakaway cord to the Dumpster.

R026510-20 Underground Storage Tank Removal

Underground Storage Tank Removal can be divided into two categories: Non-Leaking and Leaking. Prior to removing an underground storage tank, tests should be made, with the proper authorities present, to determine whether a tank has been leaking or the surrounding soil has been contaminated.

To safely remove Liquid Underground Storage Tanks:

1. Excavate to the top of the tank.
2. Disconnect all piping.
3. Open all tank vents and access ports.
4. Remove all liquids and/or sludge.
5. Purge the tank with an inert gas.
6. Provide access to the inside of the tank and clean out the interior using proper personal protective equipment (PPE).
7. Excavate soil surrounding the tank using proper PPE for on-site personnel.
8. Pull and properly dispose of the tank.
9. Clean up the site of all contaminated material.
10. Install new tanks or close the excavation.

R028213-20 Asbestos Removal Process

Asbestos removal is accomplished by a specialty contractor who understands the federal and state regulations regarding the handling and disposal of the material. The process of asbestos removal is divided into many individual steps. An accurate estimate can be calculated only after all the steps have been priced.

The steps are generally as follows:

1. Obtain an asbestos abatement plan from an industrial hygienist.
2. Monitor the air quality in and around the removal area and along the path of travel between the removal area and transport area. This establishes the background contamination.
3. Construct a two part decontamination chamber at entrance to removal area.
4. Install a HEPA filter to create a negative pressure in the removal area.
5. Install wall, floor, and ceiling protection as required by the plan, usually 2 layers of fireproof 6 mil polyethylene.
6. Industrial hygienist visually inspects work area to verify compliance with plan.
7. Provide temporary supports for conduit and piping affected by the removal process.
8. Proceed with asbestos removal and bagging process. Monitor air quality as described in Step #2. Discontinue operations when contaminate levels exceed applicable standards.
9. Document the legal disposal of materials in accordance with EPA standards.
10. Thoroughly clean removal area including all ledges, crevices, and surfaces.
11. Post abatement inspection by industrial hygienist to verify plan compliance.
12. Provide a certificate from a licensed industrial hygienist attesting that contaminate levels are within acceptable standards before returning area to regular use.

R028319-60 Lead Paint Remediation Methods

Lead paint remediation can be accomplished by the following methods:

1. Abrasive blast
2. Chemical stripping
3. Power tool cleaning with vacuum collection system
4. Encapsulation
5. Remove and replace
6. Enclosure

Each of these methods has strengths and weakness depending on the specific circumstances of the project. The following is an overview of each method.

1. **Abrasive blasting** is usually accomplished with sand or recyclable metallic blast. Before work can begin, the area must be contained to ensure the blast material with lead does not escape to the atmosphere. The use of vacuum blast greatly reduces the containment requirements. Lead abatement equipment that may be associated with this work includes a negative air machine. In addition, it is necessary to have an industrial hygienist monitor the project on a continual basis. When the work is complete, the spent blast sand with lead must be disposed of as a hazardous material. If metallic shot was used, the lead is separated from the shot and disposed of as hazardous material. Worker protection includes disposable clothing and respiratory protection.
2. **Chemical stripping** requires strong chemicals be applied to the surface to remove the lead paint. Before the work can begin, the area under/adjacent to the work area must be covered to catch the chemical and removed lead. After the chemical is applied to the painted surface it is usually covered with paper. The chemical is left in place for the specified period, then the paper with lead paint is pulled or scraped off. The process may require several chemical applications. The paper with chemicals and lead paint adhered to it, plus the containment and loose scrapings collected by a HEPA (High Efficiency Particulate Air Filter) vac, must be disposed of as a hazardous material. The chemical stripping process usually requires a neutralizing agent and several wash downs after the paint is removed. Worker protection includes a neoprene or other compatible protective clothing and respiratory protection with face shield. An industrial hygienist is required intermittently during the process.
3. **Power tool cleaning** is accomplished using shrouded needle blasting guns. The shrouding with different end configurations is held up against the surface to be cleaned. The area is blasted with hardened needles and the shroud captures the lead with a HEPA vac and deposits it in a holding tank. An industrial hygienist monitors the project, protective clothing and a respirator is required until air samples prove otherwise. When the work is complete the lead must be disposed of as a hazardous material.
4. **Encapsulation** is a method that leaves the well bonded lead paint in place after the peeling paint has been removed. Before the work can begin, the area under/adjacent to the work must be covered to catch the scrapings. The scraped surface is then washed with a detergent and rinsed. The prepared surface is covered with approximately 10 mils of paint. A reinforcing fabric can also be embedded in the paint covering. The scraped paint and containment must be disposed of as a hazardous material. Workers must wear protective clothing and respirators.
5. **Remove and replace** is an effective way to remove lead paint from windows, gypsum walls and concrete masonry surfaces. The painted materials are removed and new materials are installed. Workers should wear a respirator and tyvek suit. The demolished materials must be disposed of as hazardous waste if it fails the TCLP (Toxicity Characteristic Leachate Process) test.
6. **Enclosure** is the process that permanently seals lead painted materials in place. This process has many applications such as covering lead painted drywall with new drywall, covering exterior construction with tyvek paper then residing, or covering lead painted structural members with aluminum or plastic. The seams on all enclosing materials must be securely sealed. An industrial hygienist monitors the project, and protective clothing and a respirator is required until air samples prove otherwise.

All the processes require clearance monitoring and wipe testing as required by the hygienist.

R031113-60 Formwork Labor-Hours

Item	Hours Required				Total Hours	Multiple Use		
	Unit	Fabricate	Erect & Strip	Clean & Move	1 Use	2 Use	3 Use	4 Use
Beam and Girder, interior beams, 12″ wide	100 S.F.	6.4	8.3	1.3	16.0	13.3	12.4	12.0
Hung from steel beams		5.8	7.7	1.3	14.8	12.4	11.6	11.2
Beam sides only, 36″ high		5.8	7.2	1.3	14.3	11.9	11.1	10.7
Beam bottoms only, 24″ wide		6.6	13.0	1.3	20.9	18.1	17.2	16.7
Box out for openings		9.9	10.0	1.1	21.0	16.6	15.1	14.3
Buttress forms, to 8′ high		6.0	6.5	1.2	13.7	11.2	10.4	10.0
Centering, steel, 3/4″ rib lath			1.0		1.0			
3/8″ rib lath or slab form			0.9		0.9			
Chamfer strip or keyway	100 L.F.		1.5		1.5	1.5	1.5	1.5
Columns, fiber tube 8″ diameter			20.6		20.6			
12″			21.3		21.3			
16″			22.9		22.9			
20″			23.7		23.7			
24″			24.6		24.6			
30″			25.6		25.6			
Columns, round steel, 12″ diameter			22.0		22.0	22.0	22.0	22.0
16″			25.6		25.6	25.6	25.6	25.6
20″			30.5		30.5	30.5	30.5	30.5
24″			37.7		37.7	37.7	37.7	37.7
Columns, plywood 8″ x 8″	100 S.F.	7.0	11.0	1.2	19.2	16.2	15.2	14.7
12″ x 12″		6.0	10.5	1.2	17.7	15.2	14.4	14.0
16″ x 16″		5.9	10.0	1.2	17.1	14.7	13.8	13.4
24″ x 24″		5.8	9.8	1.2	16.8	14.4	13.6	13.2
Columns, steel framed plywood 8″ x 8″			10.0	1.0	11.0	11.0	11.0	11.0
12″ x 12″			9.3	1.0	10.3	10.3	10.3	10.3
16″ x 16″			8.5	1.0	9.5	9.5	9.5	9.5
24″ x 24″			7.8	1.0	8.8	8.8	8.8	8.8
Drop head forms, plywood		9.0	12.5	1.5	23.0	19.0	17.7	17.0
Coping forms		8.5	15.0	1.5	25.0	21.3	20.0	19.4
Culvert, box			14.5	4.3	18.8	18.8	18.8	18.8
Curb forms, 6″ to 12″ high, on grade		5.0	8.5	1.2	14.7	12.7	12.1	11.7
On elevated slabs		6.0	10.8	1.2	18.0	15.5	14.7	14.3
Edge forms to 6″ high, on grade	100 L.F.	2.0	3.5	0.6	6.1	5.6	5.4	5.3
7″ to 12″ high	100 S.F.	2.5	5.0	1.0	8.5	7.8	7.5	7.4
Equipment foundations		10.0	18.0	2.0	30.0	25.5	24.0	23.3
Flat slabs, including drops		3.5	6.0	1.2	10.7	9.5	9.0	8.8
Hung from steel		3.0	5.5	1.2	9.7	8.7	8.4	8.2
Closed deck for domes		3.0	5.8	1.2	10.0	9.0	8.7	8.5
Open deck for pans		2.2	5.3	1.0	8.5	7.9	7.7	7.6
Footings, continuous, 12″ high		3.5	3.5	1.5	8.5	7.3	6.8	6.6
Spread, 12″ high		4.7	4.2	1.6	10.5	8.7	8.0	7.7
Pile caps, square or rectangular		4.5	5.0	1.5	11.0	9.3	8.7	8.4
Grade beams, 24″ deep		2.5	5.3	1.2	9.0	8.3	8.0	7.9
Lintel or Sill forms		8.0	17.0	2.0	27.0	23.5	22.3	21.8
Spandrel beams, 12″ wide		9.0	11.2	1.3	21.5	17.5	16.2	15.5
Stairs			25.0	4.0	29.0	29.0	29.0	29.0
Trench forms in floor		4.5	14.0	1.5	20.0	18.3	17.7	17.4
Walls, Plywood, at grade, to 8′ high		5.0	6.5	1.5	13.0	11.0	9.7	9.5
8′ to 16′		7.5	8.0	1.5	17.0	13.8	12.7	12.1
16′ to 20′		9.0	10.0	1.5	20.5	16.5	15.2	14.5
Foundation walls, to 8′ high		4.5	6.5	1.0	12.0	10.3	9.7	9.4
8′ to 16′ high		5.5	7.5	1.0	14.0	11.8	11.0	10.6
Retaining wall to 12′ high, battered		6.0	8.5	1.5	16.0	13.5	12.7	12.3
Radial walls to 12′ high, smooth		8.0	9.5	2.0	19.5	16.0	14.8	14.3
2′ chords		7.0	8.0	1.5	16.5	13.5	12.5	12.0
Prefabricated modular, to 8′ high		—	4.3	1.0	5.3	5.3	5.3	5.3
Steel, to 8′ high		—	6.8	1.2	8.0	8.0	8.0	8.0
8′ to 16′ high		—	9.1	1.5	10.6	10.3	10.2	10.2
Steel framed plywood to 8′ high		—	6.8	1.2	8.0	7.5	7.3	7.2
8′ to 16′ high		—	9.3	1.2	10.5	9.5	9.2	9.0

Concrete — R0321 Reinforcing Steel

R032110-10 Reinforcing Steel Weights and Measures

Bar Designation No.**	Nominal Weight Lb./Ft.	U.S. Customary Units Nominal Dimensions*			SI Units Nominal Dimensions*			
		Diameter in.	Cross Sectional Area, in.^{2}	Perimeter in.	Nominal Weight kg/m	Diameter mm	Cross Sectional Area, cm^2	Perimeter mm
3	.376	.375	.11	1.178	.560	9.52	.71	29.9
4	.668	.500	.20	1.571	.994	12.70	1.29	39.9
5	1.043	.625	.31	1.963	1.552	15.88	2.00	49.9
6	1.502	.750	.44	2.356	2.235	19.05	2.84	59.8
7	2.044	.875	.60	2.749	3.042	22.22	3.87	69.8
8	2.670	1.000	.79	3.142	3.973	25.40	5.10	79.8
9	3.400	1.128	1.00	3.544	5.059	28.65	6.45	90.0
10	4.303	1.270	1.27	3.990	6.403	32.26	8.19	101.4
11	5.313	1.410	1.56	4.430	7.906	35.81	10.06	112.5
14	7.650	1.693	2.25	5.320	11.384	43.00	14.52	135.1
18	13.600	2.257	4.00	7.090	20.238	57.33	25.81	180.1

* The nominal dimensions of a deformed bar are equivalent to those of a plain round bar having the same weight per foot as the deformed bar.
** Bar numbers are based on the number of eighths of an inch included in the nominal diameter of the bars.

R032110-80 Shop-Fabricated Reinforcing Steel

The material prices for reinforcing, shown in the unit cost sections of the book, are for 50 tons or more of shop-fabricated reinforcing steel and include:

1. Mill base price of reinforcing steel
2. Mill grade/size/length extras
3. Mill delivery to the fabrication shop
4. Shop storage and handling
5. Shop drafting/detailing
6. Shop shearing and bending
7. Shop listing
8. Shop delivery to the job site

Both material and installation costs can be considerably higher for small jobs consisting primarily of smaller bars, while material costs may be slightly lower for larger jobs.

Concrete — R0330 Cast-In-Place Concrete

R033053-50 Industrial Chimneys

Foundation requirements in C.Y. of concrete for various sized chimneys.

Size Chimney	2 Ton Soil	3 Ton Soil	Size Chimney	2 Ton Soil	3 Ton Soil	Size Chimney	2 Ton Soil	3 Ton Soil
75′ x 3′-0″	13 C.Y.	11 C.Y.	160′ x 6′-6″	86 C.Y.	76 C.Y.	300′ x 10′-0″	325 C.Y.	245 C.Y.
85′ x 5′-6″	19	16	175′ x 7′-0″	108	95	350′ x 12′-0″	422	320
100′ x 5′-0″	24	20	200′ x 6′-0″	125	105	400′ x 14′-0″	520	400
125′ x 5′-6″	43	36	250′ x 8′-0″	230	175	500′ x 18′-0″	725	575

R033105-20 Materials for One C.Y. of Concrete

This is an approximate method of figuring quantities of cement, sand and coarse aggregate for a field mix with waste allowance included.

With crushed gravel as coarse aggregate, to determine barrels of cement required, divide 10 by total mix; that is, for 1:2:4 mix, 10 divided by 7 = 1-3/7 barrels.

If the coarse aggregate is crushed stone, use 10-1/2 instead of 10 as given for gravel.

To determine tons of sand required, multiply barrels of cement by parts of sand and then by 0.2; that is, for the 1:2:4 mix, as above, 1-3/7 x 2 x .2 = .57 tons.

Tons of crushed gravel are in the same ratio to tons of sand as parts in the mix, or 4/2 x .57 = 1.14 tons.

1 bag cement = 94#
4 bags = 1 barrel

1 C.Y. sand or crushed gravel = 2700#
1 ton sand or crushed gravel = 20 C.F.

1 C.Y. crushed stone = 2575#
1 ton crushed stone = 21 C.F.

Average carload of cement is 692 bags; of sand or gravel is 56 tons.

Do not stack stored cement over 10 bags high.

R033105-70 Placing Ready-Mixed Concrete

For ground pours allow for 5% waste when figuring quantities.

Prices in the front of the book assume normal deliveries. If deliveries are made before 8 A.M. or after 5 P.M. or on Saturday afternoons add 30%. Negotiated discounts for large volumes are not included in prices in front of book.

For the lower floors without truck access, concrete may be wheeled in rubber-tired buggies, conveyer handled, crane handled or pumped. Pumping is economical if there is top steel. Conveyers are more efficient for thick slabs.

At higher floors the rubber-tired buggies may be hoisted by a hoisting tower and wheeled to location. Placement by a conveyer is limited to three floors and is best for high-volume pours. Pumped concrete is best when building has no crane access. Concrete may be pumped directly as high as thirty-six stories using special pumping techniques. Normal maximum height is about fifteen stories.

Best pumping aggregate is screened and graded bank gravel rather than crushed stone.

Pumping downward is more difficult than pumping upward. Horizontal distance from pump to pour may increase preparation time prior to pour. Placing by cranes, either mobile, climbing or tower types, continues as the most efficient method for high-rise concrete buildings.

R033105-80 Slab on Grade

General: Ground slabs are classified on the basis of use. Thickness is generally controlled by the heaviest concentrated load supported. If load area is greater than 80 sq. in., soil bearing may be important. The base granular fill must be a uniformly compacted material of limited capillarity, such as gravel or crushed rock. Concrete is placed on this surface of the vapor barrier on top of base.

Ground slabs are either single or two course floors. Single course are widely used. Two course floors have a subsequent wear resistant topping.

Reinforcement is provided to maintain tightly closed cracks.

Control joints limit crack locations and provide for differential horizontal movement only. Isolation joints allow both horizontal and vertical differential movement.

Use of Table: Determine appropriate type of slab (A, B, C, or D) by considering type of use or amount of abrasive wear of traffic type.

Determine thickness by maximum allowable wheel load or uniform load, opposite 1st column, thickness. Increase the controlling thickness if details require, and select either plain or reinforced slab thickness and type.

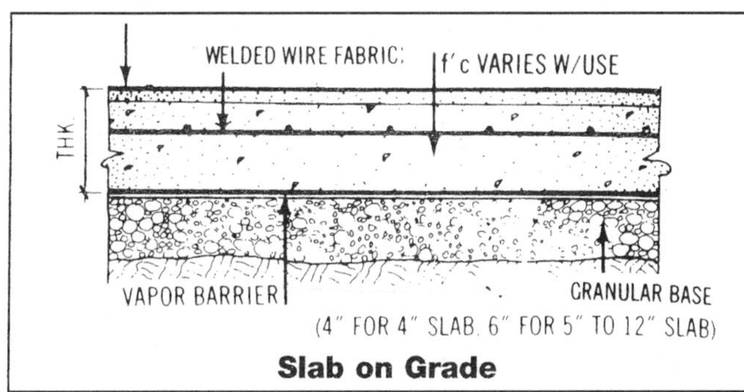

Slab on Grade

Thickness and Loading Assumptions by Type of Use

		A	B	C	D	◄ Slab I.D.
SLAB THICKNESS (IN.)	TYPE	Non	Light	Normal	Heavy	◄ Industrial
		Little	Light	Moderate	Severe	◄ Abrasion
		Foot Only	Pneumatic Wheels	Solid Rubber Wheels	Steel Tires	◄ Type of Traffic
		Load* (K)	Load* (K)	Load* (K)	Load* (K)	Max. Uniform Load to Slab ▼ (PSF)
4″	Reinf. Plain	4K				100
5″	Reinf. Plain	6K	4K			200
6″	Reinf. Plain		8K	6K	6K	500 to 800
7″	Reinf. Plain			9K	8K	1,500
8″	Reinf. Plain				11K	
10″	Reinf. Plain				14K	* Max. Wheel Load in Kips (incl. impact)
12″	Reinf. Plain					
DESIGN ASSUMPTIONS	Concrete, Chuted	f′c = 3.5 KSI	4 KSI	4.5 KSI	Slab @ 3.5 KSI	ASSUMPTIONS BY SLAB TYPE
	Toppings			1″ Integral	1″ Bonded	
	Finish	Steel Trowel	Steel Trowel	Steel Trowel	Screed & Steel Trowel	
	Compacted Granular Base	4″ deep for 4″ slab thickness 6″ deep for 5″ slab thickness & greater				ASSUMPTIONS FOR ALL SLAB TYPES
	Vapor Barrier	6 mil polyethylene				
	Forms & Joints	Allowances included				
	Rein-forcement	WWF as required ≥ 60,000 psi				

R033543-10 Polished Concrete Floors

A polished concrete floor has a glossy mirror-like appearance and is created by grinding the concrete floor with finer and finer diamond grits, similar to sanding wood, until the desired level of reflective clarity and sheen are achieved. The technical term for this type of polished concrete is bonded abrasive polished concrete. The basic piece of equipment used in the polishing process is a walk-behind planetary grinder for working large floor areas. This grinder drives diamond-impregnated abrasive discs, which progress from coarse- to fine-grit discs.

The process begins with the use of very coarse diamond segments or discs bonded in a metallic matrix. These segments are coarse enough to allow the removal of pits, blemishes, stains, and light coatings from the floor surface in preparation for final smoothing. The condition of the original concrete surface will dictate the grit coarsness of the initial grinding step which will generally end up being a three- to four-step process using ever finer grits. The purpose of this initial grinding step is to remove surface coatings and blemishes and to cut down into the cream for very fine aggregate exposure, or deeper into the fine aggregate layer just below the cream layer, or even deeper into the coarse aggregate layer. These initial grinding steps will progress up to the 100/120 grit. If wet grinding is done, a waste slurry is produced that must be removed between grit changes and disposed of properly. If dry grinding is done, a high performance vacuum will pick up the dust during grinding and collect it in bags which must be disposed of properly.

The process continues with honing the floor in a series of steps that progress from 100-grit to 400-grit diamond abrasive discs embedded in a plastic or resin matrix. At some point during, or just prior to, the honing step, one or two coats of stain or dye can be sprayed onto the surface to give color to the concrete, and two coats of densifier/hardener must be applied to the floor surface and allowed to dry. This sprayed-on densifier/hardener will penetrate about 1/8″ into the concrete to make the surface harder, denser and more abrasion-resistant.

The process ends with polishing the floor surface in a series of steps that progress from resin-impregnated 800-grit (medium polish) to 1500-grit (high polish) to 3000-grit (very high polish), depending on the desired level of reflective clarity and sheen.

The Concrete Polishing Association of America (CPAA) has defined the flooring options available when processing concrete to a desired finish. The first category is aggregate exposure, the grinding of a concrete surface with bonded abrasives, in as many abrasive grits necessary, to achieve one of the following classes:

A. Cream – very little surface cut depth; little aggregate exposure
B. Fine aggregate (salt and pepper) – surface cut depth of 1/16″; fine aggregate exposure with little or no medium aggregate exposure at random locations
C. Medium aggregate – surface cut depth of 1/8″; medium aggregate exposure with little or no large aggregate exposure at random locations
D. Large aggregate – surface cut depth of 1/4″; large aggregate exposure with little or no fine aggregate exposure at random locations

The second CPAA defined category is reflective clarity and sheen, the polishing of a concrete surface with the minimum number of bonded abrasives as indicated to achieve one of the following levels:

1. Ground – flat appearance with none to very slight diffused reflection; none to very low reflective sheen; using a minimum total of 4 grit levels up 100-grit
2. Honed – matte appearance with or without slight diffused reflection; low to medium reflective sheen; using a minimum total of 5 grit levels up to 400-grit
3. Semi-polished – objects being reflected are not quite sharp and crisp but can be easily identified; medium to high reflective sheen; using a minimum total of 6 grit levels up to 800-grit
4. Highly-polished – objects being reflected are sharp and crisp as would be seen in a mirror-like reflection; high to highest reflective sheen; using a minimum total of up to 8 grit levels up to 1500-grit or 3000-grit

The CPAA defines reflective clarity as the degree of sharpness and crispness of the reflection of overhead objects when viewed 5′ above and perpendicular to the floor surface; and reflective sheen as the degree of gloss reflected from a surface when viewed at least 20′ from and at an angle to the floor surface. These terms are relatively subjective. The final outcome depends on the internal makeup and surface condition of the original concrete floor, the experience of the floor polishing crew, and expectations of the owner. Before the grinding, honing, and polishing work commences on the main floor area, it might be beneficial to do a mock-up panel in the same floor but in an out of the way place to demonstrate the sequence of steps with increasingly fine abrasive grits and to demonstrate the final reflective clarity and reflective sheen. This mock-up panel will be within the area of, and part of, the final work.

Masonry — R0405 Common Work Results for Masonry

R040519-50 Masonry Reinforcing

Horizontal joint reinforcing helps prevent wall cracks where wall movement may occur and in many locations is required by code. Horizontal joint reinforcing is generally not considered to be structural reinforcing and an unreinforced wall may still contain joint reinforcing.

Reinforcing strips come in 10′ and 12′ lengths and in truss and ladder shapes, with and without drips. Field labor runs between 2.7 to 5.3 hours per 1000 L.F. for wall thicknesses up to 12″.

The wire meets ASTM A82 for cold drawn steel wire and the typical size is 9 ga. sides and ties with 3/16″ diameter also available. Typical finish is mill galvanized with zinc coating at .10 oz. per S.F. Class I (.40 oz. per S.F.) and Class III (.80 oz per S.F.) are also available, as is hot dipped galvanizing at 1.50 oz. per S.F.

Masonry — R0421 Clay Unit Masonry

R042110-20 Common and Face Brick

Common building brick manufactured according to ASTM C62 and facing brick manufactured according to ASTM C216 are the two standard bricks available for general building use.

Building brick is made in three grades; SW, where high resistance to damage caused by cyclic freezing is required; MW, where moderate resistance to cyclic freezing is needed; and NW, where little resistance to cyclic freezing is needed. Facing brick is made in only the two grades SW and MW. Additionally, facing brick is available in three types; FBS, for general use; FBX, for general use where a higher degree of precision and lower permissible variation in size than FBS is needed; and FBA, for general use to produce characteristic architectural effects resulting from non-uniformity in size and texture of the units.

In figuring the material cost of brickwork, an allowance of 25% mortar waste and 3% brick breakage was included. If bricks are delivered palletized with 280 to 300 per pallet, or packaged, allow only 1-1/2% for breakage. Packaged or palletized delivery is practical when a job is big enough to have a crane or other equipment available to handle a package of brick. This is so on all industrial work but not always true on small commercial buildings.

The use of buff and gray face is increasing, and there is a continuing trend to the Norman, Roman, Jumbo and SCR brick.

Common red clay brick for backup is not used that often. Concrete block is the most usual backup material with occasional use of sand lime or cement brick. Building brick is commonly used in solid walls for strength and as a fire stop.

Brick panels built on the ground and then crane erected to the upper floors have proven to be economical. This allows the work to be done under cover and without scaffolding.

R042110-50 Brick, Block & Mortar Quantities

Running Bond						For Other Bonds Standard Size Add to S.F. Quantities in Table to Left		
Number of Brick per S.F. of Wall - Single Wythe with 3/8″ Joints				C.F. of Mortar per M Bricks, Waste Included				
Type Brick	Nominal Size (incl. mortar) L H W	Modular Coursing	Number of Brick per S.F.	3/8″ Joint	1/2″ Joint	Bond Type	Description	Factor
Standard	8 x 2-2/3 x 4	3C=8″	6.75	8.1	10.3	Common	full header every fifth course	+20%
Economy	8 x 4 x 4	1C=4″	4.50	9.1	11.6		full header every sixth course	+16.7%
Engineer	8 x 3-1/5 x 4	5C=16″	5.63	8.5	10.8	English	full header every second course	+50%
Fire	9 x 2-1/2 x 4-1/2	2C=5″	6.40	550 # Fireclay	—	Flemish	alternate headers every course	+33.3%
Jumbo	12 x 4 x 6 or 8	1C=4″	3.00	22.5	29.2		every sixth course	+5.6%
Norman	12 x 2-2/3 x 4	3C=8″	4.50	11.2	14.3	Header = W x H exposed		+100%
Norwegian	12 x 3-1/5 x 4	5C=16″	3.75	11.7	14.9	Rowlock = H x W exposed		+100%
Roman	12 x 2 x 4	2C=4″	6.00	10.7	13.7	Rowlock stretcher = L x W exposed		+33.3%
SCR	12 x 2-2/3 x 6	3C=8″	4.50	21.8	28.0	Soldier = H x L exposed		—
Utility	12 x 4 x 4	1C=4″	3.00	12.3	15.7	Sailor = W x L exposed		-33.3%

Concrete Blocks Nominal Size	Approximate Weight per S.F.		Blocks per 100 S.F.	Mortar per M block, waste included	
	Standard	Lightweight		Partitions	Back up
2″ x 8″ x 16″	20 PSF	15 PSF	113	27 C.F.	36 C.F.
4″ ↓	30 ↓	20 ↓	↓	41 ↓	51 ↓
6″	42	30		56	66
8″	55	38		72	82
10″	70	47		87	97
12″	85	55		102	112

Brick & Mortar Quantities
©Brick Industry Association. 2009 Feb. Technical Notes on Brick Construction 10:
Dimensioning and Estimating Brick Masonry. Reston (VA): BIA. Table 1 Modular Brick Sizes and Table 4 Quantity Estimates for Brick Masonry.

R042210-20 Concrete Block

The material cost of special block such as corner, jamb and head block can be figured at the same price as ordinary block of same size. Labor on specials is about the same as equal-sized regular block.

Bond beam and 16″ high lintel blocks are more expensive than regular units of equal size. Lintel blocks are 8″ long and either 8″ or 16″ high.

Use of motorized mortar spreader box will speed construction of continuous walls.

Hollow non-load-bearing units are made according to ASTM C129 and hollow load-bearing units according to ASTM C90.

R050516-30 Coating Structural Steel

On field-welded jobs, the shop-applied primer coat is necessarily omitted. All painting must be done in the field and usually consists of red oxide rust inhibitive paint or an aluminum paint. The table below shows paint coverage and daily production for field painting.

See Division 05 05 13.50 for hot-dipped galvanizing and Division 09 97 13.23 for field-applied cold galvanizing and other paints and protective coatings.

See Division 05 01 10.51 for steel surface preparation treatments such as wire brushing, pressure washing and sand blasting.

Type Construction	Surface Area per Ton	Coat	One Gallon Covers		In 8 Hrs. Person Covers		Average per Ton Spray	
			Brush	Spray	Brush	Spray	Gallons	Labor-hours
Light Structural	300 S.F. to 500 S.F.	1st	500 S.F.	455 S.F.	640 S.F.	2000 S.F.	0.9 gals.	1.6 L.H.
		2nd	450	410	800	2400	1.0	1.3
		3rd	450	410	960	3200	1.0	1.0
Medium	150 S.F. to 300 S.F.	All	400	365	1600	3200	0.6	0.6
Heavy Structural	50 S.F. to 150 S.F.	1st	400	365	1920	4000	0.2	0.2
		2nd	400	365	2000	4000	0.2	0.2
		3rd	400	365	2000	4000	0.2	0.2
Weighted Average	225 S.F.	All	400	365	1350	3000	0.6	0.6

R050521-20 Welded Structural Steel

Usual weight reductions with welded design run 10% to 20% compared with bolted or riveted connections. This amounts to about the same total cost compared with bolted structures since field welding is more expensive than bolts. For normal spans of 18′ to 24′ figure 6 to 7 connections per ton.

Trusses — For welded trusses add 4% to weight of main members for connections. Up to 15% less steel can be expected in a welded truss compared to one that is shop bolted. Cost of erection is the same whether shop bolted or welded.

General — Typical electrodes for structural steel welding are E6010, E6011, E60T and E70T. Typical buildings vary between 2# to 8# of weld rod per ton of steel. Buildings utilizing continuous design require about three times as much welding as conventional welded structures. In estimating field erection by welding, it is best to use the average linear feet of weld per ton to arrive at the welding cost per ton. The type, size and position of the weld will have a direct bearing on the cost per linear foot. A typical field welder will deposit 1.8# to 2# of weld rod per hour manually. Using semiautomatic methods can increase production by as much as 50% to 75%.

R050523-10 High Strength Bolts

Common bolts (A307) are usually used in secondary connections (see Division 05 05 23.10).

High strength bolts (A325 and A490) are usually specified for primary connections such as column splices, beam and girder connections to columns, column bracing, connections for supports of operating equipment or of other live loads which produce impact or reversal of stress, and in structures carrying cranes of over 5-ton capacity.

Allow 20 field bolts per ton of steel for a 6 story office building, apartment house or light industrial building. For 6 to 12 stories allow 25 bolts per ton, and above 12 stories, 30 bolts per ton. On power stations, 20 to 25 bolts per ton are needed.

R051223-10 Structural Steel

The bare material prices for structural steel, shown in the unit cost sections of the book, are for 100 tons of shop-fabricated structural steel and include:

1. Mill base price of structural steel
2. Mill scrap/grade/size/length extras
3. Mill delivery to a metals service center (warehouse)
4. Service center storage and handling
5. Service center delivery to a fabrication shop
6. Shop storage and handling
7. Shop drafting/detailing
8. Shop fabrication
9. Shop coat of primer paint
10. Shop listing
11. Shop delivery to the job site

In unit cost sections of the book that contain items for field fabrication of steel components, the bare material cost of steel includes:

1. Mill base price of structural steel
2. Mill scrap/grade/size/length extras
3. Mill delivery to a metals service center (warehouse)
4. Service center storage and handling
5. Service center delivery to the job site

R051223-15 Structural Steel Estimating for Repair and Remodeling Projects

The correct approach to estimating structural steel is dependent upon the amount of steel required for the particular project. If the project is a sizable addition to a building, the data can be used directly from the cost data book. This is not the case however if the project requires a small amount of steel, for instance to reinforce existing roof or floor structural systems. To better understand this, please refer to the unit price line for a W16x31 beam with bolted connections. Assume your project requires the reinforcement of the structural members of the roof system of a 3 story building required for the installation a new piece of HVAC equipment. The project will need 4 pieces of W16x31 that are 30′ long. After pricing this 120 L.F. job using the unit prices for a W16x31, an analysis will reveal that the price is wholly inadequate.

The first problem is apparent if you examine the amount of steel that can be installed per day. The unit price line indicates 900 linear feet per day can be installed by a 5 man crew with an 90 ton crane. This productivity is correct for new construction but certainly is not for repair and remodeling work. Installation of new structural steel considers that each member is installed from the foundation to the roof of the structure in a planned and systematic manner with a crane having unrestricted access to all parts of the project. Additionally each connection is planned and detailed with full field access for fit-up and final bolting. The erection is planned and progresses such that interferences and conflicts with other structural members are minimized, if not completely eliminated. All of these assumptions are clearly not the case with a repair and remodeling job, and a significant decrease in the stated productivity will be observed.

A crane will certainly be needed to lift the members into the general area of the project but in most cases will not be able to place the beams into their final position. An opening in the existing roof may not be large enough to permit the beams to pass through and it may be necessary to bring them into the building through existing windows or doors. Moving the beams to the actual area where they will be installed may involve hand labor and the use of dollies. Finally, hoists and/or jacks may be needed for final positioning.

The connection of new members to existing can often be accomplished by field bolting with accurate field measurements and good planning but in many cases access to both sides of existing members is not possible and field welding becomes the only alternative. In addition to the cost of the actual welding, protection of existing finishes, systems and structure and fire protection must be considered.

New beams can never be installed tight to the existing decks or floors which they must support and the use of shims and tack welding becomes necessary. Additionally, further planning and cost is involved in assuring that existing loads are minimized during the installation and shimming process.

It is apparent that installation of structural steel as part of a repair and remodeling project involves more than simply installing the members and estimating the cost in the same manner as new construction. The best procedure for estimating the total cost is adequate planning and coordination of each process and activity that will be needed. Unit costs for the materials, labor and equipment can then be attached to each needed activity and a final, complete price can be determined.

R051223-20 Steel Estimating Quantities

One estimate on erection is that a crane can handle 35 to 60 pieces per day. Say the average is 45. With usual sizes of beams, girders, and columns, this would amount to about 20 tons per day. The type of connection greatly affects the speed of erection. Moment connections for continuous design slow down production and increase erection costs.

Short open web bar joists can be set at the rate of 75 to 80 per day, with 50 per day being the average for setting long span joists.

After main members are calculated, add the following for usual allowances: base plates 2% to 3%; column splices 4% to 5%; and miscellaneous details 4% to 5%, for a total of 10% to 13% in addition to main members.

The ratio of column to beam tonnage varies depending on type of steels used, typical spans, story heights and live loads.

It is more economical to keep the column size constant and to vary the strength of the column by using high strength steels. This also saves floor space. Buildings have recently gone as high as ten stories with 8″ high strength columns. For light columns under W8X31 lb. sections, concrete filled steel columns are economical.

High strength steels may be used in columns and beams to save floor space and to meet head room requirements. High strength steels in some sizes sometimes require long lead times.

Round, square and rectangular columns, both plain and concrete filled, are readily available and save floor area, but are higher in cost per pound than rolled columns. For high unbraced columns, tube columns may be less expensive.

Below are average minimum figures for the weights of the structural steel frame for different types of buildings using A36 steel, rolled shapes and simple joints. For economy in domes, rise to span ratio = .13. Open web joist framing systems will reduce weights by 10% to 40%. Composite design can reduce steel weight by up to 25% but additional concrete floor slab thickness may be required. Continuous design can reduce the weights up to 20%. There are many building codes with different live load requirements and different structural requirements, such as hurricane and earthquake loadings, which can alter the figures.

Structural Steel Weights per S.F. of Floor Area

Type of Building	No. of Stories	Avg. Spans	L.L. #/S.F.	Lbs. Per S.F.	Type of Building	No. of Stories	Avg. Spans	L.L. #/S.F.	Lbs. Per S.F.
Steel Frame Mfg.	1	20′x20′	40	8	Apartments	2-8	20′x20′	40	8
		30′x30′		13		9-25			14
		40′x40′		18	Office	to 10	Various	80	10
Parking garage	4	Various	80	8.5		20			18
Domes	1	200′	30	10		30			26
(Schwedler)*		300′		15		over 50			35

R051223-25 Common Structural Steel Specifications

ASTM A992 (formerly A36, then A572 Grade 50) is the all-purpose carbon grade steel widely used in building and bridge construction.

The other high-strength steels listed below may each have certain advantages over ASTM A992 stuctural carbon steel, depending on the application. They have proven to be economical choices where, due to lighter members, the reduction of dead load and the associated savings in shipping cost can be significant.

ASTM A588 atmospheric weathering, high-strength low-alloy steels can be used in the bare (uncoated) condition, where exposure to normal atmosphere causes a tightly adherant oxide to form on the surface protecting the steel from further oxidation. ASTM A242 corrosion-resistant, high-strength low-alloy steels have enhanced atmospheric corrosion resistance of at least two times that of carbon structural steels with copper, or four times that of carbon structural steels without copper. The reduction or elimination of maintenance resulting from the use of these steels often offsets their higher initial cost.

Steel Type	ASTM Designation	Minimum Yield Stress in KSI	Shapes Available
Carbon	A36	36	All structural shape groups, and plates & bars up thru 8″ thick
	A529	50	Structural shape group 1, and plates & bars up thru 2″ thick
High-Strength Low-Alloy Quenched & Self-Tempered	A913	50	All structural shape groups
		60	
		65	
		70	
High-Strength Low-Alloy Columbium-Vanadium	A572	42	All structural shape groups, and plates & bars up thru 6″ thick
		50	All structural shape groups, and plates & bars up thru 4″ thick
		55	Structural shape groups 1 & 2, and plates & bars up thru 2″ thick
		60	Structural shape groups 1 & 2, and plates & bars up thru 1-1/4″ thick
		65	Structural shape group 1, and plates & bars up thru 1-1/4″ thick
High-Strength Low-Alloy Columbium-Vanadium	A992	50	All structural shape groups
Weathering High-Strength Low-Alloy	A242	42	Structural shape groups 4 & 5, and plates & bars over 1-1/2″ up thru 4″ thick
		46	Structural shape group 3, and plates & bars over 3/4″ up thru 1-1/2″ thick
		50	Structural shape groups 1 & 2, and plates & bars up thru 3/4″ thick
Weathering High-Strength Low-Alloy	A588	42	Plates & bars over 5″ up thru 8″ thick
		46	Plates & bars over 4″ up thru 5″ thick
		50	All structural shape groups, and plates & bars up thru 4″ thick
Quenched and Tempered Low-Alloy	A852	70	Plates & bars up thru 4″ thick
Quenched and Tempered Alloy	A514	90	Plates & bars over 2-1/2″ up thru 6″ thick
		100	Plates & bars up thru 2-1/2″ thick

R051223-30 High Strength Steels

The mill price of high strength steels may be higher than A992 carbon steel but their proper use can achieve overall savings thru total reduced weights. For columns with L/r over 100, A992 steel is best; under 100, high strength steels are economical. For heavy columns, high strength steels are economical when cover plates are eliminated. There is no economy using high strength steels for clip angles or supports or for beams where deflection governs. Thinner members are more economical than thick.

The per ton erection and fabricating costs of the high strength steels will be higher than for A992 since the same number of pieces, but less weight, will be installed.

R051223-35 Common Steel Sections

The upper portion of this table shows the name, shape, common designation, and basic characteristics of commonly used steel sections. The lower portion explains how to read the designations used for the above illustrated common sections.

Shape & Designation	Name & Characteristics	Shape & Designation	Name & Characteristics
W	W Shape Parallel flange surfaces	MC	Miscellaneous Channel Infrequently rolled by some producers
S	American Standard Beam (I Beam) Sloped inner flange	L	Angle Equal or unequal legs, constant thickness
M	Miscellaneous Beams Cannot be classified as W, HP or S; infrequently rolled by some producers	T	Structural Tee Cut from W, M or S on center of web
C	American Standard Channel Sloped inner flange	HP	Bearing Pile Parallel flanges and equal flange and web thickness

Common drawing designations follow:

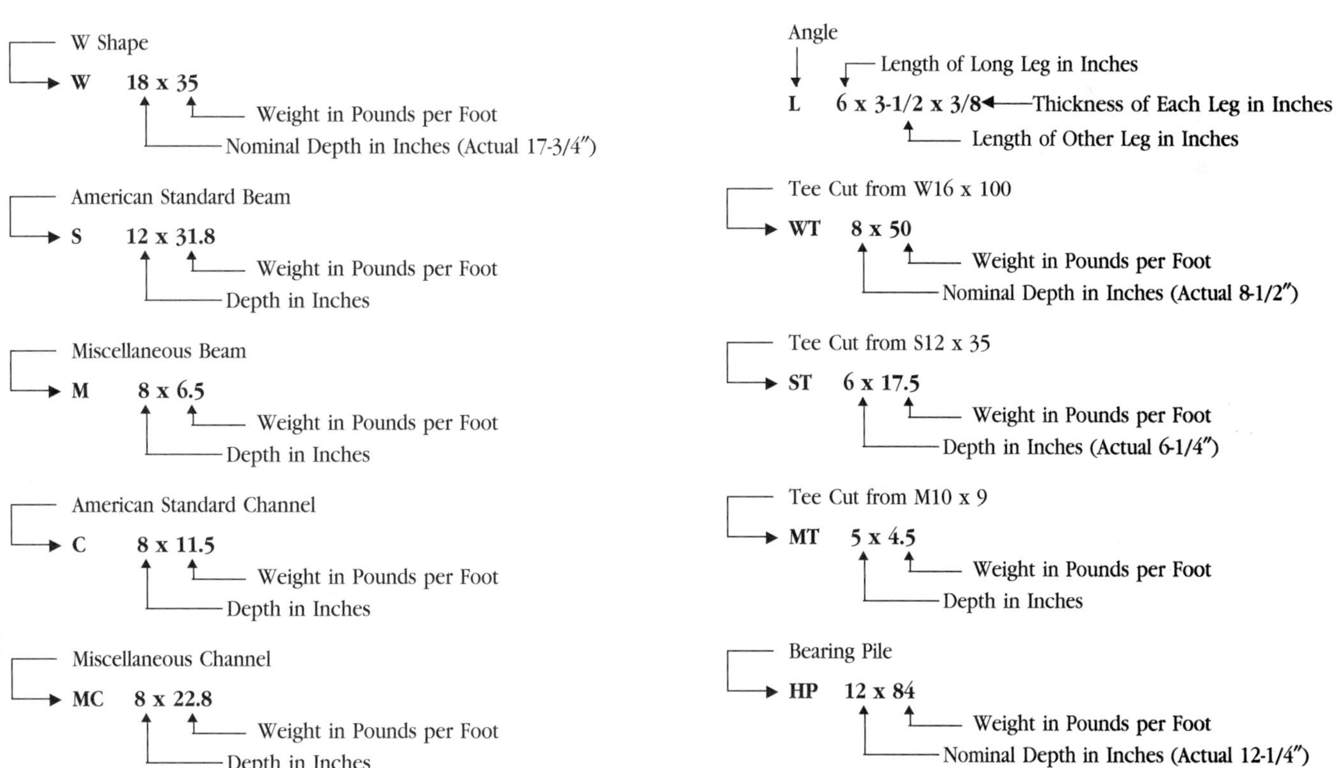

R051223-45 Installation Time for Structural Steel Building Components

The following tables show the expected average installation times for various structural steel shapes. Table A presents installation times for columns, Table B for beams, Table C for light framing and bolts, and Table D for structural steel for various project types.

Table A

Description	Labor-Hours	Unit
Columns		
Steel, Concrete Filled		
3-1/2″ Diameter	.933	Ea.
6-5/8″ Diameter	1.120	Ea.
Steel Pipe		
3″ Diameter	.933	Ea.
8″ Diameter	1.120	Ea.
12″ Diameter	1.244	Ea.
Structural Tubing		
4″ x 4″	.966	Ea.
8″ x 8″	1.120	Ea.
12″ x 8″	1.167	Ea.
W Shape 2 Tier		
W8 x 31	.052	L.F.
W8 x 67	.057	L.F.
W10 x 45	.054	L.F.
W10 x 112	.058	L.F.
W12 x 50	.054	L.F.
W12 x 190	.061	L.F.
W14 x 74	.057	L.F.
W14 x 176	.061	L.F.

Table B

Description	Labor-Hours	Unit	Labor-Hours	Unit
Beams, W Shape				
W6 x 9	.949	Ea.	.093	L.F.
W10 x 22	1.037	Ea.	.085	L.F.
W12 x 26	1.037	Ea.	.064	L.F.
W14 x 34	1.333	Ea.	.069	L.F.
W16 x 31	1.333	Ea.	.062	L.F.
W18 x 50	2.162	Ea.	.088	L.F.
W21 x 62	2.222	Ea.	.077	L.F.
W24 x 76	2.353	Ea.	.072	L.F.
W27 x 94	2.581	Ea.	.067	L.F.
W30 x 108	2.857	Ea.	.067	L.F.
W33 x 130	3.200	Ea.	.071	L.F.
W36 x 300	3.810	Ea.	.077	L.F.

Table C

Description	Labor-Hours	Unit
Light Framing		
Angles 4″ and Larger	.055	lbs.
Less than 4″	.091	lbs.
Channels 8″ and Larger	.048	lbs.
Less than 8″	.072	lbs.
Cross Bracing Angles	.055	lbs.
Rods	.034	lbs.
Hanging Lintels	.069	lbs.
High Strength Bolts in Place		
3/4″ Bolts	.070	Ea.
7/8″ Bolts	.076	Ea.

Table D

Description	Labor-Hours	Unit	Labor-Hours	Unit
Apartments, Nursing Homes, etc.				
1-2 Stories	4.211	Piece	7.767	Ton
3-6 Stories	4.444	Piece	7.921	Ton
7-15 Stories	4.923	Piece	9.014	Ton
Over 15 Stories	5.333	Piece	9.209	Ton
Offices, Hospitals, etc.				
1-2 Stories	4.211	Piece	7.767	Ton
3-6 Stories	4.741	Piece	8.889	Ton
7-15 Stories	4.923	Piece	9.014	Ton
Over 15 Stories	5.120	Piece	9.209	Ton
Industrial Buildings				
1 Story	3.478	Piece	6.202	Ton

R051223-50 Subpurlins

Bulb tee subpurlins are structural members designed to support and reinforce a variety of roof deck systems such as precast cement fiber roof deck tiles, monolithic roof deck systems, and gypsum or lightweight concrete over formboard. Other uses include interstitial service ceiling systems, wall panel systems, and joist anchoring in bond beams. See Unit Price section for pricing on a square foot basis at 32-5/8″ O.C. Maximum span is based on a 3-span condition with a total allowable vertical load of 40 psf.

R051223-80 Dimensions and Weights of Sheet Steel

Gauge No.	Approximate Thickness				Weight		
	Inches (in fractions)	Inches (in decimal parts)		Millimeters			
	Wrought Iron	Wrought Iron	Steel	Steel	per S.F. in Ounces	per S.F. in Lbs.	per Square Meter in Kg.
0000000	1/2″	.5	.4782	12.146	320	20.000	97.650
000000	15/32″	.46875	.4484	11.389	300	18.750	91.550
00000	7/16″	.4375	.4185	10.630	280	17.500	85.440
0000	13/32″	.40625	.3886	9.870	260	16.250	79.330
000	3/8″	.375	.3587	9.111	240	15.000	73.240
00	11/32″	.34375	.3288	8.352	220	13.750	67.130
0	5/16″	.3125	.2989	7.592	200	12.500	61.030
1	9/32″	.28125	.2690	6.833	180	11.250	54.930
2	17/64″	.265625	.2541	6.454	170	10.625	51.880
3	1/4″	.25	.2391	6.073	160	10.000	48.820
4	15/64″	.234375	.2242	5.695	150	9.375	45.770
5	7/32″	.21875	.2092	5.314	140	8.750	42.720
6	13/64″	.203125	.1943	4.935	130	8.125	39.670
7	3/16″	.1875	.1793	4.554	120	7.500	36.320
8	11/64″	.171875	.1644	4.176	110	6.875	33.570
9	5/32″	.15625	.1495	3.797	100	6.250	30.520
10	9/64″	.140625	.1345	3.416	90	5.625	27.460
11	1/8″	.125	.1196	3.038	80	5.000	24.410
12	7/64″	.109375	.1046	2.657	70	4.375	21.360
13	3/32″	.09375	.0897	2.278	60	3.750	18.310
14	5/64″	.078125	.0747	1.897	50	3.125	15.260
15	9/128″	.0713125	.0673	1.709	45	2.813	13.730
16	1/16″	.0625	.0598	1.519	40	2.500	12.210
17	9/160″	.05625	.0538	1.367	36	2.250	10.990
18	1/20″	.05	.0478	1.214	32	2.000	9.765
19	7/160″	.04375	.0418	1.062	28	1.750	8.544
20	3/80″	.0375	.0359	.912	24	1.500	7.324
21	11/320″	.034375	.0329	.836	22	1.375	6.713
22	1/32″	.03125	.0299	.759	20	1.250	6.103
23	9/320″	.028125	.0269	.683	18	1.125	5.490
24	1/40″	.025	.0239	.607	16	1.000	4.882
25	7/320″	.021875	.0209	.531	14	.875	4.272
26	3/160″	.01875	.0179	.455	12	.750	3.662
27	11/640″	.0171875	.0164	.417	11	.688	3.357
28	1/64″	.015625	.0149	.378	10	.625	3.052

R053100-10 Decking Descriptions

General - All Deck Products

Steel deck is made by cold forming structural grade sheet steel into a repeating pattern of parallel ribs. The strength and stiffness of the panels are the result of the ribs and the material properties of the steel. Deck lengths can be varied to suit job conditions, but because of shipping considerations, are usually less than 40 feet. Standard deck width varies with the product used but full sheets are usually 12″, 18″, 24″, 30″, or 36″. Deck is typically furnished in a standard width with the ends cut square. Any cutting for width, such as at openings or for angular fit, is done at the job site.

Deck is typically attached to the building frame with arc puddle welds, self-drilling screws, or powder or pneumatically driven pins. Sheet-to-sheet fastening is done with screws, button punching (crimping), or welds.

Composite Floor Deck

After installation and adequate fastening, floor deck serves several purposes. It (a) acts as a working platform, (b) stabilizes the frame, (c) serves as a concrete form for the slab, and (d) reinforces the slab to carry the design loads applied during the life of the building. Composite decks are distinguished by the presence of shear connector devices as part of the deck. These devices are designed to mechanically lock the concrete and deck together so that the concrete and the deck work together to carry subsequent floor loads. These shear connector devices can be rolled-in embossments, lugs, holes, or wires welded to the panels. The deck profile can also be used to interlock concrete and steel.

Composite deck finishes are either galvanized (zinc coated) or phosphatized/painted. Galvanized deck has a zinc coating on both the top and bottom surfaces. The phosphatized/painted deck has a bare (phosphatized) top surface that will come into contact with the concrete. This bare top surface can be expected to develop rust before the concrete is placed. The bottom side of the deck has a primer coat of paint.

Composite floor deck is normally installed so the panel ends do not overlap on the supporting beams. Shear lugs or panel profile shape often prevent a tight metal to metal fit if the panel ends overlap; the air gap caused by overlapping will prevent proper fusion with the structural steel supports when the panel end laps are shear stud welded.

Adequate end bearing of the deck must be obtained as shown on the drawings. If bearing is actually less in the field than shown on the drawings, further investigation is required.

Roof Deck

Roof deck is not designed to act compositely with other materials. Roof deck acts alone in transferring horizontal and vertical loads into the building frame. Roof deck rib openings are usually narrower than floor deck rib openings. This provides adequate support of rigid thermal insulation board.

Roof deck is typically installed to endlap approximately 2″ over supports. However, it can be butted (or lapped more than 2″) to solve field fit problems. Since designers frequently use the installed deck system as part of the horizontal bracing system (the deck as a diaphragm), any fastening substitution or change should be approved by the designer. Continuous perimeter support of the deck is necessary to limit edge deflection in the finished roof and may be required for diaphragm shear transfer.

Standard roof deck finishes are galvanized or primer painted. The standard factory applied paint for roof deck is a primer paint and is not intended to weather for extended periods of time. Field painting or touching up of abrasions and deterioration of the primer coat or other protective finishes is the responsibility of the contractor.

Cellular Deck

Cellular deck is made by attaching a bottom steel sheet to a roof deck or composite floor deck panel. Cellular deck can be used in the same manner as floor deck. Electrical, telephone, and data wires are easily run through the chase created between the deck panel and the bottom sheet.

When used as part of the electrical distribution system, the cellular deck must be installed so that the ribs line up and create a smooth cell transition at abutting ends. The joint that occurs at butting cell ends must be taped or otherwise sealed to prevent wet concrete from seeping into the cell. Cell interiors must be free of welding burrs, or other sharp intrusions, to prevent damage to wires.

When used as a roof deck, the bottom flat plate is usually left exposed to view. Care must be maintained during erection to keep good alignment and prevent damage.

Cellular deck is sometimes used with the flat plate on the top side to provide a flat working surface. Installation of the deck for this purpose requires special methods for attachment to the frame because the flat plate, now on the top, can prevent direct access to the deck material that is bearing on the structural steel. It may be advisable to treat the flat top surface to prevent slipping.

Cellular deck is always furnished galvanized or painted over galvanized.

Form Deck

Form deck can be any floor or roof deck product used as a concrete form. Connections to the frame are by the same methods used to anchor floor and roof deck. Welding washers are recommended when welding deck that is less than 20 gauge thickness.

Form deck is furnished galvanized, prime painted, or uncoated. Galvanized deck must be used for those roof deck systems where form deck is used to carry a lightweight insulating concrete fill.

Wood, Plastics & Comp. R0611 Wood Framing

R061110-30 Lumber Product Material Prices

The price of forest products fluctuates widely from location to location and from season to season depending upon economic conditions. The bare material prices in the unit cost sections of the book show the National Average material prices in effect Jan. 1 of this book year. It must be noted that lumber prices in general may change significantly during the year.

Availability of certain items depends upon geographic location and must be checked prior to firm-price bidding.

Wood, Plastics & Comp. R0616 Sheathing

R061636-20 Plywood

There are two types of plywood used in construction: interior, which is moisture-resistant but not waterproofed, and exterior, which is waterproofed.

The grade of the exterior surface of the plywood sheets is designated by the first letter: A, for smooth surface with patches allowed; B, for solid surface with patches and plugs allowed; C, which may be surface plugged or may have knot holes up to 1″ wide; and D, which is used only for interior type plywood and may have knot holes up to 2-1/2″ wide. "Structural Grade" is specifically designed for engineered applications such as box beams. All CC & DD grades have roof and floor spans marked on them.

Underlayment-grade plywood runs from 1/4″ to 1-1/4″ thick. Thicknesses 5/8″ and over have optional tongue and groove joints which eliminate the need for blocking the edges. Underlayment 19/32″ and over may be referred to as Sturd-i-Floor.

The price of plywood can fluctuate widely due to geographic and economic conditions.

Typical uses for various plywood grades are as follows:

AA-AD Interior — cupboards, shelving, paneling, furniture

BB Plyform — concrete form plywood

CDX — wall and roof sheathing

Structural — box beams, girders, stressed skin panels

AA-AC Exterior — fences, signs, siding, soffits, etc.

Underlayment — base for resilient floor coverings

Overlaid HDO — high density for concrete forms & highway signs

Overlaid MDO — medium density for painting, siding, soffits & signs

303 Siding — exterior siding, textured, striated, embossed, etc.

Thermal & Moist. Protec. R0752 Modified Bituminous Membrane Roofing

R075213-30 Modified Bitumen Roofing

The cost of modified bitumen roofing is highly dependent on the type of installation that is planned. Installation is based on the type of modifier used in the bitumen. The two most popular modifiers are atactic polypropylene (APP) and styrene butadiene styrene (SBS). The modifiers are added to heated bitumen during the manufacturing process to change its characteristics. A polyethylene, polyester or fiberglass reinforcing sheet is then sandwiched between layers of this bitumen. When completed, the result is a pre-assembled, built-up roof that has increased elasticity and weatherablility. Some manufacturers include a surfacing material such as ceramic or mineral granules, metal particles or sand.

The preferred method of adhering SBS-modified bitumen roofing to the substrate is with hot-mopped asphalt (much the same as built-up roofing). This installation method requires a tar kettle/pot to heat the asphalt, as well as the labor, tools and equipment necessary to distribute and spread the hot asphalt.

The alternative method for applying APP and SBS modified bitumen is as follows. A skilled installer uses a torch to melt a small pool of bitumen off the membrane. This pool must form across the entire roll for proper adhesion. The installer must unroll the roofing at a pace slow enough to melt the bitumen, but fast enough to prevent damage to the rest of the membrane.

Modified bitumen roofing provides the advantages of both built-up and single-ply roofing. Labor costs are reduced over those of built-up roofing because only a single ply is necessary. The elasticity of single-ply roofing is attained with the reinforcing sheet and polymer modifiers. Modifieds have some self-healing characteristics and because of their multi-layer construction, they offer the reliability and safety of built-up roofing.

Thermal & Moist. Protec. R0784 Firestopping

R078413-30 Firestopping

Firestopping is the sealing of structural, mechanical, electrical, and other penetrations through fire-rated assemblies. The basic components of firestop systems are safing insulation and firestop sealant on both sides of wall penetrations and the top side of floor penetrations.

Pipe penetrations are assumed to be through concrete, grout, or joint compound and can be sleeved or unsleeved. Costs for the penetrations and sleeves are not included. An annular space of 1″ is assumed. Escutcheons are not included.

Metallic pipe is assumed to be copper, aluminum, cast iron, or similar metallic material. Insulated metallic pipe is assumed to be covered with a thermal insulating jacket of varying thickness and materials.

Non-metallic pipe is assumed to be PVC, CPVC, FR Polypropylene, or similar plastic piping material. Intumescent firestop sealant or wrap strips are included. Collars on both sides of wall penetrations and a sheet metal plate on the underside of floor penetrations are included.

Ductwork is assumed to be sheet metal, stainless steel or similar metallic material. Duct penetrations are assumed to be through concrete, grout or joint compound. Costs for penetrations and sleeves are not included. An annular space of 1/2″ is assumed.

Multi-trade openings include costs for sheet metal forms, firestop mortar, wrap strips, collars, and sealants as necessary.

Structural penetrations joints are assumed to be 1/2″ or less. CMU walls are assumed to be within 1-1/2″ of metal deck. Drywall walls are assumed to be tight to the underside of metal decking.

Metal panel, glass, or curtain wall systems include a spandrel area of 5′ filled with mineral wool foil-faced insulation. Fasteners and stiffeners are included.

Openings R0813 Metal Doors

R081313-20 Steel Door Selection Guide

Standard steel doors are classified into four levels, as recommended by the Steel Door Institute in the chart below. Each of the four levels offers a range of construction models and designs, to meet architectural requirements for preference and appearance, including full flush, seamless, and stile & rail. Recommended minimum gauge requirements are also included.

For complete standard steel door construction specifications and available sizes, refer to the Steel Door Institute Technical Data Series, ANSI A250.8-98 (SDI-100), and ANSI A250.4-94 Test Procedure and Acceptance Criteria for Physical Endurance of Steel Door and Hardware Reinforcements.

Level		Model	Construction	For Full Flush or Seamless		
				Min. Gauge	Thickness (in)	Thickness (mm)
I	Standard Duty	1 2	Full Flush Seamless	20	0.032	0.8
II	Heavy Duty	1 2	Full Flush Seamless	18	0.042	1.0
III	Extra Heavy Duty	1 2 3	Full Flush Seamless *Stile & Rail	16	0.053	1.3
IV	Maximum Duty	1 2	Full Flush Seamless	14	0.067	1.6

*Stiles & rails are 16 gauge; flush panels, when specified, are 18 gauge

R087110-10 Hardware Finishes

This table describes hardware finishes used throughout the industry. It also shows the base metal and the respective symbols in the three predominate systems of identification. Many of these are used in pricing descriptions in Division Eight.

US"	BMHA*	CDN	Base	Description
US P	600	CP	Steel	Primed for Painting
US 1B	601	C1B	Steel	Bright Black Japanned
US 2C	602	C2C	Steel	Zinc Plated
US 2G	603	C2G	Steel	Zinc Plated
US 3	605	C3	Brass	Bright Brass, Clear Coated
US 4	606	C4	Brass	Satin Brass, Clear Coated
US 5	609	C5	Brass	Satin Brass, Blackened, Satin Relieved, Clear Coated
US 7	610	C7	Brass	Satin Brass, Blackened, Bright Relived, Clear Coated
US 9	611	C9	Bronze	Bright Bronze, Clear Coated
US 10	612	C10	Bronze	Satin Bronze, Clear Coated
US 10A	641	C10A	Steel	Antiqued Bronze, Oiled and Lacquered
US 10B	613	C10B	Bronze	Antiqued Bronze, Oiled
US 11	616	C11	Bronze	Satin Bronze, Blackened, Satin Relieved, Clear Coated
US 14	618	C14	Brass/Bronze	Bright Nickel Plated, Clear Coated
US 15	619	C15	Brass/Bronze	Satin Nickel, Clear Coated
US 15A	620	C15A	Brass/Bronze	Satin Nickel Plated, Blackened, Satin Relieved, Clear Coated
US 17A	621	C17A	Brass/Bronze	Nickel Plated, Blackened, Relieved, Clear Coated
US 19	622	C19	Brass/Bronze	Flat Black Coated
US 20	623	C20	Brass/Bronze	Statuary Bronze, Light
US 20A	624	C20A	Brass/Bronze	Statuary Bronze, Dark
US 26	625	C26	Brass/Bronze	Bright Chromium
US 26D	626	C26D	Brass/Bronze	Satin Chromium
US 20	627	C27	Aluminum	Satin Aluminum Clear
US 28	628	C28	Aluminum	Anodized Dull Aluminum
US 32	629	C32	Stainless Steel	Bright Stainless Steel
US 32D	630	C32D	Stainless Steel	Stainless Steel
US 3	632	C3	Steel	Bright Brass Plated, Clear Coated
US 4	633	C4	Steel	Satin Brass, Clear Coated
US 7	636	C7	Steel	Satin Brass Plated, Blackened, Bright Relieved, Clear Coated
US 9	637	C9	Steel	Bright Bronze Plated, Clear Coated
US 5	638	C5	Steel	Satin Brass Plated, Blackened, Bright Relieved, Clear Coated
US 10	639	C10	Steel	Satin Bronze Plated, Clear Coated
US 10B	640	C10B	Steel	Antique Bronze, Oiled
US 10A	641	C10A	Steel	Antiqued Bronze, Oiled and Lacquered
US 11	643	C11	Steel	Satin Bronze Plated, Blackened, Bright Relieved, Clear Coated
US 14	645	C14	Steel	Bright Nickel Plated, Clear Coated
US 15	646	C15	Steel	Satin Nickel Plated, Clear Coated
US 15A	647	C15A	Steel	Nickel Plated, Blackened, Bright Relieved, Clear Coated
US 17A	648	C17A	Steel	Nickel Plated, Blackened, Relieved, Clear Coated
US 20	649	C20	Steel	Statuary Bronze, Light
US 20A	650	C20A	Steel	Statuary Bronze, Dark
US 26	651	C26	Steel	Bright Chromium Plated
US 26D	652	C26D	Steel	Satin Chromium Plated

* - BMHA Builders Hardware Manufacturing Association
" - US Equivalent
^ - Canadian Equivalent
Japanning is imitating Asian lacquer work

R092910-10 Levels of Gypsum Drywall Finish

In the past, contract documents often used phrases such as "industry standard" and "workmanlike finish" to specify the expected quality of gypsum board wall and ceiling installations. The vagueness of these descriptions led to unacceptable work and disputes.

In order to resolve this problem, four major trade associations concerned with the manufacture, erection, finish and decoration of gypsum board wall and ceiling systems have developed an industry-wide *Recommended Levels of Gypsum Board Finish*.

The finish of gypsum board walls and ceilings for specific final decoration is dependent on a number of factors. A primary consideration is the location of the surface and the degree of decorative treatment desired. Painted and unpainted surfaces in warehouses and other areas where appearance is normally not critical may simply require the taping of wallboard joints and 'spotting' of fastener heads. Blemish-free, smooth, monolithic surfaces often intended for painted and decorated walls and ceilings in habitated structures, ranging from single-family dwellings through monumental buildings, require additional finishing prior to the application of the final decoration.

Other factors to be considered in determining the level of finish of the gypsum board surface are (1) the type of angle of surface illumination (both natural and artificial lighting), and (2) the paint and method of application or the type and finish of wallcovering specified as the final decoration. Critical lighting conditions, gloss paints, and thin wall coverings require a higher level of gypsum board finish than do heavily textured surfaces which are subsequently painted or surfaces which are to be decorated with heavy grade wall coverings.

The following descriptions were developed jointly by the Association of the Wall and Ceiling Industries-International (AWCI), Ceiling & Interior Systems Construction Association (CISCA), Gypsum Association (GA), and Painting and Decorating Contractors of America (PDCA) as a guide.

Level 0: Used in temporary construction or wherever the final decoration has not been determined. Unfinished. No taping, finishing or corner beads are required. Also could be used where non-predecorated panels will be used in demountable-type partitions that are to be painted as a final finish. .

Level 1: Frequently used in plenum areas above ceilings, in attics, in areas where the assembly would generally be concealed, or in building service corridors and other areas not normally open to public view. Some degree of sound and smoke control is provided; in some geographic areas, this level is referred to as "fire-taping," although this level of finish does not typically meet fire-resistant assembly requirements. Where a fire resistance rating is required for the gypsum board assembly, details of construction should be in accordance with reports of fire tests of assemblies that have met the requirements of the fire rating acceptable.

All joints and interior angles shall have tape embedded in joint compound. Accessories are optional at specifier discretion in corridors and other areas with pedestrian traffic. Tape and fastener heads need not be covered with joint compound. Surface shall be free of excess joint compound. Tool marks and ridges are acceptable.

Level 2: It may be specified for standard gypsum board surfaces in garages, warehouse storage, or other similar areas where surface appearance is not of primary importance.

All joints and interior angles shall have tape embedded in joint compound and shall be immediately wiped with a joint knife or trowel, leaving a thin coating of joint compound over all joints and interior angles. Fastener heads and accessories shall be covered with a coat of joint compound. Surface shall be free of excess joint compound. Tool marks and ridges are acceptable.

Level 3: Typically used in areas that are to receive heavy texture (spray or hand applied) finishes before final painting, or where commercial-grade (heavy duty) wall coverings are to be applied as the final decoration. This level of finish should not be used where smooth painted surfaces or where lighter weight wall coverings are specified. The prepared surface shall be coated with a drywall primer prior to the application of final finishes.

All joints and interior angles shall have tape embedded in joint compound and shall be immediately wiped with a joint knife or trowel, leaving a thin coating of joint compound over all joints and interior angles. One additional coat of joint compound shall be applied over all joints and interior angles. Fastener heads and accessories shall be covered with two separate coats of joint compound. All joint compounds shall be smooth and free of tool marks and ridges. The prepared surface shall be covered with a drywall primer prior to the application of the final decoration.

Level 4: This level should be used where residential grade (light duty) wall coverings, flat paints, or light textures are to be applied. The prepared surface shall be coated with a drywall primer prior to the application of final finishes. Release agents for wall coverings are specifically formulated to minimize damage if coverings are subsequently removed.

The weight, texture, and sheen level of the wall covering material selected should be taken into consideration when specifying wall coverings over this level of drywall treatment. Joints and fasteners must be sufficiently concealed if the wall covering material is lightweight, contains limited pattern, has a glossy finish, or has any combination of these features. In critical lighting areas, flat paints applied over light textures tend to reduce joint photographing. Gloss, semi-gloss, and enamel paints are not recommended over this level of finish.

All joints and interior angles shall have tape embedded in joint compound and shall be immediately wiped with a joint knife or trowel, leaving a thin coating of joint compound over all joints and interior angles. In addition, two separate coats of joint compound shall be applied over all flat joints and one separate coat of joint compound applied over interior angles. Fastener heads and accessories shall be covered with three separate coats of joint compound. All joint compounds shall be smooth and free of tool marks and ridges. The prepared surface shall be covered with a drywall primer like Sheetrock first coat prior to the application of the final decoration.

Level 5: The highest quality finish is the most effective method to provide a uniform surface and minimize the possibility of joint photographing and of fasteners showing through the final decoration. This level of finish is required where gloss, semi-gloss, or enamel is specified; when flat joints are specified over an untextured surface; or where critical lighting conditions occur. The prepared surface shall be coated with a drywall primer prior to the application of final decoration.

All joints and interior angles shall have tape embedded in joint compound and be immediately wiped with a joint knife or trowel, leaving a thin coating of joint compound over all joints and interior angles. Two separate coats of joint compound shall be applied over all flat joints and one separate coat of joint compound applied over interior angles. Fastener heads and accessories shall be covered with three separate coats of joint compound.

A thin skim coat of joint compound shall be trowel applied to the entire surface. Excess compound is immediately troweled off, leaving a film or skim coating of compound completely covering the paper. As an alternative to a skim coat, a material manufactured especially for this purpose may be applied such as Sheetrock Tuff-Hide primer surfacer. The surface must be smooth and free of tool marks and ridges. The prepared surface shall be covered with a drywall primer prior to the application of the final decoration.

R099100-10 Painting Estimating Techniques

Proper estimating methodology is needed to obtain an accurate painting estimate. There is no known reliable shortcut or square foot method. The following steps should be followed:

- List all surfaces to be painted, with an accurate quantity (area) of each. Items having similar surface condition, finish, application method and accessibility may be grouped together.
- List all the tasks required for each surface to be painted, including surface preparation, masking, and protection of adjacent surfaces. Surface preparation may include minor repairs, washing, sanding and puttying.
- Select the proper Means line for each task. Review and consider all adjustments to labor and materials for type of paint and location of work. Apply the height adjustment carefully. For instance, when applying the adjustment for work over 8' high to a wall that is 12' high, apply the adjustment only to the area between 8' and 12' high, and not to the entire wall.

When applying more than one percent (%) adjustment, apply each to the base cost of the data, rather than applying one percentage adjustment on top of the other.

When estimating the cost of painting walls and ceilings remember to add the brushwork for all cut-ins at inside corners and around windows and doors as a LF measure. One linear foot of cut-in with brush equals one square foot of painting.

All items for spray painting include the labor for roll-back.

Deduct for openings greater than 100 SF, or openings that extend from floor to ceiling and are greater than 5' wide. Do not deduct small openings.

The cost of brushes, rollers, ladders and spray equipment are considered to be part of a painting contractor's overhead, and should not be added to the estimate. The cost of rented equipment such as scaffolding and swing staging should be added to the estimate.

R099100-20 Painting

Item	Coat	One Gallon Covers			In 8 Hours a Laborer Covers			Labor-Hours per 100 S.F.		
		Brush	Roller	Spray	Brush	Roller	Spray	Brush	Roller	Spray
Paint wood siding	prime	250 S.F.	225 S.F.	290 S.F.	1150 S.F.	1300 S.F.	2275 S.F.	.695	.615	.351
	others	270	250	290	1300	1625	2600	.615	.492	.307
Paint exterior trim	prime	400	—	—	650	—	—	1.230	—	—
	1st	475	—	—	800	—	—	1.000	—	—
	2nd	520	—	—	975	—	—	.820	—	—
Paint shingle siding	prime	270	255	300	650	975	1950	1.230	.820	.410
	others	360	340	380	800	1150	2275	1.000	.695	.351
Stain shingle siding	1st	180	170	200	750	1125	2250	1.068	.711	.355
	2nd	270	250	290	900	1325	2600	.888	.603	.307
Paint brick masonry	prime	180	135	160	750	800	1800	1.066	1.000	.444
	1st	270	225	290	815	975	2275	.981	.820	.351
	2nd	340	305	360	815	1150	2925	.981	.695	.273
Paint interior plaster or drywall	prime	400	380	495	1150	2000	3250	.695	.400	.246
	others	450	425	495	1300	2300	4000	.615	.347	.200
Paint interior doors and windows	prime	400	—	—	650	—	—	1.230	—	—
	1st	425	—	—	800	—	—	1.000	—	—
	2nd	450	—	—	975	—	—	.820	—	—

Special Construction R1334 Fabricated Engineered Structures

R133419-10 Pre-Engineered Steel Buildings

These buildings are manufactured by many companies and normally erected by franchised dealers throughout the U.S. The four basic types are: Rigid Frames, Truss type, Post and Beam and the Sloped Beam type. Most popular roof slope is low pitch of 1″ in 12″. The minimum economical area of these buildings is about 3000 S.F. of floor area. Bay sizes are usually 20′ to 24′ but can go as high as 30′ with heavier girts and purlins. Eave heights are usually 12′ to 24′ with 18′ to 20′ most typical.

Material prices shown in the Unit Price section are bare costs for the building shell only and do not include floors, foundations, anchor bolts, interior finishes or utilities. Costs assume at least three bays of 24′ each, a 1″ in 12″ roof slope, and they are based on 30 psf roof load and 20 psf wind load (wind load is a function of wind speed, building height, and terrain characteristics; this should be determined by a Registered Structural Engineer) and no unusual requirements. Costs include the structural frame, 26 ga. non-insulated colored corrugated or ribbed roofing and siding panels, fasteners, closures, trim and flashing but no allowance for insulation, doors, windows, skylights, gutters or downspouts. Very large projects would generally cost less for materials than the prices shown. For roof panel substitutions and wall panel substitutions, see appropriate Unit Price sections.

Conditions at the site, weather, shape and size of the building, and labor availability will affect the erection cost of the building.

R142000-10 Freight Elevators

Capacities run from 2,000 lbs. to over 100,000 lbs. with 3,000 lbs. to 10,000 lbs. most common. Travel speeds are generally lower and control less intricate than on passenger elevators. Costs in the Unit Price sections are for hydraulic and geared elevators.

R142000-20 Elevator Selective Costs See R142000-40 for cost development.

	Passenger		Freight		Hospital	
A. Base Unit	Hydraulic	Electric	Hydraulic	Electric	Hydraulic	Electric
Capacity	1,500 lb.	2,000 lb.	2,000 lb.	4,000 lb.	4,000 lb.	4,000 lb.
Speed	100 F.P.M.	200 F.P.M.	100 F.P.M.	200 F.P.M.	100 F.P.M.	200 F.P.M.
#Stops/Travel Ft.	2/12	4/40	2/20	4/40	2/20	4/40
Push Button Oper.	Yes	Yes	Yes	Yes	Yes	Yes
Telephone Box & Wire	"	"	"	"	"	"
Emergency Lighting	"	"	No	No	"	"
Cab	Plastic Lam. Walls	Plastic Lam. Walls	Painted Steel	Painted Steel	Plastic Lam. Walls	Plastic Lam. Walls
Cove Lighting	Yes	Yes	No	No	Yes	Yes
Floor	V.C.T.	V.C.T.	Wood w/Safety Treads	Wood w/Safety Treads	V.C.T.	V.C.T.
Doors, & Speedside Slide	Yes	Yes	Yes	Yes	Yes	Yes
Gates, Manual	No	No	No	No	No	No
Signals, Lighted Buttons	Car and Hall	Car and Hall	Car and Hall	Car and Hall	Car and Hall	Car and Hall
O.H. Geared Machine	N.A.	Yes	N.A.	Yes	N.A.	Yes
Variable Voltage Contr.	"	"	N.A.	"	"	"
Emergency Alarm	Yes	"	Yes	"	Yes	"
Class "A" Loading	N.A.	N.A.	"	"	N.A.	N.A.

R142000-30 Passenger Elevators

Electric elevators are used generally but hydraulic elevators can be used for lifts up to 70′ and where large capacities are required. Hydraulic speeds are limited to 200 F.P.M. but cars are self leveling at the stops. On low rises, hydraulic installation runs about 15% less than standard electric types but on higher rises this installation cost advantage is reduced. Maintenance of hydraulic elevators is about the same as electric type but underground portion is not included in the maintenance contract.

In electric elevators there are several control systems available, the choice of which will be based upon elevator use, size, speed and cost criteria. The two types of drives are geared for low speeds and gearless for 450 F.P.M. and over.

The tables above illustrate typical installed costs of the various types of elevators available.

R142000-40 Elevator Cost Development

To price a new car or truck from the factory, you must start with the manufacturer's basic model, then add or exchange optional equipment and features. The same is true for pricing elevators.

Example:

Requirement: One-passenger elevator, five-story hydraulic, 2,500 lb. capacity, 12′ floor to floor, speed 150 F.P.M., emergency power switching and maintenance contract.

Description	Adjustment
A. Base Elevator: Hydraulic Passenger, 1500 lb. Capacity, 100 fpm, 2 Stops, Standard Finish	1 Ea.
B. Capacity Adjustment (2,500 lb.)	1 Ea.
C. Excess Travel Adjustment: 48′ Total Travel (4 x 12′) minus 12′ Base Unit Travel =	36 V.L.F.
D. Stops Adjustment: 5 Total Stops minus 2 Stops (Base Unit) =	3 Stops
E. Speed Adjustment (150 F.P.M.)	1 Ea.
F. Options:	
1. Intercom Service	1 Ea.
2. Emergency Power Switching, Automatic	1 Ea.
3. Stainless Steel Entrance Doors	5 Ea.
4. Maintenance Contract (12 Months)	1 Ea.
5. Position Indicator for main floor level (none indicated in Base Unit)	1 Ea.

Conveying Equipment — R1432 Moving Walks

R143210-20 Moving Ramps and Walks

These are a specialized form of conveyor 3′ to 6′ wide with capacities of 3,600 to 18,000 persons per hour. Maximum speed is 140 F.P.M. and normal incline is 0° to 15°.

Local codes will determine the maximum angle. Outdoor units would require additional weather protection.

Fire Suppression — R2112 Fire-Suppression Standpipes

R211226-10 Standpipe Systems

The basis for standpipe system design is National Fire Protection Association NFPA 14, however, the authority having jurisdiction should be consulted for special conditions, local requirements and approval.

Standpipe systems, properly designed and maintained, are an effective and valuable time saving aid for extinquishing fires, especially in the upper stories of tall buildings, the interior of large commercial or industrial malls, or other areas where construction features or access make the laying of temporary hose lines time consuming and/or hazardous. Standpipes are frequently installed with automatic sprinkler systems for maximum protection.

There are three general classes of service for standpipe systems:
Class I for use by fire departments and personnel with special training for heavy streams (2-1/2″ hose connections).
Class II for use by building occupants until the arrival of the fire department (1-1/2″ hose connector with hose).
Class III for use by either fire departments and trained personnel or by the building occupants (both 2-1/2″ and 1-1/2″ hose connections or one 2-1/2″ hose valve with an easily removable 2-1/2″ by 1-1/2″ adapter).

Standpipe systems are also classified by the way water is supplied to the system. The four basic types are:
Type 1: Wet standpipe system having supply valve open and water pressure maintained at all times.
Type 2: Standpipe system so arranged through the use of approved devices as to admit water to the system automatically by opening a hose valve.
Type 3: Standpipe system arranged to admit water to the system through manual operation of approved remote control devices located at each hose station.
Type 4: Dry standpipe having no permanent water supply.

R211226-20 NFPA 14 Basic Standpipe Design

Class	Design-Use	Pipe Size Minimums	Water Supply Minimums
Class I	2 1/2″ hose connection on each floor All areas within 150′ of an exit in every exit stairway Fire Department Trained Personnel	Height to 100′, 4″ dia. Heights above 100′, 6″ dia. (275′ max. except with pressure regulators 400′ max.)	For each standpipe riser 500 GPM flow For common supply pipe allow 500 GPM for first standpipe plus 250 GPM for each additional standpipe (2500 GPM max. total) 30 min. duration 65 PSI at 500 GPM
Class II	1 1/2″ hose connection with hose on each floor All areas within 130′ of hose connection measured along path of hose travel Occupant personnel	Height to 50′, 2″ dia. Height above 50′, 2 1/2″ dia.	For each standpipe riser 100 GPM flow For multiple riser common supply pipe 100 GPM 300 min. duration, 65 PSI at 100 GPM
Class III	Both of above. Class I valved connections will meet Class III with additional 2 1/2″ by 1 1/2″ adapter and 1 1/2″ hose.	Same as Class I	Same as Class I

*Note: Where 2 or more standpipes are installed in the same building or section of building they shall be interconnected at the bottom.

Combined Systems

Combined systems are systems where the risers supply both automatic sprinklers and 2-1/2″ hose connection outlets for fire department use. In such a system the sprinkler spacing pattern shall be in accordance with NFPA 13 while the risers and supply piping will be sized in accordance with NFPA 14. When the building is completely sprinklered the risers may be sized by hydraulic calculation.The minimum size riser for buildings not completely sprinklered is 6″.

The minimum water supply of a completely sprinklered, light hazard, high-rise occupancy building will be 500 GPM while the supply required for other types of completely sprinklered high-rise buildings is 1000 GPM.

General System Requirements

1. Approved valves will be provided at the riser for controlling branch lines to hose outlets.
2. A hose valve will be provided at each outlet for attachment of hose.
3. Where pressure at any standpipe outlet exceeds 100 PSI a pressure reducer must be installed to limit the pressure to 100 PSI. Note that the pressure head due to gravity in 100′ of riser is 43.4 PSI. This must be overcome by city pressure, fire pumps, or gravity tanks to provide adequate pressure at the top of the riser.
4. Each hose valve on a wet system having linen hose shall have an automatic drip connection to prevent valve leakage from entering the hose.
5. Each riser will have a valve to isolate it from the rest of the system.
6. One or more fire department connections as an auxiliary supply shall be provided for each Class I or Class III standpipe system. In buildings having two or more zones, a connection will be provided for each zone.
7. There will be no shutoff valve in the fire department connection, but a check valve will be located in the line before it joins the system.
8. All hose connections street side will be identified on a cast plate or fitting as to purpose.

R211313-10 Sprinkler Systems (Automatic)

Sprinkler systems may be classified by type as follows:

1. **Wet Pipe System.** A system employing automatic sprinklers attached to a piping system containing water and connected to a water supply so that water discharges immediately from sprinklers opened by a fire.
2. **Dry Pipe System.** A system employing automatic sprinklers attached to a piping system containing air under pressure, the release of which as from the opening of sprinklers permits the water pressure to open a valve known as a "dry pipe valve". The water then flows into the piping system and out the opened sprinklers.
3. **Pre-Action System.** A system employing automatic sprinklers attached to a piping system containing air that may or may not be under pressure, with a supplemental heat responsive system of generally more sensitive characteristics than the automatic sprinklers themselves, installed in the same areas as the sprinklers; actuation of the heat responsive system, as from a fire, opens a valve which permits water to flow into the sprinkler piping system and to be discharged from any sprinklers which may be open.
4. **Deluge System.** A system employing open sprinklers attached to a piping system connected to a water supply through a valve which is opened by the operation of a heat responsive system installed in the same areas as the sprinklers. When this valve opens, water flows into the piping system and discharges from all sprinklers attached thereto.
5. **Combined Dry Pipe and Pre-Action Sprinkler System.** A system employing automatic sprinklers attached to a piping system containing air under pressure with a supplemental heat responsive system of generally more sensitive characteristics than the automatic sprinklers themselves, installed in the same areas as the sprinklers; operation of the heat responsive system, as from a fire, actuates tripping devices which open dry pipe valves simultaneously and without loss of air pressure in the system. Operation of the heat responsive system also opens approved air exhaust valves at the end of the feed main which facilitates the filling of the system with water which usually precedes the opening of sprinklers. The heat responsive system also serves as an automatic fire alarm system.
6. **Limited Water Supply System.** A system employing automatic sprinklers and conforming to these standards but supplied by a pressure tank of limited capacity.
7. **Chemical Systems.** Systems using halon, carbon dioxide, dry chemical or high expansion foam as selected for special requirements. Agent may extinguish flames by chemically inhibiting flame propagation, suffocate flames by excluding oxygen, interrupting chemical action of oxygen uniting with fuel or sealing and cooling the combustion center.
8. **Firecycle System.** Firecycle is a fixed fire protection sprinkler system utilizing water as its extinguishing agent. It is a time delayed, recycling, preaction type which automatically shuts the water off when heat is reduced below the detector operating temperature and turns the water back on when that temperature is exceeded. The system senses a fire condition through a closed circuit electrical detector system which controls water flow to the fire automatically. Batteries supply up to 90 hour emergency power supply for system operation. The piping system is dry (until water is required) and is monitored with pressurized air. Should any leak in the system piping occur, an alarm will sound, but water will not enter the system until heat is sensed by a firecycle detector.

Area coverage sprinkler systems may be laid out and fed from the supply in any one of several patterns as shown below. It is desirable, if possible, to utilize a central feed and achieve a shorter flow path from the riser to the furthest sprinkler. This permits use of the smallest sizes of pipe possible with resulting savings.

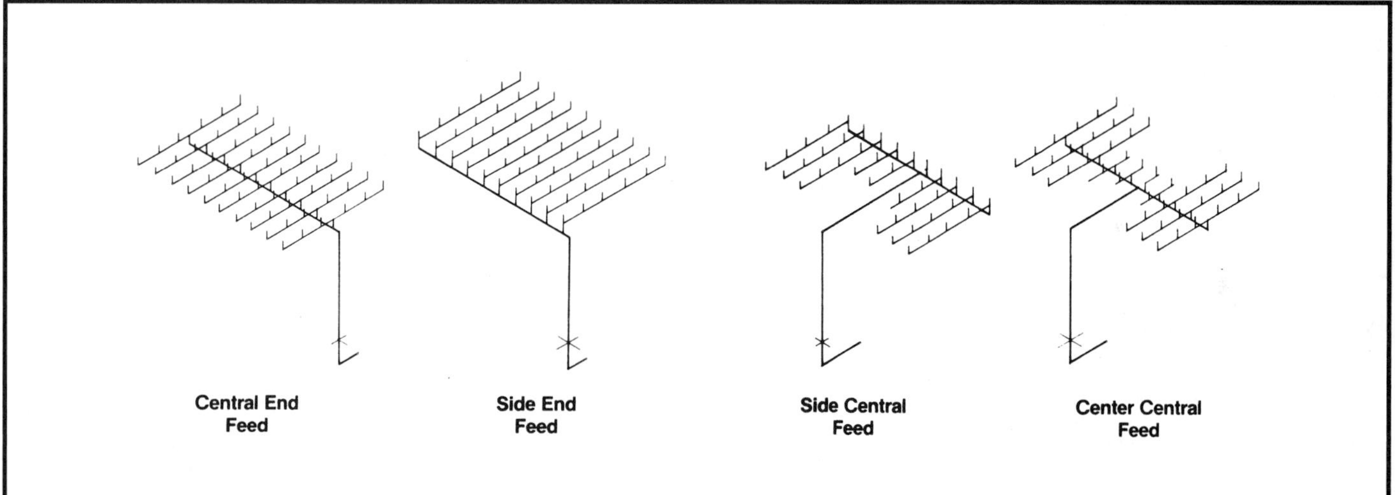

R211313-20 System Classification

System Classification

Rules for installation of sprinkler systems vary depending on the classification of occupancy falling into one of three categories as follows:

Light Hazard Occupancy

The protection area allotted per sprinkler should not exceed 225 S.F., with the maximum distance between lines and sprinklers on lines being 15′. The sprinklers do not need to be staggered. Branch lines should not exceed eight sprinklers on either side of a cross main. Each large area requiring more than 100 sprinklers and without a sub-dividing partition should be supplied by feed mains or risers sized for ordinary hazard occupancy.
Maximum system area = 52,000 S.F.

Included in this group are:

Churches
Clubs
Educational
Hospitals
Institutional
Libraries
(except large stack rooms)
Museums
Nursing Homes
Offices
Residential
Restaurants
Theaters and Auditoriums
(except stages and prosceniums)
Unused Attics

Ordinary Hazard Occupancy

The protection area allotted per sprinkler shall not exceed 130 S.F. of noncombustible ceiling and 130 S.F. of combustible ceiling. The maximum allowable distance between sprinkler lines and sprinklers on line is 15′. Sprinklers shall be staggered if the distance between heads exceeds 12′. Branch lines should not exceed eight sprinklers on either side of a cross main.
Maximum system area = 52,000 S.F.

Included in this group are:

Group 1
Automotive Parking and Showrooms
Bakeries
Beverage manufacturing
Canneries
Dairy Products Manufacturing/Processing
Electronic Plans
Glass and Glass Products Manufacturing
Laundries
Restaurant Service Areas

Group 2
Cereal Mills
Chemical Plants—Ordinary
Confectionery Products
Distilleries
Dry Cleaners
Feed Mills
Horse Stables
Leather Goods Manufacturing
Libraries—Large Stack Room Areas
Machine Shops
Metal Working
Mercantile
Paper and Pulp Mills
Paper Process Plants
Piers and Wharves
Post Offices
Printing and Publishing
Repair Garages
Stages
Textile Manufacturing
Tire Manufacturing
Tobacco Products Manufacturing
Wood Machining
Wood Product Assembly

Extra Hazard Occupancy

The protection area allotted per sprinkler shall not exceed 100 S.F. of noncombustible ceiling and 100 S.F. of combustible ceiling. The maximum allowable distance between lines and between sprinklers on lines is 12′. Sprinklers on alternate lines shall be staggered if the distance between sprinklers on lines exceeds 8′. Branch lines should not exceed six sprinklers on either side of a cross main.
Maximum system area:
Design by pipe schedule = 25,000 S.F.
Design by hydraulic calculation = 40,000 S.F.

Included in this group are:

Group 1
Aircraft hangars
Combustible Hydraulic Fluid Use Area
Die Casting
Metal Extruding
Plywood/Particle Board Manufacturing
Printing (inks with flash points < 100 degrees F)
Rubber Reclaiming, Compounding, Drying, Milling, Vulcanizing
Saw Mills
Textile Picking, Opening, Blending, Garnetting, Carding, Combing of Cotton, Synthetics, Wood Shoddy, or Burlap
Upholstering with Plastic Foams

Group 2
Asphalt Saturating
Flammable Liquids Spraying
Flow Coating
Manufactured/Modular Home Building Assemblies (where finished enclosure is present and has combustible interiors)
Open Oil Quenching
Plastics Processing
Solvent Cleaning
Varnish and Paint Dipping

Fire Suppression — R2113 Fire-Suppression Sprinkler Systems

R211313-40 Adjustment for Sprinkler/Standpipe Installations

Quality/Complexity Multiplier (For all installations)

Economy installation, add	0 to 5%
Good quality, medium complexity, add	5 to 15%
Above average quality and complexity, add	15 to 25%

Plumbing — R2201 Operation & Maintenance of Plumbing

R220102-20 Labor Adjustment Factors

Labor Adjustment Factors are provided for Divisions 21, 22, and 23 to assist the mechanical estimator account for the various complexities and special conditions of any particular project. While a single percentage has been entered on each line of Division 22 01 02.20, it should be understood that these are just suggested midpoints of ranges of values commonly used by mechanical estimators. They may be increased or decreased depending on the severity of the special conditions.

The group for "existing occupied buildings", has been the subject of requests for explanation. Actually there are two stages to this group: buildings that are existing and "finished" but unoccupied, and those that also are occupied. Buildings that are "finished" may result in higher labor costs due to the workers having to be more careful not to damage finished walls, ceilings, floors, etc. and may necessitate special protective coverings and barriers. Also corridor bends and doorways may not accommodate long pieces of pipe or larger pieces of equipment. Work above an already hung ceiling can be very time consuming. The addition of occupants may force the work to be done on premium time (nights and/or weekends), eliminate the possible use of some preferred tools such as pneumatic drivers, powder charged drivers etc. The estimator should evaluate the access to the work area and just how the work is going to be accomplished to arrive at an increase in labor costs over "normal" new construction productivity.

R220105-10 Demolition (Selective vs. Removal for Replacement)

Demolition can be divided into two basic categories.

One type of demolition involves the removal of material with no concern for its replacement. The labor-hours to estimate this work are found under "Selective Demolition" in the Fire Protection, Plumbing and HVAC Divisions. It is selective in that individual items or all the material installed as a system or trade grouping such as plumbing or heating systems are removed. This may be accomplished by the easiest way possible, such as sawing, torch cutting, or sledge hammer as well as simple unbolting.

The second type of demolition is the removal of some item for repair or replacement. This removal may involve careful draining, opening of unions, disconnecting and tagging of electrical connections, capping of pipes/ducts to prevent entry of debris or leakage of the material contained as well as transport of the item away from its in-place location to a truck/dumpster. An approximation of the time required to accomplish this type of demolition is to use half of the time indicated as necessary to install a new unit. For example; installation of a new pump might be listed as requiring 6 labor-hours so if we had to estimate the removal of the old pump we would allow an additional 3 hours for a total of 9 hours. That is, the complete replacement of a defective pump with a new pump would be estimated to take 9 labor-hours.

Plumbing — R2205 Common Work Results for Plumbing

R220523-90 Valve Selection Considerations

INTRODUCTION: In any piping application, valve performance is critical. Valves should be selected to give the best performance at the lowest cost.

The following is a list of performance characteristics generally expected of valves.

1. Stopping flow or starting it.
2. Throttling flow (Modulation).
3. Flow direction changing.
4. Checking backflow (Permitting flow in only one direction).
5. Relieving or regulating pressure.

In order to properly select the right valve, some facts must be determined.

A. What liquid or gas will flow through the valve?
B. Does the fluid contain suspended particles?
C. Does the fluid remain in liquid form at all times?
D. Which metals does fluid corrode?
E. What are the pressure and temperature limits? (As temperature and pressure rise, so will the price of the valve.)
F. Is there constant line pressure?
G. Is the valve merely an on-off valve?
H. Will checking of backflow be required?
I. Will the valve operate frequently or infrequently?

Valves are classified by design type into such classifications as Gate, Globe, Angle, Check, Ball, Butterfly and Plug. They are also classified by end connection, stem, pressure restrictions and material such as bronze, cast iron, etc. Each valve has a specific use. A quality valve used correctly will provide a lifetime of trouble-free service, but a high quality valve installed in the wrong service may require frequent attention.

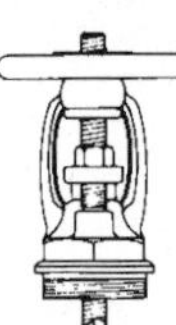

STEM TYPES
(OS & Y)—Rising Stem-Outside Screw and Yoke

Offers a visual indication of whether the valve is open or closed. Recommended where high temperatures, corrosives, and solids in the line might cause damage to inside-valve stem threads. The stem threads are engaged by the yoke bushing so the stem rises through the hand wheel as it is turned.

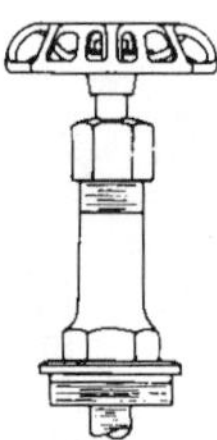

(R.S.)—Rising Stem-Inside Screw

Adequate clearance for operation must be provided because both the hand wheel and the stem rise.
The valve wedge position is indicated by the position of the stem and hand wheel.

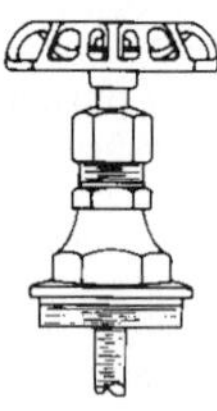

(N.R.S.)—Non-Rising Stem-Inside Screw

A minimum clearance is required for operating this type of valve. Excessive wear or damage to stem threads inside the valve may be caused by heat, corrosion, and solids. Because the hand wheel and stem do not rise, wedge position cannot be visually determined.

VALVE TYPES
Gate Valves

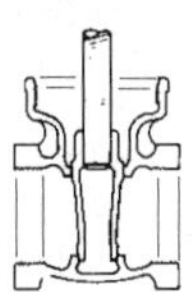

Provide full flow, minute pressure drop, minimum turbulence and minimum fluid trapped in the line.
They are normally used where operation is infrequent.

Globe Valves

Globe valves are designed for throttling and/or frequent operation with positive shut-off. Particular attention must be paid to the several types of seating materials available to avoid unnecessary wear. The seats must be compatible with the fluid in service and may be composition or metal. The configuration of the Globe valve opening causes turbulence which results in increased resistance. Most bronze Globe valves are rising stem-inside screw, but they are also available on O.S. & Y.

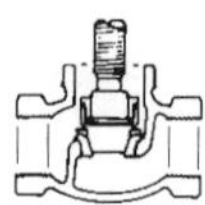

Angle Valves

The fundamental difference between the Angle valve and the Globe valve is the fluid flow through the Angle valve. It makes a 90° turn and offers less resistance to flow than the Globe valve while replacing an elbow. An Angle valve thus reduces the number of joints and installation time.

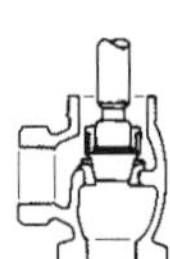

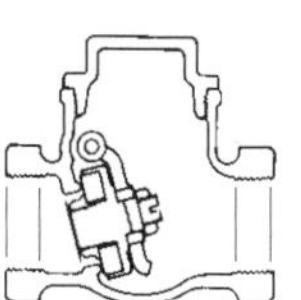

Check Valves

Check valves are designed to prevent backflow by automatically seating when the direction of fluid is reversed.
Swing Check valves are generally installed with Gate-valves, as they provide comparable full flow. Usually recommended for lines where flow velocities are low and should not be used on lines with pulsating flow. Recommended for horizontal installation, or in vertical lines only where flow is upward.

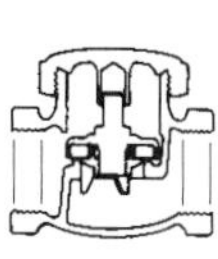

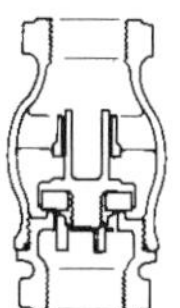

Lift Check Valves

These are commonly used with Globe and Angle valves since they have similar diaphragm seating arrangements and are recommended for preventing backflow of steam, air, gas and water, and on vapor lines with high flow velocities. For horizontal lines, horizontal lift checks should be used and vertical lift checks for vertical lines.

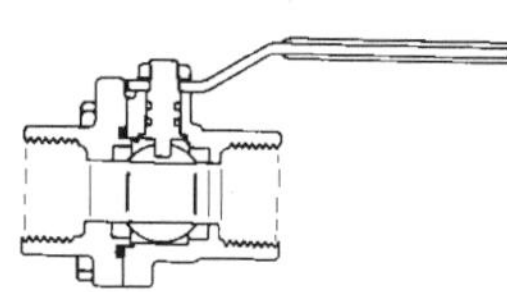

Ball Valves

Ball valves are light and easily installed, yet because of modern elastomeric seats, provide tight closure. Flow is controlled by rotating up to 90° a drilled ball which fits tightly against resilient seals. This ball seats with flow in either direction, and valve handle indicates the degree of opening. Recommended for frequent operation readily adaptable to automation, ideal for installation where space is limited.

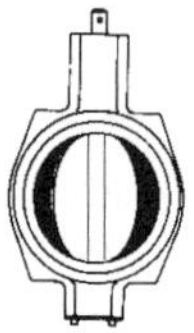

Butterfly Valves

Butterfly valves provide bubble-tight closure with excellent throttling characteristics. They can be used for full-open, closed and for throttling applications.

The Butterfly valve consists of a disc within the valve body which is controlled by a shaft. In its closed position, the valve disc seals against a resilient seat. The disc position throughout the full 90° rotation is visually indicated by the position of the operator.

A Butterfly valve is only a fraction of the weight of a Gate valve and requires no gaskets between flanges in most cases. Recommended for frequent operation and adaptable to automation where space is limited.

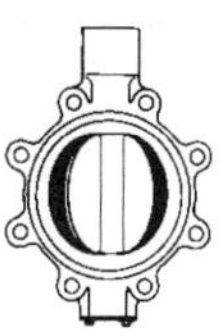

Wafer and Lug type bodies when installed between two pipe flanges, can be easily removed from the line. The pressure of the bolted flanges holds the valve in place.
Locating lugs makes installation easier.

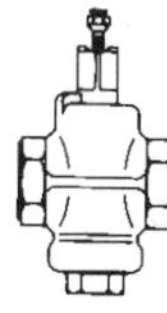

Plug Valves

Lubricated plug valves, because of the wide range of service to which they are adapted, may be classified as all purpose valves. They can be safely used at all pressure and vacuums, and at all temperatures up to the limits of available lubricants. They are the most satisfactory valves for the handling of gritty suspensions and many other destructive, erosive, corrosive and chemical solutions.

R221113-40 Plumbing Approximations for Quick Estimating

Water Control

Water Meter; Backflow Preventer, Shock Absorbers; Vacuum Breakers; Mixer. .... 10 to 15% of Fixtures

Pipe And Fittings .... 30 to 60% of Fixtures

Note: Lower percentage for compact buildings or larger buildings with plumbing in one area.
Larger percentage for large buildings with plumbing spread out.
In extreme cases pipe may be more than 100% of fixtures.
Percentages **do not** include special purpose or process piping.

Plumbing Labor

1 & 2 Story Residential .... Rough-in Labor = 80% of Materials

Apartment Buildings .... Rough-in Labor = 90 to 100% of Materials

Labor for handling and placing fixtures is approximately 25 to 30% of fixtures

Quality/Complexity Multiplier (for all installations)

Economy installation, add .... 0 to 5%

Good quality, medium complexity, add .... 5 to 15%

Above average quality and complexity, add .... 15 to 25%

R221113-50 Pipe Material Considerations

1. Malleable fittings should be used for gas service.
2. Malleable fittings are used where there are stresses/strains due to expansion and vibration.
3. Cast fittings may be broken as an aid to disassembling of heating lines frozen by long use, temperature and minerals.
4. Cast iron pipe is extensively used for underground and submerged service.
5. Type M (light wall) copper tubing is available in hard temper only and is used for nonpressure and less severe applications than K and L.
6. Type L (medium wall) copper tubing, available hard or soft for interior service.
7. Type K (heavy wall) copper tubing, available in hard or soft temper for use where conditions are severe. For underground and interior service.
8. Hard drawn tubing requires fewer hangers or supports but should not be bent. Silver brazed fittings are recommended, however soft solder is normally used.
9. Type DMV (very light wall) copper tubing designed for drainage, waste and vent plus other non-critical pressure services.

Domestic/Imported Pipe and Fittings Cost

The prices shown in this publication for steel/cast iron pipe and steel, cast iron, malleable iron fittings are based on domestic production sold at the normal trade discounts. The above listed items of foreign manufacture may be available at prices of 1/3 to 1/2 those shown. Some imported items after minor machining or finishing operations are being sold as domestic to further complicate the system.

Caution: Most pipe prices in this book also include a coupling and pipe hangers, which, for the larger sizes, can add significantly to the per-foot cost and should be taken into account when comparing "book cost" with quoted supplier's cost.

R221113-70 Piping to 10′ High

When taking off pipe, it is important to identify the different material types and joining procedures, as well as distances between supports and components required for proper support.

During the takeoff, measure through all fittings. Do not subtract the lengths of the fittings, valves, or strainers, etc. This added length plus the final rounding of the totals will compensate for nipples and waste.

When rounding off totals always increase the actual amount to correspond with manufacturer's shipping lengths.

A. Both red brass and yellow brass pipe are normally furnished in 12′ lengths, plain end. The Unit Price section includes in the linear foot costs two field threads and one coupling per 10′ length. A carbon steel clevis type hanger assembly every 10′ is also prorated into the linear foot costs, including both material and labor.

B. Cast iron soil pipe is furnished in either 5′ or 10′ lengths. For pricing purposes, the Unit Price section features 10′ lengths with a joint and a carbon steel clevis hanger assembly every 5′ prorated into the per foot costs of both material and labor.

Three methods of joining are considered: lead and oakum poured joints or push-on gasket type joints for the bell and spigot pipe and a joint clamp for the no-hub soil pipe. The labor and material costs for each of these individual joining procedures are also prorated into the linear costs per foot.

C. Copper tubing covers types K, L, M, and DWV which are furnished in 20′ lengths. Means pricing data is based on a tubing cut each length and a coupling and two soft soldered joints every 10′. A carbon steel, clevis type hanger assembly every 10′ is also prorated into the per foot costs. The prices for refrigeration tubing are for materials only. Labor for full lengths may be based on the type L labor but short cut measures in tight areas can increase the installation labor-hours from 20 to 40%.

D. Corrosion-resistant piping does not lend itself to one particular standard of hanging or support assembly due to its diversity of application and placement. The several varieties of corrosion-resistant piping do not include any material or labor costs for hanger assemblies (See the Unit Price section for appropriate selection).

E. Glass pipe is furnished in standard lengths either 5′ or 10′ long, beaded on one end. Special orders for diverse lengths beaded on both ends are also available. For pricing purposes, R.S. Means features 10′ lengths with a coupling and a carbon steel band hanger assembly every 10′ prorated into the per foot linear costs.

Glass pipe is also available with conical ends and standard lengths ranging from 6″ through 3′ in 6″ increments, then up to 10′ in 12″ increments. Special lengths can be customized for particular installation requirements.

For pricing purposes, Means has based the labor and material pricing on 10′ lengths. Included in these costs per linear foot are the prorated costs for a flanged assembly every 10′ consisting of two flanges, a gasket, two insertable seals, and the required number of bolts and nuts. A carbon steel band hanger assembly based on 10′ center lines has also been prorated into the costs per foot for labor and materials.

F. Plastic pipe of several compositions and joining methods are considered. Fiberglass reinforced pipe (FRP) is priced, based on 10′ lengths (20′ lengths are also available), with coupling and epoxy joints every 10′. FRP is furnished in both "General Service" and "High Strength." A carbon steel clevis hanger assembly, 3 for every 10′, is built into the prorated labor and material costs on a per foot basis.

The PVC and CPVC pipe schedules 40, 80 and 120, plus SDR ratings are all based on 20′ lengths with a coupling installed every 10′, as well as a carbon steel clevis hanger assembly every 3′. The PVC and ABS type DWV piping is based on 10′ lengths with solvent weld couplings every 10′, and with carbon steel clevis hanger assemblies, 3 for every 10′. The rest of the plastic piping in this section is based on flexible 100′ coils and does not include any coupling or supports.

This section ends with PVC drain and sewer piping based on 10′ lengths with bell and spigot ends and 0-ring type, push-on joints.

G. Stainless steel piping includes both weld end and threaded piping, both in the type 304 and 316 specification and in the following schedules, 5, 10, 40, 80, and 160. Although this piping is usually furnished in 20′ lengths, this cost grouping has a joint (either heli-arc butt-welded or threads and coupling) every 10′. A carbon steel clevis type hanger assembly is also included at 10′ intervals and prorated into the linear foot costs.

H. Carbon steel pipe includes both black and galvanized. This section encompasses schedules 40 (standard) and 80 (extra heavy).

Several common methods of joining steel pipe — such as thread and coupled, butt welded, and flanged (150 lb. weld neck flanges) are also included.

For estimating purposes, it is assumed that the piping is purchased in 20′ lengths and that a compatible joint is made up every 10′. These joints are prorated into the labor and material costs per linear foot. The following hanger and support assemblies every 10′ are also included: carbon steel clevis for the T & C pipe, and single rod roll type for both the welded and flanged piping. All of these hangers are oversized to accommodate pipe insulation 3/4″ thick through 5″ pipe size and 1-1/2″ thick from 6″ through 12″ pipe size.

I. Grooved joint steel pipe is priced both black and galvanized, in schedules 10, 40, and 80, furnished in 20′ lengths. This section describes two joining methods: cut groove and roll groove. The schedule 10 piping is roll-grooved, while the heavier schedules are cut-grooved. The labor and material costs are prorated into per linear foot prices, including a coupled joint every 10′, as well as a carbon steel clevis hanger assembly.

Notes:

The pipe hanger assemblies mentioned in the preceding paragraphs include the described hanger; appropriately sized steel, box-type insert and nut; plus 18″ of threaded hanger rod.

C clamps are used when the pipe is to be supported from steel shapes rather than anchored in the slab. C clamps are slightly less costly than inserts. However, to save time in estimating, it is advisable to use the given line number cost, rather than substituting a C clamp for the insert.

Add to piping labor for elevated installation:

10′ to 14.5′ high	10%	30′ to 34.5′ high	40%
15′ to 19.5′ high	20%	35′ to 39.5′ high	50%
20′ to 24.5′ high	25%	40′ and higher	55%
25′ to 29.5′ high	35%		

When using the percentage adds for elevated piping installations as shown above, bear in mind that the given heights are for the pipe supports, even though the insert, anchor, or clamp may be several feet higher than the pipe itself.

An allowance has been included in the piping installation time for testing and minor tightening of leaking joints, fittings, stuffing boxes, packing glands, etc. For extraordinary test requirements such as x-rays, prolonged pressure or demonstration tests, a percentage of the piping labor, based on the estimator's experience, must be added to the labor total. A testing service specializing in weld x-rays should be consulted for pricing if it is an estimate requirement. Equipment installation time includes start-up with associated adjustments.

R224000-10 Water Consumption Rates

Fixture Type	Water Supply Fixture Unit Value		
	Hot Water	Cold Water	Combined
Bathtub	1.0	1.0	1.4
Clothes Washer	1.0	1.0	1.4
Dishwasher	1.4	0.0	1.4
Kitchen Sink	1.0	1.0	1.4
Laundry Tub	1.0	1.0	1.4
Lavatory	0.5	0.5	0.7
Shower Stall	1.0	1.0	1.4
Water Closet (tank type)	0.0	2.2	2.2
Full Bath Group w/bathtub or shower stall	1.5	2.7	3.6
Half Bath Group w/W.C. and Lavatory	0.5	2.5	2.6
Kitchen Group w/Dishwasher and Sink	1.9	1.0	2.5
Laundry Group w/Clothes Washer and Laundry Tub	1.8	1.8	2.5
Hose bibb (sillcock)	0.0	2.5	2.5

Notes:
Typically, WSFU = 1GPM

Supply loads in the building water-distribution system shall be determined by total load on the pipe being sized, in terms of water supply fixture units (WSFU) and gallons per minute (GPM) flow rates. For fixtures not listed, choose a WSFU value of a fixture with similar flow characteristics. Water Fixture Supply Units determined the required water supply to fixtures and their service systems. Fixture units are equal to (1) cubic foot of water drained in a 1-1/4″ pipe per minute. It is not a flow rate unit but a design factor.

R224000-20 Fixture Demands in Gallons per Fixture per Hour

Table below is based on 140°F final temperature except for dishwashers in public places (*) where 180°F water is mandatory.

Supply Systems for Flush Tanks			Supply Systems for Flushometer Valves		
Load	Demand		Load	Demand	
WSFU	GPM	CU. FT.	WSFU	GPM	CU. FT.
1.0	3.0	0.041040	-	-	-
2.0	5.0	0.068400	-	-	-
3.0	6.5	0.868920	-	-	-
4.0	8.0	1.069440	-	-	-
5.0	9.4	1.256592	5.0	15	2.0052
6.0	10.7	1.430376	6.0	17.4	2.326032
7.0	11.8	1.577424	7.0	19.8	2.646364
8.0	12.8	1.711104	8.0	22.2	2.967696
9.0	13.7	1.831416	9.0	24.6	3.288528
10.0	14.6	1.951728	10.0	27	3.60936
11.0	15.4	2.058672	11.0	27.8	3.716304
12.0	16.0	2.138880	12.0	28.6	3.823248
13.0	16.5	2.205720	13.0	29.4	3.930192
14.0	17.0	2.272560	14.0	30.2	4.037136
15.0	17.5	2.339400	15.0	31	4.14408
16.0	18.0	2.906240	16.0	31.8	4.241024
17.0	18.4	2.459712	17.0	32.6	4.357968
18.0	18.8	2.513184	18.0	33.4	4.464912
19.0	19.2	2.566656	19.0	34.2	4.571856
20.0	19.6	2.620218	20.0	35.0	4.678800
25.0	21.5	2.874120	25.0	38.0	5.079840
30.0	23.3	3.114744	30.0	42.0	5.611356
35.0	24.9	3.328632	35.0	44.0	5.881920
40.0	26.3	3.515784	40.0	46.0	6.149280
45.0	27.7	3.702936	45.0	48.0	6.416640
50.0	29.1	3.890088	50.0	50.0	6.684000

Notes:

When designing a plumbing system that utilizes fixtures other than, or in addition to, water closets, use the data provided in the Supply Systems for Flush Tanks section of the above table.

To obtain the probable maximum demand, multiply the total demands for the fixtures (gal./fixture/hour) by the demand factor. The heater should have a heating capacity in gallons per hour equal to this maximum. The storage tank should have a capacity in gallons equal to the probable maximum demand multiplied by the storage capacity factor.

R224000-30 Minimum Plumbing Fixture Requirements

Classification	Occupancy	Description	Water		Lavatories		Bathtubs/Showers	Drinking Fountains	Other
			Male	Female	Male	Female			
Assembly	A-1	Theaters and other buildings for the performing arts and motion pictures	1:125	1:65	1:200			1:500	1 Service Sink
	A-2	Nightclubs, bars, taverns dance halls	1:40		1:75			1:500	1 Service Sink
		Restaurants, banquet halls, food courts	1:75		1:200			1:500	1 Service Sink
	A-3	Auditorium w/o permanent seating, art galleries, exhibition halls, museums, lecture halls, libraries, arcades & gymnasiums	1:125	1:65	1:200			1:500	1 Service Sink
		Passenger terminals and transportation facilities	1:500		1:750			1:1000	1 Service Sink
		Places of worship and other religious services	1:150	1:75	1:200			1:1000	1 Service Sink
	A-4	Indoor sporting events and activities, coliseums, arenas, skating rinks, pools, and tennis courts	1:75 for the first 1500, then 1:120 for the remainder	1:40 for the first 1520, then 1:60 for the remainder	1:200	1:150		1:1000	1 Service Sink
	A-5	Outdoor sporting events and activities, stadiums, amusement parks, bleachers, grandstands	1:75 for the first 1500, then 1:120 for the remainder	1:40 for the first 1520, then 1:60 for the remainder	1:200	1:150		1:1000	1 Service Sink
Business	B	Buildings for the transaction of business, professional services, other services involving merchandise, office buildings, banks, light industrial	1:25 for the first 50, then 1:50 for the remainder		1:40 for the first 80, then 1:80 for the remainder			1:100	1 Service Sink
Educational	E	Educational Facilities	1:50		1:50			1:100	1 Service Sink
Factory and industrial	F-1 and F-2	Structures in which occupants are engaged in work fabricating, assembly or processing of products or materials	1:100		1:100		See *International Plumbing Code*	1:400	1 Service Sink
Institutional	I-1	Residential Care	1:10		1:10		1:8	1:100	1 Service Sink
	I-2	Hospitals, ambulatory nursing home care recipient	1 per room		1 per room		1:15	1:100	1 Service Sink
		Employees, other than residential care	1:25		1:35			1:100	
		Visitors, other than residential care	1:75		1:100			1:500	
	I-3	Prisons	1 per cell		1 per cell		1:15	1:100	1 Service Sink
		Reformatories, detention and correction centers	1:15		1:15		1:15	1:100	1 Service Sink
		Employees	1:25		1:35			1:100	
	I-4	Adult and child day care	1:15		1:15		1	1:100	1 Service Sink
Mercantile	M	Retail stores, service stations, shops, salesrooms, markets and shopping centers	1:500		1:750			1:1000	1 Service Sink
Residential	R-1	Hotels, Motels, boarding houses (transient)	1 per sleeping unit		1 per sleeping unit		1 per sleeping unit		1 Service Sink
	R-2	Dormitories, fraternities, sororities and boarding houses (not transient)	1:10		1:10		1:8	1:100	1 Service Sink
		Apartment House	1 per dwelling unit		1 per dwelling unit		1 per dwelling unit		1 Kitchen sink per dwelling; 1 clothes washer connection per 20 dwellings
	R-3	1 and 2 Family Dwellings	1 per dwelling unit		1:10		1 per dwelling unit		1 Kitchen sink per dwelling; 1 clothes washer connection per dwelling
	R-3	Congregate living facilities w/<16 people	1:10		1:10		1:8	1:100	1 Service Sink
	R-4	Congregate living facilities w/<16 people	1:10		1:10		1:8	1:100	1 Service Sink
Storage	S-1 and S-2	Structures for the storage of good, warehouses, storehouses and freight depots, low and moderate hazard	1:100		1:100		See *International Plumbing Code*	1;1000	1 Service Sink

Table 2902.1

R230500-10 Subcontractors

On the unit cost pages of the R.S. Means Cost Data books, the last column is entitled "Total Incl. O&P". This is normally the cost of the installing contractor. In the HVAC Division, this is the cost of the mechanical contractor. If the particular work being estimated is to be performed by a sub to the mechanical contractor, the mechanical's profit and handling charge (usually 10%) is added to the total of the last column.

R233100-40 Steel Sheet Metal Calculator (Weight in Lb./Ft. of Length)

Gauge	26	24	22	20	18	16	Gauge	26	24	22	20	18	16
Wt.-Lb./S.F.	.906	1.156	1.406	1.656	2.156	2.656	Wt.-Lb./S.F.	.906	1.156	1.406	1.656	2.156	2.656
SMACNA Max. Dimension – Long Side		30″	54″	84″	85″ Up		SMACNA Max. Dimension – Long Side		30″	54″	84″	85″ Up	
Sum-2 sides							Sum-2 Sides						
2	.3	.40	.50	.60	.80	.90	56	9.3	12.0	14.0	16.2	21.3	25.2
3	.5	.65	.80	.90	1.1	1.4	57	9.5	12.3	14.3	16.5	21.7	25.7
4	.7	.85	1.0	1.2	1.5	1.8	58	9.7	12.5	14.5	16.8	22.0	26.1
5	.8	1.1	1.3	1.5	1.9	2.3	59	9.8	12.7	14.8	17.1	22.4	26.6
6	1.0	1.3	1.5	1.7	2.3	2.7	60	10.0	12.9	15.0	17.4	22.8	27.0
7	1.2	1.5	1.8	2.0	2.7	3.2	61	10.2	13.1	15.3	17.7	23.2	27.5
8	1.3	1.7	2.0	2.3	3.0	3.6	62	10.3	13.3	15.5	18.0	23.6	27.9
9	1.5	1.9	2.3	2.6	3.4	4.1	63	10.5	13.5	15.8	18.3	24.0	28.4
10	1.7	2.2	2.5	2.9	3.8	4.5	64	10.7	13.7	16.0	18.6	24.3	28.8
11	1.8	2.4	2.8	3.2	4.2	5.0	65	10.8	13.9	16.3	18.9	24.7	29.3
12	2.0	2.6	3.0	3.5	4.6	5.4	66	11.0	14.1	16.5	19.1	25.1	29.7
13	2.2	2.8	3.3	3.8	4.9	5.9	67	11.2	14.3	16.8	19.4	25.5	30.2
14	2.3	3.0	3.5	4.1	5.3	6.3	68	11.3	14.6	17.0	19.7	25.8	30.6
15	2.5	3.2	3.8	4.4	5.7	6.8	69	11.5	14.8	17.3	20.0	26.2	31.1
16	2.7	3.4	4.0	4.6	6.1	7.2	70	11.7	15.0	17.5	20.3	26.6	31.5
17	2.8	3.7	4.3	4.9	6.5	7.7	71	11.8	15.2	17.8	20.6	27.0	32.0
18	3.0	3.9	4.5	5.2	6.8	8.1	72	12.0	15.4	18.0	20.9	27.4	32.4
19	3.2	4.1	4.8	5.5	7.2	8.6	73	12.2	15.6	18.3	21.2	27.7	32.9
20	3.3	4.3	5.0	5.8	7.6	9.0	74	12.3	15.8	18.5	21.5	28.1	33.3
21	3.5	4.5	5.3	6.1	8.0	9.5	75	12.5	16.1	18.8	21.8	28.5	33.8
22	3.7	4.7	5.5	6.4	8.4	9.9	76	12.7	16.3	19.0	22.0	28.9	34.2
23	3.8	5.0	5.8	6.7	8.7	10.4	77	12.8	16.5	19.3	22.3	29.3	34.7
24	4.0	5.2	6.0	7.0	9.1	10.8	78	13.0	16.7	19.5	22.6	29.6	35.1
25	4.2	5.4	6.3	7.3	9.5	11.3	79	13.2	16.9	19.8	22.9	30.0	35.6
26	4.3	5.6	6.5	7.5	9.9	11.7	80	13.3	17.1	20.0	23.2	30.4	36.0
27	4.5	5.8	6.8	7.8	10.3	12.2	81	13.5	17.3	20.3	23.5	30.8	36.5
28	4.7	6.0	7.0	8.1	10.6	12.6	82	13.7	17.5	20.5	23.8	31.2	36.9
29	4.8	6.2	7.3	8.4	11.0	13.1	83	13.8	17.8	20.8	24.1	31.5	37.4
30	5.0	6.5	7.5	8.7	11.4	13.5	84	14.0	18.0	21.0	24.4	31.9	37.8
31	5.2	6.7	7.8	9.0	11.8	14.0	85	14.2	18.2	21.3	24.7	32.3	38.3
32	5.3	6.9	8.0	9.3	12.2	14.4	86	14.3	18.4	21.5	24.9	32.7	38.7
33	5.5	7.1	8.3	9.6	12.5	14.9	87	14.5	18.6	21.8	25.2	33.1	39.2
34	5.7	7.3	8.5	9.9	12.9	15.3	88	14.7	18.8	22.0	25.5	33.4	39.6
35	5.8	7.5	8.8	10.2	13.3	15.8	89	14.8	19.0	22.3	25.8	33.8	40.1
36	6.0	7.8	9.0	10.4	13.7	16.2	90	15.0	19.3	22.5	26.1	34.2	40.5
37	6.2	8.0	9.3	10.7	14.1	16.7	91	15.2	19.5	22.8	26.4	34.6	41.0
38	6.3	8.2	9.5	11.0	14.4	17.1	92	15.3	19.7	23.0	26.7	35.0	41.4
39	6.5	8.4	9.8	11.3	14.8	17.6	93	15.5	19.9	23.3	27.0	35.3	41.9
40	6.7	8.6	10.0	11.6	15.2	18.0	94	15.7	20.1	23.5	27.3	35.7	42.3
41	6.8	8.8	10.3	11.9	15.6	18.5	95	15.8	20.3	23.8	27.6	36.1	42.8
42	7.0	9.0	10.5	12.2	16.0	18.9	96	16.0	20.5	24.0	27.8	36.5	43.2
43	7.2	9.2	10.8	12.5	16.3	19.4	97	16.2	20.8	24.3	28.1	36.9	43.7
44	7.3	9.5	11.0	12.8	16.7	19.8	98	16.3	21.0	24.5	28.4	37.2	44.1
45	7.5	9.7	11.3	13.1	17.1	20.3	99	16.5	21.2	24.8	28.7	37.6	44.6
46	7.7	9.9	11.5	13.3	17.5	20.7	100	16.7	21.4	25.0	29.0	38.0	45.0
47	7.8	10.1	11.8	13.6	17.9	21.2	101	16.8	21.6	25.3	29.3	38.4	45.5
48	8.0	10.3	12.0	13.9	18.2	21.6	102	17.0	21.8	25.5	29.6	38.8	45.9
49	8.2	10.5	12.3	14.2	18.6	22.1	103	17.2	22.0	25.8	29.9	39.1	46.4
50	8.3	10.7	12.5	14.5	19.0	22.5	104	17.3	22.3	26.0	30.2	39.5	46.8
51	8.5	11.0	12.8	14.8	19.4	23.0	105	17.5	22.5	26.3	30.5	39.9	47.3
52	8.7	11.2	13.0	15.1	19.8	23.4	106	17.7	22.7	26.5	30.7	40.3	47.7
53	8.8	11.4	13.3	15.4	20.1	23.9	107	17.8	22.9	26.8	31.0	40.7	48.2
54	9.0	11.6	13.5	15.7	20.5	24.3	108	18.0	23.1	27.0	31.3	41.0	48.6
55	9.2	11.8	13.8	16.0	20.9	24.8	109	18.2	23.3	27.3	31.6	41.4	49.1
							110	18.3	23.5	27.5	31.9	41.8	49.5

Example: If duct is 34″ x 20″ x 15′ long, 34″ is greater than 30″ maximum, for 24 ga. so must be 22 ga. 34″ + 20″ = 54″ going across from 54″ find 13.5 lb. per foot. 13.5 x 15′ = 202.5 lbs. For S.F. of surface area 202.5 ÷ 1.406 = 144 S.F.

Note: Figures include an allowance for scrap.
***Do Not** use unless engineer specified. 26GA is very light and sometimes used for toilet exhaust. 16 GA is heavy plate, and mostly specified for plenums and hoods.

Heating, Ventilating & A.C. R2337 Air Outlets & Inlets

R233700-60 Diffuser Evaluation

CFM = V × An × K where V = Outlet velocity in feet per minute. An = Neck area in square feet and K = Diffuser delivery factor. An undersized diffuser for a desired CFM will produce a high velocity and noise level. When air moves past people at a velocity in excess of 25 FPM, an annoying draft is felt. An oversized diffuser will result in low velocity with poor mixing. Consideration must be given to avoid vertical stratification or horizontal areas of stagnation.

Heating, Ventilating & A.C. R2350 Central Heating Equipment

R235000-10 Heating Systems

Heating Systems

The basic function of a heating system is to bring an enclosed volume up to a desired temperature and then maintain that temperature within a reasonable range. To accomplish this, the selected system must have sufficient capacity to offset transmission losses resulting from the temperature difference on the interior and exterior of the enclosing walls in addition to losses due to cold air infiltration through cracks, crevices and around doors and windows. The amount of heat to be furnished is dependent upon the building size, construction, temperature difference, air leakage, use, shape, orientation and exposure. Air circulation is also an important consideration. Circulation will prevent stratification which could result in heat losses through uneven temperatures at various levels. For example, the most efficient use of unit heaters can usually be achieved by circulating the space volume through the total number of units once every 20 minutes or 3 times an hour. This general rule must, of course, be adapted for special cases such as large buildings with low ratios of heat transmitting surface to cubical volume. The type of occupancy of a building will have considerable bearing on the number of heat transmitting units and the location selected.
It is axiomatic, however, that the basis of any successful heating system is to provide the maximum amount of heat at the points of maximum heat loss such as exposed walls, windows, and doors. Large roof areas, wind direction, and wide doorways create problems of excessive heat loss and require special consideration and treatment.

Heat Transmission

Heat transfer is an important parameter to be considered during selection of the exterior wall style, material and window area. A high rate of transfer will permit greater heat loss during the wintertime with the resultant increase in heating energy costs and a greater rate of heat gain in the summer with proportionally greater cooling cost. Several terms are used to describe various aspects of heat transfer. However, for general estimating purposes this book lists U values for systems of construction materials. U is the "overall heat transfer coefficient." It is defined as the heat flow per hour through one square foot when the temperature difference in the air on either side of the structure wall, roof, ceiling or floor is one degree Fahrenheit. The structural segment may be a single homogeneous material or a composite.

Total heat transfer is found using the following equation:

$Q = AU(T_2 - T_1)$ where

Q = Heat flow, BTU per hour
A = Area, square feet
U = Overall heat transfer coefficient
$(T_2 - T_1)$ = Difference in temperature of air on each side of the construction component. (Also abbreviated TD)

Note that heat can flow through all surfaces of any building and this flow is in addition to heat gain or loss due to ventilation, infiltration and generation (appliances, machinery, people).

R235000-20 Heating Approximations for Quick Estimating

Oil Piping & Boiler Room Piping

Small System	20 to 30% of Boiler
Complex System with Pumps, Headers, Etc.	80 to 110% of Boiler

Breeching With Insulation:

Small	10 to 15% of Boiler
Large	15 to 25% of Boiler
Coils:	15 to 30% of Containing Unit
Balancing (Independent)	1/2% of H.V.A.C. Estimate

Quality/Complexity Adjustment: For all heating installations add these adjustments to the estimate to more closely allow for the equipment and conditions of the particular job under consideration.

Economy installation, add	0 to 5% of System
Good quality, medium complexity, add	5 to 15% of System
Above average quality and complexity, add	15 to 25% of System

R235000-30 The Basics of a Heating System

The function of a heating system is to achieve and maintain a desired temperature in a room or building by replacing the amount of heat being dissipated. There are four kinds of heating systems: hot-water, steam, warm-air and electric resistance. Each has certain essential and similar elements with the exception of electric resistance heating.

The basic elements of a heating system are:

A. A **combustion chamber** in which fuel is burned and heat transferred to a conveying medium.

B. The **"fluid"** used for conveying the heat (water, steam or air).

C. **Conductors** or pipes for transporting the fluid to specific desired locations.

D. A means of disseminating the heat, sometimes called **terminal units.**

A. The **combustion chamber** in a furnace heats air which is then distributed. This is called a warm-air system.

The combustion chamber in a boiler heats water which is either distributed as hot water or steam and this is termed a hydronic system.

The maximum allowable working pressures are limited by ASME "Code for Heating Boilers" to 15 PSI for steam and 160 PSI for hot water heating boilers, with a maximum temperature limitation of 250° F. Hot water boilers are generally rated for a working pressure of 30 PSI. High pressure boilers are governed by the ASME "Code for Power Boilers" which is used almost universally for boilers operating over 15 PSIG. High pressure boilers used for a combination of heating/process loads are usually designed for 150 PSIG.

Boiler ratings are usually indicated as either Gross or Net Output. The Gross Load is equal to the Net Load plus a piping and pickup allowance. When this allowance cannot be determined, divide the gross output rating by 1.25 for a value equal to or greater than the net heat loss requirement of the building.

B. Of the three **fluids** used, steam carries the greatest amount of heat per unit volume. This is due to the fact that it gives up its latent heat of vaporization at a temperature considerably above room temperature. Another advantage is that the pressure to produce a positive circulation is readily available. Piping conducts the steam to terminal units and returns condensate to the boiler.

The **steam system** is well adapted to large buildings because of its positive circulation, its comparatively economical installation and its ability to deliver large quantities of heat. Nearly all large office buildings, stores, hotels, and industrial buildings are so heated, in addition to many residences.

Hot water, when used as the heat carrying fluid, gives up a portion of its sensible heat and then returns to the boiler or heating apparatus for reheating. As the heat conveyed by each pound of water is about one-fiftieth of the heat conveyed by a pound of steam, it is necessary to circulate about fifty times as much water as steam by weight (although only one-thirtieth as much by volume). The hot water system is usually, although not necessarily, designed to operate at temperatures below that of the ordinary steam system and so the amount of heat transfer surface must be correspondingly greater. A temperature of 190° F to 200° F is normally the maximum. Circulation in small buildings may depend on the difference in density between hot water and the cool water returning to the boiler; circulating pumps are normally used to maintain a desired rate of flow. Pumps permit a greater degree of flexibility and better control.

In **warm-air** furnace systems, cool air is taken from one or more points in the building, passed over the combustion chamber and flue gas passages and then distributed through a duct system. A disadvantage of this system is that the ducts take up much more building volume than steam or hot water pipes. Advantages of this system are the relative ease with which humidification can be accomplished by the evaporation of water as the air circulates through the heater, and the lack of need for expensive disseminating units as the warm air simply becomes part of the interior atmosphere of the building.

C. Conductors (pipes and ducts) have been lightly treated in the discussion of conveying fluids. For more detailed information such as sizing and distribution methods, the reader is referred to technical publications such as the American Society of Heating, Refrigerating and Air-Conditioning Engineers "Handbook of Fundamentals."

D. Terminal units come in an almost infinite variety of sizes and styles, but the basic principles of operation are very limited. As previously mentioned, warm-air systems require only a simple register or diffuser to mix heated air with that present in the room. Special application items such as radiant coils and infrared heaters are available to meet particular conditions but are not usually considered for general heating needs. Most heating is accomplished by having air flow over coils or pipes containing the heat transporting medium (steam, hot-water, electricity). These units, while varied, may be separated into two general types, (1) radiator/convectors and (2) unit heaters.

Radiator/convectors may be cast, fin-tube or pipe assemblies. They may be direct, indirect, exposed, concealed or mounted within a cabinet enclosure, upright or baseboard style. These units are often collectively referred to as "radiatiors" or "radiation" although none gives off heat either entirely by radiation or by convection but rather a combination of both. The air flows over the units as a gravity "current." It is necessary to have one or more heat-emitting units in each room. The most efficient placement is low along an outside wall or under a window to counteract the cold coming into the room and achieve an even distribution.

In contrast to radiator/convectors which operate most effectively against the walls of smaller rooms, **unit heaters** utilize a fan to move air over heating coils and are very effective in locations of relatively large volume. Unit heaters, while usually suspended overhead, may be floor mounted. They also may take in fresh outside air for ventilation. The heat distributed by unit heaters may be from a remote source and conveyed by a fluid or it may be from the combustion of fuel in each individual heater. In the latter case the only piping required would be for fuel, however, a vent for the products of combustion would be necessary.

The following list gives may of the advantages of unit heaters for applications other than office or residential:

a. Large capacity so smaller number of units are required, **b.** Piping system simplified, **c.** Space saved where they are located overhead out of the way, **d.** Rapid heating directed where needed with effective wide distribution, **e.** Difference between floor and ceiling temperature reduced, **f.** Circulation of air obtained, and ventilation with introduction of fresh air possible, **g.** Heat output flexible and easily controlled.

R235616-60 Solar Heating (Space and Hot Water)

Collectors should face as close to due South as possible, however, variations of up to 20 degrees on either side of true South are acceptable. Local climate and collector type may influence the choice between east or west deviations. Obviously they should be located so they are not shaded from the sun's rays. Incline collectors at a slope of latitude minus 5 degrees for domestic hot water and latitude plus 15 degrees for space heating.

Flat plate collectors consist of a number of components as follows: Insulation to reduce heat loss through the bottom and sides of the collector. The enclosure which contains all the components in this assembly is usually weatherproof and prevents dust, wind and water from coming in contact with the absorber plate. The cover plate usually consists of one or more layers of a variety of glass or plastic and reduces the reradiation by creating an air space which traps the heat between the cover and the absorber plates.

The absorber plate must have a good thermal bond with the fluid passages. The absorber plate is usually metallic and treated with a surface coating which improves absorptivity. Black or dark paints or selective coatings are used for this purpose, and the design of this passage and plate combination helps determine a solar system's effectiveness.

Heat transfer fluid passage tubes are attached above and below or integral with an absorber plate for the purpose of transferring thermal energy from the absorber plate to a heat transfer medium. The heat exchanger is a device for transferring thermal energy from one fluid to another.

Piping and storage tanks should be well insulated to minimize heat losses.

Size domestic water heating storage tanks to hold 20 gallons of water per user, minimum, plus 10 gallons per dishwasher or washing machine. For domestic water heating an optimum collector size is approximately 3/4 square foot of area per gallon of water storage. For space heating of residences and small commercial applications the collector is commonly sized between 30% and 50% of the internal floor area. For space heating of large commercial applications, collector areas less than 30% of the internal floor area can still provide significant heat reductions.

A supplementary heat source is recommended for Northern states for December through February.

The solar energy transmission per square foot of collector surface varies greatly with the material used. Initial cost, heat transmittance and useful life are obviously interrelated.

R236000-10 Air Conditioning

General: The purpose of air conditioning is to control the environment of a space so that comfort is provided for the occupants and/or conditions are suitable for the processes or equipment contained therein. The several items which should be evaluated to define system objectives are:

Temperature Control
Humidity Control
Cleanliness
Odor, smoke and fumes
Ventilation

Efforts to control the above parameters must also include consideration of the degree or tolerance of variation, the noise level introduced, the velocity of air motion and the energy requirements to accomplish the desired results.

The variation in **temperature** and **humidity** is a function of the sensor and the controller. The controller reacts to a signal from the sensor and produces the appropriate suitable response in either the terminal unit, the conductor of the transporting medium (air, steam, chilled water, etc.), or the source (boiler, evaporating coils, etc.).

The **noise level** is a by-product of the energy supplied to moving components of the system. Those items which usually contribute the most noise are pumps, blowers, fans, compressors and diffusers. The level of noise can be partially controlled through use of vibration pads, isolators, proper sizing, shields, baffles and sound absorbing liners.

Some **air motion** is necessary to prevent stagnation and stratification. The maximum acceptable velocity varies with the degree of heating or cooling which is taking place. Most people feel air moving past them at velocities in excess of 25 FPM as an annoying draft. However, velocities up to 45 FPM may be acceptable in certain cases. Ventilation, expressed as air changes per hour and percentage of fresh air, is usually an item regulated by local codes.

Selection of the system to be used for a particular application is usually a trade-off. In some cases the building size, style, or room available for mechanical use limits the range of possibilities. Prime factors influencing the decision are first cost and total life (operating, maintenance and replacement costs). The accuracy with which each parameter is determined will be an important measure of the reliability of the decision and subsequent satisfactory operation of the installed system.

Heat delivery may be desired from an air conditioning system. Heating capability usually is added as follows: A gas fired burner or hot water/steam/electric coils may be added to the air handling unit directly and heat all air equally. For limited or localized heat requirements the water/steam/electric coils may be inserted into the duct branch supplying the cold areas. Gas fired duct furnaces are also available.

Note: When water or steam coils are used the cost of the piping and boiler must also be added. For a rough estimate use the cost per square foot of the appropriate sized hydronic system with unit heaters. This will provide a cost for the boiler and piping, and the unit heaters of the system would equate to the approximate cost of the heating coils.

R236000-20 Air Conditioning Requirements

BTU's per hour per S.F. of floor area and S.F. per ton of air conditioning.

Type of Building	BTU/Hr per S.F.	S.F. per Ton	Type of Building	BTU/Hr per S.F.	S.F. per Ton	Type of Building	BTU/Hr per S.F.	S.F. per Ton
Apartments, Individual	26	450	Dormitory, Rooms	40	300	Libraries	50	240
Corridors	22	550	Corridors	30	400	Low Rise Office, Exterior	38	320
Auditoriums & Theaters	40	300/18*	Dress Shops	43	280	Interior	33	360
Banks	50	240	Drug Stores	80	150	Medical Centers	28	425
Barber Shops	48	250	Factories	40	300	Motels	28	425
Bars & Taverns	133	90	High Rise Office—Ext. Rms.	46	263	Office (small suite)	43	280
Beauty Parlors	66	180	Interior Rooms	37	325	Post Office, Individual Office	42	285
Bowling Alleys	68	175	Hospitals, Core	43	280	Central Area	46	260
Churches	36	330/20*	Perimeter	46	260	Residences	20	600
Cocktail Lounges	68	175	Hotel, Guest Rooms	44	275	Restaurants	60	200
Computer Rooms	141	85	Corridors	30	400	Schools & Colleges	46	260
Dental Offices	52	230	Public Spaces	55	220	Shoe Stores	55	220
Dept. Stores, Basement	34	350	Industrial Plants, Offices	38	320	Shop'g. Ctrs., Supermarkets	34	350
Main Floor	40	300	General Offices	34	350	Retail Stores	48	250
Upper Floor	30	400	Plant Areas	40	300	Specialty	60	200

*Persons per ton
12,000 BTU = 1 ton of air conditioning

R236000-30 Psychrometric Table

Dewpoint or Saturation Temperature (F)

Relative humidity (%)	32	35	40	45	50	55	60	65	70	75	80	85	90	95	100
100	32	35	40	45	50	55	60	65	70	75	80	85	90	95	100
90	30	33	37	42	47	52	57	62	67	72	77	82	87	92	97
80	27	30	34	39	44	49	54	58	64	68	73	78	83	88	93
70	24	27	31	36	40	45	50	55	60	64	69	74	79	84	88
60	20	24	28	32	36	41	46	51	55	60	65	69	74	79	83
50	16	20	24	28	33	36	41	46	50	55	60	64	69	73	78
40	12	15	18	23	27	31	35	40	45	49	53	58	62	67	71
30	8	10	14	18	21	25	29	33	37	42	46	50	54	59	62
20	6	7	8	9	13	16	20	24	28	31	35	40	43	48	52
10	4	4	5	5	6	8	9	10	13	17	20	24	27	30	34

Dry bulb temperature (F) (column headings)

This table shows the relationship between RELATIVE HUMIDITY, DRY BULB TEMPERATURE AND DEWPOINT.

As an example, assume that the thermometer in a room reads 75° F, and we know that the relative humidity is 50%. The chart shows the dewpoint temperature to be 55° F. That is, any surface colder than 55° F will "sweat" or collect condensing moisture. This surface could be the outside of an uninsulated chilled water pipe in the summertime, or the inside surface of a wall or deck in the wintertime. After determining the extreme ambient parameters, the table at the left is useful in determining which surfaces need insulation or vapor barrier protection.

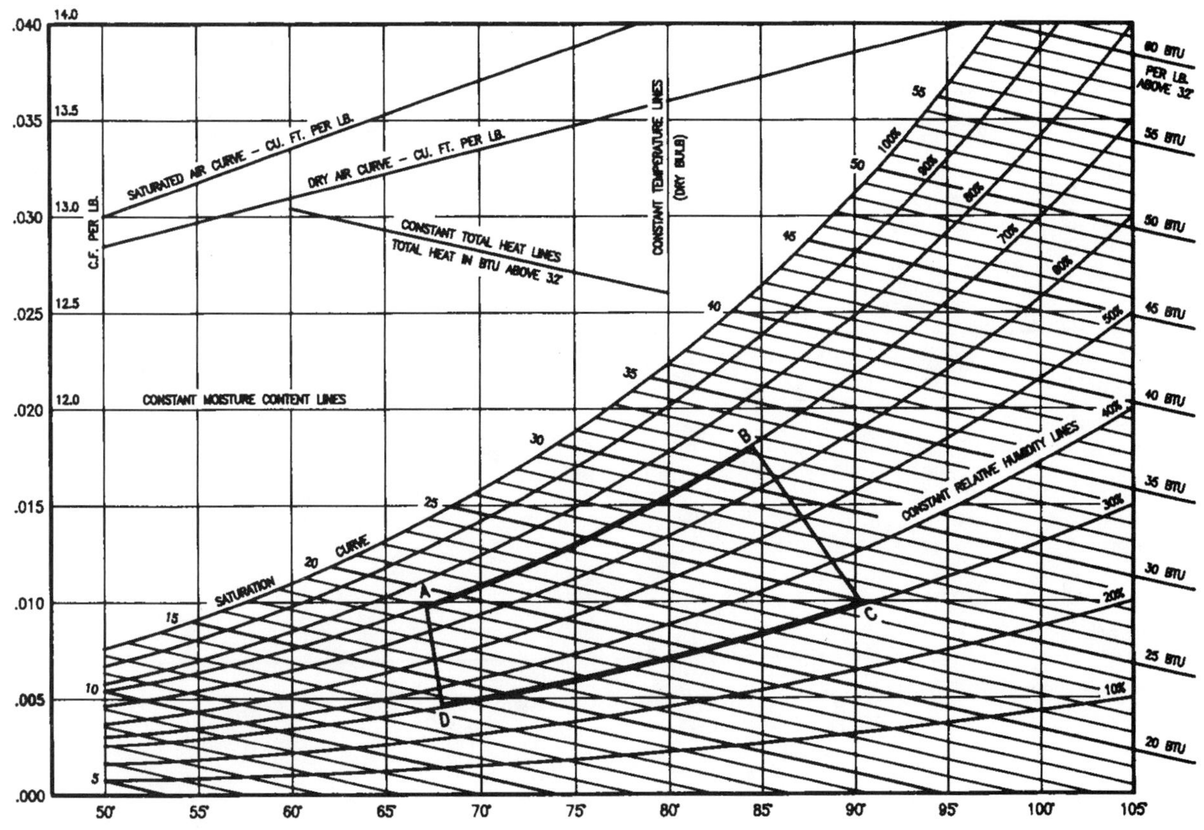

TEMPERATURE DEGREES FAHRENHEIT
TOTAL PRESSURE = 14.696 LB. PER SQ. IN. ABS.

Psychrometric chart showing different variables based on one pound of dry air. Space marked A B C D is temperature-humidity range which is most comfortable for majority of people.

R236000-90 Quality/Complexity Adjustment for Air Conditioning Systems

Economy installation, add	0 to 5%
Good quality, medium complexity, add	5 to 15%
Above average quality and complexity, add	15 to 25%

Add the above adjustments to the estimate to more closely allow for the equipment and conditions of the particular job under consideration.

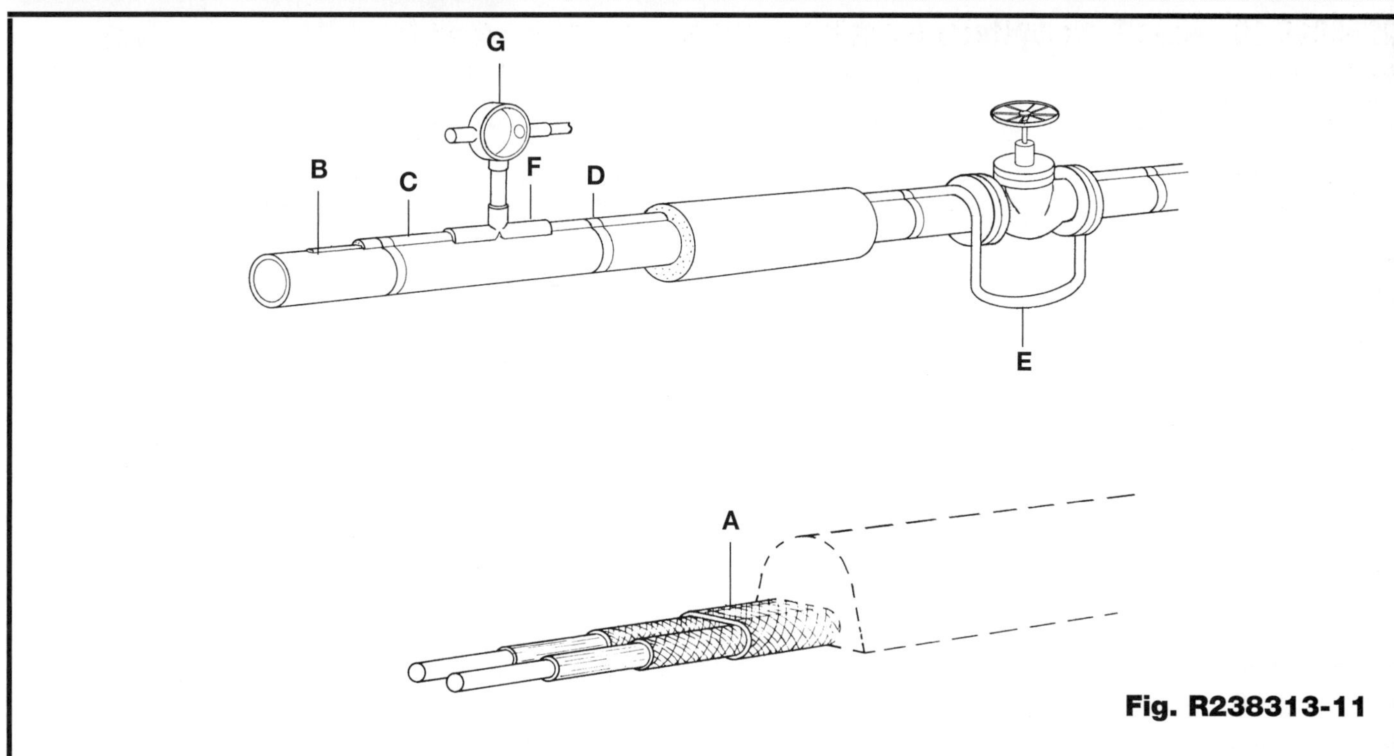

Fig. R238313-11

R238313-10 Heat Trace Systems

Before you can determine the cost of a HEAT TRACE installation the method of attachment must be established. There are (4) common methods:

1. Cable is simply attached to the pipe with polyester tape every 12′.
2. Cable is attached with a continuous cover of 2″ wide aluminum tape.
3. Cable is attached with factory extruded heat transfer cement and covered with metallic raceway with clips every 10′.
4. Cable is attached between layers of pipe insulation using either clips or polyester tape.

Example: Components for method 3 must include:

A. Heat trace cable by voltage and watts per linear foot.
B. Heat transfer cement, 1 gallon per 60 linear feet of cover.
C. Metallic raceway by size and type.
D. Raceway clips by size of pipe.

When taking off linear foot lengths of cable add the following for each valve in the system. (E)

In all of the above methods each component of the system must be priced individually.

SCREWED OR WELDED VALVE:

Size		Length
1/2″	=	6″
3/4″	=	9″
1″	=	1′ -0″
1-1/2″	=	1′ -6″
2″	=	2′
2-1/2″	=	2′ -6″
3″	=	2′ -6″
4″	=	4′ -0″
6″	=	7′ -0″
8″	=	9′ -6″
10″	=	12′ -6″
12″	=	15′ -0″
14″	=	18′ -0″
16″	=	21′ -6″
18″	=	25′ -6″
20″	=	28′ -6″
24″	=	34′ -0″
30″	=	40′ -0″

FLANGED VALVE:

Size		Length
1/2″	=	1′ -0″
3/4″	=	1′ -6″
1″	=	2′ -0″
1-1/2″	=	2′ -6″
2″	=	2′ -6″
2-1/2″	=	3′ -0″
3″	=	3′ -6″
4″	=	4′ -0″
6″	=	8′ -0″
8″	=	11′ -0″
10″	=	14′ -0″
12″	=	16′ -6″
14″	=	19′ -6″
16″	=	23′ -0″
18″	=	27′ -0″
20″	=	30′ -0″
24″	=	36′ -0″
30″	=	42′ -0″

BUTTERFLY VALVE:

Size		Length
1/2″	=	0′
3/4″	=	0′
1″	=	1′ -0″
1-1/2″	=	1′ -6″
2″	=	2′ -0″
2-1/2″	=	2′ -6″
3″	=	2′ -6″
4″	=	3′ -0″
6″	=	3′ -6″
8″	=	4′ -0″
10″	=	4′ -0″
12″	=	5′ -0″
14″	=	5′ -6″
16″	=	6′ -0″
18″	=	6′ -6″
20″	=	7′ -0″
24″	=	8′ -0″
30″	=	10′ -0″

R238313-10 Heat Trace Systems (cont.)

Add the following quantities of heat transfer cement to linear foot totals for each valve:

Nominal Valve Size	Gallons of Cement per Valve
1/2″	0.14
3/4″	0.21
1″	0.29
1-1/2″	0.36
2″	0.43
2-1/2″	0.70
3″	0.71
4″	1.00
6″	1.43
8″	1.48
10″	1.50
12″	1.60
14″	1.75
16″	2.00
18″	2.25
20″	2.50
24″	3.00
30″	3.75

The following must be added to the list of components to accurately price HEAT TRACE systems:

1. Expediter fitting and clamp fasteners (F)
2. Junction box and nipple connected to expediter fitting (G)
3. Field installed terminal blocks within junction box
4. Ground lugs
5. Piping from power source to expediter fitting
6. Controls
7. Thermostats
8. Branch wiring
9. Cable splices
10. End of cable terminations
11. Branch piping fittings and boxes

Deduct the following percentages from labor if cable lengths in the same area exceed:

150′ to 250′	10%	351′ to 500′	20%
251′ to 350′	15%	Over 500′	25%

Add the following percentages to labor for elevated installations:

15′ to 20′ high	10%	31′ to 35′ high	40%
21′ to 25′ high	20%	36′ to 40′ high	50%
26′ to 30′ high	30%	Over 40′ high	60%

R238313-20 Spiral-Wrapped Heat Trace Cable (Pitch Table)

In order to increase the amount of heat, occasionally heat trace cable is wrapped in a spiral fashion around a pipe; increasing the number of feet of heater cable per linear foot of pipe.

Engineers first determine the heat loss per foot of pipe (based on the insulating material, its thickness, and the temperature differential across it). A ratio is then calculated by the formula:

$$\text{Feet of Heat Trace per Foot of Pipe} = \frac{\text{Watts/Foot of Heat Loss}}{\text{Watts/Foot of the Cable}}$$

The linear distance between wraps (pitch) is then taken from a chart or table. Generally, the pitch is listed on a drawing leaving the estimator to calculate the total length of heat tape required. An approximation may be taken from this table.

Feet of Heat Trace Per Foot of Pipe																
	Nominal Pipe Size in Inches															
Pitch In Inches	1	1¼	1½	2	2½	3	4	6	8	10	12	14	16	18	20	24
3.5	1.80															
4	1.65															
5	1.46	1.60	1.80													
6	1.34	1.45	1.55	1.75												
7	1.25	1.35	1.43	1.57	1.75											
8	1.20	1.28	1.34	1.45	1.60	1.80										
9	1.16	1.23	1.28	1.37	1.51	1.68										
10	1.13	1.19	1.24	1.32	1.44	1.57	1.82									
15	1.06	1.08	1.10	1.15	1.21	1.29	1.42	1.78								
20	1.04	1.05	1.06	1.08	1.13	1.17	1.25	1.49	1.73							
25		1.04	1.04	1.06	1.08	1.11	1.17	1.33	1.51	1.72						
30				1.04	1.05	1.07	1.12	1.24	1.37	1.54	1.70	1.80				
35					1.06	1.06	1.09	1.17	1.28	1.42	1.54	1.64	1.78			
40						1.05	1.07	1.14	1.22	1.33	1.44	1.52	1.64	1.75		
50							1.05	1.09	1.15	1.22	1.29	1.35	1.44	1.53	1.64	1.83
60								1.06	1.11	1.16	1.21	1.25	1.31	1.39	1.46	1.62
70								1.05	1.08	1.12	1.17	1.19	1.24	1.30	1.35	1.47
80									1.06	1.09	1.13	1.15	1.19	1.24	1.28	1.38
90									1.04	1.06	1.10	1.13	1.16	1.19	1.23	1.32
100										1.05	1.08	1.10	1.13	1.15	1.19	1.23

Note: Common practice would normally limit the lower end of the table to 5% of additional heat and above 80% an engineer would likely opt for two (2) parallel cables.

R260105-30 Electrical Demolition (Removal for Replacement)

The purpose of this reference number is to provide a guide to users for electrical "removal for replacement" by applying the rule of thumb: 1/3 of new installation time (typical range from 20% to 50%) for removal. Remember to use reasonable judgment when applying the suggested percentage factor. For example:

Contractors have been requested to remove an existing fluorescent lighting fixture and replace with a new fixture utilizing energy saver lamps and electronic ballast.

In order to fully understand the extent of the project, contractors should visit the job site and estimate the time to perform the renovation work in accordance with applicable national, state and local regulation and codes.

The contractor may need to add extra labor hours to his estimate if he discovers unknown concealed conditions such as: contaminated asbestos ceiling, broken acoustical ceiling tile and need to repair, patch and touch-up paint all the damaged or disturbed areas, tasks normally assigned to general contractors. In addition, the owner could request that the contractors salvage the materials removed and turn over the materials to the owner or dispose of the materials to a reclamation station. The normal removal item is 0.5 labor hour for a lighting fixture and 1.5 labor-hours for new installation time. Revise the estimate times from 2 labor-hours work up to a minimum 4 labor-hours work for just fluorescent lighting fixture.

For removal of large concentrations of lighting fixtures in the same area, apply an "economy of scale" to reduce estimating labor-hours.

R260519-80 Undercarpet Systems

Takeoff Procedure for Power Systems: List components for each fitting type, tap, splice, and bend on your quantity takeoff sheet. Each component must be priced separately. Start at the power supply transition fittings and survey each circuit for the components needed. List the quantities of each component under a specific circuit number. Use the floor plan layout scale to get cable footage.

Reading across the list, combine the totals of each component in each circuit and list the total quantity in the last column. Calculate approximately 5% for scrap for items such as cable, top shield, tape, and spray adhesive. Also provide for final variations that may occur on-site.

Suggested guidelines are:

1. Equal amounts of cable and top shield should be priced.
2. For each roll of cable, price a set of cable splices.
3. For every 1 ft. of cable, price 2-1/2 ft. of hold-down tape.
4. For every 3 rolls of hold-down tape, price 1 can of spray adhesive.

Adjust final figures wherever possible to accommodate standard packaging of the product. This information is available from the distributor.

Each transition fitting requires:

1. 1 base
2. 1 cover
3. 1 transition block

Each floor fitting requires:

1. 1 frame/base kit
2. 1 transition block
3. 2 covers (duplex/blank)

Each tap requires:

1. 1 tap connector for each conductor
2. 1 pair insulating patches
3. 2 top shield connectors

Each splice requires:

1. 1 splice connector for each conductor
2. 1 pair insulating patches
3. 3 top shield connectors

Each cable bend requires:

1. 2 top shield connectors

Each cable dead end (outside of transition block) requires:

1. 1 pair insulating patches

Labor does not include:

1. Patching or leveling uneven floors
2. Filling in holes or removing projections from concrete slabs.
3. Sealing porous floors
4. Sweeping and vacuuming floors
5. Removal of existing carpeting
6. Carpet square cut-outs
7. Installation of carpet squares

Takeoff Procedures for Telephone Systems: After reviewing floor plans identify each transition. Number or letter each cable run from that fitting. Start at the transition fitting and survey each circuit for the components needed. List the cable type, terminations, cable length, and floor fitting type under the specific circuit number. Use the floor plan layout scale to get the cable footage. Add some extra length (next higher increment of 5 feet) to preconnectorized cable.

Transition fittings require:

1. 1 base plate
2. 1 cover
3. 1 transition block

Floor fittings require:

1. 1 frame/base kit
2. 2 covers
3. Modular jacks

Reading across the list, combine the list of components in each circuit and list the total quantity in the last column. Calculate the necessary scrap factors for such items as tape, bottom shield and spray adhesive. Also provide for final variations that may occur on-site.

Adjust final figures whenever possible to accommodate standard packaging. Check that items such as transition fittings, floor boxes, and floor fittings that are to utilize both power and telephone have been priced as combination fittings, so as to avoid duplication.

Make sure to include marking of floors and drilling of fasteners if fittings specified are not the adhesive type.

Labor does not include:

1. Conduit or raceways before transition of floor boxes
2. Telephone cable before transition boxes
3. Terminations before transition boxes
4. Floor preparation as described in power section

Be sure to include all cable folds when pricing labor.

Takeoff Procedure for Data Systems: Start at the transition fittings and take off quantities in the same manner as the telephone system, keeping in mind that data cable does not require top or bottom shields.

The data cable is simply cross-taped on the cable run to the floor fitting.

Data cable can be purchased in either bulk form in which case coaxial connector material and labor must be priced, or in preconnectorized cut lengths.

Data cable cannot be folded and must be notched at 1 inch intervals. A count of all turns must be added to the labor portion of the estimate. (Note: Some manufacturers have prenotched cable.)

Notching required:

1. 90 degree turn requires 8 notches per side
2. 180 degree turn requires 16 notches per side

Floor boxes, transition boxes, and fittings are the same as described in the power and telephone procedures.

Since undercarpet systems require special hand tools, be sure to include this cost in proportion to number of crews involved in the installation.

Job Conditions: Productivity is based on new construction in an unobstructed area. Staging area is assumed to be within 200′ of work being performed.

R260519-90 Wire

Wire quantities are taken off by either measuring each cable run or by extending the conduit and raceway quantities times the number of conductors in the raceway. Ten percent should be added for waste and tie-ins. Keep in mind that the unit of measure of wire is C.L.F. not L.F. as in raceways so the formula would read:

$$\frac{\text{(L.F. Raceway x No. of Conductors) x 1.10}}{100} = \text{C.L.F.}$$

Price per C.L.F. of wire includes:
1. Set up wire coils or spools on racks
2. Attaching wire to pull in means
3. Measuring and cutting wire
4. Pulling wire into a raceway
5. Identifying and tagging

Price does not include:
1. Connections to breakers, panelboards, or equipment
2. Splices

Job Conditions: Productivity is based on new construction to a height of 15′ using rolling staging in an unobstructed area. Material staging is assumed to be within 100′ of work being performed.

Economy of Scale: If more than three wires at a time are being pulled, deduct the following percentages from the labor of that grouping:

4-5 wires	25%
6-10 wires	30%
11-15 wires	35%
over 15 wires	40%

If a wire pull is less than 100′ in length and is interrupted several times by boxes, lighting outlets, etc., it may be necessary to add the following lengths to each wire being pulled:

Junction box to junction box	2 L.F.
Lighting panel to junction box	6 L.F.
Distribution panel to sub panel	8 L.F.
Switchboard to distribution panel	12 L.F.
Switchboard to motor control center	20 L.F.
Switchboard to cable tray	40 L.F.

Measure of Drops and Riser: It is important when taking off wire quantities to include the wire for drops to electrical equipment. If heights of electrical equipment are not clearly stated, use the following guide:

	Bottom A.F.F.	Top A.F.F.	Inside Cabinet
Safety switch to 100A	5′	6′	2′
Safety switch 400 to 600A	4′	6′	3′
100A panel 12 to 30 circuit	4′	6′	3′
42 circuit pancl	3′	6′	4′
Switch box	3′	3′6″	1′
Switchgear	0′	8′	8′
Motor control centers	0′	8′	8′
Transformers - wall mount	4′	8′	2′
Transformers - floor mount	0′	12′	4′

R260519-92 Minimum Copper and Aluminum Wire Size Allowed for Various Types of Insulation

Minimum Wire Sizes									
	Copper		Aluminum			Copper		Aluminum	
Amperes	THW THWN or XHHW	THHN XHHW *	THW XHHW	THHN XHHW *	Amperes	THW THWN or XHHW	THHN XHHW *	THW XHHW	THHN XHHW *
15A	#14	#14	#12	#12	195	3/0	2/0	250kcmil	4/0
20	#12	#12	#10	#10	200	3/0	3/0	250kcmil	4/0
25	#10	#10	#10	#10	205	4/0	3/0	250kcmil	4/0
30	#10	#10	# 8	# 8	225	4/0	3/0	300kcmil	250kcmil
40	# 8	# 8	# 8	# 8	230	4/0	4/0	300kcmil	250kcmil
45	# 8	# 8	# 6	# 8	250	250kcmil	4/0	350kcmil	300kcmil
50	# 8	# 8	# 6	# 6	255	250kcmil	4/0	400kcmil	300kcmil
55	# 6	# 8	# 4	# 6	260	300kcmil	4/0	400kcmil	350kcmil
60	# 6	# 6	# 4	# 6	270	300kcmil	250kcmil	400kcmil	350kcmil
65	# 6	# 6	# 4	# 4	280	300kcmil	250kcmil	500kcmil	350kcmil
75	# 4	# 6	# 3	# 4	285	300kcmil	250kcmil	500kcmil	400kcmil
85	# 4	# 4	# 2	# 3	290	350kcmil	250kcmil	500kcmil	400kcmil
90	# 3	# 4	# 2	# 2	305	350kcmil	300kcmil	500kcmil	400kcmil
95	# 3	# 4	# 1	# 2	310	350kcmil	300kcmil	500kcmil	500kcmil
100	# 3	# 3	# 1	# 2	320	400kcmil	300kcmil	600kcmil	500kcmil
110	# 2	# 3	1/0	# 1	335	400kcmil	350kcmil	600kcmil	500kcmil
115	# 2	# 2	1/0	# 1	340	500kcmil	350kcmil	600kcmil	500kcmil
120	# 1	# 2	1/0	1/0	350	500kcmil	350kcmil	700kcmil	500kcmil
130	# 1	# 2	2/0	1/0	375	500kcmil	400kcmil	700kcmil	600kcmil
135	1/0	# 1	2/0	1/0	380	500kcmil	400kcmil	750kcmil	600kcmil
150	1/0	# 1	3/0	2/0	385	600kcmil	500kcmil	750kcmil	600kcmil
155	2/0	1/0	3/0	3/0	420	600kcmil	500kcmil		700kcmil
170	2/0	1/0	4/0	3/0	430		500kcmil		750kcmil
175	2/0	2/0	4/0	3/0	435		600kcmil		750kcmil
180	3/0	2/0	4/0	4/0	475		600kcmil		

*Dry Locations Only

Notes:

1. Size #14 to 4/0 is in AWG units (American Wire Gauge).
2. Size 250 to 750 is in kcmil units (Thousand Circular Mils).
3. Use next higher ampere value if exact value is not listed in table.
4. For loads that operate continuously increase ampere value by 25% to obtain proper wire size.
5. Table R260519-92 has been written for estimating purpose only, based on ambient temperature of 30°C (86° F); for ambient temperature other than 30°C (86° F), ampacity correction factors will be applied.

R260533-20 Conduit To 15′ High

List conduit by quantity, size, and type. Do not deduct for lengths occupied by fittings, since this will be allowance for scrap. Example:

A. Aluminum — size
B. Rigid galvanized — size
C. Steel intermediate (IMC) — size
D. Rigid steel, plastic-coated 20 Mil. — size
E. Rigid steel, plastic-coated 40 Mil. — size
F. Electric metallic tubing (EMT) — size
G. PVC Schedule 40 — size

Types (A) thru (E) listed above contain the following per 100 L.F.:

1. (11) Threaded couplings
2. (11) Beam-type hangers
3. (2) Factory sweeps
4. (2) Fiber bushings
5. (4) Locknuts
6. (2) Field threaded pipe terminations
7. (2) Removal of concentric knockouts

Type (F) contains per 100 L.F.:

1. (11) Set screw couplings
2. (11) Beam clamps
3. (2) Field bends on 1/2″ and 3/4″ diameter
4. (2) Factory sweeps for 1″ and above
5. (2) Set screw steel connectors
6. (2) Removal of concentric knockouts

Type (G) contains per 100 L.F.:

1. (11) Field cemented couplings
2. (34) Beam clamps
3. (2) Factory sweeps
4. (2) Adapters
5. (2) Locknuts
6. (2) Removal of concentric knockouts

Labor-hours for all conduit to 15′ include:

1. Unloading by hand
2. Hauling by hand to an area up to 200′ from loading dock
3. Setup of rolling staging
4. Installation of conduit and fittings as described in Conduit models (A) thru (G)

Not included in the material and labor are:

1. Staging rental or purchase
2. Structural modifications
3. Wire
4. Junction boxes
5. Fittings in excess of those described in conduit models (A) thru (G)
6. Painting of conduit

Fittings

Only those fittings listed above are included in the linear foot totals, although they should be listed separately from conduit lengths, without prices, to ensure proper quantities for material procurement.

If the fittings required exceed the quantities included in the model conduit runs, then material and labor costs must be added to the difference. If actual needs per 100 L.F. of conduit are: (2) sweeps, (4) LBs and (1) field bend, then, (4) LBs and (1) field bend must be priced additionally.

R260533-21 Hangers

It is sometimes desirable to substitute an alternate style of hanger if the support being used is not the type described in the conduit models.

One approach is the substitution method:

1. Find the cost of the type hanger described in the conduit model.
2. Calculate the cost of the desired type hanger (it may be necessary to calculate individual components such as drilling, expansion shields, etc.).
3. Calculate the cost difference (delta) between the two types of hangers.
4. Multiply the cost delta by the number of hangers in the model.
5. Divide the total delta cost for hangers in the model by the length of the model to find the delta cost per L.F. for that model.
6. Modify the given unit costs per L.F. for the model by the delta cost per L.F.

Another approach to hanger configurations would be to start with the conduit only and add all the supports and any other items as separate lines. This procedure is most useful if the project involves racking many runs of conduit on a single hanger, for instance, a trapeze type hanger.

Example: Five (5) 2″ RGS conduits, 50 L.F. each, are to be run on trapeze hangers from one pull box to another. The run includes one 90° bend.

1. List the hangers' components to create an assembly cost for each 2′-wide trapeze.
2. List the components for the 50′ conduit run, noting that 6 trapeze supports will be required.

Job Conditions: Productivities are based on new construction to 15′ high, using scaffolding in an unobstructed area. Material storage is assumed to be within 100′ of work being performed.

Add to labor for elevated installations:

15′ to 20′ High–10%	30′ to 35 High–30%
20′ to 25′ High–20%	35′ to 40′ High–35%
25′ to 30′ High–25%	Over 40′ High–40%

Add these percentages to the L.F. labor cost, but not to fittings. Add these percentages only to quantities exceeding the different height levels, rather than the total conduit quantities.

Linear foot price for labor does not include penetrations in walls or floors and must be added to the estimate.

R260533-22 Conductors in Conduit

Table below lists maximum number of conductors for various sized conduit using THW, TW or THWN insulations.

Copper Wire Size	Conduit Size ½″			¾″			1″			1-¼″			1-½″			2″			2-½″			3″		3-½″		4″	
	TW	THW	THWN	TW	THW	THWN	TW	THW	THWN	TW	THW	THWN	TW	THW	THWN	TW	THW	THWN	TW	THW	THWN	THW	THWN	THW	THWN	THW	THWN
#14	9	6	13	15	10	24	25	16	39	44	29	69	60	40	94	99	65	154	142	93		143		192			
#12	7	4	10	12	8	18	19	13	29	35	24	51	47	32	70	78	53	114	111	76	164	117		157			
#10	5	4	6	9	6	11	15	11	18	26	19	32	36	26	44	60	43	73	85	61	104	95	160	127		163	
#8	2	1	3	4	3	5	7	5	9	12	10	16	17	13	22	28	22	36	40	32	51	49	79	66	106	85	136
#6		1	1		2	4		4	6		7	11		10	15		16	26		23	37	36	57	48	76	62	98
#4		1	1		1	2		3	4		5	7		7	9		12	16		17	22	27	35	36	47	47	60
#3		1	1		1	1		2	3		4	6		6	8		10	13		15	19	23	29	31	39	40	51
#2		1	1		1	1		2	3		4	5		5	7		9	11		13	16	20	25	27	33	34	43
#1					1	1		1	1		3	3		4	5		6	8		9	12	14	18	19	25	25	32
1/0					1	1		1	1		2	3		3	4		5	7		8	10	12	15	16	21	21	27
2/0					1	1		1	1		1	2		3	3		5	6		7	8	10	13	14	17	18	22
3/0					1	1		1	1		1	1		2	3		4	5		6	7	9	11	12	14	15	18
4/0						1		1	1		1	1		1	2		3	4		5	6	7	9	10	12	13	15
250 kcmil								1	1		1	1		1	1		2	3		4	4	6	7	8	10	10	12
300								1	1		1	1		1	1		2	3		3	4	5	6	7	8	9	11
350									1		1	1		1	1		1	2		3	3	4	5	6	7	8	9
400											1	1		1	1		1	1		2	3	4	5	5	6	7	8
500											1	1		1	1		1	1		1	2	3	4	4	5	6	7
600												1		1	1		1	1		1	1	3	3	4	4	5	5
700														1	1		1	1		1	1	2	3	3	4	4	5
750 ▼														1	1		1	1		1	1	2	2	3	3	4	4

R260533-30 Labor for Couplings and Fittings

The labor included in the unit price lines for couplings is for their installation separate from fitting. However, the labor included in the unit price lines for fittings covers their complete installation, which may include the installation of two couplings. If couplings are required to complete a fitting installation, the material only for two couplings needs to be added to the appropriate unit price lines.

Electrical — R2624 Switchboards & Panelboards

R262416-50 Load Centers and Panelboards

When pricing Load Centers list panels by size and type. List Breakers in a separate column of the "Quantity Sheet," and define by phase and ampere rating.

Material and Labor prices include breakers; for example: a 100A, 3-Wire, 102/240V, 18 circuit panel w/ main breaker, as described in the unit cost section, contains 18 single pole 20A breakers.

If you do not choose to include a full panel of single pole breakers, use the following method to adjust material and labor costs.

Example: In an 18 circuit panel only 16 single pole breakers are desired, requiring that the cost of 2 breakers be subtracted from the panel cost.

1. Go to the appropriate unit cost section of the book to find the unit prices of the given circuit breaker type.
2. Modify those costs as follows:
   Bare material price x 0.50
   Bare labor cost x 0.60.
3. Multiply those modified bare costs by 2 (breakers in this example).
4. Subtract the modified costs for the 2 breakers from the given cost of the panel.

Labor-hours for Load Center installation includes:

1. Unloading, uncrating, and handling enclosures 200′ from unloading area
2. Measuring and marking
3. Drilling (4) lead anchor type fasteners using a hammer drill
4. Mounting and leveling panel to a height of 6′
5. Preparation and termination of feeder cable to lugs or main breaker
6. Branch circuit identification
7. Lacing using tie wraps
8. Testing and load balancing
9. Marking panel directory

Not included in the material and labor are:

1. Modifications to enclosure
2. Structural supports
3. Additional lugs
4. Plywood backboards
5. Painting or lettering

Note: Knockouts are included in the price of terminating pipe runs and need not be added to the Load Center costs.

Job Conditions: Productivity is based on new construction to a height of 6′, in an unobstructed area. Material staging area is assumed to be within 100′ of work being performed.

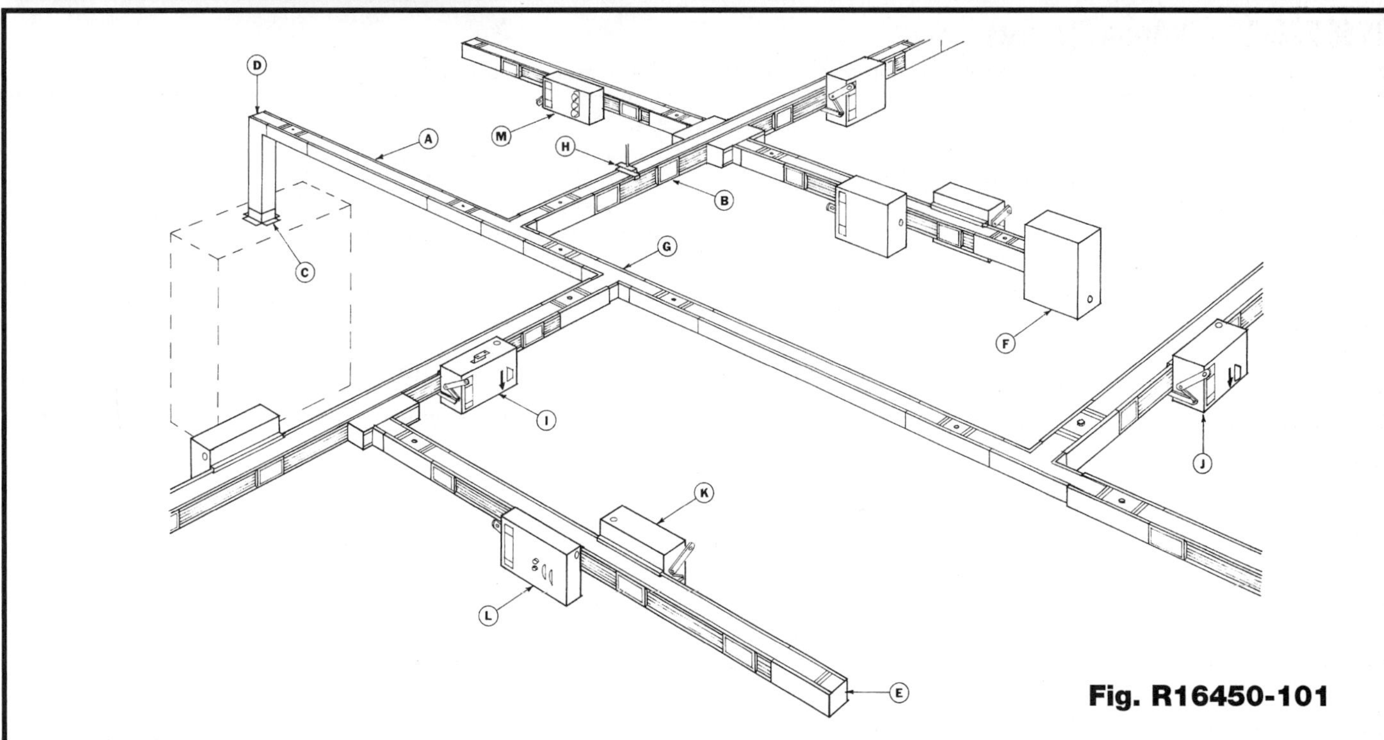

Fig. R16450-101

R262513-10 Aluminum Bus Duct

When taking off bus duct identify the system as either:

1. Aluminum
2. Copper

List straight lengths by type and size

A. Feeder — 800 A
B. Plug in — 800 A

Do not measure thru fittings as you would on conduit, since there is no allowance for scrap in bus duct systems.

If upon taking off linear foot quantities of bus duct you find your quantities are not divisible evenly by 10 ft., then the remainder must be priced as a special item and quoted from the manufactuer. Do not use the bare material cost per L.F. for these special items. You can, however, safely use the bare labor cost per L.F. for the entire length.

Identify fittings by type and ampere rating.

Example:

C. Switchboard stub 800 A

D. Elbows 800 A

E. End box 800 A

F. Cable tap box 800 A

G. Tee Fittings 800 A

H. Hangers

Plug-in Units – List separately plug-in units and identify by type and ampere rating

I. Plug-in switches 600 Volt 3 phase 60 A

J. Plug-in molded case C.B. 60 A

K. Combination starter FVNR NEMA 1

L. Combination contactor & fused switch NEMA 1

M. Combination fusible switch & lighting control 60 A

Labor-hours for feeder and plug-in sections include:

1. Unloading and uncrating
2. Hauling up to 200 ft. from loading dock
3. Measuring and marking
4. Setup of rolling staging
5. Installing hangers
6. Hanging and bolting sections
7. Aligning and leveling
8. Testing

Labor-hours do not include:

1. Modifications to existing structure for hanger supports
2. Threaded rod in excess of 2 ft.
3. Welding
4. Penetrations thru walls
5. Staging rental

Deduct the following percentages from labor only:

150 ft. to 250 ft. — 10%
251 ft. to 350 ft. — 15%
351 ft. to 500 ft. — 20%
Over 500 ft. — 25%

Deduct percentage only if runs are contained in the same area.

Example: If the job entails running 100 ft. in 5 different locations do not deduct 20%, but if the duct is being run in 1 area and the quantity is 500 ft. then you would deduct 20%.

Deduct only from straight lengths, not fittings or plug-in units.

R262726-90 Wiring Devices

Wiring devices should be priced on a separate takeoff form which includes boxes, covers, conduit and wire.

Labor-hours for devices include:

1. Stripping of wire
2. Attaching wire to device using terminators on the device itself, lugs, set screws etc.
3. Mounting of device in box

Labor-hours do not include:

1. Conduit
2. Wire
3. Boxes
4. Plates

Economy of Scale – for large concentrations of devices in the same area deduct the following percentages from labor-hours:

1	to	10	0%
11	to	25	20%
26	to	50	25%
51	to	100	30%
	over	100	35%

R262726-90 Wiring Devices (cont.)

NEMA No.	15 R	20 R	30 R	50 R	60 R
1 125V 2 Pole, 2 Wire	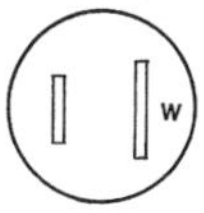				
2 250V 2 Pole, 2 Wire		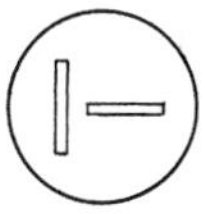	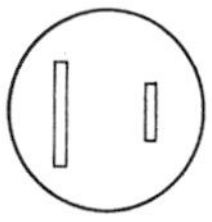		
5 125V 2 Pole, 3 Wire	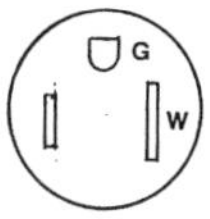	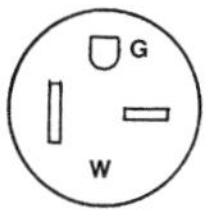		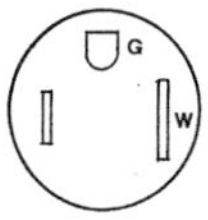	
6 250V 2 Pole, 3 Wire			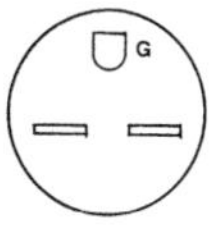	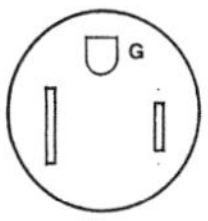	
7 277V, AC 2 Pole, 3 Wire				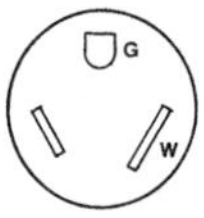	
10 125/250V 3 Pole, 3 Wire		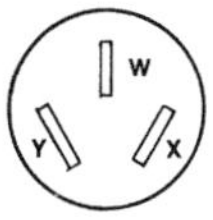	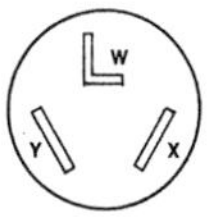	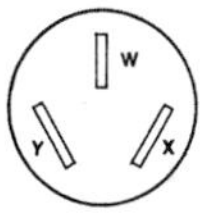	
11 3 Phase 250V 3 Pole, 3 Wire	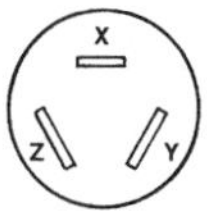	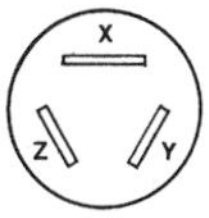	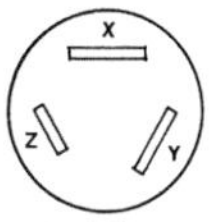	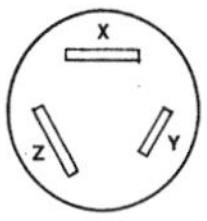	
14 125/250V 3 Pole, 4 Wire	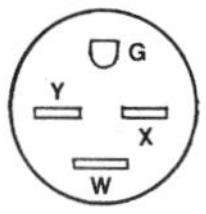	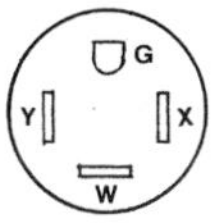	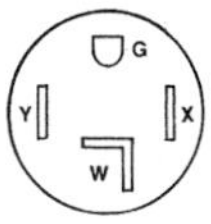	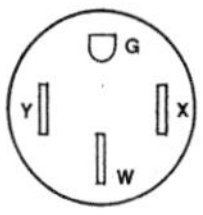	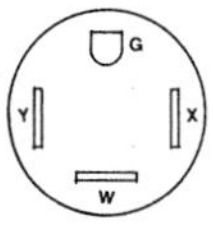
15 3 Phase 250V 3 Pole, 4 Wire	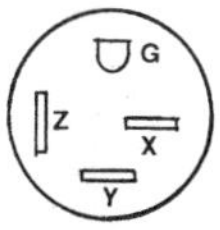		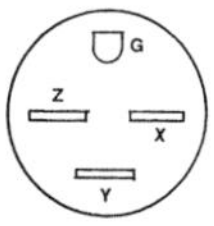	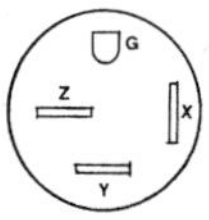	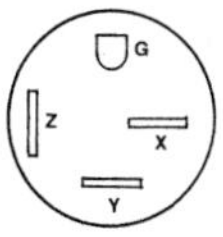
18 3 Phase 208Y/120V 4 Pole, 4 Wire	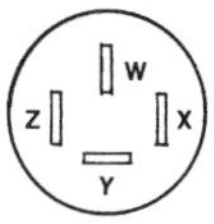	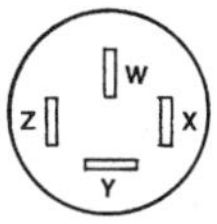	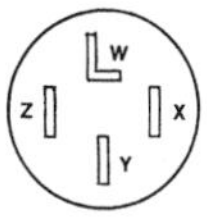	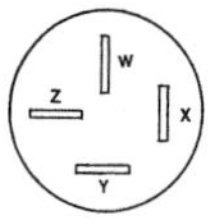	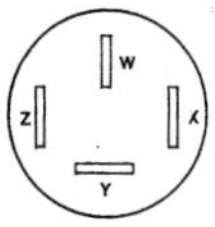

R262726-90 Wiring Devices (cont.)

NEMA No.	15 R	20 R	30 R	NEMA No.	15 R	20 R	30 R
L 1 125V 2 Pole, 2 Wire	W			**L 13** 3 Phase 600V 3 Pole, 3 Wire			X Y Z
L 2 250V 2 Pole, 2 Wire		15A		**L 14** 125/250V 3 Pole, 4 Wire		G X Y Z	G X W Z
L 5 125 V 2 Pole, 3 Wire	G W	G W	G W	**L 15** 3 Phase 250 V 3 Pole, 4 Wire		G X W Y	G X Y Z
L 6 250 V 2 Pole, 3 Wire	G X Y	G X Y	G X Y	**L 16** 3 Phase 480V 3 Pole, 4 Wire		G X Y Z	G X Y Z
L 7 227 V, AC 2 Pole, 3 Wire	G W	G W	G W	**L 17** 3 Phase 600 V 3 Pole, 4 Wire			G X Y Z
L 8 480 V 2 Pole, 3 Wire		G X Y	G X Y	**L 18** 3 Phase 208Y/120V 4 Pole, 4 Wire		W X Y Z	W X Y Z
L 9 600 V 2 Pole, 3 Wire		G X Y	G X Y	**L 19** 3 Phase 480Y/277V 4 Pole, 4 Wire		W X Y Z	W X Y Z
L 10 125 /250V 3 Pole, 3 Wire		W X Y	W X Y	**L 20** 3 Phase 600Y/347V 4 Pole, 4 Wire		W X Y Z	W X Y Z
L 11 3 Phase 250 V 3 Pole, 3 Wire		Y Z	X Y Z	**L 21** 3 Phase 208Y/120V 4 Pole, 5 Wire		W X G Y Z	W X G Y Z
L 12 3 Phase 480 V 3 Pole, 3 Wire		X Y Z	X Y Z	**L 22** 3 Phase 480Y/277V 4 Pole, 5 Wire		W X G Y Z	W X G Y Z
				L 23 3 Phase 600Y/347V 4 Pole, 5 Wire		W X G Y Z	W X G Y Z

R265113-40 Interior Lighting Fixtures

When taking off interior lighting fixtures, it is advisable to set up your quantity work sheet to conform to the lighting schedule as it appears on the print. Include the alpha-numeric code plus the symbol on your work sheet.

Take off a particular section or floor of the building and count each type of fixture before going on to another type. It would also be advantageous to include on the same work sheet the pipe, wire, fittings and circuit number associated with each type of lighting fixture. This will help you identify the costs associated with any particular lighting system and in turn make material purchases more specific as to when and how much to order under the classification of lighting.

By taking off lighting first you can get a complete "WALK THRU" of the job. This will become helpful when doing other phases of the project.

Materials for a recessed fixture include:
1. Fixture
2. Lamps
3. 6′ of jack chain
4. (2) S hooks
5. (2) Wire nuts

Labor for interior recessed fixtures include:
1. Unloading by hand
2. Hauling by hand to an area up to 200′ from loading dock
3. Uncrating
4. Layout
5. Installing fixture
6. Attaching jack chain & S hooks
7. Connecting circuit power
8. Reassembling fixture
9. Installing lamps
10. Testing

Material for surface mounted fixtures includes:
1. Fixture
2. Lamps
3. Either (4) lead type anchors, (4) toggle bolts, or (4) ceiling grid clips
4. (2) Wire nuts

Material for pendent mounted fixtures includes:
1. Fixture
2. Lamps
3. (2) Wire nuts
4. Rigid pendents as required by type of fixtures
5. Canopies as required by type of fixture

Labor hours include the following for both surface and pendent fixtures:
1. Unloading by hand
2. Hauling by hand to an area up to 200′ from loading dock
3. Uncrating
4. Layout and marking
5. Drilling (4) holes for either lead anchors or toggle bolts using a hammer drill
6. Installing fixture
7. Leveling fixture
8. Connecting circuit power
9. Installing lamps
10. Testing

Labor for surface or pendent fixtures does not include:
1. Conduit
2. Boxes or covers
3. Connectors
4. Fixture whips
5. Special support
6. Switching
7. Wire

Economy of Scale: For large concentrations of lighting fixtures in the same area deduct the following percentages from labor:

25	to	50	fixtures	15%
51	to	75	fixtures	20%
76	to	100	fixtures	25%
101	and over			30%

Job Conditions: Productivity is based on new construction in a unobstructed first floor location, using rolling staging to 15′ high.

Material staging is assumed to be within 100′ of work being performed.

Add the following percentages to labor for elevated installations:

15′	to	20′	high	10%
21′	to	25′	high	20%
26′	to	30′	high	30%
31′	to	35′	high	40%
36′	to	40′	high	50%
41′	and over			60%

R271323-40 Fiber Optics

Fiber optic systems use optical fiber such as plastic, glass, or fused silica, a transparent material, to transmit radiant power (i.e., light) for control, communication, and signaling applications. The types of fiber optic cables can be nonconductive, conductive, or composite. The composite cables contain fiber optics and current-carrying electrical conductors. The configuration for one of the fiber optic systems is as follows:

The transceiver module acts as transmitting and receiving equipment in a common house, which converts electrical energy to light energy or vice versa.

Pricing the fiber optic system is not an easy task. The performance of the whole system will affect the cost significantly. New specialized tools and techniques decrease the installing cost tremendously. In the fiber optic section of Means Electrical Cost Data, a benchmark for labor-hours and material costs is set up so that users can adjust their costs according to unique project conditions.

Units for Measure: Fiber optic cable is measured in hundred linear feet (C.L.F.) or industry units of measure - meter (m) or kilometer (km). The connectors are counted as units (EA.)

Material Units: Generally, the material costs include only the cable. All the accessories shall be priced separately.

Labor Units: The following procedures are generally included for the installation of fiber optic cables:

- Receiving
- Material handling
- Setting up pulling equipment
- Measuring and cutting cable
- Pulling cable

These additional items are listed and extended: Terminations

Takeoff Procedure: Cable should be taken off by type, size, number of fibers, and number of terminations required. List the lengths of each type of cable on the takeoff sheets. Total and add 10% for waste. Transfer the figures to a cost analysis sheet and extend.

Communications — R2715 Communications Horizontal Cabling

R271513-75 High Performance Cable

There are several categories used to describe high performance cable. The following information includes a description of categories CAT 3, 5, 5e, 6, and 7, and details classifications of frequency and specific standards. The category standards have evolved under the sponsorship of organizations such as the Telecommunication Industry Association (TIA), the Electronic Industries Alliance (EIA), the American National Standards Institute (ANSI), the International Organization for Standardization (ISO), and the International Electrotechnical Commission (IEC), all of which have catered to the increasing complexities of modern network technology. For network cabling, users must comply with national or international standards. A breakdown of these categories is as follows:

Category 3: Designed to handle frequencies up to 16 MHz.

Category 5: (TIA/EIA 568A) Designed to handle frequencies up to 100 MHz.

Category 5e: Additional transmission performance to exceed Category 5.

Category 6 (draft): Development by TIA and other international groups to handle frequencies of 250 MHz.

Category 7 (draft): Under development to handle a frequency range from 1 to 600 MHz.

Exterior Improvements — R3292 Turf & Grasses

R329219-50 Seeding

The type of grass is determined by light, shade and moisture content of soil plus intended use. Fertilizer should be disked 4″ before seeding. For steep slopes disk five tons of mulch and lay two tons of hay or straw on surface per acre after seeding. Surface mulch can be staked, lightly disked or tar emulsion sprayed. Material for mulch can be wood chips, peat moss, partially rotted hay or straw, wood fibers, and sprayed emulsions. Hemp seed blankets with fertilizer are also available. For spring seeding, watering is necessary. Late fall seeding may have to be reseeded in the spring. Hydraulic seeding, power mulching, and aerial seeding can be used on large areas.

Utilities — R3311 Water Utility Distribution Piping

R331113-80 Piping Designations

There are several systems currently in use to describe pipe and fittings. The following paragraphs will help to identify and clarify classifications of piping systems used for water distribution.

Piping may be classified by schedule. Piping schedules include 5S, 10S, 10, 20, 30, Standard, 40, 60, Extra Strong, 80, 100, 120, 140, 160, and Double Extra Strong. These schedules are dependent upon the pipe wall thickness. The wall thickness of a particular schedule may vary with pipe size.

Ductile iron pipe for water distribution is classified by Pressure Classes such as Class 150, 200, 250, 300, and 350. These classes are actually the rated water working pressure of the pipe in pounds per square inch (psi). The pipe in these pressure classes is designed to withstand the rated water working pressure plus a surge allowance of 100 psi.

The American Water Works Association (AWWA) provides standards for various types of **plastic pipe**. C-900 is the specification for polyvinyl chloride (PVC) piping used for water distribution in sizes ranging from 4″ through 12″. C-901 is the specification for polyethylene (PE) pressure pipe, tubing and fittings used for water distribution in sizes ranging from 1/2″ through 3″. C-905 is the specification for PVC piping sizes 14″ and greater.

PVC pressure-rated pipe is identified using the standard dimensional ratio (SDR) method. This method is defined by the American Society for Testing and Materials (ASTM) Standard D 2241. This pipe is available in SDR numbers 64, 41, 32.5, 26, 21, 17, and 13.5. Pipe with an SDR of 64 will have the thinnest wall while pipe with an SDR of 13.5 will have the thickest wall. When the pressure rating (PR) of a pipe is given in psi, it is based on a line supplying water at 73 degrees F.

The National Sanitation Foundation (NSF) seal of approval is applied to products that can be used with potable water. These products have been tested to ANSI/NSF Standard 14.

Valves and strainers are classified by American National Standards Institute (ANSI) Classes. These Classes are 125, 150, 200, 250, 300, 400, 600, 900, 1500 and 2500. Within each class there is an operating pressure range dependent upon temperature. Design parameters should be compared to the appropriate material dependent, pressure-temperature rating chart for accurate valve selection.

Utilities — R3371 Elec. Utility Transmission & Distribution

R337119-30 Concrete for Conduit Encasement

Table below lists C.Y. of concrete for 100 L.F. of trench. Conduits separation center to center should meet 7.5″ (N.E.C.).

Number of Conduits	1	2	3	4	6	8	9	Number of Conduits
Trench Dimension	11.5″ x 11.5″	11.5″ x 19″	11.5″ x 27″	19″ x 19″	19″ x 27″	19″ x 38″	27″ x 27″	Trench Dimension
Conduit Diameter 2.0″	3.29	5.39	7.64	8.83	12.51	17.66	17.72	Conduit Diameter 2.0″
2.5″	3.23	5.29	7.49	8.62	12.19	17.23	17.25	2.5″
3.0″	3.15	5.13	7.24	8.29	11.71	16.59	16.52	3.0″
3.5″	3.08	4.97	7.02	7.99	11.26	15.98	15.84	3.5″
4.0″	2.99	4.80	6.76	7.65	10.74	15.30	15.07	4.0″
5.0″	2.78	4.37	6.11	6.78	9.44	13.57	13.12	5.0″
6.0″	2.52	3.84	5.33	5.74	7.87	11.48	10.77	6.0″

Transportation — R3472 Railway Construction

R347216-10 Single Track R.R. Siding

The costs for a single track RR siding in the Unit Price section include the components shown in the table below.

Description of Component	Qty. per L.F. of Track	Unit
Ballast, 1-1/2″ crushed stone	.667	C.Y.
6″ x 8″ x 8′-6″ Treated timber ties, 22″ O.C.	.545	Ea.
Tie plates, 2 per tie	1.091	Ea.
Track rail	2.000	L.F.
Spikes, 6″, 4 per tie	2.182	Ea.
Splice bars w/ bolts, lock washers & nuts, @ 33′ O.C.	.061	Pair
Crew B-14 @ 57 L.F./Day	.018	Day

R347216-20 Single Track, Steel Ties, Concrete Bed

The costs for a R.R. siding with steel ties and a concrete bed in the Unit Price section include the components shown in the table below.

Description of Component	Qty. per L.F. of Track	Unit
Concrete bed, 9′ wide, 10″ thick	.278	C.Y.
Ties, W6x16 x 6′-6″ long, @ 30″ O.C.	.400	Ea.
Tie plates, 4 per tie	1.600	Ea.
Track rail	2.000	L.F.
Tie plate bolts, 1″, 8 per tie	3.200	Ea.
Splice bars w/bolts, lock washers & nuts, @ 33′ O.C.	.061	Pair
Crew B-14 @ 22 L.F./Day	.045	Day

Change Orders

Change Order Considerations

A change order is a written document, usually prepared by the design professional, and signed by the owner, the architect/engineer, and the contractor. A change order states the agreement of the parties to: an addition, deletion, or revision in the work; an adjustment in the contract sum, if any; or an adjustment in the contract time, if any. Change orders, or "extras" in the construction process occur after execution of the construction contract and impact architects/engineers, contractors, and owners.

Change orders that are properly recognized and managed can ensure orderly, professional, and profitable progress for all who are involved in the project. There are many causes for change orders and change order requests. In all cases, change orders or change order requests should be addressed promptly and in a precise and prescribed manner. The following paragraphs include information regarding change order pricing and procedures.

The Causes of Change Orders

Reasons for issuing change orders include:

- Unforeseen field conditions that require a change in the work
- Correction of design discrepancies, errors, or omissions in the contract documents
- Owner-requested changes, either by design criteria, scope of work, or project objectives
- Completion date changes for reasons unrelated to the construction process
- Changes in building code interpretations, or other public authority requirements that require a change in the work
- Changes in availability of existing or new materials and products

Procedures

Properly written contract documents must include the correct change order procedures for all parties—owners, design professionals and contractors—to follow in order to avoid costly delays and litigation.

Being "in the right" is not always a sufficient or acceptable defense. The contract provisions requiring notification and documentation must be adhered to within a defined or reasonable time frame.

The appropriate method of handling change orders is by a written proposal and acceptance by all parties involved. Prior to starting work on a project, all parties should identify their authorized agents who may sign and accept change orders, as well as any limits placed on their authority.

Time may be a critical factor when the need for a change arises. For such cases, the contractor might be directed to proceed on a "time and materials" basis, rather than wait for all paperwork to be processed—a delay that could impede progress. In this situation, the contractor must still follow the prescribed change order procedures including, but not limited to, notification and documentation.

Lack of documentation can be very costly, especially if legal judgments are to be made, and if certain field personnel are no longer available. For time and material change orders, the contractor should keep accurate daily records of all labor and material allocated to the change.

Owners or awarding authorities who do considerable and continual building construction (such as the federal government) realize the inevitability of change orders for numerous reasons, both predictable and unpredictable. As a result, the federal government, the American Institute of Architects (AIA), the Engineers Joint Contract Documents Committee (EJCDC) and other contractor, legal, and technical organizations have developed standards and procedures to be followed by all parties to achieve contract continuance and timely completion, while being financially fair to all concerned.

Pricing Change Orders

When pricing change orders, regardless of their cause, the most significant factor is when the change occurs. The need for a change may be perceived in the field or requested by the architect/engineer *before* any of the actual installation has begun, or may evolve or appear *during* construction when the item of work in question is partially installed. In the latter cases, the original sequence of construction is disrupted, along with all contiguous and supporting systems. Change orders cause the greatest impact when they occur *after* the installation has been completed and must be uncovered, or even replaced. Post-completion changes may be caused by necessary design changes, product failure, or changes in the owner's requirements that are not discovered until the building or the systems begin to function.

Specified procedures of notification and record keeping must be adhered to and enforced regardless of the stage of construction: *before, during,* or *after* installation. Some bidding documents anticipate change orders by requiring that unit prices including overhead and profit percentages—for additional as well as deductible changes—be listed. Generally these unit prices do not fully take into account the ripple effect, or impact on other trades, and should be used for general guidance only.

When pricing change orders, it is important to classify the time frame in which the change occurs. There are two basic time frames for change orders: *pre-installation change orders,* which occur before the start of construction, and *post-installation change orders,* which involve reworking after the original installation. Change orders that occur between these stages may be priced according to the extent of work completed using a combination of techniques developed for pricing *pre-* and *post-installation* changes.

Factors To Consider When Pricing Change Orders

As an estimator begins to prepare a change order, the following questions should be reviewed to determine their impact on the final price.

General

- *Is the change order work* pre-installation *or* post-installation?

  Change order work costs vary according to how much of the installation has been completed. Once workers have the project scoped in their minds, even though they have not started, it can be difficult to refocus. Consequently they may spend more than the normal amount of time understanding the change. Also, modifications to work in place, such as trimming or refitting, usually take more time than was initially estimated. The greater the amount of work in place, the more reluctant workers are to change it. Psychologically they may resent the change and as a result the rework takes longer than normal. Post-installation change order estimates must include demolition of existing work as required to accomplish the change. If the work is performed at a later time, additional obstacles, such as building finishes, may be present which must be protected. Regardless of whether the change occurs pre-installation or post-installation, attempt to isolate the identifiable factors and price them separately. For example, add shipping costs that may be required pre-installation or any demolition required post-installation. Then analyze the potential impact on productivity of psychological and/or learning curve factors and adjust the output rates accordingly. One approach is to break down the typical workday into segments and quantify the impact on each segment.

Change Order Installation Efficiency

The labor-hours expressed (for new construction) are based on average installation time, using an efficiency level. For change order situations, adjustments to this efficiency level should reflect the daily labor-hour allocation for that particular occurrence.

- *Will the change substantially delay the original completion date?*

  A significant change in the project may cause the original completion date to be extended. The extended schedule may subject the contractor to new wage rates dictated by relevant labor contracts. Project supervision and other project overhead must also be extended beyond the original completion date. The schedule extension may also put installation into a new weather season. For example, underground piping scheduled for October installation was delayed until January. As a result, frost penetrated the trench area, thereby changing the degree of difficulty of the task. Changes and delays may have a ripple effect throughout the project. This effect must be analyzed and negotiated with the owner.

- *What is the net effect of a deduct change order?*

  In most cases, change orders resulting in a deduction or credit reflect only bare costs. The contractor may retain the overhead and profit based on the original bid.

Materials

- *Will you have to pay more or less for the new material, required by the change order, than you paid for the original purchase?*

  The same material prices or discounts will usually apply to materials purchased for change orders as new construction. In some instances, however, the contractor may forfeit the advantages of competitive pricing for change orders. Consider the following example:

  A contractor purchased over $20,000 worth of fan coil units for an installation, and obtained the maximum discount. Some time later it was determined the project required an additional matching unit. The contractor has to purchase this unit from the original supplier to ensure a match. The supplier at this time may not discount the unit because of the small quantity, and the fact that he is no longer in a competitive situation. The impact of quantity on purchase can add between 0% and 25% to material prices and/or subcontractor quotes.

- *If materials have been ordered or delivered to the job site, will they be subject to a cancellation charge or restocking fee?*

  Check with the supplier to determine if ordered materials are subject to a cancellation charge. Delivered materials not used as result of a change order may be subject to a restocking fee if returned to the supplier. Common restocking charges run between 20% and 40%. Also, delivery charges to return the goods to the supplier must be added.

Labor

- *How efficient is the existing crew at the actual installation?*

  Is the same crew that performed the initial work going to do the change order? Possibly the change consists of the installation of a unit identical to one already installed; therefore, the change should take less time. Be sure to consider this potential productivity increase and modify the productivity rates accordingly.

- *If the crew size is increased, what impact will that have on supervision requirements?*

  Under most bargaining agreements or management practices, there is a point at which a working foreman is replaced by a nonworking foreman. This replacement increases project overhead by adding a nonproductive worker. If additional workers are added to accelerate the project or to perform changes while maintaining the schedule, be sure to add additional supervision time if warranted. Calculate the hours involved and the additional cost directly if possible.

- *What are the other impacts of increased crew size?*

  The larger the crew, the greater the potential for productivity to decrease. Some of the factors that cause this productivity loss are: overcrowding (producing restrictive conditions in the working space), and possibly a shortage of any special tools and equipment required. Such factors affect not only the crew working on the elements directly involved in the change order, but other crews whose movement may also be hampered. As the crew increases, check its basic composition for changes by the addition or deletion of apprentices or nonworking foreman, and quantify the potential effects of equipment shortages or other logistical factors.

- *As new crews, unfamiliar with the project, are brought onto the site, how long will it take them to become oriented to the project requirements?*

  The orientation time for a new crew to become 100% effective varies with the site and type of project. Orientation is easiest at a new construction site, and most difficult at existing, very restrictive renovation sites. The type of work also affects orientation time. When all elements of the work are exposed, such as concrete or masonry work, orientation is decreased. When the work is concealed or less visible, such as existing electrical systems, orientation takes longer. Usually orientation can be accomplished in one day or less. Costs for added orientation should be itemized and added to the total estimated cost.

- *How much actual production can be gained by working overtime?*

  Short term overtime can be used effectively to accomplish more work in a day. However, as overtime is scheduled to run beyond several weeks, studies have shown marked decreases in output. The following chart shows the effect of long term overtime on worker efficiency. If the anticipated change requires extended overtime to keep the job on schedule, these factors can be used as a guide to predict the impact on time and cost. Add project overhead, particularly supervision, that may also be incurred.

Days per Week	Hours per Day	Production Efficiency: 1 Week	2 Weeks	3 Weeks	4 Weeks	Average 4 Weeks	Payroll Cost Factors: @ 1-1/2 Times	@ 2 Times
5	8	100%	100%	100%	100%	100%	100%	100%
	9	100	100	95	90	96.25	105.6	111.1
	10	100	95	90	85	91.25	110.0	120.0
	11	95	90	75	65	81.25	113.6	127.3
	12	90	85	70	60	76.25	116.7	133.3
6	8	100	100	95	90	96.25	108.3	116.7
	9	100	95	90	85	92.50	113.0	125.9
	10	95	90	85	80	87.50	116.7	133.3
	11	95	85	70	65	78.75	119.7	139.4
	12	90	80	65	60	73.75	122.2	144.4
7	8	100	95	85	75	88.75	114.3	128.6
	9	95	90	80	70	83.75	118.3	136.5
	10	90	85	75	65	78.75	121.4	142.9
	11	85	80	65	60	72.50	124.0	148.1
	12	85	75	60	55	68.75	126.2	152.4

Effects of Overtime

Caution: Under many labor agreements, Sundays and holidays are paid at a higher premium than the normal overtime rate.

The use of long-term overtime is counterproductive on almost any construction job; that is, the longer the period of overtime, the lower the actual production rate. Numerous studies have been conducted, and while they have resulted in slightly different numbers, all reach the same conclusion. The figure above tabulates the effects of overtime work on efficiency.

As illustrated, there can be a difference between the *actual* payroll cost per hour and the *effective* cost per hour for overtime work. This is due to the reduced production efficiency with the increase in weekly hours beyond 40. This difference between actual and effective cost results from overtime work over a prolonged period. Short-term overtime work does not result in as great a reduction in efficiency and, in such cases, effective cost may not vary significantly from the actual payroll cost. As the total hours per week are increased on a regular basis, more time is lost due to fatigue, lowered morale, and an increased accident rate.

As an example, assume a project where workers are working 6 days a week, 10 hours per day. From the figure above (based on productivity studies), the average effective productive hours over a 4-week period are:

$$0.875 \times 60 = 52.5$$

Depending upon the locale and day of week, overtime hours may be paid at time and a half or double time. For time and a half, the overall (average) *actual* payroll cost (including regular and overtime hours) is determined as follows:

$$\frac{40 \text{ reg. hrs.} + (20 \text{ overtime hrs.} \times 1.5)}{60 \text{ hrs.}} = 1.167$$

Based on 60 hours, the payroll cost per hour will be 116.7% of the normal rate at 40 hours per week. However, because the effective production (efficiency) for 60 hours is reduced to the equivalent of 52.5 hours, the effective cost of overtime is calculated as follows:

For time and a half:

$$\frac{40 \text{ reg. hrs.} + (20 \text{ overtime hrs.} \times 1.5)}{52.5 \text{ hrs.}} = 1.33$$

Installed cost will be 133% of the normal rate (for labor).

Thus, when figuring overtime, the actual cost per unit of work will be higher than the apparent overtime payroll dollar increase, due to the reduced productivity of the longer workweek. These efficiency calculations are true only for those cost factors determined by hours worked. Costs that are applied weekly or monthly, such as equipment rentals, will not be similarly affected.

Equipment

- *What equipment is required to complete the change order?*

Change orders may require extending the rental period of equipment already on the job site, or the addition of special equipment brought in to accomplish the change work. In either case, the additional rental charges and operator labor charges must be added.

Summary

The preceding considerations and others you deem appropriate should be analyzed and applied to a change order estimate. The impact of each should be quantified and listed on the estimate to form an audit trail.

Change orders that are properly identified, documented, and managed help to ensure the orderly, professional and profitable progress of the work. They also minimize potential claims or disputes at the end of the project.

Estimating Tips

- The cost figures in this section were derived from approximately 11,000 projects contained in the RSMeans database of completed construction projects. They include the contractor's overhead and profit, but do not generally include architectural fees or land costs. The figures have been adjusted to January of the current year. New projects are added to our files each year, and outdated projects are discarded. For this reason, certain costs may not show a uniform annual progression. In no case are all subdivisions of a project listed.
- These projects were located throughout the U.S. and reflect a tremendous variation in square foot (S.F.) and cubic foot (C.F.) costs. This is due to differences, not only in labor and material costs, but also in individual owners' requirements. For instance, a bank in a large city would have different features than one in a rural area. This is true of all the different types of buildings analyzed. Therefore, caution should be exercised when using these square foot costs. For example, for courthouses, costs in the database are local courthouse costs and will not apply to the larger, more elaborate federal courthouses. As a general rule, the projects in the 1/4 column do not include any site work or equipment, while the projects in the 3/4 column may include both equipment and site work. The median figures do not generally include site work.
- None of the figures "go with" any others. All individual cost items were computed and tabulated separately. Thus, the sum of the median figures for plumbing, HVAC, and electrical will not normally total up to the total mechanical and electrical costs arrived at by separate analysis and tabulation of the projects.
- Each building was analyzed as to total and component costs and percentages. The figures were arranged in ascending order with the results tabulated as shown. The 1/4 column shows that 25% of the projects had lower costs and 75% had higher. The 3/4 column shows that 75% of the projects had lower costs and 25% had higher. The median column shows that 50% of the projects had lower costs and 50% had higher.
- There are two times when square foot costs are useful. The first is in the conceptual stage when no details are available. Then, square foot costs make a useful starting point. The second is after the bids are in and the costs can be worked back into their appropriate categories for information purposes. As soon as details become available in the project design, the square foot approach should be discontinued and the project priced as to its particular components. When more precision is required, or for estimating the replacement cost of specific buildings, the current edition of *RSMeans Square Foot Costs* should be used.
- In using the figures in this section, it is recommended that the median column be used for preliminary figures if no additional information is available. The median figures, when multiplied by the total city construction cost index figures (see City Cost Indexes) and then multiplied by the project size modifier at the end of this section, should present a fairly accurate base figure, which would then have to be adjusted in view of the estimator's experience, local economic conditions, code requirements, and the owner's particular requirements. There is no need to factor the percentage figures, as these should remain constant from city to city. All tabulations mentioning air conditioning had at least partial air conditioning.
- The editors of this book would greatly appreciate receiving cost figures on one or more of your recent projects, which would then be included in the averages for next year. All cost figures received will be kept confidential, except that they will be averaged with other similar projects to arrive at square foot cost figures for next year's book. See the last page of the book for details and the discount available for submitting one or more of your projects. ■

	50 17 00 | S.F. Costs		UNIT	UNIT COSTS 1/4	UNIT COSTS MEDIAN	UNIT COSTS 3/4	% OF TOTAL 1/4	% OF TOTAL MEDIAN	% OF TOTAL 3/4	
01	0010	**APARTMENTS Low Rise (1 to 3 story)**	S.F.	74.50	94	125				01
	0020	Total project cost	C.F.	6.70	8.85	10.95				
	0100	Site work	S.F.	5.45	8.70	15.30	6.05%	10.55%	13.95%	
	0500	Masonry		1.47	3.62	5.95	1.54%	3.92%	6.50%	
	1500	Finishes		7.90	10.85	13.45	9.05%	10.75%	12.85%	
	1800	Equipment		2.44	3.70	5.50	2.71%	3.99%	5.95%	
	2720	Plumbing		5.80	7.45	9.50	6.65%	8.95%	10.05%	
	2770	Heating, ventilating, air conditioning		3.70	4.56	6.70	4.20%	5.60%	7.60%	
	2900	Electrical		4.33	5.75	7.80	5.20%	6.65%	8.35%	
	3100	Total: Mechanical & Electrical	↓	15.40	19.95	24.50	16.05%	18.20%	23%	
	9000	Per apartment unit, total cost	Apt.	69,500	106,000	156,500				
	9500	Total: Mechanical & Electrical	"	13,100	20,700	27,000				
02	0010	**APARTMENTS Mid Rise (4 to 7 story)**	S.F.	99	119	147				02
	0020	Total project costs	C.F.	7.70	10.65	14.55				
	0100	Site work	S.F.	3.95	8	15.45	5.25%	6.70%	9.20%	
	0500	Masonry		4.13	9.05	12.40	5.10%	7.25%	10.50%	
	1500	Finishes		12.95	17.35	20.50	10.70%	13.50%	17.70%	
	1800	Equipment		2.77	4.30	5.65	2.54%	3.47%	4.31%	
	2500	Conveying equipment		2.24	2.76	3.33	2.05%	2.27%	2.69%	
	2720	Plumbing		5.80	9.30	9.85	5.70%	7.20%	8.95%	
	2900	Electrical		6.50	8.85	10.75	6.35%	7.20%	8.95%	
	3100	Total: Mechanical & Electrical	↓	21	26	31.50	18.25%	21%	23%	
	9000	Per apartment unit, total cost	Apt.	112,000	132,000	218,500				
	9500	Total: Mechanical & Electrical	"	21,100	24,400	25,600				
03	0010	**APARTMENTS High Rise (8 to 24 story)**	S.F.	112	129	155				03
	0020	Total project costs	C.F.	10.90	12.65	14.95				
	0100	Site work	S.F.	4.07	6.60	9.20	2.58%	4.84%	6.15%	
	0500	Masonry		6.50	11.80	14.65	4.74%	9.65%	11.05%	
	1500	Finishes		12.45	15.55	18.35	9.75%	11.80%	13.70%	
	1800	Equipment		3.61	4.44	5.85	2.78%	3.49%	4.35%	
	2500	Conveying equipment		2.55	3.87	5.25	2.23%	2.78%	3.37%	
	2720	Plumbing		7.15	9.75	11.95	6.80%	7.20%	10.45%	
	2900	Electrical		7.70	9.75	13.15	6.45%	7.65%	8.80%	
	3100	Total: Mechanical & Electrical	↓	23	29.50	35.50	17.95%	22.50%	24.50%	
	9000	Per apartment unit, total cost	Apt.	116,500	128,500	178,000				
	9500	Total: Mechanical & Electrical	"	23,200	28,800	30,500				
04	0010	**AUDITORIUMS**	S.F.	117	161	234				04
	0020	Total project costs	C.F.	7.30	10.15	14.55				
	2720	Plumbing	S.F.	6.90	10.20	12.20	5.85%	7.20%	8.70%	
	2900	Electrical		9.40	13.55	22.50	6.85%	9.50%	11.40%	
	3100	Total: Mechanical & Electrical	↓	62.50	83	101	24.50%	27.50%	31%	
05	0010	**AUTOMOTIVE SALES**	S.F.	86.50	119	146				05
	0020	Total project costs	C.F.	5.70	6.80	8.85				
	2720	Plumbing	S.F.	3.93	6.85	7.45	2.89%	6.05%	6.50%	
	2770	Heating, ventilating, air conditioning		6.05	9.25	10	4.61%	10%	10.35%	
	2900	Electrical		6.95	10.90	16	7.25%	8.80%	12.15%	
	3100	Total: Mechanical & Electrical	↓	19.35	31	37.50	17.30%	20.50%	22%	
06	0010	**BANKS**	S.F.	170	211	268				06
	0020	Total project costs	C.F.	12.10	16.45	21.50				
	0100	Site work	S.F.	19.40	30.50	43	7.90%	12.95%	17%	
	0500	Masonry		7.70	16.50	29.50	3.36%	6.90%	10.05%	
	1500	Finishes		15.15	23	28.50	5.85%	8.65%	11.70%	
	1800	Equipment		6.35	14	28.50	1.34%	5.55%	10.50%	
	2720	Plumbing		5.30	7.55	11.05	2.82%	3.90%	4.93%	
	2770	Heating, ventilating, air conditioning		10.05	13.45	17.90	4.86%	7.15%	8.50%	
	2900	Electrical		16.05	21.50	28	8.25%	10.20%	12.20%	
	3100	Total: Mechanical & Electrical	↓	38.50	51.50	62	16.55%	19.45%	24%	
	3500	See also division 11 22 00								

	50 17 00 | S.F. Costs		UNIT	UNIT COSTS 1/4	UNIT COSTS MEDIAN	UNIT COSTS 3/4	% OF TOTAL 1/4	% OF TOTAL MEDIAN	% OF TOTAL 3/4	
13	0010	**CHURCHES**	S.F.	114	145	190				13
	0020	Total project costs	C.F.	7.05	8.90	11.75				
	1800	Equipment	S.F.	1.22	3.19	6.80	.83%	2.04%	4.30%	
	2720	Plumbing		4.43	6.20	9.15	3.51%	4.96%	6.25%	
	2770	Heating, ventilating, air conditioning		10.35	13.50	19.15	7.50%	10%	12%	
	2900	Electrical		9.60	13.20	18	7.30%	8.80%	10.95%	
	3100	Total: Mechanical & Electrical		29.50	39.50	53	18.30%	22%	25%	
	3500	See also division 11 91 00								
15	0010	**CLUBS, COUNTRY**	S.F.	122	147	185				15
	0020	Total project costs	C.F.	9.85	12	16.55				
	2720	Plumbing	S.F.	7.40	10.95	25	5.60%	7.90%	10%	
	2900	Electrical		9.60	13.15	17.15	7%	8.95%	11%	
	3100	Total: Mechanical & Electrical		51	64	67.50	19%	26.50%	29.50%	
17	0010	**CLUBS, SOCIAL Fraternal**	S.F.	103	141	188				17
	0020	Total project costs	C.F.	6.10	9.25	11				
	2720	Plumbing	S.F.	6.15	7.65	11.55	5.60%	6.90%	8.55%	
	2770	Heating, ventilating, air conditioning		7.15	10.70	13.75	8.20%	9.25%	14.40%	
	2900	Electrical		7.15	12.05	13.75	5.95%	9.30%	10.55%	
	3100	Total: Mechanical & Electrical		38	41	52	21%	23%	23.50%	
18	0010	**CLUBS, Y.M.C.A.**	S.F.	134	167	252				18
	0020	Total project costs	C.F.	5.65	9.45	14.05				
	2720	Plumbing	S.F.	7.75	15.40	17.25	5.65%	7.60%	10.85%	
	2900	Electrical		10.05	13	22.50	6.45%	7.95%	8.90%	
	3100	Total: Mechanical & Electrical		41	46.50	76	20.50%	21.50%	28.50%	
19	0010	**COLLEGES Classrooms & Administration**	S.F.	118	167	216				19
	0020	Total project costs	C.F.	7.10	12.50	20				
	0500	Masonry	S.F.	9.30	17.15	21	5.10%	8.05%	10.50%	
	2720	Plumbing		5.30	12.90	24	5.10%	6.60%	8.95%	
	2900	Electrical		10.05	15.85	22	7.70%	9.85%	12%	
	3100	Total: Mechanical & Electrical		42.50	58.50	77.50	23%	28%	31.50%	
21	0010	**COLLEGES Science, Engineering, Laboratories**	S.F.	230	269	310				21
	0020	Total project costs	C.F.	13.20	19.25	22				
	1800	Equipment	S.F.	6.25	29	31.50	2%	6.45%	12.65%	
	2900	Electrical		18.95	27	41.50	7.10%	9.40%	12.10%	
	3100	Total: Mechanical & Electrical		70.50	83.50	129	28.50%	31.50%	41%	
	3500	See also division 11 53 00								
23	0010	**COLLEGES Student Unions**	S.F.	147	199	241				23
	0020	Total project costs	C.F.	8.20	10.70	13.25				
	3100	Total: Mechanical & Electrical	S.F.	55	59.50	70.50	23.50%	26%	29%	
25	0010	**COMMUNITY CENTERS**	S.F.	122	150	203				25
	0020	Total project costs	C.F.	7.90	11.30	14.65				
	1800	Equipment	S.F.	2.45	4.80	7.90	1.47%	3.01%	5.30%	
	2720	Plumbing		5.75	10.05	13.70	4.85%	7%	8.95%	
	2770	Heating, ventilating, air conditioning		8.65	13.35	19.15	6.80%	10.35%	12.90%	
	2900	Electrical		10.25	13.55	19.90	7.15%	8.90%	10.40%	
	3100	Total: Mechanical & Electrical		34	43	61.50	18.80%	23%	30%	
28	0010	**COURT HOUSES**	S.F.	175	206	284				28
	0020	Total project costs	C.F.	13.40	16	20				
	2720	Plumbing	S.F.	8.30	11.60	13.15	5.95%	7.45%	8.20%	
	2900	Electrical		18.55	21	30.50	9.05%	10.65%	12.15%	
	3100	Total: Mechanical & Electrical		52	68	74.50	22.50%	26.50%	30%	
30	0010	**DEPARTMENT STORES**	S.F.	64.50	87.50	110				30
	0020	Total project costs	C.F.	3.46	4.48	6.10				
	2720	Plumbing	S.F.	2.01	2.54	3.85	1.82%	4.21%	5.90%	
	2770	Heating, ventilating, air conditioning		4.88	9.05	13.65	8.20%	9.10%	14.80%	

50 17 00 | S.F. Costs

		50 17 00 | S.F. Costs	UNIT	UNIT COSTS 1/4	UNIT COSTS MEDIAN	UNIT COSTS 3/4	% OF TOTAL 1/4	% OF TOTAL MEDIAN	% OF TOTAL 3/4	
30	2900	Electrical	S.F.	7.40	10.15	12	9.05%	12.15%	14.95%	30
	3100	Total: Mechanical & Electrical	↓	13.05	16.65	29	13.20%	21.50%	50%	
31	0010	**DORMITORIES Low Rise (1 to 3 story)**	S.F.	122	170	212				31
	0020	Total project costs	C.F.	6.90	11.20	16.75				
	2720	Plumbing	S.F.	7.35	9.85	12.45	8.05%	9%	9.65%	
	2770	Heating, ventilating, air conditioning		7.80	9.35	12.45	4.61%	8.05%	10%	
	2900	Electrical		8.10	12.35	16.90	6.40%	8.65%	9.50%	
	3100	Total: Mechanical & Electrical	↓	42	45	70.50	22%	25%	27%	
	9000	Per bed, total cost	Bed	52,000	57,500	123,500				
32	0010	**DORMITORIES Mid Rise (4 to 8 story)**	S.F.	150	196	242				32
	0020	Total project costs	C.F.	16.55	18.20	22				
	2900	Electrical	S.F.	15.95	18.15	24.50	8.20%	10.20%	11.95%	
	3100	Total: Mechanical & Electrical	″	44.50	89	90.50	25%	30.50%	35.50%	
	9000	Per bed, total cost	Bed	21,400	48,800	282,500				
34	0010	**FACTORIES**	S.F.	57	85	131				34
	0020	Total project costs	C.F.	3.65	5.45	9.05				
	0100	Site work	S.F.	6.50	11.85	18.75	6.95%	11.45%	17.95%	
	2720	Plumbing		3.07	5.70	9.45	3.73%	6.05%	8.10%	
	2770	Heating, ventilating, air conditioning		5.95	8.55	11.55	5.25%	8.45%	11.35%	
	2900	Electrical		7.05	11.20	17.05	8.10%	10.50%	14.20%	
	3100	Total: Mechanical & Electrical	↓	20	32.50	41	21%	28.50%	35.50%	
36	0010	**FIRE STATIONS**	S.F.	113	156	213				36
	0020	Total project costs	C.F.	6.60	9.05	12.05				
	0500	Masonry	S.F.	15.40	30	38	7.50%	10.95%	15.55%	
	1140	Roofing		3.68	9.95	11.35	1.90%	4.94%	5.05%	
	1580	Painting		2.86	4.27	4.37	1.37%	1.57%	2.07%	
	1800	Equipment		1.40	2.69	4.97	.62%	1.63%	3.42%	
	2720	Plumbing		6.30	10.05	14.25	5.85%	7.35%	9.45%	
	2770	Heating, ventilating, air conditioning		6.25	10.15	15.65	5.15%	7.40%	9.40%	
	2900	Electrical		8.15	14.35	19.35	6.90%	8.60%	10.60%	
	3100	Total: Mechanical & Electrical	↓	41.50	53	60	18.40%	23%	27%	
37	0010	**FRATERNITY HOUSES & Sorority Houses**	S.F.	113	145	199				37
	0020	Total project costs	C.F.	11.25	11.70	14.10				
	2720	Plumbing	S.F.	8.55	9.75	17.90	6.80%	8%	10.85%	
	2900	Electrical	↓	7.45	16.10	19.70	6.60%	9.90%	10.65%	
38	0010	**FUNERAL HOMES**	S.F.	119	162	295				38
	0020	Total project costs	C.F.	12.15	13.55	26				
	2900	Electrical	S.F.	5.25	9.65	10.55	3.58%	4.44%	5.95%	
39	0010	**GARAGES, COMMERCIAL (Service)**	S.F.	67.50	104	144				39
	0020	Total project costs	C.F.	4.43	6.55	9.50				
	1800	Equipment	S.F.	3.80	8.55	13.30	2.21%	4.62%	6.80%	
	2720	Plumbing		4.67	7.20	13.10	5.45%	7.85%	10.65%	
	2730	Heating & ventilating		4.60	7.95	10.15	5.25%	6.85%	8.20%	
	2900	Electrical		6.40	9.75	14.10	7.15%	9.25%	10.85%	
	3100	Total: Mechanical & Electrical	↓	12.10	27	40	12.35%	17.40%	26%	
40	0010	**GARAGES, MUNICIPAL (Repair)**	S.F.	105	135	192				40
	0020	Total project costs	C.F.	6.20	7.85	13.50				
	0500	Masonry	S.F.	5.20	18.20	28	4.03%	9.15%	12.50%	
	2720	Plumbing		4.44	8.55	16.05	3.59%	6.70%	7.95%	
	2730	Heating & ventilating		7.60	11	21	6.15%	7.45%	13.50%	
	2900	Electrical		7.65	12.35	22.50	6.90%	9.40%	13%	
	3100	Total: Mechanical & Electrical	↓	34.50	56.50	68	21.50%	25.50%	35.50%	
41	0010	**GARAGES, PARKING**	S.F.	39.50	56	96.50				41
	0020	Total project costs	C.F.	3.62	4.92	7.15				

50 17 00 | S.F. Costs

		50 17 00 | S.F. Costs	UNIT	UNIT COSTS 1/4	UNIT COSTS MEDIAN	UNIT COSTS 3/4	% OF TOTAL 1/4	% OF TOTAL MEDIAN	% OF TOTAL 3/4	
41	2720	Plumbing	S.F.	.66	1.69	2.61	1.72%	2.70%	3.85%	41
	2900	Electrical		2.12	2.80	4.08	4.52%	5.40%	6.35%	
	3100	Total: Mechanical & Electrical		4.11	6.05	7.50	7%	8.90%	11.05%	
	3200									
	9000	Per car, total cost	Car	16,300	20,400	26,000				
43	0010	**GYMNASIUMS**	S.F.	108	143	196				43
	0020	Total project costs	C.F.	5.35	7.25	8.90				
	1800	Equipment	S.F.	2.55	4.78	8.60	1.76%	3.26%	6.70%	
	2720	Plumbing		5.90	7.90	10.40	4.65%	6.40%	7.75%	
	2770	Heating, ventilating, air conditioning		6.40	11.10	22.50	5.15%	9.05%	11.10%	
	2900	Electrical		8.50	11.25	14.95	6.75%	8.50%	10.30%	
	3100	Total: Mechanical & Electrical		29	40.50	49.50	19.75%	23.50%	29%	
	3500	See also division 11 66 00								
46	0010	**HOSPITALS**	S.F.	206	258	355				46
	0020	Total project costs	C.F.	15.60	19.40	28				
	1800	Equipment	S.F.	5.20	10.05	17.30	.80%	2.53%	4.80%	
	2720	Plumbing		17.70	25	32	7.60%	9.10%	10.85%	
	2770	Heating, ventilating, air conditioning		26	33.50	46.50	7.80%	12.95%	16.65%	
	2900	Electrical		22.50	30.50	45.50	10%	11.75%	14.10%	
	3100	Total: Mechanical & Electrical		66.50	92.50	137	28%	33.50%	37%	
	9000	Per bed or person, total cost	Bed	238,000	328,500	378,500				
	9900	See also division 11 71 00								
48	0010	**HOUSING For the Elderly**	S.F.	101	128	157				48
	0020	Total project costs	C.F.	7.20	10	12.80				
	0100	Site work	S.F.	7.05	10.95	16.05	5.05%	7.90%	12.10%	
	0500	Masonry		1.91	11.55	16.85	1.30%	6.05%	11%	
	1800	Equipment		2.45	3.37	5.35	1.88%	3.23%	4.43%	
	2510	Conveying systems		2.29	3.31	4.49	1.78%	2.20%	2.81%	
	2720	Plumbing		7.50	9.60	12.10	8.15%	9.55%	10.50%	
	2730	Heating, ventilating, air conditioning		3.86	5.45	8.15	3.30%	5.60%	7.25%	
	2900	Electrical		7.55	10.25	13.10	7.30%	8.50%	10.25%	
	3100	Total: Mechanical & Electrical		26	32	41	18.10%	22.50%	29%	
	9000	Per rental unit, total cost	Unit	94,000	110,000	122,500				
	9500	Total: Mechanical & Electrical	"	21,000	24,100	28,100				
50	0010	**HOUSING Public (Low Rise)**	S.F.	85	118	154				50
	0020	Total project costs	C.F.	6.75	9.45	11.75				
	0100	Site work	S.F.	10.85	15.60	25.50	8.35%	11.75%	16.50%	
	1800	Equipment		2.31	3.77	5.75	2.26%	3.03%	4.24%	
	2720	Plumbing		6.15	8.10	10.25	7.15%	9.05%	11.60%	
	2730	Heating, ventilating, air conditioning		3.08	6	6.55	4.26%	6.05%	6.45%	
	2900	Electrical		5.15	7.65	10.65	5.10%	6.55%	8.25%	
	3100	Total: Mechanical & Electrical		24.50	31.50	35	14.50%	17.55%	26.50%	
	9000	Per apartment, total cost	Apt.	93,500	106,500	133,500				
	9500	Total: Mechanical & Electrical	"	19,900	24,600	27,200				
51	0010	**ICE SKATING RINKS**	S.F.	72.50	170	187				51
	0020	Total project costs	C.F.	5.35	5.45	6.30				
	2720	Plumbing	S.F.	2.72	5.10	5.20	3.12%	3.23%	5.65%	
	2900	Electrical		7.80	11.95	12.65	6.30%	10.15%	15.05%	
52	0010	**JAILS**	S.F.	189	286	370				52
	0020	Total project costs	C.F.	19.95	28	32.50				
	1800	Equipment	S.F.	8.65	25.50	43.50	2.80%	6.95%	9.80%	
	2720	Plumbing		21.50	28.50	37.50	7%	8.90%	13.35%	
	2770	Heating, ventilating, air conditioning		19.95	26.50	51.50	7.50%	9.45%	17.75%	
	2900	Electrical		23.50	31.50	40	9.40%	11.70%	15.25%	
	3100	Total: Mechanical & Electrical		62	110	131	28%	31%	36%	
53	0010	**LIBRARIES**	S.F.	143	190	248				53
	0020	Total project costs	C.F.	9.55	12	15.30				

50 17 00 | S.F. Costs

		50 17 00 | S.F. Costs	UNIT	UNIT COSTS 1/4	UNIT COSTS MEDIAN	UNIT COSTS 3/4	% OF TOTAL 1/4	% OF TOTAL MEDIAN	% OF TOTAL 3/4	
53	0500	Masonry	S.F.	9.50	19.35	32.50	5.60%	7.15%	10.95%	53
	1800	Equipment		1.91	5.15	7.75	.28%	1.39%	4.07%	
	2720	Plumbing		5.15	7.30	10.20	3.38%	4.60%	5.70%	
	2770	Heating, ventilating, air conditioning		11.45	19.35	23.50	7.80%	10.95%	12.80%	
	2900	Electrical		14.40	19	25.50	8.40%	10.55%	12%	
	3100	Total: Mechanical & Electrical		47	57	67	21%	24%	28%	
54	0010	**LIVING, ASSISTED**	S.F.	129	154	180				54
	0020	Total project costs	C.F.	10.90	12.70	14.45				
	0500	Masonry	S.F.	3.75	4.53	5.55	2.36%	3.16%	3.86%	
	1800	Equipment		2.87	3.42	4.38	2.12%	2.45%	2.87%	
	2720	Plumbing		10.85	14.50	15.05	6.05%	8.15%	10.60%	
	2770	Heating, ventilating, air conditioning		12.85	13.45	14.75	7.95%	9.35%	9.70%	
	2900	Electrical		12.70	14.25	16.50	9%	9.95%	10.65%	
	3100	Total: Mechanical & Electrical		35	41	48	24%	28.50%	31.50%	
55	0010	**MEDICAL CLINICS**	S.F.	132	162	207				55
	0020	Total project costs	C.F.	9.65	12.50	16.60				
	1800	Equipment	S.F.	1.68	7.50	11.65	1.05%	2.94%	6.35%	
	2720	Plumbing		8.75	12.30	16.50	6.15%	8.40%	10.10%	
	2770	Heating, ventilating, air conditioning		10.45	13.70	20	6.65%	8.85%	11.35%	
	2900	Electrical		11.30	16.10	21	8.10%	10%	12.20%	
	3100	Total: Mechanical & Electrical		36	49	67.50	22%	27%	33.50%	
	3500	See also division 11 71 00								
57	0010	**MEDICAL OFFICES**	S.F.	124	154	190				57
	0020	Total project costs	C.F.	9.25	12.50	16.95				
	1800	Equipment	S.F.	4.09	8.10	11.35	.66%	4.87%	6.50%	
	2720	Plumbing		6.85	10.55	14.25	5.60%	6.80%	8.50%	
	2770	Heating, ventilating, air conditioning		8.25	11.95	15.75	6.10%	8%	9.70%	
	2900	Electrical		9.90	14.50	20.50	7.50%	9.80%	11.70%	
	3100	Total: Mechanical & Electrical		27	39.50	56	19.30%	22.50%	27.50%	
59	0010	**MOTELS**	S.F.	78	113	148				59
	0020	Total project costs	C.F.	6.95	9.30	15.25				
	2720	Plumbing	S.F.	7.90	10.10	12.05	9.45%	10.60%	12.55%	
	2770	Heating, ventilating, air conditioning		4.83	7.20	12.90	5.60%	5.60%	10%	
	2900	Electrical		7.40	9.35	11.65	7.45%	9.05%	10.45%	
	3100	Total: Mechanical & Electrical		22.50	31.50	54	18.50%	24%	25.50%	
	5000									
	9000	Per rental unit, total cost	Unit	39,800	75,500	82,000				
	9500	Total: Mechanical & Electrical	"	7,750	11,700	13,600				
60	0010	**NURSING HOMES**	S.F.	123	158	197				60
	0020	Total project costs	C.F.	9.65	12.05	16.45				
	1800	Equipment	S.F.	3.34	5.10	8.55	2%	3.62%	4.99%	
	2720	Plumbing		10.50	15.90	19.20	8.75%	10.10%	12.70%	
	2770	Heating, ventilating, air conditioning		11.05	16.80	22.50	9.70%	11.45%	11.80%	
	2900	Electrical		12.15	15.20	21	9.40%	10.60%	12.50%	
	3100	Total: Mechanical & Electrical		29	40.50	68	26%	28%	30%	
	9000	Per bed or person, total cost	Bed	54,500	68,000	88,000				
61	0010	**OFFICES Low Rise (1 to 4 story)**	S.F.	103	135	175				61
	0020	Total project costs	C.F.	7.35	10.15	13.40				
	0100	Site work	S.F.	8.30	14.45	21.50	5.95%	9.70%	13.55%	
	0500	Masonry		4	8.05	14.70	2.61%	5.40%	8.45%	
	1800	Equipment		.93	2.15	5.85	.57%	1.50%	3.42%	
	2720	Plumbing		3.69	5.70	8.35	3.66%	4.50%	6.10%	
	2770	Heating, ventilating, air conditioning		8.15	11.40	16.65	7.20%	10.30%	11.70%	
	2900	Electrical		8.45	12.10	17.20	7.45%	9.65%	11.40%	
	3100	Total: Mechanical & Electrical		23.50	32.50	48.50	18.20%	22%	27%	
62	0010	**OFFICES Mid Rise (5 to 10 story)**	S.F.	109	132	180				62
	0020	Total project costs	C.F.	7.75	9.90	14				

	50 17 00 | S.F. Costs		Unit	Unit Costs 1/4	Unit Costs Median	Unit Costs 3/4	% of Total 1/4	% of Total Median	% of Total 3/4	
62	2720	Plumbing	S.F.	3.30	5.10	7.35	2.83%	3.74%	4.50%	62
	2770	Heating, ventilating, air conditioning		8.30	11.85	18.95	7.65%	9.40%	11%	
	2900	Electrical		8.10	10.40	15.70	6.35%	7.80%	10%	
	3100	Total: Mechanical & Electrical		21	27	51.50	18.95%	21%	27.50%	
63	0010	**OFFICES High Rise (11 to 20 story)**	S.F.	134	169	208				63
	0020	Total project costs	C.F.	9.40	11.75	16.85				
	2900	Electrical	S.F.	8.15	9.95	14.80	5.80%	7.85%	10.50%	
	3100	Total: Mechanical & Electrical		26.50	35.50	59.50	16.90%	23.50%	34%	
64	0010	**POLICE STATIONS**	S.F.	161	212	270				64
	0020	Total project costs	C.F.	12.85	15.70	21.50				
	0500	Masonry	S.F.	16.10	26.50	33.50	7.80%	9.10%	11.35%	
	1800	Equipment		2.31	11.25	17.80	.98%	3.35%	6.70%	
	2720	Plumbing		9	17.95	22.50	5.65%	6.90%	10.75%	
	2770	Heating, ventilating, air conditioning		13.20	18.70	26.50	5.85%	10.55%	11.70%	
	2900	Electrical		17.60	25.50	33	9.80%	11.70%	14.50%	
	3100	Total: Mechanical & Electrical		67	70.50	94	28.50%	31.50%	32.50%	
65	0010	**POST OFFICES**	S.F.	127	157	200				65
	0020	Total project costs	C.F.	7.65	9.70	11				
	2720	Plumbing	S.F.	5.15	7.10	8.95	4.24%	5.30%	5.60%	
	2770	Heating, ventilating, air conditioning		8	11.05	12.30	6.65%	7.15%	9.35%	
	2900	Electrical		10.50	14.75	17.50	7.25%	9%	11%	
	3100	Total: Mechanical & Electrical		30.50	39.50	45	16.25%	18.80%	22%	
66	0010	**POWER PLANTS**	S.F.	880	1,175	2,150				66
	0020	Total project costs	C.F.	24.50	53	113				
	2900	Electrical	S.F.	62.50	132	197	9.30%	12.75%	21.50%	
	8100	Total: Mechanical & Electrical		155	505	1,125	32.50%	32.50%	52.50%	
67	0010	**RELIGIOUS EDUCATION**	S.F.	103	137	170				67
	0020	Total project costs	C.F.	5.70	8.15	10.20				
	2720	Plumbing	S.F.	4.28	6.05	8.60	4.40%	5.30%	7.10%	
	2770	Heating, ventilating, air conditioning		10.80	12.25	17.30	10.05%	11.45%	12.35%	
	2900	Electrical		8.15	11.65	17.10	7.70%	9.05%	10.35%	
	3100	Total: Mechanical & Electrical		35.50	45	53	22%	23.50%	26%	
69	0010	**RESEARCH Laboratories & Facilities**	S.F.	157	222	325				69
	0020	Total project costs	C.F.	11.40	22.50	27				
	1800	Equipment	S.F.	6.80	13.40	32.50	1.22%	4.80%	9.35%	
	2720	Plumbing		11.20	19.70	31.50	6.15%	8.30%	10.80%	
	2770	Heating, ventilating, air conditioning		13.80	46.50	55	7.25%	16.50%	17.50%	
	2900	Electrical		18.45	29.50	48.50	9.45%	11.15%	14.60%	
	3100	Total: Mechanical & Electrical		59	104	147	29.50%	36%	41%	
70	0010	**RESTAURANTS**	S.F.	149	192	249				70
	0020	Total project costs	C.F.	12.50	16.40	21.50				
	1800	Equipment	S.F.	8.80	23.50	35.50	6.10%	13%	15.65%	
	2720	Plumbing		11.80	14.30	18.75	6.10%	8.15%	9%	
	2770	Heating, ventilating, air conditioning		14.95	21	25	9.20%	12%	12.40%	
	2900	Electrical		15.70	19.40	25	8.35%	10.55%	11.55%	
	3100	Total: Mechanical & Electrical		48.50	52	67.50	21%	25%	29.50%	
	9000	Per seat unit, total cost	Seat	5,450	7,275	8,600				
	9500	Total: Mechanical & Electrical	"	1,375	1,825	2,150				
72	0010	**RETAIL STORES**	S.F.	69.50	93.50	124				72
	0020	Total project costs	C.F.	4.71	6.70	9.35				
	2720	Plumbing	S.F.	2.52	4.20	7.15	3.26%	4.60%	6.80%	
	2770	Heating, ventilating, air conditioning		5.45	7.45	11.20	6.75%	8.75%	10.15%	
	2900	Electrical		6.25	8.55	12.35	7.25%	9.90%	11.60%	
	3100	Total: Mechanical & Electrical		16.65	21.50	28.50	17.05%	21%	23.50%	
74	0010	**SCHOOLS Elementary**	S.F.	114	141	172				74
	0020	Total project costs	C.F.	7.40	9.50	12.25				
	0500	Masonry	S.F.	8	17.30	25.50	4.89%	10.50%	14%	
	1800	Equipment		2.63	4.96	9.40	1.83%	3.13%	4.61%	

50 17 00 | S.F. Costs

			UNIT	UNIT COSTS 1/4	UNIT COSTS MEDIAN	UNIT COSTS 3/4	% OF TOTAL 1/4	% OF TOTAL MEDIAN	% OF TOTAL 3/4
74	2720	Plumbing	S.F.	6.50	9.20	12.30	5.70%	7.15%	9.35%
	2730	Heating, ventilating, air conditioning		9.80	15.55	22.50	8.15%	10.80%	14.90%
	2900	Electrical		10.70	14.25	18.20	8.45%	10.05%	11.85%
	3100	Total: Mechanical & Electrical		39	47.50	60	25%	27.50%	30%
	9000	Per pupil, total cost	Ea.	13,000	19,300	43,300			
	9500	Total: Mechanical & Electrical	"	3,675	4,650	11,700			
76	0010	**SCHOOLS Junior High & Middle**	S.F.	117	145	177			
	0020	Total project costs	C.F.	7.40	9.60	10.75			
	0500	Masonry	S.F.	12.55	18.80	23	7.55%	11.10%	14.30%
	1800	Equipment		3.20	6	8.80	1.80%	3.03%	4.80%
	2720	Plumbing		6.80	8.40	10.40	5.30%	6.80%	7.25%
	2770	Heating, ventilating, air conditioning		13.60	16.55	29	8.90%	11.55%	14.20%
	2900	Electrical		11.90	14.70	18.60	8.05%	9.55%	10.60%
	3100	Total: Mechanical & Electrical		37.50	48	59.50	23%	25.50%	29.50%
	9000	Per pupil, total cost	Ea.	14,800	19,400	26,100			
78	0010	**SCHOOLS Senior High**	S.F.	123	150	188			
	0020	Total project costs	C.F.	7.35	10.30	17.40			
	1800	Equipment	S.F.	3.24	7.55	10.45	1.88%	2.67%	4.30%
	2720	Plumbing		6.80	10.25	18.70	5.60%	6.90%	8.30%
	2770	Heating, ventilating, air conditioning		11.75	15.95	24.50	8.95%	11.60%	15%
	2900	Electrical		12.30	16.40	24	8.70%	10.35%	12.50%
	3100	Total: Mechanical & Electrical		40.50	48	77.50	24%	26.50%	28.50%
	9000	Per pupil, total cost	Ea.	11,500	23,300	29,200			
80	0010	**SCHOOLS Vocational**	S.F.	98.50	143	177			
	0020	Total project costs	C.F.	6.10	8.80	12.15			
	0500	Masonry	S.F.	4.33	14.25	22	3.20%	4.61%	10.95%
	1800	Equipment		3.08	7.65	10.60	1.24%	3.10%	4.26%
	2720	Plumbing		6.30	9.35	13.80	5.40%	6.90%	8.55%
	2770	Heating, ventilating, air conditioning		8.80	16.40	27.50	8.60%	11.90%	14.65%
	2900	Electrical		10.25	14	19.15	8.45%	11%	13.20%
	3100	Total: Mechanical & Electrical		38.50	67.50	86.50	27.50%	29.50%	31%
	9000	Per pupil, total cost	Ea.	13,700	36,700	54,500			
83	0010	**SPORTS ARENAS**	S.F.	86	115	177			
	0020	Total project costs	C.F.	4.67	8.35	10.80			
	2720	Plumbing	S.F.	4.31	7.55	15.95	4.35%	6.35%	9.40%
	2770	Heating, ventilating, air conditioning		10.75	12.70	17.60	8.80%	10.20%	13.55%
	2900	Electrical		7.95	12.15	15.70	7.65%	9.75%	11.90%
85	0010	**SUPERMARKETS**	S.F.	79.50	92	108			
	0020	Total project costs	C.F.	4.42	5.35	8.10			
	2720	Plumbing	S.F.	4.44	5.60	6.50	5.40%	6%	7.45%
	2770	Heating, ventilating, air conditioning		6.55	8.65	10.55	8.60%	8.65%	9.60%
	2900	Electrical		9.30	11.35	13.50	10.40%	12.45%	13.60%
	3100	Total: Mechanical & Electrical		25.50	27	36.50	23.50%	27.50%	28.50%
86	0010	**SWIMMING POOLS**	S.F.	129	300	460			
	0020	Total project costs	C.F.	10.30	12.85	14			
	2720	Plumbing	S.F.	11.90	13.60	18.25	4.80%	9.70%	20.50%
	2900	Electrical		9.75	15.65	34.50	6.05%	6.95%	7.60%
	3100	Total: Mechanical & Electrical		61	80.50	106	11.15%	14.10%	23.50%
87	0010	**TELEPHONE EXCHANGES**	S.F.	160	243	315			
	0020	Total project costs	C.F.	10.65	17.05	23.50			
	2720	Plumbing	S.F.	7.20	11.15	16.30	4.52%	5.80%	6.90%
	2770	Heating, ventilating, air conditioning		16.75	33.50	42	11.80%	16.05%	18.40%
	2900	Electrical		17.40	27.50	49	10.90%	14%	17.85%
	3100	Total: Mechanical & Electrical		51.50	97.50	138	29.50%	33.50%	44.50%
91	0010	**THEATERS**	S.F.	102	138	203			
	0020	Total project costs	C.F.	4.96	7.35	10.80			

	50 17 00 | S.F. Costs		UNIT	UNIT COSTS 1/4	UNIT COSTS MEDIAN	UNIT COSTS 3/4	% OF TOTAL 1/4	% OF TOTAL MEDIAN	% OF TOTAL 3/4	
91	2720	Plumbing	S.F.	3.34	3.88	15.85	2.92%	4.70%	6.80%	91
	2770	Heating, ventilating, air conditioning		10.45	12.65	15.65	8%	12.25%	13.40%	
	2900	Electrical		9.40	12.70	26	8.05%	9.35%	12.25%	
	3100	Total: Mechanical & Electrical	↓	24	32.50	38.50	21.50%	25.50%	27.50%	
94	0010	**TOWN HALLS City Halls & Municipal Buildings**	S.F.	112	152	199				94
	0020	Total project costs	C.F.	8.10	12.60	18.45				
	2720	Plumbing	S.F.	5	9.35	16.45	4.31%	5.95%	7.95%	
	2770	Heating, ventilating, air conditioning		9.05	17.95	26.50	7.05%	9.05%	13.45%	
	2900	Electrical		11.40	15.70	21.50	8.05%	9.45%	11.65%	
	3100	Total: Mechanical & Electrical	↓	39.50	50	76.50	22%	26.50%	31%	
97	0010	**WAREHOUSES & Storage Buildings**	S.F.	44.50	63.50	95				97
	0020	Total project costs	C.F.	2.34	3.67	6.05				
	0100	Site work	S.F.	4.62	9.15	13.80	6.05%	12.95%	19.55%	
	0500	Masonry		2.54	6.35	13.70	3.60%	7.35%	12%	
	1800	Equipment		.68	1.55	8.70	.72%	1.69%	5.55%	
	2720	Plumbing		1.49	2.68	5	2.90%	4.80%	6.55%	
	2730	Heating, ventilating, air conditioning		1.70	4.80	6.45	2.41%	5%	8.90%	
	2900	Electrical		2.65	4.98	8.20	5.15%	7.20%	10.05%	
	3100	Total: Mechanical & Electrical	↓	7.40	11.35	22.50	13.30%	18.90%	26%	
99	0010	**WAREHOUSE & OFFICES Combination**	S.F.	55	73.50	101				99
	0020	Total project costs	C.F.	2.82	4.09	6.05				
	1800	Equipment	S.F.	.95	1.84	2.74	.42%	1.19%	2.40%	
	2720	Plumbing		2.13	3.70	5.50	3.74%	4.76%	6.30%	
	2770	Heating, ventilating, air conditioning		3.36	5.25	7.35	5%	5.65%	10.05%	
	2900	Electrical		3.75	5.55	8.65	5.85%	8%	10%	
	3100	Total: Mechanical & Electrical	↓	10.45	15.85	25	14.55%	19.95%	24.50%	

Square Foot Project Size Modifier

One factor that affects the S.F. cost of a particular building is the size. In general, for buildings built to the same specifications in the same locality, the larger building will have the lower S.F. cost. This is due mainly to the decreasing contribution of the exterior walls, plus the economy of scale usually achievable in larger buildings. The area conversion scale shown below will give a factor to convert costs for the typical size building to an adjusted cost for the particular project.

The square foot base size table lists the median costs, most typical project size in our accumulated data, and the range in size of the projects.

The size factor for your project is determined by dividing your project area in S.F. by the typical project size for the particular building type. With this factor, enter the area conversion scale at the appropriate size factor and determine the appropriate cost multiplier for your building size.

Example: Determine the cost per S.F. for a 100,000 S.F. Mid-rise apartment building.

$$\frac{\text{Proposed building area} = 100{,}000 \text{ S.F.}}{\text{Typical size from below} = 50{,}000 \text{ S.F.}} = 2.00$$

Enter area conversion scale at 2.0, intersect curve, read horizontally the appropriate cost multiplier of .94. Size adjusted cost becomes .94 × \$119.00 = \$112.00 based on national average costs.

Note: For size factors less than .50, the cost multiplier is 1.1
For size factors greater than 3.5, the cost multiplier is .90

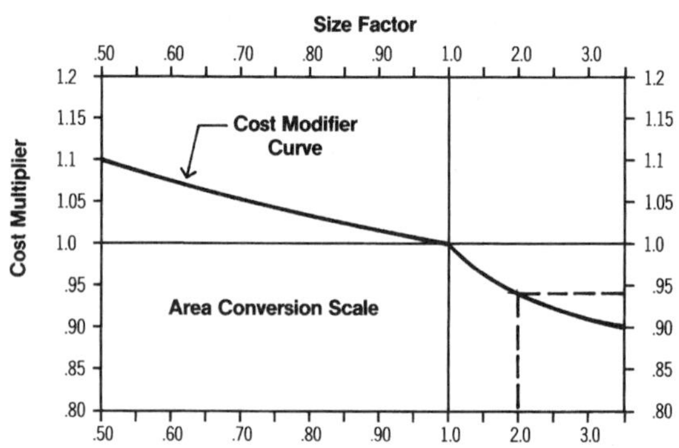

Square Foot Base Size

Building Type	Median Cost per S.F.	Typical Size Gross S.F.	Typical Range Gross S.F.	Building Type	Median Cost per S.F.	Typical Size Gross S.F.	Typical Range Gross S.F.
Apartments, Low Rise	$ 94.00	21,000	9,700 - 37,200	Jails	$ 286.00	40,000	5,500 - 145,000
Apartments, Mid Rise	119.00	50,000	32,000 - 100,000	Libraries	190.00	12,000	7,000 - 31,000
Apartments, High Rise	129.00	145,000	95,000 - 600,000	Living, Assisted	154.00	32,300	23,500 - 50,300
Auditoriums	161.00	25,000	7,600 - 39,000	Medical Clinics	162.00	7,200	4,200 - 15,700
Auto Sales	119.00	20,000	10,800 - 28,600	Medical Offices	154.00	6,000	4,000 - 15,000
Banks	211.00	4,200	2,500 - 7,500	Motels	113.00	40,000	15,800 - 120,000
Churches	145.00	17,000	2,000 - 42,000	Nursing Homes	158.00	23,000	15,000 - 37,000
Clubs, Country	147.00	6,500	4,500 - 15,000	Offices, Low Rise	135.00	20,000	5,000 - 80,000
Clubs, Social	141.00	10,000	6,000 - 13,500	Offices, Mid Rise	132.00	120,000	20,000 - 300,000
Clubs, YMCA	167.00	28,300	12,800 - 39,400	Offices, High Rise	169.00	260,000	120,000 - 800,000
Colleges (Class)	167.00	50,000	15,000 - 150,000	Police Stations	212.00	10,500	4,000 - 19,000
Colleges (Science Lab)	269.00	45,600	16,600 - 80,000	Post Offices	157.00	12,400	6,800 - 30,000
College (Student Union)	199.00	33,400	16,000 - 85,000	Power Plants	1175.00	7,500	1,000 - 20,000
Community Center	150.00	9,400	5,300 - 16,700	Religious Education	137.00	9,000	6,000 - 12,000
Court Houses	206.00	32,400	17,800 - 106,000	Research	222.00	19,000	6,300 - 45,000
Dept. Stores	87.50	90,000	44,000 - 122,000	Restaurants	192.00	4,400	2,800 - 6,000
Dormitories, Low Rise	170.00	25,000	10,000 - 95,000	Retail Stores	93.50	7,200	4,000 - 17,600
Dormitories, Mid Rise	196.00	85,000	20,000 - 200,000	Schools, Elementary	141.00	41,000	24,500 - 55,000
Factories	85.00	26,400	12,900 - 50,000	Schools, Jr. High	145.00	92,000	52,000 - 119,000
Fire Stations	156.00	5,800	4,000 - 8,700	Schools, Sr. High	150.00	101,000	50,500 - 175,000
Fraternity Houses	145.00	12,500	8,200 - 14,800	Schools, Vocational	143.00	37,000	20,500 - 82,000
Funeral Homes	162.00	10,000	4,000 - 20,000	Sports Arenas	115.00	15,000	5,000 - 40,000
Garages, Commercial	104.00	9,300	5,000 - 13,600	Supermarkets	92.00	44,000	12,000 - 60,000
Garages, Municipal	135.00	8,300	4,500 - 12,600	Swimming Pools	300.00	20,000	10,000 - 32,000
Garages, Parking	56.00	163,000	76,400 - 225,300	Telephone Exchange	243.00	4,500	1,200 - 10,600
Gymnasiums	143.00	19,200	11,600 - 41,000	Theaters	138.00	10,500	8,800 - 17,500
Hospitals	258.00	55,000	27,200 - 125,000	Town Halls	152.00	10,800	4,800 - 23,400
House (Elderly)	128.00	37,000	21,000 - 66,000	Warehouses	63.50	25,000	8,000 - 72,000
Housing (Public)	118.00	36,000	14,400 - 74,400	Warehouse & Office	73.50	25,000	8,000 - 72,000
Ice Rinks	170.00	29,000	27,200 - 33,600				

Abbreviations

A	Area Square Feet; Ampere
AAFES	Army and Air Force Exchange Service
ABS	Acrylonitrile Butadiene Stryrene; Asbestos Bonded Steel
A.C., AC	Alternating Current; Air-Conditioning; Asbestos Cement; Plywood Grade A & C
ACI	American Concrete Institute
ACR	Air Conditioning Refrigeration
ADA	Americans with Disabilities Act
AD	Plywood, Grade A & D
Addit.	Additional
Adj.	Adjustable
af	Audio-frequency
AFUE	Annual Fuel Utilization Efficiency
AGA	American Gas Association
Agg.	Aggregate
A.H., Ah	Ampere Hours
A hr.	Ampere-hour
A.H.U., AHU	Air Handling Unit
A.I.A.	American Institute of Architects
AIC	Ampere Interrupting Capacity
Allow.	Allowance
alt., alt	Alternate
Alum.	Aluminum
a.m.	Ante Meridiem
Amp.	Ampere
Anod.	Anodized
ANSI	American National Standards Institute
APA	American Plywood Association
Approx.	Approximate
Apt.	Apartment
Asb.	Asbestos
A.S.B.C.	American Standard Building Code
Asbe.	Asbestos Worker
ASCE.	American Society of Civil Engineers
A.S.H.R.A.E.	American Society of Heating, Refrig. & AC Engineers
ASME	American Society of Mechanical Engineers
ASTM	American Society for Testing and Materials
Attchmt.	Attachment
Avg., Ave.	Average
AWG	American Wire Gauge
AWWA	American Water Works Assoc.
Bbl.	Barrel
B&B, BB	Grade B and Better; Balled & Burlapped
B&S	Bell and Spigot
B.&W.	Black and White
b.c.c.	Body-centered Cubic
B.C.Y.	Bank Cubic Yards
BE	Bevel End
B.F.	Board Feet
Bg. cem.	Bag of Cement
BHP	Boiler Horsepower; Brake Horsepower
B.I.	Black Iron
bidir.	bidirectional
Bit., Bitum.	Bituminous
Bit., Conc.	Bituminous Concrete
Bk.	Backed
Bkrs.	Breakers
Bldg., bldg	Building
Blk.	Block
Bm.	Beam
Boil.	Boilermaker
bpm	Blows per Minute
BR	Bedroom
Brg.	Bearing
Brhe.	Bricklayer Helper
Bric.	Bricklayer
Brk., brk	Brick
brkt	Bracket
Brng.	Bearing
Brs.	Brass
Brz.	Bronze
Bsn.	Basin
Btr.	Better
Btu	British Thermal Unit
BTUH	BTU per Hour
Bu.	bushels
BUR	Built-up Roofing
BX	Interlocked Armored Cable
°C	degree centegrade
c	Conductivity, Copper Sweat
C	Hundred; Centigrade
C/C	Center to Center, Cedar on Cedar
C-C	Center to Center
Cab	Cabinet
Cair.	Air Tool Laborer
Cal.	caliper
Calc	Calculated
Cap.	Capacity
Carp.	Carpenter
C.B.	Circuit Breaker
C.C.A.	Chromate Copper Arsenate
C.C.F.	Hundred Cubic Feet
cd	Candela
cd/sf	Candela per Square Foot
CD	Grade of Plywood Face & Back
CDX	Plywood, Grade C & D, exterior glue
Cefi.	Cement Finisher
Cem.	Cement
CF	Hundred Feet
C.F.	Cubic Feet
CFM	Cubic Feet per Minute
CFRP	Carbon Fiber Reinforced Plastic
c.g.	Center of Gravity
CHW	Chilled Water; Commercial Hot Water
C.I., CI	Cast Iron
C.I.P., CIP	Cast in Place
Circ.	Circuit
C.L.	Carload Lot
CL	Chain Link
Clab.	Common Laborer
Clam	Common maintenance laborer
C.L.F.	Hundred Linear Feet
CLF	Current Limiting Fuse
CLP	Cross Linked Polyethylene
cm	Centimeter
CMP	Corr. Metal Pipe
CMU	Concrete Masonry Unit
CN	Change Notice
Col.	Column
CO_2	Carbon Dioxide
Comb.	Combination
comm.	Commercial, Communication
Compr.	Compressor
Conc.	Concrete
Cont., cont	Continuous; Continued, Container
Corr.	Corrugated
Cos	Cosine
Cot	Cotangent
Cov.	Cover
C/P	Cedar on Paneling
CPA	Control Point Adjustment
Cplg.	Coupling
CPM	Critical Path Method
CPVC	Chlorinated Polyvinyl Chloride
C.Pr.	Hundred Pair
CRC	Cold Rolled Channel
Creos.	Creosote
Crpt.	Carpet & Linoleum Layer
CRT	Cathode-ray Tube
CS	Carbon Steel, Constant Shear Bar Joist
Csc	Cosecant
C.S.F.	Hundred Square Feet
CSI	Construction Specifications Institute
CT	Current Transformer
CTS	Copper Tube Size
Cu	Copper, Cubic
Cu. Ft.	Cubic Foot
cw	Continuous Wave
C.W.	Cool White; Cold Water
Cwt.	100 Pounds
C.W.X.	Cool White Deluxe
C.Y.	Cubic Yard (27 cubic feet)
C.Y./Hr.	Cubic Yard per Hour
Cyl.	Cylinder
d	Penny (nail size)
D	Deep; Depth; Discharge
Dis., Disch.	Discharge
Db	Decibel
Dbl.	Double
DC	Direct Current
DDC	Direct Digital Control
Demob.	Demobilization
d.f.t.	Dry Film Thickness
d.f.u.	Drainage Fixture Units
D.H.	Double Hung
DHW	Domestic Hot Water
DI	Ductile Iron
Diag.	Diagonal
Diam., Dia	Diameter
Distrib.	Distribution
Div.	Division
Dk.	Deck
D.L.	Dead Load; Diesel
DLH	Deep Long Span Bar Joist
dlx	Deluxe
Do.	Ditto
DOP	Dioctyl Phthalate Penetration Test (Air Filters)
Dp., dp	Depth
D.P.S.T.	Double Pole, Single Throw
Dr.	Drive
DR	Dimension Ratio
Drink.	Drinking
D.S.	Double Strength
D.S.A.	Double Strength A Grade
D.S.B.	Double Strength B Grade
Dty.	Duty
DWV	Drain Waste Vent
DX	Deluxe White, Direct Expansion
dyn	Dyne
e	Eccentricity
E	Equipment Only; East; emissivity
Ea.	Each
EB	Encased Burial
Econ.	Economy
E.C.Y	Embankment Cubic Yards
EDP	Electronic Data Processing
EIFS	Exterior Insulation Finish System
E.D.R.	Equiv. Direct Radiation
Eq.	Equation
EL	elevation
Elec.	Electrician; Electrical
Elev.	Elevator; Elevating
EMT	Electrical Metallic Conduit; Thin Wall Conduit
Eng.	Engine, Engineered
EPDM	Ethylene Propylene Diene Monomer
EPS	Expanded Polystyrene
Eqhv.	Equip. Oper., Heavy
Eqlt.	Equip. Oper., Light
Eqmd.	Equip. Oper., Medium
Eqmm.	Equip. Oper., Master Mechanic
Eqol.	Equip. Oper., Oilers
Equip.	Equipment
ERW	Electric Resistance Welded

Abbreviations

E.S.	Energy Saver
Est.	Estimated
esu	Electrostatic Units
E.W.	Each Way
EWT	Entering Water Temperature
Excav.	Excavation
excl	Excluding
Exp., exp	Expansion, Exposure
Ext., ext	Exterior; Extension
Extru.	Extrusion
f.	Fiber stress
F	Fahrenheit; Female; Fill
Fab., fab	Fabricated; fabric
FBGS	Fiberglass
F.C.	Footcandles
f.c.c.	Face-centered Cubic
f'c.	Compressive Stress in Concrete; Extreme Compressive Stress
F.E.	Front End
FEP	Fluorinated Ethylene Propylene (Teflon)
F.G.	Flat Grain
F.H.A.	Federal Housing Administration
Fig.	Figure
Fin.	Finished
FIPS	Female Iron Pipe Size
Fixt.	Fixture
FJP	Finger jointed and primed
Fl. Oz.	Fluid Ounces
Flr.	Floor
FM	Frequency Modulation; Factory Mutual
Fmg.	Framing
FM/UL	Factory Mutual/Underwriters Labs
Fdn.	Foundation
FNPT	Female National Pipe Thread
Fori.	Foreman, Inside
Foro.	Foreman, Outside
Fount.	Fountain
fpm	Feet per Minute
FPT	Female Pipe Thread
Fr	Frame
F.R.	Fire Rating
FRK	Foil Reinforced Kraft
FSK	Foil/scrim/kraft
FRP	Fiberglass Reinforced Plastic
FS	Forged Steel
FSC	Cast Body; Cast Switch Box
Ft., ft	Foot; Feet
Ftng.	Fitting
Ftg.	Footing
Ft lb.	Foot Pound
Furn.	Furniture
FVNR	Full Voltage Non-Reversing
FVR	Full Voltage Reversing
FXM	Female by Male
Fy.	Minimum Yield Stress of Steel
g	Gram
G	Gauss
Ga.	Gauge
Gal., gal.	Gallon
gpm, GPM	Gallon per Minute
Galv., galv	Galvanized
GC/MS	Gas Chromatograph/Mass Spectrometer
Gen.	General
GFI	Ground Fault Interrupter
GFRC	Glass Fiber Reinforced Concrete
Glaz.	Glazier
GPD	Gallons per Day
gpf	Gallon per flush
GPH	Gallons per Hour
GPM	Gallons per Minute
GR	Grade
Gran.	Granular
Grnd.	Ground
GVW	Gross Vehicle Weight
GWB	Gypsum wall board
H	High Henry
HC	High Capacity
H.D., HD	Heavy Duty; High Density
H.D.O.	High Density Overlaid
HDPE	High density polyethelene plastic
Hdr.	Header
Hdwe.	Hardware
H.I.D., HID	High Intensity Discharge
Help.	Helper Average
HEPA	High Efficiency Particulate Air Filter
Hg	Mercury
HIC	High Interrupting Capacity
HM	Hollow Metal
HMWPE	high molecular weight polyethylene
HO	High Output
Horiz.	Horizontal
H.P., HP	Horsepower; High Pressure
H.P.F.	High Power Factor
Hr.	Hour
Hrs./Day	Hours per Day
HSC	High Short Circuit
Ht.	Height
Htg.	Heating
Htrs.	Heaters
HVAC	Heating, Ventilation & Air-Conditioning
Hvy.	Heavy
HW	Hot Water
Hyd.; Hydr.	Hydraulic
Hz	Hertz (cycles)
I.	Moment of Inertia
IBC	International Building Code
I.C.	Interrupting Capacity
ID	Inside Diameter
I.D.	Inside Dimension; Identification
I.F.	Inside Frosted
I.M.C.	Intermediate Metal Conduit
In.	Inch
Incan.	Incandescent
Incl.	Included; Including
Int.	Interior
Inst.	Installation
Insul., insul	Insulation/Insulated
I.P.	Iron Pipe
I.P.S., IPS	Iron Pipe Size
IPT	Iron Pipe Threaded
I.W.	Indirect Waste
J	Joule
J.I.C.	Joint Industrial Council
K	Thousand; Thousand Pounds; Heavy Wall Copper Tubing, Kelvin
K.A.H.	Thousand Amp. Hours
kcmil	Thousand Circular Mils
KD	Knock Down
K.D.A.T.	Kiln Dried After Treatment
kg	Kilogram
kG	Kilogauss
kgf	Kilogram Force
kHz	Kilohertz
Kip	1000 Pounds
KJ	Kiljoule
K.L.	Effective Length Factor
K.L.F.	Kips per Linear Foot
Km	Kilometer
KO	Knock Out
K.S.F.	Kips per Square Foot
K.S.I.	Kips per Square Inch
kV	Kilovolt
kVA	Kilovolt Ampere
kVAR	Kilovar (Reactance)
KW	Kilowatt
KWh	Kilowatt-hour
L	Labor Only; Length; Long; Medium Wall Copper Tubing
Lab.	Labor
lat	Latitude
Lath.	Lather
Lav.	Lavatory
lb.; #	Pound
L.B., LB	Load Bearing; L Conduit Body
L. & E.	Labor & Equipment
lb./hr.	Pounds per Hour
lb./L.F.	Pounds per Linear Foot
lbf/sq.in.	Pound-force per Square Inch
L.C.L.	Less than Carload Lot
L.C.Y.	Loose Cubic Yard
Ld.	Load
LE	Lead Equivalent
LED	Light Emitting Diode
L.F.	Linear Foot
L.F. Nose	Linear Foot of Stair Nosing
L.F. Rsr	Linear Foot of Stair Riser
Lg.	Long; Length; Large
L & H	Light and Heat
LH	Long Span Bar Joist
L.H.	Labor Hours
L.L., LL	Live Load
L.L.D.	Lamp Lumen Depreciation
lm	Lumen
lm/sf	Lumen per Square Foot
lm/W	Lumen per Watt
LOA	Length Over All
log	Logarithm
L-O-L	Lateralolet
long.	longitude
L.P., LP	Liquefied Petroleum; Low Pressure
L.P.F.	Low Power Factor
LR	Long Radius
L.S.	Lump Sum
Lt.	Light
Lt. Ga.	Light Gauge
L.T.L.	Less than Truckload Lot
Lt. Wt.	Lightweight
L.V.	Low Voltage
M	Thousand; Material; Male; Light Wall Copper Tubing
M^{2}CA	Meters Squared Contact Area
m/hr.; M.H.	Man-hour
mA	Milliampere
Mach.	Machine
Mag. Str.	Magnetic Starter
Maint.	Maintenance
Marb.	Marble Setter
Mat; Mat'l.	Material
Max.	Maximum
MBF	Thousand Board Feet
MBH	Thousand BTU's per hr.
MC	Metal Clad Cable
MCC	Motor Control Center
M.C.F.	Thousand Cubic Feet
MCFM	Thousand Cubic Feet per Minute
M.C.M.	Thousand Circular Mils
MCP	Motor Circuit Protector
MD	Medium Duty
MDF	Medium-density fibreboard
M.D.O.	Medium Density Overlaid
Med.	Medium
MF	Thousand Feet
M.F.B.M.	Thousand Feet Board Measure
Mfg.	Manufacturing
Mfrs.	Manufacturers
mg	Milligram
MGD	Million Gallons per Day
MGPH	Thousand Gallons per Hour
MH, M.H.	Manhole; Metal Halide; Man-Hour
MHz	Megahertz
Mi.	Mile
MI	Malleable Iron; Mineral Insulated
MIPS	Male Iron Pipe Size
mj	Mechanical Joint
m	Meter
mm	Millimeter
Mill.	Millwright
Min., min.	Minimum, minute

Abbreviations

Misc.	Miscellaneous	PDCA	Painting and Decorating Contractors of America	SC	Screw Cover
ml	Milliliter, Mainline	P.E., PE	Professional Engineer; Porcelain Enamel; Polyethylene; Plain End	SCFM	Standard Cubic Feet per Minute
M.L.F.	Thousand Linear Feet	P.E.C.I.	Porcelain Enamel on Cast Iron	Scaf.	Scaffold
Mo.	Month	Perf.	Perforated	Sch., Sched.	Schedule
Mobil.	Mobilization	PEX	Cross linked polyethylene	S.C.R.	Modular Brick
Mog.	Mogul Base	Ph.	Phase	S.D.	Sound Deadening
MPH	Miles per Hour	P.I.	Pressure Injected	SDR	Standard Dimension Ratio
MPT	Male Pipe Thread	Pile.	Pile Driver	S.E.	Surfaced Edge
MRGWB	Moisture Resistant Gypsum Wallboard	Pkg.	Package	Sel.	Select
MRT	Mile Round Trip	Pl.	Plate	SER, SEU	Service Entrance Cable
ms	Millisecond	Plah.	Plasterer Helper	S.F.	Square Foot
M.S.F.	Thousand Square Feet	Plas.	Plasterer	S.F.C.A.	Square Foot Contact Area
Mstz.	Mosaic & Terrazzo Worker	plf	Pounds Per Linear Foot	S.F. Flr.	Square Foot of Floor
M.S.Y.	Thousand Square Yards	Pluh.	Plumbers Helper	S.F.G.	Square Foot of Ground
Mtd., mtd., mtd	Mounted	Plum.	Plumber	S.F. Hor.	Square Foot Horizontal
Mthe.	Mosaic & Terrazzo Helper	Ply.	Plywood	SFR	Square Feet of Radiation
Mtng.	Mounting	p.m.	Post Meridiem	S.F. Shlf.	Square Foot of Shelf
Mult.	Multi; Multiply	Pntd.	Painted	S4S	Surface 4 Sides
M.V.A.	Million Volt Amperes	Pord.	Painter, Ordinary	Shee.	Sheet Metal Worker
M.V.A.R.	Million Volt Amperes Reactance	pp	Pages	Sin.	Sine
MV	Megavolt	PP, PPL	Polypropylene	Skwk.	Skilled Worker
MW	Megawatt	P.P.M.	Parts per Million	SL	Saran Lined
MXM	Male by Male	Pr.	Pair	S.L.	Slimline
MYD	Thousand Yards	P.E.S.B.	Pre-engineered Steel Building	Sldr.	Solder
N	Natural; North	Prefab.	Prefabricated	SLH	Super Long Span Bar Joist
nA	Nanoampere	Prefin.	Prefinished	S.N.	Solid Neutral
NA	Not Available; Not Applicable	Prop.	Propelled	SO	Stranded with oil resistant inside insulation
N.B.C.	National Building Code	PSF, psf	Pounds per Square Foot	S-O-L	Socketolet
NC	Normally Closed	PSI, psi	Pounds per Square Inch	sp	Standpipe
NEMA	National Electrical Manufacturers Assoc.	PSIG	Pounds per Square Inch Gauge	S.P.	Static Pressure; Single Pole; Self-Propelled
NEHB	Bolted Circuit Breaker to 600V.	PSP	Plastic Sewer Pipe	Spri.	Sprinkler Installer
NFPA	National Fire Protection Association	Pspr.	Painter, Spray	spwg	Static Pressure Water Gauge
NLB	Non-Load-Bearing	Psst.	Painter, Structural Steel	S.P.D.T.	Single Pole, Double Throw
NM	Non-Metallic Cable	P.T.	Potential Transformer	SPF	Spruce Pine Fir; Sprayed Polyurethane Foam
nm	Nanometer	P. & T.	Pressure & Temperature	S.P.S.T.	Single Pole, Single Throw
No.	Number	Ptd.	Painted	SPT	Standard Pipe Thread
NO	Normally Open	Ptns.	Partitions	Sq.	Square; 100 Square Feet
N.O.C.	Not Otherwise Classified	Pu	Ultimate Load	Sq. Hd.	Square Head
Nose.	Nosing	PVC	Polyvinyl Chloride	Sq. In.	Square Inch
NPT	National Pipe Thread	Pvmt.	Pavement	S.S.	Single Strength; Stainless Steel
NQOD	Combination Plug-on/Bolt on Circuit Breaker to 240V.	PRV	Pressure Relief Valve	S.S.B.	Single Strength B Grade
N.R.C., NRC	Noise Reduction Coefficient/ Nuclear Regulator Commission	Pwr.	Power	sst, ss	Stainless Steel
N.R.S.	Non Rising Stem	Q	Quantity Heat Flow	Sswk.	Structural Steel Worker
ns	Nanosecond	Qt.	Quart	Sswl.	Structural Steel Welder
nW	Nanowatt	Quan., Qty.	Quantity	St.; Stl.	Steel
OB	Opposing Blade	Q.C.	Quick Coupling	STC	Sound Transmission Coefficient
OC	On Center	r	Radius of Gyration	Std.	Standard
OD	Outside Diameter	R	Resistance	Stg.	Staging
O.D.	Outside Dimension	R.C.P.	Reinforced Concrete Pipe	STK	Select Tight Knot
ODS	Overhead Distribution System	Rect.	Rectangle	STP	Standard Temperature & Pressure
O.G.	Ogee	recpt.	receptacle	Stpi.	Steamfitter, Pipefitter
O.H.	Overhead	Reg.	Regular	Str.	Strength; Starter; Straight
O&P	Overhead and Profit	Reinf.	Reinforced	Strd.	Stranded
Oper.	Operator	Req'd.	Required	Struct.	Structural
Opng.	Opening	Res.	Resistant	Sty.	Story
Orna.	Ornamental	Resi.	Residential	Subj.	Subject
OSB	Oriented Strand Board	RF	Radio Frequency	Subs.	Subcontractors
OS&Y	Outside Screw and Yoke	RFID	Radio-frequency identification	Surf.	Surface
OSHA	Occupational Safety and Health Act	Rgh.	Rough	Sw.	Switch
Ovhd.	Overhead	RGS	Rigid Galvanized Steel	Swbd.	Switchboard
OWG	Oil, Water or Gas	RHW	Rubber, Heat & Water Resistant; Residential Hot Water	S.Y.	Square Yard
Oz.	Ounce	rms	Root Mean Square	Syn.	Synthetic
P.	Pole; Applied Load; Projection	Rnd.	Round	S.Y.P.	Southern Yellow Pine
p.	Page	Rodm.	Rodman	Sys.	System
Pape.	Paperhanger	Rofc.	Roofer, Composition	t.	Thickness
P.A.P.R.	Powered Air Purifying Respirator	Rofp.	Roofer, Precast	T	Temperature; Ton
PAR	Parabolic Reflector	Rohe.	Roofer Helpers (Composition)	Tan	Tangent
P.B., PB	Push Button	Rots.	Roofer, Tile & Slate	T.C.	Terra Cotta
Pc., Pcs.	Piece, Pieces	R.O.W.	Right of Way	T & C	Threaded and Coupled
P.C.	Portland Cement; Power Connector	RPM	Revolutions per Minute	T.D.	Temperature Difference
P.C.F.	Pounds per Cubic Foot	R.S.	Rapid Start	Tdd	Telecommunications Device for the Deaf
PCM	Phase Contrast Microscopy	Rsr	Riser	T.E.M.	Transmission Electron Microscopy
		RT	Round Trip	temp	Temperature, Tempered, Temporary
		S.	Suction; Single Entrance; South	TFFN	Nylon Jacketed Wire
		SBS	Styrene Butadiere Styrene		

Abbreviations

TFE	Tetrafluoroethylene (Teflon)
T. & G.	Tongue & Groove; Tar & Gravel
Th.,Thk.	Thick
Thn.	Thin
Thrded	Threaded
Tilf.	Tile Layer, Floor
Tilh.	Tile Layer, Helper
THHN	Nylon Jacketed Wire
THW.	Insulated Strand Wire
THWN	Nylon Jacketed Wire
T.L.,TL	Truckload
T.M.	Track Mounted
Tot.	Total
T-O-L	Threadolet
tmpd	Tempered
TPO	Thermoplastic Polyolefin
T.S.	Trigger Start
Tr.	Trade
Transf.	Transformer
Trhv.	Truck Driver, Heavy
Trlr	Trailer
Trlt.	Truck Driver, Light
TTY	Teletypewriter
TV	Television
T.W.	Thermoplastic Water Resistant Wire
UCI	Uniform Construction Index
UF	Underground Feeder
UGND	Underground Feeder
UHF	Ultra High Frequency
U.I.	United Inch
U.L., UL	Underwriters Laboratory
Uld.	unloading
Unfin.	Unfinished
UPS	Uninterruptible Power Supply
URD	Underground Residential Distribution
US	United States
USGBC	U.S. Green Building Council
USP	United States Primed
UTMCD	Uniform Traffic Manual For Control Devices
UTP	Unshielded Twisted Pair
V	Volt
VA	Volt Amperes
VAT	Vinyl Asbestos Tile
V.C.T.	Vinyl Composition Tile
VAV	Variable Air Volume
VC	Veneer Core
VDC	Volts Direct Current
Vent.	Ventilation
Vert.	Vertical
V.F.	Vinyl Faced
V.G.	Vertical Grain
VHF	Very High Frequency
VHO	Very High Output
Vib.	Vibrating
VLF	Vertical Linear Foot
VOC	Volitile Organic Compound
Vol.	Volume
VRP	Vinyl Reinforced Polyester
W	Wire; Watt; Wide; West
w/	With
W.C., WC	Water Column; Water Closet
W.F.	Wide Flange
W.G.	Water Gauge
Wldg.	Welding
W. Mile	Wire Mile
W-O-L	Weldolet
W.R.	Water Resistant
Wrck.	Wrecker
W.S.P.	Water, Steam, Petroleum
WT., Wt.	Weight
WWF	Welded Wire Fabric
XFER	Transfer
XFMR	Transformer
XHD	Extra Heavy Duty
XHHW, XLPE	Cross-Linked Polyethylene Wire Insulation
XLP	Cross-linked Polyethylene
Xport	Transport
Y	Wye
yd	Yard
yr	Year
Δ	Delta
%	Percent
~	Approximately
Ø	Phase; diameter
@	At
#	Pound; Number
<	Less Than
>	Greater Than
Z	zone

A

B

C

D

E

F

H

I

J

K

L

M

Q

R

T

U

V

X

Y

Z

Notes

Notes

Other RSMeans Products & Services

RSMeans—
a tradition of excellence in construction cost information and services since 1942

Table of Contents

For more information visit the RSMeans website at **www.rsmeans.com**

Unit prices according to the latest MasterFormat!

Book Selection Guide

The following table provides definitive information on the content of each cost data publication. The number of lines of data provided in each unit price or assemblies division, as well as the number of crews, is listed for each book. The presence of other elements such as reference tables, square foot models, equipment rental costs, historical cost indexes, and city cost indexes, is also indicated. You can use the table to help select the RSMeans book that has the quantity and type of information you most need in your work.

Unit Cost Divisions	Building Construction	Mechanical	Electrical	Commercial Renovation	Square Foot	Site Work Landsc.	Green Building	Interior	Concrete Masonry	Open Shop	Heavy Construction	Light Commercial	Facilities Construction	Plumbing	Residential
1	573	393	410	506		509	191	312	456	572	501	246	1042	403	178
2	777	279	85	733		993	178	399	213	776	734	481	1221	286	257
3	1690	340	230	1085		1471	987	355	2032	1689	1685	481	1788	316	388
4	957	21	0	922		720	179	612	1155	925	612	529	1172	0	441
5	1893	158	155	1090		841	1799	1095	720	1893	1037	979	1909	204	721
6	2452	18	18	2110		110	589	1528	281	2448	123	2140	2124	22	2660
7	1584	215	128	1623		581	757	532	524	1581	26	1317	1685	227	1037
8	2082	81	45	2609		265	1123	1753	105	2077	0	2193	2879	0	1526
9	1971	72	26	1767		313	449	2056	390	1911	15	1659	2220	54	1435
10	1026	17	10	642		215	27	842	158	1026	29	511	1118	235	224
11	1087	208	165	546		124	54	932	28	1071	0	230	1107	169	110
12	544	0	2	319		217	137	1654	14	532	0	352	1704	23	310
13	719	140	137	242		343	111	248	66	695	244	83	746	94	79
14	273	36	0	221		0	0	257	0	273	0	12	293	16	6
21	90	0	16	37		0	0	250	0	87	0	68	426	431	220
22	1157	7544	150	1190		1568	1067	838	20	1121	1677	857	7440	9347	712
23	1177	6971	580	928		158	902	776	38	1101	111	878	5187	1918	469
26	1354	454	10081	1021		793	611	1136	55	1349	562	1298	10030	399	617
27	72	0	297	34		13	0	71	0	72	39	52	279	0	22
28	96	58	136	71		0	21	78	0	99	0	40	151	44	25
31	1498	735	610	803		3217	289	7	1208	1443	3256	600	1560	660	611
32	822	53	8	886		4426	356	405	291	793	1841	420	1701	165	468
33	530	1079	538	252		2178	41	0	237	522	2159	128	1699	1286	154
34	107	0	47	4		190	0	0	31	62	193	0	128	0	0
35	18	0	0	0		327	0	0	0	18	442	0	84	0	0
41	60	0	0	32		7	0	22	0	60	30	0	67	14	0
44	75	79	0	0		0	0	0	0	0	0	0	75	75	0
46	23	16	0	0		274	261	0	0	23	264	0	33	33	0
48	12	0	25	0		0	25	0	0	12	17	12	12	0	12
Totals	24719	18967	13899	19673		19853	10154	16158	8022	24231	15597	15566	49880	16421	12682

Assem Div	Building Construction	Mechanical	Elec-trical	Commercial Renovation	Square Foot	Site Work Landscape	Assemblies	Green Building	Interior	Concrete Masonry	Heavy Construction	Light Commercial	Facilities Construction	Plumbing	Asm Div	Residential
A		15	0	188	150	577	598	0	0	536	571	154	24	0	1	376
B		0	0	848	2498	0	5658	56	329	1975	368	2089	174	0	2	211
C		0	0	647	926	0	1304	0	1629	146	0	816	250	0	3	588
D		1067	941	712	1859	72	2538	330	825	0	0	1345	1105	1088	4	851
E		0	0	86	260	0	300	0	5	0	0	257	5	0	5	392
F		0	0	0	114	0	114	0	0	0	0	114	3	0	6	357
G		527	447	318	249	3365	729	0	0	535	1350	205	293	677	7	307
															8	760
															9	80
															10	0
															11	0
															12	0
Totals		1609	1388	2799	6056	4014	11241	386	2788	3192	2289	4980	1854	1765		3922

Reference Section	Building Construction Costs	Mechanical	Electrical	Commercial Renovation	Square Foot	Site Work Landscape	Assem.	Green Building	Interior	Concrete Masonry	Open Shop	Heavy Construction	Light Commercial	Facilities Construction	Plumbing	Resi.
Reference Tables	yes	yes	yes	yes	no	yes	yes	yes	yes	yes	yes	yes	yes	yes	yes	yes
Models					111			25					50			28
Crews	571	571	571	550		571		571	571	571	548	571	548	550	571	548
Equipment Rental Costs	yes	yes	yes	yes		yes		yes	yes	yes	yes	yes	yes	yes	yes	yes
Historical Cost Indexes	yes	yes	yes	yes	yes	yes	yes	yes	yes	yes	yes	yes	yes	yes	yes	no
City Cost Indexes	yes	yes	yes	yes	yes	yes	yes	yes	yes	yes	yes	yes	yes	yes	yes	yes

RSMeans Building Construction Cost Data 2015

Offers you unchallenged unit price reliability in an easy-to-use format. Whether used for verifying complete, finished estimates or for periodic checks, it supplies more cost facts better and faster than any comparable source. More than 24,700 unit prices have been updated for 2015. The City Cost Indexes and Location Factors cover more than 930 areas, for indexing to any project location in North America. Order and get *RSMeans Quarterly Update Service* FREE.

$206.95 | Available Sept. 2014 | Catalog no. 60015

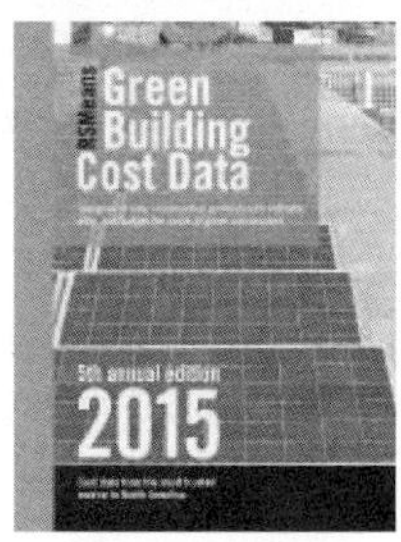

RSMeans Green Building Cost Data 2015

Estimate, plan, and budget the costs of green building for both new commercial construction and renovation work with this fifth edition of *RSMeans Green Building Cost Data*. More than 10,000 unit costs for a wide array of green building products plus assemblies costs. Easily identified cross references to LEED and Green Globes building rating systems criteria.

$170.95 | Available Nov. 2014 | Catalog no. 60555

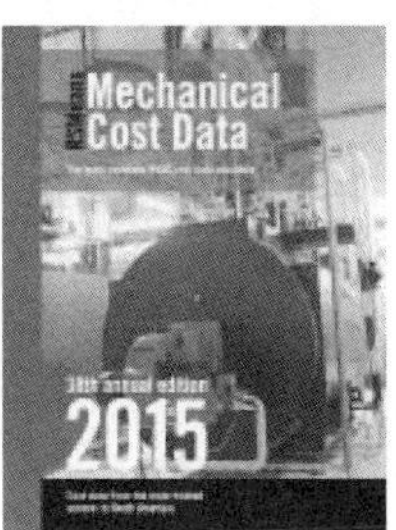

RSMeans Mechanical Cost Data 2015

Total unit and systems price guidance for mechanical construction . . . materials, parts, fittings, and complete labor cost information. Includes prices for piping, heating, air conditioning, ventilation, and all related construction.

Plus new 2015 unit costs for:

- Thousands of installed HVAC/controls, sub-assemblies and assemblies
- "On-site" Location Factors for more than 930 cities and towns in the U.S. and Canada
- Crews, labor, and equipment

$203.95 | Available Oct. 2014 | Catalog no. 60025

RSMeans Facilities Construction Cost Data 2015

For the maintenance and construction of commercial, industrial, municipal, and institutional properties. Costs are shown for new and remodeling construction and are broken down into materials, labor, equipment, and overhead and profit. Special emphasis is given to sections on mechanical, electrical, furnishings, site work, building maintenance, finish work, and demolition.

More than 49,800 unit costs, plus assemblies costs and a comprehensive Reference Section are included.

$524.95 | Available Nov. 2014 | Catalog no. 60205A

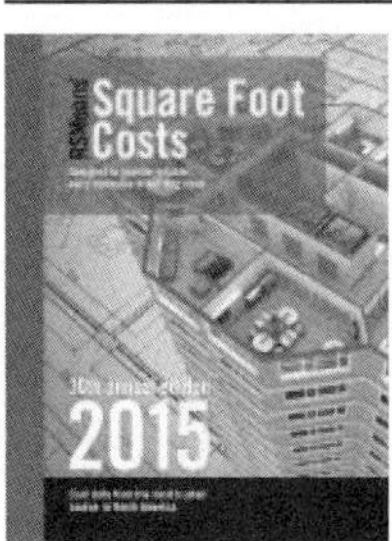

RSMeans Square Foot Costs 2015

Accurate and Easy-to-Use

- **Updated price information** based on nationwide figures from suppliers, estimators, labor experts, and contractors.
- Green building models
- Realistic graphics, offering true-to-life illustrations of building projects
- Extensive information on using square foot cost data, including sample estimates and alternate pricing methods

$218.95 | Available Oct. 2014 | Catalog no. 60055

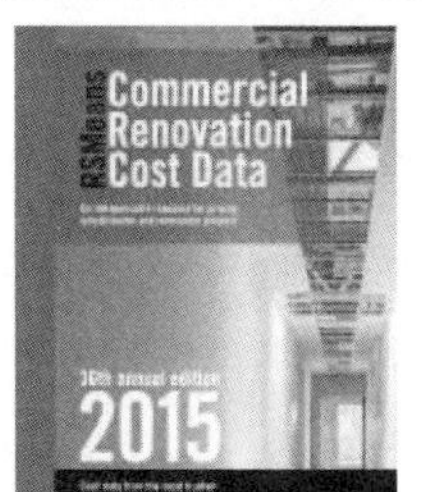

RSMeans Commercial Renovation Cost Data 2015

Commercial/Multi-family Residential

Use this valuable tool to estimate commercial and multi-family residential renovation and remodeling.

Includes: Updated costs for hundreds of unique methods, materials, and conditions that only come up in repair and remodeling, PLUS:

- Unit costs for more than 19,600 construction components
- Installed costs for more than 2,700 assemblies
- More than 930 "on-site" localization factors for the U.S. and Canada

$166.95 | Available Oct. 2014 | Catalog no. 60045

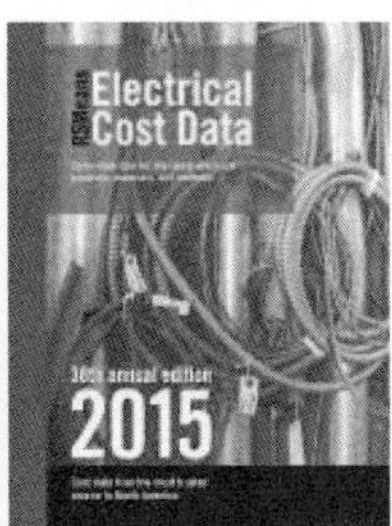

RSMeans Electrical Cost Data 2015

Pricing information for every part of electrical cost planning. More than 13,800 unit and systems costs with design tables; clear specifications and drawings; engineering guides; illustrated estimating procedures; complete labor-hour and materials costs for better scheduling and procurement; and the latest electrical products and construction methods.

- A variety of special electrical systems, including cathodic protection
- Costs for maintenance, demolition, HVAC/mechanical, specialties, equipment, and more

$208.95 | Available Oct. 2014 | Catalog no. 60035

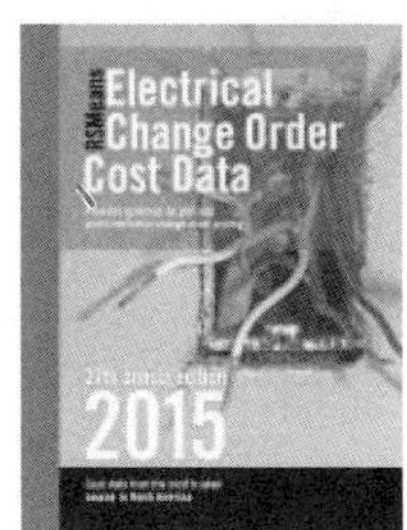

RSMeans Electrical Change Order Cost Data 2015

RSMeans Electrical Change Order Cost Data provides you with electrical unit prices exclusively for pricing change orders—based on the recent, direct experience of contractors and suppliers. Analyze and check your own change order estimates against the experience others have had doing the same work. It also covers productivity analysis, and change order cost justifications. With useful information for calculating the effects of change orders and dealing with their administration.

$199.95 | Available Dec. 2014 | Catalog no. 60235

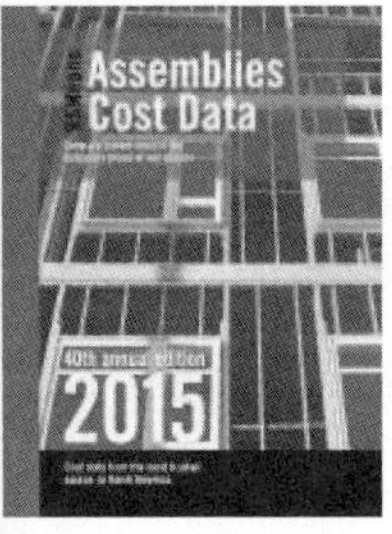

RSMeans Assemblies Cost Data 2015

RSMeans Assemblies Cost Data takes the guesswork out of preliminary or conceptual estimates. Now you don't have to try to calculate the assembled cost by working up individual component costs. We've done all the work for you.

Presents detailed illustrations, descriptions, specifications, and costs for every conceivable building assembly—over 350 types in all—arranged in the easy-to-use UNIFORMAT II system. Each illustrated "assembled" cost includes a complete grouping of materials and associated installation costs, including the installing contractor's overhead and profit.

$334.95 | Available Sept. 2014 | Catalog no. 60065

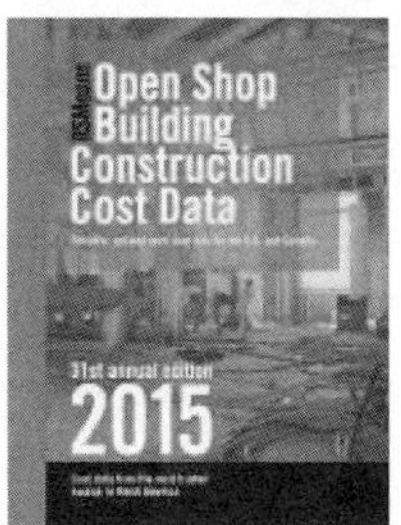

RSMeans Open Shop Building Construction Cost Data 2015

The latest costs for accurate budgeting and estimating of new commercial and residential construction . . . renovation work . . . change orders . . . cost engineering.

RSMeans Open Shop "BCCD" will assist you to:

- Develop benchmark prices for change orders.
- Plug gaps in preliminary estimates and budgets.
- Estimate complex projects.
- Substantiate invoices on contracts.
- Price ADA-related renovations.

$175.95 | Available Dec. 2014 | Catalog no. 60155

For more information visit the RSMeans website at www.rsmeans.com

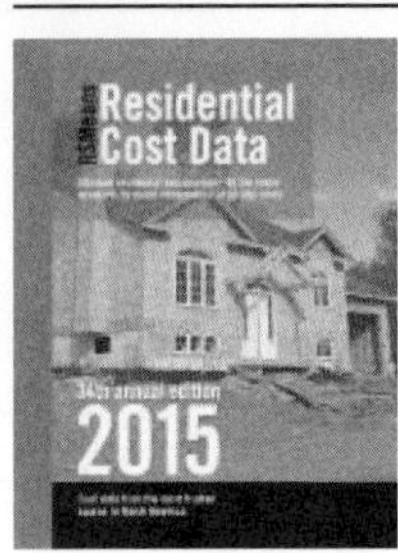

RSMeans Residential Cost Data 2015

Contains square foot costs for 28 basic home models with the look of today, plus hundreds of custom additions and modifications you can quote right off the page. Includes more than 3,900 costs for 89 residential systems. Complete with blank estimating forms, sample estimates, and step-by-step instructions.

Contains line items for cultured stone and brick, PVC trim, lumber, and TPO roofing.

$148.95 | Available Oct. 2014 | Catalog no. 60175

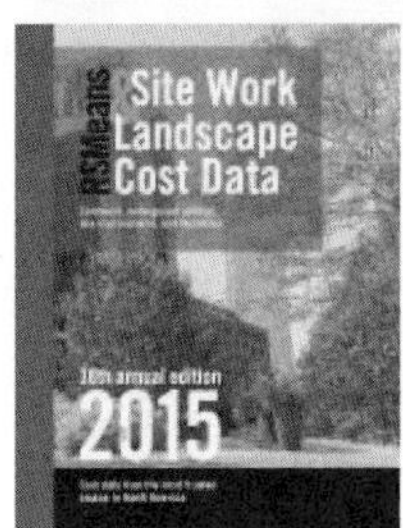

RSMeans Site Work & Landscape Cost Data 2015

Includes unit and assemblies costs for earthwork, sewerage, piped utilities, site improvements, drainage, paving, trees and shrubs, street openings/repairs, underground tanks, and more. Contains more than 60 types of assemblies costs for accurate conceptual estimates.

Includes:

- Estimating for infrastructure improvements
- Environmentally-oriented construction
- ADA-mandated handicapped access
- Hazardous waste line items

$199.95 | Available Dec. 2014 | Catalog no. 60285

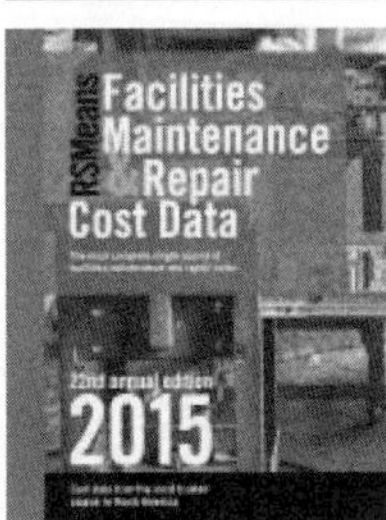

RSMeans Facilities Maintenance & Repair Cost Data 2015

RSMeans Facilities Maintenance & Repair Cost Data gives you a complete system to manage and plan your facility repair and maintenance costs and budget efficiently. Guidelines for auditing a facility and developing an annual maintenance plan. Budgeting is included, along with reference tables on cost and management, and information on frequency and productivity of maintenance operations.

The only nationally recognized source of maintenance and repair costs. Developed in cooperation with the Civil Engineering Research Laboratory (CERL) of the Army Corps of Engineers.

$461.95 | Available Nov. 2014 | Catalog no. 60305

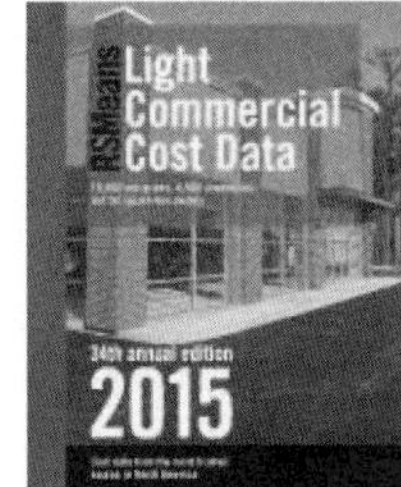

RSMeans Light Commercial Cost Data 2015

Specifically addresses the light commercial market, which is a specialized niche in the construction industry. Aids you, the owner/designer/contractor, in preparing all types of estimates—from budgets to detailed bids. Includes new advances in methods and materials.

Assemblies Section allows you to evaluate alternatives in the early stages of design/planning.

More than 15,500 unit costs ensure that you have the prices you need, when you need them.

$153.95 | Available Nov. 2014 | Catalog no. 60185

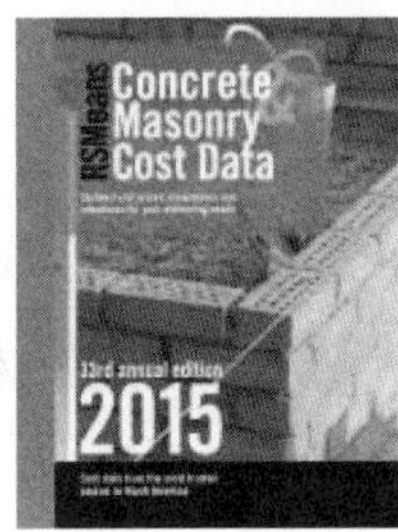

RSMeans Concrete & Masonry Cost Data 2015

Provides you with cost facts for virtually all concrete/masonry estimating needs, from complicated formwork to various sizes and face finishes of brick and block—all in great detail. The comprehensive Unit Price Section contains more than 8,000 selected entries. Also contains an Assemblies [Cost] Section, and a detailed Reference Section that supplements the cost data.

$188.95 | Available Dec. 2014 | Catalog no. 60115

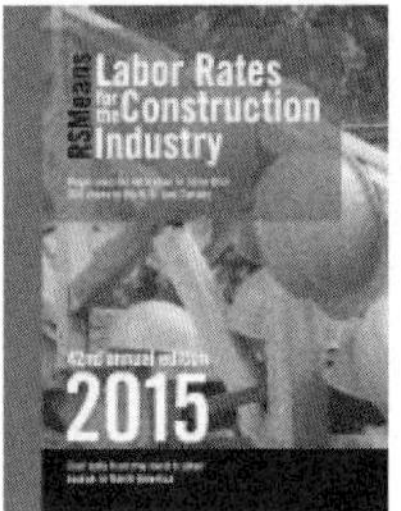

RSMeans Labor Rates for the Construction Industry 2015

Complete information for estimating labor costs, making comparisons, and negotiating wage rates by trade for more than 300 U.S. and Canadian cities. With 46 construction trades in each city, and historical wage rates included for comparison. Each city chart lists the county and is alphabetically arranged with handy visual flip tabs for quick reference.

$450.95 | Available Dec. 2014 | Catalog no. 60125

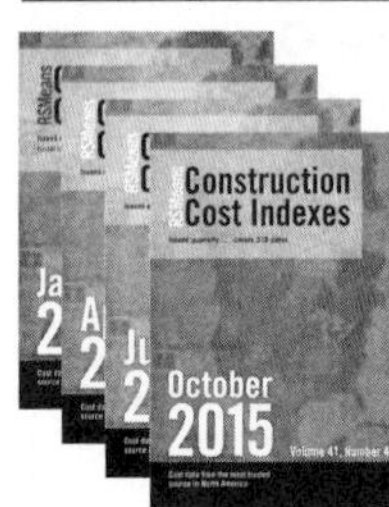

RSMeans Construction Cost Indexes 2015

What materials and labor costs will change unexpectedly this year? By how much?

- Breakdowns for 318 major cities
- National averages for 30 key cities
- Expanded five major city indexes
- Historical construction cost indexes

$359.95 per year (subscription) | Catalog no. 50145
$94.95 individual quarters | Catalog no. 60145 A,B,C,D

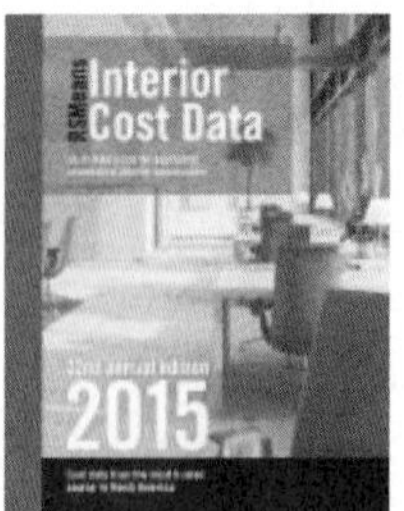

RSMeans Interior Cost Data 2015

Provides you with prices and guidance needed to make accurate interior work estimates. Contains costs on materials, equipment, hardware, custom installations, furnishings, and labor for new and remodel commercial and industrial interior construction, including updated information on office furnishings, and reference information.

$208.95 | Available Nov. 2014 | Catalog no. 60095

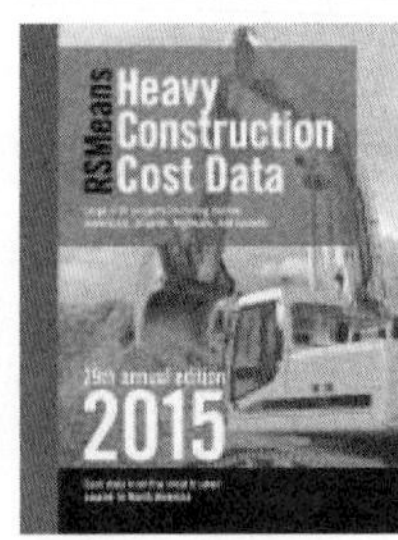

RSMeans Heavy Construction Cost Data 2015

A comprehensive guide to heavy construction costs. Includes costs for highly specialized projects such as tunnels, dams, highways, airports, and waterways. Information on labor rates, equipment, and materials costs is included. Features unit price costs, systems costs, and numerous reference tables for costs and design.

$204.95 | Available Dec. 2014 | Catalog no. 60165

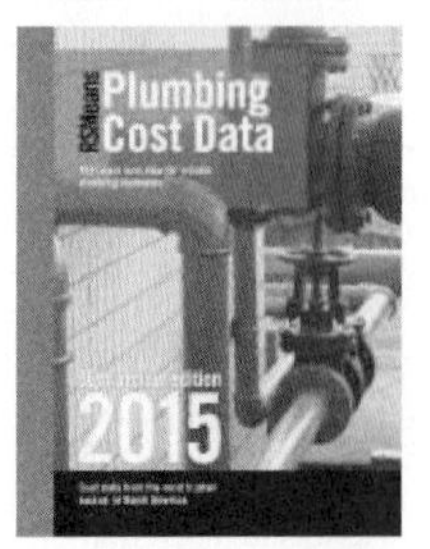

RSMeans Plumbing Cost Data 2015

Comprehensive unit prices and assemblies for plumbing, irrigation systems, commercial and residential fire protection, point-of-use water heaters, and the latest approved materials. This publication and its companion, *RSMeans Mechanical Cost Data*, provide full-range cost estimating coverage for all the mechanical trades.

Contains updated costs for potable water, radiant heat systems, and high efficiency fixtures.

$206.95 | Available Oct. 2014 | Catalog no. 60215

RSMeans Online and RSMeans CostWorks CD Comparison Matrix

Both the RSMeans CostWorks CD package and the RSMeans Online estimating tool provide the same comprehensive, up-to-date, and localized cost information to improve planning, estimating, and budgeting with RSMeans—the industry's most-trusted source of construction cost data.

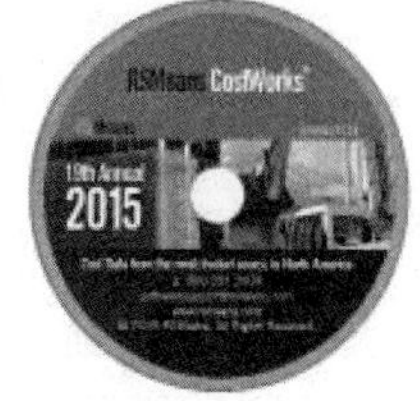

In addition to key differences noted in the comparison matrix below, both products let you do the following:

- Quickly locate costs in the searchable RSMeans database.
- Instantly adjust the data to your specific location for accurate costs anywhere in the U.S. and Canada.
- Create reliable, updated cost estimates in minutes.
- Manage, share, and export your estimates to MS Excel® with ease.
- Access a wide range of reference materials, including crew rates and national average labor rates.

Both the CD and Online options are available in a convenient one-year subscription!

Use the handy comparison matrix below to help you decide the best product for your cost estimating needs!

Why us?

For more than 70 years, RSMeans has provided the quality cost data construction professionals everywhere depend on to estimate with confidence, save time, improve decision-making, and increase profits.

Improve planning and decision-making

Back your estimates with complete, accurate, and up-to-date cost data for informed business decisions.

- Verify construction costs from third parties.
- Check validity of subcontractor proposals.
- Evaluate material and assembly alternatives.

Increase profits

Use RSMeans to estimate projects quickly and accurately, so you can gain an edge over your competition.

- Create accurate and competitive bids.
- Minimize the risk of cost overruns.
- Reduce variability.
- Gain control over costs.

The following matrix summarizes the key comparison points between RSMeans' two electronic cost data products.

	Question	CostWorks CD	RSMeans Online.com
1	How is the data presented?	• The data is organized by book title and sold at comparable book cost. • Several packaged bundles of five titles are available at a cost savings. • Available in a convenient one-year subscription	• The data is organized by book title and sold at comparable book cost. • Several packaged bundles of five titles are available at a cost savings. • RSMeans Online also offers a Complete Library title with all online data at a cost savings. • Available in a convenient one-year subscription
2	Where is the data stored?	With a typical install, the data is read off of your hard drive or off of the CD.	All data is stored on our secure server.
3	How are updates delivered?	Updates are available on a quarterly basis for download from the Costworks HomePort website.	Data is updated automatically on a quarterly basis, and users are notified when updates occur.
4	Where are my projects and estimates stored?	Upon installation, a "My CostWorks Projects" folder is created within the "My Documents" directory. All projects are defaulted to save here unless otherwise specified.	Estimates are stored on our secure server.
5	Does my IT team need to assist me in installing and running the program?	Yes, depending on the level of your permission. See #11 for system requirements.	While the product is entirely Web-based, you may need IT support with permissions for certain tasks. See #11 for system requirements.
6	Can users collaborate on projects/estimates?	Yes, you can copy project folders between computers.	You can create shared folders that can be viewed, copied, and edited by anyone within your account workgroup.

Comparison matrix continued

7	Can I create and use my own data with the RSMeans database?	Custom lines cannot be added to the CD database. With the Estimator tool, however, you can create and add custom cost lines to any estimate. You can also save custom lines as an Estimator template for use in creating new estimates.	You can create • a custom line within an estimate • a custom cost book consisting of custom unit (not assembly) cost lines, no more than once per quarter **Note** that custom data is searchable using the Include My Custom Data feature.
8	Can I enter my own trade/ labor rates and markups?	Although you cannot adjust the trade labor rates, you can add custom markups for each report template using the Estimator tool.	For each quarterly update, you can create one custom unit cost book. You can edit the labor rates by individual trade and the crew equipment costs, and also edit the associated material and equipment markups.
9	What are my reporting options?	The Estimator tool lets you create editable custom reports and report templates. Reports can also be exported to MS Excel®.	You can generate pre-formatted estimate summaries and cost lists in both PDF and Excel formats.
10	How do I find specific cost data?	You can search by CSI line number and by keyword within a division.	You can search by CSI line number and by keyword within a division. There are several additional search options, including keyword and advanced search using special characters.
11	What are the system requirements?	• Operating system – Windows XP or later • PC requirements - 256 MB RAM (recommended), 1024 x 768 High Color Monitor, 800MHz Computer (recommended), 24X CD-ROM, 250 MB Hard Disk Space	• Resolution – Recommended resolution is 1024 x 768 or higher • Compatibility - Internet Explorer version 8.x, 9.x and Mozilla Firefox version 10 and higher • iPad - iOS 5 and higher

2015 RSMeans Seminar Schedule

Note: call for exact dates and details.

Location	Dates
Seattle, WA	January and August
Dallas/Ft. Worth, TX	January
Austin, TX	February
Anchorage, AK	March and September
Las Vegas, NV	March and October
Washington, DC	April and September
Phoenix, AZ	April
Kansas City, MO	April
Toronto	May
Denver, CO	May
San Francisco, CA	June
Bethesda, MD	June
Columbus, GA	June
El Segundo, CA	August

Location	Dates
Jacksonville FL	September
Dallas, TX	September
Charleston, SC	October
Houston, TX	October
Atlanta, GA	November
Baltimore, MD	November
Orlando, FL	November
San Diego, CA	December
San Antonio, TX	December
Raleigh, NC	December

1-800-334-3509, Press 1

Professional Development

For more information visit the RSMeans website at www.rsmeans.com

eLearning Training Sessions

Learn how to use *RSMeans Online®* or *RSMeans CostWorks®* CD from the convenience of your home or office. Our eLearning training sessions let you join a training conference call and share the instructors' desktops, so you can view the presentation and step-by-step instructions on your own computer screen. The live webinars are held from 9 a.m. to 4 p.m. eastern standard time, with a one-hour break for lunch.

For these sessions, you must have a computer with high speed Internet access and a compatible Web browser. Learn more at www.rsmeansonline.com or call for a schedule: 1-800-334-3509 and press 1. Webinars are generally held on selected Wednesdays each month.

RSMeans Online®
$299 per person

RSMeans CostWorks® CD
$299 per person

RSMeans Online™ Training

Construction estimating is vital to the decision-making process at each state of every project. RSMeansOnline works the way you do. It's systematic, flexible and intuitive. In this one day class you will see how you can estimate any phase of any project faster and better.

Some of what you'll learn:

- Customizing RSMeansOnline
- Making the most of RSMeans "Circle Reference" numbers
- How to integrate your cost data
- Generate reports, exporting estimates to MS Excel, sharing, collaborating and more

Also available: RSMeans Online® training webinar

Maintenance & Repair Estimating for Facilities

This *two-day* course teaches attendees how to plan, budget, and estimate the cost of ongoing and preventive maintenance and repair for existing buildings and grounds.

Some of what you'll learn:

- The most financially favorable maintenance, repair, and replacement scheduling and estimating
- Auditing and value engineering facilities
- Preventive planning and facilities upgrading
- Determining both in-house and contract-out service costs
- Annual, asset-protecting M&R plan

Who should attend: facility managers, maintenance supervisors, buildings and grounds superintendents, plant managers, planners, estimators, and others involved in facilities planning and budgeting.

Facilities Construction Estimating

In this *two-day* course, professionals working in facilities management can get help with their daily challenges to establish budgets for all phases of a project.

Some of what you'll learn:

- Determining the full scope of a project
- Identifying the scope of risks and opportunities
- Creative solutions to estimating issues
- Organizing estimates for presentation and discussion
- Special techniques for repair/remodel and maintenance projects
- Negotiating project change orders

Who should attend: facility managers, engineers, contractors, facility tradespeople, planners, and project managers.

Practical Project Management for Construction Professionals

In this *two-day* course, acquire the essential knowledge and develop the skills to effectively and efficiently execute the day-to-day responsibilities of the construction project manager.

Covers:

- General conditions of the construction contract
- Contract modifications: change orders and construction change directives
- Negotiations with subcontractors and vendors
- Effective writing: notification and communications
- Dispute resolution: claims and liens

Who should attend: architects, engineers, owner's representatives, project managers.

Construction Cost Estimating: Concepts and Practice

This *one-day* introductory course to improve estimating skills and effectiveness starts with the details of interpreting bid documents, ending with the summary of the estimate and bid submission.

Some of what you'll learn:

- Using the plans and specifications for creating estimates
- The takeoff process—deriving all tasks with correct quantities
- Developing pricing, using various sources; how subcontractor pricing fits in
- Summarizing the estimate to arrive at the final number
- Formulas for area and cubic measure, adding waste and adjusting productivity to specific projects
- Evaluating subcontractors' proposals and prices
- Adding Insurance and bonds
- Understanding how labor costs are calculated
- Submitting bids and proposals

Who should attend: project managers, architects, engineers, owner's representatives, contractors, and anyone who's responsible for budgeting or estimating construction projects.

Mechanical & Electrical Estimating

This *two-day* course teaches attendees how to prepare more accurate and complete mechanical/electrical estimates, avoiding the pitfalls of omission and double-counting, while understanding the composition and rationale within the RSMeans mechanical/electrical database.

Some of what you'll learn:

- The unique way mechanical and electrical systems are interrelated
- M&E estimates—conceptual, planning, budgeting, and bidding stages
- Order of magnitude, square foot, assemblies, and unit price estimating
- Comparative cost analysis of equipment and design alternatives

Who should attend: architects, engineers, facilities managers, mechanical and electrical contractors, and others who need a highly reliable method for developing, understanding, and evaluating mechanical and electrical contracts.

Unit Price Estimating

This interactive *two-day* seminar teaches attendees how to interpret project information and process it into final, detailed estimates with the greatest accuracy level.

The most important credential an estimator can take to the job is the ability to visualize construction and estimate accurately.

Some of what you'll learn:

- Interpreting the design in terms of cost
- The most detailed, time-tested methodology for accurate pricing
- Key cost drivers—material, labor, equipment, staging, and subcontracts
- Understanding direct and indirect costs for accurate job cost accounting and change order management

Who should attend: corporate and government estimators and purchasers, architects, engineers, and others who need to produce accurate project estimates.

Conceptual Estimating Using RSMeans CostWorks® CD

This *two-day* class uses the leading industry data and a powerful software package to develop highly accurate conceptual estimates for your construction projects. All attendees must bring a laptop computer loaded with the current year *Square Foot Costs* and the *Assemblies Cost Data* CostWorks titles.

Some of what you'll learn:

- Introduction to conceptual estimating
- Types of conceptual estimates
- Helpful hints
- Order of magnitude estimating
- Square foot estimating
- Assemblies estimating

Who should attend: architects, engineers, contractors, construction estimators, owner's representatives, and anyone looking for an electronic method for performing square foot estimating.

RSMeans CostWorks® CD Training

This one-day course helps users become more familiar with the functionality of *RSMeans CostWorks* program. Each menu, icon, screen, and function found in the program is explained in depth. Time is devoted to hands-on estimating exercises.

Some of what you'll learn:

- Searching the database using all navigation methods
- Exporting RSMeans data to your preferred spreadsheet format
- Viewing crews, assembly components, and much more
- Automatically regionalizing the database

This training session requires you to bring a laptop computer to class.

When you register for this course you will receive an outline for your laptop requirements.

Also offering web training for RSMeans CostWorks CD!

Assessing Scope of Work for Facility Construction Estimating

This *two-day* practical training program addresses the vital importance of understanding the SCOPE of projects in order to produce accurate cost estimates for facility repair and remodeling.

Some of what you'll learn:

- Discussions of site visits, plans/specs, record drawings of facilities, and site-specific lists
- Review of CSI divisions, including means, methods, materials, and the challenges of scoping each topic
- Exercises in SCOPE identification and SCOPE writing for accurate estimating of projects
- Hands-on exercises that require SCOPE, take-off, and pricing

Who should attend: corporate and government estimators, planners, facility managers, and others who need to produce accurate project estimates.

Facilities Estimating Using RSMeans CostWorks® CD

Combines hands-on skill building with best estimating practices and real-life problems. Brings you up-to-date with key concepts, and provides tips, pointers, and guidelines to save time and avoid cost oversights and errors.

Some of what you'll learn:

- Estimating process concepts
- Customizing and adapting RSMeans cost data
- Establishing scope of work to account for all known variables
- Budget estimating: when, why, and how
- Site visits: what to look for—what you can't afford to overlook
- How to estimate repair and remodeling variables

This training session requires you to bring a laptop computer to class.

Who should attend: facility managers, architects, engineers, contractors, facility tradespeople, planners, project managers and anyone involved with JOC, SABRE, or IDIQ.

Unit Price Estimating Using RSMeans CostWorks® CD

Step-by-step instruction and practice problems to identify and track key cost drivers—material, labor, equipment, staging, and subcontractors—for each specific task. Learn the most detailed, time-tested methodology for accurately "pricing" these variables, their impact on each other and on total cost.

Some of what you'll learn:

- Unit price cost estimating
- Order of magnitude, square foot, and assemblies estimating
- Quantity takeoff
- Direct and indirect construction costs
- Development of contractor's bill rates
- How to use *RSMeans Building Construction Cost Data*

This training session requires you to bring a laptop computer to class.

Who should attend: architects, engineers, corporate and government estimators, facility managers, and government procurement staff.

Registration Information

Register early . . . Save up to $100! Register 30 days before the start date of a seminar and save $100 off your total fee. *Note: This discount can be applied only once per order. It cannot be applied to team discount registrations or any other special offer.*

How to register Register by phone today! The RSMeans toll-free number for making reservations is **1-800-334-3509, Press 1.**

Two-day seminar registration fee - $935. One-day *RSMeans CostWorks*® training registration fee - $375. To register by mail, complete the registration form and return, with your full fee, to: RSMeans Seminars, 700 Longwater Drive, Norwell, MA 02061.

Government pricing All federal government employees save off the regular seminar price. Other promotional discounts cannot be combined with the government discount.

Team discount program for two to four seminar registrations. Call for pricing: 1-800-334-3509, Press 1

Multiple course discounts When signing up for two or more courses, call for pricing.

Refund policy Cancellations will be accepted up to ten business days prior to the seminar start. There are no refunds for cancellations received later than ten working days prior to the first day of the seminar. A $150 processing fee will be applied for all cancellations. Written notice of cancellation is required. Substitutions can be made at any time before the session starts. **No-shows are subject to the full seminar fee.**

AACE approved courses Many seminars described and offered here have been approved for 14 hours (1.4 recertification credits) of credit by the AACE International Certification Board toward meeting the continuing education requirements for recertification as a Certified Cost Engineer/ Certified Cost Consultant.

AIA Continuing Education We are registered with the AIA Continuing Education System (AIA/CES) and are committed to developing quality learning activities in accordance with the CES criteria. Many seminars meet the AIA/CES criteria for Quality Level 2. AIA members may receive (14) learning units (LUs) for each two-day RSMeans course.

Daily course schedule The first day of each seminar session begins at 8:30 a.m. and ends at 4:30 p.m. The second day begins at 8:00 a.m. and ends at 4:00 p.m. Participants are urged to bring a hand-held calculator, since many actual problems will be worked out in each session.

Continental breakfast Your registration includes the cost of a continental breakfast, and a morning and afternoon refreshment break. These informal segments allow you to discuss topics of mutual interest with other seminar attendees. (You are free to make your own lunch and dinner arrangements.)

Hotel/transportation arrangements RSMeans arranges to hold a block of rooms at most host hotels. To take advantage of special group rates when making your reservation, be sure to mention that you are attending the RSMeans seminar. You are, of course, free to stay at the lodging place of your choice. **(Hotel reservations and transportation arrangements should be made directly by seminar attendees.)**

Important Class sizes are limited, so please register as soon as possible.

Note: *Pricing subject to change.*

Registration Form

ADDS-1000

Call 1-800-334-3509, Press 1 to register or FAX this form 1-800-632-6732. Visit our website: www.rsmeans.com

Please register the following people for the RSMeans construction seminars as shown here. We understand that we must make our own hotel reservations if overnight stays are necessary.

☐ Full payment of $____________________ enclosed.

☐ Bill me.

Please print name of registrant(s).

(To appear on certificate of completion)

____

____

____

____

P.O. #: ________________________

GOVERNMENT AGENCIES MUST SUPPLY PURCHASE ORDER NUMBER OR TRAINING FORM.

Firm name ________________________

Address ________________________

City/State/Zip ________________________

Telephone no. ____________ Fax no. ____________

E-mail address ________________________

Charge registration(s) to: ☐ MasterCard ☐ VISA ☐ American Express

Account no. ____________ Exp. date ____________

Cardholder's signature ________________________

Seminar name ________________________

Seminar City ________________________

Please mail check to: RSMeans Seminars, 700 Longwater Drive, Norwell, MA 02061 USA

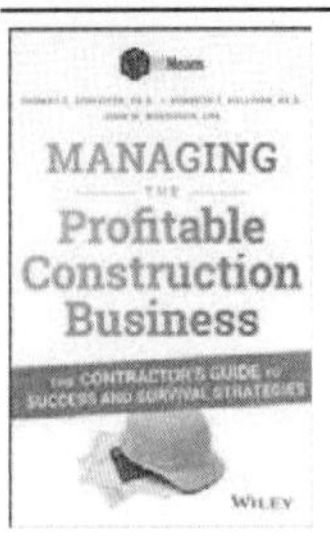

Managing the Profitable Construction Business: The Contractor's Guide to Success and Survival Strategies, 2nd Ed.

by Thomas C. Scheifler, Ph.D.

Learn from a team of construction business veterans led by Thomas C. Schleifer, who is commonly referred to as a construction business "turnaround" expert due to the number of construction companies he has rescued from financial distress. His financial acumen, combined with his practical, hands-on experience, has made him a sought-after private consultant.

$60.00 | 288 pages, hardover | Catalog no. 67370

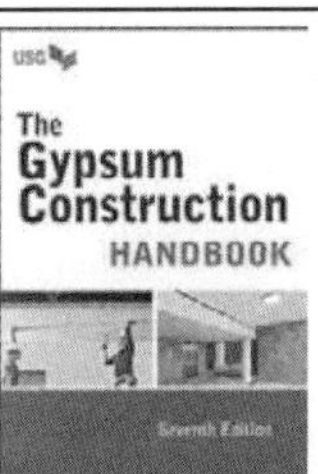

The Gypsum Construction Handbook, 7th Edition

by USG

The tried-and-true Gypsum Construction Handbook is a systematic guide to selecting and using gypsum drywall, veneer plaster, tile backers, ceilings, and conventional plaster building materials. A widely respected training text for aspiring architects and engineers, the book provides detailed product information and efficient installation methodology.

The 7th edition features updates in gypsum products, including ultralight panels, glass-mat panels, paper-faced plastic bead, and ultralightweight joint compound, and modern specialty acoustical and ceiling product guidelines. This comprehensive reference also incorporates the latest in sustainable products.

$40.00 | 576 pages, softcover | Catalog no. 67357B

Construction Project Safety

by John Schaufelberger, Ken-Yu Lin

This essential introduction to construction safety for construction management personnel takes a project-based approach to present potential hazards in construction and their mitigation or prevention.

Beginning with an introduction to Accident Prevention Programs and OSHA compliance requirements, the book integrates safety instruction into the building process by following a building project from site construction through interior finish. Reinforcing this applied approach are photographs, drawings, contract documentation, and an online 3D BIM model to help visualize the on-site scenarios.

$85.00 | 320 pages, hardover | Catalog no. 67368

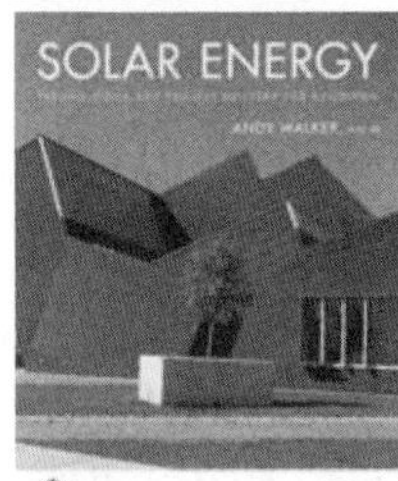

Solar Energy: Technologies & Project Delivery for Buildings

by Andy Walker

An authoritative reference on the design of solar energy systems in building projects, with applications, operating principles, and simple tools for the construction, engineering, and design professionals, the book simplifies the solar design and engineering process and provides sample documentation for the complete design of a solar energy system for buildings.

$85.00 | 320 pages, hardover | Catalog no. 67365

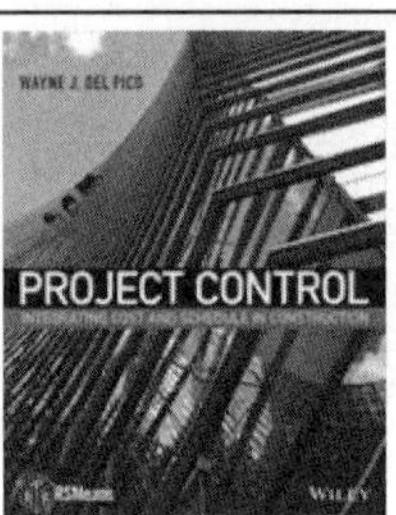

Project Control: Integrating Cost & Schedule in Construction

by Wayne J. Del Pico

Written by a seasoned professional in the field, *Project Control: Integrating Cost and Schedule in Construction* fills a void in the area of project control as applied in the construction industry today. It demonstrates how productivity models for an individual project are created, monitored, and controlled, and how corrective actions are implemented as deviations from the baseline occur.

$65.00 | 240 pages, softcover | Catalog no. 67366

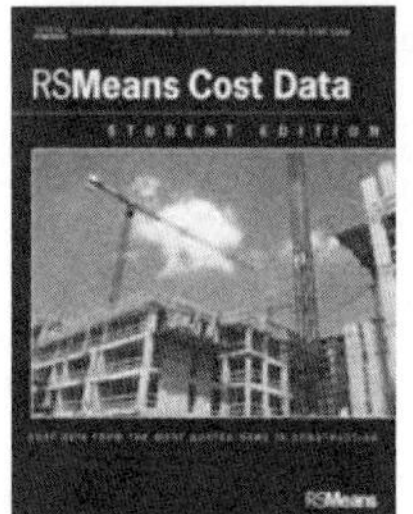

RSMeans Cost Data, Student Edition

Provides a thorough introduction to cost estimating in a print and online package. Clear explanations and example-driven. The ideal reference for students and new professionals. Features include:

- Commercial and residential construction cost data in print and online formats
- Complete how-to guidance on the essentials of cost estimating
- A supplemental website with plans, problem sets, and a full sample estimate

$110.00 | 512 pages, softcover | Catalog no. 67363

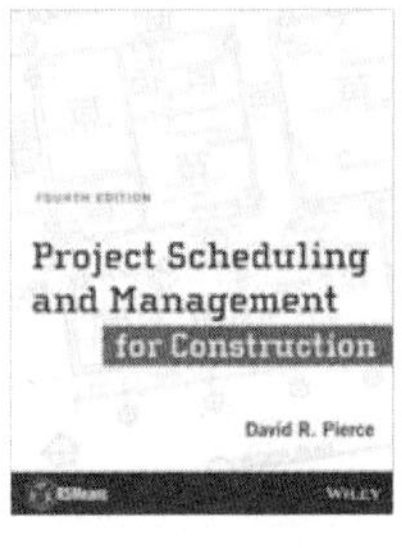

Project Scheduling and Management for Construction, 4th Edition

by David R. Pierce

First published in 1988 by RSMeans, the new edition of *Project Scheduling and Management for Construction* has been substantially revised for both professionals and students enrolled in construction management and civil engineering programs. While retaining its emphasis on developing practical, professional-level scheduling skills, the new edition is a relatable, real world case study.

$95.00 | 272 pages, softcover | Catalog no. 67367

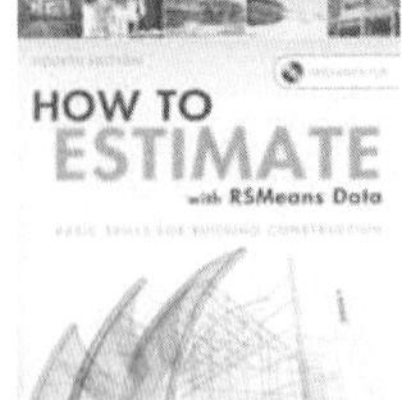

How to Estimate with Means Data & CostWorks, 4th Edition

by RSMeans and Saleh A. Mubarak, Ph.D.

This step-by-step guide takes you through all the major construction items with extensive coverage of site work, concrete and masonry, wood and metal framing, doors and windows, and other divisions. The only construction cost estimating handbook that uses the most popular source of construction data, RSMeans, this indispensible guide features access to the instructional version of *CostWorks* in electronic form, enabling you to practice techniques to solve real-world estimating problems.

$75.00 | 320 pages, softcover | Includes CostWorks CD | Catalog no. 67324C

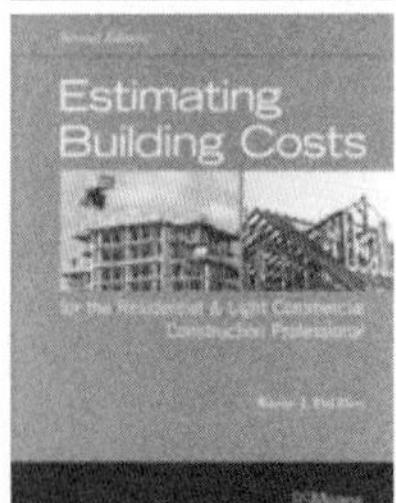

Estimating Building Costs, 2nd Edition

For the Residential & Light Commercial Construction Professional

by Wayne J. Del Pico

Explains, in detail, how to put together a reliable estimate that can be used not only for budgeting, but also for developing a schedule, managing a project, dealing with contingencies, and, ultimately, making a profit.

Completely revised and updated to reflect the CSI MasterFormat 2010 system.

$75.00 | 528 pages, softcover | Catalog no. 67343A

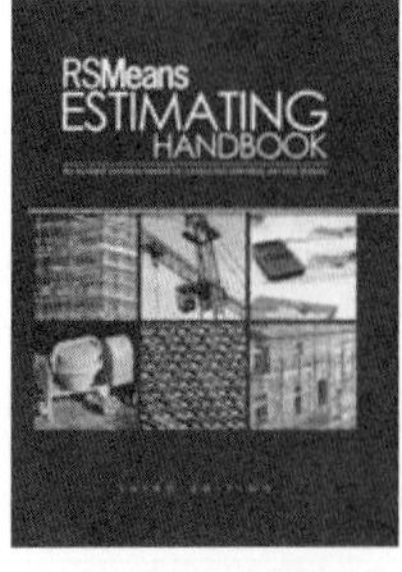

RSMeans Estimating Handbook, 3rd Edition

Widely used in the industry for tasks ranging from routine estimates to special cost analysis projects. This handbook will help construction professionals:

- Evaluate architectural plans and specifications
- Prepare accurate quantity takeoffs
- Compare design alternatives and costs
- Perform value engineering
- Double-check estimates and quotes
- Estimate change orders

$110.00 | Over 976 pages, hardcover | Catalog No. 67276B

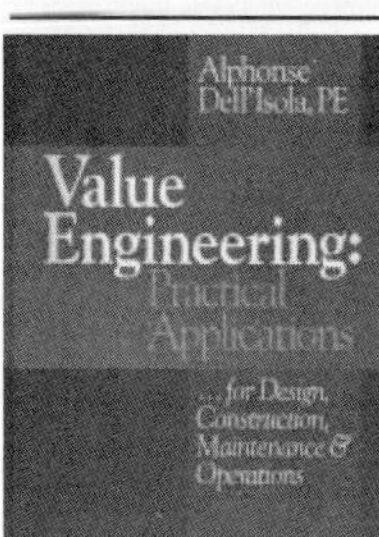

Value Engineering: Practical Applications

For Design, Construction, Maintenance & Operations

by Alphonse Dell'Isola, PE

A tool for immediate application—for engineers, architects, facility managers, owners, and contractors. Includes making the case for VE—the management briefing; integrating VE into planning, budgeting, and design; conducting life cycle costing; using VE methodology in design review and consultant selection; and case studies.

$90.00 | Over 427 pages, illustrated, softcover | Catalog no. 67319A

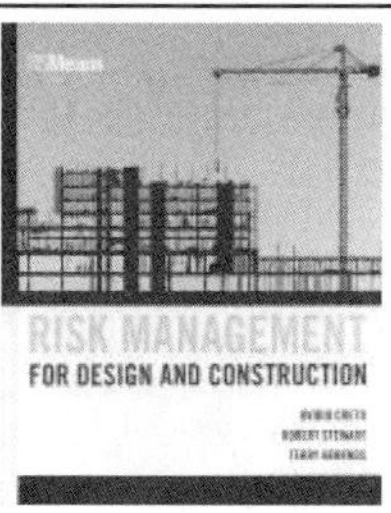

Risk Management for Design and Construction

Introduces risk as a central pillar of project management and shows how a project manager can prepare to deal with uncertainty. Includes:

- Integrated cost and schedule risk analysis
- An introduction to a ready-to-use system of analyzing a project's risks and tools to proactively manage risks
- A methodology that was developed and used by the Washington State Department of Transportation
- Case studies and examples on the proper application of principles
- Combining value analysis with risk analysis

$125.00 | Over 288 pages, softcover | Catalog no. 67359

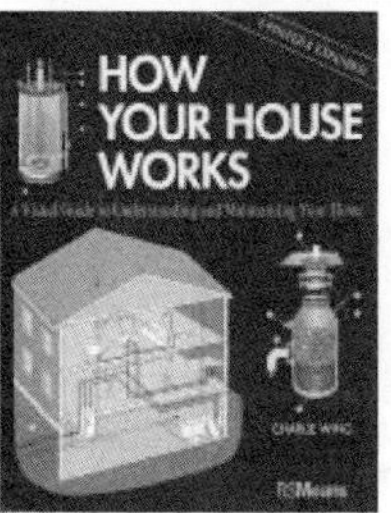

How Your House Works, 2nd Edition

by Charlie Wing

Knowledge of your home's systems helps you control repair and construction costs and makes sure the correct elements are being installed or replaced. This book uncovers the mysteries behind just about every major appliance and building element in your house. See-through, cross-section drawings in full color show you exactly how these things should be put together and how they function, including what to check if they don't work. It just might save you having to call in a professional.

$22.95 | 208 pages, softcover | Catalog no. 67351A

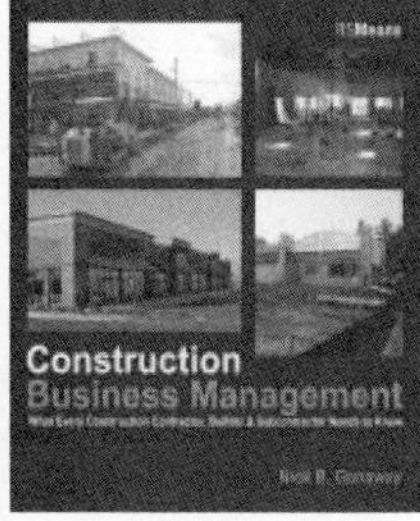

Construction Business Management

by Nick Ganaway

Only 43% of construction firms stay in business after four years. Make sure your company thrives with valuable guidance from a pro with 25 years of success as a commercial contractor. Find out what it takes to build all aspects of a business that is profitable, enjoyable, and enduring. With a bonus chapter on retail construction.

$60.00 | 201 pages, softcover | Catalog no. 67352

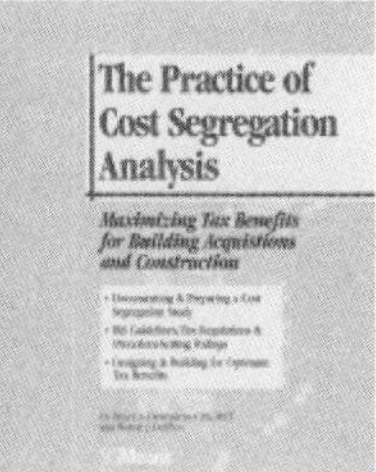

The Practice of Cost Segregation Analysis

by Bruce A. Desrosiers and Wayne J. Del Pico

This expert guide walks you through the practice of cost segregation analysis, which enables property owners to defer taxes and benefit from "accelerated cost recovery" through depreciation deductions on assets that are properly identified and classified.

With a glossary of terms, sample cost segregation estimates for various building types, key information resources, and updates via a dedicated website, this book is a critical resource for anyone involved in cost segregation analysis.

$105.00 | Over 240 pages | Catalog no. 67345

Complete Book of Framing, 2nd Edition

by Scot Simpson

This updated, easy-to-learn guide to rough carpentry and framing is written by an expert with more than thirty years of framing experience. Starting with the basics, this book begins with types of lumber, nails, and what tools are needed, followed by detailed, fully illustrated steps for framing each building element. Framer-Friendly Tips throughout the book show how to get a task done right—and more easily.

$29.95 | 368 pages, softcover | Catalog no. 67353A

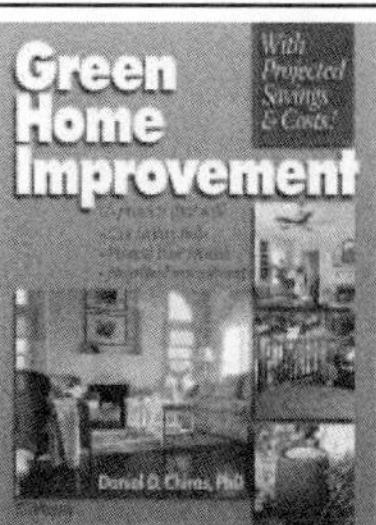

Green Home Improvement

by Daniel D. Chiras, PhD

With energy costs rising and environmental awareness increasing, people are looking to make their homes greener. This book, with 65 projects and actual costs and projected savings, helps homeowners prioritize their green improvements.

Projects range from simple water savers that cost only a few dollars, to bigger-ticket items such as HVAC systems. With color photos and cost estimates, each project compares options and describes the work involved, the benefits, and the savings.

$34.95 | 320 pages, illustrated, softcover | Catalog no. 67355

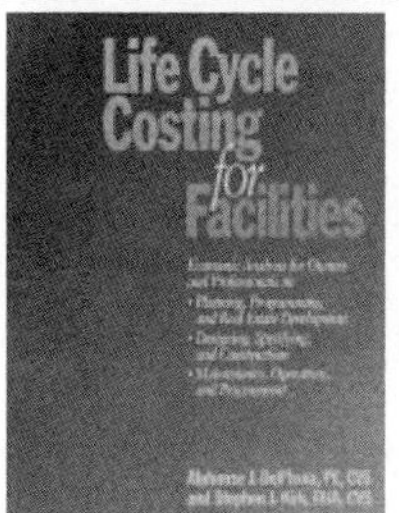

Life Cycle Costing for Facilities

by Alphonse Dell'Isola and Dr. Stephen Kirk

Guidance for achieving higher quality design and construction projects at lower costs! Cost-cutting efforts often sacrifice quality to yield the cheapest product. Life cycle costing enables building designers and owners to achieve both. The authors of this book show how LCC can work for a variety of projects—from roads to HVAC upgrades to different types of buildings.

$110.00 | 396 pages, hardcover | Catalog no. 67341

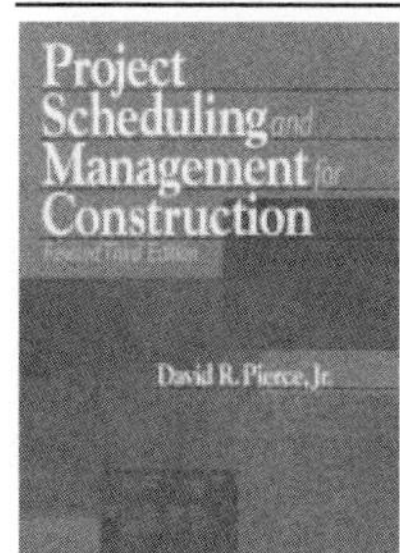

Project Scheduling and Management for Construction, 3rd Edition

by David R. Pierce, Jr.

A comprehensive, yet easy-to-follow guide to construction project scheduling and control—from vital project management principles through the latest scheduling, tracking, and controlling techniques. The author is a leading authority on scheduling, with years of field and teaching experience at leading academic institutions. Spend a few hours with this book and come away with a solid understanding of this essential management topic.

$64.95 | Over 286 pages, illustrated, hardcover | Catalog no. 67247B

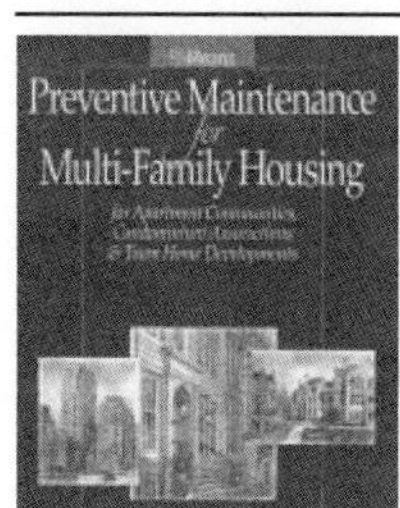

Preventive Maintenance for Multi-Family Housing

by John C. Maciha

Prepared by one of the nation's leading experts on multi-family housing.

This complete PM system for apartment and condominium communities features expert guidance, checklists for buildings and grounds maintenance tasks and their frequencies, and a dedicated website featuring customizable electronic forms. A must-have for anyone involved with multi-family housing maintenance and upkeep.

$95.00 | 290 pages | Catalog no. 67346

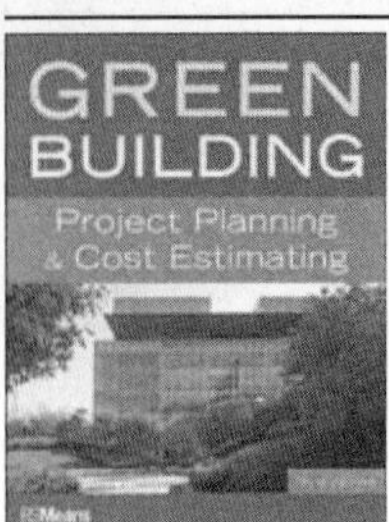

Green Building: Project Planning & Cost Estimating, 3rd Edition

Since the widely read first edition of this book, green building has gone from a growing trend to a major force in design and construction.

This new edition has been updated with the latest in green building technologies, design concepts, standards, and costs. Full-color with all new case studies—plus a new chapter on commercial real estate.

$110.00 | 480 pages, softcover | Catalog no. 67338B

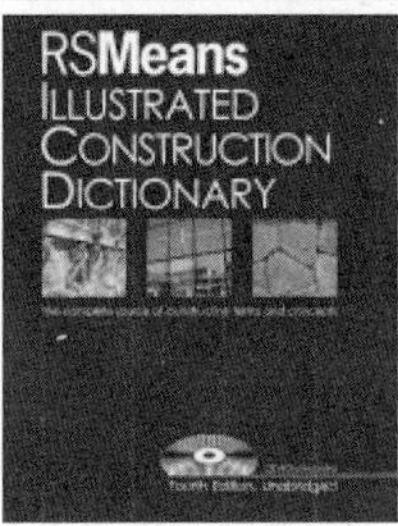

RSMeans Illustrated Construction Dictionary Unabridged 4th Edition, with CD-ROM

Long regarded as the industry's finest, *Means Illustrated Construction Dictionary* is now even better. With nearly 20,000 terms and more than 1,400 illustrations and photos, it is the clear choice for the most comprehensive and current information. The companion CD-ROM that comes with this new edition adds extra features, such as larger graphics and expanded definitions.

$110.00 | Over 880 pages, illust., hardcover | Catalog no. 67292B

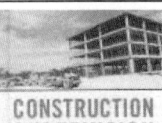

Construction Supervision

Inspires excellence with proven tactics and techniques applied by thousands of construction supervisors. Leadership guidelines carve out a practical blueprint for motivating work performance and increasing productivity through effective communication. Features:

- A unique focus on field supervision and crew management
- Coverage of supervision from the foreman to the superintendent level
- An overview of technical skills whose mastery will build confidence and success for the supervisor
- A detailed view of "soft" management and communication skills

$95.00 | 464 pages, softcover | Catalog no. 67358

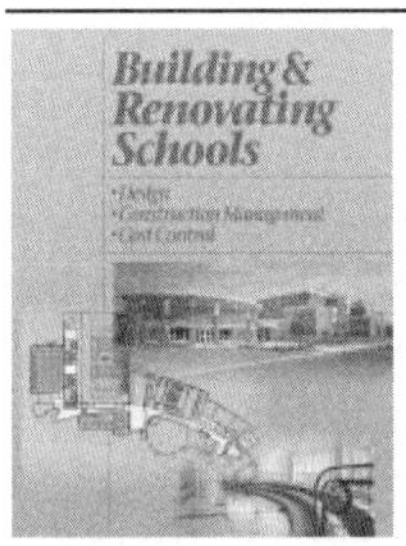

Building & Renovating Schools

This all-inclusive guide covers every step of the school construction process—from initial planning, needs assessment, and design, right through moving into the new facility. A must-have resource for anyone concerned with new school construction or renovation. With square foot cost models for elementary, middle, and high school facilities, and real-life case studies of recently completed school projects.

The contributors to this book—architects, construction project managers, contractors, and estimators who specialize in school construction—provide start-to-finish, expert guidance on the process.

$110.00 | Over 412 pages, hardcover | Catalog no. 67342

The Homeowner's Guide to Mold

By Michael Pugliese

Mold, whether caused by leaks, humidity or flooding, is a real health and financial issue—for homeowners and contractors. This full-color book explains:

- Construction and maintenance practices to prevent mold
- How to inspect for and remove mold
- Mold remediation procedures and costs
- What to do after a flood
- How to deal with insurance companies

$21.95 | 144 pages, softcover | Catalog no. 67344

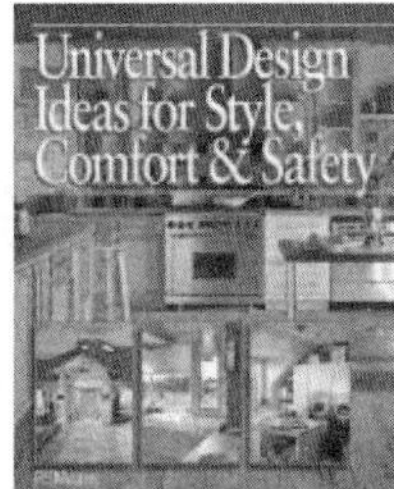

Universal Design Ideas for Style, Comfort & Safety

by RSMeans and Lexicon Consulting, Inc.

Incorporating universal design when building or remodeling helps people of any age and physical ability more fully and safely enjoy their living spaces. This book shows how universal design can be artfully blended into the most attractive homes. It discusses specialized products like adjustable countertops and chair lifts, as well as simple ways to enhance a home's safety and comfort. With color photos and expert guidance, every area of the home is covered. Includes budget estimates that give an idea how much projects will cost.

$21.95 | 160 pages, illustrated, softcover | Catalog no. 67354

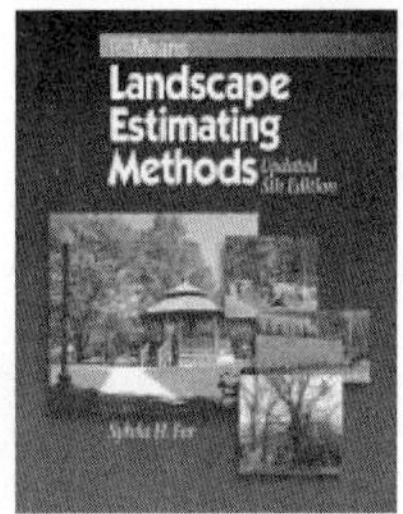

Landscape Estimating Methods, 5th Edition

Answers questions about preparing competitive landscape construction estimates, with up-to-date cost estimates and the new MasterFormat classification system. Expanded and revised to address the latest materials and methods, including new coverage on approaches to green building. Includes:

- Step-by-step explanation of the estimating process
- Sample forms and worksheets that save time and prevent errors

$75.00 | 336 pages, softcover | Catalog no. 67295C

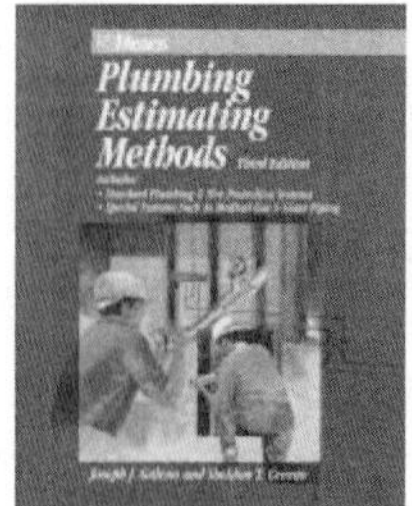

Plumbing Estimating Methods, 3rd Edition

by Joseph J. Galeno and Sheldon T. Greene

Updated and revised! This practical guide walks you through a plumbing estimate, from basic materials and installation methods through change order analysis. *Plumbing Estimating Methods* covers residential, commercial, industrial, and medical systems, and features sample takeoff and estimate forms and detailed illustrations of systems and components.

$65.00 | 381+ pages, softcover | Catalog no. 67283B

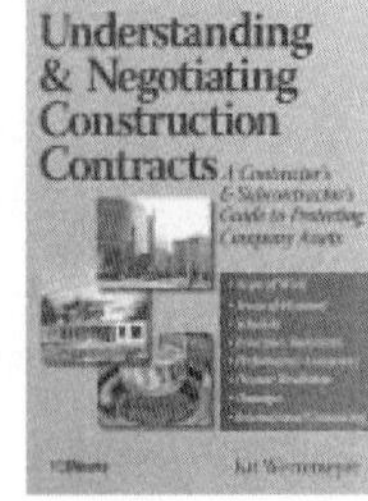

Understanding & Negotiating Construction Contracts

by Kit Werremeyer

Take advantage of the author's 30 years' experience in small-to-large (including international) construction projects. Learn how to identify, understand, and evaluate high risk terms and conditions typically found in all construction contracts—then, negotiate to lower or eliminate the risk, improve terms of payment, and reduce exposure to claims and disputes. The author avoids "legalese" and gives real-life examples from actual projects.

$80.00 | 320 pages, softcover | Catalog no. 67350

2015 Order Form

ORDER TOLL FREE 1-800-334-3509
OR FAX 1-800-632-6732

For more information visit the RSMeans website at **www.rsmeans.com**

Qty.	Book no.	COST ESTIMATING BOOKS	Unit Price	Total
	60065	Assemblies Cost Data 2015	$334.95	
	60015	Building Construction Cost Data 2015	206.95	
	60045	Commercial Renovaion Cost Data 2015	166.95	
	60115	Concrete & Masonry Cost Data 2015	188.95	
	50145	Construction Cost Indexes 2015 (subscription)	359.95	
	60145A	Construction Cost Index–January 2015	94.95	
	60145B	Construction Cost Index–April 2015	94.95	
	60145C	Construction Cost Index–July 2015	94.95	
	60145D	Construction Cost Index–October 2015	94.95	
	60345	Contr. Pricing Guide: Resid. R & R Costs 2015	39.95	
	60235	Electrical Change Order Cost Data 2015	199.95	
	60035	Electrical Cost Data 2015	208.95	
	60205A	Facilities Construction Cost Data 2015	524.95	
	60305	Facilities Maintenance & Repair Cost Data 2015	461.95	
	60555	Green Building Cost Data 2015	170.95	
	60165	Heavy Construction Cost Data 2015	204.95	
	60095	Interior Cost Data 2015	208.95	
	60125	Labor Rates for the Const. Industry 2015	450.95	
	60185	Light Commercial Cost Data 2015	153.95	
	60025	Mechanical Cost Data 2015	203.95	
	60155	Open Shop Building Const. Cost Data 2015	175.95	
	60215	Plumbing Cost Data 2015	206.95	
	60175	Residential Cost Data 2015	148.95	
	60285	Site Work & Landscape Cost Data 2015	199.95	
	60055	Square Foot Costs 2015	218.95	
	62014	Yardsticks for Costing (2014)	189.95	
	62015	Yardsticks for Costing (2015)	201.95	
		REFERENCE BOOKS		
	67329	Bldrs Essentials: Best Bus. Practices for Bldrs	45.00	
	67307	Bldrs Essentials: Plan Reading & Takeoff	50.00	
	67342	Building & Renovating Schools	110.00	
	67261A	The Building Prof. Guide to Contract Documents	75.00	
	67353A	Complete Book of Framing, 2nd Ed.	29.95	
	67146	Concrete Repair & Maintenance Illustrated	75.00	
	67352	Construction Business Management	60.00	
	67364	Construction Law	115.00	
	67358	Construction Supervision	95.00	
	67363	Cost Data Student Edition	110.00	
	67314	Cost Planning & Est. for Facil. Maint.	95.00	
	67230B	Electrical Estimating Methods, 3rd Ed.	75.00	
	67343A	Est. Bldg. Costs for Resi. & Lt. Comm., 2nd Ed.	75.00	
	67276B	Estimating Handbook, 3rd Ed.	110.00	
	67318	Facilities Operations & Engineering Reference	115.00	
	67338B	Green Building: Proj. Planning & Cost Est., 3rd Ed.	110.00	
	67355	Green Home Improvement	34.95	
	67357B	The Gypsum Construction Handbook	40.00	
	67148	Heavy Construction Handbook	110.00	
	67308E	Home Improvement Costs–Int. Projects, 9th Ed.	24.95	
	67309E	Home Improvement Costs–Ext. Projects, 9th Ed.	24.95	
	67344	Homeowner's Guide to Mold	21.95	
	67324C	How to Est.w/Means Data & CostWorks, 4th Ed.	75.00	

Qty.	Book no.	REFERENCE BOOKS (Cont.)	Unit Price	Total
	67351A	How Your House Works, 2nd Ed.	22.95	
	67282A	Illustrated Const. Dictionary, Condensed, 2nd Ed.	65.00	
	67292B	Illustrated Const. Dictionary, w/CD-ROM, 4th Ed.	110.00	
	67362	Illustrated Const. Dictionary, Student Ed.	60.00	
	67348	Job Order Contracting	130.00	
	67295C	Landscape Estimating Methods, 5th Ed.	75.00	
	67341	Life Cycle Costing for Facilities	110.00	
	67370	Managing the Profitable Construction Business	60.00	
	67294B	Mechanical Estimating Methods, 4th Ed.	75.00	
	67345	Practice of Cost Segregation Analysis	105.00	
	67366	Project Control: Integrating Cost & Sched.	95.00	
	67346	Preventive Maint. for Multi-Family Housing	95.00	
	67367	Project Sched. and Manag. for Constr., 4th Ed.	95.00	
	67322B	Resi. & Light Commercial Const. Stds., 3rd Ed.	65.00	
	67359	Risk Management for Design and Construction	125.00	
	67365	Solar Energy: Technology & Project Delivery	85.00	
	67145B	Sq. Ft. & Assem. Estimating Methods, 3rd Ed.	75.00	
	67321	Total Productive Facilities Management	85.00	
	67350	Understanding and Negotiating Const. Contracts	80.00	
	67303B	Unit Price Estimating Methods, 4th Ed.	75.00	
	67354	Universal Design	21.95	
	67319A	Value Engineering: Practical Applications	90.00	

MA residents add 6.25% state sales tax	
Shipping & Handling**	
Total (U.S. Funds)*	

Prices are subject to change and are for U.S. delivery only. *Canadian customers may call for current prices. **Shipping & handling charges: Add 8% of total order for check and credit card payments. Add 8% of total order for invoiced orders.

Send order to: ADDV-1000

Name (please print) ______________________

Company ______________________

☐ Company

☐ Home Address ______________________

____

City/State/Zip ______________________

Phone # ______________ P.O. # ______________

(Must accompany all orders being billed)

Mail to: RSMeans, 700 Longwater Drive, Norwell, MA 02061

RSMeans Project Cost Report

By filling out this report, your project data will contribute to the database that supports the RSMeans® Project Cost Square Foot Data. When you fill out this form, RSMeans will provide a $50 discount off one of the RSMeans products advertised in the preceding pages. Please complete the form including all items where you have cost data, and all the items marked (☑).

$50.00 Discount per product for each report you submit

Project Description (NEW construction only, please)

☑ Building Use (Office School . . .) ____________

☑ Address (City, State) ____________

____

☑ Total Building Area (SF) ____________

☑ Ground Floor (SF) ____________

☑ Frame (Wood, Steel . . .) ____________

☑ Exterior Wall (Brick, Tilt-up . . .) ____________

☑ Basement: (checkone) ☐ Full ☐ Partial ☐ None

☑ Number of Stories ____________

☑ Floor-to-Floor Height ____________

☑ Volume (C.F.) ____________

% Air Conditioned ____________ Tons ____________

Total Project Cost $ ____________

Owner ____________

Architect ____________

General Contractor ____________

☑ Bid Date ____________

☑ Typical Bay Size ____________

☑ Occupant Capacity ____________

☑ Labor Force: ____ % Union ____ % Non-Union

☑ Project Description (Circle one number in each line.)

1. Economy 2. Average 3. Custom 4. Luxury

1. Square 2. Rectangular 3. Irregular 4. Very Irregular

Comments ____________

____

A	☑	General Conditions	$
B	☑	Site Work	$
C	☑	Concrete	$
D	☑	Masonry	$
E	☑	Metals	$
F	☑	Wood & Plastics	$
G	☑	Thermal & Moisture Protection	$
GR		Roofing & Flashing	$
H	☑	Doors and Windows	$
J	☑	Finishes	$
JP		Painting & Wall Covering	$

K	☑	Specialties	$
L	☑	Equipment	$
M	☑	Furnishings	$
N	☑	Special Construction	$
P	☑	Conveying Systems	$
Q	☑	Mechanical	$
QP		Plumbing	$
QB		HVAC	$
R	☑	Electrical	$
S	☑	Mech./Elec. Combined	$

Please specify the RSMeans product you wish to receive.
Complete the address information.

Product Name ____________

Product Number ____________

Your Name ____________

Title ____________

Company ____________

☐ Company

☐ Home Street Address ____________

City, State, Zip ____________

Email Address ____________

Return by mail or fax 888-492-6770.

Method of Payment:

Credit Card # ____________

Expiration Date ____________

Check ____________

Purchase Order ____________

Phone Number ____________
(if you would prefer a sales call)

RSMeans
Square Foot Costs Department
700 Longwater Drive
Norwell, MA 02061
www.RSMeans.com